URANO METRIA 2000.0 Volume 3 DEEP SKY FIELD GUIDE

Cragin, Bonanno

Published by

P.O.Box 35025 • Richmond, VA 23235 USA
804-320-7016 • Fax (804) 272-5920
www.willbell.com

Uranometria and Uranometria 2000.0 are Trademarks of Willmann-Bell, Inc.
Published by Willmann-Bell, Inc.
P.O. Box 35025, Richmond, Virginia 23235

Second English Edition
First Printing, October 2001

Library of Congress Cataloging-in-Publication Data.
Uranometria 2000.0 : deep sky atlas.
p. cm.
Vol. 3 lacks subtitle.
Includes bibliographical references.
Contents: v. 1. The Northern Hemisphere to –6° / Wil Tirion, Barry Rappaport, Will Remaklus -- v. 2. The Southern Hemisphere to +6° / Wil Tirion, Barry Rappaport, Will Remaklus -- v. 3. Deep sky field guide / Murray Cragin, Emil Bonanno.
ISBN 0-943396-71-9 (v. 1) -- ISBN 0-943396-72-7 (v. 2) -- ISBN 0-943396-73-5 (v. 3)
1. Stars--Atlases. 2. Astronomy--Charts, diagrams, etc. 3. Astronomy Amateurs' manuals. 4. Astronomy Observers' manuals. I. Tirion, Wil

QB65.U78 2001
520'.22'3--dc21 2001043287

Foreword to the First Edition

> I tell you what follows because *The Deep Sky Field Guide to Uranometria 2000.0* is one of those rare frontier-busting events that will permanently alter the course of amateur astronomy. This is not a catalog in the 19th century sense. It is more like a data file of easy access. It is not built for the coffee table but to augment another frontier buster—*Uranometria 2000.0.*

There are great moments in astronomy when barred doors open and unassailable frontiers crumble so that humans can swarm into the unknown, or so that is what we like to think. Actually, these great historical breakthroughs usually get treated in one of a number of interesting (meaning human) ways.

The new experience, the new knowledge, simply gets ignored, and when historians a century later write up the event, they never mention the indifferent response. Or, if the public is receptive, the new knowledge is embraced, raised to mythic status, and few young seekers dare raise questions or do more frontier-busting. Newton did such a good job on gravity that for a couple of centuries no more work was done. Newton was the giant oak whose biochemistry killed off all the sproutings of bay oaks around his "turf.'"

But, more often than the public realizes, the new exposure leads to major conflict (in some cultures, bloody ones). When Alvarez, a Nobel winner in physics, ventured his asteroid to explain the great slaughter of species, the astronomers sat back to see the fun. The geologists fought with each other, and in some cases professors were fired merely for having the wrong departmental opinion.

Of course, other "frontier busters" came up with something that had obvious and immediate value and fared better. Foucault's silver-on-glass mirrors almost doubled the light grasp of a reflector. And most importantly, re-silvering did not change the figure of the mirror like re-polishing a metal speculum always did. Immediately, silvering took over the field.

What can we expect from the DSFG and its twin, *Uranometria 2000.0*? For one thing the constellations, useful as they were to goat herders 5000 years ago and despite the cultural job they provide for historians, are nothing but obstacles to today's fast-lane observers. The *Uranometria 2000.0* map number is now the key to the data.

With the book you are now holding in your hands and *Uranometria 2000.0*, you are about to experience the joy of having a map to 9.5 magnitude *and* the nonstellar tabular data for just that map. The working package is now *map + guide* and a *telescope*. That is all you need to begin your assault. And if a nova or comet intrudes on your dreams you can check immediately, no running back to the house to find the right catalog. This bridge, this union, will come from this symbiosis of *map-guide-telescope-observer.*

And what about the future? Add a few of the very talented, the "frontier busters" of which James Lucyk, Murray Cragin, and Barry Rappaport are current examples, and the next generation of amateur astronomers will excel and exceed anything we plan and write today. I can hardly wait.

This book is a classic before it even goes through the literary Foucault test.

WALTER SCOTT HOUSTON

Introduction

What is the *Deep Sky Field Guide* (DSFG)?

To describe the DSFG, we must first introduce *Uranometria 2000.0 Deep Sky Atlas,* the "Second Edition" of *Uranometria 2000.0,* which was first published in 1987. With the rise in popularity and availability of large amateur telescopes and sophisticated CCD cameras, the number of nonstellar deep-sky objects available to amateur observers has increased dramatically. This new version of *Uranometria 2000.0* reflects this increase by plotting more than 30,000 nonstellar deep-sky objects—a considerable increase over the first edition. Although all of these objects are named and their symbols reflect their object type on the *Uranometria* charts, and while many are plotted to reflect their actual size, little else about them can be discerned merely from their chart symbols. This is where the DSFG comes in.

The purpose of the DSFG is to provide basic catalog data for each of the more than 30,000 nonstellar objects plotted on the 220 *Uranometria 2000.0 Deep Sky Atlas* charts. This catalog data is that information which would be most useful to telescopic observers using any aperture, as well as astrophotographers and CCD imagers. Basic data includes each object's coordinates, brightness, and size. Also provided is information specific to the type of object being observed or imaged, such as alternate names, number of stars, position angle, opacity, and so on.

While readers familiar with the first edition of the DSFG will see a familiar format, there are many changes and improvements in this edition, not only in the data provided, but in how it was created. To better understand this creative process, we first need to offer a short history of modern astronomical cataloging.

The modern era of cataloging began in 1958 with the completion of The National Geographic Society–Palomar Observatory Sky Survey (POSS) which was a photographic survey of the northern hemisphere night sky taken with the 48-inch Oschin Schmidt Telescope on Palomar Mountain. The southern hemisphere was later photographed in 1975–82 using the UK 1.2-meter Schmidt Telescope at Siding Spring Observatory in Australia. While they both extend beyond the Celestial Equator, only when they are combined do we get a high-quality photographic record of the entire sky from pole to pole.

Until recently, catalogs based on these surveys were created by taking visual measurements from the photographic plates, and compiling the data by hand. This methodology made these early catalogs highly prone to human error, both in measurement and transcription. Moreover, subsequent works took this earlier data at face value, thus propagating those errors through many years and many catalogs. If a newer catalog does make a legitimate correction, it may be deemed erroneous because it does not agree with the majority of other (incorrect) ones!

Starting in the 1980s, both the POSS and the UK Schmidt Telescope Surveys were digitized (the photographic plates were scanned and converted to computer-readable files) by the Space Telescope Science Institute as part of an eight-year effort to support the Hubble Space Telescope. By the mid-1990s, this data, now called the *Digitized Sky Survey* (DSS), was compressed and made available to the general public for download via the Internet. With the availability of the DSS to anyone with a computer, it was possible to access DSS images and verify catalog data against the original source—the sky itself! However, manually downloading and inspecting images from a very large list of objects would be a daunting task, and corrections would still have to be transcribed by hand—a process both time-consuming and open to human error. What was needed was a way to more or less automate the process of inspecting DSS images and recording corrections.

The solution used in the creation of *Uranometria 2000.0 Deep Sky Atlas*, and its companion *Deep Sky Field Guide*, was provided by Emil Bonanno in the form of a highly modified version of his commercial computer program *Megastar: A Sky Atlas for Windows*. Normally this program allows a user to replace the program's artificial sky with registered images from the DSS, and overlay these images with symbols from the stellar and nonstellar databases. This allows a user not only to see the real stars, galaxies, clusters, and nebulae of the actual sky, but to see where the various catalogs place these objects relative to their true positions.

The standard program was then modified to add tools that allowed the user to move the nonstellar deep-sky object symbols, as well as to change their sizes and orientations (such as position angle for galaxies). There were special tools for use with galaxies, clusters and planetary nebulae. The end result was, where a specific nonstellar object's catalog symbol did not fall precisely on its DSS image, its symbol could be moved to the proper position. Then any changes made in an object's placement, size, orientation, etc., were automatically transferred to the DSFG/

Uranometria 2000.0 database. In addition, provisions were made that enabled the user to manually enter notes for each nonstellar object if desired—notes such as the distance and direction to nearby bright stars or nonstellar objects not plotted on the charts. Co-author Murray Cragin then spent the next two years going through the entire nonstellar database several times, refining the data as required, and adding notes.

Since the DSFG is primarily a field guide, and its information is targeted toward visual, photographic and CCD observers, other astrophysical information has been omitted, such as the absolute magnitude of clusters or the radial velocities of galaxies (things not truly needed at the telescope). And because it is possible to determine positions directly from the DSS with a high degree of accuracy, for this edition of the DSFG we have chosen to provide coordinates in the more precise format of HH MM SS.d and +/− DD MM SS. Hopefully users of digital setting circles and computer-controlled telescopes will find this increased accuracy useful, particularly in crowded fields. We believe that these positions are quite accurate for galaxies, planetary nebulae and globular clusters. For larger, less well-defined objects such as open clusters and nebulae, one should not infer that we have determined the center of mass by anything other than visual inspection of the DSS.

In the notes, stars are often referenced, giving their distance and direction, to help aid observers in finding particularly elusive objects. Two stellar databases were used for these notes, the *Guide Star Catalog* (GSC—a companion project of the DSS) and the *Hipparcos/Tycho* star catalogs. When available, the more accurate *Hipparchos/Tycho* data was used. In the latter, star brightnesses are expressed as visual (V) magnitudes and have *two* places to the right of the decimal (e.g., mag 8.34v), while the *Guide Star Catalog* magnitudes are photographic and have only *one* place to the right of the decimal (e.g., mag 13.4).

The selection of nonstellar deep-sky objects for inclusion in this edition of *Uranometria 2000.0* and the DSFG was made so as to provide a much more complete catalog than the original. Galaxies were chosen based on an approximate magnitude cutoff of 15(B) and a major axis of at least 1.5 arcminutes, with exceptions being made on an individual basis. However, all NGCs were included regardless of magnitude or size. In addition, over 1,100 planetary nebulae are plotted. The open clusters are a 1,617-object subset of a catalog prepared by Brent Archinal, comprising objects which are clearly visible on the DSS, and likely to be visible with amateur instruments. And new to this edition, we have added all confirmable dark nebulae from Barnard's *Atlas of Selected Regions of the Milky Way, Part II: Charts and Tables*. The list goes on as you will see in the sections that follow. While we may not have included everyone's favorite "challenge object," we do provide a greatly expanded database, with a much higher degree of accuracy, that should push the envelope for all observers.

What follows is a general outline of the information collected here: the sources, the uncertainties, and how the various data is relevant to observing. Insofar as possible, the best and most recent deep-sky data has been used, with many changes and additions having been made in the last months before publication based on new findings from the professional community. Some idea of the array of information collected here can be obtained by scanning the bibliography at the end of this Introduction.

The Data

Overall Arrangement

The *Deep Sky Field Guide* presents data as 220 tables—it is, fundamentally, 220 mini-catalogs, one table for each *Uranometria 2000.0* double-page chart. Each table is headed on either side by large bold numbers that correspond to its atlas chart number. Within each table, deep-sky objects are grouped into separate sections by type, starting with galaxies, then galaxy clusters, open clusters, globular clusters, bright nebulae, dark nebulae and finally planetary nebulae. Other objects—quasars, radio and X-ray sources, variable and double stars—although charted in the atlas, are not included in DSFG. Within each section of a table individual objects are listed in alphabetical order by catalog prefix (Be, IC, NGC, etc.), and ordered numerically within each prefix group. This is another break from conventional catalogs where objects are arranged in Right Ascension order. We have found that in these more compact tables, it is much easier to use names rather than the coordinates. Also, three Indexes—Messier Objects, Name, and Alternate Name—are provided to enable rapid location of an object by chart-table.

Uranometria 2000.0 "outlying objects"

A quick look at the pages of *Uranometria 2000.0* shows that many large nonstellar objects span one or more charts. For example, NGC 224 (M31) appears on Chart 30; but part of this large galaxy can also be found on adjoining Charts 44, 45 and 62. In addition to extended objects, such as the larger dark and bright nebulae, there are instances where smaller objects fall on, or partially on, the RA or Declination lines bridging two adjoining maps, with the majority of the object falling outside the chart's boundary lines (see MCG +3-59-47 in the upper right corner of Chart 81, or NGC 4556 and 4558 at Chart 71's right margin). We call these *outlying objects*. And as the numbers of these outlying objects began to add up, it became quickly apparent that to avoid costly duplication of data, and more bulk to an already large book, a decision had to be made as to when to include data for an outlying object in a table, or leave it off and let the adjoining chart's table pick it up. Our answer became the "51% rule." If 51% of an object lay within the boundary lines of a chart, data for the object was included in that chart's DSFG table. If less than 51% of the object fell within a chart, its data will be found in the table for the chart that holds the majority of the object. In the case of NGC 224 (M31) above, its index listing shows only Chart 30, so only table 30 lists this object with its data.

Thus, if you find an object only partially inside the boundary lines of a chart, and there is no line of data for it in the corresponding DSFG table, then look for it in the adjoining chart's table.

Galaxies

The DSFG contains 25,895 galaxies. These objects are complete star systems like our own Milky Way Galaxy. A typical galaxy is composed of billions of stars, while the largest can easily exceed a trillion stars. For the most part, clusters and nebulae in other galaxies are well out of visual range. Amateurs equipped with large scopes will be able to observe detail in only a few hundred of the brightest examples.

As with any deep-sky object, the most basic question asked by the observer is: "Will this galaxy be visible in my telescope under my observing conditions?" The data presented here will help you answer that question and others. A key feature of the DSFG is the "Notes" section for each chart, which draws on visual examination of the DSS and a variety of other sources to expand on and complement the numerical data.

This work includes galaxies of magnitude 15 or brighter, or with a major axis of 1.5 arcminutes or larger. As noted before, all NGC objects are included regardless of magnitude or size. Positions, dimensions and position angles were confirmed, and modified where necessary, by visual inspection of Digitized Sky Survey images. There are undoubtedly still identification, position and general data problems in the resulting DSFG galaxy database, but we believe that the procedures we have followed have greatly reduced them.

Galaxy Classification

Introduction

The DSFG provides type-classifications for almost all of its galaxies. The majority of these classifications come from the *Third Reference Catalogue of Bright Galazies* (RC3), and employ the "Revised Hubble" system that Gerard de Vaucouleurs developed and extended in the 1950s from Edwin Hubble's original system of the 1920s, and from Harlow Shapley's work during the 1930s. The Revised Hubble system is described by de Vaucouleurs in an article that appeared in *Handbuch der Physik* in 1958. Minor revisions to it were presented in the introductions to the three *Reference Catalogues of Bright Galaxies* (de Vaucouleurs and de Vaucouleurs 1964; de Vaucouleurs et al 1976; and de Vaucouleurs et al 1991). The Revised Hubble system is summarized in Table 1, adopted from RC3.

Most of the Revised Hubble types in RC3 come from the *Uppsala General Catalogue of Galaxies* (Nilson 1973), the *Southern Galaxy Catalogue* (Corwin et al 1985), the *ESO-Uppsala Survey of the ESO (B) Atlas* (Lauberts 1982), and the *South-Equatorial Galaxy Catalogue* (Corwin and Skiff 1999). Some of the Revised Hubble classifications in RC3 are transformed from the less widely used

Table 1
The Revised Morphological Galaxy Classification System

Classes	Families	Varieties	Stages	T	Type
Ellipticals		Compact		–6	cE
			Elliptical (0–6)	–5	E0
				–5	E0–1
		"cD"		–4	E^+
Lenticulars					S0
	Non-barred				SA0
	Barred				SB0
	Mixed				SAB0
		Inner ring			S(r)0
		S-shaped			S(s)0
		Mixed			S(rs)0
			Early	–3	$S0^-$
			Intermediate	–2	$S0^{\circ}$
			Late	–1	$S0^+$
Spirals	Non-barred				SA
	Barred				SB
	Mixed				SAB
		Inner ring			S(r)
		S-shaped			S(rs)
		Mixed	0/a	0	S0/a
			a	1	Sa
			ab	2	Sab
			b	3	Sb
			bc	4	Sbc
			c	5	Sc
			cd	6	Scd
			d	7	Sd
			dm	8	Sdm
			m	9	Sm
Irregulars	Non-barred				IA
	Barred				IB
	Mixed				IAB
		S-shaped			I(s)
			Non-Magellanic	90	I0
			Magellanic	10	Im
		Compact		11	cI
Peculiars				99	Pec
Peculiarities			Peculiarity		p
(All types)			Uncertain		:
			Doubtful		?
			Spindle		sp
			Outer ring		(R)
			Pseudo outer ring		(R′)

systems of Eric Holmberg (1958) and Sidney van den Bergh (1960).

Several hundred galaxies in the DSFG do not have classifications in RC3. Some of these have, however, been classified on the similar "Hubble-Sandage" system, Hubble's system as extended and revised by Allan Sandage, beginning in the 1950s. This system is fully described and beautifully illustrated in the *Carnegie Atlas of Galaxies* by Sandage and John Bedke (1994). The Hubble-Sandage system is summarized in Table 2.

Over much of their range, the two revised systems differ only in notation, so the galaxy types from either system can be used interchangeably for most objects. Both systems use terminology implying evolution: "early" type galaxies are dominated by central bulges composed of old, cool, red stars; while "late" types are dominated by extended disks of young, hot, blue stars. The terminology was introduced by Hubble as a shorthand to describe his sequence of galaxy types running from "elliptical" through "spiral" to "irregular." However, it has been clear for many years that no galaxy actually evolves along the Hubble sequence. The galaxy types are dependent on many factors including mass, rotation speed, and environment. The

Table 2
Coding for Hubble-Sandage Galaxy Classification

Classes	Sub-Classes		Other Elements	Type Codes
Ellipticals				E
Dwarf Ellipticals				d E
			Flattening	E(0–7)
(Transition)				E/SO
SO	Normal	Normal		SO(s)
	Barred	Ringed		SB0
		Inner		S0(r)
		Outer		RSO
		Partial Outer		PSO
Dwarf SO				d SO
			Disk/Bar Strength	$SB0_{[1-3]}$
			Flattening	S0[(0)–(9)]
(Transition)				SO/a
Spirals	Normal	Normal		S[a–d,m](s)
	Barred	Ringed		SB[a–d, m]
		Inner		S(r)
		Outer		RS
		Partial Outer		PS
Irregular	Normal			IM
	Barred			IBm
Blue Compact Dwarf				BCD
			Peculiarity	. . . pec
Amorphous				Amorphous
Merged				Merged
Tides				Tides
Jets				Jets
Peculiar				Peculiar

actual evolution of a single galaxy through its lifetime is thus so complex that it has not yet been fully worked out.

All of these now-traditional galaxy classifications come from photographic plates with peak spectral responses in the blue. These plates give prominence to galaxy disks that feature hot, young objects associated with star formation (O and B stars, HII regions, star clusters, and so on). Images taken with plates or detectors sensitive to longer wavelengths emphasize the cooler, older, redder stars that make up the bulges of galaxies. However, the largest amount of classification has been done on blue-sensitive plates, so virtually all of the galaxy types in the DSFG are derived from them.

Explanation of the Classification Symbols

Both systems have adopted notations based on seven classification elements of varying length, filled or not as possible (depending primarily on plate resolution), with letters, numbers, or other symbols. If a feature corresponding to one of the elements is present in a galaxy, the symbol describing that feature is written in a standard sequence. Elements missing from a particular galaxy are simply omitted, but the symbols are distinctive enough so that little confusion can arise. A "complete" classification might look like this in the Revised Hubble system: "(R)SB(s)a pec sp". In the Hubble-Sandage system, this would be expressed as "RSBa(s) pec".

Uncertainty symbols (":" and "?") were originally applied immediately following the symbols for the uncertain features. For example, if a bar was merely suspected in an edgewise galaxy, it would have been noted "SB?c". However, because limited character sets were available on early computers, classifications were further coded in the three Reference Catalogues (see the introductions to those catalogs for explanations of that coding). In that process, the uncertainty symbols were moved to the end of the coded classification. Since DSFG uses classifications decoded from RC3, the uncertainty symbols appear at the end of the classification rather than just after the symbol which is actually uncertain. Thus, an "SB?c" classification becomes "SBc?" in the DSFG.

In the following paragraphs, the notation used in the Revised Hubble (RH) system is described first (because most of the classifications in the DSFG are on this system), followed by additional explanation of the Hubble-Sandage (HS) system when needed.

- The first element is optional: if the galaxy has an outer ring, or a partial outer ring, it is noted as (R) or (R') by the RH system, and as R or PR by the HS. This element is also occasionally used to note a dwarf ("d") spheroidal galaxy in both classification systems, or a compact ("c") elliptical galaxy in the RH system. The RH system also uses "c" here for compact irregular galaxies, while the HS system codes these as "BCD" (blue compact dwarf).
- The second element codes the major class of the galaxy: "E" for elliptical; "S" for spiral or lenticular; "E/S" for a transition case; "Sph" for spheroidal (usually dwarf galaxies); "I" for irregular; and "P" for peculiar. Some classifiers use an "L" for the lenticular galaxies, but RC3 did not adopt this notation. The dwarf lenticular galaxies are called "dS0" in the HS system.
- The third element is the family: "A" for unbarred (or "ordinary"), "B" for barred, and "AB" for transition cases. The HS system omits the "A" completely; transition cases there are noted with a forward slash between two complete types. "Sa(s)/SBa(s)" is an example. For elliptical galaxies, this third element codes the flattening index, 10 x (a-b)/a, rounded to a single digit between 0 and 6 (0 and 7 in the HS system).
- The fourth element in the RH system labels the variety: "(r)" for ringed, "(s)" for S-shaped, and "(rs)" or "(sr)" for transition cases.
- The fifth element in the RH system codes the stage along the Hubble sequence with lower case letters: "a", "ab", "b", "bc", "c", "cd", "d", "dm", and "m". This element also uses "0" (zero) to distinguish the lenticular galaxies from the spirals. In the RH system, this "0" has a superscript minus sign, zero, or plus sign to indicate the stage within the lenticular class. The HS system adopts subscript numbers 1 to 3 for stages.
The RH system also has an "0/a" designation for galaxies that bridge the transition between lenticular and spiral. The zero is also used with the irregular ("I") class to flag the amorphous, irregular systems like M82 and NGC 3077. (These galaxies are simply called "Amorphous" in the HS system).

The HS system interchanges the fourth and fifth elements of RH classifications, adopting the Hubble stage as the fourth element, and using the fifth one for the variety (ringed or non-ringed) codes described above.

- The sixth element is used in the RH system primarily for indicating minor peculiarities by "pec" or sometimes just "p". The HS system uses this element for the luminosity class (discussed below) for spiral galaxies, or for the flattening index (in parentheses) for lenticulars (e.g. "(8)").
- The seventh and final element, also optional, in the RH system flags, with "sp" for "spindle", galaxies edge-on or nearly edge-on to our line of sight. This element is used in the HS system for "pec" to indicate minor peculiarities.

Summary

Galaxies are classified by assigning them to one of several major classes: elliptical, lenticular, spiral, irregular, spheroidal, or peculiar. Ellipticals are pure bulge galaxies, characterized only by a flattening index that describes their apparent shape. Lenticulars, still dominated by bulges, begin to show features associated with rotating disks. Spirals, with bulge and disk combined in a single system, are assigned stages depending on how tightly wound the arms appear, how large the bulge in the center of the galaxy is in relation to the disk, and how many knots (HII regions, star clouds, superassociations, etc.) appear along the arms. Irregulars are either of the Magellanic type (pure disk galaxies, forming an extension of the spiral sequence), or are amorphous. Over 95% of the nearby galaxies that we see on the sky can be comfortably fit within one or another of the first four classes, but some defy easy morphological binning, so they are simply called peculiar. The spheroidals are virtually all dwarf galaxies similar to the Fornax and Sculptor systems discovered by Shapley in the late 1930s.

Bars can appear among the lenticular, spiral, and Magellanic irregular galaxies. Barred lenticulars and spirals are just as common as non-barred galaxies, while barred Magellanic irregulars are much more common than non-barred irregulars.

Outer rings are most common at the lenticular/spiral transition stage. They can be seen in the later lenticular stages, and in spiral stages up to "Sbc". Galaxies of later stages "Sc" through "Sm" almost never have outer rings; the few that do are very unusual, and are probably the result of mergers or collisions. Inner rings are also seen in the same range of lenticular and spiral stages, though they are occasionally present as far along as stage "Sd".

High-resolution images reveal other interesting features in the cores of galaxies. Many nearby objects are now known to have bars, rings, and spiral features contained entirely within the central part of the bulges that are usually overexposed on survey plates.

We now think that most of the major morphological features in galaxies are the result of 1) gravitational interactions among the myriad stars that comprise them; or 2) gravitational interactions among pairs or multiple galaxies. The diffusely-distributed gas and dust present in them interacts primarily through shock waves of various origins (supernovae, spiral density waves, magnetodynamics, random motions within spiral arms, etc.). These contribute the additional features associated with star formation (HII regions, star clusters and associations, etc.) seen in spiral and irregular galaxies. Interactions of gas and dust clouds, and the dynamically active cores of galaxies, are responsible for many of the interesting features currently found by extragalactic surveys in other parts of the spectrum (gamma ray, X-ray, ultraviolet, infrared, and radio).

Magnitudes

Visual magnitudes have been determined from data in the ESO-LV, RC3, or PGC in one of five ways:

- When the RC3 gives a total V magnitude instead of its more common total B magnitude, we adopted the V_T directly.
- When B_T and $(B - V)_T$ photoelectric data exists, a total V magnitude was calculated: $V_T = B_T - (B - V)_T$. This is the usual case for galaxies with photoelectric data, which number about 3,000.
- When B_T data is absent, a total V magnitude was calculated from m_B and $(B - V)_T$. These cases are rare.
- When $(B - V)_T$ is not known it was calculated from B_T with the following color estimates based on the galaxy's type (for T, see the preceding section on Galaxy Classification):

	(B – V)
$T \leq 0.0$	1.0
$0.0 < T \leq 2.0$	0.9
$2.0 < T \leq 4.0$	0.8
$4.0 < T \leq 6.0$	0.7
$6.0 < T \leq 7.5$	0.6
$7.5 < T \leq 11.0$	0.6
T = 50	0.8
T = not given	1.0

These cases are more common than the previous one, but still relatively rare.

- When neither B_T nor $(B - V)_T$ exists, a total V magnitude was calculated from m_B using the color estimates from the table above. This is the most common case.

Uncertainties in the total magnitudes range from only a few hundredths in the best instances with photoelectric and/or CCD data (the first and second cases above) to an upper limit of one-and-a-half magnitudes for values calculated by the last, rather indirect, method.

Dimensions

Galaxy dimensions cannot be easily measured. In practice, the *perceived* diameter is limited by the luminosity of the night sky, which is about magnitude 23.0 (B) per square arcsecond at a perfectly dark site. Some catalogs determine the extent of a galaxy by generating an isophote at some chosen surface brightness value (an isophote is a contour line tracing points of equal brightness around a galaxy). For example, the *RC3* uses 25.0 B-magnitudes per square arcsecond. For most observers, this results in a size which is considerably beyond the limits of visual perception, even in a large telescope under excellent sky conditions.

Most of the dimensions shown here have been obtained by visual inspection of the DSS, and therefore may differ from those found in other catalogs and literature. But the sizes presented here, in our experience, provide a closer match to what is perceptible using amateur equipment (for both visual observation and imaging).

Surface Brightness (SB)

As a broad generalization, surface brightness is calculated by dividing a galaxy's magnitude by its area. More precisely, once the total V magnitude and dimensions are known, the mean surface brightness is computed as follows:

$$V'_{25} = V_T + \Delta V + 5\log D - 2.5\log\left(\frac{D}{d}\right) - 0.26$$

where

V_T = total V magnitude
ΔV = 0.25 or 0.33 for cD or E type galaxies (T = –6 to 4),
ΔV = 0.13 or 0.16 for $S0^-$, S0, $S0^+$ type galaxies (T = –3 to –1),
ΔV = 0.11 for all other galaxy types (T $\geq$ 0),
D = major axis in arcminutes, and
d = minor axis in arcminutes.

The value of ΔV is dependent upon the morphology of the galaxy and the magnitude isophote to which the dimensions are measured. For about 30% of the galaxies in the DSFG, our measured dimensions match those of the RC3, which indicates an isophote of 25.0 mag/arcsec^2. For those galaxies, the lower ΔV values (0.25 and 0.13) were used in the surface brightness computation. For the remainder, it was determined that our average isophote was 24.5 for elliptical galaxies, 24.7 for S0 types, and 25.0 for spirals. This corresponds to ΔV values of 0.33, 0.16 and 0.11, respectively.

The total magnitude in combination with the surface brightness provides a better indication of the visibility of a galaxy than the magnitude alone. For example, UGC 1378 (Chart 2) has a surface brightness of 14.7, meaning that it will appear as if each square arcminute of its area is as bright as a 14.7 magnitude star. (Try defocusing a 14th magnitude star until it is an arcminute across to get an idea of how faint this is!). Therefore, this galaxy will be more difficult to observe than its total magnitude (V = 12.6) might indicate. Experience suggests that objects with surface brightnesses fainter than 14.5 will be difficult to detect no matter what the total magnitude is. The "average" galaxy has a surface brightness of about 13.5, while the highest surface brightness objects have values of 12.5 or brighter.

The examples of the Andromeda Galaxy and its companions (Chart 30) serve to show that surface brightness does not depend on total magnitude. The main galaxy, M31, is a naked-eye object but has a surface brightness of 13.5, typical of mid-stage spirals. The small companion, NGC 221 = M32, is nearly five magnitudes fainter overall yet is visible in handheld binoculars because its surface brightness, 12.5, is so high. The companion NGC 205, although about the same total magnitude as M32, is much more difficult to see at a mean surface brightness of 14.0. The more distant companions NGC 185 and NGC 147 have still lower surface brightnesses, and despite being among the brightest galaxies in the sky by total magnitude, are often elusive to those seeking them from light-polluted observing sites.

Finally, we recommend that the surface brightness values presented here be regarded as qualitative indicators of relative brightness. The RC3 cites an average error of 0.33 for its surface brightness values, and we believe that our average error is somewhat greater than that. We have not normalized the density of the DSS images to achieve uniform isophotal dimensions, and many of the magnitudes used are of an unknown type. But even assuming perfect data, there can be structural variations within the same morphological type which will affect the visibility of a particular galaxy.

Position Angles (PA)

Position angles (PA) are given for noticeably elongated galaxies and were obtained directly from a visual inspection of the DSS. Obviously, position angles do not apply to circular objects. These angles, expressed in degrees, are measured from north through east. A value of 0° implies a north-south elongation. A PA of 90° indicates an east-west elongation. After a further quarter-turn counterclockwise, the object is again orientated in a PA of 180° (equivalent to 0°) again in a north-south direction. All PAs for galaxies are referred to the first half of the circle, so they do not exceed 179°.

To measure the PA of a galaxy in your eyepiece, you must first determine north as precisely as possible. The best method of doing so, especially with telescopes on altazimuth mounts (Dobsonians, etc.), is to first determine west. This is easily done by centering a bright star in the eyepiece and removing all drives to the telescope so that the star drifts to the edge of the field. The point where the star *exits* the field is *west*. North will then be 90° *counterclockwise* from the west point (*clockwise* for refractors and Schmidt-Cassegrain telescopes with star diagonals).

Knowing the position angle of a galaxy with respect to others can obviously be very helpful in identifying it in a crowded field. In a similar way, determining the PA of elongation for an "unknown" galaxy aids in identifying it later through catalog descriptions or on sky survey photographs.

DDO Luminosity Class

The range of intrinsic brightness among spiral and irregular galaxies is greater than a factor of 100 (five magnitudes). In 1960 Sidney van den Bergh, then at the University of Toronto's David Dunlap Observatory (DDO), introduced *luminosity classes* for these galaxies. This classification scheme employs the same capital Roman numerals that are used for the stellar luminosity classes: class I spirals are the bright supergiants; luminosity class V are the faint dwarfs; and classes II, III, and IV are objects of intermediate brightness.

Table 3
Examples of the DDO Luminosity Classes

DDO Class	Name	Chart No.	Mag V	Dimen.	SB	Type
I	NGC 4565	71		15.8 × 2.1		SA(s)b? sp
I–II	NGC 4725	71		5.4 × 4.3		SAB(r)ab pec
II	NGC 5012	71		2.9 × 1.7		SAB(rs)c
II–III	NGC 4826	71		10.0 × 5.4		(R)SA(rs)ab
III	NGC 4793	71		2.8 × 1.5		SAB(rs)c
III–IV	NGC 4961	71		1.6 × 1.1		SB(s)cd
IV	NGC 4455	72		2.8 × 0.8		SB(s)d? sp
IV–V	NGC 4561	71		1.5 × 1.3		SB(rs)dm
V	UGC 7673	72		1.8 × 1.5		Im

Since the above luminosity classes are assigned on the strength of the spiral arm pattern, they cannot be applied to elliptical or lenticular galaxies. Spirals with strong, well-formed arms are assigned luminosity class I. Those with weak but still traceable arms are denominated class III, and those with little or no trace of spiral structure are called class V. In practice, a trained observer can evaluate a galaxy to within half a step on this scale, so the intermediate classes I–II, II–III, III–IV, and IV–V are used as well.

There is also a loose correlation between luminosity class and surface brightness: as luminosity class increases, so does surface brightness. Most class I spirals have high surface brightnesses, while those of most dwarf galaxies of class V are low. This fact led Harold Corwin to introduce luminosity classes V-VI and VI for the extremely low-surface-brightness dwarf galaxies being found on the high-contrast, deeply-exposed IIIa-J plates used for the *ESO/SERC Southern Sky Survey* and the *Second Palomar Observatory Sky Survey*. It is not yet known, however, if these new classes correlate with absolute magnitude as do the "traditional" DDO luminosity classes. In any case, it is very doubtful that any of these low surface brightness dwarf systems can be seen visually in even large telescopes. Their average surface brightnesses are below 16th magnitude per square arcminute.

Examples of galaxies of varying luminosity classes are given in Table 3, and all are found on adjoining charts. Examining each of the galaxies on this list should demonstrate to the observer the ease in seeing those of classes I and II, and the difficulty in seeing class IV-V and V objects under nearly identical conditions.

Notes

The galaxy notes will very often identify nearby field stars, to help pinpoint the position of elusive targets, or they might point to other nearby nonstellar objects not plotted on the charts, yet seen on DSS images. These non-plotted, nonstellar objects are usually far fainter than those in the normal database; call them "challenge objects" if you will. The nonstellar objects in the notes whose names are printed in bold type are not plotted on the charts (common names and Messier names excepted), nor do they have data in the tables; however, direction and distance to them is provided in the notes. Nonstellar objects whose names are in normal type are found on the charts and appear in the tables.

Galaxy Clusters

The DSFG contains 671 galaxy clusters derived from George Abell's 1958 catalog based upon photographs from Palomar Mountain. In 1980, Abell issued a revision to this catalog, and in 1984 he died while engaged on a southern extension of it using photographs taken at the Anglo-Australian Observatory. This work was expanded upon and subsequently completed in 1989 by Harold Corwin and Ronald Olowin. The tables provided the following data:

- Name. Objects from the 1958 catalog are designated with an "A", and those from the 1989 work are designated with an "AS" prefix.
- Magnitude of 10th brightest galaxy.
- Number of galaxies.
- Diameters. Those for "A" objects are taken directly from Abell's revised catalog (1980). Similar data was not available for the "AS" clusters so the diameters were estimated based upon the distance class. For all "A" catalog objects having the same distance class, mean diameters were computed, and those values were applied to the "AS" clusters. Therefore these dimensions are indicative only, not precisely measured.

Open Clusters

The DSFG contains 1,617 objects that have been classified as open clusters, including those in the Magellanic Clouds. As will be explained later, not all of these objects are true open clusters, which are relatively young star associations scattered throughout the disk of the Galaxy. A familiar example is the Pleiades, found on Chart 78. These objects are also known as Galactic clusters, since all but those nearest the Earth are found near the plane of our Galaxy (Chart 8 shows twenty-seven).

In the very youngest cases, the remnants of the gas from which they formed are visible as nebulae in the surrounding field. In fact, a small number (78) of open clusters share a single designation with their nebulae. An example is NGC 1962 on Chart 212, which is listed as both an open cluster and a bright nebula in the respective sections, has two separate symbols on the chart, and is listed twice in the Index.

The number of stars in these objects ranges from only a few to thousands. In appearance they vary from rich, compact, easily distinguished objects, to a few stars so loosely grouped that they are hardly discernible from the surrounding star field. Many of the latter require experimentation with different apertures and magnifications to be seen. Small telescope users will probably be able to view a larger percentage of open clusters than other types of deep-sky objects.

The principal source used for data on open clusters is the book *Star Clusters* by Brent Archinal and Steven Hynes, in press, to be published by Willmann-Bell. The DSFG employs a subset of that data which was determined by a review of the entire catalog by direct inspection of the DSS images by Cragin. His criteria were general visibility and the likelihood that the object could be seen by amateurs. In addition to the cluster name, five other items of data are presented:

- Position. Archinal and Skiff directly inspected DSS images for each object taking as their starting position that specified in the primary source catalog. This work included, as necessary, identification of faint or sparse clusters or those with obviously wrong positions, using original source material. All positions are visual estimates from the DSS and are not based upon center-of-mass calculations. In all instances these should be sufficient for DSFG users to locate and identify the object.
- Integrated magnitude. Unless otherwise noted, these are V (visual) magnitudes. For Galactic objects, these values have generally been determined by Skiff, who summed the brightnesses of individual stars in clusters with published photometry. Their uncertainties are typically ±0.2 magnitude. A colon indicates a particularly uncertain value, nearly always a faint limit, due to there being photometry for only a handful of the cluster stars. Photographic blue magnitudes, coded with a "p" following the value, derive ultimately from Collinder's 1931 monograph on star clusters. A comparison with Skiff's magnitudes indicates the Collinder values are on average about half a magnitude fainter and scatter by one-half to two-thirds of a magnitude about the mean difference.

  For clusters with no magnitude determined by Skiff, earlier, much less accurate (usually photographically determined) magnitudes are given. These and other values given (from SIMBAD) should generally be assumed to be only rough values good to within 0.5 to 1.0 magnitude.

  In the Magellanic Clouds, V magnitudes have been determined via photoelectric photometry and are accurate to perhaps ±0.1.
- Diameter in arcminutes. Since these values for cluster diameters are usually approximate, and are derived from widely varying photographic materials, they should be considered rough estimations.
- An estimate of the star count within the cluster and magnitude of the brightest star. For most of the brighter objects, the star count is representative of what is visible in a modest aperture (~20 cm) telescope from a dark site. The value greatly underestimates the true population of the cluster except for the poorest ones. In other instances the count may just be the number of stars for which photometry or spectral types have been obtained.
- Type. True clusters are assemblages of stars that have a common ancestry as revealed by spectroscopy. Over time many objects cataloged as clusters have been proven to be just chance groupings or appearance groupings of unrelated stars. Many of these objects have NGC or IC designations and are well established in the literature. We have chosen to include them here. This column presents Archinal's findings as to the nature of these objects. See Table 4.

Table 4
Cluster Types

Type	Explanation
2 cl	Two open clusters ("Double Cluster".)
ass	Association (NGC 1869).
ast	Asterism (NGC 1869).
ast?	Probable asterism, but possibly a real cluster NGC 1802).
cl	Open cluster (NGC 1846).
cl?	Possible open cluster, not well differentiated from an asterism (NGC 2609).
cl??	Type unknown, but possibly an open cluster (NGC 7826).
clpn	Open cluster and planetary nebula (NGC 2818).
MW	Asterism, either a rich Milky Way star field or obvious patch of bright Milky Way (NGC 5045).
MWcl	MW type or possible real cluster (NGC 5120).
ptcl	A portion of a larger open cluster (NGC 1750 part of NGC 1746).

- Notes. These include descriptions based on a three-part system developed by Trumpler in 1930. The three elements are: relative degree of concentration and detachment from the surrounding field; range of brightness of the apparent members; and richness or population ("poor," less than 50 stars; "moderate," 50 to 100 stars; "rich," over 100 stars). Also included are comments provided by Archinal, denoted by (A), and Skiff (S), as a by-product of their efforts to locate and measure positions and sometimes sizes on the DSS. Their estimates of cluster diameters may not agree precisely with compiled catalog data because they are from a different source.

The degree of "detachment" will naturally depend on the density of the surrounding field. The visual impression can differ considerably from assessments made from sky survey photographs. It is not difficult to imagine, for instance, that the Hyades (Charts 78 and 97) would be practically invisible as a cluster were it situated along the Galactic plane in Cygnus. Observers will often have an easier time visually distinguishing an object from its field, compared to their professional counterparts examining a sky survey photograph, where a perfectly good open cluster can be utterly swamped by a dense field of background stars.

We have not seen the faintest star in any open cluster, so except for the most meager objects, the true range in brightness is always very wide. The brightness range is better defined in this classification system as the range of the brightest members and actually correlates with cluster *age*. Young clusters (less than a few tens of millions of years old) have steeply rising main sequences that are essentially vertical on the color-magnitude diagram. So the 10th brightest star, say, can be a couple of magnitudes fainter than the brightest. There will invariably be many more fainter stars than brighter ones, so the magnitude range will seem large. In an old cluster (over several hundred million years), there is usually a well-populated giant branch of about the same brightness as the brightest stars still on the main sequence. Thus, what one sees visually and photographically is a mass of similarly-bright stars, and so the range of magnitudes seems small. Both young and old clusters, however, have very faint, low-mass stars 10 or 15 magnitudes fainter than the brightest, so the real range is always large. Nevertheless, this type of classification is useful for observers.

The "richness" element should be thought of as indicative of how many stars there are in the brightest few magnitudes of a cluster. In reality, few objects have less than 100 stars, provided that very faint members are taken into account.

Globular Clusters

The DSFG contains data on 170 globular clusters, including both Milky Way and Magellanic Cloud objects. While open clusters have irregular outlines, globulars are strongly circular, and their stars are usually much more numerous, and densely packed toward the center. They are among the oldest objects to be found in galaxies, with ages near 13 billion years. The number of stars comprising them can range from 10,000 to over 100,000. While open clusters are found strictly in the Galactic plane, globulars are grouped into two regions: around the "hub" at the center of our Milky Way, and in its extended Galactic halo. All are gravitationally bound to our Galaxy even though a few are twice as far from its center as the Magellanic Clouds! Globulars are most numerous in the direction of the Galactic center (Chart 146 shows 19 of them), although many of these are greatly dimmed by interstellar dust (the non-NGC globulars listed for Chart 145 are examples).

The globular cluster data presented in this work contains the following information:

- Positions. These are from the precise measures by Shawl and White. Magellanic Cloud cluster positions are from lists by Kontizas *et al*, Welch, and the Lauberts *ESO/Uppsala Survey of the ESO (B) Atlas*.
- Integrated V (visual) magnitudes. The compilation by Webbink is the source for these, except in the case of Magellanic Cloud objects, which come from van den Bergh's 1981 summary and the work of Bica et al.
- Diameters. Since these values—given in arcminutes—are usually approximate, and are derived from widely varying photographic materials, they should be considered rough estimations.
- Concentration class. Such groupings were first assigned by Shapley and Sawyer. They assigned globulars to 12 numerical classes based on decreasing central concentration, with a value of 1 denoting the highest stellar concentration toward the center, and 12, almost no central concentration. The latter variety appears similar to faint versions of the richest open clusters. NGC 2808, found on Chart 210, is an example of a high (1) concentration, while IC 1276 on Chart 126 has the lowest (12) concentration classification.

  The Shapley-Sawyer concentration classes are correlated with central surface brightness. They are *not* generally indicative of how easily a cluster can be resolved at the eyepiece. This is instead dependent almost solely on the magnitude of the brightest cluster stars compared to the limiting magnitude of the telescope in use.
- Estimates of the V magnitude of the brightest cluster stars and the magnitude of the "horizontal branch or HB" of the cluster's color-magnitude diagram. These are useful for both visual and photographic observing. To resolve a cluster into even a few stars requires that your telescope be able to reach at least the brightest members. The Shapley-Sawyer concentration class may play a small role (but only the most concentrated clusters will be more difficult to resolve), and the visual magnitude limit must be beyond these brightest stars in any case. As one goes to fainter limits, more and more stars will become resolved. The number of stars per magnitude interval increases slowly at first in globulars but takes a sudden leap upward at the magnitude of the horizontal branch, where there are many stars at nearly the same brightness in the cluster. Visually, the cluster changes in appearance from being "partially resolved" to "well resolved" if you can see stars this faint in the telescope.

  Descriptive notes for LMC globulars are based on their appearance in the plates of the Hodge and Wright atlas.

Star Clouds

The DSFG includes 14 star clouds. These objects are composed of countless stars and often span vast stretches of the night sky. When viewed at low power the nebulosity is often overlaid with sparkling stars which often do not show up in photographs. Usually star clouds are so vast that they cannot be plotted on a large-scale atlas however Chart 164 plots 7: NGC 6360, 6415, 6421, 6437, 6455, 6476 and 6480. Another, M 24 on Chart 145 is a visually striking star cloud that contains a number of other observable objects. Because there are so few of these plotted objects we have not included them in the chart legend but have assigned a special symbol and annotated the name with "(star cloud)" to aid in their identification. The symbol sizing is the same as that used for open clusters.

Bright Nebulae

The DSFG includes 377 bright nebulae. These objects, also called diffuse or Galactic nebulae, occur in two main classes depending on their source of illumination: emission or reflection (although they occasionally are a combination of the two). Much less common is a third type of bright nebula, the supernova remnant (SNR).

Emission nebulae are clouds of dust and glowing hydrogen gas, sometimes referred to as H II ("H-two") regions. The atoms in the cloud are ionized by nearby hot stars, and when their excited electrons fall back to their previous energy state the process releases energy in the form of visible light. A well-known example is the Lagoon

Nebula, found on Charts 145 and 146. Since most of the light visible from emission objects comes from just the three lines produced by hydrogen (Hβ at 4861Å) and oxygen ([OIII] at 5007Å and 4959Å), the use of narrowband "nebula" filters can be helpful in viewing them.

Reflection nebulae have the same composition as emission objects but lack stars sufficiently hot to cause the gas comprising them to fluoresce; therefore, they shine merely by the dust in the nebulae scattering starlight (the gas does not actually reflect any light). Good examples are the nebulosities surrounding the Pleiades star cluster (Chart 78) and M78 in Orion (Chart 116). Because these objects scatter starlight of all colors, filters are not generally helpful in viewing them.

A supernova remnant is the remains of a catastrophic stellar explosion, wherein much of a star's material is ejected, often as a highly structured cloud. Examples are the Veil Nebula in Cygnus (Chart 47) and the Crab Nebula (M1) in Taurus (Chart 77). These objects have strong emission lines similar to H II regions, hence also benefit from the application of nebula filters.

For bright nebulae the following information is provided herein:

- Dimensions. These are given in arcminutes and were taken from photographs. They represent the size of the nebula's brightest parts. Nevertheless, observers will often find these values to be considerably larger than what they can detect visually, even from dark sites.
- Type. This is denoted by the letters E (emission), R (reflection), E+R, and SNR (supernova remnant). As just explained, this information is useful in deciding whether a filter will be helpful.
- Brightness. Objects are assigned to a brightness category, which is a numerical value describing an object's brightness on the POSS. An arbitrary scale was used by Beverly Lynds in her catalogue of northern nebulae and ranges from 1 to 6, where 1 is brightest and 6 is barely detectable. Brightness is almost always expressed as a range in the tables. Provided sufficient aperture is used under good skies, experience suggests most category 1 and 2 emission nebulae are within visual range. Category 3 objects will prove difficult, while those rated 4 and above are likely to be beyond visual range and thus candidates for photographers instead. When compared on this same scale, reflection nebulae are generally considered to be more difficult to observe visually. Blue light is scattered more than red light, so the reflection nebulae are more often than not comparatively blue in color. But the light is not concentrated in just a few emission lines, so filters do not help much.
- Nebular color This is presented as a numerical value from 1 to 4, again based on Lynds' system from her inspection of the blue- and red-sensitive POSS plates. It is usually expressed as a range where

  1 = brightest on blue plates
  2 = equally prominent on both blue and red plates
  3 = brightest on red plates
  4 = visible only on red plates.

  Nearly all the class 1 objects are reflection nebulae, since the scattering process makes them slightly bluer than the hot, blue stars that illuminate them. A notable exception is IC 4606 on Chart 147, which surrounds Antares: although it is a "reflection" object, it scatters the light of a red supergiant star. Class 2 objects are a mixture of emission and reflection, or more commonly, are generally overexposed emission objects. Nearly all class 3 nebulae will be emission objects. Class 4 includes weak or obscured emission objects.
- Notes. This explanatory material is largely based on photographic appearances of objects on the POSS, and the *ESO/SERC Atla*s reproductions in the *Atlas of Galactic Nebulae* by Neckel and Vehrenberg.

Dark Nebulae

The DSFG contains 367 dark nebulae. These objects are composed of clouds of gas and dust grains sufficiently dense as to become opaque. They are detectable if positioned between the observer and a bright nebula or a very dense star field, against which they will be seen silhouetted. Observationally, dark nebulae are among the most difficult deep-sky objects. The Coalsack, near the Southern Cross (Charts 198 and 209), is an example clearly visible to the unaided eye because it happens to lie in front of the rich, starry background of the Milky Way. On the other hand, the well-known Horsehead Nebula (cataloged as Barnard 33 on Chart 116) is much more typical. It is a difficult object because of its small size and because it lies in front of a rather faint background object, emission nebula IC 434. In this case the resulting contrast between the dark nebula and background is very low, making it difficult to distinguish. Dark nebulae are usually seen best with wide-field, low-power eyepieces, which provide a large, bright surrounding field. While these objects emit no light of their own, nebular filters may help increase the contrast if the background is an emission nebula.

The dark nebulae are presented here with the following information:

- Dimensions. These are given in arcminutes and were derived from photographs. As with bright nebulae, many of these objects have indistinct boundaries, and therefore the values provided should be considered approximate.
- Contrast. This is measured on a numerical scale of 1 to 6, where a value of 1 indicates the object is barely darker than the surrounding background, while those rated at 6 are the blackest and the most easily visible. Lynds found that the values corresponded fairly well with the number of magnitudes of absorption of background starlight by the nebula, so that a contrast 6 object dims starlight by about six magnitudes.

- Notes. These are primarily based on Barnard's "Catalogue of 349 Dark Objects in the Sky," found in *A Photographic Atlas of Selected Regions of the Milky Way*. Barnard's descriptions have been used (when available) because he often cataloged and described small areas within much larger complexes, while the Lynds *Catalogue of Dark Nebulae* often groups many of these smaller areas into a single, larger object. Where an object has a Lynds (LDN) designation, we provide Barnard's more detailed description, if available.

Planetary Nebulae

The DSFG encompasses 1,144 planetary nebulae. These are shells of gas thrown off by stars having approximately the Sun's mass, that are nearing toward the end of their evolutionary cycle following the red-giant stage. The shell gradually expands, until after perhaps 100,000 years it becomes undetectably thin and all that remains is the central star. Some of the earliest such objects discovered (e.g., NGC 3242 on Chart 151, and NGC 6210 on Chart 68) were found by William Herschel, who noted the resemblance of their well-defined disks to those of planets, and gave them the name "planetary nebulae", but of course they have nothing to do with planets.

The brightest planetaries have a substantial disk, typically 30 arcseconds across, but the majority listed here are stellar (or nearly so) and can be identified at the eyepiece only with a nebula filter or direct-vision prism. Chart 164 lists over 100 such objects, hardly any of which are larger than a few arcseconds in diameter. A few of the oldest and nearest objects are so distended as to be practically invisible against the background sky. Between these two extremes, planetaries come in about as many shapes as there are objects, including the aptly-named Ring Nebula on Chart 49 and the complex southern object NGC 5189 on Chart 208.

In visible light, planetary nebulae shine predominantly at the two wavelengths emitted by doubly-ionized oxygen, denoted by the symbol [OIII]. Typically, 90% of the visually-detectable light comes from the [OIII] lines at 5007Å and 4959Å in the blue-green part of the spectrum. As a result, when an [OIII] filter is placed between the eye and the eyepiece, stars are dimmed by as much as three magnitudes (the night-sky light even more), while light from a planetary passes through virtually unchanged. By rapidly passing an [OIII] filter between the eye and eyepiece and out again while examining the field, a planetary will usually appear to "blink" as the stars are dimmed, and the nebula retains it brightness. There are a few planetaries that have weak [OIII] emission (a bright example is PN G64.7+5.0, Campbell's Hydrogen Star, on Chart 48), and these are best viewed either directly or with an Hβ filter, which passes the line that is brightest in these cases.

The planetary nebula data is based on a catalog prepared by Hynes (presented in his book *Planetary Nebulae*) and the *Strasbourg-ESO Catalogue of Galactic Planetary Nebulae* and contains the following information:

- Positions. These are taken from Gordon and Kapplan.
- Name. The original "PK" system is based on Galactic coordinates to whole degrees only and within each one-degree block planetary nebulae were assigned a serial number. Thus NGC 40 = PK 120+09 1 (no decimal) is in the Galactic longitude 120° block at +9° latitude, and is the first planetary cataloged in that block. Acker et al. decided that the general improvement in positions meant that it is was possible to assign a unique name based on Galactic coordinates if the positions were given to 0.1 degree precision. Thus PN G120.0+09.8 shows that the object is at 120.0° longitude and +9.8° latitude. Since the boxes are now only 6′ on a side they are small enough that every object can have a unique name and it is not anticipated that another naming scheme will be needed in the future. These names have been adopted by modern catalogers.
- Magnitudes. These are given in two columns: photographic-blue and visual. The visual magnitudes were computed and provided by Marling, who summed the brightness of the emission lines for each object and adjusted the resulting value to account for the sensitivity of the dark-adapted eye. The uncertainties are in the range of ±0.2 magnitude. These data are far superior to early photographic estimates in blue light, most of which date from the 1930s. Since the photographic-blue magnitude does not include the visually dominant [OIII] emission, it is usually too faint by one to three magnitudes compared with the appearance to the eye.
- Diameters. Given in arcseconds, these are mostly from measures on various photographic plates. Because they have not been reduced to any uniform system, these values do not refer to any fixed brightness level of the nebulosity and hence should be considered approximate only. They usually refer to the bright, well-defined portion of the object, excluding extremely faint coronae visible on images reaching to very low light levels.
- Central star magnitudes. These are, for the very brightest stars, independent of any associated nebulosity. Although based on the most recent work, these magnitudes may, however, be subject to substantial error due to the brightness of the enveloping nebula. Brighter than about 14th magnitude, they are quite reliable to better than 0.1; fainter than this, the majority are good to within ±0.5 magnitude. Since few of these stars are brighter than magnitude 12.0 and since they are associated with nebulosity, they are most difficult to observe. Good results can be achieved at high telescopic magnifications, which have the effect of reducing the apparent brightness of the surrounding nebulosity while accentuating the central star.
- Alternate names. These are provided for all planetary nebulae which, of all deep-sky objects, suffer the most from an overabundance of alternate or secondary names. In the DSFG, as with *Uranometria 2000.0*, the name precedence is as follows: *New General Cata-*

logue (NGC), *Index Catalogue* (IC), and PN G from the *Strasbourgh/ESO Catalogue of Planetary Nebulae*. All planetaries have a PN G number assigned. Where an object has either an NGC or IC number, the alternate name will be its PN G designation. For objects with a PN G number as a primary name, the PK number is usually given as an alternate. An exhaustive cross-reference may be found in Hynes' *Planetary Nebulae*. Note also that some planetaries have Abell names and these should not be confused with similarly named galaxy clusters. Examples of these can be found in the notes for Charts 8 and 26 among others.

- Notes. These comments primarily point out two or more stars close by each planetary, giving their magnitudes, distance and direction from the object, to aid the observer in pinpointing the exact location of what is often an almost stellar target; or, on the other hand, to a planetary that might be quite large, faint and diffuse. The notes may also contain brief descriptions, where warranted, based on DSS images. Also, nonstellar objects whose names appear in bold type are not plotted on the charts, nor do they have data in the tables; however, direction and distance to these objects will be provided in the notes. Nonstellar objects whose names are in normal type will be found on the charts and will have an entry in the tables.

Abbreviations and Notation

For the most part, we have sought to avoid codes and abbreviations and wherever possible to use plain English descriptions. The following are the codes and abbreviations as used in the DSFG:

Colon(s) : denotes uncertainty
Question mark ? doubtful
Arcminutes ′ or ′.
Arcseconds ″
Directions N = north, S = south, E = west, W = west
Greater than >
Less than <
Approximately ≈
Greater than or equal to ≥
Plus or minus ±

Table 5
The Greek Alphabet

Alpha	α	Iota	ι	Rho	ρ
Beta	β	Kappa	κ	Sigma	σ
Gamma	γ	Lambda	λ	Tau	τ
Delta	δ	Mu	μ	Upsilon	υ
Epsilon	ε	Nu	ν	Phi	ϕ
Zeta	ζ	Xi	ξ	Chi	χ
Eta	η	Omicron	ο	Psi	ψ
Theta	θ	Pi	π	Omega	ω

Acknowledgments

Almost exactly 400 years ago, in 1603, a Bavarian lawyer named Johann Bayer (1572–1625) published the first *Uranometria*, which had as its "database" the observations of the Danish astronomer Tycho Brahe (1546–1601). The work you now have in your hands, in contrast, is the product of the knowledge and talents of a very large group of people, from which it has benefited in many ways—both seen and unseen. It comes at a point in time when cartographers can create a massive and reasonably uniform database, and then actually confirm the existence, size and position of the objects they plot—truly a revolution in celestial mapping.

In just the last few years there have been remarkable developments in technology which have enabled anyone with Internet access and suitable software to access and analyze the massive NASA/IPAC Extragalactic Database (NED) which is maintained by the Jet Propulsion Laboratory, California Institute of Technology, for NASA. While this alone would be a major advance over what was available just 14 years ago when the First Edition of *Uranometria 2000.0* was published, the availability of the Digitized Sky Survey created by the Space Telescope Science Institute, and operated for NASA by the Association of Universities for Research in Astronomy (AURA), has enabled us to personally verify the positions, orientations and sizes of objects—to an unprecedented level of accuracy—against the actual images upon which much of the data contained in NED was originally based.

Brian Skiff of Lowell Observatory has, from the beginning of this project, provided us with assistance on a wide variety of issues. Harold Corwin of IPAC/NED has also, over a number of years, provided data and valuable guidance—especially for galaxies and galaxy clusters. He and Steve Gottlieb were helpful in unraveling some of the tangled history of the *New General Catalog of Non-stellar Objects* and the two *Index Catalogs*.

Brent Archinal has graciously allowed us to use a subset of his Star Cluster catalog which will appear in *Star Clusters*, a book that he and Steven Hynes have in-press at Willmann-Bell.

We have also received valuable assistance from George Kepple, David Kriege, John Koester, Harold Suiter, Susan French, Glen Sanner, Larry Mitchell, Craig Crossen, Alister Ling, Richard Huziak, Jay LeBlanc, Kent Blackwell, and Leos Ondra.

While we have gone to extraordinary lengths to create the best possible large-scale atlas of the deep sky, we have no illusions—errors will be found. Whatever these errors turn out to be they are the responsibility of the authors and publisher. However, to our knowledge the First Edition of *Uranometria 2000.0* remains the only atlas ever published that has provided systematic errata which have been made widely available. That tradition will be maintained with this edition and, hopefully, those who come after us will not have to repeat them.

Bibliography

To reduce congestion on the atlas charts, we have chosen to abbreviate prefixes and remove leading zeros. In the

DSFG, however, we have used the full prefixes. In the listings below, the abbreviations used in the chart appear in brackets after the standard prefix.

General Stellar and Nonstellar Objects

Burnham, R. Jr. *Burnham's Celestial Handbook.* New York, 1978: Dover Publications, Inc.

Delporte, E. *Délimitation Scientifique des Constellations.* Cambridge, 1930: Cambridge University Press.

Dreyer, J. L. E. *New General Catalogue of Nebulae and Clusters of Stars (1888), Index Catalogue (1895), Second Index Catalogue (1908).* London, 1962: Royal Astronomical Society.

ESA. *The Hipparcos and Tycho Catalogues.* ESA SP-1200. Noordwijk, 1997: European Space Agency.

Hirshfeld, A., and R. W. Sinnott. *Sky Catalogue 2000.0,* Vol. 2. *Double Stars, Variable Stars, and Nonstellar Objects.* Cambridge, MA, 1985: Sky Publishing Corp.

Kholopov P. N., N. N. Samus, M. S. Frolov, V. P. Goranskij, N. A. Gorynya, E. A. Karitskaya, E. V. Kazarovets, N. N. Kireeva, N. P. Kukarkina, N. E. Kurochkin, G. I. Medvedeva, E. N. Pastukhova, N. B. Perova, A. S. Rastorguev, S. Yu. Shugarov, *The Combined General Catalogue of Variable Stars*, 4.1 Edition, 1998: Centre de Données astronomiques de Strasbourg and Sternberg Astronomical Institute and Institute of Astronomy (Russian Acad. Sci.), Moscow.

Kukarkin B. V., P. N. Kholopov, N. M. Artiukhina, V. P. Fedorovich, M. S. Frolov, V. P. Goranskij, N. A. Gorynya, E. A. Karitskaya, N. N. Kireeva, N. P. Kukarkina, N. E. Kurochkin, G. I. Medvedeva, N. B. Perova, G. A. Ponomareva, N. N. Samus, S. Yu Shugarov. *New Catalogue of Suspected Variable Stars*, 1982: Institute of Astronomy of Russian Academy of Sciences and Sternberg Astronomical Institute.

Kazarovets E. V., O. V Durlevich, N. N. Samus, *New Catalogue of Suspected Variable Stars. Supplement*, Version 1.0 (1998): Institute of Astronomy of Russian Academy of Sciences and Sternberg Astronomical Institute.

Neckel, T., and H. Vehrenberg. *Atlas of Galactic Nebulae.* Düsseldorf, 1985, 1987, and 1990: Treugesell-Verlag, Dr. Vehrenberg KG.

Luginbuhl, C. B., and B. A. Skiff. *Observing Handbook and Catalogue of Deep-Sky Objects.* Cambridge, 1989: Cambridge University Press.

Sinnott, R. W., ed. *The Complete New General Catalogue and Index Catalogue of Nebulae and Star Clusters by J. L. E. Dreyer.* Cambridge, MA 1988: Sky Publishing Corp. and Cambridge University Press.

Sinnott, R. W., and M. A. C. Perryman, *Millennium Star Atlas*, Vols. 1, 2, and 3. 1997: Sky Publishing Corp and European Space Agency.

Bright Nebulae

Ced Cederblad, S. "Catalogue of Bright Diffuse Galactic Nebulae," *Meddelanden fran Lunds Astronomiska Observatorium.* Ser. 2, 119 (1946).

DWB Dickel H. R., H. Wendker, and J. H. Bieritz, "The Cygnus X Region, V. Catalogue and Distances of Optically Visible H II Regions," *Astron. & Astrophysics* 1 (1969): 270–280.

Gum Gum, C. S. "A Survey of Southern H II Regions." *Memoirs of the Royal Astronomical Society* 67, 21 (1955).

IRAS Joint IRAS Science Working Group. 1988. Infrared Astronomical Satellite (IRAS), Catalogs and Atlases. Washington, D.C.: U. S. Government Printing Office. NASA RP-1190, v. 2–6.

LBN Lynds, B. T. "Catalogue of Bright Nebulae," *Astrophysical Journal Supplement Series* 12 (1965): 163.

LH Lucke, P. B., and P. W. Hodge. "A Catalogue of Stellar Associations in the Large Magellanic Cloud." *Astronomical Journal* 75, 2 (1970): 171–175.

Mi Minkowski, R. "New Emission Nebulae." *Publications of the Astronomical Society of the Pacific* 58 (1946): 305.

NGC, IC [I] Dreyer, J. L. E. *New General Catalogue of Nebulae and Clusters of Stars (1888), Index Catalogue (1895), Second Index Catalogue (1908).* London, 1962: Royal Astronomical Society.

RCW Rogers, A. W., C. T. Campbell, and J. B. Whiteoak, "A Catalogue of Hα-Emission Regions in the Southern Milky Way." *Monthly Notices of the Royal Astronomical Society* 121, 103 (1960).

Sh2 Sharpless, S. "A Catalogue of H II Regions." *Astrophysical Journal* Supp. 4, 257–279 (1959). (Also available as U.S. Naval Observatory Reprint No. 7.)

vdB van den Bergh, S. "A Search for Cepheids in Galactic Clusters." *Astrophysical Journal* 126 (1957): 323.

———. "A Study of Reflection Nebulae," *Astronomical Journal* 71 (1966): 990.

———. "Photometric and Spectroscopic Observations of Globular Clusters in the Andromeda Nebula." *Astrophysical Journal* Supp. 171, (1969): 145–174.

vdBH van den Bergh, S., and W. Herbst. "Catalogue of Southern Stars Embedded in Nebulosity." *Astronomical Journal* 80 (1975): 212.

Dark Nebulae

B Barnard, E. E. "Catalogue of 349 Dark Objects in the Sky." In *A Photographic Atlas of Selected Regions of the Milky Way.* Washington, DC, 1927: Carnegie Institution.

Be Bernes, C. "A Catalogue of Bright Nebulosities in Opaque Dust Clouds." *Astronomy & Astrophysics Supplement Series* 29, 65 (1977).

LDN Lynds, B. T. "Catalogue of Dark Nebulae." *Astrophysical Journal Supplement Series* 7, 1 (1962).

Sa Sandqvist, Aa. "More Southern Dark Dust Clouds." *Astronomy & Astrophysics* 57, 467 (1977).

SL Sandqvist, Aa., and K.P. Lindroos. "Interstellar Formaldehyde in Southern Dark Dust Clouds." *Astronomy & Astrophysics* 53, 179 (1976).

Shapley, H., and E. Lindsay. "A Catalogue of Clusters in the Large Magellanic Cloud." *Irish Astron. Journal* 6 (1963): 74–91.

Galaxies

Arp, H. C. and B. F. Madore. *A Catalogue of Southern Peculiar Galaxies and Associations.* 2 vols. Cambridge, 1987: Cambridge University Press.

Corwin, H. G., and Skiff, B. A. A South-Equatorial Galaxy Catalogue, 1999; available at http://spider.ipac.caltech.edu/staff/hgcjr

de Vaucouleurs, G. *Handbuch der Physik* 53, 275. Berlin, 1959: Springer-Verlag.

Holmberg, E. *Meddelande Lund Astr. Obs.* Series 2, no. 136 (1958).

Sandage, A. R., and J. Bedke. *The Carnegie Atlas of Galaxies.* Washington, DC, 1994: Carnegie Institution, Publ. no. 638.

A Abell, G. O. "The Distribution of Rich Clusters of Galaxies." *Astrophysical Journal Supplement Series* 3 (1958): 211.

AS Abell, G. O., H. G. Corwin, Jr., and R. P. Olowin. "A Catalogue of Rich Clusters of Galaxies." *Astrophysical Journal Supplement Series* 70 (1989): 1.

CGCG [C] Zwicky, F., E. Herzog, and P. Wild. *Catalogue of Galaxies and Clusters of Galaxies.* Vol. 1. Pasadena, 1961: California Institute of Technology.

Zwicky, F., and E. Herzog. *Catalogue of Galaxies and Clusters of Galaxies.* Vol. 2. Pasadena, 1963: California Institute of Technology.

———. *Catalogue of Galaxies and Clusters of Galaxies.* Vol. 3. Pasadena, 1966: California Institute of Technology.

———. *Catalogue of Galaxies and Clusters of Galaxies.* Vol. 4. Pasadena, 1968: California Institute of Technology.

Zwicky, F., M. Karpowicz, and C. T. Kowal. *Catalogue of Galaxies and Clusters of Galaxies.* Vol. 5. Pasadena, 1965: California Institute of Technology.

Zwicky, F., and C. T. Kowal. *Catalogue of Galaxies and Clusters of Galaxies.* Vol. 6. Pasadena, 1968: California Institute of Technology.

D, DDO van den Bergh, S. "A Preliminary Luminosity Classification of Late-Type Galaxies." *Astrophysical Journal* 131 (1960): 215.

———. "A Preliminary Luminosity Classification for Galaxies of Type Sb." *Astrophysical Journal* 131 (1960): 558.

———. "Luminosity Classifications of Dwarf Galaxies." *Astronomical Journal* 71 (1966): 922.

———. "A Reclassification of the Northern Shapley-Ames Galaxies", *Publ. David Dunlap Observatory 2*, no. 6 (1960): 159.

ESO [E], LV — Lauberts, A., E. A. Valentijn. *The Surface Photometry Catalogue of the ESO/Uppsala Galaxies.* Garching bei München, 1989: European Southern Observatory.

ESO-B — Lauberts, A. *The ESO/Uppsala Survey of the ESO(B) Atlas.* Garching bei München, 1982: European Southern Observatory.

NGC, IC [I] — Dreyer, J. L. E. *New General Catalogue of Nebulae and Clusters of Stars (1888), Index Catalogue (1895), Second Index Catalogue (1908).* London, 1962: Royal Astronomical Society.

Maffei — Maffei, P. "Infrared object in the region of IC 1805." *Publications of the Astronomical Society of the Pacific* 80 (1968): 618–621.

MCG [M] — Vorontsov-Velyaminov, B. A., and A. A. Krasnogorskaja. *Morphological Catalogue of Galaxies.* Part 1. Moscow, 1962: Moscow University Press.

Vorontsov-Velyaminov, B. A., and V. P. Archipova. *Morphological Catalogue of Galaxies.* Part 2. Moscow, 1964: Moscow University Press.

———. *Morphological Catalogue of Galaxies.* Part 3. Moscow, 1963: Moscow University Press.

———. *Morphological Catalogue of Galaxies.* Part 4. Moscow, 1968: Moscow University Press.

———. *Morphological Catalogue of Galaxies.* Part 5. Moscow, 1974: Moscow University Press.

PGC — Paturel, G., P. Fouqué, L. Bottinelli, and L. Gouguenheim. *Catalogue of Principal Galaxies.* 3 vols. Monographies de la Base de Données Extragalactiques, no. 1. Saint-Genis Laval, 1989: Observatorie de Lyon.

RC1 — de Vaucouleurs, G., and A. de Vaucouleurs. *Reference Catalogue of Bright Galaxies.* Austin, TX, 1964: University of Texas Press.

RC2 — de Vaucouleurs, G., A. de Vaucouleurs, and H. G. Corwin. *Second Reference Catalogue of Bright Galaxies.* Austin, TX, 1976: University of Texas Press.

RC3 — de Vaucouleurs, G., A. de Vaucouleurs, H. G. Corwin, R. J. Buta, G. Paturel, and P. Fouqué. *Third Reference Catalogue of Bright Galaxies.* New York, 1991: Springer-Verlag.

SGC — Corwin, H. G., A. de Vaucouleurs, and G. de Vaucouleurs. *Southern Galaxy Catalogue.* Austin, TX, 1985: University of Texas Monographs in Astronomy No. 4.

UGC [U] — Nilson, P. N. *Uppsala General Catalogue of Galaxies.* Uppsala, 1973: Uppsala Astronomical Observatory.

UGCA — ———. *Catalogue of Selected Non-UGC Galaxies*, Uppsala, 1974: Uppsala Astronomical Observatory.

Globular Clusters

AM — Arp, H. C. and B. F. Madore. *A Catalogue of Southern Peculiar Galaxies and Associations.* 2 vols. Cambridge, 1987: Cambridge University Press.

Djorg — Djorgovski, S. "Discovery of three obscured globular clusters." Part 2, Letters to the Editor. *Astrophysical Journal* 317 (June 1, 1987): L13, L14.

ESO [E] — Lauberts, A. *The ESO/Uppsala Survey of the ESO(B) Atlas.* Garching bei München, 1982: European Southern Observatory.

H60b — Hodge, P. W. "Studies of the Large Magellanic Cloud. I. The red globular clusters." *Astrophysical Journal* 131 (1960): 351–357.

HP — Haute Provence Observatory. Original reference unknown. IAU nomenclature page suggests the authors are J. Dufay, E. Berthier, and Morignat.

Liller — Liller, W. "Searches for the Optical Counterparts of the X-ray Burst Sources MXB 1728–34 and MXB 1730–33." *Astrophysical Journal* 213 (1977): L21–23.

NGC, IC [I] — Dreyer, J. L. E. *New General Catalogue of Nebulae and Clusters of Stars (1888), Index Catalogue (1895), Second Index Catalogue (1908).* London, 1962: Royal Astronomical Society.

Pal — Abell, G. O. "Globular Clusters and Planetary Nebulae discovered on the National Geographic Society-Palomar Observatory Sky Survey." *Publications of the Astronomical Society of the Pacific* 67 (1955): 258–261.

Zwicky, F. "Galaxies-The Local Group." Carnegie Institution Yearbook 58 (1959): 60–61.

Pal, Arp — Arp, H., and S. van den Bergh. "A New Faint Globular Cluster." *Publications of the Astronomical Society of the Pacific* 72 (1960): 424, 48 and plate 1.

SSWZ94 — Saurer, W., R. Seeberger, and R. Weinberger. "Berkeley 93: A Distant Star Cluster Nestled in a Dust Cloud." *Astronomical Journal* 107, no. 6 (1994): 2101–2107.

Terzan — King, I. R. "The Identity and Aliases of the Stellar System Terzan 5." *Astron. & Astrophysics* 19 (1972): 166.

Terzan, A. "Un Nouvel Amas Globulaire dans la Region du Centre de la Voie Lactee." *C.R. Acad. Sci. Ser. B* 263 (July 11, 1966): 221–222.

———. "Un Nouvel Amas Globularie dans la Region Centrale de la Galaxie." *C.R. Acad. Sci. Ser. B* 265 (Sept. 25, 1967): 734–736.

———. "Six Nouveaux Amas Stellaires (Terzan 3–8) dans la Region du Centre de la Voie Lactee et les Constellations du Scorpion et du Sagittaire." *C.R. Acad. Sci. Ser. B* 267 (Nov. 25, 1968): 1245–1248.

———. "Quatre Nouveaux Amas Stellaires dans la Direction de la Région Centrale de la Galaxie." *Astron. & Astrophysics* 12 (1971): 477–481.

———. "Erratum Four New Star Clusters in the Direction of the Central Area of the Galaxy." *Astron. & Astrophysics* 15 (1971): 336.

Ton — Pismis, P. "Nuevos Cumulos Estelares En Regiones Del Sur." *Boletin de Los Observatorios Tonantzintla y Tacubaya* 18 (Aug. 1959): 37–38.

UKS — Malkan, M., D. Kleinmann, and J. Apt. "Infrared Studies of Globular Clusters Near the Galactic Center." *Astrophysical Journal* 237 (1959): 432–437.

Open Clusters

Hodge, P. W., F. W. Wright. *The Large Magellanic Cloud.* Washington, DC, 1967: Smithsonian Press.

———. *The Small Magellanic Cloud.* Seattle, 1977: University of Washington Press.

Morel, M. *A Visual Atlas of the Large Magellanic Cloud*, privately published, 1983.

———. *A Visual Atlas of the Small Magellanic Cloud*, privately published, 1989.

———. *LMC Selected Areas*, privately published, 1990.

Ruprecht, J., B. Baláz, R. E. White. *Catalogue of Star Clusters and Associations.* Supplement 1, Part A (Introduction), Part B1 (New Data for Open Clusters), Part B2 (New Data for Associations, Globular Clusters and Extragalactic Objects). Budapest, 1981: Akadémiai Kiadó.

ADS — Aitken, R. G. *New General Catalogue of Double Stars Within 120° of the North Pole.* 2 vols. Washington, DC, 1932: Carnegie Inst. Publ. No. 417.

AL — Andrews, A. D., and E. M. Lindsay. "New Southern Clusters and Nebulous Ovals." *Irish Astron. Journal* 8 (1967): 126–127, plates 7 and 8.

AM — Arp, H. C. and B. F. Madore. *A Catalogue of Southern Peculiar Galaxies and Associations.* 2 vols. Cambridge, 1987: Cambridge University Press.

Ant — Antalová, A. "UBV Photographic Photometry of Stars in the Region AR1950: 17h03m–17h41m Decl1950: −28.8° to −33.4° III. The Catalogue and Identification Maps of Open Star Clusters: NGC 6405, NGC 6383, NGC 6374, Av 2, NGC 6416 and Hα Emission Regions: Gum 67 (Av 3), Gum 68 (Av 2 [sic, Av 1])." *Bulletin of the Astronomical Society of Czechoslovakia* 23, 2 (1972): 126–139.

Archinal — Archinal, Brent A. *The "Non-Existent" Star Clusters of the RNGC.* Portsmouth, England, 1993: The Webb Society.

Auner — Auner, G., J. Dengel, H. Hartl, and R. Weinberger. "A Ghost Image of Sirius as a Hiding Place for a New Star Cluster." *Publications of the Astronomical Society of the Pacific* 92 (1980): 422–425.

Bar — Barkhatova. "Ucenye zapiski Ural'skogo Gosudarstvennogo Universiteta." No. 22, 1958. Obtained via loan from the University of Arizona library. Photocopy now available in the U. S. Naval Observatory library.

Bas — Becker, W. and R. Fenkart. "A Catalogue of Galactic Star Clusters Observed in Three Colours." *Astron. & Astrophysics Supplement Series* 4 (1971): 241–252.

Be — Setteducati, A. F. and H. F. Weaver. "Newly Found Star Clusters." Berkeley, CA, 1960: Radio Astronomy Laboratory. (We have been unable to locate a copy of this publication. It is cited in the Lund Catalogue [with a publication date of 1962] and Ruprecht et al. [1981] (with a publication date of 1960, assumed here).

Bergeron — Archinal, B., S. J. Hynes. *Star Clusters.* In press, Willmann-Bell, Inc.

Bi — Biurakan Observatory, outside Yerevan in Armenia. The original reference is unknown.

Blanco — Blanco, V. M. "A New Galactic Star Cluster in Sculptor." *Publications of the Astronomical Society of the Pacific* 61 (1949): 183–184.

Bo — Moffat, A. F. J., and N. Vogt. "Southern Open Star Clusters IV. UBV-Hβ Photometry of 26 Clusters from Monoceros to Vela." *Astron. & Astrophysics Suppl. Ser.* 20 (1975): 85–124.

———. "Southern Open Star Clusters IV. UBV-Hβ Photometry of 20 Clusters in Carina." *Astron. & Astrophysics Suppl. Ser.* 20 (1975): 125–153.

———. "Southern Open Star Clusters VI. UBV-Hβ Photometry of 18 Clusters from Centaurus to Sagittarius." *Astron. & Astrophysics Supplement Series* 20 (1975): 155–182.

Canali — Canali, E. "An Interesting Asterism in Corvus." *The Guide Star: Newsletter of the Amateur Astronomers Association of Pittsburgh, Inc.* 31, 2 (May 1997): 7–8.

ClvdB — van den Bergh, S. "A Search for Cepheids in Galactic Clusters." *Astrophysical Journal* 126 (1957): 323.

———. " A Study of Reflection Nebulae." *Astronomical Journal* 71, 10 (Dec. 1966): 990–998.

———. "Photometric and Spectroscopic Observations of Globular Clusters in the Andromeda Nebula." *Astrophysical Journal Supp.* 171, (1969): 145–174.

Cr — Collinder, Per. On Structural Properties of Open Galactic Clusters and their Spatial Distribution. Dissertation. Annals of the Observatory of Lund, no. 2. 1931.

Cz, Ru — Alter, G., H. S. Hogg, and J. Ruprecht. "Catalogue of Star Clusters and Associations." Supp. 3. *Bulletin of the Astronomical Institutes of Czechoslovakia* 12, (1961): 2–3.

———. "Catalogue of Star Clusters and Associations." Supp. 8. *Bulletin of the Astronomical Institutes of Czechoslovakia* 17, no. 1 (1966).

Danks — Danks, A., M. Dennefield, W. Wamsteker, and P. Shaver. "Near Infrared Spectroscopy and Infrared Photometry of a New WC9 Star." *Astron. & Astrophysics* 118 (1983): 301–305.

Do — Dolidze, M. V. "On the Star Cluster Near γ Cyg." *Astr. Cirk.* 223 (1961): 11–12.

———. "Some Data of the Nebulae and Star Clusters." *Astr. Cirk.* 224 (1961): 18–22.

DoDz — Dolidze, M. V. and G. Dzimselejsvili [or Jimsheleishvili]. "New Possible Star Clusters." *Astr. Cirk.* 382 (1966): 7–8.

ESO [E] — Lauberts, A. *The ESO/Uppsala Survey of the ESO(B) Atlas.* Garching bei München, 1982: European Southern Observatory.

Fei — Feinstein, A. "A Group of Stars Around the Helium star HD 96446." *The Observatory* 84 (June 1964): 111–117.

French — French, S. C. "The Dolphin's Deep-Sky Delights." *Sky & Telescope* 93, 12 (June 1997): 106–108.

Graham — Graham, J. "The Space Distribution of the OB Stars in Car." *Astronomical Journal* 75 (1970): 703–717.

h — Dreyer, J. L. E. *The Scientific Papers of Sir William Herschel.* 2 vols. Ed. by a joint committee (mostly Dreyer) of the R.S. and R.A.S. London: Royal Society and Royal Astronomical Society. 1912.

Ha — Shapley, Harlow. *Star Clusters.* Harvard Observatory Monographs no. 2. New York, 1930: McGraw-Hill Book Co.

Haf — Haffner, H. "Neue Galaktische Sternhaufen in der Sudlichen Milchstrasse." *Z. Astrophys.* 43 (1957): 89–94.

HM — Havlen, R., and A. Moffat. "A New Cluster Containing 2 Wolf-Rayet Stars and 2 Of Stars." *Astron. & Astrophysics* 58 (1977): 351–356.

Ho — Hogg, A. R. "Catalogue of Open Clusters South of –45° Declination." *Memoirs of the Mount Stromlo Observatory* 4, no. 17 (1965).

HS66 — Hodge, P., and J. Sexton. "457 new star clusters of the Large Magellanic Cloud." *Astronomical Journal* 71 (1966): 363–368.

Kemble — Houston, W. C. "Deep-Sky Wonders." *Sky & Telescope* 60, 6 (December 1980): 546–547.

King — King, I. R. "Some New Galactic Clusters." *Harvard Bull.* 919 (1949): 41–42.

———. "A New Galactic Cluster." *Publications of the Astronomical Society of the Pacific* 73, 431 (1961): 163–164.

———. "Five New Open Clusters." *Publications of the Astronomical Society of the Pacific* 78, 460 (1966): 81–82.

———. "The Identity and Aliases of the Stellar System Terzan 5." *Astron. & Astrophysics* 19 (1972): 166.

KMHK — Kontizas, M., D. H. Morgan, D. Hatzidimitriou, and E. Kontizas. "The Cluster System of the Large Magellanic Cloud." *Astron. & Astrophysics Supplement Series* 84 (1990): 527.

Lederman — Lederman, R. D. and S. Gottlieb. "If You Don't See it ... Maybe it Isn't There." *Webb Society Quarterly Journal* 105 (July 1996): 25–31.

Lind — Lindsay, E. M. "The Cluster System of the Small Magellanic Cloud." *Monthly Notices of the Royal Astronomical Society* 118 (1958): 172–182.

Lo — Lodén, L. O. "A List of Suspected Clusters in the Southern Milky Way." *Astron. & Astrophysics Supplement Series* 10 (1973): 125–133.

———. "A Study of Some Loose Clustering in the Southern Milky Way." *Astron. & Astrophysics Supplement Series* 29 (1977): 131–50, plates 1 to 9.

———. "A Photometric Study of Two Stellar Clusterings in the Southern Milky Way (and a General Consideration on Previous and Present Data Concerning Galactic Clusterings)." *Astron. & Astrophysics Supplement Series* 44 (1977): 155–158.

———. "Photometry of Loose Clusterings in the Southern Milky Way." *Astron. & Astrophysics Supplement Series* 36 (1979): 83–93.

———. "Continued Studies of Loose Clusterings in the Southern Milky Way." *Astron. & Astrophysics Supplement Series* 38 (1979): 355–365.

———. "Concluding Observations of Loose Clusterings in the Southern Milky Way." *Astron. & Astrophysics Supplement Series* 41 (1980): 173–181.

Lynga — Lyngå, G. "Studies of the Milky Way from Centaurus to Norma. II. Open Clusters." *Medd. Lunds Ser.* 2, 140 (1964).

M — Messier, C. In the *Connaissance des Temps* for 1784. Paris, 1781.

Mayer — Lyngå, G. *Catalogue of Open Cluster Data.* 5th ed. 1987: Lund Observatory, Lund.

Mel — Melotte, P. J. "A Catalogue of Star Clusters Shown on Franklin-Adams Chart Plates." *Memoirs of the Royal Astronomical Society* 60 (1915): 175.

Moffat — Moffat, A. F. J. "Mass Loss from the M3 Supergiant HD 143183 in a Young Compact Star Cluster in Nor." *Astron. & Astrophysics* 50 (1976): 429–434.

Mrk — Markarian, B. E. "On the Classification of Open (Galactic) Stellar Clusters. II. Preliminary List of Open O-type Star Clusters." Soob. Biurakan Obs. no. 9, 1951.

NGC, IC [I] — ———, *New General Catalogue of Nebulae and Clusters of Stars (1888), Index Catalogue (1895), Second Index Catalogue (1908),* London, 1962: Royal Astronomical Society.

Pi — Pismis, P. "Un Nuevo Cumulo Galactico En Puppis." *Boletin de Los Observatorios Tonantzintla y Tacubaya* 16 (June 1957): 37–38.

PMH79 — Pismis, P., M. A. Moreno, and I. Hasse. "Internal Motions in H II Regions. VI. S 140 and the Associated CO Cloud." *Revista Mexicana de Astronomia y Astrofisca* 4, no. 4 (Aug. 1979): 331–335, plate 5.

Pool — Poole, R. M. "Credulous Clusters." *Betelgeuse* 18, 2 (Summer 1993): 3–4.

PWM78 — Pfleiderer, J., R. Weinberger, and R. M. Ross. 1978. "Some New Possible Very Red and Faint Star Clusters." In *Star Cluster Symposium—Budapest.* B. A. Balàzs, ed. Budapest, 1977: Roland Eötvös University.

Ro — Roslund, C. "Remarks on some New and some Known Galactic Clusters." *Publications of the Astronomical Society of the Pacific* 72 (1960): 205–207.

Sch Original reference to Schroeter 1 unknown.

Sh Sharpless, S. "A Catalogue of H II Regions." *Astrophysical Journal Supp.* 4 (1959): 257–279.

Skiff Luginbuhl, C. B., and B. A. Skiff. *Observing Handbook and Catalogue of Deep-Sky Objects.* Cambridge, 1990: Cambridge University Press.

Skiff, Brian. "A Galaxy Behind the Milky Way, and Other CCD Catches." *CCD Astronomy* 2, 1 (Winter 1995): 40–43.

SSWZ94 Saurer, W., R. Seeberger, and R. Weinberger. "Berkeley 93: A Distant Star Cluster Nestled in a Dust Cloud." *Astronomical Journal* 107, 6 (1994): 2101–2107.

St Stock, Jurgen. "Magnitudes and Colors for Stars in Two New Galactic Clusters." *Astrophysical Journal* 123 (1956): 258–265.

Ste Stephenson, C. B. "A Possible New Galactic Cluster Involving Delta Lyrae." *Publications of the Astronomical Society of the Pacific* 71 (1959): 145–151.

———. "A Possible New and Very Remote Galactic Cluster." *Astronomical Journal* 99, no. 6 (June 1990): 1867–1868.

Tom Tombaugh, C. "Two New Faint Galactic Star Clusters." *Publications of the Astronomical Society of the Pacific* 50 (1938): 171.

Tr Trumpler, R. J. "Preliminary Results on the Distances, Dimensions and Space Distribution of Open Star Clusters." Lick Observatory Bulletin 14, no. 420 (1930): 154–188.

Up Upgren, A. and V. Rubin. "An Old Open Cluster Near the North Galactic Pole." *Publications of the Astronomical Society of the Pacific* 77 (1965): 355–358.

vdB-Ha van den Bergh, S. and G. L. Hagen. "Uniform Survey of Clusters in the Southern Milky Way." *Astronomical Journal* 80, 1 (Jan. 1975): 11–16.

Wa FitzGerald, M. P., and A. F. J. Moffat. "Luminous Stars Beyond the Solar Circle: Investigation of a Galactic Field at l=231°." *Monthly Notices of the Royal Astronomical Society* 193 (1980): 761–774.

Moffat, A. F. J., M. P. FitzGerald, and P. D. Jackson. "The Rotation and Structure of the Galaxy Beyond the Solar Circle. I. Photometry and Spectroscopy of 276 Stars in 45 H II Regions and Other Young Stellar Groups Toward the Galactic Anticentre." *Astron. & Astrophysics Supplement Series* 38 (1979): 197–225.

We Westerlund, B. "On the Identification of a Radio Source in Carina." *Arkiv för Astronomi* 2, 39 (1960).

———. "On the Extended Infrared Source in Ara." *Astrophysical Journal Letters.* 154 (1968): L67-L68, plate L1.

WG71 Westerlund, B. and J. Glaspey. "On the Structure of the Wing of the Small Magellanic Cloud." *Astron. & Astrophysics Supplement Series* 10 (1971): 1–7.

Planetary Nebuae

Condon, J. J., Kaplan, D. L., "Planetary Nebulae in the NRAO VLA Sky Survey." *Astron. & Astrophysics Supplement Series*, 117 (1998): 361.

Hynes, S.J. *Planetary Nebulae.* Richmond VA, 1991: Willmann-Bell, Inc.

NGC, IC [I] Dreyer, J. L. E. *New General Catalogue of Nebulae and Clusters of Stars (1888), Index Catalogue (1895), Second Index Catalogue (1908).* London, 1962: Royal Astronomical Society.

PK Perek, L., and L. Kohoutek. *Catalogue of Galactic Planetary Nebulae.* Prague, 1967: Academia Publishing House of the Czecholosovak Academy of Sciences.

PN G [P] Acker A., Ochsenbein F., Stenholm B., Tylenda R., Marcout J., and Schohn C. *Strasbourg-ESO Catalogue of Galactic Planetary Nebulae.* 1992: European Southern Observatory.

Sh2 Sharpless, S. "A Catalogue of H II Regions." *Astrophysical Journal Supp.* 4 (1959): 257–279. (Also available as U.S. Naval Observatory Reprint No. 7)

Star Clouds

NGC, IC [I] Dreyer, J. L. E. *New General Catalogue of Nebulae and Clusters of Stars (1888),* Index *Catalogue (1895), Second Index Catalogue (1908).* London, 1962: Royal Astronomical Society.

1 GALAXIES 1

RA h m s	Dec ° ′ ″	Name	Mag (V)	Dim ′ Maj x min	SB	Type Class	PA	Notes
07 34 58.7	+85 32 13	IC 455	13.3	1.1 x 0.7	12.9	S0	69	Mag 11.0 star S 6′.1.
07 56 00.7	+85 09 30	IC 469	12.6	2.2 x 0.8	13.1	SAB(rs)ab:	90	Mag 8.96v star N 3′.9.
08 45 18.4	+85 44 21	IC 499	12.5	2.1 x 1.1	13.2	Sa	80	Mag 8.50v star NE 5′.7.
09 03 50.8	+85 30 03	IC 512	12.2	1.8 x 1.3	13.0	SAB(s)cd	1	Mag 7.86v star ENE 11′.5; low surface brightness **UGC 4612** W 4′.5.
02 29 21.6	+84 01 15	MCG +14-2-9	14.8	0.7 x 0.3	13.0	Sb	105	Mag 10.9 star S 3′.9.
07 32 22.6	+86 39 58	MCG +14-4-29	14.3	1.2 x 0.6	13.8		39	Mag 10.9 star E 2′.5.
08 33 40.4	+85 58 56	MCG +14-4-49	14.4	0.6 x 0.5	12.9	Sb		UGC 4348 S 2′.3.
12 00 38.1	+88 08 14	MCG +15-1-12	14.2	0.7 x 0.3	12.4	Sb	177	Mag 13.9 star N edge; mag 12.1 star ESE 2′.1.
05 02 36.8	+86 13 19	NGC 1544	13.3	1.3 x 0.9	13.3	S?	130	
07 14 20.6	+84 22 54	NGC 2268	11.5	2.7 x 1.5	12.9	SAB(r)bc II	63	
07 27 23.8	+85 45 24	NGC 2276	11.4	2.3 x 1.9	12.9	SAB(rs)c II-III	20	Located 2′.4 NE of mag 8.08v star; NGC 2300 ESE 6′.4.
07 32 23.6	+85 42 34	NGC 2300	11.0	2.8 x 2.0	12.7	SA0°	78	NGC 2276 WNW 6′.4.
11 47 11.3	+89 05 35	NGC 3172	14.1	1.1 x 0.7	13.7	Sb	39	**Polarissima Borealis**. **MCG +15-1-10** WSW 1′.6.
01 32 46.5	+85 00 47	UGC 1039	14.2	1.4 x 0.5	13.7	Sab	101	Mag 14.3 star on N edge.
01 49 17.8	+85 15 36	UGC 1198	13.8	0.8 x 0.6	13.1	E?	85	
01 58 58.8	+86 40 22	UGC 1285	13.5	1.1 x 0.9	13.4	Sbc II	179	
05 19 42.1	+84 03 08	UGC 3253	12.4	1.7 x 0.9	12.7	SB(r)b	96	
05 21 44.7	+84 28 56	UGC 3257	13.8	1.5 x 0.7	13.7	SBa	157	
05 55 45.7	+85 54 51	UGC 3336	14.3	0.7 x 0.4	12.8	SB(s)b	45	Mag 10.92v star N 1′.8.
06 31 06.0	+84 55 29	UGC 3442	13.9	1.1 x 0.3	12.5	S0/a	107	Mag 11.9 star on W end; mag 10.83v star W 0′.6.
06 47 56.3	+84 09 58	UGC 3500	13.8	1.7 x 0.4	13.2	S?	6	Mag 11.3 star E 1′.8.
06 55 00.9	+84 02 28	UGC 3521	14.3	1.0 x 0.6	13.6	S?	75	UGC 3528 NNE 2′.7.
06 56 06.1	+84 55 03	UGC 3522	14.3	2.1 x 1.1	15.1	S?	132	Stellar nucleus.
06 56 09.5	+84 04 39	UGC 3528	13.5	1.4 x 0.7	13.3	SBab	40	UGC 3521 SSE 2′.7.
07 03 22.2	+86 33 26	UGC 3536A	13.6	0.7 x 0.6	12.7	E?		**UGC 3528A** NW 1′.5.
07 17 46.7	+85 42 44	UGC 3654	14.3	0.4 x 0.3	11.8	Compact	26	Compact; star on N end; **UGC 3661** N 3′.7.
07 25 22.6	+86 12 43	UGC 3715	14.8	1.6 x 0.7	14.8	Sbc	90	
07 53 36.0	+84 37 05	UGC 3992	15.2	1.2 x 0.9	15.1	Sb	159	Mag 13.3 star superimposed S end.
07 55 44.2	+84 55 34	UGC 3993	12.8	1.6 x 1.2	13.4	S0?	35	Mag 11.8 star E 1′.1.
08 04 27.1	+84 38 30	UGC 4078	14.3	2.1 x 0.3	13.6	Sbc	82	Located between a pair of mag 12.4 and 11.4 stars N-S.
08 06 58.0	+84 45 18	UGC 4100	15.4	1.4 x 1.0	15.6	IBm	135	Irregular, low surface brightness.
08 28 30.7	+85 36 28	UGC 4297	13.6	1.6 x 0.4	12.9	Sa	83	Mag 7.40v star NW 7′.6.
08 34 00.6	+85 56 42	UGC 4348	14.2	1.5 x 1.2	14.7	SB?	21	MCG +14-4-49 N 2′.3.
08 34 44.3	+84 19 08	UGC 4396	14.4	2.0 x 0.4	14.0	SBb-c	26	Mag 11.3 star E 1′.4.
08 57 06.3	+84 48 32	UGC 4601	14.0	1.4 x 0.4	13.2		140	Mag 13.6 star SE end.
09 56 41.5	+88 10 20	UGC 5083A	16.1	2.2 x 1.9	17.5	SAdm	123	Mag 10.53v star superimposed E edge.
13 08 28.1	+84 37 50	UGC 8264	13.2	1.5 x 1.4	13.9	Pec	93	**MCG +14-6-17** NE 8′.9.
13 23 06.1	+84 30 16	UGC 8454	14.7	2.3 x 0.5	14.7	Double System	69	Double system, long, faint bridge; mag 11.4 star NW 2′.5.
13 30 58.3	+85 58 05	UGC 8615	15.4	1.5 x 0.3	14.4	Sd	28	Mag 9.72v star N 6′.0.
15 59 32.3	+86 11 11	UGC 10263	15.6	0.9 x 0.9	15.2	Dwarf Spiral		Stellar nucleus.
16 53 28.5	+86 35 25	UGC 10740	14.4	1.0 x 0.8	14.0	SB0	140	
17 19 32.4	+86 44 08	UGC 10923	13.4	1.2 x 0.6	12.9	S?	6	Double system, strongly distorted; small component on E edge is **MCG +14-8-25**.
18 06 41.2	+87 48 30	UGC 11267	15.3	1.7 x 0.3	14.4	Scd:	121	Mag 10.79v star SSW 2′.8.

OPEN CLUSTERS

RA h m s	Dec ° ′ ″	Name	Mag	Diam ′	No. ★	B ★	Type	Notes
00 47 30.0	+85 15 00	NGC 188	8.1	15	550	10.0	cl	Rich in stars; moderate brightness range; strong central concentration; detached.

2 GALAXIES 2

RA h m s	Dec ° ′ ″	Name	Mag (V)	Dim ′ Maj x min	SB	Type Class	PA	Notes
23 36 20.9	+75 38 54	IC 1502	13.5	1.2 x 0.4	12.6	$S0^+$	52	Mag 10.63v star N 5′.5.
02 07 42.4	+75 09 37	MCG +12-3-2	13.9	1.4 x 0.6	13.8	E	15	Mag 11.7 star W 1′.8.
00 40 34.9	+83 13 21	UGC 392	13.9	1.7 x 0.3	13.0	S	15	
01 06 50.8	+75 35 59	UGC 670	13.6	1.5 x 1.3	14.2	SBb?	39	Large core, numerous stars superimposed on envelope.
01 21 43.5	+78 37 39	UGC 863	15.4	1.5 x 1.5	16.2	Scd?		Stellar nucleus; mag 11.9 star W 2′.5.
01 40 36.7	+83 56 24	UGC 1148	14.4	0.9 x 0.9	14.0	S?		
01 56 18.8	+73 16 54	UGC 1378	12.6	3.5 x 2.3	14.7	(R)SB(rs)a:	11	Numerous stars superimposed on faint outer envelope; mag 9.06v star SE 5′.5.
21 56 24.3	+73 15 35	UGC 11861	13.6	3.5 x 2.6	15.8	SABdm	32	Very patchy; few superimposed stars along SE edge.
22 30 46.6	+76 30 39	UGC 12069	14.5	1.8 x 1.2	15.2	SAB(s)dm	91	Stellar nucleus, few superimposed stars and/or knots.
22 40 54.3	+75 09 51	UGC 12160	14.8	2.1 x 1.7	16.0	Scd:	11	Mag 13.4 star E edge.
22 45 38.0	+73 09 42	UGC 12182	14.6	0.8 x 0.6	13.7	S	69	
22 50 27.1	+82 52 37	UGC 12221	14.0	2.3 x 0.7	14.4	SA(s)d	95	Several superimposed stars and/or bright knots W end; mag 12.4 star W 2′.0.
23 18 29.4	+77 19 12	UGC 12504	14.6	1.3 x 0.6	14.2	Scd:	30	Mag 12.2 star E 1′.0.
00 02 21.7	+77 15 21	UGC 12921	14.7	1.3 x 0.9	14.7	SAdm-m	18	Mag 11.6 star N 1′.8.

OPEN CLUSTERS

RA h m s	Dec ° ′ ″	Name	Mag	Diam ′	No. ★	B ★	Type	Notes
02 01 05.0	+75 29 36	Be 8		5	40	14.0	cl	

PLANETARY NEBULAE

RA h m s	Dec ° ′ ″	Name	Diam ″	Mag (P)	Mag (V)	Mag cent ★	Alt Name	Notes
22 42 25.3	+80 26 28	IC 1454	38	14.8	14.:	18.8	PK 117+18.1	Mag 7.01v star E 4′.2.
01 07 12.9	+73 32 57	PN G124.0+10.7	270			16.4	PK 124+10.1	Group of 3 mag 11 stars form a right angle shape NW 4′.1; a close pair of mag 11 stars SE 3′.7.

3 GALAXIES 3

RA h m s	Dec ° ′ ″	Name	Mag (V)	Dim ′ Maj x min	SB	Type Class	PA	Notes
18 48 49.8	+73 21 48	CGCG 341-9	15.6	0.6 x 0.3	13.6		153	CGCG 341-10 and galaxy group **Hickson 85** E 6′.4.
18 50 18.6	+73 21 01	CGCG 341-10	14.4	0.4 x 0.3	12.2	E?	12	= **Hickson 85A**; **Hickson 85B** SE 0′.7; **Hickson 85C** NE 0′.9; **Hickson 85D** E 1′.2.
19 11 49.9	+83 54 17	CGCG 368-2	15.0	0.3 x 0.3	12.5	E		Almost stellar.
17 54 16.9	+73 25 23	NGC 6538	13.3	1.0 x 0.5	12.4	S	48	
18 19 46.4	+74 34 05	NGC 6643	11.1	3.8 x 1.9	13.1	SA(rs)c II	38	Very patchy, many faint knots.
18 24 07.4	+73 10 59	NGC 6654	12.0	2.6 x 2.1	13.7	(R′)SB(s)0/a	0	Bright NNE-SSW bar; mag 10.87v star W 2′.4.
18 39 26.1	+73 34 47	NGC 6654A	12.9	2.6 x 0.8	13.6	SB(s)d pec?	63	UGC 11331 NW 2′.6.
19 10 53.9	+73 24 38	NGC 6786	12.9	1.1 x 0.9	12.7	SB?	45	**UGC 11415** NE 1′.2.
18 28 06.8	+78 44 58	UGC 11270	14.0	1.5 x 1.0	14.3	Sab	5	Mag 14.3 star on E edge; mag 14.9 star on W edge.
18 33 07.1	+75 24 06	UGC 11295	15.9	1.5 x 0.3	14.9	Sdm:	138	Located between close mag 11.7 and 12.5 stars.
18 38 59.5	+73 36 35	UGC 11331	14.3	1.1 x 0.7	13.9	Sm:	33	NGC 6654A SE 2′.6.
18 39 30.8	+73 49 32	UGC 11334	15.1	1.9 x 0.3	14.3	Pec?	105	Almost stellar core; mag 11.2 star W 1′.5.
18 52 24.4	+73 11 36	UGC 11377	15.0	1.5 x 0.2	13.5	Scd:	10	Mag 9.17v star S 4′.2.
19 03 17.7	+73 02 32	UGC 11402	14.3	0.7 x 0.7	13.4	S		Mag 11.5 star E 0′.8.
20 43 30.4	+80 09 22	UGC 11635	13.0	2.6 x 1.1	14.0	Sbc	35	
21 56 24.3	+73 15 35	UGC 11861	13.6	3.5 x 2.6	15.8	SABdm	32	Very patchy; few superimposed stars along SE edge.

GALAXY CLUSTERS

RA h m s	Dec ° ′ ″	Name	Mag 10th brightest	No. Gal	Diam ′	Notes
17 54 00.0	+77 40 00	A 2296	15.9	30	16	All members anonymous, faint, stellar.
18 42 06.0	+77 42 00	A 2309	15.8	41	34	Mag 6.95v star on W edge; **CGCG 356-11** lies 10′.5 N of this star; other members anonymous, stellar.

4 GALAXIES 4

RA h m s	Dec ° ′ ″	Name	Mag (V)	Dim ′ Maj x min	SB	Type Class	PA	Notes
15 30 56.4	+82 27 19	IC 1143	13.2	0.9 x 0.9	13.1	E		Mag 10.5 star NW 5′.8.
17 21 45.1	+75 50 55	IC 4660	13.6	1.3 x 0.3	12.4	S	170	Mag 9.97v star on S edge.
14 34 34.7	+77 57 57	MCG +13-11-2	14.0	0.9 x 0.4	12.8	S0?	54	Mag 12.8 star NW 3′.0.
14 33 48.5	+80 13 50	MCG +13-11-3	14.4	1.1 x 0.8	14.1	SBc	9	
15 11 49.4	+75 18 44	MCG +13-11-12	15.5	0.6 x 0.4	13.8	Sb	144	NGC 5909, NGC 5912 N 4′.5.
16 24 10.5	+78 46 58	MCG +13-12-3	13.7	0.6 x 0.4	12.0	Irr	96	
17 26 44.9	+77 37 25	MCG +13-12-25	16.1	0.9 x 0.3	14.5	Sb		Mag 14.2 star superimposed on E end.
13 41 13.5	+83 29 53	MCG +14-6-28	14.6	0.7 x 0.5	13.3		135	
13 45 36.6	+76 49 38	NGC 5323	13.5	1.4 x 0.4	12.7	Sab	163	Very faint anonymous galaxy SW 0′.8.
13 50 12.4	+73 57 08	NGC 5344	14.5	0.5 x 0.4	12.6		84	Very small, very faint galaxy W 0′.7.
13 57 13.3	+73 37 00	NGC 5412	13.4	1.2 x 1.0	13.5	S0$^-$:	20	
13 54 24.2	+78 13 14	NGC 5452	13.3	2.0 x 1.5	14.3	SAB(s)d	120	Weak stellar nucleus; stellar galaxy **VII Zw 531** W 7′.7.
14 09 45.1	+78 36 00	NGC 5547	14.5	0.6 x 0.4	12.8		66	Mag 9.20v star SE 5′.4; located on S edge of galaxy cluster **A 1892**.
14 20 41.6	+80 07 19	NGC 5640	14.6	0.7 x 0.4	13.1	Ia	24	**CGCG 353-34** W 7′.0.
14 29 42.3	+78 51 50	NGC 5712	14.5	0.7 x 0.7	13.8	E/S0		Almost stellar nucleus; **IC 4470** NW 4′.0.
14 54 03.0	+73 07 53	NGC 5819	13.5	0.9 x 0.9	13.1	SAB(rs)bc		Located between a pair of mag 11.5 stars on NW-SE line.
14 59 31.0	+73 53 36	NGC 5836	13.9	1.2 x 1.0	13.9	SB(rs)b	45	Bright nucleus, star S of nucleus; mag 10.88v star SE 1′.7.
15 11 27.3	+75 22 58	NGC 5909	13.8	1.1 x 0.5	13.0	S?	52	NGC 5912 on E edge; MCG +13-11-12 S 4′.5.
15 11 41.4	+75 23 03	NGC 5912	13.7	1.2 x 1.1	14.1	E?	129	NGC 5909 on W edge; mag 7.97v star NE 6′.9.
15 55 26.1	+78 59 45	NGC 6068	12.8	1.1 x 0.7	12.4	SBbc?	155	NGC 6068A W 1′.9; mag 10.59v star SE 2′.7.
15 54 48.1	+78 59 04	NGC 6068A	14.0	0.9 x 0.2	12.0	S0? sp	170	
16 32 38.6	+78 11 55	NGC 6217	11.2	3.0 x 2.5	13.2	(R)SB(rs)bc II	153	Long SE-NW bar; strong dark patch NE of nucleus.
16 32 30.9	+82 32 15	NGC 6251	12.6	1.8 x 1.5	13.7	E	36	
16 32 41.3	+82 34 33	NGC 6252	14.2	0.7 x 0.3	12.4		60	Bright center, uniform envelope; NGC 6252 N 2′.3.
17 05 24.9	+75 24 18	NGC 6324	12.9	0.9 x 0.5	11.9	S?	72	Strong dark lane SE of center.
17 03 35.8	+78 37 42	NGC 6331	14.4	0.5 x 0.4	12.7	E	126	Small anonymous galaxy on NW edge; **MCG +13-12-17** E 1′.9; **UGC 10726** ENE 2′.7; **PGC 59495** SW 2′.4.
17 29 37.0	+75 42 09	NGC 6412	11.8	2.5 x 2.2	13.5	SA(s)c II	120	Patchy arms with knots.
17 30 37.0	+74 22 35	NGC 6414	14.7	1.0 x 0.5	13.8		145	
17 38 42.8	+73 30 52	NGC 6461	15.6	0.4 x 0.3	13.1		132	Mag 10.25v star W 2′.6.
17 54 16.9	+73 25 23	NGC 6538	13.3	1.0 x 0.5	12.4	S	48	
18 19 46.4	+74 34 05	NGC 6643	11.1	3.8 x 1.9	13.1	SA(rs)c II	38	Very patchy, many faint knots.
13 48 56.5	+74 14 57	UGC 8747	13.9	1.2 x 0.9	13.8	SA(rs)c	18	
14 00 39.7	+78 51 10	UGC 8964	14.8	1.5 x 0.4	14.1	Sdm	48	
14 07 17.2	+75 55 45	UGC 9052	14.5	1.0 x 0.8	14.1	SAd	63	Mag 12.8 star SW 2′.1.
14 34 08.7	+77 53 55	UGC 9413	14.0	0.9 x 0.6	13.2	Sbc	120	Very patchy.
14 53 50.0	+83 35 22	UGC 9650	13.5	1.5 x 0.4	12.8	Sa	6	
14 56 06.7	+83 31 21	UGC 9668	12.9	1.4 x 0.7	12.7	S?	82	Mag 10.61v star E 2′.5.
14 59 34.1	+81 24 01	UGC 9683	13.8	0.9 x 0.9	13.4	SB(r)a		
15 04 01.7	+77 38 02	UGC 9730	14.6	1.4 x 1.1	14.9	SB(s)d	30	Patchy.
15 07 23.0	+76 02 56	UGC 9748	13.8	1.4 x 1.0	14.0	(R)SB(r)0/a	20	E-W bar with fainter, uniform envelope N and S; mag 11.2 star SW 3′.1.
15 07 32.5	+76 08 43	UGC 9750	14.4	1.5 x 0.4	13.7	Sb III	0	Located between two close mag 11.5 stars.
15 27 15.3	+77 09 23	UGC 9874	13.6	1.0 x 0.6	13.0	E	0	
15 35 48.4	+73 26 59	UGC 9944	13.9	1.3 x 0.4	13.1	S?	174	Mag 10.35 v star SE 3′.4.
15 33 18.3	+82 13 44	UGC 9950	14.3	1.1 x 0.6	13.7	Sbc	82	Mag 11.8 star E 0′.8.
15 43 54.3	+81 48 37	UGC 10054	13.1	2.0 x 1.1	13.8	SBdm IV-V	150	Bright, narrow, elongated core.
16 03 00.0	+77 36 25	UGC 10212	14.9	1.5 x 0.3	13.9	Sbc	141	
16 04 46.0	+79 37 44	UGC 10237	14.2	0.9 x 0.8	13.7	SABcd	3	Uniform surface brightness.
16 07 30.5	+81 18 08	UGC 10280	14.0	1.4 x 0.7	13.8	SAB(s)b	170	Close pair of mag 11.6 stars and a small, faint anonymous galaxy SW 2′.5.
16 20 33.9	+73 51 53	UGC 10368	13.8	1.1 x 0.4	12.8	S0$^-$:	150	**MCG +12-15-58** S 3′.2.
16 29 05.2	+79 29 04	UGC 10446	15.2	1.7 x 0.2	13.9	Sbc	150	
16 28 18.2	+80 21 16	UGC 10447	14.4	2.2 x 0.5	14.4	Sb III	137	Located between a pair of stars, mags 10.77v and 11.5.
16 32 24.8	+76 44 20	UGC 10466	14.0	1.1 x 0.7	13.6	SABb	44	
16 30 44.3	+81 32 39	UGC 10471	14.3	1.2 x 0.9	14.2	S?	155	**MCG +14-8-8** NNW 1′.9.
16 37 17.9	+78 07 56	UGC 10509	14.2	0.9 x 0.6	13.4	SBb	105	A pair of faint, almost stellar anonymous galaxies N 0′.9.
16 47 56.9	+81 50 29	UGC 10604	14.8	1.4 x 0.3	13.7	S	152	Mag 10.40v star SW 1′.1.
16 53 29.1	+79 04 07	UGC 10632	14.9	1.5 x 0.2	13.4	Sd	29	
17 01 48.8	+79 02 13	UGC 10704	14.7	1.0 x 0.6	14.0	S	165	Mag 12.2 star N 1′.8.
17 16 05.1	+73 26 08	UGC 10803	12.1	1.2 x 0.7	12.0	E?	15	

(Continued from previous page)

4 GALAXIES 4

RA h m s	Dec ° ′ ″	Name	Mag (V)	Dim ′ Maj x min	SB	Type Class	PA	Notes
17 29 04.4	+74 15 24	UGC 10892	15.4	1.7 x 0.2	14.1	Sdm:	135	Mag 11.4 star SW 1′.5.
17 29 59.7	+77 23 45	UGC 10907	14.4	1.2 x 0.3	13.1	Sbc	39	
17 38 52.9	+74 50 13	UGC 10949	13.7	0.8 x 0.5	12.7	E	69	Mag 9.69v star NNE 3′.2.

GALAXY CLUSTERS

RA h m s	Dec ° ′ ″	Name	Mag 10th brightest	No. Gal	Diam ′	Notes
16 47 24.0	+81 33 00	A 2247	15.3	35	22	A number of UGC, MCG and PGC GX between center and E edge.
16 57 48.0	+77 00 00	A 2248	15.5	34	25	Most members anonymous, stellar/almost stellar.
17 03 42.0	+78 43 00	A 2256	15.3	88	56	
17 17 18.0	+78 00 00	A 2271	15.7	35	34	**MCG +13-12-22** ENE of center 3′.1; all other members anonymous, stellar.
17 54 00.0	+77 40 00	A 2296	15.9	30	16	All members anonymous, faint, stellar.

5 GALAXIES 5

RA h m s	Dec ° ′ ″	Name	Mag (V)	Dim ′ Maj x min	SB	Type Class	PA	Notes
12 35 17.5	+75 55 01	CGCG 352-36	15.5	0.6 x 0.6	14.3			Mag 8.46v star N 6′.3.
12 23 56.2	+73 53 35	MCG +12-12-7	14.6	0.5 x 0.5	13.0	Sb		Mag 11.8 star W 1′.8.
09 44 18.1	+76 21 05	MCG +13-7-36	14.3	0.9 x 0.4	13.1	Pec	39	Close pair of mag 11-12 star NW 3′.4.
09 50 18.9	+75 44 49	MCG +13-7-38	14.8	0.7 x 0.6	13.7	Sb	9	Mag 9.26v star SSW 2′.8.
09 56 26.2	+78 51 14	MCG +13-7-39	15.9	0.6 x 0.3	13.9	Sb	33	Mag 13.3 SW 3′.4.
10 10 28.4	+77 18 16	MCG +13-8-8	15.1	0.6 x 0.5	13.7	Sc	12	Mag 10.8 star NE 2′.3.
10 19 58.4	+79 27 05	MCG +13-8-12	14.6	0.7 x 0.6	13.5	Sbc	141	Mag 9.87v star SE 2′.7.
10 22 01.3	+77 50 12	MCG +13-8-14	14.4	0.7 x 0.4	13.1	E/S0	39	Mag 10.19v star N 4′.3.
10 30 22.4	+74 59 40	MCG +13-8-26	14.5	0.8 x 0.5	13.3	Sc	3	
10 49 19.0	+80 09 18	MCG +13-8-40	14.7	0.9 x 0.7	14.0		42	Located between a close pair of mag 14 stars.
10 48 38.7	+76 48 11	MCG +13-8-41	14.4	1.0 x 0.4	13.2		0	**UGC 5877** NW 3′.6; mag 8.88v star NNE 4′.0.
10 51 06.3	+76 47 43	MCG +13-8-43	14.7	0.9 x 0.8	14.2	Sbc	3	Mag 14.7 star N edge.
11 03 26.2	+78 54 31	MCG +13-8-54	15.5	0.6 x 0.3	13.5	Sbc	96	Close pair of mag 13 stars S 1′.3; mag 7.53v star WSW 5′.7.
11 14 44.2	+79 43 35	MCG +13-8-56	14.8	0.8 x 0.4	13.4		24	Mag 12.1 star N 0′.9.
11 24 47.0	+74 55 44	MCG +13-8-57	14.7	0.8 x 0.6	14.0	E/S0	15	Mag 10.78v star NNE 4′.3.
11 28 18.4	+77 37 43	MCG +13-8-62	14.8	0.7 x 0.2	12.5	S?	39	Mag 13.0 star NE end.
11 55 05.6	+79 25 28	MCG +13-9-6	14.4	0.7 x 0.5	13.1	S?	30	Pair with MCG +13-9-7 E.
11 55 25.1	+79 25 22	MCG +13-9-7	14.6	0.6 x 0.5	13.2	Sb	24	Pair with MCG +13-9-6 W.
12 17 30.3	+77 16 41	MCG +13-9-21	14.4	0.8 x 0.5	13.2	Sc	30	**MCG +13-9-22** SSE 1′.6.
12 20 04.4	+75 29 49	MCG +13-9-23	14.4	0.9 x 0.4	13.2	Sb	0	Mag 12.7 star N 1′.3.
12 45 38.3	+74 28 50	MCG +13-9-31	14.2	0.8 x 0.7	13.5	Spherical	15	
13 01 15.6	+80 01 43	MCG +13-9-43	14.6	0.6 x 0.4	13.1	E/S0	0	**UGC 8147** NW 3′.4.
13 06 49.3	+82 51 59	MCG +14-6-15	14.7	0.7 x 0.5	13.6	E	159	
13 20 59.4	+80 22 29	MCG +14-6-23	14.4	0.9 x 0.6	13.6		141	Mag 11.30v star N 3′.2.
13 41 13.5	+83 29 53	MCG +14-6-28	14.6	0.7 x 0.5	13.3		135	
09 43 31.4	+79 42 05	NGC 2908	14.1	0.8 x 0.8	13.5	S		Mag 10.02v star NE 7′.0.
09 43 47.3	+74 51 36	NGC 2977	12.5	1.8 x 0.8	12.7	Sb:	145	
10 05 38.7	+80 17 11	NGC 3057	13.0	2.2 x 1.3	14.0	SB(s)dm IV	5	Several strong knots; mag 9.63v star NE 9′.3.
09 56 13.1	+75 51 55	NGC 3061	12.8	1.7 x 1.5	13.6	(R′)SB(rs)c	6	Faint stellar nucleus.
10 15 32.0	+74 13 14	NGC 3144	13.4	1.2 x 0.7	13.1	SB(s)ab pec:	0	Mag 13.1 star on E edge.
10 16 54.2	+73 23 56	NGC 3147	10.6	3.9 x 3.5	13.3	SA(rs)bc II	155	
10 17 41.0	+74 20 48	NGC 3155	12.9	1.0 x 0.5	12.0	S?	41	
10 21 48.6	+74 10 34	NGC 3183	11.9	2.3 x 1.4	13.0	SB(s)bc:	161	Three superimposed stars or bright knots on N edge of envelope.
10 14 27.7	+77 49 11	NGC 3197	13.5	1.3 x 1.0	13.7	Sbc II-III	155	**Mkn 1248** N 3′.4; mag 9.98v star S 3′.5.
10 28 17.4	+79 49 23	NGC 3212	13.2	1.5 x 0.9	13.4	SB?	92	NGC 3215 on E edge.
10 28 41.3	+79 48 46	NGC 3215	13.1	1.1 x 1.0	13.1	S?	52	Bright knot or star on NE edge; NGC 3212 on W edge.
10 34 23.4	+73 45 48	NGC 3252	13.5	2.0 x 0.6	13.6	SBd? sp	35	Mag 7.58v star N 4′.3.
10 44 40.4	+76 48 33	NGC 3329	12.2	1.8 x 1.0	12.6	(R)SA(r)b:	140	Mag 9.89v star WNW 7′.7.
10 46 09.9	+73 21 12	NGC 3343	13.4	1.3 x 0.9	13.6	E	55	
10 53 54.0	+73 41 24	NGC 3403	12.2	3.0 x 1.2	13.4	SAbc: III	73	
10 59 31.5	+75 11 30	NGC 3465	13.5	1.3 x 1.1	13.8	Sab	171	Almost stellar nucleus; NGC 3500 E 8′.9.
11 01 51.6	+75 12 03	NGC 3500	13.5	1.4 x 0.7	13.4	Sab	54	NGC 3465 W 8′.9.
11 03 06.5	+75 06 55	NGC 3523	12.9	1.3 x 1.3	13.3	Sbc		Weak, almost stellar nucleus.
11 35 41.5	+73 27 06	NGC 3736	14.5	1.2 x 0.7	14.1	Sb	155	Stellar nucleus, faint star on E edge; mag 8.33v star SSW 4′.7.
11 32 32.4	+74 37 43	NGC 3752	12.9	1.7 x 0.7	12.9	Sab	155	Located 2′.3 W of mag 11.26v star.
11 49 19.4	+74 18 09	NGC 3890	13.3	0.9 x 0.9	12.9	S?		
11 42 49.3	+77 22 24	NGC 3901	13.7	1.8 x 0.8	13.9	Scd:	165	Stellar nucleus with a bright knot or star on S edge.
12 08 27.3	+76 48 13	NGC 4127	12.7	2.5 x 1.2	13.7	SAc?	140	
12 08 49.1	+74 54 15	NGC 4133	12.3	1.8 x 1.3	13.1	SABb:	125	Stellar **MCG +3-9-17** E 7′.8.
12 10 52.3	+76 07 27	NGC 4159	13.5	1.3 x 0.5	12.9	S?	35	
12 20 17.8	+75 22 12	NGC 4291	11.5	1.9 x 1.6	12.7	E	124	Located 1′.4 W of mag 10.22v star.
12 21 44.7	+75 19 20	NGC 4319	11.9	2.8 x 2.3	13.8	SB(r)ab	160	Small nucleus in bright bar; two main arms; **Mkn 205** just S of bar.
12 22 35.8	+76 10 09	NGC 4331	14.0	2.2 x 0.4	13.8	Im?	2	Large, bright patch northern half; very faint star or knot on N end.
12 23 28.8	+74 57 09	NGC 4363	13.5	1.4 x 1.4	14.1	SAb:		**CGCG 352-30** SW 8′.4.
12 24 29.9	+75 31 42	NGC 4386	11.7	2.5 x 1.3	12.8	SAB0°:	135	
12 35 45.9	+74 14 44	NGC 4572	13.9	1.6 x 0.5	13.5	S	170	Long, slightly curving bar.
12 37 24.6	+74 11 29	NGC 4589	10.7	3.2 x 2.6	13.0	E2	75	Bright nucleus with faint star E and W .
12 41 43.8	+74 25 15	NGC 4648	12.0	1.6 x 1.1	12.7	E3	70	Mag 8.08v star W 7′.3.
13 02 18.9	+75 24 13	NGC 4954	13.6	0.8 x 0.6	12.7	S0?	62	
13 35 38.8	+75 02 22	NGC 5262	13.8	1.2 x 0.7	13.4	S0$^-$:	14	**UGC 8595** ESE 1′.9.
13 38 39.4	+79 27 30	NGC 5295	14.3	0.4 x 0.4	12.4	E/S0		Very compact; mag 7.46v star SW 8′.1.
13 45 36.6	+76 49 38	NGC 5323	13.5	1.4 x 0.4	12.7	Sab	163	Very faint anonymous galaxy SW 0′.8.
13 50 12.4	+73 57 08	NGC 5344	14.5	0.5 x 0.4	12.6		84	Very small, very faint galaxy W 0′.7.
13 57 13.3	+73 37 00	NGC 5412	13.4	1.2 x 1.0	13.5	S0$^-$:	20	
13 54 24.2	+78 13 14	NGC 5452	13.3	2.0 x 1.5	14.3	SAB(s)d	120	Weak stellar nucleus; stellar galaxy **VII Zw 531** W 7′.7.
14 09 45.1	+78 36 00	NGC 5547	14.5	0.6 x 0.4	12.8		66	Mag 9.20v star SE 5′.4; located on S edge of galaxy cluster **A 1892**.
09 40 04.4	+82 06 20	UGC 5114	15.1	1.7 x 0.7	15.1	IBm:	140	Mag 10.74v star W 2′.0.
09 44 05.6	+83 48 23	UGC 5128	14.4	0.9 x 0.7	13.7	Scd:	177	Mag 8.52v star NW 3′.9.
09 46 52.1	+79 48 36	UGC 5203	14.9	2.1 x 0.2	13.8	Sd	80	Mag 10.02v star SW 4′.2.
10 04 59.5	+79 16 16	UGC 5402	13.8	1.1 x 0.7	13.3	S?	175	Located between a pair of mag 13 stars.

(Continued from previous page)

5 GALAXIES 5

RA h m s	Dec ° ′ ″	Name	Mag (V)	Dim ′ Maj x min	SB	Type Class	PA	Notes
10 23 44.8	+78 52 25	UGC 5596	13.3	0.8 x 0.7	12.8	E?	99	Elongated **MCG +13-8-17** N 3′5.
10 24 09.9	+78 37 42	UGC 5600	13.2	1.3 x 0.9	13.3	S0?	177	Bright elongated core, faint envelope; UGC 5609 SSW.
10 24 25.1	+78 36 30	UGC 5609	13.5	1.2 x 0.7	13.2	S?	9	Pair with UGC 5600 NNW.
10 31 09.9	+78 53 11	UGC 5682	14.4	1.1 x 0.7	14.0	SAB(rs)c	102	Mag 10.27v star W 5′0.
10 30 42.3	+74 13 39	UGC 5686	16.0	1.2 x 1.0	16.0	Im	140	Dwarf, extremely low surface brightness; mag 10.16v star N 2′8.
10 30 46.3	+73 53 07	UGC 5689	14.8	1.5 x 0.2	13.4	Sd	154	Close pair of mag 13.8 stars N 1′2.
10 32 19.5	+77 48 15	UGC 5701	15.4	1.5 x 0.8	15.4	Im:	0	Dwarf, low surface brightness; faint star on NE edge.
10 38 09.4	+79 21 57	UGC 5757	13.7	2.1 x 0.8	14.1	Sa?	115	Mag 11.6 star NNE 2′3.
10 40 37.4	+78 35 57	UGC 5782	14.9	0.7 x 0.7	14.0	Sb-c		
10 42 37.8	+77 29 37	UGC 5814	14.1	1.3 x 0.7	13.8	Pec	130	Stellar nucleus; mag 10.6 star S 1′9.
10 43 28.9	+80 53 27	UGC 5820	14.3	1.3 x 0.4	13.4	Sbc	75	Mag 11.2 star N 0′8.
10 44 56.5	+76 41 15	UGC 5841	14.1	1.5 x 0.8	14.1	SAB(s)c	130	Mag 13.7 star E edge; mag 11.5 star W 2′0.
10 45 56.4	+77 06 05	UGC 5854	14.4	0.8 x 0.6	13.5	Sdm:	72	
10 50 44.5	+76 56 11	UGC 5926	13.5	0.9 x 0.5	12.5	S	79	Mag 11.3 star E 2′6; UGC 5939 E 2′6.
10 51 33.0	+76 55 06	UGC 5939	14.0	0.8 x 0.3	12.3	Pec	7	Mag 11.3 star N 1′3; UGC 5926 W 2′6.
10 52 02.1	+79 09 02	UGC 5946	13.4	1.1 x 0.5	12.6	Double System	162	Double system, envelopes in contact.
11 00 26.0	+76 55 44	UGC 6065	14.6	1.0 x 0.3	13.1	Sb	99	Stellar **MCG +13-8-50** NNW 1′5; close pair of mag 11 stars WNW 2′8.
11 07 19.3	+76 41 41	UGC 6154	13.9	0.8 x 0.5	12.7	SBa	65	Mag 10.25v star SE 1′7.
11 27 59.5	+78 59 36	UGC 6456	13.7	1.5 x 0.9	13.9	Pec	0	**MCG +13-8-59** N 4′4.
11 28 56.7	+73 02 04	UGC 6473	14.1	1.5 x 0.6	13.8	Sab	20	Located between a pair of stars, mags 14.1 and 12.5.
11 45 16.0	+79 40 51	UGC 6728	13.9	0.9 x 0.6	13.1	SB0/a	160	
11 49 32.1	+75 59 46	UGC 6789	14.0	1.0 x 0.6	13.3	SBb	128	
12 04 19.9	+79 24 32	UGC 7059	14.6	1.2 x 0.7	14.3	SABb	36	**UGC 7058** S 1′7.
12 05 55.9	+77 30 18	UGC 7086	14.3	2.4 x 0.6	14.5	Sb III	72	
12 06 42.0	+80 55 11	UGC 7097	15.1	1.5 x 0.4	14.4	Sbc	82	
12 11 18.6	+74 48 29	UGC 7189	13.9	1.8 x 0.9	14.3	SBdm:	162	Bright knot or star on both N and S ends.
12 13 18.6	+75 02 02	UGC 7226	14.3	1.7 x 1.2	14.9	Sm	15	
12 13 55.4	+74 30 09	UGC 7238	15.3	1.5 x 0.2	13.8	Scd:	51	Mag 11.8 star SW edge; mag 13.8 star NE edge.
12 35 32.5	+73 40 29	UGC 7767	12.6	0.9 x 0.9	12.4	E		Mag 10.13v star S 2′7.
12 41 53.3	+75 18 23	UGC 7872	15.4	1.7 x 0.7	15.4	Im	174	Dwarf irregular, very low surface brightness; mag 8.02v star N 3′0.
12 43 38.2	+73 36 46	UGC 7908	14.6	1.5 x 0.4	13.8	Scd:	50	
12 46 18.6	+83 24 32	UGC 7956	14.4	1.5 x 1.0	14.6	Sc	54	Mag 5.81v star 32 Camelopardalis E 4′8.
12 49 59.7	+78 23 03	UGC 7995	15.9	1.8 x 1.5	16.8	Im:	99	Dwarf irregular, extremely low surface brightness.
12 54 53.0	+73 10 44	UGC 8052	13.5	1.8 x 0.5	13.2	S?	155	Mag 12.9 star S edge; mag 11.9 star E 2′0.
12 59 56.1	+73 41 30	UGC 8120	13.8	1.5 x 0.7	13.7	Sd	22	Mag 13.3 star S edge; mag 10.9 star SSW 1′9.
13 00 23.0	+80 07 21	UGC 8148	14.9	1.1 x 0.6	14.3	Sb I	61	
13 02 18.2	+78 32 29	UGC 8164	14.0	1.7 x 0.9	14.3	SAbc III	5	
13 04 10.3	+78 23 15	UGC 8183	14.5	1.5 x 0.4	13.8	Sb	121	Mag 11.9 star N edge.
13 08 35.0	+78 56 10	UGC 8245	13.9	1.5 x 0.6	13.6	Im	75	
13 08 28.1	+84 37 50	UGC 8264	13.2	1.5 x 1.4	13.9	Pec	93	**MCG +14-6-17** NE 8′9.
13 11 02.9	+78 24 45	UGC 8287	13.8	1.3 x 0.8	13.7	(R′)SB(s)a	155	Mag 12.0 star NNE 2′2.
13 23 06.1	+84 30 16	UGC 8454	14.7	2.3 x 0.5	14.7	Double System	69	Double system, long, faint bridge; mag 11.4 star NW 2′5.
13 48 56.5	+74 14 57	UGC 8747	13.9	1.2 x 0.9	13.8	SA(rs)c	18	
14 00 39.7	+78 51 10	UGC 8964	14.8	1.5 x 0.4	14.1	Sdm	48	
14 07 17.2	+75 55 45	UGC 9052	14.5	1.0 x 0.8	14.1	SAd	63	Mag 12.8 star SW 2′1.

GALAXY CLUSTERS

RA h m s	Dec ° ′ ″	Name	Mag 10th brightest	No. Gal	Diam ′	Notes
11 33 42.0	+76 13 00	A 1297	16.1	37	11	**CGCG 351-64** NW of center 1′2; all other members anonymous and stellar.
11 55 48.0	+73 28 00	A 1412	15.9	86	22	Mag 7.53v star on SW edge; all GX faint, stellar and anonymous.
12 13 48.0	+74 23 00	A 1500	15.6	44	28	Almost stellar **PGC 38861** and **38866** SW of center 8′4.

PLANETARY NEBULAE

RA h m s	Dec ° ′ ″	Name	Diam ″	Mag (P)	Mag (V)	Mag cent ★	Alt Name	Notes
12 33 06.8	+82 33 50	IC 3568	10	11.6	10.6	13.4	PK 123+34.1	

6 GALAXIES 6

RA h m s	Dec ° ′ ″	Name	Mag (V)	Dim ′ Maj x min	SB	Type Class	PA	Notes
08 53 15.8	+73 11 18	CGCG 332-23	14.6	0.4 x 0.3	12.2		21	Located 1′2 W of mag 9.58v star.
09 12 14.4	+73 35 36	CGCG 332-33	13.9	0.5 x 0.4	12.0		141	Mag 10.08v star NE 6′9.
06 19 13.3	+80 04 05	IC 440	13.3	1.7 x 0.9	13.6	SA(r)ab II-III	33	Mag 8.30v star ENE 9′2.
06 36 12.2	+82 58 05	IC 442	12.9	0.9 x 0.9	12.5	S?		Mag 8.93v star S 7′4.
06 52 12.7	+74 25 37	IC 450	13.9	0.9 x 0.5	12.9	SAB0$^+$:	130	Mag 11.1 star S 3′0; mag 9.74v star SSE 7′7.
06 52 52.7	+74 28 50	IC 451	13.8	1.3 x 1.1	14.0	SAB(r)b: III	155	Mag 10.30v star NW 9′1.
07 30 18.5	+79 52 20	IC 467	12.6	3.2 x 1.0	13.7	SAB(s)c:	80	Mag 10.33v star N 3′9.
08 40 50.3	+73 29 10	IC 511	13.4	1.6 x 0.5	13.1	S0/a:	143	Mag 6.15v star N 9′8.
08 53 42.7	+73 29 23	IC 520	11.7	1.5 x 1.3	12.3	SAB(rs)ab?	0	Mag 11.6 star WSW 5′9; low surface brightness **UGC 4634** N 2′6.
09 18 33.7	+73 45 32	IC 529	11.9	3.3 x 1.3	13.4	SA(s)c:	145	Mag 10.52v star N 4′6.
07 09 06.0	+75 21 09	IC 2174	13.5	1.0 x 0.9	13.2	(R′)SB(r)a:	42	NGC 2314 E 5′7; semi-circle of five mag 12-14 stars centered NE 2′9.
08 47 58.7	+73 32 18	IC 2389	13.4	1.6 x 0.3	12.4	SB(s)b?	126	Mag 11.8 star SE 2′2; NGC 2646 ESE 11′1.
08 25 08.6	+74 29 46	MCG +12-8-40	14.8	0.8 x 0.8	14.2	Spiral		UGC 4363 S 3′8.
09 02 47.8	+73 18 34	MCG +12-9-29	15.0	0.7 x 0.6	13.9	Sc	114	Mag 11.5 star N 2′3.
08 01 57.3	+75 56 22	MCG +13-6-15	15.0	0.7 x 0.3	13.1	Spiral	120	Mag 11.3 star SW 3′8.
08 27 58.5	+75 23 43	MCG +13-6-23	13.9	0.6 x 0.5	12.4			
09 24 12.6	+76 31 51	MCG +13-7-27	14.2	0.5 x 0.4	12.5	E?		Located between a N-S pair of mag 11 stars.
09 29 26.5	+77 37 53	MCG +13-7-28	14.8	0.6 x 0.5	13.3	Sbc	33	Mag 10.9 double star SE 2′4.
09 44 18.1	+76 21 05	MCG +13-7-36	14.3	0.9 x 0.4	13.1	Pec	39	Close pair of mag 11-12 star NW 3′4.
09 50 18.9	+75 44 49	MCG +13-7-38	14.8	0.7 x 0.6	13.7	Sb	9	Mag 9.26v star SSW 2′8.
09 56 26.2	+78 51 14	MCG +13-7-39	15.9	0.6 x 0.3	13.9	Sb	33	Mag 13.3 SW 3′4.
10 10 28.4	+77 18 16	MCG +13-8-8	15.1	0.6 x 0.5	13.7	Sc	12	Mag 10.8 star NE 2′3.
10 19 58.4	+79 27 05	MCG +13-8-12	14.6	0.7 x 0.6	13.5	Sbc	141	Mag 9.87v star SE 2′7.

GALAXIES

RA h m s	Dec ° ′ ″	Name	Mag (V)	Dim ′ Maj x min	SB	Type Class	PA	Notes
06 18 40.9	+78 21 15	NGC 2146	10.6	5.4 x 2.9	13.4	SB(s)ab pec II	131	Numerous superimposed stars on envelope on NE half.
06 23 55.6	+78 31 47	NGC 2146A	12.9	2.8 x 1.1	13.9	SAB(s)c:	30	
06 47 14.1	+74 14 10	NGC 2256	12.5	2.3 x 2.0	14.0	SAB0⁻?	72	**MCG +12-7-13** NW 8.′8.
06 47 46.6	+74 28 54	NGC 2258	11.9	2.3 x 1.5	13.1	SA(r)0°	150	Mag 11.06v star on E edge, with fainter star S of bright core and star on S edge.
07 14 20.6	+84 22 54	NGC 2268	11.5	2.7 x 1.5	12.9	SAB(r)bc II	63	
07 10 33.2	+75 19 34	NGC 2314	12.2	1.1 x 0.8	12.1	E3	25	Mag 10.7 star NNE 3.′8; IC 2174 W 5.′7.
07 27 03.5	+80 10 39	NGC 2336	10.4	7.1 x 3.9	13.9	SAB(r)bc I	178	Several dark lanes with numerous stars superimposed.
07 51 54.8	+73 00 56	NGC 2441	12.2	1.8 x 1.6	13.2	SAB(r)b: II	45	
08 15 00.1	+73 34 45	NGC 2523	11.9	3.0 x 1.8	13.6	SB(r)bc I	57	Strong E-W bar, mag 8.18v star NNE 9.′0; **UGC 4279** NE 7.′0.
08 04 08.9	+74 02 51	NGC 2523A	13.8	1.0 x 0.7	13.3	SB(s)c:	95	
08 12 56.5	+73 33 45	NGC 2523B	13.9	2.1 x 0.3	13.3	SA(s)b: sp III	92	Located 8.′8 W of NGC 2523.
08 17 43.8	+73 19 03	NGC 2523C	12.9	1.5 x 0.8	13.1	E?	95	
08 21 39.8	+73 59 17	NGC 2544	12.9	1.1 x 0.8	12.6	SB(s)a:	70	Much elongated **MCG +12-8-35** NE 1.′4.
08 24 33.4	+74 00 41	NGC 2550	12.8	1.0 x 0.4	11.7	Sb:	103	Located 2.′4 W of mag 8.45 star.
08 28 40.1	+73 44 56	NGC 2550A	12.7	1.5 x 1.4	13.4	Sc	0	**UGC 4389** SSW 6.′8; **MCG +12-8-44** SSE 7.′2; UGC 4413 NNE 8.′0.
08 24 51.0	+73 24 43	NGC 2551	12.1	1.7 x 1.1	12.6	SA(s)0/a	55	
08 37 24.4	+78 01 29	NGC 2591	12.3	3.0 x 0.6	12.7	Scd: sp	34	
08 45 35.0	+72 59 54	NGC 2630	14.2	1.4 x 0.2	12.7	Sbc	29	Mag 9.93v star NW 6.′5.
08 48 04.7	+74 05 55	NGC 2633	12.2	2.5 x 1.5	13.5	SB(s)b I-II	175	Two slender main arms; mag 10.62v star E 5.′2.
08 48 25.2	+73 58 00	NGC 2634	12.0	1.7 x 1.6	13.1	E1:	51	Pair with NGC 2634A SE 1.′9.
08 48 37.0	+73 56 17	NGC 2634A	13.6	1.8 x 0.4	13.1	SB(s)bc? sp	64	Pair with NGC 2634 NW 1.′9.
08 48 24.9	+73 40 14	NGC 2636	13.8	0.4 x 0.4	11.9	E0:		Mag 9.93v star E 6.′0.
08 50 21.7	+73 27 46	NGC 2646	12.1	1.3 x 1.3	12.5	SB(r)0°:		Mag 11.5 star S 2.′5; IC 2389 WNW 11.′1.
08 55 37.8	+78 13 18	NGC 2655	10.1	4.9 x 4.1	13.2	SAB(s)0/a	90	Large core in smooth envelope; mag 7.33v star E 9.′7.
09 08 06.0	+78 05 03	NGC 2715	11.2	4.9 x 1.5	13.2	SAB(rs)c II	22	Located on NE edge of galaxy cluster **A 719**.
09 13 23.6	+79 11 13	NGC 2732	11.9	2.1 x 0.8	12.4	S0 sp	67	Mag 13.4 star on E edge; UGC 4832 E 3.′4.
09 13 43.2	+76 28 32	NGC 2748	11.7	3.0 x 1.1	12.8	SAbc III	38	
09 43 31.4	+79 42 05	NGC 2908	14.1	0.8 x 0.8	13.5	S		Mag 10.02v star NE 7.′0.
09 38 24.7	+76 19 03	NGC 2938	13.5	1.7 x 0.9	13.8	SB(rs)cd	105	Mag 8.84v star SE 8.′1.
09 43 47.3	+74 51 36	NGC 2977	12.5	1.8 x 0.8	12.7	Sb:	145	
10 05 38.7	+80 17 11	NGC 3057	13.0	2.2 x 1.3	14.0	SB(s)dm IV	5	Several strong knots; mag 9.63v star NE 9.′3.
09 56 13.1	+75 51 55	NGC 3061	12.8	1.7 x 1.5	13.6	(R′)SB(rs)c	6	Faint stellar nucleus.
10 15 32.0	+74 13 14	NGC 3144	13.4	1.2 x 0.7	13.1	SB(s)ab pec:	0	Mag 13.1 star on E edge.
10 16 54.2	+73 23 56	NGC 3147	10.6	3.9 x 3.5	13.3	SA(rs)bc II	155	
10 17 41.0	+74 20 48	NGC 3155	12.9	1.0 x 0.5	12.0	S?	41	
10 14 27.7	+77 49 11	NGC 3197	13.5	1.3 x 1.0	13.7	Sbc II-III	155	**Mkn 1248** N 3.′4; mag 9.98v star S 3.′5.
05 47 27.7	+79 38 03	UGC 3340	13.7	1.1 x 0.4	12.7	Sab	158	Mag 7.75v star WSW 3.′9; **CGCG 348-3** SW 8.′3; **UGC 3347** SE 7.′4.
05 51 17.8	+79 41 44	UGC 3353	13.6	1.4 x 1.0	13.8	SB0	32	
05 51 26.0	+74 16 57	UGC 3357	13.9	1.5 x 0.4	13.2	S0/a	54	
05 54 06.3	+76 39 49	UGC 3364	14.8	1.0 x 1.0	14.7	SAd		Stellar nucleus, uniform envelope.
05 56 36.8	+75 18 56	UGC 3371	14.2	4.6 x 3.7	17.2	Im: V	135	Dwarf, extremely low surface brightness; mag 9.90v star W 3.′9.
05 59 08.1	+78 30 32	UGC 3373	13.2	1.4 x 0.8	13.2	SABc II	100	**UGC 3370** W 7.′4.
06 06 19.5	+83 50 19	UGC 3378	13.7	1.5 x 1.5	14.4			**UGC 3381** S 7.′4.
06 01 37.7	+73 06 57	UGC 3384	15.4	1.7 x 1.7	16.4	Sm:		Dwarf spiral, very low surface brightness; mag 11.5 star E 1.′1.
06 04 53.2	+80 08 38	UGC 3385	13.9	1.4 x 0.5	13.4	S0	168	Pair of stars, mags 11 and 12, S 2.′0.
06 09 57.0	+79 55 31	UGC 3396	14.1	0.7 x 0.7	13.4	E		**UGC 3397** SE 3.′8.
06 13 47.2	+81 04 21	UGC 3401	14.8	1.7 x 0.2	13.4	Sb II-III	20	Mag 9.95v star WSW 1.′9.
06 14 02.3	+80 00 13	UGC 3404	14.3	0.8 x 0.5	13.2	S0	6	Mag 10.73v star N 1.′8.
06 13 57.9	+80 28 33	UGC 3405	14.5	1.5 x 0.3	13.5	Sbc	127	
06 14 13.4	+79 41 19	UGC 3408	15.1	1.5 x 0.6	14.8	Triple System	12	Triple system.
06 14 30.8	+80 26 59	UGC 3410	14.2	2.1 x 0.7	14.5	Sb	118	Pair with UGC 3410 SE 2.′0.
06 15 39.1	+79 42 32	UGC 3412	15.1	1.1 x 0.5	14.3	Scd:	153	Mag 10.61v star N 1.′9.
06 17 00.7	+81 08 17	UGC 3413	12.7	1.5 x 1.0	13.0	SB(r)bc	85	
06 16 01.2	+75 56 13	UGC 3420	13.3	2.3 x 0.7	13.7	Sb II-III	152	
06 17 42.7	+78 49 22	UGC 3423	15.9	1.5 x 0.2	14.4	Sdm:	156	Mag 11.4 star on N end; mag 9.24v star S 2.′7.
06 19 22.4	+76 48 41	UGC 3431	12.4	1.8 x 1.3	13.3	E	173	Located 1.′2 SSW of a mag 7.84v star.
06 24 57.8	+82 19 05	UGC 3435	13.8	0.8 x 0.4	12.4	S?	116	
06 27 49.6	+83 18 38	UGC 3441	14.4	0.8 x 0.7	13.7	(R)SABa?	55	Elongated **CGCG 362-17** SW 10.′2.
06 31 06.0	+84 55 29	UGC 3442	13.9	1.1 x 0.3	12.5	S0/a	107	Mag 11.9 star on W end; mag 10.83v star W 0.′6.
06 24 52.9	+73 07 43	UGC 3453	13.7	1.2 x 0.7	13.4	SB?	50	
06 28 00.3	+74 18 04	UGC 3460	14.1	1.7 x 0.8	14.3	SB(s)dm:	12	Two main arms.
06 29 20.7	+74 29 31	UGC 3464	14.4	1.5 x 0.4	13.7	Sa?	42	
06 31 35.2	+74 26 00	UGC 3471	12.8	1.4 x 1.1	13.1	SA(s)c I-II	30	Mag 12.1 star on N edge; **CGCG 330-4** W 4.′5.
06 33 02.9	+75 52 02	UGC 3472	14.1	1.4 x 0.5	13.6	S0	60	
06 36 17.7	+74 20 36	UGC 3486	14.6	0.9 x 0.4	13.3	Sab	130	Mag 8.88v star S 3.′1.
06 47 56.3	+84 09 58	UGC 3500	13.8	1.7 x 0.4	13.2	S?	6	Mag 11.3 star E 1.′8.
06 55 00.9	+84 02 28	UGC 3521	14.3	1.0 x 0.6	13.6	S?	75	UGC 3528 NNE 2.′7.
06 56 06.1	+84 55 03	UGC 3522	14.3	2.1 x 1.1	15.1	S?	132	Stellar nucleus.
06 56 09.5	+84 04 39	UGC 3528	13.5	1.4 x 0.7	13.3	SBab	40	UGC 3521 SSE 2.′7.
06 54 34.4	+82 28 44	UGC 3540	14.8	1.7 x 0.4	14.2	Scd:	125	
06 53 27.5	+77 24 33	UGC 3548	13.7	0.5 x 0.4	11.8	Sa	33	Small companion N edge.
06 54 58.9	+80 57 54	UGC 3549	13.5	0.6 x 0.6	12.4	E		**MCG +13-5-32** SW 1.′4.
06 59 02.5	+80 00 12	UGC 3581	12.9	1.2 x 0.8	12.7	SAB(rs)c: I-II	100	
07 02 22.9	+80 57 08	UGC 3604	13.3	0.7 x 0.7	12.4	S		
07 06 02.5	+75 26 37	UGC 3636	14.1	1.0 x 0.3	12.7	Sab	3	Mag 6.99v star WSW 7.′0.
07 10 47.2	+80 58 17	UGC 3668	14.7	1.1 x 0.7	14.3	Scd:	3	
07 10 24.3	+74 53 25	UGC 3675	14.8	1.5 x 0.2	13.3	Scd:	21	Mag 8.87v star SSW 4.′2.
07 12 11.3	+73 28 11	UGC 3705	13.3	0.8 x 0.6	12.4	S?	66	Mag 10.8 star N 0.′7.
07 13 14.3	+73 50 35	UGC 3717	13.0	2.1 x 0.9	13.6	Sbc	103	
07 14 21.1	+73 28 23	UGC 3730	12.5	2.7 x 1.0	13.4	Ring	171	Strongly peculiar, jets plus plume.
07 22 52.7	+77 48 27	UGC 3794	15.9	1.7 x 1.1	16.4	Sm:	120	Dwarf spiral, extremely low surface brightness.
07 30 47.1	+73 37 45	UGC 3858	12.9	1.1 x 0.9	12.7	Sa	120	
07 30 48.8	+73 42 21	UGC 3859	12.7	1.5 x 1.0	13.0	Sa	33	Elongated core, smooth envelope.
07 36 40.2	+74 26 50	UGC 3906	13.8	1.5 x 0.9	14.0	S	45	Double system, contact, disrupted.
07 36 59.2	+73 42 49	UGC 3909	14.1	2.6 x 0.4	14.0	SBcd:	82	Mag 11.25v star S 2.′9.
07 39 55.6	+75 25 28	UGC 3929	14.4	1.3 x 0.8	14.3	Compact	114	
07 44 42.2	+73 49 16	UGC 3972	13.8	1.2 x 0.7	13.4	SB?	160	Mag 10.84v star SW 1.′2.
07 53 36.0	+84 37 05	UGC 3992	15.2	1.2 x 0.9	15.1	Sb	159	Mag 13.3 star superimposed S end.
07 55 44.2	+84 55 34	UGC 3993	12.8	1.6 x 1.2	13.4	S0?	35	Mag 11.8 star E 1.′1.

(Continued from previous page)

6 GALAXIES 6

RA h m s	Dec ° ′ ″	Name	Mag (V)	Dim ′ Maj x min	SB	Type Class	PA	Notes
07 49 25.5	+74 20 00	UGC 4014	13.6	1.0 x 0.5	12.7	S0$^-$:	122	Double galaxy system **MCG +12-8-12** E 3′5.
07 50 50.5	+74 21 27	UGC 4028	12.7	1.0 x 0.8	12.3	SAB(s)c?	10	Double galaxy system **MCG +12-8-12** SW 3′3.
07 52 38.4	+73 30 08	UGC 4041	12.9	1.0 x 0.5	12.1	E?	129	**UGC 4037** N 5′6.
07 54 04.4	+74 23 07	UGC 4057	12.5	2.4 x 0.7	12.9	Sa	55	
07 56 16.6	+78 00 42	UGC 4066	13.2	1.5 x 1.3	13.8	Scd:	30	
08 04 27.1	+84 38 30	UGC 4078	14.3	2.1 x 0.3	13.6	Sbc	82	Located between a pair of mag 12.4 and 11.4 stars N-S.
07 56 51.3	+73 47 10	UGC 4080	13.2	1.0 x 0.7	12.6	Sa	60	Pair of mag 13 stars NE 1′7.
08 06 58.0	+84 45 18	UGC 4100	15.4	1.4 x 1.0	15.6	IBm	135	Irregular, low surface brightness.
08 02 03.0	+73 02 37	UGC 4137	14.7	1.5 x 1.1	15.1	Sdm	133	Low surface brightness; mag 12.7 star on NE edge.
08 04 18.6	+77 49 00	UGC 4151	12.6	1.2 x 1.2	12.8	Sdm:		
08 07 06.2	+80 07 45	UGC 4173	14.7	1.9 x 0.6	14.7	Im:	145	Irregular, low surface brightness.
08 06 39.9	+73 20 36	UGC 4199	13.8	1.2 x 0.4	12.9	S?	99	**CGCG 331-27** WSW 2′3.
08 11 38.2	+76 25 18	UGC 4238	12.7	2.6 x 1.5	14.1	SBd	83	Mag 9.44v star on W edge.
08 19 06.0	+83 15 48	UGC 4262	13.1	1.7 x 0.9	13.4	Sb III	155	Double system, companion superimposed.
08 15 29.2	+78 51 44	UGC 4263	13.1	0.6 x 0.4	11.4	Pec	45	
08 17 39.3	+77 30 01	UGC 4282	14.2	1.0 x 0.5	13.3	Sbc	147	**MCG +13-6-19** S 2′1.
08 19 42.8	+79 14 03	UGC 4292	13.8	1.1 x 0.4	12.8	S0/a	84	Faint star superimposed.
08 23 35.7	+79 11 16	UGC 4328	14.6	1.3 x 0.4	13.7	SBdm:	10	Low surface brightness, pair of stars superimposed.
08 25 02.7	+74 25 59	UGC 4363	14.5	1.4 x 1.4	15.1	(R′)SB(s)d: IV-V		Very filamentary; mag 12.7 stars on E and S edges; mag 8.62v star W 1′9.
08 27 51.9	+73 31 03	UGC 4390	14.3	2.0 x 1.8	15.5	SBd	78	Mag 10.6 star W 2′2.
08 34 44.3	+84 19 08	UGC 4396	14.4	2.0 x 0.4	14.0	SBb-c	26	Mag 11.3 star E 1′4.
08 29 46.3	+73 51 25	UGC 4413	14.7	1.5 x 0.3	13.7	Scd:	105	NGC 2550A SSW 6′8; faint stellar galaxy **KUG 0825+739** SE 8′4.
08 33 41.0	+74 24 14	UGC 4448	13.4	0.8 x 0.5	12.4	E	120	
08 34 33.4	+75 09 10	UGC 4451	13.6	0.9 x 0.3	12.0	Sa-b	129	Mag 8.33v star NE 3′4; stellar Mkn SE 1′0.
08 39 54.8	+73 45 15	UGC 4502	14.7	1.5 x 0.4	14.0	SBcd:	75	Mag 10.45v star N 1′4.
08 57 06.3	+84 48 32	UGC 4601	14.0	1.4 x 0.4	13.2		140	Mag 13.6 star SE end.
08 53 13.0	+76 29 13	UGC 4623	12.7	3.2 x 0.8	13.5	Scd:	60	Mag 11.6 star superimposed on SE edge.
08 57 10.2	+79 42 12	UGC 4644	14.2	1.1 x 1.0	14.2	SAB(s)b	165	
09 02 01.9	+78 16 37	UGC 4701	14.5	2.0 x 0.9	15.0	Sd:	110	
09 03 15.3	+78 33 43	UGC 4714	12.9	1.3 x 1.0	13.1	SABb II-III	105	Mag 14.2 star on N edge.
09 04 27.2	+74 53 58	UGC 4736	15.5	1.5 x 1.2	16.0	SAdm	160	
09 14 53.3	+79 11 44	UGC 4832	14.1	0.9 x 0.4	12.8	Pec	118	Mag 10.20v star N 3′0; NGC 2732 W 3′4.
09 14 48.3	+74 13 56	UGC 4841	12.4	2.8 x 2.1	14.2	SAB(s)d	150	**CGCG 332-36** S 6′6.
09 17 14.9	+80 02 07	UGC 4852	14.5	1.0 x 0.4	13.3	Im	95	
09 18 32.2	+74 19 09	UGC 4883	12.3	1.3 x 0.7	12.0	S	27	Located near center of small galaxy cluster A 762.
09 21 36.6	+75 07 21	UGC 4937	14.3	1.0 x 0.6	13.6	Sd:	165	**MCG +13-7-22** W 6′7.
09 31 14.1	+73 48 36	UGC 5052	13.2	1.5 x 1.2	13.6	SB(s)a	120	Mag 11.56v star S edge.
09 34 36.8	+81 08 42	UGC 5068	13.4	0.8 x 0.7	12.6	S	90	
09 37 20.7	+73 23 00	UGC 5110	14.5	1.7 x 1.3	15.2	IABm	135	Irregular, low surface brightness.
09 40 04.4	+82 06 20	UGC 5114	15.1	1.7 x 0.7	15.1	IBm:	140	Mag 10.74v star W 2′0.
09 44 05.6	+83 48 23	UGC 5128	14.4	0.9 x 0.7	13.7	Scd:	177	Mag 8.52v star NW 3′9.
09 46 52.1	+79 48 36	UGC 5203	14.9	2.1 x 0.2	13.8	Sd	80	Mag 10.02v star SW 4′2.
10 04 59.5	+79 16 16	UGC 5402	13.8	1.1 x 0.7	13.3	S?	175	Located between a pair of mag 13 stars.

GALAXY CLUSTERS

RA h m s	Dec ° ′ ″	Name	Mag 10th brightest	No. Gal	Diam ′	Notes
09 19 06.0	+74 17 00	A 762	16.2	32	22	Most members faint, anonymous and stellar.

7 GALAXIES 7

RA h m s	Dec ° ′ ″	Name	Mag (V)	Dim ′ Maj x min	SB	Type Class	PA	Notes
03 45 18.4	+76 38 27	IC 334	11.3	2.5 x 2.1	13.0	S?	51	A number of stars superimposed; mag 9.10v star NE 14′3.
04 44 28.8	+75 38 19	IC 381	12.3	2.4 x 1.3	13.4	SAB(rs)bc	177	Mag 10.72v star NW 8′3.
04 57 21.8	+78 11 23	IC 391	12.7	1.1 x 1.1	12.7	SA(s)c		Mag 9.90v star NNW 5′8.
06 19 13.3	+80 04 05	IC 440	13.3	1.7 x 0.9	13.6	SA(r)ab II-III	33	Mag 8.30v star ENE 9′2.
02 07 42.4	+75 09 37	MCG +12-3-2	13.9	1.4 x 0.6	13.8	E	15	Mag 11.7 star W 1′8.
04 34 06.8	+73 17 55	MCG +12-5-6	14.6	0.9 x 0.5	13.6	Sb	27	NGC 1573 SE 4′6; mag 8.51v star W 9′1.
04 40 04.8	+74 11 06	MCG +12-5-14	14.0	1.0 x 0.6	13.5	E/S0	156	Mag 11.5 star NE 3′2.
02 29 21.6	+84 01 15	MCG +14-2-9	14.8	0.7 x 0.3	13.0	Sb	105	Mag 10.9 star S 3′9.
03 16 45.5	+80 47 35	NGC 1184	12.4	2.8 x 0.6	12.9	S0/a	168	**MCG +13-3-3** E 5′2 between a pair of mag 11.4, 12.4 stars.
04 23 26.4	+75 17 58	NGC 1530	11.4	4.6 x 2.4	13.9	SB(rs)b	24	Faint ESE-WNW bar.
04 35 04.4	+73 15 43	NGC 1573	11.7	1.9 x 1.3	12.7	E	35	UGC 3069 SW 4′3.
06 18 40.9	+78 21 15	NGC 2146	10.6	5.4 x 2.9	13.4	SB(s)ab pec II	131	Numerous superimposed stars on envelope on NE half.
01 40 36.7	+83 56 24	UGC 1148	14.4	0.9 x 0.9	14.0	S?		
01 56 18.8	+73 16 54	UGC 1378	12.6	3.5 x 2.3	14.7	(R)SB(rs)a:	11	Numerous stars superimposed on faint outer envelope; mag 9.06v star SE 5′5.
02 55 42.5	+75 09 14	UGC 2358	14.2	2.0 x 1.2	15.0	SB(r)b	25	Mag 12.4 star S edge; mag 7.91v star NE 3′6.
02 58 48.0	+75 44 45	UGC 2411	15.6	4.2 x 0.3	15.7	S?	12	Pair of stars, mags 12.5 and 13.5, along N end of E edge.
03 04 04.6	+74 26 38	UGC 2485	16.1	0.9 x 0.7	15.5	Scd:	30	Mag 10.9 star E 1′4.
03 09 18.9	+80 07 48	UGC 2519	13.6	1.4 x 0.8	13.6	Scd?	75	Two strong knots N edge.
03 19 21.1	+81 20 54	UGC 2603	14.2	1.7 x 1.5	15.1	Im	55	Dwarf irregular, very low surface brightness.
03 20 17.2	+80 14 47	UGC 2620	14.7	1.5 x 1.1	15.1	Sm:	160	
03 35 34.3	+80 05 09	UGC 2767	14.0	1.5 x 1.4	14.7	Sm?		Dwarf spiral, very low surface brightness.
03 50 36.1	+73 01 48	UGC 2865	13.9	1.4 x 0.6	13.6	SB?	14	Pair of mag 11.8 stars 2′0 S.
03 59 02.2	+79 33 22	UGC 2896	14.7	1.5 x 0.3	13.6	Scd:	54	Mag 8.73v star SE 2′7.
04 01 01.5	+74 05 00	UGC 2906	13.0	2.7 x 1.6	14.4	Sb:	5	
04 02 43.6	+78 16 45	UGC 2907	13.8	1.6 x 1.4	14.5	S?	81	Mag 10.56v star SSE 3′7.
04 33 15.5	+76 34 01	UGC 3057	14.2	1.0 x 0.8	13.8	SAB(s)b	90	Two main arms.
04 34 28.8	+73 12 14	UGC 3069	13.8	1.3 x 0.7	13.6	S0$^-$:	55	NGC 1573 NE 4′3.
04 41 54.6	+79 57 39	UGC 3101	15.3	1.2 x 0.6	14.8	SAdm:	115	Mag 11.8 star 1′6 N.
04 41 09.0	+73 40 15	UGC 3110	13.9	2.3 x 0.9	14.5	SBcd:	115	Mag 10.61v star E end.
04 49 58.3	+82 39 00	UGC 3132	14.1	1.5 x 0.6	13.8	S0/a	162	Close pair of mag 12 stars NW 1′2.
04 46 15.2	+76 25 03	UGC 3137	14.2	2.7 x 0.5	14.4	S? III-IV	75	
04 47 55.9	+74 55 40	UGC 3144	14.4	1.9 x 1.0	14.9	IBm IV-V	117	Dwarf irregular, low surface brightness.
04 48 26.5	+73 28 07	UGC 3150	13.5	1.7 x 1.2	14.1	SABbc	165	Mag 9.01v star S 1′5.
04 52 30.5	+74 29 43	UGC 3175	14.4	1.5 x 0.2	13.0	Sbc	173	

(Continued from previous page)

7 GALAXIES 7

RA h m s	Dec ° ′ ″	Name	Mag (V)	Dim ′ Maj x min	SB	Type Class	PA	Notes
04 59 55.6	+80 10 41	UGC 3197	14.0	0.9 x 0.9	13.8	E?		Mag 14.9 star on S edge; mag 9.99v star S 2′.5.
05 06 35.7	+75 35 41	UGC 3230	13.1	1.1 x 0.9	12.9	S0/a	28	**MCG +13-4-15** NW 3′.9.
05 08 54.9	+75 25 42	UGC 3241	13.4	1.4 x 0.8	13.5	S0	172	Several stars superimposed on edges.
05 19 42.1	+84 03 08	UGC 3253	12.4	1.7 x 0.9	12.7	SB(r)b	96	
05 21 44.7	+84 28 56	UGC 3257	13.8	1.5 x 0.7	13.7	SBa	157	
05 21 17.5	+76 21 21	UGC 3276	14.8	1.5 x 0.3	13.8	Sb II-III	80	Mag 13.2 star W end; mag 11.3 star E end.
05 28 23.1	+76 39 47	UGC 3302	12.9	1.2 x 1.0	12.9	SAB(r)bc: II	160	Mag 11.9 star W 0′.9.
05 33 37.5	+73 43 29	UGC 3317	14.4	1.5 x 1.1	14.8	Im V	177	Dwarf irregular, extremely low surface brightness.
05 35 27.9	+76 48 19	UGC 3318	14.1	1.5 x 1.3	14.7	Sm:	45	Dwarf spiral, very low surface brightness.
05 39 35.8	+77 18 41	UGC 3326	14.6	3.1 x 0.3	14.4	Scd:	69	
05 47 27.7	+79 38 03	UGC 3340	13.7	1.1 x 0.4	12.7	Sab	158	Mag 7.75v star WSW 3′.9; **CGCG 348-3** SW 8′.3; **UGC 3347** SE 7′.4.
05 51 17.8	+79 41 44	UGC 3353	13.6	1.4 x 1.0	13.8	SB0	32	
05 51 26.0	+74 16 57	UGC 3357	13.9	1.5 x 0.4	13.2	S0/a	54	
05 54 06.3	+76 39 49	UGC 3364	14.8	1.0 x 1.0	14.7	SAd		Stellar nucleus, uniform envelope.
05 56 36.8	+75 18 56	UGC 3371	14.2	4.6 x 3.7	17.2	Im: V	135	Dwarf, extremely low surface brightness; mag 9.90v star W 3′.9.
05 59 08.1	+78 30 32	UGC 3373	13.2	1.4 x 0.8	13.2	SABc II	100	**UGC 3370** W 7′.4.
06 06 19.5	+83 50 19	UGC 3378	13.7	1.5 x 1.5	14.4			**UGC 3381** S 7′.4.
06 01 37.7	+73 06 57	UGC 3384	15.4	1.7 x 1.7	16.4	Sm:		Dwarf spiral, very low surface brightness; mag 11.5 star E 1′.1.
06 04 53.2	+80 08 38	UGC 3385	13.9	1.4 x 0.5	13.4	S0	168	Pair of stars, mags 11 and 12, S 2′.0.
06 09 57.0	+79 55 31	UGC 3396	14.1	0.7 x 0.7	13.4	E		**UGC 3397** SE 3′.8.
06 13 47.2	+81 04 21	UGC 3401	14.8	1.7 x 0.2	13.4	Sb II-III	20	Mag 9.95v star WSW 1′.9.
06 14 02.3	+80 00 13	UGC 3404	14.3	0.8 x 0.5	13.2	S0	6	Mag 10.73v star N 1′.8.
06 13 57.9	+80 28 33	UGC 3405	14.5	1.5 x 0.3	13.5	Sbc	127	
06 14 13.4	+79 41 19	UGC 3408	15.1	1.5 x 0.6	14.8	Triple System	12	Triple system.
06 14 30.8	+80 26 59	UGC 3410	14.2	2.1 x 0.7	14.5	Sb	118	Pair with UGC 3410 SE 2′.0.
06 15 39.1	+79 42 32	UGC 3412	15.1	1.1 x 0.5	14.3	Scd:	153	Mag 10.61v star N 1′.9.
06 17 00.7	+81 08 17	UGC 3413	12.7	1.5 x 1.0	13.0	SB(r)bc	85	
06 16 01.2	+75 56 13	UGC 3420	13.3	2.3 x 0.7	13.7	Sb II-III	152	
06 17 42.7	+78 49 22	UGC 3423	15.9	1.5 x 0.2	14.4	Sdm:	156	Mag 11.4 star on N end; mag 9.24v star S 2′.7.
06 19 22.4	+76 48 41	UGC 3431	12.4	1.8 x 1.3	13.3	E	173	Located 1′.2 SSW of a mag 7.84v star.

GALAXY CLUSTERS

RA h m s	Dec ° ′ ″	Name	Mag 10th brightest	No. Gal	Diam ′	Notes
03 49 36.0	+75 11 00	A 449	16.2	56	17	All cluster members anonymous, faint and stellar.
05 00 12.0	+80 00 00	A 505	15.2	39	34	UGC 3197 near N edge; other members faint and stellar.
05 10 18.0	+73 41 00	A 527	15.7	34	22	All members anonymous, faint and stellar.

OPEN CLUSTERS

RA h m s	Dec ° ′ ″	Name	Mag	Diam ′	No. ★	B ★	Type	Notes
02 01 05.0	+75 29 36	Be 8		5	40	14.0	cl	
05 12 39.0	+73 58 00	Cr 464	4.2	120	50		cl	

GLOBULAR CLUSTERS

RA h m s	Dec ° ′ ″	Name	Total V m	B ★ V m	HB V m	Diam ′	Conc. Class Low = 12 High = 1	Notes
03 33 20.8	+79 34 57	Pal 1	13.6	16.3		2.8	12	

8 GALAXIES 8

RA h m s	Dec ° ′ ″	Name	Mag (V)	Dim ′ Maj x min	SB	Type Class	PA	Notes
22 40 36.0	+72 51 50	UGC 12154	14.7	1.5 x 0.2	13.2	Sb	24	
22 45 38.0	+73 09 42	UGC 12182	14.6	0.8 x 0.6	13.7	S	69	
22 57 01.5	+72 42 30	UGC 12263	16.9	2.8 x 1.9	18.5	Im?	165	Dwarf, very low surface brightness.

OPEN CLUSTERS

RA h m s	Dec ° ′ ″	Name	Mag	Diam ′	No. ★	B ★	Type	Notes
00 45 13.0	+64 23 30	Be 4	10.6	4	48	18.0	cl	Few stars; moderate brightness range; strong central concentration; detached. Probably not a cluster.
00 02 10.0	+67 25 00	Be 59		10	40	11.0	cl	Moderately rich in bright and faint stars with a strong central concentration; detached; involved in nebulosity.
00 48 10.0	+67 12 06	Be 61		3	15	14.0	cl	
01 01 12.0	+63 56 30	Be 62	9.3	6	50	13.0	cl	Moderately rich in stars; moderate brightness range; no central concentration; detached.
23 21 16.9	+71 47 42	Be 99		5	60	14.0	cl	
23 25 58.0	+63 47 00	Be 100		2.7	100	16.0	cl	(A) Compact rich cluster.
23 32 46.5	+64 12 16	Be 101		2	10	16.0	cl:	
00 03 27.0	+63 35 30	Be 104		3	15	16.0	cl	
01 03 07.0	+62 47 12	Cz 3	9.9	1.5	10		cl:	
00 50 00.0	+64 08 00	Do 13		10	30		cl	Moderately rich in stars; small brightness range; not well detached.
00 21 52.0	+64 23 00	King 1	19.3	9	100	13.0	cl	
23 47 36.6	+68 38 26	King 11		6	50	17.0	cl	
00 31 54.0	+63 10 00	King 14	8.5	7	186	10.0	cl	Few stars; small brightness range; no central concentration; detached. (S) Sparse.
00 43 42.0	+64 11 00	King 16	10.3	5	71	11.0	cl:	Moderately rich in stars; moderate brightness range; strong central concentration; detached. Probably not a cluster.
23 49 54.0	+62 42 00	King 21	9.6	4	20	10.0	cl	Few stars; moderate brightness range; strong central concentration; detached.
00 27 25.4	+71 23 00	NGC 110		19	100	9.7	cl?	Few stars; small brightness range; not well detached.
00 31 18.0	+63 21 12	NGC 133	9.4	3	5		cl:	Few stars; small brightness range; not well detached.
00 33 06.0	+63 18 00	NGC 146	9.1	5	132	11.6	cl	Few stars; moderate brightness range; slight central concentration; detached. (S) Sparse.
01 05 11.0	+62 01 18	NGC 358		3	25		cl?	
01 06 27.0	+62 13 42	NGC 366		4	30	10.0	cl	Moderately rich in bright and faint stars; detached.
23 36 30.0	+72 51 00	NGC 7708		30	25	7.3	ast?	

(Continued from previous page)

8 OPEN CLUSTERS 8

RA h m s	Dec ° ′ ″	Name	Mag	Diam ′	No. ★	B ★	Type	Notes
23 49 54.0	+68 01 00	NGC 7762	10.0	15	40	11.0	cl	Moderately rich in stars; moderate brightness range; slight central concentration; detached.
00 58 24.0	+68 28 00	Skiff J0058.4+6828		7	200	10.2	cl	
01 12 03.0	+62 16 24	St 3		5	8	11.0	cl	Few stars; small brightness range; not well detached.
23 43 47.0	+62 09 37	St 17		1		8.4	cl	
00 01 35.0	+64 37 24	St 18		5			cl	
00 25 16.0	+62 37 48	St 20		1		13.0	cl	

BRIGHT NEBULAE

RA h m s	Dec ° ′ ″	Name	Dim ′ Maj x min	Type	BC	Color	Notes
00 04 42.0	+67 10 00	Ced 214	50 x 40	E	2-5	3-4	Nebulosity brightest in the NE sector. The N part is broken up by absorption patches.
00 03 36.0	+68 37 00	NGC 7822	65 x 20	ESNR?	3-5	3-4	Brightest in the E part which contains two narrow bright wisps about 10′ in length along opposing borders.
22 56 48.0	+62 37 00	Sh2-155	50 x 30	E	3-5	3-4	**Cave Nebula**. The N half of this nebula, a crescent-shaped nebulosity, is its brightest part.

PLANETARY NEBULAE

RA h m s	Dec ° ′ ″	Name	Diam ″	Mag (P)	Mag (V)	Mag cent ★	Alt Name	Notes
00 13 00.9	+72 31 19	NGC 40	74	10.7	12.3	11.5	PK 120+9.1	Elongated N-S.
22 49 02.2	+67 01 40	PN G111.2+7.0	15	18.2			PK 111+6.1	Slightly elongated N-S; mag 12.1 star NE 1′.2; mag 11.5 star NW 2′.5; stellar planetary **Koh 4-58** W 6′.5.
23 12 15.7	+64 39 16	PN G112.5+3.7	5	18.6			PK 112+3.1	Mag 13.3 star S 1′.2; mag 7.33vb star N 5′.3.
23 31 51.5	+70 22 16	PN G116.2+8.5	63	14.3	14.2	21.1	PK 116+8.1	Diamond shaped, oriented NE-SW; mag 11.5 star E 1′.2; mag 10.79v star NE 2′.0; mag 7.87v star E 8′.3.
00 01 30.9	+70 42 30	PN G118.7+8.2	63	16.7		20.0	PK 118+8.2	= **Abell 86**; close pair of mag 14 stars W 1′.6; mag 9.82v star N 3′.9.
00 19 58.8	+62 59 01	PN G119.3+0.3	60	15.6	14.7	22.3	PK 119+0.1	Very slender, elongated E-W; mag 12.8 star NNW 0′.7; mag 11.19v star E 2′.3.
00 12 54.6	+69 10 23	PN G119.4+6.5	47	18.3		20.5	PK 119+6.1	Mag 14.3 star SW 2′.7.
00 40 21.6	+62 51 25	PN G121.6-0.0	38		15.4		PK 121+0.1	Elongated E-W; mag 11.4 star W 4′.4; mag 9.6 star SE 5′.5.
00 38 54.6	+66 23 47	PN G121.6+3.5	24	21.9		21.0	PK 121+3.1	Mag 12.2 star SW 3′.0; mag 13.4 star N 3′.0.
01 07 12.9	+73 32 57	PN G124.0+10.7	270			16.4	PK 124+10.1	Group of 3 mag 11 stars form a right angle shape NW 4′.1; a close pair of mag 11 stars SE 3′.7.

9 GALAXIES 9

RA h m s	Dec ° ′ ″	Name	Mag (V)	Dim ′ Maj x min	SB	Type Class	PA	Notes
20 54 52.7	+65 09 30	MCG +11-25-3	15.0	0.6 x 0.5	13.5	Spiral	138	Mag 9.60v star NW 2′.1.
20 19 39.0	+66 43 40	NGC 6911	14.3	1.9 x 1.1	15.0	SBb:	115	Almost stellar nucleus.
20 35 07.3	+64 48 09	NGC 6949	13.6	1.4 x 1.2	14.0	S?	6	Stellar nucleus.
20 37 14.6	+66 06 18	NGC 6951	10.7	3.9 x 3.2	13.2	SAB(rs)bc I-II	170	Numerous stars superimposed on envelope.
20 18 31.6	+62 41 26	UGC 11539	13.6	1.2 x 1.2	14.0	E		Bright center, uniform envelope; mag 13.4 star N edge; mag 9.67v star NW 1′.2.
20 53 11.6	+67 10 13	UGC 11648	16.5	1.5 x 0.3	15.5	Im	147	Moderately bright knot NW end; mag 10.8 star NE 1′.6.
21 05 07.3	+66 46 21	UGC 11678	13.7	1.4 x 1.4	14.3	S0?		Bright, elongated center, uniform envelope; mag 12.8 star on NNW edge.
21 09 37.0	+66 10 10	UGC 11689	13.6	1.3 x 1.0	13.7	SB(r)b	153	Mag 7.13v star E 6′.1.
21 47 16.7	+72 28 18	UGC 11818	14.1	1.2 x 0.9	14.0	S?	147	
21 56 24.3	+73 15 35	UGC 11861	13.6	3.5 x 2.6	15.8	SABdm	32	Very patchy; few superimposed stars along SE edge.
22 40 36.0	+72 51 50	UGC 12154	14.7	1.5 x 0.2	13.2	Sb	24	
22 45 38.0	+73 09 42	UGC 12182	14.6	0.8 x 0.6	13.7	S	69	

OPEN CLUSTERS

RA h m s	Dec ° ′ ″	Name	Mag	Diam ′	No. ★	B ★	Type	Notes
21 55 29.0	+63 56 24	Be 93		4.3	120	16.0	cl	
22 13 24.5	+70 15 00	ClvdB 152		25	15		cl	(A) Centered on star of nebulosity vdB 152.
21 00 36.6	+68 09 35	Cr 427	13.8	4	6		cl	(A) On W side of reflection nebula NGC 7023. Note this is not NGC 7023 itself.
21 42 00.0	+66 05 00	NGC 7129	11.5	8	10		cl	Few stars; moderate brightness range; not well detached; nebulosity involved. (A) Scattered group with bright nebula at center.
21 45 09.0	+65 47 00	NGC 7142	9.3	12	186	11.0	cl	Rich in stars; moderate brightness range; strong central concentration; detached.
21 53 40.0	+62 36 12	NGC 7160	6.1	5	61	7.0	cl	Few stars; large brightness range; strong central concentration; detached.
22 19 00.0	+63 15 00	PMH79 1		9.3	21	7.8	cl?	

BRIGHT NEBULAE

RA h m s	Dec ° ′ ″	Name	Dim ′ Maj x min	Type	BC	Color	Notes
21 01 37.0	+68 10 00	NGC 7023	10 x 8	R	1-5	1-4	Contains open cluster Cr 427.
21 42 54.0	+66 06 00	NGC 7129	7 x 7	R	1-5	1-4	Reflection nebula enveloping a compact cluster. (A) Scattered group with bright nebula at center.
21 43 48.0	+66 08 00	NGC 7133	3 x 3	R	1-5	1-4	
21 37 06.0	+68 12 00	vdB 143	7 x 6	R	2-5	1-4	Approximately diamond-shaped nebulosity with a mag 8.3 star involved.
22 13 36.0	+70 18 00	vdB 152	4 x 3	R	1-5	1-4	Nebula with a mag 8.8 star involved; a dust-ball at the edge of a small, very dense cloud.

PLANETARY NEBULAE

RA h m s	Dec ° ′ ″	Name	Diam ″	Mag (P)	Mag (V)	Mag cent ★	Alt Name	Notes
21 26 23.5	+62 53 33	NGC 7076	56	17.0		18.0	PN G101.8+8.7	Mag 14.4 star N 1′.1; mag 13.8 star N 1′.9. Brightest along E edge.
21 46 08.6	+63 47 31	NGC 7139	77	<13.	13.3	18.7	PK 104+7.1	Pair of mag 11.3 stars, on NE-SW line, SW 4′.3.
22 13 22.8	+65 53 55	PN G107.7+7.8	900			17.7	PK 107+7.1	Numerous mag 10-12 stars scattered about disc; mag 9.53v star WNW 10′.0;, and close pair of mag 10.70v, 10.06v stars NE 9′.9. from center.
22 19 33.9	+70 56 05	PN G111.0+11.6	530			15.1	PK 111+11.1	Brightest along NW edge; mag 10.29v star on SE edge.
22 49 02.2	+67 01 40	PN G111.2+7.0	15	18.2			PK 111+6.1	Slightly elongated N-S; mag 12.1 star NE 1′.2; mag 11.5 star NW 2′.5; stellar planetary **Koh 4-58** W 6′.5.

GALAXIES

RA h m s	Dec ° ′ ″	Name	Mag (V)	Dim ′ Maj x min	SB	Type Class	PA	Notes
17 56 49.9	+70 17 23	CGCG 340-27	14.7	0.3 x 0.3	12.0	Spherical		Forms an equilateral triangle with two mag 11.5 stars S.
18 48 49.8	+73 21 48	CGCG 341-9	15.6	0.6 x 0.3	13.6		153	CGCG 341-10 and galaxy group **Hickson 85** E 6′.4.
18 50 18.6	+73 21 01	CGCG 341-10	14.4	0.4 x 0.3	12.2	E?	12	= **Hickson 85A**; **Hickson 85B** SE 0′.7; **Hickson 85C** NE 0′.9; **Hickson 85D** E 1′.2.
18 33 36.0	+67 06 37	IC 4763	13.3	0.9 x 0.4	12.0	S?	102	Mag 14.6 star S edge; NGC 6677 and NGC 6679 NNW 1′.7.
17 56 35.0	+62 37 00	MCG +10-25-121	14.8	0.8 x 0.2	12.7	Sp		Located 1′.1 E of a mag 7.59v star, NE member of a bright pair.
18 27 34.3	+62 06 57	MCG +10-26-39	14.3	0.7 x 0.7	13.6	E?		Mag 12.4 star WNW 2′.6.
19 33 28.8	+62 05 15	MCG +10-28-3	14.4	0.5 x 0.4	12.5	Sb	114	Mag 8.34v star SSW 2′.2.
17 54 50.2	+62 38 41	NGC 6512	13.9	0.7 x 0.5	12.8	E?	57	
17 55 16.8	+62 40 10	NGC 6516	14.8	0.5 x 0.2	12.1	Sp	147	Mag 9.38v star NW 2′.4; a pair of stellar anonymous galaxies SE at 0′.9 and 1′.8.
17 55 48.7	+62 36 43	NGC 6521	12.9	1.6 x 1.3	13.7	E	160	
17 56 08.5	+64 16 57	NGC 6534	14.5	0.8 x 0.5	13.4	S?	24	Mag 10.40v star on N edge; spherical anonymous galaxy S 1′.2.
17 57 16.3	+64 56 17	NGC 6536	13.4	1.2 x 1.1	13.5	SB(r)b	117	Stellar nucleus.
17 54 16.9	+73 25 23	NGC 6538	13.3	1.0 x 0.5	12.4	S	48	
18 00 07.8	+66 36 54	NGC 6552	13.7	1.0 x 0.7	13.1	SB?	105	Bright NW-SE bar.
18 08 55.5	+69 04 03	NGC 6598	13.6	1.6 x 1.1	14.0	Pec	40	
18 12 57.5	+68 21 43	NGC 6621	13.6	2.1 x 0.8	14.0	Sb: pec	145	Double system.
18 12 55.3	+68 21 50	NGC 6622	13.7	0.5 x 0.4	11.8	Sa	145	
18 22 03.0	+66 36 59	NGC 6636	13.7	2.2 x 0.4	13.4	S?	179	**MCG +11-22-47** on NE edge.
18 25 28.1	+68 00 19	NGC 6650	13.9	0.4 x 0.4	11.8			
18 24 19.6	+71 36 04	NGC 6651	13.1	1.6 x 0.7	13.0	SA(r)c	30	Strong dark lane SE side.
18 24 07.4	+73 10 59	NGC 6654	12.0	2.6 x 2.1	13.7	(R′)SB(s)0/a	0	Bright NNE-SSW bar; mag 10.87v star W 2′.4.
18 39 26.1	+73 34 47	NGC 6654A	12.9	2.6 x 0.8	13.6	SB(s)d pec?	63	UGC 11331 NW 2′.6.
18 30 39.0	+67 59 09	NGC 6667	12.7	2.3 x 1.1	13.5	SABab? pec	105	Bright center offset N; very broad extension of faint material extends N and W beyond published dimensions.
18 33 09.6	+66 57 35	NGC 6676	14.4	1.6 x 0.3	13.5	Sbc	142	
18 33 30.5	+67 08 11	NGC 6677	15.1	0.4 x 0.3	12.7	Compact	63	IC 4763 SSE 1′.7; almost stellar NGC 6679 just N.
18 33 31.6	+67 08 46	NGC 6679	15.1	0.3 x 0.2	11.9	Spherical	103	Almost stellar.
18 34 50.1	+70 31 21	NGC 6690	12.5	3.8 x 1.3	14.0	Sd? sp	171	
18 55 22.0	+72 46 16	NGC 6747	14.6	0.5 x 0.3	12.4		126	
19 05 37.2	+63 56 03	NGC 6762	13.4	1.4 x 0.4	12.6	S0/a	119	
19 10 53.9	+73 24 38	NGC 6786	12.9	1.1 x 0.9	12.7	SB?	45	**UGC 11415** NE 1′.2.
19 16 41.9	+63 58 14	NGC 6789	13.2	1.3 x 1.0	13.3	Im	60	
19 37 23.1	+62 22 57	NGC 6817	14.9	0.8 x 0.5	13.8	Sbc	87	Multi-galaxy system, two bright cores.
19 41 54.7	+64 04 18	NGC 6825	14.4	0.5 x 0.4	12.5		174	
20 00 42.5	+66 13 40	NGC 6869	12.0	1.5 x 1.3	12.6	S0	90	
20 19 39.0	+66 43 40	NGC 6911	14.3	1.9 x 1.1	15.0	SBb:	115	Almost stellar nucleus.
17 54 52.8	+71 32 26	UGC 11066	14.5	1.5 x 0.7	14.4	SAB(s)dm	10	Uniform surface brightness; mag 10.8 star W 1′.1.
18 01 07.2	+67 25 33	UGC 11099	14.3	1.2 x 0.4	13.3	SBcd?	82	Very faint, small galaxy **KUG 1800+674** WNW 6′.0.
18 02 04.3	+69 42 48	UGC 11106	14.1	0.9 x 0.9	13.7	SB(r)bc		Stellar nucleus, uniform envelope; **CGCG 340-34** NE 11′.8.
18 11 33.8	+68 42 00	UGC 11165	15.3	1.5 x 0.5	14.8	Scd:	84	Uniform surface brightness; single slender arm E of center.
18 15 01.6	+71 00 25	UGC 11193	14.2	1.5 x 0.9	14.4	Im	162	Dwarf irregular, very low surface brightness.
18 18 38.8	+68 21 07	UGC 11208	14.1	1.0 x 0.9	13.8	Scd:		Stellar nucleus; star superimposed E; mag 9.71v star NW 2′.1.
18 21 56.5	+68 07 41	UGC 11222	14.0	1.0 x 0.6	13.3	SBcd:	71	
18 24 02.4	+65 18 21	UGC 11230	15.4	2.4 x 0.2	14.5	Sd	14	Mag 7.81v star SE 5′.2.
18 38 59.5	+73 36 35	UGC 11331	14.3	1.1 x 0.7	13.9	Sm:	33	NGC 6654A SE 2′.6.
18 39 30.8	+73 49 32	UGC 11334	15.1	1.9 x 0.3	14.3	Pec?	105	Almost stellar core; mag 11.2 star W 1′.5.
18 41 12.2	+70 32 51	UGC 11342	14.6	0.8 x 0.5	13.5	Sb	123	
18 47 34.5	+70 44 00	UGC 11363	13.9	1.0 x 0.7	13.4	S0	111	Faint star or small anonymous galaxy on S edge.
18 52 24.4	+73 11 36	UGC 11377	15.0	1.5 x 0.2	13.5	Scd:	10	Mag 9.17v star S 4′.2.
19 03 17.7	+73 02 32	UGC 11402	14.3	0.7 x 0.7	13.4	S		Mag 11.5 star E 0′.8.
19 08 43.5	+70 17 01	UGC 11411	15.1	0.8 x 0.6	14.2	Sc	147	NE-SW oriented row of five mag 11 stars SE 1′.5.
19 15 58.1	+72 46 25	UGC 11427	13.8	1.0 x 0.2	11.9	S	69	
19 29 57.1	+72 06 44	UGC 11455	14.3	2.6 x 0.5	14.4	Scd:	61	**MCG +12-18-2** NW 3′.4.
19 45 58.8	+67 59 14	UGC 11476	14.7	1.5 x 0.7	14.6	S	140	
19 49 24.6	+63 30 31	UGC 11487	14.5	0.8 x 0.5	13.4	SBb	114	Mag 10.97v star SW 1′.5.
19 53 02.2	+67 39 54	UGC 11496	13.3	2.1 x 2.1	14.8	Sm:		Dwarf spiral, very low surface brightness.
20 04 56.9	+62 47 48	UGC 11515	13.8	1.3 x 1.3	14.2	Sd		Mag 14.0 star on NW edge; mag 11.8 star S 2′.0.
20 13 18.5	+66 17 31	UGC 11529	15.7	1.7 x 0.5	15.4	Sbc	17	Mag 12.3 star E of center; mag 8.89v star SW 3′.8.
20 18 31.6	+62 41 26	UGC 11539	13.6	1.2 x 1.2	14.0	E		Bright center, uniform envelope; mag 13.4 star N edge; mag 9.67v star NW 1′.2.

GALAXY CLUSTERS

RA h m s	Dec ° ′ ″	Name	Mag 10th brightest	No. Gal	Diam ′	Notes
17 59 54.0	+69 13 00	A 2295	16.2	48	28	Most members anonymous, stellar/almost stellar.
18 14 48.0	+69 39 00	A 2301	15.8	34	22	**MCG +12-17-19** W of center 2′.6; all other members anonymous, stellar.
18 33 36.0	+71 01 00	A 2308	16.4	39	16	All members anonymous, faint and stellar.
18 49 42.0	+70 22 00	A 2311	16.0	60	22	All members anonymous, faint, stellar.
18 53 36.0	+68 21 00	A 2312	15.8	74	34	**CGCG 323-5** N of center 7′.6; all other members anonymous.
19 00 48.0	+69 57 00	A 2315	16.3	66	16	All members anonymous, faint and stellar.

OPEN CLUSTERS

RA h m s	Dec ° ′ ″	Name	Mag	Diam ′	No. ★	B ★	Type	Notes
18 35 00.0	+72 23 00	Kemble 2		30	17	7.0	ast	(A) "Mini-Cassiopeia" asterism.

PLANETARY NEBULAE

RA h m s	Dec ° ′ ″	Name	Diam ″	Mag (P)	Mag (V)	Mag cent ★	Alt Name	Notes
17 58 33.5	+66 37 59	NGC 6543	20	8.8	8.1	11.1	PK 96+29.1	**Cat's Eye Nebula**. Very bright central star; **IC 4677** W 1′.8, is part of the outer envelope of NGC 6543; mag 9.78v star WNW 2′.7.
18 21 52.2	+64 21 53	PN G94.0+27.4	114	14.5	14.2	15.0	PK 94+27.1	Mag 12.9 star on E edge; mag 11.40v star W 2′.1; mag 11.0 star N 2′.5.

RA h m s	Dec ° ′ ″	Name	Mag (V)	Dim ′ Maj x min	SB	Type Class	PA	Notes
17 20 25.8	+62 36 13	CGCG 300-14	14.7	0.5 x 0.5	13.1	Spherical		Mag 8.81v star W 5′3.
16 28 33.7	+69 41 32	CGCG 339-6	14.1	0.5 x 0.5	12.4			Mag 11.8 star N 1′4.
17 56 49.9	+70 17 23	CGCG 340-27	14.7	0.3 x 0.3	12.0	Spherical		Forms an equilateral triangle with two mag 11.5 stars S.
15 32 01.1	+68 14 44	IC 1129	13.1	1.1 x 0.9	12.9	Scd:	170	Mag 9.78v star SW 11′1.
15 44 08.7	+72 25 48	IC 1145	14.2	1.5 x 0.5	13.7	Sbc	168	Mag 10.42v star NNE 4′4.
15 48 22.2	+69 23 07	IC 1146	13.8	0.9 x 0.8	13.3		105	Mag 7.49v star NW 9′0; elongated **MCG +12-15-20** NNE 3′8; **MCG +12-15-18** and **UGC 10053** N 5′1.
15 50 11.8	+69 33 35	IC 1147	14.6	0.6 x 0.6	13.3			
15 52 28.6	+70 22 31	IC 1154	13.3	1.0 x 0.8	13.1	E	145	Mag 12.9 star N 1′3.
16 14 30.2	+62 32 11	IC 1210	13.1	1.6 x 0.5	12.7	Sab	168	Mag 11.1 star E 4′6.
16 16 11.9	+65 58 04	IC 1214	14.0	1.4 x 0.6	13.7	S0/a	18	Pair of faint stars superimposed N of nucleus.
16 15 35.0	+68 23 54	IC 1215	13.2	1.1 x 0.7	12.8	SB?	10	Mag 10.74v star W 9′4; IC 1216 SSE 3′4.
16 15 55.5	+68 20 59	IC 1216	14.1	1.0 x 0.9	13.8	Scd:	33	Mag 12.0 star SW 1′9; IC 1215 NNW 3′4.
16 16 37.9	+68 12 07	IC 1218	13.6	1.1 x 0.3	12.3	S?	57	Mag 11.39v star NE 2′3.
16 36 53.0	+67 37 45	IC 1225	14.5	1.4 x 0.4	13.7	SBb?	73	Mag 10.91v star N 6′6.
16 42 06.3	+65 35 11	IC 1228	13.3	1.6 x 1.5	14.1	(R′)SB(s)ab	39	Mag 13.0 star N edge; mag 12.2 star W 2′5.
16 52 03.6	+63 06 54	IC 1235	14.1	0.8 x 0.5	13.0		21	Mag 8.69v star NE 3′2.
17 01 27.9	+63 41 25	IC 1241	13.6	1.2 x 1.0	13.6	Sc	165	Mag 11.4 star NE 3′9.
17 10 12.7	+72 24 31	IC 1251	13.5	1.4 x 0.8	13.5	Scd:	71	Mag 11.04v star WNW 3′0.
17 11 33.3	+72 24 01	IC 1254	13.8	1.6 x 0.7	13.8	Sb? pec	35	Mag 10.67v star NW 4′7.
17 23 26.5	+71 15 44	IC 1261	14.4	0.9 x 0.7	14.0	E	27	**MCG +12-16-32B** on W edge.
16 27 19.9	+62 39 33	MCG +10-23-75	14.7	0.9 x 0.5	13.7	Sb	117	**UGC 10411** NE 3′9.
16 46 13.1	+62 13 46	MCG +10-24-52	14.7	0.6 x 0.6	13.4			Mag 14.8 star on S edge.
17 08 45.8	+62 02 34	MCG +10-24-101	14.9	0.8 x 0.3	13.2	S	138	Mag 11.73v star NW 2′8.
17 40 52.6	+62 36 25	MCG +10-25-81	14.9	0.7 x 0.7	14.0			Mag 10.5 star S 3′4.
17 56 35.0	+62 37 00	MCG +10-25-121	14.8	0.8 x 0.2	12.7	Sp		Located 1′1 E of a mag 7.59v star, NE member of a bright pair.
15 47 18.9	+68 22 02	MCG +11-19-15	14.1	0.7 x 0.4	12.8	E/S0	177	
15 51 27.9	+64 07 39	MCG +11-19-21	15.1	1.0 x 0.6	14.4	Sm	165	Mag 12.8 star superimposed N edge.
15 57 51.0	+64 00 18	MCG +11-19-32	14.9	0.6 x 0.6	13.7			Mag 10.81v star SE 6′1.
16 02 43.7	+66 29 31	MCG +11-20-1	14.4	0.9 x 0.7	13.8		6	Mag 12.8 star on E edge.
16 14 25.3	+64 58 58	MCG +11-20-8	13.9	1.0 x 0.3	12.4		42	Mag 13.1 star NE 1′2.
16 29 48.6	+67 22 39	MCG +11-20-16	14.5	0.8 x 0.4	13.1	S?	168	Mag 12.5 star on W edge.
16 36 14.3	+66 14 16	MCG +11-20-19	13.8	1.1 x 0.6	13.2	Sa	42	Multi-galaxy system?
16 45 25.5	+67 57 11	MCG +11-20-28	14.2	0.9 x 0.5	13.2	S?	168	
15 42 55.8	+70 49 39	MCG +12-15-13	14.7	0.6 x 0.5	13.5	E	0	Mag 10.54v star S 0′9.
15 48 52.4	+69 38 35	MCG +12-15-21	14.9	0.9 x 0.5	13.9	Spherical	123	Mag 9.35v star S 4′0.
15 49 23.4	+71 10 19	MCG +12-15-25	14.5	0.9 x 0.5	13.5	Sbc	126	Located in a small triangle of mag 14 stars.
15 58 29.3	+70 23 51	MCG +12-15-39	15.3	0.6 x 0.4	13.6	Sb	64	Mag 10.92v star E 2′3.
16 01 26.7	+70 23 01	MCG +12-15-45	15.1	0.6 x 0.3	13.1	Sab	33	Located between a pair of N-S mag 13 stars.
16 02 07.1	+70 25 00	MCG +12-15-47	13.9	1.0 x 1.0	14.0	E/S0		
16 07 15.2	+69 55 51	MCG +12-15-53	14.7	1.0 x 0.3	13.2	Sb	69	NGC 6091 SE 3′6.
16 51 39.8	+68 55 31	MCG +12-16-12A	14.1	0.6 x 0.4	12.4	S?	81	Mag 12.4 star NW 3′8.
17 24 23.5	+72 09 24	MCG +12-16-37	14.7	0.7 x 0.5	13.4		108	
15 28 00.4	+64 45 48	NGC 5949	12.0	2.2 x 1.0	12.7	SA(r)bc?	147	
15 46 31.7	+72 10 11	NGC 6011	13.5	2.0 x 0.7	13.7	Sb I-II	110	
15 51 25.0	+62 18 33	NGC 6015	11.1	5.4 x 2.1	13.6	SA(s)cd II-III	28	Patchy center, many small knots.
15 52 09.3	+64 50 23	NGC 6019	15.4	0.4 x 0.4	13.3			Located 2′6 S of mag 10.08v star.
15 53 08.1	+64 55 02	NGC 6024	14.2	0.7 x 0.6	13.1		117	**MCG +11-19-28** SE 8′4; mag 9.22v star W 6′2.
15 57 30.1	+70 41 17	NGC 6048	12.3	2.2 x 1.7	13.7	E	140	Elongated anonymous galaxy SE 2′5.
16 00 59.3	+70 36 03	NGC 6071	14.0	0.9 x 0.9	13.7	SB(s)b		Short E-W bar.
16 04 29.0	+69 39 56	NGC 6079	12.7	1.4 x 1.0	13.1	E	150	**MCG +12-15-48** W 3′4; **IC 1201** SE 7′6.
16 07 52.8	+69 54 16	NGC 6091	14.1	0.7 x 0.4	12.6		114	MCG +12-15-53 NW 3′6.
16 06 33.4	+72 29 40	NGC 6094	13.2	1.8 x 1.4	14.1	S0	120	
16 14 22.4	+63 15 38	NGC 6111	14.0	0.7 x 0.6	13.0	E/S0	21	**MCG +11-20-6** W 1′3.
16 20 57.1	+65 23 23	NGC 6140	11.3	6.3 x 4.6	14.8	SB(s)cd pec	95	Bright center, patchy arms, star on NW edge.
16 39 32.0	+66 02 20	NGC 6214	13.5	1.0 x 0.8	13.1	S	145	**MCG +11-20-23** WSW 4′7.
16 43 20.4	+70 37 52	NGC 6232	12.5	1.5 x 1.5	13.2	(R′)SB(s)a		Bright nucleus in short bar with dark patches N and S.
16 44 35.1	+70 46 52	NGC 6236	11.9	2.9 x 1.7	13.5	SAB(s)cd	15	Very patchy, many bright knots; two very bright knots S of center.
16 47 16.8	+62 08 45	NGC 6238	14.7	0.6 x 0.4	13.0	Pec	17	
16 48 04.0	+62 12 02	NGC 6244	13.5	1.5 x 0.3	12.5	SBa:	140	Very faint star N edge; mag 8.74v star E 7′8.
16 48 19.4	+62 58 32	NGC 6247	13.5	1.0 x 0.3	12.0	Pec	58	
16 46 23.2	+70 21 10	NGC 6248	13.1	3.2 x 1.2	14.4	SBd	150	Narrow, elongated center.
16 51 50.2	+63 42 54	NGC 6260	13.9	0.8 x 0.8	13.3	Sc		Mag 9.79v star ESE 2′6.
16 55 34.1	+63 14 28	NGC 6275	14.3	0.5 x 0.3	12.1	Compact	120	
16 57 24.6	+68 27 22	NGC 6288	14.5	0.8 x 0.4	13.3	E/S0	102	
16 57 45.1	+68 30 50	NGC 6289	13.8	0.8 x 0.6	13.1	E/S0	21	Mag 10.78v star N 1′4; **CGCG 321-10** NE 4′8.
17 03 36.6	+62 01 30	NGC 6297	13.6	0.7 x 0.5	12.3	S0	90	
17 05 04.6	+62 27 26	NGC 6299	14.1	0.6 x 0.6	12.8			
17 05 02.7	+68 49 38	NGC 6303	13.7	1.3 x 0.8	13.8	E	60	
17 08 59.5	+62 53 53	NGC 6317	15.1	1.0 x 0.3	13.6		51	
17 09 44.2	+62 58 23	NGC 6319	13.4	0.8 x 0.8	13.0	E/S0		
17 10 24.6	+72 18 11	NGC 6340	11.0	3.2 x 3.0	13.3	SA(s)0/a	120	Mag 11.0 star on NNW edge.
17 22 44.2	+62 09 51	NGC 6365A	14.5	1.0 x 0.9	14.2	SBcd:	137	NGC 6365B on N edge; mag 10.21v star NE 1′6.
17 22 43.5	+62 10 21	NGC 6365B	15.9	0.9 x 0.3	14.3	Sdm:	31	
17 26 31.7	+71 05 48	NGC 6395	12.3	2.4 x 0.7	12.7	Scd:	15	
17 36 06.3	+68 09 19	NGC 6419	14.6	1.3 x 0.3	13.4	Sa	131	**MCG +11-21-13** SW 5′4.
17 36 16.7	+68 03 06	NGC 6420	14.4	0.7 x 0.4	13.1	E/S0	54	**MCG +11-21-11** SW 6′1.
17 36 30.2	+68 03 29	NGC 6422	14.8	0.6 x 0.6	13.8	E/S0		Mag 8.91v star E 2′9; anonymous galaxy NW 1′1.
17 36 53.5	+68 10 13	NGC 6423	14.7	0.6 x 0.5	13.3		165	
17 36 12.4	+69 59 18	NGC 6424	14.1	0.9 x 0.6	13.4	Compact	81	**CGCG 340-14** E 9′3.
17 36 49.3	+72 05 20	NGC 6434	12.4	2.3 x 0.8	12.9	SBbc:	100	Mag 7.64v star S 2′1.
17 40 11.3	+62 38 29	NGC 6435	14.3	1.1 x 0.6	13.7	Compact	5	MCG +10-25-81 E 5′3.
17 42 32.0	+67 35 32	NGC 6456	14.8	0.8 x 0.5	13.7		45	Elongated anonymous galaxy W 1′6; mag 9.50v star and **CGCG 321-32** S 6′4.
17 42 53.0	+66 28 34	NGC 6457	14.2	1.2 x 0.9	14.2	S0$^-$:	140	Stellar galaxy **VII Zw 736** NW 6′2.
17 38 42.8	+73 30 52	NGC 6461	15.6	0.4 x 0.3	13.1		132	Mag 10.25v star W 2′6.
17 43 34.3	+67 36 12	NGC 6463	14.1	0.6 x 0.6	13.0	E		
17 44 13.0	+67 35 35	NGC 6470	15.1	1.3 x 0.2	13.5	Scd:	170	NGC 6471 E 0′5.
17 44 18.2	+67 35 27	NGC 6471	15.1	0.3 x 0.3	12.4	Scd:		Stellar; NGC 6470 W 0′5.
17 44 15.1	+67 37 11	NGC 6472	14.2	1.1 x 0.7	13.8	SBb	159	Dark lanes E and W of center; anonymous galaxy ESE 1′5.

(Continued from previous page)

GALAXIES

RA h m s	Dec ° ′ ″	Name	Mag (V)	Dim ′ Maj x min	SB	Type Class	PA	Notes
17 44 03.5	+67 37 47	NGC 6477	15.0	0.6 x 0.3	13.2	E/S0	15	Almost stellar.
17 49 20.9	+62 13 21	NGC 6488	14.5	0.6 x 0.4	12.8		63	Trio of stars NW 1′.7.
17 49 29.2	+70 08 28	NGC 6503	10.2	7.1 x 2.4	13.2	SA(s)cd III	123	Mag 8.60v star E 3′.5.
17 51 07.3	+65 31 51	NGC 6505	13.8	1.1 x 1.0	14.0	E/S0	126	
17 49 46.6	+72 01 15	NGC 6508	12.8	1.3 x 1.3	13.3	E:		
17 54 50.2	+62 38 41	NGC 6512	13.9	0.7 x 0.5	12.8	E?	57	
17 55 16.8	+62 40 10	NGC 6516	14.8	0.5 x 0.2	12.1	Sp	147	Mag 9.38v star NW 2′.4; a pair of stellar anonymous galaxies SE at 0′.9 and 1′.8.
17 55 48.7	+62 36 43	NGC 6521	12.9	1.6 x 1.3	13.7	E	160	
17 56 08.5	+64 16 57	NGC 6534	14.5	0.8 x 0.5	13.4	S?	24	Mag 10.40v star on N edge; spherical anonymous galaxy S 1′.2.
17 57 16.3	+64 56 17	NGC 6536	13.4	1.2 x 1.1	13.5	SB(r)b	117	Stellar nucleus.
17 54 16.9	+73 25 23	NGC 6538	13.3	1.0 x 0.5	12.4	S	48	
18 00 07.8	+66 36 54	NGC 6552	13.7	1.0 x 0.7	13.1	SB?	105	Bright NW-SE bar.
15 31 45.6	+67 33 19	UGC 9896	13.5	1.3 x 1:1	13.7	Scd:	145	Weak stellar nucleus.
15 35 48.4	+73 26 59	UGC 9944	13.9	1.3 x 0.4	13.1	S?	174	Mag 10.35 v star SE 3′.4.
15 40 07.3	+70 44 25	UGC 9982	14.0	1.0 x 0.9	13.7	Scd	3	
15 41 47.7	+67 15 15	UGC 9992	14.6	1.7 x 1.1	15.1	Im	147	Dwarf irregular, very low surface brightness; mag 12.3 star W 2′.4.
15 43 39.5	+67 45 36	UGC 10018	13.9	1.2 x 0.9	13.8	(R′)SB(s)bc	111	N-S bar, dark patches E and W.
15 48 17.2	+68 13 13	UGC 10057	14.1	1.5 x 0.7	14.0	(R′)SB(s)bc	92	= **Hickson 78A**; **Hickson 78B** S 1′.1; **Hickson 78C** SE 2′.7; **Hickson 78D** NE 1′.7.
15 49 32.2	+71 14 45	UGC 10072	14.1	1.7 x 1.1	14.6	SABbc	161	Two slender, slightly curving arms; **MCG +12-15-24** W 2′.5.
15 51 05.0	+67 09 29	UGC 10078	13.6	1.4 x 1.2	14.2	E	141	Almost stellar **MCG +11-19-20** N 1′.8; almost stellar **MCG +11-19-23** SE 1′.1.
15 57 08.4	+63 54 59	UGC 10115	12.8	1.3 x 0.8	12.9	E	128	Bright knot or very small galaxy S edge; stellar galaxy **Kaz 50** SW 1′.1.
15 59 33.1	+70 20 00	UGC 10142	14.5	1.5 x 0.5	14.0	SBa	13	Bright core, faint arms; mag 13.2 star N end.
16 01 25.8	+71 23 53	UGC 10162	14.4	1.6 x 0.4	13.8	SBb	63	**MCG +12-15-44** N 6′.0.
16 04 04.3	+63 42 42	UGC 10194	15.4	1.7 x 0.3	14.5	Sd:	50	Small triangle of mag 12.9 and 13.9 stars SE 2′.3.
16 07 55.5	+64 21 03	UGC 10231	15.0	1.5 x 0.9	15.1	S	125	Mag 14.7 star E end; bright knot N edge.
16 13 22.3	+65 25 06	UGC 10294	15.9	1.6 x 0.9	16.1	Sm?	5	Dwarf spiral, extremely low surface brightness.
16 13 47.5	+63 42 29	UGC 10298	13.8	1.1 x 0.3	12.4	S?	103	
16 17 15.8	+63 50 59	UGC 10334	14.1	1.5 x 0.7	14.0	IBm IV-V	150	Mag 12.3 star S 1′.7.
16 20 33.9	+73 51 53	UGC 10368	13.8	1.1 x 0.4	12.8	S0$^-$:	150	**MCG +12-15-58** S 3′.2.
16 24 43.6	+64 30 41	UGC 10383	14.4	1.4 x 0.7	14.2	Sdm:	115	
16 28 55.8	+64 12 31	UGC 10425	14.5	1.7 x 0.3	13.6	Sb II-III	155	Mag 10.0v star WNW 1′.9.
16 32 45.7	+62 40 48	UGC 10449	14.5	1.0 x 0.9	14.2	Sdm	177	Stellar nucleus, uniform envelope.
16 35 28.2	+68 32 01	UGC 10476	14.8	1.5 x 0.3	13.8	Sa	38	
16 35 44.4	+68 25 39	UGC 10478	14.6	1.5 x 0.3	13.6	Sbc	59	Mag 12.4 star E 1′.1.
16 36 47.5	+72 24 12	UGC 10497	13.5	1.1 x 0.4	12.5	S	96	Mag 8.48v star SW 2′.6.
16 37 38.0	+72 22 26	UGC 10502	12.2	2.4 x 2.0	13.8	SA(rs)c	130	Stellar nucleus, branching arms.
16 37 58.6	+71 30 41	UGC 10503	15.4	1.7 x 0.4	14.8	Im	30	Patchy.
16 41 07.1	+62 24 00	UGC 10518	13.8	1.0 x 0.8	13.4	Sbc	145	Stellar nucleus, uniform envelope; mag 7.21v star S 5′.6.
16 44 06.1	+62 05 13	UGC 10536	13.8	1.8 x 0.7	13.9	Sb II-III	123	Mag 9.23v star N 3′.8.
16 46 35.9	+62 49 26	UGC 10561	14.3	2.3 x 0.3	13.7	Sb II-III	150	Mag 8.24v star SE 2′.0; mag 9.80v star W 2′.4.
16 49 03.7	+68 45 23	UGC 10587	14.3	1.5 x 0.4	13.6	Sab	92	
16 54 33.2	+65 18 11	UGC 10614	14.7	1.1 x 0.8	14.4	Sdm	93	
16 54 51.1	+70 26 13	UGC 10622	14.7	1.5 x 0.2	13.2	Scd:	81	Mag 7.0v star NW 2′.5.
17 04 33.9	+72 26 48	UGC 10713	13.2	1.8 x 0.3	12.4	Sb II	8	
17 08 10.6	+63 37 50	UGC 10731	12.2	1.0 x 0.6	11.5	S?	33	Elongated knot of small galaxy S of core.
17 08 04.6	+69 27 53	UGC 10736	13.9	2.5 x 0.7	14.4	SABdm	155	**MCG +12-16-19** S 2′.4; mag 11.1 star on E edge.
17 14 38.9	+72 23 50	UGC 10791	14.4	1.5 x 0.8	14.4	Sm:	105	Faint star superimposed near W edge.
17 16 05.1	+73 26 08	UGC 10803	12.1	1.2 x 0.7	12.0	E?	15	
17 33 37.9	+70 50 35	UGC 10912	14.8	0.7 x 0.5	13.5	S	63	Very faint streamers extend NE and SW well beyond bright core; pair of faint stars on E edge.
17 42 14.8	+68 20 52	UGC 10963	14.3	1.4 x 0.7	14.1	SB(s)b	72	Two thin main arms.
17 47 33.0	+64 01 41	UGC 10995	14.0	1.0 x 0.4	12.8	S?	7	
17 51 58.3	+72 08 46	UGC 11038	14.3	0.9 x 0.6	13.5	Scd:	30	Weak stellar nucleus.
17 54 52.8	+71 32 26	UGC 11066	14.5	1.5 x 0.7	14.4	SAB(s)dm	10	Uniform surface brightness; mag 10.8 star W 1′.1.
18 01 07.2	+67 25 33	UGC 11099	14.3	1.2 x 0.4	13.3	SBcd?	82	Very faint, small galaxy **KUG 1800+674** WNW 6′.0.
18 02 04.3	+69 42 48	UGC 11106	14.1	0.9 x 0.9	13.7	SB(r)bc		Stellar nucleus, uniform envelope; **CGCG 340-34** NE 11′.8.

GALAXY CLUSTERS

RA h m s	Dec ° ′ ″	Name	Mag 10th brightest	No. Gal	Diam ′	Notes
17 12 30.0	+64 05 00	A 2255	15.3	102	37	A number of PGC galaxies grouped near the center.
17 59 54.0	+69 13 00	A 2295	16.2	48	28	Most members anonymous, stellar/almost stellar.

PLANETARY NEBULAE

RA h m s	Dec ° ′ ″	Name	Diam ″	Mag (P)	Mag (V)	Mag cent ★	Alt Name	Notes
17 58 33.5	+66 37 59	NGC 6543	20	8.8	8.1	11.1	PK 96+29.1	**Cat's Eye Nebula**. Very bright central star; **IC 4677** W 1′.8, is part of the outer envelope of NGC 6543; mag 9.78v star WNW 2′.7.

12 GALAXIES 12

RA h m s	Dec ° ′ ″	Name	Mag (V)	Dim ′ Maj x min	SB	Type Class	PA	Notes
15 02 27.6	+69 52 52	CGCG 337-28	14.6	0.6 x 0.4	12.9		24	Mag 12.0 star E 4′.0; mag 9.82v star WNW 9′.9.
13 49 56.6	+71 09 51	IC 954	13.7	1.1 x 0.6	13.1	Pec	91	Mag 9.35v star E 3′.2.
14 37 53.9	+69 00 49	IC 1046	14.5	0.8 x 0.4	13.1	Sc	90	Mag 9.95v star SE 6′.6.
14 39 33.6	+62 00 08	IC 1049	13.8	0.9 x 0.7	13.2	SBa	74	Very faint ENE-WSW bar with dark patch N.
14 49 21.5	+63 16 11	IC 1065	13.6	1.1 x 0.8	13.3	SB0	141	Bright center elongated E-W; faint outer envelope.
15 06 21.1	+62 58 47	IC 1100	13.4	0.8 x 0.7	12.6	S?	60	Mag 11.6 star N 1′.9.
15 12 05.1	+67 21 44	IC 1110	14.0	1.4 x 0.4	13.2	Sa	77	Mag 10.8 star SW 3′.9.
15 32 01.1	+68 14 44	IC 1129	13.1	1.1 x 0.9	12.9	Scd:	170	Mag 9.78v star SW 11′.1.
13 06 15.5	+62 00 23	MCG +10-19-27	14.8	0.6 x 0.4	13.1	Sb	120	Located 2′.3 S of mag 6.15v star.
14 35 49.4	+62 25 14	MCG +10-21-17	14.7	1.0 x 0.3	13.3		21	
14 38 09.4	+66 20 15	MCG +11-18-4	14.1	0.8 x 0.8	13.4	Sab		
14 07 05.1	+72 10 06	MCG +12-13-31	14.4	0.7 x 0.5	13.1	Irr	171	Mag 13.8 star S edge; **UGC 9093** S 2′.8.
14 11 57.8	+72 16 25	MCG +12-13-34	14.0	0.9 x 0.9	13.7			Mag 12.8 star SE edge.

(Continued from previous page)

12 GALAXIES 12

RA h m s	Dec ° ′ ″	Name	Mag (V)	Dim ′ Maj x min	SB	Type Class	PA	Notes
15 20 22.1	+69 07 32	MCG +12-14-25	15.2	0.6 x 0.3	13.1	Sc	54	Elongated **MCG +12-14-24** W edge.
15 23 59.8	+69 45 40	MCG +12-15-5	14.5	0.5 x 0.5	13.1	E/S0		
15 42 55.8	+70 49 39	MCG +12-15-13	14.7	0.6 x 0.5	13.5	E	0	Mag 10.54v star S 0′.9.
13 09 14.2	+62 10 26	NGC 5007	13.3	0.9 x 0.6	12.5	$S0^-$:	135	**MCG +10-19-44** ESE 2′.0.
13 12 19.5	+70 38 58	NGC 5034	13.3	0.9 x 0.7	12.7	S?	15	
13 22 54.8	+70 30 48	NGC 5144	13.1	1.2 x 0.9	13.0	SAc? pec	150	Bright knot S of center.
13 30 03.1	+62 30 44	NGC 5205	12.3	3.2 x 1.8	14.1	S?	166	Small, faint anonymous galaxy E 1′.6.
13 32 06.5	+62 41 59	NGC 5216	12.6	1.8 x 1.2	13.5	E0 pec	28	Member of **Keenan's System**. Bright nucleus.
13 32 10.6	+62 46 04	NGC 5218	12.3	1.8 x 1.3	13.1	SB(s)b? pec	100	Member of **Keenan's System**. Bright, wide center NE-SW; faint envelope E and W.
13 41 06.1	+67 40 20	NGC 5283	13.2	1.1 x 1.0	13.2	S0?	0	
13 46 12.1	+70 20 21	NGC 5314	13.8	0.8 x 0.5	12.7		90	
13 48 43.6	+72 35 13	NGC 5340	14.3	0.5 x 0.5	12.6			**MCG +12-13-11** NW 8′.4.
13 50 12.4	+73 57 08	NGC 5344	14.5	0.5 x 0.4	12.6		84	Very small, very faint galaxy W 0′.7.
13 57 13.3	+73 37 00	NGC 5412	13.4	1.2 x 1.0	13.5	$S0^-$:	20	
13 57 53.8	+64 54 39	NGC 5413	13.2	1.2 x 1.0	13.4	E	45	Mag 7.04v star E 3′.9.
13 56 57.0	+70 45 14	NGC 5415	14.4	0.8 x 0.5	13.3		126	**CGCG 336-35** E 10′.1.
14 05 57.5	+65 41 26	NGC 5479	14.0	0.7 x 0.5	13.0	E/S0	27	**MCG +11-17-18** and **MCG +11-17-16** NNW 1′.4.
14 19 26.9	+71 35 15	NGC 5607	13.4	0.9 x 0.8	12.9	Pec	66	
14 22 40.5	+69 35 39	NGC 5620	14.1	0.5 x 0.5	12.6	E/S0		Mag 10.4v star N 0′.8.
14 27 43.0	+69 41 39	NGC 5671	13.3	1.7 x 1.2	13.9	SB(r)b	45	Bright nucleus in short bar.
14 55 48.9	+63 54 10	NGC 5807	13.9	0.5 x 0.5	12.3	Compact		Almost stellar nucleus.
14 54 03.0	+73 07 53	NGC 5819	13.5	0.9 x 0.9	13.1	SAB(rs)bc		Located between a pair of mag 11.5 stars on NW-SE line.
14 57 47.7	+71 40 55	NGC 5832	12.1	3.7 x 2.2	14.2	SB(rs)b?	45	Diffuse center, faint envelope.
14 59 31.0	+73 53 36	NGC 5836	13.9	1.2 x 1.0	13.9	SB(rs)b	45	Bright nucleus, star S of nucleus; mag 10.88v star SE 1′.7.
15 10 38.3	+64 53 54	NGC 5881	14.5	2.1 x 1.3	15.4	SB(s)dm	80	Very patchy, elongated nucleus.
15 24 46.3	+68 43 49	NGC 5939	13.1	0.9 x 0.5	12.1	S?	35	**CGCG 338-6** SW 8′.1.
15 28 00.4	+64 45 48	NGC 5949	12.0	2.2 x 1.0	12.7	SA(r)bc?	147	
13 06 26.8	+67 42 22	UGC 8201	12.5	3.3 x 1.9	14.3	Im IV-V	99	Dwarf irregular; mag 11.8 star NNE 2′.9.
13 07 36.6	+62 12 52	UGC 8214	13.5	0.9 x 0.8	13.0	SB?	168	Mag 11.3 star SW 2′.2.
13 08 46.3	+62 16 15	UGC 8234	13.1	1.4 x 0.7	12.9	S0/a	139	UGC 8237 N 2′.4.
13 08 54.2	+62 18 19	UGC 8237	13.1	1.2 x 0.9	13.0	(R′)SBb:	148	Pair with UGC 8234 S 2′.4..
13 15 32.6	+62 07 36	UGC 8335	13.5	1.7 x 0.7	13.5	S?	120	Double system, disrupted, bridge + streamers; pair of stars on N edge, mags 13.3 and 14.2.
13 18 51.0	+72 45 56	UGC 8374	13.8	1.3 x 0.6	13.4	SAB(s)bc	9	Mag 12.1 star SW 3′.0.
13 24 34.4	+63 05 34	UGC 8436	14.3	1.0 x 0.9	14.0	Sd	6	Mag 12.9 star N 0′.9.
13 27 43.4	+62 45 59	UGC 8467	14.0	1.0 x 0.7	13.4	SBb	21	Two main arms.
13 29 32.4	+62 49 31	UGC 8491	15.5	1.4 x 0.2	14.0	Irr	165	Mag 14.4 star S 0′.9.
13 30 53.2	+62 13 23	UGC 8511	14.7	1.0 x 0.2	12.8	S	3	
13 31 34.8	+71 28 47	UGC 8525	13.3	1.0 x 0.6	12.6	$S0^-$?	150	Located between to close mag 12 stars.
13 36 02.2	+66 18 07	UGC 8604	14.4	1.5 x 1.2	14.9	SAB(s)c	69	Mag 11.9 star W 1′.6.
13 47 08.0	+72 04 10	UGC 8732	14.2	1.0 x 0.6	13.5	Sb II	126	Galaxy **IC 945** E 3′.4.
13 48 15.7	+68 05 13	UGC 8737	13.9	2.2 x 0.5	13.8	Sbc	152	
13 52 15.2	+72 43 53	UGC 8811	13.7	1.8 x 1.3	14.5	SB(s)b	12	Bright E-W bar.
13 53 03.7	+69 18 26	UGC 8823	13.6	0.9 x 0.5	12.6	S0	33	**MCG +12-13-24** E 0′.8; mag 11.8 star E 1′.5; **MCG +12-13-25** ESE 2′.4.
13 57 48.1	+63 23 51	UGC 8894	15.9	2.0 x 1.3	16.8	Sm	20	Dwarf spiral, extremely low surface brightness.
14 01 49.5	+68 52 36	UGC 8959	15.9	1.7 x 1.3	16.6	Sm:	170	Dwarf spiral, very low surface brightness.
14 06 56.5	+72 07 22	UGC 9039	14.6	1.5 x 0.2	13.2	Scd:	173	MCG +12-13-31 N 2′.8.
14 45 54.8	+68 55 47	UGC 9529	14.3	1.5 x 0.3	13.3	Sb III-IV	162	Mag 11.6 star, with stellar anonymous galaxy on E edge, NE 1′.9.
15 06 55.7	+66 10 36	UGC 9734	14.2	1.4 x 1.0	14.4	S?	66	Mag 10.7 star superimposed inside N edge.
15 08 48.5	+67 11 33	UGC 9749	10.9	30 x 19	17.8	E	70	**Ursa Minor Dwarf**. Dwarf elliptical, low surface brightness.
15 25 05.3	+66 15 11	UGC 9855	14.2	1.5 x 0.6	14.0	Im:	37	Located between a close pair of mag 13 stars.
15 31 45.6	+67 33 19	UGC 9896	13.5	1.3 x 1.1	13.7	Scd:	145	Weak stellar nucleus.
15 35 48.4	+73 26 59	UGC 9944	13.9	1.3 x 0.4	13.1	S?	174	Mag 10.35 v star SE 3′.4.
15 40 07.3	+70 44 25	UGC 9982	14.0	1.0 x 0.9	13.7	Scd	3	
15 41 47.7	+67 15 15	UGC 9992	14.6	1.7 x 1.1	15.1	Im	147	Dwarf irregular, very low surface brightness; mag 12.3 star W 2′.4.
15 43 39.5	+67 45 36	UGC 10018	13.9	1.2 x 0.9	13.8	(R′)SB(s)bc	111	N-S bar, dark patches E and W.

13 GALAXIES 13

RA h m s	Dec ° ′ ″	Name	Mag (V)	Dim ′ Maj x min	SB	Type Class	PA	Notes
11 03 04.4	+64 46 25	CGCG 314-4	13.9	0.5 x 0.5	12.2			Mag 11.9 star NW 1′.8.
11 18 36.4	+63 16 49	CGCG 314-19	14.0	0.4 x 0.4	11.9			= **Mkn 165**. Mag 9.03v star W 10′.6.
12 07 48.4	+67 23 00	CGCG 315-17	13.8	0.6 x 0.6	12.6			= **Mkn 197**. NGC 4108B SSW 9′.6.
10 51 32.9	+68 41 44	CGCG 333-59	13.9	0.9 x 0.4	12.6		120	Mag 12.1 star 1′.4 S.
11 46 45.4	+70 39 21	CGCG 334-50	13.9	0.5 x 0.5	12.2			Located 1′.9 NE of mag 10.6 star.
12 55 30.0	+70 49 44	CGCG 335-28	14.0	0.6 x 0.6	12.7			Elongated anonymous galaxy 1′.7 S; mag 11.6 star ESE 2′.0.
12 04 11.8	+62 30 04	IC 758	13.3	1.9 x 1.6	14.4	SB(rs)cd:	9	Mag 9.40v star W 7′.0.
12 55 53.5	+63 36 39	IC 836	13.7	1.4 x 0.3	12.6		73	Mag 12.3 star E end.
10 46 38.8	+72 19 11	IC 2600	14.7	0.6 x 0.4	13.0		123	**IC 2601** E 2′.7; mag 6.98v star SW 4′.1.
11 11 15.3	+62 08 26	MCG +10-16-77	14.6	0.9 x 0.3	13.0	Sb	108	Mag 12.2 star N 2′.1.
11 14 23.1	+62 04 16	MCG +10-16-90	14.4	0.8 x 0.5	13.2		51	Located between a pair of mag 12.5 and 13.6 stars.
12 54 36.2	+62 13 41	MCG +10-18-89	14.1	1.0 x 0.7	13.8	E/S0	60	Mag 11.18v star SW 5′.3.
13 06 15.5	+62 00 23	MCG +10-19-27	14.8	0.6 x 0.4	13.1	Sb	120	Located 2′.3 S of mag 6.15v star.
11 07 13.3	+65 05 59	MCG +11-14-3AB	15.1	0.7 x 0.5	13.8		177	
11 07 53.7	+66 36 15	MCG +11-14-5	13.9	1.1 x 0.7	13.5	Sc	63	Mag 9.33v star SW 4′.2.
11 10 02.2	+63 38 29	MCG +11-14-8	13.8	0.6 x 0.6	12.8	E/S0		Mag 9.68v star ENE 2′.9.
11 26 56.2	+63 25 27	MCG +11-14-25A	13.8	0.9 x 0.7	13.4	E?	33	**MCG +11-14-26** SE 8′.3.
11 36 06.6	+62 14 55	MCG +11-14-33	13.8	0.9 x 0.5	12.8	Sb	3	
12 04 51.0	+63 07 50	MCG +11-15-16	14.2	0.6 x 0.6	12.9			Mag 9.75v star SW 1′.5.
12 12 49.7	+65 48 52	MCG +11-15-36	13.9	0.8 x 0.5	12.9	E/S0	66	Mag 8.70v star N 3′.3.
12 25 53.3	+62 25 42	MCG +11-15-54	14.3	1.0 x 0.5	13.4	Sbc	60	Mag 10.91v star S 1′.6.
10 44 08.0	+70 24 17	MCG +12-10-67	14.7	1.0 x 0.5	13.8	Sb	12	**MCG +12-10-68** S 2′.2; **MCG +12-10-69** SE 2′.1; **MCG +12-10-70** E 1′.4.
11 21 06.6	+69 26 28	MCG +12-11-16	14.5	0.6 x 0.6	13.2	Sb		Mag 13.5 star W 1′.3.
11 49 54.3	+70 43 49	MCG +12-11-42	15.0	0.7 x 0.4	13.4	Sbc	177	Mag 11.4 star S 2′.7.
12 23 56.2	+73 53 35	MCG +12-12-7	14.6	0.5 x 0.5	13.0	Sb		Mag 11.8 star W 1′.8.
12 40 57.1	+71 46 16	MCG +12-12-15	14.7	0.7 x 0.3	12.9	Sb	165	MCG +12-12-16 N 2′.5.
12 41 01.1	+71 48 40	MCG +12-12-16	15.4	0.9 x 0.3	13.8	Sc	129	Mag 10.82v star W 2′.0.
10 46 09.9	+73 21 12	NGC 3343	13.4	1.3 x 0.9	13.6	E	55	
10 47 09.8	+72 50 23	NGC 3348	11.2	2.0 x 2.0	12.6	E0		**MCG +12-10-79** ESE 2′.6.

(Continued from previous page)

GALAXIES

RA h m s	Dec ° ′ ″	Name	Mag (V)	Dim ′ Maj x min	SB	Type Class	PA	Notes
10 46 35.9	+63 13 17	NGC 3359	10.6	7.2 x 4.4	14.2	SB(rs)c II	170	Several bright knots in multi-branching arms.
10 48 30.0	+72 25 30	NGC 3364	12.8	1.5 x 1.5	13.5	SAB(rs)c		**MCG +12-10-80** W 3′5; mag 6.98v star SW 14′6.
10 51 03.4	+65 46 54	NGC 3392	13.7	0.8 x 0.6	12.9	E?	120	Located 4′1 NNE of NGC 3394.
10 50 40.0	+65 43 40	NGC 3394	12.4	1.9 x 1.4	13.3	SA(rs)c	35	Mag 10.73v star E 16′3; note: **UGC 5992** lies 2′3 S of this star.
10 53 54.0	+73 41 24	NGC 3403	12.2	3.0 x 1.2	13.4	SAbc: III	73	
11 06 47.2	+72 34 07	NGC 3516	11.7	1.7 x 1.3	12.4	(R)SB(s)0°:	30	Located 2′2 SW of mag 10.51 star.
11 12 59.4	+72 52 44	NGC 3562	12.2	1.7 x 1.3	13.1	E	16	**CGCG 334-15** NE 4′9; **MCG +12-11-12** N 6′6.
11 20 12.4	+67 14 32	NGC 3622	13.2	1.2 x 0.5	12.5	S?	7	Mag 6.20v star SSE 9′4.
11 24 11.1	+69 24 47	NGC 3654	12.8	1.2 x 0.6	12.3	S?	21	
11 25 30.7	+63 26 46	NGC 3668	12.3	1.7 x 1.3	13.0	Sbc	144	Small, faint anonymous galaxy at NE edge.
11 27 41.5	+66 35 22	NGC 3682	12.5	1.7 x 1.1	13.1	SA(s)0/a:?	95	
11 35 59.4	+70 31 58	NGC 3735	11.8	4.2 x 0.8	13.0	SAc: sp II-III	131	Mag 9.61v star E 6′8.
11 35 41.5	+73 27 06	NGC 3736	14.5	1.2 x 0.7	14.1	Sb	155	Stellar nucleus, faint star on E edge; mag 8.33v star SSW 4′7.
11 46 49.8	+69 22 57	NGC 3879	13.0	2.6 x 0.5	13.1	Sdm:	130	
11 54 58.0	+69 19 48	NGC 3961	13.5	1.3 x 1.3	13.9	(R)SB(r)a:		
12 01 29.2	+69 19 28	NGC 4034	13.5	1.7 x 1.1	14.0	Scd:	5	Stellar nucleus; mag 7.42v star SSW 9′5.
12 02 11.4	+62 08 12	NGC 4041	11.3	2.7 x 2.5	13.2	SA(rs)bc: II-III	72	**MCG +10-17-132** E 7′5.
12 04 33.8	+64 26 11	NGC 4081	12.9	1.5 x 0.6	12.6	Sa?	135	Mag 10.81v star NW 4′7.
12 06 45.3	+67 09 46	NGC 4108	12.3	1.7 x 1.4	13.1	(R′)SAc:	105	Mag 9.51v star W 6′6; NGC 4108B NNE 5′0; NGC 4108A NW 7′6.
12 05 49.6	+67 15 07	NGC 4108A	13.8	1.4 x 0.5	13.3	SBbc:	7	
12 07 11.6	+67 14 09	NGC 4108B	13.7	1.3 x 1.1	13.9	SAB(s)d pec?	125	
12 08 31.0	+69 32 40	NGC 4120	13.5	1.8 x 0.4	13.0	SA(rs)cd:	166	Mag 9.43v star N 3′3.
12 07 56.6	+65 06 52	NGC 4121	13.5	0.4 x 0.4	11.6	E		Compact; located 3′6 S of NGC 4125.
12 08 06.3	+65 10 30	NGC 4125	9.7	5.8 x 3.2	12.9	E6 pec	81	NGC 4121 S 3′6; a mag 9.93v star E 2′5.
12 08 32.0	+68 46 01	NGC 4128	12.0	2.6 x 0.7	12.5	SA0: sp	67	**MCG +12-12-2** NW 1′9.
12 14 55.5	+63 46 51	NGC 4205	13.0	1.7 x 0.6	12.9	S?	28	Mag 13.1 star on W edge.
12 15 16.5	+65 59 05	NGC 4210	12.5	2.0 x 1.5	13.5	SB(r)b	105	Mag 6.74v star and UGC 7242 NW 10′2.
12 16 00.0	+66 13 49	NGC 4221	12.3	2.2 x 1.7	13.6	(R)SB(r)0^+	25	Mag 6.74v star and UGC 7242 SW 13′6.
12 16 42.3	+69 27 51	NGC 4236	9.6	21.9 x 7.2	15.0	SB(s)dm IV-V	162	Very long faint bar. Brightest emission patches in SE end 5′5 from the center.
12 16 56.1	+63 24 35	NGC 4238	13.6	1.8 x 0.5	13.4	Sd	36	
12 17 25.8	+70 48 07	NGC 4250	11.8	2.7 x 2.1	13.5	SAB(r)0^+	168	Round, compact core.
12 18 42.0	+65 53 43	NGC 4256	11.9	4.5 x 0.8	13.1	SA(s)b: sp	42	Very faint, narrow lane NE half.
12 22 47.1	+65 50 38	NGC 4332	12.2	2.1 x 1.5	13.3	SB(s)a	130	Bright elongated bar, very faint star on SE tip of bar.
12 25 18.9	+64 56 01	NGC 4391	12.7	1.1 x 1.1	12.8	SA0^-:		An E-W line of three faint stars, with a mag 10.66v star just S, lies SW 3′0.
12 27 20.5	+64 48 10	NGC 4441	12.7	3.4 x 2.5	14.9	SAB0^+ pec	9	Bright nucleus with very faint, irregular and filamentary arms N-S; very faint anonymous spherical galaxy, paired with star, 1′5 SE of center.
12 29 49.2	+64 01 55	NGC 4481	14.0	0.8 x 0.3	12.3		150	Mag 9.28v star SSE 9′0.
12 31 47.5	+64 13 58	NGC 4510	13.0	1.5 x 0.9	13.3	E	153	**MCG +11-15-62** NE 8′6.
12 32 33.7	+63 52 40	NGC 4512	13.7	2.0 x 1.4	14.7	SB(s)dm	80	Diffuse spiral.
12 32 01.4	+66 19 56	NGC 4513	13.0	1.4 x 0.9	13.1	(R)SA0°	15	Elongated **CGCG 315-43** N 3′5; stellar **CGCG 315-44** N 4′2; stellar galaxy **VII Zw 468** NNE 6′1.
12 32 47.9	+63 56 21	NGC 4521	12.2	2.5 x 0.5	12.3	S0/a	167	Mag 11 star 1′9 NW.
12 34 33.7	+63 31 30	NGC 4545	12.3	2.5 x 1.5	13.6	SB(s)cd: II	8	Mottled appearance, several bright knots.
12 47 09.2	+71 10 35	NGC 4693	13.4	2.5 x 0.5	13.5	Sd	34	Almost stellar nucleus.
12 51 12.5	+71 38 07	NGC 4749	13.5	1.8 x 0.4	13.0	Sb? sp	158	Mag 7.92v star ESE 8′5.
12 50 07.2	+72 52 26	NGC 4750	11.2	2.0 x 1.9	12.5	(R)SA(rs)ab II-III	147	Bright, diffuse center; faint outer envelope.
12 57 18.3	+70 12 12	NGC 4857	13.8	1.3 x 0.6	13.4	SABb	110	Almost stellar nucleus, faint envelope.
13 09 14.2	+62 10 26	NGC 5007	13.3	0.9 x 0.6	12.5	S0^-:	135	**MCG +10-19-44** ESE 2′0.
13 12 19.5	+70 38 58	NGC 5034	13.3	0.9 x 0.7	12.7	S?	15	
10 43 50.5	+69 44 48	UGC 5835	13.7	1.2 x 0.7	13.4	Sa?	134	**MCG +12-10-63** NNW 6′0; **UGC 5834** S 3′4.
10 44 35.0	+69 04 16	UGC 5843	14.6	1.1 x 0.3	13.2	S	9	Mag 11.5 star NW 3′6.
10 48 38.2	+66 21 39	UGC 5904	13.7	2.0 x 0.4	13.3	Sb II-III	152	Double system, contact with small companion N end.
10 49 37.7	+65 31 50	UGC 5918	14.6	2.8 x 2.8	16.6	Im: V		Dwarf, extremely low surface brightness.
10 50 24.0	+64 48 32	UGC 5932	14.2	1.5 x 0.6	13.9	Sbc	52	
10 52 04.3	+71 46 22	UGC 5955	13.1	1.2 x 1.2	13.5	E		**MCG +12-10-85** SW 5′1; **MCG +12-10-84** and mag 11.8 star SW 8′7.
10 52 36.1	+67 58 58	UGC 5979	14.5	1.7 x 1.3	15.2	Im	51	Dwarf irregular; mag 10.57v star superimposed SW end.
10 56 09.6	+72 53 42	UGC 6032	15.4	1.5 x 0.6	15.2	Sd	78	Weak, almost stellar nucleus.
11 04 48.4	+63 59 56	UGC 6133	16.1	1.1 x 1.1	16.2	Im:		Dwarf, very low surface brightness; mag 13.4 star N edge; mag 11.3 star E 1′7.
11 07 58.9	+63 55 36	UGC 6179	14.8	1.7 x 0.7	14.8	S	155	
11 12 14.7	+65 13 29	UGC 6237	15.6	0.7 x 0.3	13.8	Disturbed	69	
11 18 23.2	+65 01 49	UGC 6316	14.0	1.7 x 0.6	13.9	S?	81	Double system, consists of **MCG +11-14-14** and **MCG +11-14-15**.
11 22 07.1	+69 38 05	UGC 6378	14.1	2.2 x 0.3	13.5	Sd	142	Pair of mag 10 stars N 3′1.
11 22 16.4	+69 07 21	UGC 6381	14.2	1.5 x 1.2	14.7	Sm:	5	Dwarf spiral, very low surface brightness.
11 25 19.1	+63 43 46	UGC 6429	13.3	2.1 x 1.7	14.6	SA(rs)c I-II	24	Mag 11.6 star N 1′8.
11 26 50.3	+64 08 21	UGC 6448	14.1	1.8 x 0.9	14.5	Pec	160	
11 28 56.7	+73 02 04	UGC 6473	14.1	1.5 x 0.6	13.8	Sab	20	Located between a pair of stars, mags 14.1 and 12.5.
11 32 27.9	+62 30 24	UGC 6520	13.2	1.2 x 0.7	12.9	SB?	150	Mag 9.04v star W 3′8; **MCG +11-14-29** S 4′9.
11 33 05.1	+72 06 38	UGC 6530	14.2	1.1 x 0.6	13.6	Sb	135	
11 33 10.1	+70 23 25	UGC 6532	14.4	0.8 x 0.5	13.2	SB(r)cd:	3	**MCG +12-11-31** S 5′9.
11 33 19.7	+63 16 42	UGC 6534	12.8	2.8 x 0.6	13.2	Sd	60	Mag 12.1 star SE 3′1.
11 34 33.2	+71 32 24	UGC 6552	13.5	1.1 x 0.4	12.5	S?	46	
11 44 10.3	+71 12 56	UGC 6698	14.2	0.8 x 0.6	13.3	S	176	
11 44 29.8	+69 43 47	UGC 6711	13.3	0.7 x 0.4	11.8	S?	144	Mag 10.45v star WSW 1′5.
11 47 40.5	+69 06 13	UGC 6764	15.9	1.0 x 0.4	14.8	Sdm:	70	
12 01 43.6	+62 19 43	UGC 7009	15.2	1.5 x 0.5	14.7	Im	175	
12 02 29.4	+62 25 01	UGC 7019	14.6	1.3 x 0.7	14.3	Im	85	Uniform surface brightness; mag 11.4 star W end; mag 9.40v star NE 6′9.
12 02 37.0	+64 22 35	UGC 7020A	13.7	1.2 x 0.6	13.2	S0? I-II	100	
12 05 21.1	+63 09 22	UGC 7079	14.0	1.0 x 1.0	13.9	S		**MCG +11-15-17** and **MCG +11-15-18** NNW 1′9.
12 09 46.8	+62 16 09	UGC 7153	14.3	1.9 x 0.3	13.5	Scd:	166	
12 10 22.1	+70 30 51	UGC 7164	14.0	1.7 x 1.4	14.8	Sm:	135	Dwarf spiral, very low surface brightness; mag 11.2 star NE 2′5.
12 10 57.7	+63 54 54	UGC 7179	13.3	1.7 x 0.3	12.4	S?	135	Mag 12.8 star E 1′3.
12 14 10.3	+66 05 17	UGC 7242	13.9	1.5 x 0.7	13.8	Scd:	172	Mag 6.74v star 0′7 N.
12 24 25.3	+70 20 04	UGC 7490	12.7	3.6 x 3.6	15.3	SAm V		Dwarf spiral, low surface brightness; mag 11.1 star superimposed W of center.
12 34 06.1	+64 34 01	UGC 7730	15.4	1.2 x 0.7	15.1	Sm:	30	Dwarf spiral, extremely low surface brightness.
12 35 12.3	+72 13 54	UGC 7761	13.6	0.9 x 0.5	12.6	SB?	85	
12 35 32.5	+73 40 29	UGC 7767	12.6	0.9 x 0.9	12.4	E		Mag 10.13v star S 2′7.
12 38 09.5	+71 25 04	UGC 7809	14.5	0.8 x 0.8	13.9	SB(s)c		
12 40 57.4	+63 31 10	UGC 7848	13.7	2.4 x 0.9	14.4	SABcd	60	Stellar nucleus, smooth envelope.
12 43 38.2	+73 36 46	UGC 7908	14.6	1.5 x 0.4	13.8	Scd:	50	
12 46 01.0	+64 34 07	UGC 7941	14.0	2.9 x 0.5	14.3	Sd:	8	Mag 10.75v star SE 2′4.
12 54 53.0	+73 10 44	UGC 8052	13.5	1.8 x 0.5	13.2	S?	155	Mag 12.9 star S edge; mag 11.9 star E 2′0.

(Continued from previous page)

13 GALAXIES 13

RA h m s	Dec ° ′ ″	Name	Mag (V)	Dim ′ Maj x min	SB	Type Class	PA	Notes
12 59 56.1	+73 41 30	UGC 8120	13.8	1.5 x 0.7	13.7	Sd	22	Mag 13.3 star S edge; mag 10.9 star SSW 1'.9.
13 06 26.8	+67 42 22	UGC 8201	12.5	3.3 x 1.9	14.3	Im IV-V	99	Dwarf irregular; mag 11.8 star NNE 2'.9.
13 07 36.6	+62 12 52	UGC 8214	13.5	0.9 x 0.8	13.0	SB?	168	Mag 11.3 star SW 2'.2.
13 08 46.3	+62 16 15	UGC 8234	13.1	1.4 x 0.7	12.9	S0/a	139	UGC 8237 N 2'.4.
13 08 54.2	+62 18 19	UGC 8237	13.1	1.2 x 0.9	13.0	(R')SBb:	148	Pair with UGC 8234 S 2'.4..
13 15 32.6	+62 07 36	UGC 8335	13.5	1.7 x 0.7	13.5	S?	120	Double system, disrupted, bridge + streamers; pair of stars on N edge, mags 13.3 and 14.2.
13 18 51.0	+72 45 56	UGC 8374	13.8	1.3 x 0.6	13.4	SAB(s)bc	9	Mag 12.1 star SW 3'.0.

GALAXY CLUSTERS

RA h m s	Dec ° ′ ″	Name	Mag 10th brightest	No. Gal	Diam ′	Notes
11 48 24.0	+71 26 00	A 1382	15.9	57	34	All members faint, stellar and anonymous.
11 55 48.0	+73 28 00	A 1412	15.9	86	22	Mag 7.53v star on SW edge; all GX faint, stellar and anonymous.

14 GALAXIES 14

RA h m s	Dec ° ′ ″	Name	Mag (V)	Dim ′ Maj x min	SB	Type Class	PA	Notes
10 10 16.3	+62 54 48	CGCG 312-32	14.5	1.6 x 0.6	14.3	Sm	78	A mag 7.10v star on E end.
08 34 46.4	+70 09 44	CGCG 331-49	15.3	0.7 x 0.5	14.1		75	Very filamentary; mag 12.0 star SE 1'.8.
08 53 15.8	+73 11 18	CGCG 332-23	14.6	0.4 x 0.3	12.2		21	Located 1'.2 W of mag 9.58v star.
08 53 19.3	+68 28 15	CGCG 332-26	13.9	0.7 x 0.7	13.0			Bright, diffuse center; mag 11.8 star SE 2'.8.
09 12 14.4	+73 35 36	CGCG 332-33	13.9	0.5 x 0.4	12.0		141	Mag 10.08v star NE 6'.9.
10 51 32.9	+68 41 44	CGCG 333-59	13.9	0.9 x 0.4	12.6		120	Mag 12.1 star 1'.4 S.
08 40 50.3	+73 29 10	IC 511	13.4	1.6 x 0.5	13.1	S0/a:	143	Mag 6.15v star N 9'.8.
08 53 42.7	+73 29 23	IC 520	11.7	1.5 x 1.3	12.3	SAB(rs)ab?	0	Mag 11.6 star WSW 5'.9; low surface brightness **UGC 4634** N 2'.6.
09 18 33.7	+73 45 32	IC 529	11.9	3.3 x 1.3	13.4	SA(s)c:	145	Mag 10.52v star N 4'.6.
08 47 58.7	+73 32 18	IC 2389	13.4	1.6 x 0.3	12.4	SB(s)b?	126	Mag 11.8 star SE 2'.2; NGC 2646 ESE 11'.1.
09 21 30.1	+64 14 16	IC 2458	15.0	0.5 x 0.2	12.4	I0 pec:	21	Located on SW end of NGC 2820.
10 28 24.2	+68 25 02	IC 2574	10.4	13.2 x 5.4	14.8	SAB(s)m IV-V	41	**Coddington's Nebula**. Low surface brightness, several knots or condensations; mag 9.68v star E 11'.0 from center.
10 46 38.8	+72 19 11	IC 2600	14.7	0.6 x 0.4	13.0		123	**IC 2601** E 2'.7; mag 6.98v star SW 4'.1.
10 05 09.4	+70 33 24	MCG +10-12-19	15.6	0.5 x 0.4	13.7	Spiral	126	
08 23 42.0	+62 16 20	MCG +10-12-131	14.4	0.8 x 0.8	14.0	E/S0		Mag 14.9 star on W edge; mag 7.70v star NE 3'.9.
09 11 23.2	+62 00 58	MCG +10-13-64	14.6	0.7 x 0.5	13.4		36	Mag 10.25v star N 3'.5.
09 26 36.0	+62 24 30	MCG +10-14-8	14.0	0.8 x 0.6	13.3	E/S0	90	Mag 12.9 star SW 2'.4.
09 32 49.3	+62 20 11	MCG +10-14-19	14.6	0.6 x 0.6	13.4			Mag 11.35v star NW 4'.6.
10 31 53.5	+62 24 00	MCG +10-15-93	15.2	0.6 x 0.4	13.5	Sb	99	Mag 11.1 star E 0'.9.
08 27 51.5	+63 16 41	MCG +11-11-5	14.2	0.7 x 0.4	12.6	S0?	54	Located between a pair of mag 10.86v and 11.0 stars.
08 54 10.0	+67 54 26	MCG +11-11-31	14.4	0.8 x 0.5	13.3	Sc	150	Multi-galaxy system, small companion galaxy S edge.
09 24 59.5	+65 27 54	MCG +11-12-8A	14.4	0.8 x 0.4	13.0		3	Pair with MCG +11-12-8B N edge.
09 25 02.5	+65 28 21	MCG +11-12-8B	16.8	0.7 x 0.3	15.0			Pair with MCG +11-12-8A S edge.
09 25 00.7	+64 33 36	MCG +11-12-10	14.1	0.7 x 0.3	12.3	Sb	66	Mag 8.28v star E 4'.8; **MCG +11-12-9** S 2'.0.
10 25 46.1	+66 56 56	MCG +11-13-25	14.1	0.9 x 0.4	12.8		129	Mag 8.83v star SW 1'.2.
10 37 31.9	+65 01 03	MCG +11-13-36	14.9	0.7 x 0.4	13.4	Sb	42	Mag 12.1 star S 1'.7.
09 02 47.8	+73 18 34	MCG +12-9-29	15.0	0.7 x 0.6	13.9	Sc	114	Mag 11.5 star N 2'.3.
09 16 20.3	+69 52 53	MCG +12-9-33	14.3	0.7 x 0.6	13.2		168	Mag 10.9 star W edge.
09 18 49.1	+71 43 51	MCG +12-9-37	14.8	0.6 x 0.5	13.3	Sb	141	Mag 12.2 star SSW 2'.3.
09 38 12.2	+68 49 58	MCG +12-9-58	14.7	0.8 x 0.3	13.0		57	Mag 11.4 star superimposed SW end.
10 01 15.4	+70 59 05	MCG +12-10-13	14.7	0.6 x 0.4	13.0		141	Star on E edge.
10 05 15.6	+70 19 44	MCG +12-10-20	14.7	0.5 x 0.5	13.3	E/S0		UGC 5423 NE 2'.5; mag 10.71v star W 2'.4.
10 23 23.5	+69 52 41	MCG +12-10-30	15.1	0.7 x 0.4	13.6	SBc	66	Mag 14.1 star S 1'.3.
10 29 50.8	+69 05 01	MCG +12-10-43	14.4	0.9 x 0.6	13.6		12	Mag 13.8 star S 2'.5.
10 35 56.4	+72 36 58	MCG +12-10-54	13.7	0.8 x 0.8	13.3	E/S0		Mag 13.3 star N 1'.7.
10 44 08.0	+70 24 17	MCG +12-10-67	14.7	1.0 x 0.5	13.8	Sb	12	**MCG +12-10-68** S 2'.2; **MCG +12-10-69** SE 2'.1; **MCG +12-10-70** E 1'.4.
08 17 43.8	+73 19 03	NGC 2523C	12.9	1.5 x 0.8	13.1	E?	95	
08 21 39.8	+73 59 17	NGC 2544	12.9	1.1 x 0.8	12.6	SB(s)a:	70	Much elongated **MCG +12-8-35** NE 1'.4.
08 28 40.1	+73 44 56	NGC 2550A	12.7	1.5 x 1.4	13.4	Sc	0	**UGC 4389** SSW 6'.8; **MCG +12-8-44** SSE 7'.2; UGC 4413 NNE 8'.0.
08 24 51.0	+73 24 43	NGC 2551	12.1	1.7 x 1.1	12.6	SA(s)0/a	55	
08 42 47.8	+72 58 32	NGC 2614	12.9	2.0 x 1.6	14.0	SA(r)c:	150	Stellar nucleus with smooth envelope.
08 47 15.3	+72 59 10	NGC 2629	12.3	1.5 x 1.3	12.9	SA(r)0°:	105	Mag 13.1 star on E edge; mag 10.38v star NE 7'.3.
08 45 35.0	+72 59 54	NGC 2630	14.2	1.4 x 0.2	12.7	Sbc	29	Mag 9.93v star NW 6'.5.
08 48 25.2	+73 58 00	NGC 2634	12.0	1.7 x 1.6	13.1	E1:	51	Pair with NGC 2643A SE 1'.9.
08 48 37.0	+73 56 17	NGC 2634A	13.6	1.8 x 0.4	13.1	SB(s)bc? sp	64	Pair with NGC 2634 NW 1'.9.
08 48 24.9	+73 40 14	NGC 2636	13.8	0.4 x 0.4	11.9	E0:		Mag 9.93v star E 6'.0.
08 47 57.0	+72 53 44	NGC 2641	13.6	1.3 x 1.1	13.9	S0:	5	Very small, very faint anonymous galaxy W 2'.2.
08 50 21.7	+73 27 46	NGC 2646	12.1	1.3 x 1.3	12.5	SB(r)0°:		Mag 11.5 star S 2'.5; IC 2389 WNW 11'.1.
08 49 59.0	+70 17 56	NGC 2650	13.3	1.6 x 1.2	13.8	SB(rs)b:	55	Bright knot or star E of nucleus; mag 7.15v star E 7'.6; stellar **Mkn 95** SSW 8'.6.
09 09 58.2	+62 14 46	NGC 2742A	13.2	1.5 x 0.8	13.3	SB(s)b pec?	90	Mag 10.73v star 1'.8 WNW.
09 19 17.6	+69 12 11	NGC 2787	10.8	3.2 x 2.0	12.6	SB(r)0$^+$	111	Mag 8.51v star S 7'.7.
09 20 20.0	+64 06 12	NGC 2805	11.0	6.3 x 4.8	14.6	SAB(rs)d	140	Almost stellar nucleus; a number of stars superimposed on envelope.
09 22 05.2	+71 50 39	NGC 2810	12.2	1.7 x 1.7	13.4	E		
09 21 11.6	+64 15 03	NGC 2814	13.7	1.2 x 0.3	12.5	Sb:	179	IC 2458 on SW end.
09 21 45.4	+64 15 23	NGC 2820	12.8	4.1 x 0.5	13.4	SB(s)c pec sp	62	IC 2458 at SW end.
09 29 34.8	+62 29 28	NGC 2880	11.5	2.0 x 1.2	12.4	SB0$^-$	140	Mag 9.47v star WSW 9'.8.
09 32 53.0	+67 37 02	NGC 2892	13.1	1.4 x 1.4	13.8	E$^+$ pec:		Mag 9.80v star and **UGC 5061** W 6'.5.
09 44 09.8	+65 58 37	NGC 2909	13.1	0.6 x 0.4	11.4	Pec	78	Faint star on SW edge.
09 47 18.7	+72 59 00	NGC 2957	14.7	0.8 x 0.3	13.2	E	42	NGC 2957A on W edge.
09 47 16.3	+72 59 08	NGC 2957A	14.4	0.4 x 0.4	12.5	E1:		NGC 2957 on E edge.
09 45 08.5	+68 35 42	NGC 2959	12.8	1.3 x 1.3	13.2	(R')SAB(rs)ab pec:		NGC 2961 NE 1'.5; mag 9.44v star N 6'.6.
09 45 22.0	+68 36 31	NGC 2961	14.7	0.7 x 0.2	12.4	Sb:	137	NGC 2959 SW 1'.5.
09 47 50.4	+72 57 50	NGC 2963	13.5	1.2 x 0.6	13.0	SBab	165	Mag 9.78v star S 4'.8.
09 47 15.9	+67 54 55	NGC 2976	10.2	5.9 x 2.7	13.0	SAc pec IV	143	
09 50 21.3	+72 16 43	NGC 2985	10.4	4.6 x 3.6	13.3	(R')SA(rs)ab I	0	Mag 12.5 star superimposed E edge of core.
09 55 42.5	+72 12 04	NGC 3027	11.8	3.9 x 1.4	13.5	SB(rs)d: III	130	The small elongated galaxy **KUG 0950+724** lies 2'.7 NW of the center of NGC 3027.
09 55 38.5	+69 03 40	NGC 3031	6.9	26.9 x 14.1	13.2	SA(s)ab I-II	157	= **M 81**. Included along with NGC 3034 in **Bode's Nebula**. Low surface brightness UGC 5336 4'.5 E of center.

RA h m s	Dec ° ′ ″	Name	Mag (V)	Dim ′ Maj x min	SB	Type Class	PA	Notes
09 55 52.2	+69 40 48	NGC 3034	8.4	11.2 x 4.3	12.5	I0 sp	65	= **M 82**. Included along with NGC 3031 in **Bode's Nebula**.
10 01 55.5	+72 10 12	NGC 3065	12.5	1.7 x 1.5	13.4	SA(r)0°	21	
10 02 10.6	+72 07 28	NGC 3066	12.9	1.1 x 1.0	12.9	(R′)SAB(s)bc pec	159	
10 03 21.7	+68 44 03	NGC 3077	9.8	5.4 x 4.5	13.2	I0 pec	45	Lies 3′.8 SE of mag 7.97v star.
10 16 54.2	+73 23 56	NGC 3147	10.6	3.9 x 3.5	13.3	SA(rs)bc II	155	
10 34 23.4	+73 45 48	NGC 3252	13.5	2.0 x 0.6	13.6	SBd? sp	35	Mag 7.58v star N 4′.3.
10 32 34.8	+65 02 28	NGC 3259	12.1	1.7 x 0.9	12.4	SAB(rs)bc: III	20	Mag 10.55v star NW 3′.2.
10 33 17.5	+64 44 58	NGC 3266	12.4	1.5 x 1.3	13.0	SAB0°?	105	
10 46 09.9	+73 21 12	NGC 3343	13.4	1.3 x 0.9	13.6	E	55	
10 47 09.8	+72 50 23	NGC 3348	11.2	2.0 x 2.0	12.6	E0		**MCG +12-10-79** ESE 2′.6.
10 46 35.9	+63 13 17	NGC 3359	10.6	7.2 x 4.4	14.2	SB(rs)c II	170	Several bright knots in multi-branching arms.
10 48 30.0	+72 25 30	NGC 3364	12.8	1.5 x 1.5	13.5	SAB(rs)c		**MCG +12-10-80** W 3′.5; mag 6.98v star SW 14′.6.
10 51 03.4	+65 46 54	NGC 3392	13.7	0.8 x 0.6	12.9	E?	120	Located 4′.1 NNE of NGC 3394.
10 50 40.0	+65 43 40	NGC 3394	12.4	1.9 x 1.4	13.3	SA(rs)c	35	Mag 10.73v star E 16′.3; note: **UGC 5992** lies 2′.3 S of this star.
10 53 54.0	+73 41 24	NGC 3403	12.2	3.0 x 1.2	13.4	SAbc: III	73	
08 38 23.4	+65 07 17	PGC 24283	13.3	0.5 x 0.5	11.6	Pec		
08 17 27.0	+64 32 46	UGC 4302	14.3	1.1 x 0.9	14.1	SBb:	45	
08 19 03.9	+70 42 55	UGC 4305	10.7	8.5 x 6.0	14.8	Im IV-V	15	Dwarf irregular, low surface brightness; mag 11.4 star N end; a small triangle of stars, mags 11.4, 13.1 and 13.2, E of the center.
08 20 01.6	+62 49 50	UGC 4322	13.7	1.4 x 1.0	14.1	E	57	
08 20 19.8	+66 58 53	UGC 4323	13.1	1.8 x 1.3	14.0	E?	50	
08 20 35.1	+68 36 01	UGC 4326	13.1	1.4 x 0.6	12.8	Sbc	152	Mag 11.51v star SW 3′.1.
08 24 25.6	+65 36 17	UGC 4369	14.8	0.8 x 0.7	14.1	SBc		
08 24 54.0	+66 52 10	UGC 4376	14.0	1.0 x 0.7	13.5	S?	129	11.6 and 12.6 mag stars superimposed.
08 27 51.9	+73 31 03	UGC 4390	14.3	2.0 x 1.8	15.5	SBd	78	Mag 10.6 star W 2′.2.
08 27 35.7	+64 13 40	UGC 4398	14.1	1.1 x 0.7	13.7	SA(rs)bc	170	Mag 12.8 star on E edge.
08 29 46.3	+73 51 25	UGC 4413	14.7	1.5 x 0.3	13.7	Scd:	105	NGC 2550A SSW 6′.8; faint stellar galaxy **KUG 0825+739** SE 8′.4.
08 29 11.3	+63 20 17	UGC 4420	13.4	1.6 x 0.8	13.5	S0	129	**MCG +11-11-8** S 2′.8.
08 34 06.9	+66 10 49	UGC 4459	14.4	1.7 x 1.4	15.1	Im V	120	Dwarf irregular, very low surface brightness.
08 38 18.6	+69 01 15	UGC 4495	14.4	1.0 x 0.6	13.7	Scd:	57	
08 39 19.9	+71 42 11	UGC 4500	14.3	1.5 x 0.9	14.5	SAB(s)cd	45	Mag 11.7 star N 1′.2.
08 39 54.8	+73 45 15	UGC 4502	14.7	1.5 x 0.4	14.0	SBcd:	75	Mag 10.45v star N 1′.4.
08 41 17.7	+66 51 38	UGC 4516	14.5	1.5 x 0.3	13.5	Sdm:	123	Mag 12.4 star S 0′.5.
08 42 26.8	+70 57 59	UGC 4522	14.1	1.0 x 0.7	13.6	Scd:	159	Bright knot or star superimposed S.
08 43 41.1	+65 11 01	UGC 4536	14.2	1.7 x 1.2	14.8	SB(r)c	24	Short SE-NW bar; mag 8.83v star NW 2′.3.
08 44 07.3	+66 57 41	UGC 4539	13.8	1.0 x 0.7	13.3	Scd:	21	
08 49 00.2	+70 06 34	UGC 4593	13.0	0.6 x 0.4	11.3	S?	155	Compact; **Mkn 95** NNE 3′.9.
08 58 53.9	+66 28 04	UGC 4687	13.7	0.9 x 0.9	13.3	Pec		
09 00 12.3	+70 01 05	UGC 4697	15.3	1.5 x 0.3	14.3	SBcd:	57	Mag 10.07v star NW 1′.4.
09 03 48.4	+69 29 09	UGC 4739	14.3	0.8 x 0.5	13.2	Pec	105	Mag 13.1 star W 2′.1.
09 07 38.8	+66 34 30	UGC 4775	13.1	1.4 x 0.9	13.2	S0⁻:	0	
09 13 32.2	+70 34 54	UGC 4836	14.4	1.3 x 0.6	14.0	SB(s)ab	57	
09 18 23.3	+69 48 35	UGC 4896	13.8	1.9 x 0.5	13.6	Sab	145	Mag 7.08v star NW 5′.8.
09 21 14.1	+69 24 00	UGC 4944	15.0	1.5 x 0.2	13.5	Scd:	93	Mag 7.35v star E 11′.1.
09 21 43.1	+71 32 37	UGC 4951	14.2	0.8 x 0.6	13.3	S?	130	Stellar **CGCG 332-44** N 8′.2.
09 23 16.9	+72 03 48	UGC 4967	13.8	1.1 x 0.9	13.7	Sa	0	
09 24 02.3	+68 33 42	UGC 4981	13.9	1.1 x 0.4	12.9	Sb II	146	
09 25 12.6	+68 23 02	UGC 4998	14.4	1.6 x 0.8	14.5	Im:	80	Mag 10.7 star NNW 2′.4.
09 27 50.2	+68 24 43	UGC 5028	13.7	0.5 x 0.4	11.8	SB(s)dm pec	145	Pair with UGC 5029 E.
09 28 03.3	+68 25 10	UGC 5029	13.4	1.2 x 0.8	13.2	SAB(s)c	13	Pair with UGC 5028 W.
09 28 11.4	+64 56 03	UGC 5033	15.7	1.5 x 0.2	14.2	Scd:	167	
09 31 14.1	+73 48 36	UGC 5052	13.2	1.5 x 1.2	13.6	SB(s)a	120	Mag 11.56v star S edge.
09 37 20.7	+73 23 00	UGC 5110	14.5	1.7 x 1.3	15.2	IABm	135	Irregular, low surface brightness.
09 36 53.0	+66 47 16	UGC 5111	14.6	1.5 x 0.4	13.9	Sbc	120	Mag 11.6 star NE 3′.1.
09 40 29.1	+71 11 01	UGC 5139	12.6	4.0 x 3.3	15.3	IAB(s)m V	117	Dwarf irregular, extremely low surface brightness.
09 46 24.6	+68 57 23	UGC 5210	15.3	1.9 x 0.2	14.1	Scd:	153	
09 48 47.8	+64 09 57	UGC 5244	14.5	1.5 x 0.3	13.5	Scd:	34	Mag 10.60v star E 1′.5.
09 49 16.8	+69 25 21	UGC 5247	14.5	1.7 x 0.8	14.7	SAdm	155	Mag 10.9 star S end; mag 9.92v star W 2′.2.
09 51 40.0	+65 29 25	UGC 5277	14.0	1.7 x 1.6	14.9	SB(rs)bc	177	Mag 11.07v star W edge.
09 54 22.2	+68 19 58	UGC 5302	13.9	1.7 x 1.4	14.7	SA(s)dm	159	Mag 13.3 star superimposed S edge.
09 57 34.5	+69 02 42	UGC 5336	14.1	2.8 x 2.5	16.1	Im V	135	Dwarf irregular, extremely low surface brightness; 4′.5 E of center of M 81 (NGC 3031).
10 02 41.2	+70 44 29	UGC 5386	14.2	0.8 x 0.7	13.4	Sb II	22	Mag 8.10v star N 1′.4.
10 05 15.0	+72 12 01	UGC 5415	13.6	1.6 x 0.3	12.7	S	96	
10 05 30.6	+70 21 50	UGC 5423	14.6	0.9 x 0.6	13.8	Im	155	MCG +12-10-20 SW 2′.5.
10 07 01.0	+67 49 39	UGC 5442	14.6	1.8 x 0.9	15.0	Im:	27	Dwarf, low surface brightness.
10 08 50.1	+70 38 06	UGC 5455	13.9	1.8 x 1.8	15.0	Im		Dwarf irregular, very low surface brightness; mag 13.9 star W 1′.4.
10 12 48.4	+64 06 25	UGC 5497	15.1	0.7 x 0.6	14.0	Sd	3	Mag 13.7 star SE 0′.7.
10 14 09.6	+69 06 54	UGC 5508	14.3	1.0 x 0.8	13.9	Scd:	63	Mag 14.5 star superimposed on W end.
10 14 59.9	+65 08 22	UGC 5520	13.4	1.8 x 1.1	14.0	Scd:	100	Mag 8.57v star NW 9′.6.
10 16 08.4	+65 54 58	UGC 5530	14.2	1.3 x 0.5	13.6	S	95	**MCG +11-13-11** W 10′.3.
10 20 43.5	+65 10 18	UGC 5576	13.2	1.0 x 0.6	12.5	S?	0	Faint star or bright knot E edge.
10 24 05.7	+70 52 51	UGC 5612	12.1	3.2 x 2.1	14.0	SB(s)dm IV-V	165	Filamentary, many knots; mag 11.20v star superimposed W of center.
10 25 33.6	+70 04 33	UGC 5630	13.8	1.3 x 0.7	13.5	Double System	15	Double system, connected, disrupted.
10 30 25.0	+70 02 59	UGC 5688	13.2	3.3 x 1.9	15.1	SBm: V	145	Numerous superimposed stars; very faint, elongated anonymous galaxy SE end.
10 30 46.3	+73 53 07	UGC 5689	14.8	1.5 x 0.2	13.4	Sd	154	Close pair of mag 13.8 stars N 1′.2.
10 30 36.4	+70 37 07	UGC 5692	12.7	3.6 x 2.3	14.8	Sm: V	0	Mag 11.01v star E of center; mag 10.39v star E 5′.1.
10 31 15.5	+72 07 34	UGC 5700	13.9	1.5 x 1.2	14.4	SB?	3	
10 33 22.3	+67 36 02	UGC 5724	14.6	2.3 x 0.6	14.8	Double System	30	Double system, common envelope, in group with **MCG +11-13-29** and **MCG +11-13-31** which are just off E edge.
10 33 23.5	+64 30 10	UGC 5727	14.1	1.4 x 0.4	13.4	Sbc	29	Stellar **Mkn 145** E 11′.9.
10 37 28.8	+68 49 11	UGC 5765	14.3	1.7 x 0.3	13.4	Sbc	129	Mag 10.52v star SW 3′.9.
10 38 01.9	+64 15 59	UGC 5776	14.4	0.5 x 0.5	12.7	S?		
10 43 50.5	+69 44 48	UGC 5835	13.7	1.2 x 0.7	13.4	Sa?	134	**MCG +12-10-63** NNW 6′.0; **UGC 5834** S 3′.4.
10 44 35.0	+69 04 16	UGC 5843	14.6	1.1 x 0.3	13.2	S	9	Mag 11.5 star NW 3′.6.
10 48 38.2	+66 21 39	UGC 5904	13.7	2.0 x 0.4	13.3	Sb II-III	152	Double system, contact with small companion N end.
10 49 37.7	+65 31 50	UGC 5918	14.6	2.8 x 2.8	16.6	Im: V		Dwarf, extremely low surface brightness.
10 50 24.0	+64 48 32	UGC 5932	14.2	1.5 x 0.6	13.9	Sbc	52	
10 52 04.3	+71 46 22	UGC 5955	13.1	1.2 x 1.2	13.5	E		**MCG +12-10-85** SW 5′.1; **MCG +12-10-84** and mag 11.8 star SW 8′.7.
10 52 36.1	+67 58 58	UGC 5979	14.5	1.7 x 1.3	15.2	Im	51	Dwarf irregular; mag 10.57v star superimposed SW end.

(Continued from previous page)

14 GALAXY CLUSTERS 14

RA h m s	Dec ° ′ ″	Name	Mag 10th brightest	No. Gal	Diam ′	Notes
09 40 42.0	+66 40 00	A 834	16.3	30	13	All members are anonymous and almost stellar to stellar.

OPEN CLUSTERS

RA h m s	Dec ° ′ ″	Name	Mag	Diam ′	No. ★	B ★	Type	Notes
10 27 29.0	+66 48 00	NGC 3231		9.5	19	10.0	cl?	

15 GALAXIES 15

RA h m s	Dec ° ′ ″	Name	Mag (V)	Dim ′ Maj x min	SB	Type Class	PA	Notes
07 18 04.6	+68 20 29	CGCG 309-28	14.3	0.5 x 0.5	12.6			Mag 12.3 star SW 1′.6.
07 31 44.6	+63 14 31	CGCG 309-35	14.5	0.6 x 0.3	12.5	S?	177	= **Mkn 73**. Mag 9.13v star SSW 5′.5.
06 37 21.6	+67 51 33	IC 445	13.2	1.2 x 0.9	13.1	S0?	7	**MCG +11-9-2** SE 2′.5.
06 45 41.7	+71 20 37	IC 449	12.5	1.7 x 1.3	13.3	E:	69	Mag 8.35v star E 4′.1.
07 15 32.4	+64 55 36	IC 2179	12.4	1.1 x 1.1	12.6	E1-2		Lies 1′.4 E of mag 9.60v star.
07 29 25.3	+72 07 40	IC 2184	13.1	1.0 x 0.8	12.7	I?	0	Multiple galaxy system.
07 02 55.0	+62 46 09	MCG +10-10-21	14.0	0.8 x 0.5	12.9	SB?	114	Mag 11.5 star W 1′.6.
07 54 17.7	+62 51 30	MCG +10-11-41	14.8	0.8 x 0.6	14.1	E/S0	168	
07 36 19.7	+62 21 13	MCG +10-11-92	15.2	1.0 x 0.7	14.7	SBbc	9	Faint, slightly elongated anonymous galaxy W 0′.9.
07 37 23.8	+62 20 22	MCG +10-11-96	13.7	0.8 x 0.8	13.3	E/S0		Mag 11.6 star N 2′.4.
07 51 11.9	+62 04 50	MCG +10-11-150	13.7	0.9 x 0.6	13.1	E/S0	165	Mag 12.9 star on W edge.
07 57 22.9	+62 24 39	MCG +10-12-28	13.5	1.0 x 0.7	13.0		42	**MCG +10-12-20** W 2′.3; **MCG +10-12-27** N 1′.6.
08 23 42.0	+62 16 20	MCG +10-12-131	14.4	0.8 x 0.8	14.0	E/S0		Mag 14.9 star on W edge; mag 7.70v star NE 3′.9.
06 08 30.3	+64 43 24	MCG +11-8-12	13.7	0.8 x 0.5	12.8	E/S0	78	Mag 13.9 stars superimposed on N edge.
06 12 25.4	+67 35 57	MCG +11-8-16	14.6	0.6 x 0.6	13.3			
06 13 05.3	+64 33 40	MCG +11-8-19	14.0	1.1 x 0.7	13.6		15	Mag 11.07v star NW 3′.9.
06 14 02.4	+64 42 32	MCG +11-8-21	14.3	0.9 x 0.7	13.9	E/S0	9	
06 15 48.7	+66 50 22	MCG +11-8-25	14.3	1.1 x 0.5	13.5	S?	120	
06 16 42.2	+66 30 23	MCG +11-8-27	13.9	1.1 x 0.8	13.8	E/S0	99	
06 17 41.7	+66 36 01	MCG +11-8-30	14.8	0.9 x 0.4	13.5		15	Small, round galaxy on W edge; mag 12.3 star W 0′.7.
06 24 50.0	+65 43 50	MCG +11-8-42	14.9	0.7 x 0.4	13.4		81	Close pair of mag 14 stars WSW 1′.8.
06 25 00.9	+64 33 54	MCG +11-8-43	14.0	1.0 x 0.8	13.8	E/S0	174	Mag 9.81v star NE 6′.9.
06 25 37.5	+65 19 56	MCG +11-8-45	14.5	0.8 x 0.5	13.3	Sbc	156	Mag 13.6 star W 2′.4.
06 29 52.0	+65 12 11	MCG +11-8-50	15.8	0.7 x 0.7	14.9			Mag 13.0 star S 2′.4.
06 56 17.1	+64 56 47	MCG +11-9-21	14.6	0.7 x 0.5	13.3		123	Mag 12.6 star W 0′.6.
07 17 52.7	+63 29 07	MCG +11-9-41	13.8	1.0 x 0.5	12.9	S?	33	= **Mkn 379**; located between a close pair of E-W mag 14 stars.
07 28 51.9	+63 23 19	MCG +11-9-50	14.4	1.0 x 0.4	13.3	Sb	126	Mag 12.6 star NW 3′.0.
07 53 31.7	+63 53 46	MCG +11-10-37	13.8	0.5 x 0.5	12.4	E/S0		Mag 11.1 star W edge.
08 03 24.9	+67 09 23	MCG +11-10-52	14.0	1.0 x 0.8	13.7	Sb	54	
08 04 38.2	+65 23 29	MCG +11-10-59	14.8	0.5 x 0.5	13.1			Mag 10.37v star S 4′.2.
08 08 20.5	+64 22 52	MCG +11-10-63	14.4	0.5 x 0.5	12.7			Mag 9.27v star SE 3′.2.
08 08 51.2	+64 34 23	MCG +11-10-64	14.7	0.8 x 0.3	13.0			
08 09 44.2	+63 15 28	MCG +11-10-65	14.8	0.7 x 0.5	13.6	Sb	129	Mag 10.21v star NW 2′.8.
08 10 54.2	+67 26 55	MCG +11-10-67	15.1	1.0 x 0.5	14.2		12	
08 13 43.1	+64 20 10	MCG +11-10-69	15.3	0.6 x 0.5	14.1	E/S0	159	**MCG +11-10-70** E 1′.1.
08 27 51.5	+63 16 41	MCG +11-11-5	14.2	0.7 x 0.4	12.6	S0?	54	Located between a pair of mag 10.86v and 11.0 stars.
07 59 37.0	+71 45 10	MCG +12-8-22	14.8	0.6 x 0.6	13.6	Sb		Mag 10.88v star S 0′.9.
06 43 14.9	+65 40 37	NGC 2253	14.7	0.6 x 0.4	13.2	E/S0	12	Almost stellar.
07 16 04.0	+64 42 40	NGC 2347	12.5	1.8 x 1.3	13.2	(R′)SA(r)b: I-II	4	Mag 7.31v star N 4′.2.
07 28 56.1	+69 13 09	NGC 2366	11.1	8.1 x 3.0	14.5	IB(s)m V	25	HII region **NGC 2363** is 1′.3 W of bright knots located on SW edge of NGC 2366.
07 36 54.2	+65 36 07	NGC 2403	8.5	21.9 x 12.3	14.4	SAB(s)cd III	127	Numerous knots and superimposed stars.
07 51 54.8	+73 00 56	NGC 2441	12.2	1.8 x 1.6	13.2	SAB(r)b: II	45	
08 15 00.1	+73 34 45	NGC 2523	11.9	3.0 x 1.8	13.6	SB(r)bc I	57	Strong E-W bar, mag 8.18v star NNE 9′.0; **UGC 4279** NE 7′.0.
08 12 56.5	+73 33 45	NGC 2523B	13.9	2.1 x 0.3	13.3	SA(s)b: sp III	92	Located 8′.8 W of NGC 2523.
08 17 43.8	+73 19 03	NGC 2523C	12.9	1.5 x 0.8	13.1	E?	95	
08 21 39.8	+73 59 17	NGC 2544	12.9	1.1 x 0.8	12.6	SB(s)a:	70	Much elongated **MCG +12-8-35** NE 1′.4.
08 28 40.1	+73 44 56	NGC 2550A	12.7	1.5 x 1.4	13.4	Sc	0	**UGC 4389** SSW 6′.8; **MCG +12-8-44** SSE 7′.2; UGC 4413 NNE 8′.0.
08 24 51.0	+73 24 43	NGC 2551	12.1	1.7 x 1.1	12.6	SA(s)0/a	55	
05 52 36.2	+66 48 57	UGC 3365	14.0	2.2 x 0.3	13.4	Sa:	154	Mag 8.16v star N 3′.8; **CGCG 307-26** SE 9′.7.
05 58 26.0	+68 27 38	UGC 3379	12.9	1.5 x 1.1	13.3	(R′)SABb	115	Mag 11.9 star on S edge.
05 59 47.7	+62 09 29	UGC 3382	13.7	1.1 x 1.1	13.8	SB(rs)a		Almost stellar nucleus in short NE-SW bar, uniform envelope.
06 01 37.7	+73 06 57	UGC 3384	15.4	1.7 x 1.7	16.4	Sm:		Dwarf spiral, very low surface brightness; mag 11.5 star E 1′.1.
06 02 38.3	+65 22 17	UGC 3386	13.5	1.1 x 0.6	12.9	Sa:	43	
06 10 05.2	+67 57 00	UGC 3402	15.4	1.0 x 0.8	15.0	Sd	126	Mag 10.9 star N 3′.6.
06 10 35.5	+71 22 56	UGC 3403	13.6	2.5 x 0.7	14.1	SBcd?	27	
06 10 52.6	+64 34 00	UGC 3409	15.4	1.5 x 0.4	14.7	Sm:	10	Dwarf spiral, extremely low surface brightness.
06 10 55.0	+62 01 12	UGC 3411	14.4	1.7 x 1.1	14.9	Scd?	44	Stellar nucleus.
06 12 16.2	+64 16 06	UGC 3414	13.3	0.6 x 0.6	12.0	S?		**MCG +11-8-17** NW 2′.0; mag 11.8 star NW 2′.9.
06 13 11.8	+70 15 39	UGC 3415	15.9	1.5 x 0.2	14.4	Im?	154	Dwarf, extremely low surface brightness; mag 11.2 star S 2′.2.
06 13 39.2	+69 43 44	UGC 3416	14.0	1.5 x 0.6	13.7	Scd:	93	Located 2′.8 N of mag 9.78v star.
06 15 07.9	+71 08 10	UGC 3422	13.3	2.2 x 1.7	14.6	SAB(rs)b I-II	43	
06 14 47.7	+66 34 00	UGC 3425	14.2	2.3 x 0.3	13.6	Sb III	90	Mag 14.3 star on E end.
06 15 36.0	+71 02 10	UGC 3426	13.0	1.2 x 1.0	13.1	S0:	27	Mag 12.2 star N 1′.4.
06 15 42.2	+67 42 27	UGC 3428	15.9	1.5 x 0.4	15.2	Sdm:	77	Low surface brightness.
06 18 59.8	+63 26 40	UGC 3436	15.3	0.9 x 0.7	14.6	SAB(s)c	87	
06 19 41.6	+64 17 01	UGC 3437	15.4	1.0 x 0.6	14.7	Scd:	35	
06 20 11.7	+66 35 03	UGC 3438	14.1	1.2 x 0.6	13.6	SAB(s)b	107	Superimposed star SE and NW edges.
06 22 28.6	+66 34 42	UGC 3448	13.7	1.2 x 0.3	12.4	S0/a	45	
06 24 52.9	+73 07 43	UGC 3453	13.7	1.2 x 0.7	13.4	SB?	50	
06 25 55.8	+64 44 21	UGC 3458	13.9	2.6 x 0.7	14.4	Sb III	122	Strong dark lane almost full length; **MCG +11-8-46** SW 1′.4.
06 32 38.6	+71 33 31	UGC 3474	14.7	2.4 x 0.2	13.8	Scd:	161	Mag 11.9 star W 0′.5.
06 32 47.2	+63 40 23	UGC 3478	12.8	1.8 x 0.6	12.7	Sb II	42	Mag 11.8 star NE 1′.9.
06 39 48.5	+65 27 07	UGC 3502	13.9	1.5 x 0.6	13.7	SBb:	114	Mag 13.0 star NW 2′.4.
06 43 42.1	+65 12 21	UGC 3511	12.5	1.5 x 1.1	12.8	Scd:	135	Close pair of stars S, mags 13.4 and 14.0.
06 48 54.0	+66 15 39	UGC 3539	14.4	1.9 x 0.2	13.2	SBbc?	116	Located between a pair of mag 13.9 stars.
06 54 51.5	+70 44 51	UGC 3575	15.3	1.7 x 0.2	14.0	Scd:	166	Mag 9.89v star NE 0′.8.

(Continued from previous page)

GALAXIES

RA h m s	Dec ° ′ ″	Name	Mag (V)	Dim ′ Maj x min	SB	Type Class	PA	Notes
06 54 12.9	+65 12 08	UGC 3577	14.2	1.3 x 1.0	14.3	SB(r)b	90	Mag 11.6 star SW 0'.9.
06 55 31.0	+69 33 45	UGC 3580	11.8	3.4 x 1.9	13.7	SA(s)a pec:	3	Mag 14.0 star superimposed E of nucleus.
06 57 50.5	+63 41 33	UGC 3606	14.1	1.5 x 0.9	14.3	Sdm	12	Mag 12.6 star N 1'.4.
07 02 41.1	+70 58 08	UGC 3626	14.3	1.2 x 0.4	13.3	SBbc?	150	Mag 8.84v star SW 5'.5.
07 04 20.8	+64 01 13	UGC 3642	12.4	1.4 x 1.0	12.6	SA0°	30	**MCG +11-9-28** W 2'.3; **MCG +11-9-29** SW 2'.9.
07 05 38.7	+71 04 06	UGC 3644	13.8	2.0 x 1.2	14.6	SAB(s)d	140	Mag 11.4 star, plus chain of four fainter stars, NW edge.
07 06 35.0	+63 50 58	UGC 3660	12.7	1.7 x 0.9	13.0	SABa:	110	
07 11 24.2	+71 50 12	UGC 3697	12.9	3.0 x 0.2	12.2	Sd: pec	79	Mag 6.36v star E 12'.2.
07 11 43.0	+72 10 10	UGC 3701	13.9	1.8 x 1.8	15.0	SA(rs)cd:		Stellar nucleus.
07 12 11.3	+73 28 11	UGC 3705	13.3	0.8 x 0.6	12.4	S?	66	Mag 10.8 star N 0'.7.
07 12 32.8	+71 45 00	UGC 3714	11.9	1.8 x 1.5	12.8	S? pec	35	Bright core, uniform envelope; mag 6.36v star NE 7'.6.
07 13 14.3	+73 50 35	UGC 3717	13.0	2.1 x 0.9	13.6	Sbc	103	
07 14 21.1	+73 28 23	UGC 3730	12.5	2.7 x 1.0	13.4	Ring	171	Strongly peculiar, jets plus plume.
07 15 53.5	+67 58 56	UGC 3749	14.5	1.5 x 0.3	13.5	Scd:	21	Mag 11.6 star SE 1'.5.
07 16 53.0	+67 06 37	UGC 3764	13.2	1.2 x 0.7	12.9	SB?	120	Mag 8.32v star SW 2'.5.
07 17 53.1	+70 31 05	UGC 3771	14.0	1.1 x 0.8	13.7	SB(s)b	35	Short, bright bar; mag 11.01v star NW 1'.7.
07 22 34.5	+71 35 57	UGC 3804	12.4	1.7 x 1.2	13.0	Scd:	13	
07 26 45.6	+64 48 31	UGC 3836	14.4	1.0 x 0.6	13.7	SB(rs)cd	21	
07 28 12.1	+72 34 21	UGC 3838	13.5	0.8 x 0.3	11.8	Pec	17	Lies between a pair of N-S oriented mag 13 stars.
07 28 15.2	+63 15 16	UGC 3850	12.8	1.8 x 1.6	13.8	(R′)SAB(s)a	63	Bright core, dark patches N and S.
07 30 47.1	+73 37 45	UGC 3858	12.9	1.1 x 0.9	12.7	Sa	120	
07 30 48.8	+73 42 21	UGC 3859	12.7	1.5 x 1.0	13.0	Sa	33	Elongated core, smooth envelope.
07 30 56.9	+72 31 00	UGC 3864	13.9	1.5 x 1.0	14.2	S?	27	
07 31 27.3	+62 27 24	UGC 3886	15.3	1.0 x 1.0	15.2	SA(s)c		Mag 9.46v star W edge; mag 6.99v star N 2'.9.
07 32 56.1	+64 59 31	UGC 3893	13.8	1.0 x 0.9	13.5	SA(r)b: III	33	Mag 14.1 star on W edge.
07 33 04.8	+65 04 43	UGC 3894	13.2	1.2 x 1.2	13.6	E		Mag 13.6 star on NW edge; faint, elongated anonymous galaxy SW 2'.3.
07 34 49.4	+62 32 45	UGC 3905	13.6	0.8 x 0.6	12.7	Sab	80	
07 36 59.2	+73 42 49	UGC 3909	14.1	2.6 x 0.4	14.0	SBcd:	82	Mag 11.25v star S 2'.9.
07 37 05.2	+64 33 09	UGC 3919	15.0	1.5 x 0.2	13.5	Scd:	81	Mag 14.9 star E end, mag 14.3 star W end.
07 42 45.7	+66 15 26	UGC 3968	14.1	1.3 x 1.1	14.3	SB(r)c	60	Small dark lane SW of nucleus; mag 8.45v star W 6'.1.
07 44 42.2	+73 49 16	UGC 3972	13.8	1.2 x 0.7	13.4	SB?	160	Mag 10.84v star SW 1'.2.
07 44 59.7	+72 48 38	UGC 3975	14.1	1.4 x 1.2	14.5	SBdm	160	
07 43 55.4	+62 04 14	UGC 3978	14.1	1.2 x 0.9	14.0	SBb	55	Bright SE-NW bar, faint envelope.
07 44 31.2	+67 16 24	UGC 3979	14.0	1.7 x 0.3	13.1	Sbc	153	Mag 10.98v star NW 2'.6.
07 45 13.7	+70 01 54	UGC 3984	13.3	1.9 x 0.9	13.7	SB(s)b I-II	55	Mag 10.83v star E 2'.5.
07 45 35.9	+65 45 30	UGC 3991	14.0	1.1 x 0.5	13.2	Sc	162	
07 46 24.9	+62 29 11	UGC 4001	13.5	1.3 x 0.6	13.2	E	144	
07 47 48.1	+62 19 25	UGC 4015	14.4	1.1 x 0.6	13.8	SB(s)cd	54	Stellar **Mkn 82** NW 5'.9.
07 48 31.6	+65 20 13	UGC 4021	14.0	1.1 x 0.8	13.7	SBab	50	
07 48 44.5	+66 11 50	UGC 4023	13.1	1.6 x 0.7	13.1	S0	136	
07 48 33.4	+62 12 24	UGC 4024	14.1	1.5 x 0.4	13.4	Scd:	115	Mag 14.4 star W end.
07 52 38.4	+73 30 08	UGC 4041	12.9	1.0 x 0.5	12.1	E?	129	**UGC 4037** N 5'.6.
07 53 00.5	+72 02 27	UGC 4050	13.4	0.6 x 0.5	11.9	S?	70	**MCG +12-8-14** W 5'.0.
07 52 36.4	+62 32 47	UGC 4059	14.5	1.1 x 0.6	13.9	Scd:	50	Patchy; knot NE edge.
07 54 54.2	+72 36 39	UGC 4067	14.4	1.0 x 0.5	13.5	SBcd:	165	Located on W edge of very small Galaxy cluster **A 596**; mag 8.68v star NNW 9'.4.
07 56 51.3	+73 47 10	UGC 4080	13.2	1.0 x 0.7	12.6	Sa	60	Pair of mag 13 stars NE 1'.7.
07 56 34.8	+63 36 30	UGC 4094	13.8	1.3 x 1.0	14.0	SABab	110	Strong knot or star S edge.
07 56 54.0	+66 36 40	UGC 4095	13.8	0.8 x 0.3	12.1	S?	51	Mag 9.04v star S 3'.4.
07 57 21.6	+66 26 09	UGC 4098	13.7	0.8 x 0.5	12.6	S?	105	Mag 14.8 star on W end.
08 02 03.0	+73 02 37	UGC 4137	14.7	1.5 x 1.1	15.1	Sdm	133	Low surface brightness; mag 12.7 star on NE edge.
08 04 07.9	+62 59 01	UGC 4186	14.6	1.4 x 0.2	13.1	S	147	
08 05 06.9	+66 46 58	UGC 4195	13.6	1.9 x 1.0	14.1	SB(r)b	20	Mag 11.3 star ENE 2'.3.
08 06 39.9	+73 20 36	UGC 4199	13.8	1.2 x 0.4	12.9	S?	99	**CGCG 331-27** WSW 2'.3.
08 06 39.2	+67 02 52	UGC 4206	15.2	1.5 x 0.2	13.7	Sbc	132	
08 10 59.4	+72 47 39	UGC 4242	14.5	0.9 x 0.7	13.8	S?	165	
08 10 25.6	+67 14 35	UGC 4243	13.5	1.1 x 0.6	12.9	Sab	158	Mag 8.38v star SE 2'.8.
08 17 27.0	+64 32 46	UGC 4302	14.3	1.1 x 0.9	14.1	SBb:	45	
08 19 03.9	+70 42 55	UGC 4305	10.7	8.5 x 6.0	14.8	Im IV-V	15	Dwarf irregular, low surface brightness; mag 11.4 star N end; a small triangle of stars, mags 11.4, 13.1 and 13.2, E of the center.
08 20 01.6	+62 49 50	UGC 4322	13.7	1.4 x 1.0	14.1	E	57	
08 20 19.8	+66 58 53	UGC 4323	13.1	1.8 x 1.3	14.0	E?	50	
08 20 35.1	+68 36 01	UGC 4326	13.1	1.4 x 0.6	12.8	Sbc	152	Mag 11.51v star SW 3'.1.
08 24 25.6	+65 36 17	UGC 4369	14.8	0.8 x 0.7	14.1	SBc		
08 24 54.0	+66 52 10	UGC 4376	14.0	1.0 x 0.7	13.5	S?	129	11.6 and 12.6 mag stars superimposed.
08 27 51.9	+73 31 03	UGC 4390	14.3	2.0 x 1.8	15.5	SBd	78	Mag 10.6 star W 2'.2.
08 27 35.7	+64 13 40	UGC 4398	14.1	1.1 x 0.7	13.7	SA(rs)bc	170	Mag 12.8 star on E edge.
08 29 46.3	+73 51 25	UGC 4413	14.7	1.5 x 0.3	13.7	Scd:	105	NGC 2550A SSW 6'.8; faint stellar galaxy **KUG 0825+739** SE 8'.4.
08 29 11.3	+63 20 17	UGC 4420	13.4	1.6 x 0.8	13.5	S0	129	**MCG +11-11-8** S 2'.8.

GALAXY CLUSTERS

RA h m s	Dec ° ′ ″	Name	Mag 10th brightest	No. Gal	Diam ′	Notes
06 40 06.0	+69 41 00	A 559	15.8	33	39	All members stellar and very faint.
07 00 42.0	+69 48 00	A 564	16.2	56	34	Mag 6.63v star on SSE edge; all members faint and stellar.
07 04 30.0	+63 17 00	A 566	16.4	127	13	GX **KUG 0659+633** located 1'.4 N of cluster center; **MCG +11-9-31** NE 2'.6; all members faint and stellar.

PLANETARY NEBULAE

RA h m s	Dec ° ′ ″	Name	Diam ″	Mag (P)	Mag (V)	Mag cent ★	Alt Name	Notes
06 29 34.7	+71 05 07	PN G143.6+23.8	111			12.7	EGB 4	Very faint horseshoe shape with open end to the N; brightest along W side; mag 12.5 star on S edge; mag 14.0 star at N end.

GALAXIES

RA h m s	Dec ° ′ ″	Name	Mag (V)	Dim ′ Maj x min	SB	Type Class	PA	Notes
05 14 15.5	+62 34 28	CGCG 283-4	13.3	5.5 x 3.4	16.3	Im?	9	Very filamentary; many superimposed stars; mag 8.33v star SE 3'.4.
05 20 49.6	+66 14 43	CGCG 307-10	13.9	0.5 x 0.3	11.7		168	Mag 10.18v star NE 5'.8.
05 31 25.9	+67 43 48	CGCG 307-15	13.7	0.7 x 0.7	12.8			**CGCG 307-17** E 7'.0; mag 9.95v star SW 4'.5.
05 31 44.7	+67 36 42	CGCG 307-16	14.3	0.4 x 0.4	12.1			Mag 9.95v star NW 6'.7.
05 43 23.4	+69 25 47	CGCG 329-11	13.9	0.7 x 0.5	12.6	Sb	108	Bright nucleus slightly offset to N.
03 46 49.6	+68 05 56	IC 342	8.4	21.4 x 20.9	14.9	SAB(rs)cd I-II	168	Low surface brightness; numerous superimposed stars and bright knots.
04 07 46.6	+69 48 46	IC 356	10.0	5.9 x 3.9	13.2	SA(s)ab pec	105	Many stars superimposed; mag 8.48v star N 3'.2.
04 57 58.8	+68 19 22	IC 396	12.1	2.1 x 1.4	13.1	S?	85	Mag 9.72v star S 6'.0.
05 43 40.4	+68 56 42	MCG +11-7-13	14.1	1.1 x 0.3	12.7	S	42	
04 34 06.8	+73 17 55	MCG +12-5-6	14.6	0.9 x 0.5	13.6	Sb	27	NGC 1573 SE 4'.6; mag 8.51v star W 9'.1.
05 21 42.0	+72 19 56	MCG +12-6-2	13.7	0.7 x 0.4	12.4	E/S0	12	
05 45 38.5	+69 03 36	MCG +12-6-13	14.1	0.7 x 0.6	13.0	S?	30	Mag 11.2 star N 0'.8.
03 37 49.5	+72 34 16	NGC 1343	12.7	2.6 x 1.6	14.1	SAB(s)b: pec	80	Two faint stars superimposed E of core.
04 00 27.8	+68 34 41	NGC 1469	12.7	1.9 x 0.8	13.0	SA0$^-$:	153	Mag 10.68 star on E edge.
04 05 03.8	+70 59 49	NGC 1485	12.6	2.1 x 0.7	12.9	SAb? sp	22	
04 32 48.5	+71 53 00	NGC 1560	11.4	9.8 x 1.7	14.3	SA(s)d sp	23	Numerous superimposed stars; mag 9.56v star W 5'.6.
04 30 48.3	+64 50 55	NGC 1569	11.0	3.6 x 1.8	12.9	IBm IV-V	120	Mag 9.77v star N edge; numerous stars superimposed S and W of bright, elongated center.
04 35 04.4	+73 15 43	NGC 1573	11.7	1.9 x 1.3	12.7	E	35	UGC 3069 SW 4'.3.
05 42 04.1	+69 22 26	NGC 1961	11.0	4.6 x 3.0	13.7	SAB(rs)c II	85	Several knots, CGCG 329-11 NE 7'.4.
03 36 13.9	+67 33 52	UGC 2789	14.5	1.8 x 1.3	15.3	SB?	13	Filamentary, several knots; mag 11.6 star E 1'.4.
03 40 04.0	+71 24 16	UGC 2800	15.9	2.2 x 1.0	16.6	Im?	100	Dwarf, low surface brightness; mag 11.3 star ENE 2'.2.
03 42 34.8	+71 18 26	UGC 2813	15.9	1.7 x 0.7	15.9	Im:	35	Dwarf, very low surface brightness.
03 43 45.7	+68 18 03	UGC 2826	14.0	1.3 x 0.7	14.0	E	120	
03 48 22.4	+70 07 55	UGC 2855	12.8	4.2 x 2.0	15.0	SABc	112	Knotty; mag 11.22v star superimposed N of nucleus.
03 50 36.1	+73 01 48	UGC 2865	13.9	1.4 x 0.6	13.6	SB?	14	Pair of mag 11.8 stars 2'.0 S.
03 50 15.0	+70 05 37	UGC 2866	14.6	1.0 x 0.8	14.2	Spiral	60	
03 56 04.9	+72 55 15	UGC 2890	13.7	2.4 x 0.7	14.1	Sdm pec:	33	Mag 12.8 star superimposed NE of nucleus.
04 02 34.9	+71 42 14	UGC 2916	13.3	1.3 x 1.0	13.4	Sab	42	Mag 14.3 star superimposed inside S edge.
04 28 40.0	+70 20 48	UGC 3042	13.2	1.9 x 0.8	13.5	SAB(s)bc I-II	70	Mag 10.6 stars on NE and SW edges.
04 29 20.1	+69 31 51	UGC 3046	13.9	1.5 x 1.3	14.5	Sb	177	Mag 9.62 star SE 2'.7.
04 29 23.7	+70 25 27	UGC 3048	14.1	1.3 x 0.8	14.0	SAB(rs)bc:	104	Mag 7.76v star SE 10'.2.
04 34 28.8	+73 12 14	UGC 3069	13.8	1.3 x 0.7	13.6	S0$^-$:	55	NGC 1573 NE 4'.3.
04 36 56.8	+71 31 45	UGC 3090	14.7	1.5 x 1.1	15.1	Im:	24	Dwarf, low surface brightness, stars superimposed.
04 38 19.4	+72 16 53	UGC 3092	15.7	1.5 x 0.2	14.3	Sd	162	
04 41 09.0	+73 40 15	UGC 3110	13.9	2.3 x 0.9	14.5	SBcd:	115	Mag 10.61v star E end.
04 40 30.1	+66 38 02	UGC 3114	14.0	2.3 x 0.7	14.4	Scd:	72	Stars superimposed SW half.
04 46 36.3	+70 07 13	UGC 3143	15.0	1.7 x 0.2	13.7	Scd:	102	Located 2'.3 N of mag 8.99v star.
04 47 50.1	+72 51 30	UGC 3147	13.8	1.3 x 0.6	13.4	S?	42	**UGC 3159** NE 11'.2.
04 47 11.5	+66 03 15	UGC 3149	13.5	1.7 x 0.9	13.8	S?	110	Bright core, uniform envelope.
04 48 26.5	+73 28 07	UGC 3150	13.5	1.7 x 1.2	14.1	SABbc	165	Mag 9.01v star S 1'.5.
04 49 33.5	+69 28 30	UGC 3163	14.2	1.5 x 0.3	13.2	Sa	86	
04 49 28.0	+63 55 33	UGC 3167	15.1	1.5 x 0.2	13.7	Sbc	167	Mag 11.09v star NW 3'.1.
04 51 37.0	+67 12 57	UGC 3176	13.2	1.2 x 1.2	13.5	S0$^-$:		
04 54 07.0	+72 19 22	UGC 3182	13.2	1.3 x 0.9	13.2	S0	135	Close pair of almost stellar **MCG +12-5-24** and **MCG +12-5-25** W 11'.7.
04 54 37.9	+69 42 51	UGC 3189	14.4	2.0 x 0.6	14.4	Sd	40	Mag 11.25v star W 0'.9.
05 00 44.1	+62 14 38	UGC 3218	13.5	1.7 x 0.9	13.8	SAb	145	Numerous stars superimposed.
05 05 32.3	+67 00 37	UGC 3235	15.3	2.3 x 0.9	15.9	SAB(s)cd:	150	Mag 14.0 star NE edge.
05 08 40.8	+70 28 48	UGC 3245	13.9	1.3 x 1.1	14.1	SAB(s)c	90	Stellar nucleus; mag 7.83v star N 4'.2.
05 10 31.8	+63 09 53	UGC 3250	13.1	1.8 x 1.3	13.9	SBb	100	Mag 12.1 star NE 1'.2.
05 11 45.4	+67 29 15	UGC 3252	12.9	1.6 x 1.2	13.5	SAB(s)c	30	Mag 9.75v star NE 2'.8.
05 14 15.8	+72 20 02	UGC 3259	14.4	1.9 x 1.4	15.3	SABd	120	Stellar nucleus, uniform envelope; mag 10.0v star NW 2'.3.
05 15 46.5	+71 27 33	UGC 3267	14.1	1.5 x 1.0	14.4	Scd:	15	
05 19 10.3	+65 28 12	UGC 3277	14.2	2.2 x 0.7	14.5	Scd:	156	Faint star NE edge.
05 20 41.8	+72 32 10	UGC 3281	13.5	1.2 x 0.7	13.2	S0	150	Mag 10.09v star SE 2'.4.
05 21 23.5	+72 42 02	UGC 3284	15.3	1.0 x 0.6	14.6	SAB(s)c	113	Located between a pair of stars, mags 11.2 and 12.2.
05 27 57.5	+63 51 31	UGC 3307	13.6	0.8 x 0.5	12.7	E	79	
05 29 14.9	+67 22 15	UGC 3309	14.1	0.9 x 0.6	13.3	S?	7	
05 33 37.5	+73 43 29	UGC 3317	14.4	1.5 x 1.1	14.8	Im V	177	Dwarf irregular, extremely low surface brightness.
05 34 15.5	+70 11 24	UGC 3319	15.8	1.7 x 0.3	14.9	Scd:	127	Mag 8.66v star S 2'.0.
05 44 29.6	+69 17 55	UGC 3342	14.5	1.7 x 0.4	13.9	Scd:	42	Mag 8.74v star E 5'.9.
05 45 24.2	+72 21 22	UGC 3343	13.0	2.3 x 0.6	13.2	S?	80	Mag 11.11v star E edge.
05 44 56.7	+69 09 34	UGC 3344	13.4	1.5 x 1.1	13.8	SABbc II-III	37	Mag 9.74v star S 3'.8.
05 46 31.3	+69 02 59	UGC 3349	13.5	1.1 x 0.7	13.1	Sab	85	Mag 11.2 star and MCG +12-6-13 W 4'.8.
05 52 36.2	+66 48 57	UGC 3365	14.0	2.2 x 0.3	13.4	Sa:	154	Mag 8.16v star N 3'.8; **CGCG 307-26** SE 9'.7.
05 58 26.0	+68 27 38	UGC 3379	12.9	1.5 x 1.1	13.3	(R')SABb	115	Mag 11.9 star on S edge.
05 59 47.7	+62 09 29	UGC 3382	13.7	1.1 x 1.1	13.8	SB(rs)a		Almost stellar nucleus in short NE-SW bar, uniform envelope.
06 01 37.7	+73 06 57	UGC 3384	15.4	1.7 x 1.7	16.4	Sm:		Dwarf spiral, very low surface brightness; mag 11.5 star E 1'.1.
06 02 38.3	+65 22 17	UGC 3386	13.5	1.1 x 0.6	12.9	Sa:	43	

GALAXY CLUSTERS

RA h m s	Dec ° ′ ″	Name	Mag 10th brightest	No. Gal	Diam ′	Notes
05 10 18.0	+73 41 00	A 527	15.7	34	22	All members anonymous, faint and stellar.

OPEN CLUSTERS

RA h m s	Dec ° ′ ″	Name	Mag	Diam ′	No. ★	B ★	Type	Notes
03 39 37.0	+66 30 00	Be 10		10	50	14.0	cl	Moderately rich in stars; moderate brightness range; slight central concentration; detached.
05 12 39.0	+73 58 00	Cr 464	4.2	120	50		cl	
04 07 50.0	+62 20 00	NGC 1502	6.9	20	63	7.0	cl	Moderately rich in bright and faint stars with a strong central concentration; detached. (A) At SE end of W.S. Houston's **Kemble's Cascade**.

17 GALAXIES 17

RA h m s	Dec ° ′ ″	Name	Mag (V)	Dim ′ Maj x min	SB	Type Class	PA	Notes
03 37 49.5	+72 34 16	NGC 1343	12.7	2.6 x 1.6	14.1	SAB(s)b: pec	80	Two faint stars superimposed E of core.
01 56 18.8	+73 16 54	UGC 1378	12.6	3.5 x 2.3	14.7	(R)SB(rs)a:	11	Numerous stars superimposed on faint outer envelope; mag 9.06v star SE 5′.5.
03 08 55.5	+70 33 45	UGC 2542	14.4	1.5 x 0.2	12.9	Sb	145	Mag 10.15v star W 2′.6.
03 36 13.9	+67 33 52	UGC 2789	14.5	1.8 x 1.3	15.3	SB?	13	Filamentary, several knots; mag 11.6 star E 1′.4.
03 40 04.0	+71 24 16	UGC 2800	15.9	2.2 x 1.0	16.6	Im?	100	Dwarf, low surface brightness; mag 11.3 star ENE 2′.2.
03 42 34.8	+71 18 26	UGC 2813	15.9	1.7 x 0.7	15.9	Im:	35	Dwarf, very low surface brightness.
03 43 45.7	+68 18 03	UGC 2826	14.0	1.3 x 0.7	14.0	E	120	

OPEN CLUSTERS

RA h m s	Dec ° ′ ″	Name	Mag	Diam ′	No. ★	B ★	Type	Notes
01 47 44.0	+62 56 24	Be 5		5	15	17.0	cl	
01 54 12.0	+62 22 12	Be 7		4	18	14.0	cl	
03 39 37.0	+66 30 00	Be 10		10	50	14.0	cl	Moderately rich in stars; moderate brightness range; slight central concentration; detached.
02 19 27.0	+63 43 18	Be 63		3	12	15.0	cl	
02 21 46.0	+65 53 30	Be 64		2	30	14.0	cl	
01 45 45.0	+71 49 00	Cr 463	5.7	57	79	8.5	cl	Moderately rich in stars; moderate brightness range; no central concentration; detached.
02 02 08.0	+62 50 48	Cz 6		2	12		cl	
02 03 01.0	+62 14 48	Cz 7		4	20		cl:	
02 44 27.0	+62 19 30	Cz 13	10.4	4	56	12.8	cl	Few stars; moderate brightness range; no central concentration; detached. Probably not a cluster.
01 05 11.0	+62 01 18	NGC 358		3	25		cl?	
01 06 27.0	+62 13 42	NGC 366		4	30	10.0	cl	Moderately rich in bright and faint stars; detached.
01 29 33.0	+63 18 00	NGC 559	9.5	7	120	9.0	cl	Moderately rich in stars; small brightness range; strong central concentration; detached.
01 36 24.0	+64 32 12	NGC 609	11.0	3	77	14.0	cl	Rich in stars; large brightness range; slight central concentration; detached.
01 43 04.0	+64 02 12	NGC 637	8.2	3	55	8.0	cl	Moderately rich in stars; moderate brightness range; strong central concentration; detached.
02 23 28.1	+63 46 00	NGC 886		14	20	11.0	cl	
01 12 03.0	+62 16 24	St 3		5	8	11.0	cl	Few stars; small brightness range; not well detached.
02 04 24.0	+64 23 00	St 5		24	25	4.9	cl	Moderately rich in stars; large brightness range; no central concentration; detached.
03 12 00.0	+63 11 00	Tr 3	7.0	15	30		cl	Moderately rich in stars; moderate brightness range; no central concentration; detached. (S) Sparse.

BRIGHT NEBULAE

RA h m s	Dec ° ′ ″	Name	Dim ′ Maj x min	Type	BC	Color	Notes
02 26 30.0	+62 04 00	IC 1795	40 x 15	E	1-5	3-4	Separated from NGC 896 by a narrow dark lane aligned N-S. It is brightest at the N end of the dark lane and along the irregular N extension.
02 24 48.0	+62 01 00	NGC 896	20 x 20	E	1-5	3-4	Separated from IC 1795 by a narrow dark lane aligned N-S. It is higher in surface brightness than IC 1795 and there is a fairly prominent absorption patch in the SW part.
02 51 36.0	+67 52 00	vdB 8	4 x 1	R	1-5	1-4	Extended NW-SE; mag 8.6 star involved.

PLANETARY NEBULAE

RA h m s	Dec ° ′ ″	Name	Diam ″	Mag (P)	Mag (V)	Mag cent ★	Alt Name	Notes
01 57 36.0	+63 19 17	IC 1747	19	13.6	12.0	15.4	PK 130+1.1	Mag 12.0 star N 0′.9.
01 07 12.9	+73 32 57	PN G124.0+10.7	270			16.4	PK 124+10.1	Group of 3 mag 11 stars form a right angle shape NW 4′.1; a close pair of mag 11 stars SE 3′.7.
01 24 58.9	+65 38 33	PN G126.3+2.9	9	16.0			PK 126+3.1	Mag 12.1 star W 0′.9; mag 10.56v star W 1′.5.
01 58 35.7	+66 33 58	PN G129.5+4.5	23	20.9		20.5	PK 129+4.1	Slightly elongated E-W; mag 12.2 star N 3′.5; mag 10.84v star SE 3′.3.
02 03 41.5	+64 57 36	PN G130.4+3.1	12	16.7		20.6	PK 130+3.1	Mag 11.4 star NNW 1′.8; mag 12.1 star NNE 2′.2.
02 12 06.6	+64 09 03	PN G131.5+2.6	60	18.2		18.8	PK 131+2.1	Mag 12.7 star on E edge; mag 8.85v star W 5′.0; mag 10.96v star E 2′.4.
02 26 30.3	+65 47 50	PN G132.4+4.7	10	17.6			PK 132+4.1	Mag 10.1 star W 1′.1; mag 13.2 star SE 1′.1.
02 58 42.0	+64 30 07	PN G136.1+4.9	186	>15.5	14.3	19.6	PK 136+4.1	Mag 13.0 star on N edge; mag 10.07v star WNW 3′.6.
03 03 48.8	+64 53 28	PN G136.3+5.5	500			13.3	PK 136+5.1	Numerous faint stars on disc; mag 11.4 star NW 6′.2; close pair of mag 9.17v, 12.2 stars SW 7′.1.
03 11 02.1	+62 47 57	PN G138.1+4.1	340			12.6	PK 138+4.1	Widely spaced pair of mag 11 stars near center.

18 GALAXIES 18

RA h m s	Dec ° ′ ″	Name	Mag (V)	Dim ′ Maj x min	SB	Type Class	PA	Notes
00 20 16.7	+59 18 02	IC 10	10.4	6.3 x 5.1	14.0	IBm	135	Numerous stars superimposed; mag 11.24v star N 5′.3; mag 10.26v star on S edge.
00 39 08.5	+58 35 38	MCG +10-2-1	15.0	0.9 x 0.5	14.0	S	45	Located between a close pair of mag 11 stars.
00 46 19.4	+51 13 02	UGC 475	13.4	1.5 x 0.7	13.3	Scd:	105	Mag 10.81v star NE 2′.1.
22 59 23.0	+53 44 21	UGC 12287	14.8	1.5 x 0.7	14.7	Scd:	5	Numerous stars superimposed S half; mag 12 star E edge; mag 8.40v star S 1′.8.

OPEN CLUSTERS

RA h m s	Dec ° ′ ″	Name	Mag	Diam ′	No. ★	B ★	Type	Notes
00 09 41.0	+60 28 42	Be 1		5	10		cl	(S) Sparse.
00 25 15.0	+60 23 18	Be 2		2	30	15.0	cl	
00 00 12.0	+60 56 30	Be 58	9.7	5	39	15.0	cl	Moderately rich in stars; small brightness range; slight central concentration; detached.
00 17 42.0	+60 56 30	Be 60		3	20	14.0	cl	
23 38 42.0	+56 38 00	Be 102		5		18.0	cl	
23 45 12.0	+59 18 00	Be 103		2	20	15.0	cl	
23 04 45.3	+60 04 40	Bergeron 1		1	10		cl?	Possible cluster. About 10′ south of IC 1470. Noted 1997 by T. Bergeron.
00 07 39.0	+61 28 42	Cz 1		3	12		cl:	
00 43 39.0	+60 12 00	Cz 2		10	30		cl	
01 03 07.0	+62 47 12	Cz 3	9.9	1.5	10		cl:	
23 25 48.0	+61 18 00	Cz 43		7	15		cl	Few stars; small brightness range; no central concentration; detached.
23 21 54.0	+55 46 00	Do 46		12			ast	Few stars; small brightness range; not well detached. Probably not a cluster. (A) Scattering of bright stars.
23 57 26.5	+61 37 23	Fr 1	9.2		26	10.6	cl	
23 54 17.6	+61 43 43	Ha 21	9.0	3	6		cl	Few stars; small brightness range; not well detached.

(Continued from previous page)

OPEN CLUSTERS

RA h m s	Dec ° ′ ″	Name	Mag	Diam ′	No. ★	B ★	Type	Notes
00 52 49.0	+56 37 42	IC 1590	7.4	4		9.0	cl	(A) Involved in nebula NGC 281.
00 50 57.0	+58 11 30	King 2	19.8	4	40	17.0	cl	(A) Two thirds complete "stellar ring" visible on DSS at cluster center.
23 53 00.0	+61 57 00	King 12	9.0	3	15	10.0	cl	Few stars; small brightness range; slight central concentration; detached. Probably not a cluster.
00 10 10.0	+61 11 00	King 13		5	30	12.0	cl	
00 33 07.0	+61 51 12	King 15		3	12	18.0	cl	
23 08 18.0	+60 31 00	King 19	9.2	5	52	12.0	cl	Few stars; moderate brightness range; no central concentration; detached.
23 33 14.0	+58 28 48	King 20		5	20	13.0	cl	
23 49 54.0	+62 42 00	King 21	9.6	4	20	10.0	cl	Few stars; moderate brightness range; strong central concentration; detached.
00 21 54.0	+61 45 00	Mayer 1		8	20		cl	
23 15 14.0	+60 26 42	Mrk 50	8.5	2	39	9.8	cl	Few stars; small brightness range; no central concentration; detached; involved in nebulosity.
00 25 18.0	+61 19 18	NGC 103	9.8	5	88	11.0	cl	Moderately rich in stars; small brightness range; slight central concentration; detached.
00 29 58.0	+60 13 00	NGC 129	6.5	12	193	11.0	cl	Moderately rich in stars; moderate brightness range; no central concentration; detached.
00 31 31.0	+61 30 36	NGC 136		1.5	20	13.0	cl	Few stars; small brightness range; slight central concentration; detached.
00 39 36.0	+61 05 42	NGC 189	8.8	5	90	10.9	cl	Few stars; small brightness range; no central concentration; detached.
00 43 36.0	+61 46 00	NGC 225	7.0	15	76	9.3	cl	Few stars; small brightness range; no central concentration; detached; involved in nebulosity.
22 57 03.0	+54 18 00	NGC 7438		20	30	11.0	cl?	
23 11 04.2	+60 34 00	NGC 7510	7.9	7	75	10.0	cl	Rich in stars; large brightness range; slight central concentration; detached; involved in nebulosity.
23 24 51.0	+61 36 00	NGC 7654	6.9	16	173	11.0	cl	= **M 52**. Rich in stars; moderate brightness range; slight central concentration; detached. (A) 12′ diameter condensed central portion with 16′ outer diameter. Rich.
23 56 44.0	+61 24 00	NGC 7788	9.4	4	20		cl	Few stars; moderate brightness range; strong central concentration; detached.
23 57 28.6	+56 43 00	NGC 7789	6.7	25	583	10.0	cl	Rich in stars; moderate brightness range; slight central concentration; detached.
23 58 24.0	+61 12 30	NGC 7790	8.5	5	134	10.0	cl	Moderately rich in stars; moderate brightness range; slight central concentration; detached.
23 08 10.8	+60 52 28	PWM78 3		3	180		cl	(A) Very reddened cluster.
23 30 12.0	+60 15 30	Skiff J2330.2+6015		2.8	40		cl?	
23 30 55.0	+55 45 00	St 11		10		8.0	ast?	Few stars; moderate brightness range; not well detached. Probably not a cluster.
23 36 20.0	+52 33 00	St 12		35	9	8.2	cl	Few stars; moderate brightness range; not well detached.
23 43 47.0	+62 09 37	St 17		1		8.4	cl	
00 04 39.0	+56 04 54	St 19		2	6	8.0	cl	Few stars; small brightness range; no central concentration; detached.
00 25 16.0	+62 37 48	St 20		1		13.0	cl	
00 30 28.0	+57 55 36	St 21		4.7	10	12.0	cl	
00 39 50.0	+61 57 36	St 24	8.8	5	180	13.0	cl	Few stars; small brightness range; no central concentration; detached.

BRIGHT NEBULAE

RA h m s	Dec ° ′ ″	Name	Dim ′ Maj x min	Type	BC	Color	Notes
00 56 42.0	+61 04 00	IC 59	10 x 5	E+R	2-5	1-4	With IC 63 collectively known as **Sharpless 185**, the N part.
00 59 30.0	+60 49 00	IC 63	10 x 3	E+R	1-5	2-4	With IC 59 collectively known as **Sharpless 185**, the E part.
23 05 12.0	+60 15 00	IC 1470	1.2 x 0.8	E	2-5		A tiny patch 2′ SE of a 10th mag star.
00 52 48.0	+56 37 00	NGC 281	35 x 30	E	1-5	3-4	Somewhat reminiscent of the North America Nebula. A large dark protuberance juts in from the SW side.
23 13 30.0	+61 31 00	NGC 7538	9 x 6	E	1-5	3-4	A pair of 11th mag stars (sep. 35″, NNE-SSW) involved in nebulosity.
23 20 42.0	+61 12 00	NGC 7635	15 x 8	E	1-5	3-4	**Bubble Nebula**.
22 56 48.0	+62 37 00	Sh2-155	50 x 30	E	3-5	3-4	**Cave Nebula**. The N half of this nebula, a crescent-shaped nebulosity, is its brightest part.
23 16 06.0	+60 02 00	Sh2-157a	3 x 2	E	1-5	2-4	The brightest spot in a more extensive nebulosity which is Sh 2-157.
00 11 00.0	+58 46 00	vdB 1	5 x 5	R	2-5	1-4	Three stars involved in nebulosity; mags 8.6, 8.9, 8.9. Two faint stars NW of the cluster are in small nebulosities.

PLANETARY NEBULAE

RA h m s	Dec ° ′ ″	Name	Diam ″	Mag (P)	Mag (V)	Mag cent ★	Alt Name	Notes
22 56 19.9	+57 09 20	PN G107.7−2.2	8	14.0	14.0		PK 107−2.1	Mag 10.61v star E 0′5.
22 58 51.5	+61 57 44	PN G110.1+1.9	6				PM 1-339	Mag 11.5 star SSE 2′2; mag 9.14v star E 3′6.
23 26 14.9	+58 10 53	PN G111.8−2.8	16	14.0	11.9	13.8	PK 111−2.1	Mag 11.7 star ENE 1′0.
23 24 10.6	+60 57 27	PN G112.5−0.1	4	18.2			PK 112−0.1	Mag 12.7 star W 1′8; mag 10.94v star WNW 2′5.
23 47 43.9	+51 23 56	PN G112.9−10.2	152	14.4	13.0	18.4	PK 112−10.1	Slightly elongated N-S; brighter along E edge with mag 10.7 star on this edge; mag 11.7 star 1′9 W of center.
23 46 46.9	+54 44 38	PN G113.6−6.9	47	17.6		21.0	PK 113−6.1	Mag 11.4 star NE 2′6; mag 10.4 star WSW 2′5.
23 45 47.5	+57 03 56	PN G114.0−4.6	94	15.2	12.7	14.9	PK 114−4.1	= **Abell 82**; mag 12.9 star slightly SE of center; mag 11.19v star NW 1′7; mag 13.1 star E 1′7.
00 18 42.4	+53 52 18	PN G118.0−8.6	14	12.6	12.5	14.1	PK 118−8.1	Mag 11.6 star WSW 1′0; mag 11.0 star NNE 2′1.
00 19 58.8	+62 59 01	PN G119.3+0.3	60	15.6	14.7	22.3	PK 119+0.1	Very slender, elongated E-W; mag 12.8 star NNW 0′7; mag 11.19v star E 2′3.
00 28 15.7	+55 57 54	PN G119.6−6.7	16	13.3	12.3	19.1	PK 119−6.1	Close pair of mag 11-12 stars NW 1′9; mag 13.0 star ENE 1′8.
00 31 53.6	+57 22 33	PN G120.2−5.3	720			18.1	PK 120−5.1	Mag 7.70v star laying on the W boundary, 6′7 W of the center; mag 9.36v star E 9′1.
00 40 21.6	+62 51 25	PN G121.6−0.0	38		15.4		PK 121+0.1	Elongated E-W; mag 11.4 star W 4′4; mag 9.6 star SE 5′5.
00 45 34.8	+57 57 35	PN G122.1−4.9	36	16.3	14.1	20.0	PK 122−4.1	Mag 12.2 star NNW 2′1; mag 12.1 star SE 2′5.
01 00 53.9	+55 03 54	PN G124.3−7.7	150				PK 124−7.1	Mag 11.45v star NW 2′5; mag 11.6 star S 3′7.

GALAXIES

RA h m s	Dec ° ′ ″	Name	Mag (V)	Dim ′ Maj x min	SB	Type Class	PA	Notes
22 59 23.0	+53 44 21	UGC 12287	14.8	1.5 x 0.7	14.7	Scd:	5	Numerous stars superimposed S half; mag 12 star E edge; mag 8.40v star S 1′8.

OPEN CLUSTERS

RA h m s	Dec ° ′ ″	Name	Mag	Diam ′	No. ★	B ★	Type	Notes
21 43 38.0	+51 04 17	Bar 2	8.4	5	21	10.0	cl	
20 55 59.0	+51 04 36	Be 53		6	40	16.0	cl	Moderately rich in stars; moderate brightness range; no central concentration; detached.
21 16 57.0	+51 45 30	Be 55		5	30	14.0	cl	
21 24 46.3	+57 32 20	Be 92		2	15	15.0	cl:	
22 22 53.0	+55 52 30	Be 94	8.7	3	12	13.0	cl	Few stars; large brightness range; slight central concentration; detached; involved in nebulosity.
22 28 18.0	+59 08 00	Be 95		3	15	15.0	cl	
22 29 51.2	+55 24 22	Be 96		2	15	13.0	cl	
22 39 22.2	+59 00 58	Be 97		2	12	11.0	cl	Few stars; moderate brightness range; not well detached.
22 42 39.2	+52 24 00	Be 98		15	50	15.0	cl:	
22 39 35.0	+59 54 48	Cz 42		2	15		cl:	

(Continued from previous page)

OPEN CLUSTERS

RA h m s	Dec ° ′ ″	Name	Mag	Diam ′	No. ★	B ★	Type	Notes
21 39 00.0	+57 30 00	IC 1396	3.5	90	50	3.8	cl	Moderately rich in stars; large brightness range; not well detached; involved in nebulosity. Excluding Mu Cephei, visual mag = 4.8.
22 10 30.0	+52 50 00	IC 1434	9.0	7	40	12.0	cl	Moderately rich in stars; moderate brightness range; no central concentration; detached.
22 16 30.0	+54 03 00	IC 1442	9.1	5	104	12.0	cl	Few stars; small brightness range; no central concentration; detached.
22 15 31.0	+54 24 36	King 9		3	40	18.0	cl	
22 55 00.0	+59 10 00	King 10		4	40	11.0	cl	Moderately rich in stars; small brightness range; slight central concentration; detached.
22 52 08.0	+58 18 42	King 18		5	20	12.0	cl	
21 19 25.8	+57 35 25	NGC 7055			10		cl?	
21 30 28.0	+51 36 00	NGC 7086	8.4	12	80	10.2	cl	Moderately rich in stars; moderate brightness range; slight central concentration; detached.
21 43 52.9	+54 35 46	NGC 7127		6	12		cl	Few stars; small brightness range; not well detached.
21 43 58.0	+53 42 54	NGC 7128	9.7	4	71	11.5	cl	Moderately rich in bright and faint stars; strong central concentration; detached.
21 53 40.0	+62 36 12	NGC 7160	6.1	5	61	7.0	cl	Few stars; large brightness range; strong central concentration; detached.
22 10 26.0	+55 24 00	NGC 7226	9.6	2	83	10.8	cl	Moderately rich in stars; moderate brightness range; strong central concentration; detached.
22 12 28.0	+57 02 30	NGC 7234		3.5	8		cl?	
22 12 24.0	+57 16 24	NGC 7235	7.7	6	98	8.8	cl	Moderately rich in bright and faint stars; detached.
22 15 18.0	+54 20 00	NGC 7245	9.2	5	169	12.8	cl	Moderately rich in stars; moderate brightness range; slight central concentration; detached.
22 20 10.0	+58 06 36	NGC 7261	8.4	6	62	9.6	cl	Moderately rich in bright and faint stars; detached.
22 24 55.7	+57 50 00	NGC 7281		12	20		cl:	Few stars; moderate brightness range; not well detached. Probably not a cluster. (A) Loose scattering of stars, with 3 equal stars in E-W line at center.
22 27 52.0	+52 49 06	NGC 7295		2	20	10.0	ast?	
22 28 01.0	+52 18 48	NGC 7296	9.7	3	20	10.0	cl	Few stars; moderate brightness range; slight central concentration; detached.
22 39 44.2	+57 23 06	NGC 7352		5			cl?	
22 47 21.0	+58 08 00	NGC 7380	7.2	20	125	10.0	cl	Moderately rich in stars; moderate brightness range; no central concentration; detached; involved in nebulosity.
22 50 24.0	+52 09 00	NGC 7394		10	19	11.0	cl?	
22 54 17.0	+60 48 48	NGC 7419	13.0	6	40	10.0	cl	Moderately rich in stars; moderate brightness range; strong central concentration; detached.
22 55 06.5	+57 05 41	NGC 7423		5	40	15.0	cl	
22 55 54.0	+59 59 00	NGC 7429		15	15	11.0	cl	Few stars; moderate brightness range; not well detached.
22 57 03.0	+54 18 00	NGC 7438		20	30	11.0	cl?	

BRIGHT NEBULAE

RA h m s	Dec ° ′ ″	Name	Dim ′ Maj x min	Type	BC	Color	Notes
21 39 06.0	+57 30 00	IC 1396	170 x 140	E	3-5	3-4	Vaguely annular in appearance. It is brightest in the NW and NE sectors and there are many small absorption patches scattered throughout the object.
21 11 48.0	+59 57 00	Sh2-129	110 x 100	E	5-5	3-4	A very faint, incomplete ring of nebulosity, filamentary in structure.
22 18 42.0	+56 08 00	Sh2-132	35 x 20	E	2-5	3-4	In a very rich star field, especially to the S. The NE part considerably higher in surface brightness.
22 56 48.0	+62 37 00	Sh2-155	50 x 30	E	3-5	3-4	**Cave Nebula**. The N half of this nebula, a crescent-shaped nebulosity, is its brightest part.
21 17 30.0	+58 36 00	vdB 140	12 x 10	R	3-5	1-4	Nebulosity with a mag 6.4 star involved in the SW part.
21 37 06.0	+57 29 00	vdB 142	1 x 1	R	3-5	1-4	Nebula with a mag 8.8 star involved.

DARK NEBULAE

RA h m s	Dec ° ′ ″	Name	Dim ′ Maj x min	Opacity	Notes
21 08 13.6	+56 19 20	B 151	1 x 1	5	Very small; very dark.
21 14 29.6	+61 44 00	B 152	15 x 3	5	Small; extends SE and NW
21 21 03.3	+56 26 54	B 153	1 x 1		Black; irregular.
21 21 22.5	+56 37 00	B 154	8 x 2	5	Narrow; extends NE and SW.
21 33 41.6	+54 40 00	B 157	4 x 4	4	Round.
21 37 58.9	+56 14 00	B 160	30 x 15	4	Large, dark, irregular; extends NE and SW.
21 40 22.2	+57 49 00	B 161		6	Small, black spot.
21 41 09.2	+56 19 00	B 162	13 x 2	4	Very thin, curved, dark strip, extends N and S; a curve of small stars on the SW end.
21 42 10.1	+56 42 00	B 163	4 x 4	4	Small; very black; pointed to the S; two dark streams running from this toward the N.
21 46 31.9	+51 06 00	B 164	12 x 6	5	V-shaped vacancy.
21 48 56.1	+60 13 00	B 165	18 x 1	5	Length E and W.
21 51 05.2	+60 05 00	B 166	5 x 5	5	Round; small star on SW edge.
21 51 58.2	+60 04 00	B 167	5 x 5	5	Small; irregularly round.
21 58 52.9	+58 46 00	B 169	60 x 60	3	Elliptical black ring.
21 58 02.9	+58 58 00	B 170	26 x 4	4	Irregular black strip NE and SW.
22 01 18.8	+58 52 00	B 171	19 x 19	5	Irregular broken region.
22 07 28.9	+59 41 00	B 173	4 x 4	6	Rather definite.
22 07 19.9	+59 05 00	B 174	19 x 19	6	Narrow; irregular; elongated NE and SW.
20 58 13.1	+58 09 00	B 354	60 x 60	2	Dusky; irregular; NE and SW.
20 59 52.1	+55 34 00	B 357	30 x 30	5	Irregular; dusky.
21 06 46.9	+57 10 00	B 359	20 x 3	5	Narrow; dusky; NE and SW.
21 07 52.2	+56 30 00	B 360		5	Irregular dark marking.
21 34 04.5	+54 33 00	B 364	40 x 40	5	Region of many small dark lanes.
21 34 53.5	+56 43 00	B 365	22 x 3	4	Dark S-shaped object; N and S.
21 40 21.5	+59 34 00	B 366	10 x 10	3	Roundish; dusky.
21 44 24.5	+57 10 00	B 367	5 x 5	5	Small; dark; NW and SE.
21 50 55.6	+58 59 00	B 368	14 x 3	5	Dusky spot, NE and SW.
22 15 54.5	+56 01 00	B 369	5 x 5	3	Round; dusky.
22 34 50.7	+56 39 00	B 370			Region of narrow dark lanes.

PLANETARY NEBULAE

RA h m s	Dec ° ′ ″	Name	Diam ″	Mag (P)	Mag (V)	Mag cent ★	Alt Name	Notes
21 00 33.1	+54 32 32	NGC 7008	86	13.3	10.7	13.2	PK 93+5.2	Several faint stars on disc, short dark lane runs N-S E of central star; mag 9.51v star S edge.
21 26 23.5	+62 53 33	NGC 7076	56	17.0		18.0	PN G101.8+8.7	Mag 14.4 star N 1′.1; mag 13.8 star N 1′.9. Brightest along E edge.
22 40 20.1	+61 17 06	NGC 7354	36	12.9	12.2	16.2	PK 107+2.1	Close pair of mag 12 stars NNW 3′.3.
21 20 45.0	+51 53 25	PN G93.5+1.4	15	16.9			PK 93+1.1	Mag 12.4 star W 1′.6; mag 12.1 star E 3′.2.
21 31 50.3	+52 33 49	PN G95.2+0.7	3	16.0	15.6		PK 95+0.1	Mag 13.6 star W 0′.5; mag 6.18v star NW 4′.8.
20 56 26.7	+57 26 00	PN G95.2+7.8	73	>17.4	16.5	21.1	PK 95+7.1	Light annular appearance; mag 13.1 star SW 2′.6; mag 10.32v star NW .
21 30 00.9	+54 27 25	PN G96.3+2.3	6	16.8		14.8	PK 96+2.1	Mag 13.5 star E 0′.7; mag 11.5 star ENE 2′.3.
21 32 10.3	+55 52 42	PN G97.5+3.1	76	16.4		15.7	PK 97+3.1	= **Abell 77**; elongated ENE-WSW; mag 11.7 star E 3′.7; mag 12.3 star NW 4′.5.

(Continued from previous page)

PLANETARY NEBULAE

RA h m s	Dec ° ′ ″	Name	Diam ″	Mag (P)	Mag (V)	Mag cent ★	Alt Name	Notes
21 57 41.9	+51 41 36	PN G97.6−2.4	4	14.5	14.7	15.1	PK 97−2.1	Mag 11.9 star N 1′.3; mag 9.60v star NE 2′.8.
21 39 12.2	+55 46 02	PN G98.1+2.4	7	16.0	16.0		PK 98+2.1	Forms a small triangle with mag 12.2 star 0′.8 NW and mag 11.4 star W 0′.8; mag 7.87v star WNW 4′.0.
21 27 26.6	+57 39 05	PN G98.2+4.9	3	16.0	16.2		PK 98+4.1	Mag 11.6 star E 2′.1.
22 34 45.5	+52 26 12	PN G102.8-5.0	171	15.2	16.0	19.6	PK 102−5.1	= **Abell 80**; is elongated NNE-SSW; brightest portions along middle E and W edges; mag 12.9 star at top of bright portion on W edge.
22 26 17.3	+54 49 40	PN G102.9−2.3	117	15.8	15.0	18.7	PK 102−2.1	Central area forms a slightly flattened ring, oriented E-W, with fainter extensions NE; mag 12.7 star on NE edge; mag 12.4 star SW 1′.6. from center.
22 16 03.8	+57 28 33	PN G103.2+0.6	94	13.5	13.5	20.4	PK 103+0.1	Overall elongated SSE-NNW with brighter center more oriented NE-SW; mag 13.0 star SW 1′.1; mag 13.3 star NE 2′.1.
22 20 30.9	+57 36 18	PN G103.7+0.4	20	14.0	15.3		PK 103+0.2	Forms a triangle with mag 13.4 star W 1′.6 and mag 12.6 star SSW 1′.6.
22 20 16.8	+58 14 14	PN G104.1+1.0	2	17.2			PK 104+0.1	Stellar; mag 13.2 star NE 3′.5; mag 13.5 star WSW 2′.7.
22 32 17.6	+56 10 26	PN G104.4−1.6	26	15.1	14.8	21.2	PK 104−1.1	Oval shape, oriented E-W; mag 12.1 star S 2′.4; mag 9.91v star NE 2′.6.
22 51 39.2	+51 50 39	PN G104.8−6.7	8			12.0	PK 104−6.1	Mag 11.5 star W 1′.2; mag 10.9 star N 2′.4.
22 48 34.5	+58 29 06	PN G107.4−0.6		19.6			PK 107−0.1	Pair of N-S mag 11.9, 12.5 stars N 1′.3; mag 11.7 star W 1′.8; mag 6.32v star W 9′.3.
22 55 07.1	+56 42 30	PN G107.4−2.6	4	17.6			PK 107−2.2	Mag 10.76v star ESE 1′.1; mag 12.1 star WNW 2′.6.
22 56 19.9	+57 09 20	PN G107.7−2.2	8	14.0	14.0		PK 107−2.1	Mag 10.61v star E 0′.5.
22 58 51.5	+61 57 44	PN G110.1+1.9	6				PM 1-339	Mag 11.5 star SSE 2′.2; mag 9.14v star E 3′.6.

20 GALAXIES 20

RA h m s	Dec ° ′ ″	Name	Mag (V)	Dim ′ Maj x min	SB	Type Class	PA	Notes
20 20 35.2	+56 14 57	CGCG 282-3	14.8	0.7 x 0.6	13.7		114	Faint NNE-SSW bar with stellar nucleus.
18 57 07.3	+51 08 55	MCG +9-31-12	14.6	0.9 x 0.6	13.8		150	Mag 11.0 star N 4′.3.
19 08 20.9	+52 32 08	MCG +9-31-20	15.1	0.7 x 0.5	13.8		132	Mag 5.90v star S 6′.7.
19 08 28.9	+52 13 25	MCG +9-31-21	14.3	1.0 x 0.5	13.4	Sb	129	Mag 12.4 star and very small anonymous galaxy E 1′.0.
19 09 28.8	+52 53 35	MCG +9-31-23	14.5	0.9 x 0.7	13.9		9	Mag 11.27v star on W edge.
19 19 26.8	+52 19 51	MCG +9-31-32	14.5	1.1 x 0.6	13.9	Sbc	66	Mag 13.2 star superimposed NE of center.
19 22 23.2	+58 14 56	MCG +10-27-11	15.0	0.5 x 0.4	13.1	Sb	27	Mag 10.90v star S 2′.6.
19 24 29.1	+59 42 00	MCG +10-27-12	14.7	1.0 x 0.3	13.3	SBb	69	Mag 13.5 star S edge; small, faint anonymous galaxy E 1′.1.
19 33 28.8	+62 05 15	MCG +10-28-3	14.4	0.5 x 0.4	12.5	Sb	114	Mag 8.34v star SSW 2′.2.
20 12 44.2	+61 41 53	MCG +10-28-17	14.6	0.8 x 0.4	13.2	Spiral	18	Mag 9.85v star NW 2′.9.
20 21 30.5	+59 44 08	MCG +10-29-3	14.3	0.7 x 0.6	13.2	Spiral	111	Mag 12.5 star W 2′.4.
18 56 24.3	+52 22 38	NGC 6732	13.8	1.2 x 0.7	13.6	E?	100	Faint star on E edge of bright center.
19 00 36.3	+59 10 00	NGC 6750	13.8	1.0 x 0.6	13.1	Sc	5	
19 05 06.3	+55 43 00	NGC 6757	12.9	1.4 x 1.0	13.1	SB(r)0/a	84	Strong dark patches N and S.
19 16 10.7	+60 24 59	NGC 6787	13.9	1.0 x 0.8	13.5	SAB(s)b	102	Dark lane W of bright center.
19 21 30.8	+61 08 36	NGC 6796	12.6	1.9 x 0.4	12.2	Sbc: sp	179	
19 24 03.3	+53 37 25	NGC 6798	13.2	1.7 x 0.9	13.6	S0	141	Faint star on SE edge.
19 27 36.8	+54 22 24	NGC 6801	13.9	1.3 x 0.7	13.7	SAcd	44	
19 37 23.1	+62 22 57	NGC 6817	14.9	0.8 x 0.5	13.8	Sbc	87	Multi-galaxy system, two bright cores.
19 43 40.8	+56 06 28	NGC 6824	12.2	1.7 x 1.2	12.8	SA(s)b:	60	Large, bright, diffuse center.
19 47 07.7	+59 54 21	NGC 6829	14.1	1.6 x 0.4	13.5	Sb II-III	31	Mag 11.0 star on S edge.
19 47 57.5	+59 53 31	NGC 6831	13.3	1.4 x 1.4	13.9	S0		
20 23 33.5	+58 20 37	NGC 6916	13.5	1.8 x 1.1	14.1	SBbc	90	Stellar nucleus; mag 11.16v star 2′.3 NE.
20 34 54.9	+60 09 23	NGC 6946	8.8	11.5 x 9.8	13.8	SAB(rs)cd I-II	57	Bright nucleus, many bright knots and superimposed stars.
19 10 10.5	+60 07 31	UGC 11411A	15.1	2.0 x 0.3	14.4	Scd:	1	
19 10 46.7	+52 08 55	UGC 11412	13.8	0.9 x 0.5	12.8	Sbc	165	Mag 10.56v star N 1′.1.
19 23 50.8	+55 59 14	UGC 11435	14.2	2.0 x 0.3	13.5	Scd:	82	
19 31 08.2	+54 06 04	UGC 11453	12.0	1.7 x 1.3	12.7	Sb II	62	Strong dark lane NW of center; bright knot or companion galaxy S edge.
19 41 25.8	+51 49 15	UGC 11464	14.3	0.9 x 0.4	13.0	S	15	Mag 10.23v star W 2′.6.
19 46 41.1	+59 42 02	UGC 11475	14.5	1.5 x 0.7	14.4	SBd	85	
19 52 39.9	+57 27 35	UGC 11492	13.1	1.9 x 1.0	13.6	SAB(s)bc I	70	Strong dark patch NE of center.
19 57 46.2	+52 52 56	UGC 11502	14.4	1.5 x 0.8	14.4	Compact	174	Very knotty; bright center with moderately narrow extension N.
20 01 49.6	+57 01 33	UGC 11509	14.2	0.9 x 0.3	12.6	Sbc	172	Mag 11.37v star NE edge; mag 11.7 star SW edge.
20 02 55.9	+53 53 07	UGC 11510	14.1	1.9 x 0.6	14.1	Sd	170	Located 2′.0 E of mag 8.34v star.
20 04 56.9	+62 47 48	UGC 11515	13.8	1.3 x 1.3	14.2	Sd		Mag 14.0 star on NW edge; mag 11.8 star S 2′.0.
20 18 31.6	+62 41 26	UGC 11539	13.6	1.2 x 1.2	14.0	E		Bright center, uniform envelope; mag 13.4 star N edge; mag 9.67v star NW 1′.2.
20 24 00.8	+60 11 45	UGC 11557	13.1	2.0 x 1.6	14.2	SAB(s)dm	90	Stellar nucleus, numerous stars superimposed.
20 30 14.9	+60 26 22	UGC 11583	13.7	2.3 x 0.6	13.9	Irr	93	Uniform surface brightness; mag 13.3 star 1′.0 E.

OPEN CLUSTERS

RA h m s	Dec ° ′ ″	Name	Mag	Diam ′	No. ★	B ★	Type	Notes
20 55 59.0	+51 04 36	Be 53		6	40	16.0	cl	Moderately rich in stars; moderate brightness range; no central concentration; detached.
19 59 15.7	+56 08 00	NGC 6856			15		cl?	
20 31 30.0	+60 40 00	NGC 6939	7.8	10	301	11.9	cl	Moderately rich in stars; small brightness range; slight central concentration; detached.

DARK NEBULAE

RA h m s	Dec ° ′ ″	Name	Dim ′ Maj x min	Opacity	Notes
20 47 42.9	+59 37 40	B 148	3 x 3	5	Very small; round; indefinite.
20 49 01.8	+59 31 54	B 149	2 x 2	5	Very small; round; indefinite.
20 50 39.7	+60 18 00	B 150	60 x 3	5	Curved dark marking.
20 58 13.1	+58 09 00	B 354	60 x 60	2	Dusky; irregular; NE and SW.
20 59 52.1	+55 34 00	B 357	30 x 30	5	Irregular; dusky.

PLANETARY NEBULAE

RA h m s	Dec ° ′ ″	Name	Diam ″	Mag (P)	Mag (V)	Mag cent ★	Alt Name	Notes
21 00 33.1	+54 32 32	NGC 7008	86	13.3	10.7	13.2	PK 93+5.2	Several faint stars on disc, short dark lane runs N-S E of central star; mag 9.51v star S edge.
20 53 13.8	+53 45 42	PN G92.1+5.8	12	17.8			PK 92+5.1	Mag 12.3 star E 1′.5.
20 56 26.7	+57 26 00	PN G95.2+7.8	73	>17.4	16.5	21.1	PK 95+7.1	Light annular appearance; mag 13.1 star SW 2′.6; mag 10.32v star NW .

RA h m s	Dec ° ′ ″	Name	Mag (V)	Dim ′ Maj x min	SB	Type Class	PA	Notes
18 10 22.5	+54 13 27	CGCG 279-3	14.1	0.6 x 0.4	12.4		153	In the glare of a mag 5.97v star NNE 4′0.
17 03 14.7	+61 27 02	CGCG 299-48	14.0	0.5 x 0.2	11.4		99	Very faint anonymous galaxy NW 0′8; mag 12.4 star NE 2′0.
17 20 25.8	+62 36 13	CGCG 300-14	14.7	0.5 x 0.5	13.1	Spherical		Mag 8.81v star W 5′3.
16 56 16.2	+55 01 33	IC 1237	13.6	1.9 x 1.0	14.1	SB(s)c: I-II	20	Mag 8.73v star NW 10′1.
17 11 40.5	+59 59 46	IC 1248	13.6	0.9 x 0.8	13.1	SB(r)c	30	Mag 10.39v star N 5′3.
17 27 17.3	+58 29 05	IC 1258	13.5	0.9 x 0.7	12.9	Sab?	51	Pair of faint stars superimposed; UGC 10869 NE 2′2; stellar galaxy **Kaz 140** E 2′0.
17 38 46.0	+59 22 19	IC 1267	13.4	1.4 x 0.8	13.4	SBb I	35	Mag 11.4 star WNW 8′3; **MCG +10-25-78** SE 1′9.
18 16 14.4	+55 35 25	IC 1286	13.8	1.4 x 0.5	13.3	Sab	87	Mag 10.31v star N 2′4.
17 47 13.0	+61 26 01	IC 4669	14.1	0.7 x 0.3	12.3	SB(s)b	94	Mag 9.56v star E 4′0.
17 00 52.8	+55 05 57	MCG +9-28-15	14.8	0.6 x 0.5	13.4	Sb	120	
17 03 28.7	+55 08 52	MCG +9-28-19	14.1	0.9 x 0.5	13.1	Sc	144	
17 14 47.5	+52 23 15	MCG +9-28-30	14.5	1.1 x 0.5	13.7	Sab	9	Mag 9.58v star E 7′1.
17 15 45.4	+52 50 32	MCG +9-28-31	14.0	0.9 x 0.7	13.3	Sc	84	Mag 10.76v star NE 2′2.
17 24 08.9	+53 27 58	MCG +9-28-43	14.7	0.9 x 0.6	14.1	E/S0	6	Mag 9.49v star N 2′8.
17 24 31.0	+56 54 31	MCG +9-28-45	14.8	0.8 x 0.7	14.0		6	Faint dark patch S of nucleus.
17 37 00.1	+55 00 23	MCG +9-29-14	14.4	1.0 x 0.5	13.5	Sb	168	Mag 14.4 star on N edge; mag 10.57v star W 1′9.
17 40 16.4	+51 03 23	MCG +9-29-16	14.4	0.9 x 0.4	13.1	S?	168	**UGC 10942** S 1′7.
17 40 40.8	+51 10 49	MCG +9-29-17	15.7	1.0 x 0.6	15.0	Sbc	144	Located in a triangle of mag 13 stars.
17 44 26.9	+55 49 39	MCG +9-29-23	14.4	0.9 x 0.6	13.5		111	Mag 9.02v star S 4′6.
17 47 05.4	+54 52 52	MCG +9-29-31	14.9	0.8 x 0.5	13.8	Sc	171	
17 49 32.6	+55 30 26	MCG +9-29-36	14.9	0.7 x 0.4	13.4		144	Mag 11.2 star NE 4′0.
17 52 06.1	+52 26 46	MCG +9-29-42	14.4	0.9 x 0.4	13.2		114	
18 05 28.4	+54 01 03	MCG +9-29-52	15.5	0.7 x 0.7	14.6	Sb		
18 15 43.0	+51 47 11	MCG +9-30-9	14.4	0.8 x 0.4	13.0		177	Mag 10.85v star W edge.
18 35 58.7	+51 27 32	MCG +9-30-23	14.0	1.0 x 0.8	13.8	E/S0	87	**MCG +9-30-22** S 0′9.
18 49 36.9	+51 54 15	MCG +9-31-7	15.2	0.8 x 0.4	13.9		36	Mag 10.92v star WNW 4′7.
18 57 07.3	+51 08 55	MCG +9-31-12	14.6	0.9 x 0.6	13.8		150	Mag 11.0 star N 4′3.
16 58 42.9	+57 05 50	MCG +10-24-80	14.7	0.8 x 0.5	13.8	E/S0	45	Mag 11.9 star N 3′4.
17 00 40.0	+58 55 13	MCG +10-24-86	14.6	0.7 x 0.5	13.3		147	NGC 6291 NE 2′3.
17 01 06.4	+58 40 59	MCG +10-24-87	15.5	0.7 x 0.5	14.2	Sc	18	Mag 8.99v star W 3′8.
17 01 13.1	+60 15 01	MCG +10-24-91	15.2	0.4 x 0.3	12.7		6	Mag 9.19v star E 1′9.
17 08 45.8	+62 02 34	MCG +10-24-101	14.9	0.8 x 0.3	13.2	S	138	Mag 11.73v star NW 2′8.
17 15 54.3	+61 21 24	MCG +10-24-121	15.4	0.7 x 0.6	14.3		81	Mag 10.7 star SW 2′9.
17 16 37.4	+58 24 41	MCG +10-24-123	15.1	0.8 x 0.5	13.9	SBbc	171	Mag 10.71v star SE 2′1.
17 22 05.8	+59 06 43	MCG +10-25-11	12.9	0.8 x 0.5	12.0	E	90	Multi-galaxy system; elongated **MCG +10-25-12** SE 2′8.
17 22 41.2	+60 00 37	MCG +10-25-14	15.8	0.9 x 0.9	15.4	Ib/c		Mag 10.98v star N 3′1.
17 22 52.8	+60 07 46	MCG +10-25-17	15.2	0.5 x 0.4	13.3	Spiral	165	Slightly elongated **MCG +10-25-16** on W edge; mag 10.33v star just beyond.
17 26 26.2	+58 35 19	MCG +10-25-32	15.5	0.5 x 0.4	13.6	Sb	63	Mag 6.50v star NW 4′7.
17 28 37.5	+57 08 42	MCG +10-25-46	14.5	0.5 x 0.4	12.6		105	Mag 8.07v star S 2′2.
17 40 52.6	+62 36 25	MCG +10-25-81	14.9	0.7 x 0.7	14.0			Mag 10.5 star S 3′4.
17 50 17.2	+59 37 25	MCG +10-25-104	14.2	0.7 x 0.7	13.3	Sab		
17 56 35.0	+62 37 00	MCG +10-25-121	14.8	0.8 x 0.2	12.7	Sp		Located 1′1 E of a mag 7.59v star, NE member of a bright pair.
18 03 52.6	+61 53 38	MCG +10-26-6	15.2	0.7 x 0.3	13.3	Sc	120	Mag 8.76v star E 6′0.
18 09 50.2	+58 22 46	MCG +10-26-17	15.2	0.9 x 0.7	14.6	N galaxy?	156	
18 27 34.3	+62 06 57	MCG +10-26-39	14.3	0.7 x 0.7	13.6	E?		Mag 12.4 star WNW 2′6.
18 32 24.1	+59 52 45	MCG +10-26-43	15.0	0.6 x 0.4	13.3		54	Mag 11.6 star N 1′2.
18 45 32.0	+59 58 49	MCG +10-27-1	14.6	0.9 x 0.3	13.0	Sb	45	Mag 11.9 star E 2′4.
16 58 24.2	+58 57 20	NGC 6285	13.5	1.3 x 0.7	13.2	S0+? pec	129	Bright lens with two main arms.
16 58 32.0	+58 56 16	NGC 6286	13.3	1.3 x 1.2	13.6	Sb: pec	33	Multi-galaxy system; faint envelope extends SE of bright bar.
17 00 56.6	+58 58 09	NGC 6290	13.5	1.1 x 1.0	13.5	SBa	30	Short SE-NW bar; mag 10.12v star NW 3′0.
17 00 56.1	+58 56 12	NGC 6291	14.0	0.6 x 0.6	12.7	SBb		MCG +10-24-86 SW 2′3.
17 03 03.0	+61 02 34	NGC 6292	13.5	1.5 x 0.8	13.5	Sbc	105	Star on E end.
17 03 15.1	+60 20 15	NGC 6295	14.8	1.0 x 0.4	13.7	S	77	
17 03 36.6	+62 01 30	NGC 6297	13.6	0.7 x 0.5	12.3	S0	90	
17 05 04.6	+62 27 26	NGC 6299	14.1	0.6 x 0.6	12.8			
17 07 36.8	+60 43 46	NGC 6306	13.7	1.0 x 0.3	12.3	SB(s)ab pec:	166	Mag 7.88v star SW 10′7; note: **Mkn 892** lies 4′3 N of this star.
17 07 40.8	+60 45 03	NGC 6307	12.9	1.3 x 1.0	13.1	(R′)SB(s)0/a pec	145	Star on NW edge of bright center.
17 07 57.4	+60 59 24	NGC 6310	13.1	2.0 x 0.4	12.7	Sb: sp	69	
17 08 59.5	+62 53 53	NGC 6317	15.1	1.0 x 0.3	13.6		51	
17 09 44.2	+62 58 23	NGC 6319	13.4	0.8 x 0.8	13.0	E/S0		
17 15 22.7	+57 24 39	NGC 6338	12.3	1.5 x 1.0	12.6	S0	15	**MCG +10-24-117** N 1′3.
17 15 24.5	+57 20 57	NGC 6345	14.4	0.8 x 0.3	12.8	S/S0	33	**IC 1252** E 3′6.
17 15 24.5	+57 19 16	NGC 6346	14.3	0.7 x 0.5	13.2	E	95	**MCG +10-24-112** SW 2′3.
17 18 53.3	+52 36 54	NGC 6358	14.0	0.9 x 0.4	12.8		110	Double star NW 1′9.
17 17 52.8	+61 46 48	NGC 6359	12.6	1.2 x 0.9	12.6	SA0−:	145	
17 18 41.2	+60 36 31	NGC 6361	13.1	2.2 x 0.6	13.2	SAb: sp I-II	54	**MCG +10-25-3** SW 1′9.
17 22 44.2	+62 09 51	NGC 6365A	14.5	1.0 x 0.9	14.2	SBcd:	137	NGC 6365B on N edge; mag 10.21v star NE 1′6.
17 22 43.5	+62 10 21	NGC 6365B	15.9	0.9 x 0.3	14.3	Sdm:	31	
17 23 25.3	+56 58 25	NGC 6370	12.9	1.4 x 1.4	13.6	E		**MCG +10-25-21** NE 3′1; mag 6.53v star N 2′3.
17 24 08.1	+58 59 40	NGC 6373	13.6	1.3 x 1.0	13.8	SAB(s)c	90	**MCG +10-25-22** N 3′9.
17 25 19.2	+58 48 57	NGC 6376	15.5	0.6 x 0.3	13.5	S0/a	142	Pair with NGC 6377; mag 9.81v star SW 3′0.
17 25 22.9	+58 49 20	NGC 6377	14.9	0.6 x 0.3	12.9	Sa	58	
17 27 16.9	+60 00 47	NGC 6381	13.0	1.3 x 1.0	13.1	SA(s)c?	25	**UGC 10870** NW 1′3.
17 27 55.3	+56 52 05	NGC 6382	14.0	0.9 x 0.7	13.3		105	
17 28 01.5	+57 31 20	NGC 6385	13.1	1.3 x 1.3	13.6	SB(s)a		Mag 8.36v star W 2′8.
17 28 51.7	+52 43 22	NGC 6386	14.2	1.0 x 1.0	14.1	Sbc		**MCG +9-29-2** NW 2′6.
17 28 24.1	+57 32 42	NGC 6387	17.0	0.5 x 0.3	14.8	H II	95	
17 28 27.9	+60 05 34	NGC 6390	13.8	1.6 x 0.3	12.8	Sbc	8	**MCG +10-25-45** W 4′4; **MCG +10-25-51** E 4′2.
17 28 48.9	+58 51 02	NGC 6391	14.5	0.9 x 0.7	14.0	E/S0	90	
17 30 21.5	+59 38 21	NGC 6393	14.5	1.3 x 0.4	13.6	SBb	39	Mag 7.95v star NW 4′0.
17 30 08.4	+59 31 54	NGC 6394	15.7	0.4 x 0.4	13.6	Spiral		
17 31 50.4	+59 36 56	NGC 6399	13.7	1.2 x 0.7	13.4	S0/a	5	**MCG +10-25-58** SSW 8′5.
17 35 32.7	+60 48 46	NGC 6411	11.9	2.0 x 1.3	13.0	E	79	Bright nucleus, smooth envelope.
17 38 09.4	+58 42 52	NGC 6418	14.4	0.5 x 0.4	12.5	Sb	168	Stellar galaxy **Kaz 148** SW 2′1.
17 40 11.3	+62 38 29	NGC 6435	14.3	1.1 x 0.6	13.7	Compact	5	MCG +10-25-81 E 5′3.
17 41 13.1	+60 26 56	NGC 6436	14.0	1.3 x 0.8	13.9	Scd:	177	Stellar nucleus with trio of stars N.
17 43 46.7	+56 48 12	NGC 6449	13.8	1.0 x 0.8	13.4	Scd:	135	
17 44 56.9	+55 42 14	NGC 6454	13.6	1.0 x 0.8	13.2	S?	171	Bright center, uniform envelope; mag 9.02v star NW 5′9.
17 45 47.2	+55 46 36	NGC 6459	14.7	0.7 x 0.3	12.8	Sab	78	Located 1′6 N of mag 9.88v star.

(Continued from previous page)

GALAXIES

RA h m s	Dec ° ′ ″	Name	Mag (V)	Dim ′ Maj x min	SB	Type Class	PA	Notes
17 44 48.7	+61 54 37	NGC 6462	13.9	0.4 x 0.4	11.9	E?		
17 45 47.5	+60 53 51	NGC 6464	14.8	0.6 x 0.6	13.6	Spiral		Dark patch NE.
17 48 08.2	+51 23 52	NGC 6466	14.1	0.7 x 0.4	12.6		111	**CGCG 278-28** WNW 5′.1.
17 47 05.6	+57 18 00	NGC 6473	14.5	1.1 x 0.5	13.7	S	75	**UGC 10980** WSW 9′.5.
17 48 38.2	+51 09 24	NGC 6478	13.3	1.9 x 0.7	13.4	SAc:	37	
17 48 21.8	+54 08 57	NGC 6479	13.7	1.0 x 0.9	13.4	Sc	129	Mag 6.61v star ENE 6′.6.
17 49 20.9	+62 13 21	NGC 6488	14.5	0.6 x 0.4	12.8		63	Trio of stars NW 1′.7.
17 49 24.1	+60 04 22	NGC 6489	15.3	0.6 x 0.4	13.6	Spherical	24	**MCG +10-25-107** ESE 8′.6.
17 50 00.8	+61 31 55	NGC 6491	13.6	1.2 x 0.5	12.9	Sab	39	**UGC 11007** N 4′.7.
17 50 22.6	+61 33 33	NGC 6493	14.4	1.1 x 1.1	14.5	SAB(r)cd:		faint stellar nucleus, uniform envelope.
17 51 17.9	+59 28 14	NGC 6497	13.5	1.4 x 0.7	13.3	SB(r)b	113	Short N-S bar.
17 54 39.5	+60 49 03	NGC 6510	14.3	1.0 x 0.6	13.6	SB	30	
17 54 50.2	+62 38 41	NGC 6512	13.9	0.7 x 0.5	12.8	E?	57	
17 55 16.8	+62 40 10	NGC 6516	14.8	0.5 x 0.2	12.1	Sp	147	Mag 9.38v star NW 2′.4; a pair of stellar anonymous galaxies SE at 0′.9 and 1′.8.
17 55 48.7	+62 36 43	NGC 6521	12.9	1.6 x 1.3	13.7	E	160	
17 59 13.8	+56 13 56	NGC 6532	13.9	1.8 x 0.9	14.3	SAc	123	
17 59 39.2	+61 21 30	NGC 6542	13.3	1.3 x 0.4	12.4	S0/a: sp	98	
18 05 01.0	+56 15 45	NGC 6562	13.9	0.7 x 0.7	13.0			
18 07 00.7	+52 15 37	NGC 6566	14.5	0.6 x 0.6	13.5	E		
18 09 50.7	+61 25 19	NGC 6592	14.4	0.7 x 0.6	13.6	E/S0	126	
18 10 05.7	+61 08 00	NGC 6594	14.5	1.0 x 0.5	13.6		90	
18 11 13.7	+61 10 48	NGC 6597	14.8	0.9 x 0.4	13.7	E/S0	99	A mag 8.19v star W 3′.4.
18 11 44.4	+61 27 08	NGC 6601	14.7	0.7 x 0.3	13.1	E/S0	42	
18 12 15.1	+61 19 54	NGC 6607	15.3	0.5 x 0.5	13.7	Sb		
18 12 33.7	+61 19 54	NGC 6608	14.5	0.7 x 0.7	13.6	Sb		
18 12 29.0	+61 17 50	NGC 6609	15.1	1.0 x 0.1	12.4	Irr	42	
18 14 03.1	+61 19 07	NGC 6617	14.6	1.1 x 0.7	14.2	SA(s)d	80	Stellar nucleus.
18 33 35.2	+59 53 16	NGC 6670	15.1	1.1 x 0.6	14.5		90	Triple system.
18 37 22.4	+59 38 30	NGC 6687	14.0	1.5 x 1.2	14.5	SAd	60	Stellar nucleus.
18 39 12.1	+55 38 27	NGC 6691	12.9	1.6 x 1.5	13.7	SB(rs)bc I	48	
18 40 05.2	+59 19 59	NGC 6696	15.1	0.8 x 0.2	13.0		0	
18 43 12.9	+60 39 12	NGC 6701	12.1	1.5 x 1.3	12.7	(R′)SB(s)a	25	Bright SE-NW bar with strong dark patches NE and SW.
18 56 24.3	+52 22 38	NGC 6732	13.8	1.2 x 0.7	13.6	E?	100	Faint star on E edge of bright center.
19 00 36.3	+59 10 00	NGC 6750	13.8	1.0 x 0.6	13.1	Sc	5	
16 58 35.8	+54 52 12	UGC 10644	13.9	0.9 x 0.5	12.9		44	Mag 9.45v star S 1′.5.
16 58 27.5	+59 07 08	UGC 10646	13.8	1.4 x 0.6	13.6	E	25	
16 58 33.0	+59 45 14	UGC 10648	14.3	0.8 x 0.5	13.2	SB(s)b	137	Mag 9.16v star N 2′.7.
17 03 27.5	+59 43 29	UGC 10687	14.6	1.1 x 0.7	14.2	SBd	42	Mag 10.26v star on E edge.
17 12 15.2	+53 07 11	UGC 10758	14.2	0.8 x 0.7	13.4	S?	81	Located 1′.6 NW of mag 10.31v star.
17 13 08.9	+59 19 31	UGC 10770	13.7	1.5 x 0.6	13.4	Im pec	30	Double system, contact, disrupted.
17 16 47.9	+61 55 11	UGC 10796	13.7	1.8 x 1.1	14.3	SB(s)b	21	Stellar nucleus, faint E-W bar.
17 18 43.8	+58 08 06	UGC 10811	14.0	1.7 x 0.5	13.7	SBab?	93	Mag 10.8 star NE 2′.8.
17 20 30.4	+52 38 35	UGC 10821	14.3	1.0 x 0.8	13.9	SB(r)b	111	Mag 12.6 star on E edge.
17 20 12.6	+57 55 05	UGC 10822	9.9	50 x 31	17.9	E pec		**Draco Dwarf**. Dwarf elliptical, very low surface brightness.
17 22 49.2	+60 26 42	UGC 10830	13.5	1.1 x 0.5	12.7	Spiral?	97	Pair of mag 13 and 14 stars E 1′.9.
17 23 41.4	+52 00 39	UGC 10839	13.9	1.0 x 0.6	13.2	S	165	**UGC 10843** SE 4′.2.
17 27 26.6	+58 30 58	UGC 10869	13.1	1.1 x 1.1	13.2	S?		Double system, common envelope; **IC 1259** W edge of nucleus.
17 28 18.6	+60 56 31	UGC 10880	14.7	0.8 x 0.5	13.6	Sdm:	99	Mag 11.3 star NE 3′.2.
17 29 59.4	+60 20 59	UGC 10888	13.5	1.0 x 0.6	12.8	(R′)SB(r)b II	150	Stellar nucleus; faint, uniform envelope.
17 38 27.2	+61 02 23	UGC 10936	14.2	0.9 x 0.5	13.2	S	170	Faint star on W edge; mag 12.3 star NW 2′.1.
17 44 49.6	+55 20 58	UGC 10971	14.1	1.0 x 0.3	12.6	Sa	173	
17 46 39.7	+51 16 51	UGC 10984	14.0	1.3 x 0.6	13.5	SBab:	162	Star or knot N edge of nucleus; mag 9.66v star N 6′.2.
17 46 47.2	+59 14 41	UGC 10988	13.2	1.3 x 0.9	13.3	E	0	Bright almost stellar nucleus with faint envelope; mag 10.24v star SW 0′.9; mag 7.81v star ESE 1′.8; **UGC 10982** NW 7′.4.
17 52 06.0	+60 35 52	UGC 11028	14.6	1.4 x 0.4	13.8	Sb III	53	Mag 14 star E of center.
17 53 36.7	+58 20 58	UGC 11036	15.9	1.7 x 0.2	14.6	Sdm:	38	Mag 11.4 star W 0′.6.
18 21 26.4	+60 53 45	UGC 11215	14.8	0.8 x 0.8	14.1	Sc		Mag 12.3 star S 2′.2.
18 26 50.0	+51 08 17	UGC 11241	14.1	0.8 x 0.7	13.3	Sb III	18	Stellar nucleus, uniform envelope; mag 10.6 star SE 2′.2.
18 35 21.0	+52 43 39	UGC 11292	13.4	0.9 x 0.7	12.7	Compact	135	Mag 15.2 star superimposed N of nucleus.
18 36 20.6	+51 27 56	UGC 11298	12.5	1.4 x 0.8	12.7	E	96	MCG +9-30-23 W 3′.4.
18 42 57.7	+53 02 43	UGC 11343	14.6	0.9 x 0.6	13.8	S	66	Mag 11.01v star E 1′.2.
18 51 28.5	+58 04 55	UGC 11373	14.0	1.1 x 0.4	13.0	Sb II	146	Mag 8.38v star WNW 3′.0.

GALAXY CLUSTERS

RA h m s	Dec ° ′ ″	Name	Mag 10th brightest	No. Gal	Diam ′	Notes
18 01 18.0	+57 38 00	A 2293	16.2	38	16	All members anonymous, faint and stellar.

GALAXIES

RA h m s	Dec ° ′ ″	Name	Mag (V)	Dim ′ Maj x min	SB	Type Class	PA	Notes
15 28 04.3	+55 32 38	CGCG 274-45	14.0	0.7 x 0.3	12.1		3	Mag 13.6 star W 4′.4.
16 15 03.0	+61 46 52	CGCG 298-22	14.0	0.6 x 0.4	12.3		114	Triangle of mag 11.5-12.5 stars N 1′.9.
17 03 14.7	+61 27 02	CGCG 299-48	14.0	0.5 x 0.2	11.4		99	Very faint anonymous galaxy NW 0′.8; mag 12.4 star NE 2′.0.
15 06 54.8	+56 30 29	IC 1099	14.0	1.2 x 1.1	14.2	SAB(r)c: II	162	Mag 11.89v star NNE 7′.6.
15 06 21.1	+62 58 47	IC 1100	13.4	0.8 x 0.7	12.6	S?	60	Mag 11.6 star N 1′.9.
16 14 30.2	+62 32 11	IC 1210	13.1	1.6 x 0.5	12.7	Sab	168	Mag 11.1 star E 4′.6.
16 16 52.2	+53 00 20	IC 1211	12.7	1.0 x 0.9	12.6	E	42	Mag 10.09v star NNW 7′.6.
16 46 59.3	+58 25 19	IC 1231	12.9	2.2 x 1.0	13.6	Scd:	155	Mag 9.56v star SW 5′.3.
16 56 16.2	+55 01 33	IC 1237	13.6	1.9 x 1.0	14.1	SB(s)c: I-II	20	Mag 8.73v star NW 10′.1.
14 58 19.3	+53 14 11	MCG +9-24-60	14.4	0.9 x 0.7	13.8	S0?	0	Located between a pair of NE-SW mag 14.6, 13.6 stars.
14 59 32.9	+53 53 54	MCG +9-25-3	13.8	1.0 x 0.7	13.3	SBbc	174	Close pair of mag 7.62v and 6.83v stars S 2′.0.
15 04 40.0	+53 49 23	MCG +9-25-12	14.2	0.8 x 0.7	13.5		126	Located between a pair of SE-NW stars, mags 11.06v and 10.80v.
15 04 39.9	+54 42 52	MCG +9-25-13	14.3	1.1 x 0.9	14.1		21	
15 06 10.3	+53 24 36	MCG +9-25-15	14.7	0.7 x 0.5	13.5	Spiral	0	
15 07 45.2	+51 27 09	MCG +9-25-22	14.7	1.0 x 0.3	13.2	Sbc	147	= **Mkn 845**; mag 11.4 star ESE 2′.4.

(Continued from previous page)

GALAXIES

RA h m s	Dec ° ′ ″	Name	Mag (V)	Dim ′ Maj x min	SB	Type Class	PA	Notes
15 10 06.6	+56 22 19	MCG +9-25-29	14.9	0.9 x 0.5	13.9	Sb	69	Mag 12.7 star on S edge.
15 10 57.8	+52 27 31	MCG +9-25-31	15.1	0.8 x 0.5	14.0		33	Mag 12.9 star E 2′.0.
15 11 02.9	+52 02 55	MCG +9-25-32	15.1	0.8 x 0.6	14.1	Sc	150	
15 12 52.4	+51 23 52	MCG +9-25-36	16.1	0.7 x 0.4	14.5	Sb	105	Elongated **MCG +9-25-35** S edge.
15 14 00.4	+55 32 17	MCG +9-25-37	15.5	1.1 x 0.8	15.3		141	Mag 13.5 star N 2′.2.
15 27 59.5	+51 40 39	MCG +9-25-52	14.7	0.8 x 0.3	13.0		42	Mag 9.71v star NE 3′.4.
15 27 54.9	+54 16 10	MCG +9-25-53	14.6	1.1 x 0.2	12.8			Mag 9.39v star E 3′.8.
15 40 28.8	+53 30 42	MCG +9-26-6	13.4	1.0 x 0.8	13.0		138	**MCG +9-26-8** SE 1′.0; **MCG +9-26-9** SE 1′.6; mag 9.73v star NNE 2′.0.
15 43 40.5	+53 34 22	MCG +9-26-12	14.5	0.9 x 0.3	12.9	Sb	135	
15 48 07.0	+55 42 59	MCG +9-26-18	15.1	0.6 x 0.4	13.4	Sbc	153	Mag 11.1 star SSE 1′.4.
15 53 26.5	+52 05 13	MCG +9-26-24	15.3	0.7 x 0.5	14.0		150	Mag 12.8 star NW 3′.0.
15 54 45.6	+53 08 39	MCG +9-26-27	15.3	0.5 x 0.5	13.7	Sc		Mag 11.6 star NW 1′.5.
15 57 21.5	+54 40 14	MCG +9-26-32	14.6	0.5 x 0.4	12.9	E/S0	24	Mag 4.96v star NE 6′.1.
16 05 37.7	+55 36 50	MCG +9-26-52	14.7	1.0 x 0.6	14.0		51	Mag 11.4 star NW 2′.7.
16 05 49.3	+55 16 19	MCG +9-26-53	14.2	0.8 x 0.8	13.6			Mag 11.9 star SE 0′.9.
16 14 57.0	+55 52 21	MCG +9-27-4	14.4	0.9 x 0.7	13.7	Sc	147	Mag 10.9 star W 3′.9.
16 17 49.2	+52 12 08	MCG +9-27-12	14.5	0.7 x 0.5	13.2	Sb	30	Mag 10.76v star ENE 2′.7.
16 18 07.1	+54 24 22	MCG +9-27-13	14.9	0.7 x 0.3	13.0		144	Mag 13.0 star NW 2′.9.
16 20 51.0	+53 32 23	MCG +9-27-18	14.7	0.7 x 0.4	13.2		15	Mag 12.1 star E 1′.4.
16 22 18.1	+51 08 55	MCG +9-27-26	13.9	0.9 x 0.7	13.2	SBbc	57	
16 22 37.3	+56 35 12	MCG +9-27-27	14.8	0.9 x 0.5	13.7		144	
16 22 42.1	+53 54 28	MCG +9-27-28	15.2	0.7 x 0.7	14.2	Sbc		
16 23 18.6	+55 45 23	MCG +9-27-31	14.9	1.0 x 0.9	14.6	SBbc	45	Mag 10.82v star W 3′.5.
16 28 07.5	+52 23 04	MCG +9-27-45	14.6	0.7 x 0.3	12.8		24	Multi-galaxy system, elongated companion S end; mag 11.1 star SE 3′.0.
16 29 26.1	+55 29 15	MCG +9-27-47	15.0	0.7 x 0.3	13.2		137	NGC 6182 NE 2′.1.
16 32 36.8	+52 00 40	MCG +9-27-50	15.3	0.5 x 0.4	13.4	Sbc	21	Mag 10.23v star W 1′.7.
16 35 16.2	+54 18 40	MCG +9-27-54	16.2	0.7 x 0.4	14.6	Sbc	99	Mag 11.08v star E 3′.3.
16 36 02.8	+56 02 50	MCG +9-27-56	15.0	1.0 x 0.2	13.1	Sbc	57	Mag 11.9 star SE 1′.8.
16 37 32.5	+51 39 06	MCG +9-27-58	15.4	0.9 x 0.4	14.2		42	
16 44 16.7	+51 30 04	MCG +9-27-68	15.1	0.7 x 0.4	13.6	Sbc	132	Mag 11.7 star E 0′.8.
16 44 43.6	+51 30 38	MCG +9-27-69	14.7	0.9 x 0.4	13.4	Sbc	33	Mag 12.4 star N 2′.4.
16 47 33.4	+53 51 37	MCG +9-27-85	14.6	0.7 x 0.5	13.3	Sb	138	Mag 11.9 star N 2′.3.
16 48 23.6	+53 59 53	MCG +9-27-88	15.2	0.7 x 0.2	12.9	Sb	156	MCG +9-27-89 S 2′.1.
16 48 31.1	+53 58 06	MCG +9-27-89	14.1	0.8 x 0.8	13.5	Sc		MCG +9-27-88 N 2′.1.
16 52 52.0	+52 57 45	MCG +9-28-4	14.6	0.8 x 0.5	13.5	Sbc	18	Mag 11.9 star W 1′.3.
17 00 52.8	+55 05 57	MCG +9-28-15	14.8	0.6 x 0.5	13.4	Sb	120	
17 03 28.7	+55 08 52	MCG +9-28-19	14.1	0.9 x 0.5	13.1	Sc	144	
15 05 35.3	+58 04 17	MCG +10-21-46	14.9	0.9 x 0.4	13.6	Sc	141	
15 13 29.4	+58 30 29	MCG +10-22-6	14.3	0.7 x 0.5	13.0	S?	12	Mag 13.3 star NW 1′.9.
15 15 04.3	+61 12 10	MCG +10-22-9	14.0	1.0 x 0.3	12.5	Sb	114	Mag 11.2 star W 1′.8.
15 17 25.3	+56 39 46	MCG +10-22-10	14.3	0.7 x 0.5	13.0	Spiral	72	
15 33 58.1	+57 21 47	MCG +10-22-21	14.9	1.0 x 0.3	13.4	Sc	63	
15 34 14.3	+59 43 30	MCG +10-22-22	16.0	0.7 x 0.5	14.7		45	Mag 10.73v star S 1′.4.
15 46 24.4	+59 12 52	MCG +10-22-38	15.3	0.8 x 0.7	14.5	SBbc	12	Mag 12.4 star NE 3′.4.
15 56 40.7	+59 47 36	MCG +10-23-9	14.3	1.1 x 0.3	12.9	Sc	63	Mag 11.6 star NE 4′.2.
16 00 10.8	+58 23 07	MCG +10-23-15	14.3	0.7 x 0.3	12.6	E	135	Mag 11.00v star NNE 3′.8.
16 08 20.7	+57 56 32	MCG +10-23-23	14.2	0.8 x 0.5	13.0		135	Mag 9.89v star S 2′.6.
16 11 45.5	+60 09 50	MCG +10-23-39	14.5	0.6 x 0.4	13.0	E	36	Mag 14.0 star E edge; mag 10.56v star W 3′.2.
16 15 53.4	+60 22 53	MCG +10-23-53	14.7	1.1 x 0.2	12.9	Sbc	117	Mag 11.4 star W 1′.3.
16 16 17.9	+60 25 41	MCG +10-23-54	14.1	1.1 x 0.6	13.7	E/S0	108	
16 17 12.6	+59 47 47	MCG +10-23-55	13.4	1.1 x 0.3	12.0	Lenticular	117	Located 2′.5 N of 5.43v star Theta Draconis.
16 21 36.8	+58 41 51	MCG +10-23-68	14.1	0.6 x 0.6	12.9	SBc		Mag 13.2 star N 2′.5.
16 24 48.0	+58 05 30	MCG +10-23-72	13.9	0.9 x 0.5	12.9		63	Mag 10.78v star SSW 3′.8.
16 27 19.9	+62 39 33	MCG +10-23-75	14.7	0.9 x 0.5	13.7	Sb	117	**UGC 10411** NE 3′.9.
16 32 57.3	+60 03 58	MCG +10-23-85	16.0	1.0 x 0.3	14.5	SBc	45	
16 41 44.7	+57 50 51	MCG +10-24-32	15.2	0.7 x 0.3	13.4	S	33	NGC 6213 S 2′.3.
16 46 13.1	+62 13 46	MCG +10-24-52	14.7	0.6 x 0.6	13.4			Mag 14.8 star on S edge.
16 47 58.4	+60 48 13	MCG +10-24-58	16.1	0.5 x 0.3	13.9		21	Very faint star on E edge; mag 13.4 star E 1′.2.
16 49 28.0	+58 53 53	MCG +10-24-64	14.6	0.7 x 0.4	13.1		30	Mag 12.2 star NE 0′.7; mag 10.70v star SSE 2′.0.
16 58 42.9	+57 05 50	MCG +10-24-80	14.7	0.8 x 0.5	13.8	E/S0	45	Mag 11.9 star N 3′.4.
17 00 40.0	+58 55 13	MCG +10-24-86	14.6	0.7 x 0.5	13.3		147	NGC 6291 NE 2′.3.
17 01 06.4	+58 40 59	MCG +10-24-87	15.5	0.7 x 0.5	14.2	Sc	18	Mag 8.99v star W 3′.8.
17 01 13.1	+60 15 01	MCG +10-24-91	15.2	0.4 x 0.3	12.7		6	Mag 9.19v star E 1′.9.
14 58 40.2	+53 53 06	NGC 5820	12.4	1.7 x 1.1	13.0	S0 sp	87	Close pair of mag 7.62v, 6.83v stars E 8′.0.
14 59 00.2	+53 55 21	NGC 5821	13.6	1.4 x 0.8	13.6	S?	148	Small nucleus; NGC 5820 SW 3′.7.
15 06 03.4	+55 34 24	NGC 5862	14.8	0.5 x 0.5	13.1			Almost stellar; mag 7.70v star NW 8′.4.
15 06 29.7	+55 45 46	NGC 5866	9.9	6.4 x 2.9	12.9	SA0$^+$ sp	128	
15 12 07.2	+55 47 06	NGC 5866B	14.2	2.7 x 1.9	15.8	SAB(rs)dm:	20	
15 06 34.0	+55 28 43	NGC 5870	13.9	1.2 x 0.9	13.8	S0?	25	Diffuse core; mag 11.4 star 1′.1 E.
15 07 51.8	+54 45 09	NGC 5874	12.4	2.3 x 1.6	13.7	SAB(rs)c I-II	53	Tightly wound, strong dark lanes.
15 09 12.7	+52 31 47	NGC 5875	12.4	2.5 x 1.2	13.5	SAb: II	145	**CGCG 274-26** SSW 5′.6; stellar **Mkn 846** SW 6′.7.
15 08 33.5	+52 17 41	NGC 5875A	13.4	0.5 x 0.5	11.7	S?		**MCG +9-25-25** WNW 4′.4.
15 09 31.5	+54 30 24	NGC 5876	12.7	2.4 x 1.2	13.7	SB(r)ab:	50	Prominent N-S bar, uniform envelope.
15 09 47.1	+56 59 57	NGC 5879	11.6	4.2 x 1.3	13.3	SA(rs)bc:? II-III	6	A pair of mag 11 and 12 stars E 3′.8.
15 11 40.5	+59 48 22	NGC 5894	12.8	3.0 x 0.4	12.8	SBdm?	13	
15 15 23.3	+55 30 59	NGC 5905	11.7	4.0 x 2.6	14.1	SB(r)b I	135	Bright center; bright, narrow arm extends NW.
15 15 53.0	+56 19 57	NGC 5907	10.3	12.6 x 1.4	13.3	SA(s)c: sp II	155	Very prominent dark lane down W side.
15 16 43.3	+55 24 39	NGC 5908	11.8	3.2 x 1.2	13.1	SA(s)b: sp II-III	154	Very strong dark lane down W side.
15 33 27.9	+56 33 32	NGC 5963	12.5	3.3 x 2.6	14.7	S pec	40	Pair of stars on SE edge, extending SE from strong core.
15 34 02.0	+56 41 08	NGC 5965	11.7	5.2 x 0.7	13.0	Sb II-III	53	Strong dark lane E of center; small anonymous galaxy W 1′.2.
15 34 51.4	+56 27 02	NGC 5969	14.4	0.4 x 0.4	12.5	E/S0		Almost stellar; mag 8.16v star SW 8′.2; NGC 5971 E 6′.4.
15 35 37.2	+56 27 40	NGC 5971	13.8	1.6 x 0.6	13.6	Sa	136	NGC 5969 W 6′.4.
15 36 48.0	+59 23 50	NGC 5976	14.8	0.8 x 0.4	13.4	SA0:	119	Stellar galaxy **VII Zw 615** SW 1′.6.
15 37 53.2	+59 23 34	NGC 5981	13.0	2.8 x 0.5	13.2	Sc? sp	140	Mag 10.79v star N 3′.3.
15 38 39.8	+59 21 22	NGC 5982	11.1	2.6 x 1.9	12.9	E3	110	Bright diffuse center.
15 39 37.2	+59 19 52	NGC 5985	11.1	5.5 x 3.0	14.0	SAB(r)b I	13	Bright center, tightly wound arms.
15 39 57.7	+58 04 50	NGC 5987	11.7	4.2 x 1.3	13.4	Sb II-III	63	Two parallel dark lanes S of center; mag 10.14v star NW 1′.3.
15 41 32.8	+59 45 14	NGC 5989	13.1	0.9 x 0.9	12.7	Scd?		
15 51 25.0	+62 18 33	NGC 6015	11.1	5.4 x 2.1	13.6	SA(s)cd II-III	28	Patchy center, many small knots.
16 10 42.6	+57 27 59	NGC 6088A	12.8	0.7 x 0.3	11.2	E	138	NGC 6088B on S edge; mag 9.00v star W 6′.6.

(Continued from previous page)

GALAXIES

RA h m s	Dec ° ′ ″	Name	Mag (V)	Dim ′ Maj x min	SB	Type Class	PA	Notes
16 10 44.1	+57 27 44	NGC 6088B	13.2	0.4 x 0.3	10.7	Sb		NGC 6088A on N edge; **MCG +10-23-36** SE 8.′0.
16 11 40.8	+52 27 26	NGC 6090	13.7	2.4 x 1.0	14.5	S0?	39	Multi-galaxy system; very faint extensions/arms NE and SW; **CGCG 275-28** W 2.′5.
16 11 11.2	+61 16 04	NGC 6095	12.6	1.8 x 1.6	13.6	S0^{-}:	132	Stellar nucleus; **MCG +10-23-35** SSE 2.′9.
16 17 20.1	+61 56 17	NGC 6123	13.8	0.8 x 0.3	12.1	S0/a	4	
16 19 12.0	+57 59 00	NGC 6127	12.0	1.4 x 1.4	12.7	E		
16 19 33.5	+57 36 54	NGC 6130	13.5	1.0 x 0.7	12.9	SBbc:	25	Located 2.′7 ENE of mag 8.70v star.
16 20 59.5	+55 58 12	NGC 6136	14.7	0.9 x 0.3	13.1	Sc	99	**MCG +9-27-21** NE 4.′3.
16 21 42.2	+55 05 10	NGC 6143	13.1	1.0 x 0.9	12.9	SAB(rs)bc:	177	Mag 10.08v star SE 2.′3.
16 25 48.5	+55 21 36	NGC 6157	14.5	0.6 x 0.5	13.3	E/S0	9	Almost stellar; located just outside W edge of small galaxy cluster **A 2201**.
16 27 36.7	+59 33 40	NGC 6176	13.8	0.5 x 0.5	12.4	E/S0		
16 29 34.0	+55 31 04	NGC 6182	13.5	1.4 x 0.4	12.8	Sa	146	MCG +9-27-47 SW 2.′1.
16 31 36.9	+57 42 21	NGC 6187	14.4	0.6 x 0.6	13.1			Mag 8.36v star N 3.′4.
16 31 41.0	+59 37 35	NGC 6189	12.7	1.9 x 0.9	13.1	Scd:	20	Triangle of faint stars on S end.
16 32 06.7	+58 26 17	NGC 6190	12.6	1.4 x 1.3	13.1	Scd:	60	Bright nucleus, patchy arms.
16 35 30.8	+57 29 09	NGC 6198	13.6	1.0 x 0.7	13.2	E	90	Large diffuse center.
16 40 08.3	+58 37 02	NGC 6206	13.6	0.7 x 0.7	12.6	S0:		**MCG +10-24-19** N 3.′0.
16 41 27.7	+57 47 00	NGC 6211	12.6	1.4 x 1.0	12.9	SB(r)0° pec:	99	Large, bright center; NGC 6213 N 2.′3.
16 41 37.3	+57 48 51	NGC 6213	14.7	0.7 x 0.3	12.9	Early Spiral	57	MCG +10-24-32 N 2.′3.
16 43 04.8	+61 34 42	NGC 6223	11.9	3.5 x 2.6	14.1	Pec	96	Bright center, faint envelope.
16 43 23.1	+61 59 03	NGC 6226	13.2	0.8 x 0.4	11.8	Pec	68	Very elongated **MCG +10-24-42** SSE 1.′6; **MCG +10-24-39** WNW 3.′9.
16 47 16.8	+62 08 45	NGC 6238	14.7	0.6 x 0.4	13.0	Pec	17	
16 48 04.0	+62 12 02	NGC 6244	13.5	1.5 x 0.3	12.5	SBa:	140	Very faint star N edge; mag 8.74v star E 7.′8.
16 49 53.4	+55 32 35	NGC 6246	13.6	1.5 x 0.6	13.4	SBb?	43	
16 50 14.1	+55 23 05	NGC 6246A	13.4	2.0 x 2.0	14.8	SAB(r)c pec:		Many bright knots and/or superimposed stars; mag 9.77v star on N edge.
16 48 19.4	+62 58 32	NGC 6247	13.5	1.0 x 0.3	12.0	Pec	58	
16 52 30.1	+60 30 52	NGC 6258	13.4	0.9 x 0.7	12.9	E	70	Mag 8.49v star W 4.′4; **MCG +10-24-74** NE 4.′4.
16 58 24.2	+58 57 20	NGC 6285	13.5	1.3 x 0.7	13.2	S0^{+}? pec	129	Bright lens with two main arms.
16 58 32.0	+58 56 16	NGC 6286	13.3	1.3 x 1.2	13.6	Sb: pec	33	Multi-galaxy system; faint envelope extends SE of bright bar.
17 00 56.6	+58 58 09	NGC 6290	13.5	1.1 x 1.0	13.5	SBa	30	Short SE-NW bar; mag 10.12v star NW 3.′0.
17 00 56.1	+58 56 12	NGC 6291	14.0	0.6 x 0.6	12.7	SBb		MCG +10-24-86 SW 2.′3.
17 03 03.0	+61 02 34	NGC 6292	13.5	1.5 x 0.8	13.5	Sbc	105	Star on E end.
17 03 15.1	+60 20 15	NGC 6295	14.8	1.0 x 0.4	13.7	S	77	
17 03 36.6	+62 01 30	NGC 6297	13.6	0.7 x 0.5	12.3	S0	90	
14 57 11.5	+52 20 43	UGC 9629	13.6	1.6 x 0.7	13.6	Sa	156	
14 57 56.5	+53 47 04	UGC 9632	14.3	1.3 x 1.0	14.4	SAd	135	Very patchy, several faint knots.
14 58 09.9	+58 52 15	UGC 9638	14.5	1.4 x 0.9	14.6	Im	40	Dwarf irregular; mag 12.4 star S 1.′2; mag 11.9 star SW 3.′1.
15 01 13.7	+52 35 44	UGC 9663	14.3	1.3 x 1.1	14.6	Im: IV-V	6	Dwarf, very low, uniform surface brightness.
15 04 03.1	+52 55 45	UGC 9688	13.7	1.6 x 0.7	13.7	S0/a	9	Mag 12.5 star N 2.′5.
15 05 14.8	+51 09 28	UGC 9702	14.8	1.2 x 0.3	13.6	Sbc	168	**CGCG 274-13** NE 8.′1.
15 06 30.9	+51 54 47	UGC 9722	15.1	1.0 x 0.7	14.5	Sd	18	Mag 11.9 star W 2.′7.
15 07 48.5	+55 11 04	UGC 9737	13.8	1.2 x 0.7	13.5	S	63	Mag 12.1 star WSW 0.′8.
15 10 41.1	+55 20 59	UGC 9759	14.2	1.8 x 0.3	13.4	S?	51	Mag 8.62v star N 4.′5.
15 11 18.3	+61 07 26	UGC 9766	14.3	1.5 x 0.5	13.8	SBbc	167	**MCG +10-22-2** NW 3.′6.
15 23 51.7	+58 03 10	UGC 9837	13.3	1.8 x 1.7	14.4	SAB(s)c I-II	155	Knotty, multi-branching arms.
15 25 16.8	+53 24 25	UGC 9849	14.7	1.5 x 0.4	14.0	Scd:	82	
15 25 46.9	+52 26 41	UGC 9853	14.0	1.7 x 0.3	13.2	Sb III	93	
15 36 07.4	+59 33 58	UGC 9934	14.2	1.0 x 0.9	13.9	SBab	27	Mag 12.9 star NW 2.′7.
15 44 25.6	+55 39 55	UGC 10013	13.9	0.9 x 0.5	12.9	SBab:	12	Mag 13.3 star on W edge.
15 45 45.8	+61 33 15	UGC 10031	14.0	0.9 x 0.5	13.0	Sm:	168	
16 00 05.8	+51 47 40	UGC 10129	14.8	1.4 x 1.2	15.2	Pec	30	Bright core offset W; mag 9.58v star SW 4.′4.
16 06 04.2	+55 25 28	UGC 10214	13.7	1.5 x 0.7	13.6	SB(s)c pec	79	Dimensions do not include a very long, very faint, and narrow streamer extending NE.
16 09 25.8	+60 05 15	UGC 10247	14.8	0.8 x 0.6	13.9	SBm:	48	
16 11 30.5	+58 47 07	UGC 10271	14.1	1.3 x 0.5	13.5	SB	172	**MCG +10-23-31** S 8.′0.
16 11 47.1	+60 34 51	UGC 10279	13.6	0.9 x 0.5	12.6	Double System	177	Double galaxy system; **MCG +10-23-32** WNW 6.′6.
16 11 59.5	+61 10 35	UGC 10284	14.5	0.8 x 0.3	12.8	S0/a	51	Mag 8.80v star WNW 3.′3.
16 17 20.4	+59 19 15	UGC 10331	13.9	1.5 x 0.2	12.4	S pec	142	**MCG +10-23-59** E 0.′9.
16 22 09.6	+57 16 14	UGC 10361	13.9	1.4 x 0.7	13.7	SAB0°	60	Bright center, uniform envelope.
16 26 53.4	+51 33 16	UGC 10396	13.8	1.1 x 0.6	13.2	Sbc II	147	Strong dark patch S edge; bright knot N edge; mag 10.39v star NW 2.′0.
16 27 49.2	+56 01 37	UGC 10408	14.9	1.6 x 0.2	13.5	Sb II-III	78	
16 32 45.7	+62 40 48	UGC 10449	14.5	1.0 x 0.9	14.2	Sdm	177	Stellar nucleus, uniform envelope.
16 33 50.7	+56 07 59	UGC 10454	14.1	1.2 x 0.3	12.9	Sab	54	
16 34 15.1	+52 56 27	UGC 10456	13.9	1.4 x 0.7	13.7	Sb I-II	140	Mag 9.45v star W 2.′2.
16 38 00.4	+55 25 35	UGC 10493	13.6	1.1 x 0.8	13.4	E:	135	
16 38 59.3	+57 43 28	UGC 10500	12.9	1.3 x 1.2	13.3	S0/a	108	Mag 9.98v star SE 3.′1.
16 40 31.1	+58 05 25	UGC 10510	13.1	1.0 x 0.8	12.7	Sc II-III	33	Mag 10.19v star S 1.′7.
16 41 15.4	+51 11 03	UGC 10511	13.9	1.0 x 0.5	13.1	S0	20	**MCG +9-27-64** ESE 2.′5.
16 41 06.6	+59 19 23	UGC 10515	14.2	0.8 x 0.5	13.0	Sb II-III	10	
16 41 09.0	+61 19 35	UGC 10517	13.1	1.5 x 0.3	12.1	Sa	148	Lies between mag 11.07v star NE 2.′7 and **MCG +10-24-23** SW 2.′7.
16 41 07.1	+62 24 00	UGC 10518	13.8	1.0 x 0.8	13.4	Sbc	145	Stellar nucleus, uniform envelope; mag 7.21v star S 5.′6.
16 44 06.1	+62 05 13	UGC 10536	13.8	1.8 x 0.7	13.9	Sb II-III	123	Mag 9.23v star N 3.′8.
16 45 07.8	+61 50 24	UGC 10542	14.9	1.5 x 0.3	13.9	Sc	88	
16 45 52.8	+59 37 17	UGC 10548	13.7	1.5 x 0.7	13.6	SBb	143	Strong N-S core; mag 11.8 star W 1.′2.
16 46 35.9	+62 49 26	UGC 10561	14.3	2.3 x 0.3	13.7	Sb II-III	150	Mag 8.24v star SE 2.′0; mag 9.80v star W 2.′4.
16 49 56.0	+53 57 13	UGC 10579	13.2	1.2 x 0.7	13.0	E	12	Pair of mag 12 and 13 stars E 1.′5.
16 50 24.4	+55 50 21	UGC 10589	14.5	1.6 x 0.6	14.3	Pec	62	Mag 10.72v star N 1.′3.
16 50 56.4	+59 43 06	UGC 10590	13.4	0.9 x 0.9	13.1	SBcd?		Large, bright center with short N-S bar with dark patches E and W.
16 52 19.2	+55 54 16	UGC 10593	14.0	1.2 x 0.5	13.3	Sab	84	
16 53 34.2	+54 34 10	UGC 10601	15.2	1.2 x 0.6	14.7	Sb II-III	70	Mag 10.83v star NE 3.′2.
16 58 35.8	+54 52 12	UGC 10644	13.9	0.9 x 0.5	12.9		44	Mag 9.45v star S 1.′5.
16 58 27.5	+59 07 08	UGC 10646	13.8	1.4 x 0.6	13.6	E	25	
16 58 33.0	+59 45 14	UGC 10648	14.3	0.8 x 0.5	13.2	SB(s)b	137	Mag 9.16v star N 2.′7.
17 03 27.5	+59 43 29	UGC 10687	14.6	1.1 x 0.7	14.2	SBd	42	Mag 10.26v star on E edge.

GALAXY CLUSTERS

RA h m s	Dec ° ′ ″	Name	Mag 10th brightest	No. Gal	Diam ′	Notes
16 01 36.0	+53 52 00	A 2149	16.1	42	16	**MCG +9-26-45** W of center 2.′0; **MCG +9-26-46** N 2.′5; **MCG +9-26-47** N 5.′0; other members anonymous.

RA h m s	Dec ° ′ ″	Name	Mag (V)	Dim ′ Maj x min	SB	Type Class	PA	Notes
14 07 39.3	+54 47 36	CGCG 272-32	13.8	0.9 x 0.6	13.0	S0?	21	MCG +9-23-41 NE 9′.7.
13 05 50.9	+53 38 46	IC 847	13.9	0.3 x 0.3	11.4	E/S0		Located 1′.1 SW of NGC 4973.
13 07 36.7	+60 09 25	IC 852	13.6	1.1 x 0.9	13.4	S0⁻:	20	Mag 8.26v star W 5′.7; note: stellar **MCG +10-19-32** is 2′.0 N of this star.
13 08 41.0	+52 46 20	IC 853	13.6	1.6 x 1.3	14.2	(R′)SAB(s)ab I	33	Bright center with strong dark patches N and SE.
13 17 07.5	+57 32 21	IC 875	12.8	1.4 x 0.8	12.8	S0	150	Triangle of mag 11.3, 11.3 and 11.29v stars centered NNE 7′.3.
13 47 41.1	+56 37 17	IC 942	14.1	0.7 x 0.5	13.1	E/S0		Mag 10.73v star NE 7′.1.
14 16 31.9	+57 48 30	IC 995	14.1	1.6 x 0.4	13.5	Sdm:	147	Mag 11.5 star W 7′.3.
14 17 22.2	+57 37 49	IC 996	14.0	1.4 x 0.3	12.9	Sbc	155	Mag 10.60v star S 3′.8.
14 29 48.6	+53 57 54	IC 1027	14.5	0.8 x 0.8	13.9			Mag 10.89v star SW 6′.7.
14 39 33.6	+62 00 08	IC 1049	13.8	0.9 x 0.7	13.2	SBa	74	Very faint ENE-WSW bar with dark patch N.
14 50 46.5	+54 24 41	IC 1069	13.8	1.3 x 0.8	13.7	S0	45	Mag 11.6 star WNW 6′.6.
12 56 26.9	+51 05 04	MCG +9-21-68	14.6	0.8 x 0.3	12.9		78	Mag 9.43v star SE 0′.7; mag 11.1 star and **MCG +9-21-66** NW 2′.8.
12 57 45.0	+53 38 11	MCG +9-21-70	14.8	0.8 x 0.5	13.6		63	
12 58 38.5	+53 24 33	MCG +9-21-72	15.0	0.7 x 0.6	13.9	SBbc	33	Mag 9.63v star NW 2′.9.
12 59 03.7	+55 57 56	MCG +9-21-76	15.2	0.7 x 0.3	13.4	Sbc	150	Mag 12.7 star NE 2′.5.
13 01 11.3	+53 48 39	MCG +9-21-83	15.2	0.5 x 0.3	13.0	Sb	3	Mag 11.3 star on S edge; **MCG +9-21-85** SSE 1′.1.
13 01 17.2	+53 51 37	MCG +9-21-84	14.2	0.9 x 0.7	13.5	SBbc	111	Mag 10.88v star ESE 1′.6.
13 02 25.1	+52 12 21	MCG +9-21-87	15.7	0.8 x 0.3	14.0	Sc	156	Mag 9.61v star NNE 2′.8.
13 03 27.7	+53 31 55	MCG +9-21-94	14.8	0.9 x 0.5	13.8	Sbc	33	Mag 14.1 star E 0′.6.
13 04 17.2	+55 26 27	MCG +9-21-98	15.6	0.8 x 0.2	13.4	Sb	171	Mag 12.0 star NNW 0′.8; galaxy NPM1G +55.0166 NNW 2′.3.
13 04 24.1	+55 22 03	MCG +9-21-100	14.5	1.0 x 0.7	14.0		18	
13 05 03.1	+53 39 11	MCG +9-21-101	14.4	0.7 x 0.6	13.3	SB	102	**MCG +9-22-1** NE 1′.1; mag 11.02v star NE 2′.4.
13 05 25.8	+53 35 28	MCG +9-22-3	14.9	0.5 x 0.3	12.7	S?	21	NGC 4967 SE 2′.3.
13 07 21.9	+53 35 09	MCG +9-22-13	14.6	1.0 x 0.7	14.3	E/S0	15	Mag 8.25v star E 6′.9.
13 08 02.6	+53 56 06	MCG +9-22-16	14.5	0.8 x 0.7	13.7	Sb	66	Mag 11.1 star N 3′.8.
13 08 42.2	+52 00 58	MCG +9-22-20	13.6	1.0 x 0.8	13.2		159	UGC 8222 W 3′.8.
13 10 13.2	+52 30 58	MCG +9-22-23	14.1	0.9 x 0.9	13.7			Mag 12.4 star N 3′.0.
13 12 33.5	+52 55 28	MCG +9-22-26	15.4	0.7 x 0.4	13.9	Sbc	63	Mag 9.40v star SE 8′.5.
13 14 46.0	+53 49 11	MCG +9-22-36	14.1	1.0 x 0.3	12.6	Sab	90	Mag 9.06v star W 2′.6.
13 16 34.3	+51 25 55	MCG +9-22-41	14.6	0.7 x 0.5	13.5	E/S0	9	Located between a pair of mag 12.6, 13.0 stars; mag 11.9 and 14.5 stars superimposed on N end.
13 18 16.5	+53 57 50	MCG +9-22-44	14.6	1.0 x 0.7	14.3	E/S0	162	Mag 10.54v star N 2′.7.
13 19 47.7	+52 04 09	MCG +9-22-47	12.9	0.7 x 0.3	11.1	BCG	12	**MCG +9-22-48** SE 0′.6; **Mkn 251** ESE 2′.3.
13 22 33.0	+54 49 02	MCG +9-22-53	14.5	0.8 x 0.6	13.6		135	Mag 11.1 star N 0′.8.
13 22 59.9	+52 58 08	MCG +9-22-54	14.7	0.6 x 0.5	13.2	S?	162	= **Mkn 255**.
13 24 38.1	+52 57 31	MCG +9-22-57	14.6	1.0 x 0.9	14.3		78	Mag 13.1 star on E edge; elongated **MCG +9-22-58** ESE 2′.3.
13 27 47.4	+51 08 10	MCG +9-22-65	14.5	0.7 x 0.4	13.0		108	Mag 13.5 star E 1′.1.
13 28 00.3	+56 02 02	MCG +9-22-67	14.5	0.7 x 0.4	13.0	S?	111	MCG +9-22-68 NE 1′.2; mag 8.20v star W 4′.2.
13 28 06.7	+56 02 41	MCG +9-22-68	14.5	0.7 x 0.5	13.2	S0?	129	Mag 12.0 star on S edge; MCG +9-22-67 SW 1′.2.
13 29 46.0	+55 36 10	MCG +9-22-74	14.3	1.0 x 0.7	13.8		51	
13 33 22.4	+54 56 55	MCG +9-22-79	14.2	0.8 x 0.7	13.4		102	Mag 9.87v star NNW 2′.5.
13 35 25.5	+56 24 17	MCG +9-22-84	16.4	0.5 x 0.5	14.7	Sb		Mag 11.12v star NE 2′.9.
13 41 18.4	+52 06 06	MCG +9-22-92	15.1	0.8 x 0.5	14.0	SBb	21	Mag 9.82v star S 3′.2.
13 44 40.2	+55 59 37	MCG +9-23-3	15.3	0.6 x 0.4	13.6	Sbc	108	Mag 10.8 star E 2′.7.
13 48 13.7	+54 47 49	MCG +9-23-9	14.7	0.7 x 0.5	13.4	S?	48	Mag 10.98v star W 3′.9.
13 59 53.5	+54 47 16	MCG +9-23-23	14.9	0.7 x 0.6	13.8	Sc	39	Mag 12.1 star NW 0′.9.
14 01 16.2	+54 25 05	MCG +9-23-25	14.4	1.0 x 0.4	13.3	Sab	81	Mag 9.05v star S 3′.4.
14 08 26.4	+54 54 30	MCG +9-23-41	16.1	0.9 x 0.6	15.3	Sdm-m	33	Mag 10.77v star SW 2′.5; CGCG 272-32 SW 9′.7.
14 09 09.9	+53 49 17	MCG +9-23-42	14.8	0.8 x 0.3	13.1		174	Mag 11.53v star E 2′.8.
14 10 01.5	+54 52 52	MCG +9-23-47	14.2	1.1 x 0.9	14.0		114	**MCG +9-23-48** N 2′.4.
14 11 18.5	+55 10 51	MCG +9-23-51	14.4	1.0 x 0.7	13.9		36	Mag 10.84v star NE edge.
14 11 29.1	+55 07 16	MCG +9-23-52	14.9	0.6 x 0.4	13.2	Sc	12	
14 19 28.5	+52 08 05	MCG +9-23-61	14.8	0.6 x 0.6	13.6			
14 23 40.9	+53 54 55	MCG +9-24-4	14.9	0.7 x 0.5	13.6	Sb	3	Mag 11.8 star ENE 3′.2.
14 31 47.8	+55 56 50	MCG +9-24-10	14.5	0.8 x 0.5	13.4		159	Mag 14.7 star superimposed S edge.
14 33 10.0	+53 13 57	MCG +9-24-12	14.1	1.1 x 0.9	13.9		156	MCG +9-24-13 N 2′.0.
14 33 06.2	+53 15 53	MCG +9-24-13	14.5	0.9 x 0.9	14.1			MCG +9-24-12 S 2′.0.
14 33 52.2	+56 30 48	MCG +9-24-16	15.3	0.7 x 0.3	13.4	Sb	54	Mag 9.84v star E 2′.0.
14 34 09.9	+54 04 42	MCG +9-24-17	15.1	1.1 x 0.8	14.8	Sbc	78	
14 36 46.0	+51 27 31	MCG +9-24-22	15.6	0.7 x 0.5	14.3	Spiral	105	
14 37 35.2	+51 34 56	MCG +9-24-24	14.8	0.7 x 0.7	13.9			Located on the N end of NGC 5707; mag 6.99v star E 4′.7.
14 38 44.8	+51 19 05	MCG +9-24-26	14.6	1.4 x 1.2	15.0		135	Mag 9.40v star superimposed nearly on center.
14 38 50.8	+52 16 19	MCG +9-24-27	15.2	0.9 x 0.3	13.6	Spiral	78	
14 46 19.6	+54 26 46	MCG +9-24-34	15.0	0.6 x 0.5	13.5	SAB0°	12	Mag 9.27v star NNE 4′.9.
14 45 45.5	+51 34 44	MCG +9-24-35	13.7	0.9 x 0.7	13.1	S?	108	Very faint plume extends E; mag 8.57v star NE 3′.4.
14 46 37.2	+56 13 56	MCG +9-24-38	15.1	0.8 x 0.3	13.4	Sc	108	
14 50 02.8	+52 24 18	MCG +9-24-43	14.9	0.4 x 0.3	12.4	Compact	0	= **Mkn 826**; mag 7.15v star E 1′.7.
14 54 26.7	+53 13 00	MCG +9-24-54	14.6	0.7 x 0.3	12.8		138	**MCG +9-24-53** NW 2′.6.
14 58 19.3	+53 14 11	MCG +9-24-60	14.4	0.9 x 0.7	13.8	S0?	0	Located between a pair of NE-SW mag 14.6, 13.6 stars.
14 59 32.9	+53 53 54	MCG +9-25-3	13.8	1.0 x 0.7	13.3	SBbc	174	Close pair of mag 7.62v and 6.83v stars S 2′.0.
13 01 13.9	+58 06 41	MCG +10-19-14	15.0	0.6 x 0.4	13.3		159	
13 06 15.5	+62 00 23	MCG +10-19-27	14.8	0.6 x 0.4	13.1	Sb	120	Located 2′.3 S of mag 6.15v star.
13 06 30.9	+57 03 49	MCG +10-19-29	15.7	0.8 x 0.3	14.0		132	Mag 7.49v star S 2′.3.
13 10 54.2	+56 37 10	MCG +10-19-49	14.9	0.7 x 0.5	13.6	Sbc	144	Mag 9.52v star E 2′.0.
13 25 23.2	+59 36 45	MCG +10-19-71	13.7	0.8 x 0.6	12.7	Pair?	33	
13 26 49.9	+58 06 34	MCG +10-19-75	15.1	0.7 x 0.5	13.8		30	Mag 10.35v star NE 1′.2.
13 28 58.8	+61 54 49	MCG +10-19-77	14.7	1.0 x 0.4	13.6	Sb	36	Mag 11.26v star on E edge.
13 29 33.8	+57 45 58	MCG +10-19-79	18.5	0.7 x 0.4	17.0	SBbc	150	Mag 10.96v star NE 1′.1.
13 31 47.0	+59 10 42	MCG +10-19-83	14.2	1.1 x 0.5	13.4		105	
13 32 16.1	+60 23 36	MCG +10-19-85	15.2	0.8 x 0.4	13.8		72	Mag 11.2 star NW 5′.0.
13 32 39.9	+56 44 32	MCG +10-19-86	15.3	0.6 x 0.5	13.9	Sc	39	Mag 10.9 star SSW 1′.7.
13 34 30.1	+60 15 46	MCG +10-19-93	14.8	0.8 x 0.7	14.0		126	Mag 12.1 star N 0′.9.
13 34 36.3	+56 27 48	MCG +10-19-94	14.3	0.7 x 0.5	13.2	E/S0	15	
13 38 10.1	+60 16 15	MCG +10-19-102	14.3	1.2 x 0.8	14.1	Sb	120	Close pair **MCG +10-19-99** and **MCG +10-19-100** W 4′.7.
13 42 34.5	+60 07 42	MCG +10-20-11	14.6	0.8 x 0.6	13.7		165	
13 44 16.0	+57 55 57	MCG +10-20-16	15.1	0.5 x 0.5	13.4			Mag 9.88v star SW 2′.1.
13 44 37.0	+61 24 46	MCG +10-20-19	13.7	0.9 x 0.9	13.3			Mag 11.7 star NE 1′.2.
13 44 50.3	+59 25 24	MCG +10-20-21	15.0	0.8 x 0.5	13.9	Sbc	12	Mag 7.82v star SW 6′.3.
13 45 00.6	+61 16 47	MCG +10-20-23	14.0	0.8 x 0.5	12.9	Sc	108	**MCG +10-20-25** SE 1′.8.

RA h m s	Dec ° ′ ″	Name	Mag (V)	Dim ′ Maj x min	SB	Type Class	PA	Notes
13 49 16.4	+56 55 37	MCG +10-20-36	14.2	0.9 x 0.7	13.8	E/S0	33	
14 00 02.7	+59 15 55	MCG +10-20-59	14.2	0.7 x 0.7	13.5	E?		Mag 12.3 star N 1′.2.
14 09 54.9	+56 49 20	MCG +10-20-78	14.5	0.8 x 0.4	13.1		147	Mag 8.92v star SE 4′.8.
14 10 04.2	+58 02 20	MCG +10-20-79	14.8	0.9 x 0.7	14.1		114	
14 13 50.1	+57 46 07	MCG +10-20-84	14.9	0.9 x 0.6	14.1	Sb-c	117	On SW edge of NGC 5526.
14 34 17.0	+59 39 25	MCG +10-21-10	14.8	0.8 x 0.5	13.7		120	Mag 14.3 star NW 0′.9.
14 35 49.4	+62 25 14	MCG +10-21-17	14.7	1.0 x 0.3	13.3		21	
14 51 54.9	+60 44 49	MCG +10-21-36	15.0	0.6 x 0.4	13.3	Sbc	63	Mag 10.97v star N 1′.5.
12 56 25.2	+52 17 46	NGC 4834	14.8	0.9 x 0.3	13.2	Sbc	108	UGC 8050 W 9′.4.
13 02 57.7	+51 19 07	NGC 4938	14.3	0.8 x 0.7	13.5	S?	39	
13 05 24.7	+56 19 21	NGC 4964	13.3	1.1 x 0.6	12.7	S?	134	
13 05 36.5	+53 33 50	NGC 4967	14.2	0.7 x 0.6	13.3	E	139	MCG +9-22-3 NW 2′.3.
13 05 32.1	+53 41 04	NGC 4973	13.9	0.7 x 0.7	13.2	E/S0		Mag 11.0 star WSW 2′.6.
13 05 56.1	+53 39 33	NGC 4974	13.3	1.4 x 1.2	13.9	E/S0	129	IC 847 SW 1′.1.
13 06 04.9	+55 39 19	NGC 4977	13.3	2.0 x 2.0	14.7	SA(r)b II		Bright center, faint outer shell.
13 07 59.1	+51 55 43	NGC 4987	13.4	1.2 x 0.7	13.2	E	35	Mag 7.98v star S 8′.5.
13 09 33.4	+53 29 36	NGC 5001	13.6	1.1 x 0.4	12.6	SB	160	
13 09 14.2	+62 10 26	NGC 5007	13.3	0.9 x 0.6	12.5	S0$^-$:	135	**MCG +10-19-44** ESE 2′.0.
13 13 32.8	+51 15 30	NGC 5040	14.2	1.0 x 0.5	13.3		66	
13 20 53.1	+57 38 32	NGC 5109	12.9	1.7 x 0.5	12.6	S	153	
13 21 24.4	+57 41 43	NGC 5113	14.5	0.8 x 0.3	12.8	Sbc	42	
13 26 54.4	+52 45 10	NGC 5163	13.6	1.1 x 0.7	13.3	E	10	Mag 6.35v star E 10′.0.
13 27 11.9	+55 29 12	NGC 5164	13.7	1.0 x 0.9	13.4	SBb	27	Thin, bright bar NE-SW.
13 29 16.5	+53 04 55	NGC 5201	13.2	1.7 x 1.0	13.6	S?	145	Very faint arm extends SE then curves W outside of published dimensions.
13 29 35.6	+58 25 03	NGC 5204	11.3	5.0 x 3.0	14.1	SA(s)m IV-V	5	Very patchy, many knots.
13 30 03.1	+62 30 44	NGC 5205	12.3	3.2 x 1.8	14.1	S?	166	Small, faint anonymous galaxy E 1′.6.
13 32 06.5	+62 41 59	NGC 5216	12.6	1.8 x 1.2	13.5	E0 pec	28	Member of **Keenan's System**. Bright nucleus.
13 34 42.1	+61 59 37	NGC 5216A	13.5	1.1 x 0.3	12.1	S?	67	
13 32 10.6	+62 46 04	NGC 5218	12.3	1.8 x 1.3	13.1	SB(s)b? pec	100	Member of **Keenan's System**. Bright, wide center NE-SW; faint envelope E and W.
13 33 20.1	+51 29 27	NGC 5225	13.6	0.7 x 0.7	12.7	S?		
13 34 42.8	+51 36 49	NGC 5238	13.3	1.7 x 1.4	14.1	SAB(s)dm	160	Stellar **Mkn 1479** on SE edge of stellar nucleus.
13 36 07.3	+51 14 10	NGC 5250	13.0	1.0 x 0.9	12.7	S0	120	**MCG +9-22-83** SW 5′.9; mag 7.39v star NE 5′.8.
13 37 18.4	+57 06 30	NGC 5255	14.5	0.8 x 0.2	12.4	Sb	24	Mag 10.49v star E 1′.7.
13 41 41.0	+55 40 18	NGC 5278	13.5	1.3 x 1.0	13.7	SA(s)b? pec	57	One main arm with bright knot at end, NGC 5279 lies NE of nucleus. UGC 8671 SW 2′.8.
13 41 43.8	+55 40 23	NGC 5279	14.7	0.7 x 0.4	13.1	SB(s)a pec	29	NE of NGC 5278 nucleus.
13 45 18.3	+55 17 26	NGC 5294	14.3	0.6 x 0.5	12.8		120	
13 46 59.1	+60 58 20	NGC 5308	11.4	3.7 x 0.7	12.3	S0$^-$ sp	60	**MCG +10-20-30** S of NE end; elongated anonymous galaxy NW 1′.9.
13 49 15.4	+60 11 23	NGC 5322	10.2	5.9 x 3.9	13.6	E3-4	83	Large, bright center.
13 51 25.8	+59 51 50	NGC 5342	13.5	1.1 x 0.4	12.5	S0	152	
13 54 29.1	+54 19 50	NGC 5368	13.0	0.9 x 0.7	12.3	(R′)SABab:	10	
13 54 09.2	+60 40 40	NGC 5370	13.2	1.1 x 1.1	13.2	SB(r)0°		Bright nucleus on short bar.
13 54 46.5	+58 39 57	NGC 5372	13.3	0.6 x 0.4	11.6	S?	140	
13 55 16.1	+59 30 22	NGC 5376	12.1	2.1 x 1.3	13.1	SAB(r)b? II	70	Very faint, elongated anonymous galaxy E 3′.2; UGC 8859 N 7′.2.
13 55 34.5	+59 44 34	NGC 5379	12.9	2.2 x 0.7	13.2	S0	60	NGC 5389 E 4′.1.
13 56 06.4	+59 44 22	NGC 5389	12.0	4.1 x 1.1	13.5	SAB(r)0/a:?	3	Mag 9.09v star 4′.0 NE. NGC 5379 W 4′.1..
13 58 16.7	+59 48 51	NGC 5402	13.8	1.3 x 0.3	12.6	S?	167	Mag 10.3 star S 5′.0.
14 00 42.0	+55 09 57	NGC 5422	11.9	3.9 x 0.7	12.9	S0 sp	152	Mag 11.26v star E 2′.2.
14 00 46.0	+59 19 39	NGC 5430	11.9	2.2 x 1.1	12.7	SB(s)b II	0	Northern half brightest.
14 02 11.6	+55 48 50	NGC 5443	12.3	2.7 x 1.0	13.2	SB(s)b?	34	Mag 9.49v star N 4′.1.
14 03 12.7	+54 21 12	NGC 5457	7.9	28.8 x 26.9	14.9	SAB(rs)cd I	26	= **M 101**. **M 102** is duplicate observation of M 101. NGC 5447, NGC 5449, NGC 5450, NGC 5451, NGC 5453, NGC 5455, NGC 5458, NGC 5461, and NGC 5462 are condensations involved in M 101.
14 04 43.1	+54 53 33	NGC 5473	11.4	2.3 x 1.7	12.8	SAB(s)0$^-$:	160	Faint star E of nucleus; mag 10.71v star SW 1′.7.
14 05 01.6	+53 39 23	NGC 5474	10.8	4.7 x 4.7	14.0	SA(s)cd pec IV-V		Very bright, large nucleus offset to the N; many superimposed stars and/or knots.
14 05 12.4	+55 44 29	NGC 5475	12.6	2.0 x 0.5	12.4	Sa? sp	166	
14 05 32.8	+54 27 41	NGC 5477	14.0	1.7 x 1.3	14.7	SA(s)m IV-V	95	Stellar nucleus.
14 06 48.2	+55 01 48	NGC 5484	14.7	0.5 x 0.5	13.3	E2		Almost stellar; very faint, elongated anonymous galaxy SW 2′.4.
14 07 11.1	+55 00 04	NGC 5485	11.4	2.3 x 1.9	12.9	SA0 pec	170	Large, bright nucleus.
14 07 24.9	+55 06 06	NGC 5486	13.2	1.7 x 1.0	13.6	SA(s)m: III-IV	80	Bright nucleus, many knots.
14 09 34.1	+60 24 32	NGC 5503	15.3	0.5 x 0.2	12.6		87	Located 1′.8 NW of mag 10.95v star.
14 13 53.3	+57 46 22	NGC 5526	13.4	2.3 x 0.3	12.8	Sbc	135	MCG +10-20-84 on W edge.
14 14 54.5	+60 00 36	NGC 5540	13.9	0.7 x 0.6	13.0	E/S0	36	**MCG +10-20-88** NW 3′.0; **MCG +10-20-89** SW 2′.7.
14 17 17.0	+58 42 30	NGC 5561	14.8	1.0 x 0.5	13.8	Sm	127	
14 19 47.9	+56 43 52	NGC 5585	10.7	5.8 x 3.7	13.9	SAB(s)d IV-V	30	Bright center, many knots; mag 11.5 star NE edge; mag 9.43v star 5′.0 SE.
14 26 35.7	+51 35 03	NGC 5624	13.2	1.1 x 0.8	12.9	S?	3	Bright elongated center.
14 26 33.6	+56 34 55	NGC 5631	11.5	1.7 x 1.7	12.5	SA(s)0°		Bright center, smooth envelope.
14 30 22.8	+59 28 13	NGC 5667	12.5	1.7 x 1.1	13.0	Scd: pec	168	Bright knot S of elongated center.
14 32 05.4	+57 55 14	NGC 5678	11.3	3.3 x 1.6	13.0	SAB(rs)b II-III	5	Mag 9.43v star NNW 2′.7.
14 34 53.0	+54 28 33	NGC 5687	11.8	2.4 x 1.7	13.1	S0$^-$?	105	Nucleus lies in a triangle of superimposed stars.
14 37 30.9	+51 33 44	NGC 5707	12.5	2.6 x 0.5	12.6	Sab: sp	35	MCG +9-24-24 N at end; mag 6.99v star E 4′.7.
14 43 49.0	+53 24 00	NGC 5751	13.2	1.5 x 0.8	13.2	Scd:	55	
14 51 17.8	+58 58 45	NGC 5777	13.3	3.1 x 0.4	13.4	Sb III	144	**UGC 9570** SE 2′.7.
14 52 09.7	+55 53 55	NGC 5779	14.8	0.4 x 0.4	12.7			Mag 9.55v star ESE 6′.7.
14 53 28.0	+52 04 30	NGC 5783	12.8	2.9 x 1.8	14.4	SAB(s)c	0	Faint star N of nucleus; very faint, elongated anonymous galaxy on SW edge; NGC 5788 SW 2′.6.
14 53 17.0	+52 02 37	NGC 5788	14.7	0.5 x 0.4	12.8		39	Located 2′.6 SW of NGC 5783.
14 58 40.2	+53 53 06	NGC 5820	12.4	1.7 x 1.1	13.0	S0 sp	87	Close pair of mag 7.62v, 6.83v stars E 8′.0.
14 59 00.2	+53 55 21	NGC 5821	13.6	1.4 x 0.8	13.6	S?	148	Small nucleus; NGC 5820 SW 3′.7.
12 56 14.4	+56 52 24	UGC 8058	13.6	1.5 x 1.0	13.9	SA(rs)c? pec	10	= **Mkn 230**; very compact nucleus.
12 59 39.6	+53 20 25	UGC 8107	13.9	2.4 x 1.0	14.7	IBm:	53	Irregular; mag 11.22v star SE 2′.9.
13 02 07.7	+58 41 53	UGC 8146	13.8	3.5 x 0.5	14.3	Sd	30	Located between a pair of N-S stars, mags 12.3 and 11.02v.
13 02 40.6	+51 46 46	UGC 8151	14.7	0.8 x 0.8	14.1	Sc		
13 07 13.6	+58 08 07	UGC 8205	14.0	1.0 x 0.6	13.3	Sb II-III	105	
13 07 28.3	+52 24 13	UGC 8211	14.6	1.1 x 0.8	14.3	Scd:	60	Mag 10.49v star SE 5′.4.
13 07 36.6	+62 12 52	UGC 8214	13.5	0.9 x 0.8	13.0	SB?	168	Mag 11.3 star SW 2′.2.
13 08 17.9	+52 00 24	UGC 8222	15.1	1.0 x 0.6	14.4	SBm	0	MCG +9-22-20 E 3′.8.
13 08 04.8	+61 34 57	UGC 8226	14.2	1.1 x 0.8	13.9	SB0/a	177	Mag 13.7 star superimposed N edge.
13 08 37.1	+54 04 28	UGC 8231	13.3	1.5 x 0.5	12.8	SB?	78	Several faint knots.
13 08 46.3	+62 16 15	UGC 8234	13.1	1.4 x 0.7	12.9	S0/a	139	UGC 8237 N 2′.4.

(Continued from previous page)

23 GALAXIES 23

RA h m s	Dec ° ′ ″	Name	Mag (V)	Dim ′ Maj x min	SB	Type Class	PA	Notes
13 08 54.2	+62 18 19	UGC 8237	13.1	1.2 x 0.9	13.0	(R′)SBb:	148	Pair with UGC 8234 S 2′.4..
13 11 51.7	+60 14 49	UGC 8282	14.1	1.5 x 0.6	13.8	Scd:	92	Mag 14.9 star on N edge.
13 15 32.6	+62 07 36	UGC 8335	13.5	1.7 x 0.7	13.5	S?	120	Double system, disrupted, bridge + streamers; pair of stars on N edge, mags 13.3 and 14.2.
13 15 50.2	+56 48 29	UGC 8339	14.6	1.5 x 0.3	13.6	Scd:	95	
13 25 29.1	+57 49 16	UGC 8441	14.2	2.6 x 1.6	15.6	Im V	66	Dwarf irregular.
13 27 43.4	+62 45 59	UGC 8467	14.0	1.0 x 0.7	13.4	SBb	21	Two main arms.
13 29 32.4	+62 49 31	UGC 8491	15.5	1.4 x 0.2	14.0	Irr	165	Mag 14.4 star S 0′.9.
13 30 45.2	+54 54 34	UGC 8508	13.8	1.5 x 0.9	14.0	IAm	120	Irregular, very knotty.
13 30 53.2	+62 13 23	UGC 8511	14.7	1.0 x 0.2	12.8	S	3	
13 34 04.8	+52 42 10	UGC 8551	15.2	1.5 x 0.3	14.2	Sb II-III	18	Mag 13.3 star S 1′.7.
13 38 59.7	+51 26 06	UGC 8639	14.4	1.5 x 1.1	14.8	Im:	170	Dwarf, low, uniform surface brightness; mag 13.0 star on E edge.
13 39 11.5	+61 30 21	UGC 8649	13.9	1.3 x 1.0	14.2	E	55	
13 40 39.9	+54 19 55	UGC 8658	12.7	2.5 x 1.5	13.9	SAB(rs)c I-II	110	Two strong dark lanes N and S edges.
13 41 24.5	+55 38 32	UGC 8671	13.6	0.9 x 0.5	12.6	S?	80	**MCG +9-22-94** N 2′.3; NGC 5278 NE 2′.7.
13 42 15.8	+60 46 34	UGC 8684	13.8	1.9 x 0.4	13.4	Scd:	33	**MCG +10-20-12** NE 8′.5.
13 46 29.7	+60 22 19	UGC 8714	15.2	1.6 x 1.0	15.5	Im	3	Two faint, round anonymous galaxies SE edge.
13 48 56.2	+59 50 07	UGC 8741	14.4	1.1 x 0.9	14.2	SB	45	
13 54 32.5	+58 23 44	UGC 8836	14.8	0.8 x 0.5	13.7	SB	138	
13 54 44.4	+53 53 58	UGC 8837	13.4	4.6 x 1.2	15.1	IB(s)m sp IV-V	18	Irregular, several small knots.
13 55 35.7	+59 37 12	UGC 8859	15.9	1.5 x 0.9	16.1	Sm	160	NGC 5376 S 7′.2.
13 57 14.7	+54 06 00	UGC 8882	15.1	1.0 x 0.7	14.6		75	
13 57 42.5	+57 00 12	UGC 8892	14.4	2.0 x 1.3	15.3	Im	66	Irregular.
13 58 38.0	+60 47 47	UGC 8909	13.8	1.5 x 0.9	14.0	SABd	60	**MCG +10-20-53** SW 4′.6; mag 9.56v star E 5′.1.
14 02 49.0	+57 58 29	UGC 8970	15.3	1.5 x 0.9	15.4	SAdm	170	Weak almost stellar nucleus; mag 9.10v star NW 2′.7.
14 03 08.8	+59 25 59	UGC 8985	15.4	1.2 x 0.6	14.9	SABdm:	129	**MCG +10-20-65** W 6′.2.
14 03 20.2	+60 59 22	UGC 8988	14.9	1.5 x 0.2	13.4	Scd:	174	
14 07 08.4	+56 32 23	UGC 9032	14.2	0.9 x 0.5	13.1	Sbc	2	
14 09 19.6	+55 16 52	UGC 9060	15.0	1.0 x 0.8	14.6	Sc	39	Mag 12.5 star N 1′.1.
14 10 03.1	+54 13 07	UGC 9071	14.2	2.0 x 0.4	13.8	Scd	37	
14 19 54.7	+51 53 38	UGC 9178	14.6	1.2 x 0.8	14.4	SB?	66	Double system, bridge; mag 10.56v star S 1′.1.
14 25 26.9	+56 19 12	UGC 9245	14.0	1.9 x 1.4	14.9	SB(s)d V	5	Elongated nucleus, patchy envelope.
14 34 37.3	+59 20 15	UGC 9391	14.3	1.3 x 1.0	14.4	SBdm V	29	**MCG +10-21-12** NNE 5′.7; **MCG +10-21-15** SE 11′.9.
14 36 22.4	+58 47 35	UGC 9412	13.6	0.7 x 0.6	12.5	S?	6	
14 38 59.5	+51 07 13	UGC 9448	14.3	1.5 x 0.2	12.9	Sb II-III	88	Located between a pair of stars, mags 11.18v and 12.4.
14 39 18.9	+53 52 31	UGC 9452	14.4	2.2 x 0.7	14.7	Sm:	141	Dwarf spiral, very low surface brightness; mag 5.82v star NW 12′.9.
14 49 46.0	+60 23 54	UGC 9556	14.5	1.8 x 0.2	13.3	SBcd?	136	Pair of mag 12 stars E 1′.0.
14 57 11.5	+52 20 43	UGC 9629	13.6	1.6 x 0.7	13.6	Sa	156	
14 57 56.5	+53 47 04	UGC 9632	14.3	1.3 x 1.0	14.4	SAd	135	Very patchy, several faint knots.
14 58 09.9	+58 52 15	UGC 9638	14.5	1.4 x 0.9	14.6	Im	40	Dwarf irregular; mag 12.4 star S 1′.2; mag 11.9 star SW 3′.1.
15 01 13.7	+52 35 44	UGC 9663	14.3	1.3 x 1.1	14.6	Im: IV-V	6	Dwarf, very low, uniform surface brightness.

GALAXY CLUSTERS

RA h m s	Dec ° ′ ″	Name	Mag 10th brightest	No. Gal	Diam ′	Notes
13 36 00.0	+59 12 00	A 1767	15.7	65	20	**MCG +10-19-96** at center, other members PGC and anonymous, stellar/almost stellar.
13 43 24.0	+55 35 00	A 1783	16.3	47	16	Contains numerous GX, on E-W line from center, listed as "unknown" IC GX in some references.
14 54 06.0	+54 18 00	A 1999	15.7	68	28	Except **UGC 9589**, all member anonymous, faint and stellar.

BRIGHT NEBULAE

RA h m s	Dec ° ′ ″	Name	Dim ′ Maj x min	Type	BC	Color	Notes
14 02 28.2	+54 16 32	NGC 5447	1.0 x 0.2	E			HII region in M 101. A double, semi-detached knot aligned NNW-SSE. It lies nearly 8′ SW of M 101's center and there is a 10th mag star off the NNE end.
14 02 27.3	+54 19 49	NGC 5449	1.0 x 1.0	E			
14 02 29.5	+54 16 14	NGC 5450	1.0 x 0.5	E			
14 02 37.0	+54 21 45	NGC 5451	0.4 x 0.4	E			
14 02 56.4	+54 18 29	NGC 5453	0.3 x 0.3	E			
14 03 01.2	+54 14 27	NGC 5455	0.5 x 0.5	E			HII region in M 101, located about 6′.5 SSW of the galaxy's center; mag about 15.0v.
14 03 12.4	+54 17 55	NGC 5458	0.5 x 0.3	E			
14 03 41.0	+54 19 02	NGC 5461	1.1 x 0.4	E			HII region in M 101; 4.5′ ESE of the galaxy's center.
14 03 53.0	+54 21 54	NGC 5462	1.5 x 0.6	E			Visually, the brightest HII region in M 101, located nearly 6′ E and slightly N of center; mag about 14.0v.
14 04 29.1	+54 23 49	NGC 5471	0.3 x 0.3	E			Compact; located 11′.6 E of the center of NGC 5457 (M 101/M 102); mag about 13.7v.

24 GALAXIES 24

RA h m s	Dec ° ′ ″	Name	Mag (V)	Dim ′ Maj x min	SB	Type Class	PA	Notes
11 19 34.3	+51 30 12	Arp's Galaxy		0.2 x 0.1			90	Almost stellar; mag 10.09v star S 4′.0.
11 14 06.2	+55 42 37	CGCG 268-6	14.3	0.7 x 0.3	12.5		30	**MCG +9-19-1** WSW 9′.9; **MCG +9-19-2** NW 10′.3.
12 48 09.4	+54 01 22	CGCG 270-24	14.1	0.6 x 0.5	12.6		78	Mag 9.13v star ESE 12′.2.
12 15 40.8	+61 53 20	CGCG 292-85	14.3	0.5 x 0.4	12.6	E?	93	Mag 12.0 star SSE 6′.0; mag 12.3 star NW 7′.3.
11 26 44.4	+59 09 18	IC 691	13.9	0.8 x 0.5	12.8	I?	141	Mag 11.10v star W 1′.9; **MCG +10-16-141** E 4′.8.
11 28 30.5	+58 33 25	IC 694	11.4	0.9 x 0.5	10.4	SBm? pec	0	In contact with NGC 3690, see NGC 3690 notes.
12 04 11.8	+62 30 04	IC 758	13.3	1.9 x 1.6	14.4	SB(rs)cd:	9	Mag 9.40v star W 7′.0.
12 33 45.2	+52 15 13	IC 801	13.6	1.1 x 0.7	13.2	Sa?	58	Mag 11.9 star on S edge.
12 51 16.5	+53 41 44	IC 830	13.6	1.1 x 0.6	13.0	S?	165	Mag 10.57v star W 10′.7.
10 58 23.0	+55 37 03	MCG +9-18-59	14.8	0.8 x 0.4	13.4	Sc	66	UGC 6059 ESE 3′.1.
10 58 53.5	+55 38 05	MCG +9-18-63	14.4	0.6 x 0.6	13.2			UGC 6059 SSW 2′.6.
11 02 07.5	+54 27 13	MCG +9-18-70	15.5	0.6 x 0.5	14.1	Sb	138	
11 02 38.7	+51 10 31	MCG +9-18-72	14.8	0.9 x 0.7	14.2	Sbc	27	
11 06 38.6	+53 22 01	MCG +9-18-84	14.7	1.3 x 0.5	14.1	Sb	177	
11 10 01.5	+52 23 14	MCG +9-18-95	15.4	0.7 x 0.5	14.1	Sc	15	Located between a pair of mag 12.1, 13.2 stars.
11 12 21.9	+54 01 38	MCG +9-18-102	14.5	0.9 x 0.8	14.0	II		**MCG +9-18-99** W 6′.2.
11 13 13.1	+55 33 18	MCG +9-19-3	14.6	0.6 x 0.6	13.4			Mag 7.45v star SSE 8′.9.
11 15 24.0	+54 26 37	MCG +9-19-16	16.5	0.7 x 0.6	15.6	E/S0	171	

RA h m s	Dec ° ′ ″	Name	Mag (V)	Dim ′ Maj x min	SB	Type Class	PA	Notes
11 15 29.2	+55 27 22	MCG +9-19-18	14.3	0.9 x 0.6	13.5	SBb	39	
11 18 21.0	+53 45 10	MCG +9-19-35	15.6	0.4 x 0.4	13.5	SBb		= **Mkn 39**; **UGC 6315** on W edge.
11 19 52.8	+54 27 45	MCG +9-19-42	14.3	0.9 x 0.6	13.5	Scd:	174	Mag 13.9 star N 0′.9; almost stellar anonymous galaxy, or star, W 0′.9.
11 20 39.5	+52 37 26	MCG +9-19-44	14.3	1.1 x 0.4	13.3	Sb	95	
11 21 57.2	+52 57 44	MCG +9-19-49	13.8	0.8 x 0.6	13.1	E	156	
11 22 25.5	+53 41 18	MCG +9-19-52	14.8	0.9 x 0.3	13.3		6	Located in galaxy cluster A 1225, many anonymous galaxies W and NW.
11 22 32.5	+51 31 55	MCG +9-19-53	14.5	0.6 x 0.6	13.5	E/S0		
11 23 08.5	+51 33 26	MCG +9-19-57	15.3	0.8 x 0.3	13.6	Sc	21	
11 23 39.5	+53 49 47	MCG +9-19-64	14.1	0.4 x 0.2	11.2		90	On S edge of NGC 3656.
11 24 24.6	+51 14 04	MCG +9-19-68	14.4	0.9 x 0.8	13.9	Sbc	9	
11 25 34.9	+52 32 44	MCG +9-19-72	14.7	0.8 x 0.5	13.7	E/S0	99	
11 25 35.2	+54 23 13	MCG +9-19-73	15.2	1.1 x 0.2	13.5	S0 pec	153	Mag 8.53v star N 1′.3.
11 26 10.6	+52 07 30	MCG +9-19-75	14.0	0.9 x 0.9	13.8	E/S0		Mag 7.31v star W 2′.4.
11 30 31.8	+51 27 19	MCG +9-19-97	14.1	0.9 x 0.7	13.5	Sb	132	
11 32 23.6	+54 58 57	MCG +9-19-107	14.7	0.5 x 0.4	12.8	Sb	105	
11 32 34.3	+54 55 38	MCG +9-19-112	13.8	1.0 x 0.4	12.9	E	30	Mag 13.6 star NW 0′.9.
11 35 04.3	+54 39 33	MCG +9-19-125	14.2	0.6 x 0.6	13.0	Sc		Mag 5.83v star N 7′.6; mag 11.17v star SW 3′.0.
11 35 31.8	+54 55 49	MCG +9-19-126	14.8	0.7 x 0.3	13.2	E/S0	150	NGC 3737 NNE 1′.3; several almost stellar anonymous galaxies NW.
11 37 04.9	+55 34 52	MCG +9-19-137	16.5	0.7 x 0.5	15.2		126	
11 39 10.8	+55 39 53	MCG +9-19-148	14.1	1.3 x 0.7	13.8		90	Double galaxy system; mag 12.3 star E edge.
11 39 31.2	+55 09 32	MCG +9-19-149	14.0	0.9 x 0.4	12.7	S?	54	Mag 10.88v star SE 3′.5.
11 43 46.4	+55 02 50	MCG +9-19-165	14.3	0.7 x 0.6	13.2	SB?	171	= **Mkn 1452**; NGC 3846A E 4′.2.
11 44 20.3	+51 01 09	MCG +9-19-167	14.5	0.7 x 0.3	12.7		165	Mag 9.64v star N 4′.5.
11 44 43.5	+54 27 42	MCG +9-19-172	14.7	0.8 x 0.6	13.7		150	Mag 13.5 star N edge; large knot or small galaxy E edge.
11 45 51.8	+56 05 14	MCG +9-19-176	14.9	0.9 x 0.4	13.6	Spiral	126	Mag 10.04v star E 4′.6.
11 48 15.1	+55 36 57	MCG +9-19-196	14.8	0.5 x 0.4	12.9	S	42	Mag 12.2 star NW 0′.9.
11 54 17.5	+55 28 13	MCG +9-20-27	15.1	0.7 x 0.5	13.8	Sbc	174	Mag 10.69v star SW 3′.4.
11 58 58.8	+54 14 17	MCG +9-20-49	13.9	0.9 x 0.7	13.2	S?	123	Mag 11.20v star SW 4′.9.
12 02 11.2	+56 21 53	MCG +9-20-68	14.5	1.1 x 0.3	13.1		9	On NE boundary of galaxy cluster A 1436; **MCG +9-20-64** W 9′.3.
12 03 32.2	+51 04 39	MCG +9-20-77	15.5	0.9 x 0.2	13.5	Sc	78	
12 04 20.9	+56 20 46	MCG +9-20-80	14.7	0.6 x 0.5	13.2		69	Mag 14.4 star NE edge.
12 05 44.0	+53 11 02	MCG +9-20-91	14.9	0.9 x 0.4	13.6	Sb	108	Mag 11.3 star ESE 4′.2.
12 06 53.4	+55 10 08	MCG +9-20-96	14.7	0.7 x 0.3	12.9	Spiral	165	
12 07 09.3	+53 40 15	MCG +9-20-97	14.4	0.8 x 0.5	13.5	E/S0	27	Mag 14.9 star on Edge; mag 12.0 star E 1′.7.
12 09 50.5	+54 00 04	MCG +9-20-103	15.0	0.8 x 0.4	13.6		102	Mag 12.8 star and small, faint anonymous galaxy ENE 1′.3.
12 11 31.3	+51 51 46	MCG +9-20-108	14.9	1.1 x 0.3	13.5	Sb	114	Close pair of E-W mag 10-11 stars W 4′.6.
12 15 44.9	+54 51 25	MCG +9-20-130	14.7	1.1 x 0.6	14.1	Sbc	12	Mag 9.08v star ESE 5′.8.
12 15 47.1	+52 23 16	MCG +9-20-131	15.8	0.9 x 0.6	15.0		129	
12 17 22.7	+53 33 23	MCG +9-20-138	14.8	0.7 x 0.4	13.3		105	Mag 13.1 star W 1′.9.
12 20 35.6	+52 15 24	MCG +9-20-151	14.9	1.1 x 0.3	13.5	Sbc	84	Mag 12.2 star S 3′.2.
12 21 08.1	+51 53 46	MCG +9-20-153	14.5	0.6 x 0.4	12.8		102	Anonymous galaxy with superimposed star S 0′.8.
12 21 41.5	+55 47 12	MCG +9-20-155	14.5	0.9 x 0.8	14.0		144	Mag 10.98v star W 2′.8.
12 21 53.9	+51 02 30	MCG +9-20-156	14.6	0.8 x 0.7	13.8	Sbc	126	Close pair of mag 10 stars SW 6′.9.
12 22 07.0	+55 49 10	MCG +9-20-157	14.5	0.9 x 0.6	13.7		159	MCG +9-20-155 SW 4′.1.
12 25 45.4	+51 40 05	MCG +9-20-172	14.0	1.1 x 0.7	13.6		66	
12 26 03.8	+55 30 42	MCG +9-20-174	14.6	0.7 x 0.4	13.1	Sbc	120	Mag 9.05v star S 2′.2.
12 26 29.8	+54 10 46	MCG +9-20-176	14.4	1.1 x 0.3	13.0		126	
12 26 55.4	+52 04 30	MCG +9-20-181	14.5	0.6 x 0.4	12.8		18	Mag 11.31v star E 1′.5.
12 26 58.6	+53 41 00	MCG +9-20-183	14.8	0.8 x 0.5	13.7		12	Mag 13.3 star W 0′.5.
12 28 31.4	+53 35 46	MCG +9-21-1	14.3	1.0 x 0.7	13.8		162	Bright knot or companion galaxy N edge.
12 30 57.9	+51 20 36	MCG +9-21-8	14.8	0.6 x 0.3	12.8		66	Pair with **MCG +9-21-7** W edge.
12 30 58.9	+51 36 32	MCG +9-21-9	15.9	0.8 x 0.3	14.2	Sbc	24	
12 31 52.1	+52 23 54	MCG +9-21-11	14.3	0.8 x 0.7	13.5	S?	177	Mag 13.0 star W 1′.2.
12 42 31.5	+53 12 17	MCG +9-21-29	14.9	0.8 x 0.4	13.6	Sbc	21	Mag 13.4 star on SE edge.
12 42 53.2	+55 08 39	MCG +9-21-32	15.2	1.1 x 0.2	13.4	S?	135	Located 1′.5 E of NGC 4644.
12 45 45.1	+54 50 33	MCG +9-21-41	15.2	1.1 x 0.3	13.9		27	Forms a triangle with mag 12.2 star NE 2′.6 and mag 12.4 star SE 2′.6.
12 56 26.9	+51 05 04	MCG +9-21-68	14.6	0.8 x 0.3	12.9		78	Mag 9.43v star SE 0′.7; mag 11.1 star and **MCG +9-21-66** NW 2′.8.
12 57 45.0	+53 38 11	MCG +9-21-70	14.8	0.8 x 0.5	13.6		63	
12 58 38.5	+53 24 33	MCG +9-21-72	15.0	0.7 x 0.6	13.9	SBbc	33	Mag 9.63v star NW 2′.9.
12 59 03.7	+55 57 56	MCG +9-21-76	15.2	0.7 x 0.3	13.4	Sbc	150	Mag 12.7 star NE 2′.5.
13 01 11.3	+53 48 39	MCG +9-21-83	15.2	0.5 x 0.3	13.0	Sb	3	Mag 11.3 star on S edge; **MCG +9-21-85** SSE 1′.1.
13 01 17.2	+53 51 37	MCG +9-21-84	14.2	0.9 x 0.7	13.5	SBbc	111	Mag 10.88v star ESE 1′.6.
13 02 25.1	+52 12 21	MCG +9-21-87	15.7	0.8 x 0.3	14.0	Sc	156	Mag 9.61v star NNE 2′.8.
13 03 27.7	+53 31 55	MCG +9-21-94	14.8	0.9 x 0.5	13.8	Sbc	33	Mag 14.1 star E 0′.6.
11 00 07.6	+57 25 43	MCG +10-16-41	14.5	0.7 x 0.5	13.3		51	Mag 13.3 star W 0′.9.
11 01 00.4	+57 47 02	MCG +10-16-44	14.5	0.7 x 0.4	13.0	Sbc	126	Mag 10.65v star NW 4′.9; **MCG +10-16-43** SW 0′.9.
11 04 14.7	+57 24 07	MCG +10-16-49	14.2	0.8 x 0.5	13.1		120	Mag 8.44v star SW 4′.4.
11 05 02.0	+59 40 58	MCG +10-16-52	14.6	0.7 x 0.4	13.1		81	Mag 13.1 star E 1′.8; MCG +10-16-52 S 1′.1; **MCG +10-16-53** SE 2′.7.
11 06 49.5	+57 41 04	MCG +10-16-61	14.2	0.9 x 0.7	13.5	SB(s)c pec	39	Mag 9.89v star WSW 3′.8.
11 10 36.2	+58 46 04	MCG +10-16-71	15.0	0.5 x 0.4	13.1	Sb	117	Pair of mag 13 stars SW 1′.6.
11 11 05.3	+57 56 51	MCG +10-16-74	14.4	0.8 x 0.7	13.6	Sbc	3	Located between a pair of E-W mag 13.5 and 14.2 stars.
11 11 15.3	+62 08 26	MCG +10-16-77	14.6	0.9 x 0.3	13.0	Sb	108	Mag 12.2 star N 2′.1.
11 12 16.7	+57 04 32	MCG +10-16-81	14.4	0.5 x 0.4	12.5		111	Mag 12.1 star SW 2′.0.
11 14 11.3	+56 47 04	MCG +10-16-89	14.5	0.7 x 0.5	13.2	SB(r)bc:	63	
11 14 23.1	+62 04 16	MCG +10-16-90	14.4	0.8 x 0.5	13.2		51	Located between a pair of mag 12.5 and 13.6 stars.
11 14 41.9	+57 49 31	MCG +10-16-92	14.2	0.9 x 0.6	13.4		99	
11 18 48.4	+58 36 29	MCG +10-16-110	14.7	0.8 x 0.3	13.0	Sbc	171	
11 20 05.9	+58 06 20	MCG +10-16-118	17.1	1.0 x 0.7	16.5		150	Mag 13.8 star SSE 1′.3; mag 10.01v star NNW 3′.7.
11 20 58.7	+56 55 44	MCG +10-16-122	16.8	0.6 x 0.2	14.3	Sbc	150	MCG +10-16-123 NNW 1′.1.
11 20 54.0	+56 56 37	MCG +10-16-123	13.6	0.7 x 0.7	12.9	E/S0		MCG +10-16-122 SSE 1′.1.
11 25 29.4	+57 01 10	MCG +10-16-134	14.6	0.9 x 0.8	14.1	SBb		NGC 3674 E 7′.9.
11 28 41.2	+57 50 04	MCG +10-17-4	14.9	0.7 x 0.5	13.6	Sb	0	Strong E-W bar; close pair of mag 13 stars NE 1′.2.
11 30 17.7	+58 08 01	MCG +10-17-7	14.1	0.8 x 0.8	13.4	SA(s)m:		Mag 13.2 star NE 1′.6.
11 32 04.8	+57 26 23	MCG +10-17-10	11.9	1.9 x 1.4	12.8	SAB(rs)bc I-II	81	Patchy.
11 35 24.9	+57 38 56	MCG +10-17-19	13.9	0.8 x 0.5	12.8		105	**MCG +10-17-16** W 1′.8; **MCG +10-17-18** NNW 2′.0.
11 44 33.0	+61 31 58	MCG +10-17-57	15.1	0.7 x 0.5	13.8	Sbc	54	Mag 13.8 star SW 1′.7 with stellar anonymous galaxy half way in between.
11 44 52.1	+57 52 24	MCG +10-17-58	15.7	0.8 x 0.4	14.3		54	Mag 9.78v star SSE 4′.6; NGC 3838 NW 6′.8.
11 46 35.4	+61 44 59	MCG +10-17-64	13.9	1.0 x 0.6	13.2		81	
11 48 23.4	+59 25 45	MCG +10-17-75	15.0	0.8 x 0.3	13.3		87	NGC 3894 3′.5 E.

(Continued from previous page)

GALAXIES

RA h m s	Dec ° ′ ″	Name	Mag (V)	Dim ′ Maj x min	SB	Type Class	PA	Notes
11 48 50.3	+60 11 40	MCG +10-17-79	14.3	0.6 x 0.5	12.8		117	Mag 12.7 star SW 0′.8.
11 53 29.3	+59 41 52	MCG +10-17-97	15.1	0.7 x 0.5	13.8		66	Mag 8.20v star ESE 3′.9.
11 58 48.2	+60 03 21	MCG +10-17-120	15.1	0.8 x 0.2	13.0	Sbc	90	Mag 9.89v star S 2′.0.
11 59 05.6	+58 20 34	MCG +10-17-121	14.9	0.6 x 0.5	13.5	Sb	108	Mag 10.31v star N 3′.9.
12 06 18.9	+57 59 32	MCG +10-17-141	15.1	0.8 x 0.6	14.2	Sb	138	
12 07 12.6	+60 01 46	MCG +10-17-147	15.0	0.6 x 0.3	12.9		27	Mag 9.82v star E 3′.0.
12 12 35.1	+58 26 34	MCG +10-18-6	14.8	0.8 x 0.5	13.6		57	Mag 12.1 star on S edge.
12 18 56.6	+60 10 06	MCG +10-18-21	15.2	0.7 x 0.5	13.9		153	Mag 11.6 star NW 1′.8.
12 20 18.3	+57 52 03	MCG +10-18-27	14.4	0.8 x 0.7	13.7		141	Located 4′.3 W of mag 5.52v star 70 Ursae Majoris.
12 20 46.5	+57 47 31	MCG +10-18-28	15.1	0.6 x 0.5	13.6		171	Located 4′.3 S of mag 5.52v star 70 Ursae Majoris.
12 24 54.5	+61 03 47	MCG +10-18-44	16.6	0.7 x 0.5	15.3	Dwarf Irr	135	Mag 12.8 star NW 1′.5.
12 52 44.4	+59 15 53	MCG +10-18-87	15.1	0.8 x 0.6	14.1	Sc	90	Mag 14.0 star S 1′.7.
12 54 22.6	+58 53 42	MCG +10-18-88	14.9	0.7 x 0.5	13.6			Mag 15.1 star on SW edge; MCG +10-19-1 E 3′.6.
12 54 36.2	+62 13 41	MCG +10-18-89	14.1	1.0 x 0.7	13.8	E/S0	60	Mag 11.18v star SW 5′.3.
12 54 50.1	+58 52 55	MCG +10-19-1	14.9	1.1 x 0.9	14.8		48	MCG +10-18-88 W 3′.6.
13 01 13.9	+58 06 41	MCG +10-19-14	15.0	0.6 x 0.4	13.3		159	
11 36 06.6	+62 14 55	MCG +11-14-33	13.8	0.9 x 0.5	12.8	Sb	3	
12 25 53.3	+62 25 42	MCG +11-15-54	14.3	1.0 x 0.5	13.4	Sbc	60	Mag 10.91v star S 1′.6.
10 56 01.8	+57 06 58	NGC 3458	12.3	1.4 x 0.9	12.4	SAB0:	5	
10 58 44.9	+59 30 35	NGC 3470	13.2	1.4 x 1.2	13.6	SA(r)ab:	170	Faint, elongated anonymous galaxy S 1′.5.
10 59 09.1	+61 31 47	NGC 3471	12.5	1.7 x 0.8	12.7	Sa	14	**MCG +10-16-40** NE 2′.3.
11 01 23.3	+57 40 37	NGC 3488	12.9	1.6 x 1.0	13.2	SB(s)c:	175	Mag 13.3 star on S edge.
11 03 11.0	+56 13 16	NGC 3499	13.6	0.9 x 0.7	12.9	I0?	6	Located 14′.8 SE of mag 2.35v star Merak, Beta Ursea Majoris.
11 05 36.7	+56 31 28	NGC 3517	13.0	1.0 x 0.8	12.6	SAb:	120	**MCG +10-16-55** on N edge, elongated N-S.
11 08 40.5	+57 13 47	NGC 3530	13.9	0.7 x 0.3	12.1	S?	99	
11 10 56.9	+61 20 47	NGC 3543	14.3	1.3 x 0.3	13.2	S	8	Mag 7.38v star SE 12′.4.
11 10 56.0	+53 23 07	NGC 3549	12.1	3.2 x 1.2	13.4	SA(s)c: II-III	38	
11 11 29.6	+55 40 22	NGC 3556	10.0	8.7 x 2.2	13.1	SB(s)cd sp III-IV	80	= **M 108**.
11 15 13.1	+60 42 01	NGC 3589	13.8	1.5 x 0.8	13.8	Sd:	48	Mag 8.30v star N 4′.0.
11 16 14.3	+55 42 15	NGC 3594	13.7	1.3 x 1.1	14.0	SB0:	10	Mag 10.61v star NE 2′.6; **MCG +9-19-21** SW 2′.3.
11 18 25.4	+58 47 11	NGC 3610	10.8	2.7 x 2.3	12.8	E5:	144	Mag 7.94v star NE 8′.4.
11 18 36.7	+57 59 59	NGC 3613	10.9	3.9 x 1.9	13.1	E6	102	**MCG +10-16-111** and **MCG +10-16-112** NE 4′.6.
11 19 21.4	+57 45 27	NGC 3619	11.5	2.7 x 2.3	13.4	(R)SA(s)0$^+$:	57	
11 20 31.4	+57 46 49	NGC 3625	13.1	2.0 x 0.6	13.1	SAB(s)b:	148	**UGC 6344** SW 3′.3.
11 21 02.5	+53 10 08	NGC 3631	10.4	5.0 x 3.7	13.4	SA(s)c I-II	114	Numerous knots in multi-branching arms.
11 22 18.2	+59 04 27	NGC 3642	11.2	5.4 x 4.5	14.5	SA(r)bc: I	105	Several superimposed stars, mostly along W edge.
11 23 38.5	+53 50 31	NGC 3656	12.5	1.6 x 1.6	13.4	(R′)I0: pec		MCG +9-19-64 on S edge; star on W edge; anonymous galaxy E 2′.3.
11 23 55.6	+52 55 15	NGC 3657	12.4	1.1 x 0.7	12.0	SAB(rs)c pec	168	Compact.
11 25 26.3	+57 43 24	NGC 3669	12.4	2.2 x 0.5	12.4	SBcd: sp	153	
11 25 52.6	+60 28 44	NGC 3671	14.8	0.4 x 0.4	12.7			Almost stellar; mag 14.7 star on E edge.
11 26 26.6	+57 02 54	NGC 3674	12.3	1.9 x 0.6	12.3	S0 sp	33	MCG +10-16-134 W 7′.9.
11 27 31.8	+56 52 39	NGC 3683	12.5	1.9 x 0.7	12.6	SB(s)c? III-IV	128	Mag 10.80v star SSE 7′.9.
11 29 11.9	+57 07 59	NGC 3683A	11.9	2.3 x 1.7	13.2	SB(rs)c	75	Mag 11.0 star NE 1′.7.
11 28 33.3	+58 33 53	NGC 3690	11.4	3.2 x 2.1	13.3	IBm pec	152	Disrupted double system in contact with IC 694. Two very complex nuclei separated by 0′.3. **Arp 296** lies NE 2′.4.
11 32 35.7	+53 04 17	NGC 3718	10.8	8.1 x 4.0	14.4	SB(s)a pec	6	Very thin lane runs SE-NW across center; very faint extensions N and S of bright center.
11 33 40.8	+61 53 13	NGC 3725	13.0	1.2 x 0.9	12.9	SBc	145	Mag 9.75v star E 8′.6.
11 33 49.5	+53 07 35	NGC 3729	11.4	2.8 x 1.9	13.1	SB(r)a pec	165	Faint star just inside SW edge.
11 35 02.3	+54 51 10	NGC 3733	12.4	4.8 x 2.2	14.8	SAB(s)cd:	170	4′.0 N of a mag 5.63v star.
11 35 36.6	+54 56 51	NGC 3737	12.9	0.8 x 0.7	12.1	SB0	81	Two small, anonymous galaxies NW 1′.7.
11 35 48.3	+54 31 33	NGC 3738	11.7	2.5 x 1.9	13.3	Im III-IV	155	Located 2′.4 SW of mag 10.53v star.
11 36 12.7	+59 58 30	NGC 3740	14.0	0.9 x 0.4	12.7	S	110	
11 36 47.9	+54 17 50	NGC 3756	11.5	4.2 x 2.1	13.7	SAB(rs)bc II	177	Mag 10.55v star 3′.7 NNW.
11 37 03.1	+58 24 53	NGC 3757	12.6	1.1 x 1.1	12.6	S0?		Mag 9.75v star ENE 4′.2.
11 36 54.2	+54 49 20	NGC 3759	13.3	1.1 x 1.1	13.3	SAB0°:		**IC 2943** NW 2′.2.
11 36 58.3	+55 09 44	NGC 3759A	13.6	1.2 x 1.0	13.6	SABc pec	123	Faint, elongated anonymous galaxy S 1′.7.
11 37 23.9	+61 45 31	NGC 3762	12.6	1.9 x 0.5	12.4	Sa	167	
11 37 58.6	+59 37 00	NGC 3770	12.9	1.0 x 0.7	12.4	SBa	107	
11 39 21.8	+56 16 19	NGC 3780	11.5	3.1 x 2.5	13.5	SA(s)c: II-III	90	Numerous knots and/or superimposed stars.
11 40 07.0	+58 36 47	NGC 3795	13.2	2.1 x 0.5	13.1	S	53	
11 38 08.7	+58 45 27	NGC 3795B	12.9	0.9 x 0.9	12.8	E?		Mag 9.24v star NW 3′.7.
11 40 30.9	+60 17 57	NGC 3796	12.6	1.3 x 0.9	12.7	S?	127	
11 40 54.3	+56 12 05	NGC 3804	12.9	2.2 x 1.4	14.0	SAB(s)d	120	Several knots; mag 13.1 star on W edge.
11 41 16.1	+59 53 08	NGC 3809	12.7	1.0 x 0.8	12.4	S0	123	
11 42 45.1	+52 46 47	NGC 3824	13.6	1.3 x 0.7	13.3	SA(s)a? sp	118	
11 43 27.3	+52 42 41	NGC 3829	14.0	1.0 x 0.6	13.3	SB(s)b:	113	
11 44 04.9	+60 07 09	NGC 3835	12.4	1.9 x 0.8	12.7	Sab: sp	60	Mag 7.68v star SE 6′.6.
11 44 13.9	+57 56 52	NGC 3838	12.3	1.5 x 0.6	12.1	SA0/a?	141	MCG +10-17-58 SE 6′.8.
11 44 29.4	+55 39 06	NGC 3846	13.9	1.0 x 0.7	13.3	SAc:	135	Elongated anonymous galaxy SW 1′.3.
11 44 15.7	+55 02 10	NGC 3846A	13.2	1.9 x 1.5	14.2	SB(s)m:	40	Faint envelope extends NE of bright center; MCG +9-19-165 W 4′.2.
11 45 35.5	+55 53 11	NGC 3850	13.3	2.2 x 1.1	14.1	SB(s)c:	118	Faint, slightly elongated anonymous galaxy S 5′.4.
11 47 33.9	+55 58 00	NGC 3888	12.1	1.7 x 1.3	12.8	SAB(rs)c II	120	Stellar **MCG +9-19-183** N 4′.6.
11 47 48.5	+56 01 04	NGC 3889	14.8	0.5 x 0.3	12.6	S0	126	Faint anonymous galaxy S 1′.8.
11 48 50.1	+59 24 54	NGC 3894	11.6	2.8 x 1.7	13.3	E4-5	20	MCG +10-17-75 W 3′.5.
11 49 04.2	+59 25 57	NGC 3895	13.1	1.3 x 1.0	13.2	SB(rs)a:	116	NGC 3894 SW edge.
11 49 15.8	+56 04 59	NGC 3898	10.7	4.4 x 2.6	13.2	SA(s)ab I-II	107	**MCG +9-19-206** NE 4′.9.
11 50 39.1	+55 21 15	NGC 3913	12.6	2.6 x 2.6	14.5	(R′)SA(rs)d: III		Compact nucleus, faint, smooth envelope.
11 50 51.2	+55 08 37	NGC 3916	13.9	1.6 x 0.4	13.3	SAb: sp II-III	45	
11 50 45.7	+51 49 28	NGC 3917	11.8	5.1 x 1.3	13.7	SAcd: III-IV	77	Elongated UGC 6802 WNW 6′.3.
11 51 06.9	+55 04 43	NGC 3921	12.4	2.1 x 1.3	13.3	(R′)SA(s)0/a pec	20	Anonymous galaxy N 2′.4; another anonymous galaxy W edge; **MCG +9-19-213** W 5′.0.
11 51 13.7	+52 00 04	NGC 3931	13.4	1.1 x 0.9	13.2	SA0$^-$:	160	Mag 8.89v star E 5′.0; **MCG +9-20-18** E 8′.8.
11 53 13.6	+60 40 32	NGC 3945	10.9	5.2 x 3.5	13.9	(R)SB(rs)0$^+$	159	Several stars superimposed on faint halo.
11 53 48.7	+52 19 24	NGC 3953	10.1	6.9 x 3.5	13.4	SB(r)bc I-II	13	Numerous knots and clumps; **MCG +9-20-23** SW 10′.2.
11 54 33.6	+58 22 00	NGC 3958	13.0	1.5 x 0.7	12.9	SB(s)a	28	
11 54 58.7	+58 29 28	NGC 3963	11.9	2.8 x 2.5	13.9	SAB(rs)bc I-II	36	
11 55 45.6	+55 19 11	NGC 3972	12.3	3.9 x 1.1	13.8	SA(s)bc	120	Stellar anonymous galaxy WNW 4′.7.
11 55 53.8	+60 31 45	NGC 3975	15.3	0.7 x 0.3	13.5	SAb?	92	Located 2′.0 W of NGC 3978.
11 56 07.3	+55 23 27	NGC 3977	13.4	1.7 x 1.5	14.3	(R)SA(rs)ab:	9	Compact core in smooth, faint envelope.
11 56 10.4	+60 31 16	NGC 3978	12.7	1.6 x 1.5	13.5	SABbc: II-III	27	Mag 8.36v star E 4′.2; stellar **MCG +10-17-108** ESE 3′.9; **MCG +10-17-106** S 6′.7.

RA h m s	Dec ° ′ ″	Name	Mag (V)	Dim ′ Maj x min	SB	Type Class	PA	Notes
11 56 28.1	+55 07 29	NGC 3982	11.0	2.3 x 2.0	12.5	SAB(r)b: III	48	
11 57 35.7	+55 27 27	NGC 3990	12.5	1.4 x 0.8	12.5	S0⁻: sp	40	NGC 3998 E 3′.0.
11 57 35.7	+53 22 28	NGC 3992	9.8	7.6 x 4.6	13.5	SB(rs)bc I	68	= **M 109**. Very bright, diffuse nucleus, bright smooth bar with dark lane 1′.7 X 0′.5, three main filamentary arms with some branching.
11 57 56.5	+55 27 05	NGC 3998	10.7	2.7 x 2.2	12.5	SA(r)0°?	140	Mag 9.74v star NNW 5′.8; NGC 3990 W 3′.0.
12 01 27.6	+61 53 47	NGC 4036	10.7	4.3 x 1.7	12.7	S0⁻	85	
12 02 11.4	+62 08 12	NGC 4041	11.3	2.7 x 2.5	13.2	SA(rs)bc: II-III	72	**MCG +10-17-132** E 7′.5.
12 03 13.4	+57 53 32	NGC 4054	15.0	0.8 x 0.7	14.2		144	Triple system.
12 04 02.7	+52 35 26	NGC 4068	12.4	2.5 x 1.6	13.8	IAm	30	Bright clump S edge.
12 06 23.3	+52 42 39	NGC 4102	11.2	3.0 x 1.7	12.8	SAB(s)b? II	38	
12 09 47.8	+58 50 59	NGC 4141	13.7	1.3 x 0.9	13.7	SBcd:	75	
12 09 30.4	+53 06 15	NGC 4142	13.3	2.2 x 1.2	14.2	SB(s)d:	175	
12 10 33.7	+58 18 09	NGC 4149	13.3	1.3 x 0.3	12.1	S?	84	**MCG +10-18-3** ESE 9′.0.
12 11 33.4	+57 44 15	NGC 4161	13.0	1.1 x 0.7	12.6	S?	50	
12 12 14.7	+56 10 37	NGC 4172	13.3	1.3 x 1.1	13.6	S?	6	Star or bright knot S edge of core.
12 12 48.9	+52 54 13	NGC 4181	14.0	0.8 x 0.6	13.3	E	11	A mag 10.94v star 1′.5 S.
12 14 09.9	+54 31 39	NGC 4194	12.5	2.7 x 1.6	13.9	IBm pec	168	Very bright nucleus with much fainter extensions N and S.
12 14 18.1	+59 36 54	NGC 4195	14.2	1.6 x 1.4	14.9	SB(s)cd	117	Mag 8.26v star WNW 11′.4.
12 14 21.8	+56 00 44	NGC 4198	13.6	1.2 x 0.7	13.2	S0/a	130	Nucleus offset to SE.
12 14 48.7	+59 54 20	NGC 4199	14.3	0.5 x 0.5	12.6	Sb		Stellar; pair of faint stars on N edge; located inside W edge of galaxy cluster A 1507.
12 19 32.5	+56 44 11	NGC 4271	12.6	1.5 x 1.3	13.1	S0⁻:	55	Mag 9.72v star NE 7′.3.
12 20 12.8	+58 05 35	NGC 4284	13.5	2.5 x 1.2	14.5	Sbc	102	Mag 13.1 star E edge.
12 20 47.5	+58 05 26	NGC 4290	11.8	2.3 x 1.6	13.0	SB(rs)ab: II	90	**MCG +10-18-32** E 3′.7.
12 23 01.9	+58 26 40	NGC 4335	12.4	1.9 x 1.5	13.5	E	145	Bright nucleus with star or knot on W edge; **MCG +10-18-36** E 4′.0.
12 23 57.6	+58 22 49	NGC 4358	15.7	0.3 x 0.1	11.8	S0	42	
12 24 11.4	+58 21 36	NGC 4362	14.4	1.1 x 0.4	13.4	S?	39	
12 24 02.2	+58 23 06	NGC 4364	13.8	1.1 x 1.1	13.9	S0		Double system with NGC 4358, which is located on SW edge.
12 25 12.4	+54 30 23	NGC 4384	13.0	1.3 x 1.0	13.2	Sa	90	
12 31 22.0	+57 57 53	NGC 4500	12.5	1.6 x 1.0	12.9	SB(s)a	130	Bright nucleus on bright lens; two smooth arms encircle the center. Mag 10.66v star on E edge.
12 32 08.5	+56 28 11	NGC 4511	14.1	1.1 x 0.5	13.3	Sbc	9	
12 34 54.7	+58 54 43	NGC 4547	14.5	0.5 x 0.4	12.8	E/S0	96	Double system with NGC 4549, in common halo.
12 34 51.9	+58 55 00	NGC 4549	15.5	1.1 x 0.9	15.4		63	**MCG +10-18-68** NW 3′.1. Companion NGC 4547 on SE edge.
12 36 00.3	+54 13 14	NGC 4566	13.1	1.3 x 0.9	13.1	S	80	
12 40 01.0	+61 36 33	NGC 4605	10.3	5.8 x 2.2	12.9	SB(s)c pec III-IV	125	Bright, patchy center; several knots SE.
12 42 42.7	+55 08 40	NGC 4644	13.9	1.6 x 0.5	13.5	SBb:	53	MCG +9-21-32 E 1′.5.
12 42 52.2	+54 51 18	NGC 4646	13.4	1.0 x 0.8	13.0	S0?	12	Very faint star at NE end of bright, elongated center.
12 43 19.8	+58 57 50	NGC 4652	14.6	1.0 x 0.3	13.2	Sb	39	Close pair of mag 7.86v, 9.62v stars SSE 4′.8.
12 44 46.9	+54 52 33	NGC 4669	13.2	1.8 x 0.6	13.1	S	177	Bright elongated center with faint dark lane down middle; mag 9.74v star E 3′.0.
12 45 31.8	+54 44 11	NGC 4675	14.4	1.6 x 0.6	14.2	SBb:	97	Bright E-W bar; mag 10.88v star SE 3′.9.
12 46 39.8	+54 32 01	NGC 4686	12.6	2.0 x 0.6	12.6	Sa	3	
12 47 31.8	+54 22 23	NGC 4695	13.5	1.1 x 0.7	13.1	S	80	Very faint, very small anonymous galaxy 0′.5 SSE.
12 48 24.3	+51 09 42	NGC 4707	12.9	2.2 x 2.1	14.4	Sm: V	25	Stellar nucleus; several stars superimposed.
12 50 07.0	+52 51 00	NGC 4732	13.9	1.2 x 0.6	13.5	E	8	
12 54 37.7	+53 05 24	NGC 4801	14.2	0.8 x 0.6	13.2		138	Stellar anonymous galaxy and adjacent mag 13.9 star NNW 6′.2.
12 55 22.1	+58 20 38	NGC 4814	12.0	3.1 x 2.3	14.0	SA(s)b I-II	129	Bright core; two main filamentary arms; dark lanes.
12 56 25.2	+52 17 46	NGC 4834	14.8	0.9 x 0.3	13.2	Sbc	108	UGC 8050 W 9′.4.
13 02 57.7	+51 19 07	NGC 4938	14.3	0.8 x 0.7	13.5	S?	39	
10 58 43.8	+55 35 51	UGC 6059	14.4	1.5 x 0.7	14.3	SAB(s)bc	15	MCG +9-18-63 NE 2′.6.
11 00 39.6	+61 19 17	UGC 6080	14.9	1.9 x 0.2	13.7	Sd	129	
11 02 37.9	+59 07 32	UGC 6110	14.1	1.5 x 0.5	13.6		65	
11 06 54.8	+51 12 11	UGC 6162	13.2	2.3 x 1.1	14.1	Sd	88	Very knotty; mag 12.2 star on SE edge.
11 08 03.0	+53 36 59	UGC 6182	13.2	0.9 x 0.8	12.7	S?	90	
11 08 17.5	+53 49 19	UGC 6186	13.8	1.2 x 0.3	12.5	S	5	Mag 9.79v star SSE 1′.4.
11 09 12.8	+61 23 45	UGC 6192	14.9	0.8 x 0.4	13.5	SBb	153	Mag 10.95v star S 2′.5.
11 11 35.6	+56 31 23	UGC 6228	14.3	1.3 x 1.2	14.6	SBcd?	24	Anonymous galaxy SW 1′.5.
11 13 20.9	+59 54 30	UGC 6249	13.5	1.5 x 1.5	14.2	SAcd		**MCG +10-16-87** E 1′.2; **MCG +10-16-85** N 1′.0; **MCG +10-16-86** N 1′.9.
11 13 25.9	+53 35 42	UGC 6251	14.7	1.7 x 1.3	15.4	SABm: IV-V	60	Dwarf spiral, low surface brightness.
11 17 47.1	+51 28 29	UGC 6309	12.8	1.7 x 1.1	13.4	SB?	125	**Mkn 1445** SSW 3′.5; **Mkn 1444** SSW 5′.1.
11 19 46.1	+59 16 49	UGC 6335	13.9	1.7 x 1.5	14.8	SA(s)cd	90	
11 26 40.3	+53 44 48	UGC 6446	13.1	3.7 x 2.4	15.3	SAd	28	Mag 10.11v star S edge.
11 27 19.3	+59 37 31	UGC 6452	13.3	1.0 x 0.5	12.4	S?	30	
11 30 51.1	+60 30 05	UGC 6501	15.5	0.7 x 0.4	14.0	Sd:	96	
11 31 12.0	+51 41 54	UGC 6505	15.3	1.1 x 1.1	15.3	SABd:		
11 32 20.9	+53 54 18	UGC 6518	13.7	1.0 x 0.5	12.8	S	18	Galaxy **KUG 1129+542** NW 2′.7.
11 32 27.9	+62 30 24	UGC 6520	13.2	1.2 x 0.7	12.9	SB?	150	Mag 9.04v star W 3′.8; **MCG +11-14-29** S 4′.9.
11 32 38.0	+52 56 55	UGC 6527	15.3	1.2 x 0.6	14.8	Sa pec sp	75	Triple system, connected: E component = **Hickson 56B**, center component = **Hickson 56C**, W component = **Hickson 56D**; **Hickson 56A** E 1′.4; **Hickson 56E** SW 0′.9.
11 32 44.3	+61 49 34	UGC 6528	13.4	1.1 x 0.9	13.2	SA(rs)c	163	Stellar nucleus in uniform envelope.
11 35 00.2	+51 13 02	UGC 6555	13.3	1.3 x 1.0	13.4	S0⁻:	87	**MCG +9-19-124** and mag 11.4 star NE 1′.7.
11 35 43.5	+58 11 32	UGC 6566	14.8	1.0 x 0.7	14.3	SBm	120	Mag 7.53v star W 1′.3; **CGCG 292-7** N 8′.0.
11 36 26.8	+58 11 28	UGC 6575	13.6	2.0 x 0.4	13.2	Scd:	174	Mag 7.53v star and UGC 6566 W 6′.9.
11 39 21.1	+58 16 03	UGC 6616	13.1	2.3 x 2.1	14.7	SA(s)d:	80	Pair of bright knots SW edge.
11 42 25.9	+51 35 52	UGC 6667	13.5	3.2 x 0.5	13.8	Scd:	89	
11 43 08.8	+59 06 23	UGC 6682	13.9	1.5 x 1.3	14.5	Sm: V	100	Dwarf spiral, extremely low surface brightness.
11 45 10.4	+61 42 28	UGC 6727	13.8	0.8 x 0.3	12.1	S?	125	**MCG +10-17-54** SW 9′.7.
11 45 33.2	+58 58 39	UGC 6732	13.1	0.8 x 0.7	12.4	S0?	12	Faint, stellar galaxy **SBS 1142+592** on E edge.
11 47 23.0	+60 18 00	UGC 6762	13.0	1.0 x 1.0	12.9	S?		
11 47 40.0	+54 30 53	UGC 6766	14.8	1.5 x 0.2	13.3	Sbc	132	**MCG +9-19-186** S 1′.0.
11 47 40.1	+57 38 45	UGC 6767	13.0	0.7 x 0.6	11.9	Double System	123	Multi-galaxy system.
11 48 11.6	+54 59 27	UGC 6774	14.7	1.9 x 0.2	13.5	Scd:	155	**MCG +9-19-194** N 1′.8; mag 9.24v star SE 4′.4.
11 50 06.6	+51 51 17	UGC 6802	15.3	2.1 x 0.3	14.6	Scd:	160	Mag 11.27v star SW 2′.9.
11 50 47.7	+56 27 15	UGC 6816	13.8	1.7 x 1.3	14.5	IBm IV-V	63	Knotty; bright clump on W edge.
11 51 23.2	+53 26 32	UGC 6828	14.2	1.1 x 0.8	13.9	Sbc II-III	168	Mag 9.65v star W 6′.2; **MCG +9-20-6** and **MCG +9-20-7** SW 5′.2.
11 52 07.0	+52 06 28	UGC 6840	13.9	2.8 x 2.3	15.8	SB(rs)m IV-V	83	Dwarf spiral, low surface brightness; mag 8.89v star SSW 7′.7.
11 52 58.3	+61 12 56	UGC 6858	14.1	1.5 x 0.9	14.3	Scd:	15	Mag 11.5 star S 2′.3; mag 10.99v star W 2′.6.
11 55 23.8	+54 39 26	UGC 6894	14.8	1.6 x 0.3	13.8	Scd:	90	
11 56 17.9	+58 11 41	UGC 6912	13.7	2.2 x 1.0	14.4	S?	129	Located 1′.7 NE of mag 8.30v star.
11 56 37.8	+55 37 57	UGC 6919	14.6	1.4 x 0.4	13.8	Sdm:	90	Mag 11.7 star W 2′.7.

24 (Continued from previous page) GALAXIES 24

RA h m s	Dec ° ′ ″	Name	Mag (V)	Dim ′ Maj x min	SB	Type Class	PA	Notes
11 56 49.5	+53 09 38	UGC 6923	13.4	2.0 x 0.7	13.6	Im:	175	Mag 8.57v star S 4′.1; **UGC 6940** NE 9′.6.
11 57 25.1	+57 55 47	UGC 6931	13.7	1.5 x 1.0	14.0	SBm:	59	Mag 11.6 star SE 3′.3.
11 57 42.9	+57 33 53	UGC 6939	13.6	0.9 x 0.8	13.1	E/S0	87	**UGC 6926** SW 7′.0.
11 58 27.2	+57 35 46	UGC 6957	13.9	1.5 x 0.8	13.9	SB	57	Mag 11.6 star N 2′.3.
11 58 47.7	+53 25 29	UGC 6969	14.6	1.5 x 0.5	14.1	Im	152	Located 11′.2 E of the center of NGC 3992.
11 59 09.3	+52 42 26	UGC 6983	12.6	3.4 x 2.3	14.7	SB(rs)cd III	85	Short, elongated nucleus.
12 01 43.6	+62 19 43	UGC 7009	15.2	1.5 x 0.5	14.7	Im	175	
12 02 29.4	+62 25 01	UGC 7019	14.6	1.3 x 0.7	14.3	Im	85	Uniform surface brightness; mag 11.4 star W end; mag 9.40v star NE 6′.9.
12 04 44.0	+60 40 20	UGC 7064A	15.2	1.8 x 0.6	15.1		81	
12 04 56.6	+58 06 14	UGC 7070	14.2	1.8 x 1.4	15.1	S?	57	Double system, disrupted, bridge; mag 9.19v star W edge.
12 05 32.2	+51 30 15	UGC 7082	14.2	1.5 x 0.6	13.9	Sb II-III	53	Almost stellar anonymous galaxy S 1′.5.
12 09 45.5	+56 31 19	UGC 7144	13.5	1.0 x 0.8	13.1	Sb	81	
12 09 46.8	+62 16 09	UGC 7153	14.3	1.9 x 0.3	13.5	Scd:	166	
12 12 56.5	+52 15 51	UGC 7218	14.2	1.3 x 0.6	13.7	Im:	170	Superimposed star on E edge.
12 15 23.3	+51 20 57	UGC 7267	13.7	1.8 x 0.7	13.8	Sdm:	53	**MCG +9-20-122** WSW 10′.4.
12 19 43.4	+53 03 36	UGC 7379	15.3	1.1 x 0.6	14.7	Sd	174	
12 21 05.6	+61 05 13	UGC 7406	13.8	0.9 x 0.7	13.1	SBcd:	30	
12 26 08.4	+58 19 20	UGC 7534	13.9	3.2 x 1.9	15.7	IBm IV-V	111	Irregular; filamentary; mag 10.67v star on S edge.
12 29 08.7	+57 54 54	UGC 7618	15.1	1.5 x 0.3	14.1	Sc	45	Mag 11.4 star NE 1′.7.
12 29 45.4	+56 49 31	UGC 7635	13.5	1.2 x 0.7	13.2	S0	100	Mag 13.7 star on W end.
12 30 44.9	+57 18 04	UGC 7659	14.1	1.0 x 0.3	12.6	S?	138	
12 32 21.3	+56 39 20	UGC 7691	13.7	1.0 x 0.5	12.8	S?	147	
12 33 02.5	+56 52 16	UGC 7705	15.1	0.9 x 0.8	14.6	Sc	140	
12 43 49.0	+54 54 03	UGC 7905	13.2	1.7 x 0.9	13.5	Pec	45	Double system, disrupted; = **MCG +9-21-33**, **MCG +9-21-34**, **Mkn 220** and **Mkn 221**.
12 44 32.7	+56 09 00	UGC 7922	14.3	1.9 x 0.3	13.6	SBbc?	136	
12 46 56.3	+51 36 43	UGC 7950	14.5	1.7 x 1.2	15.1	Im	5	Elongated E-W core, uniform envelope; mag 8.04v star NW 1′.9.
12 50 31.5	+52 07 24	UGC 7993	14.4	2.1 x 0.2	13.3	Scd:	144	
12 52 03.8	+51 40 45	UGC 8012	13.6	0.7 x 0.5	12.3	S?	115	
12 54 43.4	+58 46 33	UGC 8040	13.8	1.5 x 0.2	12.3	S?	57	UGC 8046 ENE 2′.8.
12 55 04.1	+58 47 24	UGC 8046	14.6	1.0 x 0.7	14.1	SBdm:	141	UGC 8040 WSW 2′.8.
12 55 06.5	+55 08 57	UGC 8047	14.3	1.0 x 0.7	13.8	Sb-c	156	Bright knot or star SW edge.
12 55 25.0	+52 16 01	UGC 8050	13.9	1.2 x 0.9	13.8	SB(rs)bc	57	Short N-S bar with tightly wound spiral arms; NGC 4834 E 9′.4.
12 56 14.4	+56 52 24	UGC 8058	13.6	1.5 x 1.0	13.9	SA(rs)c? pec	10	= **Mkn 230**; very compact nucleus.
12 59 39.6	+53 20 25	UGC 8107	13.9	2.4 x 1.0	14.7	IBm:	53	Irregular; mag 11.22v star SE 2′.9.
13 02 07.7	+58 41 53	UGC 8146	13.8	3.5 x 0.5	14.3	Sd	30	Located between a pair of N-S stars, mags 12.3 and 11.02v.
13 02 40.6	+51 46 46	UGC 8151	14.7	0.8 x 0.8	14.1	Sc		

GALAXY CLUSTERS

RA h m s	Dec ° ′ ″	Name	Mag 10th brightest	No. Gal	Diam ′	Notes
10 58 18.0	+56 46 00	A 1132	17.0	74	10	**Ursa Major II Cluster**. Mag 16.5 **MCG +10-16-34** N of center 1′.9; all members faint and stellar.
11 18 48.0	+51 42 00	A 1218	16.0	47	17	All members anonymous and stellar.
11 21 18.0	+53 45 00	A 1225	15.4	43	15	Other than **MCG +9-19-56**, all members anonymous and stellar to almost stellar.
11 29 30.0	+54 03 00	A 1270	15.4	40	22	All members anonymous and stellar to almost stellar.
11 32 06.0	+56 01 00	A 1291	15.4	61	22	Mag 7.48v star 5′.2 N of center; Mkn 174 NW 9′.9; other members anonymous, stellar to almost stellar.
11 36 24.0	+54 57 00	A 1318	15.0	56	26	**MCG +9-19-132** N of center 2′.4; many MCG and anonymous GX throughout area.
11 47 00.0	+55 44 00	A 1377	15.0	59	20	**Usra Major I Cluster**. Mag 5.25v star 6′.6 S of center; many stellar anonymous GX grouped around center.
11 48 12.0	+54 37 00	A 1383	15.7	54	20	Almost stellar **MCG +9-19-193** N of center 2′.0; **MCG +9-19-199** SE 7′.6; most other members anonymous and stellar.
12 00 30.0	+56 15 00	A 1436	15.4	69	16	**MCG +9-20-57A** and **B** NW of center 2′.7; most other members anonymous and stellar.
12 03 36.0	+51 44 00	A 1452	15.7	46	15	**MCG +9-20-74** and **75** SW from center 1′.2; **Hickson 60** group SW 5′.7.
12 05 36.0	+51 25 00	A 1468	16.0	50	15	Other than UGC 7082, all members faint, stellar and anonymous.
12 13 24.0	+59 16 00	A 1496	16.0	58	21	Most members faint, stellar and anonymous.
12 15 48.0	+59 58 00	A 1507	15.8	39	16	**MCG +10-18-15** and **16** E of center 3′.1; most other members anonymous and stellar/almost stellar.
12 47 36.0	+55 02 00	A 1616	16.0	39	19	All members stellar, faint and anonymous.

PLANETARY NEBULAE

RA h m s	Dec ° ′ ″	Name	Diam ″	Mag (P)	Mag (V)	Mag cent ★	Alt Name	Notes
11 14 47.8	+55 01 09	NGC 3587	170	12.0	9.9	16.0	PK 148+57.1	= **M 97**, **Owl Nebula**. Faint galaxy **MCG +9-19-14** SSE 3′.8.

25 GALAXIES 25

RA h m s	Dec ° ′ ″	Name	Mag (V)	Dim ′ Maj x min	SB	Type Class	PA	Notes
09 14 57.9	+53 23 07	CGCG 264-86	14.0	0.5 x 0.4	12.1		90	Located 6′.4 S of mag 8.36v star.
09 16 45.8	+53 26 33	CGCG 264-90	14.2	0.4 x 0.2	11.3	Pec	12	Located 3′.0 NNE of a mag 9.46v star.
09 31 11.3	+52 38 04	CGCG 265-23	14.1	0.6 x 0.6	12.8			Mag 11.9 star W 0′.6.
09 59 19.1	+52 15 22	CGCG 266-3	14.1	0.4 x 0.4	12.0	S?		Mag 9.68v star ESE 5′.9.
10 01 55.1	+53 15 34	CGCG 266-9	14.0	0.8 x 0.5	12.9		177	Mag 9.43v star NNE 7′.1.
10 15 42.3	+55 39 59	CGCG 266-31	13.9	0.9 x 0.7	13.2	SB(r)b:	24	Mag 12.2 star NW 1′.0.
10 28 29.6	+53 51 55	CGCG 266-46	14.0	0.6 x 0.6	12.7			Mag 12.6 star SW 1′.5.
10 29 24.1	+55 11 10	CGCG 266-48	14.0	0.6 x 0.6	12.7			Mag 11.5 star NE 2′.8.
10 49 05.4	+52 20 00	CGCG 267-15	14.8	0.5 x 0.3	12.6	S?	174	Mag 8.49v star ESE 12′.7.
10 10 16.3	+62 54 48	CGCG 312-32	14.5	1.6 x 0.6	14.3	Sm	78	A mag 7.10v star on E end.
10 51 35.0	+55 27 54	IC 646	14.4	0.7 x 0.5	13.1	Sa?	114	Mag 8.09v star SSE 7′.7; NGC 3398 S 4′.5.
08 56 24.3	+51 52 44	MCG +9-15-51	14.2	0.6 x 0.6	13.2	E/S0		MCG +9-15-52 NE 3′.1.
08 56 37.0	+51 55 12	MCG +9-15-52	14.5	0.5 x 0.4	12.6		135	MCG +9-15-51 SW 3′.1.
09 13 55.3	+52 17 24	MCG +9-15-110	14.9	0.8 x 0.5	13.8	Sc	132	Mag 13.3 star SW 1′.5.
09 14 58.3	+51 21 37	MCG +9-15-111	13.9	1.1 x 0.5	13.1	Sc	15	
09 17 59.4	+52 44 31	MCG +9-15-116	15.0	1.2 x 0.5	14.3	Sc	66	
09 18 57.9	+51 22 11	MCG +9-15-118	14.4	1.1 x 0.3	13.0	Sb	141	Mag 11.4 star WSW 2′.8.
09 23 27.4	+54 49 44	MCG +9-16-9	15.2	0.9 x 0.6	14.3		126	Mag 10.73v star SE 4′.0.
09 24 55.9	+55 53 46	MCG +9-16-14	14.4	0.8 x 0.8	13.8	Sd		Mag 10.80v star SE 3′.4.
09 33 45.5	+51 55 09	MCG +9-16-29	15.5	0.8 x 0.4	14.1		54	Mag 12.0 star NNE 2′.3.
09 33 56.7	+51 31 41	MCG +9-16-30	14.5	1.1 x 0.8	14.2	Sb	60	Elongated anonymous galaxy N edge.

(Continued from previous page)

GALAXIES

RA h m s	Dec ° ′ ″	Name	Mag (V)	Dim ′ Maj x min	SB	Type Class	PA	Notes
09 46 08.4	+54 36 50	MCG +9-16-42	14.5	0.7 x 0.7	13.6			Mag 12.8 star NNE 2'.0.
09 46 45.4	+54 25 37	MCG +9-16-43	14.1	0.7 x 0.7	13.2	S0?		MCG +9-16-44 NE 1'.6.
09 46 53.3	+54 26 44	MCG +9-16-44	14.2	0.6 x 0.6	13.2	E?		MCG +9-16-43 SW 1'.6.
09 47 21.9	+54 28 25	MCG +9-16-48	14.5	0.6 x 0.6	13.5	E/S0		Mag 9.35v star NE 3'.6.
10 01 14.4	+55 43 04	MCG +9-17-9	13.9	1.0 x 0.4	12.8	Sb-Sc	102	NGC 3079 E 5'.7.
10 02 28.4	+53 51 53	MCG +9-17-11	14.6	0.7 x 0.3	12.7		90	Mag 14.0 star WNW 1'.7.
10 03 55.5	+54 42 53	MCG +9-17-16	14.0	1.0 x 0.9	13.7		33	Mag 12.7 star W 2'.3.
10 04 51.8	+54 34 02	MCG +9-17-19	14.5	0.7 x 0.5	13.2		78	Mag 11.9 star S 1'.3.
10 06 08.9	+52 10 15	MCG +9-17-23	14.4	0.8 x 0.5	13.3	Sc	57	
10 07 32.1	+52 53 38	MCG +9-17-26	15.4	0.8 x 0.4	14.1	SBbc	135	Mag 12.6 star W 1'.9.
10 10 01.2	+54 40 18	MCG +9-17-33	14.7	1.0 x 0.4	13.6	Sc	60	
10 14 54.7	+56 17 57	MCG +9-17-42	14.6	0.8 x 0.5	13.5		12	Pair of mag 12 stars W 1'.8.
10 17 21.2	+53 47 45	MCG +9-17-49	15.2	0.7 x 0.4	13.6	Sc	117	
10 20 28.3	+53 51 24	MCG +9-17-55	14.4	0.7 x 0.7	13.5	Sb		Mag 11.3 star S 3'.3.
10 20 46.9	+53 58 04	MCG +9-17-56	15.1	1.1 x 0.2	13.3	Sb		Located between a pair of E-W mag 11.8, 11.2 stars.
10 25 24.5	+55 31 26	MCG +9-17-64	14.5	0.7 x 0.6	13.4	SAB(s)c:	171	Star or small, round anonymous galaxy E edge.
10 25 14.9	+53 52 26	MCG +9-17-65	16.3	0.4 x 0.4	14.2	SBbc		Mag 11.17v star E 0'.9.
10 37 43.3	+55 01 38	MCG +9-18-4	15.0	0.9 x 0.7	14.4	SBbc	102	
10 44 18.8	+52 46 11	MCG +9-18-14	13.9	1.1 x 0.8	13.8	E?	90	Mag 12.8 star W 1'.8.
10 44 38.6	+56 22 10	MCG +9-18-16	13.0	0.9 x 0.4	11.7	Sa	92	Mag 11.9 star E 2'.4.
10 46 16.0	+51 04 41	MCG +9-18-24	14.9	0.7 x 0.7	14.0	Sb		Mag 12'.4 star S edge.
10 58 23.0	+55 37 03	MCG +9-18-59	14.8	0.8 x 0.4	13.4	Sc	66	UGC 6059 ESE 3'.1.
10 58 53.5	+55 38 05	MCG +9-18-63	14.4	0.6 x 0.6	13.2			UGC 6059 SSW 2'.6.
11 02 07.5	+54 27 13	MCG +9-18-70	15.5	0.6 x 0.5	14.1	Sb	138	
11 02 38.7	+51 10 31	MCG +9-18-72	14.8	0.9 x 0.7	14.2	Sbc	27	
08 58 07.9	+59 12 22	MCG +10-13-47	15.0	1.0 x 0.7	14.5	Sb	111	Mag 10.93v star E 3'.1.
09 11 23.2	+62 00 58	MCG +10-13-64	14.6	0.7 x 0.5	13.4		36	Mag 10.25v star N 3'.5.
09 13 04.2	+60 14 40	MCG +10-13-66	16.0	1.0 x 0.4	14.9	S?	117	**MCG +10-13-67** on NW edge.
09 26 36.0	+62 24 30	MCG +10-14-8	14.0	0.8 x 0.6	13.3	E/S0	90	Mag 12.9 star SW 2'.4.
09 27 50.3	+58 32 31	MCG +10-14-12	14.9	0.7 x 0.7	14.0	SBbc		
09 32 49.3	+62 20 11	MCG +10-14-19	14.6	0.6 x 0.6	13.4			Mag 11.35v star NW 4'.6.
09 33 03.8	+58 56 09	MCG +10-14-22	15.0	0.6 x 0.5	13.5	Sd	54	Mag 8.46v star SW 3'.1.
09 48 41.4	+59 15 38	MCG +10-14-39	13.6	1.0 x 0.3	12.1	Sb	15	Mag 8.40v star WNW 7'.3.
09 51 19.9	+57 56 25	MCG +10-14-44	14.8	0.7 x 0.5	13.5		108	Mag 11.4 star W 3'.1.
09 54 02.9	+58 24 07	MCG +10-14-47	14.9	0.7 x 0.3	13.0		135	Pair of mag 12-13 stars SW 2'.0.
09 59 13.3	+57 57 05	MCG +10-14-56	14.4	0.9 x 0.7	13.7		87	Mag 9.48v star WNW 5'.5.
10 03 19.2	+59 44 17	MCG +10-15-3	15.0	0.7 x 0.3	13.2	Sb	123	Mag 10.82v star N 4'.1.
10 12 54.4	+58 51 54	MCG +10-15-34	15.4	0.5 x 0.5	13.7			Mag 13.1 star N 3'.2.
10 16 18.3	+60 18 39	MCG +10-15-51	14.5	0.5 x 0.2	11.9	Sb	24	Mag 10.15v star W 1'.1.
10 21 13.2	+57 00 02	MCG +10-15-67	16.0	0.8 x 0.3	14.3	SBb	96	NGC 3206 SE 6'.3.
10 22 24.8	+57 02 07	MCG +10-15-70	15.2	0.7 x 0.4	13.6	Sbc	111	Mag 13.7 star S 0'.7; NGC 3214 E 6'.0.
10 26 16.6	+57 17 37	MCG +10-15-79	14.8	0.5 x 0.4	13.1	E/S0	168	NGC 3238 SE 5'.4.
10 30 27.8	+57 39 20	MCG +10-15-90	15.7	0.5 x 0.5	14.0	Sb		
10 31 53.5	+62 24 00	MCG +10-15-93	15.2	0.6 x 0.4	13.5	Sb	99	Mag 11.1 star E 0'.9.
10 32 32.0	+56 42 45	MCG +10-15-96	14.5	0.5 x 0.3	12.3		141	Mag 12.1 star SSE 2'.6.
10 32 59.2	+56 44 52	MCG +10-15-99	14.2	0.7 x 0.5	12.9		156	
10 33 28.7	+57 03 10	MCG +10-15-100	14.5	0.7 x 0.4	13.0	Spiral		Mag 13.9 star E 0'.9.
10 34 04.6	+59 24 10	MCG +10-15-103	14.3	0.7 x 0.7	13.6	E/S0		MCG +10-15-108 ESE 2'.7.
10 34 10.0	+61 36 26	MCG +10-15-105	14.6	0.4 x 0.4	12.7	E/S0		MCG +10-15-106 N 2'.0; pair of small, faint anonymous galaxies NW 1'.6.
10 34 14.4	+61 38 25	MCG +10-15-106	13.6	0.7 x 0.5	12.5	E/S0	147	MCG +10-15-105 S 2'.0; mag 10.99v star N 2'.2.
10 34 24.1	+59 23 13	MCG +10-15-108	14.6	0.5 x 0.5	13.2	E/S0		Mag 13.2 star E 0'.8; MCG +10-15-103 WNW 2'.7.
10 38 10.2	+58 01 01	MCG +10-15-116	15.5	0.9 x 0.5	14.5		18	
10 38 43.1	+56 33 09	MCG +10-15-118	16.2	0.6 x 0.5	14.8	Sb	15	
10 47 43.5	+59 24 41	MCG +10-16-9	14.0	0.8 x 0.6	13.3	E/S0	168	Mag 10.52v star NW 2'.8.
10 52 25.5	+59 41 05	MCG +10-16-18	14.4	0.7 x 0.5	13.1	SBc?	153	Mag 12.4 star SE 0'.9.
10 54 44.3	+56 58 56	MCG +10-16-24	15.9	0.6 x 0.2	13.5	Sm	99	Located on E edge of NGC 3445.
10 55 30.6	+57 54 08	MCG +10-16-25	15.3	1.2 x 0.5	14.6	Sc	126	Multi-galaxy system; mag 12.1 star WSW 2'.9.
11 00 07.6	+57 25 43	MCG +10-16-41	14.5	0.7 x 0.5	13.3		51	Mag 13.3 star W 0'.9.
11 01 00.4	+57 47 02	MCG +10-16-44	14.5	0.7 x 0.4	13.0	Sbc	126	Mag 10.65v star NW 4'.9; **MCG +10-16-43** SW 0'.9.
08 56 58.0	+52 04 01	NGC 2692	13.3	1.3 x 0.5	12.7	SBab:	165	Mag 9.82v star E 5'.6; UGC 4671 NW 3'.3.
08 56 59.5	+51 20 53	NGC 2693	11.9	2.6 x 1.8	13.5	E3:	160	NGC 2694 on S edge; **UGC 4679** N 7'.6; **MCG +9-15-54** NNW 9'.3.
08 56 59.4	+51 19 53	NGC 2694	14.4	1.2 x 1.2	14.8	E1		Stellar nucleus; located on S edge of NGC 2693.
08 59 06.5	+53 46 10	NGC 2701	12.3	2.2 x 1.6	13.5	SAB(rs)c: II-III	23	Mag 8.14v star E 7'.8.
08 59 48.5	+55 42 21	NGC 2710	12.9	2.0 x 1.0	13.5	SB(rs)b	125	Two main arms; mag 10.17v star SE 7'.0.
09 04 57.0	+59 55 59	NGC 2726	12.5	1.6 x 0.5	12.1	Sa?	87	
09 06 03.1	+51 44 39	NGC 2739	14.7	0.8 x 0.3	13.0	Sa	96	Located on N edge of NGC 2740.
09 06 05.1	+51 44 06	NGC 2740	14.0	0.9 x 0.7	13.4	S?	45	Located on S edge of NGC 2739; mag 7.20v star SE 12'.3.
09 07 34.4	+60 28 47	NGC 2742	11.4	3.0 x 1.5	12.9	SA(s)c: II	87	Mag 7.80v star WNW 4'.6.
09 09 58.2	+62 14 46	NGC 2742A	13.2	1.5 x 0.8	13.3	SB(s)b pec?	90	Mag 10.73v star 1'.8 WNW.
09 09 01.0	+53 50 57	NGC 2756	12.4	1.7 x 1.1	12.9	Sb II	0	
09 11 38.1	+60 02 15	NGC 2768	9.9	6.4 x 3.0	13.1	E6:	95	**UGC 4808** SW 11'.4; galaxy **VII Zw 267** E 11'.4.
09 18 35.0	+52 30 50	NGC 2800	12.8	1.4 x 0.9	13.1	E	15	Mag 13.4 star on E edge of nucleus; mag 9.10v star ESE 3'.8.
09 27 53.5	+57 22 33	NGC 2870	13.0	2.5 x 0.6	13.3	Sbc	123	
09 29 34.8	+62 29 28	NGC 2880	11.5	2.0 x 1.2	12.4	SB0⁻	140	Mag 9.47v star WSW 9'.8.
09 32 25.2	+57 28 58	NGC 2895	13.8	0.9 x 0.8	13.3	SBc	57	**MCG +10-14-17** NW 5'.3.
09 42 35.0	+58 51 07	NGC 2950	10.9	2.7 x 1.8	12.5	(R)SB(r)0°	145	Three faint stars in a line N-S just inside the W edge.
09 56 14.1	+59 18 22	NGC 3043	12.6	1.7 x 0.6	12.5	Sb: sp	81	Mag 8.18v star N 8'.1.
10 00 52.3	+55 37 06	NGC 3073	13.4	1.3 x 1.2	13.8	SAB0⁻	156	Bright round core in faint envelope; mag 7.94v star E 6'.9.
10 01 57.9	+55 40 36	NGC 3079	10.9	7.9 x 1.4	13.3	SB(s)c sp II	165	MCG +9-17-9 W 5'.7; mag 7.94v star S 6'.1.
10 04 31.7	+60 06 26	NGC 3102	13.3	0.9 x 0.9	13.0	S0⁻:		
10 15 11.7	+56 40 20	NGC 3164	13.7	0.9 x 0.7	13.0	S?	0	**MCG +10-15-44** ENE 5'.2.
10 16 22.9	+60 14 05	NGC 3168	13.4	1.0 x 0.9	13.3	E	45	Mag 10.5 star SW 3'.7; UGC 5542 NE 4'.7.
10 19 33.1	+58 12 19	NGC 3182	12.1	1.8 x 1.5	13.0	SA(r)a?	155	
10 19 42.9	+57 25 24	NGC 3188	13.7	0.8 x 0.6	12.8	(R)SB(r)ab	168	NGC 3188A on W edge.
10 19 38.3	+57 25 06	NGC 3188A	16.6	0.4 x 0.2	13.7	S?	78	
10 21 47.9	+56 55 52	NGC 3206	11.9	3.0 x 1.9	13.6	SB(s)cd	0	Close pair of strong knots or superimposed stars SW of core; MCG +10-15-67 NW 6'.3.
10 23 08.9	+57 02 21	NGC 3214	13.9	0.8 x 0.5	12.8	SA0/a:	39	MCG +10-15-70 W 6'.0.
10 23 45.7	+57 01 34	NGC 3220	13.0	1.2 x 0.6	12.5	Sb: sp	97	Galaxy **KUG 1020+571** S 6'.0.
10 25 10.3	+58 08 59	NGC 3225	12.6	2.0 x 1.0	13.2	Scd:	155	

(Continued from previous page)

GALAXIES

RA h m s	Dec ° ′ ″	Name	Mag (V)	Dim ′ Maj x min	SB	Type Class	PA	Notes
10 26 48.6	+61 16 22	NGC 3236	14.3	0.5 x 0.3	12.3	E/S0	48	Almost stellar; mag 10.91v star SW 4′.1.
10 26 43.2	+57 13 33	NGC 3238	12.9	1.4 x 1.3	13.4	SA(r)0°:	132	MCG +10-15-79 NW 5′.4; **MCG +10-15-78** SW 10′.6.
10 32 19.5	+56 04 59	NGC 3264	12.0	2.9 x 1.2	13.2	SBdm:	177	
10 36 21.6	+58 37 12	NGC 3286	13.6	0.9 x 0.6	12.9	E?	88	Mag 10.36v star SE 4′.8.
10 36 25.5	+58 33 21	NGC 3288	14.0	1.0 x 0.8	13.6	SABbc:	175	Weak stellar nucleus; mag 10.36v star ENE 3′.7.
10 38 46.0	+53 30 13	NGC 3310	10.8	3.1 x 2.4	12.8	SAB(r)bc pec II	156	Several bright knots N.
10 45 22.5	+55 57 34	NGC 3353	12.8	1.3 x 1.0	12.9	Sb? pec	60	
10 51 31.6	+55 23 26	NGC 3398	14.4	0.9 x 0.5	13.4	Sa?	78	Mag 8.09v star SE 4′.7. **MCG +9-18-41** E 2′.3.
10 51 44.3	+51 01 25	NGC 3406	12.7	1.0 x 0.7	12.4	E	42	Double system in contact, strongly disturbed.
10 52 17.9	+61 22 45	NGC 3407	13.6	1.4 x 0.7	13.4	S0⁻:	15	
10 52 11.8	+58 26 15	NGC 3408	13.5	1.0 x 0.7	12.9	Sc:	127	Mag 9.10v star S 4′.2; note: **MCG +10-16-15** lies 1′.8 W of this star.
10 51 54.0	+51 00 19	NGC 3410	14.2	0.8 x 0.5	13.1	S?	24	Stellar nucleus in smooth envelope; mag 9.84v star SW 5′.2.
10 54 48.1	+61 17 19	NGC 3435	13.2	1.9 x 1.2	13.9	Sb II	35	Mag 9.24v star NNE 7′.9.
10 53 49.8	+57 07 09	NGC 3440	13.2	2.1 x 0.5	13.1	SBb? sp	48	Faint, stellar anonymous galaxy NE 6′.2.
10 54 35.7	+56 59 19	NGC 3445	12.6	1.6 x 1.5	13.4	SAB(s)m III-IV	105	Mag 10.29v star NE 2′.1; MCG +10-16-24 on E edge.
10 54 39.2	+54 18 21	NGC 3448	12.1	5.6 x 1.8	14.4	I0	65	Bright, elongated nucleus.
10 56 01.8	+57 06 58	NGC 3458	12.3	1.4 x 0.9	12.4	SAB0:	5	
10 58 44.9	+59 30 35	NGC 3470	13.2	1.4 x 1.2	13.6	SA(r)ab:	170	Faint, elongated anonymous galaxy S 1′.5.
10 59 09.1	+61 31 47	NGC 3471	12.5	1.7 x 0.8	12.7	Sa	14	**MCG +10-16-40** NE 2′.3.
11 01 23.3	+57 40 37	NGC 3488	12.9	1.6 x 1.0	13.2	SB(s)c:	175	Mag 13.3 star on S edge.
11 03 11.0	+56 13 16	NGC 3499	13.6	0.9 x 0.7	12.9	I0?	6	Located 14′.8 SE of mag 2.35v star Merak, Beta Ursea Majoris.
08 56 42.5	+52 06 18	UGC 4671	12.7	1.5 x 1.3	13.3	S?	96	Mag 12.9 star Superimposed NE of nucleus; NGC 2692 SE 3′.3.
08 57 54.5	+59 04 55	UGC 4683	14.1	1.8 x 1.0	14.6	Im	110	Dwarf irregular, extremely low surface brightness.
08 58 10.6	+52 10 57	UGC 4690	13.9	1.1 x 0.5	13.1	Sab	120	Mag 12.0 star N 1′.2.
08 59 07.5	+53 37 57	UGC 4696	15.1	1.1 x 0.2	13.3	Scd:	135	Mag 10.88v star E 1′.5.
09 00 20.3	+52 29 39	UGC 4713	12.8	1.7 x 1.2	13.4	Sb II-III	177	Located between a mag 12.0 star E and a mag 10.97v star W.
09 00 37.4	+51 12 25	UGC 4717	14.1	1.0 x 0.5	13.2	Sab	40	
09 01 58.7	+60 09 04	UGC 4730	13.7	0.8 x 0.3	12.0	S?	93	**UGC 4727** W 1′.9; mag 10.22v star SW 1′.3.
09 04 33.7	+51 36 48	UGC 4749	13.4	0.7 x 0.7	12.5	S		Compact; mag 9.22v star W 6′.0.
09 08 23.4	+59 04 28	UGC 4788	14.6	1.5 x 0.8	14.6	SB	5	Mag 9.21v star on N edge; very small, very faint anonymous galaxy on S edge.
09 09 19.2	+54 54 40	UGC 4800	13.9	1.5 x 0.5	13.5	SB(s)cd?	120	Mag 9.45v star E 3′.0.
09 10 05.7	+54 34 47	UGC 4807	13.7	1.0 x 0.9	13.4	SA(rs)cd:	117	Pair of mag 11 stars E 1′.5.
09 11 33.6	+51 15 16	UGC 4824	13.7	2.3 x 0.6	13.8	SAd	98	Mag 9.33v star WNW 5′.0.
09 13 26.0	+52 58 52	UGC 4851	13.1	0.9 x 0.6	12.3	S0⁻:	145	Mag 9.69v star NNW 2′.4.
09 16 02.8	+52 50 30	UGC 4879	13.2	2.3 x 1.4	14.3	IAm	85	Irregular.
09 17 39.4	+52 59 28	UGC 4906	12.6	2.2 x 0.6	12.8	Sa	48	Mag 8.35v star SSW 5′.8.
09 19 34.2	+51 06 36	UGC 4932	14.6	1.7 x 0.6	14.4	Sdm:	140	Mag 11.5 star superimposed E of center.
09 22 38.1	+60 51 55	UGC 4973	14.0	0.9 x 0.8	13.5	Sbc		Mag 9.88v star S 2′.3.
09 26 16.9	+61 22 52	UGC 5013	14.3	1.5 x 0.8	14.3	Sb II-III	120	
09 28 49.8	+51 33 35	UGC 5047	15.5	1.5 x 0.2	14.0	Sdm:	160	Mag 8.81v star E 6′.5.
09 30 11.6	+55 51 08	UGC 5055	13.4	1.6 x 1.3	14.0	(R′)SB(s)b	162	Slightly elongated SE-NW core.
09 32 36.8	+51 52 19	UGC 5076	14.6	1.1 x 0.9	14.4	Im:	150	Dwarf, extremely low surface brightness; mag 3.18v star Theta Ursae Majoris S 11′.9.
09 32 50.7	+59 44 40	UGC 5077	14.1	1.5 x 0.7	14.0	SBb	77	Mag 9.12v star NE 6′.0.
09 35 59.1	+54 36 36	UGC 5106	13.9	0.9 x 0.6	13.1		95	Mag 8.89v star N 5′.4.
09 41 13.4	+61 03 38	UGC 5153	14.5	0.8 x 0.8	13.9	Sbc		Mag 11.9 star SSE 2′.1.
09 43 01.6	+58 58 24	UGC 5179	14.2	0.8 x 0.5	13.0	S?	126	Mag 9.83v star SSE 2′.9.
09 44 35.1	+55 45 48	UGC 5201	14.2	1.4 x 0.8	14.2	SAB(s)c	45	Small nucleus, filamentary arms.
09 45 12.5	+53 41 10	UGC 5207	14.0	0.7 x 0.6	12.9	SBab	55	
09 47 58.4	+54 00 52	UGC 5241	13.2	0.7 x 0.7	12.2	S?		**MCG +9-16-53** E 3′.6; mag 8.30v star SE 9′.4.
09 48 30.9	+57 58 15	UGC 5243	14.2	0.8 x 0.5	13.1	SB	99	
10 00 19.1	+54 32 16	UGC 5369	14.0	1.4 x 0.4	13.2	S	156	Stellar **Mkn 24** WSW 8′.5.
10 03 52.0	+59 26 10	UGC 5408	14.3	0.5 x 0.4	12.5	E?	156	Mag 9.71v star SW 3′.4.
10 04 41.9	+55 18 41	UGC 5421	15.9	1.8 x 0.3	15.1	Sdm:	172	Stellar nucleus; **MCG +9-17-21** and **MCG +9-17-22** ESE 11′.1.
10 04 47.8	+57 36 05	UGC 5422	14.5	1.0 x 0.9	14.2	SB(s)b	170	
10 06 00.0	+58 48 41	UGC 5435	12.6	0.9 x 0.5	11.6	E/S0	40	
10 08 12.0	+53 04 49	UGC 5459	12.6	4.0 x 0.5	13.2	SB(s)c? sp	132	Mag 8.69v star on E edge.
10 08 09.5	+51 50 31	UGC 5460	12.9	2.4 x 2.2	14.6	SB(rs)d	126	Patchy, located between a close pair of mag 12.3 and 10.86v stars.
10 09 30.9	+58 27 00	UGC 5475	14.4	1.5 x 0.7	14.3	SB(s)c	30	Mag 9.62v star NNE 6′.9.
10 10 09.5	+54 30 06	UGC 5479	13.9	1.0 x 0.8	13.5	S?	100	**MCG +9-17-36** E edge.
10 10 14.0	+58 29 15	UGC 5480	14.8	1.0 x 0.5	13.9	Sdm:	162	**CGCG 200-11** ESE 1′.6.
10 16 10.6	+58 25 37	UGC 5534	13.9	1.0 x 0.4	12.8	Sb	98	Irregular galaxy **UGC 5541** E 6′.1; stellar **MCG +10-15-45** N 3′.1.
10 16 53.4	+60 17 04	UGC 5542	13.6	1.0 x 1.0	13.6	E:		NGC 3168 SW 4′.7; **UGC 5553** ENE 9′.7.
10 17 05.5	+54 49 13	UGC 5546	14.2	1.1 x 0.7	13.7	SBb	150	Mag 9.02v star N 6′.6.
10 17 16.0	+53 27 41	UGC 5549	14.2	1.1 x 0.8	13.9	Sbc II-III	55	
10 23 47.3	+53 06 14	UGC 5615	13.3	1.1 x 0.6	12.7	Double System	171	Double system, contact; components consist of **MCG +9-17-61** and **MCG +9-17-62**.
10 24 28.1	+57 23 34	UGC 5626	14.3	1.5 x 0.6	14.0	Im	53	Irregular, low surface brightness.
10 28 56.8	+58 58 17	UGC 5673	13.9	0.7 x 0.5	12.6	SB	109	Located between two mag 12 stars on N-S line.
10 29 04.6	+54 43 05	UGC 5676	13.9	1.5 x 0.9	14.1	SBdm:	15	Mag 11.5 star N 1′.2.
10 32 31.9	+54 24 03	UGC 5720	12.8	1.1 x 0.7	12.4	Im pec:	144	
10 32 46.1	+58 51 36	UGC 5722	14.2	1.1 x 0.6	13.6	S	40	Mag 14.7 star on E edge.
10 33 48.0	+52 22 12	UGC 5733	14.0	0.7 x 0.6	12.9	SB	92	
10 34 15.1	+52 52 14	UGC 5734	13.2	1.5 x 0.2	11.7	S0/a	156	
10 44 29.6	+60 22 01	UGC 5846	15.0	1.6 x 0.9	15.2	Im V	108	Dwarf irregular, extremely low surface brightness.
10 44 23.4	+56 25 17	UGC 5848	14.2	1.9 x 1.0	14.7	Sm:	115	Dwarf spiral, very low surface brightness; MCG +9-18-16 SE 3′.7.
10 47 19.7	+54 02 19	UGC 5883	14.6	1.1 x 1.0	14.6	Im	96	Mag 12.9 star S 1′.6.
10 47 45.7	+56 05 27	UGC 5888	14.2	1.2 x 1.0	14.2	Im	42	Irregular; close pair of mag 13 stars N.
10 49 47.2	+51 53 36	UGC 5928	13.7	1.1 x 0.8	13.4	S0⁻:	99	Mag 13.7 star on NE edge; **MCG +9-18-36** SE 2′.2.
10 52 03.0	+55 36 03	UGC 5976	14.2	1.4 x 1.1	14.5	Scd:	30	Very small triangle of two mag 14 stars and a mag 10.55v star W 3′.6.
10 54 13.4	+54 17 12	UGC 6016	14.6	2.0 x 1.4	15.6	Im	30	NGC 3448 off E edge.
10 58 43.8	+55 35 51	UGC 6059	14.4	1.5 x 0.7	14.3	SAB(s)bc	15	MCG +9-18-63 NE 2′.6.
11 00 39.6	+61 19 17	UGC 6080	14.9	1.9 x 0.2	13.7	Sd	129	
11 02 37.9	+59 07 32	UGC 6110	14.1	1.5 x 0.5	13.6		65	

GALAXY CLUSTERS

RA h m s	Dec ° ′ ″	Name	Mag 10th brightest	No. Gal	Diam ′	Notes
10 58 18.0	+56 46 00	A 1132	17.0	74	10	**Ursa Major II Cluster**. Mag 16.5 **MCG +10-16-34** N of center 1′.9; all members faint and stellar.

RA h m s	Dec ° ′ ″	Name	Mag (V)	Dim ′ Maj x min	SB	Type Class	PA	Notes
07 25 19.6	+55 30 37	CGCG 261-65	14.0	0.6 x 0.5	12.5		18	Double system.
07 43 13.7	+54 58 56	CGCG 262-21	14.1	0.6 x 0.3	12.1		135	Very faint anonymous galaxy S 0′9.
07 59 25.6	+55 17 43	CGCG 262-54	14.2	1.1 x 0.4	13.2	S?	54	= **Mkn 1412**; mag 12.2 star 1′2 WNW.
08 32 28.3	+52 36 18	CGCG 263-43	13.7	0.6 x 0.6	12.4	S?		= **Mkn 91**.
08 40 33.8	+53 43 32	CGCG 263-70	14.7	0.4 x 0.4	12.5			Mag 9.25v star E 8′5.
08 02 05.3	+56 54 56	CGCG 287-31	15.0	0.5 x 0.3	13.0	E/S0	123	Mag 13.5 star W 0′7.
08 54 35.1	+57 09 59	IC 522	13.0	1.0 x 0.8	12.6	S0	165	Mag 8.32v star NNW 8′2.
07 56 14.6	+60 18 08	IC 2209	13.7	1.0 x 0.9	13.4	SB(rs)b:	145	Mag 10.92v star W 4′4; NGC 2460 NE 5′5; **MCG +10-12-22** E 4′7.
07 03 59.0	+56 29 08	MCG +9-12-22	14.9	1.0 x 0.4	13.8		66	Close pair of mag 9 stars S 2′6.
07 03 54.9	+53 39 58	MCG +9-12-23	14.7	0.7 x 0.5	13.4		33	
07 18 23.5	+52 21 46	MCG +9-12-49	14.8	1.0 x 0.8	14.4		0	Mag 11.7 star NW 1′6.
07 27 13.0	+52 56 43	MCG +9-13-1	13.9	0.8 x 0.8	13.3			
07 34 49.4	+55 57 40	MCG +9-13-17	14.0	1.1 x 0.7	13.8	E/S0	27	Mag 11.9 star NW 1′4.
07 45 51.4	+51 47 24	MCG +9-13-47	15.6	0.9 x 0.2	13.5	SBc	174	
07 46 16.0	+53 45 59	MCG +9-13-50	14.2	0.9 x 0.4	12.9	Sb	96	Mag 12.5 star WSW 2′7.
07 47 31.9	+51 11 30	MCG +9-13-56	14.0	0.8 x 0.3	12.3	Sb	123	
07 49 21.7	+52 39 56	MCG +9-13-63	14.1	0.7 x 0.7	13.4	E/S0		Mag 11.6 star NW 3′0.
07 50 03.6	+56 46 08	MCG +9-13-67	12.5	1.0 x 0.8	12.1		177	
07 50 21.3	+55 54 00	MCG +9-13-71	14.7	0.9 x 0.8	14.2	S?	171	Mag 8.83v star NW 6′8.
07 51 51.0	+55 05 11	MCG +9-13-74	14.1	1.2 x 0.9	14.0		171	
07 54 54.0	+54 07 12	MCG +9-13-88	14.4	0.7 x 0.5	13.1	S?	129	Mag 9.77v star NE 3′7.
08 13 14.1	+52 27 28	MCG +9-14-17	14.3	0.9 x 0.7	13.6	Sb	138	
08 20 21.8	+54 08 19	MCG +9-14-29	14.5	0.7 x 0.3	12.7		177	UGC 4335 E 3′1.
08 20 45.8	+56 29 26	MCG +9-14-31	13.8	0.9 x 0.6	13.2	E/S0	90	
08 27 31.0	+54 54 21	MCG +9-14-38	13.9	0.5 x 0.4	12.2	E/S0		
08 29 27.7	+55 00 55	MCG +9-14-45	14.1	0.7 x 0.4	12.6	Sc	45	
08 32 13.8	+55 51 53	MCG +9-14-50	13.5	0.9 x 0.6	12.7	S?		**MCG +9-14-51** and a mag 12.9 star E 1′2.
08 33 41.2	+54 32 55	MCG +9-14-61	14.1	1.0 x 0.7	13.5		66	**MCG +9-14-59** W 2′3.
08 39 12.8	+53 37 23	MCG +9-14-80	13.9	1.0 x 0.6	13.2	Sc	117	
08 40 15.4	+56 02 50	MCG +9-14-83	14.2	0.9 x 0.7	13.5		27	Mag 13.4 star W 0′5.
08 44 38.2	+55 01 07	MCG +9-15-11	13.7	1.0 x 0.6	13.0		105	**MCG +9-15-8** W 1′9; **MCG +9-15-13** NE 3′0.
08 46 02.4	+52 32 01	MCG +9-15-21	14.9	0.7 x 0.7	13.9			
08 50 22.8	+55 22 42	MCG +9-15-33	14.3	0.9 x 0.5	13.3		9	Mag 13.3 star WNW 1′1; small, almost stellar anonymous galaxy N 0′5.
08 56 24.3	+51 52 44	MCG +9-15-51	14.2	0.6 x 0.6	13.2	E/S0		MCG +9-15-52 NE 3′1.
08 56 37.0	+51 55 12	MCG +9-15-52	14.5	0.5 x 0.4	12.6		135	MCG +9-15-51 SW 3′1.
07 02 55.0	+62 46 09	MCG +10-10-21	14.0	0.8 x 0.5	12.9	SB?	114	Mag 11.5 star W 1′6.
07 09 51.2	+58 46 11	MCG +10-11-3	14.7	0.6 x 0.5	13.2		69	Mag 11.7 star NNE 0′8.
07 17 18.3	+59 10 17	MCG +10-11-9	15.6	0.7 x 0.7	14.7	Sbc		Mag 12.2 star N 2′4.
07 20 41.3	+58 12 25	MCG +10-11-29	15.1	0.7 x 0.4	13.6	Sb	12	Mag 12.5 star SW 1′5.
07 23 04.5	+58 05 50	MCG +10-11-33	14.2	0.9 x 0.5	13.4	E	168	UGC 3816 SSE 2′2.
07 23 55.0	+57 16 48	MCG +10-11-37	14.4	0.8 x 0.7	13.6	Sc	108	Close pair of mag 11.7 stars NNW 1′7.
07 54 17.7	+62 51 30	MCG +10-11-41	14.8	0.8 x 0.6	14.1	E/S0	168	
07 25 32.9	+60 07 28	MCG +10-11-42	16.3	0.5 x 0.4	14.4		90	
07 28 32.6	+57 44 42	MCG +10-11-53	15.0	0.7 x 0.5	13.7	SBbc	45	Mag 10.06v star NE 3′2.
07 32 33.0	+59 20 51	MCG +10-11-70	14.8	0.6 x 0.6	13.5	Sb		Mag 12.7 star ENE 0′9.
07 34 12.0	+57 37 14	MCG +10-11-82	15.5	0.8 x 0.3	13.8	Sbc	144	Mag 9.55v star ESE 2′8.
07 36 19.7	+62 21 13	MCG +10-11-92	15.2	1.0 x 0.7	14.7	SBbc	9	Faint, slightly elongated anonymous galaxy W 0′9.
07 36 33.6	+59 16 50	MCG +10-11-94	15.0	0.9 x 0.7	14.3	SBbc	90	Mag 12.9 star S 1′6.
07 37 23.8	+62 20 22	MCG +10-11-96	13.7	0.8 x 0.8	13.3	E/S0		Mag 11.6 star N 2′4.
07 39 19.7	+57 18 59	MCG +10-11-118	15.1	0.7 x 0.6	14.0	Sc	108	Mag 10.91v star N 0′7.
07 42 43.0	+60 12 29	MCG +10-11-126	15.2	0.4 x 0.4	13.1			**UGC 3971** SE 2′9; mag 10.58v star W 2′0.
07 46 08.4	+56 59 08	MCG +10-11-132	14.1	0.8 x 0.5	13.0		108	MCG +10-11-135 NE 3′2.
07 46 04.4	+57 06 01	MCG +10-11-133	15.1	0.7 x 0.7	14.2			
07 46 22.6	+57 01 43	MCG +10-11-135	14.4	1.0 x 0.4	13.3	S?	138	MCG +10-11-132 SW 3′2.
07 49 56.7	+59 38 46	MCG +10-11-146	16.7	0.7 x 0.5	15.4	Sc	18	**MCG +10-11-148** S 1′6.
07 51 11.9	+62 04 50	MCG +10-11-150	13.7	0.9 x 0.6	13.1	E/S0	165	Mag 12.9 star on W edge.
07 52 23.9	+60 44 43	MCG +10-12-1	15.0	0.9 x 0.4	13.7	Spiral	150	Mag 12.5 star SSE 1′9.
07 54 48.7	+57 47 53	MCG +10-12-12	11.2	0.7 x 0.4	9.7	SA(r)0° sp	150	
07 55 25.5	+57 21 01	MCG +10-12-14	15.1	0.8 x 0.4	13.7	Sc	123	
07 56 55.5	+58 52 38	MCG +10-12-26	14.5	0.8 x 0.6	13.6		27	Mag 10.2 star SW 3′3.
07 57 22.9	+62 24 39	MCG +10-12-28	13.5	1.0 x 0.7	13.0		42	**MCG +10-12-20** W 2′3; **MCG +10-12-27** N 1′6.
07 58 13.7	+58 31 20	MCG +10-12-36	14.6	0.5 x 0.3	12.4		138	Mag 13.1 star SE edge; mag 10.77v star W 1′3.
07 59 51.2	+56 54 49	MCG +10-12-50	14.5	1.0 x 0.6	14.0	E/S0	177	
08 02 20.2	+56 42 57	MCG +10-12-63	14.7	0.5 x 0.4	12.8		177	UGC 4164 SSW 5′1.
08 03 49.1	+61 17 20	MCG +10-12-67	14.8	0.8 x 0.4	13.4	Sb	171	UGC 4182 N 3′6.
08 06 33.2	+61 04 22	MCG +10-12-72	14.6	0.6 x 0.4	12.9	Sb	159	
08 09 11.3	+57 41 03	MCG +10-12-78	14.7	0.9 x 0.5	13.7	Sb	123	Mag 13.3 star NW 1′6.
08 09 20.2	+57 49 59	MCG +10-12-79	14.4	1.1 x 0.9	14.4	E/S0	54	Mag 8.66v star W 4′8.
08 09 52.5	+57 54 42	MCG +10-12-81	13.9	0.6 x 0.4	12.4	E	3	Mag 11.09v star SE 3′5.
08 10 07.2	+57 50 09	MCG +10-12-82	14.2	0.8 x 0.6	13.3	S?	99	Mag 11.7 star NE edge; galaxy **SBS 806+579B** SE 1′5.
08 11 03.4	+57 01 43	MCG +10-12-85	14.5	0.6 x 0.5	13.0	S?	162	
08 11 57.6	+58 42 00	MCG +10-12-87	14.8	0.7 x 0.4	13.3		108	Mag 13.1 star E 0′9.
08 13 42.8	+60 55 03	MCG +10-12-94	14.9	0.5 x 0.4	13.0		114	Mag 11.7 star SSE 2′5.
08 14 39.1	+58 01 21	MCG +10-12-100	14.3	0.9 x 0.8	13.8	SB?	30	Mag 14.5 star superimposed NE edge; mag 11.2 star SW 2′7.
08 15 25.8	+58 10 40	MCG +10-12-107	13.7	1.0 x 0.8	13.5	E	129	Stellar galaxy **SBS 811-583** E 2′2; mag 12.3 star S 1′1.
08 15 38.8	+58 17 46	MCG +10-12-108	14.4	0.7 x 0.7	13.5			UGC 4289 NNE 1′6.
08 15 50.8	+60 37 42	MCG +10-12-109	15.2	0.5 x 0.4	13.3		138	Mag 14.1 star E 0′5; mag 11.9 star SW 2′0.
08 17 36.5	+59 57 12	MCG +10-12-118	14.4	0.9 x 0.7	13.7	SBb	81	
08 18 11.0	+57 45 28	MCG +10-12-120	15.3	0.8 x 0.4	13.9	Sc	111	Mag 14.8 star W 0′7; **PGC 23275** N 1′2.
08 22 02.7	+57 18 31	MCG +10-12-130	15.2	0.7 x 0.5	13.9	SBb	123	Mag 12.4 star SE 3′0.
08 23 42.0	+62 16 20	MCG +10-12-131	14.4	0.8 x 0.8	14.0	E/S0		Mag 14.9 star on W edge; mag 7.70v star NE 3′9.
08 26 25.1	+59 53 43	MCG +10-12-137	14.5	0.8 x 0.6	13.8	E/S0	117	Mag 10.91v star ESE 3′0.
08 28 52.7	+58 12 09	MCG +10-12-141	14.6	0.7 x 0.5	13.3		96	**UGC 3861** S 3′4.
08 32 58.4	+61 48 14	MCG +10-12-143	13.9	0.7 x 0.7	12.9			Faint, elongated anonymous galaxy N 1′5.
08 33 45.5	+56 49 05	MCG +10-12-144	14.6	0.6 x 0.3	12.6	Sb	63	Mag 10.63v star superimposed E edge.
08 33 53.5	+57 32 20	MCG +10-12-145	14.3	0.7 x 0.5	13.0	Sa	141	Mag 14.9 star W edge; **MCG +10-12-146** NE 1′9.
08 41 02.7	+59 56 09	MCG +10-13-9	14.6	0.6 x 0.6	13.5	E/S0		**MCG +10-13-10** NE edge; mag 9.71v star S 2′1.
08 49 55.6	+57 07 16	MCG +10-13-20	15.6	0.6 x 0.4	13.9	Sbc	144	**MCG +10-13-19** W edge.
08 50 26.7	+60 04 53	MCG +10-13-21	15.1	0.9 x 0.3	13.5	Sb	84	Mag 10.92v star ENE 5′7.

(Continued from previous page)

GALAXIES

RA h m s	Dec ° ′ ″	Name	Mag (V)	Dim ′ Maj x min	SB	Type Class	PA	Notes
08 50 37.1	+57 21 34	MCG +10-13-24	15.2	0.7 x 0.3	13.3	Sbc	60	Mag 13.3 star 0′.7 N; mag 10.49v star SE 1′.5.
08 54 49.4	+57 40 10	MCG +10-13-32	15.1	0.9 x 0.5	14.1	Sb	126	Mag 11.4 star NE 3′.9.
08 58 07.9	+59 12 22	MCG +10-13-47	15.0	1.0 x 0.7	14.5	Sb	111	Mag 10.93v star E 3′.1.
07 43 18.4	+52 19 08	NGC 2426	13.1	1.1 x 1.1	13.3	E		Located 2′.7 NNW of mag 9.91v star.
07 43 47.6	+52 21 26	NGC 2429A	13.7	1.5 x 0.4	13.0	S?	145	Mag 11 star at S end of NGC 2429A, and SW end of NGC 2429B.
07 43 52.1	+52 20 52	NGC 2429B	16.5	0.4 x 0.2	13.6	Sbc	57	Mag 11 star at SW end.
07 45 13.3	+53 04 29	NGC 2431	13.4	0.9 x 0.9	13.0	(R′)SB(s)a:		**MCG +9-13-46** SE 5′.6; **MCG +9-13-44** SSE 6′.9.
07 48 38.9	+54 36 42	NGC 2446	12.9	1.9 x 1.0	13.5	Sb I-II	124	Stars superimposed N and E edges; mag 10.24v star SW 6′.8.
07 54 10.6	+55 29 43	NGC 2456	13.1	1.1 x 0.8	13.0	E	30	
07 54 54.8	+55 32 45	NGC 2457	15.4	0.5 x 0.2	12.7	Sbc	84	**MCG +9-13-86** W 1′.3; **MCG +9-13-89** N 3′.3.
07 55 34.9	+56 44 07	NGC 2458	15.1	0.4 x 0.3	12.6		72	Stellar; mag 10.70v star SW 3′.5; MCG +10-12-14 SE 2′.7.
07 56 53.1	+60 20 58	NGC 2460	11.8	2.5 x 1.9	13.4	SA(s)a III	40	IC 2209 SW 5′.5; **MCG +10-12-22** S 4′.5.
07 55 51.6	+56 42 38	NGC 2461	14.5	0.5 x 0.4	12.8	E/S0	150	Mag 10.70v star W 4′.8; NGC 2458 NW 2′.7.
07 56 32.4	+56 41 10	NGC 2462	13.2	0.6 x 0.4	11.5		162	**MCG +9-13-92** SW 9′.3.
07 57 12.6	+56 40 32	NGC 2463	14.2	0.6 x 0.6	12.9			Mag 10.30v star S 3′.3.
07 58 02.4	+56 21 35	NGC 2468	13.9	1.3 x 0.6	13.5	SA0:	57	Small anonymous galaxy S edge.
07 58 03.3	+56 40 46	NGC 2469	12.7	1.1 x 0.8	12.4	Sbc pec:	160	Mag 9.53v star NNE 2′.5.
07 58 41.7	+56 42 01	NGC 2472	15.3	0.5 x 0.5	13.6	Sb		Almost stellar; mag 9.53v star W 4′.4.
07 57 59.0	+52 51 24	NGC 2474	13.2	0.5 x 0.5	11.8	E0		Pair with NGC 2475.
07 58 00.8	+52 51 42	NGC 2475	13.1	0.8 x 0.8	12.6	E1		Located 2′.2 SW of a mag 9.27v star; pair with NGC 2474.
08 01 45.7	+56 33 10	NGC 2488	12.4	1.4 x 0.8	12.4	S0$^-$:	100	Mag 9.75v star W 6′.9; close pair **MCG +9-13-111** and **MCG +9-13-112** ESE 6′.8.
08 02 11.0	+56 56 28	NGC 2497	13.2	1.4 x 1.2	13.7	E?	24	Elongated **MCG +10-12-64** SE 1′.4.
08 04 06.8	+53 32 58	NGC 2505	13.2	1.2 x 0.6	12.7	SBa	0	Mag 7.04v star NNE 8′.9.
08 07 20.5	+51 07 53	NGC 2518	13.0	1.2 x 1.0	13.1	S0$^-$:	35	
08 08 49.3	+57 46 05	NGC 2521	12.8	1.2 x 0.7	12.5	SA0$^-$ pec?	45	Mag 8.66v star N 3′.6; UGC 4241 E 4′.7.
08 12 54.2	+55 40 18	NGC 2534	12.9	1.0 x 1.0	13.0	E1? pec		Mag 7.97v star S 3′.0.
08 18 58.4	+57 47 57	NGC 2549	11.2	3.9 x 1.3	12.9	SA(r)0° sp	177	Mag 9.61v star SW 8′.7; MCG +10-12-120 WSW 6′.9.
08 34 45.4	+52 42 55	NGC 2600	14.2	1.2 x 0.3	12.9	Sb II-III	78	Close pair of mag 11.8, 12.2 stars SW 1′.9.
08 35 04.3	+52 49 52	NGC 2602	14.7	0.4 x 0.2	11.8		33	Mag 10.77v star ENE 2′.6.
08 35 34.4	+52 47 16	NGC 2603	14.3	0.7 x 0.3	12.5	Sbc	39	Mag 10.77v star NNW 3′.9.
08 49 11.8	+60 13 15	NGC 2654	11.8	4.3 x 0.8	13.0	SBab: sp II-III	63	Mag 10.96v star N 4′.4.
08 47 52.1	+53 52 35	NGC 2656	13.8	1.0 x 1.0	13.7	SA0$^-$ pec:		Multi-galaxy system. Bright knot or companion galaxy W edge.
08 52 05.0	+53 37 00	NGC 2675	13.3	1.5 x 1.1	13.8	E	80	
08 53 32.9	+51 18 48	NGC 2681	10.3	3.6 x 3.3	12.8	(R′)SAB(rs)0/a	36	
08 55 34.9	+58 44 03	NGC 2685	11.3	4.6 x 2.5	13.8	(R)SB0$^+$ pec	38	**Helix Galaxy**. Mag 10.99v star just off N edge; almost stellar anonymous galaxy E 3′.0.
08 56 58.0	+52 04 01	NGC 2692	13.3	1.3 x 0.5	12.7	SBab:	165	Mag 9.82v star E 5′.6; UGC 4671 NW 3′.3.
08 56 59.5	+51 20 53	NGC 2693	11.9	2.6 x 1.8	13.5	E3:	160	NGC 2694 on S edge; **UGC 4679** N 7′.6; **MCG +9-15-54** NNW 9′.3.
08 56 59.4	+51 19 53	NGC 2694	14.4	1.2 x 1.2	14.8	E1		Stellar nucleus; located on S edge of NGC 2693.
08 59 06.5	+53 46 10	NGC 2701	12.3	2.2 x 1.6	13.5	SAB(rs)c: II-III	23	Mag 8.14v star E 7′.8.
08 59 48.5	+55 42 21	NGC 2710	12.9	2.0 x 1.0	13.5	SB(rs)b	125	Two main arms; mag 10.17v star SE 7′.0.
06 56 00.6	+55 24 00	UGC 3595	14.0	1.0 x 0.9	13.7	(R)SB(r)b:	165	
06 56 30.8	+60 39 04	UGC 3598	14.3	2.0 x 1.1	15.0	IBm	25	Irregular, very filamentary; mag 9.62v star W edge.
07 01 05.1	+51 16 08	UGC 3627	13.9	1.2 x 0.8	13.7	Sd	80	Elongated **UGC 3625** N 4′.3; stellar **PGC 20001** N 5′.5.
07 04 16.4	+54 13 19	UGC 3646	13.6	0.8 x 0.6	12.8	E	76	
07 08 23.6	+51 14 01	UGC 3684	14.0	1.4 x 0.9	14.1	(R′)SB(s)b	176	Bright SE-NW bar, faint envelope.
07 09 05.7	+61 35 44	UGC 3685	12.0	2.0 x 1.9	13.3	SB(rs)b	150	Pair of bright knots or superimposed stars SW of center.
07 10 30.7	+61 47 08	UGC 3704	13.8	1.3 x 0.3	12.6	Sa	50	Mag 8.10v star N 4′.8.
07 16 02.0	+56 49 02	UGC 3765	13.3	0.8 x 0.8	12.7	S0		**MCG +9-12-47** E 1′.3.
07 19 30.9	+59 21 18	UGC 3789	12.4	1.5 x 1.3	13.0	(R)SA(r)ab	170	**UGC 3797** E 4′.3.
07 19 18.8	+51 17 34	UGC 3792	12.8	1.7 x 1.2	13.4	SA0/a	65	
07 20 06.5	+53 00 33	UGC 3799	14.1	1.0 x 0.9	13.8	SAB(s)d		Stellar nucleus, uniform envelope.
07 23 12.4	+58 03 50	UGC 3816	12.8	1.0 x 0.7	12.3	S0	112	MCG +10-11-33 NNW 2′.2; mag 9.71v star S 2′.1.
07 24 27.8	+61 41 41	UGC 3826	13.5	3.6 x 3.2	16.0	SAB(s)d	85	Very small, weak nucleus, faint, uniform envelope.
07 24 35.9	+57 58 03	UGC 3828	12.1	2.0 x 1.2	12.9	SAB(rs)b	0	Bright core, two main arms.
07 25 14.2	+53 26 42	UGC 3832	14.1	1.2 x 0.7	13.8	SBab	107	
07 28 13.1	+58 30 25	UGC 3855	13.3	1.8 x 0.4	12.8	Sab	52	Star on E edge.
07 29 34.4	+57 06 56	UGC 3867	14.5	1.3 x 0.9	14.5	Sb	132	Multi-galaxy system; mag 10.65v star SE 2′.6.
07 29 50.0	+52 48 28	UGC 3875	14.3	1.3 x 1.1	14.5	SAB(rs)d	50	
07 30 43.1	+55 32 08	UGC 3882	14.7	1.5 x 0.6	14.4	SA(r)cd:	170	Very close pair of mag 13 stars SW edge.
07 31 07.4	+59 28 51	UGC 3885	13.1	0.9 x 0.8	12.6	S?		
07 31 27.3	+62 27 24	UGC 3886	15.3	1.0 x 1.0	15.2	SA(s)c		Mag 9.46v star W edge; mag 6.99v star N 2′.9.
07 33 19.7	+59 37 28	UGC 3897	13.1	1.4 x 1.2	13.5	SB(r)0$^+$	45	Mag 10.38v star E edge; mag 11.37v star NE edge.
07 34 49.4	+62 32 45	UGC 3905	13.6	0.8 x 0.6	12.7	Sab	80	
07 36 46.9	+55 02 26	UGC 3922	14.8	1.5 x 0.2	13.4	Scd:	164	
07 36 48.3	+55 19 16	UGC 3923	14.0	0.9 x 0.5	13.0	Sb II-III	153	
07 37 42.4	+59 18 54	UGC 3928	14.0	1.7 x 0.8	14.2	Pec	54	Mag 10.82v star E 3′.7; stellar galaxy **KUG 0734+594A** E 4′.7.
07 39 11.2	+59 09 40	UGC 3943	14.0	1.5 x 0.6	13.7	SB(s)c	108	Stellar galaxy **KUG 0735+592** S 2′.4.
07 39 49.7	+54 19 26	UGC 3948	14.6	1.0 x 0.8	14.2	SB(r)cd	159	
07 40 58.2	+55 25 36	UGC 3957	12.9	1.2 x 1.2	13.3	E		Several bright stars on the edges S and NE; mag 10.21v star S 1′.8.
07 41 32.3	+51 41 18	UGC 3963	15.9	1.5 x 0.3	14.9	Sdm:	54	Mag 12.0 star NW 2′.1.
07 43 55.4	+62 04 14	UGC 3978	14.1	1.2 x 0.9	14.0	SBb	55	Bright SE-NW bar, faint envelope.
07 43 53.5	+56 59 14	UGC 3981	13.1	1.5 x 0.9	13.3	S0	73	
07 46 24.9	+62 29 11	UGC 4001	13.5	1.3 x 0.6	13.2	E	144	
07 46 28.5	+58 57 45	UGC 4003	13.7	1.5 x 0.7	13.6	Sa:	81	
07 47 28.7	+60 56 02	UGC 4013	12.7	1.7 x 0.7	12.7	Sb II-III	130	**MCG +10-11-141** E 2′.2.
07 47 48.1	+62 19 25	UGC 4015	14.4	1.1 x 0.6	13.8	SB(s)cd	54	Stellar **Mkn 82** NW 5′.9.
07 47 58.9	+59 00 51	UGC 4020	13.4	2.0 x 0.9	13.9	SABb I-II	18	Faint galaxy **SBS 743+591B** W 2′.9; **UGC 4012** W 5′.1.
07 48 33.4	+62 12 24	UGC 4024	14.1	1.5 x 0.4	13.4	Scd:	115	Mag 14.4 star W end.
07 49 47.1	+61 21 17	UGC 4033	13.9	1.0 x 0.6	13.2	SAB(r)ab III	168	Mag 10.82v star NNW 2′.8.
07 50 08.6	+55 23 00	UGC 4035	13.1	0.8 x 0.8	12.6	E		**MCG +9-13-69** NE 1′.9.
07 50 46.3	+54 21 41	UGC 4043	14.2	2.0 x 0.3	13.5	Sd	5	
07 51 18.7	+56 54 32	UGC 4049	13.7	1.6 x 1.0	14.1	S?	16	Mag 10.91v star N 1′.9.
07 52 36.4	+62 32 47	UGC 4059	14.5	1.1 x 0.6	13.9	Scd:	50	Patchy; knot NE edge.
07 53 05.1	+55 14 28	UGC 4065	14.8	1.5 x 0.2	13.4	Sd	159	Mag 6.36v star SW 4′.5.
07 53 34.2	+51 46 47	UGC 4071	14.1	0.9 x 0.7	13.4		74	Mag 10.80v star W 1′.8.
07 53 41.5	+52 52 39	UGC 4072	13.8	1.3 x 0.5	13.2	SB(s)b	17	**MCG +9-13-78** N 2′.1.
07 54 19.0	+54 15 18	UGC 4074	13.1	1.0 x 0.9	12.9	SA(rs)cd	144	Stellar nucleus.
07 55 06.7	+55 42 12	UGC 4079	13.7	1.2 x 0.7	13.4	S pec	6	**MCG +9-13-89** S 6′.6.
07 55 19.3	+53 19 50	UGC 4085	13.5	1.1 x 0.8	13.2	S?	65	

(Continued from previous page)

GALAXIES

RA h m s	Dec ° ′ ″	Name	Mag (V)	Dim ′ Maj x min	SB	Type Class	PA	Notes
07 56 01.1	+58 25 45	UGC 4092	13.5	1.1 x 0.5	12.8	Sab	56	Mag 12 star E 1.′6.
07 58 54.4	+58 02 27	UGC 4121	14.7	2.3 x 0.8	15.2	Sm: IV-V	172	Dwarf spiral, very low surface brightness; located between a pair of mag 12.4 stars NE and SW.
07 59 01.0	+59 06 59	UGC 4122	13.3	1.6 x 1.1	13.9	E	3	**UGC 4124** N edge.
08 00 12.9	+60 17 14	UGC 4128	13.1	1.7 x 1.0	13.5	Scd:	45	
08 00 08.5	+56 21 56	UGC 4133	14.4	1.7 x 0.3	13.6	Scd:	165	**UGC 4134** on S edge.
08 01 51.0	+61 24 44	UGC 4159	13.2	0.8 x 0.5	12.1	S?	98	Low surface brightness **UGC 4153** W 2.′9.
08 02 02.5	+56 38 33	UGC 4164	14.1	1.3 x 0.6	13.7	S0/a	138	MCG +10-12-63 NE 5.′1.
08 02 32.7	+61 23 17	UGC 4169	12.9	1.5 x 0.7	12.8	Scd	140	Close pair of N-S oriented mag 13 stars E 1.′6.
08 03 49.5	+61 20 44	UGC 4182	13.6	0.8 x 0.5	12.5	Disturbed	170	MCG +10-12-67 S 3.′6; stellar galaxy **KUG 0758+614** W 6.′0.
08 04 07.9	+62 59 01	UGC 4186	14.6	1.4 x 0.2	13.1	S	147	
08 04 44.3	+61 32 07	UGC 4196	14.3	1.0 x 0.7	13.8	Sd	165	Patchy.
08 08 06.6	+55 30 55	UGC 4230	14.6	1.6 x 0.2	13.2	S	171	Mag 9.17v star SE 5.′5; **MCG +9-14-10** SE 9.′3.
08 09 24.0	+57 45 44	UGC 4241	14.0	1.0 x 0.4	12.9	Sab	146	NGC 2521 W 4.′7.
08 12 56.7	+54 58 07	UGC 4267	13.5	0.9 x 0.5	12.5	S?	35	
08 13 21.2	+57 51 00	UGC 4270	13.7	1.5 x 1.2	14.2	SAB(rs)bc	130	Mag 10.8 star N 1.′3.
08 13 57.7	+52 38 50	UGC 4277	14.2	3.5 x 0.4	14.4	Scd:	110	Faint, thin dark lane most of length; mag 11.9 star N 2.′0.
08 14 33.3	+54 47 57	UGC 4280	13.6	1.5 x 0.4	12.9	Sa	6	
08 14 43.6	+57 57 23	UGC 4280A	13.4	0.8 x 0.5	12.5	E/S0	147	Galaxy **VII Zw 220** on E edge.
08 14 41.3	+58 13 27	UGC 4281	14.3	1.3 x 0.4	13.4	Sab	75	Multi-galaxy system; **MCG +10-12-102** NNE 1.′7.
08 15 44.7	+58 19 15	UGC 4289	13.1	1.4 x 1.2	13.6	E	105	**MCG +10-12-112** and **PGC 23168** NE 2.′2; MCG +10-12-108 SW 1.′7.
08 16 18.3	+52 25 20	UGC 4293	14.5	0.7 x 0.6	13.4	S	39	
08 18 58.2	+58 15 47	UGC 4314	14.0	0.8 x 0.6	13.1	Double System	18	
08 20 01.6	+62 49 50	UGC 4322	13.7	1.4 x 1.0	14.1	E	57	
08 20 42.4	+54 07 33	UGC 4335	13.8	0.9 x 0.5	12.8	(R)SB(r)a:	125	Pair of mag 12 stars N 1.′8.
08 22 40.2	+56 19 43	UGC 4357	13.6	1.0 x 1.0	13.5	(R)SA(r)0°:		Mag 13.8 star S edge of nucleus.
08 24 31.8	+54 51 12	UGC 4380	14.0	1.1 x 1.1	14.1	Scd:		
08 27 54.0	+55 09 34	UGC 4410	14.2	1.1 x 0.5	13.4	S	0	Mag 7.91v star NW 2.′8.
08 28 02.0	+54 26 43	UGC 4415	14.4	0.9 x 0.6	13.6	SB(s)d	173	
08 29 14.7	+55 31 22	UGC 4427	14.0	1.0 x 0.4	12.9	Sbc	6	
08 30 00.1	+52 41 46	UGC 4438	13.2	0.9 x 0.7	12.5	S(r) pec:	18	
08 30 41.5	+52 17 59	UGC 4442	14.0	1.1 x 0.8	13.7	SAB(s)bc II	110	Strong dark patch SE of center.
08 31 26.7	+60 59 40	UGC 4445	14.1	1.5 x 1.5	14.8	SA(s)c		Mag 14 star on W edge; mag 9.86v star SW 3.′3.
08 33 22.0	+52 31 49	UGC 4461	13.5	1.7 x 0.5	13.2	Sbc	43	**MCG +9-14-54** SW 8.′0.
08 35 17.6	+55 49 41	UGC 4478	14.3	1.3 x 0.9	14.3	(R)SB(s)a	32	Bright bar, faint envelope.
08 37 41.2	+51 39 14	UGC 4499	12.9	2.8 x 2.2	14.7	SABdm	140	Bright "knot" S of core is **Mkn 94**.
08 39 39.6	+60 58 08	UGC 4512	14.4	1.4 x 0.7	14.2	Scd:	110	Mag 9.18v star E 6.′6.
08 39 37.7	+53 27 23	UGC 4514	13.2	2.1 x 1.0	13.8	SBcd?	70	Bright, elongated center; mag 13.4 star on NE edge.
08 40 09.4	+52 27 20	UGC 4515	13.6	1.5 x 0.6	13.3	SB(r)b:	175	Mag 12.5 star NW 1.′5.
08 41 32.1	+51 14 44	UGC 4525	14.1	1.2 x 0.7	13.7	SB(s)b	88	Mag 13.2 star 0.′9 W.
08 43 51.6	+51 59 34	UGC 4546	14.1	1.3 x 0.3	12.9	Sa	25	
08 44 21.5	+58 50 29	UGC 4549	13.4	1.5 x 1.3	14.0	Sdm?	19	
08 45 23.5	+55 07 05	UGC 4564	14.1	0.9 x 0.3	12.5	S	37	Spherical **MCG +9-15-16** on SW edge.
08 51 40.8	+51 07 07	UGC 4628	14.6	1.7 x 0.3	13.7	Scd:	74	
08 56 42.5	+52 06 18	UGC 4671	12.7	1.5 x 1.3	13.3	S?	96	Mag 12.9 star Superimposed NE of nucleus; NGC 2692 SE 3.′3.
08 57 54.5	+59 04 55	UGC 4683	14.1	1.8 x 1.0	14.6	Im	110	Dwarf irregular, extremely low surface brightness.
08 58 10.6	+52 10 57	UGC 4690	13.9	1.1 x 0.5	13.1	Sab	120	Mag 12.0 star N 1.′2.
08 59 07.5	+53 37 57	UGC 4696	15.1	1.1 x 0.2	13.3	Scd:	135	Mag 10.88v star E 1.′5.
09 00 20.3	+52 29 39	UGC 4713	12.8	1.7 x 1.2	13.4	Sb II-III	177	Located between a mag 12.0 star E and a mag 10.97v star W.
09 00 37.4	+51 12 25	UGC 4717	14.1	1.0 x 0.5	13.2	Sab	40	
09 01 58.7	+60 09 04	UGC 4730	13.7	0.8 x 0.3	12.0	S?	93	**UGC 4727** W 1.′9; mag 10.22v star SW 1.′3.

GALAXY CLUSTERS

RA h m s	Dec ° ′ ″	Name	Mag 10th brightest	No. Gal	Diam ′	Notes
07 21 24.0	+55 44 00	A 576	14.4	61	28	Numerous MCG, CGCG and PGC GX populate the center one third of the cluster diameter.
07 48 48.0	+52 04 00	A 595	15.6	45	17	A N-S line of stellar GX in the NW quadrant, and another N-S line E of the center.
08 14 36.0	+58 02 00	A 634	14.9	40	28	Many faint stellar and almost stellar MCG galaxies, particularly in N half.

PLANETARY NEBULAE

RA h m s	Dec ° ′ ″	Name	Diam ″	Mag (P)	Mag (V)	Mag cent ★	Alt Name	Notes
08 41 35.5	+58 13 49	PN G158.8+37.1	270	>14.6	13.5	17.4	PK 158+37.1	= **Abell 28**; mag 14.3 star 1.′1 W of center; mag 12.6 star E 2.′5, on E edge; mag 11.3 star S 4.′0.
07 57 51.7	+53 25 16	PN G164.8+31.1	380	14.	12.1	16.8	PK 164+31.1	Annular with brightest areas along NW and SE edges.

GALAXIES

RA h m s	Dec ° ′ ″	Name	Mag (V)	Dim ′ Maj x min	SB	Type Class	PA	Notes
05 11 20.0	+51 31 58	CGCG 258-1	15.1	0.8 x 0.4	13.8	S	150	Mag 12.8 star ENE 1.′8; mag 12.7 star WNW 2.′4.
05 32 35.9	+51 35 19	CGCG 258-9	16.1	0.8 x 0.5	14.9			Very faint stellar nucleus; mag 10.7 star N 1.′6.
05 33 27.3	+51 45 00	CGCG 258-10	15.2	0.8 x 0.5	14.1		42	Located between two mag 10.5 stars N-S.
06 04 00.2	+51 05 34	CGCG 259-8	14.7	0.4 x 0.4	12.6			Mag 10.06v star N 2.′4.
06 37 20.5	+56 22 45	CGCG 260-20	14.7	0.8 x 0.5	13.6	SBc	0	Bright E-W bar with two main N-S arms.
06 48 36.0	+51 50 23	CGCG 260-37	14.3	1.2 x 0.6	13.8		150	Bright center with faint envelope; faint star SE edge of center; mag 12.3 star NW 2.′8.
05 14 15.5	+62 34 28	CGCG 283-4	13.3	5.5 x 3.4	16.3	Im?	9	Very filamentary; many superimposed stars; mag 8.33v star SE 3.′4.
06 26 55.5	+59 04 47	IC 2166	12.4	3.0 x 2.1	14.2	SAB(s)bc	115	Mag 10.68v star W 3.′3.
06 00 49.5	+53 33 07	MCG +9-10-4	15.4	0.9 x 0.7	14.7	SBbc	126	Close pair of stars, mags 10.7 and 11.5, NE 1.′9.
06 06 42.2	+54 09 18	MCG +9-10-5	16.1	1.1 x 1.1	16.2			Several faint stars superimposed; mag 12.2 star SE 2.′3.
06 12 04.0	+51 51 48	MCG +9-11-2	15.0	0.9 x 0.6	14.2	Sc	99	Pair of mag 11.2 star NE 2.′2.
06 16 09.2	+54 32 04	MCG +9-11-8	14.4	1.0 x 0.8	14.0	Sb	129	Mag 9.1 star NW 5.′9.
06 50 22.9	+53 27 10	MCG +9-12-1	13.6	0.8 x 0.7	12.8		153	Mag 11.3 star on NW edge; mag 9.43v star E 1.′3.
06 52 44.3	+55 35 41	MCG +9-12-4	14.1	0.8 x 0.8	13.5			Faint, round anonymous galaxy E 1.′8; mag 11.05v star SE 3.′2.
06 53 36.9	+54 22 02	MCG +9-12-6	14.4	0.9 x 0.5	13.4	SBc	18	Elongated **MCG +9-12-5** NW 1.′4.
07 03 59.0	+56 29 08	MCG +9-12-22	14.9	1.0 x 0.4	13.8		66	Close pair of mag 9 stars S 2.′6.
07 03 54.9	+53 39 58	MCG +9-12-23	14.7	0.7 x 0.5	13.4		33	

(Continued from previous page)

27 GALAXIES 27

RA h m s	Dec ° ′ ″	Name	Mag (V)	Dim ′ Maj x min	SB	Type Class	PA	Notes
06 00 54.8	+60 50 18	MCG +10-9-6	14.7	0.8 x 0.4	13.3	Sc	126	
06 15 51.0	+61 04 22	MCG +10-9-14	14.7	0.8 x 0.7	13.9	Sd	162	Mag 12.3 star N 1′.8.
06 22 47.9	+57 33 14	MCG +10-9-18	14.5	0.9 x 0.6	13.7	Sbc	21	Mag 9.29v star E edge.
06 31 21.5	+58 47 02	MCG +10-10-3	14.7	0.8 x 0.7	13.9	Sc	132	Mag 11.1 star SW 4′.0.
07 02 55.0	+62 46 09	MCG +10-10-21	14.0	0.8 x 0.5	12.9	SB?	114	Mag 11.5 star W 1′.6.
06 04 34.5	+57 37 40	NGC 2128	12.6	1.5 x 1.1	13.0	S0-:	60	Mag 9.62v star N 6′.2.
06 22 34.8	+51 54 31	NGC 2208	12.8	1.7 x 1.0	13.2	S0:	110	Mag 9.86v star NE 5′.4.
06 50 08.6	+60 50 45	NGC 2273	11.7	3.6 x 2.0	13.7	SB(r)a:	50	Slightly elongated E-W core in uniform NE-SW envelope; mag 8.58v star N 5′.2.
06 46 31.8	+60 20 24	NGC 2273B	12.5	2.7 x 1.2	13.7	SB(rs)cd:	43	Mag 10.59v star on SW end.
05 00 44.1	+62 14 38	UGC 3218	13.5	1.7 x 0.9	13.8	SAb	145	Numerous stars superimposed.
05 12 29.8	+51 17 13	UGC 3260	14.1	1.3 x 0.3	12.9	S0?	116	**CGCG 258-4** NE 7′.7.
05 17 44.9	+53 33 01	UGC 3273	14.2	3.0 x 0.9	15.1	Sm	45	Numerous superimposed stars.
05 30 11.7	+55 52 16	UGC 3314	15.2	1.7 x 0.2	13.9	S?	120	Mag 12.7 star S 1′.0.
05 44 10.0	+51 11 58	UGC 3346	14.2	1.3 x 0.5	13.6	Sdm:	135	
05 45 48.2	+58 41 58	UGC 3351	14.0	1.5 x 0.3	13.0	Sab	163	Mag 10.1 star SE 3′.4.
05 47 18.5	+56 06 41	UGC 3354	13.8	1.9 x 0.6	13.8	Sab:	164	Mag 5.93v star W 6′.7.
05 55 25.5	+51 54 39	UGC 3375	13.2	2.1 x 1.0	13.9	SA(rs)c	45	Mag 10.78v star SW 1′.9.
05 59 47.7	+62 09 29	UGC 3382	13.7	1.1 x 1.1	13.8	SB(rs)a		Almost stellar nucleus in short NE-SW bar, uniform envelope.
06 04 49.8	+56 09 57	UGC 3394	13.8	1.5 x 1.2	14.3	SB?	45	Mag 10.10 star E 1′.2.
06 10 55.0	+62 01 12	UGC 3411	14.4	1.7 x 1.1	14.9	Scd?	44	Stellar nucleus.
06 16 13.4	+57 03 07	UGC 3432	14.6	1.2 x 0.3	13.3	Scd:	136	Mag 11.4 star 1′.9 NW.
06 21 31.3	+59 07 36	UGC 3445	13.3	1.4 x 0.3	12.2	S0/a	101	On W edge of UGC 3446.
06 21 39.4	+59 07 31	UGC 3446	12.9	1.4 x 1.0	13.1	S0:	150	UGC 3445 on W edge.
06 33 07.5	+53 31 07	UGC 3480	14.2	1.1 x 1.1	14.3	SAB(rs)c		Mag 10.9 star SW 3′.1.
06 34 02.0	+58 51 37	UGC 3484	14.7	1.7 x 0.2	13.4	Sbc	75	Elongated **MCG +10-10-4** W 5′.5; mag 8.37v star NW 7′.5.
06 34 09.7	+52 07 37	UGC 3488	13.9	1.0 x 0.6	13.2	S0/a	50	
06 40 07.1	+60 04 52	UGC 3504	12.3	2.5 x 1.9	13.8	SAB(s)cd	123	Many knots and/or superimposed stars.
06 49 56.1	+61 37 22	UGC 3545	15.1	1.0 x 1.0	14.9	SB?		
06 52 27.6	+57 09 40	UGC 3569	14.6	1.1 x 0.9	14.4	Sdm	159	Mag 11.2 star S 1′.2.
06 53 10.2	+57 10 39	UGC 3574	12.5	4.1 x 3.5	15.2	SA(s)cd	120	Semi-circle of mag 11-12 stars around NW edge.
06 56 00.6	+55 24 00	UGC 3595	14.0	1.0 x 0.9	13.7	(R)SB(r)b:	165	
06 56 30.8	+60 39 04	UGC 3598	14.3	2.0 x 1.1	15.0	IBm	25	Irregular, very filamentary; mag 9.62v star W edge.
07 01 05.1	+51 16 08	UGC 3627	13.9	1.2 x 0.8	13.7	Sd	80	Elongated **UGC 3625** N 4′.3; stellar **PGC 20001** N 5′.5.

OPEN CLUSTERS

RA h m s	Dec ° ′ ″	Name	Mag	Dia m ′	No. ★	B ★	Type	Notes
05 03 28.0	+52 52 00	NGC 1708		20	10	10.0	cl?	(A) 20′ diameter loose cluster which stands out in 1 degree DSS image.
05 11 02.0	+52 07 00	NGC 1790		15	8	10.0	ast?	(A) Irregular group of brighter stars.
05 44 24.0	+55 47 36	NGC 2013			8	11.0	cl??	
06 11 05.0	+51 40 36	NGC 2165		6	15	10.0	cl?	(S) Sparse group.

PLANETARY NEBULAE

RA h m s	Dec ° ′ ″	Name	Diam ″	Mag (P)	Mag (V)	Mag cent ★	Alt Name	Notes
06 43 54.9	+61 47 25	PN G153.7+22.8	140	15.9	14.5	17.4	PK 153+22.1	Mag 11.16v star S 2′.1; mag 9.32v star SSW 3′.2.
06 19 33.9	+55 36 45	PN G158.9+17.8	1200	11.2		15.3	PK 158+17.1	

28 GALAXIES 28

RA h m s	Dec ° ′ ″	Name	Mag (V)	Dim ′ Maj x min	SB	Type Class	PA	Notes
05 00 44.1	+62 14 38	UGC 3218	13.5	1.7 x 0.9	13.8	SAb	145	Numerous stars superimposed.

OPEN CLUSTERS

RA h m s	Dec ° ′ ″	Name	Mag	Diam ′	No. ★	B ★	Type	Notes
03 32 38.0	+52 39 06	Be 9		4	20	15.0	cl	
04 55 46.2	+52 49 00	Be 13		8	130	15.0	cl	
03 04 05.0	+58 44 30	Be 66		4	30	16.0	cl	(S) Stars very faint.
02 59 23.2	+60 34 00	Cr 34	6.8	24		7.9	cl:	Few stars; large brightness range; strong central concentration; detached. At east end of nebular complex. IC 1848 at west end.
03 23 13.0	+52 13 12	Cz 15		4	10		cl	(S) Sparse.
03 52 20.0	+61 58 00	Cz 17		4	15		cl:	(A) Includes two "arcs" of stars or partial "stellar rings."
04 18 51.0	+58 15 00	IC 361	11.7	7	60	14.0	cl	Rich in stars; moderate brightness range; slight central concentration; detached.
03 14 40.0	+52 42 00	King 5		6	40	13.0	cl	Moderately rich in stars; moderate brightness range; strong central concentration; detached.
03 28 00.0	+56 27 00	King 6		10	35	10.0	cl	Moderately rich in stars; moderate brightness range; slight central concentration; detached.
03 59 10.0	+51 47 00	King 7		8	80	16.0	cl	
04 19 51.0	+53 09 30	Mayer 2		3	18		cl	
03 11 41.0	+53 20 54	NGC 1220	11.8	2	15	13.0	cl	Few stars similar in brightness; strong central concentration; detached.
03 34 09.0	+51 25 12	NGC 1348		6	30	8.5	cl	Moderately rich in stars; moderate brightness range; no central concentration; detached.
03 49 27.0	+52 39 18	NGC 1444	6.6	4	57	6.8	cl	Few stars; small brightness range; not well detached. Visual mag 8.8 without brightest star.
04 04 32.0	+52 39 42	NGC 1496	9.6	3	10	12.0	cl	Few stars; moderate brightness range; no central concentration; detached.
04 07 50.0	+62 20 00	NGC 1502	6.9	20	63	7.0	cl	Moderately rich in bright and faint stars with a strong central concentration; detached. (A) At SE end of W.S. Houston's **Kemble's Cascade**.
04 15 19.0	+51 13 00	NGC 1528	6.4	18	165	10.0	cl	Moderately rich in stars; moderate brightness range; slight central concentration; detached.
05 03 28.0	+52 52 00	NGC 1708		20	10	10.0	cl?	(A) 20′ diameter loose cluster which stands out in 1 degree DSS image.
03 16 10.8	+60 07 00	St 23		29	25	7.6	cl	Few stars; large brightness range; slight central concentration; detached; involved in nebulosity. (A) Named **Pazmino's Cluster** by W. S. Houston, after John Pazmino.
03 47 44.0	+59 05 00	Tom 5	8.4	15	102	14.0	cl	Rich in stars; moderate brightness range; no central concentration; detached.

(Continued from previous page)

28 BRIGHT NEBULAE 28

RA h m s	Dec ° ′ ″	Name	Dim ′ Maj x min	Type	BC	Color	Notes
03 06 24.0	+60 41 00	IC 1871	4 x 4	E	4-5	3-4	
04 03 24.0	+51 19 00	NGC 1491	25 x 25	E	1-5	3-4	Considerably smaller (about 4′) on blue photographs than on red. Visual observers may be able to detect two strong wisps forming a narrow V-shape (open end to NNE) in the nebula's brightest part. A star about 11th mag lies just SE of this area.
03 56 06.0	+53 12 00	Sh2-205	100 x 30	E	5-5	3-4	Very faint and diffuse; of fairly uniform surface brightness. The brightest part, about 20′ across, is roughly centered on the 8th mag star at 03 51.9 +53 29.
03 29 12.0	+59 57 00	vdB 14	20 x 8	R	3-5	1-4	
03 30 06.0	+58 54 00	vdB 15	25 x 10	R	3-5	1-4	

DARK NEBULAE

RA h m s	Dec ° ′ ″	Name	Dim ′ Maj x min	Opacity	Notes
03 56 23.4	+56 07 00	B 6		4	Round; indefinite.
04 18 56.9	+55 04 00	B 9			Dark, irregular vacancy.
04 26 37.3	+55 02 00	B 11	10 x 10	3	East end of irregular lane, B8.
04 29 47.4	+54 15 00	B 12	24 x 24	5	Isolated dark spot SE of B11.
04 31 18.1	+54 54 00	B 13	11 x 11	4	Irregular.
04 37 57.6	+55 22 00	B 21	10 x 10	4	Indefinite; irregularly round.

PLANETARY NEBULAE

RA h m s	Dec ° ′ ″	Name	Diam ″	Mag (P)	Mag (V)	Mag cent ★	Alt Name	Notes
03 10 19.7	+61 19 01	IC 289	48	12.3	13.2	15.9	PK 138+2.1	Mag 10.0 star S 1′.8.
04 06 59.6	+60 55 11	NGC 1501	52	13.3	11.5	14.3	PK 144+6.1	
03 11 02.1	+62 47 57	PN G138.1+4.1	340			12.6	PK 138+4.1	Widely spaced pair of mag 11 stars near center; planetary is located in the SE corner of the large bright nebula Sh2-200.
03 36 08.1	+60 03 45	PN G142.1+3.4	12	16.1			PK 142+3.1	Mag 11.1 star N 2′.5; mag 11.1 star ENE 2′.9.
04 25 51.2	+60 07 11	PN G146.7+7.6	10		15.8	13.9	PK 146+7.1	Mag 11.10v star NE 2′.1; mag 11.1 star NW 3′.1.
03 41 43.5	+52 16 57	PN G147.4−2.3	8	13.6	13.6	16.7	PK 147−2.1	Mag 12.3 star N 3′.8; mag11.36v star S 3′.2.
04 13 15.2	+56 56 56	PN G147.8+4.1	14	14.0	13.7		PK 147+4.1	Mag 12.9 star N 1′.1; mag 11.4 star SE 2′.6.
04 20 45.2	+56 18 11	PN G149.0+4.4	0				PK 149+4.1	Mag 12.1 star NNW 2′.5; mag 14.3 star SW 2′.7.
04 13 27.4	+51 50 59	PN G151.4+0.5	14	16.9			PK 151+0.1	Mag 11.8 star E 2′.5; mag 10.95v star NNE 3′.7.

29 GALAXIES 29

RA h m s	Dec ° ′ ″	Name	Mag (V)	Dim ′ Maj x min	SB	Type Class	PA	Notes
02 36 35.0	+59 39 09	Maffei I	11.4	3.2 x 3.2	13.8	S0⁻ pec:		= **Sh 2-191**. Likely asterism **Cz 11** superimposed.
02 41 56.0	+59 36 16	Maffei II	13.7	3.2 x 3.2	16.1	SAB(rs)bc:		Small triangle of mag 11-12 stars, oriented N-S, located NW 3′.2; mag 8.50v star NW 8′.1.
02 13 16.9	+53 24 40	UGC 1699	13.8	1.7 x 0.4	13.3	S0/a	84	Stellar nucleus; mag 12.0 star E 1′.1.
02 28 30.2	+51 26 45	UGC 1930	13.5	2.7 x 2.0	15.2	SB(r)0⁺	27	Bright nucleus in SE-NW bar, numerous superimposed stars on uniform envelope.
02 55 11.1	+51 54 25	UGC 2380	15.2	1.5 x 0.7	15.1	Sb	9	Numerous faint stars superimposed.

OPEN CLUSTERS

RA h m s	Dec ° ′ ″	Name	Mag	Diam ′	No. ★	B ★	Type	Notes
02 19 28.0	+58 18 00	Bas 10	9.9	2	23	10.2	cl	Few stars; small brightness range; slight central concentration; detached.
01 47 44.0	+62 56 24	Be 5		5	15	17.0	cl	
01 51 12.0	+61 03 36	Be 6		5	20	14.0	cl	
01 54 12.0	+62 22 12	Be 7		4	18	14.0	cl	
02 39 06.0	+60 24 00	Be 65	10.2	5	41	13.0	cl	Few stars; large brightness range; slight central concentration; detached.
02 59 23.2	+60 34 00	Cr 34	6.8	24		7.9	cl:	Few stars; large brightness range; strong central concentration; detached. At east end of nebular complex. IC 1848 at west end.
01 03 07.0	+62 47 12	Cz 3	9.9	1.5	10		cl:	
01 35 36.4	+61 28 35	Cz 4		4	10		cl:	
01 55 40.0	+61 21 24	Cz 5		2	10		cl	
02 02 08.0	+62 50 48	Cz 6		2	12		cl	
02 03 01.0	+62 14 48	Cz 7		4	20		cl:	
02 32 58.0	+58 45 48	Cz 8	9.7	2	19	9.9	cl:	Few stars; moderate brightness range; slight central concentration; detached.
02 33 36.0	+59 53 00	Cz 9		10	12		cl:	
02 33 55.0	+60 10 30	Cz 10		4	15		cl:	
02 39 23.0	+54 55 06	Cz 12		3	15		cl	
02 44 27.0	+62 19 30	Cz 13	10.4	4	56	12.8	cl	Few stars; moderate brightness range; no central concentration; detached. Probably not a cluster.
01 52 22.0	+61 51 00	IC 166	11.7	8	179	17.0	cl	Moderately rich in stars; small brightness range; slight central concentration; detached.
02 32 42.0	+61 27 00	IC 1805	6.5	20	62	7.9	cl	Few stars; large brightness range; no central concentration; detached; involved in nebulosity.
02 51 11.0	+60 24 00	IC 1848	6.5	18	74	7.1	cl	Few stars; large brightness range; strong central concentration; detached; involved in nebulosity.
02 36 02.0	+59 01 00	King 4	10.5	5	44	13.0	cl	Few stars; small brightness range; no central concentration; detached.
02 29 40.0	+60 42 24	Mrk 6	7.1	6	29	8.8	cl	Few stars; small brightness range; no central concentration; detached.
01 05 11.0	+62 01 18	NGC 358		3	25		cl?	
01 06 27.0	+62 13 42	NGC 366		4	30	10.0	cl	Moderately rich in bright and faint stars; detached.
01 08 18.0	+61 35 00	NGC 381	9.3	7	50	10.0	cl	Moderately rich in stars; small brightness range; no central concentration; detached.
01 15 10.0	+60 07 30	NGC 433		4	15	9.0	cl	Few stars; moderate brightness range; no central concentration; detached.
01 15 58.0	+58 49 00	NGC 436	8.8	5	49	10.0	cl	Moderately rich in stars; moderate brightness range; strong central concentration; detached.
01 19 33.0	+58 17 00	NGC 457	6.4	20	204	8.6	cl	Rich in stars; large brightness range; slight central concentration; detached.
01 33 22.0	+60 39 30	NGC 581	7.4	6	172	9.0	cl	= **M 103**. Moderately rich in stars; moderate brightness range; slight central concentration; detached.
01 44 00.0	+61 53 00	NGC 654	6.5	6	83	10.0	cl	Rich in stars; moderate brightness range; slight central concentration; detached.
01 43 29.8	+55 53 00	NGC 657		7	40	7.1	cl	(A) Circlet of 5 stars included on SW side.
01 44 24.0	+60 40 12	NGC 659	7.9	6	186	10.0	cl	Moderately rich in stars; moderate brightness range; strong central concentration; detached.
01 46 17.0	+61 13 00	NGC 663	7.1	15	108	9.0	cl	Rich in stars; large brightness range; slight central concentration; detached.
01 58 32.0	+60 10 00	NGC 743		7	12	10.0	cl	Few stars; small brightness range; not well detached. (S) Sparse.
01 58 30.0	+55 28 30	NGC 744	7.9	5	99	10.0	cl	Few stars; small brightness range; no central concentration; detached.
02 19 04.0	+57 08 00	NGC 869	5.3	18	317	6.6	cl	**Double Cluster** with NGC 884. Rich in stars; large brightness range; strong central concentration; detached.
02 22 05.0	+57 08 00	NGC 884	6.1	18	303	6.4	cl	**Double Cluster** with NGC 869. Rich in stars; large brightness range; strong central concentration; detached.
02 33 20.0	+57 34 00	NGC 957	7.6	10	119	11.0	cl	Moderately rich in stars; moderate brightness range; no central concentration; detached.

(Continued from previous page)

29 OPEN CLUSTERS 29

RA h m s	Dec ° ′ ″	Name	Mag	Diam ′	No. ★	B ★	Type	Notes
02 42 36.0	+61 36 00	NGC 1027	6.7	15	152	9.0	cl	Moderately rich in bright and faint stars; detached; involved in nebulosity.
02 14 43.0	+59 29 00	St 2	4.4	60	166	8.2	cl	Moderately rich in stars; moderate brightness range; strong central concentration; detached. (A) Large group of bright stars.
01 12 03.0	+62 16 24	St 3		5	8	11.0	cl	Few stars; small brightness range; not well detached.
01 52 42.0	+57 04 00	St 4		12	15	11.0	cl	Few stars; small brightness range; not well detached. (S) Too sparse. (A) DSS image shows group of brighter stars in rich Milky Way.
02 29 10.7	+61 47 06	Tom 4		5.3	40	16.0	cl	
01 35 40.0	+61 17 12	Tr 1	8.1	3	112	10.0	cl	Few stars; moderate brightness range; slight central concentration; detached.
02 36 53.0	+55 55 00	Tr 2	5.9	17	109	7.4	cl	Few stars; moderate brightness range; slight central concentration; detached.

BRIGHT NEBULAE

RA h m s	Dec ° ′ ″	Name	Dim ′ Maj x min	Type	BC	Color	Notes
00 56 42.0	+61 04 00	IC 59	10 x 5	E+R	2-5	1-4	With IC 63 collectively known as **Sharpless 185**, the N part.
00 59 30.0	+60 49 00	IC 63	10 x 3	E+R	1-5	2-4	With IC 59 collectively known as **Sharpless 185**, the E part.
02 26 30.0	+62 04 00	IC 1795	40 x 15	E	1-5	3-4	Separated from NGC 896 by a narrow dark lane aligned N-S. It is brightest at the N end of the dark lane and along the irregular N extension.
02 32 42.0	+61 27 00	IC 1805	60 x 60	E	3-5	3-4	An oval ring, irregular in shape and slightly extended N-S. The ring is broken along the S perimeter and in the NW sector. Brightest in the region of the open cluster **Mel 15** and the area to the E.
02 51 12.0	+60 26 00	IC 1848	120 x 55	E	2-5	3-4	Nebulosity most extensive and brightest in the NE sector.
02 24 48.0	+62 01 00	NGC 896	20 x 20	E	1-5	3-4	Separated from IC 1795 by a narrow dark lane aligned N-S. It is higher in surface brightness than IC 1795 and there is a fairly prominent absorption patch in the SW part.

DARK NEBULAE

RA h m s	Dec ° ′ ″	Name	Dim ′ Maj x min	Opacity	Notes
02 13 00.1	+57 05 00	B 201	10 x 10		Small; 45′ W of NGC 869.

PLANETARY NEBULAE

RA h m s	Dec ° ′ ″	Name	Diam ″	Mag (P)	Mag (V)	Mag cent ★	Alt Name	Notes
01 42 19.9	+51 34 35	NGC 650-51	187	12.2	10.1	15.9	PK 130−10.1	= **M 76, Little Dumbbell**. Bright elongated "core" oriented NE-SW with fainter nebulosity extending SE-NW to give it it's "dumbbell" shape.
01 00 53.9	+55 03 54	PN G124.3−7.7	150				PK 124−7.1	Mag 11.45v star NW 2′.5; mag 11.6 star S 3′.7.
01 42 38.2	+60 10 06	PN G129.2−2.0	195				PK 129−2.1	Numerous faint stars on disc; mag 14.7 star near center; close pair of mag 11 stars on E edge, 1′.9 E of center.
01 53 03.1	+56 24 17	PN G131.4−5.4	28	14.9	14.2	18.0	PK 131−5.1	Mag 10.8 star E 0′.8; mag 13.0 star NNE 0′.6.
01 58 49.7	+52 53 47	PN G133.1−8.6	18	14.9		13.4	PK 133−8.1	Mag 11.4 star W 1′.2; mag 11.04v star SW 0′.9.
01 30 45.0	+58 24 00	Sh2-188	430			17.4	PK 128−4.1	A large right angle or crescent-shaped planetary with one arm extending N and the other extending W; brightest along it's N-S arm. Also known as **PN G128.0-4.1**, **LBN 128.04-04.12** and **LBN 633**.

30 GALAXIES 30

RA h m s	Dec ° ′ ″	Name	Mag (V)	Dim ′ Maj x min	SB	Type Class	PA	Notes
23 37 13.7	+49 33 19	CGCG 548-1	14.0	1.0 x 0.6	13.3	Spiral	54	Multi-galaxy system, strong dark lane through core; mag 10.18v star S 1′.8.
23 59 15.9	+46 53 19	IC 1525	12.2	1.9 x 1.4	13.1	SBb	20	Mag 7.36v star and MCG +8-1-19 NE 9′.5.
00 13 45.6	+48 09 02	IC 1534	13.8	0.9 x 0.4	12.6	S0	72	Mag 9.91v star SW 3′.8; **IC 1535** E 2′.0.
00 14 19.1	+48 08 36	IC 1536	14.3	0.7 x 0.5	13.1	E/S0	171	**IC 1535** W 3′.8.
00 21 59.2	+41 47 57	MCG +7-1-8	15.3	0.6 x 0.5	13.9	Sb	165	MCG +7-1-9 E 1′.4.
00 22 06.6	+41 47 31	MCG +7-1-9	15.4	0.6 x 0.5	14.0	Sc	39	MCG +7-1-8 W 1′.4.
23 24 36.2	+44 13 41	MCG +7-48-6	15.2	0.5 x 0.4	13.3	Sc	168	Mag 11.20v star SW 1′.9.
23 37 12.8	+44 09 26	MCG +7-48-14	14.9	0.5 x 0.4	13.0	Sc	27	Mag 11.9 star NNW 2′.6.
23 40 23.1	+43 15 10	MCG +7-48-16	15.3	0.7 x 0.4	13.7	Sbc	27	Mag 10.03v star NE 3′.6.
23 52 28.7	+47 23 05	MCG +8-1-4	14.5	1.1 x 0.2	12.7		66	Mag 7.86v star NW 11′.2; **MCG +8-1-3** W 2′.9.
23 52 47.6	+46 48 13	MCG +8-1-5	14.5	0.9 x 0.3	13.0	Sb	48	
23 54 01.3	+47 29 20	MCG +8-1-6	14.1	1.1 x 1.1	14.2			Close pair of stars, mags 9.98v and 10.56v, NE 2′.2.
23 57 29.8	+47 26 19	MCG +8-1-13	14.0	1.1 x 1.1	14.1	S?		Mag 10.17v star S 3′.0.
23 58 37.0	+47 56 06	MCG +8-1-15	15.7	1.3 x 1.0	15.9		129	Several superimposed stars; mag 10.37v star SW 2′.9.
00 00 13.0	+46 57 54	MCG +8-1-19	13.9	1.3 x 1.2	14.3		18	Mag 7.36v star SW 1′.6.
00 04 26.7	+47 29 23	MCG +8-1-20	14.5	1.0 x 0.4	13.4	Sb	111	
00 07 24.7	+46 59 17	MCG +8-1-23	14.4	1.0 x 0.6	13.7		129	UGC 61 and MCG +8-1-25 N 3′.0.
00 07 29.9	+47 02 20	MCG +8-1-25	14.8	0.5 x 0.3	12.6		108	Located on the E edge of UGC 61.
00 09 32.8	+48 08 04	MCG +8-1-27	14.5	1.0 x 0.5	13.6		9	Mag 10.38v star W 1′.1.
00 13 42.7	+48 41 43	MCG +8-1-29	14.6	0.9 x 0.5	13.6		84	Mag 12.4 star W 1′.6.
00 22 01.2	+49 07 56	MCG +8-1-41	14.4	1.0 x 0.5	13.4	Spiral	24	Mag 10.5 star WNW 1′.6.
00 26 51.5	+49 07 00	MCG +8-2-2	14.1	0.5 x 0.4	12.2			Mag 6.92v star SW 6′.3.
00 45 50.5	+49 40 09	MCG +8-2-12	14.6	0.7 x 0.6	13.5	SBc	120	Mag 10.54v star SE 3′.6.
23 50 01.6	+46 17 33	MCG +8-43-8	15.8	1.0 x 0.2	13.9	Sc	138	Located between a pair of mag 10 stars.
00 14 02.3	+48 14 01	NGC 48	13.6	1.4 x 0.9	13.7	SABbc pec:	15	Very faint stars on E and W edges; mag 10.44v star E 4′.6; NGC 49 E 3′.5.
00 14 22.5	+48 14 47	NGC 49	13.7	1.1 x 0.8	13.4	S0?	165	Mag 10.44v star SE 1′.8; NGC 51 E 2′.2.
00 14 35.1	+48 15 14	NGC 51	13.1	1.3 x 1.0	13.3	S0° pec:	18	Mag 10.44v star SSW 2′.1; NGC 49 W 2′.2.
00 33 12.1	+48 30 16	NGC 147	9.5	13.2 x 7.8	14.5	E5 pec	25	Numerous stars superimposed overall.
00 38 57.5	+48 20 12	NGC 185	9.2	8.0 x 7.0	13.6	E3 pec	35	Numerous superimposed stars.
00 40 19.5	+41 41 25	NGC 205	8.1	21.9 x 11.0	14.0	E5 pec	170	= **M 110**. Numerous stars superimposed.
00 42 41.7	+40 51 59	NGC 221	8.1	8.7 x 6.5	12.5	cE2	179	= **M 32**.
00 42 33.3	+41 15 20	NGC 224	3.4	190.5 x 61.7	13.5	SA(s)b I-II	35	= **M 31, Andromeda Galaxy**. Visible to the naked eye under fairly dark skies.
23 19 47.3	+42 51 05	NGC 7618	13.0	1.2 x 1.0	13.2	E	5	Mag 7.58v star N 6′.8; note: **UGC 12524** lies 3′.2 E of this star.
23 22 06.7	+40 50 43	NGC 7640	11.3	10.5 x 1.8	14.4	SB(s)c II	167	Bright nucleus bisected by short dark lane; numerous superimposed stars.
23 34 51.6	+44 18 09	NGC 7707	13.4	1.3 x 1.1	13.7	S0⁻:	45	Two star pairs superimposed.
00 06 37.1	+47 52 43	UGC 48	15.4	1.5 x 1.3	16.0	Sd	0	Stellar nucleus, several stars superimposed on faint envelope.
00 07 24.0	+47 02 25	UGC 61	12.8	1.6 x 1.4	13.5	S0	145	MCG +8-1-25 E edge.
00 07 44.1	+40 52 30	UGC 64	14.6	1.0 x 0.5	13.7		35	Mag 14.2 star on S edge.
00 09 28.5	+47 21 18	UGC 85	13.5	1.0 x 0.4	12.4	Sbc	84	Several stars superimposed on edges; mag 8.62v star ESE 7′.0.

(Continued from previous page)

GALAXIES

RA h m s	Dec ° ′ ″	Name	Mag (V)	Dim ′ Maj x min	SB	Type Class	PA	Notes
00 12 08.8	+41 45 13	UGC 112	14.2	1.2 x 1.2	14.4	Sm		Low surface brightness; mag 13.9 star on SE edge.
00 17 05.2	+42 09 35	UGC 158	14.4	1.5 x 0.4	13.7	Sb III	176	Two knots or stars N end; mag 11.9 star S 1′.6.
00 18 24.3	+48 43 52	UGC 171	13.4	0.8 x 0.5	12.3	Im	30	
00 18 27.1	+49 59 51	UGC 175	13.9	1.1 x 0.6	13.3	SABcd:	32	Faint stellar nucleus; mag 11.3 star W 1′.4.
00 18 53.3	+48 00 58	UGC 178	13.6	1.1 x 0.4	12.6	S	17	
00 19 35.0	+47 14 28	UGC 183	12.8	1.4 x 0.5	12.3	Sab	50	Located in a triangle of stars with mags 9.82v, 10.8 and 11.6.
00 20 34.3	+47 26 02	UGC 196	13.2	1.1 x 0.8	12.9	SB(s)c	144	Mag 9.93v star E 1′.6.
00 24 59.0	+43 39 43	UGC 236	13.8	0.7 x 0.6	12.8	E	90	
00 25 30.3	+45 55 15	UGC 243	13.9	1.7 x 0.4	13.3	Sb III-IV	2	Mag 11.2 star on S end; mag 10.52v star NW 2′.7.
00 26 56.8	+50 01 48	UGC 256	14.2	1.9 x 0.3	13.4	Sbc	174	Mag 11.6 star E 1′.8; mag 6.94v star SE 3′.6.
00 27 16.6	+44 03 18	UGC 263	15.9	1.5 x 0.3	14.9	Sdm	171	
00 30 31.8	+42 06 34	UGC 303	15.9	1.8 x 0.9	16.3	Im	169	Irregular, low surface brightness.
00 30 56.7	+42 11 32	UGC 306	14.7	0.9 x 0.5	13.7	Scd:	96	Mag 11.7 star W 1′.1.
00 34 07.8	+44 08 41	UGC 336	14.5	1.1 x 0.8	14.2	SAd	30	Mag 12.4 star W 0′.9.
00 37 27.9	+42 54 11	UGC 372	16.0	1.2 x 1.2	16.2	SAm IV-V		Very low surface brightness; mag 10.23v star SE 1′.3.
00 38 43.5	+41 59 49	UGC 394	14.5	2.0 x 0.6	14.5	SABdm	5	Low surface brightness.
00 43 23.5	+50 40 36	UGC 460	13.3	1.0 x 0.5	12.4	Scd:	115	Strong dark lane N edge.
00 44 41.9	+45 25 16	UGC 471	14.2	0.9 x 0.7	13.5	SA(s)bc	50	Mag 12.2 star S 0′.6.
00 46 19.4	+51 13 02	UGC 475	13.4	1.5 x 0.7	13.3	Scd:	105	Mag 10.81v star NE 2′.1.
00 47 28.3	+50 52 57	UGC 486	13.4	0.9 x 0.6	12.6	SAB(s)b II	0	Bright nucleus with dark patches E and W.
00 51 27.1	+40 43 30	UGC 522	15.2	0.8 x 0.7	14.4	Sb II-III	65	Pair of mag 12.5 stars NE 1′.5.
23 08 23.7	+46 54 27	UGC 12389	12.9	0.6 x 0.5	11.4	Compact	168	Almost stellar nucleus, faint envelope; mag 11.9 star W 0′.7.
23 10 00.4	+42 33 58	UGC 12396	14.6	1.1 x 0.9	14.4	Compact	174	
23 12 03.6	+48 48 58	UGC 12411	15.9	2.1 x 0.2	14.8	Sm:	175	Mag 10.56v star N end.
23 13 44.1	+49 40 35	UGC 12433	14.7	1.4 x 1.4	15.3	SB0/a		Short SE-NW bar; mag 13.3 star on W edge; mag 11.1 star SW 1′.9.
23 18 38.4	+42 57 27	UGC 12491	13.9	1.0 x 0.8	13.6	E:	43	Mag 9.17v star SE 2′.9.
23 18 51.5	+42 03 49	UGC 12496	14.1	0.9 x 0.2	12.1	Sa	15	Mag 11.6 star S 1′.1.
23 19 20.9	+43 58 20	UGC 12507	13.6	1.1 x 0.2	11.8	Sa	152	
23 19 54.6	+43 57 24	UGC 12517	13.0	1.4 x 1.3	13.6	E?	123	Mag 10.10v star NE 2′.6.
23 22 19.4	+50 09 32	UGC 12558	13.3	1.0 x 0.6	12.6	Scd:	0	Pair of mag 11 and 12 stars SW edge.
23 23 11.7	+43 57 28	UGC 12573	13.6	1.1 x 0.6	13.1	E:	100	**UGC 12567** W 5′.9; **UGC 12559** WNW 8′.4.
23 24 42.4	+41 20 46	UGC 12588	13.1	1.5 x 1.3	13.7	Sdm	111	Several bright knots and/or superimposed stars N of center; mag 11.0 star SE 3′.1.
23 29 59.0	+40 59 25	UGC 12632	12.1	4.6 x 3.7	15.0	Sm: V	30	Dwarf spiral, extremely low surface brightness; mag 11.08v star SW 3′.6.
23 30 07.2	+40 13 29	UGC 12634	14.8	1.0 x 0.2	12.9	Scd:	175	Triangle of two mag 12 stars and one mag 11 star NE 1′.9.
23 33 58.3	+44 40 44	UGC 12669	14.0	1.2 x 1.0	14.0	Sbc	162	Mag 8.57v star W 6′.4.
23 41 49.6	+44 58 20	UGC 12742	14.1	1.7 x 1.0	14.5	SB(s)cd:	77	Mag 12.3 star on SE edge; mag 6.41v star E 4′.5; mag 10.10v star S 2′.1.
23 43 14.8	+49 58 47	UGC 12750	13.2	1.7 x 0.7	13.2	Sa	7	Numerous stars superimposed; mag 10.8 star NE 2′.9.
23 49 10.9	+46 33 09	UGC 12795	14.0	0.9 x 0.4	12.7	S	111	Mag 10.6 star NNE 2′.2.
23 49 19.8	+47 55 05	UGC 12796	13.6	1.0 x 0.8	13.2	SB?	15	Mag 13.7 star on S edge.
23 51 03.4	+46 43 23	UGC 12809	13.8	1.2 x 0.9	13.7	Sb	12	Mag 13.6 star N of center; mag 10.52v star W 2′.7.
23 56 07.4	+49 29 52	UGC 12851	13.8	1.3 x 1.0	13.9	SBb	90	Knotty, two superimposed stars.
23 57 00.5	+47 30 55	UGC 12862	14.8	1.1 x 0.9	14.6	SB	69	Mag 11.7 star NE 3′.0; mag 9.93v star E 3′.1.
23 59 53.8	+46 53 06	UGC 12888	13.9	1.2 x 0.7	13.6	Scd:	168	Several small knots; mag 7.36v star and MCG +8-1-19 NNE 4′.2.
00 00 01.8	+47 16 28	UGC 12889	12.6	2.1 x 1.7	13.8	SB(rs)b	165	

GALAXY CLUSTERS

RA h m s	Dec ° ′ ″	Name	Mag 10th brightest	No. Gal	Diam ′	Notes
00 35 54.0	+49 41 00	A 63	16.1	33	11	Pair of mag 8.98v, 8.68v stars 4′.0 N of center; member GX faint and stellar.
00 38 30.0	+45 43 00	A 72	16.3	47	20	Mag 8.20v star, **PGC 2264** and **PGC 2267** on SW edge.

OPEN CLUSTERS

RA h m s	Dec ° ′ ″	Name	Mag	Diam ′	No. ★	B ★	Type	Notes
23 30 07.0	+49 08 00	NGC 7686	5.6	15	80	7.0	cl	Few stars; moderate brightness range; no central concentration; detached. Probably not a cluster.

STAR CLOUDS

RA h m s	Dec ° ′ ″	Name	Mag	Diam ′	No. ★	B ★	Type	Notes
00 40 36.0	+40 44 00	NGC 206		4.2				A large star cloud in the SW part of M 31, the Andromeda Galaxy. It is extended N-S and best defined on the E flank.

PLANETARY NEBULAE

RA h m s	Dec ° ′ ″	Name	Diam ″	Mag (P)	Mag (V)	Mag cent ★	Alt Name	Notes
23 25 53.9	+42 32 06	NGC 7662	37	9.2	8.3	13.2	PK 106−17.1	**Blue Snowball.** Very slightly elongated N-S; mag 13.3 star NE 0′.9.
23 22 58.1	+46 53 57	PN G107.6−13.3	10	13.9	13.6	14.7	PK 107−13.1	Close pair of mag 13 stars W 2′.2.
23 39 10.7	+48 12 30	PN G110.6−12.9	33	16.5	16.::	20.7	PK 110−12.1	Close pair of mag 10.4, 11.04v stars NE 1′.8.
23 47 43.9	+51 23 56	PN G112.9−10.2	152	14.4	13.0	18.4	PK 112−10.1	Slightly elongated N-S; brighter along E edge with mag 10.7 star on this edge; mag 11.7 star 1′.9 W of center.

GALAXIES

RA h m s	Dec ° ′ ″	Name	Mag (V)	Dim ′ Maj x min	SB	Type Class	PA	Notes
22 23 27.3	+41 10 51	CGCG 531-4	13.5	0.4 x 0.3	11.3	E	168	Mag 11.3 star ESE 2′.2.
22 26 28.4	+40 09 45	MCG +7-46-9	14.8	0.8 x 0.3	13.1	Sb	21	
22 49 39.2	+44 00 05	MCG +7-46-22	14.9	0.6 x 0.4	13.2	Spiral	30	Mag 14.2 star superimposed S end; mag 11.1 star N 2′.1.
22 58 29.4	+40 25 42	MCG +7-47-1	14.0	0.6 x 0.5	12.6	Sb	114	Pair of mag 11.3 stars N 1′.3.
23 03 23.2	+41 24 50	MCG +7-47-3	13.5	0.9 x 0.7	12.9	Compact	135	= **Mkn 925**; close pair of stars, mags 10.26v, 9.24v, SE 2′.4.
22 02 58.2	+41 03 31	NGC 7197	12.8	1.6 x 0.8	12.9	Sa	106	Stars superimposed on envelope, in star rich area.
22 10 09.2	+41 00 58	NGC 7223	12.2	1.4 x 1.1	12.5	SB(rs)bc	144	Faint, elongated anonymous galaxy on NE edge.
22 12 30.9	+45 19 42	NGC 7231	13.0	1.9 x 0.7	13.2	SBa	88	
22 16 52.8	+40 30 13	NGC 7248	12.4	1.7 x 0.9	12.7	SA0⁻:	133	
22 18 18.1	+40 33 43	NGC 7250	12.6	1.7 x 0.8	12.8	Sdm?	166	Mag 11.10v star S end.
22 25 53.8	+40 18 58	NGC 7282	13.7	2.5 x 1.0	14.5	SB(r)b	0	Stellar nucleus with many superimposed stars.
22 47 33.0	+40 14 22	NGC 7379	13.4	1.1 x 0.8	13.1	SBa	81	Mag 10.73v star NNW 2′.4.
21 43 27.1	+43 33 16	UGC 11798	14.3	1.7 x 0.3	13.4	S	137	**UGC 11797** NNW 1′.7; mag 8.49v star NW 3′.7; UGC 11801 on E edge.
21 43 35.3	+43 41 38	UGC 11799	14.5	1.5 x 1.0	14.8	SB?	40	Stellar nucleus, many superimposed stars.
21 43 34.1	+43 32 58	UGC 11801	14.4	0.7 x 0.4	12.9	S	66	UGC 11798 on W edge.
21 43 54.1	+46 37 02	UGC 11802	14.2	0.9 x 0.6	13.4	SBcd?	120	Patchy, stars superimposed, in star rich area; UGC 11806 E 3′.4.
21 44 13.1	+46 14 58	UGC 11805	14.2	1.5 x 0.4	13.5		136	Pair with UGC 11805 on SW edge.
21 44 13.4	+46 37 16	UGC 11806	13.9	1.5 x 0.6	13.6	SBb	58	Two thin arms; UGC 11802 W 3′.4.
21 45 58.7	+41 16 18	UGC 11808	15.6	2.1 x 1.5	16.7	S?	160	Small, bright core, numerous superimposed stars.
21 49 10.2	+41 56 53	UGC 11819	16.3	1.3 x 0.4	15.5	Im:	168	
21 51 41.6	+46 37 11	UGC 11836	15.2	1.5 x 0.3	14.2	Sdm:	29	Uniform surface brightness; mag 11.2 star SE 1′.4.
21 57 29.4	+41 14 52	UGC 11858	13.2	1.7 x 0.7	13.2	S0/a	47	Mag 11.0v star N 2′.0.
21 57 54.7	+42 18 20	UGC 11864	14.3	2.3 x 1.5	15.5	SB(s)dm	0	Filamentary, numerous superimposed stars.
22 03 32.8	+43 44 56	UGC 11891	13.8	3.6 x 2.8	16.2	Im	108	Uniform surface brightness, numerous superimposed stars; mag 11.7 star S of center.
22 04 30.9	+41 24 37	UGC 11897	12.7	1.3 x 1.1	12.9	SBbc:	0	Stellar nucleus, several bright knots or superimposed stars S.
22 06 16.6	+47 15 07	UGC 11909	12.4	3.0 x 0.7	13.0	S pec	3	Triangle of mag 10 and 11 stars NE 2′.1.
22 07 08.3	+44 17 53	UGC 11911	14.8	1.5 x 1.4	15.5	Compact	27	Bright core with faint outer arms.
22 08 10.8	+41 10 36	UGC 11919	13.3	1.7 x 1.1	13.8	SABbc I-II	26	Two main arms, several stars superimposed.
22 08 27.5	+48 26 27	UGC 11920	11.9	2.3 x 1.5	13.1	SB0/a	45	Bright, elongated core with numerous stars superimposed on uniform envelope.
22 09 27.5	+40 59 18	UGC 11927	13.6	1.7 x 1.1	14.1	Sb	43	Mag 9.25v star NW 3′.5.
22 10 38.3	+42 01 15	UGC 11935	13.2	1.0 x 0.9	13.0	S0:		Mag 14.1 star on N edge; mag 12.4 star on W edge; mag 10.6 star NW 1′.7.
22 11 39.1	+46 18 32	UGC 11946	13.3	1.0 x 0.7	12.8	SAB(s)c I-II	71	Stellar nucleus.
22 14 46.5	+42 10 52	UGC 11961	14.2	1.5 x 0.2	12.7	Sb	48	
22 16 49.8	+41 30 00	UGC 11973	12.1	3.0 x 0.9	13.0	SAB(s)bc	42	Tightly wound arms; mag 11.8 star on SW edge.
22 18 09.5	+45 42 22	UGC 11979	14.5	1.7 x 1.5	15.4	SBm:	36	A number of superimposed stars; mag 10.6 star SE 2′.2.
22 20 22.6	+47 42 20	UGC 11991	13.7	1.9 x 0.9	14.1	Scd: II	88	
22 22 08.5	+41 19 44	UGC 12002	14.7	1.0 x 0.7	14.3	E	18	Mag 11.23v star NE 4′.2.
22 23 12.3	+40 03 18	UGC 12016	13.2	1.1 x 0.5	12.4	S?	120	
22 25 19.5	+42 42 58	UGC 12032	14.4	2.0 x 0.3	13.7	S	108	Mag 13.3 star on W end; mag 11.4 star S 2′.0.
22 30 13.6	+42 47 24	UGC 12057	13.5	1.2 x 0.9	13.5	S0:	34	Small, bright nucleus, faint envelope with superimposed stars.
22 34 18.9	+41 19 11	UGC 12086	13.5	0.9 x 0.7	12.8	S0/a	85	
22 34 23.9	+43 48 29	UGC 12088	13.6	1.0 x 0.8	13.2	Sa	20	
22 34 44.8	+50 09 52	UGC 12095	14.7	0.9 x 0.4	13.4	Irr	72	Pair with elongated **UGC 12096** on NE edge.
22 37 43.3	+40 19 05	UGC 12125	14.9	1.9 x 0.3	14.1	Sa	3	Located between a pair of mag 12 stars.
22 49 03.9	+40 00 05	UGC 12199	13.5	1.6 x 0.8	13.6	SB(rs)b	105	Mag 11.6 star N edge; mag 11.17v star NNW 2′.1.
22 49 23.3	+40 13 54	UGC 12204	15.2	1.5 x 0.2	13.7	Sbc	28	Mag 10.98v star N end.
22 58 55.5	+40 55 58	UGC 12282	13.6	1.7 x 0.5	13.3	Sa	1	Mag 9.93v star NW 1′.5.
23 04 29.9	+44 13 13	UGC 12341	14.3	1.4 x 0.7	14.1	SBcd:	75	Pair of mag 10.48v and 12.9 stars E edge.
23 07 23.0	+43 36 09	UGC 12381	14.1	1.1 x 1.1	14.2	SB(s)c		Stellar nucleus, well defined arms; mag 11.4 star N 2′.3.
23 08 23.7	+46 54 27	UGC 12389	12.9	0.6 x 0.5	11.4	Compact	168	Almost stellar nucleus, faint envelope; mag 11.9 star W 0′.7.
23 10 00.4	+42 33 58	UGC 12396	14.6	1.1 x 0.9	14.4	Compact	174	
23 12 03.6	+48 48 58	UGC 12411	15.9	2.1 x 0.2	14.8	Sm:	175	Mag 10.56v star N end.
23 13 44.1	+49 40 35	UGC 12433	14.7	1.4 x 1.4	15.3	SB0/a		Short SE-NW bar; mag 13.3 star on W edge; mag 11.1 star SW 1′.9.

OPEN CLUSTERS

RA h m s	Dec ° ′ ″	Name	Mag	Diam ′	No. ★	B ★	Type	Notes
21 43 38.0	+51 04 17	Bar 2	8.4	5	21	10.0	cl	
21 53 30.0	+47 16 00	Cr 470	7.2	20	110	9.6	cl	This is not IC 5146, but is involved in IC 5146, the "Cocoon Nebula."
21 34 35.8	+45 59 31	NGC 7093		6	23		cl?	
22 05 08.0	+46 29 00	NGC 7209	7.7	15	98	9.0	cl	Moderately rich in stars; small brightness range; no central concentration; detached.
22 15 09.0	+49 54 00	NGC 7243	6.4	30	40	8.0	cl	Moderately rich in stars; moderate brightness range; slight central concentration; detached.

BRIGHT NEBULAE

RA h m s	Dec ° ′ ″	Name	Dim ′ Maj x min	Type	BC	Color	Notes
21 53 24.0	+47 16 00	IC 5146	10 x 10	E	1-5	2-4	**Cocoon Nebula**. Several absorption patches involved in the interior. There is a small detached patch of nebulosity, comparable in surface brightness, 10′ WSW of center.
21 43 42.0	+48 55 00	vdB 145	2 x 2	R	3-5	2-4	Nebula with a mag 7.4 star involved.

DARK NEBULAE

RA h m s	Dec ° ′ ″	Name	Dim ′ Maj x min	Opacity	Notes
21 32 08.5	+44 58 00	B 155	13 x 13	3	Round; indefinite; four small stars in a line crossing it E and W.
21 33 59.4	+45 35 00	B 156	8 x 8	3	Sharp pointed to N.
21 37 13.6	+43 24 40	B 158	3 x 3		Dark spot.
21 38 22.8	+43 14 00	B 159	25 x 25	5	Irregular partially vacant region.
21 46 31.9	+51 06 00	B 164	12 x 6	5	V-shaped vacancy.
21 53 20.7	+47 16 00	B 168	100 x 10	4	Small nebula at E end of dark lane.

(Continued from previous page)

31 PLANETARY NEBULAE 31

RA h m s	Dec ° ′ ″	Name	Diam ″	Mag (P)	Mag (V)	Mag cent ★	Alt Name	Notes
21 32 31.1	+44 35 47	IC 5117	12	13.3	11.5	16.7	PK 89−5.1	Mag 9.97v star on E edge.
22 23 55.7	+50 58 00	IC 5217	15	12.6	11.3	15.5	PK 100−5.1	Mag 10.37v star S 1′.8.
21 38 49.2	+46 00 26	PN G91.6−4.8	8	14.6			PK 91−4.1	Mag 10.99v star NW 2′.3; mag 9.64v star SE 2′.3.
21 37 01.6	+48 56 00	PN G93.3−2.4	56	13.2	13.2	19.1	PK 93−2.1	Elongated E-W; mag 12.7 star on W end; mag 10.9 star ME 2′.6.
21 35 44.1	+50 54 15	PN G94.5−0.8	6	19.7			PK 94−0.1	Mag 13.1 star SSE 0′.5.
21 43 17.8	+50 25 11	PN G95.1−2.0	8	15.8	14.3		PK 95−2.1	SE-NW line of three mag 11-12 stars centered NW 2′.3.
21 57 41.9	+51 41 36	PN G97.6−2.4	4	14.5	14.7	15.1	PK 97−2.1	Mag 11.9 star N 1′.3; mag 9.60v star NE 2′.8.
22 31 43.8	+47 48 00	PN G100.0−8.7	10	11.5	11.9	16.1	PK 100−8.1	Mag 11.6 star WNW 0′.5; mag 11.8 star ESE 1′.3.
22 51 39.2	+51 50 39	PN G104.8−6.7	8			12.0	PK 104−6.1	Mag 11.5 star W 1′.2; mag 10.9 star N 2′.4.

32 GALAXIES 32

RA h m s	Dec ° ′ ″	Name	Mag (V)	Dim ′ Maj x min	SB	Type Class	PA	Notes
19 59 04.1	+50 48 05	CGCG 257-29	13.9	0.4 x 0.3	11.5	E/S0	156	Almost stellar; **CGCG 257-28** S 0′.7.
20 02 50.1	+49 19 08	CGCG 257-35	13.7	0.4 x 0.4	11.8	E?		Mag 12.9 star W 1′.7; in star rich area.
19 56 12.9	+40 26 03	MCG +7-41-1	13.5	1.2 x 0.7	13.4	E1:	133	
19 56 54.8	+50 09 59	MCG +8-36-9	14.9	0.6 x 0.4	13.2			
19 57 01.0	+47 16 42	MCG +8-36-10	14.1	0.9 x 0.7	13.7	E	108	Mag 9.42v star 1′.0 E.
19 56 48.8	+49 53 06	UGC 11499	14.1	0.7 x 0.5	12.8	Sdm:	111	Bright nucleus, sandwiched between several faint stars; mag 11.3 star W 1′.1.
19 57 02.1	+50 10 51	UGC 11500	13.9	0.7 x 0.6	12.8	Scd:	9	MCG +8-36-9 SW 1′.5; mag 10.81v star ESE 2′.1.
19 58 42.4	+50 02 11	UGC 11503	12.9	0.6 x 0.5	11.4	S?	99	**MCG +8-36-14** SE 1′.2.
20 01 55.3	+49 09 00	UGC 11507	14.6	0.9 x 0.7	13.9	Scd:	75	Faint stellar nucleus, faint envelope; mag 10.70v star NE 1′.8.
21 30 25.1	+41 50 33	UGC 11757	14.2	1.4 x 0.9	14.3	SBab	61	Mag 10.04 star N 1′.1.

OPEN CLUSTERS

RA h m s	Dec ° ′ ″	Name	Mag	Diam ′	No. ★	B ★	Type	Notes
21 30 02.0	+48 59 00	Anon Platais		10		8.9	cl	Brightest star is V1726 Cygni, a cepheid variable, V = 8.892–9.079, period 4.237d.
20 53 42.0	+46 02 00	Bar 1		4	12		cl	Few stars; moderate brightness range; not well detached.
21 10 30.0	+46 14 00	Bas 12		4	20		cl	
21 13 11.0	+46 34 00	Bas 13		10			cl	
21 21 14.0	+44 48 11	Bas 14		5	15		cl	
20 55 59.0	+51 04 36	Be 53		6	40	16.0	cl	Moderately rich in stars; moderate brightness range; no central concentration; detached.
21 03 00.0	+40 26 12	Be 54		4	30	17.0	cl	
21 16 57.0	+51 45 30	Be 55		5	30	14.0	cl	
21 17 34.5	+41 49 36	Be 56		3	40	16.0	cl	(A) Faint.
20 24 28.4	+46 02 24	Be 89		3	15	15.0	cl	
20 35 16.0	+46 50 42	Be 90		3	20	14.0	cl	
20 18 07.0	+40 43 56	Cr 419	5.4	4.5			cl	Few stars; moderate brightness range; not well detached.
20 23 18.0	+41 42 00	Cr 421	10.1	8	22		cl	Few stars; small brightness range; no central concentration; detached; involved in nebulosity.
21 03 12.0	+44 35 00	Cr 428	8.7	10	20		cl	Few stars; small brightness range; not well detached; involved in nebulosity.
20 09 54.0	+41 22 00	Do 2		7	15		cl	
20 20 38.0	+41 21 36	Do 6		6	12		cl	(A) Involved in spectacular dark and light nebulosity.
20 24 24.0	+42 16 00	Do 8		6	20		cl	
20 25 33.4	+41 54 25	Do 9		3	5		cl	
20 26 20.0	+40 07 00	Do 10		15				
20 26 24.7	+41 25 00	Do 11		6	15		cl	
20 02 30.0	+42 06 00	Do 36		14			ast	Few stars; small brightness range; not well detached. Probably not a cluster. (A) Group of about a dozen brighter stars.
20 06 04.0	+41 11 00	Do 38		15			ast	(A) 15′ N-S, 10′ E-W, "Y" shaped group. Involved in nebulosity.
20 29 42.0	+41 43 00	Do 44		10	15		cl:	Few stars; moderate brightness range; not well detached; involved in nebulosity. Probably not a cluster.
20 05 35.0	+40 40 00	DoDz 10		24	40	7.1	cl	Few stars; moderate brightness range; not well detached; involved in nebulosity. (A) Scattered cluster in rich Milky Way and bright nebula.
20 54 43.1	+47 16 49	h 2091		6	12	8.3	cl	(A) Often incorrectly identified as NGC 6991, which it is only a part of.
20 10 47.0	+41 10 19	IC 1311	13.1	5	60	17.0	cl	Moderately rich in stars; small brightness range; strong central concentration; detached; involved in nebulosity.
21 12 07.0	+47 46 00	IC 1369	8.8	5	152	12.1	cl	Moderately rich in stars; moderate brightness range; slight central concentration; detached.
20 03 55.0	+44 20 00	NGC 6866	7.6	15	129	10.0	cl	Rich in stars; moderate brightness range; slight central concentration; detached.
20 23 12.0	+40 47 00	NGC 6910	7.4	10	66	9.6	cl	Moderately rich in bright and faint stars with a strong central concentration; detached; involved in nebulosity. Visual mag = 6.6 if bright star on NW included.
20 54 54.0	+47 25 00	NGC 6991		25	35	5.7	cl	Few stars; large brightness range; no central concentration; detached; involved in nebulosity. (A) Includes h 2091, and nebulosity IC 5076 and **vdB 137**.
20 56 22.5	+45 27 48	NGC 6996	10.0	5	40	11.0	MW	Moderately rich in stars; moderate brightness range; no central concentration; detached; involved in nebulosity. (A) Scattered group of moderately bright stars including many in a meandering E-W chain.
20 56 29.2	+44 39 00	NGC 6997	10.0	8	40		cl	
21 01 45.0	+47 20 54	NGC 7011		3.8	35		cl?	(A) Long NE opening "V"-shaped group of stars.
21 06 04.8	+41 30 00	NGC 7024		10	14	10.0	cl?	(A) Centered at W. Herschel's triangle of stars.
21 06 52.0	+50 51 00	NGC 7031	9.1	15	62	11.3	cl	Moderately rich in stars; moderate brightness range; no central concentration; detached. (A) Group of 10 brighter stars of this cluster on NE side.
21 11 12.0	+45 39 00	NGC 7039	7.6	15	185	11.3	cl	Moderately rich in stars; moderate brightness range; not well detached. (A) 7′ high by 14 wide group of stars.
21 13 08.0	+42 30 00	NGC 7044	12.0	7	60	15.0	cl	Moderately rich in stars; small brightness range; strong central concentration; detached.
21 21 48.3	+50 49 00	NGC 7058	9	10	17		cl?	
21 23 27.0	+46 22 42	NGC 7062	8.3	5	85	10.1	cl	Moderately rich in stars; moderate brightness range; slight central concentration; detached.
21 24 12.3	+48 00 41	NGC 7067	9.7	3	47	11.2	cl	Few stars; small brightness range; slight central concentration; detached.
21 26 40.0	+47 55 18	NGC 7071	10	4	15	12.0	cl?	
21 29 19.0	+47 08 00	NGC 7082	7.2	24	182	9.9	cl	Few stars; moderate brightness range; not well detached.
21 30 28.0	+51 36 00	NGC 7086	8.4	12	80	10.2	cl	Moderately rich in stars; moderate brightness range; slight central concentration; detached.
21 31 52.0	+48 26 00	NGC 7092	4.6	31	25	7.0	cl	= **M 39**. Moderately rich in stars; moderate brightness range; no central concentration; detached. (A) 25 bright stars with rich Milky Way background.
21 34 35.8	+45 59 31	NGC 7093		6	23		cl?	
20 15 08.0	+47 34 00	Om-2 Cyg		19	7		ast?	

(Continued from previous page)

BRIGHT NEBULAE

RA h m s	Dec ° ′ ″	Name	Dim ′ Maj x min	Type	BC	Color	Notes
20 26 42.0	+42 07 00	DWB 100-5-9	40 x 15	E	4-5	3-4	Very irregular in shape, mottled and of low surface brightness.
20 16 24.0	+41 49 00	IC 1318(b)	50 x 20	E	1-5	3-4	A large patch in the NNW part of the γ Cygni nebulae complex, about 2°distant from γ. It is elongated roughly N-S.
20 19 18.0	+40 44 00	IC 1318(c)	40 x 25	E	2-5	3-4	Very diffuse overall with four or five brighter condensations, the most prominent one about 30′ SE of Cr 419.
20 26 12.0	+40 30 00	IC 1318(d)	50 x 30	E	1-5	3-4	Centered approximately 1° E of γ Cygni. This and the bright patch 45′ SE are the brightest parts of the γ Cygni complex.
20 27 54.0	+40 00 00	IC 1318(e)	45 x 30	E	1-5	3-4	Centered 1.5° SE of γ Cygni. Comparable in size and brightness to the nebulosity 0.75° NW. A prominent dark lane separates them.
20 50 48.0	+44 21 00	IC 5067	25 x 10	E	1-5	3-4	Bright patch in the N part of the **Pelican Nebula**.
20 50 48.0	+42 31 00	IC 5068	40 x 30	E	2-5	3-4	
20 50 48.0	+44 11 00	IC 5070	60 x 50	E	3-5	3-4	The southern two thirds of the **Pelican Nebula**.
20 55 54.0	+47 25 00	IC 5076	9 x 7	R	1-5	1-4	The involved bright star is in the NE part of the nebula.
20 58 48.0	+44 20 00	NGC 7000	120 x 100	E	1-5	3-4	**North American Nebula**. Due to its faintness and large size, the shape giving rise to its nickname is very difficult to discern visually. The brightest portion is the area corresponding to Mexico.
20 33 54.0	+45 39 00	Sh2-112	13 x 13	E	2-5	3-4	
20 34 30.0	+46 52 00	Sh2-115	30 x 20	E	3-5	3-4	
20 24 18.0	+42 18 00	vdB 131	3 x 3	R	2-5	1-4	Nebula with a mag 9.1 star involved.
20 24 48.0	+42 23 00	vdB 132	4 x 4	R	2-5	1-4	Nebula with a star of mag 8.7 involved. A fainter, broad bridge of nebulosity, connects it to another nebula 7′ N. It is of similar size and surface brightness.

DARK NEBULAE

RA h m s	Dec ° ′ ″	Name	Dim ′ Maj x min	Opacity	Notes
21 32 08.5	+44 58 00	B 155	13 x 13	3	Round; indefinite; four small stars in a line crossing it E and W.
21 33 59.4	+45 35 00	B 156	8 x 8	3	Sharp pointed to N.
21 37 13.6	+43 24 40	B 158	3 x 3		Dark spot.
21 38 22.8	+43 14 00	B 159	25 x 25	5	Irregular partially vacant region.
20 09 30.8	+41 12 00	B 342	4.0 x 0.5	4	Small, dark marking; E and W.
20 13 26.0	+40 17 00	B 343	13 x 5	5	Elongated; NW and SE; sharpest at SE end.
20 16 12.2	+40 13 00	B 344	7 x 3	3	Dusky spot, like an arrowhead, pointed SW.
20 21 00.3	+46 34 00	B 345	15 x 2		Curved; convex to the E.
20 26 46.1	+43 45 00	B 346	10 x 4	3	Curved, black spot.
20 34 26.3	+42 06 00	B 348	60 x 3	3	Narrow; dusky; NE and SW.
20 47 24.2	+43 58 00	B 349	6 x 1		Small; curved; dusky.
20 49 06.7	+45 52 50	B 350	3 x 3	6	Small; round; dusky.
20 52 27.6	+47 24 00	B 351	3 x 25		Crooked, dusky lane, NE and SW.
20 57 10.8	+45 54 00	B 352	20 x 10	5	Large; black; sharply defined on SE side.
20 57 22.6	+45 29 00	B 353	12 x 6		Definite; dusky; N and S.
20 59 38.0	+43 11 00	B 355	5 x 5		Dark; definite; E and W lanes; like a letter V-the open end toward the E.
20 59 58.8	+46 41 00	B 356	24 x 24	5	Irregular; dusky.
21 05 40.6	+43 17 00	B 358	20 x 20		Diffuses to NE; a curve of stars along N edge.
21 12 45.8	+47 26 00	B 361	17 x 17	4	Round; irreg. extension to W for ½°±.
21 23 57.8	+50 12 00	B 362	15 x 8	5	Elongated; NE and SW.
21 24 53.8	+48 56 00	B 363	40 x 15	3	Irregular; dusky; NE and SW.
20 24 48.0	+40 10 00	LDN 889	180 x 20	4	Wide lane E of γ Cygni.
20 56 48.0	+43 52 00	LDN 935	90 x 20	4	Wide lane separating the North America and Pelican nebulae.

PLANETARY NEBULAE

RA h m s	Dec ° ′ ″	Name	Diam ″	Mag (P)	Mag (V)	Mag cent ★	Alt Name	Notes
21 32 31.1	+44 35 47	IC 5117	12	13.3	11.5	16.7	PK 89−5.1	Mag 9.97v star on E edge.
20 10 23.8	+46 27 38	NGC 6884	15	12.6	10.9	15.6	PK 82+7.1	Mag 12.1 star WNW 1′.7.
21 06 18.5	+47 51 08	NGC 7026	45	12.7	10.9	14.2	PK 89+0.1	N-S rectangular appearance; mag 9.64v star on E edge.
21 07 01.8	+42 14 07	NGC 7027	55	10.4	8.5	16.2	PK 84−3.1	Mag 11.3 star W 2′.3.
21 14 15.3	+46 17 15	NGC 7048	61	11.3	12.1	19.1	PK 88−1.1	Mag 10.19v star on S edge; mag 8.28v star WSW 3′.3.
20 12 25.8	+40 45 19	PN G77.5+3.7	6	15.5			PK 77+3.1	Mag 13.6 star S 0′.5; mag 12.2 star SE 1′.2.
20 15 22.3	+40 34 43	PN G77.7+3.1	4	17.0			PK 77+3.2	Mag 13.2 star NE 2′.1; mag 11.6 star ESE 3′.3.
20 08 43.1	+42 30 05	PN G78.6+5.2	23	16.1	15.0		PK 84+9.1	Very slightly elongated E-W; mag 13.5 star NE 0′.8; mag 8.68v star NE.
20 29 19.0	+40 15 16	PN G78.9+0.7	18	17.1			PK 78+0.1	Slightly elongated N-S, on N edge of large bright nebula IC 1318; mag 12.6 star on E edge.
20 09 02.0	+43 43 42	PN G79.6+5.8	28	14.5	13.7		PK 79+5.1	Close pair of mag 12.3, 13.3 stars W 1′.7.
20 06 55.6	+44 14 16	PN G79.9+6.4	4	16.3			PK 79+6.1	Pair of stars, mag 10.63v, 11.3, SE 1′.8; mag 10.87v star SW 1′.9.
20 04 00.2	+49 19 04	PN G84.0+9.5	16	15.2		20.8	K 3-73	Mag 13.3 star NW 0′.8; mag 9.79v star E 4′.0.
20 45 10.2	+44 39 10	PN G84.2+1.0	27				PK 84+1.1	Mag 12.4 star NNE 1′.8; mag 12.8 star W 2′.0.
21 07 39.7	+40 57 53	PN G84.2−4.2	6	17.1			PK 84−4.1	Close pair of mag 13.0, 13.7 stars E 1′.5.
20 32 23.4	+47 20 55	PN G84.9+4.4	157	15.2	14.0	18.9	PK 85+4.1	= **Abell 71**; mag 12.4 star just inside SW edge; mag 12.0 star on S edge.
20 31 52.4	+48 52 51	PN G86.1+5.4	190		15.1	18.1	PK 86+5.1	Mag 11.2 star W of center 2′.6; mag 10.9 star S of center 2′.1.
21 18 06.9	+43 48 46	PN G87.4−3.8	35				PK 87−3.1	Mag 9.88v star N 3′.3; mag 9.62v star SE 3′.7.
20 45 22.8	+50 22 40	PN G88.7+4.6	3	19.5			PK 88+4.1	Mag 11.0 star SE 2′.4.
21 19 07.5	+46 18 45	PN G89.3−2.2	14			12.1	PK 89−2.1	Mag 11.02v star SE 1′.2; mag 11.0 star S 3′.0.
21 14 07.5	+47 46 26	PN G89.8−0.6	64	14.5	14.8	19.7	PK 89−0.1	Bright portion, defined by published dimensions, slightly hourglass shaped, oriented SE-NW, with larger, fainter extensions NE-SW; mag 11.46v star ENE 1′.8.
21 10 52.6	+50 47 10	PN G91.6+1.8	32			21.0	PK 91+1.1	Rectangular shape, oriented N-S, dark patch in center; mag 10.11v star SE .
21 38 49.2	+46 00 26	PN G91.6−4.8	8	14.6			PK 91−4.1	Mag 10.99v star NW 2′.3; mag 9.64v star SE 2′.3.
21 30 51.8	+50 00 05	PN G93.3−0.9	23	16.0	14.9:	19.5	PK 93−0.1	Pair of small, dark "eyes" on face of planetary.
21 37 01.6	+48 56 00	PN G93.3−2.4	56	13.2	13.2	19.1	PK 93−2.1	Elongated E-W; mag 12.7 star on W end; mag 10.9 star ME 2′.6.
21 20 45.0	+51 53 25	PN G93.5+1.4	15	16.9			PK 93+1.1	Mag 12.4 star W 1′.6; mag 12.1 star E 3′.2.
21 35 44.1	+50 54 15	PN G94.5−0.8	6	19.7			PK 94−0.1	Mag 13.1 star SSE 0′.5.

RA h m s	Dec ° ′ ″	Name	Mag (V)	Dim ′ Maj x min	SB	Type Class	PA	Notes
19 09 21.3	+43 03 48	CGCG 229-24	13.9	0.5 x 0.3	11.7		138	Mag 12.3 star SW 1'9, in star rich area.
19 30 47.8	+41 18 13	CGCG 230-20	13.7	0.6 x 0.3	11.7		24	Mag 10.88v star NNE 0'9. CGCG 230-21 Stellar nucleus, or star; faint envelope.
19 31 28.2	+41 54 24	CGCG 230-21	14.7	0.6 x 0.6	13.4			
19 39 56.9	+50 55 25	CGCG 257-2	14.2	0.6 x 0.4	12.5		171	Mag 9.25v star SW 5'0; **MCG +8-36-1** W 8'0.
19 43 39.6	+50 32 29	CGCG 257-10	13.9	0.5 x 0.5	12.2			Star on S edge; mag 10.70v star NE 2'1.
19 50 43.9	+46 48 14	CGCG 257-16	14.5	0.5 x 0.2	11.9		48	Small triangle of mag 12-13 stars SW 2'7.
19 50 54.2	+48 03 32	CGCG 257-17	14.4	0.5 x 0.5	12.7			Almost stellar; located 1'4 SE of mag 9.54v star, in star rich area.
19 52 44.8	+48 41 07	CGCG 257-19	14.5	0.6 x 0.6	13.3			Mag 9.98v star SW 2'6.
19 59 04.1	+50 48 05	CGCG 257-29	13.9	0.4 x 0.3	11.5	E/S0	156	Almost stellar; **CGCG 257-28** S 0'7.
20 02 50.1	+49 19 08	CGCG 257-35	13.7	0.4 x 0.4	11.8	E?		Mag 12.9 star W 1'7; in star rich area.
18 33 52.8	+49 16 41	IC 1291	13.0	1.8 x 1.5	13.9	SB(s)dm?	39	Mag 9.80v star SW 4'1.
19 26 37.5	+49 45 30	IC 1301	13.7	1.2 x 0.8	13.5	Scd: I-II	155	Mag 15 stars superimposed N and S of center, in star rich area.
19 26 32.0	+50 07 29	IC 4867	13.3	1.2 x 0.7	13.0	S0	17	**MCG +8-35-11** NE 1'1; close pair of mag 8.56v and 7.40v stars NE 2'4.
18 27 39.8	+41 44 54	MCG +7-38-6	15.5	0.6 x 0.6	14.2	Sb		Mag 9.85v star NW 5'4.
18 37 04.4	+42 27 05	MCG +7-38-12	15.5	0.5 x 0.5	13.9			Mag 15.0 star on W edge; mag 11.9 star WNW 2'4.
18 49 47.6	+41 40 36	MCG +7-39-1	15.4	0.9 x 0.7	14.7	Sbc	156	Mag 11.5 star N edge.
19 00 04.5	+40 54 21	MCG +7-39-5	15.4	0.4 x 0.4	13.2			MCG +7-39-6 S edge; mag 10.72v star ENE 2'1.
19 00 05.6	+40 53 58	MCG +7-39-6	14.4	0.5 x 0.3	12.2		12	MCG +7-39-5 N edge.
19 00 03.8	+43 26 20	MCG +7-39-8	15.1	0.8 x 0.8	14.5	SBbc		Located between a pair of mag 12.2 and 10.86v stars.
19 07 08.0	+41 44 47	MCG +7-39-11	14.9	0.9 x 0.5	13.9	SBbc	99	
19 08 29.7	+41 38 03	MCG +7-39-14	15.1	0.8 x 0.4	13.7	Sbc	99	Mag 11.0 star W 3'7.
19 32 25.1	+41 54 14	MCG +7-40-10	13.9	1.1 x 1.1	14.0			Several stars superimposed; mag 13.8 star superimposed N of core.
19 41 11.8	+43 24 44	MCG +7-40-13	14.5	0.7 x 0.7	13.6	SBbc		Several stars superimposed; mag 10.92v star W 0'7.
19 43 25.7	+41 39 22	MCG +7-40-15	15.3	0.8 x 0.7	14.6		30	Several stars superimposed; mag 10.48v star NNE 0'9.
19 56 12.9	+40 26 03	MCG +7-41-1	13.5	1.2 x 0.7	13.4	E1:	133	
18 22 44.1	+47 37 08	MCG +8-33-41	14.3	0.8 x 0.7	13.5		24	Mag 13.1 star on W edge.
18 30 39.7	+48 18 24	MCG +8-34-1	14.2	1.0 x 0.5	13.3		15	Mag 8.17v star SW 1'1.
18 33 36.8	+48 14 06	MCG +8-34-2	13.9	1.1 x 0.6	13.3		138	Mag 12.9 star NE 3'3.
19 03 29.3	+45 36 50	MCG +8-34-33	14.4	0.7 x 0.3	12.5		120	
19 56 54.8	+50 09 59	MCG +8-36-9	14.9	0.6 x 0.4	13.2			
19 57 01.0	+47 16 42	MCG +8-36-10	14.1	0.9 x 0.7	13.7	E	108	Mag 9.42v star 1'0 E.
18 35 58.7	+51 27 32	MCG +9-30-23	14.0	1.0 x 0.8	13.8	E/S0	87	**MCG +9-30-22** S 0'9.
18 49 36.9	+51 54 15	MCG +9-31-7	15.2	0.8 x 0.4	13.9		36	Mag 10.92v star WNW 4'7.
18 57 07.3	+51 08 55	MCG +9-31-12	14.6	0.9 x 0.6	13.8		150	Mag 11.0 star N 4'3.
18 33 33.8	+40 02 55	NGC 6663	13.9	1.0 x 0.9	13.6	SAB(r)cd:	140	Stellar nucleus; mag 7.25v star NNW 7'7.
18 33 44.3	+42 48 01	NGC 6672	16.0	0.4 x 0.4	13.9			Located 1'0 N of mag 9.84v star.
18 37 26.6	+40 03 27	NGC 6675	12.4	1.7 x 1.3	13.2	Sbc	130	
18 40 07.1	+40 08 15	NGC 6686	14.7	0.9 x 0.8	14.2		84	UGC 11319 SW 1'8.
18 42 42.9	+40 22 01	NGC 6695	13.5	1.1 x 0.7	13.0	SBb	12	
18 46 57.5	+45 42 22	NGC 6702	12.2	1.8 x 1.3	13.2	E:	65	
18 47 18.4	+45 33 02	NGC 6703	11.3	2.5 x 2.3	13.1	$SA0^-$	0	UGC 11357 SE 9'9.
18 49 01.2	+47 39 27	NGC 6711	12.9	1.1 x 0.9	12.8	SBbc: I	90	Mag 7.65v star S 8'6.
19 01 41.3	+40 44 37	NGC 6745	13.9	1.1 x 0.5	13.1	Sm	24	Triple system; short extension N.
19 06 57.1	+50 20 54	NGC 6759	14.3	1.0 x 0.7	13.8		30	Mag 10.04v star 2'7 W.
19 08 17.2	+50 55 57	NGC 6764	11.9	2.3 x 1.3	12.9	SB(s)bc	62	Many superimposed stars; anonymous galaxy E 2'7.
19 16 47.8	+46 01 01	NGC 6783	14.4	0.6 x 0.6	13.2			Star or bright knot on SE edge; faint, elongated anonymous galaxy N 3'3.
19 20 57.4	+43 07 58	NGC 6792	12.1	2.2 x 1.3	13.1	SBb	25	Faint stars superimposed N end; mag 10.56v star NW 1'5.
18 22 15.6	+48 06 37	UGC 11217	13.8	0.9 x 0.7	13.1	Sb I-II	15	
18 24 46.2	+41 29 31	UGC 11228	13.4	1.0 x 0.6	12.7	SB0	0	Mag 13.8 star on SE edge.
18 26 00.6	+45 08 03	UGC 11235	15.1	1.6 x 0.2	13.7	SBc-Irr	99	Mag 9.68v star W 8'9.
18 26 50.0	+51 08 17	UGC 11241	14.1	0.8 x 0.7	13.3	Sb III	18	Stellar nucleus, uniform envelope; mag 10.6 star SE 2'2.
18 27 13.6	+46 55 32	UGC 11244	14.3	1.5 x 0.7	14.2	Sdm:	150	
18 28 42.1	+48 14 34	UGC 11252	14.1	0.9 x 0.5	13.1	Scd:	164	Mag 11.8 star NE edge.
18 30 35.9	+42 41 33	UGC 11262	14.6	1.5 x 0.6	14.3	Sd	48	Stellar nucleus, uniform envelope, mag 13.2 star N 1'2.
18 34 45.5	+47 02 05	UGC 11287	14.0	1.7 x 1.1	14.5	S?	15	Trio of stars SE 1'2, brightest is mag 11.5.
18 36 20.6	+51 27 56	UGC 11298	12.5	1.4 x 0.8	12.7	E	96	MCG +9-30-23 W 3'4.
18 40 00.6	+40 06 59	UGC 11319	15.1	1.1 x 0.4	14.1	S	0	Bright center, two faint, thin arms; NGC 6686 NE 1'8; mag 7.77v star W 3'4.
18 40 52.0	+45 15 42	UGC 11329	14.3	1.7 x 0.7	14.3	Sbc	17	
18 48 00.4	+45 26 28	UGC 11357	15.9	1.5 x 0.3	14.9	Sdm:	126	Mag 10.08v star SW 4'0; **MCG +8-34-21** WSW 7'3.
18 54 15.2	+48 55 40	UGC 11376	13.4	0.6 x 0.4	11.7	S	85	Small galaxy Kaz 499 N 1'9.
19 08 19.2	+43 04 04	UGC 11406	13.3	1.1 x 0.8	13.1	SB(s)bc? I-II	141	Mag 11.9 star W edge; mag 10.4 star N edge.
19 09 48.7	+41 34 32	UGC 11408	13.7	1.3 x 0.5	13.1	Sab	103	
19 15 32.5	+42 54 33	UGC 11420	14.0	1.4 x 0.3	12.9	Sb III	5	Mag 13 star on E edge.
19 15 33.4	+44 06 17	UGC 11421	13.1	1.1 x 1.1	13.2	SB0?		Bright nucleus, several stars superimposed.
19 21 08.8	+43 19 32	UGC 11430	13.3	1.0 x 1.0	13.2	SA(s)c		Mag 13.9 star superimposed NE edge.
19 27 07.7	+43 52 39	UGC 11439	14.0	0.9 x 0.5	13.0	S	159	Close pair of stars, mags 11.4 and 13.6, E 2'0.
19 29 33.3	+41 18 15	UGC 11446	13.3	1.1 x 0.8	13.0	SB	84	Brighter N-S core; mag 9.57v star SE 2'4.
19 30 58.5	+46 42 41	UGC 11450	13.8	0.9 x 0.6	13.0	SBab:	20	
19 37 22.6	+40 42 18	UGC 11459	13.4	2.0 x 1.1	14.1	Sd	12	Stellar nucleus, numerous superimposed stars; in star rich area.
19 37 51.7	+41 00 32	UGC 11460	12.9	1.0 x 0.5	12.0	$S0^-$:	108	
19 41 25.8	+51 49 15	UGC 11464	14.3	0.9 x 0.4	13.0	S	15	Mag 10.23v star W 2'6.
19 41 42.3	+50 37 55	UGC 11465	12.8	1.1 x 1.1	12.9	$SAB(s)0^-$		Almost stellar **PGC 63531** N edge; **PGC 63532** S edge; **MCG +8-36-3** S 2'3.
19 42 58.5	+45 17 51	UGC 11466	11.8	1.7 x 1.1	12.3	S?	35	Mag 10.88v star NW 1'6.
19 43 16.8	+42 03 05	UGC 11467	14.5	1.2 x 0.8	14.3	Im:	100	Low surface brightness, many superimposed stars; in star rich area.
19 43 24.9	+41 56 25	UGC 11468	14.4	0.9 x 0.9	14.0	S		Almost stellar core, uniform envelope with a few stars superimposed.
19 45 31.5	+43 08 23	UGC 11473	13.6	1.8 x 1.2	14.3	Sd	125	Numerous stars superimposed.
19 48 17.7	+46 25 41	UGC 11480	13.7	1.1 x 0.8	13.4	Scd:	165	Mag 12.1 star at N end.
19 48 59.4	+50 18 47	UGC 11485	13.1	1.3 x 0.6	12.7	(R)SBa	25	NE-SW core with dark patches N and S; mag 13.1 star E edge.
19 49 49.1	+50 41 45	UGC 11486	14.6	0.9 x 0.5	13.6	S	75	Mag 12.1 star N 0'8.
19 54 02.3	+49 56 17	UGC 11494	14.4	1.7 x 0.8	14.6	Double System	42	Double system, contact, distorted.
19 56 48.8	+49 53 06	UGC 11499	14.1	0.7 x 0.5	12.8	Sdm:	111	Bright nucleus, sandwiched between several faint stars; mag 11.3 star W 1'1.
19 57 02.1	+50 10 51	UGC 11500	13.9	0.7 x 0.6	12.8	Scd:	9	MCG +8-36-9 SW 1'5; mag 10.81v star ESE 2'1.
19 58 42.4	+50 02 11	UGC 11503	12.9	0.6 x 0.5	11.4	S?	99	**MCG +8-36-14** SE 11'2.
20 01 55.3	+49 09 00	UGC 11507	14.6	0.9 x 0.7	13.9	Scd:	75	Faint stellar nucleus, faint envelope; mag 10.70v star NE 1'8.

33

(Continued from previous page)

GALAXY CLUSTERS

RA h m s	Dec ° ′ ″	Name	Mag 10th brightest	No. Gal	Diam ′	Notes
19 20 48.0	+43 57 00	A 2319	15.4	68	15	**MCG +7-40-4** E of center 4′.1; other members PGC or anonymous GX.

OPEN CLUSTERS

RA h m s	Dec ° ′ ″	Name	Mag	Diam ′	No. ★	B ★	Type	Notes
20 02 30.0	+42 06 00	Do 36		14			ast	Few stars; small brightness range; not well detached. Probably not a cluster. (A) Group of about a dozen brighter stars.
19 37 09.6	+46 23 00	NGC 6811	6.8	15	249	11.0	cl	Moderately rich in stars; small brightness range; fairly even distribution; detached.
19 41 18.0	+40 11 00	NGC 6819	7.3	5	929	11.0	cl	Moderately rich in stars; small brightness range; strong central concentration; detached.
20 03 55.0	+44 20 00	NGC 6866	7.6	15	129	10.0	cl	Rich in stars; moderate brightness range; slight central concentration; detached.

PLANETARY NEBULAE

RA h m s	Dec ° ′ ″	Name	Diam ″	Mag (P)	Mag (V)	Mag cent ★	Alt Name	Notes
18 59 19.9	+48 27 53	NGC 6742	33	15.0	13.4	20.0	PK 78+18.1	Mag 13.5 star NE 1′.1; mag 8.84v star SW 3′.1.
19 44 48.2	+50 31 31	NGC 6826	36	9.8	8.8	10.4	PK 83+12.1	**Blinking Planetary**. Very faint halo extends out to 2 arcminutes in diameter; mag 11.1 star S 1.
19 49 46.7	+48 57 38	NGC 6833	11	13.8	12.1	14.5	PK 82+11.1	Mag 10.33v star ENE 2′.8.
19 19 10.3	+46 14 51	PN G77.6+14.7	200	14.4	13.5	17.3	PK 77+14.1	= **Abell 61**; numerous stars on disc; mag 14.7 star inside N edge; mag 12.3 star W 3′.1; mag 11.06v star SSW 4′.0.
20 04 00.2	+49 19 04	PN G84.0+9.5	16	15.2		20.8	K 3-73	Mag 13.3 star NW 0′.8; mag 9.79v star E 4′.0.

34

GALAXIES

RA h m s	Dec ° ′ ″	Name	Mag (V)	Dim ′ Maj x min	SB	Type Class	PA	Notes
17 01 29.5	+48 25 29	CGCG 252-16	14.7	0.5 x 0.5	13.0			Mag 13.8 star NW 1′.6; mag 11.2 star S 5′.9.
17 34 33.1	+49 59 14	CGCG 253-17	13.9	0.8 x 0.3	12.3		3	Located 1′.2 NNW of mag 10.23v star.
17 39 31.3	+48 21 14	CGCG 253-26	14.0	0.5 x 0.4	12.1		123	Stellar nucleus.
18 10 29.2	+49 55 17	CGCG 254-21	14.5	0.7 x 0.7	13.6			Stellar nucleus; NGC 6582 E 5′.3
17 33 02.1	+43 45 35	IC 1262	13.7	1.2 x 0.6	13.3	E	82	IC 1263 and mag 11.2 star N 4′.2.
17 33 07.4	+43 49 17	IC 1263	13.7	1.7 x 0.7	13.8	SB	168	Mag 11.2 star W 2′.6.
17 33 17.0	+43 37 44	IC 1264	14.0	1.2 x 1.1	14.2	SB0?	39	MCG +7-36-24 on E edge; mag 10.20v star E 5′.3.
17 36 39.7	+42 05 18	IC 1265	12.3	2.0 x 0.9	12.8	Sab	80	Mag 9.34v star SW 9′.2.
16 46 07.2	+42 27 34	MCG +7-34-146	14.9	0.5 x 0.4	13.0		117	Mag 9.28v star S 1′.1.
16 46 34.8	+43 01 03	MCG +7-34-147	14.6	0.6 x 0.4	12.9		81	
16 48 35.5	+41 36 00	MCG +7-34-150	14.6	0.5 x 0.5	12.9	Sc		Mag 10.18v star N 1′.7.
16 55 18.2	+44 19 24	MCG +7-35-6	14.0	0.9 x 0.3	12.4	Sb	171	Located between a pair of NE-SW mag 12 stars.
16 57 26.9	+40 44 07	MCG +7-35-9	14.4	0.9 x 0.3	12.9	Sa	144	MCG +7-35-11 and 12 ESE 2′.5.
16 57 33.6	+41 35 18	MCG +7-35-10	14.7	0.7 x 0.4	13.2	Sc	174	Mag 11.8 star N 1′.4.
16 57 41.0	+40 43 17	MCG +7-35-11	14.0	0.8 x 0.6	13.3	E	177	Pair with MCG +7-35-12 S.
16 57 40.1	+40 42 24	MCG +7-35-12	14.1	0.9 x 0.5	13.3	E	117	Pair with MCG +7-35-11 N.
16 58 50.8	+40 07 14	MCG +7-35-13	14.2	1.0 x 0.9	13.9		0	
17 02 23.6	+44 50 37	MCG +7-35-17	14.4	0.7 x 0.6	13.3		84	Mag 11.4 star superimposed on E edge.
17 02 35.9	+40 29 24	MCG +7-35-18	15.0	0.6 x 0.4	13.5	E/S0	150	Mag 11.5 star E 1′.2; mag 9.66v star SSE 3′.9.
17 02 58.2	+40 31 27	MCG +7-35-19	14.3	0.5 x 0.5	12.6			Mag 14.4 star on E edge.
17 03 09.9	+40 38 10	MCG +7-35-20	15.9	0.6 x 0.5	14.5	Sc	36	Mag 12.7 star SE 2′.7.
17 05 16.3	+41 49 34	MCG +7-35-25	14.6	0.7 x 0.5	13.5	E/S0	171	Very faint, small, slightly elongated anonymous galaxy N 1′.3; **CGCG 225-38** S 5′.7.
17 06 45.4	+42 59 56	MCG +7-35-29	14.8	0.7 x 0.4	13.3	Sb	33	Mag 13.1 star N edge.
17 06 49.9	+42 25 26	MCG +7-35-30	14.3	0.3 x 0.2	11.1		120	
17 08 25.8	+40 47 18	MCG +7-35-33	14.0	1.1 x 0.8	14.0	E/S0	33	Mag 11.9 star N 2′.3.
17 09 54.2	+40 35 46	MCG +7-35-35	14.2	0.8 x 0.3	12.5	S?	108	Mag 14.1 star N 0′.7.
17 09 52.8	+41 13 23	MCG +7-35-36	15.4	0.5 x 0.5	13.7			Mag 9.70v star N 1′.5.
17 10 05.4	+42 22 18	MCG +7-35-37	14.3	0.7 x 0.7	13.4	Sc		Mag 11.9 star N 2′.6.
17 11 44.3	+42 41 41	MCG +7-35-41	14.2	0.8 x 0.6	13.3		54	Faint star or companion galaxy N edge?
17 13 04.0	+42 23 18	MCG +7-35-47	14.8	0.7 x 0.5	13.5		144	Mag 9.77v star SW 4′.3.
17 13 40.2	+40 57 45	MCG +7-35-50	14.4	1.0 x 0.8	14.0		156	Mag 10.04v star N 2′.0.
17 14 36.6	+41 24 33	MCG +7-35-53	14.5	0.7 x 0.3	12.7	Sb	144	Small triangle of mag 13-15 stars WSW 1′.5.
17 15 15.5	+43 17 23	MCG +7-35-55	14.3	0.8 x 0.5	13.2		177	
17 15 28.3	+42 04 07	MCG +7-35-56	13.8	0.8 x 0.6	13.1	E/S0	30	Mag 9.38v star ENE 2′.9.
17 16 27.5	+43 20 34	MCG +7-35-58	14.6	0.8 x 0.3	12.9		0	
17 17 16.0	+40 34 47	MCG +7-35-61	14.4	0.6 x 0.6	13.1			Mag 9.63v star SW 1′.8.
17 17 20.0	+40 52 12	MCG +7-35-62	15.3	1.1 x 0.2	13.5	Sc	75	Located 2′.8 NE of NGC 6339.
17 20 11.8	+40 51 08	MCG +7-36-2	15.6	0.5 x 0.5	14.0	Sbc		Mag 12.1 star SW 0′.9.
17 21 21.3	+43 52 20	MCG +7-36-4	14.8	0.7 x 0.5	13.5	Sc	36	Mag 9.45v star NE 5′.5.
17 23 53.6	+41 46 33	MCG +7-36-6	15.4	0.7 x 0.5	14.1	Sbc	66	
17 24 10.4	+43 27 08	MCG +7-36-7	14.2	0.9 x 0.5	13.2	Sb	57	Mag 10.07v star E 4′.8.
17 25 01.7	+41 55 58	MCG +7-36-8	14.7	0.7 x 0.4	13.2	Sb	150	Close pair of mag 12-14 stars NNW 1′.3.
17 27 15.9	+41 08 35	MCG +7-36-11	15.1	0.7 x 0.4	13.6	Sb	111	Mag 12.5 star W 1′.8.
17 28 24.1	+41 07 56	MCG +7-36-12	15.9	0.7 x 0.4	14.4	Sc	120	Mag 9.86v star SE 4′.8.
17 29 22.2	+40 15 04	MCG +7-36-13	14.9	0.6 x 0.5	13.4		33	Mag 12.9 star SW 1′.5.
17 29 40.2	+40 16 13	MCG +7-36-14	15.1	0.5 x 0.4	13.2		0	Mag 9.47v star NE 9′.8.
17 31 24.4	+40 14 12	MCG +7-36-17	13.7	0.8 x 0.8	13.1			Mag 11.2 star W 1′.2; mag 9.41v star NE 2′.4.
17 33 20.3	+43 54 45	MCG +7-36-23	14.5	0.8 x 0.5	13.4	S	84	Faint anonymous galaxy SW 2′.7.
17 33 21.7	+43 38 02	MCG +7-36-24	14.3	0.3 x 0.3	11.5			Located on E edge of IC 1264; mag 10.20v star E 5′.3.
17 37 36.7	+43 31 11	MCG +7-36-28	14.5	0.7 x 0.5	13.2	Sb	33	Mag 11.51v star N 1′.8.
17 39 52.0	+43 18 09	MCG +7-36-29	14.3	0.6 x 0.6	13.3	E/S0		Mag 13.5 star E 0′.8.
17 40 38.3	+41 18 03	MCG +7-36-30	13.8	0.8 x 0.5	12.9	E	99	Mag 12.2 star NE 2′.4.
17 42 07.6	+44 18 19	MCG +7-36-31	14.6	0.7 x 0.6	13.5		45	Located between a pair of NE-SW stars, mags 12.4 and 11.9.
17 44 08.8	+40 52 01	MCG +7-36-33	14.3	0.5 x 0.3	12.1	S?	177	Star superimposed N end?
17 46 41.3	+41 40 20	MCG +7-36-34	14.8	0.5 x 0.5	13.2	Sbc		
17 52 53.0	+41 05 14	MCG +7-37-1	14.8	0.8 x 0.5	13.7		81	
17 58 08.2	+43 25 09	MCG +7-37-4	14.8	0.4 x 0.4	12.6			**UGC 11072** SW 2′.5; mag 6.83v star W 2′.1.
17 59 29.9	+42 30 54	MCG +7-37-10	15.1	0.9 x 0.5	14.1		144	Multi-galaxy system; mag 10.26v star SW 2′.1.
17 59 29.9	+43 23 30	MCG +7-37-11	15.7	0.7 x 0.2	13.4	Sbc	57	Mag 12.4 star E 2′.1.
18 00 04.2	+44 31 37	MCG +7-37-13	15.2	1.2 x 0.2	13.5	Sc	15	MCG +7-37-14 S 1′.3.

(Continued from previous page)

GALAXIES

RA h m s	Dec ° ′ ″	Name	Mag (V)	Dim ′ Maj x min	SB	Type Class	PA	Notes
18 00 06.9	+44 29 36	MCG +7-37-14	15.1	0.9 x 0.2	13.1	Sbc	123	MCG +7-37-13 N 1′.3; mag 8.44v star E 1′.4.
18 02 01.7	+44 28 30	MCG +7-37-17	14.6	0.8 x 0.6	13.7		30	Mag 12.1 star NE 2′.2.
18 02 40.3	+42 47 42	MCG +7-37-18	15.6	0.8 x 0.6	14.7		18	Mag 12.4 star ESE 0′.9.
18 03 50.5	+43 15 29	MCG +7-37-19	15.3	1.0 x 0.3	13.8		72	Located between a pair of E-W mag 14 stars.
18 04 39.2	+41 30 40	MCG +7-37-20	15.0	0.7 x 0.3	13.1	Sb	90	Mag 10.16v star W 5′.2.
18 04 50.6	+43 57 12	MCG +7-37-21	14.3	0.7 x 0.5	13.0		57	Mag 12.1 star E 0′.5.
18 17 26.0	+40 01 51	MCG +7-37-33	17.3	0.9 x 0.6	16.4	Sc	39	Mag 11.8 star W 0′.9.
18 17 27.2	+40 59 22	MCG +7-37-34	15.4	0.7 x 0.5	14.2	Sb	0	Mag 6.13v star SW 5′.0.
18 19 26.4	+40 58 23	MCG +7-37-38	15.3	0.7 x 0.3	13.5	Sb	150	Mag 15.1 star on W edge; mag 11.1 star NW 1′.6.
18 27 39.8	+41 44 54	MCG +7-38-6	15.5	0.6 x 0.6	14.2	Sb		Mag 9.85v star NW 5′.4.
17 00 36.4	+47 15 57	MCG +8-31-19	13.9	0.9 x 0.7	13.3		105	
17 03 08.9	+46 22 22	MCG +8-31-21	14.3	0.9 x 0.6	13.5		165	**MCG +8-31-20** W 1′.7; mag 11.00v star SSW 1′.5.
17 19 31.4	+47 42 07	MCG +8-31-44	14.4	0.8 x 0.8	13.7	Sb		
17 24 49.4	+45 27 20	MCG +8-32-5	14.0	0.8 x 0.7	13.5	E	144	**CGCG 253-7** NW 2′.8; mag 10.12v star WNW 2′.1.
17 47 18.5	+45 42 29	MCG +8-32-21	14.1	0.9 x 0.8	13.8	E/S0	54	Faint anonymous galaxy ESE 2′.1.
18 17 04.4	+48 56 31	MCG +8-33-34	15.5	0.5 x 0.5	13.9	Sc		Mag 12.3 star SE 1′.5.
18 22 44.1	+47 37 08	MCG +8-33-41	14.3	0.8 x 0.7	13.5		24	Mag 13.1 star on W edge.
16 44 16.7	+51 30 04	MCG +9-27-68	15.1	0.7 x 0.4	13.6	Sbc	132	Mag 11.7 star E 0′.8.
16 44 43.6	+51 30 38	MCG +9-27-69	14.7	0.9 x 0.4	13.4	Sbc	33	Mag 12.4 star N 2′.4.
17 40 16.4	+51 03 23	MCG +9-29-16	14.4	0.9 x 0.4	13.1	S?	168	**UGC 10942** S 1′.7.
17 40 40.8	+51 10 49	MCG +9-29-17	15.7	1.0 x 0.6	15.0	Sbc	144	Located in a triangle of mag 13 stars.
18 15 43.0	+51 47 11	MCG +9-30-9	14.4	0.8 x 0.4	13.0		177	Mag 10.85v star W edge.
16 50 06.3	+42 44 15	NGC 6239	12.4	2.6 x 1.1	13.4	SB(s)b pec? III-IV	118	Bright, elongated center, elongated dark patch N.
16 50 11.1	+45 25 11	NGC 6241	16.2	1.0 x 0.8	15.8	Sc	98	**Arp 103** WNW 7′.9.
16 59 01.4	+47 14 11	NGC 6279	13.6	1.1 x 1.0	13.6	SB0	19	Bright nucleus in short E-W bar; mag 11.45v star NE 2′.5.
16 59 26.5	+49 55 18	NGC 6283	13.5	1.1 x 1.1	13.5	S		
17 08 32.7	+42 20 22	NGC 6301	13.4	1.8 x 1.1	14.0	Scd:	115	Star on S edge.
17 10 43.7	+41 39 00	NGC 6311	13.5	1.1 x 1.1	13.7	E		Stellar **CGCG 225-56** SSW 4′.7.
17 10 48.3	+42 17 13	NGC 6312	14.2	0.7 x 0.7	13.3			
17 10 20.9	+48 19 53	NGC 6313	13.9	1.3 x 0.4	13.0	Sab	156	Close double star W 1′.4.
17 12 55.8	+40 15 55	NGC 6320	13.9	1.2 x 0.8	13.7	Sbc	103	Stellar nucleus; stars on S and NE edges.
17 13 18.2	+43 46 55	NGC 6323	13.9	1.1 x 0.4	12.9	Sab	175	**CGCG 225-70** S 5′.4.
17 14 02.4	+43 38 54	NGC 6327	14.9	0.4 x 0.4	12.7			Almost stellar.
17 14 15.2	+43 41 04	NGC 6329	12.8	1.8 x 1.8	14.1	E		Bright nucleus, uniform envelope.
17 15 03.1	+43 39 32	NGC 6332	13.6	1.6 x 0.8	13.7	Sa	45	Mag 9.30v star N 5′.8; note: **CGCG 225-87** lies 2′.5 E of this star.
17 16 16.4	+43 49 12	NGC 6336	13.6	0.9 x 0.7	12.9	SAB(s)a	170	Located 2′.4 N of mag 10.02v star.
17 17 06.8	+40 50 43	NGC 6339	12.7	2.9 x 1.7	14.3	SBd	10	Bright nucleus, patchy arms; MCG +7-35-62 NE 2′.8.
17 17 16.4	+41 03 10	NGC 6343	13.8	1.1 x 1.1	13.9			
17 18 21.3	+41 38 50	NGC 6348	14.4	0.7 x 0.6	13.3	Sb	125	Elongated anonymous galaxy NE 2′.4.
17 18 42.2	+41 41 38	NGC 6350	13.2	1.0 x 1.0	13.0	S0		Mag 8.39v star S 8′.8.
17 22 40.0	+41 06 04	NGC 6363	13.3	1.1 x 0.9	13.2	E	14	
17 36 35.8	+50 45 54	NGC 6409	13.8	0.8 x 0.6	12.9		60	
17 44 33.8	+48 06 51	NGC 6443	13.8	1.2 x 0.5	13.1	Sab	128	
17 48 08.2	+51 23 52	NGC 6466	14.1	0.7 x 0.4	12.6		111	**CGCG 278-28** WNW 5′.1.
17 48 38.2	+51 09 24	NGC 6478	13.3	1.9 x 0.7	13.4	SAc:	37	
17 57 25.1	+50 43 39	NGC 6515	13.0	1.6 x 1.0	13.5	E	10	Stellar anonymous galaxy or star E 1′.1.
17 59 14.8	+45 53 14	NGC 6524	12.8	1.3 x 1.0	13.0	S0$^-$:	155	
18 05 14.2	+46 52 53	NGC 6560	13.6	1.2 x 0.8	13.4	S pec? I-II	55	Two main arms; mag 10.36v star NE 2′.2.
18 11 01.8	+49 54 43	NGC 6582	13.9	0.6 x 0.6	12.9	E		Double system with envelopes in contact; **UGC 11149** S 2′.9
18 14 41.4	+43 16 02	NGC 6606	13.9	0.9 x 0.7	13.3	S	69	Mag 7.76v star W 6′.9.
16 44 24.5	+43 43 48	UGC 10531	15.1	0.9 x 0.6	14.3	Sbc	84	Close pair of stars, mags 11.5 and 12.2, N 2′.1.
16 47 14.9	+40 14 39	UGC 10553	14.0	1.0 x 0.5	13.1	SBab:	153	Elongated **UGC 10550** S 6′.5.
16 49 14.5	+48 37 22	UGC 10571	14.5	1.5 x 0.5	14.1	SBcd:	15	Stellar **Mkn 499**, **Mkn 500**, and dwarf **UGC 10565** NW 10′.1.
16 50 48.6	+45 24 07	UGC 10586	13.8	1.5 x 1.4	14.5	Sb II	48	Stellar anonymous galaxy SW edge; stellar anonymous galaxy and mag 12.8 star 0′.8 NE.
16 54 22.4	+41 20 08	UGC 10602	13.9	0.9 x 0.8	13.4	SBbc	31	
16 55 00.4	+43 03 26	UGC 10610	14.1	1.4 x 0.7	13.9	SB?	105	Double system, consists of **MCG +7-35-4** and **MCG +7-35-5**; narrow streamers N and S from W component.
16 57 15.7	+46 52 24	UGC 10627	14.7	1.5 x 0.3	13.7	Scd:	142	Very faint star E edge.
16 59 44.2	+42 32 21	UGC 10651	14.3	1.1 x 0.9	14.1	S?	168	
17 03 33.0	+45 27 32	UGC 10681	14.9	1.5 x 0.3	13.9	Scd:	39	Mag 10.03v star S 2′.2.
17 04 52.9	+41 51 54	UGC 10693	12.8	1.8 x 1.2	13.6	E	105	**UGC 10686** NW 7′.8.
17 05 05.7	+43 02 34	UGC 10695	13.9	1.5 x 1.2	14.6	E	114	Very small, faint, round galaxy NW edge.
17 06 23.6	+41 02 31	UGC 10707	14.5	1.1 x 1.0	14.5	SA(rs)cd	24	Mag 12.1 star on W edge; **MCG +7-35-27** W 1′.6.
17 06 52.5	+43 07 20	UGC 10710	14.2	1.5 x 0.3	13.2	Sb III	150	**CGCG 225-41** SW 2′.9.
17 08 20.0	+42 57 28	UGC 10722	13.9	1.2 x 0.8	13.9	E:	30	
17 12 48.6	+42 48 51	UGC 10759	13.6	1.0 x 1.0	13.4	Compact		Small, faint anonymous galaxy SSE 2′.6.
17 12 48.7	+44 09 36	UGC 10760	14.4	1.3 x 1.1	14.6	S	3	Stellar nucleus, faint envelope.
17 13 25.0	+42 53 52	UGC 10765	14.0	1.0 x 0.7	13.5	Scd:	123	Stellar **CGCG 225-69** SW 3′.8.
17 18 51.2	+49 52 58	UGC 10806	13.1	2.2 x 0.9	13.7	SB(s)dm	77	Very knotty.
17 19 28.7	+40 55 25	UGC 10812	14.1	1.1 x 0.5	13.3	SBb	125	Bright E-W bar, faint envelope.
17 19 21.7	+49 02 19	UGC 10814	14.1	1.1 x 0.6	13.5	SABb pec	165	A very faint, wide ribbon of material extends many arcminutes N.
17 20 57.4	+42 09 21	UGC 10820	15.2	1.5 x 0.2	13.7	Scd:	139	
17 24 24.4	+44 56 24	UGC 10845	13.6	1.0 x 0.9	13.3	Scd:	168	
17 33 38.0	+50 22 23	UGC 10908	15.4	1.7 x 0.7	15.4	SBdm?	160	Very filamentary; mag 11.12v star E 1′.3.
17 39 32.9	+49 36 03	UGC 10938	13.6	1.4 x 1.0	13.9	(R)SAB0$^+$	177	Bright nucleus, uniform envelope; mag 9.73v star E 2′.5.
17 41 12.8	+45 09 03	UGC 10946	13.8	1.5 x 0.7	13.7	SBbc	36	Mag 12.1 star N 2′.2.
17 46 39.7	+51 16 51	UGC 10984	14.0	1.3 x 0.6	13.5	SBab:	162	Star or knot N edge of nucleus; mag 9.66v star N 6′.2.
17 57 05.6	+47 44 27	UGC 11065	14.0	0.9 x 0.8	13.5	S	75	
18 04 44.2	+46 16 51	UGC 11112	14.2	1.3 x 0.7	13.9	SABb	102	**MCG +8-33-15** S 1′.6.
18 15 36.4	+43 18 21	UGC 11181	15.1	1.5 x 0.7	15.0	Multiple System	126	Multiple system, bridge + plumes; **MCG +7-37-28** E 1′.5.
18 16 10.9	+42 39 34	UGC 11185	15.0	1.5 x 1.0	15.3	Double System	51	Double system, bridge; mag 10.9 star SW edge.
18 16 18.5	+47 50 07	UGC 11186	13.7	0.8 x 0.6	12.8	S	36	Mag 11.13v star WSW 1′.7.
18 16 56.6	+40 48 11	UGC 11190	15.3	0.8 x 0.8	14.7	Scd:		Mag 11.51v star SW 2′.6.
18 18 39.1	+50 16 38	UGC 11202	13.2	1.1 x 0.9	13.1	S0	175	Mag 10.58v star N 1′.9.
18 19 01.0	+47 44 49	UGC 11204	14.7	0.8 x 0.7	13.9	Scd:	42	Mag 11.9 star E 2′.0.
18 19 44.2	+48 33 39	UGC 11206	13.6	0.8 x 0.6	12.7	S	160	Mag 9.47v star ESE 1′.9.
18 22 15.6	+48 06 37	UGC 11217	13.8	0.9 x 0.7	13.1	Sb I-II	15	
18 24 46.2	+41 29 31	UGC 11228	13.4	1.0 x 0.6	12.7	SB0	0	Mag 13.8 star on SE edge.
18 26 00.6	+45 08 03	UGC 11235	15.1	1.6 x 0.2	13.7	SBc-Irr	99	Mag 9.68v star W 8′.9.

34

(Continued from previous page)

GALAXIES

34

RA h m s	Dec ° ′ ″	Name	Mag (V)	Dim ′ Maj x min	SB	Type Class	PA	Notes
18 26 50.0	+51 08 17	UGC 11241	14.1	0.8 x 0.7	13.3	Sb III	18	Stellar nucleus, uniform envelope; mag 10.6 star SE 2'.2.
18 27 13.6	+46 55 32	UGC 11244	14.3	1.5 x 0.7	14.2	Sdm:	150	

GLOBULAR CLUSTERS

RA h m s	Dec ° ′ ″	Name	Total V m	B ★ V m	HB V m	Diam ′	Conc. Class Low = 12 High = 1	Notes
16 46 58.9	+47 31 40	NGC 6229	9.4	15.5	18.0	4.5	4	
17 17 07.3	+43 08 11	NGC 6341	6.5	12.1	15.2	14.0	4	= **M 92**.

PLANETARY NEBULAE

RA h m s	Dec ° ′ ″	Name	Diam ″	Mag (P)	Mag (V)	Mag cent ★	Alt Name	Notes
17 13 50.6	+49 16 08	PN G75.7+35.8	16	15.1	13.:	14.3	PK 75+35.1	Mag 15.2 star E 1'.5; mag 13.2 star WSW 2'.0.

35

GALAXIES

35

RA h m s	Dec ° ′ ″	Name	Mag (V)	Dim ′ Maj x min	SB	Type Class	PA	Notes
15 18 06.3	+42 44 34	CGCG 221-50	14.9	0.8 x 0.4	13.6	S0? pec	153	Multi-galaxy system?
15 55 04.5	+45 29 05	CGCG 250-18	14.2	0.9 x 0.7	13.6		177	Mag 11.1 star on E edge.
16 27 01.8	+49 32 03	CGCG 251-19	14.6	0.9 x 0.5	13.6	Pec	15	Brighter N of nucleus.
16 38 37.6	+49 08 09	CGCG 251-36	14.0	0.5 x 0.5	12.3			Located 2'.7 NE of mag 10.90v star.
15 51 21.7	+43 25 00	IC 1144	13.5	1.0 x 0.6	12.8	S0$^-$:	100	Mag 10.86v star SE 6'.1.
15 56 43.4	+48 05 42	IC 1152	13.1	1.2 x 1.2	13.4	E		Mag 10.56v star and IC 1153, MCG +8-29-27 NNE 6'.7.
15 57 03.1	+48 10 04	IC 1153	12.6	1.2 x 1.1	12.7	S0	156	Mag 10.56v star and MCG +8-29-27 NE 1'.5.
16 34 41.8	+46 23 30	IC 1221	13.8	1.2 x 1.0	13.8	Sd	0	Located 2'.9 NE of mag 12.0 star.
16 35 08.9	+46 12 46	IC 1222	13.4	1.7 x 1.3	14.2	SAB(s)c I-II	50	Mag 13.4 star W 2'.7.
16 35 42.5	+49 13 11	IC 1223	14.4	1.1 x 0.7	14.0		18	Mag 8.48v star S 8'.3.
15 35 57.1	+43 29 33	IC 4562	12.6	1.2 x 1.2	13.0	E?		**IC 4562A** NE 1'.2; mag 11.6 star W 1'.4.
15 36 27.3	+43 31 03	IC 4564	13.5	1.4 x 0.5	13.0	S?	70	IC 4566 NE 3'.1.
15 36 35.2	+43 25 28	IC 4565	14.1	0.9 x 0.5	13.1	Sc	8	Mag 8.85v star S 10'.2.
15 36 42.0	+43 32 25	IC 4566	13.3	1.6 x 1.0	13.6	Sab II	165	IC 4564 SW 3'.1.
15 37 13.3	+43 17 53	IC 4567	12.8	1.4 x 0.9	12.9	Scd?	125	Mag 8.85v star SW 7'.3; IC 4565 NW 10'.3.
15 10 36.0	+41 03 17	MCG +7-31-35	14.7	0.7 x 0.7	13.8			
15 12 25.6	+40 33 06	MCG +7-31-36	14.3	0.8 x 0.6	13.4	S?	123	Mag 11.3 star N 2'.5.
15 13 13.9	+40 32 14	MCG +7-31-39	14.3	1.1 x 0.9	14.2		87	MCG +7-31-40 S 1'.1.
15 13 15.6	+40 31 09	MCG +7-31-40	15.0	0.8 x 0.3	13.3	Sb	6	MCG +7-31-39 N 1'.1; mag 11.7 star E 0'.8.
15 13 29.5	+42 19 45	MCG +7-31-41	14.7	0.8 x 0.4	13.3	SBbc	150	
15 15 06.1	+43 08 56	MCG +7-31-47	14.6	0.9 x 0.5	13.6		120	Mag 13.2 star E 2'.3.
15 16 04.0	+43 09 34	MCG +7-31-49	15.4	0.8 x 0.5	14.2		84	UGC 9796 W 1'.5; mag 11.4 star E 0'.9.
15 17 14.2	+40 56 41	MCG +7-31-53	15.2	0.5 x 0.4	13.3	Sc	117	Pair of mag 13-14 stars WSW 1'.6.
15 24 14.2	+42 23 51	MCG +7-32-4	15.2	0.7 x 0.4	13.7	Sc	111	
15 28 09.7	+43 00 23	MCG +7-32-12	14.8	0.8 x 0.4	13.4	Sb	78	Mag 14.0 star on N edge; NGC 5934 and NGC 5935 S 4'.1.
15 29 40.4	+42 39 12	MCG +7-32-15	14.8	0.7 x 0.5	13.5		105	UGC 9873 SE 2'.3.
15 32 27.8	+41 48 40	MCG +7-32-22	15.0	0.7 x 0.6	13.9		15	Mag 10.35v star N 0'.9.
15 33 13.4	+41 04 34	MCG +7-32-26	14.6	0.9 x 0.7	14.0		156	Mag 13.2 star NNW 1'.8.
15 34 14.9	+43 14 41	MCG +7-32-27	16.0	0.8 x 0.6	15.1	SBb	9	Mag 14.6 star NW edge; mag 13.6 star ESE 1'.4.
15 36 32.4	+41 47 55	MCG +7-32-35	15.1	0.7 x 0.5	13.8	SBbc	66	
15 39 59.8	+44 27 55	MCG +7-32-43	15.1	0.8 x 0.5	14.0		129	Mag 11.4 star SW 3'.4.
15 43 24.5	+44 07 16	MCG +7-32-45	15.3	0.6 x 0.4	13.6	Sc	75	Mag 13.4 star N 2'.4.
15 44 51.4	+43 30 47	MCG +7-32-51	14.3	1.0 x 0.8	13.9		24	Mag 13.3 star E 1'.2.
15 48 26.7	+41 58 50	MCG +7-32-53	15.4	0.8 x 0.4	14.0	Sc	141	Faint anonymous galaxy W edge; mag 9.58v star ESE 4'.5.
15 49 38.3	+42 01 59	MCG +7-32-54	14.1	1.1 x 0.8	14.0	E	9	**CGCG 222-55** NE 4'.6.
15 52 00.5	+42 43 34	MCG +7-33-3	15.7	0.8 x 0.3	14.0	S+	126	Mag 11.6 star N 0'.5.
15 53 29.1	+41 34 48	MCG +7-33-6	14.4	0.8 x 0.5	13.5	E	150	Mag 11.6 star N 2'.0.
15 55 02.3	+41 34 39	MCG +7-33-11	13.1	1.1 x 0.9	12.9	S?	141	Mag 11.20v star NNE 3'.6.
15 55 47.7	+42 36 22	MCG +7-33-14	14.5	0.5 x 0.4	12.6	Sbc	153	Located 4'.0 NE of 5.75v mag star 4 Herculis.
15 56 05.3	+42 07 33	MCG +7-33-15	14.5	0.6 x 0.3	12.7	E	24	Mag 13.3 star S 1'.3.
15 56 33.2	+41 53 50	MCG +7-33-16	14.2	0.6 x 0.4	12.5	S?	147	UGC 10099 SE 1'.2.
15 57 54.6	+41 56 10	MCG +7-33-18	14.2	0.7 x 0.5	12.9	S?	135	Mag 10.66v star S 3'.6.
15 57 59.6	+40 01 59	MCG +7-33-19	14.5	0.5 x 0.3	12.3		165	Pair of mag 13.4 stars W and N edges.
15 58 47.5	+41 56 14	MCG +7-33-20	14.3	0.7 x 0.4	12.8	S?	120	Mag 11.7 star SW 3'.4.
15 59 57.3	+41 28 38	MCG +7-33-23	14.3	0.8 x 0.5	13.1	S?	15	MCG +7-33-24 E 1'.2; mag 12.0 star S 1'.9.
16 00 03.6	+41 28 43	MCG +7-33-24	14.4	0.5 x 0.3	12.2	S0?	105	MCG +7-33-23 W 1'.2.
16 00 33.3	+43 55 38	MCG +7-33-25	15.2	0.7 x 0.7	14.3			
16 01 27.9	+42 23 16	MCG +7-33-26	15.3	0.7 x 0.5	14.0	Sc	0	
16 02 16.7	+42 54 57	MCG +7-33-27	13.7	1.0 x 0.8	13.3	S?	168	Mag 11.3 star W 1'.3.
16 02 44.9	+41 11 56	MCG +7-33-30	14.7	0.5 x 0.2	12.0	S	63	Located on E edged of UGC 10152.
16 05 29.0	+42 37 38	MCG +7-33-37	14.1	0.8 x 0.5	13.0	S?	177	Mag 12.5 star NNW 2'.3.
16 07 17.8	+41 24 10	MCG +7-33-42	14.5	0.5 x 0.5	12.9	Sc		Mag 10.78v star E 4'.4.
16 10 20.6	+41 51 16	MCG +7-33-49	14.2	0.8 x 0.7	13.4	S0?	30	Mag 13.2 star E 0'.8; mag 11.19v star SE 1'.8.
16 10 52.4	+41 53 07	MCG +7-33-51	13.9	1.0 x 1.0	13.8	S?		Mag 11.7 star W 4'.1.
16 22 39.0	+43 27 51	MCG +7-34-7	14.1	0.9 x 0.7	13.4		78	
16 23 23.1	+41 38 59	MCG +7-34-12	14.4	0.8 x 0.3	12.7	S?	156	
16 23 28.2	+41 42 20	MCG +7-34-14	14.1	0.8 x 0.5	12.9	S?	36	Small companion galaxy on W edge?
16 25 44.5	+40 28 30	MCG +7-34-27	14.6	0.8 x 0.3	12.9	Sb	147	NGC 6150 NE 1'.3.
16 25 49.3	+40 20 38	MCG +7-34-28	14.1	0.6 x 0.5	12.7	Irr	21	= **Mkn 881**.
16 25 57.8	+43 57 43	MCG +7-34-30	14.3	0.7 x 0.5	13.0		120	Multi-galaxy system; anonymous galaxy W 1'.1; mag 11.12v star NNE 2'.2.
16 26 22.0	+40 54 40	MCG +7-34-31	14.5	0.8 x 0.8	13.9	S?		Mag 9.08v star E 4'.0.
16 26 27.5	+44 06 46	MCG +7-34-32	15.1	0.6 x 0.4	13.4		177	**MCG +7-34-34** NE 0'.9.
16 26 39.9	+40 28 39	MCG +7-34-33	14.0	0.8 x 0.8	13.3	SB?		Close N-S pair of mag 13-14 stars N 1'.4.
16 26 53.3	+41 15 10	MCG +7-34-35	13.7	0.9 x 0.8	13.1	SB? III-IV	117	Mag 12.9 star NW 0'.9.
16 27 37.2	+40 40 13	MCG +7-34-39	14.6	0.5 x 0.4	12.9	E	18	Small, faint anonymous galaxy W 2'.5.
16 28 14.3	+40 18 46	MCG +7-34-47	14.4	0.7 x 0.7	13.5	S?		Mag 12.8 star WNW 1'.9.

(Continued from previous page)

RA h m s	Dec ° ′ ″	Name	Mag (V)	Dim ′ Maj x min	SB	Type Class	PA	Notes
16 28 25.0	+41 10 05	MCG +7-34-51	14.3	0.8 x 0.7	13.5	Sa	33	Pair with MCG +7-34-52 SE edge; **MCG +7-34-58** E 2′.4.
16 28 27.5	+41 09 37	MCG +7-34-52	14.3	0.6 x 0.4	12.6	S0	102	Pair with MCG +7-34-51 NW; **MCG +7-34-63** SE 2′.9.
16 28 39.7	+40 07 21	MCG +7-34-61	14.3	0.7 x 0.4	12.7	SB	144	Pair with MCG +7-34-62 E edge; mag 13.6 star W 0′.9.
16 28 42.4	+40 07 22	MCG +7-34-62	14.7	0.5 x 0.4	12.8	SB	108	Pair with MCG +7-34-61 W edge.
16 28 52.2	+42 48 40	MCG +7-34-77	14.5	0.7 x 0.5	13.2	Sbc	108	Mag 14.8 star on N edge.
16 29 11.1	+40 51 24	MCG +7-34-81	14.6	0.9 x 0.3	13.1	Sb	120	MCG +7-34-82 NE 2′.7.
16 29 23.9	+40 52 26	MCG +7-34-82	14.4	0.6 x 0.6	13.1	S0?		MCG +7-34-81 SW 2′.7.
16 30 17.9	+40 35 52	MCG +7-34-89	15.0	0.5 x 0.5	13.3	S		Faint anonymous galaxy NE 1′.8.
16 30 22.0	+40 55 26	MCG +7-34-90	14.6	0.6 x 0.6	13.4	S..		**PGC 58350** on NW edge.
16 30 27.1	+40 43 27	MCG +7-34-92	14.7	1.0 x 0.3	13.5	E	33	Located 1′.1 SE of mag 7.87v star.
16 30 42.0	+40 31 43	MCG +7-34-97	15.1	0.5 x 0.5	13.4	S		NGC 6180 W 1′.7.
16 31 38.0	+41 29 34	MCG +7-34-110	14.6	0.7 x 0.6	13.5	S?	108	Mag 14.6 star on S edge.
16 31 54.9	+42 42 21	MCG +7-34-111	14.4	0.7 x 0.5	13.1	Sb	174	Mag 11.8 star on W edge.
16 36 15.6	+44 04 35	MCG +7-34-116	14.5	0.5 x 0.3	12.3	H II	51	
16 36 17.9	+44 08 05	MCG +7-34-117	13.3	0.8 x 0.6	12.4	Sab	15	Mag 10.34v star E 5′.7.
16 37 16.4	+44 25 03	MCG +7-34-119	14.5	0.7 x 0.7	13.6			Mag 13.3 star W 1′.6.
16 37 22.6	+44 20 47	MCG +7-34-120	14.6	0.4 x 0.4	12.5			MCG +7-34-123 E 2′.4.
16 37 35.8	+44 20 36	MCG +7-34-123	14.6	0.4 x 0.3	12.1		138	MCG +7-34-120 W 2′.4.
16 38 13.9	+40 08 45	MCG +7-34-125	14.1	0.9 x 0.4	12.8	Sb	168	**MCG +7-34-126** N 1′.6.
16 39 39.0	+43 17 38	MCG +7-34-128	15.2	0.6 x 0.4	13.5	Sbc	135	Mag 12.4 star SW 1′.5.
16 41 37.5	+40 09 45	MCG +7-34-135	14.3	1.2 x 0.6	13.8	Sb	57	Mag 10.15v star SSW 4′.0.
16 42 08.6	+40 16 34	MCG +7-34-138	14.7	0.9 x 0.9	14.3	Sbc		
16 42 56.7	+43 59 46	MCG +7-34-140	14.9	1.0 x 0.2	13.0		96	
16 46 07.2	+42 27 34	MCG +7-34-146	14.9	0.5 x 0.4	13.0		117	Mag 9.28v star S 1′.1.
16 46 34.8	+43 01 03	MCG +7-34-147	14.6	0.6 x 0.4	12.9		81	
16 48 35.5	+41 36 00	MCG +7-34-150	14.6	0.5 x 0.5	12.9	Sc		Mag 10.18v star N 1′.7.
15 09 30.6	+46 29 35	MCG +8-28-3	14.2	0.5 x 0.5	12.5			Mag 8.76v star W 7′.9.
15 10 24.2	+45 09 55	MCG +8-28-4	14.5	1.0 x 0.7	13.9	Sd	57	Mag 9.94v star N 3′.1.
15 20 41.5	+49 06 21	MCG +8-28-19	14.2	0.8 x 0.4	12.9		27	**PGC 54757** NE 1′.6.
15 20 52.4	+48 39 37	MCG +8-28-20	13.8	0.7 x 0.5	12.8	E/S0	99	Mag 12.0 star NW 2′.7.
15 25 11.4	+47 39 29	MCG +8-28-28	14.6	0.7 x 0.5	13.3	SBb	168	Mag 11.6 star N 1′.7.
15 26 39.3	+50 19 06	MCG +8-28-31	14.8	1.1 x 0.8	14.5	Sbc	84	Close pair of mag 12-13 stars NE 2′.6.
15 31 21.5	+47 01 23	MCG +8-28-36	15.1	0.9 x 0.3	13.5	SBc	99	Mag 9.82v star N edge; **UGC 9883** NW 2′.4.
15 32 21.7	+48 24 18	MCG +8-28-37	15.1	0.7 x 0.6	14.0	Sbc	21	
15 40 04.7	+45 56 51	MCG +8-28-42	15.2	1.0 x 0.3	13.7	Sbc		Mag 12.3 star E 1′.0.
15 41 58.0	+48 33 22	MCG +8-29-4	14.2	0.6 x 0.6	12.9			**MCG +8-29-1** WSW 2′.0.
15 42 18.0	+48 12 54	MCG +8-29-5	14.8	0.5 x 0.4	12.9		6	Mag 11.6 star SE 3′.0.
15 44 27.1	+47 00 23	MCG +8-29-8	14.3	1.0 x 0.8	13.9		69	
15 49 07.9	+44 41 55	MCG +8-29-12	14.6	0.9 x 0.7	13.9		162	
15 50 39.1	+50 27 36	MCG +8-29-13	14.9	0.6 x 0.4	13.2	Sc	108	Mag 11.6 star SW 5′.0.
15 54 42.2	+45 20 37	MCG +8-29-18	15.4	0.7 x 0.6	14.3		96	**MCG +8-29-19** SE edge; mag 10.99v star N 2′.7.
15 57 09.4	+48 11 08	MCG +8-29-27	14.3	0.6 x 0.4	12.6	SB?	93	Mag 10.56v star S edge; IC 1153 SW 1′.5.
16 00 43.0	+48 33 26	MCG +8-29-31	13.9	0.9 x 0.8	13.4	S?	54	Mag 8.37v star E 4′.9.
16 15 09.4	+46 46 25	MCG +8-30-1	13.9	1.1 x 1.0	13.9	IB?	30	
16 24 58.0	+50 18 10	MCG +8-30-10	14.6	0.6 x 0.5	13.2		12	
16 31 25.7	+46 19 25	MCG +8-30-20	14.8	0.8 x 0.8	14.1			Mag 10.09v star NW 4′.9.
16 32 57.4	+50 24 05	MCG +8-30-24	13.3	0.8 x 0.8	12.7	S?		Mag 14.9 star SE edge.
16 33 14.1	+50 23 53	MCG +8-30-25	14.2	0.8 x 0.6	13.5	E/S0	3	MCG +8-30-26 E 1′.3.
16 33 21.6	+50 24 16	MCG +8-30-26	14.7	0.6 x 0.5	13.3		63	Multi-galaxy system, elongated companion W edge; MCG +8-30-25 W 1′.3.
16 33 41.2	+50 22 30	MCG +8-30-27	15.1	0.7 x 0.5	13.8	S	18	MCG +8-30-26, MCG +8-30-25 WNW 3′.8.
16 36 06.7	+50 24 05	MCG +8-30-35	14.4	1.0 x 0.8	14.0		48	Mag 12.8 star NE 3′.0.
16 37 30.9	+49 28 10	MCG +8-30-39	14.8	0.4 x 0.3	12.4	Sc	117	**UGC 10487** SE 2′.7.
15 12 17.9	+50 33 25	MCG +9-25-33	14.7	0.9 x 0.4	13.5		87	
15 12 52.4	+51 23 52	MCG +9-25-36	16.1	0.7 x 0.4	14.5	Sb	105	Elongated **MCG +9-25-35** S edge.
15 21 49.7	+50 30 55	MCG +9-25-45	15.8	0.8 x 0.3	14.1	Sc	171	Mag 11.6 star W edge.
15 27 59.5	+51 40 39	MCG +9-25-52	14.7	0.8 x 0.3	13.0		42	Mag 9.71v star NE 3′.4.
15 22 18.1	+51 08 55	MCG +9-27-26	13.9	0.9 x 0.7	13.2	SBbc	57	
16 37 32.5	+51 39 06	MCG +9-27-58	15.4	0.9 x 0.4	14.2		42	
16 44 16.7	+51 30 04	MCG +9-27-68	15.1	0.7 x 0.4	13.6	Sbc	132	Mag 11.7 star E 0′.8.
16 44 43.6	+51 30 38	MCG +9-27-69	14.7	0.9 x 0.4	13.4	Sbc	33	Mag 12.4 star N 2′.4.
15 12 45.5	+41 14 00	NGC 5886	14.1	0.9 x 0.4	13.1	E/S0	86	Faint star N edge; mag 9.07v star SE 2′.9.
15 13 07.4	+41 15 55	NGC 5888	13.4	1.3 x 0.8	13.3	SB(s)bc	158	Weak dark lane W of center.
15 13 15.8	+41 19 38	NGC 5889	15.4	0.9 x 0.4	14.2	SBb?	39	Mag 9.28v star ENE 9′.5.
15 13 34.3	+41 57 28	NGC 5893	13.1	1.3 x 1.1	13.4	SB(r)b	30	Strong dark lane E of center.
15 13 50.0	+42 00 26	NGC 5895	14.2	0.9 x 0.2	12.2	Sc	21	NGC 5896 N 1′.0.
15 13 50.8	+42 01 24	NGC 5896	16.1	0.3 x 0.2	12.9		76	Almost stellar, NGC 5895 S 1′.0.
15 15 03.4	+42 02 54	NGC 5899	11.7	3.2 x 1.2	13.0	SAB(rs)c II	18	Several dark patches.
15 15 05.1	+42 12 38	NGC 5900	14.0	1.7 x 0.5	13.6	Sb: sp III	131	Strong dark lane NE.
15 14 22.0	+50 19 46	NGC 5902	13.2	1.1 x 1.0	13.2		105	
15 18 43.9	+41 51 55	NGC 5914A	14.4	0.8 x 0.4	13.0	Sb	153	NGC 5914B N 1′.5.
15 18 45.3	+41 53 24	NGC 5914B	16.0	0.6 x 0.6	14.7			Multi-galaxy system; elongated companion galaxy extends S from S edge.
15 19 25.9	+45 52 48	NGC 5918	13.2	1.9 x 0.8	13.5	Scd:	85	Mag 8.89v star WNW 9′.9; **MCG +8-28-16** S 5′.5.
15 21 14.4	+41 43 32	NGC 5923	13.1	1.8 x 1.8	14.2	SAB(s)bc I-II		Stellar nucleus.
15 26 08.1	+41 40 26	NGC 5929	14.1	2.1 x 0.7	14.4	Sab: pec	160	Multi-galaxy system.
15 26 06.1	+41 40 11	NGC 5930	12.6	0.9 x 0.8	12.1	SAB(rs)b pec	37	Multi-galaxy system.
15 26 48.2	+48 36 53	NGC 5932	14.3	0.8 x 0.8	13.7			Almost stellar NGC 5933 E 2′.2.
15 27 01.6	+48 36 46	NGC 5933	14.8	0.3 x 0.2	11.6		33	
15 28 12.7	+42 55 44	NGC 5934	13.9	1.3 x 0.6	13.5	S?	2	Close pair of mag 7.48v, 9.77v stars SW 6′.4.
15 28 16.7	+42 56 36	NGC 5935	14.8	0.9 x 0.5	13.8	Sb	27	NGC 5934 SW 1′.1.
15 29 44.1	+42 46 40	NGC 5943	13.2	1.3 x 1.3	13.7	S0?		
15 29 44.8	+42 55 08	NGC 5945	12.8	2.9 x 1.8	14.4	SB(rs)ab	99	N-S bar with dark patches E and W.
15 30 36.8	+42 43 03	NGC 5947	13.7	1.2 x 1.2	13.9	SBbc		
15 31 30.8	+40 25 51	NGC 5950	13.7	1.5 x 0.8	13.8	Sb II	37	Strong dark lane NW.
15 44 21.6	+41 05 07	NGC 5992	13.7	1.0 x 0.6	13.0	S	175	Two main arms.
15 44 27.6	+41 07 11	NGC 5993	13.1	1.2 x 0.9	13.0	SB(r)b:	140	Multi-galaxy system; main arm S.
15 52 52.9	+40 38 50	NGC 6013	13.6	1.3 x 0.8	13.5	SBb	174	Mag 8.31v star NE 8′.0.
16 24 54.2	+41 03 00	NGC 6138	14.7	0.9 x 0.2	12.7	Sbc	102	Faint galaxy **I Zw 148** SW 3′.3.
16 23 06.3	+40 51 30	NGC 6141		0.3 x 0.3				
16 25 02.5	+40 56 43	NGC 6145	14.2	0.8 x 0.4	12.8	S..	0	

(Continued from previous page)

GALAXIES

RA h m s	Dec ° ′ ″	Name	Mag (V)	Dim ′ Maj x min	SB	Type Class	PA	Notes
16 25 10.2	+40 53 33	NGC 6146	12.5	1.3 x 1.0	12.8	E?	75	
16 25 05.8	+40 55 43	NGC 6147	15.1	0.4 x 0.4	13.0	Sb		NGC 6145 1.′2 NW.
16 25 50.1	+40 29 15	NGC 6150	14.0	1.3 x 0.5	13.6	E?	61	MCG +7-34-27 SW 1.′3.
16 25 30.7	+49 50 22	NGC 6154	12.7	2.1 x 2.0	14.1	SB(r)a	0	Bright bar SE-NW with dark patches NE and SW.
16 26 08.3	+48 22 02	NGC 6155	12.3	1.3 x 0.9	12.3	S?	145	Pair of stars SW 3.′1, brightest is mag 9.41v.
16 27 25.3	+42 40 45	NGC 6159	14.2	1.4 x 1.0	14.4	S0⁻:	142	**MCG +7-34-40** SE 3.′5.
16 27 41.0	+40 55 33	NGC 6160	13.2	1.8 x 1.5	14.3	E	65	**PGC 58191** on S edge; pair of stars N of nucleus.
16 29 44.8	+40 48 41	NGC 6173	12.1	1.9 x 1.4	13.2	E	140	Stellar **PGC 58358** NE 1.′5.
16 29 47.8	+40 52 15	NGC 6174	14.5	0.4 x 0.4	12.4			Stellar **PGC 58350** on NW edge.
16 29 57.7	+40 37 45	NGC 6175	13.7	1.3 x 0.8	13.6		92	Double system.
16 30 34.0	+40 32 16	NGC 6180	14.1	0.9 x 0.6	13.5	E?	9	MCG +7-34-97 E 1.′7.
16 31 34.6	+40 33 53	NGC 6184	14.0	0.8 x 0.5	12.9	Sb	128	Mag 8.59v star S 7.′0.
16 50 06.3	+42 44 15	NGC 6239	12.4	2.6 x 1.1	13.4	SB(s)b pec? III-IV	118	Bright, elongated center, elongated dark patch N.
16 50 11.1	+45 25 11	NGC 6241	16.2	1.0 x 0.8	15.8	Sc	98	**Arp 103** WNW 7.′9.
15 11 13.4	+46 09 02	UGC 9761	13.0	1.2 x 1.2	13.2	SB(r)0/a		**MCG +8-28-6** NE 8.′6.
15 14 01.2	+44 35 20	UGC 9780	14.6	2.2 x 0.4	14.3	Scd:	148	
15 15 56.2	+43 10 00	UGC 9796	15.1	1.5 x 0.4	14.4	S?	17	Short E-W core, faint arms; MCG +7-31-49 E 1.′5; mag 11.4 star E 2.′3.
15 19 15.1	+44 55 01	UGC 9815	14.9	1.6 x 0.4	14.2	Sdm:	5	Pair of mag 13 star E 2.′0.
15 23 56.0	+46 35 54	UGC 9835	13.8	1.0 x 0.8	13.4	Sb II-III	96	Mag 11.8 star and close by faint, elongated anonymous galaxy SE 2.′3.
15 24 49.3	+42 10 17	UGC 9840	15.1	0.9 x 0.6	14.3	Sc-Irr	66	
15 26 29.9	+41 17 31	UGC 9856	14.7	2.3 x 0.3	14.2	Sd	152	Mag 15.2 star on E edge.
15 26 31.0	+41 43 31	UGC 9857	15.4	1.7 x 0.6	15.3	IBm	136	Irregular; mag 13.0 star NE 2.′2.
15 26 40.6	+40 33 47	UGC 9858	12.8	4.9 x 1.0	14.4	SABbc	83	Mag 8.74v star N 2.′0.
15 29 50.5	+42 37 43	UGC 9873	14.7	1.5 x 0.3	13.7	Sc	123	MCG +7-32-15 NW 2.′3.
15 32 52.2	+41 11 25	UGC 9892	14.4	1.5 x 0.3	13.4	Sb II-III	101	
15 35 45.9	+44 50 05	UGC 9924	13.6	0.8 x 0.6	12.6	SB?	126	
15 36 45.2	+44 14 09	UGC 9936	14.5	1.5 x 0.8	14.6	Sm V	165	Dwarf spiral, knotty.
15 39 08.3	+43 51 50	UGC 9959	13.5	1.8 x 1.1	14.1	SB(s)b	130	Mag 12.2 star superimposed N edge; mag 8.66v star W edge.
15 43 47.3	+43 47 23	UGC 9997	14.7	0.7 x 0.5	13.4	Sb-c	171	
15 43 55.1	+44 18 20	UGC 10001	14.5	1.5 x 1.0	14.8	S	111	
15 44 45.7	+46 04 40	UGC 10010	14.4	1.5 x 0.6	14.1	Im:	5	Mag 14.0 star on E edge.
15 51 12.9	+47 15 15	UGC 10070	13.2	1.3 x 0.6	12.8	S?	111	
15 54 04.9	+41 37 05	UGC 10087	14.4	1.0 x 0.9	14.1	Scd:		Anonymous galaxy W edge.
15 55 43.5	+47 51 59	UGC 10097	12.9	1.4 x 1.0	13.2	S0	118	Bright, diffuse center, smooth envelope.
15 56 36.5	+41 52 46	UGC 10099	14.0	0.4 x 0.4	11.9	S?		MCG +7-33-16 NW 1.′2.
15 57 21.1	+47 09 58	UGC 10109	13.9	1.2 x 0.7	13.5	SAB(r)b	147	Mag 12.2 star E 2.′0.
15 58 40.4	+48 40 56	UGC 10118	13.2	1.0 x 0.7	12.7	S0⁻:	97	
16 00 05.8	+51 47 40	UGC 10129	14.8	1.4 x 1.2	15.2	Pec	30	Bright core offset W; mag 9.58v star SW 4.′4.
16 02 42.6	+41 11 48	UGC 10152	13.6	0.6 x 0.4	11.9	Double System	6	MCG +7-33-30 on E edge.
16 02 50.7	+47 13 26	UGC 10156	13.7	1.1 x 0.4	12.7	Double System	81	Double system, bridge; mag 6.47v star WNW 2.′2.
16 03 31.8	+49 20 12	UGC 10168	13.1	1.9 x 1.3	13.9	(R)SAB0/a	172	Bright, diffuse core, almost surrounded by dark ring, uniform outer envelope.
16 05 45.9	+41 20 36	UGC 10200	13.0	0.6 x 0.6	11.7	S?		**MCG +7-33-40** S 1.′6.
16 09 48.4	+42 20 01	UGC 10241	14.3	0.9 x 0.8	13.7	Sbc	135	
16 12 18.7	+49 23 53	UGC 10278	13.9	1.0 x 0.5	13.0	Sab	170	Mag 8.43v star NNE 5.′0.
16 16 18.0	+47 02 37	UGC 10310	13.2	2.7 x 2.3	15.0	SB(s)m IV-V	21	Numerous knots; one bright knot E edge.
16 17 30.8	+46 05 27	UGC 10325	13.6	0.6 x 0.3	11.6	Double System	108	Mag 13.3 star SW 0.′5; almost stellar anonymous galaxy SE 1.′1.
16 18 11.9	+40 05 38	UGC 10330	14.3	0.8 x 0.6	13.4	SAB(s)b	114	Single, faint, thin arms extend N and S.
16 20 54.0	+40 06 14	UGC 10349	13.6	1.4 x 0.5	13.0	SBab	65	Stellar galaxy **KUG 1618+402** WNW 8.′4.
16 21 22.2	+40 48 33	UGC 10354	14.6	1.2 x 0.8	14.4	SAB(s)c	81	Large, strong core, faint envelope.
16 22 00.4	+40 26 52	UGC 10357	14.1	0.9 x 0.7	13.4	S?	81	
16 23 39.2	+50 58 08	UGC 10374	13.5	1.2 x 0.7	13.2	(R)SB(r)a	129	Mag 14.3 star on NE edge.
16 26 53.4	+51 33 16	UGC 10396	13.8	1.1 x 0.6	13.2	Sbc II	147	Strong dark patch S edge; bright knot N edge; mag 10.39v star NW 2.′0.
16 28 28.3	+41 12 59	UGC 10407	13.7	0.7 x 0.6	12.6	S?	12	MCG +7-34-51 and MCG +7-34-52 S 2.′9.
16 29 00.4	+41 16 59	UGC 10415	13.8	0.9 x 0.8	13.3	SABb	144	Bright E-W center; faint anonymous galaxy S 1.′7.
16 30 19.2	+41 06 07	UGC 10427	14.2	1.4 x 1.2	14.6	SBcd:		Bright center, filamentary arms.
16 30 27.0	+41 29 01	UGC 10430	14.1	0.9 x 0.8	13.6	SB(rs)bc	111	Mag 9.48v star E 5.′4.
16 31 04.2	+41 09 19	UGC 10436	13.8	1.3 x 1.3	14.3	Sc		Bright center; branching arms.
16 31 07.6	+43 20 52	UGC 10437	13.8	2.0 x 1.7	15.0	S?	174	Bright nucleus, patchy arms.
16 34 46.0	+45 19 24	UGC 10457	14.3	1.0 x 0.8	13.9	Scd:	153	Mag 12.6 star S 2.′4.
16 35 07.8	+40 59 25	UGC 10459	15.1	1.5 x 0.2	13.6	Scd:	120	Mag 7.95v star SE 10.′0.
16 37 30.2	+44 08 37	UGC 10480	13.6	1.2 x 0.8	13.5	E?	45	Mag 15.3 star superimposed S edge.
16 37 34.4	+50 20 43	UGC 10486	12.9	1.2 x 1.0	13.0	S0⁻:	90	Mag 11.7 star N 2.′1.
16 41 15.4	+51 11 03	UGC 10511	13.9	1.0 x 0.5	13.1	S0	20	**MCG +9-27-64** ESE 2.′5.
16 44 24.5	+43 43 48	UGC 10531	15.1	0.9 x 0.6	14.3	Sbc	84	Close pair of stars, mags 11.5 and 12.2, N 2.′1.
16 47 14.9	+40 14 39	UGC 10553	14.0	1.0 x 0.5	13.1	SBab:	153	Elongated **UGC 10550** S 6.′5.
16 49 14.5	+48 37 22	UGC 10571	14.5	1.5 x 0.5	14.1	SBcd:	15	Stellar **Mkn 499**, **Mkn 500**, and dwarf **UGC 10565** NW 10.′1.
16 50 48.6	+45 24 07	UGC 10586	13.8	1.5 x 1.4	14.5	Sb II	48	Stellar anonymous galaxy SW edge; stellar anonymous galaxy and mag 12.8 star 0.′8 NE.

GALAXY CLUSTERS

RA h m s	Dec ° ′ ″	Name	Mag 10th brightest	No. Gal	Diam ′	Notes
16 14 06.0	+49 07 00	A 2169	15.9	45	16	All members anonymous, faint and stellar.
16 21 00.0	+50 11 00	A 2184	15.9	31	8	All members anonymous, faint and stellar/almost stellar.
16 28 12.0	+40 54 00	A 2197	13.9	73	90	General scattering of UGC, MCG and PGC GX mixed with anonymous GX.

GLOBULAR CLUSTERS

RA h m s	Dec ° ′ ″	Name	Total V m	B ★ V m	HB V m	Diam ′	Conc. Class Low = 12 High = 1	Notes
16 46 58.9	+47 31 40	NGC 6229	9.4	15.5	18.0	4.5	4	

PLANETARY NEBULAE

RA h m s	Dec ° ′ ″	Name	Diam ″	Mag (P)	Mag (V)	Mag cent ★	Alt Name	Notes
16 04 26.6	+40 40 59	NGC 6058	40	13.3	12.9	13.9	PK 64+48.1	Slightly elongated, PA 135 degrees; mag 9.32v star NW 4.′8.

RA h m s	Dec ° ′ ″	Name	Mag (V)	Dim ′ Maj x min	SB	Type Class	PA	Notes
14 54 39.2	+42 01 22	CGCG 221-10	14.0	0.7 x 0.7	13.1			Mag 11.13v star ESE 3′.5.
14 15 21.5	+49 19 33	CGCG 247-15	14.4	0.5 x 0.5	12.7	Compact		= **Mkn 672**.
13 36 01.0	+49 57 44	IC 902	13.7	2.2 x 0.4	13.5	Sb II-III	162	Mag 10.59v star NNW 1′.8.
13 51 47.5	+50 58 40	IC 951	13.6	1.2 x 1.2	13.8	Sd		Mag 10.79v star NW 10′.2.
14 33 16.6	+41 38 58	IC 1028	14.1	1.1 x 0.5	13.3	S	15	Star on E edge.
14 32 27.5	+49 54 10	IC 1029	11.3	2.8 x 0.5	11.5	SAb: sp II-III	152	Mag 10.03v star E 3′.7.
14 45 48.9	+50 23 39	IC 1056	13.3	1.8 x 1.3	14.1	Sb II	29	Mag 7.77v star E 6′.6.
15 03 45.9	+42 41 58	IC 1090	14.6	0.6 x 0.4	12.9		75	Mag 10.30v star W 4′.9.
13 33 30.0	+40 31 45	MCG +7-28-32	14.7	0.7 x 0.5	13.4		51	
13 35 50.1	+40 16 14	MCG +7-28-35	14.9	0.8 x 0.5	13.8	Sbc	99	
13 40 02.7	+40 25 14	MCG +7-28-47	14.7	0.8 x 0.3	13.0	Sb	48	Mag 11.09v star W 3′.3.
13 43 40.5	+41 41 42	MCG +7-28-57	15.1	0.7 x 0.5	13.8	Sbc	30	Mag 9.07v star E 7′.2.
13 46 50.3	+42 39 00	MCG +7-28-64	14.6	0.7 x 0.5	13.3	SBc	48	
13 48 45.1	+41 42 33	MCG +7-28-71	14.4	0.6 x 0.4	12.7		3	Mag 10.19v star N 3′.1.
13 50 13.8	+42 44 45	MCG +7-28-75	15.0	0.8 x 0.6	14.1	Sc	69	Mag 13.2 star SW 0′.6.
13 50 41.4	+40 16 42	MCG +7-28-78	14.4	1.1 x 0.8	14.1		60	
13 51 25.3	+40 12 46	MCG +7-29-2	14.5	0.8 x 0.8	13.8	S?		= **Mkn 462**. Mag 13.4 star N 2′.4.
13 57 00.7	+43 21 41	MCG +7-29-24	14.3	0.9 x 0.7	13.9	E/S0	150	Mag 12.7 star NE 1′.2.
13 59 43.3	+40 23 15	MCG +7-29-27	16.7	0.7 x 0.5	15.4	Spiral	93	Located on NW edged of UGC 8917.
14 05 46.2	+43 32 41	MCG +7-29-46	15.8	0.7 x 0.7	14.9	S?		
14 20 40.4	+40 18 54	MCG +7-29-62	15.2	0.6 x 0.4	13.5		156	Mag 11.2 star NNW 2′.3.
14 20 55.2	+40 07 12	MCG +7-30-1	14.8	0.8 x 0.6	13.8		36	
14 33 12.8	+41 19 12	MCG +7-30-24	14.9	0.7 x 0.3	13.1	Sc	15	Mag 11.1 star NE 2′.3.
14 33 48.6	+40 05 40	MCG +7-30-28	14.0	0.7 x 0.5	12.7	S?	18	On N edge of UGC 9376.
14 37 58.6	+41 20 30	MCG +7-30-43	15.1	0.8 x 0.5	14.0	Sb	6	Located between a pair of N-S mag 14.5 stars.
14 43 54.8	+43 34 47	MCG +7-30-54	14.6	0.9 x 0.3	13.0	S?	84	Mag 13.7 star N 1′.0.
14 54 47.0	+40 47 03	MCG +7-31-7	14.6	1.0 x 0.5	13.7		177	Mag 10.82v star SW 4′.4.
14 55 27.5	+41 15 26	MCG +7-31-10	14.2	0.8 x 0.5	13.2	E/S0	33	Close pair of stars, mags 10.09v and 11.14v, E 3′.0.
14 57 28.7	+40 59 19	MCG +7-31-12	15.1	0.8 x 0.4	13.8		12	Mag 11.5 star WSW 3′.3.
14 58 33.7	+43 52 35	MCG +7-31-13	14.8	0.6 x 0.6	13.5			Mag 12.4 star E 2′.8.
14 59 00.7	+43 24 24	MCG +7-31-15	14.5	0.7 x 0.5	13.2	Sb	60	Mag 9.93v star S 2′.3.
15 00 23.8	+43 16 54	MCG +7-31-17	14.6	0.5 x 0.5	13.0			MCG +7-31-18 N edge.
15 00 25.0	+43 17 14	MCG +7-31-18	15.4	0.5 x 0.3	13.2		111	MCG +7-31-17 S edge.
15 02 43.3	+43 16 26	MCG +7-31-21	15.1	0.8 x 0.3	13.4		108	MCG +7-31-22 S 2′.0.
15 02 42.4	+43 14 29	MCG +7-31-22	15.4	0.4 x 0.4	13.2	Sb		MCG +7-31-21 N 2′.0.
15 05 28.6	+43 06 34	MCG +7-31-29	15.2	0.7 x 0.5	13.9	SBb	90	Mag 9.85v star NE 2′.4.
15 10 36.0	+41 03 17	MCG +7-31-35	14.7	0.7 x 0.7	13.8			
15 12 25.6	+40 33 06	MCG +7-31-36	14.3	0.8 x 0.6	13.4	S?	123	Mag 11.3 star N 2′.5.
15 13 13.9	+40 32 14	MCG +7-31-39	14.3	1.1 x 0.9	14.2		87	MCG +7-31-40 S 1′.1.
15 13 15.6	+40 31 09	MCG +7-31-40	15.0	0.8 x 0.3	13.3	Sb	6	MCG +7-31-39 N 1′.1; mag 11.7 star E 0′.8.
15 13 29.5	+42 19 45	MCG +7-31-41	14.7	0.8 x 0.4	13.3	SBbc	150	
15 15 06.1	+43 08 56	MCG +7-31-47	14.6	0.9 x 0.5	13.6		120	Mag 13.2 star E 2′.3.
13 33 23.3	+49 06 09	MCG +8-25-18	16.0	0.8 x 0.5	14.9	Dwarf Irr	3	Mag 10.45v star SE 3′.2.
13 49 02.2	+49 24 33	MCG +8-25-43	16.4	0.6 x 0.2	14.0	Sb	51	Mag 8.21v star E 2′.1.
14 12 36.1	+46 12 16	MCG +8-26-15	15.0	1.1 x 0.5	14.2		102	Mag 14.9 star NE edge; mag 9.64v star W 4′.3.
14 19 26.8	+45 50 32	MCG +8-26-20	14.0	0.8 x 0.7	13.3		102	**CGCG 247-19** N 1′.3.
14 21 37.2	+50 23 25	MCG +8-26-22	14.3	1.0 x 0.4	13.1		141	NGC 5602 NE 9′.4.
14 22 12.3	+49 50 31	MCG +8-26-28	15.1	0.9 x 0.7	14.5	Sbc		
14 26 02.8	+49 31 15	MCG +8-26-31	15.0	1.0 x 0.7	14.5	SBbc	30	
14 41 17.2	+44 28 43	MCG +8-27-18	14.0	0.7 x 0.7	13.1	S?		Mag 12.0 star E 1′.2; UGC 9476 NE 3′.4.
14 53 14.9	+46 29 07	MCG +8-27-27	14.9	0.9 x 0.3	13.3	Sb	162	MCG +8-27-28 E 1′.7; mag 10.23v star NW 3′.8.
14 53 24.5	+46 29 04	MCG +8-27-28	15.5	0.5 x 0.4	13.6	Sa	54	MCG +8-27-27 W 1′.7; mag 12.8 star NE 1′.7.
14 55 41.6	+45 32 26	MCG +8-27-31	14.5	1.1 x 0.5	13.7		66	Located between a mag 8.88v star ENE 10′.4 and close pair of mag 9.72v, 9.76v stars E 9′.2.
14 57 29.1	+45 50 41	MCG +8-27-40	14.2	1.3 x 1.0	14.3		135	Mag 9.53v star SE 4′.0.
14 58 28.0	+48 29 29	MCG +8-27-41	14.5	0.8 x 0.7	13.9	E	48	Pair with MCG +8-27-43 SE.
14 58 32.4	+48 29 00	MCG +8-27-43	14.7	0.9 x 0.7	14.3	E	171	Pair with MCG +8-27-41 NW; **CGCG 248-37** E 2′.0.
15 00 45.4	+49 58 55	MCG +8-27-52	13.7	0.4 x 0.3	11.2	Sb	81	Located S edge of NGC 5828.
15 01 33.5	+49 06 41	MCG +8-27-55	14.3	0.8 x 0.5	13.2		174	Mag 12.2 star NW 1′.7.
15 03 56.4	+48 32 59	MCG +8-27-60	13.9	1.2 x 0.8	13.7		123	**MCG +8-27-59** N 1′.6.
15 04 24.8	+49 23 52	MCG +8-27-62	14.2	0.8 x 0.6	13.3	Spiral	162	**MCG +8-27-61** S 1′.8.
15 09 30.6	+46 29 35	MCG +8-28-3	14.2	0.5 x 0.5	12.5			Mag 8.76v star W 7′.9.
15 10 24.2	+45 09 55	MCG +8-28-4	14.5	1.0 x 0.7	13.9	Sd	57	Mag 9.94v star N 3′.1.
14 36 46.0	+51 27 31	MCG +9-24-22	15.6	0.7 x 0.5	14.3	Spiral	105	
14 37 35.2	+51 34 56	MCG +9-24-24	14.8	0.7 x 0.7	13.9			Located on the N end of NGC 5707; mag 6.99v star E 4′.7.
14 38 44.8	+51 19 05	MCG +9-24-26	14.6	1.4 x 1.2	15.0		135	Mag 9.40v star superimposed nearly on center.
14 45 45.5	+51 34 44	MCG +9-24-35	13.7	0.9 x 0.7	13.1	S?	108	Very faint plume extends E; mag 8.57v star NE 3′.4.
15 07 45.2	+51 27 09	MCG +9-25-22	14.7	1.0 x 0.3	13.2	Sbc	147	= **Mkn 845**; mag 11.4 star ESE 2′.4.
15 12 17.9	+50 33 25	MCG +9-25-33	14.7	0.9 x 0.4	13.5		87	
15 12 52.4	+51 23 52	MCG +9-25-36	16.1	0.7 x 0.4	14.5	Sb	105	Elongated **MCG +9-25-35** S edge.
13 32 48.5	+41 52 21	NGC 5214	13.6	1.1 x 0.8	13.3	Scd:	140	Mag 10.66v star NW 2′.7; **MCG +7-28-29** on SW edge.
13 33 20.1	+51 29 27	NGC 5225	13.6	0.7 x 0.7	12.7	S?		
13 34 02.8	+47 54 52	NGC 5229	13.7	3.3 x 0.6	14.3	SB(s)d? sp	167	Mag 9.47v star NNW 6′.9.
13 34 42.8	+51 36 49	NGC 5238	13.3	1.7 x 1.4	14.1	SAB(s)dm	160	Stellar **Mkn 1479** on SE edge of stellar nucleus.
13 36 07.3	+51 14 10	NGC 5250	13.0	1.0 x 0.9	12.7	S0	120	**MCG +9-22-83** SW 5′.9; mag 7.39v star NE 5′.8.
13 38 17.6	+48 16 34	NGC 5256	13.2	1.2 x 1.1	13.4	Pec	30	Pair of compact galaxies.
13 45 08.6	+41 30 08	NGC 5289	13.0	1.9 x 0.6	13.0	(R)SABab: sp	100	**MCG +7-28-60** S 7′.2.
13 45 19.1	+41 42 45	NGC 5290	12.5	3.5 x 0.9	13.6	Sbc: sp	95	
13 46 18.8	+43 51 05	NGC 5296	14.4	1.0 x 0.6	13.7	$S0^{+}$: I	177	Stellar anonymous galaxy S end.
13 46 24.2	+43 52 11	NGC 5297	11.8	5.6 x 1.3	13.8	SAB(s)c: sp I-II	148	Triangle of faint stars on SE end; mag 9.46v star NE 2′.0.
13 46 24.9	+46 06 18	NGC 5301	12.7	4.2 x 0.9	14.0	SA(s)bc: sp II-III	151	Pair of stars flank SE end.
13 50 20.3	+41 21 58	NGC 5320	12.1	3.4 x 1.7	13.9	SAB(rs)c:	18	
13 52 09.7	+43 14 30	NGC 5336	12.8	1.4 x 1.0	13.0	Scd:	109	
13 53 21.1	+40 21 51	NGC 5350	11.3	3.2 x 2.3	13.4	SB(r)b I-II	40	Mag 6.48v star SW 2′.9.
13 53 27.2	+40 16 52	NGC 5353	11.0	2.2 x 1.1	11.8	S0 sp	145	NGC 5354 on N edge.
13 53 26.7	+40 18 11	NGC 5354	11.4	1.4 x 1.3	11.9	S0 sp	72	NGC 5353 on S edge.
13 53 45.5	+40 20 18	NGC 5355	13.1	1.2 x 0.7	12.8	S0?	35	Mag 9.16v star N 5′.0.
13 54 00.5	+40 16 38	NGC 5358	13.6	1.1 x 0.4	12.5	S0/a	138	Mag 13.0, 13.6 double star SW 1′.1.
13 54 53.8	+41 18 47	NGC 5362	12.3	2.3 x 1.0	13.1	Sb? pec	88	Very faint elongated N-S oriented anonymous galaxy at midpoint on S edge.
13 55 40.3	+40 27 45	NGC 5371	10.6	4.4 x 3.5	13.4	SAB(rs)bc I	8	Mag 8.95v star on NE edge.

(Continued from previous page)

GALAXIES

RA h m s	Dec ° ′ ″	Name	Mag (V)	Dim ′ Maj x min	SB	Type Class	PA	Notes
13 56 17.0	+47 14 05	NGC 5377	11.3	4.7 x 2.4	13.8	(R)SB(s)a	23	Bright elongated center, dark patches N and S.
13 57 05.2	+41 50 49	NGC 5383	11.4	3.2 x 2.7	13.6	(R′)SB(rs)b: pec II	85	Bright, diffuse nucleus, several knots SE.
13 57 04.6	+46 15 50	NGC 5391	15.8	1.0 x 0.3	14.3	Sc	87	
14 00 54.2	+40 59 13	NGC 5410	13.1	1.5 x 0.8	13.2	SB?	66	**UGC 8932** 1′.2 N.
14 00 48.0	+48 26 37	NGC 5425	13.6	1.9 x 0.5	13.4	Sd	127	
14 01 57.7	+46 18 41	NGC 5439	13.9	1.1 x 0.4	12.8	Sa	9	Mag 9.63v star E 6′.2.
14 02 50.5	+49 10 22	NGC 5448	11.0	4.0 x 1.8	13.0	(R)SAB(r)a	115	Bright center, dark patch W of nucleus.
14 06 21.4	+50 43 30	NGC 5480	12.1	1.7 x 1.1	12.6	SA(s)c: III	0	Several dark patches; NGC 5481 E 3′.2.
14 06 41.4	+50 43 23	NGC 5481	12.3	1.8 x 1.5	13.3	E^+	115	Strong diffuse core, faint envelope; NGC 5480 W 3′.2.
14 10 15.1	+48 32 46	NGC 5500	13.3	1.0 x 0.9	13.2	E	129	Mag 9.42v star SW 5′.9.
14 12 22.6	+50 20 50	NGC 5520	12.4	2.0 x 1.1	13.2	Sb II-III	63	Small, faint anonymous galaxy NE 2′.3.
14 22 28.4	+40 19 13	NGC 5598	13.0	1.5 x 1.0	13.3	S0	50	Mag 9.48v star W 6′.0.
14 22 53.3	+40 18 35	NGC 5601	14.6	0.8 x 0.3	12.9	Sb	0	NGC 5598 W 4′.9.
14 22 18.9	+50 30 06	NGC 5602	12.7	1.4 x 0.8	12.7	Sa	166	MCG +8-26-22 SW 9′.4.
14 23 01.7	+40 22 37	NGC 5603	13.0	1.1 x 1.1	13.1	S0		Bright center with faint outer envelope; UGC 9216 N 2′.6.
14 23 17.8	+41 46 29	NGC 5608	13.3	2.2 x 1.2	14.2	Im:	95	Bright nucleus.
14 26 12.1	+48 33 46	NGC 5622	13.2	1.7 x 1.0	13.6	Sb II	90	
14 26 35.7	+51 35 03	NGC 5624	13.2	1.1 x 0.8	12.9	S?	3	Bright elongated center.
14 27 36.5	+41 15 25	NGC 5630	13.0	2.2 x 0.7	13.3	Sdm:	92	
14 27 28.3	+46 08 51	NGC 5633	12.4	2.0 x 1.2	13.2	(R)SA(rs)b II-III	10	
14 29 49.5	+49 37 21	NGC 5660	11.9	2.8 x 2.5	13.9	SAB(rs)c II	90	**MCG +8-26-38** WNW 2′.7.
14 31 31.8	+49 57 20	NGC 5673	12.1	2.5 x 0.6	12.4	SBc? sp	136	Star on NW end; mag 11.3 star W 4′.3.
14 32 46.3	+49 27 20	NGC 5676	11.2	4.0 x 1.9	13.2	SA(rs)bc II	47	
14 34 45.2	+48 40 11	NGC 5682	14.1	1.7 x 0.6	14.0	SB(s)b	127	NGC 5683 SE end.
14 34 52.5	+48 39 38	NGC 5683	14.8	0.5 x 0.5	13.2	SB(s)0/a?		Stellar, on SE end of NGC 5682.
14 35 29.7	+48 44 29	NGC 5689	11.9	3.5 x 1.0	13.1	SB(s)0/a:	85	Bright central bulge.
14 36 11.2	+48 35 08	NGC 5693	13.5	1.8 x 1.5	14.4	SB(rs)d	0	Weak almost stellar nucleus, bright knot or star S edge.
14 36 57.1	+41 49 41	NGC 5696	13.0	2.0 x 1.5	14.0	Sbc	45	
14 36 32.1	+41 41 09	NGC 5697	13.8	1.1 x 0.7	13.4	S?	21	Mag 10.95v star W 3′.6.
14 37 02.0	+48 32 41	NGC 5700	14.5	1.0 x 0.4	13.4	SB	32	Mag 14.4 star on SE edge; located on SW edge of small galaxy cluster **A 1948**.
14 37 30.9	+51 33 44	NGC 5707	12.5	2.6 x 0.5	12.6	Sab: sp	35	MCG +9-24-24 N at end; mag 6.99v star E 4′.7.
14 38 16.2	+40 27 24	NGC 5708	13.3	1.6 x 0.6	13.1	Sdm:	177	
14 38 11.9	+46 38 19	NGC 5714	13.4	3.2 x 0.4	13.5	Scd:	82	Mag 11.2 star N 1′.0.
14 38 37.6	+46 39 46	NGC 5717	14.4	1.0 x 0.7	13.9		36	NGC 5722 E 2′.8.
14 38 33.3	+50 48 56	NGC 5720	13.4	2.1 x 1.4	14.4	Sb II	140	
14 38 52.9	+46 40 28	NGC 5721	15.9	0.3 x 0.2	12.7		140	NGC 5722 SE 0′.6.
14 38 34.5	+46 39 56	NGC 5722	16.0	0.6 x 0.6	14.7			NGC 5721 NW 0′.6.
14 38 58.0	+46 41 22	NGC 5723	14.6	0.8 x 0.3	13.1	E/S0	3	Small, elongated nucleus, faint envelope.
14 39 02.1	+46 41 32	NGC 5724	17.0	0.1 x 0.1	11.9		112	Almost stellar. NGC 5723 W 0′.7.
14 39 52.3	+42 44 34	NGC 5730	14.0	1.8 x 0.4	13.5	Im:	88	Mag 10.69v star NW 3′.0.
14 40 09.5	+42 46 45	NGC 5731	13.2	1.6 x 0.4	12.6	S?	116	Mag 10.69v star W 5′.6.
14 42 29.0	+41 50 29	NGC 5739	12.1	2.3 x 2.1	13.6	$SAB(r)0^+$:	0	Bright knot or star NE edge.
14 49 34.4	+47 22 36	NGC 5767	14.0	0.9 x 0.7	13.4	SBab:	150	
14 51 39.0	+40 35 56	NGC 5772	12.8	2.1 x 1.3	13.8	SA(r)b: II-III	35	
14 54 16.7	+42 33 30	NGC 5784	12.4	1.9 x 1.8	13.6	S0	0	**UGC 9585** W 4′.5; **UGC 9583** SW 10′.0.
14 55 15.5	+42 30 23	NGC 5787	13.2	1.0 x 0.9	13.0	S?	0	Slightly elongated core in smooth envelope.
14 55 53.7	+49 43 33	NGC 5794	13.5	1.0 x 1.0	13.3	S?		Bright diffuse center; faint envelope.
14 56 19.3	+49 23 53	NGC 5795	13.9	1.6 x 0.3	12.9	S?	64	Star on NE end.
14 56 23.9	+49 41 45	NGC 5797	12.8	1.6 x 1.0	13.2	S0/a	110	Mag 5.64v star S 4′.2.
14 57 06.6	+49 40 07	NGC 5804	13.1	1.2 x 1.1	13.3	SB(s)b	0	Bright nucleus with knot E; NGC 5805 S 2′.6.
14 57 11.6	+49 37 43	NGC 5805	15.0	0.4 x 0.3	12.6		140	NGC 5804 N 2′.6.
14 58 58.5	+49 49 18	NGC 5818	13.7	1.2 x 0.9	13.6	S0	170	Mag 8.79v star E 7′.4; located on W edge of small galaxy cluster **A 2011**.
15 00 46.0	+49 59 34	NGC 5828	14.9	0.8 x 0.6	14.0	S	50	MCG +8-27-52 on S edge.
15 01 50.9	+47 52 31	NGC 5830	14.2	1.0 x 0.7	13.7	Sb II-III	170	Mag 9.09v star N 3′.9.
15 02 25.5	+48 52 40	NGC 5835	14.4	1.1 x 0.9	14.2	Sa	160	Star NW of nucleus.
15 06 33.8	+42 38 27	NGC 5860	13.2	0.8 x 0.8	12.6	S?		Multi-galaxy system.
15 12 45.5	+41 14 00	NGC 5886	14.1	0.9 x 0.4	13.1	E/S0	86	Faint star N edge; mag 9.07v star SE 2′.9.
15 13 07.4	+41 15 55	NGC 5888	13.4	1.3 x 0.8	13.3	SB(s)bc	158	Weak dark lane W of center.
15 13 15.8	+41 19 38	NGC 5889	15.4	0.9 x 0.4	14.2	SBb?	39	Mag 9.28v star ENE 9′.5.
15 13 34.3	+41 57 28	NGC 5893	13.1	1.3 x 1.1	13.4	SB(r)b	30	Strong dark lane E of center.
15 13 50.0	+42 00 26	NGC 5895	14.2	0.9 x 0.2	12.2	Sc	21	NGC 5896 N 1′.0.
15 13 50.8	+42 01 24	NGC 5896	16.1	0.3 x 0.2	12.9		76	Almost stellar, NGC 5895 S 1′.0.
15 15 03.4	+42 02 54	NGC 5899	11.7	3.2 x 1.2	13.0	SAB(rs)c II	18	Several dark patches.
15 15 05.1	+42 12 38	NGC 5900	14.0	1.7 x 0.5	13.6	Sb: sp III	131	Strong dark lane NE.
15 14 22.0	+50 19 46	NGC 5902	13.2	1.1 x 1.0	13.2		105	
13 35 42.6	+45 55 42	UGC 8588	14.2	1.2 x 1.0	14.2	Sm: V	42	Dwarf spiral, very low, uniform surface brightness.
13 36 15.4	+46 11 58	UGC 8597	13.8	1.7 x 1.4	14.5	SB(s)d V	16	Knotty, short, bright E-W bar.
13 36 49.7	+44 52 48	UGC 8611	14.3	1.5 x 1.4	15.0	SAB(s)d	105	Pair of mag 14.4 stars E edge.
13 38 59.7	+51 26 06	UGC 8639	14.4	1.5 x 1.1	14.8	Im:	170	Dwarf, low, uniform surface brightness; mag 13.0 star on E edge.
13 39 29.0	+46 00 54	UGC 8642	15.2	1.6 x 0.2	13.8	Sd	108	
13 39 53.3	+40 44 27	UGC 8651	14.1	2.8 x 1.4	15.4	Im IV-V	80	Dwarf irregular; prominent clump NE edge.
13 40 30.1	+42 59 31	UGC 8656	14.5	1.2 x 0.4	13.6	S	45	
13 41 42.4	+40 52 26	UGC 8670	14.4	1.2 x 0.9	14.3	SABd	90	Mag 13.4 star SW 1′.7.
13 43 30.6	+43 27 47	UGC 8688	13.9	0.9 x 0.6	13.1	Pec	72	Mag 12.6 star on NW edge.
13 45 27.3	+47 55 21	UGC 8702	14.2	1.5 x 0.7	14.1	Sd	70	Mag 12.8 star S 1′.7.
13 47 49.0	+40 29 16	UGC 8726	14.4	2.1 x 0.5	14.3	Sd	127	Mag 10.59v star NW 1′.0.
13 48 38.9	+43 24 37	UGC 8733	13.4	2.4 x 1.4	14.6	SBcd:	12	Very knotty; mag 12.5 star W 2′.2.
13 50 36.4	+42 32 26	UGC 8756	13.6	2.1 x 0.6	13.7	Sab	81	Mag 7.00v star W 5′.1.
13 52 48.0	+43 49 38	UGC 8798	15.4	1.5 x 0.7	15.3	Sdm	147	Mag 13.4 star superimposed E edge.
13 55 07.9	+40 10 02	UGC 8841	13.9	1.8 x 1.0	14.4	SBb	115	Large core, two main arms.
13 56 58.0	+45 58 18	UGC 8876	13.5	1.1 x 0.4	12.5	S0/a	18	
13 59 47.9	+40 22 54	UGC 8917	14.4	1.6 x 0.3	13.5	Scd:	123	MCG +7-29-27 NW end.
14 09 21.7	+49 02 18	UGC 9056	13.7	1.0 x 0.3	12.2	S?	143	
14 10 13.6	+48 18 50	UGC 9068	15.4	1.5 x 0.9	15.6	SAdm	160	Mag 14 star on E edge; mag 11.6 star NE 2′.7.
14 11 27.1	+50 12 30	UGC 9083	14.3	1.1 x 0.7	13.8	Sdm	75	Very knotty; mag 12.4 star N 1′.5.
14 12 28.3	+45 41 26	UGC 9098	13.2	1.1 x 0.5	12.4	Double System	53	Double system, contact.
14 13 40.0	+43 51 56	UGC 9105	14.2	1.1 x 0.9	14.0	S	145	Mag 13.3 star W 2′.6.
14 15 08.0	+45 35 42	UGC 9125	13.9	1.1 x 0.5	13.1	S	120	Mag 6.51v star E 9′.6.
14 19 54.7	+51 53 38	UGC 9178	14.6	1.2 x 0.8	14.4	SB?	66	Double system, bridge; mag 10.56v star S 1′.1.

(Continued from previous page)

36 GALAXIES 36

RA h m s	Dec ° ′ ″	Name	Mag (V)	Dim ′ Maj x min	SB	Type Class	PA	Notes
14 22 32.0	+45 23 01	UGC 9211	14.4	2.3 x 1.9	15.9	Im: V	102	Dwarf irregular, very low surface brightness; mag 11.2 star N edge.
14 22 56.5	+40 24 59	UGC 9216	13.9	1.3 x 1.3	14.3	SAcd		Stellar nucleus; NGC 5603 S 2′6; mag 12.1 star W 2′2.
14 23 45.1	+40 01 23	UGC 9223	13.7	1.0 x 0.4	12.6	S0	171	
14 24 43.1	+44 31 34	UGC 9240	12.9	1.8 x 1.8	14.0	IAm IV		Irregular, low surface brightness.
14 29 23.7	+45 04 19	UGC 9314	14.2	1.2 x 0.4	13.3	S	37	Mag 14.3 star on SW end.
14 29 50.6	+44 26 40	UGC 9324	14.3	2.4 x 1.4	15.5	SBm V	102	Filamentary.
14 33 46.8	+40 04 50	UGC 9376	13.8	1.5 x 0.8	13.8	S?	147	Pair with MCG +7-30-28 N edge, disturbed.
14 33 58.7	+40 14 38	UGC 9379	14.4	1.5 x 0.8	14.4	Scd:	145	Several small knots.
14 34 52.8	+40 44 49	UGC 9386	13.8	1.1 x 0.6	13.2	SBab	140	
14 37 13.4	+43 41 48	UGC 9422	14.6	1.8 x 0.2	13.3	Scd:	161	Mag 9.77v star NW 1′0.
14 38 00.2	+40 06 20	UGC 9429	13.9	1.0 x 0.7	13.4	S	90	Mag 12.9 star E 1′8; very small, very faint anonymous galaxy S 0′8.
14 38 49.6	+41 00 26	UGC 9441	13.8	1.1 x 0.4	12.8	Sa	33	
14 38 59.5	+51 07 13	UGC 9448	14.3	1.5 x 0.2	12.9	Sb II-III	88	Located between a pair of stars, mags 11.18v and 12.4.
14 41 32.2	+44 30 49	UGC 9476	13.0	1.5 x 0.9	13.2	SAB(rs)c	137	MCG +8-27-18 SW 3′4.
14 49 01.2	+42 27 51	UGC 9542	14.4	1.8 x 0.6	14.3	Sd	32	Mag 13.6 star on SW edge.
14 50 39.2	+42 44 28	UGC 9559	13.8	1.1 x 0.4	12.7	S?	147	Faint, stellar anonymous galaxy SE edge; **Mkn 828** NE 1′8.
14 51 46.1	+43 38 36	UGC 9567	14.2	1.0 x 0.6	13.5	Im:	134	Mag 9.34v star WNW 5′8.
14 51 59.7	+43 43 16	UGC 9569	13.5	1.5 x 1.3	14.1	SB(rs)d	153	Patchy; mag 10.26v star SE 3′3.
14 55 09.1	+43 49 06	UGC 9598	14.1	1.5 x 0.6	13.8	Scd:	124	
14 56 04.6	+45 24 18	UGC 9612	14.1	1.2 x 0.9	14.0	SAB(s)c	75	Located 1′6 W of mag 9.18v star.
14 58 35.9	+44 52 57	UGC 9639	13.5	1.1 x 0.9	13.4	Sab	24	Bright core, fainter envelope.
14 59 03.0	+41 26 47	UGC 9641	14.0	1.0 x 0.8	13.6	S	85	
15 01 09.9	+44 41 50	UGC 9660	13.6	1.0 x 0.5	12.7	S?	80	
15 01 33.0	+48 19 04	UGC 9665	14.0	1.5 x 0.3	13.0	Sbc	142	**UGC 9657** WNW 7′7.
15 03 50.7	+42 06 53	UGC 9684	13.6	1.5 x 0.7	13.5	SBab	105	Much elongated **UGC 9681** W 2′0.
15 04 37.8	+40 22 18	UGC 9691	14.0	1.1 x 0.9	13.8	SBbc	58	Mag 10.53v star SE 3′5.
15 05 14.8	+51 09 28	UGC 9702	14.8	1.2 x 0.3	13.6	Sbc	168	**CGCG 274-13** NE 8′1.
15 06 30.9	+51 54 47	UGC 9722	15.1	1.0 x 0.7	14.5	Sd	18	Mag 11.9 star W 2′7.
15 11 13.4	+46 09 02	UGC 9761	13.0	1.2 x 1.2	13.2	SB(r)0/a		**MCG +8-28-6** NE 8′6.
15 14 01.2	+44 35 20	UGC 9780	14.6	2.2 x 0.4	14.3	Scd:	148	
15 15 56.2	+43 10 00	UGC 9796	15.1	1.5 x 0.4	14.4	S?	17	Short E-W core, faint arms; MCG +7-31-49 E 1′5; mag 11.4 star E 2′3.

GALAXY CLUSTERS

RA h m s	Dec ° ′ ″	Name	Mag 10th brightest	No. Gal	Diam ′	Notes
14 22 06.0	+48 33 00	A 1904	15.6	83	16	**MCG +8-26-24** N of center 1′4; **MCG +8-26-25** S 3′7; most other members anonymous, stellar/almost stellar.

37 GALAXIES 37

RA h m s	Dec ° ′ ″	Name	Mag (V)	Dim ′ Maj x min	SB	Type Class	PA	Notes
12 11 56.0	+40 39 13	CGCG 215-50	14.6	0.7 x 0.5	13.3		93	Galaxy pair.
12 32 34.9	+45 46 00	CGCG 244-27	14.1	0.4 x 0.4	12.0			= **Mkn 215**.
11 58 34.6	+42 44 03	IC 749	12.4	2.3 x 1.9	13.8	SAB(rs)cd III	150	Many faint knots and/or superimposed stars; mag 8.96v star SW 3′1.
11 58 52.3	+42 43 23	IC 750	11.9	2.6 x 1.2	13.0	Sab: sp	43	IC 749 W 3′4; mag 8.96v star WSW 5′5.
11 58 52.3	+42 34 04	IC 751	14.3	1.3 x 0.4	13.4	Sb? sp	30	Mag 8.96v star NW 9′2; **IC 752** E 4′2; IC 749, IC 750 N 9′4.
13 36 01.0	+49 57 44	IC 902	13.7	2.2 x 0.4	13.5	Sb II-III	162	Mag 10.59v star NNW 1′8.
12 44 03.3	+41 10 05	IC 3713	14.7	0.7 x 0.5	13.4		132	Pair of stars mags 12.6 and 10.85v N 2′1.
12 44 42.8	+40 40 40	IC 3726	14.7	1.4 x 0.3	13.6	Scd:	111	Mag 9.71v star NE 13′0; **IC 3723** NNW 4′2; **UGC 7904** WNW 9′2.
12 47 28.1	+40 33 57	IC 3783	14.7	0.8 x 0.6	13.7		120	**IC 3778** W 5′3.
12 48 58.9	+40 35 42	IC 3810	14.7	1.0 x 0.5	13.8	SBc	156	Mag 14.4 star on E edge.
12 52 39.5	+40 06 09	IC 3850	14.8	0.8 x 0.7	14.0		45	Mag 10.96v star W 1′0.
13 00 52.6	+40 35 00	IC 4060	14.3	0.8 x 0.6	13.4	Sb	141	Mag 13.1 star W 2′3; mag 10.53v star W 5′9.
13 02 04.9	+40 24 30	IC 4100	14.0	1.4 x 0.9	14.1	Scd:	110	Mag 11.18v star NNW 10′1.
13 28 33.6	+46 55 34	IC 4263	14.5	2.0 x 0.4	14.1	SB(s)d: sp	105	Mag 9.13v star SW 14′4.
11 58 57.0	+44 11 32	MCG +7-25-12	15.4	1.1 x 0.5	14.6	Sc	168	
12 05 15.6	+43 10 08	MCG +7-25-19A	13.8	1.4 x 1.1	14.4	E/S0	24	UGC 7069 WSW 3′3; mag 8.65v star E 3′2.
12 06 07.8	+40 09 01	MCG +7-25-21	14.4	0.8 x 0.7	13.6		120	Mag 13.8 star W 1′5.
12 09 15.4	+41 36 36	MCG +7-25-31	14.4	1.4 x 1.1	14.7		168	Mag 13.6 star E 0′9.
12 09 26.2	+44 04 21	MCG +7-25-34	15.3	0.4 x 0.4	13.2	Sb		Stellar; NGC 4137 NW 1′8.
12 10 05.3	+43 26 33	MCG +7-25-43	15.1	0.8 x 0.3	13.4	Sb	150	Mag 13.4 star WSW 1′8.
12 10 59.0	+43 14 56	MCG +7-25-47	14.7	0.8 x 0.8	14.1	Sb		Mag 10.80v star NW 4′0.
12 18 57.3	+43 36 41	MCG +7-25-57	14.7	0.8 x 0.4	13.3	Sb	24	Mag 12.8 star SE 1′6.
12 21 57.2	+42 18 44	MCG +7-26-2	15.1	0.6 x 0.6	13.9	Sc		
12 27 38.2	+40 09 34	MCG +7-26-7	16.9	0.7 x 0.6	15.8	Sc	21	
12 28 52.2	+42 10 40	MCG +7-26-11	15.7	0.8 x 0.5	14.6	Dwarf Irr	150	Mag 10.78v star NW 5′6.
12 30 24.2	+42 54 04	MCG +7-26-12	15.8	0.8 x 0.4	14.4	Dwarf Irr?	114	Mag 12.8 star E 2′7.
12 31 00.4	+42 31 34	MCG +7-26-15	15.0	0.6 x 0.3	13.0		99	Mag 10.64v star NE 3′0.
12 34 03.3	+43 24 58	MCG +7-26-27	14.4	0.9 x 0.3	12.9	Sb	165	Mag 10.78v star SE 4′2.
12 43 39.0	+44 05 37	MCG +7-26-43	15.1	0.5 x 0.3	12.9		144	Pair with MCG +7-26-44 N edge.
12 43 39.3	+44 06 03	MCG +7-26-44	15.3	0.6 x 0.4	13.6		96	Pair with MCG +7-26-43 S edge.
12 49 35.8	+41 41 40	MCG +7-26-56	15.2	0.9 x 0.6	14.4	Sb	33	Mag 11.22v star N 2′4.
13 04 21.9	+43 48 34	MCG +7-27-24	14.7	0.8 x 0.4	13.3	Sbc	153	Mag 11.3 star WSW 2′5.
13 04 58.0	+43 33 11	MCG +7-27-26	13.4	0.9 x 0.9	13.0			Pair with MCG +7-27-28 E edge; **MCG +7-27-27** NNE 1′9.
13 05 02.7	+43 33 11	MCG +7-27-28	14.8	0.6 x 0.3	12.8		111	Pair with **MCG +7-27-27** W edge.
13 09 14.2	+43 20 30	MCG +7-27-34	14.6	0.8 x 0.6	13.7		132	
13 10 42.3	+44 01 28	MCG +7-27-38	14.4	0.6 x 0.3	12.4		45	Mag 12.4 star on NW edge.
13 12 51.8	+40 32 33	MCG +7-27-45	14.1	0.7 x 0.5	12.8		69	Mag 14.8 star SW end; mag 9.01v star S 2′5.
13 12 58.7	+41 47 11	MCG +7-27-46	15.3	1.0 x 0.9	15.0	Im:	129	Uniform surface brightness; mag 12.7 star W 2′1.
13 16 06.4	+41 30 00	MCG +7-27-56	15.0	0.5 x 0.3	12.8		159	MCG +7-27-57 E 1′6.
13 16 14.8	+41 29 38	MCG +7-27-57	14.9	0.7 x 0.6	13.8	Sc	60	MCG +7-27-56 W 1′6.
13 21 57.5	+44 15 45	MCG +7-28-4	15.3	0.8 x 0.4	13.9	Sc	156	Very close pair of mag 11-12 stars W 1′1.
13 23 48.6	+43 18 01	MCG +7-28-6	14.7	0.8 x 0.5	13.6		162	Close pair of mag 11.4 stars W 5′9.
13 24 32.9	+42 45 50	MCG +7-28-8	14.6	0.7 x 0.7	13.7			
13 27 04.6	+41 57 24	MCG +7-28-13	14.6	0.8 x 0.5	13.5		105	Mag 13.6 star W 1′3.
13 29 07.6	+41 17 07	MCG +7-28-18	15.2	0.6 x 0.4	13.5	Spiral	168	
13 29 09.8	+41 43 46	MCG +7-28-20	14.9	0.5 x 0.5	13.3	Sbc		

(Continued from previous page)

RA h m s	Dec ° ′ ″	Name	Mag (V)	Dim ′ Maj x min	SB	Type Class	PA	Notes
13 29 56.8	+41 37 37	MCG +7-28-22	15.0	0.9 x 0.9	14.7	Sc		
13 30 25.0	+39 59 52	MCG +7-28-23	14.9	0.7 x 0.4	13.4	Spiral	57	Pair of E-W mag 13-14 stars N 1′.8.
13 33 30.0	+40 31 45	MCG +7-28-32	14.7	0.7 x 0.5	13.4		51	
13 35 50.1	+40 16 14	MCG +7-28-35	14.9	0.8 x 0.5	13.8	Sbc	99	
12 01 09.2	+47 25 23	MCG +8-22-54	15.5	0.7 x 0.3	13.6	Sc	117	Mag 11.1 star SSE 3′.7.
12 09 14.1	+47 03 29	MCG +8-22-73	14.0	1.0 x 0.8	13.6	(R)SAB0° pec	33	= **Mkn 198**.
12 09 30.6	+48 53 58	MCG +8-22-75	14.5	1.0 x 0.4	13.4	Sc	57	Mag 10.50v star S 1′.1.
12 09 48.7	+47 10 05	MCG +8-22-76	14.4	0.8 x 0.4	13.0	Sb	135	MCG +8-22-80 E 5′.4.
12 10 20.2	+47 09 53	MCG +8-22-80	15.0	0.7 x 0.5	13.7		168	MCG +8-22-76 W 5′.4.
12 12 50.2	+46 46 23	MCG +8-22-84	13.9	0.8 x 0.8	13.3	S?		
12 20 40.7	+46 15 17	MCG +8-23-7	14.9	0.5 x 0.5	13.4	E2		NGC 4288 N 2′.4.
12 25 29.2	+47 16 23	MCG +8-23-24	14.2	0.9 x 0.7	13.5	Sbc	117	Stellar **Mkn 208** NNE 4′.2.
12 25 33.8	+46 03 59	MCG +8-23-25	14.6	0.5 x 0.4	12.7		117	Pair with MCG +8-23-26 E edge.
12 25 36.8	+46 03 50	MCG +8-23-26	15.2	0.5 x 0.4	13.3		171	Pair with MCG +8-23-25 W edge.
12 25 49.6	+50 05 37	MCG +8-23-29	14.8	0.9 x 0.8	14.3	Sbc	39	
12 25 46.5	+45 17 22	MCG +8-23-31	14.4	0.8 x 0.5	13.2		126	UGC 7525 SE 1′.3.
12 25 47.3	+45 54 32	MCG +8-23-32	14.7	0.8 x 0.4	13.3	Sc	92	**UGC 7530** SE 2′.9; NGC 4392 SW 6′.2.
12 26 16.9	+48 29 37	MCG +8-23-35	14.3	0.9 x 0.6	13.4	S?	90	Mag 11.8 star W 1′.1.
12 28 25.4	+44 37 55	MCG +8-23-40	14.4	0.8 x 0.7	13.6		18	Mag 13.0 star E 1′.3; UGC 7617 E 7′.4.
12 32 01.2	+49 37 48	MCG +8-23-49	15.7	0.9 x 0.5	14.7	Sd	84	Mag 10.69v star NNE 3′.2.
12 32 17.5	+47 20 13	MCG +8-23-51	14.1	1.3 x 1.0	14.3		36	
12 34 46.8	+47 45 31	MCG +8-23-61	14.3	1.2 x 0.8	14.1	S0?	90	Faint, round anonymous galaxy W 1′.8.
12 36 14.9	+47 53 16	MCG +8-23-65	14.8	0.8 x 0.5	13.7	Sb	93	**MCG +8-23-64** NW 2′.3.
12 36 51.4	+45 39 02	MCG +8-23-67	14.5	0.7 x 0.7	13.6	Sc		Mag 11.7 star N 3′.1.
12 37 13.8	+49 26 52	MCG +8-23-69	13.9	0.9 x 0.8	13.4	Sc	6	
12 43 02.1	+48 34 43	MCG +8-23-80	14.8	0.6 x 0.5	13.3	Sc	57	Mag 13.0 star S 1′.8.
12 57 35.8	+48 29 10	MCG +8-24-14	14.6	0.8 x 0.7	13.8	Sc	171	
13 02 29.9	+49 57 25	MCG +8-24-26	15.1	0.5 x 0.5	13.4	Sbc		
13 03 10.5	+47 23 34	MCG +8-24-28	14.5	1.0 x 0.9	14.3		90	**MCG +8-24-27** NNW 6′.2.
13 03 11.5	+48 47 04	MCG +8-24-29	15.1	0.8 x 0.7	14.4	SBb	99	Mag 13.2 star E 2′.5.
13 03 53.8	+47 23 06	MCG +8-24-31	14.8	0.8 x 0.8	14.2	SBbc		Mag 10.71v star S 2′.7.
13 04 20.6	+45 03 21	MCG +8-24-33	15.0	0.7 x 0.4	13.4		126	
13 04 23.1	+47 35 45	MCG +8-24-34	15.1	0.6 x 0.6	13.8	SBc		Mag 13.1 star NNW 2′.1.
13 04 55.9	+47 30 09	MCG +8-24-35	14.5	1.0 x 0.6	13.8	SBc	51	
13 05 41.9	+49 25 16	MCG +8-24-36	14.3	1.3 x 0.7	14.1		81	Mag 9.26v star N 3′.0.
13 06 19.9	+49 18 17	MCG +8-24-40	15.8	0.8 x 0.6	14.9	SBc	57	
13 06 26.9	+45 17 20	MCG +8-24-41	14.6	0.9 x 0.6	13.8		153	Mag 5.64v star W 6′.2.
13 07 27.2	+45 09 12	MCG +8-24-45	15.3	0.8 x 0.3	13.6	Sc	30	Mag 11.9 star NNW 2′.2.
13 09 38.8	+45 16 09	MCG +8-24-56	15.4	0.7 x 0.4	13.8	Sbc	171	Mag 11.23v star NE 1′.7.
13 11 17.9	+47 32 47	MCG +8-24-68	15.5	0.5 x 0.5	13.9			UGC 8272 E 1′.5.
13 11 31.1	+46 20 29	MCG +8-24-70	14.6	0.8 x 0.6	13.7	Sc	84	Mag 11.5 star E 1′.2.
13 12 17.3	+44 50 20	MCG +8-24-86	13.9	0.8 x 0.8	13.5	E		**UGC 8281** and **MCG +8-24-79** SW 3′.0.
13 12 56.7	+47 27 21	MCG +8-24-89	14.6	0.9 x 0.6	13.8	Sbc	21	
13 17 17.5	+47 20 31	MCG +8-24-99	15.3	0.5 x 0.4	13.4	Sc	27	
13 17 30.0	+47 47 00	MCG +8-24-101	14.8	1.1 x 0.7	14.4		132	
13 19 46.7	+48 12 20	MCG +8-24-108	14.7	0.7 x 0.7	13.8	Sb		Mag 12.2 star W 1′.6.
13 19 46.4	+47 42 55	MCG +8-24-109	14.5	0.8 x 0.6	13.6	Sc	108	Mag 8.59v star N 3′.7.
13 31 37.8	+46 09 02	MCG +8-25-17	14.2	0.9 x 0.8	13.6		6	
13 33 23.3	+49 06 09	MCG +8-25-18	16.0	0.8 x 0.5	14.9	Dwarf Irr	3	Mag 10.45v star SE 3′.2.
12 03 32.2	+51 04 39	MCG +9-20-77	15.5	0.9 x 0.2	13.5	Sc	78	
12 11 31.3	+51 51 46	MCG +9-20-108	14.9	1.1 x 0.3	13.5	Sb	114	Close pair of E-W mag 10-11 stars W 4′.6.
12 13 27.2	+50 42 35	MCG +9-20-116	15.6	0.6 x 0.5	14.2		163	Located 2′.0 S of NGC 4187.
12 21 08.1	+51 53 46	MCG +9-20-153	14.5	0.6 x 0.4	12.8		102	Anonymous galaxy with superimposed star S 0′.8.
12 21 53.9	+51 02 30	MCG +9-20-156	14.6	0.8 x 0.7	13.8	Sbc	126	Close pair of mag 10 stars SW 6′.9.
12 25 45.4	+51 40 05	MCG +9-20-172	14.0	1.1 x 0.7	13.6		66	
12 30 57.9	+51 20 36	MCG +9-21-8	14.8	0.6 x 0.3	12.8		66	Pair with **MCG +9-21-7** W edge.
12 30 58.9	+51 36 32	MCG +9-21-9	15.9	0.8 x 0.3	14.2	Sbc	24	
12 46 23.2	+50 53 56	MCG +9-21-42	16.7	0.5 x 0.5	15.1	SBbc		Mag 10.35v star W 5′.4.
12 46 26.9	+50 47 30	MCG +9-21-43	14.7	0.5 x 0.7	13.4	Sb	105	Mag 12.2 star SW 2′.9.
12 46 42.1	+50 43 47	MCG +9-21-46	15.4	0.8 x 0.3	13.8	Sb	18	MCG +9-21-43 NW 4′.4.
12 56 26.9	+51 05 04	MCG +9-21-68	14.6	0.8 x 0.3	12.9		78	Mag 9.43v star SE 0′.7; mag 11.1 star and **MCG +9-21-66** NW 2′.8.
13 10 08.3	+50 30 37	MCG +9-22-24	15.5	0.8 x 0.4	14.1	SBbc	123	
13 14 26.2	+50 58 45	MCG +9-22-35	14.8	0.7 x 0.7	13.9			Mag 10.50v star SW 2′.8.
13 16 34.3	+51 25 55	MCG +9-22-41	14.6	0.7 x 0.5	13.5	E/S0	9	Located between a pair of mag 12.6, 13.0 stars; mag 11.9 and 14.5 stars superimposed on N end.
13 27 47.4	+51 08 10	MCG +9-22-65	14.5	0.7 x 0.4	13.0		108	Mag 13.5 star E 1′.1.
11 56 41.8	+48 20 02	NGC 3985	12.5	1.3 x 0.8	12.4	SB(s)m:	73	**MCG +8-22-42** and **MCG +8-22-43** SW 9′.8.
11 58 06.8	+47 20 06	NGC 4001	15.3	0.6 x 0.3	13.5	E4; cand. dwarf	162	Almost stellar; NGC 4010 SE 6′.9.
11 58 36.8	+47 15 31	NGC 4010	12.6	4.3 x 0.8	13.8	SB(s)d: sp	66	NGC 4001 NW 6′.9.
11 58 30.7	+43 56 47	NGC 4013	11.2	5.2 x 1.0	12.9	Sb sp III	66	Moderately dark lane extends almost full length of galaxy.
11 59 25.4	+50 57 41	NGC 4026	10.8	5.2 x 1.3	12.7	S0 sp	178	UGC 6956 W 9′.8; mag 9.43v star NNE 7′.3.
12 02 51.0	+48 38 10	NGC 4047	12.2	1.6 x 1.3	12.8	(R)SA(rs)b: II	105	
12 03 11.1	+44 31 59	NGC 4051	10.2	5.2 x 3.9	13.3	SAB(rs)bc II	135	Several bright knots; arms branching mostly NE of nucleus.
12 05 22.7	+50 21 11	NGC 4085	12.4	2.8 x 0.8	13.1	SAB(s)c:? III-IV	78	Mag 8.27v star SW 5′.6.
12 05 33.5	+50 32 14	NGC 4088	10.6	5.8 x 2.2	13.2	SAB(rs)bc II-III	43	**MCG +9-20-92** SE 5′.1.
12 06 00.7	+47 28 36	NGC 4096	10.9	6.6 x 1.8	13.4	SAB(rs)c II-III	20	
12 06 08.7	+49 34 55	NGC 4100	11.2	5.4 x 1.8	13.5	(R′)SA(rs)bc I-II	167	Mag 8.88v star NNW 7′.2.
12 06 51.2	+42 59 44	NGC 4109	13.9	0.7 x 0.6	12.8	Sa?	156	**MCG +7-25-25** ESE 2′.4; anonymous galaxy E 0′.6.
12 07 03.0	+43 04 02	NGC 4111	10.7	4.6 x 1.0	12.3	SA(r)0$^+$: sp	150	Located 3′.7 SW of a mag 8.07v star.
12 07 45.7	+43 07 33	NGC 4117	13.0	1.8 x 0.9	13.4	S0°:	18	NGC 4118 E 1′.4.
12 07 52.8	+43 06 41	NGC 4118	14.6	0.7 x 0.4	13.1	S0$^+$?	150	Almost stellar; NGC 4117 W 1′.4.
12 09 09.0	+44 00 11	NGC 4135	13.9	1.0 x 0.6	13.2	SAB(s)bc?	87	
12 09 17.6	+44 05 22	NGC 4137	14.1	1.1 x 0.8	13.8	SB(rs)c?	54	Narrow arms for open "S" shape. Companion at 0′.55 W, 0′.1 X 0′.1, at tip of spiral arm.
12 09 29.7	+43 41 00	NGC 4138	11.3	2.6 x 1.7	12.8	SA(r)0$^+$	150	Mag 11.18v star 2′.2 N.
12 09 35.9	+42 32 02	NGC 4143	10.7	2.3 x 1.4	11.8	SAB(s)0°	144	Mag 8.47v star SW 4′.9.
12 09 59.3	+46 27 23	NGC 4144	11.6	6.0 x 1.3	13.7	SAB(s)cd? sp III	104	
12 11 04.2	+50 29 06	NGC 4157	11.4	6.8 x 1.1	13.4	SAB(s)b? sp II	66	Mag 10.36v star at SW end; mag 8.05v star NW 4′.3.
12 13 16.7	+43 41 57	NGC 4183	12.3	5.2 x 0.8	13.7	SA(s)cd? sp III-IV	166	Bright in center, no definite nucleus, knotty filamentary arms.

RA h m s	Dec ° ′ ″	Name	Mag (V)	Dim ′ Maj x min	SB	Type Class	PA	Notes
12 13 29.3	+50 44 28	NGC 4187	13.2	1.3 x 1.0	13.5	E	145	**MCG +9-20-114** W 3′.3; MCG +9-20-116 S 1′.9. note: several small anonymous galaxies 0′.4 E of MCG +9-20-116.
12 15 51.8	+47 05 27	NGC 4217	11.2	5.2 x 1.5	13.3	Sb sp	50	Dark lane extends full length of galaxy along center line. Mag 9.01v star N 1′.8 from NE half of galaxy.
12 15 46.3	+48 07 54	NGC 4218	12.5	1.2 x 0.7	12.2	Sa?	142	Located 2′.5 WNW of mag 8.87v star.
12 16 11.9	+47 53 04	NGC 4220	11.4	3.9 x 1.4	13.1	SA(r)0$^+$	141	
12 16 26.6	+47 01 29	NGC 4226	13.5	1.0 x 0.5	12.6	Sa pec?	127	NGC 4217 NW 7′.2.
12 16 49.0	+47 27 26	NGC 4231	13.3	1.2 x 1.1	13.4	SA0$^+$ pec?	33	Compact nucleus and smooth envelope, NGC 4232 on S edge.
12 16 48.9	+47 26 19	NGC 4232	13.6	1.4 x 0.7	13.5	SBb pec:	155	NGC 4231 on N edge.
12 17 30.2	+45 37 06	NGC 4242	10.8	5.0 x 3.8	13.9	SAB(s)dm III-IV	25	Numerous small, bright knots.
12 17 50.8	+47 24 34	NGC 4248	12.5	3.0 x 1.1	13.7	I0? sp	108	Bright, elongated nucleus with very faint star W end.
12 18 58.1	+47 18 07	NGC 4258	8.4	18.6 x 7.2	13.6	SAB(s)bc II-III	150	= **M 106**.
12 20 38.0	+46 17 33	NGC 4288	12.9	2.1 x 1.6	14.0	SB(s)dm IV-V	130	Bright, elongated nucleus; MCG +8-23-7 S 2′.4.
12 23 27.4	+46 59 33	NGC 4346	11.2	3.3 x 1.3	12.7	S0 sp	99	
12 23 58.8	+48 46 47	NGC 4357	12.4	3.6 x 1.3	13.9	SAbc II	77	Mag 8.38v star NW 10′.0.
12 25 35.3	+45 41 07	NGC 4389	11.7	2.6 x 1.3	12.9	SB(rs)bc pec: IV	105	Bright, narrow bar cross by faint dust lane on E end.
12 25 18.9	+45 50 50	NGC 4392	13.4	0.7 x 0.6	12.4		72	MCG +8-23-32 NE 6′.2; faint, stellar anonymous galaxy S 5′.4.
12 28 12.3	+44 05 47	NGC 4449	9.6	6.2 x 4.4	13.0	IBm IV	45	Large. Bright central area with numerous bright knots and patches.
12 28 45.5	+44 51 52	NGC 4460	11.3	4.0 x 1.2	12.8	SB(s)0$^+$? sp	40	Mag 7.39v star SW 8′.2.
12 30 31.3	+41 41 58	NGC 4485	11.9	2.3 x 1.6	13.2	IB(s)m pec III-IV	15	Numerous bright knots/patches S of center.
12 30 35.8	+41 38 26	NGC 4490	9.8	6.3 x 3.1	12.9	SB(s)d pec III	125	Bright core with a number of bright knots and emission patches, especially NW of the center.
12 34 59.7	+50 51 01	NGC 4537	14.9	1.0 x 0.3	13.5		87	NGC 4542 SSW 3′.2.
12 34 49.0	+50 48 17	NGC 4542	13.9	1.0 x 0.5	13.0	S	28	NGC 4537 NNE 3′.2.
12 41 05.9	+50 23 26	NGC 4617	13.2	3.0 x 0.5	13.5	Sb II-III	179	Mag 9.33v star NE 2′.8.
12 41 32.3	+41 08 51	NGC 4618	10.8	4.2 x 3.4	13.5	SB(rs)m II-III	25	Large bright center with many knots E and S; mag 11.00v star 3′.6 S.
12 41 52.5	+41 16 19	NGC 4625	12.4	1.6 x 1.4	13.1	SAB(rs)m pec	132	Brightest in N half; moderate dark patch S of bright area.
12 43 36.6	+41 01 05	NGC 4655	13.9	1.0 x 1.0	13.8			Round, compact nucleus in faint envelope.
12 48 46.5	+41 55 16	NGC 4704	13.7	1.0 x 0.9	13.5	SB(rs)bc pec I	105	Short bar with arms forming a slightly compressed "S."
12 48 24.3	+51 09 42	NGC 4707	12.9	2.2 x 2.1	14.4	Sm: V	25	Stellar nucleus; several stars superimposed.
12 50 52.7	+41 07 07	NGC 4736	8.2	14.4 x 12.1	13.7	(R)SA(r)ab II	117	= **M 94**. Extremely bright nucleus in bright inner ring.
12 50 59.8	+47 40 15	NGC 4741	13.7	1.3 x 0.9	13.7	Scd:	165	Very faint, very small anonymous galaxy N 1′.5.
12 54 37.4	+46 31 47	NGC 4800	11.5	1.6 x 1.2	12.1	SA(rs)b III	25	
12 56 49.0	+48 17 51	NGC 4837	13.8	1.3 x 0.6	13.4	Pec	70	A connected pair; **MCG +8-24-12** ENE 2′.0.
12 59 56.4	+47 12 19	NGC 4901	14.5	1.0 x 1.0	14.4			Mag 8.09v star WSW 9′.5.
13 00 55.6	+47 13 22	NGC 4917	13.8	1.4 x 1.0	14.0	(R′)SB(s)b	160	Stellar nucleus in short SE-NW bar.
13 02 37.7	+50 26 19	NGC 4932	13.6	1.7 x 1.3	14.3	SA(r)c	0	Mag 10.74v star S 1′.6; very faint anonymous galaxy W 2′.3.
13 02 57.7	+51 19 07	NGC 4938	14.3	0.8 x 0.7	13.5	S?	39	
13 05 52.1	+41 43 17	NGC 4963	13.3	0.8 x 0.8	12.7	S?		Mag 8.78v star SW 5′.3.
13 08 12.2	+41 40 35	NGC 4985	13.7	1.3 x 1.1	14.0	S0	135	
13 07 59.1	+51 55 43	NGC 4987	13.4	1.2 x 0.7	13.2	E	35	Mag 7.98v star S 8′.5.
13 08 10.2	+50 39 51	NGC 4998	14.5	0.9 x 0.8	14.0	Sc	48	Very small companion galaxy, or extension of NGC 4998?, S edge.
13 08 37.9	+43 44 15	NGC 5003	14.8	1.1 x 0.7	14.3	Sa	72	
13 10 46.7	+50 05 33	NGC 5009	14.5	1.1 x 0.7	14.1	SBb	75	
13 12 06.2	+46 11 46	NGC 5021	13.4	1.5 x 0.7	13.3	SBb	78	Located on E edge of galaxy cluster **A 1697**.
13 12 11.2	+44 02 03	NGC 5023	12.3	6.0 x 0.8	13.9	Scd: sp	28	**MCG +7-27-41** NW 9′.7.
13 12 37.6	+47 03 47	NGC 5029	13.1	1.7 x 1.1	13.7	E	150	**MCG +8-24-88** S 7′.9; mag 9.49v star E 7′.1.
13 13 32.8	+51 15 30	NGC 5040	14.2	1.0 x 0.5	13.3		66	
13 15 49.3	+42 01 50	NGC 5055	8.6	12.6 x 7.2	13.3	SA(rs)bc II-III	105	= **M 63**, **Sunflower Galaxy**. Very bright nucleus. Mag 9.30v star 3′.6 WNW of center.
13 19 38.0	+40 23 09	NGC 5093	13.7	1.4 x 0.7	13.5	Sa	143	
13 20 30.2	+43 05 00	NGC 5103	12.7	1.4 x 1.0	12.9	S?	143	Located 1′.8 S of mag 8.12v star.
13 23 10.7	+43 05 09	NGC 5123	12.8	1.3 x 1.1	13.1	Scd:	174	Patchy, branching arms.
13 25 14.2	+43 15 59	NGC 5145	12.4	2.0 x 1.8	13.7	S?	90	
13 28 10.2	+46 40 16	NGC 5169	13.5	2.3 x 0.9	14.1	SB(rs)b:	103	Stellar nucleus with bright knot, superimposed star or small galaxy on W edge.
13 28 25.4	+46 35 27	NGC 5173	12.2	1.0 x 0.9	12.2	E0:	117	Mag 10.47v star S 5′.5; note: **MCG +8-25-6** lies 1′.0 E of this star.
13 29 53.2	+47 11 09	NGC 5194	8.4	11.2 x 6.9	12.9	SA(s)bc pec I-II	28	= **M 51**, **Whirlpool Galaxy**. Galaxy **IC 4278** NE 6′.8.
13 29 59.0	+47 15 53	NGC 5195	9.5	5.8 x 4.6	13.0	I0 pec	79	**IC 4277** NE 4′.0.
13 30 11.3	+46 40 16	NGC 5198	11.8	2.1 x 1.8	13.2	E1-2:	0	
13 32 48.5	+41 52 21	NGC 5214	13.6	1.1 x 0.8	13.3	Scd:	140	Mag 10.66v star NW 2′.7; **MCG +7-28-29** on SW edge.
13 33 20.1	+51 29 27	NGC 5225	13.6	0.7 x 0.7	12.7	S?		
13 34 02.8	+47 54 52	NGC 5229	13.7	3.3 x 0.6	14.3	SB(s)d? sp	167	Mag 9.47v star NNW 6′.9.
13 34 42.8	+51 36 49	NGC 5238	13.3	1.7 x 1.4	14.1	SAB(s)dm	160	Stellar **Mkn 1479** on SE edge of stellar nucleus.
13 36 07.3	+51 14 10	NGC 5250	13.0	1.0 x 0.9	12.7	S0	120	**MCG +9-22-83** SW 5′.9; mag 7.39v star NE 5′.8.
13 38 17.6	+48 16 34	NGC 5256	13.2	1.2 x 1.1	13.4	Pec	30	Pair of compact galaxies.
11 56 28.4	+50 25 44	UGC 6917	12.5	3.8 x 2.3	14.7	SBm	130	Mag 11.4 star superimposed W of center.
11 56 52.6	+50 49 01	UGC 6922	13.2	1.5 x 1.2	13.7	S?	48	
11 57 17.2	+49 16 57	UGC 6930	12.1	3.4 x 2.9	14.4	SAB(s)d	30	Mag 10.94v star NW 2′.2.
11 58 25.9	+50 55 03	UGC 6956	14.5	2.5 x 2.3	16.3	SB(s)m V	27	Dwarf spiral, low surface brightness; NGC 4026 E 9′.8.
12 00 18.9	+50 39 11	UGC 6992	14.1	1.4 x 0.7	13.9	Sdm:	63	
12 02 36.6	+41 03 13	UGC 7020	13.2	2.4 x 1.8	14.7	SA(rs)bc I-II	171	Two main arms, strong dark lane W of center.
12 04 58.2	+43 08 56	UGC 7069	14.8	1.5 x 0.3	13.8	SBcd:	70	MCG +7-25-19A ENE 3′.3.
12 05 32.2	+51 30 15	UGC 7082	14.2	1.5 x 0.6	13.9	Sb II-III	53	Almost stellar anonymous galaxy S 1′.5.
12 05 59.8	+43 09 06	UGC 7089	13.8	3.2 x 0.9	14.8	Sdm:	36	Mag 8.65v star W 4′.7.
12 08 55.1	+41 44 25	UGC 7129	13.4	1.2 x 0.7	13.1	Sab	75	
12 10 55.2	+50 17 14	UGC 7176	15.8	1.5 x 0.5	15.3	Im: V	80	Dwarf, very low surface brightness.
12 12 40.5	+40 33 43	UGC 7213	14.4	1.1 x 1.0	14.4	SBd	51	Mag 10.75v star ESE 2′.2.
12 15 23.3	+51 20 57	UGC 7267	13.7	1.8 x 0.7	13.8	Sdm:	53	**MCG +9-20-122** WSW 10′.4.
12 15 33.4	+43 25 59	UGC 7271	14.4	2.1 x 0.7	14.7	SBd:	160	Mag 13.4 star SW 3′.3.
12 16 42.1	+46 04 43	UGC 7301	14.9	1.9 x 0.3	14.1	Sd	82	
12 17 28.9	+46 49 37	UGC 7325	14.2	1.5 x 0.6	13.9	Sab	145	
12 18 12.4	+44 09 59	UGC 7340	13.3	1.4 x 0.7	13.1	S?	5	= **Mkn 203**; double system with long, extremely faint bridge, S component a short E-W bar.
12 19 12.4	+49 21 14	UGC 7358	13.6	1.5 x 0.5	13.1	Scd:	37	
12 19 32.1	+49 48 56	UGC 7367	12.8	1.5 x 0.9	12.9	S0/a	93	
12 21 15.6	+45 48 49	UGC 7408	13.1	2.5 x 1.5	14.4	IAm IV-V	106	Dwarf irregular, very low surface brightness; double system **UGC 7391** NW 11′.7.
12 21 39.4	+40 50 53	UGC 7416	12.9	1.6 x 1.4	13.6	SB(r)b I	3	Strong dark patches N and S of nucleus.
12 22 23.9	+45 03 28	UGC 7444	13.7	1.3 x 0.6	13.3	Sa	44	
12 24 14.0	+44 56 17	UGC 7486	15.1	0.8 x 0.8	14.5	Sc		
12 25 53.6	+45 16 49	UGC 7525	14.4	1.1 x 0.9	14.2	SBcd:	39	
12 26 56.8	+48 16 36	UGC 7560	13.7	0.7 x 0.6	12.6	SB	120	
12 27 41.0	+43 29 48	UGC 7577	12.3	4.2 x 2.8	14.9	Im IV-V	130	Dwarf irregular, low, uniform surface brightness; mag 10.39v star W 3′.3.

(Continued from previous page)

37 GALAXIES 37

RA h m s	Dec ° ′ ″	Name	Mag (V)	Dim ′ Maj x min	SB	Type Class	PA	Notes
12 28 14.6	+44 27 08	UGC 7593	13.9	1.1 x 0.8	13.6	Double System	162	Double system, connected.
12 28 45.1	+43 13 43	UGC 7608	13.3	3.6 x 3.2	15.8	Im V	63	Dwarf irregular, low surface brightness.
12 29 06.3	+44 39 16	UGC 7617	14.7	2.3 x 0.3	14.1	Sd	68	MCG +8-23-40 W 7′.4.
12 29 53.5	+47 31 50	UGC 7639	13.7	2.5 x 1.5	15.0	Im	153	
12 32 27.2	+42 42 17	UGC 7690	12.6	2.1 x 1.7	13.9	Im:	20	Mag 11.2 star SW 2′.5.
12 36 22.3	+40 00 18	UGC 7774	14.2	3.4 x 0.5	14.6	Sd	99	Mag 13.8 star S edge; mag 12.1 star E 4′.1.
12 39 33.8	+47 37 22	UGC 7823	14.2	1.0 x 0.5	13.3	Scd:	75	Located between a pair of mag 10.9v and 10.2v stars.
12 44 06.8	+45 00 20	UGC 7910	14.2	1.1 x 0.8	13.9	SB?	21	
12 46 56.3	+51 36 43	UGC 7950	14.5	1.7 x 1.2	15.1	Im	5	Elongated E-W core, uniform envelope; mag 8.04v star NW 1′.9.
12 52 03.8	+51 40 45	UGC 8012	13.6	0.7 x 0.5	12.3	S?	115	
13 00 30.6	+48 18 29	UGC 8119	14.8	1.5 x 0.4	14.1	S	46	
13 02 40.6	+51 46 46	UGC 8151	14.7	0.8 x 0.8	14.1	Sc		
13 03 16.2	+50 37 14	UGC 8162	14.5	1.5 x 1.3	15.1	Scd:	123	
13 05 46.5	+46 27 44	UGC 8189	14.8	1.5 x 0.2	13.3	Sbc	133	
13 08 03.2	+46 49 37	UGC 8215	16.0	0.9 x 0.7	15.3	Im	72	
13 08 19.2	+45 21 41	UGC 8225	13.3	1.1 x 0.9	13.2	S0	33	
13 11 27.4	+47 32 58	UGC 8272	15.1	1.5 x 1.0	15.4	(R′)SB(rs)c	26	MCG +8-24-68 W 1′.6.
13 13 10.1	+50 34 37	UGC 8304	13.7	2.1 x 0.5	13.6	Sb II-III	17	
13 13 53.8	+42 12 28	UGC 8313	14.0	1.7 x 0.5	13.7	SB(s)c? sp	35	Mag 12.8 star W 1′.1.
13 14 26.9	+45 55 20	UGC 8320	12.4	3.2 x 1.3	13.8	IBm IV-V	150	Irregular.
13 15 15.7	+44 24 23	UGC 8327	14.0	1.4 x 0.7	13.8	Double System	92	= **Mkn 248**; Double system, two bridges; mag 12.8 star N 1′.7.
13 15 30.4	+47 29 50	UGC 8331	14.2	2.8 x 0.9	15.0	IAm V	140	Uniform surface brightness.
13 18 45.5	+41 56 55	UGC 8365	13.8	2.3 x 1.4	14.9	SB(s)d IV-V	115	
13 21 36.6	+42 17 06	UGC 8400	13.6	0.7 x 0.6	12.5	SB?	42	
13 29 38.8	+45 23 16	UGC 8489	14.2	2.1 x 0.6	14.3	SABdm IV-V	106	
13 30 38.9	+49 08 12	UGC 8506	16.4	1.6 x 0.2	15.1	Scd:	162	
13 35 42.6	+45 55 42	UGC 8588	14.2	1.2 x 1.0	14.2	Sm: V	42	Dwarf spiral, very low, uniform surface brightness.
13 36 15.4	+46 11 58	UGC 8597	13.8	1.7 x 1.4	14.5	SB(s)d V	16	Knotty, short, bright E-W bar.
13 36 49.7	+44 52 48	UGC 8611	14.3	1.5 x 1.4	15.0	SAB(s)d	105	Pair of mag 14.4 stars E edge.
13 38 59.7	+51 26 06	UGC 8639	14.4	1.5 x 1.1	14.8	Im:	170	Dwarf, low, uniform surface brightness; mag 13.0 star on E edge.
13 39 29.0	+46 00 54	UGC 8642	15.2	1.6 x 0.2	13.8	Sd	108	
13 39 53.3	+40 44 27	UGC 8651	14.1	2.8 x 1.4	15.4	Im IV-V	80	Dwarf irregular; prominent clump NE edge.

GALAXY CLUSTERS

RA h m s	Dec ° ′ ″	Name	Mag 10th brightest	No. Gal	Diam ′	Notes
12 03 36.0	+51 44 00	A 1452	15.7	46	15	**MCG +9-20-74** and **75** SW from center 1′.2; **Hickson 60** group SW 5′.7.
12 05 36.0	+51 25 00	A 1468	16.0	50	15	Other than UGC 7082, all members faint, stellar and anonymous.

38 GALAXIES 38

RA h m s	Dec ° ′ ″	Name	Mag (V)	Dim ′ Maj x min	SB	Type Class	PA	Notes
11 19 34.3	+51 30 12	Arp's Galaxy		0.2 x 0.1			90	Almost stellar; mag 10.09v star S 4′.0.
10 38 06.7	+44 50 38	CGCG 241-2	14.5	0.7 x 0.3	12.6		99	Located 2′.3 S of a mag 10.51v star.
10 48 19.5	+50 01 19	CGCG 241-13	14.0	0.9 x 0.3	12.4	S0?	90	CGCG 241-15 E 5′.3.
10 48 52.7	+50 02 09	CGCG 241-15	14.1	0.7 x 0.6	13.0	SB?	63	= **Mkn 152**. NW-SE bar; CGCG 241-13 W 5′.3.
10 54 43.2	+46 33 11	CGCG 241-36	14.6	0.9 x 0.3	13.0	S?	105	Mag 10.71v star SW 8′.1.
11 01 49.7	+44 26 52	CGCG 241-60	14.5	0.7 x 0.6	13.4			Galaxy pair. Mag 12.1 star SSW 2′.7.
11 05 08.1	+44 44 47	CGCG 241-74	14.0	0.6 x 0.3	12.0		18	= **Mkn 162**.
11 11 06.5	+43 37 58	IC 674	13.4	1.7 x 0.5	13.1	Sab	120	Mag 10.81v star S 2′.4.
11 24 17.5	+47 50 49	IC 687	14.3	0.8 x 0.8	13.7	Sb		Mag 10.94v star NW 9′.5.
11 32 56.6	+50 14 26	IC 705	14.5	0.8 x 0.4	13.1		27	Mag 10.46v star NW 10′.0; **UGC 6535** NNE 4′.5.
11 33 59.4	+49 03 44	IC 708	13.0	1.4 x 0.9	13.2	E	174	Mag 11.4 star NW 3′.7; IC 709 SE 2′.8.
11 34 14.8	+49 02 32	IC 709	13.9	0.6 x 0.6	12.9	E?		Almost stellar; IC 708 NW 2′.8.
11 34 46.6	+48 57 21	IC 711	14.1	0.5 x 0.5	12.7	E?		Mag 10.59v star E 4′.1; **MCG +8-21-60** SW 1′.0.
11 34 49.4	+49 04 39	IC 712	13.8	1.1 x 0.7	13.3	S?	98	Mag 8.81v star N 1′.9; near center of galaxy cluster A 1314.
11 45 18.6	+49 34 15	IC 731	15.1	0.5 x 0.3	12.9			
11 58 34.6	+42 44 03	IC 749	12.4	2.3 x 1.9	13.8	SAB(rs)cd III	150	Many faint knots and/or superimposed stars; mag 8.96v star SW 3′.1.
11 58 52.3	+42 43 23	IC 750	11.9	2.6 x 1.2	13.0	Sab: sp	43	IC 749 W 3′.4; mag 8.96v star WSW 5′.5.
11 58 52.3	+42 34 04	IC 751	14.3	1.3 x 0.4	13.4	Sb? sp	30	Mag 8.96v star NW 9′.2; **IC 752** E 4′.2; IC 749, IC 750 N 9′.4.
10 24 22.7	+41 42 22	MCG +7-22-1	13.8	0.8 x 0.6	13.1	E/S0	24	Bright star on W edge.
10 29 44.7	+40 29 36	MCG +7-22-10	15.1	0.9 x 0.3	13.6	Spiral	6	Mag 7.02v star ESE 4′.0.
10 30 11.4	+43 21 36	MCG +7-22-12	14.4	1.0 x 0.5	13.5	Sb	30	Large knot or companion galaxy E edge.
10 30 23.4	+43 43 25	MCG +7-22-14	14.7	0.9 x 0.2	12.7		6	Mag 12.4 star E 3′.5.
10 30 38.6	+44 00 43	MCG +7-22-15	14.4	0.9 x 0.7	13.8		153	
10 32 38.8	+43 59 27	MCG +7-22-21	14.9	1.0 x 0.7	14.3	Sbc	42	Large knot E edge; mag 13.0 star NE 1′.7.
10 32 49.9	+43 10 24	MCG +7-22-22	15.4	0.6 x 0.3	13.4	Spiral	48	Mag 12.3 star S 1′.4.
10 33 00.5	+42 55 42	MCG +7-22-23	14.7	0.6 x 0.6	13.5			Mag 10.86v star W 1′.7.
10 33 50.5	+40 17 27	MCG +7-22-25	14.6	1.0 x 0.5	13.7	Sb	102	Mag 12.3 star N 3′.1.
10 35 11.4	+42 46 12	MCG +7-22-27	14.5	0.9 x 0.6	13.7		108	Mag 12.4 star S 2′.8.
10 35 21.8	+40 52 28	MCG +7-22-29	14.5	0.8 x 0.7	13.7		12	Mag 7.79v star W 1′.8.
10 37 31.8	+43 39 15	MCG +7-22-34	14.4	0.9 x 0.3	12.9	Sab	6	Mag 9.83v star W 7′.0.
10 41 44.3	+40 02 26	MCG +7-22-44	14.6	0.8 x 0.7	13.8	Spiral	66	
10 42 53.6	+40 08 19	MCG +7-22-46	14.5	0.7 x 0.3	12.7	Sb	18	Mag 8.89v star NE 6′.1.
10 45 49.5	+43 54 08	MCG +7-22-59	15.4	0.6 x 0.5	13.9	Sc	51	Forms a triangle with a pair of mag 12.5 stars 2′.0 N and W.
10 54 56.3	+41 29 53	MCG +7-23-1	15.0	0.9 x 0.6	14.2		129	Mag 12.1 star WNW 3′.3.
10 56 16.0	+42 19 56	MCG +7-23-3	14.6	0.9 x 0.4	13.3		36	Mag 14.3 star SE edge.
10 58 43.4	+43 19 13	MCG +7-23-9	15.6	0.5 x 0.5	14.0	Spiral		
10 59 54.8	+43 08 49	MCG +7-23-10	16.2	0.6 x 0.5	14.7	Sb	96	Mag 10.60v star N 1′.2.
11 03 11.2	+41 42 15	MCG +7-23-15	14.7	0.9 x 0.7	14.2	E/S0	168	
11 03 55.5	+43 29 01	MCG +7-23-18	14.7	0.5 x 0.4	12.8	Sb	159	Mag 8.8 star on E edge.
11 03 53.7	+40 50 57	MCG +7-23-19	14.8	0.6 x 0.5	13.3	Ring	94	**Mayall's Object**. Multi-galaxy system.
11 12 11.2	+41 02 12	MCG +7-23-32	15.6	0.7 x 0.5	14.3	Sb	24	Mag 13.3 star SW 2′.5.
11 21 16.8	+40 20 43	MCG +7-23-40	14.7	0.7 x 0.6	13.6		171	Mag 8.08v star SE 5′.6.
11 26 10.9	+44 09 09	MCG +7-24-5	14.8	0.9 x 0.3	13.2	Sc	162	Mag 10.65v star W 2′.4.
11 28 18.5	+41 26 32	MCG +7-24-10	15.7	0.7 x 0.5	14.4		87	Mag 10.71v star W 4′.1.

(Continued from previous page)

GALAXIES

RA h m s	Dec ° ′ ″	Name	Mag (V)	Dim ′ Maj x min	SB	Type Class	PA	Notes
11 30 21.4	+44 09 51	MCG +7-24-16	14.9	0.7 x 0.5	13.6	Sc	87	Small, faint anonymous galaxy on N edge.
11 32 52.5	+43 30 54	MCG +7-24-21	14.3	0.6 x 0.5	12.8	S?	33	Mag 13.5 star N 1′7.
11 33 16.6	+43 47 42	MCG +7-24-23	15.6	0.4 x 0.4	13.6	E/S0		Pair of stars, mags 14.5 and 14.9, S 1′3.
11 39 06.2	+43 14 49	MCG +7-24-27	14.8	0.7 x 0.4	13.2	Sb	120	Mag 10.07v star SW 3′4.
11 41 11.1	+43 26 36	MCG +7-24-29	14.3	0.9 x 0.7	13.7	Sbc	42	
11 58 57.0	+44 11 32	MCG +7-25-12	15.4	1.1 x 0.5	14.6	Sc	168	
10 27 04.7	+46 03 24	MCG +8-19-32	15.1	0.6 x 0.6	13.9	Sb		Mag 10.85v star NW 2′5.
10 27 52.2	+46 00 21	MCG +8-19-35	15.2	0.8 x 0.4	13.8	Sc	123	Mag 11.3 star WNW 3′0.
10 34 53.5	+45 16 07	MCG +8-19-39	13.9	0.9 x 0.9	13.7	E/S0		Mag 11.7 star S 2′3.
10 40 03.0	+49 33 03	MCG +8-20-12	15.1	0.8 x 0.5	14.0	Sb	21	Mag 11.8 star W 2′3.
10 45 16.3	+49 52 44	MCG +8-20-17	14.6	0.8 x 0.4	13.2		72	Mag 14.1 star SSW 0′9.
10 46 00.3	+45 55 38	MCG +8-20-18	15.3	0.8 x 0.5	14.1	Sbc	159	Mag 12.9 star SE 0′9.
10 47 18.0	+47 25 17	MCG +8-20-22	14.7	1.1 x 0.8	14.4	Sbc	114	Mag 12.9 star NE 0′7.
10 48 09.5	+48 19 50	MCG +8-20-24	14.3	0.7 x 0.6	13.2		159	
10 50 34.3	+48 41 34	MCG +8-20-31	14.6	0.8 x 0.8	13.9	Sc		
10 50 45.3	+46 50 28	MCG +8-20-32	14.6	0.9 x 0.3	13.0	Sbc	177	Mag 10.77v star N 1′5.
10 53 46.7	+46 13 06	MCG +8-20-38	14.4	0.7 x 0.7	13.5			Mag 12.8 star NW 1′3.
10 54 00.7	+49 36 15	MCG +8-20-40	13.8	1.0 x 0.5	12.9	Sb	108	Mag 10.90v star superimposed S edge; UGC 6013 N 3′4.
10 54 38.2	+47 28 11	MCG +8-20-45	14.1	0.7 x 0.7	13.2			MCG +8-20-45 ESE 1′3; mag 9.34v star NW 1′8.
10 56 11.7	+49 56 16	MCG +8-20-52	14.1	0.8 x 0.5	13.1	E/S0	126	
10 58 33.6	+46 16 02	MCG +8-20-56	14.5	0.8 x 0.8	13.9	Sb		
10 59 53.3	+50 00 51	MCG +8-20-61	13.8	0.8 x 0.8	13.2	S?		UGC 6071 NE 2′7.
11 02 23.9	+50 05 38	MCG +8-20-71	14.9	0.7 x 0.3	13.1		63	Very close pair of mag 13 stars N 1′1.
11 04 04.6	+45 15 43	MCG +8-20-77	13.6	0.9 x 0.8	13.1		3	**MCG +8-20-78** N 2′6.
11 06 49.3	+45 57 50	MCG +8-20-87	14.1	0.9 x 0.8	13.8	E/S0	114	Mag 13.3 star E 1′5.
11 07 19.6	+46 22 59	MCG +8-20-89	14.3	0.8 x 0.6	13.3		30	
11 08 31.9	+47 20 07	MCG +8-20-92	14.8	1.0 x 0.3	13.3	Sc	66	
11 11 52.4	+45 32 26	MCG +8-21-1	14.7	0.9 x 0.5	13.7	Sb	150	
11 12 27.0	+45 15 55	MCG +8-21-2	14.8	0.5 x 0.4	12.9		27	Close pair of mag 14 stars S 2′0.
11 17 08.6	+46 56 45	MCG +8-21-12	14.5	0.8 x 0.3	12.8	Sb	168	**MCG +8-21-11** S 5′1.
11 17 41.6	+47 05 05	MCG +8-21-13	14.7	0.7 x 0.4	13.2			Mag 10.33v star S 1′3.
11 19 17.7	+47 38 53	MCG +8-21-16	13.4	1.1 x 1.1	13.5			Mag 10.89v star W 2′8.
11 27 35.9	+47 22 46	MCG +8-21-37	14.9	0.7 x 0.5	13.6		60	
11 27 55.0	+47 23 45	MCG +8-21-38	14.1	0.7 x 0.5	12.8		33	Mag 10.67v star NE 3′3.
11 28 54.0	+48 31 46	MCG +8-21-39	14.8	0.7 x 0.6	13.7		174	
11 30 01.7	+44 41 04	MCG +8-21-42	14.9	0.5 x 0.5	13.2	Sb		Mag 9.54v star S 1′7.
11 30 41.0	+47 01 03	MCG +8-21-44	14.4	1.1 x 0.6	13.8	Sc	57	Mag 11.16v star W 1′0.
11 31 52.1	+48 49 32	MCG +8-21-45	15.1	0.8 x 0.2	13.0	Sb	48	Mag 9.12v star E 2′1.
11 32 09.6	+49 00 37	MCG +8-21-46	14.1	0.8 x 0.4	12.7		132	**MCG +8-21-47** S 1′0.
11 34 28.5	+46 21 38	MCG +8-21-58	14.4	0.6 x 0.6	13.3	E/S0		Mag 11.5 star S 1′3.
11 34 48.5	+46 59 22	MCG +8-21-61	14.9	1.1 x 0.3	13.6	Sb	102	
11 35 43.1	+49 44 13	MCG +8-21-67	14.6	0.7 x 0.4	13.1		30	Mag 12.3 star W edge; mag 9.24v star SSW 2′0.
11 36 03.9	+49 27 01	MCG +8-21-69	15.0	0.6 x 0.4	13.3		123	Mag 10.49v star W 3′2.
11 36 30.7	+49 07 51	MCG +8-21-72	14.1	0.7 x 0.7	13.2	S?		**MCG +8-21-64** W 10′0.
11 36 36.8	+49 03 44	MCG +8-21-74	14.5	0.9 x 0.8	14.0	S?	51	Mag 12.0 star ENE 2′6.
11 37 16.2	+45 45 25	MCG +8-21-75	14.9	0.8 x 0.6	14.0		165	Small, faint anonymous galaxy SE edge.
11 40 10.5	+50 17 54	MCG +8-21-88	15.4	0.5 x 0.3	13.2		144	Star superimposed SE end? Mag 11.9 star W 1′5.
11 40 47.5	+46 32 29	MCG +8-21-90	14.5	0.5 x 0.5	13.1	E/S0		
11 41 22.1	+46 23 38	MCG +8-21-92	14.8	0.8 x 0.4	13.5		123	
11 43 04.8	+48 23 55	MCG +8-21-93	14.0	0.9 x 0.3	12.5	Sb	12	Mag 11.3 star SE 3′7.
11 49 42.6	+49 55 46	MCG +8-22-13	14.8	0.7 x 0.5	13.5		30	
11 51 57.3	+46 49 02	MCG +8-22-21	14.8	0.6 x 0.4	13.1		123	
11 54 01.6	+44 38 33	MCG +8-22-32	15.3	1.1 x 0.3	13.9	Sc	171	Mag 10.45v star S edge.
12 01 09.2	+47 25 23	MCG +8-22-54	15.5	0.7 x 0.3	13.6	Sc	117	Mag 11.1 star SSE 3′7.
10 46 16.0	+51 04 41	MCG +9-18-24	14.9	0.7 x 0.7	14.0	Sb		Mag 12′4 star S edge.
11 01 37.8	+50 49 53	MCG +9-18-69	15.0	1.1 x 0.3	13.7	Sbc	102	
11 02 38.7	+51 10 31	MCG +9-18-72	14.8	0.9 x 0.7	14.2	Sbc	27	
11 22 04.0	+50 48 52	MCG +9-19-51	14.5	0.7 x 0.6	13.4	Sb	24	Mag 11.3 star NE 4′3.
11 22 32.5	+51 31 55	MCG +9-19-53	14.5	0.6 x 0.6	13.5	E/S0		
11 23 08.5	+51 33 26	MCG +9-19-57	15.3	0.8 x 0.3	13.6	Sc	21	
11 24 24.6	+51 14 04	MCG +9-19-68	14.4	0.9 x 0.8	13.9	Sbc	9	
11 30 31.8	+51 27 19	MCG +9-19-97	14.1	0.9 x 0.7	13.5	Sb	132	
11 44 20.3	+51 01 09	MCG +9-19-167	14.5	0.7 x 0.3	12.7		165	Mag 9.64v star N 4′5.
11 50 31.3	+50 32 33	MCG +9-20-2	15.1	0.8 x 0.2	13.0		174	UGC 6811 S; UGC 6812 N 2′2.
12 03 32.2	+51 04 39	MCG +9-20-77	15.5	0.9 x 0.2	13.5	Sc	78	
10 20 31.7	+43 01 16	NGC 3202	13.2	1.2 x 0.8	13.0	SB(r)a	20	Mag 12.0 star NE 2′8.
10 20 50.0	+42 58 17	NGC 3205	13.3	1.4 x 1.1	13.6	S?	166	Very faint star NE edge of nucleus.
10 21 00.8	+42 59 04	NGC 3207	13.3	1.3 x 0.8	13.2	S?	97	
10 37 12.2	+50 07 11	NGC 3298	13.9	0.9 x 0.6	13.3	E/S0	138	
10 39 09.6	+41 41 09	NGC 3319	11.1	6.2 x 3.4	14.2	SB(rs)cd II-III	37	Very elongated nucleus.
10 39 36.5	+47 23 53	NGC 3320	12.3	2.2 x 1.0	13.0	Scd: III	20	Very faint star on S end.
10 48 01.3	+43 11 08	NGC 3374	13.7	1.2 x 0.9	13.6	SBc	142	**CGCG 212-55** S 2′2.
10 51 44.3	+51 01 25	NGC 3406	12.7	1.0 x 0.7	12.4	E	42	Double system in contact, strongly disturbed.
10 51 54.0	+51 00 19	NGC 3410	14.2	0.8 x 0.5	13.1	S?	24	Stellar nucleus in smooth envelope; mag 9.84v star SW 5′2.
10 51 42.6	+43 42 41	NGC 3415	12.6	2.1 x 1.1	13.4	SA0^{+}:	10	Mag 13.0 star on S edge, part of a small triangle of stars extending SE; NGC 2416 N 3′5.
10 51 48.5	+43 45 49	NGC 3416	14.5	0.6 x 0.2	12.0	S?	27	NGC 3415 S 3′5.
10 57 31.2	+40 56 45	NGC 3468	13.0	1.3 x 0.7	12.8	S0	8	Very faint, small anonymous galaxy SW 1′8.
10 59 27.5	+46 07 21	NGC 3478	12.9	2.6 x 1.2	13.9	SB(rs)bc I-II	132	**MCG +8-20-58** S 2′1.
11 13 44.8	+48 16 21	NGC 3577	13.4	1.4 x 1.1	13.7	SB(r)a	179	**MCG +8-21-4** NW 2′8.
11 14 10.9	+48 19 06	NGC 3583	11.1	2.8 x 1.8	12.7	SB(s)b II	125	Mag 10.10v star NE 6′6; **MCG +8-21-4** W 5′7.
11 15 25.3	+47 26 48	NGC 3595	12.1	1.6 x 0.7	12.2	E?	176	Located 2′0 S of mag 7.56v star.
11 15 52.2	+41 35 23	NGC 3600	11.7	4.1 x 0.9	13.0	Sa?	3	Narrow extension S of nucleus; very small, faint anonymous galaxy WNW 2′7.
11 18 22.1	+45 44 56	NGC 3614	11.6	4.6 x 2.6	14.1	SAB(r)c I-II	104	NGC 3614A SSW 2′5.
11 18 11.8	+45 43 00	NGC 3614A	12.0	0.7 x 0.7	11.1	SB(s)m		
11 26 07.5	+43 35 00	NGC 3675	10.2	5.9 x 3.1	13.2	SA(s)b II	178	Mag 12.9 star superimposed inside SW edge.
11 26 16.7	+46 58 30	NGC 3677	12.3	1.9 x 1.5	13.3	(R′)SA(r)0/a:	130	**CGCG 242-36** NE 2′7.
11 33 21.1	+47 01 47	NGC 3726	10.4	6.2 x 4.3	13.8	SAB(r)c I-II	10	Many knots and clumps in multi-branching arms.
11 36 06.2	+45 17 05	NGC 3741	14.0	1.4 x 0.8	14.0	Im	5	
11 37 44.1	+47 53 32	NGC 3769	11.8	3.1 x 1.0	12.8	SB(r)b: II-III	152	NGC 3769A on SE edge.

(Continued from previous page)

GALAXIES

RA h m s	Dec ° ′ ″	Name	Mag (V)	Dim ′ Maj x min	SB	Type Class	PA	Notes
11 37 50.7	+47 52 54	NGC 3769A	14.1	1.0 x 0.4	13.0	SBm pec: III	113	Located on SE edge of NGC 3769.
11 39 20.7	+46 30 49	NGC 3782	12.4	1.7 x 1.1	12.9	SAB(s)cd: IV-V	0	Faint star on S end.
11 41 17.1	+47 41 14	NGC 3811	12.3	2.4 x 1.5	13.5	SB(r)cd: III	160	Bright portion offset NW.
11 45 56.7	+50 11 58	NGC 3870	13.0	1.0 x 0.8	12.7	S0?	25	
11 46 07.3	+47 29 39	NGC 3877	11.0	5.5 x 1.3	13.0	SA(s)c: II-III	35	A mag 9.87v star 3'.7 NW.
11 48 38.8	+48 42 41	NGC 3893	10.5	4.5 x 2.8	13.1	SAB(rs)c: I	165	**MCG +8-22-9** NE 4'.9.
11 48 56.5	+48 40 26	NGC 3896	12.9	1.4 x 1.0	13.1	SB0/a: pec	126	Faint star N edge.
11 49 39.8	+48 25 25	NGC 3906	12.9	1.9 x 1.7	14.0	SB(s)d	0	Strong E-W bar, smooth envelope.
11 50 45.7	+51 49 28	NGC 3917	11.8	5.1 x 1.3	13.7	SAcd: III-IV	77	Elongated UGC 6802 WNW 6'.3.
11 51 13.2	+50 09 20	NGC 3922	12.8	1.7 x 0.8	12.9	S0/a	38	Mag 9.26v star NE 8'.4; **CGCG 269-11** NE 10'.8.
11 52 39.0	+50 02 14	NGC 3924	14.4	1.7 x 1.6	15.3	Sm:	105	Mag 7.07v star SSW 7'.5.
11 51 47.8	+48 41 01	NGC 3928	12.6	1.5 x 1.5	13.3	SA(s)b?		**Miniature Spiral**.
11 51 13.7	+52 00 04	NGC 3931	13.4	1.1 x 0.9	13.2	SA0$^-$:	160	Mag 8.89v star E 5'.0; **MCG +9-20-18** E 8'.8.
11 52 29.3	+48 27 27	NGC 3932	14.5	1.1 x 0.5	13.7	Sbc	48	
11 52 49.5	+44 07 06	NGC 3938	10.4	5.4 x 4.9	13.8	SA(s)c I	0	Multi-branching arms with many knots; mag 8.92v star NW 8'.6.
11 53 41.3	+47 51 25	NGC 3949	11.1	2.9 x 1.7	12.7	SA(s)bc: III-IV	120	NGC 3950 N 1'.5.
11 53 41.5	+47 53 02	NGC 3950	15.7	0.3 x 0.3	13.1	E0; cand. dwarf		Located 1'.5 N of the center of NGC 3949.
11 56 41.8	+48 20 02	NGC 3985	12.5	1.3 x 0.8	12.4	SB(s)m:	73	**MCG +8-22-42** and **MCG +8-22-43** SW 9'.8.
11 58 06.8	+47 20 06	NGC 4001	15.3	0.6 x 0.3	13.5	E4; cand. dwarf	162	Almost stellar; NGC 4010 SE 6'.9.
11 58 36.8	+47 15 31	NGC 4010	12.6	4.3 x 0.8	13.8	SB(s)d: sp	66	NGC 4001 NW 6'.9.
11 58 30.7	+43 56 47	NGC 4013	11.2	5.2 x 1.0	12.9	Sb sp III	66	Moderately dark lane extends almost full length of galaxy.
11 59 25.4	+50 57 41	NGC 4026	10.8	5.2 x 1.3	12.7	S0 sp	178	UGC 6956 W 9'.8; mag 9.43v star NNE 7'.3.
12 02 51.0	+48 38 10	NGC 4047	12.2	1.6 x 1.3	12.8	(R)SA(rs)b: II	105	
12 03 11.1	+44 31 59	NGC 4051	10.2	5.2 x 3.9	13.3	SAB(rs)bc II	135	Several bright knots; arms branching mostly NE of nucleus.
10 22 40.8	+46 14 20	UGC 5604	13.0	2.4 x 1.2	14.0	Sc I-II	47	
10 22 46.0	+48 38 15	UGC 5605	14.6	0.9 x 0.4	13.3	S	48	
10 26 51.8	+41 53 19	UGC 5657	14.5	1.5 x 0.2	13.0	S	87	Mag 13.6 star on S edge; mag 9.62v star S 4'.2.
10 30 09.8	+44 07 20	UGC 5698	15.2	1.7 x 0.2	13.9	Sbc	127	Located between a close pair of mags 12.3 and 12.9 stars.
10 31 14.4	+43 08 16	UGC 5707	13.4	2.4 x 1.8	14.8	SAB(s)cd	155	Stellar nucleus; mag 12.4 star on E edge.
10 31 45.2	+46 40 18	UGC 5714	13.9	1.0 x 0.7	13.4	SB(s)c	85	Forms a triangle with a mag 12.8 star 1'.6SE and a mag 12.5 star 1'.7 S.
10 34 45.8	+50 46 08	UGC 5740	14.5	2.0 x 1.2	15.3	SABm	140	Dwarf spiral, low surface brightness; mag 11.1 star W edge.
10 35 04.9	+46 33 37	UGC 5744	13.4	0.8 x 0.7	12.6	S0?	110	
10 35 08.6	+45 05 03	UGC 5746	13.2	1.5 x 1.0	13.5	SBab	40	
10 36 12.9	+42 39 31	UGC 5754	15.6	0.8 x 0.5	14.5	Sc	126	Mag 14.1 star NW 1'.2.
10 37 19.3	+43 35 16	UGC 5771	13.3	1.5 x 1.1	13.7	S0/a	60	Mag 9.83v star NW 5'.9.
10 39 26.9	+47 56 50	UGC 5791	13.6	1.5 x 0.5	13.1	S?	43	Mag 10.27v star E 2'.0.
10 39 47.1	+47 55 52	UGC 5798	13.9	0.9 x 0.2	11.9	S?	45	Mag 10.27v star NW 1'.8.
10 43 25.6	+40 46 23	UGC 5838	13.3	1.2 x 0.5	12.6	SBb	130	
10 44 03.7	+43 54 08	UGC 5845	13.9	1.0 x 0.7	13.4		160	
10 45 02.7	+43 42 14	UGC 5859	13.8	1.4 x 0.7	13.6	SBa	25	Bright core, faint envelope.
10 46 09.6	+49 32 36	UGC 5872	14.2	1.5 x 0.6	13.9	Sbc	74	Mag 8.50v star S 8'.1.
10 48 53.8	+46 43 14	UGC 5917	14.3	1.1 x 0.5	13.5	Im	163	
10 49 47.2	+51 53 36	UGC 5928	13.7	1.1 x 0.8	13.4	S0$^-$:	99	Mag 13.7 star on NE edge; **MCG +9-18-36** SE 2'.2.
10 50 21.6	+41 27 56	UGC 5941	13.6	0.9 x 0.7	12.9	S?	168	Bright knot or star N end.
10 51 18.3	+44 34 15	UGC 5953	13.0	0.7 x 0.4	11.5	Pec	85	Mag 10.75v star S 2'.2.
10 52 47.9	+49 36 40	UGC 5991	13.5	1.1 x 0.7	13.1	S0$^-$	155	Mag 10.99v star SE 3'.4.
10 52 57.5	+40 22 42	UGC 5996	15.6	0.6 x 0.6	14.3	Sdm		Mag 13.0 star on N edge.
10 53 09.5	+50 16 59	UGC 5998	13.9	0.8 x 0.3	12.2	Pec?	110	Disrupted pair? Mag 12.4 star S 0'.5.
10 53 36.0	+50 46 21	UGC 6008	14.5	1.7 x 0.2	13.2	SBb	154	Pair of stars, mags 11.8 and 12.2, W and SW 2'.3.
10 53 59.8	+49 39 34	UGC 6013	13.0	0.9 x 0.7	12.3	S0	45	MCG +8-20-40 S 3'.4.
10 54 02.8	+46 01 36	UGC 6015	14.1	0.9 x 0.7	13.5	Scd:	0	**UGC 6005** W 6'.1.
10 55 02.6	+49 43 36	UGC 6029	13.2	0.9 x 0.7	12.6	Pec	35	**MCG +8-20-47** S 7'.1.
10 56 09.8	+47 23 31	UGC 6038	14.3	1.1 x 1.1	14.4	Sb II		
10 59 59.6	+50 03 21	UGC 6071	13.1	1.4 x 0.8	13.3	E	135	MCG +8-20-61 SW 2'.7.
10 59 58.5	+50 54 09	UGC 6074	13.3	1.6 x 0.9	13.5	S?	90	Bright N-S core; mag 8.14v star W 1'.8.
11 00 11.5	+45 44 16	UGC 6075	13.4	1.7 x 0.7	13.5	Sb II-III	107	
11 00 09.7	+45 54 59	UGC 6076	13.5	0.8 x 0.6	12.8	E?	150	Mag 10.60v star E 2'.0.
11 01 34.1	+45 39 15	UGC 6100	13.4	0.8 x 0.6	12.5	Sa?	10	Mag 9.09v star WNW 2'.1.
11 01 55.7	+47 05 38	UGC 6101	15.2	1.5 x 0.2	13.7	Sbc	64	
11 01 59.1	+45 13 40	UGC 6103	13.5	0.9 x 0.7	12.8	Pec	48	Mag 12.8 star W 1'.4.
11 02 11.2	+45 53 17	UGC 6106	13.6	1.5 x 0.6	13.3	SBb?	160	
11 02 35.0	+50 34 55	UGC 6109	14.2	1.3 x 1.3	14.6	SAB(r)c		Mag 14.3 star SW of nucleus.
11 02 57.0	+50 39 47	UGC 6114	13.6	1.4 x 0.8	13.6	Double System	12	Double system, contact.
11 03 09.8	+50 12 20	UGC 6117	13.5	0.9 x 0.6	12.7	SBa	122	Mag 10.05v star NW 2'.8.
11 03 38.6	+45 10 49	UGC 6125	13.9	1.3 x 0.5	13.3	(R′)SB(s)ab	33	Faint anonymous galaxy SW 0'.9.
11 04 03.5	+49 49 18	UGC 6127	13.7	0.9 x 0.8	13.2	Scd:	164	Mag 11.6 star on S edge; **MCG +8-20-75** SW 0'.9.
11 04 18.2	+50 01 58	UGC 6129	14.3	0.6 x 0.6	13.0	Sbc		Mag 10.48v star NE 2'.5.
11 04 28.0	+44 01 48	UGC 6131	13.4	1.3 x 0.8	13.3	S0	80	
11 04 37.1	+45 07 29	UGC 6135	12.4	0.9 x 0.9	12.0	S?		**MCG +8-20-80** W 1'.7; mag 9.10v star NW 2'.2.
11 04 36.7	+45 59 06	UGC 6136	13.8	1.7 x 0.5	13.5	S?	7	**CGCG 241-70** NNW 8'.5.
11 06 04.5	+43 19 49	UGC 6149	13.7	0.9 x 0.3	12.1	S	110	
11 06 49.5	+43 43 22	UGC 6161	13.4	2.8 x 1.5	14.8	SBdm	40	
11 06 54.8	+51 12 11	UGC 6162	13.2	2.3 x 1.1	14.1	Sd	88	Very knotty; mag 12.2 star on SE edge.
11 07 04.4	+45 49 20	UGC 6165	14.3	1.7 x 0.3	13.4	Sab	57	Mag 10.57v star S 2'.1.
11 08 30.0	+45 07 47	UGC 6187	13.4	0.6 x 0.6	12.1	S?		
11 09 43.7	+46 48 50	UGC 6201	13.6	0.8 x 0.6	12.7	S?	110	
11 09 58.9	+46 05 44	UGC 6205	14.3	1.7 x 1.3	15.0	Sm:	50	Dwarf spiral, very low surface brightness.
11 11 27.5	+47 02 06	UGC 6227	13.3	0.8 x 0.8	12.7	S0$^-$:		**CGCG 241-92** SW 7'.6.
11 11 26.0	+43 54 25	UGC 6232	14.2	1.7 x 0.3	13.3	Sa	150	
11 13 41.2	+47 34 40	UGC 6255	13.1	0.9 x 0.6	12.3	S?	44	Located near N edge of small galaxy cluster **A 1202**.
11 14 33.4	+43 14 08	UGC 6266	14.2	1.2 x 0.8	14.0	Sm	55	Mag 12.5 star S edge.
11 17 47.1	+51 28 29	UGC 6309	12.8	1.7 x 1.1	13.4	SB?	125	**Mkn 1445** SSW 3'.5; **Mkn 1444** SSW 5'.1.
11 22 02.3	+50 35 37	UGC 6380	13.5	1.5 x 0.7	13.4	S0/a	58	Mag 11.1 star NW 2'.6.
11 23 23.3	+50 53 34	UGC 6399	13.6	3.0 x 0.8	14.4	Sm:	142	
11 24 06.1	+45 48 19	UGC 6410	13.9	1.1 x 0.8	13.6	SABc	5	Mag 10.72v star on N edge.
11 27 22.0	+40 00 47	UGC 6455	14.7	1.7 x 0.7	14.7	S	83	Mag 12.6 star W 2'.6.
11 29 11.2	+42 31 34	UGC 6489	14.8	1.5 x 0.7	14.7	S	55	Mag 11.6 star NE 2'.5.
11 31 12.0	+51 41 54	UGC 6505	15.3	1.1 x 1.1	15.3	SABd:		
11 32 45.1	+40 50 32	UGC 6529	14.3	1.0 x 0.7	13.8	Sbc	72	

(Continued from previous page)

38 GALAXIES 38

RA h m s	Dec ° ′ ″	Name	Mag (V)	Dim ′ Maj x min	SB	Type Class	PA	Notes
11 33 29.4	+49 14 19	UGC 6541	14.2	1.2 x 0.7	13.8	Im	133	**UGC 6538** NNW 3′3.
11 35 00.2	+51 13 02	UGC 6555	13.3	1.3 x 1.0	13.4	S0$^-$:	87	**MCG +9-19-124** and mag 11.4 star NE 1′7.
11 36 31.2	+47 49 06	UGC 6576	14.0	1.4 x 1.4	14.6	SB0		
11 38 19.7	+47 26 13	UGC 6606	14.0	1.7 x 0.9	14.3	Double System	81	Double system, connected, disrupted.
11 38 51.5	+43 09 51	UGC 6611	14.5	1.5 x 0.4	13.8	S	9	Mag 10.07v star N 3′0.
11 40 05.9	+45 56 32	UGC 6628	12.6	3.2 x 3.2	15.0	SAm		Filamentary with several knots; mag 11.9 star superimposed SE of center.
11 42 25.9	+51 35 52	UGC 6667	13.5	3.2 x 0.5	13.8	Scd:	89	
11 44 24.9	+48 50 07	UGC 6713	14.4	1.4 x 1.1	14.7	Sm	123	Dwarf spiral, extremely low surface brightness.
11 45 09.3	+49 43 04	UGC 6726	13.6	1.0 x 0.5	12.7	S0$^-$:	107	
11 46 57.5	+50 42 08	UGC 6755	13.8	1.5 x 0.7	13.7	S	60	
11 48 00.6	+49 48 27	UGC 6773	14.2	1.7 x 0.7	14.2	Im	168	Located 1′9 SE of mag 7.12v star.
11 48 35.7	+43 43 16	UGC 6776	15.8	1.5 x 0.2	14.3	Scd:	30	**UGC 6768** W 8′6; **CGCG 214-26** W 12′8.
11 50 06.6	+51 51 17	UGC 6802	15.3	2.1 x 0.3	14.6	Scd:	160	Mag 11.27v star SW 2′9.
11 50 12.4	+42 04 28	UGC 6805	13.6	0.5 x 0.5	12.1	E?		Compact.
11 50 34.9	+50 31 40	UGC 6811	13.8	1.1 x 0.4	12.8	S0/a	12	MCG +9-20-2 and UGC 6811 S 2′4.
11 50 35.4	+50 34 46	UGC 6812	17.0	1.5 x 0.4	16.3	Sdm:	54	MCG +9-20-2 N 1′0.
11 50 46.7	+45 48 26	UGC 6818	13.4	1.8 x 0.7	13.5	SB?	78	Mag 12.1 star SSE 2′1.
11 53 39.8	+43 27 31	UGC 6865	13.8	1.5 x 0.6	13.5	S?	35	Double system connected.
11 55 38.4	+43 02 40	UGC 6901	13.5	1.3 x 0.6	13.0	S?	32	
11 56 28.4	+50 25 44	UGC 6917	12.5	3.8 x 2.3	14.7	SBm	130	Mag 11.4 star superimposed W of center.
11 56 52.6	+50 49 01	UGC 6922	13.2	1.5 x 1.2	13.7	S?	48	
11 57 17.2	+49 16 57	UGC 6930	12.1	3.4 x 2.9	14.4	SAB(s)d	30	Mag 10.94v star NW 2′2.
11 58 25.9	+50 55 03	UGC 6956	14.5	2.5 x 2.3	16.3	SB(s)m V	27	Dwarf spiral, low surface brightness; NGC 4026 E 9′8.
12 00 18.9	+50 39 11	UGC 6992	14.1	1.4 x 0.7	13.9	Sdm:	63	
12 02 36.6	+41 03 13	UGC 7020	13.2	2.4 x 1.8	14.7	SA(rs)bc I-II	171	Two main arms, strong dark lane W of center.

GALAXY CLUSTERS

RA h m s	Dec ° ′ ″	Name	Mag 10th brightest	No. Gal	Diam ′	Notes
10 32 06.0	+40 12 00	A 1035	15.4	94	18	**MCG +7-22-19** N of center 4′7; most other members anonymous and stellar.
11 18 48.0	+51 42 00	A 1218	16.0	47	17	All members anonymous and stellar.
11 34 48.0	+49 02 00	A 1314	13.9	44	28	Many MCG, PGC and anonymous GX scattered throughout cluster area.
12 03 36.0	+51 44 00	A 1452	15.7	46	15	**MCG +9-20-74** and **75** SW from center 1′2; **Hickson 60** group SW 5′7.

39 GALAXIES 39

RA h m s	Dec ° ′ ″	Name	Mag (V)	Dim ′ Maj x min	SB	Type Class	PA	Notes
10 15 49.4	+43 47 11	CGCG 211-34	13.9	0.7 x 0.4	12.4	S?	162	= **Mkn 139**; mag 10.75v star WNW 5′2.
08 46 20.1	+47 09 11	CGCG 237-18	13.9	0.8 x 0.4	12.6	S0	3	Multi-galaxy system.
08 48 24.8	+47 17 18	CGCG 237-21	14.2	0.5 x 0.3	12.0	S?	108	Mag 9.50v star W 5′8.
10 16 28.4	+45 19 14	CGCG 240-22	14.5	0.5 x 0.3	12.3		9	Very faint, elongated anonymous galaxy NE 2′6.
09 26 05.2	+50 26 16	CGCG 265-17	14.1	0.6 x 0.3	12.1		69	Mag 12.4 star SE 2′4.
10 12 48.7	+43 08 43	IC 598	12.9	1.4 x 0.5	12.4	S0/a	8	Mag 10.95v star WSW 6′4.
08 45 05.7	+40 33 32	MCG +7-18-44	14.8	0.9 x 0.3	13.2	Spiral	150	Mag 13.4 star SW 1′9.
08 45 41.7	+41 16 56	MCG +7-18-45	14.1	1.0 x 0.6	13.4		144	Mag 10.7 star E 1′7.
08 48 59.8	+41 00 08	MCG +7-18-50	14.1	0.8 x 0.7	13.3		12	
08 49 31.0	+42 16 25	MCG +7-18-52	15.1	0.8 x 0.2	13.0		126	Mag 11.55v star SW 0′6.
08 49 45.5	+40 14 45	MCG +7-18-54	15.3	0.8 x 0.4	13.9	Spiral	30	Mag 10.9 star S 2′0.
08 50 11.4	+40 15 18	MCG +7-18-55	14.0	0.8 x 0.6	13.1		81	Mag 13.0 star SW 1′6.
08 50 37.9	+40 15 01	MCG +7-18-57	14.7	0.7 x 0.4	13.2	Sc	39	Pair of stars, mags 13.3 and 10.68v, E 0′9.
08 52 41.7	+40 41 02	MCG +7-18-59	14.4	0.9 x 0.6	13.6	Spiral	141	Mag 11.1 star S 2′0.
08 54 40.6	+42 31 34	MCG +7-18-63	14.3	1.1 x 0.6	13.7	Sb	117	Mag 10.31v star ENE 4′6.
09 02 52.5	+40 19 29	MCG +7-19-13	15.1	0.5 x 0.5	13.5	Sbc		
09 03 02.3	+40 25 57	MCG +7-19-14	14.5	0.6 x 0.6	13.2	Spiral		**MCG +7-19-15** ENE 3′9.
09 04 43.5	+42 19 10	MCG +7-19-20	14.2	0.8 x 0.7	13.5		174	
09 04 48.1	+40 35 28	MCG +7-19-21	15.4	0.7 x 0.7	14.4			Mag 9.27v star W 3′8.
09 04 50.6	+42 15 20	MCG +7-19-22	15.0	0.8 x 0.6	14.0	Spiral	72	Located between a pair of SE-NE mag 13-14 stars.
09 07 01.7	+41 16 43	MCG +7-19-30	14.9	0.9 x 0.4	13.6	Sb	27	Located between a pair of E-W stars, mags 12.6 and 13.7.
09 07 06.9	+42 35 42	MCG +7-19-31	15.3	0.6 x 0.5	13.8	Spiral	27	Mag 11.18v star S 2′7.
09 14 13.2	+43 08 02	MCG +7-19-37	14.3	0.7 x 0.5	13.0	Spiral	171	
09 15 44.8	+42 15 57	MCG +7-19-44	14.1	0.9 x 0.3	12.5	Sab	66	
09 16 28.3	+42 08 17	MCG +7-19-46	14.5	0.7 x 0.5	13.5	E/S0	87	Almost stellar; mag 10.19v star S 8′9.
09 16 36.2	+41 42 54	MCG +7-19-50	14.4	0.7 x 0.7	13.5			Mag 10.39v star on E edge.
09 16 46.8	+43 42 41	MCG +7-19-51	14.3	0.8 x 0.6	13.4		36	**MCG +7-19-52** NE 0′8; **MCG +7-19-53** E 1′1.
09 17 41.7	+44 02 16	MCG +7-19-57	15.2	0.6 x 0.4	13.5	Spiral	15	Mag 11.99v star NE 2′8.
09 23 41.2	+40 49 59	MCG +7-19-67	15.0	0.9 x 0.3	13.4	Sb	105	Mag 9.84v star SE 3′8.
09 24 14.0	+40 38 11	MCG +7-20-1	15.3	0.5 x 0.4	13.4		3	MCG +7-20-2 SE 1′7.
09 24 22.0	+40 37 24	MCG +7-20-2	15.0	0.9 x 0.3	13.4	Sb	126	MCG +7-20-1 NW 1′7; mag 13.1 star S 1′2.
09 24 54.1	+42 09 31	MCG +7-20-4	14.8	0.6 x 0.5	13.3		120	Mag 11.9 star on NE edge.
09 24 56.6	+40 23 57	MCG +7-20-6	15.0	0.7 x 0.3	13.1	Spiral	132	Mag 10.6 star NE 5′1.
09 25 39.7	+40 38 46	MCG +7-20-9	15.0	0.7 x 0.3	13.2	Spiral	159	
09 27 07.6	+42 34 14	MCG +7-20-11	14.6	0.8 x 0.7	13.9	Sm	9	Mag 11.3 star S edge.
09 32 37.2	+41 46 16	MCG +7-20-19	15.0	0.9 x 0.9	14.6			Mag 11.6 star ESE 2′7.
09 32 51.8	+42 03 30	MCG +7-20-20	15.1	0.8 x 0.2	13.0		150	
09 37 08.2	+41 37 33	MCG +7-20-25	14.7	0.6 x 0.4	13.0		39	Mag 12.5 star SW 1′8.
09 37 09.1	+42 58 15	MCG +7-20-26	15.2	0.5 x 0.4	13.3		33	Mag 13.3 star on N edge.
09 38 04.0	+42 58 25	MCG +7-20-27	15.0	0.7 x 0.4	13.5		54	
09 37 45.4	+43 05 59	MCG +7-20-30	14.2	0.8 x 0.4	12.8	Sb	69	Mag 9.15v star W 3′1.
09 40 30.2	+40 05 13	MCG +7-20-31	14.5	0.8 x 0.5	13.4		12	Mag 9.20v star ESE 3′8.
09 41 30.0	+41 23 03	MCG +7-20-32	15.3	0.9 x 0.3	13.7	Spiral	6	Mag 11.2 star NE 1′8.
09 42 35.1	+41 49 33	MCG +7-20-33	16.2	0.7 x 0.7	15.2	Spiral		Forms a triangle with a mag 12.7 star SE 3′3 and a mag 12.8 star SW 2′7.
09 43 47.7	+42 28 21	MCG +7-20-37	14.4	0.9 x 0.7	13.7		129	MCG +7-20-40 SE 3′2; mag 11.7 star NE 2′9.
09 43 59.3	+42 25 57	MCG +7-20-40	14.0	1.0 x 0.8	13.8	E/S0	108	**MCG +7-20-43** E 1′5; MCG +7-20-37 NW 3′2.
09 49 21.5	+43 55 03	MCG +7-20-57	14.7	0.7 x 0.3	12.9	Spiral	126	NGC 3006 N 6′5; mag 8.45v star NE 11′6.
09 49 20.4	+43 26 31	MCG +7-20-58	15.0	0.7 x 0.7	14.1			
09 50 33.2	+44 18 49	MCG +7-20-66	13.2	0.6 x 0.4	11.5		45	NGC 3010 NE 0′6; NGC 3009 SW 4′1.

RA h m s	Dec ° ′ ″	Name	Mag (V)	Dim ′ Maj x min	SB	Type Class	PA	Notes
09 50 39.4	+44 19 49	MCG +7-20-67	14.2	0.5 x 0.4	12.3	Compact	30	= **Mkn 1237**; NGC 3010 SW 0.′9.
09 50 37.1	+43 45 31	MCG +7-20-68	14.0	0.8 x 0.5	12.9	S?	48	
09 52 28.3	+42 49 07	MCG +7-20-72	15.9	0.7 x 0.6	14.9	Spiral	153	Mag 13.9 star SE 0.′7.
09 55 24.7	+41 09 16	MCG +7-21-2	14.4	0.8 x 0.7	13.7		156	**MCG +7-21-1** W 2.′4; mag 11.3 star N 2.′0.
10 16 15.7	+41 09 57	MCG +7-21-35	14.9	0.8 x 0.3	13.2	Spiral	63	Mag 12.6 star SSE 2.′5.
10 24 22.7	+41 42 22	MCG +7-22-1	13.8	0.8 x 0.6	13.1	E/S0	24	Bright star on W edge.
08 59 52.1	+49 46 10	MCG +8-17-9	15.2	0.6 x 0.4	13.5	Sc	51	Mag 8.94v star S 2.′6.
09 00 58.5	+47 47 40	MCG +8-17-14	14.7	1.0 x 0.3	13.3		6	Mag 14.2 star W edge; mag 10.97v star NE 1.′7.
09 06 50.0	+48 46 19	MCG +8-17-27	15.3	0.7 x 0.5	14.0	Sbc	78	
09 07 56.7	+49 35 46	MCG +8-17-34	14.0	0.4 x 0.4	12.1	E/S0		Pair with MCG +8-17-35 N edge.
09 07 57.7	+49 36 08	MCG +8-17-35	14.1	0.4 x 0.3	11.6		18	Pair with MCG +8-17-34 S edge.
09 08 53.3	+49 45 02	MCG +8-17-40	15.1	0.8 x 0.2	12.9	Sbc	165	**MCG +8-17-39** W 1.′4; **MCG +8-17-42** E 1.′4.
09 10 12.1	+50 24 02	MCG +8-17-48	13.8	0.9 x 0.7	13.1	S?	170	Located half way between NGC 2771 E and NGC 2767 W.
09 12 14.5	+49 45 38	MCG +8-17-55	14.5	0.8 x 0.4	13.1		105	Mag 13.8 star S 0.′9.
09 16 47.8	+49 46 46	MCG +8-17-69	14.4	0.9 x 0.9	14.0			Mag 13.1 star SE 0.′9.
09 17 01.6	+50 02 40	MCG +8-17-70	14.8	0.9 x 0.7	14.1	Sc	105	Multi-galaxy system.
09 21 10.9	+45 53 14	MCG +8-17-84	14.1	0.5 x 0.5	12.4	S?		Mag 8.16v star S 1.′4.
09 22 30.9	+50 25 55	MCG +8-17-87	12.1	1.7 x 0.8	12.3		72	Bright core offset W, bright knot E, double system?
09 27 55.0	+44 30 21	MCG +8-17-102	14.4	0.7 x 0.4	12.9	Spiral	147	
09 27 58.7	+47 53 17	MCG +8-17-103	15.1	1.1 x 0.3	13.8		99	Mag 10.89v star WNW 2.′0.
09 30 13.3	+49 29 18	MCG +8-17-109	13.9	1.3 x 0.8	14.0	E/S0	33	Mag 11.2 star W 4.′3.
09 31 06.6	+48 03 01	MCG +8-17-110	14.5	1.4 x 0.8	14.7	E/S0	30	Mag 10.79v star WNW 8.′9.
09 33 52.2	+46 51 46	MCG +8-18-7	13.7	0.8 x 0.6	13.0	E/S0	72	Mag 6.55v star NE 5.′2.
09 36 37.3	+48 28 25	MCG +8-18-12	14.3	0.8 x 0.5	13.1	Sc	60	Pair with MCG +8-18-13 W.
09 36 31.5	+48 28 09	MCG +8-18-13	14.4	0.8 x 0.4	13.0	Sb		Pair with MCG +8-18-12 E; mag 13.1 star W 1.′5.
09 37 24.9	+48 30 53	MCG +8-18-14	14.5	0.9 x 0.4	13.3	Spiral	60	Close pair of mag 11 stars E 0.′9; mag 9.23v star E 1.′9.
09 38 48.6	+48 34 18	MCG +8-18-16	13.8	0.7 x 0.7	12.9			Elongated **MCG +8-18-16A** S edge.
09 41 37.8	+48 40 10	MCG +8-18-23	16.1	0.5 x 0.5	14.4	Spiral		UGC 5172 E 2.′3.
09 42 02.8	+48 05 33	MCG +8-18-25	17.4	0.8 x 0.2	15.3	Sc	9	**CGCG 239-24** N 3.′8; stellar galaxy **Mkn 1420** S 5.′0.
09 46 15.2	+47 08 02	MCG +8-18-29	15.3	0.8 x 0.4	13.9	Spiral	90	
09 48 33.5	+47 22 03	MCG +8-18-32	15.3	0.5 x 0.4	13.4	Spiral		
09 48 42.1	+47 20 41	MCG +8-18-33	14.8	0.8 x 0.2	12.7	Spiral	144	
09 49 02.3	+47 02 05	MCG +8-18-34	15.0	0.7 x 0.6	14.1	E/S0	42	
09 49 14.4	+45 36 03	MCG +8-18-35	15.2	0.6 x 0.5	13.7	Spiral	9	Mag 12.4 star ESE 1.′5.
09 51 03.8	+47 10 00	MCG +8-18-40	14.3	0.7 x 0.3	12.5	Spiral	90	Mag 13.7 star SE 1.′4.
09 53 40.1	+46 55 13	MCG +8-18-42	15.5	0.5 x 0.5	13.8	Spiral		Mag 10.20v star NNE 2.′9.
10 02 33.5	+44 47 14	MCG +8-18-56	14.6	0.8 x 0.5	13.5	Spiral	102	
10 03 03.4	+45 35 50	MCG +8-18-57	14.6	1.0 x 0.4	13.5		45	Mag 9.26v star SE 3.′9.
10 03 28.9	+46 37 37	MCG +8-18-59	14.4	1.3 x 0.8	14.3	Spiral	120	
10 03 23.3	+48 21 54	MCG +8-18-60	14.1	0.8 x 0.5	12.9	S?	30	Located between a pair of NW-SE mag 13.8 stars.
10 04 04.9	+45 02 08	MCG +8-18-62	14.8	0.8 x 0.5	13.6	Spiral	138	
10 04 53.1	+44 38 07	MCG +8-18-64	15.0	0.7 x 0.6	13.9		141	
10 11 08.0	+46 54 52	MCG +8-19-8	13.9	1.2 x 0.8	13.9	E/S0		**MCG +8-19-6** W 9.′9.
10 13 26.5	+50 26 41	MCG +8-19-11	15.1	0.5 x 0.3	12.9		15	Sometimes misidentified as **NGC 3148**.
10 27 04.7	+46 03 24	MCG +8-19-32	15.1	0.6 x 0.6	13.9	Sb		Mag 10.85v star NW 2.′5.
10 27 52.2	+46 00 21	MCG +8-19-35	15.2	0.8 x 0.4	13.8	Sc	123	Mag 11.3 star WNW 3.′0.
08 56 24.3	+51 52 44	MCG +9-15-51	14.2	0.6 x 0.6	13.2	E/S0		MCG +9-15-52 NE 3.′1.
08 56 37.0	+51 55 12	MCG +9-15-52	14.5	0.5 x 0.4	12.6		135	MCG +9-15-51 SW 3.′1.
09 14 58.3	+51 21 37	MCG +9-15-111	13.9	1.1 x 0.5	13.1	Sc	15	
09 18 57.9	+51 22 11	MCG +9-15-118	14.4	1.1 x 0.3	13.0	Sb	141	Mag 11.4 star WSW 2.′8.
09 23 08.4	+50 50 06	MCG +9-16-8	13.9	1.0 x 0.8	13.5		30	Mag 9.53v star S 3.′5.
09 28 32.6	+50 47 33	MCG +9-16-22	14.2	1.0 x 0.5	13.3	Sc	171	Mag 12.9 star N 3.′4.
09 33 45.5	+51 55 09	MCG +9-16-29	15.5	0.8 x 0.4	14.1		54	Mag 12.0 star NNE 2.′3.
09 33 56.7	+51 31 41	MCG +9-16-30	14.5	1.1 x 0.8	14.2	Sb	60	Elongated anonymous galaxy N edge.
08 51 35.7	+47 33 28	NGC 2676	13.1	1.2 x 1.2	13.4	S0:		lies 1.′9 SW of mag 10.57v star. Small anonymous galaxy SE 3.′0.
08 53 32.9	+51 18 48	NGC 2681	10.3	3.6 x 3.3	12.8	(R′)SAB(rs)0/a	36	
08 54 53.8	+49 09 39	NGC 2684	12.9	0.9 x 0.8	12.4	S?	27	NGC 2787A and NGC 2787B E 1.′8.
08 54 59.1	+49 08 30	NGC 2686A	14.8	0.5 x 0.3	12.6		54	On W edge of NGC 2886B.
08 55 00.6	+49 08 30	NGC 2686B	16.2	0.3 x 0.2	13.0		39	On E edge of NGC 2886A.
08 55 04.9	+49 09 19	NGC 2687A	17.1	0.3 x 0.1	13.1		39	On W edge of NGC 2687B.
08 55 06.2	+49 09 19	NGC 2687B	14.8	0.4 x 0.2	11.9		27	On E edge of NGC 2687A.
08 55 11.6	+49 07 18	NGC 2688	15.8	0.5 x 0.3	13.6		96	Mag 10.72v star S 2.′8.
08 55 15.7	+49 09 02	NGC 2689	16.3	0.3 x 0.2	13.1		102	Anonymous galaxy ESE 2.′3.
08 56 59.5	+51 20 53	NGC 2693	11.9	2.6 x 1.8	13.5	E3:	160	NGC 2694 on S edge; **UGC 4679** N 7.′6; **MCG +9-15-54** NNW 9.′3.
08 56 59.4	+51 19 53	NGC 2694	14.4	1.2 x 1.2	14.8	E1		Stellar nucleus; located on S edge of NGC 2693.
08 59 30.0	+44 54 47	NGC 2712	12.1	2.9 x 1.6	13.6	SB(r)b: I	178	
09 06 03.1	+51 44 39	NGC 2739	14.7	0.8 x 0.3	13.0	Sa	96	Located on N edge of NGC 2740.
09 06 05.1	+51 44 06	NGC 2740	14.0	0.9 x 0.7	13.4	S?	45	Located on S edge of NGC 2739; mag 7.20v star SE 12.′3.
09 07 58.6	+41 42 29	NGC 2755	13.3	1.2 x 0.8	13.1	S?	130	
09 09 52.7	+50 24 29	NGC 2762	15.3	0.5 x 0.3	13.1		81	Almost stellar; located 0.′7 S of NGC 2767.
09 09 54.5	+50 25 07	NGC 2767	14.0	0.5 x 0.4	12.1	S?	0	NGC 2762 S 0.′7.
09 10 32.1	+50 26 01	NGC 2769	13.0	1.7 x 0.4	12.4	Sa	146	NGC 2771 S 3.′4.
09 10 39.7	+50 22 44	NGC 2771	12.7	2.2 x 1.9	14.1	(R′)SB(r)ab	137	Strong nucleus in NE-SW bar; NGC 2769 N 3.′4.
09 12 14.4	+44 57 16	NGC 2776	11.6	3.0 x 2.7	13.7	SAB(rs)c I-II	0	Numerous knots.
09 14 09.1	+40 06 54	NGC 2782	11.6	3.5 x 2.6	13.9	SAB(rs)a pec	74	The bright portion of the galaxy is offset to the W. A faint, wide "arm" extends E then curves N to fill out the published size for this galaxy.
09 15 15.1	+40 55 06	NGC 2785	14.1	1.5 x 0.5	13.7	Im?	120	
09 17 23.1	+41 59 59	NGC 2798	12.3	2.6 x 1.0	13.2	SB(s)a pec	160	Mag 8.35v star NE 8.′8; NGC 2799 on E edge; UGC 4904 S 5.′4.
09 17 31.2	+41 59 35	NGC 2799	13.7	1.9 x 0.5	13.5	SB(s)m?	125	Located on E edge of NGC 2798.
09 22 02.4	+50 58 39	NGC 2841	9.2	8.1 x 3.5	12.7	SA(r)b: I	147	Located 4.′6 W of mag 8.49v star.
09 21 48.1	+40 09 01	NGC 2844	12.9	1.5 x 0.7	12.9	SA(r)a:	13	Mag 7.67v star NE 8.′0.
09 23 14.8	+40 09 47	NGC 2852	13.2	1.0 x 0.6	12.5	SAB(r)a?	165	Mag 10.3 star W 3.′1.
09 23 17.5	+40 11 55	NGC 2853	13.3	1.0 x 0.6	12.6	SB0:	25	Mag 14.9 star on SE edge; mag 7.67v star W 9.′7.
09 24 03.2	+49 12 15	NGC 2854	13.0	1.7 x 0.6	12.9	SB(s)b	50	Mag 10.32v star N 3.′1.
09 24 16.5	+49 14 54	NGC 2856	13.2	1.1 x 0.5	12.4	S?	134	Mag 10.32v star W 3.′5.
09 24 38.0	+49 21 27	NGC 2857	12.3	2.2 x 2.0	13.7	SA(s)c I-II	138	Stellar nucleus; **CGCG 238-51** ENE 3.′9.
09 24 53.3	+41 03 35	NGC 2860	13.7	1.4 x 0.6	13.4	SBa	108	Mag 9.00v star N 5.′8.
09 48 43.7	+44 04 52	NGC 2998	12.5	2.9 x 1.3	13.8	SAB(rs)c I-II	53	Mag 10.7 star S 3.′0.
09 49 15.0	+44 07 49	NGC 3005	14.9	1.1 x 0.2	13.1	Sb	150	Mag 8.89v star N 3.′8.

RA h m s	Dec ° ′ ″	Name	Mag (V)	Dim ′ Maj x min	SB	Type Class	PA	Notes
09 49 17.3	+44 01 30	NGC 3006	14.9	0.7 x 0.2	12.6	Spiral	81	NGC 2998 NW 7′0; MCG +7-20-57 S 6′5.
09 49 34.3	+44 06 07	NGC 3008	14.5	0.7 x 0.4	13.0	Sb	135	Mag 8.45v star SE 10′1.
09 50 11.1	+44 17 44	NGC 3009	13.6	1.0 x 0.8	13.2	S?	161	Stellar nucleus; MCG +7-20-66 NE 4′1.
09 50 34.8	+44 19 22	NGC 3010	13.3	0.5 x 0.3	11.1		129	MCG +7-20-66 S edge; MCG +7-20-6 SW 0′6.
10 03 57.4	+40 45 26	NGC 3104	13.0	3.3 x 2.2	15.0	IAB(s)m	35	Faint star and knot? inline S of nucleus.
10 06 07.3	+47 15 44	NGC 3111	13.0	0.9 x 0.8	12.5	S0$^-$:	144	Very faint anonymous galaxy N 2′0.
10 10 54.3	+45 57 00	NGC 3135	13.5	1.0 x 0.6	12.8	S?	90	
10 17 57.3	+41 06 52	NGC 3179	13.1	1.9 x 0.5	12.9	S0	48	Mag 13.7 star on SW edge.
10 18 17.0	+41 25 20	NGC 3184	9.8	7.4 x 6.9	13.9	SAB(rs)cd II-III	135	Many knots and superimposed stars cover this spiral galaxy.
10 19 05.3	+46 27 18	NGC 3191	13.3	0.8 x 0.6	12.4	SB(s)bc pec	5	
10 19 54.5	+45 32 58	NGC 3198	10.3	8.5 x 3.3	13.8	SB(rs)c II	35	Several strong knots; mag 11.2 star on N edge.
10 20 31.7	+43 01 16	NGC 3202	13.2	1.2 x 0.8	13.0	SB(r)a	20	Mag 12.0 star NE 2′8.
10 20 50.0	+42 58 17	NGC 3205	13.3	1.4 x 1.1	13.6	S?	166	Very faint star NE edge of nucleus.
10 21 00.8	+42 59 04	NGC 3207	13.3	1.3 x 0.8	13.2	S?	97	
08 44 06.0	+49 47 36	UGC 4551	12.5	1.7 x 0.6	12.4	S0?	113	Located between a pair of mag 12 stars N and S .
08 44 15.3	+41 43 00	UGC 4556	13.5	0.7 x 0.3	11.7	I?	21	
08 44 55.2	+47 44 44	UGC 4562	14.7	1.2 x 0.7	14.4	SB(s)c	127	
08 46 14.5	+41 34 48	UGC 4578	14.1	1.2 x 0.7	13.7	Sb II-III	65	
08 46 33.6	+48 25 44	UGC 4580	13.9	0.9 x 0.5	12.9	Sbc	7	
08 47 22.7	+49 33 30	UGC 4587	12.8	1.7 x 1.0	13.2	S0?	8	Large, diffuse core, smooth envelope; mag 11.3 star SW 2′8.
08 48 07.0	+41 51 29	UGC 4598	15.2	0.8 x 0.8	14.5	Sc		Mag 12.3 star NE 3′0.
08 49 01.2	+41 11 54	UGC 4609	14.5	0.6 x 0.6	13.2	Pec		Compact; mag 12.4 star E 1′4; faint, almost stellar anonymous galaxy SE 2′7.
08 50 20.4	+41 17 18	UGC 4622	14.2	1.0 x 0.8	13.8	SAd	147	Stellar nucleus, smooth envelope.
08 51 40.8	+51 07 07	UGC 4628	14.6	1.7 x 0.3	13.7	Scd:	74	
08 51 54.6	+40 49 58	UGC 4635	14.3	1.2 x 0.8	14.1	Pec	36	Very faint plume extends N and then E from bright core.
08 52 56.5	+42 24 49	UGC 4642	13.4	0.8 x 0.7	12.6	S?	72	Faint star W end.
08 53 35.9	+45 20 09	UGC 4648	16.0	2.2 x 1.4	17.0	Im:	141	Dwarf, almost stellar nucleus with low surface brightness envelope.
08 53 38.1	+40 43 44	UGC 4649	14.0	0.8 x 0.7	13.2		25	Faint star E of nucleus.
08 54 40.8	+47 06 15	UGC 4659	14.4	1.8 x 0.6	14.3	SAdm:	115	Mag 12.8 star W 2′5.
08 57 41.7	+43 07 33	UGC 4686	13.8	0.7 x 0.4	12.3	S?	78	Mag 11.6 star NW 2′1.
08 58 50.5	+41 34 46	UGC 4700	13.6	0.7 x 0.7	12.7	Sb II-III		Mag 9.88 star on NE edge.
08 59 34.9	+45 55 33	UGC 4709	13.9	0.7 x 0.5	12.6	S?	31	Stellar **MCG +8-17-8** E 2′5.
09 00 37.4	+51 12 25	UGC 4717	14.1	1.0 x 0.5	13.2	Sab	40	
09 00 38.1	+50 40 41	UGC 4719	14.2	2.1 x 0.3	13.5	Sc	95	Very faint anonymous galaxy E 0′9; mag 10.61v star S 2′3.
09 04 33.7	+51 36 48	UGC 4749	13.4	0.7 x 0.7	12.5	S		Compact; mag 9.22v star W 6′0.
09 04 33.6	+45 17 21	UGC 4753	15.0	1.5 x 0.2	13.5	Sd	33	A close pair of mag 14.1 stars W 0′9.
09 05 54.6	+47 10 44	UGC 4765	13.6	0.7 x 0.5	12.3	Pec	15	**MCG +8-17-24** S 1′0.
09 07 11.2	+50 42 43	UGC 4778	13.4	1.2 x 1.2	13.7	Sb II		Strong N-S dark lane W edge.
09 07 19.7	+46 13 18	UGC 4784	15.1	1.5 x 0.7	15.0	S	75	Mag 10.64v star WNW 1′8; **MCG +8-17-30** WSW 1′7; **MCG +8-17-29** WSW 2′5.
09 08 42.6	+44 48 37	UGC 4798	14.2	1.0 x 0.8	13.8	SAcd	159	
09 09 30.6	+45 57 07	UGC 4805	13.7	1.2 x 0.9	13.6	S	115	
09 10 08.0	+50 03 26	UGC 4812	13.5	0.8 x 0.6	12.6	Sc	123	**PGC 25848** SE 0′8; mag 9.95v star SE 2′5.
09 11 33.6	+51 15 16	UGC 4824	13.7	2.3 x 0.6	13.8	SAd	98	Mag 9.33v star WNW 5′0.
09 11 39.8	+46 38 20	UGC 4829	13.6	0.7 x 0.6	12.5	S?	21	Mag 10.94v star SW 1′9.
09 13 02.2	+49 38 22	UGC 4844	13.1	1.5 x 1.1	13.5	SABbc:	170	Pair of stars, mag 11.6 and 12.3, S and SE.
09 14 43.3	+40 52 49	UGC 4867	14.1	1.7 x 1.3	14.8	SAB(s)d	85	Mag 12.3 star N edge; mottled envelope; **UGC 4863** SSW 7′1.
09 14 51.7	+48 35 35	UGC 4868	15.6	1.1 x 0.5	14.8	SBm	36	
09 14 55.7	+46 54 09	UGC 4870	13.3	1.0 x 0.5	12.4	S II	142	
09 15 01.6	+40 02 11	UGC 4872	14.8	1.7 x 0.2	13.5	SBb	12	Mag 10.61v star WNW 2′5.
09 15 16.1	+48 39 56	UGC 4874	15.9	1.2 x 1.0	15.9	SBm	36	Mag 15.1 star superimposed E of center; mag 10.62v star SW 2′4.
09 15 55.5	+44 19 42	UGC 4881	14.0	0.9 x 0.8	13.5	Double System		Double system, contact, disrupted.
09 16 25.4	+42 59 26	UGC 4882	14.4	1.1 x 0.4	13.4	Spiral	81	Elongated **MCG +7-19-48** N 0′9.
09 16 51.3	+45 41 59	UGC 4892	13.9	1.1 x 0.6	13.3	Pec	63	Eruptive?
09 17 21.8	+41 54 35	UGC 4904	14.1	1.0 x 0.5	13.2	SB	137	NGC 2798 and NGC 2799 N 5′4.
09 18 16.0	+47 56 52	UGC 4917	14.6	0.6 x 0.6	13.3	Scd:		Mag 10.64v star NNE 3′7.
09 18 14.0	+45 39 05	UGC 4919	13.7	1.4 x 0.9	13.8	SAB(rs)c II	14	
09 18 35.7	+47 52 10	UGC 4922	13.3	3.3 x 1.5	14.9	SAm	55	Almost stellar nucleus.
09 18 54.5	+50 01 13	UGC 4927	13.5	1.0 x 1.0	13.4	S0$^-$?		
09 19 12.5	+48 58 02	UGC 4930	14.6	1.1 x 0.7	14.1	Scd:	114	Mag 12.4 star E edge.
09 19 34.2	+51 06 36	UGC 4932	14.6	1.7 x 0.6	14.4	Sdm:	140	Mag 11.5 star superimposed E of center.
09 21 02.3	+49 36 06	UGC 4958	14.1	1.4 x 0.4	13.3	SB	122	
09 23 03.2	+44 33 13	UGC 4982	13.6	1.2 x 0.3	12.3	Sdm:	4	Located between two mag 11.5 stars.
09 26 39.9	+45 50 50	UGC 5026	13.1	1.1 x 0.8	12.9	S0	15	Bright, elongated core.
09 28 10.4	+44 39 49	UGC 5045	13.5	1.1 x 0.9	13.3	SAB(r)c	141	
09 28 49.8	+51 33 35	UGC 5047	15.5	1.5 x 0.2	14.0	Sdm:	160	Mag 8.81v star E 6′5.
09 32 36.8	+51 52 19	UGC 5076	14.6	1.1 x 0.9	14.4	Im:	150	Dwarf, extremely low surface brightness; mag 3.18v star Theta Ursae Majoris S 11′9.
09 34 12.5	+46 42 26	UGC 5090	14.7	0.7 x 0.7	13.7	SABc		
09 33 50.6	+44 15 19	UGC 5091	14.3	1.6 x 0.5	13.9	Sdm:	75	Mag 12.3 star SE 3′1.
09 38 23.4	+43 30 33	UGC 5133	13.7	1.1 x 0.3	12.4	S0	146	
09 38 36.3	+43 10 34	UGC 5135	14.4	0.8 x 0.6	13.5	S	120	Stellar nucleus; mag 9.53v star NE 1′6.
09 39 34.4	+48 25 15	UGC 5145	14.2	1.7 x 0.4	13.6	S	46	**MCG +8-18-17** SW 8′1.
09 40 27.2	+48 20 13	UGC 5151	13.0	1.0 x 0.7	12.5	I?	6	
09 40 58.1	+47 37 13	UGC 5157	13.6	1.7 x 0.4	13.0	S	18	Mag 9.63v star SSW 4′5; **UGC 5150** SSW 7′9.
09 41 46.6	+46 58 25	UGC 5169	14.6	0.9 x 0.9	14.2	Spiral		Mag 11.2 star NW 4′6.
09 41 52.2	+48 40 12	UGC 5172	14.3	1.7 x 1.7	15.3	SA(s)m		MCG +8-18-23 W 2′3; mag 10.26v star ESE 2′1.
09 43 06.0	+41 05 32	UGC 5187	13.2	1.0 x 0.7	12.7	SB?	126	
09 43 57.7	+41 41 14	UGC 5198	13.9	1.0 x 0.6	13.2	SBa	135	
09 43 57.4	+42 40 22	UGC 5199	14.1	1.5 x 0.6	13.8	Sab	66	**MCG +7-20-38** WSW 1′2; **MCG +7-20-41** S 2′4.
09 46 28.8	+45 45 07	UGC 5225	13.8	1.0 x 1.0	13.6	S?		Bright center, uniform envelope.
09 47 20.2	+46 36 37	UGC 5237	14.0	1.5 x 0.7	13.9	Scd:	155	Close pair of stars, mags 9.48v and 10.14v, SE 2′8.
09 52 04.2	+40 51 30	UGC 5290	13.8	1.0 x 0.6	13.1	S0/a?	0	Mag 10.90v star NW 1′5.
09 52 54.7	+42 50 41	UGC 5295	13.3	2.1 x 0.9	13.8	SAB(s)b	150	Very thin, very faint anonymous galaxy NW 2′8.
09 57 35.9	+45 13 45	UGC 5345	13.7	1.6 x 0.3	12.7	S?	63	= **Hickson 41A**. **UGC 5346** = **Hickson 41B** N 2′0; **Hickson 41C** W 1′5; **Hickson 41D** E 2′6; **MCG +8-18-45** SW 1′4.
09 58 53.2	+47 44 10	UGC 5354	13.2	1.4 x 0.8	13.2	SB?	78	Mag 9.61v star N 1′6.
09 59 09.0	+50 59 14	UGC 5356	14.2	1.1 x 0.5	13.4	SB(rs)bc	58	Mag 9.01v star on E edge.
10 00 01.9	+46 14 16	UGC 5368	13.9	1.0 x 0.9	13.6		108	
10 07 19.6	+47 00 21	UGC 5451	13.5	1.8 x 0.9	13.9	Im	103	Mag 12.3 star N 1′8.
10 08 09.5	+51 50 31	UGC 5460	12.9	2.4 x 2.2	14.6	SB(rs)d	126	Patchy, located between a close pair of mag 12.3 and 10.86v stars.

(Continued from previous page)

39 GALAXIES 39

RA h m s	Dec ° ′ ″	Name	Mag (V)	Dim ′ Maj x min	SB	Type Class	PA	Notes
10 09 14.5	+40 21 02	UGC 5476	13.5	0.8 x 0.8	12.9	SABb		Bright, diffuse nucleus, faint envelope; mag 11.10v star on NE edge.
10 16 59.1	+49 37 37	UGC 5545	13.5	0.9 x 0.7	12.8	S?	130	Mag 11.9 star NNE 1′.7.
10 22 40.8	+46 14 20	UGC 5604	13.0	2.4 x 1.2	14.0	Sc I-II	47	
10 22 46.0	+48 38 15	UGC 5605	14.6	0.9 x 0.4	13.3	S	48	
10 26 51.8	+41 53 19	UGC 5657	14.5	1.5 x 0.2	13.0	S	87	Mag 13.6 star on S edge; mag 9.62v star S 4′.2.

GALAXY CLUSTERS

RA h m s	Dec ° ′ ″	Name	Mag 10th brightest	No. Gal	Diam ′	Notes
09 12 48.0	+47 42 00	A 757	15.6	32	22	Mostly stellar anonymous GX extending E to W through the center.

40 GALAXIES 40

RA h m s	Dec ° ′ ″	Name	Mag (V)	Dim ′ Maj x min	SB	Type Class	PA	Notes
07 49 22.0	+43 45 58	CGCG 206-29	13.8	0.6 x 0.6	12.5	Spiral		Mag 11.8 star N 1′.9.
07 13 38.3	+50 01 49	CGCG 234-105	14.2	0.5 x 0.3	12.0		108	**CGCG 234-101** W 9′.0; mag 8.94v star SSW 6′.7.
07 54 52.0	+45 49 18	CGCG 236-13	14.8	0.5 x 0.5	13.2	Spiral		Mag 10.68v star W 3′.7; mag 10.6 star NE 4′.7.
08 46 20.1	+47 09 11	CGCG 237-18	13.9	0.8 x 0.4	12.6	S0	3	Multi-galaxy system.
08 48 24.8	+47 17 18	CGCG 237-21	14.2	0.5 x 0.3	12.0	S?	108	Mag 9.50v star W 5′.8.
07 10 34.3	+50 07 05	IC 458	13.5	0.9 x 0.4	12.3	S0$^-$:	170	**IC 461** SE 2′.8; mag 10.11v star NW 2′.3.
07 11 04.8	+50 08 12	IC 464	13.8	0.8 x 0.4	12.5	E?	57	NGC 2340 N 2′.5; mag 10.11v star W 6′.8.
07 11 33.7	+50 14 52	IC 465	13.7	0.9 x 0.9	13.3	S?		Mag 9.11v star NE 7′.5.
07 43 36.4	+49 40 04	IC 471	13.3	0.6 x 0.6	12.2	E		Mag 10.59v star E 5′.6.
07 43 50.1	+49 36 51	IC 472	13.4	1.6 x 1.0	13.8	SAB(rs)b	167	Mag 9.77v star SE 4′.7.
08 13 59.0	+45 44 44	IC 2233	12.6	4.7 x 0.5	13.4	SB(s)d: sp	172	Mag 10.13v star on E edge.
07 22 11.7	+40 28 08	MCG +7-15-15	14.4	0.9 x 0.4	13.1	SBb	123	
07 41 12.6	+42 44 57	MCG +7-16-10	14.1	0.8 x 0.5	12.9		171	
07 59 46.3	+42 15 24	MCG +7-17-5	15.4	0.6 x 0.6	14.2	SBbc		Mag 8.91v star NW 4′.9.
08 05 05.5	+40 10 28	MCG +7-17-12	14.7	0.8 x 0.7	14.0		6	Mag 10.7 star WNW 4′.7.
08 13 48.8	+41 38 56	MCG +7-17-23	15.4	0.5 x 0.3	13.2	Spiral	129	Mag 9.16v star S 2′.4.
08 14 30.2	+43 08 34	MCG +7-17-24	14.6	0.7 x 0.5	13.5	E/S0	84	Mag 11.8 star E 1′.9.
08 15 12.8	+41 12 52	MCG +7-17-26	14.8	0.8 x 0.5	13.7		69	Mag 8.01v star SSW 5′.0.
08 25 09.4	+42 17 56	MCG +7-17-30	15.6	0.8 x 0.6	14.7	Spiral	102	Mag 10.78v star NE 1′.4.
08 26 00.7	+40 58 50	MCG +7-18-1	13.9	0.9 x 0.7	13.2		165	Mag 10.5 star NW 2′.4.
08 27 26.9	+41 29 36	MCG +7-18-2	15.4	0.5 x 0.4	13.5	Spiral	21	Mag 10.5 star WSW 2′.7.
08 28 25.7	+40 37 46	MCG +7-18-5	14.7	0.7 x 0.4	13.1		117	Mag 12.4 star N 1′.4.
08 30 32.8	+42 59 26	MCG +7-18-11	15.1	0.7 x 0.3	13.3	Spiral	165	Mag 12.1 star NE 3′.9.
08 32 20.6	+41 21 28	MCG +7-18-16	14.5	0.7 x 0.7	13.6			Mag 12.6 star W 1′.1.
08 32 27.1	+41 59 30	MCG +7-18-17	15.4	1.0 x 0.5	14.5	Spiral	147	N-S line of three stars, mags 11.7, 12.5 and 13.5, NE 1′.4.
08 32 29.4	+44 11 07	MCG +7-18-18	15.2	0.9 x 0.3	13.7	Spiral	126	Mag 12.5 star ESE 2′.6.
08 33 12.6	+41 05 39	MCG +7-18-18A	14.2	1.1 x 0.8	13.9		162	Mag 13.2 star WNW 2′.2.
08 33 57.5	+41 24 19	MCG +7-18-24	15.1	0.5 x 0.3	12.9	Sb	150	**UGC 4471** WNW 3′.0.
08 34 46.5	+41 12 44	MCG +7-18-25	15.2	0.7 x 0.5	13.9	Sc	15	Mag 12.7 star N 2′.5.
08 34 55.2	+41 24 05	MCG +7-18-26	14.3	0.6 x 0.4	12.8	E/S0	132	Mag 9.93v star NW 2′.9.
08 36 42.8	+41 28 18	MCG +7-18-28	15.3	0.5 x 0.5	13.6			Mag 12.4 star N 1′.8.
08 37 06.7	+41 27 23	MCG +7-18-29	14.3	0.6 x 0.4	12.6	Spiral	153	Mag 13.5 star W 1′.9.
08 41 41.2	+40 39 24	MCG +7-18-36	13.9	0.6 x 0.5	12.7	E	18	Pair with MCG +7-18-37 S on edge.
08 41 40.9	+40 39 01	MCG +7-18-37	14.2	0.3 x 0.3	11.4	Sb		Pair with MCG +7-18-36 N edge.
08 42 23.7	+41 54 30	MCG +7-18-38	14.8	0.7 x 0.5	13.5	Sb	81	Forms a triangle with a mag 11.6 star NW 2′.4 and a mag 11.9 star SW 2′.0.
08 42 41.9	+42 48 45	MCG +7-18-39	15.2	0.8 x 0.4	13.8	Spiral	120	Mag 9.43v star NE 2′.9.
08 45 05.7	+40 33 32	MCG +7-18-44	14.8	0.9 x 0.3	13.2	Spiral	150	Mag 13.4 star SW 1′.9.
08 45 41.7	+41 16 56	MCG +7-18-45	14.1	1.0 x 0.6	13.4		144	Mag 10.7 star E 1′.7.
08 48 59.8	+41 00 08	MCG +7-18-50	14.1	0.8 x 0.7	13.3		12	
08 49 31.0	+42 16 25	MCG +7-18-52	15.1	0.8 x 0.2	13.0		126	Mag 11.55v star SW 0′.6.
08 49 45.5	+40 14 45	MCG +7-18-54	15.3	0.8 x 0.4	13.9	Spiral	30	Mag 10.9 star S 2′.0.
08 50 11.4	+40 15 18	MCG +7-18-55	14.0	0.8 x 0.6	13.1		81	Mag 13.0 star SW 1′.6.
08 50 37.9	+40 15 01	MCG +7-18-57	14.7	0.7 x 0.4	13.2	Sc	39	Pair of stars, mags 13.3 and 10.68v, E 0′.9.
07 07 59.5	+48 39 55	MCG +8-13-61	13.5	1.1 x 0.9	13.3	S?	51	Mag 9.17v star NW 3′.5.
07 09 03.7	+48 34 22	MCG +8-13-72	15.1	0.5 x 0.4	13.2	Sc	137	NGC 2329 N 2′.5.
07 09 13.7	+46 05 54	MCG +8-13-74	14.2	0.8 x 0.8	13.8	E/S0		Mag 12.0 star S 1′.8.
07 10 58.2	+49 01 18	MCG +8-13-91	14.4	1.1 x 0.3	13.1	Sbc	99	**MCG +8-13-99** NE 6′.2.
07 10 47.9	+48 13 26	MCG +8-13-93	15.4	0.9 x 0.2	13.4	Sc	21	Mag 12.3 star W 1′.7.
07 12 04.8	+50 35 32	MCG +8-13-102	14.0	0.7 x 0.4	12.5		42	
07 12 51.3	+49 00 15	MCG +8-13-105	14.3	1.3 x 0.4	13.4	Sc	153	Mag 10.11v star E 4′.8.
07 12 56.3	+49 45 50	MCG +8-13-106	14.0	0.7 x 0.5	12.8		171	Close pair of mag 13-14 stars W 1′.2.
07 13 54.1	+50 23 53	MCG +8-13-109	13.9	0.6 x 0.5	12.4		45	
07 20 33.7	+49 05 17	MCG +8-14-6	15.0	0.8 x 0.5	13.8	Sd	105	
07 39 34.8	+49 21 19	MCG +8-14-30	14.2	0.8 x 0.5	13.1	Sab	165	Mag 12.6 star WSW 1′.4.
07 40 14.7	+49 16 56	MCG +8-14-32	14.6	0.8 x 0.4	13.2	Sb	12	**CGCG 235-25** N 1′.7.
07 42 32.3	+49 11 26	MCG +8-14-34	14.7	0.9 x 0.4	13.4		0	
07 45 33.3	+45 46 19	MCG +8-14-38	14.5	0.7 x 0.7	13.5			
07 51 13.4	+50 11 34	MCG +8-15-5	14.3	0.7 x 0.3	12.5	Sab	168	UGC 4051 SE 1′.1.
07 56 44.2	+45 52 36	MCG +8-15-24	14.6	0.6 x 0.5	13.2		174	MCG +8-15-26 NE 0′.9; mag 11.4 star N 1′.3.
07 56 47.6	+45 53 12	MCG +8-15-26	14.3	0.6 x 0.4	12.6	Sc	6	Mag 11.4 star NW 1′.1; MCG +8-15-24 SW 0′.9.
07 58 46.3	+50 27 34	MCG +8-15-32	14.8	0.7 x 0.3	13.0		3	Mag 11.1 star N 3′.2.
08 00 15.8	+45 13 57	MCG +8-15-35	14.3	0.8 x 0.3	12.6	Spiral	48	Located between a pair of N-S mag 13 stars.
08 15 17.2	+46 04 27	MCG +8-15-56	14.9	0.8 x 0.8	14.3	Spiral		Elongated NE-SW core; mag 9.43v star NNE 2′.3.
08 30 02.0	+49 41 09	MCG +8-16-8	15.0	0.8 x 0.5	13.8	Sc	150	Mag 11.1 star NW 2′.0.
08 34 22.7	+48 05 16	MCG +8-16-13	14.2	1.0 x 0.7	13.7		57	Star or small galaxy W edge; almost stellar anonymous galaxy S 0′.9; another anonymous galaxy SSW 2′.5.
08 36 11.8	+50 25 05	MCG +8-16-16	14.2	1.5 x 0.4	13.5	S	57	Mag 15.2 star SW edge.
08 36 45.9	+48 41 57	MCG +8-16-19	15.4	0.8 x 0.5	14.3	Sb	150	Mag 11.00v star NE 3′.6.
08 37 49.0	+47 21 01	MCG +8-16-22	14.5	0.9 x 0.3	13.0	Spiral	75	
08 39 09.3	+45 07 46	MCG +8-16-23	14.7	0.8 x 0.8	14.0	Spiral		Mag 11.2 star E 10′.2.
07 45 51.4	+51 47 24	MCG +9-13-47	15.6	0.9 x 0.2	13.5	SBc	174	
07 47 31.9	+51 11 30	MCG +9-13-56	14.0	0.8 x 0.3	12.3	Sb	123	

(Continued from previous page)

GALAXIES

RA h m s	Dec ° ′ ″	Name	Mag (V)	Dim ′ Maj x min	SB	Type Class	PA	Notes
07 08 11.1	+50 40 52	NGC 2326	13.1	1.9 x 1.8	14.3	SB(rs)b	120	Mag 10.14v star NE 5'.1; NGC 2326A SE 4'.8.
07 08 34.3	+50 37 51	NGC 2326A	14.7	1.0 x 0.6	13.9	SA(s)m:	15	Stellar nucleus; NGC 2326 NW 4'.8.
07 09 08.2	+48 36 52	NGC 2329	12.5	1.3 x 1.1	12.7	S0$^-$:	175	Small, faint anonymous galaxy 2'.0 SW.
07 09 28.5	+50 09 06	NGC 2330	14.8	0.4 x 0.3	12.6	E	30	NGC 2332 N 2'.1.
07 09 34.4	+50 10 56	NGC 2332	12.8	1.5 x 0.8	12.9	S0:	60	NGC 2330 S 2'.1.
07 10 14.2	+44 27 19	NGC 2337	12.4	2.1 x 1.4	13.4	IBm	120	Mag 9.42v star SSW 6'.4; **UGC 3698** SW 10'.9.
07 11 10.8	+50 10 25	NGC 2340	11.7	1.8 x 1.2	12.5	E	84	IC 464 S 2'.5; mag 10.11v star WSW 8'.0.
07 12 28.8	+47 09 59	NGC 2344	12.0	1.5 x 1.2	12.5	SA(rs)c: II	165	
08 01 53.4	+50 44 11	NGC 2500	11.6	2.9 x 2.6	13.7	SB(rs)d III-IV	48	Many bright knots.
08 07 20.5	+51 07 53	NGC 2518	13.0	1.2 x 1.0	13.1	S0$^-$:	35	
08 13 14.8	+45 59 23	NGC 2537	11.7	1.7 x 1.5	12.6	SB(s)m pec III-IV	162	**Bear Paw Galaxy**. Mag 8.45v star W 6'.5; NGC 2537A E 4'.5.
08 13 41.2	+45 59 35	NGC 2537A	15.4	0.6 x 0.6	14.1	SB(s)c		NGC 2537 W 4'.5; **MCG +8-15-53** E 6'.4.
08 14 40.3	+49 03 42	NGC 2541	11.8	6.3 x 3.2	14.9	SA(s)cd III-IV	165	Many bright knots and/or superimposed stars.
08 19 19.8	+50 00 16	NGC 2552	12.1	3.5 x 2.0	14.1	SA(s)m? IV-V	57	Numerous superimposed stars and/or knots on faint envelope; mag 10.55v star NE 3'.3.
08 43 38.2	+50 12 19	NGC 2639	11.7	1.8 x 1.1	12.2	(R)SA(r)a:?	140	
08 51 35.7	+47 33 28	NGC 2676	13.1	1.2 x 1.2	13.4	S0:		lies 1'.9 SW of mag 10.57v star. Small anonymous galaxy SE 3'.0.
07 08 13.9	+46 06 55	UGC 3683	12.8	2.0 x 1.3	13.7	S0	50	Mag 13.6 star W edge.
07 08 23.6	+51 14 01	UGC 3684	14.0	1.4 x 0.9	14.1	(R')SB(s)b	176	Bright SE-NW bar, faint envelope.
07 09 23.2	+48 38 05	UGC 3696	12.8	1.2 x 0.6	12.4	E?	77	NGC 2329 SW 2'.8.
07 09 54.6	+47 54 34	UGC 3706	13.8	1.0 x 0.5	12.9	Double System	156	Double system, contact, disrupted; **MCG +8-13-81** W 2'.3.
07 11 27.2	+48 14 21	UGC 3724	13.5	1.5 x 0.8	13.5	SBb	55	Brighter core offset NE; mag 13.5 star N edge.
07 11 41.8	+49 51 41	UGC 3725	13.0	1.4 x 0.8	13.0	S0$^-$:	140	Mag 15.4 star N of nucleus; mag 13 star on W edge.
07 15 04.9	+50 32 06	UGC 3758	13.3	0.7 x 0.7	12.4	Compact		**MCG +8-13-112** NE 2'.4.
07 17 58.4	+40 59 03	UGC 3781	14.0	1.5 x 0.8	14.0	Double System	51	Double system, bridge + plume; mag 7.44v star SW 4'.1.
07 19 18.8	+51 17 34	UGC 3792	12.8	1.7 x 1.2	13.4	SA0/a	65	
07 22 18.6	+49 17 28	UGC 3812	13.4	0.7 x 0.7	12.6	E		Very slender **UGC 3814** SE 1'.4.
07 22 44.5	+45 06 24	UGC 3817	14.6	1.8 x 0.9	15.0	Im:	162	Dwarf, very low surface brightness.
07 23 33.1	+41 26 04	UGC 3825	14.1	1.3 x 1.2	14.4	SAB(s)bc	120	Mag 7.85v star S 4'.3.
07 24 59.8	+49 29 31	UGC 3831	12.8	1.1 x 0.9	12.7	SB(s)bc	85	Mag 9.48v star E 6'.6.
07 26 35.5	+43 17 43	UGC 3844	13.0	1.7 x 1.1	13.7	E?	5	Mag 6.66v star SE 4'.7.
07 26 42.8	+47 05 39	UGC 3845	12.8	1.5 x 1.0	13.1	SB(s)bc	176	
07 28 17.4	+40 46 09	UGC 3860	14.5	1.7 x 1.1	15.1	Im V	30	Dwarf irregular, very low surface brightness.
07 28 54.3	+49 08 16	UGC 3863	13.1	1.2 x 0.6	12.6	(R')SBa	76	Diffuse core; with strong dark patches E and W.
07 29 16.6	+42 16 44	UGC 3871	14.0	0.9 x 0.6	13.2	SAB(s)c	45	
07 37 37.2	+41 56 47	UGC 3933	13.5	1.2 x 0.8	13.3	SAB(s)bc II	0	Mag 9.56v star SE 2'.0.
07 39 45.1	+48 44 28	UGC 3949	13.8	1.1 x 1.1	13.9	SB(s)d II-III		Stellar nucleus.
07 41 32.3	+51 41 18	UGC 3963	15.9	1.5 x 0.3	14.9	Sdm:	54	Mag 12.0 star NW 2'.1.
07 41 26.1	+40 06 36	UGC 3966	13.5	2.0 x 1.8	14.7	Im V	141	Dwarf irregular, very low surface brightness.
07 42 33.0	+49 48 32	UGC 3973	13.3	1.3 x 1.3	13.7	SBb		Mag 11.0 star E 2'.5.
07 42 40.8	+40 55 43	UGC 3976	14.5	0.7 x 0.5	13.2	S	36	Mag 12.5 star W 0'.5.
07 44 38.6	+40 21 58	UGC 3997	13.8	1.0 x 0.8	13.4	Im?	30	Mag 12.1 star N 1'.6.
07 45 07.3	+46 04 21	UGC 4000	13.9	1.5 x 0.5	13.4	SB?b	40	
07 46 25.1	+48 19 58	UGC 4007	13.8	1.0 x 0.4	12.6	SBa	129	Pair of stars, mags 11.6 and 12.8, N 0'.5.
07 46 38.0	+44 47 24	UGC 4008	13.3	1.0 x 0.5	12.4	S0/a	171	Very slender anonymous galaxy N 0'.9.
07 47 02.0	+41 32 07	UGC 4018	13.8	0.8 x 0.7	13.2	E:	24	Anonymous galaxy on S edge.
07 51 17.7	+50 10 43	UGC 4051	13.6	0.9 x 0.6	12.9	E?	24	MCG +8-15-5 NW 1'.1.
07 51 20.1	+50 14 08	UGC 4052	13.2	1.3 x 0.8	13.1	Double System	90	Double system, contact; MCG +8-15-5 SSW 2'.7.
07 51 35.9	+42 52 45	UGC 4056	14.1	1.0 x 0.6	13.4	SAB(s)c	27	
07 52 57.2	+40 10 28	UGC 4068	16.0	1.5 x 0.2	14.5	Scd:	8	Mag 9.03v star NW 1'.1.
07 53 26.5	+50 13 42	UGC 4070	13.8	0.9 x 0.7	13.2	SAB(r)a:	135	**MCG +8-15-17** E 2'.5; **MCG +8-15-15** SW 2'.5; mag 10.83v star NE 1'.6.
07 53 34.2	+51 46 47	UGC 4071	14.1	0.9 x 0.7	13.4		74	Mag 10.80v star W 1'.8.
07 54 50.6	+50 02 15	UGC 4082	13.2	1.0 x 1.0	13.3	E:		
07 55 06.1	+43 29 11	UGC 4087	14.6	1.1 x 0.7	14.2	Sd	108	Mag 11.1 star superimposed E end; mag 9.17v star ESE 1'.7.
07 57 01.8	+49 34 03	UGC 4107	13.0	1.5 x 1.4	13.7	SA(rs)c	120	Mag 8.23v star WNW 2'.5.
07 59 54.4	+47 24 46	UGC 4136	14.1	1.4 x 0.3	13.0	Sa	142	
08 00 23.5	+42 11 18	UGC 4148	14.6	2.3 x 0.3	14.0	Sd:	10	
08 02 42.9	+40 40 41	UGC 4176	14.3	1.7 x 0.5	14.0	SBd	112	
08 03 24.0	+41 54 54	UGC 4188	13.5	1.5 x 0.7	13.4	S0	115	Mag 12.7 star and very faint, very small anonymous galaxy E edge.
08 04 31.2	+40 12 20	UGC 4200	13.7	0.9 x 0.6	12.9	(R')SB(s)bc I	26	Located between two mag 10.7 stars; **CGCG 207-25** W 2'.5.
08 05 26.6	+46 42 25	UGC 4209	13.7	1.1 x 0.6	13.1	S0$^-$:	80	Small, faint anonymous galaxy WNW 1'.2.
08 07 21.4	+40 23 53	UGC 4226	14.0	1.9 x 1.4	14.9	SA(r)cd	35	Close trio of stars, mags 12-14, S edge.
08 10 05.6	+46 11 31	UGC 4250	14.2	1.4 x 1.2	14.6	Scd:	132	Large core with faint arms.
08 10 00.0	+40 06 11	UGC 4252	14.3	1.9 x 0.4	13.9	Scd:	17	Uniform surface brightness; mag 12.5 star N 2'.6.
08 10 47.7	+46 54 43	UGC 4258	14.8	1.5 x 0.5	14.3	Sd	122	
08 11 07.6	+46 27 41	UGC 4260	13.5	1.6 x 1.3	14.2	Im: IV-V	48	Brightest portion NE edge.
08 18 13.8	+49 15 07	UGC 4309	14.5	1.1 x 0.8	14.2	S	51	
08 26 05.0	+45 58 09	UGC 4393	12.4	2.3 x 1.6	13.7	SB?	45	Elongated core; mag 9.52v star N 4'.8.
08 28 28.3	+41 51 17	UGC 4426	14.6	1.8 x 0.9	14.9	Im: V	0	Dwarf, extremely low surface brightness, uniform surface brightness; mag 15.1 star on E edge.
08 31 25.9	+40 57 23	UGC 4449	14.8	1.0 x 0.8	14.4	Pec		
08 33 17.8	+41 15 36	UGC 4465	13.9	1.1 x 0.7	13.4	Sa	160	Very faint, small anonymous galaxy E 0'.9.
08 33 30.7	+41 31 30	UGC 4468	13.4	1.5 x 0.9	13.6	S0	165	Two stars superimposed E edge; mag 10.77v star N 2'.0.
08 37 26.6	+40 02 04	UGC 4498	13.8	1.0 x 0.6	13.1	SBa	13	Mag 11.0 star S 1'.3.
08 37 41.2	+51 39 14	UGC 4499	12.9	2.8 x 2.2	14.7	SABdm	140	Bright "knot" S of core is **Mkn 94**.
08 38 44.5	+43 32 49	UGC 4507	14.1	1.7 x 0.4	13.5	Sb	43	
08 39 00.9	+40 20 43	UGC 4513	14.1	0.9 x 0.9	13.7	Compact		
08 41 32.1	+51 14 44	UGC 4525	14.1	1.2 x 0.7	13.7	SB(s)b	88	Mag 13.2 star 0'.9 W.
08 43 21.6	+45 44 09	UGC 4543	13.7	2.1 x 1.4	14.7	SAdm	4	Patchy; mag 10.21v star NE 1'.8.
08 43 51.6	+51 59 34	UGC 4546	14.1	1.3 x 0.3	12.9	Sa	25	
08 44 06.0	+49 47 36	UGC 4551	12.5	1.7 x 0.6	12.4	S0?	113	Located between a pair of mag 12 stars N and S .
08 44 15.3	+41 43 00	UGC 4556	13.5	0.7 x 0.3	11.7	I?	21	
08 44 55.2	+47 44 44	UGC 4562	14.7	1.2 x 0.7	14.4	SB(s)c	127	
08 46 14.5	+41 34 48	UGC 4578	14.1	1.2 x 0.7	13.7	Sb II-III	65	
08 46 33.6	+48 25 44	UGC 4580	13.9	0.9 x 0.5	12.9	Sbc	7	
08 47 22.7	+49 33 30	UGC 4587	12.8	1.7 x 1.0	13.2	S0?	8	Large, diffuse core, smooth envelope; mag 11.3 star SW 2'.8.
08 48 07.0	+41 51 29	UGC 4598	15.2	0.8 x 0.8	14.5	Sc		Mag 12.3 star NE 3'.0.
08 49 01.2	+41 11 54	UGC 4609	14.5	0.6 x 0.6	13.2	Pec		Compact; mag 12.4 star E 1'.4; faint, almost stellar anonymous galaxy SE 2'.7.
08 50 20.4	+41 17 18	UGC 4622	14.2	1.0 x 0.8	13.8	SAd	147	Stellar nucleus, smooth envelope.
08 51 40.8	+51 07 07	UGC 4628	14.6	1.7 x 0.3	13.7	Scd:	74	
08 51 54.6	+40 49 58	UGC 4635	14.3	1.2 x 0.8	14.1	Pec	36	Very faint plume extends N and then E from bright core.

(Continued from previous page)

40 GALAXY CLUSTERS 40

RA h m s	Dec ° ′ ″	Name	Mag 10th brightest	No. Gal	Diam ′	Notes
07 09 12.0	+48 37 00	A 569	13.8	36	26	Many faint UGC, MCG and CGCG GX, along with anonymous GX.
07 28 12.0	+41 57 00	A 582	16.4	34	11	All members faint, stellar and anonymous.

41 GALAXIES 41

RA h m s	Dec ° ′ ″	Name	Mag (V)	Dim ′ Maj x min	SB	Type Class	PA	Notes
06 19 19.2	+42 37 07	CGCG 203-6	14.0	0.4 x 0.3	11.5	Sbc	150	Stellar nucleus; mag 11.7 star E 1:0.
06 13 11.0	+46 29 42	CGCG 233-2	14.7	0.4 x 0.3	12.3	S	3	Mag 13 star 0:8 S; mag 12.5 star NW 1:7.
06 16 04.2	+47 09 26	CGCG 233-5	15.2	0.3 x 0.3	12.4			Stellar; mag 10.71v star SW 2:6.
06 20 38.3	+50 06 28	CGCG 233-11	14.8	0.5 x 0.4	12.9		147	Mag 11.1 star SSE 4:3; mag 10.73v star NW 5:4.
06 56 43.5	+48 54 02	CGCG 234-31	14.0	0.8 x 0.3	12.3	Sab	129	SE end sit in a triangle of mag 14 stars; mag 12.1 star NW 2:5.
06 59 02.0	+45 26 37	CGCG 234-39	14.4	0.6 x 0.5	13.2	E/S0	141	Mag 14 star SE end.
07 13 38.3	+50 01 49	CGCG 234-105	14.2	0.5 x 0.3	12.0		108	**CGCG 234-101** W 9:0; mag 8.94v star SSW 6:7.
05 32 35.9	+51 35 19	CGCG 258-9	16.1	0.8 x 0.5	14.9			Very faint stellar nucleus; mag 10.7 star N 1:6.
05 33 27.3	+51 45 00	CGCG 258-10	15.2	0.8 x 0.5	14.1		42	Located between two mag 10.5 stars N-S.
06 04 00.2	+51 05 34	CGCG 259-8	14.7	0.4 x 0.4	12.6			Mag 10.06v star N 2:4.
06 48 36.0	+51 50 23	CGCG 260-37	14.3	1.2 x 0.6	13.8		150	Bright center with faint envelope; faint star SE edge of center; mag 12.3 star NW 2:8.
07 10 34.3	+50 07 05	IC 458	13.5	0.9 x 0.4	12.3	S0$^-$:	170	**IC 461** SE 2:8; mag 10.11v star NW 2:3.
07 11 04.8	+50 08 12	IC 464	13.8	0.8 x 0.4	12.5	E?	57	NGC 2340 N 2:5; mag 10.11v star W 6:8.
07 11 33.7	+50 14 52	IC 465	13.7	0.9 x 0.9	13.3	S?		Mag 9.11v star NE 7:5.
05 58 36.1	+40 31 13	MCG +7-13-3	14.8	0.7 x 0.5	13.5	S		
06 47 03.0	+40 01 18	MCG +7-14-15	14.3	0.4 x 0.3	11.8		60	Mag 10.4 star E 1:9.
06 49 34.0	+44 25 50	MCG +7-14-16	14.4	0.9 x 0.5	13.4	S?	6	
06 53 11.3	+44 04 18	MCG +7-14-18	14.2	0.9 x 0.5	13.2		153	Mag 13.7 star superimposed S edge.
06 55 10.2	+41 00 07	MCG +7-14-19	13.9	0.7 x 0.7	13.1	E		Several faint stars on edges; UGC 3593 S 2:7.
06 55 50.8	+40 41 35	MCG +7-15-2	14.1	0.6 x 0.5	12.7	S?	15	Close pair of mag 11 stars ESE 2:3.
07 00 07.3	+40 23 15	MCG +7-15-4	14.8	0.5 x 0.4	12.9	Sbc	33	Mag 11.3 star WNW 2:4.
07 07 13.0	+44 48 59	MCG +7-15-6	14.6	1.0 x 0.4	13.4	Sb	141	UGC 3679 SE 3:0; **UGC 3673** NW 3:0.
05 52 12.5	+46 50 22	MCG +8-11-10	15.2	0.9 x 0.3	13.6	Sb		
05 56 17.1	+48 32 37	MCG +8-11-12	13.9	0.9 x 0.9	13.5	S?		
06 16 01.2	+47 05 55	MCG +8-12-9	14.8	0.3 x 0.3	12.0			
06 21 16.0	+45 12 45	MCG +8-12-18	15.0	0.8 x 0.3	13.3	Sc	153	Mag 13.3 star E 1:2.
06 24 55.4	+49 30 29	MCG +8-12-24	14.0	0.8 x 0.5	12.9	S?	21	Mag 13.7 star superimposed on E edge.
06 41 05.9	+50 09 18	MCG +8-12-33	14.5	0.9 x 0.4	13.2	S?	69	Mag 9.37v star SW 2:9; **MCG +8-12-34** NE 7:8.
06 45 46.6	+46 02 53	MCG +8-13-7	14.5	1.0 x 0.5	13.6	Sb	144	Mag 11.6 star SW 1:1.
06 45 52.0	+45 27 21	MCG +8-13-8	14.2	1.2 x 0.8	14.0		162	
06 49 47.9	+48 30 11	MCG +8-13-13	14.7	0.8 x 0.6	13.7	Sb	153	
07 02 52.3	+50 35 12	MCG +8-13-46	15.2	0.5 x 0.5	13.5			Mag 12.1 star on NE edge; pair of mag 12 stars NW 0:7.
07 07 32.7	+48 54 02	MCG +8-13-59	14.9	0.7 x 0.4	13.3	Sb	129	
07 07 59.5	+48 39 55	MCG +8-13-61	13.5	1.1 x 0.9	13.3	S?	51	Mag 9.17v star NW 3:5.
07 09 03.7	+48 34 22	MCG +8-13-72	15.1	0.5 x 0.4	13.2	Sc	137	NGC 2329 N 2:5.
07 09 13.7	+46 05 54	MCG +8-13-74	14.2	0.8 x 0.8	13.8	E/S0		Mag 12.0 star S 1:8.
07 10 58.2	+49 01 18	MCG +8-13-91	14.4	1.1 x 0.3	13.1	Sbc	99	**MCG +8-13-99** NE 6:2.
07 10 47.9	+48 13 26	MCG +8-13-93	15.4	0.9 x 0.2	13.4	Sc	21	Mag 12.3 star W 1:7.
07 12 04.8	+50 35 32	MCG +8-13-102	14.0	0.7 x 0.4	12.5		42	
07 12 51.3	+49 00 15	MCG +8-13-105	14.3	1.3 x 0.4	13.4	Sc	153	Mag 10.11v star E 4:8.
07 12 56.3	+49 45 50	MCG +8-13-106	14.0	0.7 x 0.5	12.8		171	Close pair of mag 13-14 stars W 1:2.
07 13 54.1	+50 23 53	MCG +8-13-109	13.9	0.6 x 0.5	12.4		45	
06 12 04.0	+51 51 48	MCG +9-11-2	15.0	0.9 x 0.6	14.2	Sc	99	Pair of mag 11.2 star NE 2:2.
06 22 34.8	+51 54 31	NGC 2208	12.8	1.7 x 1.0	13.2	S0:	110	Mag 9.86v star NE 5:4.
06 56 17.5	+45 29 34	NGC 2303	12.6	1.5 x 1.5	13.4	E		
06 58 37.7	+45 12 38	NGC 2308	13.2	1.8 x 1.2	13.9	Sab	170	Mag 4.90v star 16 Lyncis SW 12:7.
07 02 33.0	+50 35 24	NGC 2315	13.6	1.3 x 0.4	12.7	S0/a	116	Mag 8.18v star N 3:7; MCG +8-13-46 E 3:0.
07 05 42.3	+50 34 48	NGC 2320	11.9	1.4 x 0.8	12.0	E	140	Lies 1:5 SW of mag 9.34v star.
07 05 59.0	+50 45 19	NGC 2321	13.6	1.4 x 1.1	13.9	SBa	123	Almost stellar nucleus, faint, smooth envelope.
07 06 00.5	+50 30 37	NGC 2322	13.8	1.1 x 0.4	12.7	SBa:	136	NGC 2320 and mag 9.34v star NNW 5:1.
07 08 11.1	+50 40 52	NGC 2326	13.1	1.9 x 1.8	14.3	SB(rs)b	120	Mag 10.14v star NE 5:1; NGC 2326A SE 4:8.
07 08 34.3	+50 37 51	NGC 2326A	14.7	1.0 x 0.6	13.9	SA(s)m:	15	Stellar nucleus; NGC 2326 NW 4:8.
07 09 08.2	+48 36 52	NGC 2329	12.5	1.3 x 1.1	12.7	S0$^-$:	175	Small, faint anonymous galaxy 2:0 SW.
07 09 28.5	+50 09 06	NGC 2330	14.8	0.4 x 0.3	12.6	E	30	NGC 2332 N 2:1.
07 09 34.4	+50 10 56	NGC 2332	12.8	1.5 x 0.8	12.9	S0:	60	NGC 2330 S 2:1.
07 10 14.2	+44 27 19	NGC 2337	12.4	2.1 x 1.4	13.4	IBm	120	Mag 9.42v star SSW 6:4; **UGC 3698** SW 10:9.
07 11 10.8	+50 10 25	NGC 2340	11.7	1.8 x 1.2	12.5	E	84	IC 464 S 2:5; mag 10.11v star WSW 8:0.
07 12 28.8	+47 09 59	NGC 2344	12.0	1.5 x 1.2	12.5	SA(rs)c: II	165	
05 35 19.0	+40 53 12	UGC 3325	14.6	1.1 x 0.8	14.3	Sbc	30	Many superimposed stars; mag 13.1 star S edge.
05 44 10.0	+51 11 58	UGC 3346	14.2	1.3 x 0.5	13.6	Sdm:	135	
05 47 15.0	+50 52 12	UGC 3355	13.5	1.1 x 1.1	13.6	S0:		Several stars superimposed; mag 10.6 star ENE 1:9.
05 52 12.7	+41 47 10	UGC 3369	16.5	1.5 x 0.6	16.3	Sdm:	165	Mag 10.9 star N 1:8.
05 54 53.4	+46 26 23	UGC 3374	14.1	2.6 x 2.3	15.9	SB?	90	Bright N-S core; superimposed stars W side of core.
05 55 25.5	+51 54 39	UGC 3375	13.2	2.1 x 1.0	13.9	SA(rs)c	45	Mag 10.78v star SW 1:9.
06 09 08.3	+42 05 07	UGC 3407	12.8	1.2 x 0.8	12.6	Sa	18	Bright core, uniform envelope; mag 10.69v star on S edge.
06 12 26.1	+44 26 17	UGC 3418	13.5	1.0 x 0.8	13.1	SB(rs)b	138	Short, bright N-S bar.
06 27 18.0	+49 36 49	UGC 3467	13.4	1.4 x 1.0	13.6	Sab	20	Bright core, uniform envelope.
06 31 22.0	+50 05 35	UGC 3477	14.8	1.9 x 0.4	14.4	Sd	0	Mag 12.1 star W 2:1.
06 32 34.9	+40 12 26	UGC 3481	14.3	0.8 x 0.2	12.2	I?	96	Very close pair of mag 14.1 stars E 0:8.
06 33 39.6	+40 41 15	UGC 3487	14.2	1.0 x 0.7	13.7	Sd	110	**MCG +7-14-5** S 1:2.
06 35 26.7	+48 50 36	UGC 3493	13.6	1.5 x 0.7	13.5	S?	97	
06 40 32.8	+50 06 16	UGC 3506	13.4	1.6 x 1.2	13.9	S0$^-$:	108	Mag 9.37v star NE 3:1.
06 41 32.6	+40 10 06	UGC 3510	14.1	1.1 x 0.8	13.8	S?	135	Very slender **MCG +7-14-9** NW 1:1.
06 42 42.7	+42 24 57	UGC 3512	16.1	0.9 x 0.4	14.8	Im	111	Dwarf, very low surface brightness.
06 45 03.6	+40 24 49	UGC 3525	13.4	1.1 x 0.8	13.3	E?	70	
06 46 12.4	+43 47 33	UGC 3532	14.2	1.4 x 0.5	13.7	SB(s)b	94	UGC 3535 NE 2:7.
06 46 23.5	+43 49 20	UGC 3535	13.6	1.0 x 0.8	13.2	S?	54	Mag 11.9 star N 1:5.
06 47 32.8	+47 39 17	UGC 3538	13.8	1.5 x 0.7	13.7	SB(s)a	0	Mag 11.9 star W 1:1.
06 50 26.0	+43 03 09	UGC 3554	13.8	1.1 x 0.6	13.2	SBb?	148	Mag 12.8 star W 1:2.
06 51 20.3	+49 53 06	UGC 3561	13.8	0.7 x 0.4	12.3	S?	77	
06 51 47.5	+48 29 46	UGC 3567	13.2	1.0 x 0.8	12.8	S0	103	Mag 11.3 star WSW 2:3.

(Continued from previous page)

41 GALAXIES 41

RA h m s	Dec ° ′ ″	Name	Mag (V)	Dim ′ Maj x min	SB	Type Class	PA	Notes
06 51 59.0	+50 23 52	UGC 3568	14.8	1.5 x 0.2	13.3	Sbc	15	
06 53 07.1	+50 02 03	UGC 3576	13.5	1.5 x 0.7	13.5	SB(s)b	128	
06 54 32.8	+45 42 12	UGC 3588	13.6	1.7 x 0.5	13.3	S0	30	Faint star E edge of nucleus; mag 6.32v star NE 10'.4.
06 55 13.8	+40 20 17	UGC 3592	14.1	0.9 x 0.5	13.1	(R)SB(s)a	6	Mag 10.84v star 0'.6 W.
06 55 14.4	+40 57 34	UGC 3593	13.7	1.6 x 0.8	13.8	SABc:	40	MCG +7-14-19 N 2'.7.
06 55 45.8	+47 55 47	UGC 3597	14.3	1.9 x 0.3	13.5	S	102	
06 55 49.6	+39 59 57	UGC 3601	13.9	0.5 x 0.4	12.0	S?	29	Mag 8.50v star NNW 7'.5.
06 57 34.6	+46 24 10	UGC 3608	12.7	1.0 x 0.8	12.3	S?	20	
06 58 16.8	+45 03 21	UGC 3614	13.5	1.3 x 1.0	13.6	SBab	177	Mag 4.90v star NW 7'.3; mag 10.83v star NE 1'.7.
07 01 05.1	+51 16 08	UGC 3627	13.9	1.2 x 0.8	13.7	Sd	80	Elongated **UGC 3625** N 4'.3; stellar **PGC 20001** N 5'.5.
07 03 02.8	+49 25 23	UGC 3638	13.4	1.0 x 0.6	12.7	SBab:	48	
07 04 01.0	+50 40 46	UGC 3645	14.6	1.5 x 0.2	13.1	S	33	
07 04 58.0	+46 08 25	UGC 3655	13.8	0.5 x 0.4	11.9	S?	9	Mag 15 star on S edge.
07 07 28.0	+44 47 24	UGC 3679	14.3	1.0 x 0.6	13.6	S?	145	MCG +7-15-6 NW 3'.0.
07 08 13.9	+46 06 55	UGC 3683	12.8	2.0 x 1.3	13.7	S0	50	Mag 13.6 star W edge.
07 08 23.6	+51 14 01	UGC 3684	14.0	1.4 x 0.9	14.1	(R')SB(s)b	176	Bright SE-NW bar, faint envelope.
07 09 23.2	+48 38 05	UGC 3696	12.8	1.2 x 0.6	12.4	E?	77	NGC 2329 SW 2'.8.
07 09 54.6	+47 54 34	UGC 3706	13.8	1.0 x 0.5	12.9	Double System	156	Double system, contact, disrupted; **MCG +8-13-81** W 2'.3.
07 11 27.2	+48 14 21	UGC 3724	13.5	1.5 x 0.8	13.5	SBb	55	Brighter core offset NE; mag 13.5 star N edge.
07 11 41.8	+49 51 41	UGC 3725	13.0	1.4 x 0.8	13.0	S0-:	140	Mag 15.4 star N of nucleus; mag 13 star on W edge.
07 15 04.9	+50 32 06	UGC 3758	13.3	0.7 x 0.7	12.4	Compact		**MCG +8-13-112** NE 2'.4.

GALAXY CLUSTERS

RA h m s	Dec ° ′ ″	Name	Mag 10th brightest	No. Gal	Diam ′	Notes
06 12 36.0	+48 36 00	A 553	15.3	30	13	Many stellar anonymous GX concentrated near center of cluster; mag 6.17v star 11'.8 NW of center.
07 09 12.0	+48 37 00	A 569	13.8	36	26	Many faint UGC, MCG and CGCG GX, along with anonymous GX.

OPEN CLUSTERS

RA h m s	Dec ° ′ ″	Name	Mag	Diam ′	No. ★	B ★	Type	Notes
06 02 34.0	+49 52 00	NGC 2126	10.2	6	40	13.0	cl	Moderately rich in stars; moderate brightness range; no central concentration; detached.
06 11 05.0	+51 40 36	NGC 2165		6	15	10.0	cl?	(S) Sparse group.
06 48 18.0	+41 05 00	NGC 2281	5.4	25	119	8.0	cl	Moderately rich in bright and faint stars with a strong central concentration; detached.

PLANETARY NEBULAE

RA h m s	Dec ° ′ ″	Name	Diam ″	Mag (P)	Mag (V)	Mag cent ★	Alt Name	Notes
05 56 24.0	+46 06 15	IC 2149	34	11.2	10.6	11.5	PK 166+10.1	Mag 10.6 star NW 4'.2.
06 34 07.6	+44 46 37	NGC 2242	22	15.1	15.0	17.6	PN G170.3+15.8	Mag 112.9 star NE 1'.4.

42 GALAXIES 42

RA h m s	Dec ° ′ ″	Name	Mag (V)	Dim ′ Maj x min	SB	Type Class	PA	Notes
05 11 20.0	+51 31 58	CGCG 258-1	15.1	0.8 x 0.4	13.8	S	150	Mag 12.8 star ENE 1'.8; mag 12.7 star WNW 2'.4.
05 32 35.9	+51 35 19	CGCG 258-9	16.1	0.8 x 0.5	14.9			Very faint stellar nucleus; mag 10.7 star N 1'.6.
05 33 27.3	+51 45 00	CGCG 258-10	15.2	0.8 x 0.5	14.1		42	Located between two mag 10.5 stars N-S.
03 58 07.8	+43 33 00	MCG +7-9-1	14.0	1.0 x 0.5	13.1		72	Mag 12.0 star N 1'.8.
05 14 05.2	+50 00 32	MCG +8-10-1	14.3	0.4 x 0.4	12.4	E/S0		Close pair of stars, mags 11.2 and 12.2, NNW 1'.0.
05 28 45.1	+49 52 57	MCG +8-10-7	14.6	0.8 x 0.7	13.8	S	144	
03 58 59.5	+43 20 51	UGC 2908	14.5	1.1 x 0.7	14.1	S?	106	Mag 10.7 star SE 2'.7.
04 38 28.6	+44 02 12	UGC 3108	14.5	1.5 x 1.1	14.9	S?	99	Numerous superimposed stars, in star rich area.
05 12 29.8	+51 17 13	UGC 3260	14.1	1.3 x 0.3	12.9	S0?	116	**CGCG 258-4** NE 7'.7.
05 35 19.0	+40 53 12	UGC 3325	14.6	1.1 x 0.8	14.3	Sbc	30	Many superimposed stars; mag 13.1 star S edge.

OPEN CLUSTERS

RA h m s	Dec ° ′ ″	Name	Mag	Diam ′	No. ★	B ★	Type	Notes
04 20 27.0	+44 55 12	Be 11	10.4	5	35	15.0	cl	Moderately rich in stars; moderate brightness range; slight central concentration; detached.
04 44 24.0	+42 41 12	Be 12		4	30	16.0	cl	
04 59 44.0	+43 28 54	Be 14		6	30	16.0	cl	
05 02 05.0	+44 30 48	Be 15		5	35	15.0	cl	
05 22 12.0	+45 28 00	Be 18		12	300	16.0	cl	Moderately rich in stars; small brightness range; fairly even distribution; detached.
04 37 50.0	+50 46 00	Be 67		10	30	15.0	cl:	(S) Sparse.
04 44 30.0	+42 04 00	Be 68	9.8	12	113	13.7	cl:	Moderately rich in stars; small brightness range; fairly even distribution; detached.
05 25 48.0	+41 57 00	Be 70		7	40	15.0	cl:	
05 20 53.0	+41 05 00	Cr 62	4.2	28			cl	Few stars; large brightness range; not well detached.
03 59 10.0	+51 47 00	King 7		8	80	16.0	cl	
04 09 55.0	+49 31 00	NGC 1513	8.4	12	50	11.0	cl	Moderately rich in stars; small brightness range; slight central concentration; detached. (S) Sparse.
04 15 19.0	+51 13 00	NGC 1528	6.4	18	165	10.0	cl	Moderately rich in stars; moderate brightness range; slight central concentration; detached.
04 20 57.0	+50 15 00	NGC 1545	6.2	12	65	9.0	cl	Few stars; moderate brightness range; not well detached.
04 31 39.0	+43 45 00	NGC 1582	7.0	24	20	9.0	cl	Few stars; moderate brightness range; not well detached. (A) Scattering of bright stars on DSS image.
04 34 53.0	+45 16 18	NGC 1605	10.7	5	40	12.5	cl	Moderately rich in stars; small brightness range; no central concentration; detached.
04 40 37.0	+50 27 42	NGC 1624	11.8	3	12		cl	About a dozen stars of mag 12 and fainter are involved in the nebulosity.
04 51 06.0	+43 41 00	NGC 1664	7.6	18	101	10.0	cl	Few stars; small brightness range; no central concentration; detached.
05 03 33.0	+49 29 30	NGC 1724		1			ast	
05 11 40.0	+47 41 42	NGC 1798	10.0	5	50	13.0	cl	Moderately rich in stars; small brightness range; a strong central concentration; detached.
05 25 55.0	+46 29 24	NGC 1883	12.0	5	30	14.0	cl	Moderately rich in stars; small brightness range; slight central concentration; detached.
04 46 35.0	+44 42 36	Ru 148	9.5	4	112	14.0	cl:	Few stars; small brightness range; not well detached. (S) Sparse group.
04 58 11.0	+43 01 00	Skiff J0458.2+4301		4		10.5	cl?	(S) Dubious.

(Continued from previous page)

BRIGHT NEBULAE

RA h m s	Dec ° ′ ″	Name	Dim ′ Maj x min	Type	BC	Color	Notes
04 03 24.0	+51 19 00	NGC 1491	25 x 25	E	1-5	3-4	Considerably smaller (about 4′) on blue photographs than on red. Visual observers may be able to detect two strong wisps forming a narrow V-shape (open end to NNE) in the nebula's brightest part. A star about 11th mag lies just SE of this area.
04 40 24.0	+50 27 00	NGC 1624	5 x 5	E	1-5	3-4	
05 27 18.0	+42 59 00	Sh2-224	20 x 3	SNR-E	5-5	4-4	An incomplete oval ring. The brightest part is a slightly convex crescent along the N circumference. This very narrow strip about 12′ long is convex toward a mag 8 star a short distance NW.

DARK NEBULAE

RA h m s	Dec ° ′ ″	Name	Dim ′ Maj x min	Opacity	Notes
04 31 56.0	+46 37 00	B 15	10 x 15	5	Elliptical.
04 32 26.0	+46 36 00	B 16		5	Very small; elongated N and S; close to SE edge of B15.
04 32 30.6	+46 31 00	B 17		5	Very small; elongated N and S; close to E edge of B15.
04 37 04.5	+50 59 00	B 20	60 x 60	2	In S part of larger, relatively vacant area.
04 52 04.1	+46 01 00	B 25	8 x 8	3	Irregularly round.

PLANETARY NEBULAE

RA h m s	Dec ° ′ ″	Name	Diam ″	Mag (P)	Mag (V)	Mag cent ★	Alt Name	Notes
04 13 27.4	+51 50 59	PN G151.4+0.5	14	16.9			PK 151+0.1	Mag 11.8 star E 2′.5; mag 10.95v star NNE 3′.7.
04 15 54.6	+48 49 38	PN G153.7−1.4	2	18.9			PK 153−1.1	Very close pair of mag 13.5, 14.0 stars S 0′.6.
04 46 42.9	+44 28 00	PN G160.5−0.5	92			20.8	PK 160−0.1	Mag 13.3 star 0′.3 NE of center.
04 54 31.2	+42 16 40	PN G163.1−0.8	115			21.0	PK 163−0.1	Close pair of stars, mags 11.2, 12.7, on S edge.

GALAXIES

RA h m s	Dec ° ′ ″	Name	Mag (V)	Dim ′ Maj x min	SB	Type Class	PA	Notes
02 20 04.1	+41 16 25	CGCG 538-43	14.3	0.5 x 0.3	12.1	Irr	168	Forms a large triangle with a mag 10.12v star WNW 8′.1 and mag 9.94v star NNW 7′.9.
02 31 21.9	+43 27 54	CGCG 539-37	14.1	0.8 x 0.3	12.4		156	Mag 10.30v star N 1′.6; mag 10.17v star WSW 2′.3.
02 59 01.3	+42 20 42	CGCG 540-19	14.1	0.6 x 0.4	12.4		87	Mag 12.9 star on N edge.
03 13 29.0	+41 04 26	CGCG 540-59	14.1	0.8 x 0.2	12.0		0	Mag 11.5 star E 3′.0.
03 14 08.7	+42 44 55	CGCG 540-62	13.4	0.8 x 0.5	12.3		3	Fairly bright anonymous galaxy ENE 4′.5; **UGC 2591** W 11′.2.
03 17 03.7	+41 37 58	CGCG 540-79	14.1	0.8 x 0.4	12.7		171	Mag 13.5 star on S edge; mag 11.0 star 1′.3 W; **PGC 12203** 2′.1 NE.
03 28 27.7	+40 09 13	CGCG 541-11	14.2	0.8 x 0.2	12.1	Sbc	84	Located 1′.1 WSW of mag 7.03v star.
02 56 39.1	+47 31 31	CGCG 554-16	13.9	0.7 x 0.4	12.6	E/S0	24	Mag 11.01v star 0′.6 S.
02 58 05.4	+46 29 51	CGCG 554-17	14.3	0.5 x 0.4	12.6	E	147	Mag 9.13v star WNW 6′.8.
02 49 45.5	+46 58 33	IC 257	12.6	1.4 x 0.8	12.6	S0⁻:	155	Mag 8.72v star W 12′.6.
02 49 41.0	+41 03 14	IC 258	14.3	0.7 x 0.5	13.0	SB0	126	IC 259 on E edge.
02 49 46.1	+41 03 04	IC 259	14.1	1.4 x 1.2	14.5	SB0?	165	Bright N-S bar, faint outer envelope, few superimposed stars; IC 258 on W edge.
02 51 00.8	+46 57 19	IC 260	13.1	1.0 x 0.7	12.8	E:	175	Mag 11.8 star on W edge; mag 9.41v star N 10′.4.
02 51 43.7	+42 49 38	IC 262	13.2	1.6 x 1.4	13.9	SB0	62	Mag 10.04v star N 2′.5.
02 54 44.1	+41 39 16	IC 265	14.6	0.7 x 0.7	13.9	E		Mag 9.66v star SE 4′.5; **PGC 10962** W 3′.1.
02 55 04.8	+42 15 45	IC 266	14.8	0.7 x 0.4	13.3		126	Several bright knots and/or superimposed stars.
03 06 10.3	+42 22 16	IC 284	11.5	3.8 x 1.2	13.0	SAdm	13	Galaxy **V Zw 319** on W edge just below mid-point.
03 07 33.0	+42 23 13	IC 288	13.9	1.1 x 0.3	12.5	S?	42	Mag 9.98v star NE 6′.2.
03 10 13.0	+40 45 52	IC 292	13.5	1.2 x 0.6	13.0	Sdm:	75	Mag 10.13v star N 2′.4.
03 10 56.2	+41 08 11	IC 293	14.0	0.8 x 0.6	13.2	E	90	Mag 8.12v star NE 6′.0.
03 11 03.3	+40 37 19	IC 294	13.9	2.1 x 1.5	15.0	(R)SB(rs)0/a:	123	Mag 11.29v star S 4′.5.
03 14 47.9	+42 13 16	IC 301	13.2	0.8 x 0.8	12.8	E		Mag 9.37v star N 8′.5.
03 16 06.4	+40 48 14	IC 309	13.5	0.9 x 0.9	13.1	SA(s)0°		Located between a pair of mag 11 stars.
03 16 43.0	+41 19 30	IC 310	12.7	1.1 x 1.1	12.8	SA(r)0°:		**UGC 2626** NE 3′.7.
03 16 46.8	+40 00 11	IC 311	14.1	0.9 x 0.6	13.3	S?	113	Mag 9.49v star ENE 6′.2.
03 18 08.4	+41 45 14	IC 312	13.4	1.2 x 0.6	13.0	E:	125	**PGC 12288** SE 1′.9.
03 21 20.1	+41 55 45	IC 316	14.2	1.2 x 0.7	13.9	S?	64	Very faint envelope, several stars superimposed; very small, elongated galaxy **MCG +7-7-74** E edge of nucleus.
03 25 59.3	+40 47 18	IC 320	13.7	1.3 x 1.0	13.8	SB(rs)ab	48	Several stars superimposed SW and W of nucleus.
02 28 46.2	+45 58 13	IC 1799	13.7	1.2 x 0.4	12.8	S?	34	Mag 6.71v star ENE 11′.0.
03 19 34.5	+41 34 47	IC 1907	14.2	0.9 x 0.6	13.6	E?	88	Mag 11.7 star W 1′.9; **PGC 12430** NE 2′.7.
02 20 35.0	+41 34 22	MCG +7-5-39	15.0	0.6 x 0.6	13.8	U/N		Mag 10.59v star W 1′.8.
02 21 23.7	+42 52 31	MCG +7-5-42	14.6	0.7 x 0.3	12.8	Irr	129	Mag 11.41v star SW 0′.8; **MCG +7-5-41** WSW 2′.2.
02 21 59.7	+44 22 03	MCG +7-5-43	14.7	0.6 x 0.3	12.9	E/S0	99	Mag 10.20v star N 1′.5.
02 24 01.9	+41 59 45	MCG +7-6-6	13.7	0.9 x 0.7	13.3	E/U	147	Mag 10.8 star superimposed N end; mag 12.0 star W edge; mag 6.66v star N 6′.4.
02 26 23.8	+42 51 35	MCG +7-6-18	14.8	0.6 x 0.4	13.1		111	Star superimposed? Mag 10.73v star W 3′.7.
02 26 46.5	+41 50 02	MCG +7-6-20	14.0	1.0 x 0.5	13.1	S?	147	Located between a close pair of mag 11.4 stars.
02 27 37.0	+42 00 24	MCG +7-6-23	14.3	0.4 x 0.3	11.9	S?	141	Mag 8.72v star ENE 0′.8.
02 34 03.8	+42 40 29	MCG +7-6-34	15.2	0.7 x 0.6	14.1		135	Mag 13.3 star NW edge.
02 39 36.7	+42 00 53	MCG +7-6-53	14.4	0.7 x 0.5	13.1		57	Mag 9.71v star E 2′.6; close pair of mag 12 stars W edge.
02 49 48.1	+41 27 43	MCG +7-6-74	13.7	0.6 x 0.5	12.2	S0?	21	Faint, elongated anonymous galaxy SW edge; mag 12.1 star SSW 1′.6.
02 54 16.7	+42 43 30	MCG +7-7-1	14.7	0.7 x 0.5	13.4	Spiral	153	Mag 10.88v star E 1′.1.
02 54 25.3	+41 34 32	MCG +7-7-3	12.2	0.3 x 0.2	9.2	E/S0	45	Superimposed on NGC 1129 on SW edge of core.
02 54 27.6	+41 30 45	MCG +7-7-5	16.1	0.3 x 0.2	13.1	E/S0	6	Located 4′.0 S of the center of NGC 1129.
02 54 39.3	+42 09 35	MCG +7-7-7	14.3	0.8 x 0.4	12.9		129	Bright star S 1′.0.
02 54 44.8	+41 31 40	MCG +7-7-8	14.1	1.0 x 0.6	13.4		75	NGC 1129 NW 4′.2.
03 20 22.2	+41 38 25	MCG +7-7-70	15.6	0.9 x 0.7	15.0	SBR(pec)	57	NGC 1281 W 3′.1.
03 20 25.3	+42 32 49	MCG +7-7-71	13.6	0.9 x 0.7	13.2	E/S0?	27	Several faint stars superimposed on edges, in star rich area.
03 25 24.8	+40 32 11	MCG +7-8-4	14.9	0.5 x 0.5	13.3	SBR/R		Mag 10.73v star superimposed on E edge.
03 29 11.7	+41 23 18	MCG +7-8-13	14.2	1.0 x 0.5	13.3		15	
03 31 01.2	+43 14 49	MCG +7-8-20	13.9	0.8 x 0.5	12.8	S?	87	Mag 11.52v star S 0′.9.
03 33 22.5	+40 09 46	MCG +7-8-25	14.6	0.8 x 0.8	14.0			Mag 11.4 star ESE 2′.7.
03 39 37.2	+41 04 58	MCG +7-8-28	14.4	0.9 x 0.7	13.7		90	Mag 11.2 star NE 2′.6.
03 50 47.8	+42 16 51	MCG +7-8-35	14.5	0.8 x 0.4	13.1	S	63	
03 58 07.8	+43 33 00	MCG +7-9-1	14.0	1.0 x 0.5	13.1		72	Mag 12.0 star N 1′.8.
02 22 05.1	+50 38 15	MCG +8-5-5	14.4	1.1 x 1.1	14.5			Mag 11.7 star superimposed NE edge.
02 29 10.4	+48 25 51	MCG +8-5-14	14.6	0.5 x 0.4	12.7	SB	3	Mag 13.3 star W edge; mag 10.84v star NE 4′.3.

(Continued from previous page)

GALAXIES

RA h m s	Dec ° ′ ″	Name	Mag (V)	Dim ′ Maj x min	SB	Type Class	PA	Notes
02 22 33.8	+42 21 01	NGC 891	9.9	11.7 x 1.6	13.0	SA(s)b? sp III	22	dark lane runs almost full length of galaxy.
02 23 20.3	+41 57 09	NGC 898	12.9	1.9 x 0.5	12.7	Sab sp	170	MCG +7-6-6 E 8'.1.
02 25 16.4	+42 05 21	NGC 906	12.9	1.6 x 1.4	13.6	SBab II	143	Two main arms; mag 9.19v star SW 4'.7.
02 25 23.0	+42 02 07	NGC 909	13.3	0.9 x 0.9	13.0	E		Located 2'.3 NE of mag 9.82v star.
02 25 26.7	+41 49 26	NGC 910	12.2	1.6 x 1.6	13.3	E^{+}		Mag 9.61v star W 5'.3; **CGCG 539-18** N 4'.5.
02 25 42.5	+41 57 18	NGC 911	12.7	1.3 x 0.7	12.7	E	115	Located 2'.2 S of mag 9.29v star.
02 25 42.7	+41 46 35	NGC 912	14.1	0.8 x 0.7	13.5	E	153	Compact; NGC 913 N 1'.5; **CGCG 539-15** SW 4'.8.
02 25 44.7	+41 47 52	NGC 913	15.0	0.7 x 0.3	13.4	E	24	NGC 912 S 1'.5.
02 26 05.2	+42 08 36	NGC 914	13.0	1.5 x 1.1	13.4	SA(s)c	117	
02 27 52.2	+45 56 48	NGC 920	13.9	1.5 x 1.1	14.3	(R′)SB(s)ab	154	A triangle of mag 10.17v, 10.98v, 10.98v stars centered SW 1'.1.
02 27 34.7	+41 58 41	NGC 923	13.7	0.8 x 0.6	12.7	Sb: III	107	MCG +7-6-23 and a mag 8.72v star lie 2'.3 NNE.
02 29 17.4	+45 54 39	NGC 933	13.9	1.3 x 0.9	13.9	S?	35	Mag 6.71v star NE 8'.8.
02 29 28.1	+42 15 01	NGC 937	14.2	1.1 x 0.5	13.4	SBcd?	117	Mag 10.41v star E 5'.3; stellar anonymous galaxy SW 6'.1.
02 30 38.5	+42 13 57	NGC 946	13.2	1.4 x 1.0	13.4	S0:	65	Mag 9.05v star S 8'.6; stellar anonymous galaxy NNW 6'.4.
02 35 18.8	+40 55 35	NGC 980	13.0	1.7 x 0.9	13.4	S0	110	Very elongated **UGC 2068** SE 2'.4.
02 35 24.9	+40 52 10	NGC 982	12.5	1.5 x 0.6	12.2	Sa	132	Much elongated **UGC 2068** N 1'.5; **MCG +7-6-40** E 2'.2; and a mag 10.27v star 2'.5 SW.
02 38 31.9	+41 31 41	NGC 995	13.4	1.3 x 0.6	13.0	S0	35	Small, faint, elongated anonymous galaxy E 3'.0.
02 38 39.9	+41 38 49	NGC 996	13.0	1.2 x 1.2	13.5	E		Small, faint anonymous galaxy and mag 14.1 star S 2'.9.
02 38 47.6	+41 40 14	NGC 999	13.5	1.2 x 0.8	13.3	(R′)SAB(s)a	48	Mag 11.4 star 1'.0 NE; bright knot, star on SW edge.
02 38 49.9	+41 27 35	NGC 1000	14.6	0.7 x 0.7	13.6			Almost stellar, NGC 995 NW 5'.3.
02 39 12.7	+41 40 17	NGC 1001	14.8	0.9 x 0.5	13.8	Sbc	132	Knot or star on NW end.
02 39 17.1	+40 52 18	NGC 1003	11.4	4.3 x 1.3	13.2	SA(s)cd III	97	Mag 9.88v star on S edge; pair of small, faint anonymous galaxies S 6'.6.
02 39 27.9	+41 29 33	NGC 1005	13.8	1.0 x 0.8	13.6	E/S0	65	Close pair of mag 11.6, 12.1 stars W 3'.2.
02 43 12.7	+41 29 59	NGC 1053	12.9	1.4 x 0.7	12.7	S0	40	Small anonymous galaxy N 0'.7.
02 46 03.0	+40 05 36	NGC 1077A	16.3	0.5 x 0.4	14.4	SBb	30	NGC 1077B on W edge.
02 46 00.6	+40 05 23	NGC 1077B	13.6	1.0 x 0.8	13.2	Sb	165	NGC 1077A on E edge. Mag 10.9 star W 4'.5.
02 47 56.4	+41 14 47	NGC 1086	12.8	1.5 x 1.0	13.1	Scd:	35	Mag 9.89v star NE 3'.5.
02 50 40.7	+41 40 17	NGC 1106	12.3	1.3 x 1.0	12.5	$SA0^{+}$	27	Mag 8.75v star E 3'.0; faint anonymous galaxy SE 3'.6; **UGC 2330** SE 9'.5.
02 52 51.4	+42 12 14	NGC 1122	12.1	1.7 x 1.2	12.7	SABb	40	Three bright knots or superimposed stars; **UGC 2354** N 2'.4.
02 54 27.4	+41 34 43	NGC 1129	12.4	2.9 x 2.1	14.4	E	90	Almost stellar MCG +7-7-3 0'.5 superimposed SW edge of core.
02 54 24.5	+41 36 19	NGC 1130	15.1	0.5 x 0.3	12.9	compact	36	Located 1'.5 N of the center of NGC 1129.
02 54 34.2	+41 33 30	NGC 1131	14.6	0.4 x 0.4	12.5	E/S0		Located on SE edge of NGC 1129.
02 56 36.8	+43 02 48	NGC 1138	12.8	1.1 x 1.1	12.9	SB0		
03 00 46.7	+43 09 45	NGC 1159	13.5	0.5 x 0.4	11.6	Spiral HSB	123	Mag 7.70v star W 6'.9.
03 01 13.1	+44 57 16	NGC 1160	12.8	1.5 x 0.7	12.7	Scd:	50	Located 3'.7 N of NGC 1161.
03 01 14.0	+44 53 46	NGC 1161	11.0	2.8 x 2.0	12.7	S0	20	A pair of mag 9.15v and 9.85v stars on W edge.
03 02 00.0	+42 35 04	NGC 1164	13.1	1.1 x 1.0	13.1	(R′)SAB(s)ab	145	Small, slightly elongated core; mag 8.67v star NNW 8'.3.
03 03 34.6	+46 23 08	NGC 1169	11.3	4.0 x 2.6	13.6	SAB(r)b I-II	28	Several stars superimposed; pair of mag 13 stars and a small anonymous galaxy NE 3'.2.
03 03 58.8	+43 23 59	NGC 1171	12.3	1.9 x 1.0	12.8	Scd:	147	Mag 9.28v star N 8'.0.
03 04 32.5	+42 20 21	NGC 1175	12.9	1.9 x 0.5	12.7	$SA(r)0^{+}$	153	Almost stellar NGC 1177 N 1'.7; mag 7.94v star NW 10'.4.
03 04 37.3	+42 21 43	NGC 1177	14.6	0.4 x 0.4	12.5	S?		NGC 1175 S 1'.7.
03 05 30.7	+42 50 07	NGC 1186	11.4	3.2 x 1.0	12.5	SB(r)bc:	122	Mag 10.17v star W 3'.7.
03 06 13.3	+41 50 53	NGC 1198	12.5	1.4 x 0.8	12.5	$S0^{-}$:	120	Mag 9.38v star S 5'.9; **CGCG 540-36** 2'.5 SW of this star.
03 11 13.7	+41 21 49	NGC 1224	13.7	1.0 x 0.8	13.4	$S0^{-}$:	0	Mag 8.73v star NE 9'.7.
03 15 21.1	+41 21 22	NGC 1250	12.8	2.1 x 0.6	12.9	$S0^{0}$: sp	159	Stellar **PGC 12103** S 6'.5.
03 16 26.2	+41 31 46	NGC 1257	13.7	1.3 x 0.2	12.1	Sa	68	Mag 10.48 star on N edge.
03 17 17.4	+41 23 04	NGC 1259	14.3	0.7 x 0.7	13.4	S0		**PGC 12225** ESE 2'.3.
03 17 27.4	+41 24 18	NGC 1260	13.3	1.1 x 0.5	12.5	S0/a: sp	74	**PGC 12230** ENE 1'.0; **PGC 12225** S 1'.9.
03 17 59.7	+41 31 13	NGC 1264	14.1	1.0 x 0.7	13.6	SBa-b	30	**PGC 12263** S 1'.5.
03 18 15.7	+41 51 25	NGC 1265	12.1	1.8 x 1.5	13.2	E^{+}	165	Almost stellar galaxy **PGC 12293** on W edge.
03 18 44.8	+41 28 03	NGC 1267	13.1	0.8 x 0.8	12.7	E^{+}:		**CGCG 540-89** W 1'.8.
03 18 45.3	+41 29 17	NGC 1268	13.4	1.0 x 0.6	12.7	SAB(rs)b: II-III	120	**CGCG 540-89** WSW 1'.9.
03 18 58.2	+41 28 09	NGC 1270	13.1	1.0 x 0.8	12.9	E:	15	**PGC 12358** E 1'.2.
03 19 11.4	+41 21 10	NGC 1271	13.9	0.7 x 0.3	12.1	SB0?	127	Stellar **PGC 12386** N 5'.0.
03 19 21.3	+41 29 25	NGC 1272	11.8	2.0 x 1.9	13.2	E^{+}	0	Mag 10.9star E 2'.9; **PGC 12409** E 3'.0.
03 19 27.0	+41 32 23	NGC 1273	13.2	0.8 x 0.7	12.4	$SA(r)0^{0}$?	144	**PGC 12354** W 4'.6.
03 19 40.7	+41 32 51	NGC 1274	14.0	0.8 x 0.3	12.4	E3	38	Mag 11.8 star E 1'.4; NGC 1275 SE 2'.6.
03 19 48.2	+41 30 44	NGC 1275	11.9	2.2 x 1.7	13.2	Pec	110	**Perseus A**. **PGC 12448** SE 2'.8; **PGC 12441** NE 1'.5.
03 19 52.5	+41 33 00	NGC 1276		0.5 x 0.2			175	
03 19 51.5	+41 34 22	NGC 1277	13.5	0.8 x 0.3	11.9	$S0^{+}$: pec	87	Located on N edge of NGC 1278.
03 19 54.2	+41 33 46	NGC 1278	12.4	1.3 x 1.0	12.8	E pec:	102	NGC 1277 on N edge; galaxy NGC 1276 S edge; galaxy **V Zw 339** ESE 1'.4.
03 20 06.3	+41 37 45	NGC 1281	13.3	0.9 x 0.4	12.3	E5	66	Mag 10.51v star WSW 1'.0.
03 20 12.1	+41 21 57	NGC 1282	12.9	1.2 x 0.7	12.8	E:	25	**PGC 12465** S 1'.1; **PGC 12487** SE 1'.4; **PGC 12497** ENE 1'.7.
03 20 15.7	+41 23 52	NGC 1283	13.6	0.8 x 0.6	12.8	E1:	73	Located 2'.1 N of NGC 1282.
03 21 36.6	+41 23 30	NGC 1293	13.4	0.9 x 0.9	13.2	E0		**CGCG 540-113** NNW 5'.0; **CGCG 540-115** N 6'.7.
03 21 40.3	+41 21 35	NGC 12943	13.2	1.1 x 0.8	13.0	$SA0^{-}$?	0	Stellar **PGC 12544** W 9'.6.
03 30 01.4	+41 49 55	NGC 1334	13.2	1.5 x 0.7	13.1	S?	115	
03 30 19.6	+41 34 19	NGC 1335	13.8	1.2 x 0.5	13.1	$S0^{-}$:	174	Mag 9.03v star S 2'.9.
02 20 23.1	+40 47 31	UGC 1796	14.5	1.1 x 1.1	14.5	SAB(s)dm		Stellar nucleus.
02 21 06.6	+48 57 33	UGC 1802	13.3	0.9 x 0.8	13.0	E	150	Mag 12.0 star W 1'.4.
02 21 13.7	+42 45 45	UGC 1807	15.9	1.5 x 1.5	16.6	Im:		Dwarf, very low surface brightness; mag 8.79v star W 1'.9.
02 22 15.8	+43 32 47	UGC 1827	14.8	0.6 x 0.5	13.3	Sdm:	0	Mag 14.1 star on N edge; mag 8.35v star NW 2'.5.
02 22 31.6	+47 50 56	UGC 1830	12.7	2.8 x 2.7	14.7	SB0/a	39	Small 1'.2 X 0'.9 core with very faint arms extending to published dimensions; mag 9.43v star NW 3'.3.
02 22 31.2	+43 03 55	UGC 1832	14.3	1.2 x 0.7	14.0	Sa	151	
02 22 58.6	+43 00 40	UGC 1837	13.8	0.8 x 0.6	12.9	S0	39	UGC 1841 SE 2'.7.
02 23 09.8	+41 22 12	UGC 1840	13.2	1.7 x 1.6	14.1	Pec	25	Published dimensions contain undocumented anonymous companion galaxy E side, dark patch (loop) NW; mag 5.80v star WNW 4'.0.
02 23 11.5	+42 59 27	UGC 1841	14.0	3.3 x 2.3	16.2	E	126	Bright core, smooth envelope; mag 8.48v star N edge; UGC 1837 NW 2'.7..
02 24 20.0	+43 15 43	UGC 1854	14.7	1.4 x 0.3	13.6	S	45	**UGC 1849** NW 2'.6.
02 24 29.7	+40 52 10	UGC 1855	13.9	1.0 x 0.6	13.2	SB(s)a	88	Mag 10.49v star ENE 2'.5.
02 24 47.4	+41 40 48	UGC 1858	14.8	1.3 x 0.9	14.8	SB	3	SE-NW bar; mag 12.1 star N 1'.1.
02 24 44.3	+42 37 21	UGC 1859	12.9	1.5 x 0.8	13.2	E?	47	Mag 10.9 star NW 2'.1.
02 25 07.7	+41 51 06	UGC 1866	13.9	1.0 x 0.5	13.0	SBa	30	Bright knot or star superimposed N end; mag 9.61v star SW 2'.7.
02 25 15.6	+45 27 04	UGC 1867	14.4	1.9 x 0.3	13.6	Scd	57	Mag 11.1 star NE 2'.3.
02 26 38.6	+50 02 40	UGC 1893	14.1	1.2 x 0.7	13.8	SB0/a	33	
02 27 25.6	+41 03 33	UGC 1914	13.7	1.0 x 0.8	13.5	E	81	Mag 7.99v star SE 0'.9.
02 28 30.2	+51 26 45	UGC 1930	13.5	2.7 x 2.0	15.2	$SB(r)0^{+}$	27	Bright nucleus in SE-NW bar, numerous superimposed stars on uniform envelope.
02 29 23.0	+47 29 27	UGC 1957	14.4	1.5 x 0.2	12.9	Scd:	175	Mag 11.5 star W 0'.8.
02 31 13.2	+43 20 21	UGC 1987	14.3	0.4 x 0.3	12.1	E?	14	

RA h m s	Dec ° ′ ″	Name	Mag (V)	Dim ′ Maj x min	SB	Type Class	PA	Notes
02 31 14.2	+40 23 24	UGC 1988	13.9	1.0 x 0.4	12.7	Sab	123	Mag 12.5 star SE 1′.1.
02 31 18.7	+42 52 37	UGC 1989	15.1	1.0 x 1.0	14.9	S		**UGC 1992** NW 1′.4.
02 32 22.8	+42 11 54	UGC 2001	13.4	1.5 x 1.0	13.7	Sab	35	
02 33 43.2	+40 31 44	UGC 2034	13.2	2.5 x 1.9	14.8	Im IV-V	170	Dwarf irregular, low surface brightness; mag 12.1 star E of center.
02 33 58.7	+44 20 39	UGC 2035	13.1	1.5 x 1.0	13.4	Sb II	125	Mag 10.9 star E 2′.0.
02 34 23.9	+45 00 07	UGC 2043	13.9	0.8 x 0.8	13.3	SB(s)c		
02 34 38.6	+44 54 45	UGC 2050	14.1	1.0 x 0.6	13.4	SAB(s)c	2	Filamentary, numerous faint stars superimposed.
02 35 03.3	+41 21 24	UGC 2060	13.7	1.2 x 0.5	13.0	SBab	177	
02 36 04.6	+42 25 15	UGC 2073	12.8	1.8 x 1.3	13.6	S0?	105	Mag 9.89v star W 2′.7.
02 36 02.6	+42 36 47	UGC 2074	14.1	0.7 x 0.5	12.8		130	
02 37 37.7	+42 38 05	UGC 2101	14.2	1.9 x 0.2	13.0	Sb III	162	**UGC 2108** SSE 5′.9; small, faint anonymous galaxy N 1′.9.
02 38 05.3	+41 47 23	UGC 2111	13.8	2.1 x 0.4	13.4	Sab	119	Mag 11.7 star NE 2′.6.
02 38 47.2	+40 41 53	UGC 2126	14.3	1.0 x 0.8	13.9	SABdm?	0	Mag 13.3 star on W edge.
02 39 52.0	+43 05 49	UGC 2146	17.1	1.0 x 0.7	16.6	Im:	153	Dwarf, stellar nucleus, low surface brightness; mag 9.90v star NE 2′.7.
02 40 47.2	+43 49 03	UGC 2161	13.6	1.5 x 1.5	14.4	SAB0°:		Trio of mag 10-11 stars SE 2′.1.
02 42 11.0	+43 21 18	UGC 2172	13.7	1.1 x 0.7	13.3	Irr	30	Mag 11.9 star N edge.
02 42 16.8	+42 23 32	UGC 2175	14.3	0.9 x 0.7	13.6	SAB(s)bc	108	Pair of widely spaced mag 10.8 stars W 1′.3; mag 12.3 star on W edge.
02 43 11.0	+40 25 42	UGC 2185	12.8	2.6 x 0.6	13.2	Scd:	144	Very knotty; **MCG +7-6-57** NW 3′.1.
02 43 07.4	+41 03 49	UGC 2186	13.4	1.4 x 0.7	13.2	S0/a	142	Mag 10.81v star E 1′.9; mag 11.49v star SE 2′.9.
02 43 26.8	+41 24 21	UGC 2194	13.6	1.5 x 0.5	13.2	SBb:	84	Mag 12.0 star SW 2′.5.
02 43 31.6	+44 21 34	UGC 2196	14.7	0.8 x 0.8	14.1	S		
02 44 52.9	+41 59 18	UGC 2215	14.0	0.9 x 0.3	12.4	Sb	3	
02 45 54.3	+42 48 40	UGC 2227	13.2	1.9 x 1.5	14.2	SB(r)b	144	Short NE-SW bar; mag 10.53v star SW 2′.4.
02 46 20.5	+44 57 07	UGC 2233	14.2	1.5 x 0.3	13.2	Sb III-IV	108	Mag 11.6 star N 0′.8.
02 46 56.9	+48 11 35	UGC 2240	13.2	1.2 x 0.6	12.7	SB0	5	
02 47 23.9	+45 31 53	UGC 2249	13.7	1.7 x 1.0	14.2	SAB0+	43	Located between a close pair of mag 10.6 stars, one star on W edge.
02 47 40.9	+40 29 40	UGC 2256	13.8	1.3 x 0.7	13.5	S0	85	Several stars superimposed in N-S line over nucleus; mag 12.9 star on E end.
02 48 17.6	+50 48 01	UGC 2261	13.6	1.5 x 0.9	13.9	E	70	UGC 2270 SE 3′.6.
02 48 15.6	+44 39 31	UGC 2269	14.4	0.9 x 0.5	13.4	Im:	100	Mag 10.4 star S 2′.5.
02 48 31.2	+50 45 11	UGC 2270	13.4	1.7 x 1.5	14.3	Sb:	96	Mag 13.3 star superimposed NW of nucleus; mag 10.76v star E 2′.0.
02 48 49.8	+40 40 48	UGC 2277	14.6	1.0 x 1.0	14.4	Compact		
02 48 53.9	+41 46 37	UGC 2280	13.8	2.2 x 0.7	14.1	SAbc II-III	44	Mag 11.9 star superimposed E edge.
02 50 19.5	+47 30 36	UGC 2314	14.4	0.8 x 0.6	13.5	Im:	36	Pair of mag 10.8 stars NW 1′.6.
02 50 24.2	+46 41 51	UGC 2317	13.7	1.5 x 0.7	13.6	Sa	14	Pair of stars, mags 10.81v and 12.0, E edge.
02 52 20.0	+44 03 54	UGC 2342	14.3	1.0 x 0.9	14.0			= **Mkn 1060**.
02 52 40.6	+41 23 46	UGC 2350	13.8	0.8 x 0.6	12.9	Sb III-IV	117	**UGC 2349** S 4′.3; **CGCG 539-118** N 9′.0.
02 52 44.4	+46 56 14	UGC 2351	13.0	1.5 x 1.1	13.4	SB(s)b	120	Small nucleus in short, weak bar; mag 5.87v star SW 12′.3.
02 53 34.9	+41 53 06	UGC 2361	13.3	1.3 x 0.8	13.2	SAB(s)b	69	**CGCG 540-10** E 13′.0.
02 54 26.2	+42 38 58	UGC 2370	14.8	2.8 x 0.2	14.0	Sdm?	127	Razor thin; mag 7.69v star NE 3′.8.
02 55 11.1	+51 54 25	UGC 2380	15.2	1.5 x 0.7	15.1	Sb	9	Numerous faint stars superimposed.
02 59 21.7	+47 04 53	UGC 2449	13.2	1.1 x 0.9	13.0	S?	129	Bright core offset to SE of center.
03 00 36.5	+49 02 33	UGC 2459	13.8	2.7 x 0.2	13.0	Sdm:	62	
03 00 37.4	+40 15 04	UGC 2463	13.8	2.0 x 0.9	14.3	SABm	95	Mag 10.64v star WNW 2′.7.
03 00 59.6	+43 01 00	UGC 2470	13.7	1.3 x 0.3	12.5	Sa	169	Very small, very faint anonymous galaxy on W edge.
03 01 10.2	+41 23 45	UGC 2473	14.3	1.0 x 0.4	13.2	Scd?	36	Located at the center of the small galaxy cluster **A 404**; **PGC 11417** NE 7′.1.
03 02 07.0	+41 35 36	UGC 2495	13.4	1.0 x 0.7	12.9	S0	105	Mag 11.55v star W 1′.8; mag 8.68v star N 2′.1.
03 02 44.4	+41 37 33	UGC 2500	13.9	1.3 x 0.6	13.5	SB(s)a	165	Mag 9.5 star SW 2′.8.
03 03 36.3	+46 56 39	UGC 2504	14.6	1.5 x 0.6	14.3	Pec	137	Mag 11.5 star SE end.
03 05 55.4	+41 35 30	UGC 2528	13.4	1.7 x 0.7	13.4	S0/a	12	Mag 12.2 star N end.
03 06 32.8	+41 29 09	UGC 2534	13.9	1.2 x 1.0	14.2	E?	33	
03 07 05.0	+46 37 12	UGC 2537	13.2	0.9 x 0.9	12.8	SA(s)cd		Pair of bright knots or stars E edge.
03 07 09.1	+41 44 57	UGC 2538	14.3	1.4 x 0.5	13.8	SBa	134	Mag 13.3 star SE end of nucleus; mag 11.9 star on S edge.
03 09 18.6	+42 58 21	UGC 2559	13.0	1.6 x 0.5	12.6	S0	44	Mag 11.2 star E 1′.2.
03 10 14.6	+42 13 16	UGC 2568	13.7	1.3 x 0.5	13.1	S0	163	**UGC 2564** SSW 3′.0.
03 10 56.1	+45 01 02	UGC 2573	14.0	1.4 x 0.4	13.2	SBb	82	Stellar nucleus; mag 7.54v star W 3′.2.
03 13 29.4	+44 07 50	UGC 2596	13.0	2.0 x 0.9	13.5	SAm:	17	Two bright knots or stars N end.
03 14 08.5	+41 17 31	UGC 2598	13.4	1.0 x 0.5	12.5	S0:	25	Galaxy **V Zw 331** SSW 2′.9.
03 15 01.6	+42 02 04	UGC 2608	12.9	0.9 x 0.8	12.4	(R′)SB(s)b	61	Bright E-W bar, strong dark patches N and S; mag 10.69v star SW 1′.7.
03 15 15.0	+41 58 53	UGC 2612	14.2	0.7 x 0.7	13.3	Scd:		Several stars superimposed.
03 15 18.5	+42 41 46	UGC 2614	13.3	1.5 x 0.4	12.6	S0/a	90	Mag 10.8 star superimposed W end.
03 16 00.9	+40 53 13	UGC 2617	13.2	2.2 x 0.6	13.4	SAB(s)d	176	Numerous knots and/or superimposed stars; mag 10.6 star WNW 2′.0.
03 16 00.9	+42 04 30	UGC 2618	13.6	1.1 x 0.5	12.8	Sab	167	Faint, stellar anonymous galaxy NE 1′.2.
03 17 52.3	+43 18 13	UGC 2640	13.4	0.9 x 0.5	12.4	SBb	70	Close pair of mag 12.4 stars NW 1′.5.
03 18 43.3	+42 17 55	UGC 2654	13.3	1.5 x 0.4	12.6	S?	3	Pair of very faint, stellar anonymous galaxies E edge.
03 18 45.4	+43 14 20	UGC 2655	12.9	1.9 x 0.6	12.9	SAB(s)d	175	Several stars superimposed.
03 18 53.6	+40 35 42	UGC 2659	13.8	1.2 x 0.3	12.5	Sbc	65	Mag 9.76v star E 2′.4.
03 20 01.7	+41 15 04	UGC 2673	13.6	1.5 x 0.9	13.8	S0:	135	Mag 10.5 star N 0′.9.
03 21 01.3	+40 47 49	UGC 2686	14.5	1.1 x 0.5	13.7	SABa	43	Mag 10.37v star N 1′.9.
03 21 27.7	+40 48 05	UGC 2689	13.9	1.2 x 0.6	13.5	S0?	127	Several faint stars superimposed; mag 11.1 star NW 2′.9.
03 22 03.0	+40 51 47	UGC 2698	12.9	1.0 x 0.7	12.5	E	105	Mag 10.51v star on E edge.
03 22 53.7	+42 33 13	UGC 2700	14.0	1.7 x 0.5	13.7	SBb?	132	
03 23 49.1	+40 33 26	UGC 2708	13.8	1.3 x 1.0	14.0	S0	24	Very faint, elongated anonymous galaxy N 1′.5.
03 24 36.6	+40 41 24	UGC 2717	13.3	0.7 x 0.6	12.4	E		Mag 7.92v star N 4′.5; elongated **UGC 2715** NNW 6′.4.
03 25 15.7	+42 40 21	UGC 2723	15.2	1.1 x 0.7	14.8	SB(r)c	0	Stellar nucleus, smooth envelope; mag 13.0 star on W edge.
03 25 29.9	+41 14 24	UGC 2725	13.8	0.9 x 0.7	13.2	S0	162	Mag 6.51v star W 4′.0.
03 25 52.2	+40 44 56	UGC 2730	14.1	1.5 x 0.3	13.1	Sb III	127	IC 320 N 2′.7.
03 26 03.2	+41 15 06	UGC 2733	13.5	1.0 x 0.6	13.0	E?	57	Mag 14.9 star on W edge.
03 26 27.3	+40 30 25	UGC 2736	13.5	1.7 x 0.3	12.6	Sab	69	Mag 8.06v star S 2′.9.
03 27 40.1	+40 53 51	UGC 2742	14.1	0.8 x 0.6	13.2	SB(s)bc	117	
03 29 04.1	+40 49 24	UGC 2752	13.8	1.4 x 0.5	13.3	S0:	83	Mag 10.14v star SW 2′.5.
03 29 35.2	+40 52 18	UGC 2756	14.1	1.4 x 0.7	13.9	Double System	12	Double system, contact; consists of **MCG +7-8-15** and **MCG +7-8-16**.
03 30 56.7	+40 48 33	UGC 2766	15.1	0.5 x 0.5	13.4	S0		
03 32 07.6	+47 47 33	UGC 2773	14.1	1.1 x 0.8	13.8	Double System	123	Numerous faint stars superimposed.
03 32 05.8	+41 35 25	UGC 2775	14.8	1.4 x 0.8	14.8	SAB(s)cd:	138	Mag 11.8 star NW 3′.1.
03 36 27.9	+48 30 13	UGC 2794	12.9	1.5 x 1.5	13.6	S?		
03 38 10.1	+40 59 00	UGC 2798	13.2	2.1 x 0.7	13.5	SABbc	70	Mag 10.38v star NW 2′.2.
03 43 54.6	+40 00 56	UGC 2837	13.4	1.7 x 0.8	13.6	SB(rs)b	155	Patchy with E-W bar; mag 12.3 star S 1′.3.
03 44 58.7	+45 57 59	UGC 2844	13.8	1.0 x 0.8	13.6	E:	42	Mag 7.71v star NNW 4′.5; elongated **MCG +8-7-11** S 7′.5.
03 45 33.2	+44 51 17	UGC 2849	14.9	1.4 x 0.9	15.0	Scd:	73	Stellar nucleus; mag 11.8 star S 0′.8.

(Continued from previous page)

GALAXIES

RA h m s	Dec ° ′ ″	Name	Mag (V)	Dim ′ Maj x min	SB	Type Class	PA	Notes
03 45 44.0	+46 40 24	UGC 2851	14.2	1.6 x 0.8	14.3	Sb: III	125	Mag 10.69v star on NE edge.
03 47 22.9	+45 58 07	UGC 2858	15.3	1.4 x 1.0	15.5	Sdm	160	Numerous stars superimposed, in star rich area.
03 47 18.5	+40 51 37	UGC 2859	14.8	1.1 x 1.1	14.8	Scd:		Mag 121.2 star WSW 2'.6.
03 58 59.5	+43 20 51	UGC 2908	14.5	1.1 x 0.7	14.1	S?	106	Mag 10.7 star SE 2'.7.

GALAXY CLUSTERS

RA h m s	Dec ° ′ ″	Name	Mag 10th brightest	No. Gal	Diam ′	Notes
02 25 48.0	+41 52 00	A 347	13.3	32	56	Numerous almost stellar and larger catalog and anonymous GX.
02 43 36.0	+41 50 00	A 372	15.7	40	17	Mag 8.51v star NW of center 3'.0; note: **UGC 2195** lies 4'.3 N of this star.
03 18 36.0	+41 30 00	A 426	12.5	88	190	**Perseus Cluster**. Numerous stellar and almost stellar PGC and anonymous GX populate this cluster.

OPEN CLUSTERS

RA h m s	Dec ° ′ ″	Name	Mag	Diam ′	No. ★	B ★	Type	Notes
03 59 10.0	+51 47 00	King 7		8	80	16.0	cl	
02 27 18.0	+42 19 00	Lederman 1		14			ast	(A) "Black Widow spider" asterism of GSC stars. Noted 1997 by R. Lederman.
03 24 19.0	+49 52 00	Mel 20	2.3	300	50	3.0	cl	
02 32 14.9	+44 39 00	NGC 956	8.9	9	30	9.0	cl	Moderately rich in stars; moderate brightness range; not well detached. (S) Sparse field. (A) DSS image shows 4 bright stars with some fainter background stars.
02 42 08.0	+42 45 00	NGC 1039	5.2	25	60	9.0	cl	= **M 34**. Rich in stars; large brightness range; slight central concentration; detached.
03 05 56.0	+44 23 00	NGC 1193	12.6	3	40	14.0	cl	Moderately rich in stars; moderate brightness range; strong central concentration; detached.
03 14 42.0	+47 14 00	NGC 1245	8.4	10	200	12.0	cl	Rich in stars; moderate brightness range; slight central concentration; detached.
03 34 09.0	+51 25 12	NGC 1348		6	30	8.5	cl	Moderately rich in stars; moderate brightness range; no central concentration; detached.

BRIGHT NEBULAE

RA h m s	Dec ° ′ ″	Name	Dim ′ Maj x min	Type	BC	Color	Notes
03 31 12.0	+43 54 00	GK-N1901	1 x 1	SNR			**Nova Persei 1901**, a supernova remnant (= **MCG +7-8-22**). The nebulosity is approximately centered around the involved star and has a very clumpy appearance; brightest and most concentrated along the SW perimeter.
04 03 24.0	+51 19 00	NGC 1491	25 x 25	E	1-5	3-4	Considerably smaller (about 4′) on blue photographs than on red. Visual observers may be able to detect two strong wisps forming a narrow V-shape (open end to NNE) in the nebula's brightest part. A star about 11th mag lies just SE of this area.

PLANETARY NEBULAE

RA h m s	Dec ° ′ ″	Name	Diam ″	Mag (P)	Mag (V)	Mag cent ★	Alt Name	Notes
02 52 14.8	+50 35 54	PN G141.7−7.8	127	>16.		21.4	PK 141−7.1	Lightly annular, brightest portions along SE and NW edges; mag 14.1 star near center.
02 45 23.8	+42 33 03	PN G144.3−15.5	25	16.7	14.4	19.9	PK 144−15.1	Elongated galaxy **CGCG 539-91** WNW 0'.8; mag 10.96v star SE 1'.6; mag 8.69v star SE 3'.1.
03 27 15.5	+45 24 20	PN G149.4−9.2	540			17.0	PK 149−9.1	Mag 10.25v star near center; mag 9.69v star E 6'.9; mag 10.5 star W 6'.0 and mag 10.29v star WSW 6'.0.
03 49 05.9	+50 00 14	PN G149.7−3.3	780			16.5	PK 149−3.1	Small trapezoid shape of four mag 11 stars slightly SE of center; mag 10.30v star on WNW edge, 6'.0 from center.

GALAXIES

RA h m s	Dec ° ′ ″	Name	Mag (V)	Dim ′ Maj x min	SB	Type Class	PA	Notes
02 19 13.8	+41 42 58	CGCG 538-41	14.2	0.5 x 0.3	12.0		171	Mag 12.1 star N 0'.5.
02 20 04.1	+41 16 25	CGCG 538-43	14.3	0.5 x 0.3	12.1	Irr	168	Forms a large triangle with a mag 10.12v star WNW 8'.1 and mag 9.94v star NNW 7'.9.
01 20 31.1	+50 08 38	CGCG 551-15	14.0	0.7 x 0.4	12.5		111	Mag 6.57v star E 13'.1.
01 40 41.8	+45 50 01	CGCG 552-5	14.2	0.7 x 0.4	12.7		3	Located 1'.8 SE of mag 9.03v star.
01 00 55.5	+47 40 53	IC 65	12.9	3.0 x 0.9	13.9	SAB(s)bc	155	Mag 7.75v star E 8'.6.
00 53 38.8	+41 06 00	MCG +7-3-3	14.6	0.9 x 0.4	13.3		84	Mag 11.7 star NW 0'.9.
00 53 58.8	+40 24 49	MCG +7-3-4	14.9	0.7 x 0.4	13.3	Sc	117	
00 54 35.9	+42 16 29	MCG +7-3-5	14.5	0.6 x 0.5	13.0	SB?	21	Mag 12.3 star E 2'.6.
00 55 11.4	+42 13 13	MCG +7-3-6	15.0	0.7 x 0.5	13.7	Sc	105	Mag 8.27v star N 1'.8.
00 56 38.6	+42 39 54	MCG +7-3-8	15.4	0.6 x 0.4	13.7	Sc	132	Mag 10.32v star ESE 3'.3.
00 57 42.4	+43 42 03	MCG +7-3-11	14.5	1.0 x 0.6	13.8	Sb I	72	Mag 11.19v star NW 2'.7.
01 06 52.8	+44 17 02	MCG +7-3-18	15.0	0.8 x 0.6	14.0		135	
01 12 59.0	+40 01 15	MCG +7-3-22	15.0	0.7 x 0.4	13.4	Sc	6	Mag 12.1 star S 1'.5.
01 17 05.7	+40 58 22	MCG +7-3-25	15.3	0.7 x 0.2	13.0	Sb	36	Mag 11.14v star N 0'.7.
01 21 01.4	+40 26 45	MCG +7-3-29	14.1	0.8 x 0.3	12.4	Sc	123	
01 21 09.2	+40 28 15	MCG +7-3-31	14.9	0.7 x 0.4	13.3	Sc	21	NGC 477 NE 2'.4.
02 13 59.8	+41 52 34	MCG +7-5-27	14.5	1.1 x 0.4	13.4	S0	135	Mag 12.1 star N 1'.4.
02 20 35.0	+41 34 22	MCG +7-5-39	15.0	0.6 x 0.6	13.8	U/N		Mag 10.59v star W 1'.8.
02 21 23.7	+42 52 31	MCG +7-5-42	14.6	0.7 x 0.3	12.8	Irr	129	Mag 11.41v star SW 0'.8; **MCG +7-5-41** WSW 2'.2.
02 21 59.7	+44 22 03	MCG +7-5-43	14.7	0.6 x 0.3	12.9	E/S0	99	Mag 10.20v star N 1'.5.
02 24 01.9	+41 59 45	MCG +7-6-6	13.7	0.9 x 0.7	13.3	E/U	147	Mag 10.8 star superimposed N end; mag 12.0 star W edge; mag 6.66v star N 6'.4.
02 26 23.8	+42 51 35	MCG +7-6-18	14.8	0.6 x 0.4	13.1		111	Star superimposed? Mag 10.73v star W 3'.7.
02 26 46.5	+41 50 02	MCG +7-6-20	14.0	1.0 x 0.5	13.1	S?	147	Located between a close pair of mag 11.4 stars.
02 27 37.0	+42 00 24	MCG +7-6-23	14.3	0.4 x 0.3	11.9	S?	141	Mag 8.72v star ENE 0'.8.
00 45 50.5	+49 40 09	MCG +8-2-12	14.6	0.7 x 0.6	13.5	SBc	120	Mag 10.54v star SE 3'.6.
00 58 39.8	+50 02 39	MCG +8-2-22	14.0	2.1 x 0.6	14.1		18	Several stars superimposed S end.
01 01 45.3	+47 54 05	MCG +8-3-6	14.0	0.5 x 0.3	12.0	E/S0	63	Mag 13.5 star superimposed S edge.
01 16 04.5	+46 43 50	MCG +8-3-11	14.8	0.9 x 0.4	13.5	Sb	18	UGC 813 and UGC 816 ENE 2'.2.
01 18 11.7	+49 58 28	MCG +8-3-14	15.2	0.7 x 0.7	14.3	Sbc		Mag 12.3 star S 2'.9.
01 26 38.0	+48 23 34	MCG +8-3-22	14.1	0.6 x 0.5	12.6	S?	111	Mag 9.50v star E 1'.7.
01 32 35.4	+49 24 14	MCG +8-3-29	14.3	1.0 x 0.5	13.4	SBbc	63	Mag 10.90v star N 0'.8.
02 04 32.0	+45 46 20	MCG +8-4-17	14.9	0.9 x 0.6	14.1	SBc	90	UGC 1562 W edge.
02 22 05.1	+50 38 15	MCG +8-5-5	14.4	1.1 x 1.1	14.5			Mag 11.7 star superimposed NE edge.
00 52 04.4	+47 32 58	NGC 278	10.8	2.1 x 2.0	12.2	SAB(rs)b II-III	12	Mag 8.82v star N 2'.8.
00 57 39.3	+43 48 04	NGC 317A	14.0	1.4 x 1.3	14.6	S?	111	NGC 317B on S edge.
00 57 41.3	+43 47 28	NGC 317B	13.4	1.0 x 0.5	12.4	SB?	105	MCG +7-3-11 S 5'.4.

(Continued from previous page)

GALAXIES

RA h m s	Dec ° ′ ″	Name	Mag (V)	Dim ′ Maj x min	SB	Type Class	PA	Notes
01 21 20.2	+40 29 16	NGC 477	13.0	1.5 x 0.9	13.2	SAB(s)c	150	Three faint stars in a row across S end on E-W line.
01 28 29.5	+48 23 14	NGC 562	13.3	1.3 x 1.0	13.4	SA(rs)c I	20	Close pair of mag 8.49v, 10.1 stars S 4′.0.
01 30 49.5	+41 15 22	NGC 573	13.2	0.4 x 0.4	11.1	S?		Compact; mag 9.55v star NW 4′.6.
01 33 41.1	+44 55 45	NGC 590	12.9	2.6 x 1.3	14.1	SBa	150	**CGCG 537-12** NW 2′.3; mag 10.24v star NE 2′.0.
01 35 02.4	+41 14 56	NGC 605	12.9	2.2 x 1.1	13.7	S0	145	
01 37 00.1	+42 19 17	NGC 620	13.8	0.8 x 0.7	13.0	Pec	3	
01 57 50.9	+44 55 01	NGC 746	12.9	1.7 x 1.0	13.3	Im	90	
02 06 51.7	+44 34 17	NGC 812	11.3	2.2 x 0.8	11.8	S pec	160	
02 12 12.5	+44 34 05	NGC 846	12.1	1.9 x 1.7	13.2	SB(rs)ab	140	Mag 9.2 star N 5′.6.
02 22 33.8	+42 21 01	NGC 891	9.9	11.7 x 1.6	13.0	SA(s)b? sp III	22	dark lane runs almost full length of galaxy.
02 23 20.3	+41 57 09	NGC 898	12.9	1.9 x 0.5	12.7	Sab sp	170	MCG +7-6-6 E 8′.1.
02 25 16.4	+42 05 21	NGC 906	12.9	1.6 x 1.4	13.6	SBab II	143	Two main arms; mag 9.19v star SW 4′.7.
02 25 23.0	+42 02 07	NGC 909	13.3	0.9 x 0.9	13.0	E		Located 2′.3 NE of mag 9.82v star.
02 25 26.7	+41 49 26	NGC 910	12.2	1.6 x 1.6	13.3	E+		Mag 9.61v star W 5′.3; **CGCG 539-18** N 4′.5.
02 25 42.5	+41 57 18	NGC 911	12.7	1.3 x 0.7	12.7	E	115	Located 2′.2 S of mag 9.29v star.
02 25 42.7	+41 46 35	NGC 912	14.1	0.8 x 0.7	13.5	E	153	Compact; NGC 913 N 1′.5; **CGCG 539-15** SW 4′.8.
02 25 44.7	+41 47 52	NGC 913	15.0	0.7 x 0.3	13.4	E	24	NGC 912 S 1′.5.
02 26 05.2	+42 08 36	NGC 914	13.0	1.5 x 1.1	13.4	SA(s)c	117	
02 27 52.2	+45 56 48	NGC 920	13.9	1.5 x 1.1	14.3	(R′)SB(s)ab	154	A triangle of mag 10.17v, 10.98v, 10.98v stars centered SW 1′.1.
02 27 34.7	+41 58 41	NGC 923	13.7	0.8 x 0.6	12.7	Sb: III	107	MCG +7-6-23 and a mag 8.72v star lie 2′.3 NNE.
00 44 41.9	+45 25 16	UGC 471	14.2	0.9 x 0.7	13.5	SA(s)bc	50	Mag 12.2 star S 0′.6.
00 46 19.4	+51 13 02	UGC 475	13.4	1.5 x 0.7	13.3	Scd:	105	Mag 10.81v star NE 2′.1.
00 47 28.3	+50 52 57	UGC 486	13.4	0.9 x 0.6	12.6	SAB(s)b II	0	Bright nucleus with dark patches E and W.
00 51 27.1	+40 43 30	UGC 522	15.2	0.8 x 0.7	14.4	Sb II-III	65	Pair of mag 12.5 stars NE 1′.5.
00 52 53.8	+41 58 10	UGC 539	13.8	0.8 x 0.7	13.0	Scd:	162	Mag 12.9 star on W edge.
00 56 27.1	+50 37 19	UGC 576	13.9	0.9 x 0.5	12.9	SABcd:	125	Patchy; mag 10.96v star SW 1′.7.
00 58 24.2	+48 39 40	UGC 600	13.2	1.5 x 1.1	13.6	SAB(s)b	135	Few knots or superimposed stars; mag 13.4 star E edge.
00 59 02.3	+48 01 07	UGC 608	14.3	1.8 x 0.8	14.5	SABdm	133	Several stars superimposed, in star rich area.
01 00 28.3	+47 59 43	UGC 622	13.5	1.1 x 0.6	12.9	Scd:	160	Mag 10.67v star NW 3′.1.
01 04 01.5	+41 50 33	UGC 655	13.8	2.8 x 2.8	15.9	Sm		
01 04 51.6	+42 12 40	UGC 665	14.5	0.8 x 0.6	13.6	SB	24	Mag 12.2 star NW edge; mag 10.64v star NE 1′.2.
01 10 10.5	+43 06 30	UGC 725	13.7	2.0 x 0.4	13.3	SBcd?	43	Mag 12.4 star W 2′.0.
01 10 28.9	+43 17 13	UGC 728	13.7	1.6 x 1.1	14.2	SAB(s)c	92	Knotty; two main arms.
01 10 43.5	+49 36 09	UGC 731	14.2	2.2 x 2.0	15.7	Im: V	66	Dwarf irregular, very low surface brightness; located between a mag 11.37 star NW and mag 11.4 star SE.
01 11 36.1	+49 07 12	UGC 746	13.0	1.5 x 0.9	13.4	E	100	Close pair of mag 12.4 stars on S edge.
01 13 21.1	+50 38 40	UGC 761	15.0	1.0 x 0.9	15.0	E?	117	Faint star superimposed W of core.
01 14 10.7	+42 14 25	UGC 777	14.4	0.9 x 0.9	14.0	Scd:		
01 14 26.3	+42 33 22	UGC 783	13.7	1.5 x 0.9	13.9	SAB(rs)c II	153	
01 16 16.6	+46 44 21	UGC 813	13.9	1.2 x 0.5	13.2	S?	110	Strong knot E of nucleus; MCG +8-3-11 WSW 2′.2; UGC 816 NE edge.
01 16 20.4	+46 44 51	UGC 816	13.3	1.7 x 0.8	13.5	S?	87	Pair with UGC 813 S edge; mag 10.12v star N 1′.5.
01 17 39.5	+43 38 48	UGC 826	13.8	1.1 x 0.7	13.3	Sb:	165	Dark patch N of center; close pair of mag 11 stars NW 1′.5.
01 18 59.5	+44 17 10	UGC 836	14.5	0.8 x 0.8	14.1	E:		Small, faint, round anonymous galaxy E 1′.8.
01 21 08.1	+42 51 02	UGC 880	14.1	1.0 x 0.8	13.7	S	60	Bright knot or star SW of nucleus; knot on S edge.
01 21 56.1	+50 02 48	UGC 902	13.4	1.0 x 0.6	12.9	E:	117	Mag 6.57v star N 4′.6; **MCG +8-3-17** E 3′.6; **MCG +8-3-18** E 6′.2.
01 27 48.9	+48 49 08	UGC 1035	14.0	1.2 x 0.4	13.1	Sb II-III	17	Mag 11.5 star N edge.
01 28 06.2	+49 14 28	UGC 1042	13.3	1.3 x 0.5	12.7	S0	95	
01 29 47.2	+45 35 57	UGC 1068	12.9	1.6 x 1.0	13.2	Sc	30	Line of four mag 12 and 13 stars extend N from E edge.
01 30 00.2	+40 58 28	UGC 1070	13.2	1.9 x 1.4	14.2	Scd:	55	
01 32 35.7	+41 59 10	UGC 1101	13.9	0.9 x 0.7	13.3	S0	18	Bright center, faint envelope, several stars superimposed; pair of mag 12 and 13 stars S 1′.3.
01 35 31.7	+47 32 58	UGC 1132	14.1	1.2 x 0.5	13.4	Sdm	13	Mag 10.8 star S 2′.5.
01 36 21.4	+47 23 19	UGC 1142	14.8	0.9 x 0.8	14.3	Compact	90	Mag 11.5 star SW 0′.9.
01 38 15.5	+41 39 12	UGC 1162	13.8	1.5 x 1.5	14.5	SB(r)b		Short N-S bar; mag 11.4 star NE 3′.3.
01 39 18.3	+48 45 50	UGC 1168	14.0	1.2 x 0.7	13.7	S? III	85	Mag 3.59v star 51 Andromedae SW 15′.3.
01 40 35.4	+43 51 40	UGC 1179	14.7	0.9 x 0.7	14.0	Compact	3	Close pair of mag 12 stars S 1′.1.
01 41 13.6	+45 00 04	UGC 1185	14.1	0.9 x 0.6	13.3	Compact	135	
01 41 48.1	+49 02 24	UGC 1186	14.8	1.5 x 1.2	15.3	Pec	90	Bright, elongated core, uniform envelope with few superimposed stars; mag 11.9 star W edge.
01 50 52.7	+48 21 04	UGC 1303	13.9	1.1 x 1.0	13.9	SBa	99	Mag 9.94v star E edge.
01 52 10.9	+42 55 01	UGC 1331	14.0	0.9 x 0.5	12.9	Scd:	20	Mag 10.66v star ENE 1′.9.
01 52 18.3	+48 05 15	UGC 1332	13.6	1.3 x 1.2	14.2	E	95	Mag 14.1 star E edge; mag 12.4 star W 1′.0.
01 53 36.4	+43 57 56	UGC 1355	13.2	1.1 x 0.9	13.0	SAB(s)b	60	
01 55 30.8	+47 57 15	UGC 1389	13.5	1.6 x 1.4	14.4	E	36	Several stars superimposed around core; mag 8.90v star SW 2′.2.
01 55 42.0	+46 48 06	UGC 1394	13.4	2.1 x 1.2	14.3	Sa	70	Several faint stars superimposed; mag 10.99v star N edge.
01 56 57.0	+40 20 27	UGC 1418	13.2	1.7 x 1.1	13.7	S0	50	Mag 10.05v star E 2′.2.
01 58 34.7	+44 34 30	UGC 1447	14.1	0.9 x 0.7	13.5	SABcd	150	Bright knot or star E edge.
02 00 58.2	+46 08 28	UGC 1492	13.7	1.2 x 0.4	12.8	S0/a	3	Chain of five mag 11-12 stars SW 4′.8.
02 01 09.3	+50 30 23	UGC 1493A	13.4	0.6 x 0.5	12.2	E	90	
02 01 32.9	+45 00 14	UGC 1504	13.7	1.0 x 0.6	13.0	S0/a	155	Several stars superimposed, in star rich area.
02 01 51.5	+44 53 33	UGC 1508	14.7	1.8 x 0.4	14.2	S0	7	**MCG +7-5-8** N 2′.4.
02 04 10.8	+47 58 30	UGC 1552	13.1	1.0 x 0.7	12.6	SB(r)0°	0	Mag 10.40v star NE 1′.2.
02 04 23.9	+45 46 29	UGC 1562	14.0	1.4 x 0.7	14.0	E	105	Galaxy pair, contact. MCG +8-4-17 E 1′.4.
02 04 36.5	+47 56 16	UGC 1563	13.5	1.3 x 1.3	14.1	E:		Two stars, mags 10.7 and 10.9, SE 1′.6.
02 06 02.0	+45 11 34	UGC 1585	13.5	0.9 x 0.6	12.7	SAB(s)ab	165	**CGCG 538-15** W 2′.8.
02 07 42.9	+45 37 22	UGC 1607	13.0	1.8 x 0.5	12.7	SB(s)ab	78	Mag 14.4 star on S edge.
02 07 56.1	+43 35 17	UGC 1612	14.4	0.8 x 0.5	13.3	SAc	55	Stellar nucleus; mag 10.5 star E 1′.3.
02 08 21.4	+41 28 43	UGC 1626	13.4	1.4 x 1.2	13.8	SAB(rs)c	114	Small, faint anonymous galaxy on S edge.
02 08 57.2	+47 13 09	UGC 1634	13.3	2.1 x 2.1	14.8	SAB(r)cd		Knotty, multi-branching arms; mag 10.94v star NW 3′.2.
02 10 14.9	+41 31 15	UGC 1661	13.8	1.1 x 0.7	13.6	E:	12	Mag 12.7 star on W edge; very slender **UGC 1656** N 2′.4.
02 12 15.6	+42 22 45	UGC 1692	15.7	0.7 x 0.6	14.6	Sbc	18	Mag 10.50v star N 1′.4.
02 13 16.6	+41 14 32	UGC 1704	14.6	1.1 x 0.4	13.5	Sd	35	Close pair of stars, mags 11.8 and 12.3, SW edge.
02 15 21.1	+49 50 34	UGC 1728	15.9	1.0 x 0.2	14.0	Sd	37	Mag 11.9 star N 2′.3.
02 18 56.7	+40 33 49	UGC 1780	14.7	1.6 x 0.4	14.0	IBm:	158	Mag 11.1 star ENE 1′.3.
02 19 15.1	+42 44 02	UGC 1782	17.1	1.5 x 0.4	16.4	SBdm: IV	63	Mag 11.6 star SW 2′.3.
02 20 23.1	+40 47 31	UGC 1796	14.5	1.1 x 1.1	14.5	SAB(s)dm		Stellar nucleus.
02 21 06.6	+48 57 33	UGC 1802	13.3	0.9 x 0.8	13.0	E	150	Mag 12.0 star W 1′.4.
02 21 13.7	+42 45 45	UGC 1807	15.9	1.5 x 1.5	16.6	Im:		Dwarf, very low surface brightness; mag 8.79v star W 1′.9.
02 22 15.8	+43 32 47	UGC 1827	14.8	0.6 x 0.5	13.3	Sdm:	0	Mag 14.1 star on N edge; mag 8.35v star NW 2′.5.
02 22 31.6	+47 50 56	UGC 1830	12.7	2.8 x 2.7	14.7	SB0/a	39	Small 1′.2 X 0′.9 core with very faint arms extending to published dimensions; mag 9.43v star NW 3′.3.
02 22 31.2	+43 03 55	UGC 1832	14.3	1.2 x 0.7	14.0	Sa	151	

(Continued from previous page)

GALAXIES

RA h m s	Dec ° ′ ″	Name	Mag (V)	Dim ′ Maj x min	SB	Type Class	PA	Notes
02 22 58.6	+43 00 40	UGC 1837	13.8	0.8 x 0.6	12.9	S0	39	UGC 1841 SE 2.′7.
02 23 09.8	+41 22 12	UGC 1840	13.2	1.7 x 1.6	14.1	Pec	25	Published dimensions contain undocumented anonymous companion galaxy E side, dark patch (loop) NW; mag 5.80v star WNW 4.′0.
02 23 11.5	+42 59 27	UGC 1841	14.0	3.3 x 2.3	16.2	E	126	Bright core, smooth envelope; mag 8.48v star N edge; UGC 1837 NW 2.′7..
02 24 20.0	+43 15 43	UGC 1854	14.7	1.4 x 0.3	13.6	S	45	**UGC 1849** NW 2.′6.
02 24 29.7	+40 52 10	UGC 1855	13.9	1.0 x 0.6	13.2	SB(s)a	88	Mag 10.49v star ENE 2.′5.
02 24 47.4	+41 40 48	UGC 1858	14.8	1.3 x 0.9	14.8	SB	3	SE-NW bar; mag 12.1 star N 1.′1.
02 24 44.3	+42 37 21	UGC 1859	12.9	1.5 x 0.8	13.2	E?	47	Mag 10.9 star NW 2.′1.
02 25 07.7	+41 51 06	UGC 1866	13.9	1.0 x 0.5	13.0	SBa	30	Bright knot or star superimposed N end; mag 9.61v star SW 2.′7.
02 25 15.6	+45 27 04	UGC 1867	14.4	1.9 x 0.3	13.6	Scd	57	Mag 11.1 star NE 2.′3.
02 26 38.6	+50 02 40	UGC 1893	14.1	1.2 x 0.7	13.8	SB0/a	33	
02 27 25.6	+41 03 33	UGC 1914	13.7	1.0 x 0.8	13.5	E	81	Mag 7.99v star SE 0.′9.

GALAXY CLUSTERS

RA h m s	Dec ° ′ ″	Name	Mag 10th brightest	No. Gal	Diam ′	Notes
01 57 06.0	+41 21 00	A 276	16.3	39	8	Mag 7.58v star on E edge; all cluster members faint and stellar.
02 25 48.0	+41 52 00	A 347	13.3	32	56	Numerous almost stellar and larger catalog and anonymous GX.

OPEN CLUSTERS

RA h m s	Dec ° ′ ″	Name	Mag	Diam ′	No. ★	B ★	Type	Notes
02 27 18.0	+42 19 00	Lederman 1		14			ast	(A) "Black Widow spider" asterism of GSC stars. Noted 1997 by R. Lederman.

PLANETARY NEBULAE

RA h m s	Dec ° ′ ″	Name	Diam ″	Mag (P)	Mag (V)	Mag cent ★	Alt Name	Notes
01 42 19.9	+51 34 35	NGC 650-51	187	12.2	10.1	15.9	PK 130−10.1	= **M 76**, **Little Dumbbell**. Bright elongated "core" oriented NE-SW with fainter nebulosity extending SE-NW to give it it's "dumbbell" shape.
01 37 19.6	+50 28 11	PN G130.3−11.7	9	14.1	14.1	16.2	PK 130−11.1	Mag 11.5 star E 0.′8; mag 12.6 star W 1.′2.

GALAXIES

RA h m s	Dec ° ′ ″	Name	Mag (V)	Dim ′ Maj x min	SB	Type Class	PA	Notes
00 35 31.3	+36 30 31	Andromeda III	14.0	2.0 x 2.0	15.6	E?		
23 21 38.8	+33 29 01	CGCG 497-7	14.1	0.5 x 0.5	12.4			Mag 10.93v star E 2.′7; UGC 12538 SW 7.′8.
23 37 39.6	+30 07 47	CGCG 497-42	14.9	0.3 x 0.2	11.9	E?		**Barbon's Galaxy**, (= **Mkn 328**); mag 8.80v star SW 8.′1.
00 01 36.3	+33 33 42	CGCG 499-38	14.4	0.4 x 0.3	11.9		15	Mag 12.7 star N 1.′2; mag 9.00v star E 7.′6.
00 03 09.1	+31 02 08	CGCG 499-40	14.1	0.7 x 0.3	12.3		135	Mag 12.8 star SE 0.′8.
00 16 49.1	+29 37 07	CGCG 499-95	13.9	0.6 x 0.6	12.6			Mag 10.43v star SE 6.′9.
00 17 02.8	+29 56 28	CGCG 499-98	13.9	0.9 x 0.7	13.2		123	8.76v star NNW 5.′2.
00 23 40.7	+30 33 32	CGCG 500-13	14.4	0.8 x 0.6	13.5	S0	3	Close pair of mag 11 stars N 4.′7; mag 11.7 star W 3.′7.
00 24 37.1	+29 17 11	CGCG 500-15	13.9	0.6 x 0.3	11.9		129	Mag 10.03v star N 5.′0.
00 29 07.2	+31 00 06	CGCG 500-32	14.1	0.8 x 0.5	13.0		95	Mag 12.1 star W 1.′2.
00 29 25.5	+30 33 25	CGCG 500-33	15.3	0.6 x 0.4	13.7	Sbc	141	= **Mkn 551**; double system; forms triangle with mag 9.90v star N 2.′5; mag 10.5 star NE 2.′7.
00 35 13.0	+29 47 56	CGCG 500-48	14.1	0.5 x 0.3	11.9		123	Mag 11.8 star SW 4.′3.
00 39 45.0	+33 09 27	CGCG 500-65	14.0	0.8 x 0.7	13.2		57	Mag 8.73v star W 8.′9.
00 42 48.0	+29 55 18	CGCG 500-74	14.2	0.5 x 0.4	12.5	E	120	Mag 12.4 star NW 1.′7.
00 42 22.0	+29 38 27	IC 43	13.2	1.5 x 1.2	13.7	SABc II-III	117	Mag 10.4 star SE 3.′4; **IC 45** N 3.′4.
00 24 27.9	+38 11 03	IC 1550	13.9	0.9 x 0.6	13.1	S0	159	Mag 13.8 star on N edge.
23 47 15.3	+32 46 56	IC 5355	13.7	1.0 x 0.5	12.8	SBcd:	10	Mag 12.2 star W 2.′0.
00 00 09.4	+32 44 16	IC 5370	14.1	0.6 x 0.5	12.7	S0/a	117	Mag 9.60v star SSE 9.′1; **IC 5369** SW 4.′4.
00 00 29.3	+32 46 51	IC 5373	14.7	0.6 x 0.4	13.0	Sb	150	Multi-galaxy system; mag 9.07v star NE 6.′0; **IC 5371** W 2.′7.
00 01 19.8	+34 31 31	IC 5376	13.7	2.0 x 0.4	13.3	Sab	4	Mag 9.77v star S 8.′1.
23 55 21.7	+30 23 04	MCG +5-1-3	14.6	0.9 x 0.3	13.0	S?	102	MCG +5-1-4 N edge.
23 55 23.4	+30 23 32	MCG +5-1-4	14.2	0.5 x 0.4	12.3	Sc	60	Mag 12.8 star NE 0.′7; MCG +5-1-3 S edge.
23 56 16.2	+29 24 30	MCG +5-1-8	14.5	0.7 x 0.5	13.2	S?	57	**MCG +5-1-9** S 1.′2; **UGC 12850** SW 2.′8.
00 02 50.2	+31 29 05	MCG +5-1-27	14.3	0.7 x 0.7	13.4	S0?		Mag 14.9 star superimposed W edge; mag 12.1 star E 1.′7.
00 05 29.4	+32 31 33	MCG +5-1-31	13.9	0.6 x 0.6	12.7	S?		Mag 11.0 star SW 0.′8.
00 13 44.8	+30 11 39	MCG +5-1-55	13.8	0.8 x 0.4	12.5	S?	60	Mag 14.0 star superimposed on NE end.
00 16 32.9	+30 20 44	MCG +5-1-58	13.8	1.1 x 0.8	13.6	S0	6	**CGCG 499-94** NW 2.′4.
00 17 17.4	+30 12 32	MCG +5-1-61	14.4	0.7 x 0.5	13.1	S?	102	Mag 10.94v star S 2.′1.
00 17 59.1	+29 28 09	MCG +5-1-63	14.0	0.7 x 0.7	13.1			A N-S pair of mag 14 stars S 1.′4.
00 21 13.3	+30 28 26	MCG +5-2-1	14.3	0.6 x 0.4	12.6	S0?	72	MCG +5-2-2 N 2.′0; mag 11.5 star N 1.′4.
00 21 16.1	+30 30 19	MCG +5-2-2	14.1	0.9 x 0.7	13.5	SB?	132	MCG +5-2-1 S 2.′0; mag 11.5 star SW 1.′3.
00 21 12.1	+29 38 42	MCG +5-2-3	14.6	0.7 x 0.5	13.3		78	Mag 11.5 star W 1.′7.
00 22 16.2	+30 14 49	MCG +5-2-5	14.4	0.9 x 0.5	13.4		72	Mag 12.7 star on S edge; faint, elongated anonymous galaxy NE 2.′4.
00 22 45.8	+29 56 19	MCG +5-2-8	14.6	0.8 x 0.5	13.4	S?	165	Mag 14.4 star on NE edge.
00 25 13.7	+30 02 14	MCG +5-2-11	14.3	1.0 x 0.8	13.9	SB?	78	Knotty; mag 9.90v star S 1.′6.
00 32 44.6	+30 45 00	MCG +5-2-23	14.1	1.4 x 0.7	13.9	S?	153	**CGCG 500-43** S 6.′3.
00 36 18.4	+32 54 10	MCG +5-2-28	14.2	0.9 x 0.5	13.2	SB?	18	
00 38 28.4	+29 37 27	MCG +5-2-31	14.3	1.0 x 0.4	13.1	S?	24	Numerous small, faint anonymous galaxies E and S.
23 16 43.8	+29 35 12	MCG +5-54-56	14.8	0.8 x 0.3	13.1	S?	18	Mag 14.5 star SW end.
23 17 46.1	+29 01 35	MCG +5-54-60	14.7	0.5 x 0.5	13.0			UGC 12482 W 2.′7.
23 18 15.4	+29 13 49	MCG +5-54-61	14.2	0.9 x 0.9	14.0	E		Mag 14.3 star SW edge.
23 20 01.3	+32 55 20	MCG +5-55-1	14.7	0.9 x 0.2	12.7	S?	99	Mag 10.7 star N 2.′0.
23 22 58.4	+29 14 02	MCG +5-55-7	14.7	0.6 x 0.6	13.4	Spherical		UGC 12566 SSW 6.′4.
23 23 22.1	+29 25 43	MCG +5-55-9	14.6	0.4 x 0.4	12.5	SB?		Mag 13.4 star SW 0.′7.
23 23 34.4	+29 10 54	MCG +5-55-10	13.8	0.8 x 0.6	12.9	S0	39	Mag 12.9 star NE 3.′3.
23 24 59.6	+29 47 22	MCG +5-55-12	15.3	0.8 x 0.7	14.5	SB?	39	**MCG +5-55-13** E 0.′9.
23 26 35.9	+32 50 45	MCG +5-55-17	14.6	0.9 x 0.4	13.3	SB?	81	Mag 11.3 star NW 3.′8.
23 28 01.2	+32 09 43	MCG +5-55-21	14.3	0.9 x 0.3	12.7	S?	75	Mag 11.7 star SSW 2.′1.

(Continued from previous page)

GALAXIES

RA h m s	Dec ° ′ ″	Name	Mag (V)	Dim ′ Maj x min	SB	Type Class	PA	Notes
23 28 10.9	+32 28 19	MCG +5-55-22	14.3	0.8 x 0.5	13.2	S0	21	Mag 11.3 star NNW 3'.6.
23 28 49.2	+29 44 04	MCG +5-55-24	14.4	0.7 x 0.4	12.8	Sbc	53	UGC 12625 NE 4'.5.
23 29 53.7	+32 05 04	MCG +5-55-27	14.0	0.7 x 0.3	12.2	S0?	15	Mag 11.3 star SE 3'.4.
23 29 55.0	+32 38 54	MCG +5-55-28	14.2	0.9 x 0.3	12.6	S?	24	Mag 13.1 star NE 1'.5.
23 30 21.4	+32 53 01	MCG +5-55-30	14.5	0.7 x 0.5	13.2	S?	36	Mag 13.1 star N end; mag 11.2 star W 2'.2.
23 37 16.7	+31 47 59	MCG +5-55-41	14.0	0.5 x 0.4	12.1		126	Mag 10.80v star ENE 0'.8.
23 39 08.6	+32 05 06	MCG +5-55-45	13.8	0.7 x 0.5	12.6	S0/a	18	Mag 9.80v star N edge.
23 42 30.1	+30 35 48	MCG +5-55-48	15.4	0.9 x 0.3	13.8	Sbc	162	Mag 13.5 star W 2'.4.
23 43 56.9	+29 07 54	MCG +5-55-50	15.3	0.7 x 0.3	13.5	Sb	27	Mag 11.4 star 0'.6 SW; **MCG +5-55-52** ESE 1'.9; MCG +5-55-51 S 1'.0.
23 43 57.8	+29 06 54	MCG +5-55-51	16.4	0.7 x 0.3	14.5	Sb	111	MCG +5-55-50 and mag 11.4 star NW 0'.8.
23 49 47.8	+30 02 18	MCG +5-56-11	14.4	0.5 x 0.4	12.5	S?	0	Mag 8.72v star NE 2'.5.
23 53 45.8	+29 15 19	MCG +5-56-20	14.7	0.7 x 0.4	13.1	SB?	120	**MCG +5-56-19** W 1'.5.
23 59 53.8	+35 25 03	MCG +6-1-2	15.3	0.7 x 0.4	13.8	Sc	156	
00 01 03.7	+34 39 10	MCG +6-1-4	14.3	0.7 x 0.6	13.2	SAB a:	126	Small, elongated **MCG +6-1-3** W 1'.6.
00 01 15.8	+36 58 10	MCG +6-1-6	15.2	0.4 x 0.4	13.1			
00 02 26.0	+33 56 56	MCG +6-1-8	15.3	0.6 x 0.4	13.6	Sb	102	Mag 11.11v star E 2'.8.
00 03 32.2	+37 20 13	MCG +6-1-9	15.3	0.7 x 0.3	13.5	Sb	171	
00 07 06.7	+35 49 04	MCG +6-1-12	15.0	0.9 x 0.8	14.5		111	Located between two N-S mag 11 stars.
00 10 41.7	+35 51 03	MCG +6-1-16	15.9	0.9 x 0.5	14.9	Spiral	15	Mag 11.14v star N 3'.9.
00 13 42.9	+36 37 07	MCG +6-1-18	15.0	0.6 x 0.4	13.6	E/S0	57	Mag 9.35v star W 3'.0.
00 16 34.0	+36 11 51	MCG +6-1-20	14.9	0.8 x 0.5	13.7	Sb	174	Mag 11.6 star W 2'.4.
00 17 04.5	+35 16 18	MCG +6-1-22	15.0	0.5 x 0.4	13.1	Sbc	21	
00 17 12.7	+35 53 00	MCG +6-1-23	15.7	0.8 x 0.5	14.6	Sc	12	Mag 9.28v star SE 3'.7.
00 21 05.0	+34 30 52	MCG +6-1-26	13.4	0.6 x 0.4	11.7		156	Mag 11.7 star NW 1'.9.
00 21 31.4	+38 04 35	MCG +6-1-28	16.3	0.6 x 0.2	13.8	Sb	126	MCG +6-1-30 N 1'.1; **MCG +6-1-29** WSW 1'.4.
00 21 34.2	+38 05 33	MCG +6-1-30	14.6	0.8 x 0.5	13.5	Sb	138	MCG +6-1-28 S 1'.1; mag 11.4 star N 2'.5.
00 21 55.4	+39 33 56	MCG +6-2-1	15.2	0.7 x 0.5	14.0	Sbc	3	Mag 9.92v star superimposed N edge.
00 39 30.1	+36 20 46	MCG +6-2-12	14.1	0.9 x 0.5	13.1	S0/a	60	Mag 14.9 star on W edge; mag 11.3 star NW 4'.7.
00 41 54.0	+37 06 01	MCG +6-2-14	15.5	0.7 x 0.6	14.5	Sc	117	Mag 10.24 star W 2'.3.
23 27 18.9	+37 13 15	MCG +6-51-4	14.1	0.8 x 0.7	13.3	Sbc	48	Mag 12.3 star NW 1'.7.
23 28 05.1	+34 51 35	MCG +6-51-5	14.8	0.8 x 0.4	13.4	Sb	9	
23 46 38.9	+39 23 58	MCG +6-52-3	14.8	0.7 x 0.5	13.5		69	Mag 10.3 star SW 2'.7.
00 07 49.7	+35 21 38	NGC 5	13.3	1.1 x 0.7	13.1	E:	115	
00 08 42.3	+37 26 53	NGC 11	13.7	1.5 x 0.3	12.7	Sa	111	Mag 10.2 star N 2'.8.
00 08 47.8	+33 25 59	NGC 13	13.2	2.5 x 0.6	13.5	(R)Sab: III	53	Mag 8.46v star NNW 8'.1.
00 09 32.8	+33 18 29	NGC 20	13.0	1.5 x 1.5	13.8	S0$^-$:		Mag 12.0 star E of core; mag 10.24v star E 2'.4.
00 10 41.0	+32 59 00	NGC 19	13.2	1.2 x 0.6	12.7	SB(r)bc	42	
00 10 33.0	+28 59 44	NGC 27	13.5	1.3 x 0.6	13.1	S?	117	Forms triangle with UGC 95 W and mag 10.51v star S.
00 10 47.2	+33 21 05	NGC 29	12.7	1.5 x 0.8	12.8	SAB(s)bc:	154	**NGC 21** (now nonexistent) was a duplicate observation of NGC 29.
00 12 19.0	+31 03 38	NGC 39	13.5	1.1 x 1.0	13.5	SA(rs)c	120	Weak stellar nucleus in uniform envelope.
00 13 01.0	+30 54 55	NGC 43	12.6	1.6 x 1.5	13.4	SB0	15	Mag 13.4 star on N edge; mag 10.94v star SW 3'.9.
00 18 15.0	+30 03 44	NGC 67	14.2	0.4 x 0.2	11.5	E:	51	Anonymous galaxy SW 0'.8.
00 18 18.5	+30 04 16	NGC 68	12.9	1.2 x 1.0	13.0	SA0$^-$	0	Located on SW edge of NGC 70.
00 18 20.6	+30 02 21	NGC 69	14.8	0.4 x 0.3	12.3	SB(s)0$^-$	33	Compact; NGC 72 E 1'.7.
00 18 22.8	+30 04 48	NGC 70	13.5	1.4 x 1.2	13.9	SA(rs)c III	0	Pair of faint stars superimposed N and S of nucleus.
00 18 23.6	+30 03 44	NGC 71	13.2	1.2 x 0.9	13.1	SA0$^-$ pec:	129	Located on S edge of NGC 70.
00 18 28.6	+30 02 28	NGC 72	13.5	1.1 x 0.9	13.3	SB(rs)ab	15	NGC 72A E 1'.3.
00 18 34.4	+30 02 09	NGC 72A	14.7	0.3 x 0.3	12.0	E3:		Anonymous galaxy on E edge.
00 18 49.5	+30 03 40	NGC 74	14.8	0.8 x 0.3	13.1	Sb	131	Located 5'.9 E of NGC 70; faint, stellar anonymous galaxy ENE 6'.8.
00 19 37.9	+29 56 01	NGC 76	13.1	1.0 x 0.9	12.8	S?	65	**MCG +5-1-73** E 1'.2.
00 22 29.9	+29 44 44	NGC 97	12.3	1.3 x 1.3	12.9	E:		Mag 9.64v star N 4'.2; **CGCG 500-11** NE 10'.9.
00 25 59.9	+29 12 40	NGC 108	12.1	2.3 x 1.8	13.5	(R)SB(r)0$^+$	153	Mag 9.33v star NNW 7'.1.
00 26 49.0	+31 42 09	NGC 112	13.6	1.0 x 0.5	12.7	S?	105	Mag 11.2 star N 3'.5.
00 31 20.6	+30 47 24	NGC 140	13.2	1.5 x 1.3	13.8	Scd:	54	
00 33 50.3	+30 43 25	NGC 149	13.7	1.2 x 0.7	13.4	S0:	155	
00 38 23.2	+29 28 20	NGC 181	14.9	0.6 x 0.2	12.4	S?	147	Located 2'.7 S of NGC 183.
00 38 29.5	+29 30 40	NGC 183	12.7	1.7 x 1.1	13.5	E	130	Very small anonymous galaxy 2'.5 NE; **CGCG 500-61** E 4'.8.
00 38 36.0	+29 26 51	NGC 184	14.7	0.7 x 0.2	12.4	S0/a	6	Located 8'.3 due N of mag 4.35v star Epsilon Andromedae.
00 41 44.8	+36 21 30	NGC 218	14.1	0.9 x 0.7	13.5	S?	126	Faint star on S edge; mag 9.23v star E 9'.1.
00 42 41.7	+40 51 59	NGC 221	8.1	8.7 x 6.5	12.5	cE2	179	= **M 32**.
00 42 54.3	+32 34 47	NGC 226	13.4	0.9 x 0.6	12.6	S?	105	Mag 13.4 star on S edge; mag 7.66v star W 10'.6.
00 43 36.8	+30 35 10	NGC 233	12.4	1.2 x 1.2	12.9	E?		
23 22 06.7	+40 50 43	NGC 7640	11.3	10.5 x 1.8	14.4	SB(s)c II	167	Bright nucleus bisected by short dark lane; numerous superimposed stars.
23 28 35.2	+32 24 53	NGC 7680	12.6	1.9 x 1.9	13.8	S0$^-$:		Bright center, uniform envelope with some superimposed stars.
23 40 33.7	+29 11 14	NGC 7729	13.5	1.9 x 0.6	13.5	Sa	7	
23 46 58.7	+29 27 31	NGC 7752	14.3	0.9 x 0.5	13.3	I0:	107	Located on SW edge of NGC 7753.
23 47 04.7	+29 28 58	NGC 7753	12.0	3.3 x 2.1	14.0	SAB(rs)bc I	50	Bright nucleus, western arm extends S towards NGC 7752.
23 49 11.9	+30 58 57	NGC 7760	13.4	0.9 x 0.9	13.3	E?		Faint star SW edge of nucleus.
23 52 10.1	+31 16 35	NGC 7773	13.4	1.2 x 1.2	13.7	SBbc		
23 59 20.4	+31 17 05	NGC 7799	15.7	1.1 x 0.2	13.9	Scd:	16	
00 01 26.8	+31 26 00	NGC 7805	13.3	1.0 x 0.7	12.8	SAB0°: pec	45	
00 01 30.0	+31 26 30	NGC 7806	13.5	1.2 x 0.9	13.4	SA(rs)bc? pec	20	**MCG +5-1-26** E 0'.9; NGC 7805 on SW edge.
00 04 24.8	+31 28 20	NGC 7819	13.5	1.5 x 1.2	14.0	SB(s)b II	109	Bright nucleus in short NE-SW bar.
00 07 19.7	+32 36 33	NGC 7831	12.8	1.7 x 0.4	12.2	Sb: sp III	38	Mag 9.46v star S 2'.2.
00 08 01.7	+33 04 14	NGC 7836	13.7	0.9 x 0.5	12.7	I?	133	
00 03 20.5	+29 47 48	UGC 12	14.7	0.7 x 0.5	13.4	Scd:	105	
00 07 51.0	+35 57 56	UGC 63	14.7	0.7 x 0.4	13.2	Im	39	Mag 11.8 star S 1'.5.
00 07 44.1	+40 52 30	UGC 64	14.6	1.0 x 0.5	13.7		35	Mag 14.2 star on S edge.
00 10 16.9	+30 50 59	UGC 93	14.2	2.0 x 1.4	15.2	SAdm	60	Stellar nucleus; mag 10.02v star S 1'.8.
00 11 01.1	+30 03 08	UGC 102	13.5	1.5 x 1.2	14.0	SABa	85	Bright, slightly elongated core, uniform envelope; mag 9.65v star N 1'.5.
00 12 54.9	+33 21 37	UGC 117	14.1	1.4 x 0.8	14.1	SAB(s)cd	110	Mag 10.27v star SW 1'.1.
00 13 50.7	+35 59 35	UGC 128	15.9	1.9 x 1.2	16.6	Sdm IV-V	65	Very low surface brightness.
00 13 57.1	+30 52 54	UGC 130	13.8	0.6 x 0.4	12.1	S?	163	Almost stellar nucleus; mag 13.2 star and very faint anonymous galaxy N 3'.4.
00 15 43.7	+29 39 58	UGC 147	13.6	1.5 x 0.2	12.1	S0/a	156	Located between a pair of mag 12.0 and 13.0 stars; very faint, small anonymous galaxy on E edge.
00 16 28.7	+29 55 23	UGC 152	14.7	1.6 x 0.3	13.7	Scd:	78	Lies between a mag 9.55v star S 4'.7 and a mag 8.76v star NNE 8'.2.
00 17 03.6	+34 29 56	UGC 160	13.7	0.9 x 0.8	13.1	SAdm IV-V	108	Mag 12.7 star E edge; mag 13.5 star N edge.
00 17 35.9	+30 12 15	UGC 166	14.2	1.7 x 0.7	14.2	Scd:	18	Mag 8.73v star and **CGCG 499-101** SSE 3'.9.
00 22 26.5	+29 30 13	UGC 215	13.8	1.3 x 0.8	13.6	SBab	135	Mag 12.7 stars on NW end and E edge.
00 24 38.7	+33 15 23	UGC 232	13.7	1.5 x 1.0	13.9	SB(r)a	45	Bright N-S oriented center, faint envelope.

(Continued from previous page)

45 GALAXIES 45

RA h m s	Dec ° ′ ″	Name	Mag (V)	Dim ′ Maj x min	SB	Type Class	PA	Notes
00 25 03.4	+31 20 38	UGC 238	13.4	1.8 x 0.5	13.1	S?	178	Mag 12.6 star on SE edge.
00 27 16.4	+39 47 28	UGC 262	14.1	1.2 x 0.6	13.6	Pec	67	Brighter E of center.
00 28 15.3	+30 48 11	UGC 279	13.3	1.7 x 0.5	13.0	SB?	118	Mag 10.09v star SW 2′8.
00 28 34.3	+33 16 19	UGC 284	14.0	1.5 x 0.8	14.1	SAB(s)c	23	Weak stellar nucleus; **CGCG 500-25** W 4′2.
00 29 53.4	+31 23 35	UGC 299	14.3	1.5 x 0.7	14.2	Sc	74	Stellar nucleus; mag 10.05v star E 4′2; **UGC 294** W 6′0.
00 31 53.0	+37 40 39	UGC 318	15.0	0.6 x 0.5	13.6	SBc	159	Close pair of stars, mags 10.9 and 11.1, W 2′3.
00 32 12.1	+31 40 56	UGC 319	13.9	1.5 x 0.9	14.1	Sbc II	135	Mag 10.8 star SW 1′6.
00 33 42.1	+39 32 39	UGC 330	13.7	1.1 x 0.4	12.7	S0	140	Mag 9.17v star NE 4′0.
00 33 54.6	+31 27 03	UGC 334	14.6	1.6 x 1.6	15.5	Sm V		Dwarf spiral, extremely low surface brightness.
00 34 56.0	+31 56 52	UGC 346	14.0	1.1 x 0.5	13.2	Sbc	128	
00 35 55.7	+31 52 13	UGC 355	13.7	1.2 x 0.4	12.7	SBab	123	Mag 10.78v star NW 2′4.
00 37 21.3	+29 08 54	UGC 371	14.4	1.7 x 0.3	13.5	Scd:	45	
00 38 12.4	+30 53 27	UGC 381	14.5	0.8 x 0.6	13.6	Sb	5	Located 14′6 W of mag 3.27v star Delta Andomedae.
00 38 22.3	+32 38 15	UGC 384	13.8	1.2 x 0.8	13.6	SAB(rs)d	174	Stellar nucleus.
00 38 31.9	+30 17 25	UGC 388	15.3	1.2 x 0.6	14.8	Scd:	3	
00 39 18.4	+29 39 26	UGC 400	14.1	0.7 x 0.6	13.0	S?	75	Elongated **UGC 412** N 6′7.
00 40 41.4	+30 09 39	UGC 431	14.3	0.5 x 0.4	12.4	Sb	54	Mag 11.2 star NW 2′1.
00 41 03.7	+31 43 54	UGC 433	13.9	1.7 x 0.3	13.0	Scd:	74	
00 41 52.0	+32 59 28	UGC 442	14.3	1.5 x 0.5	13.9	SA(rs)d	175	
00 42 04.8	+36 48 13	UGC 444	13.1	1.0 x 0.7	12.6	S?	163	
00 43 50.8	+32 51 12	UGC 465	13.5	1.7 x 0.4	13.0	SBa	178	
23 16 43.6	+33 59 41	UGC 12474	13.3	1.1 x 0.4	12.3	Sa	88	Mag 8.38v star S 6′8.
23 17 02.9	+30 20 05	UGC 12476	13.0	1.4 x 0.7	12.8	S0/a	93	Mag 7.69v star NNW 2′6.
23 17 33.7	+29 01 08	UGC 12482	13.0	1.4 x 1.2	13.5	E:	39	MCG +5-54-60 E 2′8; **CGCG 496-75** NNW 2′9.
23 20 31.6	+29 18 22	UGC 12530	13.8	1.1 x 1.1	13.8	SBcd:		Mag 12.4 star E 2′4.
23 21 10.0	+33 23 56	UGC 12538	13.9	0.9 x 0.3	12.3	S?	20	CGCG 497-7 NE 7′8.
23 22 28.7	+29 10 46	UGC 12557	14.1	2.1 x 0.4	13.8	S?	9	Mag 10.45v star W 9′2.
23 22 46.0	+29 08 16	UGC 12566	13.6	1.7 x 1.3	14.3	Sab	140	Bright center, prominent dark patches SE and NW, faint envelope.
23 23 10.3	+32 31 35	UGC 12570	14.0	0.5 x 0.4	12.1	Pec	168	Mag 10.7 star NW 0′8.
23 29 07.1	+29 46 29	UGC 12625	14.2	1.5 x 0.2	12.8	Sb III	176	Mag 10.02v star NNW 3′0.
23 29 59.0	+40 59 25	UGC 12632	12.1	4.6 x 3.7	15.0	Sm: V	30	Dwarf spiral, extremely low surface brightness; mag 11.08v star SW 3′6.
23 30 07.2	+40 13 29	UGC 12634	14.8	1.0 x 0.2	12.9	Scd:	175	Triangle of two mag 12 stars and one mag 11 star NE 1′9.
23 30 27.0	+30 13 16	UGC 12639	13.7	1.2 x 0.8	13.5	SB(s)b	171	Mag 12.8 star NE 2′3.
23 31 29.5	+32 28 50	UGC 12645	14.1	0.9 x 0.5	13.1	S?	97	Dark patch E of center.
23 32 02.5	+32 25 23	UGC 12650	14.7	1.6 x 0.2	13.3	Scd:	159	
23 32 10.5	+35 23 11	UGC 12651	15.9	0.9 x 0.4	14.6	Sdm:	123	Mag 7.98v star SE 4′1.
23 32 43.6	+29 27 37	UGC 12657	13.2	1.9 x 0.6	13.3	S0	140	Large, diffuse core, uniform envelope.
23 33 41.2	+32 23 02	UGC 12666	13.7	1.5 x 0.6	13.4	Scd:	132	Mag 12.1 star on NW end; mag 11.46v star 0′8 N of this star.
23 33 49.7	+30 03 37	UGC 12667	12.8	1.5 x 0.9	13.0	Scd:	142	**UGC 12665** SW 2′6.
23 34 23.1	+34 37 29	UGC 12672	13.6	1.0 x 0.8	13.2	S?	18	Mag 12.4 star NE edge.
23 35 43.8	+32 23 08	UGC 12693	14.7	2.5 x 0.2	13.8	SBcd?	3	Mag 9.54v star NE 6′8.
23 37 59.5	+31 59 40	UGC 12711	13.4	0.6 x 0.6	12.1	Sb:		Mag 10.6 star NE 2′2.
23 38 14.4	+30 42 27	UGC 12713	14.4	1.1 x 0.5	13.6	S0/a	60	
23 38 13.2	+32 20 05	UGC 12714	14.7	1.7 x 0.2	13.4	Scd:	168	Pair of mag 13 stars S 1′2.
23 41 55.5	+30 34 51	UGC 12741	13.7	0.9 x 0.3	12.1	Sa	102	
23 46 12.2	+33 22 14	UGC 12776	12.9	2.4 x 1.8	14.3	SB(rs)b	141	Brighter N-S bar; mag 11.6 star W edge.
23 50 33.9	+28 59 52	UGC 12803	14.0	1.3 x 1.0	14.2	S?	126	Mag 12.1 star on E edge.
23 55 41.8	+31 53 57	UGC 12845	14.1	1.7 x 1.5	15.0	Sd	61	Stellar nucleus with uniform envelope; mag 12.8 star NE edge.
23 57 06.5	+29 50 18	UGC 12861	14.2	0.9 x 0.5	13.2	Sab	140	Mag 10.9 star E 1′6; mag 10.7 star N 2′5.
23 57 24.0	+30 59 29	UGC 12864	13.5	1.7 x 0.9	13.8	SBb	110	Brighter E-W bar, uniform envelope.
23 58 29.1	+32 13 48	UGC 12869	14.2	0.9 x 0.9	13.8	SBcd:		
00 01 14.3	+34 40 33	UGC 12904	14.2	1.1 x 0.9	14.0	SBab	36	Disturbed.

GALAXY CLUSTERS

RA h m s	Dec ° ′ ″	Name	Mag 10th brightest	No. Gal	Diam ′	Notes
00 37 48.0	+29 35 00	A 71	15.5	30	19	Band of numerous, faint, stellar anonymous GX extend SE to NW.

STAR CLOUDS

RA h m s	Dec ° ′ ″	Name	Mag	Diam ′	No. ★	B ★	Type	Notes
00 40 36.0	+40 44 00	NGC 206		4.2				A large star cloud in the SW part of M 31, the Andromeda Galaxy. It is extended N-S and best defined on the E flank.

PLANETARY NEBULAE

RA h m s	Dec ° ′ ″	Name	Diam ″	Mag (P)	Mag (V)	Mag cent ★	Alt Name	Notes
23 35 53.6	+30 28 02	PN G104.2−29.6	320	15.1	12.1	16.1	PK 104−29.1	= **Jones 1**; slight brightening along NW and S edges; mag 11.00v star NE 8′0; mag 11.35v star W 9′0.

46 GALAXIES 46

RA h m s	Dec ° ′ ″	Name	Mag (V)	Dim ′ Maj x min	SB	Type Class	PA	Notes
21 57 12.0	+30 43 47	CGCG 493-15	14.4	0.5 x 0.4	12.7	E	144	**CGCG 493-16** ESE 2′6.
23 10 11.2	+29 54 57	CGCG 496-50	13.8	0.7 x 0.6	12.7		162	Multi-galaxy system.
23 21 38.8	+33 29 01	CGCG 497-7	14.1	0.5 x 0.5	12.4			Mag 10.93v star E 2′7; UGC 12538 SW 7′8.
22 10 14.9	+36 39 18	CGCG 513-10	15.0	0.8 x 0.5	13.8	Sc	69	Group of 10 mag 10-12 stars, including a mag 10.49v star, W 5′5.
22 28 34.5	+33 59 54	CGCG 514-36	14.2	0.6 x 0.5	12.9	E	174	Mag 10.4 star1′6 N.
22 38 30.7	+34 14 19	CGCG 514-79	13.9	0.6 x 0.4	12.2		144	**UGC 12132** NE 5′7, also several very small, faint anonymous galaxies in that area.
23 11 05.4	+29 38 33	IC 1473	12.9	2.0 x 1.0	13.5	S0	176	Mag 11.04v star SW 7′5.
22 11 12.1	+38 55 35	IC 5180	13.3	1.0 x 0.8	13.0	E:	120	Mag 8.67v star NE 10′6; **MCG +6-48-14** SE 3′2; **MCG +6-48-13** S 2′2.
22 11 06.5	+29 36 31	MCG +5-52-2	14.6	0.6 x 0.5	13.1	S?		MCG +5-52-3 N 2′2; mag 8.51v star NW 2′2.
22 11 08.3	+29 38 38	MCG +5-52-3	15.5	0.5 x 0.5	14.1	E		MCG +5-52-2 S 2′2; mag 8.51v star W 2′8.
22 14 38.5	+32 57 15	MCG +5-52-4	14.7	0.8 x 0.3	13.0	Sc	39	Close pair of mag 11-12 stars S 2′6.

(Continued from previous page)

GALAXIES

RA h m s	Dec ° ′ ″	Name	Mag (V)	Dim ′ Maj x min	SB	Type Class	PA	Notes
22 22 46.2	+29 46 36	MCG +5-52-13	14.2	0.8 x 0.5	13.1	SB?	39	Mag 12.6 star 1′.2 S.
22 27 11.6	+31 30 22	MCG +5-53-1	14.5	1.1 x 0.2	12.7	S?	141	Trio of mag 12-13 stars ENE 1′.1.
22 36 03.1	+32 32 18	MCG +5-53-7	14.5	0.9 x 0.7	13.9	SBbc	90	Filamentary; mag 13.6 star on SE edge; mag 8.73v star E 1′.1 ENE.
22 49 33.3	+32 22 00	MCG +5-53-17	14.3	0.6 x 0.5	12.8	S0	9	Mag 10.2 star N 3′.5.
22 51 03.7	+31 22 25	MCG +5-53-21	14.6	0.4 x 0.3	12.4	E	159	Located on E edge of UGC 12214; **MCG +5-53-22** ENE 2′.4.
22 53 29.6	+31 38 39	MCG +5-54-3	14.2	0.5 x 0.5	12.6	SBcd:		= **Mkn 922**; mag 11.4 star W 1′.9; **UGC 12231** S 1′.7.
22 54 20.4	+32 29 08	MCG +5-54-8	13.9	0.5 x 0.4	12.2	E:	60	Almost stellar galaxy **IV Zw 123B** ESE 0.9, at mid point to UGC 12242.
22 54 45.2	+32 12 45	MCG +5-54-12	14.8	0.9 x 0.3	13.2	Sbc	51	
22 54 42.6	+32 19 10	MCG +5-54-13	15.9	0.8 x 0.6	15.2	E	129	Faint anonymous galaxy SW 0′.9; mag 9.61v star W 4′.3.
22 54 38.1	+32 24 26	MCG +5-54-14	15.0	0.6 x 0.4	13.3	Sc	75	Mag 12.1 star SW 0′.5.
22 54 52.3	+32 47 37	MCG +5-54-15	14.9	0.6 x 0.4	13.2	Sd	141	Sandwiched between a pair of mag 10.29v and 12.3 stars.
22 55 07.8	+31 18 29	MCG +5-54-16	14.2	0.9 x 0.4	13.0	S?	69	Faint, slightly elongated anonymous galaxy W 0′.9.
22 55 06.4	+32 11 01	MCG +5-54-17	13.9	0.7 x 0.6	13.0	E$_p$/shells	0	Mag 10.1 star WNW 2′.8.
22 55 55.5	+31 40 14	MCG +5-54-19	14.0	0.9 x 0.5	12.9	S?	24	Mag 11.7 star W 3′.0.
23 00 09.6	+31 22 28	MCG +5-54-23	14.6	0.9 x 0.3	13.0	S?	102	UGC 12297 N 2′.1.
23 00 38.8	+30 44 29	MCG +5-54-25	13.7	1.1 x 0.5	13.0	S0/a	120	Mag 12.0 star superimposed N edge.
23 09 29.9	+29 28 57	MCG +5-54-39	14.8	1.0 x 0.2	12.9	S?	102	Mag 12.7 star N 2′.1.
23 10 04.8	+30 13 29	MCG +5-54-41	13.9	0.6 x 0.4	12.2	Sc	156	Mag 11.6 star on S edge; mag 11.33v star E 2′.0.
23 12 38.6	+33 41 56	MCG +5-54-47	14.4	0.9 x 0.3	12.8	S?	48	Mag 13.2 star SE 1′.0.
23 16 43.8	+29 35 12	MCG +5-54-56	14.8	0.8 x 0.3	13.1	S?	18	Mag 14.5 star SW end.
23 17 46.1	+29 01 35	MCG +5-54-60	14.7	0.5 x 0.5	13.0			UGC 12482 W 2′.7.
23 18 15.4	+29 13 49	MCG +5-54-61	14.2	0.9 x 0.9	14.0	E		Mag 14.3 star SW edge.
23 20 01.3	+32 55 20	MCG +5-55-1	14.7	0.9 x 0.2	12.7	S?	99	Mag 10.7 star N 2′.0.
23 22 58.4	+29 14 02	MCG +5-55-7	14.7	0.6 x 0.6	13.4	Spherical		UGC 12566 SSW 6′.4.
23 23 22.1	+29 25 43	MCG +5-55-9	14.6	0.4 x 0.4	12.5	SB?		Mag 13.4 star SW 0′.7.
23 23 34.4	+29 10 54	MCG +5-55-10	13.8	0.8 x 0.6	12.9	S0	39	Mag 12.9 star NE 3′.3.
22 09 43.1	+38 11 45	MCG +6-48-10	14.8	0.8 x 0.6	13.8	Sc	21	Mag 12.3 star NNE 1′.2.
22 13 49.3	+37 01 32	MCG +6-48-20	14.0	0.6 x 0.5	12.6	S0	24	Close N-S pair of mag 10-11 stars NNW 1′.0.
22 16 12.2	+37 28 25	MCG +6-49-1	14.6	0.8 x 0.4	13.2		108	Mag 11.8 star W 1′.3.
22 25 28.3	+39 15 23	MCG +6-49-15	14.1	0.8 x 0.7	13.3	Sb	111	Mag 9.81v star SE 3′.2.
22 26 40.4	+35 03 25	MCG +6-49-17	14.6	0.7 x 0.5	13.5	E?	75	
22 28 44.6	+39 05 03	MCG +6-49-21	14.3	0.7 x 0.5	13.0	Sbc	150	Mag 10.3 star E 1′.3.
22 30 41.2	+39 17 27	MCG +6-49-27	14.4	0.7 x 0.5	13.1		24	Several faint stars on S edge; mag 10.1 star NW 4′.0.
22 32 39.2	+37 13 16	MCG +6-49-31	15.2	0.8 x 0.5	14.0	Sb	57	
22 34 59.9	+37 11 56	MCG +6-49-36	15.3	0.7 x 0.6	14.2	Ia	144	Mag 12.4 star N 1′.5.
22 36 22.9	+34 32 34	MCG +6-49-44	15.4	0.8 x 0.3	13.7	Sb	39	**PGC 69291** SE 3′.4.
22 37 14.0	+38 38 46	MCG +6-49-48	14.5	0.9 x 0.4	13.3	Sb	63	NGC 7330 SSW 6′.9.
22 38 25.4	+35 21 58	MCG +6-49-56	15.1	1.0 x 0.3	13.6	S0/a	24	MCG +6-49-61 NE 1′.2; MCG +6-49-60 and UGC 12127 SSE 12′.1.
22 38 34.9	+35 20 26	MCG +6-49-60	13.8	0.4 x 0.4	11.7			UGC 12127 SW.
22 38 30.4	+35 22 38	MCG +6-49-61	14.9	0.9 x 0.3	13.3		147	Star on W edge; MCG +6-49-56 SW 1′.2; MCG +6-49-60 and UGC 12127 S 2′.2.
22 38 58.7	+35 26 16	MCG +6-49-66	14.2	0.5 x 0.4	12.4	S0	3	Located between a pair of E-W mag 12.4, 12.9 stars.
22 39 21.9	+35 57 48	MCG +6-49-67	14.6	0.8 x 0.6	13.7	Compact	9	Mag 12.1 star NE 1′.9.
22 39 45.6	+34 22 45	MCG +6-49-68	14.2	0.6 x 0.6	12.9	S? IV		Mag 13.6 star on E edge.
22 40 17.8	+38 25 35	MCG +6-49-70	14.9	0.5 x 0.3	12.7	Sc	159	
22 41 23.0	+34 37 17	MCG +6-49-72	13.7	0.9 x 0.6	13.0	S0	126	Mag 10.61v star NE 4′.5.
22 41 47.1	+36 06 13	MCG +6-49-76	14.5	0.8 x 0.7	13.7	Sc	9	Mag 14.1 star superimposed NE of core; mag 9.80v star SE 0′.8.
22 41 48.8	+37 43 05	MCG +6-49-77	14.9	0.7 x 0.6	13.8	Sc	33	Mag 9.34v star SSE 2′.0.
22 53 24.9	+37 10 27	MCG +6-50-7	13.8	0.9 x 0.7	13.2	S0	6	Mag 12.2 star NW 0′.8.
22 55 14.9	+36 40 16	MCG +6-50-10	13.9	0.8 x 0.6	13.0	Sc		Mag 10.09v star E 2′.7.
22 55 44.1	+39 16 51	MCG +6-50-11	14.3	0.8 x 0.6	13.4		111	Mag 10.17v star W 2′.6.
23 02 31.5	+38 42 49	MCG +6-50-18	14.6	0.9 x 0.4	13.3	Sb	9	Mag 12.3 star E 1′.6.
23 03 44.1	+34 58 57	MCG +6-50-20	14.0	0.7 x 0.7	13.0	S?		Mag 12.7 star NW 0′.9.
23 04 56.3	+36 01 14	MCG +6-50-21	14.5	0.8 x 0.4	13.1	Sc	72	
23 07 20.5	+36 21 41	MCG +6-50-25	14.2	0.8 x 0.4	12.8	S0	171	Mag 12.1 star NNE 2′.6.
22 26 28.4	+40 09 45	MCG +7-46-9	14.8	0.8 x 0.3	13.1	Sb	21	
22 58 29.4	+40 25 42	MCG +7-47-1	14.0	0.6 x 0.5	12.6	Sb	114	Pair of mag 11.3 stars N 1′.3.
22 07 52.4	+31 21 33	NGC 7217	10.1	3.9 x 3.2	12.7	(R)SA(r)ab II-III	83	Many faint superimposed stars.
22 11 31.5	+38 43 16	NGC 7227	13.5	1.3 x 0.6	13.1	S0	8	Mag 10.77v star on SW edge.
22 11 49.0	+38 41 59	NGC 7228	13.5	2.1 x 1.2	14.4	(R′)SB(s)a	156	Two main arms.
22 15 22.7	+37 16 49	NGC 7240	14.2	0.6 x 0.6	13.0	S0$^-$:		**IC 1441** NW 1′.4; **IC 5192** WSW 2′.1.
22 15 39.6	+37 17 55	NGC 7242	13.2	2.3 x 1.7	14.6	E$^+$:	30	Bright nucleus, many superimposed stars.
22 16 52.8	+40 30 13	NGC 7248	12.4	1.7 x 0.9	12.7	SA0$^-$:	133	
22 18 18.1	+40 33 43	NGC 7250	12.6	1.7 x 0.8	12.8	Sdm?	166	Mag 11.10v star S end.
22 19 26.8	+29 23 35	NGC 7253A	13.2	1.7 x 0.8	13.4	SB?	116	Double system with NGC 7253B SE edge.
22 19 29.6	+29 23 07	NGC 7253B	14.5	1.6 x 0.5	14.1	S?	59	
22 21 45.3	+36 21 00	NGC 7263	14.6	0.8 x 0.7	13.8	S?	60	
22 22 13.5	+36 23 11	NGC 7264	13.8	2.2 x 0.3	13.2	Sb III	57	
22 22 27.5	+36 12 39	NGC 7265	12.2	2.4 x 1.9	13.7	S0$^-$	170	UGC 12007 E 2′.8; small, elongated anonymous galaxy NW 2′.5.
22 23 47.8	+32 24 07	NGC 7270	14.0	1.0 x 0.6	13.3	S?	93	
22 23 57.7	+32 21 58	NGC 7271	15.6	0.5 x 0.3	13.4		117	Located 2′.1 W of mag 10.46v star.
22 24 09.2	+36 11 58	NGC 7273	13.8	0.8 x 0.5	12.7	S0	6	Mag 11.26v star W 2′.1.
22 24 11.1	+36 07 32	NGC 7274	12.8	1.5 x 1.5	13.6	E		
22 24 17.3	+32 26 43	NGC 7275	14.3	0.9 x 0.2	12.3	Sa	37	Very close mag 10.69v star W 2′.0; **MCG +5-52-18** SW 2′.7.
22 24 14.4	+36 05 15	NGC 7276	13.9	0.9 x 0.9	13.8	E		
22 25 53.8	+40 18 58	NGC 7282	13.7	2.5 x 1.0	14.5	SB(r)b	0	Stellar nucleus with many superimposed stars.
22 27 50.7	+29 05 42	NGC 7286	12.5	1.7 x 0.7	12.5	S0/a	98	
22 28 26.0	+30 17 34	NGC 7292	12.5	2.1 x 1.7	13.8	IBm IV	117	Elongated nucleus, a number of superimposed stars.
22 31 32.0	+30 57 31	NGC 7303	12.7	1.5 x 1.2	13.2	S?	125	SE half far brighter.
22 35 32.0	+34 48 10	NGC 7315	12.5	1.6 x 1.6	13.4	S0		Mag 10.36v star NE 3′.2.
22 35 52.1	+33 56 37	NGC 7317	13.6	0.4 x 0.4	11.7	E4		Member of **Stephan's Quintet** with NGC 7318A, NGC 7318B, NGC 7319 and NGC 7320. = **Hickson 92E**.
22 35 56.7	+33 57 57	NGC 7318A	13.4	0.8 x 0.8	13.0	E2 pec		Member of **Stephan's Quintet** with NGC 7317, NGC 7318B, NGC 7319 and NGC 7320. = **Hickson 92D**.
22 35 58.7	+33 57 56	NGC 7318B	13.1	1.4 x 0.9	13.2	SB(s)bc pec	99	Member of **Stephan's Quintet** with NGC 7317, NGC 7318A, NGC 7319 and NGC 7320. = **Hickson 92B**.
22 36 04.1	+33 58 25	NGC 7319	13.1	1.5 x 1.1	13.5	SB(s)bc pec	52	Member of **Stephan's Quintet** with NGC 7317, NGC 7318A, NGC 7318B and NGC 7320. = **Hickson 92C**.
22 36 03.4	+33 56 56	NGC 7320	12.6	2.2 x 1.1	13.4	SA(s)d	132	Member of **Stephan's Quintet** with NGC 7317, NGC 7318A, NGC 7318B, and NGC 7319. = **Hickson 92A**.

RA h m s	Dec ° ′ ″	Name	Mag (V)	Dim ′ Maj x min	SB	Type Class	PA	Notes
22 36 20.6	+33 59 03	NGC 7320C	15.5	0.6 x 0.4	13.8	(R)SAB(s)0/a	172	= **Hickson 92E**. Located 2′4 W of NGC 7320.
22 36 56.3	+38 32 51	NGC 7330	12.2	1.4 x 1.4	13.0	E		Mag 10.51v star W 5′5; MCG +6-49-48 NNE 6′9.
22 37 04.9	+34 24 49	NGC 7331	9.5	10.5 x 3.5	13.2	SA(s)b I-II	171	Many dark lanes, knots and superimposed stars.
22 37 19.5	+34 26 54	NGC 7335	13.4	1.3 x 0.6	13.0	SA(rs)0$^+$	151	
22 37 22.0	+34 28 51	NGC 7336	14.5	0.8 x 0.4	13.1		139	
22 37 26.8	+34 22 24	NGC 7337	14.4	1.1 x 0.7	14.0	SB(rs)b	177	Faint star SE of nucleus.
22 37 44.4	+34 24 32	NGC 7340	13.7	0.9 x 0.6	13.0	E?	162	
22 38 13.1	+35 29 56	NGC 7342	13.9	1.3 x 1.3	14.3	SBa		Bright nucleus.
22 38 38.0	+34 04 18	NGC 7343	13.5	1.0 x 0.8	13.1	(R′)SB(s)bc: I-II	160	
22 38 44.9	+35 32 23	NGC 7345	14.3	1.2 x 0.3	13.0	Sa	39	Located 1′7 N of mag 9.59v star.
22 42 02.4	+30 42 28	NGC 7356	14.0	1.1 x 0.5	13.3	Sbc	76	
22 42 24.0	+30 10 16	NGC 7357	14.0	1.5 x 0.7	13.9	Sb II-III	120	Located 8′3 SW of mag 2.94v "Matar", Eta Pegasi.
22 43 18.3	+33 59 54	NGC 7363	13.8	1.1 x 0.9	13.6	SAB(s)d:	91	Stellar nucleus; mag 7.81v star SE 7′9.
22 44 12.2	+34 21 05	NGC 7369	13.7	1.4 x 1.1	14.1	S?	26	Bright, stellar nucleus.
22 47 33.0	+40 14 22	NGC 7379	13.4	1.1 x 0.8	13.1	SBa	81	Mag 10.73v star NNW 2′4.
22 51 03.1	+37 05 14	NGC 7395	13.8	1.2 x 1.1	14.0	S0?	123	
22 53 21.6	+32 07 39	NGC 7407	13.1	2.0 x 0.9	13.6	Sbc	152	
22 55 32.1	+29 48 19	NGC 7420	13.9	0.8 x 0.6	13.0	S?	54	Mag 9.57v star W 4′7.
22 56 02.7	+36 21 38	NGC 7426	12.3	1.7 x 1.4	13.2	E	72	Mag 5.74v star W 3′7.
22 58 09.9	+29 13 37	NGC 7439	14.0	1.1 x 0.7	13.6	SB0	150	
22 58 32.9	+35 48 08	NGC 7440	13.5	1.4 x 1.1	13.8	SB(r)a II-III	84	Bright nucleus.
22 59 22.5	+39 06 24	NGC 7445	14.6	0.7 x 0.2	12.5	E/S0	88	Located 2′0 NW of NGC 7446.
22 59 29.1	+39 04 54	NGC 7446	13.8	0.8 x 0.8	13.2			Anonymous galaxy W 2′9.
22 59 37.7	+39 08 40	NGC 7449	14.0	1.0 x 0.8	13.7	E:	130	NGC 7445 and NGC 7446 SW 3′7.
23 00 59.9	+30 08 41	NGC 7457	11.2	4.3 x 2.3	13.6	SA(rs)0$^-$?	130	Mag 10.71v star E 2′8; mag 10.60v star NE 3′4.
23 03 57.1	+30 09 34	NGC 7473	13.7	1.1 x 0.5	12.9	SB0	45	
23 06 05.0	+34 06 29	NGC 7485	13.1	1.2 x 0.6	12.7	S0	146	Pair of faint, stellar anonymous galaxies S 2′0.
23 07 25.0	+32 22 27	NGC 7490	12.3	2.8 x 2.6	14.3	Sbc I	60	Bright nucleus, several superimposed stars.
23 12 21.0	+31 07 31	NGC 7512	12.6	1.5 x 1.0	13.2	E	30	Stellar **Mkn 927** NE 1′6.
23 12 25.9	+34 52 54	NGC 7514	12.6	1.4 x 0.9	12.7	S?	132	Mag 9.39v star N 3′2.
23 22 06.7	+40 50 43	NGC 7640	11.3	10.5 x 1.8	14.4	SB(s)c II	167	Bright nucleus bisected by short dark lane; numerous superimposed stars.
21 57 44.4	+38 55 54	UGC 11862	15.9	1.7 x 0.2	14.6	Sdm:	132	Mag 10.30v star S 2′0.
22 03 33.0	+38 33 18	UGC 11890	13.6	0.9 x 0.4	12.3	Sab	40	
22 03 52.1	+35 59 22	UGC 11892	13.7	0.9 x 0.7	13.1	S0$^-$	120	
22 04 06.8	+35 56 18	UGC 11893	15.4	2.1 x 0.3	14.7	Sd	118	Mag 12.3 star N 1′8.
22 04 22.5	+39 44 31	UGC 11895	13.4	1.3 x 0.8	13.3	Sb	20	Strong dark lanes SE and NW edges; mag 11.37v star 0′8 S.
22 07 23.6	+38 44 54	UGC 11912	15.1	1.7 x 0.7	15.1	S	10	Mag 11.5 star SW 1′2.
22 09 27.5	+40 59 18	UGC 11927	13.6	1.7 x 1.1	14.1	Sb	43	Mag 9.25v star NW 3′5.
22 09 37.6	+39 16 55	UGC 11929	13.0	1.0 x 0.7	12.5	S0	53	**CGCG 513-6** S 4′5.
22 12 23.2	+39 16 41	UGC 11949	13.8	1.0 x 0.2	11.9	Sab	0	Mag 10.22v star ESE 2′7.
22 12 31.7	+38 40 52	UGC 11950	12.9	1.3 x 0.9	13.1	E:	40	Star superimposed NE of nucleus; faint, elongated anonymous galaxy N 1′0.
22 13 49.2	+39 14 12	UGC 11955	13.7	1.0 x 0.6	13.0	Scd:	50	Bright center, two very faint arms.
22 17 15.9	+33 30 11	UGC 11974	13.7	1.1 x 0.9	13.5	SBb	45	Mag 7.71v star NW 6′1.
22 17 19.7	+35 34 17	UGC 11975	14.2	1.5 x 1.1	14.6	S0/a	126	Mag 11.3 star on W edge of core.
22 18 31.8	+29 14 36	UGC 11981	14.2	1.3 x 1.1	14.4	SA(s)c	170	Stellar nucleus; mag 11.7 star E 1′3.
22 20 52.7	+33 17 44	UGC 11994	14.0	2.3 x 0.3	13.5	Sbc	122	Mag 9.77v star NE 1′9.
22 21 09.1	+36 35 27	UGC 11995	14.0	1.5 x 0.2	12.6	Sa	141	Mag 10.8 star NW 3′3.
22 22 29.6	+37 06 27	UGC 12006	13.4	1.4 x 0.7	13.2	SB0/a:	150	
22 22 40.9	+36 11 43	UGC 12007	14.5	1.2 x 1.0	14.5	Sb	53	Mag 10.20v star superimposed NE edge; NGC 7265 W 2′8.
22 22 40.3	+37 58 36	UGC 12009	12.9	1.4 x 0.7	12.7	S?	176	
22 23 01.9	+30 55 25	UGC 12011	13.1	0.7 x 0.5	11.8	Pair	93	Pair of compacts, disrupted; mag 10.59v star NW 1′7.
22 23 12.3	+40 03 18	UGC 12016	13.2	1.1 x 0.5	12.4	S?	120	
22 23 37.2	+30 51 28	UGC 12018	13.9	1.7 x 0.3	13.0	SBb?	34	Mag 10.8 star SW 3′8.
22 23 47.4	+35 23 16	UGC 12020	14.1	1.3 x 1.0	14.2	SBa	9	Bright core, uniform envelope.
22 24 02.5	+33 26 11	UGC 12022	14.0	1.5 x 0.5	13.5	SB?	96	
22 26 36.9	+36 43 41	UGC 12037	14.1	1.4 x 1.2	14.5	SB(r)b	161	Bright center, faint envelope; mag 8.08 star N 2′7.
22 26 48.7	+35 31 06	UGC 12039	13.0	1.8 x 1.4	13.9	SB(r)b	0	NW-SE bar, two main arms.
22 27 06.0	+36 21 39	UGC 12040	13.3	1.4 x 1.0	13.5	(R)SB0°	178	Mag 10.36v star N 2′0.
22 27 48.4	+38 35 13	UGC 12044	14.1	1.7 x 0.5	13.8	Sab	81	Mag 10.9 star SE.
22 29 16.8	+37 45 04	UGC 12051	14.1	0.9 x 0.8	13.6	(R′)SAB(s)bc:	60	Two main arms form "S" shape.
22 29 52.7	+36 43 10	UGC 12056	14.1	1.5 x 0.9	14.3	Sa	160	Mag 11.14v star superimposed S edge.
22 30 34.0	+33 49 11	UGC 12060	14.3	1.5 x 0.8	14.3	IBm	138	Mag 12.2 star SE 1′2.
22 31 11.0	+35 22 29	UGC 12063	14.4	1.6 x 0.5	14.0	Sab	147	
22 31 20.7	+39 21 27	UGC 12064	13.6	0.5 x 0.4	11.7	S0$^-$:	12	Brightest of three galaxies; companions N 0′6 and WNW 1′6.
22 32 24.5	+30 50 04	UGC 12071	14.2	1.2 x 0.4	13.2	SBb	29	Two main arms; mag 11.2 star N 1′9.
22 32 33.3	+39 12 56	UGC 12073	13.6	2.0 x 0.7	13.8	SB(s)b	100	UGC 12075 E 3′0.
22 32 49.0	+39 12 37	UGC 12075	13.9	1.5 x 1.1	14.3	SBcd:	39	**MCG +6-49-34** on E edge; mag 10.4 star SE 2′6.
22 34 10.8	+32 51 19	UGC 12082	13.5	2.6 x 2.3	15.3	Sm V	9	Dwarf spiral, low surface brightness; mag 11.1 star NE 3′3.
22 37 40.5	+34 50 43	UGC 12121	13.8	1.1 x 0.5	13.0	S0	70	
22 37 43.3	+40 19 05	UGC 12125	14.9	1.9 x 0.3	14.1	Sa	3	Located between a pair of mag 12 stars.
22 38 29.5	+35 19 44	UGC 12127	13.4	1.3 x 1.2	13.9	E	9	MCG +6-49-56 and MCG +6-49-61 N 2′5; MCG +6-49-60 NE 1′3, **MCG +6-49-62** E 1′7.
22 38 44.6	+37 35 49	UGC 12131	14.1	0.7 x 0.6	13.0	Sb II	54	
22 39 49.3	+38 12 55	UGC 12137	12.8	1.7 x 1.5	13.6	SAB(s)bc I-II	25	N-S bar with two main arms, several stars superimposed S edge.
22 40 25.7	+31 50 29	UGC 12143	14.0	1.5 x 0.5	13.6	Sab	64	Mag 8.07v star S 3′5.
22 41 07.8	+32 10 07	UGC 12149	13.8	0.8 x 0.7	13.0	(R′)SB(s)a		Mag 8.50v star NE 4′0; **CGCG 495-9** SW 11′8.
22 41 12.4	+34 14 53	UGC 12150	13.9	1.0 x 0.5	13.0	SB0/a	31	
22 41 26.5	+39 17 29	UGC 12156	13.9	1.0 x 0.7	13.4	SB(s)c	36	Faint stellar nucleus.
22 41 55.1	+34 55 07	UGC 12157	14.6	1.9 x 0.5	14.4	S	102	Pair of faint stars, or bright knot and star E end; mag 10.7 star SW 1′3.
22 42 12.3	+33 12 05	UGC 12161	14.1	0.8 x 0.6	13.1	S?	30	Located 1′1 SE of mag 7.83v star.
22 42 39.5	+29 43 27	UGC 12163	14.0	0.7 x 0.4	12.4	SB	118	Located between a pair of mag 8.95v stars.
22 42 50.2	+30 30 10	UGC 12164	14.3	1.5 x 0.7	14.2	Sb II-III	132	
22 43 51.8	+38 22 35	UGC 12173	12.8	1.9 x 1.2	13.5	SAB(rs)c	80	
22 44 45.8	+33 27 35	UGC 12177	12.7	0.9 x 0.9	12.3	S?		Bright star superimposed inside W edge; mag 12.0 star NE 1′1 with a very faint anonymous galaxy SE of this star.
22 45 04.0	+33 59 46	UGC 12179	13.2	1.3 x 1.3	13.6	S0		**Mkn 920** NE 5′0.
22 46 09.4	+38 03 08	UGC 12181	13.5	1.0 x 0.8	13.1	SABcd: I	155	Two main arms; mag 10.9 star NW 1′0.
22 47 25.1	+31 22 25	UGC 12185	13.6	1.7 x 0.8	13.8	SB(s)ab	150	Bright core; mag 11.0 star W 2′1.
22 47 37.6	+39 52 38	UGC 12188	13.5	0.8 x 0.5	12.4	S	150	Double system, contact; round companion on NE edge.
22 49 03.9	+40 00 05	UGC 12199	13.5	1.6 x 0.8	13.6	SB(rs)b	105	Mag 11.6 star N edge; mag 11.17v star NNW 2′1.

(Continued from previous page)

46 GALAXIES 46

RA h m s	Dec ° ′ ″	Name	Mag (V)	Dim ′ Maj x min	SB	Type Class	PA	Notes
22 49 09.6	+34 59 31	UGC 12201	14.0	1.5 x 0.3	13.0	Sab	48	Mag 13.3 star N edge.
22 49 23.3	+40 13 54	UGC 12204	15.2	1.5 x 0.2	13.7	Sbc	28	Mag 10.98v star N end.
22 49 32.7	+33 21 35	UGC 12206	13.8	2.8 x 1.6	15.3	SB(rs)a	137	Triangle of stars, mags 10.6, 11.9 and 13.9, superimposed S half.
22 50 11.6	+31 22 39	UGC 12210	14.1	1.2 x 1.0	14.2	SB(r)a		N-S bar with dark patches E and W. Faint anonymous galaxy 1′.7.
22 50 30.3	+29 08 16	UGC 12212	15.4	2.0 x 1.0	16.0	Sm:	84	Dwarf spiral, very low surface brightness; mag 10.29v star WNW 1′.9.
22 51 00.9	+31 22 28	UGC 12214	13.0	1.1 x 0.8	12.8	S0?	160	MCG +5-53-21 on E edge.
22 50 56.3	+34 51 23	UGC 12215	14.4	1.4 x 0.9	14.5	Sd	65	Bright center, uniform envelope, stars, mag 14.7, superimposed N and S edges.
22 51 21.9	+32 20 57	UGC 12218	14.3	0.9 x 0.4	13.0	Im?	22	
22 53 41.0	+33 42 29	UGC 12234	12.7	1.2 x 0.7	12.3	Sbc	130	
22 54 04.2	+32 22 23	UGC 12235	13.8	1.1 x 0.7	13.4	S0	45	Mag 9.61v star SE 4′.7.
22 53 57.4	+36 14 28	UGC 12236	15.0	1.3 x 1.3	15.4	SAm		Stellar nucleus; mag 10.59v star SE 2′.7.
22 54 16.9	+32 15 04	UGC 12238	13.5	1.2 x 0.9	13.5	SAB0°	70	Mag 10.30v star on N edge.
22 54 25.7	+32 27 05	UGC 12242	13.1	1.1 x 1.0	13.3	E:	117	MCG +5-54-8 NNW 2′.4.
22 55 43.4	+31 46 17	UGC 12252	15.4	1.5 x 0.2	13.9	Sd	173	Mag 10.65v star S 1′.7.
22 56 31.9	+37 44 18	UGC 12260	14.8	1.5 x 0.3	13.8	Sd	21	Mag 11.6 star N 1′.8.
22 58 55.5	+40 55 58	UGC 12282	13.6	1.7 x 0.5	13.3	Sa	1	Mag 9.93v star NW 1′.5.
23 00 09.3	+31 24 32	UGC 12297	13.8	1.1 x 0.5	13.0	S0?	132	Elongated anonymous galaxy NE 0′.8; MCG +5-54-23 S 2′.1.
23 00 14.5	+39 14 07	UGC 12298	14.0	1.4 x 0.4	13.2	SBab	79	
23 01 25.0	+30 14 22	UGC 12311	14.3	1.5 x 0.4	13.6	S?	143	An arc of three mag 12-13 stars SE 2′.6.
23 02 33.2	+32 35 42	UGC 12323	13.8	1.0 x 0.9	13.5	Sc I-II	21	
23 05 49.2	+31 05 14	UGC 12356	13.7	0.9 x 0.9	13.3	Compact		Elongated **UGC 12357** S 1′.5.
23 06 15.4	+31 53 03	UGC 12362	14.2	1.5 x 0.6	13.9	SB(s)b	118	
23 07 01.0	+35 46 36	UGC 12372	13.6	1.1 x 0.8	13.3	S?	51	Mag 6.39v star S 8′.4.
23 07 49.3	+30 19 01	UGC 12385	14.5	1.0 x 0.8	14.1	SBcd:	162	Four mag 12-14 stars superimposed.
23 09 31.1	+32 40 28	UGC 12394	13.7	0.9 x 0.7	13.1	SBab	45	
23 11 50.4	+31 01 16	UGC 12410	15.0	1.7 x 0.2	13.7	Scd:	48	Mag 7.28v star NW 13′.6.
23 13 43.5	+29 00 29	UGC 12430	14.2	2.3 x 0.2	13.2	Sd	164	**UGC 12427** SW 5′.4.
23 14 29.5	+31 32 55	UGC 12444	13.5	0.9 x 0.9	13.1	S?		Mag 11.1 star N 2′.6; mag 11.9 star E 1′.9.
23 16 43.6	+33 59 41	UGC 12474	13.3	1.1 x 0.4	12.3	Sa	88	Mag 8.38v star S 6′.8.
23 17 02.9	+30 20 05	UGC 12476	13.0	1.4 x 0.7	12.8	S0/a	93	Mag 7.69v star NNW 2′.6.
23 17 33.7	+29 01 08	UGC 12482	13.0	1.4 x 1.2	13.5	E:	39	MCG +5-54-60 E 2′.8; **CGCG 496-75** NNW 2′.9.
23 20 31.6	+29 18 22	UGC 12530	13.8	1.1 x 1.1	13.8	SBcd:		Mag 12.4 star E 2′.4.
23 21 10.0	+33 23 56	UGC 12538	13.9	0.9 x 0.3	12.3	S?	20	CGCG 497-7 NE 7′.8.
23 22 28.7	+29 10 46	UGC 12557	14.1	2.1 x 0.4	13.8	S?	9	Mag 10.45v star W 9′.2.
23 22 46.0	+29 08 16	UGC 12566	13.6	1.7 x 1.3	14.3	Sab	140	Bright center, prominent dark patches SE and NW, faint envelope.
23 23 10.3	+32 31 35	UGC 12570	14.0	0.5 x 0.4	12.1	Pec	168	Mag 10.7 star NW 0′.8.

GALAXY CLUSTERS

RA h m s	Dec ° ′ ″	Name	Mag 10th brightest	No. Gal	Diam ′	Notes
22 54 48.0	+29 28 00	A 2503	16.4	47	8	All members anonymous, faint, stellar.

47 GALAXIES 47

RA h m s	Dec ° ′ ″	Name	Mag (V)	Dim ′ Maj x min	SB	Type Class	PA	Notes
21 48 29.7	+29 23 46	CGCG 493-8	14.1	0.7 x 0.3	12.3		66	Mag 9.14v star SW 7′.3.
21 57 12.0	+30 43 47	CGCG 493-15	14.4	0.5 x 0.4	12.7	E	144	**CGCG 493-16** ESE 2′.6.
21 35 32.8	+35 23 53	IC 1392	11.5	1.1 x 0.7	11.1	S0⁻:	75	Located between close N-S mag 12.0, 12.4 stars.
21 46 31.0	+38 28 44	MCG +6-47-6	14.5	0.7 x 0.7	13.6			MCG +6-47-7 NE 2′.7.
21 46 43.5	+38 29 55	MCG +6-47-7	14.7	0.5 x 0.5	13.1	Sb		MCG +6-47-6 SW 2′.7.
21 03 33.7	+29 53 50	NGC 7013	11.3	4.0 x 1.4	13.1	SA(r)0/a	157	Mag 9.88v star on N edge; many superimposed stars.
21 26 06.3	+30 29 36	UGC 11743	14.0	1.1 x 0.7	13.6	S?	45	Patchy; mag 8.77v star NE 4′.8.
21 28 51.4	+31 50 12	UGC 11753	14.1	0.7 x 0.7	13.2	SB(r)b		Mag 11.4 star SE 1′.1.
21 35 48.3	+35 21 03	UGC 11775	13.1	1.0 x 0.4	12.0	S0	50	IC 1392 and pair of mag 12.0, 12.4 stars NW 4′.2.
21 36 39.3	+35 41 40	UGC 11781	12.1	1.4 x 1.0	12.3	SAB0°	75	Published dimensions include only the bright core, very faint halo extends much further out.
21 52 44.9	+38 56 07	UGC 11841	15.9	2.8 x 0.2	15.1	Sdm:	66	Pair of mag 11 stars N 1′.4.
21 57 44.4	+38 55 54	UGC 11862	15.9	1.7 x 0.2	14.6	Sdm:	132	Mag 10.30v star S 2′.0.
22 03 33.0	+38 33 18	UGC 11890	13.6	0.9 x 0.4	12.3	Sab	40	
22 03 52.1	+35 59 22	UGC 11892	13.7	0.9 x 0.7	13.1	S0⁻	120	

OPEN CLUSTERS

RA h m s	Dec ° ′ ″	Name	Mag	Diam ′	No. ★	B ★	Type	Notes
21 03 00.0	+40 26 12	Be 54		4	30	17.0	cl	
21 09 00.0	+37 36 00	Do 45		18	35		cl	Moderately rich in stars; moderate brightness range; no central concentration; detached; involved in nebulosity. (A) Scattering of a dozen or so bright stars over an 18′ field.
20 41 44.2	+36 37 15	Do 47		5	15		cl	
20 51 04.0	+35 54 30	DoDz 11		5	12		cl	Few stars; moderate brightness range; not well detached.
21 10 57.0	+33 46 00	NGC 7037		7	10	12.0	cl?	(A) Slight condensation of stars on DSS image.
21 24 30.0	+36 30 00	NGC 7063	7.0	9	66	8.9	cl	Few stars; small brightness range; no central concentration; detached.
20 41 41.2	+35 33 00	Ru 173		40	20	8.0	cl	Few stars; moderate brightness range; slight central concentration; detached.
20 43 28.3	+37 01 49	Ru 174		2	10	14.0	cl	
20 45 24.0	+35 31 00	Ru 175		15	30	11.0	cl	Moderately rich in stars; moderate brightness range; not well detached.

(Continued from previous page)

47 BRIGHT NEBULAE 47

RA h m s	Dec ° ′ ″	Name	Dim ′ Maj x min	Type	BC	Color	Notes
20 56 12.0	+31 04 00	IC 1340	25 x 20	E	3-5		
20 45 42.0	+30 43 00	NGC 6960	70 x 6	SNR-E	2-5		Extreme edge outlined on map but not labeled: see Map 120. The westernmost part of the **Veil Nebula**. With NGC 6992, it is the brightest part. The entire nebulous complex is collectively known under several common names, e.g., **Cirrus**, **Filamentary**, **Network** or **Cygnus Loop**.
20 50 48.0	+31 52 00	NGC 6974	6 x 4	SNR-E			Part of the **Veil Nebula** complex, S-shaped; semi-detached from NGC 6979.
20 51 00.0	+32 09 00	NGC 6979	7 x 3	SNR-E			Northernmost bright patch in the **Veil Nebula**; semi-detached from NGC 6974.
20 56 24.0	+31 43 00	NGC 6992	60 x 8	SNR-E	2-5		Part of the **Veil Nebula** complex.
20 57 06.0	+31 13 00	NGC 6995	12 x 12	SNR-E	2-5		Part of the **Veil Nebula** complex.

PLANETARY NEBULAE

RA h m s	Dec ° ′ ″	Name	Diam ″	Mag (P)	Mag (V)	Mag cent ★	Alt Name	Notes
21 02 18.7	+36 41 40	PK 80−6.1	37	13.5				**Egg Nebula**.
20 42 16.5	+37 40 22	PN G78.3−2.7	20	16.0			PK 78−2.1	Mag 12.3 star SE 0′.9; mag 12.2 star NE 1′.9.
21 35 29.5	+31 41 44	PN G81.2−14.9	107	16.0	13.4	13.2	PK 81−14.1	= **Abell 78**; mag 12.4 star ENE 1′.6; mag 13.2 star NW 1′.3; mag 7.45v star NW 7′.9.
21 22 15.5	+38 07 13	PN G83.9−8.4	10	15.4	15.5:	15.7	PK 83−8.1	Mag 11.9 star NW edge; mag 10.26v star ESE 2′.7.
21 07 39.7	+40 57 53	PN G84.2−4.2	6	17.1			PK 84−4.1	Close pair of mag 13.0, 13.7 stars E 1′.5.
21 33 08.2	+39 38 12	PN G86.5−8.8	32	12.7	12.0	17.3	PK 86−8.1	Rectangular shape, oriented SE-NW; mag 13.9 star on SE end; mag 11.5 star W 2′.0.

48 GALAXIES 48

RA h m s	Dec ° ′ ″	Name	Mag (V)	Dim ′ Maj x min	SB	Type Class	PA	Notes
19 19 33.6	+37 24 17	CGCG 202-8	15.9	0.4 x 0.4	13.8			Bright center or star? Very faint envelope, closely surrounded by stars in star rich area.
19 30 52.8	+35 47 06	IC 1302	13.4	0.9 x 0.5	12.4	SAB(s)c?	45	Core surrounded by four superimposed stars.
19 31 30.3	+35 52 31	IC 1303	14.3	1.1 x 0.6	13.7	SAc:	106	Three faint stars superimposed in a row along S edge.
19 31 22.9	+37 02 18	MCG +6-43-3	14.3	0.5 x 0.5	12.6			Mag 10.61v star ESE 1′.8.
19 56 12.9	+40 26 03	MCG +7-41-1	13.5	1.2 x 0.7	13.4	E1:	133	
19 18 21.7	+34 50 12	UGC 11426	13.5	0.8 x 0.6	12.6	S	55	Mag 9.85v star S 1′.2.
19 20 28.6	+30 49 31	UGC 11428	13.8	1.1 x 0.8	13.5	Scd:	168	Numerous stars superimposed; in star rich area.
19 23 49.9	+34 47 32	UGC 11433	14.7	0.9 x 0.2	12.7	S	25	Mag 12.4 star SW 0′.8; in star rich area.
19 30 33.4	+35 46 36	UGC 11448	14.9	1.1 x 0.7	14.5	Scd:	57	Pair of mag 11 stars SW 2′.2.
19 32 39.6	+37 52 28	UGC 11456	14.6	1.3 x 0.5	14.0	S	37	Mag 10.74v star SE 1′.7.
19 37 22.6	+40 42 18	UGC 11459	13.4	2.0 x 1.1	14.1	Sd	12	Stellar nucleus, numerous superimposed stars; in star rich area.

OPEN CLUSTERS

RA h m s	Dec ° ′ ″	Name	Mag	Diam ′	No. ★	B ★	Type	Notes
20 02 23.6	+35 18 35	ADS 13292		1	8	9.4	cl?	(S) Likely cluster. Involved in Sh2-101.
19 59 30.2	+34 38 34	Be 49		3	20	16.0	cl	
20 11 53.0	+34 24 30	Be 51		2	20	15.0	cl	
20 04 42.2	+33 54 25	Be 84		2	20	16.0	cl	
20 18 47.0	+37 45 18	Be 85		6	40	15.0	cl	
20 20 21.2	+38 42 00	Be 86	7.9	7	30	13.0	cl	Moderately rich in stars; moderate brightness range; not well detached; involved in nebulosity.
20 21 35.0	+37 24 00	Be 87		10	30	13.0	cl	Moderately rich in stars; moderate brightness range; no central concentration; detached.
20 07 30.0	+35 41 00	Bi 1		10	15		cl	Few stars; large brightness range; not well detached; involved in nebulosity. (A)Rich field around bright star, but no obvious cluster on DSS image.
20 09 14.3	+35 29 00	Bi 2	6.3	20	78	16.0	cl	Few stars; moderate brightness range; no central concentration; detached.
20 17 42.2	+39 21 00	ClvdB 130	9.3	7	15	10.3	cl	Few stars; moderate brightness range; slight central concentration; detached; involved in nebulosity.
20 18 07.0	+40 43 56	Cr 419	5.4	4.5			cl	Few stars; moderate brightness range; not well detached.
20 08 12.0	+36 33 00	Do 1		6	10		cl:	(A) Mostly E-W scattering of brighter stars in rich Milky Way field.
20 15 27.0	+36 49 00	Do 3		7	40		cl	Moderately rich in stars; moderate brightness range; no central concentration; detached; involved in nebulosity. (A) Condensation in Milky Way.
20 17 46.6	+36 45 00	Do 4		9	20		cl:	(A) Possible cluster involved with nebula Sh2-104.
20 20 29.0	+39 22 00	Do 5		10	30		cl	(A) Loose cluster of bright stars involved in nebulosity.
20 26 20.0	+40 07 00	Do 10		15				
20 03 00.0	+37 41 00	Do 37		8			ast	(A) Group of about 10 brighter stars.
20 16 12.0	+37 55 00	Do 39		12	40		cl	Moderately rich in stars; moderate brightness range; not well detached; involved in nebulosity. (A) Scattering of bright stars with many fainter stars in background and very rich Milky Way field (or nebulosity).
20 18 12.0	+37 51 00	Do 40		5	12		cl	Few stars; small brightness range; no central concentration; detached; involved in nebulosity.
20 19 12.0	+37 45 00	Do 41		11			cl	Few stars; small brightness range; not well detached. (A) About 9 bright stars, with Be 85 on W side.
20 19 39.7	+38 08 00	Do 42		8	20		cl	Few stars; small brightness range; no central concentration; detached; involved in nebulosity. (A) Has prominent chain of 10 stars NNE to SSW.
20 21 42.0	+39 57 00	Do 43		20			ast	Few stars; moderate brightness range; not well detached. Probably not a cluster. (A) Scattering of stars, including 5 brighter ones on south side. Surrounded by bright nebulae.
20 41 44.2	+36 37 15	Do 47		5	15		cl	
20 05 35.0	+40 40 00	DoDz 10		24	40	7.1	cl	Few stars; moderate brightness range; not well detached; involved in nebulosity. (A) Scattered cluster in rich Milky Way and bright nebula.
20 10 03.0	+34 57 42	IC 1310		3	12	14.0	cl	
20 16 31.7	+37 39 00	IC 4996	7.3	7	56	8.0	cl	Few stars; large brightness range; slight central concentration; detached; involved in nebulosity.
19 20 52.7	+37 46 00	NGC 6791	9.5	10	380	15.0	cl	Rich in stars; moderate brightness range; strong central concentration; detached. (A) Very rich cluster - appears as a sparse globular cluster on DSS image.
19 41 18.0	+40 11 00	NGC 6819	7.3	5	929	11.0	cl	Moderately rich in stars; small brightness range; strong central concentration; detached.
19 52 13.0	+29 24 30	NGC 6834	7.8	6	128	11.0	cl	Moderately rich in stars; moderate brightness range; slight central concentration; detached.
19 56 28.1	+32 20 55	NGC 6846	14.2	0.8	40	12.8	cl	Few stars; small brightness range; not well detached.
20 06 27.0	+35 47 00	NGC 6871	5.2	30	66	6.8	cl	Few stars; moderate brightness range; slight central concentration; detached; involved in nebulosity. (A) About a dozen bright stars over a 30′ field of rich Milky Way. Dark nebulae on SW side.
20 07 48.5	+38 14 00	NGC 6874	7.7	7	20	10.0	cl?	
20 11 18.0	+35 51 00	NGC 6883	8.0	35	30		cl	Moderately rich in stars; moderate brightness range; not well detached; involved in nebulosity.
20 23 12.0	+40 47 00	NGC 6910	7.4	10	66	9.6	cl	Moderately rich in bright and faint stars with a strong central concentration; detached; involved in nebulosity. Visual mag = 6.6 if bright star on NW included.
20 24 06.0	+38 30 00	NGC 6913	6.6	10	81	9.0	cl	= **M 29**. Moderately rich in bright and faint stars; detached; involved in nebulosity. (A) 30 bright stars, about 100 stars total.
20 04 52.0	+29 12 54	Ro 4	10.0	6	30		cl	(A) 3 patches of bright nebulosity on W side, including IC 4954/55.

(Continued from previous page)

OPEN CLUSTERS

RA h m s	Dec ° ′ ″	Name	Mag	Diam ′	No. ★	B ★	Type	Notes
20 10 00.0	+33 46 00	Ro 5		50	15		cl	Few stars; moderate brightness range; not well detached; involved in nebulosity. (A) Scattering of bright stars over 50′ rich Milky Way field.
20 28 49.0	+39 20 00	Ro 6		24	30		cl	(A) Scattering of bright stars in strong nebulosity. Identification difficult.
20 11 31.1	+35 38 23	Ru 172		5	20	12.0	cl	(A) Slight condensation of stars in rich Milky Way and nebulosity.
20 41 41.2	+35 33 00	Ru 173		40	20	8.0	cl	Few stars; moderate brightness range; slight central concentration; detached.
20 43 28.3	+37 01 49	Ru 174		2	10	14.0	cl	
19 42 24.0	+38 39 00	SkiffJ1942.3+3839		3	12		cl?	(A) Possibly as large as 50 stars over 10′ diameter.
19 51 01.9	+32 14 35	SSWZ94 6		1			cl	

GLOBULAR CLUSTERS

RA h m s	Dec ° ′ ″	Name	Total V m	B ★ V m	HB V m	Diam ′	Conc. Class Low = 12 High = 1	Notes
19 16 35.5	+30 11 05	NGC 6779	8.4	13.2	16.3	8.8	10	= **M 56**.

BRIGHT NEBULAE

RA h m s	Dec ° ′ ″	Name	Dim ′ Maj x min	Type	BC	Color	Notes
20 02 48.0	+36 58 00	Ced 174	15 x 5	E	3-5	3-4	
20 14 18.0	+39 54 00	IC 1318(a)	40 30	E	4-5	3-4	Centered about 1.6° W, slightly S of γ Cygni. It has an undefined outline, blending well with the surrounding Milky Way field; some filamentary structure in the S part and a conspicuous dark patch, B 343 in the N end.
20 19 18.0	+40 44 00	IC 1318(c)	40 x 25	E	2-5	3-4	Very diffuse overall with four or five brighter condensations, the most prominent one about 30' SE of Cr 419.
20 26 12.0	+40 30 00	IC 1318(d)	50 x 30	E	1-5	3-4	Centered approximately 1° E of γ Cygni. This and the bright patch 45' SE are the brightest parts of the γ Cygni complex.
20 27 54.0	+40 00 00	IC 1318(e)	45 x 30	E	1-5	3-4	Centered 1.5° SE of γ Cygni. Comparable in size and brightness to the nebulosity 0.75° NW. A prominent dark lane separates them.
20 04 48.0	+29 15 00	IC 4954/55	3 x 3	R	2-5	1-4	A pair of detached nebulosities aligned NW-SE; distance between centers about 3′.5. The SE component is largest and about 3' in diameter; the NW component appears to be double, aligned nearly N-S.
19 36 18.0	+29 33 00	Mi 92	0.2 x 0.1	R	2-5		**Footprint Nebula**.
20 01 54.0	+33 31 00	NGC 6857	0.6 x 0.6				
20 12 00.0	+38 21 00	NGC 6888	18 x 13	E	1-5	3-4	**Crescent Nebula**. Irregular oval ring, broken along most of the E perimeter.
20 00 00.0	+35 17 00	Sh2-101	18 x 10	E	1-5	3-4	A crown-shaped nebula with many faint stars involved; extended NNE-SSW. It is brightest in the SE sector and along the E perimeter.
20 17 48.0	+36 44 00	Sh2-104	7 x 7	E	2-5	3-4	
20 19 06.0	+39 21 00	Sh2-108	110 x 60	E	3-5	3-4	The SW part of the γ Cygni nebular complex.
20 04 36.0	+32 15 00	vdB 128	8 x 8	R	3-5	2-4	Nebula with a mag 5.6 star involved.
20 30 42.0	+36 56 00	vdB 133	10 x 10	R	2-5	1-4	The involved star, mag 6.2, is in the SE part of the nebula.

DARK NEBULAE

RA h m s	Dec ° ′ ″	Name	Dim ′ Maj x min	Opacity	Notes
19 58 41.4	+35 20 00	B 144	30 x 30	1	**Fish on the Platter**. Large, semi-vacant region.
20 02 49.3	+37 41 00	B 145	35 x 6	4	Sharply defined E and W.
20 03 29.0	+36 01 00	B 146	1 x 1	6	Very small.
20 06 50.3	+35 23 00	B 147	20 x 8	5	Very narrow sinuous dark lane E and W.
19 50 01.3	+34 17 00	B 341	30 x 1	3	Narrow, dark lane; E and W.
20 13 26.0	+40 17 00	B 343	13 x 5	5	Elongated; NW and SE; sharpest at SE end.
20 16 12.2	+40 13 00	B 344	7 x 3	3	Dusky spot, like an arrowhead, pointed SW.
20 28 27.8	+39 55 00	B 347	4.0 x 0.7	4	Dark streak in nebulous cloud.
20 24 48.0	+40 10 00	LDN 889	180 x 20	4	Wide lane E of γ Cygni.

PLANETARY NEBULAE

RA h m s	Dec ° ′ ″	Name	Diam ″	Mag (P)	Mag (V)	Mag cent ★	Alt Name	Notes
19 55 02.4	+29 17 17	NGC 6842	57	13.6	13.1	15.9	PK 65+0.1	Numerous stars on disc; mag 8.72v star NE 6′.3.
20 10 52.5	+37 24 41	NGC 6881	16	14.3	13.9	18.3	PK 74+2.1	Slightly elongated, or egg shaped, with a PA of 146 degrees; mag 11.1 star E 1′.9; pair of mag 8.49v and 8.90v stars NW 3′.9.
20 16 24.0	+30 33 51	NGC 6894	55	14.4	12.3	18.1	PK 69−2.1	Hint of annularity; small, faint dark patch in center.
19 34 45.2	+30 31 01	PN G64.7+5.0	35	9.6	11.3	12.5	PK 64+5.1	**Campbell's Hydrogen Star**. Mag 11.6 star E 1′.9; mag 10.14v star NE 3′.3.
19 51 00.9	+31 02 27	PN G66.9+2.2	12				PK 66+2.1	Close pair of mag 11 stars ENE 3′.0; mag 11.07v star W 3′.3.
20 03 11.6	+30 32 32	PN G67.9−0.2	1	18.6			PK 67−0.1	Triangle of mag 12.5-13 stars centered NE 1′.3; mag 12.4 star W 2′.8.
20 13 58.0	+29 33 54	PN G68.3−2.7	9				PK 68−2.1	Mag 12.8 star NNE 0′.9; mag 12.2 star SW 1′.2.
19 59 18.5	+31 54 34	PN G68.6+1.1	22	14.1	14.7:	21.1	PK 68+1.2	Appears slightly elongated SE-NW, possibly by a star on SE edge; mag 14.0 star N 0′.6.
19 56 34.3	+32 22 11	PN G68.7+1.9	3	15.9			PK 68+1.1	Open cluster NGC 6846 SW 1′.9; mag 12.5 star N 0′.3.
19 51 52.9	+32 59 17	PN G68.7+3.0	5	14.7			PK 68+3.1	Mag 11.6 star W 0′.9; mag 9.18v star SE 3′.3.
20 04 44.1	+31 27 20	PN G68.8−0.0	40		16.0	21.0	PK 68−0.1	Rectangular shape, elongated SSE-NNW; mag 13.6 star on S edge.
19 54 00.8	+33 22 12	PN G69.2+2.8	3	18.9			PK 69+2.1	Mag 13.6 star E 0′.7; mag 11.9 star WSW 2′.6, in star rich area.
19 50 00.3	+33 45 53	PN G69.2+3.8	42	16.4			PK 69+3.1	Rectangular shape, oriented N-S, small dark patch in center; close group of three mag 10.15v, 10.22v and 11.4 stars W 1′.4.
20 21 58.6	+29 59 18	PN G69.6−3.9	14	18.6	16.::		PK 69−3.1	Close pair of mag 12.4 stars NNW 2′.8; mag 12.8 stars S 2′.3.
20 06 56.2	+32 16 32	PN G69.7−0.0	10				PK 69+0.1	Mag 12.3 star SE 0′.7; mag 8.78v star ENE 1′.5.
20 21 03.9	+32 29 21	PN G71.6−2.3	11	14.6	14.5	15.7	PK 71−2.1	Mag 11.2 star NNE 1′.3; mag 11.4 star NW 2′.4.
20 12 47.9	+34 20 31	PN G72.1+0.1	8	15.8			PK 72+0.1	Mag 12.3 star S 1′.9; mag 9.81v star NNE 1′.6.
20 25 05.0	+33 34 48	PN G73.0−2.4	4	17.2			PK 73−2.1	Mag 13.3 star NNW 2′.3; mag 12.3 star S 2′.5.
20 04 16.4	+39 35 30	PN G75.6+4.3	28	16.0			PK 75+4.1	Several faint stars on disc; mag 11.7 star E 3′.1; mag 11.5 star WNW 3′.3.
20 19 58.4	+38 23 59	PN G76.3+1.1	25	>20.1		21.0	PK 76+1.1	= **Abell 69**; annular; close pair of mag 11.2, 12.4 stars S 2′.0; mag 12.5 star E 0′.9.
20 17 15.6	+38 50 21	PN G76.4+1.8	6	18.9			PK 76+1.2	Small group of four mag 12-13 stars N 0′.5; mag 11.8 star on SE edge.
20 12 25.8	+40 45 19	PN G77.5+3.7	6	15.5			PK 77+3.1	Mag 13.6 star S 0′.5; mag 12.2 star SE 1′.2.
20 15 22.3	+40 34 43	PN G77.7+3.1	4	17.0			PK 77+3.2	Mag 13.2 star NE 2′.1; mag 11.6 star ESE 3′.3.
20 42 16.5	+37 40 22	PN G78.3−2.7	20	16.0			PK 78−2.1	Mag 12.3 star SE 0′.9; mag 12.2 star NE 1′.9.
20 29 19.0	+40 15 16	PN G78.9+0.7	18	17.1			PK 78+0.1	Slightly elongated N-S, on N edge of large bright nebula IC 1318; mag 12.6 star on E edge.

RA h m s	Dec ° ′ ″	Name	Mag (V)	Dim ′ Maj x min	SB	Type Class	PA	Notes
18 30 57.6	+29 52 32	CGCG 172-40	14.3	0.7 x 0.4	12.8		9	Uniform surface brightness; mag 12.7 star on W edge.
18 39 48.0	+32 51 40	CGCG 173-21	15.4	0.4 x 0.3	12.9		0	MCG +5-44-9 S 3′4.
19 19 33.6	+37 24 17	CGCG 202-8	15.9	0.4 x 0.4	13.8			Bright center or star? Very faint envelope, closely surrounded by stars in star rich area.
18 10 27.3	+31 00 05	IC 1277	13.4	1.2 x 0.9	13.3	Scd:	34	Mag 11.01v star NNE 5′3.
18 11 15.2	+36 00 28	IC 1279	13.5	2.6 x 0.6	13.8	Sb II	159	Mag 10.95v star E 2′9; note: **IC 1281** is 1′7 E of this star.
18 29 22.5	+39 42 47	IC 1288	13.4	1.1 x 0.7	13.0	SBa	0	Mag 9.12v star NNW 4′7.
18 30 02.5	+39 57 46	IC 1289	15.0	0.7 x 0.3	13.2		132	Mag 8.82v star WSW 5′9.
18 53 18.9	+33 03 56	IC 1296	14.0	1.1 x 0.8	13.7	SBbc	80	Very filamentary; mag 10.4 star WSW 1′7.
17 56 46.9	+31 16 59	MCG +5-42-17	15.0	0.7 x 0.5	13.7	Sb	90	Pair of mag 13 stars W 0′9.
18 02 57.2	+29 18 35	MCG +5-42-29	14.8	0.6 x 0.6	13.5	Sc		**CGCG 171-49** N 4′0; UGC 11098 W 1′9.
17 57 53.4	+31 49 21	MCG +5-42-41		0.8 x 0.6			0	
18 23 04.2	+29 54 02	MCG +5-43-16	14.1	0.9 x 0.9	13.7			Few stars superimposed; mag 10.5 star NW 3′0.
18 25 30.8	+32 38 03	MCG +5-43-19	15.8	0.6 x 0.4	14.1	SBb	174	
18 25 38.5	+31 59 58	MCG +5-43-20	14.6	0.4 x 0.4	12.4			Mag 10.8 star on N edge.
18 32 06.2	+31 02 32	MCG +5-44-2	14.0	0.4 x 0.4	11.9			**UGC 11264** NW 5′2.
18 39 51.3	+32 48 22	MCG +5-44-9	13.6	0.8 x 0.7	12.9		39	Mag 11.6 star NE edge; CGCG 173-21 N 3′4.
17 58 54.0	+36 58 46	MCG +6-39-29	14.8	0.6 x 0.5	13.3		156	Mag 9.58v star W 1′7.
17 59 46.4	+36 40 31	MCG +6-39-30	14.9	0.8 x 0.4	13.5	Sb	27	Mag 9.10v star NE 4′5.
18 05 11.7	+34 42 14	MCG +6-40-2	14.7	0.9 x 0.7	14.0	Sd	93	Bright knot or superimposed star N of center.
18 05 45.6	+34 45 02	MCG +6-40-3	14.2	0.6 x 0.4	12.5	S	138	Mag 10.78v star SSE 3′8.
18 07 49.0	+35 52 58	MCG +6-40-6	14.1	0.6 x 0.5	12.7	S	111	
18 08 22.9	+34 58 37	MCG +6-40-7	15.0	0.9 x 0.4	13.7	Sa	177	Mag 9.58v star N 2′5.
18 20 33.7	+38 11 07	MCG +6-40-13	14.2	0.7 x 0.4	12.9	E/S0	18	Mag 11.8 star WNW 2′5.
18 20 45.0	+38 08 33	MCG +6-40-15	14.4	0.7 x 0.5	13.1		123	MCG +6-40-16 E 0′9.
18 20 49.8	+38 08 30	MCG +6-40-16	14.6	0.7 x 0.2	12.3	S?	9	MCG +6-40-15 W 0′9.
18 22 04.1	+36 37 14	MCG +6-40-17	14.3	0.7 x 0.5	13.0	S?	90	Forms a triangle with a pair of mag 13 stars NE 3′0 and SE 2′8.
18 35 16.8	+38 27 25	MCG +6-41-5	14.7	0.4 x 0.4	12.8	E/S0		
18 35 57.2	+38 08 03	MCG +6-41-6	13.8	0.6 x 0.6	12.6			**CGCG 201-15** NE 2′9; mag 10.64v star ESE 2′4.
18 39 58.0	+37 49 43	MCG +6-41-12	15.0	0.6 x 0.6	13.8			Mag 9.80v star SW 2′4.
18 41 14.9	+35 02 52	MCG +6-41-17	14.1	0.7 x 0.4	12.5		120	Mag 12.0 star E 0′9.
18 43 03.2	+37 23 55	MCG +6-41-20	14.3	0.7 x 0.4	12.8		9	Mag 8.63v star NW 3′7.
18 49 02.0	+34 27 39	MCG +6-41-21	14.7	0.6 x 0.4	13.0	Sbc	18	Mag 8.91v star W 2′8.
18 57 37.6	+38 00 28	MCG +6-41-24	12.6	1.1 x 0.9	12.7	E?	60	Several stars superimposed; mag 10.8 star E 2′6.
18 01 17.8	+39 42 26	MCG +7-37-16	14.6	0.8 x 0.5	13.5	Sb	159	Mag 9.81v star N 3′0.
18 17 26.0	+40 01 51	MCG +7-37-33	17.3	0.9 x 0.6	16.4	Sc	39	Mag 11.8 star W 0′9.
18 17 27.2	+40 59 22	MCG +7-37-34	15.4	0.7 x 0.5	14.2	Sb	0	Mag 6.13v star SW 5′0.
18 18 06.1	+39 49 31	MCG +7-37-36	14.0	0.7 x 0.5	12.7	S?	21	**CGCG 227-25** SE 2′3.
18 19 26.4	+40 58 23	MCG +7-37-38	15.3	0.7 x 0.3	13.5	Sb	150	Mag 15.1 star on W edge; mag 11.1 star NW 1′6.
18 55 05.7	+39 31 39	MCG +7-39-3	14.7	0.7 x 0.7	13.8			Mag 10.69v star W 2′1.
18 55 53.2	+39 56 53	MCG +7-39-4	14.1	0.6 x 0.4	12.4		60	Mag 9.70v star on E edge.
19 00 04.5	+40 54 21	MCG +7-39-5	15.4	0.4 x 0.4	13.2			MCG +7-39-6 S edge; mag 10.72v star ENE 2′1.
19 00 05.6	+40 53 58	MCG +7-39-6	14.4	0.5 x 0.3	12.2		12	MCG +7-39-5 N edge.
17 56 05.8	+33 12 29	NGC 6504	12.6	2.2 x 0.5	12.6	S	94	Large central bulge, located between two mag 11 stars.
18 10 57.5	+31 06 55	NGC 6575	12.7	1.8 x 1.3	13.6	E	65	Mag 11.01v star SW 4′2.
18 12 21.7	+39 37 57	NGC 6585	12.9	1.9 x 0.4	12.5	S?	50	Mag 10.84v star NE 2′4; mag 11.05v star S 2′3.
18 16 10.8	+36 04 42	NGC 6612	14.2	0.7 x 0.7	13.3			Faint star NE of stellar nucleus.
18 28 08.0	+34 18 11	NGC 6640	13.5	1.1 x 0.8	13.2	Scd:	153	
18 29 38.8	+39 51 56	NGC 6646	12.6	1.6 x 1.3	13.2	Sa	63	Mag 8.82v star N 3′8.
18 33 01.6	+34 03 37	NGC 6657	13.5	1.0 x 0.6	12.8	SB	138	**CGCG 201-7** and mag 11.27v star W 6′6.
18 34 11.2	+32 03 51	NGC 6662	13.7	1.6 x 0.5	13.3	SBab?	20	Mag 8.11v star NW 5′2.
18 33 33.8	+40 02 55	NGC 6663	13.9	1.0 x 0.9	13.6	SAB(r)cd:	140	Stellar nucleus; mag 7.25v star NNW 7′7.
18 34 30.0	+30 43 14	NGC 6665	13.9	1.1 x 0.6	13.3	Sc	30	**MCG +5-44-5** SE 9′8.
18 37 26.6	+40 03 27	NGC 6675	12.4	1.7 x 1.3	13.2	Sbc	130	
18 39 58.6	+39 58 54	NGC 6685	13.4	1.1 x 0.9	13.3	S0$^-$:	30	**IC 4772** N 2′8.
18 40 07.1	+40 08 15	NGC 6686	14.7	0.9 x 0.8	14.2		84	UGC 11319 SW 1′8.
18 40 40.1	+36 17 18	NGC 6688	12.6	1.6 x 1.3	13.3	SA0$^+$	75	Bright nucleus, numerous stars superimposed on shell.
18 41 41.7	+34 50 36	NGC 6692	14.1	1.0 x 0.7	13.8	E/S0	110	Located 3′1 SW of mag 9.60v star.
18 42 42.9	+40 22 01	NGC 6695	13.5	1.1 x 0.7	13.0	SBb	12	
18 46 04.2	+32 16 45	NGC 6700	13.1	1.4 x 1.0	13.3	SB(rs)c	115	Very patchy; mag 8.91v star NE 2′8.
18 50 44.5	+33 57 33	NGC 6713	13.7	0.4 x 0.4	11.6	Pec		
19 01 41.3	+40 44 37	NGC 6745	13.9	1.1 x 0.5	13.1	Sm	24	Triple system; short extension N.
17 56 55.0	+32 38 09	UGC 11058	12.6	1.5 x 1.1	13.0	SB(s)b I-II	45	Several stars superimposed SW edge, one bright star or knot E edge; mag 9.06v star S 2′1.
17 58 51.9	+34 00 22	UGC 11076	13.9	1.2 x 0.5	13.2	Sbc	63	Pair of mag 11.5 stars N 1′9.
18 00 19.1	+34 38 24	UGC 11087	13.7	0.6 x 0.5	12.2	S?	135	Mag 10.47v star SW 2′2.
18 02 48.8	+29 18 23	UGC 11098	14.7	1.0 x 0.5	13.8	S	132	Mag 10.7 star on W edge.
18 07 27.9	+35 33 49	UGC 11124	12.7	2.4 x 2.1	14.3	SB(s)cd	132	Very filamentary; mag 11.03 v star superimposed W of NE-SW bar.
18 09 10.9	+30 19 35	UGC 11129	13.7	1.2 x 0.7	13.3	Sa	83	Mag 10.85v star NE 1′8.
18 09 25.9	+38 47 43	UGC 11132	14.6	2.0 x 0.3	13.9	Sb II-III	145	
18 11 09.3	+39 10 13	UGC 11140	14.3	2.3 x 0.4	14.1	Scd:	31	Almost stellar nucleus; pair of mag 13 stars W edge.
18 12 05.6	+29 09 21	UGC 11151	14.0	1.5 x 0.3	13.0	S	0	Mag 9.87v star NW 2′2.
18 12 25.6	+30 12 03	UGC 11157	14.4	1.7 x 0.4	13.8	Sd:	150	Mag 11.21v star NE of center 0′8.
18 13 01.6	+29 41 44	UGC 11162	14.4	1.4 x 0.7	14.2	S	115	
18 13 18.7	+33 49 47	UGC 11163	13.8	1.1 x 0.9	13.6	B	6	**UGC 11167** S 5′5.
18 14 09.8	+30 40 23	UGC 11171	13.9	1.0 x 0.5	13.0	S	153	
18 16 56.6	+40 48 11	UGC 11190	15.3	0.8 x 0.8	14.7	Scd:		Mag 11.51v star SW 2′6.
18 17 57.0	+30 38 59	UGC 11195	13.9	1.1 x 0.5	13.1	SB?	114	Multi-galaxy system.
18 18 29.1	+36 19 13	UGC 11199	13.5	1.6 x 0.8	13.6	Sa	165	Mag 11.13v star SE 2′9.
18 23 17.8	+30 15 10	UGC 11219	14.4	0.9 x 0.4	13.1	Double System	129	Small nucleus with two faint stars W; single faint arm extends SE.
18 25 44.4	+32 10 41	UGC 11233	13.6	0.7 x 0.6	12.5	S0	54	Almost stellar; mag 9.20v star NE 4′3.
18 31 35.0	+33 56 12	UGC 11265	13.2	1.0 x 1.0	13.0	S?		Mag 8.44v star SE 4′1; faint, elongated anonymous galaxy NE 5′5; **CGCG 201-7** and mag 11.27v star NE 13′9.
18 32 07.1	+37 36 45	UGC 11268	13.8	1.3 x 0.5	13.2	Sbc	169	
18 33 39.7	+32 08 21	UGC 11275	15.8	1.5 x 0.2	14.3	Scd:	25	Mag 8.11v star E 3′4; **CGCG 173-8** and **CGCG 173-9** NE 7′7.
18 33 43.8	+38 37 03	UGC 11278	14.0	1.0 x 0.4	12.9	S0	57	
18 34 04.9	+38 35 49	UGC 11281	14.0	1.5 x 0.3	13.0	Sb III	6	Two faint stars superimposed S of center; mag 13.8 star on S end; mag 9.22v star S 3′7.
18 36 04.9	+30 50 22	UGC 11291	14.0	1.9 x 0.4	13.6	Sd:	116	Uniform surface brightness; mag 11.2 star E end.
18 36 59.8	+30 36 58	UGC 11296	14.2	0.9 x 0.6	13.4	S	40	**CGCG 173-18** E 2′8; mag 7.94v star N 6′1.
18 37 32.2	+33 16 50	UGC 11303	13.9	1.2 x 0.7	13.6	Sb II	118	
18 37 31.6	+36 51 20	UGC 11304	13.6	1.4 x 0.6	13.3	Sa	128	

(Continued from previous page)

49 GALAXIES 49

RA h m s	Dec ° ′ ″	Name	Mag (V)	Dim ′ Maj x min	SB	Type Class	PA	Notes
18 38 42.6	+37 02 49	UGC 11311	14.5	1.5 x 0.5	14.0	S	131	
18 38 50.1	+37 57 02	UGC 11312	13.8	1.2 x 0.6	13.3	S0	176	
18 39 19.5	+38 56 24	UGC 11313	14.4	2.1 x 0.3	13.7	Sb III	23	Mag 11.5 star S 2′2.
18 40 00.6	+40 06 59	UGC 11319	15.1	1.1 x 0.4	14.1	S	0	Bright center, two faint, thin arms; NGC 6686 NE 1′8; mag 7.77v star W 3′4.
18 40 44.2	+36 09 47	UGC 11325	13.6	2.1 x 0.5	13.5	Sb III	136	**UGC 11322** SW 3′5.
18 42 26.3	+32 22 32	UGC 11333	14.5	1.3 x 0.5	13.9	S	151	Elongated center, faint envelope NW.
18 42 37.3	+34 28 13	UGC 11336	14.6	0.9 x 0.7	13.9	Scd:	51	Filamentary.
18 42 41.9	+35 37 31	UGC 11338	13.6	0.5 x 0.4	11.7	Pec		
18 50 47.3	+35 20 51	UGC 11367	15.1	0.5 x 0.2	12.4	SB	60	Stellar **PGC 62490** NE 0′8.
18 52 33.5	+33 49 36	UGC 11372	14.1	1.2 x 0.3	12.8	S	154	
18 56 51.1	+36 37 22	UGC 11380	12.8	1.5 x 0.6	12.5	Sab	172	Faint dark patch S of center; mag 10.61v star SE 3′1.
19 03 49.2	+33 50 39	UGC 11397	12.8	1.1 x 0.6	12.2	SBa	95	
19 04 57.5	+34 27 50	UGC 11399	13.8	1.5 x 0.3	12.8	S	124	Located N of a pair of mag 10.9 stars.
19 07 03.8	+29 00 21	UGC 11404	13.2	1.9 x 1.3	14.0	Sbc II	60	Many stars superimposed, bright knot NE edge; mag 10.70v star NW 1′5.
19 18 21.7	+34 50 12	UGC 11426	13.5	0.8 x 0.6	12.6	S	55	Mag 9.85v star S 1′2.
19 20 28.6	+30 49 31	UGC 11428	13.8	1.1 x 0.8	13.5	Scd:	168	Numerous stars superimposed; in star rich area.
19 23 49.9	+34 47 32	UGC 11433	14.7	0.9 x 0.2	12.7	S	25	Mag 12.4 star SW 0′8; in star rich area.

OPEN CLUSTERS

RA h m s	Dec ° ′ ″	Name	Mag	Diam ′	No. ★	B ★	Type	Notes
18 08 48.0	+31 32 00	DoDz 9		28	15		cl	Few stars; moderate brightness range; no central concentration; detached.
19 01 26.7	+29 17 00	NGC 6743		8	35	8.2	cl?	
19 20 52.7	+37 46 00	NGC 6791	9.5	10	380	15.0	cl	Rich in stars; moderate brightness range; strong central concentration; detached. (A) Very rich cluster - appears as a sparse globular cluster on DSS image.
18 54 30.6	+36 54 00	Ste 1	3.8	40	77	4.3	cl:	Few stars; large brightness range; not well detached.

GLOBULAR CLUSTERS

RA h m s	Dec ° ′ ″	Name	Total V m	B ★ V m	HB V m	Diam ′	Conc. Class Low = 12 High = 1	Notes
19 16 35.5	+30 11 05	NGC 6779	8.4	13.2	16.3	8.8	10	= **M 56**.

PLANETARY NEBULAE

RA h m s	Dec ° ′ ″	Name	Diam ″	Mag (P)	Mag (V)	Mag cent ★	Alt Name	Notes
18 53 35.2	+33 01 44	NGC 6720	76	9.7	8.8	15.2	PK 63+13.1	= **M 57**, **Ring Nebula**. Galaxy IC 1296 NW 4′2.
19 11 06.8	+30 32 39	NGC 6765	40	13.1	12.9	16.0	PK 62+9.1	Brightest portion appears elongated NE-SW with secondary bright patch on E edge; close pair of mag 9.68v and 11.8 stars W 3′8.
18 50 02.3	+35 14 33	PN G64.9+15.5	28	12.8	13.3		PK 64+15.1	Slightly elongated E-W; mag 11.1 star S 2′0.
19 00 26.6	+38 21 06	PN G68.7+14.8	2	13.7			PK 68+14.1	Mag 10.30v star N 0′7; mag 12.2 star S 2′2.

50 GALAXIES 50

RA h m s	Dec ° ′ ″	Name	Mag (V)	Dim ′ Maj x min	SB	Type Class	PA	Notes
16 51 03.8	+30 39 44	CGCG 169-7	15.0	0.6 x 0.4	13.4		24	Mag 10.03v star W 8′3.
16 56 22.0	+29 19 50	CGCG 169-12	16.0	0.5 x 0.3	13.7		102	Mag 13.3 star S 1′1.
17 15 10.7	+29 15 44	CGCG 170-5	14.9	0.6 x 0.6	13.6			Almost stellar nucleus, faint outer envelope.
17 16 54.9	+31 31 36	CGCG 170-10	14.9	0.9 x 0.4	13.6		138	Close pair of mag 7.06v, 8.62v stars E 6′8; **CGCG 170-11** N 3′4.
17 17 59.6	+33 20 00	CGCG 198-32	14.0	0.5 x 0.3	11.8		90	Almost stellar.
17 40 05.7	+33 48 34	CGCG 199-5	14.3	0.6 x 0.3	12.3		102	Double system, small companion at E end.
17 52 36.8	+37 44 48	CGCG 199-26	14.0	0.7 x 0.3	12.2		21	Located 1′8 S of mag 11.16v star; = **Mkn 1119**.
17 10 34.0	+36 18 10	IC 1244	13.9	1.0 x 1.0	13.7	Sb		Mag 11.2 star ENE 2′1.
17 12 37.0	+38 01 10	IC 1245	13.7	1.7 x 0.9	14.1	S0	125	Mag 10.25v star E 3′9.
17 14 55.3	+35 31 11	IC 1249	14.0	0.6 x 0.6	12.7			Mag 10.06v star S 6′6.
16 37 47.4	+36 06 51	IC 4614	14.3	0.8 x 0.6	13.4	S0/a	83	Mag 15.4 star on N edge; NGC 6169 SE 2′7.
16 42 07.7	+36 41 00	IC 4617		1.2 x 0.4			32	
16 58 07.4	+29 54 33	MCG +5-40-14	14.3	0.7 x 0.4	12.8	S?	0	
16 59 28.1	+32 21 27	MCG +5-40-21	15.2	0.5 x 0.5	13.6			Mag 11.9 star NE 3′3.
17 00 59.6	+29 25 07	MCG +5-40-25	15.3	0.6 x 0.5	13.9		90	**Mkn 504** ESE 1′4; mag 9.28v star N 2′8.
17 06 54.9	+30 16 11	MCG +5-40-38	14.2	0.9 x 0.9	13.8	Sb		
17 08 33.5	+31 48 19	MCG +5-40-43	13.9	0.7 x 0.7	13.0	SBc		
17 10 12.4	+30 18 31	MCG +5-40-47	14.1	0.9 x 0.7	13.4	Sbc		
17 52 40.1	+29 48 15	MCG +5-42-7	14.8	0.3 x 0.3	12.0			Almost stellar; 1′3 SE of NGC 6486.
17 56 46.9	+31 16 59	MCG +5-42-17	15.0	0.7 x 0.5	13.7	Sb	90	Pair of mag 13 stars W 0′9.
18 02 57.2	+29 18 35	MCG +5-42-29	14.8	0.6 x 0.6	13.5	Sc		**CGCG 171-49** N 4′0; UGC 11098 W 1′9.
17 57 53.4	+31 49 21	MCG +5-42-41		0.8 x 0.6			0	
16 39 54.9	+37 10 44	MCG +6-37-1	14.8	0.7 x 0.3	13.0	Spiral	36	MCG +6-37-2 E 2′4.
16 40 06.7	+37 11 24	MCG +6-37-2	14.2	0.8 x 0.7	13.4	S0	87	MCG +6-37-1 W 2′4.
16 41 00.2	+33 46 19	MCG +6-37-4	14.3	0.9 x 0.4	13.0	S0/a	66	Mag 10.56v star SW 2′8.
16 48 42.3	+35 56 34	MCG +6-37-11	14.8	0.9 x 0.4	13.5	Sc	27	Mag 7.42v star S 1′2.
16 58 55.4	+36 31 16	MCG +6-37-18	14.3	0.9 x 0.4	13.1	SBab	60	Mag 13.0 star NW 1′6.
17 00 01.8	+38 48 55	MCG +6-37-19	14.9	1.1 x 0.7	14.4	Sb	123	Large N-S core; mag 10.65v star W 2′7.
17 00 07.2	+37 50 19	MCG +6-37-20	14.9	0.7 x 0.6	13.8		159	Mag 8.46v star W 3′4.
17 03 27.9	+36 04 17	MCG +6-37-23	14.8	1.0 x 0.6	14.1	Sb	177	Mag 9.27v star E 3′3.
17 06 41.8	+35 52 06	MCG +6-37-29	14.2	0.9 x 0.6	13.4		111	Mag 13.4 star W 2′3.
17 07 06.7	+34 14 03	MCG +6-37-31	14.8	0.7 x 0.5	13.5		6	Mag 10.8 star and MCG +6-37-32 N 1′6.
17 07 10.1	+34 16 08	MCG +6-37-32	14.8	0.8 x 0.5	13.8	E	18	Mag 10.8 star SW 1′0.
17 08 00.4	+36 20 36	MCG +6-38-1	14.8	0.9 x 0.5	13.8		147	
17 11 51.5	+37 33 31	MCG +6-38-4	14.6	0.6 x 0.5	13.2	Sc	150	
17 12 28.6	+35 52 59	MCG +6-38-5	14.0	0.8 x 0.7	13.3		48	Mag 12.3 star S edge.
17 17 16.2	+38 53 47	MCG +6-38-12	14.6	0.5 x 0.4	12.7		90	Faint, slightly elongated anonymous galaxy WSW 1′6.
17 19 12.0	+34 12 00	MCG +6-38-18	14.0	0.8 x 0.4	12.8	E/S0	60	**MCG +6-38-15** S edge; **MCG +6-38-14** W 1′4; **MCG +6-38-13** W 1′8.

(Continued from previous page)

GALAXIES

RA h m s	Dec ° ′ ″	Name	Mag (V)	Dim ′ Maj x min	SB	Type Class	PA	Notes
17 22 06.1	+35 20 02	MCG +6-38-19	15.2	0.9 x 0.6	14.3	Sbc	168	
17 26 21.9	+35 16 06	MCG +6-38-21	14.3	1.1 x 0.7	13.9		132	Mag 12.8 star NE 1′1.
17 27 05.2	+37 51 03	MCG +6-38-22	14.6	0.9 x 0.7	13.9		135	Mag 9.81v star SW 0′8.
17 36 39.4	+35 57 30	MCG +6-39-1	14.7	0.5 x 0.5	13.0			Mag 8.96v star E 2′0.
17 38 57.7	+33 25 47	MCG +6-39-5	14.9	0.7 x 0.4	13.4		9	Mag 10.15v star E 1′7.
17 40 06.7	+37 41 04	MCG +6-39-6	14.6	1.0 x 0.7	14.1		141	Pair of mag 13 stars S 1′3.
17 40 32.1	+35 38 43	MCG +6-39-9	14.0	0.8 x 0.6	13.3	E/S0	21	MCG +6-39-10 N edge.
17 40 34.1	+35 39 11	MCG +6-39-10	15.0	0.6 x 0.4	13.4		120	MCG +6-39-9 S edge.
17 40 55.1	+38 43 53	MCG +6-39-11	14.4	0.7 x 0.6	13.3		174	Several faint, stellar galaxies S.
17 43 37.2	+33 04 54	MCG +6-39-12	14.9	0.5 x 0.4	13.0		12	MCG +6-39-13 N 0′9.
17 43 38.7	+33 05 45	MCG +6-39-13	15.3	0.4 x 0.3	12.8		132	Mag 12.2 star on E edge.
17 43 46.3	+33 05 28	MCG +6-39-14	15.2	0.6 x 0.4	13.5	Sc		MCG +6-39-13 W 1′6; mag 10.23v star NE 2′7.
17 44 15.4	+38 04 47	MCG +6-39-16	14.7	0.5 x 0.5	13.0			Located between a pair of N-S mag 12 stars.
17 48 26.0	+34 04 13	MCG +6-39-21	15.0	0.4 x 0.3	12.6		3	UGC 10990 NW 2′5.
17 51 42.7	+37 24 20	MCG +6-39-23	15.8	0.7 x 0.5	14.6	Sc	75	Mag 14.8 star on N edge; very slender **UGC 11015** S 1′0.
17 58 54.0	+36 58 46	MCG +6-39-29	14.8	0.6 x 0.5	13.3		156	Mag 9.58v star W 1′7.
17 59 46.4	+36 40 31	MCG +6-39-30	14.9	0.8 x 0.4	13.5	Sb	27	Mag 9.10v star NE 4′5.
16 38 13.9	+40 08 45	MCG +7-34-125	14.1	0.9 x 0.4	12.8	Sb	168	**MCG +7-34-126** N 1′6.
16 40 28.1	+39 19 10	MCG +7-34-130	13.8	0.9 x 0.7	13.2	Compact	12	Mag 11.3 star S 1′3.
16 40 29.8	+39 14 04	MCG +7-34-131	14.9	0.8 x 0.3	13.2	Sc	48	Faint anonymous galaxy of similar size and shape ESE 1′7.
16 41 15.1	+39 45 07	MCG +7-34-132	14.7	0.8 x 0.3	13.0	Sb	141	Pair with MCG +7-34-133 W.
16 41 10.0	+39 45 20	MCG +7-34-133	14.6	0.8 x 0.5	13.5	Sb	108	Pair with MCG +7-34-132 E; forms a triangle with a pair of mag 12 stars W.
16 41 37.5	+40 09 45	MCG +7-34-135	14.3	1.2 x 0.6	13.8	Sb	57	Mag 10.15v star SSW 4′0.
16 41 45.3	+39 38 34	MCG +7-34-136	14.4	0.7 x 0.4	12.9	Sb	114	Mag 9.66v star W 1′2.
16 42 08.6	+40 16 34	MCG +7-34-138	14.7	0.9 x 0.9	14.3	Sbc		
16 42 11.8	+39 27 18	MCG +7-34-139	15.1	0.5 x 0.5	13.4	Sc		Located between an E-W pair of mag 13 stars.
16 44 11.0	+39 47 16	MCG +7-34-143	14.4	0.8 x 0.5	13.3	Sc	72	Mag 12.3 star WSW 1′9.
16 55 47.1	+39 18 34	MCG +7-35-7	15.0	0.9 x 0.6	14.2	Sb	42	
16 57 26.9	+40 44 07	MCG +7-35-9	14.4	0.9 x 0.3	12.9	Sa	144	MCG +7-35-11 and 12 ESE 2′5.
16 57 41.0	+40 43 17	MCG +7-35-11	14.0	0.8 x 0.6	13.3	E	177	Pair with MCG +7-35-12 S.
16 57 40.1	+40 42 24	MCG +7-35-12	14.1	0.9 x 0.5	13.3	E	117	Pair with MCG +7-35-11 N.
16 58 50.8	+40 07 14	MCG +7-35-13	14.2	1.0 x 0.9	13.9		0	
17 01 00.5	+39 33 55	MCG +7-35-15	13.7	0.9 x 0.7	13.1		174	Mag 8.02v star SW 1′8.
17 02 35.9	+40 29 24	MCG +7-35-18	15.0	0.6 x 0.4	13.5	E/S0	150	Mag 11.5 star E 1′2; mag 9.66v star SSE 3′9.
17 02 58.2	+40 31 27	MCG +7-35-19	14.3	0.5 x 0.5	12.6			Mag 14.4 star on E edge.
17 03 09.9	+40 38 10	MCG +7-35-20	15.9	0.6 x 0.5	14.5	Sc	36	Mag 12.7 star SE 2′7.
17 05 57.5	+39 52 26	MCG +7-35-26	14.9	0.8 x 0.8	14.2	Sbc		Mag 10.66v star N 2′3.
17 08 25.8	+40 47 18	MCG +7-35-33	14.0	1.1 x 0.8	14.0	E/S0	33	Mag 11.9 star N 2′3.
17 09 54.2	+40 35 46	MCG +7-35-35	14.2	0.8 x 0.3	12.5	S?	108	Mag 14.1 star N 0′7.
17 10 34.9	+39 55 31	MCG +7-35-38	14.2	1.0 x 0.8	13.8		33	Mag 10.40v star SSE 2′3.
17 12 12.5	+39 44 24	MCG +7-35-42	15.0	0.6 x 0.6	13.7	S?		Forms a triangle with a mag 13.2 star NW 2′2 and a mag 12.4 star SW 1′7.
17 13 40.2	+40 57 45	MCG +7-35-50	14.4	1.0 x 0.8	14.0		156	Mag 10.04v star N 2′0.
17 14 21.8	+39 04 11	MCG +7-35-52	14.8	0.7 x 0.5	13.5	Sb	18	Mag 8.86v star N 2′8.
17 17 16.0	+40 34 47	MCG +7-35-61	14.4	0.6 x 0.6	13.1			Mag 9.63v star SW 1′8.
17 17 20.0	+40 52 12	MCG +7-35-62	15.3	1.1 x 0.2	13.5	Sc	75	Located 2′8 NE of NGC 6339.
17 20 16.0	+39 15 42	MCG +7-36-1	13.5	0.7 x 0.5	12.5	E/S0	21	
17 20 11.8	+40 51 08	MCG +7-36-2	15.6	0.5 x 0.5	14.0	Sbc		Mag 12.1 star SW 0′9.
17 29 22.2	+40 15 04	MCG +7-36-13	14.9	0.6 x 0.5	13.4		33	Mag 12.9 star SW 1′5.
17 29 40.2	+40 16 13	MCG +7-36-14	15.1	0.5 x 0.4	13.2		0	Mag 9.47v star NE 9′8.
17 31 24.4	+40 14 12	MCG +7-36-17	13.7	0.8 x 0.8	13.1			Mag 11.2 star W 1′2; mag 9.41v star NE 2′4.
17 33 11.9	+39 19 05	MCG +7-36-19	15.2	0.5 x 0.3	13.0	Sb	99	Mag 11.02v star N 4′2.
17 42 27.7	+39 19 02	MCG +7-36-32	14.4	0.7 x 0.4	12.9	Sb	117	Mag 10.1 star W 3′5.
17 44 08.8	+40 52 01	MCG +7-36-33	14.3	0.5 x 0.3	12.1	S?	177	Star superimposed N end?
18 01 17.8	+39 42 26	MCG +7-37-16	14.6	0.8 x 0.5	13.5	Sb	159	Mag 9.81v star N 3′0.
16 36 37.2	+36 11 58	NGC 6194	13.8	1.0 x 0.8	13.4	Compact	105	
16 36 32.5	+39 01 41	NGC 6195	13.0	1.5 x 1.1	13.4	Sb I	45	Strong dark lane NW.
16 37 54.0	+36 04 22	NGC 6196	12.9	2.0 x 1.2	13.7	SAB0⁻: pec	140	IC 4614 2 NW 2′7.
16 37 59.8	+35 59 39	NGC 6197	14.6	1.3 x 0.5	14.0		37	
16 43 04.1	+36 49 58	NGC 6207	11.6	3.0 x 1.3	13.0	SA(s)c III	15	Many knots and superimposed stars.
16 43 23.1	+39 48 18	NGC 6212	14.2	0.7 x 0.4	12.6	S?	105	
16 54 46.6	+36 30 04	NGC 6255	12.7	3.6 x 1.5	14.4	SBcd:	85	Dark patch S of center; bright condensation or star E end.
16 56 03.7	+39 38 41	NGC 6257	15.1	0.8 x 0.3	13.4		123	
16 59 20.7	+29 56 43	NGC 6274	13.8	0.6 x 0.5	12.3	Double System	23	NGC 6274A on S edge.
16 59 21.8	+29 56 21	NGC 6274A		0.7 x 0.2			123	NGC 6274 on N edge.
17 00 47.2	+29 49 13	NGC 6282	14.4	0.7 x 0.5	13.1	SBbc	36	**UGC 10654** SSW 6′3.
17 12 55.8	+40 15 55	NGC 6320	13.9	1.2 x 0.8	13.7	Sbc	103	Stellar nucleus; stars on S and NE edges.
17 15 44.5	+29 24 17	NGC 6330	14.0	1.4 x 0.5	13.5	SBb	160	Strong N-S bar.
17 17 06.8	+40 50 43	NGC 6339	12.7	2.9 x 1.7	14.3	SBd	10	Bright nucleus, patchy arms; MCG +7-35-62 NE 2′8.
17 19 06.5	+36 03 36	NGC 6349	14.3	0.8 x 0.2	12.2	S?	81	**CGCG 198-34** SW 5′8.
17 24 27.6	+29 23 24	NGC 6364	12.9	1.5 x 1.2	13.4	S0	2	Three faint stars on E edge.
17 25 09.1	+37 45 36	NGC 6367	14.2	0.8 x 0.7	13.5		156	Almost stellar nucleus; mag 7.68v star ENE 4′4.
17 43 56.4	+36 47 59	NGC 6433	13.3	2.0 x 0.5	13.1	Sb III	163	**CGCG 199-14** N 4′2.
17 46 07.7	+35 34 10	NGC 6446	15.2	0.7 x 0.6	14.1	Sa	9	Faint stars N and S of bright center.
17 46 17.3	+35 34 19	NGC 6447	12.8	1.6 x 0.9	13.1	SB?	145	Two main arms, dark patches E and W of center.
17 51 52.9	+31 27 41	NGC 6485	12.9	1.5 x 1.4	13.6	Sbc I	24	Mag 9.57v star ENE 2′9.
17 52 35.3	+29 49 03	NGC 6486	14.3	0.8 x 0.8	13.9	E/S0		Stellar **MCG −5-42-7** SE 1′3; NGC 6487 NE.
17 52 41.9	+29 50 16	NGC 6487	11.9	1.8 x 1.8	13.1	E		NGC 6486 SW 1′9.
17 56 05.8	+33 12 29	NGC 6504	12.6	2.2 x 0.5	12.6	S	94	Large central bulge, located between two mag 11 stars.
16 36 54.4	+36 25 25	UGC 10473	13.9	1.5 x 0.4	13.2	SBa	170	**CGCG 196-84** NE 2′3; mag 11.5 star NNW 2′3.
16 37 34.7	+37 17 06	UGC 10477	14.3	1.7 x 0.3	13.4	S?	26	Located between a close pair of stars, mags 9.54v and 10.75v.
16 40 22.8	+33 40 46	UGC 10504	14.0	1.0 x 0.5	13.1	Sab	169	
16 41 47.1	+39 58 59	UGC 10512	13.8	0.8 x 0.8	13.4	E		Star or almost stellar galaxy on NE edge.
16 44 33.7	+32 13 02	UGC 10529	14.0	1.0 x 0.8	13.6	Sbc	24	Row of three mag 14 and 15 stars along S edge.
16 46 57.0	+34 09 19	UGC 10547	13.7	1.4 x 0.7	13.6	SB(s)b	10	Mag 11.2 star NE 2′9; almost stellar anonymous galaxy SSE 2′3.
16 47 14.9	+40 14 39	UGC 10553	14.0	1.0 x 0.5	13.1	SBab:	153	Elongated **UGC 10550** S 6′5.
16 49 02.1	+36 11 32	UGC 10566	14.3	1.9 x 0.5	14.1	S	73	Two slender, slightly curving arms; UGC 10567 NE 2′4.
16 49 09.5	+36 13 23	UGC 10567	14.3	0.8 x 0.7	13.5	Compact	28	UGC 10566 SW 2′4.
16 53 52.4	+39 45 37	UGC 10599	13.3	1.1 x 0.9	13.3	E?	160	Bright, diffuse core, uniform envelope.
16 58 01.8	+38 13 09	UGC 10635	12.6	0.6 x 0.6	11.3	S?		Almost stellar, stellar **PGC 59357** E 6′9.

(Continued from previous page)

50 GALAXIES 50

RA h m s	Dec ° ′ ″	Name	Mag (V)	Dim ′ Maj x min	SB	Type Class	PA	Notes
17 01 30.4	+30 10 02	UGC 10662	13.4	1.5 x 0.9	13.7	E	75	Very small, very faint superimposed galaxy SE edge; mag 12.2 star N 1′2.
17 01 29.2	+30 19 28	UGC 10663	13.6	1.2 x 1.0	13.6	SAB(r)b	48	Faint star NW edge.
17 02 16.1	+30 42 09	UGC 10668	13.5	0.8 x 0.6	12.6	S0		**UGC 10666** N 2′2.
17 03 01.7	+29 52 22	UGC 10673	14.5	1.5 x 0.5	14.1		62	Almost stellar nucleus, uniform envelope; faint star SW end.
17 03 15.3	+31 27 18	UGC 10675	14.5	0.7 x 0.4	13.0	S?	36	Almost stellar core NE, faint extension SW.
17 03 30.6	+36 25 04	UGC 10677	13.6	0.8 x 0.6	12.7	Pec	111	
17 03 53.0	+31 29 55	UGC 10679	14.4	1.5 x 0.3	13.4	Sb III	52	Mag 11.1 star NE 3′0.
17 04 43.4	+34 33 28	UGC 10688	14.3	1.4 x 1.1	14.6	Scd:	78	
17 06 28.5	+38 21 45	UGC 10706	14.2	1.1 x 0.4	13.2	Sbc	32	
17 06 46.6	+35 39 26	UGC 10708	13.9	1.0 x 0.3	12.4	S	125	Mag 11.1 star N 1′2.
17 07 19.6	+31 26 19	UGC 10712	14.1	0.9 x 0.7	13.4	(R)SABa	60	Mag 10.34v star SW 3′3.
17 07 25.3	+30 13 31	UGC 10714	14.6	1.5 x 0.2	13.2	Sbc	159	Mag 11.9 star E 1′8.
17 07 27.5	+31 14 51	UGC 10715	13.7	0.9 x 0.7	13.1	SBa	37	Mag 13.3 star on W edge.
17 09 42.6	+36 24 52	UGC 10732	14.5	0.9 x 0.6	13.6	Scd:	45	Mag 11.4 star SE 1′9.
17 10 05.4	+32 29 58	UGC 10733	13.8	1.0 x 0.5	13.1	E	116	Mag 10.59v star N 0′7.
17 12 13.9	+30 10 06	UGC 10749	14.1	1.1 x 0.6	13.5	SBb	22	
17 12 35.4	+34 41 31	UGC 10753	14.8	0.9 x 0.3	13.2	Scd:	102	
17 16 26.0	+37 20 41	UGC 10785	14.0	1.5 x 0.5	13.5	S	108	
17 19 02.4	+32 16 11	UGC 10801	15.4	1.5 x 0.6	15.1	Sd	7	Stellar nucleus, uniform envelope.
17 19 28.7	+40 55 25	UGC 10812	14.1	1.1 x 0.5	13.3	SBb	125	Bright E-W bar, faint envelope.
17 31 40.2	+32 13 52	UGC 10890	15.1	1.7 x 0.2	13.8	Sd	88	Almost stellar central bulge; mag 12.0 star S of E end.
17 38 16.0	+35 32 02	UGC 10929	13.7	1.1 x 0.8	13.5	Sab	178	Bright center, uniform envelope.
17 38 35.8	+39 12 46	UGC 10933	14.5	1.5 x 0.4	13.8	Pec	12	Brightest N, faint streamer S; mag 12.5 star on N end.
17 45 14.4	+38 54 55	UGC 10969	14.7	1.6 x 0.3	13.8	S	105	Mag 8.21v star N 4′6.
17 46 27.9	+30 42 10	UGC 10976	13.5	0.8 x 0.5	12.4	SB	39	Elongated **UGC 10977** on E edge.
17 48 17.9	+34 06 06	UGC 10990	14.3	1.7 x 0.4	13.7	Sab	177	MCG +6-39-21 SE 2′5; mag 11.3 star NE 1′8.
17 49 26.8	+36 08 40	UGC 11000	13.5	0.8 x 0.5	12.4	S?	156	Mag 11.8 star N edge.
17 52 08.8	+29 51 44	UGC 11017	13.9	1.2 x 0.7	13.6	SBdm:	165	Mag 10.04v star NE 2′9.
17 54 07.1	+30 41 41	UGC 11031	14.4	0.8 x 0.8	13.8	Im:		Knotty, several stars superimposed.
17 54 30.2	+32 53 11	UGC 11035	13.0	1.5 x 1.0	13.3	SB?	141	Strongly peculiar; mag 10.49v star W 1′3.
17 54 51.8	+34 46 35	UGC 11041	13.3	1.0 x 0.6	12.6	Sab	65	
17 55 21.0	+31 50 48	UGC 11045	13.8	1.2 x 0.9	13.7	Sa	120	Bright center; mag 12.1 star E 1′4.
17 55 43.4	+34 35 14	UGC 11050	14.8	1.5 x 0.2	13.3	Sbc	56	Mag 14.4 star on w edge.
17 56 55.0	+32 38 09	UGC 11058	12.6	1.5 x 1.1	13.0	SB(s)b I-II	45	Several stars superimposed SW edge, one bright star or knot E edge; mag 9.06v star S 2′1.
17 58 51.9	+34 00 22	UGC 11076	13.9	1.2 x 0.5	13.2	Sbc	63	Pair of mag 11.5 stars N 1′9.
18 00 19.1	+34 38 24	UGC 11087	13.7	0.6 x 0.5	12.2	S?	135	Mag 10.47v star SW 2′2.
18 02 48.8	+29 18 23	UGC 11098	14.7	1.0 x 0.5	13.8	S	132	Mag 10.7 star on W edge.

GALAXY CLUSTERS

RA h m s	Dec ° ′ ″	Name	Mag 10th brightest	No. Gal	Diam ′	Notes
16 59 42.0	+32 32 00	A 2241	15.6	30	22	Several PGC GX, most members stellar/almost stellar.
17 09 42.0	+34 27 00	A 2249	15.4	39	16	All members anonymous, stellar/almost stellar.

OPEN CLUSTERS

RA h m s	Dec ° ′ ″	Name	Mag	Diam ′	No. ★	B ★	Type	Notes
16 45 28.6	+38 21 21	DoDz 6		3.5	5	8.8	cl?	Few stars; moderate brightness range; not well detached.

GLOBULAR CLUSTERS

RA h m s	Dec ° ′ ″	Name	Total V m	B ★ V m	HB V m	Diam ′	Conc. Class Low = 12 High = 1	Notes
16 41 41.5	+36 27 37	NGC 6205	5.8	11.9	15.0	20.0	5	= **M 13, Great Hercules Cluster**. The galaxy IC 4617, mag about 15v and 1.1 x 0.4 in PA 29° lies nearly 15′ NNE. It can be found nearly midway between the cluster and the galaxy NGC 6207, slightly west of a line between them.

PLANETARY NEBULAE

RA h m s	Dec ° ′ ″	Name	Diam ″	Mag (P)	Mag (V)	Mag cent ★	Alt Name	Notes
16 40 18.2	+38 42 19	PN G61.9+41.3	10	13.3		15.4	PK 61+41.1	Mag 14.4 star SW 0′7; mag 14.7 star N 2′9.

51 GALAXIES 51

RA h m s	Dec ° ′ ″	Name	Mag (V)	Dim ′ Maj x min	SB	Type Class	PA	Notes
15 25 39.5	+29 36 09	CGCG 165-54	14.6	0.6 x 0.4	13.1	E	21	Mag 14.7 star on S end.
15 38 43.5	+29 24 49	CGCG 166-23	13.4	0.9 x 0.4	12.1	SBc	0	**CGCG 166-24** NE 1′6.
15 54 11.7	+30 09 17	CGCG 167-1	13.9	0.4 x 0.3	11.7	E	57	Mag 14.4 star N 0′7.
16 14 11.0	+29 51 04	CGCG 167-55	13.9	0.5 x 0.4	12.1	S0	15	**CGCG 167-56** NNE 8′1; mag 11.31v star W 7′4.
16 22 35.1	+38 21 57	CGCG 196-50	14.3	0.3 x 0.3	11.6	S0		Almost stellar; = **Mkn 879**..
15 18 31.2	+32 23 30	IC 4539	15.3	0.4 x 0.4	13.2	SBb		Mag 10.67v star NE 4′0.
15 22 05.9	+33 08 54	IC 4542	14.1	0.8 x 0.5	13.0	S0/a	123	
15 29 14.9	+32 49 28	IC 4549	14.6	0.6 x 0.2	12.1	Sbc	99	Mag 11.1 star W 7′2.
16 33 49.9	+39 15 46	IC 4610	13.8	0.6 x 0.6	12.5	S?		Galaxy **KUG 1632+393A** NW 0′9; faint, elongated anonymous galaxy W 2′1.
16 37 47.4	+36 06 51	IC 4614	14.3	0.8 x 0.6	13.4	S0/a	83	Mag 15.4 star on N edge; NGC 6169 SE 2′7.
16 42 07.7	+36 41 00	IC 4617		1.2 x 0.4			32	
15 46 30.1	+32 07 00	MCG +5-37-23	14.5	0.8 x 0.4	13.1	SBb	15	
16 02 50.9	+32 08 37	MCG +5-38-12	14.6	0.4 x 0.4	12.5	Sab		Mag 10.91v star N 3′0.
16 13 04.9	+30 54 05	MCG +5-38-39	15.5	0.6 x 0.4	14.0	E	57	Mag 15.0 star N edge; mag 14.1 star SW edge; **MCG +5-38-41** E 2′1.
16 18 18.7	+31 36 09	MCG +5-38-52	15.5	0.5 x 0.5	13.9	Sbc		
15 16 53.9	+38 14 27	MCG +6-34-1	16.0	0.5 x 0.3	13.8		126	Mag 10.10v star NW edge.
15 18 01.6	+38 12 59	MCG +6-34-2	14.6	0.6 x 0.5	13.1	S?	6	Mag 12.0 star N 1′7.

RA h m s	Dec ° ′ ″	Name	Mag (V)	Dim ′ Maj x min	SB	Type Class	PA	Notes
15 20 24.6	+32 51 31	MCG +6-34-4	15.2	0.7 x 0.4	13.7		99	MCG +6-34-5 S 0.'9.
15 20 26.4	+32 50 41	MCG +6-34-5	15.4	0.7 x 0.3	13.6	Sb	87	MCG +6-34-4 N 0.'9.
15 27 57.3	+34 57 58	MCG +6-34-10	15.1	0.8 x 0.2	13.0	Sc	132	Mag 13.4 star E edge; mag 12.1 star W 2.'0.
15 31 39.0	+35 16 46	MCG +6-34-13	14.9	0.5 x 0.5	13.2			
15 33 07.3	+35 03 36	MCG +6-34-14	15.2	0.5 x 0.4	13.3	Sc	108	Mag 11.00v star S 2.'6.
15 33 20.5	+37 58 10	MCG +6-34-15	16.1	0.6 x 0.4	14.4		3	Mag 13.2 star N 2.'3.
15 34 00.5	+32 49 05	MCG +6-34-16	14.7	0.5 x 0.5	13.0	Sbc		Mag 12.2 star WNW 3.'0.
15 35 34.0	+36 14 34	MCG +6-34-18	15.1	0.7 x 0.4	13.5	SBbc	135	Mag 8.63v star S 2.'2.
15 37 24.9	+33 40 18	MCG +6-34-22	14.3	0.8 x 0.6	13.3	SBb	108	Mag 12.1 star on SW edge.
15 37 29.1	+36 14 36	MCG +6-34-23	14.7	0.8 x 0.5	13.6		135	Mag 11.5 star S 1.'3; mag 8.99v star NE 3.'9.
15 38 38.8	+36 57 30	MCG +6-34-24	14.3	0.6 x 0.4	12.6	Sa		Mag 12.4 star W 2.'7.
15 39 08.9	+32 24 24	MCG +6-34-25	15.4	0.6 x 0.4	13.7	SBbc	36	
15 39 46.0	+34 01 58	MCG +6-34-26	15.2	0.5 x 0.4	13.3	Sb	51	
15 40 37.8	+35 46 15	MCG +6-34-27	15.2	0.6 x 0.4	13.5	Sbc	147	Mag 11.6 star NNW 4.'6.
15 40 56.2	+32 47 00	MCG +6-34-28	14.1	0.8 x 0.5	12.9	SBab	36	Mag 8.62v star SW 2.'5.
15 44 18.2	+34 41 42	MCG +6-35-1	14.6	0.8 x 0.5	13.4		105	Possible star superimposed E end? Faint, round anonymous galaxy SW 2.'2.
15 46 34.3	+33 13 19	MCG +6-35-5	14.4	0.9 x 0.9	14.0	Sb		
15 50 57.9	+34 06 07	MCG +6-35-7	14.7	0.6 x 0.5	13.2		108	Mag 14.6 star on N edge.
15 50 58.5	+33 52 00	MCG +6-35-8	14.9	0.7 x 0.7	13.9	Sb		Mag 11.3 star NE 5.'1.
15 52 12.1	+34 05 33	MCG +6-35-11	15.5	0.6 x 0.3	13.5		18	Pair with MCG +6-35-12 S edge.
15 52 10.9	+34 05 01	MCG +6-35-12	14.4	0.5 x 0.5	12.8			Pair with MCG +6-35-11 N edge.
15 55 28.3	+34 10 39	MCG +6-35-15	15.2	0.5 x 0.3	13.0	Sc	24	
15 59 16.4	+35 43 29	MCG +6-35-18	14.9	0.6 x 0.4	13.2	Sc	171	
16 01 58.5	+36 42 06	MCG +6-35-24	14.5	0.8 x 0.5	13.4	SBa	99	Mag 11.4 star N 3.'8.
16 02 23.7	+36 06 32	MCG +6-35-25	14.9	0.7 x 0.7	14.0	S0		
16 02 54.1	+33 08 03	MCG +6-35-27	14.9	0.7 x 0.6	13.8	S0	9	Mag 12.0 star SW 2.'5.
16 04 45.6	+34 37 11	MCG +6-35-29	14.3	0.6 x 0.5	12.9	Sa	39	Almost stellar **MCG +6-35-30** on N edge; MCG +6-35-31 SE 2.'0.
16 04 56.6	+34 35 44	MCG +6-35-31	14.6	0.5 x 0.4	12.7	SBa	102	MCG +6-35-29 NW 2.'0; mag 12.9 star W 1.'3.
16 05 34.5	+32 39 35	MCG +6-35-32	15.0	0.7 x 0.5	13.7	Sc	153	Distorted; mag 11.1 star S 2.'4.
16 06 40.9	+37 05 46	MCG +6-35-34	14.9	0.8 x 0.5	13.8	Sbc	60	Mag 13.1 star NE 1.'3.
16 08 16.6	+36 05 34	MCG +6-35-36	14.7	0.8 x 0.5	13.5		18	Mag 12.5 star NE 0.'8.
16 09 24.1	+32 59 57	MCG +6-35-40	14.9	0.5 x 0.4	13.0	Spiral	54	Mag 10.08v star NE 5.'0.
16 10 41.5	+36 51 38	MCG +6-35-41	14.7	0.8 x 0.5	13.6	Compact	177	Mag 11.23v star on W edge.
16 11 12.6	+35 56 17	MCG +6-35-44	14.3	0.7 x 0.4	12.7	Compact, jet	126	Elongated **MCG +6-35-45** on NE edge.
16 11 45.8	+37 27 38	MCG +6-35-46	15.6	0.8 x 0.4	14.2	Sbc	144	Mag 11.42v star S 2.'8.
16 13 01.6	+37 17 14	MCG +6-36-3	14.9	0.9 x 0.6	14.1	Sc	177	
16 13 08.0	+36 35 13	MCG +6-36-4	14.9	0.8 x 0.5	13.7	Spiral	123	Mag 9.54v star NNW 3.'3.
16 13 49.3	+38 09 04	MCG +6-36-5	14.3	0.9 x 0.7	13.9	E/S0	147	Mag 11.5 star S 2.'0.
16 14 06.4	+34 52 11	MCG +6-36-6	14.6	0.4 x 0.4	12.5	S0		Mag 10.76v star SW 2.'4.
16 15 52.7	+36 59 46	MCG +6-36-8	14.5	0.7 x 0.6	13.5	Compact	81	Mag 11.39v star N 0.'8.
16 16 01.4	+37 55 05	MCG +6-36-9	14.9	0.6 x 0.3	12.9	Spiral	153	**MCG +6-36-10** N 1.'3.
16 16 50.0	+35 42 04	MCG +6-36-12	14.6	0.6 x 0.5	13.3	E2	111	Located between NGC 6104 3.'9 W and a mag 8.28v star E 4.'4.
16 18 33.0	+35 29 09	MCG +6-36-20	15.0	0.6 x 0.5	13.5	Compact	87	Mag 10.54v star WSW 4.'9.
16 19 22.3	+35 34 29	MCG +6-36-23	15.1	0.9 x 0.6	14.3	Spiral	156	Mag 13.8 star ESE 1.'8.
16 19 21.8	+37 04 06	MCG +6-36-24	15.8	0.6 x 0.3	13.7	Spiral	117	Located half way between NGC 6117 and a mag 12.1 star 2.'8 SE.
16 20 11.0	+35 07 49	MCG +6-36-31	14.5	0.8 x 0.4	13.1	Sc	60	Mag 11.4 star on N edge.
16 21 09.7	+34 45 29	MCG +6-36-33	14.4	0.7 x 0.6	13.4	Compact	79	**CGCG 196-43** W 2.'0.
16 21 11.0	+36 04 04	MCG +6-36-34	15.2	1.0 x 0.7	14.6	SBb	30	= **Mkn 878**.
16 21 34.1	+38 00 28	MCG +6-36-36	15.0	0.8 x 0.6	14.1	S0/a	150	NGC 6129 ESE 2.'0.
16 22 59.8	+37 56 54	MCG +6-36-38	14.3	0.7 x 0.6	13.3	S0/a	81	Located on N edge of NGC 6137.
16 23 07.9	+35 50 14	MCG +6-36-40	14.4	0.9 x 0.5	13.4	Compact	63	
16 24 48.8	+37 30 59	MCG +6-36-42	15.6	0.9 x 0.8	15.1	Spiral	135	Pair of mag 12-13 stars N 2.'1.
16 26 30.4	+35 40 17	MCG +6-36-43	14.8	0.8 x 0.3	13.1	Spiral	39	Mag 13.3 star E 1.'3.
16 28 20.6	+33 10 22	MCG +6-36-45	14.9	0.7 x 0.4	13.3	Sc	75	Mag 13.0 star N 2.'4.
16 32 48.8	+37 21 24	MCG +6-36-50	14.5	0.9 x 0.4	13.3	Spiral	147	Mag 11.11v star S 1.'6.
16 33 09.6	+34 55 33	MCG +6-36-51	15.1	0.5 x 0.5	13.5	S?		Mag 13.2 star W 2.'3.
16 33 49.5	+33 19 40	MCG +6-36-53	14.7	0.8 x 0.7	13.9	Spiral	165	Mag 12.4 star NW 1.'6.
16 39 54.9	+37 10 44	MCG +6-37-1	14.8	0.7 x 0.3	13.0	Spiral	36	MCG +6-37-2 E 2.'4.
16 40 06.7	+37 11 24	MCG +6-37-2	14.2	0.8 x 0.7	13.4	S0	87	MCG +6-37-1 W 2.'4.
16 41 00.2	+33 46 19	MCG +6-37-4	14.3	0.9 x 0.4	13.0	S0/a	66	Mag 10.56v star SW 2.'8.
15 17 01.8	+39 41 44	MCG +7-31-52	15.2	0.8 x 0.6	14.2	Sb	60	Mag 10.56v star W 5.'0.
15 17 14.2	+40 56 41	MCG +7-31-53	15.2	0.5 x 0.4	13.3	Sc	117	Pair of mag 13-14 stars WSW 1.'6.
15 29 16.1	+39 37 39	MCG +7-32-14	14.5	0.7 x 0.5	13.2	Sb	129	
15 42 15.3	+39 58 59	MCG +7-32-44	14.2	0.8 x 0.8	13.6			
15 53 25.3	+39 23 07	MCG +7-33-5	15.3	0.9 x 0.8	14.8	Sbc	54	
15 57 59.6	+40 01 59	MCG +7-33-19	14.5	0.5 x 0.3	12.3		165	Pair of mag 13.4 stars W and N edges.
16 21 05.2	+39 55 01	MCG +7-34-2	14.6	0.6 x 0.5	13.1	S?	57	Mag 13.0 star N 2.'4.
16 22 00.4	+38 54 50	MCG +7-34-5	15.4	0.5 x 0.3	13.2	Sc	36	Located 1.'9 SE of NGC 6131.
16 23 08.0	+39 18 46	MCG +7-34-10	15.4	0.7 x 0.3	13.6	Spiral	51	
16 23 28.1	+39 11 27	MCG +7-34-13	14.8	0.5 x 0.5	13.2	S		MCG +7-34-16 NE 2.'1; mag 8.89v star W 2.'6.
16 23 32.7	+39 07 18	MCG +7-34-15	13.8	1.0 x 1.0	13.7	S?		
16 23 38.1	+39 12 07	MCG +7-34-16	15.3	0.4 x 0.3	12.9	S?	36	MCG +7-34-13 SW 2.'1; slightly elongated anonymous galaxy NW 1.'3.
16 24 37.3	+39 07 36	MCG +7-34-18	14.9	1.0 x 0.3	13.4	S	12	**MCG +7-34-17** NNW 6.'3.
16 24 47.5	+39 44 02	MCG +7-34-19	14.4	1.0 x 0.5	13.5	SB	33	A mag 14.0 star on both the E and W edges.
16 25 41.9	+39 36 00	MCG +7-34-26	14.4	0.8 x 0.4	13.0	S?	27	Mag 9.52v star S 1.'2.
16 25 44.5	+40 28 30	MCG +7-34-27	14.6	0.8 x 0.3	12.9	Sb	147	NGC 6150 NE 1.'3.
16 25 49.3	+40 20 38	MCG +7-34-28	14.1	0.6 x 0.5	12.7	Irr	21	= **Mkn 881**.
16 26 22.0	+40 54 40	MCG +7-34-31	14.5	0.8 x 0.8	13.9	S?		Mag 9.08v star E 4.'0.
16 26 39.9	+40 28 39	MCG +7-34-33	14.0	0.8 x 0.8	13.3	SB?		Close N-S pair of mag 13-14 stars N 1.'4.
16 27 22.4	+39 06 31	MCG +7-34-36	13.9	0.8 x 0.7	13.1	S	150	**Mkn 882** E 2.'5.
16 27 28.9	+38 52 05	MCG +7-34-37	14.5	0.7 x 0.5	13.2	SBa	129	Mag 8.1v star WSW 5.'6.
16 27 37.2	+40 40 13	MCG +7-34-39	14.6	0.5 x 0.4	12.9	E	18	Small, faint anonymous galaxy W 2.'5.
16 27 55.3	+39 15 28	MCG +7-34-43	14.7	0.8 x 0.6	13.7	S0	3	MCG +7-34-44 on SE edge; round anonymous galaxy N 1.'4.
16 27 57.7	+39 15 08	MCG +7-34-44	13.8	0.4 x 0.2	11.1	E	132	Located on SE edge of MCG +7-34-43.
16 27 59.1	+39 18 09	MCG +7-34-45	14.9	0.4 x 0.4	12.8	S?		Mag 9.80v star W 2.'7; MCG +7-34-43 S 2.'7 with an anonymous galaxy located at the half way point.
16 28 14.3	+40 18 46	MCG +7-34-47	14.4	0.7 x 0.7	13.5	S?		Mag 12.8 star WNW 1.'9.
16 28 38.8	+39 05 14	MCG +7-34-57	13.4	0.4 x 0.3	10.9	Sc	62	Mag 9.48v star S edge.
16 28 39.7	+40 07 21	MCG +7-34-61	14.3	0.7 x 0.4	12.7	SB	144	Pair with MCG +7-34-62 E edge; mag 13.6 star W 0.'9.
16 28 42.4	+40 07 22	MCG +7-34-62	14.7	0.5 x 0.4	12.8	SB	108	Pair with MCG +7-34-61 W edge.

(Continued from previous page)

GALAXIES

RA h m s	Dec ° ′ ″	Name	Mag (V)	Dim ′ Maj x min	SB	Type Class	PA	Notes
16 28 43.0	+39 23 16	MCG +7-34-67	14.6	0.3 x 0.2	11.4	S0	75	Almost stellar, mag 13.6 star SW 2′4.
16 28 50.3	+39 50 02	MCG +7-34-75	14.5	0.7 x 0.5	13.2	S?	132	Faint, almost stellar anonymous galaxy SE 2′5.
16 29 07.3	+39 29 42	MCG +7-34-79	15.1	0.4 x 0.4	13.2	E		The NGC 6166 group NW 5′4.
16 29 11.1	+40 51 24	MCG +7-34-81	14.6	0.9 x 0.3	13.1	Sb	120	MCG +7-34-82 NE 2′7.
16 29 23.9	+40 52 26	MCG +7-34-82	14.4	0.6 x 0.6	13.1	S0?		MCG +7-34-81 SW 2′7.
16 30 17.9	+40 35 52	MCG +7-34-89	15.0	0.5 x 0.5	13.3	S		Faint anonymous galaxy NE 1′8.
16 30 22.0	+40 55 26	MCG +7-34-90	14.6	0.6 x 0.6	13.4	S..		**PGC 58350** on NW edge.
16 30 27.1	+40 43 27	MCG +7-34-92	14.7	1.0 x 0.3	13.5	E	33	Located 1′1 SE of mag 7.87v star.
16 30 42.0	+40 31 43	MCG +7-34-97	15.1	0.5 x 0.5	13.4	S		NGC 6180 W 1′7.
16 30 54.2	+39 50 48	MCG +7-34-102	14.4	0.9 x 0.5	13.4	S0?	96	MCG +7-34-105 E 1′9.
16 31 02.9	+39 47 32	MCG +7-34-104	15.0	1.0 x 1.0	15.0	E		MCG +7-34-105 N 2′6; MCG +7-34-102 NW 3′6.
16 31 03.7	+39 50 15	MCG +7-34-105	14.4	0.9 x 0.6	13.6	S?	102	MCG +7-34-102 W 1′9; MCG +7-34-104 S 2′6.
16 38 13.9	+40 08 45	MCG +7-34-125	14.1	0.9 x 0.4	12.8	Sb	168	**MCG +7-34-126** N 1′6.
16 40 28.1	+39 19 10	MCG +7-34-130	13.8	0.9 x 0.7	13.2	Compact	12	Mag 11.3 star S 1′3.
16 40 29.8	+39 14 04	MCG +7-34-131	14.9	0.8 x 0.3	13.2	Sc	48	Faint anonymous galaxy of similar size and shape ESE 1′7.
16 41 15.1	+39 45 07	MCG +7-34-132	14.7	0.8 x 0.3	13.0	Sb	141	Pair with MCG +7-34-133 W.
16 41 10.0	+39 45 20	MCG +7-34-133	14.6	0.8 x 0.5	13.5	Sb	108	Pair with MCG +7-34-132 E; forms a triangle with a pair of mag 12 stars W.
16 41 37.5	+40 09 45	MCG +7-34-135	14.3	1.2 x 0.6	13.8	Sb	57	Mag 10.15v star SSW 4′0.
16 41 45.3	+39 38 34	MCG +7-34-136	14.4	0.7 x 0.4	12.9	Sb	114	Mag 9.66v star W 1′2.
16 42 08.6	+40 16 34	MCG +7-34-138	14.7	0.9 x 0.9	14.3	Sbc		
16 42 11.8	+39 27 18	MCG +7-34-139	15.1	0.5 x 0.5	13.4	Sc		Located between an E-W pair of mag 13 stars.
15 22 02.1	+31 13 56	NGC 5924	14.6	0.7 x 0.2	12.3	S0?	12	**PGC 54863** E 2′1.
15 31 30.8	+40 25 51	NGC 5950	13.7	1.5 x 0.8	13.8	Sb II	37	Strong dark lane NW.
15 35 16.4	+30 51 47	NGC 5961	14.1	0.8 x 0.3	12.4	S?	97	**UGC 9920** S 3′8.
15 35 52.4	+39 46 07	NGC 5966	12.2	1.8 x 1.1	12.9	E	90	Located 2′8 SW of mag 6.79v star.
15 39 02.5	+31 45 31	NGC 5974	14.2	0.6 x 0.3	12.2	S?	110	Mag 8.37v star SSE 5′0; stellar **CGCG 166-27** N 3′9.
15 52 52.9	+40 38 50	NGC 6013	13.6	1.3 x 0.8	13.5	SBb	174	Mag 8.31v star NE 8′0.
16 02 40.5	+37 21 35	NGC 6038	13.5	1.1 x 1.1	13.6	Sc		
16 07 41.7	+38 55 49	NGC 6069	14.3	0.8 x 0.8	13.6			Faint anonymous galaxy N 1′8.
16 12 35.4	+29 21 54	NGC 6085	13.0	1.5 x 1.2	13.5	Sa	165	Bright nucleus; mag 10.97v star E 1′8.
16 12 35.4	+29 29 02	NGC 6086	12.8	1.7 x 1.2	13.5	E	12	Star on NW edge; mag 10.71v star SW 2′8.
16 12 40.9	+33 02 07	NGC 6089	13.8	0.9 x 0.6	13.0	Compact	42	Multi-galaxy system; star and small companion NE edge.
16 14 26.2	+35 06 33	NGC 6097	13.9	1.3 x 0.7	13.7	S0/a	156	
16 15 44.5	+31 57 48	NGC 6103	13.9	0.7 x 0.5	12.6	S?	80	
16 16 30.8	+35 42 29	NGC 6104	13.3	0.8 x 0.7	12.5	S?	45	Compact; MCG +6-36-12 E 3′9.
16 17 09.4	+34 52 44	NGC 6105	14.4	0.6 x 0.5	12.9	S?	138	Located 2′6 SW of NGC 6107.
16 17 20.2	+34 54 06	NGC 6107	13.8	1.5 x 1.1	14.3	E:	40	Mag 8.90v star on N edge.
16 17 25.7	+35 08 08	NGC 6108	14.4	1.1 x 1.0	14.4	S?	124	Short bar SW-NW; **PGC 57737** E 2′0.
16 17 40.7	+35 00 15	NGC 6109	12.7	1.0 x 1.0	12.6	S?		Small, faint anonymous galaxy S 2′4.
16 17 44.2	+35 05 10	NGC 6110	14.8	0.8 x 0.4	13.5	Sab	105	**PGC 57742** SW 2′0.
16 18 00.6	+35 06 36	NGC 6112	14.0	1.0 x 0.8	13.6	S?	69	Faint, elongated anonymous galaxy NNW 3′3.
16 18 23.7	+35 10 28	NGC 6114	14.3	1.2 x 0.6	13.8	I?	100	Faint stars on E and W edges.
16 18 54.7	+35 09 15	NGC 6116	14.3	2.2 x 0.9	14.9	Sb II	17	Strong dark lane E of center.
16 19 18.3	+37 05 43	NGC 6117	13.6	1.2 x 1.2	13.8	SA(s)bc		MCG +6-36-24 S 1′7; mag 9.67v star N 2′6.
16 19 41.9	+37 48 22	NGC 6119	15.3	0.7 x 0.5	14.0	Spiral	87	Star on S edge.
16 19 48.2	+37 46 28	NGC 6120	13.8	1.0 x 0.7	13.3	Pec	18	Located 2′3 NNE of mag 10.34v star.
16 20 09.6	+37 47 53	NGC 6122	14.6	0.8 x 0.2	12.5		156	**MCG +6-36-30** W 1′3.
16 21 28.1	+36 22 35	NGC 6126	13.6	1.0 x 1.0	13.5	S?		Located 1′6 S of mag 9.97v star.
16 21 43.2	+37 59 44	NGC 6129	14.0	0.8 x 0.8	13.5	S0		MCG +6-36-36 WNW 2′0.
16 21 52.3	+38 56 02	NGC 6131	13.3	1.0 x 1.0	13.1	SABcd:		MCG +7-34-5 SE 1′9.
16 23 03.3	+37 55 22	NGC 6137	12.4	1.9 x 1.2	13.3	E	175	MCG +6-36-38 NNW 1′7.
16 23 06.3	+40 51 30	NGC 6141		0.3 x 0.3				
16 23 21.1	+37 15 29	NGC 6142	13.8	1.9 x 0.5	13.6	Sb III	165	**CGCG 196-55** S 2′6.
16 25 02.5	+40 56 43	NGC 6145	14.2	0.8 x 0.4	12.8	S..	0	
16 25 10.2	+40 53 33	NGC 6146	12.5	1.3 x 1.0	12.8	E?	75	
16 25 05.8	+40 55 43	NGC 6147	15.1	0.4 x 0.4	13.0	Sb		NGC 6145 1′2 NW.
16 25 50.1	+40 29 15	NGC 6150	14.0	1.3 x 0.5	13.6	E?	61	MCG +7-34-27 SW 1′3.
16 27 40.9	+39 22 56	NGC 6158	13.7	0.9 x 0.6	13.1	E?	81	**PGC 58195** on SE edge; mag 10.20v star SE 2′1.
16 27 41.0	+40 55 33	NGC 6160	13.2	1.8 x 1.5	14.3	E	65	**PGC 58191** on S edge; pair of stars N of nucleus.
16 28 20.7	+32 48 38	NGC 6161	14.7	0.7 x 0.3	12.9	S?	159	= **Hickson 82C**. **Hickson 82D** W 0′8.
16 28 22.5	+32 50 54	NGC 6162	13.6	0.9 x 0.7	13.0	S0	30	= **Hickson 82A**.
16 28 28.0	+32 50 47	NGC 6163	14.4	0.8 x 0.5	13.2	SB0?	54	= **Hickson 82B**.
16 28 38.5	+39 33 02	NGC 6166	11.8	1.9 x 1.4	12.8	E+2 pec	35	Small, faint anonymous galaxies on W and S edges.
16 28 31.1	+39 31 13	NGC 6166A	14.4	0.4 x 0.2	11.5	S0	120	
16 28 53.3	+39 33 33	NGC 6166B	14.9	0.4 x 0.3	12.4	Spiral	6	Located 2′9 E of NGC 6166.
16 28 23.5	+39 34 09	NGC 6166C	13.3	0.4 x 0.4	11.3	E		Located 3′0 WNW of NGC 6166.
16 28 39.5	+39 31 03	NGC 6166D		0.5 x 0.5				**MCG +7-34-70** E 1′0.
16 29 44.8	+40 48 41	NGC 6173	12.1	1.9 x 1.4	13.2	E	140	Stellar **PGC 58358** NE 1′5.
16 29 47.8	+40 52 15	NGC 6174	14.5	0.4 x 0.4	12.4			Stellar **PGC 58350** on NW edge.
16 29 57.7	+40 37 45	NGC 6175	13.7	1.3 x 0.8	13.6		92	Double system.
16 30 38.8	+35 03 20	NGC 6177	13.6	1.7 x 1.2	14.2	(R′)SB(s)b	10	Strong dark lanes E and W of elongated center.
16 30 47.2	+35 06 05	NGC 6179	15.5	0.4 x 0.4	13.4			Almost stellar.
16 30 34.0	+40 32 16	NGC 6180	14.1	0.9 x 0.6	13.5	E?	9	MCG +7-34-97 E 1′7.
16 31 34.6	+40 33 53	NGC 6184	14.0	0.8 x 0.5	12.9	Sb	128	Mag 8.59v star S 7′0.
16 33 17.9	+35 20 33	NGC 6185	13.4	1.2 x 0.9	13.4	S?	0	Star on N end.
16 36 37.2	+36 11 58	NGC 6194	13.8	1.0 x 0.8	13.4	Compact	105	
16 36 32.5	+39 01 41	NGC 6195	13.0	1.5 x 1.1	13.4	Sb I	45	Strong dark lane NW.
16 37 54.0	+36 04 22	NGC 6196	12.9	2.0 x 1.2	13.7	SAB0−: pec	140	IC 4614 2 NW 2′7.
16 37 59.8	+35 59 39	NGC 6197	14.6	1.3 x 0.5	14.0		37	
16 43 04.1	+36 49 58	NGC 6207	11.6	3.0 x 1.3	13.0	SA(s)c III	15	Many knots and superimposed stars.
16 43 23.1	+39 48 18	NGC 6212	14.2	0.7 x 0.4	12.6	S?	105	
15 18 01.1	+30 41 21	UGC 9809	14.0	1.2 x 0.9	13.9	SB(r)c	111	Stellar nucleus, uniform envelope; mag 10.97v star S edge.
15 22 44.9	+29 46 07	UGC 9831	13.9	0.6 x 0.6	12.6	Sa		Mag 12.6 star N 1′1.
15 23 56.9	+38 07 17	UGC 9834	14.0	0.9 x 0.9	13.7	S0		Mag 9.24v star NE 3′7.
15 25 05.7	+37 57 36	UGC 9842	14.3	1.5 x 0.5	13.8	SBb	67	
15 26 40.6	+40 33 47	UGC 9858	12.8	4.9 x 1.0	14.4	SABbc	83	Mag 8.74v star N 2′0.
15 34 45.0	+31 03 14	UGC 9910	15.4	1.7 x 1.4	16.2	Sm	165	Bright core, patchy envelope; low surface brightness.
15 35 53.8	+38 40 33	UGC 9922	13.5	0.9 x 0.4	12.2	S?	168	Double system, contact.
15 46 45.5	+31 00 39	UGC 10034	13.7	1.2 x 0.9	13.6	SAB(r)b	36	Two main arms; bright nucleus in short, bright bar.

(Continued from previous page)

51 GALAXIES 51

RA h m s	Dec ° ′ ″	Name	Mag (V)	Dim ′ Maj x min	SB	Type Class	PA	Notes
15 59 46.7	+37 02 11	UGC 10065	14.2	0.9 x 0.6	13.4	Sb	160	
15 57 28.0	+30 03 35	UGC 10104	13.5	2.0 x 1.7	14.6	SA(rs)bc I-II	27	Stellar nucleus; mag 12.5 star on NW edge; mag 9.75v star NW 2′.3.
15 59 09.7	+35 01 44	UGC 10120	13.8	1.4 x 1.0	14.0	SB(r)b	141	Bright nucleus with bright NE-SW oriented bar; faint envelope NW-SE.
15 59 28.9	+39 49 38	UGC 10122	14.1	1.3 x 1.1	14.3	Sa	24	Mag 11.0 star NNW edge.
16 03 02.2	+39 38 44	UGC 10155	13.7	0.9 x 0.8	13.2	Scd:	66	
16 03 52.3	+39 58 58	UGC 10166	13.9	1.0 x 0.7	13.4	SB	159	
16 06 40.1	+30 05 54	UGC 10205	13.4	1.5 x 0.8	13.5	Sa	132	Faint dark lane through bright core.
16 06 42.5	+31 53 40	UGC 10207	14.5	1.7 x 0.7	14.5	SB(s)bc:	157	Two main arms; **UGC 10208** S 1′.8.
16 08 58.3	+36 36 33	UGC 10227	14.6	2.0 x 0.2	13.5	SBcd?	169	Mag 4.73v star Coronae Borealis S 7′.2.
16 09 43.1	+30 27 09	UGC 10234	13.6	1.2 x 1.0	13.7	$S0^+$	15	Bright center with faint envelope; mag 11.1 star W 1′.2.
16 11 07.7	+38 15 04	UGC 10257	14.0	1.7 x 0.3	13.1	Sbc	165	Mag 12.7 star superimposed S edge.
16 11 45.4	+29 44 42	UGC 10259	15.1	1.8 x 0.6	15.0		171	Chain of five galaxies, plumes.
16 11 58.5	+29 50 18	UGC 10262	13.8	1.0 x 1.0	13.7	S0		An anonymous galaxy SW 0′.8, and a pair of anonymous galaxies NW 1′.8.
16 17 31.9	+31 11 38	UGC 10312	13.8	0.9 x 0.6	13.0	S?	110	Mag 13.0 star W 2′.1.
16 18 11.9	+40 05 38	UGC 10330	14.3	0.8 x 0.6	13.4	SAB(s)b	114	Single, faint, thin arms extend N and S.
16 19 41.6	+36 05 15	UGC 10342	14.2	1.1 x 0.8	14.0	(R)SABb	129	Strong NE-SW core; mag 10.82v star E 1′.8.
16 19 45.1	+38 14 03	UGC 10344	14.5	1.3 x 0.9	14.5	S	25	Faint star or knot W of nucleus.
16 20 54.0	+40 06 14	UGC 10349	13.6	1.4 x 0.5	13.0	SBab	65	Stellar galaxy **KUG 1618+402** WNW 8′.4.
16 21 22.2	+40 48 33	UGC 10354	14.6	1.2 x 0.8	14.4	SAB(s)c	81	Large, strong core, faint envelope.
16 22 00.4	+40 26 52	UGC 10357	14.1	0.9 x 0.7	13.4	S?	81	
16 22 55.5	+39 47 30	UGC 10362	14.2	1.5 x 0.7	14.1	SB(r)b	65	Faint, slightly elongated anonymous galaxy SE 2′.7.
16 23 16.3	+39 55 04	UGC 10367	13.3	1.4 x 1.2	13.7	SB(r)b	99	Short, bright N-S bar.
16 24 12.3	+30 09 44	UGC 10372	14.2	1.0 x 0.6	13.5	SAB(rs)b: III	170	
16 25 26.2	+39 52 12	UGC 10381	13.7	1.2 x 0.7	13.4	S0/a	95	**CGCG 224-23** NE 10′.9.
16 28 11.5	+39 49 16	UGC 10404	14.1	1.4 x 1.0	14.3	SB?	160	Strong dark lane W of center.
16 29 51.2	+39 45 55	UGC 10420	13.8	1.3 x 1.0	13.9	SB(r)b	155	Small N-S bar; mag 13.9 star N 2′.9.
16 30 33.4	+39 49 50	UGC 10429	14.1	0.9 x 0.9	13.7	S?		
16 35 07.8	+40 59 25	UGC 10459	15.1	1.5 x 0.2	13.6	Scd:	120	Mag 7.95v star SE 10′.0.
16 36 54.4	+36 25 25	UGC 10473	13.9	1.5 x 0.4	13.2	SBa	170	**CGCG 196-84** NE 2′.3; mag 11.5 star NNW 2′.3.
16 37 34.7	+37 17 06	UGC 10477	14.3	1.7 x 0.3	13.4	S?	26	Located between a close pair of stars, mags 9.54v and 10.75v.
16 40 22.8	+33 40 46	UGC 10504	14.0	1.0 x 0.5	13.1	Sab	169	
16 41 47.1	+39 58 59	UGC 10512	13.8	0.8 x 0.8	13.4	E		Star or almost stellar galaxy on NE edge.

GALAXY CLUSTERS

RA h m s	Dec ° ′ ″	Name	Mag 10th brightest	No. Gal	Diam ′	Notes
15 21 18.0	+30 39 00	A 2061	15.7	71	20	**CGCG 165-41** at center; other members PGC and anonymous, most stellar; A 2067 NW.
15 23 12.0	+30 54 00	A 2067	15.7	58	13	All members PGC or anonymous, most stellar.
15 33 18.0	+31 08 00	A 2092	15.7	55	16	Mag 4.16v star on N edge; mostly PGC and anonymous GX along N-S centerline.
15 45 00.0	+36 03 00	A 2124	15.6	50	13	Most members grouped NW of center, this area is also the smaller **A 2122**.
16 12 30.0	+29 32 00	A 2162	13.7	37	56	Large concentration of GX around center, another near NW edge.
16 20 24.0	+29 54 00	A 2175	16.2	61	16	All members anonymous, faint and stellar.
16 28 12.0	+40 54 00	A 2197	13.9	73	90	General scattering of UGC, MCG and PGC GX mixed with anonymous GX.
16 28 36.0	+39 31 00	A 2199	13.9	88	90	Many UGC, MCG and PGC GX, heavy N-S concentration towards the center.

OPEN CLUSTERS

RA h m s	Dec ° ′ ″	Name	Mag	Diam ′	No. ★	B ★	Type	Notes
16 27 24.0	+38 04 00	DoDz 5		27			ast	Few stars; small brightness range; no central concentration; detached.

GLOBULAR CLUSTERS

RA h m s	Dec ° ′ ″	Name	Total V m	B ★ V m	HB V m	Diam ′	Conc. Class Low = 12 High = 1	Notes
16 41 41.5	+36 27 37	NGC 6205	5.8	11.9	15.0	20.0	5	= **M 13, Great Hercules Cluster**. The galaxy IC 4617, mag about 15v and 1.1 x 0.4 in PA 29° lies nearly 15′ NNE. It can be found nearly midway between the cluster and the galaxy NGC 6207, slightly west of a line between them.

PLANETARY NEBULAE

RA h m s	Dec ° ′ ″	Name	Diam ″	Mag (P)	Mag (V)	Mag cent ★	Alt Name	Notes
16 04 26.6	+40 40 59	NGC 6058	40	13.3	12.9	13.9	PK 64+48.1	Slightly elongated, PA 135°; mag 9.32v star NW 4′.8.
16 40 18.2	+38 42 19	PN G61.9+41.3	10	13.3		15.4	PK 61+41.1	Mag 14.4 star SW 0′.7; mag 14.7 star N 2′.9.

52 GALAXIES 52

RA h m s	Dec ° ′ ″	Name	Mag (V)	Dim ′ Maj x min	SB	Type Class	PA	Notes
13 56 09.0	+29 40 11	CGCG 162-29	15.0	0.7 x 0.5	13.7	Sc	54	Very close pair of mag 11 and 13 stars NW 9′.7.
14 17 11.4	+30 13 52	CGCG 163-14	15.2	0.6 x 0.5	13.8	Sc	144	Mag 14.7 star E 2′.2.
15 04 58.0	+31 40 23	CGCG 165-13	16.1	0.5 x 0.5	14.5			Mag 10.25v star WNW 7′.8.
14 15 49.2	+39 47 49	IC 990	14.7	0.6 x 0.4	13.0		51	Mag 6.38v star SE 7′.5; mag 10.60v star SW 3′.1.
14 27 09.7	+30 56 51	IC 1012	13.5	1.2 x 0.7	13.2	S?	105	Very faint star SE edge of core; mag 13.1 star W 3′.5.
14 00 43.8	+31 53 34	IC 4357	13.9	1.1 x 0.5	13.1	S?	70	Mag 11.56v star N 1′.5.
14 04 10.2	+33 20 13	IC 4370	15.2	1.4 x 0.3	14.1	Sa	138	= **Hickson 70A**; **Hickson 70D** on N edge; **Hickson 70C** is MCG +6-31-65 ESE 2′.4.
14 04 10.9	+33 18 25	IC 4371	14.2	0.8 x 0.6	13.3	S0?	42	= **Hickson 70B**; **Hickson 70E** = **IC 4369** NW 1′.3; **Hickson 70F** NW 1′.9; **Hickson 70G** NW 2′.7.
14 10 02.3	+37 32 57	IC 4380	14.6	0.5 x 0.5	12.9	Spiral		Mag 8.49v star ESE 7′.2.
14 16 26.0	+39 38 59	IC 4394		0.9 x 0.3			15	Mag 6.38v star N 5′.6.
14 18 17.0	+31 39 11	IC 4403	13.9	1.3 x 0.5	13.3	S?	132	Mag 9.20v star WSW 10′.5.
14 21 33.2	+31 35 05	IC 4409	14.4	0.7 x 0.3	12.6	S?	36	
14 25 59.2	+30 28 22	IC 4422	13.9	0.6 x 0.5	12.7	E?	96	Mag 11.9 star W 2′.4.
14 27 24.4	+37 28 16	IC 4435	14.2	0.7 x 0.6	13.4	E/S0	45	Mag 8.03v star NE 10′.3.
14 29 01.4	+37 27 45	IC 4446	15.2	0.9 x 0.4	14.0	Sbc	114	Mag 8.03v star NW 11′.8.

(Continued from previous page)

GALAXIES

RA h m s	Dec ° ′ ″	Name	Mag (V)	Dim ′ Maj x min	SB	Type Class	PA	Notes
14 29 18.0	+30 49 55	IC 4447	13.4	1.2 x 0.8	13.2	S0/a	166	Bright center with strong dark patches N and S; mag 12.5 star W edge.
14 34 36.8	+30 16 44	IC 4460	14.9	0.5 x 0.3	12.7	Sa	171	Stellar; mag 11.1 star W 4'2.
14 43 54.5	+33 24 19	IC 4496	14.0	1.0 x 0.8	13.6	S0	36	Mag 11.3 star NNW 5'7.
14 44 35.6	+37 28 55	IC 4500	14.7	0.5 x 0.3	12.5		84	Mag 10.46v star NE 7'0.
14 46 33.4	+33 24 29	IC 4505	13.7	1.1 x 0.9	13.6	S0⁻:	174	**IC 4506** ESE 1'4.
14 51 07.0	+37 29 40	IC 4515	14.9	0.7 x 0.5	13.6		60	Mag 11.05v star E 1'9.
14 55 07.1	+33 43 29	IC 4520	15.3	0.4 x 0.4	13.2	Sb		Mag 8.63v star WSW 14'1.
15 18 31.2	+32 23 30	IC 4539	15.3	0.4 x 0.4	13.2	SBb		Mag 10.67v star NE 4'0.
15 22 05.9	+33 08 54	IC 4542	14.1	0.8 x 0.5	13.0	S0/a	123	
14 00 45.9	+30 04 32	MCG +5-33-39	14.2	0.7 x 0.4	12.7	S?	78	
14 01 04.1	+29 31 35	MCG +5-33-41	14.4	0.6 x 0.5	13.0	Sc	90	
14 05 12.9	+29 25 02	MCG +5-33-47	15.3	0.9 x 0.2	13.3	Sbc	168	Pair with MCG +5-33-48 E; mag 10.37v star SW 1'3.
14 05 15.4	+29 25 27	MCG +5-33-48	14.6	0.6 x 0.5	13.2		69	Pair with MCG +5-33-47 W.
14 13 43.9	+29 25 44	MCG +5-34-3	14.3	0.8 x 0.3	12.6	Sab	174	Mag 15.2 star superimposed N end.
14 45 27.9	+30 25 42	MCG +5-35-11	14.6	0.8 x 0.4	13.3	S0	27	
15 10 08.9	+31 53 15	MCG +5-36-10	14.3	1.0 x 0.7	13.8	S?	3	
13 57 15.1	+34 30 36	MCG +6-31-29	14.8	0.8 x 0.4	13.4	Spiral	90	Mag 8.58v star S 4'9.
13 58 07.4	+32 38 30	MCG +6-31-31	14.6	0.8 x 0.5	13.4		132	Mag 7.16v star SW 3'3.
13 58 41.0	+34 31 26	MCG +6-31-35	14.8	0.7 x 0.5	13.6	Spiral	147	Mag 14.5 star on E edge.
13 58 41.9	+35 05 20	MCG +6-31-36	14.9	0.9 x 0.4	13.6	Sb	165	
13 58 54.6	+35 15 18	MCG +6-31-37	14.4	0.5 x 0.5	12.7			
13 59 11.3	+34 04 16	MCG +6-31-38	14.6	0.7 x 0.7	13.7	Spiral		
13 59 57.4	+38 12 02	MCG +6-31-42	14.7	0.8 x 0.4	13.3	Spiral	48	Located 1'7 NE of NGC 5403.
14 01 14.9	+37 52 55	MCG +6-31-43	14.4	0.9 x 0.5	13.4	Sab	81	Mag 12.3 star NW 4'1.
14 01 24.2	+36 47 57	MCG +6-31-44	14.3	0.6 x 0.6	13.1	S?		= **Mkn 465**.
14 01 46.7	+34 40 02	MCG +6-31-47	13.8	0.9 x 0.8	13.4	S0	36	
14 02 20.6	+32 26 53	MCG +6-31-49	15.0	0.9 x 0.2	13.0	Sb	135	Mag 12.0 star NW 3'2; **CGCG 191-34** W 5'2.
14 04 15.7	+35 48 01	MCG +6-31-62	14.4	0.6 x 0.4	12.9	E/S0	33	Mag 11.33v star NW 1'4.
14 04 20.4	+33 19 13	MCG +6-31-65	14.3	0.9 x 0.2	12.2	S?	45	= **Hickson 70C**. IC 4370 and IC 4371 W 1'9.
14 04 25.7	+35 36 40	MCG +6-31-66	15.6	0.7 x 0.6	14.5	Spiral	13	Stellar nucleus.
14 06 49.0	+33 46 18	MCG +6-31-70	14.7	0.9 x 0.3	13.2	Sb	150	Mag 11.3 star W 4'1.
14 09 12.1	+35 22 46	MCG +6-31-74	17.1	0.7 x 0.5	15.8	Sd	75	
14 11 28.9	+35 53 49	MCG +6-31-77	14.6	0.5 x 0.5	12.9			Mag 12.6 star NW 2'9.
14 14 51.4	+36 47 20	MCG +6-31-82	14.7	0.7 x 0.4	13.1	Spiral	30	Mag 9.99v star ESE 4'5.
14 15 16.3	+34 20 51	MCG +6-31-84	14.9	0.8 x 0.5	13.7		135	Mag 15.4 star on SE end; mag 10.88v star WNW 2'3.
14 15 38.4	+36 22 28	MCG +6-31-86	14.5	0.5 x 0.4	12.7	Compact	30	Mag 10.21v star WSW 3'1.
14 15 44.9	+36 10 39	MCG +6-31-87	15.0	0.5 x 0.4	13.1	Spiral	6	NGC 5529 NW 3'0.
14 17 11.8	+35 24 30	MCG +6-31-92	14.9	0.8 x 0.3	13.2	Sbc	15	Pair of mag 10 stars NW 4'6.
14 19 15.0	+35 20 18	MCG +6-31-95	14.4	0.6 x 0.5	13.1	E/S0	81	Mag 14.6 star NE edge.
14 21 55.7	+34 48 20	MCG +6-32-8	14.8	0.9 x 0.3	13.3	Sb	24	Mag 14.3 star W 1'2.
14 22 12.1	+35 47 50	MCG +6-32-9	15.4	0.4 x 0.3	13.0		117	Mag 12.8 star WSW 1'3.
14 23 04.5	+38 08 28	MCG +6-32-17	14.4	0.6 x 0.6	13.2			
14 24 12.2	+35 08 44	MCG +6-32-24	14.7	0.6 x 0.4	13.0		150	**UGC 9238**, **UGC 9233** and **UGC 9235** NE 10'1.
14 24 51.4	+38 15 12	MCG +6-32-31	14.7	0.9 x 0.4	13.5	SBb	150	Mag 12.2 star NE 2'7.
14 28 24.4	+32 23 34	MCG +6-32-39	14.5	0.9 x 0.7	13.8	Spiral	81	Mag 11.05v star E 3'2.
14 28 21.2	+35 24 17	MCG +6-32-40	14.4	0.9 x 0.8	13.9	SBab	57	
14 29 50.4	+35 34 09	MCG +6-32-47	14.9	0.7 x 0.2	12.6	Sa	12	Pair with UGC 9315 NW.
14 29 49.8	+34 36 07	MCG +6-32-48	16.1	0.8 x 0.6	15.1	Sc	168	
14 30 24.7	+32 56 12	MCG +6-32-51	15.3	0.6 x 0.4	13.7	SBb	117	Mag 14.4 star N edge; pair with MCG +6-32-52 S.
14 30 25.9	+32 55 30	MCG +6-32-52	15.6	0.4 x 0.2	12.7	Sb	126	Pair with MCG +6-32-51 N.
14 31 03.1	+35 31 08	MCG +6-32-54	16.3	0.9 x 0.4	15.0	Sc	51	Mag 13.1 star WNW 0'9; mag 9.37v star W 8'6.
14 31 21.4	+35 37 19	MCG +6-32-56	14.9	0.8 x 0.3	13.2	Sb	102	
14 31 25.5	+33 13 48	MCG +6-32-57	14.9	0.9 x 0.3	13.3	Spiral	87	
14 31 56.2	+33 38 30	MCG +6-32-58A	14.4	0.9 x 0.3	12.9	Sb	111	**MCG +6-32-58** on S edge; mag 13.1 star on NE edge.
14 32 46.3	+36 42 25	MCG +6-32-63	14.7	0.7 x 0.5	13.4	Spiral	174	Mag 14.3 star N 1'3.
14 33 03.2	+34 44 41	MCG +6-32-64	15.3	0.8 x 0.3	13.6	Sbc	162	**UGC 9367** E 3'2.
14 33 30.4	+34 33 36	MCG +6-32-66	13.8	0.4 x 0.4	11.7			Mag 12.1 star NE 1'6.
14 33 39.6	+35 18 42	MCG +6-32-68	14.8	0.6 x 0.6	13.5			
14 33 42.2	+34 59 04	MCG +6-32-69	14.6	0.8 x 0.5	13.5	Sb	66	
14 35 18.4	+35 07 07	MCG +6-32-70	13.3	0.6 x 0.6	12.1			Mag 15.0 star on N edge.
14 35 55.7	+36 43 06	MCG +6-32-74	14.4	0.7 x 0.5	13.2	S0/a	174	UGC 9387 NW 7'4.
14 36 50.9	+34 17 30	MCG +6-32-76	14.9	0.4 x 0.4	12.8			Mag 12.4 star SW 1'7.
14 38 07.6	+36 27 23	MCG +6-32-78	15.3	0.4 x 0.4	13.2			Mag 9.53v star S 2'2; NGC 5695 NW 11'3.
14 38 06.6	+37 00 32	MCG +6-32-79	14.8	0.8 x 0.6	13.9		138	Pair of mag 14.6 stars on E and N edges.
14 39 08.0	+38 08 02	MCG +6-32-80	14.8	0.5 x 0.5	13.1	Sb		Close pair of mag 14.4 stars E 1'7.
14 39 25.6	+36 32 47	MCG +6-32-82	16.1	0.5 x 0.4	14.2	Sc	116	Mag 8.54v star S 2'2.
14 41 40.3	+36 44 17	MCG +6-32-84	15.0	0.5 x 0.5	13.3	Sc		Mag 7.28v star E 12'0.
14 41 40.1	+36 38 46	MCG +6-32-85	15.0	0.8 x 0.4	13.6	Sc	129	Small triangle of mag 14-15 stars E 1'0; pair of mag 9.39v, 8.66v stars SE 5'7.
14 42 32.7	+33 44 46	MCG +6-32-87	14.8	0.7 x 0.5	13.5	Sc	165	
14 42 48.0	+37 52 18	MCG +6-32-89	15.3	0.9 x 0.3	13.7	Sc	126	Mag 11.3 star SW 3'5.
14 53 06.0	+36 55 25	MCG +6-33-5	14.9	0.7 x 0.5	13.6		24	**MCG +6-33-6** SE 1'1.
14 55 28.3	+32 50 21	MCG +6-33-9	15.6	0.4 x 0.3	13.2		24	**MCG +6-33-11** and **UGC 9603** SE 0'9; mag 7.22v star SW 2'2.
15 02 37.4	+33 10 17	MCG +6-33-17	15.8	0.7 x 0.3	13.9	Sb	90	
15 04 29.8	+35 53 23	MCG +6-33-19	15.3	1.0 x 0.4	14.2		66	Galaxy or star, or both, NE end.
15 04 32.9	+35 57 52	MCG +6-33-20	13.3	1.0 x 0.8	13.1	E?	81	Mag 11.01v star N 2'4.
15 08 05.0	+34 23 12	MCG +6-33-22	15.8	0.8 x 0.3	14.1	S?	48	Very faint plume extends N.
15 16 53.9	+38 14 27	MCG +6-34-1	16.0	0.5 x 0.3	13.8		126	Mag 10.10v star NW edge.
15 18 01.6	+38 12 59	MCG +6-34-2	14.6	0.6 x 0.5	13.1	S?	6	Mag 12.0 star N 1'7.
15 20 24.6	+32 51 31	MCG +6-34-4	15.2	0.7 x 0.4	13.7		99	MCG +6-34-5 S 0'9.
15 20 26.4	+32 50 41	MCG +6-34-5	15.4	0.7 x 0.3	13.6	Sb	87	MCG +6-34-4 N 0'9.
13 59 43.3	+40 23 15	MCG +7-29-27	16.7	0.7 x 0.5	15.4	Spiral	93	Located on NW edged of UGC 8917.
14 01 35.1	+38 41 08	MCG +7-29-37	13.7	0.8 x 0.6	12.7	S?	80	Mag 11.11v star on E edge.
14 02 48.5	+39 07 34	MCG +7-29-41	15.9	0.4 x 0.2	13.0		91	Located on SE edge of UGC 8962.
14 20 40.4	+40 18 54	MCG +7-29-62	15.2	0.6 x 0.4	13.5		156	Mag 11.2 star NNW 2'3.
14 20 55.2	+40 07 12	MCG +7-30-1	14.8	0.8 x 0.6	13.8		36	
14 22 32.7	+39 35 04	MCG +7-30-5	14.6	0.8 x 0.4	13.3		30	Mag 12.6 star SW 2'6.
14 26 03.2	+38 46 36	MCG +7-30-12	15.0	0.9 x 0.5	14.0	SBbc	141	Pair of mag 13.4, 13.6 stars S 1'1.
14 31 27.8	+39 15 31	MCG +7-30-23	14.4	0.8 x 0.4	13.0	S?	81	Mag 13.8 star NW 1'7.
14 33 48.6	+40 05 40	MCG +7-30-28	14.0	0.7 x 0.5	12.7	S?	18	On N edge of UGC 9376.

(Continued from previous page)

RA h m s	Dec ° ′ ″	Name	Mag (V)	Dim ′ Maj x min	SB	Type Class	PA	Notes
14 36 42.8	+39 56 35	MCG +7-30-34	14.1	0.6 x 0.5	12.6		126	Mag 12.2 star NE edge.
14 36 46.5	+39 57 56	MCG +7-30-35	15.6	0.7 x 0.2	13.3	Sc	24	Mag 12.2 star and MCG +7-30-34 SW 1′.0.
14 48 53.9	+38 46 04	MCG +7-30-65	14.7	0.9 x 0.2	12.7	Sbc	60	Mag 9.82v star S 1′.4.
14 54 47.0	+40 47 03	MCG +7-31-7	14.6	1.0 x 0.5	13.7		177	Mag 10.82v star SW 4′.4.
14 57 28.7	+40 59 19	MCG +7-31-12	15.1	0.8 x 0.4	13.8		12	Mag 11.5 star WSW 3′.3.
15 05 29.7	+39 56 36	MCG +7-31-28	14.5	0.8 x 0.5	13.3		111	**MCG +7-31-27** N 1′.1; mag 9.46v star W 1′.9.
15 08 53.5	+39 19 17	MCG +7-31-34	15.2	0.8 x 0.3	13.5	Sb	60	
15 12 25.6	+40 33 06	MCG +7-31-36	14.3	0.8 x 0.6	13.4	S?	123	Mag 11.3 star N 2′.5.
15 13 13.9	+40 32 14	MCG +7-31-39	14.3	1.1 x 0.9	14.2		87	MCG +7-31-40 S 1′.1.
15 13 15.6	+40 31 09	MCG +7-31-40	15.0	0.8 x 0.3	13.3	Sb	6	MCG +7-31-39 N 1′.1; mag 11.7 star E 0′.8.
15 17 01.8	+39 41 44	MCG +7-31-52	15.2	0.8 x 0.6	14.2	Sb	60	Mag 10.56v star W 5′.0.
15 17 14.2	+40 56 41	MCG +7-31-53	15.2	0.5 x 0.4	13.3	Sc	117	Pair of mag 13-14 stars WSW 1′.6.
13 56 56.0	+29 09 53	NGC 5375	11.5	3.2 x 2.8	13.7	SB(r)ab	0	Dark patches E and W of bright center.
13 56 50.9	+37 47 49	NGC 5378	12.5	2.6 x 2.1	14.2	(R′)SB(r)a	90	Strong nucleus in short, N-S bar; mag 9.06v star E 4′.9.
13 56 56.7	+37 36 37	NGC 5380	12.3	1.7 x 1.7	13.3	SA0⁻		Compact core, smooth envelope.
13 58 33.5	+37 27 11	NGC 5394	13.0	1.9 x 1.3	13.8	SB(s)b pec I-II	135	Extremely bright nucleus or star; two main arms, one connecting with NGC 5395 S.
13 58 37.3	+37 25 34	NGC 5395	11.4	2.9 x 1.5	12.8	SA(s)b pec I-II	2	Strong dark lanes W side. Connected to NGC 5394 N.
13 59 31.4	+34 46 23	NGC 5399	13.9	1.2 x 0.3	12.6	S?	88	
13 59 43.3	+36 14 15	NGC 5401	13.7	1.5 x 0.3	12.7	Sa	81	
13 59 50.9	+38 10 55	NGC 5403	13.6	3.1 x 0.9	14.6	SB(s)b: sp III	145	Strong dark lane through central lens.
14 00 19.9	+38 54 51	NGC 5406	12.3	1.9 x 1.4	13.2	SAB(rs)bc I-II	120	Mag 6.63v star N 6′.9; **UGC 8913** SW 12′.1.
14 00 50.3	+39 09 21	NGC 5407	13.2	1.1 x 0.6	12.8	E?	100	Mag 6.63v star SSW 9′.3; mag 9.23v star W 3′.7.
14 00 54.2	+40 59 13	NGC 5410	13.1	1.5 x 0.8	13.2	SB?	66	**UGC 8932** 1′.2 N.
14 01 42.0	+33 49 31	NGC 5421	14.1	1.4 x 1.0	14.3	SB?	164	Double system in contact; very faint anonymous galaxy S 0′.4.
14 02 36.1	+32 30 35	NGC 5433	13.5	1.6 x 0.4	12.9	Sdm:	3	MCG +6-31-49 SW 5′.0.
14 03 00.5	+34 45 23	NGC 5440	12.3	2.2 x 1.0	13.0	Sa	50	Star on SW end.
14 03 12.2	+34 41 02	NGC 5441	15.6	0.6 x 0.6	14.3	Spiral		Stellar nucleus.
14 03 24.2	+35 07 56	NGC 5444	11.9	2.4 x 2.1	13.6	E⁺:	90	
14 03 31.4	+35 01 27	NGC 5445	13.0	1.5 x 0.5	12.6	S0?	27	Mag 13.6 star on S edge.
14 10 31.7	+38 53 38	NGC 5497	14.1	1.2 x 0.7	13.7	SB(s)b	66	**MCG +7-29-47** W 2′.0.
14 10 47.6	+35 54 47	NGC 5499	13.6	0.9 x 0.6	12.8	S?	150	Mag 11.23v star S 2′.8.
14 12 41.2	+30 51 16	NGC 5512	14.2	0.6 x 0.5	13.0	E	78	Almost stellar.
14 12 38.0	+39 18 35	NGC 5515	12.9	1.3 x 0.7	12.6	Sab	108	
14 12 51.2	+35 42 40	NGC 5517	13.8	1.2 x 0.8	13.6	S0/a	125	Mag 8.08v star S 6′.8; elongated galaxy **KUG 1410+359** W 8′.6.
14 14 27.2	+36 24 15	NGC 5524	13.8	0.8 x 0.7	13.1	Spiral	150	**UGC 9123** NE 8′.0.
14 15 18.8	+36 12 08	NGC 5527	15.1	0.9 x 0.7	14.4		179	Uniform surface brightness.
14 15 33.3	+36 13 42	NGC 5529	11.9	6.2 x 0.8	13.5	Sc: sp	115	Faint dark lane N of center; MCG +6-31-87 SE 3′.0; NGC 5527 SW 3′.3.
14 16 07.6	+35 20 38	NGC 5533	11.8	3.1 x 1.9	13.6	SA(rs)ab I-II	30	
14 16 23.8	+39 30 05	NGC 5536	13.5	1.2 x 1.0	13.6	SBa	144	Bright N-S bar; mag 7.49v star NW 14′.0.
14 16 31.9	+39 35 15	NGC 5541	12.9	0.9 x 0.7	12.2	S?	12	Mag 6.38v star N 9′.6; IC 4394 N 3′.9.
14 17 02.6	+36 34 19	NGC 5544	13.0	1.1 x 1.0	12.9	(R)SB(rs)0/a	62	NGC 5545 overlapping NE edge.
14 17 06.0	+36 34 37	NGC 5545	14.1	1.0 x 0.3	12.6	SA(s)bc:	58	Bright core offset SW end.
14 18 25.7	+36 29 35	NGC 5557	11.0	2.3 x 1.9	12.6	E1	105	Faint star S of bright nucleus.
14 19 17.6	+35 08 15	NGC 5567	13.7	1.1 x 0.9	13.6	S0/a	61	Faint anonymous galaxy SE 1′.6.
14 19 21.4	+35 05 28	NGC 5568	14.7	0.8 x 0.6	13.7	Sc	96	Small bright nucleus.
14 19 35.4	+36 08 26	NGC 5572	14.2	0.9 x 0.7	13.5	Sb II	170	
14 20 26.2	+35 11 13	NGC 5579	13.6	1.9 x 1.4	14.5	SABcd	165	Bright nucleus, two main arms S.
14 20 43.2	+39 41 36	NGC 5582	11.6	2.8 x 1.7	13.3	E	25	
14 21 25.2	+35 16 13	NGC 5589	13.3	1.1 x 1.1	13.3	SBa		**MCG +6-32-4** WNW 3′.8; mag 9.71v star W 6′.0.
14 21 38.3	+35 12 20	NGC 5590	12.3	1.8 x 1.8	13.5	S0		Bright nucleus.
14 22 28.6	+37 07 18	NGC 5596	13.5	1.1 x 0.8	13.2	S0	100	**MCG +6-32-16** E 6′.2.
14 22 28.4	+40 19 13	NGC 5598	13.0	1.5 x 1.0	13.3	S0	50	Mag 9.48v star W 6′.0.
14 22 53.3	+40 18 35	NGC 5601	14.6	0.8 x 0.3	12.9	Sb	0	NGC 5598 W 4′.9.
14 23 01.7	+40 22 37	NGC 5603	13.0	1.1 x 1.1	13.1	S0		Bright center with faint outer envelope; UGC 9216 N 2′.6.
14 23 48.5	+34 50 32	NGC 5609		0.4 x 0.3			99	Stellar.
14 24 04.7	+33 02 49	NGC 5611	12.6	1.3 x 0.6	12.2	S0	63	**UGC 9232** SSE 7′.6.
14 24 06.1	+34 53 28	NGC 5613	14.9	0.5 x 0.4	13.0	(R)SAB(r)0⁺	18	Almost stellar, located 2′.0 N of NGC 5614.
14 24 07.3	+34 51 33	NGC 5614	11.7	2.5 x 2.2	13.4	SA(r)ab pec	130	Faint star on NW edge of bright center.
14 24 20.6	+36 27 43	NGC 5616	13.8	2.1 x 0.4	13.5	Sbc	157	Faint, elongated anonymous galaxy SW 5′.0; mag 9.83v star S 6′.8.
14 27 08.6	+33 15 08	NGC 5623	12.5	1.6 x 1.1	13.1	E	17	Mag 9.33v star S 4′.4; round anonymous galaxy with faint star on S edge ESE 9′.7.
14 27 02.1	+39 57 28	NGC 5625	13.8	1.0 x 0.8	13.4	Sbc	78	
14 28 46.6	+30 24 45	NGC 5639	13.5	1.4 x 0.9	13.6	Scd:	98	Stellar nucleus.
14 29 13.9	+30 01 33	NGC 5642	12.6	1.8 x 1.3	13.6	E	130	Star SE of nucleus.
14 29 34.1	+35 27 44	NGC 5646	14.2	1.5 x 0.4	13.5	SBb:	81	Mag 10.21v star E 1′.9.
14 30 10.4	+31 12 54	NGC 5653	12.2	1.7 x 1.3	12.9	(R′)SA(rs)b III-IV	125	
14 30 01.6	+36 21 34	NGC 5654	13.0	1.5 x 1.0	13.3	S?	145	Double nucleus or bright knot?
14 30 25.0	+35 19 11	NGC 5656	11.8	1.9 x 1.5	12.8	Sab	50	Elongated anonymous galaxy S 2′.4.
14 30 43.5	+29 10 49	NGC 5657	13.3	1.9 x 0.8	13.6	SBb:	163	Two thin arms form an open "S."
14 32 38.4	+31 40 13	NGC 5672	13.5	0.9 x 0.6	12.7	Sb? sp	50	**MCG +5-34-67** S 1′.4.
14 32 39.6	+36 18 07	NGC 5675	12.8	2.8 x 1.0	13.8	S?	128	UGC 9350 W 7′.9.
14 35 50.3	+36 32 31	NGC 5684	12.7	1.5 x 1.3	13.3	S0	105	Diffuse core, smooth envelope.
14 36 15.4	+29 54 31	NGC 5685	13.3	1.2 x 1.2	13.7	E:		
14 36 02.7	+36 30 10	NGC 5686	14.4	0.6 x 0.6	13.1	Sa		Compact.
14 37 22.4	+36 34 04	NGC 5695	12.8	1.5 x 1.1	13.2	S?	150	MCG +6-32-78 SE 11′.3.
14 37 14.5	+38 27 14	NGC 5698	13.0	1.9 x 0.9	13.5	SBb	70	Dark patches E and W of center; mag 10.76v star SW 2′.6.
14 38 42.5	+30 27 53	NGC 5706	14.8	0.4 x 0.3	12.3		75	Star N edge.
14 38 16.2	+40 27 24	NGC 5708	13.3	1.6 x 0.6	13.1	Sdm:	177	
14 38 50.1	+30 26 32	NGC 5709	13.6	1.6 x 0.4	12.9	SBa	105	Almost stellar NGC 5706 NW 2′.2.
14 40 25.8	+33 59 21	NGC 5727	13.6	2.2 x 1.2	14.5	SABdm	135	Very patchy.
14 40 38.8	+38 38 11	NGC 5732	13.5	1.3 x 0.7	13.2	Sbc	40	
14 45 14.2	+38 43 41	NGC 5752	14.0	0.5 x 0.2	11.4	Sb	120	On W edge of NGC 5754.
14 45 18.9	+38 48 20	NGC 5753	15.9	0.6 x 0.5	14.5		153	Bright nucleus; uniform envelope; NGC 5755 SSE 1′.9.
14 45 19.9	+38 43 56	NGC 5754	13.0	2.0 x 1.8	14.3	SB(rs)b	33	Bright bar, strong dark lane W side of center. NGC 5752 on W edge.
14 45 24.4	+38 46 47	NGC 5755	14.2	1.3 x 1.0	14.3	SB?	15	Faint single arms N and S.
14 52 14.3	+29 50 42	NGC 5771	13.6	0.8 x 0.7	12.9	E?	153	NGC 5773 SE 4′.2.
14 51 39.0	+40 35 56	NGC 5772	12.8	2.1 x 1.3	13.8	SA(r)b: II-III	35	
14 52 30.5	+29 48 26	NGC 5773	13.6	0.9 x 0.9	13.2	S?		NGC 5771 NW 4′.2.
14 56 36.0	+30 14 04	NGC 5789	13.6	0.9 x 0.8	13.1	Sdm	135	Several bright knots.
14 57 38.0	+29 58 06	NGC 5798	13.0	1.4 x 1.0	13.3	Im:	42	Patchy; pair of stars on W edge.

(Continued from previous page)

52 GALAXIES 52

RA h m s	Dec ° ′ ″	Name	Mag (V)	Dim ′ Maj x min	SB	Type Class	PA	Notes
15 05 53.3	+39 31 19	NGC 5853	14.1	1.4 x 0.8	14.1	SB	150	
15 22 02.1	+31 13 56	NGC 5924	14.6	0.7 x 0.2	12.3	S0?	12	**PGC 54863** E 2′1.
13 56 09.0	+30 05 17	UGC 8856	14.7	2.0 x 0.7	14.9	Double System	33	Double system, bridge, streamer, consists of **MCG +5-33-21** and **MCG +5-33-22**.
13 56 04.4	+38 18 11	UGC 8858	14.5	1.5 x 0.6	14.2	Sdm:	53	
13 59 47.9	+40 22 54	UGC 8917	14.4	1.6 x 0.3	13.5	Scd:	123	MCG +7-29-27 NW end.
14 00 15.9	+38 30 08	UGC 8923	14.1	1.5 x 0.4	13.3	S0/a	143	
14 01 49.2	+37 00 30	UGC 8945	13.5	1.0 x 0.7	12.9	S0/a	175	
14 02 43.2	+39 10 05	UGC 8960	14.3	1.5 x 0.3	13.3	S	178	UGC 8962 and MCG +7-29-41 S.
14 02 45.7	+39 08 06	UGC 8962	14.8	1.1 x 1.1	14.9	SB		MCG +7-29-41 on SE edge; UGC 8960 N.
14 03 17.0	+38 31 48	UGC 8975	13.4	1.0 x 0.5	12.5	S?	22	
14 03 37.0	+39 03 07	UGC 8980	13.5	1.0 x 0.9	13.3	(R′)SB(s)b	155	Bright center, dark patches N and S.
14 03 47.4	+35 44 28	UGC 8984	13.6	1.2 x 0.3	12.3	S?	35	
14 04 36.8	+29 11 53	UGC 8999	14.2	0.9 x 0.7	13.5	Scd:	5	**MCG +5-33-44** S 1′5.
14 04 37.0	+35 32 41	UGC 9003	14.0	1.4 x 0.7	13.8	S0	25	**CGCG 191-51** NE 5′8.
14 05 28.3	+30 45 59	UGC 9012	14.0	0.8 x 0.5	12.9	S0	90	**MCG +5-33-51** SW 1′8.
14 06 15.8	+35 47 41	UGC 9022	13.9	1.5 x 1.2	14.4	Sdm	93	
14 07 55.5	+29 52 22	UGC 9035	13.6	1.3 x 1.0	13.7	SB(rs)b	141	
14 08 14.6	+35 44 11	UGC 9042	15.9	0.9 x 0.9	15.5	Sd:		Mag 11.5 star SW 3′0.
14 08 48.1	+35 36 44	UGC 9045	14.5	1.1 x 0.7	14.1	S0/a	111	Mag 14.1 star NE edge; mag 11.12v star NNE 2′9.
14 08 53.1	+33 31 57	UGC 9048	14.6	1.9 x 0.4	14.1	Sbc	60	
14 11 38.6	+39 38 29	UGC 9081	13.0	1.5 x 1.0	13.3	Sa:	60	Located between a pair of stars, mags 12.3 and 11.7.
14 11 53.5	+38 11 37	UGC 9088	13.4	2.3 x 0.7	13.8	S0/a	65	
14 13 57.4	+31 33 53	UGC 9107	14.1	0.7 x 0.5	12.8	SBb	84	
14 14 14.7	+35 25 23	UGC 9113	14.4	1.8 x 0.5	14.1	S?	58	
14 21 52.8	+39 58 41	UGC 9203	14.0	0.9 x 0.8	13.5	(R′)SB(s)b	33	Bright NE-SW bar; mag 9.38v star S 4′1.
14 22 46.6	+37 59 41	UGC 9213	14.2	1.8 x 0.7	14.3	Sb II-III	160	Very faint, elongated anonymous galaxy N edge; mag 10.70v star S 2′8.
14 22 55.5	+32 50 59	UGC 9214	13.8	0.9 x 0.7	13.1	SBa	23	Stellar **Mkn 679** E 6′6.
14 22 56.5	+40 24 59	UGC 9216	13.9	1.3 x 1.3	14.3	SAcd		Stellar nucleus; NGC 5603 S 2′6; mag 12.1 star W 2′2.
14 23 42.7	+34 00 28	UGC 9221	13.5	0.8 x 0.6	12.6	S?	12	Elongated **UGC 9222** on NE edge.
14 23 45.1	+40 01 23	UGC 9223	13.7	1.0 x 0.4	12.6	S0	171	
14 25 20.5	+32 28 53	UGC 9241	13.6	0.5 x 0.5	11.9	S?		
14 25 20.5	+39 32 18	UGC 9242	13.5	4.9 x 0.3	13.7	Sd	71	
14 25 33.3	+33 50 48	UGC 9243	14.4	1.2 x 0.5	13.7	Scd:	147	Uniform surface brightness; mag 11.7 star SW 1′3.
14 26 58.4	+31 30 56	UGC 9253	13.8	1.7 x 0.4	13.2	Sbc	48	Mag 13.6 star S edge; close pair of stars, mags 11 and 12, NE 1′8.
14 27 10.0	+35 55 15	UGC 9262	14.0	1.5 x 0.2	12.5	S	51	Very faint, stellar anonymous galaxy SW 1′5.
14 28 23.5	+33 15 10	UGC 9284	14.0	1.5 x 0.3	12.9	Sa?	137	Mag 10.98v star S 2′6.
14 28 36.9	+38 59 56	UGC 9291	13.0	2.6 x 1.4	14.3	Sd	115	Located 2′8 W of mag 9.18v star.
14 29 47.8	+35 34 24	UGC 9315	14.5	0.8 x 0.2	12.4	S	69	Pair with MCG +6-32-47 SE.
14 29 58.3	+36 52 23	UGC 9320	14.4	1.0 x 1.0	14.3	SBdm		Mag 9.07v star SW 2′9.
14 32 01.1	+36 18 17	UGC 9350	14.4	1.0 x 0.3	12.9	S	27	Mag 11.4 star on NE edge; **MCG +6-32-59** W 0′6.
14 33 37.2	+35 57 55	UGC 9372	14.4	0.9 x 0.6	13.6	SBb	108	Mag 9.08v star N 5′3.
14 33 46.8	+40 04 50	UGC 9376	13.8	1.5 x 0.8	13.8	S?	147	Pair with MCG +7-30-28 N edge, disturbed.
14 33 58.7	+40 14 38	UGC 9379	14.4	1.5 x 0.8	14.4	Scd:	145	Several small knots.
14 34 52.8	+40 44 49	UGC 9386	13.8	1.1 x 0.6	13.2	SBab	140	
14 35 34.0	+36 49 08	UGC 9387	14.1	1.9 x 0.5	13.9	S0	30	MCG +6-32-74 SE 7′4.
14 37 51.0	+30 28 52	UGC 9425	14.1	1.1 x 0.6	13.5	S?	3	Double system, connected; long, slender, curving plumes.
14 38 00.2	+40 06 20	UGC 9429	13.9	1.0 x 0.7	13.4	S	90	Mag 12.9 star E 1′8; very small, very faint anonymous galaxy S 0′8.
14 41 33.0	+38 51 06	UGC 9473	12.7	1.4 x 1.2	13.1	S0:	118	
14 45 28.0	+31 25 59	UGC 9504	14.4	1.3 x 0.7	14.1	Double System	9	Extremely faint **UGC 9506** SE 1′5.
14 45 45.5	+32 37 47	UGC 9510	14.8	1.6 x 0.3	13.8	Sbc	57	**CGCG 192-60** E 9′5.
14 46 21.8	+32 46 46	UGC 9518	13.3	1.2 x 0.8	13.3	E	6	Bright core, uniform envelope.
14 46 21.0	+34 22 10	UGC 9519	13.4	0.8 x 0.7	12.7	S0:	75	Mag 12.5 star NW 1′0.
14 48 26.9	+34 59 52	UGC 9537	13.8	2.3 x 0.5	13.8	Sb III-IV	140	
14 50 56.5	+35 34 13	UGC 9560	14.6	0.8 x 0.3	12.9	Pec	58	
14 51 14.5	+35 32 27	UGC 9562	13.9	1.1 x 1.1	14.0	Pec		Strongly peculiar, brighter jets NE-SW, faint extended envelope NW-SE; mag 10.45v star SE 2′6.
14 54 11.7	+30 12 28	UGC 9588	14.0	0.8 x 0.6	13.1	S?	54	Patchy, brighter core offset NE.
14 56 51.3	+38 45 22	UGC 9623	13.2	1.0 x 0.6	12.6	S0	173	Very faint, stellar anonymous galaxy SW 1′1.
14 59 34.9	+32 50 28	UGC 9647	14.5	1.5 x 1.0	14.8	Sm:	95	Dwarf spiral, low surface brightness.
15 04 37.8	+40 22 18	UGC 9691	14.0	1.1 x 0.9	13.8	SBbc	58	Mag 10.53v star SE 3′5.
15 06 39.2	+37 17 03	UGC 9716	15.1	0.8 x 0.6	14.2	Sb-c	120	Mag 13.5 star on W edge.
15 18 01.1	+30 41 21	UGC 9809	14.0	1.2 x 0.9	13.9	SB(r)c	111	Stellar nucleus, uniform envelope; mag 10.97v star S edge.
15 22 44.9	+29 46 07	UGC 9831	13.9	0.6 x 0.6	12.6	Sa		Mag 12.6 star N 1′1.
15 23 56.9	+38 07 17	UGC 9834	14.0	0.9 x 0.9	13.7	S0		Mag 9.24v star NE 3′7.

GALAXY CLUSTERS

RA h m s	Dec ° ′ ″	Name	Mag 10th brightest	No. Gal	Diam ′	Notes
15 21 18.0	+30 39 00	A 2061	15.7	71	20	**CGCG 165-41** at center; other members PGC and anonymous, most stellar; A 2067 NW.
15 23 12.0	+30 54 00	A 2067	15.7	58	13	All members PGC or anonymous, most stellar.

53 GALAXIES 53

RA h m s	Dec ° ′ ″	Name	Mag (V)	Dim ′ Maj x min	SB	Type Class	PA	Notes
12 56 44.1	+30 43 05	CGCG 160-30	14.0	0.6 x 0.5	12.6	S0	168	Mag 11.4 star N 1′3; MCG +5-31-22 E 3′0.
13 09 16.2	+29 21 57	CGCG 160-151	14.8	0.4 x 0.4	12.7	S?		Mag 8.54v star W 8′2.
13 15 47.2	+31 50 43	CGCG 160-172	14.1	0.9 x 0.4	12.8	S?	66	Mag 13.9 star on E edge.
13 16 20.6	+30 40 37	CGCG 160-175	14.4	0.7 x 0.4	12.9	S?	24	Very close pair of mag 13-14 stars S 1′0; mag 12.7 star E 2′5.
13 18 31.8	+31 20 13	CGCG 160-185	13.8	0.5 x 0.4	12.1	E	3	Anonymous galaxy W 1′6; anonymous galaxy S 2′3.
13 18 59.8	+30 46 48	CGCG 160-190	14.6	0.8 x 0.5	13.7	E	177	**CGCG 161-9** N 2′2.
13 20 14.8	+30 59 12	CGCG 161-19	14.1	0.7 x 0.4	12.6		39	Bright center.
13 39 03.5	+32 09 16	CGCG 161-103	15.6	0.6 x 0.5	14.2	Irr	95	Mag 13.8 star W 2′1; mag 9.91v star ESE 5′3.
13 56 09.0	+29 40 11	CGCG 162-29	15.0	0.7 x 0.5	13.7	Sc	54	Very close pair of mag 11 and 13 stars NW 9′7.
13 24 20.5	+36 35 42	CGCG 190-4	14.0	0.5 x 0.3	11.8		141	= **Mkn 451**.
12 46 44.7	+29 44 05	IC 818	14.6	1.0 x 0.3	13.1	S?	48	Mag 10.48v star SW 5′1.
12 47 26.4	+29 47 13	IC 821	13.8	1.1 x 1.1	13.9	SAB(s)bc I		Mag 10.48v star SW 5′1.

RA h m s	Dec ° ′ ″	Name	Mag (V)	Dim ′ Maj x min	SB	Type Class	PA	Notes
12 51 20.0	+31 03 31	IC 826	14.0	0.6 x 0.6	12.8	S?		Mag 10.98v star S 4′.2.
13 00 39.9	+29 01 11	IC 842	13.9	1.2 x 0.5	13.2	S?	57	Stellar nucleus; mag 13.6 star S 3′.9.
13 01 33.7	+29 07 47	IC 843	13.6	1.2 x 0.5	12.9	S0	131	Mag 11.09v star NE 5′.4.
13 20 36.3	+34 08 07	IC 883	13.8	1.4 x 0.7	13.6	Im: pec	141	Double system? Faint jets extend out SE and SW from bright center.
12 42 15.4	+38 30 15	IC 3687	13.5	2.5 x 1.5	14.7	IAB(s)m IV-V	9	Mag 8.32v star SW 12′.1.
12 44 42.8	+40 40 40	IC 3726	14.7	1.4 x 0.3	13.6	Scd:	111	Mag 9.71v star NE 13′.0; **IC 3723** NNW 4′.2; **UGC 7904** WNW 9′.2.
12 47 28.1	+40 33 57	IC 3783	14.7	0.8 x 0.6	13.7		120	**IC 3778** W 5′.3.
12 48 58.9	+40 35 42	IC 3810	14.7	1.0 x 0.5	13.8	SBc	156	Mag 14.4 star on E edge.
12 49 28.7	+37 13 47	IC 3816	14.4	0.5 x 0.5	12.7			Mag 10.65v star SW 5′.9.
12 52 39.5	+40 06 09	IC 3850	14.8	0.8 x 0.7	14.0		45	Mag 10.96v star W 1′.0.
12 55 06.1	+39 13 20	IC 3892	14.4	1.3 x 0.6	14.0	S?	175	Mag 10.21v star N 2′.4; IC 3895 S 1′.3.
12 55 09.3	+39 12 09	IC 3895	14.5	0.8 x 0.8	13.9	S?		Pair with IC 3892 N 1′.3.
12 55 45.6	+36 17 33	IC 3904	15.1	0.5 x 0.4	13.2	Sa	72	Almost stellar.
12 59 13.1	+35 51 12	IC 3966	14.8	0.6 x 0.4	13.1	Spiral	3	MCG +6-29-6 N 3′.4.
12 59 12.9	+36 07 41	IC 3967	15.2	0.5 x 0.3	13.0	S0	36	Mag 10.58v star E 3′.4.
12 59 15.7	+38 52 55	IC 3975	14.4	0.8 x 0.5	13.3		30	Almost stellar galaxy E 4′.4, possibly **IC 4001**?
12 59 39.4	+38 48 54	IC 4003	14.3	0.7 x 0.7	13.4			**IC 4004** ESE 0′.8.
13 00 16.1	+36 15 12	IC 4028	14.4	0.7 x 0.6	13.3	SABc:	144	Mag 10.84v star S 5′.2.
13 00 42.8	+36 20 42	IC 4049	14.0	0.8 x 0.8	13.4	S0?		Mag 14.4 star on NW edge; mag 11.01v star WSW 11′.5.
13 00 44.5	+39 45 09	IC 4056	15.2	1.0 x 0.9	14.9	SBb-c		Very narrow extension southward from W end, with small knot at end; mag 10.37v star S 3′.1.
13 00 52.6	+40 35 00	IC 4060	14.3	0.8 x 0.6	13.4	Sb	141	Mag 13.1 star W 2′.3; mag 10.53v star W 5′.9.
13 01 06.8	+39 50 28	IC 4064	13.1	1.7 x 1.4	13.9	S0	27	Mag 10.26v star 1′.2 E; **IC 4062** NW 1′.8.
13 01 11.1	+39 44 38	IC 4065	14.6	0.8 x 0.5	13.5		177	Mag 10.37v star WSW 7′.0; IC 4056 W 5′.1.
13 01 42.8	+36 38 52	IC 4086	14.6	0.8 x 0.6	13.7	Scd:	12	Mag 9.34v star WNW 5′.1.
13 01 43.7	+29 02 38	IC 4088	13.8	1.5 x 0.5	13.4	Sab	89	Mag 8.89v star ESE 5′.6; small, faint anonymous galaxy S 2′.6.
13 02 04.9	+40 24 30	IC 4100	14.0	1.4 x 0.9	14.1	Scd:	110	Mag 11.18v star NNW 10′.1.
13 02 19.1	+38 00 59	IC 4103	14.9	0.8 x 0.5	13.8		75	Mag 7.82v star W 11′.8.
13 02 31.5	+38 28 40	IC 4108	14.8	0.8 x 0.4	13.4		129	Close pair of E-W oriented mag 12.0, 13.0 stars SE 2′.2.
13 02 48.8	+37 13 21	IC 4112	15.1	0.4 x 0.4	13.0	Spiral		Mag 8.11v star NNW 9′.1.
13 02 51.8	+38 17 31	IC 4118	14.8	0.5 x 0.3	12.6		114	IC 4123 ENE 3′.1.
13 03 05.9	+38 18 47	IC 4123	15.0	0.5 x 0.5	13.3	Sb		Mag 12.9 star NW 1′.7; IC 4118 WSW 3′.1.
13 03 17.5	+38 02 46	IC 4127	15.3	0.6 x 0.4	13.6		51	Almost stellar; mag 10.86v star ESE 9′.5.
13 04 57.2	+39 55 27	IC 4165	14.8	0.7 x 0.7	13.8	Sc		Mag 9.36v star SW 10′.1.
13 05 18.8	+31 26 32	IC 4166	14.1	0.9 x 0.5	13.1	SBab?	0	Mag 10.78v star on SW edge.
13 05 49.2	+37 36 05	IC 4182	11.4	4.8 x 4.1	14.5	SA(s)m IV-V	90	Numerous bright knots and/or superimposed stars.
13 05 59.7	+36 17 53	IC 4187	14.9	0.3 x 0.3	12.4	E/S0		Pair of mag 12.5 and 14.0 stars S 0′.8; mag 9.70v star SW 11′.7; IC 4188 N 1′.9.
13 06 02.5	+36 19 38	IC 4188	16.2	0.5 x 0.4	14.3	Spiral	72	Mag 9.65v star ESE 12′.0; IC 4187 S 1′.9.
13 06 03.8	+35 58 45	IC 4189	13.5	1.3 x 0.9	13.5	Scd:	0	Mag 11.1 star SE 5′.7; **IC 4178** NW 5′.0.
13 07 20.6	+35 47 14	IC 4199	14.1	0.5 x 0.5	12.4	Spiral		Mag 10.79 star ENE 2′.0.
13 07 51.3	+35 50 01	IC 4201	14.6	0.5 x 0.5	12.9			Mag 10.79v star WSW 4′.8.
13 12 11.2	+35 40 13	IC 4213	13.3	2.5 x 0.5	13.4	Scd:	174	Mag 11.13v star SE 8′.9.
13 29 21.1	+37 37 20	IC 4269	15.1	0.8 x 0.6	14.2		111	Mag 9.24v star E 4′.8; located near center of galaxy cluster A 1749, many faint, stellar galaxies nearby.
13 29 21.6	+37 24 43	IC 4271	14.2	0.8 x 0.5	13.0		9	Multi-galaxy system; smaller, round component S.
13 35 35.8	+33 28 48	IC 4302	14.9	1.4 x 0.2	13.4	Scd:	126	**IC 4300** SSW 4′.2; **IC 4301** S 6′.4.
13 35 57.9	+33 25 42	IC 4304	14.1	1.2 x 0.4	13.1	Sab	42	IC 4305 N 2′.7; **IC 4306** E 4′.6.
13 35 58.5	+33 28 26	IC 4305	13.8	0.7 x 0.6	12.9	E5	171	IC 4304 S 2′.7.
13 50 43.0	+39 42 29	IC 4336	14.0	1.3 x 0.4	13.2	SBb	158	Mag 7.49v star E 12′.4.
13 53 33.6	+37 23 08	IC 4340	14.0	1.0 x 0.7	13.4	S?	45	**CGCG 191-4** W 7′.1.
13 53 34.7	+37 31 19	IC 4341	13.9	0.7 x 0.7	13.0	S?		Mag 11.7 star NW 1′.5; **UGC 8795** W 9′.3.
14 00 43.8	+31 53 34	IC 4357	13.9	1.1 x 0.5	13.1	S?	70	Mag 11.56v star N 1′.5.
12 43 14.0	+31 05 04	MCG +5-30-64	13.7	0.9 x 0.9	13.3	S?		Mag 13.3 star NW 2′.6.
12 45 28.2	+30 37 29	MCG +5-30-74	14.3	0.7 x 0.3	12.5		108	Mag 14.2 star on S edge.
12 48 05.9	+29 26 37	MCG +5-30-87	16.6	0.8 x 0.5	15.4	Sc	39	
12 53 03.1	+32 06 23	MCG +5-30-115	13.9	0.8 x 0.6	13.0	S?	114	Mag 12.8 star E 1′.0.
12 56 58.0	+30 42 53	MCG +5-31-22	13.5	0.7 x 0.7	12.8	E?		Round anonymous galaxy S 0′.9; CGCG 160-30 W 3′.0.
12 58 10.4	+32 00 57	MCG +5-31-42	14.6	0.8 x 0.4	13.2	S?	153	
12 58 18.4	+29 07 37	MCG +5-31-45	14.3	0.9 x 0.5	13.3	S0	24	Multi-galaxy system.
12 59 45.5	+32 02 37	MCG +5-31-67	13.6	0.9 x 0.9	13.3	SB?		Mag 7.91v star NE 4′.3.
13 02 04.4	+29 15 06	MCG +5-31-106	14.5	0.9 x 0.3	12.9	S?	159	Mag 11.09v star SW 4′.6.
13 05 59.8	+29 16 41	MCG +5-31-127	14.0	0.8 x 0.6	13.1	S0?	21	Mag 12.4 star S 2′.6.
13 08 55.2	+29 02 24	MCG +5-31-139	15.3	0.8 x 0.4	13.9	S?	90	Mag 13.4 star on W edge.
13 10 20.9	+31 26 37	MCG +5-31-147	14.0	0.7 x 0.7	13.3	E		Mag 13.9 star on NE edge.
13 11 36.7	+31 30 24	MCG +5-31-151	13.3	1.1 x 0.8	13.0	S?	24	Mag 10.79v star ESE 2′.7.
13 13 20.8	+30 33 30	MCG +5-31-158	15.5	0.9 x 0.4	14.2		54	
13 16 15.9	+30 15 50	MCG +5-31-167	13.7	0.4 x 0.4	11.8	E3		Mag 11.6 star SE 2′.8.
13 20 21.6	+31 30 52	MCG +5-32-1	13.8	0.9 x 0.8	13.3	S?	60	
13 21 18.8	+31 23 16	MCG +5-32-4	14.3	0.9 x 0.5	13.3	S?	21	**CGCG 160-206** W 5′.8.
13 25 57.3	+31 37 08	MCG +5-32-17	14.1	0.8 x 0.6	13.1	S?	21	
13 30 39.2	+31 16 58	MCG +5-32-35	15.4	0.6 x 0.4	13.7	S?	42	UGC 8502 on W edge.
13 38 43.2	+31 16 14	MCG +5-32-49	14.0	0.5 x 0.5	12.6	E1		
13 38 59.1	+31 21 54	MCG +5-32-50	14.4	0.9 x 0.3	12.9	S0/a	120	
13 55 10.7	+31 16 53	MCG +5-33-18	14.6	0.6 x 0.5	13.2	Sc	42	
14 00 45.9	+30 04 32	MCG +5-33-39	14.2	0.7 x 0.4	12.7	S?	78	
14 01 04.1	+29 31 35	MCG +5-33-41	14.4	0.6 x 0.5	13.0	Sc	90	
12 36 26.1	+34 34 03	MCG +6-28-12	14.6	0.6 x 0.4	12.9	Spiral	153	Mag 13.5 star SE 1′.3.
12 43 07.2	+32 29 24	MCG +6-28-22	15.2	0.7 x 0.5	13.9	Sc	81	Mag 11.2 star superimposed on N edge.
12 54 52.3	+35 23 37	MCG +6-28-42	13.6	0.7 x 0.6	12.5	SBb	150	**MCG +6-28-41** NW 1′.4.
12 56 55.8	+32 26 51	MCG +6-28-44	15.0	0.8 x 0.4	13.6	Sc?	108	Double system?
12 57 33.8	+35 31 25	MCG +6-29-1	15.6	0.8 x 0.3	13.9	Sc	153	Mag 12.4 star N 2′.3.
12 59 12.3	+35 54 34	MCG +6-29-6	14.4	0.6 x 0.6	13.1	S0/a		IC 3966 S 3′.4.
13 00 02.2	+33 26 11	MCG +6-29-10	14.3	0.7 x 0.4	12.7		36	= **Mkn 235**.
13 00 18.1	+34 56 34	MCG +6-29-11	15.3	0.6 x 0.5	13.8		9	Mag 7.63v star SW 2′.6.
13 01 27.4	+34 45 22	MCG +6-29-15	15.5	0.6 x 0.5	14.0	Sbc	18	
13 03 16.6	+32 43 14	MCG +6-29-21	15.2	0.9 x 0.5	14.2		117	
13 05 28.9	+33 50 58	MCG +6-29-28	15.2	0.8 x 0.3	13.5	Spiral	177	Mag 12.1 star N 1′.3.
13 07 10.6	+34 17 52	MCG +6-29-37	14.6	1.0 x 0.3	13.1	Sb	129	Mag 13.4 star SE 1′.0.
13 07 55.1	+34 05 10	MCG +6-29-43	14.2	0.7 x 0.6	13.3	E	85	Mag 13.2 star E 0′.9.
13 09 36.6	+34 52 14	MCG +6-29-46	14.9	0.8 x 0.4	13.5	S?	75	**UGC 8242** N 6′.8.

(Continued from previous page)

GALAXIES

RA h m s	Dec ° ′ ″	Name	Mag (V)	Dim ′ Maj x min	SB	Type Class	PA	Notes
13 11 57.6	+34 21 53	MCG +6-29-56	13.8	0.9 x 0.9	13.4	S?		Stellar galaxy **KUG 1309+345A** WSW 5′.6.
13 12 29.6	+34 03 18	MCG +6-29-58	15.0	0.9 x 0.4	13.7	Spiral	168	MCG +6-29-59 E 2′.1.
13 12 39.4	+34 03 51	MCG +6-29-59	15.9	0.7 x 0.5	14.6	Spiral	129	MCG +6-29-58 W 2′.1.
13 13 43.7	+33 58 46	MCG +6-29-63	15.6	0.6 x 0.6	14.3	Spiral		Faint anonymous galaxy SE 4′.6.
13 14 54.8	+34 42 07	MCG +6-29-66	14.2	0.7 x 0.6	13.3	E/S0	72	
13 14 51.0	+38 02 14	MCG +6-29-67	14.9	0.7 x 0.5	13.6	Sc	108	Mag 12.4 star SW 2′.0.
13 16 12.3	+34 03 45	MCG +6-29-70	15.2	0.6 x 0.3	13.2	Sbc	12	Mag 10.77v star E 1′.9.
13 16 16.5	+35 45 42	MCG +6-29-72	15.0	0.5 x 0.3	12.8	Spiral	15	
13 16 48.6	+33 58 43	MCG +6-29-73	14.5	0.8 x 0.6	13.6		51	**UGC 8352** N 7′.8.
13 17 04.9	+37 57 05	MCG +6-29-75	15.4	0.7 x 0.6	14.3	Sc	0	Mag 11.3 star ENE 2′.2.
13 20 31.8	+33 17 26	MCG +6-29-79	14.1	0.8 x 0.5	12.9	Sa	156	Mag 11.9 star NW 2′.9.
13 20 50.6	+36 36 47	MCG +6-29-80	14.9	0.6 x 0.5	13.5	Sc	54	Mag 8.40v star NW 4′.4.
13 21 02.8	+33 20 19	MCG +6-29-81	14.5	0.7 x 0.5	13.4	E6	36	Mag 10.13v star N 0′.9.
13 21 55.4	+35 21 32	MCG +6-29-82	14.8	0.5 x 0.5	13.1	Irr		
13 22 03.2	+35 30 47	MCG +6-29-83	14.1	0.8 x 0.3	12.5	S0/a	108	
13 22 40.7	+32 53 20	MCG +6-29-84	14.7	0.6 x 0.4	13.0	Sa	144	Mag 14.4 star on S edge.
13 24 43.8	+32 32 23	MCG +6-30-3	14.5	0.7 x 0.7	13.6	SBa(sr)		Galaxy **NGP9 F269-1984430** NNW 1′.6.
13 25 38.7	+33 40 47	MCG +6-30-7	14.3	0.9 x 0.7	13.6	SBa(s)	39	
13 25 49.3	+34 29 25	MCG +6-30-9	15.3	0.6 x 0.5	13.9	Spiral	42	
13 28 04.7	+34 18 39	MCG +6-30-13	14.6	0.8 x 0.5	13.4	Sbc	36	UGC 8461 N 1′.3.
13 29 36.4	+34 36 06	MCG +6-30-17	14.9	0.7 x 0.3	13.0	Sb	36	Mag 13.2 star ENE 1′.9.
13 29 41.1	+36 54 14	MCG +6-30-18	14.8	0.6 x 0.4	13.1	Spiral	60	
13 30 36.9	+34 55 01	MCG +6-30-21	14.4	0.9 x 0.4	13.1		90	NGC 5199 S 5′.3.
13 31 05.6	+34 41 04	MCG +6-30-26	14.2	0.8 x 0.6	13.3		153	Mag 11.2 star NW 3′.1.
13 31 30.3	+33 32 20	MCG +6-30-27	14.8	0.7 x 0.6	13.7	SBa(r)	30	Mag 12.1 star S 2′.1.
13 32 57.7	+32 36 18	MCG +6-30-29	14.2	0.6 x 0.6	13.1	E1		MCG +6-30-30 ESE 1′.9.
13 33 06.1	+32 35 25	MCG +6-30-30	14.7	0.6 x 0.4	13.0	Sb	63	MCG +6-30-29 WNW 1′.9.
13 33 10.3	+32 41 04	MCG +6-30-31	14.9	0.9 x 0.3	13.3	Sa	156	Mag 11.3 star E 2′.4.
13 33 02.9	+37 11 42	MCG +6-30-32	14.4	1.1 x 0.8	14.1	Spiral	9	Small, faint, round anonymous galaxy SE edge.
13 33 06.7	+33 09 04	MCG +6-30-34	14.7	0.8 x 0.5	13.5	Sb	24	Mag 11.8 star E 1′.7.
13 33 13.3	+33 06 36	MCG +6-30-35	14.5	1.0 x 0.3	13.0	Sb	156	
13 33 42.4	+36 19 00	MCG +6-30-38	15.5	0.6 x 0.6	14.2	Spiral		
13 34 31.6	+35 14 32	MCG +6-30-41	14.0	0.6 x 0.3	12.0	Sc	12	
13 34 32.5	+36 02 49	MCG +6-30-42	16.0	0.9 x 0.3	14.4	Sc	72	Mag 13.0 star N 0′.9; mag 8.75v star S 3′.8.
13 34 40.4	+32 57 01	MCG +6-30-44	14.4	0.9 x 0.8	14.1	E1	42	Mag 13.3 star NE 0′.9.
13 34 51.5	+34 03 17	MCG +6-30-45	15.1	0.6 x 0.2	12.6	S?	80	UGC 8561 ESE 1′.4.
13 35 27.1	+33 43 27	MCG +6-30-49	14.4	0.8 x 0.6	13.7	E2	15	Mag 11.46v star E 2′.4.
13 36 00.6	+34 49 03	MCG +6-30-57	14.5	0.9 x 0.5	13.5		153	Mag 11.7 star NE 0′.8; Slightly elongated anonymous galaxy WSW 4′.3.
13 38 41.8	+32 18 28	MCG +6-30-64	14.9	0.6 x 0.6	13.7	Sbc		Mag 12.2 star S 0′.9.
13 39 40.4	+35 01 21	MCG +6-30-65	14.9	0.6 x 0.5	13.4	Spiral	156	Close pair of mag 13 stars ENE 2′.3.
13 39 47.5	+33 40 10	MCG +6-30-66	15.2	0.6 x 0.4	13.5	SBa(s)	9	MCG +6-30-67 N 1′.1.
13 39 48.5	+33 41 15	MCG +6-30-67	14.9	0.9 x 0.6	14.3	E6	24	MCG +6-30-66 S 1′.1.
13 41 10.5	+37 01 06	MCG +6-30-70	15.7	0.8 x 0.3	14.0	Spiral	102	Mag 12.0 star E 2′.6.
13 41 20.0	+36 56 03	MCG +6-30-71	14.7	1.1 x 0.5	13.9		15	Mag 9.92v star S 5′.3.
13 42 06.1	+37 02 25	MCG +6-30-73	14.0	0.9 x 0.7	13.4	Sbc	96	Mag 12.9 star N 2′.5.
13 45 35.2	+35 36 38	MCG +6-30-79	14.2	1.0 x 0.6	13.5	S0?	27	Elongated anonymous galaxy SE 8′.0.
13 46 11.7	+36 35 45	MCG +6-30-80	14.1	0.8 x 0.6	13.2	S?	36	Mag 15.0 star on SW edge.
13 47 10.6	+34 05 16	MCG +6-30-84	14.6	1.1 x 0.4	13.5		89	Mag 8.66v star S 6′.0.
13 48 34.8	+37 06 45	MCG +6-30-89	13.9	1.0 x 0.9	13.6	S0?	90	Mag 11.6 star SW 3′.7.
13 49 35.2	+35 15 05	MCG +6-30-91	15.2	0.7 x 0.5	13.9	Spiral	6	UGC 8739 W 4′.3.
13 50 10.0	+38 13 08	MCG +6-30-93	14.2	0.6 x 0.6	13.2	E		
13 50 20.7	+35 28 29	MCG +6-30-94	14.0	0.7 x 0.6	12.9	S0	168	Mag 12.1 star ESE 2′.4.
13 51 00.4	+36 57 12	MCG +6-30-104	14.0	1.1 x 0.7	13.5	S0/a	12	Small, round, faint anonymous galaxy SW edge.
13 51 28.2	+34 33 57	MCG +6-30-106	14.6	1.0 x 0.6	13.9		30	Mag 5.90v star NW 7′.0.
13 53 44.0	+33 13 20	MCG +6-31-15	13.8	0.9 x 0.7	13.2		129	Part of multi-galaxy system; **UGC 8817** on NW edge; **MCG +6-31-13** and **MCG +6-31-16** inline SW on SW edge.
13 54 02.9	+37 22 49	MCG +6-31-17	14.5	0.8 x 0.4	13.1		105	MCG +6-31-18 N 2′.2.
13 54 03.8	+37 24 57	MCG +6-31-18	14.9	0.9 x 0.6	14.1		105	MCG +6-31-17 S 2′.2.
13 55 13.0	+32 51 56	MCG +6-31-25	14.7	0.8 x 0.3	13.0	Spiral	108	**MCG +6-31-24** NW 2′.8.
13 57 15.1	+34 30 36	MCG +6-31-29	14.8	0.8 x 0.4	13.4	Spiral	90	Mag 8.58v star S 4′.9.
13 58 07.4	+32 38 30	MCG +6-31-31	14.6	0.8 x 0.5	13.4		132	Mag 7.16v star SW 3′.3.
13 58 41.0	+34 31 26	MCG +6-31-35	14.8	0.7 x 0.5	13.6	Spiral	147	Mag 14.5 star on E edge.
13 58 41.9	+35 05 20	MCG +6-31-36	14.9	0.9 x 0.4	13.6	Sb	165	
13 58 54.6	+35 15 18	MCG +6-31-37	14.4	0.5 x 0.5	12.7			
13 59 11.3	+34 04 16	MCG +6-31-38	14.6	0.7 x 0.7	13.7	Spiral		
13 59 57.4	+38 12 02	MCG +6-31-42	14.7	0.8 x 0.4	13.3	Spiral	48	Located 1′.7 NE of NGC 5403.
14 01 14.9	+37 52 55	MCG +6-31-43	14.4	0.9 x 0.5	13.4	Sab	81	Mag 12.3 star NW 4′.1.
14 01 24.2	+36 47 57	MCG +6-31-44	14.3	0.6 x 0.6	13.1	S?		= **Mkn 465**.
14 01 46.7	+34 40 02	MCG +6-31-47	13.8	0.9 x 0.8	13.4	S0	36	
14 02 20.6	+32 26 53	MCG +6-31-49	15.0	0.9 x 0.2	13.0	Sb	135	Mag 12.0 star NW 3′.2; **CGCG 191-34** W 5′.2.
12 37 15.7	+39 28 57	MCG +7-26-32	14.6	0.8 x 0.6	13.7		159	Mag 12.3 star SW 4′.7.
13 12 51.8	+40 32 33	MCG +7-27-45	14.1	0.7 x 0.5	12.8		69	Mag 14.8 star SW end; mag 9.01v star S 2′.5.
13 14 07.9	+38 16 40	MCG +7-27-49	13.9	1.0 x 0.9	13.6	Sbc	150	Small **MCG +7-27-50** on N edge; mag 11.42v star NNW 1′.7.
13 28 48.9	+38 44 41	MCG +7-28-15	15.5	0.7 x 0.4	13.9	Spiral	96	Mag 10.74v star SW 2′.5.
13 29 57.1	+38 24 17	MCG +7-28-21	14.6	0.8 x 0.4	13.2	Sc	129	
13 30 25.0	+39 59 52	MCG +7-28-23	14.9	0.7 x 0.4	13.4	Spiral	57	Pair of E-W mag 13-14 stars N 1′.8.
13 33 30.0	+40 31 45	MCG +7-28-32	14.7	0.7 x 0.5	13.4		51	
13 35 50.1	+40 16 14	MCG +7-28-35	14.9	0.8 x 0.5	13.8	Sbc	99	
13 40 02.7	+40 25 14	MCG +7-28-47	14.7	0.8 x 0.3	13.0	Sb	48	Mag 11.09v star W 3′.3.
13 40 15.3	+38 52 08	MCG +7-28-48	14.4	1.0 x 0.4	13.3		92	Mag 12.6 star N 2′.6.
13 47 45.8	+38 15 31	MCG +7-28-66	14.5	0.8 x 0.4	13.1	S?	108	NGC 5303 N 2′.6.
13 50 41.4	+40 16 42	MCG +7-28-78	14.4	1.1 x 0.8	14.1		60	
13 51 25.3	+40 12 46	MCG +7-29-2	14.5	0.8 x 0.8	13.8	S?		= **Mkn 462**. Mag 13.4 star N 2′.4.
13 54 03.8	+39 59 07	MCG +7-29-14	15.3	0.7 x 0.7	14.4	Spiral		Mag 11.6 star ESE 2′.3.
13 59 43.3	+40 23 15	MCG +7-29-27	16.7	0.7 x 0.5	15.4	Spiral	93	Located on NW edged of UGC 8917.
14 01 35.1	+38 41 08	MCG +7-29-37	13.7	0.8 x 0.6	12.7	S?	80	Mag 11.11v star on E edge.
14 02 48.5	+39 07 34	MCG +7-29-41	15.9	0.4 x 0.2	13.0		91	Located on SE edge of UGC 8962.
12 38 04.6	+33 27 31	NGC 4583	13.4	1.1 x 0.9	13.3	SBa	91	**MCG +6-28-13** WSW 4′.7.
12 41 44.5	+35 03 43	NGC 4619	12.7	1.3 x 1.3	13.1	SB(r)b pec?		Located 1′.9 NW of mag 9.14v star.

(Continued from previous page)

GALAXIES

RA h m s	Dec ° ′ ″	Name	Mag (V)	Dim ′ Maj x min	SB	Type Class	PA	Notes
12 41 59.8	+32 34 22	NGC 4627	12.4	1.7 x 1.0	13.1	E4 pec	26	Located 2′.4 N of the center of NGC 4631.
12 42 04.1	+32 32 12	NGC 4631	9.2	12.8 x 2.4	12.8	SB(s)d sp III	84	Many bright and dark patches entire length.
12 43 54.8	+32 09 23	NGC 4656	10.5	10.0 x 1.8	13.5	SB(s)m pec IV	37	Brightest part of galaxy is NE half; several bright patches. NGC 4657 on NE end.
12 44 11.4	+32 12 20	NGC 4657	10.9	1.3 x 0.6	10.5	SBm IV	160	Bright center.
12 44 26.2	+37 07 16	NGC 4662	12.7	1.9 x 1.6	13.8	SB(rs)bc III-IV	55	Stellar nucleus in a short bar; tightly wound arms.
12 46 10.3	+30 44 39	NGC 4676A	13.1	2.3 x 0.7	13.5	S0 pec?	0	**The Mice**. Interacting pair with NGC 4676B. Bright nucleus with long, narrow arm extending N.
12 46 11.7	+30 42 36	NGC 4676B	13.8	2.2 x 0.8	14.2	SB(s)0/a pec	2	**The Mice**. Interacting pair with NGC 4676A. Bright nucleus with broad, faint arm extending S.
12 47 23.8	+35 21 07	NGC 4687	13.2	0.7 x 0.6	12.3	E1	118	Compact.
12 48 45.8	+35 19 56	NGC 4711	13.4	1.5 x 0.9	13.6	Sb:	40	Mag 8.03v star 6′.3.
12 50 08.7	+33 09 30	NGC 4719	13.2	1.4 x 1.1	13.5	SB(s)b	15	Bright nucleus in bright, slender N-S bar.
12 50 53.1	+34 09 22	NGC 4737	14.3	0.9 x 0.6	13.4	Sab	50	
12 53 06.9	+36 49 01	NGC 4774	14.3	0.6 x 0.4	12.6	Ring:	90	Double system with faint bridge to small companion N.
12 57 47.8	+36 22 13	NGC 4846	13.6	1.3 x 0.6	13.2	S?	62	
12 59 02.3	+34 51 53	NGC 4861	12.3	4.2 x 1.6	14.2	SB(s)m:	15	Filamentary appearance; very bright clump S end.
12 59 08.8	+37 18 29	NGC 4868	12.2	1.6 x 1.5	13.0	SAab? II	90	Bright center with faint star E.
12 59 17.9	+37 02 48	NGC 4870	14.6	0.7 x 0.3	12.8		0	Lies between a pair of mag 11.3v stars.
12 59 59.6	+37 11 35	NGC 4893	14.3	0.5 x 0.5	12.6	Sb		**MCG +6-29-9** on S edge.
13 00 43.3	+37 18 51	NGC 4914	11.6	3.5 x 1.9	13.6	E^+	155	NGC 4916 N 4′.6.
13 00 40.3	+37 23 29	NGC 4916	16.3	0.6 x 0.2	13.9		8	Mag 9.58v star E 5′.2; NGC 4914 S 4′.6.
13 01 24.8	+29 18 36	NGC 4922	13.0	1.3 x 1.0	13.2	I0 pec	129	Double system in common halo.
13 04 18.1	+29 01 41	NGC 4949	14.9	0.8 x 0.4	13.5	SB(s)b pec sp	108	Anonymous galaxy SW 1′.9.
13 04 58.5	+29 07 17	NGC 4952	12.4	1.8 x 1.1	13.1	E	23	Elongated galaxy **KUG 1303+292** SE 10′.1.
13 05 01.0	+35 10 43	NGC 4956	12.4	1.5 x 1.5	13.1	S0		
13 05 41.3	+33 10 42	NGC 4959	14.5	0.8 x 0.8	13.8	Sa		
13 06 17.3	+29 03 44	NGC 4966	13.3	1.0 x 0.5	12.4	S I	143	Located 2′.5 NE of mag 6.52v star; small companion galaxy or star at NW edge.
13 08 24.5	+35 12 18	NGC 4986	13.2	1.7 x 0.9	13.5	SB(r)b	70	Stellar nucleus or star? Star immediately S of nucleus.
13 10 38.0	+36 37 59	NGC 5002	13.8	1.7 x 1.0	14.2	SBm:	173	
13 11 01.6	+29 38 10	NGC 5004	12.9	1.4 x 1.1	13.2	S0	170	Bright center with uniform outer halo.
13 11 01.9	+29 34 41	NGC 5004A	13.7	1.4 x 0.8	13.7	SBab	172	Bright nucleus in N-S bar; star on SE edge.
13 10 55.7	+37 03 26	NGC 5005	9.8	5.8 x 2.8	12.7	SAB(rs)bc II	65	Several strong dark lanes.
13 11 31.0	+36 16 53	NGC 5014	12.9	1.7 x 0.6	12.7	Sa? sp	102	
13 12 44.6	+31 48 31	NGC 5025	13.4	2.0 x 0.6	13.5	Sb II-III	57	Mag 14 star on NE end.
13 13 27.9	+36 35 41	NGC 5033	10.2	10.7 x 5.0	14.4	SA(s)c I-II	170	Bright nucleus, several filamentary, branching arms.
13 14 32.6	+30 42 15	NGC 5041	13.4	1.7 x 1.5	14.2	Scd:	150	Bright nucleus.
13 15 34.9	+29 40 33	NGC 5052	13.2	1.5 x 1.0	13.5	S0/a	160	Three faint stars SE 1′.5 along with small, faint anonymous galaxy.
13 16 12.4	+30 57 01	NGC 5056	13.1	1.7 x 1.0	13.6	Scd:	0	Mag 8.67v star S 3′.3.
13 16 27.9	+31 01 50	NGC 5057	13.0	1.3 x 1.2	13.4	S0	177	Bright, diffuse center, faint outer envelope.
13 17 30.8	+31 05 30	NGC 5065	13.6	1.3 x 0.8	13.5	Sd	85	**CGCG 160-180** SE 2′.8.
13 18 26.0	+31 28 04	NGC 5074	14.0	0.9 x 0.8	13.5	SAb pec?	60	**CGCG 161-5** NNE 5′.4.
13 19 03.0	+39 35 22	NGC 5083	14.2	1.2 x 1.1	14.4	SB(r)cd	130	
13 19 39.2	+30 15 21	NGC 5089	13.1	1.7 x 0.9	13.4	S?	120	Elongated center flanked by two very faint stars.
13 19 38.0	+40 23 09	NGC 5093	13.7	1.4 x 0.7	13.5	Sa	143	
13 20 08.7	+33 05 16	NGC 5096	13.9	0.7 x 0.7	13.2	E0+E0+E1		Multi-galaxy system, 3 components; two anonymous galaxies NE 0′.8.
13 20 14.7	+33 08 36	NGC 5098A	14.1	0.7 x 0.7	13.2			Separated from NGC 5098B E by 0′.8; mag 11.14v star 2′.0 S.
13 20 18.0	+33 08 39	NGC 5098B	14.6	0.6 x 0.6	13.3			NGC 5098A W 0′.8.
13 21 24.5	+38 32 21	NGC 5107	13.2	1.7 x 0.5	12.9	SB(s)d? sp	128	
13 21 55.6	+38 44 04	NGC 5112	12.1	4.0 x 2.8	14.6	SB(rs)cd II	130	Brightest on NE half with bright knots.
13 23 45.1	+31 33 55	NGC 5127	11.9	2.8 x 2.2	13.9	E pec	75	Bright nucleus, uniform envelope.
13 23 56.8	+30 59 15	NGC 5131	13.5	2.1 x 0.3	12.9	Sa	81	**IC 4239** ESE 6′.4; **IC 4240** E 6′.6; **IC 4238** S 3′.3.
13 24 51.5	+36 22 43	NGC 5141	12.8	1.3 x 1.0	13.0	S0	80	Faint galaxy **KUG 1322+365** S 6′.2.
13 25 01.2	+36 24 00	NGC 5142	13.3	1.0 x 0.7	12.8	S0	5	**UGC 8440** E 7′.2; NGC 5143 N 2′.4.
13 25 01.3	+36 26 16	NGC 5143	15.8	0.7 x 0.4	14.2	Spiral	97	NGC 5142 S 2′.4.
13 26 09.2	+35 56 01	NGC 5149	12.9	1.5 x 0.9	13.1	SBbc	155	Moderately bright nucleus on short bar; star N end.
13 26 28.5	+36 00 36	NGC 5154	13.8	1.4 x 1.2	14.2	Scd:	56	Stellar nucleus.
13 27 16.8	+32 01 48	NGC 5157	13.3	1.3 x 0.9	13.3	SAB(r)a	111	Short bar SE-NW.
13 28 14.9	+32 01 55	NGC 5166	13.5	2.3 x 0.4	13.2	Sb II-III	67	Very faint dark lane lengthwise through middle.
13 29 48.3	+31 07 48	NGC 5187	13.4	1.1 x 0.7	13.0	S?	48	Almost stellar galaxy **KUG 1327+312** SSE 7′.8.
13 30 42.7	+34 49 51	NGC 5199	13.6	0.9 x 0.9	13.3	Compact		MCG +6-30-21 N 5′.3.
13 34 25.2	+34 41 25	NGC 5223	13.0	1.5 x 1.3	13.7	E	168	Star and very small, round galaxy SW 0′.8.
13 34 35.1	+34 46 39	NGC 5228	13.3	1.0 x 0.9	13.1	$S0^-$:	9	**UGC 8547** W 5′.2.
13 35 13.4	+34 40 38	NGC 5233	13.9	1.4 x 0.7	13.7	Sab	80	
13 35 55.3	+35 35 18	NGC 5240	13.1	1.9 x 1.4	14.0	SB(s)cd?	60	
13 36 15.4	+38 20 36	NGC 5243	13.2	1.5 x 0.4	12.5	S	126	
13 39 24.8	+30 59 25	NGC 5259	14.2	1.1 x 0.7	14.0	E0	108	Star or very small anonymous galaxy NW of nucleus.
13 40 09.2	+36 51 42	NGC 5265	13.9	0.7 x 0.6	12.8	Irr	66	
13 40 40.1	+38 47 39	NGC 5267	13.5	1.4 x 0.5	13.0	SBb	56	Mag 9.94v star NE 10′.0.
13 41 42.4	+30 07 32	NGC 5271	14.1	1.1 x 0.9	13.9	S?	168	
13 42 08.3	+35 39 14	NGC 5273	11.6	2.8 x 2.5	13.6	SA(s)0°	10	NGC 5276 SE 3′.2.
13 42 23.3	+29 50 48	NGC 5274	14.6	0.4 x 0.4	12.7	E		Pair of small anonymous galaxies close N and NW; **UGC 8682** E 3′.0.
13 42 23.6	+29 49 29	NGC 5275	14.2	0.7 x 0.7	13.2	S?		Very small, elongated N-S anonymous galaxy on W edge.
13 42 22.0	+35 37 26	NGC 5276	13.8	1.0 x 0.6	13.1	SAB(s)b	153	Located 3′.2 SE of NGC 5273.
13 42 38.4	+29 57 14	NGC 5277	14.5	0.7 x 0.6	13.4	S?	138	
13 42 55.6	+29 52 06	NGC 5280	13.6	0.8 x 0.8	13.1	E?		**MCG +5-32-73** SW 0′.9.
13 43 24.7	+30 04 04	NGC 5282	13.3	1.1 x 0.8	13.0	S?	96	A mag 10.43V star 2′.1 NNW.
13 44 59.7	+29 48 51	NGC 5287	14.8	0.5 x 0.2	12.1		21	Very small anonymous galaxy on E edge.
13 47 45.6	+38 18 13	NGC 5303	12.6	0.9 x 0.4	11.3	Pec	83	MCG +7-28-66 S 2′.6.
13 47 55.7	+37 49 34	NGC 5305	13.6	1.5 x 1.1	14.0	SB(r)b	30	Mag 6.99v star NW 6′.3; **UGC 8724** S 4′.7.
13 48 56.2	+39 59 07	NGC 5311	12.3	2.6 x 2.2	14.0	S0/a	110	
13 49 50.5	+33 37 16	NGC 5312	13.9	0.8 x 0.5	12.7	S0/a	36	Mag 8.39v star NW 9′.6.
13 49 44.2	+39 59 07	NGC 5313	12.0	1.9 x 1.1	12.7	Sb? II	43	Faint dark lane S of nucleus.
13 50 36.1	+33 42 15	NGC 5318	12.9	1.5 x 0.9	13.1	S0?	165	Almost stellar anonymous galaxy N edge, 0′.9 from center; **MCG +6-30-97** N 1′.9.
13 50 40.8	+33 45 38	NGC 5319	15.5	0.6 x 0.2	13.0	Spiral	66	**MCG +6-30-97** SW 3′.3; a pair of anonymous galaxies NW 3′.0.
13 50 43.6	+33 37 58	NGC 5321	14.0	0.7 x 0.6	12.9	S0/a	86	Short N-S bar.
13 50 54.2	+38 16 24	NGC 5325	15.3	0.9 x 0.8	14.8	SB(s)m:	91	**MCG +7-28-81** S 2′.0.
13 50 50.5	+39 34 26	NGC 5326	11.9	2.2 x 1.1	12.7	SAa:	137	Faint dark lane NE side of elongated center; mag 7.49v star ENE 12′.2.
13 52 23.1	+39 41 19	NGC 5337	12.6	1.7 x 0.8	12.7	S?	20	Mag 7.49v star W 7′.0; **CGCG 219-13** W 7′.0.
13 52 32.2	+37 48 58	NGC 5341	13.3	1.3 x 0.5	12.6	S?	164	Mag 11.18v star NE 2′.0.
13 53 02.0	+39 34 49	NGC 5346	13.8	2.0 x 0.8	14.2	Scd:	158	Faint, stellar galaxy **KUG 1350+397** SW 10′.6.
13 53 18.0	+33 29 26	NGC 5347	12.6	1.7 x 1.3	13.4	(R′)SB(rs)ab II	130	Bright center with bar.

(Continued from previous page)

GALAXIES

RA h m s	Dec ° ′ ″	Name	Mag (V)	Dim ′ Maj x min	SB	Type Class	PA	Notes
13 53 13.2	+37 52 58	NGC 5349	14.0	1.8 x 0.4	13.5	SBb	76	NGC 5351 NE 3′.5.
13 53 21.1	+40 21 51	NGC 5350	11.3	3.2 x 2.3	13.4	SB(r)b I-II	40	Mag 6.48v star SW 2′.9.
13 53 27.9	+37 54 54	NGC 5351	12.1	3.0 x 1.5	13.5	SA(r)b: I-II	100	Mag 9.94v star E 5′.6; NGC 5349 SW 3′.5.
13 53 38.6	+36 07 59	NGC 5352	13.0	1.2 x 1.0	13.1	S0$^-$:	58	Almost stellar nucleus.
13 53 27.2	+40 16 52	NGC 5353	11.0	2.2 x 1.1	11.8	S0 sp	145	NGC 5354 on N edge.
13 53 26.7	+40 18 11	NGC 5354	11.4	1.4 x 1.3	11.9	S0 sp	72	NGC 5353 on S edge.
13 53 45.5	+40 20 18	NGC 5355	13.1	1.2 x 0.7	12.8	S0?	35	Mag 9.16v star N 5′.0.
13 54 00.5	+40 16 38	NGC 5358	13.6	1.1 x 0.4	12.5	S0/a	138	Mag 13.0, 13.6 double star SW 1′.1.
13 54 35.6	+38 26 56	NGC 5361	13.9	0.8 x 0.5	12.7	S?	63	
13 55 40.3	+40 27 45	NGC 5371	10.6	4.4 x 3.5	13.4	SAB(rs)bc I	8	Mag 8.95v star on NE edge.
13 56 56.0	+29 09 53	NGC 5375	11.5	3.2 x 2.8	13.7	SB(r)ab	0	Dark patches E and W of bright center.
13 56 50.9	+37 47 49	NGC 5378	12.5	2.6 x 2.1	14.2	(R′)SB(r)a	90	Strong nucleus in short, N-S bar; mag 9.06v star E 4′.9.
13 56 56.7	+37 36 37	NGC 5380	12.3	1.7 x 1.7	13.3	SA0$^-$		Compact core, smooth envelope.
13 58 33.5	+37 27 11	NGC 5394	13.0	1.9 x 1.3	13.8	SB(s)b pec I-II	135	Extremely bright nucleus or star; two main arms, one connecting with NGC 5395 S.
13 58 37.3	+37 25 34	NGC 5395	11.4	2.9 x 1.5	12.8	SA(s)b pec I-II	2	Strong dark lanes W side. Connected to NGC 5394 N.
13 59 31.4	+34 46 23	NGC 5399	13.9	1.2 x 0.3	12.6	S?	88	
13 59 43.3	+36 14 15	NGC 5401	13.7	1.5 x 0.3	12.7	Sa	81	
13 59 50.9	+38 10 55	NGC 5403	13.6	3.1 x 0.9	14.6	SB(s)b: sp III	145	Strong dark lane through central lens.
14 00 19.9	+38 54 51	NGC 5406	12.3	1.9 x 1.4	13.2	SAB(rs)bc I-II	120	Mag 6.63v star N 6′.9; **UGC 8913** SW 12′.1.
14 00 50.3	+39 09 21	NGC 5407	13.2	1.1 x 0.6	12.8	E?	100	Mag 6.63v star SSW 9′.3; mag 9.23v star W 3′.7.
14 00 54.2	+40 59 13	NGC 5410	13.1	1.5 x 0.8	13.2	SB?	66	**UGC 8932** 1′.2 N.
14 01 42.0	+33 49 31	NGC 5421	14.1	1.4 x 1.0	14.3	SB?	164	Double system in contact; very faint anonymous galaxy S 0′.4.
14 02 36.1	+32 30 35	NGC 5433	13.5	1.6 x 0.4	12.9	Sdm:	3	MCG +6-31-49 SW 5′.0.
14 03 00.5	+34 45 23	NGC 5440	12.3	2.2 x 1.0	13.0	Sa	50	Star on SW end.
14 03 12.2	+34 41 02	NGC 5441	15.6	0.6 x 0.6	14.3	Spiral		Stellar nucleus.
14 03 24.2	+35 07 56	NGC 5444	11.9	2.4 x 2.1	13.6	E$^+$:	90	
14 03 31.4	+35 01 27	NGC 5445	13.0	1.5 x 0.5	12.6	S0?	27	Mag 13.6 star on S edge.
12 36 22.3	+40 00 18	UGC 7774	14.2	3.4 x 0.5	14.6	Sd	99	Mag 13.8 star S edge; mag 12.1 star E 4′.1.
12 37 55.8	+34 01 19	UGC 7799	14.5	1.5 x 0.3	13.5	SB?	123	**MCG +6-28-16** W 0′.9; **MCG +6-28-14** E 1′.1.
12 38 49.0	+31 59 03	UGC 7811	13.6	1.3 x 0.6	13.2	S?	111	Mag 11.2 star S 2′.3.
12 38 48.0	+32 05 32	UGC 7812	13.3	0.9 x 0.7	12.6	S?	39	Faint dark patch NW edge.
12 38 57.2	+38 05 22	UGC 7816	14.5	1.7 x 1.1	15.0	Pec	66	Strongly peculiar, streamers; **PGC 42288** N edge; mag 14.4 star S edge.
12 40 56.3	+29 27 56	UGC 7836	14.1	1.5 x 0.5	13.7	Scd:	42	Stellar **CGCG 159-53** NNE 5′.1.
12 44 25.2	+34 23 12	UGC 7916	14.4	2.4 x 1.9	15.9	Im V	177	Dwarf irregular, very low surface brightness; close pair of stars, mags 9.43v and 10.10v, N edge.
12 46 59.8	+36 28 37	UGC 7949	14.8	1.9 x 1.1	15.5	Im: V	45	Dwarf, very low, uniform surface brightness; mag 11.4 star E edge.
12 49 37.0	+30 50 41	UGC 7978	13.8	1.2 x 0.7	13.5	Scd:	8	**MCG +5-30-92** NW 8′.6; **UGC 7981** S 5′.6; **MCG +5-30-99** E 11′.7.
12 51 38.0	+31 21 05	UGC 8004	13.9	1.7 x 0.6	13.8	Sd	5	
12 54 02.5	+29 36 11	UGC 8025	14.0	1.8 x 0.4	13.5	Sb III	74	**MCG +5-30-117** W 2′.9.
12 57 11.6	+29 02 36	UGC 8069	14.2	1.4 x 0.6	13.8	SB?	21	Elongated core offset to W; **CGCG 160-35** WNW 2′.4.
12 57 50.0	+29 39 13	UGC 8076	14.5	1.2 x 0.7	14.2	SABd:	89	
13 01 29.6	+36 21 42	UGC 8139	15.6	1.0 x 0.7	15.1	S	159	Located between a close pair of E-W mag 14 stars.
13 05 14.3	+31 59 56	UGC 8179	13.8	1.5 x 0.5	13.4	Sb II	15	Round anonymous galaxy S 1′.7; mag 8.66v star NW 2′.5.
13 05 25.0	+32 53 59	UGC 8181	15.4	1.5 x 0.4	14.7	Sdm:	80	**CGCG 189-19** SE 5′.9.
13 06 45.1	+35 06 00	UGC 8199	13.9	0.9 x 0.4	12.6	S?	142	**UGC 8200** NE 2′.9.
13 07 24.4	+32 51 44	UGC 8203	15.1	1.5 x 0.2	13.6	Scd:	163	**UGC 8210** S 6′.0.
13 07 31.0	+36 22 41	UGC 8207	15.1	0.4 x 0.4	13.0	SBc		
13 10 05.0	+34 10 50	UGC 8246	13.9	3.0 x 0.8	14.7	SB(s)cd	83	
13 10 20.3	+32 28 56	UGC 8250	14.8	1.5 x 0.2	13.3	Scd:	13	
13 11 22.6	+34 25 04	UGC 8266	14.1	1.4 x 0.9	14.2	S0$^-$:	126	Mag 8.31v star NNE 4′.7.
13 12 58.3	+31 15 24	UGC 8294	14.1	1.0 x 0.7	13.5	Scd:	110	Mag 12.4 star S 1′.2.
13 13 13.4	+33 59 02	UGC 8299	14.1	0.9 x 0.6	13.3	SBb	30	Located between a pair of mag 13 stars.
13 13 17.6	+36 12 50	UGC 8303	13.1	2.2 x 2.2	14.7	IAB(s)m IV-V		Dwarf irregular, very low surface brightness.
13 14 30.9	+35 23 12	UGC 8318	14.3	2.0 x 1.0	14.9	SB(s)cd?	70	
13 14 48.2	+34 52 48	UGC 8323	13.8	1.0 x 0.7	13.2	Im?	101	Two knots, or stars, W edge.
13 16 03.7	+35 02 33	UGC 8338	13.8	1.1 x 0.4	12.7	Sbc	83	Mag 11.6 star W 2′.7.
13 19 56.6	+30 07 06	UGC 8377	13.6	0.9 x 0.8	13.3	E?	90	
13 21 13.2	+31 13 11	UGC 8392	13.7	1.5 x 0.9	13.9	Scd:	105	Mag 14.8 star E edge.
13 21 40.5	+31 21 01	UGC 8397	13.8	1.3 x 0.6	13.3	Sbc	13	Mag 14.6 star superimposed SW edge.
13 21 45.5	+31 14 10	UGC 8399	13.9	1.0 x 0.6	13.2	SB(r)b	46	
13 24 20.3	+32 50 50	UGC 8431	14.4	1.0 x 0.6	13.7	S0/a	102	Mag 11.1 star E 3′.5.
13 26 54.0	+32 11 39	UGC 8451	13.7	1.7 x 0.9	14.0	SAc II-III	65	Stellar galaxy **KUG 1324+325** NW 8′.3.
13 28 03.9	+34 19 56	UGC 8461	14.2	1.3 x 1.0	14.4	Pec	108	Stellar nucleus, smooth envelope; MCG +6-30-13 S; small, faint anonymous galaxy NW 2′.5.
13 28 26.1	+30 48 54	UGC 8466	14.7	1.2 x 0.6	14.2	Sbc	125	
13 28 52.2	+38 34 41	UGC 8471	14.1	1.0 x 0.4	13.0	S	21	**MCG +7-28-17** E 0′.8.
13 29 39.4	+29 46 13	UGC 8483	14.0	0.9 x 0.9	13.6	S?		
13 30 06.7	+31 23 15	UGC 8492	14.0	0.9 x 0.8	13.5	S?	121	
13 30 16.4	+31 19 59	UGC 8496	13.6	1.2 x 0.7	13.3	S?	72	Disrupted multiple system.
13 30 26.0	+31 37 14	UGC 8498	12.8	2.6 x 0.9	13.6	Sb II-III	3	
13 30 36.6	+31 17 06	UGC 8502	13.7	0.5 x 0.4	11.8	Double System	24	Double system, in contact with MCG +5-32-35 on E edge.
13 31 17.6	+29 22 02	UGC 8510	14.0	1.1 x 1.0	14.0	Sbc	111	
13 33 29.2	+33 02 29	UGC 8539	13.5	1.3 x 0.5	12.9	S?	2	Mag 11.3 star and stellar galaxy **KUG 1330+332** SW 6′.9.
13 34 14.9	+31 25 26	UGC 8548	14.1	1.2 x 0.6	13.6	SB?	6	Slender extension S, fainter N.
13 34 15.1	+37 12 23	UGC 8554	14.8	1.5 x 0.3	13.8	Sbc	139	Mag 4.94v star E 6′.7.
13 34 55.3	+31 23 32	UGC 8560	13.6	1.2 x 0.9	13.5	Sb II	162	Two main arms.
13 34 57.4	+34 02 37	UGC 8561	13.2	1.1 x 0.9	13.0	Scd:	131	MCG +6-30-45 WNW 1′.4.
13 34 57.6	+38 27 26	UGC 8564	13.4	1.7 x 0.7	13.5	S0	25	Mag 12.8 star SW 1′.9.
13 35 49.9	+34 59 56	UGC 8583	13.6	1.1 x 0.7	13.1	SBab	155	Mag 7.90v star W 4′.6.
13 36 42.0	+34 44 45	UGC 8600	14.5	1.5 x 0.2	13.1	Sb II-III	56	Faint, stellar anonymous galaxy N 6′.0.
13 36 54.3	+32 05 42	UGC 8605	15.7	1.7 x 0.7	15.8	Sm:	156	Dwarf spiral, extremely low surface brightness; another dwarf spiral, **UGC 8602**, W 2′.1.
13 36 59.3	+33 34 16	UGC 8609	15.2	1.1 x 0.9	15.0	SABcd:	57	Stellar nucleus; faint, small anonymous galaxy SSW 5′.4.
13 37 31.0	+38 37 12	UGC 8619	14.2	1.0 x 0.5	13.3	S	15	
13 37 40.1	+39 09 13	UGC 8621	13.4	0.7 x 0.7	12.4	S?		Close pair of mag 7.83v, 9.11v stars ENE 3′.7.
13 38 13.1	+32 49 20	UGC 8627	14.0	1.0 x 0.8	13.6	Sbc	120	
13 38 27.1	+33 06 58	UGC 8630	13.4	1.6 x 0.5	13.0	S?	95	Located between two mag 12.3 stars, N-S.
13 39 53.3	+40 44 27	UGC 8651	14.1	2.8 x 1.4	15.4	Im IV-V	80	Dwarf irregular; prominent clump NE edge.
13 41 42.4	+40 52 26	UGC 8670	14.4	1.2 x 0.9	14.3	SABd	90	Mag 13.4 star SW 1′.7.
13 42 32.8	+39 39 27	UGC 8683	14.6	1.3 x 0.8	14.5	Im V	129	Mag 12.3 star NW 2′.7.
13 43 09.1	+30 20 12	UGC 8685	13.4	1.2 x 1.0	13.4	SB(r)bc	29	Elongated **CGCG 161-128** SW 9′.6.

(Continued from previous page)

53 GALAXIES 53

RA h m s	Dec ° ′ ″	Name	Mag (V)	Dim ′ Maj x min	SB	Type Class	PA	Notes
13 44 28.3	+35 11 36	UGC 8693	13.6	1.2 x 0.4	12.7	S?	167	Close pair of mag 13 stars NW 1′.5.
13 47 01.2	+33 53 38	UGC 8713	14.7	1.7 x 0.3	13.8	SB(s)d: sp	87	Pair with UGC 8715 SE 1′.5..
13 47 07.1	+33 52 46	UGC 8715	13.9	1.3 x 1.1	14.1	SB(s)d	159	Strong N-S bar with dark patches E and W; UGC 8713 NW 1′.5.
13 47 17.8	+34 08 54	UGC 8718	13.7	0.7 x 0.6	12.6	S	125	Very faint, small, elongated anonymous galaxy E 2′.3.
13 47 49.0	+40 29 16	UGC 8726	14.4	2.1 x 0.5	14.3	Sd	127	Mag 10.59v star NW 1′.0.
13 49 04.6	+39 29 52	UGC 8736	13.4	1.3 x 0.6	13.0	S?	175	Mag 9.02v star NNW 3′.0.
13 49 14.1	+35 15 24	UGC 8739	13.7	1.8 x 0.4	13.2	SB?	122	Two faint, round anonymous galaxies N 2′.0.
13 49 37.7	+38 55 11	UGC 8742	14.5	1.5 x 1.3	15.1	Im	30	Dwarf irregular, very low surface brightness; mag 11.38v star NNE 3′.0.
13 50 39.1	+35 02 14	UGC 8754	13.6	1.0 x 1.0	13.5	S0⁻:		**UGC 8752** N 5′.8.
13 50 50.4	+38 00 59	UGC 8760	14.0	2.1 x 0.6	14.1	Im IV-V	33	Dwarf irregular, very low surface brightness.
13 52 07.0	+38 03 59	UGC 8778	14.1	1.2 x 0.3	12.8	S?	117	
13 52 35.1	+38 42 19	UGC 8793	15.4	1.7 x 0.5	15.1	Sd	46	
13 53 14.6	+38 13 36	UGC 8806	14.1	1.8 x 0.5	13.8	Sb III	80	
13 54 13.1	+31 05 47	UGC 8824	14.8	0.9 x 0.5	13.8	Pec	40	Bright core offset to N.
13 54 05.1	+33 35 12	UGC 8825	13.8	0.6 x 0.4	12.1	Compact	156	Compact, plumes; **PGC 49397** SW edge.
13 54 20.2	+32 55 47	UGC 8829	13.7	1.1 x 0.8	13.4	SBa	80	Bright, diffuse center; faint anonymous galaxy SW 2′.6.
13 54 48.8	+35 50 17	UGC 8833	14.5	0.9 x 0.8	14.0	Im:	164	
13 55 07.9	+40 10 02	UGC 8841	13.9	1.8 x 1.0	14.4	SBb	115	Large core, two main arms.
13 55 38.1	+39 42 58	UGC 8851	15.9	1.5 x 0.5	15.4	Sdm:	138	
13 55 53.4	+37 11 45	UGC 8854	13.7	1.0 x 0.8	13.3	S0	115	
13 56 09.0	+30 05 17	UGC 8856	14.7	2.0 x 0.7	14.9	Double System	33	Double system, bridge, streamer, consists of **MCG +5-33-21** and **MCG +5-33-22**.
13 56 04.4	+38 18 11	UGC 8858	14.5	1.5 x 0.6	14.2	Sdm:	53	
13 59 47.9	+40 22 54	UGC 8917	14.4	1.6 x 0.3	13.5	Scd:	123	MCG +7-29-27 NW end.
14 00 15.9	+38 30 08	UGC 8923	14.1	1.5 x 0.4	13.3	S0/a	143	
14 01 49.2	+37 00 30	UGC 8945	13.5	1.0 x 0.7	12.9	S0/a	175	
14 02 43.2	+39 10 05	UGC 8960	14.3	1.5 x 0.3	13.3	S	178	UGC 8962 and MCG +7-29-41 S.
14 02 45.7	+39 08 06	UGC 8962	14.8	1.1 x 1.1	14.9	SB		MCG +7-29-41 on SE edge; UGC 8960 N.
14 03 17.0	+38 31 48	UGC 8975	13.4	1.0 x 0.5	12.5	S?	22	
14 03 37.0	+39 03 07	UGC 8980	13.5	1.0 x 0.9	13.3	(R′)SB(s)b	155	Bright center, dark patches N and S.
14 03 47.4	+35 44 28	UGC 8984	13.6	1.2 x 0.3	12.3	S?	35	

GALAXY CLUSTERS

RA h m s	Dec ° ′ ″	Name	Mag 10th brightest	No. Gal	Diam ′	Notes
13 11 24.0	+39 12 00	A 1691	15.4	64	19	**MCG +7-27-39** NW of center 3′.4; most other members anonymous, stellar/almost stellar.
13 29 30.0	+37 37 00	A 1749	16.0	55	18	Most prominent members spread SE and NW from center.
13 44 30.0	+29 50 00	A 1781	15.4	41	34	**MCG +5-32-78** N of center 3′.2; **UGC 8692** N 4′.4; except NGC 5287, most other members anonymous.
13 48 18.0	+32 17 00	A 1793	16.4	54	16	All members faint, stellar and anonymous.
13 53 30.0	+35 31 00	A 1813	16.0	32	18	**CGCG 191-5** NW of center 5′.4; other members anonymous and stellar.

54 GALAXIES 54

RA h m s	Dec ° ′ ″	Name	Mag (V)	Dim ′ Maj x min	SB	Type Class	PA	Notes
11 29 14.6	+31 55 39	CGCG 156-84	15.7	0.5 x 0.4	13.8	Spiral	33	Pair of mag 11 stars ENE 9′.4.
11 44 11.1	+37 11 09	CGCG 186-30	14.4	0.6 x 0.5	12.9	S?	60	= **Mkn 428**.
11 53 07.0	+35 01 20	CGCG 186-63	14.1	0.4 x 0.3	11.6	Irr	90	Stellar galaxy **Mkn 641** SW 11′.1; **CGCG 186-64** NE 11′.3.
12 03 00.3	+39 25 31	CGCG 215-17	14.2	0.5 x 0.3	12.0	S?	42	= **Mkn 43**; multi-galaxy system **KUG 1200+397A** W 2′.0.
12 11 56.0	+40 39 13	CGCG 215-50	14.6	0.7 x 0.5	13.3		93	Galaxy pair.
12 19 38.7	+29 52 57	IC 779	13.8	0.9 x 0.9	13.4	S?		Mag 10.7 star S 3′.8.
11 21 23.1	+34 21 20	IC 2738	14.4	0.7 x 0.7	13.5	S?		Mag 9.46v star N 3′.5; **IC 2735** W 4′.0.
11 21 42.9	+34 21 45	IC 2744	14.5	0.5 x 0.5	12.8	S0?		Mag 9.46v star NW 5′.5; note: **CGCG 185-43** 2′.8 N of this star.
11 22 03.2	+34 18 50	IC 2751	14.9	0.9 x 0.3	13.4	Sb	72	Mag 10.24v star SW 5′.6; faint, round anonymous galaxy SE 2′.0.
11 28 59.0	+38 51 00	IC 2861	14.3	0.9 x 0.7	13.8	E/S0	99	Mag 12.3 star S 2′.6.
11 33 29.9	+34 18 56	IC 2928	13.7	1.0 x 0.7	13.2	Sbc	143	Mag 9.94v star WSW 6′.8; **IC 2925** SW 4′.6; small anonymous galaxy SW 2′.0; **IC 2933** E 8′.9.
11 37 31.2	+31 21 42	IC 2947	14.0	0.7 x 0.4	12.5	S?	12	Mag 9.97v star W 3′.2.
11 41 37.9	+37 59 28	IC 2950	14.2	0.6 x 0.5	12.7	Sc		Mag 10.33v star E 5′.4.
11 44 25.9	+33 21 16	IC 2953	14.0	1.2 x 1.0	14.0	(R′)SB(r)b	72	Mag 9.82v star S 8′.4. **IC 2952** W 1′.8; almost stellar anonymous galaxy ENE 1′.5.
11 45 37.0	+31 17 56	IC 2957	14.1	0.8 x 0.4	12.7	S0/a	24	Mag 11.01v star WSW 3′.9.
11 50 55.2	+30 50 59	IC 2967	13.7	1.0 x 0.6	13.2	E?	15	Mag 10.96v star E 9′.0.
11 53 50.8	+33 21 51	IC 2973	13.7	1.4 x 0.8	13.7	SB(s)d:	125	Mag 10.96v star W 9′.5.
11 56 54.4	+32 09 30	IC 2979	13.5	0.8 x 0.7	12.7	SB0	171	Located 3′.1 SSE of mag 7.97v star.
11 59 12.8	+30 43 49	IC 2985	14.2	1.0 x 0.5	13.3	SB?	142	Mag 9.78v star NW 4′.9. **IC 2984** WSW 2′.3.
11 59 49.7	+30 50 38	IC 2986	13.9	0.7 x 0.6	12.8	S0?	0	Mag 1.4 star NNE 4′.4.
12 05 16.0	+30 51 18	IC 2992	14.1	0.6 x 0.5	12.6	S0?	0	mag 8.87v star SW 2′.5.
12 06 16.8	+33 31 29	IC 3001	14.9	0.8 x 0.3	13.2		72	Mag 7.91v star NE 5′.3.
12 07 04.3	+33 22 54	IC 3002	15.6	0.5 x 0.5	14.0	Sc		Mag 8.08v star NNW 4′.3.
12 07 32.8	+32 48 43	IC 3003	14.3	0.8 x 0.6	13.4	S0	138	Mag 10.3 star NNW 5′.2.
12 08 37.2	+38 49 52	IC 3014	13.5	1.2 x 0.8	13.4	SB?	69	Bright N-S oriented center, faint envelope E-W.
12 10 02.4	+38 44 24	IC 3022	13.5	0.9 x 0.9	13.3	E?		Mag 10.24v star NE 1′.7.
12 25 56.5	+30 50 32	IC 3330	14.0	1.2 x 0.6	13.5	SAB(r)ab:	100	Mag 11.02v star W 2′.0.
12 42 15.4	+38 30 15	IC 3687	13.5	2.5 x 1.5	14.7	IAB(s)m IV-V	9	Mag 8.32v star SW 12′.1.
11 18 10.0	+30 24 25	MCG +5-27-49	14.6	0.8 x 0.5	13.4	SABb	33	
11 31 55.5	+31 52 52	MCG +5-27-86	15.0	0.8 x 0.8	14.4	Spiral		
11 48 46.1	+29 38 27	MCG +5-28-32	14.9	0.8 x 0.4	13.6		12	Mag 11.9 star N 0′.7.
12 00 16.2	+31 13 28	MCG +5-28-74	14.2	0.8 x 0.4	12.8	S?	36	Mag 12.6 star N 2′.4.
12 05 50.8	+31 02 50	MCG +5-29-12	16.4	0.4 x 0.2	13.5	Sb	177	The center of elongated UGC 7085A NW 1′.4.
12 09 09.7	+29 16 23	MCG +5-29-24	15.8	0.9 x 0.2	13.8	Spiral	9	NGC 4132 SW 2′.3.
12 20 35.4	+30 47 49	MCG +5-29-66	14.3	0.9 x 0.7	13.7	IB?	18	
12 30 13.4	+30 23 09	MCG +5-30-9	15.7	0.9 x 0.4	14.5		3	
12 32 56.9	+32 08 31	MCG +5-30-17	15.7	0.8 x 0.3	14.0	Spiral	81	NGC 4509 SSE 3′.6.
12 43 14.0	+31 05 04	MCG +5-30-64	13.7	0.9 x 0.9	13.3	S?		Mag 13.3 star NW 2′.6.
11 16 33.1	+35 18 19	MCG +6-25-32	14.9	0.8 x 0.4	13.5	SBa	75	Mag 13.5 star S 1′.7.
11 17 00.4	+32 30 21	MCG +6-25-36	15.0	0.7 x 0.4	13.5		54	Mag 12.1 star WNW 2′.3.

(Continued from previous page)

RA h m s	Dec ° ′ ″	Name	Mag (V)	Dim ′ Maj x min	SB	Type Class	PA	Notes
11 17 00.1	+32 35 48	MCG +6-25-37	14.5	0.7 x 0.5	13.4	E/S0	51	Mag 13.7 star W 2′3.
11 18 54.7	+36 40 26	MCG +6-25-42	15.5	0.5 x 0.4	13.6	Spiral?	159	
11 19 17.1	+34 07 40	MCG +6-25-43	14.6	0.9 x 0.5	13.7	S0	15	Mag 9.36v star SSE 4′5.
11 21 22.8	+33 43 38	MCG +6-25-50	14.9	0.8 x 0.4	13.5		147	
11 21 30.7	+33 57 27	MCG +6-25-51	14.4	0.5 x 0.5	12.8	S0/a		
11 23 10.9	+34 29 19	MCG +6-25-59	14.8	0.9 x 0.2	12.8	Sb	15	UGC 6397 W 1′8.
11 23 24.7	+33 49 40	MCG +6-25-61	14.8	0.4 x 0.4	12.7	S0		Mag 12.6 star NE 0′5; mag 11.15v star S 1′4.
11 24 11.7	+34 02 46	MCG +6-25-62	14.7	0.7 x 0.5	13.4		111	Mag 11.03v star W 5′3.
11 25 11.5	+32 15 28	MCG +6-25-63	14.2	0.9 x 0.7	13.5	Sb	90	**MCG +6-25-64** N 4′9; **UGC 6434** NE 6′5.
11 25 29.3	+32 14 39	MCG +6-25-65	15.8	0.4 x 0.3	13.4	Spiral	165	Located between a pair of NE-SW oriented stars, mags 11.1 and 10.65v.
11 25 49.9	+33 39 42	MCG +6-25-68	16.0	0.5 x 0.5	14.3	Irr		Faint, slightly elongated anonymous galaxy S 1′5.
11 26 26.8	+35 19 55	MCG +6-25-70	15.2	0.5 x 0.4	13.3	Sd	87	**MCG +6-25-69** W 2′0 with an anonymous galaxy on it's N edge.
11 26 38.0	+35 25 24	MCG +6-25-71	14.5	0.8 x 0.5	13.3	Sa	138	Mag 11.8 star on SE end; mag 11.02v star E 1′4.
11 26 48.6	+35 14 57	MCG +6-25-72	13.9	0.6 x 0.4	12.2	S0?	9	= **Mkn 423**; pair of mag 12.4 stars N 2′2.
11 26 54.0	+33 07 05	MCG +6-25-73	15.5	0.8 x 0.5	14.4	Spiral	117	Mag 7.24v star E 3′8.
11 27 33.9	+36 03 36	MCG +6-25-74	14.2	0.9 x 0.3	12.6	S?	60	**MCG +6-25-75** N 0′5.
11 31 07.3	+35 35 19	MCG +6-25-79A	15.0	0.9 x 0.3	13.4	Spiral	30	Mag 11.4 star NW 2′7.
11 34 15.3	+33 18 45	MCG +6-26-3	14.9	0.8 x 0.3	13.2	Spiral	33	
11 34 27.6	+33 10 42	MCG +6-26-5	14.1	1.1 x 0.7	13.7	S0?	75	Mag 14.4 star W 1′4; mag 10.89v star NE 5′3.
11 38 36.4	+33 52 02	MCG +6-26-10	13.7	0.8 x 0.7	13.2	E	130	UGC 6610 SSE 4′2.
11 39 13.9	+33 55 48	MCG +6-26-12	14.4	0.8 x 0.6	13.5	S?	21	Forms a triangle with a pair of mag 12.8 stars NE 4′6 and SE 4′6.
11 39 45.1	+37 00 00	MCG +6-26-13	15.4	0.9 x 0.3	13.8		123	Mag 10.46v star SE 4′3.
11 40 35.3	+36 41 30	MCG +6-26-15	14.7	0.6 x 0.6	13.5	Sc		Pair of small, faint anonymous galaxies S and SE 1′3.
11 40 49.1	+35 12 16	MCG +6-26-16	14.3	0.9 x 0.5	13.3	S?	6	
11 41 18.2	+35 43 49	MCG +6-26-18	14.8	0.8 x 0.8	14.1			Mag 13.9 star N 2′5.
11 42 26.5	+35 48 42	MCG +6-26-22	15.1	0.9 x 0.7	14.4	Sbc	147	
11 44 27.3	+32 40 32	MCG +6-26-26	14.5	0.8 x 0.6	13.6	S0/a	0	
11 46 43.7	+34 59 53	MCG +6-26-35	15.2	0.5 x 0.5	13.5	Spiral		Located between a close pair of stars, mags 13.9 and 14.7.
11 48 04.8	+37 26 32	MCG +6-26-38	14.5	0.8 x 0.6	13.6	S0	87	Multi-galaxy system **MCG +6-26-39** NE 0′8; mag 11.4 star SW 2′7.
11 51 40.0	+36 35 15	MCG +6-26-44	14.9	0.9 x 0.6	14.1	Sa	150	**UGC 6836** E 1′6; mag 8.98v star W 1′0.
11 52 34.8	+37 48 05	MCG +6-26-50	13.9	0.7 x 0.3	12.1	Sb	21	
11 55 30.8	+37 24 22	MCG +6-26-56	14.7	0.9 x 0.6	13.9	S0/a	141	Mag 13.0 star S 1′1.
11 59 15.1	+34 34 39	MCG +6-26-65	13.8	0.9 x 0.5	12.8	S0/a	129	MCG +6-26-66 N 1′2.
11 59 17.1	+34 35 45	MCG +6-26-66	14.3	0.5 x 0.3	12.1	Sb	150	MCG +6-26-65 S 1′2.
11 59 28.7	+34 53 33	MCG +6-26-67	14.4	1.1 x 0.4	13.4	S?	63	Mag 13.9 star W 1′8.
12 05 40.1	+36 44 53	MCG +6-27-2	14.9	1.0 x 0.5	14.0		141	
12 06 21.1	+37 00 50	MCG +6-27-8	14.9	1.1 x 0.4	13.9	Sbc	66	Mag 11.7 star E 1′7.
12 09 47.1	+36 27 43	MCG +6-27-16	14.5	0.9 x 0.5	13.5	SBa	60	MCG +6-27-17 SE 2′5.
12 09 56.4	+36 26 02	MCG +6-27-17	15.7	0.9 x 0.4	14.5	Spiral	9	MCG +6-27-16 NW 2′5.
12 10 51.7	+36 37 23	MCG +6-27-20	14.2	1.2 x 0.9	14.3	E/S0	3	Faint, oval anonymous galaxy WSW 2′5.
12 11 17.2	+32 35 36	MCG +6-27-22	15.2	0.7 x 0.4	13.7	S?	66	
12 12 04.6	+32 44 04	MCG +6-27-25	15.0	0.9 x 0.4	13.8	Spiral	144	Mag 11.05v star SE 2′1.
12 12 39.5	+34 42 11	MCG +6-27-28	15.4	0.9 x 0.5	14.4	Sd	138	Pair with UGC 7212 S edge.
12 14 06.0	+34 31 31	MCG +6-27-31	16.2	0.8 x 0.8	15.6	Spiral		
12 14 01.6	+35 01 10	MCG +6-27-32	16.3	0.9 x 0.3	14.7	Spiral	63	
12 14 14.2	+32 25 12	MCG +6-27-33	14.4	0.8 x 0.5	13.3	S0	25	UGC 7243 E 2′0; **MCG +6-27-35** NE 2′2.
12 14 16.3	+34 37 08	MCG +6-27-34	14.6	0.9 x 0.7	14.0	S0	33	Mag 10.39v star N 6′1.
12 15 11.1	+34 36 42	MCG +6-27-41	14.4	0.8 x 0.6	13.7	E	69	Mag 13.2 star N 1′3.
12 19 43.9	+35 30 24	MCG +6-27-46	15.1	0.9 x 0.3	13.5		24	Mag 9.71v star W 4′8.
12 20 37.3	+34 26 28	MCG +6-27-48	14.9	0.9 x 0.6	14.1	Sb	123	
12 22 21.8	+35 09 13	MCG +6-27-50	15.4	1.1 x 0.3	14.1	Sbc	51	
12 23 46.7	+36 09 06	MCG +6-27-52	14.8	0.7 x 0.5	13.6	SBa	27	
12 26 55.3	+37 54 29	MCG +6-27-54	14.0	1.2 x 0.9	13.9	SBa	132	
12 31 34.8	+37 58 45	MCG +6-28-5	14.4	0.5 x 0.3	12.2	SB?	24	
12 32 08.4	+34 53 07	MCG +6-28-6	14.8	0.7 x 0.6	13.7	SBb	177	Mag 12.5 star S 2′1.
12 35 11.8	+34 07 17	MCG +6-28-11	14.7	0.8 x 0.6	13.7	Sc	30	Mag 8.51v star SE 6′8.
12 36 26.1	+34 34 03	MCG +6-28-12	14.6	0.6 x 0.4	12.9	Spiral	153	Mag 13.5 star SE 1′3.
12 43 07.2	+32 29 24	MCG +6-28-22	15.2	0.7 x 0.5	13.9	Sc	81	Mag 11.2 star superimposed on N edge.
11 21 16.8	+40 20 43	MCG +7-23-40	14.7	0.7 x 0.6	13.6		171	Mag 8.08v star SE 5′6.
11 21 45.2	+38 27 06	MCG +7-23-41	15.2	0.8 x 0.4	13.8		27	Mag 11.9 star W 1′7.
11 23 01.4	+38 31 08	MCG +7-24-1	15.7	0.5 x 0.3	13.5		15	Mag 10.09v star WNW 2′4.
11 26 46.3	+39 15 58	MCG +7-24-6	13.9	0.9 x 0.6	13.1	S?	51	Mag 12.1 star SW 1′4.
11 30 33.5	+38 14 26	MCG +7-24-17	14.9	0.7 x 0.4	13.3	Spiral	6	Mag 12.6 star N 2′3.
11 37 16.1	+39 16 05	MCG +7-24-24	15.2	0.7 x 0.3	13.3	Sb	81	Pair with MCG +7-24-25 N; mag 10.53v star SE 3′4.
11 37 16.8	+39 17 02	MCG +7-24-25	15.1	0.8 x 0.3	13.4	Sc	36	Pair with MCG +7-24-24 S.
11 48 20.0	+38 44 54	MCG +7-24-32	14.8	0.7 x 0.5	13.6	Spiral	6	
11 55 47.8	+39 13 48	MCG +7-25-5	14.9	0.7 x 0.5	13.6		159	Mag 12.0 star N 3′8.
11 57 25.4	+39 45 45	MCG +7-25-7	15.2	1.2 x 1.0	15.2		111	Mag 12.8 star W 2′9.
11 59 50.7	+39 36 06	MCG +7-25-13	13.9	0.9 x 0.5	12.9	Spiral	63	
12 02 45.4	+39 05 44	MCG +7-25-15	15.2	0.7 x 0.6	14.1	Spiral	30	
12 06 07.8	+40 09 01	MCG +7-25-21	14.4	0.8 x 0.7	13.6		120	Mag 13.8 star W 1′5.
12 21 02.9	+39 52 00	MCG +7-25-59	14.9	0.6 x 0.5	13.5	Sa	87	MCG +7-25-60 and MCG +7-25-61 N 1′5.
12 21 02.1	+39 53 57	MCG +7-25-60	14.3	0.7 x 0.6	13.4	E	78	MCG +7-25-61 SE 0′9.
12 21 06.3	+39 53 31	MCG +7-25-61	14.9	0.7 x 0.4	13.4	Sa	36	**MCG +7-27-60** NW 0′9.
12 24 42.6	+38 22 37	MCG +7-26-5	14.9	0.7 x 0.4	13.3		12	Mag 8.02v star S 3′6.
12 27 38.2	+40 09 34	MCG +7-26-7	16.9	0.7 x 0.6	15.8	Sc	21	
12 33 50.4	+39 31 10	MCG +7-26-23	15.2	0.9 x 0.6	14.4	Sc	9	
12 33 53.1	+39 37 36	MCG +7-26-24	16.7	0.8 x 0.5	15.6	Sc	24	
12 34 46.5	+39 35 16	MCG +7-26-28	15.5	0.5 x 0.5	13.8			Mag 11.1 star S 2′5.
12 35 52.0	+38 22 32	MCG +7-26-30	14.8	0.9 x 0.4	13.6	S	54	
12 37 15.7	+39 28 57	MCG +7-26-32	14.6	0.8 x 0.6	13.7		159	Mag 12.3 star SW 4′7.
11 22 31.3	+39 52 36	NGC 3648	12.6	1.3 x 0.8	12.5	S0	75	Mag 9.71v star NW 7′0.
11 22 39.0	+37 46 00	NGC 3652	12.2	2.6 x 0.7	12.7	Scd?	150	
11 23 58.3	+38 33 45	NGC 3658	12.2	1.6 x 1.5	13.0	SA(r)0°:	27	
11 24 43.6	+38 45 50	NGC 3665	10.8	4.3 x 3.3	13.6	SA(s)0°	18	
11 28 00.6	+29 30 40	NGC 3687	12.0	1.9 x 1.9	13.3	(R′)SAB(r)bc? I-II		Bright knots or stars E and S of core; mag 9.75v star SW 6′6.
11 28 54.2	+35 24 50	NGC 3694	13.0	1.0 x 0.8	12.6	S?	120	Moderately compact; mag 8.48v star WSW 12′7.
11 29 17.3	+35 34 33	NGC 3695	13.9	0.9 x 0.7	13.2	S	145	Mag 9.24v star NE 6′9.
11 29 38.6	+35 30 52	NGC 3700	14.0	1.0 x 0.7	13.4	(R′)SB(r)ab	1	Very faint, very small anonymous galaxy NE 1′5; mag 9.24v star N 7′6.

RA h m s	Dec ° ′ ″	Name	Mag (V)	Dim ′ Maj x min	SB	Type Class	PA	Notes
11 36 34.1	+36 24 32	NGC 3755	12.8	3.2 x 1.4	14.3	SAB(rs)c pec	133	
11 39 42.8	+31 54 33	NGC 3786	12.3	2.2 x 1.3	13.3	SAB(rs)a pec	77	On S edge of NGC 3788.
11 39 44.8	+31 55 51	NGC 3788	12.6	2.1 x 0.7	12.8	SAB(rs)ab pec	178	Very bright, diffuse nucleus.
11 39 27.0	+31 51 14	NGC 3793	14.9	0.7 x 0.4	13.4	Sc	107	Very faint star on S edge.
11 41 18.9	+36 32 50	NGC 3813	11.7	2.2 x 1.1	12.5	SA(rs)b: III	87	Surrounded by five faint stars.
11 44 14.0	+33 30 50	NGC 3847	13.3	1.1 x 1.1	13.5	E		**CGCG 186-29** NW 2.′5; an anonymous galaxy lies immediately N of CGCG 186-29.
11 44 45.1	+33 19 13	NGC 3855	14.8	0.6 x 0.6	13.6			Mag 9.82v star SW 8.′9; **IC 729** E 7.′0.
11 46 10.1	+33 06 27	NGC 3871	14.8	1.4 x 0.5	14.2	S	102	Elongated anonymous galaxy ESE 2.′4; mag 9.77v star SE 3.′6.
11 46 17.8	+33 12 15	NGC 3878	12.8	0.5 x 0.5	11.4	E		Almost stellar; galaxy **KUG 1143+334** W 4.′6.
11 46 22.3	+33 09 41	NGC 3880	13.9	0.7 x 0.7	12.9	S0?		Faint galaxy **KUG 1144+334** E 6.′3.
11 46 34.4	+33 06 21	NGC 3881	13.9	0.7 x 0.7	13.1	S0/a		Stellar anonymous galaxy SW 1.′3; mag 9.77v star SW 4.′0.
11 48 03.4	+30 21 34	NGC 3891	12.4	2.0 x 1.7	13.6	Sbc	70	
11 48 59.6	+35 00 59	NGC 3897	12.9	1.9 x 1.9	14.2	Sbc I		Mag 5.73v star SE 10.′1.
11 51 45.4	+38 00 52	NGC 3930	12.4	3.2 x 2.4	14.5	SAB(s)c	30	
11 52 24.0	+32 24 13	NGC 3935	13.3	1.0 x 0.5	12.4	S?	114	
11 52 55.1	+36 59 05	NGC 3941	10.3	3.5 x 2.3	12.5	SB(s)0°	10	
11 55 42.5	+32 11 20	NGC 3966	14.6	0.4 x 0.3	12.1	Compact	171	Almost stellar.
11 55 36.5	+29 59 44	NGC 3971	12.7	1.4 x 1.2	13.1	S0	30	**UGC 6905** SE 3.′9.
11 57 51.8	+29 02 22	NGC 3984	13.5	1.2 x 1.1	13.7	SB(rs)b	153	Pair of faint stars on SW edge.
11 56 44.4	+32 01 15	NGC 3986	12.7	3.1 x 0.7	13.4	S0 sp	110	**IC 2978** W 4.′6.
11 57 30.6	+32 20 10	NGC 3991	13.1	1.4 x 0.4	12.3	Im pec sp	33	Mag 11.8 star NE end.
11 57 36.8	+32 16 41	NGC 3994	12.7	1.0 x 0.6	12.0	SA(r)c pec?	10	W of NGC 3995; mag 6.43v star E 6.′4.
11 57 44.8	+32 17 39	NGC 3995	12.4	2.8 x 1.0	13.4	SAm pec III-IV	33	Strong (0 II) emission. Small bright nucleus off center in cardioid-shaped lens or inner ring.
11 58 56.6	+30 24 42	NGC 4020	12.7	2.1 x 0.9	13.2	SBd? sp	15	
11 59 10.1	+37 47 28	NGC 4025	13.5	2.5 x 1.4	14.8	SB(s)cd IV	31	Very thin N-S bar; mag 7.84v star NW 7.′1.
12 00 31.3	+31 56 47	NGC 4031	14.4	0.7 x 0.4	12.9	S?	60	UGC 6997 SE 5.′8.
12 04 03.9	+31 53 43	NGC 4062	11.1	4.1 x 1.7	13.1	SA(s)c II-III	100	
12 06 02.5	+36 51 49	NGC 4097	13.4	1.2 x 0.7	13.1	S0	98	Located 1.′5 N of mag 11.29v star.
12 07 08.6	+32 59 44	NGC 4122	14.7	0.8 x 0.5	13.6	Spiral	68	Stellar nucleus.
12 08 47.3	+29 18 12	NGC 4131	13.3	1.3 x 0.7	13.0	S	73	Mag 11.36v star S 3.′7.
12 09 01.4	+29 14 59	NGC 4132	14.0	1.1 x 0.4	12.9	S	21	MCG +5-29-24 NE 2.′3.
12 09 10.3	+29 10 36	NGC 4134	12.8	2.2 x 0.9	13.4	Sb	150	
12 09 17.9	+29 55 38	NGC 4136	11.0	4.0 x 3.7	13.8	SAB(r)c II	90	Numerous knots surround bright nucleus.
12 10 01.9	+39 52 59	NGC 4145	11.3	5.9 x 4.3	14.6	SAB(rs)d II	100	Chains of HII regions in the arms.
12 10 08.0	+35 52 37	NGC 4148	13.3	1.5 x 1.0	13.6	$S0^+$	165	
12 10 33.8	+30 24 05	NGC 4150	11.6	2.3 x 1.6	12.9	SA(r)0°?	147	
12 10 32.2	+39 24 20	NGC 4151	10.8	6.3 x 4.5	14.3	(R′)SAB(rs)ab:	146	Mag 8.18v star W 7.′6; **UGC 7188** E 8.′5.
12 10 49.5	+39 28 19	NGC 4156	13.2	1.4 x 1.1	13.5	SB(rs)b I	48	Faint, stellar anonymous galaxy N 2.′1.
12 12 08.9	+36 10 12	NGC 4163	14.0	1.8 x 1.6	15.0	IAm	0	Brightest portion offset S.
12 12 18.8	+29 10 45	NGC 4169	12.2	1.8 x 0.9	12.6	S0	153	Member of **The Box**. 2.′0 S of NGC 4173.
12 12 21.0	+29 12 35	NGC 4173	13.0	5.0 x 0.7	14.2	SBd:	134	Uniform surface brightness.
12 12 27.0	+29 08 54	NGC 4174	13.4	0.6 x 0.3	11.4	S?	50	Member of **The Box**. 1.′5 SW of NGC 4175.
12 12 31.2	+29 10 02	NGC 4175	13.3	1.8 x 0.4	12.8	S	130	Member of **The Box**. Located on SE end of NGC4173; NGC 4174 SW 1.′5.
12 13 44.8	+36 38 06	NGC 4190	13.3	1.7 x 1.5	14.2	Im pec IV-V	24	bright elongated core, faint outer envelope.
12 15 05.2	+33 11 48	NGC 4203	10.9	3.4 x 3.2	13.3	$SAB0^-$:	10	Very bright nucleus, a N-S line of four faint stars W of nucleus; a mag 8.16v star NNW 3.′8.
12 15 39.0	+36 19 43	NGC 4214	9.8	8.0 x 6.6	13.9	IAB(s)m III-IV	144	Bright nucleus with numerous knots and clumps overall.
12 16 33.8	+33 31 18	NGC 4227	12.7	1.5 x 0.9	12.9	SAB0°:	70	Mag 14.3 star on N edge; **UGC 7295** S 4.′4.
12 16 38.7	+33 33 39	NGC 4229	13.3	1.3 x 0.9	13.4	S?	12	Almost stellar nucleus, smooth envelope.
12 17 29.3	+37 48 21	NGC 4244	10.4	16.6 x 1.9	14.0	SA(s)cd: sp IV	48	Bright patch near SW end; short, faint lane near center; several stars along NW edge.
12 17 36.8	+29 36 33	NGC 4245	11.4	2.9 x 2.2	13.3	SB(r)0/a:	153	
12 18 26.6	+29 48 43	NGC 4253	13.1	1.0 x 0.8	12.7	(R′)SB(s)a:	54	
12 19 47.5	+30 20 19	NGC 4272	13.1	1.7 x 1.3	14.0	E	120	Compact core; stellar **CGCG 158-63** and mag 13.0 star W 5.′9.
12 19 50.3	+29 36 52	NGC 4274	10.4	6.8 x 2.5	13.3	(R)SB(r)ab II-III	102	Center has almost a "Saturn" like appearance with bright nucleus surrounded by brighter ring; fainter outer envelope.
12 20 06.8	+29 16 49	NGC 4278	10.2	3.8 x 3.8	13.1	E1-2		Large, bright center.
12 20 21.0	+29 18 38	NGC 4283	12.1	1.5 x 1.5	12.9	E0		On NE edge of NGC 4278.
12 20 42.3	+29 20 46	NGC 4286	13.1	1.6 x 1.0	13.5	SA(r)0/a:	150	Weak, almost stellar nucleus in smooth envelope.
12 21 56.8	+30 04 28	NGC 4308	13.4	0.8 x 0.7	12.8	E:	30	**UGC 7438** E 5.′1.
12 22 26.5	+29 12 29	NGC 4310	12.2	2.2 x 1.2	13.1	(R′)SAB(r)0^+?	161	
12 22 31.8	+29 53 45	NGC 4314	10.6	4.2 x 3.7	13.4	SB(rs)a	69	Bright, moderately large nucleus with bright bar extending NW-SE.
12 24 11.9	+31 31 14	NGC 4359	12.7	3.5 x 0.8	13.7	SB(rs)c? sp	108	
12 24 36.3	+39 22 59	NGC 4369	11.7	2.1 x 2.0	13.1	(R)SA(rs)a	127	
12 25 50.2	+33 32 42	NGC 4395	10.2	13.2 x 11.0	15.4	SA(s)m: IV	147	Very low surface brightness. Many knots and condensations. **NGC 4399**, **NGC 4400** and **NGC 4401** are HII condensations in disc.
12 26 27.3	+31 13 17	NGC 4414	10.1	4.4 x 3.0	12.8	SA(rs)c? II-III	155	Many filamentary arms with dark lanes and many branches. Group of small, faint anonymous galaxies NE 4.′5.
12 31 22.9	+29 08 10	NGC 4495	13.2	1.6 x 0.8	13.3	Sab	130	Bright, elongated center.
12 33 06.9	+32 05 30	NGC 4509	13.5	0.8 x 0.5	12.4	Sab pec?	155	Mag 9.55v star SE 4.′6; MCG +5-30-17 NNW 3.′6.
12 32 43.0	+29 42 42	NGC 4514	13.2	1.2 x 0.9	13.1	Sbc	51	Bright, stellar nucleus.
12 33 51.3	+30 16 40	NGC 4525	12.2	2.6 x 1.3	13.4	Scd:	47	
12 34 05.7	+35 31 06	NGC 4534	12.3	2.6 x 2.1	14.0	SA(s)dm:	107	
12 38 04.6	+33 27 31	NGC 4583	13.4	1.1 x 0.9	13.3	SBa	91	**MCG +6-28-13** WSW 4.′7.
12 41 44.5	+35 03 43	NGC 4619	12.7	1.3 x 1.3	13.1	SB(r)b pec?		Located 1.′9 NW of mag 9.14v star.
12 41 59.8	+32 34 22	NGC 4627	12.4	1.7 x 1.0	13.1	E4 pec	26	Located 2.′4 N of the center of NGC 4631.
12 42 04.1	+32 32 12	NGC 4631	9.2	12.8 x 2.4	12.8	SB(s)d sp III	84	Many bright and dark patches entire length.
12 43 54.8	+32 09 23	NGC 4656	10.5	10.0 x 1.8	13.5	SB(s)m pec IV	37	Brightest part of galaxy is NE half; several bright patches. NGC 4657 on NE end.
11 16 28.0	+29 19 35	UGC 6292	13.6	1.2 x 0.7	13.3	SB(s)b	66	Faint anonymous galaxy SE 1.′4; mag 10.84v star SE 3.′1; **PGC 34386** S 1.′6; **PGC 34394** S 2.′4.
11 17 03.4	+36 08 26	UGC 6298	14.1	1.0 x 0.5	13.2	S?	75	Bright, narrow E-W core.
11 17 33.8	+36 03 54	UGC 6303	14.0	1.2 x 1.0	14.0	Scd:	98	Star or bright knot N of nucleus.
11 17 40.7	+38 02 59	UGC 6307	14.2	1.3 x 0.5	13.6	Sdm:	165	
11 18 11.8	+30 23 41	UGC 6314	14.2	1.1 x 0.9	14.0	S?	6	
11 18 32.2	+37 56 50	UGC 6326	15.0	0.9 x 0.9	14.6	Sd-dm		Mag 10.46v star E 2.′3.
11 19 54.5	+33 05 23	UGC 6337	14.1	0.8 x 0.6	13.1	S?	108	Mag 3.49v star Nu Ursae Majoris W 17.′9.
11 20 00.8	+36 06 00	UGC 6338	14.2	1.0 x 0.5	13.3	Scd:	150	
11 20 40.2	+31 13 17	UGC 6355	14.0	1.9 x 0.3	13.3	Sd	102	Mag 13.3 star W end; almost stellar anonymous galaxy S 0.′8; small, faint anonymous galaxy SE 2.′6.
11 21 04.9	+31 15 06	UGC 6367	13.7	0.8 x 0.7	12.9	SB(s)ab	20	
11 22 53.6	+34 20 27	UGC 6393	13.9	0.9 x 0.6	13.1	SBa	40	Bright SE-NW bar.

GALAXIES

RA h m s	Dec ° ′ ″	Name	Mag (V)	Dim ′ Maj x min	SB	Type Class	PA	Notes
11 22 56.7	+34 06 41	UGC 6394	13.3	1.1 x 1.0	13.4	E:	60	Quartet of stellar anonymous galaxies NE 2′3.
11 23 02.1	+34 29 51	UGC 6397	14.0	1.7 x 0.3	13.2	Sab	0	MCG +6-25-59 E 1′8.
11 25 32.3	+38 03 34	UGC 6433	13.8	1.3 x 0.6	13.4	Pec	79	Strong knot E end.
11 27 19.0	+38 39 49	UGC 6454	15.1	1.5 x 0.3	14.1	Scd:	34	Mag 12.7 star W 2′0.
11 27 22.0	+40 00 47	UGC 6455	14.7	1.7 x 0.7	14.7	S	83	Mag 12.6 star W 2′6.
11 29 24.1	+34 52 15	UGC 6491	14.4	1.8 x 0.9	14.8	Sdm	0	
11 30 09.8	+38 37 11	UGC 6497	14.7	2.1 x 0.3	14.1	Sd:	86	Mag 10.62v star SSE 2′9.
11 30 11.2	+35 52 08	UGC 6499	15.3	1.5 x 0.6	15.0	Im:	5	Mag 9.87v star SE 2′8.
11 31 44.7	+34 19 56	UGC 6512	14.1	1.1 x 0.5	13.3	S?	150	Mag 11.3 star SW 2′4.
11 32 02.4	+36 41 49	UGC 6517	13.1	1.8 x 1.1	13.7	Sbc III	31	Located between a pair of stars, mags 13 and 12.2.
11 32 38.6	+35 19 41	UGC 6526	13.4	1.7 x 0.4	12.8	S?	84	
11 32 45.1	+40 50 32	UGC 6529	14.3	1.0 x 0.7	13.8	Sbc	72	
11 32 49.1	+39 05 05	UGC 6531	14.3	1.8 x 1.8	15.4	SB(r)dm		Mag 9.21v star W 3′2.
11 33 44.1	+32 37 59	UGC 6545	13.8	1.3 x 0.4	12.9	S?	133	Faint star S edge; mag 11.3 star N 1′8.
11 34 07.5	+36 40 57	UGC 6551	15.4	1.6 x 0.2	14.0	Sd	111	Mag 6.38v star N 8′2.
11 35 49.9	+35 20 03	UGC 6570	13.5	1.1 x 0.5	12.7	S0/a	123	Mag 12.5 star NE 2′1.
11 38 02.3	+35 12 13	UGC 6603	14.0	2.1 x 0.5	13.9	Scd:	78	Mag 14.0 star N 2′0.
11 38 44.1	+33 48 21	UGC 6610	14.3	2.1 x 0.4	14.0	Scd:	13	Mag 10.16v star E 2′7.
11 40 34.9	+36 07 39	UGC 6639	14.4	1.1 x 0.6	13.8	S0/a	174	Mag 13.4 star S 1′2.
11 42 00.7	+32 32 52	UGC 6659	14.3	0.6 x 0.5	12.9	Sc	171	
11 42 18.2	+30 13 46	UGC 6664	14.7	1.5 x 0.2	13.2	Scd:	102	
11 44 41.5	+35 58 03	UGC 6716	13.8	1.4 x 1.3	14.3	Sbc II-III	45	Almost stellar nucleus; mag 7.01v star ESE 11′7.
11 47 08.1	+29 34 37	UGC 6761	14.2	1.1 x 0.5	13.4	S0/a	97	
11 48 38.0	+32 38 29	UGC 6777	14.2	0.5 x 0.5	12.7	E?		
11 49 23.4	+39 46 15	UGC 6792	13.9	2.4 x 0.4	13.7	Scd:	172	N end sits in a triangle of mag 11.6, 12.8 and 13.4 stars.
11 50 52.2	+38 52 40	UGC 6817	13.1	3.4 x 1.8	14.9	Im V	65	Close pair of mag 121 stars on W edge; mag 10.77v star N 1′9.
11 51 28.2	+35 26 00	UGC 6827	13.4	1.1 x 0.5	12.6	S0/a	127	
11 52 47.5	+29 19 41	UGC 6853	13.7	1.2 x 0.6	13.2	S0	35	**UGC 6868** ESE 13′2.
11 55 25.1	+33 07 36	UGC 6892	14.4	1.3 x 0.4	13.6	S?	77	
11 55 24.5	+39 13 27	UGC 6893	14.7	1.0 x 0.9	14.4	Sd	3	
11 55 39.6	+31 31 06	UGC 6900	14.1	1.8 x 1.1	14.7	Im: V	115	Uniform surface brightness; mag 13.4 star on N edge.
11 56 28.4	+39 44 33	UGC 6916	13.8	1.7 x 0.4	13.2	SBbc:	34	Stellar galaxy **KUG 1153+400A** W 4′7.
11 57 08.6	+30 23 29	UGC 6927	13.4	0.9 x 0.7	12.7	S0	90	
11 57 19.8	+36 24 52	UGC 6929	14.3	1.1 x 0.9	14.1	Scd:	60	
11 57 54.8	+36 23 31	UGC 6945	14.2	1.2 x 0.8	14.0	S?	117	
11 58 28.2	+38 04 28	UGC 6955	13.3	3.9 x 1.7	15.2	IB(s)m: V	70	Dwarf irregular; extremely low surface brightness.
11 59 31.9	+30 09 18	UGC 6987	14.1	1.2 x 0.7	13.8	SB(s)b	10	Bright N-S bar.
12 00 50.4	+31 52 41	UGC 6997	14.5	0.8 x 0.6	13.5	Sm:	96	Uniform surface brightness; mag 9.93v star NW 1′5.
12 01 33.2	+33 20 22	UGC 7007	15.2	1.7 x 1.2	15.8	Sm:	90	Dwarf spiral, low surface brightness.
12 02 03.2	+29 50 55	UGC 7012	13.3	1.9 x 1.0	13.8	Scd:	8	
12 02 22.4	+29 51 40	UGC 7017	13.6	1.7 x 0.4	13.1	Sb III	70	
12 03 21.5	+29 25 09	UGC 7031	14.3	1.5 x 0.3	13.3	Sbc	57	
12 04 43.4	+31 10 38	UGC 7064	13.3	0.9 x 0.9	12.9	S?		**CGCG 158-11** N 0′9; **CGCG 158-10** S 1′1; mag 9.97v star S 2′7.
12 05 15.0	+38 14 08	UGC 7071	14.3	1.0 x 0.8	13.9	Sb II-III	144	
12 05 38.1	+33 05 52	UGC 7084	14.0	1.5 x 0.9	14.2	S	70	
12 05 45.1	+31 03 51	UGC 7085A	14.1	2.6 x 1.0	15.1	Sab	6	= **Arp 97**; multiple galaxy system, long bridges; MCG +5-29-12 E 1′3.
12 06 33.5	+39 14 13	UGC 7098	13.6	1.1 x 0.7	13.1	S?	89	Very faint, elongated anonymous galaxy NE 1′1.
12 07 18.6	+36 39 21	UGC 7105	14.0	1.2 x 0.7	13.6	SBa	160	
12 08 42.5	+36 48 13	UGC 7125	13.6	4.3 x 0.8	14.8	Sm	85	
12 09 11.7	+30 54 23	UGC 7131	14.5	1.7 x 0.5	14.2	Sdm:	25	
12 09 09.8	+31 34 10	UGC 7132	13.0	1.2 x 1.2	13.3	E		
12 09 51.0	+38 13 06	UGC 7145	14.4	1.5 x 0.5	13.9	Scd:	153	
12 10 08.7	+39 03 08	UGC 7159	14.6	0.8 x 0.8	14.0	Scd:		
12 10 53.9	+39 45 22	UGC 7175	14.4	1.9 x 0.5	14.2	Sdm:	75	Faint, small, round anonymous galaxy S edge.
12 11 21.0	+35 50 39	UGC 7187	14.1	1.2 x 0.5	13.4	SBa:	102	
12 11 32.0	+29 05 19	UGC 7190	13.7	1.1 x 0.9	13.5	S0?	30	Almost stellar nucleus; mag 15.6 star on NW edge; mag 9.51v star W 4′1.
12 12 19.3	+37 00 46	UGC 7207	15.3	2.3 x 1.1	16.2	Im	120	Dwarf irregular, very low surface brightness; mag 9.22v star SW 2′1.
12 12 28.0	+39 06 37	UGC 7208	13.5	1.3 x 1.2	13.8	S?	71	Almost stellar galaxy **KUG 1210+393** S 2′2.
12 12 39.9	+34 41 25	UGC 7212	14.7	1.3 x 0.2	13.1	Scd:	42	Pair with MCG +6-27-28 N edge.
12 12 40.5	+40 33 43	UGC 7213	14.4	1.1 x 1.0	14.4	SBd	51	Mag 10.75v star ESE 2′2.
12 14 23.1	+32 24 51	UGC 7243	13.4	0.9 x 0.6	12.7	S0	25	MCG +6-27-33 W 2′0; **MCG +6-27-35** N 2′2.
12 15 02.8	+35 57 31	UGC 7257	13.5	1.2 x 0.8	13.3	Sdm:	160	Knot N end; mag 12.1 star NNW 2′5.
12 20 16.0	+33 39 41	UGC 7388	15.1	1.2 x 0.8	14.9	SB	39	
12 20 27.8	+31 10 15	UGC 7395	14.1	0.5 x 0.4	12.2	S?	159	
12 21 39.4	+40 50 53	UGC 7416	12.9	1.6 x 1.4	13.6	SB(r)b I	3	Strong dark patches N and S of nucleus.
12 22 02.7	+32 05 31	UGC 7428	13.6	1.2 x 1.1	13.8	Im:	12	Faint core offset N; mag 9.98v star E 1′4.
12 27 05.3	+37 08 33	UGC 7559	13.9	3.2 x 2.0	15.7	IBm V	135	Dwarf irregular, very low surface brightness.
12 28 30.8	+32 32 49	UGC 7598	14.5	0.7 x 0.7	13.6	SBcd:		
12 28 28.4	+37 14 02	UGC 7599	14.5	1.8 x 0.9	14.9	Sm V	135	Mag 14.1 star S edge.
12 28 38.7	+35 43 03	UGC 7605	14.4	1.5 x 1.0	14.7	Im	30	Dwarf irregular, very low surface brightness.
12 31 58.1	+29 42 32	UGC 7673	14.8	1.8 x 1.5	15.7	Im V	44	Dwarf irregular, low surface brightness.
12 32 00.6	+39 49 55	UGC 7678	12.9	1.5 x 0.9	13.1	SB?	90	
12 32 29.1	+39 35 22	UGC 7689	14.8	1.7 x 0.2	13.5	Scd:	64	**MCG +7-26-22** SE 5′0, between a close pair of mag 9.50v stars.
12 32 53.2	+31 32 30	UGC 7698	12.6	6.5 x 4.7	16.2	Im V	177	Dwarf irregular, extremely low surface brightness.
12 32 48.4	+37 37 19	UGC 7699	12.5	3.6 x 1.0	13.7	SBcd:	32	Mag 12.4 star E 1′6.
12 34 00.8	+39 01 11	UGC 7719	14.4	1.7 x 0.6	14.3	Sdm	163	
12 35 09.0	+29 44 39	UGC 7750	13.9	1.5 x 0.6	13.6	S?	144	
12 36 22.3	+40 00 18	UGC 7774	14.2	3.4 x 0.5	14.6	Sd	99	Mag 13.8 star S edge; mag 12.1 star E 4′1.
12 37 55.8	+34 01 19	UGC 7799	14.5	1.5 x 0.3	13.5	SB?	123	**MCG +6-28-16** W 0′9; **MCG +6-28-14** E 1′1.
12 38 49.0	+31 59 03	UGC 7811	13.6	1.3 x 0.6	13.2	S?	111	Mag 11.2 star S 2′3.
12 38 48.0	+32 05 32	UGC 7812	13.3	0.9 x 0.7	12.6	S?	39	Faint dark patch NW edge.
12 38 57.2	+38 05 22	UGC 7816	14.5	1.7 x 1.1	15.0	Pec	66	Strongly peculiar, streamers; **PGC 42288** N edge; mag 14.4 star S edge.
12 40 56.3	+29 27 56	UGC 7836	14.1	1.5 x 0.5	13.7	Scd:	42	Stellar **CGCG 159-53** NNE 5′1.

(Continued from previous page)

GALAXY CLUSTERS

RA h m s	Dec ° ′ ″	Name	Mag 10th brightest	No. Gal	Diam ′	Notes
11 16 30.0	+29 15 00	A 1213	14.5	51	22	Main concentration of brighter PGC and anonymous GX near center of cluster area.
11 21 30.0	+34 19 00	A 1228	13.8	50	50	Many anonymous GX along with scattering of IC, MCG and UGC GX.
11 26 06.0	+35 19 00	A 1257	15.0	42	16	Many members concentrated N and NE of cluster center.
11 30 00.0	+36 40 00	A 1275	15.7	45	16	Mkn 424 NE of center 7′0, 2′7 N of a mag 10.78v star; other members anonymous and stellar.
11 39 24.0	+32 24 00	A 1336	16.0	46	21	All members anonymous, faint and stellar.
11 44 24.0	+30 54 00	A 1365	15.7	51	15	**Mkn 1453** SE of center 9′4, 2′8 N of mag 9.66v star; all other members anonymous and stellar.

OPEN CLUSTERS

RA h m s	Dec ° ′ ″	Name	Mag	Diam ′	No. ★	B ★	Type	Notes
12 35 00.0	+36 23 00	Up 1		18	10		ast	

55

GALAXIES

55

RA h m s	Dec ° ′ ″	Name	Mag (V)	Dim ′ Maj x min	SB	Type Class	PA	Notes
11 09 12.9	+31 32 46	CGCG 155-78	14.6	0.9 x 0.6	13.8		54	Pair of mag 13-14 stars ENE 5′1.
11 13 26.9	+30 36 25	CGCG 156-27	14.2	0.7 x 0.6	13.1	S?	33	Mag 15.2 star superimposed on NE edge.
09 57 16.1	+33 58 32	IC 2521	15.0	0.5 x 0.5	13.4	Spiral		
09 57 32.8	+33 37 09	IC 2524	14.0	0.6 x 0.4	12.3	S?	63	Mag 10.55v star S 2′9.
10 01 31.0	+37 12 14	IC 2530	14.3	0.6 x 0.6	13.0	S?		UGC 5391 NE 3′3.
10 04 31.9	+38 00 17	IC 2535	13.6	0.8 x 0.6	12.7	S?	90	Mag 6.77v star W 7′6.
10 07 50.8	+34 18 54	IC 2542	13.9	1.0 x 0.7	13.4	Sb	177	Mag 7.56v star ESE 11′8.
10 16 06.2	+38 06 31	IC 2557	14.8	0.4 x 0.3	12.3		93	Stellar, mag 9.77v star W 4′1.
10 19 08.6	+34 40 22	IC 2561	14.0	1.1 x 0.6	13.4	S?	14	Mag 13.8 star on N edge; mag 10.68v star SW 3′1.
10 22 19.5	+36 34 54	IC 2566	14.0	1.0 x 0.6	13.3	S?	3	Mag 10.09v star WSW 7′9; IC 2568 ENE 2′4.
10 22 30.1	+36 35 54	IC 2568	14.1	1.2 x 0.6	13.6	SBa	97	Bright, elongated center, faint outer envelope; IC 2566 WSW 2′4.
10 28 01.5	+32 45 47	IC 2577	15.1	0.7 x 0.3	13.2	Sa	102	Mag 12.5 star SW 0′7.
10 36 38.8	+35 03 10	IC 2591	13.5	1.4 x 0.7	13.3	S?	128	Mag 8.86v star NW 5′3.
10 49 24.6	+32 46 19	IC 2604	14.1	1.3 x 0.9	14.1	SB(s)m pec?	40	Mag 12.3 star S 1′1; mag 10.63v star SW 8′8.
10 50 17.8	+37 57 20	IC 2606	14.0	0.7 x 0.5	12.7	S?	120	IC 2607 N 2′3.
10 50 19.0	+37 59 36	IC 2607	14.8	0.7 x 0.4	13.2	SBab	66	Mag 10.59v star N 2′0.
11 02 05.8	+38 47 17	IC 2616	14.8	0.3 x 0.2	11.6			Stellar; mag 10.18v star E 13′0.
11 02 08.0	+38 39 55	IC 2617	14.4	0.9 x 0.5	13.4			Mag 10.69v star S 5′6.
11 02 23.9	+38 30 19	IC 2620	14.2	1.0 x 0.9	14.0	S?	141	Mag 10.74v star E 4′4.
11 21 23.1	+34 21 20	IC 2738	14.4	0.7 x 0.7	13.5	S?		Mag 9.46v star N 3′5; **IC 2735** W 4′0.
11 21 42.9	+34 21 45	IC 2744	14.5	0.5 x 0.5	12.8	S0?		Mag 9.46v star NW 5′5; note: **CGCG 185-43** 2′8 N of this star.
11 22 03.2	+34 18 50	IC 2751	14.9	0.9 x 0.3	13.4	Sb	72	Mag 10.24v star SW 5′6; faint, round anonymous galaxy SE 2′0.
09 57 59.7	+32 14 22	MCG +5-24-5	14.9	0.7 x 0.4	13.4	SBc	165	= **Mkn 412**; mag 12.4 star ESE 0′7.
10 10 03.4	+32 04 07	MCG +5-24-22	14.6	0.9 x 0.4	13.3	Sab	3	
10 26 11.2	+32 05 58	MCG +5-25-10	14.5	0.8 x 0.5	13.6	E	160	
10 39 13.3	+31 06 48	MCG +5-25-30	15.8	0.8 x 0.5	14.6		9	Mag 13.2 star S 1′6.
10 45 24.4	+29 13 07	MCG +5-26-2	14.4	0.6 x 0.5	12.9	Sb	135	Mag 12.1 star SE 2′9.
10 47 43.5	+30 37 30	MCG +5-26-7	14.8	1.0 x 0.5	13.9		84	Galaxy pair; **MCG +5-26-10** ENE 2′2.
11 02 57.5	+31 23 31	MCG +5-26-37	14.2	0.5 x 0.4	12.3	Sb	15	Much fainter anonymous galaxy, of same size, SE 2′0.
11 04 58.3	+29 08 17	MCG +5-26-46	15.4	0.6 x 0.4	13.7		153	= **Mkn 36**.
11 04 58.4	+30 01 38	MCG +5-26-47	14.1	0.9 x 0.7	13.5	SBbc	6	**MCG +5-26-48** S 4′7.
11 05 46.2	+30 31 37	MCG +5-26-51	15.3	0.6 x 0.5	13.8		168	Located between a pair of stars, mags 11.6 and 12.4.
11 10 39.1	+31 39 20	MCG +5-27-5	14.3	0.7 x 0.7	13.6	E		
11 18 10.0	+30 24 25	MCG +5-27-49	14.6	0.8 x 0.5	13.4	SABb	33	
09 56 22.3	+37 01 12	MCG +6-22-34	14.9	0.8 x 0.2	12.8	Spiral	120	
09 57 22.0	+36 04 49	MCG +6-22-36	14.8	0.6 x 0.4	13.1		54	Mag 11.6 star W edge.
09 57 31.5	+36 05 29	MCG +6-22-37	15.5	0.4 x 0.3	13.1	S0	63	MCG +6-22-38 S 1′5; MCG +6-22-36 WSW 2′1.
09 57 33.7	+36 04 04	MCG +6-22-38	13.6	0.8 x 0.3	11.9		144	MCG +6-22-37 N 1′5; MCG +6-22-36 W 2′4.
09 57 44.4	+36 42 46	MCG +6-22-40	15.2	0.7 x 0.4	13.7		90	Mag 11.12v star NE 4′7.
09 57 53.0	+36 20 51	MCG +6-22-42	14.0	0.9 x 0.6	13.4	E/S0	177	Mag 9.63v star SW 9′9.
09 58 09.2	+36 07 28	MCG +6-22-44	14.9	0.8 x 0.3	13.3		147	Mag 13.4 star WSW 1′0.
09 58 31.9	+36 15 13	MCG +6-22-45	15.0	0.6 x 0.6	13.8	Sb		Mag 12.4 star NW 2′7.
09 59 52.6	+37 54 29	MCG +6-22-48	15.7	0.6 x 0.6	14.4	S?		Mag 11.4 star SE 4′8.
10 01 12.5	+36 51 05	MCG +6-22-51	14.2	0.8 x 0.7	13.4	S0/a	30	Mag 11.7 star W 1′4.
10 01 26.6	+36 40 15	MCG +6-22-52	14.9	0.7 x 0.5	13.6	Sb	150	Almost stellar nucleus; mag 10.54v star NE 4′8.
10 02 34.5	+34 56 09	MCG +6-22-57	14.7	0.8 x 0.7	14.0		147	Faint, slightly elongated anonymous galaxy E 1′9.
10 03 32.0	+33 45 21	MCG +6-22-61	14.9	0.6 x 0.5	13.5	Spiral	90	Mag 12.6 star W 2′2.
10 03 30.1	+37 27 31	MCG +6-22-62	15.7	0.5 x 0.3	13.5	S0	0	MCG +6-22-63 SE 1′2.
10 03 34.9	+37 26 45	MCG +6-22-63	14.7	0.7 x 0.5	13.5	S0	141	MCG +6-22-62 NW 1′2.
10 04 23.8	+37 49 03	MCG +6-22-64	14.9	0.6 x 0.4	13.2	Sbc	51	
10 04 30.3	+37 03 54	MCG +6-22-66	15.8	0.5 x 0.3	13.6		51	MCG +6-22-67 NE 1′8; mag 11.7 star NNW 1′0.
10 04 38.4	+37 04 36	MCG +6-22-67	14.6	0.4 x 0.4	12.5	Spiral		MCG +6-22-66 SW 1′8.
10 04 54.5	+37 21 29	MCG +6-22-68	14.1	0.9 x 0.6	13.3	Sb	111	Mag 14.0 star N edge.
10 04 50.8	+35 06 57	MCG +6-22-69	15.1	0.7 x 0.7	14.1			MCG +6-22-70 NE 2′3; mag 9.00v star W 3′5.
10 04 57.9	+35 08 39	MCG +6-22-70	14.8	0.8 x 0.6	13.8	Sb	120	MCG +6-22-69 SW 2′3.
10 06 40.6	+38 10 07	MCG +6-22-72	15.2	0.9 x 0.8	14.7	Sbc	33	Mag 14.3 star SW edge.
10 07 08.3	+34 59 19	MCG +6-22-73	14.7	0.8 x 0.4	13.3	Sbc	21	Mag 10.50v star SE 3′3.
10 13 05.9	+35 16 55	MCG +6-23-1	14.9	0.5 x 0.3	12.7	Sb?	99	Located in a small triangle of mag 12.9, 13.8 and 14.2 stars; MCG +6-23-2 N 1′9.
10 13 07.2	+35 18 46	MCG +6-23-2	15.3	0.8 x 0.3	13.6	Sb	75	MCG +6-23-1 S 1′9.
10 23 36.1	+37 38 13	MCG +6-23-10	15.5	0.7 x 0.6	14.4	Spiral	3	MCG +6-23-12 E 1′6.
10 23 44.1	+37 38 26	MCG +6-23-12	15.5	0.7 x 0.4	13.9		117	MCG +6-23-10 W 1′6.
10 25 23.4	+37 26 45	MCG +6-23-14	14.7	0.6 x 0.6	13.7	E		
10 25 52.1	+35 57 27	MCG +6-23-15	14.9	0.9 x 0.7	14.4	E/S0	27	
10 26 14.5	+37 27 33	MCG +6-23-16	15.0	0.8 x 0.5	13.9	Spiral	75	
10 43 11.3	+37 16 22	MCG +6-24-7	14.6	0.7 x 0.5	13.4		141	
10 47 04.8	+37 33 39	MCG +6-24-13	15.6	0.9 x 0.5	14.5	Sb	150	Mag 10.50v star SW 3′9.
10 50 10.8	+38 09 52	MCG +6-24-20	14.3	0.8 x 0.8	13.7	Sbc		Mag 11.3 and 12.8 double star N 0′9.
10 50 19.3	+33 37 39	MCG +6-24-22	14.6	0.8 x 0.4	13.2	S0	162	
10 52 58.9	+37 36 49	MCG +6-24-30	14.1	0.7 x 0.7	13.1	Sa		**CGCG 184-32** N 7′7.
10 53 13.4	+37 17 43	MCG +6-24-32	15.1	0.7 x 0.7	14.2	Spiral		Faint, elongated anonymous galaxy E edge.

RA h m s	Dec ° ′ ″	Name	Mag (V)	Dim ′ Maj x min	SB	Type Class	PA	Notes
10 53 38.3	+34 01 56	MCG +6-24-34	14.7	0.9 x 0.4	13.4	Spiral	66	Mag 11.7 star E 1′.5.
10 57 48.0	+36 15 43	MCG +6-24-38	15.8	0.9 x 0.4	14.6	NE	54	
10 57 42.6	+37 39 16	MCG +6-24-39	14.3	0.9 x 0.6	13.5	S0	150	Mag 9.11v star W 2′.9.
11 00 10.2	+35 19 18	MCG +6-24-42	14.2	0.9 x 0.4	13.0	S0/a	111	Mag 11.6 star E 2′.8.
11 02 51.6	+36 46 57	MCG +6-24-44	14.4	0.9 x 0.6	13.6	SBab	45	
11 05 09.2	+38 04 09	MCG +6-24-47	14.6	0.7 x 0.4	13.1	Sbc	132	**CGCG 185-3** SE 4′.2.
11 05 42.5	+35 07 02	MCG +6-25-1	14.6	0.6 x 0.4	12.9	S?	156	Mag 12.0 star E 3′.4.
11 06 06.6	+34 41 43	MCG +6-25-3	14.5	0.5 x 0.5	12.9			
11 06 58.0	+35 33 02	MCG +6-25-5	14.4	0.6 x 0.5	12.9		84	Faint star on E edge.
11 07 59.7	+36 52 14	MCG +6-25-6	13.7	0.9 x 0.8	13.1	SB?	30	Mag 11.7 star S 2′.9.
11 08 28.5	+33 38 49	MCG +6-25-8	15.4	0.7 x 0.4	13.9		6	Mag 14.1 star N edge; mag 10.95v star W 4′.6.
11 08 41.3	+36 09 37	MCG +6-25-9	14.2	0.6 x 0.6	13.1	E/S0		Mag 7.33v star S 9′.0.
11 09 49.2	+35 25 49	MCG +6-25-12	15.4	0.5 x 0.5	13.7			Mag 12.1 star NW 2′.4.
11 09 52.5	+37 00 04	MCG +6-25-14	14.5	0.7 x 0.3	12.7	Sab	36	NGC 3542 S 3′.3.
11 09 58.5	+37 08 07	MCG +6-25-15	14.6	0.8 x 0.4	13.2	SBa	27	Mag 10.87v star NE 3′.5.
11 11 18.2	+35 23 05	MCG +6-25-18	14.2	0.8 x 0.4	12.8	SBb	69	
11 13 31.1	+37 32 43	MCG +6-25-21	15.4	0.8 x 0.3	13.7	Sbc	42	
11 13 31.4	+35 53 14	MCG +6-25-22	15.5	0.6 x 0.3	13.4	Spiral	156	
11 14 15.3	+34 09 12	MCG +6-25-25	14.6	0.7 x 0.5	13.3	S0/a	21	
11 15 07.4	+36 32 22	MCG +6-25-29	15.3	0.7 x 0.3	13.5	Sb	72	
11 15 35.3	+36 30 28	MCG +6-25-31	14.6	0.5 x 0.5	13.0	S0/a		Faint star on W edge; mag 12.6 star W 2′.1.
11 16 33.1	+35 18 19	MCG +6-25-32	14.9	0.8 x 0.4	13.5	SBa	75	Mag 13.5 star S 1′.7.
11 17 00.4	+32 30 21	MCG +6-25-36	15.0	0.7 x 0.4	13.5		54	Mag 12.1 star WNW 2′.3.
11 17 00.1	+32 35 48	MCG +6-25-37	14.5	0.7 x 0.5	13.4	E/S0	51	Mag 13.7 star W 2′.3.
11 18 54.7	+36 40 26	MCG +6-25-42	15.5	0.5 x 0.4	13.6	Spiral?	159	
11 19 17.1	+34 07 40	MCG +6-25-43	14.6	0.9 x 0.5	13.7	S0	15	Mag 9.36v star SSE 4′.5.
11 21 22.8	+33 43 38	MCG +6-25-50	14.9	0.8 x 0.4	13.5		147	
11 21 30.7	+33 57 27	MCG +6-25-51	14.4	0.5 x 0.5	12.8	S0/a		
11 23 10.9	+34 29 19	MCG +6-25-59	14.8	0.9 x 0.2	12.8	Sb	15	UGC 6397 W 1′.8.
11 23 24.7	+33 49 40	MCG +6-25-61	14.8	0.4 x 0.4	12.7	S0		Mag 12.6 star NE 0′.5; mag 11.15v star S 1′.4.
10 07 23.1	+38 58 15	MCG +7-21-10	15.3	1.0 x 0.5	14.4	Sc	3	Mag 12.2 star E 1′.9.
10 12 43.0	+39 22 25	MCG +7-21-15	14.4	0.8 x 0.5	13.3	SBb	117	Mag 10.10v star N 0′.8; mag 9.98v star E 2′.0.
10 13 48.0	+38 40 27	MCG +7-21-19	15.0	0.5 x 0.4	13.1	SBb	90	NGC 3159 SSE 1′.5.
10 14 29.8	+39 30 02	MCG +7-21-29	14.5	0.8 x 0.6	13.6	S0	24	Mag 12.6 star on S edge.
10 14 45.5	+38 58 53	MCG +7-21-30	14.5	0.7 x 0.7	13.6			Mag 11.64v star SSE 1′.2.
10 15 01.7	+39 39 48	MCG +7-21-32	14.8	0.6 x 0.4	13.1	SBb	150	Mag 10.55v star SW 5′.0.
10 21 26.6	+38 17 46	MCG +7-21-44	14.9	0.6 x 0.4	13.2	Spiral	99	Elongated **MCG +7-21-45** NE 1′.1.
10 29 44.7	+40 29 36	MCG +7-22-10	15.1	0.9 x 0.3	13.6	Spiral	6	Mag 7.02v star ESE 4′.0.
10 33 50.5	+40 17 27	MCG +7-22-25	14.6	1.0 x 0.5	13.7	Sb	102	Mag 12.3 star N 3′.1.
10 35 21.8	+40 52 28	MCG +7-22-29	14.5	0.8 x 0.7	13.7		12	Mag 7.79v star W 1′.8.
10 38 06.7	+39 10 36	MCG +7-22-35	14.0	0.8 x 0.4	12.7	S?	72	
10 40 00.6	+39 07 19	MCG +7-22-39	13.8	0.8 x 0.5	12.6	SB?	144	Mag 13.5 star on W edge; **MCG +7-22-40** ESE 1′.3.
10 40 44.5	+39 04 26	MCG +7-22-43	14.2	0.6 x 0.4	12.5	Spiral	105	**MCG +7-22-42** SW 1′.1.
10 41 44.3	+40 02 26	MCG +7-22-44	14.6	0.8 x 0.7	13.8	Spiral	66	
10 42 53.6	+40 08 19	MCG +7-22-46	14.5	0.7 x 0.3	12.7	Sb	18	Mag 8.89v star NE 6′.1.
10 43 10.3	+39 02 19	MCG +7-22-47	14.3	0.5 x 0.4	12.4		171	Mag 10.71v star E 3′.7.
10 44 03.8	+38 39 08	MCG +7-22-50	15.0	0.9 x 0.3	13.4	Sbc	33	
10 44 33.7	+39 09 26	MCG +7-22-52	15.0	1.2 x 0.2	13.3	Sc	96	Mag 11.38v star NNW 4′.0.
10 44 39.2	+38 45 32	MCG +7-22-53	13.9	0.9 x 0.8	13.4	S?	174	
10 44 43.2	+38 18 36	MCG +7-22-54	14.0	0.7 x 0.5	12.8		45	Mag 10.7 star on E edge.
10 45 25.0	+39 09 49	MCG +7-22-57	14.0	0.7 x 0.7	13.1			MCG +7-22-58 on N edge; mag 10.51v star SE 3′.0.
10 45 24.7	+39 10 26	MCG +7-22-58	15.1	0.8 x 0.2	13.0	Sb	168	MCG +7-22-57 S edge.
10 46 33.2	+39 58 57	MCG +7-22-60	14.7	0.5 x 0.5	13.0			Mag 10.29v star SW 3′.4.
10 47 26.8	+38 56 10	MCG +7-22-61	14.7	0.5 x 0.5	13.0	Sbc		
10 47 38.2	+38 55 26	MCG +7-22-62	14.6	0.5 x 0.3	12.4	Sb	36	On S edge of UGC 5893.
10 48 28.2	+38 17 42	MCG +7-22-69	15.4	0.5 x 0.4	13.5	Spiral	24	Mag 12.2 star SW 2′.2.
10 53 41.9	+38 46 14	MCG +7-22-76	15.2	0.7 x 0.4	13.6	SBb	117	
10 58 28.6	+39 14 09	MCG +7-23-8	15.2	0.7 x 0.2	12.9	Sc	168	
11 03 53.7	+40 50 57	MCG +7-23-19	14.8	0.6 x 0.5	13.3	Ring	94	**Mayall's Object**. Multi-galaxy system.
11 04 42.7	+38 14 03	MCG +7-23-21	15.8	0.3 x 0.3	13.0	S		Mag 7.50v star N 0′.8; **UGC 6140** S 0′.8.
11 11 40.2	+39 22 25	MCG +7-23-29	14.8	0.9 x 0.3	13.3	Sb	165	Mag 11.7 star WSW 2′.7 with small, faint anonymous galaxy half way to the star.
11 14 35.0	+39 29 21	MCG +7-23-36	15.9	0.5 x 0.4	14.0	Sb	74	
11 21 16.8	+40 20 43	MCG +7-23-40	14.7	0.7 x 0.6	13.6		171	Mag 8.08v star SE 5′.6.
11 21 45.2	+38 27 06	MCG +7-23-41	15.2	0.8 x 0.4	13.8		27	Mag 11.9 star W 1′.7.
11 23 01.4	+38 31 08	MCG +7-24-1	15.7	0.5 x 0.3	13.5		15	Mag 10.09v star WNW 2′.4.
09 58 21.9	+32 22 09	NGC 3067	12.1	2.5 x 0.9	12.8	SAB(s)ab? III	105	Mag 9.77v star E 3′.9.
09 58 53.0	+31 37 11	NGC 3071	14.3	0.6 x 0.5	12.9	S0	177	Stellar **Mkn 413** NE 6′.9; mag 9.50v star NE 9′.4.
09 59 41.1	+35 23 34	NGC 3074	12.7	2.3 x 1.9	14.1	SAB(rs)c II	166	Several bright knots and/or superimposed stars near E and W edges.
10 02 30.9	+32 42 52	NGC 3099	14.0	0.4 x 0.4	11.9			**MCG +6-22-58** ESE 1′.3.
10 03 57.4	+40 45 26	NGC 3104	13.0	3.3 x 2.2	15.0	IAB(s)m	35	Faint star and knot? inline S of nucleus.
10 04 05.4	+31 11 06	NGC 3106	12.4	1.8 x 1.8	13.5	S0		Small bright nucleus in smooth, faint envelope.
10 06 45.2	+31 05 50	NGC 3116	14.5	0.4 x 0.4	12.6	E		Almost stellar; mag 8.12v star NW 10′.7.
10 07 11.7	+33 01 41	NGC 3118	13.5	2.5 x 0.4	13.4	Sbc	41	**UGC 5446** SW 9′.7.
10 08 20.8	+31 51 44	NGC 3126	12.8	2.8 x 0.5	13.0	Sb III	123	
10 13 26.5	+38 39 25	NGC 3150	14.6	0.8 x 0.7	13.8	Sb	173	
10 13 29.1	+38 37 09	NGC 3151	13.8	0.8 x 0.5	12.7	SA0:	168	Very small and faint anonymous galaxy, and star, E 0′.9.
10 13 34.0	+38 50 35	NGC 3152	14.2	0.8 x 0.7	13.5	(R′)SB0°:	75	
10 13 50.6	+38 45 51	NGC 3158	11.9	2.0 x 1.8	13.3	E3:	0	**MCG +7-21-25** ESE 2′.3; **MCG +7-21-27** E 3′.3.
10 13 53.0	+38 39 14	NGC 3159	13.6	1.2 x 0.9	13.8	E2 pec:	156	
10 13 54.8	+38 50 38	NGC 3160	14.1	1.3 x 0.3	13.0	S	140	Mag 11.5 star N 2′.4; NGC 3152 W 4′.1.
10 13 59.2	+38 39 24	NGC 3161	13.5	0.8 x 0.6	12.8	E2	2	Located between NGC 3163 and NGC 3159.
10 14 07.0	+38 39 07	NGC 3163	13.3	1.2 x 1.1	13.5	SA0$^-$:	30	Bright knot or star E edge of nucleus.
10 25 43.3	+39 38 45	NGC 3237	13.0	1.3 x 1.3	13.4	(R)SAB0°		Small, faint anonymous galaxy NW 2′.3.
10 29 19.8	+29 29 24	NGC 3254	11.7	4.5 x 1.4	13.6	SA(s)bc II	46	Close pair of mag 9.36v, 9.92v stars E 6′.2.
10 36 16.0	+37 19 28	NGC 3294	11.8	3.5 x 1.8	13.6	SA(s)c I-II	122	Very small, anonymous galaxy NE 3′.9.
10 37 37.8	+37 27 20	NGC 3304	13.4	1.7 x 0.6	13.3	SB(s)a?	158	Faint galaxy **KUG 1034+377** W 6′.6.
10 41 31.1	+37 18 46	NGC 3334	12.8	1.1 x 1.0	12.8	S0?	0	
10 44 22.9	+30 43 29	NGC 3350	14.3	0.6 x 0.6	13.1	S0		Pair of mag 9 -10 stars S 1′.5.
10 48 24.2	+34 42 34	NGC 3381	11.8	2.0 x 1.7	13.0	SB pec	60	Mag 9.62v star SW 9′.5.

(Continued from previous page)

RA h m s	Dec ° ′ ″	Name	Mag (V)	Dim ′ Maj x min	SB	Type Class	PA	Notes
10 49 49.8	+32 58 50	NGC 3395	12.1	2.1 x 1.1	12.8	SAB(rs)cd pec: III	50	Asymmetrical or disrupted appearance; located on S edge of NGC 3396 with some apparent overlap.
10 49 55.1	+32 59 25	NGC 3396	12.0	3.1 x 1.1	13.2	IBm pec	100	NGC 3395 overlapping on S edge.
10 51 20.6	+32 45 58	NGC 3413	12.2	1.7 x 0.8	12.4	S0 sp	178	Mag 7.69v star SE 12′.3; note: **PGC 32631** lies 3′.9 E of this star.
10 51 47.2	+32 53 57	NGC 3424	12.4	2.8 x 0.8	13.1	SB(s)b:?	112	Mag 12.3 star on S edge; NGC 3430 NE 5′.9.
10 52 11.6	+32 57 00	NGC 3430	11.5	4.0 x 2.2	13.8	SAB(rs)c II	30	Close pair of mag 7.31v, 9.35v stars NE 7′.3.
10 52 32.1	+36 37 24	NGC 3432	11.3	6.8 x 1.5	13.6	SB(s)m sp III	38	A mag 9.20v star lies NE of center 6′.6; UGC 5983 W of S end.
10 53 08.1	+33 54 34	NGC 3442	13.4	0.6 x 0.5	11.9	Sa?	30	Mag 10.67v star E 6′.8.
10 57 31.2	+40 56 45	NGC 3468	13.0	1.3 x 0.7	12.8	S0	8	Very faint, small anonymous galaxy SW 1′.8.
11 09 15.8	+36 01 13	NGC 3540	13.3	1.4 x 1.2	13.7	SB0	63	Mag 7.33v star W 8′.2.
11 09 55.6	+36 56 46	NGC 3542	14.2	0.8 x 0.3	12.5	S?	47	MCG +6-25-14 N 3′.3.
11 10 13.4	+36 57 57	NGC 3545A	13.8	0.5 x 0.5	12.4	E		Nuclei with NGC 3545B separated by only 0′.25.
11 10 12.2	+36 57 52	NGC 3545B	13.8	0.5 x 0.5	12.4	E		On W edge of NGC 3545A.
11 12 08.2	+35 27 04	NGC 3569	13.3	1.1 x 1.0	13.3	SA0°	78	Small, faint anonymous galaxy 1′.8 SW.
11 22 31.3	+39 52 36	NGC 3648	12.6	1.3 x 0.8	12.5	S0	75	Mag 9.71v star NW 7′.0.
11 22 39.0	+37 46 00	NGC 3652	12.2	2.6 x 0.7	12.7	Scd?	150	
11 23 58.3	+38 33 45	NGC 3658	12.2	1.6 x 1.5	13.0	SA(r)0°:	27	
09 58 06.5	+37 17 34	UGC 5349	13.6	2.4 x 0.7	14.0	Sdm:	35	Mag 13.8 star NE end.
09 59 25.9	+30 44 43	UGC 5364	12.6	5.0 x 3.2	15.5	IBm V	102	Dwarf Irregular, very low surface brightness; mag 10.8 star NE 3′.4.
10 01 03.2	+36 37 06	UGC 5382	13.8	0.9 x 0.5	12.8	Sab	65	
10 01 37.7	+39 37 36	UGC 5389	14.9	1.8 x 0.2	13.6	Scd:	122	Mag 12.2 star S 1′.1.
10 01 41.3	+37 14 54	UGC 5391	13.9	2.1 x 0.8	14.3	Sm	170	Many knots; IC 2530 SW 3′.3.
10 01 42.3	+33 08 05	UGC 5393	13.9	2.1 x 1.2	14.8	SB(r)dm:	114	Mag 12.9 star NE 2′.1.
10 01 47.8	+36 29 53	UGC 5394	15.8	1.5 x 0.2	14.3	Scd:	56	Mag 8.13v star W 3′.1.
10 04 40.6	+29 21 50	UGC 5427	14.0	1.2 x 0.8	13.8	Sdm:	120	
10 09 01.9	+32 29 30	UGC 5474	13.9	1.1 x 1.0	13.9	SAB(rs)cd	84	Short N-S bar.
10 09 14.5	+40 21 02	UGC 5476	13.5	0.8 x 0.8	12.9	SABb		Bright, diffuse nucleus, faint envelope; mag 11.10v star on NE edge.
10 09 31.7	+30 09 00	UGC 5478	14.2	1.7 x 1.5	15.1	Im V	153	Dwarf irregular, very low surface brightness.
10 09 51.5	+30 19 12	UGC 5481	13.9	1.5 x 0.5	13.4	Sa:	91	Bright, elongated core; close pair of mag 12.9 and 13.7 stars N edge.
10 11 30.4	+30 47 25	UGC 5490	13.7	1.2 x 0.6	13.2	Scd:	10	Uniform surface brightness; mag 9.12v star W edge.
10 14 10.9	+39 27 10	UGC 5518	15.9	1.4 x 1.0	16.1	Im	60	Dwarf irregular, extremely low surface brightness; mag 10.57v star SE 2′.9.
10 16 22.0	+37 46 47	UGC 5540	13.9	1.5 x 0.3	12.9	Scd?	111	Stellar galaxy **KUG 1013-381** N 8′.0.
10 18 48.6	+38 28 12	UGC 5563	13.7	1.1 x 0.4	12.6	S?	3	**CGCG 211-41** SSE 6′.5.
10 20 04.2	+38 36 57	UGC 5577	13.4	1.0 x 0.9	13.1	S?	100	**CGCG 211-45** SE 9′.7.
10 23 41.8	+33 46 22	UGC 5622	14.3	0.8 x 0.6	13.4	SAB(s)bc	165	Mag 5.52v star 28 Leonis Minoris 6′.6 SE.
10 23 48.8	+33 48 28	UGC 5623	15.8	0.7 x 0.5	14.5	Scd:	126	Mag 5.52v star 28 Leonis Minoris 6′.8 SE.
10 26 49.8	+34 55 08	UGC 5656	14.0	1.4 x 0.7	13.8	SAB(s)bc	135	
10 34 29.9	+35 15 21	UGC 5738	13.3	0.9 x 0.5	12.3	S?	30	
10 36 33.8	+38 26 18	UGC 5759	13.2	1.0 x 0.8	12.9	S0⁻:	25	
10 36 43.5	+31 32 51	UGC 5764	14.7	1.8 x 1.1	15.3	IB(s)m: V	60	Dwarf irregular, very low surface brightness.
10 40 17.9	+38 29 12	UGC 5804	14.0	1.3 x 0.8	13.9	SABb II-III	110	Mag 13.4 star S edge.
10 40 38.4	+37 19 59	UGC 5806	14.8	0.8 x 0.8	14.2	Scd:		Mag 10.86v star W 2′.8.
10 41 08.7	+36 22 19	UGC 5813	14.2	1.7 x 0.7	14.2	Scd:	35	
10 41 50.9	+38 42 56	UGC 5819	13.7	1.5 x 0.3	12.7	Sbc	120	Very faint, elongated anonymous galaxy NW end; another very faint, elongated anonymous galaxy E 1′.2.
10 42 41.7	+34 27 00	UGC 5829	13.5	4.5 x 4.0	16.5	Im V	3	Dwarf irregular, low surface brightness, a few knots; mag 11.1 star N 4′.5.
10 43 25.6	+40 46 23	UGC 5838	13.3	1.2 x 0.5	12.6	SBb	130	
10 43 29.8	+39 41 14	UGC 5839	13.8	1.1 x 0.3	12.5	S0/a	147	
10 45 06.4	+38 59 00	UGC 5861	14.8	1.6 x 0.2	13.4	Scd:	147	
10 45 46.7	+37 12 41	UGC 5868	13.9	1.3 x 1.1	14.1	Sb II-III	24	**UGC 5871** SE 2′.2.
10 45 59.5	+34 57 50	UGC 5870	13.1	1.0 x 1.0	13.0	S0?		Multi-galaxy system.
10 47 05.8	+30 03 12	UGC 5885	14.1	1.5 x 0.6	13.8	Sb II-III	138	Mag 14.1 star superimposed N edge.
10 47 40.4	+38 55 53	UGC 5893	13.1	1.0 x 0.7	12.8	E?	108	MCG +7-22-62 on S edge; **MCG +7-22-65** ESE 1′.5; **MCG +7-22-64** N 1′.3.
10 47 50.5	+33 43 45	UGC 5898	15.4	1.5 x 0.4	14.7	Sdm:	158	**CGCG 184-15** E 4′.4.
10 48 27.9	+38 23 50	UGC 5910	13.4	1.9 x 1.6	14.5	SB(r)b	123	Small nucleus in short, weak SE-NW bar.
10 49 59.0	+31 54 33	UGC 5934	14.1	1.7 x 1.0	14.5	SB(s)dm	30	Several bright knots or superimposed stars.
10 50 07.2	+36 20 31	UGC 5936	13.2	1.2 x 0.9	13.2	(R)SA0⁺:	83	Mag 12.2 star SW 2′.9.
10 52 17.0	+36 35 36	UGC 5983	14.7	0.9 x 0.7	14.0	Dwarf	105	NGC 3432 W.
10 52 16.0	+30 03 45	UGC 5984	13.7	2.4 x 1.8	15.1	S?	45	Double system, strongly distorted, consists of **MCG +5-26-24** and **MCG +5-26-25**.
10 52 38.5	+34 29 00	UGC 5990	14.1	1.4 x 0.4	13.4	Sab	14	
10 52 57.5	+40 22 42	UGC 5996	15.6	0.6 x 0.6	14.3	Sdm		Mag 13.0 star on N edge.
10 53 10.3	+37 25 29	UGC 6002	14.3	1.1 x 0.9	14.1	SB(s)b		
10 55 00.7	+29 32 34	UGC 6031	14.8	1.1 x 0.7	14.4	S	141	Mag 12.4 star SSE 2′.0.
10 55 55.3	+36 51 41	UGC 6036	13.7	1.5 x 0.3	12.7	Sab	101	Mag 9.74v star NNW 3′.2.
10 59 46.3	+33 23 32	UGC 6070	12.8	0.5 x 0.5	11.1	S?		
11 00 52.4	+38 06 15	UGC 6089	14.4	0.9 x 0.7	13.7	Scd:	138	
11 03 25.6	+39 15 12	UGC 6121	14.4	1.5 x 0.4	13.7	S?	97	
11 04 27.5	+38 12 28	UGC 6132	12.8	0.7 x 0.6	11.7	S?	104	Located 2′.2 S of mag 6.03v star.
11 05 04.7	+35 21 55	UGC 6143	14.4	1.5 x 0.4	13.7	Sb: III	12	Faint, small, elongated anonymous galaxy S 4′.9.
11 06 07.1	+29 55 53	UGC 6152	14.2	1.2 x 0.5	13.5	SBb	154	Mag 14.4 star S edge.
11 07 59.1	+35 27 46	UGC 6183	13.9	1.1 x 0.6	13.3	SABb	5	Faint, elongated anonymous galaxy SSW 4′.0.
11 09 26.0	+29 34 07	UGC 6198	13.6	1.0 x 0.7	13.1	S0?	121	
11 11 00.0	+29 20 10	UGC 6220	14.9	0.7 x 0.3	13.1	Double System	66	**PGC 33971** SW 0′.9; mag 11.05v star E 1′.1.
11 14 37.1	+30 18 49	UGC 6271	13.7	1.1 x 0.5	12.9	Sa	52	
11 14 51.2	+35 30 08	UGC 6273	13.5	1.5 x 0.3	12.5	Sab	29	Mag 9.77v star S 5′.1.
11 14 56.6	+33 49 30	UGC 6274	13.9	0.9 x 0.7	13.3	Scd:	145	
11 15 10.2	+31 02 01	UGC 6276	12.8	1.5 x 1.1	13.3	S0	9	
11 15 22.0	+35 30 03	UGC 6279	13.8	1.3 x 0.6	13.4	Sa	7	**MCG +6-25-27** on S edge.
11 16 28.0	+29 19 35	UGC 6292	13.6	1.2 x 0.7	13.3	SB(s)b	66	Faint anonymous galaxy SE 1′.4; mag 10.84v star SE 3′.1; **PGC 34386** S 1′.6; **PGC 34394** S 2′.4.
11 17 03.4	+36 08 26	UGC 6298	14.1	1.0 x 0.5	13.2	S?	75	Bright, narrow E-W core.
11 17 33.8	+36 03 54	UGC 6303	14.0	1.2 x 1.0	14.0	Scd:	98	Star or bright knot N of nucleus.
11 17 40.7	+38 02 59	UGC 6307	14.2	1.3 x 0.5	13.6	Sdm:	165	
11 18 11.8	+30 23 41	UGC 6314	14.2	1.1 x 0.9	14.0	S?	6	
11 18 32.2	+37 56 50	UGC 6326	15.0	0.9 x 0.9	14.6	Sd-dm		Mag 10.46v star E 2′.3.
11 19 54.5	+33 05 23	UGC 6337	14.1	0.8 x 0.6	13.1	S?	108	Mag 3.49v star Nu Ursae Majoris W 17′.9.
11 20 00.8	+36 06 00	UGC 6338	14.2	1.0 x 0.5	13.3	Scd:	150	
11 20 40.2	+31 13 17	UGC 6355	14.0	1.9 x 0.3	13.3	Sd	102	Mag 13.3 star W end; almost stellar anonymous galaxy S 0′.8; small, faint anonymous galaxy SE 2′.6.
11 21 04.9	+31 15 06	UGC 6367	13.7	0.8 x 0.7	12.9	SB(s)ab	20	

(Continued from previous page)

GALAXIES

RA h m s	Dec ° ′ ″	Name	Mag (V)	Dim ′ Maj x min	SB	Type Class	PA	Notes
11 22 53.6	+34 20 27	UGC 6393	13.9	0.9 x 0.6	13.1	SBa	40	Bright SE-NW bar.
11 22 56.7	+34 06 41	UGC 6394	13.3	1.1 x 1.0	13.4	E:	60	Quartet of stellar anonymous galaxies NE 2′.3.
11 23 02.1	+34 29 51	UGC 6397	14.0	1.7 x 0.3	13.2	Sab	0	MCG +6-25-59 E 1′.8.

GALAXY CLUSTERS

RA h m s	Dec ° ′ ″	Name	Mag 10th brightest	No. Gal	Diam ′	Notes
10 32 06.0	+40 12 00	A 1035	15.4	94	18	**MCG +7-22-19** N of center 4′.7; most other members anonymous and stellar.
10 48 30.0	+31 28 00	A 1097	16.0	38	34	All members faint, anonymous and stellar.
11 11 42.0	+39 34 00	A 1187	15.6	55	20	Most members anonymous and stellar to almost stellar.
11 16 30.0	+29 15 00	A 1213	14.5	51	22	Main concentration of brighter PGC and anonymous GX near center of cluster area.
11 21 30.0	+34 19 00	A 1228	13.8	50	50	Many anonymous GX along with scattering of IC, MCG and UGC GX.

GALAXIES

RA h m s	Dec ° ′ ″	Name	Mag (V)	Dim ′ Maj x min	SB	Type Class	PA	Notes
08 40 02.3	+29 48 59	CGCG 150-14	15.0	0.5 x 0.4	13.1	Radio galaxy	36	Mag 10.59v star ENE 7′.6.
08 52 08.3	+29 41 35	CGCG 150-43	15.3	0.4 x 0.4	13.2			Located 2′.9 NE of mag 12.0 star.
08 56 29.2	+29 40 01	CGCG 150-48	15.4	0.4 x 0.4	13.2			Located 1′.4 SSE of mag 12.6 star.
08 56 47.7	+29 55 05	CGCG 150-50	15.5	0.5 x 0.3	13.3		114	Mag 11.0 star NNW 8′.1; **CGCG 150-52** ESE 9′.5.
09 00 13.4	+31 59 52	CGCG 150-56	13.9	0.6 x 0.3	11.9	SBa	162	Located 2′.6 SSE of mag 11.9 star.
09 28 11.4	+30 13 55	CGCG 152-10	15.6	0.4 x 0.4	13.4			
09 50 04.1	+31 44 41	CGCG 152-70	14.0	0.5 x 0.3	11.8	Sa	147	Mag 14.5 star SE 0′.6.
09 48 05.2	+32 52 54	CGCG 182-20	14.0	0.6 x 0.4	12.3		171	= **Mkn 408**.
09 09 41.9	+37 36 02	IC 527	13.3	1.2 x 0.9	13.2	S?	66	Mag 9.15v star E 11′.3; **CGCG 180-50** N 6′.6.
09 45 00.2	+29 27 08	IC 558	13.6	0.8 x 0.8	13.0	S?		Mag 9.92v star NNE 9′.0.
08 38 34.0	+30 47 53	IC 2387	14.0	1.1 x 0.5	13.2	Scd:	18	Located 2′.9 SE of mag 9.87v star.
08 47 59.2	+38 04 07	IC 2400	14.3	0.7 x 0.3	12.5	Spiral	111	Mag 7.40v star W 14′.0.
08 48 10.4	+37 45 16	IC 2401	13.8	1.2 x 0.7	13.5	S0:	101	Mag 9.73v star E 10′.1.
08 48 42.8	+37 13 03	IC 2405	14.5	0.8 x 0.5	13.4	Spiral	3	Mag 10.10v star SW 2′.5.
08 54 21.6	+32 40 47	IC 2421	13.3	2.0 x 1.5	14.3	SA(rs)c I	147	Mag 7.40v star NE 8′.7. MCG +6-20-13A SSE 3′.2.
09 03 14.7	+30 35 28	IC 2428	13.8	1.9 x 0.4	13.3	Scd:	75	Mag 10.30v star ENE 8′.3.
09 07 16.0	+37 12 55	IC 2434	13.4	1.5 x 0.7	13.4	SB?	13	Mag 8.18v star SE 10′.2.
09 08 38.4	+32 35 34	IC 2439	13.9	1.2 x 0.4	13.0	S0?	30	Mag 6.48v star SW 7′.8.
09 12 51.0	+30 12 44	IC 2444	13.8	0.6 x 0.6	12.5	S?		Mag 8.68v star NW 8′.3.
09 13 12.5	+31 48 23	IC 2445	14.2	0.7 x 0.4	12.6	SB?	16	Mag 11.2 star W 10′.6.
09 19 58.0	+37 11 29	IC 2461	14.0	2.3 x 0.4	13.8	Sb II-III	143	Mag 9.52v star SSW 4′.4.
09 24 52.8	+38 21 03	IC 2467	13.9	0.7 x 0.7	13.0			IC 2468 ESE 1′.8.
09 25 01.5	+38 20 37	IC 2468	15.1	0.4 x 0.4	13.0			MCG +7-20-8 N 2′.5.
09 27 23.4	+30 26 27	IC 2473	13.0	1.6 x 1.3	13.6	SB(r)bc	18	Mag 9.04v star SSW 10′.4.
09 27 52.9	+29 59 07	IC 2476	12.9	1.2 x 0.9	12.8	S0⁻:	21	Mag 11.13v star W 3′.8; **IC 2479** E 2′.5; **IC 2478** NNE 3′.6; **CGCG 152-2** W 5′.9; **CGCG 152-3** WNW 6′.5.
09 33 03.6	+29 55 40	IC 2490	13.4	1.3 x 1.0	13.5	Sbc III	175	Mag 10.14v star W 7′.0.
09 35 14.3	+34 43 51	IC 2491	13.7	1.1 x 0.8	13.5	S0?	75	Mag 7.63v star NE 6′.9.
09 36 17.7	+37 21 47	IC 2493	13.9	0.6 x 0.4	12.3	S0	0	Mag 13.1 star NW 1′.4.
09 42 23.3	+36 20 56	IC 2500	14.3	1.1 x 0.5	13.5	Sab	165	Mag 10.72v star W 9′.6.
09 54 48.5	+37 41 11	IC 2516	14.1	1.0 x 0.8	13.7	S0/a	18	Mag 10.92v star E 8′.1.
09 57 16.1	+33 58 32	IC 2521	15.0	0.5 x 0.5	13.4	Spiral		
09 57 32.8	+33 37 09	IC 2524	14.0	0.6 x 0.4	12.3	S?	63	Mag 10.55v star S 2′.9.
10 01 31.0	+37 12 14	IC 2530	14.3	0.6 x 0.6	13.0	S?		UGC 5391 NE 3′.3.
08 36 01.6	+30 16 00	MCG +5-20-29	14.9	0.4 x 0.3	12.6	E		
09 13 37.3	+29 59 56	MCG +5-22-20	14.5	0.4 x 0.2	11.6	S?	66	Superimposed on NGC 2783.
09 15 03.6	+29 16 09	MCG +5-22-25	14.6	0.9 x 0.7	14.0	SBc	159	Located between a pair of NE-SW mag 12.8, 13.8 stars.
09 15 28.1	+31 48 23	MCG +5-22-27	14.1	0.7 x 0.5	12.9	SBb	81	Mag 15.1 star S edge; mag 11.1 star E edge.
09 34 41.1	+32 04 05	MCG +5-23-17	14.2	0.8 x 0.4	12.8	S?		Mag 13.5 star W 1′.2.
09 46 34.8	+30 39 10	MCG +5-23-37	14.3	0.8 x 0.5	13.2	SBb	174	Faint, elongated anonymous galaxy WNW 1′.4.
09 53 31.2	+30 06 47	MCG +5-24-1	16.2	0.7 x 0.7	15.3			Mag 11.4 star NW 4′.0.
09 57 59.7	+32 14 22	MCG +5-24-5	14.9	0.7 x 0.4	13.4	SBc	165	= **Mkn 412**; mag 12.4 star ESE 0′.7.
08 40 36.4	+36 54 13	MCG +6-19-13	14.3	0.5 x 0.3	12.1		81	**UGC 4520** E 2′.0.
08 42 57.5	+37 32 47	MCG +6-19-17	15.4	0.6 x 0.4	13.7	Sbc	6	Mag 13.2 star NW 3′.1.
08 45 27.5	+34 25 08	MCG +6-19-20	13.9	0.7 x 0.4	12.4	Spiral	111	Mag 10.72v star E 2′.9.
08 46 17.9	+37 41 46	MCG +6-19-22	15.5	0.5 x 0.5	13.8	Sc		Mag 9.94v star N 2′.0.
08 48 12.5	+36 46 47	MCG +6-20-3	14.8	0.6 x 0.4	13.1	Irr	18	Mag 11.42v star W 2′.2.
08 48 03.4	+37 39 41	MCG +6-20-4	15.3	0.5 x 0.4	13.4	Sbc	27	Mag 11.2 star E 1′.9.
08 54 29.6	+32 38 05	MCG +6-20-13A	14.6	0.7 x 0.3	12.8	S?	42	Mag 7.40v star NE 8′.8;mag 14.7 star on S edge; IC 2421 NNW 3′.2.
09 01 14.3	+34 48 36	MCG +6-20-20	15.2	0.8 x 0.7	14.4		9	Star superimposed E edge?
09 03 09.1	+34 05 42	MCG +6-20-21	15.3	0.5 x 0.4	13.4	Spiral	12	
09 07 15.4	+37 30 18	MCG +6-20-26	14.2	0.8 x 0.6	13.3	S0 ·	147	Elongated **MCG +6-20-27** on E edge.
09 08 26.4	+35 37 59	MCG +6-20-30	15.4	0.4 x 0.4	13.3	Spiral		MCG +6-20-31 SSE 1′.4.
09 08 29.9	+35 36 52	MCG +6-20-31	14.5	0.6 x 0.4	12.9	S0	177	Mag 9.18v star SE 3′.6.
09 09 19.6	+33 07 18	MCG +6-20-36	15.7	0.7 x 0.3	13.8	Double	106	NGC 2770 E 3′.0.
09 09 33.4	+32 30 24	MCG +6-20-37	14.6	0.8 x 0.6	13.7	Sc	15	Mag 10.9 star W 1′.9.
09 12 36.2	+35 47 19	MCG +6-20-46	15.0	0.6 x 0.5	13.5		55	
09 13 45.3	+34 50 13	MCG +6-20-48	14.7	0.7 x 0.4	13.2	S?	54	Mag 13.1 star ENE 0′.8.
09 14 27.6	+35 48 16	MCG +6-20-49	14.4	0.9 x 0.7	13.7		162	Elongated **UGC 4865** S 5′.4.
09 14 26.3	+36 06 40	MCG +6-20-50	13.9	0.9 x 0.7	13.3	SB?	33	**UGC 4866** E 2′.4.
09 18 18.9	+33 33 24	MCG +6-21-5	14.7	0.7 x 0.6	13.6		78	
09 19 36.2	+33 25 40	MCG +6-21-12	14.5	0.7 x 0.7	13.6	S?		Mag 11.0 star W 4′.6.
09 20 00.9	+33 00 09	MCG +6-21-18	14.8	0.6 x 0.6	13.5			Mag 10.62v star N 2′.7.
09 20 45.9	+33 42 15	MCG +6-21-24	15.1	0.6 x 0.4	13.4	S?	126	Mag 12.1 star NNW 2′.1.
09 23 32.9	+33 45 16	MCG +6-21-29	14.4	0.9 x 0.7	13.8	S0	141	Mag 14.4 star NE 3′.2.
09 24 46.4	+33 46 44	MCG +6-21-32	14.5	0.8 x 0.7	13.7	S0/a	66	MCG +6-21-33 SE 3′.1; mag 14.4 star W 0′.9.
09 24 55.3	+33 44 11	MCG +6-21-33	15.1	0.8 x 0.6	14.2	SBb	48	MCG +6-21-32 NW 3′.1.
09 25 57.8	+34 37 47	MCG +6-21-36	15.9	0.6 x 0.5	14.4	Sc	10	UGC 5020 N 1′.6.
09 27 25.0	+33 41 55	MCG +6-21-38	14.6	0.8 x 0.4	13.2	Sc	165	Mag 12.8 star NW 2′.9.
09 29 38.5	+35 20 26	MCG +6-21-40	14.7	0.8 x 0.4	13.3		99	Mag 11.4 star W 1′.4 with MCG +6-21-41 just N of this star.

(Continued from previous page)

GALAXIES

RA h m s	Dec ° ′ ″	Name	Mag (V)	Dim ′ Maj x min	SB	Type Class	PA	Notes
09 29 31.5	+35 21 44	MCG +6-21-41	16.0	0.8 x 0.4	14.6	Sb	153	Mag 11.4 star on S edge.
09 30 03.4	+34 13 25	MCG +6-21-42	14.8	0.6 x 0.6	13.7	E		A mag 14.5 star paired with a faint anonymous galaxy E 1′.2.
09 30 15.6	+36 41 31	MCG +6-21-43	14.5	0.8 x 0.4	13.1	SBab	84	
09 31 55.2	+35 33 15	MCG +6-21-44	14.6	0.7 x 0.6	13.5	Sa	51	Forms a triangle with a mag 11.1 star NE 2′.2 and a mag 10.00v star SE 2′.6.
09 32 05.5	+35 32 16	MCG +6-21-45	14.6	0.7 x 0.3	12.8	Sb	132	Mag 11.17v star on S edge; mag 10.00v star S 1′.0.
09 33 16.8	+33 54 17	MCG +6-21-46	14.3	0.7 x 0.6	13.2	Sbc	171	Mag 9.61v star NW 1′.3.
09 33 34.3	+33 38 53	MCG +6-21-48	14.1	0.6 x 0.6	13.1	E		**MCG +6-21-49** SSE 3′.5.
09 33 54.4	+34 03 44	MCG +6-21-50	15.3	0.9 x 0.3	13.7	Spiral	21	Mag 11.27v star SW 3′.1.
09 34 00.9	+33 59 28	MCG +6-21-51	13.8	0.9 x 0.7	13.3	E	171	Mag 13.7 star w 1′.1.
09 35 15.3	+33 59 05	MCG +6-21-52	14.4	1.0 x 0.7	13.9	S0/a	144	Mag 14.3 star NE 3′.0.
09 37 02.6	+35 42 22	MCG +6-21-58	15.7	0.8 x 0.4	14.3	Spiral	72	Mag 7.74v star W 4′.9.
09 37 25.7	+35 14 23	MCG +6-21-61	15.0	0.5 x 0.3	12.7	Sc	33	**CGCG 181-72** E 2′.6.
09 38 54.1	+34 17 59	MCG +6-21-64	14.2	0.8 x 0.5	13.1	Sab	66	Mag 13.1 star SE 0′.9; mag 11.9 star SE 3′.4.
09 39 17.4	+36 33 42	MCG +6-21-68	14.4	0.8 x 0.7	13.7	S?	75	MCG +6-21-69 NE 1′.2; mag 14.1 star W 1′.2.
09 39 22.4	+36 34 26	MCG +6-21-69	14.6	0.7 x 0.5	13.3	S?	30	MCG +6-21-68 SE 1′.2.
09 39 28.8	+33 50 53	MCG +6-21-70	14.5	0.7 x 0.4	12.9	Spiral	141	Mag 12.9 star SE 3′.3.
09 43 26.5	+36 14 13	MCG +6-22-4	15.4	0.4 x 0.4	13.2	Sc		Almost stellar, located 1′.6 E of NGC 2965.
09 44 31.3	+32 48 42	MCG +6-22-6	14.3	0.9 x 0.4	13.1	S0	81	
09 44 52.4	+37 36 13	MCG +6-22-7	14.6	1.1 x 0.3	13.2	Spiral	21	Mag 11.9 star WSW 1′.4.
09 45 52.3	+34 41 06	MCG +6-22-9	14.0	0.8 x 0.3	12.3	S?	126	Mag 10.7 star NW 1′.2.
09 46 20.1	+34 47 41	MCG +6-22-10	14.9	0.8 x 0.3	13.2	Spiral	9	Faint anonymous galaxy E 1′.4.
09 47 23.2	+34 46 59	MCG +6-22-11	14.7	0.9 x 0.7	14.0	Spiral	132	
09 47 41.2	+35 03 37	MCG +6-22-12	14.0	1.1 x 0.4	13.0	Sab	144	Mag 11.9 star NW 3′.3.
09 49 36.6	+34 26 13	MCG +6-22-16	15.7	0.9 x 0.3	14.1	Spiral	99	Mag 10.9 star ENE 2′.6.
09 51 23.5	+35 45 17	MCG +6-22-22	15.7	0.5 x 0.4	13.8	Spiral	105	
09 53 18.7	+36 05 07	MCG +6-22-24	13.9	0.8 x 0.7	13.1	S?	3	Mag 12.3 star N 2′.1.
09 54 12.8	+34 08 29	MCG +6-22-26	15.7	0.7 x 0.5	14.4	SBc	99	Mag 10.4 star NW 5′.3.
09 55 43.3	+37 00 32	MCG +6-22-30	14.6	0.8 x 0.3	12.9	SBb	78	Mag 11.9 star N 0′.6.
09 55 52.5	+35 57 55	MCG +6-22-32	15.2	0.9 x 0.7	14.5	SBbc	48	Strong NE-SW core, two main arms.
09 56 22.3	+37 01 12	MCG +6-22-34	14.9	0.8 x 0.2	12.8	Spiral	120	
09 57 22.0	+36 04 49	MCG +6-22-36	14.8	0.6 x 0.4	13.1		54	Mag 11.6 star W edge.
09 57 31.5	+36 05 29	MCG +6-22-37	15.5	0.4 x 0.3	13.1	S0	63	MCG +6-22-38 S 1′.5; MCG +6-22-36 WSW 2′.1.
09 57 33.7	+36 04 04	MCG +6-22-38	13.6	0.8 x 0.3	11.9		144	MCG +6-22-37 N 1′.5; MCG +6-22-36 W 2′.4.
09 57 44.4	+36 42 46	MCG +6-22-40	15.2	0.7 x 0.4	13.7		90	Mag 11.12v star NE 4′.7.
09 57 53.0	+36 20 51	MCG +6-22-42	14.0	0.9 x 0.6	13.4	E/S0	177	Mag 9.63v star SW 9′.9.
09 58 09.2	+36 07 28	MCG +6-22-44	14.9	0.8 x 0.3	13.3		147	Mag 13.4 star WSW 1′.0.
09 58 31.9	+36 15 13	MCG +6-22-45	15.0	0.6 x 0.6	13.8	Sb		Mag 12.4 star NW 2′.7.
09 59 52.6	+37 54 29	MCG +6-22-48	15.7	0.6 x 0.6	14.4	S?		Mag 11.4 star SE 4′.8.
10 01 12.5	+36 51 05	MCG +6-22-51	14.2	0.8 x 0.7	13.4	S0/a	30	Mag 11.7 star W 1′.4.
10 01 26.6	+36 40 15	MCG +6-22-52	14.9	0.7 x 0.5	13.6	Sb	150	Almost stellar nucleus; mag 10.54v star NE 4′.8.
10 02 34.5	+34 56 09	MCG +6-22-57	14.7	0.8 x 0.7	14.0		147	Faint, slightly elongated anonymous galaxy E 1′.9.
10 03 32.0	+33 45 21	MCG +6-22-61	14.9	0.6 x 0.5	13.5	Spiral	90	Mag 12.6 star W 2′.2.
10 03 30.1	+37 27 31	MCG +6-22-62	15.7	0.5 x 0.3	13.5	S0	0	MCG +6-22-63 SE 1′.2.
10 03 34.9	+37 26 45	MCG +6-22-63	14.7	0.7 x 0.5	13.5	S0	141	MCG +6-22-62 NW 1′.2.
08 37 57.7	+39 53 44	MCG +7-18-31	14.7	1.0 x 0.2	12.8		84	Triangle of mag 13-14 stars W and NW 1′.4.
08 41 41.2	+40 39 24	MCG +7-18-36	13.9	0.6 x 0.5	12.7	E	18	Pair with MCG +7-18-37 S on edge.
08 41 40.9	+40 39 01	MCG +7-18-37	14.2	0.3 x 0.3	11.4	Sb		Pair with MCG +7-18-36 N edge.
08 45 05.7	+40 33 32	MCG +7-18-44	14.8	0.9 x 0.3	13.2	Spiral	150	Mag 13.4 star SW 1′.9.
08 47 45.7	+39 32 10	MCG +7-18-47	14.5	1.0 x 0.5	13.6		57	Mag 8.04v star NE 4′.8.
08 48 59.8	+41 00 08	MCG +7-18-50	14.1	0.8 x 0.7	13.3		12	
08 49 31.0	+39 12 35	MCG +7-18-53	14.7	0.9 x 0.7	14.0	Spiral	93	Mag 9.81v star NE 6′.5.
08 49 45.5	+40 14 45	MCG +7-18-54	15.3	0.8 x 0.4	13.9	Spiral	30	Mag 10.9 star S 2′.0.
08 50 11.4	+40 15 18	MCG +7-18-55	14.0	0.8 x 0.6	13.1		81	Mag 13.0 star SW 1′.6.
08 50 37.9	+40 15 01	MCG +7-18-57	14.7	0.7 x 0.4	13.2	Sc	39	Pair of stars, mags 13.3 and 10.68v, E 0′.9.
08 52 41.7	+40 41 02	MCG +7-18-59	14.4	0.9 x 0.6	13.6	Spiral	141	Mag 11.1 star S 2′.0.
08 56 32.8	+39 08 24	MCG +7-19-4	14.1	0.9 x 0.7	13.4	Sb	105	
09 02 52.5	+40 19 29	MCG +7-19-13	15.1	0.5 x 0.5	13.5	Sbc		
09 03 02.3	+40 25 57	MCG +7-19-14	14.5	0.6 x 0.6	13.2	Spiral		**MCG +7-19-15** ENE 3′.9.
09 04 48.1	+40 35 28	MCG +7-19-21	15.4	0.7 x 0.7	14.4			Mag 9.27v star W 3′.8.
09 21 01.8	+39 09 23	MCG +7-19-62	14.8	0.7 x 0.7	13.9	Sb		Mag 10.03v star SW 2′.5.
09 23 41.2	+40 49 59	MCG +7-19-67	15.0	0.9 x 0.3	13.4	Sb	105	Mag 9.84v star SE 3′.8.
09 24 14.0	+40 38 11	MCG +7-20-1	15.3	0.5 x 0.4	13.4		3	MCG +7-20-2 SE 1′.7.
09 24 22.0	+40 37 24	MCG +7-20-2	15.0	0.9 x 0.3	13.4	Sb	126	MCG +7-20-1 NW 1′.7; mag 13.1 star S 1′.2.
09 24 56.6	+40 23 57	MCG +7-20-6	15.0	0.7 x 0.3	13.1	Spiral	132	Mag 10.6 star NE 5′.1.
09 25 04.4	+38 23 03	MCG +7-20-8	15.2	0.6 x 0.4	13.5	SBb		Mag 12.4 star superimposed S edge; mag 11.15v star NNE 1′.1.
09 25 39.7	+40 38 46	MCG +7-20-9	15.0	0.7 x 0.3	13.2	Spiral	159	
09 40 30.2	+40 05 13	MCG +7-20-31	14.5	0.8 x 0.5	13.4		12	Mag 9.20v star ESE 3′.8.
09 43 17.0	+38 52 39	MCG +7-20-35	14.5	0.7 x 0.7	13.6			Mag 11.3 star WNW 1′.6.
09 44 44.3	+38 47 19	MCG +7-20-45	14.6	0.4 x 0.3	12.2		81	Mag 13.1 star S 2′.1.
09 47 50.7	+39 32 00	MCG +7-20-50	14.5	0.6 x 0.4	12.8	Spiral	114	Mag 11.7 star SW 2′.0.
09 49 53.2	+38 54 23	MCG +7-20-61	14.5	0.5 x 0.4	12.6		135	Mag 10.05v star W 2′.1.
08 42 25.8	+37 13 12	NGC 2638	12.8	1.7 x 0.6	12.6	S0/a	72	
08 44 08.4	+34 43 00	NGC 2649	12.3	1.6 x 1.5	13.1	SAB(rs)bc:	90	Very bright knot or superimposed star on N edge.
08 49 22.6	+36 42 35	NGC 2668	13.8	1.5 x 0.9	14.0	Sab	155	Tow main N-S arms; mag 9.40vb star N 7′.4.
08 51 32.9	+30 51 52	NGC 2679	12.6	1.6 x 1.3	13.3	SB0:	3	Mag 10.10v star N 5′.1.
08 52 40.2	+33 25 01	NGC 2683	9.8	9.3 x 2.1	12.8	SA(rs)b II-III	44	
08 54 46.5	+39 32 18	NGC 2691	13.1	0.9 x 0.5	12.1	Sa?	165	
08 56 47.9	+39 22 54	NGC 2704	13.4	1.0 x 1.0	13.2	SB(r)ab		**UGC 4689** NE 12′.4.
09 00 15.6	+35 43 40	NGC 2719	13.1	1.3 x 0.3	11.9	Im pec?	133	Located on N edge of NGC 2719A.
09 00 15.8	+35 43 10	NGC 2719A	13.9	0.6 x 0.5	12.4	Im pec:	135	Located on S edge of NGC 2719.
09 01 01.7	+35 45 38	NGC 2724	13.6	1.7 x 1.5	14.5	SAB(s)c	2	Almost stellar nucleus; mag 10.36v star SW 3′.0.
09 05 59.7	+35 22 40	NGC 2746	13.1	1.6 x 1.5	13.9	SB(rs)a	123	Almost stellar nucleus in short N-S bar; mag 12.3 star on N edge.
09 08 37.6	+37 37 16	NGC 2759	13.0	1.2 x 0.9	13.0	S0$^-$:	50	**MCG +6-20-29** SSW 5′.7; **UGC 4799** S 7′.4; **MCG +6-20-35** SSE 9′.3.
09 08 47.8	+29 51 52	NGC 2766	13.6	1.3 x 0.5	13.0	Sab	132	
09 09 34.1	+33 07 16	NGC 2770	12.2	3.8 x 1.0	13.5	SA(s)c:	148	MCG +6-20-36 W 3′.0.
09 12 24.4	+35 01 40	NGC 2778	12.4	1.4 x 1.0	12.8	E	40	Mag 9.41v star W 8′.1; note: **UGC 4834** is 3′.9 S of this star.
09 12 28.4	+35 03 11	NGC 2779	15.0	0.7 x 0.6	13.9	SBa	161	Located 1′.9 N of NGC 2778.
09 12 44.5	+34 55 33	NGC 2780	13.4	0.9 x 0.7	12.7	SB?	150	Close pair of mag 12.8, 13.7 stars W 1′.9.

(Continued from previous page)

GALAXIES

RA h m s	Dec ° ′ ″	Name	Mag (V)	Dim ′ Maj x min	SB	Type Class	PA	Notes
09 14 09.1	+40 06 54	NGC 2782	11.6	3.5 x 2.6	13.9	SAB(rs)a pec	74	The bright portion of the galaxy is offset to the W. A faint, wide "arm" extends E then curves N to fill out the published size for this galaxy.
09 13 39.6	+29 59 35	NGC 2783	12.7	2.1 x 1.5	13.9	E	168	The small galaxy MCG +5-22-20 lies NW of the center and within NGC 2783's envelope. UGC 4856 lies slightly further W.
09 15 15.1	+40 55 06	NGC 2785	14.1	1.5 x 0.5	13.7	Im?	120	
09 14 59.8	+29 43 46	NGC 2789	12.2	1.5 x 1.3	12.8	S0/a	20	Bright core, faint envelope; **CGCG 151-30** NW 7′.8.
09 16 46.4	+34 25 54	NGC 2793	13.0	1.3 x 1.1	13.2	SB(s)m pec	63	Small anonymous galaxy 1′.5 NW.
09 16 41.8	+30 54 57	NGC 2796	13.5	1.2 x 0.8	13.3	Sa?	80	Small anonymous galaxies lie 0′.9 ESE and 1′.1 SW.
09 19 17.3	+34 00 27	NGC 2823	14.6	0.9 x 0.5	13.5	SBa	30	Lies 1′.1 NNE of mag 10.44v star.
09 19 22.5	+33 44 34	NGC 2825	14.4	0.9 x 0.4	13.1	Sa: sp	83	Small, faint anonymous galaxy N 3′.1.
09 19 24.4	+33 37 26	NGC 2826	13.7	1.5 x 0.3	12.7	S0$^+$: sp	143	NGC 2829 NE 2′.0.
09 19 19.1	+33 52 53	NGC 2827	14.8	0.8 x 0.3	13.1	Sb	6	Mag 10.34v star NNE 4′.8.
09 19 34.8	+33 53 17	NGC 2828	15.0	0.5 x 0.3	12.8		48	Almost stellar; mag 10.34v star N 4′.1.
09 19 30.3	+33 38 52	NGC 2829	16.0	0.3 x 0.2	12.8		9	Almost stellar; small triangle of faint stars on E edge, star closest to E edge could be companion galaxy?
09 19 41.6	+33 44 17	NGC 2830	14.3	1.3 x 0.3	13.1	SB0/a: sp	112	On the W edge of NGC 2832.
09 19 45.4	+33 44 40	NGC 2831	13.3	0.5 x 0.5	11.9	E0		Located on SW edge of NGC 2832's bright core.
09 19 46.9	+33 44 56	NGC 2832	11.9	3.0 x 2.0	13.8	E$^+$2:	160	NGC 2831 on SW edge of core; NGC 2830 on W.
09 19 58.0	+33 55 37	NGC 2833	14.5	0.9 x 0.3	12.9		168	Mag 10.34v star W 6′.1.
09 20 02.5	+33 42 38	NGC 2834	14.5	0.6 x 0.5	13.3	E/S0	65	Almost stellar; mag 10.4 star W 2′.2.
09 20 43.1	+39 18 54	NGC 2838	13.7	0.9 x 0.9	13.6	E/S0		
09 20 36.4	+33 39 02	NGC 2839	14.2	0.9 x 0.9	14.1	E?		Almost stellar; MCG +6-21-24 NNE 3′.7.
09 20 52.8	+35 22 06	NGC 2840	13.8	1.0 x 0.9	13.5	SB(rs)bc	110	Mag 11.8 star on NW edge.
09 21 48.1	+40 09 01	NGC 2844	12.9	1.5 x 0.7	12.9	SA(r)a:	13	Mag 7.67v star NE 8′.0.
09 23 14.8	+40 09 47	NGC 2852	13.2	1.0 x 0.6	12.5	SAB(r)a?	165	Mag 10.3 star W 3′.1.
09 23 17.5	+40 11 55	NGC 2853	13.3	1.0 x 0.6	12.6	SB0:	25	Mag 14.9 star on SE edge; mag 7.67v star W 9′.7.
09 24 18.3	+34 30 53	NGC 2859	10.9	4.6 x 4.1	14.0	(R)SB(r)0$^+$	85	Annular around bright core.
09 30 17.0	+29 32 24	NGC 2893	13.2	1.1 x 1.0	13.2	(R)SB0/a	79	3′.1 SW of mag 9.15v star.
09 35 44.2	+31 42 19	NGC 2918	12.6	1.4 x 1.0	12.9	E	65	Almost stellar **CGCG 152-35** NE 10′.8.
09 36 53.2	+37 41 38	NGC 2922	14.0	1.2 x 0.5	13.3	Im?	103	Mag 9.23v star N 6′.5.
09 37 31.3	+32 50 26	NGC 2926	13.5	1.0 x 0.8	13.1	S?	120	Mag 9.56v star N 4′.4.
09 39 08.0	+34 00 22	NGC 2942	12.6	1.9 x 1.5	13.6	SA(s)c: I-II	165	Stellar nucleus; mag 10.7 star SSE 5′.3; mag 9.14v star SW 12′.0.
09 39 17.5	+32 18 38	NGC 2944	14.0	1.0 x 0.4	12.9	SB(s)c pec?	96	Small galaxy **KUG 0936+325B** located S of E end of NGC 2944.
09 41 16.6	+35 52 55	NGC 2955	12.9	1.7 x 0.9	13.2	(R′)SA(r)b II	162	
09 42 54.3	+31 50 48	NGC 2964	11.3	2.9 x 1.6	12.8	SAB(r)bc: II-III	97	**Mkn 404** is located E of center; mag 10.51v star S 4′.9.
09 43 19.0	+36 14 53	NGC 2965	13.4	1.2 x 0.9	13.4	S0	64	Mag 11.21v star NNE 7′.4; MCG +6-22-4 E 1′.6; stellar **CGCG 181-84** WSW 5′.5.
09 43 12.1	+31 55 40	NGC 2968	11.7	2.3 x 1.6	13.0	I0	45	Star or small galaxy superimposed on SW edge.
09 43 31.2	+31 58 35	NGC 2970	13.6	0.6 x 0.5	12.3	E1:	54	Mag 11.7 star W 2′.9.
09 43 46.2	+36 10 46	NGC 2971	14.0	1.1 x 0.8	13.7	SB(r)b	135	Mag 11.5 star W 6′.8; NGC 2965 and MCG +6-22-4 NW 6′.8.
09 44 56.6	+31 05 49	NGC 2981	13.6	1.2 x 1.0	13.6	SAB(rs)bc:	77	Mag 12.4 star SE 1′.1.
09 48 35.3	+33 25 13	NGC 3003	11.9	5.8 x 1.3	13.9	Sbc? III-IV	79	
09 49 41.2	+32 13 13	NGC 3011	13.3	0.9 x 0.8	12.8	S0	69	A mag 10.23v star lies 2′.5 NE.
09 49 52.1	+34 42 49	NGC 3012	13.5	1.0 x 1.0	13.6	E:		Faint, stellar anonymous galaxy W 3′.6.
09 50 09.5	+33 34 09	NGC 3013	14.9	0.8 x 0.5	13.7		91	Located 2′.7 SE of a mag 7.90v star.
09 50 57.4	+33 33 12	NGC 3021	12.1	1.6 x 0.9	12.4	SA(rs)bc: II	110	A mag 11.37v star at the SE end.
09 52 08.2	+29 14 08	NGC 3032	12.5	1.7 x 1.3	13.3	SAB(r)0°	95	Located between a mag 9.08v star 1′.8 N and a mag 10.96v star 1′.8 S.
09 58 21.9	+32 22 09	NGC 3067	12.1	2.5 x 0.9	12.8	SAB(s)ab? III	105	Mag 9.77v star E 3′.9.
09 58 53.0	+31 37 11	NGC 3071	14.3	0.6 x 0.5	12.9	S0	177	Stellar **Mkn 413** NE 6′.9; mag 9.50v star NE 9′.4.
09 59 41.1	+35 23 34	NGC 3074	12.7	2.3 x 1.9	14.1	SAB(rs)c II	166	Several bright knots and/or superimposed stars near E and W edges.
10 02 30.9	+32 42 52	NGC 3099	14.0	0.4 x 0.4	11.9			**MCG +6-22-58** ESE 1′.3.
10 03 57.4	+40 45 26	NGC 3104	13.0	3.3 x 2.2	15.0	IAB(s)m	35	Faint star and knot? inline S of nucleus.
08 37 26.6	+40 02 04	UGC 4498	13.8	1.0 x 0.6	13.1	SBa	13	Mag 11.0 star S 1′.3.
08 39 00.9	+40 20 43	UGC 4513	14.1	0.9 x 0.9	13.7	Compact		
08 41 53.2	+32 52 02	UGC 4531	13.8	1.1 x 0.4	12.7	SBb	37	Mag 11.0 star W 2′.6.
08 44 09.0	+33 30 59	UGC 4558	14.3	1.7 x 0.3	13.4	Sbc	14	
08 44 07.6	+30 07 09	UGC 4559	13.2	2.8 x 0.6	13.6	Sab	50	
08 45 38.0	+36 56 01	UGC 4572	13.2	0.6 x 0.6	11.9	S?		
08 48 50.5	+29 52 11	UGC 4611	14.8	1.7 x 0.2	13.5	Sd	104	
08 49 16.6	+36 07 08	UGC 4614	13.5	0.7 x 0.7	12.6	S?		Mag 10.73v star W 1′.6.
08 49 27.2	+29 31 12	UGC 4617	14.4	1.5 x 0.6	14.1	SAd	57	Located between a pair of stars, mags 11.3 and 12.4.
08 50 11.7	+35 04 36	UGC 4621	13.0	1.2 x 0.7	12.7	S?	140	Located NW of a mag 7.96v star.
08 51 54.6	+40 49 58	UGC 4635	14.3	1.2 x 0.8	14.1	Pec	36	Very faint plume extends N and then E from bright core.
08 53 38.1	+40 43 44	UGC 4649	14.0	0.8 x 0.7	13.2		25	Faint star E of nucleus.
08 53 37.5	+39 08 06	UGC 4650	13.4	1.7 x 0.7	13.4	Sab	90	Bright, elongated core; mag 14.3 star W end.
08 53 54.7	+35 08 45	UGC 4653	13.7	1.1 x 0.7	13.3	SB(s)b	3	Triple system, contact, plume NW; mag 9.88v star S 1′.5.
08 54 24.2	+34 33 22	UGC 4660	15.7	1.2 x 1.1	15.8	Sm:	132	
08 58 45.3	+39 30 33	UGC 4699	13.3	0.6 x 0.6	12.2	E?		Compact; located on NNW edge of small galaxy cluster **A 727**.
08 58 51.3	+38 48 32	UGC 4702	13.5	1.4 x 1.4	14.1	S0?		
08 59 02.5	+39 12 25	UGC 4704	14.4	3.7 x 0.4	14.7	Sdm:	115	Mag 11.1 star SE end.
09 05 45.2	+36 21 17	UGC 4767	13.1	1.3 x 1.1	13.4	S0	25	Bright core, smooth envelope; mag 9.42v star NE 3′.6.
09 06 40.0	+34 37 17	UGC 4777	14.3	2.2 x 0.4	14.0	Im:	140	Uniform surface brightness; mag 9.21v star SE 6′.1.
09 07 34.8	+33 16 26	UGC 4787	13.6	2.1 x 0.5	13.5	Sdm	6	
09 11 35.5	+32 51 02	UGC 4831	14.4	1.1 x 1.1	14.5	SABd		
09 12 09.5	+35 31 55	UGC 4837	14.5	1.9 x 1.0	15.1	Sm? V	160	Mag 7.55v star SW 4′.2.
09 13 32.8	+29 59 59	UGC 4856	14.3	1.5 x 0.2	12.8	Sb III	77	NGC 2783 and MCG +5-22-20 E end.
09 13 48.9	+29 10 24	UGC 4860	14.1	0.9 x 0.3	12.5	Double System	17	Very faint anonymous galaxy E 1′.2.
09 14 43.3	+40 52 49	UGC 4867	14.1	1.7 x 1.3	14.8	SAB(s)d	85	Mag 12.3 star N edge; mottled envelope; **UGC 4863** SSW 7′.1.
09 14 34.3	+30 08 25	UGC 4869	13.0	2.0 x 0.8	13.4	S0?	35	**MCG +5-22-23** NE 2′.7.
09 14 56.5	+39 15 42	UGC 4871	15.2	1.7 x 0.6	15.0	SB(s)m V	20	Mag 15.1 star NE edge; mag 12.4 star NW 2′.5.
09 15 01.6	+40 02 11	UGC 4872	14.8	1.7 x 0.2	13.5	SBb	12	Mag 10.61v star WNW 2′.5.
09 15 26.5	+31 24 03	UGC 4878	15.3	1.7 x 0.7	15.3	SBm	25	
09 16 33.4	+39 52 18	UGC 4889	14.1	1.5 x 1.4	14.8	S	66	Mag 11.6 star NE 2′.1.
09 18 35.8	+34 33 09	UGC 4926	14.4	1.5 x 0.3	13.4	S? III	154	Mag 11.6 star NE 2′.4; small, faint anonymous galaxy SW 0′.7.
09 20 08.9	+39 09 46	UGC 4950	14.1	0.9 x 0.5	13.1	Sa	62	Mag 10.26v star W 0′.9.
09 20 25.8	+36 14 44	UGC 4953	15.4	1.5 x 0.4	14.7	Sd:	122	Mag 11.4 star SE 2′.9.
09 21 45.6	+39 31 28	UGC 4970	14.9	1.5 x 0.2	13.5	Scd:	105	Mag 6.66v star NW 10′.2.
09 21 51.6	+33 24 05	UGC 4972	13.2	1.5 x 1.0	13.5	S0:	0	Mag 12.5 star NE 3′.0.
09 22 10.5	+33 50 57	UGC 4974	13.3	1.5 x 1.2	13.8	S0:	0	Faint, slightly elongated anonymous galaxy WNW 1′.8.
09 23 15.2	+34 44 02	UGC 4988	14.7	1.2 x 0.9	14.6	SABm	51	Mag 12.0 star NW 3′.5.
09 25 19.1	+34 06 47	UGC 5011	15.3	1.0 x 1.0	15.2	SBcd:		

(Continued from previous page)

56 GALAXIES 56

RA h m s	Dec ° ′ ″	Name	Mag (V)	Dim ′ Maj x min	SB	Type Class	PA	Notes
09 25 48.0	+34 16 38	UGC 5015	14.3	1.8 x 1.7	15.4	SABdm	20	Weak stellar nucleus, faint envelope; mag 10.2 star SW 4′.7.
09 26 01.4	+34 39 12	UGC 5020	14.3	2.1 x 0.5	14.2	Scd:	79	MCG +6-21-36 S 1′.6.
09 31 13.0	+30 01 15	UGC 5070	13.6	1.0 x 0.4	12.4	S?	45	Mag 12.4 star E 1′.6.
09 31 53.0	+29 47 33	UGC 5074	13.7	0.9 x 0.7	13.1	Sa?	36	
09 33 25.6	+34 02 50	UGC 5088	14.2	1.3 x 0.9	14.3	S0	126	Almost stellar nucleus.
09 35 26.3	+29 48 44	UGC 5108	13.6	1.3 x 0.6	13.2	SBab	134	
09 37 12.7	+38 05 28	UGC 5119	13.7	0.6 x 0.6	12.4	S0?		
09 39 25.5	+32 21 48	UGC 5146	15.2	1.1 x 0.8	14.9	SAB0°: pec	75	Double system, contact, disrupted.
09 43 02.3	+37 49 21	UGC 5184	14.0	1.1 x 0.7	13.6	SBb	79	Bright E-W bar, faint envelope.
09 43 33.0	+39 24 52	UGC 5193	13.8	1.1 x 0.8	13.7	E:	9	
09 45 13.4	+39 26 17	UGC 5212	15.1	0.9 x 0.4	13.8	Sc	81	Mag 12.9 star SW 2′.3.
09 50 22.1	+31 29 18	UGC 5272	14.1	2.0 x 0.7	14.3	Im IV-V	115	Bright knot or superimposed star NW end and S edge; mag 12.7 star SSE 2′.6.
09 50 42.4	+30 29 31	UGC 5276	13.6	1.0 x 0.6	12.9	S?	125	Mag 10.11v star SW 3′.5.
09 51 01.4	+30 35 38	UGC 5281	14.2	1.9 x 0.3	13.4	Double System	164	Double system, bridge; consists of **MCG +5-23-44**, **45**.
09 51 09.9	+33 07 48	UGC 5282	14.8	1.1 x 0.6	14.2	Sm:	54	Uniform surface brightness.
09 51 28.2	+32 56 34	UGC 5287	13.7	1.7 x 0.9	14.1	SB(s)cd	171	**UGC 5294** NE 12′.7.
09 52 04.2	+40 51 30	UGC 5290	13.8	1.0 x 0.6	13.1	S0/a?	0	Mag 10.90v star NW 1′.5.
09 55 25.2	+33 15 43	UGC 5326	13.8	1.0 x 0.8	13.4	Im	99	Very faint, stellar anonymous galaxy NE 2′.7.
09 58 06.5	+37 17 34	UGC 5349	13.6	2.4 x 0.7	14.0	Sdm:	35	Mag 13.8 star NE end.
09 59 25.9	+30 44 43	UGC 5364	12.6	5.0 x 3.2	15.5	IBm V	102	Dwarf Irregular, very low surface brightness; mag 10.8 star NE 3′.4.
10 01 03.2	+36 37 06	UGC 5382	13.8	0.9 x 0.5	12.8	Sab	65	
10 01 37.7	+39 37 36	UGC 5389	14.9	1.8 x 0.2	13.6	Scd:	122	Mag 12.2 star S 1′.1.
10 01 41.3	+37 14 54	UGC 5391	13.9	2.1 x 0.8	14.3	Sm	170	Many knots; IC 2530 SW 3′.3.
10 01 42.3	+33 08 05	UGC 5393	13.9	2.1 x 1.2	14.8	SB(r)dm:	114	Mag 12.9 star NE 2′.1.
10 01 47.8	+36 29 53	UGC 5394	15.8	1.5 x 0.2	14.3	Scd:	56	Mag 8.13v star W 3′.1.

GALAXY CLUSTERS

RA h m s	Dec ° ′ ″	Name	Mag 10th brightest	No. Gal	Diam ′	Notes
09 19 48.0	+33 46 00	A 779	13.8	32	50	Majority of members are almost stellar to stellar.

57 GALAXIES 57

RA h m s	Dec ° ′ ″	Name	Mag (V)	Dim ′ Maj x min	SB	Type Class	PA	Notes
07 21 05.8	+30 45 08	CGCG 147-10	15.9	0.5 x 0.5	14.3	Sc		Mag 9.81v stars WNW 9′.5; mag 10.34v star E 5′.9.
07 21 58.7	+29 29 47	CGCG 147-13	15.4	0.5 x 0.4	13.5	Sm	18	Mag 8.87v star NE 4′.8.
07 31 50.5	+30 32 04	CGCG 147-31	15.3	0.5 x 0.4	13.4		18	Mag 13.2 star NE 0′.7; mag 11.2 star WSW 2′.1.
07 38 32.7	+29 11 06	CGCG 147-48	15.4	0.4 x 0.2	12.5		168	Stellar.
07 39 52.2	+29 44 10	CGCG 147-51	15.9	0.4 x 0.4	13.8			Mag 11.8 star S 2′.9; mag 12.4 star W 3′.6.
07 45 43.6	+32 06 03	CGCG 148-9	15.3	0.9 x 0.4	14.0		6	Mag 9.74v star NNW 7′.0.
07 47 20.6	+29 24 42	CGCG 148-19	15.2	0.5 x 0.3	12.9		117	Mag 10.6 star E 2′.5.
07 47 29.5	+31 12 49	CGCG 148-21	15.3	0.5 x 0.4	13.4			Mag 12.1 star S 2′.9; mag 12.0 star NNW 4′.1.
08 08 04.2	+29 57 13	CGCG 148-115	15.6	0.4 x 0.3	13.1		18	Mag 10.77v star WNW 6′.4.
08 23 19.8	+29 27 31	CGCG 149-19	15.9	0.5 x 0.3	13.7		66	Mag 13.2 star on W end.
08 40 02.3	+29 48 59	CGCG 150-14	15.0	0.5 x 0.4	13.1	Radio galaxy	36	Mag 10.59v star ENE 7′.6.
08 07 07.4	+36 13 58	CGCG 178-28	14.1	0.7 x 0.4	12.8	E/S0	24	Mag 7.99v star E 8′.4.
07 47 09.3	+30 29 17	IC 475	14.5	0.8 x 0.5	13.4	S?	141	Mag 10.31v star SE 5′.5.
07 23 15.9	+32 29 42	IC 2185	14.4	0.6 x 0.4	12.7	S?	129	Pair of mag 10.66v, 10.72v stars SSW 4′.1.
07 29 54.3	+37 27 04	IC 2190	14.0	0.9 x 0.5	13.0	SBbc:	21	With three mag 12 stars, it is the E point of a N-S oriented four point, diamond shaped asterism.
07 33 23.8	+31 28 58	IC 2193	13.4	1.5 x 0.7	13.3	Sb II	87	Mag 13.8 star on N edge; mag 11.1 star SE 3′.5.
07 33 40.2	+31 19 59	IC 2194	13.9	0.9 x 0.3	12.3	Sb	48	Mag 8.91v star SE 10′.8.
07 34 09.7	+31 24 20	IC 2196	12.7	1.4 x 1.1	13.2	E	150	Mag 8.91v star S 10′.7.
07 34 55.6	+31 16 30	IC 2199	13.2	1.1 x 0.6	12.6	SB?	25	Mag 8.91v star WSW 7′.8.
07 36 17.1	+33 07 19	IC 2201	14.0	1.3 x 0.3	12.8	Sa	67	Mag 9.54v star NNE 6′.2.
07 40 33.7	+34 13 48	IC 2203	13.6	1.2 x 0.9	13.5	SBcd	162	Mag 10.32v star W 7′.2.
07 41 18.1	+34 13 53	IC 2204	15.0	1.1 x 0.9	14.8	(R)SB(r)ab	51	Bright center with strong dark patches NE and SW; mag 9.75v star SSE 1′.4.
07 49 51.3	+33 57 38	IC 2207	14.5	2.0 x 0.3	13.8	Scd:	124	Mag 9.27v star SW 10′.1.
07 57 45.7	+32 33 26	IC 2211	13.7	0.9 x 0.5	12.7	Sa	140	Mag 7.13v star NNW 6′.5.
07 58 57.3	+32 36 40	IC 2212	14.5	0.5 x 0.5	12.9			Almost stellar; mag 8.30v star NE 10′.0.
07 59 54.1	+33 17 23	IC 2214	13.6	0.8 x 0.7	12.8	SBab	51	Bright N-S bar; star on NE edge.
08 05 28.3	+35 56 49	IC 2225	13.9	0.9 x 0.7	13.2	S0/a	84	
08 28 23.8	+30 22 48	IC 2376	13.4	0.3 x 0.3	10.6			Numerous anonymous galaxies within 2′.5 N, E and S; mag 10.10v star WSW 11′.4.
08 28 22.2	+30 26 31	IC 2378	13.6	0.4 x 0.4	11.5	S?		Mag 10.48v star N 7′.1; **IC 2380** E 2′.2; **CGCG 147-27** and 149-28 N 1′.4; **IC 2374** SSE 2′.0.
08 35 10.8	+37 15 50	IC 2385	14.3	0.7 x 0.4	12.8	Spiral	18	Mag 12.8 star W 0′.7; mag 13.3 star NNE 0′.9.
08 38 34.0	+30 47 53	IC 2387	14.0	1.1 x 0.5	13.2	Scd:	18	Located 2′.9 SE of mag 9.87v star.
07 21 05.3	+29 20 15	MCG +5-18-7	14.1	0.9 x 0.6	13.3	Sb	51	Mag 10.7 star W 4′.5.
07 26 55.5	+30 08 46	MCG +5-18-13	15.3	0.5 x 0.4	13.4	Sb	141	
07 39 06.1	+29 09 35	MCG +5-18-26	14.7	0.6 x 0.6	13.5	Sb		Mag 10.04v star NW 2′.1.
07 49 35.6	+30 24 56	MCG +5-19-18	14.2	0.7 x 0.4	12.7		174	Mag 12.3 star SE 3′.2.
08 00 47.1	+29 44 51	MCG +5-19-30	14.6	0.7 x 0.4	13.0	SBc	129	Mag 11.3 star E 2′.0.
08 03 15.1	+30 47 41	MCG +5-19-36	14.0	0.8 x 0.5	12.9	S?	48	Mag 14.2 star on S edge.
08 04 23.9	+29 30 49	MCG +5-19-37	15.3	0.6 x 0.2	12.8	Sb	96	Pair with MCG +5-19-37 S.
08 04 25.0	+29 30 22	MCG +5-19-38	14.6	0.8 x 0.2	12.5	S0	92	Pair with MCG +5-19-37 N; mag 9.63v star S 1′.9.
08 12 05.8	+29 24 14	MCG +5-20-3	14.8	0.9 x 0.3	13.2	S0/a	15	Mag 11.5 star 0′.7 W; faint anonymous galaxy SW 1′.3.
08 29 30.1	+31 40 28	MCG +5-20-20	14.4	0.7 x 0.3	12.6	S?	81	Mag 11.1 star SE 2′.6.
08 35 08.7	+29 42 33	MCG +5-20-26	14.0	0.7 x 0.7	13.1	S?		
08 35 32.8	+30 32 01	MCG +5-20-28	14.6	0.5 x 0.3	12.4	S?	69	Mag 11.05v star E 1′.9.
08 36 01.6	+30 16 00	MCG +5-20-29	14.9	0.4 x 0.3	12.6	E		
07 16 04.2	+37 25 09	MCG +6-16-32	14.8	0.7 x 0.7	13.8	Sb		Mag 11.0 star NNW 1′.8.
07 22 15.6	+38 20 42	MCG +6-16-38	15.1	0.5 x 0.4	13.2		30	Close pair of mag 10 stars SE 2′.0.
07 28 49.8	+35 32 52	MCG +6-17-9	14.1	0.6 x 0.4	12.4	S?	78	
07 52 32.5	+33 28 52	MCG +6-18-1	15.3	0.7 x 0.5	14.0	Spiral	78	Mag 10.51v star SW 5′.1.
07 57 04.1	+33 04 56	MCG +6-18-2	13.8	0.7 x 0.7	12.9			Mag 14.9 star superimposed N edge; mag 11.8 star S 1′.3.
07 59 57.0	+35 48 51	MCG +6-18-6	14.1	0.8 x 0.8	13.4	Sb		Mag 10.9 star N 1′.6.
08 02 39.3	+34 46 32	MCG +6-18-8	14.5	0.8 x 0.6	13.6	S?	123	Located between a pair of N-S mag 12.7 and 13.0 stars.
08 03 29.4	+33 27 43	MCG +6-18-9	13.9	0.7 x 0.7	13.0	S0?		

GALAXIES

RA h m s	Dec ° ′ ″	Name	Mag (V)	Dim ′ Maj x min	SB	Type Class	PA	Notes
08 05 28.9	+37 33 27	MCG +6-18-11	14.4	0.6 x 0.4	12.7	S0/a	24	Mag 15.1 star W edge; **MCG +6-18-10** and mag 12.1 star S 1′.2.
08 19 22.0	+35 02 46	MCG +6-19-1	14.0	0.8 x 0.3	12.3	S?	108	Mag 7.17v star due E 2′.0.
08 22 36.1	+37 06 01	MCG +6-19-2	14.6	0.8 x 0.3	12.9	Sbc	144	
08 24 11.6	+37 51 10	MCG +6-19-3	14.7	0.5 x 0.4	12.8	Spiral	24	Mag 14.4 star N edge.
08 31 24.4	+35 51 32	MCG +6-19-8	15.0	0.7 x 0.3	13.2	Sc	120	Mag 11.1 star ESE 2′.9.
08 40 36.4	+36 54 13	MCG +6-19-13	14.3	0.5 x 0.3	12.1		81	**UGC 4520** E 2′.0.
08 42 57.5	+37 32 47	MCG +6-19-17	15.4	0.6 x 0.4	13.7	Sbc	6	Mag 13.2 star NW 3′.1.
07 22 11.7	+40 28 08	MCG +7-15-15	14.4	0.9 x 0.4	13.1	SBb	123	
07 46 18.9	+39 04 01	MCG +7-16-15	14.8	1.0 x 0.6	14.1	Sbc	48	Mag 12.0 star on S edge.
07 53 16.2	+39 02 57	MCG +7-16-22	14.4	0.8 x 0.8	13.8			
07 54 24.9	+39 22 16	MCG +7-16-23	14.0	0.7 x 0.5	12.7	S?	147	Located on the N side of a small triangle of mag 13 stars.
07 55 25.4	+39 11 08	MCG +7-17-1	14.4	0.8 x 0.7	13.6	S?	168	Mag 9.95v star SW 5′.4.
08 05 05.5	+40 10 28	MCG +7-17-12	14.7	0.8 x 0.7	14.0		6	Mag 10.7 star WNW 4′.7.
08 08 27.1	+39 57 49	MCG +7-17-17	14.0	0.7 x 0.3	12.2	S?	162	Mag 8.89v star S 1′.2.
08 12 00.8	+39 00 44	MCG +7-17-21	15.1	0.7 x 0.4	13.6	SBb	141	Mag 10.62v star N 3′.0.
08 17 49.6	+38 47 42	MCG +7-17-27	14.4	1.0 x 0.4	13.2	SBb	6	Located 1′.2 W of a mag 7.86v star.
08 18 42.0	+39 08 29	MCG +7-17-28	14.9	0.9 x 0.6	14.1		36	Mag 11.0 star S 2′.5.
08 21 10.1	+39 15 49	MCG +7-17-29	14.1	0.7 x 0.5	12.8		45	
08 26 00.7	+40 58 50	MCG +7-18-1	13.9	0.9 x 0.7	13.2		165	Mag 10.5 star NW 2′.4.
08 28 25.7	+40 37 46	MCG +7-18-5	14.7	0.7 x 0.4	13.1		117	Mag 12.4 star N 1′.4.
08 29 00.5	+39 01 11	MCG +7-18-8	14.8	0.7 x 0.5	13.5		147	Faint anonymous galaxy ESE 1′.7.
08 33 54.8	+39 47 52	MCG +7-18-23	15.0	0.7 x 0.5	13.8	Sd	135	
08 37 57.7	+39 53 44	MCG +7-18-31	14.7	1.0 x 0.2	12.8		84	Triangle of mag 13-14 stars W and NW 1′.4.
08 41 41.2	+40 39 24	MCG +7-18-36	13.9	0.6 x 0.5	12.7	E	18	Pair with MCG +7-18-37 S on edge.
08 41 40.9	+40 39 01	MCG +7-18-37	14.2	0.3 x 0.3	11.4	Sb		Pair with MCG +7-18-36 N edge.
07 26 37.0	+33 49 23	NGC 2373	13.8	0.6 x 0.5	12.3	S?	159	Mag 12.6 star on NE edge.
07 27 09.7	+33 49 54	NGC 2375	13.6	1.3 x 1.0	13.8	SB(s)b	170	Strong E-W core in faint N-S envelope.
07 27 26.4	+33 48 37	NGC 2379	13.5	0.9 x 0.8	13.0	SA0:	127	Mag 10.97v star NE 3′.5.
07 28 28.3	+33 50 17	NGC 2385	13.9	1.0 x 0.7	13.4	Sb	54	Mag 9.18v star NNW 6′.6.
07 28 58.1	+36 52 48	NGC 2387	15.2	0.6 x 0.5	13.7		39	Almost stellar; mag 9.46v star NE 6′.9.
07 28 53.4	+33 49 07	NGC 2388	13.8	1.0 x 0.6	13.1	S?	65	NGC 2389 NE 3′.4.
07 29 04.7	+33 51 39	NGC 2389	12.9	2.0 x 1.4	13.9	SAB(rs)c	83	Mag 11.17v star N 1′.6; NGC 2388 SW 3′.4.
07 30 04.9	+34 01 36	NGC 2393	14.0	1.2 x 0.8	13.8	Sc	103	
07 35 02.1	+32 49 15	NGC 2410	13.0	2.5 x 0.7	13.4	SBb?	31	
07 36 56.7	+35 14 28	NGC 2415	12.4	0.9 x 0.9	12.0	Im?		1′.8 SW of mag 9.52v star.
07 40 39.2	+39 13 57	NGC 2424	12.6	3.8 x 0.5	13.2	SB(r)b: sp II-III	81	Mag 7.97v star E 6′.7.
07 44 13.8	+31 39 04	NGC 2435	12.8	2.1 x 0.5	12.7	Sa	36	Mag 14.3 star on NE edge; mag 10.27v star WSW 4′.3.
07 46 53.0	+39 01 54	NGC 2444	13.2	1.6 x 0.9	13.5	Ring A	27	Pair with NGC 2445 S edge.
07 46 55.1	+39 00 39	NGC 2445	13.3	1.7 x 1.3	14.0	Ring B	18	Multi-galaxy system. Multiple bright knots or galactic cores. Pair with NGC 2444 N.
07 56 45.4	+39 55 40	NGC 2476	12.6	1.4 x 0.8	12.8	E?	153	
07 58 28.1	+37 47 11	NGC 2484	13.0	0.9 x 0.8	12.6	S0:	145	Mag 10.57v star E 7′.0.
08 00 23.8	+39 49 47	NGC 2493	12.0	1.9 x 1.9	13.2	SB0		NGC 2495 E 2′.0; mag 7.59v star NE 7′.9.
08 00 33.3	+39 50 25	NGC 2495	15.2	0.4 x 0.2	12.3	I?	5	Located 2′.0 E of NGC 2493; **CGCG 207-17** NE 2′.4.
08 08 09.7	+39 09 24	NGC 2524	12.7	1.4 x 1.0	12.9	S0/a	125	
08 07 25.1	+39 11 41	NGC 2528	12.6	1.5 x 1.5	13.3	SAB(rs)b I		Mag 10.85v star NE 6′.0.
08 10 15.3	+33 57 25	NGC 2532	12.4	1.9 x 1.4	13.3	SAB(rs)c	10	Numerous faint knots.
08 12 57.5	+36 15 11	NGC 2543	11.9	2.3 x 1.3	12.9	SB(s)b	45	Arms form tight "S" shape.
08 33 23.3	+29 32 16	NGC 2604	12.3	1.7 x 1.7	13.3	SB(rs)cd		**MCG +5-20-23** SE 3′.6.
08 42 25.8	+37 13 12	NGC 2638	12.8	1.7 x 0.6	12.6	S0/a	72	
07 16 13.9	+32 56 14	UGC 3774	14.3	1.6 x 0.3	13.3	Sab	60	
07 16 38.6	+33 59 15	UGC 3776	13.4	1.7 x 0.4	12.8	S?	66	
07 16 43.0	+29 51 13	UGC 3777	13.9	1.8 x 0.3	13.1	Scd:	152	Mag 10.8 star W edge; mag 9.06v star NW 3′.1.
07 17 28.6	+34 04 40	UGC 3780	14.0	1.1 x 0.3	12.6	S?	59	Mag 10.09v star NW 2′.8; **UGC 3779** S 6′.3.
07 17 58.4	+40 59 03	UGC 3781	14.0	1.5 x 0.8	14.0	Double System	51	Double system, bridge + plume; mag 7.44v star SW 4′.1.
07 18 26.0	+31 33 36	UGC 3788	14.1	1.3 x 0.4	13.2	S	155	Pair of mag 10 stars S 1′.9.
07 21 43.0	+35 44 17	UGC 3811	14.3	1.3 x 0.9	14.3	Sd	50	
07 23 13.0	+37 27 33	UGC 3822	13.9	1.4 x 0.7	13.7	SB(rs)bc	21	Strong NE-SW bar.
07 23 43.8	+33 26 34	UGC 3829	12.9	1.3 x 1.3	13.3	S?		Double system, contact, distorted.
07 24 47.3	+32 48 12	UGC 3833	14.1	0.8 x 0.8	13.5	S?		Faint envelope; mag 11.27v star SE 2′.3.
07 25 33.8	+36 41 09	UGC 3837	14.2	0.9 x 0.5	13.2	S	105	
07 26 42.1	+37 23 15	UGC 3849	13.7	1.3 x 1.0	13.8	Double System	141	Double system, strongly distorted.
07 28 17.4	+40 46 09	UGC 3860	14.5	1.7 x 1.1	15.1	Im V	30	Dwarf irregular, very low surface brightness.
07 29 44.9	+33 41 21	UGC 3879	14.4	2.1 x 0.3	13.8	Sdm:	103	
07 30 24.8	+36 06 43	UGC 3887	14.2	1.2 x 1.0	14.2	SB(s)cd	131	Slender SE-NW bar, filamentary arms.
07 33 31.9	+30 33 44	UGC 3904	13.0	1.3 x 1.3	13.4	(R)SB(r)0/a		Short, bright, N-S bar.
07 34 51.2	+33 32 17	UGC 3913	14.6	1.5 x 0.5	14.1	S?	160	Mag 9.47v star NW 3′.2.
07 37 35.5	+35 36 15	UGC 3937	13.2	1.9 x 0.5	13.0	SB?	151	Dwarf spiral **UGC 3934** N 3′.4; **MCG +6-17-22** NW 5′.7.
07 38 36.4	+37 38 00	UGC 3944	13.9	1.8 x 0.8	14.1	Scd:	124	Mag 13.1 star NW end.
07 41 26.1	+40 06 36	UGC 3966	13.5	2.0 x 1.8	14.7	Im V	141	Dwarf irregular, very low surface brightness.
07 42 40.8	+40 55 43	UGC 3976	14.5	0.7 x 0.5	13.2	S	36	Mag 12.5 star W 0′.5.
07 44 08.6	+29 14 51	UGC 3995	12.4	2.3 x 1.0	13.1	S pec	85	Mag 11.0 star S 1′.8.
07 44 38.6	+40 21 58	UGC 3997	13.8	1.0 x 0.8	13.4	Im?	30	Mag 12.1 star N 1′.6.
07 48 19.0	+34 19 54	UGC 4029	13.6	2.5 x 0.5	13.7	SBbc:	63	
07 50 00.2	+30 01 27	UGC 4042	13.9	1.1 x 0.7	13.5	(R)SB(r)b:	52	Mag 8.76v star S 5′.3; **UGC 4039** SSW 6′.3.
07 50 09.1	+30 43 54	UGC 4047	13.3	1.6 x 0.9	13.5	SBb I	147	**MCG +5-19-16** WNW 2′.1.
07 51 08.1	+34 03 18	UGC 4055	14.6	1.1 x 0.8	14.3	Sd	123	Mag 10.14v star W 3′.6.
07 52 57.2	+40 10 28	UGC 4068	16.0	1.5 x 0.2	14.5	Scd:	8	Mag 9.03v star NW 1′.1.
07 59 12.0	+31 48 22	UGC 4131	14.2	0.8 x 0.6	13.3	SABc	140	Mag 11.4 star N 1′.8.
07 59 13.0	+32 54 43	UGC 4132	12.9	2.0 x 0.5	12.7	Sbc	28	**CGCG 178-12** SW 6′.8.
08 02 42.9	+40 40 41	UGC 4176	14.3	1.7 x 0.5	14.0	SBd	112	
08 04 31.2	+40 12 20	UGC 4200	13.7	0.9 x 0.6	12.9	(R′)SB(s)bc I	26	Located between two mag 10.7 stars; **CGCG 207-25** W 2′.5.
08 06 43.0	+39 05 25	UGC 4219	14.0	2.5 x 1.8	15.5	SA(rs)b II	150	Moderately bright N-S core, faint envelope.
08 07 21.4	+40 23 53	UGC 4226	14.0	1.9 x 1.4	14.9	SA(r)cd	35	Close trio of stars, mags 12-14, S edge.
08 07 41.1	+39 00 12	UGC 4229	13.7	0.6 x 0.5	12.2	S?	90	Mag 9.49v star S 2′.9.
08 10 00.0	+40 06 11	UGC 4252	14.3	1.9 x 0.4	13.9	Scd:	17	Uniform surface brightness; mag 12.5 star N 2′.6.
08 10 56.5	+36 49 41	UGC 4261	13.9	0.9 x 0.4	12.6	S?	75	Plume plus jet; mag 10.94v star W 1′.2.
08 14 22.4	+39 15 03	UGC 4283	14.0	1.2 x 0.8	13.8	SB(s)b	112	Bright N-S bar, faint envelope.
08 17 37.3	+35 26 40	UGC 4306	14.2	1.2 x 0.7	13.8	S?	135	Mag 11.9 star N 1′.0.
08 28 54.8	+34 39 00	UGC 4434	13.8	0.9 x 0.4	12.6	S?	105	

(Continued from previous page)

57 GALAXIES 57

RA h m s	Dec ° ′ ″	Name	Mag (V)	Dim ′ Maj x min	SB	Type Class	PA	Notes
08 31 25.9	+40 57 23	UGC 4449	14.8	1.0 x 0.8	14.4	Pec		
08 37 26.6	+40 02 04	UGC 4498	13.8	1.0 x 0.6	13.1	SBa	13	Mag 11.0 star S 1′.3.
08 39 00.9	+40 20 43	UGC 4513	14.1	0.9 x 0.9	13.7	Compact		
08 41 53.2	+32 52 02	UGC 4531	13.8	1.1 x 0.4	12.7	SBb	37	Mag 11.0 star W 2′.6.

GALAXY CLUSTERS

RA h m s	Dec ° ′ ″	Name	Mag 10th brightest	No. Gal	Diam ′	Notes
07 53 18.0	+29 21 00	A 602	15.8	32	22	All members anonymous, faint and stellar.
08 10 06.0	+35 13 00	A 628	15.9	43	9	Mag 6.66v star on N edge; all member GX stellar.
08 28 30.0	+30 25 00	A 671	14.9	38	21	See notes for IC 2376, IC 2378.

GLOBULAR CLUSTERS

RA h m s	Dec ° ′ ″	Name	Total V m	B ★ V m	HB V m	Diam ′	Conc. Class Low = 12 High = 1	Notes
07 38 08.5	+38 52 55	NGC 2419	10.3	17.3	20.2	4.6	2	**Intergalactic Wanderer**. Shapley originally named this the **Intergalactic Tramp**. Probably the most distant of the Milky Way's globular star clusters.

PLANETARY NEBULAE

RA h m s	Dec ° ′ ″	Name	Diam ″	Mag (P)	Mag (V)	Mag cent ★	Alt Name	Notes
07 25 34.8	+29 29 22	NGC 2371-72	62	13.0	11.2	14.8	PK 189+19.1	Very faint patches of nebulosity separated from main body NW and SE; mag11.2 star NW 4′.4.

58 GALAXIES 58

RA h m s	Dec ° ′ ″	Name	Mag (V)	Dim ′ Maj x min	SB	Type Class	PA	Notes
06 45 35.1	+29 13 16	CGCG 145-6	14.4	0.7 x 0.6	13.3	S	33	Several stars superimposed; pair of mag 11 stars S; mag 10.47v star NW 1′.7.
06 51 54.6	+32 34 19	CGCG 146-3	15.4	0.4 x 0.2	12.5		15	Staggered N-S line of five mag 10-12 star W 5′.7.
07 21 05.8	+30 45 08	CGCG 147-10	15.9	0.5 x 0.5	14.3	Sc		Mag 9.81v stars WNW 9′.5; mag 10.34v star E 5′.9.
07 21 58.7	+29 29 47	CGCG 147-13	15.4	0.5 x 0.4	13.5	Sm	18	Mag 8.87v star NE 4′.8.
06 59 49.9	+35 27 28	IC 2175	13.9	1.6 x 0.8	14.0	SBc:	66	Mag 8.60v star N 1′.4.
07 07 31.9	+32 28 11	IC 2176	14.1	0.9 x 0.4	12.9		6	**IC 2178** NNE 2′.9.
07 23 15.9	+32 29 42	IC 2185	14.4	0.6 x 0.4	12.7	S?	129	Pair of mag 10.66v, 10.72v stars SSW 4′.1.
06 41 30.6	+32 50 00	MCG +5-16-1	13.7	1.0 x 0.7	13.2		171	
06 49 55.2	+29 31 21	MCG +5-16-7	14.4	0.8 x 0.7	13.7	SB0/a	120	
07 08 00.2	+31 39 44	MCG +5-17-12	13.9	1.0 x 0.5	13.0	Sb	90	Pair of galaxies at right angles; mag 11.5 star NNE 2′.3.
07 21 05.3	+29 20 15	MCG +5-18-7	14.1	0.9 x 0.6	13.3	Sb	51	Mag 10.7 star W 4′.5.
06 28 12.5	+35 27 58	MCG +6-15-1	15.0	0.5 x 0.4	13.1	S	111	Mag 10.52v star N 0′.9.
06 31 55.5	+35 33 07	MCG +6-15-2	14.5	0.7 x 0.5	13.2	S	39	Mag 10.84v star S 2′.6.
06 51 23.8	+36 54 44	MCG +6-15-15	14.7	0.7 x 0.4	13.1		120	
06 55 27.8	+33 16 47	MCG +6-16-1	13.9	0.7 x 0.4	12.4	S?	117	Forms a small triangle with a pair of mag 12-13 stars NE and NW.
06 58 23.2	+35 50 22	MCG +6-16-4	13.8	0.8 x 0.6	13.1	E	21	Mag 11.5 star N 0′.9.
07 01 53.3	+37 43 18	MCG +6-16-8	14.9	0.8 x 0.6	13.9	Spiral	9	Mag 14.6 star N edge; mag 9.30v star E 3′.7.
07 03 22.4	+33 46 42	MCG +6-16-11	16.0	0.6 x 0.3	13.9	SBb	72	Pair with MCG +6-16-12 E.
07 03 26.3	+33 46 42	MCG +6-16-12	13.3	0.6 x 0.6	12.1			Pair with MCG +6-16-11 W.
07 03 51.2	+37 56 12	MCG +6-16-13	15.9	0.6 x 0.6	14.6			MCG +6-16-14 SE 2′.9.
07 04 03.7	+37 54 39	MCG +6-16-14	14.3	0.6 x 0.5	12.8		171	Mag 12.0 star superimposed on SE edge.
07 04 34.1	+37 30 43	MCG +6-16-16	14.4	0.8 x 0.4	13.0	Sb	105	Mag 6.63v star NW 4′.6.
07 10 14.8	+33 39 49	MCG +6-16-23	15.2	0.8 x 0.5	14.1		81	Mag 13.1 star N 0′.7.
07 14 52.0	+36 04 44	MCG +6-16-29	14.6	0.7 x 0.4	13.0		105	Mag 10.9 star N 2′.2.
07 14 53.1	+35 20 21	MCG +6-16-30	15.5	0.4 x 0.4	13.4			Compact; mag 13.1 star SW 0′.5.
07 15 22.3	+36 02 16	MCG +6-16-31	14.7	0.9 x 0.3	13.1		144	Mag 10.47v star SW 1′.5.
07 16 04.2	+37 25 09	MCG +6-16-32	14.8	0.7 x 0.7	13.8	Sb		Mag 11.0 star NNW 1′.8.
07 22 15.6	+38 20 42	MCG +6-16-38	15.1	0.5 x 0.4	13.2		30	Close pair of mag 10 stars SE 2′.0.
05 58 36.1	+40 31 13	MCG +7-13-3	14.8	0.7 x 0.5	13.5	S		
06 19 46.0	+39 01 59	MCG +7-13-11	15.8	0.7 x 0.7	14.9			
06 36 08.5	+39 24 49	MCG +7-14-6	14.1	0.5 x 0.5	12.4	SB?		Mag 12.5 star S 0′.5.
06 47 03.0	+40 01 18	MCG +7-14-15	14.3	0.4 x 0.3	11.8		60	Mag 10.4 star E 1′.9.
06 55 10.2	+41 00 07	MCG +7-14-19	13.9	0.7 x 0.7	13.1	E		Several faint stars on edges; UGC 3593 S 2′.7.
06 55 50.8	+40 41 35	MCG +7-15-2	14.1	0.6 x 0.5	12.7	S?	15	Close pair of mag 11 stars ESE 2′.3.
07 00 07.3	+40 23 15	MCG +7-15-4	14.8	0.5 x 0.4	12.9	Sbc	33	Mag 11.3 star WNW 2′.4.
07 09 24.9	+39 06 10	MCG +7-15-8	14.4	0.7 x 0.4	12.8	Sc	120	Mag 12.3 star NW 2′.0.
07 22 11.7	+40 28 08	MCG +7-15-15	14.4	0.9 x 0.4	13.1	SBb	123	
06 47 17.4	+33 33 59	NGC 2274	12.1	1.2 x 1.1	12.5	E	169	NGC 2275 N 2′.0; **UGC 3544** E 10′.6.
06 47 18.0	+33 35 56	NGC 2275	13.2	1.3 x 1.0	13.3	S?	157	Almost stellar nucleus, bright knot or star S of core; UGC 3537 W 6′.8.
06 50 52.1	+33 27 41	NGC 2288	14.4	0.4 x 0.2	11.5		93	Almost stellar; located 1′.0 S of NGC 2289.
06 50 53.7	+33 28 40	NGC 2289	13.2	1.1 x 0.7	12.8	S0	92	NGC 2288 S 1′.0.
06 50 57.0	+33 26 15	NGC 2290	13.2	1.3 x 0.7	12.9	(R)SAa:	50	Faint stars NE and SW of core.
06 50 58.6	+33 31 30	NGC 2291	13.2	1.0 x 0.8	12.8	SA0°:	126	Stellar nucleus, smooth, faint envelope; NGC 2294 E 2′.7.
06 51 11.5	+33 31 35	NGC 2294	13.8	0.9 x 0.3	12.5	E6?	6	NGC 2291 W 2′.7.
07 08 21.4	+35 10 10	NGC 2333	13.3	1.0 x 0.7	12.7	Sa	35	Located on NE edge of galaxy cluster A 568; small group of stellar galaxies SW 10′.4.
06 02 05.4	+36 06 16	UGC 3390	14.6	1.8 x 1.1	15.2	SABdm	35	Numerous stars superimposed right up to elongated core.
06 30 29.0	+39 30 13	UGC 3475	13.9	1.7 x 1.0	14.3	Sm:	85	Dwarf spiral, very low surface brightness; mag 11.8 star on N edge.
06 30 28.3	+33 18 00	UGC 3476	14.3	1.0 x 0.3	12.8	Im:	63	Uniform surface brightness.
06 32 23.6	+35 11 21	UGC 3479	14.0	1.1 x 0.3	12.6	Sab	179	Mag 6.79v star NNE 7′.0.
06 32 34.9	+40 12 26	UGC 3481	14.3	0.8 x 0.2	12.2	I?	96	Very close pair of mag 14.1 stars E 0′.8.
06 33 39.6	+40 41 15	UGC 3487	14.2	1.0 x 0.7	13.7	Sd	110	**MCG +7-14-5** S 1′.2.
06 36 51.8	+37 34 05	UGC 3499	13.9	1.1 x 0.7	13.5	S	24	Patchy, two superimposed stars S of nucleus; mag 10.4 star W 1′.3.
06 41 32.6	+40 10 06	UGC 3510	14.1	1.1 x 0.8	13.8	S?	135	Very slender **MCG +7-14-9** NW 1′.1.
06 45 03.6	+40 24 49	UGC 3525	13.4	1.1 x 0.8	13.3	E?	70	
06 45 26.2	+34 29 09	UGC 3529	13.8	0.9 x 0.4	12.5	S?	165	
06 46 03.6	+29 20 51	UGC 3536	13.0	1.2 x 0.6	12.5	S0	142	Mag 8.47v star SW 4′.4.

(Continued from previous page)

58 GALAXIES 58

RA h m s	Dec ° ′ ″	Name	Mag (V)	Dim ′ Maj x min	SB	Type Class	PA	Notes
06 46 45.5	+33 37 05	UGC 3537	14.0	0.8 x 0.7	13.2	SBcd:	16	NGC 2274 and NGC 2275 E 6′.8; mag 9.39v star W 6′.0.
06 51 40.7	+29 04 17	UGC 3571	13.5	1.1 x 0.7	13.1	Scd:	105	Stellar nucleus; mag 11.4 star NW 1′.3.
06 54 31.5	+30 04 12	UGC 3590	14.1	0.9 x 0.4	12.8	SBab	45	
06 55 13.8	+40 20 17	UGC 3592	14.1	0.9 x 0.5	13.1	(R)SB(s)a	6	Mag 10.84v star 0′.6 W.
06 55 14.4	+40 57 34	UGC 3593	13.7	1.6 x 0.8	13.8	SABc:	40	MCG +7-14-19 N 2′.7.
06 55 35.7	+39 45 51	UGC 3596	12.8	1.1 x 1.0	12.8	S0?	132	
06 55 49.6	+39 59 57	UGC 3601	13.9	0.5 x 0.4	12.0	S?	29	Mag 8.50v star NNW 7′.5.
06 57 39.8	+39 05 13	UGC 3612	13.5	0.9 x 0.4	12.2	SB(s)b	25	
06 57 59.1	+35 44 01	UGC 3615	14.4	1.5 x 0.3	13.4	Sab	33	Very faint anonymous galaxy on W edge of mag 14.5 star NE 2′.9.
06 58 27.7	+34 12 37	UGC 3619	14.3	0.7 x 0.6	13.2	SBc	30	
07 01 41.9	+29 52 58	UGC 3631	14.8	1.3 x 0.6	14.4	SB	107	Two main arms.
07 04 01.4	+29 14 35	UGC 3649	13.8	0.9 x 0.5	12.8	Sa	63	**MCG +5-17-9** E 1′.4.
07 05 33.7	+33 49 39	UGC 3664	14.4	0.9 x 0.5	13.4	Scd:	72	Mag 8.80v star SW 6′.6.
07 08 30.4	+29 52 41	UGC 3692	13.9	1.5 x 0.6	13.7	S0?	13	Mag 12.0 star W 1′.4.
07 08 37.1	+32 40 52	UGC 3694	14.0	1.5 x 0.7	13.9	Sd	95	Located between a pair of stars, mags 10.5 and 10.9.
07 09 21.4	+36 17 02	UGC 3703	14.2	1.1 x 1.1	14.3	SA(s)c		Weak stellar nucleus; **CGCG 176-21** NNE 6′.6.
07 10 42.2	+34 25 15	UGC 3723	13.1	1.6 x 0.8	13.3	SB0	3	Bright knot or star S edge.
07 11 23.1	+30 10 00	UGC 3728	14.2	0.7 x 0.4	12.7	SBb:	140	Mag 4.39v star Tau Geminorum NW 5′.6.
07 11 38.6	+29 09 55	UGC 3731	14.1	1.1 x 0.4	13.1	Sdm:	158	
07 11 56.3	+33 05 17	UGC 3735	15.4	1.1 x 0.8	15.1	Sd	66	Mag 7.15v star W 7′.8.
07 13 28.0	+35 05 47	UGC 3742	13.5	1.4 x 0.7	13.3	Sdm	72	Mag 14.7 star E end; mag 11.4 star S 1′.5.
07 13 36.7	+35 39 20	UGC 3743	15.3	1.5 x 1.1	15.7	SAB(s)cd	168	Mag 11.1 star N end; mag 8.71v star NW 2′.7.
07 14 04.0	+35 16 43	UGC 3752	13.9	0.6 x 0.4	12.2	S?	78	Mag 11.1 star S 0′.7.
07 16 13.9	+32 56 14	UGC 3774	14.3	1.6 x 0.3	13.3	Sab	60	
07 16 38.6	+33 59 15	UGC 3776	13.4	1.7 x 0.4	12.8	S?	66	
07 16 43.0	+29 51 13	UGC 3777	13.9	1.8 x 0.3	13.1	Scd:	152	Mag 10.8 star W edge; mag 9.06v star NW 3′.1.
07 17 28.6	+34 04 40	UGC 3780	14.0	1.1 x 0.3	12.6	S?	59	Mag 10.09v star NW 2′.8; **UGC 3779** S 6′.3.
07 17 58.4	+40 59 03	UGC 3781	14.0	1.5 x 0.8	14.0	Double System	51	Double system, bridge + plume; mag 7.44v star SW 4′.1.
07 18 26.0	+31 33 36	UGC 3788	14.1	1.3 x 0.4	13.2	S	155	Pair of mag 10 stars S 1′.9.
07 21 43.0	+35 44 17	UGC 3811	14.3	1.3 x 0.9	14.3	Sd	50	
07 23 13.0	+37 27 33	UGC 3822	13.9	1.4 x 0.7	13.7	SB(rs)bc	21	Strong NE-SW bar.
07 23 43.8	+33 26 34	UGC 3829	12.9	1.3 x 1.3	13.3	S?		Double system, contact, distorted.

GALAXY CLUSTERS

RA h m s	Dec ° ′ ″	Name	Mag 10th brightest	No. Gal	Diam ′	Notes
07 07 36.0	+35 02 00	A 568	15.4	36	17	**Gemini Cluster**. Members faint and stellar, see notes for NGC 2333.

OPEN CLUSTERS

RA h m s	Dec ° ′ ″	Name	Mag	Diam ′	No. ★	B ★	Type	Notes
06 15 18.0	+39 51 18	NGC 2192	10.9	5	45	14.0	cl	Moderately rich in stars; moderate brightness range; slight central concentration; detached.

BRIGHT NEBULAE

RA h m s	Dec ° ′ ″	Name	Dim ′ Maj x min	Type	BC	Color	Notes
06 04 06.0	+30 15 00	Sh2-241	2 x 2	E+R	2-5	3-4	Brightest across a 2′ area. A very diffuse and very faint fan-shaped region of nebulosity extending out by about 7′, tracing an arc from a point due S to nearly due W where it abruptly stops.

GALAXIES

RA h m s	Dec ° ′ ″	Name	Mag (V)	Dim ′ Maj x min	SB	Type Class	PA	Notes
05 58 36.1	+40 31 13	MCG +7-13-3	14.8	0.7 x 0.5	13.5	S		
04 56 15.0	+30 03 09	UGC 3205	14.0	2.1 x 0.6	14.1	Sab	44	
05 35 19.0	+40 53 12	UGC 3325	14.6	1.1 x 0.8	14.3	Sbc	30	Many superimposed stars; mag 13.1 star S edge.
06 02 05.4	+36 06 16	UGC 3390	14.6	1.8 x 1.1	15.2	SABdm	35	Numerous stars superimposed right up to elongated core.

59 OPEN CLUSTERS 59

RA h m s	Dec ° ′ ″	Name	Mag	Diam ′	No. ★	B ★	Type	Notes
05 48 54.0	+30 11 06	Bas 4	9.1	5	134	12.2	cl	Few stars; small brightness range; slight central concentration; detached. (S) Low contrast against Milky Way.
05 20 30.0	+30 35 00	Be 17		8	100	16.0	cl	Moderately rich in stars; small brightness range; fairly even distribution; detached.
05 24 03.0	+29 34 12	Be 19	11.4	4	150	15.0	cl	Moderately rich in stars; small brightness range; slight central concentration; detached.
05 24 21.0	+32 36 24	Be 69		3	30	14.0	cl	
05 40 57.0	+32 16 00	Be 71		5	30	15.0	cl	
05 20 32.5	+39 33 00	Cz 20		36	30		cl	Few stars; small brightness range; not well detached. (A) 36′ diameter group of about 30 bright stars, including NGC 1857 and **Iss 206**.
05 26 40.0	+36 00 00	Cz 21		8	40		cl:	
05 04 36.0	+34 50 00	Do 15		18			ast	Few stars; small brightness range; not well detached. Probably not a cluster.
05 14 36.0	+32 43 00	Do 16		6	10		cl	Few stars; moderate brightness range; not well detached; involved in nebulosity.
05 24 06.0	+33 17 00	Do 18		6	15		cl	Few stars; small brightness range; not well detached; involved in nebulosity.
05 28 33.0	+33 40 36	Do 20		5	10		cl:	Few stars; large brightness range; not well detached. Probably not a cluster. (S) Sparse.
05 49 17.0	+33 37 36	King 8	11.2	4	198	15.0	cl	Moderately rich in stars; moderate brightness range; slight central concentration; detached.
05 08 24.0	+39 05 00	King 17		5	25	14.0	cl	
05 18 10.5	+33 22 00	Mel 31		135	35	4.5	cl?	
05 08 06.0	+37 01 00	NGC 1778	7.7	8	112	10.1	cl	Few stars; moderate brightness range; no central concentration; detached.
05 20 06.0	+39 21 00	NGC 1857	7.0	10	40	11.0	cl	Moderately rich in bright and faint stars; strong central concentration; detached.
05 22 46.0	+33 25 00	NGC 1893	7.5	25	270	9.3	cl	Rich in stars; large brightness range; slight central concentration; detached; involved in nebulosity. (A) Involved with nebula IC 410.
05 25 41.9	+29 20 00	NGC 1896		20	25	8.5	cl?	
05 28 05.0	+35 19 30	NGC 1907	8.2	5	113	11.0	cl	Moderately rich in stars; small brightness range; strong central concentration; detached; involved in nebulosity.

(Continued from previous page)

OPEN CLUSTERS

RA h m s	Dec ° ′ ″	Name	Mag	Diam ′	No. ★	B ★	Type	Notes
05 28 43.0	+35 51 00	NGC 1912	6.4	15	160	8.0	cl	= **M 38**. Rich in stars; moderate brightness range; slight central concentration; detached.
05 31 26.0	+34 14 42	NGC 1931	10.1	6	20		cl	Several stars involved in nebulosity.
05 36 18.0	+34 08 00	NGC 1960	6.0	10	60	9.0	cl	= **M 36**. Rich in stars; large brightness range; strong central concentration; detached.
05 52 19.0	+32 33 00	NGC 2099	5.6	15	1842	11.0	cl	= **M 37**. Rich in stars; moderate brightness range; strong central concentration; detached.
05 07 13.8	+30 50 34	Skiff J0507.2+3050		6		10.8	cl?	
05 28 07.0	+34 25 00	St 8		15	40	9.0	cl	Moderately rich in bright and faint stars with a strong central concentration; detached; involved in nebulosity (IC 417). (A) Nebulosity is IC 417.
05 39 00.0	+37 52 00	St 10		25	15		cl	Few stars; moderate brightness range; not well detached. (S) Sparse group.

GLOBULAR CLUSTERS

RA h m s	Dec ° ′ ″	Name	Total V m	B ★ V m	HB V m	Diam ′	Conc. Class Low = 12 High = 1	Notes
04 46 05.9	+31 22 51	Pal 2	13.0	18.8	21.7	2.2	9	

BRIGHT NEBULAE

RA h m s	Dec ° ′ ″	Name	Dim ′ Maj x min	Type	BC	Color	Notes
05 16 12.0	+34 16 00	IC 405	30 x 20	E+R	2-5	2-4	**Flaming Star Nebula**. This object has an entirely different appearance on red and blue photos. Brightest in the area around AE Aurigae.
05 22 36.0	+33 31 00	IC 410	40 x 30	E	2-5	3-4	N side brightest. The central and S part largely obscured by absorption matter.
05 28 06.0	+34 26 00	IC 417	13 x 10	E	2-5	3-4	
05 18 12.0	+37 36 00	IC 2120	1 x 1				
05 31 24.0	+34 15 00	NGC 1931	4 x 4	E+R	1-5	3-4	
05 37 42.0	+32 00 00	NGC 1985					
05 39 24.0	+35 56 00	Sh2-231	10 x 5	E	4-5	3-4	Very diffuse and faint, even surface brightness nebula; extended N-S.
05 41 06.0	+35 52 00	Sh2-235	7 x 5	E	2-5	3-4	
04 48 24.0	+29 47 00	vdB 29	7 x 5	R	3-5	1-4	The mag 6.5 involved star at the E side of a diffuse nebulosity.
04 55 42.0	+30 33 00	vdB 31	8 x 5	R	3-5	1-4	Nebula with a mag 6.8 star involved.

DARK NEBULAE

RA h m s	Dec ° ′ ″	Name	Dim ′ Maj x min	Opacity	Notes
04 40 33.4	+29 53 00	B 23	5 x 5	5	Sharply pointed to the SE.
04 42 53.3	+29 44 00	B 24	8 x 8	5	Sharply pointed to the S.
04 54 38.2	+30 37 00	B 26	5 x 5	6	Irregular.
04 55 08.0	+30 33 00	B 27	5 x 5	6	Irregular.
04 55 52.4	+30 38 00	B 28	4 x 4	6	Irregular.
05 06 23.4	+31 35 00	B 29	10 x 10	6	Round; indefinite.
05 43 30.7	+32 39 00	B 34	20 x 20	4	Round; starless; indefinite.
04 44 00.8	+31 44 00	B 221	45 x 45		Partly vacant region.
05 08 23.8	+32 10 00	B 222	10 x 10		Round; indefinite.
05 36 34.6	+33 42 00	B 226	17 x 17		Dark spot; S of M 36.

PLANETARY NEBULAE

RA h m s	Dec ° ′ ″	Name	Diam ″	Mag (P)	Mag (V)	Mag cent ★	Alt Name	Notes
04 39 48.0	+36 45 39	PN G165.5−6.5	13	13.7		18.5	PK 165−6.1	Mag 14.4 star NNE 1′.9; mag 14.3 star W 3′.5.
04 42 53.9	+36 06 50	PN G166.4−6.5	12	17.9		17.0	PK 166−6.1	Mag 10.38v star N 1′.5; mag 13.5 star E 2′.3.
05 06 38.4	+39 08 09	PN G167.0−0.9	60	16.6		20.2	PK 167−0.1	Mag 12.0 star on SE edge; mag 10.23v star SE 2′.3.
04 36 37.4	+33 39 27	PN G167.4−9.1	13	15.4		15.3	PK 167−9.1	Mag 14.1 star SW 0′.6; mag 12.3 star SW 2′.4.
05 41 22.2	+39 15 06	PN G170.7+4.6	0	16.5			PK 170+4.1	Mag 14.2 star SW 0′.8; mag 12.4 star SW 3′.0.
05 42 34.2	+36 09 06	PN G173.5+3.2	20	18.1		19.9	PK 175+6.1	Pair of mag 8.93v and 9.73v stars E 1′.4; mag 8.06v star N 3′.0.
05 40 53.2	+35 42 18	PN G173.7+2.7	30				PK 173+2.1	Second bright patch of nebulosity, adjacent to star, S 0′.9; mag 10.62v star NW 3′.5.
05 07 08.1	+30 49 26	PN G173.7−5.8	130	13.7		18.8	PK 173−5.1	Irregularly round and very diffuse in a moderately rich field of faint stars. A faint pair of stars oriented roughly E-W (sep. about 15″) at N edge; another fainter pair of similar separation, aligned roughly N-S near S edge.

60 GALAXIES 60

RA h m s	Dec ° ′ ″	Name	Mag (V)	Dim ′ Maj x min	SB	Type Class	PA	Notes
03 28 27.7	+40 09 13	CGCG 541-11	14.2	0.8 x 0.2	12.1	Sbc	84	Located 1′.1 WSW of mag 7.03v star.
03 16 06.4	+40 48 14	IC 309	13.5	0.9 x 0.9	13.1	SA(s)0°		Located between a pair of mag 11 stars.
03 16 46.8	+40 00 11	IC 311	14.1	0.9 x 0.6	13.3	S?	113	Mag 9.49v star ENE 6′.2.
03 25 59.3	+40 47 18	IC 320	13.7	1.3 x 1.0	13.8	SB(rs)ab	48	Several stars superimposed SW and W of nucleus.
04 06 39.7	+37 06 53	IC 2027	14.1	0.4 x 0.4	12.2	E		Mag 9.13v star E 13′.3.
03 53 16.5	+32 17 51	MCG +5-10-1	15.0	0.7 x 0.4	13.4	Scd	108	UGC 2888 N 2′.0.
04 03 55.7	+30 50 23	MCG +5-10-5	14.8	0.7 x 0.7	14.1	E		Pair of mag 10-11 stars S 1′.9.
03 20 53.9	+37 52 01	MCG +6-8-15	14.8	0.7 x 0.5	13.6	SBc	27	Mag 10.77v star WSW 2′.4.
03 31 08.4	+39 37 42	MCG +6-8-27	13.9	0.9 x 0.5	12.9		51	Mag 8.63v star NW 1′.3.
03 42 28.6	+38 58 00	MCG +6-9-4	14.9	0.5 x 0.3	12.7	S	141	Mag 11.7 star WSW 0′.9.
03 49 17.5	+37 17 29	MCG +6-9-9	15.2	0.5 x 0.5	13.6	Sd		Mag 10.14v star NW edge.
03 51 34.3	+37 04 46	MCG +6-9-10	14.6	0.5 x 0.4	12.8	S0	168	Mag 10.06v star W edge.
03 25 24.8	+40 32 11	MCG +7-8-4	14.9	0.5 x 0.5	13.3	SBR/R		Mag 10.73v star superimposed on E edge.
03 27 59.4	+39 54 13	MCG +7-8-11	13.7	0.8 x 0.5	12.6	S?	114	Mag 8.75v star NE 6′.4.
03 29 53.9	+39 50 18	MCG +7-8-17	15.1	0.7 x 0.5	13.8		99	Star or large knot W edge; mag 1.8 star W 1′.8.
03 33 22.5	+40 09 46	MCG +7-8-25	14.6	0.8 x 0.8	14.0			Mag 11.4 star ESE 2′.7.
03 53 32.0	+32 29 34	NGC 1465	13.7	1.7 x 0.5	13.4	S0/a	165	
03 16 00.9	+40 53 13	UGC 2617	13.2	2.2 x 0.6	13.4	SAB(s)d	176	Numerous knots and/or superimposed stars; mag 10.6 star WNW 2′.0.
03 16 28.2	+35 03 56	UGC 2623	14.2	1.8 x 1.4	15.1	SB(s)d	120	Mag 7.59v star W 5′.5.
03 16 59.4	+31 34 02	UGC 2627	13.8	1.5 x 1.3	14.4	SA(s)c II-III	78	Weak dark patch E of nucleus.

(Continued from previous page)

GALAXIES

RA h m s	Dec ° ′ ″	Name	Mag (V)	Dim ′ Maj x min	SB	Type Class	PA	Notes
03 17 07.2	+31 34 58	UGC 2629	14.6	0.8 x 0.8	13.9	SBcd:		Uniform surface brightness.
03 17 22.7	+36 34 05	UGC 2633	15.4	1.5 x 0.2	14.0	Scd:	81	Mag 13.1 star S 0′.8.
03 17 31.6	+37 02 47	UGC 2636	15.9	1.7 x 0.2	14.5	Scd:	10	**UGC 2630** NNW 6′.8.
03 18 38.5	+37 36 28	UGC 2653	15.1	1.3 x 0.5	14.5	Sd	155	Mag 10.49v star W 1′.3.
03 18 53.6	+40 35 42	UGC 2659	13.8	1.2 x 0.3	12.5	Sbc	65	Mag 9.76v star E 2′.4.
03 19 12.0	+39 26 32	UGC 2661	13.6	1.3 x 0.3	12.4	S0	147	
03 20 34.5	+37 29 40	UGC 2678	14.7	1.0 x 0.8	14.3	SBb:	51	Stellar nucleus; mag 11.8 star WNW 1′.9.
03 20 51.8	+38 15 16	UGC 2685	13.5	2.0 x 1.5	14.5	SAB(s)b	9	Bright E-W core, faint envelope.
03 21 01.3	+40 47 49	UGC 2686	14.5	1.1 x 0.5	13.7	SABa	43	Mag 10.37v star N 1′.9.
03 21 27.7	+40 48 05	UGC 2689	13.9	1.2 x 0.6	13.5	S0?	127	Several faint stars superimposed; mag 11.1 star NW 2′.9.
03 22 03.0	+40 51 47	UGC 2698	12.9	1.0 x 0.7	12.5	E	105	Mag 10.51v star on E edge.
03 22 55.3	+37 01 54	UGC 2702	14.6	2.2 x 1.8	15.9		36	Stellar nucleus, smooth envelope; mag 9.25v star SW 1′.6.
03 23 42.9	+36 55 07	UGC 2706	14.7	0.8 x 0.7	14.0	Sdm:	81	Mag 11.3 star WSW 1′.8.
03 23 49.1	+40 33 26	UGC 2708	13.8	1.3 x 1.0	14.0	S0	24	Very faint, elongated anonymous galaxy N 1′.5.
03 23 53.8	+38 40 36	UGC 2709	13.5	1.9 x 0.7	13.7	SABb	3	
03 23 58.5	+37 45 18	UGC 2710	13.5	1.1 x 0.8	13.3	S0	6	Bright knot, or star, SE of nucleus; several faint stars superimposed on S edge.
03 24 36.6	+40 41 24	UGC 2717	13.3	0.7 x 0.6	12.4	E		Mag 7.92v star N 4′.5; elongated **UGC 2715** NNW 6′.4.
03 25 52.2	+40 44 56	UGC 2730	14.1	1.5 x 0.3	13.1	Sb III	127	IC 320 N 2′.7.
03 26 27.3	+40 30 25	UGC 2736	13.5	1.7 x 0.3	12.6	Sab	69	Mag 8.06v star S 2′.9.
03 27 40.1	+40 53 51	UGC 2742	14.1	0.8 x 0.6	13.2	SB(s)bc	117	
03 28 31.6	+36 46 33	UGC 2746	14.4	1.1 x 0.8	14.2	$S0^-$:	172	Close stellar pair **MCG +6-8-23** and **MCG +6-8-24** S 5′.5.
03 28 46.6	+36 33 20	UGC 2750	15.0	1.5 x 0.3	14.0	Sbc	156	Mag 14 star on W edge; close pair of mag 13 and 14 stars E edge.
03 29 04.1	+40 49 24	UGC 2752	13.8	1.4 x 0.5	13.3	S0:	83	Mag 10.14v star SW 2′.5.
03 29 24.0	+39 47 29	UGC 2755	14.0	0.7 x 0.7	13.1	S?		Trio of mag 12-13 stars NW 1′.2. Very faint, very small anonymous galaxy on N edge.
03 29 35.2	+40 52 18	UGC 2756	14.1	1.4 x 0.7	13.9	Double System	12	Double system, contact; consists of **MCG +7-8-15** and **MCG +7-8-16**.
03 30 56.7	+40 48 33	UGC 2766	15.1	0.5 x 0.5	13.4	S0		
03 31 18.4	+35 28 00	UGC 2770	13.7	1.5 x 1.4	14.4	SB(rs)b	95	Bright, narrow E-W bar.
03 31 23.6	+39 44 29	UGC 2771	13.5	1.7 x 0.5	13.2	S0	12	Mag 8.63v star and MCG +6-8-27 SSW 7′.0.
03 33 13.3	+36 11 03	UGC 2780	15.2	1.5 x 1.0	15.5	S?	27	Stellar nucleus.
03 34 18.6	+39 21 22	UGC 2783	12.7	1.1 x 1.1	13.0	E?		Mag 8.09v star SW 4′.3.
03 34 19.6	+39 32 42	UGC 2784	12.8	1.2 x 0.9	12.8	S0:	160	**CGCG 525-43** N 2′.9.
03 38 10.1	+40 59 00	UGC 2798	13.2	2.1 x 0.7	13.5	SABbc	70	Mag 10.38v star NW 2′.2.
03 38 56.8	+30 15 16	UGC 2807	16.6	1.1 x 0.6	16.0	Sm:	161	Mag 12.0 star NW 1′.6.
03 39 37.3	+38 41 23	UGC 2808	14.0	1.0 x 0.6	13.3	SAcd	160	Mag 11.5 star N 1′.5.
03 40 08.2	+39 36 26	UGC 2810	13.6	0.9 x 0.7	12.9	SA(rs)c	114	Faint stellar nucleus, branching arms.
03 41 32.5	+37 13 29	UGC 2817	15.0	1.8 x 0.5	14.7	S?	126	
03 41 44.5	+39 20 06	UGC 2818	13.6	1.0 x 0.8	13.2	S	143	Mag 14.4 star on S edge, plus several knots.
03 42 24.4	+39 14 36	UGC 2828	13.7	0.9 x 0.8	13.2	SB(rs)bc	33	Mag 11.2 star NW 2′.6.
03 43 57.0	+39 17 41	UGC 2836	12.4	1.2 x 1.0	12.5	$S0^-$:	162	Mag 10.56v star NE 1′.8.
03 43 54.6	+40 00 56	UGC 2837	13.4	1.7 x 0.8	13.6	SB(rs)b	155	Patchy with E-W bar; mag 12.3 star S 1′.3.
03 46 58.5	+38 38 01	UGC 2857	13.5	1.1 x 0.7	13.1	SBbc	35	Several faint stars superimposed.
03 47 18.5	+40 51 37	UGC 2859	14.8	1.1 x 1.1	14.8	Scd:		Mag 12.2 star WSW 2′.6.
03 47 39.7	+39 20 47	UGC 2861	13.9	1.5 x 0.8	14.0	SA(r)b	50	Faint star superimposed S of nucleus.
03 48 32.7	+35 08 58	UGC 2868	15.1	1.7 x 0.2	13.8	Sbc	140	Planetary nebula IC 351 SW 13′.7.
03 51 03.9	+36 53 47	UGC 2877	16.3	0.8 x 0.5	15.1	SBdm:	10	
03 52 17.0	+36 14 12	UGC 2881	13.3	1.5 x 1.0	13.6	$(R)SAB0^+$	148	Close pair of stars, mags 10.74v and 11.2, SE 2′.3.
03 52 35.8	+34 41 03	UGC 2882	17.1	1.7 x 0.7	17.2	Sdm: IV	94	Faint core, low surface brightness.
03 53 02.3	+35 35 20	UGC 2885	12.8	3.2 x 1.6	14.4	SA(rs)c	40	Mag 10.60v star superimposed NE end; mag 8.67v star SE 4′.5.
03 53 04.0	+34 57 29	UGC 2886	13.3	0.8 x 0.7	12.5	Sm: IV-V	9	Low surface brightness; another low surface brightness, **UGC 2887**, SE 1′.0.
03 53 19.2	+32 19 47	UGC 2888	14.1	1.2 x 0.6	13.6	S?	20	MCG +5-10-1 S 2′.0.
03 53 37.5	+37 15 51	UGC 2889	14.0	1.5 x 0.7	13.9	SABbc:	100	Faint, almost stellar anonymous galaxy SE 2′.0.
03 56 38.5	+34 51 00	UGC 2901	14.8	1.1 x 1.1	14.8	S?		
04 00 48.0	+35 00 40	UGC 2920	14.0	2.1 x 0.3	13.3	Scd:	63	Mag 10.4 star S 1′.2.
04 04 37.0	+33 17 23	UGC 2944	13.2	1.1 x 0.5	12.4	S?	96	Mag 10.14v star E 2′.6.
04 04 39.7	+33 48 29	UGC 2945	13.7	1.4 x 1.0	13.9	S0/a	60	
04 05 06.3	+31 01 34	UGC 2950	15.2	1.7 x 0.3	14.4	S?	167	Several bright knots or stars N end.
04 05 55.0	+37 03 48	UGC 2952	14.2	1.1 x 0.6	13.6	S	33	Mag 10.62v star W 9′.1.
04 06 28.5	+31 16 56	UGC 2956	14.6	1.3 x 0.4	13.7	Scd:	124	
04 09 43.4	+37 00 35	UGC 2971	14.2	1.5 x 0.9	14.4	S	143	Stellar nucleus, uniform envelope.
04 09 54.1	+36 31 01	UGC 2972	15.2	1.5 x 0.9	15.4		170	small nucleus, uniform envelope.
04 12 15.6	+34 54 55	UGC 2978	14.7	1.5 x 0.4	14.0	SB?	117	Mag 11.2 star NE 1′.9.
04 13 56.1	+29 09 24	UGC 2989	14.4	1.5 x 0.7	14.3	SB?	30	Almost stellar nucleus, uniform envelope.
04 14 00.6	+36 50 50	UGC 2991	15.7	1.7 x 1.1	16.2	SB(r)b	107	Mag 11.6 star superimposed near center.
04 21 19.8	+36 45 34	UGC 3016	15.7	1.5 x 0.3	14.7	Scd:	91	
04 21 52.1	+36 07 36	UGC 3021	14.2	1.5 x 1.1	14.7	E	35	Pair of stars or bright knots S of center.
04 24 35.0	+33 52 28	UGC 3028	14.8	1.5 x 0.7	14.7	S?	143	Mag 11.2 star E 2′.6.
04 27 16.2	+32 37 31	UGC 3047	14.9	1.5 x 1.3	15.5	SBcd:	30	Stellar nucleus.
04 28 26.4	+39 37 27	UGC 3052	14.1	1.5 x 1.0	14.4	SB?	10	Numerous stars superimposed; mag 9.95v star N 2′.2.
04 32 21.7	+33 13 06	UGC 3078	14.7	1.5 x 0.4	14.0	Sbc	9	

OPEN CLUSTERS

RA h m s	Dec ° ′ ″	Name	Mag	Diam ′	No. ★	B ★	Type	Notes
04 27 48.0	+30 57 00	Cz 18		6	15		cl	(A) Faint galaxy **UGC 3050** on W side.
03 44 34.0	+32 10 00	IC 348	7.3	8	21	10.0	cl	Few stars; moderate brightness range; no central concentration; detached. (A) The IC 348 name applies to both the cluster and nebulosity.
03 31 40.0	+37 23 00	NGC 1342	6.7	17	99	8.0	cl	Moderately rich in stars; moderate brightness range; no central concentration; detached.
04 21 20.0	+36 55 00	NGC 1548		30			ast	Few stars; small brightness range; not well detached. (S) Small sparse group.

BRIGHT NEBULAE

RA h m s	Dec ° ′ ″	Name	Dim ′ Maj x min	Type	BC	Color	Notes
03 44 30.0	+32 17 00	IC 348	7 x 7	R	2-5	2-4	Nebula with an 8th mag star involved. (A) The IC 348 name applies to both the cluster and nebulosity.
03 29 18.0	+31 25 00	NGC 1333	6 x 3	R	3-5	3-4	Extended NNE-SSW; brightest at the ends.
04 00 42.0	+36 37 00	NGC 1499	160 x 40	E	1-5	3-4	**California Nebula**. Reportedly visible to the unaided eye when viewed through an **O-III** filter.
04 30 12.0	+35 16 00	NGC 1579	12 x 8	R	1-5	1-4	Irregular in shape and extended roughly N-S. Brightest and concentrated in the central region; several stars involved.

(Continued from previous page)

60 BRIGHT NEBULAE 60

RA h m s	Dec ° ′ ″	Name	Dim ′ Maj x min	Type	BC	Color	Notes
03 28 18.0	+29 48 00	vdB 16	4.5 x 4.5	R	3-5	2-4	Nebulosity with a mag 9.1 star involved.
03 49 36.0	+38 59 00	vdB 24	5 x 3	R	3-5	2-4	Brightest part of vdB 24 lies immediately S of a mag 8.8 star. This 1′ patch is cometary in shape and fans away from the star.

DARK NEBULAE

RA h m s	Dec ° ′ ″	Name	Dim ′ Maj x min	Opacity	Notes
03 32 57.4	+31 10 00	B 1	30 x 30	4	Large, indefinite.
03 33 31.3	+32 19 00	B 2	20 x 20	4	Indefinite; elongated SE and NW.
03 40 01.9	+31 59 00	B 3	20 x 20	5	Irregular, dark space in nebula.
03 44 02.4	+31 48 00	B 4		5	Very large; indefinite.
03 47 53.1	+32 53 00	B 5	60 x 60	5	Indefinite; elongated NE and SW; Eta Persei near NE side.
04 40 33.4	+29 53 00	B 23	5 x 5	5	Sharply pointed to the SE.
04 42 53.3	+29 44 00	B 24	8 x 8	5	Sharply pointed to the S.
03 25 38.7	+30 17 00	B 202	33 x 12	4	Elongated NW and SE.
03 25 50.3	+30 47 00	B 203		4	Elongated E and W.
03 28 29.2	+30 11 00	B 204	14 x 14	5	Irregular.
03 28 32.0	+31 06 00	B 205	15 x 2	5	Two dark strips, N and S.
03 29 09.4	+30 11 00	B 206	5 x 5	5	
04 34 51.5	+29 36 00	B 219	55 x 3	5	Partly vacant space; NE and SW.
04 44 00.8	+31 44 00	B 221	45 x 45		Partly vacant region.

PLANETARY NEBULAE

RA h m s	Dec ° ′ ″	Name	Diam ″	Mag (P)	Mag (V)	Mag cent ★	Alt Name	Notes
03 47 33.1	+35 02 45	IC 351	18	12.4	11.9	15.8	PK 159−15.1	Mag 9.6v star SE 3'.4.
03 56 22.1	+33 52 27	IC 2003	20	12.6	11.4	15.0	PK 161−14.1	Mag 13.4 star on SW edge; close pair of stars, mags 10.04v and 10.9, NE 2'.9.
04 09 16.9	+30 46 34	NGC 1514	132	10.	10.9	9.4	PK 165−15.1	Strong dark patch S of central star.
03 45 26.6	+37 48 53	PN G156.9−13.3	34			17.7	PK 156−13.1	Mag 14.4 star on SW edge; mag 12.1 star SSE 1'.3.
04 39 48.0	+36 45 39	PN G165.5−6.5	13	13.7		18.5	PK 165−6.1	Mag 14.4 star NNE 1'.9; mag 14.3 star W 3'.5.
04 42 53.9	+36 06 50	PN G166.4−6.5	12	17.9		17.0	PK 166−6.1	Mag 10.38v star N 1'.5; mag 13.5 star E 2'.3.
04 36 37.4	+33 39 27	PN G167.4−9.1	13	15.4		15.3	PK 167−9.1	Mag 14.1 star SW 0'.6; mag 12.3 star SW 2'.4.

61 GALAXIES 61

RA h m s	Dec ° ′ ″	Name	Mag (V)	Dim ′ Maj x min	SB	Type Class	PA	Notes
02 10 38.7	+31 20 09	CGCG 504-22	14.3	0.7 x 0.4	13.0	E	12	Elongated **CGCG 504-23** SE 4'.8.
02 30 33.3	+30 52 21	CGCG 504-101	14.2	0.5 x 0.4	12.3	I?	177	Mag 11.1 star SE 1'.5.
02 34 35.8	+29 57 23	CGCG 505-17	15.4	0.5 x 0.3	13.2		99	Galaxy **V Zw 256** on N edge.
02 36 23.9	+31 42 38	CGCG 505-19	14.0	0.6 x 0.3	12.0		177	Mag 6.07v star SSE 7'.4; note that **UGC 2087** is 2'.5 W of this star.
02 14 56.0	+36 35 00	CGCG 523-2	15.1	1.0 x 0.4	13.9	Spiral	9	Located 2'.7 NNE of mag 10.09v star.
02 46 15.1	+39 21 27	CGCG 524-13	14.6	0.5 x 0.4	12.7		114	Mag 9.09v star ESE 7'.1.
03 00 06.0	+39 29 51	CGCG 524-41	13.9	0.5 x 0.4	12.0		162	Located 2'.8 S of mag 9.65v star.
03 07 46.5	+38 22 15	CGCG 524-54	14.6	0.7 x 0.5	13.3		3	Mag 11.8 star N 1'.0; NGC 1207 E 5'.6.
01 58 55.1	+36 40 28	IC 178	13.3	1.1 x 0.9	13.1	Sab	161	Mag 7.30v star SSW 4'.6.
02 00 11.7	+38 01 17	IC 179	12.6	1.6 x 1.2	13.3	E	110	Mag 10.67v star NE 2'.5.
02 36 28.0	+38 58 21	IC 239	11.1	4.6 x 4.2	14.2	SAB(rs)cd	3	Mag 9.6v star N edge, mag 8.6v star S edge.
03 01 30.5	+37 45 55	IC 278	13.2	1.1 x 1.1	13.4	E:		Mag 9.56v star N 3'.8.
03 10 13.0	+40 45 52	IC 292	13.5	1.2 x 0.6	13.0	Sdm:	75	Mag 10.13v star N 2'.4.
03 11 03.3	+40 37 19	IC 294	13.9	2.1 x 1.5	15.0	(R)SB(rs)0/a:	123	Mag 11.29v star S 4'.5.
03 15 01.4	+37 52 52	IC 304	13.8	1.2 x 0.7	13.5	Sb II-III	25	Knotty; IC 305 S 1'.4.
03 15 03.9	+37 51 33	IC 305	14.1	0.8 x 0.6	13.3	E	45	Very small, very faint, elongated anonymous galaxy on E edge.
03 16 06.4	+40 48 14	IC 309	13.5	0.9 x 0.9	13.1	SA(s)0°		Located between a pair of mag 11 stars.
03 16 46.8	+40 00 11	IC 311	14.1	0.9 x 0.6	13.3	S?	113	Mag 9.49v star ENE 6'.2.
02 16 12.8	+32 38 56	IC 1784	13.1	1.4 x 0.7	12.9	SA(rs)bc pec:	88	Two main arms; IC 1785 NE 2'.0.
02 16 21.2	+32 39 57	IC 1785	14.5	0.9 x 0.4	13.5	E	144	IC 1784 SW 2'.0; mag 11.2 star S 5'.8.
02 17 51.2	+32 23 41	IC 1789	13.7	2.2 x 0.4	13.4	Sa:	27	Mag 11.0 star N 5'.8.
02 19 01.2	+34 27 41	IC 1792	13.6	1.1 x 0.7	13.1	Compact	33	Mag 7.72v star W 4'.6.
02 21 32.4	+32 32 37	IC 1793	13.8	1.5 x 0.5	13.4	Sab	34	Mag 10.63v star NE 2'.6.
02 34 20.2	+32 25 45	IC 1815	12.9	1.4 x 1.1	13.2	SB0	141	Two small, elongated galaxies ESE and SE 1'.5.
02 38 37.0	+32 04 10	IC 1823	13.5	2.1 x 2.0	14.9	SB(r)c	72	Mag 10.42v star E 8'.4.
03 06 22.0	+36 00 50	IC 1874	13.7	0.9 x 0.5	12.7	S0/a	96	Bright N-S oriented center.
03 15 55.2	+37 09 12	IC 1900	14.0	0.5 x 0.4	12.1	S0	96	**IC 1901** ESE 3'.0.
01 56 50.3	+31 42 10	MCG +5-5-29	13.9	1.0 x 0.6	13.2	S?	144	= **Mkn 1167**.
02 04 53.2	+32 16 40	MCG +5-5-52	14.4	0.8 x 0.6	13.4	S?	138	Located between a close pair of mag 10-11 stars.
02 07 15.2	+32 57 08	MCG +5-6-7	14.0	0.6 x 0.6	12.7	SB?		MCG +5-6-8 E 2'.1.
02 07 25.5	+32 57 09	MCG +5-6-8	13.9	0.6 x 0.5	12.5	S?	0	MCG +5-6-7 W 2'.1.
02 10 59.5	+32 43 45	MCG +5-6-12	14.8	1.0 x 0.3	13.4	S?	75	**UGC 1671** S 1'.9.
02 12 06.5	+31 35 10	MCG +5-6-14	14.2	0.7 x 0.5	12.9	S?	45	Mag 9.84v star S 1'.4.
02 23 19.1	+32 11 18	MCG +5-6-35	14.4	0.5 x 0.4	12.6	S?	6	MCG +5-6-36 NE 0'.8.
02 23 22.1	+32 11 47	MCG +5-6-36	14.1	0.6 x 0.3	12.1	S?	69	MCG +5-6-35 SW 0'.8.
02 32 54.2	+30 44 39	MCG +5-7-5	14.3	0.8 x 0.7	13.5		15	Mag 11.7 star SE 1'.6.
02 43 33.0	+32 15 07	MCG +5-7-39	14.3	0.9 x 0.6	13.5		72	Mag 8.71v star E 4'.9.
03 01 45.6	+29 12 34	MCG +5-8-2	14.0	0.7 x 0.7	13.3	E/S0		Mag 9.38v star W edge.
01 56 40.0	+35 35 32	MCG +6-5-61	15.4	0.9 x 0.4	14.1	Sc	138	Mag 8.65v star N 3'.3.
01 56 58.0	+35 56 44	MCG +6-5-63	14.8	0.9 x 0.4	13.6	Spiral	150	Mag 10.20v star NW 3'.9.
01 58 35.4	+38 43 05	MCG +6-5-69	14.9	0.7 x 0.4	13.3	Sb	132	Mag 13.4 star NW edge.
01 59 21.9	+36 49 35	MCG +6-5-73	14.0	0.7 x 0.5	12.9	E?	159	**PGC 7512** W 2'.6; mag 12.3 star N 2'.1.
01 59 58.9	+36 35 25	MCG +6-5-74	15.3	0.7 x 0.4	13.7	Irr	156	Star superimposed S end; mag 12.0 star WSW 2'.1.
02 04 45.7	+35 59 32	MCG +6-5-80	15.5	0.5 x 0.4	13.9	E/S0	177	Mag 11.2 star N 1'.4.
02 05 17.4	+34 47 17	MCG +6-5-83	14.8	0.9 x 0.5	13.7	Double	150	Mag 12.3 star E 2'.0.
02 09 58.7	+35 43 12	MCG +6-5-90	14.6	0.6 x 0.5	13.2	Sb	24	**MCG +6-5-89** NW 2'.8.
02 14 08.7	+35 43 05	MCG +6-5-106	14.4	0.7 x 0.6	13.4	S0	42	Close pair of mag 13-14 stars N 1'.9.
02 15 16.9	+33 46 21	MCG +6-6-1	15.0	0.6 x 0.6	13.8	Spiral		Mag 11.7 star SW 3'.1.
02 15 55.1	+33 48 35	MCG +6-6-4	14.4	1.1 x 0.6	13.8	Spiral	72	

(Continued from previous page)

GALAXIES

RA h m s	Dec ° ′ ″	Name	Mag (V)	Dim ′ Maj x min	SB	Type Class	PA	Notes
02 16 50.9	+37 21 05	MCG +6-6-5	15.6	0.4 x 0.4	13.5	Sb		Star or large knot on S edge; mag 11.8 star NNE 1'.7.
02 17 49.6	+35 48 23	MCG +6-6-7	14.7	0.7 x 0.5	13.4	Spiral	33	**UGC 1765** SE 3'.4.
02 18 18.6	+37 27 49	MCG +6-6-12	15.3	0.5 x 0.4	13.4	Spiral	90	Mag 10.49v star ENE 1'.5.
02 19 40.2	+37 05 19	MCG +6-6-20	14.8	0.6 x 0.3	12.8	Sb	168	**UGC 1786** NW 2'.3.
02 20 24.4	+36 59 39	MCG +6-6-21	14.1	1.1 x 0.7	13.9	E/S0	27	Mag 12.3 star NE 2'.1.
02 24 55.1	+35 17 30	MCG +6-6-27	15.9	0.7 x 0.4	14.4	Sbc	102	
02 25 27.5	+37 10 25	MCG +6-6-29	14.0	0.7 x 0.6	13.1	E?	93	UGC 1882 NE 5'.0; mag 7.34v star SW 4'.0.
02 29 01.2	+38 05 49	MCG +6-6-43	14.5	0.9 x 0.8	14.0	S0/a	84	Close pair of mag 13-14 stars S 1'.3.
02 30 27.9	+38 21 46	MCG +6-6-47	14.9	0.7 x 0.4	13.3	Sb	177	
02 33 04.4	+35 30 27	MCG +6-6-53	15.3	0.5 x 0.5	13.7	Spiral		Mag 14.5 star E edge; mag 12.5 star SSE 1'.9; NGC 959 W 8'.2.
02 34 13.9	+34 28 00	MCG +6-6-57	15.2	0.6 x 0.3	13.2		60	NGC 968 WNW 1'.8.
02 34 36.9	+34 52 07	MCG +6-6-59	14.7	0.8 x 0.3	13.0	Irr	90	Located between a pair of NE-SW mag 11 stars.
02 40 14.4	+38 20 46	MCG +6-6-72	14.5	0.6 x 0.4	12.8		123	Mag 9.66v star E 2'.0; an anonymous galaxy on the S edge of this star.
02 42 34.9	+36 32 42	MCG +6-6-77	15.3	0.6 x 0.6	14.0			Mag 10.47v star N 4'.0.
02 51 37.4	+37 50 36	MCG +6-7-15	14.9	0.7 x 0.7	14.0	Spiral		Mag 12.2 star SSE 1'.1.
02 54 57.3	+39 15 22	MCG +6-7-19	15.0	0.6 x 0.4	13.3		12	Mag 9.66v star SW 2'.1.
02 59 01.0	+33 54 33	MCG +6-7-24	15.8	0.5 x 0.2	13.1		69	Located 1'.1 W of UGC 2448.
03 10 30.7	+35 04 59	MCG +6-7-46	14.6	0.8 x 0.4	13.2	S0	63	Mag 8.95v star SSE 5'.6.
03 10 59.5	+38 39 30	MCG +6-7-47	15.2	0.8 x 0.2	13.1		60	
03 11 39.6	+35 22 22	MCG +6-8-2	14.3	0.7 x 0.3	12.7	E?	75	Mag 10.8 star E 2'.6.
03 20 53.9	+37 52 01	MCG +6-8-15	14.8	0.7 x 0.5	13.6	SBc	27	Mag 10.77v star WSW 2'.4.
01 56 27.7	+36 48 09	NGC 732	13.5	1.4 x 1.0	13.7	S0	28	UGC 1416 NNE 6'.3.
01 56 24.9	+33 03 51	NGC 733	15.2	0.6 x 0.3	13.2		26	NGC 736 ESE 3'.6; **CGCG 503-53** NW 6'.2.
01 56 38.5	+34 10 30	NGC 735	13.3	1.7 x 0.6	13.1	Sb I	138	**PGC 7275** NW 1'.5. Mag 10.17v star SW 1'.4.
01 56 41.1	+33 02 36	NGC 736	12.2	1.5 x 1.5	13.0	E^+:		Line of three faint stars N of nucleus; NGC 738 on NE edge.
01 56 45.7	+33 03 30	NGC 738	14.9	0.5 x 0.3	12.7	S0	155	Located on NE edge of NGC 736.
01 56 54.8	+33 15 58	NGC 739	13.9	0.9 x 0.6	13.1	S0?	127	Almost stellar core, faint, smooth envelope.
01 56 54.9	+33 00 54	NGC 740	14.0	1.6 x 0.4	13.3	SBb?	137	Mag 10.17v star E 1'.3.
01 57 32.4	+33 12 39	NGC 750	11.9	1.7 x 1.3	12.7	E pec	162	Pair with NGC 751 on S edge; mag 9.13v star ESE 4'.8.
01 57 33.0	+33 12 07	NGC 751	12.7	1.2 x 1.2	13.2	E pec		Pair with NGC 750 on N edge.
01 57 42.3	+35 54 51	NGC 753	12.3	3.0 x 1.9	14.0	SAB(rs)bc I-II	125	several faint stars on face near edges.
01 57 50.6	+36 20 32	NGC 759	12.7	1.4 x 1.4	13.5	E		Chain of faint stars starts E of nucleus and extends S and then SE; **UGC 1434** SSW 5'.8.
01 57 49.6	+33 22 37	NGC 761	13.5	1.5 x 0.5	13.0	SBa:	143	
01 59 36.0	+30 54 31	NGC 769	12.9	0.9 x 0.6	12.1	S?	73	Very faint, elongated anonymous galaxy 1'.1 S.
02 00 14.9	+31 25 45	NGC 777	11.4	2.5 x 2.0	13.2	E1	155	Mag 8.72v star SE 6'.3.
02 00 19.6	+31 18 43	NGC 778	13.2	1.1 x 0.5	12.4	S0:	150	Lies 3'.0 SW of mag 8.72v star.
02 01 06.5	+31 52 58	NGC 783	12.1	1.6 x 1.4	12.9	Sc	35	Bright knots or superimposed stars on E and W sides of core; **UGC 1499** N 4'.7.
02 01 40.1	+31 49 33	NGC 785	13.0	1.7 x 0.9	13.3	$S0^-$:	83	Mag 8.92v star WSW 7'.2; note: **IC 1766** lies 1'.7 S of this star.
02 02 26.1	+32 04 13	NGC 789	13.5	1.2 x 0.8	13.3	S?	3	Mag 8.13v star WNW 9'.2.
02 03 28.1	+38 06 59	NGC 797	12.7	1.6 x 1.3	13.3	SAB(s)a	65	Star of very small galaxy SW edge.
02 03 19.6	+32 04 38	NGC 798	13.5	1.2 x 0.5	12.9	E	137	Mag 9.86v star NE 6'.5.
02 03 44.9	+38 15 33	NGC 801	13.1	3.2 x 0.7	13.8	Sc	150	2'.3 E of mag 10.69v star.
02 04 02.2	+30 49 56	NGC 804	13.7	1.4 x 0.3	12.6	S0	7	
02 08 08.9	+29 15 21	NGC 816	14.3	0.4 x 0.4	12.2	Compact		Compact.
02 08 44.8	+38 46 34	NGC 818	12.5	3.0 x 1.0	13.6	SABc:	113	
02 08 34.4	+29 13 58	NGC 819	13.5	0.7 x 0.5	12.2	S?	10	Mag 11.3 star N 4'.0.
02 09 25.0	+30 44 24	NGC 826	13.9	0.8 x 0.7	13.1		60	Multi-galaxy system; small companion N of bright nucleus.
02 10 09.7	+39 11 20	NGC 828	12.3	2.5 x 1.6	13.6	Sa: pec	141	Slight "hourglass" shape, faint lane down W side of nucleus.
02 11 01.4	+37 39 56	NGC 834	13.1	1.1 x 0.5	12.3	S?	20	
02 11 17.4	+37 29 48	NGC 841	12.6	1.8 x 1.0	13.1	(R')SAB(s)ab	135	Mag 8.11v star S 8'.9.
02 12 19.9	+37 28 34	NGC 845	13.5	1.7 x 0.4	12.9	Sb II	149	Mag 8.44v star S 6'.6.
02 15 00.1	+30 46 43	NGC 860	14.3	0.7 x 0.5	13.2	E	138	**PGC 8613** SSE 1'.9; **KUG 0212+305** SSE 2'.5.
02 15 51.1	+35 54 45	NGC 861	13.8	1.5 x 0.5	13.4	Sb III	38	Mag 12.6 star on S end.
02 22 00.7	+33 15 54	NGC 890	11.2	2.5 x 1.7	12.7	$SAB(r)0^-$?	54	
02 27 16.9	+33 34 44	NGC 925	10.1	10.5 x 5.9	14.5	SAB(s)d II-III	117	Many superimposed stars and relatively bright knots.
02 28 13.9	+31 18 37	NGC 931	12.8	3.9 x 0.8	13.9	SAbc II	78	Very small galaxy on N edge, above nucleus.
02 29 27.5	+31 38 28	NGC 940	12.4	1.2 x 1.0	12.5	S0:	9	
02 30 48.9	+37 08 08	NGC 949	11.8	3.0 x 1.6	13.3	SA(rs)b:? III-IV	148	Bright knots or superimposed stars N and S of core; faint, stellar anonymous galaxy N 3'.9.
02 31 09.7	+29 35 21	NGC 953	13.5	1.3 x 1.3	14.1	E		
02 32 24.2	+35 29 40	NGC 959	12.4	2.3 x 1.4	13.5	Sdm:	65	Uniform surface brightness; MCG +6-6-53 E 8'.2.
02 34 06.1	+34 28 46	NGC 968	12.2	2.7 x 1.5	13.8	E	60	MCG +6-6-57 ESE 1'.8.
02 34 08.1	+32 56 47	NGC 969	12.3	1.5 x 1.2	12.8	S0	3	**CGCG 505-8** NW 7'.5.
02 34 11.7	+32 58 36	NGC 970	14.8	0.7 x 0.2	12.5		60	Double system or superimposed star on W edge; now nonexistent NGC 972 is a mag 15.1 star E 1'.0.
02 34 13.7	+29 18 33	NGC 972	11.4	3.4 x 1.7	13.2	Sab	152	Pair of stars 1'.6 WSW, brightest of the pair is mag 9.91v.
02 34 19.8	+32 30 14	NGC 973	12.8	3.7 x 0.5	13.3	Sb III	48	Moderately dark lane runs almost full length of galaxy. Mag 7.64v star SW 4'.4; anonymous galaxy NE 3'.8.
02 34 26.0	+32 57 14	NGC 974	12.7	1.7 x 1.2	13.3	SAB(rs)b:	63	Very faint, broad extension W, and turning sharply N beyond published dimensions.
02 34 47.1	+32 50 48	NGC 978A	12.9	1.3 x 1.0	13.1	$S0^-$:	62	Attached to NGC 978B 0'.35 SSE; mag 8.12v star E 5'.9.
02 34 48.2	+32 50 32	NGC 978B	14.4	0.6 x 0.3	12.4	S0	166	Located on S edge of NGC 978A.
02 35 18.8	+40 55 35	NGC 980	13.0	1.7 x 0.9	13.4	S0	110	Very elongated **UGC 2068** SE 2'.4.
02 35 24.9	+40 52 10	NGC 982	12.5	1.5 x 0.6	12.2	Sa	132	Much elongated **UGC 2068** N 1'.5; **MCG +7-6-40** E 2'.2; and a mag 10.27v star 2'.5 SW.
02 36 49.6	+33 19 33	NGC 987	12.4	1.6 x 1.1	12.9	SB0/a	39	
02 38 55.8	+34 37 18	NGC 1002	13.1	1.3 x 0.7	12.9	SB(r)b:	146	Mag 12.5 star N 1'.1; mag 10.52v star W 8'.5.
02 39 17.1	+40 52 18	NGC 1003	11.4	4.3 x 1.3	13.2	SA(s)cd III	97	Mag 9.88v star on S edge; pair of small, faint anonymous galaxies S 6'.6.
02 39 14.8	+30 09 02	NGC 1012	12.0	2.5 x 1.1	12.9	S0/a?	24	
02 40 24.3	+39 03 46	NGC 1023	9.3	7.4 x 2.5	12.4	$SB(rs)0^-$	87	NGC 1023A located 2'.4 E of center.
02 40 36.8	+39 03 22	NGC 1023A	13.6	1.5 x 0.9	13.8	IB?	14	Located 2'.4 E of center of NGC 1023.
02 42 35.8	+34 45 49	NGC 1050	12.6	1.7 x 1.1	13.1	(R')SB(s)a	113	Close pair of mag 9.9, 10.80v stars W 5'.9.
02 43 03.2	+32 29 25	NGC 1057	14.2	1.2 x 0.8	14.0	S0	111	NGC 1061 SE 3'.1.
02 43 30.2	+37 20 25	NGC 1058	11.2	2.5 x 2.5	13.0	SA(rs)c III		Several bright knots; mag 10.23v star NW 8'.0.
02 43 14.9	+32 25 26	NGC 1060	11.8	2.3 x 1.7	13.2	$S0^-$:	75	Very small, very faint anonymous galaxy on SE edge.
02 43 15.8	+32 28 00	NGC 1061	14.1	0.9 x 0.6	13.3	I?	42	Located 2'.7 N of NGC 1060; NGC 1057 NW 3'.1.
02 43 49.9	+32 28 27	NGC 1066	13.3	1.7 x 1.4	14.2	E:	57	Mag 7.36v star SSE 6'.8; **UGC 2202** S 5'.1.
02 43 50.7	+32 30 36	NGC 1067	13.7	1.1 x 1.0	13.6	SAB(s)c I-II	165	Faint, anonymous galaxy NE 2'.3.
02 46 03.0	+40 05 36	NGC 1077A	16.3	0.5 x 0.4	14.4	SBb	30	NGC 1077B on W edge.
02 46 00.6	+40 05 23	NGC 1077B	13.6	1.0 x 0.8	13.2	Sb	165	NGC 1077A on E edge. Mag 10.9 star W 4'.5.
02 48 16.2	+34 25 09	NGC 1093	13.1	1.4 x 1.0	13.3	SABab?	97	Mag 9.81v star NNW 4'.2.
03 01 42.5	+35 12 21	NGC 1167	12.4	3.3 x 2.3	14.5	$SA0^-$	73	Strong, round core, uniform envelope; mag 10.55v star S 4'.1.

(Continued from previous page)

GALAXIES

RA h m s	Dec ° ′ ″	Name	Mag (V)	Dim ′ Maj x min	SB	Type Class	PA	Notes
03 08 15.9	+38 22 52	NGC 1207	12.6	2.2 x 1.5	13.7	SA(rs)b II	123	CGCG 524-54 W 5′.6; mag 8.56v star E 5′.8.
03 09 17.3	+38 38 57	NGC 1213	14.4	1.8 x 1.4	15.3	SA(s)dm	54	Faint stellar nucleus with several bright knots; mag 9.57v star E 9′.1.
03 11 05.5	+35 23 09	NGC 1226	12.9	1.6 x 1.3	13.7	E	95	**UGC 2579** NE 6′.2.
03 11 07.8	+35 19 29	NGC 1227	14.2	1.1 x 1.0	14.2	(R)SB(s)0$^+$	63	Almost stellar nucleus, faint envelope; mag 10.33v star W 4′.6.
03 12 33.1	+39 19 03	NGC 1233	13.2	1.8 x 0.6	13.1	Sb III	27	
01 55 58.7	+37 07 41	UGC 1398	13.9	1.0 x 1.0	13.7	Scd?		Mag 14 stars on E and N edges; mag 6.28v star W 12′.1.
01 56 04.5	+36 07 51	UGC 1400	13.0	2.3 x 0.3	12.4	Sb III	156	Mag 9.07v star N 2′.6.
01 56 44.1	+36 23 00	UGC 1415	13.7	1.2 x 0.3	12.4	S0/a	1	
01 56 46.4	+36 53 09	UGC 1416	13.8	1.0 x 0.6	13.1	S?	65	Mag 11.7 star NW 2′.0; NGC 732 SSW 6′.3.
01 56 57.0	+40 20 27	UGC 1418	13.2	1.7 x 1.1	13.7	S0	50	Mag 10.05v star E 2′.2.
01 57 06.8	+32 47 15	UGC 1422	13.4	1.0 x 0.4	12.3	S?	90	
01 59 06.8	+36 03 43	UGC 1459	14.6	1.5 x 0.3	13.6	Scd:	106	Mag 12.2 star N 2′.2.
01 59 04.9	+36 15 29	UGC 1460	13.7	1.6 x 0.9	14.0	Sa	150	Triangle of stars superimposed E, W and S edges; mag 11.9 star NE 1′.0.
01 59 42.6	+32 04 56	UGC 1470	15.9	1.5 x 0.2	14.4	Sdm:	138	Mag 11.0 star NW 3′.1.
02 00 11.3	+37 36 06	UGC 1474	14.0	0.8 x 0.7	13.2	SB(s)dm	21	Located 2′.2 SW of mag 8.24v star.
02 00 55.3	+38 12 37	UGC 1493	13.0	2.1 x 0.7	13.3	SBab?	87	A line of stars, mags 12.2, 13.7 and 13.0, plus a small, faint anonymous galaxy NNE 1′.0.
02 01 13.7	+30 21 27	UGC 1502	14.6	1.8 x 0.9	15.0	Im:	141	Dwarf, low surface brightness.
02 01 20.0	+33 19 46	UGC 1503	13.4	0.8 x 0.7	12.8	E	75	Mag 8.68v star SE 2′.3.
02 05 26.7	+31 10 28	UGC 1577	12.9	2.2 x 1.5	14.0	SBbc	58	Strong SE-NW bar; two stars superimposed inside N edge.
02 05 33.8	+34 52 53	UGC 1581	14.5	1.7 x 0.6	14.3	Sdm:	160	**CGCG 522-109** N 2′.3.
02 05 39.5	+39 50 20	UGC 1582	13.7	1.5 x 0.5	13.2	Sd	33	Located 1′.1 N of mag 9.01v star.
02 06 03.9	+29 47 32	UGC 1590	12.7	1.9 x 1.4	13.6	S0$^-$:	130	Mag 11.8 star NW 1′.7; faint anonymous galaxy NW 2′.8.
02 06 12.5	+29 58 02	UGC 1591	13.8	1.5 x 0.3	12.7	S?	152	Small, round galaxy **V Zw 182** on N end.
02 06 30.2	+29 59 32	UGC 1596	13.5	1.1 x 0.6	13.0	SAB0°	125	Bright nucleus; mag 13.4 star on SE end.
02 07 17.9	+37 05 49	UGC 1604	13.8	1.5 x 0.9	14.0	S0?	51	
02 09 10.2	+31 59 37	UGC 1641	13.7	1.1 x 0.8	13.4	SA(s)dm	78	Stellar nucleus, bright knot NE edge.
02 09 26.6	+37 15 30	UGC 1650	16.4	2.0 x 0.2	15.2	Scd?	28	Razor thin; mag 11.8 star SW 2′.5.
02 09 40.2	+35 47 39	UGC 1651	12.8	2.0 x 1.1	13.8	E?	108	Double system, includes E component **MCG +6-5-88A**.
02 10 09.2	+36 42 18	UGC 1654	14.2	0.9 x 0.5	13.2	Sc	39	Mag 10.08v star NE 3′.0.
02 10 18.6	+38 35 34	UGC 1660	14.6	0.8 x 0.5	13.5	Sd	27	Mag 12.3 star S edge; mag 9.64v star NNE 2′.6.
02 10 48.8	+34 59 01	UGC 1666	15.1	0.7 x 0.7	14.2	Disturbed		**MCG +6-5-97** S 0′.9.
02 10 41.9	+35 11 53	UGC 1668	14.8	1.1 x 1.0	14.8	Pec	20	Bright, diffuse center.
02 11 11.1	+38 45 23	UGC 1674	14.3	1.0 x 0.5	13.4	Sd	60	Short, narrow extension SW end leading to small knot or star.
02 11 35.3	+31 30 32	UGC 1682	14.0	1.0 x 0.4	12.8	Scd:	107	Uniform surface brightness; located 2′.4 ESE of mag 6.23v star.
02 11 38.7	+34 02 37	UGC 1685	14.1	1.1 x 0.3	12.7	Sab	72	Mag 10.40v star N 1′.5.
02 11 57.5	+29 18 46	UGC 1690	14.8	1.5 x 0.2	13.3	S?	103	Pair of stars, mags 12.2 and 11.3, NE 2′.0.
02 12 13.0	+39 13 58	UGC 1691	13.2	1.8 x 1.0	13.7	S0	48	Bright core, slightly offset to the NE, two stars superimposed; mag 10.62v star N 3′.1.
02 12 28.2	+29 51 23	UGC 1696	14.5	0.9 x 0.6	13.7	Sdm		
02 12 46.0	+36 18 09	UGC 1701	13.9	1.2 x 0.9	13.8	(R′)Sb: II-III	102	Bright center, faint envelope.
02 12 54.9	+32 48 52	UGC 1703	14.7	0.9 x 0.8	14.2	Im:	6	Dwarf, extremely low surface brightness.
02 14 34.2	+37 24 27	UGC 1721	13.2	1.8 x 1.5	14.1	SB(rs)bc	132	Large, diffuse core; mag 14.8 star NW edge.
02 14 50.6	+31 28 11	UGC 1726	13.8	1.7 x 0.4	13.3	Sbc	63	Located between a pair of mag 12.1 and 12.8 stars.
02 15 04.3	+32 43 25	UGC 1729	14.4	1.2 x 0.7	14.1	SAB(s)cd	44	**UGC 1730** ESE 3′.6; mag 8.70v star WNW 8′.1.
02 15 38.2	+35 31 20	UGC 1735	12.8	1.1 x 0.8	12.6	S0$^-$:	54	Almost stellar galaxy **VI ZC 198** NE 2′.5.
02 16 23.5	+31 59 56	UGC 1750	13.7	1.3 x 0.4	12.8	Sab	71	Elongated **UGC 1734** SW 13′.9.
02 17 23.1	+38 24 46	UGC 1757	13.2	1.1 x 0.4	12.2	S?	87	Mag 9.98v star W 2′.9.
02 18 05.2	+38 04 24	UGC 1767	13.4	1.1 x 0.9	13.2	Im	113	Patchy; mag 9.08v star W 1′.9.
02 18 11.3	+37 05 42	UGC 1769	13.2	1.0 x 0.5	12.3	Sbc	123	Mag 7.83v star ESE 5′.1; **CGCG 523-16** ENE 6′.4; **CGCG 523-7** W 10′.5.
02 18 27.4	+38 01 22	UGC 1772	13.0	0.8 x 0.5	11.9	I?	143	**CGCG 523-11** SW 5′.1.
02 18 43.0	+35 27 44	UGC 1776	14.1	1.3 x 0.8	14.0	SB(s)b	149	Strong NE-SW bar with dark patches N and S.
02 18 51.5	+33 43 25	UGC 1778	13.8	1.0 x 0.6	13.1	SAdm:	178	**CGCG 523-19** NNE 7′.1.
02 18 56.7	+40 33 49	UGC 1780	14.7	1.6 x 0.4	14.0	IBm:	158	Mag 11.1 star ENE 1′.3.
02 19 38.9	+37 56 04	UGC 1787	14.0	1.2 x 0.5	13.3	Sdm:	117	Mag 9.94v star E 3′.2.
02 19 41.4	+36 37 34	UGC 1788	14.2	0.9 x 0.6	13.6	E:	21	**UGC 1784** WNW 7′.0.
02 19 52.9	+29 02 08	UGC 1792	13.3	2.2 x 1.2	14.2	SAB(r)c	0	Mag 10.6 star NE 3′.1.
02 20 23.1	+40 47 31	UGC 1796	14.5	1.1 x 1.1	14.5	SAB(s)dm		Stellar nucleus.
02 20 29.5	+35 12 14	UGC 1800	15.1	0.9 x 0.5	14.1	S	160	Mag 12.5 star NW 1′.0.
02 20 58.0	+38 39 26	UGC 1804	14.5	1.5 x 0.7	14.4	SAm?	95	Mag 10.50v star on N edge.
02 20 59.3	+32 50 21	UGC 1805	13.6	0.9 x 0.3	12.0	Sc	164	Stellar **Mkn 1032** S 7′.9.
02 21 28.1	+39 22 18	UGC 1810	12.6	1.9 x 1.3	13.5	SA(s)b pec	50	Very long, narrow arm along E edge; pair with **UGC 1813** SE; mag 9.32v star E 3′.1.
02 21 46.2	+33 01 14	UGC 1820	14.4	1.7 x 0.3	13.5	Scd:	124	Close pair of mag 12.3 and 13.3 stars NE 3′.0.
02 22 03.6	+32 14 07	UGC 1825	14.2	0.9 x 0.6	13.3	SA(s)c	130	
02 22 05.3	+33 56 38	UGC 1826	13.8	1.0 x 0.4	12.6	SB?	45	Bright N-S center, two, thin, main arms.
02 22 22.5	+33 44 53	UGC 1829	14.8	1.0 x 0.7	14.3	Irr	132	Mag 10.92v star SE 3′.4.
02 24 29.7	+40 52 10	UGC 1855	13.9	1.0 x 0.6	13.2	SB(s)a	88	Mag 10.49v star ENE 2′.5.
02 24 31.6	+31 36 55	UGC 1856	14.2	2.0 x 0.3	13.5	Sd	123	Pair of mag 10.5v stars along NE edge.
02 24 34.2	+33 10 04	UGC 1857	14.5	1.4 x 0.6	14.2	Sd:	44	
02 25 00.1	+36 02 11	UGC 1865	13.8	3.2 x 2.6	15.9	Sm: V	80	Dwarf spiral, very low surface brightness.
02 25 38.4	+36 57 52	UGC 1877	13.2	1.2 x 1.0	13.4	E	48	Mag 7.29v star NE 3′.7.
02 25 45.4	+37 13 53	UGC 1882	13.8	1.2 x 0.9	13.8	Sc	51	Mag 8.61v star ESE 2′.4.
02 26 00.6	+39 28 13	UGC 1886	11.9	3.7 x 2.0	14.0	SAB(rs)bc II	35	Mag 11.8 star on E edge and mag 11.8 on W edge; mag 10.75v star NE 3′.1.
02 26 03.0	+30 10 21	UGC 1889	13.7	1.1 x 0.9	13.6	SB0	162	
02 26 07.9	+31 54 42	UGC 1890	13.3	2.3 x 1.2	14.3	Sab	55	Located 2′.0 N of mag 8.16v star.
02 26 20.8	+30 26 41	UGC 1896	13.7	1.1 x 0.6	13.1	SA(r)0°	5	**CGCG 504-82** N 8′.6.
02 26 57.3	+35 10 54	UGC 1910	13.8	1.6 x 0.4	13.1	S?	175	Mag 12.6 star on E edge.
02 27 38.5	+36 09 02	UGC 1919	13.8	1.5 x 1.0	14.1	SB(s)b	45	Faint star N of nucleus, mag 14.6 star on NE edge.
02 27 49.8	+31 43 40	UGC 1924	14.5	1.7 x 0.2	13.2	Scd:	3	Mag 12.6 star N 1′.7.
02 28 35.2	+37 57 09	UGC 1941	15.1	1.2 x 0.9	15.0		108	Small triangle of stars, mags 14.1 and 12.6, W 2′.2.
02 29 26.5	+31 28 15	UGC 1963	13.4	1.1 x 0.9	13.2	SAB(rs)ab II	53	Bright center, mottled surface.
02 30 06.7	+38 06 01	UGC 1972	13.8	1.1 x 0.6	13.4	E?	9	Mag 8.84v star NE 3′.0.
02 30 21.6	+35 19 34	UGC 1976	13.9	1.5 x 0.7	13.8	Sb II-III	113	Mag 13.8 star on W edge.
02 30 33.9	+32 10 30	UGC 1980	13.6	1.0 x 0.4	12.5	S?	147	
02 31 14.2	+40 23 24	UGC 1988	13.9	1.0 x 0.4	12.7	Sab	123	Mag 12.5 star SE 1′.1.
02 31 40.2	+39 22 40	UGC 1993	13.3	2.2 x 0.4	13.0	Sb III	140	
02 32 26.4	+39 24 23	UGC 2003	14.1	1.0 x 0.3	12.6	S	140	Mag 8.14v star S 2′.3.
02 32 38.1	+31 33 50	UGC 2011	14.3	1.2 x 0.7	14.0	SAB(r)ab III	104	Mag 12.0 star on N edge; mag 10.6 star N 2′.6.
02 32 54.1	+38 40 55	UGC 2014	15.3	1.9 x 0.9	15.8	Im: V	176	Dwarf, low surface brightness; mag 10.78v star SE 3′.5.
02 32 55.3	+34 51 40	UGC 2015	14.6	0.8 x 0.8	13.9	S0?		
02 33 17.1	+32 44 47	UGC 2022	14.2	0.8 x 0.8	13.7	E:		

(Continued from previous page)

GALAXIES

RA h m s	Dec ° ′ ″	Name	Mag (V)	Dim ′ Maj x min	SB	Type Class	PA	Notes
02 33 18.1	+33 29 25	UGC 2023	13.3	2.8 x 2.6	15.3	Im: V	144	Dwarf, low surface brightness; N-S line of four mag 12 stars W edge.
02 33 42.7	+37 39 59	UGC 2033	13.7	1.5 x 0.7	13.6	SB(s)b	154	Mag 12.4 star on W edge.
02 33 43.2	+40 31 44	UGC 2034	13.2	2.5 x 1.9	14.8	Im IV-V	170	Dwarf irregular, low surface brightness; mag 12.1 star E of center.
02 34 29.1	+29 45 02	UGC 2053	14.6	1.9 x 0.9	15.1	Im V	34	Dwarf irregular, low uniform surface brightness.
02 34 33.2	+33 56 33	UGC 2054	14.3	1.3 x 0.7	14.0	Sdm:	80	Located 1′.6 N of mag 8.48v star.
02 35 22.5	+37 29 12	UGC 2065	14.0	1.0 x 1.0	13.8	Sm		UGC 2067 NE 2′.5.
02 35 29.5	+37 31 08	UGC 2067	13.7	2.3 x 0.5	13.7	Sab	158	UGC 2065 SW 2′.5.
02 35 37.6	+37 38 20	UGC 2069	12.4	2.3 x 1.4	13.5	SAB(s)d III-IV	65	Patchy, several superimposed stars and bright knots.
02 36 11.2	+34 35 44	UGC 2077	14.5	0.7 x 0.7	13.6	Scd:		
02 36 30.4	+32 42 54	UGC 2083	14.2	1.5 x 0.2	12.8	Sbc	49	
02 36 37.8	+34 36 26	UGC 2090	14.3	0.9 x 0.9	13.9	SAB(s)b II-III		
02 36 51.5	+36 06 41	UGC 2094	12.8	1.4 x 1.1	13.1	SB(rs)c	162	Mag 10.39v star N edge; mag 9.99v star NNW 2′.3.
02 37 39.9	+34 25 55	UGC 2105	13.5	1.5 x 1.2	14.0	(R′)SB(s)a: III	60	Located between a pair of mag 12 stars.
02 37 58.4	+34 14 25	UGC 2109	13.5	1.7 x 1.5	14.4	SAB(s)d	141	Several bright knots.
02 38 10.9	+30 50 50	UGC 2116	13.9	1.2 x 1.0	13.9	Sdm IV	3	Mag 7.37v star SE 4′.1.
02 38 27.6	+29 45 34	UGC 2122	13.6	1.0 x 1.0	13.5	SAB(s)c I		Strong dark patch E of nucleus.
02 38 47.2	+40 41 53	UGC 2126	14.3	1.0 x 0.8	13.9	SABdm?	0	Mag 13.3 star on W edge.
02 38 40.2	+33 26 43	UGC 2131	14.6	1.5 x 0.6	14.3	SB(r)d	93	Located 1′.9 NW of mag 7.27v star.
02 39 36.8	+36 04 48	UGC 2143	13.5	0.5 x 0.5	11.8	I?		Mag 12.4 star E 1′.5; mag 11.3 star SE 1′.8.
02 40 19.2	+32 15 40	UGC 2156	13.3	1.7 x 1.3	14.1	SA(r)c	153	Mag 7.44v star NNE 3′.9.
02 40 25.2	+38 33 44	UGC 2157	14.1	1.8 x 0.5	13.8	Sdm:	39	Mag 10.00v star SE 1′.9.
02 41 15.6	+38 44 35	UGC 2165	14.4	1.5 x 0.7	14.3		160	Mag 11.4 star on W edge.
02 41 36.2	+37 13 31	UGC 2169	14.6	1.2 x 0.7	14.3	S0	50	
02 42 05.8	+32 22 42	UGC 2174	13.8	2.1 x 2.0	15.2	SAB(s)c	45	Stellar nucleus, branching arms; mag 9.04v star NE 2′.4.
02 42 33.6	+35 12 08	UGC 2179	13.9	1.5 x 1.5	14.6	SB(s)cd		Pair of mag 11 stars NW 2′.8.
02 42 40.3	+39 31 59	UGC 2180	13.8	1.1 x 0.4	12.8	S?	40	Mag 10.5 star NE 2′.1.
02 43 11.0	+40 25 42	UGC 2185	12.8	2.6 x 0.6	13.2	Scd:	144	Very knotty; **MCG +7-6-57** NW 3′.1.
02 43 26.1	+31 28 14	UGC 2197	14.4	1.4 x 0.9	14.5	Scd:	160	Mag 14.6 star SW edge.
02 43 44.5	+32 29 42	UGC 2201	14.9	1.4 x 0.3	13.8	Sd	101	Located 1′.7 NNW of NGC 1066.
02 43 55.1	+33 10 03	UGC 2205	14.3	1.1 x 0.7	13.9	SABc:	30	
02 43 52.5	+33 20 53	UGC 2206	13.9	0.9 x 0.5	12.9	S?	105	Mag 12.3 star W 2′.0; mag 11.5 star NW 1′.7.
02 44 33.8	+37 58 38	UGC 2213	14.5	1.0 x 0.9	14.2	Sb III	108	Mag 8.92v star NNW 1′.2.
02 44 58.0	+32 42 21	UGC 2218	14.6	0.8 x 0.8	14.0	Sb		Mag 10.34v star on E edge.
02 44 57.9	+30 22 39	UGC 2221	17.3	0.9 x 0.2	15.3	Sdm?	80	
02 45 09.8	+32 59 19	UGC 2222	13.6	1.4 x 0.6	13.4	E?	96	**UGC 2225** on E end.
02 45 14.4	+35 11 15	UGC 2223	14.1	1.1 x 0.3	12.8	Scd:	47	Mag 10.27v star NW 1′.3.
02 46 35.8	+32 26 56	UGC 2239	14.7	1.7 x 0.3	13.8	SBbc?	13	
02 46 58.0	+39 00 47	UGC 2243	14.0	1.3 x 0.4	13.2	SBbc	100	Mag 11.0 star WNW 2′.9.
02 47 40.9	+40 29 40	UGC 2256	13.8	1.3 x 0.7	13.5	S0	85	Several stars superimposed in N-S line over nucleus; mag 12.9 star on E end.
02 47 55.4	+37 32 20	UGC 2259	13.2	2.6 x 2.0	14.8	SB(s)dm III-IV	155	**MCG +6-7-10** just off N edge; mag 10.10v star just off NE edge.
02 48 49.8	+40 40 48	UGC 2277	14.6	1.0 x 1.0	14.4	Compact		
02 49 45.5	+38 04 07	UGC 2305	14.3	1.1 x 0.7	13.9	SABcd	20	Patchy; bright knot W edge; mag 11.7 star E 1′.7.
02 50 59.7	+37 27 58	UGC 2328	12.5	1.2 x 0.9	12.6	E	70	Pair of stars, mags 11.0 and 12.4, N 1′.8.
02 53 48.8	+39 34 39	UGC 2363	13.7	1.5 x 1.0	14.0	Compact	156	Bright core, uniform envelope; mag 11.09v star N 1′.3.
02 54 22.3	+31 17 28	UGC 2376	15.2	1.2 x 1.0	15.3	Scd:	36	Stellar nucleus, smooth envelope.
02 55 46.5	+33 46 01	UGC 2392	14.1	1.8 x 0.5	13.8	Scd?	12	Mag 13.4 star S edge.
02 58 03.5	+35 15 30	UGC 2435	13.4	2.0 x 1.4	14.4	SA(s)cd	145	Brighter nucleus, several small knots.
02 59 05.4	+33 54 48	UGC 2448	14.7	0.9 x 0.3	13.1	S?	99	Mag 8.30v star ESE 7′.7; MCG +6-7-24 E 1′.1.
02 59 58.7	+36 49 13	UGC 2456	12.6	2.0 x 1.5	13.7	(R)SB(s)0$^+$	72	Diffuse SE-NW core, faint envelope.
03 00 37.4	+40 15 04	UGC 2463	13.8	2.0 x 0.9	14.3	SABm	95	Mag 10.64v star WNW 2′.7.
03 00 37.5	+35 10 06	UGC 2465	13.9	1.2 x 0.3	12.6	Sa:	144	Mag 8.43v star W 3′.9.
03 00 37.0	+35 37 44	UGC 2466	15.7	1.5 x 0.5	15.2	IAm	172	Mag 11.2 star on S edge; mag 8.78v star N 3′.1.
03 01 37.1	+31 49 08	UGC 2483	15.1	1.0 x 0.4	14.0	S?	137	Mag 10.73v star SE 2′.3; mag 11.1 star NE 2′.4.
03 01 53.8	+35 43 59	UGC 2491	13.3	1.4 x 1.2	13.7	SB(r)0/a	0	Bright center in N-S bar; the strongly disrupted double system **UGC 2493** NE 2′.1.
03 02 04.0	+36 05 57	UGC 2494	13.8	1.0 x 0.5	12.9	S?	29	Stellar anonymous galaxy NE 1′.7.
03 02 08.1	+29 06 22	UGC 2497	13.8	3.1 x 0.7	14.5	Sdm	70	Almost stellar galaxy **V Zw 312** NNE 2′.4.
03 05 44.6	+36 46 56	UGC 2526	12.5	3.6 x 0.5	13.0	Sb III	136	Mag 6.99v star NW end.
03 07 00.6	+36 10 02	UGC 2540	14.6	1.7 x 0.2	13.3	Sab	69	Mag 12.5 star S 1′.0.
03 07 23.5	+37 50 09	UGC 2543	14.0	1.9 x 0.5	13.8	SAdm	105	Located 1′.6 N of mag 7.93v star.
03 07 37.9	+39 16 14	UGC 2546	14.5	0.4 x 0.4	12.4	(R)SB(r)b?		
03 08 15.8	+36 26 55	UGC 2550	14.0	1.5 x 0.7	13.9	Scd:	95	Uniform surface brightness.
03 14 40.6	+39 37 03	UGC 2604	13.7	1.3 x 0.9	13.7	SAB(s)c	140	Mag 10.6 star 2′.4 E.
03 16 00.9	+40 53 13	UGC 2617	13.2	2.2 x 0.6	13.4	SAB(s)d	176	Numerous knots and/or superimposed stars; mag 10.6 star WNW 2′.0.
03 16 28.2	+35 03 56	UGC 2623	14.2	1.8 x 1.4	15.1	SB(s)d	120	Mag 7.59v star W 5′.5.
03 16 59.4	+31 34 02	UGC 2627	13.8	1.5 x 1.3	14.4	SA(s)c II-III	78	Weak dark patch E of nucleus.
03 17 07.2	+31 34 58	UGC 2629	14.6	0.8 x 0.8	13.9	SBcd:		Uniform surface brightness.
03 17 22.7	+36 34 05	UGC 2633	15.4	1.5 x 0.2	14.0	Scd:	81	Mag 13.1 star S 0′.8.
03 17 31.6	+37 02 47	UGC 2636	15.9	1.7 x 0.2	14.5	Scd:	10	**UGC 2630** NNW 6′.8.
03 18 38.5	+37 36 28	UGC 2653	15.1	1.3 x 0.5	14.5	Sd	155	Mag 10.49v star W 1′.3.
03 18 53.6	+40 35 42	UGC 2659	13.8	1.2 x 0.3	12.5	Sbc	65	Mag 9.76v star E 2′.4.
03 19 12.0	+39 26 32	UGC 2661	13.6	1.3 x 0.3	12.4	S0	147	
03 20 34.5	+37 29 40	UGC 2678	14.7	1.0 x 0.8	14.3	SBb:	51	Stellar nucleus; mag 11.8 star WNW 1′.9.
03 20 51.8	+38 15 16	UGC 2685	13.5	2.0 x 1.5	14.5	SAB(s)b	9	Bright E-W core, faint envelope.
03 21 01.3	+40 47 49	UGC 2686	14.5	1.1 x 0.5	13.7	SABa	43	Mag 10.37v star N 1′.9.
03 21 27.7	+40 48 05	UGC 2689	13.9	1.2 x 0.6	13.5	S0?	127	Several faint stars superimposed; mag 11.1 star NW 2′.9.
03 22 03.0	+40 51 47	UGC 2698	12.9	1.0 x 0.7	12.5	E	105	Mag 10.51v star on E edge.
03 22 55.3	+37 01 54	UGC 2702	14.6	2.2 x 1.8	15.9		36	Stellar nucleus, smooth envelope; mag 9.25v star SW 1′.6.
03 23 42.9	+36 55 07	UGC 2706	14.7	0.8 x 0.7	14.0	Sdm:	81	Mag 11.3 star WSW 1′.8.
03 23 49.1	+40 33 26	UGC 2708	13.8	1.3 x 1.0	14.0	S0	24	Very faint, elongated anonymous galaxy N 1′.5.
03 23 53.8	+38 40 36	UGC 2709	13.5	1.9 x 0.7	13.7	SABb	3	
03 23 58.5	+37 45 18	UGC 2710	13.5	1.1 x 0.8	13.3	S0	6	Bright knot, or star, SE of nucleus; several faint stars superimposed on S edge.

GALAXY CLUSTERS

RA h m s	Dec ° ′ ″	Name	Mag 10th brightest	No. Gal	Diam ′	Notes
01 57 18.0	+32 13 00	A 278	15.6	38	19	Few PGC and numerous anonymous GX in N-S band along center line.

(Continued from previous page)

GALAXY CLUSTERS

RA h m s	Dec ° ′ ″	Name	Mag 10th brightest	No. Gal	Diam ′	Notes
02 45 48.0	+36 51 00	A 376	15.4	36	34	**UGC 2232** NE of center 4.′8; numerous stellar anonymous GX populate entire diameter of A 376.
03 01 42.0	+35 49 00	A 407	14.7	46	39	Majority of galaxies in a narrow band along the N-S centerline of cluster.

OPEN CLUSTERS

RA h m s	Dec ° ′ ″	Name	Mag	Diam ′	No. ★	B ★	Type	Notes
01 57 35.0	+37 50 00	NGC 752	5.7	75	77	8.0	cl	Rich in stars; moderate brightness range; slight central concentration; detached.

GALAXIES

RA h m s	Dec ° ′ ″	Name	Mag (V)	Dim ′ Maj x min	SB	Type Class	PA	Notes
00 45 41.5	+38 02 09	Andromeda I	12.6	4.0 x 3.0	15.4	E3 pec?	156	
01 16 26.3	+33 25 37	Andromeda II	12.5	2.0 x 2.0	14.1	E?		
00 39 45.0	+33 09 27	CGCG 500-65	14.0	0.8 x 0.7	13.2		57	Mag 8.73v star W 8.′9.
00 42 48.0	+29 55 18	CGCG 500-74	14.2	0.5 x 0.4	12.5	E	120	Mag 12.4 star NW 1.′7.
00 44 08.0	+30 21 00	CGCG 500-80	14.3	0.6 x 0.4	12.7	S0	69	Close pair of mag 10.7, 12.3 stars WSW 3.′1.
00 47 31.7	+31 39 19	CGCG 501-15	14.5	0.9 x 0.4	13.3	S?	117	Mag 13.8 star E 0.′7; mag 10.90v star WSW 5.′9.
00 59 32.3	+31 47 50	CGCG 501-56	14.1	0.5 x 0.4	12.2		48	Mag 7.78v star NE 6.′4; CGCG 501-58 NE 4.′8.
00 59 53.4	+31 49 34	CGCG 501-58	13.8	0.3 x 0.3	11.1	SA0		Located 1.′9 SW of a mag 7.78v star.
01 21 05.7	+33 22 42	CGCG 502-43	14.4	0.4 x 0.4	12.5	E		Mag 9.84v star NW 3.′7; note: **CGCG 502-39** on N edge of this star.
01 21 17.5	+33 05 24	CGCG 502-44	14.6	0.4 x 0.4	12.6	E		Mag 12.3 star 1.′0 S.
01 26 12.5	+33 24 18	CGCG 502-84	14.0	0.9 x 0.3	12.5	S0	15	Mag 10.24v star ESE 8.′1.
01 02 31.1	+34 16 31	CGCG 520-8	14.3	0.6 x 0.2	11.8	Sb	147	Mag 8.85v star E 9.′3.
01 20 28.4	+38 09 17	CGCG 521-2	14.0	0.6 x 0.3	12.0	S?	168	Mag 12.4 star SE 1.′0; mag 11.2 star NNW 3.′1.
01 44 38.4	+34 39 33	CGCG 521-74	14.7	0.6 x 0.3	12.7	Spiral		Mag 10.02v star W 9.′3.
01 46 27.4	+34 55 32	CGCG 522-2	14.0	0.7 x 0.4	12.5	Sb	18	Mag 13.1 star SW 1.′1; mag 8.83v star WSW 12.′8.
01 50 41.2	+33 44 23	CGCG 522-19	14.4	0.4 x 0.3	11.9		177	Mag 8.71v star E 12.′0.
01 52 09.1	+39 22 52	CGCG 522-28	14.0	0.5 x 0.4	12.1		69	Mag 10.56v star ENE 7.′0; mag 10.39v star SW 9.′3.
00 42 22.0	+29 38 27	IC 43	13.2	1.5 x 1.2	13.7	SABc II-III	117	Mag 10.4 star SE 3.′4; **IC 45** N 3.′4.
01 00 32.6	+30 47 46	IC 66	14.1	1.2 x 0.5	13.4	Sa	125	Mag 10.64v star NW 6.′7.
01 01 23.8	+31 02 27	IC 69	13.7	0.9 x 0.8	13.2	S?	66	Mag 8.42v star NE 4.′1.
01 20 33.3	+29 36 58	IC 96	13.9	0.9 x 0.3	12.3	S?	48	Mag 11.2 star WNW 7.′8; IC 1672 N 5.′1.
01 55 10.5	+35 16 56	IC 171	12.2	1.4 x 1.0	12.7	E?	132	Mag 11.1 star on E edge; galaxy **KUG 0152+349** SSE 6.′4.
01 58 55.1	+36 40 28	IC 178	13.3	1.1 x 0.9	13.1	Sab	161	Mag 7.30v star SSW 4.′6.
02 00 11.7	+38 01 17	IC 179	12.6	1.6 x 1.2	13.3	E	110	Mag 10.67v star NE 2.′5.
01 07 22.6	+33 04 00	IC 1619	14.6	0.7 x 0.5	13.3	Sb	102	Located between a close pair of mag 12 stars N and S.
01 11 37.5	+33 21 14	IC 1636	14.4	0.5 x 0.3	12.3	S0	90	Mag 10.9 star NW 4.′6.
01 12 21.9	+33 21 50	IC 1638	14.1	0.7 x 0.7	13.4	S0		Mag 8.50v star E 8.′1.
01 13 14.6	+38 53 06	IC 1647	15.0	0.8 x 0.3	13.3	Sc	39	Mag 10.47v star SW 3.′5.
01 13 42.2	+33 13 05	IC 1648	14.4	0.6 x 0.4	12.7	S0	132	Mag 8.34v star W 5.′4.
01 14 56.5	+31 56 52	IC 1652	13.5	1.3 x 0.3	12.3	S0/a	169	Mag 9.17v star NNW 5.′4.
01 15 12.3	+30 11 42	IC 1654	13.2	1.3 x 1.2	13.5	(R)SB(r)a	42	Mag 9.26v star NW 3.′3.
01 16 06.2	+30 20 55	IC 1659	13.1	1.2 x 0.9	13.3	E	20	Mag 9.45v star NW 10.′7.
01 19 53.6	+32 28 00	IC 1666	13.5	1.1 x 1.0	13.5	Scd:	72	Mag 9.39v star WNW 9.′3.
01 20 38.2	+29 41 54	IC 1672	13.0	1.3 x 1.0	13.1	S?	140	Mag 11.2 star WSW 8.′6; IC 96 S 5.′1.
01 20 46.5	+33 02 38	IC 1673	14.1	0.3 x 0.3	11.5	E		Mag 9.79v star SW 5.′9.
01 21 00.1	+34 14 53	IC 1675	13.4	0.8 x 0.3	11.7		69	Mag 10.46v star E 7.′1.
01 21 07.5	+33 12 56	IC 1677	14.1	0.9 x 0.5	13.0	S?	120	Knotty, several superimposed stars and/or bright knots.
01 21 44.8	+33 29 35	IC 1679	14.8	0.7 x 0.5	13.5	Sa	54	NGC 483 NE 2.′9.
01 21 51.4	+33 16 58	IC 1680	14.4	0.7 x 0.5	13.2	S0	90	Mag 9.83v star S 1.′3.
01 22 13.4	+33 15 34	IC 1682	14.3	0.9 x 0.4	13.0	Sb	126	Mag 8.82v star SE 2.′1.
01 22 39.2	+34 26 11	IC 1683	13.3	1.3 x 0.6	12.9	S?	177	Mag 10.15v star W 9.′5.
01 23 19.3	+33 16 37	IC 1687	13.7	0.4 x 0.2	10.8		3	Mag 7.61v star NW 1.′7.
01 23 48.0	+33 03 19	IC 1689	13.8	1.0 x 0.6	13.1	S0?	165	Mag 9.21v star E 11.′4.
01 23 49.6	+33 09 21	IC 1690	13.9	0.7 x 0.3	12.1	S0	135	Mag 11.8 star SE 1.′5.
01 38 27.0	+33 21 54	IC 1718	14.4	0.5 x 0.3	12.2	Compact	129	Faint star on SW edge; mag 11.01v star W 2.′5.
01 50 48.1	+35 55 58	IC 1732	14.0	1.5 x 0.4	13.3	S?	62	Mag 7.12v star S 5.′4.
01 50 43.0	+33 04 55	IC 1733	13.0	1.2 x 1.0	13.2	E:	50	Mag 10.76v star N 2.′1; IC 1735 E 2.′0.
01 50 51.8	+33 05 32	IC 1735	16.2	0.7 x 0.5	15.0	SB	149	Located on W edge of galaxy cluster A260; mag 10.76v star NW 2.′2; IC 1733 W 2.′0.
00 36 18.4	+32 54 10	MCG +5-2-28	14.2	0.9 x 0.5	13.2	SB?	18	
00 38 28.4	+29 37 27	MCG +5-2-31	14.3	1.0 x 0.4	13.1	S?	24	Numerous small, faint anonymous galaxies E and S.
00 46 24.5	+29 37 25	MCG +5-2-45	14.4	0.9 x 0.6	13.6	SB?	12	Mag 9.43v star SW 2.′4.
00 46 34.2	+29 43 28	MCG +5-2-47	14.0	0.8 x 0.6	13.0	Sc	96	Mag 11.3 star NE 3.′6.
00 47 10.5	+32 41 35	MCG +5-3-3	14.3	0.5 x 0.4	12.4	S?	21	Mag 13.7 star on N edge; UGC 484 W 3.′2; mag 9.4 star E 1.′8.
00 47 50.2	+29 57 31	MCG +5-3-6	14.5	0.5 x 0.5	12.9	S0		MCG +5-3-7 E 2.′6.
00 48 02.4	+29 57 23	MCG +5-3-7	14.5	0.7 x 0.5	13.2	S0	138	MCG +5-3-6 W 2.′6.
00 57 41.8	+33 21 04	MCG +5-3-29	14.3	0.8 x 0.4	12.9	S?	177	Mag 12.1 star WSW 1.′6.
00 59 39.8	+29 20 13	MCG +5-3-32	14.1	0.7 x 0.6	13.2	E?	27	
01 00 39.8	+30 07 26	MCG +5-3-35	14.4	0.7 x 0.5	13.1	S0?	16	Mag 10.00v star NW 2.′6.
01 01 13.1	+30 09 07	MCG +5-3-39	14.2	0.5 x 0.5	12.5	S?		UGC 632 S 1.′4.
01 01 52.9	+30 08 44	MCG +5-3-42	14.3	0.8 x 0.5	13.2	S0	147	Mag 8.62v star E 4.′5.
01 08 52.5	+32 05 57	MCG +5-3-66	14.1	0.6 x 0.5	12.7	S?	28	Mag 11.3 star N 3.′7.
01 09 42.6	+32 27 04	MCG +5-3-72	14.5	0.8 x 0.5	13.3	Sab	60	UGC 724 SE 6.′2.
01 09 56.7	+30 28 02	MCG +5-3-74	14.7	0.8 x 0.6	13.9	E/S0	87	N-S line of three mag 13 stars E 1.′3.
01 16 32.5	+32 53 31	MCG +5-4-12	15.9	0.7 x 0.7	14.9			Pair of mag 13.4 stars E 1.′6.
01 18 40.6	+32 15 25	MCG +5-4-14	14.4	0.8 x 0.6	13.5	S?	132	
01 21 34.8	+33 36 00	MCG +5-4-26	14.4	0.9 x 0.6	13.7	S0/a	18	
01 23 58.7	+33 18 45	MCG +5-4-48	14.3	0.5 x 0.4	12.4	S?	36	Very faint, small anonymous galaxy N 1.′1.
01 24 49.8	+32 14 00	MCG +5-4-55	14.2	0.7 x 0.5	12.9	S?	54	Mag 11.0 star SSE 1.′9.
01 25 04.7	+29 25 50	MCG +5-4-56	15.4	0.6 x 0.6	14.2			Mag 10.07v star W 3.′6.
01 26 33.8	+31 36 57	MCG +5-4-59	14.0	0.8 x 0.6	13.1	SAB(rs)bc: I	111	Mag 10.7 star N 3.′0.
01 32 20.8	+32 45 27	MCG +5-4-68	15.0	0.8 x 0.8	14.4			Mag 10.33v star ESE 2.′0.
01 34 14.1	+32 02 29	MCG +5-4-70	14.4	0.8 x 0.7	13.6	Sb	27	Mag 11.00v star S 2.′1.
01 34 21.5	+32 38 23	MCG +5-4-71	15.1	0.9 x 0.5	14.1	Spiral	6	Mag 13.6 star W 1.′3.
01 35 10.7	+32 42 26	MCG +5-4-72	14.8	0.7 x 0.5	13.5		126	Mag 7.6 star NW 4.′9.
01 42 51.8	+31 28 37	MCG +5-5-2	14.6	0.7 x 0.5	13.3	S?	162	Mag 9.87v star SW 3.′4.

(Continued from previous page)

RA h m s	Dec ° ′ ″	Name	Mag (V)	Dim ′ Maj x min	SB	Type Class	PA	Notes
01 46 54.7	+31 52 10	MCG +5-5-9	14.7	0.9 x 0.6	14.1	E/S0	99	
01 46 56.6	+32 06 34	MCG +5-5-10	13.9	0.8 x 0.4	12.6	S?	9	
01 56 50.3	+31 42 10	MCG +5-5-29	13.9	1.0 x 0.6	13.2	S?	144	= **Mkn 1167**.
00 39 30.1	+36 20 46	MCG +6-2-12	14.1	0.9 x 0.5	13.1	S0/a	60	Mag 14.9 star on W edge; mag 11.3 star NW 4′.7.
00 41 54.0	+37 06 01	MCG +6-2-14	15.5	0.7 x 0.6	14.5	Sc	117	Mag 10.24 star W 2′.3.
00 46 38.8	+36 19 43	MCG +6-2-17	14.5	1.1 x 0.6	13.9	S?	174	UGC 480 W 1′.5.
00 46 38.5	+37 26 22	MCG +6-2-18	13.6	1.6 x 1.2	14.2		50	Mag 11.3 star W 5′.0.
00 47 53.0	+38 25 10	MCG +6-2-19	14.2	0.9 x 0.8	13.7		39	
00 56 39.7	+34 35 14	MCG +6-3-2	15.8	0.5 x 0.4	13.9	Sbc	48	**MCG +6-3-1** NW 1′.7.
00 59 20.9	+39 18 57	MCG +6-3-4	15.1	0.7 x 0.5	13.8		141	Close pair of stars, mags 11.5 and 9.71v, E 2′.5.
01 02 55.4	+37 40 20	MCG +6-3-7	15.3	0.7 x 0.3	13.4	Spiral	159	Mag 10.5 star S 2′.2.
01 05 47.5	+34 59 18	MCG +6-3-9	15.2	0.6 x 0.3	13.2	Sbc	117	Mag 11.0 star NE 2′.5.
01 06 24.6	+33 46 09	MCG +6-3-10	15.0	0.7 x 0.5	13.7	Sb	81	
01 07 35.8	+34 01 54	MCG +6-3-12	15.3	0.5 x 0.3	13.1	Spiral	144	Mag 7.99v star SW 3′.0.
01 07 47.3	+33 48 43	MCG +6-3-13	15.7	0.6 x 0.4	14.0	Spiral	6	Located between a SE-NW pair of mag 12-13 stars.
01 08 46.4	+38 49 00	MCG +6-3-16	14.8	0.8 x 0.8	14.2			Two faint stars superimposed W edge; mag 9.68v star NE 3′.1.
01 08 58.3	+37 29 08	MCG +6-3-17	15.6	0.9 x 0.9	15.2	Sbc		Mag 11.5 star N 2′.7.
01 12 49.6	+37 01 23	MCG +6-3-22	14.6	0.6 x 0.4	13.0	S0	6	Pair of mag 9 stars S 1′.2.
01 14 25.4	+34 03 27	MCG +6-3-25	16.0	0.6 x 0.4	14.3	SBc	96	Trio of mag 12-13 stars NW 2′.4.
01 15 09.8	+33 47 25	MCG +6-3-26	15.4	0.9 x 0.5	14.4	Spiral	153	Mag 8.58v star S 2′.4.
01 20 56.9	+34 59 10	MCG +6-4-2	14.7	0.6 x 0.6	13.5	SBb		Mag 11.8 star N 1′.9.
01 21 17.4	+36 41 09	MCG +6-4-5	15.3	0.9 x 0.3	13.7	Sbc	18	Mag 11.3 star SSW 3′.5.
01 22 54.4	+39 15 23	MCG +6-4-9	13.6	0.8 x 0.7	12.9	Sb	81	Mag 9.50v star N 1′.8.
01 24 01.6	+33 51 57	MCG +6-4-14	15.1	0.7 x 0.5	13.8	Sd	48	NGC 512 N 2′.5.
01 32 04.8	+39 11 42	MCG +6-4-36	14.2	0.8 x 0.8	13.6			Mag 12.9 star N edge; mag 10.8 star SW 3′.5.
01 35 38.5	+35 27 59	MCG +6-4-42	14.5	0.9 x 0.6	13.7	S0	36	Mag 14.2 star on W edge.
01 36 09.8	+35 55 07	MCG +6-4-44	15.3	0.8 x 0.3	13.6	Sb	120	Close pair of mag 13-14 stars NW 1′.1.
01 38 06.4	+36 56 43	MCG +6-4-46	14.5	0.8 x 0.4	13.1	Sb	108	
01 39 33.0	+35 09 26	MCG +6-4-50	14.5	0.9 x 0.3	12.9	S?	18	Mag 11.8 star NNW 1′.5.
01 39 57.4	+34 24 25	MCG +6-4-51	15.1	0.6 x 0.4	13.4	Spiral	110	A short SE-NW line of three mag 8.09v, 9.82c and 10.81v stars SE 7′.3.
01 40 21.8	+37 12 28	MCG +6-4-52	14.9	0.7 x 0.3	13.1	Sb	90	
01 41 15.7	+34 48 38	MCG +6-4-55	13.9	0.9 x 0.6	13.1	SB?	168	Mag 10.24v star NW 3′.4.
01 47 43.8	+35 01 19	MCG +6-5-5	13.9	0.9 x 0.9	13.5	S?		
01 49 24.5	+34 58 25	MCG +6-5-10	15.5	0.5 x 0.4	13.6	S0	39	**MCG +6-5-8** W 1′.3.
01 51 18.7	+34 52 20	MCG +6-5-18	14.6	0.8 x 0.5	13.5	S0a	111	**UGC 1316** S 1′.5.
01 51 45.6	+36 08 00	MCG +6-5-21	15.2	0.8 x 0.5	14.1	Spiral	175	Mag 13.8 star W edge.
01 52 02.8	+36 07 45	MCG +6-5-24	15.3	0.8 x 0.6	14.4	S+	12	NGC 700 SE 2′.9.
01 53 50.2	+36 47 13	MCG +6-5-38	14.8	0.6 x 0.6	13.5	Spiral		Mag 13.2 star W 2′.4.
01 53 50.3	+36 20 58	MCG +6-5-40	14.3	0.7 x 0.7	13.6	E?		Small, faint, elongated anonymous galaxy E edge; almost stellar galaxy **VI Zw 95** SW.
01 54 58.1	+35 25 11	MCG +6-5-45	14.5	0.7 x 0.5	13.2	Spiral	168	= **Mkn 1010**; **CGCG 522-65** NE 4′.0.
01 54 59.0	+36 52 06	MCG +6-5-47	14.7	0.5 x 0.5	13.0	S0		Mag 10.9 star NE 1′.1.
01 55 02.0	+36 55 08	MCG +6-5-48	15.2	0.9 x 0.5	14.2	Sc	60	Located between a pair of N-S stars, mags 10.72v and 10.9; UGC 1385 W 1′.6.
01 55 39.9	+37 02 16	MCG +6-5-53	15.1	0.9 x 0.3	13.5	S..	105	Mag 6.28v star NW 10′.0.
01 56 40.0	+35 35 32	MCG +6-5-61	15.4	0.9 x 0.4	14.1	Sc	138	Mag 8.65v star N 3′.3.
01 56 58.0	+35 56 44	MCG +6-5-63	14.8	0.9 x 0.4	13.6	Spiral	150	Mag 10.20v star NW 3′.9.
01 58 35.4	+38 43 05	MCG +6-5-69	14.9	0.7 x 0.4	13.3	Sb	132	Mag 13.4 star NW edge.
01 59 21.9	+36 49 35	MCG +6-5-73	14.0	0.7 x 0.5	12.9	E?	159	**PGC 7512** W 2′.6; mag 12.3 star N 2′.1.
01 59 58.9	+36 35 25	MCG +6-5-74	15.3	0.7 x 0.4	13.7	Irr	156	Star superimposed S end; mag 12.0 star WSW 2′.1.
00 52 39.9	+39 53 04	MCG +7-3-1	15.1	0.6 x 0.6	13.9			
00 53 58.8	+40 24 49	MCG +7-3-4	14.9	0.7 x 0.4	13.3	Sc	117	
01 12 59.0	+40 01 15	MCG +7-3-22	15.0	0.7 x 0.4	13.4	Sc	6	Mag 12.1 star S 1′.5.
01 17 05.7	+40 58 22	MCG +7-3-25	15.3	0.7 x 0.2	13.0	Sb	36	Mag 11.14v star N 0′.7.
01 21 01.4	+40 26 45	MCG +7-3-29	14.1	0.8 x 0.3	12.4	Sc	123	
01 21 09.2	+40 28 15	MCG +7-3-31	14.9	0.7 x 0.4	13.3	Sc	21	NGC 477 NE 2′.4.
00 38 23.2	+29 28 20	NGC 181	14.9	0.6 x 0.2	12.4	S?	147	Located 2′.7 S of NGC 183.
00 38 29.5	+29 30 40	NGC 183	12.7	1.7 x 1.1	13.5	E	130	Very small anonymous galaxy 2′.5 NE; **CGCG 500-61** E 4′.8.
00 38 36.0	+29 26 51	NGC 184	14.7	0.7 x 0.2	12.4	S0/a	6	Located 8′.3 due N of mag 4.35v star Epsilon Andromedae.
00 41 44.8	+36 21 30	NGC 218	14.1	0.9 x 0.7	13.5	S?	126	Faint star on S edge; mag 9.23v star E 9′.1.
00 42 41.7	+40 51 59	NGC 221	8.1	8.7 x 6.5	12.5	cE2	179	= **M 32**.
00 42 54.3	+32 34 47	NGC 226	13.4	0.9 x 0.6	12.6	S?	105	Mag 13.4 star on S edge; mag 7.66v star W 10′.6.
00 43 36.8	+30 35 10	NGC 233	12.4	1.2 x 1.2	12.9	E?		
00 46 00.9	+29 57 34	NGC 243	13.7	0.8 x 0.4	12.3	S0?	149	Mag 10.47v star 1′.5 W.
00 48 47.2	+31 57 22	NGC 262	13.1	1.1 x 1.1	13.1	SA(s)0/a:		Almost stellar nucleus with smooth envelope.
00 49 48.1	+32 16 36	NGC 266	11.6	3.0 x 2.9	13.8	SB(rs)ab	99	Strong E-W bar; mag 8.13v star S 3′.8.
00 52 42.2	+30 38 20	NGC 282	13.4	1.0 x 0.9	13.4	E	50	Mag 8.04v star S 2′.0.
00 53 28.4	+32 28 56	NGC 287	14.0	0.7 x 0.5	12.7	S0/a	21	
00 55 07.6	+31 32 32	NGC 295	12.6	2.2 x 1.0	13.3	SBb:	164	Mag 9.78v star at SE edge.
00 55 21.8	+31 40 36	NGC 296	15.0	1.1 x 0.3	13.6	Scd:	148	UGC 566 N 3′.2; UGC 567 N 4′.4.
00 57 32.9	+30 16 47	NGC 311	13.0	1.5 x 0.6	12.8	S0	120	NGC 315 NNE 5′.4.
00 57 48.9	+30 21 04	NGC 315	11.2	3.2 x 2.5	13.4	E$^+$:	43	
00 58 05.4	+30 25 28	NGC 318	14.4	0.6 x 0.4	12.7	S0	10	Located 5′.7 NW of NGC 315.
01 00 36.6	+30 40 06	NGC 338	12.8	1.9 x 0.6	12.8	Sab	109	Mag 9.97v star W 6′.0; mag 9.05v star E 10′.3.
01 06 58.5	+32 18 28	NGC 373	14.9	0.4 x 0.4	13.0	E		Faint star on NW edge.
01 07 05.8	+32 47 39	NGC 374	13.4	1.3 x 0.5	12.7	S0/a	175	Mag 10.7 star WNW 5′.5.
01 07 06.2	+32 20 50	NGC 375	14.5	0.5 x 0.5	13.0	E2:		Almost stellar; stellar anonymous galaxy E 0′.9.
01 07 15.8	+32 31 11	NGC 379	12.9	1.4 x 0.8	12.9	S0	0	Several very faint, small anonymous galaxies N 4′.0.
01 07 17.7	+32 28 57	NGC 380	12.5	1.3 x 1.3	13.2	E2		Bright core in smooth envelope.
01 07 24.0	+32 24 12	NGC 382	13.2	0.7 x 0.7	12.4	E:		Almost stellar nucleus; located on S edge of NGC 383.
01 07 25.1	+32 24 41	NGC 383	12.4	1.4 x 1.4	13.0	SA0$^-$:		NGC 382 on S edge.
01 07 25.1	+32 17 33	NGC 384	13.1	1.1 x 0.9	13.1	E3	135	NGC 385 N 1′.6.
01 07 27.2	+32 19 11	NGC 385	13.0	1.1 x 1.0	12.9	SA0$^-$:	147	NGC 384 S 1′.6.
01 07 31.4	+32 21 40	NGC 386	14.3	0.5 x 0.4	12.6	E3:	9	Located 3′.3 S of NGC 383.
01 07 33.2	+32 23 26	NGC 387	15.5	0.4 x 0.4	13.6	E		Stellar nucleus, faint envelope.
01 07 47.3	+32 18 32	NGC 388	14.3	0.6 x 0.3	12.5	E3:	165	Stellar nucleus.
01 08 30.1	+39 41 44	NGC 389	13.8	1.3 x 0.4	13.0	S0	54	Mag 10.9 star on NE edge; **CGCG 520-16** W 9′.0.
01 08 13.0	+32 27 13	NGC 390	15.2	0.6 x 0.3	13.2		63	
01 08 23.5	+33 07 59	NGC 392	12.7	1.2 x 0.9	12.7	S0$^-$:	50	NGC 394 on N edge.
01 08 37.1	+39 38 38	NGC 393	12.5	1.7 x 1.4	13.3	S0$^-$:	29	
01 08 26.1	+33 08 50	NGC 394	13.8	0.7 x 0.4	12.3	S0	129	NGC 392 on S edge.

GALAXIES

RA h m s	Dec ° ′ ″	Name	Mag (V)	Dim ′ Maj x min	SB	Type Class	PA	Notes
01 08 31.3	+33 06 30	NGC 397	14.8	0.7 x 0.5	13.7	E	54	Almost stellar nucleus in faint envelope.
01 08 53.8	+32 30 52	NGC 398	14.5	0.8 x 0.5	13.4	S0	144	Mag 10.63v star NE 7′.3.
01 08 59.3	+32 38 00	NGC 399	13.5	1.1 x 0.8	13.3	SBa:	40	Small, faint, elongated anonymous galaxy E 9′.5.
01 09 14.3	+32 45 05	NGC 403	12.5	1.9 x 0.6	12.5	S0/a:	86	**MCG +5-3-71** SE 2′.2, on N edge of a mag 11.03v star.
01 09 27.0	+35 43 02	NGC 404	10.3	3.5 x 3.5	12.9	SA(s)0$^-$:		In the glare of mag 2.08v Mirach, Beta Andromedae, SE 6′.7.
01 10 36.6	+33 07 32	NGC 407	13.4	1.7 x 0.4	12.8	S0/a: sp	0	Faint galaxy **KUG 0107+327** SSE 5′.9.
01 10 59.1	+33 09 02	NGC 410	11.5	2.4 x 1.3	12.7	E$^+$:	30	**CGCG 501-119** N 4′.8; PGC 4221 SSW 2′.9; now nonexistent **NGC 408** is a single star W 1′.6.
01 11 17.5	+33 06 46	NGC 414	13.8	0.8 x 0.6	12.9	Double System	35	Double system.
01 12 09.7	+32 07 23	NGC 420	12.1	2.0 x 1.8	13.4	S0:	174	Close double star mag 7.00v, 8.26v, ESE 9′.3; **CGCG 501-128** S 7′.2.
01 13 03.0	+38 46 06	NGC 425	12.7	1.0 x 0.8	12.3	S?	54	Mag 12.5 star on NW edge; mag 10.47v star N 5′.3.
01 14 04.4	+33 42 11	NGC 431	12.9	1.4 x 0.9	13.0	SB0	20	Mag 10.85v star NE 2′.5; stellar galaxy **KUG 0111+335** N 6′.9.
01 15 07.7	+33 22 37	NGC 443	13.1	0.8 x 0.7	12.3	S:	57	
01 15 49.8	+31 04 49	NGC 444	14.3	1.9 x 0.4	13.8	Sd	157	Mag 10.41v star ESE 2′.8.
01 15 37.8	+33 03 56	NGC 447	14.0	2.2 x 2.2	15.5	(R)SB(rs)0/a:		Several stars E and S of the elongated nucleus. A mag 10.49v star 2′.7 NE.
01 16 07.3	+33 05 18	NGC 449	14.2	0.6 x 0.5	12.8	(R′)S?	71	Located 2′.9 SW of a mag 6.05v star.
01 16 12.4	+33 03 48	NGC 451	14.0	0.7 x 0.5	12.7	S?	36	Located 3′.3 SSW of a mag 6.05v star.
01 16 15.2	+31 02 00	NGC 452	12.6	2.5 x 0.8	13.2	SBab	43	Mag 10.41v star NW 3′.2; stellar galaxy **IV Zw 43** SW 4′.2.
01 19 48.6	+32 46 02	NGC 468	14.3	0.8 x 0.5	13.2	S0/a	171	Mag 9.68v star ESE 5′.8.
01 20 28.8	+32 42 29	NGC 472	14.1	1.2 x 1.0	14.2	Sb	120	Mag 9.68v star NW 3′.5.
01 21 20.2	+40 29 16	NGC 477	13.0	1.5 x 0.9	13.2	SAB(s)c	150	Three faint stars in a row across S end on E-W line.
01 21 56.4	+33 31 14	NGC 483	13.2	0.7 x 0.7	12.3	S?		Faint anonymous galaxy ESE 1′.1; mag 10.23v star ESE 2′.4; IC 1679 SW 2′.9.
01 22 55.3	+33 10 23	NGC 494	12.9	2.0 x 0.8	13.2	Sab	100	Small, faint anonymous galaxy NE 2′.4; **CGCG 502-55** SW 6′.2.
01 22 56.2	+33 28 13	NGC 495	12.9	1.3 x 0.8	12.8	(R′)SB(s)0/a pec:	170	Mag 10.52v star SW 6′.7.
01 23 11.8	+33 31 39	NGC 496	13.3	1.6 x 0.9	13.5	Sbc	28	Very faint, elongated anonymous galaxy N 2′.1.
01 23 11.4	+33 29 19	NGC 498	14.3	0.5 x 0.5	12.6	S0		Almost stellar; located 1′.8 N of NGC 499.
01 23 11.6	+33 27 36	NGC 499	12.1	1.6 x 1.3	12.8	S0$^-$	84	Mag 11.35v star E 3′.5; NGC 498 N 1′.8.
01 23 22.6	+33 25 57	NGC 501	14.5	0.5 x 0.5	13.0	E0		A mag 11.35v star 1′.8 NE; NGC 499 NW.
01 23 28.6	+33 19 53	NGC 503	14.1	0.7 x 0.5	13.0	E?	27	Mag 7.61v star SW 4′.4.
01 23 27.9	+33 12 11	NGC 504	13.0	1.7 x 0.4	12.4	S0	47	Mag 7.61v star NNW 6′.0.
01 23 40.0	+33 15 19	NGC 507	11.2	2.5 x 2.5	13.1	SA(r)0°		Mag 7.61v star NW 6′.1; NGC 508 on N edge.
01 23 40.7	+33 16 50	NGC 508	13.1	1.1 x 1.1	13.4	E0:		Located on N edge of NGC 507.
01 23 59.9	+33 54 26	NGC 512	13.2	1.6 x 0.4	12.6	Sab	116	MCG +6-4-14 S 2′.5.
01 24 26.9	+33 47 56	NGC 513	13.0	0.9 x 0.6	12.2	S?	75	Faint, round anonymous galaxy S 1′.9.
01 24 38.6	+33 28 19	NGC 515	13.0	1.4 x 1.1	13.4	S0	126	Star or knot on NW edge of core; mag 11.18v star N 3′.9.
01 24 43.9	+33 25 44	NGC 517	12.4	1.4 x 0.5	11.9	S0	20	Small, faint anonymous galaxy WSW 3′.9.
01 25 22.3	+34 01 19	NGC 523	12.7	3.0 x 0.8	13.5	Pec	108	Very faint extensions E and W of bright, elongated core fill out published dimensions.
01 25 34.0	+33 40 17	NGC 528	12.5	1.2 x 0.8	12.4	S0	64	
01 25 40.5	+34 42 47	NGC 529	12.1	2.4 x 2.1	13.8	S0$^-$:	160	Mag 9.33v star S 4′.5.
01 26 18.9	+34 45 14	NGC 531	13.8	1.9 x 0.5	13.6	SB0/a:	34	Almost stellar nucleus, faint envelope; mag 13.4 star on E edge.
01 26 21.5	+34 42 12	NGC 536	12.4	3.3 x 1.2	13.7	SB(r)b II	68	Faint, broad extensions NE and SW from bright core; mag 6.34v star S 7′.9.
01 26 31.1	+34 40 27	NGC 542	14.8	1.0 x 0.3	13.3	S?	146	NGC 536 N 2′.6; mag 8.08v star ESE 6′.8.
01 27 40.6	+37 10 58	NGC 551	12.7	1.8 x 0.8	12.9	SBbc	140	
01 28 18.8	+34 18 31	NGC 561	12.9	1.6 x 1.5	13.7	(R)SB(s)a	45	
01 29 03.2	+32 19 55	NGC 566	13.5	1.6 x 0.4	12.9	S0	178	Mag 6′.4 star NNW 8′.7.
01 29 56.3	+32 30 02	NGC 571	13.7	1.3 x 1.3	14.1	S?		Stellar nucleus in faint envelope; mag 9.37v star N 5′.6.
01 31 45.8	+33 36 49	NGC 579	13.3	1.5 x 1.3	13.9	Scd:	150	
01 31 58.3	+33 28 33	NGC 582	13.2	2.2 x 0.6	13.3	SB?	58	Star on SW end.
01 32 33.5	+35 21 31	NGC 587	12.8	2.2 x 0.8	13.2	SAB(s)b	67	
01 33 31.6	+35 40 05	NGC 591	12.9	1.3 x 1.0	13.0	(R′)SB0/a	5	Mag 6.96v star SE 5′.9; note: faint stellar galaxies **KUG 0131+353** and **KUG 0131+353A** lie 4′.6 E of this star.
01 33 51.3	+30 39 54	NGC 598	5.7	70.8 x 41.7	14.2	SA(s)cd II-III	23	= **M 33**, **Pinwheel Galaxy**. A difficult object in small telescopes due to its low surface brightness. Numerous HII regions visible in large aperture telescopes.
01 35 28.3	+33 39 22	NGC 608	13.3	1.9 x 1.5	14.3	S?	32	
01 35 52.4	+33 40 51	NGC 614	12.7	1.4 x 1.4	13.3	S0?		Very faint outer envelope surrounds bright core.
01 36 49.1	+35 30 41	NGC 621	12.7	1.2 x 1.0	12.8	SB0	24	Lies 2′.0 SW of mag 9.82v star.
01 38 18.6	+35 21 54	NGC 634	13.0	2.1 x 0.6	13.1	Sa	167	2.2' NE of mag 7.62v star.
01 42 25.8	+35 38 17	NGC 653	13.4	1.5 x 0.2	12.0	Sab	39	
01 44 35.5	+37 41 45	NGC 662	13.0	0.8 x 0.5	11.8	S pec	20	3′.0 NE of a mag 7.19v star.
01 46 06.2	+34 22 25	NGC 666	13.4	0.7 x 0.5	12.1	S?	80	Compact; mag 10.07v star SW 8′.3; **CGCG 521-77** SW 11′.4.
01 46 22.7	+36 27 39	NGC 668	13.1	1.8 x 1.2	13.7	Sb II-III	30	Mag 10.53v star 2′.4 E.
01 47 16.4	+35 33 49	NGC 669	12.3	3.2 x 0.6	12.9	Sab	36	
01 49 43.8	+35 47 04	NGC 679	12.3	1.9 x 1.9	13.6	S0$^-$:		Large bright center, uniform envelope; mag 7.12v star E 14′.6.
01 50 33.3	+36 22 11	NGC 687	12.3	1.4 x 1.4	12.9	S0		Almost stellar galaxy **VI Zw 60** SE 2′.8.
01 50 44.4	+35 16 59	NGC 688	12.7	2.5 x 1.5	14.0	(R′)SAB(rs)b	145	**UGC 1299** N 5′.1.
01 52 12.9	+36 05 49	NGC 700	14.4	1.2 x 0.3	13.1	S0	7	Stellar nucleus; pair of faint stars on S end; **CGCG 522-33** E 4′.2; mag 9.23v star SW 4′.8.
01 52 39.8	+36 10 15	NGC 703	13.3	0.9 x 0.7	12.7	S0$^-$:	50	Located on NW edge of NGC 708.
01 52 38.0	+36 07 28	NGC 704	12.8	0.6 x 0.5	11.4	S0	90	Double system, small companion S of nucleus. **CGCG 522-33** SW 1′.2.
01 52 41.6	+36 08 36	NGC 705	13.6	1.3 x 0.3	12.5	S0/a	114	Located on SW edge of NGC 708.
01 52 46.6	+36 09 06	NGC 708	12.7	3.0 x 2.5	14.8	E	35	Several stars superimposed on envelope; NGC 703 on NW edge; NGC 705 on SW edge.
01 52 50.9	+36 13 23	NGC 709	14.3	1.1 x 0.5	13.5	S0	135	Lies 2′.1 SE of mag 9.81v star.
01 52 54.0	+36 03 09	NGC 710	13.7	1.3 x 1.0	13.8	Scd:	45	Almost stellar nucleus; stellar **PGC 6974** N 3′.5.
01 53 08.5	+36 49 09	NGC 712	12.8	1.0 x 0.6	12.1	S0	85	Mag 10.24v star 2′.4 W.
01 53 29.6	+36 13 15	NGC 714	13.1	1.5 x 0.4	12.4	S0/a	109	Stellar galaxy **VI Zw 95** N 6′.1.
01 53 55.1	+36 13 44	NGC 717	13.9	1.5 x 0.3	12.9	S0/a	117	Mag 9.18v star E 4′.2.
01 54 45.5	+39 22 56	NGC 721	13.5	1.7 x 1.0	13.9	SB(rs)bc	135	Stellar nucleus, smooth envelope.
01 56 27.7	+36 48 09	NGC 732	13.5	1.4 x 1.0	13.7	S0	28	UGC 1416 NNE 6′.3.
01 56 24.9	+33 03 51	NGC 733	15.2	0.6 x 0.3	13.2		26	NGC 736 ESE 3′.6; **CGCG 503-53** NW 6′.2.
01 56 38.5	+34 10 30	NGC 735	13.3	1.7 x 0.6	13.1	Sb I	138	**PGC 7275** NW 1′.5. Mag 10.17v star SW 1′.4.
01 56 41.1	+33 02 36	NGC 736	12.2	1.5 x 1.5	13.0	E$^+$:		Line of three faint stars N of nucleus; NGC 738 on NE edge.
01 56 45.7	+33 03 30	NGC 738	14.9	0.5 x 0.3	12.7	S0	155	Located on NE edge of NGC 736.
01 56 54.8	+33 15 58	NGC 739	13.9	0.9 x 0.6	13.1	S0?	127	Almost stellar core, faint, smooth envelope.
01 56 54.9	+33 00 54	NGC 740	14.0	1.6 x 0.4	13.3	SBb?	137	Mag 10.17v star E 1′.3.
01 57 32.4	+33 12 39	NGC 750	11.9	1.7 x 1.3	12.7	E pec	162	Pair with NGC 751 on S edge; mag 9.13v star ESE 4′.8.
01 57 33.0	+33 12 07	NGC 751	12.7	1.2 x 1.2	13.2	E pec		Pair with NGC 750 on N edge.
01 57 42.3	+35 54 51	NGC 753	12.3	3.0 x 1.9	14.0	SAB(rs)bc I-II	125	several faint stars on face near edges.
01 57 50.6	+36 20 32	NGC 759	12.7	1.4 x 1.4	13.5	E		Chain of faint stars starts E of nucleus and extends S and then SE; **UGC 1434** SSW 5′.8.
01 57 49.6	+33 22 37	NGC 761	13.5	1.5 x 0.5	13.0	SBa:	143	
01 59 36.0	+30 54 31	NGC 769	12.9	0.9 x 0.6	12.1	S?	73	Very faint, elongated anonymous galaxy 1′.1 S.
02 00 14.9	+31 25 45	NGC 777	11.4	2.5 x 2.0	13.2	E1	155	Mag 8.72v star SE 6′.3.

(Continued from previous page)

GALAXIES

RA h m s	Dec ° ′ ″	Name	Mag (V)	Dim ′ Maj x min	SB	Type Class	PA	Notes
02 00 19.6	+31 18 43	NGC 778	13.2	1.1 x 0.5	12.4	S0:	150	Lies 3′0 SW of mag 8.72v star.
02 01 06.5	+31 52 58	NGC 783	12.1	1.6 x 1.4	12.9	Sc	35	Bright knots or superimposed stars on E and W sides of core; **UGC 1499** N 4′7.
02 01 40.1	+31 49 33	NGC 785	13.0	1.7 x 0.9	13.3	S0⁻:	83	Mag 8.92v star WSW 7′2; note: **IC 1766** lies 1′7 S of this star.
02 02 26.1	+32 04 13	NGC 789	13.5	1.2 x 0.8	13.3	S?	3	Mag 8.13v star WNW 9′2.
02 03 28.1	+38 06 59	NGC 797	12.7	1.6 x 1.3	13.3	SAB(s)a	65	Star of very small galaxy SW edge.
02 03 19.6	+32 04 38	NGC 798	13.5	1.2 x 0.5	12.9	E	137	Mag 9.86v star NE 6′5.
02 03 44.9	+38 15 33	NGC 801	13.1	3.2 x 0.7	13.8	Sc	150	2′3 E of mag 10.69v star.
00 37 21.3	+29 08 54	UGC 371	14.4	1.7 x 0.3	13.5	Scd:	45	
00 38 12.4	+30 53 27	UGC 381	14.5	0.8 x 0.6	13.6	Sb	5	Located 14′6 W of mag 3.27v star Delta Andomedae.
00 38 22.3	+32 38 15	UGC 384	13.8	1.2 x 0.8	13.6	SAB(rs)d	174	Stellar nucleus.
00 38 31.9	+30 17 25	UGC 388	15.3	1.2 x 0.6	14.8	Scd:	3	
00 39 18.4	+29 39 26	UGC 400	14.1	0.7 x 0.6	13.0	S?	75	Elongated **UGC 412** N 6′7.
00 40 41.4	+30 09 39	UGC 431	14.3	0.5 x 0.4	12.4	Sb	54	Mag 11.2 star NW 2′1.
00 41 03.7	+31 43 54	UGC 433	13.9	1.7 x 0.3	13.0	Scd:	74	
00 41 52.0	+32 59 28	UGC 442	14.3	1.5 x 0.5	13.9	SA(rs)d	175	
00 42 04.8	+36 48 13	UGC 444	13.1	1.0 x 0.7	12.6	S?	163	
00 43 50.8	+32 51 12	UGC 465	13.5	1.7 x 0.4	13.0	SBa	178	
00 46 26.0	+30 14 15	UGC 478	13.7	1.5 x 0.2	12.3	Sa	116	Mag 10.91v star NE 3′4.
00 46 31.7	+36 19 39	UGC 480	12.6	1.6 x 1.1	13.1	S?	177	Pair with MCG +6-2-17 1′5 E, disrupted.
00 46 56.2	+32 40 26	UGC 484	13.1	2.2 x 0.8	13.5	(R′)SB(s)b	25	Elongated core with dark patches E and W; mag 13.2 star on E edge.
00 47 08.3	+30 20 30	UGC 485	14.1	2.0 x 0.2	12.9	Scd:	179	Mag 10.7 star S 2′1.
00 50 09.9	+31 43 51	UGC 511	14.6	1.7 x 0.3	13.7	Scd:	105	Located between a pair of mag 12.3 stars.
00 50 49.6	+30 09 48	UGC 518	14.2	0.8 x 0.7	13.4	SB(r)c	90	Mag 10.21v star E 2′6.
00 51 27.1	+40 43 30	UGC 522	15.2	0.8 x 0.7	14.4	Sb II-III	65	Pair of mag 12.5 stars NE 1′5.
00 51 35.0	+29 24 04	UGC 524	13.6	0.8 x 0.8	13.0	(R′)SB(s)b		Mag 12.0 star on SE edge; mag 14.8 star on W edge.
00 51 34.5	+29 42 58	UGC 525	14.4	1.5 x 0.9	14.6	SB?	150	Stellar nucleus, faint envelope.
00 52 12.4	+29 40 34	UGC 529	13.6	1.3 x 0.4	12.7	S0/a	90	Mag 9.4 double star W 2′8.
00 52 58.4	+29 01 55	UGC 540	13.7	0.7 x 0.5	12.4	S?	137	Faint star on E edge; mag 8.82v star NNW 2′8.
00 53 26.7	+29 16 12	UGC 542	13.3	2.1 x 0.3	12.7	S?	160	**UGC 536** SW 12′0.
00 54 23.2	+31 39 50	UGC 548	13.8	0.9 x 0.8	13.3	Scd:	96	Stellar nucleus, smooth envelope; mag 10.04v star W 5′3.
00 54 50.4	+29 14 44	UGC 556	14.3	1.0 x 0.4	13.1	S?	100	
00 54 46.9	+31 21 53	UGC 557	14.6	1.0 x 0.5	13.7	SB?	37	
00 55 18.4	+31 43 45	UGC 566	14.5	1.1 x 0.8	14.2	SAdm	31	Stellar nucleus, uniform envelope; UGC 567 N 1′7.
00 55 19.5	+31 44 54	UGC 567	13.8	1.0 x 0.4	12.7	S0	62	UGC 566 S 1′7.
00 56 21.3	+39 49 31	UGC 578	14.0	1.5 x 0.6	13.7	Sb	2	Mag 13 stars close N, E and S.
00 57 50.0	+31 29 03	UGC 598	13.6	1.7 x 0.5	13.2	S0/a	29	
00 58 23.3	+36 43 49	UGC 602	14.1	1.4 x 1.0	14.3	SAB(s)c I-II	138	
00 59 36.2	+35 33 35	UGC 614	13.3	1.5 x 0.9	13.4	SB(s)d	70	**Mkn 967** SW 3′0.
01 01 01.9	+29 36 04	UGC 629	13.9	0.8 x 0.6	12.9	Scd:	158	
01 01 11.4	+30 07 50	UGC 632	13.8	1.0 x 0.6	13.1	SBab	163	MCG +5-3-39 N 1′4.
01 01 21.2	+31 30 29	UGC 633	14.1	1.5 x 0.3	13.0	Sb II-III	9	
01 03 26.2	+32 14 11	UGC 646	13.9	1.6 x 0.7	13.9	SB?	105	**MCG +5-3-44**, with mag 11.6 star on N edge, N 5′0.
01 05 19.1	+31 40 55	UGC 669	14.7	1.5 x 0.4	14.0	Scd:	125	
01 06 09.6	+31 24 21	UGC 673	15.1	1.1 x 0.5	14.3	SAc? III-IV	33	
01 07 32.9	+39 23 58	UGC 690	12.7	1.5 x 1.3	13.3	Scd:	84	Mag 10.7 star NE 1′7.
01 07 37.1	+32 56 21	UGC 692	14.1	1.0 x 1.0	14.0	SB(s)cd		Stellar nucleus; mag 13.1 star on S edge.
01 08 05.0	+33 27 09	UGC 697	13.8	1.0 x 0.5	12.9	SB?	85	Full length, narrow, bright bar.
01 09 14.2	+32 09 03	UGC 714	13.7	1.1 x 0.8	13.4	SAc	10	**CGCG 501-107** ENE 4′3.
01 09 59.3	+32 22 05	UGC 724	12.7	1.7 x 1.2	13.3	S	24	**CGCG 501-114** E 4′6; MCG +5-3-72 NW 6′2.
01 10 44.3	+33 33 29	UGC 732	14.0	1.5 x 0.8	14.1	SA(r)d	80	Mag 11.5 star N 1′1.
01 11 17.3	+31 44 20	UGC 742	14.4	1.0 x 0.8	14.0	Sdm	147	
01 11 18.6	+31 53 15	UGC 743	13.7	1.3 x 0.7	13.4	Sa	8	Mag 13.3 star NW edge.
01 11 43.4	+35 16 29	UGC 748	13.9	1.3 x 0.4	13.1	S?	79	
01 12 39.0	+38 30 29	UGC 755	13.6	1.4 x 1.4	14.2	SAB(rs)c		Star superimposed E of stellar nucleus.
01 12 43.1	+32 58 10	UGC 756	13.9	0.9 x 0.7	13.3	Scd:	120	Mag 9.31v star E 5′2.
01 15 51.9	+33 48 35	UGC 809	14.7	1.4 x 0.3	13.6	Scd:	23	
01 16 04.2	+37 38 54	UGC 811	14.4	0.9 x 0.4	13.2	S0/a	162	
01 16 26.3	+39 03 09	UGC 822	14.0	1.2 x 0.6	13.5	SBa	4	Pair of mag 13 stars E and NE 2′0.
01 18 10.2	+38 26 31	UGC 831	13.9	0.7 x 0.5	12.6	S?	177	
01 18 40.2	+31 02 09	UGC 835	13.9	1.0 x 0.7	13.4	Scd:	75	Uniform surface brightness.
01 19 10.4	+33 01 49	UGC 841	14.1	1.5 x 0.3	13.1	Sbc	54	**CGCG 502-31** SE 10.6; **IC 1668** NNW 9′2.
01 20 13.3	+33 30 21	UGC 862	13.5	0.8 x 0.8	13.1	E		Mag 13.6 star S 2′1; mag 12.6 star SW 3′1.
01 21 03.2	+33 53 53	UGC 878	13.4	1.0 x 0.7	13.0	S0?	95	Mag 13.4 star on W end; mag 10.46v star SW 1′0.
01 21 37.3	+32 36 17	UGC 901	13.7	0.7 x 0.2	11.4	Sb	145	Mag 14.4 star on W edge; mag 8.64v star NE 2′5.
01 22 01.7	+37 24 06	UGC 909	13.5	1.5 x 1.0	13.8	Sd	45	Strong dark lane W of center.
01 22 10.0	+32 12 54	UGC 911	14.1	0.8 x 0.6	13.1	SBb	29	Bright SE-NW bar.
01 22 15.1	+34 40 07	UGC 913	13.9	0.6 x 0.5	12.4	S?	53	
01 23 28.3	+30 47 03	UGC 934	13.6	2.0 x 0.6	13.6	S?	132	Very faint, small anonymous galaxy SW 0′8; mag 12.0 star SW 2′8.
01 23 37.8	+32 37 44	UGC 937	14.1	0.9 x 0.7	13.4	S?	42	Mag 11.4 star SSE 2′6.
01 23 38.1	+34 34 06	UGC 940	14.6	0.9 x 0.5	13.6	SA(s)c	76	Stellar nucleus, uniform envelope; **Mkn 988** W 1′6.
01 24 45.3	+32 09 54	UGC 959	13.5	0.9 x 0.5	12.5	Sa:	63	Very faint plume curving off NE end; mag 11.0 star NE 2′9.
01 25 15.8	+34 21 31	UGC 975	14.1	0.8 x 0.5	13.0	S	115	Mag 13.4 star on NE edge.
01 25 31.8	+32 08 11	UGC 987	13.4	2.1 x 0.6	13.5	Sa	32	Mag 11.3 star E 2′2.
01 26 57.7	+39 01 05	UGC 1022	13.9	1.0 x 0.7	13.4	S	110	
01 27 35.2	+31 33 14	UGC 1033	13.5	2.8 x 0.5	13.7	Scd:	133	
01 28 00.9	+32 02 00	UGC 1045	13.8	1.2 x 0.4	12.8	S?	143	Mag 6.78v star SW 5′6.
01 28 05.5	+38 12 53	UGC 1046	14.2	1.0 x 1.0	14.0	SAm		Uniform surface brightness.
01 29 12.5	+39 25 33	UGC 1059	13.5	0.6 x 0.5	12.0	SB?	4	**CGCG 521-37** N 7′9; **CGCG 521-33** NW 10′7; mag 9.43v star NW 4′8.
01 30 00.2	+40 58 28	UGC 1070	13.2	1.9 x 1.4	14.2	Scd:	55	
01 31 34.7	+34 46 55	UGC 1086	14.1	1.0 x 0.3	12.6	S?	38	Mag 9.29v star SW 1′7.
01 31 37.7	+36 49 59	UGC 1088	14.5	1.1 x 0.9	14.3	Scd:	90	Stellar nucleus, uniform envelope; mag 13.6 star on N edge.
01 31 47.0	+38 42 59	UGC 1090	13.8	1.2 x 1.1	13.9	Sb III	21	Small, bright nucleus; mag 7.27v star NW 5′0.
01 32 13.5	+32 06 10	UGC 1095	14.1	1.0 x 0.5	13.2	S?	102	Small bright nucleus with two faint, thin arms; **MCG +5-4-66** W 2′0.
01 34 49.0	+34 02 04	UGC 1125	13.4	0.7 x 0.6	12.3	S	147	**CGCG 521-48** S 6′8.
01 35 16.6	+34 28 19	UGC 1131	14.4	1.0 x 0.8	14.0	SABdm:	12	Stellar nucleus; mag 13.8 star on NW edge.
01 36 32.7	+39 55 20	UGC 1145	14.1	1.2 x 0.6	13.6	SBb	157	
01 36 56.4	+31 59 05	UGC 1152	13.7	1.1 x 0.5	12.9	Sb II-III	177	Very faint, small, elongated anonymous galaxy W 1′6.
01 38 03.4	+32 29 38	UGC 1160	14.9	1.5 x 0.2	13.4	Scd:	102	
01 38 14.1	+36 36 31	UGC 1161	14.1	0.8 x 0.6	13.2	Compact		Mag 11.4 star WNW 2′2.

(Continued from previous page)

GALAXIES

RA h m s	Dec ° ′ ″	Name	Mag (V)	Dim ′ Maj x min	SB	Type Class	PA	Notes
01 38 34.6	+34 59 28	UGC 1166	13.1	1.5 x 0.5	12.6	S0	69	
01 40 28.0	+34 37 27	UGC 1178	14.0	1.8 x 0.3	13.2	Scd:	55	Mag 12 and 14 stars superimposed SW end; mag 12.9 star on NE edge.
01 44 08.3	+34 23 10	UGC 1212	13.4	1.3 x 1.1	13.7	Sb	104	Bright core, uniform envelope.
01 44 09.2	+31 19 13	UGC 1213	13.6	1.0 x 1.0	13.5	S0		
01 44 38.4	+38 12 08	UGC 1221	14.0	0.9 x 0.5	12.9	Sbc	145	
01 45 51.7	+35 06 35	UGC 1234	13.9	1.1 x 0.7	13.5	SAB(s)c	170	Knot or star N of nucleus; mag 11.3 star SW 1′.1.
01 47 30.2	+36 01 59	UGC 1251	14.0	0.9 x 0.4	12.8	S?	44	
01 48 07.2	+36 27 08	UGC 1257	14.0	1.1 x 0.5	13.2	Sab	107	
01 49 05.9	+34 58 55	UGC 1269	14.1	0.9 x 0.4	12.9	S0⁻:	108	**MCG +6-5-8** ESE 2′.6.
01 49 15.7	+35 04 23	UGC 1272	13.2	1.5 x 0.9	13.4	S0:	23	Bright core, uniform envelope; mag 11.6 star NE 2′.3.
01 49 25.8	+35 27 08	UGC 1277	13.2	1.6 x 0.9	13.5	S0/a	75	Mag 8.14v star WSW 2′.1.
01 49 31.6	+32 35 16	UGC 1281	12.3	3.9 x 0.6	13.0	Sdm	38	Faint, elongated anonymous galaxy 3′.0 NE of center; **CGCG 503-27** E 0′.9.
01 50 15.6	+33 29 41	UGC 1295	14.3	0.9 x 0.9	13.9	S0		**CGCG 503-30** NNE 8′.7.
01 50 47.0	+32 32 43	UGC 1306	13.6	1.2 x 1.1	13.8	(R)SAB0°	67	Stellar nucleus, uniform envelope; faint anonymous galaxies NE 1′.4; NW 2′.4, and S 2′.2.
01 50 51.3	+36 16 31	UGC 1308	12.8	2.0 x 2.0	14.3	E		Two bright stars or knots adjacent to nucleus; mag 9.76v star SE 1′.8.
01 51 01.2	+29 48 06	UGC 1311	16.7	1.5 x 0.9	16.9	Im:	20	Dwarf, low surface brightness.
01 51 23.7	+33 01 48	UGC 1318	14.0	1.0 x 0.7	13.6	E:	144	Mag 10.5 star S 6′.3.
01 51 29.2	+36 03 56	UGC 1319	13.9	0.9 x 0.7	13.2	S?	155	Mag 9.23v star E 4′.5.
01 51 48.4	+30 32 34	UGC 1327	13.6	1.2 x 0.7	13.3	S0⁻:	100	Small, faint anonymous galaxy E 2′.1.
01 52 21.8	+35 47 42	UGC 1338	14.2	0.8 x 0.7	13.4	Sb II	63	Mag 7.79v star SW 11′.5; note: **MCG +6-5-22** lies 2′.5 E of this star.
01 52 24.9	+35 51 20	UGC 1339	13.7	1.1 x 0.8	13.4	SB(r)0⁺	24	Bright knot or star NE of nucleus.
01 52 30.5	+31 59 06	UGC 1341	14.5	0.9 x 0.7	13.8	SAB(s)d	3	Uniform surface brightness.
01 52 34.9	+36 30 02	UGC 1344	12.7	1.5 x 0.7	12.6	(R)SBa	34	Mag 10.50v star S edge; mag 10.40v star N 1′.2.
01 52 46.1	+36 37 06	UGC 1347	12.9	1.0 x 0.9	12.6	SAB(rs)c I-II	36	Mag 11.69v star SE edge; mag 10.37v star N 1′.5.
01 52 57.6	+36 30 44	UGC 1350	13.2	1.4 x 1.1	13.5	SB(r)b	40	Several stars superimposed SW edge.
01 53 23.1	+36 57 16	UGC 1353	13.1	1.0 x 0.7	12.7	S0⁻:	104	Forms a triangle with mag 12 stars W and NW.
01 53 42.3	+29 55 56	UGC 1359	13.4	0.9 x 0.8	12.9	SB?	6	
01 53 50.7	+36 33 50	UGC 1361	15.0	0.8 x 0.6	14.1	Scd:	147	Two stars superimposed; mag 11.8 star on S edge.
01 54 20.1	+36 37 45	UGC 1366	13.9	1.5 x 0.4	13.2	SBcd:	140	Mag 10.90v star NW 3′.3; mag 12.1 star SE 1′.8.
01 54 54.0	+36 55 03	UGC 1385	13.3	0.7 x 0.6	12.2	(R)SB0/a	170	MCG +6-5-48 E 1′.6.
01 55 11.0	+36 15 36	UGC 1387	14.6	0.8 x 0.4	13.2	Sdm:	171	Anonymous galaxy SW 1′.9.
01 55 53.8	+32 59 20	UGC 1397	13.8	0.9 x 0.4	12.6	S0	32	Mag 9.49v star SE 7′.1; faint, stellar anonymous galaxy NNE 2′.7.
01 55 58.7	+37 07 41	UGC 1398	13.9	1.0 x 1.0	13.7	Scd?		Mag 14 stars on E and N edges; mag 6.28v star W 12′.1.
01 56 04.5	+36 07 51	UGC 1400	13.0	2.3 x 0.3	12.4	Sb III	156	Mag 9.07v star N 2′.6.
01 56 44.1	+36 23 00	UGC 1415	13.7	1.2 x 0.3	12.4	S0/a	1	
01 56 46.4	+36 53 09	UGC 1416	13.8	1.0 x 0.6	13.1	S?	65	Mag 11.7 star NW 2′.0; NGC 732 SSW 6′.3.
01 56 57.0	+40 20 27	UGC 1418	13.2	1.7 x 1.1	13.7	S0	50	Mag 10.05v star E 2′.2.
01 57 06.8	+32 47 15	UGC 1422	13.4	1.0 x 0.4	12.3	S?	90	
01 59 06.8	+36 03 43	UGC 1459	14.6	1.5 x 0.3	13.6	Scd:	106	Mag 12.2 star N 2′.2.
01 59 04.9	+36 15 29	UGC 1460	13.7	1.6 x 0.9	14.0	Sa	150	Triangle of stars superimposed E, W and S edges; mag 11.9 star NE 1′.0.
01 59 42.6	+32 04 56	UGC 1470	15.9	1.5 x 0.2	14.4	Sdm:	138	Mag 11.0 star NW 3′.1.
02 00 11.3	+37 36 06	UGC 1474	14.0	0.8 x 0.7	13.2	SB(s)dm	21	Located 2′.2 SW of mag 8.24v star.
02 00 55.3	+38 12 37	UGC 1493	13.0	2.1 x 0.7	13.3	SBab?	87	A line of stars, mags 12.2, 13.7 and 13.0, plus a small, faint anonymous galaxy NNE 1′.0.
02 01 13.7	+30 21 27	UGC 1502	14.6	1.8 x 0.9	15.0	Im:	141	Dwarf, low surface brightness.
02 01 20.0	+33 19 46	UGC 1503	13.4	0.8 x 0.7	12.8	E	75	Mag 8.68v star SE 2′.3.
01 45 15.6	+32 03 42	UGC 12227	15.8	0.6 x 0.5	14.4	SAB(s)d		**UGC 1224** N 3′.9.

GALAXY CLUSTERS

RA h m s	Dec ° ′ ″	Name	Mag 10th brightest	No. Gal	Diam ′	Notes
00 37 48.0	+29 35 00	A 71	15.5	30	19	Band of numerous, faint, stellar anonymous GX extend SE to NW.
01 14 48.0	+37 24 00	A 161	16.4	41	11	Mag 8.56v star 19′.2 NW of center; stellar **KUG 0112+371** W of center 5′.3.
01 15 24.0	+32 35 00	A 165	17.5	37	13	All members faint and stellar.
01 19 48.0	+35 48 00	A 174	15.8	36	16	All members faint and stellar.
01 51 54.0	+33 09 00	A 260	15.8	51	20	**MCG +5-5-21** NW of center 5′.9; numerous anonymous and PGC GX on W boundary and beyond.
01 52 48.0	+36 08 00	A 262	13.3	40	100	Numerous catalog and anonymous GX spread over entire area of A 262.
01 57 18.0	+32 13 00	A 278	15.6	38	19	Few PGC and numerous anonymous GX in N-S band along center line.

STAR CLOUDS

RA h m s	Dec ° ′ ″	Name	Mag	Diam ′	No. ★	B ★	Type	Notes
00 40 36.0	+40 44 00	NGC 206		4.2				A large star cloud in the SW part of M 31, the Andromeda Galaxy. It is extended N-S and best defined on the E flank.
01 33 12.0	+30 39 00	NGC 592		0.7				A knot involved in the galaxy M 33. It lies 9′ E of M 33's center.

BRIGHT NEBULAE

RA h m s	Dec ° ′ ″	Name	Dim ′ Maj x min	Type	BC	Color	Notes
01 32 42.0	+30 40 00	NGC 588	0.7 x 0.4	E			This knot lies in the outlying regions of M 33, 14′ E of the galaxy's center. The knot NGC 592 is 5′.7 due E.
01 33 30.0	+30 42 00	NGC 595	0.7 x 0.5	E			A knot involved in the galaxy M 33. It can be found 5′ WNW of M 33's center.
01 34 30.0	+30 48 00	NGC 604	1.0 x 0.7	E			Located 12′ NE of the galaxy's center, this is the brightest HII region in M 33. It contains several stars of 16th mag and fainter.

OPEN CLUSTERS

RA h m s	Dec ° ′ ″	Name	Mag	Diam ′	No. ★	B ★	Type	Notes
00 51 26.0	+35 49 18	NGC 272	8.5	5	8	9.0	ast	(S) Asterism, position is for L-shaped circlet.
01 57 35.0	+37 50 00	NGC 752	5.7	75	77	8.0	cl	Rich in stars; moderate brightness range; slight central concentration; detached.

RA h m s	Dec ° ′ ″	Name	Mag (V)	Dim ′ Maj x min	SB	Type Class	PA	Notes
23 28 29.1	+19 51 48	CGCG 454-71	14.0	0.5 x 0.3	11.8		84	Located 1′.8 SE of mag 6.68v star.
23 57 27.8	+18 18 37	CGCG 456-5	14.7	0.5 x 0.4	12.9	Compact	54	Mag 7.88v star S 3′.8.
23 57 37.8	+18 11 25	CGCG 456-6	15.0	0.4 x 0.4	12.9			Bright center, faint envelope; mag 10.63v star NW 1′.7; mag 7.88v star NNW 3′.9.
00 11 43.2	+20 58 27	CGCG 456-37	14.1	0.5 x 0.5	12.5	S0		= **Mkn 337**. Mag 10.56v star SE 9′.4.
23 17 30.6	+22 07 32	CGCG 475-59	14.3	0.6 x 0.4	12.6	Sc	72	Low, uniform surface brightness.
23 24 25.9	+25 23 05	CGCG 476-30	14.2	0.5 x 0.3	12.0	S?	72	Mag 7.39v star NE 8′.0.
23 25 42.8	+22 54 15	CGCG 476-34	14.3	0.7 x 0.3	12.5	S?	39	Mag 13.5 star on W edge; mag 11.6 star NW 2′.3.
23 30 09.7	+25 31 47	CGCG 476-55	14.4	0.5 x 0.3	12.2	Im	75	Mag 13 star on N edge.
23 33 12.8	+26 52 14	CGCG 476-66	14.3	0.8 x 0.7	13.5		138	Double system? Small, round galaxy N edge.
23 38 50.7	+27 16 03	CGCG 476-96	13.8	0.9 x 0.6	13.2	E	129	**CGCG 476-95** SSW 3′.5; numerous faint anonymous galaxies in surrounding area.
23 42 56.8	+27 28 13	CGCG 476-120	13.7	0.6 x 0.6	12.5	S0/a		Mag 9.82v star WNW 3′.2.
23 43 17.2	+27 21 28	CGCG 476-122	14.7	0.8 x 0.5	13.8	E	48	
23 55 49.3	+25 30 27	CGCG 478-2	15.4	0.9 x 0.7	14.8	Pec	111	Galaxy pair; = **CGCG 477-3**; **MCG +4-1-4** SE 1′.1.
00 04 44.6	+26 49 55	CGCG 478-21	13.7	0.5 x 0.5	12.0			Bright E-W core.
00 39 25.3	+25 15 12	CGCG 479-54	14.1	0.6 x 0.3	12.1			Mag 11.9 star NE 1′.3.
00 16 49.1	+29 37 07	CGCG 499-95	13.9	0.6 x 0.6	12.6			Mag 10.43v star SE 6′.9.
00 16 56.3	+27 53 22	CGCG 499-97	14.3	0.8 x 0.5	13.2		12	Small, faint, elongated anonymous galaxy NE edge; UGC 157 S 2′.5.
00 17 02.8	+29 56 28	CGCG 499-98	13.9	0.9 x 0.7	13.2		123	8.76v star NNW 5′.2.
00 24 37.1	+29 17 11	CGCG 500-15	13.9	0.6 x 0.3	11.9		129	Mag 10.03v star N 5′.0.
00 35 13.0	+29 47 56	CGCG 500-48	14.1	0.5 x 0.3	11.9		123	Mag 11.8 star SW 4′.3.
00 42 48.0	+29 55 18	CGCG 500-74	14.2	0.5 x 0.4	12.5	E	120	Mag 12.4 star NW 1′.7.
00 13 27.1	+17 29 11	IC 4	13.1	1.0 x 0.9	12.9	S?	10	Mag 10.10v star NNE 8′.3.
00 42 22.0	+29 38 27	IC 43	13.2	1.5 x 1.2	13.7	SABc II-III	117	Mag 10.4 star SE 3′.4; **IC 45** N 3′.4.
00 42 58.0	+27 15 14	IC 46	13.9	0.7 x 0.4	12.3	S0?	87	Mag 8.99v star SE 6′.1.
00 19 48.8	+23 46 17	IC 1540	14.1	1.1 x 0.4	13.1	SBb	27	Mag 9.97v star SE 14′.0.
00 20 41.6	+22 35 32	IC 1542	14.1	0.7 x 0.5	12.8	S?	78	Several small, bright patches.
00 20 55.7	+21 51 58	IC 1543	13.4	0.7 x 0.7	12.5	S?		Mag 9.15v star SW 4′.3.
00 21 17.5	+23 05 26	IC 1544	13.7	1.4 x 0.9	13.8	SAB(s)c	150	Stellar nucleus; mag 9.24v star NE 2′.0.
00 21 29.2	+22 30 16	IC 1546	14.7	0.9 x 0.3	13.1	S?	129	Very small, very faint stellar anonymous galaxy NW 1′.6.
00 36 52.3	+23 59 04	IC 1559	14.0	0.6 x 0.2	11.5	SAB0 pec: II-III	10	On S edge of NGC 169; mag 6.17v star ENE 3′.7.
23 16 00.8	+25 33 21	IC 5298	14.0	0.7 x 0.6	12.9	S?	108	Mag 8.71v star E 8′.3.
23 20 58.5	+19 19 04	IC 5312	14.4	0.4 x 0.3	12.0	Sb	99	Faint, round anonymous galaxy SE edge; IC 5314 E 2′.4.
23 21 08.6	+19 18 38	IC 5314	14.7	0.5 x 0.5	13.0			Mag 11.05v star S 0′.9.
23 21 18.2	+25 23 05	IC 5315	13.5	0.9 x 0.6	12.8	E?	138	Trio of mag 12, 13 and 14 stars N 1′.6.
23 23 28.7	+21 09 46	IC 5317	14.1	1.0 x 0.7	13.6	S0	21	Mag 11.9 star SSW 5′.3.
23 33 09.8	+21 14 13	IC 5329	14.7	1.7 x 0.2	13.4	Scd:	108	Mag 11.0 star N 5′.3; mag 11.0 star E 5′.6.
23 33 24.9	+21 07 49	IC 5331	14.1	1.2 x 0.3	12.8	Sab	18	Mag 12.5 star on SE edge; mag 11.5 star E 1′.6.
23 36 30.5	+21 08 41	IC 5338	13.7	1.1 x 0.6	13.3	E?	30	**IC 5337** W 1′.3; **CGCG 455-28** SE 3′.4.
23 38 38.9	+27 00 38	IC 5341	14.6	0.5 x 0.5	13.1	E:		**CGCG 476-90** SW 2′.9.
23 38 36.4	+27 01 42	IC 5342	14.4	0.7 x 0.3	12.7	E:	171	Located 1′.5 E of NGC 7720.
00 15 52.5	+19 46 26	MCG +3-1-33	14.1	0.7 x 0.4	12.6	S0?	21	Mag 11.01v star N 4′.1.
00 19 58.4	+19 24 04	MCG +3-2-1	14.8	0.8 x 0.3	13.1	Sc	54	Mag 10.33v star W edge.
00 20 51.7	+21 32 10	MCG +3-2-5	14.8	0.8 x 0.7	14.1	S0	81	Mag 11.8 star SE 2′.9.
00 37 50.6	+17 22 57	MCG +3-2-19	15.5	0.7 x 0.4	14.0	Sc	123	Mag 10.92v star NW 0′.9.
00 39 48.6	+21 12 53	MCG +3-2-22	14.4	0.6 x 0.4	12.7	S?	27	**MCG +3-2-21** NW 1′.8; mag 5.88v star 54 Piscium WNW 6′.6.
23 17 21.7	+18 25 23	MCG +3-59-26	15.5	0.7 x 0.5	14.2	Spiral	174	Mag 6.72v star SSE 8′.0; **CGCG 454-28** E 4′.5.
23 17 21.2	+18 03 20	MCG +3-59-27	15.2	0.5 x 0.5	13.6	S0/a		Mag 11.0 star SW 2′.7.
23 20 48.4	+18 54 16	MCG +3-59-40	14.4	0.8 x 0.3	12.8	S?	81	Almost stellar **MCG +3-59-39** NW 7′.9.
23 23 01.2	+17 44 27	MCG +3-59-46	14.7	0.6 x 0.6	13.4			Faint star SW edge.
23 23 04.1	+18 00 55	MCG +3-59-47	14.6	0.7 x 0.4	13.0	S?	54	
23 23 20.5	+20 35 21	MCG +3-59-48	14.6	1.0 x 1.0	14.4			MCG +3-59-49 S 1′.4; **MCG +3-59-50** E 1′.2.
23 23 21.7	+20 33 58	MCG +3-59-49	14.4	0.5 x 0.3	12.2		120	MCG +3-59-48 N 1′.4.
23 23 32.8	+19 35 56	MCG +3-59-51	15.0	0.6 x 0.5	13.5	Spiral	3	**UGC 12574** SE 2′.6.
23 27 04.2	+17 48 50	MCG +3-59-57	14.3	1.0 x 1.0	14.1	S0?		Short line of three mag 15 stars extend from SE edge.
23 28 05.1	+18 31 49	MCG +3-59-58	13.8	0.8 x 0.8	13.1	S?		Located 1′.9 W of a mag 7.43v star.
23 28 24.2	+18 27 14	MCG +3-59-59	15.4	0.6 x 0.4	13.7	Sc	129	Mag 11.10v star W 2′.6.
23 28 30.3	+17 16 36	MCG +3-59-60	13.7	0.7 x 0.4	12.2	S0°: sp	117	**CGCG 454-73** E 1′.6.
23 29 37.4	+19 22 45	MCG +3-59-64	14.3	0.9 x 0.9	13.9	SB?		Mag 14.4 start on W edge; mag 13.1 star NE 0′.8.
23 35 31.2	+18 18 35	MCG +3-60-9	14.7	0.7 x 0.6	13.6	Im		Mag 9.76v star SE 5′.3.
23 43 06.0	+18 39 22	MCG +3-60-21	14.3	0.5 x 0.3	12.1	Sc	105	**MCG +3-60-22** N edge.
23 44 35.2	+17 45 59	MCG +3-60-23	15.2	0.9 x 0.3	13.6	Spiral	54	Close pair of stars, mags 12.9 and 14.4, SW 1′.6.
23 45 08.7	+19 54 03	MCG +3-60-24	14.4	0.7 x 0.5	13.1	Irr	120	Mag 10.02v star S 4′.6.
23 48 50.6	+17 46 41	MCG +3-60-28	15.5	0.5 x 0.5	13.8	Sc		Mag 14.2 star superimposed on S edge.
23 50 33.4	+19 05 45	MCG +3-60-29	15.2	0.8 x 0.7	14.4	Sc	114	
23 51 14.1	+20 13 43	MCG +3-60-31	15.6	0.4 x 0.3	13.1	Sbc	27	Mag 11.7 star N 2′.3.
23 51 26.9	+20 35 08	MCG +3-60-36	14.0	0.8 x 0.5	12.8	S?	162	= **Mkn 331**. **UGC 12812** WSW 1′.9.
00 05 00.2	+22 07 54	MCG +4-1-19	15.0	0.6 x 0.4	13.3		90	Mag 9.95v star W 2′.0.
00 05 52.5	+22 32 08	MCG +4-1-20	15.1	0.7 x 0.3	13.3	Sc	177	**UGC 41** S 2′.8; mag 10.94v star NNW 2′.6.
00 12 56.0	+23 42 32	MCG +4-1-40	15.1	0.8 x 0.5	13.9	Sc	156	
00 14 12.9	+22 45 59	MCG +4-1-43	14.7	0.8 x 0.6	13.8		84	Mag 9.23v star S 2′.2.
00 14 55.0	+26 19 52	MCG +4-1-44	13.9	0.7 x 0.5	12.7	S?	111	
00 16 57.3	+22 31 19	MCG +4-1-46	14.4	0.9 x 0.6	13.8	E/S0	33	
00 17 59.8	+24 33 41	MCG +4-1-48	13.8	0.9 x 0.6	13.0	S?	42	Mag 10.53v star W 2′.6.
00 21 33.7	+22 35 27	MCG +4-2-10	14.9	0.8 x 0.3	13.2		114	NGC 86 S 2′.4.
00 26 02.7	+21 47 42	MCG +4-2-19	15.3	0.8 x 0.7	14.5	Sd	0	NGC 109 E 2′.8; mag 10.67v star NW 1′.6.
00 28 28.5	+23 29 47	MCG +4-2-26	15.0	0.8 x 0.5	13.8		129	**CGCG 479-33** SW 4′.4.
00 33 35.9	+23 23 54	MCG +4-2-30	14.4	0.9 x 0.4	13.2	S?	135	
00 37 51.1	+26 19 17	MCG +4-2-39	15.0	0.9 x 0.7	14.4		0	Mag 9.21v star WNW 0′.9.
00 40 24.9	+24 26 35	MCG +4-2-43	14.3	0.7 x 0.7	13.6	E/S0		Mag 10.9 star SW 2′.3.
00 42 23.3	+25 33 45	MCG +4-2-45	14.7	0.8 x 0.7	13.9		60	**CGCG 479-60** W 5′.1.
00 42 27.3	+25 37 09	MCG +4-2-46	15.4	0.7 x 0.6	14.3		63	
00 42 49.2	+23 29 28	MCG +4-2-47	14.9	0.8 x 0.4	13.5	S?	48	NGC 228 NE 1′.4.
00 43 42.0	+23 28 34	MCG +4-2-50	14.4	0.9 x 0.6	13.6		105	NGC 229 W 8′.7.
23 15 59.9	+27 27 10	MCG +4-54-39	14.4	0.6 x 0.4	12.7	S?	153	Mag 9.10v star NW 2′.6.
23 19 14.5	+26 03 18	MCG +4-55-2	14.2	0.9 x 0.7	13.6	Spiral	3	Mag 10.84v star SE 4′.5.
23 32 02.8	+23 54 35	MCG +4-55-23	14.8	1.2 x 0.8	14.6		18	Multi-galaxy system; mag 6.40v star SE 7′.3.
23 32 23.5	+23 30 41	MCG +4-55-24	14.5	0.7 x 0.7	13.6	S?		Faint, almost stellar anonymous galaxy SW 2′.6.
23 33 26.1	+25 39 00	MCG +4-55-26	14.2	0.7 x 0.4	12.7	S?	39	Close pair of mag 14.4, 13.5 stars N 1′.7.
23 33 45.6	+25 47 08	MCG +4-55-28	13.8	0.7 x 0.7	12.9	S?		Close pair of mag 121 stars SW 1′.8.

RA h m s	Dec ° ′ ″	Name	Mag (V)	Dim ′ Maj x min	SB	Type Class	PA	Notes
23 37 26.1	+27 04 15	MCG +4-55-32	14.9	0.5 x 0.4	13.0	S0	54	Very faint anonymous galaxy S edge; mag 11.1 star ESE 4′.3.
23 41 06.5	+25 10 07	MCG +4-55-44	14.4	0.9 x 0.4	13.1	S?	111	
23 41 16.2	+25 33 02	MCG +4-55-45	13.9	0.9 x 0.5	12.9	S?	54	
23 42 38.5	+27 05 24	MCG +4-55-47	14.3	0.5 x 0.3	12.1	S?	6	NGC 7737 SE 2′.9.
23 44 26.7	+27 36 06	MCG +4-56-1	14.7	0.6 x 0.6	13.5			Mag 8.60v star SW 2′.7.
23 45 47.8	+27 35 48	MCG +4-56-6	14.9	0.8 x 0.6	13.9	Sd	21	
23 48 19.3	+24 05 22	MCG +4-56-9	14.5	0.7 x 0.6	13.5	S?	15	Close pair of mag 11.8 and 13.1 stars W 2′.1.
23 48 49.7	+22 42 25	MCG +4-56-10	15.5	0.8 x 0.6	14.6	Spiral	132	Star or large knot on NW end.
23 50 47.5	+27 17 14	MCG +4-56-14	14.1	0.7 x 0.5	12.8	Spiral	177	Mag 13.2 star SE 2′.6.
23 51 00.0	+27 13 07	MCG +4-56-19	14.7	1.1 x 0.3	13.3	S?	147	Slightly elongated **PGC 72606** N 1′.2; round **PGC 72608** N 2′.4.
23 56 16.2	+29 24 30	MCG +5-1-8	14.5	0.7 x 0.5	13.2	S?	57	**MCG +5-1-9** S 1′.2; **UGC 12850** SW 2′.8.
23 58 28.1	+28 35 11	MCG +5-1-14	14.9	0.7 x 0.5	13.6	Sc	21	Mag 12.8 star N 2′.4.
23 58 28.5	+28 02 00	MCG +5-1-15	14.2	1.0 x 0.8	13.9	S0	6	Mag 12.0 star NW 3′.9.
23 58 32.7	+28 50 19	MCG +5-1-17	15.0	0.8 x 0.5	13.9		117	Mag 8.10v star W 2′.1.
00 00 44.1	+28 24 03	MCG +5-1-21	14.8	0.7 x 0.5	13.5	S?	99	= **Hickson 99C**. UGC 12899 on E edge.
00 10 43.4	+28 33 53	MCG +5-1-47	14.3	0.6 x 0.6	13.0	SB?		Mag 10.58v star N edge.
00 13 55.8	+28 45 43	MCG +5-1-56	14.5	0.9 x 0.8	14.0	S?	0	Mag 10.75v star S 1′.9.
00 17 59.1	+29 28 09	MCG +5-1-63	14.0	0.7 x 0.7	13.1			A N-S pair of mag 14 stars S 1′.4.
00 21 12.1	+29 38 42	MCG +5-2-3	14.6	0.7 x 0.5	13.3		78	Mag 11.5 star W 1′.7.
00 22 45.8	+29 56 19	MCG +5-2-8	14.6	0.8 x 0.5	13.4	S?	165	Mag 14.4 star on NE edge.
00 38 28.4	+29 37 27	MCG +5-2-31	14.3	1.0 x 0.4	13.1	S?	24	Numerous small, faint anonymous galaxies E and S.
23 16 43.8	+29 35 12	MCG +5-54-56	14.8	0.8 x 0.3	13.1	S?	18	Mag 14.5 star SW end.
23 17 18.0	+28 35 42	MCG +5-54-58	14.4	0.7 x 0.5	13.1	Sc	147	Mag 7.91v star N edge.
23 17 46.1	+29 01 35	MCG +5-54-60	14.7	0.5 x 0.5	13.0			UGC 12482 W 2′.7.
23 18 15.4	+29 13 49	MCG +5-54-61	14.2	0.9 x 0.9	14.0	E		Mag 14.3 star SW edge.
23 22 58.4	+29 14 02	MCG +5-55-7	14.7	0.6 x 0.6	13.4	Spherical		UGC 12566 SSW 6′.4.
23 23 22.1	+29 25 43	MCG +5-55-9	14.6	0.4 x 0.4	12.5	SB?		Mag 13.4 star SW 0′.7.
23 23 34.4	+29 10 54	MCG +5-55-10	13.8	0.8 x 0.6	12.9	S0	39	Mag 12.9 star NE 3′.3.
23 24 59.6	+29 47 22	MCG +5-55-12	15.3	0.8 x 0.7	14.5	SB?	39	**MCG +5-55-13** E 0′.9.
23 25 22.3	+28 51 20	MCG +5-55-14	13.9	0.9 x 0.5	12.9	S?	114	Mag 13.4 star on W edge.
23 28 49.2	+29 44 04	MCG +5-55-24	14.4	0.7 x 0.4	12.8	Sbc	53	UGC 12625 NE 4′.5.
23 43 56.9	+29 07 54	MCG +5-55-50	15.3	0.7 x 0.3	13.5	Sb	27	Mag 11.4 star 0′.6 SW; **MCG +5-55-52** ESE 1′.9; MCG +5-55-51 S 1′.0.
23 43 57.8	+29 06 54	MCG +5-55-51	16.4	0.7 x 0.3	14.5	Sb	111	MCG +5-55-50 and mag 11.4 star NW 0′.8.
23 47 24.6	+28 23 36	MCG +5-56-7	14.2	0.9 x 0.4	12.9	S?	6	Mag 8.12v star E 4′.8.
23 53 45.8	+29 15 19	MCG +5-56-20	14.7	0.7 x 0.4	13.1	SB?	120	**MCG +5-56-19** W 1′.5.
00 07 16.0	+27 42 28	NGC 1	12.9	1.7 x 1.1	13.4	SA(s)b: II-III	120	NGC 2 S 1′.8; mag 10.57v star NNW 5′.8.
00 07 17.3	+27 40 39	NGC 2	14.2	1.1 x 0.6	13.6	Sab	112	Mag 13.1 star W 1′.1; NGC 1 N 1′.8.
00 08 54.6	+23 48 59	NGC 9	13.5	1.0 x 0.5	12.6	Sb: pec	155	Mag 9.42v star E 5′.9.
00 09 02.5	+21 37 26	NGC 15	13.8	1.0 x 0.6	13.1	Sa	30	Mag 9.41v star W 6′.5.
00 09 04.4	+27 43 46	NGC 16	12.0	1.8 x 1.0	12.5	SAB0⁻ sp	16	**MCG +5-1-35** N 3′.2.
00 09 48.3	+27 49 55	NGC 22	13.6	1.3 x 1.0	13.8	Sb	160	Mag 10.34v star N 2′.5.
00 09 53.5	+25 55 24	NGC 23	12.0	2.1 x 1.3	13.0	SB(s)a	8	Bright core oriented SE-NW in faint N-S envelope.
00 10 27.1	+25 49 47	NGC 26	12.7	1.9 x 1.4	13.6	SA(rs)ab II-III	100	Bright center slightly offset to W.
00 10 33.0	+28 59 44	NGC 27	13.5	1.3 x 0.6	13.1	S?	117	Forms triangle with UGC 95 W and mag 10.51v star S.
00 12 48.0	+22 01 19	NGC 41	13.7	0.8 x 0.5	12.6	S?	123	Stellar galaxy **KUG 0009+216** SW 7′.2.
00 12 56.4	+22 05 59	NGC 42	13.8	1.1 x 0.6	13.2	S0⁻:	115	Very faint anonymous galaxy ENE 1′.1.
00 14 40.3	+18 34 54	NGC 52	13.4	2.1 x 0.4	13.1	S?	127	Mag 10.24v star E 7′.3.
00 15 30.9	+17 19 43	NGC 57	11.6	2.2 x 1.9	13.2	E	40	
00 19 37.9	+29 56 01	NGC 76	13.1	1.0 x 0.9	12.8	S?	65	**MCG +5-1-73** E 1′.2.
00 21 03.1	+22 33 57	NGC 79	14.0	0.6 x 0.6	13.0	E		Compact.
00 21 10.9	+22 21 23	NGC 80	12.1	1.6 x 1.6	13.0	SA0⁻:		Small, faint anonymous galaxy 1′.3 NW; another small, faint anonymous galaxy SE 0′.9; NGC 81 N 1′.7.
00 21 13.4	+22 22 56	NGC 81	15.7	0.3 x 0.2	12.5		84	Almost stellar; located 1′.7 N of NGC 80.
00 21 22.6	+22 25 57	NGC 83	12.5	1.5 x 1.5	13.4	E		Mag 11.2 star on S edge; pair of mag 11.1, 11.2 stars E 1′.3.
00 21 25.6	+22 30 43	NGC 85	14.8	0.7 x 0.5	13.5	S0	146	
00 21 28.7	+22 33 17	NGC 86	14.8	0.8 x 0.2	12.6	Sbc	9	Mag 12.3 star at S end.
00 21 51.6	+22 23 57	NGC 90	13.7	1.9 x 1.0	14.2	SAB(s)c pec I	132	Almost stellar nucleus, two main strongly curving arms; now nonexistent **NGC 91** is a faint star S 1′.9.
00 22 03.4	+22 24 31	NGC 93	13.3	1.4 x 0.6	12.9	S?	48	Very faint, elongated anonymous galaxy N 1′.4.
00 22 13.7	+22 28 58	NGC 94	14.6	0.6 x 0.3	12.6	S0	30	Anonymous companion S 0′.6.
00 22 17.9	+22 32 44	NGC 96	14.6	0.6 x 0.6	13.3			Almost stellar; bright knot or star S of core.
00 22 29.9	+29 44 44	NGC 97	12.3	1.3 x 1.3	12.9	E:		Mag 9.64v star N 4′.2; **CGCG 500-11** NE 10′.9.
00 25 59.9	+29 12 40	NGC 108	12.1	2.3 x 1.8	13.5	(R)SB(r)0⁺	153	Mag 9.33v star NNW 7′.1.
00 26 14.5	+21 48 27	NGC 109	13.7	1.2 x 0.9	13.6	SB(r)a	77	MCG +4-2-19 W 2′.8.
00 36 04.0	+23 57 28	NGC 160	12.7	2.3 x 1.2	13.7	(R)SA0⁺ pec	45	Mag 7.25v star N 4′.2.
00 35 57.9	+24 02 14	NGC 162	15.6	0.9 x 0.2	13.6	Sbc	119	Mag 7.25v star E 2′.8.
00 36 52.5	+23 59 28	NGC 169	13.2	2.6 x 0.6	13.6	SA(s)ab: sp II-III	88	Mag 6.17v star ENE 3′.7; IC 1559 on S edge.
00 38 23.2	+29 28 20	NGC 181	14.9	0.6 x 0.2	12.4	S?	147	Located 2′.7 S of NGC 183.
00 38 29.5	+29 30 40	NGC 183	12.7	1.7 x 1.1	13.5	E	130	Very small anonymous galaxy 2′.5 NE; **CGCG 500-61** E 4′.8.
00 38 36.0	+29 26 51	NGC 184	14.7	0.7 x 0.2	12.4	S0/a	6	Located 8′.3 due N of mag 4.35v star Epsilon Andromedae.
00 41 28.2	+25 30 01	NGC 214	12.3	1.9 x 1.4	13.2	SAB(r)c I	35	Mag 9.91v star NNE 7′.5.
00 42 54.5	+23 30 13	NGC 228	13.7	1.2 x 1.0	13.7	(R)SB(r)ab	126	MCG +4-2-47 SW 1′.4.
00 43 04.8	+23 30 32	NGC 229	14.1	0.9 x 0.3	12.5	Sab	96	MCG +4-2-50 E 8′.7.
23 16 24.9	+24 29 50	NGC 7568	13.6	0.9 x 0.6	12.8	S?	120	
23 16 50.4	+18 28 57	NGC 7572	14.4	0.9 x 0.3	12.9	S0/a	162	**CGCG 454-20** S 4′.5; **UGC 12471** NW 7′.1.
23 17 12.0	+18 42 01	NGC 7578A	13.3	0.8 x 0.8	12.8	S0° pec		= **Hickson 94B**; stellar nucleus.
23 17 13.6	+18 42 27	NGC 7578B	14.0	0.4 x 0.4	12.0	E1:		= **Hickson 94A**; **Hickson 94C** NE 2′.2; **Hickson 94D** NE 0′.5; **Hickson 94E** NNE 1′.3; **Hickson 94F** NE 2′.2; **Hickson 94G** NE 2′.9.
23 17 57.8	+18 45 06	NGC 7588	14.8	0.5 x 0.3	12.6		99	Located 2′.1 NW of a mag 9.04v star.
23 18 30.2	+18 41 23	NGC 7597	14.0	1.0 x 1.0	13.9	S?		Mag 9.79v star NNW 2′.9 with anonymous galaxy W of star.
23 18 33.3	+18 44 59	NGC 7598	14.9	0.5 x 0.5	13.5	E		Located 2′.2 NE of mag 9.79 star.
23 18 43.6	+18 41 53	NGC 7602	14.4	0.6 x 0.6	13.2	S?		Almost stellar; mag 8.03v star S 3′.2.
23 20 05.8	+24 13 13	NGC 7620	13.0	1.1 x 1.1	13.1	Scd:		
23 20 22.9	+27 18 57	NGC 7624	13.1	1.0 x 0.7	12.6	Scd:	30	Star or bright knot on S edge.
23 20 30.1	+17 13 32	NGC 7625	12.1	1.6 x 1.4	12.8	SA(rs)a pec	60	Pair of strong dark patches SW of center.
23 20 54.9	+25 53 54	NGC 7628	12.7	1.1 x 0.9	12.7	E	117	
23 25 48.7	+27 01 45	NGC 7660	12.7	1.4 x 1.1	13.1	E	35	Very faint anonymous spiral galaxy E 2′.2; elongated anonymous galaxy SE 1′.7.
23 26 40.8	+25 04 39	NGC 7664	12.7	2.6 x 1.5	14.0	Sc:	90	Bright center offset to the N; faint envelope.
23 27 41.3	+23 35 18	NGC 7673	12.8	1.3 x 1.2	13.1	(R′)SAc? pec	57	Multi-galaxy system.

(Continued from previous page)

RA h m s	Dec ° ′ ″	Name	Mag (V)	Dim ′ Maj x min	SB	Type Class	PA	Notes
23 28 06.1	+23 31 51	NGC 7677	13.2	1.6 x 1.0	13.6	SAB(r)bc:	35	Multi-galaxy system; mag 8.82v star N 2'.8.
23 28 28.2	+22 25 15	NGC 7678	11.8	2.3 x 1.7	13.2	SAB(rs)c I-II	5	Bright, knotty arms.
23 28 54.9	+17 18 35	NGC 7681	14.4	1.6 x 1.4	15.2	S0°: sp	42	Star or bright knot NE edge of core.
23 31 05.7	+21 24 40	NGC 7688	14.0	0.4 x 0.4	11.9	S0		**CGCG 455-3** NW 2'.5.
23 34 01.7	+24 56 40	NGC 7698	13.3	1.0 x 0.8	13.0	S0	170	Elongated nucleus, smooth envelope.
23 35 51.8	+23 37 08	NGC 7712	12.7	0.9 x 0.8	12.4	E?	120	
23 38 04.9	+25 43 10	NGC 7718	14.1	1.1 x 0.8	13.8	S?	151	**MCG +4-55-33** NW 3'.3.
23 38 29.5	+27 01 51	NGC 7720	12.4	1.6 x 1.3	13.1	E+ pec:	18	IC 5342 E 1'.6; IC 5341 SE 2'.4. Several stellar, anonymous galaxies in area.
23 39 11.9	+27 06 52	NGC 7726	14.2	1.5 x 0.5	13.8	SB(s)b	60	A mag 10.84v star, with an anonymous galaxy immediately N, lies SW 2'.6.
23 40 01.0	+27 08 00	NGC 7728	13.1	1.0 x 0.8	12.9	E	75	Mag 9.68v star SW 2'.4.
23 40 33.7	+29 11 14	NGC 7729	13.5	1.9 x 0.6	13.5	Sa	7	
23 42 17.4	+26 13 52	NGC 7735	13.6	1.3 x 0.9	13.7	E	90	Star on NE edge.
23 42 46.5	+27 03 05	NGC 7737	13.8	1.1 x 0.5	13.0	S0/a	147	MCG +4-55-47 NW 2'.9.
23 43 32.6	+27 18 40	NGC 7740	14.0	0.9 x 0.5	13.1	S0	140	
23 43 53.8	+26 04 30	NGC 7741	11.3	4.4 x 3.0	14.0	SB(s)cd II-III	170	Short, bright E-W bar, knotty, patchy arms; mag 9.84v star on N edge.
23 44 45.8	+25 54 29	NGC 7745	14.2	0.7 x 0.7	13.4	E		Stellar anonymous galaxy ESE 2'.1.
23 45 32.4	+27 21 36	NGC 7747	13.6	1.5 x 0.5	13.1	(R')SB(s)b	36	Elongated center, thin dark lanes E and W .
23 46 58.7	+29 27 31	NGC 7752	14.3	0.9 x 0.5	13.3	I0:	107	Located on SW edge of NGC 7753.
23 47 04.7	+29 28 58	NGC 7753	12.0	3.3 x 2.1	14.0	SAB(rs)bc I	50	Bright nucleus, western arm extends S towards NGC 7752.
23 50 52.2	+27 09 59	NGC 7765	14.6	0.7 x 0.7	13.6	SB?		Faint, anonymous galaxy S 1'.7.
23 50 56.0	+27 07 31	NGC 7766	15.5	0.5 x 0.2	12.9	S?	36	Located 1'.4 SSW of center of NGC 7768.
23 50 56.5	+27 05 12	NGC 7767	13.5	1.1 x 0.2	11.7	S0/a	142	Mag 12.8 star on W edge.
23 50 58.5	+27 08 49	NGC 7768	12.3	1.6 x 1.3	13.1	E	60	Mag 10.98v star E 2'.6.
23 51 03.9	+20 08 59	NGC 7769	12.0	3.2 x 2.7	14.2	(R)SA(rs)b II	170	Bright, patchy center with very faint outer envelope.
23 51 22.7	+20 05 43	NGC 7770	13.8	0.7 x 0.4	12.3	S0/a?	17	On S edge of NGC 7771.
23 51 24.7	+20 06 44	NGC 7771	12.3	2.5 x 1.0	13.1	SB(s)a	68	Strong E-W dark lane S of center; NGC 7770 on S edge.
23 52 24.4	+28 46 23	NGC 7775	13.3	1.0 x 0.8	12.9	Scd:	20	Patchy, star or bright knot S edge.
23 53 12.7	+28 16 57	NGC 7777	13.3	1.2 x 0.8	13.1	S0	48	
23 55 23.2	+21 48 33	NGC 7784	14.4	0.5 x 0.5	12.8			Stellar.
23 55 21.8	+21 35 16	NGC 7786	13.2	1.0 x 0.7	12.7	S?	2	
23 59 25.9	+20 44 59	NGC 7798	12.4	1.4 x 1.3	12.9	S	51	Strong dark lanes around center.
00 03 58.5	+20 44 59	NGC 7817	11.8	3.5 x 0.9	12.9	SAbc: sp	45	Mag 7.49v star SW 9'.1.
00 02 46.6	+18 53 09	UGC 3	13.5	1.4 x 0.6	13.1	SBa	84	Mag 8.58v star SSW 4'.0; **CGCG 456-22** E 3'.9.
00 03 10.1	+21 57 31	UGC 6	14.0	0.8 x 0.7	13.2	Pec	117	Bright core with one arm E.
00 03 21.9	+22 06 03	UGC 11	14.4	0.9 x 0.6	13.6	S?	42	Mag 14.0 star W 0'.9.
00 03 20.5	+29 47 48	UGC 12	14.7	0.7 x 0.5	13.4	Scd:	105	
00 03 29.3	+27 21 04	UGC 13	13.7	1.1 x 0.9	13.5	(R')SB(s)0/a	24	Bright E-W bar, faint envelope.
00 03 35.0	+23 11 56	UGC 14	12.9	1.9 x 1.1	13.5	S?	32	
00 04 33.8	+28 18 04	UGC 29	13.6	1.2 x 0.8	13.7	E:	9	
00 04 52.1	+17 11 31	UGC 31	14.2	1.0 x 0.7	13.7	IAm	55	
00 05 48.4	+27 26 55	UGC 40	14.0	1.2 x 0.7	13.7	SB?	112	Strong nucleus with dark patches E and W.
00 06 21.8	+17 26 00	UGC 46	13.7	0.8 x 0.8	13.1	Spiral		Double system, bridge, consists of **MCG +3-1-23** and **MCG +3-1-24**.
00 06 40.4	+26 09 12	UGC 50	14.1	0.9 x 0.3	12.5	Sab	10	Mag 11.27v star SE 2'.7.
00 06 50.4	+19 19 18	UGC 53	14.3	0.5 x 0.4	12.4	SABbc	171	
00 08 10.2	+27 00 11	UGC 68	13.5	0.7 x 0.6	12.4	SB?	145	
00 08 10.9	+27 31 40	UGC 69	13.8	0.9 x 0.7	13.1	Scd:	42	
00 09 04.4	+25 37 06	UGC 79	14.4	1.4 x 1.0	14.6	Scd:	85	Stellar nucleus, uniform envelope.
00 09 46.5	+28 20 25	UGC 87	13.6	1.1 x 0.7	13.3	E?	138	Bright knot or star N of nucleus; mag 15.3 star on SW edge; faint, almost anonymous galaxy SW 1'.4.
00 10 02.1	+28 12 35	UGC 92	13.7	1.3 x 0.9	13.7	Sb II	54	Very faint anonymous galaxy NE 1'.3.
00 10 26.5	+28 59 18	UGC 95	14.8	1.7 x 0.2	13.5	Scd:	7	Forms a triangle with NGC 27 E and a mag 10.51v star SE 1'.8.
00 11 14.4	+28 54 19	UGC 105	13.7	1.2 x 1.2	13.9	(R')SAB(s)0+		Short, moderately bright N-S bar; stellar galaxy **KUG 0008+287** N 4'.8.
00 11 45.1	+28 29 53	UGC 108	14.0	1.0 x 0.5	13.1	SBb:	51	Mag 7.52v star S 4'.8.
00 12 01.4	+26 23 33	UGC 110	14.3	1.0 x 0.6	13.6	SA(s)cd	145	
00 12 15.7	+22 19 14	UGC 113	13.9	1.1 x 0.3	12.6	Sa	105	Galaxy **KUG 0009+220A** 1'.6 W.
00 13 17.4	+17 01 46	UGC 122	14.6	2.4 x 0.4	14.4	Im	109	
00 13 56.7	+26 58 02	UGC 127	15.8	1.2 x 0.3	14.5	SBcd:	126	Mag 12.3 star NW 2'.3.
00 15 43.9	+27 26 37	UGC 146	14.1	1.1 x 0.4	13.1	S0	65	
00 15 43.7	+29 39 58	UGC 147	13.6	1.5 x 0.2	12.1	S0/a	156	Located between a pair of mag 12.0 and 13.0 stars; very faint, small anonymous galaxy on E edge.
00 16 28.7	+29 55 23	UGC 152	14.7	1.6 x 0.3	13.7	Scd:	78	Lies between a mag 9.55v star S 4'.7 and a mag 8.76v star NNE 8'.2.
00 16 53.9	+27 50 51	UGC 157	14.0	1.0 x 0.8	13.7	SBb:	15	CGCG 499-97 N 2'.6.
00 17 23.9	+18 05 05	UGC 164	14.0	1.7 x 0.6	13.8	SBbc	65	Mag 12.1 star W 2'.7.
00 18 18.7	+19 23 35	UGC 169	13.8	1.0 x 0.7	13.3	SBcd?	3	Mag 8.95v star on N edge.
00 19 00.6	+23 28 31	UGC 179	14.3	0.8 x 0.4	12.9	Scd:	3	
00 20 48.8	+20 01 50	UGC 197	13.8	0.9 x 0.7	13.1	SABa	12	
00 21 08.3	+27 12 46	UGC 202	13.3	1.2 x 0.9	13.2	(R)SB0°	22	Mag 10.72v star W 1'.7.
00 22 05.1	+23 44 10	UGC 210	14.1	1.0 x 0.4	12.9	Sb III	19	
00 22 26.5	+29 30 13	UGC 215	13.8	1.3 x 0.8	13.6	SBab	135	Mag 12.7 stars on NW end and E edge.
00 23 10.8	+27 25 54	UGC 221	13.8	0.6 x 0.4	12.1	S?	164	
00 23 15.6	+20 16 03	UGC 223	15.9	1.5 x 1.5	16.6	Im:		Dwarf, very low surface brightness.
00 23 49.7	+26 55 17	UGC 227	13.8	0.9 x 0.6	13.0	Compact	145	Mag 10.43v star NE 1'.6.
00 23 56.8	+24 18 17	UGC 228	13.9	1.0 x 0.6	13.2	Sbc	43	
00 25 28.7	+20 14 15	UGC 242	13.3	1.4 x 1.0	13.5	SABd	111	Very faint anonymous galaxy N 2'.2.
00 25 30.0	+24 48 35	UGC 244	14.0	1.0 x 0.5	13.1	SBbc?	154	
00 26 06.8	+25 43 25	UGC 248	14.0	1.7 x 0.9	14.3	S?	48	= **Hickson 1A**. **Hickson 1B** is bright "knot" S of nucleus; **Hickson 1C** w 2'.9; **Hickson 1D** w 1'.8.
00 27 17.4	+24 10 17	UGC 261	13.6	0.9 x 0.8	13.1	S?	0	Brighter southern half.
00 27 14.5	+20 04 17	UGC 265	14.0	0.8 x 0.5	12.9	S?	117	
00 28 21.3	+27 22 09	UGC 278	14.3	1.6 x 0.3	13.4	SBbc	6	
00 31 17.7	+28 59 30	UGC 310	14.1	1.1 x 0.3	12.8	Scd:	30	Mag 10.41 star N 2'.4.
00 32 32.4	+23 23 44	UGC 321	14.6	1.5 x 0.3	13.6	SBcd?	149	Mag 14.5 star superimposed S end.
00 36 10.0	+25 48 27	UGC 360	15.1	0.8 x 0.5	14.0	I?	18	**CGCG 479-42** SW 10'.3.
00 36 45.1	+21 33 56	UGC 364	13.6	1.2 x 0.4	12.6	S?	58	**CGCG 457-21** N 2'.5.
00 37 05.4	+25 41 57	UGC 367	13.9	1.0 x 1.0	14.0	E?		In contact with **MCG +4-3-37** on SE edge; mag 10.6 star on NW edge.
00 37 21.3	+29 08 54	UGC 371	14.4	1.7 x 0.3	13.5	Scd:	45	
00 37 43.8	+25 38 23	UGC 375	14.1	0.7 x 0.7	13.2	Compact		
00 38 39.8	+17 24 14	UGC 393	15.2	1.8 x 0.4	14.7	Sdm:	170	Mag 12.0 star S 2'.2.
00 38 57.4	+25 38 18	UGC 398	13.4	1.1 x 0.9	13.3	SA(r)a	25	**CGCG 479-50** N 5'.3; stellar galaxy **IV Zw 27** S 5'.3.
00 39 18.4	+29 39 26	UGC 400	14.1	0.7 x 0.6	13.0	S?	75	Elongated **UGC 412** N 6'.7.

(Continued from previous page)

GALAXIES

RA h m s	Dec ° ′ ″	Name	Mag (V)	Dim ′ Maj x min	SB	Type Class	PA	Notes
00 39 29.6	+25 38 35	UGC 411	13.2	1.1 x 0.7	13.0	E?	100	Mag 9.10v star N 5.′5.
00 40 10.9	+22 42 52	UGC 425	13.8	0.6 x 0.6	12.5	SB?		Interacting pair?
23 16 21.7	+28 29 21	UGC 12470	14.2	1.0 x 0.7	13.6	Sa	135	
23 17 33.7	+29 01 08	UGC 12482	13.0	1.4 x 1.2	13.5	E:	39	MCG +5-54-60 E 2.′8; **CGCG 496-75** NNW 2.′9.
23 18 38.8	+25 13 54	UGC 12490	13.6	0.9 x 0.6	12.8	SBa	75	**CGCG 475-61** N 2.′1; mag 9.28v star SE 1.′9.
23 19 51.1	+26 15 44	UGC 12515	12.9	1.3 x 1.0	13.1	S0	55	Located 0.′9 S of mag 9.40v star.
23 20 31.6	+29 18 22	UGC 12530	13.8	1.1 x 1.1	13.8	SBcd:		Mag 12.4 star E 2.′4.
23 22 28.7	+29 10 46	UGC 12557	14.1	2.1 x 0.4	13.8	S?	9	Mag 10.45v star W 9.′2.
23 22 46.0	+29 08 16	UGC 12566	13.6	1.7 x 1.3	14.3	Sab	140	Bright center, prominent dark patches SE and NW, faint envelope.
23 24 50.2	+26 38 42	UGC 12587	14.4	1.2 x 0.9	14.4	Sb III	129	Mag 12.1 star SE 1.′2.
23 25 21.8	+28 29 39	UGC 12591	12.9	1.5 x 0.7	12.8	S0/a	58	Full length dark lane ; mag 9.97v star S 1.′6.
23 28 45.2	+22 14 05	UGC 12619	14.3	1.5 x 0.9	14.5	SAB(s)dm	173	Very knotty; mag 11.6 star N 2.′5.
23 29 07.1	+29 46 29	UGC 12625	14.2	1.5 x 0.2	12.8	Sb III	176	Mag 10.02v star NNW 3.′0.
23 30 01.3	+27 05 11	UGC 12631	14.1	1.2 x 0.3	12.9	Sb III	150	
23 30 19.2	+27 08 00	UGC 12636	14.2	0.9 x 0.8	13.8	S0	141	Mag 13.6 star on W edge; **MCG +4-55-21** S 2.′3.
23 31 39.1	+25 56 42	UGC 12646	13.2	1.8 x 1.5	14.1	SB(r)b	45	Bright SE-NW bar, uniform envelope; mag 9.53v star E 2.′9.
23 32 29.1	+23 55 53	UGC 12655	12.8	1.5 x 0.9	13.0	S0	135	Mag 6.40v star S 5.′4.
23 32 43.6	+29 27 37	UGC 12657	13.2	1.9 x 0.6	13.3	S0	140	Large, diffuse core, uniform envelope.
23 33 31.3	+24 01 09	UGC 12663	15.1	0.9 x 0.9	14.7	S?		Mag 11.8 star W 1.′3.
23 34 54.7	+18 04 13	UGC 12681	13.9	1.2 x 0.7	13.6	SB?	15	
23 34 53.8	+18 13 41	UGC 12682	13.5	1.5 x 1.2	14.0	Im	40	Mag 9.14v star SW 8.′8.
23 36 01.3	+23 02 14	UGC 12696	14.0	1.0 x 0.9	13.8	SBbc	51	Mag 15.3 star superimposed SW of nucleus.
23 36 41.0	+17 30 52	UGC 12707	13.6	1.5 x 1.0	13.9	SB?	40	
23 37 33.3	+17 59 52	UGC 12710	13.8	1.4 x 0.8	13.7	Im?	140	Bright core offset SW.
23 39 36.0	+21 54 53	UGC 12725	14.0	1.0 x 0.4	12.9	S0	10	
23 40 38.9	+20 26 26	UGC 12731	14.7	1.5 x 0.2	13.2	Sb	108	Mag 8.34v star SSE 5.′3; located on W edge of very small galaxy cluster **A 2643**.
23 40 40.1	+26 14 08	UGC 12732	13.2	3.3 x 1.5	14.8	Sm:	174	Mag 15 star superimposed on E edge; mag 14.8 star superimposed SW edge.
23 40 46.9	+26 50 05	UGC 12733	13.8	1.0 x 0.8	13.5	E	0	Anonymous galaxy E 0.′9.
23 43 02.6	+19 25 19	UGC 12747	14.3	1.0 x 0.6	13.6	Im:	110	Brightest W.
23 43 49.8	+28 20 18	UGC 12755	13.7	1.4 x 0.6	13.4	SB?	102	Mag 11.9 star W 1.′4.
23 44 38.9	+22 00 34	UGC 12764	14.6	0.9 x 0.7	13.9	Scd:	9	Mag 14.0 star N edge.
23 48 04.4	+17 28 29	UGC 12784	13.9	1.0 x 0.9	13.6	SAB(s)bc	36	Mag 11.07v star N 2.′2.
23 48 01.0	+27 22 29	UGC 12785	13.9	1.2 x 0.4	13.0	S0	100	**CGCG 477-6** NW 5.′0.
23 48 49.6	+26 13 12	UGC 12791	15.0	1.5 x 0.6	14.7	Im: IV-V	83	Dwarf irregular, very low surface brightness; mag 8.27v star N 5.′8.
23 49 02.1	+26 47 20	UGC 12792	13.8	1.1 x 0.5	13.0	SBb	49	
23 50 33.9	+28 59 52	UGC 12803	14.0	1.3 x 1.0	14.2	S?	126	Mag 12.1 star on E edge.
23 50 36.8	+24 33 19	UGC 12804	14.1	0.8 x 0.7	13.6	E	141	
23 52 44.2	+27 07 56	UGC 12823	14.0	0.9 x 0.7	13.4	SB0	174	
23 53 56.7	+28 29 28	UGC 12835	13.3	1.0 x 0.8	13.0	E?	0	Mag 10.01v star S; note: faint, small elongated anonymous galaxy 0.′8 W of this star.
23 54 29.8	+28 18 28	UGC 12839	13.8	1.1 x 0.9	13.6	S0	18	Mag 12.7 star S of nucleus; mag 11.6 star W 2.′0.
23 54 30.1	+28 52 13	UGC 12840	13.1	1.2 x 1.0	13.1	(R)SAB(s)0°	15	Bright N-S center, uniform envelope.
23 55 30.6	+17 55 15	UGC 12843	13.2	2.8 x 1.2	14.4	SABdm	27	Mag 10.31v star superimposed almost on center.
23 55 39.5	+19 30 56	UGC 12844	14.4	1.5 x 0.5	13.9	Scd:	138	Mag 8.83v star N 6.′7.
23 56 37.4	+27 06 14	UGC 12855	13.5	1.3 x 0.6	13.1	S0/a	170	Elongated **CGCG 478-4** N 5.′8.
23 57 06.5	+29 50 18	UGC 12861	14.2	0.9 x 0.5	13.2	Sab	140	Mag 10.9 star E 1.′6; mag 10.7 star N 2.′5.
23 58 31.8	+26 12 51	UGC 12873	14.5	1.2 x 0.8	14.3	Sm	108	Mag 10.67v star S 1.′4.
23 59 01.3	+18 50 00	UGC 12879	14.0	0.6 x 0.4	12.3	S?	51	
23 59 30.8	+18 12 08	UGC 12886	13.7	0.7 x 0.5	12.5	Sbc	101	
00 00 28.2	+17 13 08	UGC 12893	14.0	1.8 x 1.7	15.1	SAdm	90	Almost stellar nucleus, uniform envelope.
00 00 31.7	+26 19 29	UGC 12896	13.8	1.0 x 0.9	13.5	S	105	**MCG +4-1-9** S 1.′2.
00 00 38.0	+28 23 00	UGC 12897	13.9	1.1 x 0.4	12.9	Sab	5	= **Hickson 99A**. UGC 12899 = **Hickson 99B** 2.′2 NE; MCG +5-1-21 = **Hickson 99C** 1.′7 NE; **Hickson 99D** ESE 1.′7; **Hickson 99E** ESE 1.′3.
00 00 47.1	+28 24 05	UGC 12899	13.8	0.6 x 0.6	12.5	S0?		= **Hickson 99B**.
00 00 56.3	+20 20 13	UGC 12900	14.8	1.8 x 0.2	13.5	Scd:	111	Mag 10.95v star NW 3.′3.
00 00 59.0	+28 54 39	UGC 12901	13.8	1.4 x 0.5	13.3	SBb	48	Mag 11.3 star NW 2.′7.
00 01 38.5	+23 28 56	UGC 12914	12.4	2.5 x 1.2	13.4	(R)S(r)cd: pec	160	Pair with UGC 12915 NE, disrupted.
00 01 41.6	+23 29 46	UGC 12915	13.0	1.4 x 0.5	12.5	S?	137	Pair with UGC 12914 SW; pair of mag 13 stars NE 1.′6.

GALAXY CLUSTERS

RA h m s	Dec ° ′ ″	Name	Mag 10th brightest	No. Gal	Diam ′	Notes
00 20 30.0	+28 37 00	A 21	16.2	56	13	Only faint, stellar GX seen on DSS image.
00 28 54.0	+17 34 00	A 43	15.9	37	20	Mag 8.41v star on E edge; member galaxies faint and stellar.
00 37 48.0	+29 35 00	A 71	15.5	30	19	Band of numerous, faint, stellar anonymous GX extend SE to NW.
00 39 48.0	+21 15 00	A 75	15.5	42	13	Mag 5.88v star 54 Piscium 6.′0 W of center.
23 18 24.0	+18 44 00	A 2572	15.3	32	28	Galaxy group **Hickson 94** on E edge.
23 33 48.0	+23 00 00	A 2618	15.9	35	15	All members anonymous, stellar/almost stellar.
23 34 54.0	+27 25 00	A 2622	15.9	41	18	**PGC 71807** SE of center 3.′2; all other members anonymous.
23 36 18.0	+20 31 00	A 2625	15.6	45	15	All members anonymous; stellar/almost stellar.
23 36 30.0	+21 09 00	A 2626	15.2	47	13	See notes for IC 5338; all other members anonymous, stellar/almost stellar.
23 38 18.0	+27 01 00	A 2634	13.8	52	22	Tight concentration near center, other members evenly distributed over cluster area.
23 50 54.0	+27 08 00	A 2666	13.8	34	78	Numerous galaxies around NGC 7768.

GALAXIES

RA h m s	Dec ° ′ ″	Name	Mag (V)	Dim ′ Maj x min	SB	Type Class	PA	Notes
22 59 36.0	+17 58 14	CGCG 453-38	14.7	0.5 x 0.3	12.5		153	**CGCG 453-39** NE 4.′3; **CGCG 453-41** SE 7.′3.
23 05 27.3	+21 09 38	CGCG 453-65	14.0	0.5 x 0.4	12.1	Sbc	57	Mag 10.21v star SSW 1.′4.
23 08 50.8	+20 04 09	CGCG 454-1	15.1	0.5 x 0.5	13.4	Sc		Mag 9.53v star ESE 5.′4.
23 10 16.8	+26 37 13	CGCG 475-41	14.4	0.5 x 0.4	12.5		117	Mag 11.4 star NE 2.′3.
23 17 30.6	+22 07 32	CGCG 475-59	14.3	0.6 x 0.4	12.6	Sc	72	Low, uniform surface brightness.
23 10 11.2	+29 54 57	CGCG 496-50	13.8	0.7 x 0.6	12.7		162	Multi-galaxy system.
22 02 31.5	+19 44 56	IC 1420	13.2	1.2 x 0.8	13.0	SB?	90	Mag 10.25v star SSE 6.′2.
23 09 06.7	+17 15 29	IC 1472	14.2	1.0 x 0.5	13.3	S0/a	57	Mag 10.26v star SW 4.′5.
23 11 05.4	+29 38 33	IC 1473	12.9	2.0 x 1.0	13.5	S0	176	Mag 11.04v star SW 7.′5.
22 34 00.7	+23 20 18	IC 5231	14.5	1.0 x 0.9	14.3		69	Mag 10.93v star NW 9.′5.

(Continued from previous page)

GALAXIES

RA h m s	Dec ° ′ ″	Name	Mag (V)	Dim ′ Maj x min	SB	Type Class	PA	Notes
22 36 33.0	+25 45 47	IC 5233	13.8	1.0 x 0.8	13.4	S?	19	Mag 12.4 star on NE edge.
22 41 15.4	+23 24 22	IC 5242	13.7	0.8 x 0.8	13.1	S?		Mag 13.6 star on N edge; strong dark lane SE of center.
22 41 24.8	+23 22 26	IC 5243	14.5	0.7 x 0.6	13.4	S?	30	Pair of stars, mags 9.42v and 10.75v, SW 2′.5.
22 51 31.8	+23 04 47	IC 5258	12.8	1.2 x 0.9	12.8	S0⁻:	105	Mag 10.15v star S 5′.1.
22 58 27.7	+18 55 06	IC 5274	13.8	0.8 x 0.8	13.2	S0:		Mag 9.44v star W 12′.2; **IC 5276** SSE 6′.5.
23 02 48.2	+21 52 27	IC 5282	14.3	1.3 x 0.6	13.9	Sd	173	Mag 11.1 star E 4′.4.
23 06 46.4	+19 07 17	IC 5284	13.9	1.0 x 0.3	12.5	Sa: sp	141	Mag 7.45v star SW 11′.1.
23 06 59.0	+22 56 11	IC 5285	12.6	1.6 x 1.2	13.1	S?	100	Mag 10.4 star S 1′.6.
23 16 00.8	+25 33 21	IC 5298	14.0	0.7 x 0.6	12.9	S?	108	Mag 8.71v star E 8′.3.
23 20 58.5	+19 19 04	IC 5312	14.4	0.4 x 0.3	12.0	Sb	99	Faint, round anonymous galaxy SE edge; IC 5314 E 2′.4.
23 21 08.6	+19 18 38	IC 5314	14.7	0.5 x 0.5	13.0			Mag 11.05v star S 0′.9.
23 21 18.2	+25 23 05	IC 5315	13.5	0.9 x 0.6	12.8	E?	138	Trio of mag 12, 13 and 14 stars N 1′.6.
23 23 28.7	+21 09 46	IC 5317	14.1	1.0 x 0.7	13.6	S0	21	Mag 11.9 star SSW 5′.3.
22 06 38.2	+17 27 39	MCG +3-56-9	14.5	0.7 x 0.4	13.0	Sbc	3	Mag 12.1 star SE 0′.8.
22 09 13.0	+20 23 17	MCG +3-56-13	15.3	0.9 x 0.9	15.0	Sc		Mag 10.05v star NW 3′.6.
22 24 23.8	+18 04 13	MCG +3-57-2	15.1	0.7 x 0.3	13.3	Sbc	126	**MCG +3-57-4** E 2′.3.
22 28 27.5	+19 30 27	MCG +3-57-10	14.3	0.8 x 0.5	13.1	S?	24	Mag 11.7 star NW 1′.2.
22 35 29.0	+19 40 42	MCG +3-57-19	14.4	0.7 x 0.3	12.8	E	0	Mag 7.08v star NE 4′.9.
22 36 41.2	+19 31 21	MCG +3-57-22	15.2	0.8 x 0.2	13.1	Spiral	33	Mag 11.2 star N 2′.6.
22 36 42.1	+19 38 52	MCG +3-57-23	14.8	0.7 x 0.3	13.0	Sbc	132	Mag 9.57v star SE 3′.1.
22 37 39.1	+20 00 56	MCG +3-57-28	14.4	0.8 x 0.2	12.3	S?	123	
22 38 18.4	+18 35 12	MCG +3-57-29	14.0	1.0 x 0.7	13.5	SBa	3	Small line of three mag 13-14 stars NE 1′.3.
22 41 56.0	+20 15 38	MCG +3-57-31	14.3	0.5 x 0.3	12.1	I0?	24	Faint anonymous galaxy NE 0′.9; mag 9.44v star SE 3′.2.
22 44 50.6	+19 08 02	MCG +3-58-1	14.9	0.7 x 0.4	13.4	SBc	162	Mag 10.02v star WSW 3′.3.
22 48 48.8	+18 33 47	MCG +3-58-5	14.3	0.7 x 0.4	12.7	S?	147	Mag 9.01v star SE 2′.5.
22 49 30.6	+17 25 58	MCG +3-58-7	14.5	0.8 x 0.5	13.4	SB?	3	Mag 11.3 star NW 3′.7.
22 51 04.5	+18 51 19	MCG +3-58-8	15.3	0.8 x 0.3	13.6	Sbc	30	Pair of stars, mags 11.8 and 13.6, S 1′.7.
22 55 32.5	+19 17 30	MCG +3-58-13	15.3	0.7 x 0.3	13.5	Sb	96	Faint star N edge.
23 08 11.3	+18 06 07	MCG +3-59-1	15.8	0.6 x 0.3	13.7	Sb	120	Mag 11.1 star SW 3′.2.
23 11 38.0	+18 46 22	MCG +3-59-9	14.5	0.7 x 0.3	12.7	Sb	45	
23 13 07.1	+18 03 21	MCG +3-59-11	15.1	0.5 x 0.4	13.2	Sb	156	Mag 12.9 star superimposed N end.
23 17 21.7	+18 25 23	MCG +3-59-26	15.5	0.7 x 0.5	14.2	Spiral	174	Mag 6.72v star SSE 8′.0; **CGCG 454-28** E 4′.5.
23 17 21.2	+18 03 20	MCG +3-59-27	15.2	0.5 x 0.5	13.6	S0/a		Mag 11.0 star SW 2′.7.
23 20 48.4	+18 54 16	MCG +3-59-40	14.4	0.8 x 0.3	12.8	S?	81	Almost stellar **MCG +3-59-39** NW 7′.9.
23 23 01.2	+17 44 27	MCG +3-59-46	14.7	0.6 x 0.6	13.4			Faint star SW edge.
23 23 04.1	+18 00 55	MCG +3-59-47	14.6	0.7 x 0.4	13.0	S?	54	
23 23 20.5	+20 35 21	MCG +3-59-48	14.6	1.0 x 1.0	14.4			MCG +3-59-49 S 1′.4; **MCG +3-59-50** E 1′.2.
23 23 21.7	+20 33 58	MCG +3-59-49	14.4	0.5 x 0.3	12.2		120	MCG +3-59-48 N 1′.4.
23 23 32.8	+19 35 56	MCG +3-59-51	15.0	0.6 x 0.5	13.5	Spiral	3	**UGC 12574** SE 2′.6.
21 58 49.5	+25 30 03	MCG +4-51-15	15.3	0.5 x 0.5	13.7			Mag 10.9 star WSW 2′.8.
21 59 29.9	+23 42 58	MCG +4-52-1	14.6	0.8 x 0.3	12.9	S?	6	Close pair of mag 10.9, 12.3 stars ESE 2′.0.
22 05 09.0	+26 36 24	MCG +4-52-3	14.1	0.7 x 0.3	12.2	S?	135	Mag 5.75v star N 4′.1.
22 12 26.1	+27 12 26	MCG +4-52-6	14.7	0.8 x 0.2	12.6	Sb	72	Mag 8.63v star W 1′.1.
22 17 13.2	+25 12 45	MCG +4-52-7	14.5	0.8 x 0.5	13.4	S?	0	Mag 10.34v star NW 3′.0.
22 21 29.1	+25 15 41	MCG +4-52-9	14.7	0.5 x 0.5	13.1	S?		Mag 12.2 star NE 2′.3.
22 21 47.8	+25 19 10	MCG +4-52-10	14.8	0.9 x 0.2	12.8	S?	165	Mag 11.6 star SE 2′.2.
22 22 32.8	+25 17 53	MCG +4-52-11	14.8	0.5 x 0.5	13.1			Small group of four mag 13-14 stars N 1′.6.
22 44 22.1	+25 06 39	MCG +4-53-14	14.5	0.7 x 0.5	13.2	Compact	105	Mag 11.9 star N 1′.9.
22 58 12.9	+25 13 13	MCG +4-54-9	14.5	0.6 x 0.6	13.2	Compact		Mag 12.5 star NW 1′.2; small, faint anonymous galaxy just S of this star.
23 03 10.4	+23 17 39	MCG +4-54-20	14.4	0.5 x 0.5	13.0	E?		Mag 8.75v star SE 1′.9.
23 04 17.4	+22 32 23	MCG +4-54-22	14.4	0.9 x 0.5	13.4	SB?	54	Stellar **Mkn 315** NW 6′.0.
23 04 28.2	+27 21 24	MCG +4-54-23	13.8	0.8 x 0.4	12.4		123	Mag 14.8 star superimposed SE edge; **CGCG 475-34** E 6′.6.
23 07 25.6	+22 48 33	MCG +4-54-27	14.8	0.8 x 0.3	13.1	S?	87	Mag 15.1 star on W end; mag 13.2 star SW 1′.3.
23 13 58.3	+25 26 32	MCG +4-54-34	14.5	0.5 x 0.3	12.3	Sb	126	Mag 8.10v star NE 0′.9; **MCG +4-54-33** NNW 0′.5.
23 15 59.9	+27 27 10	MCG +4-54-39	14.4	0.6 x 0.4	12.7	S?	153	Mag 9.10v star NW 2′.6.
23 19 14.5	+26 03 18	MCG +4-55-2	14.2	0.9 x 0.7	13.6	Spiral	3	Mag 10.84v star SE 4′.5.
22 11 06.5	+29 36 31	MCG +5-52-2	14.6	0.6 x 0.5	13.1	S?		MCG +5-52-3 N 2′.2; mag 8.51v star NW 2′.2.
22 11 08.3	+29 38 38	MCG +5-52-3	15.5	0.5 x 0.5	14.1	E		MCG +5-52-2 S 2′.2; mag 8.51v star W 2′.8.
22 16 11.8	+27 54 16	MCG +5-52-6	14.6	0.7 x 0.3	12.7	SB?	15	Mag 12.9 star N end; mag 10.2 star SE 1′.6.
22 22 46.2	+29 46 36	MCG +5-52-13	14.2	0.8 x 0.5	13.1	SB?	39	Mag 12.6 star 1′.2 S.
22 56 21.5	+28 26 32	MCG +5-54-20	14.2	0.9 x 0.7	13.6	S?	108	Mag 10.60v star NE 1′.5.
23 05 46.1	+28 41 51	MCG +5-54-31	14.3	1.3 x 0.7	14.0	SB?	78	
23 08 04.0	+28 34 07	MCG +5-54-38	14.7	0.6 x 0.5	13.2	S?	156	Small triangle of mag 12-13 stars ESE 1′.4.
23 09 29.9	+29 28 57	MCG +5-54-39	14.8	1.0 x 0.2	12.9	S?	102	Mag 12.7 star N 2′.1.
23 15 10.5	+28 57 03	MCG +5-54-53	14.5	1.1 x 0.5	13.7	Sbc	165	
23 16 43.8	+29 35 12	MCG +5-54-56	14.8	0.8 x 0.3	13.1	S?	18	Mag 14.5 star SW end.
23 17 18.0	+28 35 42	MCG +5-54-58	14.4	0.7 x 0.5	13.1	Sc	147	Mag 7.91v star N edge.
23 17 46.1	+29 01 35	MCG +5-54-60	14.7	0.5 x 0.5	13.0			UGC 12482 W 2′.7.
23 18 15.4	+29 13 49	MCG +5-54-61	14.2	0.9 x 0.9	14.0	E		Mag 14.3 star SW edge.
23 22 58.4	+29 14 02	MCG +5-55-7	14.7	0.6 x 0.6	13.4	Spherical		UGC 12566 SSW 6′.4.
23 23 22.1	+29 25 43	MCG +5-55-9	14.6	0.4 x 0.4	12.5	SB?		Mag 13.4 star SW 0′.7.
23 23 34.4	+29 10 54	MCG +5-55-10	13.8	0.8 x 0.6	12.9	S0	39	Mag 12.9 star NE 3′.3.
22 00 41.1	+17 44 13	NGC 7177	11.2	3.1 x 2.0	13.0	SAB(r)b I-II	93	
22 11 35.5	+25 51 53	NGC 7224	13.2	1.6 x 1.0	13.7	E	110	
22 15 49.7	+19 13 53	NGC 7241	12.6	3.4 x 1.1	13.9	SB(s)bc? pec	20	UGC 11964 W 5′.0.
22 19 26.8	+29 23 35	NGC 7253A	13.2	1.7 x 0.8	13.4	SB?	116	Double system with NGC 7253B SE edge.
22 19 29.6	+29 23 07	NGC 7253B	14.5	1.6 x 0.5	14.1	S?	59	
22 28 32.8	+17 28 14	NGC 7283	14.4	0.9 x 0.4	13.1	S?	9	Mag 10.43v star W 2′.7.
22 27 50.7	+29 05 42	NGC 7286	12.5	1.7 x 0.7	12.5	S0/a	98	
22 28 26.4	+17 08 49	NGC 7290	13.3	1.6 x 1.0	13.7	SA(r)bc	161	
22 35 56.5	+20 19 20	NGC 7316	13.0	1.1 x 0.9	12.9	S	60	Located 3′.3 NE of a mag 6.61v star.
22 36 28.2	+21 37 19	NGC 7321	12.9	1.6 x 1.1	13.4	SB(r)b	12	Galaxy **Kaz 227** S 1′.7.
22 36 53.7	+19 08 33	NGC 7323	12.9	1.4 x 1.1	13.3	Sb	170	NGC 7324 E 1′.7.
22 37 01.0	+19 08 49	NGC 7324	13.9	1.0 x 0.8	13.7	E	168	Star on SE edge.
22 37 24.9	+23 47 51	NGC 7332	11.1	4.1 x 1.1	12.6	S0 pec sp	155	Mag 10.9 star on S edge.
22 37 47.2	+23 47 12	NGC 7339	12.2	3.0 x 0.7	12.9	SAB(s)bc:?	93	
22 46 32.0	+21 04 59	NGC 7375	13.7	1.0 x 0.7	13.2	S(r)0	60	Bright nucleus.
22 53 48.2	+20 12 37	NGC 7409	14.9	0.5 x 0.4	13.2	E	162	**CGCG 453-19** E 4′.7.
22 54 34.9	+20 14 07	NGC 7411	13.4	0.9 x 0.9	13.2	E:		NGC 7415 and mag 10.7 star NW 2′.6.

(Continued from previous page)

RA h m s	Dec ° ′ ″	Name	Mag (V)	Dim ′ Maj x min	SB	Type Class	PA	Notes
22 54 24.5	+20 15 06	NGC 7415	15.5	0.9 x 0.2	13.5	Sab	36	Mag 10.7 star NE 1′0.
22 55 32.1	+29 48 19	NGC 7420	13.9	0.8 x 0.6	13.0	S?	54	Mag 9.57v star W 4′7.
22 57 51.7	+26 09 41	NGC 7431	14.9	0.5 x 0.2	12.2	S?	48	Located 1′5 NW of NGC 7436's bright nucleus.
22 57 56.1	+26 08 58	NGC 7433	14.0	0.7 x 0.2	11.8	S0/a	99	On W edge of NGC 7436's bright nucleus.
22 57 54.5	+26 08 19	NGC 7435	14.2	1.3 x 0.6	13.8	SB(s)a	120	On SW edge of NGC 7436. Bright N-S bar with faint envelope SE-NW.
22 57 57.8	+26 09 00	NGC 7436	12.6	2.0 x 2.0	14.1	E		NGC 7433 on W edge of bright nucleus.
22 58 09.9	+29 13 37	NGC 7439	14.0	1.1 x 0.7	13.6	SB0	150	
23 02 03.5	+27 03 07	NGC 7466	13.5	1.5 x 0.5	13.1	Sb II-III	26	
23 04 04.5	+20 03 59	NGC 7474	14.5	0.5 x 0.5	12.9			
23 04 10.9	+20 04 52	NGC 7475	13.5	1.1 x 0.9	13.3	Double System	54	Multi-galaxy system; **CGCG 453-57** and a mag 10.42v star N 6′6.
23 06 50.6	+28 10 45	NGC 7487	13.5	2.0 x 1.8	14.8	Compact	141	Bright center, uniform envelope.
23 07 32.6	+22 59 52	NGC 7489	13.4	2.1 x 1.1	14.1	Sd	170	Group of six faint stars on W edge; mag 9.28v star NNE 4′3.
23 09 04.8	+18 10 56	NGC 7497	12.2	4.9 x 1.1	13.9	SB(s)d	48	
23 12 52.0	+20 14 52	NGC 7516	13.3	1.1 x 1.0	13.2	S0	110	Bright nucleus, smooth envelope.
23 13 41.9	+24 54 06	NGC 7527	13.3	1.3 x 0.9	13.5	E	165	**UGC 12432** E 3′8.
23 14 29.5	+23 41 02	NGC 7539	12.5	1.5 x 1.0	12.8	S0	165	Bright, diffuse center, smooth envelope.
23 14 34.7	+28 19 36	NGC 7543	13.1	1.1 x 0.9	13.0	S?	140	**CGCG 496-62** N 7′5.
23 15 03.5	+18 58 23	NGC 7547	13.7	1.0 x 0.4	12.6	(R′)SAB(s)0/a: pec	98	= **Hickson 93C**.
23 15 11.2	+25 16 55	NGC 7548	13.3	1.1 x 0.9	13.2	(R)SAB(r)0°:	15	Stellar nucleus.
23 15 17.0	+19 02 27	NGC 7549	13.0	2.8 x 0.8	13.8	SB(s)cd pec	8	= **Hickson 93B**; multi-galaxy system.
23 15 16.1	+18 57 39	NGC 7550	12.2	1.4 x 1.2	12.6	SA0⁻	171	= **Hickson 93A**. Bright center, smooth envelope.
23 15 33.2	+19 02 51	NGC 7553	14.7	0.4 x 0.4	12.8	S0		= **Hickson 93D**.
23 15 38.3	+18 55 10	NGC 7558	14.9	0.4 x 0.4	12.9	E?		= **Hickson 93E**. Stellar nucleus.
23 16 24.9	+24 29 50	NGC 7568	13.6	0.9 x 0.6	12.8	S?	120	
23 16 50.4	+18 28 57	NGC 7572	14.4	0.9 x 0.3	12.9	S0/a	162	**CGCG 454-20** S 4′5; **UGC 12471** NW 7′1.
23 17 12.0	+18 42 01	NGC 7578A	13.3	0.8 x 0.8	12.8	S0° pec		= **Hickson 94B**; stellar nucleus.
23 17 13.6	+18 42 27	NGC 7578B	14.0	0.4 x 0.4	12.0	E1:		= **Hickson 94A**; **Hickson 94C** NE 2′2; **Hickson 94D** NE 0′5; **Hickson 94E** NNE 1′3; **Hickson 94F** NE 2′2; **Hickson 94G** NE 2′9.
23 17 57.8	+18 45 06	NGC 7588	14.8	0.5 x 0.3	12.6		99	Located 2′1 NW of a mag 9.04v star.
23 18 30.2	+18 41 23	NGC 7597	14.0	1.0 x 1.0	13.9	S?		Mag 9.79v star NNW 2′9 with anonymous galaxy W of star.
23 18 33.3	+18 44 59	NGC 7598	14.9	0.5 x 0.5	13.5	E		Located 2′2 NE of mag 9.79 star.
23 18 43.6	+18 41 53	NGC 7602	14.4	0.6 x 0.6	13.2	S?		Almost stellar; mag 8.03v star S 3′2.
23 20 05.8	+24 13 13	NGC 7620	13.0	1.1 x 1.1	13.1	Scd:		
23 20 22.9	+27 18 57	NGC 7624	13.1	1.0 x 0.7	12.6	Scd:	30	Star or bright knot on S edge.
23 20 30.1	+17 13 32	NGC 7625	12.1	1.6 x 1.4	12.8	SA(rs)a pec	60	Pair of strong dark patches SW of center.
23 20 54.9	+25 53 54	NGC 7628	12.7	1.1 x 0.9	12.7	E	117	
21 55 59.3	+27 53 56	UGC 11852	13.8	1.5 x 0.7	13.7	SBa?	15	Elongated core with smooth envelope.
21 57 59.9	+24 16 00	UGC 11860	14.8	0.9 x 0.4	13.6	Sdm	133	
21 59 04.7	+18 10 42	UGC 11868	13.4	2.2 x 1.7	14.7	SB(s)m	78	Small, moderately bright center, faint envelope.
22 02 23.3	+18 18 52	UGC 11878	13.9	1.4 x 1.2	14.3	SB?	57	Small, bright core with one large, faint arm recurving S, then SW under the bright core.
22 05 53.9	+20 37 55	UGC 11905	13.7	1.7 x 0.7	13.8	S?	12	Long, narrow streamer extends S; mag 9.86v star SE 2′2.
22 09 15.4	+21 31 07	UGC 11924	13.7	1.5 x 0.9	13.9	Sd	120	
22 12 01.0	+17 54 17	UGC 11944	15.4	2.1 x 0.7	15.7	Im:	28	Dwarf irregular, extremely low surface brightness; mag 11.6 star S 2′1.
22 15 28.8	+19 13 10	UGC 11964	14.3	1.9 x 0.2	13.1	Scd:	45	Mag 10.41v star SE 2′3.
22 17 39.8	+28 16 52	UGC 11978	14.6	1.0 x 0.8	14.2	SAB(r)cd:	69	Mag 10.43v star W edge.
22 18 31.8	+29 14 36	UGC 11981	14.2	1.3 x 1.1	14.4	SA(s)c	170	Stellar nucleus; mag 11.7 star E 1′3.
22 23 25.5	+19 50 55	UGC 12015	14.4	1.3 x 1.1	14.6	SABd	45	Strong core, disrupted or filamentary envelope.
22 26 36.5	+25 06 30	UGC 12036	13.7	1.1 x 0.9	13.5	S0⁻:	165	Mag 14.0 star on SW edge.
22 26 54.5	+19 32 52	UGC 12038	14.2	1.1 x 0.3	12.8	Sb	55	Mag 11.9 star E 1′7.
22 29 03.3	+19 06 00	UGC 12050	13.9	0.9 x 0.5	12.9	S?	82	Located 3′2 SE of mag 10.61v star.
22 30 44.0	+22 32 29	UGC 12059	14.0	1.1 x 0.8	13.7	SBb	10	Mag 13.3 star on NW edge; mag 9.09v star W 3′0.
22 31 50.9	+19 41 34	UGC 12066	13.5	1.1 x 0.9	13.3	S pec	51	Double system, contact, disrupted; small, round **Mkn 305** on NW edge.
22 31 46.5	+20 36 09	UGC 12067	14.1	0.9 x 0.5	13.1	Sa	47	**CGCG 452-21** N 6′8.
22 34 26.8	+25 01 49	UGC 12084	14.0	1.5 x 1.1	14.4	SABb	120	
22 34 49.3	+18 38 18	UGC 12093	14.2	1.2 x 0.6	13.7	Scd:	114	Mag 10.84v star WSW 2′1.
22 36 46.7	+19 22 55	UGC 12107	13.8	1.3 x 0.7	13.5	S0	135	Mag 10.94v star S 2′6.
22 37 53.5	+25 20 19	UGC 12124	14.0	1.0 x 0.5	13.1	S0?	113	
22 42 10.6	+19 59 48	UGC 12158	13.9	1.2 x 1.1	14.1	Sb	87	Stellar nucleus.
22 42 39.5	+29 43 27	UGC 12163	14.0	0.7 x 0.4	12.4	SB	118	Located between a pair of mag 8.95v stars.
22 48 06.6	+28 17 36	UGC 12190	14.7	2.0 x 0.2	13.5	Sc	172	Mag 10.9 star E 1′7.
22 48 40.9	+27 36 43	UGC 12191	13.8	1.4 x 0.7	13.6	S?	2	Bright core, faint envelope; UGC 12193 S 1′8.
22 48 44.2	+27 34 58	UGC 12193	13.1	1.3 x 1.0	13.3	SB(s)bc	93	UGC 12191 N 1′8; **CGCG 474-32** E 2′3.
22 49 20.5	+19 17 21	UGC 12200	13.9	0.9 x 0.5	12.8	Sb	25	Stellar **Mkn 1125** N 8′2.
22 50 30.3	+29 08 16	UGC 12212	15.4	2.0 x 1.0	16.0	Sm:	84	Dwarf spiral, very low surface brightness; mag 10.29v star WNW 1′9.
22 53 45.0	+25 50 36	UGC 12233	14.4	0.7 x 0.5	13.1	SAB(s)b	84	Mag 12.2 star NE 3′9.
22 56 31.4	+17 47 01	UGC 12258	14.0	1.5 x 0.3	13.0	Sa	85	Mag 10.6 star NE 3′3; mag 11.3 star NNW 2′8.
22 56 27.1	+18 42 05	UGC 12259	15.1	1.5 x 0.3	14.1	SB?	7	Located between a close pair of mag 13.9 and 13.6 stars.
22 57 36.0	+19 47 21	UGC 12265	13.6	0.7 x 0.3	11.8	Double System	84	
22 58 17.5	+25 46 13	UGC 12272	13.9	0.9 x 0.7	13.3	Sa	75	
22 58 50.1	+20 17 51	UGC 12278	14.3	0.9 x 0.6	13.5	Scd:	175	Uniform surface brightness; mag 11.4 star SW 3′3.
22 59 19.8	+24 06 21	UGC 12283	14.3	0.9 x 0.9	14.0	SBab		Mag 8.96v star S 4′6.
22 59 41.5	+24 04 28	UGC 12289	15.4	1.2 x 1.0	15.4	Sd	117	Mag 8.96v star WSW 6′6.
22 59 48.4	+26 18 05	UGC 12291	14.5	0.9 x 0.6	13.6	Scd:	66	Located on NE edge of small galaxy cluster **A 2515**; **CGCG 475-16** S 6′1.
22 59 51.6	+26 01 56	UGC 12293	14.3	0.9 x 0.7	13.6	Sa	12	Mag 12.5 star NW 0′9.
23 00 49.1	+26 44 27	UGC 12303	13.6	1.5 x 1.2	14.1	Sb II	90	Mag 8.31v star NE 4′4.
23 01 19.8	+28 24 01	UGC 12310	14.0	0.9 x 0.8	13.5	S?		
23 02 46.6	+22 05 41	UGC 12326	14.5	0.9 x 0.6	13.7	Scd:	81	
23 03 18.5	+19 57 19	UGC 12333	14.7	1.2 x 0.4	13.8	SBb	156	Mag 10.34v star S 2′1; **CGCG 453-55** SE 2′5.
23 03 58.1	+27 18 08	UGC 12334	13.9	1.0 x 0.9	13.7	SB0	15	
23 05 00.0	+18 42 10	UGC 12344	14.1	2.1 x 0.8	14.5	SBd:	177	Very thin core; mag 8.79v star E 2′5.
23 05 12.2	+18 52 05	UGC 12347	14.5	0.9 x 0.6	13.7	Im:	12	Irregular.
23 05 45.5	+18 59 06	UGC 12351	14.5	1.0 x 0.5	13.6	Im:	38	Mag 12.2 star NE 2′3.
23 05 40.5	+27 39 58	UGC 12352	13.5	1.4 x 0.8	13.5	S?	166	
23 08 58.3	+23 21 58	UGC 12390	14.7	0.8 x 0.6	13.7	S0	96	
23 10 35.8	+21 42 39	UGC 12400	14.0	1.3 x 1.3	14.4	SBa		Mag 11.0 star W 3′8.
23 12 19.9	+28 43 20	UGC 12413	14.2	1.1 x 1.0	14.1	SB0?	78	Mag 11.7 star NE 2′7.
23 13 10.0	+24 14 40	UGC 12425	14.1	1.5 x 0.2	12.6	Sab	172	Mag 9.93v star W 2′1.
23 13 43.5	+29 00 29	UGC 12430	14.2	2.3 x 0.2	13.2	Sd	164	**UGC 12427** SW 5′4.
23 16 21.7	+28 29 21	UGC 12470	14.2	1.0 x 0.7	13.6	Sa	135	

(Continued from previous page)

64 GALAXIES 64

RA h m s	Dec ° ′ ″	Name	Mag (V)	Dim ′ Maj x min	SB	Type Class	PA	Notes
23 17 33.7	+29 01 08	UGC 12482	13.0	1.4 x 1.2	13.5	E:	39	MCG +5-54-60 E 2′8; **CGCG 496-75** NNW 2′9.
23 18 38.8	+25 13 54	UGC 12490	13.6	0.9 x 0.6	12.8	SBa	75	**CGCG 475-61** N 2′1; mag 9.28v star SE 1′9.
23 19 51.1	+26 15 44	UGC 12515	12.9	1.3 x 1.0	13.1	S0	55	Located 0′9 S of mag 9.40v star.
23 20 31.6	+29 18 22	UGC 12530	13.8	1.1 x 1.1	13.8	SBcd:		Mag 12.4 star E 2′4.
23 22 28.7	+29 10 46	UGC 12557	14.1	2.1 x 0.4	13.8	S?	9	Mag 10.45v star W 9′2.
23 22 46.0	+29 08 16	UGC 12566	13.6	1.7 x 1.3	14.3	Sab	140	Bright center, prominent dark patches SE and NW, faint envelope.

GALAXY CLUSTERS

RA h m s	Dec ° ′ ″	Name	Mag 10th brightest	No. Gal	Diam ′	Notes
22 54 48.0	+29 28 00	A 2503	16.4	47	8	All members anonymous, faint, stellar.
23 18 24.0	+18 44 00	A 2572	15.3	32	28	Galaxy group **Hickson 94** on E edge.

65 GALAXIES 65

RA h m s	Dec ° ′ ″	Name	Mag (V)	Dim ′ Maj x min	SB	Type Class	PA	Notes
21 07 26.8	+18 03 08	CGCG 449-2	14.1	0.6 x 0.3	12.1		90	Faint, elongated anonymous galaxy S edge; mag 10.63v star W 1′7.
21 18 33.2	+22 44 41	CGCG 471-4	13.9	0.5 x 0.4	12.0		51	Mag 8.89v star SE 1′5.
21 48 29.7	+29 23 46	CGCG 493-8	14.1	0.7 x 0.3	12.3		66	Mag 9.14v star SW 7′3.
22 02 31.5	+19 44 56	IC 1420	13.2	1.2 x 0.8	13.0	SB?	90	Mag 10.25v star SSE 6′2.
21 21 29.2	+21 14 32	IC 5104	13.4	1.6 x 0.4	12.7	SBab?	173	Mag 7.89v star N 4′5.
20 46 10.5	+18 56 05	MCG +3-53-1	14.3	0.4 x 0.3	11.8	Spiral	120	Mag 12.3 star W 2′4.
21 05 23.7	+19 44 51	MCG +3-53-14	14.7	0.7 x 0.5	13.4	SBc	111	Mag 11.1 star N 2′0.
21 08 21.3	+18 11 58	MCG +3-54-2	13.8	0.7 x 0.7	13.1	E		UGC 11683 E 1′4.
21 14 37.7	+19 00 17	MCG +3-54-5	14.3	0.6 x 0.4	12.6	Sb		
21 23 33.3	+17 34 58	MCG +3-54-9	14.8	0.5 x 0.3	12.6	Sb	156	Double system? Small, faint companion galaxy SE edge.
21 23 50.2	+17 34 13	MCG +3-54-10	14.7	0.8 x 0.3	13.0	Sb	159	Located between a pair of E-W oriented mag 12.5 stars.
20 45 37.9	+22 44 34	MCG +4-49-1	14.6	0.4 x 0.3	12.1			Mag 8.63v star NW 3′0.
20 58 28.0	+23 59 30	MCG +4-49-6	14.5	0.5 x 0.4	12.6		24	Mag 12.3 star N 3′0.
21 16 53.0	+24 12 13	MCG +4-50-4	14.1	0.6 x 0.4	12.4	S0	30	Mag 12.4 star NW 0′8; located within the boundaries of the planetary nebula PN G72.7-17.1, near the N edge.
21 18 59.1	+25 25 49	MCG +4-50-7	14.2	1.0 x 0.3	12.7	S?	81	Close pair of mag 11 star E 1′4.
21 51 29.7	+22 50 45	MCG +4-51-7	15.9	0.7 x 0.3	14.1	Sc		
21 51 35.2	+22 39 02	MCG +4-51-8	14.8	0.7 x 0.2	12.5	S?	90	Mag 9.37v star E 1′8.
21 53 09.5	+22 27 31	MCG +4-51-10	14.3	0.5 x 0.5	12.6			Mag 12-13 double star SE 2′7.
21 58 49.5	+25 30 03	MCG +4-51-15	15.3	0.5 x 0.5	13.7			Mag 10.9 star WSW 2′8.
21 59 29.9	+23 42 58	MCG +4-52-1	14.6	0.8 x 0.3	12.9	S?	6	Close pair of mag 10.9, 12.3 stars ESE 2′0.
21 00 42.4	+17 48 15	NGC 7003	13.0	1.1 x 0.8	12.7	Sbc	120	Very patchy.
21 03 33.7	+29 53 50	NGC 7013	11.3	4.0 x 1.4	13.1	SA(r)0/a	157	Mag 9.88v star on N edge; many superimposed stars.
21 18 33.2	+26 26 48	NGC 7052	12.4	2.1 x 1.1	13.4	E	64	
21 21 07.6	+23 05 06	NGC 7053	13.1	1.4 x 1.3	13.6	S?	27	Star on SW edge; small, faint anonymous galaxy SSW 2′4.
21 22 07.3	+18 39 56	NGC 7056	12.9	1.0 x 0.9	12.7	SBb	57	
21 30 02.2	+26 43 09	NGC 7080	12.3	1.8 x 1.7	13.4	SB(r)b I-II	50	Bright nucleus, many superimposed stars.
21 42 40.6	+28 56 48	NGC 7116	13.4	1.1 x 0.4	12.4	S?	105	
21 48 13.1	+22 09 32	NGC 7137	12.4	1.6 x 1.4	13.1	SAB(rs)c III	36	Dark patches E and W of nucleus.
22 00 41.1	+17 44 13	NGC 7177	11.2	3.1 x 2.0	13.0	SAB(r)b I-II	93	
20 39 25.5	+27 15 02	UGC 11608	9.6	0.3 x 0.3	6.8	Pair		
20 41 40.4	+19 12 01	UGC 11615	14.4	1.5 x 0.5	13.9	Im:	153	Many superimposed stars S half; in star rich area; mag 11.5 star NE 2′5.
20 51 26.0	+18 58 02	UGC 11643	13.9	1.0 x 0.4	12.8	SBb	65	**MCG +3-53-4** WNW 1′1.
20 57 15.4	+25 58 03	UGC 11651	13.8	3.2 x 0.9	14.8	Sdm	165	Very knotty, many superimposed stars.
20 57 29.6	+18 47 47	UGC 11653	14.5	1.2 x 0.7	14.2	Sd:	6	Bright nucleus, faint envelope; three stars in N-S row along E side; mag 11.5 star SE 0′9.
21 05 50.2	+18 28 02	UGC 11676	13.9	0.9 x 0.4	12.6	S?	2	Mag 10.32v star NE 2′8.
21 08 22.5	+17 49 15	UGC 11682	13.7	1.5 x 0.5	13.2	S?	170	Mag 10.59v star N 2′6.
21 08 26.8	+18 11 29	UGC 11683	14.5	1.0 x 0.2	12.6	Scd:	168	MCG +3-54-2 W 1′4.
21 14 31.5	+26 44 06	UGC 11707	13.9	3.3 x 1.9	15.7	SAdm	55	Small, moderately bright core, numerous stars superimposed; mag 11.3 star on NE edge.
21 14 54.9	+25 51 50	UGC 11709	15.9	0.9 x 0.9	15.5	Im:		Stellar nucleus, uniform envelope.
21 18 35.4	+19 43 05	UGC 11717	14.2	1.3 x 0.5	13.6	S	48	Mag 10.35v star N 0′7.
21 19 38.3	+21 58 13	UGC 11722	14.1	1.5 x 0.4	13.4	S?	110	Mag 6.29v star NE 8′8.
21 21 07.0	+26 15 59	UGC 11728	14.4	0.8 x 0.8	13.8	Sd		Mag 7.32v star W 6′5.
21 28 48.5	+20 30 35	UGC 11749	13.4	1.1 x 0.7	13.0	SBb	162	Stellar nucleus; mag 12.8 star superimposed S end.
21 29 31.5	+27 19 14	UGC 11754	14.2	1.9 x 1.6	15.3	SABcd	150	Very patchy, a number of stars superimposed.
21 34 22.6	+26 21 10	UGC 11769	14.4	1.5 x 0.6	14.1	Sb	142	Mag 8.89v star NW edge.
21 40 04.3	+25 02 25	UGC 11787	14.5	0.5 x 0.4	12.6	Sa	162	Mag 8.79v star SSE 3′2.
21 40 26.9	+25 09 53	UGC 11788	14.8	1.2 x 0.7	14.5	S	108	Double system, connected, distorted; mag 10.8 star SW 0′8.
21 41 59.1	+22 42 48	UGC 11793	13.8	1.3 x 1.2	14.1	SB(s)b		bright nucleus in NE-SW bar.
21 51 07.0	+25 51 47	UGC 11830	13.8	1.3 x 1.3	14.2	Sb II		Bright knot or superimposed star W of center.
21 51 44.0	+25 15 18	UGC 11834	13.5	1.0 x 0.7	12.9	S?	129	Star superimposed S.
21 52 35.8	+28 18 21	UGC 11838	14.7	1.9 x 0.2	13.5	Sd	23	
21 55 40.2	+24 53 49	UGC 11849	13.8	1.5 x 0.9	13.9	SBcd:	73	Mag 7.88v star N 8′6.
21 55 59.3	+27 53 56	UGC 11852	13.8	1.5 x 0.7	13.7	SBa?	15	Elongated core with smooth envelope.
21 57 59.9	+24 16 00	UGC 11860	14.8	0.9 x 0.4	13.6	Sdm	133	
21 59 04.7	+18 10 42	UGC 11868	13.4	2.2 x 1.7	14.7	SB(s)m	78	Small, moderately bright center, faint envelope.
22 02 23.3	+18 18 52	UGC 11878	13.9	1.4 x 1.2	14.3	SB?	57	Small, bright core with one large, faint arm recurving S, then SW under the bright core.

PLANETARY NEBULAE

RA h m s	Dec ° ′ ″	Name	Diam ″	Mag (P)	Mag (V)	Mag cent ★	Alt Name	Notes
21 16 52.3	+24 08 52	PN G72.7−17.1	830	>12.2		17.1	PK 72−17.1	= **Abell 74**. A mag 11.28v star lies at the center, the galaxy MCG +4-50-4 lies 3′6 N of this star.

GALAXIES

RA h m s	Dec ° ′ ″	Name	Mag (V)	Dim ′ Maj x min	SB	Type Class	PA	Notes
20 29 02.9	+18 22 31	CGCG 447-5	14.1	0.6 x 0.4	12.4	Sc	30	Low surface brightness; few stars superimposed, in star rich area.
20 28 28.8	+25 43 26	NGC 6921	13.4	0.9 x 0.2	11.4	SA(r)0/a:	141	**MCG +4-48-2** NE 1′.5.
20 30 49.5	+20 17 44	UGC 11582	13.7	1.5 x 1.0	14.0	Sdm:	30	Numerous stars superimposed.
20 39 25.5	+27 15 02	UGC 11608	9.6	0.3 x 0.3	6.8	Pair		
20 41 40.4	+19 12 01	UGC 11615	14.4	1.5 x 0.5	13.9	Im:	153	Many superimposed stars S half; in star rich area; mag 11.5 star NE 2′.5.

OPEN CLUSTERS

RA h m s	Dec ° ′ ″	Name	Mag	Diam ′	No. ★	B ★	Type	Notes
19 17 17.0	+19 32 42	Be 44		2	30	16.0	cl	
19 28 27.3	+17 21 57	Be 47		3	20	16.0	cl	
20 14 30.0	+28 56 36	Be 52		2	40	18.0	cl	
20 01 25.1	+28 38 36	Be 83		2	20	17.0	cl:	
19 26 12.0	+20 06 00	Cr 399	3.6	90	40	5.2	ast	**Brocchi's Cluster** or **Coathanger**. Moderately rich in stars; large brightness range; no central concentration; detached. Not a true cluster.
20 11 35.0	+26 32 00	Cr 416	5.6	8	40		cl	Part of, or associated with, NGC 6885.
19 42 38.0	+21 09 12	Cz 40		4	30		cl	
19 50 50.8	+25 18 00	Cz 41		8	30		cl:	
19 53 10.3	+18 21 00	Ha 20	7.7	8	28	8.9	cl	Few stars; moderate brightness range; not well detached.
19 23 12.6	+22 09 00	NGC 6793		7	15		cl	Few stars; moderate brightness range; no central concentration; detached.
19 27 00.0	+25 05 36	NGC 6800		5	20	10.0	cl	Few stars; small brightness range; not well detached.
19 30 35.0	+20 15 48	NGC 6802	8.8	5	201	14.0	cl	Moderately rich in stars; small brightness range; strong central concentration; detached.
19 43 10.0	+23 18 00	NGC 6823	7.1	7	79	8.8	cl	Moderately rich in bright and faint stars with a strong central concentration; detached; involved in nebulosity.
19 48 53.2	+21 12 58	NGC 6827		4	30	13.0	cl	
19 51 00.0	+23 06 00	NGC 6830	7.9	6	82	10.0	cl	Few stars; moderate brightness range; slight central concentration; detached.
19 52 13.0	+29 24 30	NGC 6834	7.8	6	128	11.0	cl	Moderately rich in stars; moderate brightness range; slight central concentration; detached.
19 54 19.0	+17 53 18	NGC 6839		4	12		ast?	
20 12 01.0	+26 29 00	NGC 6885	8.1	20	34	5.9	cl	Moderately rich in stars; moderate brightness range; no central concentration; detached. (A) Cr 416 on northwest side.
20 34 32.3	+28 17 00	NGC 6940	6.3	25	170	11.0	cl	Rich in stars; moderate brightness range; no central concentration; detached.
19 45 06.0	+17 35 00	Ro 1		3	15		cl	(A) Condensation in rich Milky Way.
19 45 24.4	+23 57 00	Ro 2		45	20	7.1	cl	
19 58 41.1	+20 31 25	Ro 3		5	20		cl	
20 04 52.0	+29 12 54	Ro 4	10.0	6	30		cl	(A) 3 patches of bright nebulosity on W side, including IC 4954/55.

GLOBULAR CLUSTERS

RA h m s	Dec ° ′ ″	Name	Total V m	B ★ V m	HB V m	Diam ′	Conc. Class Low = 12 High = 1	Notes
19 53 46.1	+18 46 42	NGC 6838	8.4	12.1	14.5	7.2		= **M 71**.
19 18 02.1	+18 34 18	Pal 10	13.2	18	19.4	4.0	12	

BRIGHT NEBULAE

RA h m s	Dec ° ′ ″	Name	Dim ′ Maj x min	Type	BC	Color	Notes
20 04 48.0	+29 15 00	IC 4954/55	3 x 3	R	2-5	1-4	A pair of detached nebulosities aligned NW-SE; distance between centers about 3′.5. The SE component is largest and about 3′ in diameter; the NW component appears to be double, aligned nearly N-S.
19 36 18.0	+29 33 00	Mi 92	0.2 x 0.1	R	2-5		**Footprint Nebula**.
19 40 24.0	+27 18 00	NGC 6813	3 x 3	E	3-5	3-4	In an exceedingly rich star field.
19 43 06.0	+23 17 00	NGC 6820	40 x 30	E	3-5	3-4	Has a very mottled appearance. In the N central section of NGC 6820 there is a very narrow, prominent dark lane about 5′ in length, aligned roughly E-W. There are several small absorption patches in the NE section.
19 30 18.0	+18 16 00	Sh2-82	7 x 7	E+R	3-5	3-4	A larger, roundish nebula with a faint central star. It is accompanied by a smaller, semi-detached nebulosity due N.
19 49 00.0	+18 24 00	Sh2-84	6 x 3	E	3-5	3-4	A broad, V-shaped nebulosity. The surrounding star field is exceedingly rich.
19 46 00.0	+25 20 00	Sh2-88	18 x 6	E	3-5	3-4	Two bright patches or knots 11′ SE of center. Each of these knots, aligned ENE-WSW, is about 2′ across; distance between centers about 2′.4.
19 49 18.0	+26 52 00	Sh2-90	8 x 3	E	2-5	3-4	A broad C-shaped nebulosity with the convex side to the W.
19 26 06.0	+22 43 00	vdB 126	7 x 5	R	2-5	1-4	Nebula with a mag 8.3 star involved.

PLANETARY NEBULAE

RA h m s	Dec ° ′ ″	Name	Diam ″	Mag (P)	Mag (V)	Mag cent ★	Alt Name	Notes
19 55 02.4	+29 17 17	NGC 6842	57	13.6	13.1	15.9	PK 65+0.1	Numerous stars on disc; mag 8.72v star NE 6′.3.
19 59 36.1	+22 43 13	NGC 6853	402	7.6	7.4	13.9	PK 60−3.1	= **M 27**, **Dumbbell Nebula**. Bright and large; hourglass shape is easily visible in small telescopes.
20 12 43.0	+19 59 20	NGC 6886	10	12.2	11.4	18.0	PK 60−7.2	Mag 10.7 star S 0′.8; mag 10.21v star E 1′.5.
20 22 23.0	+20 06 16	NGC 6905	72	11.9	11.1	15.7	PK 61−9.1	Overall shape elongated NNW-SSE, bright center approximately 40 arcseconds in diameter; mag 10.39v star N edge; mag 12.0 star S edge.
19 19 18.8	+17 11 45	PN G51.3+1.8	20			15.1	PK 51+1.1	Forms a triangle with a mag 11.8 star SW 2′.2 and a mag 12.6 star WNW 2′.5.
19 19 02.8	+19 02 18	PN G52.9+2.7	2	17.4			PK 52+2.1	Mag 14.3 star E 0′.6; mag 9.40v star .
19 35 18.5	+17 12 58	PN G53.2−1.5	4				PK 53−1.1	Close pair of mag 11.12v, 11.5 stars E 2′.0; mag 12.9 star NW 1′.9.
19 18 40.1	+19 34 26	PN G53.3+3.0	87	17.2		21.1	PK 53+3.1	= **Abell 59**; bright arc N edge; mag 11.07v star SE 2′.0; mag 6.51v star NE 2′.9.
19 42 10.4	+17 05 11	PN G53.8−3.0	40	17.1		14.6	PK 53−3.1	= **Abell 63**; mag 11.5 star WSW 1′.5; mag 11.4 star SE 2′.0.
19 41 34.1	+17 45 14	PN G54.4−2.5	10	16.5		17.7	PK 54−2.1	Mag 12.1 star W 0′.9; mag 12.6 star NNW 1′.0.
19 40 26.0	+18 49 10	PN G55.1−1.8	3	18.5			PK 55−1.1	Small triangle of mag 11 stars E 1′.5; mag 12.8 star W 1′.0.
19 23 25.0	+21 07 59	PN G55.2+2.8	5	16.9	15.8		PK 55+2.1	
19 23 46.9	+21 06 36	PN G55.3+2.7	11	16.3			PK 55+2.2	Mag 11.9 star NW 1′.3.
19 36 25.7	+19 42 32	PN G55.5−0.5	4	14.0	13.9		PK 55−0.1	Mag 13.9 star SE 0′.3; mag 12.6 star NE 1′.3.
19 26 37.8	+21 09 24	PN G55.6+2.1	5			16.7	PK 55+2.3	Located 2′.5 W of a mag 8.09v star.
19 27 44.1	+21 30 00	PN G56.0+2.0	5	18.4			PK 56+2.1	Mag 13.0 star NE 2′.5; mag 10.7 star SW 4′.4.
19 39 36.0	+20 19 03	PN G56.4−0.9	3				PK 56−0.1	Mag 11.1 star W 2′.1; mag 12.5 star SE 1′.6.
20 02 36.4	+17 36 50	PN G56.8−6.9	15	14.7	14.4		PK 56−6.1	Mag 12.9 star SW 0′.6; mag 11.1 star N 2′.8.
19 45 22.2	+21 20 02	PN G57.9−1.5	5	17.2			PK 57−1.1	Mag 11.4 star N 2′.2; mag 11.5 star NE 2′.6.
20 01 42.1	+19 54 37	PN G58.6−5.5	150			17.4	PK 58−5.1	Disc dense with stars; mag 12.8 star 0′.8 N of center; mag 8.73v star S of center 2′.7.
19 17 05.9	+25 37 33	PN G58.6+6.1	37	17.5	14.7	17.6	PK 58+6.1	= **Abell 57**; SE-NW line of three mag 12 stars SW 2′.0; mag 10.84v star SE 1′.2.
19 36 21.9	+23 39 46	PN G58.9+1.3	4	16.4			PK 58+1.1	Mag 10.51v star S 0′.6; mag 9.51v star S 1′.7; mag 8.02v star N 1′.7.

(Continued from previous page)

66 PLANETARY NEBULAE 66

RA h m s	Dec ° ′ ″	Name	Diam ″	Mag (P)	Mag (V)	Mag cent ★	Alt Name	Notes
19 48 26.4	+22 08 34	PN G59.0−1.7	12	16.0	14.8	11.6	PK 59−1.1	Close pair of mag 12 stars SW 1′.6; mag 11.1 star N 1′.5.
19 24 02.9	+25 18 47	PN G59.0+4.6	10	15.8			PK 59+4.1	Mag 12.7 star W 0′.9; mag 11.2 star NNE 1′.2.
19 33 46.8	+24 32 26	PN G59.4+2.3	3	16.0	16.2		PK 59+2.1	Mag 13.2 star NE 1′.5.
19 35 54.6	+24 54 47	PN G59.9+2.0	3				PK 59+2.2	Mag 13.2 star NE 2′.4; mag 11.30v star NW 4′.3.
20 00 10.5	+21 42 54	PN G60.0−4.3	38	16.6	15.2	13.2	PK 60−4.1	= **Abell 68**; mag 10.6 star S 0′.9; mag 10.4 star ESE 1′.8.
20 11 56.1	+20 20 04	PN G60.3−7.3	36	16.2	16.0	11.3	PK 60−7.1	Mag 12.3 star NW 2′.8.
19 38 52.1	+25 05 35	PN G60.4+1.5	9			18.6	PM 1−310	Mag 12.9 star NW 1′.1; mag 11.25v star NE 1′.8.
19 46 15.7	+24 11 02	PN G60.5−0.3	7	19.4			PK 60−0.1	Close pair of mag 11-12 stars N 1′.8; mag 10.24v star E 2′.6.
19 38 08.5	+25 15 38	PN G60.5+1.8	3	16.5			PK 60+1.1	Mag 10.46v star N 1′.5.
19 32 57.8	+26 52 41	PN G61.3+3.6	35	16.9			PK 61+3.1	Elongated, very slender, oriented E-W; mag 12.2 star N 2′.4; mag 12.6 star S 2′.5.
19 39 43.5	+26 29 30	PN G61.8+2.1	10	16.2			PK 61+2.1	Mag 13.7 star E 0′.7.
19 50 28.6	+25 54 27	PN G62.4−0.2	29	16.5			PK 62−0.1	Elongated E-W; mag 12.3 star S 1′.0; mag 10.8 star E 1′.9.
20 04 58.7	+25 26 36	PN G63.8−3.3	4				PK 63−3.1	Mag 11.8 star S 0′.9; mag 10.33v star E 0′.7.
20 03 22.6	+27 00 53	PN G64.9−2.1	6	16.8	16.0		PK 64−2.1	Mag 12.6 star N 1′.4; mag 12.1 star ESE 3′.5.
20 09 04.7	+26 26 55	PN G65.1−3.5	24			21.0	PK 65−3.1	Forms a triangle with mag 9.95v star NW 3′.8 and mag 10.61v star SW 4′.3; small dark patch near W of center.
20 17 21.5	+25 21 44	PN G65.2−5.6	24	14.9	14.8		PK 65−5.1	Pair of mag 11 stars W 1′.8; close pair of mag 11.5 stars SE 1′.6.
20 19 38.3	+27 00 08	PN G66.9−5.2	5	13.5			PK 66−5.1	Mag 13.4 star N 1′.7; mag 9.31v star SE 5′.7.
20 13 58.0	+29 33 54	PN G68.3−2.7	9				PK 68−2.1	Mag 12.8 star NNE 0′.9; mag 12.2 star SW 1′.2.
20 21 58.6	+29 59 18	PN G69.6−3.9	14	18.6	16.::		PK 69−3.1	Close pair of mag 12.4 stars NNW 2′.8; mag 12.8 stars S 2′.3.

67 GALAXIES 67

RA h m s	Dec ° ′ ″	Name	Mag (V)	Dim ′ Maj x min	SB	Type Class	PA	Notes
18 09 26.7	+19 07 00	CGCG 113-24	14.2	0.7 x 0.5	12.9		108	Mag 9.39v star SE 9′.8.
18 12 14.1	+21 53 03	CGCG 142-19	14.3	1.0 x 0.8	13.9		33	Bright elongated core, faint outer envelope.
18 40 27.7	+23 48 17	CGCG 143-14	14.7	0.8 x 0.4	13.3		162	Mag 8.83v star NW 10′.2; note: **CGCG 143-11** lies 4′.0 NE of this star.
18 08 59.2	+28 15 35	CGCG 172-4	15.4	0.3 x 0.3	12.6			Mag 13.2 star W 1′.1; mag 11.4 star SW 2′.2.
18 30 57.6	+29 52 32	CGCG 172-40	14.3	0.7 x 0.4	12.8		9	Uniform surface brightness; mag 12.7 star on W edge.
18 36 39.7	+19 43 43	MCG +3-47-8	12.4	1.3 x 1.3	13.0	E0?		Numerous stars superimposed along S edge.
18 38 26.2	+17 11 47	MCG +3-47-10	12.9	0.6 x 0.6	11.7	SA0^{-}		
17 58 07.0	+21 16 16	MCG +4-42-17	14.1	0.5 x 0.3	11.9		108	Mag 8.03v star N 3′.1.
18 01 33.5	+26 15 09	MCG +4-42-22	13.6	0.7 x 0.4	12.2	E/S0	108	Mag 14.6 star N edge.
18 06 09.0	+23 27 56	MCG +4-43-4	14.4	0.9 x 0.5	13.4	Sbc	63	
18 24 48.7	+23 04 00	MCG +4-43-31	14.0	0.9 x 0.9	13.6			Located between a close pair of mag 11-12 stars.
18 26 09.8	+23 18 50	MCG +4-43-33	14.1	0.9 x 0.4	12.9	Sbc	0	
18 02 57.2	+29 18 35	MCG +5-42-29	14.8	0.6 x 0.6	13.5	Sc		**CGCG 171-49** N 4′.0; UGC 11098 W 1′.9.
18 12 51.5	+28 12 49	MCG +5-43-9	15.8	1.1 x 0.3	14.5	Sbc	102	Pair of mag 8.23v, 8.09v stars NE 2′.9.
18 23 04.2	+29 54 02	MCG +5-43-16	14.1	0.9 x 0.9	13.7			Few stars superimposed; mag 10.5 star NW 3′.0.
19 10 03.1	+28 57 12	MCG +5-45-5	14.2	1.0 x 0.9	13.9	Spiral	120	Mag 10.65v star W 2′.3.
17 55 59.6	+18 20 16	NGC 6500	12.2	2.2 x 1.6	13.4	SAab:	50	Mag 7.43v star E 5′.9; NGC 6501 on N edge.
17 56 03.7	+18 22 21	NGC 6501	12.0	2.0 x 1.8	13.3	SA0^{+}:	54	
17 59 34.3	+24 53 12	NGC 6513	13.3	1.2 x 0.8	13.1	SB0:	40	
17 59 43.9	+28 51 58	NGC 6518	13.9	0.4 x 0.4	11.8			**MCG +5-42-25** E 1′.2.
18 01 46.5	+19 43 42	NGC 6527	13.4	1.4 x 1.0	13.6	Sa:	150	
18 05 10.1	+25 13 54	NGC 6547	13.6	1.5 x 0.4	12.9	S0	133	
18 05 59.1	+18 35 16	NGC 6548	11.7	3.0 x 2.8	13.9	SB0	60	Bright N-S bar with dark patches E and W.
18 05 49.4	+18 32 15	NGC 6549	13.8	1.4 x 0.4	13.0	S?	53	
18 07 49.1	+17 36 15	NGC 6555	12.4	2.0 x 1.5	13.4	SAB(rs)c	110	Patchy, many knots.
18 10 49.4	+21 14 15	NGC 6571	14.4	0.4 x 0.4	12.5	E/S0		**MCG +4-43-5** W 3′.7.
18 11 48.1	+21 25 38	NGC 6576	14.7	0.6 x 0.6	13.5			Mag 10.24v star NE 2′.4.
18 12 01.2	+21 27 49	NGC 6577	12.6	1.5 x 1.3	13.4	E	6	Mag 10.24v star SW 1′.4; **CGCG 142-15** NW 3′.1.
18 12 31.9	+21 25 13	NGC 6579	13.8	0.4 x 0.4	11.6			On SW edge of NGC 6580.
18 12 33.7	+21 25 32	NGC 6580	13.3	1.3 x 0.7	13.2	E/S0	126	NGC 6579 on SW edge; very small, elongated anonymous galaxy on S edge.
18 12 47.7	+25 24 40	NGC 6581	14.9	0.6 x 0.3	12.9		57	Located 1′.8 SW of mag 10.08v star.
18 13 38.4	+21 05 26	NGC 6586	13.7	0.9 x 0.5	12.7	S	105	
18 13 51.1	+18 49 30	NGC 6587	12.9	1.3 x 1.1	13.2	SAB0^{-}:	21	
18 14 03.7	+22 17 00	NGC 6593	14.3	0.9 x 0.6	13.7	E/S0	162	
18 15 42.9	+24 54 43	NGC 6599	12.6	1.3 x 1.2	13.0	S0	69	Bright center, several stars superimposed on envelope.
18 16 34.4	+25 02 36	NGC 6602	13.8	0.9 x 0.7	13.1	S	0	
18 17 41.1	+22 14 18	NGC 6616	13.8	1.4 x 0.6	13.5	Sab	59	Strong dark lane along center line.
18 18 55.8	+23 39 19	NGC 6619	13.0	1.2 x 1.1	13.3	E?	102	Several faint stars superimposed on envelope.
18 19 43.0	+23 42 31	NGC 6623	13.0	1.3 x 1.2	13.5	E	155	**MCG +4-43-28** NE 0′.9; stellar **MCG +4-43-27** S of NGC 6623's nucleus.
18 22 22.0	+23 28 42	NGC 6628	12.9	1.9 x 1.3	13.8	S0?	87	E-W dark lane N of center.
18 25 03.3	+27 32 06	NGC 6632	12.1	3.0 x 1.4	13.5	SA(rs)bc III	155	Bright center, strong dark lanes.
18 28 57.4	+22 54 09	NGC 6641	13.4	0.9 x 0.7	12.7	S?	102	
18 33 55.8	+22 53 15	NGC 6658	12.9	1.7 x 0.4	12.4	S0 sp	5	
18 34 36.7	+22 54 33	NGC 6661	12.1	1.7 x 1.1	12.6	SA(s)0/a	145	
18 37 54.9	+22 04 36	NGC 6669	15.1	0.9 x 0.9	14.7	SBdm:		Star superimposed center; mag 9.90v star NNE 2′.5; very star rich area.
18 37 26.4	+26 25 00	NGC 6671	13.1	1.5 x 1.3	13.7	S	27	Very large, bright center flanked by a pair of stars.
18 38 33.8	+25 22 28	NGC 6674	12.2	4.0 x 2.2	14.4	SB(r)b	143	Bright nucleus, many stars superimposed.
18 39 44.0	+22 19 02	NGC 6680	14.6	0.7 x 0.4	13.1		45	**UGC 11310** SW 10′.8.
18 45 15.3	+25 30 47	NGC 6697	12.7	1.2 x 1.0	12.9	E	48	Bright center; mag 9.84v star E 1′.8.
18 50 34.2	+26 50 18	NGC 6710	13.1	1.7 x 1.0	13.5	SA0^{+}?	40	Bright center, many superimposed stars.
19 00 50.5	+28 46 13	NGC 6740	13.9	0.9 x 0.8	13.3	S	30	Stellar nucleus, several faint stars superimposed, in star rich area.
17 57 10.3	+27 57 43	UGC 11060	14.0	1.4 x 0.4	13.2	Sa	130	Mag 10.30v star N 2′.1.
17 57 35.5	+23 01 36	UGC 11063	14.0	1.0 x 1.0	13.9	SAB(s)c		Mag 12.4 star NE edge.
17 57 40.9	+27 50 04	UGC 11064	12.9	1.8 x 1.8	14.0	Scd:		Several stars superimposed S edge; mag 8.2 double star NE 1′.3.
17 58 05.0	+28 14 37	UGC 11068	13.6	1.3 x 1.1	13.8	S?	90	Stellar nucleus.
17 58 14.3	+27 15 45	UGC 11070	14.2	1.0 x 0.7	13.6	SB(r)b	75	Mag 10.55v star N 2′.6.
18 00 05.7	+26 22 00	UGC 11082	13.4	0.8 x 0.8	12.8	S0		**UGC 11080** SW 1′.9.
18 00 43.2	+28 42 44	UGC 11090	13.6	1.0 x 0.7	13.1	Scd:	20	Mag 7.37v star W 5′.9; **UGC 11086** NW 6′.2.
18 02 24.5	+26 02 35	UGC 11097	13.9	1.3 x 0.4	13.0	S?	157	
18 02 48.8	+29 18 23	UGC 11098	14.7	1.0 x 0.5	13.8	S	132	Mag 10.7 star on W edge.
18 04 36.1	+21 38 10	UGC 11105	13.6	2.3 x 1.7	14.9	Sdm	66	Stellar nucleus, numerous superimposed stars, faint branching arms.
18 05 04.9	+17 15 59	UGC 11107	14.1	0.8 x 0.8	13.5	SB(r)bc:		Bright N-S bar.
18 05 30.8	+23 16 19	UGC 11113	13.6	1.6 x 1.2	14.2	SABd	177	Strong E-W bar, fainter arms.
18 07 02.2	+20 29 15	UGC 11120	14.7	1.6 x 0.3	13.8	Im:	112	

(Continued from previous page)

GALAXIES

RA h m s	Dec ° ′ ″	Name	Mag (V)	Dim ′ Maj x min	SB	Type Class	PA	Notes
18 07 38.1	+28 29 09	UGC 11123	14.3	0.8 x 0.6	13.4	Sdm	78	
18 08 39.8	+28 03 47	UGC 11127	14.3	1.0 x 0.4	13.2	S?	12	Very narrow extension northward for 0′.5.
18 11 38.6	+25 39 22	UGC 11142	14.7	1.5 x 0.2	13.3	Scd:	14	Mag 6.78v star S 5′.9.
18 12 07.0	+25 35 43	UGC 11150	14.6	1.5 x 0.2	13.1	S?	7	Mag 6.78v star W 8′.1; **IC 1280** NNE 4′.8.
18 12 05.6	+29 09 21	UGC 11151	14.0	1.5 x 0.3	13.0	S	0	Mag 9.87v star NW 2′.2.
18 12 32.5	+18 35 57	UGC 11152	13.5	1.9 x 1.0	14.0	SBdm	140	Bright N-S bar, numerous stars superimposed; mag 10.87v star SE 2′.0.
18 12 31.9	+25 26 07	UGC 11155	13.9	1.1 x 0.8	13.6	Scd?	36	**IC 4697** SW 1′.3.
18 12 37.1	+25 32 10	UGC 11156	13.3	1.0 x 1.0	13.3	E?		Mag 10.29v star SE 2′.4.
18 13 01.6	+29 41 44	UGC 11162	14.4	1.4 x 0.7	14.2	S	115	
18 14 01.8	+21 30 29	UGC 11170	14.7	1.3 x 1.1	14.9		93	Strong, narrow dark N-S lane E of center, bright elongated knot E of lane.
18 15 47.0	+25 22 34	UGC 11179	14.3	1.0 x 0.7	13.8	Scd:	10	Pair of mag 10.5v stars SW 2′.7.
18 17 57.0	+26 45 25	UGC 11194	13.3	1.5 x 0.9	13.4	SBbc	112	Knotty, strong dark lane S of center.
18 18 22.8	+21 17 32	UGC 11197	13.3	1.1 x 1.1	13.4	S0		Numerous stars superimposed.
18 25 38.7	+22 19 35	UGC 11229	13.8	0.9 x 0.5	12.8	S0/a	74	Mag 10.64v star WNW 1′.4.
18 27 17.3	+24 53 06	UGC 11237	14.0	1.3 x 0.8	13.9	Scd:	101	Filamentary, numerous stars superimposed.
18 28 24.0	+22 44 11	UGC 11246	14.0	1.2 x 0.4	13.1	Sab	152	
18 31 06.8	+22 24 32	UGC 11261	14.3	0.9 x 0.5	13.3	Sdm:	150	Mag 10.5 star NE 0′.9.
18 35 14.4	+22 30 00	UGC 11285	14.1	1.5 x 0.4	13.4	Sdm:	42	
18 35 47.6	+22 28 15	UGC 11289	13.2	1.5 x 0.9	13.4	S?	0	Numerous stars superimposed; mag 12.3 star N edge.
18 36 37.5	+19 55 16	UGC 11294	12.8	1.3 x 0.9	13.0	E0?	40	Strong dark lane SW of center.
18 37 54.9	+17 31 55	UGC 11301	14.6	2.2 x 0.2	13.6	S	110	
18 37 55.4	+27 47 32	UGC 11307	14.1	1.4 x 0.4	13.3	SBcd:	50	Mag 11.7 star on S edge.
18 40 10.6	+21 29 35	UGC 11314	14.4	0.9 x 0.4	13.1	Im:	91	Mag 8.27v star S 4′.9.
18 40 19.8	+24 11 58	UGC 11315	12.6	2.1 x 1.0	13.3	Sb II-III	25	Numerous stars superimposed.
18 40 48.2	+23 41 00	UGC 11320	14.1	1.7 x 0.2	12.8	Sbc	89	Mag 13.4 star S 1′.1; mag 12.0 star NNE 2′.4.
18 40 57.5	+23 05 20	UGC 11323	13.7	1.5 x 0.6	13.4	SB(s)bc	85	
18 43 11.4	+18 43 38	UGC 11337	12.3	1.5 x 1.4	13.0	(R)SBa		Large core, uniform envelope with numerous stars superimposed; mag 11.9 star on W edge.
18 44 15.4	+24 08 34	UGC 11344	12.7	2.0 x 1.0	13.3	Sab	160	Mag 11.8 star S of core; mag 10.85v star NW edge.
18 44 16.9	+25 15 21	UGC 11346	13.7	1.0 x 0.7	13.2	SBb	77	Very slender core, faint, uniform envelope.
18 46 16.7	+22 36 46	UGC 11350	13.4	1.1 x 0.8	13.1	Scd:	3	Stellar nucleus, branching arms.
18 47 44.2	+23 20 48	UGC 11353	13.6	1.1 x 0.3	12.2	S0/a	153	
18 47 56.9	+22 56 32	UGC 11355	13.3	1.5 x 0.6	13.0	S?	126	Numerous stars superimposed; mag 12.3 star NW end.
18 51 30.0	+26 29 00	UGC 11368	13.3	1.5 x 0.5	12.8	S0/a	142	Mag 13.2 star on N edge.
18 51 37.9	+23 38 01	UGC 11369	13.2	1.0 x 0.5	12.3	SBa:	158	
18 51 43.8	+26 33 18	UGC 11370	14.0	1.4 x 0.4	13.2	Sdm:	48	Uniform envelope, few stars superimposed; mag 8.24v star E 9′.8.
18 51 56.1	+26 29 10	UGC 11371	13.4	1.7 x 0.5	13.1	S?	44	Mag 12.2 star E of center; mag 8.99v star S 5′.2.
18 54 23.6	+24 39 06	UGC 11375	13.4	1.1 x 0.4	12.4	Scd:	153	
18 56 43.8	+25 14 12	UGC 11379	13.8	0.9 x 0.4	12.5	Scd:	93	Numerous stars superimposed, in star rich area.
19 02 45.3	+27 18 18	UGC 11393	14.8	1.1 x 0.9	14.6	Scd:	126	Several stars superimposed around center; mag 10.03v star S 2′.3.
19 03 36.6	+27 36 20	UGC 11394	14.4	1.8 x 0.2	13.1	Scd:	37	Mag 10.8 star W 0′.9.
19 07 03.8	+29 00 21	UGC 11404	13.2	1.9 x 1.3	14.0	Sbc II	60	Many stars superimposed, bright knot NE edge; mag 10.70v star NW 1′.5.

OPEN CLUSTERS

RA h m s	Dec ° ′ ″	Name	Mag	Diam ′	No. ★	B ★	Type	Notes
19 17 17.0	+19 32 42	Be 44		2	30	16.0	cl	
19 01 26.7	+29 17 00	NGC 6743		8	35	8.2	cl?	
19 23 12.6	+22 09 00	NGC 6793		7	15		cl	Few stars; moderate brightness range; no central concentration; detached.

GLOBULAR CLUSTERS

RA h m s	Dec ° ′ ″	Name	Total V m	B ★ V m	HB V m	Diam ′	Conc. Class Low = 12 High = 1	Notes
19 18 02.1	+18 34 18	Pal 10	13.2	18	19.4	4.0	12	

PLANETARY NEBULAE

RA h m s	Dec ° ′ ″	Name	Diam ″	Mag (P)	Mag (V)	Mag cent ★	Alt Name	Notes
19 04 32.4	+17 57 08	PN G50.4+5.2	37	16.5		18.4	PK 50+5.1	= **Abell 52**; close pair of mag 13 stars E 1′.4; mag 12.5 star N 1′.3.
19 14 59.4	+17 22 47	PN G51.0+2.8	9	17.0		13.4	PK 51+2.1	Mag 12.4 star W 0′.5; mag 11.8 star N 1′.5.
19 14 04.1	+17 31 32	PN G51.0+3.0	5	15.9	15.0		PK 51+3.1	Mag 11.7 star N 1′.5; mag 12.2 star S 1′.8.
19 19 18.8	+17 11 45	PN G51.3+1.8	20			15.1	PK 51+1.1	Forms a triangle with a mag 11.8 star SW 2′.2 and a mag 12.6 star WNW 2′.5.
18 49 47.6	+20 50 36	PN G51.4+9.6	13	12.2	11.4	13.3	PK 51+9.1	Close pair of mag 10.92v, 12.8 stars SW 1′.9; mag 13.2 star NE 2′.5.
19 03 37.4	+19 21 21	PN G51.5+6.1	45	16.6	15.2	19.2	PK 51+6.1	Mag 12.2 star W 3′.0; mag 12.5 star S 2′.8.
18 59 03.8	+20 37 00	PN G52.2+7.6	30	13.6	14.7		PK 52+7.1	Three faint stars in a NE-SW row across disc; mag 11.7 star S 1′.7; mag 10.88v star WSW 2′.7 and 6.75v star W 5′.2.
19 19 02.8	+19 02 18	PN G52.9+2.7	2	17.4			PK 52+2.1	Mag 14.3 star E 0′.6; mag 9.40v star .
19 18 40.1	+19 34 26	PN G53.3+3.0	87	17.2		21.1	PK 53+3.1	= **Abell 59**; bright arc N edge; mag 11.07v star SE 2′.0; mag 6.51v star NE 2′.9.
19 23 25.0	+21 07 59	PN G55.2+2.8	5	16.9	15.8		PK 55+2.1	
19 23 46.9	+21 06 36	PN G55.3+2.7	11	16.3			PK 55+2.2	Mag 11.9 star NW 1′.3.
19 08 39.6	+22 58 59	PN G55.3+6.6	56	17.1			PK 55+6.1	= **Abell 54**; tight trio of mag 12-13 stars NW 1′.3; mag 12.4 star E 1′.8.
18 31 18.6	+26 56 11	PN G55.4+16.0	63	15.6	14.3	14.9	PK 55+16.1	= **Abell 46**; mag 13.8 star NW 2′.0; mag 14.2 star E 2′.4.
19 17 05.9	+25 37 33	PN G58.6+6.1	37	17.5	14.7	17.6	PK 58+6.1	= **Abell 57**; SE-NW line of three mag 12 stars SW 2′.0; mag 10.84v star SE 1′.2.
19 14 30.2	+28 40 43	PN G61.0+8.0	16	14.9	14.3	17.2	PK 61+8.1	Mag 11.6 star E 2′.4; mag 8.10v star W 2′.7.

68 68

GALAXIES

RA h m s	Dec ° ′ ″	Name	Mag (V)	Dim ′ Maj x min	SB	Type Class	PA	Notes
16 37 29.5	+17 55 55	CGCG 109-35	14.9	0.7 x 0.4	13.3	Irr	0	Mag 11.7 star N 2′.1.
16 48 09.9	+18 01 11	CGCG 110-9	15.2	0.5 x 0.4	13.3	SBbc	27	Mag 11.9 star S 4′.7.
16 48 30.1	+19 18 16	CGCG 110-10	15.0	0.6 x 0.3	13.0	SBb	108	Pair of stars, mags 12.0 and 12.8, S 1′.1.
16 53 54.0	+20 35 49	CGCG 110-18	15.2	0.4 x 0.4	13.0			Located between a pair of stars mags 11.1 and 12.3.
17 47 42.2	+18 16 16	CGCG 112-41	15.6	0.8 x 0.8	14.9			Stellar nucleus with faint star S; faint outer envelope.

RA h m s	Dec ° ′ ″	Name	Mag (V)	Dim ′ Maj x min	SB	Type Class	PA	Notes
16 39 30.8	+21 18 58	CGCG 138-50	14.2	0.8 x 0.5	13.0	S?	66	Located in a small triangle of mag 12-14 stars; mag 8.18v star SE 6′.6.
16 44 28.0	+26 04 09	CGCG 138-65	15.5	0.4 x 0.4	13.3			Mag 13.3 star NE 0′.4.
16 45 14.9	+22 08 14	CGCG 138-69	14.8	0.6 x 0.4	13.1	Sab	24	Small triangle of mag 10-12 stars NW 6′.0; mag 11.0 star W 5′.9.
16 56 15.6	+20 58 32	CGCG 139-22	15.0	0.7 x 0.5	13.7		69	Mag 14.2 star SE 2′.1; mag 13.7 star W 1′.8.
17 27 01.6	+24 36 03	CGCG 140-25	15.4	0.6 x 0.4	13.7		24	Mag 11.7 star W 3′.7.
17 45 37.3	+22 13 19	CGCG 141-8	14.9	0.6 x 0.6	13.6			Patchy appearance.
16 52 24.2	+28 08 20	CGCG 169-8	15.1	0.5 x 0.5	13.4			Mag 10.9 star W 5′.9; mag 9.79v star E 7′.1.
16 56 22.0	+29 19 50	CGCG 169-12	16.0	0.5 x 0.3	13.7		102	Mag 13.3 star S 1′.1.
17 15 10.7	+29 15 44	CGCG 170-5	14.9	0.6 x 0.6	13.6			Almost stellar nucleus, faint outer envelope.
16 58 29.9	+20 02 27	IC 1236	13.6	1.0 x 0.8	13.2	SAB(s)c I	51	Mag 7.70v star N 6′.4.
17 23 47.2	+26 29 09	IC 1256	13.2	1.6 x 1.1	13.7	Sb I-II	97	Mag 10.30v star SW 2′.1.
17 50 39.3	+17 12 29	IC 1268	14.7	0.6 x 0.5	13.3	Sc	117	Mag 10.28v star SE 9′.5; stellar **CGCG 112-59** SE 5′.5.
17 52 05.9	+21 34 08	IC 1269	12.8	1.7 x 1.3	13.5	Sbc I	125	Mag 10.2 star S 4′.2.
16 51 33.5	+17 26 55	IC 4624	14.8	0.4 x 0.4	12.6			Low surface brightness; mag 10.00v star WSW 2′.6.
16 55 09.6	+26 39 32	IC 4630	13.6	0.8 x 0.5	12.5	S?	6	Bright core comprises northern half; very faint, narrow extension ENE of bright core.
16 49 44.1	+17 51 48	MCG +3-13-5		0.6 x 0.5			12	
16 57 54.8	+20 54 03	MCG +3-43-9	14.9	0.5 x 0.4	13.0	Sb		Mag 12.2 star ENE 2′.1.
17 35 33.8	+20 47 46	MCG +3-45-3	15.2	0.8 x 0.6	14.3	Sb	0	Mag 10.03v star N 0′.9.
17 37 26.6	+19 33 35	MCG +3-45-5	14.1	0.8 x 0.5	13.0		63	Mag 9.32v star N 2′.8.
17 39 49.5	+19 47 53	MCG +3-45-8	14.3	0.5 x 0.5	12.9	E		
17 48 12.5	+17 37 29	MCG +3-45-25	14.2	0.5 x 0.5	12.5	Sb		Mag 10.54v star NNE 1′.4; MCG +3-45-26 NNW 1′.8.
17 48 10.5	+17 39 15	MCG +3-45-26	14.3	0.3 x 0.2	11.1		159	**MCG +3-45-24** W 4′.2.
17 49 30.2	+18 33 52	MCG +3-45-32	14.4	0.4 x 0.3	12.1	S0	45	Mag 10.28v star SW 3′.5.
16 40 14.7	+23 01 17	MCG +4-39-18	15.3	0.5 x 0.5	13.6	Sbc		
17 14 30.7	+23 03 37	MCG +4-41-1	14.5	0.4 x 0.3	12.0		108	
17 29 25.2	+24 52 54	MCG +4-41-18	14.1	0.9 x 0.7	13.4		108	Mag 12.3 star W 1′.3; mag 9.72v star SE 3′.4.
17 42 02.8	+25 37 10	MCG +4-42-2	14.5	0.6 x 0.4	12.8		9	Mag 10.54v star N 2′.8.
17 58 07.0	+21 16 16	MCG +4-42-17	14.1	0.5 x 0.3	11.9		108	Mag 8.03v star N 3′.1.
18 01 33.5	+26 15 09	MCG +4-42-22	13.6	0.7 x 0.4	12.2	E/S0	108	Mag 14.6 star N edge.
16 50 02.5	+28 10 27	MCG +5-40-1	15.6	0.6 x 0.5	14.1	Sc	171	
16 56 43.4	+27 17 15	MCG +5-40-7	14.9	0.8 x 0.4	13.5			Mag 14.3 star superimposed N edge.
16 58 07.4	+29 54 33	MCG +5-40-14	14.3	0.7 x 0.4	12.8	S?	0	
16 58 31.6	+27 35 07	MCG +5-40-15	15.2	0.7 x 0.7	14.3	SBb		Stellar nucleus; mag 8.00v star NE 4′.9.
17 00 45.0	+27 49 37	MCG +5-40-23	14.3	0.5 x 0.4	12.4	S	177	Faint anonymous galaxy N 1′.3; mag 10.90v star W 1′.5.
17 00 53.7	+27 43 26	MCG +5-40-24	14.3	0.4 x 0.4	12.1	SB0		Mag 12.1 star SE 2′.4.
17 00 59.6	+29 25 07	MCG +5-40-25	15.3	0.6 x 0.5	13.9		90	**Mkn 504** ESE 1′.4; mag 9.28v star N 2′.8.
17 26 51.5	+28 55 18	MCG +5-41-15	15.3	0.5 x 0.3	13.0	Sb	99	Mag 11.27v star W 1′.8.
17 52 40.1	+29 48 15	MCG +5-42-7	14.8	0.3 x 0.3	12.0			Almost stellar; 1′.3 SE of NGC 6486.
17 53 09.6	+27 30 29	MCG +5-42-10	16.3	0.7 x 0.7	15.4	Sbc		Stellar nucleus, uniform envelope; mag 11.5 star W 0′.9.
17 54 45.5	+27 30 39	MCG +5-42-14	14.9	1.0 x 0.6	14.2		132	
18 02 57.2	+29 18 35	MCG +5-42-29	14.8	0.6 x 0.6	13.5	Sc		**CGCG 171-49** N 4′.0; UGC 11098 W 1′.9.
16 40 14.5	+23 45 53	NGC 6201	14.6	0.3 x 0.2	11.4		21	Almost stellar; 3′.0 WSW of NGC 6203.
16 40 27.5	+23 46 27	NGC 6203	14.4	0.6 x 0.6	13.1			Mag 10.22v star NW 2′.2.
16 48 02.6	+26 12 42	NGC 6228	14.0	0.9 x 0.5	13.0	S?	130	Very strong dark patch W of nucleus.
16 50 15.8	+23 34 46	NGC 6233	13.3	1.4 x 1.0	13.5	S0	24	
16 52 26.4	+23 19 57	NGC 6243	14.1	1.1 x 0.4	13.1	Sa	154	**CGCG 139-14** SW 3′.5; **MCG +4-40-5** SE 7′.1.
16 56 30.5	+27 58 36	NGC 6261	14.0	1.4 x 0.5	13.4	S0/a	88	
16 56 43.2	+27 49 17	NGC 6263	13.7	0.9 x 0.9	13.4	E		Mag 9.34v star N 2′.6.
16 57 16.0	+27 50 58	NGC 6264	14.5	0.7 x 0.5	13.2	S?	15	
16 57 29.1	+27 50 39	NGC 6265	14.2	0.9 x 0.6	13.4	S0	23	
16 58 08.9	+22 59 04	NGC 6267	13.1	1.3 x 1.0	13.3	SB(r)bc	35	Bright nucleus; mag 10.84v star NW 2′.1.
16 57 58.1	+27 51 15	NGC 6269	12.2	2.0 x 1.6	13.5	E	80	
16 58 44.1	+27 51 32	NGC 6270	13.3	0.5 x 0.5	11.9	E/S0		
16 58 50.7	+27 57 52	NGC 6271	14.1	0.6 x 0.6	12.8	S0?		
16 58 58.4	+27 55 48	NGC 6272	14.5	0.5 x 0.2	11.9	Spiral	167	Small, faint anonymous galaxy, SE 1′.4.
16 59 20.7	+29 56 43	NGC 6274	13.8	0.6 x 0.5	12.3	Double System	23	NGC 6274A on S edge.
16 59 21.8	+29 56 21	NGC 6274A		0.7 x 0.2			123	NGC 6274 on N edge.
17 00 45.1	+23 02 37	NGC 6276	14.6	0.4 x 0.3	12.2	S0	123	Almost stellar.
17 00 50.5	+23 00 37	NGC 6278	12.4	2.0 x 1.2	13.2	S0	130	NGC 6276 NW 2′.3.
17 00 47.2	+29 49 13	NGC 6282	14.4	0.7 x 0.5	13.1	SBbc	36	**UGC 10654** SSW 6′.3.
17 11 59.7	+23 22 49	NGC 6308	13.4	1.2 x 1.1	13.5	SAB(rs)c:	150	
17 12 38.8	+23 16 15	NGC 6314	13.0	1.4 x 0.7	12.8	SA(s)a: sp	175	Dark lane W side; mag 10.20v star ENE 2′.7.
17 12 46.1	+23 13 23	NGC 6315	13.2	0.8 x 0.6	12.3	SB(s)c:	36	Short E-W bar.
17 14 24.2	+20 18 50	NGC 6321	13.4	1.1 x 1.0	13.4	SB(rs)bc	63	Almost stellar anonymous galaxy NE 2′.5.
17 15 44.5	+29 24 17	NGC 6330	14.0	1.4 x 0.5	13.5	SBb	160	Strong N-S bar.
17 24 27.6	+29 23 24	NGC 6364	12.9	1.5 x 1.2	13.4	S0	2	Three faint stars on E edge.
17 27 20.8	+26 30 15	NGC 6371	14.3	0.8 x 0.3	12.6	Sb	162	
17 27 31.9	+26 28 31	NGC 6372	12.9	1.7 x 1.1	13.4	Sb? pec I	90	Bright nucleus, stars or knots on E edge..
17 38 47.4	+18 52 39	NGC 6408	12.7	1.6 x 1.4	13.4	SB(rs)a	132	Dark patches E and W of nucleus.
17 41 47.9	+23 40 21	NGC 6417	13.1	1.4 x 1.2	13.5	SB(r)b:	45	Mag 6.96v star N 5′.0.
17 43 38.6	+25 29 38	NGC 6427	13.3	1.6 x 0.6	13.1	S0⁻:	36	
17 44 05.4	+25 21 03	NGC 6429	13.1	1.9 x 0.6	13.1	SBa	23	Several stars along E edge, strong dark patch N.
17 46 51.6	+20 45 36	NGC 6442	12.6	1.9 x 1.5	13.7	E	120	Several faint stars superimposed around bright center.
17 47 58.6	+20 50 14	NGC 6452	14.4	0.5 x 0.5	12.7			Mag 9.17v star N 5′.0.
17 49 11.1	+20 48 12	NGC 6458	13.4	1.3 x 0.9	13.4	S0	155	
17 49 30.2	+20 45 44	NGC 6460	13.1	1.9 x 1.1	13.8	Scd:	157	
17 50 40.1	+17 32 15	NGC 6467	12.6	2.6 x 1.7	14.1	S	77	
17 51 49.0	+23 04 19	NGC 6482	11.4	2.0 x 1.7	12.8	E:	70	Mag 10.65v star NW 2′.4.
17 51 47.2	+24 29 02	NGC 6484	12.3	1.9 x 1.9	13.6	Sb: II		Two main, knotty arms.
17 52 35.3	+29 49 03	NGC 6486	14.3	0.8 x 0.8	13.9	E/S0		Stellar **MCG −5-42-7** SE 1′.3; NGC 6487 NE.
17 52 41.9	+29 50 16	NGC 6487	11.9	1.8 x 1.8	13.1	E		NGC 6486 SW 1′.9.
17 54 30.5	+18 22 32	NGC 6490	13.5	1.1 x 0.9	13.3	S0⁻?	115	**CGCG 112-69** N 4′.4.
17 54 50.8	+18 19 37	NGC 6495	12.2	2.0 x 1.8	13.5	E	69	Bright center, many superimposed stars on faint envelope; **UGC 11037** SE 5′.4.
17 55 59.6	+18 20 16	NGC 6500	12.2	2.2 x 1.6	13.4	SAab:	50	Mag 7.43v star E 5′.9; NGC 6501 on N edge.
17 56 03.7	+18 22 21	NGC 6501	12.0	2.0 x 1.8	13.3	SA0⁺:	54	
17 59 34.3	+24 53 12	NGC 6513	13.3	1.2 x 0.8	13.1	SB0:	40	
17 59 43.9	+28 51 58	NGC 6518	13.9	0.4 x 0.4	11.8			**MCG +5-42-25** E 1′.2.
18 01 46.5	+19 43 42	NGC 6527	13.4	1.4 x 1.0	13.6	Sa:	150	
16 38 49.6	+17 21 09	UGC 10490	14.2	1.0 x 0.8	13.8	Sd	113	Faint stellar nucleus, uniform envelope.

(Continued from previous page)

GALAXIES

RA h m s	Dec ° ′ ″	Name	Mag (V)	Dim ′ Maj x min	SB	Type Class	PA	Notes
16 42 23.9	+25 04 59	UGC 10514	14.2	1.6 x 0.5	13.8	Sdm:	163	Mag 10.73v star ESE 1′.4.
16 44 49.6	+22 31 15	UGC 10528	12.2	1.9 x 1.1	12.9	SA0$^+$	65	
16 46 43.5	+26 56 45	UGC 10543	14.1	0.9 x 0.9	13.7	SB(r)a		Uniform envelope; mag 8.63v star SE 0′.9.
16 47 26.3	+21 07 28	UGC 10549	14.4	1.0 x 0.8	14.0	Im	138	Very filamentary; mag 11.9 star SE 1′.0; mag 10.91v star SE 2′.7.
17 00 14.7	+23 06 24	UGC 10650	14.4	1.3 x 0.3	13.2	Im?	30	Bright knot or faint star S end.
17 00 15.1	+27 34 56	UGC 10653	14.0	1.1 x 0.7	13.6	SBb	132	
17 00 51.6	+28 01 17	UGC 10658	14.6	1.1 x 0.6	14.0	S0/a	15	Moderately bright, round anonymous galaxy W 1′.7.
17 03 21.0	+24 56 06	UGC 10672	13.9	1.8 x 0.3	13.1	S0/a	82	Very faint anonymous galaxy W 1′.4.
17 03 01.7	+29 52 22	UGC 10673	14.5	1.5 x 0.5	14.1		62	Almost stellar nucleus, uniform envelope; faint star SW end.
17 05 28.1	+23 09 10	UGC 10692	14.1	2.0 x 0.4	13.7	Sb III-IV	179	**MCG +4-40-16** W 6′.2.
17 06 36.5	+24 46 28	UGC 10702	14.5	1.0 x 0.6	13.8	Disturbed	177	
17 08 05.4	+26 22 39	UGC 10717	14.4	0.9 x 0.7	13.7	Sd	100	
17 08 25.6	+25 31 00	UGC 10721	13.8	1.2 x 0.5	13.0	Scd?	110	Mag 11.5 star W 2′.6.
17 09 25.2	+22 12 51	UGC 10728	14.0	1.5 x 0.7	13.9	Sb II-III	55	Mag 11.9 star on E edge.
17 16 57.3	+21 36 50	UGC 10787	13.7	1.0 x 0.3	12.2	SB?	31	Anonymous galaxy W 0′.9.
17 20 11.4	+23 50 23	UGC 10813	13.6	0.8 x 0.8	13.0			Mag 12.1 star S 0′.9.
17 23 26.3	+23 38 31	UGC 10828	13.9	1.2 x 0.8	13.8	S0?	144	Very small galaxy or bright knot S edge.
17 24 28.7	+24 58 18	UGC 10831	14.0	1.0 x 0.8	13.6	Sd	12	UGC 10837 E 3′.8.
17 24 45.5	+24 58 14	UGC 10837	13.4	1.8 x 0.8	13.6	SBbc	145	Bright elongated core, uniform envelope; mag 12.1 star E 1′.7.
17 24 50.9	+23 44 08	UGC 10840	13.4	0.9 x 0.6	12.8	E	51	
17 28 37.5	+26 33 37	UGC 10866	14.1	1.0 x 0.8	13.7	Scd:	154	Stellar nucleus, faint envelope.
17 29 56.3	+24 53 00	UGC 10879	13.9	0.9 x 0.7	13.2	Scd:	100	Mag 10.64v star W 1′.6.
17 33 04.0	+27 34 26	UGC 10894	14.0	1.1 x 0.5	13.2	S?	167	
17 33 35.9	+20 46 01	UGC 10899	13.2	1.7 x 1.0	13.6	S?	174	Numerous stars superimposed, four brightest on SW edge.
17 34 06.6	+25 20 35	UGC 10905	13.3	1.4 x 0.8	13.3	S0/a	176	Mag 8.35v star N 3′.0.
17 34 38.0	+25 34 06	UGC 10909	14.3	1.0 x 0.6	13.6	Scd:	155	Filamentary; several superimposed stars.
17 37 33.1	+17 32 00	UGC 10919	13.4	1.3 x 0.7	13.1	S?	7	Numerous knots.
17 38 27.3	+24 57 10	UGC 10926	14.2	1.1 x 0.5	13.4	SB	135	Pair of mag 14 stars SE edge.
17 38 42.0	+21 40 28	UGC 10928	13.9	1.5 x 0.6	13.6	Sb II	111	
17 41 18.4	+17 22 02	UGC 10941	14.5	1.1 x 0.5	13.7	S	7	
17 41 51.5	+18 17 19	UGC 10944	13.4	1.0 x 0.8	13.2	E	170	Mag 9.30v star NE 2′.8.
17 45 14.4	+18 08 16	UGC 10966	13.5	1.7 x 0.5	13.2	Sab	97	knot w of elongated core; mag 13.7 star E end.
17 46 22.0	+26 32 36	UGC 10972	13.3	2.4 x 0.6	13.5	Scd:	57	Small triangle of mag 13-15 stars on NW edge.
17 47 08.4	+20 51 30	UGC 10979	14.0	0.9 x 0.8	13.4	Sc	33	Very patchy; mag 10.81v star on S edge.
17 49 48.0	+25 47 42	UGC 10999	14.0	0.9 x 0.5	13.0	Sb	160	
17 52 08.8	+29 51 44	UGC 11017	13.9	1.2 x 0.7	13.6	SBdm:	165	Mag 10.04v star NE 2′.9.
17 52 54.6	+23 12 38	UGC 11024	14.1	1.5 x 0.3	13.1	S?	143	Sits in a rectangle of stars with mags of 11.2, 12.2, 12.5 and 12.6.
17 52 57.8	+27 40 25	UGC 11025	13.1	1.1 x 0.9	13.0	S0	0	
17 53 16.7	+24 34 28	UGC 11027	13.7	1.5 x 1.0	14.0	Im:	72	Several superimposed stars; one large bright knot, or galaxy?, near SW edge.
17 53 51.8	+24 27 59	UGC 11029	13.7	1.1 x 0.8	13.4	SBd	51	Several small knots NE.
17 55 07.3	+18 32 16	UGC 11039	14.4	0.9 x 0.7	13.7	Im		Stellar nucleus, brighter E.
17 55 17.8	+26 22 16	UGC 11042	13.5	1.1 x 1.1	13.6	Sa?		Very small, very faint companion galaxies on NE edge and SW edge.
17 55 22.9	+25 25 49	UGC 11043	14.6	1.7 x 1.2	15.2	Triple System	120	Triple system, extensive halo.
17 55 35.4	+18 55 23	UGC 11044	14.1	0.8 x 0.6	13.2	SB?	168	**UGC 11052** SE 12′.1.
17 55 31.4	+28 49 37	UGC 11046	13.4	1.2 x 0.8	13.3	(R′)SB(s)a	126	Bright E-W bar, dark patches N and S.
17 57 10.3	+27 57 43	UGC 11060	14.0	1.4 x 0.4	13.2	Sa	130	Mag 10.30v star N 2′.1.
17 57 35.5	+23 01 36	UGC 11063	14.0	1.0 x 1.0	13.9	SAB(s)c		Mag 12.4 star NE edge.
17 57 40.9	+27 50 04	UGC 11064	12.9	1.8 x 1.8	14.0	Scd:		Several stars superimposed S edge; mag 8.2 double star NE 1′.3.
17 58 05.0	+28 14 37	UGC 11068	13.6	1.3 x 1.1	13.8	S?	90	Stellar nucleus.
17 58 14.3	+27 15 45	UGC 11070	14.2	1.0 x 0.7	13.6	SB(r)b	75	Mag 10.55v star N 2′.6.
18 00 05.7	+26 22 00	UGC 11082	13.4	0.8 x 0.8	12.8	S0		**UGC 11080** SW 1′.9.
18 00 43.2	+28 42 44	UGC 11090	13.6	1.0 x 0.7	13.1	Scd:	20	Mag 7.37v star W 5′.9; **UGC 11086** NW 6′.2.
18 02 24.5	+26 02 35	UGC 11097	13.9	1.3 x 0.4	13.0	S?	157	
18 02 48.8	+29 18 23	UGC 11098	14.7	1.0 x 0.5	13.8	S	132	Mag 10.7 star on W edge.

OPEN CLUSTERS

RA h m s	Dec ° ′ ″	Name	Mag	Diam ′	No. ★	B ★	Type	Notes
17 26 24.0	+24 12 00	DoDz 8	6.8	14	6	8.3	ast	Few stars; moderate brightness range; not well detached.

PLANETARY NEBULAE

RA h m s	Dec ° ′ ″	Name	Diam ″	Mag (P)	Mag (V)	Mag cent ★	Alt Name	Notes
16 44 29.7	+23 47 58	NGC 6210	21	9.3	8.8	12.6	PK 43+37.1	Strong, narrow extensions both N and S; mag 9.46v star NE 4′.5.
17 42 36.6	+21 27 01	PN G45.6+24.3	47	15.5	15.1	16.4	PK 45+24.1	Mag 11.9 star SE 2′.2; mag 11.5 star NW 3′.0.
17 44 56.7	+27 20 05	PN G51.9+25.8	43			20.3	PK 51+25.1	Mag 12.6 star E 1′.4; mag 10.53v star W 2′.4.
17 54 23.1	+27 59 56	PN G53.3+24.0	13	12.2	12.1	17.6	PK 53+24.1	Mag 10.21v star S 2′.6; mag 11.6 star NW 3′.0.

69

GALAXIES

69

RA h m s	Dec ° ′ ″	Name	Mag (V)	Dim ′ Maj x min	SB	Type Class	PA	Notes
15 16 40.0	+19 05 30	CGCG 106-26	14.2	0.5 x 0.4	12.3	S?	33	= **Mkn 688**; between very close mag 12.6 and 13.9 stars.
15 54 06.0	+18 38 50	CGCG 107-53	13.9	0.8 x 0.6	12.9	SAb:	144	Mag 8.16v star N 7′.8; mag 8.47v star NW 8′.6; **MCG +3-40-57** WSW 7′.3; UGC 10084 E 4′.4.
16 00 47.3	+18 04 39	CGCG 108-46	14.8	0.6 x 0.4	13.1	SB(s)b:	48	Mag 12.8 star W 1′.5.
16 06 02.0	+18 06 40	CGCG 108-138	15.5	0.4 x 0.3	13.1	SAB(s)bc	24	Mag 11.54v star NE 3′.2; almost stellar **PGC 57119** S 1′.5.
16 11 05.5	+18 29 55	CGCG 108-165	14.7	0.5 x 0.4	12.8	S	150	Mag 11.6 star W 2′.1.
16 37 29.5	+17 55 55	CGCG 109-35	14.9	0.7 x 0.4	13.3	Irr	0	Mag 11.7 star N 2′.1.
15 33 40.1	+24 24 14	CGCG 136-12	14.7	0.5 x 0.5	13.1			Mag 14.0 star E 1′.3; mag 13.8 star NW 1′.5.
15 38 04.6	+26 16 51	CGCG 136-36	15.2	0.3 x 0.3	12.4			Almost stellar. **CGCG 136-35** S 1′.2; close pair of mag 11.4 stars S 3′.6.
15 38 35.5	+26 13 16	CGCG 136-37	15.4	0.6 x 0.5	14.0		177	Small right angle shaped asterism of three mag 13.5 stars S 2′.9.
15 39 35.2	+23 11 54	CGCG 136-43	16.5	0.5 x 0.4	14.6	SBc	0	Mag 10.09v star SSW 6′.8.
15 40 08.9	+21 30 55	CGCG 136-50	14.7	0.7 x 0.6	13.6	SBb	153	Mag 11.1 star S 1′.3; NGC 5975 SW 3′.7.
15 42 46.0	+24 59 09	CGCG 136-62	14.3	0.8 x 0.6	13.4		54	Mag 13.8 star SW 1′.6.
15 49 28.3	+21 02 00	CGCG 136-88	14.2	0.6 x 0.4	12.5	Sa	33	**MCG +4-37-39** S 2′.3.

(Continued from previous page)

GALAXIES

RA h m s	Dec ° ′ ″	Name	Mag (V)	Dim ′ Maj x min	SB	Type Class	PA	Notes
15 51 26.3	+20 57 04	CGCG 136-99	14.9	0.5 x 0.5	13.2	Sc		Mag 4.74v Rho Serpentis NW 2′9.
16 09 37.5	+20 24 56	CGCG 137-47	15.4	0.5 x 0.5	13.8	SB		Low surface brightness; mag 11.1 star NE 2′2.
16 12 12.8	+25 33 38	CGCG 137-59	15.0	0.5 x 0.3	12.8		144	Mag 10.19v star SE 6′9; note: **CGCG 137-61** 1′0 W of this star.
16 12 18.0	+25 59 28	CGCG 137-60	15.0	0.5 x 0.4	13.1		171	Mag 11.2 star SW 4′5.
16 21 54.9	+25 37 30	CGCG 138-2	14.4	0.6 x 0.4	12.7	S0/a	108	Mag 13.2 star W 2′4; mag 10.20v star S 6′4.
16 24 09.2	+25 48 56	CGCG 138-7	15.9	0.4 x 0.4	13.8			Forms a triangle with mag 14.2 star W 1′6 and mag 12.7 star NNW 1′9.
16 27 18.9	+22 01 59	CGCG 138-12	15.8	0.6 x 0.4	14.1		57	Mag 10.86v star NE 2′1; **CGCG 138-13** N 2′3.
16 28 36.2	+24 33 17	CGCG 138-16	14.8	0.6 x 0.4	13.1	Sc	18	Mag 12.1 star NNW 0′8.
16 28 47.2	+21 29 34	CGCG 138-17	14.9	0.8 x 0.6	13.9		156	Mag 9.91v star SSE 4′7.
16 39 30.8	+21 18 58	CGCG 138-50	14.2	0.8 x 0.5	13.0	S?	66	Located in a small triangle of mag 12-14 stars; mag 8.18v star SE 6′6.
15 25 39.5	+29 36 09	CGCG 165-54	14.6	0.6 x 0.4	13.1	E	21	Mag 14.7 star on S end.
15 30 16.4	+27 05 50	CGCG 166-5	13.9	0.6 x 0.6	12.7	S0		**CGCG 166-4** NNW 2′4; spherical anonymous galaxy SW 2′0.
15 38 43.5	+29 24 49	CGCG 166-23	13.4	0.9 x 0.4	12.1	SBc	0	**CGCG 166-24** NE 1′6.
15 50 34.4	+28 38 17	CGCG 166-64	13.9	0.8 x 0.7	13.2	S0	114	Mag 12.1 star on SW edge.
16 14 11.0	+29 51 04	CGCG 167-55	13.9	0.5 x 0.4	12.1	S0	15	**CGCG 167-56** NNE 8′1; mag 11.31v star W 7′4.
16 16 25.1	+27 14 45	CGCG 167-64	14.3	0.6 x 0.4	12.6	Sab	108	Mag 11.2 star NE 2′0.
15 30 00.9	+23 38 14	IC 1124	14.0	1.0 x 0.4	12.9	S?	74	Mag 7.61v star N 7′4.
15 40 06.8	+20 40 49	IC 1132	13.4	1.2 x 1.0	13.5	SA(rs)c	36	Mag 10.80v star E 8′1.
15 45 34.8	+17 41 58	IC 1135	14.7	0.8 x 0.3	13.0	Sb	66	Mag 10.9 star NE 6′1.
15 50 26.0	+18 08 17	IC 1142	13.5	1.4 x 1.1	13.8	Scd:	144	Mag 10.21v star NE 6′4.
15 58 32.4	+17 26 24	IC 1151	12.9	2.5 x 0.8	13.6	SB(rs)c	28	Mag 9.37v star E 5′8.
16 00 37.5	+19 43 20	IC 1156	13.4	0.9 x 0.7	13.0	E	0	Faint star S of nucleus.
16 05 33.5	+17 36 12	IC 1178	14.1	1.2 x 1.0	14.2	S0$^-$ pec?	36	Bright nucleus offset to SW; **IC 1181** on S edge.
16 05 37.0	+17 48 04	IC 1182	14.2	1.0 x 0.5	13.3	SA0$^+$ pec	81	NGC 6054 SSW 2′4.
16 05 38.2	+17 46 01	IC 1183	14.2	0.8 x 0.4	12.9	SAB0$^-$	66	Small companion NW end; NGC 6054 W 1′8.
15 34 56.6	+23 30 08	IC 4553	13.2	1.8 x 1.7	14.3	S?	144	= **IC 1127** = **Arp 220**; strongly peculiar, double nucleus; symbiotic pair?
15 35 22.6	+25 17 47	IC 4556	13.9	0.8 x 0.5	12.8		171	Mag 9.92v star W 3′2; **IC 4558** and **IC 4559** NE 6′7.
15 40 48.3	+28 17 30	IC 4569	14.0	0.9 x 0.6	13.2	S0?	132	Mag 10.81v star NE 11′2; **MCG +5-37-12** NNW 5′0.
15 41 22.8	+28 13 44	IC 4570	14.1	1.0 x 0.8	13.7	SABcd:	63	Mag 10.81v star N 10′6; **IC 4574** E 8′1.
15 41 54.2	+28 07 59	IC 4572	13.7	1.1 x 0.8	13.4	SBab	69	Small, round anonymous galaxy on S edge; mag 9.78v star NE 11′6.
15 42 35.6	+23 40 09	IC 4576	13.9	1.0 x 0.6	13.2		63	Mag 9.87v star SW 12′5.
15 42 51.7	+23 46 19	IC 4579	14.9	0.7 x 0.5	13.6		66	**IC 4577** NW 1′8.
15 45 39.5	+28 05 20	IC 4582	14.0	1.4 x 0.4	13.3	S?	171	Mag 10.94v star NNE 4′4.
15 46 22.0	+23 48 29	IC 4583	14.3	1.0 x 0.3	12.9	S?	36	Mag 11.1 star NE 6′1.
15 36 36.3	+17 20 15	MCG +3-40-12	14.7	0.7 x 0.4	13.2	Sa	96	
15 39 32.9	+17 25 52	MCG +3-40-18	14.7	0.5 x 0.4	12.8	S0	126	Mag 12.3 star NE 2′4; this star has a faint, elongated anonymous galaxy on it's E edge.
15 54 10.0	+19 06 22	MCG +3-40-60	14.3	0.6 x 0.4	12.6	SB(r)0/a:	159	Stellar galaxy **KUG 1552+191** SSE 5′0.
15 56 33.8	+20 03 06	MCG +3-41-4	13.7	0.8 x 0.6	12.8	SA0/a	24	Faint, elongated anonymous galaxy N 1′4; another faint, elongated anonymous galaxy W 1′3.
16 01 06.9	+19 26 53	MCG +3-41-33	13.7	0.9 x 0.8	13.3	SAB0$^-$	153	Mag 10.47v star W 3′2.
16 02 04.3	+17 04 32	MCG +3-41-47	13.8	0.7 x 0.5	12.7	E	141	Mag 13.2 star NW 1′5.
16 09 31.7	+18 15 06	MCG +3-41-137	15.0	0.7 x 0.5	13.7	S	3	Mag 12.8 star W 1′9.
16 10 51.3	+17 03 23	MCG +3-41-141	14.4	0.9 x 0.4	13.2	Sb	108	Mag 11.3 star ESE 1′5.
16 30 47.3	+20 21 18	MCG +3-42-14	14.8	0.8 x 0.6	13.9	SBb	168	
16 31 42.8	+18 38 56	MCG +3-42-18	15.1	0.6 x 0.4	13.4	Sd	15	
16 32 04.0	+20 45 51	MCG +3-42-19	16.4	0.9 x 0.4	15.1		138	Close pair of mag 11.6 stars E 1′7.
15 20 11.0	+25 43 22	MCG +4-36-38	13.8	1.0 x 0.8	13.4			
15 29 03.8	+25 27 30	MCG +4-36-46	14.1	0.8 x 0.5	12.9	SB?	171	
15 33 48.9	+21 08 08	MCG +4-37-3	14.2	0.6 x 0.4	12.5		6	Mag 10.9 star WNW 2′1.
15 34 04.2	+22 56 00	MCG +4-37-4	14.3	0.8 x 0.4	12.9	Sb	117	Mag 14.0 star on E edge.
15 37 45.3	+22 25 37	MCG +4-37-14	14.3	0.7 x 0.3	12.4	S	153	
15 50 00.1	+24 48 03	MCG +4-37-42	14.2	0.9 x 0.6	13.6	E/S0	78	Small, faint, elongated N-S anonymous galaxy on W edge; mag 12.1 star NNW 2′3.
15 50 38.6	+20 22 51	MCG +4-37-44	15.8	0.9 x 0.4	14.5	S	60	Multi-galaxy system with small companion galaxy N edge; mag 7.60v star SW.
15 54 02.9	+23 07 49	MCG +4-37-55	14.2	0.8 x 0.3	12.5	S?	21	Mag 12.0 star SE 1′7.
15 56 33.9	+21 17 19	MCG +4-38-1	13.6	0.8 x 0.6	12.9	E/S0	24	Mag 7.97v star SW 3′1.
16 01 29.0	+22 25 41	MCG +4-38-14	14.3	1.1 x 0.7	13.9		18	
16 02 30.5	+21 07 13	MCG +4-38-15	13.8	0.6 x 0.5	12.6	E	9	Mag 10.62v star E 2′9.
16 15 25.6	+26 06 36	MCG +4-38-43	14.4	0.7 x 0.5	13.2		48	
16 17 49.6	+20 41 31	MCG +4-38-44	15.2	0.7 x 0.4	13.7	SBc	54	
16 17 57.0	+22 56 42	MCG +4-38-45	13.8	0.9 x 0.4	12.8	S0	168	
16 27 24.9	+21 36 39	MCG +4-39-3	14.3	0.8 x 0.5	13.1		63	Mag 11.2 star on N edge; **CGCG 138-11** NNW 2′6.
16 32 17.8	+21 22 20	MCG +4-39-13	14.3	0.7 x 0.5	13.0	Irr	3	
16 40 14.7	+23 01 17	MCG +4-39-18	15.3	0.5 x 0.5	13.6	Sbc		
15 24 26.5	+26 25 41	MCG +5-36-27	14.5	0.6 x 0.4	12.8		126	Mag 13.7 star W edge.
15 43 46.2	+28 24 52	MCG +5-37-17	14.3	0.5 x 0.4	12.4	S?	114	
15 56 31.6	+28 26 14	MCG +5-38-3	14.4	0.7 x 0.7	13.4	SBb		
15 58 43.8	+26 48 59	MCG +5-38-6	14.0	0.9 x 0.6	13.4	E?	81	Mag 8.54v star NW 3′6.
16 09 37.7	+28 03 05	MCG +5-38-19	14.2	0.6 x 0.6	12.9	Sc		Mag 11.4 star N 3′1.
16 14 42.9	+27 15 33	MCG +5-38-42	14.4	0.6 x 0.4	12.7	SBb	81	Mag 12.9 star W 2′4.
15 19 24.7	+20 53 47	NGC 5910	13.6	0.7 x 0.7	12.9	E?		= **Hickson 74A**. **Hickson 74B** on S edge; **Hickson 74C** on NE edge; **Hickson 74D** ESE 1′8; **Hickson 74E** NE 1′0.
15 26 03.0	+18 04 25	NGC 5928	12.2	2.2 x 1.6	13.5	S0	105	Mag 7.88v star N 6′1.
15 34 49.0	+28 39 19	NGC 5958	12.7	1.0 x 1.0	12.5	S?		
15 38 54.3	+17 01 35	NGC 5972	13.6	1.2 x 0.8	13.4	S0/a	5	
15 39 58.0	+21 28 13	NGC 5975	14.2	1.0 x 0.3	12.7	S?	171	Mag 8.42v star SE 8′2; CGCG 136-50 NE 3′7.
15 40 33.6	+17 07 42	NGC 5977	13.4	1.2 x 1.0	13.5	SB0	155	Almost stellar nucleus; **MCG +3-40-24** E 4′0.
15 45 16.9	+24 37 46	NGC 5991	14.0	1.0 x 0.9	13.7		126	**MCG +4-37-30** S 7′5.
15 46 53.4	+17 52 19	NGC 5994	15.4	0.5 x 0.3	13.2	SB?	87	Located SW end of NGC 5996.
15 46 57.9	+17 52 52	NGC 5996	12.8	1.8 x 1.0	13.3	S?	33	Multi-galaxy system. Bright, narrow extension S. NGC 5994 W of S end.
15 47 46.0	+28 38 31	NGC 6001	13.6	1.0 x 1.0	13.4	Sc		NGC 6002 SW 1′1.
15 47 42.1	+28 37 48	NGC 6002	16.8	0.4 x 0.2	13.9	Sb	48	Stellar; located 1′1 SW of NGC 6001.
15 49 25.7	+19 01 54	NGC 6003	13.4	0.9 x 0.8	12.9	S0?	126	
15 50 22.8	+18 56 17	NGC 6004	12.3	1.9 x 1.7	13.4	SAB(rs)bc	105	Strong dark patch NW of nucleus.
15 52 55.9	+21 06 02	NGC 6008A	12.9	1.4 x 1.3	13.4	SB(r)b	63	
15 53 08.3	+21 04 26	NGC 6008B	14.3	0.7 x 0.4	12.8		168	Located 3′3 SE of NGC 6008A.
15 55 55.0	+26 57 57	NGC 6016	14.3	1.0 x 0.5	13.4	Scd:	26	Mag 7.97v star N 5′4.
15 57 08.2	+22 24 13	NGC 6020	12.7	1.4 x 1.0	13.1	E	140	
15 59 11.9	+20 44 51	NGC 6027	14.3	0.5 x 0.2	11.7	S0 pec	0	Member of **Seyfert's Sextet**.
15 59 11.2	+20 45 13	NGC 6027A	13.9	0.8 x 0.8	13.3	Sa pec		Member of **Seyfert's Sextet**.
15 59 10.9	+20 45 43	NGC 6027B	14.3	0.4 x 0.3	11.9	S0 pec	42	Member of **Seyfert's Sextet**.

(Continued from previous page)

GALAXIES

RA h m s	Dec ° ′ ″	Name	Mag (V)	Dim ′ Maj x min	SB	Type Class	PA	Notes
15 59 14.4	+20 45 52	NGC 6027C	15.9	0.7 x 0.4	14.4	SB(s)c? sp	56	Member of **Seyfert's Sextet**.
15 59 13.0	+20 45 32	NGC 6027D	15.6	0.2 x 0.2	11.9	S?		Member of **Seyfert's Sextet**.
15 59 12.6	+20 45 46	NGC 6027E	13.5	1.0 x 0.7	12.9	S0?	84	Member of **Seyfert's Sextet**.
16 01 28.7	+19 21 34	NGC 6028	13.5	1.3 x 1.2	13.8	(R)SA0$^+$:	30	Bright center is surrounded by dark ring.
16 01 51.4	+17 57 23	NGC 6030	12.8	1.1 x 0.8	12.5	S0°	37	Elongated anonymous galaxies N 2′3 and W 2′4.
16 03 01.2	+20 57 21	NGC 6032	13.5	1.6 x 0.7	13.4	SB(rs)b:	0	
16 03 32.1	+17 11 55	NGC 6034	13.5	1.1 x 0.8	13.4	E$^+$	54	
16 03 24.2	+20 53 29	NGC 6035	13.5	0.9 x 0.9	13.1	SAB(rs)c		Bright nucleus, dark lane E.
16 04 26.5	+17 44 53	NGC 6040	14.2	1.4 x 0.5	13.6	SAB(s)c	42	NGC 6040B on S edge.
16 04 26.5	+17 44 28	NGC 6040B	14.0	0.8 x 0.8	13.4	SA0$^+$ pec		On S edge of NGC 6040.
16 04 35.9	+17 43 19	NGC 6041A	13.3	1.2 x 1.1	13.5	SB0°	36	Stellar NGC 6041B SW of nucleus; **IC 1170** W 0′9.
16 04 35.0	+17 43 00	NGC 6041B	15.6	0.3 x 0.3	12.9	SB0$^-$		Stellar; located SW edge of NGC 6041A core.
16 04 39.7	+17 42 01	NGC 6042	13.9	0.8 x 0.7	13.2	SA0$^-$	60	Located 1′7 NE of mag 10.72v star.
16 05 01.5	+17 46 27	NGC 6043A	14.3	0.5 x 0.5	12.7	SAB0$^-$		
16 04 59.7	+17 52 13	NGC 6044	14.3	0.6 x 0.6	13.0	SA0°		Located 2′9 SW of mag 10.91v star.
16 05 07.9	+17 45 25	NGC 6045	13.9	1.3 x 0.3	12.8	SB(s)c sp	79	Faint, elongated, anonymous galaxy E end.
16 05 09.1	+17 43 46	NGC 6047	13.5	0.8 x 0.8	13.1	E$^+$		Star on NW edge; **UGC 10190** SE 4′6.
16 05 23.4	+17 45 26	NGC 6050	14.7	0.8 x 0.5	13.6	SA(s)c	132	Double galaxy, small companion SW edge.
16 04 56.8	+23 55 59	NGC 6051	13.1	1.6 x 1.1	13.7	E	165	Star on S edge. **PGC 57025** SE 2′0; **PGC 57014** N 2′3; **PGC 57010** N 2′0.
16 05 12.9	+20 32 31	NGC 6052	13.0	0.9 x 0.7	12.3	Sc	171	Multi-galaxy system, components on E-W line; mag 10.94v star S 2′5.
16 05 30.7	+17 46 03	NGC 6054	15.2	0.7 x 0.4	13.7	(R′)SAB(s)b	21	IC 1182 NE 2′5; IC 1183 E 1′8.
16 05 32.7	+18 09 34	NGC 6055	13.7	1.0 x 0.6	13.0	(R)SAB0$^+$	40	Stellar **PGC 57057** WSW 2′1.
16 05 31.4	+17 57 46	NGC 6056	13.9	0.9 x 0.5	12.9	SB(s)0$^+$	56	**PGC 57052** W 2′2; **PGC 57070** S 2′1.
16 05 39.6	+18 09 47	NGC 6057	14.7	0.6 x 0.5	13.4	E	159	Almost stellar.
16 05 52.0	+21 29 03	NGC 6060	13.1	2.0 x 1.1	13.8	SAB(rs)c	105	Strong dark patches E of nucleus.
16 06 16.0	+18 14 58	NGC 6061	13.6	1.0 x 0.8	13.2	SA0$^-$	66	**IC 1189** S 4′0; **IC 1191** and **PGC 57154** ENE 3′3.
16 06 22.9	+19 46 43	NGC 6062	13.6	1.2 x 0.9	13.6	SB(rs)bc	10	Pair of faint stars on W edge; NGC 6062B SW 1′3.
16 06 19.1	+19 45 47	NGC 6062B	15.3	0.4 x 0.4	13.2	Sc		Mag 9.04v star SW 3′7.
16 11 22.7	+23 57 55	NGC 6075	13.9	0.9 x 0.7	13.5	E/S0	90	
16 11 13.4	+26 52 20	NGC 6076	14.0	1.0 x 0.5	13.1		63	Mag 7.53v star NW 1′7.
16 11 14.1	+26 55 22	NGC 6077	13.3	1.2 x 1.1	13.6	E	60	Mag 7.53v star SW 2′3.
16 14 16.8	+17 45 24	NGC 6084	13.9	1.0 x 0.6	13.2	Sa sp	30	Mag 9.98v star SW 1′9.
16 12 35.4	+29 21 54	NGC 6085	13.0	1.5 x 1.2	13.5	Sa	165	Bright nucleus; mag 10.97v star E 1′8.
16 12 35.4	+29 29 02	NGC 6086	12.8	1.7 x 1.2	13.5	E	12	Star on NW edge; mag 10.71v star SW 2′8.
16 12 47.0	+27 59 11	NGC 6092	13.2	1.4 x 1.2	13.8	E	10	
16 14 46.9	+26 33 29	NGC 6096	14.3	0.9 x 0.4	13.0	S?	122	Mag 10v star SW 2′2.
16 15 35.6	+19 27 09	NGC 6098	13.3	1.0 x 0.7	13.0	E$^+$	141	NGC 6099 on NW edge.
16 15 34.3	+19 27 42	NGC 6099	13.4	0.9 x 0.9	13.3	E$^+$		NGC 6098 on SE edge.
16 15 37.0	+28 09 30	NGC 6102	13.9	1.2 x 0.8	13.7	S?	70	
16 27 04.0	+24 05 33	NGC 6148	16.1	0.5 x 0.3	13.8		60	
16 27 24.2	+19 35 49	NGC 6149	13.5	1.1 x 0.8	13.2	S0	22	Mag 8.99v star S 3′8.
16 31 21.4	+20 11 04	NGC 6168	14.2	1.4 x 0.3	13.1	Sdm:	111	Star on E end. Mag 10.65v star S 1′7.
16 32 21.1	+19 49 36	NGC 6181	11.9	2.5 x 1.1	12.8	SAB(rs)c I-II	175	Mag 11.04 star W 2′6.
16 34 25.6	+21 32 23	NGC 6186	12.9	1.7 x 1.3	13.6	(R′)SB(s)a	60	Large condensation S of bright nucleus.
16 40 14.5	+23 45 53	NGC 6201	14.6	0.3 x 0.2	11.4		21	Almost stellar; 3′0 WSW of NGC 6203.
16 40 27.5	+23 46 27	NGC 6203	14.4	0.6 x 0.6	13.1			Mag 10.22v star NW 2′2.
15 21 23.7	+28 33 36	UGC 9820	13.8	0.6 x 0.4	12.1	S0/a	110	Almost stellar.
15 21 45.5	+23 09 06	UGC 9825	14.1	1.0 x 0.8	13.7	SA(rs)c	171	Uniform surface brightness; mag 11.6 star W 0′8.
15 22 31.0	+19 15 36	UGC 9828	15.0	1.5 x 0.2	13.5	Scd:	178	
15 22 44.9	+29 46 07	UGC 9831	13.9	0.6 x 0.6	12.6	Sa		Mag 12.6 star N 1′1.
15 23 52.8	+23 32 47	UGC 9833	14.3	1.1 x 0.9	14.1	SAm	18	Uniform surface brightness.
15 25 34.0	+18 16 35	UGC 9841	13.5	2.3 x 0.4	13.3	Sbc	55	Mag 11.4 star W 1′8.
15 25 39.8	+20 47 10	UGC 9843	14.1	1.0 x 0.6	13.4	S?	155	Distorted.
15 27 02.2	+26 42 16	UGC 9859	14.0	0.7 x 0.7	13.0	I?		Very faint star NE of nucleus.
15 30 47.3	+23 03 53	UGC 9875	13.9	1.5 x 1.5	14.6	Sm:		Dwarf elliptical, very low surface brightness.
15 35 19.1	+20 50 30	UGC 9917	14.1	1.0 x 0.5	13.2	SB?	55	
15 36 27.9	+22 29 59	UGC 9927	13.7	1.0 x 0.9	13.4	SB0	171	Brighter E-W bar, uniform envelope.
15 39 16.8	+24 27 19	UGC 9954	14.3	1.0 x 0.5	13.3	Sab	13	
15 39 39.1	+21 46 54	UGC 9958	13.2	1.2 x 1.1	13.4	S0$^-$:	105	**CGCG 136-41** SW 4′7; **CGCG 136-47** NE 4′8.
15 42 42.0	+22 52 51	UGC 9984	14.0	1.1 x 0.7	13.6	SB(r)b	156	
15 44 23.0	+25 19 38	UGC 9999	13.3	1.4 x 0.8	13.3	S0	35	Almost stellar anonymous galaxy SW 1′9.
15 45 44.6	+20 33 37	UGC 10020	12.9	2.0 x 2.0	14.3	Sd		Bright center, patchy envelope.
15 47 36.4	+26 03 45	UGC 10035	14.0	0.8 x 0.5	12.8	S?	102	**IC 1138** NE 12′2.
15 48 41.3	+21 52 10	UGC 10043	14.0	2.2 x 0.3	13.4	Sbc	151	**MCG +4-37-35** SE 2′7.
15 49 45.0	+18 31 35	UGC 10050	12.8	1.3 x 0.7	12.6	S0	154	Mag 10.7 star S 0′8.
15 49 58.5	+20 48 18	UGC 10052	13.5	1.2 x 0.7	13.2	Double System	84	Double galaxy system.
15 50 50.9	+22 14 14	UGC 10059	14.3	1.1 x 0.7	13.9	Scd:	115	Stellar nucleus; mag 7.85v star NNE 5′8.
15 51 07.1	+20 12 06	UGC 10060	14.1	1.1 x 0.7	13.7	SB(s)b	114	Bright N-S bar with dark patches E and W.
15 51 11.8	+21 56 31	UGC 10062	14.0	1.3 x 0.6	13.6	Scd:	156	
15 52 20.8	+24 37 34	UGC 10073	13.8	1.1 x 0.6	13.2	SABbc	147	Mag 11.6 star S 1′9.
15 52 47.9	+24 23 21	UGC 10074	14.3	1.0 x 0.6	13.6	Scd:	85	**CGCG 136-104** WSW 11′5.
15 54 24.5	+18 39 06	UGC 10084	13.4	1.2 x 0.7	13.1	S0$^+$?	126	CGCG 107-53 W 4′4; mag 8.16v star NNW 7′7.
15 54 26.7	+18 31 26	UGC 10085	14.3	0.9 x 0.6	13.5	SAcd:	65	Mag 1.9 star S 4′9; UGC 10084 N 7′6.
15 55 55.5	+17 09 47	UGC 10093	14.2	1.3 x 0.9	14.2	SAB(s)c	111	Mag 13.0 star WNW 0′9.
15 59 46.0	+18 47 59	UGC 10121	13.9	1.2 x 0.6	13.4	SAB(rs)bc	175	Stellar **Mkn 293** S 4′1.
16 00 23.7	+20 50 55	UGC 10127	13.3	1.7 x 0.7	13.3	Sb	75	Three stars superimposed: mag 10.9 NE edge; mags 13.7 and 12.1 W end.
16 01 40.3	+21 21 11	UGC 10138	13.9	1.3 x 0.6	13.5	Sa	178	
16 03 00.2	+27 00 35	UGC 10151	13.9	0.9 x 0.7	13.2	Scd:	170	Several branching arms.
16 03 19.6	+20 38 10	UGC 10153	13.8	1.2 x 0.8	13.6	SB(r)b	128	
16 03 54.4	+25 00 37	UGC 10160	13.9	1.5 x 0.4	13.2	SBa	93	Mag 12.4 star 0′5 N.
16 05 52.2	+18 13 14	UGC 10195	14.7	1.5 x 0.4	14.0	(R′)SAB(s)b	120	**MCG +3-41-115** SE 1′4; **MCG +3-41-107** NW 4′8.
16 07 07.1	+22 03 35	UGC 10211	14.0	1.1 x 0.4	13.0	Sab	13	
16 09 51.4	+22 36 53	UGC 10223	15.1	1.6 x 0.2	13.7	SBb	57	
16 09 51.1	+20 10 34	UGC 10232	15.2	1.6 x 0.2	13.8	Sd: sp	132	
16 10 05.1	+22 39 02	UGC 10236	14.9	1.6 x 0.2	13.5	Sb	15	**CGCG 137-51** E 11′1; mag 10.40v star W 7′1.
16 10 29.7	+19 57 11	UGC 10243	14.1	0.8 x 0.7	13.4	SB(rs)c:	9	Mag 10.8 star NW 3′1.
16 10 59.0	+26 20 50	UGC 10250	15.2	0.9 x 0.7	14.6	S?	93	Brighter N half; mag 8.17v star N 3′1.
16 11 45.4	+29 44 42	UGC 10259	15.1	1.8 x 0.6	15.0		171	Chain of five galaxies, plumes.
16 11 57.9	+20 55 22	UGC 10260	14.5	1.0 x 0.6	13.8	S	150	Almost stellar nucleus; mag 8.53v star W 7′7.

(Continued from previous page)

69 GALAXIES 69

RA h m s	Dec ° ′ ″	Name	Mag (V)	Dim ′ Maj x min	SB	Type Class	PA	Notes
16 11 58.5	+29 50 18	UGC 10262	13.8	1.0 x 1.0	13.7	S0		An anonymous galaxy SW 0′.8, and a pair of anonymous galaxies NW 1′.8.
16 12 44.7	+28 17 15	UGC 10273	14.3	1.5 x 0.2	12.8	S?	159	Faint, elongated anonymous galaxy E edge; **CGCG 167-43** NW 3′.0.
16 15 28.9	+18 54 13	UGC 10297	14.3	1.9 x 0.3	13.5	Sc? sp	1	Mag 5.71v star 16 Herculis S 6′.2.
16 18 05.8	+21 33 31	UGC 10321	14.6	1.8 x 0.6	14.5		3	UGC 10321 Quintuple group with **MCG +4-38-47** located E 0′.9.
16 21 28.4	+28 38 23	UGC 10351	14.4	0.9 x 0.5	13.4	Sdm:	40	
16 28 54.2	+17 53 23	UGC 10405	14.5	1.0 x 0.8	14.1	Scd:	162	Trio of stars, mags 11.9, 12.1 and 13.1, SE 1′.7.
16 29 03.6	+25 42 47	UGC 10410	13.9	0.8 x 0.6	13.1	E:	90	
16 29 27.1	+21 20 16	UGC 10413	13.6	1.9 x 0.5	13.4	Scd:	173	Mag 2.78v star "Komephoros", Beta Herculis, 13′.8 NE; mag 12.3 star NW 1′.5.
16 31 21.6	+22 41 49	UGC 10435	14.1	0.8 x 0.6	13.2	SB(r)b:	88	
16 33 47.8	+28 59 05	UGC 10445	12.6	1.9 x 1.5	13.6	Scd?	145	
16 34 57.0	+25 41 31	UGC 10455	14.0	1.0 x 0.9	13.7	SB(r)b	126	Stellar nucleus in short bar.
16 38 49.6	+17 21 09	UGC 10490	14.2	1.0 x 0.8	13.8	Sd	113	Faint stellar nucleus, uniform envelope.
16 42 23.9	+25 04 59	UGC 10514	14.2	1.6 x 0.5	13.8	Sdm:	163	Mag 10.73v star ESE 1′.4.

GALAXY CLUSTERS

RA h m s	Dec ° ′ ″	Name	Mag 10th brightest	No. Gal	Diam ′	Notes
15 22 42.0	+27 43 00	A 2065	15.6	109	22	**Corona Cluster**. Particularly dense immediately W and S of center.
15 28 06.0	+28 52 00	A 2079	15.4	57	18	Mag 3.66v star on N edge; **UGC 9861** S of this star 10′.6; most other members PGC and anonymous GX.
15 32 42.0	+28 00 00	A 2089	15.8	70	16	PGC and anonymous members, most stellar.
15 39 48.0	+21 46 00	A 2107	15.7	51	28	Except UGC 9958, most members stellar.
15 40 06.0	+17 53 00	A 2108	15.7	45	13	All members anonymous, faint and stellar/almost stellar.
15 58 18.0	+27 13 00	A 2142	16.0	89	22	Almost stellar **PGC 56527** and **PGC 56515** near center; most other members anonymous and stellar.
16 03 18.0	+25 27 00	A 2148	15.4	41	16	Most members anonymous, stellar/almost stellar.
16 05 12.0	+17 44 00	A 2151	13.8	87	56	**Hercules Galaxy Cluster**. Somewhat strong concentration at center; many GX spread over whole area of cluster.
16 12 30.0	+29 32 00	A 2162	13.7	37	56	Large concentration of GX around center, another near NW edge.
16 16 48.0	+23 10 00	A 2170	15.9	38	16	All members anonymous, faint and stellar.
16 20 24.0	+29 54 00	A 2175	16.2	61	16	All members anonymous, faint and stellar.

PLANETARY NEBULAE

RA h m s	Dec ° ′ ″	Name	Diam ″	Mag (P)	Mag (V)	Mag cent ★	Alt Name	Notes
16 27 33.8	+27 54 33	PN G47.0+42.4	174	13.7	13.0	15.6	PK 47+42.1	= **Abell 39**; mag 14.0 star W 1′.8 just outside edge; small, very faint anonymous galaxy 0′.9 WNW of center; another small, very faint anonymous galaxy on S edge, between two faint stars.

70 GALAXIES 70

RA h m s	Dec ° ′ ″	Name	Mag (V)	Dim ′ Maj x min	SB	Type Class	PA	Notes
14 09 04.1	+17 45 56	CGCG 103-86	14.4	0.8 x 0.4	13.0	S?	90	Mag 7.69v star SSW 8′.9; mag 11.8 star SW 3′.2.
14 34 31.7	+19 44 28	CGCG 104-44	14.4	0.8 x 0.3	12.7	S?	147	Mag 11.1 star W 7′.3.
15 16 40.0	+19 05 30	CGCG 106-26	14.2	0.5 x 0.4	12.3	S?	33	= **Mkn 688**; between very close mag 12.6 and 13.9 stars.
13 58 25.7	+24 09 22	CGCG 132-64	14.5	0.7 x 0.3	12.6		171	Located 1′.7 NE of mag 10.67v star.
14 00 34.4	+25 19 05	CGCG 132-69	14.7	0.6 x 0.5	13.3	Sm	105	Mag 13.5 star E 2′.3; mag 13.1 star W 3′.6.
14 11 29.0	+24 02 17	CGCG 133-4	15.1	0.4 x 0.4	13.0	Spiral		Mag 13.4 star N 2′.0; stellar galaxy **KUG 1409+243** NNE 2′.9.
14 44 06.5	+26 01 10	CGCG 134-24	14.6	0.6 x 0.4	12.9		171	Mag 11.2 star NNW 4′.7.
14 47 20.7	+23 57 00	CGCG 134-33	14.7	0.7 x 0.4	13.2		123	Mag 10.14v star W 5′.2.
14 50 05.4	+21 32 27	CGCG 134-38	14.9	0.5 x 0.4	13.0		90	Mag 14.9 star WNW 1′.5; mag 10.31v star NW 9′.5.
14 56 16.3	+20 31 16	CGCG 134-57	14.4	0.9 x 0.5	13.4		15	Mag 8.74v star W 6′.3.
14 59 52.2	+20 57 09	CGCG 134-63	14.8	0.6 x 0.5	13.3		81	Mag 14.4 star on N edge.
15 06 32.9	+21 42 11	CGCG 135-13	14.2	0.9 x 0.6	13.3	Sab	102	Mag 12.5, 13.4 double star NNW 3′.0.
15 14 57.5	+25 53 41	CGCG 135-36	14.6	0.6 x 0.5	13.1		0	Pair of N-S oriented mag 11.4, 11.1 stars E 6′.6.
13 56 09.0	+29 40 11	CGCG 162-29	15.0	0.7 x 0.5	13.7	Sc	54	Very close pair of mag 11 and 13 stars NW 9′.7.
14 17 52.6	+26 45 23	CGCG 163-17	15.2	0.9 x 0.5	14.2	Spiral	159	Mag 8.18v star NE 4′.2; IC 4395 NW 9′.3.
14 21 25.1	+27 07 13	CGCG 163-32	15.1	0.7 x 0.4	13.5	Sc	135	Mag 13.1 star on SW edge.
14 31 04.8	+28 17 11	CGCG 163-70	14.3	0.7 x 0.5	13.0	S?	150	= **Mkn 864**.
14 58 55.8	+26 44 46	CGCG 164-49	14.6	0.6 x 0.5	13.1	SBbc	9	Mag 7.55v star WNW 10′.5.
13 55 59.5	+17 30 21	IC 960	14.0	1.4 x 0.7	13.8	Disrupted Spiral	24	Multi-galaxy system.
14 09 59.2	+17 41 46	IC 982	13.0	1.0 x 1.0	12.9	SA0+		Located on SW edge of IC 983.
14 10 04.5	+17 43 57	IC 983	11.7	3.9 x 2.6	14.1	SB(r)bc	126	8.99v mag star on SE end; IC 982 SW 2′.6.
14 10 07.7	+18 21 48	IC 984	13.5	1.9 x 0.5	13.3	Sb II	35	Mag 9.18v star WSW 5′.7.
14 19 32.7	+17 52 29	IC 999	13.9	0.8 x 0.4	12.5	S0?	142	IC 1000 SE 2′.2.
14 19 40.3	+17 51 16	IC 1000	13.7	0.9 x 0.5	12.7	S0	23	IC 999 NW 2′.2; elongated **UGC 9171** ESE 3′.0.
14 28 07.3	+25 52 05	IC 1017	13.7	1.0 x 0.5	12.8	S0?	129	NGC 5629 SE 2′.4.
14 29 17.2	+20 39 15	IC 1021	13.8	1.1 x 0.8	13.5	SBa	127	Mag 7.30v star NW 8′.2.
14 38 25.4	+18 11 00	IC 1037	14.1	0.8 x 0.5	12.9		108	Mag 5.91v star N 7′.4; **IC 1036** S 4′.5.
14 44 07.1	+18 00 41	IC 1050	14.5	0.7 x 0.4	13.0	Sb	33	Mag 10.17v star SW 7′.5.
14 51 17.7	+18 41 12	IC 1062	14.5	0.4 x 0.3	12.3	E/S0	102	Mag 7.59v star S 2′.3.
14 54 49.5	+18 06 22	IC 1075	13.8	1.1 x 0.5	13.0	SBb	156	IC 1076 SSE 4′.8; **CGCG 105-73** E 5′.4.
14 54 59.7	+18 02 12	IC 1076	13.6	1.0 x 0.5	12.7	S?	14	Mag 10.65v star SSW 7′.1.
15 02 43.4	+17 15 10	IC 1085	13.8	0.9 x 0.7	13.1		27	Mag 9.40v star W 6′.4.
15 08 24.1	+19 12 27	IC 1096	14.3	0.8 x 0.8	13.7			Located between a mag 11.4 star 1′.3 SE and a mag 10.1 star W 1′.0. **MCG +3-39-8** S 1′.1.
15 08 31.3	+19 11 00	IC 1097	13.9	1.1 x 0.6	13.3	S?	52	Mag 11.4 star NW 0′.9; **UGC 9738** NE 8′.8.
13 58 06.3	+28 25 19	IC 4355	14.5	0.5 x 0.3	12.3	Sa	0	Almost stellar; mag 10.57v star SSE 7′.0.
14 10 57.3	+25 29 47	IC 4381	13.7	1.4 x 1.0	13.9	Scd:	150	= **Hickson 71A**. **IC 4382** (**Hickson 71B**) NE 1′.8; **Hickson 71C** ESE 2′.0; **Hickson 71D** SE 3′.4.
14 11 56.1	+27 06 49	IC 4384	13.9	1.1 x 0.6	13.3	S0/a	164	Mag 11.24v star E 11′.1.
14 17 21.3	+26 51 21	IC 4395	13.8	1.2 x 0.8	13.6	S?	21	Nucleus offset to W; very small, elongated E-W, anonymous galaxy on W edge; mag 11.3 star S 2′.0.
14 17 58.9	+26 24 48	IC 4397	13.2	1.1 x 0.9	13.1	S?	165	Mag 11.8 star N 4′.6.
14 18 23.9	+26 23 07	IC 4399	14.1	1.1 x 0.7	13.7	Sb III-IV	140	Mag 15.3 star S of nucleus; mag 11.0 star E 2′.6.
14 19 16.4	+26 17 53	IC 4405	13.8	0.9 x 0.6	13.0	S?	120	**MCG +5-34-20** ENE 1′.2.
14 24 53.8	+17 02 16	IC 4417	14.0	0.8 x 0.7	13.4	E/S0	156	Stellar; mag 11.4 star W 1′.9.
14 25 27.3	+25 31 34	IC 4418	14.3	0.6 x 0.5	12.8		96	Mag 9.30v star NW 12′.9.
14 28 34.6	+17 20 04	IC 4438	15.8	0.5 x 0.4	13.9	Sbc	84	**IC 4440** E 6′.0.
14 28 45.3	+28 57 50	IC 4442	14.0	1.0 x 0.5	13.1	SBa	13	Located 11′.0 NNW of NGC 5641.

RA h m s	Dec ° ′ ″	Name	Mag (V)	Dim ′ Maj x min	SB	Type Class	PA	Notes
14 32 27.7	+27 25 36	IC 4452	14.4	0.5 x 0.5	12.9	E?		Mag 9.37v star E 11.′0.
14 37 20.9	+18 14 54	IC 4469	14.9	1.6 x 0.2	13.5	Scd:	110	Mag 5.91v star E 12.′9.
14 38 23.4	+23 19 57	IC 4475	13.8	0.9 x 0.8	13.3		174	Multi-galaxy system; two small, faint companions; spherical galaxy on NE edge, elongated galaxy on NW edge.
14 38 45.9	+28 30 19	IC 4479	13.4	1.3 x 1.3	13.8	Scd:		Stellar nucleus; stellar **IC 4477** SW 3.′7.
14 40 12.6	+18 56 33	IC 4482	15.1	0.6 x 0.4	13.4		9	Mag 9.49v star E 11.′6.
14 44 20.9	+28 32 59	IC 4497	13.9	0.5 x 0.5	12.2	S0?		Stellar; mag 12.0 star SW 7.′4.
14 45 00.9	+26 18 02	IC 4498	14.4	0.5 x 0.5	12.7	S0?		Mag 10.86v star SSW 3.′7.
14 50 55.5	+27 34 41	IC 4514	14.1	0.9 x 0.6	13.3	SB(s)ab	125	Short, bright bar.
15 04 30.5	+27 47 31	IC 4533	13.8	1.0 x 0.9	13.5	Sa	160	Strong dark lane W.
14 15 15.2	+18 07 38	MCG +3-36-93	13.8	0.6 x 0.5	12.3		9	Mag 8.76v star W 7.′2.
14 33 24.6	+20 00 17	MCG +3-37-24	14.0	0.8 x 0.7	13.2	Sb	135	Strong NE-SW bar.
14 49 02.7	+18 06 37	MCG +3-38-27	14.9	0.6 x 0.3	12.9		105	Mag 13.8 star NE 0.′9.
14 51 09.8	+18 34 30	MCG +3-38-38	14.6	0.8 x 0.4	13.2	Sb	156	Mag 13.7 star SE 1.′4; mag 7.59v star N 4.′7.
15 04 31.7	+17 18 53	MCG +3-38-81	13.6	0.9 x 0.9	13.2	SBbc		
15 04 41.4	+19 30 16	MCG +3-38-82	15.7	0.7 x 0.5	14.4	SBc	90	
15 04 37.6	+17 41 11	MCG +3-38-83	15.0	0.9 x 0.5	14.0	Sbc	153	Mag 11.2 star NW edge; **MCG +3-39-79** W 1.′9; **MCG +3-38-80** WSW 2.′0.
15 08 45.3	+18 56 01	MCG +3-39-11	14.4	0.7 x 0.5	13.1	Sb	9	Mag 13.9 star NW 2.′4.
15 12 31.6	+19 09 27	MCG +3-39-14	13.4	0.9 x 0.9	13.0			Mag 7.24v star SE 5.′1.
13 58 23.1	+22 53 12	MCG +4-33-37	14.6	0.7 x 0.2	12.3	S0?	6	
14 02 06.0	+26 03 32	MCG +4-33-39	14.3	0.9 x 0.4	13.1	Sb	177	Mag 11.8 star NE 2.′0.
14 04 52.7	+21 37 57	MCG +4-33-40	14.1	0.5 x 0.3	11.9	Compact	132	Mag 12.6 star N 3.′9.
14 11 49.1	+21 22 31	MCG +4-33-44	14.2	0.9 x 0.4	12.9	Sbc	24	
14 42 18.0	+25 30 31	MCG +4-35-5	14.2	0.6 x 0.5	12.8	Sb	60	
14 43 41.7	+23 01 07	MCG +4-35-6	14.1	0.8 x 0.4	12.7	Sa	90	
15 04 35.3	+25 54 10	MCG +4-36-2	14.4	0.5 x 0.5	12.8	Sbc		
15 06 05.7	+24 23 27	MCG +4-36-10	14.9	0.5 x 0.5	13.2	Sbc		
15 07 01.1	+25 45 17	MCG +4-36-16	14.4	0.7 x 0.5	13.1		6	
15 10 53.6	+20 56 03	MCG +4-36-23	15.4	0.5 x 0.4	13.5	Sd	177	
15 20 11.0	+25 43 22	MCG +4-36-38	13.8	1.0 x 0.8	13.4			
14 00 32.7	+28 39 36	MCG +5-33-38	14.1	0.7 x 0.7	13.2	Sc		
14 01 04.1	+29 31 35	MCG +5-33-41	14.4	0.6 x 0.5	13.0	Sc	90	
14 05 12.9	+29 25 02	MCG +5-33-47	15.3	0.9 x 0.2	13.3	Sbc	168	Pair with MCG +5-33-48 E; mag 10.37v star SW 1.′3.
14 05 15.4	+29 25 27	MCG +5-33-48	14.6	0.6 x 0.5	13.2		69	Pair with MCG +5-33-47 W.
14 13 43.9	+29 25 44	MCG +5-34-3	14.3	0.8 x 0.3	12.6	Sab	174	Mag 15.2 star superimposed N end.
14 20 39.7	+26 51 47	MCG +5-34-23	14.3	0.9 x 0.8	13.8	Sa	93	Mag 13.2 star E 2.′3.
14 49 28.0	+27 46 50	MCG +5-35-18	14.0	1.0 x 0.8	13.6	S0		Mag 11.4 star SW 1.′6.
15 09 06.9	+26 30 34	MCG +5-36-9	14.0	0.7 x 0.5	12.7	S0	105	Mag 14.1 star NE edge.
13 56 56.0	+29 09 53	NGC 5375	11.5	3.2 x 2.8	13.7	SB(r)ab	0	Dark patches E and W of bright center.
14 09 57.4	+17 32 45	NGC 5490	12.1	2.4 x 1.9	13.7	E	5	
14 10 06.9	+17 36 56	NGC 5490C	13.9	1.1 x 0.7	13.5	SB(s)bc		
14 10 35.3	+19 36 42	NGC 5492	12.8	1.6 x 0.4	12.2	Sb pec?	150	
14 11 04.4	+25 41 51	NGC 5498	13.6	1.0 x 0.8	13.2	S0⁻:	120	
14 12 29.0	+24 38 08	NGC 5508	13.1	1.1 x 0.8	12.9	S0	141	Mag 9.98v star NE 4.′1.
14 12 39.6	+20 23 13	NGC 5509	14.1	0.9 x 0.6	13.5	E/S0	100	
14 13 08.7	+20 24 59	NGC 5513	12.6	1.9 x 1.1	13.3	S0	115	**MCG +4-34-4** SW 1.′4; small anonymous galaxy E 1.′2.
14 13 47.8	+20 50 52	NGC 5518	14.0	0.7 x 0.7	13.3	E/S0		Mag 10.20v star NW 1.′2.
14 14 52.1	+25 19 02	NGC 5523	12.1	4.6 x 1.3	13.8	SA(s)cd: II-III	93	
14 17 59.7	+25 08 08	NGC 5548	12.6	1.4 x 1.3	13.1	(R′)SA(s)0/a	110	Strong curved dark lane E of center.
14 18 30.0	+26 17 15	NGC 5553	14.1	1.3 x 0.3	13.0	Sa:	88	Mag 11.2 star W 6.′2.
14 19 12.6	+24 47 53	NGC 5559	14.0	1.4 x 0.4	13.2	SBb	67	
14 21 16.3	+23 28 47	NGC 5581	14.1	0.9 x 0.7	13.7	E/S0	162	**Mkn 678** ENE 8.′5; **UGC 9186** NNW 9.′0.
14 23 10.3	+26 15 55	NGC 5594	14.2	1.1 x 0.6	13.5		144	Faint star W of nucleus; mag 10.45v star SSE 1.′9.
14 24 22.8	+24 36 48	NGC 5610	13.2	2.0 x 0.7	13.4	SB(s)ab	114	Bright E-W bar.
14 28 25.9	+17 55 27	NGC 5628	13.3	1.1 x 0.7	13.0	E	175	
14 28 16.5	+25 50 57	NGC 5629	12.1	1.8 x 1.8	13.2	S0		**IC 1018** SW 1.′3; IC 1017 NW 2.′4.
14 28 31.7	+27 24 30	NGC 5635	12.6	2.3 x 1.1	13.4	S pec	65	Bright nucleus.
14 28 59.5	+23 11 30	NGC 5637	13.9	0.9 x 0.5	12.9	S?	7	Located on the NW edge of galaxy cluster **A 1921**; **CGCG 133-74** NE 9.′7.
14 29 17.0	+28 49 12	NGC 5641	12.2	2.5 x 1.3	13.3	(R′)SAB(r)ab I	158	Strong dark lane E; mag 9.98v star S 9.′6.
14 30 43.5	+29 10 49	NGC 5657	13.3	1.9 x 0.8	13.6	SBb:	163	Two thin arms form an open "S."
14 31 05.9	+25 21 15	NGC 5659	13.9	1.6 x 0.4	13.3	Sb II	43	UGC 9340 N 8.′2.
14 34 12.7	+25 28 05	NGC 5677	13.8	0.9 x 0.7	13.1	S?	135	Mag 10.34v star NW 2.′1.
14 36 15.4	+29 54 31	NGC 5685	13.3	1.2 x 1.2	13.7	E:		
14 38 55.3	+20 30 22	NGC 5702	13.3	1.1 x 0.8	13.1	S0	150	
14 39 16.3	+20 02 33	NGC 5710	13.0	1.2 x 1.1	13.3	E	153	Faint, elongated anonymous galaxy N 2.′1.
14 39 22.9	+19 59 22	NGC 5711	14.0	1.0 x 0.6	13.3	S?	70	Double star NW 1.′0.
14 42 33.3	+28 43 33	NGC 5735	12.3	2.4 x 1.9	13.8	SB(rs)bc	40	**CGCG 164-14** S 4.′3.
14 43 11.9	+18 52 48	NGC 5737	13.5	1.3 x 0.8	13.4	SBb	170	Mag 9.83v star E 4.′6.
14 45 05.3	+21 54 55	NGC 5748	14.6	0.8 x 0.7	13.8		150	Mag 9.51v star NE 2.′6.
14 47 42.1	+18 30 08	NGC 5760	13.3	1.5 x 0.7	13.2	Sa	96	**IC 4507** S 2.′9.
14 52 14.3	+29 50 42	NGC 5771	13.6	0.8 x 0.7	12.9	E?	153	NGC 5773 SE 4.′2.
14 52 30.5	+29 48 26	NGC 5773	13.6	0.9 x 0.9	13.2	S?		NGC 5771 NW 4.′2.
14 54 31.5	+18 38 29	NGC 5778	13.8	1.2 x 0.9	13.9	E?	10	Faint star W of nucleus; several faint anonymous galaxies close by N and S.
14 54 22.5	+28 56 24	NGC 5780	14.1	0.9 x 0.4	12.8	S?	138	Mag 10.14v star NNE 6.′1; **CGCG 164-42** N 7.′8.
14 57 38.0	+29 58 06	NGC 5798	13.0	1.4 x 1.0	13.3	Im:	42	Patchy; pair of stars on W edge.
15 01 53.6	+25 57 51	NGC 5827	13.0	1.1 x 0.9	12.8	Sab pec:	135	Bright knot on NW edge.
15 02 41.9	+23 20 01	NGC 5829	13.4	1.8 x 1.5	14.3	SA(s)c I-II	45	= **Hickson 73A**. **IC 4526** = **Hickson 73B** NW 1.′3; **Hickson 73C** NE 2.′4; **Hickson 73D** NW 3.′2; **Hickson 73E** W 2.′5.
15 04 52.0	+21 04 06	NGC 5842	14.3	0.4 x 0.4	12.2			Almost stellar.
15 07 27.1	+19 35 52	NGC 5857	13.1	1.2 x 0.6	12.6	SB(s)b	137	NGC 5859 E 1.′3.
15 07 35.1	+19 34 55	NGC 5859	12.4	2.9 x 0.9	13.3	SB(s)bc	136	Strong dark lanes N and S of center; NGC 5857 just beyond W edge.
15 19 24.7	+20 53 47	NGC 5910	13.6	0.7 x 0.7	12.9	E?		= **Hickson 74A**. **Hickson 74B** on S edge; **Hickson 74C** on NE edge; **Hickson 74D** ESE 1.′8; **Hickson 74E** NE 1.′0.
13 56 02.9	+18 22 17	UGC 8850	13.7	0.6 x 0.4	12.0	S?		Faint streamer extends SW; mag 11.8 star N 1.′2.
13 57 15.4	+24 15 26	UGC 8873	14.1	1.6 x 0.4	13.5	S?	27	
13 57 39.9	+25 46 28	UGC 8879	14.8	1.5 x 0.2	13.3	Scd:	163	
13 58 03.5	+20 24 02	UGC 8887	14.1	1.1 x 0.5	13.3	Sab	122	
13 58 02.3	+20 37 30	UGC 8888	14.5	1.7 x 0.8	14.7	Quad System	132	Quadruple system; bridges? At center of galaxy cluster A 1825.
13 58 05.6	+21 47 48	UGC 8889	13.8	1.5 x 0.5	13.3	S?	30	Mag 5.76v star SE 9.′9; located just inside N edge of small galaxy cluster **A 1827**.

(Continued from previous page)

GALAXIES

RA h m s	Dec ° ′ ″	Name	Mag (V)	Dim ′ Maj x min	SB	Type Class	PA	Notes
13 58 51.1	+26 06 22	UGC 8904	15.1	0.7 x 0.6	14.0		78	Mag 11.5 star NW 1′.6.
13 59 39.2	+28 03 36	UGC 8911	14.0	0.9 x 0.7	13.3	Scd:	96	Disturbed; round, almost stellar anonymous galaxy N 2′.0.
14 03 03.9	+28 01 59	UGC 8961	14.0	1.7 x 0.3	13.1	S?	75	Mag 9.99v star SE 2′.6.
14 04 36.8	+29 11 53	UGC 8999	14.2	0.9 x 0.7	13.5	Scd:	5	**MCG +5-33-44** S 1′.5.
14 06 40.7	+22 04 12	UGC 9024	15.1	2.0 x 2.0	16.5	S?		Stellar nucleus, faint envelope; mag 7.42v star NW 8′.6.
14 07 55.5	+29 52 22	UGC 9035	13.6	1.3 x 1.0	13.7	SB(rs)b	141	
14 11 17.9	+17 30 23	UGC 9078	13.2	1.4 x 1.1	13.5	(R′)SB(s)a	63	N-S bar, smooth envelope.
14 12 16.8	+18 17 56	UGC 9087	13.5	1.0 x 0.7	13.0	S0?	30	**MCG +3-36-80** NNW 2′.7.
14 12 24.5	+18 24 40	UGC 9090	13.6	1.0 x 0.8	13.2	S0⁻:	70	Mag 11.6 star SE 2′.5.
14 13 16.3	+27 00 28	UGC 9101	13.2	1.9 x 1.0	13.7	Sb III	145	
14 15 56.7	+23 03 16	UGC 9128	14.0	1.7 x 1.5	14.9	Im	63	Dwarf irregular, very low surface brightness.
14 16 47.0	+23 00 10	UGC 9138	14.5	1.9 x 0.2	13.3	Scd:	171	Pair of mag 13 star W 0′.7.
14 18 18.4	+28 58 03	UGC 9155	14.1	0.7 x 0.5	12.8	SAB(rs)c	33	Stellar nucleus; galaxy **IC 4398** SSW 7′.0.
14 18 42.9	+21 49 03	UGC 9164	13.9	1.0 x 0.5	13.0	S?		Very patchy, strongly peculiar.
14 19 36.5	+17 38 38	UGC 9167	14.2	1.4 x 0.4	13.4	Sa	48	Mag 9.49v star on S edge.
14 20 46.1	+21 56 08	UGC 9182	13.6	2.4 x 0.5	13.6	Sd	123	Mag 8.41v star E 3′.8; **CGCG 133-37** NE 9′.6.
14 23 03.9	+18 16 44	UGC 9212	13.6	1.1 x 0.6	13.0	S0	120	
14 24 46.8	+26 08 22	UGC 9234	13.7	1.9 x 0.9	14.1	S?	162	Small bright core, faint arms.
14 28 02.3	+21 18 08	UGC 9274	13.4	1.3 x 0.8	13.3	SB?	44	Bright knot of star NE end; mag 10.99v star SW 2′.5.
14 28 41.4	+21 20 24	UGC 9282	14.1	1.5 x 0.9	14.3	Im:	110	Dwarf, low surface brightness.
14 28 57.0	+25 33 12	UGC 9294	13.8	1.2 x 0.7	13.5	SAB(s)bc	5	Pair of mag 12.8 stars NE 1′.4.
14 30 18.0	+21 44 47	UGC 9316	13.8	1.8 x 0.5	13.6	S0?	28	
14 30 11.3	+27 31 53	UGC 9317	13.7	1.0 x 1.0	13.6	SB(rs)c		
14 30 20.1	+23 03 42	UGC 9322	13.9	1.1 x 0.9	13.8	SAB(s)bc	65	**Mkn 683** SW 3′.0.
14 31 00.4	+25 29 22	UGC 9340	14.3	0.8 x 0.7	13.5	SBd	90	**MCG +4-34-42** W 6′.4; NGC 5659 S 8′.2.
14 35 45.7	+24 43 29	UGC 9396	13.7	0.9 x 0.8	13.2	SBa		N-S bar; very faint, almost stellar anonymous galaxy N 1′.7.
14 36 09.2	+21 47 32	UGC 9401	13.4	1.5 x 0.6	13.2	S?	85	Close pair of mag 12.5 stars NW 0′.5.
14 37 28.2	+25 45 53	UGC 9418	13.7	1.0 x 1.0	13.5	(R)SBa		Bright E-W core with dark patches N and S.
14 39 22.6	+17 00 45	UGC 9444	13.2	1.1 x 1.0	13.2	S0?	150	Line of three very faint stars SE along SE edge.
14 42 22.5	+22 19 55	UGC 9480	13.8	0.9 x 0.5	12.8	SB0/a	59	**MCG +4-35-4** and in contact companion NE 2′.0; almost stellar anonymous galaxy SW 1′.8.
14 45 26.4	+19 27 56	UGC 9503	14.3	1.6 x 0.5	13.9	Sb II-III	88	Mag 11.4 star ESE 2′.3.
14 49 36.0	+25 22 49	UGC 9544	13.4	1.8 x 1.3	14.2	S?	147	Bright nucleus, envelope slightly brighter SE; small knot E edge.
14 51 09.9	+17 11 16	UGC 9558	14.5	1.2 x 0.6	14.0	Scd: I-II	170	Stellar nucleus, two main arms.
14 53 46.8	+20 06 56	UGC 9578	13.9	1.0 x 0.9	13.6	SBb	129	Stellar nucleus.
14 54 44.9	+24 05 45	UGC 9594	14.1	1.0 x 0.7	13.5	SBcd:	78	**CGCG 134-51** SSW 5′.1.
14 54 48.6	+25 47 29	UGC 9596	14.2	0.9 x 0.6	13.4	Sbc	0	Elongated **CGCG 134-53** S 5′.2.
14 57 00.7	+24 36 45	UGC 9618	13.4	1.5 x 0.6	13.1	Double System	9	Double system, contact; consists of **MCG +4-35-18** and **MCG +4-35-19**.
14 57 11.2	+19 41 50	UGC 9620	13.5	1.1 x 0.5	12.8	S?	121	Mag 9.65v star W 5′.6.
14 57 21.1	+19 40 15	UGC 9622	13.2	1.2 x 0.7	12.8	S?	178	UGC 9620 NW 2′.9.
14 59 04.2	+19 35 11	UGC 9635	14.1	0.8 x 0.6	13.2	Sab	20	Mag 10.19v star S 1′.2.
14 59 34.5	+27 06 55	UGC 9644	13.3	1.3 x 1.3	13.7	SB(r)a		Bright N-S bar.
15 02 41.7	+19 49 33	UGC 9672	16.5	1.2 x 0.7	16.2	Sm	35	Several knots and/or superimposes stars NE end; mag 11.7 star NE 2′.8.
15 06 09.5	+25 46 56	UGC 9705	14.1	0.6 x 0.3	12.1	S0/a	70	
15 06 41.8	+23 38 29	UGC 9713	13.2	1.5 x 0.7	13.1	SB0	160	
15 07 03.9	+21 12 35	UGC 9718	14.1	1.3 x 0.5	13.5	Sb	96	
15 12 02.2	+21 17 51	UGC 9763	14.3	1.8 x 0.4	13.8	S?	15	Peculiar, long streamers N; mag 12.1 star WNW 2′.8.
15 13 28.0	+25 12 22	UGC 9770	14.1	1.1 x 0.7	13.7	SABb	126	Mag 12.9 star N 0′.6.
15 14 15.2	+20 28 42	UGC 9777	13.5	1.5 x 0.9	13.7	S?	150	Mag 7.97v star N 5′.9.
15 21 23.7	+28 33 36	UGC 9820	13.8	0.6 x 0.4	12.1	S0/a	110	Almost stellar.
15 21 45.5	+23 09 06	UGC 9825	14.1	1.0 x 0.8	13.7	SA(rs)c	171	Uniform surface brightness; mag 11.6 star W 0′.8.
15 22 31.0	+19 15 36	UGC 9828	15.0	1.5 x 0.2	13.5	Scd:	178	
15 22 44.9	+29 46 07	UGC 9831	13.9	0.6 x 0.6	12.6	Sa		Mag 12.6 star N 1′.1.
15 23 52.8	+23 32 47	UGC 9833	14.3	1.1 x 0.9	14.1	SAm	18	Uniform surface brightness.

GALAXY CLUSTERS

RA h m s	Dec ° ′ ″	Name	Mag 10th brightest	No. Gal	Diam ′	Notes
13 58 00.0	+20 39 00	A 1825	15.7	49	13	Except UGC 8888, all members anonymous and stellar/almost stellar.
13 59 12.0	+27 59 00	A 1831	15.4	67	26	**MCG +5-33-33** at center; **MCG +5-33-32** N 2′.4; large concentration of anonymous GX SW of center.
14 11 42.0	+28 08 00	A 1873	16.3	41	19	All members anonymous, faint and stellar.
14 21 24.0	+17 41 00	A 1899	16.0	33	22	**IC 1004** W of center 5′.4; **MCG +3-37-8** NE 6′.5; all others anonymous and stellar.
14 31 00.0	+25 39 00	A 1927	16.0	50	20	Most members anonymous, stellar/almost stellar.
14 54 30.0	+18 37 00	A 1991	15.4	60	22	Small concentration of GX N of center; larger concentration SSW of center.
14 58 42.0	+27 49 00	A 2005	16.0	105	16	All members anonymous, faint and stellar.
15 03 00.0	+27 11 00	A 2019	16.3	38	16	All members anonymous, faint and stellar.
15 04 18.0	+28 25 00	A 2022	15.6	50	21	**CGCG 165-8** at center; heaviest concentration of GX N and NW of center.
15 11 30.0	+18 03 00	A 2036	16.0	39	13	**CGCG 106-19** S of center 4′.7; all other members faint, anonymous and stellar.
15 22 42.0	+27 43 00	A 2065	15.6	109	22	**Corona Cluster**. Particularly dense immediately W and S of center.

GLOBULAR CLUSTERS

RA h m s	Dec ° ′ ″	Name	Total V m	B ★ V m	HB V m	Diam ′	Conc. Class Low = 12 High = 1	Notes
14 05 27.3	+28 32 04	NGC 5466	9.2	13.8	16.6	9.0	12	

GALAXIES

RA h m s	Dec ° ′ ″	Name	Mag (V)	Dim ′ Maj x min	SB	Type Class	PA	Notes
13 27 13.9	+19 48 12	CGCG 101-51	15.1	0.4 x 0.3	12.9	E	54	UGC 8448 NW 11′.6.
13 43 39.3	+19 34 01	CGCG 102-41	15.0	0.5 x 0.5	13.3			**CGCG 102-48** E 8′.6.
13 46 52.5	+17 41 55	CGCG 102-58	14.7	0.5 x 0.5	13.1	Spiral		Mag 11.5 star W 4′.8; stellar galaxy **KUG 1344+178** SW 6′.5.
13 50 30.2	+19 17 47	CGCG 102-67	14.7	0.7 x 0.5	13.4	Sb	123	Mag 14.2 star on E edge.
13 22 26.4	+21 25 30	CGCG 131-4	14.4	0.7 x 0.5	13.1	S?	78	= **Mkn 659**; mag 12.1 star WNW 2′.4; IC 885 S 6′.7.
13 46 36.5	+22 41 34	CGCG 132-9	14.1	0.5 x 0.4	12.2		33	**CGCG 132-7** NW 1′.5; bright star or galaxy E 1′.2.
13 53 23.9	+23 02 50	CGCG 132-36	14.7	0.5 x 0.4	13.0	E		Mag 13.9 star ESE 1′.2.

(Continued from previous page)

GALAXIES

RA h m s	Dec ° ′ ″	Name	Mag (V)	Dim ′ Maj x min	SB	Type Class	PA	Notes
13 58 25.7	+24 09 22	CGCG 132-64	14.5	0.7 x 0.3	12.6		171	Located 1′.7 NE of mag 10.67v star.
14 00 34.4	+25 19 05	CGCG 132-69	14.7	0.6 x 0.5	13.3	Sm	105	Mag 13.5 star E 2′.3; mag 13.1 star W 3′.6.
12 57 26.0	+27 32 43	CGCG 160-40	13.9	0.4 x 0.3	11.5	S0	147	Located 2′.8 N of NGC 4839.
12 59 05.4	+27 38 37	CGCG 160-73	14.6	0.5 x 0.5	13.0	SBa		= **Mkn 58**; **PGC 44562** SW 1′.8.
13 07 13.3	+28 02 45	CGCG 160-141	14.8	0.5 x 0.5	13.2	S?		Mag 9.41v star W 9′.4.
13 09 16.2	+29 21 57	CGCG 160-151	14.8	0.4 x 0.4	12.7	S?		Mag 8.54v star W 8′.2.
13 35 43.8	+27 24 33	CGCG 161-85	15.5	0.5 x 0.5	13.8	SBc		Forms a triangle with a mag 11.1 star N 2′.8 and a mag 12.3 star W 2′.3.
13 44 45.1	+27 08 11	CGCG 161-138	15.1	0.5 x 0.5	13.4	S+		Mag 10.02v star NNE 4′.1; mag 9.79v star E 7′.0.
13 56 09.0	+29 40 11	CGCG 162-29	15.0	0.7 x 0.5	13.7	Sc	54	Very close pair of mag 11 and 13 stars NW 9′.7.
12 45 12.0	+23 02 06	IC 813	13.7	0.9 x 0.8	13.2	S?	36	Mag 11.1 star WNW 3′.9.
12 46 44.7	+29 44 05	IC 818	14.6	1.0 x 0.3	13.1	S?	48	Mag 10.48v star SW 5′.1.
12 47 26.4	+29 47 13	IC 821	13.8	1.1 x 1.1	13.9	SAB(s)bc I		Mag 10.48v star SW 5′.1.
12 53 59.2	+26 26 33	IC 832	13.9	0.5 x 0.5	12.5	E?		Close pair of mag 11.8 star NW 2′.2.
12 56 52.3	+26 29 12	IC 835	14.2	0.6 x 0.6	12.9	S?		Mag 9.98v star W 10′.9; **IC 837** E 8′.9.
12 58 13.8	+26 25 32	IC 838	15.1	0.5 x 0.5	13.5			Stellar nucleus; NGC 4849 S 1′.8.
13 00 39.9	+29 01 11	IC 842	13.9	1.2 x 0.5	13.2	S?	57	Stellar nucleus; mag 13.6 star S 3′.9.
13 01 33.7	+29 07 47	IC 843	13.6	1.2 x 0.5	12.9	S0	131	Mag 11.09v star NE 5′.4.
13 08 34.3	+21 02 58	IC 851	14.1	1.0 x 0.4	12.9	S?	150	Mag 12.5 star S 4′.6.
13 13 50.2	+17 04 29	IC 857	13.8	1.0 x 0.8	13.4	SB(s)b	110	Mag 12.4 star SW 9′.4; mag 13.1 star E 8′.0.
13 14 51.9	+17 13 36	IC 858	12.7	1.4 x 0.9	12.9	S0?	100	**IC 859** E 1′.3.
13 15 03.5	+24 37 05	IC 860	14.5	0.8 x 0.5	13.4	S?	12	Mag 11.07v star E 5′.4.
13 16 15.4	+20 02 47	IC 862	14.2	0.4 x 0.4	12.3	E		Mag 8.82v star NE 4′.8.
13 17 19.8	+20 38 18	IC 867	13.9	1.4 x 1.0	14.1	SAB(s)c	30	**IC 866** N 3′.3; **IC 864** NNW 4′.1; **IC 868** SE 2′.6; **IC 870** SE 3′.5.
13 22 31.0	+21 18 58	IC 885	13.5	0.8 x 0.8	13.1	E		CGCG 131-4 N 6′.7.
13 32 04.8	+17 02 52	IC 894	14.2	1.1 x 0.5	13.4	Sb	78	Mag 8.64v star NW 6′.0.
13 40 03.0	+23 08 33	IC 905	14.2	0.6 x 0.6	13.0			Mag 11.7 star WNW 2′.8.
13 40 51.3	+24 28 20	IC 909	14.3	0.4 x 0.4	12.1	Compact		MCG +4-32-26 E 6′.3.
13 45 16.2	+23 13 08	IC 933	13.3	1.2 x 0.8	13.2	S0	155	Faint, elongated anonymous galaxy NW 2′.2.
13 55 59.5	+17 30 21	IC 960	14.0	1.4 x 0.7	13.8	Disrupted Spiral	24	Multi-galaxy system.
12 36 03.6	+26 59 13	IC 3559	14.5	0.6 x 0.2	12.0	S?	75	Almost stellar.
12 36 38.0	+24 25 38	IC 3581	14.6	0.9 x 0.4	13.3	S?	51	Mag 11.0 star NW 1′.3.
12 36 30.5	+26 11 58	IC 3582	14.6	0.6 x 0.3	12.6	Sab?	39	Mag 10.7 star NW 10′.7; note: **IC 3543** is 1′.5 NW and **IC 3546** is 3′.0 S of this star.
12 36 40.2	+26 49 46	IC 3585	13.4	1.1 x 0.9	13.2	SA(s)0°:	126	Bright nucleus; mag 13.1 star S 0′.8.
12 36 48.7	+27 32 50	IC 3587	15.1	1.4 x 0.2	13.5	Scd:	122	Mag 10.69v star S 4′.8.
12 37 21.2	+28 12 27	IC 3598	13.8	1.5 x 0.4	13.1	SA(r)ab:	140	Mag 9.89v star E 2′.5.
12 40 36.4	+26 30 21	IC 3644	14.6	0.9 x 0.2	12.6	S?	15	**IC 3646** N 1′.3.
12 40 52.9	+26 43 41	IC 3651	13.2	1.0 x 1.0	13.1	S0		Mag 9.74v star E 7′.5; **MCG +5-30-57** E 4′.8.
12 42 54.0	+20 59 21	IC 3692	13.8	1.0 x 0.7	13.3	SBa:	97	Mag 10.45v star W 3′.0.
12 44 53.1	+18 45 17	IC 3721	14.1	0.9 x 0.3	12.5	S?	136	Mag 9.47v star E 13′.6.
12 45 19.0	+21 32 09	IC 3736	16.0	0.6 x 0.4	14.3	S?	165	Mag 10.76v star SE 3′.1.
12 45 44.8	+19 10 37	IC 3745	13.6	0.7 x 0.7	12.7	S0		Mag 10.7 star W 6′.2.
12 54 48.5	+19 10 36	IC 3881	13.0	3.7 x 0.9	14.2	SBcd:	24	Mag 10.28v star S 12′.1.
12 55 41.4	+27 14 57	IC 3900	14.0	0.9 x 0.4	12.7	SB0:	177	Mag 10.98v star S 3′.0.
12 59 39.2	+28 53 42	IC 3990	14.4	1.2 x 0.4	13.5	S?	27	**IC 3991** and a mag 11.9 star N 1′.9.
13 00 38.0	+28 03 21	IC 4040	14.8	1.0 x 0.3	13.4	Sdm:	153	IC 4045 NE 3′.1; NGC 4908 ESE 3′.0.
13 00 48.7	+28 05 23	IC 4045	13.9	0.6 x 0.4	12.5	E4	115	Very faint, elongated **PGC 44815** NW 1′.1.
13 00 54.6	+28 00 25	IC 4051	13.2	1.2 x 0.9	13.3	E0	105	IC 4051 SSE 2′.2.
13 01 43.7	+29 02 38	IC 4088	13.8	1.5 x 0.5	13.4	Sab	89	Mag 8.89v star ESE 5′.6; small, faint anonymous galaxy S 2′.6.
13 03 46.6	+19 16 17	IC 4130	16.0	1.0 x 0.6	15.5	cD	69	Mag 9.99v star NE 9′.7.
13 08 31.7	+24 42 00	IC 4202	14.5	1.7 x 0.3	13.6	Sbc	143	Mag 6.82v star NW 10′.3.
13 16 16.9	+25 24 16	IC 4215	14.2	1.5 x 0.2	12.7	Sab	45	Mag 11.13v star N 2′.1.
13 23 00.1	+27 06 54	IC 4234	14.0	0.7 x 0.5	12.8	S?	27	Mag 9.23v star W 11′.0.
13 38 25.2	+26 44 32	IC 4314	13.7	0.6 x 0.6	12.4	S?		**IC 4313** NW 1′.4.
13 49 52.6	+25 11 26	IC 4332	14.2	0.8 x 0.8	13.6	S		Almost stellar; mag 10.15v star E 12′.1; note: UGC 8753 is 2′.9 S of this star.
13 54 55.7	+25 07 18	IC 4343	14.0	1.0 x 0.5	13.4	E/S0	108	Mag 9.14v star NW 8′.5; note: **IC 4342** is 3′.3 S of this star.
13 55 12.6	+25 01 12	IC 4344	13.9	0.8 x 0.6	13.0	SB	66	Mag 11.2 star W 3′.7; IC 4345 N 1′.8.
13 55 13.5	+25 03 03	IC 4345	14.9	0.9 x 0.9	14.7	E?		= **Hickson 69B**; **UGC 8842** = **Hickson 69A** E 4′.0; note: **Hickson 69D** on E edge of UGC 8842; **Hickson 69C** E 4′.8.
13 58 06.3	+28 25 19	IC 4355	14.5	0.5 x 0.3	12.3	Sa	0	Almost stellar; mag 10.57v star SSE 7′.0.
13 10 00.6	+19 43 03	MCG +3-34-3	14.5	0.3 x 0.3	12.0	E?		
13 25 10.4	+17 03 09	MCG +3-34-28	14.8	0.7 x 0.4	13.3	Sc	135	Pair of mag 12 stars 1′.3 N and NE.
13 32 26.9	+18 08 37	MCG +3-35-3	14.9	0.3 x 0.3	12.2			
13 36 53.4	+17 23 10	MCG +3-35-13	13.4	1.0 x 0.8	13.0	S0?	108	MCG +3-35-14 N edge; mag 11.30v star W 2′.4.
13 36 54.5	+17 23 56	MCG +3-35-14	13.9	0.6 x 0.4	12.4	E	90	MCG +3-35-13 S edge.
13 43 35.0	+18 03 22	MCG +3-35-19	14.4	1.0 x 1.0	14.5	Ep		**CGCG 102-42** SE 3′.4.
13 49 05.6	+17 14 14	MCG +3-35-27	15.8	0.5 x 0.5	14.1			
13 53 10.2	+17 20 00	MCG +3-35-35	14.3	0.3 x 0.3	11.5	Compact		Almost stellar.
12 38 07.7	+22 41 54	MCG +4-30-10	16.3	0.8 x 0.8	15.7	Sm		Mag 10.53v star S 3′.6; mag 6.38v star E 12′.7.
13 05 16.2	+25 57 24	MCG +4-31-5	14.3	0.9 x 0.7	13.7	S?	21	
13 06 15.1	+25 27 37	MCG +4-31-6	14.7	0.6 x 0.4	13.0	S?	141	Mag 11.3 star NW 3′.3.
13 21 52.0	+22 25 43	MCG +4-32-2	13.9	1.0 x 0.6	13.2	S0	132	Elongated anonymous galaxy parallel along E edge.
13 32 30.0	+22 36 05	MCG +4-32-12	15.4	0.5 x 0.4	13.6	Sbc	90	Mag 13.1 star N 1′.9.
13 34 54.2	+26 12 19	MCG +4-32-13	14.4	0.8 x 0.7	13.6	S0	18	Mag 11.4 star NE 3′.8.
13 34 57.4	+22 34 55	MCG +4-32-14	14.8	0.7 x 0.7	13.9			Mag 11.26v star N 1′.8.
13 38 57.3	+24 58 46	MCG +4-32-17	14.4	0.9 x 0.5	13.4	S	84	Mag 12.7 star N 0′.8.
13 39 51.6	+22 38 50	MCG +4-32-19	15.5	0.9 x 0.4	14.3	SBc	126	
13 41 18.1	+24 29 46	MCG +4-32-26	14.6	0.9 x 0.5	13.6	(R)SB(r)a?	87	Stellar nucleus; IC 909 W 6′.3.
13 50 30.8	+24 57 46	MCG +4-33-9	14.2	0.7 x 0.6	13.3	E	177	MCG +4-33-10 N edge.
13 50 31.0	+24 58 30	MCG +4-33-10	14.2	0.6 x 0.6	13.2	E		MCG +4-33-9 S edge.
13 53 16.7	+21 16 05	MCG +4-33-19	14.5	0.9 x 0.5	13.5		90	
13 54 07.2	+21 58 26	MCG +4-33-20	15.3	0.7 x 0.5	14.0	SBb	51	Mag 11.45v star E 2′.7.
13 58 23.1	+22 53 12	MCG +4-33-37	14.6	0.7 x 0.2	12.3	S0?	6	
14 02 06.0	+26 03 32	MCG +4-33-39	14.3	0.9 x 0.4	13.1	Sb	177	Mag 11.8 star NE 2′.0.
12 44 38.3	+28 28 19	MCG +5-30-70	14.2	0.7 x 0.4	12.7	S:	114	
12 46 55.4	+26 33 48	MCG +5-30-79	14.5	0.9 x 0.5	13.7	E+:	27	= **Mkn 1335**.
12 48 05.9	+29 26 37	MCG +5-30-87	16.6	0.8 x 0.5	15.4	Sc	39	
12 49 42.6	+26 53 31	MCG +5-30-94	13.8	0.9 x 0.7	13.4	E1:	144	
12 50 54.1	+27 50 27	MCG +5-30-101	14.8	0.7 x 0.7	14.1	E0		Mag 12.2 star E 1′.9.
12 51 43.9	+27 57 42	MCG +5-30-108	14.8	0.6 x 0.5	13.6	E/S0	171	

(Continued from previous page)

GALAXIES

RA h m s	Dec ° ′ ″	Name	Mag (V)	Dim ′ Maj x min	SB	Type Class	PA	Notes
12 52 09.1	+27 28 37	MCG +5-30-109	14.6	0.8 x 0.6	13.8	E/S0	96	Mag 4.93v star 31 Comae Berenices NW 7′.1.
12 52 16.4	+27 31 56	MCG +5-30-111	15.9	0.7 x 0.4	14.4	Spiral	132	Mag 4.93v star 31 Comae Berenices W 7′.6.
12 53 16.1	+27 05 41	MCG +5-30-116	13.9	0.9 x 0.7	13.3	S?	141	
12 56 51.6	+26 53 52	MCG +5-31-18	14.6	0.7 x 0.6	13.7	E	3	Almost stellar, faint envelope.
12 57 09.6	+27 27 57	MCG +5-31-23	14.5	0.6 x 0.6	13.2	S?		Located 3′.7 SW of NGC 4839.
12 57 54.5	+27 29 20	MCG +5-31-36	15.6	0.6 x 0.5	14.1	SBb	53	Pair with MCG +5-31-37 E 1′.5.
12 58 01.5	+27 29 20	MCG +5-31-37	14.9	0.7 x 0.5	13.6	SBa	87	Pair with MCG +5-31-36 W 1′.5.
12 58 18.4	+29 07 37	MCG +5-31-45	14.3	0.9 x 0.5	13.3	S0	24	Multi-galaxy system.
12 58 30.4	+28 00 52	MCG +5-31-46	14.1	0.9 x 0.9	13.7	S0?		
12 59 47.5	+27 42 35	MCG +5-31-74	15.9	1.1 x 0.4	14.9	S0/a	165	
13 00 54.3	+27 46 58	MCG +5-31-94	15.3	0.4 x 0.3	12.9	SA0	42	Located on SW edge of NGC 4911.
13 00 52.2	+28 21 57	MCG +5-31-95	13.8	0.7 x 0.4	12.5	E?	18	Faint, stellar anonymous galaxy E 3′.8.
13 02 04.4	+29 15 06	MCG +5-31-106	14.5	0.9 x 0.3	12.9	S?	159	Mag 11.09v star SW 4′.6.
13 05 59.8	+29 16 41	MCG +5-31-127	14.0	0.8 x 0.6	13.1	S0?	21	Mag 12.4 star S 2′.6.
13 06 36.5	+27 52 22	MCG +5-31-132	15.3	0.9 x 0.5	14.2	S?	126	Faint, elongated anonymous galaxy NE 1′.2.
13 06 37.9	+28 50 57	MCG +5-31-133	14.7	0.9 x 0.5	13.7	S?	177	Mag 10.84v star N 1′.7.
13 08 55.2	+29 02 24	MCG +5-31-139	15.3	0.8 x 0.4	13.9	S?	90	Mag 13.4 star on W edge.
13 08 57.7	+28 43 59	MCG +5-31-140	15.1	0.8 x 0.3	13.6	E	18	Mag 9.04v star SE 4′.4.
13 08 57.3	+28 16 50	MCG +5-31-141	15.7	0.7 x 0.5	14.4	SBbc	27	NGC 4983 WNW 7′.0; faint, elongated anonymous galaxy ENE 5′.7.
13 12 07.9	+27 19 49	MCG +5-31-152	14.4	0.9 x 0.7	13.9	E0	123	Faint, round anonymous galaxy on S edge.
13 12 17.9	+26 41 06	MCG +5-31-153	13.9	1.0 x 0.5	13.0	S?	108	Located between a mag 10.47v star N and mag 11.7 star S.
13 12 59.5	+27 08 26	MCG +5-31-154	14.7	0.9 x 0.7	14.2	E?	36	Multi-galaxy system, star superimposed on NE edge.
13 39 44.3	+27 46 34	MCG +5-32-56	14.5	0.7 x 0.6	13.4	S?	45	Mag 13.2 star N 1′.3.
13 41 45.3	+27 00 16	MCG +5-32-62	14.1	0.8 x 0.5	13.0	S?	174	Mag 11.4 star S 1′.9.
13 52 22.6	+27 41 20	MCG +5-33-11	14.5	0.6 x 0.4	12.8	Sbc	99	Mag 11.1 star SE 4′.4.
13 53 44.9	+28 36 22	MCG +5-33-13	15.8	0.7 x 0.5	14.5	S?	105	Mag 9.72v star NW 3′.4.
13 55 25.0	+26 47 37	MCG +5-33-19	15.1	0.8 x 0.5	13.9		48	Forms triangle with mag 11.34V star SE 1′.3 and MCG +5-33-20 E 1′.4.
13 55 31.1	+26 47 49	MCG +5-33-20	15.4	0.9 x 0.5	14.4	SBbc	138	Forms triangle with mag 11.34v star SW 1′.3 and MCG +5-33-19 W 1′.4.
14 00 32.7	+28 39 36	MCG +5-33-38	14.1	0.7 x 0.7	13.2	Sc		
14 01 04.1	+29 31 35	MCG +5-33-41	14.4	0.6 x 0.5	13.0	Sc	90	
12 36 08.4	+19 19 24	NGC 4561	12.5	1.5 x 1.3	13.0	SB(rs)dm IV-V	30	Very bright condensation on E side of center.
12 36 12.9	+26 56 25	NGC 4563	14.7	0.5 x 0.3	12.5		97	
12 36 21.5	+25 59 07	NGC 4565	9.6	15.8 x 2.1	13.2	SA(s)b? sp I	136	Very strong and complex dark lane runs entire length of galaxy; modest central bulge.
12 38 13.4	+28 56 09	NGC 4585	14.1	0.8 x 0.5	13.0	S?	114	Mag 10.27v star E 5′.7.
12 41 29.0	+26 05 17	NGC 4613	15.2	0.5 x 0.5	13.6	Sa		Very faint star or knot NW edge.
12 41 31.6	+26 02 31	NGC 4614	13.3	1.1 x 0.9	13.1	SB0/a	175	Very short E-W bar.
12 41 37.6	+26 04 19	NGC 4615	13.1	1.6 x 0.7	13.1	Scd:	125	Two main arms form an elongated open "S."
12 42 39.2	+19 56 42	NGC 4635	12.6	2.0 x 1.4	13.6	SAB(s)d III	170	Many small knots overall.
12 45 16.9	+27 07 28	NGC 4670	12.7	1.4 x 1.1	13.0	SB(s)0/a pec:	90	Mag 9.13v star E 4′.4.
12 45 34.6	+27 03 37	NGC 4673	12.9	1.0 x 0.9	12.7	E1-2	170	Mag 9.13v star N 4′.2.
12 47 11.6	+19 27 52	NGC 4685	12.6	2.0 x 1.2	13.4	S0⁻:	158	
12 47 55.4	+27 13 20	NGC 4692	12.6	1.3 x 1.3	13.1	E⁺:		Just N of a triangle of mag 14 stars.
12 49 01.7	+27 10 42	NGC 4702	15.7	0.7 x 0.4	14.1	Irr	0	Lies 1′.4 SE of mag 10.9 star.
12 49 34.2	+25 28 11	NGC 4712	12.8	2.5 x 1.1	13.7	SA(s)bc III	160	Small, elongated nucleus with bright knot or star on N edge; NGC 4725 E 11′.9.
12 49 57.9	+27 49 19	NGC 4715	13.1	1.6 x 1.1	13.6	SA0⁺	20	Bright nucleus.
12 50 20.0	+27 19 23	NGC 4721	14.5	0.8 x 0.3	12.8	S0?	114	**MCG +5-30-102** ENE 10′.0; mag 9.90v star E 12′.4.
12 50 26.7	+25 30 12	NGC 4725	9.4	10.7 x 7.6	14.0	SAB(r)ab pec I-II	35	Bright nucleus; tightly wound spiral arms; many superimposed stars around bright core.
12 50 28.1	+27 26 05	NGC 4728	13.5	1.0 x 0.7	13.2	E1	108	Elongated **UGC 7992** ESE 2′.2.
12 51 02.0	+28 55 38	NGC 4735	14.6	0.6 x 0.5	13.1	S?	105	Mag 9.08v star SE 5′.8; very faint, elongated anonymous galaxy NW 5′.4.
12 51 08.8	+28 47 12	NGC 4738	13.4	2.1 x 0.3	12.8	Scd?	34	Mag 9.08v star NNE 4′.8.
12 51 26.2	+27 25 17	NGC 4745	15.1	0.8 x 0.8	14.5	S0?		Elongated N-S **MCG +5-30-105** NW 1′.6.
12 51 45.8	+25 46 31	NGC 4747	12.4	3.5 x 1.2	13.8	SBcd? sp pec	30	Short, bright, narrow bar with knot at S end.
12 54 05.6	+27 04 06	NGC 4787	14.4	1.2 x 0.3	13.2	S0/a	2	Located 3′.0 W of NGC 4789.
12 54 16.0	+27 18 14	NGC 4788	14.4	0.8 x 0.4	13.0	S?	141	Mag 10.5 star 1′.3 W.
12 54 19.2	+27 03 56	NGC 4789	12.1	1.9 x 1.5	13.1	SA0:	171	Almost stellar nucleus; mag 10.41v star on N edge.
12 54 05.5	+27 08 49	NGC 4789A	13.6	1.8 x 1.3	14.4	IB(s)m IV-V	35	Uniform surface brightness.
12 54 41.1	+28 56 16	NGC 4793	11.6	2.8 x 1.5	13.0	SAB(rs)c III	50	Very faint anonymous galaxy 1′.3 SE; mag 10.11v star N 1′.8.
12 54 55.3	+27 24 42	NGC 4798	13.2	1.2 x 0.9	13.1	S0⁻:	30	Stellar **PGC 43965** S 4′.7.
12 55 34.0	+27 56 30	NGC 4805		0.3 x 0.2			108	Stellar.
12 55 29.2	+27 31 14	NGC 4807	13.5	1.0 x 0.8	13.1	SAB0⁻ pec:	21	NGC 4807A N 1′.5.
12 55 30.6	+27 32 38	NGC 4807A	15.2	0.5 x 0.2	12.6		123	NGC 4807 S 1′.5; **MCG +5-31-8** N 6′.7.
12 56 12.2	+27 44 40	NGC 4816	12.8	1.3 x 1.1	13.1	S0⁻:	84	**CGCG 160-23** E 1′.7.
12 56 27.8	+26 59 14	NGC 4819	13.2	1.2 x 0.8	13.0	(R′)SAB(r)a:	160	NGC 4821 S 1′.9; **MCG +5-31-19** NNE 8′.0.
12 56 29.3	+26 57 24	NGC 4821	14.5	0.5 x 0.3	12.4	E4:	9	Almost stellar; NGC 4819 N 1′.9.
12 56 34.4	+27 32 18	NGC 4824	15.3	0.5 x 0.4	13.6	SA0	141	Small, faint anonymous galaxy 1′.9 E.
12 56 43.9	+21 40 54	NGC 4826	8.5	10.0 x 5.4	12.7	(R)SA(rs)ab II-III	115	= **M 64, Blackeye Galaxy**. Extremely bright nucleus party hidden by very small, dark lane.
12 56 43.6	+27 10 42	NGC 4827	12.9	1.4 x 1.1	13.2	S0⁻	48	**MCG +5-31-19** S 5′.2.
12 56 43.1	+28 01 11	NGC 4828	14.2	0.7 x 0.7	13.3	S?		Almost stellar anonymous galaxy 2′.0 NNE.
12 57 24.4	+27 29 50	NGC 4839	12.1	4.0 x 1.9	14.3	E⁺	65	Bright nucleus with very small knot or star SE. CGCG 160-40 N 2′.9; NGC 4842A and NGC 4842B E 2′.6; MCG +5-31-23 SW 3′.7.
12 57 33.0	+27 36 36	NGC 4840	13.7	0.7 x 0.7	12.9	E1:		Compact; stellar **PGC 44382** E 4′.9.
12 57 32.1	+28 28 36	NGC 4841A	12.8	1.6 x 1.0	13.4	E⁺ pec	124	Double system with NGC 4841B, common envelope.
12 57 34.1	+28 28 54	NGC 4841B	12.6	1.0 x 0.7	12.3	E pec	134	Very elongated anonymous galaxy WNW 2′.1.
12 57 36.0	+27 29 34	NGC 4842A	14.0	0.4 x 0.4	12.0	E0		NGC 4839 W 2′.5; NGC 4842B on S edge.
12 57 36.2	+27 29 01	NGC 4842B	15.1	0.3 x 0.2	12.0	E3:	42	NGC 4839 NE 3′.4; NGC 4842A on N edge.
12 58 05.7	+28 14 33	NGC 4848	13.7	1.6 x 0.5	13.3	SBab: sp	158	Mag 6.94v star NE 7′.7; stellar **CGCG 160-49** SW 5′.2.
12 58 12.8	+26 23 43	NGC 4849	12.9	1.7 x 1.3	13.6	S0°:	175	**IC 383** N 1′.9.
12 58 21.8	+27 58 03	NGC 4850	14.2	0.7 x 0.5	12.9	SA0	63	
12 58 21.7	+28 08 49	NGC 4851	15.0	0.5 x 0.3	12.8	SBa	112	Double system with **IC 839** which sits on it's N edge; Galaxy **AGC 221210** NW 2′.7; **CGCG 160-57** SW 1′.3.
12 58 35.4	+27 35 46	NGC 4853	13.6	0.8 x 0.7	12.8	(R′)SA0⁻?	81	Compact; mag 8.54v star S 7′.3.
12 58 47.6	+27 40 28	NGC 4854	13.9	1.1 x 0.8	13.7	SB0	57	Stellar **PGC 44469** W 3′.6.
12 59 02.2	+28 06 55	NGC 4858	15.2	0.5 x 0.4	13.3	SBb	36	See notes for NGC 4860.
12 59 02.0	+26 48 55	NGC 4859	13.6	1.4 x 0.8	13.6	S0/a	95	
12 59 04.1	+28 07 23	NGC 4860	13.5	1.0 x 0.8	13.3	E2:	126	NGC 4858 on S edge; faint anonymous galaxy W 2′.3.
12 59 13.2	+27 58 32	NGC 4864	13.6	0.7 x 0.5	12.4	E2	129	NGC 4867 SE 0′.6; **IC 3955** NW 2′.0.
12 59 20.1	+28 05 01	NGC 4865	13.7	0.9 x 0.5	12.8	E6	123	Located 3′.0 NW of mag 7.17v star; **MCG +5-31-63** SW 1′.5; **PGC 44574** S 0′.6; and **PGC 44609** E 1′.9.
12 59 15.4	+27 58 12	NGC 4867	14.5	0.5 x 0.4	12.8	E3	36	Several faint galaxies E 1′.8 to 2′.0.

(Continued from previous page)

GALAXIES

RA h m s	Dec ° ′ ″	Name	Mag (V)	Dim ′ Maj x min	SB	Type Class	PA	Notes
12 59 23.5	+27 54 39	NGC 4869	13.8	0.8 x 0.7	13.1	E3	69	Mag 14.4 star on NW edge; **PGC 44585** S 0′9; **PGC 44581** S 1′7.
12 59 30.2	+27 57 22	NGC 4871	14.1	0.7 x 0.4	12.6	SA0	177	On W edge of NGC 4874.
12 59 34.2	+27 56 47	NGC 4872	14.4	0.6 x 0.4	12.8	SB0	123	On S edge of NGC 4874.
12 59 32.9	+27 58 59	NGC 4873	14.1	0.7 x 0.5	12.9	SA0	105	A trio of PGC galaxies lie ENE 1′7.
12 59 36.1	+27 57 34	NGC 4874	11.7	1.9 x 1.9	13.1	E$^+$0		**PGC 44651** and **PGC 44644** 1′0 E; NGC 4872 SSW 0′9; NGC 4871 W 1′3; **IC 3998** E 2′6.
12 59 38.0	+27 54 22	NGC 4875	14.7	0.4 x 0.3	12.2	SA0	123	**IC 3973** SW 2′0.
12 59 44.6	+27 54 44	NGC 4876	14.4	0.5 x 0.4	12.6	E5	18	Very faint star NE of nucleus. **IC 3973** SE 2′2; **PGC 44585** S 0′9; **PGC 44581** SSW 1′6.
12 59 57.9	+28 14 45	NGC 4881	13.6	1.0 x 1.0	13.6	E$^+$		Very faint **PGC 44691** E 1′0; mag 8.15v star W 5′5.
12 59 56.2	+28 02 02	NGC 4883	14.4	0.6 x 0.5	12.9	SB0	97	
13 00 04.6	+27 59 13	NGC 4886	13.9	0.6 x 0.6	12.8	E0		Located 0′9 SSW of the bright core of NGC 4874.
13 00 08.0	+27 58 31	NGC 4889	11.5	2.9 x 1.9	13.3	E$^+$4	80	NGC 4886 on NW edge; **PGC 44708** on W edge of core; **IC 4011** N 1′8.
13 00 03.8	+26 53 53	NGC 4892	14.0	1.3 x 0.3	12.9	S	13	
13 00 16.6	+27 58 00	NGC 4894	15.2	0.5 x 0.2	12.6	SA0	27	**CGCG 160-233** NW 0′9; **PGC 44666** ENE 1′5.
13 00 18.0	+28 12 04	NGC 4895	13.2	1.8 x 0.6	13.2	SA0 pec sp	153	NGC 4895A SW 2′7.
13 00 09.2	+28 10 10	NGC 4895A	15.0	0.7 x 0.4	13.7	E	99	Faint anonymous spiral with stellar nucleus S 0′7.
13 00 31.0	+28 20 48	NGC 4896	13.9	1.0 x 0.6	13.2	S0$^-$ pec:	5	**CGCG 160-89** E 3′0.
13 00 17.9	+27 57 19	NGC 4898	13.5	0.6 x 0.4	12.0	E pec	90	Double system. **PGC 44741** S 1′1; **PGC 44763** W 2′2; **PGC 44771** NE 2′5.
13 00 39.9	+27 55 25	NGC 4906	14.1	0.5 x 0.5	12.7	E3		**PGC 44821** E 1′7; **PGC 44809** and **IC 4042** N 2′7.
13 00 48.8	+28 09 28	NGC 4907	13.6	1.1 x 1.0	13.5	SB(r)b	42	NGC 4895A SW 2′7.
13 00 51.5	+28 02 30	NGC 4908	13.6	0.8 x 0.6	12.9	E5	49	IC 4051 S 2′2; small galaxy or knot N edge.
13 00 56.0	+27 47 22	NGC 4911	12.8	1.2 x 1.1	12.9	SAB(r)bc	127	MCG +5-31-94 on SW edge.
13 01 17.7	+27 48 28	NGC 4919	14.1	1.1 x 0.7	13.6	(R′)SA(r)0°:	140	**MCG +5-31-96** WNW 1′9; **PGC 44879** W 1′6.
13 01 26.1	+27 53 08	NGC 4921	12.2	2.5 x 2.2	13.9	SB(rs)ab	165	Bright nucleus; NGC 4923 SE 2′6.
13 01 24.8	+29 18 36	NGC 4922	13.0	1.3 x 1.0	13.2	I0 pec	129	Double system in common halo.
13 01 32.0	+27 50 46	NGC 4923	13.7	0.8 x 0.8	13.1	(R′)SA(r)0$^-$?		NGC 4921 NW 2′6.
13 01 54.0	+27 37 27	NGC 4926	13.0	1.2 x 1.1	13.2	SA0$^-$:	57	Faint, elongated anonymous galaxy SW 1′7; spherical anonymous galaxy NE 2′3, not to be confused with NGC 4926A 1′6 E.
13 02 08.2	+27 38 54	NGC 4926A	14.2	0.6 x 0.5	12.8	S0 pec?	90	Spherical anonymous galaxy W 1′6.
13 01 57.7	+28 00 15	NGC 4927	13.7	0.8 x 0.6	12.8	SA0$^-$:	15	Faint star on N edge.
13 02 44.5	+28 02 41	NGC 4929	13.4	0.8 x 0.8	13.0	E1:		Compact core; **IC 4106** N 4′4.
13 03 00.8	+28 01 53	NGC 4931	13.5	1.7 x 0.7	13.5	S0 sp	78	**IC 4111** NNW 2′5.
13 03 16.3	+28 01 45	NGC 4934	14.4	1.1 x 0.3	13.0	S?	104	Faint, elongated anonymous galaxy ESE 2′1.
13 03 45.2	+28 04 59	NGC 4943	14.4	0.7 x 0.5	13.2	SAB0	111	Small, faint, elongated anonymous galaxy SE 2′1; mag 10.78v star N 2′5.
13 03 49.9	+28 11 06	NGC 4944	12.9	1.7 x 0.6	12.7	S0/a? sp	89	Double system with anonymous companion at W end; another, much smaller galaxy on S edge.
13 04 18.1	+29 01 41	NGC 4949	14.9	0.8 x 0.4	13.5	SB(s)b pec sp	108	Anonymous galaxy SW 1′9.
13 04 58.5	+29 07 17	NGC 4952	12.4	1.8 x 1.1	13.1	E	23	Elongated galaxy **KUG 1303+292** SE 10′1.
13 05 12.6	+27 34 10	NGC 4957	13.0	1.2 x 1.0	13.2	E3	100	Bright, diffuse nucleus.
13 05 53.3	+27 33 06	NGC 4960	15.6	0.6 x 0.4	13.9		153	Almost stellar.
13 05 47.9	+27 44 01	NGC 4961	13.6	1.6 x 1.1	14.0	SB(s)cd III-IV	100	Bright nucleus with several surrounding knots.
13 06 17.3	+29 03 44	NGC 4966	13.3	1.0 x 0.5	12.4	S I	143	Located 2′5 NE of mag 6.52v star; small companion galaxy or star at NW edge.
13 06 55.2	+28 32 51	NGC 4971	13.5	1.0 x 1.0	13.4	S0?		Very close pair of mag 10.90v, 10.94v stars N 3′1.
13 07 50.5	+18 24 52	NGC 4978	13.1	1.5 x 0.8	13.2	S0/a	142	
13 07 42.9	+24 48 37	NGC 4979	13.9	1.0 x 0.7	13.4	SB?	100	Bright nucleus in short E-W bar; mag 6.82v star E 4′6.
13 08 27.7	+28 19 11	NGC 4983	14.0	1.1 x 0.7	13.5	S?	123	MCG +5-31-141 ESE 7′0.
13 09 47.3	+28 54 17	NGC 5000	13.2	1.7 x 1.4	14.0	SB(rs)bc	0	Bright, long E-W bar; bright condensation at tip of arm S; anonymous galaxy on E edge.
13 11 01.6	+29 38 10	NGC 5004	12.9	1.4 x 1.1	13.2	S0	170	Bright center with uniform outer halo.
13 11 01.9	+29 34 41	NGC 5004A	13.7	1.4 x 0.8	13.7	SBab	172	Bright nucleus in N-S bar; star on SE edge.
13 11 37.0	+22 54 53	NGC 5012	12.2	2.9 x 1.7	13.8	SAB(rs)c II	10	
13 12 06.8	+24 05 39	NGC 5016	12.8	1.7 x 1.3	13.5	SAB(rs)c II-III	50	
13 13 27.1	+27 48 08	NGC 5032	12.8	2.1 x 1.1	13.6	SB(r)b	22	**MCG +5-31-159** S 2′4.
13 15 34.9	+29 40 33	NGC 5052	13.2	1.5 x 1.0	13.5	S0/a	160	Three faint stars SE 1′5 along with small, faint anonymous galaxy.
13 19 08.1	+28 30 22	NGC 5081	13.0	2.2 x 0.8	13.5	SBb	103	Elongated core E-W; several faint stars on W end.
13 19 51.6	+23 00 00	NGC 5092	13.3	1.0 x 1.0	13.3	E		
13 22 55.5	+26 58 49	NGC 5116	12.7	2.0 x 0.7	12.9	SB(s)c: II	40	Mag 9.23v star NW 14′6.
13 22 56.6	+28 19 00	NGC 5117	13.2	2.2 x 1.0	13.9	SBcd?	154	Short bar with moderately dark lanes either side.
13 27 47.0	+17 46 45	NGC 5158	12.8	1.3 x 1.2	13.1	SBab?	126	Mag 8.98v star NE 9′7.
13 29 19.0	+17 03 08	NGC 5172	11.9	3.3 x 1.7	13.6	SAB(rs)bc: I-II	103	**CGCG 101-59** ESE 6′8.
13 30 38.6	+18 08 05	NGC 5190	13.2	1.0 x 0.8	12.8	SBb:	159	
13 34 05.9	+17 51 24	NGC 5217	12.6	1.5 x 1.4	13.4	E	36	**CGCG 102-20** E 3′3; **CGCG 102-17** NW 3′8.
13 37 24.8	+27 25 08	NGC 5251	13.9	0.7 x 0.7	12.9	S?		
13 39 55.6	+28 23 58	NGC 5263	13.4	1.6 x 0.4	12.7	S?	26	Mag 9.64v star S 3′1.
13 42 23.3	+29 50 48	NGC 5274	14.6	0.4 x 0.4	12.7	E		Pair of small anonymous galaxies close N and NW; **UGC 8682** E 3′0.
13 42 23.6	+29 49 29	NGC 5275	14.2	0.7 x 0.7	13.2	S?		Very small, elongated N-S anonymous galaxy on W edge.
13 42 38.4	+29 57 14	NGC 5277	14.5	0.7 x 0.6	13.4	S?	138	
13 42 55.6	+29 52 06	NGC 5280	13.6	0.8 x 0.8	13.1	E?		**MCG +5-32-73** SW 0′9.
13 44 59.7	+29 48 51	NGC 5287	14.8	0.5 x 0.2	12.1		21	Very small anonymous galaxy on E edge.
13 56 56.0	+29 09 53	NGC 5375	11.5	3.2 x 2.8	13.7	SB(r)ab	0	Dark patches E and W of bright center.
12 40 56.3	+29 27 56	UGC 7836	14.1	1.5 x 0.5	13.7	Scd:	42	Stellar **CGCG 159-53** NNE 5′1.
12 43 05.3	+27 42 49	UGC 7890	14.0	0.7 x 0.4	12.5	I?	24	
12 47 11.6	+26 42 42	UGC 7955	14.8	1.5 x 0.2	13.3	Scd:	30	
12 47 42.2	+26 58 50	UGC 7959	13.8	1.2 x 0.3	12.5	S0 sp	63	
12 52 36.3	+26 44 55	UGC 8013	14.3	1.6 x 0.5	13.9	S?	91	Mag 7.65v star ENE 6′4.
12 54 02.5	+29 36 11	UGC 8025	14.0	1.8 x 0.4	13.5	Sb III	74	**MCG +5-30-117** W 2′9.
12 57 11.6	+29 02 36	UGC 8069	14.2	1.4 x 0.6	13.8	SB?	21	Elongated core offset to W; **CGCG 160-35** WNW 2′4.
12 57 50.0	+29 39 13	UGC 8076	14.5	1.2 x 0.7	14.2	SABd:	89	
12 58 02.6	+26 51 35	UGC 8080	14.0	1.0 x 0.5	13.1	S0/a	56	Mag 12.3 star E edge.
13 00 50.0	+27 24 17	UGC 8122	14.0	1.2 x 0.7	13.7	S0/a	160	Mag 13.3 star E 1′0.
13 08 54.3	+28 10 59	UGC 8229	13.6	1.5 x 1.1	14.0	SB(r)b	60	Mag 12.1 star SE 1′8.
13 09 52.5	+28 22 57	UGC 8244	15.4	1.3 x 0.6	15.0	Sd	75	Mag 11.9 star and anonymous galaxy WNW 5′3.
13 10 27.3	+18 26 15	UGC 8248	14.1	1.0 x 0.4	13.0	S?	101	
13 12 42.8	+22 49 55	UGC 8290	13.6	1.5 x 1.1	14.0	Sm:	48	Large knot or clump SW end.
13 16 42.0	+21 58 48	UGC 8343	14.0	1.2 x 0.7	13.7	S?	30	NE-SW bar.
13 17 45.3	+27 34 11	UGC 8359	13.8	1.5 x 0.6	13.5	Sab	112	
13 18 46.7	+27 43 58	UGC 8363	14.3	1.2 x 0.8	14.1	IBm	119	Very patchy.
13 23 00.8	+23 18 24	UGC 8409	14.4	2.4 x 1.0	15.2	SAdm	80	
13 26 40.0	+19 56 40	UGC 8448	13.8	1.0 x 0.8	13.4	Sbc I-II	45	**MCG +3-34-34** NNE 2′9; CGCG 101-51 SE 11′6.
13 29 39.4	+29 46 13	UGC 8483	14.0	0.9 x 0.9	13.6	S?		
13 30 59.3	+19 26 12	UGC 8507	13.3	1.5 x 0.9	13.5	Im?	12	
13 31 17.6	+29 22 02	UGC 8510	14.0	1.1 x 1.0	14.0	Sbc	111	

(Continued from previous page)

71 GALAXIES 71

RA h m s	Dec ° ′ ″	Name	Mag (V)	Dim ′ Maj x min	SB	Type Class	PA	Notes
13 31 52.5	+20 00 06	UGC 8516	13.3	1.0 x 0.7	12.8	Scd:	30	
13 33 29.2	+17 28 14	UGC 8535	13.6	1.2 x 0.6	13.1	S0/a	97	**CGCG 102-12** NW 9′.7.
13 36 40.8	+20 11 58	UGC 8598	14.1	1.5 x 0.2	12.7	Sbc	169	Mag 12.9 star W 2′.1.
13 39 10.9	+28 57 31	UGC 8636	13.9	0.9 x 0.8	13.4	Sbc	156	
13 39 19.2	+24 46 30	UGC 8638	13.9	1.2 x 0.8	13.7	Im	70	Many knots and/or superimposed stars along S half.
13 45 45.8	+22 05 24	UGC 8701	14.1	1.7 x 0.3	13.2	S?	26	**CGCG 132-1** S 2′.4; mag 10.60v star N edge.
13 46 09.0	+21 53 39	UGC 8703	14.1	1.0 x 0.5	13.2	S?	30	
13 46 31.9	+20 50 46	UGC 8705	14.0	1.0 x 0.5	13.1	Scd:	75	
13 51 05.5	+25 05 39	UGC 8753	14.7	1.1 x 0.7	14.3	Sbc	36	
13 51 01.0	+21 59 29	UGC 8759	13.8	0.9 x 0.7	13.2	S0	108	Mag 15.3 star on SW edge.
13 51 00.8	+24 05 24	UGC 8762	16.1	1.5 x 1.0	16.4	SB(s)d	165	Elongated nucleus.
13 52 14.1	+25 05 46	UGC 8763	13.7	0.7 x 0.4	12.2	Double System	144	
13 52 22.8	+21 32 21	UGC 8781	13.4	1.5 x 0.7	13.3	SBb	160	**CGCG 132-28** N 2′.9; **CGCG 132-29** NNE 7′.3.
13 52 37.4	+24 44 55	UGC 8788	13.6	1.0 x 0.7	13.1	S0/a	35	Mag 12.3 star SE 2′.7.
13 52 57.7	+20 55 00	UGC 8794	14.3	1.7 x 0.6	14.2	Sb II-III	68	Located between a pair of mag 13.4 stars.
13 55 25.0	+17 47 43	UGC 8839	13.0	3.0 x 2.3	14.9	Im V	114	Dwarf irregular, extremely low surface brightness.
13 56 02.9	+18 22 17	UGC 8850	13.7	0.6 x 0.4	12.0	S?		Faint streamer extends SW; mag 11.8 star N 1′.2.
13 57 15.4	+24 15 26	UGC 8873	14.1	1.6 x 0.4	13.5	S?	27	
13 57 39.9	+25 46 28	UGC 8879	14.8	1.5 x 0.2	13.3	Scd:	163	
13 58 03.5	+20 24 02	UGC 8887	14.1	1.1 x 0.5	13.3	Sab	122	
13 58 02.3	+20 37 30	UGC 8888	14.5	1.7 x 0.8	14.7	Quad System	132	Quadruple system; bridges? At center of galaxy cluster A 1825.
13 58 05.6	+21 47 48	UGC 8889	13.8	1.5 x 0.5	13.3	S?	30	Mag 5.76v star SE 9′.9; located just inside N edge of small galaxy cluster **A 1827**.
13 58 51.1	+26 06 22	UGC 8904	15.1	0.7 x 0.6	14.0		78	Mag 11.5 star NW 1′.6.
13 59 39.2	+28 03 36	UGC 8911	14.0	0.9 x 0.7	13.3	Scd:	96	Disturbed; round, almost stellar anonymous galaxy N 2′.0.
14 03 03.9	+28 01 59	UGC 8961	14.0	1.7 x 0.3	13.1	S?	75	Mag 9.99v star SE 2′.6.

GALAXY CLUSTERS

RA h m s	Dec ° ′ ″	Name	Mag 10th brightest	No. Gal	Diam ′	Notes
12 54 42.0	+18 59 00	A 1638	16.0	33	15	Many possible IC galaxies, listed as "unknown" in some references, SW of center.
12 59 48.0	+27 58 00	A 1656	13.5	106	220	**Coma Cluster**. Highest concentration at center; many GX extend beyond cluster boundaries NE and SW.
13 41 54.0	+26 21 00	A 1775	15.7	92	34	**UGC 8669** NW of center 1′.2; most other members PGC and anonymous.
13 44 30.0	+29 50 00	A 1781	15.4	41	34	**MCG +5-32-78** N of center 3′.2; **UGC 8692** N 4′.4; except NGC 5287, most other members anonymous.
13 49 00.0	+26 35 00	A 1795	16.0	115	16	**MCG +5-33-5** near center; most other members faint, stellar and anonymous.
13 49 42.0	+28 04 00	A 1800	15.4	40	16	Elongated **UGC 8738** NW of center 4′.8; **UGC 8748** NE of center 12′.9.
13 58 00.0	+20 39 00	A 1825	15.7	49	13	Except UGC 8888, all members anonymous and stellar/almost stellar.
13 59 12.0	+27 59 00	A 1831	15.4	67	26	**MCG +5-33-33** at center; **MCG +5-33-32** N 2′.4; large concentration of anonymous GX SW of center.

GLOBULAR CLUSTERS

RA h m s	Dec ° ′ ″	Name	Total V m	B ★ V m	HB V m	Diam ′	Conc. Class Low = 12 High = 1	Notes
13 12 55.3	+18 10 09	NGC 5024	7.7	13.8	16.9	13.0	5	= **M 53**.
13 16 27.0	+17 41 53	NGC 5053	9.0	13.8	16.7	10.0	11	
13 42 11.2	+28 22 32	NGC 5272	6.3	12.7	15.6	18.0	6	= **M 3**.

PLANETARY NEBULAE

RA h m s	Dec ° ′ ″	Name	Diam ″	Mag (P)	Mag (V)	Mag cent ★	Alt Name	Notes
12 59 27.9	+27 38 08	PN G49.3+88.1	6	16.0	15.0	19.5	PK 49+88.1	Very faint, small anonymous galaxy SW 3′.2; mag 15.3 star W 3′.1; other faint galaxies in the area.
12 55 33.8	+25 53 30	PN G339.9+88.4	525			14.9	PK 339+88.1	Mag 8.86v star at center; mag 11.3 star S 3′.9.

72 GALAXIES 72

RA h m s	Dec ° ′ ″	Name	Mag (V)	Dim ′ Maj x min	SB	Type Class	PA	Notes
11 33 50.5	+20 01 33	CGCG 97-11	15.4	0.3 x 0.3	12.6	SAb		Mag 13.6 star E 0′.6; mag 13.4 star S 1′.6.
11 38 36.1	+17 49 20	CGCG 97-35	15.3	0.3 x 0.3	12.5	S0		Mag 11.4 star on SE edge.
11 38 51.1	+19 36 03	CGCG 97-36	15.3	0.5 x 0.4	13.4	S0/a	171	**MCG +3-30-30** ESE 8′.8; mag 9.58v star SSE 8′.7.
11 45 44.8	+20 01 51	CGCG 97-138	13.8	0.6 x 0.5	12.3	Irr	141	**CGCG 97-133** and in-contact anonymous companion galaxy W 6′.5.
12 19 46.8	+17 33 53	CGCG 99-16	14.5	0.3 x 0.3	11.8	S		Mag 10.9 star SW 11′.0.
11 34 50.8	+25 31 47	CGCG 126-101	14.1	1.1 x 0.3	12.8		78	Mag 11.4 star W 5′.4.
11 46 46.8	+21 16 13	CGCG 127-55	13.9	0.5 x 0.4	12.1	SAa	3	= **Mkn 127-55**; forms a triangle with a pair of mag 12.5 stars N 2′.6 and NNE 2′.9.
12 05 09.6	+22 00 26	CGCG 128-16	14.2	0.4 x 0.3	11.8	Sm	132	Located 1′.8 S of mag 10.8 star.
11 29 15.5	+20 34 55	IC 700	13.1	0.8 x 0.4	11.7	S0?	59	A chain of four galaxies; IC 700 = **Hickson 54A** with the other three members being **Hicksons 54B**, **54C** and **54D**; Mag 9.97v star SW 4′.9.
11 31 00.7	+20 28 03	IC 701	14.7	0.6 x 0.5	13.2	SB(rs)dm pec	105	Faint, narrow extension NE.
11 33 44.7	+21 22 46	IC 707	13.8	0.5 x 0.4	11.9	S?	18	Mag 9.14v star NNW 13′.1.
11 51 31.2	+23 51 43	IC 739	13.8	1.1 x 0.7	13.3	SBab:	150	Mag 10.92v star E 2′.3.
11 51 02.3	+20 47 57	IC 742	13.9	1.2 x 1.1	14.0	SBab		Mag 10.4 star SW 4′.4; **CGCG 127-67** NW 8′.4.
11 55 35.1	+25 53 21	IC 746	13.9	1.2 x 0.5	13.2	S?	169	Mag 9.33v star SW 11′.1.
12 08 12.0	+25 45 23	IC 762	14.3	0.8 x 0.5	13.2	S?	144	**IC 763** N 3′.5; mag 12.0 star SSE 2′.4.
12 19 23.6	+28 18 31	IC 777	13.5	1.1 x 0.7	13.0	SB?	149	Mag 10.62v star N 4′.4; **CGCG 158-74** NE 8′.0.
12 19 38.7	+29 52 57	IC 779	13.8	0.9 x 0.9	13.4	S?		Mag 10.7 star S 3′.8.
12 19 58.5	+25 46 17	IC 780	12.9	1.1 x 0.8	12.6	S0⁻:	7	Mag 9.00v star E 10′.6.
12 26 59.5	+22 38 22	IC 791	13.1	1.1 x 1.1	13.2	(R)SB(r)a		Mag 9.84v star W 8′.0.
11 18 05.3	+17 38 54	IC 2703	14.4	0.6 x 0.6	13.1			Mag 10.33v star WSW 12′.6.
11 43 24.6	+19 44 56	IC 2951	13.6	1.4 x 0.7	13.4	Sa	80	**UGC 6683** W 1′.9.
11 45 17.5	+26 46 01	IC 2956	13.8	1.3 x 0.8	13.6	SAB(s)bc	68	Mag 9.50v star ENE 11′.3; stellar **CGCG 157-28** NW 4′.7.
12 15 55.0	+23 35 41	IC 3075	14.5	0.9 x 0.6	13.7	S..	60	Mag 10.42v star NW 7′.3.
12 18 21.4	+25 13 01	IC 3122	13.5	1.4 x 0.8	13.5	Sb II-III	150	Mag 9.91v star SW 9′.0.
12 19 05.6	+27 17 50	IC 3143	15.7	0.8 x 0.5	14.6	Spiral	150	Mag 7.47v star W 12′.0.
12 20 04.7	+27 58 29	IC 3165	13.3	1.9 x 1.1	14.0	SBb:	5	Mag 10.26v star S 4′.1; **IC 3168** SE 4′.4.
12 20 24.2	+25 33 35	IC 3171	13.7	0.8 x 0.7	13.1	E?	60	Mag 11.8 star NE 4′.8.

(Continued from previous page)

GALAXIES

RA h m s	Dec ° ′ ″	Name	Mag (V)	Dim ′ Maj x min	SB	Type Class	PA	Notes
12 20 55.9	+24 40 03	IC 3186	15.0	0.6 x 0.4	13.3	Sb	171	Mag 11.36v star WSW 11'.6.
12 21 45.8	+25 52 59	IC 3203	14.7	1.5 x 0.2	13.3	Sb	145	Mag 4.80v star 12 Comae Berenices E 10'.3.
12 22 10.9	+26 03 03	IC 3215	14.4	1.8 x 0.5	14.1	Sdm	92	Mag 11.3 star WNW 9'.3.
12 23 11.2	+27 45 53	IC 3243	15.1	0.7 x 0.4	13.6	S	54	Mag 11.01v star SE 8'.8; **IC 3230** W 7'.1.
12 23 50.5	+28 11 53	IC 3263	14.1	0.7 x 0.6	13.0	S?	111	Mag 10.77v star NNE 10'.4; stellar IC 3263 N 5'.5.
12 25 18.1	+26 42 50	IC 3308	14.8	1.3 x 0.2	13.2	Sdm:	65	Mag 11.4 star SW 1'.6; mag 8.33v star NE 8'.4.
12 25 20.3	+28 22 52	IC 3309	13.7	1.3 x 1.1	13.9	S?	85	Very filamentary.
12 27 50.3	+26 59 38	IC 3376	12.9	1.7 x 1.3	13.6	(R)SBa	60	Mag 9.18v star W 8'.5; note: **IC 3367** is 4'.2 S of this star.
12 28 27.3	+18 24 51	IC 3391	13.4	1.1 x 0.9	13.2	Scd:	63	Mag 8.56v star SW 7'.2.
12 29 03.8	+27 46 41	IC 3407	13.9	1.1 x 0.7	13.4	SB?	145	Two main arms; strong dark lane W of center.
12 29 38.6	+26 13 48	IC 3421	15.6	0.7 x 0.5	14.3	Sb	33	Mag 8.07v star SSE 7'.9.
12 30 29.9	+19 40 23	IC 3436	13.7	0.7 x 0.7	12.8	S0		Bright SE-NW bar.
12 31 04.6	+28 51 08	IC 3441	14.7	0.6 x 0.4	13.0		141	Stellar anonymous galaxy WNW 1'.8; IC 3451 E 4'.3.
12 31 24.4	+28 51 20	IC 3451	14.6	0.5 x 0.4	12.7			IC 3441 W 4'.3.
12 34 17.5	+27 27 07	IC 3516	15.7	1.4 x 0.2	14.1	Sbc	72	Mag 9.07v star E 8'.1.
12 34 49.3	+17 48 51	IC 3530	14.0	1.2 x 0.9	13.9	S0	162	Mag 11.6 star SW 9'.0.
12 35 12.6	+26 32 00	IC 3536	15.8	0.9 x 0.2	13.8	Spiral	156	Located 6'.4 W of NGC 4555.
12 36 03.6	+26 59 13	IC 3559	14.5	0.6 x 0.2	12.0	S?	75	Almost stellar.
12 36 38.0	+24 25 38	IC 3581	14.6	0.9 x 0.4	13.3	S?	51	Mag 11.0 star NW 1'.3.
12 36 30.5	+26 11 58	IC 3582	14.6	0.6 x 0.3	12.6	Sab?	39	Mag 10.7 star NW 10'.7; note: **IC 3543** is 1'.5 NW and **IC 3546** is 3'.0 S of this star.
12 36 40.2	+26 49 46	IC 3585	13.4	1.1 x 0.9	13.2	SA(s)0°:	126	Bright nucleus; mag 13.1 star S 0'.8.
12 36 48.7	+27 32 50	IC 3587	15.1	1.4 x 0.2	13.5	Scd:	122	Mag 10.69v star S 4'.8.
12 37 21.2	+28 12 27	IC 3598	13.8	1.5 x 0.4	13.1	SA(r)ab:	140	Mag 9.89v star E 2'.5.
12 40 36.4	+26 30 21	IC 3644	14.6	0.9 x 0.2	12.6	S?	15	**IC 3646** N 1'.3.
12 40 52.9	+26 43 41	IC 3651	13.2	1.0 x 1.0	13.1	S0		Mag 9.74v star E 7'.5; **MCG +5-30-57** E 4'.8.
12 42 54.0	+20 59 21	IC 3692	13.8	1.0 x 0.7	13.3	SBa:	97	Mag 10.45v star W 3'.0.
11 16 35.3	+18 07 02	MCG +3-29-18	14.3	0.9 x 0.3	12.9	E/S0	171	NGC 3608 ENE 6'.0.
11 17 38.3	+17 49 05	MCG +3-29-24	14.5	1.3 x 0.8	14.4	SB0/p	165	Almost stellar nucleus; very faint, small anonymous galaxy SSE 5'.1.
11 17 58.1	+17 26 30	MCG +3-29-25	14.4	0.5 x 0.3	12.2		159	
11 18 06.2	+18 47 52	MCG +3-29-27	15.5	0.5 x 0.4	13.6	Sc	30	Mag 11.15v star N 1'.8.
11 20 50.7	+19 22 26	MCG +3-29-34	13.9	0.9 x 0.8	13.4	Sc	111	Close pair of mag 12.6 stars NE 2'.8; **MCG +3-29-33** SW 2'.8.
11 28 27.6	+19 11 03	MCG +3-29-55	14.4	0.8 x 0.6	13.5	Sbc	171	
11 29 13.3	+19 46 20	MCG +3-29-58	13.9	0.8 x 0.4	12.5	S?	33	Mag 13.1 star NE 1'.8.
11 31 02.8	+18 23 26	MCG +3-29-60	15.7	0.5 x 0.3	13.5		168	Mag 5.55v star W 8'.1.
11 31 03.9	+20 14 04	MCG +3-29-61	14.8	1.0 x 0.7	14.2	SB	129	Large NE-SW oriented core; mag 12.4 star NW 3'.3.
11 36 28.4	+19 48 39	MCG +3-30-14	14.2	0.5 x 0.5	12.7	E/SA0		
11 36 54.4	+19 59 49	MCG +3-30-21	14.1	0.7 x 0.4	12.6	S?	33	Almost stellar **MCG +3-30-18** NW 0'.8; UGC 6583 S 1'.6.
11 37 36.0	+20 09 47	MCG +3-30-26	14.6	0.7 x 0.4	13.1	S?	90	
11 40 16.2	+17 27 26	MCG +3-30-38	13.7	0.9 x 0.7	13.1	S?	144	
11 41 43.3	+17 00 26	MCG +3-30-44A	15.2	0.9 x 0.3	13.7	Sc	165	Mag 13.4 star on NE edge.
11 42 12.9	+18 24 14	MCG +3-30-49	15.2	0.8 x 0.8	14.6	S0		
11 42 24.6	+20 07 09	MCG +3-30-51	13.8	1.1 x 0.6	13.2	S?	105	**MCG +3-30-48** SW 2'.6.
11 42 45.1	+20 01 52	MCG +3-30-55	14.1	0.9 x 0.4	12.9	S?	120	Stellar **CGCG 97-74** NE 4'.7.
11 43 59.8	+19 46 47	MCG +3-30-71	14.6	0.8 x 0.4	13.3	S?	14	
11 44 30.6	+20 04 37	MCG +3-30-83	14.5	0.9 x 0.2	12.5	S?	159	**CGCG 97-111**, with small companion on S edge, NNW 2'.2.
11 45 07.1	+19 57 57	MCG +3-30-94	15.2	0.6 x 0.2	12.7	Sc	117	Located on SE edge of NGC 3861.
11 45 14.9	+19 50 43	MCG +3-30-98	13.9	0.7 x 0.5	12.9	E	99	Mag 7.53v star N 3'.1.
11 46 03.2	+19 26 13	MCG +3-30-108	15.1	0.4 x 0.4	13.0	S0		Stellar.
11 47 15.3	+19 10 35	MCG +3-30-113	14.9	0.5 x 0.5	13.3	SAa:		
11 48 03.6	+20 00 21	MCG +3-30-115	13.9	0.8 x 0.5	12.7	S0?	30	Mag 7.38v star W 8'.4.
12 01 44.4	+17 54 01	MCG +3-31-15	14.0	0.8 x 0.6	13.1	Sb	138	**CGCG 98-26** E 4'.9.
12 03 31.1	+17 08 56	MCG +3-31-28	14.4	1.0 x 0.5	13.5		66	Mag 11.1 star W 4'.9.
12 03 23.6	+20 01 15	MCG +3-31-29	14.1	0.5 x 0.4	12.2	S?	21	Mag 10.44v star E 0'.8.
12 07 09.7	+16 59 40	MCG +3-31-41	13.6	0.8 x 0.6	12.7	SBb?	162	**MCG +3-31-42** SSE 1'.7.
12 09 32.9	+17 00 48	MCG +3-31-49	13.6	0.9 x 0.7	13.0	S?	33	
12 11 15.2	+17 53 13	MCG +3-31-61	13.7	0.5 x 0.5	12.1	S?		**MCG +3-31-56** and stellar **Mkn 758** W 7'.9.
12 17 27.3	+17 38 59	MCG +3-31-94	13.8	0.8 x 0.5	12.8	E/S0	135	Mag 11.1 star S 2'.7.
12 20 14.5	+17 20 45	MCG +3-32-3	13.9	0.5 x 0.4	12.0	Sc(r)II	174	Mag 8.52v star SW 4'.6.
11 18 59.1	+22 53 11	MCG +4-27-17	14.4	0.7 x 0.4	12.8	SBc	138	Compact; UGC 6333 E 4'.3.
11 22 02.2	+25 55 16	MCG +4-27-25	15.3	0.5 x 0.5	13.6			
11 25 33.4	+22 49 05	MCG +4-27-34	13.9	0.9 x 0.7	13.3	SB(s)b I-II	84	Mag 9.60v star NE 3'.5.
11 26 18.7	+21 05 43	MCG +4-27-36	14.5	0.8 x 0.4	13.1	SB?	123	= **Hickson 52A**. **Hickson 52B** N 0'.8; **Hickson 52C** S 2'.4; **Hickson 52D** SW 0'.7.
11 28 34.5	+21 00 35	MCG +4-27-39	14.6	0.8 x 0.6	13.6		159	Mag 13.5 star W 2'.0.
11 28 58.8	+22 05 28	MCG +4-27-46	15.0	0.4 x 0.3	12.6	Sb	123	
11 31 15.9	+23 02 20	MCG +4-27-53	15.3	0.7 x 0.3	13.4	SBb	117	Mag 11.7 star E edge.
11 32 01.3	+25 38 40	MCG +4-27-57	14.6	0.5 x 0.5	13.0			
11 36 15.4	+22 25 57	MCG +4-27-72	14.6	0.8 x 0.3	12.9	Sbc	171	
11 52 20.0	+21 06 07	MCG +4-27-78		0.5 x 0.5				Close pair of mag 13 stars S 1'.3.
11 39 44.9	+22 41 08	MCG +4-28-16	14.1	0.7 x 0.7	13.2	S?		Mag 11.26v star E 2'.9.
11 41 56.2	+20 19 10	MCG +4-28-29	14.7	0.6 x 0.4	13.0	SBbc	39	Bright star on E edge; NGC 3821 E 3'.1.
11 42 08.5	+25 58 24	MCG +4-28-31	15.4	0.7 x 0.2	13.1	Sb	24	MCG +4-28-32 NNE 3'.7; mag 8.39v star SW 4'.7.
11 42 16.8	+26 01 39	MCG +4-28-32	15.0	0.6 x 0.3	12.9	Sbc	60	
11 42 29.1	+21 35 11	MCG +4-28-33	15.1	0.7 x 0.3	13.3	Sbc	51	
11 43 08.4	+24 00 14	MCG +4-28-36	15.4	0.9 x 0.5	14.4	Sbc	18	**UGC 6681** S 3'.9; mag 7.40v star E 8'.8.
11 44 34.6	+25 25 16	MCG +4-28-45	14.2	0.9 x 0.7	13.5	Sbc	63	Mag 13.6 star NE 3'.4.
11 45 23.9	+20 19 29	MCG +4-28-47	14.5	1.1 x 0.7	14.1	S0/a	75	Faint, slightly elongated anonymous galaxy SSE 2'.1.
11 45 27.7	+20 48 22	MCG +4-28-48	14.0	0.7 x 0.5	12.9	E	84	Mag 10.68v star WNW 2'.7.
11 50 44.4	+21 06 31	MCG +4-28-62	15.3	0.8 x 0.6	14.4		0	Galaxy **Mkn 1461** NE 3'.2.
11 50 56.2	+21 58 24	MCG +4-28-65	15.0	0.8 x 0.5	13.9	Spiral	144	
11 51 14.7	+21 00 05	MCG +4-28-70	14.6	0.6 x 0.3	12.8	E	51	= **Mkn 1463**; mag 11.9 star W 2'.1.
11 51 59.9	+21 06 28	MCG +4-28-77	14.0	0.8 x 0.6	13.0	S?	129	
11 53 17.0	+23 27 50	MCG +4-28-87	14.2	0.7 x 0.4	12.6	S?	90	Mag 11.1 star W 1'.9.
11 53 57.6	+25 41 06	MCG +4-28-92	13.8	0.7 x 0.6	12.9	E	48	Mag 10.46v star NE 1'.6.
11 59 36.4	+21 14 54	MCG +4-28-115	13.7	0.5 x 0.5	12.2	E		Mag 10.39v star W 1'.0.
11 59 46.3	+21 17 38	MCG +4-28-116	15.6	0.5 x 0.4	13.7	Spiral		
11 59 49.5	+21 26 51	MCG +4-28-117	15.9	0.8 x 0.5	14.8	Spiral	171	Mag 10.8 star E 3'.3.
12 01 38.7	+22 39 34	MCG +4-28-121	14.4	0.9 x 0.6	13.6		30	**UGC 7010** SSE 8'.3; **MCG +4-28-120** SW 6'.6.
12 02 26.6	+23 49 07	MCG +4-28-123	14.0	0.6 x 0.6	13.0	E		
12 02 54.0	+21 38 37	MCG +4-29-2	14.4	0.5 x 0.5	12.7	Spiral		

(Continued from previous page)

RA h m s	Dec ° ′ ″	Name	Mag (V)	Dim ′ Maj x min	SB	Type Class	PA	Notes
12 03 27.3	+22 12 34	MCG +4-29-4	14.1	0.8 x 0.6	13.2	I?	45	
12 04 56.2	+21 14 22	MCG +4-29-12	14.3	0.8 x 0.8	13.7	Sb		Almost stellar; NGC 4084 E 4′.7.
12 04 55.9	+21 25 29	MCG +4-29-13	14.3	0.9 x 0.9	13.9			
12 08 05.2	+25 35 23	MCG +4-29-32	13.8	0.5 x 0.5	12.1	S?		
12 08 55.5	+25 11 33	MCG +4-29-37	14.6	0.4 x 0.4	12.5	S?		
12 09 17.9	+24 58 00	MCG +4-29-38	14.4	0.9 x 0.6	13.6	S?	161	**UGC 7124** W 7′.3.
12 10 34.7	+25 55 38	MCG +4-29-44	14.0	1.1 x 0.6	13.4	SB?	141	Mag 9.24v star W 6′.2.
12 10 34.4	+25 25 43	MCG +4-29-45	14.3	0.4 x 0.2	11.6	E/S0	129	
12 14 26.6	+24 10 58	MCG +4-29-48	14.4	0.6 x 0.6	13.2	S?		
12 14 41.5	+23 51 09	MCG +4-29-50	15.1	0.8 x 0.6	14.2	Sc	129	Mag 12.3 star NW 2′.3.
12 24 15.0	+20 17 14	MCG +4-29-69	14.5	0.7 x 0.4	13.0		33	Multi-galaxy system? Mag 10.87v star E 1′.8.
12 38 07.7	+22 41 54	MCG +4-30-10	16.3	0.8 x 0.8	15.7	Sm		Mag 10.53v star S 3′.6; mag 6.38v star E 12′.7.
11 18 17.9	+28 13 29	MCG +5-27-52	14.0	0.8 x 0.5	12.9	SB?	114	UGC 6322 N 2′.1; elongated anonymous galaxy SSE 1′.3.
11 24 48.5	+28 45 06	MCG +5-27-68	15.4	0.7 x 0.5	14.1	Spiral	9	Mag 9.08v star NW 4′.4.
11 26 12.7	+27 11 52	MCG +5-27-70	16.0	0.5 x 0.4	14.1		48	
11 28 03.2	+27 23 06	MCG +5-27-74	14.6	0.8 x 0.6	13.7	SBb	153	
11 29 23.8	+28 32 34	MCG +5-27-79	14.2	0.8 x 0.5	13.1	S0	165	
11 36 30.8	+26 51 39	MCG +5-27-89	14.9	0.7 x 0.7	14.0	Scd		Mag 10.3 star NW edge.
11 38 53.3	+26 18 34	MCG +5-28-3	16.6	0.7 x 0.3	14.7	Sd	111	Mag 13.5 star E 1′.0; mag 11.9 star S 2′.6.
11 40 39.2	+28 51 51	MCG +5-28-11	14.8	0.8 x 0.5	13.6	Sc	54	
11 40 47.2	+27 53 58	MCG +5-28-13	16.0	0.8 x 0.6	15.0	S..	60	
11 41 26.0	+27 51 35	MCG +5-28-15	14.3	0.8 x 0.5	13.2	S0/a	120	
11 43 00.6	+27 23 50	MCG +5-28-20	14.5	0.9 x 0.6	13.7	S0	138	
11 48 46.1	+29 38 27	MCG +5-28-32	14.9	0.8 x 0.4	13.6		12	Mag 11.9 star N 0′.7.
11 51 35.0	+27 38 20	MCG +5-28-40	14.6	0.8 x 0.6	13.6	S0/a	36	Pair with MCG +5-28-41 NW edge.
11 51 32.1	+27 38 46	MCG +5-28-41	15.2	0.7 x 0.2	12.9	S0	120	Pair with MCG +5-28-40 SE edge.
11 55 02.7	+27 17 54	MCG +5-28-44	14.3	0.8 x 0.4	13.0	S?	90	
11 58 55.6	+28 23 59	MCG +5-28-67	14.4	0.8 x 0.6	13.7	E	12	
11 59 40.2	+26 32 44	MCG +5-28-71	14.1	0.8 x 0.7	13.3	S?	165	
12 04 04.8	+28 58 55	MCG +5-29-3	15.4	0.5 x 0.4	13.8	E?	144	
12 06 31.0	+28 08 12	MCG +5-29-15	14.4	0.5 x 0.5	13.0	E		NGC 4104 2′.8 NE.
12 07 22.8	+27 51 03	MCG +5-29-17	15.3	0.8 x 0.4	13.9	S+	102	Mag 10.07v star SW 6′.3.
12 09 09.7	+29 16 23	MCG +5-29-24	15.8	0.9 x 0.2	13.8	Spiral	9	NGC 4132 SW 2′.3.
12 12 24.5	+28 48 57	MCG +5-29-35	14.2	0.9 x 0.4	12.9	S?	27	Mag 13.9 star E 0′.5; close pair of mag 12.4 stars NW 1′.6.
12 15 54.7	+26 39 43	MCG +5-29-45	13.9	0.8 x 0.5	12.8	SB?	99	
11 16 46.5	+18 01 00	NGC 3605	12.3	1.5 x 1.0	12.7	E4-5	12	Located on SW edge of NGC 3607.
11 16 54.6	+18 03 07	NGC 3607	9.9	4.7 x 3.9	12.9	SA(s)0°:	120	Elongated MCG +3-29-18 NW 6′.0; NGC 3605 on SW edge.
11 16 59.1	+18 08 54	NGC 3608	10.8	3.2 x 2.6	13.1	E2	75	NGC 3607 SE 6′.2; MCG +3-29-18 WSW 6′.0.
11 17 50.8	+26 37 32	NGC 3609	13.1	1.2 x 1.0	13.2	Sab	50	**MCG +5-27-44** N 1′.9.
11 18 14.7	+26 37 13	NGC 3612	14.1	1.0 x 0.8	13.7	Sdm:	160	Weak stellar nucleus; mag 10.95v star NW 3′.9.
11 18 06.8	+23 23 48	NGC 3615	12.8	1.4 x 0.9	13.1	E	40	Stellar **CGCG 126-22** NE 3′.2.
11 18 32.5	+23 28 07	NGC 3618	13.6	0.9 x 0.8	13.1	SABb: III	175	Weak stellar nucleus; **MCG +4-27-16** SSE 9′.3.
11 20 03.5	+18 21 23	NGC 3626	11.0	2.7 x 1.9	12.6	(R)SA(rs)0^+	163	Elongated **UGC 6341** S 5′.9.
11 20 31.9	+26 57 50	NGC 3629	12.1	2.3 x 1.6	13.3	SA(s)cd: III	30	
11 21 35.7	+18 27 31	NGC 3639	13.7	0.7 x 0.6	12.6	S?	39	
11 21 43.1	+20 10 04	NGC 3646	11.1	3.9 x 2.2	13.3	Ring I-II	50	Weak almost stellar nucleus; strong dark area SW of nucleus; NGC 3649 E 7′.8.
11 22 14.8	+20 12 30	NGC 3649	13.7	1.4 x 0.6	13.3	SB(s)a	140	Pair of very faint stars S edge; NGC 3646 W 7′.8.
11 22 35.3	+20 42 13	NGC 3650	13.9	1.7 x 0.3	13.0	SAb sp III	54	Located 2′.7 SE of a mag 11.17v star.
11 22 26.4	+24 17 56	NGC 3651	13.2	1.1 x 1.1	13.4	E		= **Hickson 51A**. **Hickson 51B** = **IC 2759** W 2′.8. **Hickson 51D** E 1′.0. **Hickson 51E** 3′.2 WNW. **Hickson 51F** just S of nucleus. **Hickson 51G** on SE edge.
11 22 30.2	+24 16 45	NGC 3653	13.7	0.6 x 0.4	12.0	S0?	79	= **Hickson 51C**.
11 23 45.1	+17 49 02	NGC 3659	12.2	2.1 x 1.1	12.9	SB(s)m? III-IV	60	
11 24 49.6	+23 56 42	NGC 3670	13.5	1.3 x 0.7	13.2	SB0/a	35	Stellar **CGCG 126-47** NNW 3′.7.
11 26 15.8	+27 52 00	NGC 3678	13.6	0.8 x 0.8	13.0	Sbc		
11 27 11.3	+17 01 45	NGC 3684	11.4	3.1 x 2.1	13.3	SA(rs)bc II-III	130	Mag 10.61v star E 7′.8; located on S edge of small galaxy cluster **A 1264**.
11 27 44.1	+17 13 20	NGC 3686	11.3	3.2 x 2.5	13.4	SB(s)bc II	15	Many small knots overall.
11 28 00.6	+29 30 40	NGC 3687	12.0	1.9 x 1.9	13.3	(R′)SAB(r)bc? I-II		Bright knots or stars E and S of core; mag 9.75v star SW 6′.6.
11 28 10.8	+25 39 37	NGC 3689	12.3	1.7 x 1.1	12.9	SAB(rs)c II	97	
11 28 50.4	+20 47 41	NGC 3697	13.1	2.3 x 0.7	13.4	SABb	93	**Integral Sign Galaxy**, = **Hickson 53A**. **Hickson 53b** SSE 3′.3; **Hickson 53C** SSE 4′.1; **Hickson 53D** E 4′.1, located N of a mag 11.8 star.
11 29 29.1	+24 05 35	NGC 3701	13.0	1.9 x 0.9	13.4	Sbc	145	**MCG +4-27-50** NE 11′.0.
11 31 07.0	+22 46 02	NGC 3710	13.1	1.0 x 0.8	12.8	E	105	Moderately compact core; mag 7.83v star NE 4′.4.
11 31 09.3	+28 34 00	NGC 3712	14.0	2.4 x 1.2	15.0	SB?	160	Very slender nucleus; mag 6.76v star S 7′.7.
11 31 42.2	+28 09 11	NGC 3713	13.2	1.2 x 0.8	13.0	S0^-:	125	**CGCG 156-92** and a mag 11.4 star NW 5′.4.
11 31 53.9	+28 21 27	NGC 3714	14.2	1.3 x 0.8	14.1	I?	68	
11 33 15.8	+24 26 48	NGC 3728	13.0	2.0 x 1.5	14.0	Sb II-III	25	
11 35 37.6	+25 05 20	NGC 3739	14.4	1.1 x 0.3	13.1	Sbc	17	
11 35 57.5	+21 43 18	NGC 3743	15.2	0.5 x 0.5	13.7	E/S0		Stellar; mag 9.95v star SE 1′.4.
11 35 57.9	+23 00 41	NGC 3744	15.0	0.6 x 0.3	13.0		9	
11 37 44.5	+22 01 13	NGC 3745	15.2	0.3 x 0.2	12.0	SB(s)0^-:	102	Member of **Copeland's Septet**.
11 37 43.8	+22 00 33	NGC 3746	14.2	1.0 x 0.5	13.3	SB(r)b	126	Member of **Copeland's Septet**.
11 37 49.2	+22 01 30	NGC 3748	14.8	0.5 x 0.3	12.6	SB0°? sp	132	Member of **Copeland's Septet**.
11 37 51.8	+21 58 25	NGC 3750	13.9	0.5 x 0.4	12.1	SAB0^-?	132	= **Hickson 57C**. Member of **Copeland's Septet**.
11 37 54.0	+21 56 07	NGC 3751	13.9	0.5 x 0.2	11.3	S0^- pec?	9	= **Hickson 57F**. Member of **Copeland's Septet**.
11 37 53.9	+21 58 49	NGC 3753	13.6	1.7 x 0.5	13.3	Sab? sp pec	117	= **Hickson 57A**. Member of **Copeland's Septet**.
11 37 55.0	+21 59 07	NGC 3754	14.3	0.5 x 0.4	12.4	SBb? pec	18	= **Hickson 57D**. Member of **Copeland's Septet**.
11 36 29.2	+21 35 47	NGC 3758	14.3	0.5 x 0.5	12.6	S?		Located 2′.6 W of mag 9.70v star.
11 36 44.2	+22 59 32	NGC 3761	13.9	0.9 x 0.8	13.6	E	63	Stellar nucleus.
11 36 54.2	+17 53 20	NGC 3764	14.7	0.9 x 0.7	14.0	S?	123	Almost stellar nucleus; faint extension NW; NGC 3768 SE 5′.5.
11 37 04.3	+24 05 45	NGC 3765	14.1	0.8 x 0.6	13.2	S?	57	
11 37 14.5	+17 50 20	NGC 3768	12.4	1.6 x 0.9	12.7	S0	155	NGC 3764 NW 5′.5.
11 37 48.5	+22 41 32	NGC 3772	13.5	1.1 x 0.6	12.9	SBa	16	
11 39 03.7	+26 21 40	NGC 3781	13.8	0.6 x 0.5	12.3	S?	30	Compact.
11 39 29.7	+26 18 35	NGC 3784	14.4	0.9 x 0.3	12.8	S?	135	Faint, elongated anonymous galaxy W 2′.1.
11 39 32.8	+26 18 07	NGC 3785	14.2	1.0 x 0.4	13.0	S0	25	Mag 10.4 star W 1′.5.
11 39 38.0	+20 27 15	NGC 3787	13.7	0.6 x 0.4	12.1	S0	21	CGCG WSW 5′.4.
11 39 47.2	+17 42 44	NGC 3790	13.9	1.2 x 0.3	12.6	S0/a	154	**MCG +3-30-35** E 4′.5.
11 40 14.0	+24 41 47	NGC 3798	12.1	2.3 x 1.2	13.1	SB0	60	Mag 9.85v star S 4′.7.
11 40 17.0	+17 43 40	NGC 3801	12.0	2.5 x 1.6	13.4	S0?	120	**MCG +3-30-35** SW 3′.3; NGC 3802 N 2′.2.

(Continued from previous page)

GALAXIES

RA h m s	Dec ° ′ ″	Name	Mag (V)	Dim ′ Maj x min	SB	Type Class	PA	Notes
11 40 19.0	+17 45 54	NGC 3802	13.3	1.1 x 0.3	12.0	S	85	NGC 3803 N 2′.1.
11 40 17.4	+17 48 04	NGC 3803		0.4 x 0.4				Stellar.
11 40 41.8	+20 20 31	NGC 3805	12.7	1.4 x 1.1	13.0	$S0^-$:	60	
11 40 46.7	+17 47 45	NGC 3806	13.6	2.2 x 1.8	14.9	SABb	169	Almost stellar nucleus; mag 9.21v star S 5′.0.
11 40 44.2	+22 25 46	NGC 3808	13.4	1.7 x 0.9	13.8	SAB(rs)c: pec	0	Double system, interacting with NGC 3808A, bridge; NGC 3808A on N end.
11 40 44.7	+22 26 46	NGC 3808A	14.6	0.7 x 0.3	12.7	I0? pec	57	NGC 3808 on S edge.
11 41 07.9	+24 49 20	NGC 3812	12.4	1.7 x 1.6	13.4	E	63	Located 1′.7 NW of mag 8.41v star.
11 41 27.7	+24 48 16	NGC 3814	14.7	0.9 x 0.3	13.1	S0	174	Located between NGC 3815 E 2′.5 and a mag 8.41 star W 3′.1.
11 41 39.4	+24 48 01	NGC 3815	13.0	1.7 x 0.9	13.3	Sab	72	NGC 3814 W 2′.5.
11 41 48.0	+20 06 11	NGC 3816	12.5	1.9 x 1.1	13.2	S0	70	**CGCG 97-63** ESE 7′.3; **CGCG 97-62** SE 9′.9.
11 42 09.2	+20 18 56	NGC 3821	12.9	1.4 x 1.3	13.4	(R)SAB(s)ab	165	Very faint outer ring.
11 42 32.9	+26 29 20	NGC 3826	13.4	0.9 x 0.7	12.9	E	65	**UGC 6677** NE 5′.3.
11 42 36.3	+18 50 43	NGC 3827	13.3	0.9 x 0.7	12.6	S?	65	Mag 9.12v star S 4′.4.
11 43 11.9	+26 33 32	NGC 3830	15.2	0.9 x 0.4	14.0	S0	15	Stellar NGC 3830A N 0′.9.
11 43 10.9	+26 34 25	NGC 3830A	14.6	0.4 x 0.3	12.1		171	Stellar; NGC 3830 S 0′.9.
11 43 31.5	+22 43 33	NGC 3832	13.0	1.9 x 1.6	14.0	SB(rs)bc	120	Short N-S bar; mag 11.4 star NW 2′.2.
11 43 37.8	+19 05 26	NGC 3834	13.5	1.4 x 1.0	13.7	S?	129	Compact core, faint envelope; mag 9.23v star NE 8′.1.
11 43 56.4	+19 53 40	NGC 3837	13.3	0.8 x 0.8	12.8	E		**MCG +3-30-76**, **MCG +3-30-79** and **MCG +3-30-78** Clustered together SE 6′.7.
11 43 59.0	+20 04 36	NGC 3840	13.8	1.1 x 0.8	13.6	Sa	67	Stellar, faint anonymous galaxy NNE 1′.7.
11 44 02.3	+19 58 15	NGC 3841	13.6	0.7 x 0.7	12.7	S?		Almost stellar.
11 44 02.3	+19 56 57	NGC 3842	11.8	1.4 x 1.0	12.2	E	5	**CGCG 97-90** SW 1′.4; anonymous galaxy W 1′.1; UGC 6697 W 3′.4.
11 44 00.9	+20 01 44	NGC 3844	13.9	1.2 x 0.3	12.6	S0/a	28	Small, faint, elongated anonymous galaxy W 3′.4.
11 44 05.7	+19 59 44	NGC 3845	14.0	0.8 x 0.4	12.6	S?	135	NGC 3844 NNW 2′.2.
11 44 20.4	+19 58 50	NGC 3851	14.7	0.6 x 0.4	13.2	E	71	Located 2′.2 E of mag 11.38v star.
11 44 50.3	+19 31 55	NGC 3857	14.1	0.9 x 0.4	12.8	S0?	42	Located 3′.2 NW of mag 8.50v star.
11 44 52.4	+19 27 13	NGC 3859	14.1	1.2 x 0.3	12.9	S?	58	Located 3′.5 SW of mag 8.50v star; note: **CGCG 97-123** lies 1′.4 W of this star.
11 44 49.4	+19 47 41	NGC 3860	13.4	1.1 x 0.5	12.6	S?	38	**CGCG 97-114** S 1′.2; stellar **MCG +3-30-92** SE 1′.8; **CGCG 97-113** S 2′.3.
11 45 04.0	+19 58 25	NGC 3861	12.7	2.3 x 1.3	13.7	(R′)SAB(r)b III	77	Elongated MCG +3-30-94 on SE edge.
11 45 05.1	+19 36 20	NGC 3862	12.7	1.5 x 1.5	13.5	E		Galaxy **IC 2955** located on N edge.
11 45 15.7	+19 23 28	NGC 3864	14.2	0.7 x 0.5	12.9	S?	66	NGC 3867 E 3′.4.
11 45 29.6	+19 23 58	NGC 3867	13.2	1.5 x 0.6	12.9	S?	173	NGC 3864 W 3′.4; stellar anonymous galaxy NE 1′.7.
11 45 30.1	+19 26 38	NGC 3868	14.4	0.8 x 0.3	12.7	S?	80	Mag 12.1 star N 2′.1; NGC 3867 S 2′.8.
11 45 46.2	+19 46 21	NGC 3873	12.9	1.0 x 1.0	12.9	E		Stellar anonymous galaxy W 1′.2; NGC 3875 on SE edge.
11 45 49.6	+19 45 59	NGC 3875	13.9	1.2 x 0.3	12.6	S0/a	87	Located on SE edge of NGC 3873; **CGCG 97-143** ENE 4′.2.
11 46 47.2	+20 40 33	NGC 3883	12.7	3.0 x 2.4	14.6	SA(rs)b II	159	Bright stellar nucleus in uniform envelope; mag 9.41v star S 5′.4.
11 46 12.2	+20 23 28	NGC 3884	12.6	2.1 x 1.3	13.5	SA(r)0/a	10	**IC 732** and an anonymous galaxy NW 4′.0; mag 7.68v star S 4′.4.
11 47 05.7	+19 50 12	NGC 3886	13.1	1.2 x 0.9	13.1	$S0^-$:	132	**MCG +3-30-114** NE 10′.0.
11 49 09.4	+27 01 22	NGC 3900	11.4	3.2 x 1.7	13.1	$SA(r)0^+$	2	
11 49 18.8	+26 07 18	NGC 3902	12.8	1.6 x 1.3	13.5	SAB(s)bc:	85	
11 49 59.3	+21 20 02	NGC 3910	12.8	1.6 x 1.2	13.4	$S0^-$:	150	Faint star on NNW edge.
11 49 22.2	+24 56 18	NGC 3911	13.9	1.1 x 0.8	13.6	SB?	110	Star on W edge.
11 50 04.4	+26 28 48	NGC 3912	12.4	1.5 x 0.9	12.6	SAB(s)b? pec III	5	
11 50 41.6	+20 00 53	NGC 3919	13.3	0.9 x 0.9	13.1	E		**CGCG 97-162** NE 2′.9.
11 50 06.0	+24 55 13	NGC 3920	13.2	1.0 x 1.0	13.1	S?		
11 51 21.1	+21 53 19	NGC 3925	14.5	0.7 x 0.5	13.2	SAB0/a	3	Mag 10.5 star SW 4′.5; **MCG +4-28-67** SSW 7′.1.
11 51 26.5	+22 01 36	NGC 3926A	14.3	0.5 x 0.4	12.6	E/S0	117	Double system with NGC 3926B, in common halo.
11 51 28.3	+22 01 30	NGC 3926B	14.2	0.9 x 0.9	14.0	E/S0		**MCG +4-28-75** NE 2′.6.
11 51 42.7	+21 00 07	NGC 3929	14.0	0.6 x 0.4	12.4	E?	80	Mag 13.3 star on S edge.
11 52 42.7	+20 37 53	NGC 3937	12.5	1.8 x 1.6	13.5	$S0^-$:	24	**CGCG 127-85** W 2′.9.
11 52 46.4	+20 59 22	NGC 3940	12.8	1.7 x 1.6	13.9	E	99	Mag 10.12v star E 5′.4; NGC 3946 E 8′.3.
11 52 56.5	+20 28 44	NGC 3943	13.4	1.1 x 1.1	13.4	S?		Mag 7.27v star SW 7′.3.
11 53 05.1	+26 12 24	NGC 3944	12.9	1.4 x 1.1	13.2	$S0^-$:	25	
11 53 20.6	+21 01 18	NGC 3946	14.3	0.6 x 0.5	12.9	S0/a	90	Mag 10.12v star W 3′.0.
11 53 20.5	+20 45 06	NGC 3947	13.2	1.4 x 1.2	13.6	(R)SB(rs)b II	90	Stellar **CGCG 127-92** SSW 6′.1.
11 53 41.2	+23 22 56	NGC 3951	13.6	1.1 x 0.5	12.8	S?	172	Stellar galaxy **KUG 1151+236** SE 5′.8.
11 53 41.7	+20 52 54	NGC 3954	13.7	0.7 x 0.7	12.9	E?		Elongated **UGC 6873** S 8′.6.
11 54 53.4	+28 15 44	NGC 3964	14.0	1.0 x 0.8	13.6	S?	76	Compact core, smooth envelope; mag 12.0 star on N edge; **MCG +5-28-46** NE 8′.0.
11 55 36.5	+29 59 44	NGC 3971	12.7	1.4 x 1.2	13.1	S0	30	**UGC 6905** SE 3′.9.
11 56 23.8	+23 52 02	NGC 3983	14.0	1.3 x 0.4	13.1	S0/a	114	
11 57 51.8	+29 02 22	NGC 3984	13.5	1.2 x 1.1	13.7	SB(rs)b	153	Pair of faint stars on SW edge.
11 57 20.8	+25 11 42	NGC 3987	12.9	2.2 x 0.4	12.6	Sb III	58	Mag 10.72v star and NGC 3989 N 2′.3.
11 57 24.2	+27 52 39	NGC 3988	13.3	1.0 x 0.8	13.1	E	42	Almost stellar nucleus in smooth envelope.
11 57 26.8	+25 13 59	NGC 3989	14.8	0.6 x 0.3	12.8	S?	123	Located 1′.3 E of mag 10.72v star.
11 57 37.9	+25 14 25	NGC 3993	13.7	1.8 x 0.5	13.4	Sb: sp	141	Mag 12.0 star on NW end.
11 57 48.5	+25 16 14	NGC 3997	13.4	1.7 x 0.6	13.3	SBb pec	141	Mag 11.7 star on E edge; mag 12.0 star on W edge.
11 57 56.6	+25 04 05	NGC 3999	14.8	0.4 x 0.3	12.3	S0	81	Almost stellar; mag 11.2 star W 1′.5.
11 57 57.0	+25 08 41	NGC 4000	14.6	1.0 x 0.2	12.7	S?	3	Mag 8.27v star E 1′.7.
11 57 59.5	+23 12 07	NGC 4002	14.1	1.1 x 0.6	13.5	S0?	111	Mag 9.37v star E 4′.0.
11 57 59.1	+23 07 31	NGC 4003	13.3	1.5 x 0.9	13.5	SB0	151	Mag 9.37v star NE 5′.5.
11 58 05.3	+27 52 19	NGC 4004	13.7	2.0 x 0.7	13.9	Pec	2	Small bright core N end with slender arm extending S; mag 11.4 star and **IC 2982** W 3′.9.
11 58 10.4	+25 07 20	NGC 4005	13.1	1.2 x 0.7	12.8	S?	92	Mag 8.27v star WNW 1′.8.
11 58 16.9	+28 11 31	NGC 4008	12.0	2.5 x 1.3	13.3	E5	167	UGC 6968 NE 8′.5.
11 58 28.2	+25 12 50	NGC 4009	15.5	0.4 x 0.2	12.6		65	Almost stellar.
11 58 25.5	+25 05 51	NGC 4011	14.9	0.5 x 0.2	12.3	S0	39	Located 1′.7 SE of mag 8.27v star.
11 58 42.6	+25 02 11	NGC 4015	12.8	2.0 x 1.5	13.9	Double System	131	Elongated **MCG +4-28-110** on N edge of core.
11 58 29.0	+27 31 46	NGC 4016	13.2	1.5 x 0.8	13.2	SBdm:	175	Faint stars or knots N and SE of core.
11 58 45.7	+27 27 10	NGC 4017	12.2	1.8 x 1.4	13.1	SABbc	72	Arms form an open "S" shape.
11 58 40.6	+25 19 01	NGC 4018	13.8	1.7 x 0.3	12.9	Sab	163	
11 59 02.7	+25 04 59	NGC 4021	14.8	0.4 x 0.4	12.9	E		Compact.
11 59 00.9	+25 13 20	NGC 4022	13.0	1.2 x 1.2	13.3	SAB0°:		NGC 4009 W 7′.4.
11 59 05.4	+24 59 16	NGC 4023	13.7	0.9 x 0.7	13.0	S?	25	Elongated **MCG +4-28-110** Extends from NGC 4015's nucleus NE; **PGC 37703** just off E edge.
12 00 33.0	+20 04 25	NGC 4032	12.3	1.6 x 1.2	12.8	Im: III	171	
12 02 05.4	+17 49 24	NGC 4040	13.3	1.9 x 1.3	14.3	E	145	**CGCG 98-27** S 8′.0.
12 02 50.2	+18 00 53	NGC 4048	13.7	0.6 x 0.5	12.2	S?	92	
12 02 54.7	+18 45 09	NGC 4049	13.4	0.9 x 0.6	12.5	I?	52	
12 03 11.6	+19 43 42	NGC 4053	13.6	1.2 x 0.5	12.9	S0/a	109	
12 03 57.9	+20 18 41	NGC 4056	15.5	0.5 x 0.5	13.8			Stellar nucleus.
12 05 15.2	+20 18 35	NGC 4057	15.4	0.5 x 0.2	12.8		80	Almost stellar; NGC 4090 E 3′.0.

(Continued from previous page)

GALAXIES

RA h m s	Dec ° ′ ″	Name	Mag (V)	Dim ′ Maj x min	SB	Type Class	PA	Notes
12 04 01.0	+20 20 12	NGC 4060	14.7	0.5 x 0.4	12.8	S0	93	Located 2.′1 W of NGC 4066.
12 04 01.7	+20 13 54	NGC 4061	13.1	1.2 x 0.9	13.2	E:	171	NGC 4065 on E edge.
12 04 11.0	+18 26 40	NGC 4064	11.5	4.4 x 1.7	13.5	SB(s)a: pec	150	
12 04 06.2	+20 14 03	NGC 4065	12.6	1.1 x 1.0	12.7	E	129	NGC 4061 on W edge.
12 04 09.6	+20 20 50	NGC 4066	12.9	1.2 x 1.2	13.3	E		Compact core, faint envelope; NGC 4069 SSW 1.′7; NGC 4060 W 2.′1.
12 04 06.1	+20 19 23	NGC 4069		0.4 x 0.3			36	Almost stellar, very faint star S edge.
12 04 11.4	+20 24 37	NGC 4070	13.1	1.0 x 1.0	13.1	E		Stellar anonymous galaxy SW 1.′3.
12 04 13.9	+20 12 32	NGC 4072	15.4	0.6 x 0.2	12.9	S0	30	Very elongated **UGC 7049** SW 1.′8.
12 04 29.7	+20 18 54	NGC 4074	14.5	1.0 x 0.6	13.8	S0: pec	111	
12 04 32.8	+20 12 18	NGC 4076	13.4	0.9 x 0.8	12.9	S?	75	Elongated N-S anonymous galaxy NE 2.′2.
12 04 52.0	+26 59 32	NGC 4080	13.7	1.4 x 0.7	13.5	Im?	122	
12 05 15.4	+21 12 51	NGC 4084	14.5	0.6 x 0.6	13.2	S?		MCG +4-29-12 W 4.′7.
12 05 29.5	+20 14 44	NGC 4086	13.6	0.9 x 0.7	13.0	S0	73	Mag 8.79v star W 6.′1.
12 05 37.4	+20 33 18	NGC 4089	13.7	0.8 x 0.8	13.2	E?		NGC 4091 on E edge.
12 05 28.0	+20 18 33	NGC 4090	13.9	1.2 x 0.5	13.2	Sab	38	NGC 4057 W 3.′0.
12 05 40.3	+20 33 19	NGC 4091	14.5	1.0 x 0.3	13.0	S	43	NGC 4089 on W edge.
12 05 50.3	+20 28 40	NGC 4092	13.3	1.0 x 1.0	13.1	S?		Mag 11.7 star on NW edge.
12 05 51.5	+20 31 15	NGC 4093	14.3	0.7 x 0.5	13.2	E?	45	Almost stellar.
12 05 54.4	+20 34 21	NGC 4095	13.5	1.3 x 1.0	13.9	E?	153	Compact core, faint envelope.
12 06 03.8	+20 36 23	NGC 4098	13.1	1.2 x 0.9	13.1	S?	156	Multi-galaxy system.
12 06 10.8	+25 33 24	NGC 4101	13.5	1.2 x 0.9	13.5	S0/a	60	Mag 9.64v star NNW 7.′3.
12 06 39.1	+28 10 30	NGC 4104	12.1	2.6 x 1.5	13.4	S0	35	MCG +5-29-15 SW 2.′8.
12 07 03.6	+18 31 53	NGC 4110	13.7	1.3 x 0.7	13.4	SBb	128	Mag 12.1 star N 7.′2.
12 08 47.3	+29 18 12	NGC 4131	13.3	1.3 x 0.7	13.0	S	73	Mag 11.36v star S 3.′7.
12 09 01.4	+29 14 59	NGC 4132	14.0	1.1 x 0.4	12.9	S	21	MCG +5-29-24 NE 2.′3.
12 09 10.3	+29 10 36	NGC 4134	12.8	2.2 x 0.9	13.4	Sb	150	
12 09 17.9	+29 55 38	NGC 4136	11.0	4.0 x 3.7	13.8	SAB(r)c II	90	Numerous knots surround bright nucleus.
12 10 18.3	+26 25 49	NGC 4146	12.7	1.4 x 1.3	13.2	(R)SAB(s)ab:	66	
12 10 45.6	+19 02 28	NGC 4155	13.3	1.1 x 1.0	13.4	E	81	**MCG +3-31-63** NE 9.′1.
12 11 10.2	+20 10 31	NGC 4158	12.1	1.9 x 1.7	13.2	SA(r)b:	78	Mag 11.4 star E 1.′6.
12 11 52.5	+24 07 25	NGC 4162	12.2	2.3 x 1.4	13.4	(R)SA(rs)bc II	174	Located 2.′6 NE of mag 10.51v star.
12 12 09.6	+17 45 23	NGC 4166	13.1	1.2 x 1.0	13.1	SB0	20	Stellar galaxy **KUG 1209+179** SW 10.′4.
12 12 18.8	+29 10 45	NGC 4169	12.2	1.8 x 0.9	12.6	S0	153	Member of **The Box**. 2.′0 S of NGC 4173.
12 12 21.0	+29 12 35	NGC 4173	13.0	5.0 x 0.7	14.2	SBd:	134	Uniform surface brightness.
12 12 27.0	+29 08 54	NGC 4174	13.4	0.6 x 0.3	11.4	S?	50	Member of **The Box**. 1.′5 SW of NGC 4175.
12 12 31.2	+29 10 02	NGC 4175	13.3	1.8 x 0.4	12.8	S	130	Member of **The Box**. Located on SE end of NGC4173; NGC 4174 SW 1.′5.
12 13 22.4	+28 30 38	NGC 4185	12.1	2.6 x 1.9	13.7	Sbc	165	Stellar **PGC 39016** ENE 3.′9; mag 7.68v star NNW 8.′5.
12 14 29.7	+28 25 23	NGC 4196	12.8	1.6 x 1.2	13.4	S0?	60	Very faint jet extending NE.
12 15 14.1	+20 39 34	NGC 4204	12.3	3.6 x 2.9	14.7	SB(s)dm	36	Long, bright and asymmetrical nucleus extends across narrow axis of galaxy.
12 15 35.9	+28 10 39	NGC 4211	14.1	1.2 x 0.8	13.9	S0/a pec	105	Connected double system with NGC 4211A, bridge; separation 0.′55.
12 15 37.6	+28 09 48	NGC 4211A	15.3	1.4 x 0.3	14.2	S0/a pec	3	Small, bright nucleus at N end of galaxy, on edge of NGC 4211. Remainder of NGC 4211A extends due S.
12 15 37.6	+23 58 54	NGC 4213	12.5	1.7 x 1.7	13.7	E		Mag 4.93v star 7 Comae Berenices E 10.′1; stellar C 772 W 5.′3; UGC 7266 N 7.′0.
12 17 36.8	+29 36 33	NGC 4245	11.4	2.9 x 2.2	13.3	SB(r)0/a:	153	
12 18 08.9	+28 10 30	NGC 4251	10.7	3.6 x 1.5	12.4	SB0? sp	100	Bright, diffuse nucleus with narrow bar.
12 18 26.6	+29 48 43	NGC 4253	13.1	1.0 x 0.8	12.7	(R′)SB(s)a:	54	
12 19 50.3	+29 36 52	NGC 4274	10.4	6.8 x 2.5	13.3	(R)SB(r)ab II-III	102	Center has almost a "Saturn" like appearance with bright nucleus surrounded by brighter ring; fainter outer envelope.
12 19 52.6	+27 37 13	NGC 4275	13.2	1.1 x 1.0	13.2	S?	80	Very faint, small, elongated anonymous galaxy S 1.′1.
12 20 06.8	+29 16 49	NGC 4278	10.2	3.8 x 3.8	13.1	E1-2		Large, bright center.
12 20 21.0	+29 18 38	NGC 4283	12.1	1.5 x 1.5	12.9	E0		On NE edge of NGC 4278.
12 20 42.3	+29 20 46	NGC 4286	13.1	1.6 x 1.0	13.5	SA(r)0/a:	150	Weak, almost stellar nucleus in smooth envelope.
12 21 13.2	+18 23 03	NGC 4293	10.4	5.6 x 2.6	13.1	(R)SB(s)0/a	72	Bright, elongated bar with hint of dark lane.
12 21 09.6	+28 09 53	NGC 4295	13.7	0.9 x 0.6	12.9	S?	165	Compact.
12 22 26.5	+29 12 29	NGC 4310	12.2	2.2 x 1.2	13.1	(R′)SAB(r)0^+?	161	
12 22 31.8	+29 53 45	NGC 4314	10.6	4.2 x 3.7	13.4	SB(rs)a	69	Bright, moderately large nucleus with bright bar extending NW-SE.
12 23 29.7	+19 25 37	NGC 4336	12.5	2.0 x 0.9	13.0	SB0/a	162	Mag 9.59v star SSE 3.′5.
12 23 14.0	+28 53 34	NGC 4338	14.7	2.2 x 0.3	14.0	Sd	175	**IC 3222** WSW 12.′4.
12 23 37.5	+17 32 29	NGC 4344	12.3	1.7 x 1.6	13.3	SB0?	90	Faint galaxy **VCC 696** E 7.′5.
12 25 00.5	+28 33 31	NGC 4375	12.8	1.4 x 1.2	13.2	SB(r)ab pec:	5	Small, faint, round anonymous galaxy NNW 1.′6.
12 25 24.1	+18 11 18	NGC 4382	9.1	7.1 x 5.5	13.0	SA(s)0^+ pec	3	= **M 85**. A mag 10.44v star is located on the SE edge. The galaxy **MCG +3-32-28** is located on the S edge.
12 25 51.5	+27 33 44	NGC 4393	12.1	3.2 x 2.9	14.4	SABd	12	Bright center with numerous bright knots and/or superimposed stars.
12 25 55.5	+18 12 50	NGC 4394	10.9	3.6 x 3.2	13.4	(R)SB(r)b II	141	Bright nucleus in short, bright bar; faint dust lanes.
12 26 17.4	+27 52 15	NGC 4408	13.9	0.8 x 0.6	13.0	S0	30	Compact.
12 28 15.9	+28 37 11	NGC 4448	11.1	3.9 x 1.4	12.8	SB(r)ab II	94	**UGC 7597** NE 4.′2; **MCG +5-29-88** W 4.′8; an anonymous galaxy lies NNW 4.′5.
12 28 29.8	+17 05 00	NGC 4450	10.1	5.2 x 3.9	13.2	SA(s)ab I-II	175	Smooth tightly wound arms; a mag 9.12v star lies 4.′0 SW.
12 28 44.0	+22 49 13	NGC 4455	12.3	2.8 x 0.8	13.1	SB(s)d? sp IV	16	Many very small, faint knots.
12 29 47.7	+27 14 35	NGC 4475	13.6	1.7 x 1.0	14.0	SAbc I-II	5	Almost stellar nucleus.
12 31 24.3	+25 46 30	NGC 4494	9.8	4.8 x 3.5	12.9	E1-2	171	Bright center; small, elongated anonymous galaxy W 4.′6, possibly **IC 3455**?
12 31 22.9	+29 08 10	NGC 4495	13.2	1.6 x 0.8	13.3	Sab	130	Bright, elongated center.
12 32 43.0	+29 42 42	NGC 4514	13.2	1.2 x 0.9	13.1	Sbc	51	Bright, stellar nucleus.
12 31 42.6	+20 28 55	NGC 4529	14.1	0.8 x 0.6	13.2		147	
12 34 34.6	+18 12 07	NGC 4539	12.0	3.3 x 1.3	13.5	SB(s)a: sp	95	
12 35 41.4	+26 31 24	NGC 4555	12.1	1.9 x 1.4	13.2	E	125	Mag 11.4 star NNE 8.′5; IC 3536 W 6.′4.
12 35 45.9	+26 54 29	NGC 4556	13.1	1.2 x 1.0	13.3	E^+:	80	Mag 14.5 star on S edge; **IC 3561** E 4.′2.
12 35 52.6	+26 59 29	NGC 4558	14.7	0.8 x 0.6	13.7		19	**MCG +5-30-31** SE 2.′0.
12 35 57.7	+27 57 52	NGC 4559	10.0	10.7 x 4.4	14.0	SAB(rs)cd II-III	147	Several filamentary branching arms. **IC 3550**, **IC 3551**, **IC 3552**, **IC 3554**, **IC 3555**, **IC 3563** and **IC 3564** are condensations in this galaxy.
12 36 08.4	+19 19 24	NGC 4561	12.5	1.5 x 1.3	13.0	SB(rs)dm IV-V	30	Very bright condensation on E side of center.
12 35 35.4	+25 51 04	NGC 4562	13.4	2.5 x 0.8	14.0	SB(s)dm: sp	48	
12 36 12.9	+26 56 25	NGC 4563	14.7	0.5 x 0.3	12.5		97	
12 36 21.5	+25 59 07	NGC 4565	9.6	15.8 x 2.1	13.2	SA(s)b? sp I	136	Very strong and complex dark lane runs entire length of galaxy; modest central bulge.
12 38 13.4	+28 56 09	NGC 4585	14.1	0.8 x 0.5	13.0	S?	114	Mag 10.27v star E 5.′7.
12 41 29.0	+26 05 17	NGC 4613	15.2	0.5 x 0.5	13.6	Sa		Very faint star or knot NW edge.
12 41 31.6	+26 02 31	NGC 4614	13.3	1.1 x 0.9	13.1	SB0/a	175	Very short E-W bar.
12 41 37.6	+26 04 19	NGC 4615	13.1	1.6 x 0.7	13.1	Scd:	125	Two main arms form an elongated open "S."
12 42 39.2	+19 56 42	NGC 4635	12.6	2.0 x 1.4	13.6	SAB(s)d III	170	Many small knots overall.

RA h m s	Dec ° ′ ″	Name	Mag (V)	Dim ′ Maj x min	SB	Type Class	PA	Notes
11 16 28.0	+29 19 35	UGC 6292	13.6	1.2 x 0.7	13.3	SB(s)b	66	Faint anonymous galaxy SE 1′.4; mag 10.84v star SE 3′.1; **PGC 34386** S 1′.6; **PGC 34394** S 2′.4.
11 16 51.0	+17 47 50	UGC 6296	13.6	1.4 x 0.4	12.8	S?	166	Mag 11.0 star SSW 5′.1.
11 17 22.4	+22 20 17	UGC 6301	14.5	1.5 x 0.3	13.5	Scd:	139	
11 18 17.9	+18 50 46	UGC 6320	13.1	1.0 x 0.9	12.8	S?	9	Mag 11.15v star SW 2′.8.
11 18 16.8	+28 15 37	UGC 6322	14.4	1.3 x 1.0	14.6	$SA0^{+}$: II	18	MCG +5-27-52 S 2′.1.
11 18 22.2	+18 44 17	UGC 6324	13.4	1.4 x 0.5	12.9	S0	174	Mag 14.6 star W edge; mag 12.0 star N 2′.1.
11 18 28.3	+25 19 26	UGC 6325	14.2	1.5 x 0.5	13.7	Scd?	10	Mag 9.51v star ESE 2′.8.
11 19 16.8	+20 48 49	UGC 6332	13.2	1.5 x 1.2	13.7	(R)SBa	130	Short, bright NE-SW bar; dark patches NW and SE.
11 19 17.2	+22 52 52	UGC 6333	14.4	1.1 x 0.5	13.6	Sdm:	120	Stellar nucleus; MCG +4-27-17 W 4′.3.
11 19 30.3	+28 39 13	UGC 6334	12.6	2.5 x 1.0	13.4	(R)S0/a:	82	Pair of mag 12 and 11.6 stars off E end; bright core offset on S edge.
11 24 05.2	+24 36 52	UGC 6414	15.2	1.7 x 0.2	13.9	Sbc	51	
11 24 10.5	+27 00 49	UGC 6415	13.9	2.0 x 0.5	13.8	Sab	20	
11 24 26.1	+27 27 23	UGC 6421	14.2	1.1 x 0.7	13.8	Sdm:	98	
11 24 45.1	+23 36 50	UGC 6425	13.8	1.1 x 0.8	13.5	SAbc III	153	
11 25 46.5	+18 56 22	UGC 6437	14.1	0.8 x 0.5	13.0	Sbc	35	
11 28 13.0	+21 59 48	UGC 6465	13.7	1.1 x 0.8	13.4	SB?	120	Very patchy, several knots.
11 28 36.2	+23 24 17	UGC 6476	13.6	1.5 x 1.0	13.9	Sbc II-III	111	
11 29 02.5	+17 13 54	UGC 6483	14.4	1.9 x 0.2	13.2	Scd:	68	Mag 9.88v star E 2′.3.
11 29 45.4	+22 07 31	UGC 6495	13.9	0.8 x 0.6	13.0	SAB(s)b: I	90	
11 31 21.4	+26 17 41	UGC 6508	14.2	0.7 x 0.5	12.9	Sb II-III	90	Mag 12.1 star WNW 1′.8.
11 31 22.8	+23 06 52	UGC 6509	14.8	1.7 x 0.2	13.5	Sd	79	
11 32 21.7	+28 02 53	UGC 6522	13.1	1.4 x 1.4	13.7	S0?		Very faint, almost stellar anonymous galaxy NE edge; **MCG +5-27-88** NNE 4′.2.
11 32 41.6	+20 26 18	UGC 6525	13.4	1.3 x 1.2	13.7	(R)SB(r)b:	135	
11 33 42.3	+23 24 44	UGC 6544	14.0	1.1 x 0.5	13.3	Sbc	95	
11 33 55.7	+23 38 37	UGC 6548	14.0	1.0 x 0.3	12.6	S0	36	Very small, very faint anonymous galaxy S edge; mag 10.89v star ESE 1′.9.
11 36 54.4	+19 58 10	UGC 6583	13.6	0.6 x 0.3	11.5	Pec	5	MCG +3-30-21 and **MCG +3-30-18** N 1′.6.
11 37 22.3	+25 44 26	UGC 6593	14.0	0.8 x 0.5	12.9	S?	0	Mag 11.3 star W 0′.6.
11 38 25.3	+20 44 27	UGC 6607	14.5	1.1 x 0.6	13.9	Sd	75	**CGCG 127-10** WNW 6′.8.
11 38 29.6	+20 31 36	UGC 6609	13.3	1.2 x 0.8	13.4	E	100	**MCG +4-28-14** W 1′.5.
11 39 14.9	+17 08 36	UGC 6614	13.5	1.6 x 1.3	14.1	(R)SA(r)a?	171	Bright core, uniform envelope.
11 39 47.7	+19 55 58	UGC 6625	13.6	0.7 x 0.6	12.5	S?	55	
11 40 11.6	+17 18 39	UGC 6631	13.8	0.8 x 0.5	12.7	S?	117	
11 40 25.0	+28 22 23	UGC 6637	14.1	0.8 x 0.4	12.7	S?	72	
11 40 56.5	+25 46 49	UGC 6645	13.5	1.5 x 1.2	14.0	SAB(r)b III	0	Almost stellar nucleus; mag 8.14v star W 9′.0.
11 42 29.4	+18 19 56	UGC 6670	12.9	2.8 x 0.8	13.6	IBm	153	Located between a pair of stars, mags 10.9 and 11.7.
11 42 39.6	+24 49 19	UGC 6674	14.0	1.1 x 0.6	13.4	Scd:	153	
11 43 20.0	+18 11 25	UGC 6687	15.4	1.5 x 0.3	14.4	Sdm:	25	Uniform surface brightness.
11 43 48.9	+19 58 08	UGC 6697	13.6	1.5 x 0.3	12.6	Im:	137	NGC 3842 E 3′.4.
11 44 47.0	+20 07 26	UGC 6719	13.6	1.1 x 0.7	13.2	S?	30	Very faint, small anonymous galaxy NE 5′.0.
11 45 06.0	+20 26 12	UGC 6725	12.9	1.2 x 0.7	12.6	S0	40	
11 45 55.0	+17 11 32	UGC 6741	15.1	1.5 x 0.3	14.1	Double System	172	
11 45 55.7	+21 01 31	UGC 6743	13.7	1.4 x 1.4	14.3	SABbc		Stellar nucleus.
11 46 42.6	+23 57 48	UGC 6751	14.2	1.3 x 0.4	13.3	Sbc	97	Mag 9.65v star N 0′.9.
11 47 08.1	+29 34 37	UGC 6761	14.2	1.1 x 0.5	13.4	S0/a	97	
11 48 57.3	+23 50 14	UGC 6782	14.4	1.8 x 1.8	15.6	Im V		Dwarf irregular, very low surface brightness; mag 11.3 star SW 3′.0.
11 49 23.7	+26 44 23	UGC 6791	14.4	2.2 x 0.4	14.1	Sd	0	
11 50 20.1	+25 57 41	UGC 6806	13.2	1.8 x 0.6	13.1	S pec	45	Low surface brightness dwarf **UGC 6807** NE 2′.0.
11 51 01.2	+20 23 55	UGC 6821	13.8	1.2 x 1.1	14.0	SAB(rs)bc:	66	**UGC 6820** NNW 6′.3.
11 52 36.9	+23 34 56	UGC 6846	13.0	1.0 x 0.8	12.7	E?	24	Mag 12.3 star on W edge.
11 52 38.1	+24 18 26	UGC 6847	14.2	1.5 x 0.3	13.2	Sbc	151	
11 52 47.5	+29 19 41	UGC 6853	13.7	1.2 x 0.6	13.2	S0	35	**UGC 6868** ESE 13′.2.
11 53 11.2	+25 26 12	UGC 6861	14.2	0.9 x 0.6	13.4	S	127	
11 53 59.8	+20 34 20	UGC 6876	13.8	0.9 x 0.7	13.2	Sab	30	Mag 9.12v star S 4′.9.
11 54 45.3	+20 03 17	UGC 6881	15.3	1.5 x 0.7	15.2	Im:	125	Dwarf, low surface brightness; mag 12.3 star SE 1′.3.
11 54 58.6	+26 12 06	UGC 6883	13.9	0.8 x 0.8	13.3	SAB(s)cd:		
11 55 16.8	+17 29 12	UGC 6891	14.6	1.5 x 0.3	13.6	Sab	110	
11 56 15.1	+17 01 46	UGC 6913	14.1	1.3 x 0.6	13.7	Sb	0	Mag 11.5 star SW 2′.7.
11 58 44.6	+28 17 22	UGC 6968	13.2	2.8 x 0.9	14.1	S?	85	Bright almost stellar core with two very narrow arms.
11 59 02.3	+21 48 27	UGC 6976	13.1	1.1 x 0.9	13.0	S0:	45	Mag 9.48v star SE 4′.2.
12 02 03.2	+29 50 55	UGC 7012	13.3	1.9 x 1.0	13.8	Scd:	8	
12 02 22.4	+29 51 40	UGC 7017	13.6	1.7 x 0.4	13.1	Sb III	70	
12 03 21.5	+29 25 09	UGC 7031	14.3	1.5 x 0.3	13.3	Sbc	57	
12 03 53.7	+25 26 00	UGC 7040	13.5	1.7 x 1.0	13.9	SBdm	170	Mag 11.46v star E 2′.5.
12 05 13.0	+28 46 49	UGC 7072	13.3	1.4 x 1.0	13.5	SAdm	27	
12 05 23.3	+17 53 05	UGC 7073	13.4	1.0 x 0.7	12.8	SB(rs)c:	115	UGC 7074 N 2′.1.
12 05 23.4	+17 55 09	UGC 7074	13.8	0.9 x 0.2	11.8	Sa	98	
12 06 44.5	+17 42 50	UGC 7100	13.7	1.7 x 0.3	12.8	S?	8	
12 08 05.7	+25 14 12	UGC 7115	13.3	1.0 x 1.0	13.3	E		Bright core, diffuse envelope; mag 12.6 star NNW 1′.2.
12 09 19.8	+18 59 49	UGC 7133	14.3	1.5 x 0.9	14.5	SABd	170	Mag 12.4 star SE 3′.2.
12 09 27.8	+22 06 14	UGC 7137	14.8	0.8 x 0.5	13.7	Double System	72	Considered double system with bridge to **MCG +4-29-40** S 1′.3; mag 10.29v star NW 3′.5.
12 09 28.9	+26 13 35	UGC 7138	13.7	1.7 x 1.1	14.2	Sd	85	Faint anonymous galaxy WSW 1′.9.
12 09 39.6	+23 17 19	UGC 7141	14.3	1.5 x 0.3	13.3	S?	146	
12 09 46.7	+25 01 33	UGC 7143	13.8	0.9 x 0.4	12.6	S?	107	
12 10 15.2	+25 18 33	UGC 7157	13.6	1.1 x 0.5	12.9	S0	79	Mag 9.40v star NE 2′.6; **CGCG 128-46** W 4′.6.
12 10 37.0	+18 49 42	UGC 7170	14.3	2.8 x 0.3	13.9	Scd:	12	
12 11 22.1	+18 00 59	UGC 7186	14.9	1.5 x 0.3	13.9	Sm: IV-V	83	Close pair of stars, mags 12.6 and 10.63v, S 2′.4.
12 11 32.0	+29 05 19	UGC 7190	13.7	1.1 x 0.9	13.5	S0?	30	Almost stellar nucleus; mag 15.6 star on NW edge; mag 9.51v star W 4′.1.
12 13 00.2	+25 16 55	UGC 7217	14.6	1.5 x 0.2	13.1	S?	1	Mag 14.5 star W edge.
12 13 13.8	+28 50 06	UGC 7221	13.9	1.0 x 0.5	13.0	S?	65	
12 13 18.4	+21 38 02	UGC 7224	13.4	1.3 x 1.0	13.7	E	18	
12 15 25.2	+19 17 30	UGC 7263	13.8	1.0 x 0.8	13.4	SBbc?	130	
12 15 28.9	+24 05 31	UGC 7266	14.0	1.0 x 0.6	13.3	S0	40	NGC 4213 S 7′.0.
12 15 34.5	+21 49 57	UGC 7270	14.0	1.5 x 0.4	13.3	S?	162	Mag 11.03v star SSE 2′.3.
12 15 59.2	+27 26 29	UGC 7286	13.9	1.4 x 0.4	13.2	Sab	102	Strong dark patch w of nucleus.
12 16 09.1	+28 07 43	UGC 7287	14.4	1.0 x 0.6	13.7	SBcd:	81	Mag 8.26v star S 5′.0; stellar **Mkn 765** SE 8′.3.
12 16 43.4	+28 43 48	UGC 7300	14.7	1.4 x 0.9	14.8	Im V	132	Dwarf irregular, very low, uniform surface brightness.
12 17 36.3	+22 32 29	UGC 7321	13.4	5.0 x 0.3	13.7	Sd	82	
12 17 53.7	+17 26 31	UGC 7331	13.6	1.5 x 0.7	13.5	(R′)SB(s)b	0	Strong dark patches E and W of elongated core.

(Continued from previous page)

GALAXIES

RA h m s	Dec ° ′ ″	Name	Mag (V)	Dim ′ Maj x min	SB	Type Class	PA	Notes
12 18 41.9	+17 43 07	UGC 7346	14.6	2.2 x 2.2	16.2	S?		
12 19 13.3	+22 25 49	UGC 7357	13.6	1.5 x 1.4	14.3	SAB(s)c	63	Stellar nucleus; mag 8.31v star N 6'.2.
12 20 49.0	+17 29 12	UGC 7399A	13.6	1.7 x 1.0	14.0	S0	85	**CGCG 99-22** NNE 10'.7.
12 27 41.9	+28 41 52	UGC 7576	15.0	1.3 x 0.5	14.4	S?	51	"Double system." Two very faint, narrow extensions, or arms, NE and SW; mag 12.1 star WNW 1'.5.
12 31 58.1	+29 42 32	UGC 7673	14.8	1.8 x 1.5	15.7	Im V	44	Dwarf irregular, low surface brightness.
12 32 51.7	+20 11 00	UGC 7697	14.3	1.9 x 0.4	13.9	Scd:	99	Stellar **Mkn 771** W 11'.5.
12 35 09.0	+29 44 39	UGC 7750	13.9	1.5 x 0.6	13.6	S?	144	
12 40 56.3	+29 27 56	UGC 7836	14.1	1.5 x 0.5	13.7	Scd:	42	Stellar **CGCG 159-53** NNE 5'.1.
12 43 05.3	+27 42 49	UGC 7890	14.0	0.7 x 0.4	12.5	I?	24	

GALAXY CLUSTERS

RA h m s	Dec ° ′ ″	Name	Mag 10th brightest	No. Gal	Diam ′	Notes
11 16 30.0	+29 15 00	A 1213	14.5	51	22	Main concentration of brighter PGC and anonymous GX near center of cluster area.
11 27 54.0	+26 51 00	A 1267	15.4	37	31	**MCG +5-27-72** SE of center 3'.5; **MCG +5-27-75** SE 7'.7; **CGCG 156-77** N 6'.8.
11 44 30.0	+19 50 00	A 1367	13.5	117	100	A mix of brighter catalog GX and stellar anonymous GX.

OPEN CLUSTERS

RA h m s	Dec ° ′ ″	Name	Mag	Diam ′	No. ★	B ★	Type	Notes
12 25 06.0	+26 07 00	Mel 111	1.8	300	273	5.0	cl	

GLOBULAR CLUSTERS

RA h m s	Dec ° ′ ″	Name	Total V m	B ★ V m	HB V m	Diam ′	Conc. Class Low = 12 High = 1	Notes
12 10 06.2	+18 32 31	NGC 4147	10.4	14.5	16.9	4.4	6	
11 29 16.8	+28 58 25	Pal 4	14.2	18.0	20.8	1.3	12	

GALAXIES

RA h m s	Dec ° ′ ″	Name	Mag (V)	Dim ′ Maj x min	SB	Type Class	PA	Notes
10 25 01.0	+17 17 30	CGCG 94-39	14.5	0.7 x 0.6	13.4	Sb	90	**CGCG 94-43** SE 1'.7; **MCG +3-27-26** SE 3'.3; **MCG +3-27-24** SSE 3'.8.
10 37 25.4	+18 04 02	CGCG 94-98	14.2	0.3 x 0.3	11.4			Located 1'.1 N of mag 10.19v star.
10 42 05.6	+18 14 56	CGCG 94-105	14.3	0.8 x 0.5	13.1		51	Mag 15.3 star on W edge.
10 51 50.3	+18 44 45	CGCG 95-51	14.2	0.5 x 0.5	12.5			Mag 14.2 star superimposed SW of center; mag 13.1 star just off SW edge; mag 8.71v stars ESE 1'.7.
10 55 44.6	+17 00 14	CGCG 95-70	14.3	0.5 x 0.4	12.4		0	Mag 11.7 star WSW 3'.3.
11 04 32.2	+17 07 38	CGCG 95-104	14.2	0.8 x 0.5	13.1		24	Mag 13.3 star E 0'.6; mag 10.2 star NE 4'.4.
09 57 30.5	+23 36 29	CGCG 123-9	15.6	0.7 x 0.4	14.0		15	Mag 14.5 star E 1'.5; mag 10.50v star W 5'.6.
10 15 50.5	+20 39 01	CGCG 123-31	14.8	0.7 x 0.4	13.2		30	Mag 12.6 star SE 2'.9.
10 41 50.3	+21 18 40	CGCG 124-54	14.0	0.7 x 0.6	13.0	S?	30	= **Mkn 725**; UGC 5822 S 6'.4.
10 45 52.3	+25 57 04	CGCG 125-5	14.1	0.8 x 0.6	13.2		150	**UGC 5874** SE 4'.5; mag 9.35v star ENE 6'.4.
11 07 35.9	+22 51 05	CGCG 125-30	14.2	0.8 x 0.5	13.1		60	Mag 8.17v star NNE 8'.4.
11 15 43.7	+22 05 06	CGCG 126-14	15.2	0.5 x 0.5	13.5			Close pair of mag 9.69v and 9.24v stars NNW 9'.1.
10 29 44.2	+27 15 12	CGCG 154-21	15.1	0.7 x 0.7	14.2			Mag 8.17v star NE 9'.6.
11 02 23.0	+26 54 13	CGCG 155-46	15.0	0.4 x 0.4	12.8			Mag 13.4 star E 1'.3; mag 10.85v star W 5'.1.
11 14 17.0	+28 33 26	CGCG 156-31	14.0	0.7 x 0.5	12.7	S0?	57	Mag 12.7 star SE 2'.6.
09 59 00.2	+17 49 00	IC 582	14.0	0.9 x 0.9	13.6	S		Elongated **IC 583** E 1'.1.
10 26 28.3	+20 13 40	IC 610	13.9	1.9 x 0.3	13.2	Sbc sp II	29	Mag 9.06v star NW 4'.9.
10 48 08.2	+18 11 16	IC 642	12.6	1.3 x 1.3	13.3	E?		Mag 9.76v star NW 13'.5.
09 56 20.3	+27 13 38	IC 2520	13.8	0.7 x 0.6	12.7	S?	90	Mag 10.22v star ESE 1'.8.
10 10 28.2	+27 57 12	IC 2550	13.6	1.0 x 0.8	13.2	SAB(rs)b: III	129	Mag 11.7 star on SE edge.
10 10 40.5	+24 24 48	IC 2551	13.6	1.2 x 1.0	13.6	S?	33	Mag 10.22v star NNW 3'.3.
10 31 10.5	+26 03 15	IC 2583	14.7	0.8 x 0.5	13.6		110	UGC 5713 SE 7'.7.
10 36 16.6	+26 57 45	IC 2590	13.3	1.1 x 1.1	13.3	S0		**MCG +5-25-23** SW 1'.4, with star in between.
10 39 42.3	+26 43 37	IC 2598	14.1	0.6 x 0.3	12.1	Sa	144	Mag 8.29v star S 5'.5.
11 18 05.3	+17 38 54	IC 2703	14.4	0.6 x 0.6	13.1			Mag 10.33v star WSW 12'.6.
10 20 56.2	+20 09 18	MCG +3-27-1	15.8	0.6 x 0.4	14.1	Sm	85	Mag 10.24v star NNE 7'.5.
10 23 10.4	+17 57 44	MCG +3-27-14	13.6	0.5 x 0.4	11.7	Compact	33	= **Mkn 630**.
10 25 18.1	+20 00 41	MCG +3-27-28	14.4	0.7 x 0.7	13.5	Sbc		**CGCG 94-50** E 10'.1.
10 28 12.0	+18 36 22	MCG +3-27-46	14.6	1.9 x 1.0	15.2	SBc_p	6	Mag 8.30v star W 3'.3.
10 34 08.6	+19 42 14	MCG +3-27-61	14.0	1.0 x 0.8	13.7	Sb	30	
10 34 53.6	+17 45 31	MCG +3-27-63	15.7	0.5 x 0.4	13.8		45	Mag 11.2 star WSW 7'.2.
10 44 23.1	+18 29 12	MCG +3-28-2	14.0	0.9 x 0.7	13.3	Sb	36	Mag 12.8 star SW 2'.4.
10 53 30.4	+18 55 50	MCG +3-28-29	14.0	0.7 x 0.6	12.9		81	Mag 12.1 star SSW 2'.2.
11 00 03.9	+17 25 25	MCG +3-28-45	14.0	1.0 x 0.5	13.1		123	Mag 9.31v star N 5'.1.
11 02 59.6	+17 19 45	MCG +3-28-52	14.2	0.7 x 0.4	12.6	Sb	132	**CGCG 95-102** E 8'.9.
11 03 38.6	+19 19 39	MCG +3-28-54	14.2	0.9 x 0.5	13.2	Sc	162	
11 05 32.6	+17 38 21	MCG +3-28-56	14.1	1.0 x 0.4	13.0	Sb	18	Mag 9.38v star NW 2'.6.
11 16 35.3	+18 07 02	MCG +3-29-18	14.3	0.9 x 0.3	12.9	E/S0	171	NGC 3608 ENE 6'.0.
11 17 38.3	+17 49 05	MCG +3-29-24	14.5	1.3 x 0.8	14.4	SB0/p	165	Almost stellar nucleus; very faint, small anonymous galaxy SSE 5'.1.
11 17 58.1	+17 26 30	MCG +3-29-25	14.4	0.5 x 0.3	12.2		159	
11 18 06.2	+18 47 52	MCG +3-29-27	15.5	0.5 x 0.4	13.6	Sc	30	Mag 11.15v star N 1'.8.
11 20 50.7	+19 22 26	MCG +3-29-34	13.9	0.9 x 0.8	13.4	Sc	111	Close pair of mag 12.6 stars NE 2'.8; **MCG +3-29-33** SW 2'.8.
10 16 33.3	+22 05 26	MCG +4-24-22	14.1	1.1 x 0.9	13.9		138	Mag 10.06v star NE 5'.3.
10 23 18.7	+22 23 25	MCG +4-25-14	14.9	1.0 x 0.5	14.0	Sbc	120	Mag 13.4 star S 0'.9.
10 27 41.6	+24 16 17	MCG +4-25-19	15.9	0.5 x 0.5	14.2	Sb		Mag 14.5 double star NE 2'.4.
10 31 18.8	+25 51 11	MCG +4-25-28	14.1	0.9 x 0.3	12.5		24	Faint anonymous galaxy W 0'.8; mag 11.2 star W 4'.3.
10 36 20.6	+21 01 22	MCG +4-25-34	14.0	0.8 x 0.4	12.6		3	Mag 15.2 star on E edge; mag 12.9 star N 1'.6.
10 43 51.0	+21 28 08	MCG +4-25-47	15.5	0.7 x 0.5	14.3		81	Mag 9.27v star NW 2'.0.

(Continued from previous page)

GALAXIES

RA h m s	Dec ° ′ ″	Name	Mag (V)	Dim ′ Maj x min	SB	Type Class	PA	Notes
10 44 30.0	+21 42 56	MCG +4-25-49	14.3	0.8 x 0.3	12.6	Sb	84	
10 44 36.6	+21 39 58	MCG +4-25-50	15.0	0.7 x 0.3	13.1	Sb	36	
10 48 35.5	+22 14 59	MCG +4-26-7	14.7	0.8 x 0.7	13.9	Sb	69	**MCG +4-26-10** SE 3′0.
10 48 44.4	+26 03 14	MCG +4-26-9	14.8	0.3 x 0.3	12.0	Compact		Mag 13.4 star NW 3′3.
10 49 27.1	+21 46 50	MCG +4-26-12	14.3	1.0 x 0.8	13.9			
10 50 14.1	+21 08 24	MCG +4-26-13	15.3	0.7 x 0.3	13.4	Sc	117	
10 54 05.9	+20 38 37	MCG +4-26-18	15.9	1.1 x 1.1	16.0	Im		
10 54 29.5	+21 09 47	MCG +4-26-20	14.3	0.9 x 0.7	13.7	Sb	138	**UGC 6020** and mag 10.5 star SW 4′8.
10 58 28.5	+24 22 25	MCG +4-26-23	14.1	0.8 x 0.7	13.4	Sc	114	
11 02 24.3	+22 15 49	MCG +4-26-26	15.2	0.6 x 0.3	13.2		90	Mag 10.57v star N 1′6.
11 05 01.8	+24 48 44	MCG +4-26-27	15.1	0.8 x 0.3	13.4	Sbc	99	
11 06 13.3	+24 21 17	MCG +4-26-27A	14.4	0.7 x 0.5	13.1	Sb	138	Mag 13.1 star NW 1′3.
11 07 32.0	+23 22 41	MCG +4-26-32	14.3	0.8 x 0.6	13.3		30	Mag 6.48v star SSE 3′7.
11 12 31.1	+23 16 31	MCG +4-27-1	14.3	0.7 x 0.7	13.4			Mag 15.3 star on NE edge; **MCG +4-27-2** NE 3′2.
11 18 59.1	+22 53 11	MCG +4-27-17	14.4	0.7 x 0.4	12.8	SBc	138	Compact; UGC 6333 E 4′3.
11 22 02.2	+25 55 16	MCG +4-27-25	15.3	0.5 x 0.5	13.6			
10 06 18.0	+28 56 37	MCG +5-24-11	14.1	0.5 x 0.5	12.4	S?		Mag 14.1 star N 1′7.
10 25 47.8	+26 34 13	MCG +5-25-9	13.8	0.9 x 0.6	13.0	S?	114	
10 28 52.2	+26 47 34	MCG +5-25-14	15.8	0.9 x 0.8	15.3	Sb	18	
10 28 58.7	+27 38 55	MCG +5-25-16	15.8	1.0 x 0.2	13.9	Sc	120	Mag 12.0 star on N edge; **MCG +5-25-17** NNE 2′9.
10 37 30.5	+27 20 09	MCG +5-25-28	15.3	0.4 x 0.3	12.8		165	**MCG +5-25-27** NW 2′3.
10 41 01.9	+28 38 09	MCG +5-25-31	15.7	0.7 x 0.6	14.6	Sbc	135	Mag 12.4 star E 1′3.
10 45 24.4	+29 13 07	MCG +5-26-2	14.4	0.6 x 0.5	12.9	Sb	135	Mag 12.1 star SE 2′9.
10 45 49.7	+27 37 10	MCG +5-26-3	13.9	0.7 x 0.5	12.6	S0?	27	= **Mkn 726**; mag 12.4 star SW 2′5.
10 46 25.2	+28 28 45	MCG +5-26-4	15.2	0.8 x 0.4	13.8	Sbc	126	
10 48 12.4	+26 36 09	MCG +5-26-13	14.6	0.7 x 0.3	12.7	Sb	177	UGC 5912 E 3′5.
10 53 56.5	+26 54 30	MCG +5-26-27	14.9	0.5 x 0.5	13.3			Mag 11.6 star S 2′3.
10 54 58.8	+28 01 03	MCG +5-26-29	15.5	0.8 x 0.3	13.8	Sb	78	Mag 13-15 double star E 1′6.
10 55 33.8	+27 04 52	MCG +5-26-31	15.8	0.7 x 0.5	14.6	Sb	141	
11 04 58.3	+29 08 17	MCG +5-26-46	15.4	0.6 x 0.4	13.7		153	= **Mkn 36**.
11 07 10.1	+28 25 30	MCG +5-26-58	15.2	0.6 x 0.5	14.0	E	108	Stellar nucleus, faint envelope, mag 9.26v star SW 3′9.
11 09 41.9	+28 15 08	MCG +5-26-68	15.9	0.5 x 0.4	14.0	Spiral	24	Mag 9.98v star NW 4′4.
11 10 38.5	+28 19 01	MCG +5-27-1	13.5	0.8 x 0.8	12.8	S0?		
11 12 50.2	+28 04 17	MCG +5-27-25	15.7	0.6 x 0.5	14.2	Spiral	111	
11 18 17.9	+28 13 29	MCG +5-27-52	14.0	0.8 x 0.5	12.9	SB?	114	UGC 6322 N 2′1; elongated anonymous galaxy SSE 1′3.
09 58 39.6	+28 52 27	NGC 3068	14.3	1.1 x 0.9	14.2	S0⁻: pec	53	Double system with companion at 0′6 SW.
10 01 08.3	+22 24 20	NGC 3088A	13.8	0.5 x 0.3	11.6		69	NGC 3088B on SE edge; **UGC 5381** SW 6′7.
10 01 10.2	+22 23 59	NGC 3088B	15.4	0.5 x 0.1	12.0	Sbc	138	NGC 3088A on NW edge.
10 02 17.0	+24 42 39	NGC 3098	12.0	2.3 x 0.5	12.0	S0 sp	90	
10 08 36.5	+18 13 52	NGC 3131	13.0	2.4 x 0.7	13.4	SBb:	54	
10 13 01.4	+17 02 03	NGC 3154	13.5	0.9 x 0.4	12.2	Sb	124	Lies 2′1 NW of mag 8.77v star.
10 13 31.5	+22 44 12	NGC 3162	11.6	3.0 x 2.5	13.7	SAB(rs)bc II	33	Mag 13.4 star on E edge; **MCG +4-24-18** W 9′0.
10 16 34.2	+21 07 23	NGC 3177	12.4	1.4 x 1.2	12.8	SA(rs)b II-III	135	
10 17 38.6	+21 41 08	NGC 3185	12.2	2.3 x 1.6	13.4	(R)SB(r)a	130	Mag 14.1 star on W edge; mag 10.46v star E 4′6.
10 17 48.1	+21 52 25	NGC 3187	13.4	3.6 x 1.6	15.2	SB(s)c pec	57	Very faint arms extending NE and SW form an open "S" shape.
10 18 05.9	+21 49 59	NGC 3190	11.1	4.4 x 1.5	13.0	SA(s)a pec sp	125	Faint, dark lane runs the length of the galaxy.
10 18 25.0	+21 53 34	NGC 3193	10.9	2.0 x 2.0	12.5	E2		Mag 9.63v star sits on N edge.
10 18 49.1	+27 40 05	NGC 3196	15.7	0.4 x 0.2	12.8		115	Faint, stellar anonymous galaxy NE 2′5.
10 20 10.9	+27 48 58	NGC 3204	13.5	1.3 x 0.9	13.6	SAB(r)b	110	**CGCG 154-2** WNW 10′0.
10 20 38.5	+25 30 16	NGC 3209	12.7	1.3 x 1.1	13.1	E	80	Almost stellar **MCG +4-25-4** E 4′7.
10 21 17.5	+19 39 04	NGC 3213	13.5	1.1 x 0.9	13.3	Sbc:	133	Mag 8.05v star SW 9′8.
10 21 41.2	+23 55 23	NGC 3216	13.4	1.3 x 1.0	13.6	E:	0	Elongated anonymous galaxy NNW 2′6.
10 22 20.2	+21 34 12	NGC 3221	13.1	3.2 x 0.7	13.8	SB(s)cd: sp III-IV	167	Mag 9.82v star SW 9′2; galaxy **II Zw 45** S 3′2.
10 22 34.5	+19 53 13	NGC 3222	12.8	1.3 x 1.1	13.0	SB0:	51	Very faint star on S edge; **CGCG 94-22** SSE 6′5.
10 23 27.4	+19 53 56	NGC 3226	11.4	2.8 x 2.0	13.3	E2: pec	29	On N edge of NGC 3227.
10 23 31.1	+19 51 40	NGC 3227	10.3	4.1 x 3.9	13.1	SAB(s)a pec II-III	150	Very faint envelope extends beyond bright central core to fill out published dimensions; NGC 3326 on N edge.
10 24 24.5	+28 01 41	NGC 3232	14.3	0.8 x 0.8	13.7	S?		**MCG +5-25-5** N 1′2; mag 10.26v star SW 2′6.
10 24 59.3	+28 01 25	NGC 3235	13.3	1.2 x 0.9	13.2	S0⁻:	76	**IC 2572** NNE 4′6.
10 25 05.2	+17 09 01	NGC 3239	11.3	5.0 x 3.3	14.2	IB(s)m pec	81	Very irregular appearance; a mag 9.96v star superimposed on the center; **MCG +3-27-27** E 3′2 from center.
10 27 18.5	+28 30 20	NGC 3245	10.8	3.2 x 1.8	12.6	SA(r)0°:?	177	
10 27 01.2	+28 38 22	NGC 3245A	13.9	3.3 x 0.3	13.7	SB(s)b sp	150	
10 27 45.2	+22 50 50	NGC 3248	12.4	2.5 x 1.1	13.3	S0	135	
10 29 16.7	+26 05 54	NGC 3251	13.4	2.0 x 0.4	13.0	SB?	55	Small, faint, elongated anonymous galaxy SE 2′3.
10 29 19.8	+29 29 24	NGC 3254	11.7	4.5 x 1.4	13.6	SA(s)bc II	46	Close pair of mag 9.36v, 9.92v stars E 6′2.
10 31 07.0	+28 47 46	NGC 3265	12.9	1.0 x 0.6	12.4	E:	73	
10 31 30.0	+24 52 04	NGC 3270	13.1	3.2 x 0.8	13.9	SAB(r)b: II	10	
10 32 17.4	+27 40 06	NGC 3274	12.8	2.0 x 1.0	13.4	SABd? IV-V	100	Very small, faint anonymous galaxy on NW edge.
10 32 55.4	+28 30 41	NGC 3277	11.7	1.9 x 1.7	12.8	SA(r)ab II	170	Large round core in smooth envelope.
10 34 47.4	+21 38 59	NGC 3287	12.3	2.1 x 1.0	12.9	SB(s)d III-IV	20	Mag 7.40v star SW 6′4.
10 36 55.7	+21 52 52	NGC 3301	11.4	3.5 x 1.0	12.6	(R′)SB(rs)0/a	52	
10 37 02.2	+18 07 57	NGC 3303	13.7	2.3 x 1.7	15.0	Pec	159	A small bright nucleus with a faint, broad band of material extending S and even fainter material filling the area E giving the larger published size.
10 39 39.0	+25 19 20	NGC 3323	13.4	1.2 x 0.7	13.0	SB?	174	
10 39 58.0	+24 05 28	NGC 3327	13.4	1.1 x 0.9	13.2	SA(r)b: II-III	85	Mag 14.1 star on W edge.
10 43 31.3	+24 55 23	NGC 3344	9.9	7.1 x 6.5	13.9	(R)SAB(r)bc I-II	18	Many bright knots overall; a mag 10.22v star 1′6 E of center.
10 44 15.0	+22 22 18	NGC 3352	12.5	1.6 x 1.2	13.1	S0	0	
10 45 09.5	+22 04 44	NGC 3363	13.4	1.3 x 0.8	13.3	S?	0	
10 47 03.9	+17 16 27	NGC 3370	11.6	2.6 x 1.5	12.9	SA(s)c II	148	
10 48 12.1	+28 36 05	NGC 3380	12.5	1.7 x 1.3	13.2	(R′)SBa?	12	
10 50 45.4	+28 28 07	NGC 3400	13.2	1.3 x 0.8	13.1	SB(s)a:	100	Mag 9.53v star ENE 6′3.
10 51 16.2	+27 58 30	NGC 3414	11.0	3.5 x 2.6	13.3	S0 pec	12	Stellar anonymous galaxy on N edge, 2′0 from center.
10 51 23.8	+28 06 42	NGC 3418	13.2	1.4 x 1.1	13.5	SAB(s)0/a:	75	Mag 9.08v star W 12′8.
10 51 41.8	+18 28 51	NGC 3426	13.2	1.3 x 1.0	13.3	S?	114	
10 52 36.0	+22 56 03	NGC 3437	12.1	2.5 x 0.8	12.7	SAB(rs)c: III	122	
10 53 00.4	+17 34 23	NGC 3443	13.1	2.8 x 1.4	14.4	SAd	145	Small, weak nucleus in faint, uniform envelope.
10 54 20.9	+27 14 19	NGC 3451	13.0	1.7 x 0.8	13.2	Sd	50	
10 54 29.5	+17 20 34	NGC 3454	13.5	2.1 x 0.4	13.1	SB(s)c? sp	116	Mag 11 star 1′6 S.

GALAXIES

RA h m s	Dec ° ′ ″	Name	Mag (V)	Dim ′ Maj x min	SB	Type Class	PA	Notes
10 54 30.7	+17 17 04	NGC 3455	12.0	2.6 x 2.0	13.7	(R′)SAB(rs)b II-III	62	Very faint arms encircle brighter, elongated core.
10 54 48.8	+17 37 10	NGC 3457	12.7	0.9 x 0.9	12.3	S?		Faint anonymous galaxy E 4′.7, near original **IC 656** position.
10 58 05.1	+17 07 20	NGC 3473	13.5	1.2 x 1.0	13.6	SBb:	40	Mag 12.8 star on N edge; mag 9.46v star NE 6′.7; faint, elongated anonymous galaxy N 1′.2.
10 58 08.8	+17 05 41	NGC 3474	13.9	0.8 x 0.7	13.2	S?	138	NGC 3474 N 1′.9.
10 58 25.3	+24 13 35	NGC 3475	13.1	1.7 x 1.1	13.6	Sa	65	**Mkn 419** S 1′.9; anonymous galaxy SSW 2′.4.
11 00 24.0	+28 58 30	NGC 3486	10.5	7.1 x 5.2	14.3	SAB(r)c II	80	Numerous knots and superimposed stars.
11 00 46.7	+17 35 12	NGC 3487	13.9	0.9 x 0.4	12.6	Sb? sp	153	
11 01 28.0	+27 43 06	NGC 3493	14.3	1.1 x 0.3	12.9	S?	84	Stellar galaxy **KUG 1058+280** N 5′.8.
11 02 47.7	+17 59 30	NGC 3501	12.9	3.9 x 0.5	13.4	Scd:	27	Mag 9.07v star NW 7′.2.
11 03 11.4	+27 58 19	NGC 3504	11.1	2.7 x 2.1	12.9	(R)SAB(s)ab I-II	159	
11 03 25.9	+18 08 07	NGC 3507	10.9	3.4 x 2.9	13.3	SB(s)b	110	Bright knot or superimposed star NE of center.
11 03 43.5	+28 53 11	NGC 3510	12.1	4.0 x 0.8	13.2	SB(s)m sp	163	Mag 7.06v star W 7′.6; stellar galaxy **KUG 1101+290** S 6′.4.
11 04 03.0	+28 02 11	NGC 3512	12.3	1.6 x 1.5	13.1	SAB(rs)c II	132	
11 04 37.5	+28 13 39	NGC 3515	13.9	0.9 x 0.7	13.2	SAc:	55	Mag 9.58v star N 8′.5.
11 06 40.3	+20 05 07	NGC 3522	13.1	1.2 x 0.7	12.9	E	117	
11 07 18.3	+28 31 40	NGC 3527	13.7	1.0 x 0.9	13.5	(R)SB(r)ab:	159	UGC 6166 NNW 4′.5; stellar **Mkn 1283** E 7′.6.
11 08 57.7	+26 35 42	NGC 3534	14.0	1.1 x 0.4	13.0	Sb	170	
11 08 51.2	+28 28 31	NGC 3536	13.8	0.9 x 0.7	13.1	SB?	155	Stellar nucleus; faint, stellar galaxy **KUG 1105+286** SW 8′.8.
11 09 08.9	+28 40 20	NGC 3539	14.6	1.2 x 0.3	13.3	S?	0	Faint, elongated anonymous galaxy and a short line of three mag 13 stars ESE 6′.1.
11 10 38.5	+28 46 03	NGC 3550	13.3	1.0 x 1.0	13.2	Pec		**CGCG 156-2** S 2′.7; a mag 11.14v star E 1′.5; and a pair of stellar anonymous galaxies just off the SE edge.
11 10 42.9	+28 41 34	NGC 3552	14.5	0.7 x 0.7	13.8	E		Pair of faint, stellar anonymous galaxies SW 0′.8.
11 10 39.9	+28 42 24	NGC 3553	16.0	0.5 x 0.4	14.3	E	17	A pair of stellar anonymous galaxies inline SW at 0′.7 and 1′.1.
11 10 47.9	+28 39 37	NGC 3554	14.2	0.5 x 0.5	12.8	E		An anonymous galaxy NE 2′.6.
11 09 44.5	+21 45 29	NGC 3555	12.8	1.8 x 1.7	14.0	E?	30	Compact nucleus; **CGCG 125-34** NNE 3′.4.
11 10 55.8	+28 32 38	NGC 3558	13.8	1.0 x 0.9	13.5	S?	138	An anonymous galaxy 1′.1 N.
11 11 13.1	+28 41 45	NGC 3561	13.8	0.8 x 0.8	13.2	SA(r)a pec		**MCG +5-27-12** SE 1′.7.
11 11 13.0	+28 42 41	NGC 3561A	13.3	0.9 x 0.9	12.9	S0°: pec		Pair with NGC 3561 S edge; small, faint anonymous galaxy NE edge.
11 11 25.3	+26 57 45	NGC 3563	13.7	1.0 x 0.7	13.3	SB0:	174	NGC 3563A on E edge; mag 9.53v star N 3′.4.
11 11 23.8	+26 57 41	NGC 3563A	15.0	0.5 x 0.3	12.8	SB:0	168	Located on E edge of NGC 3563.
11 12 03.4	+27 35 18	NGC 3570	13.5	1.0 x 1.0	13.4	S0		**MCG +5-27-20** due S 1′.4.
11 12 08.1	+27 34 28	NGC 3574	14.9	0.4 x 0.4	12.8	S?		**MCG +5-27-21** S 0′.4.
11 14 02.7	+20 23 13	NGC 3588	14.5	0.6 x 0.6	13.2	Sb		Double system, in contact.
11 14 27.6	+17 15 29	NGC 3592	13.7	1.8 x 0.6	13.6	Sc? sp	120	
11 15 11.8	+17 15 41	NGC 3598	12.3	1.7 x 1.3	13.0	S0⁻:	35	Mag 14.0 star on N edge.
11 15 27.1	+18 06 35	NGC 3599	12.0	2.7 x 2.1	13.7	SA0:	99	
11 15 48.5	+17 24 52	NGC 3602	15.0	1.0 x 0.3	13.5	Sb	48	Mag 10.05v star SSW 3′.4; small, faint anonymous galaxy NW 6′.0.
11 16 46.5	+18 01 00	NGC 3605	12.3	1.5 x 1.0	12.7	E4-5	12	Located on SW edge of NGC 3607.
11 16 54.6	+18 03 07	NGC 3607	9.9	4.7 x 3.9	12.9	SA(s)0°:	120	Elongated MCG +3-29-18 NW 6′.0; NGC 3605 on SW edge.
11 16 59.1	+18 08 54	NGC 3608	10.8	3.2 x 2.6	13.1	E2	75	NGC 3607 SE 6′.2; MCG +3-29-18 WSW 6′.0.
11 17 50.8	+26 37 32	NGC 3609	13.1	1.2 x 1.0	13.2	Sab	50	**MCG +5-27-44** N 1′.9.
11 18 14.7	+26 37 13	NGC 3612	14.1	1.0 x 0.8	13.7	Sdm:	160	Weak stellar nucleus; mag 10.95v star NW 3′.9.
11 18 06.8	+23 23 48	NGC 3615	12.8	1.4 x 0.9	13.1	E	40	Stellar **CGCG 126-22** NE 3′.2.
11 18 32.5	+23 28 07	NGC 3618	13.6	0.9 x 0.8	13.1	SABb: III	175	Weak stellar nucleus; **MCG +4-27-16** SSE 9′.3.
11 20 03.5	+18 21 23	NGC 3626	11.0	2.7 x 1.9	12.6	(R)SA(rs)0⁺	163	Elongated **UGC 6341** S 5′.9.
11 20 31.9	+26 57 50	NGC 3629	12.1	2.3 x 1.6	13.3	SA(s)cd: III	30	
11 21 35.7	+18 27 31	NGC 3639	13.7	0.7 x 0.6	12.6	S?	39	
11 21 43.1	+20 10 04	NGC 3646	11.1	3.9 x 2.2	13.3	Ring I-II	50	Weak almost stellar nucleus; strong dark area SW of nucleus; NGC 3649 E 7′.8.
11 22 14.8	+20 12 30	NGC 3649	13.7	1.4 x 0.6	13.3	SB(s)a	140	Pair of very faint stars S edge; NGC 3646 W 7′.8.
11 22 35.3	+20 42 13	NGC 3650	13.9	1.7 x 0.3	13.0	SAb sp III	54	Located 2′.7 SE of a mag 11.17v star.
11 22 26.4	+24 17 56	NGC 3651	13.2	1.1 x 1.1	13.4	E		= **Hickson 51A**. **Hickson 51B** = **IC 2759** W 2′.8. **Hickson 51D** E 1′.0. **Hickson 51E** 3′.2 WNW. **Hickson 51F** just S of nucleus. **Hickson 51G** on SE edge.
11 22 30.2	+24 16 45	NGC 3653	13.7	0.6 x 0.4	12.0	S0?	79	= **Hickson 51C**.
11 23 45.1	+17 49 02	NGC 3659	12.2	2.1 x 1.1	12.9	SB(s)m? III-IV	60	
09 56 22.7	+20 28 46	UGC 5339	13.5	1.0 x 0.6	12.8	Double System	150	Double system, contact; **MCG +4-24-3** NW 2′.0.
09 56 45.9	+28 49 43	UGC 5340	14.5	2.4 x 1.0	15.3	Im pec:	7	Dwarf peculiar, low, fairly uniform surface brightness.
09 56 35.7	+20 38 37	UGC 5341	14.3	2.6 x 0.2	13.5	Scd:	57	
09 58 52.0	+19 12 50	UGC 5359	14.1	1.2 x 0.4	13.1	Sbc	97	
10 02 35.7	+19 10 37	UGC 5403	13.5	1.2 x 0.4	12.5	S0/a	80	Mag 8.54v star SW 3′.4.
10 03 59.1	+22 16 30	UGC 5420	13.5	1.0 x 0.7	13.0	S0:	81	Anonymous companion on S edge.
10 04 40.6	+29 21 50	UGC 5427	14.0	1.2 x 0.8	13.8	Sdm:	120	
10 04 59.0	+21 32 14	UGC 5431	15.2	1.5 x 0.2	13.8	Scd:	48	Mag 8.72v star W 5′.3.
10 05 13.4	+21 27 17	UGC 5434	13.5	1.0 x 0.8	13.1	SAB(s)b	25	Mag 10.04v star E 6′.0.
10 05 21.6	+19 17 10	UGC 5436	13.4	1.4 x 0.6	13.1	S?	130	**MCG +3-26-22** on S edge
10 08 12.9	+18 42 23	UGC 5467	13.2	0.8 x 0.8	12.6	S0?		Mag 8.82v star N 3′.5.
10 10 39.9	+20 04 12	UGC 5489	12.7	2.0 x 1.0	13.3	Sa	0	Mag 13.1 star SE 1′.7.
10 12 03.7	+23 05 09	UGC 5498	14.4	1.5 x 0.2	12.9	Sa:	63	Knot or very small superimposed galaxy NE end.
10 12 17.6	+27 51 43	UGC 5499	13.1	2.8 x 0.6	13.5	SBb:	42	Mag 8.39v star W 9′.3.
10 13 32.4	+20 10 30	UGC 5509	14.3	1.5 x 0.2	12.9	Sb III	66	Mag 13.3 star W 1′.5.
10 13 42.7	+18 07 33	UGC 5514	15.0	1.5 x 0.3	13.9	Scd:	141	Located between a pair of mag 13.9 and 12.5 stars; **MCG +3-26-43** E 1′.4.
10 14 21.6	+22 07 26	UGC 5524	14.6	1.9 x 0.4	14.1	Scd:	43	Mag 12.2 star W 1′.2.
10 17 26.0	+17 05 47	UGC 5552	14.1	1.5 x 0.3	13.1	Sbc	32	
10 20 57.2	+25 21 55	UGC 5588	14.1	0.5 x 0.4	12.2	S?	30	Mag 10.36v star SW 1′.2; **UGC 5583** W 5′.1.
10 22 14.1	+20 35 24	UGC 5598	14.6	1.5 x 0.3	13.6	S	35	
10 23 36.1	+28 18 43	UGC 5621	14.1	1.0 x 0.7	13.6	SAB(s)bc	90	Stellar nucleus.
10 25 20.8	+26 27 31	UGC 5638	13.8	0.6 x 0.4	12.1	S0	108	
10 25 26.1	+17 15 40	UGC 5639	13.9	1.2 x 0.8	13.7	SABc	138	CGCG 94-39 WNW 6′.2 (see notes for CGCG 94-39 for other galaxies in area).
10 26 25.6	+17 30 39	UGC 5651	13.3	1.7 x 1.2	13.9	SABcd:	18	Mag 9.58v star S 4′.3.
10 28 20.8	+22 34 19	UGC 5672	13.6	2.1 x 0.6	13.7	S?	158	
10 28 30.2	+19 33 45	UGC 5675	14.2	1.8 x 1.4	15.1	Sm:	162	Dwarf spiral, low surface brightness.
10 28 52.7	+26 20 08	UGC 5679	13.9	1.5 x 0.6	13.6	S?	104	Stellar nucleus.
10 29 16.4	+23 43 42	UGC 5683	14.6	1.0 x 0.4	13.5	Double System	147	Stellar nucleus, smooth envelope; mag 9.29v star NNW 8′.2.
10 29 29.2	+19 37 22	UGC 5690	13.6	0.7 x 0.6	12.5	SB?	21	
10 29 28.4	+20 59 28	UGC 5691	14.7	0.5 x 0.5	13.0	S		Almost stellar.
10 29 51.7	+19 49 58	UGC 5696	13.6	1.3 x 0.7	13.3	Sb II	39	**MCG +3-27-51** S 4′.1.
10 31 16.1	+19 22 58	UGC 5709	14.2	0.9 x 0.6	13.4	Sd:	115	**CGCG 94-83** and a mag 11.0 star NE 7′.3.
10 31 31.8	+24 07 20	UGC 5710	15.1	1.7 x 0.5	14.8	S	70	Line of three stars E, mags 7.93v, 11.2 and 12.1.
10 31 38.8	+25 59 02	UGC 5713	13.8	1.7 x 0.5	13.5	Sbc	5	IC 2583 NW 7′.7.
10 35 39.3	+28 33 54	UGC 5749	14.2	0.5 x 0.4	12.3	S?	101	

(Continued from previous page)

GALAXIES

RA h m s	Dec ° ′ ″	Name	Mag (V)	Dim ′ Maj x min	SB	Type Class	PA	Notes
10 35 46.3	+21 02 53	UGC 5751	13.6	1.9 x 0.4	13.2	S?	172	**CGCG 124-43** NE 2′1.
10 39 52.5	+21 50 42	UGC 5801	14.8	0.6 x 0.5	13.3		144	Mag 8.83v star E 6′2.
10 40 24.6	+21 37 10	UGC 5805	14.0	1.1 x 0.7	13.6	SBdm	90	Mag 12.0 star S 1′9.
10 41 52.8	+21 15 11	UGC 5822	13.8	1.1 x 0.5	13.0	(R)SB(r)a	171	CGCG 124-54 N 6′4; **CGCG 124-53** SSW 4′5.
10 42 11.2	+23 44 48	UGC 5825	14.0	1.1 x 0.5	13.2	Sa?	92	
10 42 38.0	+23 57 03	UGC 5830	14.0	1.2 x 0.7	13.7	SAB(r)b: II-III	30	
10 43 06.0	+20 25 23	UGC 5833	13.7	1.2 x 0.5	13.1	S0	148	
10 43 56.0	+28 08 50	UGC 5844	15.4	1.5 x 0.3	14.4	Sd	121	Close pair of mag 13 stars W end.
10 44 37.1	+26 10 54	UGC 5855	14.0	1.6 x 0.4	13.4	S?	96	
10 46 42.9	+25 55 53	UGC 5881	14.0	1.1 x 0.5	13.2	Sa	55	Mag 9.14v star E 5′6.
10 47 02.9	+26 32 33	UGC 5884	13.5	1.1 x 0.7	13.1	SA(s)b: II-III	97	
10 47 39.6	+26 17 43	UGC 5894	13.4	1.5 x 0.7	13.3	SAB(s)ab	155	**CGCG 155-13** SE 4′0.
10 48 01.1	+28 14 48	UGC 5903	13.8	1.1 x 0.6	13.2	SAB(s)ab	18	
10 48 27.4	+26 35 04	UGC 5912	13.2	1.5 x 1.5	14.0	SA(s)c II		MCG +5-26-13 W 3′5.
10 48 39.3	+21 44 16	UGC 5916	15.3	1.1 x 0.8	15.0	SABcd:	48	Mag 11.5 star SW 2′9.
10 49 12.4	+27 55 34	UGC 5921	14.4	1.5 x 0.4	13.7	Sdm:	161	
10 49 12.2	+22 00 59	UGC 5924	13.9	1.5 x 0.5	13.4	Sa	52	Located on S edge of galaxy cluster A 1100.
10 50 24.4	+17 33 47	UGC 5945	13.8	1.7 x 0.7	13.8	IBm	95	Elongated core offset E.
10 50 30.6	+19 38 43	UGC 5947	14.5	1.3 x 0.7	14.2	Im pec:	25	Mag 12.1 star S 1′9.
10 51 15.9	+27 50 54	UGC 5958	14.5	1.5 x 0.3	13.5	Sbc	179	Mag 10.37v star W 7′3.
10 52 31.8	+19 47 31	UGC 5989	13.7	1.6 x 0.5	13.3	Im?	126	Mag 12.3 star S 1′6.
10 55 00.7	+29 32 34	UGC 6031	14.8	1.1 x 0.7	14.4	S	141	Mag 12.4 star SSE 2′0.
10 55 29.3	+17 08 29	UGC 6035	13.7	1.1 x 0.9	13.5	IBm	81	Bright knot or star W end.
11 01 48.6	+28 41 16	UGC 6102	14.3	1.0 x 0.7	13.8	Im	140	Mag 11.9 star NW 2′9.
11 04 39.9	+27 43 25	UGC 6138	13.9	1.3 x 0.9	13.9	Sm:	57	Mag 8.60v star N 2′6.
11 05 56.2	+19 49 36	UGC 6151	14.4	1.7 x 1.7	15.4	Sm: V		Dwarf spiral, low surface brightness.
11 06 07.1	+29 55 53	UGC 6152	14.2	1.2 x 0.5	13.5	SBb	154	Mag 14.4 star S edge.
11 06 25.3	+17 30 28	UGC 6157	13.3	1.6 x 1.4	14.0	SA(s)dm	24	Several bright knots; close pair of stars, mags 10.8 and 12.1, S 1′3.
11 06 51.1	+23 00 57	UGC 6163	13.7	1.0 x 0.7	13.2	Sa	90	**UGC 6164** E 1′9.
11 07 08.7	+28 35 32	UGC 6166	13.9	1.0 x 0.7	13.4	Sbc	106	NGC 3527 SSE 4′5; close pair **UGC 6160** and **MCG +5-26-55** NNW 8′7.
11 07 09.9	+18 34 08	UGC 6171	13.8	2.3 x 0.6	14.0	IBm:	68	Irregular, slightly brighter core.
11 07 19.5	+23 29 01	UGC 6173	13.5	0.9 x 0.6	12.7	S?	160	
11 07 20.7	+18 25 51	UGC 6175	13.7	1.3 x 1.0	13.9	SAB(r)0$^+$ pec	48	Double system, disrupted, plumes extending E and W.
11 07 24.9	+21 39 26	UGC 6176	13.6	1.2 x 0.5	12.9	SB0	20	Mag 10.9 star SW 2′0.
11 07 46.4	+19 32 53	UGC 6181	14.1	1.0 x 0.8	13.7	Im:	39	Mag 13.4 star superimposed NE end.
11 08 55.8	+26 36 35	UGC 6190	14.3	1.7 x 0.6	14.2	S?	85	
11 09 00.8	+22 55 41	UGC 6194	13.0	1.2 x 0.9	12.9	S?	42	Mag 8.93v star W 1′4.
11 09 26.0	+29 34 07	UGC 6198	13.6	1.0 x 0.7	13.1	S0?	121	
11 09 51.6	+24 15 43	UGC 6204	14.0	1.0 x 0.5	13.1	Sb? pec	177	Contact with UGC 6207 SE, distorted.
11 09 54.8	+24 15 24	UGC 6207	14.2	1.5 x 0.3	13.2	Sb? pec sp	60	Pair with UGC 6204 NW edge.
11 10 57.1	+19 10 54	UGC 6219	14.0	1.5 x 0.4	13.3	S?	83	**MCG +3-29-6** ENE 1′6.
11 11 00.0	+29 20 10	UGC 6220	14.9	0.7 x 0.3	13.1	Double System	66	**PGC 33971** SW 0′9; mag 11.05v star E 1′1.
11 12 50.5	+23 15 19	UGC 6246	14.8	1.5 x 0.3	13.8	Scd:	42	MCG +4-27-1 W 4′7.
11 12 52.7	+27 26 31	UGC 6247	13.6	1.4 x 0.6	13.3	Sab	120	Mag 10.87v star SW 2′0.
11 13 10.5	+27 49 01	UGC 6250	13.2	1.1 x 0.9	13.3	E?	132	Elongated **CGCG 156-28** NE 5′9.
11 13 19.2	+25 51 46	UGC 6252	14.2	0.9 x 0.7	13.5	Scd:	25	
11 13 27.4	+22 09 39	UGC 6253	12.0	10.1 x 9.0	17.0	E0	101	**Leo II**, dwarf elliptical.
11 13 49.7	+21 30 56	UGC 6258	14.3	1.9 x 0.5	14.1	Im	175	Mag 13.3 star E 1′4.
11 16 28.0	+29 19 35	UGC 6292	13.6	1.2 x 0.7	13.3	SB(s)b	66	Faint anonymous galaxy SE 1′4; mag 10.84v star SE 3′1; **PGC 34386** S 1′6; **PGC 34394** S 2′4.
11 16 51.0	+17 47 50	UGC 6296	13.6	1.4 x 0.4	12.8	S?	166	Mag 11.0 star SSW 5′1.
11 17 22.4	+22 20 17	UGC 6301	14.5	1.5 x 0.3	13.5	Scd:	139	
11 18 17.9	+18 50 46	UGC 6320	13.1	1.0 x 0.9	12.8	S?	9	Mag 11.15v star SW 2′8.
11 18 16.8	+28 15 37	UGC 6322	14.4	1.3 x 1.0	14.6	SA0$^+$: II	18	MCG +5-27-52 S 2′1.
11 18 22.2	+18 44 17	UGC 6324	13.4	1.4 x 0.5	12.9	S0	174	Mag 14.6 star W edge; mag 12.0 star N 2′1.
11 18 28.3	+25 19 26	UGC 6325	14.2	1.5 x 0.5	13.7	Scd?	10	Mag 9.51v star ESE 2′8.
11 19 16.8	+20 48 49	UGC 6332	13.2	1.5 x 1.2	13.7	(R)SBa	130	Short, bright NE-SW bar; dark patches NW and SE.
11 19 17.2	+22 52 52	UGC 6333	14.4	1.1 x 0.5	13.6	Sdm:	120	Stellar nucleus; MCG +4-27-17 W 4′3.
11 19 30.3	+28 39 13	UGC 6334	12.6	2.5 x 1.0	13.4	(R)S0/a:	82	Pair of mag 12 and 11.6 stars off E end; bright core offset on S edge.

GALAXY CLUSTERS

RA h m s	Dec ° ′ ″	Name	Mag 10th brightest	No. Gal	Diam ′	Notes
10 48 54.0	+22 14 00	A 1100	15.7	35	30	Most members anonymous and stellar to almost stellar.
11 09 30.0	+21 41 00	A 1177	15.7	32	15	Most members stellar to almost stellar.
11 10 48.0	+28 40 00	A 1185	14.3	52	28	**Leo A Cluster**. Brighter galaxies in center one third of cluster diameter.
11 16 30.0	+29 15 00	A 1213	14.5	51	22	Main concentration of brighter PGC and anonymous GX near center of cluster area.

GALAXIES

RA h m s	Dec ° ′ ″	Name	Mag (V)	Dim ′ Maj x min	SB	Type Class	PA	Notes
08 46 10.1	+19 21 09	CGCG 90-4	14.6	0.7 x 0.4	13.1		12	Mag 9.46v star S 4′1.
08 54 08.7	+17 25 43	CGCG 90-33	14.6	0.6 x 0.4	12.9		123	Star on NW end.
09 17 12.3	+20 09 10	CGCG 91-55	14.0	0.5 x 0.3	11.8		129	Mag 12.0 star 1′4 NE.
09 24 55.7	+19 08 22	CGCG 91-91	15.2	0.6 x 0.4	13.5		105	Mag 10.36v star SW 7′6.
09 46 04.6	+19 26 29	CGCG 92-46	16.6	0.5 x 0.5	14.9	SBbc		Very close pair of mag 14.6, 14.7 stars SW 1′0; mag 10.10v star W 11′6.
09 51 30.2	+17 15 28	CGCG 92-58	16.4	0.8 x 0.3	14.7		153	Mag 9.27v star NE 7′0.
09 52 04.0	+19 39 00	CGCG 92-61	15.1	0.8 x 0.3	13.4		135	Close pair of mag 12.4, 13.9 stars WSW 7′0; mag 12.4 star ENE 8′0.
09 52 12.7	+17 35 32	CGCG 92-62	14.7	0.6 x 0.5	13.2		48	Mag 12.4 star 1′8 NE.
09 21 08.5	+21 38 28	CGCG 121-61	14.8	0.5 x 0.4	12.9		141	Mag 9.10v star WNW 5′0.
09 29 40.3	+22 57 38	CGCG 122-6	14.4	0.4 x 0.2	11.5		156	Almost stellar.
09 31 15.4	+25 15 08	CGCG 122-13	14.4	0.6 x 0.4	12.7		39	Mag 14.2 star E 1′6; mag 10.04v star WNW 7′2.
09 37 42.1	+21 17 25	CGCG 122-37	15.1	0.5 x 0.4	13.2		54	Mag 9.81v star SW 6′3.
09 42 04.0	+26 02 38	CGCG 122-56	15.8	0.3 x 0.3	13.0			Mag 13.9 star NE 0′9.
09 46 07.6	+21 33 49	CGCG 122-73	14.8	0.5 x 0.3	12.6		75	**CGCG 122-72** NNW 3′9; mag 9.66v star NE 9′6.
09 46 49.6	+26 01 28	CGCG 122-80	15.4	0.4 x 0.3	12.9		75	**CGCG 122-77** S 5′6; mag 9.80v star W 9′1.

(Continued from previous page)

RA h m s	Dec ° ′ ″	Name	Mag (V)	Dim ′ Maj x min	SB	Type Class	PA	Notes
09 47 40.6	+21 38 13	CGCG 122-85	15.4	0.5 x 0.3	13.2		171	**CGCG 122-86** NNE 2′.0.
09 57 30.5	+23 36 29	CGCG 123-9	15.6	0.7 x 0.4	14.0		15	Mag 14.5 star E 1′.5; mag 10.50v star W 5′.6.
08 36 47.1	+27 19 55	CGCG 150-5	15.6	0.6 x 0.4	13.9			Mag 14.4 star on N edge.
08 40 02.3	+29 48 59	CGCG 150-14	15.0	0.5 x 0.4	13.1	Radio galaxy	36	Mag 10.59v star ENE 7′.6.
08 47 07.2	+27 52 20	CGCG 150-30	15.5	0.5 x 0.5	13.8			Mag 11.8 star NNE 2′.8.
08 52 08.3	+29 41 35	CGCG 150-43	15.3	0.4 x 0.4	13.2			Located 2′.9 NE of mag 12.0 star.
08 56 29.2	+29 40 01	CGCG 150-48	15.4	0.4 x 0.4	13.2			Located 1′.4 SSE of mag 12.6 star.
08 56 47.7	+29 55 05	CGCG 150-50	15.5	0.5 x 0.3	13.3		114	Mag 11.0 star NNW 8′.1; **CGCG 150-52** ESE 9′.5.
09 36 05.5	+24 56 49	IC 545	14.4	0.5 x 0.4	12.5	Sab	3	Mag 13.9 star on E edge.
09 45 00.2	+29 27 08	IC 558	13.6	0.8 x 0.8	13.0	S?		Mag 9.92v star NNE 9′.0.
09 59 00.2	+17 49 00	IC 582	14.0	0.9 x 0.9	13.6	S		Elongated **IC 583** E 1′.1.
08 44 30.9	+18 17 08	IC 2392	14.8	0.6 x 0.5	13.4	Sc	6	Mag 3.93v star Delta Cancri S 8′.4.
08 46 49.1	+28 10 18	IC 2393	13.2	1.3 x 0.9	13.3	E	20	Mag 6.60v star E 4′.0.
08 47 07.0	+28 14 11	IC 2394	13.8	1.5 x 0.7	13.7	SB(s)b	105	UGC 4591 W 2′.0; mag 6.60v star S 4′.5.
08 46 44.6	+17 45 14	IC 2398	14.9	0.5 x 0.3	12.6		9	Mag 12.5 star NE 0′.5; mag 10.89v star SSE 2′.1.
08 47 49.8	+18 54 39	IC 2399	15.2	0.4 x 0.4	13.0			Mag 11.8 star S 6′.3; stellar **CGCG 90-14** ENE 8′.0.
08 48 04.7	+17 42 07	IC 2406	13.2	1.1 x 0.6	12.6	S0/a	173	Mag 8.46v star SE 5′.2; **IC 2407** S 5′.5.
08 48 24.7	+18 19 49	IC 2409	13.5	1.0 x 0.7	13.0	SAB(s)a	165	Mag 11.1 star NNE 1′.2.
08 54 47.2	+20 13 10	IC 2423	13.7	1.0 x 0.8	13.3	SAB(s)b	130	Mag 8.63v star W 10′.9; stellar **IC 2422** W 5′.4.
09 04 22.8	+27 57 10	IC 2430	13.5	1.0 x 0.5	12.6	S0/a	43	Mag 6.68v star SW 4′.4.
09 06 49.8	+26 16 29	IC 2435	14.3	1.0 x 0.4	13.3	E	120	Mag 11.2 star W 3′.4.
09 11 30.8	+28 49 35	IC 2443	14.1	0.7 x 0.6	13.0	S?	162	Mag 7.89v star ENE 6′.4.
09 13 31.4	+28 57 08	IC 2446	13.8	1.4 x 0.6	13.4	Sa	148	Mag 10.7 star W 6′.0.
09 16 01.8	+17 49 11	IC 2454	13.2	0.9 x 0.6	12.4	Sa?		NGC 2797 SE 7′.4.
09 27 52.9	+29 59 07	IC 2476	12.9	1.2 x 0.9	12.8	S0⁻:	21	Mag 11.13v star W 3′.8; **IC 2479** E 2′.5; **IC 2478** NNE 3′.6; **CGCG 152-2** W 5′.9; **CGCG 152-3** WNW 6′.5.
09 30 17.5	+26 38 26	IC 2486	14.0	1.0 x 0.7	13.5	SAB(rs)bc	129	Mag 9.59v star NNW 5′.1.
09 30 09.2	+20 05 24	IC 2487	13.4	1.8 x 0.4	12.9	Sb: sp III	164	Mag 9.04v star WNW 9′.3.
09 33 03.6	+29 55 40	IC 2490	13.4	1.3 x 1.0	13.5	Sbc III	175	Mag 10.14v star W 7′.0.
09 38 07.4	+28 03 23	IC 2495	13.9	0.8 x 0.5	12.8	S?	15	Mag 7.86v star SE 5′.4.
09 56 20.3	+27 13 38	IC 2520	13.8	0.7 x 0.6	12.7	S?	90	Mag 10.22v star ESE 1′.8.
08 38 23.9	+17 37 50	MCG +3-22-20	14.3	0.7 x 0.5	13.0	S0?	39	Located between a close pair of mag 14.8 stars.
08 50 34.6	+19 22 06	MCG +3-23-13	14.2	0.8 x 0.3	12.5		27	Mag 8.80v star SSW 7′.3; **CGCG 90-22** S 7′.6.
08 55 52.6	+18 09 22	MCG +3-23-19	14.5	0.6 x 0.3	12.5		21	Mag 12.1 star N 2′.2.
08 57 54.0	+17 05 23	MCG +3-23-21	14.7	0.7 x 0.3	12.9	Sc	111	Mag 8.64v star N 5′.7.
08 57 52.7	+20 07 18	MCG +3-23-22	14.1	0.5 x 0.5	12.5			**MCG +3-23-23** E edge.
09 04 44.9	+18 48 41	MCG +3-23-32	14.3	0.7 x 0.5	13.0	Sb	9	
09 09 35.0	+18 36 55	MCG +3-24-2	13.8	0.5 x 0.5	12.4	E		Located between a pair of stars, mags 11.8 and 12.1.
09 09 49.6	+19 51 27	MCG +3-24-3	15.6	0.6 x 0.6	14.3	Sc		
09 16 34.4	+17 55 06	MCG +3-24-24	14.3	0.9 x 0.6	13.5		129	Mag 12.8 star WSW 1′.4.
09 16 53.7	+17 27 24	MCG +3-24-29	14.5	1.3 x 0.8	14.4		6	Stellar nucleus, smooth envelope.
09 18 10.6	+18 46 34	MCG +3-24-38	15.2	0.8 x 0.4	13.8	Sbc	102	Mag 7.88v star E 1′.3.
09 20 32.6	+17 42 07	MCG +3-24-46	14.5	0.9 x 0.6	13.7		39	
09 21 18.7	+18 12 48	MCG +3-24-51	13.9	0.9 x 0.5	12.8	SB?	51	**MCG +3-24-52** NE 1′.2.
09 24 39.4	+17 39 44	MCG +3-24-55	14.8	0.4 x 0.2	11.9		171	Located 3′.0 S of mag 7.11v star.
09 36 14.4	+17 29 03	MCG +3-25-4	14.1	0.7 x 0.5	12.8		6	
09 37 31.3	+19 35 17	MCG +3-25-7	14.6	0.9 x 0.7	14.0	Sb	99	Mag 13.5 star NE 3′.4.
08 38 06.8	+25 16 22	MCG +4-21-7	14.4	0.5 x 0.5	12.7			
08 41 10.0	+20 53 54	MCG +4-21-13	14.9	0.5 x 0.5	13.2	Sm		
08 54 40.7	+20 34 57	MCG +4-21-25	14.2	0.8 x 0.3	12.5		63	Mag 14.7 star NE end; mag 10.5 star SW 1′.8.
09 13 22.6	+22 51 56	MCG +4-22-27	14.8	0.7 x 0.6	13.7	S?		
09 20 45.9	+22 04 38	MCG +4-22-32	14.1	1.0 x 0.5	13.4	E/S0	123	
09 26 21.7	+22 07 51	MCG +4-22-52	14.1	0.9 x 0.4	12.8		24	
08 36 15.4	+28 03 34	MCG +5-21-1	14.0	0.8 x 0.6	13.1	SBc	168	Mag 13.6 star on E edge.
08 46 15.1	+27 20 45	MCG +5-21-5	14.6	0.8 x 0.3	12.9	S?	33	
09 04 33.1	+28 20 58	MCG +5-22-6	16.2	0.6 x 0.4	14.5	SBbc	165	Mag 10.09v star ENE 4′.1.
09 13 37.3	+29 59 56	MCG +5-22-20	14.5	0.4 x 0.2	11.6	S?	66	Superimposed on NGC 2783.
09 15 03.6	+29 16 09	MCG +5-22-25	14.6	0.9 x 0.7	14.0	SBc	159	Located between a pair of NE-SW mag 12.8, 13.8 stars.
09 17 09.7	+27 20 51	MCG +5-22-34	15.4	0.6 x 0.3	13.7	E/S0	99	Mag 6.78v star NE 7′.3; **MCG +5-22-30** W 6′.1.
08 37 32.7	+28 42 12	NGC 2619	12.4	1.8 x 1.1	13.0	Sbc	35	
08 37 28.4	+24 56 45	NGC 2620	13.5	2.0 x 0.5	13.4	S?	93	NGC 2621 N 3′.9.
08 37 37.1	+24 59 58	NGC 2621	14.8	0.8 x 0.4	13.4		171	Mag 8.85v star NW 8′.5.
08 38 10.9	+24 53 40	NGC 2622	14.1	0.8 x 0.4	12.7	S?	33	**CGCG 120-11** SW 1′.0.
08 38 24.3	+25 45 13	NGC 2623	13.4	2.4 x 0.7	13.8	Pec	60	Multi-galaxy system; long streamers extend from bright center NE and SW.
08 38 09.7	+19 43 30	NGC 2624	14.1	0.6 x 0.5	12.6		27	Mag 7.75v star W 7′.7.
08 38 23.2	+19 42 56	NGC 2625	15.0	0.4 x 0.4	12.9			**CGCG 89-56** S 7′.3.
08 40 22.8	+23 32 21	NGC 2628	13.3	1.1 x 1.1	13.3	SAB(r)c?		Mag 9.34v star S 5′.2; **CGCG 120-19** SW 6′.8.
08 41 13.5	+19 41 24	NGC 2637	15.4	0.8 x 0.5	14.2		51	Mag 6.82 star 42 Cancri W 7′.3.
08 41 51.8	+19 42 07	NGC 2643	14.9	0.7 x 0.4	13.3		21	Pair of E-W oriented mag 8.47v, 9.60v stars SW 4′.0.
08 42 43.1	+19 39 00	NGC 2647	14.3	0.8 x 0.5	13.2		18	Almost stellar; mag 9.73v star S 4′.2.
08 48 27.4	+19 01 06	NGC 2667	14.0	0.7 x 0.2	11.7	Sb	78	Elongated galaxy **IC 2411** NNE 1′.6.
08 49 21.7	+19 04 30	NGC 2672	11.7	3.0 x 2.8	14.0	E1-2	117	NGC 2673 on E edge of core.
08 49 24.2	+19 04 23	NGC 2673	13.5	1.2 x 1.2	13.9	E0 pec		Almost stellar nucleus. Located on E edge of core of NGC 2672.
08 50 01.4	+19 00 35	NGC 2677	14.6	0.7 x 0.4	13.1	Sb	171	
08 57 23.5	+17 17 18	NGC 2711	13.7	0.9 x 0.6	12.9	SB?	170	
09 02 38.9	+25 56 01	NGC 2735	13.3	1.2 x 0.4	12.3	SAB(rs)b? pec	94	NGC 2735A on NE edge; mag 7.28v star S 4′.9.
09 02 42.0	+25 56 15	NGC 2735A	15.3	0.4 x 0.2	12.4	Im: pec		Located on NE edge of NGC 2735.
09 03 59.7	+21 54 23	NGC 2737	14.1	0.9 x 0.4	12.9	Sab	61	NGC 2738 N 3′.9.
09 04 00.6	+21 58 06	NGC 2738	13.1	1.4 x 0.6	12.8	S?	55	Mag 9.87v star E 6′.7; NGC 2737 S 3′.9.
09 04 54.4	+25 00 15	NGC 2743	13.6	1.1 x 0.8	13.4	Sdm:	105	
09 04 38.8	+18 27 46	NGC 2744	13.5	1.7 x 1.1	14.0	SB(s)ab: pec	120	Double galaxy system. Small companion 0′.4 S of center of galaxy.
09 04 39.3	+18 15 24	NGC 2745	14.6	0.4 x 0.3	12.1		0	Almost stellar; mag 9.20v star WNW 4′.3.
09 05 18.4	+18 26 31	NGC 2747	14.9	0.6 x 0.4	13.2		170	Mag 10.9 star SE 2′.3.
09 05 21.3	+18 18 44	NGC 2749	11.8	1.7 x 1.4	12.7	E3	69	Several superimposed stars along E edge.
09 05 48.4	+25 26 02	NGC 2750	11.9	2.2 x 1.9	13.3	SABc	81	Many bright knots northern half.
09 05 32.2	+18 15 45	NGC 2751	14.3	0.8 x 0.6	13.3	Sbc	141	Mag 13.0 star on N edge; NGC 2749 NW 4′.0.
09 05 43.0	+18 20 20	NGC 2752	13.7	1.9 x 0.4	13.3	SBb: sp II-III	58	Mag 9.8 star NE 1′.7.
09 07 08.5	+25 20 28	NGC 2753	14.5	0.8 x 0.5	13.4		15	
09 07 30.7	+18 26 02	NGC 2761	14.3	0.6 x 0.4	12.6	Sm	150	

(Continued from previous page)

GALAXIES

RA h m s	Dec ° ′ ″	Name	Mag (V)	Dim ′ Maj x min	SB	Type Class	PA	Notes
09 08 17.5	+21 26 35	NGC 2764	12.9	1.5 x 0.9	13.1	S0:	15	Mag 9.27v star SE 6′.1.
09 08 47.8	+29 51 52	NGC 2766	13.6	1.3 x 0.5	13.0	Sab	132	
09 10 39.8	+18 41 45	NGC 2774	13.7	0.9 x 0.9	13.4			Almost stellar; bright knot or star W of core.
09 13 39.6	+29 59 35	NGC 2783	12.7	2.1 x 1.5	13.9	E	168	The small galaxy MCG +5-22-20 lies NW of the center and within NGC 2783's envelope. UGC 4856 lies slightly further W.
09 14 59.8	+29 43 46	NGC 2789	12.2	1.5 x 1.3	12.8	S0/a	20	Bright core, faint envelope; **CGCG 151-30** NW 7′.8.
09 15 02.8	+19 41 49	NGC 2790	14.5	0.6 x 0.4	12.8	Sc	51	Located 1′.3 SE of a mag 11.02v star.
09 15 02.1	+17 35 27	NGC 2791	14.6	0.8 x 0.3	12.9		156	Pair of faint stars on E edge.
09 16 02.0	+17 35 23	NGC 2794	13.2	1.2 x 1.2	13.4	SB?		Mag 8.64v star S 6′.5; several anonymous stellar galaxies SSE 3′.8.
09 16 04.0	+17 37 41	NGC 2795	12.8	1.4 x 1.0	13.2	E	170	Small, faint anonymous galaxy on S edge. Another anonymous galaxy lies 2′.0 NW; **MCG +3-24-22** lies 2′.4 NE.
09 16 21.8	+17 43 34	NGC 2797	14.0	0.7 x 0.6	12.9	Pec	21	IC 2454 NW 7′.4; mag 11.9 star W 2′.9; **MCG +3-24-21** W 3′.4.
09 16 44.1	+19 56 05	NGC 2801	14.0	1.1 x 1.0	14.0	SA(s)c	62	Mag 9.87v star NE 4′.4; faint, elongated anonymous galaxy W 4′.1; stellar **CGCG 91-45** S 5′.9.
09 16 41.5	+18 57 48	NGC 2802	14.3	0.9 x 0.6	13.5	S?	132	
09 16 43.9	+18 57 16	NGC 2803	14.0	1.1 x 0.9	13.8	S?	45	
09 16 50.2	+20 11 53	NGC 2804	12.9	1.4 x 1.2	13.3	S0	60	Faint, stellar anonymous galaxy S 2′.0.
09 17 00.6	+20 02 11	NGC 2806	15.5	0.7 x 0.4	14.0	Sb	153	NGC 2807 SW 1′.0; mag 9.87v star S 3′.3.
09 16 57.7	+20 01 43	NGC 2807	14.3	0.5 x 0.4	12.4		81	NGC 2806 NE 1′.0.
09 17 06.9	+20 04 09	NGC 2809	13.0	1.5 x 1.3	13.6	S0	171	Almost stellar anonymous galaxy NW 1′.6; another SE 2′.9.
09 17 40.8	+19 55 07	NGC 2812	15.2	0.5 x 0.1	11.8	S0/Sb	150	NGC 2813 SE 1′.9.
09 17 45.4	+19 54 23	NGC 2813	13.5	1.3 x 1.1	13.8	S0	145	A mag 9.22v star SW 2′.0; NGC 2812 NW 1′.9; **CGCG 91-58** SW 4′.3.
09 19 02.2	+26 16 11	NGC 2824	13.3	0.9 x 0.6	12.5	S0	160	Located 3′.1 WNW of mag 6.60v star.
09 20 28.7	+18 55 34	NGC 2843	15.5	0.4 x 0.2	12.7		57	Lies 2′.6 NW of mag 10.32 star, and has faint star on N edge.
09 24 54.7	+26 46 28	NGC 2862	12.9	2.5 x 0.5	13.0	S?	114	Mag 8.59v star S 6′.0.
09 27 18.6	+23 01 10	NGC 2885	13.9	0.7 x 0.4	12.4	S0	70	
09 30 17.0	+29 32 24	NGC 2893	13.2	1.1 x 1.0	13.2	(R)SB0/a	79	3′.1 SW of mag 9.15v star.
09 30 16.8	+23 39 47	NGC 2896	13.9	0.9 x 0.9	13.5			
09 32 10.2	+21 29 54	NGC 2903	9.0	12.6 x 6.0	13.6	SAB(rs)bc I-II	17	**UGC 5086** ESE 9′.3.
09 34 57.8	+21 42 18	NGC 2916	12.1	2.5 x 1.7	13.5	SA(rs)b?	20	**CGCG 122-22** S 5′.0.
09 37 15.2	+23 35 23	NGC 2927	12.9	1.3 x 1.0	13.0	SAB(rs)b	155	
09 37 29.8	+23 09 39	NGC 2929	13.8	1.2 x 0.3	12.5	S?	144	Mag 7.13v star SW 10′.9; **MCG +4-23-21** and 22 E 6′.1; NGC 2930 N 2′.6.
09 37 32.7	+23 12 11	NGC 2930	14.2	0.7 x 0.4	12.7	S?	138	NGC 2929 S 2′.6; NGC 2931 N 2′.6.
09 37 37.9	+23 14 26	NGC 2931	14.2	0.8 x 0.6	13.2	Sb	69	NGC 2930 S 2′.6.
09 37 55.1	+17 00 51	NGC 2933	15.3	1.0 x 0.4	14.1	Sbc	30	Mag 9.93v star S 1′.7; **CGCG 92-13** NW 7′.9.
09 38 24.3	+17 02 40	NGC 2941	15.0	0.8 x 0.5	13.9		163	Stellar nucleus in smooth envelope.
09 38 32.6	+17 01 52	NGC 2943	12.4	2.2 x 1.2	13.5	E	130	Stellar **MCG +3-25-12** E 1′.6.
09 39 01.7	+17 01 32	NGC 2946	14.0	1.2 x 0.4	13.1	SB?	13	Stellar nucleus.
09 46 47.9	+22 00 39	NGC 2988	14.6	0.8 x 0.3	12.9	SBbc	30	Located open W edge of NGC 2991.
09 46 50.2	+22 00 48	NGC 2991	12.6	1.4 x 1.1	13.0	S0	138	NGC 2988 on W edge.
09 47 16.1	+22 05 21	NGC 2994	13.1	1.3 x 1.0	13.2	S0	125	
09 50 54.3	+28 32 59	NGC 3026	12.9	2.7 x 0.8	13.6	Im	82	
09 52 08.2	+29 14 08	NGC 3032	12.5	1.7 x 1.3	13.3	SAB(r)0°	95	Located between a mag 9.08v star 1′.8 N and a mag 10.96v star 1′.8 S.
09 53 05.1	+19 25 57	NGC 3040	13.8	0.7 x 0.5	12.5	Sb	171	A spherical anonymous galaxy S 1′.1; an elongated anonymous galaxy NW 0′.7.
09 58 39.6	+28 52 27	NGC 3068	14.3	1.1 x 0.9	14.2	S0⁻: pec	53	Double system with companion at 0′.6 SW.
10 01 08.3	+22 24 20	NGC 3088A	13.8	0.5 x 0.3	11.6		69	NGC 3088B on SE edge; **UGC 5381** SW 6′.7.
10 01 10.2	+22 23 59	NGC 3088B	15.4	0.5 x 0.1	12.0	Sbc	138	NGC 3088A on NW edge.
10 02 17.0	+24 42 39	NGC 3098	12.0	2.3 x 0.5	12.0	S0 sp	90	
08 37 43.0	+20 30 15	UGC 4504	14.1	1.2 x 0.6	13.6	Sd	25	Uniform surface brightness.
08 40 54.0	+19 21 15	UGC 4526	13.9	1.5 x 0.4	13.2	Sab	54	Mag 9.50v star SW 3′.2.
08 41 41.4	+18 51 33	UGC 4532	14.7	1.5 x 0.2	13.2	Sbc	34	Located in a small triangle of mag 13 stars.
08 42 52.6	+25 04 10	UGC 4542	14.1	1.1 x 0.9	13.9	SAm	135	Uniform surface brightness.
08 46 35.7	+19 01 07	UGC 4588	14.3	1.5 x 0.3	13.3	Scd:	159	Mag 12.83v star SW 2′.3.
08 46 58.1	+28 14 15	UGC 4591	14.9	1.5 x 0.2	13.4	Scd:	18	IC 2394 E 2′.0; mag 6.60v star S 5′.0.
08 46 58.0	+21 42 46	UGC 4592	14.0	1.5 x 0.5	13.5	SB?	131	Mag 11.6 star NE 2′.5.
08 47 08.7	+19 37 46	UGC 4596	13.3	1.2 x 1.0	13.4	(R)SA(r)0⁺	155	Bright center, faint envelope.
08 47 40.2	+25 53 37	UGC 4597	13.9	1.0 x 0.5	13.0	S?	70	
08 47 55.9	+25 49 53	UGC 4602	14.0	1.2 x 0.5	13.3	Sab	130	
08 48 50.5	+29 52 11	UGC 4611	14.8	1.7 x 0.2	13.5	Sd	104	
08 49 27.2	+29 31 12	UGC 4617	14.4	1.5 x 0.6	14.1	SAd	57	Located between a pair of stars, mags 11.3 and 12.4.
08 51 15.9	+19 20 29	UGC 4631	13.4	0.9 x 0.8	12.9	S0	45	MCG +3-23-13 W 9′.9.
08 52 40.5	+21 25 21	UGC 4643	13.1	1.3 x 0.9	13.1	SA(s)bc	96	
08 55 06.7	+18 56 03	UGC 4669	13.9	1.5 x 1.3	14.5	Sdm:	90	Stellar nucleus with superimposed star or bright knot W of nucleus; smooth envelope.
08 58 32.9	+28 16 01	UGC 4698	13.3	1.5 x 0.7	13.2	SBab	55	
09 00 23.8	+25 36 38	UGC 4722	14.6	1.5 x 0.2	13.1	Sdm:	32	Mag 10.76v star W 2′.5.
09 00 42.1	+17 37 14	UGC 4729	13.8	0.9 x 0.7	13.1	SB(s)cd:	156	Bright E-W bar; mag 11.01v star SE 3′.3.
09 02 06.2	+23 23 09	UGC 4740	13.9	1.1 x 0.5	13.1	S?	166	Mag 10.7 star SW 2′.1.
09 02 44.0	+25 25 14	UGC 4746	14.0	1.5 x 0.6	13.7	Sbc	147	
09 04 47.3	+22 01 40	UGC 4758	14.0	1.3 x 0.6	13.6	SBb	15	Mag 14.3 star on NW edge.
09 04 56.2	+17 27 17	UGC 4761	13.2	0.9 x 0.8	12.9	E	72	**MCG +3-23-33** W 1′.2; **MCG +3-23-29** W 10′.0.
09 06 00.3	+18 45 52	UGC 4773	13.7	1.6 x 0.4	13.1	S?	69	Mag 10.24v star S 1′.9.
09 06 39.7	+19 20 14	UGC 4780	13.5	1.5 x 1.2	14.0	SABdm	60	
09 09 20.3	+20 41 51	UGC 4809	13.9	1.6 x 0.6	13.7	SBcd?	105	**MCG +4-22-18** SSW 7′.8.
09 11 09.7	+19 40 02	UGC 4828	15.0	1.5 x 0.2	13.6	Scd:	6	Mag 13.3 star W 1′.4.
09 13 01.3	+20 21 49	UGC 4853	13.9	1.1 x 0.9	13.7	SBcd:	25	
09 13 32.8	+29 59 59	UGC 4856	14.3	1.5 x 0.2	12.8	Sb III	77	NGC 2783 and MCG +5-22-20 E end.
09 13 22.6	+19 22 11	UGC 4858	14.9	1.3 x 0.8	14.8	Im V	69	Dwarf irregular, very low surface brightness, mag 14.5 star on S edge; mag 8.99v star SW 10′.6.
09 13 48.9	+29 10 24	UGC 4860	14.1	0.9 x 0.3	12.5	Double System	17	Very faint anonymous galaxy E 1′.2.
09 16 49.6	+27 29 21	UGC 4895	14.1	1.2 x 0.4	13.1	Sb II-III	163	Mag 6.78v star ESE 11′.0.
09 17 05.2	+25 25 45	UGC 4902	12.8	1.4 x 0.8	12.9	S0?	157	
09 17 29.5	+25 58 00	UGC 4912	13.9	0.7 x 0.4	12.4	S0?	39	
09 18 20.0	+17 45 09	UGC 4925	14.7	1.5 x 0.2	13.2	Sd	78	Mag 9.50v star SW 2′.5.
09 19 37.9	+27 27 23	UGC 4940	13.7	1.2 x 0.7	13.4	SABab	58	Mag 9.0v star SE 2′.5.
09 22 53.1	+21 58 32	UGC 4985	13.7	1.5 x 0.6	13.5	SAB(s)b	177	Located 2′.0 NW of mag 9.83v star.
09 23 35.4	+24 45 37	UGC 4994	14.0	1.0 x 0.4	12.9	Sab	33	**CGCG 121-77** NE 4′.5; **CGCG 121-81** SE 10′.0.
09 24 23.3	+28 17 26	UGC 5002	14.1	0.9 x 0.5	13.1	SB?	176	Mag 9.82v star NE 2′.4.
09 24 29.6	+22 16 31	UGC 5005	14.5	1.5 x 1.2	15.0	Im	33	Dwarf irregular, very low surface brightness.
09 24 44.6	+20 01 43	UGC 5009	14.1	1.0 x 0.8	13.7	Scd:	150	
09 26 01.3	+19 22 58	UGC 5023	13.8	0.7 x 0.6	12.7	S?	57	Mag 13.3 star on NE edge.
09 27 10.4	+21 35 40	UGC 5035	13.9	1.0 x 0.9	13.6	(R)SB(s)a	172	

(Continued from previous page)

74 GALAXIES 74

RA h m s	Dec ° ′ ″	Name	Mag (V)	Dim ′ Maj x min	SB	Type Class	PA	Notes
09 27 36.1	+28 47 53	UGC 5040	13.4	2.0 x 1.9	14.7	Im?	126	Stellar nucleus, uniform envelope; mag 13.7 star S edge.
09 31 53.0	+29 47 33	UGC 5074	13.7	0.9 x 0.7	13.1	Sa?	36	
09 32 58.8	+27 29 57	UGC 5084	13.8	1.0 x 0.6	13.1	S?	15	Pair of mag 12 stars SE 1′.9.
09 34 14.3	+24 13 33	UGC 5094	14.0	0.6 x 0.4	12.4	S0	150	Mag 9.70v star W 3′.4; **CGCG 122-19** E 3′.0.
09 35 26.3	+29 48 44	UGC 5108	13.6	1.3 x 0.6	13.2	SBab	134	
09 37 09.3	+19 50 10	UGC 5123	13.9	0.6 x 0.5	12.4	S?	155	**CGCG 92-14** SE 9′.7.
09 37 58.2	+25 29 40	UGC 5129	13.2	1.7 x 0.7	13.2	Sa	103	
09 42 55.2	+28 58 54	UGC 5185	13.6	1.7 x 0.6	13.5	Sc	168	Stellar galaxy **KUG 0939+292** NW 1′.4; **CGCG 152-55** WSW 6′.4.
09 43 08.4	+21 10 54	UGC 5192	13.4	1.9 x 0.6	13.4	SBb:	90	
09 45 49.8	+28 28 20	UGC 5219	15.9	0.9 x 0.7	15.2	SAB(s)d	123	Mag 13.9 star NE 3′.1.
09 47 49.2	+23 43 20	UGC 5246	14.2	0.8 x 0.6	13.3	S	140	Mag 6.49v star SW 7′.5.
09 53 56.5	+23 22 57	UGC 5313	14.0	0.7 x 0.6	12.9	S?	130	
09 54 17.8	+23 17 15	UGC 5320	13.5	1.8 x 0.7	13.6	SB(r)cd	102	
09 55 35.4	+19 17 39	UGC 5330	13.9	1.8 x 1.2	14.6	Pec	102	
09 56 22.7	+20 28 46	UGC 5339	13.5	1.0 x 0.6	12.8	Double System	150	Double system, contact; **MCG +4-24-3** NW 2′.0.
09 56 45.9	+28 49 43	UGC 5340	14.5	2.4 x 1.0	15.3	Im pec:	7	Dwarf peculiar, low, fairly uniform surface brightness.
09 56 35.7	+20 38 37	UGC 5341	14.3	2.6 x 0.2	13.5	Scd:	57	
09 58 52.0	+19 12 50	UGC 5359	14.1	1.2 x 0.4	13.1	Sbc	97	
10 02 35.7	+19 10 37	UGC 5403	13.5	1.2 x 0.4	12.5	S0/a	80	Mag 8.54v star SW 3′.4.
10 03 59.1	+22 16 30	UGC 5420	13.5	1.0 x 0.7	13.0	S0:	81	Anonymous companion on S edge.

GALAXY CLUSTERS

RA h m s	Dec ° ′ ″	Name	Mag 10th brightest	No. Gal	Diam ′	Notes
08 40 54.0	+26 44 00	A 692	16.2	38	20	All members anonymous, faint and stellar.

OPEN CLUSTERS

RA h m s	Dec ° ′ ″	Name	Mag	Diam ′	No. ★	B ★	Type	Notes
08 40 22.0	+19 40 00	NGC 2632	3.1	70	161	6.0	cl	= **M 44**, **Beehive**, **Praesepe**. Moderately rich in bright and faint stars; detached.

PLANETARY NEBULAE

RA h m s	Dec ° ′ ″	Name	Diam ″	Mag (P)	Mag (V)	Mag cent ★	Alt Name	Notes
08 46 53.6	+17 52 44	PN G208.5+33.2	127	15.6		14.3	PK 208+33.1	Mag 13.8 star near the center; mag 14.2 star on N edge.

75 GALAXIES 75

RA h m s	Dec ° ′ ″	Name	Mag (V)	Dim ′ Maj x min	SB	Type Class	PA	Notes
07 32 05.8	+17 48 50	CGCG 87-4	14.8	0.4 x 0.3	12.3		51	Low surface brightness; mag 11.7 star NE 0′.9.
07 39 29.2	+18 10 45	CGCG 87-25	15.0	0.7 x 0.5	13.7		105	Mag 12.9 star W 2′.3; mag 12.0 star NNE 3′.2.
08 24 14.4	+17 19 53	CGCG 89-16	14.8	0.4 x 0.4	12.9	E		= **Mkn 387**. Mag 9.73v star SSE 4′.4; mag 8.83v star S 7′.9.
07 18 25.1	+25 56 15	CGCG 117-6	15.0	0.6 x 0.4	13.3		90	Mag 7.97v star E 6′.7.
07 23 39.8	+22 38 38	CGCG 117-33	15.3	0.5 x 0.4	13.4		90	Mag 11.5 star W 3′.4.
07 27 57.8	+26 25 10	CGCG 117-41	15.1	0.4 x 0.4	13.0			Located in an E-W rectangle of eight mag 12-14 stars.
07 40 42.7	+23 37 31	CGCG 117-65	14.6	0.4 x 0.4	12.4			Almost stellar.
07 45 25.5	+23 59 20	CGCG 118-3	15.2	0.5 x 0.4	13.3		63	Mag 10.2 star NW 5′.2.
07 46 15.9	+24 05 51	CGCG 118-5	15.4	0.6 x 0.3	13.4		147	Mag 10.25v star NE 5′.4; mag 10.6 star SW 5′.7.
07 50 40.7	+23 29 36	CGCG 118-14	14.7	0.6 x 0.5	13.2		15	Mag 13.8 star W 1′.2.
07 53 45.4	+24 15 07	CGCG 118-18	15.6	0.6 x 0.3	13.6		150	Mag 10.8 star on N end.
07 54 30.9	+25 48 53	CGCG 118-21	15.0	0.4 x 0.4	12.8			Mag 10.90v star NE 3′.2.
08 00 28.6	+21 08 08	CGCG 118-40	15.1	0.5 x 0.3	12.9	Sc	174	Located between a mag 8.71v star SE 8′.4 and a mag 9.72v star NW 8′.8.
08 19 13.1	+20 45 23	CGCG 119-51	13.3	0.8 x 0.5	12.2	Sb	138	Stellar nucleus.
07 21 58.7	+29 29 47	CGCG 147-13	15.4	0.5 x 0.4	13.5	Sm	18	Mag 8.87v star NE 4′.8.
07 38 32.7	+29 11 06	CGCG 147-48	15.4	0.4 x 0.2	12.5		168	Stellar.
07 39 52.2	+29 44 10	CGCG 147-51	15.9	0.4 x 0.4	13.8			Mag 11.8 star S 2′.9; mag 12.4 star W 3′.6.
07 44 57.5	+28 55 35	CGCG 148-6	14.9	0.6 x 0.5	13.5		165	Mag 11.5 star E 4′.4.
07 47 20.6	+29 24 42	CGCG 148-19	15.2	0.5 x 0.3	12.9		117	Mag 10.6 star E 2′.5.
07 49 44.4	+27 51 09	CGCG 148-39	15.2	0.5 x 0.3	13.0		81	Forms a triangle with a mag 12.9 star N 1′.9 and a mag 12.6 star E 2′.3; **CGCG 148-45** E 7′.2.
08 08 04.2	+29 57 13	CGCG 148-115	15.6	0.4 x 0.3	13.1		18	Mag 10.77v star WNW 6′.4.
08 09 20.2	+28 00 42	CGCG 148-117	13.5	0.5 x 0.5	11.8	Sc		Mag 15.4 star on S edge.
08 23 19.8	+29 27 31	CGCG 149-19	15.9	0.5 x 0.3	13.7		66	Mag 13.2 star on W end.
08 28 56.5	+27 44 45	CGCG 149-34	15.8	0.4 x 0.3	13.3	Irr	75	Very faint star N edge.
08 33 42.7	+27 42 39	CGCG 149-50	14.8	0.6 x 0.4	13.1	Spiral	33	Mag 10.96v star NE 2′.4.
08 36 47.1	+27 19 55	CGCG 150-5	15.6	0.6 x 0.4	13.9			Mag 14.4 star on N edge.
08 40 02.3	+29 48 59	CGCG 150-14	15.0	0.5 x 0.4	13.1	Radio galaxy	36	Mag 10.59v star ENE 7′.6.
07 54 22.6	+27 00 28	IC 479	14.7	0.6 x 0.5	13.2	SBc	171	Mag 8.22v star NE 11′.1; **CGCG 148-57** SW 5′.3.
07 55 23.3	+26 44 33	IC 480	14.2	1.7 x 0.3	13.4	Sbc	168	Mag 10.4 star NNW 2′.4.
07 59 47.3	+25 21 22	IC 482	14.5	0.6 x 0.5	13.0		156	Mag 9.97v star ESE 8′.3.
08 00 19.9	+26 42 05	IC 485	14.5	1.2 x 0.3	13.2	Sa:	153	Mag 10.55v star WNW 9′.1; **IC 484** SW 4′.6.
08 00 21.2	+26 36 45	IC 486	13.7	1.0 x 0.7	13.2	SBa	139	Mag 9.64v star SW 10′.7.
08 05 38.7	+26 10 02	IC 492	13.5	1.1 x 0.9	13.3	SB(s)bc:	138	Mag 8.14v star S 3′.3.
08 07 27.6	+25 07 59	IC 493	14.1	0.8 x 0.5	12.9	Sb	24	Mag 11.5 star E 7′.2.
08 09 45.0	+25 52 47	IC 496	15.8	0.8 x 0.5	14.7	Sb	108	Galaxy pair; slightly elongated component W.
08 28 22.4	+25 07 28	IC 508	14.0	0.7 x 0.7	13.1	S?		Bright, elongated center.
08 32 03.5	+24 00 35	IC 509	13.0	1.2 x 1.1	13.1	SA(rs)c I-II	0	Stellar nucleus; mag 5.70v star NW 8′.6.
07 22 43.2	+21 30 45	IC 2186	13.9	0.8 x 0.5	12.8		111	IC 2188 S 1′.8.
07 22 43.4	+21 28 56	IC 2188	13.9	0.4 x 0.4	11.8			Mag 11.9 star on N edge.
08 00 50.0	+27 29 59	IC 2217	14.0	0.6 x 0.5	12.5	S?	80	Mag 11.9 star W 2′.2.
08 02 36.8	+27 26 15	IC 2219	13.7	1.4 x 0.6	13.4	Sc	175	Pair of stars, mags 11.6 and 10.04v, NW 2′.5.
08 14 06.8	+23 51 58	IC 2239	13.6	1.0 x 0.7	13.1	S0?	168	Star on E edge.
08 16 04.9	+23 07 59	IC 2248	14.4	1.1 x 0.8	14.2	Sb	0	Bright NW-SW oriented center, faint envelope N and S.

GALAXIES

RA h m s	Dec ° ′ ″	Name	Mag (V)	Dim ′ Maj x min	SB	Type Class	PA	Notes
08 16 34.0	+21 24 32	IC 2253	13.9	0.7 x 0.5	12.8	E?	168	Mag 9.15v star SW 13′.7.
08 18 01.8	+24 44 02	IC 2267	14.1	2.1 x 0.3	13.5	SBcd?	153	Mag 7.51v star NW 8′.3.
08 19 32.2	+21 23 36	IC 2293	14.0	0.9 x 0.7	13.4	S?	123	Mag 7.86v star NW 10′.6; NGC 2557 NW 5′.7.
08 22 20.0	+19 24 52	IC 2329	14.0	2.1 x 0.4	13.7	Sdm:	117	Mag 8.74v star NW 8′.4.
08 23 34.3	+21 20 50	IC 2338	14.4	1.0 x 0.5	13.4	SAB(s)cd pec	72	**IC 2339** on S edge; mag 9.57v star ENE 5′.2.
08 23 29.9	+18 44 57	IC 2340	13.9	0.7 x 0.5	12.6		171	
08 23 41.5	+21 26 04	IC 2341	13.6	1.3 x 0.6	13.2	S0^{-}:	1	Mag 9.57v star SE 4′.8.
08 25 44.7	+27 52 23	IC 2361	13.9	1.2 x 0.4	13.0	S?	78	Mag 5.58v star 22 Cancri E 9′.5.
08 25 45.5	+19 26 54	IC 2363	13.9	0.8 x 0.8	13.3	SB(s)bc: I-II		Bright N-S bar with strong dark patches E and W.
08 26 18.1	+27 50 20	IC 2365	13.4	1.1 x 0.8	13.1	S0	45	Mag 5.58v star 22 Cancri 3′.8 NE.
08 26 48.9	+20 21 50	IC 2373	14.7	0.9 x 0.9	14.3	Scd?		Mag 9.96v star NW 4′.8.
08 28 46.2	+22 03 08	IC 2382	14.2	0.6 x 0.4	12.5	Sb	168	Star on N end.
07 27 22.1	+19 38 21	MCG +3-19-13	13.9	0.5 x 0.4	12.0	S?	18	MCG +3-19-14 S 1′.0.
07 27 24.0	+19 37 27	MCG +3-19-14	14.1	0.6 x 0.6	12.8	S?		MCG +3-19-13 N 1′.0.
07 29 28.0	+20 03 15	MCG +3-19-18	14.1	0.8 x 0.4	12.7	Sb	147	Mag 12.4 star N 1′.1.
07 31 12.3	+18 14 35	MCG +3-19-20	14.2	0.8 x 0.4	12.8	Sb	18	Mag 14.3 star N 0′.9.
07 35 56.8	+18 02 55	MCG +3-20-7	15.0	0.6 x 0.5	13.6	Sc	45	Mag 9.19v star N 4′.3.
07 56 41.7	+17 59 25	MCG +3-21-2	15.0	0.4 x 0.3	12.5		57	Mag 12.2 star N 0′.9 with faint, elongated anonymous galaxy in between them; mag 9.49v star SE 2′.8.
08 06 52.1	+18 44 16	MCG +3-21-15	14.3	0.7 x 0.4	12.7		90	Mag 9.45v star NE 1′.3.
08 23 48.0	+17 59 29	MCG +3-22-8	15.8	0.5 x 0.3	13.8	E/S0	171	
08 30 54.1	+20 14 52	MCG +3-22-16	15.0	0.8 x 0.8	14.3			Mag 11.5 star NE 1′.1.
08 32 40.5	+19 30 38	MCG +3-22-18	16.4	0.5 x 0.5	14.7	S		Mag 10.22v star W 3′.3.
08 38 23.9	+17 37 50	MCG +3-22-20	14.3	0.7 x 0.5	13.0	S0?	39	Located between a close pair of mag 14.8 stars.
07 20 53.9	+23 19 02	MCG +4-18-5	14.2	1.0 x 0.5	13.3			
07 21 34.2	+23 32 54	MCG +4-18-7	13.3	0.9 x 0.9	12.9			Located between a close pair of E-W mag 11-12 stars.
07 22 48.7	+26 31 41	MCG +4-18-9	13.8	0.5 x 0.4	11.9		12	
07 23 11.7	+23 39 46	MCG +4-18-14	14.7	0.8 x 0.8	14.1	Sb		
07 48 53.3	+21 44 50	MCG +4-19-3	14.2	0.8 x 0.4	13.1	E/S0	27	
07 53 45.0	+21 02 55	MCG +4-19-7	14.4	0.8 x 0.5	13.2	Sb	36	
07 59 53.5	+23 23 25	MCG +4-19-17	14.3	1.0 x 0.5	13.4	Sbc	9	Mag 12.7 star W 2′.0.
08 20 49.4	+22 39 23	MCG +4-20-34	14.0	0.6 x 0.4	12.3	S?	168	Located between a pair of N-S mag 10.71v and 11.8 stars.
08 35 25.0	+23 31 29	MCG +4-20-69	13.6	0.9 x 0.8	13.1	S?	156	Mag 10.9 star NE 4′.1.
08 38 06.8	+25 16 22	MCG +4-21-7	14.4	0.5 x 0.5	12.7			
08 41 10.0	+20 53 54	MCG +4-21-13	14.9	0.5 x 0.5	13.2	Sm		
07 16 24.5	+27 37 08	MCG +5-17-19	15.3	0.5 x 0.4	13.4		51	
07 21 05.3	+29 20 15	MCG +5-18-7	14.1	0.9 x 0.6	13.3	Sb	51	Mag 10.7 star W 4′.5.
07 25 01.5	+27 19 27	MCG +5-18-11	13.7	0.9 x 0.7	13.0	S?	156	Mag 12.4 star S 0′.9.
07 37 31.6	+28 50 12	MCG +5-18-24	15.7	0.9 x 0.6	14.9		45	Faint stellar nucleus, uniform envelope; mag 11.3 star E 1′.8.
07 37 47.4	+28 38 40	MCG +5-18-25	14.5	0.7 x 0.7	13.6			
07 39 06.1	+29 09 35	MCG +5-18-26	14.7	0.6 x 0.6	13.5	Sb		Mag 10.04v star NW 2′.1.
07 47 04.9	+27 56 47	MCG +5-19-4	13.3	1.0 x 0.7	12.7	Sbc I-II	33	Mag 13.3 star on W edge; mag 11.1 star E 1′.5.
08 00 47.1	+29 44 51	MCG +5-19-30	14.6	0.7 x 0.4	13.0	SBc	129	Mag 11.3 star E 2′.0.
08 02 04.2	+27 13 32	MCG +5-19-33	14.7	0.8 x 0.4	13.3		3	
08 04 23.9	+29 30 49	MCG +5-19-37	15.3	0.6 x 0.2	12.8	Sb	96	Pair with MCG +5-19-37 S.
08 04 25.0	+29 30 22	MCG +5-19-38	14.6	0.8 x 0.2	12.5	S0	92	Pair with MCG +5-19-37 N; mag 9.63v star S 1′.9.
08 12 05.8	+29 24 14	MCG +5-20-3	14.8	0.9 x 0.3	13.2	S0/a	15	Mag 11.5 star 0′.7 W; faint anonymous galaxy SW 1′.3.
08 18 07.6	+26 38 01	MCG +5-20-7	14.4	0.8 x 0.6	13.5	SBa	51	Mag 10.32v star ENE 3′.8.
08 23 29.9	+27 08 11	MCG +5-20-11	14.2	0.6 x 0.5	12.7	S?		
08 35 08.7	+29 42 33	MCG +5-20-26	14.0	0.7 x 0.7	13.1	S?		
08 36 15.4	+28 03 34	MCG +5-21-1	14.0	0.8 x 0.6	13.1	SBc	168	Mag 13.6 star on E edge.
07 17 40.7	+23 21 24	NGC 2357	13.3	3.5 x 0.5	13.8	Scd:	122	
07 22 22.7	+22 05 01	NGC 2365	12.4	2.8 x 1.4	13.7	SABa	173	Faint envelope extends beyond brighter, elongated core.
07 25 01.6	+23 46 59	NGC 2370	13.7	0.9 x 0.5	12.7	SB?	43	
07 26 36.0	+23 04 23	NGC 2376	14.4	0.6 x 0.6	13.2	Sbc		
07 30 16.3	+24 29 11	NGC 2398	13.9	0.5 x 0.3	11.7	Sb	150	**MCG +4-18-22** WNW 0′.7.
07 32 14.0	+25 54 21	NGC 2405	14.1	0.7 x 0.4	12.5		105	Star or very bright knot E of nucleus; mag 9.56v star ESE 6′.9.
07 31 47.7	+18 17 17	NGC 2406	14.2	0.7 x 0.6	13.4	E/S0	51	NGC 2407 NNE 3′.5.
07 31 56.7	+18 19 57	NGC 2407	13.4	1.1 x 0.9	13.2	S0^{-}:	42	**MCG +3-20-2** E 2′.4; mag 7.94v star ENE 7′.5.
07 34 36.3	+18 16 53	NGC 2411	13.8	0.9 x 0.6	13.2	E?	50	Anonymous elongated galaxy ENE 1′.0.
07 36 38.1	+17 53 11	NGC 2418	12.2	1.8 x 1.8	13.4	E		Faint material fans outward NE from bright center.
07 47 20.3	+26 55 46	NGC 2449	13.4	1.3 x 0.6	12.9	Sab	137	Galaxy **IC 476** NW 1′.5.
07 47 32.4	+27 01 12	NGC 2450	14.7	0.9 x 0.2	12.6	Sbc	156	
07 57 10.5	+23 46 47	NGC 2480	13.8	1.3 x 0.7	13.5	SB?	160	
07 57 13.7	+23 45 58	NGC 2481	13.0	1.4 x 0.5	12.5	S?	18	
07 57 56.8	+25 09 38	NGC 2486	13.3	1.7 x 0.9	13.6	Sa	97	NGC 2487 E 5′.6.
07 58 20.5	+25 08 54	NGC 2487	12.5	2.6 x 2.1	14.2	SBb	115	Mag 10.45v star S edge; NGC 2486 W 5′.6.
07 59 17.9	+27 04 38	NGC 2490	14.6	0.5 x 0.4	12.7		45	Located 1′.4 S of mag 9.45v star; at the center of galaxy cluster A 610.
07 59 29.8	+27 01 32	NGC 2492	12.7	1.0 x 1.0	12.6	S0^{-}:		Mag 9.45v star NW 5′.3.
07 59 38.9	+24 58 54	NGC 2498	13.4	1.1 x 0.8	13.1	SBa:	113	Mag 8.69v star SSW 4′.8.
08 00 36.7	+22 23 59	NGC 2503	13.7	1.0 x 1.0	13.5	SAB(rs)bc		Located 1′.9 NW of mag 10.43v star.
08 03 07.9	+23 23 29	NGC 2512	13.1	1.4 x 0.9	13.2	SBb	113	
08 06 13.6	+17 42 24	NGC 2522	13.8	1.1 x 0.4	12.8	S0/a	32	
08 07 55.6	+17 49 03	NGC 2530	13.6	1.4 x 1.0	13.9	SB(s)d	170	Mag 13.7 star on N edge; **UGC 4232** SW 9′.0.
08 11 13.5	+25 12 13	NGC 2535	12.8	2.3 x 1.2	13.7	SA(r)c pec I	12	Forms an open "S" shape; NGC 2536 on S edge.
08 11 16.0	+25 10 43	NGC 2536	14.1	1.0 x 0.5	13.2	SB(rs)c pec	49	Almost stellar core, on S edge of NGC 2535.
08 12 46.6	+26 21 38	NGC 2540	13.5	1.3 x 0.9	13.5	SB(rs)cd:	128	
08 14 14.2	+21 21 21	NGC 2545	12.4	2.0 x 1.1	13.1	(R)SB(r)ab I-II	170	
08 17 34.9	+20 54 06	NGC 2553	13.9	1.7 x 0.8	14.1	S?	72	Slightly elongated, almost stellar N-S core in faint envelope.
08 17 53.8	+23 28 17	NGC 2554	12.0	3.2 x 2.3	14.0	S0/a	147	Very faint outer envelope. Almost stellar **CGCG 119-32** W 1′.4.
08 19 00.9	+20 56 12	NGC 2556	14.5	0.6 x 0.3	12.5	S0	138	Mag 8.13v star E 10′.2; **CGCG 119-53** NNE 8′.6.
08 19 10.8	+21 26 08	NGC 2557	13.2	1.2 x 1.0	13.3	SB0	55	Mag 7.86v star NW 5′.1; IC 2293 SE 5′.7.
08 19 12.7	+20 30 39	NGC 2558	13.0	1.7 x 1.3	13.7	SAB(rs)ab	160	Mag 10.70v star W 4′.8.
08 19 52.3	+20 59 02	NGC 2560	13.3	1.4 x 0.3	12.2	S0/a	93	Located 1′.6 E of mag 10.39 star. There is a small anonymous galaxy 0′.5 SW of this same star.
08 20 23.6	+21 07 54	NGC 2562	12.9	1.0 x 0.7	12.3	S0/a:	3	**MCG +4-20-29** SSW 4′.7.
08 20 35.6	+21 04 03	NGC 2563	12.2	2.1 x 1.5	13.3	S0^{0}:	80	Brightest member of the **Cancer Cluster**. Stellar galaxy **PGC 23406** lies 1′.2 NE.
08 19 48.5	+22 01 45	NGC 2565	12.6	1.7 x 0.9	12.9	(R′)SBbc: II	167	**CGCG 119-56** NW 1′.8.
08 21 21.2	+20 52 02	NGC 2569	14.3	0.6 x 0.5	12.9	E?	117	NGC 2570 N 2′.6; mag 9.93v star SW 6′.3.
08 21 22.6	+20 54 36	NGC 2570	14.5	1.1 x 0.6	13.9	Sab	75	Stellar nucleus, faint envelope; NGC 2569 S 2′.6.

(Continued from previous page)

RA h m s	Dec ° ′ ″	Name	Mag (V)	Dim ′ Maj x min	SB	Type Class	PA	Notes
08 21 24.8	+19 08 51	NGC 2572	13.8	1.3 x 0.5	13.2	Sa?	133	
08 22 45.1	+24 17 50	NGC 2575	12.7	1.7 x 1.3	13.4	SA(rs)cd:	145	Bright knots or superimposed stars E and S of almost stellar nucleus.
08 22 57.8	+25 44 17	NGC 2576	14.3	1.7 x 0.3	13.4	Sb III	41	Mag 8.47v star W 3'2.
08 22 43.6	+22 33 12	NGC 2577	12.4	1.8 x 1.1	13.0	S0⁻:	105	**UGC 4345** NE 9'5; **UGC 4361** NW 12'7.
08 24 31.1	+18 35 44	NGC 2581	13.4	1.1 x 0.8	13.1	SB?	10	Mag 9.95v star S 4'2; stellar **CGCG 89-21** E 9'3.
08 25 12.2	+20 20 04	NGC 2582	13.0	1.2 x 1.2	13.3	(R')SAB(s)ab I		
08 27 08.1	+25 58 10	NGC 2592	12.3	1.7 x 1.4	13.2	E	45	NGC 2594 SSE 5'9.
08 26 48.0	+17 22 28	NGC 2593	13.9	1.0 x 0.5	13.0	S0/a	172	Almost stellar **CGCG 89-26** W 7'5.
08 27 17.1	+25 52 43	NGC 2594	14.1	0.7 x 0.4	12.6		27	Located 0'5 NNW of a mag 11.40v star; NGC 2592 N 5'9.
08 27 42.0	+21 28 41	NGC 2595	12.3	3.2 x 2.4	14.3	SAB(rs)c	45	Mag 9.01v stars lies 2'2 SW of center.
08 27 26.8	+17 17 01	NGC 2596	13.5	1.5 x 0.6	13.2	Sb I	65	
08 30 02.7	+21 29 19	NGC 2598	13.6	1.3 x 0.6	13.2	SBa?	3	
08 32 11.4	+22 33 38	NGC 2599	12.2	1.9 x 1.7	13.4	SAa	105	Faint galaxy **KUG 0829+227B** E 8'2.
08 33 23.3	+29 32 16	NGC 2604	12.3	1.7 x 1.7	13.3	SB(rs)cd		**MCG +5-20-23** SE 3'6.
08 33 56.9	+26 58 19	NGC 2607	13.7	1.1 x 1.0	13.7	S?	57	Stellar nucleus, faint envelope.
08 35 17.0	+28 28 22	NGC 2608	12.3	1.9 x 1.2	13.0	SB(s)b: II	60	Mag 8.53v star W 8'9.
08 35 29.3	+25 01 36	NGC 2611	14.4	0.7 x 0.4	12.9		42	**CGCG 119-130** NE 10'1.
08 37 32.7	+28 42 12	NGC 2619	12.4	1.8 x 1.1	13.0	Sbc	35	
08 37 28.4	+24 56 45	NGC 2620	13.5	2.0 x 0.5	13.4	S?	93	NGC 2621 N 3'9.
08 37 37.1	+24 59 58	NGC 2621	14.8	0.8 x 0.4	13.4		171	Mag 8.85v star NW 8'5.
08 38 10.9	+24 53 40	NGC 2622	14.1	0.8 x 0.4	12.7	S?	33	**CGCG 120-11** SW 1'0.
08 38 24.3	+25 45 13	NGC 2623	13.4	2.4 x 0.7	13.8	Pec	60	Multi-galaxy system; long streamers extend from bright center NE and SW.
08 38 09.7	+19 43 30	NGC 2624	14.1	0.6 x 0.5	12.6		27	Mag 7.75v star W 7'7.
08 38 23.2	+19 42 56	NGC 2625	15.0	0.4 x 0.4	12.9			**CGCG 89-56** S 7'3.
08 40 22.8	+23 32 21	NGC 2628	13.3	1.1 x 1.1	13.3	SAB(r)c?		Mag 9.34v star S 5'2; **CGCG 120-19** SW 6'8.
08 41 13.5	+19 41 24	NGC 2637	15.4	0.8 x 0.5	14.2		51	Mag 6.82 star 42 Cancri W 7'3.
08 41 51.8	+19 42 07	NGC 2643	14.9	0.7 x 0.4	13.3		21	Pair of E-W oriented mag 8.47v, 9.60v stars SW 4'0.
08 42 43.1	+19 39 00	NGC 2647	14.3	0.8 x 0.5	13.2		18	Almost stellar; mag 9.73v star S 4'2.
07 16 43.0	+29 51 13	UGC 3777	13.9	1.8 x 0.3	13.2	Scd:	152	Mag 10.8 star W edge; mag 9.06v star NW 3'1.
07 19 58.5	+22 05 28	UGC 3803	13.9	0.7 x 0.5	12.6	Sa:	15	Mag 3.54v star Delta Geminorum SW 6'8.
07 20 01.0	+17 56 41	UGC 3805	13.8	1.2 x 0.4	12.9	S?	147	Mag 11.9 star W 1'5.
07 20 22.8	+22 54 25	UGC 3806	14.1	0.9 x 0.8	13.6	SBcd:	117	Mag 9.90v star E 2'4.
07 21 02.0	+25 10 45	UGC 3808	14.0	1.8 x 0.9	14.4	SB(s)d:	50	Pair of mag 10.5 stars NW 1'5.
07 22 20.0	+17 17 12	UGC 3820	14.2	1.5 x 0.3	13.2	Scd:	100	Located between a pair of stars N-S, mags 10.61v and 9.92v.
07 22 47.7	+18 55 35	UGC 3823	13.9	1.1 x 0.6	13.3	Sb II-III	91	Mag 11.6 star W end; mag 11.4 star S 1'1.
07 22 51.1	+22 35 18	UGC 3824	13.4	0.9 x 0.8	12.9	S0	21	Compact; mag 7.84v star WNW 9'7.
07 23 07.6	+22 12 27	UGC 3827	13.6	1.0 x 0.8	13.2	SB?	45	Mag 13.2 star on SE edge.
07 25 21.1	+19 10 35	UGC 3840	13.0	1.1 x 1.1	13.2	E		**CGCG 86-24** W 5'8.
07 25 37.5	+19 07 38	UGC 3842	13.4	1.1 x 0.8	13.2	S0	120	Bright knot or star NW edge.
07 27 47.8	+20 23 52	UGC 3862	14.2	1.5 x 0.4	13.5	S?	13	Mag 11.4 star SE 2'8.
07 28 38.8	+24 28 48	UGC 3869	14.2	1.0 x 0.4	13.1	S0?	25	Mag 10.73v star SW 1'6.
07 28 54.2	+20 35 31	UGC 3873	13.8	2.1 x 0.5	13.7	Scd:	138	Located between a close pair of stars, mags 11.3 and 11.5.
07 29 06.0	+26 54 56	UGC 3874	13.6	1.6 x 1.1	14.1	S?	156	Peculiar, plumes.
07 29 17.5	+27 54 02	UGC 3876	13.1	2.2 x 1.1	13.9	SA(s)d	2	Mag 5.01v star 65 Geminorum E 7'0.
07 33 09.5	+19 11 58	UGC 3903	13.6	1.5 x 0.6	13.4	(R')SAB(rs)b	174	Mag 13.7 star N edge.
07 34 07.2	+23 05 41	UGC 3911	13.7	1.1 x 0.8	13.4	S0	160	
07 35 19.0	+19 02 45	UGC 3920	13.2	1.3 x 1.2	13.5	SABb II-III	36	Bright nucleus; mag 10.66v star on NW edge.
07 37 03.0	+22 21 04	UGC 3932	14.1	1.5 x 0.3	13.1	Sab	150	Mag 6.97v star E 8'6.
07 37 19.6	+19 58 31	UGC 3939	13.7	1.1 x 0.9	13.5	S0	25	
07 40 22.8	+23 16 26	UGC 3960	13.3	1.3 x 1.1	13.6	E	50	Mag 8.99v star S 3'0.
07 44 08.6	+29 14 51	UGC 3995	12.4	2.3 x 1.0	13.1	S pec	85	Mag 11.0 star S 1'8.
07 48 10.9	+28 13 45	UGC 4030	13.5	0.5 x 0.2	10.9	Double System		Almost stellar; stellar **Mkn 1203** NNW 6'4; stellar **CGCG 148-23** SW 4'6.
07 48 19.4	+23 14 20	UGC 4031	14.2	0.9 x 0.6	13.4	Scd:	165	Star on S edge; mag 6.20v star SSE 6'7.
07 49 51.3	+18 49 45	UGC 4044	13.9	1.4 x 0.7	13.7	Scd:	147	
07 50 56.2	+23 53 47	UGC 4054	13.5	2.0 x 0.5	13.4	Sb II-III	175	
07 55 48.6	+24 42 21	UGC 4099	13.1	1.7 x 1.3	13.8	SB(r)b	100	Mag 7.69v star SE 3'7.
07 56 25.5	+27 00 44	UGC 4105	13.8	1.0 x 0.6	13.1	S?	85	Mag 10 star S 2'0; pair of mag 10 stars NW 2'3.
07 59 22.4	+18 06 37	UGC 4140	14.2	1.6 x 0.2	12.9	Sbc	141	
08 04 48.3	+20 41 39	UGC 4207	13.4	1.5 x 0.9	13.6	Sbc?	125	Bright knot or star N edge.
08 08 46.0	+18 11 39	UGC 4245	13.9	1.5 x 0.4	13.2	SBb	110	**MCG +3-21-22** S 6'6.
08 10 10.8	+24 53 26	UGC 4257	14.5	2.1 x 0.2	13.4	Scd:	33	**IC 497** NW 2'2; **MCG +4-20-3** S 0'9; **CGCG 119-4** S 2'5.
08 12 01.3	+19 21 46	UGC 4269	13.7	1.1 x 0.5	12.9	S?	90	Mag 11 star S 0'8.
08 14 16.9	+18 26 26	UGC 4286	13.4	1.4 x 0.5	12.9	S?	42	Elongated **CGCG 88-52** W 1'3.
08 15 59.2	+23 11 53	UGC 4299	13.2	1.8 x 0.3	12.4	Sbc	150	Mag 13.7 star on NW edge.
08 16 00.8	+27 04 31	UGC 4300	13.4	1.3 x 1.0	13.5	Scd:	90	Close pair of mag 12.7 stars SE 2'0.
08 16 02.3	+28 37 25	UGC 4301	13.8	1.2 x 0.5	13.1	Sbc	146	
08 16 28.3	+23 48 36	UGC 4304	14.1	1.5 x 0.3	13.1	S0	168	Mag 14.5 star S end; mag 11.3 star NE 2'7.
08 17 25.9	+21 41 00	UGC 4308	12.8	2.0 x 1.6	13.9	SB(rs)c I	110	Mag 9.51v star SSW 2'6.
08 18 29.4	+20 45 37	UGC 4324	14.1	1.5 x 0.5	13.6	Sab: sp	27	**CGCG 119-40** NW 1'7.
08 19 01.8	+21 11 09	UGC 4329	13.6	1.9 x 1.3	14.4	SA(r)cd	125	Mag 9.52v star E 2'4.
08 19 37.9	+21 06 51	UGC 4332	13.9	1.5 x 0.5	13.4	S?	54	Located in a small triangle of mag 12 stars.
08 20 12.6	+26 01 19	UGC 4340	14.2	1.2 x 0.7	13.9	SABd:	120	Weak stellar nucleus, smooth envelope; mag 11.6 star SSE 2'7.
08 20 09.9	+27 05 32	UGC 4341	13.8	1.0 x 0.5	12.9	S0/a	43	Mag 5.13v star N 7'6.
08 20 16.6	+20 52 30	UGC 4344	14.2	1.5 x 1.5	14.9	SAdm		Stellar nucleus; mag 5.80v star S 7'7.
08 20 40.8	+25 54 15	UGC 4346	13.9	0.8 x 0.7	13.1	SB(s)ab	48	
08 22 24.8	+25 30 29	UGC 4364	13.8	1.0 x 1.0	13.6	SABab:		
08 22 58.7	+27 42 23	UGC 4373	14.3	0.8 x 0.5	13.2	SBcd:	55	Uniform surface brightness; mag 10.7 star NE 1'1.
08 23 11.2	+22 39 52	UGC 4375	12.1	2.5 x 1.7	13.5	SABc:	0	Several stars Along S edge; bright star superimposed E of nucleus.
08 24 01.6	+21 01 35	UGC 4386	13.2	1.7 x 0.5	12.9	Sb III	21	**MCG +4-20-47** S 3'4.
08 25 47.6	+28 07 06	UGC 4395	14.6	1.7 x 0.2	13.3	Scd:	154	Mag 10.6 star NW 2'7.
08 26 05.7	+21 40 04	UGC 4400	15.7	1.5 x 0.2	14.2	Scd:	5	Mag 11.8 star W 2'7.
08 26 36.1	+22 56 58	UGC 4406	13.6	1.2 x 1.0	13.7	(R)SB0°	125	Small, bright center, faint envelope.
08 27 06.0	+21 38 41	UGC 4414	13.2	1.2 x 1.1	13.3	(R')SB(s)0/a	111	Short N-S bar with dark patches E and W.
08 27 16.7	+22 52 37	UGC 4416	12.9	2.3 x 0.9	13.6	SB(s)b	165	Mag 11.9 star Superimposed S of center; mag 11.2 star W 1'9; **MCG +4-20-57** SW 3'0.
08 28 14.4	+28 03 22	UGC 4425	13.9	1.3 x 0.4	13.1	SB?	130	
08 28 31.0	+17 27 55	UGC 4433	13.4	1.0 x 0.7	12.9	S pec	15	Mag 9.83v star W 2'0.
08 30 01.5	+17 15 34	UGC 4444	13.4	1.5 x 0.9	13.6	SB(s)cd?	125	Several bright knots, or superimposed stars, around core.
08 31 57.6	+19 12 41	UGC 4457	13.5	1.5 x 0.8	13.5	SAB(rs)c pec	125	Two main arms; mag 11.5 star NW 2'0.
08 33 00.2	+26 00 48	UGC 4464	14.6	1.5 x 0.3	13.6	SBcd?	110	Small, faint anonymous galaxy S 1'4.

(Continued from previous page)

75 GALAXIES 75

RA h m s	Dec ° ′ ″	Name	Mag (V)	Dim ′ Maj x min	SB	Type Class	PA	Notes
08 37 43.0	+20 30 15	UGC 4504	14.1	1.2 x 0.6	13.6	Sd	25	Uniform surface brightness.
08 40 54.0	+19 21 15	UGC 4526	13.9	1.5 x 0.4	13.2	Sab	54	Mag 9.50v star SW 3′.2.
08 41 41.4	+18 51 33	UGC 4532	14.7	1.5 x 0.2	13.2	Sbc	34	Located in a small triangle of mag 13 stars.
08 42 52.6	+25 04 10	UGC 4542	14.1	1.1 x 0.9	13.9	SAm	135	Uniform surface brightness.

GALAXY CLUSTERS

RA h m s	Dec ° ′ ″	Name	Mag 10th brightest	No. Gal	Diam ′	Notes
07 53 18.0	+29 21 00	A 602	15.8	32	22	All members anonymous, faint and stellar.
07 59 18.0	+27 06 00	A 610	16.4	46	12	Most members faint and stellar.
08 40 54.0	+26 44 00	A 692	16.2	38	20	All members anonymous, faint and stellar.

OPEN CLUSTERS

RA h m s	Dec ° ′ ″	Name	Mag	Diam ′	No. ★	B ★	Type	Notes
07 38 24.0	+21 34 24	NGC 2420	8.3	6	304	11.0	cl	Moderately rich in stars; small brightness range; strong central concentration; detached.
08 40 22.0	+19 40 00	NGC 2632	3.1	70	161	6.0	cl	= **M 44**, **Beehive**, **Praesepe**. Moderately rich in bright and faint stars; detached.

PLANETARY NEBULAE

RA h m s	Dec ° ′ ″	Name	Diam ″	Mag (P)	Mag (V)	Mag cent ★	Alt Name	Notes
07 25 34.8	+29 29 22	NGC 2371-72	62	13.0	11.2	14.8	PK 189+19.1	Very faint patches of nebulosity separated from main body NW and SE; mag11.2 star NW 4′.4.
07 29 11.0	+20 54 39	NGC 2392	54	9.9	9.1	10.5	PK 197+17.1	**Eskimo Nebula**. Located 1′.7 S of mag 8.26v star.

76 GALAXIES 76

RA h m s	Dec ° ′ ″	Name	Mag (V)	Dim ′ Maj x min	SB	Type Class	PA	Notes
07 02 59.2	+18 32 14	CGCG 85-23	15.3	0.3 x 0.3	12.5			MCG +3-18-6 S 3′.2.
06 38 38.2	+26 30 19	CGCG 115-6	14.5	0.7 x 0.5	13.2		105	Mag 8.78v star on SE edge.
06 47 34.9	+26 44 20	CGCG 115-14	14.4	0.7 x 0.4	12.9		21	Mag 9.54v star W 2′.8.
06 53 11.7	+22 18 39	CGCG 116-2	15.0	0.5 x 0.5	13.4	S		Forms small triangle with mag 13.8 star WSW 0′.6 and mag 12.9 star SSE 0′.7.
06 53 20.5	+25 17 49	CGCG 116-3	15.8	0.3 x 0.3	13.0			Almost stellar, in star rich area; located 3′.3 ESE of mag 7.40v star.
07 08 03.4	+22 11 57	CGCG 116-24	16.0	0.5 x 0.3	13.8		117	Located 1′.8 SW of mag 10.18v star.
07 18 25.1	+25 56 15	CGCG 117-6	15.0	0.6 x 0.4	13.3		90	Mag 7.97v star E 6′.7.
07 23 39.8	+22 38 38	CGCG 117-33	15.3	0.5 x 0.4	13.4		90	Mag 11.5 star W 3′.4.
06 45 35.1	+29 13 16	CGCG 145-6	14.4	0.7 x 0.6	13.3	S	33	Several stars superimposed; pair of mag 11 stars S; mag 10.47v star NW 1′.7.
07 12 59.1	+27 46 26	CGCG 146-40	14.4	0.4 x 0.4	12.2	Sb		Mag 8.08v star ENE 1′.7.
07 21 58.7	+29 29 47	CGCG 147-13	15.4	0.5 x 0.4	13.5	Sm	18	Mag 8.87v star NE 4′.8.
07 11 19.7	+26 22 15	IC 2180	13.6	0.8 x 0.8	13.0	Sab		Located 3′.1 SW of mag 6.70v star.
07 13 10.4	+18 59 42	IC 2181	13.6	0.9 x 0.5	12.6	Sab	140	Mag 7.96v star S 4′.5.
07 22 43.2	+21 30 45	IC 2186	13.9	0.8 x 0.5	12.8		111	IC 2188 S 1′.8.
07 22 43.4	+21 28 56	IC 2188	13.9	0.4 x 0.4	11.8			Mag 11.9 star on N edge.
07 01 08.1	+19 37 26	MCG +3-18-5	14.0	0.9 x 0.2	12.0	Sab	6	Mag 8.50v star on N edge.
07 03 03.8	+18 29 08	MCG +3-18-6	14.1	0.4 x 0.3	11.6		156	Located between a close pair of mag 11.4 and 13.0 stars; CGCG 85-23 N 3′.2.
07 03 34.6	+18 57 22	MCG +3-18-7	14.5	0.8 x 0.3	12.8	Sb	45	
07 04 28.5	+18 35 25	MCG +3-18-9	13.9	0.9 x 0.9	13.7	E/S0		Several stars superimposed; mag 7.31v star NW 6′.6; UGC 3656 S 3′.0.
06 49 48.5	+25 41 16	MCG +4-16-1	14.8	0.5 x 0.3	12.6		24	Mag 10.51v star on W edge.
06 49 56.2	+25 42 15	MCG +4-16-2	14.3	0.8 x 0.2	12.2	S	99	
06 49 57.3	+25 38 21	MCG +4-16-3	14.6	0.5 x 0.4	12.7	Sm	135	Located on W edge of UGC 3555.
06 50 03.8	+25 40 17	MCG +4-16-5	14.2	0.4 x 0.3	11.7	Sb	21	UGC 3555 S 2′.5.
06 53 03.4	+23 08 06	MCG +4-17-1	13.8	0.9 x 0.8	13.3		162	Group of eight mag 12-14 stars form an arc on E side.
07 04 18.6	+23 54 03	MCG +4-17-6	14.0	0.9 x 0.7	13.3		102	
07 13 30.5	+22 47 03	MCG +4-17-10	15.5	0.5 x 0.4	13.6		78	Mag 13 star on E edge; mag 10.1 star WNW 2′.3.
07 20 53.9	+23 19 02	MCG +4-18-5	14.2	1.0 x 0.5	13.3			
07 21 34.2	+23 32 54	MCG +4-18-7	13.3	0.9 x 0.9	12.9			Located between a close pair of E-W mag 11-12 stars.
07 22 48.7	+26 31 41	MCG +4-18-9	13.8	0.5 x 0.4	11.9		12	
07 23 11.7	+23 39 46	MCG +4-18-14	14.7	0.8 x 0.8	14.1	Sb		
06 49 55.2	+29 31 21	MCG +5-16-7	14.4	0.8 x 0.7	13.7	SB0/a	120	
06 52 02.4	+27 27 35	MCG +5-16-10	14.0	0.8 x 0.5	12.9	S0?	36	UGC 3573 WNW 3′.4.
06 52 57.0	+27 37 57	MCG +5-17-2	14.5	0.5 x 0.5	12.8	Sbc		Galaxy pair?
07 02 02.3	+27 49 44	MCG +5-17-7	14.7	0.5 x 0.4	12.8	S		
07 06 01.4	+28 17 44	MCG +5-17-10	14.2	0.7 x 0.4	12.7	S?	135	Mag 13.5 star on NW end.
07 16 24.5	+27 37 08	MCG +5-17-19	15.3	0.5 x 0.4	13.4		51	
07 21 05.3	+29 20 15	MCG +5-18-7	14.1	0.9 x 0.6	13.3	Sb	51	Mag 10.7 star W 4′.5.
07 08 20.6	+18 46 45	NGC 2339	11.8	2.7 x 2.0	13.5	SAB(rs)bc II-III	175	Numerous stars superimposed; **CGCG 85-37** W 8′.1.
07 09 12.1	+20 36 06	NGC 2341	13.2	0.8 x 0.8	12.6	Pec		Faint star S of elongated center.
07 09 18.3	+20 38 09	NGC 2342	12.6	1.4 x 1.3	13.1	S pec	126	
07 17 40.7	+23 21 24	NGC 2357	13.3	3.5 x 0.5	13.8	Scd:	122	
07 22 22.7	+22 05 01	NGC 2365	12.4	2.8 x 1.4	13.7	SABa	173	Faint envelope extends beyond brighter, elongated core.
06 20 34.1	+27 58 42	UGC 3447	13.9	0.9 x 0.5	13.1	E:	81	Mag 11.3 star NE 1′.6.
06 21 12.7	+27 51 24	UGC 3450	14.5	0.9 x 0.5	13.5	SBb	162	Located between a close pair of mag 10.5 and 11.3 stars.
06 33 33.8	+21 02 09	UGC 3489	14.7	2.0 x 0.3	14.0	Sbc	123	Mag 8.19v star ENE 3′.2.
06 38 01.9	+22 39 04	UGC 3503	14.5	1.7 x 0.4	13.9	Sdm:	115	Few stars superimposed; mag 7.35v star NW 6′.6.
06 43 42.5	+28 27 13	UGC 3518	13.9	1.0 x 0.3	12.4	S0/a	164	Mag 8.25v star S 3′.3.
06 45 30.3	+25 49 50	UGC 3531	13.2	1.3 x 0.8	13.1	SABbc	150	Bright knot or star N of nucleus.
06 45 41.3	+22 25 45	UGC 3534	15.7	1.5 x 0.6	15.5	S?	114	Mag 10.34v star S 2′.1.
06 46 03.6	+29 20 51	UGC 3536	13.0	1.2 x 0.6	12.5	S0	142	Mag 8.47v star SW 4′.4.
06 49 51.9	+28 22 15	UGC 3552	13.8	1.0 x 0.5	12.9	SBcd:	75	Mag 8.16v star SW 4′.0.
06 49 43.5	+20 04 52	UGC 3553	14.6	1.0 x 0.4	13.4	Scd:	158	Mag 10.87v star S 0′.9.
06 49 59.7	+25 37 58	UGC 3555	13.5	1.2 x 0.8	13.3	SAB(rs)bc	60	MCG +4-16-3 on W edge; MCG +4-16-5 NNE 2′.4.
06 50 15.0	+20 08 24	UGC 3558	13.7	1.0 x 0.9	13.6	E	130	Mag 12.7 star on E edge, in star rich area.
06 51 40.7	+29 04 17	UGC 3571	13.5	1.1 x 0.7	13.1	Scd:	105	Stellar nucleus; mag 11.4 star NW 1′.3.

(Continued from previous page)

GALAXIES

RA h m s	Dec ° ′ ″	Name	Mag (V)	Dim ′ Maj x min	SB	Type Class	PA	Notes
06 51 48.0	+27 28 50	UGC 3573	13.8	1.8 x 0.3	13.0	Sb	140	Mag 9.26v star NE 4′.6; MCG +5-16-10 ESE 3′.4.
06 53 12.0	+27 04 48	UGC 3584	14.1	1.5 x 0.2	12.6	Sbc	14	Mag 11.5 star on E edge; mag 10.57v star W 2′.1.
06 53 34.1	+27 18 31	UGC 3585	14.0	0.7 x 0.5	12.7	Scd:	0	Two faint stars superimposed; mag 11.6 star E 1′.1.
06 53 55.1	+19 17 58	UGC 3587	12.9	2.8 x 0.8	13.7	S?	107	
06 54 49.0	+23 30 05	UGC 3594	13.5	1.0 x 0.6	12.8	SAB(s)b	137	
06 55 11.7	+24 13 47	UGC 3599	13.0	1.0 x 0.8	12.6	S?	177	**CGCG 116-6** N 2′.1.
06 57 11.7	+20 26 12	UGC 3611	13.3	0.9 x 0.3	11.7	S0/a	65	Pair of stars, mags 11.7 and 10.36v, NW 2′.8.
07 01 41.9	+29 52 58	UGC 3631	14.8	1.3 x 0.6	14.4	SB	107	Two main arms.
07 01 41.3	+17 10 53	UGC 3635	14.5	0.7 x 0.4	13.0	Sdm:	115	Small, bright nucleus offset S.
07 02 26.2	+19 58 09	UGC 3639	14.1	0.9 x 0.2	12.1	S?	93	
07 04 01.4	+29 14 35	UGC 3649	13.8	0.9 x 0.5	12.8	Sa	63	**MCG +5-17-9** E 1′.4.
07 03 51.2	+22 22 10	UGC 3652	13.2	1.9 x 0.6	13.2	Scd:	115	
07 04 23.8	+18 32 32	UGC 3656	13.9	1.0 x 0.4	12.8	S0/a	174	MCG +3-18-9 N 3′.0.
07 04 40.6	+17 35 03	UGC 3658	14.9	1.8 x 0.7	15.0	Im:	65	Dwarf irregular, very low surface brightness.
07 06 45.9	+25 27 00	UGC 3674	14.3	1.0 x 0.5	13.4	Scd:	101	Mag 11.1 star N 0′.8.
07 06 40.3	+23 53 32	UGC 3676	13.9	1.0 x 0.5	13.0	Sb III	108	Mag 13.5 star on S edge; mag 9.96v star NE 3′.3.
07 06 48.4	+19 22 56	UGC 3678	14.5	0.9 x 0.7	13.8	S	0	
07 08 30.4	+29 52 41	UGC 3692	13.9	1.5 x 0.6	13.7	S0?	13	Mag 12.0 star W 1′.4.
07 09 10.7	+28 41 38	UGC 3702	14.3	0.8 x 0.7	13.5	Sd	30	Very patchy.
07 11 06.7	+25 54 55	UGC 3726	13.9	1.0 x 0.4	12.8	S?	110	Mag 12.2 star NE edge; mag 12.7 star W end.
07 11 38.6	+29 09 55	UGC 3731	14.1	1.1 x 0.4	13.1	Sdm:	158	
07 13 27.2	+27 30 46	UGC 3745	13.5	0.9 x 0.6	12.7	SBb:	99	
07 13 54.2	+23 04 52	UGC 3751	13.8	1.3 x 0.4	12.9	S?	30	
07 13 56.1	+23 14 24	UGC 3753	13.8	1.4 x 0.6	13.5	Sa:	35	
07 15 29.5	+23 25 38	UGC 3770	14.3	1.2 x 0.7	14.0	Im	27	Mag 9.91v star NE 1′.5.
07 16 43.0	+29 51 13	UGC 3777	13.9	1.8 x 0.3	13.1	Scd:	152	Mag 10.8 star W edge; mag 9.06v star NW 3′.1.
07 19 58.5	+22 05 28	UGC 3803	13.9	0.7 x 0.5	12.6	Sa:	15	Mag 3.54v star Delta Geminorum SW 6′.8.
07 20 01.0	+17 56 41	UGC 3805	13.8	1.2 x 0.4	12.9	S?	147	Mag 11.9 star W 1′.5.
07 20 22.8	+22 54 25	UGC 3806	14.1	0.9 x 0.8	13.6	SBcd:	117	Mag 9.90v star E 2′.4.
07 21 02.0	+25 10 45	UGC 3808	14.0	1.8 x 0.9	14.4	SB(s)d:	50	Pair of mag 10.5 stars NW 1′.5.
07 22 20.0	+17 17 12	UGC 3820	14.2	1.5 x 0.3	13.2	Scd:	100	Located between a pair of stars N-S, mags 10.61v and 9.92v.
07 22 47.7	+18 55 35	UGC 3823	13.9	1.1 x 0.6	13.3	Sb II-III	91	Mag 11.6 star W end; mag 11.4 star S 1′.1.
07 22 51.1	+22 35 18	UGC 3824	13.4	0.9 x 0.8	12.9	S0	21	Compact; mag 7.84v star WNW 9′.7.
07 23 07.6	+22 12 27	UGC 3827	13.6	1.0 x 0.8	13.2	SB?	45	Mag 13.2 star on SE edge.

OPEN CLUSTERS

RA h m s	Dec ° ′ ″	Name	Mag	Diam ′	No. ★	B ★	Type	Notes
05 58 10.0	+21 57 36	Bas 11b	8.9	3	70	11.5	cl	Few stars; small brightness range; no central concentration; detached.
06 33 15.0	+20 31 48	Be 23		4	25	15.0	cl	
06 25 24.9	+19 46 00	Bo 1	7.9	26	8	8.4	cl	
06 18 00.0	+23 38 00	Cr 89	5.7	60	15		cl	Few stars; moderate brightness range; not well detached; involved in nebulosity.
06 04 50.6	+24 09 37	IC 2156		3.5	40		cl?	(A) IC 2157 6′ to south.
06 04 49.8	+24 03 21	IC 2157	8.4	5	56	12.0	cl	Few stars; small brightness range; slight central concentration; detached. (A) IC 2156 6′ to north.
06 01 06.0	+23 19 24	NGC 2129	6.7	6	73	10.0	cl	Moderately rich in bright and faint stars with a strong central concentration; detached. Visual mag without the two brightest stars = 8.2.
06 07 26.0	+24 05 48	NGC 2158	8.6	5	973	15.0	cl	Rich in stars; large brightness range; slight central concentration; detached.
06 09 00.0	+24 21 00	NGC 2168	5.1	25	434	8.0	cl	= **M 35**. Rich in stars; large brightness range; no central concentration; detached. (S) Very large.
06 10 53.0	+20 36 36	NGC 2175.1		5	20		cl	Rich in stars; large brightness range; no central concentration; detached; involved in nebulosity. (A) Also known (perhaps preferably) as **C0607+206**.
06 24 28.0	+19 21 36	NGC 2218		1.2	20	11.5	ast	(A) Triangular asterism.
06 43 20.0	+26 58 12	NGC 2266	9.5	5	50	11.0	cl	Moderately rich in stars; moderate brightness range; slight central concentration; detached.
06 55 12.0	+17 59 18	NGC 2304	10.0	3	30		cl	Moderately rich in stars; small brightness range; slight central concentration; detached.
07 07 01.0	+27 16 00	NGC 2331	8.5	19	30	9.0	cl	Moderately rich in stars; moderate brightness range; not well detached. (A) Includes 1′ oval asterism of 7 stars on SE edge.
06 19 22.0	+18 33 00	Skiff J0619.3+1832		20	50		cl?	(A) Three clumps of stars over a 20′ area.

BRIGHT NEBULAE

RA h m s	Dec ° ′ ″	Name	Dim Maj x min	Type	BC	Color	Notes
06 16 54.0	+22 47 00	IC 443	50 x 40	SNR-E	2-5		The NE part is brightest and has a crescent shape whose convex side is sharply defined. The entire object is intricately structured and very filamentary in appearance.
06 20 24.0	+23 16 00	IC 444	8 x 4	R	3-5	1-4	
06 13 06.0	+17 58 00	IC 2162	3 x 3	E	2-5	3-4	Similar in size, shape and surface brightness to Sh2-257.
06 07 48.0	+18 40 00	NGC 2163	3 x 2	R	2-5	1-4	A double system, dumbbell in shape and extended N-S. Each part is roughly 1′ across and centered on a 12th mag central star. The N part is slightly larger and brighter.
06 09 42.0	+20 30 00	NGC 2174	40 x 30	E	1-5	3-4	A mottled nebulosity surrounding a mag 7.5 star; several other stars involved. The brightest spot is a small, oval patch about 1′ across. It lies 4′ E and slightly N of the mag 7.5 star.
06 08 30.0	+21 37 00	Sh2-247	10 x 10	E	3-5	4-4	
06 12 48.0	+17 58 00	Sh2-257	3 x 3	E	2-5	3-4	This roundish nebula is quite bright and well defined.

DARK NEBULAE

RA h m s	Dec ° ′ ″	Name	Dim ′ Maj x min	Opacity	Notes
06 07 23.9	+19 39 00	B 227	12 x 12		Round; one or two faint stars in it.

PLANETARY NEBULAE

RA h m s	Dec ° ′ ″	Name	Diam ″	Mag (P)	Mag (V)	Mag cent ★	Alt Name	Notes
06 16 11.1	+28 22 17	PN G183.8+5.5	135				PK 183+5.1	Mag 11.9 star NW 2′.2; mag 11.3 star SSW 3′.6.
05 58 45.5	+25 18 42	PN G184.6+0.6	2	17.3			PK 184+0.1	Mag 8.88v star W 2′.7; mag 8.84v star S 3′.7.
06 13 55.0	+26 52 57	PN G184.8+4.4	3	18.1			PK 184+4.1	Mag 11.6 star SW 0′.9, this star is part of a small triangle of stars with two mag 12 stars.
06 37 20.9	+24 00 37	PN G189.8+7.7	37	13.4	13.0	19.6	PK 189+7.1	Elongated NNW-SSE; mag 10.21v star NW 1′.2.

(Continued from previous page)

76 PLANETARY NEBULAE 76

RA h m s	Dec ° ′ ″	Name	Diam ″	Mag (P)	Mag (V)	Mag cent ★	Alt Name	Notes
06 40 09.5	+21 24 50	PN G192.5+7.2	76			19.0	PK 192+7.1	Close pair of mag 12-13 stars E 2'.4; mag 10.26v star SW 1'.7.
06 25 57.4	+17 47 26	PN G194.2+2.5	14	12.4	11.7	17.8	PK 194+2.1	Mag 12.5 star on S edge; mag 11.7 star SSE 1'.7.

77 GALAXIES 77

RA h m s	Dec ° ′ ″	Name	Mag (V)	Dim ′ Maj x min	SB	Type Class	PA	Notes
04 36 02.0	+19 57 01	NGC 1615	13.9	1.2 x 0.7	13.6	SA0-:	115	Forms a triangle with mag 7.24v star W 5'.3 and mag 6.34v star SW 6'.1.
04 36 29.6	+20 36 15	UGC 3099	14.4	1.5 x 0.3	13.4	S	8	
04 40 39.0	+17 08 26	UGC 3129	15.9	2.3 x 1.2	16.9	Im:	45	Dwarf, very low surface brightness.
04 42 07.8	+20 02 55	UGC 3135	15.4	1.1 x 0.8	15.1	SBcd:	68	Stellar nucleus; mag 9.24v star NW 5'.9.
04 46 29.7	+18 27 41	UGC 3157	14.1	1.5 x 1.1	14.5	SB?	165	
04 47 27.7	+23 58 58	UGC 3165	14.2	1.6 x 1.1	14.6	IAm	132	Stellar nucleus with star superimposed on W edge.
05 11 32.8	+17 03 22	UGC 3261	14.1	1.2 x 0.7	13.8	Sdm:	55	Two bright knots or superimposed stars S of nucleus; mag 11.3 star W 0'.9.
05 41 55.5	+18 29 29	UGC 3341	14.6	1.6 x 1.6	15.5	SBab		Two stars on N edge.
05 49 31.7	+17 41 08	UGC 3363	14.1	1.5 x 0.7	14.0	S0/a	35	Mag 10.46v star NE 2'.3.

OPEN CLUSTERS

RA h m s	Dec ° ′ ″	Name	Mag	Diam ′	No. ★	B ★	Type	Notes
05 58 10.0	+21 57 36	Bas 11b	8.9	3	70	11.5	cl	Few stars; small brightness range; no central concentration; detached.
05 24 03.0	+29 34 12	Be 19	11.4	4	150	15.0	cl	Moderately rich in stars; small brightness range; slight central concentration; detached.
05 51 46.0	+21 48 36	Be 21	11.1	6	194	6.0	cl	Moderately rich in stars; moderate brightness range; strong central concentration; detached.
05 50 12.0	+22 15 00	Be 72		2	15	15.0	cl	
04 56 56.0	+28 48 00	Cz 19		15	50		cl	Moderately rich in stars; small brightness range; slight central concentration; detached. (A) DSS image shows two condensations to cluster, on NW and SE.
05 50 08.0	+28 53 36	Cz 23		5			cl	
05 55 26.0	+20 52 42	Cz 24		5	35		cl:	
05 33 30.0	+26 31 00	DoDz 3		10	10		cl	Few stars; moderate brightness range; not well detached. (S) Sparse.
05 35 54.0	+25 57 00	DoDz 4		25	15		cl	Few stars; moderate brightness range; not well detached. (S) Sparse group.
04 45 42.2	+19 07 00	NGC 1647	6.4	40	200	9.0	cl	Rich in stars; moderate brightness range; slight central concentration; detached.
05 03 36.0	+23 49 00	NGC 1746	6.1	40	20	8.0	cl	Few stars; moderate brightness range; no central concentration; detached. NGC 1746 = **NGC 1750**.
05 04 24.0	+23 47 00	NGC 1758		40	20		cl	(A) Probably clump on E side of NGC 1746.
05 10 13.0	+24 08 00	NGC 1802		20	25	9.2	ast?	(A) This may also be a larger 12′ x 60′ (elongated N-S) group of some 80 stars.
05 25 41.9	+29 20 00	NGC 1896		20	25	8.5	cl?	
05 38 14.9	+25 49 00	NGC 1996		22			cl?	(S) Perhaps large group? (A) Rich scattering of stars on DSS.
05 43 08.0	+20 08 00	NGC 2026		10	35	9.0	cl?	(S) Sparse.
06 01 06.0	+23 19 24	NGC 2129	6.7	6	73	10.0	cl	Moderately rich in bright and faint stars with a strong central concentration; detached. Visual mag without the two brightest stars = 8.2.

BRIGHT NEBULAE

RA h m s	Dec ° ′ ″	Name	Dim ′ Maj x min	Type	BC	Color	Notes
04 40 00.0	+25 44 00	IC 2087	4 x 4	R	3-5	3-4	Faint and diffuse; in a starless field. This object is also catalogued as Barnard 22.
05 34 30.0	+22 01 00	NGC 1952	6 x 4	SNR-E	1-5	3-4	= **M 1**, **Crab Nebula** or **Taurus A**. Supernova of 1054.
05 39 06.0	+28 00 00	Sh2-240	200 x 180	SNR-E	4-5	3-4	**Simeis 147**. Very wispy or filamentary in structure.
04 48 24.0	+29 47 00	vdB 29	7 x 5	R	3-5	1-4	The mag 6.5 involved star at the E side of a diffuse nebulosity.

DARK NEBULAE

RA h m s	Dec ° ′ ″	Name	Dim ′ Maj x min	Opacity	Notes
04 39 59.0	+25 45 00	B 14	3 x 3		Very small, bright nebula.
04 38 39.9	+26 03 00	B 22	120 x 120	4	**Taurus Dark Cloud**. Irregular; unequally dark; extended SE and NW.
04 40 33.4	+29 53 00	B 23	5 x 5	5	Sharply pointed to the SE.
04 42 53.3	+29 44 00	B 24	8 x 8	5	Sharply pointed to the S.
04 41 30.0	+26 00 00	B 220	7 x 7		Round; feebler extension runs NW for 1°.

PLANETARY NEBULAE

RA h m s	Dec ° ′ ″	Name	Diam ″	Mag (P)	Mag (V)	Mag cent ★	Alt Name	Notes
04 37 23.6	+25 02 36	PN G174.2−14.6	25	15.3	15.1	18.6	PK 174−14.1	Mag 12.8 star S 2'.4; mag 14.2 star NNW 1'.8.
05 31 36.0	+28 58 39	PN G178.3−2.5	10	16.1			PK 178−2.1	Mag 11.0 star S 1'.2; mag 10.6 star NW 3'.1.
05 52 48.2	+28 05 57	PN G181.5+0.9	65	15.8		21.1	PK 181+0.1	Mag 10.9 star S 2'.3; mag 12.6 star NW 1'.5.
05 46 50.1	+24 22 00	PN G184.0−2.1	6		14.6		PK 184−2.1	Mag 12.0 star SW 1'.3; close pair of mag 13 stars N 1'.6.
05 58 45.5	+25 18 42	PN G184.6+0.6	2	17.3			PK 184+0.1	Mag 8.88v star W 2'.7; mag 8.84v star S 3'.7.

78 GALAXIES 78

RA h m s	Dec ° ′ ″	Name	Mag (V)	Dim ′ Maj x min	SB	Type Class	PA	Notes
03 38 56.7	+20 24 03	CGCG 465-3	15.4	0.5 x 0.3	13.2	SBc	33	Located 1'.9 ENE of mag 9.33v star.
03 58 12.5	+18 35 03	CGCG 466-3	14.7	0.6 x 0.4	13.0		177	
03 53 46.2	+19 58 27	IC 355	14.9	0.5 x 0.5	13.3	Sc		
04 03 44.0	+22 09 27	IC 357	13.2	1.2 x 1.0	13.3	SB(s)ab	175	Mag 8.91v star W 11'.8.
04 03 43.0	+19 53 38	IC 358	14.1	1.2 x 0.3	12.8	S0	64	Mag 8.70v star NE 7'.3.
04 12 28.3	+27 42 06	IC 359	13.9	1.3 x 1.3	14.3	S0-:		Numerous faint stars superimposed on faint envelope.
03 40 45.2	+17 44 27	IC 1977	13.5	1.4 x 0.8	13.5	SBb	177	Mag 10.43v star SSW 7'.1.
03 36 50.5	+21 07 34	MCG +3-10-1	14.9	0.8 x 0.8	14.3	Sb		Mag 12.8 star S 1'.9.
03 38 17.7	+19 23 10	MCG +3-10-2	14.9	0.6 x 0.6	13.6			Mag 7.70v star SW 2'.2.
04 00 33.2	+20 19 43	MCG +3-11-3	15.2	0.8 x 0.5	14.1	SBc	9	

(Continued from previous page)

GALAXIES

RA h m s	Dec ° ′ ″	Name	Mag (V)	Dim ′ Maj x min	SB	Type Class	PA	Notes
04 01 26.8	+23 14 58	MCG +4-10-4	14.4	0.8 x 0.4	13.1	S?	138	Mag 13.7 star N edge; pair of mag 11 stars N 1'.6; mag 6.88v star SW 7'.5.
04 02 10.6	+21 57 06	MCG +4-10-9	14.6	0.6 x 0.6	13.4			Mag 12.4 star ENE 2'.4.
04 02 30.6	+21 53 22	MCG +4-10-11	14.4	0.8 x 0.3	12.7		114	Mag 8.91v star E 5'.9.
04 02 39.5	+21 33 37	MCG +4-10-12	14.0	0.7 x 0.7	13.1	S0		Mag 11.9 star ENE 1'.4.
04 02 55.6	+21 29 00	MCG +4-10-14	14.7	0.9 x 0.3	13.2	S?	150	
04 03 29.9	+26 21 44	MCG +4-10-15	14.0	0.8 x 0.4	12.7	SB?	66	Close pair of mag 11-12 star NE 2'.6.
04 03 52.7	+24 23 49	MCG +4-10-17	14.7	0.6 x 0.4	13.0	S?	117	Mag 8.95v star W 0'.9.
04 08 02.8	+23 17 34	MCG +4-10-24	13.7	0.6 x 0.6	12.7	E?		Mag 11.3 star SE 1'.6.
04 14 14.8	+24 39 38	MCG +4-10-28	14.6	0.6 x 0.6	13.3	S?		
04 02 06.9	+23 07 56	NGC 1497	13.1	1.5 x 0.7	13.0	S0	60	UGC 2927 W 5'.9.
04 05 47.6	+25 24 30	NGC 1508	14.5	0.7 x 0.6	13.4	Sc	24	**CGCG 487- 20** N 5'.5; mag 8.87v star W 12'.1.
04 19 02.1	+26 49 36	NGC 1539	14.8	0.5 x 0.5	13.1			
04 36 02.0	+19 57 01	NGC 1615	13.9	1.2 x 0.7	13.6	SA0$^-$:	115	Forms a triangle with mag 7.24v star W 5'.3 and mag 6.34v star SW 6'.1.
03 20 23.5	+17 17 47	UGC 2684	14.6	1.8 x 0.9	15.0	Im?	111	Dwarf, low surface brightness.
03 24 07.7	+17 45 13	UGC 2716	14.2	1.8 x 0.9	14.6	Sdm:	73	**MCG +3-9-13** W 1'.7.
03 37 33.5	+23 17 36	UGC 2797	15.1	0.5 x 0.5	13.4			
03 39 33.2	+19 47 00	UGC 2809	15.7	1.3 x 0.8	15.6	Im?	156	
03 43 57.1	+22 39 39	UGC 2840	15.4	1.4 x 0.9	15.5	Im?	174	Bright portion on E edge.
03 53 36.8	+19 06 18	UGC 2892	14.5	1.4 x 1.2	14.9	SB(rs)bc	69	
04 01 40.6	+23 06 42	UGC 2927	13.7	2.2 x 1.4	14.8	(R)SB(s)a	75	NGC 1497 E 5'.9.
04 01 41.0	+23 12 19	UGC 2928	14.3	1.0 x 0.6	13.6	(R′)SAB(s)a	150	Mag 6.88v star W 10'.1.
04 02 18.3	+25 48 56	UGC 2931	13.9	0.9 x 0.9	13.5	SA(s)c		Mag 14.4 star on NE edge.
04 04 10.0	+22 07 55	UGC 2942	14.8	1.3 x 0.3	13.6	S?	66	**UGC 2943** on S edge; mag 4.35v star 37 Tauri SE 7'.9.
04 05 02.4	+25 15 54	UGC 2949	13.7	1.1 x 0.5	12.9	SBab:	119	Mag 9.35v star NW 6'.3.
04 06 53.0	+22 51 42	UGC 2958	14.2	1.3 x 0.3	13.0	Sb	20	
04 08 57.5	+27 11 57	UGC 2964	14.4	1.2 x 0.9	14.3	SAcd	72	
04 09 08.9	+17 06 39	UGC 2968	14.8	1.8 x 0.5	14.5	Sb: III-IV	28	Mag 9.08v star NNE 4'.5.
04 11 24.0	+26 52 37	UGC 2976	13.7	1.7 x 0.7	13.7	Scd:	149	Mag 11.3 star superimposed on W edge.
04 13 39.0	+25 28 57	UGC 2988	14.1	2.7 x 0.5	14.3	Sb:	5	
04 13 56.1	+29 09 24	UGC 2989	14.4	1.5 x 0.7	14.3	SB?	30	Almost stellar nucleus, uniform envelope.
04 19 06.1	+26 10 46	UGC 3009	15.0	1.7 x 0.4	14.5	Scd:	107	Mag 8.75v star NE 5'.4.
04 28 09.9	+21 39 15	UGC 3053	14.1	0.9 x 0.7	13.4	Scd:	170	Located 2'.9 NE of mag 5.71v star.
04 36 29.6	+20 36 15	UGC 3099	14.4	1.5 x 0.3	13.4	S	8	
04 40 39.0	+17 08 26	UGC 3129	15.9	2.3 x 1.2	16.9	Im:	45	Dwarf, very low surface brightness.
04 42 07.8	+20 02 55	UGC 3135	15.4	1.1 x 0.8	15.1	SBcd:	68	Stellar nucleus; mag 9.24v star NW 5'.9.

GALAXY CLUSTERS

RA h m s	Dec ° ′ ″	Name	Mag 10th brightest	No. Gal	Diam ′	Notes
03 41 42.0	+23 29 00	A 450	16.4	40	25	Few stellar anonymous GX in S half, most others very faint.

OPEN CLUSTERS

RA h m s	Dec ° ′ ″	Name	Mag	Diam ′	No. ★	B ★	Type	Notes
04 06 41.0	+27 34 00	Do 14		10	18		cl	Few stars; small brightness range; no central concentration; detached.
03 47 29.0	+24 06 00	M 45	1.5	120	100	3.0	cl	**Pleiades**. Rich in stars; large brightness range; strong central concentration; detached; involved in nebulosity (S) Center star is Alcyone.

BRIGHT NEBULAE

RA h m s	Dec ° ′ ″	Name	Dim ′ Maj x min	Type	BC	Color	Notes
04 27 06.0	+26 07 00	Ced 33	5 x 2	R	4-5	3-4	Very faint and diffuse
04 27 18.0	+23 00 00	Ced 34	10 x 6	R	2-5	1-4	
03 46 18.0	+23 56 00	IC 349	26 x 26	R	1-5	1-4	Small knot in **vdB 22**. A small patch of nebulosity about 35' south-southeast the star Merope (23 Tauri) in the Pleiades cluster.
03 55 00.0	+25 29 00	IC 353	180 x 30	R	3-5	1-4	(see IC 1995).
03 50 18.0	+25 35 00	IC 1995	135 x 135	R	3-5	1-4	A part of the larger nebula known as **LBN 774** which includes IC 353. This section, the E part of LBN 774, is more filamentary in structure than IC 353.
04 40 00.0	+25 44 00	IC 2087	4 x 4	R	3-5	3-4	Faint and diffuse; in a starless field. This object is also catalogued as Barnard 22.
03 45 48.0	+24 22 00	NGC 1432	26 x 26	R	2-5	1-4	= **vdB 21**. Brighter part of nebulosity surrounding the star Maia (20 Tauri) in the Pleiades cluster.
03 46 06.0	+23 47 00	NGC 1435	30 x 30	R	1-5	1-4	**Tempel's Nebula**, **Merope Nebula**. Condensation of nebulosity S of and semi-detached (in shorter exposures) from IC 349 which is a small knot in **vdB 22**.
04 21 48.0	+19 32 00	NGC 1554/55	1 x 1	R	2-5	3-4	**Hind's Variable Nebula**. Involved with the irregular variable star T Tauri. NGC 1554 lies 0'.75 W of the star but according to early observers it was SW of T Tauri. A much smaller nebulosity (apparently NGC 1555) surrounds the star itself. NGC 1555 also known as **Struveis Lost Nebula**.
03 28 18.0	+29 48 00	vdB 16	4.5 x 4.5	R	3-5	2-4	Nebulosity with a mag 9.1 star involved.
03 44 54.0	+24 07 00	vdB 20	11 x 11	R	3-5	1-4	Nebulosity surrounding the star Electra (17 Tauri) in the Pleiades cluster.
03 47 30.0	+24 06 00	vdB 23	17 x 17	R	3-5	1-4	Nebulosity surrounding the star Alcyone (η) (25 Tauri) in the Pleiades cluster.

DARK NEBULAE

RA h m s	Dec ° ′ ″	Name	Dim ′ Maj x min	Opacity	Notes
04 17 25.0	+28 34 00	B 7		5	Large, irregular, with brighter condensation (B10) in SE part.
04 18 41.2	+28 17 00	B 10	8 x 8	5	The brightest part of B7.
04 39 59.0	+25 45 00	B 14	3 x 3		Very small, bright nebula.
04 31 13.3	+24 21 00	B 18	60 x 60	5	Group of dark spots.
04 33 39.9	+26 16 00	B 19	60 x 60		Large; indefinite.
04 38 39.9	+26 03 00	B 22	120 x 120	4	**Taurus Dark Cloud**. Irregular; unequally dark; extended SE and NW.
04 40 33.4	+29 53 00	B 23	5 x 5	5	Sharply pointed to the SE.
04 42 53.3	+29 44 00	B 24	8 x 8	5	Sharply pointed to the S.
04 04 35.4	+26 21 00	B 207			Small black spot, elongated nearly N and S.
04 11 32.9	+25 10 00	B 208			Darker spot in dark lane.
04 12 23.3	+28 20 00	B 209			Center of a broad extension from B7, toward the W.

(Continued from previous page)

DARK NEBULAE

RA h m s	Dec ° ′ ″	Name	Dim ′ Maj x min	Opacity	Notes
04 15 33.3	+25 04 00	B 210		5	Darker spot in dark lane.
04 17 12.4	+27 49 00	B 211		5	North end of dark lane running SE from B7.
04 19 14.6	+25 18 00	B 212		5	Darker spot in dark lane.
04 21 10.6	+27 03 00	B 213			Dark spot in lane.
04 21 55.7	+28 33 00	B 214	5 x 5		Round.
04 23 34.4	+25 03 00	B 215		5	Darker spot in dark lane.
04 23 59.6	+26 38 00	B 216			Dark spot in lane.
04 27 38.5	+26 07 00	B 217			Southeast end of dark lane running from B7.
04 08 06.0	+26 20 00	B 218	15 x 15		Triangular; extension to the E.
04 34 51.5	+29 36 00	B 219	55 x 3	5	Partly vacant space; NE and SW.
04 41 30.0	+26 00 00	B 220	7 x 7		Round; feebler extension runs NW for 1°.

PLANETARY NEBULAE

RA h m s	Dec ° ′ ″	Name	Diam ″	Mag (P)	Mag (V)	Mag cent ★	Alt Name	Notes
03 53 36.4	+19 29 39	PN G171.3−25.8	38	13.9	15.1	17.2	PK 171−25.1	Mag 10.65v star N 2′.3.
04 37 23.6	+25 02 36	PN G174.2−14.6	25	15.3	15.1	18.6	PK 174−14.1	Mag 12.8 star S 2′.4; mag 14.2 star NNW 1′.8.

GALAXIES

RA h m s	Dec ° ′ ″	Name	Mag (V)	Dim ′ Maj x min	SB	Type Class	PA	Notes
01 58 13.6	+21 20 59	CGCG 461-14	14.5	0.6 x 0.4	12.9	S0	141	Mag 8.31v star W 5′.1.
01 58 19.1	+20 57 41	CGCG 461-15	14.7	0.4 x 0.4	12.8	S0		Line of three mag 10 stars N 7′.6; mag 12.4 star SSW 3′.3.
02 16 11.6	+17 58 13	CGCG 461-70	14.8	0.6 x 0.5	13.3	Sc	171	Mag 11.6 star W 4′.3.
02 32 53.8	+19 15 28	CGCG 462-23	13.9	0.8 x 0.6	12.9	Sbc	84	Mag 15.3 star on S edge; mag 12.9 star S 1′.0.
01 59 15.8	+24 24 57	CGCG 482-35	13.9	0.7 x 0.2	11.6		78	Between two mag 12 stars.
02 02 39.7	+27 34 24	CGCG 482-55	14.0	0.8 x 0.2	11.9	S0/a	21	Located 2′.1 SE of mag 9.91v star.
02 04 18.7	+28 39 16	CGCG 504-2	14.4	0.9 x 0.6	13.6		102	Mag 13.8 star on W end.
02 22 40.6	+28 27 52	CGCG 504-69	14.0	0.6 x 0.5	12.6	S?	165	Mag 10.88v star W 1′.6.
02 34 35.8	+29 57 23	CGCG 505-17	15.4	0.5 x 0.3	13.2		99	Galaxy **V Zw 256** on N edge.
02 01 30.9	+26 28 50	IC 187	12.9	2.0 x 0.7	13.1	SBa	70	Double system? Small elongated companion E end; IC 188 NE 5′.2.
02 01 46.3	+26 32 46	IC 188	13.8	0.7 x 0.4	12.3	S?	45	IC 187 SW 5′.2.
02 01 52.9	+23 33 01	IC 189	13.9	0.7 x 0.6	12.8	SB?	147	NE-SW bar with strong dark patches N and S; stellar galaxy **V Zw 167** SE 2′.8.
02 02 07.3	+23 32 55	IC 190	14.1	0.6 x 0.4	12.6	E?	105	Stellar galaxy **V Zw 167** WSW 2′.8; mag 9.84v star S 2′.5.
02 22 40.8	+28 15 24	IC 221	13.0	1.4 x 1.0	13.2	Sc	9	Mag 8.37v star SE 10′.9.
02 27 46.0	+28 12 28	IC 226	14.2	1.7 x 1.4	15.0	S?	150	Mag 9.65v star E 1′.8.
02 28 03.7	+28 10 27	IC 227	14.5	1.3 x 0.8	14.6	E	60	Mag 8.62v star NNE 4′.5.
02 32 50.9	+20 38 24	IC 235	14.4	0.6 x 0.4	12.7	Pec	3	Small triangle of mag 10.37v, 11.8, 12.4 stars centered W 10′.5.
02 41 25.6	+17 48 41	IC 248	13.4	1.0 x 0.6	12.7	Sa:	145	Mag 7.63v star ESE 9′.3.
01 56 09.1	+17 38 26	IC 1748	13.7	1.0 x 0.6	13.0	SABbc	130	Mag 8.54v star SE 11′.9; **CGCG 460-51** S 6′.6; **UGC 1417** NE 9′.1.
01 57 19.4	+28 35 19	IC 1753	13.9	0.4 x 0.4	12.0	E?		**IC 1752** NW 1′.7.
02 00 23.5	+24 34 46	IC 1764	13.3	1.2 x 0.8	13.1	SB(r)b	174	Mag 11.04v star NW 1′.8.
02 28 12.8	+19 34 59	IC 1801	13.8	1.6 x 0.6	13.6	SBb:	40	NGC 935 on N edge.
02 29 14.1	+23 04 56	IC 1803	13.4	1.3 x 1.1	13.6	S0?	120	Small, faint, elongated galaxy on E edge.
02 29 35.2	+22 56 33	IC 1806	15.1	0.7 x 0.5	14.1	E	135	Mag 9.30v star SW 7′.7.
02 30 31.1	+22 56 56	IC 1807	14.7	0.4 x 0.4	12.8	E		Mag 11.0 star S 2′.1.
02 31 40.4	+22 55 02	IC 1809	13.9	0.8 x 0.6	13.0	(R′)SB(s)ab	128	Mag 9.11v star SE 8′.0.
02 41 57.8	+19 01 43	IC 1832	14.1	0.7 x 0.6	13.1	S0	30	Mag 10.4 star E 9′.7.
02 45 36.3	+18 55 42	IC 1841	15.1	0.7 x 0.5	13.8	SBb	51	Mag 8.28v star SE 11′.7.
02 49 20.8	+19 18 11	IC 1854	13.9	0.5 x 0.4	12.1	S0	162	Mag 10.90v star W 9′.8.
02 53 07.2	+25 29 21	IC 1861	13.3	1.2 x 0.8	13.2	SA0°	150	Mag 11.5 star on E edge; mag 11.1 star SE 2′.4.
03 09 58.4	+19 12 26	IC 1890	14.6	0.7 x 0.7	13.7	S0		Located between a mag 9.55v star SW 3′.5 and a mag 9.48v star NE 3′.3.
01 56 52.8	+19 35 39	MCG +3-6-3	15.5	0.6 x 0.5	14.1	Sc	3	
02 01 34.9	+18 38 38	MCG +3-6-17	16.0	0.6 x 0.5	14.5	SBc	177	Mag 13.2 star S 1′.6.
02 01 44.3	+17 55 39	MCG +3-6-18	15.3	0.7 x 0.5	14.0	Sb	147	Mag 7.28v star NW 3′.7.
02 05 56.7	+18 26 50	MCG +3-6-32	15.9	0.7 x 0.4	14.3	Sc	30	Mag 13.9 star S 1′.8.
02 09 28.5	+19 46 30	MCG +3-6-37	16.2	1.1 x 0.6	15.6		21	Mag 13.1 star SW 0′.9.
02 10 21.2	+21 23 06	MCG +3-6-40	14.1	0.6 x 0.6	12.9	S0		Pair of mag 14.1 stars S 1′.3.
02 10 57.3	+19 08 22	MCG +3-6-41	15.5	0.6 x 0.6	14.3	Sc		Located between a pair of NE-SW mag 12-13 stars.
02 11 03.0	+17 12 01	MCG +3-6-42	16.1	0.8 x 0.4	14.7	Sc	39	Mag 14.9 star N end; mag 11.1 star S 1′.7.
02 12 46.0	+19 11 07	MCG +3-6-44	15.6	0.7 x 0.5	14.3	SBc	36	Mag 10.8 star S 1′.7.
02 14 28.7	+17 11 47	MCG +3-6-47	16.5	1.1 x 0.8	16.2		129	Mag 13.8 star SE edge.
02 15 18.1	+18 00 20	MCG +3-6-49	14.2	1.0 x 0.9	14.1	E/S0	0	Mag 10.89v star S 2′.8; mag 8.76v star SSE 5′.5.
02 15 50.0	+20 41 40	MCG +3-6-50	15.6	0.6 x 0.4	13.9	Sb	42	Mag 9.90v star E 2′.7.
02 22 25.3	+17 40 07	MCG +3-7-5	14.6	0.9 x 0.4	13.4	Sbc	129	Mag 13.0 star SW 2′.9.
02 22 53.4	+17 17 57	MCG +3-7-6	15.1	0.5 x 0.5	13.4	Sc		Mag 11.5 star SW 2′.4.
02 26 56.4	+20 32 01	MCG +3-7-13	14.4	0.8 x 0.7	13.6	SB?	45	NGC 924 SW 3′.1.
02 31 25.7	+20 39 58	MCG +3-7-19	14.7	0.6 x 0.5	13.2		153	Mag 9.70v star E 2′.6.
02 31 28.4	+18 46 08	MCG +3-7-20	14.8	1.0 x 0.2	12.9	Sc	153	Mag 9.29v star NE 1′.9.
02 32 40.5	+18 27 56	MCG +3-7-23	14.7	0.7 x 0.3	12.9	Sb	126	Mag 9.20v star W 1′.9.
02 33 16.6	+21 29 49	MCG +3-7-24	14.8	0.7 x 0.5	13.5		123	
02 33 33.2	+21 28 54	MCG +3-7-26	14.8	0.7 x 0.3	13.0		123	
02 35 47.3	+20 24 11	MCG +3-7-31	15.0	0.6 x 0.5	13.6	SBc	72	Mag 10.28v star W 4′.9.
02 37 02.1	+20 25 40	MCG +3-7-32	14.4	0.7 x 0.7	13.5			Mag 12.9 star WNW 1′.4.
02 37 20.7	+20 24 53	MCG +3-7-36	14.7	0.7 x 0.5	13.4	SBbc	144	Mag 11.5 star NW 2′.1.
02 40 14.0	+17 52 57	MCG +3-7-40	14.1	0.8 x 0.4	12.7	SBb	21	Located between a close pair of mag 12 stars.
02 40 50.1	+21 36 32	MCG +3-7-42	14.6	1.1 x 0.3	13.2	S?	141	Mag 12.3 star SE 3′.4.
02 42 29.2	+18 09 51	MCG +3-7-47	15.1	0.6 x 0.4	13.4	Sc	168	Mag 10.00v star NE 3′.8. NGC 1054 NW 4′.5.
02 47 52.0	+17 16 27	MCG +3-8-11	15.2	0.3 x 0.3	12.4			Compact; mag 8.97v star WNW 2′.8.
02 50 20.0	+19 06 38	MCG +3-8-21	13.9	0.7 x 0.3	12.1	S?	150	Mag 14.9 star E edge; mag 9.61v star NW 2′.9.
03 00 10.6	+18 54 09	MCG +3-8-41	15.2	0.6 x 0.5	13.8	Sc	42	Mag 10.59v star E 1′.2.
03 07 58.4	+17 46 40	MCG +3-9-1	15.2	0.8 x 0.5	14.0		21	Mag 13.4 star S end.
03 12 26.6	+19 14 43	MCG +3-9-5	14.2	1.0 x 0.5	13.3	S0/a	81	
01 59 38.2	+27 25 59	MCG +4-5-27	14.3	0.8 x 0.4	12.9	S?	75	Mag 10.6 star W 0′.9.
02 00 04.5	+23 45 48	MCG +4-5-31	14.5	0.9 x 0.7	13.9		30	Mag 9.34v star N 4′.5.
02 01 08.0	+26 50 37	MCG +4-5-36	13.9	0.5 x 0.5	12.2			

RA h m s	Dec ° ′ ″	Name	Mag (V)	Dim ′ Maj x min	SB	Type Class	PA	Notes
02 03 41.2	+26 16 34	MCG +4-5-46	14.1	0.9 x 0.5	13.1	Sc	6	UGC 1549 NW 2′.6.
02 03 51.6	+25 55 30	MCG +4-5-47	13.5	0.8 x 0.6	12.8	E/S0	12	Located 2′.7 E of mag 5.69v star 10 Arietis.
02 05 20.5	+25 06 16	MCG +4-6-1	14.5	0.8 x 0.4	13.1	S?	24	Mag 11.1 star WSW 2′.3.
02 05 43.7	+24 13 58	MCG +4-6-2	14.3	1.0 x 0.7	13.7	SB?	147	
02 11 55.5	+22 35 29	MCG +4-6-10	14.3	0.8 x 0.5	13.1		33	Mag 11.1 star NW 2′.6.
02 13 06.2	+22 51 20	MCG +4-6-11	14.4	0.8 x 0.5	13.2		42	Mag 10.1 star NE 1′.4.
02 16 18.1	+23 38 30	MCG +4-6-15	14.5	0.9 x 0.5	13.5	SB?	45	Mag 10.7 star N 3′.0.
02 19 20.4	+27 11 35	MCG +4-6-16	16.6	0.8 x 0.6	15.7		30	
02 23 03.2	+25 26 03	MCG +4-6-19	14.8	0.8 x 0.5	13.7		30	Mag 9.45v star N 7′.8.
02 23 52.1	+25 32 30	MCG +4-6-22	14.0	0.7 x 0.6	12.9	S?	0	**CGCG 483-22** SW 6′.4.
02 24 43.6	+25 31 52	MCG +4-6-26	14.7	0.5 x 0.5	13.0	SB?		UGC 1860 N 1′.8; mag 7.80v star SE 4′.9.
02 25 10.1	+23 51 03	MCG +4-6-28	14.7	0.5 x 0.3	12.5	S?	60	Mag 9.49v star SW 2′.8.
02 25 16.5	+24 15 51	MCG +4-6-29	14.2	0.9 x 0.4	12.9	S?	129	Mag 10.9 star NE 1′.7.
02 25 38.5	+23 00 02	MCG +4-6-31	14.5	0.7 x 0.4	13.0	S?	27	
02 25 39.7	+27 17 41	MCG +4-6-32	15.1	0.7 x 0.6	14.0	Sd	3	Stellar nucleus, smooth envelope.
02 25 55.9	+24 51 22	MCG +4-6-36	14.6	0.6 x 0.5	13.1	S?	0	Faint, elongated anonymous galaxy S 2′.1.
02 26 54.2	+25 01 55	MCG +4-6-42	14.1	0.9 x 0.4	12.9	S?	9	
02 26 54.7	+23 47 57	MCG +4-6-43	15.1	0.7 x 0.5	13.8		126	
02 27 26.4	+23 05 32	MCG +4-6-44	15.0	0.8 x 0.3	13.3	S?	138	Mag 10.6 star W 1′.1; **MCG +4-6-47** E 1′.4; **MCG +4-6-48** SE 2′.6.
02 27 42.4	+26 13 33	MCG +4-6-49	13.9	0.8 x 0.7	13.1	S?	30	Mag 11.2 star SSE 3′.4; mag 9.20v star N 4′.8.
02 29 50.0	+23 06 28	MCG +4-6-60	14.0	0.9 x 0.6	13.4	E	117	**MCG +4-6-58** NNW 1′.5; mag 11.7 star SE 1′.3.
02 30 16.7	+23 09 09	MCG +4-6-61	14.6	0.6 x 0.6	13.5	E		Mag 12.3 star WSW 1′.5.
02 30 53.1	+25 24 25	MCG +4-7-1	14.8	0.6 x 0.3	12.7	S?	39	Mag 10.7 star SSE 2′.7.
02 30 50.7	+26 11 36	MCG +4-7-2	14.7	0.7 x 0.4	13.2	S?	117	Mag 11.3 star WSW 1′.3.
02 31 34.7	+26 09 56	MCG +4-7-3	15.5	0.6 x 0.5	14.0	Sd	3	
02 32 00.5	+23 45 19	MCG +4-7-5	14.3	0.8 x 0.7	13.5	S?	90	Mag 12.6 star NE 1′.7.
02 32 54.5	+25 05 33	MCG +4-7-6	14.1	0.6 x 0.5	12.6	S?	3	Mag 10.1 star SE 1′.7.
02 32 56.3	+21 53 34	MCG +4-7-8	13.9	1.2 x 0.9	13.8		156	Mag 12.2 star NNW 3′.3.
02 35 49.7	+24 27 25	MCG +4-7-13	14.4	0.7 x 0.6	13.3		66	
02 36 15.3	+27 10 43	MCG +4-7-15	14.4	0.5 x 0.5	12.9	E		Mag 8.57v star SW 1′.4.
02 36 59.1	+25 26 31	MCG +4-7-17	13.9	1.2 x 0.7	13.6		132	UGC 2082 W 9′.7.
02 37 38.0	+25 46 10	MCG +4-7-19	15.3	0.7 x 0.3	13.4	Sb	36	Mag 10.39v star E 2′.7.
02 50 50.7	+22 46 18	MCG +4-7-26	14.5	0.9 x 0.3	12.9	S0?	129	Mag 9.8 star NE 1′.8.
02 57 59.1	+25 25 25	MCG +4-8-1	14.7	0.7 x 0.3	12.8	S?	48	**MCG +4-8-2** NE 1′.9.
02 58 10.0	+24 02 01	MCG +4-8-3	14.5	0.7 x 0.4	13.0	S?	15	Close N-S pair of mag 12.2 stars WNW 2′.1.
02 58 18.3	+25 26 53	MCG +4-8-4	14.6	0.6 x 0.5	13.1	SB?	66	Close pair of stars, mags 10.71v and 11.9, N 1′.7.
03 12 55.1	+22 31 17	MCG +4-8-12	14.7	0.6 x 0.4	13.0	S?	15	Mag 9.97v star E 2′.3.
02 46 01.2	+28 01 37	MCG +5-7-50	14.6	0.9 x 0.3	13.0	S?	123	Mag 11.8 star NW 2′.7.
03 01 45.6	+29 12 34	MCG +5-8-2	14.0	0.7 x 0.7	13.3	E/S0		Mag 9.38v star W edge.
01 58 48.1	+24 53 28	NGC 765	12.8	2.8 x 2.8	14.9	SAB(rs)bc		Small, elongated core in smooth envelope; mag 7.34v star SE 8′.0.
01 59 13.7	+18 57 14	NGC 770	12.8	0.9 x 0.7	12.3	E3:	15	Located 3′.6 S of the center of NGC 772; stellar **PGC 7509** W 2′.0; stellar **PGC 7493** W 3′.6.
01 59 20.1	+19 00 26	NGC 772	10.3	7.2 x 4.3	13.9	SA(s)b I	130	Faint, small anonymous galaxy E 5′.2.
01 59 54.6	+23 38 36	NGC 776	12.4	1.7 x 1.7	13.4	SAB(rs)b		**IC 181** NE 2′.0; **IC 180** SE 2′.8.
02 00 35.2	+28 13 31	NGC 780	13.8	1.5 x 0.9	14.0	Sb	176	2′.7 NE of mag 9.71v star.
02 01 16.6	+28 50 16	NGC 784	11.7	6.6 x 1.5	14.1	SBdm: sp	0	Several faint knots in elongated core; almost stellar galaxy **V Zw 166** S 6′.5.
02 02 29.4	+18 22 22	NGC 794	12.7	1.3 x 1.1	13.0	S0⁻:	45	Close pair of faint, stellar anonymous galaxies SSE 4′.2.
02 04 29.6	+28 48 44	NGC 805	13.5	1.1 x 0.7	13.1	SB0	115	Mag 13.3 star on S edge; mag 13.9 star on W edge.
02 04 55.6	+28 59 16	NGC 807	12.5	1.8 x 1.3	13.4	E	145	Mag 10.26v star SW 1′.7.
02 08 08.9	+29 15 21	NGC 816	14.3	0.4 x 0.4	12.2	Compact		Compact.
02 07 33.7	+17 12 09	NGC 817	13.3	0.7 x 0.3	11.5	S?	27	
02 08 34.4	+29 13 58	NGC 819	13.5	0.7 x 0.5	12.2	S?	10	Mag 11.3 star N 4′.0.
02 14 03.7	+27 52 34	NGC 855	12.6	2.6 x 0.8	13.4	E	63	Mag 9.45v star NW 7′.6; stellar **CGCG 504-31** NW 12′.9.
02 16 15.3	+28 36 01	NGC 865	13.2	1.5 x 0.4	12.5	S?	159	
02 23 32.2	+26 30 42	NGC 900	13.7	1.1 x 0.7	13.3	S0	30	
02 23 34.2	+26 33 26	NGC 901	14.8	0.4 x 0.4	12.6			
02 24 05.6	+27 20 33	NGC 904	13.6	1.2 x 0.9	13.6	E	130	**UGC 1852** NW 1′.3.
02 25 45.9	+27 13 11	NGC 915	14.2	0.7 x 0.7	13.3			Located 1′.3 S of NGC 916.
02 25 47.8	+27 14 29	NGC 916	15.3	1.0 x 0.4	14.2		0	Anonymous galaxy 3′.3 NE.
02 25 50.9	+18 29 45	NGC 918	12.2	3.5 x 2.0	14.2	SAB(rs)c: I-II	158	Star or knot on N edge of small nucleus; mag 11.26v star S 3′.0.
02 26 16.9	+27 12 40	NGC 919	14.5	1.2 x 0.3	13.2	Sab	138	Mag 10.89v star SE 3′.8.
02 26 46.9	+20 29 49	NGC 924	12.4	2.3 x 1.3	13.4	S0	53	**MCG +3-3-13** NE 3′.1.
02 27 41.1	+27 13 13	NGC 928	14.0	0.7 x 0.3	12.1	S?	48	Located 2′.6 N of mag 10.18v star.
02 27 54.6	+20 19 56	NGC 932	12.4	1.9 x 1.6	13.5	SAa	42	
02 28 11.2	+19 35 52	NGC 935	12.9	1.7 x 1.1	13.5	Scd:	155	IC 1803 on S edge; mag 10.3 star SW 1′.3.
02 28 33.4	+20 16 59	NGC 938	12.4	1.6 x 1.2	13.1	E	100	Strong core in smooth envelope; star or knot on SE edge.
02 31 09.7	+29 35 21	NGC 953	13.5	1.3 x 1.3	14.1	E		
02 32 40.0	+28 04 13	NGC 962	12.9	1.4 x 1.2	13.6	E	175	Mag 9.48v star E 6′.1; faint, elongated anonymous galaxy NE 4′.2.
02 34 13.7	+29 18 33	NGC 972	11.4	3.4 x 1.7	13.2	Sab	152	Pair of stars 1′.6 WSW, brightest of the pair is mag 9.91v.
02 34 00.0	+20 58 37	NGC 976	12.4	1.5 x 1.2	12.9	SA(rs)c: II	162	Mag 9.58v star W 10′.6; small, faint, elongated anonymous galaxy N 5′.9.
02 34 43.1	+23 24 47	NGC 984	12.8	2.0 x 1.2	13.7	SA0⁺	120	Mag 8.17v star NNW 6′.3.
02 37 25.6	+21 06 03	NGC 992	12.8	0.9 x 0.7	12.1	S?	7	Stellar **Mkn 369** N 2′.7; mag 9.18v star N 4′.3.
02 39 50.7	+18 01 30	NGC 1030	13.3	1.6 x 0.7	13.3	S?	8	Faint, small anonymous galaxy ENE 4′.4.
02 40 29.2	+19 17 49	NGC 1036	13.2	1.4 x 1.0	13.4	Pec?	2	
02 42 15.8	+18 13 01	NGC 1054	13.7	0.9 x 0.5	12.7	S?	33	MCG +3-7-47 SE 4′.5 mag 11.98v star NE 2′.8.
02 42 48.6	+28 34 27	NGC 1056	12.4	2.3 x 1.1	13.3	Sa:	160	
02 59 42.3	+25 14 14	NGC 1156	11.7	2.6 x 1.7	13.2	IB(s)m IV-V	34	Numerous knots; mag 11.3 star N edge; mag 8.48v star E 6′.4.
01 57 26.8	+17 13 12	UGC 1432	13.8	1.0 x 0.6	13.1	Sbc	65	
01 57 29.8	+19 55 06	UGC 1433	14.7	0.5 x 0.5	13.0	Scd:		**MCG +3-6-6** and **UGC 1436** SE 2′.6.
01 58 14.1	+19 05 58	UGC 1445	13.7	1.0 x 0.5	12.8	S0	93	Almost stellar nucleus.
01 58 30.1	+25 21 35	UGC 1451	13.4	1.6 x 0.7	13.3	SB?	121	Mag 10.63v star E 1′.7.
01 58 45.7	+24 38 39	UGC 1453	14.4	1.5 x 1.3	15.0	Im:		Dwarf, low surface brightness.
01 59 09.5	+25 23 10	UGC 1462	14.3	1.4 x 1.0	14.5	SBcd:	65	Mag 12.7 stars superimposed S edge.
02 00 15.1	+24 15 07	UGC 1478	13.8	0.9 x 0.8	13.3	SB(rs)c	96	**CGCG 482-38** NW 5′.7.
02 00 19.3	+24 28 27	UGC 1479	14.0	1.2 x 0.4	13.1	S?	176	Mag 13.3 star on NE edge; mag 9.72v star W 1′.5.
02 00 23.8	+21 06 27	UGC 1485	13.5	1.2 x 0.7	13.2	S?	25	Mag 11.1 star SE 2′.7.
02 00 32.3	+21 17 12	UGC 1490	13.7	0.6 x 0.3	11.7	S?	85	**CGCG 461-20** W 3′.1.
02 00 47.2	+17 42 49	UGC 1495	14.5	0.8 x 0.6	13.6	Scd:	39	
02 02 00.0	+21 05 44	UGC 1514	13.3	0.7 x 0.5	12.1	S?	6	
02 02 18.9	+19 04 01	UGC 1518	14.7	1.5 x 0.6	14.4		100	Located near center of galaxy cluster **A 292**; mag 8.48v star E 8′.9; stellar anonymous galaxy S 1′.4.

(Continued from previous page)

GALAXIES

RA h m s	Dec ° ′ ″	Name	Mag (V)	Dim ′ Maj x min	SB	Type Class	PA	Notes
02 02 16.8	+19 10 47	UGC 1519	14.7	1.5 x 0.6	14.5	Sdm:	40	Located near N edge of galaxy cluster **A 292**; mag 9.16v star W 5′.0.
02 02 34.6	+17 10 01	UGC 1531	14.1	1.2 x 0.4	13.1	Sb II	114	
02 02 48.1	+26 34 50	UGC 1533	14.4	0.8 x 0.6	13.4	SABcd:	141	Mag 8.45v star ESE 9′.5.
02 03 20.3	+18 37 44	UGC 1546	13.7	1.0 x 0.9	13.5	SAB(s)c II	18	
02 03 20.6	+22 02 59	UGC 1547	13.9	1.9 x 1.5	14.9	IBm V	18	Irregular, brightest S one third, low surface brightness.
02 03 32.8	+26 18 23	UGC 1549	13.8	0.8 x 0.3	12.1	S0/a	36	MCG +4-5-46 SE 2′.5.
02 03 37.6	+24 04 28	UGC 1551	12.6	2.3 x 1.7	13.9	SB?	135	Many bright knots and/or superimposed stars.
02 03 56.4	+19 40 04	UGC 1560	13.8	0.8 x 0.7	13.0	(R′)SB(s)b:	38	Bright E-W bar; **MCG +3-6-29** SW 1′.3.
02 04 05.3	+24 12 25	UGC 1561	13.9	1.0 x 0.8	13.5	Im	109	Several knots and/or superimposed stars center; mag 10.6 star NW 2′.3.
02 04 32.1	+27 55 31	UGC 1565	14.5	1.3 x 0.5	13.9	SBdm	4	
02 06 03.9	+29 47 32	UGC 1590	12.7	1.9 x 1.4	13.6	$S0^-$:	130	Mag 11.8 star NW 1′.7; faint anonymous galaxy NW 2′.8.
02 06 12.5	+29 58 02	UGC 1591	13.8	1.5 x 0.3	12.7	S?	152	Small, round galaxy **V Zw 182** on N end.
02 06 30.2	+29 59 32	UGC 1596	13.5	1.1 x 0.6	13.0	SAB0°	125	Bright nucleus; mag 13.4 star on SE end.
02 09 14.4	+25 34 14	UGC 1648	13.6	0.9 x 0.5	12.6	S0?	75	Mag 10.10v star W 1′.5.
02 09 32.7	+21 14 48	UGC 1652	14.1	1.3 x 0.5	13.5	Sd	30	
02 11 57.5	+29 18 46	UGC 1690	14.8	1.5 x 0.2	13.3	S?	103	Pair of stars, mags 12.2 and 11.3, NE 2′.0.
02 12 28.2	+29 51 23	UGC 1696	14.5	0.9 x 0.6	13.7	Sdm		
02 13 34.0	+25 51 18	UGC 1706	14.0	1.2 x 0.5	13.3	Scd:	154	
02 15 17.0	+18 19 03	UGC 1731	13.5	1.0 x 0.6	12.9	E?	24	
02 15 20.8	+22 00 18	UGC 1733	14.8	1.5 x 0.2	13.3	Scd:	128	Mag 12.4 star N 0′.8; mag 10.8 star NE 2′.4.
02 15 44.2	+25 12 22	UGC 1739	14.1	0.4 x 1.2	13.1	S?	124	
02 16 21.0	+24 53 15	UGC 1752	15.6	1.5 x 1.2	16.1	SA(s)cd	60	Bright core, uniform envelope. mag 11.2 star NW 2′.6.
02 19 52.9	+29 02 08	UGC 1792	13.3	2.2 x 1.2	14.2	SAB(r)c	0	Mag 10.6 star NE 3′.1.
02 21 06.5	+23 36 02	UGC 1808	13.7	0.9 x 0.8	13.2	Sb II	54	Mag 13.9 star on N edge.
02 21 25.3	+25 25 20	UGC 1812	14.0	1.1 x 0.5	13.2	Scd:	20	Mag 12.6 star on N edge.
02 22 23.5	+28 42 53	UGC 1833	14.7	1.2 x 0.8	14.5	SB(s)d	123	Small, moderately bright NE-SW bar; **UGC 1818** NW 2′.8.
02 23 39.4	+27 09 30	UGC 1844	14.6	0.9 x 0.7	13.9	SBdm	165	Stellar nucleus; dwarf spiral **UGC 1850** SE 4′.9.
02 24 42.2	+25 33 37	UGC 1860	13.6	1.4 x 0.8	13.6	SB(rs)b	30	MCG +4-6-26 S 1′.8.
02 25 04.7	+22 13 01	UGC 1871	14.0	0.9 x 0.6	13.2	S?	37	
02 25 35.1	+26 44 36	UGC 1881	14.0	1.0 x 0.9	13.7	Sb III-IV	93	Small, almost stellar companion galaxy S edge, connected.
02 25 50.6	+27 24 45	UGC 1885	14.5	0.9 x 0.8	14.0	SB(r)b	138	Mag 11.5 star E 1′.5.
02 26 11.2	+26 03 20	UGC 1891	13.9	0.9 x 0.9	13.8	E		**CGCG 483-47** N 3′.0.
02 26 12.5	+27 36 09	UGC 1892	14.1	1.0 x 0.8	13.7	SB(r)b	171	
02 26 26.1	+27 39 11	UGC 1899	13.9	0.9 x 0.8	13.4	S?	6	Mag 11.7 star N 1′.0.
02 27 27.4	+24 15 33	UGC 1917	14.3	0.9 x 0.7	13.7	(R′)SB(s)ab	96	
02 27 32.6	+25 40 05	UGC 1918	13.7	1.2 x 0.5	13.0	SBab	118	Stellar nucleus, mag 14.1 star SE end; mag 8.71v star SW 1′.6.
02 27 46.5	+26 35 23	UGC 1921	13.7	0.7 x 0.5	12.4	SB(s)b	138	Mag 11.0 star E 1′.8.
02 28 22.2	+23 12 53	UGC 1938	14.1	1.4 x 0.3	13.0	Sbc	154	Mag 9.63v star NNW 6′.1.
02 28 17.6	+26 18 42	UGC 1939	13.8	1.1 x 0.7	13.4	SBa	110	Mag 10.27v star W 3′.2.
02 28 54.3	+25 20 40	UGC 1955	14.2	1.5 x 0.7	14.1	S?	165	Mag 11.7 star SW 1′.2.
02 29 26.4	+20 13 00	UGC 1965	14.4	1.5 x 0.4	13.7	S?	92	Mag 10.25v star N 5′.9.
02 29 54.1	+25 15 25	UGC 1970	14.4	2.0 x 0.3	13.6	Scd:	22	Mag 5.89v star E 8′.8.
02 30 00.1	+28 37 57	UGC 1971	13.8	1.0 x 0.6	13.1	S0?	159	**CGCG 504-96** SW 5′.3; **MCG +5-6-53** NE 9′.3.
02 31 52.5	+19 09 11	UGC 1999	13.9	2.8 x 0.5	14.1	Scd:	96	Mag 8.43v star N 8′.6.
02 32 41.5	+21 01 55	UGC 2012	13.9	1.1 x 0.8	13.6	Sbc	131	Faint, almost stellar anonymous galaxy NW 2′.5.
02 32 44.7	+28 50 21	UGC 2017	13.6	2.3 x 1.9	15.1	Im	84	Dwarf irregular, low surface brightness.
02 32 54.2	+23 19 37	UGC 2020	14.8	1.5 x 0.2	13.3	Scd:	109	
02 33 14.3	+25 30 19	UGC 2025	15.0	1.5 x 0.2	13.5	Scd:	33	Located between a pair of mag 11 stars E and W.
02 33 18.3	+22 23 33	UGC 2028	13.8	1.4 x 0.7	13.6	Scd:	104	
02 33 24.5	+20 16 17	UGC 2031	13.9	1.2 x 0.8	13.8	S0	10	
02 34 29.1	+29 45 02	UGC 2053	14.6	1.9 x 0.9	15.1	Im V	34	Dwarf irregular, low uniform surface brightness.
02 35 10.2	+20 51 03	UGC 2064	13.5	2.1 x 1.5	14.6	SAB(s)bc	165	Strong N-S bar.
02 36 09.8	+23 53 59	UGC 2079	13.9	1.7 x 0.7	13.9	SAB(s)c I	157	Two thin arms.
02 36 17.2	+25 25 11	UGC 2082	13.0	4.4 x 0.8	14.2	Scd:	133	Mag 10.67v star NW 5′.9; mag 9.61v star N 7′.0.
02 37 27.5	+23 18 03	UGC 2104	14.5	0.9 x 0.6	13.7	Scd:	57	
02 38 27.6	+29 45 34	UGC 2122	13.6	1.0 x 1.0	13.5	SAB(s)c I		Strong dark patch E of nucleus.
02 38 52.0	+27 50 51	UGC 2134	13.4	1.7 x 0.7	13.4	Sb II-III	105	Mag 8.61v star N 4′.2.
02 39 06.2	+18 23 02	UGC 2140	14.8	1.7 x 0.9	15.1	IB(s)m pec	150	= **Hickson 18B**. Strongly disrupted; UGC 2140A SE edge; mag 10.71v star E 2′.6.
02 39 09.5	+18 22 00	UGC 2140A	14.5	0.9 x 0.2	12.5	SB0/a: sp	123	= **Hickson 18A**. UGC 2140 NW.
02 39 57.9	+28 19 29	UGC 2151	13.8	1.8 x 0.6	13.7	SAc	111	
02 47 51.8	+23 24 13	UGC 2267	14.3	1.0 x 0.7	13.8	(R)SB(r)b	165	
02 48 04.3	+27 06 10	UGC 2272	13.8	1.8 x 0.3	13.0	Sbc	65	Located 2′.3 NNW of mag 7.56v star.
02 48 51.8	+20 51 42	UGC 2286	14.5	1.0 x 0.8	14.1	Compact	108	
02 49 09.2	+23 01 00	UGC 2290	13.9	1.5 x 1.5	14.6	Sm:		Mag 10.6 star W 2′.2.
02 49 09.8	+18 20 02	UGC 2296	12.2	0.8 x 0.8	11.6	S?		Mag 12.4 star NW 3′.5.
02 49 23.3	+17 39 55	UGC 2303	13.4	1.2 x 1.2	13.6	SABb II-III		
02 50 38.9	+20 40 55	UGC 2327	13.9	1.1 x 0.8	13.6	Sb: II-III	65	
02 57 07.6	+17 30 49	UGC 2424	13.4	2.3 x 0.6	13.6	Sbc	158	
02 58 35.8	+25 16 54	UGC 2442	14.3	0.9 x 0.9	13.9	S?		
02 59 55.0	+24 13 33	UGC 2457	14.4	0.9 x 0.9	14.0	SAcd		
03 01 30.1	+17 51 07	UGC 2486	13.7	1.1 x 0.8	13.4	SA(r)bc	54	Stellar nucleus; mag 9.31v star N 5′.9.
03 02 08.1	+29 06 22	UGC 2497	13.8	3.1 x 0.7	14.5	Sdm	70	Almost stellar galaxy **V Zw 312** NNE 2′.4.
03 02 12.3	+17 20 44	UGC 2498	14.4	1.5 x 0.7	14.3	SABcd	65	Located between a pair of mag 12.4 stars.
03 03 31.3	+27 41 10	UGC 2506	14.1	1.1 x 0.9	13.9	Compact	30	Mag 7.54v start S 7′.0; stellar galaxy **V Zw 313** SSW 8′.2.
03 05 46.8	+22 12 11	UGC 2530	15.9	1.7 x 0.8	16.1	Sdm:	37	Mag 13.4 star E edge.
03 08 30.9	+20 46 15	UGC 2553	13.9	1.0 x 0.7	13.4	SB(rs)b	132	
03 09 37.0	+18 30 05	UGC 2563	14.6	1.5 x 0.2	13.2	Scd:	18	
03 10 25.8	+20 23 06	UGC 2570	15.6	1.5 x 0.3	14.6	S?	117	
03 20 23.5	+17 17 47	UGC 2684	14.6	1.8 x 0.9	15.0	Im?	111	Dwarf, low surface brightness.

OPEN CLUSTERS

RA h m s	Dec ° ′ ″	Name	Mag	Diam ′	No. ★	B ★	Type	Notes
02 47 31.0	+17 15 00	DoDz 1		10	12		cl	Few stars; moderate brightness range; no central concentration; detached.

RA h m s	Dec ° ′ ″	Name	Mag (V)	Dim ′ Maj x min	SB	Type Class	PA	Notes
01 09 29.8	+19 51 50	CGCG 459-10	14.9	0.7 x 0.2	12.6	Sc	78	Mag 10.1 star W 4′.7; mag 8.21v star E 8′.3.
01 19 37.1	+18 26 37	CGCG 459-29	14.9	0.7 x 0.3	13.0	Sc	123	Mag 11.1 star W 3′.7.
01 21 42.2	+18 34 02	CGCG 459-40	15.2	0.6 x 0.5	13.8	Sc	3	Mag 6.96v star NE 10′.8.
01 30 16.3	+20 35 45	CGCG 459-66	15.6	0.7 x 0.7	14.7			Multi-galaxy system. Consists of three distinct members, closely grouped, forming a triangle.
01 40 42.0	+20 50 18	CGCG 460-18	14.9	0.6 x 0.5	13.5	Sb	39	Mag 13.8 star NNE 1′.0.
01 58 13.6	+21 20 59	CGCG 461-14	14.5	0.6 x 0.4	12.9	S0	141	Mag 8.31v star W 5′.1.
01 58 19.1	+20 57 41	CGCG 461-15	14.7	0.4 x 0.4	12.8	S0		Line of three mag 10 stars N 7′.6; mag 12.4 star SSW 3′.3.
00 39 25.3	+25 15 12	CGCG 479-54	14.1	0.6 x 0.3	12.1			Mag 11.9 star NE 1′.3.
01 59 15.8	+24 24 57	CGCG 482-35	13.9	0.7 x 0.2	11.6		78	Between two mag 12 stars.
02 02 39.7	+27 34 24	CGCG 482-55	14.0	0.8 x 0.2	11.9	S0/a	21	Located 2′.1 SE of mag 9.91v star.
00 42 48.0	+29 55 18	CGCG 500-74	14.2	0.5 x 0.4	12.5	E	120	Mag 12.4 star NW 1′.7.
00 48 08.7	+28 36 48	CGCG 501-18	14.5	0.6 x 0.3	12.6	S0	117	Mag 9.50v star N 6′.9.
00 42 22.0	+29 38 27	IC 43	13.2	1.5 x 1.2	13.7	SABc II-III	117	Mag 10.4 star SE 3′.4; **IC 45** N 3′.4.
00 42 58.0	+27 15 14	IC 46	13.9	0.7 x 0.4	12.3	S0?	87	Mag 8.99v star SE 6′.1.
00 59 24.5	+27 03 31	IC 64	14.6	1.3 x 0.9	14.9	E/S0	147	Mag 10.44v star E 6′.5; **MCG +4-3-30** W 4′.7.
01 20 33.3	+29 36 58	IC 96	13.9	0.9 x 0.3	12.3	S?	48	Mag 11.2 star WNW 7′.8; IC 1672 N 5′.1.
01 26 54.5	+19 12 50	IC 115	14.0	0.5 x 0.5	12.4	S0?		Mag 5.50v star 94 Pisicum NW 3′.5; near center of galaxy cluster A 195.
01 49 14.7	+20 42 32	IC 163	13.0	1.8 x 0.9	13.4	SBdm	95	Mag 11.5 star W 8′.2.
01 51 08.9	+21 54 48	IC 167	13.1	2.9 x 1.8	14.8	SAB(s)c	95	Two main arms; mag 10.5 star N 3′.9; NGC 694 NNW 5′.5.
02 01 30.9	+26 28 50	IC 187	12.9	2.0 x 0.7	13.1	SBa	70	Double system? Small elongated companion E end; IC 188 NE 5′.2.
02 01 46.3	+26 32 46	IC 188	13.8	0.7 x 0.4	12.3	S?	45	IC 187 SW 5′.2.
02 01 52.9	+23 33 01	IC 189	13.9	0.7 x 0.6	12.8	SB?	147	NE-SW bar with strong dark patches N and S; stellar galaxy **V Zw 167** SE 2′.8.
02 02 07.3	+23 32 55	IC 190	14.1	0.6 x 0.4	12.6	E?	105	Stellar galaxy **V Zw 167** WSW 2′.8; mag 9.84v star S 2′.5.
00 36 52.3	+23 59 04	IC 1559	14.0	0.6 x 0.2	11.5	SAB0 pec: II-III	10	On S edge of NGC 169; mag 6.17v star ENE 3′.7.
00 47 10.4	+23 04 25	IC 1583	14.9	0.7 x 0.4	13.3		21	Mag 11.04v star WNW 6′.7.
00 47 18.6	+27 49 36	IC 1584	13.6	1.6 x 1.4	14.3	SB?	90	Mag 9.91v star ENE 4′.9.
00 47 14.3	+23 03 13	IC 1585	14.6	0.9 x 0.5	13.6		108	IC 1583 NNE 1′.5; **MCG +4-3-3** E 1′.2.
00 47 56.5	+22 22 19	IC 1586	14.0	0.5 x 0.4	12.1		0	Stellar; mag 12.2 star W 6′.9.
00 54 42.9	+21 31 20	IC 1596	13.8	1.8 x 0.6	13.7	S?	120	Mag 11.3 star W 1′.3; mag 10.97v star N 4′.6.
01 20 38.2	+29 41 54	IC 1672	13.0	1.3 x 1.0	13.1	S?	140	Mag 11.2 star WSW 8′.6; IC 96 S 5′.1.
01 25 50.5	+18 10 59	IC 1701	14.4	1.0 x 0.8	14.1		99	Mag 8.57v star E 9′.8.
01 30 55.1	+17 11 17	IC 1711	13.6	2.6 x 0.5	13.7	Sb III	43	Mag 7.58v star E 5′.0.
01 44 49.9	+21 52 54	IC 1725	14.3	0.9 x 0.7	13.6	SB?	123	Bright NE-SW bar; faint star on SE edge.
01 47 30.5	+27 19 49	IC 1727	11.5	5.7 x 2.4	14.2	SB(s)m III-IV	150	Mag 10.3 star S 3′.6; mag 9.86v star SW 7′.0; NGC 672 NE 8′.0.
01 50 12.5	+27 11 43	IC 1731	13.3	1.5 x 1.0	13.6	SAB(s)c:	140	Mag 9.86v star S 4′.0; the center of open cluster Cr 21 S 7′.1.
01 50 53.3	+18 18 08	IC 1736	14.1	0.4 x 1.3	13.2	Sbc	120	Mag 8.83v star N 4′.6.
01 53 14.2	+22 43 14	IC 1742	14.2	0.7 x 0.4	12.7	S?	36	Mag 11.48v star NNW 8′.1.
01 56 09.1	+17 38 26	IC 1748	13.7	1.0 x 0.6	13.0	SABbc	130	Mag 8.54v star SE 11′.9; **CGCG 460-51** S 6′.6; **UGC 1417** NE 9′.1.
01 57 19.4	+28 35 19	IC 1753	13.9	0.4 x 0.4	12.0	E?		**IC 1752** NW 1′.7.
02 00 23.5	+24 34 46	IC 1764	13.3	1.2 x 0.8	13.1	SB(r)b	174	Mag 11.04v star NW 1′.8.
00 37 50.6	+17 22 57	MCG +3-2-19	15.5	0.7 x 0.4	14.0	Sc	123	Mag 10.92v star NW 0′.9.
00 39 48.6	+21 12 53	MCG +3-2-22	14.4	0.6 x 0.4	12.7	S?	27	**MCG +3-2-21** NW 1′.8; mag 5.88v star 54 Piscium WNW 6′.6.
00 44 50.2	+20 06 29	MCG +3-3-1	14.6	0.6 x 0.6	13.4	S0		Close pair of stars, mags 12.8 and 15.1, SW 1′.1.
00 53 56.5	+20 23 33	MCG +3-3-6	15.4	0.5 x 0.4	13.5	Sc	33	Mag 11.2 star SW 0′.9.
00 56 14.8	+19 58 40	MCG +3-3-9	14.8	0.9 x 0.8	14.3	SBb	105	Mag 9.28v star E 4′.7.
01 07 12.9	+17 59 38	MCG +3-3-16	15.3	0.8 x 0.4	13.9		63	Mag 12.7 star SE 2′.1.
01 17 42.4	+19 57 44	MCG +3-4-15	14.8	0.5 x 0.5	13.1			Mag 13.2 star E 1′.9.
01 19 57.3	+21 22 49	MCG +3-4-21	14.8	0.7 x 0.3	13.0	Sbc	33	Close pair of stars, mags 12.2 and 13.2, N 1′.6.
01 24 43.9	+17 00 14	MCG +3-4-34	15.7	0.4 x 0.4	13.6			Mag 13.2 star SW 0′.9.
01 26 34.4	+19 20 24	MCG +3-4-37	15.3	0.8 x 0.8	14.7			Mag 9.44v star S 2′.5.
01 28 51.0	+18 59 13	MCG +3-4-45	14.3	1.0 x 0.7	13.8	S0/a	30	
01 28 58.3	+19 33 54	MCG +3-4-46	14.9	0.5 x 0.5	13.2	Sc		Mag 10.65v star SE 3′.0.
01 29 11.0	+19 42 48	MCG +3-4-47	15.1	0.8 x 0.3	13.4	Sc	132	
01 31 49.0	+18 35 47	MCG +3-5-1	14.3	0.7 x 0.5	13.1	SB0/LSB arms	90	Mag 11.4 star NE 0′.9.
01 33 43.1	+18 33 09	MCG +3-5-9	15.9	0.9 x 0.7	15.2	Sc		
01 43 56.5	+17 03 44	MCG +3-5-13	14.4	0.4 x 0.4	12.3	cl pec:		= **Mkn 360**; mag 13.1 star WSW 1′.9.
01 54 44.3	+18 02 49	MCG +3-5-31	14.7	0.7 x 0.6	13.6			
01 56 52.8	+19 35 39	MCG +3-6-3	15.5	0.6 x 0.5	14.1	Sc	3	
02 01 34.9	+18 38 38	MCG +3-6-17	16.0	0.6 x 0.5	14.5	SBc	177	Mag 13.2 star S 1′.6.
02 01 44.3	+17 55 39	MCG +3-6-18	15.3	0.7 x 0.5	14.0	Sb	147	Mag 7.28v star NW 3′.7.
00 37 51.1	+26 19 17	MCG +4-2-39	15.0	0.9 x 0.7	14.4		0	Mag 9.21v star WNW 0′.9.
00 40 24.9	+24 26 35	MCG +4-2-43	14.3	0.7 x 0.7	13.6	E/S0		Mag 10.9 star SW 2′.3.
00 42 23.3	+25 33 45	MCG +4-2-45	14.7	0.8 x 0.7	13.9		60	**CGCG 479-60** W 5′.1.
00 42 27.3	+25 37 09	MCG +4-2-46	15.4	0.7 x 0.6	14.3		63	
00 42 49.2	+23 29 28	MCG +4-2-47	14.9	0.8 x 0.4	13.5	S?	48	NGC 228 NE 1′.4.
00 43 42.0	+23 28 34	MCG +4-2-50	14.4	0.9 x 0.6	13.6		105	NGC 229 W 8′.7.
00 45 54.3	+25 15 55	MCG +4-2-53	14.4	0.8 x 0.4	13.0	S?	27	Mag 9.93v star N 2′.7.
00 49 34.2	+23 34 44	MCG +4-3-8	14.3	0.8 x 0.5	13.1	S?	141	= **Hickson 8A**. **Hickson 8B** N 0′.9; **Hickson 8C** NE edge; **Hickson 8D** SE edge; small, unnamed anonymous galaxy SW edge.
00 50 13.0	+24 29 50	MCG +4-3-11	14.6	0.8 x 0.5	13.4	SB?	69	MCG +4-3-12 NE 3′.2.
00 50 25.8	+24 31 15	MCG +4-3-12	14.3	0.6 x 0.4	12.6	S	105	Faint anonymous galaxy W 2′.0.
00 54 41.6	+24 52 10	MCG +4-3-16	14.5	0.8 x 0.5	13.3	S?	177	Close pair of mag 12-13 stars N 1′.7.
00 55 48.8	+24 08 51	MCG +4-3-17	15.0	1.0 x 0.5	14.1	Spiral	81	NGC 304 ESE 4′.0.
00 56 22.3	+25 36 55	MCG +4-3-20	14.6	0.9 x 0.8	14.3	E/S0	126	
00 56 41.9	+27 01 37	MCG +4-3-21	15.5	0.7 x 0.3	13.7	Sbc	159	**UGC 585** SE 1′.4.
00 58 27.4	+23 32 28	MCG +4-3-26	14.3	0.9 x 0.8	13.8	SB?	114	Mag 12.0 star SW 2′.7.
00 58 29.2	+22 19 29	MCG +4-3-27	15.3	0.7 x 0.5	14.0			Mag 8.92v star N 5′.8.
01 00 28.1	+27 01 27	MCG +4-3-32	14.9	0.6 x 0.6	13.7			Mag 13.5 star superimposed E edge; mag 11.8 star SW 2′.7.
01 00 34.2	+27 05 52	MCG +4-3-33	14.3	0.7 x 0.6	13.2		21	Mag 14.0 star NE 0′.7.
01 04 52.3	+22 18 52	MCG +4-3-39	14.5	0.7 x 0.4	13.0	SB?	156	Mag 11.2 star E 1′.1.
01 05 01.9	+22 05 01	MCG +4-3-40	14.4	0.8 x 0.6	13.5	SB?	99	
01 06 12.1	+25 33 03	MCG +4-3-42	14.7	1.0 x 0.3	13.2	S?	9	Mag 12.0 star SW 2′.8.
01 06 28.6	+24 52 10	MCG +4-3-43	14.9	0.7 x 0.5	13.6		51	Mag 9.50v star N 2′.2.
01 42 13.7	+27 40 22	MCG +4-5-1	14.9	0.8 x 0.7	14.1	Sb	57	Mag 10.48v star N 2′.8.
01 43 08.1	+27 45 00	MCG +4-5-3	14.0	1.1 x 0.8	13.7	S?	138	Faint, slightly elongated anonymous galaxy N 2′.6.
01 46 04.3	+23 39 51	MCG +4-5-6	14.8	0.8 x 0.7	14.0		9	
01 46 04.0	+23 27 28	MCG +4-5-7	15.1	0.5 x 0.3	12.9		9	Mag 9.77v star S 1′.0.
01 46 08.8	+23 26 18	MCG +4-5-8	14.9	0.4 x 0.4	12.8			Mag 9.77v star W 1′.4.
01 47 47.7	+25 34 20	MCG +4-5-10	14.3	0.6 x 0.4	12.6	SB(r)c+	45	Mag 11.4 star W 1′.9.

RA h m s	Dec ° ′ ″	Name	Mag (V)	Dim ′ Maj x min	SB	Type Class	PA	Notes
01 59 38.2	+27 25 59	MCG +4-5-27	14.3	0.8 x 0.4	12.9	S?	75	Mag 10.6 star W 0′.9.
02 00 04.5	+23 45 48	MCG +4-5-31	14.5	0.9 x 0.7	13.9		30	Mag 9.34v star N 4′.5.
02 01 08.0	+26 50 37	MCG +4-5-36	13.9	0.5 x 0.5	12.2			
02 03 41.2	+26 16 34	MCG +4-5-46	14.1	0.9 x 0.5	13.1	Sc	6	UGC 1549 NW 2′.6.
02 03 51.6	+25 55 30	MCG +4-5-47	13.5	0.8 x 0.6	12.8	E/S0	12	Located 2′.7 E of mag 5.69v star 10 Arietis.
00 38 28.4	+29 37 27	MCG +5-2-31	14.3	1.0 x 0.4	13.1	S?	24	Numerous small, faint anonymous galaxies E and S.
00 46 24.5	+29 37 25	MCG +5-2-45	14.4	0.9 x 0.6	13.6	SB?	12	Mag 9.43v star SW 2′.4.
00 46 34.2	+29 43 28	MCG +5-2-47	14.0	0.8 x 0.6	13.0	Sc	96	Mag 11.3 star NE 3′.6.
00 47 50.2	+29 57 31	MCG +5-3-6	14.5	0.5 x 0.5	12.9	S0		MCG +5-3-7 E 2′.6.
00 48 02.4	+29 57 23	MCG +5-3-7	14.5	0.7 x 0.5	13.2	S0	138	MCG +5-3-6 W 2′.6.
00 59 39.8	+29 20 13	MCG +5-3-32	14.1	0.7 x 0.6	13.2	E?	27	
01 22 17.5	+28 47 54	MCG +5-4-31	14.6	1.2 x 0.4	13.7	S?	36	Mag 8.25v star NE 3′.7.
01 22 54.8	+28 50 06	MCG +5-4-33	14.4	0.9 x 0.3	12.9	I?	102	Mag 11.7 star NW 0′.8.
01 23 56.9	+28 37 56	MCG +5-4-49	15.0	1.1 x 0.3	13.6	Compact	30	
01 25 04.7	+29 25 50	MCG +5-4-56	15.4	0.6 x 0.6	14.2			Mag 10.07v star W 3′.6.
01 41 33.2	+28 20 21	MCG +5-5-1	15.6	0.7 x 0.7	14.7	Sc		Mag 12.2 star on NE edge.
01 46 58.1	+28 45 20	MCG +5-5-11	13.9	0.9 x 0.7	13.2	S0?	3	
00 36 04.0	+23 57 28	NGC 160	12.7	2.3 x 1.2	13.7	(R)SA0$^+$ pec	45	Mag 7.25v star N 4′.2.
00 36 52.5	+23 59 28	NGC 169	13.2	2.6 x 0.6	13.6	SA(s)ab: sp II-III	88	Mag 6.17v star ENE 3′.7; IC 1559 on S edge.
00 38 23.2	+29 28 20	NGC 181	14.9	0.6 x 0.2	12.4	S?	147	Located 2′.7 S of NGC 183.
00 38 29.5	+29 30 40	NGC 183	12.7	1.7 x 1.1	13.5	E	130	Very small anonymous galaxy 2′.5 NE; **CGCG 500-61** E 4′.8.
00 38 36.0	+29 26 51	NGC 184	14.7	0.7 x 0.2	12.4	S0/a	6	Located 8′.3 due N of mag 4.35v star Epsilon Andromedae.
00 41 28.2	+25 30 01	NGC 214	12.3	1.9 x 1.4	13.2	SAB(r)c I	35	Mag 9.91v star NNE 7′.5.
00 42 54.5	+23 30 13	NGC 228	13.7	1.2 x 1.0	13.7	(R)SB(r)ab	126	MCG +4-2-47 SW 1′.4.
00 43 04.8	+23 30 32	NGC 229	14.1	0.9 x 0.3	12.5	Sab	96	MCG +4-2-50 E 8′.7.
00 46 00.9	+29 57 34	NGC 243	13.7	0.8 x 0.4	12.3	S0?	149	Mag 10.47v star 1′.5 W.
00 47 53.9	+19 35 46	NGC 251	13.2	2.4 x 1.7	14.6	Sc	96	Pair of mag 12.4, 12.9 stars on E and N edges; mag 6.12v star 59 Piscium W 9′.6.
00 48 01.6	+27 37 29	NGC 252	12.4	1.5 x 1.1	12.8	(R)SA(r)0$^+$:	80	Mag 10.16v star W 4′.1.
00 48 13.0	+27 39 26	NGC 258	14.2	0.5 x 0.4	12.3	Sb	108	Almost stellar; mag 11.05v star W 0′.8.
00 48 35.1	+27 41 33	NGC 260	13.5	0.8 x 0.8	12.9	Scd pec:		Mag 10.65v star NE 4′.6.
00 52 30.8	+24 20 56	NGC 280	13.3	1.7 x 1.1	13.9	SB?	95	Faint star on E end.
00 56 06.0	+24 07 38	NGC 304	13.1	1.1 x 0.7	12.7	S?	175	**MCG +4-3-178** WNW 4′.0.
00 58 22.9	+26 51 53	NGC 326	13.2	1.4 x 1.4	14.0	E$^+$4: pec		Mag 7.39v star S 5′.0; **MCG +4-3-24** WNW 4′.5.
01 03 16.4	+22 20 34	NGC 354	13.5	0.9 x 0.5	12.5	SB pec	29	Mag 13.6 star on W edge; mag 9.41v star S 3′.4.
01 18 08.2	+17 33 41	NGC 459	14.4	0.7 x 0.6	13.3	Sbc	33	Stellar nucleus; mag 10.74v star E 4′.7.
01 30 46.7	+21 26 23	NGC 575	12.8	1.7 x 1.6	13.7	SB(rs)c	66	Stellar nucleus; star or bright knot SE of core.
01 34 50.3	+21 25 03	NGC 606	13.4	1.4 x 1.2	13.8	SB(r)c	114	2′.8 S of mag 9.15v star.
01 42 27.3	+26 08 33	NGC 656	12.4	1.5 x 1.3	12.9	SB0	35	Slightly elongated N-S core, smooth envelope; mag 10.4 star NW 1′.6; mag 9.57v star W 7′.8.
01 44 14.7	+28 42 22	NGC 661	12.2	1.7 x 1.4	13.1	E$^+$:	60	
01 47 24.8	+27 53 11	NGC 670	12.7	2.0 x 1.0	13.3	SA0	172	Mag 8.45v star E 7′.8.
01 47 54.0	+27 25 58	NGC 672	10.9	6.0 x 2.4	13.6	SB(s)cd III	65	IC 1727 SW 8′.0; mag 9.75v star NW 9′.0.
01 49 24.8	+21 59 49	NGC 678	12.2	4.5 x 0.8	13.5	SB(s)b: sp III	78	
01 49 47.2	+21 58 15	NGC 680	11.9	1.9 x 1.5	13.0	E$^+$ pec:	156	Mag 10.4 star E 3′.6; small, faint anonymous galaxy NE 3′.5.
01 50 14.5	+27 38 42	NGC 684	12.4	3.2 x 0.6	12.9	Sb sp II-III	87	Mag 8.45v star WNW 8′.9.
01 50 41.7	+21 45 33	NGC 691	11.4	3.5 x 2.6	13.7	SA(rs)bc	95	Very close pair of mag 9 stars on NE edge.
01 50 58.4	+21 59 48	NGC 694	13.7	0.7 x 0.6	12.7	S0? pec	160	Mag 10.5 star ESE 2′.3; IC 167 SSE 5′.5.
01 51 14.3	+22 34 56	NGC 695	12.8	0.8 x 0.7	12.1	S0? pec	40	Mag 12.9 star on W edge; pair of faint, stellar anonymous galaxies for a triangle with a mag 12.3 star 6′.3 NNE.
01 51 17.6	+22 21 25	NGC 697	12.0	4.0 x 1.3	13.7	SAB(r)c:	105	
01 52 27.8	+17 30 45	NGC 711	13.1	1.6 x 0.8	13.3	S0	15	**UGC 1335** W 6′.5.
01 53 38.7	+19 50 27	NGC 719	13.2	1.4 x 1.1	13.6	S0?	150	**UGC 1357** NNW 7′.9.
01 54 47.1	+20 41 52	NGC 722	13.4	1.7 x 0.5	13.1	S?	138	Located 6′.8 S of mag 2.66v Sharatan, Beta Arietis.
01 58 48.1	+24 53 28	NGC 765	12.8	2.8 x 2.8	14.9	SAB(rs)bc		Small, elongated core in smooth envelope; mag 7.34v star SE 8′.0.
01 59 13.7	+18 57 14	NGC 770	12.8	0.9 x 0.7	12.3	E3:	15	Located 3′.6 S of the center of NGC 772; stellar **PGC 7509** W 2′.0; stellar **PGC 7493** W 3′.6.
01 59 20.1	+19 00 26	NGC 772	10.3	7.2 x 4.3	13.9	SA(s)b I	130	Faint, small anonymous galaxy E 5′.2.
01 59 54.6	+23 38 36	NGC 776	12.4	1.7 x 1.7	13.4	SAB(rs)b		**IC 181** NE 2′.0; **IC 180** SE 2′.8.
02 00 35.2	+28 13 31	NGC 780	13.8	1.5 x 0.9	14.0	Sb	176	2′.7 NE of mag 9.71v star.
02 01 16.6	+28 50 16	NGC 784	11.7	6.6 x 1.5	14.1	SBdm: sp	0	Several faint knots in elongated core; almost stellar galaxy **V Zw 166** S 6′.5.
02 02 29.4	+18 22 22	NGC 794	12.7	1.3 x 1.1	13.0	S0$^-$:	45	Close pair of faint, stellar anonymous galaxies SSE 4′.2.
00 36 10.0	+25 48 27	UGC 360	15.1	0.8 x 0.5	14.0	I?	18	**CGCG 479-42** SW 10′.3.
00 36 45.1	+21 33 56	UGC 364	13.6	1.2 x 0.4	12.6	S?	58	**CGCG 457-21** N 2′.5.
00 37 05.4	+25 41 57	UGC 367	13.9	1.0 x 1.0	14.0	E?		In contact with **MCG +4-3-37** on SE edge; mag 10.6 star on NW edge.
00 37 21.3	+29 08 54	UGC 371	14.4	1.7 x 0.3	13.5	Scd:	45	
00 37 43.8	+25 38 23	UGC 375	14.1	0.7 x 0.7	13.2	Compact		
00 38 39.8	+17 24 14	UGC 393	15.2	1.8 x 0.4	14.7	Sdm:	170	Mag 12.0 star S 2′.2.
00 38 57.4	+25 38 18	UGC 398	13.4	1.1 x 0.9	13.3	SA(r)a	25	**CGCG 479-50** N 5′.3; stellar galaxy **IV Zw 27** S 5′.3.
00 39 18.4	+29 39 26	UGC 400	14.1	0.7 x 0.6	13.0	S?	75	Elongated **UGC 412** N 6′.7.
00 39 29.6	+25 38 35	UGC 411	13.2	1.1 x 0.7	13.0	E?	100	Mag 9.10v star N 5′.5.
00 40 10.9	+22 42 52	UGC 425	13.8	0.6 x 0.6	12.5	SB?		Interacting pair?
00 44 03.9	+26 12 24	UGC 469	14.2	1.5 x 0.3	13.2	S?	131	Mag 10.9 star WNW 2′.6.
00 46 13.2	+19 29 19	UGC 477	14.1	2.8 x 0.6	14.5	Sdm:	167	A pair of faint, almost stellar anonymous galaxies NE 1′.6; a single anonymous galaxy WNW 1′.3.
00 49 02.4	+28 13 01	UGC 501	14.5	1.7 x 0.2	13.1	Scd:	110	
00 49 38.8	+22 55 58	UGC 506	13.4	1.5 x 1.2	13.8	S?	177	Several very faint stars superimposed.
00 49 58.7	+21 43 15	UGC 510	14.4	1.2 x 0.3	13.1	Scd:	20	Mag 12 star N 1′.1.
00 51 35.0	+29 24 04	UGC 524	13.6	0.8 x 0.8	13.0	(R′)SB(s)b		Mag 12.0 star on SE edge; mag 14.8 star on W edge.
00 51 34.5	+29 42 58	UGC 525	14.4	1.5 x 0.9	14.6	SB?	150	Stellar nucleus, faint envelope.
00 52 12.4	+29 40 34	UGC 529	13.6	1.3 x 0.4	12.7	S0/a	90	Mag 9.4 double star W 2′.8.
00 52 58.4	+29 01 55	UGC 540	13.7	0.7 x 0.5	12.4	S?	137	Faint star on E edge; mag 8.82v star NNW 2′.8.
00 53 26.7	+29 16 12	UGC 542	13.3	2.1 x 0.3	12.7	S?	160	**UGC 536** SW 12′.0.
00 54 50.4	+29 14 44	UGC 556	14.3	1.0 x 0.4	13.1	S?	100	
00 57 19.8	+23 53 22	UGC 591	14.1	0.7 x 0.4	12.6	S?	165	Located between two stars, mags 11.6 and 12.6.
00 58 34.3	+22 19 04	UGC 605	15.1	1.1 x 0.7	14.7	S	108	Two very thin, curving arms; MCG +4-3-27 W 1′.3.
00 59 01.1	+23 51 06	UGC 612	13.7	0.9 x 0.4	12.5	S?	90	Mag 9.98v star NW 2′.9.
00 59 39.1	+18 50 34	UGC 616	14.1	1.1 x 0.9	13.9	(R′)SB(r)b	76	Strong dark patch E of center; mag 11.9 star on S edge.
00 59 49.5	+18 00 24	UGC 617	14.0	0.8 x 0.7	13.2	SBa	33	Bright E-W bar.
01 00 51.9	+19 28 36	UGC 628	16.6	1.4 x 0.7	16.5	Sm:	147	Dwarf spiral, low surface brightness.
01 01 01.9	+29 36 04	UGC 629	13.9	0.8 x 0.6	12.9	Scd:	158	
01 01 42.2	+24 03 28	UGC 636	13.3	1.5 x 1.2	13.9	E	110	Mag 12.4 star N 1′.4; mag 12.1 star E 1′.6.
01 03 05.6	+24 58 17	UGC 643	14.6	1.0 x 0.7	14.0	Scd:	6	

(Continued from previous page)

GALAXIES

RA h m s	Dec ° ′ ″	Name	Mag (V)	Dim ′ Maj x min	SB	Type Class	PA	Notes
01 03 59.3	+21 13 29	UGC 654	14.1	1.5 x 1.5	14.8	Compact		Mag 9.80v star NNW 1′.4.
01 08 04.2	+21 07 13	UGC 696	13.7	1.1 x 0.7	13.3	S0⁻:	144	Mag 11.6 star WSW 0′.7.
01 09 56.1	+20 46 21	UGC 723	14.9	1.7 x 0.2	13.5	Scd:	165	
01 12 00.7	+17 18 30	UGC 751	14.2	1.2 x 1.0	14.5	E/S0	63	Mag 13.5 star on E edge; almost stellar anonymous galaxy W 1′.9; larger anonymous galaxy S 2′.4.
01 15 19.5	+28 29 21	UGC 800	14.2	0.9 x 0.8	13.7	Scd:	107	Bright center, faint envelope.
01 19 17.3	+21 45 44	UGC 845	13.8	1.1 x 0.8	13.5	Sb II-III	40	Located on W edge of a small triangle of mag 13-14 stars.
01 19 33.2	+19 42 40	UGC 851	15.5	0.8 x 0.6	14.5	(R′)SB(s)bc	171	Short SE-NW bar.
01 21 00.9	+17 04 25	UGC 883	14.3	1.1 x 0.8	14.0	Sdm IV-V	36	Uniform surface brightness; close pair of stars, mags 12.5 and 13.5, N 1′.
01 21 36.9	+20 55 07	UGC 899	14.2	0.9 x 0.4	12.9	Sbc	63	Bright knot or star S edge.
01 21 35.2	+23 45 34	UGC 900	14.9	1.1 x 0.9	14.7	Sb	27	Mag 13.1 star superimposed NE edge; **UGC 905** NE 3′.0.
01 21 47.8	+17 35 33	UGC 903	13.6	1.8 x 0.4	13.1	S?	52	
01 21 50.0	+18 16 00	UGC 904	14.0	1.1 x 0.6	13.4	(R′)SB(s)a	48	Mag 11.9 star NW 1′.4.
01 26 44.2	+17 15 50	UGC 1020	13.9	1.2 x 0.3	12.7	S?	52	
01 27 05.5	+18 35 53	UGC 1025	13.7	0.8 x 0.8	13.1	S0/a		Bright N-S center, uniform envelope; mag 11.0 star NE 3′.0.
01 27 32.7	+19 10 39	UGC 1032	13.5	0.7 x 0.6	12.4	Pec	10	Near E side of galaxy cluster A 195.
01 30 07.7	+25 51 50	UGC 1073	13.8	1.7 x 1.0	14.2	Sm: V	163	Dwarf spiral, very low surface brightness; mag 7.24v star NE 7′.8.
01 30 35.6	+19 36 30	UGC 1077	13.9	1.0 x 0.6	13.3	S0⁻:	50	Mag 13.8 star on SE edge.
01 31 51.4	+17 33 51	UGC 1093	14.0	1.2 x 0.7	13.7	Sb	169	Brightest of three; first companion NW 0′.9; second companion ESE 1′.5.
01 32 16.1	+21 24 37	UGC 1098	13.8	1.0 x 0.5	12.9	SB?	90	
01 32 28.8	+19 38 09	UGC 1099	14.4	1.0 x 1.0	14.3	Compact		Almost stellar nucleus, faint envelope.
01 32 42.7	+18 19 01	UGC 1104	13.6	1.1 x 0.6	13.0	Im	5	Bright, diffuse center.
01 33 28.3	+17 24 26	UGC 1113	15.3	0.8 x 0.8	14.7	Scd:		
01 33 50.4	+17 13 48	UGC 1119	15.9	1.1 x 0.6	15.3	Sdm	125	Stellar nucleus; mag 12.2 star SE 2′.3.
01 37 17.1	+28 53 24	UGC 1154	13.5	0.8 x 0.7	12.7	S?	129	
01 42 33.2	+18 18 21	UGC 1197	14.2	1.8 x 0.6	14.2	Im	56	
01 44 20.7	+17 28 39	UGC 1219	12.8	1.3 x 0.6	12.3	SB?	102	
01 45 21.4	+28 43 19	UGC 1228	14.1	1.7 x 0.7	14.2	Sdm:	25	Located 3′.3 E of mag 8.04v star.
01 45 32.7	+25 31 15	UGC 1230	14.9	2.1 x 1.8	16.2	Sm:	174	Dwarf spiral, low surface brightness.
01 45 41.6	+28 48 19	UGC 1233	13.9	0.9 x 0.5	12.9	Scd:	110	Stellar galaxy **V Zw 103** ESE 6′.3.
01 46 45.8	+18 34 27	UGC 1243	13.6	1.3 x 0.6	13.2	S0?	135	Mag 13.3 star NNW 2′.1.
01 48 46.6	+20 15 53	UGC 1265	13.6	1.2 x 0.5	12.9	SB?	170	Multi-galaxy system; mag 9.30v star NW 4′.5.
01 51 01.2	+29 48 06	UGC 1311	16.7	1.5 x 0.9	16.9	Im:	20	Dwarf, low surface brightness.
01 51 36.2	+19 06 11	UGC 1324	13.5	1.3 x 0.8	13.4	SBb	90	
01 51 52.0	+17 03 05	UGC 1328	13.6	1.5 x 1.0	13.9	SAB(rs)c:	125	Located 2′.0 SW of mag 8.20v star.
01 51 47.4	+18 11 13	UGC 1329	15.1	1.8 x 0.5	14.8	SBcd:	128	Low surface brightness; mag 6.61v star N 6′.2.
01 53 42.3	+29 55 56	UGC 1359	13.4	0.9 x 0.8	12.9	SB?	6	
01 54 13.6	+18 06 02	UGC 1369	13.8	1.1 x 0.8	13.5	Scd:	65	
01 54 31.4	+17 40 24	UGC 1372	14.0	0.9 x 0.4	12.8	S?	14	Mag 11.0 star SE 1′.2; mag 10.0 star S 2′.6.
01 54 32.9	+20 03 52	UGC 1375	13.9	1.2 x 1.1	14.0	SBcd:	21	
01 54 47.7	+18 08 34	UGC 1384	17.5	1.5 x 0.6	17.2	SBd:	8	
01 55 36.8	+21 19 00	UGC 1396	13.8	1.2 x 0.5	13.1	S0	94	**UGC 1393** SW 3′.3.
01 55 53.0	+18 02 38	UGC 1399	13.7	0.8 x 0.5	12.6	S0	2	**CGCG 461-3** SSW 3′.7.
01 57 26.8	+17 13 12	UGC 1432	13.8	1.0 x 0.6	13.1	Sbc	65	
01 57 29.8	+19 55 06	UGC 1433	14.7	0.5 x 0.5	13.0	Scd:		**MCG +3-6-6** and **UGC 1436** SE 2′.6.
01 58 14.1	+19 05 58	UGC 1445	13.7	1.0 x 0.5	12.8	S0	93	Almost stellar nucleus.
01 58 30.1	+25 21 35	UGC 1451	13.4	1.6 x 0.7	13.3	SB?	121	Mag 10.63v star E 1′.7.
01 58 45.7	+24 38 39	UGC 1453	14.4	1.5 x 1.3	15.0	Im:		Dwarf, low surface brightness.
01 59 09.5	+25 23 10	UGC 1462	14.3	1.4 x 1.0	14.5	SBcd:	65	Mag 12.7 stars superimposed S edge.
02 00 15.1	+24 15 07	UGC 1478	13.8	0.9 x 0.8	13.3	SB(rs)c	96	**CGCG 482-38** NW 5′.7.
02 00 19.3	+24 28 27	UGC 1479	14.0	1.2 x 0.4	13.1	S?	176	Mag 13.3 star on NE edge; mag 9.72v star W 1′.5.
02 00 23.8	+21 06 27	UGC 1485	13.5	1.2 x 0.7	13.2	S?	25	Mag 11.1 star SE 2′.7.
02 00 32.3	+21 17 12	UGC 1490	13.7	0.6 x 0.3	11.7	S?	85	**CGCG 461-20** W 3′.1.
02 00 47.2	+17 42 49	UGC 1495	14.5	0.8 x 0.6	13.6	Scd:	39	
02 02 00.0	+21 05 44	UGC 1514	13.3	0.7 x 0.5	12.1	S?	6	
02 02 18.9	+19 04 01	UGC 1518	14.7	1.5 x 0.6	14.4		100	Located near center of galaxy cluster **A 292**; mag 8.48v star E 8′.9; stellar anonymous galaxy S 1′.4.
02 02 16.8	+19 10 47	UGC 1519	14.7	1.5 x 0.6	14.5	Sdm:	40	Located near N edge of galaxy cluster **A 292**; mag 9.16v star W 5′.0.
02 02 34.6	+17 10 01	UGC 1531	14.1	1.2 x 0.4	13.1	Sb II	114	
02 02 48.1	+26 34 50	UGC 1533	14.4	0.8 x 0.6	13.4	SABcd:	141	Mag 8.45v star ESE 9′.5.
02 03 20.3	+18 37 44	UGC 1546	13.7	1.0 x 0.9	13.5	SAB(s)c II	18	
02 03 20.6	+22 02 59	UGC 1547	13.9	1.9 x 1.5	14.9	IBm V	18	Irregular, brightest S one third, low surface brightness.
02 03 32.8	+26 18 23	UGC 1549	13.8	0.8 x 0.3	12.1	S0/a	36	MCG +4-5-46 SE 2′.5.
02 03 37.6	+24 04 28	UGC 1551	12.6	2.3 x 1.7	13.9	SB?	135	Many bright knots and/or superimposed stars.
02 03 56.4	+19 40 04	UGC 1560	13.8	0.8 x 0.7	13.0	(R′)SB(s)b:	38	Bright E-W bar; **MCG +3-6-29** SW 1′.3.

GALAXY CLUSTERS

RA h m s	Dec ° ′ ″	Name	Mag 10th brightest	No. Gal	Diam ′	Notes
00 37 48.0	+29 35 00	A 71	15.5	30	19	Band of numerous, faint, stellar anonymous GX extend SE to NW.
00 39 48.0	+21 15 00	A 75	15.5	42	13	Mag 5.88v star 54 Piscium 6′.0 W of center.
00 49 48.0	+24 31 00	A 104	15.9	50	15	Most cluster members faint and stellar to almost stellar.
01 11 00.0	+17 39 00	A 154	15.6	66	13	**IC 1634** and **PGC 4329** at center with numerous PGC and anonymous GX mostly S of center.
01 21 48.0	+19 28 00	A 179	15.3	31	28	Stellar **PCG 4990** NE of center 10′.6; slightly elongated **UGC 918** SE 16′.9.
01 26 54.0	+19 10 00	A 195	15.3	32	13	See IC 115 notes.
01 38 54.0	+18 53 00	A 225	15.9	51	15	All members faint and stellar.

OPEN CLUSTERS

RA h m s	Dec ° ′ ″	Name	Mag	Diam ′	No. ★	B ★	Type	Notes
01 50 12.0	+27 05 00	Cr 21	8.2	7	20		ast	Few stars; large brightness range; no central concentration; detached.(S) Not a true cluster.

RA h m s	Dec ° ′ ″	Name	Mag (V)	Dim ′ Maj x min	SB	Type Class	PA	Notes
23 33 06.8	+09 45 04	CGCG 407-5	14.2	0.8 x 0.5	13.1			Very faint anonymous galaxy S edge.
23 37 44.1	+08 04 54	CGCG 407-27	14.8	0.6 x 0.5	13.3		159	Faint stellar nucleus, faint outer envelope.
23 53 16.0	+07 25 31	CGCG 407-68	14.6	0.4 x 0.4	12.5			Mag 9.41v star NW 12′.6.
00 20 02.7	+06 27 52	CGCG 409-1	14.3	0.6 x 0.6	13.0	Sa		Galaxy pair, components N-S; mag 11.5 star on E edge.
00 34 16.7	+05 32 49	CGCG 409-37	14.0	0.4 x 0.4	11.9	Sb		Mag 13.5 star on SE edge.
23 33 27.2	+14 20 02	CGCG 432-2	14.4	0.4 x 0.3	11.9		114	Mag 11.9 star S 3′.0; mag 9.78v star W 9′.8.
00 13 27.1	+17 29 11	IC 4	13.1	1.0 x 0.9	12.9	S?	10	Mag 10.10v star NNE 8′.3.
00 18 53.2	+10 35 38	IC 7	13.9	0.7 x 0.6	13.0	E	0	Mag 14.6 star S of nucleus.
00 20 20.1	+07 41 59	IC 13	13.9	1.4 x 0.4	13.2	Sbc	163	Mag 9.78v star S 7′.7.
00 34 24.5	+12 16 05	IC 31	14.3	1.6 x 0.3	13.4	Sa	89	Mag 10.75v star S 4′.2.
00 35 36.4	+09 07 27	IC 34	12.6	2.4 x 0.8	13.2	SB(r)a	156	Mag 11.3 star SSW 9′.6; **UGC 353** S 7′.6.
00 37 39.8	+10 21 27	IC 35	14.0	0.8 x 0.7	13.2	Scd:	33	Mag 10.15v star NE 1′.6.
23 45 55.1	+12 03 47	IC 1508	13.3	2.0 x 0.5	13.1	Sdm:	168	Mag 10.9 star N 6′.7.
23 53 29.6	+11 19 01	IC 1513	14.1	0.9 x 0.3	12.5	S?	107	Mag 10.94v star NE 9′.7.
00 01 31.5	+11 20 44	IC 1526	14.4	0.7 x 0.5	13.1	Sc	129	Mag 1.9 star WNW 3′.1.
00 22 49.9	+06 57 50	IC 1549	13.7	0.8 x 0.8	13.1	S?		Stellar; mag 8.57v star NW 12′.8.
00 27 35.5	+08 52 38	IC 1551	13.4	2.5 x 1.2	14.4	S?	15	Bright, round core with two faint, slightly curving arms.
00 39 05.3	+06 01 10	IC 1564	13.9	1.1 x 0.5	13.1	SABbc:	83	Mag 8.74v star W 7′.8.
00 39 26.3	+06 44 00	IC 1565	13.4	1.0 x 1.0	13.3	S?		Mag 10.61v star SE 3′.4.
00 39 33.5	+06 48 51	IC 1566	13.7	0.8 x 0.7	13.1	E/S0	126	Mag 11.8 star SSW 3′.0; IC 1565 SSW 5′.2.
00 39 56.0	+06 50 55	IC 1568	13.9	0.8 x 0.8	13.3	S0		Mag 10.24v star and UGC 429 NE 8′.7.
00 02 37.7	+16 38 50	IC 5378	16.1	1.6 x 1.4	16.9	SBc	177	
00 03 11.1	+15 57 55	IC 5381	13.8	1.4 x 0.4	13.0	Sab? sp	54	Mag 9.64v star NW 8′.3.
00 02 40.1	+08 44 09	MCG +1-1-13	14.9	1.0 x 0.3	13.5	S?	21	Mag 10.28v star W 7′.7.
00 03 23.1	+05 42 14	MCG +1-1-16	14.9	0.7 x 0.7	14.0	SBc		Mag 11.1 star NW 3′.3.
00 06 14.3	+08 53 12	MCG +1-1-29	14.5	0.8 x 0.4	13.1	S?		Mag 12.7 star W 2′.0.
00 18 38.3	+07 31 00	MCG +1-1-49	14.9	0.6 x 0.5	13.4		99	
00 22 49.9	+06 49 00	MCG +1-2-4	14.6	0.9 x 0.3	13.0	Sc	45	Mag 9.14v star WSW 11′.9.
00 25 25.0	+06 42 19	MCG +1-2-8	15.0	0.7 x 0.4	13.5	Sc	162	Mag 10.74v star W 1′.9.
00 25 37.1	+07 28 27	MCG +1-2-9	14.9	0.8 x 0.5	13.7		72	
00 28 26.8	+05 00 10	MCG +1-2-14	14.8	0.6 x 0.4	13.1	SB0?	3	Mag 12.6 star SW 2′.1.
00 30 28.7	+05 51 39	MCG +1-2-15	14.3	0.7 x 0.4	12.8		141	
00 31 18.8	+08 28 30	MCG +1-2-18	13.6	0.8 x 0.5	12.4	S?	9	Pair with UGC 312 E.
00 32 58.8	+05 17 07	MCG +1-2-23	15.0	0.6 x 0.5	13.6	Sb	12	MCG +1-2-24 S 1′.5.
00 33 00.9	+05 15 45	MCG +1-2-24	15.1	0.7 x 0.5	13.8	Sb	39	MCG +1-2-23 N 1′.1
00 33 25.5	+07 51 21	MCG +1-2-26	14.4	0.8 x 0.3	12.7	S?	15	**UGC 327** NW 3′.8.
00 35 55.9	+08 43 00	MCG +1-2-34	14.1	0.5 x 0.5	12.4			Mag 11.8 star SE 1′.7.
00 36 17.0	+07 50 02	MCG +1-2-36	15.5	0.8 x 0.6	14.6		90	
00 38 54.8	+07 03 20	MCG +1-2-42	14.8	0.3 x 0.3	12.0	S0?		= **Hickson 5B**.
23 21 05.3	+08 06 07	MCG +1-59-58	14.6	1.1 x 0.5	13.8	SB?	90	Double system?
23 21 13.2	+07 21 59	MCG +1-59-59	13.5	0.9 x 0.4	12.3	SB?	30	
23 21 41.0	+08 59 22	MCG +1-59-63	14.5	0.9 x 0.2	12.5	S?	78	A very close pair of mag 12.8 and 11.6 stars on N edge.
23 21 46.2	+09 20 28	MCG +1-59-65	14.8	0.7 x 0.5	13.5	Spiral	132	Mag 8.96v star NW 9′.8; stellar **CGCG 406-92** SE 4′.5; **UGC 12551** SE 7′.1.
23 25 01.8	+08 15 51	MCG +1-59-77	14.4	0.7 x 0.3	12.5	Sb	138	Mag 10.66v star E 2′.3.
23 25 42.9	+05 04 35	MCG +1-59-78	15.4	0.5 x 0.5	13.7			
23 25 48.0	+08 02 21	MCG +1-59-79	14.3	0.7 x 0.5	13.0	S?	75	Mag 11.4 star E 1′.8.
23 28 01.3	+07 51 29	MCG +1-59-82	14.7	0.9 x 0.4	13.4	S?	6	Mag 13.5 star S 1′.2.
23 28 15.3	+09 36 45	MCG +1-59-84	14.4	0.5 x 0.3	12.2	Sb	30	Mag 10.9 star on E edge.
23 28 58.6	+08 54 35	MCG +1-59-85	14.8	0.7 x 0.4	13.2	Sc	0	
23 32 34.7	+06 50 58	MCG +1-60-1	14.1	1.0 x 0.3	12.6		60	Mag 13.3 star S 1′.1.
23 38 03.3	+07 28 11	MCG +1-60-13	15.0	0.7 x 0.3	13.1		102	
23 38 14.8	+07 48 27	MCG +1-60-14	14.2	0.9 x 0.5	13.2	S0/a	57	
23 39 05.5	+07 48 48	MCG +1-60-18	15.1	0.7 x 0.4	13.6	Interacting?	156	Mag 10.84v star NW 1′.6.
23 43 59.2	+08 19 50	MCG +1-60-23	14.5	0.7 x 0.7	13.5	SB?		
23 48 57.7	+06 07 58	MCG +1-60-38	14.6	0.5 x 0.5	12.9	SB(r)cd: pec		Mag 13.8 star N 1′.1.
23 50 50.6	+06 08 55	MCG +1-60-39	13.3	0.7 x 0.5	12.3	E/S0	99	Mag 14.8 star on SW edge.
23 53 22.6	+08 08 55	MCG +1-60-44	15.6	0.9 x 0.6	14.8	SBc	114	Mag 13.6 star on W edge; NGC 7780 SE 3′.0.
00 01 13.6	+13 08 35	MCG +2-1-9	15.0	0.9 x 0.4	13.7	S?	75	= **Hickson 100C**.
00 01 26.2	+13 06 44	MCG +2-1-12	14.4	0.8 x 0.4	13.0	S?	129	= **Hickson 100B**; NGC 7803 W.
00 01 34.1	+15 04 52	MCG +2-1-14	16.4	0.8 x 0.5	15.2	SBbc	3	Mag 12.7 star W 2′.7.
00 03 36.0	+10 36 14	MCG +2-1-16	14.0	0.5 x 0.5	12.3			Mag 13.0 star NW 2′.5.
00 18 12.1	+13 11 30	MCG +2-1-31	14.8	1.0 x 0.3	13.3	Sb	147	
00 26 53.4	+11 34 21	MCG +2-2-10	14.6	0.8 x 0.3	12.9	S?	138	UGC 260 E 2′.3.
00 29 39.5	+11 36 04	MCG +2-2-13	15.2	0.7 x 0.6	14.1	SBc	51	
00 36 09.8	+12 38 28	MCG +2-2-22	14.3	0.7 x 0.5	13.0	S?	24	Mag 13.2 star SSE 1′.9.
23 20 29.9	+12 04 51	MCG +2-59-26	15.3	0.9 x 0.6	14.5		174	
23 25 10.3	+12 18 19	MCG +2-59-39	14.6	0.7 x 0.3	12.8		141	Mag 12.4 star E 0′.9.
23 26 54.1	+15 43 27	MCG +2-59-42	14.7	0.9 x 0.8	14.2		126	Mag 10.46v star NE 2′.8.
23 28 34.2	+12 11 12	MCG +2-59-47	15.4	0.9 x 0.3	13.8	Spiral	54	
23 29 05.0	+10 31 26	MCG +2-59-49	14.9	0.8 x 0.5	13.8	Spiral	33	Faint star N edge.
23 44 29.7	+11 40 46	MCG +2-60-12	14.8	0.4 x 0.3	12.3			MCG +2-60-13 E 1′.5.
23 44 35.6	+11 41 09	MCG +2-60-13	14.2	0.8 x 0.6	13.3		171	MCG +2-60-12 W 1′.5.
23 47 09.2	+15 35 48	MCG +2-60-17	14.5	0.7 x 0.4	13.0	H II	66	Small triangle of mag 12-14 stars N 1′.1.
23 47 03.8	+14 50 29	MCG +2-60-18	14.6	0.7 x 0.7	13.7			Mag 14.3 star superimposed W edge.
23 51 53.1	+13 41 42	MCG +2-60-21	14.9	0.5 x 0.4	13.0	Spiral	72	Close pair of mag 9.15v, 10.42v stars NW 7′.6.
23 55 16.1	+14 22 31	MCG +2-60-25	14.4	1.1 x 0.3	13.0	S?	168	
00 37 50.6	+17 22 57	MCG +3-2-19	15.5	0.7 x 0.4	14.0	Sc	123	Mag 10.92v star NW 0′.9.
23 23 01.2	+17 44 27	MCG +3-59-46	14.7	0.6 x 0.6	13.4			Faint star SW edge.
23 23 51.6	+16 38 38	MCG +3-59-53	15.4	0.8 x 0.5	14.3	Sc	75	Almost stellar **PGC 71322** NNE 2′.3.
23 27 04.2	+17 48 50	MCG +3-59-57	14.3	1.0 x 1.0	14.1	S0?		Short line of three mag 15 stars extend from SE edge.
23 28 30.3	+17 16 36	MCG +3-59-60	13.7	0.7 x 0.4	12.2	S0°: sp	117	**CGCG 454-73** E 1′.6.
23 35 49.3	+16 22 02	MCG +3-60-11	14.3	0.8 x 0.6	13.4	S0	90	Mag 13.4 star E 2′.4.
23 44 35.2	+17 45 59	MCG +3-60-23	15.2	0.9 x 0.3	13.6	Spiral	54	Close pair of stars, mags 12.9 and 14.4, SW 1′.6.
23 46 46.2	+16 02 35	MCG +3-60-25	14.4	0.8 x 0.6	13.5	S0	129	
23 48 50.6	+17 46 41	MCG +3-60-28	15.5	0.5 x 0.5	13.8	Sc		Mag 14.2 star superimposed on S edge.
00 07 16.8	+08 18 05	NGC 3	13.3	0.9 x 0.6	12.6	S0?	111	Mag 11.7 star SW 1′.3; mag 9.57v star NE 6′.6.
00 07 24.6	+08 22 24	NGC 4	15.9	0.4 x 0.2	13.0		35	
00 08 46.3	+15 48 52	NGC 14	12.1	2.8 x 2.1	13.9	(R)IB(s)m pec	25	
00 11 22.3	+06 23 20	NGC 36	13.1	2.2 x 1.1	14.0	SAB(rs)b	21	Elongated **MCG +1-1-44** E 0′.5 .

(Continued from previous page)

GALAXIES

RA h m s	Dec ° ′ ″	Name	Mag (V)	Dim ′ Maj x min	SB	Type Class	PA	Notes
00 15 30.9	+17 19 43	NGC 57	11.6	2.2 x 1.9	13.2	E	40	
00 17 45.5	+11 26 57	NGC 63	11.7	1.7 x 1.1	12.3	S pec	108	
00 19 26.4	+06 26 56	NGC 75	13.2	1.1 x 1.1	13.3	S0		Almost stellar nucleus, smooth envelope.
00 22 13.7	+10 29 24	NGC 95	12.5	1.5 x 1.1	12.9	SAB(rs)c pec II	90	
00 23 59.9	+15 46 09	NGC 99	13.7	1.0 x 0.9	13.4	Scd:	42	
00 24 01.9	+16 28 58	NGC 100	13.3	4.2 x 0.5	13.9	Scd: sp	56	Mag 8.13v star SE 9.′9.
00 25 16.8	+12 53 02	NGC 105	13.2	1.1 x 0.7	12.8	SAab:	167	Mag 10.74v star W 6.′5.
00 30 58.2	+10 12 28	NGC 137	12.8	1.2 x 1.2	13.1	S0		
00 30 59.3	+05 09 36	NGC 138	13.7	1.3 x 0.6	13.2	Sa:	175	Mag 12.2 star N 1.′3.
00 31 06.5	+05 04 41	NGC 139	14.4	0.7 x 0.6	13.3	SB?	171	Located 5.′2 S of NGC 138.
00 31 17.5	+05 10 46	NGC 141	14.5	0.8 x 0.6	13.6		90	Mag 10.75v star NNW 7.′3.
00 37 57.4	+08 38 09	NGC 180	12.9	1.9 x 1.3	13.7	SB(rs)bc	160	Superimposed star or bright knot at NW end of elongated core; mag 8.04v star E 6.′7.
00 38 54.7	+07 03 44	NGC 190	14.0	0.9 x 0.7	13.4	Sab	135	= **Hickson 5A**. **Hickson 5B** on S edge; **Hickson 5C** NNW 0.′8; **Hickson 5D** S 0.′9.
23 20 09.1	+08 09 54	NGC 7617	13.8	0.9 x 0.7	13.2	SA0°:	42	Located 2.′8 SW of NGC 7619.
23 20 14.5	+08 12 18	NGC 7619	11.1	2.5 x 2.3	12.9	E	30	Member of **Pegasus I Cluster**, Bright center, uniform envelope; **UGC 12510** WNW 9.′5.
23 20 24.8	+08 21 56	NGC 7621	14.7	0.7 x 0.2	12.4		177	Located 2.′2 SW of NGC 7623.
23 20 30.1	+08 23 45	NGC 7623	12.9	1.2 x 0.9	12.8	SA0°:	165	Faint, almost stellar anonymous galaxy E 2.′0.
23 20 30.1	+17 13 32	NGC 7625	12.1	1.6 x 1.4	12.8	SA(rs)a pec	60	Pair of strong dark patches SW of center.
23 20 42.6	+08 13 00	NGC 7626	11.1	2.6 x 2.3	13.0	E pec:	9	Member of **Pegasus I Cluster**, Bright core with uniform envelope.
23 21 16.4	+11 23 48	NGC 7630	14.3	1.1 x 0.4	13.3	S?	162	
23 21 26.7	+08 13 03	NGC 7631	13.1	1.8 x 0.7	13.2	SA(r)b:	79	Faint anonymous galaxy NE 3.′5.
23 21 42.0	+08 53 08	NGC 7634	12.6	1.2 x 0.9	12.5	SB0	95	Bright center in E-W lens, uniform envelope.
23 22 33.2	+11 19 47	NGC 7638	14.9	0.6 x 0.6	13.7			Multi-galaxy system.
23 22 48.4	+11 22 19	NGC 7639	14.6	0.6 x 0.5	13.4	E/S0	123	**IC 1484** WNW 2.′2.
23 22 30.9	+11 53 30	NGC 7641	13.9	1.7 x 0.5	13.6	Sa	135	**UGC 12562** SSE 8.′1.
23 22 50.4	+11 59 20	NGC 7643	13.2	1.4 x 0.7	13.0	S?	45	Bright elongated center.
23 24 32.7	+13 58 52	NGC 7644	13.5	1.0 x 0.2	11.6		114	Multi-galaxy system.
23 23 57.4	+16 46 35	NGC 7647	13.6	1.4 x 1.0	13.9	E	14	**PGC 71317** WSW 2.′2; **PGC 71326** S 1.′6; **PGC 71337** NE 2.′2; **PGC 71331** N 2.′1.
23 23 54.1	+09 40 05	NGC 7648	13.0	1.6 x 1.0	13.4	S0	85	Faint star on E end and W end.
23 24 20.1	+14 38 46	NGC 7649	14.0	1.3 x 0.9	14.1	E	80	The immediate area abounds in small, faint anonymous galaxies. Stellar anonymous galaxy NE 0.′9; another SW 2.′5.
23 24 26.0	+13 58 10	NGC 7651	15.3	0.8 x 0.5	14.4	E/S0	18	Faint anonymous galaxy E 4.′0, possibly **IC 5319**?
23 24 49.3	+15 16 32	NGC 7653	12.7	1.6 x 1.4	13.5	Sb	132	**UGC 12590** SE 7.′0.
23 25 55.7	+14 12 34	NGC 7659	14.0	0.9 x 0.4	12.8	S0/a	110	Faint, almost stellar anonymous galaxy E 6.′3.
23 27 19.5	+12 28 02	NGC 7671	12.8	1.4 x 0.8	12.8	SA0:	138	Located 2.′2 E of mag 10.67v star.
23 27 31.5	+12 23 03	NGC 7672	13.9	0.9 x 0.7	13.2	Sb	36	Dark lanes NE and SW of bright center.
23 27 56.7	+08 46 42	NGC 7674	13.2	1.1 x 1.0	13.2	SA(r)bc pec	150	= **Hickson 96A**; **Hickson 96C** NE 0.′6; **Hickson 96D** SE 1.′1.
23 28 06.0	+08 46 04	NGC 7675	14.9	0.6 x 0.4	13.2	SAB(s)0⁻:	35	= **Hickson 96B**.
23 28 54.9	+17 18 35	NGC 7681	14.4	1.6 x 1.4	15.2	S0°: sp	42	Star or bright knot NE edge of core.
23 29 03.8	+11 26 42	NGC 7683	12.5	1.9 x 1.0	13.1	S0	140	Mag 8.91v star NW 5.′9.
23 32 24.5	+15 50 56	NGC 7691	12.9	2.1 x 1.6	14.1	SAB(rs)bc	175	Mag 10.80v star S 1.′4.
23 34 46.9	+16 04 30	NGC 7703	13.4	2.2 x 0.5	13.4	S0	147	
23 35 39.6	+15 18 04	NGC 7711	12.2	2.6 x 1.3	13.3	S0	95	Very faint, broad extension E beyond published dimensions.
23 38 41.2	+15 57 15	NGC 7722	12.4	1.7 x 1.4	13.2	S0/a	150	Strong N-S dark lane partly obscures nucleus.
23 44 15.7	+10 46 01	NGC 7742	11.6	1.7 x 1.7	12.6	SA(r)b		Bright, diffuse center, smooth envelope.
23 44 20.9	+09 55 57	NGC 7743	11.5	3.0 x 2.6	13.6	(R)SB(s)0⁺	80	
23 46 58.5	+06 51 41	NGC 7751	12.9	1.0 x 1.0	12.8	S?		
23 52 11.2	+11 28 08	NGC 7774	13.1	1.0 x 0.6	12.6	E	93	Multi-galaxy system; E component has bright nucleus.
23 53 19.9	+07 52 13	NGC 7778	12.7	1.0 x 1.0	12.7	E		Bright nucleus in bright, round core; uniform envelope.
23 53 26.9	+07 52 32	NGC 7779	12.7	1.4 x 1.1	13.0	(R′)SA0/a:	10	
23 53 32.1	+08 07 04	NGC 7780	13.9	1.0 x 0.5	13.0	Sab	3	MCG +1-60-44 NW 3.′0.
23 53 46.0	+07 51 38	NGC 7781	13.9	0.8 x 0.2	11.8	S?	13	Mag 14.8 star W edge; mag 12.2 star SW 2.′4.
23 53 54.0	+07 58 08	NGC 7782	12.2	2.4 x 1.3	13.3	SA(s)b I-II	1	
23 55 18.8	+05 54 59	NGC 7785	11.6	2.5 x 1.3	12.9	E5-6	143	Mag 10.67v star SE 2.′9.
23 58 03.7	+16 29 55	NGC 7792	14.1	1.0 x 0.8	13.7	Sb	3	Multi-galaxy system, small companion S.
23 58 34.1	+10 43 42	NGC 7794	12.6	1.3 x 1.0	12.8	S?	0	Bright, stellar nucleus.
23 59 37.0	+14 48 28	NGC 7800	12.5	2.3 x 1.6	13.7	Im?	42	Thin, bright center, filamentary envelope.
00 01 00.5	+06 14 30	NGC 7802	13.5	1.1 x 0.6	12.9	S0	51	**UGC 12903** N 5.′9.
00 01 20.0	+13 06 38	NGC 7803	13.1	1.0 x 0.6	12.4	S0/a	82	= **Hickson 100A**; MCG +2-1-12 = **Hickson 100B** E 1.′5; MCG +2-1-9 = **Hickson 100C** NW 2.′5; **MCG +2-1-10** = **Hickson 100D** W 2.′2.
00 02 19.4	+12 58 15	NGC 7810	13.0	1.0 x 0.7	12.5	S0	80	Faint stars on E and W edges.
00 03 14.8	+16 08 43	NGC 7814	10.6	5.5 x 2.3	13.2	SA(s)ab: sp	135	Strong, thin dark lane entire length of galaxy.
00 03 49.0	+07 28 43	NGC 7816	12.8	1.7 x 1.5	13.7	Sbc	171	Strong nucleus in smooth envelope.
00 04 09.0	+07 22 48	NGC 7818	14.0	1.0 x 1.0	13.8	Scd:		
00 04 31.1	+05 11 56	NGC 7820	12.9	1.3 x 0.6	12.5	S0/a	165	
00 05 06.3	+06 55 12	NGC 7824	13.2	1.6 x 1.2	13.7	Sab	145	Mag 10.76v star NW 2.′1.
00 05 27.1	+05 10 36	NGC 7825	13.7	1.1 x 0.5	12.9	SB(s)b	27	**CGCG 408-27** W 5.′3.
00 05 27.8	+05 13 18	NGC 7827	13.9	1.2 x 0.9	13.9	SB0	36	
00 06 38.0	+08 22 03	NGC 7834	14.3	1.1 x 0.9	14.1	Scd:	18	Stellar nucleus; mag 13.1 star on NW edge.
00 06 46.9	+08 25 32	NGC 7835	14.6	0.5 x 0.2	12.0	Sb	162	Mag 12.5 star W 3.′5.
00 06 51.4	+08 21 05	NGC 7837	15.6	0.5 x 0.3	13.3	Sb	171	NGC 7838 on E edge.
00 06 54.1	+08 21 03	NGC 7838	14.6	0.7 x 0.3	12.7	Sb	93	NGC 7837 on W edge.
00 03 20.6	+08 37 05	UGC 10	14.7	0.9 x 0.8	14.2	Scd:	3	Elongated anonymous galaxy on SW edge.
00 03 43.1	+15 13 06	UGC 17	14.2	2.8 x 1.9	15.9	Sm: V	171	Dwarf, very low surface brightness.
00 04 13.0	+10 47 25	UGC 23	13.9	1.0 x 0.7	13.4	SB(rs)b	18	Mag 12.4 star SW 1.′5.
00 04 29.6	+05 50 42	UGC 27	14.0	2.0 x 1.1	14.7	Scd:	135	
00 04 52.1	+17 11 31	UGC 31	14.2	1.0 x 0.7	13.7	IAm	55	
00 04 57.9	+05 07 21	UGC 33	14.0	0.9 x 0.6	13.2	SB0/a	165	Mag 12.2 star W 1.′6.
00 05 09.3	+06 15 32	UGC 35	14.0	1.5 x 1.4	14.7	Sm		Dwarf spiral, extremely low surface brightness.
00 05 14.1	+06 46 19	UGC 36	13.5	1.5 x 0.5	13.1	SA(r)a:	18	
00 06 21.8	+17 26 00	UGC 46	13.7	0.8 x 0.8	13.1	Spiral		Double system, bridge, consists of **MCG +3-1-23** and **MCG +3-1-24**.
00 06 40.5	+05 06 47	UGC 51	14.2	0.7 x 0.4	12.7	SBbc	30	
00 06 49.6	+08 37 42	UGC 52	13.6	1.7 x 1.7	14.6	SAc		Mag 10.81v star NW 2.′3.
00 08 06.8	+09 43 01	UGC 66	13.7	1.1 x 0.6	13.1	SABb	85	
00 08 14.9	+07 46 47	UGC 67	14.2	1.2 x 0.4	13.2	Sab	24	
00 09 04.5	+10 55 08	UGC 81	14.5	1.4 x 0.3	13.4	Sb III-IV	45	
00 10 40.8	+13 42 34	UGC 99	14.0	2.8 x 2.8	16.1	Sm		Dwarf spiral, very low surface brightness; mag 12.2 star NW edge.
00 13 03.0	+14 24 32	UGC 119	13.3	0.8 x 0.5	12.1	S?	78	
00 13 17.4	+17 01 46	UGC 122	14.6	2.4 x 0.4	14.4	Im	109	

(Continued from previous page)

GALAXIES

RA h m s	Dec ° ′ ″	Name	Mag (V)	Dim ′ Maj x min	SB	Type Class	PA	Notes
00 14 00.8	+12 57 50	UGC 132	14.3	1.7 x 0.5	14.0	SABdm	14	Very low surface brightness **UGC 134** S 4′.1.
00 15 15.3	+05 53 13	UGC 143	15.1	0.7 x 0.5	13.8	SB(r)a	126	Mag 6.98v star SE 6′.3.
00 15 51.4	+16 05 19	UGC 148	13.2	1.7 x 0.5	12.9	S?	98	
00 16 14.8	+10 19 53	UGC 151	13.3	1.0 x 0.8	12.9	S0?	123	Mag 6.51v star SE 9′.3.
00 16 44.3	+07 04 35	UGC 155	13.6	1.5 x 0.4	12.9	S?	3	
00 16 48.2	+12 20 43	UGC 156	13.9	2.1 x 1.1	14.7	Im	179	Dwarf irregular, low surface brightness.
00 20 05.3	+10 52 38	UGC 191	13.4	1.6 x 1.1	13.9	Sm V	150	Dwarf spiral, low surface brightness; mag 14.0 star S of center.
00 23 48.3	+14 41 00	UGC 226	13.9	0.8 x 0.4	12.5	S?	4	
00 24 42.9	+14 49 27	UGC 233	13.6	0.6 x 0.6	12.3	Pec		
00 25 10.4	+06 29 26	UGC 240	14.1	0.9 x 0.6	13.3	SAB(rs)b	66	Very faint, small, elongated anonymous galaxy NE 1′.1.
00 26 10.5	+13 39 12	UGC 249	14.2	0.9 x 0.8	13.7	Sdm:	30	
00 26 07.1	+16 26 09	UGC 250	14.4	1.1 x 0.8	14.1	SBcd:	143	
00 26 26.5	+06 16 54	UGC 253	14.1	1.1 x 0.6	13.5	SAB(s)b	30	Mag 10.7 star N edge.
00 27 02.8	+11 34 51	UGC 260	13.0	2.5 x 0.4	12.9	Scd:	21	MCG +2-2-10 W 2′.3.
00 28 54.4	+10 10 26	UGC 287	14.5	0.8 x 0.7	13.7	SAd	132	Mag 6.05v star W 8′.4.
00 29 07.9	+15 54 02	UGC 290	17.4	1.9 x 0.2	16.2	Im?	136	Mag 8.42v star N 5′.9; note: elongated **UGC 289** lies 2′.1 S of this star.
00 30 34.1	+13 21 54	UGC 305	14.1	1.1 x 0.8	13.8	SAc	171	
00 31 23.9	+08 28 04	UGC 312	13.0	1.5 x 0.7	12.9	SB?	7	Knotty; MCG +1-2-18 W 1′.4.
00 31 26.1	+06 12 21	UGC 313	13.2	1.1 x 0.5	12.4	S?	10	
00 31 28.8	+08 23 59	UGC 314	14.1	1.1 x 0.7	13.7	S?	174	**UGC 315** E 2′.5.
00 33 47.3	+07 14 52	UGC 331	14.1	0.9 x 0.6	13.3	SBa	176	UGC 335 NE 2′.7.
00 33 56.5	+07 16 20	UGC 335	13.2	2.0 x 0.8	13.8	E	143	Double system, contact; UGC 331 SW 2′.7.
00 37 58.0	+05 08 50	UGC 379	15.3	1.5 x 0.2	13.8	Scd:	120	
00 38 24.7	+13 29 09	UGC 385	14.1	2.1 x 0.7	14.4	S?	65	Almost stellar nucleus; stellar **Mkn 342** NNW 3′.1.
00 38 23.7	+15 02 22	UGC 386	13.6	1.2 x 0.6	13.2	S0	170	Almost stellar nucleus, smooth envelope.
00 38 39.8	+17 24 14	UGC 393	15.2	1.8 x 0.4	14.7	Sdm:	170	Mag 12.0 star S 2′.2.
00 39 37.8	+08 57 54	UGC 418	14.1	1.9 x 0.2	12.9	Sb III	98	Mag 11.6 star S 1′.5; UGC 422 N 2′.7.
00 39 43.2	+09 00 16	UGC 422	13.7	0.8 x 0.7	12.9	SB?	171	UGC 418 SSW 2′.7.
23 20 02.9	+15 57 07	UGC 12519	13.4	1.3 x 0.5	12.7	SB?	158	
23 20 16.8	+08 00 18	UGC 12522	14.8	1.5 x 1.4	15.4	Sm:	6	Dwarf spiral, very low surface brightness; mag 10.23v star NE 2′.2.
23 21 45.1	+09 04 42	UGC 12544	14.0	1.1 x 1.0	13.9	IB(s)m	129	Mag 10.01v star N 1′.9.
23 21 51.6	+05 00 25	UGC 12547	14.3	1.2 x 0.6	13.8	SB?	151	Two main arms form open "S" shape; **PGC 71224** E 5′.5.
23 21 57.2	+05 02 09	UGC 12548	13.8	1.2 x 0.3	12.5	Sa?	118	**UGC 12555** NE 10′.4.
23 22 04.6	+13 02 07	UGC 12552	14.4	1.7 x 0.2	13.1	Sab	169	Mag 11.1 star SW 3′.4.
23 22 13.5	+09 23 00	UGC 12553	15.3	2.2 x 1.3	16.3	Im:	126	Dwarf, very low surface brightness; mag 11.8 star W 2′.6.
23 22 58.5	+08 59 40	UGC 12561	15.2	1.5 x 0.4	14.5	Sdm:	175	
23 23 22.5	+13 19 07	UGC 12571	13.9	1.8 x 1.0	14.4	SB?	89	
23 24 34.3	+09 16 00	UGC 12581	14.1	1.2 x 0.9	14.1	SBab	135	Short NE-SW bar.
23 24 31.4	+16 52 02	UGC 12582	14.2	0.8 x 0.6	13.3	SB?	171	Small group of five stellar PGC galaxies W 9′.0.
23 24 39.4	+08 25 29	UGC 12585	14.1	1.5 x 1.4	14.8	Sdm:	141	
23 27 10.2	+15 01 02	UGC 12601	15.4	1.6 x 0.9	15.6	SB(s)dm:	60	Faint plume or extension NE.
23 28 36.3	+14 44 24	UGC 12613	12.6	4.6 x 2.8	15.2	Im V	120	**Pegasus Dwarf**. Dwarf irregular, very low surface brightness; almost stellar galaxy **IV Zw 152** SE 3′.9.
23 32 30.8	+14 48 57	UGC 12653	14.0	1.5 x 0.5	13.5	S?	103	
23 34 48.1	+16 51 07	UGC 12677	14.3	0.6 x 0.4	12.6	Sbc	3	**CGCG 455-15** WSW 4′.8.
23 35 17.6	+12 55 25	UGC 12687	13.5	1.4 x 0.7	13.3	SBbc	113	Bright N-S core, faint envelope.
23 35 26.3	+07 19 19	UGC 12688	13.4	1.5 x 0.4	12.7	I?	88	Knot E end; close pair of mag 14 stars N 1′.1.
23 35 32.3	+05 12 55	UGC 12689	13.5	1.5 x 0.3	12.5	Sab	149	
23 36 41.0	+17 30 52	UGC 12707	13.6	1.5 x 1.0	13.9	SB?	40	
23 37 33.3	+17 59 52	UGC 12710	13.8	1.4 x 0.8	13.7	Im?	140	Bright core offset SW.
23 38 36.8	+05 25 57	UGC 12717	14.2	1.1 x 0.8	13.9	Scd:	141	Stellar nucleus; elongated **UGC 12720** NE 9′.0.
23 43 16.8	+08 42 38	UGC 12749	14.5	1.5 x 0.4	13.8	S?	34	Mag 11.9 star E 1′.8.
23 43 28.7	+13 12 32	UGC 12753	13.6	1.6 x 1.2	14.2	SAB(r)bc	9	Bright knot NE edge.
23 43 57.3	+11 30 51	UGC 12756	14.4	0.8 x 0.5	13.3	SB(r)b	153	Mag 11.5 star S 2′.2.
23 45 10.4	+07 02 26	UGC 12767	13.3	1.8 x 1.5	14.2	SB(rs)b	60	Bright knot or star S edge of nucleus.
23 48 04.4	+17 28 29	UGC 12784	13.9	1.0 x 0.9	13.6	SAB(s)bc	36	Mag 11.07v star N 2′.2.
23 49 15.0	+11 48 10	UGC 12793	14.4	0.9 x 0.7	13.8	Scd:	36	Mag 11.4 star SW 2′.3.
23 50 19.2	+10 45 25	UGC 12800	14.1	1.3 x 0.5	13.5	SBa	9	Flanked by a pair of mag 13.8 stars E and W.
23 52 11.1	+08 23 39	UGC 12818	13.8	1.5 x 0.7	13.7	S?	24	
23 52 36.5	+14 33 06	UGC 12822	13.4	1.0 x 0.7	12.9	$S0^-$:	145	
23 55 30.6	+17 55 15	UGC 12843	13.2	2.8 x 1.2	14.4	SABdm	27	Mag 10.31v star superimposed almost on center.
23 56 42.3	+13 46 34	UGC 12854	13.4	0.8 x 0.5	12.3	S?	163	**CGCG 433-3** N 5′.8.
23 56 45.0	+16 48 47	UGC 12856	13.9	2.1 x 0.8	14.3	IB(s)m	12	Irregular, disturbed.
23 56 58.1	+10 49 25	UGC 12860	14.1	1.5 x 0.4	13.4	Sb	96	Mag 10.80v star NW 1′.9; mag 10.89v star S 1′.2.
23 58 30.6	+09 58 23	UGC 12871	14.0	1.0 x 0.7	13.5	Scd:	0	Stellar nucleus, uniform envelope; mag 8.34v star S 9′.0.
00 00 07.1	+08 16 43	UGC 12890	14.0	1.0 x 0.7	13.7	E	18	Slightly smaller companion galaxy NE edge; mag 9.34v star W 4′.0.
00 00 28.2	+17 13 08	UGC 12893	14.0	1.8 x 1.7	15.1	SAdm	90	Almost stellar nucleus, uniform envelope.

GALAXY CLUSTERS

RA h m s	Dec ° ′ ″	Name	Mag 10th brightest	No. Gal	Diam ′	Notes
00 28 54.0	+17 34 00	A 43	15.9	37	20	Mag 8.41v star on E edge; member galaxies faint and stellar.
00 39 48.0	+06 46 00	A 76	15.0	42	28	Numerous stellar anonymous GX over entire area.
23 24 00.0	+16 49 00	A 2589	15.3	40	15	Numerous PGC GX on N-S line through center.
23 24 30.0	+14 38 00	A 2593	15.1	42	28	Numerous anonymous GX N and NW of center.
23 37 30.0	+15 49 00	A 2630	15.2	31	22	Numerous anonymous GX from SE edge in staggered line to NNW edge.
23 44 48.0	+09 08 00	A 2657	14.9	51	30	Galaxies extend in wide E-W band through center.
23 50 48.0	+06 06 00	A 2665	15.8	34	28	MCG +1-60-39 N of center 3′.0; other members anonymous, stellar.
23 55 36.0	+11 25 00	A 2675	16.4	60	25	All members anonymous, faint and stellar.

OPEN CLUSTERS

RA h m s	Dec ° ′ ″	Name	Mag	Diam ′	No. ★	B ★	Type	Notes
23 51 48.0	+16 15 00	NGC 7772		5	10		cl:	Few stars; small brightness range; no central concentration; detached.

RA h m s	Dec ° ′ ″	Name	Mag (V)	Dim ′ Maj x min	SB	Type Class	PA	Notes
22 21 41.3	+06 55 14	CGCG 404-4	15.9	0.4 x 0.4	13.7			Mag 10.13v star N 2′.7.
22 24 55.5	+09 30 56	CGCG 404-13	14.2	0.6 x 0.4	12.5			Triangle of mag 10-11 stars NNW 7′.0.
23 06 26.2	+08 24 08	CGCG 405-33	15.0	0.4 x 0.4	12.9			Mag 9.77v star NNE 3′.6.
23 10 50.8	+09 08 46	CGCG 406-13	13.8	0.9 x 0.8	13.5	E/S0	27	Called a galaxy pair, it has two distinct members with their centers separated.
23 17 39.1	+08 15 08	CGCG 406-45	15.9	0.6 x 0.3	13.9	Sc	129	Located 3′.1 NE of a mag 9.12v star.
23 18 50.0	+07 02 41	CGCG 406-58	13.6	0.5 x 0.4	11.7		120	Elongated anonymous galaxy NE 2′.3.
22 15 59.7	+14 07 28	CGCG 428-63	13.9	0.7 x 0.6	12.8	Sc	0	Mag 11.8 star SE 1′.2; mag 9.59v star WNW 4′.3.
22 17 12.4	+14 14 17	CGCG 428-65	14.7	0.3 x 0.3	12.1	E		= **Mkn 304**.
22 41 22.1	+09 44 32	CGCG 429-22	14.9	0.6 x 0.4	13.4	E	114	Mag 10.8 star W 7′.8; mag 10.47v star NNE 7′.2.
22 59 36.0	+17 58 14	CGCG 453-38	14.7	0.5 x 0.3	12.5		153	**CGCG 453-39** NE 4′.3; **CGCG 453-41** SE 7′.3.
22 58 34.4	+15 10 20	IC 1461	14.2	0.6 x 0.4	12.5	S?	144	Mag 8.63v star NW 8′.2.
23 09 06.7	+17 15 29	IC 1472	14.2	1.0 x 0.5	13.3	S0/a	57	Mag 10.26v star SW 4′.5.
23 12 51.3	+05 48 20	IC 1474	13.9	1.0 x 0.5	13.0	Scd:	150	Mag 8.86v star SE 7′.2.
23 18 14.1	+10 17 58	IC 1478	13.7	1.4 x 0.9	13.8	Sb	30	IC 5305 W 2′.0.
23 19 25.6	+05 54 17	IC 1481	13.5	0.8 x 0.7	12.7	S?	42	Lies 1′.4 NW of a mag 8.43v star.
22 11 34.2	+11 47 45	IC 5177	13.9	1.6 x 0.8	14.0	SBb	25	Mag 12.3 star on N edge; mag 7.99v star ESE 9′.1.
23 18 06.3	+10 17 57	IC 5305	14.5	0.6 x 0.4	12.8		141	IC 1478 E 2′.0.
23 19 11.7	+08 06 36	IC 5309	13.7	1.4 x 0.6	13.4	Sb	23	Close pair of N-S oriented mag 9.18v, 9.6 stars N 7′.2.
22 09 39.1	+07 09 48	MCG +1-56-16	15.4	0.5 x 0.3	13.2		69	Mag 11.3 star SSW 2′.7.
22 09 51.9	+09 34 35	MCG +1-56-17	15.3	0.4 x 0.2	12.4			Mag 9.08v star S 2′.7.
22 17 25.1	+09 12 46	MCG +1-56-21	14.5	0.8 x 0.5	13.4	Sbc	177	Mag 11.3 star N 3′.1.
22 47 32.6	+05 28 32	MCG +1-58-6	15.2	0.5 x 0.3	12.9	Spiral	126	Mag 11.6 star E 1′.9.
23 01 16.3	+09 35 53	MCG +1-58-22	13.7	0.7 x 0.4	12.3	E	117	Forms the NW corner of a square with three mag 12-13 stars.
23 03 36.1	+09 17 22	MCG +1-58-27	14.4	1.0 x 0.6	13.7	SBd	33	Mag 13.6 star on E edge; mag 11.4 star E 2′.5.
23 04 44.8	+07 48 10	MCG +1-58-28	14.8	0.8 x 0.2	12.6	S?	81	Mag 8.41v star N 5′.5.
23 07 12.6	+08 45 02	MCG +1-58-32	14.3	0.9 x 0.3	12.7	S?	54	Mag 11.4 star N edge.
23 09 52.5	+07 30 55	MCG +1-59-2	14.8	0.8 x 0.6	13.9	SB?	45	Strong SE-NW bar.
23 14 34.4	+06 33 20	MCG +1-59-15	15.0	0.9 x 0.3	13.4	Sb	30	Pair of mag 14-15 stars S 0′.9.
23 17 05.6	+07 07 19	MCG +1-59-27	14.9	0.9 x 0.7	14.3	S?	123	Mag 9.28v star E 2′.6.
23 17 21.0	+05 39 36	MCG +1-59-28	14.2	0.8 x 0.4	12.8	S?	108	Pair with MCG +1-59-29 S 1′.0.
23 17 21.8	+05 38 38	MCG +1-59-29	14.3	0.4 x 0.4	12.2	S0?		Pair with MCG +1-59-28 N 1′.0
23 18 51.8	+06 53 41	MCG +1-59-41	15.5	0.5 x 0.3	13.3		126	UGC 12494 S 1′.2.
23 19 12.4	+09 17 22	MCG +1-59-43	14.6	0.7 x 0.3	12.8		108	NGC 7601 SW 7′.0; mag 8.87v star NNW 9′.3.
23 21 05.3	+08 06 07	MCG +1-59-58	14.6	1.1 x 0.5	13.8	SB?	90	Double system?
23 21 13.2	+07 21 59	MCG +1-59-59	13.5	0.9 x 0.4	12.3	SB?	30	
23 21 41.0	+08 59 22	MCG +1-59-63	14.5	0.9 x 0.2	12.5	S?	78	A very close pair of mag 12.8 and 11.6 stars on N edge.
23 21 46.2	+09 20 28	MCG +1-59-65	14.8	0.7 x 0.5	13.5	Spiral	132	Mag 8.96v star NW 9′.8; stellar **CGCG 406-92** SE 4′.5; **UGC 12551** SE 7′.1.
23 25 01.8	+08 15 51	MCG +1-59-77	14.4	0.7 x 0.3	12.5	Sb	138	Mag 10.66v star E 2′.3.
23 25 42.9	+05 04 35	MCG +1-59-78	15.4	0.5 x 0.5	13.7			
23 25 48.0	+08 02 21	MCG +1-59-79	14.3	0.7 x 0.5	13.0	S?	75	Mag 11.4 star E 1′.8.
23 28 01.3	+07 51 29	MCG +1-59-82	14.7	0.9 x 0.4	13.4	S?	6	Mag 13.5 star S 1′.2.
22 14 06.3	+13 29 10	MCG +2-56-21	14.5	0.6 x 0.6	13.2			Mag 11.7 star NE edge.
22 14 10.3	+13 16 44	MCG +2-56-22	15.2	0.6 x 0.4	13.5	Sc	21	Close pair of mag 11-12 stars N 2′.2.
22 15 03.3	+15 24 02	MCG +2-56-25	14.0	0.9 x 0.8	13.5	S?	123	Mag 10.32v star S 1′.4.
22 44 09.3	+10 04 11	MCG +2-58-1	14.4	0.7 x 0.6	13.5	E	177	MCG +2-58-2 S 2′.6.
22 44 11.5	+10 01 42	MCG +2-58-2	14.4	0.9 x 0.4	13.2	S?	150	MCG +2-58-3 S 2′.4.
22 44 09.8	+09 59 19	MCG +2-58-3	14.4	0.7 x 0.7	13.5	S0		Pair of stars, mags 12.1 and 12.6, SW 0′.9.
22 46 38.8	+11 11 47	MCG +2-58-9	15.3	0.7 x 0.4	13.8	Sc	3	Mag 14.5 star SE edge; mag 12.2 star NNE 1′.9.
22 49 59.8	+12 03 26	MCG +2-58-16	14.4	0.9 x 0.7	13.7	S?	42	Mag 10.9 star SW edge.
22 50 51.8	+12 33 30	MCG +2-58-25	15.2	0.9 x 0.5	14.2	Spiral	30	
22 50 59.3	+12 03 19	MCG +2-58-26	14.4	0.7 x 0.4	12.8	S?	45	
22 50 59.8	+11 13 56	MCG +2-58-28	14.9	0.9 x 0.7	14.2	Spiral	33	Faint, almost stellar anonymous galaxy NE 2′.6.
22 51 56.1	+15 40 42	MCG +2-58-29	14.6	0.8 x 0.6	13.7	Sb	114	Mag 13.3 star N 0′.9.
22 54 39.2	+15 45 09	MCG +2-58-33	14.3	0.7 x 0.4	12.7	S?	117	Uniform surface brightness; mag 9.76v star SE 8′.4.
22 57 09.2	+13 11 15	MCG +2-58-39	14.9	0.5 x 0.5	13.3	SBbc		
22 58 31.9	+10 43 52	MCG +2-58-42	14.2	1.0 x 0.6	13.4		36	Mag 11.8 star NE 1′.3.
22 59 35.8	+14 05 20	MCG +2-58-46	14.4	0.9 x 0.3	12.8	S?	63	Mag 14.4 star N edge.
23 00 02.4	+15 40 39	MCG +2-58-47	17.4	0.6 x 0.5	15.9		54	Mag 12.8 star N edge.
23 00 16.5	+13 35 37	MCG +2-58-48	15.2	0.9 x 0.3	13.7	Spiral	33	Mag 6.70v star SSE 8′.4.
23 00 46.2	+13 37 05	MCG +2-58-50	14.6	0.8 x 0.5	13.4	Sc	165	Mag 12.2 star WNW 4′.2; mag 6.70v star SSW 9′.7.
23 01 19.0	+10 22 16	MCG +2-58-54	15.4	0.7 x 0.5	14.1	Sbc	129	Mag 12.2 star NW 1′.9.
23 02 40.4	+13 19 42	MCG +2-58-58	14.6	0.7 x 0.4	13.1		18	
23 03 44.3	+10 39 54	MCG +2-58-59	14.4	0.9 x 0.5	13.4	S?	165	Mag 12.4 star SW 1′.3.
23 05 19.5	+14 10 10	MCG +2-58-61	15.2	0.9 x 0.4	13.9	Sb	15	Mag 11.4 star NE 1′.9.
23 06 27.0	+11 32 59	MCG +2-58-64	15.4	0.6 x 0.5	13.9	SBbc	27	Mag 12.5 star SE 2′.2.
23 08 32.4	+13 12 35	MCG +2-59-2	14.8	1.0 x 0.3	13.3	S?	54	Mag 11.2 star W 3′.6.
23 14 59.7	+14 59 15	MCG +2-59-12	13.6	0.8 x 0.7	12.8	SBb	12	Mag 12.4 star S 1′.7.
23 20 29.9	+12 04 51	MCG +2-59-26	15.3	0.9 x 0.6	14.5		174	
23 25 10.3	+12 18 19	MCG +2-59-39	14.6	0.7 x 0.3	12.8		141	Mag 12.4 star E 0′.9.
23 26 54.1	+15 43 27	MCG +2-59-42	14.7	0.9 x 0.8	14.2		126	Mag 10.46v star NE 2′.8.
22 09 59.2	+16 29 25	MCG +3-56-17	14.9	0.8 x 0.8	14.3	Sc		Mag 10.08v star NW 1′.2.
22 26 45.4	+16 10 59	MCG +3-57-6	14.7	0.9 x 0.7	14.0	Im	123	Knotty, several superimposed stars; mag 14.1 star on NE edge.
22 28 38.0	+16 56 00	MCG +3-57-11	14.5	0.8 x 0.8	13.8			Dwarf irregular **UGC 12049** NNE 5′.5.
22 46 18.7	+15 53 03	MCG +3-58-2	15.3	1.0 x 0.7	14.8	Sc	42	Mag 10.6 star NW 1′.4.
22 49 30.6	+17 25 58	MCG +3-58-7	14.5	0.8 x 0.5	13.4	SB?	3	Mag 11.3 star NW 3′.7.
23 16 23.7	+15 51 32	MCG +3-59-20	14.7	0.4 x 0.4	12.6	S?		Mag 13.0 star W 2′.0; NGC 7576 0′.9 WSW of this star.
23 16 38.6	+15 53 47	MCG +3-59-22	14.6	0.8 x 0.3	12.9	S?	27	Mag 14.4 star NE 2′.5.
23 17 43.0	+16 42 33	MCG +3-59-29	15.3	0.5 x 0.3	13.2	S0		Mag 8.03v star W 1′.8; MCG +3-59-30 E 1′.4.
23 17 48.8	+16 42 37	MCG +3-59-30	15.1	0.6 x 0.4	13.4		54	MCG +3-59-29 W 1′.4.
23 23 01.2	+17 44 27	MCG +3-59-46	14.7	0.6 x 0.6	13.4			Faint star SW edge.
23 23 51.6	+16 38 38	MCG +3-59-53	15.4	0.8 x 0.5	14.3	Sc	75	Almost stellar **PGC 71322** NNE 2′.3.
23 27 04.2	+17 48 50	MCG +3-59-57	14.3	1.0 x 1.0	14.1	S0?		Short line of three mag 15 stars extend from SE edge.
22 14 44.9	+13 50 46	NGC 7236	13.6	0.6 x 0.6	12.4	SA0⁻		**CGCG 428-57** SSW 5′.5.
22 14 47.0	+13 50 27	NGC 7237	13.1	0.6 x 0.6	11.9	SA0⁻		**CGCG 428-59** S 7′.6.
22 16 26.8	+16 28 16	NGC 7244	13.8	0.8 x 0.4	12.4	S?	174	
22 24 31.7	+16 35 19	NGC 7272	13.6	0.9 x 0.8	13.1	SBa	39	Anonymous galaxy 0′.9 S.
22 26 27.8	+16 08 53	NGC 7280	12.1	2.2 x 1.5	13.3	SAB(r)0⁺	78	
22 28 32.8	+17 28 14	NGC 7283	14.4	0.9 x 0.4	13.1	S?	9	Mag 10.43v star W 2′.7.
22 28 26.4	+17 08 49	NGC 7290	13.3	1.6 x 1.0	13.7	SA(r)bc	161	

(Continued from previous page)

GALAXIES

RA h m s	Dec ° ′ ″	Name	Mag (V)	Dim ′ Maj x min	SB	Type Class	PA	Notes
22 28 29.5	+16 46 58	NGC 7291	13.1	1.8 x 1.7	14.2	S0	54	Star and pair of small anonymous galaxies SE 2′1.
22 32 14.0	+11 42 42	NGC 7305	14.1	0.7 x 0.6	13.2	E/S0	15	
22 34 06.6	+05 34 13	NGC 7311	12.5	1.6 x 0.8	12.6	Sab	10	
22 34 35.0	+05 49 00	NGC 7312	13.4	1.4 x 0.8	13.4	SB(s)b	83	Faint nucleus in short bar.
22 37 29.3	+10 31 55	NGC 7328	13.1	2.0 x 0.7	13.3	Sab	88	
22 39 35.6	+11 04 57	NGC 7346	14.6	0.6 x 0.4	13.1	E	48	
22 39 56.0	+11 01 39	NGC 7347	13.7	1.5 x 0.3	12.7	S?	133	
22 40 36.2	+11 54 23	NGC 7348	13.8	1.1 x 0.6	13.2	Scd:	12	
22 39 37.5	+11 46 10	NGC 7353	13.5	2.0 x 0.7	13.7	Sbc II	145	Mag 11.3 star SE 1′7.
22 43 49.4	+08 42 18	NGC 7362	12.7	1.1 x 0.8	12.5	E:	175	Faint anonymous galaxies 1′6 NNE and 1′7 S.
22 44 26.7	+10 46 50	NGC 7366	14.3	0.4 x 0.4	12.1			Almost stellar.
22 45 37.3	+11 03 25	NGC 7370	15.3	0.6 x 0.4	13.6		132	
22 45 46.1	+11 07 48	NGC 7372	13.5	1.0 x 0.9	13.2	S?	78	Located 4′6 SW of a mag 7.22v star.
22 46 01.0	+10 51 11	NGC 7374	13.8	0.9 x 0.7	13.1	S?	93	**MCG +2-58-6** NNW 0′9.
22 49 35.8	+11 33 20	NGC 7383	13.7	0.8 x 0.7	12.9	SB0	174	NGC 7384 E 2′5.
22 49 46.0	+11 33 06	NGC 7384	15.3	0.6 x 0.3	13.3		60	NGC 7383 W 2′5.
22 49 54.6	+11 36 31	NGC 7385	12.0	1.5 x 1.3	12.7	E pec:	36	Mag 11.5 star on NW edge; faint, elongated anonymous galaxy W 3′7.
22 50 02.3	+11 41 50	NGC 7386	12.3	1.8 x 1.1	12.9	SA0:	141	Stellar NGC 7388 NE 2′2.
22 50 17.9	+11 38 14	NGC 7387	14.0	0.7 x 0.5	12.7	S0$^-$:	48	**MCG +2-58-23** S 0′5.
22 50 08.6	+11 43 22	NGC 7388	16.7	0.3 x 0.2	13.5		63	Appears stellar.
22 50 16.2	+11 33 59	NGC 7389	13.9	1.4 x 0.9	14.0	SB0	144	
22 50 19.5	+11 31 51	NGC 7390	14.2	0.9 x 0.7	13.6	S0?	177	Mag 9.35v star 3′4 SSE.
22 52 57.3	+12 35 36	NGC 7405	15.3	0.4 x 0.3	12.8	Sc		
22 55 03.1	+13 13 13	NGC 7413	14.1	1.0 x 0.7	13.6	S0$^-$ pec:	81	
22 55 24.6	+13 14 49	NGC 7414	16.0	0.5 x 0.2	13.3		174	
22 57 09.9	+08 30 20	NGC 7427	15.1	0.7 x 0.7	14.2	Compact		Bright nucleus.
22 57 29.7	+08 47 37	NGC 7430	14.3	0.6 x 0.3	12.3		60	
22 58 02.0	+13 08 03	NGC 7432	13.3	1.5 x 1.2	13.9	E	40	
22 58 10.1	+14 18 34	NGC 7437	13.3	1.8 x 1.8	14.4	SAB(rs)d		Very patchy; many bright knots and/or superimposed stars.
22 59 26.7	+15 32 54	NGC 7442	13.3	1.1 x 1.1	13.3	SAc:		
23 00 03.9	+15 58 50	NGC 7448	11.7	2.7 x 1.2	12.8	SA(rs)bc II-III	170	Pair of bright knots 0′6 N of center.
23 00 41.0	+08 27 59	NGC 7451	14.0	1.0 x 0.5	13.1	SBbc	67	
23 01 00.0	+06 44 57	NGC 7452	14.5	0.6 x 0.3	12.5	Double System	57	Galaxy pair.
23 01 06.7	+16 23 17	NGC 7454	11.8	2.2 x 1.6	13.2	E4	150	Star on NW edge; anonymous galaxy SW 1′7.
23 00 41.0	+07 18 10	NGC 7455	14.3	0.6 x 0.4	12.6	Sa	174	Mag 11.8 star NE 1′2.
23 01 48.4	+15 34 56	NGC 7461	13.3	0.9 x 0.7	12.7	SB0	150	Bright bar; mag 11.28v star SW 2′3.
23 01 52.2	+15 58 52	NGC 7463	13.2	2.6 x 0.6	13.6	SABb: pec	90	NGC 7464 on S edge; mag 8.22v star SW 2′5.
23 01 53.8	+15 58 22	NGC 7464	13.3	0.5 x 0.5	11.8	E1 pec:		Located S edge of NGC 7463.
23 02 00.9	+15 57 52	NGC 7465	12.6	2.2 x 1.8	13.9	(R′)SB(s)0°:	42	Bright N-S bar, very faint envelope; **UGC 12321** NE 5′8.
23 02 27.5	+15 33 11	NGC 7467	14.5	0.6 x 0.5	13.3	E/S0	33	
23 02 59.2	+16 36 11	NGC 7468	13.7	0.9 x 0.6	13.0	E3: pec	15	Stellar **IC 1465** SW 1′8.
23 04 53.9	+16 40 29	NGC 7468A	14.1	1.5 x 0.6	13.8	SB?	147	Multiple-galaxy system.
23 03 15.6	+08 52 25	NGC 7469	12.3	1.5 x 1.1	12.7	(R′)SAB(rs)a	125	**IC 5283** NE 1′3.
23 04 56.4	+12 19 15	NGC 7479	10.9	4.1 x 3.1	13.5	SB(s)c I-II	25	Two main knotty arms.
23 08 57.3	+12 02 51	NGC 7495	13.1	1.8 x 1.7	14.1	SAB(s)c II	5	Mag 9.06v star SE 6′2.
23 10 22.4	+07 34 50	NGC 7499	12.8	1.1 x 0.7	12.4	SA(s)0°:	10	Bright member of **Pegasus II**. **MCG +1-59-3** W 4′8.
23 10 29.9	+11 00 46	NGC 7500	13.3	2.1 x 1.1	14.0	S0	125	Bright center, smooth envelope.
23 10 30.5	+07 35 17	NGC 7501	13.4	0.5 x 0.5	12.0	E1:		Bright member of **Pegasus II**. Almost stellar; very faint anonymous galaxy on E edge.
23 10 42.5	+07 34 03	NGC 7503	13.2	0.8 x 0.8	12.8	E2:		Bright member of **Pegasus II**. NGC 7501 E 2′1.
23 11 00.8	+13 37 51	NGC 7505	14.7	0.6 x 0.3	12.7		111	
23 11 49.1	+12 56 25	NGC 7508	14.8	1.0 x 0.3	13.3	S	160	
23 12 21.5	+14 36 32	NGC 7509	13.4	1.1 x 1.1	13.4			
23 12 26.3	+13 43 34	NGC 7511	13.9	1.0 x 0.5	13.0	S?	133	
23 12 48.7	+12 40 42	NGC 7515	12.4	1.7 x 1.6	13.3	S?	15	
23 13 12.9	+06 19 16	NGC 7518	13.4	1.4 x 1.0	13.6	(R)SAB(r)a	126	UGC 12423 N 6′4.
23 13 11.3	+10 46 16	NGC 7519	14.0	1.2 x 1.0	14.0	Sb:	165	Mag 10.08v star W 6′5; note: **UGC 12416** lies 2′8 S of this star.
23 13 34.7	+13 59 10	NGC 7523	14.8	1.1 x 0.3	13.4		3	Faint star on N edge.
23 13 40.5	+14 01 19	NGC 7525	14.2	0.6 x 0.6	13.2	E		Multi-galaxy system, small elongated companion on N edge.
23 14 20.3	+10 13 53	NGC 7528	15.1	0.5 x 0.5	13.4			Mag 9.63v star SSW 5′5.
23 14 03.2	+08 59 32	NGC 7529	14.1	0.9 x 0.8	13.6	S?	75	
23 14 12.7	+13 34 55	NGC 7535	13.7	1.5 x 1.5	14.4	Sd		Very patchy with stellar nucleus.
23 14 13.1	+13 25 34	NGC 7536	13.4	1.9 x 0.7	13.6	SBbc	56	
23 14 36.1	+15 56 58	NGC 7540	14.7	0.6 x 0.4	13.0		147	Faint, stellar anonymous galaxy on E edge.
23 14 41.7	+10 38 33	NGC 7542	14.5	0.6 x 0.4	12.8		117	Mag 9.27v star N 5′3.
23 15 39.8	+06 42 28	NGC 7557	14.1	0.6 x 0.6	12.9	S?		Mag 9.43v star N 6′1.
23 15 46.6	+13 17 25	NGC 7559A	15.4	1.0 x 0.8	15.1	S0$^-$:	67	Almost stellar NGC 7559B N 0′5.
23 15 46.1	+13 17 48	NGC 7559B	13.7	0.4 x 0.3	11.3	S0$^-$:	36	Located on N edge of NGC 7559A.
23 15 57.4	+06 41 12	NGC 7562	11.6	2.2 x 1.5	12.8	E2-3	83	NGC 7562A SSE 2′3.
23 16 01.3	+06 39 04	NGC 7562A	14.8	1.5 x 0.4	14.1	Sdm: sp	0	Small, elongated nucleus, smooth envelope; small, faint anonymous galaxy S 1′5.
23 15 55.9	+13 11 47	NGC 7563	12.8	1.9 x 1.0	13.4	SBa	155	Bright center on E-W bar; mag 10.53v star E 2′2.
23 15 38.5	+07 18 37	NGC 7564	14.6	0.6 x 0.4	12.9		132	
23 16 11.1	+15 51 00	NGC 7567	14.5	0.9 x 0.3	12.9	S?	76	Mag 10.56v star NW 2′9.
23 16 44.7	+13 28 58	NGC 7570	13.2	1.5 x 0.9	13.4	SBa	30	Bright nucleus in NW-SW bar, several faint knots.
23 17 17.2	+07 21 52	NGC 7577		0.3 x 0.2			48	Stellar.
23 17 55.3	+09 26 55	NGC 7579	14.1	0.4 x 0.3	11.6		39	Mag 7.61v star N 3′4.
23 17 36.6	+14 00 03	NGC 7580	13.7	0.8 x 0.6	12.7	S?	45	
23 17 52.8	+07 22 46	NGC 7583	13.8	0.7 x 0.7	12.9	S?		Faint anonymous galaxy S 1′8.
23 17 53.1	+09 26 01	NGC 7584	14.4	0.4 x 0.4	12.2	S0?		**MCG +1-59-32** W 3′5.
23 17 55.6	+08 35 03	NGC 7586	16.3	0.5 x 0.5	14.6			
23 17 59.2	+09 40 46	NGC 7587	13.9	1.3 x 0.4	13.0	SBab: sp	123	**CGCG 406-51** S 0′9.
23 18 16.2	+06 35 12	NGC 7591	13.0	1.9 x 0.8	13.3	SBbc	145	Strong dark patches SE and NW of bright center.
23 17 57.2	+11 20 55	NGC 7593	13.6	1.0 x 0.6	12.9	S?	104	
23 18 11.4	+10 14 44	NGC 7594	13.1	0.7 x 0.3	11.2	S?	84	Very faint, elongated N-S, anonymous galaxy on W edge.
23 18 47.2	+09 13 58	NGC 7601	14.0	1.1 x 0.9	13.8	SAB(s)c:	96	Almost stellar nucleus; mag 12.0 star N 1′4; MCG +1-59-43 NE 7′0.
23 17 52.1	+07 25 45	NGC 7604	14.5	0.5 x 0.3	12.4	E	90	Faint anonymous galaxy E 1′5.
23 19 15.2	+08 20 56	NGC 7608	14.2	1.5 x 0.4	13.5	S?	20	Mag 9.68v star ESE 5′5; **UGC 12510** SE 7′8.
23 19 30.6	+09 30 19	NGC 7609	14.1	1.3 x 1.1	14.3	Pec	135	Double galaxy system; the NW component is **Hickson 95A**, the SE component is **Hickson 95C**; **Hickson 95B** is elongated galaxy SE 1′1; **Hickson 95D** is elongated galaxy SW; star NW 5′0.
23 19 41.3	+10 11 06	NGC 7610	13.0	2.5 x 1.9	14.5	Scd: I-II	45	

(Continued from previous page)

GALAXIES

RA h m s	Dec ° ′ ″	Name	Mag (V)	Dim ′ Maj x min	SB	Type Class	PA	Notes
23 19 36.8	+08 03 44	NGC 7611	12.5	1.5 x 0.6	12.3	SB0+:	139	Member of **Pegasus I Cluster**, Mag 6.91v star SSE 5.′5; IC 5309 WNW 6.′8.
23 19 44.3	+08 34 35	NGC 7612	12.8	1.6 x 0.8	12.9	S0	2	**CGCG 406-69** ESE 2.′7.
23 19 54.7	+08 23 55	NGC 7615	14.3	0.9 x 0.5	13.3	Sb?	152	
23 19 27.2	+10 09 11	NGC 7616	16.0	0.5 x 0.3	13.8	Scd:	0	Almost stellar.
23 20 09.1	+08 09 54	NGC 7617	13.8	0.9 x 0.7	13.2	SA0°:	42	Located 2.′8 SW of NGC 7619.
23 20 14.5	+08 12 18	NGC 7619	11.1	2.5 x 2.3	12.9	E	30	Member of **Pegasus I Cluster**, Bright center, uniform envelope; **UGC 12510** WNW 9.′5.
23 20 24.8	+08 21 56	NGC 7621	14.7	0.7 x 0.2	12.4		177	Located 2.′2 SW of NGC 7623.
23 20 30.1	+08 23 45	NGC 7623	12.9	1.2 x 0.9	12.8	SA0°:	165	Faint, almost stellar anonymous galaxy E 2.′0.
23 20 30.1	+17 13 32	NGC 7625	12.1	1.6 x 1.4	12.8	SA(rs)a pec	60	Pair of strong dark patches SW of center.
23 20 42.6	+08 13 00	NGC 7626	11.1	2.6 x 2.3	13.0	E pec:	9	Member of **Pegasus I Cluster**, Bright core with uniform envelope.
23 21 16.4	+11 23 48	NGC 7630	14.3	1.1 x 0.4	13.3	S?	162	
23 21 26.7	+08 13 03	NGC 7631	13.1	1.8 x 0.7	13.2	SA(r)b:	79	Faint anonymous galaxy NE 3.′5.
23 21 42.0	+08 53 08	NGC 7634	12.6	1.2 x 0.9	12.5	SB0	95	Bright center in E-W lens, uniform envelope.
23 22 33.2	+11 19 47	NGC 7638	14.9	0.6 x 0.6	13.7			Multi-galaxy system.
23 22 48.4	+11 22 19	NGC 7639	14.6	0.6 x 0.5	13.4	E/S0	123	**IC 1484** WNW 2.′2.
23 22 30.9	+11 53 30	NGC 7641	13.9	1.7 x 0.5	13.6	Sa	135	**UGC 12562** SSE 8.′1.
23 22 50.4	+11 59 20	NGC 7643	13.2	1.4 x 0.7	13.0	S?	45	Bright elongated center.
23 24 32.7	+13 58 52	NGC 7644	13.5	1.0 x 0.2	11.6		114	Multi-galaxy system.
23 23 57.4	+16 46 35	NGC 7647	13.6	1.4 x 1.0	13.9	E	14	**PGC 71317** WSW 2.′2; **PGC 71326** S 1.′6; **PGC 71337** NE 2.′2; **PGC 71331** N 2.′1.
23 23 54.1	+09 40 05	NGC 7648	13.0	1.6 x 1.0	13.4	S0	85	Faint star on E end and W end.
23 24 20.1	+14 38 46	NGC 7649	14.0	1.3 x 0.9	14.1	E	80	The immediate area abounds in small, faint anonymous galaxies. Stellar anonymous galaxy NE 0.′9; another SW 2.′5.
23 24 26.0	+13 58 10	NGC 7651	15.3	0.8 x 0.5	14.4	E/S0	18	Faint anonymous galaxy E 4.′0, possibly **IC 5319**?
23 24 49.3	+15 16 32	NGC 7653	12.7	1.6 x 1.4	13.5	Sb	132	**UGC 12590** SE 7.′0.
23 25 55.7	+14 12 34	NGC 7659	14.0	0.9 x 0.4	12.8	S0/a	110	Faint, almost stellar anonymous galaxy E 6.′3.
23 27 19.5	+12 28 02	NGC 7671	12.8	1.4 x 0.8	12.8	SA0:	138	Located 2.′2 E of mag 10.67v star.
23 27 31.5	+12 23 03	NGC 7672	13.9	0.9 x 0.7	13.2	Sb	36	Dark lanes NE and SW of bright center.
23 27 56.7	+08 46 42	NGC 7674	13.2	1.1 x 1.0	13.2	SA(r)bc pec	150	= **Hickson 96A**; **Hickson 96C** NE 0.′6; **Hickson 96D** SE 1.′1.
22 09 15.4	+14 21 39	UGC 11921	13.8	1.7 x 0.7	13.8	IBm	127	
22 09 50.7	+16 53 32	UGC 11928	14.4	0.9 x 0.6	13.5	SB?	15	Strong E-W core.
22 12 01.0	+17 54 17	UGC 11944	15.4	2.1 x 0.7	15.7	Im:	28	Dwarf irregular, extremely low surface brightness; mag 11.6 star S 2.′1.
22 12 18.0	+11 29 40	UGC 11947	13.8	1.0 x 1.0	13.7	SAB(s)b		Bright nucleus in faint bar; faint arms N and S; mag 13.7 star SE edge.
22 12 22.3	+14 01 24	UGC 11948	14.0	0.9 x 0.7	13.4	SBb	22	
22 13 38.9	+14 13 06	UGC 11952	14.8	1.5 x 0.2	13.3	SBcd?	143	**CGCG 428-47** and a close mag 10.3 star S 6.′1.
22 20 08.0	+09 01 02	UGC 11987	14.3	1.5 x 0.3	13.3	SB?	36	
22 23 35.5	+11 51 37	UGC 12017	14.0	1.2 x 0.9	13.9	SB(s)b	146	**MCG +2-57-2** SE 2.′8.
22 24 11.6	+06 00 13	UGC 12021	14.2	1.1 x 0.5	13.4	Sb	100	Mag 8.13v star WSW 8.′7.
22 24 16.4	+05 21 37	UGC 12023	14.4	0.8 x 0.6	13.4	SB(r)b:	150	Mag 12.2 star NW 3.′9.
22 29 32.3	+07 43 28	UGC 12054	14.0	1.5 x 0.2	12.6	S?	46	
22 33 06.2	+08 05 54	UGC 12074	13.5	0.7 x 0.5	12.2	S?	145	**MCG +1-57-7** NW 11.′0.
22 39 32.2	+08 36 44	UGC 12133	14.7	1.7 x 0.2	13.4	Scd:	44	**UGC 12139** E 11.′6.
22 40 17.1	+08 03 08	UGC 12138	13.3	0.8 x 0.7	12.6	SBa	3	E-W bar; mag 11.8 star WSW 1.′3.
22 45 08.4	+06 25 52	UGC 12178	13.0	3.0 x 1.6	14.6	SAB(s)dm	10	Mag 8.37v star 2.′1 SE of center.
22 47 06.4	+11 36 38	UGC 12184	15.2	1.0 x 0.9	15.0	Sd	48	Mag 8.83v star SE 3.′9.
22 48 58.2	+07 13 02	UGC 12196	14.3	1.1 x 0.4	13.3	Sa	153	**MCG +1-58-8** N 1.′1.
22 49 29.2	+11 18 07	UGC 12202	13.6	1.0 x 0.7	13.1	Sa	135	**MCG +2-58-11** SW 10.′2.
22 51 03.8	+07 17 49	UGC 12213	15.4	1.8 x 1.7	16.5	SABdm:	150	Core slightly offset SW of center.
22 52 32.8	+11 39 39	UGC 12222	14.0	1.1 x 0.7	13.5	Sb	175	Mag 9.48v star W 2.′9.
22 52 38.3	+06 05 36	UGC 12224	13.8	1.6 x 1.5	14.6	Scd:	15	Stellar nucleus; mag 8.56v star S 4.′4.
22 54 19.9	+11 46 56	UGC 12237	14.5	1.5 x 0.3	13.5	Sbc	92	**UGC 12243** SE 7.′2.
22 55 35.9	+12 47 21	UGC 12250	13.3	1.5 x 0.9	13.4	SBb	13	Strong dark patches E and W of elongated core.
22 56 01.9	+12 45 58	UGC 12253	15.1	1.5 x 0.2	13.6	Sb	145	**MCG +2-58-38** NW 2.′2 with mag 10.15v star directly N of it.
22 56 12.9	+05 23 00	UGC 12255	14.2	0.5 x 0.5	12.5	S?		
22 56 31.4	+17 47 01	UGC 12258	14.0	1.5 x 0.3	13.0	Sa	85	Mag 10.6 star NE 3.′3; mag 11.3 star NNW 2.′8.
22 58 00.6	+06 04 10	UGC 12266	13.7	1.1 x 0.9	13.5	SBab	173	Mag 9.63v star NW 3.′1.
22 59 12.6	+13 36 16	UGC 12281	14.2	3.2 x 0.2	13.6	Sdm:	30	Pair of mag 12.9 and 13.7 stars very close to W edge.
23 01 12.0	+12 43 24	UGC 12307	15.4	1.5 x 0.3	14.4	Im	156	**UGC 12300** WNW 9.′2.
23 01 18.7	+14 20 26	UGC 12308	14.1	2.3 x 0.5	14.1	Scd:	122	Mag 10.63v star S 1.′9.
23 05 19.8	+16 51 59	UGC 12350	13.9	2.8 x 0.9	14.8	Sm	95	Mag 11.7 star S edge.
23 06 04.7	+14 52 01	UGC 12359	14.0	1.5 x 1.5	14.8	S?		Very faint anonymous galaxy with stellar nucleus NW 2.′3.
23 08 30.5	+12 49 50	UGC 12388	15.4	1.7 x 0.7	15.4	Sd:	51	
23 11 51.5	+09 30 20	UGC 12407	13.2	0.8 x 0.5	12.0	S?	125	
23 13 14.0	+06 25 31	UGC 12423	13.6	3.3 x 0.4	13.7	Sc:	145	Mag 10.8 star W 2.′4; NGC 7518 S 6.′4.
23 14 45.4	+05 24 53	UGC 12451	14.9	1.5 x 0.4	14.2	Im	155	Mag 10.53v star NW 2.′1.
23 15 09.2	+09 40 44	UGC 12454	14.0	1.3 x 0.5	13.4	S0	150	Almost stellar galaxy **PGC 70822** on S edge.
23 15 53.2	+05 07 41	UGC 12461	13.9	1.1 x 1.1	14.0	S0?		
23 16 44.9	+08 54 19	UGC 12472	13.4	1.0 x 0.7	12.9	S0?	88	
23 18 52.7	+06 52 35	UGC 12494	14.6	1.4 x 0.4	13.8	SABd	30	MCG +1-59-41 N 1.′2.
23 18 48.8	+16 37 36	UGC 12495	14.2	1.3 x 0.7	13.9	Sdm:	100	Mag 11.5 star E 2.′2.
23 19 30.3	+16 04 28	UGC 12506	14.5	2.6 x 0.2	13.7	Scd:	81	
23 20 02.9	+15 57 07	UGC 12519	13.4	1.3 x 0.5	12.7	SB?	158	
23 20 16.8	+08 00 18	UGC 12522	14.8	1.5 x 1.4	15.4	Sm:	6	Dwarf spiral, very low surface brightness; mag 10.23v star NE 2.′2.
23 21 45.1	+09 04 42	UGC 12544	14.0	1.1 x 1.0	13.9	IB(s)m	129	Mag 10.01v star N 1.′9.
23 21 51.6	+05 00 25	UGC 12547	14.3	1.2 x 0.6	13.8	SB?	151	Two main arms form open "S" shape; **PGC 71224** E 5.′5.
23 21 57.2	+05 02 09	UGC 12548	13.8	1.2 x 0.3	12.5	Sa?	118	**UGC 12555** NE 10.′4.
23 22 04.6	+13 02 07	UGC 12552	14.4	1.7 x 0.2	13.1	Sab	169	Mag 11.1 star SW 3.′4.
23 22 13.5	+09 23 00	UGC 12553	15.3	2.2 x 1.3	16.3	Im:	126	Dwarf, very low surface brightness; mag 11.8 star W 2.′6.
23 22 58.5	+08 59 40	UGC 12561	15.2	1.5 x 0.4	14.5	Sdm:	175	
23 23 22.5	+13 19 07	UGC 12571	13.9	1.8 x 1.0	14.4	SB?	89	
23 24 34.3	+09 16 00	UGC 12581	14.1	1.2 x 0.9	14.1	SBab	135	Short NE-SW bar.
23 24 31.4	+16 52 02	UGC 12582	14.2	0.8 x 0.6	13.3	SB?	171	Small group of five stellar PGC galaxies W 9.′0.
23 24 39.4	+08 25 29	UGC 12585	14.1	1.5 x 1.4	14.8	Sdm:	141	
23 27 10.2	+15 01 02	UGC 12601	15.4	1.6 x 0.9	15.6	SB(s)dm:	60	Faint plume or extension NE.

(Continued from previous page)

82 GALAXY CLUSTERS 82

RA h m s	Dec ° ′ ″	Name	Mag 10th brightest	No. Gal	Diam ′	Notes
23 24 00.0	+16 49 00	A 2589	15.3	40	15	Numerous PGC GX on N-S line through center.
23 24 30.0	+14 38 00	A 2593	15.1	42	28	Numerous anonymous GX N and NW of center.

GLOBULAR CLUSTERS

RA h m s	Dec ° ′ ″	Name	Total V m	B ★ V m	HB V m	Diam ′	Conc. Class Low = 12 High = 1	Notes
23 06 44.4	+12 46 19	Pal 13	13.8	17.0	17.7	0.7	12	

83 GALAXIES 83

RA h m s	Dec ° ′ ″	Name	Mag (V)	Dim ′ Maj x min	SB	Type Class	PA	Notes
20 56 56.7	+06 49 00	CGCG 400-12	14.1	0.5 x 0.3	12.1	E	15	Mag 10.43v star ENE 3′.1.
21 00 13.6	+09 33 30	CGCG 400-17	15.6	0.5 x 0.3	13.6	E	33	UGC 11659 NW 2′.5.
21 05 40.5	+09 13 59	CGCG 400-27	14.3	0.6 x 0.5	12.8		165	Mag 10.84v star N 3′.5; mag 8.05 v star SW 6′.5.
21 56 58.5	+06 39 23	CGCG 403-5	15.2	0.5 x 0.5	13.5			= **Mkn 517**; elongated anonymous galaxy SW 0′.8.
21 02 53.9	+15 23 28	CGCG 425-36	14.0	0.5 x 0.3	11.8		75	Mag 11.5 star NE 2′.6.
21 15 09.1	+12 39 29	CGCG 426-27	15.4	0.7 x 0.5	14.1		156	Multi-galaxy system; mag 15.4 star W edge.
21 16 53.2	+13 16 08	CGCG 426-33	14.2	0.5 x 0.3	12.0		12	Mag 14.9 star SE 0′.4; mag 11.9 star SSW 2′.7.
21 37 31.0	+14 35 06	CGCG 427-7	14.2	0.6 x 0.6	12.9			Mag 11.3 star S 2′.4; **CGCG 427-8** E 7′.3.
21 45 10.6	+10 06 03	CGCG 427-19	14.4	0.6 x 0.3	12.4	Sb	96	Mag 13.6 star E 1′.3; mag 14.4 star NNW 0′.8.
22 06 38.2	+11 37 57	CGCG 428-29	15.2	0.4 x 0.2	12.3		54	Mag 12.2 star SE 3′.1; mag 6.84v star NE 10′.0.
22 15 59.7	+14 07 28	CGCG 428-63	13.9	0.7 x 0.6	12.8	Sc	0	Mag 11.8 star SE 1′.2; mag 9.59v star WNW 4′.3.
21 03 07.1	+16 01 49	CGCG 448-33	14.2	0.6 x 0.3	12.2	Sc	150	**CGCG 448-35** E 8′.8; mag 6.88v star SW 6′.1.
21 08 43.2	+12 28 57	IC 1359	13.9	1.2 x 0.4	12.9	S?	175	Mag 11.4 star NE 6′.1.
21 11 29.8	+05 03 20	IC 1361	14.2	1.0 x 1.0	14.0	S?		Bright nucleus offset SW with very faint envelope NE.
21 40 13.1	+14 37 56	IC 1394	13.8	1.0 x 0.9	13.5		177	Mag 8.98v star SW 8′.4.
21 45 51.5	+09 28 29	IC 1398	14.6	0.8 x 0.5	13.5	Sc	81	Mag 9.51v star ENE 2′.3.
21 58 18.2	+08 25 22	IC 1414	14.0	0.5 x 0.5	12.3			**CGCG 403-10** ENE 5′.6; **CGCG 403-7** SSW 3′.3.
22 03 35.3	+15 06 23	IC 1427	13.7	1.2 x 0.9	13.8	E	105	Mag 10.6 star W 3′.3.
21 54 23.2	+15 09 19	IC 5145	13.4	1.6 x 0.9	13.6	Sab	170	Mag 8.5v star NE 8′.5.
22 11 34.2	+11 47 45	IC 5177	13.9	1.6 x 0.8	14.0	SBb	25	Mag 12.3 star on N edge; mag 7.99v star ESE 9′.1.
21 18 52.7	+05 50 55	MCG +1-54-5	14.4	0.8 x 0.3	12.8	S?	6	
21 26 05.5	+06 42 40	MCG +1-54-10	14.7	0.5 x 0.5	13.0			Mag 11.5 star SE 0′.9.
21 34 09.2	+08 46 42	MCG +1-55-4	14.3	0.5 x 0.3	12.1		90	**MCG +1-55-3** WNW 1′.6; mag 10.81v star W 2′.0.
21 36 06.9	+07 56 21	MCG +1-55-5	14.0	0.6 x 0.3	12.0		21	Mag 14.9 star on N end; located between a pair of mag 11-12 stars.
21 45 59.2	+07 52 08	MCG +1-55-10	14.7	0.9 x 0.3	13.1	Sbc	162	Mag 11.0 star N 3′.5.
21 51 02.8	+06 14 32	MCG +1-55-13	14.4	0.5 x 0.4	12.5	S?	138	Mag 10.21v star W 2′.5.
21 55 43.7	+05 48 24	MCG +1-56-1	14.6	0.6 x 0.6	13.3	Sa		
21 58 37.1	+07 26 23	MCG +1-56-4	14.0	0.9 x 0.6	13.2		147	Located between a pair of mag 11-12 stars.
21 59 34.3	+09 24 15	MCG +1-56-6	14.5	0.8 x 0.7	13.7	Sc	117	
22 04 43.8	+09 14 58	MCG +1-56-11	15.3	0.6 x 0.4	13.6		18	Mag 10.27v star NE 2′.3.
22 07 08.6	+06 29 08	MCG +1-56-13	14.6	0.8 x 0.6	13.7		75	Mag 13.1 star NW 2′.0.
22 07 40.3	+08 55 06	MCG +1-56-14	14.8	1.1 x 0.8	14.5	SBc	21	
22 09 39.1	+07 09 48	MCG +1-56-16	15.4	0.5 x 0.3	13.2		69	Mag 11.3 star SSW 2′.7.
22 09 51.9	+09 34 35	MCG +1-56-17	15.3	0.4 x 0.2	12.4			Mag 9.08v star S 2′.7.
20 57 09.3	+13 31 27	MCG +2-53-6	15.3	1.0 x 0.2	13.4	Sc	129	Mag 11.4 star SW 1′.3.
20 57 45.1	+13 59 34	MCG +2-53-7	14.2	0.8 x 0.8	13.5			Mag 9.47v star NNW 2′.2.
20 58 06.2	+13 30 52	MCG +2-53-8	15.4	0.7 x 0.3	13.5	Sb	102	
20 58 03.8	+11 03 10	MCG +2-53-9	14.9	0.8 x 0.8	14.3	Sc		Stars superimposed N and W.
20 59 06.7	+11 08 43	MCG +2-53-10	15.4	0.7 x 0.7	14.5	SBbc		Weak stellar nucleus, faint envelope.
20 59 27.5	+13 56 08	MCG +2-53-11	14.6	0.9 x 0.7	13.9	SBb	141	Mag 11.8 star N 2′.2.
21 12 01.2	+11 29 44	MCG +2-54-5	14.5	1.0 x 0.6	13.8		144	Strong N-S bar; mag 11.1 star SW 2′.2.
21 12 15.9	+13 00 59	MCG +2-54-6	13.4	0.9 x 0.5	12.4			Faint, stellar anonymous galaxy W edge.
21 12 42.2	+13 58 51	MCG +2-54-12	13.9	0.9 x 0.9	13.5			Faint star superimposed SW edge.
21 16 32.9	+11 29 40	MCG +2-54-16	14.8	1.1 x 0.6	14.2	Sbc	78	Mag 12.2 star SW 1′.8.
21 16 47.0	+15 03 26	MCG +2-54-17	14.5	0.5 x 0.4	12.6	SBbc	36	Pair with MCG +2-54-18 S edge.
21 16 48.3	+15 02 53	MCG +2-54-18	14.5	0.5 x 0.4	12.6		153	Pair with MCG +2-54-17 N edge.
21 21 48.9	+10 39 20	MCG +2-54-21	14.4	0.9 x 0.6	13.6		138	Mag 10.9 star SE 3′.0.
21 25 01.1	+12 26 25	MCG +2-54-24	15.0	0.8 x 0.3	13.3	Sb	3	Mag 11.6 star W 1′.0.
21 39 43.6	+10 20 16	MCG +2-55-5	14.4	0.7 x 0.7	13.4			Almost stellar **PGC 67122** on S edge.
21 42 26.9	+13 53 57	MCG +2-55-8	15.7	1.0 x 0.5	14.8		96	
21 52 52.6	+15 34 12	MCG +2-55-22	15.4	0.9 x 0.7	14.8	Sbc	48	Mag 9.58v star SW 6′.5.
21 53 16.1	+15 36 55	MCG +2-55-23	14.7	1.1 x 0.3	13.3	Sb	30	Mag 9.27v star NNE 5′.6.
21 53 19.8	+15 33 06	MCG +2-55-24	14.9	0.4 x 0.3	12.5		177	MCG +2-55-25 SE 1′.7.
21 53 23.6	+15 31 40	MCG +2-55-25	14.4	0.4 x 0.4	12.3			Mag 12.1 star W 0′.8; MCG +2-55-24 NW 1′.7.
21 53 46.7	+10 03 05	MCG +2-55-26	14.7	0.8 x 0.6	13.8	Sc	156	Mag 14.7 star SW edge.
21 59 42.8	+10 19 20	MCG +2-56-3	14.7	0.5 x 0.5	13.0	S?		Mag 10.4 star ESE 2′.4.
21 59 53.4	+13 34 46	MCG +2-56-4	14.3	0.7 x 0.7	13.3	SB?		Stellar nucleus; mag 8.81v star N 5′.2.
22 01 38.7	+09 58 57	MCG +2-56-6	14.5	1.0 x 0.4	13.3		156	Mag 11.3 star W edge.
22 14 06.3	+13 29 10	MCG +2-56-21	14.5	0.6 x 0.6	13.2			Mag 11.7 star NE edge.
22 14 10.3	+13 16 44	MCG +2-56-22	15.2	0.6 x 0.4	13.5	Sc	21	Close pair of mag 11-12 stars N 2′.2.
22 15 03.3	+15 24 02	MCG +2-56-25	14.0	0.9 x 0.8	13.5	S?	123	Mag 10.32v star S 1′.4.
21 00 51.3	+16 18 13	MCG +3-53-9	14.2	0.8 x 0.6	13.5	E1:	66	Pair of mag 11.5 stars NW 2′.6.
21 23 33.3	+17 34 58	MCG +3-54-9	14.8	0.5 x 0.3	12.6	Sb	156	Double system? Small, faint companion galaxy SE edge.
21 23 50.2	+17 34 13	MCG +3-54-10	14.7	0.8 x 0.3	13.0	Sb	159	Located between a pair of E-W oriented mag 12.5 stars.
21 24 11.5	+15 59 16	MCG +3-54-11	14.5	0.8 x 0.8	13.9	Sa		Mag 10.42v star NW 2′.8.
22 06 38.2	+17 27 39	MCG +3-56-9	14.5	0.7 x 0.4	13.0	Sbc	3	Mag 12.1 star SE 0′.8.
22 09 59.2	+16 29 25	MCG +3-56-17	14.9	0.8 x 0.8	14.3	Sc		Mag 10.08v star NW 1′.2.
21 00 42.4	+17 48 15	NGC 7003	13.0	1.1 x 0.8	12.7	Sbc	120	Very patchy.
21 05 37.4	+11 24 53	NGC 7015	12.5	1.9 x 1.6	13.6	Sbc	165	Many branching arms and superimposed stars.
21 07 47.5	+16 20 06	NGC 7025	12.8	1.9 x 1.3	13.6	Sa	33	Mag 9.72v star on W edge.
21 09 36.5	+15 07 29	NGC 7033	14.2	0.8 x 0.5	13.1	S?	23	
21 09 38.3	+15 09 00	NGC 7034	13.8	1.1 x 0.6	13.3	E	125	NGC 7033 S 1′.7; mag 10.75v star WNW 1′.4.
21 13 16.5	+08 51 52	NGC 7040	14.0	1.0 x 0.8	13.6	S?	150	Multi-galaxy system, small, faint companion on S edge; **MCG +1-54-3** SW 8′.7.
21 13 45.7	+13 34 29	NGC 7042	12.0	2.0 x 1.8	13.2	Sb	140	

(Continued from previous page)

GALAXIES

RA h m s	Dec ° ′ ″	Name	Mag (V)	Dim ′ Maj x min	SB	Type Class	PA	Notes
21 14 04.2	+13 37 36	NGC 7043	13.7	1.3 x 1.0	13.8	SBa:	135	
21 26 13.7	+14 10 56	NGC 7066	13.9	0.9 x 0.9	13.5	S?		
21 26 32.3	+12 11 04	NGC 7068	14.0	0.9 x 0.3	12.4	Sc	165	Mag 9.16v star N 1'.2.
21 29 39.0	+06 40 53	NGC 7074	15.3	0.9 x 0.4	14.1		111	Mag 6.41v star S 6'.2.
21 32 25.3	+06 34 49	NGC 7085	14.5	1.2 x 0.7	14.2	Sc	147	
21 39 34.7	+08 52 35	NGC 7100	13.9	0.7 x 0.7	13.0			
21 39 27.2	+08 53 46	NGC 7101	15.9	0.5 x 0.3	13.7		114	
21 39 44.3	+06 17 13	NGC 7102	13.5	1.7 x 1.1	14.0	SB(rs)b?	153	
21 42 22.9	+12 29 51	NGC 7112	14.3	1.1 x 0.4	13.2	Sab	87	Located 1'.9 SW of mag 10.66v star.
21 42 26.7	+12 34 06	NGC 7113	14.2	1.4 x 1.4	15.0	E		Mag 9.24v star on W edge.
21 47 16.6	+10 14 30	NGC 7132	14.2	1.1 x 0.7	13.8	Sc	114	Mag 10.87v star W 1'.1.
21 49 01.2	+12 30 43	NGC 7138	14.2	1.2 x 0.5	13.4	SBa	177	
21 56 25.6	+13 33 42	NGC 7159	14.3	0.6 x 0.5	12.8		168	
22 00 41.1	+17 44 13	NGC 7177	11.2	3.1 x 2.0	13.0	SAB(r)b I-II	93	
22 03 06.6	+11 11 55	NGC 7190	13.8	1.0 x 0.5	12.9	SB0	66	
22 03 30.9	+12 38 11	NGC 7194	13.1	1.1 x 0.9	13.1	E	20	NGC 7195 N 1'.4; **CGCG 428-21** WNW 2'.1.
22 03 30.3	+12 39 37	NGC 7195	14.7	0.5 x 0.3	12.5		81	
22 05 41.1	+16 47 04	NGC 7206	13.3	1.0 x 0.7	12.8	S0:	147	NGC 7207 SE 1'.6; **UGC 11902** S 3'.9.
22 05 45.8	+16 46 01	NGC 7207		0.5 x 0.2			93	
22 07 02.2	+10 14 04	NGC 7212	13.9	1.3 x 0.6	13.5	S?	33	Multi-galaxy system.
22 14 44.9	+13 50 46	NGC 7236	13.6	0.6 x 0.6	12.4	SA0⁻		**CGCG 428-57** SSW 5'.5.
22 14 47.0	+13 50 27	NGC 7237	13.1	0.6 x 0.6	11.9	SA0⁻		**CGCG 428-59** S 7'.6.
21 00 04.8	+09 34 51	UGC 11659	13.9	1.3 x 0.6	13.4	S?	135	Large core with two short arms; CGCG 400-17 SE 2'.6.
21 00 26.3	+16 51 35	UGC 11661	13.7	1.1 x 0.9	13.5	Sbc	115	
21 04 33.8	+09 39 26	UGC 11671	13.5	1.1 x 0.8	13.2	Double System	75	Double system, contact, disrupted; mag 10.33v star on N edge.
21 04 29.0	+16 05 02	UGC 11672	14.2	1.6 x 1.1	14.7	Double System	81	Double system.
21 05 55.6	+07 40 36	UGC 11675	14.1	0.9 x 0.3	12.5	Sa	46	**MCG +1-53-11** NW 5'.0.
21 08 22.5	+17 49 15	UGC 11682	13.7	1.5 x 0.5	13.2	S?	170	Mag 10.59v star N 2'.6.
21 11 52.0	+11 16 32	UGC 11694	14.3	1.7 x 1.0	14.7		45	Star superimposed on core? Mag 11.8 star NW 1'.7.
21 12 06.9	+13 00 01	UGC 11698	14.0	0.9 x 0.6	13.2	Sb:	15	Mag 7.75v star SW 5'.7.
21 12 20.5	+12 36 37	UGC 11699	14.5	1.1 x 1.1	14.5	SBd:		Several stars superimposed.
21 12 24.9	+11 24 29	UGC 11700	13.9	1.2 x 0.7	13.5	SBb	13	Mag 15.4 star superimposed on NW edge.
21 14 22.2	+15 14 35	UGC 11706	14.6	1.5 x 0.2	13.2	Sb	55	Small triangle of mag 13.5 stars N 1'.8.
21 17 28.3	+15 33 55	UGC 11713	14.3	1.5 x 0.5	13.8	SBcd:	16	**CGCG 426-35** S 6'.4.
21 18 53.3	+15 40 52	UGC 11719	14.6	1.5 x 0.2	13.2	Sab	162	Mag 14.8 star on e edge; mag 10.7 star NW 2'.6; mag 9.67v star SW 2'.4.
21 19 14.7	+06 01 09	UGC 11720	13.8	1.0 x 0.7	13.3	SB(rs)cd	130	
21 19 27.0	+14 03 23	UGC 11721	14.6	1.3 x 0.2	13.0	Im:	162	
21 22 29.4	+15 16 55	UGC 11735	13.5	1.0 x 0.8	13.2	E?		**MCG +2-54-23** NE 10'.8.
21 26 14.4	+09 47 52	UGC 11740	13.6	1.5 x 0.5	13.1	S?	122	Located between a pair of stars, mags 11.4 and 12.4.
21 28 59.6	+11 22 53	UGC 11751	13.8	1.0 x 0.6	13.1	SB	162	Very faint bridge extends to companion **PGC 66835** 0'.9 S. **PGC 66835** has a bright E-W bar and faint envelope.
21 30 57.9	+13 59 10	UGC 11758	14.4	1.5 x 0.2	13.0	S?	60	
21 33 33.6	+08 46 21	UGC 11765	14.2	1.1 x 0.6	13.6	Sb	80	
21 38 09.3	+08 57 35	UGC 11782	13.9	2.3 x 1.3	14.9	SB(s)m	35	elongated core uniform envelope; mag 12.6 star SW edge.
21 42 12.8	+05 36 50	UGC 11792	14.8	1.5 x 0.2	13.3	Scd:	160	Located 4'.0 S of mag 5.31v star 7 Pegasi.
21 44 28.8	+14 53 55	UGC 11803	14.3	0.9 x 0.8	13.8	Sbc	177	Stellar nucleus, strong dark patch W side.
21 49 28.8	+14 13 48	UGC 11820	14.2	2.2 x 2.2	15.8	Sm		Dwarf spiral, extremely low surface brightness.
21 49 35.7	+15 47 25	UGC 11821	14.1	1.7 x 0.3	13.2	Sa	79	Mag 6.85v star SW 3'.7.
21 54 45.3	+06 25 19	UGC 11846	14.2	1.0 x 0.5	13.3	Sbc	42	
21 55 59.4	+05 54 10	UGC 11851	13.8	1.0 x 0.7	13.3	Sa	24	Small, elongated E-W galaxy on N edge; mag 10.7 star S 1'.1.
21 58 36.1	+12 02 17	UGC 11865	13.6	0.5 x 0.5	11.9	I?		
22 00 41.6	+10 33 02	UGC 11871	13.6	0.9 x 0.5	12.6	S?	172	Double system, brighter N half; mag 13.9 star superimposed SE edge.
22 04 47.3	+15 47 37	UGC 11896	14.4	0.8 x 0.5	13.2	Sbc	111	Close pair of mag 13 stars SW 1'.4; mag 7.89v star NW 5'.6.
22 06 35.2	+16 03 15	UGC 11908	13.5	1.2 x 0.8	13.3	S?	96	
22 09 15.4	+14 21 39	UGC 11921	13.8	1.7 x 0.7	13.8	IBm	127	
22 09 50.7	+16 53 32	UGC 11928	14.4	0.9 x 0.6	13.5	SB?	15	Strong E-W core.
22 12 01.0	+17 54 17	UGC 11944	15.4	2.1 x 0.7	15.7	Im:	28	Dwarf irregular, extremely low surface brightness; mag 11.6 star S 2'.1.
22 12 18.0	+11 29 40	UGC 11947	13.8	1.0 x 1.0	13.7	SAB(s)b		Bright nucleus in faint bar; faint arms N and S; mag 13.7 star SE edge.
22 12 22.3	+14 01 24	UGC 11948	14.0	0.9 x 0.7	13.4	SBb	22	
22 13 38.9	+14 13 06	UGC 11952	14.8	1.5 x 0.2	13.3	SBcd?	143	**CGCG 428-47** and a close mag 10.3 star S 6'.1.

OPEN CLUSTERS

RA h m s	Dec ° ′ ″	Name	Mag	Diam ′	No. ★	B ★	Type	Notes
21 07 22.0	+16 18 00	French 1		13		8.8	ast?	(A) "Toadstool" shaped, likely asterism. Includes galaxy NGC 7025. Noted 1997 by S. French.
22 02 59.0	+10 49 00	NGC 7193		20	25	9.0	cl?	(A) J. Herschel's no. 2145, an obvious group and possible real cluster. Galaxy **IC 5160** N of center 6'.8.

GLOBULAR CLUSTERS

RA h m s	Dec ° ′ ″	Name	Total V m	B ★ V m	HB V m	Diam ′	Conc. Class Low = 12 High = 1	Notes
21 01 29.5	+16 11 15	NGC 7006	10.6	15.6	18.8	3.6	1	
21 29 58.3	+12 10 01	NGC 7078	6.3	12.6	15.9	18.0	4	= **M 15**. Contains the planetary nebula PN G65.0−27.3 (Pease 1).

PLANETARY NEBULAE

RA h m s	Dec ° ′ ″	Name	Diam ″	Mag (P)	Mag (V)	Mag cent ★	Alt Name	Notes
21 36 53.0	+12 47 19	NGC 7094	94	13.6	13.4	13.6	PK 66−28.1	Prominent central star, slightly brighter along W edge.
21 29 59.4	+12 10 26	PN G65.0−27.3	1	14.9	15.5	14.9	PK 65−27.1	= **Pease 1** or **Köstner 648**; located approximately 0'.6 NE of the center of M 15; a specialized finder chart, like the HST image of M 15, is needed to properly identify this very small planetary.

RA h m s	Dec ° ′ ″	Name	Mag (V)	Dim ′ Maj x min	SB	Type Class	PA	Notes
19 56 14.3	+09 23 50	CGCG 398-2	14.2	0.6 x 0.6	12.9			Mag 10.47v star superimposed on NW end with mag 9.12v star on NW edge forming close double stars.
19 56 39.9	+07 50 37	CGCG 398-3	13.9	0.4 x 0.4	11.8			Mag 11.8 star E 1′.0.
20 39 22.7	+08 54 42	CGCG 399-26	14.8	0.6 x 0.5	13.3		30	Located between two mag 11 stars, in a star rich area.
20 41 50.1	+05 35 37	CGCG 399-28	14.1	0.5 x 0.4	12.2		36	Mag 13.6 star SW edge; mag 10.02v star S 3′.9.
20 56 56.7	+06 49 00	CGCG 400-12	14.1	0.5 x 0.3	12.1	E	15	Mag 10.43v star ENE 3′.1.
21 00 13.6	+09 33 30	CGCG 400-17	15.6	0.5 x 0.3	13.6	E	33	UGC 11659 NW 2′.5.
19 59 44.0	+14 45 52	CGCG 423-2	15.8	0.8 x 0.6	14.8	Spiral	150	Very patchy, several stars superimposed, in star rich area.
20 22 31.2	+10 08 58	CGCG 424-2	13.8	0.7 x 0.7	12.9			Mag 10.7 star SE 0′.6.
20 37 04.5	+12 39 00	CGCG 424-27	14.4	0.6 x 0.4	12.7		159	Mag 10.80v star S 2′.3.
20 38 45.0	+09 43 30	CGCG 424-33	14.2	0.6 x 0.3	12.2		0	Close pair of mag 9.6, 10.94v stars E 4′.4.
20 47 23.5	+13 05 07	CGCG 425-7	13.8	0.5 x 0.5	12.1			Close pair of mag 9.18v, 9.98v stars E 3′.5.
21 02 53.9	+15 23 28	CGCG 425-36	14.0	0.5 x 0.3	11.8		75	Mag 11.5 star NE 2′.6.
20 24 05.3	+15 45 22	CGCG 447-3	14.3	0.7 x 0.3	12.5	Spiral		Galaxy pair.
21 03 07.1	+16 01 49	CGCG 448-33	14.2	0.6 x 0.3	12.2	Sc	150	**CGCG 448-35** E 8′.8; mag 6.88v star SW 6′.1.
20 34 37.6	+09 08 06	MCG +1-52-13	15.3	1.0 x 0.7	14.7		60	
20 25 30.2	+10 34 18	MCG +2-52-6	14.6	1.0 x 0.7	14.0	Sbc	93	Close pair of mag 11.4 stars E 2′.3.
20 26 42.5	+10 41 41	MCG +2-52-7	14.3	0.5 x 0.4	12.4	Sb	171	
20 38 38.3	+12 10 32	MCG +2-52-25	14.4	0.9 x 0.9	14.0			Numerous stars superimposed; mag 12.5 star NE edge; in star rich area.
20 57 09.3	+13 31 27	MCG +2-53-6	15.3	1.0 x 0.2	13.4	Sc	129	Mag 11.4 star SW 1′.3.
20 57 45.1	+13 59 34	MCG +2-53-7	14.2	0.8 x 0.8	13.5			Mag 9.47v star NNW 2′.2.
20 58 06.2	+13 30 52	MCG +2-53-8	15.4	0.7 x 0.3	13.5	Sb	102	
20 58 03.8	+11 03 10	MCG +2-53-9	14.9	0.8 x 0.8	14.3	Sc		Stars superimposed N and W.
20 59 06.7	+11 08 43	MCG +2-53-10	15.4	0.7 x 0.7	14.5	SBbc		Weak stellar nucleus, faint envelope.
20 59 27.5	+13 56 08	MCG +2-53-11	14.6	0.9 x 0.7	13.9	SBb	141	Mag 11.8 star N 2′.2.
21 00 51.3	+16 18 13	MCG +3-53-9	14.2	0.8 x 0.6	13.5	E1:	66	Pair of mag 11.5 stars NW 2′.6.
20 22 21.5	+06 25 44	NGC 6901	13.7	1.4 x 0.5	13.1	SB(r)ab	63	
20 23 34.1	+06 26 40	NGC 6906	12.3	1.7 x 0.8	12.5	SB(rs)bc:	36	
20 27 28.4	+08 05 50	NGC 6917	13.9	1.4 x 1.0	14.2	S	40	
20 32 38.2	+09 54 59	NGC 6927	14.5	0.5 x 0.2	11.9	S0: sp	9	NGC 6927A S 2′.0.
20 32 36.7	+09 52 59	NGC 6927A	14.5	0.4 x 0.2	11.7	E6:	15	NGC 6927 N 2′.0.
20 32 50.7	+09 55 36	NGC 6928	12.2	2.0 x 0.6	12.3	SB(s)ab	106	Star or bright knot on N edge.
20 32 58.7	+09 52 34	NGC 6930	12.8	1.3 x 0.5	12.1	SB(s)ab? sp	8	Companion galaxy on N edge; anonymous galaxy SE 1′.7.
20 38 24.0	+06 59 45	NGC 6944	13.8	1.5 x 0.6	13.5	S0$^-$:	57	Bright nucleus; 2′.6 S of mag 8.35v star.
20 38 11.4	+06 54 09	NGC 6944A	14.1	0.9 x 0.7	13.5	SB(rs)d pec:	72	stellar nucleus, uniform envelope; 1′.6 NW of mag 9.09v star.
20 43 53.6	+12 30 43	NGC 6956	12.3	1.9 x 1.9	13.6	SBb		Trio of stars superimposed E edge.
20 48 27.6	+07 44 20	NGC 6969	14.0	1.1 x 0.3	12.6	Sa	15	Mag 7.10v star S 8′.9.
20 49 23.9	+05 59 40	NGC 6971	13.7	1.1 x 0.9	13.5	Sb	60	
20 49 59.2	+09 53 55	NGC 6972	13.3	1.2 x 0.6	12.7	S0/a	143	**CGCG 425-10** N 3′.8; **CGCG 425-14** SE 8′.8.
20 55 49.0	+10 30 26	NGC 6988	14.6	0.5 x 0.5	13.0			
21 00 42.4	+17 48 15	NGC 7003	13.0	1.1 x 0.8	12.7	Sbc	120	Very patchy.
19 49 20.1	+07 09 30	UGC 11482	14.5	1.8 x 0.7	14.6	Im:	97	Irregular, no defined core, many superimposed stars.
19 54 58.4	+05 52 19	UGC 11493	13.3	2.1 x 1.5	14.4	Scd:	170	Very filamentary, many superimposed stars; in star rich area.
19 57 15.1	+05 53 23	UGC 11498	12.7	3.1 x 1.0	13.8	SBb:	75	Numerous superimposed stars, a few small knots; in star rich area.
20 04 19.0	+07 24 49	UGC 11511	12.9	1.0 x 1.0	12.8	Scd:		Filamentary; numerous stars superimposed; mag 11.07v star E 2′.2.
20 04 34.6	+12 44 20	UGC 11512	13.5	1.0 x 0.5	12.6	SB?	166	Sits in a triangle of mag 10.8 and 11 stars.
20 11 49.5	+05 45 45	UGC 11522	13.7	1.0 x 0.5	12.8	Sbc	3	**CGCG 398-20** N 4′.1.
20 12 01.1	+05 30 57	UGC 11523	14.1	0.9 x 0.9	13.7	Scd:		Small core with stellar nucleus, offset to the w, faint envelope E and S.
20 12 03.9	+05 45 46	UGC 11524	13.4	1.0 x 0.9	13.1	SA(r)c: I	90	Strong dark lane between bright core and single N-S arm E; mag 10.6 star E 0′.9.
20 13 35.8	+07 16 56	UGC 11526	13.8	1.5 x 0.7	13.7	Sa	20	Line of three stars proceeding NE from E edge, mags 10.8, 12.3, and 12.4.
20 16 43.4	+07 23 47	UGC 11532	13.6	1.0 x 0.9	13.4	S0:	48	Several stars superimposed S edge.
20 22 39.0	+09 35 03	UGC 11543	14.1	1.1 x 0.6	13.5	Sb	52	Bright nucleus, located between a mag 11.2 star 2′.2 NW and a mag 10.4 star 2′.1 SE; **CGCG 399-5** SE 6′.7.
20 24 36.6	+06 54 38	UGC 11551	14.0	1.5 x 0.4	13.3	Sbc	109	
20 24 35.7	+12 26 36	UGC 11552	13.8	1.9 x 0.5	13.6	Sab	16	Prominent dark lane down E side; **UGC 11550** SW 2′.4.
20 25 12.2	+05 15 45	UGC 11555	13.3	1.7 x 0.8	13.5	SA(s)bc I	21	Two main arms; located between a mag 11.3 star E and a mag 12.0 star W.
20 27 38.7	+10 45 25	UGC 11564	14.0	1.5 x 0.2	12.5	S?	99	Double system? Bridge?
20 28 18.4	+10 45 17	UGC 11568	13.8	2.1 x 0.6	13.9	Scd?	37	Mag 10.99v star NW 3′.1.
20 29 01.5	+10 40 46	UGC 11571	14.1	1.9 x 0.5	13.9	SBcd:	18	**UGC 11569** SW 4′.9.
20 29 09.6	+10 44 36	UGC 11572	13.5	1.0 x 0.9	13.4	E	30	Very faint anonymous galaxy W 1′.6.
20 30 42.8	+09 11 21	UGC 11578	15.9	1.4 x 1.4	16.5	Sdm		Many superimposed stars.
20 32 19.3	+11 22 04	UGC 11587	13.3	1.8 x 0.5	13.1	S0	166	**MCG +2-52-13** N 2′.1.
20 34 44.9	+07 59 09	UGC 11593	13.8	0.9 x 0.3	12.2	S?	114	Double system, in contact with round companion NW end; companion is **MCG +1-52-14**.
20 36 34.1	+11 29 41	UGC 11599	12.7	2.3 x 1.7	14.0	Sb	95	Bright core, branching arms; mag 11.4 star N 1′.5.
20 37 55.0	+10 37 51	UGC 11602	13.6	1.3 x 1.0	13.7	S0	0	Large, bright, diffuse center, many superimposed stars, in star rich area.
20 39 20.8	+10 48 19	UGC 11605	13.6	1.2 x 1.1	13.9	E	87	Mag 10.7 star on S edge.
20 40 05.8	+07 15 35	UGC 11610	13.8	1.1 x 0.6	13.2	Sbc	138	Stellar nucleus, uniform envelope; mag 10.7 star W 1′.1.
20 40 31.9	+14 16 30	UGC 11611	14.3	1.2 x 0.5	13.6	SBcd:	55	Stellar nucleus, faint envelope, dark lane N of center.
20 44 09.8	+12 25 01	UGC 11620	13.7	0.6 x 0.4	12.0	S?	25	Pair of stars on SE edge, mag 12.2 and 13.9.
20 44 26.8	+12 29 49	UGC 11623	13.9	1.0 x 0.8	13.6	SB(r)a	40	Bright NE-SW bar with dark lanes E and W.
20 48 17.2	+16 43 23	UGC 11634	13.8	1.2 x 0.7	13.5	Double System	63	Double system, contact; elongated component, plus mag 11.9 star, N.
20 49 36.8	+16 51 54	UGC 11638	13.0	1.9 x 1.1	13.7	SBb:	135	Triangle of two mag 12 stars and one mag 13 star superimposed.
20 49 51.6	+06 13 11	UGC 11639	13.7	1.1 x 0.9	13.5	Compact	57	Small, bright core, faint uniform envelope.
20 53 02.4	+07 09 31	UGC 11644	14.1	1.3 x 0.4	13.3	SBb	174	Mag 10.34v star NE 0′.9.
21 00 04.8	+09 34 51	UGC 11659	13.9	1.3 x 0.6	13.4	S?	135	Large core with two short arms; CGCG 400-17 SE 2′.6.
21 00 26.3	+16 51 35	UGC 11661	13.7	1.1 x 0.9	13.5	Sbc	115	

OPEN CLUSTERS

RA h m s	Dec ° ′ ″	Name	Mag	Diam ′	No. ★	B ★	Type	Notes
19 53 28.8	+11 40 57	NGC 6837	12	3	20		cl	
19 54 19.0	+17 53 18	NGC 6839		4	12		ast?	
19 55 16.0	+12 07 30	NGC 6840		4	20	10.0	cl?	
19 56 04.6	+12 09 02	NGC 6843		4	<20	10.0	cl?	
20 03 05.2	+11 16 00	NGC 6858		10	12	10.0	cl?	
20 41 10.5	+16 39 00	NGC 6950		14	60		cl?	(A) Scattered group of 50-60 faint stars.

(Continued from previous page)

84

OPEN CLUSTERS

RA h m s	Dec ° ′ ″	Name	Mag	Diam ′	No. ★	B ★	Type	Notes
19 45 06.0	+17 35 00	Ro 1		3	15		cl	(A) Condensation in rich Milky Way.

GLOBULAR CLUSTERS

RA h m s	Dec ° ′ ″	Name	Total V m	B ★ V m	HB V m	Diam ′	Conc. Class Low = 12 High = 1	Notes
20 34 11.6	+07 24 15	NGC 6934	8.9	13.8	17.1	7.1	8	
21 01 29.5	+16 11 15	NGC 7006	10.6	15.6	18.8	3.6	1	

DARK NEBULAE

RA h m s	Dec ° ′ ″	Name	Dim ′ Maj x min	Opacity	Notes
19 44 02.6	+08 18 00	B 339	60 x 60	2	Broken, dusky region.
19 48 44.4	+11 25 00	B 340	7 x 7	5	Irregular; curved.

PLANETARY NEBULAE

RA h m s	Dec ° ′ ″	Name	Diam ″	Mag (P)	Mag (V)	Mag cent ★	Alt Name	Notes
20 20 08.8	+16 43 52	IC 4997	13	11.6	10.5	14.4	PK 58−10.1	Bright central star; mag 10.05 star SW 1.′1.
20 10 26.8	+16 55 19	NGC 6879	9	13.0	12.5	14.8	PK 57−8.1	Mag 11.8 star SW 1.′4; mag 8.95v star E 5.′3.
20 15 09.0	+12 42 12	NGC 6891	21	11.7	10.5	12.4	PK 54−12.1	Mag 12.0 star W 0.′9; mag 11.3 star E 3.′6.
20 02 36.4	+17 36 50	PN G56.8−6.9	15	14.7	14.4		PK 56−6.1	Mag 12.9 star SW 0.′6; mag 11.1 star N 2.′8.
20 50 02.1	+13 33 29	PN G59.7−18.7	127	14.6	13.8	16.1	PK 59−18.1	= **Abell 72**; mag 8.61v star W 1.′8; mag 11.6 star E 1.′6; galaxy **MCG +2-53-5** S 1.′9.

85

GALAXIES

RA h m s	Dec ° ′ ″	Name	Mag (V)	Dim ′ Maj x min	SB	Type Class	PA	Notes
18 38 26.2	+17 11 47	MCG +3-47-10	12.9	0.6 x 0.6	11.7	SA0-		
18 36 41.4	+10 26 18	UGC 11293	15.3	1.0 x 0.5	14.4	Spiral	80	Several stars superimposed, in star rich area.
18 37 54.9	+17 31 55	UGC 11301	14.6	2.2 x 0.2	13.6	S	110	
18 41 34.8	+08 08 22	UGC 11326	14.3	1.8 x 0.8	14.6		120	Several stars superimposed, very filamentary, bright knot or star W edge.
19 49 20.1	+07 09 30	UGC 11482	14.5	1.8 x 0.7	14.6	Im:	97	Irregular, no defined core, many superimposed stars.

OPEN CLUSTERS

RA h m s	Dec ° ′ ″	Name	Mag	Diam ′	No. ★	B ★	Type	Notes
18 54 49.4	+05 32 56	Archinal 1		1.5	24	13.4	cl	
19 15 30.0	+11 16 30	Be 43		5	35	15.0	cl	
19 19 06.0	+15 43 18	Be 45		2	20	15.0	cl	
19 28 27.3	+17 21 57	Be 47		3	20	16.0	cl	
19 11 20.3	+13 06 42	Be 82		2.5	20	14.0	cl	
19 25 24.0	+11 40 00	Do 35		7	10		ast?	
18 38 54.0	+05 26 00	IC 4756	4.6	40	466	8.0	cl	Rich in stars; large brightness range; slight central concentration; detached. (A) Also known as the **Graff Cluster** or **Graff 1**.
19 24 30.0	+13 42 00	King 25		5	40		cl	
19 29 00.0	+14 52 00	King 26		2	15		cl	
18 51 30.0	+10 20 00	NGC 6709	6.7	15	111	9.0	cl	Moderately rich in stars; moderate brightness range; not well detached. (A) Very irregular. No obvious center on DSS image.
18 56 45.0	+10 26 47	NGC 6724	10	3	10	12.0	cl?	
19 01 24.0	+11 36 00	NGC 6738	8.3	15			cl	Few stars; moderate brightness range; not well detached.
18 55 02.7	+10 47 21	Pool J1855.0+1047		6	40		cl	(A) Involved in star cloud **Poole J1857.0+1027**.
19 45 06.0	+17 35 00	Ro 1		3	15		cl	(A) Condensation in rich Milky Way.

DARK NEBULAE

RA h m s	Dec ° ′ ″	Name	Dim ′ Maj x min	Opacity	Notes
19 19 49.1	+05 14 00	B 140	60 x 60	3	Semi-vacant region.
19 39 41.3	+10 31 00	B 142	40 x 40	6	With B 143 **The Double Dark Nebula** = **The Triple Cave Nebula**. Large; irregular; extended E and W.
19 41 25.1	+11 00 00	B 143	60 x 60	6	With B 142 **The Double Dark Nebula** = **The Triple Cave Nebula**. The outline of a square with the W side missing.
19 19 33.6	+07 34 00	B 330	30 x 30	4	Dark; round.
19 26 03.8	+07 35 00	B 331	60 x 3	1	Dusky, narrow lane NW.
19 28 00.6	+08 45 00	B 332		1	Area of dark lanes.
19 28 55.3	+10 40 00	B 333	60 x 60	3	Area of irregular dark lanes.
19 35 06.0	+12 19 20	B 334	3 x 3	4	Small, dark marking.
19 36 54.1	+07 37 00	B 335	6 x 6	6	Small; very black.
19 36 41.1	+12 20 37	B 336	2 x 1	5	Dark; in a slightly larger vacant space.
19 37 00.9	+12 23 40	B 337	3 x 3	4	Dark; with narrow extension for 17′ to NW.
19 43 02.8	+07 28 00	B 338	8 x 8	3	Dusky; in S part of B339.
19 44 02.6	+08 18 00	B 339	60 x 60	2	Broken, dusky region.
19 48 44.4	+11 25 00	B 340	7 x 7	5	Irregular; curved.
19 20 54.0	+11 16 00	LDN 673	55 x 15	6	
19 21 48.0	+12 26 00	LDN 684	50 x 10	5	

PLANETARY NEBULAE

RA h m s	Dec ° ′ ″	Name	Diam ″	Mag (P)	Mag (V)	Mag cent ★	Alt Name	Notes
19 18 28.2	+06 32 15	NGC 6781	114	11.8	11.4	16.7	PK 41−2.1	Brightest portion annular, with a small, faint extension N giving it an overall rectangular look.
19 31 16.6	+10 03 20	NGC 6803	10	11.3	11.4	15.2	PK 46−4.1	Almost stellar; mag 10.53v star N 1.′8.

(Continued from previous page)

85 PLANETARY NEBULAE 85

RA h m s	Dec ° ′ ″	Name	Diam ″	Mag (P)	Mag (V)	Mag cent ★	Alt Name	Notes
19 31 35.2	+09 13 31	NGC 6804	66	12.2	12.0	14.3	PK 45−4.1	The bright, annular center is 35 arcseconds in diameter, with fainter extensions is rectangular, oriented NNW-SSE.
19 34 33.7	+05 41 01	NGC 6807	8	13.8	12.0	16.3	PK 42−6.1	Stellar; mag 10.13v star NNE 1′5.
18 54 56.9	+06 02 41	PN G38.7+1.9	304				PK 38+2.1	Numerous faint stars on disc; a triangle of mag 9.31v, 9.54v, and 11.4 stars is centered SW 8′3.
18 56 18.3	+07 07 22	PN G39.8+2.1	15				PK 39+2.1	Mag 13.2 star S 1′3; mag 7.35v star W 4′4.
19 06 46.1	+06 23 50	PN G40.3−0.4	31	16.9	15.5	20.9	PK 40−0.1	= **Abell 53**; mag 12.9 star N 1′6.
19 16 27.8	+05 13 17	PN G40.4−3.1	3	15.7			PK 40−3.1	Close pair of mag 13 stars ENE 2′8; mag 13.7 star NNW 1′3.
19 09 13.6	+07 05 41	PN G41.2−0.6	18	18.9			PK 41−0.1	Close pair of mag 13-14 stars E 2′2; mag 11.1 star NW 3′4.
18 51 41.6	+09 54 50	PN G41.8+4.4	8	18.0			PK 41+4.1	Mag 12.9 star ENE 0′5; mag 9.28v star S 2′5.
18 48 33.0	+10 35 48	PN G42.0+5.4	9			16.2	PK 42+5.1	Mag 11.4 star NE 0′5; mag 12.4 star W 2′0.
19 21 00.8	+07 36 50	PN G43.0−3.0	12	14.8	15.3		PK 43−3.1	Mag 10.99v star NW 4′1; mag 14.1 star NE 2′2.
18 56 33.7	+10 52 08	PN G43.1+3.8	10	16.6		14.8	PK 43+3.1	Mag 12.2 star SW 2′9; mag 9.47v star SE 4′9.
19 02 17.8	+10 17 32	PN G43.3+2.2	14			18.5	PK 43+2.1	Close pair of mag 11.7, 11.2 stars NE 1′3; mag 11.6 star SE 1′7.
18 53 01.7	+12 15 58	PN G44.0+5.2	10	17.7			PK 44+5.1	Mag 11.3 star NW 4′8; mag 11.9 star S 4′0.
18 50 46.9	+12 37 30	PN G44.1+5.8		18.1		16.5	PK 44+5.2	Mag 11.3 star ESE 0′6; mag 13.1 star SW 0′6.
19 32 39.6	+07 27 50	PN G44.3−5.6	12	15.9		17.3	PK 44−5.1	Mag 11.8 star NW 0′8; mag 13.8 star ESE 0′7.
18 34 03.1	+14 49 17	PN G44.3+10.4	135				PK 44+10.1	Numerous stars on disc; mag 12.2 star near inside S edge; mag 11.9 star NW of center 1′6; mag 10.9 star E of center 2′8.
19 24 22.3	+09 53 56	PN G45.4−2.7	14	12.7	12.7	14.6	PK 45−2.1	Mag 11.4 star W 1′3; mag 12.7 star S 1′1.
19 09 26.7	+12 00 41	PN G45.6+1.5					PK 45+1.1	Stellar; mag 12.5 star ENE 3′2; mag 13.2 star SSE 3′0.
19 22 26.7	+10 41 20	PN G45.9−1.9	1	20.5			PK 45−1.1	Stellar; mag 11.1 star SSW 1′7; mag 14.2 star NE 2′0.
19 27 45.0	+10 24 18	PN G46.3−3.1	12	14.7	14.0		PK 46−3.1	Mag 10.33v star ESE 1′1; mag 12.3 star ENE 1′0.
19 06 23.0	+13 44 43	PN G46.8+2.9	10			17.0	PK 46+2.1	Mag 13.9 star W 0′8; mag 10.10v star NE 2′4.
19 03 10.3	+14 06 56	PN G46.8+3.8	600			17.7	PK 46+3.1	Mag 9.31v star just inside N edge; mag 9.13v star just inside SSE edge.
19 02 40.4	+14 28 49	PN G47.1+4.1	7	17.6		16.9	PK 47+4.1	Mag 11.7 star NNW 0′9; mag 12.3 star SW 1′5.
19 33 18.3	+10 37 04	PN G47.1−4.2	161	14.8		18.8	PK 47−4.1	= **Abell 62**; mag 11.6 star just inside NNE edge; mag 12.1 star just inside S edge and mag 10.15v star NW 1′6, just outside NW edge.
19 28 14.5	+12 19 35	PN G48.0−2.3	10	14.8	14.4		PK 48−2.1	Mag 11.5 star N 1′2; mag 10.30v star NE 2′9.
19 15 30.7	+14 03 48	PN G48.1+1.1	1	17.3			PK 48+1.2	Stellar; pair of mag 13-14 stars SSW 0′7; mag 11.9 star E 1′3.
19 04 51.5	+15 47 35	PN G48.5+4.2	3	16.5			PK 48+4.2	Very close pair of mag 12 stars E 0′7; mag 7.01v star S 3′4.
19 26 26.7	+13 19 33	PN G48.7−1.5	40				PK 48−1.1	Bright strip along W edge; mag 12.5 star N 0′9; mag 11.0 star W 2′
19 13 38.5	+14 59 18	PN G48.7+1.9	4	16.3	15.0		PK 48+1.1	Close pair of mag 13 stars SSE 1′7; mag 12.3 star N 2′6; mag 7.26v star SW 5′7.
19 12 06.0	+15 09 02	PN G48.7+2.3	6	18.6			PK 48+2.1	Mag 12.9 star N 2′5; mag 12.0 star ESE 3′7.
19 13 05.5	+15 46 42	PN G49.4+2.4	35	18.6			PK 49+2.1	Brighter, elongated E-W with faint haze extending N; mag 12.4 star W 2′4.
19 11 30.9	+16 51 35	PN G50.1+3.3	81			11.1	PK 50+3.1	Irregular shape with dark patches E and W of center; mag 10.6 star S 1′8.
19 30 16.7	+14 47 19	PN G50.4−1.6	1				PK 50−1.1	Stellar; mag 13.0 star ESE 2′8.
19 04 32.4	+17 57 08	PN G50.4+5.2	37	16.5		18.4	PK 50+5.1	= **Abell 52**; close pair of mag 13 stars E 1′4; mag 12.5 star N 1′3.
19 14 59.4	+17 22 47	PN G51.0+2.8	9	17.0		13.4	PK 51+2.1	Mag 12.4 star W 0′5; mag 11.8 star N 1′5.
19 14 04.1	+17 31 32	PN G51.0+3.0	5	15.9	15.0		PK 51+3.1	Mag 11.7 star N 1′5; mag 12.2 star S 1′8.
19 42 03.6	+13 50 35	PN G51.0−4.5	20	14.4	13.3	18.1	PK 51−4.1	Very slightly elongated NE-SW; mag 12.1 star SE 0′7.
19 19 18.8	+17 11 45	PN G51.3+1.8	20			15.1	PK 51+1.1	Forms a triangle with a mag 11.8 star SW 2′2 and a mag 12.6 star WNW 2′5.
19 41 09.4	+14 56 57	PN G51.9−3.8	12	13.9	14.0	14.5	PK 51−3.1	Mag 11.3 star N 1′1.
19 42 18.8	+15 09 06	PN G52.2−4.0	5	12.8	12.9	18.1	PK 52−4.1	Close pair of mag 12.9 stars NNW 1′7; pair of mag 12.4, 10.4 stars E 4′2.
19 39 09.9	+15 56 45	PN G52.5−2.9	8	12.6	11.8	14.1	PK 52−2.2	Mag 11.0 star N 3′6; mag 11.4 star WSW 2′9.
19 39 16.0	+16 20 46	PN G52.9−2.7	5	18.7			PK 52−2.1	Mag 11.4 star WNW 2′7; mag 11.6 star SW 2′4.
19 35 18.5	+17 12 58	PN G53.2−1.5	4				PK 53−1.1	Close pair of mag 11.12v, 11.5 stars E 2′0; mag 12.9 star NW 1′9.
19 42 10.4	+17 05 11	PN G53.8−3.0	40	17.1		14.6	PK 53−3.1	= **Abell 63**; mag 11.5 star WSW 1′5; mag 11.4 star SE 2′0.
19 41 34.1	+17 45 14	PN G54.4−2.5	10	16.5		17.7	PK 54−2.1	Mag 12.1 star W 0′9; mag 12.6 star NNW 1′0.

86 GALAXIES 86

RA h m s	Dec ° ′ ″	Name	Mag (V)	Dim ′ Maj x min	SB	Type Class	PA	Notes
17 26 52.9	+07 07 41	CGCG 54-27	15.1	0.4 x 0.3	12.6		141	Mag 13.6 star N 0′5; mag 11.2 star SSE 3′7.
17 30 40.7	+06 23 30	CGCG 55-2	14.6	0.7 x 0.3	12.8		39	Elongated anonymous galaxy W 1′6; mag 11.4 star SW 2′5.
17 31 54.3	+06 30 01	CGCG 55-4	15.3	0.4 x 0.3	12.8		21	Star superimposed; **CGCG 55-6** NE 2′1; **CGCG 55-3** S 1′1; **CGCG 55-5** ESE 2′0.
17 50 33.3	+05 40 14	CGCG 55-20	15.6	0.7 x 0.3	13.7	S	33	Mag 9.71v star W 7′2.
17 55 33.5	+06 11 31	CGCG 56-1	14.6	0.4 x 0.3	12.1		108	Low surface brightness.
17 56 21.9	+05 43 23	CGCG 56-2	14.8	0.4 x 0.3	12.3		141	Mag 12.4 star N 0′8.
18 04 15.7	+07 02 24	CGCG 56-8	14.7	0.8 x 0.2	12.5	S	141	Mag 9.91v star W 2′7.
17 38 08.3	+13 53 08	CGCG 83-11	15.0	0.4 x 0.4	12.9			Located in the middle of a triangle consisting of two mag 11 stars and one (check, seems truncated)
17 54 25.4	+15 37 05	CGCG 113-1	15.5	0.3 x 0.3	12.7			Forms an equilateral triangle with a pair of mag 11.4 stars N.
17 23 05.4	+12 41 43	IC 1255	13.4	1.1 x 0.5	12.6	S	12	Mag 10.7 star SSW 3′5.
17 50 39.3	+17 12 29	IC 1268	14.7	0.6 x 0.5	13.3	Sc	117	Mag 10.28v star SE 9′5; stellar **CGCG 112-59** SE 5′5.
18 08 12.0	+11 42 38	IC 4688	13.1	1.5 x 1.1	13.5	Scd:	164	Mag 10.66v star NE 7′2; almost stellar **IC 4691** NE 10′9.
17 35 19.0	+06 10 03	MCG +1-45-2	14.2	0.7 x 0.4	12.6		24	Located between a pair of mag 11 stars; in star rich area.
18 04 22.6	+07 16 47	MCG +1-46-4	14.0	0.9 x 0.7	13.6	E	108	Several superimposed stars; mag 10.16v star NE 3′3.
17 50 29.0	+14 23 39	MCG +2-45-5	13.9	0.7 x 0.2	11.6	Sb	165	Mag 9.23v star SE 3′8.
17 54 03.2	+14 51 57	MCG +2-45-7	15.1	0.7 x 0.7	14.2			Few stars superimposed S edge; mag 9.61v star SSE 2′7.
18 15 58.3	+13 47 07	MCG +2-46-12	14.1	0.6 x 0.4	12.4	Spiral	105	Uniform surface brightness.
17 48 12.5	+17 37 29	MCG +3-45-25	14.2	0.5 x 0.5	12.5	Sb		Mag 10.54v star NNE 1′4; MCG +3-45-26 NNW 1′8.
17 48 10.5	+17 39 15	MCG +3-45-26	14.3	0.3 x 0.2	11.1		159	**MCG +3-45-24** W 4′2.
17 58 16.3	+15 15 14	MCG +3-46-6	14.2	0.7 x 0.7	13.3			
18 14 18.5	+15 30 15	MCG +3-46-21	13.9	1.1 x 0.6	13.3		30	Close pair of mag 13 stars NE edge.
18 38 26.2	+17 11 47	MCG +3-47-10	12.9	0.6 x 0.6	11.7	SA0$^-$		
17 27 11.9	+11 32 40	NGC 6368	12.3	3.5 x 0.6	13.0	Sb II	42	Bright center, several stars superimposed.
17 29 22.0	+16 12 25	NGC 6375	13.9	1.6 x 1.6	14.9	E		UGC 10872 SW 3′1.
17 30 42.0	+06 16 57	NGC 6378	13.5	1.3 x 0.9	13.5	S?	5	Mag 11.12v star S 1′0.
17 30 35.0	+16 17 20	NGC 6379	12.9	1.1 x 1.0	12.8	Scd: I-II	33	Mag 9.36v star W 3′1.
17 32 24.2	+07 03 37	NGC 6384	10.4	6.2 x 4.1	13.8	SAB(r)bc I	30	Elongated nucleus; many superimposed stars.
17 32 39.6	+16 24 07	NGC 6389	12.1	2.8 x 1.9	13.7	Sbc II-III	130	Bright nucleus, superimposed stars.
17 50 40.1	+17 32 15	NGC 6467	12.6	2.6 x 1.7	14.1	S	77	
17 59 25.6	+06 17 10	NGC 6509	12.5	1.6 x 1.2	13.1	Sd	105	Numerous superimposed stars, in star rich area.
18 07 49.1	+17 36 15	NGC 6555	12.4	2.0 x 1.5	13.4	SAB(rs)c	110	Patchy, many knots.
18 11 07.5	+14 05 34	NGC 6570	12.7	1.8 x 1.1	13.3	SB(rs)m:	30	One main arm NE.
18 11 51.3	+14 58 52	NGC 6574	12.0	1.4 x 1.1	12.3	SAB(rs)bc: II-III	160	
18 18 33.6	+13 15 54	NGC 6615	13.1	1.3 x 0.9	13.1	SB0$^+$:	165	Bright center, faint envelope with many superimposed stars.
18 22 38.9	+15 41 54	NGC 6627	13.3	1.3 x 1.1	13.5	(R′)SB(s)b	70	two main arms.

86 (Continued from previous page) GALAXIES 86

RA h m s	Dec ° ′ ″	Name	Mag (V)	Dim ′ Maj x min	SB	Type Class	PA	Notes
18 27 37.2	+14 49 10	NGC 6635	13.4	1.0 x 0.9	13.1	S0 pec:	42	
17 25 25.2	+10 16 41	UGC 10844	13.9	1.0 x 0.5	13.1	S0	67	
17 28 08.9	+07 25 18	UGC 10862	12.7	2.3 x 1.9	14.2	SB(rs)c	126	NE-SW bar, numerous superimposed stars; mag 10.49v star NE 1′.1.
17 28 19.5	+14 10 07	UGC 10864	13.3	1.1 x 0.5	12.5	S0	175	
17 29 00.3	+06 19 50	UGC 10868	14.3	1.9 x 0.3	13.5	Double System	21	Mag 12.1 star E 1′.8.
17 29 16.1	+16 09 40	UGC 10872	15.9	1.6 x 0.3	15.0	Sdm:	5	Mag 11.3 star on S end; NGC 6375 NW 3′.1.
17 36 37.0	+15 17 52	UGC 10913	14.2	1.0 x 0.9	13.9	SBd		Low surface brightness; mag 13.3 star on E edge.
17 37 33.4	+11 07 17	UGC 10918	13.3	1.1 x 0.9	13.2	E	170	Bright knot of star NE of nucleus; mag 10.36v star SW 1′.4.
17 37 33.1	+17 32 00	UGC 10919	13.4	1.3 x 0.7	13.1	S?	7	Numerous knots.
17 41 18.4	+17 22 02	UGC 10941	14.5	1.1 x 0.5	13.7	S	7	
17 42 50.8	+16 12 48	UGC 10950	14.2	0.8 x 0.8	13.6	Double System		Elongated **MCG +3-45-16** on SE edge.
17 47 39.4	+09 32 46	UGC 10981	14.2	0.9 x 0.4	12.9	Sbc	57	
17 48 04.7	+14 44 27	UGC 10985	13.2	1.3 x 0.9	13.2	SB(s)dm	110	Located 2′.1 E of mag 7.88v star.
17 50 13.8	+14 17 11	UGC 11001	13.8	1.3 x 0.5	13.2	Sdm:	133	Mag 9.23v star NE 7′.1.
17 57 04.7	+12 14 21	UGC 11055	13.5	0.9 x 0.7	12.8	SB?	105	Mag 10.41 star NE 2′.9.
17 57 14.6	+12 10 44	UGC 11057	12.9	1.8 x 0.7	13.0	Scd:	90	Very knotty; mag 11.2 star on S edge.
17 58 22.1	+09 41 20	UGC 11067	13.5	1.0 x 0.9	13.2	S?	15	Bright center, dark patches N and S; several superimposed stars.
17 58 59.4	+10 32 16	UGC 11073	13.6	1.1 x 0.5	12.8	Sbc	157	
17 59 07.6	+07 08 32	UGC 11074	13.3	2.2 x 1.0	14.0	Sd:	142	Numerous stars superimposed, in star rich area; mag 11.6 star NW 2′.2.
18 01 51.7	+06 58 04	UGC 11093	13.0	3.1 x 0.5	13.3	Scd:	4	Numerous stars superimposed; mag 12.2 star N of center.
18 05 04.9	+17 15 59	UGC 11107	14.1	0.8 x 0.8	13.5	SB(r)bc:		Bright N-S bar.
18 08 06.9	+09 43 17	UGC 11122	14.0	0.9 x 0.8	13.5	Scd:	129	Very faint envelope, stars superimposed, in star rich area.
18 11 48.8	+12 05 15	UGC 11141	13.1	1.8 x 1.3	13.9	Sdm:	120	Short, elongated core, numerous stars superimposed on faint halo; mag 13.9 star inside NW edge.
18 13 59.6	+13 16 29	UGC 11168	13.1	1.3 x 1.0	13.2	Sdm:	96	Mag 9.80v star on S edge; mag 9.71v star NNW 2′.4.
18 16 02.8	+06 44 48	UGC 11177	12.9	1.4 x 1.3	13.4	SB(s)b	162	Bright E-W bar, faint arms, numerous superimposed stars, in star rich area.
18 22 53.1	+12 25 45	UGC 11214	14.3	1.2 x 1.1	14.5	Scd:	10	Numerous stars superimpose along E and SE edges; in star rich area.
18 28 01.5	+16 07 29	UGC 11242	14.1	1.1 x 0.5	13.3	Sd	88	Weak stellar nucleus.
18 36 41.4	+10 26 18	UGC 11293	15.3	1.0 x 0.5	14.4	Spiral	80	Several stars superimposed, in star rich area.
18 37 54.9	+17 31 55	UGC 11301	14.6	2.2 x 0.2	13.6	S	110	

OPEN CLUSTERS

RA h m s	Dec ° ′ ″	Name	Mag	Diam ′	No. ★	B ★	Type	Notes
17 46 12.0	+05 43 00	IC 4665	4.2	70	57	6.0	cl	Moderately rich in stars; moderate brightness range; no central concentration; detached.
18 38 54.0	+05 26 00	IC 4756	4.6	40	466	8.0	cl	Rich in stars; large brightness range; slight central concentration; detached. (A) Also known as the **Graff Cluster** or **Graff 1**.
18 27 15.0	+06 30 00	NGC 6633	4.6	20	159	8.0	cl	Moderately rich in stars; moderate brightness range; no central concentration; detached.

PLANETARY NEBULAE

RA h m s	Dec ° ′ ″	Name	Diam ″	Mag (P)	Mag (V)	Mag cent ★	Alt Name	Notes
18 12 06.4	+06 51 11	NGC 6572	15	9.0	8.1	13.1	PK 34+11.1	Powerful central star; bluish-green color is intense in large-aperture telescopes; mag 9.36v star E 3′.5.
17 53 32.2	+10 37 25	PN G36.0+17.6	80	14.7	14.6	14.7	PK 36+17.1	Faint stars E and S edges; mag 9.31v star NW 3′.9; mag 11.43v star SE 2′.9.
18 17 34.2	+10 09 01	PN G38.2+12.0	16	12.4		12.5	PK 38+12.1	Mag 11.9 star N 1′.1; mag 13.0 star S 1′.1.
18 27 48.4	+14 29 05	PN G43.3+11.6	10	14.0	13.9	15.5	PK 43+11.1	Close pair of mag 12.1, 12.3 stars ENE 3′.2.
18 34 03.1	+14 49 17	PN G44.3+10.4	135				PK 44+10.1	Numerous stars on disc; mag 12.2 star near inside S edge; mag 11.9 star NW of center 1′.6; mag 10.9 star E of center 2′.8.

87 GALAXIES 87

RA h m s	Dec ° ′ ″	Name	Mag (V)	Dim ′ Maj x min	SB	Type Class	PA	Notes
16 19 48.1	+05 09 41	CGCG 52-4	14.1	0.7 x 0.4	12.8	E	24	Mag 13.4 star N 2′.4; mag 12.3 star W 4′.8.
16 27 02.5	+07 33 40	CGCG 52-19	14.4	0.8 x 0.4	13.0		69	Pair of mag 12 stars WNW 0′.8; mag 10.6 star NE 2′.6.
16 27 40.7	+05 05 15	CGCG 52-21	15.4	0.7 x 0.5	14.2	SBbc	33	Mag 13.9 star on E edge and a mag 15.4 star on W edge.
16 29 43.2	+06 18 29	CGCG 52-29	15.5	0.3 x 0.3	12.7			Mag 11.2 star NE 2′.7.
16 31 04.4	+05 09 28	CGCG 52-38	14.1	0.9 x 0.3	12.5		177	**CGCG 52-39** SE 8′.5.
16 34 47.5	+06 43 55	CGCG 52-44	14.8	0.6 x 0.4	13.1		111	Low, uniform surface brightness.
16 37 42.4	+07 15 44	CGCG 52-51	14.9	0.5 x 0.3	12.7		33	Located 2′.1 SE of mag 9.06v star.
16 44 09.3	+07 26 42	CGCG 53-2	14.5	0.4 x 0.4	12.4			Mag 8.34v star NW 6′.4.
16 45 27.0	+08 49 51	CGCG 53-4	16.4	0.7 x 0.5	15.2	SBbc	114	Pair of mag 12 stars NW 2′.0.
17 15 08.8	+08 27 20	CGCG 54-11	14.5	0.5 x 0.4	12.6		24	**CGCG 54-10** SW 2′.2.
17 15 11.8	+08 25 34	CGCG 54-13	14.4	0.7 x 0.4	12.9		27	Forms a triangle with **CGCG 54-10** W 2′.8 and CGCG 54-11 WNW 1′.9.
17 26 52.9	+07 07 41	CGCG 54-27	15.1	0.4 x 0.3	12.6		141	Mag 13.6 star N 0′.5; mag 11.2 star SSE 3′.7.
16 09 08.6	+12 05 00	CGCG 79-62	14.3	0.6 x 0.4	12.6		156	Mag 9.38v star ENE 4′.7.
16 14 25.0	+10 30 49	CGCG 79-85	14.4	0.7 x 0.3	12.6	S	72	Mag 11.1 star N 1′.6.
16 17 35.2	+13 29 28	CGCG 79-89	14.4	0.7 x 0.6	13.3	Sb	45	Mag 10.76v star SW 0′.9.
16 22 58.2	+14 36 09	CGCG 80-17	14.6	0.7 x 0.6	13.5	SBa	81	Mag 13.8 star on W edge.
16 23 41.7	+09 44 39	CGCG 80-19	14.3	0.6 x 0.4	12.6		3	Elongated **CGCG 80-18** S 2′.1; **CGCG 80-21** N 2′.7.
16 26 55.8	+10 21 10	CGCG 80-32	16.4	0.4 x 0.3	13.9		3	Mag 8.55v star NW 12′.4.
16 32 58.5	+11 43 15	CGCG 80-46	14.3	0.9 x 0.3	12.7		108	Faint star on N edge.
17 04 58.3	+09 55 49	CGCG 81-18	14.7	0.6 x 0.5	13.2	Sb	177	Close pair of mag 11.2, 13.1 stars SW 3′.5; mag 13.6 star NE 0′.7.
17 13 11.8	+13 03 24	CGCG 82-15	15.5	0.6 x 0.4	13.8		129	Mag 9.71v star E 9′.4.
16 37 29.5	+17 55 55	CGCG 109-35	14.9	0.7 x 0.4	13.3	Irr	0	Mag 11.7 star N 2′.1.
16 49 50.9	+16 37 43	CGCG 110-12	14.8	0.6 x 0.4	13.1		150	Mag 11.8 star S 1′.3.
16 08 17.5	+07 32 17	IC 1197	13.4	2.9 x 0.5	13.7	Scd:	56	Mag 10.09v star SW 7′.3.
16 08 36.6	+12 19 47	IC 1198	14.3	0.8 x 0.3	12.6	S?	105	Mag 7.29v star WNW 11′.9.
16 10 34.6	+10 02 24	IC 1199	13.6	1.3 x 0.5	13.0	S?	157	Lies 1′.9 W of mag 7,53v star.
16 14 16.0	+09 32 11	IC 1205	13.7	0.5 x 0.5	12.1	Sa		Located 2′.3 E of mag 8.47v star.
16 15 13.1	+11 17 45	IC 1206	13.6	1.2 x 0.8	13.4	Sab	2	**CGCG 79-86** SW 1′.3.
16 18 39.7	+15 33 26	IC 1209	13.3	1.1 x 0.8	13.1	S0$^-$	5	Close pair of mag 10.05v, 8.68v stars SW 2′.7.
17 23 05.4	+12 41 43	IC 1255	13.4	1.1 x 0.5	12.6	S	12	Mag 10.7 star SSW 3′.5.
16 50 51.2	+08 46 59	IC 4621	13.8	0.9 x 0.7	13.1	Sab	122	Mag 11.7 star NW 1′.2.
16 51 33.5	+17 26 55	IC 4624	14.8	0.4 x 0.4	12.6			Low surface brightness; mag 10.00v star WSW 2′.6.
16 24 53.7	+06 04 50	MCG +1-42-2	14.5	0.8 x 0.7	13.7	Sc	171	
16 29 38.3	+08 27 01	MCG +1-42-5	14.0	0.9 x 0.9	13.8	E		

(Continued from previous page)

GALAXIES

RA h m s	Dec ° ′ ″	Name	Mag (V)	Dim ′ Maj x min	SB	Type Class	PA	Notes
17 06 10.4	+07 46 52	MCG +1-43-9	14.2	0.6 x 0.5	12.8		9	
16 08 32.8	+09 36 22	MCG +2-41-10	13.9	0.7 x 0.7	13.2	E		
16 10 56.3	+13 45 07	MCG +2-41-14	13.9	1.1 x 0.7	13.4	S0	156	Mag 11.7 star SE edge.
16 12 06.0	+14 12 05	MCG +2-41-18	14.9	0.3 x 0.2	11.7			On S edge of NGC 6078.
16 25 11.2	+11 52 48	MCG +2-42-3	14.8	0.9 x 0.6	14.0		159	
16 39 26.2	+11 12 39	MCG +2-42-7	14.8	0.9 x 0.3	13.2	Sb	150	
16 48 14.8	+13 54 14	MCG +2-43-2	15.0	1.1 x 0.7	14.5	SBbc	9	Stellar nucleus in narrow, elongated core.
17 08 55.1	+11 27 55	MCG +2-44-1	14.1	0.8 x 0.6	13.2		84	Mag 10.59v star NE 1′.6.
16 49 44.1	+17 51 48	MCG +3-13-5		0.6 x 0.5			12	
16 10 51.3	+17 03 23	MCG +3-41-141	14.4	0.9 x 0.4	13.2	Sb	108	Mag 11.3 star ESE 1′.5.
16 25 38.1	+16 27 17	MCG +3-42-8	14.2	0.5 x 0.5	12.7	E		Mag 10.8 star NE 3′.0.
16 40 03.4	+15 53 01	MCG +3-42-26	13.8	0.5 x 0.4	12.0	S0	90	Mag 12.3 star NW 3′.0.
17 15 34.3	+16 19 31	MCG +3-44-3	14.7	0.7 x 0.5	13.4	SBa	111	
16 10 10.9	+16 41 55	NGC 6073	13.5	1.3 x 0.7	13.3	Sc:	130	Located 2′.6 NE of mag 10.43v star.
16 11 17.4	+14 15 30	NGC 6074	14.4	0.4 x 0.4	12.3			**MCG +2-41-15** on SW edge.
16 12 05.6	+14 12 29	NGC 6078	15.2	0.5 x 0.5	13.8	E		Stellar MCG +2-41-18 on S edge.
16 12 57.0	+09 51 58	NGC 6081	13.1	1.8 x 0.6	13.0	S0	131	Very faint dark lane NW .
16 13 12.8	+14 11 04	NGC 6083	14.6	0.9 x 0.5	13.6	Sb	42	**CGCG 79-79** NNW 7′.4.
16 14 16.8	+17 45 24	NGC 6084	13.9	1.0 x 0.6	13.2	Sa sp	30	Mag 9.98v star SW 1′.9.
16 18 47.1	+07 24 38	NGC 6106	12.2	2.5 x 1.4	13.5	SA(s)c II-III	140	Patchy with several knots; mag 7.77v star S 9′.5.
16 19 10.7	+14 07 54	NGC 6113	14.1	1.0 x 0.6	13.4	S0	147	**MCG +2-41-25** N 1′.2.
16 23 38.8	+11 47 11	NGC 6132	13.6	1.5 x 0.5	13.2	Sab	127	Mag 9.20v star S 7′.8.
16 46 22.5	+09 02 11	NGC 6219	14.2	0.7 x 0.7	13.3			
16 48 18.5	+06 18 42	NGC 6224	13.5	1.0 x 1.0	13.3	S0^{-}?		Faint anonymous galaxy S 2′.3; mag 10.68v star 1′.6 N.
16 48 21.6	+06 13 20	NGC 6225	13.8	1.0 x 0.7	13.2	S0^{-}?	156	
17 01 57.5	+06 39 54	NGC 6280	14.6	0.4 x 0.2	11.7		144	Mag 11.2 star ENE 2′.2.
17 19 54.7	+16 39 35	NGC 6347	13.7	1.2 x 0.7	13.3	SBb	100	
17 27 11.9	+11 32 40	NGC 6368	12.3	3.5 x 0.6	13.0	Sb II	42	Bright center, several stars superimposed.
16 09 20.8	+08 45 44	UGC 10225	13.2	1.7 x 1.6	14.1	SBdm	18	Bright, elongated core, several knots S.
16 10 26.6	+12 18 20	UGC 10238	13.6	1.5 x 0.7	13.5	(R′)SB(s)b	70	Large E-W bar, faint envelope.
16 10 26.0	+12 49 27	UGC 10239	13.8	0.7 x 0.6	12.7	SBa	87	
16 11 17.1	+13 51 57	UGC 10249	14.0	1.0 x 0.5	13.1	SBab	76	
16 14 03.4	+14 16 53	UGC 10287	13.3	1.5 x 1.1	13.7	SBb	15	Two main arms; mag 13.2 star on E edge; **CGCG 79-82** S 8′.3.
16 19 28.6	+07 16 41	UGC 10337	14.0	1.2 x 0.4	13.1	Sb III	66	Mag 7.98v star S 6′.3.
16 23 11.4	+16 55 55	UGC 10360	13.4	1.0 x 1.0	13.3	S0		Bright center, faint envelope; mag 9.53v star SE 2′.5.
16 25 49.8	+16 34 32	UGC 10380	14.4	1.5 x 0.3	13.4	Sb III	108	Mag 12.6 star N edge; mag 11.20v star NNW 2′.5.
16 27 14.0	+12 58 46	UGC 10387	14.3	0.9 x 0.5	13.3	SBcd:	23	Mag 10.77v star SE 1′.7.
16 27 03.0	+16 22 54	UGC 10388	13.8	1.5 x 0.6	13.5	Sa	130	Mag 12.0 star N 1′.1.
16 28 54.2	+17 53 23	UGC 10405	14.5	1.0 x 0.8	14.1	Scd:	162	Trio of stars, mags 11.9, 12.1 and 13.1, SE 1′.7.
16 29 36.2	+15 39 27	UGC 10412	13.8	1.0 x 0.5	12.9	S0	78	
16 29 47.4	+08 38 38	UGC 10414	13.8	1.0 x 0.8	13.4	SB0/a:	84	Pair of mag 13 and 14 stars S 1′.4.
16 30 20.4	+08 37 41	UGC 10416	13.8	1.2 x 0.9	13.7	SAB(s)bc	144	Stellar nucleus, branching arms.
16 36 19.5	+10 21 51	UGC 10463	14.2	0.8 x 0.5	13.1	SBb	66	Mag 11.2 star NE 1′.8.
16 38 49.6	+17 21 09	UGC 10490	14.2	1.0 x 0.8	13.8	Sd	113	Faint stellar nucleus, uniform envelope.
16 50 00.0	+09 26 08	UGC 10569	14.5	1.5 x 0.2	13.0	Sb III-IV	137	Located 0′.9 NE of mag 6.92v star.
17 01 26.8	+06 55 34	UGC 10659	13.8	1.1 x 0.6	13.2	S?	32	Mag 11.8 star S 0′.7.
17 03 39.9	+09 17 53	UGC 10674	14.6	1.5 x 0.3	13.6	S	163	Mag 12.1 star W edge.
17 04 50.8	+12 55 27	UGC 10685	13.6	1.2 x 0.5	12.9	S?	119	
17 06 19.2	+10 22 25	UGC 10699	13.8	0.7 x 0.6	12.7	SB?	132	Mag 11.02v star NE 2′.1.
17 07 11.8	+11 25 13	UGC 10705	13.8	0.8 x 0.6	12.8	SBa	24	**CGCG 82-5** E 6′.1.
17 08 39.0	+09 35 12	UGC 10720	14.1	1.0 x 0.7	13.6	Scd:	90	Faint, stellar nucleus.
17 11 05.7	+05 51 04	UGC 10738	14.3	1.7 x 0.3	13.4	Sbc	49	Mag 9.60v star N 2′.6.
17 11 30.9	+07 59 37	UGC 10743	13.8	1.1 x 0.5	13.0	Sa?	113	Mag 6.34v star SSE 6′.9.
17 16 05.0	+06 50 28	UGC 10775	13.6	1.2 x 1.0	13.6	SBa	90	Several stars superimposed; mag 8.00v star W 3′.8.
17 16 24.0	+06 33 32	UGC 10778	13.3	1.0 x 0.8	13.0	S0^{-}?	108	Close pair of mag 11.8 stars N 2′.2.
17 16 30.4	+06 25 51	UGC 10779	13.3	1.8 x 0.6	13.2	Sb II-III	70	Mag 11.6 star SE 0′.9.
17 17 44.7	+07 36 40	UGC 10789	14.8	1.5 x 0.4	14.1	Double System	171	Double system, connected.
17 18 54.3	+08 26 29	UGC 10797	14.1	1.0 x 0.3	12.7	SBcd:	16	
17 19 33.2	+11 12 24	UGC 10802	13.6	1.0 x 1.0	13.5	S0		Mag 11.4 star on NW edge.
17 19 51.4	+14 23 59	UGC 10805	14.4	1.7 x 1.5	15.2	SBm V	171	Small, elongated core, numerous superimposed stars.
17 25 25.2	+10 16 41	UGC 10844	13.9	1.0 x 0.5	13.1	S0	67	

GALAXY CLUSTERS

RA h m s	Dec ° ′ ″	Name	Mag 10th brightest	No. Gal	Diam ′	Notes
16 11 36.0	+16 58 00	A 2159	15.9	34	16	Most members anonymous, faint and stellar.

OPEN CLUSTERS

RA h m s	Dec ° ′ ″	Name	Mag	Diam ′	No. ★	B ★	Type	Notes
17 11 23.0	+15 28 36	DoDz 7		6	6		cl	Few stars; small brightness range; not well detached. (A) Confirmed as 6′ diameter group of 6 stars.

GLOBULAR CLUSTERS

RA h m s	Dec ° ′ ″	Name	Total V m	B ★ V m	HB V m	Diam ′	Conc. Class Low = 12 High = 1	Notes
16 11 00.3	+14 57 49	Pal 14	14.7	17.6	20.0	2.5		

BRIGHT NEBULAE

RA h m s	Dec ° ′ ″	Name	Dim ′ Maj x min	Type	BC	Color	Notes
17 19 00.0	+06 05 00	vdB 111	12 x 8	R	3-5	1-4	Nebula with a mag 6.5 star involved; slightly extended N-S.

(Continued from previous page)

87 PLANETARY NEBULAE 87

RA h m s	Dec ° ′ ″	Name	Diam ″	Mag (P)	Mag (V)	Mag cent ★	Alt Name	Notes
16 11 44.5	+12 04 17	IC 4593	42	10.9	10.7	11.2	PK 25+40.1	Mag 9.42v star NW 5'.2.

88 GALAXIES 88

RA h m s	Dec ° ′ ″	Name	Mag (V)	Dim ′ Maj x min	SB	Type Class	PA	Notes
15 04 41.8	+06 20 35	CGCG 48-100	15.6	0.4 x 0.3	13.2		36	**CGCG 48-102** SE 1'.1.
15 06 34.5	+05 13 24	CGCG 49-1	14.6	0.9 x 0.4	13.3		75	CGCG 49-3 NNE 6'.6.
15 06 43.5	+05 19 35	CGCG 49-3	14.4	0.5 x 0.5	12.8			CGCG 49-1 SSW 6'.6; mag 10.9 star NNE 3'.6.
15 10 30.4	+06 29 26	CGCG 49-21	14.8	0.7 x 0.4	13.3		36	Mag 11.1 star WNW 6'.5.
15 11 31.4	+07 15 05	CGCG 49-33	13.7	0.5 x 0.5	12.1	S0		Mag 7.90v star NW 8'.8.
15 11 56.5	+05 33 34	CGCG 49-39	14.0	0.8 x 0.7	13.2		174	Mag 11.9 star NNE 3'.9; many very faint galaxies in the surrounding area.
15 13 02.5	+07 17 30	CGCG 49-46	14.7	0.6 x 0.4	13.0	Sab	36	**CGCG 49-57** SE 4'.8; mag 8.70v star NNW 7'.3; **UGC 9767** NNW 9'.2.
15 15 10.8	+05 34 36	CGCG 49-75	14.8	0.5 x 0.3	12.5		84	Mag 12.3 star WSW 0'.8.
15 18 43.7	+05 53 52	CGCG 49-115	15.2	0.5 x 0.3	12.9		30	Mag 12.2 star E 1'.8.
15 22 05.2	+05 51 14	CGCG 49-148	14.1	0.6 x 0.4	12.4	Sab	57	= **Mkn 851**. **CGCG 49-149** SE 1'.9; mag 9.99v star NNW 2'.5.
15 22 21.9	+05 23 27	CGCG 49-150	14.3	0.6 x 0.4	12.6			Located between two N-S mag 10 stars.
15 31 59.8	+07 27 02	CGCG 50-15	15.8	0.7 x 0.7	14.8			faint stellar nucleus with faint envelope; mag 12.3 star NNE 2'.3.
15 32 45.4	+08 15 16	CGCG 50-22	14.8	0.3 x 0.3	12.0			Mag 11.21v star N 6'.0.
15 33 47.9	+05 17 27	CGCG 50-25	14.8	0.8 x 0.6	13.9	S0	0	Mag 10.9 star N 0'.7; **UGC 9894** SE 1'.3.
15 39 02.6	+05 50 24	CGCG 50-59	15.3	0.8 x 0.5	14.2		153	Strong dark lane W.
15 40 54.1	+05 08 30	CGCG 50-68	14.8	0.5 x 0.2	12.1		75	Mag 10.17v star NE 10'.0.
15 50 24.9	+06 46 15	CGCG 50-113	15.2	0.6 x 0.5	13.7		129	Mag 13.3 star SE 0'.8; mag 12.0 star N 4'.4.
15 55 35.4	+06 27 46	CGCG 51-5	14.7	0.3 x 0.3	11.9			Located 3'.1 NW of mag 10.7 star.
14 57 31.1	+08 23 24	CGCG 76-111	14.3	0.5 x 0.3	12.1		126	Located between two stars, mags 12 and 13, on a SE-NW line.
15 01 38.9	+10 25 11	CGCG 76-133	14.1	0.7 x 0.6	13.0		66	= **Mkn 1494**. Mag 8.63v star SW 4'.9.
15 07 09.7	+09 38 06	CGCG 77-11	14.8	0.8 x 0.6	13.8		54	Pair of N-S oriented mag 11.3, 11.4 stars W 3'.3; mag 10.55v star E 5'.4.
15 09 52.7	+13 15 23	CGCG 77-25	14.4	0.5 x 0.5	12.8			Mag 12.1 star N 4'.1.
15 10 05.4	+09 43 19	CGCG 77-26	14.3	0.6 x 0.6	13.0			Spherical anonymous galaxy SW 2'.5; mag 12.4 star W 2'.0.
15 16 11.1	+12 44 20	CGCG 77-55	14.3	0.8 x 0.5	13.2		3	**CGCG 77-56** E 5'.3; mag 10.45v star SW 9'.6.
15 17 01.3	+13 06 09	CGCG 77-59	15.0	0.4 x 0.3	12.6			Mag 13.2 star on E edge.
15 17 57.2	+12 57 50	CGCG 77-63	15.0	0.5 x 0.4	13.1		174	MCG +2-39-13 E 5'.8; mag 12.4 star S 1'.6.
15 19 32.7	+09 57 33	CGCG 77-74	13.9	0.8 x 0.5	12.8	Sa	3	Mag 10.4 star W 3'.3.
15 20 32.3	+11 59 23	CGCG 77-78	15.0	0.7 x 0.4	13.5		57	Two main arms.
15 25 21.5	+13 43 47	CGCG 77-125	14.5	0.4 x 0.3	12.0		24	Mag 12.7 star SE 1'.6.
15 26 29.6	+08 35 26	CGCG 77-130	14.3	0.4 x 0.4	12.2			Mag 12.7 star N 2'.4; mag 8.48v star SW 7'.1.
15 46 08.8	+08 47 30	CGCG 78-64	14.2	0.8 x 0.3	12.5		21	Mag 7.83v stars SW 12'.9.
15 49 29.0	+11 56 09	CGCG 78-80	14.5	1.0 x 0.6	13.8		9	Bright center with strong dark patches N and S.
15 55 47.2	+11 37 26	CGCG 79-8	15.2	0.6 x 0.4	13.5		33	Mag 8.49v star E 5'.5.
15 55 53.6	+11 58 39	CGCG 79-9	14.3	1.0 x 0.2	12.4		120	Mag 10.6 star SW 6'.9.
15 59 30.0	+09 48 57	CGCG 79-18	14.9	0.4 x 0.4	12.8			Mag 12.1 stars SE 1'.5 and WSW 2'.5.
16 04 34.4	+12 10 51	CGCG 79-37	14.3	0.4 x 0.2	11.4		9	Faint star on S end; very faint, elongated anonymous galaxy S 1'.7.
16 04 48.1	+14 07 40	CGCG 79-40	14.4	0.6 x 0.4	12.7	Sb	27	Mag 11.4 star S 2'.8; **CGCG 79-42** E 2'.5.
16 06 45.3	+11 53 13	CGCG 79-47	16.0	1.1 x 0.2	14.2		123	Mag 13.8 star W 1'.3; mag 11.6 star NE 4'.5.
16 07 08.3	+14 15 54	CGCG 79-48	15.0	0.6 x 0.5	13.5	Spiral	78	Mag 10.22v star S 3'.6.
16 09 08.6	+12 05 00	CGCG 79-62	14.3	0.6 x 0.4	12.6		156	Mag 9.38v star ENE 4'.7.
16 14 25.0	+10 30 49	CGCG 79-85	14.4	0.7 x 0.3	12.6	S	72	Mag 11.1 star N 1'.6.
15 00 55.4	+16 56 47	CGCG 105-91	14.7	0.5 x 0.4	12.8	Compact	0	= **Mkn 1391**. Close pair of mag 12.9, 13.8 stars W 4'.9.
15 21 29.1	+15 08 14	CGCG 106-33	14.2	0.8 x 0.3	12.5	Sb	15	Mag 9.79 star E 7'.1.
16 01 21.2	+16 40 33	CGCG 108-59	14.0	0.4 x 0.4	12.1	E		Mag 10.92v star NW 3'.2; **MCG +3-41-40** NNE 5'.2.
16 02 58.8	+15 50 30	CGCG 108-81	14.1	0.8 x 0.4	12.7	(R')SB(s)0/a	153	**MCG +3-41-59** E 3'.3.
16 05 22.7	+16 11 52	CGCG 108-117	14.8	0.4 x 0.3	12.6	E	138	Elongated anonymous galaxy S 1'.4.
14 56 29.2	+09 21 15	IC 1078	13.5	1.1 x 0.9	13.3	Sa	0	IC 1079 NE 2'.0.
14 56 36.1	+09 22 10	IC 1079	12.9	1.7 x 1.0	13.5	E	83	Bright nucleus with bright pseudo ring; IC 1078 SW 2'.0.
14 58 52.6	+07 00 23	IC 1082	14.4	0.7 x 0.5	13.1		33	Mag 10.78v star E 2'.4.
15 02 43.4	+17 15 10	IC 1085	13.8	0.9 x 0.7	13.1		27	Mag 9.40v star W 6'.4.
15 07 36.1	+09 21 31	IC 1092	14.2	0.7 x 0.5	12.9		69	
15 07 35.7	+14 32 48	IC 1093	13.8	1.0 x 0.7	13.2	SABbc	120	Multi-galaxy system **IC 1094** N 5'.0.
15 10 56.1	+05 44 42	IC 1101	13.7	1.1 x 0.6	13.2	$S0^-$:	23	IC 1101 at center; surrounding PGC and anonymous GX mostly stellar.
15 18 15.2	+12 29 16	IC 1113	14.4	0.7 x 0.6	13.3	Sb	123	Mag 11.6 star NE 1'.1.
15 21 55.2	+08 25 24	IC 1116	12.9	1.2 x 0.8	12.7	S0?	174	Stellar **CGCG 77-91** ESE 8'.4.
15 24 59.6	+13 26 39	IC 1118	14.3	1.0 x 0.6	13.6		84	Mag 10.11v star SSW 3'.7.
15 38 51.7	+12 04 49	IC 1131	13.8	0.5 x 0.4	12.1	E1:	153	Mag 10.37v star NNE 3'.9; NGC 5970 NW 8'.3.
15 41 12.1	+15 34 25	IC 1133	14.1	1.2 x 0.4	13.2	Scd:	130	Mag 10.77v star W 2'.1.
15 45 34.8	+17 41 58	IC 1135	14.7	0.8 x 0.3	13.0	Sb	66	Mag 10.9 star NE 6'.1.
15 49 47.0	+12 23 55	IC 1141	13.9	0.7 x 0.7	13.0	SB?		Mag 6.69v star N 10'.1.
15 58 08.1	+12 04 14	IC 1149	13.5	1.2 x 1.0	13.6	Sbc	160	Mag 10.7v star SE 5'.2.
15 58 32.4	+17 26 24	IC 1151	12.9	2.5 x 0.8	13.6	SB(rs)c	28	Mag 9.37v star E 5'.8.
16 00 35.8	+15 41 06	IC 1155	14.3	0.9 x 0.7	13.6	SAB(rs)bc	9	Mag 7.93v star SW 3'.1.
16 02 08.4	+15 41 39	IC 1165	14.2	0.8 x 0.5	13.1	$SA0^-$ pec	156	Galaxy pair (**IC 1165A**, **IC 1165B**); **MCG +3-41-50** N 1'.0; **MCG +3-41-46** SW 1'.9.
16 04 13.5	+13 44 37	IC 1169	13.3	1.0 x 0.6	12.6	S?	10	Mag 10.01v star N 6'.7.
16 05 26.8	+15 01 29	IC 1174	13.4	0.9 x 0.7	12.8	SA0/a:	50	Mag 12.2 star ENE 1'.6.
16 05 33.5	+17 36 12	IC 1178	14.1	1.2 x 1.0	14.2	$S0^-$ pec?	36	Bright nucleus offset to SW; **IC 1181** on S edge.
16 05 37.0	+17 48 04	IC 1182	14.2	1.0 x 0.5	13.3	$SA0^+$ pec	81	NGC 6054 SSW 2'.4.
16 05 38.2	+17 46 01	IC 1183	14.2	0.8 x 0.4	12.9	$SAB0^-$	66	Small companion NW end; NGC 6054 W 1'.8.
16 07 58.3	+10 46 44	IC 1196	13.7	1.0 x 0.5	12.8	Sa	3	Mag 9.34v star SE 2'.4.
16 08 17.5	+07 32 17	IC 1197	13.4	2.9 x 0.5	13.7	Scd:	56	Mag 10.09v star SW 7'.3.
16 08 36.6	+12 19 47	IC 1198	14.3	0.8 x 0.3	12.6	S?	105	Mag 7.29v star WNW 11'.9.
16 10 34.6	+10 02 24	IC 1199	13.6	1.3 x 0.5	13.0	S?	157	Lies 1'.9 W of mag 7,53v star.
16 14 16.0	+09 32 11	IC 1205	13.7	0.5 x 0.5	12.1	Sa		Located 2'.3 E of mag 8.47v star.
16 15 13.1	+11 17 45	IC 1206	13.6	1.2 x 0.8	13.4	Sab	2	**CGCG 79-86** SW 1'.3.
14 58 26.7	+08 12 29	MCG +1-38-24	14.4	0.3 x 0.3	11.7	S0		Mag 10.43v star SE 3'.0.
15 05 01.1	+05 47 46	MCG +1-38-28	14.3	1.0 x 0.4	13.2	Sb	21	
15 10 18.6	+07 42 16	MCG +1-39-3	14.5	0.8 x 0.4	13.1	Sb	69	**CGCG 49-18** SW 5'.0.
15 15 31.7	+06 22 10	MCG +1-39-10	15.1	0.5 x 0.5	13.4	Sb		Mag 6.46v star NE 7'.9.
15 17 12.8	+07 01 39	MCG +1-39-13	15.0	0.4 x 0.4	12.9	S0		Faint anonymous galaxy SE 0'.6; mag 12.1 star NE 0'.8.
15 17 23.8	+07 53 27	MCG +1-39-14	14.1	0.6 x 0.6	12.8			**CGCG 49-102** SE 2'.3.
15 18 08.5	+05 18 36	MCG +1-39-16	14.6	0.7 x 0.7	13.7	Sc		

(Continued from previous page)

GALAXIES

RA h m s	Dec ° ′ ″	Name	Mag (V)	Dim ′ Maj x min	SB	Type Class	PA	Notes
15 39 05.4	+05 34 14	MCG +1-40-10	14.3	0.5 x 0.5	12.8	E		
16 02 11.7	+07 05 09	MCG +1-41-6	14.0	1.2 x 0.6	13.5	S0	12	Mag 11.2 star SW 1′1.
14 56 02.6	+09 28 48	MCG +2-38-24	15.4	0.5 x 0.2	12.7		33	Located between a pair of mag 13.5 stars SE and NW.
14 57 17.9	+11 40 07	MCG +2-38-29	14.2	0.8 x 0.4	12.8	Sb	126	Mag 15.7 star on S edge; located between two mag 13 stars.
14 57 38.0	+09 08 44	MCG +2-38-30	14.6	0.8 x 0.4	13.2	Sa	12	Mag 12.5 star W 1′8.
14 59 38.3	+13 12 51	MCG +2-38-31	14.3	0.5 x 0.4	12.4	S	36	**MCG +2-38-32** N 0′8; mag 9.08v star NNW 1′6.
15 02 16.5	+11 55 02	MCG +2-38-33	14.1	0.9 x 0.3	12.5	Sa		Mag 11.8 star W 2′2.
15 05 56.2	+12 43 43	MCG +2-38-38	14.8	0.4 x 0.2	12.0	S0	45	
15 05 55.0	+12 44 39	MCG +2-38-39	14.8	0.4 x 0.2	11.9	S0	126	
15 06 42.3	+12 40 32	MCG +2-38-42	14.8	0.9 x 0.7	14.1	SBc	105	Mag 13.1 star ESE 1′3.
15 08 12.6	+10 16 42	MCG +2-39-1	14.0	0.8 x 0.5	12.9	S0	177	Mag 10.75v star SE 2′2.
15 11 33.0	+09 36 45	MCG +2-39-6	13.9	0.8 x 0.6	13.0	Sab	15	**CGCG 77-36** N 8′2.
15 14 36.1	+08 20 00	MCG +2-39-8	14.4	0.9 x 0.3	12.9	S0	153	Stellar **CGCG 49-69** SW 4′8.
15 18 00.2	+13 31 34	MCG +2-39-10	14.1	0.9 x 0.6	13.3	Sb	72	
15 18 21.1	+12 58 13	MCG +2-39-13	14.8	0.6 x 0.4	13.1	Pec	63	MCG +2-39-13A ESE 2′4; CGCG 77-63 W 5′8.
15 18 26.6	+12 56 14	MCG +2-39-13A	14.5	1.1 x 0.5	13.7	S0a	156	MCG +2-39-13 WNW 2′4.
15 22 07.9	+13 55 38	MCG +2-39-18	14.2	0.8 x 0.7	13.4	S	81	
15 23 05.3	+08 36 35	MCG +2-39-20	13.7	1.1 x 0.9	13.5	S?	36	Stellar **CGCG 77-96** SW edge; **MCG +2-39-23** NE 2′0; **MCG +2-39-24** NE 3′0; **MCG +2-39-22** SW 3′2.
15 23 35.3	+09 20 44	MCG +2-39-28	13.7	0.5 x 0.4	12.0	E?	66	**CGCG 77-116** NE 2′8.
15 39 50.8	+14 11 11	MCG +2-40-9	14.0	0.6 x 0.6	12.7			**CGCG 78-42** SW 5′2.
15 40 53.4	+09 44 59	MCG +2-40-10	13.7	0.5 x 0.5	12.0	Sa		
15 51 45.2	+12 44 33	MCG +2-40-15	14.4	0.9 x 0.4	13.1	Spiral	78	
16 00 55.5	+12 40 44	MCG +2-41-2	14.9	0.7 x 0.4	13.4	Spiral	66	Almost stellar; stellar galaxy **KUG 1559+128** E 8′4.
16 01 36.3	+12 21 34	MCG +2-41-3	13.7	0.9 x 0.3	12.2	S0	114	
16 05 38.1	+10 02 05	MCG +2-41-6	14.6	0.9 x 0.4	13.3	Sbc	174	Pair of mag 11-12 stars SW 2′0.
16 08 32.8	+09 36 22	MCG +2-41-10	13.9	0.7 x 0.7	13.2	E		
16 10 56.3	+13 45 07	MCG +2-41-14	13.9	1.1 x 0.7	13.4	S0	156	Mag 11.7 star SE edge.
16 12 06.0	+14 12 05	MCG +2-41-18	14.9	0.3 x 0.2	11.7			On S edge of NGC 6078.
15 01 56.1	+16 58 33	MCG +3-38-71	15.3	1.0 x 0.8	14.9	Sbc	0	Stellar nucleus, smooth envelope.
15 04 31.7	+17 18 53	MCG +3-38-81	13.6	0.9 x 0.9	13.2	SBbc		
15 04 37.6	+17 41 11	MCG +3-38-83	15.0	0.9 x 0.5	14.0	Sbc	153	Mag 11.2 star NW edge; **MCG +3-39-79** W 1′9; **MCG +3-38-80** WSW 2′0.
15 04 40.3	+16 09 04	MCG +3-38-84	14.9	0.5 x 0.5	13.2	SBb		
15 07 23.3	+16 22 38	MCG +3-39-1	15.2	0.6 x 0.6	13.9	Sb		
15 17 54.5	+16 18 40	MCG +3-39-19	14.8	0.6 x 0.6	13.5	Irr		Mag 11.6 star ESE 8′4.
15 22 55.8	+14 59 53	MCG +3-39-23	14.1	0.8 x 0.8	13.5	Sab		Strong E-W bar; mag 12.5 star NW 3′0.
15 31 39.3	+15 35 33	MCG +3-40-1	15.9	0.5 x 0.5	14.2	Sbc		
15 35 52.5	+14 31 01	MCG +3-40-9	14.3	0.5 x 0.4	12.5	S0	129	Mag 11.7 E edge.
15 36 36.3	+17 20 15	MCG +3-40-12	14.7	0.7 x 0.4	13.2	Sa	96	
15 39 32.9	+17 25 52	MCG +3-40-18	14.7	0.5 x 0.4	12.8	S0	126	Mag 12.3 star NE 2′4; this star has a faint, elongated anonymous galaxy on it's E edge.
15 56 23.6	+16 31 23	MCG +3-41-3	13.3	1.2 x 0.8	13.1	SA0$^+$	33	Located between a pair of mag 11.5 stars.
16 00 53.8	+16 20 06	MCG +3-41-28	14.4	0.5 x 0.4	12.5	SA(s)b	114	Pair with MCG +3-41-29 N.
16 00 54.6	+16 20 45	MCG +3-41-29	15.2	0.8 x 0.5	14.0	SAB(s)c	93	Pair with MCG +3-41-28 S.
16 02 04.3	+17 04 32	MCG +3-41-47	13.8	0.7 x 0.5	12.7	E	141	Mag 13.2 star NW 1′5.
16 03 33.1	+16 19 20	MCG +3-41-61	13.6	0.9 x 0.7	13.0	SA0°	69	**CGCG 108-87** E 2′5.
16 03 53.8	+14 38 09	MCG +3-41-64	14.3	0.8 x 0.7	13.5	SAB(s)c	0	
16 04 50.0	+16 35 07	MCG +3-41-83	13.8	0.8 x 0.8	13.4	E		
16 05 29.2	+16 26 09	MCG +3-41-95	14.3	0.8 x 0.8	13.8	E		**UGC 10187** NW 0′8.
16 06 25.8	+15 41 35	MCG +3-41-124	14.5	0.6 x 0.4	13.0	E	93	UGC 10201 on S edge; numerous stellar anonymous galaxies W and S.
16 10 51.3	+17 03 23	MCG +3-41-141	14.4	0.9 x 0.4	13.2	Sb	108	Mag 11.3 star ESE 1′5.
14 57 35.8	+08 17 04	NGC 5790	13.6	1.1 x 0.9	13.5	(R)SA0/a	77	Pair of mag 9 stars SE 3′0.
15 04 40.6	+12 37 57	NGC 5837	14.0	1.0 x 0.6	13.3	S	25	
15 06 22.4	+06 22 48	NGC 5847	14.9	0.7 x 0.4	13.3	Sb	165	Located 2′9 NE of mag 10.37V star.
15 06 53.3	+12 51 29	NGC 5851	14.1	1.0 x 0.3	12.7	S?	43	**MCG +2-38-43** WSW 1′7.
15 06 56.5	+12 50 50	NGC 5852	13.6	1.1 x 0.6	13.0	S?	123	Star SE of nucleus.
15 21 52.0	+07 42 32	NGC 5920	13.6	1.1 x 0.6	13.0	S0:	105	**PGC 54838** N 1′6.
15 21 56.7	+05 04 15	NGC 5921	10.8	4.9 x 4.0	13.9	SB(r)bc I-II	130	Bright nucleus, many branching arms, numerous knots.
15 23 25.0	+12 42 54	NGC 5926	13.6	0.8 x 0.6	12.6	S?	90	Mag 10.11v star NW 2′5.
15 29 29.7	+07 34 24	NGC 5931	14.0	0.9 x 0.4	12.8	S0	48	**CGCG 49-173** W 8′9.
15 30 00.8	+12 59 22	NGC 5936	12.5	1.4 x 1.3	13.0	SB(rs)b I-II	36	Dark patch S.
15 31 18.1	+07 27 27	NGC 5940	13.4	0.8 x 0.8	12.8	SBab		Located on N edge of small galaxy cluster **A 2085**; **CGCG 50-12** E 4′9.
15 31 36.9	+07 18 40	NGC 5941	14.4	0.5 x 0.5	12.8	S0?		= **Hickson 76C**. **Hickson 76B** NNE 1′8; **Hickson 76F** NNE 1′6; **Hickson 76G** N 2′3.
15 31 42.3	+07 17 10	NGC 5942	14.8	0.4 x 0.4	12.9	E?		= **Hickson 76D**.
15 31 47.8	+07 18 23	NGC 5944	15.0	0.6 x 0.3	13.0	S?	111	= **Hickson 76A**. **Hickson 76E** 0′7 E.
15 33 43.1	+15 00 14	NGC 5951	12.7	3.5 x 0.8	13.6	SBc: sp	5	Mag 9.11v star NE 7′0.
15 34 32.5	+15 11 36	NGC 5953	12.3	1.6 x 1.3	13.0	SAa: pec	169	Star SW of bright center; NGC 5953 on NE edge; **UGC 9902** S 3′6.
15 34 35.2	+15 12 10	NGC 5954	12.2	1.3 x 0.6	11.8	SAB(rs)cd: pec	5	NGC 5953 on SW edge.
15 35 12.6	+05 03 43	NGC 5955	14.9	0.9 x 0.6	14.1	Sb-Sbc	6	**MCG +1-40-6A** NNE 2′5.
15 34 58.5	+11 45 01	NGC 5956	12.3	1.6 x 1.6	13.2	Scd?		Star or bright knot on E edge.
15 35 23.2	+12 02 50	NGC 5957	11.7	2.8 x 2.6	13.7	(R′)SAB(r)b	90	Bright nucleus in faint bar.
15 36 18.6	+05 39 52	NGC 5960	14.3	0.7 x 0.6	13.2	Sb	114	
15 36 31.5	+16 36 27	NGC 5962	11.3	3.0 x 2.1	13.2	SA(r)c II-III	110	Many small knots.
15 37 36.0	+05 58 24	NGC 5964	11.9	4.2 x 3.2	14.6	SB(rs)d	145	Elongated nucleus; many superimposed stars and/or knots.
15 38 30.0	+12 11 10	NGC 5970	11.5	2.9 x 1.9	13.2	SB(r)c II-III	88	Mag 7.44v star NNE 5′3; IC 1131 SE 8′3.
15 38 54.3	+17 01 35	NGC 5972	13.6	1.2 x 0.8	13.4	S0/a	5	
15 40 33.6	+17 07 42	NGC 5977	13.4	1.2 x 1.0	13.5	SB0	155	Almost stellar nucleus; **MCG +3-40-24** E 4′0.
15 41 30.4	+15 47 14	NGC 5980	12.7	1.9 x 0.7	12.8	S	13	Faint dark lane W.
15 42 45.7	+08 14 28	NGC 5983	13.4	1.0 x 1.0	13.4	E		Mag 10.58v star NE 2′9.
15 42 53.3	+14 13 53	NGC 5984	12.5	2.9 x 0.8	13.2	SB(rs)d: III	144	
15 44 33.8	+10 17 35	NGC 5988	13.8	1.2 x 1.0	13.8	Scd:	115	
15 46 53.4	+17 52 19	NGC 5994	15.4	0.5 x 0.3	13.2	SB?	87	Located SW end of NGC 5996.
15 46 57.9	+17 52 52	NGC 5996	12.8	1.8 x 1.0	13.3	S?	33	Multi-galaxy system. Bright, narrow extension S. NGC 5994 W of S end.
15 47 27.7	+08 19 13	NGC 5997	15.2	0.3 x 0.3	12.4			Almost stellar.
15 53 02.8	+12 00 15	NGC 6006	15.1	0.6 x 0.4	13.4	Pair	162	
15 53 23.2	+11 57 32	NGC 6007	13.2	1.7 x 1.2	13.8	SABbc: I	65	
15 53 24.2	+12 03 28	NGC 6009	14.5	0.5 x 0.2	11.9		168	
15 54 13.9	+14 36 09	NGC 6012	12.0	2.1 x 1.5	13.1	(R)SB(r)ab:	153	Long SE-NW bar; mag 9.42v star NE; mag 10.71v star S.
15 55 57.5	+05 55 55	NGC 6014	12.2	1.7 x 1.6	13.2	S0	171	Pair of stars or bright knots N edge; mag 7.38v star E 10′9.
15 57 15.5	+05 59 52	NGC 6017	13.2	0.8 x 0.7	12.4	S?	140	

RA h m s	Dec ° ′ ″	Name	Mag (V)	Dim ′ Maj x min	SB	Type Class	PA	Notes
15 57 29.7	+15 52 19	NGC 6018	13.4	1.4 x 0.7	13.3	(R)SAB(s)0$^+$:	69	Bright core, fainter outer envelope.
15 57 30.8	+15 57 25	NGC 6021	13.1	1.4 x 0.8	13.2	E$^+$	160	Several faint stars superimposed.
15 57 47.8	+16 16 55	NGC 6022	14.7	0.7 x 0.5	13.4	(R′)SB(s)bc	75	Short N-S bar, dark patches E and W.
15 57 49.8	+16 18 36	NGC 6023	13.1	1.4 x 1.0	13.5	E	70	Bright knot on E edge; very small anonymous galaxy on W edge.
16 01 58.9	+12 34 25	NGC 6029	14.7	1.2 x 0.7	14.4	Compact	9	Pair of faint stars N of nucleus.
16 01 51.4	+17 57 23	NGC 6030	12.8	1.1 x 0.8	12.5	S0°	37	Elongated anonymous galaxies N 2'.3 and W 2'.4.
16 03 32.1	+17 11 55	NGC 6034	13.5	1.1 x 0.8	13.4	E$^+$	54	
16 04 26.5	+17 44 53	NGC 6040	14.2	1.4 x 0.5	13.6	SAB(s)c	42	NGC 6040B on S edge.
16 04 26.5	+17 44 28	NGC 6040B	14.0	0.8 x 0.8	13.4	SA0$^+$ pec		On S edge of NGC 6040.
16 04 35.9	+17 43 19	NGC 6041A	13.3	1.2 x 1.1	13.5	SB0°	36	Stellar **NGC** 6041A SW of nucleus; **IC 1170** W 0'.9.
16 04 35.0	+17 43 00	NGC 6041B	15.6	0.3 x 0.3	12.9	SB0$^-$		Stellar; located SW edge of **NGC 6041** core.
16 04 39.7	+17 42 01	NGC 6042	13.9	0.8 x 0.7	13.2	SA0$^-$	60	Located 1'.7 NE of mag 10.72v star.
16 05 01.5	+17 46 27	NGC 6043A	14.3	0.5 x 0.5	12.7	SAB0$^-$		
16 04 59.7	+17 52 13	NGC 6044	14.3	0.6 x 0.6	13.0	SA0°		Located 2'.9 SW of mag 10.91v star.
16 05 07.9	+17 45 25	NGC 6045	13.9	1.3 x 0.3	12.8	SB(s)c sp	79	Faint, elongated, anonymous galaxy E end.
16 05 09.1	+17 43 46	NGC 6047	13.5	0.8 x 0.8	13.1	E$^+$		Star on NW edge; **UGC 10190** SE 4'.6.
16 05 23.4	+17 45 26	NGC 6050	14.7	0.8 x 0.5	13.6	SA(s)c	132	Double galaxy, small companion SW edge.
16 05 30.7	+17 46 03	NGC 6054	15.2	0.7 x 0.4	13.7	(R′)SAB(s)b	21	IC 1182 NE 2'.5; IC 1183 E 1'.8.
16 05 31.4	+17 57 46	NGC 6056	13.9	0.9 x 0.5	12.9	SB(s)0$^+$	56	**PGC 57052** W 2'.2; **PGC 57070** S 2'.1.
16 07 13.0	+07 58 43	NGC 6063	13.1	1.7 x 0.9	13.4	Scd:	159	Stellar nucleus in smooth envelope.
16 07 22.9	+13 53 16	NGC 6065	14.0	0.8 x 0.6	13.1	S0	6	**CGCG 79-52** SE 3'.0.
16 07 35.4	+13 56 38	NGC 6066	14.0	0.7 x 0.6	12.9		135	**CGCG 79-56** E 7'.3.
16 10 10.9	+16 41 55	NGC 6073	13.5	1.3 x 0.7	13.3	Sc:	130	Located 2'.6 NE of mag 10.43v star.
16 11 17.4	+14 15 30	NGC 6074	14.4	0.4 x 0.4	12.3			**MCG +2-41-15** on SW edge.
16 12 05.6	+14 12 29	NGC 6078	15.2	0.5 x 0.5	13.8	E		Stellar MCG +2-41-18 on S edge.
16 12 57.0	+09 51 58	NGC 6081	13.1	1.8 x 0.6	13.0	S0	131	Very faint dark lane NW .
16 13 12.8	+14 11 04	NGC 6083	14.6	0.9 x 0.5	13.6	Sb	42	**CGCG 79-79** NNW 7'.4.
16 14 16.8	+17 45 24	NGC 6084	13.9	1.0 x 0.6	13.2	Sa sp	30	Mag 9.98v star SW 1'.9.
14 56 47.8	+09 30 34	UGC 9614	14.7	1.5 x 1.3	15.3	Im:	5	Dwarf, very low surface brightness.
14 56 53.2	+09 16 15	UGC 9616	13.5	0.8 x 0.6	12.5	Sab	159	Mag 12.1 star N 3'.2.
14 57 44.9	+06 37 34	UGC 9625	13.5	1.5 x 0.5	13.1	SBab:	178	
14 59 27.6	+16 38 42	UGC 9640	13.0	1.3 x 1.1	13.4	E	110	Elongated anonymous galaxy NE 1'.9.
15 00 56.7	+11 31 00	UGC 9654	13.8	1.0 x 0.7	13.3	Sab	72	**CGCG 76-128** W 6'.1.
15 02 19.9	+05 38 46	UGC 9667	13.1	1.8 x 0.9	13.5	S0/a	87	**CGCG 48-93** pair with mag 14.1 star SSE 6'.6.
15 03 22.1	+10 39 19	UGC 9675	13.7	1.1 x 0.4	12.7	SB?	90	**CGCG 76-140** SSE 3'.7.
15 04 47.4	+08 10 14	UGC 9685	14.0	0.9 x 0.7	13.3	Scd:	109	Mag 9.85v star W 3'.5.
15 05 02.6	+07 50 03	UGC 9689	14.0	0.9 x 0.7	13.3	Sbc	30	Faint stellar nucleus, uniform envelope; mag 10.83v star SE 1'.6.
15 05 30.6	+08 31 25	UGC 9696	13.6	0.8 x 0.7	12.8	Sa	159	
15 06 32.5	+09 26 58	UGC 9708	13.6	1.0 x 0.9	13.3	Sb II	168	Elongated **UGC 9711** on SE edge.
15 06 45.2	+12 33 40	UGC 9712	14.3	1.4 x 0.3	13.2	Scd:	82	
15 07 24.4	+12 34 46	UGC 9721	13.9	0.9 x 0.7	13.2	S?	95	Almost stellar nucleus, uniform envelope; mag 11.9 star E 2'.1.
15 11 13.2	+10 27 01	UGC 9755	13.6	1.0 x 0.8	13.2	Scd:	35	Very slender **MCG +2-39-5** on N edge.
15 11 37.2	+05 23 48	UGC 9757	13.8	1.4 x 0.6	13.5	Sbc III	60	Mag 11.1 star SW 2'.9.
15 11 28.1	+13 29 00	UGC 9758	13.2	1.3 x 1.3	13.6	S0		Mag 12.3 star on S edge.
15 14 49.3	+05 32 50	UGC 9781	13.8	1.3 x 1.3	14.2	Sbc		Mag 12.5 star superimposed W of nucleus.
15 16 10.3	+10 30 21	UGC 9794	12.9	2.8 x 0.8	13.6	SBdm:	42	Irregular, disrupted; mag 12.2 star N 1'.7.
15 16 41.9	+07 23 47	UGC 9798	14.0	1.0 x 0.5	13.1	S0/a	168	Mag 14.1 star on N edge.
15 16 44.5	+07 01 17	UGC 9799	13.0	1.8 x 0.9	13.5	E	30	**CGCG 49-89** S 1'.1; **PGC 54521** W 0'.8; **CGCG 49-91** N 2'.0.
15 17 59.0	+10 53 42	UGC 9807	13.8	1.3 x 0.5	13.2	Sab	105	Mag 12.2 star WSW 1'.2.
15 18 08.8	+13 49 40	UGC 9808	13.6	1.1 x 1.0	13.6	SABb II-III	65	Mag 12.4 star superimposed E edge.
15 19 25.4	+11 03 17	UGC 9814	14.1	1.0 x 0.4	12.9	Sdm:	45	
15 21 40.4	+08 24 52	UGC 9821	13.3	1.1 x 1.1	13.4	SB(r)b		Faint N-S bar; mag 8.75v star N 2'.1.
15 25 12.3	+07 09 14	UGC 9838	14.1	1.3 x 0.4	13.2	Sab	13	**CGCG 49-160** SE 5'.4.
15 25 52.7	+07 49 11	UGC 9844	13.7	1.5 x 0.5	13.2	SB(s)b	67	Bright NE-SW bar.
15 26 05.5	+09 12 13	UGC 9845	14.7	1.5 x 0.3	13.7	Scd:	114	Mag 9.73v star E 2'.1.
15 26 02.6	+16 19 18	UGC 9846	13.9	0.9 x 0.8	13.4	SABc I	103	moderately dark patches E and W of nucleus.
15 29 25.8	+05 05 03	UGC 9864	13.5	1.7 x 0.6	13.4	S0?	25	**CGCG 49-175** SW 2'.9.
15 33 51.9	+11 00 36	UGC 9897	13.8	0.8 x 0.6	12.8	SB(r)a	75	
15 34 08.1	+14 23 44	UGC 9900	13.9	1.0 x 0.9	13.6	SAdm:	3	Uniform surface brightness.
15 34 27.0	+12 16 13	UGC 9901	14.3	1.5 x 0.2	12.8	Sbc	178	
15 35 10.5	+16 32 56	UGC 9912	13.4	1.5 x 1.4	14.1	SBdm	75	
15 35 39.7	+12 36 19	UGC 9919	14.6	1.5 x 0.3	13.6	Scd:	171	
15 36 31.9	+16 26 21	UGC 9925	14.0	1.3 x 0.6	13.6	Scd:	10	Uniform surface brightness.
15 38 22.1	+12 57 36	UGC 9941	13.7	1.5 x 1.4	14.4	Im:	159	Dwarf, low surface brightness.
15 39 20.0	+15 23 02	UGC 9951	14.1	1.4 x 0.7	13.9	Sd	132	Mag 7.97v star W 8'.6.
15 40 22.5	+07 16 57	UGC 9964	13.4	1.2 x 1.0	13.4	(R)SB(r)0/a	102	Brighter N-S bar, faint envelope.
15 41 50.5	+05 51 09	UGC 9976	14.0	1.0 x 0.6	13.3	SB(r)b	144	Stellar nucleus.
15 42 12.0	+06 38 38	UGC 9978	14.1	1.2 x 0.5	13.4	Scd:	35	Very knotty.
15 43 24.1	+14 26 09	UGC 9991	14.0	1.5 x 0.5	13.5	Scd:	70	
15 44 29.3	+11 32 55	UGC 9996	14.6	1.5 x 0.2	13.1	Sb III	12	Pair of stars, mags 11.9 and 12.7, S 2'.0.
15 45 44.2	+12 30 35	UGC 10014	13.9	1.2 x 1.0	13.9	Im	24	Low surface brightness dwarf.
15 46 09.4	+06 53 51	UGC 10023	14.2	1.0 x 0.6	13.4	Im	78	Patchy.
15 46 54.3	+05 53 27	UGC 10029	13.4	1.3 x 0.7	13.1	S0	100	**CGCG 50-104** ENE 5'.6.
15 48 12.8	+11 16 43	UGC 10037	13.8	1.1 x 0.8	13.5	(R′)SB(r)b:	36	bright nucleus, moderately bright SE-NW bar.
15 49 01.2	+05 11 08	UGC 10041	12.9	2.8 x 1.6	14.4	SABdm	165	Several bright knots, mag 12.7 star on NE edge; very faint, elongated anonymous galaxy NW edge.
15 48 57.1	+07 13 19	UGC 10042	14.2	1.5 x 0.2	12.7	Scd:	169	Mag 11.4 star NW edge.
15 51 15.2	+16 19 38	UGC 10061	13.5	1.9 x 1.2	14.3	Im V	27	Very filamentary; mag 11.7 star E edge; triangle of one mag 13.5 and two mag 14.7 stars superimposed.
15 52 03.7	+12 53 54	UGC 10068	13.5	1.0 x 0.8	13.1	S0	100	
15 53 05.9	+13 14 08	UGC 10077	14.0	1.0 x 0.5	13.1	Sbc	158	Faint, uniform envelope; located 2'.7 NW of mag 6.09v star.
15 55 55.5	+17 09 47	UGC 10093	14.2	1.3 x 0.9	14.2	SAB(s)c	111	Mag 13.0 star WNW 0'.9.
15 58 13.2	+13 10 03	UGC 10111	15.1	1.7 x 0.2	13.8	Pec?	37	Faint galaxy **KUG 1556+132** SE 7'.6.
16 01 20.0	+08 50 10	UGC 10130	13.6	1.3 x 0.7	13.3	Sa	170	Mag 8.99v star NE 2'.0.
16 01 21.2	+16 18 25	UGC 10134	14.5	0.9 x 0.7	13.8	SB(rs)ab	129	Faint N-S bar.
16 01 49.8	+08 08 52	UGC 10137	13.9	1.2 x 0.6	13.4	Sc	40	Mag 14.7 star on SW edge.
16 02 17.1	+15 58 26	UGC 10143	13.1	2.0 x 1.1	13.9	E$^+$	10	**MCG +3-41-52** S 2'.4; pair of anonymous galaxies S 3'.7; stellar **PGC 56782** N 1'.7.
16 02 19.9	+16 20 44	UGC 10144	13.1	1.0 x 1.0	13.2	E		**MCG +3-41-56** N 1'.3; mag 10.08v star SE 1'.9.
16 03 05.2	+05 06 26	UGC 10146	14.5	0.7 x 0.6	13.4	Sd:	117	Faint stellar nucleus, uniform envelope.
16 03 16.4	+05 38 24	UGC 10147	13.8	0.9 x 0.6	13.0	Double System	81	Small anonymous galaxy N 0'.7.

(Continued from previous page)

88 GALAXIES 88

RA h m s	Dec ° ′ ″	Name	Mag (V)	Dim ′ Maj x min	SB	Type Class	PA	Notes
16 03 55.5	+11 43 42	UGC 10158	13.6	1.1 x 0.7	13.2	S0	20	Mag 11.7 star W 2'.3.
16 04 28.2	+14 46 52	UGC 10164	13.7	1.0 x 0.7	13.2	S0⁻:	24	Stellar nucleus, uniform envelope; UGC 10169 N 2'.4.
16 04 31.9	+14 49 07	UGC 10169	13.7	0.7 x 0.4	12.3	S0 pec	42	Long, faint streamer extends NE; UGC 10164 and mag 11.3 star S 1'.6.
16 05 02.2	+13 42 05	UGC 10176	14.6	1.6 x 0.2	13.2	Scd:	66	Mag 13.1 star N 1'.2; mag 10.4 star SW 10'.7.
16 06 25.5	+15 41 07	UGC 10201	13.3	0.6 x 0.6	12.3	E⁺0		MCG +3-41-124 on N edge; **MCG +3-41-130** SE 10'.7.
16 06 42.3	+16 19 11	UGC 10204	13.0	1.0 x 0.6	12.3	SAB(s)0/a	93	Mag 7.85 star NE 2'.8.
16 07 25.2	+10 25 29	UGC 10213	13.5	1.0 x 0.8	13.1	S?	120	**CGCG 79-49** S 3'.7.
16 09 20.8	+08 45 44	UGC 10225	13.2	1.7 x 1.6	14.1	SBdm	18	Bright, elongated core, several knots S.
16 10 26.6	+12 18 20	UGC 10238	13.6	1.5 x 0.7	13.5	(R')SB(s)b	70	Large E-W bar, faint envelope.
16 10 26.0	+12 49 27	UGC 10239	13.8	0.7 x 0.6	12.7	SBa	87	
16 11 17.1	+13 51 57	UGC 10249	14.0	1.0 x 0.5	13.1	SBab	76	
16 14 03.4	+14 16 53	UGC 10287	13.3	1.5 x 1.1	13.7	SBb	15	Two main arms; mag 13.2 star on E edge; **CGCG 79-82** S 8'.3.

GALAXY CLUSTERS

RA h m s	Dec ° ′ ″	Name	Mag 10th brightest	No. Gal	Diam ′	Notes
15 03 42.0	+07 55 00	A 2020	16.0	47	26	All members anonymous, faint and stellar.
15 09 36.0	+07 31 00	A 2028	15.7	50	19	All members stellar to almost stellar.
15 11 00.0	+05 45 00	A 2029	16.0	82	16	
15 11 30.0	+06 21 00	A 2033	15.7	40	18	**UGC 9756** and **PGC 54209** at center; all other members anonymous, faint and stellar.
15 12 48.0	+07 25 00	A 2040	15.7	52	25	Mag 8.70v star at center; **UGC 9767** 2'.2 N of this star; most members anonymous and stellar.
15 16 48.0	+07 00 00	A 2052	15.0	41	22	UGC 9799 at center; many CGCG, PGC and anonymous GX overall.
15 18 48.0	+06 12 00	A 2055	16.0	40	19	All member anonymous, faint and stellar.
15 23 00.0	+08 38 00	A 2063	15.1	63	22	MCG +2-39-20 at center; many CGCG and MCG galaxies in this cluster.
15 40 06.0	+17 53 00	A 2108	15.7	45	13	All members anonymous, faint and stellar/almost stellar.
16 02 18.0	+15 53 00	A 2147	13.8	52	39	Line of numerous anonymous GX stretch from UGC 10143 S to IC 1165.
16 05 12.0	+17 44 00	A 2151	13.8	87	56	**Hercules Galaxy Cluster**. Somewhat strong concentration at center; many GX spread over whole area of cluster.
16 05 24.0	+16 26 00	A 2152	13.8	60	37	Many UGC, MCG and CGCG galaxies in this cluster.
16 11 36.0	+16 58 00	A 2159	15.9	34	16	Most members anonymous, faint and stellar.

GLOBULAR CLUSTERS

RA h m s	Dec ° ′ ″	Name	Total V m	B ★ V m	HB V m	Diam ′	Conc. Class Low = 12 High = 1	Notes
16 11 00.3	+14 57 49	Pal 14	14.7	17.6	20.0	2.5		

PLANETARY NEBULAE

RA h m s	Dec ° ′ ″	Name	Diam ″	Mag (P)	Mag (V)	Mag cent ★	Alt Name	Notes
16 11 44.5	+12 04 17	IC 4593	42	10.9	10.7	11.2	PK 25+40.1	Mag 9.42v star NW 5'.2.

89 GALAXIES 89

RA h m s	Dec ° ′ ″	Name	Mag (V)	Dim ′ Maj x min	SB	Type Class	PA	Notes
13 52 20.4	+07 47 55	CGCG 45-125	14.3	0.4 x 0.3	11.8		165	Elongated **CGCG 45-126** NE 4'.6.
13 55 28.8	+06 35 43	CGCG 46-2	15.1	0.4 x 0.2	12.2	Compact	3	= **Mkn 1366**.
13 55 45.0	+08 11 16	CGCG 46-5	14.1	0.7 x 0.3	12.3		147	**CGCG 46-6** E 2'.9; mag 9.49v star SW 4'.7.
13 56 47.7	+05 31 32	CGCG 46-12	14.5	0.3 x 0.2	11.3		177	Mag 13.3 star NW 1'.9.
13 56 55.7	+05 09 01	CGCG 46-13	15.1	0.5 x 0.4	13.2		126	NGC 5373 NNE 6'.8; NGC 5364 SW 13'.6.
13 57 47.7	+07 23 44	CGCG 46-17	14.1	0.8 x 0.7	13.3	SBbc		**CGCG 46-19** NE 4'.1; mag 7.37v star NE 5'.4.
13 58 20.3	+07 13 30	CGCG 46-25	14.3	0.8 x 0.6	13.3	SBc	177	UGC 8896 E 4'.6; **UGC 8881** SW 6'.6.
14 00 18.6	+05 02 38	CGCG 46-33	14.7	0.6 x 0.3	12.6		36	Mag 12.0 star W 5'.6; mag 9.25v star SW 10'.0.
14 08 11.4	+05 48 37	CGCG 46-56	14.7	0.7 x 0.4	13.1		12	Mag 9.61v star E 7'.3.
14 17 25.2	+08 02 57	CGCG 46-81	14.6	0.3 x 0.3	11.8			In galaxy cluster A 1890, many faint anonymous galaxies WSW.
14 18 22.7	+08 05 49	CGCG 46-91	16.4	0.4 x 0.4	14.2			NGC 5535 WNW 12'.1.
14 20 39.9	+05 25 35	CGCG 47-14	14.8	0.8 x 0.3	13.1		96	Pair of E-W mag 10.31v, 10.61v stars E 4'.8; CGCG 47-15 N 3'.6.
14 20 42.4	+05 29 06	CGCG 47-15	15.3	0.6 x 0.2	12.9		90	CGCG 47-14 S 3'.6.
14 25 55.7	+06 27 20	CGCG 47-37	14.3	0.4 x 0.2	11.4		81	Located 1'.1 S of mag 8.74v star.
14 26 02.4	+08 06 38	CGCG 47-39	14.9	0.6 x 0.6	13.6			Slightly elongated center, faint envelope; mag 13.1 star NW 0'.9.
14 27 35.8	+08 07 26	CGCG 47-50	15.1	0.8 x 0.3	13.4		81	Mag 13.4 star SW 1'.6.
14 28 21.5	+05 02 28	CGCG 47-56	15.0	0.9 x 0.3	13.4		117	Mag 10.66v star NW 6'.7.
14 29 06.2	+07 50 43	CGCG 47-60	15.2	0.4 x 0.4	13.0			Forms a triangle with mag 12.1 star NW 3'.7 and mag 11.1 stars WSW 6'.4.
14 36 51.3	+05 57 38	CGCG 47-116	15.4	0.4 x 0.3	13.0		9	Mag 11.4 star SE 1'.7.
14 41 07.9	+06 02 32	CGCG 47-142	15.1	0.8 x 0.4	13.7		144	UGC 9471 SE 10'.3.
14 48 45.9	+08 05 59	CGCG 48-19	15.4	0.5 x 0.5	13.7	Sa		Mag 11.1 star SW 4'.8.
14 50 14.5	+06 30 26	CGCG 48-22	15.0	0.5 x 0.5	13.4			Mag 12.1 star SW 6'.7.
14 52 38.0	+06 15 34	CGCG 48-43	15.0	0.7 x 0.4	13.5		177	Mag 13.5 star S 0'.7; mag 11.2 star S 7'.6.
14 53 22.4	+08 12 17	CGCG 48-55	14.7	0.8 x 0.3	13.0	S0	12	Mag 14.4 star SE 1'.0; mag 10.71v star E 7'.2.
14 55 21.3	+05 17 50	CGCG 48-70	15.2	0.5 x 0.3	13.0		90	Mag 10.8 star E 2'.2.
13 46 10.0	+09 17 29	CGCG 73-71	14.9	0.6 x 0.3	12.8		63	Mag 11.08v star 1'.6 SE.
13 46 50.3	+11 37 12	CGCG 73-72	14.3	0.7 x 0.3	12.5	Sab	162	= **Mkn 1360**; **CGCG 73-73** E 0'.9.
13 56 55.4	+14 08 28	CGCG 74-13	15.2	0.9 x 0.5	14.2	Spiral	0	Mag 12.6 star NW 7'.1.
13 58 22.0	+08 15 14	CGCG 74-20	14.3	0.7 x 0.5	13.0		0	Mag 13.9 star with small, elongated anonymous galaxy NW 2'.3.
13 59 42.8	+12 44 11	CGCG 74-28	14.5	0.7 x 0.5	13.4	E	81	Mag 12.0 star NNW 2'.8; UGC 8907 NW 5'.0.
14 00 04.4	+12 06 38	CGCG 74-30	14.1	1.0 x 0.4	13.0		114	**CGCG 74-29** SSE 4'.1; mag 10.69v star E 7'.5.
14 01 44.1	+09 52 03	CGCG 74-43	14.3	0.9 x 0.4	13.0		105	Mag 11.4 star N 0'.9.
14 03 04.6	+08 56 48	CGCG 74-64	14.2	0.5 x 0.5	12.5			MCG +2-36-21 SSE 3'.7.
14 04 11.9	+12 42 23	CGCG 74-76	14.8	0.7 x 0.5	13.5		27	Mag 7.01v star WNW 4'.2.
14 04 09.1	+13 04 01	CGCG 74-77	14.6	0.6 x 0.3	12.6	Spiral	120	Mag 11.2 star SE 1'.8.
14 04 20.8	+12 00 55	CGCG 74-78	14.2	0.7 x 0.3	12.4		63	Very faint, elongated, anonymous galaxy SW 1'.5.
14 05 16.5	+10 25 50	CGCG 74-94	13.9	0.8 x 0.4	12.5		132	Mag 12.9 star W 5'.9; small "check mark" asterism of four mag 13-14 stars N 4'.0.
14 06 06.7	+11 47 16	CGCG 74-103	15.4	0.9 x 0.5	14.4	Spiral	72	**CGCG 74-97** WNW 8'.8; mag 9.63v star SW 12'.1.
14 09 46.9	+11 35 02	CGCG 74-124	14.7	1.0 x 0.7	14.1		57	Mag 11.3 star ENE 6'.0.

(Continued from previous page)

GALAXIES

RA h m s	Dec ° ′ ″	Name	Mag (V)	Dim ′ Maj x min	SB	Type Class	PA	Notes
14 10 04.2	+10 09 52	CGCG 74-126	14.2	0.4 x 0.4	12.0			Pair of mag 7.90v, 7.94v stars NW 9'.2.
14 10 41.5	+13 33 26	CGCG 74-129	13.7	0.8 x 0.5	12.5		93	Mag 9.26v star E 8'.0.
14 11 34.0	+08 21 24	CGCG 74-131	14.2	0.6 x 0.3	12.2		174	Mag 10.55v star NNW 3'.0.
14 12 21.1	+08 30 30	CGCG 74-135	14.3	1.0 x 0.4	13.2		171	Mag 8.48v star S 8'.6.
14 12 29.9	+08 38 50	CGCG 74-136	14.2	0.9 x 0.6	13.5	S0	135	Mag 13.4 star W 1'.1; UGC 9084 W 4'.3.
14 17 39.7	+12 57 29	CGCG 74-159	14.2	0.9 x 0.5	13.2		93	Mag 13.8 star SW 1'.2; mag 11.6 star W 9'.2.
14 26 58.2	+12 53 52	CGCG 75-34	14.7	0.9 x 0.4	13.4		27	Mag 11.7 star ESE 2'.9; mag 11.5 star S 3'.7.
14 28 01.2	+11 41 24	CGCG 75-41	14.7	0.6 x 0.4	13.0		90	Mag 13.6 star NE 1'.3; mag 13.4 star W 2'.3.
14 29 50.8	+13 47 56	CGCG 75-56	15.3	0.8 x 0.2	13.2	Spiral	144	Mag 10.17v star NE 6'.9.
14 34 11.7	+12 42 49	CGCG 75-74	14.2	0.8 x 0.6	13.3		45	**CGCG 75-73** S 1'.3.
14 35 23.9	+09 27 53	CGCG 75-81	15.0	0.7 x 0.3	13.2	Sa	33	Mag 13.0 star N 1'.3.
14 40 45.7	+08 27 01	CGCG 75-108	14.6	0.9 x 0.5	13.6		162	Knot or very small galaxy S edge.
14 41 20.4	+10 47 48	CGCG 75-110	14.5	1.1 x 0.6	13.9		174	Mag 112.3 star S 3'.4; mag 9.42v star NE 10'.5.
14 43 50.4	+11 36 33	CGCG 76-10	14.9	0.7 x 0.5	13.7		171	Mag 12.6 star NW 2'.3.
14 45 09.9	+09 16 25	CGCG 76-16	14.8	0.5 x 0.5	13.2			Mag 8.33v star NW 8'.5.
14 46 50.3	+13 50 49	CGCG 76-34	14.1	0.5 x 0.3	11.9	S0a	150	Mag 9.92v star SW 7'.1.
14 47 07.2	+12 26 44	CGCG 76-41	14.4	0.5 x 0.3	12.2	Compact	75	Mag 9.71v star SW 6'.2.
14 47 02.2	+13 40 04	CGCG 76-43	14.6	0.5 x 0.5	12.9			NGC 5758 ESE 1'.7.
14 48 30.0	+10 16 18	CGCG 76-58	15.2	0.6 x 0.2	12.8		132	Mag 13.9 star E 1'.6; mag 8.44v star ESE 11'.3.
14 51 23.8	+09 35 30	CGCG 76-77	15.0	0.5 x 0.4	13.1		6	Mag 7.51v star NNW 9'.4.
14 51 34.3	+10 44 03	CGCG 76-78	14.9	0.9 x 0.5	13.9		18	Mag 11.3 star NNE 9'.8; mag 10.7 star NW 11'.9.
14 52 16.6	+12 17 08	CGCG 76-80	15.0	0.4 x 0.4	12.9			Mag 11.3 star NNE 7'.2.
14 52 55.8	+12 44 26	CGCG 76-84	15.4	0.5 x 0.4	13.5		3	Mag 12.5 star S 2'.1.
14 54 06.1	+12 29 32	CGCG 76-91	15.6	0.9 x 0.3	14.0	Spiral	129	Mag 10.26v star NE 3'.2.
14 54 30.0	+10 19 08	CGCG 76-92	14.1	0.6 x 0.4	12.4	Sab	60	CGCG 76-92 S 0'.8.
14 55 29.3	+08 48 25	CGCG 76-97	14.8	0.7 x 0.5	13.5		9	Mag 9.92v star E 6'.5.
14 57 31.1	+08 23 24	CGCG 76-111	14.3	0.5 x 0.3	12.1		126	Located between two stars, mags 12 and 13, on a SE-NW line.
15 01 38.9	+10 25 11	CGCG 76-133	14.1	0.7 x 0.6	13.0		66	= **Mkn 1494**. Mag 8.63v star SW 4'.9.
13 46 52.5	+17 41 55	CGCG 102-58	14.7	0.5 x 0.5	13.1	Spiral		Mag 11.5 star W 4'.8; stellar galaxy **KUG 1344+178** SW 6'.5.
14 03 46.4	+14 52 04	CGCG 103-55	14.1	0.4 x 0.3	11.7	Compact	135	= **Mkn 802**.
14 04 50.2	+16 16 52	CGCG 103-65	15.1	0.7 x 0.6	14.0	S0	57	Mag 11.3 star NE 2'.8.
14 09 04.1	+17 45 56	CGCG 103-86	14.4	0.8 x 0.4	13.0	S?	90	Mag 7.69v star SSW 8'.9; mag 11.8 star SW 3'.2.
14 09 11.9	+14 33 49	CGCG 103-87	15.6	0.7 x 0.7	14.7			Mag 12.1 star W 2'.8.
14 13 33.5	+16 07 34	CGCG 103-120	15.3	0.6 x 0.5	13.8	Sc	63	Mag 9.38v star N 3'.7.
14 40 28.0	+14 49 34	CGCG 104-65	14.4	0.5 x 0.5	12.7			Mag 13.4 star E 1'.2.
14 49 04.9	+16 44 06	CGCG 105-34	14.4	0.8 x 0.6	13.5	Sbc	87	Bright N-S bar.
15 00 55.4	+16 56 47	CGCG 105-91	14.7	0.5 x 0.4	12.8	Compact	0	= **Mkn 1391**. Close pair of mag 12.9, 13.8 stars W 4'.9.
13 51 31.2	+14 05 29	IC 944	13.4	1.6 x 0.5	13.0	Sa	108	**MCG +2-35-20** N 1'.2; mag 8.14v star ESE 7'.5.
13 52 08.6	+14 06 55	IC 946	13.4	0.8 x 0.6	12.5	Sa	110	Mag 10.6 star SE 1'.1; mag 8.14v star S 6'.2.
13 52 26.7	+14 05 27	IC 948	13.1	1.3 x 0.7	13.0	E	152	Mag 8.14v star SW 8'.3.
13 56 03.3	+13 30 24	IC 959	12.9	1.7 x 1.0	13.4	Sa	0	Mag 10.7 star NNE 4'.4.
13 55 59.5	+17 30 21	IC 960	14.0	1.4 x 0.7	13.8	Disrupted Spiral	24	Multi-galaxy system.
13 57 13.3	+12 01 14	IC 962	13.3	0.8 x 0.8	12.7	S?		MCG +2-36-2 S 1'.5; mag 7.74v star SE 8'.6.
13 58 14.1	+05 24 29	IC 966	13.1	0.8 x 0.7	12.4	S0	162	Mag 10.34v star NE 9'.0.
13 58 22.9	+14 27 23	IC 967	14.0	1.0 x 0.4	12.8	Spiral	21	Mag 7.35v star NE 9'.3.
14 09 32.5	+14 49 55	IC 979	13.6	1.0 x 0.6	12.9	SBab	172	Mag 11.6 star NE 2'.2.
14 09 59.2	+17 41 46	IC 982	13.0	1.0 x 1.0	12.9	SA0⁺		Located on SW edge of IC 983.
14 10 04.5	+17 43 57	IC 983	11.7	3.9 x 2.6	14.1	SB(r)bc	126	8.99v mag star on SE end; IC 982 SW 2'.6.
14 18 22.8	+11 11 39	IC 994	13.9	1.3 x 0.5	13.3	Sa	13	**IC 993** NW 1'.7.
14 19 32.7	+17 52 29	IC 999	13.9	0.8 x 0.4	12.5	S0?	142	IC 1000 SE 2'.2.
14 19 40.3	+17 51 16	IC 1000	13.7	0.9 x 0.5	12.7	S0	23	IC 999 NW 2'.2; elongated **UGC 9171** ESE 3'.0.
14 28 17.8	+13 46 46	IC 1014	12.4	2.7 x 2.0	14.1	Sdm	90	Mag 10.33v star NE 6'.1.
14 38 10.2	+09 20 06	IC 1035	14.3	0.4 x 0.4	12.2			Mag 15.3 star on NE edge.
14 41 29.1	+09 25 49	IC 1044	14.1	0.9 x 0.5	13.1		0	Mag 7.82v star NNE 6'.9; **CGCG 75-112** ESE 7'.0.
14 45 43.3	+16 56 46	IC 1053	14.3	0.5 x 0.3	12.1	Sb	33	Mag 4.60v star Omicron Bootis W 7'.0.
14 56 29.2	+09 21 15	IC 1078	13.5	1.1 x 0.9	13.3	Sa	0	IC 1079 NE 2'.0.
14 56 36.1	+09 22 10	IC 1079	12.9	1.7 x 1.0	13.5	E	83	Bright nucleus with bright pseudo ring; IC 1078 SW 2'.0.
14 58 52.6	+07 00 23	IC 1082	14.4	0.7 x 0.5	13.1		33	Mag 10.78v star E 2'.4.
15 02 43.4	+17 15 10	IC 1085	13.8	0.9 x 0.7	13.1		27	Mag 9.40v star W 6'.4.
14 24 53.8	+17 02 16	IC 4417	14.0	0.8 x 0.7	13.4	E/S0	156	Stellar; mag 11.4 star W 1'.9.
14 28 34.6	+17 20 04	IC 4438	15.8	0.5 x 0.4	13.9	Sbc	84	**IC 4440** E 6'.0.
14 39 12.8	+15 52 36	IC 4478	14.8	0.6 x 0.4	13.1		24	Mag 10.14v star W 7'.0.
14 40 19.6	+16 41 07	IC 4483	13.9	1.5 x 0.5	13.4	Sb II-III	26	Mag 11.4 star W 9'.9; stellar galaxy **Mkn 821** E 5'.2.
14 46 39.5	+16 08 47	IC 4503	14.0	0.8 x 0.4	12.6		0	Mag 9.57v star NW 6'.8.
14 54 23.6	+16 21 20	IC 4516	13.2	1.0 x 0.7	12.9	E	162	Mag 11.1 star W 11'.0.
13 48 05.1	+07 23 29	MCG +1-35-37	13.8	0.5 x 0.5	12.1	Sab		Small, faint companion anonymous galaxy SE edge; UGC 8728 E 1'.9.
14 13 45.7	+08 13 08	MCG +1-36-24	13.9	0.4 x 0.2	11.0	Sb	6	Pair with **MCG +1-36-57** E edge; **UGC 9103** W 0'.9.
14 34 49.0	+08 09 53	MCG +1-37-33	14.5	0.9 x 0.6	13.7		120	
14 42 39.0	+05 21 05	MCG +1-37-50	14.7	0.8 x 0.4	13.3	Sb	33	
14 58 26.7	+08 12 29	MCG +1-38-24	14.4	0.3 x 0.3	11.7	S0		Mag 10.43v star SE 3'.0.
13 49 23.9	+08 30 26	MCG +2-35-18	14.1	0.9 x 0.4	12.9		3	Mag 6.61v star SE 7'.9.
13 52 19.8	+13 34 31	MCG +2-35-22	14.5	0.9 x 0.7	13.9	SBb	75	Mag 12.0 star N 3'.3.
13 54 04.7	+13 35 36	MCG +2-35-24	13.3	1.1 x 1.0	13.3	S?	72	Mag 12.0 star NW 1'.4.
13 57 14.1	+11 59 49	MCG +2-36-2	14.1	0.9 x 0.3	12.5	S?	168	**MCG +2-36-4** SE 1'.8; IC 192 N 1'.5; mag 7.74v star ESE 7'.9.
14 01 52.4	+08 47 09	MCG +2-36-10	13.8	0.7 x 0.7	12.8	S?		Mag 9.78v star SE 2'.1.
14 03 12.8	+08 53 45	MCG +2-36-21	14.1	0.9 x 0.4	12.8		162	CGCG 74-64 NNW 3'.7; **CGCG 74-61** W 4'.7.
14 03 31.3	+09 13 12	MCG +2-36-23	14.1	0.7 x 0.5	12.8		150	
14 04 38.2	+11 38 42	MCG +2-36-30	13.8	0.7 x 0.5	12.5		126	Mag 8.48v star S 3'.4.
14 04 49.4	+11 23 41	MCG +2-36-33	14.1	0.7 x 0.6	13.0		18	
14 06 01.3	+12 46 57	MCG +2-36-39	13.9	0.7 x 0.5	12.7	SB?	0	= **Mkn 804**.
14 07 00.3	+10 27 40	MCG +2-36-41	14.0	0.7 x 0.5	12.7	Sb	27	
14 07 08.3	+12 35 15	MCG +2-36-42	13.9	0.9 x 0.3	12.3	Sab	132	Mag 10.59v star SW 1'.3.
14 10 43.8	+08 59 46	MCG +2-36-45	13.9	1.0 x 0.7	13.4	Sb	171	
14 12 58.5	+09 55 13	MCG +2-36-49	14.1	0.9 x 0.5	13.1		105	Located between a pair of mag 13 stars.
14 13 20.1	+08 51 04	MCG +2-36-52	14.1	1.2 x 0.9	14.1		36	
14 13 11.9	+13 00 09	MCG +2-36-53	13.4	1.1 x 0.8	13.1	S?	123	
14 13 27.4	+08 58 49	MCG +2-36-54	14.3	1.1 x 0.5	13.5	Sc	156	Mag 8.09v star E 5'.8.
14 13 46.7	+08 13 11	MCG +2-36-57	15.5	0.4 x 0.3	13.0		129	Pair with MCG +1-36-24 W edge.
14 27 14.6	+11 20 16	MCG +2-37-8	14.0	1.2 x 0.7	13.7	SBa	93	

(Continued from previous page)

GALAXIES

RA h m s	Dec ° ′ ″	Name	Mag (V)	Dim ′ Maj x min	SB	Type Class	PA	Notes
14 27 34.7	+11 19 48	MCG +2-37-9	13.8	1.1 x 0.5	13.1	SAB0$^+$	153	Mag 12.5 star SSE 1′.3.
14 28 22.5	+11 25 02	MCG +2-37-11	14.0	1.1 x 0.7	13.6		90	Small, faint, elongated anonymous galaxy S edge; NGC 5627 SE 3′.7.
14 46 27.9	+11 30 22	MCG +2-38-5	14.1	0.6 x 0.3	12.1	S0	117	Mag 8.41v star E 7′.5.
14 46 46.4	+11 34 10	MCG +2-38-7	14.6	0.7 x 0.4	13.1	SB	138	UGC 9521 ENE 3′.6.
14 46 53.6	+11 37 31	MCG +2-38-8	14.5	0.5 x 0.3	12.4	S0	18	UGC 9521 SE 2′.6.
14 51 28.1	+09 19 16	MCG +2-38-20	14.7	0.3 x 0.3	11.9	S0:		UGC 9561 N 0′.9; mag 11.7 star NE 2′.5.
14 54 58.6	+11 41 54	MCG +2-38-21	14.2	0.9 x 0.3	12.6		15	
14 56 02.6	+09 28 48	MCG +2-38-24	15.4	0.5 x 0.2	12.7		33	Located between a pair of mag 13.5 stars SE and NW.
14 57 17.9	+11 40 07	MCG +2-38-29	14.2	0.8 x 0.4	12.8	Sb	126	Mag 15.7 star on S edge; located between two mag 13 stars.
14 57 38.0	+09 08 44	MCG +2-38-30	14.6	0.8 x 0.4	13.2	Sa	12	Mag 12.5 star W 1′.8.
14 59 38.3	+13 12 51	MCG +2-38-31	14.3	0.5 x 0.4	12.4	S	36	**MCG +2-38-32** N 0′.8; mag 9.08v star NNW 1′.6.
15 02 16.5	+11 55 02	MCG +2-38-33	14.1	0.9 x 0.3	12.5	Sa		Mag 11.8 star W 2′.2.
13 45 48.1	+15 31 53	MCG +3-35-22	15.4	0.5 x 0.5	13.7	Sbc		
13 48 14.6	+15 25 40	MCG +3-35-26	15.5	0.7 x 0.4	14.0		36	Multi-galaxy system.
13 49 05.6	+17 14 14	MCG +3-35-27	15.8	0.5 x 0.5	14.1			
13 53 05.4	+15 50 38	MCG +3-35-33	14.5	0.9 x 0.4	13.2		30	Mag 13.2 star S 1′.9.
13 53 09.8	+14 39 19	MCG +3-35-34	14.5	0.5 x 0.3	12.3		105	Double system, small companion on N edge; mag 8.83v star NNE 2′.4.
13 53 10.2	+17 20 00	MCG +3-35-35	14.3	0.3 x 0.3	11.5	Compact		Almost stellar.
13 56 52.3	+15 42 05	MCG +3-36-7	14.1	0.8 x 0.7	13.3		66	Mag 11.42 star N 3′.2.
13 58 55.5	+15 38 09	MCG +3-36-17	14.2	0.8 x 0.5	13.0	S?	137	Mag 13.4 star N 2′.9.
14 00 02.6	+15 55 13	MCG +3-36-25	14.6	0.6 x 0.4	12.9	Spiral	3	
14 04 27.4	+16 18 11	MCG +3-36-38	14.0	0.5 x 0.4	12.3	E	39	UGC 8987 NW 3′.1.
14 05 04.2	+16 34 55	MCG +3-36-44	14.0	0.7 x 0.6	12.9		39	Mag 13.0 star S 1′.3.
14 06 22.3	+16 28 58	MCG +3-36-47	13.8	0.8 x 0.3	12.1		15	Mag 15.1 star on SE edge; mag 9.35v star SW 2′.8.
14 07 22.2	+15 04 36	MCG +3-36-50	13.9	0.9 x 0.5	12.9	S?	60	Mag 11.6 star SE 2′.4.
14 07 34.2	+16 06 18	MCG +3-36-52	14.5	0.9 x 0.5	13.5	Sb	120	Mag 8.95v star SE 5′.3.
14 08 22.6	+16 55 44	MCG +3-36-54	16.0	0.5 x 0.4	14.1		90	Mag 13.0 star W 1′.8.
14 09 54.8	+15 10 37	MCG +3-36-63	14.2	0.9 x 0.4	13.0		0	
14 10 24.4	+16 06 41	MCG +3-36-72	14.5	0.7 x 0.4	12.9	Irr	78	Mag 13.6 star NNW 1′.2.
14 10 24.2	+16 10 55	MCG +3-36-73	15.5	0.9 x 0.4	14.2	Sb	168	
14 49 29.7	+16 33 34	MCG +3-38-30	14.5	1.1 x 0.3	13.1	Sab	81	Mag 11.1 star NE 1′.9.
14 50 58.3	+16 06 20	MCG +3-38-35	14.7	0.5 x 0.5	13.1	Sc		
14 52 29.6	+16 27 57	MCG +3-38-43	14.9	0.5 x 0.5	13.3	Sab		
14 52 55.5	+16 42 09	MCG +3-38-44	13.3	1.0 x 1.0	13.1	S?		**PGC 53162** SE edge; **MCG +3-38-45** N 1′.5; **CGCG 105-54** W 2′.2.
14 55 20.0	+16 30 20	MCG +3-38-57	14.6	0.4 x 0.4	12.5	Sa/0		**MCG +3-38-56** N 2′.3; mag 11.1 star S 1′.1.
15 01 56.1	+16 58 33	MCG +3-38-71	15.3	1.0 x 0.8	14.9	Sbc	0	Stellar nucleus, smooth envelope.
13 46 52.7	+16 16 23	NGC 5293	13.1	1.9 x 1.5	14.1	SA(r)c I-II	120	Very patchy.
13 52 07.8	+16 58 09	NGC 5332	12.9	0.9 x 0.9	12.6	S0$^-$:		**CGCG 102-69** WSW 3′.8.
13 53 26.8	+05 12 23	NGC 5338	12.4	2.0 x 0.9	12.9	SB0:	97	Bright nucleus; near center of galaxy cluster A 1809; numerous anonymous galaxies in area.
13 54 11.2	+05 13 40	NGC 5348	13.1	3.5 x 0.5	13.6	SBbc: sp	177	On E edge of galaxy cluster A 1809.
13 54 58.2	+05 19 56	NGC 5356	12.6	3.1 x 0.9	13.6	SABbc: sp	15	
13 56 07.0	+05 15 19	NGC 5363	10.1	4.1 x 2.6	12.5	I0?	135	Located 3′.9 SW of mag 8.06v star.
13 56 12.2	+05 00 58	NGC 5364	10.5	6.8 x 4.4	14.1	SA(rs)bc pec I	30	**Leo III**. Bright nucleus, smooth spiral arms.
13 57 07.5	+05 15 05	NGC 5373	14.9	0.5 x 0.4	13.0		174	Mag 9.21v star E 4′.0; CGCG 46-13 SSW 6′.8.
13 57 29.4	+06 05 50	NGC 5374	12.5	1.7 x 1.5	13.3	SB(r)bc? III	54	Bright nucleus; mag 10.78v star 1′.2 W.
13 58 15.1	+06 15 31	NGC 5382	12.6	0.9 x 0.6	11.8	S0	25	
13 58 12.8	+06 31 01	NGC 5384	13.1	1.5 x 0.8	13.1	S0	65	Mag 9.33v star NW 3′.1.
13 58 22.2	+06 20 17	NGC 5386	13.2	1.0 x 0.4	12.1	S0/a	51	Star or bright knot SW end.
13 58 24.6	+06 04 07	NGC 5387	13.9	1.8 x 0.3	13.1	Sbc: sp	22	
14 01 09.4	+07 42 06	NGC 5405	13.5	0.8 x 0.8	12.8	S?		
14 01 46.0	+09 29 22	NGC 5409	13.3	1.7 x 1.1	13.8	(R′)SAB(s)b	50	**CGCG 74-48** NE 3′.8.
14 01 59.3	+08 56 14	NGC 5411	13.3	1.4 x 0.8	13.3	S0$^-$:	140	Very small, very faint anonymous galaxy on NW edge.
14 02 03.6	+09 55 45	NGC 5414	12.4	1.0 x 0.8	12.0	S?	160	A pair of anonymous galaxies N and NW 1′.8.
14 02 11.3	+09 26 22	NGC 5416	13.3	1.4 x 0.8	13.2	Scd:	110	
14 02 13.1	+08 02 09	NGC 5417	13.0	1.5 x 0.6	12.7	Sa	120	
14 02 17.5	+07 40 59	NGC 5418	13.5	1.2 x 0.5	12.8	SB?	44	
14 02 48.7	+09 20 30	NGC 5423	12.8	1.5 x 0.9	13.0	S0$^-$:	75	**MCG +2-36-18** E 1′.7; **MCG +2-36-16** W 1′.6.
14 02 55.7	+09 25 13	NGC 5424	13.1	1.6 x 1.3	13.7	S0	110	Extremely faint anonymous galaxy N 3′.1.
14 03 07.2	+09 21 46	NGC 5431	13.9	0.8 x 0.6	12.9	S?	51	Weak almost stellar nucleus; mag 10.21v star E 4′.4.
14 03 23.2	+09 26 51	NGC 5434	13.2	1.5 x 1.4	13.9	SAc	33	UGC 8967 on N edge.
14 03 41.2	+09 34 22	NGC 5436	13.9	1.3 x 0.5	13.2	S0/a	126	Mag 8.92v star WSW 5′.3; NGC 5438 NNE 2′.8.
14 03 47.3	+09 31 25	NGC 5437	14.3	0.9 x 0.4	13.0	Sb	0	Stellar anonymous galaxy N 1′.8.
14 03 48.0	+09 36 37	NGC 5438	13.6	0.8 x 0.8	13.2	E/S0		Triangle of faint stars on N edge.
14 05 17.8	+09 38 14	NGC 5446	15.6	0.6 x 0.4	13.9		42	
14 04 45.8	+14 22 54	NGC 5454	12.7	1.5 x 0.9	13.0	S0	110	Faint star NE of nucleus; mag 10.87v star W 2′.2.
14 04 58.9	+11 52 19	NGC 5456	12.9	1.2 x 1.0	12.9	S0	175	**CGCG 74-97** E 8′.6.
14 05 00.1	+13 07 56	NGC 5459	13.1	1.1 x 1.0	13.1	S0	10	
14 06 10.5	+09 21 10	NGC 5463	13.2	1.2 x 0.5	12.5	S?	49	Star or galaxy on NE end.
14 06 31.5	+06 01 41	NGC 5470	13.4	2.5 x 0.4	13.2	Sb III	63	
14 08 30.6	+08 55 53	NGC 5482	12.9	1.2 x 0.6	12.5	S0	88	**CGCG 74-121** SE 5′.6.
14 09 43.9	+08 04 08	NGC 5487	14.0	0.9 x 0.5	13.0	SBbc	65	Mag 8.92v star N 3′.0.
14 09 57.4	+17 32 45	NGC 5490	12.1	2.4 x 1.9	13.7	E	5	
14 10 06.9	+17 36 56	NGC 5490C	13.9	1.1 x 0.7	13.5	SB(s)bc		
14 10 57.3	+06 21 51	NGC 5491	12.9	1.4 x 0.8	12.9	S?	78	Very small galaxy or star on N edge.
14 12 15.8	+15 50 30	NGC 5504	13.0	1.3 x 1.1	13.2	SAB(s)bc	130	**MCG +3-36-79** NW 1′.7; **IC 4383** N 2′.3; **MCG +3-36-84** ENE 2′.1.
14 12 31.5	+13 18 17	NGC 5505	13.2	1.0 x 0.8	12.8	SBa	130	Located 2′.6 N of mag 10.56v star.
14 13 08.7	+08 37 08	NGC 5511	14.1	0.7 x 0.7	13.2	Sm		Mag 10.74v star E 2′.8; **MCG +2-36-50** NW 1′.1.
14 13 40.7	+07 39 31	NGC 5514	12.8	2.2 x 1.1	13.6	S?	105	Multi-galaxy system; long, faint extension E.
14 14 20.9	+07 30 57	NGC 5519	13.1	1.6 x 1.0	13.5	Sa	66	Bright knot or star E edge of nucleus.
14 14 50.3	+15 08 47	NGC 5522	13.5	1.9 x 0.4	13.0	Sb III	50	Mag 11.28v star SW 2′.0.
14 15 39.3	+14 16 58	NGC 5525	12.8	1.4 x 0.9	13.0	S0	23	Faint dark lane along centerline.
14 16 20.0	+08 17 35	NGC 5528	13.9	1.4 x 0.7	13.8	Sb	21	Faint star NE of nucleus.
14 16 43.3	+10 53 05	NGC 5531	13.5	0.9 x 0.9	13.1	S0		Star or bright knot S of center.
14 16 53.0	+10 48 27	NGC 5532	11.9	1.6 x 1.6	12.8	S0		Small elongated galaxy S of center; a pair of stellar anonymous galaxies SE 2′.1.
14 17 37.7	+08 10 44	NGC 5535	13.7	0.8 x 0.7	12.9		48	In galaxy cluster A 1890, many small, faint anonymous galaxies in surrounding area.
14 17 37.2	+07 03 15	NGC 5537	14.3	1.0 x 0.5	13.4	Sb	36	
14 17 42.4	+07 28 32	NGC 5538	14.7	0.8 x 0.3	13.0		63	
14 17 53.3	+07 33 27	NGC 5542	14.2	0.7 x 0.5	12.9	Sb	174	
14 18 04.1	+07 39 15	NGC 5543	14.6	0.7 x 0.3	12.7		132	Faint, stellar anonymous galaxy NE 4′.7.

(Continued from previous page)

GALAXIES

RA h m s	Dec ° ′ ″	Name	Mag (V)	Dim ′ Maj x min	SB	Type Class	PA	Notes
14 18 09.1	+07 33 47	NGC 5546	12.3	1.3 x 1.1	12.6	E	3	
14 18 38.9	+07 22 34	NGC 5549	12.9	1.6 x 0.8	13.0	S0	120	Faint, elongated anonymous galaxy SW 1'.6; mag 9.04v star E 7'.2.
14 18 28.0	+12 52 59	NGC 5550	13.2	1.2 x 0.8	13.0	S?	100	Mag 14.3 star on SW edge; mag 11.13v star WNW 3'.4.
14 18 54.9	+05 27 04	NGC 5551	14.2	0.6 x 0.4	12.5		111	
14 19 04.0	+07 01 52	NGC 5552	14.5	0.8 x 0.4	13.1		177	Faint star on S end.
14 19 15.0	+07 01 13	NGC 5554	14.6	0.7 x 0.7	13.7			
14 20 11.2	+10 15 44	NGC 5562	13.6	0.7 x 0.7	12.7	S?		Compact.
14 20 13.2	+07 03 16	NGC 5563	14.6	0.8 x 0.4	13.2		81	Stellar nucleus.
14 19 30.3	+07 27 17	NGC 5570	15.8	0.5 x 0.4	13.9		42	Mag 9.04v star SW 7'.2.
14 20 41.6	+06 54 22	NGC 5573	14.5	1.3 x 0.3	13.3		105	Bright nucleus.
14 20 59.6	+06 12 07	NGC 5575	13.3	1.0 x 0.9	13.0	S0	96	
14 21 40.5	+13 13 55	NGC 5583	13.5	0.8 x 0.6	12.6	S?	80	Mag 10.9 star NW 1'.0.
14 22 10.9	+13 55 01	NGC 5587	12.5	2.6 x 0.8	13.2	S0/a	162	Mag 8.80v star S 5'.1.
14 22 34.0	+13 43 02	NGC 5591	14.6	1.3 x 0.7	14.4	S?	75	Multi-galaxy system.
14 23 50.9	+06 34 27	NGC 5599	13.6	1.4 x 0.5	13.1	Sb III	166	Stellar appearing **PGC 51412** and **PGC 51413** NW 2'.3.
14 23 49.5	+14 38 21	NGC 5600	12.1	1.4 x 1.4	12.7	Sc pec		Strong dark patch N of nucleus.
14 28 34.5	+11 22 37	NGC 5627	12.9	1.7 x 1.0	13.4	S0	120	**MCG +2-37-14** E 1'.7.
14 28 25.9	+17 55 27	NGC 5628	13.3	1.1 x 0.7	13.0	E	175	
14 30 25.6	+11 55 38	NGC 5644	12.5	1.4 x 1.4	13.1	S0		A mag 10.75v star SW 1'.7.
14 30 39.6	+07 16 30	NGC 5645	12.5	2.4 x 1.5	13.8	SB(s)d III	80	Bright knot or small galaxy SE of center.
14 30 36.1	+11 52 34	NGC 5647	14.6	1.1 x 0.3	13.2	Sa	0	Mag 10.43v star SW 1'.2; **MCG +2-37-18** E 0'.7; mag 13.3 star E 1'.3.
14 30 32.7	+14 01 21	NGC 5648	13.9	1.1 x 0.8	13.6	Sb	172	= **NGC 5649**.
14 31 01.4	+05 58 40	NGC 5652	12.5	2.0 x 1.4	13.4	SABbc II	117	Diffuse nucleus; mag 10.7 star N 5'.8.
14 30 50.8	+13 58 04	NGC 5655	13.2	1.1 x 0.9	13.1	Scd?	165	
14 31 57.5	+06 15 02	NGC 5661	13.3	1.5 x 0.6	13.0	SBb:	23	Very faint galaxy **VIII Zw 451** on SW edge.
14 32 25.6	+08 04 44	NGC 5665	12.0	1.9 x 1.3	12.8	SAB(rs)c pec? IV-V	145	
14 21 30.0	+05 04 22	NGC 5665A	14.0	1.2 x 0.5	13.3	Sbc:	40	Almost stellar nucleus in smooth envelope, star or knot on S edge.
14 33 09.2	+10 30 36	NGC 5666	12.9	0.9 x 0.8	12.4	S?	155	Large, diffuse center.
14 32 44.0	+09 53 27	NGC 5669	11.3	4.0 x 2.8	13.8	SAB(rs)cd IV	50	many bright knots.
14 33 52.1	+05 27 28	NGC 5674	13.0	1.1 x 1.0	13.0	SABc	30	**CGCG 47-99** N 7'.7.
14 35 06.4	+05 21 21	NGC 5679A	13.7	0.8 x 0.5	12.5	Sc	54	Located W edge NGC 5679B; star or knot N edge.
14 35 08.6	+05 21 31	NGC 5679B	13.6	1.2 x 0.7	13.2	Sb	127	NGC 5679A W edge, NGC 5679C SE edge.
14 35 11.1	+05 21 12	NGC 5679C		0.3 x 0.2			10	Located SE end of NGC 5679B.
14 35 43.0	+08 18 00	NGC 5681	13.9	0.9 x 0.7	13.3	S	5	
14 39 11.1	+05 21 48	NGC 5701	10.9	4.3 x 4.1	13.8	(R)SB(rs)0/a	90	Bright N-S bar with strong dark patches E and W.
14 43 30.8	+11 12 09	NGC 5736	14.3	1.1 x 0.7	13.8	Sa	108	**CGCG 76-9** NE 2'.7.
14 44 20.7	+12 07 54	NGC 5747	13.7	0.8 x 0.8	13.0	Sb		**CGCG 76-15** N 7'.7.
14 47 08.6	+13 39 14	NGC 5758	13.5	1.2 x 0.6	13.0	S0⁻:	147	CGCG 76-43 WNW 1'.7; mag 9.35v star E 2'.9.
14 47 14.2	+13 27 35	NGC 5759	14.4	1.4 x 0.6	14.0	Double System	150	Multi-galaxy system. Small companion located N at the end of a faint extension from the S component.
14 48 42.7	+12 27 26	NGC 5762	12.8	1.5 x 1.3	13.4	S?	140	Compact core in faint envelope.
14 48 58.7	+12 29 21	NGC 5763	14.4	0.4 x 0.4	12.3			Stellar.
14 50 51.0	+05 07 10	NGC 5765A	13.9	0.8 x 0.4	12.5	Sb:	82	Double system with NGC 5765B S.
14 50 51.6	+05 06 51	NGC 5765B	14.6	0.8 x 0.6	13.7	Sa-b	82	Double system with NGC 5765A N.
14 52 41.8	+07 55 52	NGC 5769	14.4	0.6 x 0.6	13.3	E		**CGCG 48-48** NE 1'.9.
14 55 55.3	+11 51 43	NGC 5782	14.0	0.7 x 0.7	13.1	S0		
14 57 35.8	+08 17 04	NGC 5790	13.6	1.1 x 0.9	13.5	(R)SA0/a	77	Pair of mag 9 stars SE 3'.0.
13 44 23.2	+05 47 30	UGC 8689	13.0	1.6 x 1.1	13.5	S0	155	**CGCG 45-88** NE 5'.1.
13 46 51.1	+07 23 16	UGC 8708	13.8	1.1 x 0.6	13.2	Im?	33	Irregular or peculiar; mag 12.0 star SE 2'.4.
13 48 12.6	+07 23 37	UGC 8728	13.5	1.1 x 0.6	12.9	SB?	79	MCG +1-35-37 W 1'.9.
13 52 03.0	+09 31 51	UGC 8769	13.5	1.3 x 0.8	13.4	S0⁻:	50	Pair of faint stars on W edge; several very faint, very small anonymous galaxies in immediate area.
13 54 06.3	+05 21 18	UGC 8818	13.9	0.9 x 0.8	13.4	Sbc	177	Large knot or small galaxy NE edge.
13 54 31.2	+15 02 37	UGC 8827	13.4	1.2 x 1.2	13.7	(R)SB0°		Bright core, smooth envelope.
13 55 25.0	+17 47 43	UGC 8839	13.0	3.0 x 2.3	14.9	Im V	114	Dwarf irregular, extremely low surface brightness.
13 56 34.5	+10 11 23	UGC 8861	15.3	1.1 x 0.6	14.7	Sdm	30	Weak nucleus, faint envelope.
13 57 18.8	+15 27 26	UGC 8872	13.0	0.8 x 0.8	12.4	S0?		Mag 9.79v star NNW 4'.1.
13 57 49.6	+09 52 22	UGC 8878	13.5	1.1 x 0.9	13.3	SBb	172	Bright E-W bar with dark patch S.
13 58 04.8	+15 18 48	UGC 8883	13.4	1.0 x 0.8	13.0	S?	18	
13 58 38.8	+07 12 59	UGC 8896	14.2	1.5 x 0.2	12.7	S?	69	Mag 8.93v star NE 6'.6; **CGCG 46-30** ENE 5'.3; CGCG 46-25 W 4'.6.
13 59 02.7	+15 33 58	UGC 8902	13.8	1.2 x 0.4	12.9	Sb: II-III	155	Located 1'.2 NE of mag 8.12v star; **MCG +3-36-20** NE 1'.0; **MCG +3-36-21** ESE 1'.5.
13 59 22.6	+05 32 20	UGC 8906	14.2	1.5 x 0.5	13.7	Sd	111	Mag 11.9 star SSE 2'.1.
13 59 26.0	+12 47 15	UGC 8907	14.3	1.5 x 0.9	14.5		15	Mag 12.0 star E 2'.7; CGCG 74-28 SE 5'.0.
14 00 14.6	+08 58 02	UGC 8918	14.3	1.5 x 0.3	13.3	S?	33	Stellar **Mkn 798** SE 5'.7.
14 02 33.1	+09 04 50	UGC 8948	14.6	1.3 x 0.7	14.4	SBb	42	**CGCG 74-56** N 2'.4; mag 11.00v star N 1'.3.
14 03 31.7	+06 46 08	UGC 8966	14.2	1.0 x 0.5	13.3	Scd:	25	
14 03 27.3	+09 28 03	UGC 8967	13.9	1.7 x 0.3	13.0	Sbc	72	Located on N edge of NGC 5434.
14 03 39.2	+11 22 39	UGC 8972	13.8	1.3 x 0.6	13.4	Sb III	157	Pair with UGC 8973 N, contact.
14 03 38.5	+11 23 16	UGC 8973	13.8	1.1 x 0.5	13.0	Sb	133	Pair with UGC 8972 S, contact.
14 03 48.7	+15 24 06	UGC 8977	14.0	1.1 x 0.6	13.4	S?	99	Located between two stars, mags 10.09v and 11.57v.
14 04 15.9	+16 19 47	UGC 8987	14.1	2.1 x 0.3	13.4	SBm	96	Brightest on E end; MCG +3-36-38 SE 2'.6; stellar galaxy **KUG 1401+165** W 7'.6.
14 04 47.4	+08 48 02	UGC 8995	13.3	2.1 x 1.1	14.1	SABdm	3	
14 04 43.3	+14 16 49	UGC 8996	14.6	1.5 x 0.2	13.1	Scd:	107	Mag 9.62v star W 5'.9.
14 04 53.7	+12 43 18	UGC 9002	13.5	1.5 x 0.7	13.4	SB?	9	**MCG +2-36-35** S edge; mag 11.2 star W 2'.7.
14 05 06.2	+09 20 19	UGC 9007	14.3	0.8 x 0.7	13.5	Sdm	70	Stellar nucleus, uniform envelope.
14 05 02.0	+11 00 37	UGC 9008	14.0	1.0 x 1.0	13.9	SAd		Stellar **CGCG 74-80** WSW 6'.3.
14 05 56.5	+09 01 31	UGC 9015	13.7	1.1 x 0.8	13.4	S?	15	Stellar nucleus, uniform envelope; mag 14.4 star E 0'.8; mag 11.2 star N 1'.8.
14 06 51.0	+09 19 16	UGC 9023	14.3	1.1 x 0.5	13.5	Scd:	145	
14 06 59.2	+12 33 38	UGC 9025	14.0	1.0 x 0.6	13.3	SAB(s)bc	75	Mag 12.9 star on SE edge; mag 10.59v star NE 1'.5; MCG +2-36-42 NE 2'.8.
14 08 29.4	+07 03 27	UGC 9037	13.4	1.7 x 0.8	13.6	Scd:	160	Mag 8.82v star SW 4'.6; **CGCG 46-55** W 8'.5.
14 08 33.0	+11 48 56	UGC 9041	13.6	1.1 x 0.7	13.2	SBa	50	**CGCG 74-120** and mag 12.4 star N 7'.7.
14 08 41.1	+16 05 48	UGC 9043	13.3	1.5 x 0.9	13.5	S0:	90	Mag 14.9 star on E edge.
14 09 02.5	+14 18 57	UGC 9044	13.6	1.0 x 0.5	12.7	S?	165	Mag 11.5 star SW 2'.9.
14 09 52.1	+14 52 20	UGC 9055	13.9	0.9 x 0.6	13.1	SBb	38	Mag 11.3 star ESE 1'.8.
14 10 45.4	+15 12 34	UGC 9067	13.5	1.5 x 0.7	13.4	Sab	12	Located 2'.5 W of mag 8.31v star.
14 11 17.9	+17 30 23	UGC 9078	13.2	1.4 x 1.1	13.5	(R′)SB(s)a	63	N-S bar, smooth envelope.
14 12 12.8	+08 39 43	UGC 9084	13.6	1.0 x 0.8	13.2	Sb III	39	Mag 11.4 star NNE 4'.2; CGCG 74-136 E 4'.3.
14 12 30.0	+11 14 23	UGC 9089	14.0	1.0 x 0.4	12.9	S	52	
14 13 42.0	+12 30 16	UGC 9104	13.5	0.9 x 0.7	12.8	S?	72	Mag 8.27v star NW 5'.6.

(Continued from previous page)

89 GALAXIES 89

RA h m s	Dec ° ′ ″	Name	Mag (V)	Dim ′ Maj x min	SB	Type Class	PA	Notes
14 14 13.2	+15 37 16	UGC 9110	13.0	1.8 x 0.6	13.0	SBb	17	**CGCG 103-124** E 3′4; **MCG +3-36-87** SSW 8′2.
14 14 52.2	+14 07 32	UGC 9117	12.9	1.8 x 0.9	13.3	Sa	7	Almost stellar **PGC 50887** W 0′8.
14 15 11.9	+15 44 28	UGC 9121	13.7	1.5 x 0.7	13.6	Sbc	140	Mag 12.0 star E 1′3.
14 15 40.6	+16 33 00	UGC 9126	14.4	1.7 x 1.5	15.3	Im: V	6	Dwarf, very low surface brightness.
14 16 35.2	+09 59 01	UGC 9134	14.1	1.1 x 0.5	13.3	Scd:	20	Faint nucleus, uniform envelope; mag 11.4 star W 1′3; mag 10.56v star SW 1′9.
14 18 48.3	+10 50 40	UGC 9162	14.2	1.1 x 0.6	13.6	SBb	135	Two slender arms.
14 19 36.5	+17 38 38	UGC 9167	14.2	1.4 x 0.4	13.4	Sa	48	Mag 9.49v star on S edge.
14 19 45.7	+09 21 55	UGC 9169	12.9	4.2 x 0.8	14.1	Im	53	
14 20 30.6	+10 25 54	UGC 9177	13.3	1.4 x 1.2	13.7	Scd:	57	Stellar nucleus.
14 22 27.4	+15 05 03	UGC 9206	14.1	1.1 x 0.6	13.5	Pec	165	Mag 13.8 star on N end.
14 24 24.3	+08 16 31	UGC 9225	14.4	1.0 x 0.5	13.5	Sm	0	Mag 5.94v star WSW 6′1.
14 26 08.5	+05 14 15	UGC 9244	14.0	1.5 x 0.8	14.0	SBbc	6	Very faint, small, elongated anonymous galaxy N edge.
14 26 30.5	+05 58 35	UGC 9246	14.0	1.2 x 0.4	13.1	S?	152	Located 3′8 NW of mag 7.20v star.
14 27 00.4	+08 41 01	UGC 9249	14.1	2.1 x 0.3	13.5	Sd:	86	Mag 10.00v star N 2′7.
14 27 28.6	+11 02 23	UGC 9259	14.1	0.9 x 0.5	13.1	S?	165	
14 27 49.8	+11 33 33	UGC 9267	13.4	1.0 x 0.7	12.9	SBb	60	Mag 13.4 star N 3′9; mag 14.0 star S 2′9.
14 28 10.8	+13 33 02	UGC 9273	14.3	1.0 x 0.5	13.4	Im:	10	**CGCG 75-49** SE 8′4.
14 28 57.0	+11 11 49	UGC 9286	14.0	1.0 x 0.5	13.1	SBb	146	Mag 12.1 star N edge.
14 28 58.3	+13 51 42	UGC 9288	12.9	1.1 x 1.1	13.0	S0		Mag 13.6 star N 1′4; mag 10.33v star WSW 4′8.
14 31 11.7	+05 18 24	UGC 9338	14.3	1.0 x 0.4	13.1	Sb II-III	49	Very faint, elongated anonymous galaxy NW 3′1.
14 31 18.1	+07 56 40	UGC 9339	13.8	1.0 x 0.5	12.9	S?	88	
14 32 53.6	+11 35 43	UGC 9356	13.0	1.5 x 0.7	12.9	S?	105	Mag 12.5 star NNW 2′4.
14 33 21.7	+06 52 37	UGC 9364	13.8	1.7 x 0.7	13.8	SBdm	20	Mag 11.4 star SW 3′4.
14 34 11.1	+10 12 44	UGC 9374	13.9	1.0 x 0.6	13.2	S?	132	
14 35 22.7	+05 16 34	UGC 9385	16.0	1.5 x 1.4	16.7	Im:	105	Dwarf, extremely low surface brightness.
14 35 33.2	+12 54 30	UGC 9389	13.2	2.0 x 0.6	13.2	SBb	48	Mag 11.4 star SW 1′9.
14 35 39.9	+13 10 11	UGC 9394	13.4	2.1 x 0.6	13.5	Scd?	46	
14 36 20.3	+05 19 49	UGC 9400	13.5	1.1 x 0.8	13.2	S0	30	
14 37 15.7	+10 00 20	UGC 9411	13.8	0.9 x 0.6	12.9	Sab	65	Short bar offset NE; **CGCG 75-94** SE 1′5.
14 39 31.9	+09 14 04	UGC 9443	13.9	1.0 x 0.4	12.8	S?	80	Mag 12.0 star NE 0′9; mag 11.7 star E 2′0.
14 39 22.6	+17 00 45	UGC 9444	13.2	1.1 x 1.0	13.2	S0?	150	Line of three very faint stars SE along SE edge.
14 40 23.9	+06 18 25	UGC 9454	13.8	1.0 x 0.8	13.4	Scd:	60	
14 41 43.0	+05 57 08	UGC 9471	13.8	1.5 x 0.5	13.3	SBb	162	CGCG 47-142 NW 10′3.
14 41 58.5	+08 27 40	UGC 9474	13.7	1.0 x 0.8	13.3	Scd:	60	Short plumes extend N and S.
14 44 12.4	+07 56 55	UGC 9492	14.0	1.6 x 0.6	13.8	Sa	50	Brighter elongated core, faint envelope.
14 45 21.5	+07 51 45	UGC 9500	12.7	2.8 x 2.8	14.8	Sm: V		Dwarf spiral, very low surface brightness.
14 46 09.4	+13 01 44	UGC 9513	14.5	1.0 x 0.4	13.4		146	Mag 15.3 star NW end; UGC 9515 E 3′0.
14 46 21.3	+13 01 13	UGC 9515	13.5	1.0 x 0.7	13.0	S?	131	Mag 11.7 star SE 2′9; UGC 9515 W 3′0.
14 46 40.7	+12 36 16	UGC 9517	13.9	0.9 x 0.5	12.8	Sb: II-III	50	
14 47 00.2	+11 35 31	UGC 9521	13.5	1.2 x 0.7	13.2	SBa:	0	UGC 9523 E 1′5; MCG +2-38-8 NW 2′6.
14 47 06.3	+11 35 35	UGC 9523	13.4	0.9 x 0.8	12.9	S0/a	120	**CGCG 76-40** E 1′3.
14 47 42.5	+09 39 34	UGC 9530	14.0	0.9 x 0.6	13.2	S?		Almost stellar **CGCG 76-46** WSW 2′1.
14 48 36.6	+11 19 31	UGC 9534	13.7	1.0 x 0.7	13.3	E?	140	**CGCG 76-59** S 3′9.
14 50 48.7	+10 07 07	UGC 9555	14.8	1.5 x 0.6	14.5	Triple System	153	Triple system, envelopes in contact; consists of **MCG +2-38-16**, **MCG +2-38-17**, and **MCG +2-38-18**.
14 51 09.9	+17 11 16	UGC 9558	14.5	1.2 x 0.6	14.0	Scd: I-II	170	Stellar nucleus, two main arms.
14 51 29.4	+09 20 03	UGC 9561	14.0	0.7 x 0.4	12.5	S?	90	Almost stellar MCG +2-38-20 S 0′9; mag 11.7 star NE 1′7.
14 56 47.8	+09 30 34	UGC 9614	14.7	1.5 x 1.3	15.3	Im:	5	Dwarf, very low surface brightness.
14 56 53.2	+09 16 15	UGC 9616	13.5	0.8 x 0.6	12.5	Sab	159	Mag 12.1 star N 3′2.
14 57 44.9	+06 37 34	UGC 9625	13.5	1.5 x 0.5	13.1	SBab:	178	
14 59 27.6	+16 38 42	UGC 9640	13.0	1.3 x 1.1	13.4	E	110	Elongated anonymous galaxy NE 1′9.
15 00 56.7	+11 31 00	UGC 9654	13.8	1.0 x 0.7	13.3	Sab	72	**CGCG 76-128** W 6′1.
15 02 19.9	+05 38 46	UGC 9667	13.1	1.8 x 0.9	13.5	S0/a	87	**CGCG 48-93** pair with mag 14.1 star SSE 6′6.
15 03 22.1	+10 39 19	UGC 9675	13.7	1.1 x 0.4	12.7	SB?	90	**CGCG 76-140** SSE 3′7.

GALAXY CLUSTERS

RA h m s	Dec ° ′ ″	Name	Mag 10th brightest	No. Gal	Diam ′	Notes
13 53 18.0	+05 09 00	A 1809	15.8	78	16	Large concentration of anonymous, stellar GX SW of NGC 5338.
14 17 36.0	+08 11 00	A 1890	15.5	37	20	Many anonymous, stellar members.
14 21 24.0	+17 41 00	A 1899	16.0	33	22	**IC 1004** W of center 5′4; **MCG +3-37-8** NE 6′5; all others anonymous and stellar.
14 26 54.0	+16 40 00	A 1913	16.0	53	28	**CGCG 104-27** WNW of center 3′7; several other CGCG galaxies near NW edge of A 1913.
14 52 42.0	+16 44 00	A 1983	15.4	51	28	Large concentration of GX extending SE from center.
15 03 42.0	+07 55 00	A 2020	16.0	47	26	All members anonymous, faint and stellar.

90 GALAXIES 90

RA h m s	Dec ° ′ ″	Name	Mag (V)	Dim ′ Maj x min	SB	Type Class	PA	Notes
12 33 51.4	+08 01 24	CGCG 42-153	14.6	0.5 x 0.3	12.4	Sbc(r)II	27	Mag 10.67v star E 6′4.
12 52 02.0	+06 33 02	CGCG 43-48	14.7	0.6 x 0.4	13.0		162	Mag 11.0 star W 5′1.
13 04 04.9	+07 54 26	CGCG 43-127	13.7	0.8 x 0.6	13.0	E	159	**CGCG 43-128** E 2′4.
13 08 02.9	+06 28 49	CGCG 44-7	14.4	0.6 x 0.3	12.3		111	Mag 9.58v star WNW 9′4; UGC 8204 SSW 10′4.
13 13 44.7	+06 59 28	CGCG 44-33	14.4	0.5 x 0.2	11.8	S	111	CGCG 44-36 SE 3′0; mag 12.8 star W 3′6.
13 13 54.2	+06 57 38	CGCG 44-36	14.5	0.3 x 0.2	11.3	S	90	**CGCG 44-35** W 1′2; CGCG 44-33 NW 3′0.
13 19 11.5	+07 25 48	CGCG 44-67	14.7	0.8 x 0.4	13.3		72	**CGCG 44-68** ENE 5′4; mag 11.6 star W 4′2.
13 33 21.3	+06 19 06	CGCG 45-18	14.3	0.7 x 0.5	13.0		39	Mag 12.9 star W 2′3; mag 10.97v star NW 6′7.
13 34 06.9	+05 06 07	CGCG 45-26	15.1	0.5 x 0.3	12.9		174	Located 3′5 SE of mag 10.10v star.
13 41 47.9	+07 36 57	CGCG 45-74	15.4	0.3 x 0.3	12.7			A trio of mag 10 stars SW 2′1.
13 02 52.4	+10 05 38	CGCG 71-100	14.5	1.0 x 0.6	13.8	Sb	30	Mag 15 star on S edge; mag 11.3 star ESE 2′5.
13 09 15.1	+08 31 53	CGCG 72-7	14.3	1.1 x 0.5	13.5		15	Mag 10.8 star NE 2′8.
13 11 05.1	+08 48 26	CGCG 72-13	14.2	0.8 x 0.5	13.1		132	Located 2′0 E of mag 9.01v star.
13 11 31.6	+08 44 36	CGCG 72-15	14.3	0.9 x 0.5	13.3	Sb	129	Mag 9.92v star S 3′8.
13 11 48.9	+10 36 06	CGCG 72-18	14.3	0.9 x 0.4	13.0		69	Mag 11.46v star WNW 6′2.
13 12 32.8	+11 43 42	CGCG 72-23	14.3	0.7 x 0.5	13.0		30	Mag 12.9 star W 3′0; mag 9.65v star NE 7′2.
13 16 21.5	+13 30 05	CGCG 72-41	14.1	1.0 x 0.8	13.7		90	Mag 10.85v star S 4′9; **CGCG 72-40** SW 7′4.
13 31 24.9	+09 57 29	CGCG 73-10	14.9	0.5 x 0.5	13.2			

(Continued from previous page)

RA h m s	Dec ° ′ ″	Name	Mag (V)	Dim ′ Maj x min	SB	Type Class	PA	Notes
13 34 13.9	+09 38 06	CGCG 73-29	14.0	0.7 x 0.5	12.7		153	
13 46 10.0	+09 17 29	CGCG 73-71	14.9	0.6 x 0.3	12.8		63	Mag 11.08v star 1′6 SE.
13 46 50.3	+11 37 12	CGCG 73-72	14.3	0.7 x 0.3	12.5	Sab	162	= **Mkn 1360**; **CGCG 73-73** E 0′9.
13 21 55.4	+14 19 56	CGCG 101-31	15.0	0.8 x 0.4	13.6		27	Mag 10.8 star W 8′9.
13 46 52.5	+17 41 55	CGCG 102-58	14.7	0.5 x 0.5	13.1	Spiral		Mag 11.5 star W 4′8; stellar galaxy **KUG 1344+178** SW 6′5.
12 33 56.8	+15 21 13	IC 800	13.4	1.5 x 1.1	13.8	SB(rs)c pec?	157	Mag 12.0 star ENE 8′7.
12 42 09.1	+12 35 47	IC 810	13.5	1.6 x 0.5	13.1	S0?	166	Close pair of N-S oriented mag 11.4., 12.4 stars ESE 5′3.
12 46 46.4	+09 50 59	IC 816	14.0	1.1 x 0.8	13.7	SB?	35	Bright NE-SW bar; **IC 817** E 2′6.
12 51 55.2	+16 16 53	IC 827	14.0	0.9 x 0.4	12.8	S?	104	Mag 11.7 star NNE 5′4.
12 58 42.1	+10 36 58	IC 840	13.6	0.8 x 0.6	12.7	SBa:	30	Mag 8.06v star ENE 10′4.
13 13 50.2	+17 04 29	IC 857	13.8	1.0 x 0.8	13.4	SB(s)b	110	Mag 12.4 star SW 9′4; mag 13.1 star E 8′0.
13 14 51.9	+17 13 36	IC 858	12.7	1.4 x 0.9	12.9	S0?	100	**IC 859** E 1′3.
13 19 56.4	+15 50 59	IC 881	13.8	1.6 x 0.4	13.1	Sa	11	IC 882 NNE 3′9.
13 20 07.0	+15 53 50	IC 882	13.8	0.8 x 0.6	12.8	S?	60	Mag 10.07v star NE 6′7; IC 881 SSW 3′9.
13 32 04.8	+17 02 52	IC 894	14.2	1.1 x 0.5	13.4	Sb	78	Mag 8.64v star NW 6′0.
13 34 43.3	+09 20 08	IC 900	12.9	1.6 x 1.0	13.3	SAB(s)bc I-II	27	Mag 12.1 star NW 1′1; mag 9.71v star SE 9′4.
13 35 42.4	+13 19 48	IC 901	14.6	0.8 x 0.6	13.6	Sab	117	**UGC 8576** E 1′3.
13 51 31.2	+14 05 29	IC 944	13.4	1.6 x 0.5	13.0	Sa	108	**MCG +2-35-20** N 1′2; mag 8.14v star ESE 7′5.
12 32 14.4	+10 15 02	IC 3468	13.1	1.4 x 1.2	13.7	E	167	Mag 8.55v star W 9′1.
12 32 23.4	+11 15 45	IC 3470	13.3	0.6 x 0.6	12.2	E?		Compact; located 6′8 NNE of NGC 4503.
12 32 41.2	+12 46 19	IC 3475	13.1	1.7 x 1.6	14.2	E:	86	Low surface brightness; mag 11.7 star ESE 8′3.
12 32 41.9	+14 02 56	IC 3476	12.7	2.0 x 1.4	13.7	IB(s)m:	30	Mag 9.79v star ESE 8′2.
12 32 44.3	+14 11 44	IC 3478	13.3	1.1 x 0.9	13.1	SAB0:	105	Close pair of mag 12.0, 12.1 stars N 6′3.
12 32 52.3	+11 24 13	IC 3481	13.3	0.7 x 0.7	12.5	$SAB0^-$: pec		Almost stellar **IC 3481A** SE 1′4; mag 10.08 star SE 5′4; note: **IC 3483** on S edge of this star.
12 33 14.2	+12 51 24	IC 3486	14.1	0.7 x 0.5	13.0	E?	45	**IC 3492** E 1′4.
12 33 13.5	+09 23 47	IC 3487	14.3	0.7 x 0.4	13.0	E6:	84	Mag 11.9 star E 2′0; mag 9.79v star NE 12′5.
12 33 13.8	+12 14 43	IC 3489	14.4	0.6 x 0.6	13.1	S?		Mag 10.43v star WSW 12′0.
12 33 45.3	+10 59 42	IC 3499	13.3	1.4 x 0.5	12.7	S0/a	125	Mag 7.60v star SE 9′7.
12 33 51.8	+13 19 20	IC 3501	13.9	0.7 x 0.7	13.2	d:E1		Mag 10.92v star NW 3′7.
12 34 14.9	+11 04 12	IC 3510	14.0	1.0 x 0.7	13.4	S?	0	Mag 13.8 star on NE edge; mag 7.60v star S 9′1.
12 34 30.9	+09 09 14	IC 3517	14.8	1.0 x 0.6	14.1	Sdm:	15	Mag 12.1 star NW 7′1; NGC 4522 W 12′8.
12 34 31.5	+09 37 24	IC 3518	14.2	1.2 x 0.7	13.9	Im:	33	Mag 10.01v star NE 2′8.
12 34 39.7	+07 09 32	IC 3521	13.2	1.4 x 0.9	13.3	IBm	27	Mag 11.8 star W 6′8.
12 34 45.9	+15 13 13	IC 3522	14.7	1.4 x 0.6	14.4	IBm: sp V	95	Mag 12.01v star NW 6′0.
12 34 49.3	+17 48 51	IC 3530	14.0	1.2 x 0.9	13.9	S0	162	Mag 11.6 star SW 9′0.
12 35 27.3	+12 45 00	IC 3540	14.0	0.7 x 0.7	13.1	S0?		Mag 10.04v star ESE 9′0.
12 36 37.7	+06 37 17	IC 3576	13.5	2.3 x 2.1	15.0	Sm IV-V	30	Mag 10.26v star ESE 9′9.
12 36 43.8	+13 15 33	IC 3583	12.7	2.2 x 1.1	13.5	IBm	0	Mag 11.5 star on E edge; located 6′1 N of the center of NGC 4569 (M 90).
12 36 54.8	+12 31 12	IC 3586	13.6	1.1 x 1.0	13.6	dS0:	90	Stellar nucleus; mag 10.80v star ESE 4′9.
12 37 03.1	+06 55 31	IC 3591	13.8	1.2 x 0.6	13.3	IBm	48	Multi-galaxy system.
12 38 37.3	+10 28 34	IC 3608	13.8	3.1 x 0.4	13.9	Sb: III-IV	95	Mag 11.10v star N 2′5.
12 39 04.3	+13 21 49	IC 3611	13.3	1.4 x 0.8	13.3	S?	137	Pair of E-W oriented mag 9.70v, 9.91v stars S 7′1.
12 39 25.0	+07 57 56	IC 3617	14.2	1.4 x 0.7	14.0	Im IV-V	65	Mag 11.2 star NE 5′4.
12 39 48.2	+12 58 25	IC 3631	13.7	0.9 x 0.6	12.8	S?	90	Mag 11.28v star W 5′1.
12 40 13.6	+12 52 22	IC 3635	14.3	0.7 x 0.6	13.2	Scd?	168	Mag 7.90v star SE 12′0.
12 40 19.6	+14 42 55	IC 3637	13.7	2.2 x 1.0	14.5	dS0(6) pec:	18	Low surface brightness, located between a close pair of E- W oriented mag 13.1, 12.7 stars.
12 40 16.5	+10 31 05	IC 3638	14.3	0.9 x 0.7	13.7	Sbc(r)II	120	Mag 9.96v star SW 6′4.
12 40 53.4	+10 28 32	IC 3647	13.9	1.5 x 0.9	14.1	Im:	140	
12 40 58.7	+11 11 01	IC 3652	13.7	0.8 x 0.8	13.2	E		Almost stellar.
12 41 15.9	+11 23 10	IC 3653	13.8	0.5 x 0.5	12.3	E?		Mag 12.4 star S 2′5.
12 41 20.6	+14 41 59	IC 3658	14.0	1.3 x 0.6	13.8	E?	81	Mag 10.72v star SSW 4′7.
12 41 46.6	+11 29 14	IC 3665	14.5	1.1 x 0.7	14.0	Im:	72	Located 10′1 SSW of NGC 4621 (M 59).
12 42 08.8	+11 45 12	IC 3672	13.5	1.0 x 1.0	13.5	E		Mag 10.8 star S 1′3; NGC 4621 (M 59) S 6′6.
12 42 36.2	+10 33 51	IC 3686	14.3	1.0 x 0.4	13.2	Sc(s)II	171	Mag 10.68v star SSE 4′4.
12 43 45.8	+10 46 13	IC 3704	14.0	1.2 x 0.3	12.8	Sbc	43	Faint galaxies **VCC 1983** SE 2′4; **VCC 1975** SW 2′7.
12 44 45.3	+12 21 02	IC 3718	13.2	2.7 x 0.8	13.9	S	72	Mag 12.4 star N 4′4.
12 44 47.3	+12 03 52	IC 3720	14.0	1.3 x 0.7	14.0	E?	125	Mag 11.4 star WSW 7′9; note: very faint galaxy **VCC 1995** on N edge of this star.
12 45 05.7	+10 54 00	IC 3727	13.9	1.0 x 0.8	13.5	Scd:	162	Mag 13.5 star on NW edge.
12 45 20.6	+13 41 31	IC 3735	13.8	1.0 x 0.6	13.3	E?	162	Mag 8.33v star NE 10′7.
12 45 31.7	+13 19 52	IC 3742	13.2	1.7 x 0.8	13.3	SB	45	Galaxy **VCC 2028** E 1′3.
12 46 15.6	+08 20 51	IC 3754	13.7	1.3 x 0.6	13.3	Sab	128	Mag 10.9 star and MCG +2-33-12 NNW 8′2.
12 47 15.2	+10 12 10	IC 3773	12.9	1.9 x 0.7	13.2	E	20	Mag 8.89v star ESE 10′8.
12 48 55.5	+14 54 28	IC 3806	13.6	1.5 x 0.5	13.2	Sa?	177	Mag 12.5 star S 5′9.
12 35 18.4	+06 32 34	MCG +1-32-108	15.3	0.8 x 0.2	13.2	S0(8)	120	
12 36 34.9	+08 03 13	MCG +1-32-111	14.2	1.0 x 0.8	13.8	S?	39	
12 41 23.7	+06 40 37	MCG +1-32-133	14.2	0.6 x 0.3	12.2	$S0_1(6)$	18	
12 56 20.4	+05 28 35	MCG +1-33-29	14.5	0.5 x 0.4	12.6	Sc		
12 58 31.9	+08 13 00	MCG +1-33-33	14.5	1.0 x 0.4	13.3	Sab	0	
13 48 05.1	+07 23 29	MCG +1-35-37	13.8	0.5 x 0.5	12.1	Sab		Small, faint companion anonymous galaxy SE edge; UGC 8728 E 1′9.
12 37 41.5	+08 33 30	MCG +2-32-161	14.0	1.4 x 0.6	13.7	S?	126	
12 46 04.7	+08 28 39	MCG +2-33-12	14.7	0.3 x 0.3	11.9	S0?		Mag 10.9 star W 0′9.
12 58 18.9	+13 23 27	MCG +2-33-38	14.1	0.9 x 0.3	12.5	Sb	15	
12 59 12.4	+08 58 48	MCG +2-33-44	15.4	2.4 x 1.3	16.4		84	True multi-galaxy system; dimensions enclose eight, very small, very faint individual galaxies.
13 00 33.3	+10 07 49	MCG +2-33-49	14.2	0.7 x 0.4	12.7			
13 39 18.4	+12 43 26	MCG +2-35-16	15.0	0.8 x 0.3	13.3	Sbc	159	
13 49 23.9	+08 30 26	MCG +2-35-18	14.1	0.9 x 0.4	12.9		3	Mag 6.61v star SE 7′9.
12 40 19.8	+15 56 05	MCG +3-32-82	13.6	0.8 x 0.6	12.7	$S0_1(6)$	177	Mag 11.3 star SSE 3′8.
13 02 43.5	+15 30 23	MCG +3-33-22	13.8	0.6 x 0.5	12.3	S?	162	Mag 10.9 star SE 2′9.
13 24 48.6	+16 08 41	MCG +3-34-26	14.0	0.6 x 0.4	12.4	Compact E	90	Mag 13.0 star NE 2′6.
13 25 10.4	+17 03 09	MCG +3-34-28	14.8	0.7 x 0.4	13.3	Sc	135	Pair of mag 12 stars 1′3 N and NE.
13 28 28.8	+16 26 46	MCG +3-34-40	15.4	0.6 x 0.3	13.3		135	Compact.
13 32 56.8	+16 38 14	MCG +3-35-6	13.8	0.8 x 0.8	13.2			Mag 11.34v star on SW edge.
13 36 53.4	+17 23 10	MCG +3-35-13	13.4	1.0 x 0.8	13.0	S0?	108	MCG +3-35-14 N edge; mag 11.30v star W 2′4.
13 36 54.5	+17 23 56	MCG +3-35-14	13.9	0.6 x 0.4	12.4	E	90	MCG +3-35-13 S edge.
13 45 48.1	+15 31 53	MCG +3-35-22	15.4	0.5 x 0.5	13.7	Sbc		
13 48 14.6	+15 25 40	MCG +3-35-26	15.5	0.7 x 0.4	14.0		36	Multi-galaxy system.
13 49 05.6	+17 14 14	MCG +3-35-27	15.8	0.5 x 0.5	14.1			
12 31 59.5	+14 25 15	NGC 4501	9.6	6.9 x 3.7	13.0	SA(rs)b I-II	140	= **M 88**.
12 32 03.2	+16 41 14	NGC 4502	13.9	1.1 x 0.6	13.3	Scd:	40	

(Continued from previous page)

RA h m s	Dec ° ′ ″	Name	Mag (V)	Dim ′ Maj x min	SB	Type Class	PA	Notes
12 32 06.1	+11 10 34	NGC 4503	11.1	3.5 x 1.7	12.9	SB0$^-$:	12	Mag 11.6 star S 2'.8.
12 32 10.7	+13 25 10	NGC 4506	12.7	1.6 x 1.1	13.2	Sa pec?	110	Located 2'.2 E of a mag 10.86v star.
12 33 05.0	+16 15 55	NGC 4515	12.3	1.3 x 1.1	12.6	S0$^-$:	9	**CGCG 99-82** W 10'.5.
12 33 07.7	+14 34 30	NGC 4516	12.8	1.7 x 1.0	13.2	SB(rs)ab?	0	Long, narrow open "S" shaped bar.
12 33 11.9	+07 51 08	NGC 4518	13.8	1.1 x 0.5	13.1	SB$0_{2/3}$(r)/a	177	**MCG +1-32-37** E 2'.2.
12 33 10.6	+07 50 00	NGC 4518B		0.7 x 0.2			36	Located on S edge of NGC 4518.
12 33 30.1	+08 39 15	NGC 4519	11.8	3.2 x 2.5	13.9	SB(rs)d II-III	145	Bright nucleus, knotty filamentary arms; NGC 4519A NW 2'.5.
12 33 24.9	+08 41 24	NGC 4519A	15.2	0.5 x 0.2	12.5	dS0?	47	Locate 2'.5 NW of NGC 4519.
12 33 39.6	+09 10 16	NGC 4522	12.3	3.7 x 1.0	13.6	SB(s)cd: sp	33	Mag 12.1 star ENE 7'.2; IC 3517 E 12'.8.
12 33 47.9	+15 10 03	NGC 4523	14.1	2.0 x 1.9	15.4	SAB(s)m V	45	Disrupted system, low surface brightness; mag 9.90v star SSW 4'.6.
12 34 02.5	+07 42 03	NGC 4526	9.7	7.2 x 2.4	12.7	SAB(s)0°:	113	Located between a mag 6.75v star 7'.2 E and a mag 6.96v star 7'.5 W.
12 34 06.2	+11 19 17	NGC 4528	12.1	1.7 x 1.0	12.5	S0°:	5	Mag 8.19v star E 7'.8; stellar galaxy **VCC 1512** SW 8'.4.
12 34 15.9	+13 04 32	NGC 4531	11.4	3.1 x 2.0	13.3	SB0$^+$:	155	
12 34 19.5	+06 27 54	NGC 4532	11.9	2.8 x 1.1	13.0	IBm III-IV	163	Long, irregular bar offset N; faint star on E edge.
12 34 19.6	+08 11 47	NGC 4535	10.0	7.1 x 5.0	13.7	SAB(s)c I-II	0	Bright nucleus in short bar, two main arms with many branches.
12 34 51.0	+15 33 05	NGC 4540	11.7	1.9 x 1.5	12.7	SAB(rs)cd IV	40	Galaxy **IC 3520** NE 1'.4.
12 35 20.4	+06 06 53	NGC 4543	13.6	0.9 x 0.6	13.0	E3	9	
12 35 26.2	+14 29 45	NGC 4548	10.2	5.4 x 4.3	13.4	SB(rs)b I-II	150	= **M 91**. Bright diffuse nucleus in strong bar.
12 35 30.7	+12 13 16	NGC 4550	11.7	3.3 x 0.9	12.7	SB0°: sp	178	NGC 4551 NNE 3'.0.
12 35 37.8	+12 15 50	NGC 4551	12.0	1.8 x 1.4	13.0	E:	70	NGC 4550 SSW 3'.0.
12 35 40.0	+12 33 22	NGC 4552	9.8	3.5 x 3.5	12.5	E0-1		= **M 89**. Faint star on E edge.
12 36 27.0	+11 26 21	NGC 4564	11.1	3.5 x 1.5	12.9	E	47	Located between a pair of mag 11.9 stars NW and SE.
12 36 32.4	+11 15 34	NGC 4567	11.3	3.0 x 2.0	13.1	SA(rs)bc III	85	Member of **Siamese Twins**. Contact pair with NGC 4568; small bright nucleus.
12 36 33.9	+11 14 07	NGC 4568	10.8	4.6 x 2.0	13.1	SA(rs)bc III	23	Member of **Siamese Twins**. Interacting pair with NGC 4567. Many dark lanes and knots.
12 36 49.8	+13 09 44	NGC 4569	9.5	9.5 x 4.4	13.4	SAB(rs)ab I-II	26	= **M 90**. Several very small, very faint anonymous galaxies along SW edge; IC 3583 N 6'.1.
12 36 53.1	+07 14 51	NGC 4570	10.9	3.8 x 1.1	12.3	S0 sp	159	Faint, stellar anonymous galaxy N 5'.2.
12 36 56.3	+14 13 02	NGC 4571	11.3	3.6 x 3.2	13.8	SA(r)d III	55	Many faint knots; mag 8.91v star 2'.5 NE; faint galaxy **VCC 1682** W 4'.8.
12 37 30.5	+09 33 18	NGC 4578	11.5	3.3 x 2.5	13.7	SA(r)0°:	35	
12 37 43.6	+11 49 08	NGC 4579	9.7	5.9 x 4.7	13.1	SAB(rs)b II	95	= **M 58**. Small, bright diffuse nucleus in smooth lens with dark lanes.
12 37 48.2	+05 22 04	NGC 4580	11.8	2.1 x 1.6	13.0	SAB(rs)a pec II-III	165	
12 38 17.9	+13 06 32	NGC 4584	12.9	1.4 x 1.0	13.1	SAB(s)a?	5	
12 38 45.6	+06 46 00	NGC 4588	14.3	1.3 x 0.5	13.7	Scd:	57	
12 39 12.6	+06 00 47	NGC 4591	13.1	1.6 x 0.8	13.2	S?	37	
12 39 51.5	+15 17 51	NGC 4595	12.1	1.7 x 1.1	12.6	SAB(rs)b? III	110	
12 39 55.8	+10 10 36	NGC 4596	10.4	4.0 x 3.0	13.0	SB(r)0$^+$	135	Very bright, diffuse nucleus in strong bar; mag 10.83v star SE 3'.4.
12 40 12.1	+08 22 59	NGC 4598	12.7	1.4 x 1.1	13.1	SB0	114	Bright nucleus in short N-S bar.
12 40 57.6	+11 54 44	NGC 4606	11.8	3.2 x 1.6	13.5	SB(s)a:	33	Two mag 14 stars in line S of center.
12 41 12.4	+11 53 09	NGC 4607	12.8	2.9 x 0.7	13.4	SBb? sp	2	Mag 11.2 star N 4'.0.
12 41 13.1	+10 09 21	NGC 4608	11.0	3.2 x 2.7	13.2	SB(r)0°	120	Large diffuse nucleus on short, strong bar; mag 12.9 star on W edge.
12 41 25.5	+13 43 44	NGC 4611	14.3	1.2 x 0.3	13.0	S?	126	UGC 7857 ENE 7'.4.
12 41 32.9	+07 18 50	NGC 4612	10.9	2.5 x 1.9	12.5	(R)SAB0°	145	Bright center; mag 10.44v star on E edge.
12 41 59.5	+12 56 31	NGC 4620	12.2	1.8 x 1.5	13.1	S0	40	Bright, compact nucleus in faint envelope.
12 42 02.7	+11 38 49	NGC 4621	9.6	5.4 x 3.7	12.9	E5	165	= **M 59**. Bright diffuse nucleus; IC 3672 N 6'.6; IC 3665 SSW 10'.1.
12 42 10.8	+07 40 37	NGC 4623	12.3	2.2 x 0.7	12.7	SB0$^+$: sp	176	
12 42 37.1	+14 21 19	NGC 4633	13.1	2.1 x 0.9	13.7	SAB(s)dm:	33	Very faint, stellar galaxy **VCC 1922** WNW 2'.7.
12 42 41.1	+14 17 38	NGC 4634	12.4	2.9 x 0.9	13.3	SBcd: sp	156	Very long, narrow and bright center. Galaxy **VCC 1944** E 2'.3.
12 42 54.1	+11 26 17	NGC 4637	13.9	1.2 x 0.6	13.4	S0?	98	Faint stellar nucleus; faint star W of nucleus.
12 42 47.4	+11 26 31	NGC 4638	11.2	2.2 x 1.4	12.3	S0$^-$ sp	125	Bright, elongated center; smooth outer envelope.
12 42 52.4	+13 15 26	NGC 4639	11.5	2.8 x 1.9	13.2	SAB(rs)bc II-III	123	Bright nucleus; bright patches N and S of nucleus; faint star SE of center.
12 42 57.8	+12 17 13	NGC 4640	13.5	1.4 x 0.8	13.4	SB?	45	Faint galaxy **VCC 1942** lies NW 2'.1 with star on W edge.
12 43 07.7	+12 03 02	NGC 4641	13.2	1.2 x 0.9	13.2	S0	170	Stellar nucleus; mag 12.6 star SE 1'.0.
12 43 32.5	+11 34 55	NGC 4647	11.3	2.9 x 2.3	13.2	SAB(rs)c III-IV	117	Small bright nucleus, many knotty arms; pair with NGC 4649 (M 60).
12 43 40.0	+11 33 05	NGC 4649	8.8	7.4 x 6.0	12.9	E2	105	= **M 60**. NGC 4647 NW 2'.5.
12 43 41.7	+16 23 33	NGC 4651	10.8	4.0 x 2.6	13.2	SA(rs)c II	80	Bright center, several filamentary, patchy arms.
12 43 57.9	+13 07 19	NGC 4654	10.5	4.9 x 2.8	13.2	SAB(rs)cd II	128	Patchy appearance and several bright knots NW half of galaxy.
12 44 29.5	+13 29 52	NGC 4659	12.1	1.7 x 1.3	12.8	S0/a	173	Bright, diffuse nucleus; lies 1'.4 NNE of a mag 9.83v star.
12 44 32.1	+11 11 23	NGC 4660	11.2	2.2 x 1.6	12.6	E:	100	Very faint galaxy **VCC 2002** NE 0'.9.
12 47 45.4	+13 45 40	NGC 4689	10.9	4.3 x 3.5	13.7	SA(rs)bc II-III	165	Bright, complex central region, many filamentary arms with dark lanes. Faint galaxy **VCC 2062** SE 4'.9.
12 48 15.0	+10 58 58	NGC 4694	11.4	3.2 x 1.5	13.0	SB0 pec	140	Bright elongated bar, smooth envelope.
12 48 23.2	+08 29 10	NGC 4698	10.6	4.0 x 2.5	12.9	SA(s)ab II	170	Bright nucleus, faint dark lane down W side.
12 49 39.0	+15 09 55	NGC 4710	11.0	4.9 x 1.2	12.8	SA(r)0$^+$? sp	27	Short, faint, dark lane over nucleus.
12 49 57.3	+05 18 38	NGC 4713	11.7	2.7 x 1.7	13.2	SAB(rs)d III	100	Many bright knots and patches overall.
12 51 06.8	+10 54 42	NGC 4733	11.9	2.2 x 2.2	13.6	E$^+$:		Bright nucleus; faint star 0'.9 W of nucleus.
12 51 55.1	+12 04 59	NGC 4746	12.6	2.1 x 0.6	12.7	Sb: sp	120	
12 52 17.2	+11 18 42	NGC 4754	10.6	4.6 x 2.5	13.1	SB(r)0$^-$:	23	Large, bright nucleus.
12 52 44.5	+15 50 45	NGC 4758	13.0	3.1 x 0.7	13.6	Im:	160	
12 52 55.3	+11 13 39	NGC 4762	10.3	8.7 x 1.7	13.1	SB(r)0°? sp	29	Located between a pair of mag 9 stars, this edge on galaxy has a bright central "bar" measuring 3'.6 in length.
12 53 51.0	+09 42 37	NGC 4779	12.4	2.1 x 1.8	13.7	SB(rs)bc	70	Strong nucleus on bright N-S bar.
12 55 02.4	+08 03 57	NGC 4795	12.1	2.3 x 1.6	13.4	(R')SB(r)a pec:	118	Star or very bright patch 0.5 E of center.
12 55 33.8	+08 14 24	NGC 4803	14.1	0.6 x 0.4	12.4	Compact	6	Located 0'.8 NNW of mag 9.83v star.
12 59 27.1	+14 10 16	NGC 4866	11.2	6.3 x 1.3	13.4	SA(r)0$^+$: sp	87	Very faint dark lanes in central lens.
13 00 10.6	+12 28 59	NGC 4880	11.4	3.2 x 2.3	13.5	SA(r)0$^+$:	165	Bright nucleus, smooth outer shell.
13 03 21.4	+14 22 35	NGC 4935	13.0	1.2 x 1.1	13.2	(R')SAB(s)b	75	Bright, slightly elongated nucleus.
13 07 02.9	+13 38 10	NGC 4969	13.9	0.7 x 0.5	12.9	E/S0	96	Multi-galaxy system.
13 09 05.7	+11 38 06	NGC 4992	13.4	1.2 x 0.7	13.1	Sa	10	Pair of faint stars on E edge.
13 12 40.1	+12 35 59	NGC 5020	11.7	3.2 x 2.7	13.9	SAB(rs)bc	85	Bright nucleus, two main branching arms with several knots.
13 13 21.0	+06 03 41	NGC 5027	13.4	1.2 x 1.1	13.6	SB(r)b:	63	Stellar nucleus in smooth envelope.
13 16 52.4	+12 32 52	NGC 5058	13.5	0.6 x 0.6	12.3	S?		Multi-galaxy system.
13 16 58.6	+07 50 41	NGC 5059	15.1	0.9 x 0.2	13.1	Sc	8	
13 17 16.2	+06 02 12	NGC 5060	13.3	1.1 x 0.8	13.0	SB(s)b	64	Bright nucleus with strong NE-SW bar.
13 18 37.3	+07 56 08	NGC 5071	15.1	0.9 x 0.7	14.5		144	Almost stellar nucleus; **CGCG 44-63** N 4'.0.
13 19 06.3	+07 49 49	NGC 5075	14.6	0.6 x 0.6	13.3			**IC 4223** SW 3'.5.
13 19 19.3	+08 25 47	NGC 5080	13.6	1.3 x 1.1	14.0	E/S0	99	Mag 7.07v star ENE 9'.9.
13 20 59.5	+08 58 41	NGC 5100	14.3	0.9 x 0.5	13.3	Sc	21	Very faint galaxy **VIII Zw 295** NW 0'.7.
13 23 00.4	+13 57 03	NGC 5115	13.7	1.4 x 0.7	13.5	SBcd:	97	**CGCG 72-60** E 9'.0.
13 23 27.4	+06 23 34	NGC 5118	13.7	0.8 x 0.8	13.0	Scd:		**CGCG 44-79** ENE 2'.5.
13 24 00.7	+09 42 34	NGC 5125	12.4	1.7 x 1.1	13.0	Sb II-III	170	Bright nucleus.
13 24 10.2	+13 58 35	NGC 5129	12.1	1.7 x 1.4	13.0	E	10	Mag 10.33v star E 1'.7; mag 9.65v star SE 2'.8.

(Continued from previous page)

GALAXIES

RA h m s	Dec ° ′ ″	Name	Mag (V)	Dim ′ Maj x min	SB	Type Class	PA	Notes
13 24 28.9	+14 05 33	NGC 5132	12.9	1.3 x 0.9	13.0	(R)SB(r)0$^+$:	54	Large nucleus in short, broad bar.
13 24 51.5	+13 44 15	NGC 5136	13.8	0.7 x 0.6	12.7		78	Compact.
13 24 52.5	+14 04 38	NGC 5137	15.1	0.8 x 0.4	13.7		117	Stellar nucleus.
13 26 40.8	+16 52 23	NGC 5151	13.9	0.8 x 0.8	13.2			Mag 8.02v star N 4'.6.
13 27 47.0	+17 46 45	NGC 5158	12.8	1.3 x 1.2	13.1	SBab?	126	Mag 8.98v star NE 9'.7.
13 29 13.6	+11 16 33	NGC 5162	14.2	1.0 x 0.3	12.7	S0	45	Stellar **CGCG 72-86** S 4'.3.
13 28 39.2	+11 23 13	NGC 5165	13.7	0.9 x 0.4	12.4		174	Mag 8.57v star NE 8'.4.
13 28 40.2	+12 42 40	NGC 5167	13.8	0.9 x 0.9	13.4	Sc		Stellar nucleus.
13 29 21.5	+11 44 04	NGC 5171	12.8	1.1 x 0.8	12.5	S0$^-$:	10	Star NW of nucleus; very small, faint anonymous galaxy SE 0'.9.
13 29 19.0	+17 03 08	NGC 5172	11.9	3.3 x 1.7	13.6	SAB(rs)bc: I-II	103	**CGCG 101-59** ESE 6'.8.
13 29 25.9	+11 00 23	NGC 5174	13.0	2.7 x 1.4	14.3	Scd:	160	Star superimposed S of center.
13 29 25.0	+11 46 53	NGC 5176	14.4	0.5 x 0.5	13.0	E		Faint, elongated anonymous galaxy SE 0'.9; NGC 5177 N.
13 29 24.2	+11 47 47	NGC 5177	14.6	0.7 x 0.2	12.3	Sb	141	Almost stellar anonymous galaxy N 0'.5; another small anonymous galaxy NE 0'.9.
13 29 29.3	+11 37 27	NGC 5178	13.8	1.2 x 0.7	13.4	S(r)0/a	95	Elongated anonymous galaxy NE 1'.4; **CGCG 72-88** on E edge of a star NW 2'.5.
13 29 31.0	+11 44 45	NGC 5179	14.2	0.7 x 0.5	12.9	Sb	45	Very faint star or knot NE of core.
13 29 26.9	+16 49 32	NGC 5180	13.0	1.4 x 1.0	13.2	S0?	25	Faint star SE of nucleus; mag 7.44v star NE 6'.1.
13 29 41.9	+13 18 13	NGC 5181	13.9	0.9 x 0.8	13.4	S0	54	Mag 9.38v star SW 5'.0.
13 30 02.3	+13 24 58	NGC 5185	13.3	1.9 x 0.7	13.5	Sb II-III	58	Line of four mag 13-14 stars SW 3'.1.
13 30 03.8	+12 10 31	NGC 5186	15.5	0.5 x 0.5	13.8			Stellar; **CGCG 72-100** W 5'.2.
13 30 47.4	+11 12 02	NGC 5191	14.1	0.9 x 0.7	13.4	Sb	84	**CGCG 73-5** NE 3'.1.
13 32 14.2	+13 53 27	NGC 5207	13.2	1.7 x 1.0	13.6	SAB(r)b II-III	140	Star on NW end.
13 32 27.9	+07 19 03	NGC 5208	13.1	1.7 x 0.6	13.0	S0	162	**CGCG 45-8** SE 1'.7.
13 32 42.7	+07 19 38	NGC 5209	13.0	1.2 x 1.0	13.2	E		**CGCG 45-11** E 3'.0; **CGCG 45-12** NNE 4'.8; **CGCG 45-14** SE 4'.0.
13 32 49.3	+07 10 09	NGC 5210	12.9	1.4 x 1.2	13.4	Sa	15	Bright nucleus.
13 34 05.9	+17 51 24	NGC 5217	12.6	1.5 x 1.4	13.4	E	36	**CGCG 102-20** E 3'.3; **CGCG 102-17** NW 3'.8.
13 34 56.0	+13 49 55	NGC 5221	13.0	1.6 x 0.8	13.1	Sb: II	104	Very wide, very faint material extends W far beyond published dimensions.
13 34 56.2	+13 44 29	NGC 5222	13.1	1.6 x 1.1	13.8	E	33	Bright knot or small galaxy E of nucleus.
13 35 08.9	+06 28 49	NGC 5224	14.0	0.7 x 0.7	13.1			Located 2'.3 SW of mag 9.18v star.
13 35 03.7	+13 55 19	NGC 5226	15.8	0.4 x 0.2	12.9		24	Stellar.
13 35 32.1	+13 40 30	NGC 5230	12.1	1.9 x 1.9	13.3	SA(s)c I-II		Branching, patchy arms, many knots.
13 36 01.8	+06 35 07	NGC 5235	14.2	1.2 x 0.5	13.5	SB	120	**CGCG 45-35** S 3'.5.
13 36 26.2	+07 22 10	NGC 5239	12.8	1.8 x 1.2	13.5	SB(rs)bc	13	Bright center with two main arms.
13 37 31.9	+08 53 02	NGC 5248	10.3	6.2 x 4.5	13.8	SAB(rs)bc I-II	122	Bright nucleus and center, faint outer envelope.
13 37 37.5	+15 58 20	NGC 5249	12.9	1.5 x 1.1	13.4	S0?	170	
13 40 16.2	+05 04 33	NGC 5261	14.4	0.8 x 0.5	13.3		141	Mag 9.80v star W 4'.2.
13 46 52.7	+16 16 23	NGC 5293	13.1	1.9 x 1.5	14.1	SA(r)c I-II	120	Very patchy.
12 32 32.4	+08 02 38	UGC 7688	14.2	1.2 x 0.9	14.1	Im:	42	Mag 11.3 star on E edge.
12 34 45.3	+06 17 55	UGC 7739	14.0	1.2 x 0.8	13.8	Im V	153	Dwarf irregular, low, uniform surface brightness.
12 37 45.4	+07 06 08	UGC 7795	14.1	1.2 x 0.9	14.0	Im V	9	Patchy, several knots; mag 10.35v star NE 4'.7.
12 38 20.7	+07 53 25	UGC 7802	14.3	1.7 x 0.3	13.4	Scd:	56	Mag 12.6 star N 0'.9.
12 41 54.4	+13 46 18	UGC 7857	14.1	1.6 x 0.9	14.3	Sd	60	Mag 15.2 star on W edge; NGC 4611 WSW 7'.4.
12 46 38.5	+09 18 22	UGC 7942	14.3	0.9 x 0.7	13.6	SABd:	9	Mag 7.80v star SW 3'.1.
12 46 45.7	+05 57 14	UGC 7943	13.1	2.3 x 1.9	14.6	Scd:	30	Mag 6.31v star E 4'.1.
12 49 39.6	+12 37 30	UGC 7979	14.5	1.5 x 0.3	13.5	Sdm:	38	
12 52 53.6	+09 59 20	UGC 8015	13.3	1.7 x 1.0	13.7	Sab	60	Pair of mag 12 stars close S.
12 54 44.1	+13 14 14	UGC 8032	13.2	2.5 x 0.6	13.5	S	167	
12 55 10.4	+07 55 05	UGC 8042	13.9	1.3 x 0.8	13.8	S?	170	Stellar nucleus, uniform envelope; mag 9.59v star S 2'.0.
12 55 23.6	+07 54 34	UGC 8045	14.3	1.0 x 0.7	13.8	IBm:	0	Knots and/or superimposed stars E and W edges.
12 56 19.9	+10 11 37	UGC 8056	14.3	1.2 x 0.7	13.9	SB(s)d:	162	Strong dark lane NW of center.
12 58 09.1	+14 51 34	UGC 8081	14.9	1.8 x 0.7	15.0	IAB(s)m: IV	21	Dwarf, low, uniform, surface brightness; mag 12.8 star SW 3'.3.
12 58 18.3	+14 33 15	UGC 8085	13.8	2.3 x 0.6	14.0	SBcd?	115	
12 58 40.0	+14 13 02	UGC 8091	14.4	1.1 x 0.7	13.9	Im V	36	Patchy, several knots.
12 58 50.9	+09 39 13	UGC 8093	14.0	1.2 x 0.6	13.5	S	151	**CGCG 71-84** WSW 6'.3; **UGC 8090** SW 8'.4.
13 00 25.0	+13 40 11	UGC 8114	13.9	1.5 x 0.6	13.6	S?	80	
13 03 14.7	+07 48 03	UGC 8155	12.4	3.0 x 2.6	14.5	S?	162	Almost stellar nucleus, strong knot S edge.
13 04 15.2	+09 13 23	UGC 8170	13.7	1.2 x 1.0	13.8	Scd:	3	
13 06 18.3	+10 22 34	UGC 8192	13.1	1.3 x 1.1	13.3	Sab	102	Mag 13.1 star N 2'.5.
13 07 39.8	+06 20 07	UGC 8204	13.2	0.8 x 0.6	12.3	(R)SAB(r)0$^+$:	100	Two stellar points of light; double system, double star, or superimposed stars? CGCG 44-7 NNE 10'.4.
13 09 23.4	+06 59 45	UGC 8233	13.7	1.6 x 0.6	13.5	Scd:	138	Mag 7.76v star on E edge.
13 10 44.0	+11 42 26	UGC 8253	16.0	1.4 x 1.0	16.2	Sm	3	Dwarf spiral, low surface brightness; mag 13.5 star NW 1'.1.
13 10 56.4	+11 28 38	UGC 8255	13.1	1.5 x 1.1	13.5	Scd:	95	Weak stellar nucleus.
13 12 33.5	+07 11 03	UGC 8285	14.1	1.8 x 0.5	13.8	Sdm:	55	
13 13 10.1	+09 41 27	UGC 8296	14.0	1.0 x 0.3	12.5	S0/a	178	
13 13 19.5	+10 11 36	UGC 8298	14.0	0.8 x 0.4	12.6	Im	103	Uniform surface brightness; **CGCG 72-27** N 7'.7.
13 13 34.3	+15 59 22	UGC 8306	14.2	1.5 x 0.3	13.1	Sb III	37	
13 15 01.2	+12 43 33	UGC 8322	14.0	1.2 x 0.3	12.7	Sa?	37	
13 17 01.7	+06 21 24	UGC 8349	13.5	1.3 x 1.0	13.6	SAB(rs)b	135	Stellar **CGCG 44-54** NE 4'.4.
13 19 33.4	+16 59 11	UGC 8370	13.8	0.9 x 0.8	13.3	Sbc	45	
13 20 32.3	+05 24 23	UGC 8382	14.2	1.0 x 0.5	13.3	IBm	140	Uniform surface brightness, **CGCG 44-74** NE 5'.1; mag 9.74v star S 3'.8.
13 20 27.9	+14 32 00	UGC 8383	13.9	0.9 x 0.6	13.1	Sb II-III	0	Mag 11.16v star on NW edge.
13 20 37.7	+09 47 15	UGC 8385	13.6	1.8 x 0.9	14.0	SABm IV-V	102	
13 24 12.8	+15 35 38	UGC 8425	14.6	1.5 x 0.2	13.1	S?	147	**MCG +3-34-24** S 1'.5.
13 24 35.1	+06 31 43	UGC 8427	13.8	1.4 x 0.7	13.6	S	15	Mag 11.9 star E 1'.1.
13 27 01.3	+10 03 20	UGC 8450	14.4	0.8 x 0.6	13.5	Im	150	Dwarf irregular, very low, uniform surface brightness.
13 27 09.9	+15 05 41	UGC 8452	13.4	0.7 x 0.6	12.4	S0	66	Located 2'.8 ENE from a mag 8.98v star.
13 33 14.1	+05 45 53	UGC 8532	14.4	1.1 x 0.9	14.2	S	150	Disrupted? Numerous bright knots SE of core, on SE edge.
13 33 29.2	+17 28 14	UGC 8535	13.6	1.2 x 0.6	13.1	S0/a	97	**CGCG 102-12** NW 9'.7.
13 35 14.8	+10 41 19	UGC 8563	14.1	1.0 x 0.8	13.7	S0?	9	Double system, contact; published UGC dimensions encompasses **UGC 8562** S.
13 35 38.1	+06 11 33	UGC 8569	13.9	0.9 x 0.7	13.3	Sab	155	Mag 12.6 star NW 1'.1.
13 35 37.8	+07 32 41	UGC 8572	13.9	1.5 x 0.5	13.5	Sb	57	Mag 14.3 star NW 0'.8.
13 35 45.6	+08 58 07	UGC 8575	14.0	2.4 x 0.5	14.0	Im	30	
13 36 13.5	+10 28 37	UGC 8585	13.8	1.1 x 0.9	13.6	SB(r)cd?	114	Bright center, faint envelope; mag 12.1 star SSW 1'.2.
13 36 47.2	+06 29 47	UGC 8596	14.2	1.0 x 0.6	13.5	S	155	Mag 9.54v star N 7'.5.
13 37 23.4	+06 26 07	UGC 8613	14.4	1.4 x 0.4	13.6	SB	107	Disturbed. Small, faint anonymous galaxy SE 1'.9; **UGC 8610** N 3'.1; **MCG +1-35-19** N 4'.4.
13 37 26.8	+07 38 45	UGC 8614	12.6	3.6 x 1.8	14.5	Im IV-V	0	Dwarf spiral, low surface brightness; mag 10.85v star Superimposed NW of center.
13 37 43.1	+05 14 32	UGC 8617	14.1	1.0 x 0.6	13.4	SB	27	Very faint, almost stellar anonymous galaxy E 2'.1.
13 38 23.6	+06 53 14	UGC 8626	13.5	1.6 x 0.9	13.7	SBbc	168	Two main arms; bright knot N edge.
13 41 05.3	+05 06 20	UGC 8657	13.5	1.5 x 1.1	13.9	Sbc	30	Stellar nucleus; **CGCG 45-71** SE 5'.8.
13 41 39.3	+05 01 55	UGC 8663	13.3	1.0 x 0.9	13.1	S0$^-$:	51	Bright nucleus, uniform envelope; mag 9.93v star S 2'.0; **CGCG 45-71** W 4'.2.

(Continued from previous page)

90 GALAXIES 90

RA h m s	Dec ° ′ ″	Name	Mag (V)	Dim ′ Maj x min	SB	Type Class	PA	Notes
13 44 23.2	+05 47 30	UGC 8689	13.0	1.6 x 1.1	13.5	S0	155	**CGCG 45-88** NE 5′.1.
13 46 51.1	+07 23 16	UGC 8708	13.8	1.1 x 0.6	13.2	Im?	33	Irregular or peculiar; mag 12.0 star SE 2′.4.
13 48 12.6	+07 23 37	UGC 8728	13.5	1.1 x 0.6	12.9	SB?	79	MCG +1-35-37 W 1′.9.

GLOBULAR CLUSTERS

RA h m s	Dec ° ′ ″	Name	Total V m	B ★ V m	HB V m	Diam ′	Conc. Class Low = 12 High = 1	Notes
13 16 27.0	+17 41 53	NGC 5053	9.0	13.8	16.7	10.0	11	

91 GALAXIES 91

RA h m s	Dec ° ′ ″	Name	Mag (V)	Dim ′ Maj x min	SB	Type Class	PA	Notes
11 23 20.3	+05 51 05	CGCG 39-162	14.3	0.4 x 0.4	12.2			Mag 11.8 star WNW 5′.4.
11 24 08.7	+06 12 51	CGCG 39-167	13.9	0.8 x 0.4	12.5	(R)SB(r)0/a	111	Mag 8.85v star SE 11′.4.
11 46 10.4	+07 30 58	CGCG 40-42	13.7	0.4 x 0.4	11.6	(R′)SB(r)dm: III-IV		Pair of E-W oriented mag 13-14 stars E 3′.8.
11 51 05.8	+06 45 59	CGCG 40-51	16.2	0.6 x 0.4	14.5		54	Mag 11.19v star W 3′.1.
11 51 33.2	+05 06 04	CGCG 40-53	13.9	0.5 x 0.5	12.5	E		Mag 12.9 star W 2′.1.
11 51 40.4	+06 42 24	CGCG 40-54	14.4	0.6 x 0.3	12.4		171	Forms a triangle with mag 13 and 14 stars 1′.1 E.
12 14 02.1	+06 43 20	CGCG 41-50	14.6	0.5 x 0.2	11.9	Sa?	24	Mag 12.1 star S 2′.8.
12 19 12.5	+06 22 50	CGCG 42-9	14.7	0.7 x 0.3	12.9	S0$_1$(6)	9	Close pair of mag 14.4, 14.7 stars N 0′.8; **CGCG 42-11** and **CGCG 42-10** S 4′.7.
12 19 27.8	+05 02 47	CGCG 42-18	14.5	0.6 x 0.3	12.7	E7	78	= **Mkn 1321**. Mag 10.17v star SE 4′.8.
12 28 14.2	+07 36 25	CGCG 42-120	14.7	0.4 x 0.4	12.6	Sb pec		Mag 12.6 star SSE 3′.1; mag 10.81v star N 7′.2.
12 33 51.4	+08 01 24	CGCG 42-153	14.6	0.5 x 0.3	12.4	Sbc(r)II	27	Mag 10.67v star E 6′.4.
11 50 42.5	+10 31 13	CGCG 68-76	15.0	0.7 x 0.3	13.1		138	Mag 10.37v star W 10′.2; MCG +2-30-42 NE 9′.7.
11 52 36.5	+13 11 51	CGCG 68-81	14.8	0.6 x 0.4	13.1		57	Mag 12.4 star SW 7′.0.
12 01 27.6	+14 02 01	CGCG 69-29	14.5	0.4 x 0.3	12.1			= **Mkn 756**. Located 1′.9 ENE of mag 10.13v star.
12 16 33.7	+13 01 51	CGCG 69-120	14.3	0.5 x 0.4	12.6	E/S0	171	Located 12′.0 SE of the center of NGC 4216.
11 38 36.1	+17 49 20	CGCG 97-35	15.3	0.3 x 0.3	12.5	S0		Mag 11.4 star on SE edge.
11 51 33.8	+16 11 03	CGCG 97-167	14.2	0.4 x 0.2	11.5	E	141	Mag 10.66v star NE 3′.3.
11 54 08.8	+14 52 50	CGCG 97-179	14.8	0.3 x 0.3	12.0			Lies between a pair of stars, mags 10.9 and 12.1, on NW-SW line.
12 19 46.8	+17 33 53	CGCG 99-16	14.5	0.3 x 0.3	11.8	S		Mag 10.9 star SW 11′.0.
11 25 53.6	+09 59 13	IC 692	13.6	0.7 x 0.5	12.3	S?	125	Mag 9.3 star NW 7′.4.
11 28 39.8	+09 05 56	IC 696	13.6	1.0 x 0.9	13.3	SBdm:	144	IC 2857 W 2′.1.
11 29 03.8	+09 06 45	IC 698	13.4	1.0 x 0.5	12.5	S0?	147	Mag 11.24v star SE 4′.4; **IC 2867** SW 1′.5.
11 29 06.6	+08 59 13	IC 699	13.9	1.2 x 0.4	12.9	Sb: III	12	Mag 10.51v star SE 6′.0.
11 39 53.0	+08 52 27	IC 718	13.9	1.1 x 0.6	13.3	Im:	4	Mag 8.96v star NW 8′.1.
11 40 18.5	+09 00 32	IC 719	13.1	1.3 x 0.4	12.3	S0?	52	Mag 10.40v star N 8′.0.
11 42 22.5	+08 45 59	IC 720	13.9	1.1 x 0.6	13.3	S?	0	multi-galaxy system.
11 42 43.8	+08 58 24	IC 722	13.6	1.0 x 0.5	12.7	S?	66	Mag 11.6 star N 2′.0; **CGCG 68-38** S 6′.1.
11 43 34.3	+08 56 26	IC 724	12.5	2.3 x 0.9	13.1	Sa	60	Stellar **CGCG 68-41** W 3′.9.
11 44 28.6	+10 46 59	IC 727	14.1	1.6 x 0.2	12.8	Sb III	161	**MCG +2-30-26** E 2′.0; mag 8.09v star SW 9′.2.
11 48 27.9	+12 43 35	IC 736	13.8	0.6 x 0.5	12.6	E?	111	= **Hickson 59A**; **Hickson 59B** 2′.2 SW ; **Hickson 59C** 1′.8 SE; **Hickson 59D** = **IC 737** W 0′.7; **Hickson 59E** NW 2′.7.
11 57 26.8	+07 27 37	IC 748	14.4	0.6 x 0.5	13.0		81	Mag 10.08v star NE 5′.3.
12 01 10.5	+14 06 15	IC 755	13.2	2.4 x 0.3	12.7	SBb? sp	145	Mag 10.13v star SSE 5′.5; note: CGCG 69-29 (= **Mkn 756**) is 1′.9 ENE of this star.
12 11 02.8	+12 06 10	IC 767	13.8	0.8 x 0.6	13.0	E?	72	Almost stellar, mag 12.0 star N 2′.3; mag 10.11v star NNW 11′.8.
12 11 47.7	+12 08 33	IC 768	13.8	1.4 x 0.7	13.6	Scd:	115	Mag 11.1 star SE 1′.5; mag 11.3 star NW 4′.6.
12 12 32.3	+12 07 25	IC 769	12.6	2.6 x 1.7	14.0	SA(rs)bc	43	Stellar nucleus, mag 9.77v star SE 13′.7.
12 15 13.5	+13 11 02	IC 771	13.9	0.6 x 0.4	12.2	SB?	98	Mag 10.68v star S 2′.8; NGC 4216 E 10′.2.
12 18 08.2	+06 08 20	IC 773	13.5	0.8 x 0.7	12.8	SB0$^+$?	0	Mag 9.46v star S 4′.4.
12 18 53.7	+12 54 41	IC 775	13.3	1.1 x 0.8	13.1	SB0$^-$ pec:	18	Mag 10.39v star E 2′.1; stellar **CGCG 69-140** W 4′.2.
12 19 03.0	+08 51 25	IC 776	13.7	1.8 x 0.8	13.9	Sdm	98	**IC 3134** N 6′.5.
12 20 03.4	+14 57 39	IC 781	13.3	1.0 x 0.8	13.0	dS0(4),N	45	Mag 9.98v star W 2′.0.
12 21 36.9	+05 45 54	IC 782	13.8	1.3 x 0.6	13.4	SB0/SBa(r)	60	Mag 13.2 star on S edge; mag 10.76v star E 3′.9.
12 21 38.9	+15 44 42	IC 783	13.8	1.2 x 0.8	13.6	SAB(rs)0/a?	141	Stellar nucleus; mag 9.40v star NW 6′.9.
12 22 19.8	+15 43 59	IC 783A	14.5	0.5 x 0.5	12.9	SB(r)0?		Mag 9.74v star SE 6′.2 NGC 4321 (M 100) NE 10′.1.
12 26 20.6	+07 27 34	IC 789	13.9	1.1 x 0.6	13.3	S0?	131	Almost stellar; mag 10.71v star E 6′.7.
12 26 35.7	+09 02 04	IC 790	14.6	0.6 x 0.4	12.9	S0?	93	**MCG +2-32-54** ENE 2′.3.
12 27 08.9	+16 19 30	IC 792	14.0	1.6 x 0.5	13.6	Sb III	59	Mag 9.26v star W 10′.8; stellar **CGCG 99-53** SW 5′.4.
12 28 08.7	+12 05 36	IC 794	13.1	1.3 x 1.0	13.3	E$^+$:	110	Galaxy **VCC 1068** SW 1′.1; **PGC 40978** SE 1′.3.
12 29 26.4	+16 24 12	IC 796	13.1	1.3 x 0.6	12.7	S0/a	145	Mag 11.36v star SSW 5′.8.
12 31 54.8	+15 07 24	IC 797	12.9	1.3 x 0.9	12.9	SBcd?	108	Mag 10.42v star NE 12′.4.
12 33 56.8	+15 21 13	IC 800	13.4	1.5 x 1.1	13.8	SB(rs)c pec?	157	Mag 12.0 star ENE 8′.7.
11 22 02.2	+08 23 34	IC 2757	14.0	0.7 x 0.5	12.7	SB(s)a I-II		Mag 12.3 star SW 3′.8.
11 22 18.9	+13 03 50	IC 2763	14.7	1.3 x 0.2	13.0	Scd?	96	Mag 10.44v star SW 9′.9; **IC 2767** NE 1′.3.
11 22 40.1	+13 19 46	IC 2776	15.3	0.8 x 0.4	13.9	Sc	168	Mag 13.2 star SE 4′.4.
11 22 55.3	+13 26 23	IC 2782	13.9	1.0 x 0.8	13.5	Sdm:	0	Mag 11.6 star ESE 11′.0; **IC 2785** and **IC 2786** ESE 5′.9.
11 23 19.1	+13 37 46	IC 2787	14.1	0.7 x 0.7	13.2	Scd?		Mag 13.0 star W 4′.2.
11 24 55.8	+13 13 19	IC 2804	14.2	0.9 x 0.3	12.6	Sc	12	Mag 9.69v star NE 7′.2.
11 26 34.3	+11 26 21	IC 2822	13.6	1.5 x 0.7	13.5	SBbc	115	Mag 10.89v star ESE 7′.5.
11 27 06.3	+13 14 16	IC 2826	14.9	0.6 x 0.6	13.7			Mag 10.95v star E 5′.6.
11 27 11.5	+08 43 51	IC 2828	13.5	0.9 x 0.6	12.7	S0?	51	Mag 6.73v star WSW 13′.8.
11 27 21.6	+07 48 45	IC 2830	15.5	0.5 x 0.5	13.8			Mag 8.93v star NE 7′.6.
11 28 00.7	+11 09 25	IC 2846	14.5	0.9 x 0.7	13.8	Sp	114	Mag 10.91v star NNE 3′.7.
11 28 13.1	+09 03 41	IC 2850	14.4	0.7 x 0.3	12.6	S?	126	Mag 14.1 star on SE end; mag 10.34v star SSW 3′.2.
11 28 14.9	+09 08 45	IC 2853	13.7	1.0 x 0.6	13.0	SBab	0	Strong dark patches N and S of nucleus; mag 10.55v star N 0′.9.
11 28 31.4	+09 06 05	IC 2857	14.5	1.9 x 0.2	13.3	Scd:	161	IC 696 E 2′.1.
11 29 20.8	+08 36 07	IC 2871	14.3	0.8 x 0.6	13.4		84	Mag 8.25v star WSW 13′.7.
11 29 27.9	+13 13 08	IC 2873	14.6	0.7 x 0.4	13.1		3	Mag 12.6 star N 3′.2.
11 34 19.7	+13 19 12	IC 2934	15.3	0.7 x 0.3	13.4	Sb	159	Located 2′.5 W of mag 8.99v star.
11 36 10.1	+10 03 19	IC 2941	13.7	0.9 x 0.8	13.2	S?	72	Strong dark lane N of center.
11 37 04.5	+12 55 36	IC 2945	14.0	1.0 x 0.8	13.8	E/S0		Almost stellar; mag 7.38v star E 13′.8.
12 04 38.7	+11 02 58	IC 2990	15.1	0.8 x 0.3	13.4		15	Mag 7.91v star SE 13′.1.
12 07 51.9	+13 34 32	IC 3008	14.0	0.8 x 0.4	12.7	S?	24	Mag 7.66v star SW 5′.3.
12 09 22.4	+13 59 31	IC 3019	13.2	1.4 x 1.3	13.8	E	123	Mag 11.8 star S 5′.1.
12 09 54.5	+13 02 58	IC 3021	14.3	0.9 x 0.6	13.5	Sm	0	Mag 10.97v star NE 1′.5.

RA h m s	Dec ° ′ ″	Name	Mag (V)	Dim ′ Maj x min	SB	Type Class	PA	Notes
12 10 02.0	+14 21 57	IC 3023	14.7	1.5 x 0.6	14.4	Im:	141	Mag 11.8 star ENE 7'.0.
12 10 41.8	+13 19 49	IC 3029	13.8	1.4 x 0.4	13.1	SB?	28	Mag 11.07v star E 8'.1.
12 11 10.0	+13 35 12	IC 3033	14.3	1.1 x 0.6	13.7	Sdm?	2	Located 2'.3 WNW of mag 9.28v star.
12 12 15.0	+12 29 17	IC 3036	13.7	1.5 x 1.2	14.2	Sm	177	Mag 14.5 star on E edge.
12 12 32.5	+12 18 40	IC 3039	14.5	0.9 x 0.3	12.9	S?	15	Mag 11.09v star SW 2'.7.
12 12 48.7	+13 58 34	IC 3044	13.5	1.9 x 0.8	13.8	SB(s)cd pec:	68	Mag 9.49v star W 12'.8.
12 13 08.0	+12 55 03	IC 3046	14.4	1.3 x 0.4	13.5	S?	132	Mag 9.61v star W 4'.2; **IC 3047** N 5'.0.
12 14 55.1	+13 27 34	IC 3059	14.2	1.7 x 1.3	14.9	Im: V	0	IC 3059 E 5'.2.
12 15 02.2	+12 32 47	IC 3060	13.9	0.7 x 0.4	12.4	Sab	0	Mag 10.75vstar WSW 9'.4.
12 15 04.4	+14 01 45	IC 3061	13.6	2.2 x 0.4	13.3	SBc? sp	122	NGC 4212 SE 11'.2.
12 15 05.6	+13 35 41	IC 3062	13.9	0.8 x 0.6	12.9	S?	6	**IC 3073** E 7'.6.
12 15 06.9	+12 00 55	IC 3063	13.9	0.9 x 0.5	12.9	Sa	20	Bright nucleus.
12 15 12.5	+14 25 58	IC 3065	13.6	0.9 x 0.6	12.8	S0?	0	Mag 10.70v star N 7'.0.
12 15 46.2	+10 41 55	IC 3074	14.1	2.3 x 0.3	13.6	SB(s)dm: sp	160	**CGCG 69-118** NE 10'.8.
12 15 56.3	+14 25 54	IC 3077	14.1	0.9 x 0.7	13.4	Sd	0	Mag 11.0 star WSW 5'.0; **MCG +3-31-88** NE 8'.9.
12 16 00.1	+12 41 11	IC 3078	14.5	0.5 x 0.5	12.9	Sb(r) I;		**IC 3081** E 2'.2.
12 16 29.1	+14 00 43	IC 3091	14.0	1.3 x 0.6	13.6	S0	126	Mag 13.7 star on N edge; mag 9.88v star N 8'.3.
12 16 56.2	+13 37 28	IC 3094	13.9	0.5 x 0.5	12.2	S?		Mag 11.45v star SW 13'.2.
12 16 52.7	+14 30 49	IC 3096	14.3	1.4 x 0.5	13.7	dS0?	96	Mag 8.17v star ESE 11'.5; **MCG +3-31-88** W 5'.7.
12 17 09.4	+12 27 14	IC 3099	14.2	1.9 x 0.3	13.5	Scd?	172	Mag 9.01v star NW 7'.1.
12 17 05.4	+12 17 21	IC 3100	13.9	1.5 x 0.5	13.4	S0?	58	Mag 12.1 star W 2'.5.
12 17 33.5	+12 23 10	IC 3105	14.1	1.9 x 0.6	14.1	Im:	27	Mag 10.19v star NNE 10'.0.
12 17 46.7	+10 50 41	IC 3107	13.3	1.4 x 0.8	13.3	S?	133	**CGCG 69-136** NE 5'.8; planetary nebula **Koh 2-4** (**PK 275+72.1**) NE 7'.7; note: **PGC 39496** is on E edge of this planetary.
12 17 44.1	+13 10 12	IC 3109	14.7	0.5 x 0.5	13.1	Sc(s)I		Very faint galaxy **VCC 261** W 2'.1.
12 18 00.0	+06 39 11	IC 3115	13.1	1.7 x 1.4	13.8	SB(s)cd	132	Mag 7.95v star WSW 8'.7; **CGCG 41-75** SE 4'.7.
12 18 11.1	+09 29 57	IC 3118	14.1	1.5 x 0.7	14.0	Im:	170	Mag 11.5 star S 3'.0.
12 18 35.3	+11 52 11	IC 3127	15.3	0.6 x 0.5	13.8	Sbc(s)I-II	177	Mag 9.38v star S 4'.8.
12 18 50.9	+07 51 40	IC 3131	13.8	1.4 x 1.4	14.4	S0/a		Mag 11.31v star S 5'.5.
12 19 22.0	+07 52 13	IC 3148	14.1	0.7 x 0.7	13.2	SB?		Mag 11.31v star SW 11'.1; **IC 3150** S 4'.6.
12 19 33.0	+09 24 47	IC 3151	14.2	0.9 x 0.4	13.0	SBa(r)I	126	Mag 11.6 star SW 1'.7; mag 9.02v star NNW 3'.8.
12 19 37.0	+05 23 48	IC 3153	14.8	0.5 x 0.5	13.2	Sc(r)I-II		Located in the center of a triangle formed by NGC 4259, NGC 4270 and NGC 4273.
12 19 45.4	+06 00 19	IC 3155	14.1	1.2 x 0.6	13.6	S0 sp	35	On SW edge of NGC 4269.
12 19 44.3	+09 08 50	IC 3156	14.6	0.7 x 0.6	13.5	SBc(s):	39	Mag 9.48v star E 10'.8.
12 20 18.7	+09 32 41	IC 3167	13.7	1.4 x 0.7	13.6	dSB0(3),N:	90	Mag 10.03v star NW 6'.7.
12 20 26.7	+09 25 23	IC 3170	14.3	0.4 x 0.4	12.1	Sbc(s)I-II		Mag 9.36v star ESE 10'.4.
12 20 55.1	+11 00 29	IC 3188	14.4	0.7 x 0.4	12.8	S?	51	Mag 10.52v star NW 5'.2.
12 22 07.4	+08 59 25	IC 3211	14.5	0.8 x 0.8	13.9	Sd		Located 3'.4 S of NGC 4307.
12 22 19.6	+06 55 37	IC 3218	14.7	0.4 x 0.4	12.8	E?		Mag 9.61v star SE 12'.4.
12 22 38.9	+06 40 33	IC 3225	13.9	1.8 x 0.6	13.8	Sdm:	40	**IC 3229** E 3'.5.
12 23 44.1	+12 28 35	IC 3258	13.1	1.6 x 1.4	13.8	IB(s)m pec:	109	Mag 10.32v star N 6'.6.
12 23 48.6	+07 11 10	IC 3259	13.5	1.7 x 0.9	13.8	SAB(s)dm?	15	Mag 10.54v star W 3'.2.
12 24 05.7	+07 02 27	IC 3267	13.4	1.2 x 1.2	13.7	SA(s)cd		Stellar nucleus; either a very small companion galaxy, or star, on E edge.
12 24 07.5	+06 36 24	IC 3268	13.3	0.9 x 0.8	12.8	Pec	48	Mag 12.8 star on E edge; mag 10.28v star E 8'.9.
12 24 14.0	+07 57 08	IC 3271	13.8	1.0 x 1.0	13.7	SABc		NGC 4353 S 10'.6.
12 25 15.2	+12 42 50	IC 3303	13.8	0.9 x 0.6	13.2	E$^+$?	73	Located 8'.3 NW of NGC 4388.
12 25 33.2	+12 15 37	IC 3311	14.3	1.7 x 0.3	13.4	Sdm:	135	Very faint galaxy **VCC 823** NNE 3'.6.
12 25 36.5	+15 49 46	IC 3313	14.2	0.8 x 0.8	13.5	Sa		**MCG +3-32-32** N 1'.1.
12 25 54.4	+07 33 12	IC 3322	13.5	2.3 x 0.5	13.5	SAB(s)cd: sp	156	Mag 10.28v star W 4'.6.
12 25 42.6	+07 13 04	IC 3322A	12.9	3.5 x 0.4	13.1	SB(s)cd: sp	157	Mag 10.19v star NW 5'.6.
12 25 58.0	+10 03 12	IC 3328	13.7	0.7 x 0.7	13.0	E?		Located 9'.1 E of NGC 4380.
12 26 05.5	+11 48 39	IC 3331	14.4	1.0 x 0.6	13.7	S?	48	Mag 8.81v star NW 12'.3; note: **IC 3305** is 5'.2 S of this star.
12 26 32.4	+13 34 40	IC 3344	14.6	0.8 x 0.3	13.1	E?	48	Mag 11.4 star N 8'.6.
12 26 50.8	+13 10 30	IC 3355	14.9	1.2 x 0.5	14.2	Im IV-V	172	Mag 10.57v star N 8'.4.
12 26 50.7	+11 33 33	IC 3356	14.6	1.5 x 0.9	14.8	Im	90	Low surface brightness; mag 13.5 star on W edge; mag 10.91v star E 9'.2.
12 26 54.5	+11 39 48	IC 3358	13.3	1.2 x 0.8	13.3	E?	120	Mag 10.91v star SE 9'.6.
12 27 03.2	+12 33 39	IC 3363	14.3	1.3 x 0.4	13.6	E?	126	Mag 12.9 star W 3'.4; NGC 4413 WNW 8'.2.
12 27 11.4	+15 53 48	IC 3365	13.8	1.7 x 0.7	13.8	Im	72	Mag 111.4 star S 2'.3; **MCG +3-32-42** N 4'.9; **IC 3369** N 7'.8.
12 27 22.3	+10 52 00	IC 3371	14.7	1.7 x 0.2	13.4	Scd:	55	Star on E edge.
12 28 15.0	+11 47 21	IC 3381	13.4	1.3 x 0.8	13.5	E$^+$:	110	Located 2'.3 S of mag 7.89v star.
12 28 43.3	+14 59 57	IC 3392	12.2	2.3 x 0.9	12.8	SAb:	40	Mag 11.6 star W 2'.5.
12 28 41.8	+12 54 53	IC 3393	13.9	1.3 x 0.4	13.0	S0$^+$:	129	Mag 11.5 star NE 6'.2; **IC 3388** SSW 6'.6.
12 29 22.5	+11 26 01	IC 3413	13.7	1.1 x 0.7	13.4	E?	160	Mag 8.53v star NNE 4'.3.
12 29 28.9	+06 46 14	IC 3414	13.3	1.5 x 0.9	13.5	SABdm?	35	Mag 8.04v star NW 8'.2.
12 29 43.3	+11 24 01	IC 3418	13.8	1.5 x 1.0	14.1	IBm: V	48	Mag 8.53v star NNW 6'.0.
12 29 56.4	+10 36 55	IC 3425	13.6	1.6 x 0.7	13.6	S0?	35	Mag 11.6 star NE 3'.2.
12 30 27.8	+14 09 34	IC 3432	14.6	0.7 x 0.4	13.0	S?	54	Mag 10.36v star W 11'.9; NGC 4474 SW 9'.9.
12 31 51.4	+12 39 24	IC 3457	13.9	1.0 x 0.7	13.5	E3:	135	Very faint galaxy **VCC 1399** SE 3'.2.
12 31 56.0	+12 10 26	IC 3459	14.2	1.1 x 0.8	13.9	Im:	159	Mag 8.07v star SW 4'.9.
12 32 14.4	+10 15 02	IC 3468	13.1	1.4 x 1.2	13.7	E	167	Mag 8.55v star W 9'.1.
12 32 23.4	+11 15 45	IC 3470	13.3	0.6 x 0.6	12.2	E?		Compact; located 6'.8 NNE of NGC 4503.
12 32 41.2	+12 46 19	IC 3475	13.1	1.7 x 1.6	14.2	E:	86	Low surface brightness; mag 11.7 star ESE 8'.3.
12 32 41.9	+14 02 56	IC 3476	12.7	2.0 x 1.4	13.7	IB(s)m:	30	Mag 9.79v star ESE 8'.2.
12 32 44.3	+14 11 44	IC 3478	13.3	1.1 x 0.9	13.1	SAB0:	105	Close pair of mag 12.0, 12.1 stars N 6'.3.
12 32 52.3	+11 24 13	IC 3481	13.3	0.7 x 0.7	12.5	SAB0$^-$: pec		Almost stellar **IC 3481A** SE 1'.4; mag 10.08 star SE 5'.4; note: **IC 3483** on S edge of this star.
12 33 14.2	+12 51 24	IC 3486	14.1	0.7 x 0.5	13.0	E?	45	**IC 3492** E 1'.4.
12 33 13.5	+09 23 47	IC 3487	14.3	0.7 x 0.4	13.0	E6:	84	Mag 11.9 star E 2'.0; mag 9.79v star NE 12'.5.
12 33 13.8	+12 14 43	IC 3489	14.4	0.6 x 0.6	13.1	S?		Mag 10.43v star WSW 12'.0.
12 33 45.3	+10 59 42	IC 3499	13.3	1.4 x 0.5	12.7	S0/a	125	Mag 7.60v star SE 9'.7.
12 33 51.8	+13 19 20	IC 3501	13.9	0.7 x 0.7	13.2	d:E1		Mag 10.92v star NW 3'.7.
12 34 14.9	+11 04 12	IC 3510	14.0	1.0 x 0.7	13.4	S?	0	Mag 13.8 star on NE edge; mag 7.60v star S 9'.1.
12 34 30.9	+09 09 14	IC 3517	14.8	1.0 x 0.6	14.1	Sdm:	15	Mag 12.1 star NW 7'.1; NGC 4522 W 12'.8.
12 34 31.5	+09 37 24	IC 3518	14.2	1.2 x 0.7	13.9	Im:	33	Mag 10.01v star NE 2'.8.
12 34 39.7	+07 09 32	IC 3521	13.2	1.4 x 0.9	13.3	IBm	27	Mag 11.8 star W 6'.8.
12 34 45.9	+15 13 13	IC 3522	14.7	1.4 x 0.6	14.4	IBm: sp V	95	Mag 12.01v star NW 6'.0.
12 34 49.3	+17 48 51	IC 3530	14.0	1.2 x 0.9	13.9	S0	162	Mag 11.6 star SW 9'.0.
12 35 27.3	+12 45 00	IC 3540	14.0	0.7 x 0.7	13.1	S0?		Mag 10.04v star ESE 9'.0.
12 36 37.7	+06 37 12	IC 3576	13.5	2.3 x 2.1	15.0	Sm IV-V	30	Mag 10.26v star ESE 9'.9.

(Continued from previous page)

GALAXIES

RA h m s	Dec ° ′ ″	Name	Mag (V)	Dim ′ Maj x min	SB	Type Class	PA	Notes
12 36 43.8	+13 15 33	IC 3583	12.7	2.2 x 1.1	13.5	IBm	0	Mag 11.5 star on E edge; located 6′1 N of the center of NGC 4569 (M 90).
12 36 54.8	+12 31 12	IC 3586	13.6	1.1 x 1.0	13.6	dS0:	90	Stellar nucleus; mag 10.80v star ESE 4′9.
12 37 03.1	+06 55 31	IC 3591	13.8	1.2 x 0.6	13.3	IBm	48	Multi-galaxy system.
12 38 37.3	+10 28 34	IC 3608	13.8	3.1 x 0.4	13.9	Sb: III-IV	95	Mag 11.10v star N 2′5.
12 39 04.3	+13 21 49	IC 3611	13.3	1.4 x 0.8	13.3	S?	137	Pair of E-W oriented mag 9.70v, 9.91v stars S 7′1.
12 39 25.0	+07 57 56	IC 3617	14.2	1.4 x 0.7	14.0	Im IV-V	65	Mag 11.2 star NE 5′4.
12 39 48.2	+12 58 25	IC 3631	13.7	0.9 x 0.6	12.8	S?	90	Mag 11.28v star W 5′1.
11 21 10.1	+05 21 47	MCG +1-29-35	13.3	1.0 x 1.0	13.3	E		
11 27 46.1	+06 08 37	MCG +1-29-42	14.2	0.6 x 0.5	12.7	SA(rs)b III-IV	57	Mag 11.9 star S 6′9.
11 31 30.0	+06 03 55	MCG +1-29-48	14.0	0.7 x 0.5	12.7	(R′)SAB(r)ab	114	Star W edge.
11 36 39.7	+06 17 28	MCG +1-30-2	13.7	0.7 x 0.4	12.2	SA(r)0° pec	24	Mag 7.14v star E 9′6.
11 39 31.2	+07 45 45	MCG +1-30-7	14.2	0.8 x 0.5	13.1	SB(r)d III-IV	33	
11 44 22.4	+08 10 28	MCG +1-30-12	14.3	0.9 x 0.4	13.0	SB?	54	Mag 13.0 star SE 1′4.
11 49 52.3	+06 39 58	MCG +1-30-16	13.9	1.0 x 0.5	13.1	SA(r)0°:	78	Mag 12.9 star E 2′4.
11 52 01.8	+06 24 31	MCG +1-30-18	13.9	0.6 x 0.6	12.7	SB(r)0⁻:		
11 56 05.0	+05 31 10	MCG +1-31-2	15.4	0.3 x 0.3	12.6	SBb		
11 58 02.8	+05 54 45	MCG +1-31-7	15.0	0.2 x 0.2	11.3			
12 01 44.3	+05 49 16	MCG +1-31-9	14.2	0.8 x 0.5	13.1	SB?	102	
12 01 42.8	+06 26 57	MCG +1-31-10	14.2	1.0 x 0.7	13.7		54	Mag 7.86v star NW 5′0.
12 02 54.6	+05 36 44	MCG +1-31-13	15.5	0.5 x 0.3	13.3	SBb		
12 04 04.4	+06 54 33	MCG +1-31-19	14.7	0.6 x 0.4	13.0	Sc pec	123	
12 03 56.0	+07 17 50	MCG +1-31-20	14.9	0.7 x 0.2	12.7	Sc	111	
12 15 18.4	+05 45 35	MCG +1-31-30	14.5	0.5 x 0.5	13.1	E0 pec?		NGC 4197 W 10′3.
12 19 22.2	+05 54 38	MCG +1-31-53	14.5	0.6 x 0.4	12.8		171	Galaxy **VCC 336** SSW 2′4.
12 19 51.0	+06 59 29	MCG +1-32-6	14.7	0.6 x 0.5	13.2	SBc(s)II	105	Mag 10.51v star W 3′0.
12 24 03.2	+05 10 55	MCG +1-32-41	13.9	0.9 x 0.8	13.4	SB?	138	
12 25 52.8	+05 48 29	MCG +1-32-56	13.8	0.8 x 0.2	11.7	IB?	30	
12 28 59.8	+07 51 02	MCG +1-32-76	14.4	0.8 x 0.5	13.2	Sb(r)II	168	
12 35 18.4	+06 32 34	MCG +1-32-108	15.3	0.8 x 0.2	13.2	S0(8)	120	
12 36 34.9	+08 03 13	MCG +1-32-111	14.2	1.0 x 0.8	13.8	S?	39	
11 40 41.6	+12 34 47	MCG +2-30-9	15.5	0.4 x 0.3	13.0	Sc	141	
11 42 16.8	+14 03 58	MCG +2-30-17	14.8	0.7 x 0.5	13.8	E/S0	3	
11 43 29.6	+10 20 41	MCG +2-30-21	14.8	0.6 x 0.4	13.1	Sc	3	Mag 12.0 star SE 2′7; mag 8.78v star E 8′4.
11 45 34.7	+12 12 20	MCG +2-30-31	16.0	0.5 x 0.5	14.3	Sab		
11 46 12.7	+10 35 41	MCG +2-30-35	13.6	0.9 x 0.7	13.0	I?	63	Mag 12.4 star SW 3′5.
11 51 10.3	+10 38 04	MCG +2-30-42	15.3	0.7 x 0.2	13.0	Sb	108	Mag 13.6 star SW 2′3; CGCG 68-76 SW 9′7.
11 59 34.0	+13 53 11	MCG +2-31-7	14.1	0.8 x 0.4	12.7	S?	39	Stellar **CGCG 69-11** S 6′5.
12 00 29.8	+11 59 33	MCG +2-31-10	14.7	0.4 x 0.3	12.2		126	
12 03 36.0	+08 23 17	MCG +2-31-18	14.0	0.5 x 0.4	12.2	S0	120	**MCG +2-31-17** SW 2′2.
12 04 19.4	+08 48 34	MCG +2-31-20	15.4	0.7 x 0.4	13.9	Sc	54	Mag 12.8 star N 2′4.
12 05 36.4	+08 59 15	MCG +2-31-27	14.4	0.7 x 0.3	12.6	S?	136	UGC 7085 E 1′5.
12 23 44.5	+09 20 20	MCG +2-32-22	14.6	0.8 x 0.2	12.5	Sb	114	
12 24 08.1	+09 23 04	MCG +2-32-25	14.4	0.4 x 0.4	12.3	$S0_1(0)$		
12 37 41.5	+08 33 30	MCG +2-32-161	14.0	1.4 x 0.6	13.7	S?	126	
11 34 21.3	+15 39 36	MCG +3-30-7	14.8	0.7 x 0.3	12.9	Irr	156	Mag 7.3 star WSW 4′1.
11 35 00.1	+15 41 07	MCG +3-30-11	14.4	0.5 x 0.5	12.7	Sc		
11 36 16.8	+15 28 14	MCG +3-30-15	14.9	0.6 x 0.5	13.5	Spiral	111	**MCG +3-30-16** N 1′7.
11 39 56.3	+16 57 13	MCG +3-30-33	15.0	0.5 x 0.2	12.3		6	= **Mkn 745**; **CGCG 97-42** NW 2′2.
11 40 16.2	+17 27 26	MCG +3-30-38	13.7	0.9 x 0.7	13.1	S?	144	
11 41 43.3	+17 00 26	MCG +3-30-44A	15.2	0.9 x 0.3	13.7	Sc	165	Mag 13.4 star on NE edge.
11 42 42.0	+15 21 24	MCG +3-30-56	15.1	0.5 x 0.5	13.4	Sbc		
11 44 20.1	+15 06 18	MCG +3-30-80	15.4	0.6 x 0.5	14.0	Spiral	27	
11 49 20.2	+15 24 35	MCG +3-30-117	14.9	0.5 x 0.5	13.2			
11 54 20.3	+16 14 27	MCG +3-30-125	14.7	0.7 x 0.4	13.1	Sb	90	
11 58 47.2	+15 42 54	MCG +3-31-7	14.1	0.8 x 0.5	12.9		174	Mag 12.3 star SW 2′0.
12 01 44.4	+17 54 01	MCG +3-31-15	14.0	0.8 x 0.6	13.1	Sb	138	**CGCG 98-26** E 4′9.
12 03 20.6	+16 30 31	MCG +3-31-25	13.9	0.6 x 0.5	12.4		33	UGC 7032 SE 2′2.
12 03 31.1	+17 08 56	MCG +3-31-28	14.4	1.0 x 0.5	13.5		66	Mag 11.1 star W 4′9.
12 03 36.1	+16 03 15	MCG +3-31-30	13.9	0.7 x 0.4	12.4	Sb	108	
12 07 09.7	+16 59 40	MCG +3-31-41	13.6	0.8 x 0.6	12.7	SBb?	162	**MCG +3-31-42** SSE 1′7.
12 09 32.9	+17 00 48	MCG +3-31-49	13.6	0.9 x 0.7	13.0	S?	33	
12 11 15.2	+17 53 13	MCG +3-31-61	13.7	0.5 x 0.5	12.1	S?		**MCG +3-31-56** and stellar **Mkn 758** W 7′9.
12 11 38.2	+16 28 39	MCG +3-31-64	14.2	0.7 x 0.5	12.9	Compact	150	Mag 10.72v star NE 4′4.
12 13 09.2	+16 17 47	MCG +3-31-75	13.9	0.7 x 0.4	12.4	SA0	135	
12 17 27.3	+17 38 59	MCG +3-31-94	13.8	0.8 x 0.5	12.8	E/S0	135	Mag 11.1 star S 2′7.
12 20 14.5	+17 20 45	MCG +3-32-3	13.9	0.5 x 0.4	12.0	Sc(r)II	174	Mag 8.52v star SW 4′6.
11 20 15.4	+12 58 59	NGC 3627	8.9	9.1 x 4.2	12.7	SAB(s)b II	173	= **M 66**. Many moderately dark lanes on either side of the bright center. A mag 9.79v star on NW edge. Member of **Leo Triplet** (**"Trio in Leo"**) with M 65 and NGC 3628.
11 20 18.2	+13 35 06	NGC 3628	9.5	14.8 x 3.0	13.4	Sb pec sp III	104	Prominent dark lane runs almost full length down the center of the galaxy. Member of **Leo Triplet** (**"Trio in Leo"**) with M 65 and M 66.
11 22 54.6	+16 35 20	NGC 3655	11.7	1.5 x 1.0	11.9	SA(s)c: III-IV	30	
11 23 45.1	+17 49 02	NGC 3659	12.2	2.1 x 1.1	12.9	SB(s)m? III-IV	60	
11 24 25.8	+11 20 28	NGC 3666	12.0	4.4 x 1.0	13.5	SA(rs)c: III	100	Mag 5.79v star NE 9′6.
11 26 29.7	+16 51 48	NGC 3681	11.2	2.0 x 2.0	12.5	SAB(r)bc I-II		
11 27 11.3	+17 01 45	NGC 3684	11.4	3.1 x 2.1	13.3	SA(rs)bc II-III	130	Mag 10.61v star E 7′8; located on S edge of small galaxy cluster **A 1264**.
11 27 44.1	+17 13 20	NGC 3686	11.3	3.2 x 2.5	13.4	SB(s)bc II	15	Many small knots overall.
11 28 09.4	+16 55 14	NGC 3691	11.8	1.3 x 1.0	11.9	SBb?	15	
11 28 24.0	+09 24 28	NGC 3692	12.1	3.2 x 0.7	12.9	Sb I-II	95	Mag 8.60v star NW 10′3.
11 30 07.1	+09 16 39	NGC 3705	11.1	4.9 x 2.0	13.4	SAB(r)ab II	122	Mag 9.28v star NE 9′7; note: **IC 2887** lies 1′2 SW of this star.
11 34 11.8	+12 30 44	NGC 3731	13.9	1.0 x 0.9	13.7	E	50	
11 36 54.2	+17 53 20	NGC 3764	14.7	0.9 x 0.7	14.0	S?	123	Almost stellar nucleus; faint extension NW; NGC 3768 SE 5′5.
11 37 15.6	+16 52 36	NGC 3767	13.4	1.0 x 0.9	13.2	(R)SB(r)0°?	75	Short E-W bar.
11 37 14.5	+17 50 20	NGC 3768	12.4	1.6 x 0.9	12.7	S0	155	NGC 3764 NW 5′5.
11 38 13.0	+12 06 44	NGC 3773	12.0	1.2 x 1.0	12.1	SA0:	165	Compact core in uniform envelope.
11 39 47.2	+17 42 44	NGC 3790	13.9	1.2 x 0.3	12.6	S0/a	154	**MCG +3-30-35** E 4′5.
11 40 09.5	+15 19 37	NGC 3799	13.9	0.7 x 0.4	12.4	SB(s)b: pec	114	Located on S edge of NGC 3800.
11 40 13.3	+15 20 28	NGC 3800	12.7	2.0 x 0.6	12.7	SAB(rs)b: pec	52	NGC 3799 on S edge.
11 40 17.0	+17 43 40	NGC 3801	12.0	2.5 x 1.6	13.4	S0?	120	**MCG +3-30-35** SW 3′3; NGC 3802 N 2′2.
11 40 19.0	+17 45 54	NGC 3802	13.3	1.1 x 0.3	12.0	S	85	NGC 3803 N 2′1.

(Continued from previous page)

RA h m s	Dec ° ′ ″	Name	Mag (V)	Dim ′ Maj x min	SB	Type Class	PA	Notes
11 40 17.4	+17 48 04	NGC 3803		0.4 x 0.4				Stellar.
11 40 46.7	+17 47 45	NGC 3806	13.6	2.2 x 1.8	14.9	SABb	169	Almost stellar nucleus; mag 9.21v star S 5′.0.
11 40 58.6	+11 28 10	NGC 3810	10.8	4.3 x 3.0	13.4	SA(rs)c I-II	15	Many knots and/or superimposed stars.
11 41 53.0	+10 18 17	NGC 3817	13.3	1.0 x 0.9	13.1	SB0/a	140	Mag 10.99v star N 2′.8 faint, stellar anonymous galaxy S 1′.9.
11 42 05.9	+10 21 02	NGC 3819	13.8	0.6 x 0.5	12.6	E?	136	NGC 3820 N 2′.0.
11 42 04.9	+10 22 59	NGC 3820	14.5	0.5 x 0.4	12.7	S?	35	NGC 3919 S 2′.0.
11 42 11.1	+10 16 38	NGC 3822	13.1	1.4 x 0.8	13.1	S0?	178	NGC 3825 E 3′.1.
11 42 23.5	+10 15 50	NGC 3825	13.0	1.3 x 1.0	13.2	SBa	160	Small, faint, stellar anonymous galaxy NE 1′.8; NGC 3822 W 3′.1.
11 42 58.4	+16 29 13	NGC 3828	14.8	0.8 x 0.5	13.7		30	UGC 6686 E 5′.9.
11 43 28.9	+10 09 39	NGC 3833	13.5	1.4 x 0.7	13.3	Sc	27	Mag 9.44v star ESE 12′.9; NGC 3848 NNE 6′.0.
11 43 54.2	+10 47 04	NGC 3839	13.5	1.0 x 0.5	12.6	Sdm:	87	Mag 8.09v star S 6′.1; **CGCG 68-42** W 7′.8.
11 43 54.7	+07 55 32	NGC 3843	13.5	1.0 x 0.4	12.4	S0/a	42	
11 43 41.8	+10 14 41	NGC 3848	14.2	0.7 x 0.4	12.6	Sc	114	Mag 8.76v star NE 7′.7.
11 44 28.4	+16 33 29	NGC 3853	12.4	1.7 x 1.0	13.0	E	140	**CGCG 97-126** E 7′.2.
11 45 06.0	+08 28 12	NGC 3863	12.9	2.8 x 0.6	13.3	Sbc	75	Mag 9.73v star WSW 8′.9; note: **CGCG 68-49** lies 3′.2 SSW of this star.
11 45 45.7	+10 49 30	NGC 3869	12.8	1.9 x 0.5	12.5	Sa	135	
11 45 49.2	+13 45 59	NGC 3872	11.7	1.9 x 1.2	12.6	E5	24	Mag 8.30v star ESE 11′.1; **UGC 6747** NE 9′.3.
11 45 26.6	+09 09 37	NGC 3876	12.8	1.1 x 0.7	12.4	Sab:	105	UGC 6734 SE 3′.4.
11 49 52.7	+12 11 07	NGC 3908	15.0	0.4 x 0.4	12.9			Located 3′.3 SE of mag 10.55v star.
11 50 32.5	+06 34 02	NGC 3914	13.2	1.1 x 0.6	12.6	(R′)SB(rs)b II-III	40	
11 52 02.3	+16 48 35	NGC 3933	13.6	1.1 x 0.6	12.9	S?	83	Much elongated **UGC 6835** NW 2′.8.
11 52 12.4	+16 51 05	NGC 3934	13.6	1.1 x 1.0	13.6	S?	66	Mag 11.31v star NW 6′.0.
11 55 28.8	+11 58 04	NGC 3968	11.8	2.7 x 1.9	13.4	SAB(rs)bc	10	NGC 3973 NE 2′.6.
11 55 37.1	+11 59 45	NGC 3973	15.0	0.6 x 0.3	13.0		12	Located 0′.7 NNW of a mag 9.84v star.
11 55 56.9	+06 44 50	NGC 3976	11.5	3.8 x 1.2	13.0	SAB(s)b I-II	53	**MCG +1-31-1A** SSE 4′.5.
11 57 46.1	+14 17 46	NGC 3996	13.5	0.9 x 0.7	12.8	S?	50	
11 58 27.5	+10 01 18	NGC 4012	13.4	1.9 x 0.6	13.4	Sb III	153	
11 58 35.8	+16 10 36	NGC 4014	12.3	1.4 x 0.9	12.4	S0/a	120	
12 00 03.4	+08 10 50	NGC 4029	13.5	1.2 x 0.7	13.2	Sb:	150	
12 01 23.6	+13 24 01	NGC 4037	11.9	2.5 x 2.0	13.5	SB(rs)b: III	9	Almost stellar nucleus; mag 8.96v star E 5′.3.
12 02 05.4	+17 49 24	NGC 4040	13.3	1.9 x 1.3	14.3	E	145	**CGCG 98-27** S 8′.0.
12 04 11.6	+10 51 14	NGC 4067	12.5	1.4 x 1.0	12.7	SA(s)b: I-II	35	
12 04 47.8	+10 35 45	NGC 4078	13.2	1.3 x 0.5	12.6	S0?	18	
12 05 11.5	+10 40 13	NGC 4082	14.8	0.9 x 0.3	13.3		69	Galaxy **IC 2991** S 1′.9.
12 05 14.2	+10 36 46	NGC 4083	14.6	0.8 x 0.5	13.5		33	Galaxy **IC 2991** N 1′.7; mag 10.27v star SSE 2′.2.
12 08 10.1	+10 22 45	NGC 4124	11.4	4.3 x 1.4	13.2	SA(r)0^+	114	
12 08 37.4	+16 08 34	NGC 4126	13.4	1.0 x 0.8	13.0	S0	0	
12 10 37.3	+16 02 01	NGC 4152	12.2	2.2 x 1.7	13.5	SAB(rs)c II	115	Stellar **MCG +3-31-53** on SE edge.
12 12 05.5	+13 12 17	NGC 4164	14.7	0.6 x 0.3	12.9	E3	111	Stellar, located 2′.9 W of NGC 4168.
12 12 11.8	+13 14 48	NGC 4165	13.5	1.2 x 0.8	13.3	SAB(r)a:? III-IV	160	Stellar nucleus; Located 2′.7 NNW of NGC 4168 and 2′.6 SE of a mag 10.13v star.
12 12 09.6	+17 45 23	NGC 4166	13.1	1.2 x 1.0	13.1	SB0	20	Stellar galaxy **KUG 1209+179** SW 10′.4.
12 12 17.3	+13 12 14	NGC 4168	11.2	2.8 x 2.3	13.2	E2	120	Large core in smooth envelope.
12 12 46.6	+10 52 07	NGC 4178	11.4	5.1 x 1.8	13.7	SB(rs)dm II	30	Three strong clumps or knots S end; mag 7.55v star E 7′.1.
12 13 03.0	+07 02 20	NGC 4180	12.6	1.6 x 0.6	12.4	Sab:	22	
12 14 06.6	+14 43 32	NGC 4186	13.8	1.0 x 0.8	13.4	SA(s)ab:	60	Located near the center of galaxy cluster **A 1499**; **UGC 7223** W 10′.2.
12 13 47.5	+13 25 38	NGC 4189	11.7	2.4 x 1.7	13.1	SAB(rs)cd? II-III	85	Several bright knots E side.
12 13 50.2	+07 12 03	NGC 4191	12.8	1.1 x 0.8	12.6	S0	5	
12 13 48.8	+14 53 52	NGC 4192	10.1	9.8 x 2.8	13.6	SAB(s)ab II	155	= **M 98**.
12 13 53.8	+13 10 21	NGC 4193	12.3	2.2 x 1.0	13.0	SAB(s)c:? III	93	
12 14 38.7	+05 48 21	NGC 4197	12.8	3.4 x 0.6	13.5	Sd	36	MCG +1-31-30 E 10′.3.
12 14 44.4	+12 10 49	NGC 4200	13.0	1.5 x 0.8	13.1	S0	98	Mag 8.00v star W 10′.0.
12 15 16.8	+13 01 21	NGC 4206	12.2	6.2 x 1.0	14.0	SA(s)bc:	0	Extremely faint galaxy **VCC 132** W 3′.2.
12 15 30.2	+09 35 10	NGC 4207	12.6	1.6 x 0.8	12.7	S?	124	
12 15 39.3	+13 54 08	NGC 4212	11.2	3.2 x 1.9	13.0	SAc: III	75	Galaxy **VCC 154** NNW 2′.7.
12 15 54.4	+06 24 00	NGC 4215	12.1	1.9 x 0.7	12.2	SA(r)0^+: sp	174	
12 15 54.3	+13 09 02	NGC 4216	10.0	8.1 x 1.8	12.8	SAB(s)b: II	19	Prominent dark lane runs almost full length of galaxy on E side of center line. **CGCG 69-113** N 4′.0 from galaxy center.
12 16 22.3	+13 18 22	NGC 4222	13.3	3.3 x 0.5	13.7	Sd: sp	56	Mag 8.84v star E 4′.6.
12 16 33.9	+07 27 41	NGC 4224	11.8	2.6 x 1.0	12.7	SA(s)a: sp	57	Strong dust lane length of bright nucleus on NW side.
12 17 07.7	+07 37 22	NGC 4233	11.9	2.4 x 1.1	12.8	S0°	177	Mag 8.92v star ENE 9′.1.
12 17 10.1	+07 11 30	NGC 4235	11.6	4.2 x 0.9	12.9	SA(s)a sp	48	
12 17 11.5	+15 19 25	NGC 4237	11.6	2.1 x 1.3	12.6	SAB(rs)bc III	108	**MCG +31 87** SW 12′.3.
12 17 14.9	+16 31 50	NGC 4239	12.8	1.4 x 0.7	12.9	E	120	A mag 10.31v star lies 2′.1 NE.
12 17 25.7	+06 41 25	NGC 4241	11.9	2.6 x 1.3	13.1	SA(s)0^+:	128	Mag 7.95v star S 5′.4.
12 17 58.2	+07 11 09	NGC 4246	12.7	2.4 x 1.3	13.8	SA(s)c II	83	Bright stellar nucleus, two knotty arms with some branching; NGC 4247 N 5′.2.
12 17 58.0	+07 16 25	NGC 4247	13.5	0.8 x 0.7	12.7	(R)SAB(s)ab pec?	48	NGC 4246 S 5′.2.
12 17 59.4	+05 35 52	NGC 4249	14.0	0.6 x 0.6	12.8	$S0_1(0)$		Almost stellar nucleus; **MCG +3-31-43** N 5′.4.
12 18 31.0	+05 33 35	NGC 4252	14.0	1.5 x 0.4	13.3	Sb? sp	48	Small, faint galaxy **VCC 286** N 3′.5.
12 18 48.1	+14 25 16	NGC 4254	9.9	5.4 x 4.7	13.2	SA(s)c I-II	51	= **M 99**. Very bright nucleus, many branching arms NE, one massive arm W, curving to the north. Numerous knots.
12 19 06.4	+05 43 32	NGC 4257	14.0	1.5 x 0.5	13.5	Sab: sp	78	Very faint galaxy **VCC 321** N 0′.9.
12 19 22.3	+05 22 36	NGC 4259	13.6	1.1 x 0.4	12.5	S0 sp	143	Mag 14.8 star on NE edge.
12 19 22.3	+06 05 55	NGC 4260	11.8	2.7 x 1.3	13.0	SB(s)a	58	Broad bright bar; **CGCG 42-7** W 5′.5; **IC 3136** NW 8′.1.
12 19 23.2	+05 49 30	NGC 4261	10.4	4.1 x 3.6	13.3	E2-3	160	Galaxy **VCC 344** on S edge; **VCC 336** 3′.4 NNW; NGC 4264 ENE 3′.5.
12 19 30.8	+14 52 38	NGC 4262	11.6	1.9 x 1.7	12.7	SB(s)0^-?	120	Mag 9.98v star and IC 781 NE 8′.3.
12 19 35.8	+05 50 51	NGC 4264	12.8	1.0 x 0.8	12.4	SB(rs)0^+	123	Center of NGC 4261 WSW 3′.5.
12 19 41.7	+05 32 16	NGC 4266	13.7	2.0 x 0.4	13.3	SB(s)a? sp	76	Mag 9.21v star on N edge.
12 19 45.5	+12 47 52	NGC 4267	10.9	3.2 x 3.0	13.3	SB(s)0^-?	33	Very bright, diffuse nucleus; mag 10.59v star WSW 11′.7.
12 19 47.2	+05 17 02	NGC 4268	12.8	1.6 x 0.6	12.6	SB0/a: sp	48	Mag 11.6 star S 3′.7; located on E edge of galaxy cluster **A 1516**.
12 19 49.2	+06 00 52	NGC 4269	12.9	1.1 x 0.8	12.7	$S0^+$	137	Located 1′.6 S of mag 7.71v star. IC 3155 on SW edge.
12 19 49.8	+05 27 49	NGC 4270	12.2	2.0 x 0.9	12.7	S0	110	Mag 9.12v star and NGC 4266 N 5′.4.
12 19 55.6	+05 20 27	NGC 4273	11.9	2.3 x 1.5	13.1	SB(s)c II-III	10	NGC 4277 E 1′.9.
12 20 07.4	+07 41 27	NGC 4276	12.5	1.6 x 1.4	13.2	SB?	3	Stellar **CGCG 42-27** NW 4′.2; **IC 3150** NW 11′.6; stellar **CGCG 42-30** N 7′.4.
12 20 03.7	+05 20 28	NGC 4277	13.4	1.0 x 0.9	13.1	SAB(rs)0/a:	141	Located 1′.9 E of NGC 4273.
12 20 21.4	+05 23 15	NGC 4281	11.3	3.0 x 1.5	12.8	$S0^+$: sp	88	NGC 4277 and NGC 4273 WSW 6′.0.
12 20 24.2	+05 34 20	NGC 4282	14.0	1.2 x 0.6	13.5	S0?	105	Mag 7.29v star NW 9′.2.
12 20 48.5	+05 38 23	NGC 4287	14.3	1.2 x 0.2	12.6	S(on-edge)	72	Mag 14.0 star on N edge; mag 7.29v star W 13′.1.
12 21 17.6	+11 30 40	NGC 4294	12.1	3.2 x 1.2	13.4	SB(s)cd III	155	Bright knot or star NW edge; faint galaxy **VCC 453** NNW 5′.9.
12 21 28.3	+06 39 08	NGC 4296	12.7	1.4 x 0.7	12.5	S0	15	Star or bright knot S end; double system with NGC 4297 N.

GALAXIES

RA h m s	Dec ° ′ ″	Name	Mag (V)	Dim ′ Maj x min	SB	Type Class	PA	Notes
12 21 27.5	+06 40 17	NGC 4297	14.6	0.6 x 0.3	12.7	S0	168	Double system with NGC 4296 S.
12 21 32.7	+14 36 21	NGC 4298	11.3	3.2 x 1.8	13.1	SA(rs)c III-IV	140	Many faint knots.
12 21 40.5	+11 30 03	NGC 4299	12.5	1.7 x 1.6	13.4	SAB(s)dm: III-IV	26	Mottled, patchy center.
12 21 41.3	+05 23 02	NGC 4300	12.9	1.5 x 0.6	12.7	Sa	42	Mag 6.46v star ESE 13'.3.
12 21 42.3	+14 35 59	NGC 4302	11.6	5.5 x 1.0	13.3	Sc: sp	178	Strong, dark lane runs almost full length down the center of the galaxy.
12 22 03.8	+12 44 27	NGC 4305	12.6	2.2 x 1.2	13.5	SA(r)a	32	Almost stellar core in smooth envelope.
12 22 04.3	+12 47 13	NGC 4306	12.6	1.6 x 1.3	13.3	SB(s)0°:	140	Compact core, faint envelope.
12 22 05.9	+09 02 42	NGC 4307	12.0	3.6 x 0.8	13.0	Sb sp II-III	24	Smooth, uniform appearance. IC 3211 S 3'.2.
12 22 12.7	+07 08 34	NGC 4309	12.7	1.8 x 0.9	13.1	SAB(r)0$^+$	85	Galaxy **VCC 538** N 1'.5.
12 22 31.3	+15 32 07	NGC 4312	11.7	4.6 x 1.1	13.3	SA(rs)ab: sp III	170	Galaxies **VCC 583** SE 3'.9; **VCC 554** SW 4'.3.
12 22 38.7	+11 47 57	NGC 4313	11.6	4.0 x 1.0	13.0	SA(rs)ab: sp	143	Galaxy **VCC 560** W 1'.6.
12 22 41.9	+09 20 00	NGC 4316	12.9	2.5 x 0.5	13.0	Scd?	113	
12 22 43.3	+08 11 50	NGC 4318	13.3	0.7 x 0.5	12.2	E?	65	Compact; mag 9.54v star N 4'.4.
12 22 57.6	+10 32 36	NGC 4320	13.9	1.1 x 0.7	13.4	S?	170	Almost stellar nucleus; faint galaxy **VCC 546** WNW 10.3.
12 22 55.9	+15 49 12	NGC 4321	9.4	7.4 x 6.3	13.4	SAB(s)bc I	30	= **M 100**. Very bright nucleus, patchy spiral arms; IC 783A SW 10'.1.
12 23 01.8	+15 54 21	NGC 4322	13.9	0.9 x 0.6	13.2	SB(r)0°:	140	Diffuse. Galaxy **VCC 619** NNE 1'.8.
12 23 06.2	+05 15 02	NGC 4324	11.6	2.8 x 1.2	12.8	SA(r)0$^+$	53	Mag 6.46v star NW 8'.9.
12 23 06.8	+10 37 17	NGC 4325	13.3	1.0 x 0.7	12.9	E?	9	Stellar **PGC 40237** NE 7'.3.
12 23 11.6	+06 04 20	NGC 4326	13.3	1.5 x 1.1	13.7	SAB(r)ab:	108	Bright, almost stellar nucleus.
12 23 20.1	+15 49 14	NGC 4328	13.0	1.3 x 0.9	13.1	SA0$^-$:	90	Galaxy **VCC 636** N 2'.9.
12 23 17.2	+11 22 03	NGC 4330	12.4	4.5 x 0.9	13.8	Scd?	59	Mag 9.57v star NW 4'.3; mag 10.28v star N 4'.7.
12 23 22.4	+06 02 27	NGC 4333	13.6	0.8 x 0.7	12.8	SB(s)ab	171	Bright, narrow E-W bar.
12 23 23.9	+07 28 22	NGC 4334	13.0	2.3 x 1.1	13.8	SB(s)ab	135	Very small, bright nucleus on a fairly bright, thin bar; star on S edge of galaxy.
12 23 35.1	+06 04 53	NGC 4339	11.4	1.9 x 1.7	12.7	E0	0	Mag 11.2 star S 1'.5.
12 23 35.5	+16 43 18	NGC 4340	11.2	3.0 x 2.0	13.0	SB(r)0$^+$	102	Strong core in weak N-S oriented bar.
12 23 53.6	+07 06 22	NGC 4341	13.2	1.7 x 0.6	13.2	SAB(s)0° sp	96	Mag 10.54v star NW 6'.0; IC 3259 NNW 4'.9.
12 23 39.1	+07 03 15	NGC 4342	12.5	1.2 x 0.6	12.0	S0$^-$ sp	168	IC 3267 E 6'.6.
12 23 38.8	+06 57 13	NGC 4343	12.1	2.5 x 0.7	12.5	SA(rs)b: III	133	Mag 9.61v star SW 10'.7.
12 23 37.5	+17 32 29	NGC 4344	12.3	1.7 x 1.6	13.3	SB0?	90	Faint galaxy **VCC 696** E 7'.5.
12 23 57.8	+16 41 35	NGC 4350	11.0	3.0 x 1.4	12.4	SA0 sp	28	
12 24 01.5	+12 12 16	NGC 4351	12.6	2.0 x 1.3	13.5	SB(rs)ab pec:	80	
12 24 05.1	+11 13 04	NGC 4352	12.6	2.1 x 1.0	13.3	SA0: sp	102	Mag 9.06v star ESE 6'.4.
12 24 00.2	+07 47 06	NGC 4353	13.6	0.9 x 0.6	12.8	IBm:	81	Mag 13.4 star N 1'.2. IC 3271 N 10'.6.
12 24 14.8	+08 32 07	NGC 4356	13.3	2.8 x 0.5	13.5	Scd:	40	
12 24 21.6	+09 17 32	NGC 4360	12.3	1.4 x 1.2	12.9	E	145	
12 24 28.1	+07 19 01	NGC 4365	9.6	6.9 x 5.0	13.4	E3	40	Very bright center. NGC 4366 lies 5'.7 NE of center.
12 24 47.1	+07 21 10	NGC 4366	14.3	0.9 x 0.6	13.7	E	51	
12 24 55.1	+07 26 41	NGC 4370	12.6	1.4 x 0.7	12.4	Sa sp	83	Faint, stellar galaxy **VCC 762** NE 4'.1.
12 24 55.5	+11 42 15	NGC 4371	10.8	4.0 x 2.2	13.0	SB(r)0$^+$	95	
12 25 03.7	+12 53 08	NGC 4374	9.1	6.5 x 5.6	13.0	E1	135	= **M 84**.
12 25 18.1	+05 44 26	NGC 4376	13.3	1.4 x 0.9	13.4	Im	157	Pair of faint stars or bright knots on W side.
12 25 12.2	+14 45 40	NGC 4377	11.9	1.7 x 1.4	12.7	SA0$^-$	177	Two very small, very faint galaxies are visible through the envelope of NGC 4377. The first is just NE of the nucleus; the second is W of the nucleus, mag 8.90v star ESE 3'.4.
12 25 14.9	+15 36 26	NGC 4379	11.7	1.9 x 1.6	12.8	S0$^-$ pec:	105	Very faint galaxy **VCC 761** W 3'.4.
12 25 22.2	+10 01 00	NGC 4380	11.7	3.5 x 1.9	13.6	SA(rs)b:?	153	
12 25 25.4	+16 28 09	NGC 4383	12.1	1.9 x 1.0	12.7	Sa? pec	28	**UGC 7504** S 2'.5.
12 25 41.7	+12 48 37	NGC 4387	12.1	1.8 x 1.1	12.9	E	140	Located 10'.1 SE of NGC 4374 (M 84).
12 25 47.4	+12 39 42	NGC 4388	11.0	5.6 x 1.3	13.0	SA(s)b: sp II-III	92	Bright central lens with two short dust lanes; IC 3303 NW 8'.3.
12 25 50.9	+10 27 33	NGC 4390	12.6	1.7 x 1.3	13.3	SAB(s)c:	95	Bright, patchy center.
12 25 59.0	+15 40 16	NGC 4396	12.6	3.3 x 1.0	13.7	SAd: sp	125	Brighter patches SE half, star or bright knot NW end.
12 26 07.0	+13 06 43	NGC 4402	11.7	3.9 x 1.1	13.2	Sb sp	90	Strong dark lane from center stretching E; mag 10.05v star W 12'.4.
12 26 07.2	+16 10 50	NGC 4405	12.0	1.8 x 1.1	12.6	SA(rs)0/a:	20	Mag 8.20v star W 9'.5; note: **IC 787** lies 2'.9 S of this star.
12 26 12.1	+12 56 45	NGC 4406	8.9	8.9 x 5.8	13.2	E3	130	= **M 86**. Small, faint galaxy **VCC 882** 1'.3 NE of bright center of M 86. **VCC 902** on edge 4'.2 SE of center.
12 26 29.9	+09 01 05	NGC 4410A	12.8	0.9 x 0.8	12.3	Sab? pec	114	Galaxy IC 790 NE 1'.7.
12 26 28.2	+09 01 11	NGC 4410B	13.9	0.7 x 0.6	12.8	S0? pec	168	Close pair with NGC 4410A, common envelope.
12 26 29.2	+08 52 13	NGC 4411A	12.7	2.0 x 1.9	14.0	SB(rs)c	30	Bright stellar nucleus, several faint superimposed stars.
12 26 47.2	+08 53 02	NGC 4411B	12.3	2.5 x 2.5	14.2	SAB(s)cd		Bright almost stellar nucleus, complex branching arms.
12 26 32.3	+12 36 37	NGC 4413	12.3	2.3 x 1.5	13.4	(R′)SB(rs)ab:	60	Mag 10.89v star N 3'.0.
12 26 40.5	+08 26 10	NGC 4415	12.1	1.3 x 1.2	12.4	S0/a	0	Faint galaxy **VCC 888** SW 7'.4.
12 26 47.0	+07 55 09	NGC 4416	12.4	1.7 x 1.5	13.3	SB(rs)cd:	108	Mag 7.94v star SW 3'.7; mag 8.06v star NE 11'.8; note: **CGCG 42-113** lies 3'.0 S of this star.
12 26 50.5	+09 35 03	NGC 4417	11.1	3.4 x 1.3	12.6	SB0: sp	49	Elongated nucleus; very small, faint anonymous galaxy SE 2'.7.
12 26 56.5	+15 02 57	NGC 4419	11.2	3.3 x 1.1	12.4	SB(s)a sp	133	Faint dark lane NE side of center.
12 27 02.6	+15 27 46	NGC 4421	11.6	2.7 x 2.0	13.3	SB(s)0/a	20	Bright, elongated nucleus; mag 9.79v star NW 2'.4.
12 27 09.2	+05 52 53	NGC 4423	13.4	2.3 x 0.4	13.2	Sdm:	18	
12 27 11.9	+09 25 11	NGC 4424	11.7	3.6 x 1.8	13.5	SB(s)a:	95	Very small galaxy, paired with a star E, 0'.6 S of center; possibly **IC 3366**?
12 27 13.3	+12 44 01	NGC 4425	11.8	3.0 x 1.0	12.9	SB0$^+$: sp	27	Long bright core with faint halo, very faint star S end.
12 27 26.2	+11 06 26	NGC 4429	10.0	5.6 x 2.6	12.8	SA(r)0$^+$	99	Bright nucleus with moderately dark patches E and W; mag 9.07v star N 2'.0.
12 27 26.1	+06 15 51	NGC 4430	12.0	2.3 x 2.0	13.5	SB(rs)b:	51	Faint, narrow lane N of nucleus with bright knot and star N of lane.
12 27 27.4	+12 17 26	NGC 4431	12.9	1.7 x 1.1	13.4	SA(r)0	177	Mag 9.16v star SSE 5'.3.
12 27 33.0	+06 13 59	NGC 4432	14.0	1.0 x 0.8	13.6	Sbc	6	Almost stellar nucleus; mag 7.71v star S 5'.0.
12 27 36.6	+08 09 15	NGC 4434	12.1	1.4 x 1.4	12.9	E		Mag 8.06v star S 8'.7; UGC 7580 SE 5'.9.
12 27 40.3	+13 04 44	NGC 4435	10.8	2.8 x 2.0	12.5	SB(s)0°	13	Bright, elongated bar; located 4'.3 N of NGC 4438.
12 27 41.1	+12 18 59	NGC 4436	13.0	1.5 x 0.7	12.9	S0	115	Star on NW end.
12 27 45.3	+13 00 30	NGC 4438	10.2	8.5 x 3.2	13.6	SA(s)0/a pec:	27	**The Eyes**. Bright, irregular center; very faint galaxy **VCC 1040** 1'.4 S of center.
12 27 53.5	+12 17 33	NGC 4440	11.7	1.9 x 1.5	12.7	SB(rs)a	3	Very bright nucleus in short bar; outer arms form "S" shape.
12 28 03.8	+09 48 16	NGC 4442	10.4	4.6 x 1.8	12.6	SB(s)0°	87	**CGCG 70-101** SE 3'.2.
12 28 16.1	+09 26 11	NGC 4445	12.8	2.6 x 0.5	13.0	Sab: sp	106	**MCG +2-32-85** E 9'.7.
12 28 06.8	+13 54 42	NGC 4446	13.9	1.1 x 0.8	13.6	Scd:	106	Weak, almost stellar nucleus; NGC 4447 ESE 1'.6.
12 28 12.6	+13 53 57	NGC 4447	14.0	0.9 x 0.7	13.3	SB0?	105	NGC 4446 WNW 1'.6.
12 28 29.8	+17 05 00	NGC 4450	10.1	5.2 x 3.9	13.2	SA(s)ab I-II	175	Smooth tightly wound arms; a mag 9.12v star lies 4'.0 SW.
12 28 40.6	+09 15 28	NGC 4451	12.5	1.5 x 1.0	12.8	Sbc:	162	
12 28 43.2	+11 45 12	NGC 4452	12.0	2.8 x 0.6	12.4	S0?	32	Very bright, thin, elongated central core.
12 28 47.1	+06 30 40	NGC 4453	14.9	0.6 x 0.4	13.2	S?	153	Double system.
12 28 57.6	+13 14 30	NGC 4458	12.1	1.7 x 1.6	13.1	E0-1	45	NW of NGC 4461; a mag 10.95v star lies 2'.2 E.
12 29 00.3	+13 58 43	NGC 4459	10.4	3.5 x 2.7	12.7	SA(r)0$^+$	110	Faint galaxy **VCC 1163** 2'.1 NE; mag 8.75v star SE 2'.1.
12 29 03.1	+13 10 59	NGC 4461	11.2	3.5 x 1.4	12.8	SB(s)0$^+$:	9	Mag 10.95v star N 3'.9; very faint galaxy **VCC 1101** W 9'.7.
12 29 21.3	+08 09 20	NGC 4464	12.5	1.1 x 0.8	12.2	S?	171	Located 11'.3 NNW from the center of NGC 4472 (M 49).
12 29 23.6	+08 01 30	NGC 4465	14.5	0.5 x 0.3	12.3	Sc	95	Located 5'.7 W of the center of NGC 4472 (M 49).

(Continued from previous page)

RA h m s	Dec ° ′ ″	Name	Mag (V)	Dim ′ Maj x min	SB	Type Class	PA	Notes
12 29 30.5	+07 41 45	NGC 4466	13.5	1.1 x 0.3	12.2	Sab? sp	101	
12 29 30.3	+07 59 30	NGC 4467	13.8	0.5 x 0.5	12.4	E2		Star W 0'.5.
12 29 31.0	+14 02 56	NGC 4468	12.8	1.4 x 1.1	13.1	SA0$^-$?	73	Small companion or star? ENE of nucleus.
12 29 28.5	+08 45 04	NGC 4469	11.2	3.8 x 1.3	12.8	SB(s)0/a? sp	89	**MCG +2-32-86** NW 4'.9.
12 29 37.9	+07 49 25	NGC 4470	12.1	1.3 x 0.9	12.1	Sa?	0	Very faint galaxy **VCC 1195** W 1'.4.
12 29 46.8	+07 59 56	NGC 4472	8.4	10.2 x 8.3	13.2	E2	155	= **M 49**. Brightest galaxy in the Virgo Cluster Galaxy **VCC 1203** SSW 4'.8.
12 29 48.5	+13 25 48	NGC 4473	10.2	4.5 x 2.5	12.8	E5	94	
12 29 53.7	+14 04 07	NGC 4474	11.5	2.4 x 1.4	12.7	S0 pec:	80	Bright nucleus on elongated bar; mag 10.36v star NW 5'.2.
12 29 59.1	+12 20 55	NGC 4476	12.2	1.7 x 1.2	12.8	SA(r)0$^-$:	25	Located 4'.7 W of NGC 4478.
12 30 02.0	+13 38 11	NGC 4477	10.4	3.8 x 3.5	13.1	SB(s)0:?	9	Elongated core, smooth outer envelope; NGC 4479 SE 5'.3.
12 30 17.2	+12 19 42	NGC 4478	11.4	1.9 x 1.6	12.6	E2	140	Located 7'.8 W of NGC 4486 (M 87); stellar **PGC 41285** N 4'.2.
12 30 18.1	+13 34 40	NGC 4479	12.4	1.5 x 1.3	13.0	SB(s)0°:?	24	Located 5'.3 SE of the center of NGC 4477.
12 30 10.4	+10 46 46	NGC 4482	12.7	1.7 x 1.0	13.2	E	145	**IC 3416** W 8'.7.
12 30 40.7	+09 00 58	NGC 4483	12.2	1.6 x 0.9	12.5	SB(s)0$^+$:	65	Mag 10.14v star W 6'.5; **IC 3430** NW 7'.1.
12 30 50.0	+12 23 24	NGC 4486	8.6	8.3 x 6.6	13.0	E$^+$0-1 pec	159	= **M 87**, = **Virgo A**. **UGC 7652** SW of center 1'.7; anonymous galaxy SW of center 2'.4; **IC 3443** SE 7'.4.
12 30 57.9	+12 16 17	NGC 4486A	12.3	1.0 x 0.8	12.2	E2	4	Located 7'.2 S of the center of NGC 4486 (M 87). 11th magnitude star superimposed on a 14th magnitude galaxy.
12 30 32.0	+12 29 22	NGC 4486B	13.4	0.6 x 0.6	12.2	cE0		Compact; mag 8.63v star E 4'.2.
12 30 51.7	+08 21 31	NGC 4488	12.2	4.1 x 1.7	14.1	SB(s)0/a pec:	176	Wide bar extending SW-NE; small nucleus; single arm extending N and curving NE; second arm extending S and curving SW.
12 30 52.5	+16 45 32	NGC 4489	12.0	1.7 x 1.5	13.0	E	165	Small, bright nucleus, faint star NW of nucleus.
12 30 57.2	+11 29 01	NGC 4491	12.6	1.7 x 0.9	12.9	SB(s)a:	148	**IC 3446** E 6'.4.
12 30 59.7	+08 04 38	NGC 4492	12.6	1.7 x 1.6	13.5	SA(s)a?	90	Star on NE edge; faint star superimposed, or a small, faint galaxy seen through the envelope W of the nucleus.
12 31 32.6	+11 37 28	NGC 4497	12.5	2.0 x 0.9	13.0	SAB(s)0$^+$:	65	Very faint galaxy **VCC 1366** S 1'.3.
12 31 39.0	+16 51 09	NGC 4498	12.2	2.9 x 1.1	13.3	SAB(s)d	133	Bright elongated bar with faint lane S; very small, faint galaxy **VCC 1385** SE 2'.0.
12 31 59.5	+14 25 15	NGC 4501	9.6	6.9 x 3.7	13.0	SA(rs)b I-II	140	= **M 88**.
12 32 03.2	+16 41 14	NGC 4502	13.9	1.1 x 0.6	13.3	Scd:	40	
12 32 06.1	+11 10 34	NGC 4503	11.1	3.5 x 1.7	12.9	SB0$^-$:	12	Mag 11.6 star S 2'.8.
12 32 10.7	+13 25 10	NGC 4506	12.7	1.6 x 1.1	13.2	Sa pec?	110	Located 2'.2 E of a mag 10.86v star.
12 33 05.0	+16 15 55	NGC 4515	12.3	1.3 x 1.1	12.6	S0$^-$:	9	**CGCG 99-82** W 10'.5.
12 33 07.7	+14 34 30	NGC 4516	12.8	1.7 x 1.0	13.2	SB(rs)ab?	0	Long, narrow open "S" shaped bar.
12 33 11.9	+07 51 08	NGC 4518	13.8	1.1 x 0.5	13.1	SB$0_{2/3}$(r)/a	177	**MCG +1-32-37** E 2'.2.
12 33 10.6	+07 50 00	NGC 4518B		0.7 x 0.2			36	Located on S edge of NGC 4518.
12 33 30.1	+08 39 15	NGC 4519	11.8	3.2 x 2.5	13.9	SB(rs)d II-III	145	Bright nucleus, knotty filamentary arms; NGC 4519A NW 2'.5.
12 33 24.9	+08 41 24	NGC 4519A	15.2	0.5 x 0.2	12.5	dS0?	47	Locate 2'.5 NW of NGC 4519.
12 33 39.6	+09 10 16	NGC 4522	12.3	3.7 x 1.0	13.6	SB(s)cd: sp	33	Mag 12.1 star ENE 7'.2; IC 3517 E 12'.8.
12 33 47.9	+15 10 03	NGC 4523	14.1	2.0 x 1.9	15.4	SAB(s)m V	45	Disrupted system, low surface brightness; mag 9.90v star SSW 4'.6.
12 34 02.5	+07 42 03	NGC 4526	9.7	7.2 x 2.4	12.7	SAB(s)0°:	113	Located between a mag 6.75v star 7'.2 E and a mag 6.96v star 7'.5 W.
12 34 06.2	+11 19 17	NGC 4528	12.1	1.7 x 1.0	12.5	S0°:	5	Mag 8.19v star E 7'.8; stellar galaxy **VCC 1512** SW 8'.4.
12 34 15.9	+13 04 32	NGC 4531	11.4	3.1 x 2.0	13.3	SB0$^+$:	155	
12 34 19.5	+06 27 54	NGC 4532	11.9	2.8 x 1.1	13.0	IBm III-IV	163	Long, irregular bar offset N; faint star on E edge.
12 34 19.6	+08 11 47	NGC 4535	10.0	7.1 x 5.0	13.7	SAB(s)c I-II	0	Bright nucleus in short bar, two main arms with many branches.
12 34 51.0	+15 33 05	NGC 4540	11.7	1.9 x 1.5	12.7	SAB(rs)cd IV	40	Galaxy **IC 3520** NE 1'.4.
12 35 20.4	+06 06 53	NGC 4543	13.6	0.9 x 0.6	13.0	E3	9	
12 35 26.2	+14 29 45	NGC 4548	10.2	5.4 x 4.3	13.4	SB(rs)b I-II	150	= **M 91**. Bright diffuse nucleus in strong bar.
12 35 30.7	+12 13 16	NGC 4550	11.7	3.3 x 0.9	12.7	SB0°: sp	178	NGC 4551 NNE 3'.0.
12 35 37.8	+12 15 50	NGC 4551	12.0	1.8 x 1.4	13.0	E:	70	NGC 4550 SSW 3'.0.
12 35 40.0	+12 33 22	NGC 4552	9.8	3.5 x 3.5	12.5	E0-1		= **M 89**. Faint star on E edge.
12 36 27.0	+11 26 21	NGC 4564	11.1	3.5 x 1.5	12.9	E	47	Located between a pair of mag 11.9 stars NW and SE.
12 36 32.4	+11 15 34	NGC 4567	11.3	3.0 x 2.0	13.1	SA(rs)bc III	85	Member of **Siamese Twins**. Contact pair with NGC 4568; small bright nucleus.
12 36 33.9	+11 14 07	NGC 4568	10.8	4.6 x 2.0	13.1	SA(rs)bc III	23	Member of **Siamese Twins**. Interacting pair with NGC 4567. Many dark lanes and knots.
12 36 49.8	+13 09 44	NGC 4569	9.5	9.5 x 4.4	13.4	SAB(rs)ab I-II	26	= **M 90**. Several very small, very faint anonymous galaxies along SW edge; IC 3583 N 6'.1.
12 36 53.1	+07 14 51	NGC 4570	10.9	3.8 x 1.1	12.3	S0 sp	159	Faint, stellar anonymous galaxy N 5'.2.
12 36 56.3	+14 13 02	NGC 4571	11.3	3.6 x 3.2	13.8	SA(r)d III	55	Many faint knots; mag 8.91v star 2'.5 NE; faint galaxy **VCC 1682** W 4'.8.
12 37 30.5	+09 33 18	NGC 4578	11.5	3.3 x 2.5	13.7	SA(r)0°:	35	
12 37 43.6	+11 49 08	NGC 4579	9.7	5.9 x 4.7	13.1	SAB(rs)b II	95	= **M 58**. Small, bright diffuse nucleus in smooth lens with dark lanes.
12 37 48.2	+05 22 04	NGC 4580	11.8	2.1 x 1.6	13.0	SAB(rs)a pec II-III	165	
12 38 17.9	+13 06 32	NGC 4584	12.9	1.4 x 1.0	13.1	SAB(s)a?	5	
12 38 45.6	+06 46 00	NGC 4588	14.3	1.3 x 0.5	13.7	Scd:	57	
12 39 12.6	+06 00 47	NGC 4591	13.1	1.6 x 0.8	13.2	S?	37	
12 39 51.5	+15 17 51	NGC 4595	12.1	1.7 x 1.1	12.6	SAB(rs)b? III	110	
12 39 55.8	+10 10 36	NGC 4596	10.4	4.0 x 3.0	13.0	SB(r)0$^+$	135	Very bright, diffuse nucleus in strong bar; mag 10.83v star SE 3'.4.
11 27 47.3	+07 59 19	UGC 6462	14.1	1.5 x 0.2	12.7	S?	3	Stellar **CGCG 39-188** E 3'.4; elongated **CGCG 39-190** E 5'.9.
11 29 02.5	+17 13 54	UGC 6483	14.4	1.9 x 0.2	13.2	Scd:	68	Mag 9.88v star E 2'.3.
11 34 55.4	+16 06 54	UGC 6556	14.8	1.5 x 0.3	13.8	Sd	90	
11 35 08.3	+15 57 30	UGC 6559	15.1	1.5 x 0.2	13.7	Scd:	176	Mag 12.6 star E 0'.5.
11 37 02.0	+15 34 14	UGC 6586	13.8	2.0 x 0.9	14.3	SABbc	175	Stellar nucleus; **Mkn 742** SE 8'.6.
11 37 13.7	+15 25 55	UGC 6588	13.9	1.3 x 0.5	13.3	Sbc	144	**Mkn 742** NE 5'.3.
11 37 37.0	+16 33 23	UGC 6594	13.9	2.5 x 0.3	13.4	Sd	134	
11 39 14.9	+17 08 36	UGC 6614	13.5	1.6 x 1.3	14.1	(R)SA(r)a?	171	Bright core, uniform envelope.
11 39 17.6	+09 57 44	UGC 6617	13.6	0.9 x 0.5	12.6	S0?	159	Very small, compact anonymous galaxy S edge.
11 40 11.6	+17 18 39	UGC 6631	13.8	0.8 x 0.5	12.7	S?	117	
11 41 03.8	+10 13 26	UGC 6647	14.0	1.2 x 0.9	13.9	Sd:	36	Mag 11.9 star N 0'.9.
11 41 39.7	+15 57 55	UGC 6653	13.7	1.6 x 0.4	13.0	SB0/a?	12	UGC 6655 E 2'.7.
11 41 50.8	+15 58 27	UGC 6655	14.2	0.3 x 0.3	11.7	E?		Almost stellar.
11 42 20.5	+16 00 39	UGC 6666	13.7	1.0 x 0.4	12.6	S?	14	**MCG +3-30-52** N 5'.0.
11 42 18.5	+14 59 40	UGC 6669	15.1	1.5 x 0.7	15.0	Im	81	Dwarf irregular, very low surface brightness; pair of stars, mags 12.2 and 14.1, E 1'.7.
11 43 22.7	+16 29 05	UGC 6686	14.2	2.6 x 0.3	13.8	Sb III	51	NGC 3828 W 5'.9.
11 44 45.9	+09 12 46	UGC 6717	15.3	1.5 x 1.5	16.0	SBm:		Dwarf spiral, low surface brightness; mag 14.0 star on S edge.
11 45 35.9	+09 07 05	UGC 6734	14.3	1.5 x 0.3	13.3	Sb III	33	NGC 3876 NW 3'.4.
11 45 48.5	+10 28 36	UGC 6740	13.7	1.0 x 0.5	12.8	S?	80	
11 45 55.0	+17 11 32	UGC 6741	15.1	1.5 x 0.3	14.1	Double System	172	
11 46 46.5	+14 32 00	UGC 6753	14.4	1.0 x 0.5	13.5	SB	81	
11 47 06.4	+13 42 24	UGC 6758	12.8	1.8 x 1.7	13.9	S?	69	Mag 13.6 star on S edge.
11 48 12.9	+13 12 30	UGC 6775	14.1	1.3 x 0.7	13.8	S	175	
11 49 24.3	+16 38 30	UGC 6794	14.2	1.5 x 0.5	13.7	SBdm:	69	Mag 9.15v star SE 3'.7.
11 50 12.5	+06 59 52	UGC 6804	14.9	1.7 x 0.3	14.0	Sd: sp	104	Very faint star NE edge.

(Continued from previous page)

91 GALAXIES 91

RA h m s	Dec ° ′ ″	Name	Mag (V)	Dim ′ Maj x min	SB	Type Class	PA	Notes
11 51 34.6	+15 28 28	UGC 6831	13.4	0.8 x 0.6	12.5	S0	75	Mag 10.54v star W 1′.4.
11 53 46.5	+10 24 08	UGC 6871	13.2	0.8 x 0.6	12.3	SB0	131	
11 55 11.6	+06 10 07	UGC 6886	13.4	1.2 x 0.8	13.2	SB(rs)bc pec I-II	80	Mag 12.3 star N 1′.8.
11 55 16.8	+17 29 12	UGC 6891	14.6	1.5 x 0.3	13.6	Sab	110	
11 56 06.1	+15 07 56	UGC 6911	14.2	1.6 x 0.4	13.6	Sbc	21	
11 56 15.1	+17 01 46	UGC 6913	14.1	1.3 x 0.6	13.7	Sb	0	Mag 11.5 star SW 2′.7.
12 01 21.7	+14 27 01	UGC 7003	14.4	1.7 x 1.3	15.1	Im:	165	Dwarf, low surface brightness.
12 02 24.0	+14 50 38	UGC 7016	13.7	1.5 x 0.3	12.7	SBab:	137	
12 03 27.4	+16 29 04	UGC 7032	13.5	0.5 x 0.5	11.8	S?		MCG +3-31-25 NW 2′.2.
12 04 08.9	+13 21 13	UGC 7043	15.1	0.9 x 0.3	13.5	Pec	49	
12 05 23.3	+17 53 05	UGC 7073	13.4	1.0 x 0.7	12.8	SB(rs)c:	115	UGC 7074 N 2′.1.
12 05 23.4	+17 55 09	UGC 7074	13.8	0.9 x 0.2	11.8	Sa	98	
12 05 42.5	+08 59 22	UGC 7085	13.6	1.7 x 0.6	13.4	S pec	67	MCG +2-31-27 W 1′.5.
12 06 44.5	+17 42 50	UGC 7100	13.7	1.7 x 0.3	12.8	S?	8	
12 07 25.2	+16 53 15	UGC 7109	14.6	1.6 x 0.2	13.2	S	57	Mag 13.4 star on S edge.
12 11 55.0	+16 13 50	UGC 7194	14.0	1.0 x 1.0	13.9	Scd:		
12 11 59.4	+15 24 04	UGC 7196	14.2	1.6 x 0.4	13.6	S?	54	**CGCG 98-97** E 6′.3; **UGC 7210** SE 10′.9; **MCG +3-31-71** NE 6′.9.
12 13 38.0	+16 07 10	UGC 7230	13.9	1.5 x 0.6	13.6	SB(s)d pec	54	Multiple system, strongly disrupted; **MCG +3-31-74** WSW 7′.6.
12 14 09.4	+07 46 27	UGC 7239	13.2	2.3 x 1.9	14.7	Im:	123	Dwarf, low surface brightness.
12 14 38.0	+12 48 46	UGC 7249	14.7	1.6 x 0.6	14.5	Im IV-V	57	
12 17 04.3	+10 00 17	UGC 7307	15.0	1.6 x 1.1	15.4	Im:	96	Dwarf, very low surface brightness; mag 12.9 star S 1′.2.
12 17 37.8	+16 43 35	UGC 7327	14.9	1.7 x 0.2	13.6	Sd	91	Mag 11.8 star WNW 2′.0.
12 17 53.7	+17 26 31	UGC 7331	13.6	1.5 x 0.7	13.5	(R′)SB(s)b	0	Strong dark patches E and W of elongated core.
12 18 41.9	+17 43 07	UGC 7346	14.6	2.2 x 2.2	16.2	S?		
12 19 03.6	+13 58 51	UGC 7355	14.5	1.5 x 0.9	14.7	Double System	90	Double system, contact.
12 20 01.5	+08 36 24	UGC 7383	13.7	0.9 x 0.6	12.8	Sab	130	**MCG +2-32-6** NE 3′.0; mag 9.87v star S 2′.3.
12 20 49.0	+17 29 12	UGC 7399A	13.6	1.7 x 1.0	14.0	S0	85	**CGCG 99-22** NNE 10′.7.
12 21 55.7	+06 27 01	UGC 7423	15.0	1.0 x 0.6	14.2	Sm:	5	
12 21 57.3	+08 40 24	UGC 7424	14.3	0.9 x 0.7	13.6	Im:	99	
12 22 19.6	+14 45 41	UGC 7436	14.3	1.7 x 0.9	14.6	S	126	Small, very faint anonymous galaxy SE 1′.4.
12 27 11.2	+07 15 46	UGC 7557	12.8	3.0 x 2.4	14.8	Sm	162	Dwarf spiral, low uniform surface brightness; pair of stars, mags 10.07v and 8.21v, SE edge.
12 27 29.2	+07 38 34	UGC 7567	14.6	1.0 x 0.6	13.9	Im:	55	
12 27 54.9	+08 05 22	UGC 7580	14.3	0.8 x 0.5	13.2	S0(4)	30	Almost stellar; mag 8.06v star SW 8′.0.
12 28 19.0	+08 43 38	UGC 7590	13.2	1.2 x 0.5	12.5	S?	168	Bright, small knot S end; mag 12.4 star NE edge.
12 28 34.0	+08 38 22	UGC 7596	14.1	1.3 x 0.6	13.7	Im:	131	Small group of four stellar anonymous galaxies centered E 6′.8.
12 30 00.8	+07 55 41	UGC 7636	14.2	0.8 x 0.4	12.8	Im	15	Located on the SE edge of NGC 4467 (M 49); mag 12.1 star E 0′.8.
12 32 32.4	+08 02 38	UGC 7688	14.2	1.2 x 0.9	14.1	Im:	42	Mag 11.3 star on E edge.
12 34 45.3	+06 17 55	UGC 7739	14.0	1.2 x 0.8	13.8	Im V	153	Dwarf irregular, low, uniform surface brightness.
12 37 45.4	+07 06 08	UGC 7795	14.1	1.2 x 0.9	14.0	Im V	9	Patchy, several knots; mag 10.35v star NE 4′.7.
12 38 20.7	+07 53 25	UGC 7802	14.3	1.7 x 0.3	13.4	Scd:	56	Mag 12.6 star N 0′.9.

GALAXY CLUSTERS

RA h m s	Dec ° ′ ″	Name	Mag 10th brightest	No. Gal	Diam ′	Notes
11 43 36.0	+07 29 00	A 1362	16.0	44	13	All members stellar to almost stellar; **CGCG 40-35** 12′.2 E of center, on E edge, only catalog GX in cluster.
11 49 36.0	+12 15 00	A 1390	16.0	31	16	NGC 3908 only catalog GX, other members faint, stellar and anonymous.
12 08 00.0	+14 57 00	A 1474	16.0	70	16	Most members faint, stellar and anonymous.
12 27 24.0	+08 50 00	A 1541	16.0	58	12	Prominent anonymous GX E of center 1′.6; most other members stellar/almost stellar.

92 GALAXIES 92

RA h m s	Dec ° ′ ″	Name	Mag (V)	Dim ′ Maj x min	SB	Type Class	PA	Notes
10 10 40.6	+07 34 34	CGCG 36-49	14.6	0.4 x 0.4	12.5			Mag 8.62v star NNE 7′.9.
10 11 09.5	+07 46 46	CGCG 36-51	15.5	0.7 x 0.3	13.6		144	Mag 13.1 star NW 0′.7.
10 11 19.7	+05 53 45	CGCG 36-52	14.7	0.6 x 0.6	13.5			Mag 15.2 star 0′.5 S; mag 11.4 star N 5′.0.
10 14 18.5	+06 30 43	CGCG 36-67	15.3	0.5 x 0.5	13.7			Mag 11.8 star SE 3′.0.
10 15 16.8	+05 23 07	CGCG 36-72	14.8	0.6 x 0.3	12.8		27	Pair of mag 10.89v, 10.69v stars N 8′.2.
10 17 36.8	+06 23 33	CGCG 36-84	14.8	0.6 x 0.3	12.7		66	Mag 8.40v star SW 2′.5.
10 20 27.8	+08 02 19	CGCG 37-5	16.5	0.5 x 0.4	14.6		33	Uniform surface brightness; mag 12.5 star W 1′.4.
10 21 29.5	+05 56 26	CGCG 37-11	14.2	0.6 x 0.5	12.7		30	Triangle of stars, mags 12, 13 and 14, SW 1′.2.
10 24 18.1	+07 13 29	CGCG 37-24	16.1	0.6 x 0.3	14.0		33	Mag 12.3 star SW 1′.8.
10 24 43.6	+06 26 00	CGCG 37-28	13.5	0.6 x 0.6	12.3	E/S0		**CGCG 37-27** S 1′.3.
10 26 44.5	+08 10 50	CGCG 37-38	15.1	0.7 x 0.3	13.3		0	CGCG 37-39 SE 3′.4.
10 26 56.3	+08 09 04	CGCG 37-39	15.5	0.4 x 0.4	13.3			CGCG 37-38 NW 3′.4.
10 28 24.8	+06 14 15	CGCG 37-43	14.8	0.7 x 0.3	13.0		48	Low, uniform surface brightness.
10 30 20.0	+08 03 36	CGCG 37-55	14.3	0.6 x 0.4	12.6		117	Mag 11.8 star NNW 2′.5.
10 31 51.6	+07 58 06	CGCG 37-67	14.2	0.7 x 0.5	12.9		174	Mag 10.24v star W 5′.2.
10 33 34.6	+07 48 20	CGCG 37-78	14.3	0.7 x 0.5	13.0		54	Faint anonymous galaxy S 0′.9; mag 11.0 star E 1′.5.
10 34 39.5	+06 38 26	CGCG 37-81	14.3	0.4 x 0.4	12.2			Forms a triangle with CGCG 37-82 N and a mag 12.8 star NE.
10 34 37.5	+06 40 08	CGCG 37-82	14.4	0.4 x 0.2	11.5		12	Forms a triangle with CGCG 37-81 S and a mag 12.8 star E.
10 35 42.4	+05 36 55	CGCG 37-87	15.6	0.6 x 0.3	13.6		90	Mag 12.5 star NE 5′.0.
10 55 12.7	+05 51 41	CGCG 38-48	15.4	0.4 x 0.3	12.9		78	Mag 7.40v star W 6′.7.
10 56 46.2	+05 39 35	CGCG 38-58	15.1	0.6 x 0.4	13.4		33	Mag 9.0v star SSE 2′.0.
10 57 50.8	+08 02 35	CGCG 38-64	14.2	0.5 x 0.4	12.3		177	Mag 10.27v star S 2′.4.
10 58 12.4	+07 34 15	CGCG 38-68	14.0	0.5 x 0.4	12.2	SB(r)0°:	0	Mag 9.40v star SW 10′.0.
10 59 33.3	+07 41 00	CGCG 38-78	13.3	0.6 x 0.4	11.7	SA(r)0°:		**CGCG 38-82** ESE 13′.0.
11 02 43.9	+05 22 49	CGCG 38-96	13.9	0.3 x 0.3	11.4	E		Mag 12.1 star NNW 3′.5.
11 03 21.4	+06 50 27	CGCG 38-103	13.9	0.5 x 0.3	11.8	SAB(r)0^-:	174	CGCG 38-105 NNE 2′.3.
11 03 21.4	+06 55 18	CGCG 38-104	14.1	0.6 x 0.4	12.4	SA(r)a:		Mag 9.11v star NNE 7′.3; CGCG 38-105 SSE 3′.3.
11 03 26.9	+06 52 19	CGCG 38-105	13.8	0.6 x 0.3	11.8	S0^-	177	A faint, anonymous galaxy E 2′.2; CGCG 38-104 NNW 3′.3; CGCG 38-103 SSW 2′.3.
11 04 11.2	+07 04 07	CGCG 38-107	15.0	0.7 x 0.4	13.5		126	Mag 9.11v star WSW 8′.4.
11 05 05.3	+07 56 37	CGCG 38-118	15.1	0.6 x 0.4	13.4		135	Mag 14.1 star E 1′.2; mag 12.5 star W 5′.8.
11 08 31.6	+05 49 29	CGCG 39-9	14.5	0.5 x 0.4	12.6		30	Mag 14.4 star W 2′.7.
11 08 55.3	+06 18 27	CGCG 39-14	14.3	0.5 x 0.4	12.4	SAB(s)c pec: II-III	159	Located 2′.2 SW of CGCG 39-16.
11 09 03.0	+05 01 51	CGCG 39-15	16.6	0.4 x 0.4	14.4			A mag 9.32v star NW 2′.0; a mag 10.67v star E 2′.3.

(Continued from previous page)

RA h m s	Dec ° ′ ″	Name	Mag (V)	Dim ′ Maj x min	SB	Type Class	PA	Notes
11 09 01.6	+06 19 57	CGCG 39-16	14.1	0.5 x 0.4	12.2	SB(s)bc pec:	60	Located 1′2 SW of a mag 10.14v star.
11 09 41.5	+07 14 39	CGCG 39-22	16.6	0.8 x 0.7	15.8		15	Elongated **UGC 6208** E 3′9; faint, round **CGCG 39-30** ENE 7′0.
11 10 23.6	+05 00 17	CGCG 39-35	15.1	0.5 x 0.4	13.2		90	CGCG 39-43 NE 8′9; **CGCG 39-24** W 7′9.
11 10 53.1	+05 05 23	CGCG 39-43	16.2	0.5 x 0.4	14.3		150	Mag 10.09v star E 6′9; CGCG 39-95 SW 8′9.
11 11 32.7	+06 26 23	CGCG 39-59	13.9	0.7 x 0.4	12.6	E	63	Mag 11.0 star E 4′2.
11 13 43.6	+06 41 14	CGCG 39-80	15.9	0.8 x 0.2	13.7		66	IC 678 SE 8′7; **CGCG 39-71** WSW 9′9.
11 17 41.2	+07 17 12	CGCG 39-104	13.9	0.6 x 0.3	12.1	E	114	Mag 10.7 star SW 8′4.
11 18 45.6	+07 29 32	CGCG 39-113	14.0	0.4 x 0.3	11.6	S0 pec	114	Located 2′2 SW of NGC 3624.
11 23 20.3	+05 51 05	CGCG 39-162	14.3	0.4 x 0.4	12.2			Mag 11.8 star WNW 5′4.
11 24 08.7	+06 12 51	CGCG 39-167	13.9	0.8 x 0.4	12.5	(R)SB(r)0/a	111	Mag 8.85v star SE 11′4.
10 11 00.3	+09 50 45	CGCG 64-85	14.8	0.9 x 0.7	14.1		81	Stellar nucleus in short, bright bar; mag 11.23v star NW 1′2.
10 17 12.8	+13 11 27	CGCG 64-101	15.0	0.7 x 0.3	13.2		150	Mag 12.3 star NW 2′5.
10 18 56.1	+13 09 26	CGCG 64-106	14.6	0.7 x 0.4	13.0		105	**CGCG 64-105** NW 6′0; **CGCG 64-103** NW 10′7.
10 23 38.3	+13 57 25	CGCG 65-18	14.3	0.8 x 0.5	13.2		108	Mag 10.31v star W 2′9; mag 8.73v star SW 4′4.
10 24 22.4	+13 40 26	CGCG 65-24	14.2	0.6 x 0.3	12.2		171	UGC 5627 SSW 6′9; mag 8.86v star SE 7′9; note **CGCG 65-27** is 5′4 E of this star.
10 36 19.8	+13 26 47	CGCG 65-62	14.6	0.7 x 0.6	13.5	SB:b:	24	Very faint dwarf galaxy **UGC 5758** W 1′7.
10 37 05.5	+12 46 09	CGCG 65-69	15.6	0.9 x 0.9	15.3	Sc		Mag 7.79v star N 5′9; NGC 3306 S 7′0.
10 37 52.5	+10 45 26	CGCG 65-73	14.6	0.6 x 0.6	13.3			Mag 10.38v star ESE 3′2; mag 9.81v star WNW 7′8.
10 39 45.6	+11 38 49	CGCG 65-78	14.8	0.7 x 0.5	13.6	Sb	129	Mag 11.9 star WSW 2′8.
10 40 28.7	+13 37 19	CGCG 65-82	14.9	0.9 x 0.6	14.1		21	Mag 9.75v star S 3′6.
10 43 30.1	+12 05 18	CGCG 66-1	14.1	0.6 x 0.4	12.4	Sa	123	
10 48 34.1	+12 58 54	CGCG 66-23	14.2	1.0 x 0.4	13.1		111	Mag 10.41v star SSW 7′8; CGCG 66-28 E 6′4.
10 48 58.9	+13 00 55	CGCG 66-28	14.2	0.7 x 0.3	12.4		9	CGCG 66-23 W 6′4; mag 8.99v star SSE 11′6; mag 9.61v star ESE 10′4.
10 50 30.7	+12 30 04	CGCG 66-36	14.2	0.8 x 0.6	13.2	SBb	111	A mag 11.8 star NNE 1′1.
10 51 59.3	+08 56 36	CGCG 66-47	15.4	0.7 x 0.4	13.9	Sc	6	Mag 8.16v star N 7′1; mag 8.80v star S 5′6.
10 54 54.5	+10 02 44	CGCG 66-62	15.0	0.6 x 0.5	13.6	Sbc	159	Mag 9.90v star W 6′3; mag 9.78v star SE 6′5; MCG +2-28-26 SW 8′8.
10 57 11.3	+08 27 04	CGCG 66-68	14.2	0.5 x 0.5	12.6			Mag 13.9 star on N edge.
10 57 36.5	+08 20 17	CGCG 66-71	15.3	0.7 x 0.5	14.0		30	MCG +2-28-31 S 2′7.
11 01 01.8	+11 02 47	CGCG 66-94	13.9	0.4 x 0.2	11.1	S0/a	168	= **Mkn 728**. **PGC 33202** SW 3′2.
11 02 21.4	+08 29 14	CGCG 66-104	14.0	0.8 x 0.6	13.1	(R)SAB0$^-$	0	Mag 9.12v star N 7′2.
11 04 04.4	+13 51 38	CGCG 66-107	14.2	0.6 x 0.3	12.2		45	Mag 11.1 star on N edge.
11 08 09.4	+13 02 34	CGCG 67-7	14.5	0.8 x 0.3	12.8		141	Mag 10.60v star S 4′9.
11 08 19.6	+10 02 22	CGCG 67-9	14.8	0.7 x 0.3	12.9	Sb	9	Mag 14.3 star on N end.
11 09 52.4	+09 31 07	CGCG 67-18	13.9	0.4 x 0.4	11.8	SA(r)0$^-$		Mag 11.6 star NNE 2′7; mag 11.1 star NW 4′8.
10 08 19.9	+16 53 52	CGCG 93-57	14.2	0.5 x 0.5	12.6			Mag 3.52v star Eta Leonis SW 16′5; mag 15.0 star N 1′9.
10 08 40.2	+16 47 10	CGCG 93-59	14.7	0.6 x 0.4	13.0		135	Mag 3.52v star Eta Leonis W 19′3; mag 12.6 star E 2′8.
10 25 01.0	+17 17 30	CGCG 94-39	14.5	0.7 x 0.6	13.4	Sb	90	**CGCG 94-43** SE 1′7; **MCG +3-27-26** SE 3′3; **MCG +3-27-24** SSE 3′8.
10 26 54.8	+16 07 23	CGCG 94-56	14.3	0.9 x 0.8	13.8			**UGC 5655** W 4′5.
10 55 44.6	+17 00 14	CGCG 95-70	14.3	0.5 x 0.4	12.4		0	Mag 11.7 star WSW 3′3.
11 01 12.7	+16 43 24	CGCG 95-91	14.4	0.8 x 0.5	13.3		15	Mag 9.80v star W 6′8; **UGC 6083** WSW 11′8; UGC 6104 SE 11′5.
11 04 32.2	+17 07 38	CGCG 95-104	14.2	0.8 x 0.5	13.1		24	Mag 13.3 star E 0′6; mag 10.2 star NE 4′4.
11 05 00.8	+14 53 38	CGCG 95-106	14.4	0.6 x 0.3	12.4	Sa	168	Mag 12.6 star on E edge.
11 18 45.3	+15 10 09	CGCG 96-28	14.7	0.8 x 0.5	13.5		156	Mag 14.5 star on W edge.
10 09 38.2	+10 59 57	IC 595	14.8	0.7 x 0.5	13.6		165	Almost stellar.
10 10 31.4	+10 02 27	IC 596	14.9	0.9 x 0.3	13.3	Sb	15	Mag 9.88v star E 11′3.
10 18 19.9	+07 02 52	IC 602	13.2	0.8 x 0.5	12.1	S:	177	**IC 601** SW 1′3.
10 24 08.5	+16 44 30	IC 607	13.2	1.3 x 0.9	13.2	SB(rs)bc	114	**CGCG 94-33** ESE 6′0; mag 11.4 star WSW 10′9.
10 27 07.8	+11 00 36	IC 613	14.4	0.8 x 0.8	14.0	E		**IC 612** N 2′7. Numerous small, anonymous galaxies in surrounding area within a few arcminutes.
10 32 47.8	+15 51 36	IC 616	13.5	1.0 x 1.0	13.4	Scd:		Mag 10.8 star WNW 5′6.
10 37 36.1	+05 36 11	IC 628	13.7	1.1 x 0.8	13.4	SAB(s)ab	118	Elongated **UGC 5779** N 1′7.
10 40 55.0	+05 59 27	IC 634	14.2	1.2 x 0.4	13.3	Pec?	116	Mag 10.91v star N 7′1.
10 41 45.3	+15 38 37	IC 635	14.0	1.6 x 0.3	13.1	S?	5	Mag 10.76v star SE 9′3.
10 45 52.1	+16 55 45	IC 639	14.0	1.1 x 0.3	12.6	Sb	0	Mag 9.24v star ESE 10′3.
10 51 00.4	+12 17 09	IC 648	14.4	0.8 x 0.5	13.2	Sc	147	Mag 8.08v star W 3′7.
10 58 16.2	+08 14 25	IC 658	13.4	0.7 x 0.5	12.3	E^+ pec	57	Mag 10.71v star SW 7′0.
11 00 45.4	+10 33 07	IC 664	13.1	1.3 x 0.9	13.1	S0?	40	Mag 10.59v star SSW 4′1; stellar **PGC 33197** N 2′6; double system **PGC 33210** NE 4′6.
11 07 16.6	+06 18 06	IC 669	13.1	1.2 x 0.7	12.8	S0$^-$	165	Close pair of E-W oriented mag 10.03v, 10.42v stars N 3′7.
11 07 28.7	+06 42 51	IC 670	13.3	1.1 x 0.9	13.3	E^+	51	Mag 7.75v star W 9′2.
11 12 40.0	+09 03 24	IC 676	11.8	2.1 x 1.3	12.8	(R)SB(r)0$^+$	10	Mag 10.09v star W 9′4.
11 13 56.7	+12 18 04	IC 677	12.9	1.5 x 0.6	12.7	Sbc II-III	45	Mag 9.50v star WSW 10′0.
11 14 06.4	+06 34 38	IC 678	13.9	0.7 x 0.5	12.8	E	147	CGCG 39-80 NW 8′7.
11 25 53.6	+09 59 13	IC 692	13.6	0.7 x 0.5	12.3	S?	125	Mag 9.3 star NW 7′4.
11 11 38.0	+12 07 14	IC 2628	14.9	0.8 x 0.6	13.9		168	Mag 13.1 star E 0′8.
11 13 28.4	+10 29 07	IC 2634	14.5	0.7 x 0.5	13.2	SAB(s)b II-III	114	Located 3′4 NE of mag 7.47v star.
11 13 49.9	+09 35 09	IC 2637	13.0	0.8 x 0.8	12.5	E^+ pec		Mag 12.1 star W 6′0.
11 13 51.9	+10 33 44	IC 2638	13.8	1.1 x 0.6	13.2	(R)SB(r)0$^+$	98	Bright, elongated center, faint envelope.
11 14 46.5	+11 07 36	IC 2649	14.5	0.5 x 0.5	12.8			Mag 12.2 star N 3′9.
11 15 43.9	+13 46 53	IC 2666	13.7	1.0 x 0.7	13.1		0	Mag 10.26v star W 8′1.
11 16 04.2	+10 09 41	IC 2672	14.0	0.9 x 0.7	13.3	SAB(rs)c: III	35	Mag 11.7 star E 1′3.
11 16 08.4	+11 02 52	IC 2674	14.1	1.0 x 0.5	13.2	Scd:	20	Mag 9.19v star SW 13′2.
11 16 36.2	+11 11 59	IC 2681	15.3	0.4 x 0.4	13.1			Almost stellar.
11 17 01.1	+13 05 55	IC 2684	14.5	0.5 x 0.5	12.9			Mag 10.47v star S 1′9.
11 18 05.3	+17 38 54	IC 2703	14.4	0.6 x 0.6	13.1			Mag 10.33v star WSW 12′6.
11 18 34.7	+12 42 38	IC 2708	14.5	0.5 x 0.5	12.8			Mag 13.0 star SW 2′6.
11 22 02.2	+08 23 34	IC 2757	14.0	0.7 x 0.5	12.7	SB(s)a I-II		Mag 12.3 star SW 3′8.
11 22 18.9	+13 03 50	IC 2763	14.7	1.3 x 0.2	13.0	Scd?	96	Mag 10.44v star SW 9′9; **IC 2767** NE 1′3.
11 22 40.1	+13 19 46	IC 2776	15.3	0.8 x 0.4	13.9	Sc	168	Mag 13.2 star SE 4′4.
11 22 55.3	+13 26 23	IC 2782	13.9	1.0 x 0.8	13.5	Sdm:	0	Mag 11.6 star ESE 11′0; **IC 2785** and **IC 2786** ESE 5′9.
11 23 19.1	+13 37 46	IC 2787	14.1	0.7 x 0.7	13.2	Scd?		Mag 13.0 star W 4′2.
11 24 55.8	+13 13 19	IC 2804	14.2	0.9 x 0.3	12.6	Sc	12	Mag 9.69v star NE 7′2.
11 26 34.3	+11 26 21	IC 2822	13.6	1.5 x 0.7	13.5	SBbc	115	Mag 10.89v star ESE 7′5.
11 27 06.3	+13 14 16	IC 2826	14.9	0.6 x 0.6	13.7			Mag 10.95v star E 5′6.
11 27 11.5	+08 43 51	IC 2828	13.5	0.9 x 0.6	12.7	S0?	51	Mag 6.73v star WSW 13′8.
11 27 21.6	+07 48 45	IC 2830	15.5	0.5 x 0.5	13.8			Mag 8.93v star NE 7′6.
11 28 00.7	+11 09 25	IC 2846	14.5	0.9 x 0.7	13.8	Sp	114	Mag 10.91v star NNE 3′7.
10 09 46.4	+05 10 19	MCG +1-26-16	14.7	0.9 x 0.7	14.0	Sbc	81	
10 10 37.2	+05 08 58	MCG +1-26-17	16.6	0.7 x 0.5	15.3		15	Mag 9.49v star SE 2′4.
10 15 53.4	+06 57 47	MCG +1-26-28	14.1	0.5 x 0.5	12.4			Small companion galaxy SW edge; mag 9.31v star SW 3′6.

RA h m s	Dec ° ′ ″	Name	Mag (V)	Dim ′ Maj x min	SB	Type Class	PA	Notes
10 21 12.8	+08 06 45	MCG +1-27-2	14.1	0.9 x 0.6	13.3	Sbc	156	
10 21 36.5	+06 01 30	MCG +1-27-3	14.5	0.7 x 0.5	13.2	Sb	51	Pair of mag 11-12 stars NE 1′.8.
10 36 35.8	+05 54 33	MCG +1-27-20	13.8	0.7 x 0.4	12.3		108	Close pair of mag 14 stars WNW 2′.2.
10 41 13.0	+06 22 09	MCG +1-27-26	14.2	0.8 x 0.6	13.5	E	41	
10 42 19.7	+06 12 00	MCG +1-27-29	14.3	0.7 x 0.3	12.5	Sc	138	
10 51 33.7	+08 09 53	MCG +1-28-14	15.8	2.2 x 1.1	16.6		159	Mag 13.8 star E 2′.2; NGC 3427 N 8′.3.
10 52 27.5	+08 05 37	MCG +1-28-16	14.1	1.2 x 0.6	13.6		24	= **Mkn 1266**.
10 56 09.2	+06 10 20	MCG +1-28-22	13.8	0.3 x 0.2	10.6	Compact	75	= **Mkn 1271**. Mag 5.98v star 56 Leonis WNW 2′.1.
10 58 23.3	+06 42 52	MCG +1-28-25	14.9	0.5 x 0.5	13.2			Star on E edge; mag 13.5 star E 1′.2.
11 02 22.2	+06 04 36	MCG +1-28-28	14.3	0.7 x 0.4	12.8	SB(s)b: II	45	
11 11 20.7	+05 49 46	MCG +1-29-12	13.8	0.8 x 0.6	12.8	S0?	132	Almost stellar; NGC 3567 on NW edge.
11 13 00.3	+07 51 36	MCG +1-29-15	15.0	0.9 x 0.3	13.4	Sbc	147	
11 13 08.0	+05 04 23	MCG +1-29-16	14.1	0.9 x 0.6	13.2	SAB(s)c II-III	21	
11 13 12.6	+05 11 48	MCG +1-29-17	15.2	0.9 x 0.9	14.8	SBc		**CGCG 39-70** NE 2′.8.
11 21 10.1	+05 21 47	MCG +1-29-35	13.3	1.0 x 1.0	13.3	E		
11 27 46.1	+06 08 37	MCG +1-29-42	14.2	0.6 x 0.5	12.7	SA(rs)b III-IV	57	Mag 11.9 star S 6′.9.
10 10 00.2	+11 54 55	MCG +2-26-29	15.0	0.9 x 0.7	14.4	Sb	6	
10 14 34.8	+11 51 53	MCG +2-26-33	15.0	0.4 x 0.4	12.8			Line of three mag 14 stars, on NE-SW track, E 0′.9.
10 24 52.5	+09 35 38	MCG +2-27-9	15.0	0.7 x 0.4	13.5	Sb	39	Located between a pair of mag 12.8 stars; **MCG +2-27-10** S 1′.5.
10 25 48.7	+13 43 38	MCG +2-27-13	15.1	0.5 x 0.5	13.4	S0?		= **Hickson 47B**; UGC 5644 SW edge.
10 25 47.9	+13 44 56	MCG +2-27-14	15.9	0.5 x 0.4	14.0	Sd	168	= **Hickson 47D**; **Hickson 47C** on NE edge; mag 10.95v star NE 1′.4.
10 31 31.3	+12 29 41	MCG +2-27-23	14.9	0.8 x 0.4	13.6	Sbc	18	
10 32 17.0	+12 03 18	MCG +2-27-24	14.2	0.7 x 0.4	12.7	S?	99	Mag 14.4 star W 2′.0.
10 38 08.2	+10 22 53	MCG +2-27-33	14.4	0.8 x 0.6	13.7	E/S0	147	
10 45 51.5	+09 43 19	MCG +2-28-4	14.2	0.5 x 0.3	12.0	Sb	135	
10 47 53.8	+10 53 45	MCG +2-28-10	14.8	1.0 x 0.7	14.3		84	
10 52 09.8	+12 47 43	MCG +2-28-24	14.2	0.7 x 0.3	12.4	Sb	39	Mag 14.5 star N 1′.1.
10 54 27.8	+09 56 57	MCG +2-28-26	14.8	0.6 x 0.5	13.3		126	Mag 9.90v star N 7′.2; CGCG 66-62 NE 8′.8.
10 55 31.9	+08 50 47	MCG +2-28-27	14.3	0.7 x 0.5	13.0		156	Mag 12.1 star SW 3′.3.
10 56 32.9	+09 55 58	MCG +2-28-29	14.4	1.0 x 0.7	13.9	Sab	171	
10 57 32.0	+08 17 55	MCG +2-28-31	14.6	0.7 x 0.6	13.5	Sc	105	Faint star S edge; CGCG 66-71 N 2′.7.
11 00 37.4	+11 24 52	MCG +2-28-40	14.5	0.9 x 0.3	12.9	Sb	102	Mag 11.09v star N 3′.0.
11 01 11.2	+12 28 26	MCG +2-28-46	14.3	1.0 x 0.6	13.6	SBc	21	
11 05 58.5	+08 21 27	MCG +2-28-49	14.2	0.9 x 0.3	12.6	(R′)SB(r)a: sp	0	Close pair of mag 13.9 stars E 1′.3.
11 06 51.2	+14 12 09	MCG +2-28-51	14.5	0.8 x 0.5	13.3		30	Faint, small, elongated anonymous galaxy W 0′.9; faint anonymous galaxy N 1′.9.
11 07 41.8	+13 01 31	MCG +2-29-1	14.1	0.7 x 0.4	12.6	Sc	33	Mag 7.69v star W 3′.8.
11 08 01.1	+14 06 11	MCG +2-29-2	16.0	0.9 x 0.6	15.1		150	Mag 10.36v star W 5′.0.
11 08 18.4	+13 13 25	MCG +2-29-3	14.6	0.6 x 0.6	13.3			Small, faint anonymous galaxy SSE 2′.3.
11 08 22.1	+13 51 19	MCG +2-29-4	13.8	0.7 x 0.5	12.5		54	
11 09 23.2	+10 50 00	MCG +2-29-5	14.0	0.7 x 0.7	13.0			Mag 10.66v star S 3′.2.
10 08 04.4	+14 48 09	MCG +3-26-30	14.6	1.0 x 0.8	14.2	Sbc	9	Stellar nucleus, mag 11.3 star W 7′.3.
10 10 38.5	+16 41 03	MCG +3-26-36	14.3	0.6 x 0.4	12.6		144	= **Mkn 1247**.
10 14 00.9	+14 29 55	MCG +3-26-44	14.1	1.1 x 0.9	13.9		42	**MCG +3-26-46** N 1′.7; **MCG +3-26-45** N 2′.1.
10 14 56.5	+16 52 41	MCG +3-26-48	14.1	0.9 x 0.3	12.5		117	Mag 8.35v star ESE 4′.0.
10 21 08.8	+16 17 56	MCG +3-27-2	14.4	0.7 x 0.5	13.1		129	Located between a pair of mag 14.0 and 14.7 stars.
10 21 21.5	+16 06 52	MCG +3-27-3	14.6	0.6 x 0.6	13.3			Mag 13.9 star N 1′.1.
10 23 10.4	+17 57 44	MCG +3-27-14	13.6	0.5 x 0.4	11.7	Compact	33	= **Mkn 630**.
10 27 14.8	+16 02 57	MCG +3-27-41	14.4	0.9 x 0.7	14.0	E	105	**CGCG 94-59** NW 2′.7; **MCG +3-27-38** SW 3′.1.
10 27 51.6	+16 09 03	MCG +3-27-45	14.3	0.7 x 0.6	13.2	Sc	33	Mag 13.3 star E 1′.9.
10 30 21.0	+15 11 30	MCG +3-27-53	14.5	0.4 x 0.3	12.0			Located between a pair of mag 12.8 stars; mag 10.66v star E 2′.5.
10 34 53.6	+17 45 31	MCG +3-27-63	15.7	0.5 x 0.4	13.8		45	Mag 11.2 star WSW 7′.2.
10 36 10.0	+16 12 53	MCG +3-27-65	15.7	0.8 x 0.8	15.1	Sbc		
10 42 36.4	+15 44 53	MCG +3-27-75	13.8	0.8 x 0.6	12.8	S?	6	Elongated **UGC 5828** W 1′.5.
10 45 22.4	+16 19 44	MCG +3-28-5	14.0	0.7 x 0.3	12.2	Sa	177	Mag 8.25v star W 3′.8.
10 49 14.2	+15 48 31	MCG +3-28-11	14.4	0.9 x 0.5	13.4	Sc	54	Mag 9.18v star NE 4′.5.
10 51 49.7	+16 00 01	MCG +3-28-22	15.0	0.9 x 0.9	14.6	IIB pec		
10 53 13.2	+16 14 04	MCG +3-28-26	13.7	0.8 x 0.6	12.8		0	Mag 11.6 star S 2′.4.
10 55 47.5	+14 52 00	MCG +3-28-36	14.3	0.7 x 0.7	13.4	Sb		
10 56 34.9	+16 47 24	MCG +3-28-38	14.5	0.4 x 0.3	12.0		120	
11 00 03.9	+17 25 25	MCG +3-28-45	14.0	1.0 x 0.5	13.1		123	Mag 9.31v star N 5′.1.
11 02 59.6	+17 19 45	MCG +3-28-52	14.2	0.7 x 0.4	12.6	Sb	132	**CGCG 95-102** E 8′.9.
11 05 32.6	+17 38 21	MCG +3-28-56	14.1	1.0 x 0.4	13.0	Sb	18	Mag 9.38v star NW 2′.6.
11 17 38.3	+17 49 05	MCG +3-29-24	14.5	1.3 x 0.8	14.4	SB0/p	165	Almost stellar nucleus; very faint, small anonymous galaxy SSE 5′.1.
11 17 58.1	+17 26 30	MCG +3-29-25	14.4	0.5 x 0.3	12.2		159	
10 08 12.4	+09 58 38	NGC 3130	13.4	1.1 x 0.6	12.8	S0/a	30	Located in the glare of mag 4.37v star 31 Leo 4′.6 WNW.
10 12 29.6	+12 22 41	NGC 3134	13.7	0.9 x 0.2	11.7		51	
10 12 50.6	+12 39 58	NGC 3153	12.7	2.1 x 0.9	13.2	Scd:	170	
10 13 01.4	+17 02 03	NGC 3154	13.5	0.9 x 0.4	12.2	Sb	124	Lies 2′.1 NW of mag 8.77v star.
10 17 38.0	+06 58 12	NGC 3186	15.1	0.7 x 0.7	14.2			Very small, elongated nucleus in faint envelope.
10 23 32.8	+10 57 35	NGC 3217	14.5	0.5 x 0.3	12.3	S?	30	= **IC 606**. Almost stellar; mag 11.6 star SE 2′.7.
10 23 43.8	+12 34 02	NGC 3230	12.8	2.3 x 1.1	13.7	S0	115	Pair of stars S of nucleus; **UGC 5625** N 4′.1; **CGCG 65-19** N 8′.9.
10 25 05.2	+17 09 01	NGC 3239	11.3	5.0 x 3.3	14.2	IB(s)m pec	81	Very irregular appearance; a mag 9.96v star superimposed on the center; **MCG +3-27-27** E 3′.2 from center.
10 28 27.4	+12 42 12	NGC 3253	13.6	1.2 x 1.1	13.7	SAB(rs)bc II	45	
10 34 42.2	+11 12 02	NGC 3279	13.4	2.9 x 0.3	13.0	Sd	152	Mag 10.49v star N 3′.6; UGC 5737 W 11′.9.
10 36 23.9	+12 42 23	NGC 3299	12.7	2.2 x 1.7	14.0	SAB(s)dm	3	Uniform surface brightness.
10 36 38.5	+14 10 16	NGC 3300	12.1	1.9 x 1.0	12.7	SAB(r)0°:?	173	Mag 8.22v star S 7′.7.
10 37 10.3	+12 39 08	NGC 3306	13.4	1.3 x 0.5	12.8	SB(s)m?	141	CGCG 65-69 N 7′.0.
10 39 32.0	+05 06 25	NGC 3326	13.7	0.6 x 0.6	12.4	Sa		Mag 10.23v star N 5′.5; note: **PGC 31711** is 1′.1 S of this star.
10 40 28.5	+09 10 53	NGC 3332	12.3	1.4 x 1.4	12.9	(R)SA0$^-$		Mag 9.88v star NE 8′.1.
10 42 07.5	+13 44 45	NGC 3338	11.1	5.9 x 3.6	14.2	SA(s)c I-II	100	A mag 8.96v star on W end.
10 42 31.5	+05 02 41	NGC 3341	14.0	1.2 x 0.4	13.1	Pec	24	Faint star on SW edge.
10 43 38.8	+14 52 20	NGC 3346	11.7	2.9 x 2.5	13.7	SB(rs)cd II-III	90	Weak stellar nucleus.
10 43 50.5	+06 45 47	NGC 3349	14.5	0.5 x 0.5	12.8	Sc		
10 43 57.4	+11 42 15	NGC 3351	9.7	7.4 x 5.0	13.5	SB(r)b II	13	= **M 95**.
10 44 12.4	+06 45 31	NGC 3356	13.3	1.7 x 0.8	13.5	Sbc I-II	102	Mag 10.65v star S 2′.8.
10 44 20.8	+14 05 01	NGC 3357	12.7	1.4 x 1.3	13.3	E$^+$:	90	Very faint anonymous galaxies 2′.0 NW and W.
10 44 51.7	+06 35 49	NGC 3362	12.8	1.4 x 1.1	13.1	SABc II	54	Mag 9.58v star E 3′.8.
10 46 35.2	+13 45 08	NGC 3367	11.5	2.5 x 2.2	13.2	SB(rs)c I-II	57	Mag 8.89v star SE 9′.9.

(Continued from previous page)

GALAXIES

RA h m s	Dec ° ′ ″	Name	Mag (V)	Dim ′ Maj x min	SB	Type Class	PA	Notes
10 46 45.9	+11 49 12	NGC 3368	9.3	7.6 x 5.2	13.1	SAB(rs)ab II	176	= **M 96**. Almost stellar anonymous galaxy SW edge; several more along E.
10 47 03.9	+17 16 27	NGC 3370	11.6	2.6 x 1.5	12.9	SA(s)c II	148	
10 47 26.6	+06 02 52	NGC 3376	13.9	0.8 x 0.4	12.5	S?	167	
10 47 42.4	+13 59 09	NGC 3377	10.4	5.0 x 3.0	13.3	E5-6	35	NGC 3377A NW 7′1.
10 47 22.3	+14 04 13	NGC 3377A	13.6	2.2 x 2.1	15.1	SAB(s)m V	90	Uniform surface brightness; mag 14.7 star superimposed E of center.
10 47 49.8	+12 34 55	NGC 3379	9.3	5.4 x 4.8	12.8	E1	71	= **M 105**.
10 48 16.9	+12 37 44	NGC 3384	9.9	5.5 x 2.5	12.6	SB(s)0$^-$:	53	
10 48 12.1	+04 59 53	NGC 3386	13.8	0.5 x 0.5	12.4	E		Compact; NGC 3385 S 4′3.
10 48 29.0	+12 31 57	NGC 3389	11.9	2.8 x 1.3	13.2	SA(s)c II-III	112	Numerous knots and clumps.
10 48 56.5	+14 13 06	NGC 3391	13.0	1.0 x 0.5	12.1	S?	35	Faint stars superimposed on NE and W edges; **CGCG 66-26** S 5′8.
10 49 27.7	+16 13 02	NGC 3399	12.9	0.8 x 0.8	12.3	S0?		Located on W edge of galaxy cluster **A 1108**.
10 49 43.9	+16 14 21	NGC 3405	13.4	0.8 x 0.7	12.7	S0	49	Double system in contact, companion is **MCG +3-28-15**; **CGCG 95-34** and companion N 2′2.
10 50 53.5	+13 24 42	NGC 3412	10.5	3.6 x 2.0	12.6	SB(s)0°	155	Mag 14.2 star superimposed N edge; UGC 5944 SW 11′7.
10 51 01.8	+08 28 23	NGC 3417	14.6	0.7 x 0.4	13.1		81	
10 51 18.0	+13 56 46	NGC 3419	12.5	1.2 x 1.0	12.5	(R)SAB(r)0$^+$	115	NGC 3419A N 4′7.
10 51 20.1	+14 01 19	NGC 3419A	14.1	1.8 x 0.3	13.3	SB(s)b: sp	137	Mag 11.6 star E 1′6; NGC 3419 S 4′7.
10 51 14.6	+05 50 21	NGC 3423	11.1	3.8 x 3.2	13.7	SA(s)cd II-III	10	Many knots and superimposed stars.
10 51 25.6	+08 34 03	NGC 3425	13.1	1.0 x 1.0	13.0	S0		Lies 2′4 NW of mag 10.07v star. Possible small, faint anonymous galaxy on S edge.
10 51 26.4	+08 17 52	NGC 3427	13.2	1.1 x 0.5	12.4	S0/a	77	MCG +1-28-14 S 8′3.
10 51 29.6	+09 16 43	NGC 3428	13.1	1.5 x 0.7	13.1	SAB(s)b	170	Mag 7.81v star SW 5′1.
10 50 57.7	+09 15 51	NGC 3429		0.7 x 0.4			87	Almost stellar; mag 7.81v star SE 4′6.
10 52 04.0	+10 08 55	NGC 3433	11.6	3.5 x 3.0	14.0	SA(s)c I-II	50	Three small, elongated anonymous galaxies; one 2′0 E, two 4′2 SW.
10 52 26.1	+10 32 50	NGC 3438	13.4	0.8 x 0.8	12.8	S?		**CGCG 66-51** S 3′4, may be a double system or associated with a star.
10 52 25.9	+08 33 23	NGC 3439	14.3	0.6 x 0.4	12.6		150	Almost stellar.
10 52 31.1	+07 13 29	NGC 3441	13.6	0.7 x 0.4	12.1	S?	5	Mag 8.73v star E 4′9.
10 53 00.4	+17 34 23	NGC 3443	13.1	2.8 x 1.4	14.4	SAd	145	Small, weak nucleus in faint, uniform envelope.
10 52 59.5	+10 12 41	NGC 3444	14.7	1.0 x 0.2	12.8	Sbc: sp	19	Mag 9.63v star E 6′2.
10 53 24.3	+16 46 26	NGC 3447	12.5	3.7 x 2.1	14.6	SAB(s)m pec	0	NGC 3347A on E edge; **PGC 32713** NE 7′8.
10 53 29.9	+16 47 06	NGC 3447A	12.5	1.5 x 0.8	12.5	IB(s)m pec	110	NGC 3347, and companion NGC 3447A, lie 3′3 NE of a mag 9.89 star.
10 54 29.5	+17 20 34	NGC 3454	13.5	2.1 x 0.4	13.1	SB(s)c? sp	116	Mag 11 star 1′6 S.
10 54 30.7	+17 17 04	NGC 3455	12.0	2.6 x 2.0	13.7	(R′)SAB(rs)b II-III	62	Very faint arms encircle brighter, elongated core.
10 54 48.8	+17 37 10	NGC 3457	12.7	0.9 x 0.9	12.3	S?		Faint anonymous galaxy E 4′7, near original **IC 656** position.
10 55 21.0	+07 41 46	NGC 3462	12.2	1.7 x 1.2	12.8	S0:	60	Mag 8.30v star E 8′9.
10 56 15.5	+09 45 14	NGC 3466	13.6	1.0 x 0.6	12.9	SB(s)b:	55	Mag 11.4 star W 3′0; **CGCG 66-64** W 7′0.
10 56 44.2	+09 45 28	NGC 3467	13.4	0.8 x 0.8	12.9	S0		Mag 11.4 star NW 3′6.
10 58 05.1	+17 07 20	NGC 3473	13.5	1.2 x 1.0	13.6	SBb:	40	Mag 12.8 star on N edge; mag 9.46v star NE 6′7; faint, elongated anonymous galaxy N 1′2.
10 58 08.8	+17 05 41	NGC 3474	13.9	0.8 x 0.7	13.2	S?	138	NGC 3474 N 1′9.
10 58 07.7	+09 16 33	NGC 3476	13.8	0.8 x 0.7	13.2	E	117	
10 58 12.8	+09 13 01	NGC 3477	14.8	0.9 x 0.3	13.2	S0/a	72	
11 00 02.4	+14 50 24	NGC 3485	11.8	2.3 x 2.0	13.3	SB(r)b: II	105	**CGCG 95-88** E 6′1.
11 00 46.7	+17 35 12	NGC 3487	13.9	0.9 x 0.4	12.6	Sb? sp	153	
11 00 18.4	+13 54 02	NGC 3489	10.3	3.5 x 2.0	12.3	SAB(rs)0$^+$	70	
10 59 54.4	+09 21 42	NGC 3490	13.8	0.5 x 0.5	12.4	E		Almost stellar.
11 00 35.5	+12 09 39	NGC 3491	13.2	0.9 x 0.9	12.9	S0$^-$:		Mag 7.93v star SE 10′1.
11 00 57.2	+10 30 15	NGC 3492	13.2	1.1 x 0.8	12.9	S?	100	Multi-galaxy system; mag 10.59v star W 4′8; **IC 666** ESE 4′5; **IC 663** SW 6′4.
11 02 47.7	+17 59 30	NGC 3501	12.9	3.9 x 0.5	13.4	Scd:	27	Mag 9.07v star NW 7′2.
11 03 13.0	+11 04 38	NGC 3506	12.5	1.2 x 1.1	12.6	Sc: I-II	45	Low surface brightness dwarf spiral **UGC 6122** NE 5′4.
11 06 32.1	+11 23 01	NGC 3524	12.8	1.6 x 0.5	12.4	S0/a	14	**CGCG 66-111** lies just W of a star, 2′1 NW of NGC 3524.
11 06 56.4	+07 10 20	NGC 3526	13.2	1.9 x 0.4	12.7	SAc pec sp	55	
11 09 56.0	+10 43 12	NGC 3547	12.8	1.9 x 0.9	13.2	Sb: III-IV	7	Mag 9.91v star SSW 3′9.
11 10 44.9	+12 00 55	NGC 3559	12.8	1.3 x 0.9	12.8	S pec	55	Located on E edge of small galaxy cluster **A 1183**.
11 11 18.7	+05 50 08	NGC 3567	13.3	0.9 x 0.7	12.6	S0 pec	132	MCG +1-29-12 on SE edge.
11 14 27.6	+17 15 29	NGC 3592	13.7	1.8 x 0.6	13.6	Sc? sp	120	
11 14 36.5	+12 49 00	NGC 3593	10.9	5.2 x 1.9	13.3	SA(s)0/a:	92	Located within the boundaries of the small galaxy cluster **A 1209**.
11 15 06.1	+14 47 10	NGC 3596	11.3	4.0 x 3.8	14.1	SAB(rs)c II-III	138	Moderate dark lanes E and W of center.
11 15 11.8	+17 15 41	NGC 3598	12.3	1.7 x 1.3	13.0	S0$^-$:	35	Mag 14.0 star on N edge.
11 15 33.2	+05 06 53	NGC 3601	13.8	0.5 x 0.5	12.1	SB(r)ab?		
11 15 48.5	+17 24 52	NGC 3602	15.0	1.0 x 0.3	13.5	Sb	48	Mag 10.05v star SSW 3′4; small, faint anonymous galaxy NW 6′0.
11 18 55.7	+13 05 37	NGC 3623	9.3	9.8 x 2.9	12.8	SAB(rs)a II	174	= **M 65**. Prominent N-S dark lane, 3′9 in length, E side of center. Member of **Leo Triplet ("Trio in Leo")** with M 66 and NGC 3628.
11 18 51.0	+07 31 15	NGC 3624	13.9	0.9 x 0.6	13.0	SB(r)b I-II	162	CGCG 39-113 SW 2′1.
11 20 15.4	+12 58 59	NGC 3627	8.9	9.1 x 4.2	12.7	SAB(s)b II	173	= **M 66**. Many moderately dark lanes on either side of the bright center. A mag 9.79v star on NW edge. Member of **Leo Triplet ("Trio in Leo")** with M 65 and NGC 3628.
11 20 18.2	+13 35 06	NGC 3628	9.5	14.8 x 3.0	13.4	Sb pec sp III	104	Prominent dark lane runs almost full length down the center of the galaxy. Member of **Leo Triplet ("Trio in Leo")** with M 65 and M 66.
11 22 54.6	+16 35 20	NGC 3655	11.7	1.5 x 1.0	11.9	SA(s)c: III-IV	30	
11 23 45.1	+17 49 02	NGC 3659	12.2	2.1 x 1.1	12.9	SB(s)m? III-IV	60	
11 24 25.8	+11 20 28	NGC 3666	12.0	4.4 x 1.0	13.5	SA(rs)c: III	100	Mag 5.79v star NE 9′6.
11 26 29.7	+16 51 48	NGC 3681	11.2	2.0 x 2.0	12.5	SAB(r)bc I-II		
11 27 11.3	+17 01 45	NGC 3684	11.4	3.1 x 2.1	13.3	SA(rs)bc II-III	130	Mag 10.61v star E 7′8; located on S edge of small galaxy cluster **A 1264**.
11 27 44.1	+17 13 20	NGC 3686	11.3	3.2 x 2.5	13.4	SB(s)bc II	15	Many small knots overall.
10 08 28.5	+12 18 18	UGC 5470	10.5	9.8 x 7.4	15.2	E3	80	**Leo I**. Low surface brightness; IC 591 W 14′6.
10 09 12.6	+15 00 18	UGC 5477	12.7	1.3 x 1.3	13.3	E		
10 11 53.4	+16 26 24	UGC 5495	13.6	2.8 x 0.3	13.3	Scd:	101	
10 13 58.9	+07 01 26	UGC 5522	12.6	2.8 x 1.7	14.1	Sd	145	Bright core, smooth envelope; mag 10.9 star SW 3′0.
10 14 28.6	+15 54 07	UGC 5526	13.9	0.9 x 0.4	12.6	S	98	
10 15 42.1	+07 19 39	UGC 5537	14.3	2.3 x 0.3	13.8	Sd	147	
10 16 40.3	+16 58 19	UGC 5547	13.6	1.1 x 0.7	13.2	S0:	125	
10 17 26.0	+17 05 47	UGC 5552	14.1	1.5 x 0.3	13.1	Sbc	32	
10 19 35.1	+06 19 34	UGC 5573	13.4	1.6 x 0.9	13.7	Scd:	129	Mag 11.2 star NW 2′2.
10 21 38.3	+12 34 31	UGC 5595	13.6	0.8 x 0.6	12.7	Sbc	36	Mag 10.66v star NE 1′6.
10 23 21.9	+09 56 15	UGC 5616	15.1	1.5 x 0.2	13.7	Scd:	138	
10 24 08.8	+13 34 20	UGC 5627	13.9	0.9 x 0.5	12.9	S	65	**CGCG 65-23** N 2′7; CGCG 65-24 NNE 6′9.
10 24 40.2	+14 45 25	UGC 5633	13.7	2.4 x 1.6	15.0	SBdm IV-V	175	Mag 12.6 star SW edge.
10 25 26.1	+17 15 40	UGC 5639	13.9	1.2 x 0.8	13.7	SABc	138	CGCG 94-39 WNW 6′2 (see notes for CGCG 94-39 for other galaxies in area).
10 25 41.8	+11 44 18	UGC 5642	13.9	1.5 x 0.3	12.9	Sbc	98	
10 25 46.1	+13 42 59	UGC 5644	14.3	1.0 x 0.6	13.5	SA(r):	12	= **Hickson 47A**.
10 25 53.1	+14 21 46	UGC 5646	12.8	2.3 x 0.7	13.2	S	165	
10 26 25.6	+17 30 39	UGC 5651	13.3	1.7 x 1.2	13.9	SABcd:	18	Mag 9.58v star S 4′3.

(Continued from previous page)

92 GALAXIES 92

RA h m s	Dec ° ′ ″	Name	Mag (V)	Dim ′ Maj x min	SB	Type Class	PA	Notes
10 29 15.5	+06 07 37	UGC 5687	14.1	1.9 x 0.3	13.3	Scd:	111	Mag 7.49v star N 7′.9.
10 29 46.8	+13 01 03	UGC 5695	13.8	1.3 x 0.5	13.1	S?	96	
10 30 04.8	+05 02 49	UGC 5702	14.3	0.9 x 0.5	13.3	S	100	
10 33 50.1	+12 52 39	UGC 5735	13.9	0.9 x 0.7	13.3	Sbc	159	
10 33 53.4	+11 12 21	UGC 5737	14.5	0.9 x 0.7	13.8	Scd:	101	NGC 3279 E 11′.9.
10 34 20.0	+13 45 10	UGC 5739	13.9	1.5 x 0.6	13.6	Im?	152	Bright, elongated center, faint envelope.
10 36 21.5	+13 42 40	UGC 5760	13.5	1.5 x 0.5	13.0	S?	3	Mag 7.77v star N 6′.4.
10 38 46.9	+05 41 46	UGC 5788	13.6	1.2 x 0.7	13.3	Sb III	10	Mag 7.57v star NNW 2′.5.
10 40 31.3	+12 17 37	UGC 5808	13.3	1.1 x 0.9	13.1	SA(s)b II-III	90	Mag 12.5 star N 1′.1.
10 41 15.6	+06 21 38	UGC 5818	13.9	1.1 x 0.7	13.5	Scd:	137	Faint stellar nucleus, uniform envelope; MCG +1-27-26 NW edge.
10 42 49.2	+13 27 28	UGC 5832	12.8	1.3 x 1.1	13.1	SB?	101	Mag 10.9 star N 1′.8.
10 44 49.1	+15 57 04	UGC 5858	14.0	1.1 x 1.1	14.1	Sdm		Mag 11.4 star NE 3′.1.
10 45 42.6	+11 20 36	UGC 5869	13.5	1.4 x 0.9	13.6	SAB(rs)b: III-IV	93	
10 47 29.6	+07 14 59	UGC 5892	13.5	1.2 x 0.9	13.5	SBb	150	**CGCG 38-11** and **CGCG 38-12** (N of the two) W 5′.7.
10 47 41.5	+11 04 35	UGC 5897	12.7	2.6 x 0.9	13.5	SAc	75	
10 49 07.7	+06 54 58	UGC 5923	13.5	1.3 x 0.5	12.9	S0/a?	173	
10 50 19.2	+13 16 16	UGC 5944	14.2	1.0 x 1.0	14.1	Im		NGC 3412 NE 11′.7.
10 50 24.4	+17 33 47	UGC 5945	13.8	1.7 x 0.7	13.8	IBm	95	Elongated core offset E.
10 50 49.2	+15 19 26	UGC 5950	14.6	1.2 x 0.6	14.1	Im?	36	Pair of dwarfs, extremely low surface brightness.
10 52 33.4	+10 01 09	UGC 5994	15.1	1.5 x 0.2	13.6	Scd:	142	Located 10′.8 SE of NGC 3433.
10 52 59.2	+07 37 09	UGC 5999	15.1	1.5 x 1.3	15.7	Im V	141	Dwarf irregular, low surface brightness.
10 53 42.7	+09 43 38	UGC 6014	13.8	1.2 x 0.8	13.6	SABdm:	70	Mag 10.16v star W 4′.7.
10 55 29.3	+17 08 29	UGC 6035	13.7	1.1 x 0.9	13.5	IBm	81	Bright knot or star W end.
10 56 15.4	+15 13 23	UGC 6043	14.3	1.2 x 0.5	13.6	Scd:	70	
10 56 51.2	+06 54 21	UGC 6046	13.9	0.9 x 0.6	13.1	SBcd:	60	**Ant Galaxy**. **CGCG 38-59** S 3′.4.
10 57 31.1	+05 41 51	UGC 6049	14.0	1.1 x 0.7	13.5	Sab	67	
10 58 04.2	+06 02 42	UGC 6053	13.8	1.0 x 0.9	13.5	SA(s)c: II	90	Stellar nucleus; **CGCG 38-63** WNW 6′.1.
10 58 37.5	+09 03 01	UGC 6062	12.8	1.4 x 0.7	12.6	SAB(r)0	25	Mag 12.9 star S 1′.8.
10 58 59.3	+06 31 20	UGC 6066	14.1	1.5 x 0.3	13.1	SAab: sp	38	
10 59 14.8	+05 17 28	UGC 6068	13.7	1.3 x 0.6	13.3	$SA0^+$ pec?	5	
10 59 44.6	+10 04 13	UGC 6072	13.5	0.9 x 0.8	13.0	(R′)SA(r)ab		**PGC 33129** NE 2′.8.
11 00 04.5	+12 14 02	UGC 6078	14.0	1.1 x 0.5	13.2	Sbc	126	Mag 10.50v star S 3′.6.
11 00 18.6	+10 02 49	UGC 6081	14.1	1.0 x 0.6	13.4	Sc pec	90	Double system, disrupted.
11 00 48.1	+10 43 38	UGC 6093	14.0	1.2 x 0.8	13.8	SAB(rs)bc II	0	**PGC 33210** S 7′.7.
11 01 51.2	+16 36 25	UGC 6104	13.7	1.6 x 0.5	13.3	Sbc	51	CGCG 95-91 NW 11′.5; very close pair of mag 14.1 stars W 2′.8.
11 02 35.2	+16 44 05	UGC 6112	13.3	2.1 x 0.7	13.6	Sd?	123	On E edge of galaxy cluster A 1145; mag 8.25v star N 9′.1.
11 04 09.1	+08 22 01	UGC 6130	13.7	1.0 x 0.8	13.3	SAB(rs)cd II-III	36	Stellar nucleus.
11 04 34.5	+16 03 40	UGC 6137	13.0	1.2 x 0.7	12.7	S0?	105	Bright nucleus; mag 9.76v star S 1′.9.
11 04 39.1	+05 11 57	UGC 6141	13.8	1.1 x 0.6	13.2	SAB(r)b: I-II	13	Bright core, faint envelope; mag 12.1 star on E edge.
11 06 25.3	+17 30 28	UGC 6157	13.3	1.6 x 1.4	14.0	SA(s)dm	24	Several bright knots; close pair of stars, mags 10.8 and 12.1, S 1′.3.
11 07 04.4	+07 48 09	UGC 6168	14.1	1.3 x 0.3	12.9	(R′)SB(r)bc: III	122	
11 07 03.4	+12 03 37	UGC 6169	13.7	1.7 x 0.3	12.8	Sb: III	0	
11 07 56.7	+08 00 05	UGC 6185	13.5	1.5 x 0.8	13.6	SAB(rs)dm III	160	Located on E edge of small galaxy cluster **A 1170**.
11 09 49.3	+12 46 12	UGC 6206	13.3	0.6 x 0.4	11.6	Double System	27	
11 10 01.6	+05 18 13	UGC 6210	13.5	1.5 x 0.7	13.4	SAB(s)bc II-III	15	Stellar nucleus; stellar **CGCG 39-21** NW 6′.1.
11 11 18.0	+11 33 11	UGC 6231	14.4	0.9 x 0.5	13.4		55	Bright N-S core; mag 10.6 star SE 2′.6.
11 11 28.2	+06 54 24	UGC 6233	13.8	1.0 x 0.3	12.3	SA(r)0/a:	3	Stellar **CGCG 39-57** N 6′.2.
11 12 52.1	+10 11 55	UGC 6248	14.6	1.5 x 1.2	15.1	IAB(s)m V-VI	156	Dwarf irregular, very low surface brightness.
11 16 51.0	+17 47 50	UGC 6296	13.6	1.4 x 0.4	12.8	S?	166	Mag 11.0 star SSW 5′.1.
11 17 17.0	+16 19 34	UGC 6300	13.8	1.2 x 0.7	13.6	E	71	
11 18 00.4	+07 50 41	UGC 6312	13.8	1.1 x 0.5	13.0	SBa pec sp	45	Mag 12.3 star SW end.
11 27 47.3	+07 59 19	UGC 6462	14.1	1.5 x 0.2	12.7	S?	3	Stellar **CGCG 39-188** E 3′.4; elongated **CGCG 39-190** E 5′.9.

GALAXY CLUSTERS

RA h m s	Dec ° ′ ″	Name	Mag 10th brightest	No. Gal	Diam ′	Notes
10 23 24.0	+12 50 00	A 999	15.6	33	22	**MCG +2-27-5** and **MCG +2-27-4** at center; most other GX anonymous and stellar; see NGC 3230 notes.
10 27 00.0	+10 58 00	A 1016	15.4	37	22	A number of anonymous and CGCG galaxies extend from center to SW edge.
10 27 48.0	+10 24 00	A 1020	16.0	68	22	**Leo Cluster**. All members anonymous, faint and stellar.
10 54 00.0	+16 51 00	A 1126	16.0	55	17	**PGC 32713** near center, see notes for NGC 3447; most cluster members stellar.
11 00 54.0	+10 32 00	A 1142	15.4	35	17	NGC 3492 and IC 664 near center; most other members stellar to almost stellar.
11 01 30.0	+16 43 00	A 1145	15.7	34	17	Most members anonymous, faint and stellar.
11 03 00.0	+07 37 00	A 1149	16.0	34	17	All members anonymous and stellar to almost stellar.

93 GALAXIES 93

RA h m s	Dec ° ′ ″	Name	Mag (V)	Dim ′ Maj x min	SB	Type Class	PA	Notes
09 01 09.2	+06 06 27	CGCG 33-43	14.4	0.7 x 0.3	12.6		6	Mag 10.5 star S 2′.4.
09 16 25.3	+07 05 42	CGCG 34-20	14.6	0.4 x 0.3	12.1		120	Mag 10.10v star NE 1′.7.
09 31 52.2	+07 34 49	CGCG 35-3	14.7	0.4 x 0.4	12.6			Mag 11.9 star E 2′.6.
09 39 30.2	+06 26 12	CGCG 35-23	14.3	0.8 x 0.2	12.2			Mag 11 star W 0′.8; MCG +1-25-8 SW 2′.7.
09 45 12.7	+07 07 33	CGCG 35-44	15.5	0.3 x 0.3	12.7			Mag 10.26v star S 3′.3; mag 9.01v star W 7′.2.
09 46 42.8	+06 40 57	CGCG 35-55	14.8	0.4 x 0.4	12.7			Located 8′.3 E of mag 5.80v star.
09 47 35.2	+06 48 47	CGCG 35-57	14.7	0.5 x 0.5	13.0			Uniform surface brightness; 0′.9 SW of mag 8.68v star.
09 53 02.8	+07 50 25	CGCG 35-75	15.0	0.5 x 0.3	12.8		138	UGC 5304 NE 2′.0.
09 54 11.8	+07 07 37	CGCG 35-82	14.9	0.5 x 0.3	12.7		27	Mag 9.89v star W 4′.7.
10 04 08.8	+06 30 36	CGCG 36-27	13.7	0.3 x 0.2	10.5		6	Stellar; mag 10.01v star W 8′.9.
10 10 40.6	+07 34 34	CGCG 36-49	14.6	0.4 x 0.4	12.5			Mag 8.62v star NNE 7′.9.
10 11 09.5	+07 46 46	CGCG 36-51	15.5	0.7 x 0.3	13.6		144	Mag 13.1 star NW 0′.7.
10 11 19.7	+05 53 45	CGCG 36-52	14.7	0.6 x 0.6	13.5			Mag 15.2 star 0′.5 S; mag 11.4 star N 5′.0.
10 14 18.5	+06 30 43	CGCG 36-67	15.3	0.5 x 0.5	13.7			Mag 11.8 star SE 3′.0.
10 15 16.8	+05 23 07	CGCG 36-72	14.8	0.6 x 0.3	12.8		27	Pair of mag 10.89v, 10.69v stars N 8′.2.
08 58 05.9	+12 59 28	CGCG 61-31	14.5	0.8 x 0.6	13.5		48	Mag 8.87v star E 12′.5.
08 58 43.3	+12 40 33	CGCG 61-32	15.0	0.6 x 0.4	13.3		21	Mag 9.24v star ENE 5′.8.
09 02 04.6	+11 37 58	CGCG 61-44	14.8	1.0 x 0.3	13.4		30	Mag 14.0 star on NE edge; mag 10.4 star SW 6′.7.

(Continued from previous page)

GALAXIES

RA h m s	Dec ° ′ ″	Name	Mag (V)	Dim ′ Maj x min	SB	Type Class	PA	Notes
09 20 50.8	+13 11 27	CGCG 62-22	14.2	1.0 x 0.3	12.7		72	Located between a pair of mag 12 stars on a NW-SE line.
09 24 57.5	+11 32 26	CGCG 62-29	14.3	0.6 x 0.4	12.6		9	NGC 2872 et al. SE 12′.8.
09 25 15.4	+10 53 09	CGCG 62-30	14.5	0.8 x 0.4	13.1		42	Mag 8.92v star NW 8′.3.
09 32 52.4	+12 17 08	CGCG 63-6	15.9	0.7 x 0.5	14.6		132	Mag 11.1 star NNE 5′.7; mag 11.42v star SW 6′.9.
09 37 37.3	+12 37 37	CGCG 63-21	14.3	0.6 x 0.3	12.3		171	Mag 11.9 star E 2′.4; mag 6.73v star NE 11′.5.
09 38 53.6	+09 44 54	CGCG 63-27	15.3	0.7 x 0.4	13.8		165	Triangle of mag 8.95v, 9.65v and 10.50v stars centered SW 9′.3; NGC 2940 5′.5 W of this triangle.
09 40 31.0	+13 32 28	CGCG 63-32	14.7	0.5 x 0.5	13.1			Mag 13.2 star E 2′.4; mag 9.58v star NW 12′.2.
09 40 43.2	+13 19 37	CGCG 63-34	14.0	0.9 x 0.3	12.4			Mag 12.1 star N 4′.1; mag 10.38v star SE 9′.8.
09 46 22.6	+09 09 53	CGCG 63-61	15.0	0.7 x 0.4	13.5		150	Mag 13.9 star N 0′.7; **CGCG 63-65** NE 8′.1.
09 46 38.5	+13 39 31	CGCG 63-62	14.3	0.6 x 0.3	12.3		93	Located 2′.8 E of mag 9.22v star.
09 49 51.6	+10 25 50	CGCG 63-79	14.2	0.7 x 0.2	11.9		6	Loose group of four mag 12-13 stars SE 3′.8; mag 12.9 star WNW 3′.4.
09 50 17.9	+13 22 07	CGCG 63-83	14.7	0.6 x 0.4	13.0		42	Mag 10.15v star N 3′.2.
09 57 47.5	+10 01 36	CGCG 64-9	15.5	0.6 x 0.2	13.1		171	Mag 10.37v star E 5′.1; note **MCG +2-26-7** is 1′.7 S of this star.
09 59 29.4	+11 12 00	CGCG 64-18	15.0	0.6 x 0.4	13.3		33	Mag 11.7 star E 2′.1.
09 59 42.5	+09 57 04	CGCG 64-19	15.1	0.7 x 0.4	13.6		156	Mag 11.1 star W 2′.3.
09 59 37.7	+10 19 53	CGCG 64-20	15.0	0.9 x 0.6	14.1		57	Mag 11.4 star E 6′.1; IC 584 E 8′.2.
10 00 07.5	+13 32 58	CGCG 64-26	15.4	0.4 x 0.2	12.5		36	Mag 13.3 star E 0′.8; mag 10.56v star SW 7′.3.
10 00 20.9	+11 20 05	CGCG 64-27	14.2	0.5 x 0.3	12.0		162	Mag 12.1 star NW 5′.0.
10 00 26.0	+12 52 12	CGCG 64-28	15.0	0.3 x 0.3	12.2	Ia		Forms a triangle with an E-W oriented pair of mag 12.8, 13.2 stars N 1′.9.
10 06 17.8	+13 58 21	CGCG 64-58	14.7	0.8 x 0.5	13.6		42	Faint, round anonymous galaxy N 1′.8.
10 06 28.9	+12 52 11	CGCG 64-60	15.7	0.9 x 0.6	14.9		141	Patchy appearance.
10 11 00.3	+09 50 45	CGCG 64-85	14.8	0.9 x 0.7	14.1		81	Stellar nucleus in short, bright bar; mag 11.23v star NW 1′.2.
09 04 58.3	+15 50 18	CGCG 90-67	15.0	0.6 x 0.4	13.3	Sm	54	Mag 14.8 star on S edge.
09 20 11.7	+16 31 46	CGCG 91-72	15.5	0.5 x 0.3	13.3		21	Mag 11.2 star E 4′.7.
09 43 55.0	+14 48 07	CGCG 92-35	15.9	0.9 x 0.3	14.3		147	Mag 12.1 star SSE 4′.5; mag 9.93v star NNE 6′.8.
09 44 36.0	+15 03 12	CGCG 92-37	15.4	0.8 x 0.3	13.7	Sbc	9	Mag 10.29v star SE 11′.9.
09 46 03.6	+16 29 07	CGCG 92-45	14.4	0.6 x 0.4	12.7		93	Mag 9.18v star W 3′.4.
09 51 30.2	+17 15 28	CGCG 92-58	16.4	0.8 x 0.3	14.7		153	Mag 9.27v star NE 7′.0.
09 51 54.0	+14 55 54	CGCG 92-59	14.6	0.8 x 0.5	13.5	Sm	84	Low, uniform surface brightness; mag 14.6 star on N edge.
09 52 12.7	+17 35 32	CGCG 92-62	14.7	0.6 x 0.5	13.2		48	Mag 12.4 star 1′.8 NE.
09 55 08.5	+14 17 24	CGCG 92-72	14.2	0.6 x 0.2	11.7	Sbc	156	**CGCG 92-73** NE 1′.4.
09 58 53.7	+15 22 44	CGCG 93-13	14.7	0.7 x 0.4	13.2		90	Mag 14.7 star on SE edge; mag 11.3 star N 4′.4.
09 59 04.0	+15 39 49	CGCG 93-15	14.3	0.5 x 0.4	12.4		30	Faint, elongated anonymous galaxy NW 2′.3.
09 59 22.6	+15 54 49	CGCG 93-18	15.1	0.6 x 0.4	13.4		42	Mag 10.32v star N 3′.6.
10 00 43.3	+15 16 58	CGCG 93-19	15.5	0.8 x 0.4	14.1		51	Strong dark lane along SE edge.
10 08 19.9	+16 53 52	CGCG 93-57	14.2	0.5 x 0.5	12.6			Mag 3.52v star Eta Leonis SW 16′.5; mag 15.0 star N 1′.9.
10 08 40.2	+16 47 10	CGCG 93-59	14.7	0.6 x 0.4	13.0		135	Mag 3.52v star Eta Leonis W 19′.3; mag 12.6 star E 2′.8.
09 02 40.7	+10 50 28	IC 526	14.4	0.8 x 0.5	13.0	Sb	48	Mag 9.36v star E 6′.7.
09 09 22.7	+15 47 49	IC 528	14.1	1.5 x 0.8	14.1	Sab III	163	= **Hickson 36A**. Mag 8.85v star NE 1′.8; **Hickson 36B** NE 1′.1; **Hickson 36C** on NW edge; **Hickson 36D** on SW edge.
09 15 17.1	+11 53 09	IC 530	13.1	1.8 x 0.4	12.6	Sab	87	Mag 9.81v star N 3′.7.
09 41 00.3	+06 56 05	IC 551	13.6	0.8 x 0.5	12.5	S?	168	Mag 10.19v star NE 11′.8.
09 41 16.6	+10 38 46	IC 552	13.4	1.0 x 0.5	12.6	S0	1	Mag 10.5 star SW 8′.0; stellar **CGCG 63-42** E 3′.7.
09 41 57.1	+12 17 44	IC 555	13.3	1.3 x 0.5	12.7	S0	18	Mag 10.76v star E 3′.6; stellar galaxy **Mkn 1422** S 7′.0.
09 44 02.6	+10 59 17	IC 557	13.9	0.8 x 0.4	12.5	Sb	42	Mag 11.3 star S 6′.1; NGC 2984 NW 7′.0.
09 44 44.0	+09 36 51	IC 559	14.2	0.7 x 0.4	12.7	Sb	90	Mag 10.42v star E 1′.8.
09 47 50.1	+15 51 03	IC 565	15.2	1.6 x 0.3	14.2	Scd?	50	Mag 10.13v star SE 5′.8.
09 51 08.4	+15 43 48	IC 568	13.5	1.7 x 1.1	14.0	SB(rs)b	33	Bright E-W bar; **IC 570** E 10′.4.
09 51 28.2	+10 55 08	IC 569	14.1	0.6 x 0.4	12.6	E	162	Mag 8.30v star WSW 8′.1.
09 52 31.7	+15 46 28	IC 571	15.1	1.0 x 0.6	14.4		0	IC 572 N 3′.1.
09 52 33.1	+15 49 33	IC 572	14.8	0.5 x 0.5	13.1			Mag 10.87v star NNW 5′.7; IC 571 S 3′.1.
09 56 03.9	+10 29 52	IC 577	14.0	0.6 x 0.6	12.7	S?		Mag 14.6 star on S edge; mag 10.14v star NE 14′.2.
09 56 16.3	+10 29 07	IC 578	13.9	1.2 x 0.4	12.9	SBa:	72	Mag 10.14v star NE 11′.9.
09 58 11.8	+15 56 47	IC 581	14.1	0.9 x 0.6	13.3	SBab:	135	Mag 10.41v star NNE 6′.6.
09 59 00.2	+17 49 00	IC 582	14.0	0.9 x 0.9	13.6	S		Elongated **IC 583** E 1′.1.
09 59 05.2	+10 21 36	IC 584	14.6	0.4 x 0.4	12.5			Star S edge; mag 7.12v star NW 7′.6; CGCG 64-20 E 8′.2.
09 59 44.2	+12 59 16	IC 585	13.4	1.3 x 0.8	13.3	S0⁻:	129	Mag 10.26v star S 4′.6; NGC 3080 NNE 4′.4.
10 07 27.7	+12 16 27	IC 591	13.2	1.0 x 0.7	12.7	S?	170	Faint, elongated anonymous galaxy S 1′.9; low surface brightness **Leo I** (UGC 5470) E 14′.6.
10 09 38.2	+10 59 57	IC 595	14.8	0.7 x 0.5	13.6		165	Almost stellar.
10 10 31.4	+10 02 27	IC 596	14.9	0.9 x 0.3	13.3	Sb	15	Mag 9.88v star E 11′.3.
09 04 35.0	+14 35 35	IC 2431	14.0	0.6 x 0.4	12.3		54	Multi-galaxy system; strong dark lane along centerline NE-SW.
09 16 01.8	+17 49 11	IC 2454	13.2	0.9 x 0.6	12.4	Sa?		NGC 2797 SE 7′.4.
08 59 49.1	+05 03 36	MCG +1-23-16	14.1	0.6 x 0.4	12.4	S0?	3	Faint star on E edge; mag 12.3 star NE 2′.4.
09 14 37.5	+08 06 58	MCG +1-24-8	14.0	0.8 x 0.6	13.1		105	Mag 11.6 star NE edge; mag 9.92v star SW 2′.1.
09 18 58.1	+05 53 15	MCG +1-24-11	14.0	0.5 x 0.5	12.3			**CGCG 34-26** E 1′.6.
09 24 01.1	+05 12 33	MCG +1-24-19	14.7	0.9 x 0.7	14.0	Sbc	171	Mag 12.2 star S 1′.5.
09 34 44.8	+06 25 30	MCG +1-25-3	14.7	0.7 x 0.5	13.4	Sb	78	
09 39 23.3	+06 24 03	MCG +1-25-8	13.9	0.9 x 0.6	13.1	S?	69	Mag 11.0 star, with CGCG 35-23 immediately E of it, NNE 2′.3.
09 42 31.5	+07 05 54	MCG +1-25-15	14.1	0.7 x 0.5	12.8	Sm	99	Mag 12.4 star NW 1′.7.
09 57 21.1	+07 11 16	MCG +1-26-4	14.7	0.4 x 0.4	12.6	S?		
10 09 46.4	+05 10 19	MCG +1-26-16	14.7	0.9 x 0.7	14.0	Sbc	81	
10 10 37.2	+05 08 58	MCG +1-26-17	16.6	0.7 x 0.5	15.3		15	Mag 9.49v star SE 2′.4.
10 15 53.4	+06 57 47	MCG +1-26-28	14.1	0.5 x 0.5	12.4			Small companion galaxy SW edge; mag 9.31v star SW 3′.6.
08 56 25.0	+12 24 45	MCG +2-23-11	14.8	0.8 x 0.7	14.0	Sc	3	
09 01 24.7	+10 11 58	MCG +2-23-19	14.9	0.7 x 0.6	13.8	Sb	33	
09 02 41.5	+13 06 26	MCG +2-23-23	14.9	0.9 x 0.2	12.9	Sb	135	Close pair of mag 13-14 stars NW 2′.8.
09 03 11.3	+13 37 55	MCG +2-23-24	13.7	0.6 x 0.6	12.5			Mag 12.8 star NW 1′.0.
09 03 23.5	+13 30 38	MCG +2-23-25	14.6	0.7 x 0.5	13.3	Sb	48	
09 04 37.1	+13 33 12	MCG +2-23-26	14.3	0.9 x 0.3	12.7	S?	60	Mag 11,8 star SE 3′.1; MCG +2-23-27 E 3′.2.
09 04 50.0	+13 33 42	MCG +2-23-27	13.7	0.9 x 0.7	13.1	S?	0	Mag 10.84v star N 3′.2.
09 18 06.5	+13 28 18	MCG +2-24-5	14.0	0.7 x 0.5	12.7	Sb	99	Mag 12.0 star N 1′.5.
09 20 56.0	+10 41 58	MCG +2-24-7	15.4	0.7 x 0.3	13.6	Sc	3	Mag 12.4 star SW 1′.7.
09 28 40.7	+12 37 01	MCG +2-24-15	14.3	0.9 x 0.4	13.0	Sb	141	
09 34 55.7	+11 40 47	MCG +2-25-8	14.0	1.0 x 0.6	13.3		132	
09 35 21.8	+13 32 51	MCG +2-25-9	14.8	0.9 x 0.5	13.8	Sc	111	Strong NE-SW bar; mag 11.1 star ESE 2′.0.
09 36 26.0	+11 19 45	MCG +2-25-10	15.0	1.1 x 0.4	14.0	Irr	3	Mag 8.12v star N 6′.3.
09 40 30.4	+11 29 46	MCG +2-25-14	13.8	0.9 x 0.5	12.8		24	Galaxy pair.
09 41 47.0	+11 28 54	MCG +2-25-19	14.1	0.7 x 0.7	13.2			Mag 14.3 star superimposed E edge; mag 12.7 star N 1′.5.
09 42 53.3	+09 29 34	MCG +2-25-21	12.9	0.9 x 0.7	12.3	Sb	18	Part of irregular multiple system, connected with UGC 5189 SE.

GALAXIES

RA h m s	Dec ° ′ ″	Name	Mag (V)	Dim ′ Maj x min	SB	Type Class	PA	Notes
09 42 54.4	+12 23 09	MCG +2-25-23	13.7	0.8 x 0.3	12.0	Sab	144	
09 44 23.5	+11 13 48	MCG +2-25-28	14.8	0.8 x 0.8	14.2			Mag 10.70v star WNW 3′2.
09 46 46.0	+13 31 46	MCG +2-25-32	14.9	0.6 x 0.6	13.6	Sbc		
09 46 49.0	+09 44 07	MCG +2-25-34	13.7	0.7 x 0.5	12.4		150	
09 47 22.9	+08 34 06	MCG +2-25-35	14.0	0.4 x 0.3	11.5		1	Trio of bright stars W, brightest is W 1′5, mag 7.01v.
09 47 51.5	+08 49 47	MCG +2-25-36	14.1	0.8 x 0.4	12.7	Sd	45	Mag 11.8 star WNW 2′3.
09 48 41.0	+09 00 30	MCG +2-25-37	14.4	0.7 x 0.6	13.3	Sb	144	Mag 9.59v star SE 3′6.
09 49 36.9	+09 00 15	MCG +2-25-39	13.8	0.9 x 0.2	11.8		78	
09 49 52.7	+09 09 36	MCG +2-25-42	13.9	0.5 x 0.2	11.2	Sb	30	Pair with UGC 5270 on E edge.
09 51 41.8	+12 26 39	MCG +2-25-49	14.1	0.5 x 0.4	12.2		174	Located between a N-S pair of mag 12 stars.
09 52 36.4	+09 52 52	MCG +2-25-51	14.0	0.7 x 0.5	12.7		87	Faint, slightly elongated anonymous galaxy W 1′2.
09 53 15.2	+11 14 49	MCG +2-25-52	14.7	0.7 x 0.2	12.4		177	Located between a pair of mag 12-13 stars on NE-SW line.
09 53 51.5	+13 36 50	MCG +2-25-53	14.2	0.7 x 0.4	12.7	Sb	39	Mag 12.9 star N 2′7.
09 54 02.3	+10 36 26	MCG +2-25-54	13.9	0.9 x 0.6	13.1		141	Mag 12.5 star W 1′3; small, faint anonymous galaxy E 1′4.
09 56 35.2	+10 05 36	MCG +2-26-3	14.9	0.9 x 0.7	14.2		57	
09 56 34.9	+11 09 46	MCG +2-26-4	13.7	0.7 x 0.6	12.6	S0?	153	= **Mkn 1241**; faint anonymous galaxy NE 2′6.
09 58 36.4	+13 15 15	MCG +2-26-8	14.4	0.7 x 0.4	12.9		162	Mag 10.47 star SW 1′9.
09 59 08.8	+13 03 30	MCG +2-26-11	15.7	1.1 x 0.2	13.9	Sb	162	Pair with MCG +2-26-12 NE edge; several small, faint anonymous galaxies SW 1′7.
09 59 09.1	+13 03 52	MCG +2-26-12	15.4	0.4 x 0.3	12.9	Sb	63	Pair with MCG +2-26-11 SW edge.
09 59 43.4	+11 39 39	MCG +2-26-13	14.3	1.1 x 0.7	13.9		30	Mag 9.01v star S 3′2.
10 00 57.2	+11 10 53	MCG +2-26-16	15.8	0.7 x 0.7	14.8			
10 01 03.4	+11 27 41	MCG +2-26-17	15.2	0.5 x 0.5	13.5	Sb		
10 04 50.0	+11 54 49	MCG +2-26-23	15.3	0.7 x 0.3	13.4		114	
10 10 00.2	+11 54 55	MCG +2-26-29	15.0	0.9 x 0.7	14.4	Sb	6	
10 14 34.8	+11 51 53	MCG +2-26-33	15.0	0.4 x 0.4	12.8			Line of three mag 14 stars, on NE-SW track, E 0′9.
08 57 54.0	+17 05 23	MCG +3-23-21	14.7	0.7 x 0.3	12.9	Sc	111	Mag 8.64v star N 5′7.
09 13 43.7	+16 51 44	MCG +3-24-13	14.3	0.8 x 0.5	13.1	Sc	57	Mag 11.08v star NE 5′1.
09 15 43.4	+15 13 15	MCG +3-24-17	15.1	0.5 x 0.4	13.2			
09 16 34.4	+17 55 06	MCG +3-24-24	14.3	0.9 x 0.6	13.5		129	Mag 12.8 star WSW 1′4.
09 16 53.7	+17 27 24	MCG +3-24-29	14.5	1.3 x 0.8	14.4		6	Stellar nucleus, smooth envelope.
09 17 38.5	+16 26 23	MCG +3-24-36	14.2	0.9 x 0.6	13.4		36	Mag 13.7 star N 0′9.
09 18 42.6	+16 28 52	MCG +3-24-44	15.1	0.7 x 0.3	13.3	Sc	24	Mag 11.52v star NW 3′1.
09 20 32.6	+17 42 07	MCG +3-24-46	14.5	0.9 x 0.6	13.7		39	
09 24 39.4	+17 39 44	MCG +3-24-55	14.8	0.4 x 0.2	11.9		171	Located 3′0 S of mag 7.11v star.
09 33 32.1	+14 35 02	MCG +3-25-1	14.1	0.5 x 0.4	12.2		84	
09 34 29.8	+16 33 36	MCG +3-25-3	16.2	0.5 x 0.4	14.3		57	
09 36 14.4	+17 29 03	MCG +3-25-4	14.1	0.7 x 0.5	12.8		6	
09 44 46.8	+14 53 28	MCG +3-25-23	14.2	0.8 x 0.4	12.8	Sb	6	Double system? Companion galaxy SE edge.
09 47 20.4	+14 43 24	MCG +3-25-27	14.1	0.6 x 0.3	12.1		147	Faint star on NE edge.
09 49 05.1	+13 52 45	MCG +3-25-38	14.5	0.9 x 0.4	13.2		141	
09 57 47.9	+15 07 28	MCG +3-26-6	14.0	0.9 x 0.9	13.7	Sb		Mag 9.01v star S 4′4.
09 57 53.7	+14 55 29	MCG +3-26-7	14.2	1.2 x 0.8	14.0		9	
10 08 04.4	+14 48 09	MCG +3-26-30	14.6	1.0 x 0.8	14.2	Sbc	9	Stellar nucleus, mag 11.3 star W 7′3.
10 10 38.5	+16 41 03	MCG +3-26-36	14.3	0.6 x 0.4	12.6		144	= **Mkn 1247**.
10 14 00.9	+14 29 55	MCG +3-26-44	14.1	1.1 x 0.9	13.9		42	**MCG +3-26-46** N 1′7; **MCG +3-26-45** N 2′1.
10 14 56.5	+16 52 41	MCG +3-26-48	14.1	0.9 x 0.3	12.5		117	Mag 8.35v star ESE 4′0.
08 57 23.5	+17 17 18	NGC 2711	13.7	0.9 x 0.6	12.9	SB?	170	
08 58 50.5	+06 17 34	NGC 2718	11.8	2.1 x 2.1	13.3	(R′)SAB(s)ab		Fainter arms form an "S" shape within the published size.
08 59 08.1	+11 08 54	NGC 2720	12.8	1.1 x 1.1	12.9	$S0^-$:		
09 01 03.3	+11 05 48	NGC 2725	13.5	0.7 x 0.6	12.4	S?	54	Mag 10.55v star N 2′5.
09 01 40.8	+11 04 57	NGC 2728	13.6	1.1 x 0.8	13.4	Sb III-IV	60	Mag 9.95v star SE 4′5.
09 02 15.7	+16 50 18	NGC 2730	12.9	1.7 x 1.3	13.6	SBdm:	80	Mag 13.9 star on S edge.
09 02 08.3	+08 18 02	NGC 2731	13.6	0.8 x 0.5	12.5	S?	70	Mag 12.8 star on N edge.
09 03 01.7	+16 51 45	NGC 2734	16.8	0.6 x 0.5	15.4		84	Almost stellar.
09 09 44.2	+07 10 22	NGC 2773	14.1	0.7 x 0.3	12.3	S?	83	Very faint, small, elongated anonymous galaxy SE 3′1.
09 10 20.4	+07 02 13	NGC 2775	10.1	4.3 x 3.3	12.9	SA(r)ab	155	Mag 11.21v star and small, faint anonymous galaxy SW 7′4.
09 10 41.8	+07 12 21	NGC 2777	13.3	0.9 x 0.6	12.5	Sab?	165	Very small, faint anonymous galaxy WSW 3′5.
09 15 02.1	+17 35 27	NGC 2791	14.6	0.8 x 0.3	12.9		156	Pair of faint stars on E edge.
09 16 02.0	+17 35 23	NGC 2794	13.2	1.2 x 1.2	13.4	SB?		Mag 8.64v star S 6′5; several anonymous stellar galaxies SSE 3′8.
09 16 04.0	+17 37 41	NGC 2795	12.8	1.4 x 1.0	13.2	E	170	Small, faint anonymous galaxy on S edge. Another anonymous galaxy lies 2′0 NW; **MCG +3-24-22** lies 2′4 NE.
09 16 21.8	+17 43 34	NGC 2797	14.0	0.7 x 0.6	12.9	Pec	21	IC 2454 NW 7′4; mag 11.9 star W 2′9; **MCG +3-24-21** W 3′4.
09 18 09.4	+16 11 50	NGC 2819	12.8	1.4 x 1.2	13.3	E	60	Mag 9.55v star N 7′6; **Mkn 704** NNE 7′5.
09 24 15.4	+05 56 27	NGC 2864	14.6	0.8 x 0.7	13.8	Sc	12	
09 25 42.7	+11 25 58	NGC 2872	11.9	2.1 x 1.8	13.3	E2	171	NGC 2874 on E edge; CGCG 62-29 NW 12′8.
09 25 48.6	+11 27 12	NGC 2873	15.4	0.7 x 0.3	13.6	Sb	117	NGC 2874 S 1′8.
09 25 47.3	+11 25 26	NGC 2874	12.5	2.4 x 0.7	12.9	SB(r)bc	43	NGC 2872 on W edge; NGC 2873 N 1′8.
09 26 36.3	+07 57 15	NGC 2882	12.7	1.5 x 0.8	12.7	S?	80	
09 29 30.2	+07 43 03	NGC 2894	12.4	1.9 x 1.0	12.9	Sa	27	2′8 NNE of mag 8.73v star. Chain of four faint stars extends across center of galaxy E-W, and curves SW.
09 32 06.0	+08 26 29	NGC 2906	12.7	1.4 x 0.9	12.8	Scd:	75	
09 33 45.9	+10 09 14	NGC 2911	11.5	4.1 x 3.0	14.1	SA(s)0: pec	140	NGC 2912 lies 1′3 E of center.
09 33 51.4	+10 09 33	NGC 2912	16.6	0.3 x 0.3	13.9	Im		Lies 1′3 E of center of NGC 2911.
09 34 02.8	+09 28 42	NGC 2913	13.2	1.1 x 0.7	12.7	S?	140	
09 34 02.8	+10 06 30	NGC 2914	13.2	1.0 x 0.7	12.6	SB(s)ab	12	Located 5′0 SE of NGC 2911; mag 10.9 star WNW 1′7.
09 34 47.4	+10 17 06	NGC 2919	12.9	1.7 x 0.6	12.7	SAB(r)b:	159	
09 36 03.8	+16 45 35	NGC 2923	15.6	0.5 x 0.4	13.7		15	Stellar.
09 37 10.1	+16 58 38	NGC 2928	14.7	1.1 x 0.6	14.1		42	Stellar nucleus; mag 10.35v star SE 3′7.
09 37 55.1	+17 00 51	NGC 2933	15.3	1.0 x 0.4	14.1	Sbc	30	Mag 9.93v star S 1′7; **CGCG 92-13** NW 7′9.
09 38 08.3	+09 31 16	NGC 2939	12.4	2.4 x 0.6	12.7	Sbc	157	NGC 2940 N 5′8.
09 38 05.2	+09 36 58	NGC 2940	13.6	0.9 x 0.7	13.0	S0?	90	Triangle of mag 8.95v, 9.65v and 10.50v stars centered 5′5 E.
09 38 24.3	+17 02 40	NGC 2941	15.0	0.8 x 0.5	13.9		163	Stellar nucleus in smooth envelope.
09 38 32.6	+17 01 52	NGC 2943	12.4	2.2 x 1.2	13.5	E	130	Stellar **MCG +3-25-12** E 1′6.
09 39 01.7	+17 01 32	NGC 2946	14.0	1.2 x 0.4	13.1	SB?	13	Stellar nucleus.
09 38 59.3	+06 57 20	NGC 2948	12.9	1.4 x 0.9	13.0	SBbc	7	Mag 9.96v star SW 3′9.
09 39 52.1	+16 55 05	NGC 2949	14.5	0.7 x 0.6	13.4	SBbc	171	Almost stellar nucleus; **CGCG 92-25** S 7′9.
09 40 24.1	+14 55 21	NGC 2954	12.4	1.2 x 0.9	12.5	E	160	
09 40 41.6	+11 53 16	NGC 2958	13.2	1.2 x 0.7	12.8	S(r)bc	10	
09 40 54.0	+05 09 54	NGC 2962	11.9	2.6 x 1.9	13.5	$(R)SAB(rs)0^+$	3	

(Continued from previous page)

GALAXIES

RA h m s	Dec ° ′ ″	Name	Mag (V)	Dim ′ Maj x min	SB	Type Class	PA	Notes
09 43 40.5	+11 03 39	NGC 2984	13.5	0.7 x 0.7	12.6	S0?		Mag 10.6 star NNW 7′.3; IC 557 SE 7′.0.
09 46 17.4	+05 42 31	NGC 2990	12.7	1.3 x 0.7	12.5	Sc: II	85	**CGCG 35-52**, and smaller anonymous companion, SE 1′.5.
09 49 50.8	+12 41 43	NGC 3016	12.9	1.2 x 0.9	12.9	Sb III	70	Stellar **CGCG 63-75** SW 3′.1; mag 9.80v star NW 6′.4.
09 50 07.1	+12 44 43	NGC 3019	15.2	0.8 x 0.4	13.8	Sbc	45	Almost stellar core; mag 10.57v star SE 5′.9.
09 50 06.4	+12 48 46	NGC 3020	11.9	3.0 x 1.6	13.5	SB(r)cd:	99	Strong E-W dark lane N side of core; mag 9.80v star W 8′.4.
09 50 27.1	+12 46 00	NGC 3024	13.1	2.1 x 0.5	13.0	Sc: sp	125	Mag 10.57v star S 5′.3.
09 53 07.3	+16 40 38	NGC 3041	11.5	3.7 x 2.4	13.8	SAB(rs)c II-III	95	Almost stellar nucleus, several stars superimposed.
09 54 56.4	+16 27 20	NGC 3048	15.4	0.7 x 0.3	13.6		123	An anonymous galaxy extends N from the SE end of the galaxy; NGC 3053 E 9′.1.
09 54 49.5	+09 16 16	NGC 3049	12.1	2.2 x 1.4	13.2	SB(rs)ab	25	
09 55 33.7	+16 25 56	NGC 3053	12.7	1.8 x 0.8	13.0	SBa?	134	NGC 3048 W 9′.1.
09 56 19.3	+16 49 48	NGC 3060	13.0	2.2 x 0.6	13.2	Sb III	78	UGC 5343 E 6′.5.
09 57 56.8	+10 25 56	NGC 3069	14.2	0.7 x 0.3	12.3	S0?	165	Mag 7.12v star E 12′.2.
09 58 07.0	+10 21 33	NGC 3070	12.3	1.4 x 1.4	13.0	E		Mag 7.12v star NE 11′.1.
09 58 56.4	+14 25 08	NGC 3075	13.6	1.2 x 0.8	13.4	Sc	135	
09 59 55.9	+13 02 37	NGC 3080	13.4	0.9 x 0.8	12.9	Sa	69	Mag 10.47v star SE 4′.2; IC 585 SSW 4′.4.
10 01 26.0	+15 46 08	NGC 3094	12.3	1.6 x 1.2	12.9	SB(s)a	78	A mag 11.09v star sits on the S edge.
10 04 22.5	+13 37 18	NGC 3107	13.4	0.7 x 0.6	12.3	Sbc:	140	Located 1′.8 NW of mag 7.79v star.
10 06 47.9	+14 18 47	NGC 3119	14.4	0.5 x 0.5	12.7			Elongated anonymous galaxy SW 1′.9.
10 06 52.0	+14 22 26	NGC 3121	12.6	1.7 x 1.4	13.5	E	20	A very faint, small anonymous galaxy on S edge.
10 08 12.4	+09 58 38	NGC 3130	13.4	1.1 x 0.6	12.8	S0/a	30	Located in the glare of mag 4.37v star 31 Leo 4′.6 WNW.
10 12 29.6	+12 22 41	NGC 3134	13.7	0.9 x 0.2	11.7		51	
10 12 50.6	+12 39 58	NGC 3153	12.7	2.1 x 0.9	13.2	Scd:	170	
10 13 01.4	+17 02 03	NGC 3154	13.5	0.9 x 0.4	12.2	Sb	124	Lies 2′.1 NW of mag 8.77v star.
08 56 24.2	+13 10 54	UGC 4677	13.6	1.2 x 0.8	13.4	Scd:	120	Mag 11.3 star WNW 1′.2.
08 57 01.5	+13 11 50	UGC 4685	13.3	1.3 x 0.6	12.9	S?	150	Mag 9.54v star NNW 6′.6.
08 57 52.3	+12 29 42	UGC 4694	14.1	0.9 x 0.5	13.1	SBcd:	100	Filamentary; mag 9.93v star E 2′.5; mag 8.96v star SW 3′.6.
09 00 08.0	+16 55 23	UGC 4721	14.1	0.8 x 0.6	13.2	SBcd:	39	
09 00 42.1	+17 37 14	UGC 4729	13.8	0.9 x 0.7	13.1	SB(s)cd:	156	Bright E-W bar; mag 11.01v star SE 3′.3.
09 01 00.2	+10 37 00	UGC 4731	13.0	1.2 x 0.9	12.9	S0/a	72	Mag 10.9 star SSW 2′.7.
09 04 56.2	+17 27 17	UGC 4761	13.2	0.9 x 0.8	12.9	E	72	**MCG +3-23-33** W 1′.2; **MCG +3-23-29** W 10′.0.
09 06 34.2	+06 18 13	UGC 4781	13.6	1.4 x 0.6	13.3	Scd:	127	Mag 9.78v star S 2′.6.
09 08 10.4	+05 55 32	UGC 4797	13.9	2.0 x 1.5	15.0	Sm: V	0	Mag 12.8 star on W edge.
09 11 01.9	+13 24 48	UGC 4827	14.0	1.3 x 0.4	13.1	S?	80	
09 12 26.0	+09 57 19	UGC 4845	13.9	1.5 x 0.5	13.4	SBd:	98	Mag 11.1 star N 0′.9.
09 13 35.8	+12 26 27	UGC 4861	13.3	0.9 x 0.5	12.3	Sa?	77	
09 14 12.6	+16 44 32	UGC 4864	12.8	1.7 x 1.7	13.8	SA(r)ab II-III		Strong nucleus, uniform envelope; mag 10.95v star W 4′.7.
09 14 41.2	+15 27 42	UGC 4873	13.3	1.1 x 0.8	13.0	S?	45	Close pair of stars, mags 11 and 13, S 1′.6.
09 15 55.0	+10 07 59	UGC 4884	13.6	0.9 x 0.5	12.6	Sab	35	Stellar **CGCG 62-12** NE 8′.9.
09 16 10.7	+06 19 44	UGC 4890	13.9	0.9 x 0.5	12.9	SBcd:	55	
09 16 39.8	+07 15 58	UGC 4900	13.6	1.1 x 0.7	13.2	Sbc	122	Mag 12.0 star NNE 2′.7.
09 16 58.0	+16 18 11	UGC 4907	14.1	0.5 x 0.5	12.4	Scd:		
09 18 20.0	+17 45 09	UGC 4925	14.7	1.5 x 0.2	13.2	Sd	78	Mag 9.50v star SW 2′.5.
09 18 40.6	+13 49 43	UGC 4931	13.7	1.0 x 0.7	13.1	S0/a	165	Mag 11.5 star on N edge; mag 10.46v star S 2′.0.
09 19 33.0	+05 52 56	UGC 4946	13.9	0.9 x 0.5	12.9	SB(r)b	30	
09 20 13.2	+08 47 40	UGC 4957	14.7	1.7 x 0.2	13.4	Sd:	152	Mag 10.8 star on E edge.
09 20 32.1	+07 04 24	UGC 4959	13.9	0.7 x 0.5	12.6	Sbc	42	
09 20 40.8	+15 06 02	UGC 4962	13.5	1.3 x 1.2	13.8	SBa	153	Small, elongated nucleus, uniform envelope; **CGCG 91-76** NE 2′.0.
09 24 14.6	+11 06 38	UGC 5003	14.2	1.0 x 0.5	13.3	Scd:	30	Mag 14.2 star on S edge.
09 26 03.4	+12 44 00	UGC 5025	14.3	0.7 x 0.5	13.1	S0?	70	Bright core, uniform envelope; mag 10.3 star NE 1′.6.
09 27 44.0	+12 17 15	UGC 5044	14.4	1.0 x 1.0	14.2	S?		Double system, = **Hickson 38B** and **C**; **Hickson 38A** WSW 2′.5; **Hickson 38D** WNW 2′.5.
09 30 36.0	+06 07 21	UGC 5069	14.0	1.0 x 0.9	13.8	S0	15	
09 32 21.0	+10 48 32	UGC 5082	13.9	1.1 x 0.3	12.5	S	127	Mag 11.6 star on S edge; mag 7.88v star S 2′.0.
09 34 00.2	+10 01 45	UGC 5093	14.0	1.0 x 0.4	12.9	S	153	**CGCG 63-12** SE 9′.4.
09 34 38.7	+05 50 30	UGC 5100	13.6	1.5 x 0.9	13.8	SB(s)b	30	Bright NE-SW bar; mag 12.2 star NE end.
09 35 07.2	+05 07 08	UGC 5107	13.9	2.0 x 0.4	13.5	SBd	47	Mag 11.3 star S 1′.4.
09 40 50.0	+11 33 12	UGC 5164	14.3	1.1 x 0.2	12.5	Sdm:	21	Double system, contact, distorted; small companion S edge; mag 8.77v star E 5′.2.
09 41 32.2	+11 24 46	UGC 5173	13.9	2.0 x 0.3	13.2	Sb III	131	**CGCG 63-43** N 3′.5.
09 41 53.9	+11 37 58	UGC 5177	14.3	1.0 x 0.4	13.2	S?	165	Mag 12.6 star on SE edge; mag 10.42v star S 0′.7.
09 42 56.6	+09 28 25	UGC 5189	13.2	1.8 x 0.7	13.3	Im?	120	Irregular multiple system with MCG +2-25-21 NW, connected.
09 43 51.2	+09 39 20	UGC 5204	14.0	0.8 x 0.8	13.4	Sb		Mag 10.31v star SSW 5′.9.
09 44 57.3	+16 42 27	UGC 5213	13.1	1.1 x 1.0	13.1	SB(rs)a	63	
09 45 14.4	+09 06 36	UGC 5215	12.8	1.5 x 0.8	12.9	Sbc III	23	
09 45 22.7	+09 46 04	UGC 5216	13.7	1.4 x 0.7	13.5	SABd:	165	
09 45 30.4	+06 22 36	UGC 5218	13.6	1.8 x 0.9	14.0	SABd	125	Almost stellar nucleus.
09 46 45.3	+13 46 22	UGC 5232	13.8	1.2 x 0.7	13.5	S?	0	
09 46 49.5	+16 02 35	UGC 5234	13.5	1.5 x 1.0	13.8	S(r)c	115	Mag 11.2 star superimposed E end.
09 49 26.1	+14 39 25	UGC 5258	14.7	1.6 x 0.2	13.3	S?	132	Mag 8.06v star SW 5′.6.
09 49 52.9	+09 05 42	UGC 5267	14.1	1.5 x 0.2	12.6	Sb III	44	UGC 5270 and MCG +2-25-42 N 4′.3.
09 49 54.8	+09 09 46	UGC 5270	13.7	1.0 x 0.6	13.0	S0	6	Mag 14.1 star SW of nucleus; MCG +2-25-42 on W edge; UGC 5267 S 4′.3.
09 50 11.4	+16 17 12	UGC 5274	13.6	0.9 x 0.8	13.1	Scd:	90	
09 51 06.1	+09 00 31	UGC 5286	13.0	1.9 x 1.2	13.7	SAd	35	
09 51 17.1	+07 49 38	UGC 5288	13.5	1.1 x 0.7	13.1	Sdm:	155	
09 51 53.5	+12 56 38	UGC 5291	13.4	1.2 x 0.7	13.1	$S0^-$:	0	
09 53 08.8	+07 51 48	UGC 5304	13.4	0.8 x 0.7	12.7	S?	147	Distorted, bridge to small companion N edge; CGCG 35-75 SW 2′.0.
09 53 26.0	+07 52 00	UGC 5308	13.5	0.8 x 0.7	12.7	S?	168	Bright nucleus offset S.
09 54 46.2	+13 22 25	UGC 5324	14.4	0.8 x 0.3	12.7	S	160	**Mkn 711** NE 7′.1.
09 56 42.6	+15 38 13	UGC 5342	13.6	1.0 x 0.5	12.7	S?	15	
09 56 45.5	+16 48 33	UGC 5343	14.2	1.4 x 0.5	13.7	Sdm	114	Stellar nucleus, faint envelope; NGC 3060 W 6′.5.
09 57 03.5	+15 34 39	UGC 5344	13.9	1.0 x 0.9	13.6	SABd	6	Stellar nucleus, faint envelope.
09 58 47.3	+11 23 16	UGC 5358	13.4	1.3 x 0.6	13.0	SB(s)b	77	Mag 9.26v star NE 4′.2.
10 00 00.0	+05 19 42	UGC 5373	11.3	5.5 x 3.7	14.5	IB(s)m IV-V	110	**Sextans B**. Dwarf irregular, low surface brightness.
10 01 05.4	+16 56 16	UGC 5385	13.8	1.0 x 0.7	13.3	Scd:	111	**CGCG 93-20** NNW 7′.5.
10 01 35.2	+12 45 52	UGC 5395	13.5	1.0 x 0.7	13.0	$S0^-$:	31	
10 01 40.6	+10 45 15	UGC 5396	13.7	1.5 x 0.4	13.0	Scd:	156	Mag 9.96v star NE 2′.9.
10 02 13.8	+13 41 45	UGC 5400	13.2	0.9 x 0.7	12.5	S0	160	**CGCG 64-39** S 6′.7.
10 03 03.7	+10 44 42	UGC 5409	14.0	1.2 x 0.3	12.7	S?	165	
10 04 53.7	+05 03 44	UGC 5432	13.1	0.8 x 0.7	12.5	E:		Mag 7.35v star S 3′.9.
10 07 07.1	+15 58 59	UGC 5453	14.0	1.0 x 0.7	13.5	Im	80	**CGCG 93-44** WSW 7′.4.
10 07 11.0	+12 39 02	UGC 5454	14.1	0.9 x 0.6	13.3	SABdm:	40	

(Continued from previous page)

93 GALAXIES 93

RA h m s	Dec ° ′ ″	Name	Mag (V)	Dim ′ Maj x min	SB	Type Class	PA	Notes
10 07 19.6	+10 21 48	UGC 5456	13.3	1.5 x 0.7	13.2	I0?	148	Located between a pair of stars, mags 12.1 and 13.0.
10 08 28.5	+12 18 18	UGC 5470	10.5	9.8 x 7.4	15.2	E3	80	**Leo I**. Low surface brightness; IC 591 W 14.′6.
10 09 12.6	+15 00 18	UGC 5477	12.7	1.3 x 1.3	13.3	E		
10 11 53.4	+16 26 24	UGC 5495	13.6	2.8 x 0.3	13.3	Scd:	101	
10 13 58.9	+07 01 26	UGC 5522	12.6	2.8 x 1.7	14.1	Sd	145	Bright core, smooth envelope; mag 10.9 star SW 3.′0.
10 14 28.6	+15 54 07	UGC 5526	13.9	0.9 x 0.4	12.6	S	98	
10 15 42.1	+07 19 39	UGC 5537	14.3	2.3 x 0.3	13.8	Sd	147	

BRIGHT NEBULAE

RA h m s	Dec ° ′ ″	Name	Dim ′ Maj x min	Type	BC	Color	Notes
09 39 54.0	+11 58 53	IRAS 09371+1212		R			**Frosty Leo Nebula**.

PLANETARY NEBULAE

RA h m s	Dec ° ′ ″	Name	Diam ″	Mag (P)	Mag (V)	Mag cent ★	Alt Name	Notes
09 52 59.1	+13 44 33	PN G221.5+46.3	720	10.4		16.0	PK 221+46.1	Mag 12.6 star W of center1.′7; mag 11.3 star WNW of center 4.′3.

94 GALAXIES 94

RA h m s	Dec ° ′ ″	Name	Mag (V)	Dim ′ Maj x min	SB	Type Class	PA	Notes
07 46 32.7	+06 51 49	CGCG 30-16	16.2	0.3 x 0.3	13.5			Low surface brightness; mag 10.53v star W 2.′0.
07 48 56.0	+06 20 03	CGCG 30-20	15.6	0.5 x 0.5	13.9			Bright nucleus, faint envelope; mag 13.1 star N edge; bracketed by faint stars; in star rich area.
07 52 59.0	+05 29 46	CGCG 30-24	14.2	0.7 x 0.5	12.9		51	Mag 12.7 star N 0.′6.
08 00 01.5	+08 00 18	CGCG 31-16	14.3	0.5 x 0.3	12.1		129	Mag 12.2 star NW 2.′7.
08 00 28.0	+07 48 52	CGCG 31-18	14.3	0.5 x 0.2	11.6	Sb	6	Mag 8.20v star NE 7.′9.
08 01 38.5	+06 37 22	CGCG 31-24	14.3	0.7 x 0.4	12.8		153	Mag 12.5 star NW 1.′1.
08 04 13.3	+07 22 32	CGCG 31-37	15.3	0.4 x 0.2	12.4		141	Close, faint double stars at SW edge, mag 14.1 star NE 1.′1.
08 11 05.0	+06 17 53	CGCG 31-65	14.3	0.5 x 0.5	12.6			Uniform, low surface brightness; forms a triangle with two mag 12 stars W.
08 23 20.8	+07 40 16	CGCG 32-14	15.6	0.3 x 0.3	12.8			N-S pair of mag 12 stars N 2.′0.
08 25 16.1	+06 46 02	CGCG 32-23	15.0	0.6 x 0.3	13.0		75	Mag 11.0v star SW 2.′8.
08 25 35.9	+05 06 56	CGCG 32-24	14.6	0.4 x 0.4	12.5			Very close pair of mag 14 stars E 0.′8.
08 38 47.0	+07 28 47	CGCG 32-46	15.3	0.5 x 0.3	13.1		123	**CGCG 32-49** SE 7.′1; mag 8.02v star NE 8.′6.
09 01 09.2	+06 06 27	CGCG 33-43	14.4	0.7 x 0.3	12.6		6	Mag 10.5 star S 2.′4.
07 46 09.3	+10 39 31	CGCG 58-33	15.6	0.7 x 0.3	13.8		144	Mag 5.25v star 11 CMi N 6.′8.
07 51 17.8	+10 49 56	CGCG 58-43	14.4	0.8 x 0.5	13.3		33	Close pair of N-S oriented mag 11 stars N 4.′0; **CGCG 58-46** SSE 4.′1.
07 51 23.7	+10 30 36	CGCG 58-45	15.0	0.6 x 0.3	13.0		111	Mag 12.1 star WSW 4.′4.
07 54 57.8	+14 27 17	CGCG 58-69	14.2	0.7 x 0.4	12.7		141	Mag 12.6 star on E edge.
07 56 00.6	+14 11 42	CGCG 59-1	14.6	0.7 x 0.3	12.8		132	Mag 12.8 star N 0.′7; mag 8.93v star N 8.′5.
08 01 39.6	+10 43 55	CGCG 59-17	16.2	0.8 x 0.3	14.5	Sb	174	Mag 10.01v star SE 1.′7.
08 02 38.0	+09 29 57	CGCG 59-27	14.3	0.5 x 0.3	12.1		168	Located 6.′1 NNE of NGC 2513; NGC 2510 W 6.′7.
08 03 56.4	+08 41 56	CGCG 59-32	13.9	0.4 x 0.4	11.8			= **Mkn 1208**. Pair of mag 12.4 and 14.3 stars NW 0.′9; mag 10.13v star ENE 3.′2; **CGCG 59-38** ENE 4.′9.
08 03 59.1	+08 32 05	CGCG 59-33	14.8	0.6 x 0.3	12.8			Mag 10.8 star N 3.′4; mag 11.0 star E 2.′8.
08 04 00.1	+10 00 32	CGCG 59-34	14.4	0.6 x 0.5	12.9			= **Mkn 1209**. UGC 4198 SW 5.′2.
08 04 06.3	+09 48 48	CGCG 59-36	14.4	0.8 x 0.5	13.3		39	Close pair of mag 12.2, 12.4 stars S 3.′0.
08 04 16.1	+08 38 41	CGCG 59-37	16.2	0.5 x 0.5	14.5			Mag 10.13v star and **CGCG 59-38** N 5.′0.
08 05 40.7	+10 42 24	CGCG 59-42	14.8	0.6 x 0.3	12.8		144	12.7 mag star on SW edge; mag 10.53v star SE 1.′2.
08 06 13.8	+11 47 32	CGCG 59-45	15.4	0.5 x 0.5	13.8	Sm		Mag 10.69v star NW 1.′3; mag 11.3 star SW 1.′1.
08 09 04.0	+11 21 09	CGCG 59-49	14.5	0.5 x 0.3	12.3		51	Mag 11.1 star E 2.′5.
08 11 25.7	+08 53 35	CGCG 59-50	14.9	0.4 x 0.4	12.8			**CGCG 59-51** N 2.′9; mag 10.95v star E 3.′4.
08 17 14.2	+12 59 26	CGCG 59-57	14.8	0.8 x 0.3	13.1		105	**UGC 4317** SE 10.′9.
08 22 55.8	+13 37 20	CGCG 60-3	14.7	0.7 x 0.4	13.2		36	Mag 12.0 star NW 1.′9; mag 9.23v star NW 9.′5.
08 23 50.2	+11 32 58	CGCG 60-5	14.6	0.5 x 0.5	13.0	S0		**CGCG 60-8** SE 4.′3; mag 9.60v star NW 8.′2.
08 25 01.6	+13 32 10	CGCG 60-10	14.2	0.8 x 0.5	13.1		162	Bright NE-SW bar.
08 26 20.6	+14 14 03	CGCG 60-13	15.1	0.5 x 0.5	13.5			E-W oriented pair of mag 11.8 stars SSW 5.′0.
08 35 40.2	+12 33 32	CGCG 60-26	15.2	0.7 x 0.3	13.3			Mag 9.98v star N 6.′3.
08 37 39.2	+12 01 58	CGCG 60-30	15.2	0.7 x 0.4	13.6	Sb	12	Mag 8.95v star N 6.′6.
08 44 30.2	+11 10 56	CGCG 61-2	14.6	0.3 x 0.3	11.8			N-S oriented pair of mag 10.6, 11.0 stars NNW 4.′9.
08 52 57.2	+09 18 59	CGCG 61-17	14.4	0.7 x 0.3	12.5	Sc	114	Mag 8.61v star S 7.′8.
08 58 05.9	+12 59 28	CGCG 61-31	14.5	0.8 x 0.6	13.5		48	Mag 8.87v star E 12.′5.
08 58 43.3	+12 40 33	CGCG 61-32	15.0	0.6 x 0.4	13.3		21	Mag 9.24v star ENE 5.′8.
09 02 04.6	+11 37 58	CGCG 61-44	14.8	1.0 x 0.3	13.4		30	Mag 14.0 star on NE edge; mag 10.4 star SW 6.′7.
07 58 01.4	+14 42 06	CGCG 88-9	14.5	0.6 x 0.6	13.3			Located 2.′5 N of mag 10.07v star.
08 24 14.4	+17 19 53	CGCG 89-16	14.8	0.4 x 0.4	12.9	E		= **Mkn 387**. Mag 9.73v star SSE 4.′4; mag 8.83v star S 7.′9.
08 54 08.7	+17 25 43	CGCG 90-33	14.6	0.6 x 0.4	12.9		123	Star on NW end.
08 09 30.2	+05 16 45	IC 498	13.8	0.9 x 0.8	13.3	S	54	
08 53 11.4	+09 08 53	IC 523	13.1	0.9 x 0.7	12.4	Sab	99	Mag 8.61v star NW 3.′5.
09 02 40.7	+10 50 28	IC 526	14.4	0.8 x 0.4	13.0	Sb	48	Mag 9.36v star E 6.′7.
08 06 11.2	+12 32 35	IC 2226	13.4	0.9 x 0.7	12.7	SABab II	150	Mag 10.1 star E 3.′6.
08 11 01.8	+05 05 10	IC 2231	14.0	1.0 x 1.0	14.1	E:		Mag 13.1 star on W edge; mag 9.88v star WNW 7.′1.
08 46 44.6	+17 45 14	IC 2398	14.9	0.5 x 0.3	12.6		9	Mag 12.5 star NE 0.′5; mag 10.89v star SSE 2.′1.
08 48 04.7	+17 42 07	IC 2406	13.2	1.1 x 0.6	12.6	S0/a	173	Mag 8.46v star SE 5.′2; **IC 2407** S 5.′5.
07 45 30.8	+07 55 51	MCG +1-20-5	13.9	0.8 x 0.4	12.5	S?	48	Mag 12.9 star NE 0.′9.
07 47 45.8	+06 47 58	MCG +1-20-7	14.5	0.9 x 0.4	13.3		144	Mag 13.6 star SE end; mag 11.5 star NE 1.′7.
07 52 46.0	+06 29 46	MCG +1-20-8	14.0	0.5 x 0.5	12.4			**CGCG 30-26** E 8.′6; **CGCG 30-27** ENE 11.′1.
08 00 05.2	+06 42 06	MCG +1-21-5	14.2	0.8 x 0.5	13.1		102	Mag 11.5 star on E edge.
08 10 20.5	+07 56 07	MCG +1-21-16	14.4	0.7 x 0.4	12.8		75	Mag 12.0 star N 2.′6.
08 10 44.1	+07 28 27	MCG +1-21-17	14.2	0.6 x 0.4	12.5		168	
08 38 02.4	+07 53 28	MCG +1-22-14	15.4	0.7 x 0.3	13.5		114	Double system or star on edge?
08 38 43.6	+07 48 23	MCG +1-22-15	14.8	0.4 x 0.4	12.6			Mag 11.4 star S 0.′9.
08 55 31.8	+07 35 51	MCG +1-23-4	14.0	0.7 x 0.4	12.5		6	Mag 11.6 star N 2.′0.
08 59 49.1	+05 03 36	MCG +1-23-16	14.1	0.6 x 0.4	12.4	S0?	3	Faint star on E edge; mag 12.3 star NE 2.′4.
07 46 38.6	+09 13 10	MCG +2-20-8	14.5	0.9 x 0.4	13.2		60	

(Continued from previous page)

GALAXIES

RA h m s	Dec ° ′ ″	Name	Mag (V)	Dim ′ Maj x min	SB	Type Class	PA	Notes
08 00 37.6	+13 41 50	MCG +2-21-2	14.1	0.6 x 0.4	12.6	E/S0	96	**CGCG 59-12** S 2′3.
08 01 54.3	+09 37 31	MCG +2-21-5	14.2	0.8 x 0.7	13.5	SBbc	60	Mag 8.07v star W 3′1.
08 01 55.5	+11 43 12	MCG +2-21-6	14.2	0.7 x 0.3	12.4		66	
08 02 24.7	+09 35 46	MCG +2-21-10	13.6	0.6 x 0.4	12.1	E/S0	114	Mag 8.73v star SE edge.
08 23 51.4	+09 50 51	MCG +2-22-1	14.7	0.6 x 0.4	13.0	Sb	90	
08 36 27.8	+10 38 39	MCG +2-22-2	14.7	0.8 x 0.6	13.7	Sc	141	Mag 14.9 star on W edge; mag 12.8 star SW 0′8.
08 40 03.7	+11 00 35	MCG +2-22-3	14.1	0.7 x 0.3	12.3	Sb	15	Mag 8.75v star N 2′6.
08 42 48.3	+14 15 52	MCG +2-22-6	14.5	0.9 x 0.2	12.5	S?	102	Located 1′3 E of SE edge of NGC 2648.
08 46 01.4	+12 47 07	MCG +2-23-5	14.2	0.9 x 0.3	12.6	S?	171	UGC 4582 W 1′5.
08 56 25.0	+12 24 45	MCG +2-23-11	14.8	0.8 x 0.7	14.0	Sc	3	
09 01 24.7	+10 11 58	MCG +2-23-19	14.9	0.7 x 0.6	13.8	Sb	33	
09 02 41.5	+13 06 26	MCG +2-23-23	14.9	0.9 x 0.2	12.9	Sb	135	Close pair of mag 13-14 stars NW 2′8.
09 03 11.3	+13 37 55	MCG +2-23-24	13.7	0.6 x 0.6	12.5			Mag 12.8 star NW 1′0.
09 03 23.5	+13 30 38	MCG +2-23-25	14.6	0.7 x 0.5	13.3	Sb	48	
07 56 40.5	+16 55 12	MCG +3-21-1	14.8	0.5 x 0.4	12.9		96	Mag 12.2 star E 0′9.
07 56 41.7	+17 59 25	MCG +3-21-2	15.0	0.4 x 0.3	12.5		57	Mag 12.2 star N 0′9 with faint, elongated anonymous galaxy in between them; mag 9.49v star SE 2′8.
08 00 36.7	+15 41 02	MCG +3-21-7	14.4	0.8 x 0.3	12.7	S?	39	Mag 9.99v star NW 1′8.
08 07 16.6	+14 57 03	MCG +3-21-18	14.1	1.0 x 1.0	13.9			Mag 11.0 star on NW edge; **CGCG 88-37** E 5′9.
08 23 48.0	+17 59 29	MCG +3-22-8	15.8	0.5 x 0.3	13.8	E/S0	171	
08 38 23.9	+17 37 50	MCG +3-22-20	14.3	0.7 x 0.5	13.0	S0?	39	Located between a close pair of mag 14.8 stars.
08 57 54.0	+17 05 23	MCG +3-23-21	14.7	0.7 x 0.3	12.9	Sc	111	Mag 8.64v star N 5′7.
07 50 35.0	+16 22 07	NGC 2454	13.8	0.9 x 0.4	12.6		100	
07 56 48.8	+07 28 37	NGC 2485	12.2	1.6 x 1.4	12.9	Sa	153	Very small anonymous galaxy on W edge.
07 58 27.5	+07 58 58	NGC 2491	14.8	0.4 x 0.3	12.3		93	Almost stellar; mag 10.9 star W 2′9; NGC 2469 NE 3′8.
07 58 37.3	+08 01 42	NGC 2496	12.9	1.1 x 0.9	13.0	E	167	Mag 8.48v star ESE 9′1; NGC 2491 SW 3′8.
07 58 51.8	+07 29 34	NGC 2499	14.0	0.8 x 0.5	12.9	SBc	18	Located 1′8 N of mag 10.14v star.
07 59 52.4	+05 36 25	NGC 2504	14.0	0.5 x 0.4	12.1	S?	96	Mag 8.34v star SW 8′8.
08 01 37.3	+15 42 36	NGC 2507	12.2	2.5 x 1.8	13.7	S0/a pec	126	A quartet of stars surround the bright center; very faint outer envelope.
08 01 57.2	+08 33 04	NGC 2508	12.7	1.4 x 1.1	13.2	E?	130	
08 02 10.7	+09 29 07	NGC 2510	13.4	1.0 x 0.7	12.9	S0:	111	CGCG 59-27 E 6′7.
08 02 15.1	+09 23 36	NGC 2511	14.2	0.9 x 0.4	12.9	S?	120	Pair of very small, anonymous galaxies SW 2′1.
08 02 24.8	+09 24 50	NGC 2513	11.6	2.5 x 2.0	13.3	E	170	NGC 2511 SW 2′6.
08 02 49.6	+15 48 27	NGC 2514	13.4	1.3 x 1.2	13.8	SB(s)bc I	21	Mag 7.80v star NNW 7′2.
08 06 13.6	+17 42 24	NGC 2522	13.8	1.1 x 0.4	12.8	S0/a	32	
08 06 58.7	+08 00 09	NGC 2526	13.8	1.2 x 0.7	13.5	S?	131	Galaxy **IC 2228** NE 3′3.
08 07 55.6	+17 49 03	NGC 2530	13.6	1.4 x 1.0	13.9	SB(s)d	170	Mag 13.7 star on N edge; **UGC 4232** SW 9′0.
08 26 48.0	+17 22 28	NGC 2593	13.9	1.0 x 0.5	13.0	S0/a	172	Almost stellar **CGCG 89-26** W 7′5.
08 27 26.8	+17 17 01	NGC 2596	13.5	1.5 x 0.6	13.2	Sb I	65	
08 42 40.1	+14 17 03	NGC 2648	11.8	3.2 x 1.0	13.0	Sa	148	Lies 1′4 W of mag 11.16 star. MCG +2-22-6 lies 1′3 S of this same star.
08 43 55.1	+11 46 17	NGC 2651	15.2	0.6 x 0.5	13.7	Sc	114	
08 45 15.8	+09 38 42	NGC 2657	13.0	1.0 x 0.9	12.7	SA(rs)d:	153	
08 45 59.5	+12 37 09	NGC 2661	12.8	1.4 x 1.3	13.3	Scd:	60	Mottled appearance; mag 11.25v star on W edge.
08 57 23.5	+17 17 18	NGC 2711	13.7	0.9 x 0.6	12.9	SB?	170	
08 58 50.5	+06 17 34	NGC 2718	11.8	2.1 x 2.1	13.3	(R′)SAB(s)ab		Fainter arms form an "S" shape within the published size.
08 59 08.1	+11 08 54	NGC 2720	12.8	1.1 x 1.1	12.9	S0⁻:		
09 01 03.3	+11 05 48	NGC 2725	13.5	0.7 x 0.6	12.4	S?	54	Mag 10.55v star N 2′5.
09 01 40.8	+11 04 57	NGC 2728	13.6	1.1 x 0.8	13.4	Sb III-IV	60	Mag 9.95v star SE 4′5.
09 02 15.7	+16 50 18	NGC 2730	12.9	1.7 x 1.3	13.6	SBdm:	80	Mag 13.9 star on S edge.
09 02 08.3	+08 18 02	NGC 2731	13.6	0.8 x 0.5	12.5	S?	70	Mag 12.8 star on N edge.
09 03 01.7	+16 51 45	NGC 2734	16.8	0.6 x 0.5	15.4		84	Almost stellar.
07 45 10.8	+07 55 51	UGC 4005	13.7	1.8 x 0.5	13.4	S?	15	Mag 12.5 star on W edge.
07 45 14.0	+11 03 45	UGC 4006	13.7	0.9 x 0.6	12.9	S?	165	Located on E edge of very small galaxy cluster **A 594**.
07 46 56.0	+07 17 45	UGC 4025	13.8	1.1 x 0.4	12.8	SBb:	37	
07 51 26.1	+14 01 10	UGC 4060	13.8	1.1 x 0.3	12.4	S?	104	Mag 11.2 star S 0′8.
07 53 29.2	+14 36 40	UGC 4077	14.0	1.1 x 0.9	13.8	SB(rs)d:	36	
07 54 32.2	+16 48 28	UGC 4090	13.9	0.9 x 0.7	13.2	S?	57	Double system, companion superimposed; pair of mag 10 stars NE 1′2.
07 56 16.8	+11 39 40	UGC 4109	13.7	0.9 x 0.8	13.2	SB(r)b	60	Mag 15.3 star on W edge; mag 10.54v star NE 1′5.
07 57 02.1	+14 23 24	UGC 4115	13.7	1.8 x 1.0	14.2	IAm	145	Mag 13.7 star E of center.
07 59 23.6	+16 25 13	UGC 4139	13.2	1.1 x 1.0	13.1	SA(s)c II	63	Mag 12.3 star on W edge.
07 59 40.2	+15 23 14	UGC 4145	13.2	1.5 x 0.7	13.1	Sa	140	Mag 7.97v star WSW 4′4.
08 00 06.4	+13 09 05	UGC 4154	14.0	1.5 x 0.7	13.9	Sd	147	Mag 12.6 star W 0′8.
08 01 23.2	+15 22 05	UGC 4170	13.2	1.1 x 0.9	13.2	E:	152	Lies between a pair of mag 11 and 12 stars.
08 01 31.0	+09 42 23	UGC 4171	13.6	2.2 x 0.4	13.3	Sb III	113	**CGCG 59-21** E 7′1.
08 01 58.5	+15 03 24	UGC 4175	13.9	0.7 x 0.6	12.9	SB0?	99	
08 02 21.8	+06 52 41	UGC 4183	13.1	1.8 x 1.1	13.7	SB(r)ab	73	Mag 12.7 star N edge.
08 02 57.4	+16 17 55	UGC 4190	13.5	1.2 x 0.4	12.6	S0:	32	**MCG +3-21-13** N 2′1.
08 03 25.5	+10 03 01	UGC 4197	13.6	1.7 x 0.3	12.7	Sb: III	133	**CGCG 59-30** E 4′5.
08 03 43.3	+09 57 35	UGC 4198	13.2	1.5 x 1.1	13.6	S0?	155	Mag 10.50v star N 2′4; mag 7.52v star W 7′1.
08 04 05.9	+05 06 47	UGC 4203	13.4	0.6 x 0.6	12.2	S?		Mag 11.9 star SE 2′4.
08 04 46.2	+10 46 35	UGC 4211	13.9	1.3 x 0.6	13.5	Sdm:	33	Small, bright knot S end.
08 05 34.2	+10 23 33	UGC 4215	13.5	1.1 x 0.9	13.3	S?	30	Stellar nucleus, faint envelope; mag 10.60v star NW 2′9.
08 05 50.9	+12 28 47	UGC 4216	14.0	0.9 x 0.5	13.0	Sab	150	Mag 9.76v star W 3′8.
08 06 47.8	+05 18 33	UGC 4228	12.4	2.0 x 1.3	13.3	S0	145	
08 07 58.2	+05 12 22	UGC 4239	14.3	0.6 x 0.6	13.0	S		Mag 13.4 star NE 0′9.
08 08 06.1	+14 50 16	UGC 4240	14.3	1.6 x 0.3	13.3	S?	0	Mag 8.26v star E 4′1.
08 08 29.0	+11 10 38	UGC 4244	13.9	0.9 x 0.6	13.0	Sab	55	Mag 9.81v star N 3′7.
08 12 47.5	+09 23 10	UGC 4276	14.0	1.0 x 0.7	13.5	(R′)SBa:	147	
08 20 46.1	+16 38 44	UGC 4350	14.2	1.1 x 0.6	13.6	(R′)SB(s)b	126	Two main arms; mag 10.86v star N 1′8.
08 23 52.1	+14 45 10	UGC 4385	13.6	0.8 x 0.7	12.8	I?	177	
08 28 31.0	+17 27 55	UGC 4433	13.4	1.0 x 0.7	12.9	S pec	15	Mag 9.83v star W 2′0.
08 30 01.5	+17 15 34	UGC 4444	13.4	1.5 x 0.9	13.6	SB(s)cd?	125	Several bright knots, or superimposed stars, around core.
08 31 10.5	+09 36 16	UGC 4452	13.3	0.7 x 0.5	12.1	S0?	138	**CGCG 60-19** and mag 10.63v star S 6′2.
08 42 35.4	+10 35 08	UGC 4540	13.2	1.6 x 1.1	13.7	SBdm	165	Mag 10.63v star SE 2′5.
08 42 56.2	+13 38 27	UGC 4545	13.2	1.0 x 0.6	12.5	SB?	90	
08 43 16.0	+13 05 08	UGC 4550	13.5	2.6 x 0.4	13.4	Sb III-IV	4	
08 43 21.3	+10 43 34	UGC 4552	13.8	1.0 x 0.6	13.1	Sb III	40	
08 44 43.8	+10 28 20	UGC 4568	14.1	1.5 x 0.4	13.3	Im?	80	
08 45 55.2	+12 46 56	UGC 4582	14.1	0.9 x 0.3	12.5	S?	150	MCG +2-23-5 E 1′5.

(Continued from previous page)

94 GALAXIES 94

RA h m s	Dec ° ′ ″	Name	Mag (V)	Dim ′ Maj x min	SB	Type Class	PA	Notes
08 46 40.0	+13 12 47	UGC 4590	14.1	1.3 x 0.4	13.2	S	20	
08 46 45.3	+06 57 33	UGC 4594	13.9	0.9 x 0.5	12.9	Sab	70	
08 47 41.9	+13 25 07	UGC 4599	12.6	2.1 x 2.1	14.1	(R)SA0°		Bright core, faint envelope; mag 9.63v star NNE 2′.2.
08 51 56.8	+16 56 39	UGC 4639	12.8	1.4 x 1.3	13.4	S0?	72	Bright core, smooth envelope; mag 12.0 star NE 1′.4.
08 55 55.6	+13 13 48	UGC 4670	12.7	1.6 x 1.2	13.2	S0:	42	Mag 12.0 star NW 1′.8.
08 56 24.2	+13 10 54	UGC 4677	13.6	1.2 x 0.8	13.4	Scd:	120	Mag 11.3 star WNW 1′.2.
08 57 01.5	+13 11 50	UGC 4685	13.3	1.3 x 0.6	12.9	S?	150	Mag 9.54v star NNW 6′.6.
08 57 52.3	+12 29 42	UGC 4694	14.1	0.9 x 0.5	13.1	SBcd:	100	Filamentary; mag 9.93v star E 2′.5; mag 8.96v star SW 3′.6.
09 00 08.0	+16 55 23	UGC 4721	14.1	0.8 x 0.6	13.2	SBcd:	39	
09 00 42.1	+17 37 14	UGC 4729	13.8	0.9 x 0.7	13.1	SB(s)cd:	156	Bright E-W bar; mag 11.01v star SE 3′.3.
09 01 00.2	+10 37 00	UGC 4731	13.0	1.2 x 0.9	12.9	S0/a	72	Mag 10.9 star SSW 2′.7.

OPEN CLUSTERS

RA h m s	Dec ° ′ ″	Name	Mag	Diam ′	No. ★	B ★	Type	Notes
08 51 24.0	+11 49 00	NGC 2682	6.9	25	324	9.0	cl	= **M 67**. Rich in stars; large brightness range; slight central concentration; detached.

PLANETARY NEBULAE

RA h m s	Dec ° ′ ″	Name	Diam ″	Mag (P)	Mag (V)	Mag cent ★	Alt Name	Notes
08 46 53.6	+17 52 44	PN G208.5+33.2	127	15.6		14.3	PK 208+33.1	Mag 13.8 star near the center; mag 14.2 star on N edge.
07 55 11.2	+09 33 10	PN G211.4+18.4	94			16.6	PK 211+18.1	Located in a small triangle of mag 12 stars.
08 11 12.8	+10 57 17	PN G211.9+22.6	170			14.2	PK 211+22.1	Mag 10.4 star N 3′.3; mag 10.17v star SW 4′.3; mag 10.5 star SE 3′.3.
08 54 13.2	+08 53 59	PN G219.1+31.2	970	12.2	12.0	15.5	PK 219+31.1	Nebulosity strongest along S and SE edges; mag 10.07v star ESE of center 1′.8; mag 9.93v star W of center 6′.6.

95 GALAXIES 95

RA h m s	Dec ° ′ ″	Name	Mag (V)	Dim ′ Maj x min	SB	Type Class	PA	Notes
07 10 49.9	+07 52 36	CGCG 29-3	14.2	0.3 x 0.3	11.5			Stellar; mag 10.43v star SW 0′.9.
07 24 47.6	+07 11 57	CGCG 29-25	14.4	0.5 x 0.2	11.8		96	Mag 14 star just off S edge; mag 11.5 star SW 1′.3.
07 30 03.1	+05 48 13	CGCG 29-29	14.4	0.5 x 0.3	12.2		27	Close pair of mag 13 stars NW 1′.8; mag 9.48v star ESE 4′.2.
07 37 05.8	+07 59 21	CGCG 30-5	15.2	0.4 x 0.4	13.1			Mag 10.9 star E 1′.2.
07 40 05.9	+05 34 56	CGCG 30-8	16.0	0.6 x 0.4	14.3		135	Mag 10.8 star E 3′.7; CGCG 30-9 NNE 4′.2.
07 40 13.7	+05 38 41	CGCG 30-9	15.7	0.4 x 0.2	12.8		6	CGCG 30-8 SSW 4′.2.
07 42 14.1	+06 42 14	CGCG 30-11	14.6	0.3 x 0.3	11.8			Mag 11.5 star NNW 1′.7.
07 46 32.7	+06 51 49	CGCG 30-16	16.2	0.3 x 0.3	13.5			Low surface brightness; mag 10.53v star W 2′.0.
07 48 56.0	+06 20 03	CGCG 30-20	15.6	0.5 x 0.5	13.9			Bright nucleus, faint envelope; mag 13.1 star N edge; bracketed by faint stars; in star rich area.
07 30 52.4	+13 03 17	CGCG 57-16	14.4	0.8 x 0.6	13.4		24	**CGCG 57-17** NE 3′.1.
07 31 34.2	+13 08 25	CGCG 57-19	14.3	0.6 x 0.6	13.0			UGC 3892 SW 7′.6.
07 37 29.7	+09 01 31	CGCG 58-16	15.0	0.5 x 0.3	12.8		129	Mag 11.8 star SW 1′.5; mag 12.3 star E 2′.8.
07 46 09.3	+10 39 31	CGCG 58-33	15.6	0.7 x 0.3	13.8		144	Mag 5.25v star 11 CMi N 6′.8.
07 51 17.8	+10 49 56	CGCG 58-43	14.4	0.8 x 0.5	13.3		33	Close pair of N-S oriented mag 11 stars N 4′.0; **CGCG 58-46** SSE 4′.1.
07 51 23.7	+10 30 36	CGCG 58-45	15.0	0.6 x 0.3	13.0		111	Mag 12.1 star WSW 4′.4.
06 52 12.7	+15 14 10	CGCG 85-11	14.2	0.7 x 0.3	12.4	S0?	21	UGC 3578 ENE 2′.1.
07 32 05.8	+17 48 50	CGCG 87-4	14.8	0.4 x 0.3	12.3		51	Low surface brightness; mag 11.7 star NE 0′.9.
07 38 27.6	+16 51 19	CGCG 87-21	15.4	0.7 x 0.5	14.1		33	Two main arms; mag 10.24v star W 1′.3.
06 51 06.5	+12 55 14	IC 454	13.4	1.2 x 0.7	13.1	SBab	140	Mag 10.65v star E 6′.1.
07 10 40.9	+06 27 15	MCG +1-19-1	14.4	0.7 x 0.5	13.1	S/I	39	Mag 12.6 star NE edge; two other mag 12-13 star close, in line NE.
07 12 43.8	+06 33 54	MCG +1-19-2	14.5	0.9 x 0.7	13.8		69	Mag 9.39v star NNW 8′.7.
07 16 52.0	+07 46 24	MCG +1-19-3	14.1	0.7 x 0.3	12.3	S	42	Mag 11.5 star N 1′.9.
07 27 46.3	+06 27 51	MCG +1-19-5	14.9	0.7 x 0.7	14.0			Mag 14.0 star E edge; mag 10.27v star NE 3′.1.
07 36 30.6	+06 21 14	MCG +1-20-2	15.4	0.6 x 0.4	13.7		147	Stellar nucleus, mag 13.1 star S edge.
07 45 30.8	+07 55 51	MCG +1-20-5	13.9	0.8 x 0.4	12.5	S?	48	Mag 12.9 star NE 0′.9.
07 47 45.8	+06 47 58	MCG +1-20-7	14.5	0.9 x 0.4	13.3		144	Mag 13.6 star SE end; mag 11.5 star NE 1′.7.
06 58 01.6	+12 39 59	MCG +2-18-3	13.4	0.4 x 0.4	11.3	S		Pair of mag 13 stars W 0′.9.
06 58 37.2	+12 39 58	MCG +2-18-4	14.3	0.7 x 0.7	13.3	Sb		Several stars superimposed S.
07 35 43.3	+11 42 33	MCG +2-20-3	14.5	0.9 x 0.4	13.2		147	Mag 12.1 star W 1′.4; **CGCG 58-4** W 3′.5.
07 46 38.6	+09 13 10	MCG +2-20-8	14.5	0.9 x 0.4	13.2		60	
07 13 12.2	+12 15 54	NGC 2350	12.3	1.3 x 0.7	12.0	S0/a	110	**UGC 3754** NE 9′.5.
07 30 46.6	+09 38 44	NGC 2402	13.9	0.8 x 0.8	13.4	Double System		Located 2′.8 NNW of mag 8.88v star. Anonymous galaxy lies on NE edge.
07 35 41.2	+11 36 44	NGC 2416	13.4	1.0 x 0.7	12.9	Scd:	110	Dwarf spiral UGC 3924 S 5′.5.
07 36 38.1	+17 53 11	NGC 2418	12.2	1.8 x 1.8	13.4	E		Faint material fans outward NE from bright center.
07 42 57.0	+09 20 52	NGC 2433	15.3	0.5 x 0.4	13.4	Sab	39	Almost stellar; **CGCG 58-30** E 6′.5.
07 50 35.0	+16 22 07	NGC 2454	13.8	0.9 x 0.4	12.6		100	
06 52 20.6	+15 14 58	UGC 3578	13.0	1.5 x 0.7	12.9	SB(s)ab	20	Several superimposed stars; CGCG 85-11 WSW 2′.1.
06 55 27.0	+15 56 03	UGC 3602	13.3	1.5 x 1.5	14.0	SAdm:		Numerous stars superimposed.
06 57 15.1	+13 32 11	UGC 3613	15.9	1.5 x 0.5	15.4	SBdm	85	Mag 9.89v star WNW 2′.7.
06 59 04.7	+14 17 41	UGC 3621	14.1	1.5 x 1.4	14.8	Im:	109	
07 01 36.8	+14 08 05	UGC 3634	14.6	1.0 x 0.7	14.1	SB(r)a	135	
07 01 41.3	+17 10 53	UGC 3635	14.5	0.7 x 0.4	13.0	Sdm:	115	Small, bright nucleus offset S.
07 02 21.6	+11 14 21	UGC 3641	14.6	1.3 x 1.3	15.1	Sb		
07 04 40.6	+17 35 03	UGC 3658	14.9	1.8 x 0.7	15.0	Im:	65	Dwarf irregular, very low surface brightness.
07 07 54.7	+13 05 12	UGC 3688	14.4	1.0 x 0.4	13.3	Sb III	43	Mag 9.81v star SW 3′.3.
07 08 01.3	+15 10 45	UGC 3691	11.9	2.2 x 1.0	12.6	SAcd	65	Knotty; mag 9.90v star NW edge.
07 09 02.5	+06 55 41	UGC 3707	14.0	0.9 x 0.7	13.3	Sab	45	
07 12 15.2	+07 14 26	UGC 3738	14.3	1.0 x 0.5	13.4	S	30	Pair of mag 14 stars N edge; mag 10.58v star E 1′.1.
07 13 51.7	+10 31 15	UGC 3755	13.5	1.7 x 1.0	13.9	Im	160	
07 14 50.2	+16 58 55	UGC 3766	14.1	1.1 x 0.5	13.3	Scd:	145	Mag 12.0 star E 2′.5.
07 14 54.3	+06 35 43	UGC 3768	14.4	0.9 x 0.2	12.4	S	152	Mag 9.35v star E 0′.9.
07 15 35.3	+15 08 35	UGC 3772	13.3	1.0 x 0.7	12.8	S?	40	Stellar nucleus with star superimposed E.
07 15 52.7	+12 06 51	UGC 3775	14.9	1.4 x 1.4	15.5	Sm:		Dwarf spiral, very low surface brightness.
07 17 46.5	+07 57 09	UGC 3785	13.8	1.1 x 0.4	12.8	Sa:	9	Anonymous galaxy and mag 11.2 star on E edge.
07 20 01.0	+17 56 41	UGC 3805	13.8	1.2 x 0.4	12.9	S?	147	Mag 11.9 star W 1′.5.

(Continued from previous page)

GALAXIES

RA h m s	Dec ° ′ ″	Name	Mag (V)	Dim ′ Maj x min	SB	Type Class	PA	Notes
07 21 27.1	+13 15 46	UGC 3813	13.7	1.3 x 0.6	13.3	Im:	90	Faint envelope.
07 22 00.7	+05 08 49	UGC 3819	14.9	1.0 x 0.3	13.4	Im:	130	
07 22 20.0	+17 17 12	UGC 3820	14.2	1.5 x 0.3	13.2	Scd:	100	Located between a pair of stars N-S, mags 10.61v and 9.92v.
07 25 10.5	+09 30 58	UGC 3839	13.3	1.0 x 0.9	13.0	SBb	27	Stellar nucleus or superimposed star?
07 29 17.0	+14 25 08	UGC 3877	13.5	1.0 x 0.5	12.6	S0	48	Mag 14 star on NE end; mag 8.22v star S 2.′9.
07 31 09.9	+13 03 31	UGC 3892	13.9	1.5 x 0.5	13.4	Scd?	78	**CGCG 57-17** NW 3.′1.
07 32 27.7	+10 54 17	UGC 3900	13.5	0.9 x 0.7	12.8	SBb	150	
07 35 38.3	+11 31 20	UGC 3924	14.1	1.0 x 0.9	13.8	Sm:		Dwarf spiral, extremely low surface brightness; NGC 2416 N 5.′5; mag 10.67v star W 4.′1.
07 37 03.0	+13 36 02	UGC 3936	13.0	1.1 x 0.9	12.8	SBbc	100	
07 37 08.9	+09 54 39	UGC 3938	13.7	0.8 x 0.7	12.9	Sc	51	**CGCG 58-14** E 2.′1; **CGCG 58-10** W 5.′0.
07 37 23.1	+14 04 41	UGC 3941	14.2	1.5 x 0.4	13.5	SBcd:	74	Mag 11.6 star WSW 2.′8.
07 39 25.4	+08 53 53	UGC 3955	14.5	1.5 x 0.2	13.0	Sd:	71	Mag 7.21v star W 5.′2.
07 40 29.9	+13 52 14	UGC 3962	13.2	2.1 x 0.6	13.3	SABb	133	Mag 6.20v star SE 7.′4.
07 41 54.3	+16 48 08	UGC 3974	13.1	3.1 x 2.9	15.4	IB(s)m V	105	Dwarf irregular, extremely low surface brightness.
07 45 10.8	+07 55 51	UGC 4005	13.7	1.8 x 0.5	13.4	S?	15	Mag 12.5 star on W edge.
07 45 14.0	+11 03 45	UGC 4006	13.7	0.9 x 0.6	12.9	S?	165	Located on E edge of very small galaxy cluster **A 594**.
07 46 56.0	+07 17 45	UGC 4025	13.8	1.1 x 0.4	12.8	SBb:	37	
07 51 26.1	+14 01 10	UGC 4060	13.8	1.1 x 0.3	12.4	S?	104	Mag 11.2 star S 0.′8.

GALAXY CLUSTERS

RA h m s	Dec ° ′ ″	Name	Mag 10th brightest	No. Gal	Diam ′	Notes
07 42 36.0	+09 21 00	A 592	15.0	53	15	Most members faint and stellar.

OPEN CLUSTERS

RA h m s	Dec ° ′ ″	Name	Mag	Diam ′	No. ★	B ★	Type	Notes
06 36 19.0	+08 20 24	Bas 7	8.5	5	139	10.4	cl	Few stars; small brightness range; not well detached.
06 34 12.0	+08 05 00	Bas 8		30	50		cl	
06 53 04.0	+16 55 42	Be 29		2	20	15.0	cl	
07 23 38.0	+05 22 24	Be 78		5	12	16.0	cl	
06 57 37.0	+08 17 18	Bi 7		4	30	14.0	cl	
06 58 07.0	+06 25 54	Bi 8		5	70	14.0	cl	Rich in stars; moderate brightness range; slight central concentration; detached.
06 51 21.0	+05 46 06	Bi 11		2	25	15.0	cl	
06 50 16.0	+05 43 42	Bi 12		4	30	17.0	cl:	
06 37 06.0	+05 58 00	Cr 106	4.6	35	20		cl	Few stars; moderate brightness range; not well detached.
06 38 42.0	+06 55 00	Cr 111	7.0	3.2			cl	Probably not a cluster.
07 30 02.0	+11 57 00	Do 26		23			cl	Few stars; small brightness range; not well detached. Probably not a cluster. (A) Faint scattered group with very bright 6 Cmi just NW of center.
06 34 39.0	+08 22 00	NGC 2251	7.3	10	92	9.1	cl	Moderately rich in stars; moderate brightness range; no central concentration; detached. (S) Sparse.
06 34 19.8	+05 19 00	NGC 2252	7.7	18	30	9.0	cl	Moderately rich in stars; moderate brightness range; no central concentration; detached; involved in nebulosity. (A) DSS image shows obvious N-S stream of bright stars, with large bright nebula involved on SW side.
06 35 46.6	+07 40 15	NGC 2254	9.1	6	93	12.0	cl	Moderately rich in stars; small brightness range; strong central concentration; detached.
06 38 33.3	+10 52 57	NGC 2259	10.8	3.5	25	14.0	cl	Few stars; small brightness range; slight central concentration; detached; involved in nebulosity.
06 40 59.0	+09 54 00	NGC 2264	4.1	40	40	5.0	cl	**Christmas Tree Cluster**. Moderately rich in stars; large brightness range; no central concentration; detached; involved in nebulosity.
06 55 12.0	+17 59 18	NGC 2304	10.0	3	30		cl	Moderately rich in stars; small brightness range; slight central concentration; detached.
07 17 00.0	+13 45 00	NGC 2355	9.7	8	40	13.0	cl	Moderately rich in stars; moderate brightness range; slight central concentration; detached.
07 27 13.0	+13 37 00	NGC 2395	8.0	15	53	10.0	cl	Moderately rich in stars; moderate brightness range; not well detached. (S) Sparse irregular group.
06 36 31.0	+09 29 00	Tr 5	10.9	15	150	17.0	cl	Moderately rich in stars; small brightness range; fairly even distribution; detached; involved in nebulosity.

BRIGHT NEBULAE

RA h m s	Dec ° ′ ″	Name	Dim ′ Maj x min	Type	BC	Color	Notes
06 32 42.0	+07 19 00	IC 448	15 x 10	R	3-5	1-4	
06 32 18.0	+05 03 00	NGC 2237-9, 46	80 x 60	E	1-5	3-4	**Rosette Nebula**. A large, bright and mottled nebulosity. This roundish nebula is involved in a sparse cluster of naked eye stars. The NW sector contains many tiny absorption patches.
06 32 42.0	+10 10 00	NGC 2245	5 x 4	R	1-5	2-4	Comet-shaped nebula with a mag 11 star involved.
06 33 12.0	+10 20 00	NGC 2247	6 x 6	R	1-5	2-4	A mag 8.5 star involved in the N part.
06 39 12.0	+08 44 00	NGC 2261	3.5 x 1.5	E+R	1-5	2-4	**Hubble's Variable Nebula**. Fan-shaped object with the apex at the S end.
06 41 06.0	+09 53 00	NGC 2264	35 x 15	E	1-5	3-4	A complex region of bright and dark nebulosity. It is densest in the area around the star 15 Monocerotis and 40′ SSE, a region which contains the dark nebula feature known as the **Cone Nebula** (**Lyons 1613**).

DARK NEBULAE

RA h m s	Dec ° ′ ″	Name	Dim ′ Maj x min	Opacity	Notes
06 32 55.0	+10 29 00	B 37	180 x 150	4	Irregular, semi-vacant region.
06 33 41.8	+11 05 00	B 38	60 x 60	4	Irregular vacancy. This seems to refer to a portion of B37.
06 38 02.5	+10 20 00	B 39		5	Small, sharply defined, elongated spot in the nebulosity NW of 15 Monocerotis.

PLANETARY NEBULAE

RA h m s	Dec ° ′ ″	Name	Diam ″	Mag (P)	Mag (V)	Mag cent ★	Alt Name	Notes
06 59 57.0	+14 36 35	PN G200.7+8.4	69	17.0			PK 200+8.1	Slightly elongated ENE-WSW; mag 12.4 star in center.
06 39 55.9	+11 06 29	PN G201.7+2.5	3	15.4			PK 201+2.1	Mag 13.1 star N 0.′3; mag 12.4 star on W edge.
06 52 27.9	+09 57 40	PN G204.1+4.7	415	12.5		15.0	PK 204+4.1	Appears partially annular with brightening along the E and S edges the SE and NE edges remain fairly clear. Mag 13.3 star near the center and a mag 11.1 star S of the center.
07 29 04.5	+13 14 55	PN G205.1+14.2	615	11.3	10.3:	15.9	PK 205+14.1	**Medusa Nebula**. Brighter in the eastern half; mag 10.9 star ESE 7.′7; mag 10.31v star.

GALAXIES

RA h m s	Dec ° ′ ″	Name	Mag (V)	Dim ′ Maj x min	SB	Type Class	PA	Notes
05 20 12.5	+05 50 10	CGCG 421-30	14.5	0.8 x 0.4	13.1	Sb	18	Mag 9.75v star SSW 4′.1.
05 23 38.2	+05 34 07	MCG +1-14-36	15.1	0.6 x 0.5	13.7		36	Mag 11.7 star SW 1′.7.
06 25 19.8	+07 31 30	MCG +1-17-1	13.6	0.9 x 0.9	13.2	S		Numerous stars superimposed; in star rich area.
05 50 05.4	+13 37 34	MCG +2-15-2	15.7	0.9 x 0.2	13.7	S	57	Located between a pair of mag 11.4 stars.
05 21 45.9	+06 41 18	NGC 1875	13.7	0.8 x 0.8	13.0	S0?		= **Hickson 34A**. Other Hickson galaxies extend in a line SE from edge of NGC 1875; **Hickson 34D** 0′.5; **Hickson 34C** 0′.8; **Hickson 34B** 1′.2.
05 57 27.0	+11 56 56	NGC 2119	14.0	1.2 x 0.8	13.9	E	145	Mag 9.70v star N 2′.2.
05 20 40.7	+08 48 29	UGC 3293	14.9	1.7 x 0.8	15.0	Scd:	131	Very knotty; mag 11.2 star N 0′.9.
05 49 31.7	+17 41 08	UGC 3363	14.1	1.5 x 0.7	14.0	S0/a	35	Mag 10.46v star NE 2′.3.
05 54 40.8	+15 10 11	UGC 3376	14.6	1.0 x 0.7	14.1	SB?	170	Patchy, elongated nucleus, several stars superimposed.

OPEN CLUSTERS

RA h m s	Dec ° ′ ″	Name	Mag	Diam ′	No. ★	B ★	Type	Notes
06 36 19.0	+08 20 24	Bas 7	8.5	5	139	10.4	cl	Few stars; small brightness range; not well detached.
06 34 12.0	+08 05 00	Bas 8		30	50		cl	
05 58 27.0	+07 45 24	Be 22		1.5	20	15.0	cl	
05 35 00.0	+09 56 00	Cr 69	2.8	70	20		cl	(S) = Lamda Orionis cluster.
05 48 40.0	+07 21 30	Cr 74	14.4	6	40		cl	
06 22 54.0	+05 07 00	Cr 92	8.6	11			cl	Probably not a cluster.
06 30 51.2	+09 57 00	Cr 95		27	10		cl	Few stars; moderate brightness range; no central concentration; detached; involved in nebulosity.
06 30 58.1	+05 50 00	Cr 97	5.4	25	15		cl	Few stars; moderate brightness range; not well detached.
06 37 06.0	+05 58 00	Cr 106	4.6	35	20		cl	Few stars; moderate brightness range; not well detached.
06 38 42.0	+06 55 00	Cr 111	7.0	3.2			cl	Probably not a cluster.
05 22 24.0	+07 07 00	Do 17		12			ast	Few stars; moderate brightness range; not well detached. Probably not a cluster. (S) Only five bright stars.
05 23 42.0	+08 10 00	Do 19		23			cl	Few stars; small brightness range; not well detached.
05 27 24.0	+07 04 00	Do 21		12			cl	Few stars; moderate brightness range; not well detached.
05 23 12.0	+11 27 00	DoDz 2		10	12		cl	Few stars; moderate brightness range; no central concentration; detached. (S) Sparse group.
06 02 56.0	+10 27 00	NGC 2141	9.4	10	365	15.0	cl	Rich in stars; moderate brightness range; strong central concentration; detached.
06 03 07.4	+05 44 00	NGC 2143		11	12	10.0	cl?	(A) sparse.
06 08 25.0	+13 57 54	NGC 2169	5.9	6	30	6.9	cl	Moderately rich in stars; large brightness range; no central concentration; detached.
06 12 08.0	+05 27 30	NGC 2186	8.7	5	30	12.0	cl	Moderately rich in stars; moderate brightness range; slight central concentration; detached.
06 13 46.0	+12 48 00	NGC 2194	8.5	9	194	13.0	cl	Rich in stars; moderate brightness range; slight central concentration; detached.
06 16 51.0	+06 00 00	NGC 2202		7	12	9.0	cl?	(S) Sparse field.
06 27 32.0	+12 39 00	NGC 2224		20	20		cl?	(A) Loose group.
06 29 29.2	+16 43 00	NGC 2234		35	50	11.0	cl?	(A) Large, sparse cluster.
06 29 40.0	+06 50 00	NGC 2236	8.5	8	243	12.0	cl	Moderately rich in stars; moderate brightness range; slight central concentration; detached.
06 34 39.0	+08 22 00	NGC 2251	7.3	10	92	9.1	cl	Moderately rich in stars; moderate brightness range; no central concentration; detached. (S) Sparse.
06 34 19.8	+05 19 00	NGC 2252	7.7	18	30	9.0	cl	Moderately rich in stars; moderate brightness range; no central concentration; detached; involved in nebulosity. (A) DSS image shows obvious N-S stream of bright stars, with large bright nebula involved on SW side.
06 35 46.6	+07 40 15	NGC 2254	9.1	6	93	12.0	cl	Moderately rich in stars; small brightness range; strong central concentration; detached.
06 38 33.3	+10 52 57	NGC 2259	10.8	3.5	25	14.0	cl	Few stars; small brightness range; slight central concentration; detached; involved in nebulosity.
06 14 49.0	+12 52 00	Skiff J0614.8+1252		5	20		cl	
06 36 31.0	+09 29 00	Tr 5	10.9	15	150	17.0	cl	Moderately rich in stars; small brightness range; fairly even distribution; detached; involved in nebulosity.

BRIGHT NEBULAE

RA h m s	Dec ° ′ ″	Name	Dim ′ Maj x min	Type	BC	Color	Notes
05 45 18.0	+09 04 00	Ced 59	3 x 2	E+R	1-5	3-4	
06 31 00.0	+10 27 00	IC 446	5 x 4	R	3-5	1-4	Nebula with a faint central star involved. At the N edge of dark nebula LDN 1607 (not plotted).
06 32 42.0	+07 19 00	IC 448	15 x 10	R	3-5	1-4	
06 13 06.0	+17 58 00	IC 2162	3 x 3	E	2-5	3-4	Similar in size, shape and surface brightness to Sh2-257.
06 31 12.0	+09 54 00	IC 2169	25 x 20	R	3-5	1-4	There is a relatively large absorption patch in the NE sector.
06 32 18.0	+05 03 00	NGC 2237-9, 46	80 x 60	E	1-5	3-4	**Rosette Nebula**. A large, bright and mottled nebulosity. This roundish nebula is involved in a sparse cluster of naked eye stars. The NW sector contains many tiny absorption patches.
06 32 42.0	+10 10 00	NGC 2245	5 x 4	R	1-5	2-4	Comet-shaped nebula with a mag 11 star involved.
06 33 12.0	+10 20 00	NGC 2247	6 x 6	R	1-5	2-4	A mag 8.5 star involved in the N part.
06 39 12.0	+08 44 00	NGC 2261	3.5 x 1.5	E+R	1-5	2-4	**Hubble's Variable Nebula**. Fan-shaped object with the apex at the S end.
06 12 48.0	+17 58 00	Sh2-257	3 x 3	E	2-5	3-4	This roundish nebula is quite bright and well defined.
06 08 54.0	+15 49 00	Sh2-261	30 x 25	E	3-5	3-4	Very mottled, roundish nebulosity. The central region is of lower surface brightness, the southern 1/3 brightest.
05 35 00.0	+10 00 00	Sh2-264	270 x 240	E	4-5	3-4	
05 21 42.0	+08 27 00	vdB 38	30 x 25	E+R	3-5	2-4	A mag 5.8 star on NE edge of irregular nebula.

DARK NEBULAE

RA h m s	Dec ° ′ ″	Name	Dim ′ Maj x min	Opacity	Notes
05 30 16.5	+12 46 00	B 30	67 x 67	4	Large dark area with few stars.
05 32 01.5	+12 46 00	B 31	30 x 30	4	Extended NE and SW; the E and darkest part of B30.
05 32 08.5	+12 26 00	B 32		3	Dark projection from S end of B31 to the E.
05 45 30.7	+09 03 00	B 35	20 x 10	2	Elongated E and W with an extension SE; three small stars N of it.
05 49 45.9	+07 25 00	B 36	120 x 120	4	Irregular narrow dark lane NE and SW. The NE end connects brokenly with B35.
06 32 55.0	+10 29 00	B 37	180 x 150	4	Irregular, semi-vacant region.
06 33 41.8	+11 05 00	B 38	60 x 60	4	Irregular vacancy. This seems to refer to a portion of B37.
06 38 02.5	+10 20 00	B 39		5	Small, sharply defined, elongated spot in the nebulosity NW of 15 Monocerotis.
05 21 28.1	+08 20 00	B 223	8 x 8		Fan-shaped nebula.
05 23 54.9	+10 37 00	B 224	20 x 20		Indefinite.
05 28 58.0	+11 36 00	B 225	35 x 5		South end of extension running S from B30.

(Continued from previous page)

96 PLANETARY NEBULAE 96

RA h m s	Dec ° ′ ″	Name	Diam ″	Mag (P)	Mag (V)	Mag cent ★	Alt Name	Notes
05 42 06.2	+09 05 13	NGC 2022	39	12.4	11.6	15.8	PK 196−10.1	Located between a pair of NE-SW mag 11.6 and 12.1 stars.
05 40 45.1	+12 21 21	PN G193.6−9.5	28	13.9	13.9		PK 193−9.1	Mag 12.0 star NE 3′.1.
06 25 57.4	+17 47 26	PN G194.2+2.5	14	12.4	11.7	17.8	PK 194+2.1	Mag 12.5 star on S edge; mag 11.7 star SSE 1′.7.
05 31 45.8	+06 55 57	PN G197.2−14.2	34	15.2	14.7	20.2	PK 197−14.1	Mag 12.1 star SSW 1′.4; mag 12.2 star NE 2′.1.
05 59 24.5	+10 41 40	PN G197.4−6.4	925			17.4	PK 197−6.1	Row of three mag 10-11 stars, oriented E-W, 7′.3 N of center, along N edge.
06 11 08.7	+11 46 45	PN G197.8−3.3	33	>18.2		15.2	PK 197−3.1	Mag 11.8 star S 2′.3; close pair of mag 13 stars NNW 2′.0.
06 02 20.0	+09 39 15	PN G198.6−6.3	37	13.9	12.0	19.7	PK 198−6.1	Located 0′.9 WNW of mag 4.13v star Mu Orionis.
06 39 55.9	+11 06 29	PN G201.7+2.5	3	15.4			PK 201+2.1	Mag 13.1 star N 0′.3; mag 12.4 star on W edge.
06 14 33.6	+07 34 29	PN G201.9−4.6	47			21.1	PK 201−4.1	Slightly elongated ESE-WNW with a bright patch at each end; small triangle of mag 12 to 13 stars SW 4′.1.
06 23 55.0	+05 30 11	PN G204.8−3.5	11				PK 204−3.1	Mag 0.02v star SW 3′.0; mag 10.2 star E 6′.1.

97 GALAXIES 97

RA h m s	Dec ° ′ ″	Name	Mag (V)	Dim ′ Maj x min	SB	Type Class	PA	Notes
04 31 13.9	+08 30 40	CGCG 419-15	14.4	0.7 x 0.6	13.3	Spiral	33	Mag 12.9 star 1′.3 NE.
04 32 00.0	+06 46 40	CGCG 419-17	14.1	0.6 x 0.4	12.4		12	Located 2′.0 N of mag 12.2 star.
04 35 41.3	+08 15 07	CGCG 419-20	14.2	0.9 x 0.7	13.5		84	Mag 9.99v star E 7′.1.
04 53 27.0	+07 08 11	CGCG 420-14	14.0	0.5 x 0.2	11.4		174	Mag 12.4 star SW 1′.8.
05 20 12.5	+05 50 10	CGCG 421-30	14.5	0.8 x 0.4	13.1	Sb	18	Mag 9.75v star SSW 4′.1.
04 13 11.5	+06 33 27	MCG +1-11-14	15.1	0.7 x 0.4	13.6		51	Mag 9.08v star N 2′.6.
04 25 37.8	+07 18 50	MCG +1-12-4	15.4	0.8 x 0.5	14.3	Spiral	147	Mag 13.6 star NE 0′.8.
04 34 44.4	+08 08 36	MCG +1-12-10	14.1	0.9 x 0.5	13.1	S?	30	Pair with MCG +1-12-11 NNE 1′.5.
04 34 46.8	+08 10 03	MCG +1-12-11	14.0	0.7 x 0.5	12.9	E	21	Pair with MCG +1-12-10 SSW 1′.5.
05 12 08.6	+05 27 39	MCG +1-14-3	14.3	0.7 x 0.6	13.2	SB?	30	Mag 10.40v star W 1′.5.
05 14 53.1	+06 15 00	MCG +1-14-5	15.6	0.9 x 0.3	14.1	Sc	36	Mag 8.32v star NE 1′.5.
05 15 15.5	+06 15 54	MCG +1-14-7	14.0	1.0 x 0.6	13.3	Sa/0	21	Wide pair of mag 12.0 stars S 1′.3.
05 15 37.0	+07 11 51	MCG +1-14-9	14.7	0.9 x 0.4	13.5	S?	102	
05 15 53.7	+07 10 06	MCG +1-14-10	14.4	0.7 x 0.5	13.1	S?	135	Mag 10.58v star NE 3′.4.
05 17 23.0	+06 52 36	MCG +1-14-22	14.8	0.8 x 0.8	14.2	Sa		Mag 12.5 star SSE 2′.5.
05 23 38.2	+05 34 07	MCG +1-14-36	15.1	0.6 x 0.5	13.7		36	Mag 11.7 star SW 1′.7.
04 09 11.8	+08 38 55	NGC 1517	13.4	1.0 x 0.9	13.1	Scd:	0	Mag 10.8 star SE 0′.9.
04 31 10.2	+07 37 47	NGC 1590	13.7	0.9 x 0.7	13.0	Pec	108	
04 40 09.2	+07 20 59	NGC 1633	13.5	1.2 x 1.0	13.5	SAB(s)ab	51	Located on NW edge of bright nebula Sh2-250; NGC 1634 on S edge; mag 8.24v star S 5′.3.
04 40 09.9	+07 20 17	NGC 1634	14.1	0.5 x 0.5	12.6	E		See notes for NGC 1633.
05 11 46.3	+05 11 59	NGC 1819	12.4	1.3 x 1.0	12.6	SB0	120	NE-SW core in SE-NW bar.
05 21 45.9	+06 41 18	NGC 1875	13.7	0.8 x 0.8	13.0	S0?		= **Hickson 34A**. Other Hickson galaxies extend in a line SE from edge of NGC 1875; **Hickson 34D** 0′.5; **Hickson 34C** 0′.8; **Hickson 34B** 1′.2.
04 09 08.9	+17 06 39	UGC 2968	14.8	1.8 x 0.5	14.5	Sb: III-IV	28	Mag 9.08v star NNE 4′.5.
04 13 12.7	+13 25 10	UGC 2984	15.4	1.7 x 1.4	16.2	SBdm:	168	
04 16 05.0	+08 10 48	UGC 2997	13.8	1.1 x 0.5	13.0	S0/a	178	
04 18 58.2	+05 26 01	UGC 3010	14.3	1.0 x 0.6	13.6	SBcd:	132	Mag 9.91v star NW 1′.7.
04 24 40.1	+07 13 06	UGC 3035	13.3	1.3 x 0.8	13.2	SB0	11	Mag 11.4 star N 1′.4.
04 30 10.8	+06 56 30	UGC 3061	13.7	1.3 x 1.3	14.1	S?		Multiple galaxy system, interaction.
04 30 58.3	+05 32 28	UGC 3066	13.9	1.4 x 1.0	14.1	SAB(r)d:	110	
04 31 01.5	+08 05 09	UGC 3067	13.7	0.8 x 0.8	13.3	E?		Mag 8.95v star NE 4′.4; note: **UGC 3075** lies 3′.7 ENE of this star.
04 31 01.1	+05 54 35	UGC 3068	13.2	1.5 x 1.0	13.5	Double System	102	Double or triple system, distorted.
04 32 48.4	+10 23 27	UGC 3084	14.9	1.7 x 0.4	14.3	Scd:	158	
04 33 11.2	+05 21 14	UGC 3087	13.3	0.8 x 0.6	12.4	S0:	130	**BW Tauri**. = **3C 120.** Has a variable Seyfert nucleus.
04 33 54.1	+16 54 43	UGC 3089	14.3	1.2 x 0.9	14.2	Sbc	69	
04 39 51.4	+07 03 19	UGC 3122	13.5	2.2 x 1.2	14.4	SAB(rs)c	25	
04 40 39.0	+17 08 26	UGC 3129	15.9	2.3 x 1.2	16.9	Im:	45	Dwarf, very low surface brightness.
04 48 00.3	+08 42 40	UGC 3172	15.5	1.5 x 0.2	14.0	Sd	138	Close pair of mag 11.6 stars SW 3′.2.
04 50 24.3	+08 42 44	UGC 3180	14.1	0.9 x 0.6	13.3	SB	170	
04 50 37.3	+06 00 27	UGC 3181	13.2	2.1 x 0.7	13.5	SBb	88	
04 51 49.6	+08 50 35	UGC 3188	13.9	1.0 x 0.5	13.0	S?	85	Located between a mag 9.80v star W and a mag 10.18v star SW.
04 59 21.0	+05 37 03	UGC 3224	13.0	1.5 x 1.2	13.5	Sb II-III	15	Stellar nucleus, uniform envelope.
05 03 24.5	+16 24 14	UGC 3234	14.3	1.8 x 1.8	15.5	Im: V		Dwarf irregular, low surface brightness.
05 06 38.2	+08 40 24	UGC 3247	14.0	0.8 x 0.8	13.4	S?		
05 11 32.8	+17 03 22	UGC 3261	14.1	1.2 x 0.7	13.8	Sdm:	55	Two bright knots or superimposed stars S of nucleus; mag 11.3 star W 0′.9.
05 14 07.0	+06 31 10	UGC 3269	14.1	0.9 x 0.5	13.1	Sbc	175	
05 16 37.2	+06 26 37	UGC 3274	13.6	1.3 x 0.5	13.0	S?	177	UGC 3274 consists of a chain of 5 or 6 galaxies. Mag 10.0v star NE 2′.4; **MCG +1-14-12** NW 2′.9; **CGCG 421-16** SW 3′.2.
05 16 46.2	+06 37 19	UGC 3275	15.1	1.7 x 0.4	14.5	S?	37	Mag 12.3 star 2′.6 NW.
05 17 37.5	+06 47 58	UGC 3282	13.8	0.9 x 0.7	13.2	SBcd?	15	
05 20 40.7	+08 48 29	UGC 3293	14.9	1.7 x 0.8	15.0	Scd:	131	Very knotty; mag 11.2 star N 0′.9.

GALAXY CLUSTERS

RA h m s	Dec ° ′ ″	Name	Mag 10th brightest	No. Gal	Diam ′	Notes
04 59 54.0	+05 26 00	A 526	16.4	71	13	Stellar **PGC 16522** NNE of center 3′.8; other members anonymous, faint and stellar.
05 16 36.0	+06 27 00	A 539	14.4	50	17	Two UGC galaxies near center, remaining cluster GX almost stellar to stellar and faint.

OPEN CLUSTERS

RA h m s	Dec ° ′ ″	Name	Mag	Diam ′	No. ★	B ★	Type	Notes
05 22 24.0	+07 07 00	Do 17		12			ast	Few stars; moderate brightness range; not well detached. Probably not a cluster. (S) Only five bright stars.
05 23 42.0	+08 10 00	Do 19		23			cl	Few stars; small brightness range; not well detached.
05 27 24.0	+07 04 00	Do 21		12			cl	Few stars; moderate brightness range; not well detached.
05 23 12.0	+11 27 00	DoDz 2		10	12		cl	Few stars; moderate brightness range; no central concentration; detached. (S) Sparse group.
04 26 54.0	+15 52 00	Hyades	.5	330	380	4.0	cl	Moderately rich in bright and faint stars; detached. (A) Famous moving cluster.
04 48 29.0	+10 56 00	NGC 1662	6.4	12	59	9.0	cl	Moderately rich in bright and faint stars; detached.
04 49 24.3	+13 08 00	NGC 1663		9	30		cl?	Few stars; moderate brightness range; not well detached.
04 58 21.1	+08 14 20	NGC 1707		0.8	5	13.5	ast?	

(Continued from previous page)

97 OPEN CLUSTERS 97

RA h m s	Dec ° ′ ″	Name	Mag	Diam ′	No. ★	B ★	Type	Notes
05 10 46.0	+16 31 00	NGC 1807	7.0	12	37	9.0	cl	Few stars; moderate brightness range; slight central concentration; detached. Probably not a cluster.
05 12 27.0	+16 41 00	NGC 1817	7.7	20	283	9.0	cl	Rich in stars; moderate brightness range; not well detached. (S) Sparse.

BRIGHT NEBULAE

RA h m s	Dec ° ′ ″	Name	Dim ′ Maj x min	Type	BC	Color	Notes
04 13 36.0	+10 13 00	vdB 26	6 x 6	R	3-5	1-4	Nebula with a mag 7.2 star involved.
05 18 12.0	+13 24 00	vdB 37	7 x 4	R	2-5	4-4	Nebula with a mag 8.2 star involved in the SE part.
05 21 42.0	+08 27 00	vdB 38	30 x 25	E+R	3-5	2-4	A mag 5.8 star on NE edge of irregular nebula.

DARK NEBULAE

RA h m s	Dec ° ′ ″	Name	Dim ′ Maj x min	Opacity	Notes
05 21 28.1	+08 20 00	B 223	8 x 8		Fan-shaped nebula.
05 23 54.9	+10 37 00	B 224	20 x 20		Indefinite.

PLANETARY NEBULAE

RA h m s	Dec ° ′ ″	Name	Diam ″	Mag (P)	Mag (V)	Mag cent ★	Alt Name	Notes
05 05 34.3	+10 42 23	PN G190.3−17.7	26	12.9	11.9	14.4	PK 190−17.1	Elongated N-S; mag 12.9 star W 1.′8.

98 GALAXIES 98

RA h m s	Dec ° ′ ″	Name	Mag (V)	Dim ′ Maj x min	SB	Type Class	PA	Notes
02 57 33.9	+05 58 33	CGCG 415-40	14.0	0.6 x 0.4	12.4	S0	150	**PGC 11171** W 2.′5; **PGC 11160** W 3.′9.
02 59 16.2	+06 07 56	CGCG 415-50	14.6	0.4 x 0.3	12.2	S0/a	30	Mag 14.4 star SE 0.′7; mag 10.9 star NNE 5.′6; faint anonymous galaxy NE 0.′6.
03 08 55.7	+11 54 02	CGCG 441-2	15.2	0.6 x 0.6	14.0			**CGCG 441-1** W 9.′4.
03 24 00.3	+16 10 56	CGCG 464-14	14.9	0.6 x 0.4	13.2	Sm	108	Mag 10.49v star WSW 5.′7.
03 03 52.9	+09 36 42	IC 1873	14.3	1.1 x 0.7	13.9	S0?	24	Mag 9.37v star NNE 3.′8.
03 35 33.2	+05 03 57	IC 1956	14.5	1.6 x 0.3	13.5	SBbc:	34	Mag 10.6 star W 5.′0.
03 40 45.2	+17 44 27	IC 1977	13.5	1.4 x 0.8	13.5	SBb	177	Mag 10.43v star SSW 7.′1.
03 54 30.5	+10 42 24	IC 2002	13.8	1.3 x 1.0	13.9	SBa	0	Mag 10.83v star W 6.′6.
02 57 41.8	+06 01 26	MCG +1-8-27	12.7	1.2 x 0.8	12.7	E0:	0	**PGC 11193** NE 2.′8; **PGC 11184** WNW 1.′7.
02 57 45.8	+05 57 05	MCG +1-8-28	15.5	0.9 x 0.4	14.2	Scd	150	**PGC 11195** N 1.′4; **PGC 11196** N 2.′0; mag 12.4 star N 2.′3.
02 58 24.7	+06 35 29	MCG +1-8-30	14.2	1.4 x 1.1	14.6	S0	6	Large, bright core.
02 59 52.3	+06 31 55	MCG +1-8-33	14.4	0.7 x 0.5	13.1	S?	150	Mag 14.2 star on S edge; faint, round anonymous galaxy S 1.′2.
03 00 00.2	+05 58 03	MCG +1-8-34	15.2	0.5 x 0.5	13.5	Sbc		Mag 12.4 star W 1.′1.
03 00 08.8	+05 48 13	MCG +1-8-35	14.5	0.5 x 0.5	12.9	Sa/0		
03 00 09.7	+05 42 40	MCG +1-8-36	15.8	0.5 x 0.4	13.9	Sbc	0	UGC 2469 E 2.′7.
03 05 13.8	+05 51 10	MCG +1-8-40	14.5	0.6 x 0.5	13.0	Sbc	12	Mag 12.9 star N 2.′3.
03 05 35.0	+06 07 58	MCG +1-8-41	14.3	0.4 x 0.3	11.8		132	Located 1.′3 NW of a mag 7.69v star.
03 43 53.8	+06 31 45	MCG +1-10-8	14.4	0.9 x 0.5	13.4	Sb	81	Mag 12.5 star N 2.′0.
03 56 58.0	+08 30 44	MCG +1-11-1	14.3	0.9 x 0.7	13.6		33	Located between a N-S pair of mag 13-14 stars.
03 59 24.0	+06 40 43	MCG +1-11-3	14.3	0.6 x 0.6	13.1	S0		**MCG +1-11-4A** N 1.′2; UGC 2914 W.
03 59 31.2	+06 36 46	MCG +1-11-4	15.4	0.4 x 0.4	13.5	E/S0:		Mag 13.3 star NW 0.′8.
04 06 49.0	+06 56 26	MCG +1-11-11	14.4	1.0 x 0.5	13.5		144	Mag 13.0 star W 1.′5.
04 13 11.5	+06 33 27	MCG +1-11-14	15.1	0.7 x 0.4	13.6		51	Mag 9.08v star N 2.′6.
02 57 29.7	+10 27 22	MCG +2-8-37	14.6	1.0 x 0.7	14.1		108	Mag 12.6 star NE edge; UGC 2430 and bright star NW.
02 58 14.0	+11 07 57	MCG +2-8-41	14.6	0.8 x 0.6	13.6		153	**MCG +2-8-42** SSE 1.′0.
02 59 47.7	+15 17 32	MCG +2-8-45	15.2	0.6 x 0.5	13.7		102	Mag 11.7 star E 2.′7.
03 03 51.0	+12 04 10	MCG +2-8-48	14.0	0.9 x 0.7	13.3		147	Stellar nucleus, faint envelope.
03 17 04.5	+15 29 05	MCG +2-9-2	13.9	0.8 x 0.3	12.2	S?	48	Mag 10.47v star NE 2.′4.
03 27 05.1	+12 33 06	MCG +2-9-7	14.8	0.9 x 0.5	13.8		111	MCG +2-9-8 on N edge.
03 27 04.8	+12 33 22	MCG +2-9-8	14.4	0.1 x 0.1	9.2			Stellar; located on N edge of MCG +2-9-7.
02 56 32.9	+15 54 34	MCG +3-8-32	15.1	0.4 x 0.4	13.2	E/S0		UGC 2413 NW edge.
02 57 31.4	+16 38 04	MCG +3-8-35	15.4	0.6 x 0.3	13.4	Sc		
02 57 37.6	+16 04 02	MCG +3-8-36	14.8	0.7 x 0.4	13.5	E/S0	66	Almost stellar; stellar **MCG +3-8-34** WNW 3.′9; mag 8.47v star ESE 10.′1.
03 07 58.4	+17 46 40	MCG +3-9-1	15.2	0.8 x 0.5	14.0		21	Mag 13.4 star S end.
03 00 35.0	+11 50 35	NGC 1166	14.0	1.0 x 1.0	13.8	Sab		Mag 7.6 star N 8.′3.
03 00 47.1	+11 46 16	NGC 1168	14.2	1.2 x 0.7	13.9	SAB(rs)b: II-III	18	Mag 9.48v star SE 5.′4.
03 11 28.0	+10 48 28	NGC 1236	14.8	0.9 x 0.6	14.0		27	
03 39 13.5	+15 49 12	NGC 1384	14.5	0.9 x 0.5	13.4	S?	135	Mag 8.64v star NE 3.′6; several stellar anonymous galaxies in the immediate area.
03 50 23.5	+06 58 22	NGC 1462	14.2	0.9 x 0.5	13.2	S?	48	
04 09 11.8	+08 38 55	NGC 1517	13.4	1.0 x 0.9	13.1	Scd:	0	Mag 10.8 star SE 0.′9.
02 56 28.8	+15 54 57	UGC 2413	14.3	0.9 x 0.7	13.6	Compact	70	MCG +3-8-32 SE 1.′0.
02 56 43.0	+07 19 59	UGC 2419	13.5	1.2 x 1.2	13.7	(R′)SB(s)a		Bright NE-SW center, faint envelope.
02 57 07.6	+17 30 49	UGC 2424	13.4	2.3 x 0.6	13.6	Sbc	158	
02 57 09.3	+05 19 15	UGC 2426	14.1	0.9 x 0.6	13.2	Sb III	65	Mag 10.89v star NE 2.′7.
02 57 26.1	+10 28 42	UGC 2430	14.1	0.8 x 0.6	13.1	Sa	90	MCG +2-8-37 SE 1.′6; mag 8.86v star W 1.′0.
02 57 32.9	+08 06 00	UGC 2434	15.3	0.9 x 0.9	14.9	SA(s)cd		
02 57 53.9	+11 10 58	UGC 2437	14.1	1.5 x 1.0	14.4	Sd	33	
02 57 53.3	+13 01 53	UGC 2438	14.4	0.9 x 0.9	14.2	cD		Located at the center of galaxy cluster A 399.
02 58 29.8	+06 18 20	UGC 2444	14.2	0.8 x 0.6	13.3	SB?	45	Faint, almost stellar anonymous galaxy NNE 2.′6.
02 59 15.1	+16 02 11	UGC 2453	14.3	1.5 x 1.2	14.8	SAB(s)b II-III	126	
03 00 20.5	+05 42 42	UGC 2469	14.3	1.1 x 0.7	13.9	Pec?	24	MCG +1-8-36 W 2.′7.
03 01 30.1	+17 51 07	UGC 2486	13.7	1.1 x 0.8	13.4	SA(r)bc	54	Stellar nucleus; mag 9.31v star N 5.′9.
03 02 12.3	+17 20 44	UGC 2498	14.4	1.5 x 0.7	14.3	SABcd	65	Located between a pair of mag 12.4 stars.
03 14 10.2	+16 29 06	UGC 2602	13.8	1.3 x 0.9	13.8	SAB(s)c	12	**MCG +3-9-9** NNE 3.′0.
03 15 54.2	+09 09 34	UGC 2622	15.2	1.5 x 0.9	15.4	SB(s)b	75	
03 19 43.2	+09 45 06	UGC 2674	13.8	0.9 x 0.6	13.1	S0?	120	
03 20 23.5	+17 17 47	UGC 2684	14.6	1.8 x 0.9	15.0	Im?	111	Dwarf, low surface brightness.
03 21 13.7	+07 26 34	UGC 2695	15.8	1.5 x 0.3	14.8	Scd?	22	

(Continued from previous page)

98 GALAXIES 98

RA h m s	Dec ° ′ ″	Name	Mag (V)	Dim ′ Maj x min	SB	Type Class	PA	Notes
03 22 34.3	+14 54 46	UGC 2703	14.5	1.5 x 0.4	13.8	Sb III-IV	27	Stellar **CGCG 441-10** N 2′8.
03 23 46.6	+06 33 40	UGC 2712	14.3	1.2 x 1.2	14.5	SBdm:		Uniform surface brightness.
03 24 07.7	+17 45 13	UGC 2716	14.2	1.8 x 0.9	14.6	Sdm:	73	**MCG +3-9-13** W 1′7.
03 26 45.5	+07 42 48	UGC 2740	14.0	1.4 x 0.7	13.8	Sb: III-IV	64	
03 41 35.7	+16 01 11	UGC 2823	15.1	1.5 x 0.3	14.1	Sb: III-IV	8	**UGC 2830** E 7′0.
03 41 51.0	+08 09 31	UGC 2829	13.9	1.0 x 0.3	12.4	S0	171	
03 44 56.0	+05 54 16	UGC 2852	15.4	1.5 x 0.2	13.9	Sdm:	120	
03 47 39.0	+13 15 20	UGC 2862	14.1	1.7 x 1.0	14.5	(R′)SAB(s)a:	30	
03 48 46.5	+13 08 20	UGC 2871	14.2	1.2 x 0.9	14.2	Sb II-III	144	Stellar nucleus.
03 54 07.6	+15 59 29	UGC 2894	14.4	0.9 x 0.6	13.6	Scd:		
03 56 58.0	+16 29 02	UGC 2904	14.0	1.5 x 1.2	14.5	Sb: III	132	**UGC 2905** N 2′3.
03 59 17.3	+06 41 09	UGC 2914	13.4	1.1 x 0.8	13.2	SB0	24	MCG +1-11-3 E 1′7; **MCG +1-11-4A** ENE 2′1.
04 09 08.9	+17 06 39	UGC 2968	14.8	1.8 x 0.5	14.5	Sb: III-IV	28	Mag 9.08v star NNE 4′5.
04 13 12.7	+13 25 10	UGC 2984	15.4	1.7 x 1.4	16.2	SBdm:	168	

GALAXY CLUSTERS

RA h m s	Dec ° ′ ″	Name	Mag 10th brightest	No. Gal	Diam ′	Notes
02 57 00.0	+15 57 00	A 397	15.1	35	50	**UGC 2420** NW of center 3′3; **CGCG 463-33** W of center 15′4.
02 57 54.0	+13 00 00	A 399	15.6	57	9	With UGC 2438 at its center, most other members are stellar PGC and anonymous GX spread N-S along centerline.
02 57 36.0	+06 01 00	A 400	13.9	58	56	Numerous PGC and anonymous GX spread throughout the cluster boundaries.
02 58 54.0	+13 34 00	A 401	15.6	90	19	Mag 7.31v star near center; **UGC 2450** SE of this star 3′1; other stellar PGC and anonymous GX in surrounding area.

BRIGHT NEBULAE

RA h m s	Dec ° ′ ″	Name	Dim ′ Maj x min	Type	BC	Color	Notes
04 13 36.0	+10 13 00	vdB 26	6 x 6	R	3-5	1-4	Nebula with a mag 7.2 star involved.

99 GALAXIES 99

RA h m s	Dec ° ′ ″	Name	Mag (V)	Dim ′ Maj x min	SB	Type Class	PA	Notes
02 41 34.9	+07 11 13	CGCG 414-40	13.8	0.6 x 0.3	11.8		99	Faint star or stellar galaxy 0′4 S; mag 13.3 star SW 1′3.
02 44 00.4	+05 25 45	CGCG 415-4	14.1	0.6 x 0.5	12.6	Sc		**CGCG 415-5** NW 0′7.
02 44 43.4	+05 20 01	CGCG 415-6	14.1	0.6 x 0.4	12.4		129	Mag 14.8 star NW 1′0.
02 57 33.9	+05 58 33	CGCG 415-40	14.0	0.6 x 0.4	12.4	S0	150	**PGC 11171** W 2′5; **PGC 11160** W 3′9.
02 59 16.2	+06 07 56	CGCG 415-50	14.6	0.4 x 0.3	12.2	S0/a	30	Mag 14.4 star SE 0′7; mag 10.9 star NNE 5′6; faint anonymous galaxy NE 0′6.
01 56 10.0	+10 47 42	CGCG 438-2	14.7	0.6 x 0.3	12.7		45	Mag 11.9 star N 0′4.
02 11 33.2	+13 55 02	CGCG 438-40	14.0	0.8 x 0.7	13.2		105	= **Mkn 366**; a very faint arc of material extends W.
02 16 11.6	+17 58 13	CGCG 461-70	14.8	0.6 x 0.5	13.3	Sc	171	Mag 11.6 star W 4′3.
01 45 16.5	+10 38 56	IC 154	14.0	1.4 x 0.2	12.4	Sb III	66	Mag 9.31v star N 6′5.
01 45 29.2	+10 33 07	IC 156	13.5	1.5 x 1.2	14.0	S?	72	Mag 9.44v star E 2′6.
01 48 43.8	+10 30 26	IC 161	13.8	0.7 x 0.5	12.5	S?	65	IC 162 ENE 2′5.
01 48 53.3	+10 31 15	IC 162	12.7	1.4 x 1.4	13.3	S0		MCG +2-5-39 E edge.
01 59 51.7	+07 24 37	IC 182	13.7	0.9 x 0.6	12.9	SBb	33	
02 02 32.5	+16 00 49	IC 192	13.5	0.8 x 0.6	12.6	S0	156	Mag 10.94v star WSW 9′5.
02 02 30.8	+11 05 31	IC 193	13.4	1.5 x 1.3	14.0	SA(rs)c	165	Mag 6.52v star ESE 8′0.
02 03 44.7	+14 42 33	IC 195	13.0	1.5 x 0.5	12.5	SAB0°	126	Mag 10.40v star W 2′4; IC 196 N 2′1.
02 03 49.9	+14 44 20	IC 196	12.9	2.8 x 1.4	14.2	SBb:	5	Two slender arms; IC 195 S 2′1.
02 06 03.2	+09 17 41	IC 198	13.8	1.0 x 0.6	13.1	S	43	Mag 12.0 star NNW 4′8; IC 199 SE 5′7; mag 9.68v star N 9′0.
02 06 19.4	+09 13 36	IC 199	14.0	1.4 x 0.7	13.8	Sab	25	Stellar nucleus; IC 198 NW 5′7; mag 10.56v star ENE 9′5.
02 07 28.6	+09 10 04	IC 202	14.3	1.4 x 0.3	13.2	Sb III-IV	132	Mag 10.39v star S 4′0.
02 08 27.8	+06 23 40	IC 208	13.4	1.3 x 1.3	13.8	SAbc III		Stellar nucleus. NGC 825 S 4′5; mag 9.66v star SE 8′5.
02 14 04.3	+16 27 18	IC 213	13.8	1.9 x 1.5	14.8	SAB(rs)b III-IV	150	Mag 14.9 star SE of center; mag 11.4 star W 11′2.
02 14 05.8	+05 10 33	IC 214	14.2	0.9 x 0.5	13.2	I?	162	Mag 11.4 star S 5′4.
02 22 48.0	+11 38 15	IC 222	14.5	0.8 x 0.5	13.3	Sc	24	Mag 8.90v star S 3′9.
02 35 22.8	+12 50 15	IC 238	12.8	1.2 x 0.6	12.3	S0	30	Mag 11.3 star N 3′3.
02 41 25.6	+17 48 41	IC 248	13.4	1.0 x 0.6	12.7	Sa:	145	Mag 7.63v star ESE 9′3.
02 53 50.2	+12 51 00	IC 267	13.0	1.6 x 0.7	13.0	(R′)SB(s)b	156	Mag 8.51v star NE 12′8.
01 56 09.1	+17 38 26	IC 1748	13.7	1.0 x 0.6	13.0	SABbc	130	Mag 8.54v star SE 11′9; **CGCG 460-51** S 6′6; **UGC 1417** NE 9′1.
01 56 11.2	+06 44 38	IC 1749	13.6	0.7 x 0.5	12.3	S0:	149	Mag 9.46v star NNE 5′5.
01 57 09.9	+14 32 58	IC 1755	13.8	1.4 x 0.3	12.7	Sa	154	Mag 8.85v star W 6′7.
02 02 14.5	+09 58 48	IC 1770	13.8	1.0 x 0.7	13.3	S0	63	A mag 12.3 star and IC 1771 on S edge; mag 9.91v star E 6′5.
02 02 16.1	+09 58 06	IC 1771	14.2	0.5 x 0.5	12.6	S0		Mag 9.91v star E 6′5; IC 1770 on N edge.
02 02 42.9	+07 44 45	IC 1772	13.6	0.7 x 0.7	12.7			Bright, almost stellar nucleus in faint envelope; mag 8.19v star N 2′7.
02 03 59.0	+15 19 03	IC 1774	13.9	1.6 x 1.2	14.5	SAB(s)d	128	Mag 9.77v star SE 10′7.
02 05 17.7	+13 30 18	IC 1775	15.0	0.7 x 0.7	14.1	SBbc		Mag 10.48v star E 5′9.
02 05 15.1	+06 06 19	IC 1776	13.2	1.3 x 1.1	13.4	SB(s)d	159	Mag 11.4 star WSW 2′3.
02 06 51.2	+14 43 16	IC 1780	14.9	0.5 x 0.3	12.7	Sb	9	Mag 10.49v star N 2′0.
02 17 41.4	+12 28 15	IC 1791	13.3	1.0 x 1.0	13.1	S0		**IC 1790** N 2′5; mag 10.51v star NW 4′6.
02 21 30.4	+15 45 38	IC 1794	13.8	0.8 x 0.7	13.0	S0?	108	Almost stellar; mag 12.7 star SW 2′3.
02 33 50.5	+11 12 08	IC 1817	15.8	0.7 x 0.7	14.9	Sm		Galaxy pair, larger component E; mag 10.69 star W 1′0.
02 38 55.6	+09 05 52	IC 1825	13.9	1.3 x 0.9	13.9	Scd:	15	Mag 10.33v star W 9′8.
02 47 43.7	+13 15 16	IC 1846	14.4	0.6 x 0.5	13.0	Compact	150	Mag 11.6 star WSW 2′5.
02 47 44.7	+09 21 21	IC 1849	14.4	0.9 x 0.6	13.8	E	90	Mag 7.64v star SW 6′1.
02 49 39.0	+14 37 08	IC 1857	14.0	0.8 x 0.5	12.9	S?	150	Mag 11.9 star N 1′9.
02 55 20.2	+08 49 43	IC 1865	13.9	1.5 x 1.1	14.3	Sbc	90	**IC 1863** SW 7′8; mag 8.83v star SW 10′0.
02 55 52.3	+09 18 43	IC 1867	14.1	1.5 x 1.2	14.6		15	Mag 9.75v star W 7′3.
03 03 52.9	+09 36 42	IC 1873	14.3	1.1 x 0.7	13.9	S0?	24	Mag 9.37v star NNE 3′8.
01 54 42.8	+05 25 29	MCG +1-5-47	14.5	0.8 x 0.5	13.4	Spiral	135	**UGC 1373** W 4′6.
01 56 36.9	+05 48 11	MCG +1-6-9	14.8	0.6 x 0.4	13.1		96	
01 58 36.3	+09 17 02	MCG +1-6-18	15.3	0.8 x 0.3	13.6		150	Mag 13.5 star SE 0′7.
01 58 44.0	+08 17 20	MCG +1-6-20	15.2	0.8 x 0.6	14.3			NGC 766 N 3′5.
01 58 53.2	+05 25 00	MCG +1-6-23	14.9	0.6 x 0.5	13.4	Sab	51	Mag 11.4 star NW 3′9.
01 58 50.9	+05 21 33	MCG +1-6-24	14.6	0.6 x 0.3	12.6	Sb	90	Mag 10.9 star S 1′1.

(Continued from previous page)

RA h m s	Dec ° ′ ″	Name	Mag (V)	Dim ′ Maj x min	SB	Type Class	PA	Notes
01 59 00.4	+06 31 19	MCG +1-6-25	14.5	0.7 x 0.4	13.0		84	Located between a pair of mag 12.4 and 12.9 stars.
02 01 32.4	+05 54 56	MCG +1-6-30	15.0	0.7 x 0.5	13.8		156	mag 12.9 star NW 2.′4.
02 14 46.4	+07 27 12	MCG +1-6-60	15.0	0.7 x 0.6	13.9		174	Mag 12.2 star ENE 2.′3.
02 22 38.7	+05 32 28	MCG +1-7-3	14.7	0.7 x 0.6	13.6	Sbc	132	Located between a pair of NW-SE mag 14.4 stars.
02 24 12.3	+05 22 20	MCG +1-7-4	14.3	0.9 x 0.3	12.7	S?	48	Mag 14.3 star SW 1.′7.
02 24 17.6	+06 28 50	MCG +1-7-5	14.3	0.8 x 0.6	13.4	Sb	42	Mag 9.40v star NW 4.′0.
02 36 24.1	+07 31 58	MCG +1-7-13	14.6	0.8 x 0.4	13.2	Sb	72	Mag 14.9 star at NW edge.
02 40 38.4	+08 50 55	MCG +1-7-20	14.6	1.1 x 0.5	13.8		174	Mag 12.8 star NE edge; mag 11.83v star E 2.′3.
02 42 52.5	+07 35 51	MCG +1-7-25	13.9	0.9 x 0.7	13.2	S?	3	= **Mkn 596**.
02 48 53.5	+08 24 18	MCG +1-8-5	16.6	1.0 x 0.5	15.7		3	Mag 14.1 star NNE 0.′9.
02 53 31.3	+06 28 22	MCG +1-8-11A	14.4	0.7 x 0.5	13.1	Sc	144	Mag 13.1 star WSW 2.′4; mag 7.82v star W 12.′9.
02 57 41.8	+06 01 26	MCG +1-8-27	12.7	1.2 x 0.8	12.7	E0:	0	**PGC 11193** NE 2.′8; **PGC 11184** WNW 1.′7.
02 57 45.8	+05 57 05	MCG +1-8-28	15.5	0.9 x 0.4	14.2	Scd	150	**PGC 11195** N 1.′4; **PGC 11196** N 2.′0; mag 12.4 star N 2.′3.
02 58 24.7	+06 35 29	MCG +1-8-30	14.2	1.4 x 1.1	14.6	S0	6	Large, bright core.
02 59 52.3	+06 31 55	MCG +1-8-33	14.4	0.7 x 0.5	13.1	S?	150	Mag 14.2 star on S edge; faint, round anonymous galaxy S 1.′2.
03 00 00.2	+05 58 03	MCG +1-8-34	15.2	0.5 x 0.5	13.5	Sbc		Mag 12.4 star W 1.′1.
03 00 08.8	+05 48 13	MCG +1-8-35	14.5	0.5 x 0.5	12.9	Sa/0		
03 00 09.7	+05 42 40	MCG +1-8-36	15.8	0.5 x 0.4	13.9	Sbc	0	UGC 2469 E 2.′7.
01 44 29.7	+12 25 44	MCG +2-5-18	14.8	0.8 x 0.4	13.4	Sbc	42	Mag 9.42v star NE 4.′5.
01 45 15.4	+10 20 49	MCG +2-5-24	15.4	0.8 x 0.4	14.1	Sb	132	NGC 665 NW 6.′5; **UGC 1226** WNW 3.′6.
01 48 57.0	+10 30 44	MCG +2-5-39	15.4	0.7 x 0.2	13.1	Sb	165	IC 162 W edge.
01 59 49.9	+09 53 32	MCG +2-6-9	15.1	0.7 x 0.7	14.2	Sd		Mag 11.0 star WSW 4.′2.
02 03 59.2	+14 18 33	MCG +2-6-21	15.2	1.3 x 0.9	15.2	Sb	45	Strong N-S bar; close pair of mag 13 stars SW 2.′1.
02 04 51.0	+11 25 21	MCG +2-6-22	14.4	0.8 x 0.5	13.2	SB?	36	Located 1.′2 NW of a mag 8.80v star.
02 11 14.5	+14 07 14	MCG +2-6-40	14.3	0.7 x 0.4	12.8	Sb	90	Spiral dwarf **UGC 1693** E 11.′9; dwarf **UGC 1689** SE 12.′7.
02 12 04.6	+15 35 18	MCG +2-6-45	14.5	0.9 x 0.4	13.2		117	
02 12 38.8	+14 21 56	MCG +2-6-46	14.2	0.8 x 0.6	13.2		147	Mag 10.8 star N 2.′8.
02 16 36.9	+12 58 28	MCG +2-6-50	15.4	0.7 x 0.6	14.3	SBbc	30	Mag 12.1 star N 0.′8.
02 25 13.1	+15 03 27	MCG +2-7-5	14.8	0.7 x 0.5	13.5	Sc	36	Mag 9.94v star W 2.′1.
02 28 21.7	+10 22 54	MCG +2-7-10	14.4	0.8 x 0.6	13.4		150	
02 29 46.5	+10 11 54	MCG +2-7-12	17.1	0.7 x 0.4	15.6		105	Mag 9.13v star SE 6.′1.
02 47 19.0	+15 23 14	MCG +2-8-5	15.1	0.6 x 0.4	13.4		30	Mag 12.8 star NNE 1.′9.
02 50 08.9	+12 50 35	MCG +2-8-14	14.8	0.7 x 0.5	13.5	compact	0	**UGC 2320** N 2.′1.
02 51 13.0	+13 11 29	MCG +2-8-20	13.9	0.5 x 0.4	12.2	E	174	Triple system; NGC 1117 S edge.
02 53 32.2	+12 38 41	MCG +2-8-26	14.2	0.9 x 0.9	13.8			Mag 14.1 star W 2.′5.
02 54 02.2	+11 49 43	MCG +2-8-31	14.8	0.7 x 0.7	13.9			Star or bright knot SE edge.
02 57 29.7	+10 27 22	MCG +2-8-37	14.6	1.0 x 0.7	14.1		108	Mag 12.6 star NE edge; UGC 2430 and bright star NW.
02 58 14.0	+11 07 57	MCG +2-8-41	14.6	0.8 x 0.6	13.6		153	**MCG +2-8-42** SSE 1.′0.
02 59 47.7	+15 17 32	MCG +2-8-45	15.2	0.6 x 0.5	13.7		102	Mag 11.7 star E 2.′7.
03 03 51.0	+12 04 10	MCG +2-8-48	14.0	0.9 x 0.7	13.3		147	Stellar nucleus, faint envelope.
01 57 36.3	+16 46 21	MCG +3-6-8	13.7	1.0 x 0.6	13.0	S?	36	
02 00 28.6	+16 33 34	MCG +3-6-13	14.7	0.9 x 0.6	13.9	Sc	90	Mag 13.4 star 0.′7 N.
02 01 44.3	+17 55 39	MCG +3-6-18	15.3	0.7 x 0.5	14.0	Sb	147	Mag 7.28v star NW 3.′7.
02 11 03.0	+17 12 01	MCG +3-6-42	16.1	0.8 x 0.4	14.7	Sc	39	Mag 14.9 star N end; mag 11.1 star S 1.′7.
02 14 28.7	+17 11 47	MCG +3-6-47	16.5	1.1 x 0.8	16.2		129	Mag 13.8 star SE edge.
02 15 18.1	+18 00 20	MCG +3-6-49	14.2	1.0 x 0.9	14.1	E/S0	0	Mag 10.89v star S 2.′8; mag 8.76v star SSE 5.′5.
02 20 52.5	+16 21 12	MCG +3-7-1	14.8	0.7 x 0.5	13.5	Sbc	81	Dwarf galaxy **UGC 1819** E 10.′6.
02 22 25.3	+17 40 07	MCG +3-7-5	14.6	0.9 x 0.4	13.4	Sbc	129	Mag 13.0 star SW 2.′9.
02 22 53.4	+17 17 57	MCG +3-7-6	15.1	0.5 x 0.5	13.4	Sc		Mag 11.5 star SW 2.′4.
02 24 51.2	+16 10 33	MCG +3-7-8	14.8	0.8 x 0.5	13.6		48	Mag 13.1 star S 1.′5.
02 40 14.0	+17 52 57	MCG +3-7-40	14.1	0.8 x 0.4	12.7	SBb	21	Located between a close pair of mag 12 stars.
02 43 26.6	+16 40 10	MCG +3-7-49	14.4	0.4 x 0.4	12.3	S?		Pair with MCG +3-7-50 S edge.
02 43 27.0	+16 39 42	MCG +3-7-50	15.3	0.6 x 0.4	13.6	Sc	30	Pair with MCG +3-7-49 N edge.
02 45 50.2	+16 25 21	MCG +3-8-6	15.1	0.6 x 0.4	13.4	Sc	96	Mag 12.9 star NE 1.′8.
02 46 03.4	+15 51 02	MCG +3-8-7	14.8	0.8 x 0.3	13.1		9	= **Mkn 597**; multi-galaxy system.
02 47 52.0	+17 16 27	MCG +3-8-11	15.2	0.3 x 0.3	12.4			Compact; mag 8.97v star WNW 2.′8.
02 54 34.8	+15 50 34	MCG +3-8-27	15.4	0.6 x 0.4	13.7	Scd	0	Mag 10.4 star NNW 3.′9.
02 55 17.5	+16 09 42	MCG +3-8-29	14.6	0.6 x 0.4	12.9		120	Mag 10.13v star NE 3.′9.
02 56 32.9	+15 54 34	MCG +3-8-32	15.1	0.4 x 0.4	13.2	E/S0		UGC 2413 NW edge.
02 57 31.4	+16 38 04	MCG +3-8-35	15.4	0.6 x 0.3	13.4	Sc		
02 57 37.6	+16 04 02	MCG +3-8-36	14.8	0.7 x 0.4	13.5	E/S0	66	Almost stellar; stellar **MCG +3-8-34** WNW 3.′9; mag 8.47v star ESE 10.′1.
01 44 56.1	+10 25 24	NGC 665	12.1	2.4 x 1.6	13.5	(R)S0°?	125	Mag 10.23v star W 6.′4; **MCG +2-5-22** NE 2.′0 from center.
01 46 59.1	+13 07 29	NGC 671	13.3	1.5 x 0.5	12.8	S?	55	
01 48 22.5	+11 31 15	NGC 673	12.6	2.1 x 1.6	13.8	SAB(s)c	0	Two main N-S branching arms; mag 10.13v star ENE 3.′3.
01 49 08.7	+13 03 31	NGC 675	14.5	1.0 x 0.4	13.4	S	96	Located on W edge of NGC 677.
01 48 57.4	+05 54 28	NGC 676	11.9	4.0 x 1.0	13.3	S0/a: sp	172	
01 49 14.0	+13 03 19	NGC 677	12.2	2.0 x 2.0	13.7	E		Compact core, faint, smooth envelope; **Mkn 1166** NW 2.′8; NGC 675 on W edge.
01 49 46.8	+11 42 07	NGC 683	13.6	1.0 x 1.0	13.4	S?		Elongated **MCG +2-5-48** sits on E edge.
01 50 30.8	+06 08 40	NGC 693	12.4	2.1 x 1.0	13.1	S0/a?	106	Lies 1.′4 W of mag 10.38v star.
01 51 50.8	+06 17 45	NGC 706	12.5	1.9 x 1.4	13.4	Sbc?	147	Mag 12.7 star on N edge.
01 52 27.8	+17 30 45	NGC 711	13.1	1.6 x 0.8	13.3	S0	15	**UGC 1335** W 6.′5.
01 52 59.5	+12 42 26	NGC 716	12.9	1.8 x 0.7	13.0	SBa:	57	**MCG +1-6-21** 0.′9 SW of nucleus.
01 56 20.9	+05 37 39	NGC 741	11.1	3.0 x 2.9	13.5	E0:	90	NGC 742 0.′8 E of nucleus. **IC 1751** NW 1.′5; **MCG +1-6-5** SW 3.′3.
01 56 24.4	+05 37 33	NGC 742	14.3	0.3 x 0.3	11.7	cE0:		Located 0.′8 E of nucleus of NGC 741.
01 58 42.0	+08 20 48	NGC 766	12.7	2.0 x 2.0	14.2	E		Small, round core, smooth envelope; **MCG +1-6-21** E 2.′9; MCG +1-6-20 S 3.′5.
01 59 34.8	+14 00 31	NGC 774	13.0	1.5 x 1.2	13.5	S0	165	UGC 1468 S 4.′0.
02 00 08.9	+12 39 20	NGC 781	13.1	1.5 x 0.4	12.4	S?	13	
02 01 24.8	+15 38 45	NGC 786	13.4	0.7 x 0.6	12.3	S?	138	Compact.
02 01 44.3	+08 30 01	NGC 791	13.1	1.6 x 1.6	14.1	E		**UGC 1513** SE 1.′6; **MCG +1-6-29** SW 2.′4.
02 02 15.4	+15 42 42	NGC 792	13.1	1.7 x 1.0	13.6	S0	130	Mag 10.8 star E 1.′4; stellar anonymous galaxy SSW 3.′2, on W edge of mag 12.0 star.
02 03 44.7	+16 01 52	NGC 803	12.6	3.0 x 1.3	13.9	SA(s)c: sp	8	Mag 10.9 star on W edge.
02 05 28.6	+13 15 05	NGC 810	13.2	1.7 x 1.3	14.0	E?	25	Stellar anonymous galaxy W 2.′1; mag 7.88v star NE 7.′2.
02 07 33.7	+17 12 09	NGC 817	13.3	0.7 x 0.3	11.5	S?	27	
02 08 24.8	+14 20 57	NGC 820	12.8	1.3 x 0.8	12.7	Sb II-III	72	Mag 9.35v star SE 5.′8.
02 08 21.2	+10 59 37	NGC 821	10.7	2.6 x 1.6	12.2	E6?	25	A mag 9.26v star sits on the NW edge.
02 08 32.3	+06 19 23	NGC 825	13.2	2.0 x 0.5	13.0	Sa	53	Mag 9.66v star E 5.′5; IC 208 N 4.′5.
02 08 56.4	+07 58 12	NGC 827	12.8	2.0 x 0.6	12.8	S?	85	Mag 9.40v star SW 6.′9.
02 09 34.5	+06 05 46	NGC 831	14.5	0.6 x 0.4	12.8	Spiral	102	Mag 6.98v star S 7.′3; **UGC 1647** NW 9.′0.
02 10 16.0	+07 50 42	NGC 840	13.4	1.8 x 1.0	13.9	SB(r)b	76	Almost stellar nucleus in short E-W bar, smooth envelope.

(Continued from previous page)

GALAXIES

RA h m s	Dec ° ′ ″	Name	Mag (V)	Dim ′ Maj x min	SB	Type Class	PA	Notes
02 10 14.4	+06 02 58	NGC 844	14.1	0.4 x 0.4	12.0			Almost stellar; mag 7.39v star NE 5′.4.
02 15 27.6	+06 00 10	NGC 864	10.9	4.7 x 3.5	13.7	SAB(rs)c II-III	20	A mag 10.74v star lies 0′.8 E of stellar nucleus.
02 17 09.4	+14 31 20	NGC 870	15.5	0.3 x 0.3	12.7			Stellar; mag 8.62v star SSE 4′.0.
02 17 10.8	+14 32 47	NGC 871	13.6	1.2 x 0.5	12.9	SB(s)c:	4	**UGC 1761** NE 4′.3.
02 17 53.2	+14 31 15	NGC 876	14.7	2.1 x 0.4	14.4	SAc: sp	20	Located on SW edge of NGC 877.
02 17 58.9	+14 32 40	NGC 877	11.9	2.4 x 1.8	13.3	SAB(rs)bc I-II	140	Strong dark lane W side of core; mag 7.68v star S 4′.7.
02 19 40.0	+15 48 50	NGC 882	13.6	1.2 x 0.6	13.1	S0	82	
02 26 37.1	+12 09 17	NGC 927	13.4	1.2 x 1.2	13.7	SB(r)c I		Very small, very faint anonymous galaxy W 1′.7.
02 33 22.7	+09 36 04	NGC 975	13.1	1.1 x 0.8	12.8	S0/a	0	Mag 10.0 star NNW 2′.2; stellar **CGCG 414-16** NNW 6′.1.
02 36 18.4	+11 38 31	NGC 990	12.5	1.8 x 1.5	13.5	E	42	Mag 10.04v star S 4′.0; mag 9.79v star E 5′.8.
02 37 14.6	+07 18 17	NGC 997	14.0	1.1 x 1.1	14.3	E		Located 1′.4 NE of mag 9.50 star; NGC 998 N 1′.9.
02 37 16.5	+07 20 07	NGC 998	14.7	0.8 x 0.5	13.6		177	NGC 997 S 1′.9.
02 39 11.7	+10 50 51	NGC 1024	12.1	3.9 x 1.4	13.8	(R′)SA(r)ab III	155	Mag 11.5 star on E edge; NGC 1028 E 6′.3.
02 39 19.3	+06 32 38	NGC 1026	12.6	1.7 x 1.3	13.3	S0	114	
02 39 37.2	+10 50 37	NGC 1028	14.8	0.9 x 0.3	13.3	Sa	9	Located 6′.3 E of NGC 1024; NGC 1029 S 3′.1.
02 39 36.7	+10 47 35	NGC 1029	13.1	1.4 x 0.4	12.3	S0/a	70	NGC 1028 N 3′.1.
02 41 12.9	+08 43 09	NGC 1044	13.2	0.7 x 0.7	12.3	S0⁻ pec:		**MCG +1-7-22** WNW 2′.0; NGC 1046 SSE 0′.8.
02 41 14.8	+08 42 26	NGC 1046	13.8	0.2 x 0.2	10.2	S0⁻:		Located 0′.8 SSE of NGC 1044.
02 47 04.1	+16 11 57	NGC 1088	14.3	1.0 x 0.5	13.4	S0/a	99	Mag 9.37v star NE 9′.9; stellar galaxy **IC 255** N 5′.4.
02 49 19.6	+08 05 34	NGC 1107	12.2	1.8 x 1.3	13.0	S0	140	
02 49 00.4	+13 13 26	NGC 1109	13.8	0.9 x 0.6	13.0	SAB(s)c	3	Almost stellar anonymous galaxy NW 5′.5, possibly **IC 1850**?
02 49 50.7	+13 08 44	NGC 1111	14.7	0.6 x 0.5	13.2		159	Mag 11.3 star NNE 2′.3.
02 50 25.4	+13 15 59	NGC 1115	14.7	0.7 x 0.2	12.4	Sb	12	Mag 10.34v star W 3′.2.
02 50 35.7	+13 20 03	NGC 1116	14.3	1.3 x 0.3	13.1	Sab	27	NGC 115 SSW 4′.9.
02 51 13.2	+13 11 08	NGC 1117	14.5	0.7 x 0.5	13.4	E	33	Triple system, MCG +2-8-20 N edge.
02 52 51.9	+13 15 21	NGC 1127	14.4	0.7 x 0.5	13.1	(R)SB(r)ab:	39	
02 53 41.2	+13 00 56	NGC 1134	12.2	2.5 x 0.9	12.9	S?	148	UGC 2362 W 7′.2.
03 00 35.0	+11 50 35	NGC 1166	14.0	1.0 x 1.0	13.8	Sab		Mag 7.6 star N 8′.3.
03 00 47.1	+11 46 16	NGC 1168	14.2	1.2 x 0.7	13.9	SAB(rs)b: II-III	18	Mag 9.48v star SE 5′.4.
01 44 20.7	+17 28 39	UGC 1219	12.8	1.3 x 0.6	12.3	SB?	102	
01 45 55.0	+09 49 22	UGC 1237	14.1	1.5 x 0.5	13.7	Sab	45	Elongated anonymous galaxy NNE 2′.2; mag 10.57v star E 3′.0.
01 46 10.3	+06 25 51	UGC 1239	14.1	0.9 x 0.5	13.1	Sdm:	155	
01 46 37.1	+14 41 28	UGC 1242	14.6	1.0 x 0.7	14.0	Sm:	141	
01 46 50.8	+11 24 04	UGC 1245	14.1	1.3 x 1.0	14.2	S	150	Bright center, uniform envelope; very faint anonymous galaxy with stellar nucleus NE 2′.6.
01 47 00.7	+12 24 17	UGC 1246	13.9	1.2 x 0.9	13.9	IAm	21	Numerous faint and bright knots overall.
01 47 30.0	+12 06 15	UGC 1253	14.2	1.2 x 0.5	13.5	S?	12	
01 48 33.3	+12 36 49	UGC 1260	13.1	1.0 x 0.7	12.6	(R′)SB(s)a	126	Mag 10.4 star NW 2′.5.
01 48 33.6	+13 25 49	UGC 1261	14.5	1.5 x 0.4	13.8	Scd:	92	**UGC 1279** ESE 13′.0.
01 49 00.1	+13 12 35	UGC 1271	13.1	1.7 x 0.9	13.4	SB0	113	
01 49 11.7	+12 51 10	UGC 1274	13.9	1.5 x 0.4	13.2	Sa:	108	Mag 13.9 star W 1′.3.
01 49 29.9	+12 30 32	UGC 1282	13.1	1.5 x 0.7	13.0	S0/a	55	
01 50 03.2	+11 48 45	UGC 1290	14.1	0.7 x 0.5	12.8	Sbc	140	
01 51 02.3	+13 17 45	UGC 1312	13.9	1.5 x 0.5	13.4	S?	93	Mag 11.9 star NE 1′.6.
01 51 29.9	+13 07 52	UGC 1322	14.1	1.2 x 1.1	14.3	SA(s)c	159	Two very faint, almost stellar anonymous galaxies outside NE and E edges.
01 51 37.1	+08 15 22	UGC 1325	12.6	1.8 x 1.8	13.8	E		Bright core, uniform envelope; UGC 1326 N 2′.7.
01 51 40.2	+08 17 59	UGC 1326	13.6	0.8 x 0.8	13.2	E:		UGC 1325 S 2′.7.
01 51 52.0	+17 03 05	UGC 1328	13.6	1.5 x 1.0	13.9	SAB(rs)c:	125	Located 2′.0 SW of mag 8.20v star.
01 54 13.3	+07 53 01	UGC 1368	13.8	1.5 x 0.4	13.1	Sab	53	
01 54 31.4	+17 40 24	UGC 1372	14.0	0.9 x 0.4	12.8	S?	14	Mag 11.0 star SE 1′.2; mag 10.0 star S 2′.6.
01 54 41.0	+06 02 02	UGC 1383	14.9	1.5 x 0.2	13.4	Scd:	36	
01 55 15.9	+10 00 48	UGC 1391	14.7	1.5 x 0.2	13.2	Scd:	0	Mag 11.1 star NE 2′.2.
01 55 22.0	+06 36 41	UGC 1395	13.4	1.2 x 0.9	13.3	SA(rs)b III	150	
01 56 43.5	+15 00 45	UGC 1420	13.8	1.0 x 0.5	12.9	S?	3	Mag 8.99v star S 7′.2; note: **CGCG 438-5** lies 1′.8 SE of this star.
01 56 52.1	+05 46 28	UGC 1425	13.6	1.0 x 0.7	13.1	E/S0	174	**MCG +1-6-11** E 1′.1.
01 56 58.8	+06 02 12	UGC 1427	13.5	1.0 x 0.6	12.8	S0	65	
01 57 26.8	+17 13 12	UGC 1432	13.8	1.0 x 0.6	13.1	Sbc	65	
01 57 21.7	+05 36 32	UGC 1435	13.9	0.8 x 0.5	12.7	S?	40	
01 58 16.3	+14 17 46	UGC 1450	14.7	0.8 x 0.7	13.9		132	
01 59 31.1	+13 56 36	UGC 1468	15.4	1.5 x 0.9	15.6	Sd	145	Stellar nucleus, faint envelope; NGC 774 N 4′.0.
02 00 47.2	+17 42 49	UGC 1495	14.5	0.8 x 0.6	13.6	Scd:	39	
02 01 00.0	+15 11 20	UGC 1496	13.2	1.6 x 0.8	13.4	S0	165	
02 00 59.0	+08 18 35	UGC 1498	13.4	0.9 x 0.5	12.4	S	26	
02 01 05.8	+06 31 18	UGC 1505	13.5	1.0 x 0.4	12.4	S0	100	
02 01 51.3	+16 15 27	UGC 1512	13.5	1.2 x 0.5	12.8	S0/a	53	
02 02 34.6	+17 10 01	UGC 1531	14.1	1.2 x 0.4	13.1	Sb II	114	
02 03 10.6	+05 39 21	UGC 1545	14.0	0.9 x 0.3	12.4	S0/a	124	
02 03 55.1	+11 57 43	UGC 1558	15.4	1.3 x 1.1	15.6	SABm:	117	Mag 11.9 star SW 2′.5.
02 04 43.9	+08 32 31	UGC 1572	13.4	1.3 x 0.4	12.5	S?	62	Mag 13.3 star on SW end.
02 05 14.0	+09 55 29	UGC 1580	14.2	1.5 x 0.4	13.4	S?	137	Mag 10.9 star N 1′.8.
02 05 26.8	+13 19 36	UGC 1584	15.3	1.5 x 0.5	14.8	Scd:	78	Mag 7.88v star E 5′.1.
02 05 37.3	+06 46 17	UGC 1587	13.7	1.0 x 0.4	12.6	S?	24	Mag 11.1 star SE 1′.6.
02 06 05.9	+13 17 06	UGC 1593	14.4	0.8 x 0.7	13.7	SAB(rs)c	18	Mag 7.88v star NW 5′.7.
02 07 01.4	+08 07 48	UGC 1605	14.5	1.4 x 0.5	14.0		60	Knot or clump SW end.
02 07 43.3	+15 21 57	UGC 1614	14.4	1.5 x 0.2	12.9	S0/a	87	
02 07 57.2	+16 13 08	UGC 1622	14.2	1.1 x 0.4	13.1	Scd:	165	Mag 8.81v star SE 1′.7.
02 08 24.7	+14 58 20	UGC 1630	13.4	1.1 x 0.5	12.6	S?	43	Mag 10.8 star N edge; mag 11.1 star SE 0′.8.
02 09 04.6	+05 06 39	UGC 1646	13.8	2.0 x 0.5	13.7	SBbc	173	Mag 9.52 star W 1′.3.
02 09 12.3	+06 36 01	UGC 1649	14.6	1.2 x 1.2	14.8	SA(s)dm		
02 09 56.8	+16 01 55	UGC 1659	13.9	1.5 x 0.6	13.6	SBcd:	38	Pair of stars, mags 10.79v and 9.96v, E 2′.5.
02 10 08.9	+07 38 43	UGC 1663	13.9	1.3 x 0.7	13.7	Scd:	13	
02 10 37.7	+05 52 10	UGC 1669	13.5	0.7 x 0.6	12.4	S	173	
02 10 44.0	+06 45 30	UGC 1670	14.3	2.3 x 2.3	15.9	Sm: V		Dwarf spiral, very low surface brightness.
02 11 24.0	+15 57 16	UGC 1684	13.2	1.6 x 1.6	14.1	(R)SAB0⁺		Bright core with dark ring around it; mag 8.46v star W 1′.4.
02 11 39.2	+14 17 55	UGC 1687	13.9	1.3 x 1.0	14.0	S0:	0	Mag 12.8 star SE edge; **MCG +2-6-41** W 2′.0.
02 12 02.5	+09 30 59	UGC 1694	13.1	1.5 x 1.5	13.8	SAB(s)dm		Weak almost stellar nucleus; mag 8.82v star NW 5′.3.
02 12 42.5	+13 58 07	UGC 1702	14.7	0.8 x 0.8	14.1	SAd?		Located 3′.0 N of a mag 7.13v star.
02 14 23.5	+07 51 38	UGC 1724	14.4	1.1 x 0.9	14.2	SAB(s)c	90	Stellar nucleus.
02 15 45.8	+15 23 40	UGC 1742	14.5	1.0 x 0.7	14.0	SABcd	51	
02 18 14.8	+13 12 08	UGC 1773	14.2	1.5 x 0.3	13.2	SB?	48	Mag 11.8 star SW 2′.3.

99 (Continued from previous page) GALAXIES 99

RA h m s	Dec ° ′ ″	Name	Mag (V)	Dim ′ Maj x min	SB	Type Class	PA	Notes
02 18 26.5	+05 39 12	UGC 1775	12.9	1.3 x 1.2	13.2	S?	141	**UGC 1774** S 6′0.
02 20 29.3	+06 48 36	UGC 1803	14.1	1.9 x 0.4	13.7	SB(s)m: sp	57	Mag 12.3 star 2′0 NE.
02 21 23.8	+16 33 59	UGC 1814	12.8	2.5 x 1.2	13.8	SAB(s)bc	157	Several bright knots W and S edges.
02 21 31.1	+14 11 55	UGC 1817	14.0	2.4 x 0.4	13.8	Sd	163	Mag 10.28v star SW 2′9.
02 25 22.9	+11 05 21	UGC 1879	14.3	1.0 x 0.8	13.9	SABc	126	Stellar nucleus, uniform envelope.
02 25 30.0	+11 28 19	UGC 1883	14.5	2.1 x 0.5	14.4	Sd:	157	
02 26 13.9	+12 27 47	UGC 1897	15.4	0.7 x 0.7	14.5	Sd		
02 28 27.3	+15 36 27	UGC 1946	15.4	1.7 x 0.2	14.1	Sd	137	Mag 12.9 star W 1′1.
02 29 23.6	+10 10 37	UGC 1966	13.6	1.5 x 1.0	13.9	(R′)SABab III	40	Almost stellar nucleus, smooth envelope; mag 11.3 star E 1′5.
02 30 24.5	+08 41 54	UGC 1982	15.9	0.9 x 0.7	15.2	Im	111	
02 32 14.4	+09 43 10	UGC 2007	13.8	1.1 x 0.8	13.5	SB0	99	Short N-S bar, uniform envelope.
02 32 52.1	+05 42 00	UGC 2021	13.5	0.7 x 0.6	12.4	S0	45	Mag 10.61v star NW 2′9.
02 33 54.0	+09 37 52	UGC 2041	14.0	1.0 x 0.6	13.3	Sab	161	Mag 10.7 star SW 2′2.
02 35 39.3	+06 16 19	UGC 2075	13.6	1.0 x 0.9	13.3	S0/a	159	
02 36 31.6	+07 18 30	UGC 2092	14.4	3.0 x 0.3	14.1	Scd:	32	Mag 11.9 star E 1′5.
02 40 55.4	+08 35 40	UGC 2167	14.2	1.2 x 0.4	13.3	SBcd:	140	Mag 7.24v star SE 8′0.
02 41 54.2	+05 56 50	UGC 2176	13.7	1.1 x 0.4	12.6	Sb II-III	14	
02 46 17.5	+13 05 46	UGC 2238	14.0	1.2 x 0.8	13.8	Im?	165	
02 47 02.6	+15 25 16	UGC 2252	13.9	1.0 x 1.0	13.8	Compact		
02 47 02.3	+08 39 05	UGC 2255	13.4	0.8 x 0.4	12.0	S0/a	85	
02 48 41.8	+14 18 44	UGC 2282	13.7	1.0 x 0.8	13.3	SB(rs)bc	150	Mag 11.6 star W 2′5.
02 48 34.5	+06 31 17	UGC 2285	14.4	1.0 x 0.3	13.0	Sdm:	85	Mag 8.37v star NE 5′1.
02 49 23.3	+17 39 55	UGC 2303	13.4	1.2 x 1.2	13.6	SABb II-III		
02 50 13.3	+08 10 04	UGC 2323	14.6	0.7 x 0.4	13.0	S(r)c		
02 50 49.0	+16 00 26	UGC 2329	13.8	0.8 x 0.6	12.9	Sab	177	**CGCG 463-26** NW 4′5.
02 51 12.0	+08 04 32	UGC 2336	14.4	1.7 x 1.1	14.9	SB(s)cd	165	Stellar nucleus.
02 51 23.6	+16 02 28	UGC 2339	15.9	1.0 x 0.7	15.4	Sdm	45	Mag 11.4 star NE 3′0.
02 52 12.7	+13 59 29	UGC 2348	13.6	0.7 x 0.6	12.5	S0?	0	Elongated **UGC 2344** SSW 5′9.
02 53 11.7	+13 01 53	UGC 2362	14.3	1.5 x 1.1	14.7	Im	109	Mag 9.48v star SW 5′3; NGC 1134 E 7′2.
02 53 21.6	+06 32 21	UGC 2364	15.0	1.5 x 0.5	14.5	S?	103	Mag 7.82v star SW 11′1.
02 53 41.2	+06 15 55	UGC 2367	13.8	2.0 x 0.7	14.0	S0	145	Elongated **UGC 2375** E 5′6.
02 54 01.8	+14 58 22	UGC 2369	13.7	1.0 x 0.8	13.3	Double System	24	Double system, contact, distorted; consists of **MCG +2-8-29** and **MCG +2-8-30**.
02 53 58.6	+05 59 22	UGC 2372	14.4	0.9 x 0.8	13.9	SAB(rs)cd:		
02 54 23.8	+11 44 48	UGC 2378	15.1	1.5 x 0.2	13.6	Scd:	177	
02 55 17.1	+12 13 28	UGC 2387	14.5	1.5 x 0.5	14.1	Scd:	12	
02 55 47.0	+06 13 04	UGC 2399	14.0	1.1 x 1.1	14.1	SAB(s)cd		Mag 11.1 star NW 2′2; mag 7.30v star SE 4′9.
02 55 57.7	+06 29 36	UGC 2405	14.0	1.7 x 0.6	13.9	Scd:	167	
02 56 28.8	+15 54 57	UGC 2413	14.3	0.9 x 0.7	13.6	Compact	70	MCG +3-8-32 SE 1′0.
02 56 43.0	+07 19 59	UGC 2419	13.5	1.2 x 1.2	13.7	(R′)SB(s)a		Bright NE-SW center, faint envelope.
02 57 07.6	+17 30 49	UGC 2424	13.4	2.3 x 0.6	13.6	Sbc	158	
02 57 09.3	+05 19 15	UGC 2426	14.1	0.9 x 0.6	13.2	Sb III	65	Mag 10.89v star NE 2′7.
02 57 26.1	+10 28 42	UGC 2430	14.1	0.8 x 0.6	13.1	Sa	90	MCG +2-8-37 SE 1′6; mag 8.86v star W 1′0.
02 57 32.9	+08 06 00	UGC 2434	15.3	0.9 x 0.9	14.9	SA(s)cd		
02 57 53.9	+11 10 58	UGC 2437	14.1	1.5 x 1.0	14.4	Sd	33	
02 57 53.3	+13 01 53	UGC 2438	14.4	0.9 x 0.9	14.2	cD		Located at the center of galaxy cluster A 399.
02 58 29.8	+06 18 20	UGC 2444	14.2	0.8 x 0.6	13.3	SB?	45	Faint, almost stellar anonymous galaxy NNE 2′6.
02 59 15.1	+16 02 11	UGC 2453	14.3	1.5 x 1.2	14.8	SAB(s)b II-III	126	
03 00 20.5	+05 42 42	UGC 2469	14.3	1.1 x 0.7	13.9	Pec?	24	MCG +1-8-36 W 2′7.
03 01 30.1	+17 51 07	UGC 2486	13.7	1.1 x 0.8	13.4	SA(r)bc	54	Stellar nucleus; mag 9.31v star N 5′9.
03 02 12.3	+17 20 44	UGC 2498	14.4	1.5 x 0.7	14.3	SABcd	65	Located between a pair of mag 12.4 stars.

GALAXY CLUSTERS

RA h m s	Dec ° ′ ″	Name	Mag 10th brightest	No. Gal	Diam ′	Notes
01 44 06.0	+06 23 00	A 245	16.4	40	16	All members faint and stellar.
01 44 42.0	+05 48 00	A 246	16.4	56	25	Difficult to see many GX, even on DSS image.
02 57 00.0	+15 57 00	A 397	15.1	35	50	**UGC 2420** NW of center 3′3; **CGCG 463-33** W of center 15′4.
02 57 54.0	+13 00 00	A 399	15.6	57	9	With UGC 2438 at its center, most other members are stellar PGC and anonymous GX spread N-S along centerline.
02 57 36.0	+06 01 00	A 400	13.9	58	56	Numerous PGC and anonymous GX spread throughout the cluster boundaries.
02 58 54.0	+13 34 00	A 401	15.6	90	19	Mag 7.31v star near center; **UGC 2450** SE of this star 3′1; other stellar PGC and anonymous GX in surrounding area.

OPEN CLUSTERS

RA h m s	Dec ° ′ ″	Name	Mag	Diam ′	No. ★	B ★	Type	Notes
02 47 31.0	+17 15 00	DoDz 1		10	12		cl	Few stars; moderate brightness range; no central concentration; detached.

100 GALAXIES 100

RA h m s	Dec ° ′ ″	Name	Mag (V)	Dim ′ Maj x min	SB	Type Class	PA	Notes
00 34 16.7	+05 32 49	CGCG 409-37	14.0	0.4 x 0.4	11.9	Sb		Mag 13.5 star on SE edge.
00 44 32.9	+08 36 04	CGCG 410-1	14.1	0.5 x 0.3	11.9		30	Mag 10.73v star ESE 10′9.
00 44 48.5	+05 08 05	CGCG 410-2	15.0	0.6 x 0.3	13.0		63	Mag 8.22v star ENE 4′9.
01 15 53.5	+05 19 28	CGCG 411-14	14.1	0.8 x 0.3	12.4		12	Mag 13.6 star E 1′2.
01 06 35.7	+10 31 15	CGCG 435-40	15.3	0.6 x 0.4	13.6	Sb	18	Mag 10.21v star SE 6′9.
01 15 54.3	+13 21 07	CGCG 436-16	14.1	0.6 x 0.4	12.4		45	Mag 12.0 star N 1′6.
01 21 30.8	+14 30 15	CGCG 436-34	14.1	0.8 x 0.3	12.4		84	Mag 9.36v star ENE 12′8.
01 12 32.9	+16 32 39	CGCG 459-16	14.9	0.7 x 0.7	13.9	Sc		Galaxy pair, components N-S; **CGCG 459-15** WSW 2′5.
01 27 00.7	+15 58 48	CGCG 459-55	15.0	0.5 x 0.3	12.8	Sc	6	Mag 12.7 star SE 0′9.
01 29 45.6	+16 55 31	CGCG 459-64	14.9	0.6 x 0.5	13.4	Sc		Mag 10.96v star 1′2 W.
00 34 24.5	+12 16 05	IC 31	14.3	1.6 x 0.3	13.4	Sa	89	Mag 10.75v star S 4′2.
00 35 36.4	+09 07 27	IC 34	12.6	2.4 x 0.8	13.2	SB(r)a	156	Mag 11.3 star SSW 9′6; **UGC 353** S 7′6.
00 37 39.8	+10 21 27	IC 35	14.0	0.8 x 0.7	13.2	Scd:	33	Mag 10.15v star NE 1′6.
00 50 41.0	+10 36 00	IC 53	13.9	0.5 x 0.5	12.3	S0$^-$:		Almost stellar, mag 11.9 star NW 2′1.

(Continued from previous page)

GALAXIES

RA h m s	Dec ° ′ ″	Name	Mag (V)	Dim ′ Maj x min	SB	Type Class	PA	Notes
00 51 42.4	+07 43 03	IC 55	14.1	0.8 x 0.5	13.0	Sab	177	Mag 13.7 star on E edge; mag 10.83v star N 3′.9.
00 54 48.5	+11 50 27	IC 57	14.1	0.7 x 0.7	13.2	S0?		Mag 11.27v star W 5′.0.
00 57 07.2	+07 30 20	IC 61	14.4	1.2 x 0.8	14.2	Compact	174	Mag 9.05v star NW 3′.9.
00 58 44.0	+11 48 27	IC 62	14.2	0.8 x 0.5	13.1	S	25	Mag 10.8 star WSW 7′.3.
01 07 11.6	+10 50 12	IC 75	13.8	1.2 x 0.9	13.7	Sb III	30	
01 24 08.6	+09 55 48	IC 101	13.8	1.4 x 0.6	13.5	S?	127	Mag 9.29v star SW 12′.7; **IC 102** SE 5′.1.
01 25 13.5	+14 52 16	IC 107	14.0	0.8 x 0.6	13.1	S?	100	Mag 9.97v star SW 3′.6.
01 26 03.0	+11 26 36	IC 112	13.7	0.8 x 0.4	12.3	Sdm:	128	Mag 10.05v star SW 8′.9.
01 26 22.6	+09 54 38	IC 114	14.1	1.3 x 0.6	13.7	S0	150	Mag 10.53v star SE 8′.9.
01 45 16.5	+10 38 56	IC 154	14.0	1.4 x 0.2	12.4	Sb III	66	Mag 9.31v star N 6′.5.
01 45 29.2	+10 33 07	IC 156	13.5	1.5 x 1.2	14.0	S?	72	Mag 9.44v star E 2′.6.
01 48 43.8	+10 30 26	IC 161	13.8	0.7 x 0.5	12.5	S?	65	IC 162 ENE 2′.5.
01 48 53.3	+10 31 15	IC 162	12.7	1.4 x 1.4	13.3	S0		MCG +2-5-39 E edge.
00 39 05.3	+06 01 10	IC 1564	13.9	1.1 x 0.5	13.1	SABbc:	83	Mag 8.74v star W 7′.8.
00 39 26.3	+06 44 00	IC 1565	13.4	1.0 x 1.0	13.3	S?		Mag 10.61v star SE 3′.4.
00 39 33.5	+06 48 51	IC 1566	13.7	0.8 x 0.7	13.1	E/S0	126	Mag 11.8 star SSW 3′.0; IC 1565 SSW 5′.2.
00 39 56.0	+06 50 55	IC 1568	13.9	0.8 x 0.8	13.3	S0		Mag 10.24v star and UGC 429 NE 8′.7.
00 40 28.2	+06 43 07	IC 1569	14.6	0.4 x 0.4	12.5	S0		Mag 10.7 star NE 2′.7.
00 53 27.0	+05 46 11	IC 1592	14.1	0.9 x 0.6	13.3	S?	165	Anonymous galaxy W 1′.4.
00 54 41.8	+05 46 27	IC 1598	13.9	1.1 x 0.5	13.1	Sa	2	Mag 7.94v star W 11′.1.
01 07 14.3	+13 57 15	IC 1620	14.3	1.0 x 0.8	13.9	SABbc	90	MCG +2-3-33 W 1′.1; mag 7.23v star SW 10′.2.
01 12 27.5	+15 44 58	IC 1645	15.5	0.4 x 0.4	13.6	E/S0		Mag 11.15v star SE 8′.6; **IC 1646** SE 4′.7.
01 25 07.7	+08 41 54	IC 1695	14.0	0.7 x 0.7	13.1	S?		Located at the center of galaxy cluster A 193, numerous very faint anonymous galaxies surrounding area; **UGC 967** NW 7′.5.
01 25 22.2	+14 50 17	IC 1698	13.4	1.3 x 0.4	12.6	S0	117	IC 1700 N 1′.7; mag 9.97v star W 5′.5.
01 25 24.8	+14 51 50	IC 1700	12.8	1.0 x 0.7	12.5	E	27	Mag 13.8 star SW edge; IC 1698 S 1′.4; IC 107 W 2′.8.
01 25 56.2	+16 36 02	IC 1702	13.5	1.2 x 0.8	13.3	Scd:	170	Mag 12.8 star on N edge; mag 10.77v star NE 7′.7.
01 27 09.5	+14 46 32	IC 1704	13.2	1.0 x 0.8	12.8	S pec?	165	Mag 10.82v star NE 3′.0; Almost stellar **IC 1706** NE 5′.9.
01 30 55.1	+17 11 17	IC 1711	13.6	2.6 x 0.5	13.7	Sb III	43	Mag 7.58v star E 5′.0.
01 33 34.8	+12 35 06	IC 1715	13.9	0.9 x 0.6	13.1	Im	100	Mag 11.2 star N 8′.5.
01 41 24.5	+08 31 28	IC 1721	13.4	1.0 x 0.5	12.5	S?	94	Mag 12.0 star E 1′.5; mag 9.93v star NW 8′.4.
01 43 14.1	+08 53 17	IC 1723	13.0	3.1 x 0.6	13.5	Sb II	29	Mag 8.58v star E 12′.6.
00 32 58.8	+05 17 07	MCG +1-2-23	15.0	0.6 x 0.5	13.6	Sb	12	MCG +1-2-24 S 1′.5.
00 33 00.9	+05 15 45	MCG +1-2-24	15.1	0.7 x 0.5	13.8	Sb	39	MCG +1-2-23 N 1′.1
00 33 25.5	+07 51 21	MCG +1-2-26	14.4	0.8 x 0.3	12.7	S?	15	**UGC 327** NW 3′.8.
00 35 55.9	+08 43 00	MCG +1-2-34	14.1	0.5 x 0.5	12.4			Mag 11.8 star SE 1′.7.
00 36 17.0	+07 50 02	MCG +1-2-36	15.5	0.8 x 0.6	14.6		90	
00 38 54.8	+07 03 20	MCG +1-2-42	14.8	0.3 x 0.3	12.0	S0?		= **Hickson 5B**.
00 41 47.1	+08 23 23	MCG +1-2-54	14.7	0.5 x 0.5	13.0			**MCG +1-2-55** SE 3′.2.
00 57 02.8	+07 27 37	MCG +1-3-8	15.4	0.7 x 0.5	14.1	Sbc	30	IC 61 N 2′.9; mag 9.29v star SW 3′.7.
01 03 14.8	+08 49 29	MCG +1-3-13	14.5	1.0 x 0.3	13.1	S?	174	Mag 9.30v star SE 3′.7.
01 12 44.5	+08 43 12	MCG +1-4-2	14.8	0.4 x 0.4	12.7			Mag 10.64v star SW 1′.3.
01 13 55.0	+07 25 30	MCG +1-4-4	15.0	0.8 x 0.4	13.6	Sc	90	Mag 11.3 star NE 3′.1.
01 15 03.7	+05 22 38	MCG +1-4-7	14.1	1.1 x 0.4	13.1	S?	156	Mag 10.7 star E 6′.5.
01 15 53.5	+05 04 54	MCG +1-4-10	15.7	0.7 x 0.3	13.8	Sbc	120	Mag 14.1 star NW 0′.5; mag 10.51v star NE 4′.6.
01 16 13.0	+05 09 58	MCG +1-4-13	15.9	0.4 x 0.4	13.7			Located 3′.8 E of NGC 455.
01 21 05.2	+05 45 41	MCG +1-4-27	15.6	0.9 x 0.3	14.0	Sb	108	
01 22 54.3	+08 51 17	MCG +1-4-42	15.0	1.0 x 0.7	14.5	SBC	90	Mag 11.36v star WNW 3′.1.
01 23 19.4	+07 47 47	MCG +1-4-44	14.9	0.9 x 0.4	13.7	Sm	144	Uniform surface brightness.
01 23 50.1	+07 47 09	MCG +1-4-47	14.7	1.0 x 0.5	13.8		126	Multi-galaxy system, two nuclei; **UGC 964** ESE 11′.8.
01 26 33.7	+09 03 21	MCG +1-4-60	14.9	0.8 x 0.6	14.0	SBb	33	
01 32 49.5	+05 44 22	MCG +1-5-4	15.1	0.7 x 0.3	13.3	Sc	60	Mag 10.28v star SE 2′.9.
01 37 07.0	+05 41 34	MCG +1-5-8	15.0	0.7 x 0.4	13.5	Sc	36	Mag 8.18v star WSW 3′.9.
01 37 40.0	+06 55 12	MCG +1-5-12	14.7	0.6 x 0.6	13.5	SBc		Mag 12.9 star NW 2′.2.
01 41 13.9	+08 28 45	MCG +1-5-18	15.8	0.7 x 0.7	14.8	SBc		Mag 11.4 star SW 3′.9; IC 1721 NNE 3′.9.
01 41 41.0	+05 51 10	MCG +1-5-20	15.1	0.9 x 0.2	13.1	Sb	96	Mag 12.1 star WNW 1′.1.
00 36 09.8	+12 38 28	MCG +2-2-22	14.3	0.7 x 0.5	13.0	S?	24	Mag 13.2 star SSE 1′.9.
00 47 30.5	+15 41 49	MCG +2-3-2	15.4	0.8 x 0.5	14.3	Sb	36	Mag 13.5 star SE 2′.3.
00 47 35.2	+14 09 16	MCG +2-3-3	15.7	0.9 x 0.4	14.4	Sc	78	Mag 12.9 star on N edge.
00 51 01.8	+11 31 32	MCG +2-3-7	15.8	0.7 x 0.4	14.3	Sc	81	Mag 14.0 star W end.
00 55 16.7	+12 19 23	MCG +2-3-12	15.3	0.7 x 0.7	14.4	Sdm		
00 55 26.7	+12 06 11	MCG +2-3-13	15.6	0.5 x 0.5	14.0	Sc		UGC 571 E 7′.9.
01 03 20.5	+14 01 49	MCG +2-3-27	15.3	0.7 x 0.3	13.4	Sc	25	UGC 644 W.
01 05 42.1	+14 44 01	MCG +2-3-31	15.5	0.7 x 0.4	14.0	Sb	96	
01 07 10.0	+13 57 11	MCG +2-3-33	15.6	0.7 x 0.3	13.7	Sb	33	IC 1620 E 1′.1; mag 7.23v star SW 9′.1.
01 19 24.6	+12 27 40	MCG +2-4-22	13.9	1.0 x 0.7	13.6	E?	9	= **Mkn 983**; UGC 849 S edge.
01 20 02.7	+14 21 39	MCG +2-4-25	14.2	0.6 x 0.3	12.2	Sb	165	Mag 11.9 star E edge.
01 20 49.6	+11 53 39	MCG +2-4-28	14.2	0.7 x 0.7	13.5	E/S0		Mag 12.9 star NE 0′.8; mag 11.4 star E 2′.4.
01 21 09.4	+15 41 40	MCG +2-4-29	14.6	0.8 x 0.4	13.2	SBc: IV	171	Mag 10.6 star 0′.9 W.
01 21 47.3	+11 43 12	MCG +2-4-31	14.9	0.7 x 0.6	13.8	SBbc	162	Faint anonymous galaxy N 2′.2.
01 25 22.3	+10 13 05	MCG +2-4-43	15.1	0.8 x 0.3	13.4			MCG +2-4-45 and MCG +2-4-46 S 2′.2.
01 25 23.4	+10 10 54	MCG +2-4-45	15.1	0.4 x 0.4	13.0			Pair with MCG +2-4-46 S edge.
01 25 24.5	+10 10 22	MCG +2-4-46	15.3	0.7 x 0.5	14.0		9	Pair with MCG +2-4-45 N edge.
01 38 54.9	+15 01 16	MCG +2-5-8	14.6	1.0 x 0.7	14.1		105	Anonymous galaxy NNW 1′.8.
01 44 29.7	+12 25 44	MCG +2-5-18	14.8	0.8 x 0.4	13.4	Sbc	42	Mag 9.42v star NE 4′.5.
01 45 15.4	+10 20 49	MCG +2-5-24	15.4	0.8 x 0.4	14.1	Sb	132	NGC 665 NW 6′.5; **UGC 1226** WNW 3′.6.
01 48 57.0	+10 30 44	MCG +2-5-39	15.4	0.7 x 0.2	13.1	Sb	165	IC 162 W edge.
00 37 50.6	+17 22 57	MCG +3-2-19	15.5	0.7 x 0.4	14.0	Sc	123	Mag 10.92v star NW 0′.9.
00 54 45.3	+16 26 12	MCG +3-3-8	14.7	0.8 x 0.5	13.6		162	Mag 9.38v star NNE 8′.6.
01 07 12.9	+17 59 38	MCG +3-3-16	15.3	0.8 x 0.4	13.9		63	Mag 12.7 star SE 2′.1.
01 16 44.0	+16 23 58	MCG +3-4-13	13.8	1.3 x 0.8	13.7	S?	141	Mag 12.2 star NW 2′.6.
01 24 43.9	+17 00 14	MCG +3-4-34	15.7	0.4 x 0.4	13.6			Mag 13.2 star SW 0′.9.
01 43 56.5	+17 03 44	MCG +3-5-13	14.4	0.4 x 0.4	12.3	cI pec:		= **Mkn 360**; mag 13.1 star WSW 1′.9.
00 37 57.4	+08 38 09	NGC 180	12.9	1.9 x 1.3	13.7	SB(rs)bc	160	Superimposed star or bright knot at NW end of elongated core; mag 8.04v star E 6′.7.
00 38 54.7	+07 03 44	NGC 190	14.0	0.9 x 0.7	13.4	Sab	135	= **Hickson 5A**. **Hickson 5B** on S edge; **Hickson 5C** NNW 0′.8; **Hickson 5D** S 0′.9.
00 41 09.9	+16 28 09	NGC 213	13.3	1.5 x 1.2	13.8	SB(rs)a	102	Bright knot or superimposed star E of core.
00 43 32.3	+14 20 31	NGC 234	12.5	1.5 x 1.2	13.0	SAB(rs)c	54	
00 45 02.1	+06 06 45	NGC 240	13.5	1.2 x 1.2	13.7	S0/a		Mag 9.54v star W 5′.8.
00 47 16.2	+07 54 32	NGC 250	13.6	1.0 x 0.6	12.9	S0/a	153	

(Continued from previous page)

GALAXIES

RA h m s	Dec ° ′ ″	Name	Mag (V)	Dim ′ Maj x min	SB	Type Class	PA	Notes
00 48 01.7	+08 17 46	NGC 257	12.6	1.9 x 1.3	13.4	Scd:	105	
00 58 49.3	+07 06 38	NGC 332	14.0	1.0 x 0.6	13.3	Compact	165	
01 14 22.4	+05 55 37	NGC 437	12.8	1.3 x 1.0	12.9	S0/a	130	
01 15 57.7	+05 10 44	NGC 455	12.7	1.9 x 1.2	13.4	S?	153	MCG +1-4-13 E 3′.8; mag 10.51v star SSE 2′.3.
01 18 08.2	+17 33 41	NGC 459	14.4	0.7 x 0.6	13.3	Sbc	33	Stellar nucleus; mag 10.74v star E 4′.7.
01 18 58.2	+16 19 35	NGC 463	14.1	1.2 x 0.4	13.2	S0:	4	
01 19 33.0	+14 52 13	NGC 469	14.1	0.6 x 0.4	12.4		165	Located at the center of galaxy cluster **A 175**; mag 8.41v star S 5′.1.
01 19 59.7	+14 47 09	NGC 471	13.3	1.0 x 0.7	12.8	S0	75	Multi-galaxy system; small companion lies just S of larger nucleus.
01 19 55.2	+16 32 40	NGC 473	12.5	1.7 x 1.1	13.0	SAB(r)0/a:	153	Mag 9.28v star SE 5′.1.
01 20 02.2	+14 51 36	NGC 475	15.0	0.4 x 0.4	12.8			Stellar.
01 20 19.9	+16 01 11	NGC 476	14.4	0.7 x 0.5	13.1	S0	102	Almost stellar nucleus.
01 21 27.8	+07 01 02	NGC 485	13.1	1.7 x 0.5	12.8	S	3	Mag 8.82v star SSW 3′.7.
01 22 10.6	+05 24 41	NGC 486	14.4	0.4 x 0.4	12.3	Sb		NGC 492 NE 1′.0.
01 21 46.8	+05 15 13	NGC 488	10.3	5.2 x 3.9	13.4	SA(r)b I	15	Mag 10.26v star on SSW edge; mag 11.5 star superimposed SE of core.
01 21 53.8	+09 12 24	NGC 489	12.7	1.7 x 0.4	12.1	S?	120	**UGC 896** WSW 6′.9; **CGCG 411-38** NE 10′.3.
01 22 02.9	+05 21 59	NGC 490	14.4	0.7 x 0.6	13.3	S?	9	Faint, stellar anonymous galaxy NW 8′.0.
01 22 13.6	+05 25 01	NGC 492	16.2	0.7 x 0.6	15.1	SBbc	120	Very small, faint anonymous galaxy NW 4′.9; NGC 486 SW 1′.0.
01 22 39.4	+05 23 12	NGC 500	14.2	0.8 x 0.6	13.5	E/S0	102	Mag 10.9 star N 1′.4.
01 22 55.4	+09 02 55	NGC 502	12.8	1.1 x 1.0	12.8	SA(r)0°	66	Faint, stellar anonymous galaxy W 4′.3; stellar **Mkn 568** W 7′.1.
01 22 57.0	+09 28 05	NGC 505	13.8	1.0 x 0.7	13.3	S0	57	Almost stellar nucleus in faint envelope.
01 23 23.9	+09 25 58	NGC 509	13.4	1.6 x 0.6	13.2	S0?	85	Mag 9.67v star SE 7′.6.
01 23 30.7	+11 17 26	NGC 511	13.7	1.1 x 1.1	14.0	E:		Several faint stars on SW inside edge.
01 24 04.0	+12 55 01	NGC 514	11.7	3.5 x 2.8	14.0	SAB(rs)c II	110	A mag 9.46v star lies E 3′.0.
01 24 08.1	+09 33 03	NGC 516	13.1	1.4 x 0.5	12.6	S0	44	Mag 10.21v star NE 6′.5.
01 24 17.8	+09 19 46	NGC 518	13.3	1.7 x 0.6	13.1	Sa: sp	98	Mag 10.7v star lies 2′.4 NE.
01 24 46.1	+09 59 45	NGC 522	12.9	2.7 x 0.4	12.9	Sbc: sp	33	IC 101 SW 9′.9.
01 24 47.7	+09 32 20	NGC 524	10.3	2.8 x 2.8	12.4	SA(rs)0$^+$		Large core in smooth halo; mag 10.08v star SW 6′.3.
01 24 53.0	+09 42 10	NGC 525	13.2	1.5 x 0.6	13.0	S0	8	**UGC 969** NNW 3′.9.
01 25 17.4	+09 15 57	NGC 532	12.9	3.1 x 1.0	14.0	Sab? sp	28	
01 29 07.3	+11 07 48	NGC 569	13.7	1.0 x 0.5	12.8	S?	163	**UGC 1065** NE 1′.1.
01 36 41.1	+15 47 16	NGC 628	9.4	10.5 x 9.5	14.2	SA(s)c I	25	= **M 74**. Many knots and superimposed stars.
01 36 47.2	+05 50 05	NGC 631	13.3	1.0 x 0.8	13.1	E	123	Compact nucleus, uniform envelope.
01 37 17.6	+05 52 39	NGC 632	12.4	1.0 x 0.8	12.0	S0?	170	Superimposed star or bright knot on NE edge of core.
01 39 37.8	+07 14 13	NGC 638	13.8	0.8 x 0.5	12.7	S?	20	
01 40 09.0	+05 43 29	NGC 645	12.6	2.6 x 1.2	13.7	SBb:	125	Lies 3′.2 SE of mag 9.40v star.
01 40 43.3	+07 58 57	NGC 652	13.7	1.0 x 0.6	12.9	S?	55	
01 42 09.6	+12 36 06	NGC 658	12.5	3.0 x 1.6	14.0	Sb II-III	23	
01 43 02.0	+13 38 38	NGC 660	11.2	8.3 x 3.2	14.6	SB(s)a pec	170	Light dust lane extends the length of the elongated NE-SW nucleus.
01 44 56.1	+10 25 24	NGC 665	12.1	2.4 x 1.6	13.5	(R)S0°?	125	Mag 10.23v star W 6′.4; **MCG +2-5-22** NE 2′.0 from center.
01 46 59.1	+13 07 29	NGC 671	13.3	1.5 x 0.5	12.8	S?	55	
01 48 22.5	+11 31 15	NGC 673	12.6	2.1 x 1.6	13.8	SAB(s)c	0	Two main N-S branching arms; mag 10.13v star ENE 3′.3.
01 49 08.7	+13 03 31	NGC 675	14.5	1.0 x 0.4	13.4	S	96	Located on W edge of NGC 677.
01 48 57.4	+05 54 28	NGC 676	11.9	4.0 x 1.0	13.3	S0/a: sp	172	
01 49 14.0	+13 03 19	NGC 677	12.2	2.0 x 2.0	13.7	E		Compact core, faint, smooth envelope; **Mkn 1166** NW 2′.8; NGC 675 on W edge.
01 49 46.8	+11 42 07	NGC 683	13.6	1.0 x 1.0	13.4	S?		Elongated **MCG +2-5-48** sits on E edge.
01 50 30.8	+06 08 40	NGC 693	12.4	2.1 x 1.0	13.1	S0/a?	106	Lies 1′.4 W of mag 10.38v star.
01 51 50.8	+06 17 45	NGC 706	12.5	1.9 x 1.4	13.4	Sbc?	147	Mag 12.7 star on N edge.
00 33 47.3	+07 14 52	UGC 331	14.1	0.9 x 0.6	13.3	SBa	176	UGC 335 NE 2′.7.
00 33 56.5	+07 16 20	UGC 335	13.2	2.0 x 0.8	13.8	E	143	Double system, contact; UGC 331 SW 2′.7.
00 37 58.0	+05 08 50	UGC 379	15.3	1.5 x 0.2	13.8	Scd:	120	
00 38 24.7	+13 29 09	UGC 385	14.1	2.1 x 0.7	14.4	S?	65	Almost stellar nucleus; stellar **Mkn 342** NNW 3′.1.
00 38 23.7	+15 02 22	UGC 386	13.6	1.2 x 0.6	13.2	S0	170	Almost stellar nucleus, smooth envelope.
00 38 39.8	+17 24 14	UGC 393	15.2	1.8 x 0.4	14.7	Sdm:	170	Mag 12.0 star S 2′.2.
00 39 37.8	+08 57 54	UGC 418	14.1	1.9 x 0.2	12.9	Sb III	98	Mag 11.6 star S 1′.5; UGC 422 N 2′.7.
00 39 43.2	+09 00 16	UGC 422	13.7	0.8 x 0.7	12.9	SB?	171	UGC 418 SSW 2′.7.
00 40 30.6	+06 55 00	UGC 429	13.9	1.2 x 0.7	13.6	S0:	120	Mag 10.24v star NW 2′.9.
00 47 19.4	+14 42 09	UGC 488	14.0	0.6 x 0.5	12.6	Sab	90	
00 48 48.3	+11 22 07	UGC 500	14.2	0.8 x 0.7	13.4	SABbc	0	
00 50 03.0	+07 54 59	UGC 512	14.2	0.8 x 0.4	12.8	Sdm:	135	Mag 14.7 star NE edge.
00 51 12.2	+12 01 25	UGC 521	14.4	0.6 x 0.4	12.7	Im	50	Low surface brightness **UGC 515** NW 8′.4.
00 52 28.0	+14 31 04	UGC 533	14.3	1.2 x 0.4	13.3	Scd:	120	Mag 12.0 star N 2′.2.
00 53 35.0	+12 41 33	UGC 545	14.0	0.5 x 0.5	12.3	S?		Almost stellar nucleus.
00 54 49.4	+10 32 13	UGC 558	14.1	1.4 x 0.5	13.6	S0:	37	
00 55 58.7	+12 07 24	UGC 571	14.2	0.9 x 0.4	12.9	SB(s)b	177	Stellar nucleus; MCG +2-3-13 W 7′.9.
00 56 02.2	+14 13 46	UGC 572	14.4	0.8 x 0.5	13.3	Scd:	140	Faint star or knot on SE edge.
00 57 39.9	+08 05 31	UGC 596	14.7	0.6 x 0.6	13.4	Compact		
00 58 25.5	+11 33 24	UGC 603	14.6	0.6 x 0.4	12.9	S?	159	Disrupted by small companion on NE edge; **CGCG 435-28** NE 2′.0.
00 58 45.1	+12 44 52	UGC 607	14.3	0.8 x 0.7	13.5	Scd?	21	Strong knot N edge.
00 58 51.2	+12 58 21	UGC 610	12.9	1.4 x 1.1	13.5	E	69	Close pair of very faint, small anonymous galaxies NNW 1′.5.
00 59 40.3	+15 19 49	UGC 615	13.3	0.8 x 0.5	12.2	SABab	23	Mag 13.8 star N edge.
01 01 00.7	+13 28 04	UGC 627	13.9	1.0 x 0.6	13.2	S?	115	
01 01 02.6	+09 43 28	UGC 631	15.3	1.5 x 0.2	13.8	Scd:	137	Mag 12.7 star NW 1′.2.
01 01 25.2	+07 37 33	UGC 634	14.5	1.6 x 1.0	14.9	SABm: V	35	Dwarf spiral, extremely low surface brightness.
01 02 15.5	+09 05 53	UGC 640	14.4	0.7 x 0.7	13.5	Sb III		Mag 13.1 star S 1′.6.
01 03 18.0	+14 02 12	UGC 644	13.6	0.8 x 0.3	11.9	S Irr	60	MCG +2-3-27 SE 0′.7.
01 06 25.7	+14 34 14	UGC 677	14.1	0.7 x 0.5	12.8	S?	114	Triangle of faint stars on NE edge.
01 07 22.5	+16 41 03	UGC 685	13.6	1.7 x 1.2	14.3	SAm	120	
01 08 25.3	+06 28 57	UGC 705	15.3	1.0 x 0.7	14.7	SB(r)c	129	
01 08 24.5	+08 20 53	UGC 706	14.0	1.0 x 0.7	13.5	SB(r)ab	21	**UGC 702** NNW 4′.0.
01 08 34.5	+16 53 57	UGC 708	14.0	0.8 x 0.5	12.9	SBcd:	48	Mag 12.8 star S 1′.6.
01 09 22.0	+14 20 31	UGC 717	13.6	1.5 x 0.7	13.5	SBb	160	Almost stellar anonymous galaxy SE 1′.1; UGC 719 NE 2′.1.
01 09 28.9	+14 21 45	UGC 719	14.1	0.7 x 0.5	12.8	SBb	24	UGC 717 SW 2′.1; very small, very faint, elongated anonymous galaxy S edge.
01 09 45.6	+13 18 29	UGC 722	15.4	1.7 x 1.1	15.9	SB(s)dm	150	**UGC 716** SW 10′.7; mag 8.10v star W 8′.4.
01 10 03.2	+13 58 39	UGC 727	15.6	0.8 x 0.8	15.0	Multiple System		Stellar nucleus; mag 12.1 star E 2′.0.
01 10 39.1	+16 35 49	UGC 733	14.6	0.7 x 0.5	13.3	SB(s)b	57	Mag 13.6 star W edge; pair of stars, mags 10.44v and 10.36v, NE 2′.6.
01 11 08.6	+08 46 31	UGC 741	13.9	0.8 x 0.8	13.5	E:		
01 12 00.7	+17 18 30	UGC 751	14.2	1.2 x 1.0	14.5	E/S0	63	Mag 13.5 star on E edge; almost stellar anonymous galaxy W 1′.9; larger anonymous galaxy S 2′.4.
01 13 51.5	+13 16 16	UGC 774	13.9	0.8 x 0.5	12.7	S?	86	Double galaxy, connected.
01 14 14.6	+15 53 51	UGC 785	13.7	1.2 x 0.8	13.5	(R)SAB0°:	155	

(Continued from previous page)

GALAXIES

RA h m s	Dec ° ′ ″	Name	Mag (V)	Dim ′ Maj x min	SB	Type Class	PA	Notes
01 15 23.0	+08 05 56	UGC 803	14.5	1.7 x 1.4	15.3	SAdm	18	Mag 8.95v star SW 5′.2.
01 15 38.2	+09 18 22	UGC 808	13.5	2.2 x 1.5	14.6	SA(r)c	60	
01 16 59.4	+13 01 27	UGC 824	14.0	1.2 x 0.2	12.3	Sa	73	
01 18 08.9	+11 22 53	UGC 833	13.1	1.9 x 1.1	13.7	SB(r)cd	50	Mag 12.7 star NW 1′.4.
01 18 56.8	+05 33 22	UGC 834	14.4	0.7 x 0.7	13.5	SAB(s)bc		Mag 13.5 star SW 0′.9.
01 18 46.1	+14 59 32	UGC 838	13.4	0.7 x 0.5	12.2	S?	171	Close pair of mag 12.0, 12.6 stars NE 6′.2.
01 19 24.3	+12 26 42	UGC 849	14.1	1.2 x 0.6	13.6	Sdm:	126	Irregular, distorted; MCG +2-4-22 N.
01 19 37.3	+08 10 22	UGC 855	13.8	0.9 x 0.8	13.3	SB(rs)b	33	
01 19 59.2	+12 55 27	UGC 860	14.4	0.8 x 0.8	13.8	Scd:		
01 20 24.2	+05 49 40	UGC 871	15.5	1.8 x 1.6	16.5	Im	162	Dwarf irregular, very low surface brightness.
01 20 55.5	+06 34 28	UGC 882	14.5	1.3 x 0.8	14.4	Im	85	Almost uniform surface brightness; **MCG +1-4-29** SSE 8′.9.
01 21 00.9	+17 04 25	UGC 883	14.3	1.1 x 0.8	14.0	Sdm IV-V	36	Uniform surface brightness; close pair of stars, mags 12.5 and 13.5, N 1′.
01 21 08.6	+05 08 21	UGC 887	14.3	0.7 x 0.6	13.2	Sbc	132	Mag 10.66v star on NW edge.
01 21 18.7	+12 24 41	UGC 891	14.1	2.4 x 0.8	14.6	SABm: IV-V	47	
01 21 28.8	+15 56 45	UGC 897	15.2	0.7 x 0.7	14.3	Sbc		Stellar nucleus.
01 21 47.8	+17 35 33	UGC 903	13.6	1.8 x 0.4	13.1	S?	52	
01 21 58.3	+15 47 35	UGC 910	13.9	1.0 x 0.9	13.7	SA(s)c	0	Located 1′.0 W of mag 7.35v star.
01 23 23.6	+15 12 49	UGC 933	14.2	0.9 x 0.8	13.6	SBa	48	
01 24 34.2	+16 32 08	UGC 958	14.5	1.7 x 0.3	13.6	SBcd:	26	
01 25 23.8	+14 31 11	UGC 985	14.1	1.7 x 0.5	13.8	Sb III	19	Mag 13.3 star N 1′.8.
01 25 29.7	+07 33 40	UGC 989	15.4	1.5 x 1.2	15.9	SAB(s)m	15	
01 25 35.8	+07 59 24	UGC 993	14.7	0.9 x 0.4	13.4	Double System		Double system, disrupted.
01 26 23.7	+06 16 35	UGC 1014	14.1	1.0 x 1.0	14.0	Sm		Uniform surface brightness; mag 11.3 star W 0′.9.
01 26 44.2	+17 15 50	UGC 1020	13.9	1.2 x 0.3	12.7	S?	52	
01 26 48.8	+12 01 25	UGC 1023	14.1	0.9 x 0.4	12.9	S0	60	Mag 11.1 star SE 1′.8.
01 26 55.0	+13 02 38	UGC 1024	14.7	0.6 x 0.4	13.0		96	
01 27 12.9	+13 36 14	UGC 1026	13.9	1.8 x 1.4	14.8	Sm:	135	Mag 12.9 star SW 1′.2.
01 27 41.5	+12 13 04	UGC 1041	14.7	1.0 x 0.6	14.0		30	Dimensions include four small galaxies, part of a chain of six extending N.
01 28 47.7	+16 41 18	UGC 1056	14.2	0.8 x 0.5	13.1	Im:	48	Very faint star on S edge.
01 28 53.3	+13 47 33	UGC 1057	13.8	1.5 x 0.5	13.4	Sbc	153	Mag 9.06v star N 2′.7.
01 30 58.3	+14 06 29	UGC 1083	13.6	1.5 x 0.7	13.5	Sb III	125	
01 31 26.7	+14 16 37	UGC 1087	13.8	1.2 x 1.2	14.0	SA(rs)c		
01 31 51.4	+17 33 51	UGC 1093	14.0	1.2 x 0.7	13.7	Sb	169	Brightest of three; first companion NW 0′.9; second companion ESE 1′.5.
01 33 17.8	+13 19 54	UGC 1110	14.2	1.8 x 0.5	13.9	Scd:	174	
01 33 28.3	+17 24 26	UGC 1113	15.3	0.8 x 0.8	14.7	Scd:		
01 33 50.4	+17 13 48	UGC 1119	15.9	1.1 x 0.6	15.3	Sdm	125	Stellar nucleus; mag 12.2 star SE 2′.3.
01 34 52.7	+12 04 37	UGC 1129	13.8	1.0 x 0.6	13.2	S0:	83	
01 35 31.5	+11 57 09	UGC 1139	13.9	0.8 x 0.6	12.9	SB?	0	
01 36 04.5	+11 41 24	UGC 1144	15.3	1.5 x 0.6	15.0	Scd:	3	Mag 11.0 star NW 1′.6.
01 38 21.0	+07 32 02	UGC 1167	13.3	1.7 x 1.3	14.0	SA(rs)cd	120	Mag 9.65v star SE 4′.6.
01 40 10.2	+15 54 26	UGC 1176	13.8	4.6 x 3.7	16.8	Im V	33	Dwarf irregular, extremely low surface brightness.
01 42 06.4	+07 39 46	UGC 1191	15.1	0.7 x 0.7	14.2	Double System		Double system; **MCG +1-5-22** SW 0′.9.
01 42 27.2	+13 58 37	UGC 1195	12.8	3.2 x 1.0	14.0	Im:	50	
01 42 48.5	+13 09 18	UGC 1200	13.2	1.6 x 0.9	13.5	IBm:	170	Stellar anonymous galaxy N 1′.5.
01 42 53.7	+05 50 10	UGC 1203	15.6	1.5 x 0.7	15.5		35	
01 43 21.1	+11 54 09	UGC 1206	14.4	1.7 x 0.8	14.6		30	
01 43 45.0	+12 09 45	UGC 1209	14.1	1.6 x 0.7	14.1	Sdm:	50	Bright, elongated center, faint envelope NW-SW.
01 43 55.3	+13 48 23	UGC 1211	17.0	2.4 x 1.4	18.2	Im:	18	Dwarf, low surface brightness; mag 9.97v star W 1′.5.
01 44 20.7	+17 28 39	UGC 1219	12.8	1.3 x 0.6	12.3	SB?	102	
01 45 55.0	+09 49 22	UGC 1237	14.1	1.5 x 0.5	13.7	Sab	45	Elongated anonymous galaxy NNE 2′.2; mag 10.57v star E 3′.0.
01 46 10.3	+06 25 51	UGC 1239	14.1	0.9 x 0.5	13.1	Sdm:	155	
01 46 37.1	+14 41 28	UGC 1242	14.6	1.0 x 0.7	14.0	Sm:	141	
01 46 50.8	+11 24 04	UGC 1245	14.1	1.3 x 1.0	14.2	S	150	Bright center, uniform envelope; very faint anonymous galaxy with stellar nucleus NE 2′.6.
01 47 00.7	+12 24 17	UGC 1246	13.9	1.2 x 0.9	13.9	IAm	21	Numerous faint and bright knots overall.
01 47 30.0	+12 06 15	UGC 1253	14.2	1.2 x 0.5	13.5	S?	12	
01 48 33.3	+12 36 49	UGC 1260	13.1	1.0 x 0.7	12.6	(R′)SB(s)a	126	Mag 10.4 star NW 2′.5.
01 48 33.6	+13 25 49	UGC 1261	14.5	1.5 x 0.4	13.8	Scd:	92	**UGC 1279** ESE 13′.0.
01 49 00.1	+13 12 35	UGC 1271	13.1	1.7 x 0.9	13.4	SB0	113	
01 49 11.7	+12 51 10	UGC 1274	13.9	1.5 x 0.4	13.2	Sa:	108	Mag 13.9 star W 1′.3.
01 49 29.9	+12 30 32	UGC 1282	13.1	1.5 x 0.7	13.0	S0/a	55	
01 50 03.2	+11 48 45	UGC 1290	14.1	0.7 x 0.5	12.8	Sbc	140	
01 51 02.3	+13 17 45	UGC 1312	13.9	1.5 x 0.5	13.4	S?	93	Mag 11.9 star NE 1′.6.
01 51 29.9	+13 07 52	UGC 1322	14.1	1.2 x 1.1	14.3	SA(s)c	159	Two very faint, almost stellar anonymous galaxies outside NE and E edges.
01 51 37.1	+08 15 22	UGC 1325	12.6	1.8 x 1.8	13.8	E		Bright core, uniform envelope; UGC 1326 N 2′.7.
01 51 40.2	+08 17 59	UGC 1326	13.6	0.8 x 0.8	13.2	E:		UGC 1325 S 2′.7.
01 51 52.0	+17 03 05	UGC 1328	13.6	1.5 x 1.0	13.9	SAB(rs)c:	125	Located 2′.0 SW of mag 8.20v star.

GALAXY CLUSTERS

RA h m s	Dec ° ′ ″	Name	Mag 10th brightest	No. Gal	Diam ′	Notes
00 39 48.0	+06 46 00	A 76	15.0	42	28	Numerous stellar anonymous GX over entire area.
01 11 00.0	+17 39 00	A 154	15.6	66	13	**IC 1634** and **PGC 4329** at center with numerous PGC and anonymous GX mostly S of center.
01 11 48.0	+16 52 00	A 158	15.9	46	17	Two faint, almost stellar anonymous GX at center as seen on DSS image; all other members stellar.
01 12 54.0	+15 30 00	A 160	15.7	34	22	**MCG +2-4-10** at cluster center; **MCG +2-4-11** ENE 4′.2; **CGCG 436-11** NNW 3′.6.
01 16 48.0	+16 15 00	A 171	15.9	42	29	**MCG +3-4-14** E of center 7′.3.
01 25 06.0	+08 41 00	A 193	16.0	58	16	Most members stellar, see notes for IC 1695.
01 41 54.0	+07 37 00	A 240	15.6	43	17	Numerous stellar anonymous GX tightly packed around the center, filling about one third the diameter of A 240.
01 44 06.0	+06 23 00	A 245	16.4	40	16	All members faint and stellar.
01 44 42.0	+05 48 00	A 246	16.4	56	25	Difficult to see many GX, even on DSS image.

PLANETARY NEBULAE

RA h m s	Dec ° ′ ″	Name	Diam ″	Mag (P)	Mag (V)	Mag cent ★	Alt Name	Notes
00 59 53.7	+15 44 00	PN G125.9−47.0	261			12.1	PK 125−47.1	Slightly elongated NNE-SSW; mag 11.89v star on E edge; mag 12.5 star W 4′.0.

RA h m s	Dec ° ′ ″	Name	Mag (V)	Dim ′ Maj x min	SB	Type Class	PA	Notes
23 32 14.5	+02 24 29	CGCG 381-2	14.2	0.3 x 0.3	11.6	E?		= **Mkn 536**. Mag 9.36v star S 3′.3.
23 53 18.8	+03 43 00	CGCG 381-57	15.0	0.4 x 0.3	12.5		147	Mag 11.7 star ENE 2′.6.
00 16 39.1	+01 35 53	CGCG 382-39	14.3	0.5 x 0.2	11.7	Sc	81	Located 0′.7 S of the center of MCG +0-1-55.
23 30 43.2	+03 58 00	CGCG 406-122	14.8	0.3 x 0.3	12.0	Compact		Almost stellar.
23 40 51.0	+04 35 15	CGCG 407-40	14.0	0.8 x 0.4	12.6		147	Mag 12.1 star SE 2′.0.
00 34 16.7	+05 32 49	CGCG 409-37	14.0	0.4 x 0.4	11.9	Sb		Mag 13.5 star on SE edge.
00 12 06.1	−00 24 54	IC 3	13.8	0.8 x 0.4	12.7	E/S0	55	Mag 10.07v star E 7′.1.
00 18 54.9	−03 16 36	IC 6	13.5	0.8 x 0.8	13.1	E		Almost stellar; IC 8 NNE 3′.8.
00 19 02.8	−03 13 23	IC 8	14.4	0.8 x 0.4	13.0	Sc	138	Mag 11.6 star N 3′.9; IC 6 SSW 3′.8.
00 28 30.0	+02 38 53	IC 17	15.1	0.5 x 0.5	13.4	S II		Mag 9.60v star E 10′.6.
00 29 10.6	−00 09 48	IC 21	15.2	0.5 x 0.4	13.3	Sab		Mag 10.13v star ENE 7′.6.
00 31 12.1	−00 24 25	IC 25	14.3	1.0 x 0.9	14.1	S0?	39	Mag 9.10v star W 10′.5.
00 34 10.8	−02 10 38	IC 29	15.1	0.7 x 0.7	14.1			Mag 10.74v star SW 6′.3; IC 30 N 5′.7.
00 34 14.8	−02 05 05	IC 30	15.9	0.7 x 0.4	14.3		21	IC 29 S 5′.7.
00 35 01.7	−02 08 32	IC 32	14.9	0.4 x 0.4	12.8			Pair with IC 33 E 0′.9.
00 35 05.1	−02 08 18	IC 33	14.5	0.7 x 0.5	13.3	S0		Pair with IC 32 W 0′.9; mag 9.51v star NE 8′.1.
00 39 21.4	+02 27 20	IC 40	14.4	1.1 x 0.5	13.6	S	13	Mag 10.36v star NW 9′.7.
23 20 49.7	+01 44 16	IC 1482	14.4	0.6 x 0.4	12.7		123	Almost stellar; mag 10.6 star NE 8′.0.
23 30 36.2	−03 02 25	IC 1492	13.3	1.0 x 0.8	12.9	SB0$^+$?	36	Mag 9.58v star NW 5′.6.
23 30 53.5	−02 56 05	IC 1496	13.4	1.7 x 1.3	14.1	(R)SB(rs)0/a	85	Mag 9.58v star WSW 9′.6.
23 31 53.7	−05 00 29	IC 1498	13.0	1.8 x 0.6	12.9	(R′)SB(r)a:	11	Mag 9.43v star WSW 9′.5.
23 34 40.1	−03 09 12	IC 1501	13.8	1.5 x 0.7	13.7	SAB(s)bc pec: II-III	136	Mag 10.02v star SW 11′.5.
23 38 27.0	+04 47 57	IC 1503	13.7	1.0 x 0.5	12.8	Sdm:	65	Mag 112.4 star S 3′.5.
23 41 19.5	+04 01 00	IC 1504	13.5	1.7 x 0.5	13.1	Sb	88	Mag 11.4 star N 6′.8.
23 41 37.0	−03 33 55	IC 1505	13.7	0.9 x 0.7	13.3	E	156	Mag 9.71v star ESE 9′.3.
23 44 48.5	+04 44 05	IC 1506	14.5	0.7 x 0.4	13.0		126	Mag 11.6 star SE 2′.2.
23 45 33.2	+01 41 19	IC 1507	13.6	1.4 x 0.6	13.2	S0°	134	Mag 8.29v star E 3′.8.
23 50 32.9	+02 04 21	IC 1510	14.7	0.7 x 0.6	13.6		156	Mag 11.08v star NNW 2′.7.
23 56 04.0	−00 59 20	IC 1515	13.5	1.0 x 0.8	13.2	(R′)SB(rs)ab	13	Mag 12.2 star on W edge; mag 9.04v star NE 6′.8; IC 1516 N 4′.4.
23 56 07.3	−00 55 02	IC 1516	13.1	1.7 x 1.6	14.1	Sbc pec II	65	Mag 9.04v star E 5′.2; IC 1515 S 4′.4.
23 56 18.8	−00 18 21	IC 1517	13.8	0.8 x 0.8	13.4	E		Mag 10.39v star W 5′.4.
23 59 03.5	+01 43 10	IC 1522	15.5	1.0 x 0.3	14.1	Sbc	12	Mag 10.00v star WNW 6′.5.
23 59 10.7	−04 07 43	IC 1524	14.1	1.7 x 0.7	14.1	SA(rs)b pec: II	80	Mag 9.92v star W 8′.9; MCG −1-1-12 S 3′.8.
00 02 21.8	+04 05 24	IC 1527	14.6	1.3 x 0.7	14.4	Sb	147	Mag 13.5 star on S edge; mag 12.3 star NW 7′.2.
23 34 36.5	−04 32 06	IC 5334	13.3	1.8 x 0.6	13.3	(R′)	127	Mag 11.6 star SW 9′.6.
23 47 19.0	−02 18 50	IC 5351	13.6	0.5 x 0.4	12.0	E?	177	= **Hickson 97D**.
23 47 24.0	−02 21 05	IC 5356	14.1	0.8 x 0.4	12.7	S?	33	= **Hickson 97C**.
23 47 23.1	−02 18 05	IC 5357	12.9	0.9 x 0.5	12.0	SAB0$^-$ pec:	150	= **Hickson 97A**; **Hickson 97E** N 1′.3; mag 10.44v star E 2′.3; UGC 5359 = **Hickson 97B** is E 3′.8.
00 01 04.5	+04 30 00	IC 5374	14.3	0.5 x 0.4	12.4	S?	18	Mag 10.58v star SW 9′.5; IC 5375 N 2′.4.
00 01 04.8	+04 32 24	IC 5375	14.3	0.8 x 0.3	12.6	S?	0	IC 5374 S 2′.4.
23 56 02.7	−01 29 06	MCG +0-1-3	14.6	0.7 x 0.7	13.7	S0		
23 56 13.8	−00 32 25	MCG +0-1-7	14.8	0.8 x 0.4	13.4		21	
23 59 22.2	+01 52 32	MCG +0-1-13	15.0	1.3 x 1.1	15.2		111	Strong, slightly elongated core, smooth envelope.
00 00 04.4	−00 04 59	MCG +0-1-14	15.1	0.2 x 0.2	11.6	E		Mag 10.37v star NW 1′.6.
00 00 07.8	−00 02 26	MCG +0-1-15	14.9	0.3 x 0.3	12.4	E		Mag 10.67v star S 1′.5.
00 00 12.9	+01 07 08	MCG +0-1-16	14.4	0.9 x 0.5	13.4	Irr pec	147	Mag 9.3 star SW 3′.7.
00 00 35.6	−01 45 49	MCG +0-1-17	14.1	0.8 x 0.5	13.0		141	
00 03 17.4	+03 37 17	MCG +0-1-22	14.5	0.9 x 0.3	12.9	Sb	147	
00 04 07.2	−00 51 46	MCG +0-1-23	15.4	1.3 x 0.8	15.3	Sd	36	
00 04 48.9	−01 29 57	MCG +0-1-24	14.2	0.6 x 0.4	12.5	S?	21	
00 04 47.2	−01 34 17	MCG +0-1-25	14.2	0.9 x 0.4	12.9	S?	57	
00 10 39.4	−00 03 10	MCG +0-1-34	14.8	0.7 x 0.7	13.9	Sc		
00 11 06.3	+02 40 39	MCG +0-1-35	14.5	0.8 x 0.7	13.7	S?	30	Mag 8.92v star SE 2′.4.
00 11 39.7	−00 28 27	MCG +0-1-36	14.6	0.4 x 0.4	12.5			Close N-S oriented pair of mag 12.0, 11.30v stars N 4′.7.
00 14 55.1	+00 15 04	MCG +0-1-46	14.7	0.4 x 0.4	12.6	Sb		Compact; mag 10.31v star SW 1′.5.
00 15 21.5	+02 33 10	MCG +0-1-47	15.9	0.7 x 0.4	14.4		162	Mag 11.00v star N edge.
00 16 36.9	+01 32 01	MCG +0-1-53	15.1	0.6 x 0.4	13.4	Sbc	108	MCG +0-1-55 and CGCG 382-39 N 4′.6.
00 16 39.7	+01 36 38	MCG +0-1-55	14.3	1.1 x 1.1	14.4	Sc(r)		CGCG 382-39 S edge; MCG +0-1-53 S 4′.6.
00 16 54.9	−00 05 16	MCG +0-1-56	14.9	1.0 x 0.7	14.3	Sb	177	
00 17 08.8	−00 57 28	MCG +0-1-57	15.0	0.8 x 0.5	13.8		150	
00 17 25.9	−00 58 44	MCG +0-1-58	14.1	1.1 x 0.9	14.0	Sb	144	
00 17 55.4	+00 20 06	MCG +0-1-59	15.0	0.8 x 0.6	14.0		39	Mag 14.8 star SW edge; mag 8.82v star N 2′.5.
00 19 41.8	+00 22 31	MCG +0-2-1	15.7	0.9 x 0.2	13.7	Sc	138	
00 19 47.4	+00 35 23	MCG +0-2-2	15.4	0.6 x 0.3	13.4		165	Mag 12.1 star ESE 1′.9.
00 21 33.9	+03 35 36	MCG +0-2-11	15.6	0.9 x 0.2	13.6	Sb	3	Mag 11.4 star SW 2′.0.
00 22 21.0	−00 56 03	MCG +0-2-13	14.9	0.8 x 0.7	14.1	Sab		
00 23 21.2	+01 50 44	MCG +0-2-16	15.1	0.7 x 0.4	13.5	Sc	159	Mag 14.1 star SE edge.
00 24 19.1	−01 38 18	MCG +0-2-20	14.3	0.6 x 0.6	13.0			
00 24 55.5	+00 31 45	MCG +0-2-23	15.0	0.6 x 0.4	13.4	SBbc	24	
00 24 54.2	−01 25 29	MCG +0-2-24	15.6	0.5 x 0.3	13.4		132	
00 25 15.2	−01 06 44	MCG +0-2-25	14.5	1.1 x 1.1	14.6			
00 27 40.3	+00 30 53	MCG +0-2-34	15.2	0.5 x 0.5	13.6	Sc		
00 28 38.2	+02 57 13	MCG +0-2-45	15.1	0.9 x 0.2	13.1	Sbc: sp	57	Mag 11.03v star NW 2′.5.
00 28 48.1	+03 25 46	MCG +0-2-47	14.3	0.6 x 0.4	12.6		159	
00 29 28.5	−00 12 45	MCG +0-2-54	14.9	0.8 x 0.5	13.8	Sa		Mag 7.83v star SE 8′.6; **MCG +0-2-61** ESE 5′.2.
00 29 38.5	+00 24 35	MCG +0-2-57	15.0	0.7 x 0.4	13.5	Spiral	21	
00 29 48.1	+01 51 14	MCG +0-2-58	15.2	0.6 x 0.6	13.9	Sd		Mag 11.5 star NE 2′.5.
00 29 47.1	+00 09 36	MCG +0-2-60	16.2	0.5 x 0.3	14.0	SBa	147	Small, faint anonymous galaxy NW 0′.8.
00 32 21.7	−01 45 35	MCG +0-2-65	15.0	0.6 x 0.4	13.3	Sb	84	Mag 7.07v star W 10′.5.
00 32 42.2	−01 49 04	MCG +0-2-68	15.1	0.7 x 0.4	13.6		12	
00 33 14.9	+00 12 01	MCG +0-2-69	15.6	0.7 x 0.4	14.1	Spiral	42	
00 34 42.6	+02 25 28	MCG +0-2-78	15.0	0.7 x 0.5	13.7		24	
00 36 25.0	+02 25 29	MCG +0-2-90	15.2	1.0 x 0.4	14.0	Sbc	15	Mag 14.4 star on NW edge.
00 37 35.8	+00 16 49	MCG +0-2-94	14.4	0.7 x 0.5	13.1	S?	111	= **Mkn 955**.
00 39 20.0	−02 08 16	MCG +0-2-108	14.8	0.9 x 0.6	13.9	Sbc	66	
00 40 00.7	−01 43 05	MCG +0-2-117	14.9	0.6 x 0.6	13.6			
23 20 07.4	−00 21 00	MCG +0-59-25	15.4	0.8 x 0.4	14.0	Sm	39	

RA h m s	Dec ° ′ ″	Name	Mag (V)	Dim ′ Maj x min	SB	Type Class	PA	Notes
23 20 29.1	−01 00 10	MCG +0-59-28	14.1	0.5 x 0.5	12.4			
23 21 39.1	+01 26 43	MCG +0-59-32	15.0	0.9 x 0.6	14.2		135	NGC 7629 SW 5′.6.
23 21 51.8	−00 41 27	MCG +0-59-33	14.3	0.7 x 0.3	12.4	Sb	165	
23 22 05.4	+01 28 50	MCG +0-59-34	15.0	0.5 x 0.5	13.4	Sbc		Almost stellar core.
23 26 16.6	−01 58 28	MCG +0-59-43	14.7	0.6 x 0.5	13.2		162	
23 34 50.7	−01 12 21	MCG +0-60-11	15.3	0.5 x 0.3	13.1	Sd	165	Mag 13.1 star SW 2′.7.
23 35 25.7	+01 03 32	MCG +0-60-13	17.0	0.8 x 0.6	16.0		33	Mag 9.33v star S 5′.3.
23 36 06.9	−01 53 44	MCG +0-60-16	14.1	1.2 x 0.5	13.4	Sab	78	Mag 13.1 star N 2′.5.
23 37 07.8	−01 21 20	MCG +0-60-20	15.0	0.8 x 0.3	13.4	Sab	21	
23 37 13.9	−01 41 02	MCG +0-60-21	14.6	0.8 x 0.6	13.7	Sc	174	Located between two N-S mag 13 stars.
23 37 43.2	−00 31 03	MCG +0-60-23	14.7	0.9 x 0.4	13.5	Sb	54	
23 38 44.0	−01 28 19	MCG +0-60-25	15.0	0.9 x 0.4	13.7	Sb	150	Mag 12.9 star SW 0′.9.
23 38 51.9	+01 47 17	MCG +0-60-26	14.7	0.6 x 0.5	13.3		63	
23 39 25.0	−01 26 59	MCG +0-60-27	15.1	0.9 x 0.6	14.3	Irr	9	
23 40 05.4	−00 37 47	MCG +0-60-28	16.5	0.9 x 0.4	15.2	Sb	156	
23 40 20.0	+00 05 58	MCG +0-60-30	15.6	0.8 x 0.3	13.9	Sb	69	Mag 9.85v star SSW 3′.2.
23 40 29.8	+02 46 26	MCG +0-60-31	14.4	0.9 x 0.4	13.1	Sc	150	
23 40 33.0	−00 33 02	MCG +0-60-32	15.3	0.8 x 0.7	14.6	Sc		
23 41 07.9	−01 02 16	MCG +0-60-33	14.6	0.9 x 0.3	13.0	Sb	108	
23 42 41.4	+01 57 54	MCG +0-60-36	14.9	0.7 x 0.7	14.0	Sa		Mag 11.4 star N 1′.7.
23 43 57.8	+02 44 44	MCG +0-60-37	14.2	0.9 x 0.7	13.6	S?	21	= **Mkn 539**; mag 9.61v star S 4′.0.
23 44 22.5	+02 48 49	MCG +0-60-40	14.8	0.9 x 0.4	13.6	Sb	120	
23 45 05.1	−00 16 06	MCG +0-60-41	15.4	1.1 x 0.2	13.6	Sb	18	
23 47 05.5	+03 15 59	MCG +0-60-46	14.7	0.7 x 0.5	13.5		144	Mag 11.4 star SE 2′.0.
23 47 13.2	−00 28 52	MCG +0-60-48	15.1	0.9 x 0.7	14.5	SBc	30	Stellar **Mkn 540** NW 3′.8.
23 47 21.1	+01 55 59	MCG +0-60-49	14.1	0.7 x 0.7	13.1	SBc		
23 48 18.0	+02 20 29	MCG +0-60-50	14.8	0.8 x 0.3	13.1	Sbc	99	Mag 12.5 star SW edge.
23 48 29.9	−01 31 09	MCG +0-60-51	15.0	0.8 x 0.3	13.3	Sab	78	Mag 13.3 star SE 2′.5.
23 50 47.0	−02 11 18	MCG +0-60-54	14.6	0.9 x 0.7	13.9		33	Mag 13.7 star W 1′.8.
23 53 49.5	−00 58 27	MCG +0-60-57	14.6	0.7 x 0.2	12.3	Sb	30	
00 03 23.1	+05 42 14	MCG +1-1-16	14.9	0.7 x 0.7	14.0	SBc		Mag 11.1 star NW 3′.3.
00 06 47.8	+04 56 12	MCG +1-1-33	15.2	0.6 x 0.4	13.6		21	
00 26 28.9	+04 32 33	MCG +1-2-12	15.5	0.7 x 0.4	14.0	Sc	123	
00 28 26.8	+05 00 10	MCG +1-2-14	14.8	0.6 x 0.4	13.1	SB0?	3	Mag 12.6 star SW 2′.1.
00 30 28.7	+05 51 39	MCG +1-2-15	14.3	0.7 x 0.4	12.8		141	
00 31 09.5	+04 30 09	MCG +1-2-17	14.6	0.8 x 0.3	12.9	S?	60	Mag 11.9 star W 2′.2.
00 32 58.8	+05 17 07	MCG +1-2-23	15.0	0.6 x 0.5	13.6	Sb	12	MCG +1-2-24 S 1′.5.
00 33 00.9	+05 15 45	MCG +1-2-24	15.1	0.7 x 0.5	13.8	Sb	39	MCG +1-2-23 N 1′.1
00 37 55.7	+04 54 40	MCG +1-2-38	14.5	1.1 x 0.2	12.7	S?	174	Mag 12.8 star S 2′.8.
23 25 42.9	+05 04 35	MCG +1-59-78	15.4	0.5 x 0.5	13.7			
23 37 01.4	+04 56 09	MCG +1-60-9	14.6	0.7 x 0.5	13.3		66	
23 37 20.9	+04 54 16	MCG +1-60-11	14.1	1.0 x 0.6	13.4		6	Very slender, faint anonymous galaxy NE 1′.9.
23 55 01.7	−03 44 48	MCG −1-1-1	14.6	0.5 x 0.4	12.7	Sc	24	
23 55 13.7	−03 44 49	MCG −1-1-2	13.8	1.2 x 0.7	13.4	SB(rs)b: II-III	24	
23 55 41.9	−03 45 32	MCG −1-1-4	14.8	1.0 x 0.2	13.1	E	108	
23 56 46.2	−02 38 53	MCG −1-1-5	14.0	1.0 x 0.4	12.9	(R′)SB(r)a	18	Mag 9.08v star NE 4′.9.
23 56 49.4	−03 21 01	MCG −1-1-6	13.9	1.1 x 0.6	13.3	SAB(rs)c: II-III	123	
23 56 49.0	−03 23 12	MCG −1-1-7	14.3	0.8 x 0.4	12.9	Sc	165	
23 57 28.2	−02 54 31	MCG −1-1-8	13.6	0.9 x 0.5	12.6	SA(s)cd: III-IV	153	Weak stellar nucleus, smooth envelope.
23 58 57.4	−02 15 01	MCG −1-1-10	14.4	1.5 x 0.5	13.9	SB(s)bc: II	141	Mag 8.34v star NE 15′.3.
23 59 11.0	−04 11 32	MCG −1-1-12	14.1	1.4 x 0.3	13.0	SA(rs)b pec: II	72	Mag 10.88v star WSW 3′.2; IC 1524 N 3′.8.
00 00 05.1	−05 12 34	MCG −1-1-15	14.7	0.6 x 0.6	13.4	Sbc		
00 00 11.4	−05 09 33	MCG −1-1-17	14.2	0.8 x 0.4	12.8	Sbc	33	
00 00 21.5	−02 36 46	MCG −1-1-20	14.1	0.9 x 0.4	13.1	E$^+$:	123	Star at NW end.
00 00 48.5	−05 34 49	MCG −1-1-21	15.5	0.7 x 0.4	13.9	S0/a	90	
00 02 35.0	−03 42 40	MCG −1-1-24	13.7	0.9 x 0.5	12.7	SB(s)bc? I-II	177	
00 02 48.9	−03 36 22	MCG −1-1-26	14.4	0.9 x 0.3	12.8	Sbc	27	**MCG −1-1-25** SW 2′.9.
00 05 07.8	−03 48 25	MCG −1-1-31	15.1	0.9 x 0.2	13.1	Sb	105	Mag 7.59v star ESE 5′.4.
00 08 35.9	−05 13 01	MCG −1-1-36	14.3	1.1 x 0.5	13.5	(R′)Sa pec:	165	
00 08 47.9	−02 24 45	MCG −1-1-38	14.7	1.1 x 0.2	12.9		117	Mag 6.21v star W 9′.4; MCG −1-1-39 NE 3′.6.
00 09 00.3	−02 22 59	MCG −1-1-39	14.7	0.4 x 0.4	12.5			
00 10 10.1	−04 42 40	MCG −1-1-43	14.0	0.6 x 0.6	12.8	SBa/b		= **Mkn 937**.
00 10 36.8	−04 50 37	MCG −1-1-46	14.7	0.5 x 0.5	13.0	Sc		
00 12 30.0	−03 17 17	MCG −1-1-48	15.3	0.8 x 0.4	14.0		126	Mag 7.40v star SE 8′.4.
00 12 57.3	−05 31 29	MCG −1-1-49	15.1	0.7 x 0.4	13.6	SBb		
00 13 34.4	−05 05 35	MCG −1-1-52	13.8	1.2 x 0.6	13.3	SA(rs)b pec?	168	
00 13 47.9	−04 28 32	MCG −1-1-53	14.1	1.1 x 0.4	13.1	S?	144	
00 16 50.9	−05 16 08	MCG −1-1-64	14.0	1.5 x 0.5	13.6	SB(s)a pec:	42	
00 17 06.4	−04 36 40	MCG −1-1-66	13.9	1.3 x 0.3	12.7	S0$^+$: sp	126	
00 17 09.8	−03 42 50	MCG −1-1-67	14.2	0.7 x 0.7	13.5	E?		
00 17 36.4	−04 37 37	MCG −1-1-70	14.2	0.9 x 0.6	13.4	Sc	138	
00 18 46.0	−03 24 31	MCG −1-1-74	15.4	1.1 x 0.5	14.6	SBbc	33	Close pair of mag 10.42v, 12.2 stars SE 8′.2.
00 19 26.1	−04 05 01	MCG −1-1-77	13.6	1.1 x 0.9	13.4	SBc	177	
00 20 02.9	−04 03 09	MCG −1-2-2	14.2	0.9 x 0.9	13.9	Sbc		
00 20 42.1	−04 26 02	MCG −1-2-5	13.6	1.2 x 0.5	12.9	SA0°?	144	
00 22 10.1	−04 20 51	MCG −1-2-6	14.1	1.1 x 0.7	13.7	SA(r)bc? II-III	90	
00 24 28.3	−03 51 24	MCG −1-2-9	14.4	0.9 x 0.3	12.8		3	= **Mkn 944**.
00 25 47.9	−02 17 06	MCG −1-2-11	13.9	2.4 x 0.2	13.0	Sbc pec sp	15	Mag 11.07v star NNW 5′.4.
00 26 17.2	−04 29 32	MCG −1-2-14	13.5	1.5 x 0.5	13.1	SB(s)a pec:	123	Mag 11.5 star on E edge.
00 29 48.8	−05 06 57	MCG −1-2-22	14.3	1.0 x 0.8	13.9	Sbc	138	
00 31 07.9	−03 04 10	MCG −1-2-24	14.7	0.9 x 0.9	14.4			
00 34 01.2	−03 10 00	MCG −1-2-31	13.5	1.4 x 0.5	13.0	S0°:	159	Mag 9.95v star SW 2′.6.
00 35 36.2	−04 44 00	MCG −1-2-38	14.3	1.4 x 0.6	14.0	Sbc pec? III	85	
00 36 09.2	−03 31 18	MCG −1-2-40	14.0	0.9 x 0.7	13.5	E	18	
23 21 03.0	−04 53 40	MCG −1-59-21	14.3	1.1 x 0.5	13.5	SAB(s)m IV-V	105	
23 28 25.0	−02 48 04	MCG −1-59-24	14.0	1.3 x 0.7	13.8	SABc I-II	60	**MCG −1-59-25** NE 1′.2.
23 30 32.4	−02 27 43	MCG −1-59-27	14.0	1.9 x 1.3	14.9	SB(r)b pec: II-III	57	Bright nucleus in short E-W bar.
23 36 29.9	−04 54 34	MCG −1-60-11	14.1	1.7 x 1.3	14.9	SB(r)d III	162	MCG −1-60-12 SSE 8′.8.
23 36 45.2	−05 02 20	MCG −1-60-12	13.7	0.5 x 0.5	12.3	E		Mag 9.35v star ESE 13′.5; MCG −1-60-11 NNW 8′.8.

(Continued from previous page)

GALAXIES

RA h m s	Dec ° ′ ″	Name	Mag (V)	Dim ′ Maj x min	SB	Type Class	PA	Notes
23 37 10.8	−03 43 31	MCG −1-60-14	13.3	0.7 x 0.5	12.2	E?	156	Mag 12.3 star on SW edge.
23 37 52.3	−03 29 28	MCG −1-60-15	13.5	0.9 x 0.3	12.0	SAB(rs)0$^+$?	60	
23 38 48.6	−05 44 52	MCG −1-60-16	14.5	1.7 x 0.7	14.5	SB(s)m: IV-V	54	Mag 7.66v star NE 19′.4.
23 41 47.8	−03 40 00	MCG −1-60-21	13.6	1.5 x 0.3	12.6	Sc? pec sp	36	= **Mkn 933**. Mag 9.71v star NE 7′.5; MCG −1-60-22 NE 4′.7.
23 42 01.2	−03 36 58	MCG −1-60-22	14.0	0.7 x 0.3	12.2	Sb pec:	108	Multi-galaxy system; mag 9.71v star E 3′.2.
23 42 01.0	−04 13 03	MCG −1-60-23	13.4	1.1 x 0.7	13.0	(R′)SB(r)b: II-III	54	
23 44 55.9	−04 16 31	MCG −1-60-27	14.4	0.7 x 0.3	12.6	Sc	123	
23 49 01.5	−02 17 34	MCG −1-60-43	14.3	0.7 x 0.4	12.8	Sc	114	Faint, elongated anonymous galaxy E 4′.8.
23 53 03.8	−03 48 10	MCG −1-60-45	14.1	1.0 x 0.8	13.7	Sc	162	
00 08 45.0	+04 36 44	NGC 12	13.1	1.1 x 1.0	13.1	SAB(rs)c: II-III	137	Stellar nucleus, smooth envelope.
00 11 47.0	−05 35 14	NGC 38	13.3	1.1 x 1.1	13.3	(R)SAa:		Strong, round core in smooth envelope.
00 15 58.4	−00 18 14	NGC 60	14.1	1.3 x 1.1	14.3	SA(r)cd pec II-III	155	Stellar nucleus.
00 20 25.9	+00 49 31	NGC 78A	13.7	1.1 x 0.7	13.3	SB(r)0/a:	80	Mag 9.82v star WNW 7′.4; NGC 78B on NE edge.
00 20 27.6	+00 49 56	NGC 78B	13.5	1.0 x 0.7	13.0	S0° pec?	134	NGC 78A on SW edge.
00 24 43.8	−05 08 56	NGC 106	13.7	0.8 x 0.6	12.8	Sc	100	
00 26 54.8	−02 30 03	NGC 113	13.1	1.0 x 1.0	13.0	SA0$^-$:		
00 26 58.3	−01 47 13	NGC 114	13.8	1.0 x 0.7	13.3	SB(rs)0:	165	NGC 118 E 4′.5.
00 27 11.1	+01 19 59	NGC 117	14.3	0.9 x 0.5	13.3	S0$^+$: sp	100	Mag 9.41v star NE 5′.2; stellar **UGC 257** NNW 7′.9.
00 27 16.4	−01 46 49	NGC 118	14.0	0.7 x 0.6	12.9	I0?	40	Compact; NGC 114 W 4′.5.
00 27 30.2	−01 30 51	NGC 120	13.4	1.5 x 0.5	13.0	SB0°:	73	Mag 8.57v star SE 12′.0.
00 27 52.5	−01 48 37	NGC 124	13.0	1.4 x 0.8	12.9	SA(s)c I-II	168	Mag 8.57v star N 7′.8.
00 28 50.2	+02 50 18	NGC 125	12.1	1.2 x 1.2	12.4	(R)SA0$^+$ pec:		Close pair of mag 12.1, 12.2 stars on S edge.
00 29 08.2	+02 48 36	NGC 126	14.2	0.7 x 0.4	12.7	SB0°?	110	Located 3′.6 SSW of NGC 128.
00 29 12.4	+02 52 20	NGC 127	14.8	0.8 x 0.5	13.6	SA0°:	70	On W edge of NGC 128.
00 29 14.9	+02 51 58	NGC 128	11.8	2.5 x 0.7	12.2	S0 pec sp	1	NGC 130 on E edge; NGC 127 on W edge.
00 29 18.6	+02 52 09	NGC 130	14.4	0.7 x 0.4	12.9	SA0$^-$:	52	Located on E edged of NGC 128.
00 30 10.3	+02 05 32	NGC 132	12.6	1.9 x 1.4	13.6	SAB(s)bc I-II	49	
00 30 59.3	+05 09 36	NGC 138	13.7	1.3 x 0.6	13.2	Sa:	175	Mag 12.2 star N 1′.3.
00 31 06.5	+05 04 41	NGC 139	14.4	0.7 x 0.6	13.3	SB?	171	Located 5′.2 S of NGC 138.
00 31 17.5	+05 10 46	NGC 141	14.5	0.8 x 0.6	13.6		90	Mag 10.75v star NNW 7′.3.
00 31 45.6	−05 09 16	NGC 145	12.7	1.8 x 1.6	13.7	SB(s)dm	57	Mag 8.60v star E 5′.6.
00 35 33.9	−02 50 56	NGC 161	13.4	1.3 x 0.8	13.3	S0°:	27	**MCG −1-2-37** S 1′.6.
00 36 32.9	+02 44 58	NGC 164	15.0	0.4 x 0.4	12.9			Almost stellar; mag 9.78v star W 6′.9.
00 36 46.0	+01 53 08	NGC 170	14.4	0.6 x 0.3	12.4	S0$^-$?	79	Located 1′.9 N of mag 9.28v star.
00 37 12.4	+01 56 30	NGC 173	13.0	3.2 x 2.6	15.2	SA(rs)c II	90	
00 38 12.3	+02 43 39	NGC 182	12.4	2.0 x 1.5	13.5	(R′)SAB(rs)a pec:	75	Very small, faint anonymous galaxies 1′.5 SW and 1′.9 SE.
00 38 25.4	+03 09 56	NGC 186	13.4	1.4 x 0.8	13.4	SB0$^+$ pec:	23	Pair of very faint, very small anonymous galaxies SW 2′.7.
00 39 13.3	+00 51 53	NGC 192	12.6	1.9 x 0.9	13.1	(R′)SB(r)a:	167	Mag 8.78v star NW 8′.0; NGC 197 and NGC 196 N 2′.6.
00 39 18.6	+03 19 51	NGC 193	12.2	1.4 x 1.2	12.6	SAB(s)0$^-$:	55	Mag 13.1 star on SW edge; mag 9.87v star ESE 2′.5.
00 39 18.5	+03 02 17	NGC 194	12.2	1.5 x 1.4	13.0	E	30	Mag 7.35v star N 5′.9.
00 39 17.9	+00 54 42	NGC 196	12.9	1.1 x 0.8	12.6	SB0 pec:	3	NGC 197 S 1′.2.
00 39 18.9	+00 53 29	NGC 197	14.1	0.7 x 0.7	13.3	SB0° pec		Almost stellar nucleus.
00 39 23.0	+02 47 51	NGC 198	13.2	1.2 x 1.2	13.4	SA(r)c II		Mag 9.92v star S 5′.3.
00 39 33.2	+03 08 16	NGC 199	13.6	1.2 x 0.7	13.3	SA0° pec:	160	Mag 7.35v star W 5′.1.
00 39 34.7	+02 53 15	NGC 200	12.6	1.9 x 1.0	13.2	SB(s)bc I	161	Elongated anonymous galaxy SW 2′.7.
00 39 34.7	+00 51 36	NGC 201	12.9	1.8 x 1.4	13.8	SAB(r)c II	155	Stellar nucleus, multi-branching arms.
00 39 40.1	+03 32 05	NGC 202	14.3	1.0 x 0.4	13.2	SB(s)0$^+$?	153	NGC 203 S 5′.5; mag 11.5 star SW 3′.2; mag 7.63v star N 7′.4.
00 39 39.5	+03 26 31	NGC 203	14.0	0.9 x 0.3	12.4	S0° pec:	85	Mag 11.5 star NNW 4′.2; NGC 202 N 5′.5.
00 39 44.3	+03 17 56	NGC 204	12.9	1.2 x 1.1	13.1	SA(s)0° pec:	30	Mag 9.87v star W 4′.2.
23 21 19.5	+01 24 09	NGC 7629	13.8	1.0 x 0.8	13.4	SB0°?	179	Mag 11.4 star W 8′.8; MCG +0-59-32 NE 5′.6.
23 22 53.4	+01 26 33	NGC 7642	13.7	0.5 x 0.5	12.1	Sa pec?		**MCG +0-59-36** SE 2′.8.
23 26 45.2	−04 57 58	NGC 7663	16.2	0.6 x 0.3	14.2	Irr III:	166	Mag 9.21v star SW 2′.3.
23 24 22.9	−00 06 32	NGC 7667	13.9	1.6 x 1.1	14.4	SB(s)m pec III-IV	100	
23 28 47.7	+03 30 37	NGC 7679	12.9	1.7 x 0.9	13.2	SB0 pec:	90	Multi-galaxy system; bright center offset to W.
23 29 03.8	+03 32 00	NGC 7682	13.2	1.2 x 1.1	13.3	SB(r)ab	153	Bright nucleus in NNW-SSE bar.
23 30 32.2	+00 04 52	NGC 7684	13.6	1.4 x 0.4	12.8	S0$^+$	21	
23 30 33.4	+03 54 08	NGC 7685	13.2	1.9 x 1.4	14.1	SAB(s)c:	170	Almost stellar nucleus, patchy arms.
23 30 54.5	+03 32 44	NGC 7687	13.4	1.3 x 1.0	13.6	SAB(rs)0° pec?	75	Bright center, uniform envelope.
23 32 46.7	−05 35 48	NGC 7692	14.8	0.5 x 0.4	12.9	Irr	100	
23 33 10.6	−01 17 31	NGC 7693	13.4	1.2 x 0.8	13.2	SB(r)0$^+$:	156	Extremely faint anonymous spiral galaxy with stellar nucleus on S edge.
23 33 16.6	−02 42 11	NGC 7694	13.4	1.6 x 0.9	13.6	Im pec: III	80	Almost stellar NGC 7695 S 1′.1.
23 33 15.1	−02 43 15	NGC 7695	15.1	0.5 x 0.3	12.9		81	
23 33 50.3	+04 52 12	NGC 7696	13.9	1.3 x 0.6	13.5		96	**Mkn 537** NE 10′.6.
23 34 26.9	−02 53 56	NGC 7699	15.0	0.7 x 0.5	13.7	Sa	109	
23 34 30.1	−02 57 12	NGC 7700	13.3	1.3 x 1.0	13.4	S0$^+$ sp	154	
23 34 31.5	−02 51 16	NGC 7701	13.8	1.6 x 0.4	13.2	S0$^+$ sp	155	
23 35 01.3	+04 53 50	NGC 7704	13.4	1.1 x 0.9	13.3	S0$^-$:	67	Small, faint anonymous galaxy WSW 1′.9.
23 35 02.7	+04 48 11	NGC 7705	14.4	0.6 x 0.5	13.0	S0	30	Almost stellar nucleus.
23 35 10.5	+04 57 49	NGC 7706	13.2	1.2 x 0.8	13.1	S0	120	**Mkn 537** W 11′.7.
23 35 46.1	−02 52 53	NGC 7710	13.9	1.4 x 0.5	13.4	S0 pec sp	135	
23 36 14.4	+02 09 21	NGC 7714	12.5	1.9 x 1.4	13.4	SB(s)b: pec	79	Multi-galaxy system; one faint arm extends W and curves NW; the other arm appears to blend with NGC 7715 E.
23 36 21.7	+02 09 22	NGC 7715	14.2	2.6 x 0.5	14.3	Im pec sp	73	Multi-galaxy system; mag 5.70v star 16 Psc S 3′.3.
23 36 31.4	+00 17 48	NGC 7716	12.1	2.1 x 1.8	13.4	SAB(r)b: I-II	35	Located 2′.0 N of a mag 10.17v star.
23 39 14.8	−04 32 22	NGC 7725	13.8	0.8 x 0.7	13.0	S0	123	
23 41 29.2	+03 44 21	NGC 7731	13.5	1.4 x 1.1	13.8	(R)SBa pec:	95	Bright center, faint envelope; mag 11.3 star E 1′.2; NGC 7732 on SE edge.
23 41 34.0	+03 43 29	NGC 7732	13.8	1.9 x 0.6	13.8	Scd pec sp	96	Located 1′.5 S of mag 11.3 star, NGC 7731 NW edge.
23 44 01.8	+00 30 54	NGC 7738	13.1	2.0 x 1.5	14.1	SB(rs)b II	80	Strong bar NE-SW with dark patches E and W.
23 44 30.0	+00 19 16	NGC 7739	13.6	1.1 x 0.9	13.7	E$^+$:	90	
23 45 20.2	−01 41 05	NGC 7746	13.1	1.4 x 1.1	13.4	S0° pec:	154	
23 46 37.9	+03 47 56	NGC 7750	12.9	1.6 x 0.8	13.0	(R′)SB(rs)c pec: II	171	
23 48 41.8	+04 10 33	NGC 7756	13.1	0.4 x 0.2	10.2		21	Located 0′.9 WNW of center of NGC 7757.
23 48 45.4	+04 10 14	NGC 7757	12.7	2.5 x 1.8	14.2	SA(rs)c	115	NGC 7756 lies 0′.9 WNW of center.
23 54 11.0	+00 22 47	NGC 7783A/B	13.1	1.3 x 0.6	12.7		120	Multi-galaxy system; = **Hickson 98A** NW end; **Hickson 98B** SE end; **Hickson 98D** N 0′.7; mag 9.39v star N 1′.8.
23 54 13.7	+00 21 22	NGC 7783C	15.4	0.2 x 0.2	11.7	E		= **Hickson 98C**.
23 55 18.8	+05 54 59	NGC 7785	11.6	2.5 x 1.3	12.9	E5-6	143	Mag 10.67v star SE 2′.9.
23 56 07.9	+00 32 55	NGC 7787	14.2	1.8 x 0.5	13.9	(R′)SB(rs)0/a:	104	Bright nucleus, filamentary arms.
23 58 58.8	+03 38 01	NGC 7797	13.7	1.0 x 0.9	13.4	Sbc	13	

(Continued from previous page)

GALAXIES

RA h m s	Dec ° ′ ″	Name	Mag (V)	Dim ′ Maj x min	SB	Type Class	PA	Notes
00 02 09.4	+02 56 25	NGC 7809	14.5	0.5 x 0.4	12.6	Im?	95	
00 02 26.5	+03 21 06	NGC 7811	14.5	0.4 x 0.4	12.3	cIm??		Mag 8.56v star E 8′.2.
00 04 31.1	+05 11 56	NGC 7820	12.9	1.3 x 0.6	12.5	S0/a	165	
00 05 27.1	+05 10 36	NGC 7825	13.7	1.1 x 0.5	12.9	SB(s)b	27	**CGCG 408-27** W 5′.3.
00 05 27.8	+05 13 18	NGC 7827	13.9	1.2 x 0.9	13.9	SB0	36	
00 06 28.6	−03 42 57	NGC 7832	13.2	1.9 x 1.0	13.9	E^+	20	**MCG −1-1-35** E 5′.3; mag 8.25v star NE 8′.4.
00 02 57.1	+04 12 29	UGC 4	14.3	0.7 x 0.4	12.8	Sbc	158	Mag 11.2 star NNW 2′.7.
00 03 05.6	−01 54 55	UGC 5	13.2	1.4 x 0.6	12.8	SABbc	45	Mag 10.79v star E 2′.9.
00 04 29.6	+05 50 42	UGC 27	14.0	2.0 x 1.1	14.7	Scd:	135	
00 04 57.9	+05 07 21	UGC 33	14.0	0.9 x 0.6	13.2	SB0/a	165	Mag 12.2 star W 1′.6.
00 06 40.5	+05 06 47	UGC 51	14.2	0.7 x 0.4	12.7	SBbc	30	
00 09 51.9	+04 40 29	UGC 88	14.3	0.6 x 0.4	12.6	S0/a	6	Mag 15.4 star on S edge.
00 13 56.0	+03 42 50	UGC 129	14.2	1.2 x 1.0	14.2	(R′)SB(s)ab		Mag 10.51v star N 1′.5.
00 14 31.9	−00 44 15	UGC 139	13.6	2.2 x 1.2	14.5	SAB(s)c? II	82	
00 15 15.3	+05 53 13	UGC 143	15.1	0.7 x 0.5	13.8	SB(r)a	126	Mag 6.98v star SE 6′.3.
00 22 23.2	−01 18 11	UGC 212	14.1	1.5 x 0.6	13.8	SB	27	Galaxy **KUG 0019-016** S 3′.5.
00 23 37.7	−00 30 35	UGC 224	14.5	0.5 x 0.3	12.3	Disrupted	3	Multi-galaxy system, includes **MCG +0-2-18** and **MCG +0-2-19**; mag 8.61v star S 2′.0.
00 27 50.0	−01 12 02	UGC 272	14.4	1.2 x 0.5	13.7	SAB(s)d: III	130	
00 27 58.5	+02 30 24	UGC 275	13.6	1.7 x 0.8	13.8	SAB(rs)b pec? II	140	**UGC 281** E 6′.0.
00 28 13.8	+03 14 00	UGC 277	14.4	1.1 x 0.9	14.2	Scd:	160	
00 28 18.3	+03 22 59	UGC 282	14.6	1.2 x 1.2	14.8	SAB(s)cd: II-III		In contact with UGC 283 E edge.
00 28 21.8	+03 23 22	UGC 283	14.4	1.0 x 0.7	13.9	SAB(rs)d II-III	168	On E edge of UGC 282.
00 33 22.0	−01 07 20	UGC 328	15.4	1.4 x 0.8	15.4	SB(rs)dm IV-V	144	Close pair of stars, mags 12.9 and 13.9, SW 1′.7.
00 33 36.2	+02 40 53	UGC 329	14.2	1.7 x 0.8	14.4	SAB(s)cd III	170	
00 34 33.0	+03 36 08	UGC 342	14.5	1.1 x 0.9	14.3	SABc	57	Mag 12.3 star WSW 1′.4.
00 35 27.2	+02 56 02	UGC 348	14.6	1.0 x 0.7	14.0	Sdm:	33	Mag 12.7 star E 2′.1.
00 36 13.0	+01 42 43	UGC 358	13.9	0.9 x 0.9	13.5	S?		
00 37 58.0	+05 08 50	UGC 379	15.3	1.5 x 0.2	13.8	Scd:	120	
00 39 18.6	+03 57 10	UGC 402	13.4	1.1 x 0.7	13.0	S0	26	**UGC 416** ESE 3′.7.
23 20 11.5	−01 51 26	UGC 12521	13.4	1.5 x 0.9	13.6	SA(s)cd: III	167	**CGCG 380-29** W 11′.7; **CGCG 380-31** W 10′.1.
23 21 51.6	+05 00 25	UGC 12547	14.3	1.2 x 0.6	13.8	SB?	151	Two main arms form open "S" shape; **PGC 71224** E 5′.5.
23 21 57.2	+05 02 09	UGC 12548	13.8	1.2 x 0.3	12.5	Sa?	118	**UGC 12555** NE 10′.4.
23 29 22.1	+03 23 21	UGC 12628	14.2	1.6 x 1.2	14.8	SB(rs)c: II-III	78	
23 30 25.7	+00 09 22	UGC 12635	14.3	1.2 x 1.2	14.5	SA(rs)d: IV-V		Almost stellar nucleus, smooth envelope; mag 12.2 star SW 2′.5.
23 31 53.5	−02 09 25	UGC 12648	14.2	0.9 x 0.8	13.7	Irr	15	
23 33 29.0	−01 59 03	UGC 12661	14.1	1.1 x 0.3	12.7	Sab	147	Elongated **MCG +0-60-6** NW 1′.2.
23 35 14.1	+00 02 30	UGC 12685	15.4	1.2 x 1.1	15.5	SB(s)d pec: III-IV:	27	Located between a pair of mag 13.0 and 14.4 stars.
23 35 32.3	+05 12 55	UGC 12689	13.5	1.5 x 0.3	12.5	Sab	149	
23 35 39.8	+01 11 51	UGC 12690	14.9	1.8 x 1.8	16.0	SB(s)m: V-VI		Dwarf, very low surface brightness; mag 6.47v star NNW 7′.5.
23 37 24.0	+00 23 32	UGC 12709	13.9	2.8 x 1.7	15.5	SAB(s)m V	145	Patchy, low surface brightness.
23 38 36.8	+05 25 57	UGC 12717	14.2	1.1 x 0.8	13.9	Scd:	141	Stellar nucleus; elongated **UGC 12720** NE 9′.0.
23 40 21.0	+01 14 45	UGC 12729	13.9	1.2 x 0.5	13.2	S0/a	74	
23 41 46.4	−01 21 09	UGC 12739	13.3	0.8 x 0.7	12.5	S?	3	**CGCG 381-27** NW 1′.8.
23 44 12.8	+02 16 37	UGC 12758	13.9	1.2 x 0.9	13.9	Sb	155	Mag 15 star on S edge.
23 46 03.4	−01 53 28	UGC 12774	13.5	1.1 x 0.9	13.3	(R′)SB(rs)a:	90	
23 51 06.4	+01 03 24	UGC 12810	13.6	1.7 x 0.7	13.6	(R′)SAB(r)bc pec: II	52	
23 51 50.5	+03 04 59	UGC 12816	13.8	1.4 x 0.7	13.6	SAB(rs)cd: III	147	Mag 5.58v star 22 Piscium S 9′.4; stellar galaxy **UM 10** W 4′.9.
23 54 26.6	−01 56 01	UGC 12838	13.8	0.9 x 0.8	13.3	Scd:	117	Mag 7.59v star E 3′.1.
23 55 52.0	+00 33 27	UGC 12847	14.8	1.0 x 1.0	14.6			
23 56 47.4	+01 21 12	UGC 12857	13.7	2.1 x 0.5	13.6	Sbc: sp	34	
23 59 19.6	+04 45 38	UGC 12881	14.6	1.7 x 0.3	13.8	Sdm:	160	Mag 9.14v star SW 3′.2.

GALAXY CLUSTERS

RA h m s	Dec ° ′ ″	Name	Mag 10th brightest	No. Gal	Diam ′	Notes
23 44 54.0	−04 06 00	A 2656	16.2	35	28	MCG −1-60-27 on S edge; other members anonymous, stellar.
00 03 54.0	+02 03 00	A 2700	16.0	59	16	All members anonymous, faint and stellar/almost stellar.

GALAXIES

RA h m s	Dec ° ′ ″	Name	Mag (V)	Dim ′ Maj x min	SB	Type Class	PA	Notes
23 12 20.3	−01 32 00	CGCG 380-9	14.6	0.7 x 0.6	13.5		177	Mag 12.1 star on E edge; MCG +0-59-6 SW 4′.4.
23 14 28.9	+04 11 30	CGCG 406-27	14.5	1.0 x 0.5	13.6		177	Mag 11.8 star W 6′.3.
23 15 29.6	+04 54 00	CGCG 406-34	14.2	0.8 x 0.6	13.3		45	Mag 10.6 star W 4′.2.
22 15 45.1	+02 03 54	IC 1437	13.5	1.0 x 0.9	13.2	(R′)SA(r)0/a	72	Mag 8.16v star SSW 7′.6.
22 29 59.7	−05 07 14	IC 1447	12.8	1.4 x 0.8	12.8	SA(r)b? II	108	Mag 9.34v star N 1′.5.
22 53 46.1	+01 22 15	IC 1455	13.8	0.9 x 0.5	12.8	SBa	30	Forms a triangle with a pair of mag 11.7 stars SE and SW.
22 57 04.1	+04 40 36	IC 1460	14.4	0.5 x 0.5	12.8	E2 pec?		Almost stellar; mag 11.4 star E 6′.8.
23 03 39.2	−02 46 35	IC 1466	13.4	0.4 x 0.4	11.3	(R?)SA(r)0		Almost stellar; mag 12.0 star W 9′.7.
23 04 49.8	−03 13 53	IC 1467	14.2	0.8 x 0.3	12.5	Sc	3	Mag 12.0 star SW 7′.0; IC 1468 ENE 4′.7.
23 05 07.5	−03 12 18	IC 1468	14.2	1.1 x 0.5	13.5	$(R')SAB0^+$ pec:	156	IC 1467 WSW 4′.7.
23 12 51.3	+05 48 20	IC 1474	13.9	1.0 x 0.5	13.0	Scd:	150	Mag 8.86v star SE 7′.2.
23 19 25.6	+05 54 17	IC 1481	13.5	0.8 x 0.7	12.7	S?	42	Lies 1′.4 NW of a mag 8.43v star.
23 20 49.7	+01 44 16	IC 1482	14.4	0.6 x 0.4	12.7		123	Almost stellar; mag 10.6 star NE 8′.0.
22 41 38.7	+02 38 20	IC 5241	13.9	1.0 x 0.7	13.4	Scd pec?	21	Mag 10.75v star ENE 8′.2.
23 09 20.3	+00 45 20	IC 5287	13.9	1.0 x 0.9	13.7	(R′)SB(r)b II	138	Bright nucleus with short NE-SW bar.
22 09 06.3	+02 00 51	MCG +0-56-10	14.9	0.8 x 0.7	14.1		0	Core offset E; mag 12.3 star NW 2′.8.
22 10 29.0	−00 01 14	MCG +0-56-11	15.8	0.9 x 0.4	14.5		132	
22 11 53.0	+00 06 28	MCG +0-56-13	14.2	0.9 x 0.6	13.4	SB(rs)bc I-II	24	Mag 10.03v star E 3′.1.
22 11 59.4	−00 15 11	MCG +0-56-14	15.2	0.7 x 0.7	14.3			
22 14 20.8	+01 01 05	MCG +0-56-15	14.6	0.9 x 0.5	13.6		57	Located between a pair of stars, mags 13.9 and 12.5.
22 24 09.2	+02 43 37	MCG +0-57-1	14.7	0.9 x 0.4	13.5	Sbc	90	Mag 10.25v star SW 3′.6.
22 28 51.7	+01 53 49	MCG +0-57-2	14.2	0.8 x 0.5	13.0	S?	111	
22 39 18.1	−01 42 06	MCG +0-57-6	16.3	0.5 x 0.4	14.4	Sm	93	Star superimposed? mag 10.00v star SE 3′.6.
22 42 57.8	−02 04 36	MCG +0-57-9	14.7	0.8 x 0.5	13.6		90	Mag 11.9 star NNW 1′.3.
22 49 42.2	+03 07 08	MCG +0-58-4	14.9	0.9 x 0.8	14.4		147	

RA h m s	Dec ° ′ ″	Name	Mag (V)	Dim ′ Maj x min	SB	Type Class	PA	Notes
22 55 16.8	+02 14 44	MCG +0-58-12	14.6	1.0 x 0.4	13.4	S?	129	Mag 12.6 star NE 3.′0.
22 56 36.5	+01 24 28	MCG +0-58-13	16.0	0.8 x 0.3	14.3	Sc	63	Mag 9.36v star NNE 2.′9.
22 59 37.4	+02 20 37	MCG +0-58-18	15.2	1.1 x 0.5	14.5	S?	69	Mag 10.9 star W 1.′8.
23 01 48.4	+01 45 03	MCG +0-58-22	16.0	0.9 x 0.4	14.8	Sbc	99	Mag 11.9 star SE 2.′0.
23 03 21.7	+01 53 11	MCG +0-58-23	15.7	0.8 x 0.7	14.9	Scd	57	Mag 10.40v star ENE 8.′4.
23 03 38.4	+01 50 57	MCG +0-58-24	14.8	0.7 x 0.5	13.5		90	Filamentary; mag 13.2 star SE 2.′5.
23 05 55.5	+02 59 19	MCG +0-58-31	15.0	0.8 x 0.6	14.0	Spiral	21	NGC 7482 NW 6.′0.
23 07 18.4	+02 11 30	MCG +0-58-34	14.3	1.0 x 0.6	13.5	SB?	117	Faint star on S edge.
23 12 05.6	−01 34 25	MCG +0-59-6	15.1	0.7 x 0.3	13.2	Sb	27	Mag 9.51v star SW 8.′4.
23 12 11.0	+03 37 46	MCG +0-59-7	14.0	0.9 x 0.7	13.4	S0?	3	Located between a pair of mag 13 stars.
23 15 43.3	+00 25 01	MCG +0-59-13	14.1	0.7 x 0.4	12.6	Sb	177	
23 19 02.1	+02 50 59	MCG +0-59-22	14.9	0.7 x 0.4	13.4	Spiral	162	
23 20 07.4	−00 21 00	MCG +0-59-25	15.4	0.8 x 0.4	14.0	Sm	39	
23 20 29.1	−01 00 10	MCG +0-59-28	14.1	0.5 x 0.5	12.4			
23 21 39.1	+01 26 43	MCG +0-59-32	15.0	0.9 x 0.6	14.2		135	NGC 7629 SW 5.′6.
23 21 51.8	−00 41 27	MCG +0-59-33	14.3	0.7 x 0.3	12.4	Sb	165	
23 22 05.4	+01 28 50	MCG +0-59-34	15.0	0.5 x 0.5	13.4	Sbc		Almost stellar core.
23 26 16.6	−01 58 28	MCG +0-59-43	14.7	0.6 x 0.5	13.2		162	
22 47 32.6	+05 28 32	MCG +1-58-6	15.2	0.5 x 0.3	12.9	Spiral	126	Mag 11.6 star E 1.′9.
23 11 19.6	+04 58 49	MCG +1-59-9	14.2	0.7 x 0.4	12.7	SB?	24	Located 1.′9 SE of a mag 6.76v star.
23 17 21.0	+05 39 36	MCG +1-59-28	14.2	0.8 x 0.4	12.8	S?	108	Pair with MCG +1-59-29 S 1.′0.
23 17 21.8	+05 38 38	MCG +1-59-29	14.3	0.4 x 0.4	12.2	S0?		Pair with MCG +1-59-28 N 1.′0
23 25 42.9	+05 04 35	MCG +1-59-78	15.4	0.5 x 0.5	13.7			
22 16 44.9	−04 43 47	MCG −1-56-3	13.6	0.7 x 0.7	12.9	E/S0		
22 18 47.1	−03 29 44	MCG −1-56-5	13.6	0.9 x 0.9	13.2	SBd		
22 23 39.3	−03 25 51	MCG −1-57-4	12.3	1.3 x 1.0	12.4	S0°	90	Between two stars.
22 23 52.8	−05 32 57	MCG −1-57-5	14.3	0.7 x 0.4	12.8	SBd	171	
22 24 11.6	−03 28 58	MCG −1-57-7	14.0	1.7 x 1.7	15.0	SB(s)d: III-IV		Knotty, branching arms.
22 24 32.8	−04 45 10	MCG −1-57-8	14.4	1.1 x 0.6	13.7	SBbc	153	Mag 5.79v star SW 8.′2.
22 26 16.7	−04 20 28	MCG −1-57-9	14.1	0.8 x 0.8	13.4	Sc		
22 34 54.7	−04 42 01	MCG −1-57-15	15.2	1.7 x 0.7	15.2	IBm sp V	107	Mag 15.0 star W 2.′5.
22 36 35.2	−02 54 27	MCG −1-57-16	12.9	2.2 x 1.9	14.3	SA(s)m IV-V	25	Mag 11.09v star NNE 3.′6; mag 8.00v star W 15.′7.
22 38 57.0	−05 51 12	MCG −1-57-18	12.6	2.4 x 0.8	13.1	SAB(s)dm? III-IV	78	Close pair of mag 8.1 stars superimposed W end.
22 40 15.4	−02 25 31	MCG −1-57-21	13.1	2.2 x 1.3	14.1	SB(s)bc pec II	81	Mag 10.75v star W 13.′3.
22 42 05.5	−02 47 01	MCG −1-57-23	14.1	1.1 x 0.7	13.6	(R)SB(r)a? V-VI	54	Mag 9.30v star SSW 3.′6.
22 43 04.5	−03 47 14	MCG −1-57-24	15.5	1.3 x 0.2	13.9	S?	21	
22 57 13.8	−03 09 46	MCG −1-58-8	14.7	0.4 x 0.4	12.6	S0		Mag 7.74v star E 3.′0.
22 58 02.3	−03 46 13	MCG −1-58-9	13.2	1.0 x 0.8	12.8	(R′)SA(s)bc: pec	45	= **Arp 314**. Mag 8.47v star NNW 7.′4.
22 58 07.7	−03 47 21	MCG −1-58-10	13.4	1.0 x 0.9	13.2	SB(rs)cd: pec	24	Several bright knots. Star on E edge.
23 04 23.5	−04 52 30	MCG −1-58-16	14.3	1.3 x 0.3	13.2	S?	27	Mag 6.70v star NW 8.′3.
23 07 25.7	−05 09 28	MCG −1-58-22	14.2	1.3 x 0.4	13.4	S?	0	
23 10 58.8	−04 31 24	MCG −1-59-3	14.9	0.5 x 0.5	13.3	Irr		
23 17 43.2	−03 30 19	MCG −1-59-13	14.7	0.6 x 0.3	12.6	Scd	114	Mag 10.01v star NNE 9.′1.
23 18 01.8	−03 41 49	MCG −1-59-14	14.0	1.5 x 1.3	14.6	SB?		Mag 10.44v star W 8.′1.
23 18 17.5	−03 37 36	MCG −1-59-16	14.2	1.1 x 0.3	12.9	S?	36	Mag 9.42v star E 12.′8.
23 18 43.6	−03 00 23	MCG −1-59-18	14.4	0.9 x 0.5	13.4	S0	105	
23 21 03.0	−04 53 40	MCG −1-59-21	14.3	1.1 x 0.5	13.5	SAB(s)m IV-V	105	
22 08 34.5	+00 30 42	NGC 7215	13.9	1.1 x 0.5	13.1	S0°?	88	
22 10 51.6	+02 06 21	NGC 7222	13.8	1.2 x 1.2	14.0	SB(r)b II-III		Stellar nucleus in short bar.
22 15 01.5	−05 03 13	NGC 7239	13.8	1.1 x 0.7	13.4	SAB0$^-$	75	
22 22 36.6	−04 07 18	NGC 7260	12.7	1.9 x 1.4	13.6	SAB(rs)bc I-II	33	
22 23 59.0	−04 04 25	NGC 7266	13.4	0.8 x 0.6	12.4	Sa	95	
22 28 15.0	−02 53 05	NGC 7288	13.0	2.3 x 1.5	14.2	S0/a pec sp	92	
22 34 06.6	+05 34 13	NGC 7311	12.5	1.6 x 0.8	12.6	Sab	10	
22 34 35.0	+05 49 00	NGC 7312	13.4	1.4 x 0.8	13.4	SB(s)b	83	Faint nucleus in short bar.
22 39 36.3	−04 09 33	NGC 7344	13.7	1.5 x 0.8	13.8	(R′)SB(r)ab I-II	20	
22 41 27.0	−04 26 43	NGC 7351	12.9	1.8 x 1.0	13.4	SAB0°:	177	
22 43 34.1	+04 09 01	NGC 7360	13.7	0.7 x 0.3	11.9	S?	153	
22 44 24.2	−00 09 43	NGC 7364	12.6	1.7 x 1.0	13.0	S0/a pec:	65	
22 44 34.7	+03 38 43	NGC 7367	13.8	1.5 x 0.4	13.1	Sab:	128	
22 46 19.4	+03 12 33	NGC 7373	13.6	1.3 x 0.5	13.0	SAB(s)0°?	16	
22 47 17.5	+03 38 43	NGC 7376	14.4	0.6 x 0.4	12.7	SB0$^+$:	142	
22 50 36.2	−01 32 42	NGC 7391	12.0	1.7 x 1.5	13.0	E:	70	Bright center, smooth envelope; faint star on N edge.
22 51 37.7	−05 33 31	NGC 7393	12.9	1.9 x 0.9	13.4	SB(rs)c pec	88	Brightest part offset to E. Dark patch E of nucleus.
22 52 22.5	+01 05 30	NGC 7396	12.9	2.4 x 1.6	14.2	Sa pec sp	103	Dark lane along centerline; anonymous galaxy SW 3.′0.
22 52 46.7	+01 07 56	NGC 7397	14.2	0.7 x 0.5	12.9	SB(rs)0/a	159	
22 52 49.3	+01 12 05	NGC 7398	13.6	1.2 x 0.8	13.4	SA(r)a:	75	Mag 9.76v star NE 6.′1.
22 52 58.5	+01 08 34	NGC 7401	14.9	0.9 x 0.6	14.1	Sa	84	NGC 7402 E 1.′5.
22 53 04.6	+01 08 40	NGC 7402	14.5	0.6 x 0.4	12.8	S0 pec:	50	
22 55 41.3	−05 29 43	NGC 7416	12.4	3.2 x 0.7	13.1	SB(r)b II-III:	110	Mag 8.97v star S 6.′0.
22 56 12.8	+03 55 30	NGC 7422	13.4	0.9 x 0.8	12.9	SB(s)b: II	140	
22 57 19.3	−01 02 58	NGC 7428	12.5	2.4 x 1.3	13.6	(R)SAB(r)a pec:	160	Bright center with smooth envelope.
22 58 21.5	−01 11 04	NGC 7434	15.1	0.6 x 0.3	13.1	E/S0	62	
23 01 28.7	+01 45 10	NGC 7458	12.5	1.4 x 1.2	13.1	E	15	Bright center with uniform envelope.
23 01 43.0	+02 15 44	NGC 7460	13.0	1.4 x 1.0	13.2	SB(s)b pec:	37	
23 04 56.6	+02 34 37	NGC 7478	15.4	0.4 x 0.4	13.3	Sa		Almost stellar; mag 11.7 star NW 1.′0.
23 05 13.8	+02 32 56	NGC 7480	14.0	1.3 x 0.3	12.9	S0/a: sp	102	Mag 9.84v star W 5.′6.
23 05 38.7	+03 03 30	NGC 7482	13.6	0.7 x 0.5	12.5	E$^+$	41	MCG +0-58-31 SE 6.′0.
23 05 48.4	+03 32 40	NGC 7483	13.0	1.6 x 1.1	13.5	(R′)SAB(rs)a	110	Bright center, smooth envelope.
23 07 49.0	+00 56 23	NGC 7488	13.8	0.7 x 0.5	12.5	S0$^-$ pec:	24	
23 08 05.9	−05 57 59	NGC 7491	13.8	0.9 x 0.6	13.0	Sbc	172	Located 2.′5 N of mag 11.20v star.
23 11 41.1	−02 09 36	NGC 7506	12.9	1.7 x 1.1	13.4	(R′)SB(r)0$^+$:	102	Bright diffuse center, smooth envelope.
23 13 13.9	−02 06 02	NGC 7517	14.4	0.6 x 0.3	12.4	E$^+$?	148	Mag 9.57v star N 2.′3.
23 13 35.5	−01 43 54	NGC 7521	13.9	0.7 x 0.6	12.8	SB0°?	168	
23 13 46.7	−01 43 50	NGC 7524	15.1	0.9 x 0.3	13.6	S0/a	172	Small, faint slightly elongated anonymous galaxy E 6.′7.
23 14 12.0	−02 46 50	NGC 7530	14.6	0.7 x 0.4	13.1	S0/a	120	
23 14 22.3	−02 43 42	NGC 7532	13.8	1.4 x 0.5	13.3	S0$^+$ pec:	150	Mag 10.68v star 2.′2 S.
23 14 22.1	−02 02 01	NGC 7533	15.1	0.7 x 0.3	13.3	S0/a	134	Very faint, elongated anonymous galaxy S 3.′7.
23 14 26.7	−02 41 58	NGC 7534	13.8	1.0 x 0.6	13.1	IB(s)m III	12	Very faint star on S edge; mag 7.14v star NW 6.′3.

(Continued from previous page)

102 GALAXIES 102

RA h m s	Dec ° ′ ″	Name	Mag (V)	Dim ′ Maj x min	SB	Type Class	PA	Notes
23 14 34.6	+04 29 55	NGC 7537	13.2	2.2 x 0.6	13.4	SAbc:	79	
23 14 43.5	+04 32 03	NGC 7541	11.7	3.5 x 1.2	13.1	SB(rs)bc: pec II	102	Strong dark patch W of center.
23 14 57.1	−02 11 59	NGC 7544	15.1	0.8 x 0.3	13.4	S0	58	
23 15 05.6	−02 19 29	NGC 7546	15.1	1.0 x 0.8	14.7	SAB(r)bc: II-III	5	**MCG −1-59-8** N 2'0.
23 15 41.4	−02 22 45	NGC 7554	14.9	0.4 x 0.4	12.9	E		Stellar; located 0'8 W of NGC 7556's center.
23 15 44.5	−02 22 54	NGC 7556	12.7	2.5 x 1.6	14.1	S0⁻	123	NGC 7554 located 0'8 W of center; anonymous galaxy on SE edge.
23 16 37.7	−02 19 52	NGC 7566	13.3	1.3 x 0.7	13.0	(R′)SB(rs)a pec?	115	Star or bright knot on E edge and W edge.
23 17 22.8	−04 43 39	NGC 7576	12.9	1.2 x 1.0	12.9	SA(r)0⁺	165	
23 18 01.2	−04 39 01	NGC 7585	11.4	2.3 x 2.0	13.0	(R′)SA(s)0⁺ pec	105	Large bright center with smooth envelope.
23 18 15.9	+00 15 36	NGC 7589	14.1	1.1 x 0.7	13.7	SAB(rs)a:	102	Stellar nucleus; double star E 1'2.
23 18 22.2	−04 24 57	NGC 7592	13.5	1.3 x 1.1	13.7	S0⁺ pec:	57	Double system, bright extension N.
23 18 56.5	+00 14 39	NGC 7603	13.2	1.5 x 1.0	13.5	SA(rs)b: pec	165	Stellar **PGC 71041** SE 0'9.
23 21 19.5	+01 24 09	NGC 7629	13.8	1.0 x 0.8	13.4	SB0°?	179	Mag 11.4 star W 8'8; MCG +0-59-32 NE 5'6.
23 22 53.4	+01 26 33	NGC 7642	13.7	0.5 x 0.5	12.1	Sa pec?		**MCG +0-59-36** SE 2'8.
23 26 45.2	−04 57 58	NGC 7663	16.2	0.6 x 0.3	14.2	Irr III:	166	Mag 9.21v star SW 2'3.
23 24 22.9	−00 06 32	NGC 7667	13.9	1.6 x 1.1	14.4	SB(s)m pec III-IV	100	
22 08 09.4	+04 41 22	UGC 11915	13.8	0.7 x 0.6	12.7	S pec IV	39	Very knotty; possible small, round anonymous companion galaxy S 0'9.
22 24 11.6	+06 00 13	UGC 12021	14.2	1.1 x 0.5	13.4	Sb	100	Mag 8.13v star WSW 8'7.
22 24 16.4	+05 21 37	UGC 12023	14.4	0.8 x 0.6	13.4	SB(r)b:	150	Mag 12.2 star NW 3'9.
22 41 33.8	+00 23 55	UGC 12151	14.6	2.8 x 1.9	16.2	IB(s)m: V	0	Dwarf, low surface brightness.
22 50 06.2	−01 02 59	UGC 12208	14.2	1.0 x 0.7	13.7	SBd	70	
22 56 12.9	+05 23 00	UGC 12255	14.2	0.5 x 0.5	12.5	S?		
22 58 21.1	+02 17 51	UGC 12271	13.7	1.5 x 1.4	14.4	SAB(rs)c I-II	72	Almost stellar nucleus.
23 00 23.7	+01 37 33	UGC 12295	14.3	1.2 x 0.8	14.1	SAB(s)cd: II-III	95	Stellar nucleus, uniform envelope.
23 04 06.0	+01 54 25	UGC 12336	13.7	1.0 x 0.9	13.5	S0⁻:	0	Bright center, uniform envelope; mag 10.40v star WNW 3'4.
23 05 12.6	+00 50 01	UGC 12346	14.3	1.5 x 1.2	14.8	(R′)SB(rs)bc?	39	
23 06 31.5	+00 10 15	UGC 12363	13.7	1.1 x 0.5	12.9	S	30	
23 14 33.2	+00 14 07	UGC 12446	13.5	0.5 x 0.5	11.8	S0?		
23 14 45.4	+05 24 53	UGC 12451	14.9	1.5 x 0.4	14.2	Im	155	Mag 10.53v star NW 2'1.
23 15 02.9	+01 26 04	UGC 12452	14.8	1.8 x 0.2	13.5	Sd: sp	137	Mag 6.76v star SE 12'1.
23 15 53.2	+05 07 41	UGC 12461	13.9	1.1 x 1.1	14.0	S0?		
23 15 59.5	−01 50 50	UGC 12466	14.0	1.0 x 0.7	13.5	SBb	148	
23 16 58.8	+03 42 35	UGC 12475	14.6	1.5 x 0.2	13.2	Sb: sp	81	Mag 10.93v star W 2'5.
23 17 25.8	−01 35 16	UGC 12479	14.2	1.1 x 0.3	12.9	Sa	33	Mag 8.01v star NE 4'9.
23 18 53.9	−01 03 39	UGC 12492	13.5	1.1 x 0.8	13.2	S0	98	
23 19 37.5	+01 28 41	UGC 12508	13.6	0.9 x 0.4	12.3	S	135	
23 20 11.5	−01 51 26	UGC 12521	13.4	1.5 x 0.9	13.6	SA(s)cd: III	167	**CGCG 380-29** W 11'7; **CGCG 380-31** W 10'1.
23 21 51.6	+05 00 25	UGC 12547	14.3	1.2 x 0.6	13.8	SB?	151	Two main arms form open "S" shape; **PGC 71224** E 5'5.
23 21 57.2	+05 02 09	UGC 12548	13.8	1.2 x 0.3	12.5	Sa?	118	**UGC 12555** NE 10'4.

GALAXY CLUSTERS

RA h m s	Dec ° ′ ″	Name	Mag 10th brightest	No. Gal	Diam ′	Notes
22 23 54.0	−01 35 00	A 2440	16.0	32	15	All members anonymous, stellar/almost stellar.
22 35 48.0	+01 28 00	A 2457	16.0	53	22	All members anonymous; stellar/almost stellar.

103 GALAXIES 103

RA h m s	Dec ° ′ ″	Name	Mag (V)	Dim ′ Maj x min	SB	Type Class	PA	Notes
21 10 48.5	−01 58 58	CGCG 375-3	15.1	0.4 x 0.4	12.9			Mag 9.61v star SW 2'6; mag 12.5 star E 1'1.
21 13 32.7	−01 22 49	CGCG 375-14	14.7	0.5 x 0.4	12.8		81	Mag 10.81v star E 5'1.
21 14 00.6	+00 32 04	CGCG 375-16	15.5	0.4 x 0.3	13.1		174	Mag 9.97v star N 8'0.
21 20 59.0	+03 40 18	CGCG 375-30	14.1	0.5 x 0.5	12.4			Anonymous galaxy S 1'2.
21 23 24.4	+02 11 31	CGCG 375-33	14.8	0.3 x 0.3	12.1			Mag 8.22v star WNW 6'9.
21 31 48.4	+03 35 16	CGCG 376-3	14.0	0.9 x 0.8	13.5		69	Mag 13.2 star E of elongated nucleus.
21 32 35.0	+02 41 38	CGCG 376-7	13.9	0.5 x 0.4	12.0		27	Mag 10.07v star SE 5'6.
21 40 02.4	+01 12 24	CGCG 376-21	15.7	0.4 x 0.3	13.3		15	Anonymous galaxy 0'6 N; mag 8.05v star SE 2'3.
21 48 28.2	+01 52 32	CGCG 376-33	14.5	0.6 x 0.6	13.3			Mag 10.9 star SW 3'2.
21 51 20.4	−01 39 00	CGCG 376-41	14.5	0.5 x 0.5	13.0	E		Mag 9.57v star N 3'7.
22 04 17.8	+04 40 00	CGCG 403-19	14.2	0.6 x 0.4	12.7	E;BLLac	108	Sits in a triangle of mag 13 stars.
21 11 29.8	+05 03 20	IC 1361	14.2	1.0 x 1.0	14.0	S?		Bright nucleus offset SW with very faint envelope NE.
21 13 24.9	+02 46 14	IC 1364	13.8	1.0 x 0.6	13.1		130	Mag 9.37v star SE 2'7.
21 13 55.8	+02 33 51	IC 1365	13.7	1.1 x 0.7	13.5	E⁺	57	Bright knot or star W of core; mag 12.3 star E 1'7.
21 14 12.7	+02 10 36	IC 1368	13.4	1.1 x 0.4	12.4	Sa? sp	48	Mag 9.96v star W 9'7.
21 15 14.3	+02 11 28	IC 1370	14.6	0.4 x 0.3	12.1		5	Mag 9.38v star N 3'5.
21 20 15.8	−04 52 35	IC 1371	13.8	0.9 x 0.7	13.2	S0⁻ pec:	15	Close pair of mag 12.1, 10.77v stars W 4'5.
21 20 37.4	+01 05 31	IC 1373	14.5	0.3 x 0.3	11.8			Almost stellar; UGC 11724 S 2'8.
21 20 59.8	+03 59 06	IC 1375	14.2	1.0 x 0.7	13.7		138	Very faint star N of nucleus.
21 25 26.8	+04 18 49	IC 1377	14.3	0.8 x 0.7	13.5	Sb	120	Mag 10.37v star NE 5'1.
21 27 33.6	−01 11 19	IC 1381	14.6	0.8 x 0.5	13.4		57	Mag 10.86v star N 6'1; note: **IC 1383** is 2'3 E of this star.
21 27 53.1	−01 22 09	IC 1384	14.7	0.7 x 0.5	13.4	Irr		Mag 12.4 star ENE 8'5.
21 28 51.2	−01 04 14	IC 1385	15.0	0.4 x 0.4	12.9	SBcd		Mag 12.5 star W 8'6.
21 29 34.5	−01 21 05	IC 1387	14.5	0.4 x 0.3	12.0		6	Mag 12.0 star NNE 4'9.
21 32 24.9	−01 51 45	IC 1390	14.5	0.5 x 0.4	12.6		162	Mag 9.16v star S 6'2.
21 46 59.5	+01 42 49	IC 1401	13.8	1.8 x 0.7	13.9	SAB(r)bc: I-II	175	Mag 10.48v star NW 11'0.
21 50 49.8	+02 01 12	IC 1405	13.8	0.9 x 0.7	13.2	SBab	115	Mag 10.36v star SW 6'3; stellar **IC 1406** SE 4'3.
21 56 00.8	−01 31 01	IC 1411	13.4	0.9 x 0.5	12.5	E	31	Mag 11.9 star SSE 8'0; note: **CGCG 377-2** is 1'6 W of this star.
22 02 00.0	+04 23 05	IC 1418	14.0	0.9 x 0.9	13.6	S?		Mag 8.31v star N 5'1.
22 03 12.9	+04 17 46	IC 1423	13.9	1.0 x 0.5	13.0	SBb	35	Mag 8.35v star NE 8'1.
22 03 24.5	+02 35 39	IC 1425	14.3	0.4 x 0.4	12.1			**IC 1422** W 6'1.
22 15 45.1	+02 03 54	IC 1437	13.5	1.0 x 0.9	13.2	(R′)SA(r)0/a	72	Mag 8.16v star SSW 7'6.
21 11 30.6	−02 01 56	IC 5090	13.5	1.2 x 0.5	12.8	Sa pec?	26	Mag 12.1 star and compact galaxy MCG +0-54-2 W 5'1.
21 28 10.9	+02 28 23	IC 5111	14.4	0.7 x 0.6	13.3	Sa	27	Mag 10.49v star N 3'4.
20 56 41.0	+00 40 24	MCG +0-53-10	15.3	0.7 x 0.7	14.4	Sbc		Mag 13.3 star NW edge; mag 11.8 star SW 1'5.
20 56 58.7	+02 17 07	MCG +0-53-11	15.3	0.9 x 0.7	14.7	Sc	27	Stellar nucleus, smooth envelope; **CGCG 374-33**, located between two close mag 12 stars, E 2'2.
20 59 36.6	+02 10 00	MCG +0-53-13	14.2	1.1 x 0.7	13.7	Sb	12	Mag 11.9 star S 1'8.

GALAXIES

RA h m s	Dec ° ′ ″	Name	Mag (V)	Dim ′ Maj x min	SB	Type Class	PA	Notes
21 11 07.7	−02 02 21	MCG +0-54-2	14.1	0.5 x 0.5	12.7	E		Compact; mag 12.1 star E 1′.1.
21 13 13.1	−00 22 01	MCG +0-54-5	14.7	0.7 x 0.3	12.8	Sbc	45	
21 27 40.2	+03 14 46	MCG +0-54-18	15.8	0.9 x 0.8	15.3	Sb	42	
21 29 13.6	−00 17 39	MCG +0-54-25	14.5	0.5 x 0.4	12.6		60	
21 31 00.0	−00 00 03	MCG +0-54-29	14.7	0.8 x 0.6	13.8	S0/a	102	
21 31 43.0	+00 21 28	MCG +0-55-1	14.9	0.7 x 0.3	13.0		30	Mag 11.1 star ESE 2′.5.
21 32 41.8	−00 07 40	MCG +0-55-5	14.5	0.9 x 0.4	13.3		9	
21 43 37.7	+02 38 06	MCG +0-55-12	14.6	1.2 x 0.3	13.3	Sb	21	Mag 12.0 star E 2′.5.
21 54 18.8	+00 21 15	MCG +0-55-28	14.9	0.6 x 0.4	13.2	Sb	0	Mag 10.43v star S 2′.8.
21 56 15.0	+02 10 20	MCG +0-56-2	14.3	0.8 x 0.4	12.9	Sc	96	
22 07 56.6	+00 22 11	MCG +0-56-9	14.8	0.8 x 0.6	13.8	SBab	6	
22 09 06.3	+02 00 51	MCG +0-56-10	14.9	0.8 x 0.7	14.1		0	Core offset E; mag 12.3 star NW 2′.8.
22 10 29.0	−00 01 14	MCG +0-56-11	15.8	0.9 x 0.4	14.5		132	
22 11 53.0	+00 06 28	MCG +0-56-13	14.2	0.9 x 0.6	13.4	SB(rs)bc I-II	24	Mag 10.03v star E 3′.1.
22 11 59.4	−00 15 11	MCG +0-56-14	15.2	0.7 x 0.7	14.3			
22 14 20.8	+01 01 05	MCG +0-56-15	14.6	0.9 x 0.5	13.6		57	Located between a pair of stars, mags 13.9 and 12.5.
21 18 52.7	+05 50 55	MCG +1-54-5	14.4	0.8 x 0.3	12.8	S?	6	
21 53 05.6	+04 17 58	MCG +1-55-15	14.6	0.9 x 0.2	12.6	S?	72	Mag 12.3 star NW 3′.4.
21 55 43.7	+05 48 24	MCG +1-56-1	14.6	0.6 x 0.6	13.3	Sa		
20 56 23.2	−03 54 30	MCG −1-53-18	13.2	1.0 x 0.9	13.0	SBbc	3	Mag 10.94v star W 1′.1.
20 56 33.3	−03 49 46	MCG −1-53-19	13.8	0.7 x 0.7	12.8			
21 02 14.9	−05 06 47	MCG −1-53-21	14.9	0.7 x 0.5	13.7	SBb	114	
21 10 05.2	−03 42 07	MCG −1-54-3	13.6	1.9 x 1.0	14.2	SB(rs)bc: II	36	Mag 10.40v star SE 5′.9.
21 20 00.9	−03 55 23	MCG −1-54-12	14.5	0.9 x 0.6	13.6	S?	51	= **Hickson 89A**. **Hickson 89C** and **Hickson 89D** ENE 1′.8. **Hickson 89B** ENE 4′.8.
21 26 00.0	−03 48 37	MCG −1-54-16	15.2	1.3 x 0.2	13.6	S?	33	
21 31 00.8	−03 59 03	MCG −1-54-20	14.6	0.8 x 0.7	13.9	S0/a	126	
21 32 09.8	−05 43 43	MCG −1-55-1	14.3	0.8 x 0.8	13.6	Sbc		Mag 2.89v star "Sadalsuud", Beta Aquarii NW 13′.0.
21 42 04.8	−04 56 40	MCG −1-55-3	14.1	0.8 x 0.5	12.9	Sbc	129	
21 47 41.5	−04 09 46	MCG −1-55-11	13.8	1.1 x 0.3	12.5	(R)SB(s)b pec	174	
21 01 07.6	−00 11 48	NGC 7001	12.9	1.4 x 1.1	13.3	SAB(rs)ab:	162	
21 14 56.3	+02 50 04	NGC 7046	13.1	1.9 x 1.3	13.9	SB(rs)cd II-III	115	Bright nucleus.
21 16 27.7	−00 49 36	NGC 7047	13.4	1.2 x 0.7	13.0	SAB(r)b: II-III:	107	Mag 8.51v star S 5′.5.
21 28 05.9	−01 38 49	NGC 7069	13.4	1.3 x 0.9	13.4	SAB(s)0$^-$?	20	
21 29 59.6	+02 24 48	NGC 7077	13.1	0.8 x 0.7	12.4	S0$^-$ pec?	37	Mag 787v star NE 4′.6.
21 31 24.2	+02 29 27	NGC 7081	12.7	1.3 x 1.3	13.1	Sb pec?		
21 44 52.6	−03 37 13	NGC 7121	13.4	1.1 x 0.5	12.6	SB(r)bc? II	12	
21 51 47.3	+03 01 01	NGC 7146	14.3	0.9 x 0.6	13.4	SAB(s)ab:	80	Stellar galaxy **II Zw 152** ESE 2′.6; mag 10.42v star N 2′.6.
21 51 58.4	+03 04 18	NGC 7147	13.5	1.1 x 0.9	13.3	SB(s)0$^+$:	5	**CGCG 376-46** SE 1′.6; an anonymous galaxy lies just S of CGCG 376-46.
21 52 23.6	+03 25 33	NGC 7148	14.3	0.5 x 0.4	12.4		129	Multi-galaxy system; stellar **PGC 67535** on NW edge.
21 52 11.6	+03 18 01	NGC 7149	13.2	1.3 x 0.9	13.3	E	25	
21 54 33.6	+02 56 34	NGC 7156	12.5	1.6 x 1.4	13.2	SAB(rs)cd: I-II	96	
21 56 23.7	+01 21 47	NGC 7164	14.2	0.9 x 0.7	13.6	SAB(rs)0$^+$ pec:	118	
22 01 26.4	−05 26 00	NGC 7170	13.8	1.2 x 0.8	13.6	SA0$^-$?	158	
22 01 43.6	−01 57 38	NGC 7181	14.0	1.0 x 0.9	13.7	S0° pec:	95	Bright center, faint envelope.
22 01 51.5	−02 11 50	NGC 7182	14.4	0.9 x 0.5	13.4	SB(rs)0$^+$:	71	
22 03 16.1	+00 34 16	NGC 7189	13.5	1.0 x 0.7	12.9	SB(rs)b pec:	115	Bright N-S bar.
22 05 14.2	−00 38 51	NGC 7198	13.3	1.5 x 1.0	13.6	S0 pec	5	
22 08 34.5	+00 30 42	NGC 7215	13.9	1.1 x 0.5	13.1	S0°?	88	
22 10 51.6	+02 06 21	NGC 7222	13.8	1.2 x 1.2	14.0	SB(r)b II-III		Stellar nucleus in short bar.
22 15 01.5	−05 03 13	NGC 7239	13.8	1.1 x 0.7	13.4	SAB0$^-$	75	
20 59 47.8	−01 52 53	UGC 11657	13.6	2.4 x 1.1	14.5	Pec	21	Double system; published dimensions combine **UGC 11658** N with UGC 11657 S. UGC 11657 has several superimposed stars; **UGC 11658** a large plume N.
21 07 42.8	+03 52 25	UGC 11680	13.8	2.1 x 0.7	14.1	Scd:	70	Double system; one component is small, round and compact, at the end of a long bridge NE of larger bright core.
21 12 10.5	−01 28 33	UGC 11695	14.2	1.3 x 0.7	13.9	SA(r)b pec: I-II	115	A very faint bridge extends to a small companion galaxy S 1′.2.
21 17 46.3	+02 22 05	UGC 11714	14.6	0.9 x 0.6	13.8	Scd:	39	
21 20 17.4	−01 41 05	UGC 11723	13.9	1.7 x 0.2	12.6	Sb sp	33	Mag 8.62v star SE 4′.4.
21 20 38.5	+01 02 45	UGC 11724	14.3	1.5 x 0.3	13.3	SBcd: II	145	Uniform surface brightness; mag 12.5 star SE 1′.3; IC 1373 N 2′.8.
21 31 39.9	+02 27 05	UGC 11760	13.4	1.0 x 0.8	13.0	SAB(s)bc pec? II	145	Stellar nucleus, faint, uniform envelope, a pair of stars, mags 12.4 and 11.3, N 1′.5.
21 41 00.0	+01 20 04	UGC 11789	13.9	0.9 x 0.4	12.6	Sb	105	Stellar nucleus; mag 7.91v star SE 2′.4.
21 41 29.8	+00 53 40	UGC 11790	14.1	1.5 x 1.0	14.4	SA(rs)d: III	160	Stellar nucleus, smooth envelope.
21 42 12.8	+05 36 50	UGC 11792	14.8	1.5 x 0.2	13.3	Scd:	160	Located 4′.0 S of mag 5.31v star 7 Pegasi.
21 48 26.6	−01 40 22	UGC 11814	13.5	0.8 x 0.5	12.4	I?	160	Mag 12.4 star S 1′.2.
21 49 07.4	+00 26 49	UGC 11816	13.6	1.1 x 0.8	13.3	SB(rs)c: III	102	
21 51 05.3	+02 46 23	UGC 11828	14.1	0.9 x 0.8	13.6	Sab		**CGCG 376-40** S 4′.4; a close group of four stellar PGC galaxies centered NW 5′.8.
21 55 59.4	+05 54 10	UGC 11851	13.8	1.0 x 0.7	13.3	Sa	24	Small, elongated E-W galaxy on N edge; mag 10.7 star S 1′.1.
21 56 18.7	−01 10 23	UGC 11853	14.2	1.0 x 0.7	13.7	SB(s)c II-III	75	Faint, elongated center, very faint envelope; pair of stars, mags 12.0 and 12.4, E 1′.8.
21 58 07.4	+01 00 32	UGC 11859	14.4	2.3 x 0.2	13.4	Sbc: sp	63	
22 06 20.5	+02 20 56	UGC 11907	13.7	1.0 x 0.6	13.0	S0/a	110	Mag 6.51v star NNE 7′.1.
22 08 09.4	+04 41 22	UGC 11915	13.8	0.7 x 0.6	12.7	S pec IV	39	Very knotty; possible small, round anonymous companion galaxy S 0′.9.

GALAXY CLUSTERS

RA h m s	Dec ° ′ ″	Name	Mag 10th brightest	No. Gal	Diam ′	Notes
22 05 24.0	−05 35 00	A 2415	15.9	40	13	All members anonymous; several prominent GX near center.

GLOBULAR CLUSTERS

RA h m s	Dec ° ′ ″	Name	Total V m	B ★ V m	HB V m	Diam ′	Conc. Class Low = 12 High = 1	Notes
21 33 27.0	−00 49 12	NGC 7089	6.6	13.1	16.1	16.0	2	= **M 2**. Visual observers may be able to detect a curving dark lane in the NE sector of this cluster.

RA h m s	Dec ° ′ ″	Name	Mag (V)	Dim ′ Maj x min	SB	Type Class	PA	Notes
19 59 29.9	+01 05 48	CGCG 372-6	14.3	0.5 x 0.3	12.1		117	Very close pair of mag 11.6, 12.5 stars E 1'.7.
20 36 09.2	+01 44 03	CGCG 373-40	14.4	0.6 x 0.4	12.7		159	Forms triangle with mag 11.5 star E 3'.0 and mag 11.5 star NE 3'.0.
20 45 49.5	+00 10 37	CGCG 374-9	13.9	0.5 x 0.3	11.7		48	**UGC 11626** E 10'.6; mag 7.57v star NW 14'.8.
20 55 22.3	+02 21 13	CGCG 374-29	14.4	0.4 x 0.4	12.2			Mag 12.1 star S edge.
20 22 36.8	+03 55 48	CGCG 399-3	14.4	0.8 x 0.5	13.3		90	Mag 14.4 star superimposed on W end.
20 29 42.3	+03 43 07	CGCG 399-13	14.2	0.4 x 0.2	11.3		114	Mag 11.3 star W 3'.3.
20 31 20.8	+04 36 24	CGCG 399-16	14.5	0.8 x 0.4	13.1		168	Mag 10.67v star NNW 2'.7.
20 41 50.1	+05 35 37	CGCG 399-28	14.1	0.5 x 0.4	12.2		36	Mag 13.6 star SW edge; mag 10.02v star S 3'.9.
20 23 15.7	+00 39 49	IC 1317	13.8	0.6 x 0.5	12.5	E⁺?	80	Mag 10.19v star ESE 5'.2.
20 26 25.8	+02 54 32	IC 1320	13.6	1.0 x 0.6	12.9	SB(s)b?	81	Mag 6.64v star NW 3'.8.
20 45 15.0	−05 37 22	IC 5050	13.5	1.3 x 0.9	13.5	SAB(r)b: II-III	75	Mag 10.58v star SE 4'.8.
19 57 39.5	−00 55 05	MCG +0-51-2	14.5	0.9 x 0.3	12.9		108	
20 02 39.7	+01 51 59	MCG +0-51-6	14.2	0.9 x 0.8	13.9	E/S0	12	Numerous stars superimposed, in star rich area.
20 06 45.6	−01 44 30	MCG +0-51-7	14.8	0.7 x 0.7	13.9			Small, faint, elongated anonymous galaxy W 1'.9.
20 17 32.7	+00 00 05	MCG +0-51-12	14.8	1.0 x 0.6	14.1	SBc	45	Mag 15.4 star on W edge; mag 13.8 star W 0'.8.
20 29 25.2	−00 10 24	MCG +0-52-16	13.9	0.9 x 0.4	12.6	Sab	105	
20 29 49.2	−02 01 47	MCG +0-52-17	14.8	0.8 x 0.2	12.6	Sb	135	Mag 8.91v star S 2'.7.
20 30 00.7	−00 51 22	MCG +0-52-19	14.5	0.6 x 0.4	12.8	S?	174	**MCG +0-52-22** E 5'.7.
20 30 30.9	+01 51 54	MCG +0-52-23	13.8	0.8 x 0.3	12.1	Sb	0	Mag 11.0 star N 1'.3.
20 31 13.5	+00 16 22	MCG +0-52-28	15.1	1.1 x 0.2	13.3	Sc	33	Mag 11.2 star W 1'.9.
20 35 30.6	−02 45 57	MCG +0-52-38	13.6	1.9 x 0.5	13.4	SAab pec sp	69	Mag 11.9 star WNW 7'.8; **MCG +0-52-37** NW 7'.1.
20 35 48.8	−02 39 18	MCG +0-52-40	14.4	0.9 x 0.4	13.1		168	Located between a pair of E- W oriented mag 12-13 stars.
20 39 26.5	+02 01 05	MCG +0-52-43	14.4	0.8 x 0.3	12.7	Sb	27	**CGCG 373-45** N 4'.2; mag 7.83v star N 7'.9.
20 46 15.8	−00 13 19	MCG +0-53-2	14.1	0.7 x 0.7	13.2	S?		Mag 6.76v star S 7'.9.
20 52 22.5	+00 04 28	MCG +0-53-8	14.0	1.0 x 0.8	13.6	Sa	111	Mag 12.5 star SW edge.
20 56 41.0	+00 40 24	MCG +0-53-10	15.3	0.7 x 0.7	14.4	Sbc		Mag 13.3 star NW edge; mag 11.8 star SW 1'.5.
20 56 58.7	+02 17 07	MCG +0-53-11	15.3	0.9 x 0.7	14.7	Sc	27	Stellar nucleus, smooth envelope; **CGCG 374-33**, located between two close mag 12 stars, E 2'.2.
20 59 36.6	+02 10 00	MCG +0-53-13	14.2	1.1 x 0.7	13.7	Sb	12	Mag 11.9 star S 1'.8.
20 28 40.5	+04 20 42	MCG +1-52-9	14.1	0.7 x 0.3	12.2	Sb	6	Stars superimposed N and S ends; mag 9.15v star SE 1'.4.
20 15 48.9	−02 54 07	MCG −1-51-1	14.1	1.9 x 0.5	13.9	SB(rs)bc? II-III	156	Mag 14.3 star E edge; mag 12.1 star NE 2'.2.
20 26 26.0	−05 04 11	MCG −1-52-2	14.2	0.5 x 0.4	12.3		3	
20 26 39.8	−05 04 31	MCG −1-52-3	14.3	0.9 x 0.3	12.7	S?	96	
20 38 24.0	−04 07 05	MCG −1-52-14	13.9	1.3 x 0.8	13.8	(R)SA(s)0/a?	96	MCG −1-52-14 W 1'.0.
20 38 54.4	−05 38 25	MCG −1-52-16	12.9	2.7 x 0.8	13.6	(R′)Sa:	24	Mag 12.8 star SW edge.
20 43 20.7	−04 30 02	MCG −1-52-18	14.6	0.4 x 0.4	12.5	S0/a		Mag 9.03v star NNE 3'.3.
20 43 46.1	−03 12 51	MCG −1-53-2	13.5	0.7 x 0.4	12.0	S0	75	**MCG −1-53-1** N 0'.3, **MCG −1-53-3** S 0'.5.
20 45 14.5	−05 29 05	MCG −1-53-4	13.3	0.8 x 0.6	12.3		120	Mag 10.53v star ESE 7'.0.
20 45 41.2	−04 57 04	MCG −1-53-6	14.0	1.2 x 0.7	13.7	(R′)Sab	87	
20 45 40.0	−05 35 26	MCG −1-53-7	14.6	0.6 x 0.5	13.2	SBbc	174	Mag 13.6 star on N edge; mag 10.53v star N 4'.2.
20 46 21.1	−02 48 46	MCG −1-53-8	13.3	1.1 x 0.6	12.7	S?	42	= **Mkn 896**. Mag 8.25v star NE 9'.4.
20 48 12.2	−03 52 24	MCG −1-53-9	13.9	1.1 x 0.8	13.6	SB(r)bc: II	114	
20 48 14.9	−03 53 01	MCG −1-53-10	15.2	0.9 x 0.4	13.9	SAB(s)c? II-III	3	
20 48 13.6	−05 47 47	MCG −1-53-11	14.9	0.9 x 0.5	13.9	SB?	171	
20 56 23.2	−03 54 30	MCG −1-53-18	13.2	1.0 x 0.9	13.0	SBbc	3	Mag 10.94v star W 1'.1.
20 56 33.3	−03 49 46	MCG −1-53-19	13.8	0.7 x 0.7	12.8			
21 02 14.9	−05 06 47	MCG −1-53-21	14.9	0.7 x 0.5	13.7	SBb	114	
20 21 35.1	−02 34 09	NGC 6900	13.5	0.8 x 0.6	12.6	Sb	178	
20 27 46.2	−03 04 38	NGC 6915	12.2	1.4 x 1.0	12.4	(R′)Sab?	80	Large, diffuse center; mag 10.58v star 2'.1 NW.
20 29 52.8	−02 11 26	NGC 6922	13.5	1.3 x 1.0	13.7	SA(rs)c pec:	150	Dark lane N.
20 33 06.1	−02 01 46	NGC 6926	12.4	1.9 x 1.3	13.3	SB(s)bc pec	0	
20 33 21.7	−02 02 16	NGC 6929	13.4	0.9 x 0.7	12.8	SA0⁺ pec:	77	
20 36 23.6	−04 37 08	NGC 6941	12.8	2.0 x 1.4	13.8	SAB(rs)b II	115	
20 39 00.6	−04 58 21	NGC 6945	13.4	1.7 x 0.9	13.8	S0⁻	126	Located 2'.9 SSW of a mag 6.46v star.
20 44 03.3	+03 12 32	NGC 6954	13.2	1.0 x 0.6	12.5	SA0⁺?	68	
20 44 18.1	+02 35 41	NGC 6955	13.6	1.4 x 1.3	14.1	SAB(r)bc II	30	Bright, almost stellar nucleus.
20 44 47.6	+02 34 49	NGC 6957	14.4	0.5 x 0.4	12.5	Scd? III	171	
20 47 07.3	+00 25 45	NGC 6959	13.7	0.7 x 0.3	11.9	(R)S0° pec:	57	Anonymous galaxy WNW 1'.1.
20 47 10.6	+00 21 47	NGC 6961	13.7	0.6 x 0.5	12.4	E⁺ pec?	134	**CGCG 374-11** W 8'.2.
20 47 19.0	+00 19 15	NGC 6962	12.1	2.9 x 2.3	14.1	SAB(r)ab	75	NGC 6964 on SE edge.
20 47 24.3	+00 18 00	NGC 6964	13.0	1.7 x 1.3	13.8	E⁺ pec:	171	Very faint, very small anonymous galaxy on NE edge.
20 47 20.5	+00 29 01	NGC 6965	14.0	0.6 x 0.4	12.4	S0°:	66	Very small, very faint anonymous galaxy or star on S edge.
20 47 34.1	+00 24 39	NGC 6967	13.1	1.0 x 0.6	12.4	SB(rs)0⁺?	105	Mag 10.76v star on E edge.
20 49 23.9	+05 59 40	NGC 6971	13.7	1.1 x 0.9	13.5	Sb	60	
20 52 13.1	−05 47 55	NGC 6975	14.9	0.9 x 0.2	12.9	S?	68	
20 52 26.0	−05 46 23	NGC 6976	14.0	0.8 x 0.7	13.2	SAB(r)bc? II	5	
20 52 29.8	−05 44 47	NGC 6977	13.2	1.2 x 0.9	13.2	SB(r)a pec:	162	
20 52 35.5	−05 42 42	NGC 6978	13.3	1.5 x 0.7	13.2	Sb II	125	
21 01 07.6	−00 11 48	NGC 7001	12.9	1.4 x 1.1	13.3	SAB(rs)ab:	162	
19 49 20.5	+04 09 11	UGC 11481	14.2	1.0 x 0.3	12.7	S	127	
19 51 27.1	+03 33 37	UGC 11488	13.7	1.2 x 0.9	13.6	Sbc	25	Faint envelope, numerous stars superimposed.
19 52 02.3	+04 46 41	UGC 11489	13.8	1.1 x 1.0	13.8	Scd:	39	Numerous stars superimposed, in star rich area.
19 54 58.4	+05 52 19	UGC 11493	13.3	2.1 x 1.5	14.4	Scd:	170	Very filamentary, many superimposed stars; in star rich area.
19 55 38.0	+02 10 52	UGC 11497	13.5	1.1 x 0.9	13.3	SBb	86	Bright N-S bar, faint envelope.
19 57 15.1	+05 53 23	UGC 11498	12.7	3.1 x 1.0	13.8	SBb:	75	Numerous superimposed stars, a few small knots; in star rich area.
19 58 37.1	+02 36 10	UGC 11501	13.1	1.4 x 1.1	13.4	SAB(rs)c	0	Stellar nucleus with mag 14.2 star NW.
20 02 28.1	+01 38 49	UGC 11505	13.9	1.7 x 0.6	13.8	Sb:	163	Stellar nucleus, stars superimposed; mag 10.7 star NW 2'.8.
20 10 53.8	+02 08 10	UGC 11521	13.6	1.0 x 0.7	13.1	S0	135	Bright center; small, very faint envelope.
20 11 49.5	+05 45 45	UGC 11522	13.7	1.0 x 0.5	12.8	Sbc	3	**CGCG 398-20** N 4'.1.
20 12 01.1	+05 30 57	UGC 11523	14.1	0.9 x 0.9	13.7	Scd:		Small core with stellar nucleus, offset to the w, faint envelope E and S.
20 12 03.9	+05 45 46	UGC 11524	13.4	1.0 x 0.9	13.1	SA(r)c: I	90	Strong dark lane between bright core and single N-S arm E; mag 10.6 star E 0'.9.
20 12 18.5	+01 57 25	UGC 11525	14.3	1.5 x 0.2	12.9	Sab: sp	53	Mag 11.35v star S 1'.9.
20 13 54.5	−01 09 30	UGC 11527	13.4	1.7 x 0.3	12.5	SB(s)ab?	21	
20 18 38.2	−00 08 58	UGC 11537	13.6	2.1 x 0.6	13.7	Sc:	82	
20 22 26.1	+00 17 32	UGC 11541	15.9	1.7 x 0.8	16.1	SB(s)d: III-IV	140	Mag 13.4 star NW edge; mag 11.8 star NW 2'.5.
20 25 12.2	+05 15 45	UGC 11555	13.3	1.7 x 0.8	13.5	SA(s)bc I	21	Two main arms; located between a mag 11.3 star E and a mag 12.0 star W.
20 26 09.6	+01 09 16	UGC 11559	13.1	1.3 x 1.0	13.3	S0	174	**MCG +0-52-8** S 3'.5.
20 26 38.1	+02 38 23	UGC 11561	13.4	1.0 x 0.8	13.1	S0?	153	Bright core, uniform envelope; mag 11.8 star NW 0'.9; mag 12.4 star E 1'.1.
20 26 42.1	+02 41 41	UGC 11562	13.5	1.1 x 1.0	13.4	SA(s)bc: I-II	120	Bright nucleus, moderately dark lane S; mag 9.96v star N 1'.4.

(Continued from previous page)

GALAXIES

RA h m s	Dec ° ′ ″	Name	Mag (V)	Dim ′ Maj x min	SB	Type Class	PA	Notes
20 28 12.1	+00 17 16	UGC 11566	13.3	0.6 x 0.5	11.8		55	Forms a triangle with a mag 11.8 star SW and a mag 12.4 star NW.
20 28 09.7	+00 28 20	UGC 11567	14.2	0.8 x 0.8	13.6	SB?		Several stars superimposed; **CGCG 373-15** N 1′.1.
20 30 07.9	+03 03 01	UGC 11575	14.0	1.5 x 0.5	13.5	SB(s)d pec: II-III	122	Numerous stars superimposed.
20 30 21.5	−01 00 31	UGC 11576	14.2	1.1 x 1.0	14.2	S0/a	39	**MCG +0-52-24** E 2′.4.
20 30 35.4	+01 22 31	UGC 11577	13.4	1.5 x 1.3	14.0	SB(r)cd pec II-III	45	Mag 11.3 star SW 2′.3.
20 31 01.5	−01 56 11	UGC 11581	14.0	1.2 x 0.7	13.7	SB(rs)0/a:	40	**MCG +0-52-30** ESE 6′.9.
20 31 49.5	+01 32 35	UGC 11584	14.7	1.7 x 0.2	13.3	Scd: sp	53	Close pair of mag 14 stars SW end.
20 32 16.0	−02 15 00	UGC 11585	13.4	1.7 x 1.4	14.2	(R′)SAB(rs)bc pec: II	63	Mag 13.7 star S edge.
20 33 18.7	+01 45 27	UGC 11591	14.4	1.1 x 0.5	13.6	S	143	Three bright knots and/or superimposed stars.
20 35 04.6	+01 56 06	UGC 11595	14.4	1.9 x 0.2	13.2	Sbc: sp	12	
20 39 44.8	+02 01 50	UGC 11607	13.9	0.9 x 0.3	12.4	S?	28	Mag 9.23v star S 3′.4.
20 44 37.4	−01 43 15	UGC 11622	14.2	1.0 x 0.4	13.0	Sb pec sp	48	
20 48 00.5	−00 10 42	UGC 11631	14.4	1.0 x 0.2	12.5	Im:	63	Mag 8.46v star E 3′.6.
20 55 27.7	−01 13 35	UGC 11649	13.1	1.4 x 1.1	13.5	SB(r)a pec	70	Bright core with dark patches E and W.
20 59 47.8	−01 52 53	UGC 11657	13.6	2.4 x 1.1	14.5	Pec	21	Double system; published dimensions combine **UGC 11658** N with UGC 11657 S. UGC 11657 has several superimposed stars; **UGC 11658** a large plume N.

PLANETARY NEBULAE

RA h m s	Dec ° ′ ″	Name	Diam ″	Mag (P)	Mag (V)	Mag cent ★	Alt Name	Notes
20 00 39.2	+01 43 41	NGC 6852	28	12.8	12.6:	17.9	PK 42−14.1	Mag 12.8 star W 0′.5; mag 7.44v star WSW 4′.6.
19 58 27.0	+03 03 00	PN G43.5−13.4	67	16.0		19.3	PK 43−13.1	= **Abell 67**; mag 13.8 star NNW 1′.8; mag 12.9 stars SE 2′.4.

GALAXIES

RA h m s	Dec ° ′ ″	Name	Mag (V)	Dim ′ Maj x min	SB	Type Class	PA	Notes
19 43 26.3	−01 10 37	MCG +0-50-1	17.2	0.8 x 0.3	15.5	Sbc	171	
19 49 20.5	+04 09 11	UGC 11481	14.2	1.0 x 0.3	12.7	S	127	
19 51 27.1	+03 33 37	UGC 11488	13.7	1.2 x 0.9	13.6	Sbc	25	Faint envelope, numerous stars superimposed.

OPEN CLUSTERS

RA h m s	Dec ° ′ ″	Name	Mag	Diam ′	No. ★	B ★	Type	Notes
18 44 19.1	−04 55 53	AL 5		2.2			cl	Faint, circular.
18 54 49.4	+05 32 56	Archinal 1		1.5	24	13.4	cl	
18 48 06.0	−05 51 07	Bas 1	8.9	5.5	94	12.6	cl	Few stars; small brightness range; not well detached. Known as **Apriamasvili** cluster by discovery priority.
18 44 57.5	−01 09 00	Be 79		7	60	15.0	cl	
18 54 20.6	−01 13 12	Be 80		3	20	15.0	cl	
19 38 21.7	+00 20 44	Cr 401	7.0	1			cl	Moderately rich in stars; moderate brightness range; not well detached.
18 49 47.0	+04 58 00	Cz 38		10	80		cl	Rich in stars; moderate brightness range; strong central concentration; detached.
18 40 30.0	−04 05 00	Do 32		8	40		cl	Moderately rich in stars; moderate brightness range; slight central concentration; detached. (A) Nothing obvious in 24′ field.
18 38 54.0	+05 26 00	IC 4756	4.6	40	466	8.0	cl	Rich in stars; large brightness range; slight central concentration; detached. (A) Also known as the **Graff Cluster** or **Graff 1**.
18 50 45.8	−05 12 18	NGC 6704	9.2	6	71	12.0	cl	Moderately rich in stars; moderate brightness range; strong central concentration; detached.
19 00 45.3	−00 27 00	NGC 6735		8	35	12.0	cl	
19 07 49.0	+04 16 00	NGC 6755	7.5	15	157	11.0	cl	Rich in stars; moderate brightness range; slight central concentration; detached.
19 08 43.0	+04 42 18	NGC 6756	10.6	4	40	13.0	cl	Moderately rich in stars; small brightness range; strong central concentration; detached.
19 16 30.0	−00 55 00	NGC 6775		13	10		cl?	(A) Straggling group of stars running NW to SE, as J. Herschel and Smyth describe it.
18 43 00.5	−04 13 31	Tr 35	9.2	6	65	11.4	cl	Moderately rich in stars; moderate brightness range; strong central concentration; detached.

STAR CLOUDS

RA h m s	Dec ° ′ ″	Name	Mag	Diam ′	No. ★	B ★	Type	Notes
18 39 37.6	−04 45 35	NGC 6682		47			MW	

GLOBULAR CLUSTERS

RA h m s	Dec ° ′ ″	Name	Total V m	B ★ V m	HB V m	Diam ′	Conc. Class Low = 12 High = 1	Notes
19 05 15.3	+01 54 03	NGC 6749	12.4	16.5	19.7	4.0		
19 11 12.1	+01 01 50	NGC 6760	9.0	15.6	17.5	9.6	9	

DARK NEBULAE

RA h m s	Dec ° ′ ″	Name	Dim ′ Maj x min	Opacity	Notes
18 47 19.4	−04 32 00	B 104	16 x 1	5	Small, definite, caret-shaped object.
18 48 50.0	−05 04 46	B 106	2 x 2	6	Free of stars.
18 49 30.8	−05 01 00	B 107	5 x 5	6	Irregular; free of stars.
18 50 08.1	−04 48 00	B 110	11 x 11	6	Irregular; free of stars; a small star near the W edge.
18 50 38.6	−04 57 00	B 111	120 x 120	3	Region full of dark structures.
18 51 24.8	−04 19 00	B 113	11 x 11	5	Irregular; small star in W part.
18 53 42.2	−04 51 00	B 117a	7 x 7		Black; irregular.
18 54 36.4	−04 33 00	B 119		2	Very small.
18 54 39.2	−05 11 00	B 119a	30 x 30	3	Irregular dark region; liberally sprinkled with stars in its NW half; several narrow dark lanes in SE part.
18 54 53.5	−04 36 00	B 120		2	Very small; E side bounded by a curve of very small stars.
18 55 25.6	−04 37 00	B 121		2	Small dusky spot; 12 mag star in center.
18 56 49.0	−04 45 00	B 122	4 x 4	5	Small; narrow extension 4′ N.
18 57 39.8	−04 43 00	B 123	90 x 90	5	Roundish; a narrow spur 4′–5′ long from its NE edge; a narrow lane 18′ long and 1½′ wide from its W side to the NW.
18 57 41.8	−04 21 10	B 124	3 x 3		Small dark spot.
18 58 21.9	−04 23 00	B 125	9 x 2		Dark; E and W; fairly well defined on N border; some faint stars in it.

105 (Continued from previous page) DARK NEBULAE 105

RA h m s	Dec ° ′ ″	Name	Dim ′ Maj x min	Opacity	Notes
18 59 02.3	−04 32 00	B 126	8 x 8	4	Dusky; round; rather definite.
19 01 31.9	−05 27 00	B 127	4 x 4	5	Irregular.
19 01 40.4	−04 34 00	B 128	10 x 10		Irregular; dusky; fairly well defined.
19 02 04.5	−05 18 00	B 129	5 x 5	5	Very black; sharply defined.
19 01 56.2	−05 34 00	B 130	7 x 7	5	Dusky; not well defined.
19 03 05.8	−04 22 14	B 131	2 x 4	5	Black spot.
19 04 28.0	−04 26 00	B 132		6	Dark; fairly well defined.
19 07 32.5	−03 55 00	B 135	13 x 13	6	Dusky spot.
19 08 49.7	−04 00 00	B 136	8 x 8	6	Dusky spot; magnitude 10 star near middle.
19 16 00.2	−01 20 00	B 137		3	Projection near S end of B138.
19 16 23.9	+00 13 00	B 138	180 x 10	2	Great curved, semi-vacant lane.
19 18 00.4	−01 25 00	B 139	10 x 2	5	Narrow black spot; magnitude 10 star on SE edge.
19 19 49.1	+05 14 00	B 140	60 x 60	3	Semi-vacant region.
19 20 11.3	+01 54 00	B 141	20 x 20	1	Semi-vacant region; N of the N end of B138.
18 41 50.5	−02 08 00	B 316	6 x 6	4	Dark.
18 51 58.0	−01 16 00	B 319	7 x 1		Thin, curved, dark marking.
18 52 47.1	−05 51 00	B 320	15 x 15	4	Irregular.
18 55 47.1	−04 27 32	B 322	2 x 2		Very small; at NE end of the "crescent" B111.
18 57 34.1	−03 25 00	B 323	17 x 8		A dark parallelogram, N and S.
18 59 22.9	−03 00 00	B 324		3	Definite, dusky loop.
18 59 54.0	−04 04 00	B 325	15 x 3		Irregular; dusky; NW and SE.
19 03 25.5	−00 23 00	B 326	25 x 3	2	Narrow; E and W.
19 04 26.9	−05 08 00	B 327	30 x 3	3	Curved, dusky lane; N and S.
19 04 49.4	−04 15 00	B 328	4 x 4	6	Small; black.
19 06 59.5	+03 12 00	B 329	6 x 6	2	Dusky spot in star cloud.
18 38 36.0	−01 47 00	LDN 557		6	Irregular shape.
18 52 36.0	−01 56 00	LDN 582	60 x 10	5	
18 57 30.0	+01 04 00	LDN 617	50 x 10	5	Irregular shape.

PLANETARY NEBULAE

RA h m s	Dec ° ′ ″	Name	Diam ″	Mag (P)	Mag (V)	Mag cent ★	Alt Name	Notes
19 02 37.1	−00 26 57	NGC 6741	8	10.8	11.5	20.3	PK 33−2.1	
19 05 55.5	−05 59 33	NGC 6751	26	12.5	11.9	15.4	PK 29−5.1	Mag 11.3 star E 2′0.
19 14 36.4	−02 42 27	NGC 6772	86	14.2	12.7	18.6	PK 33−6.1	Appears slightly elongated N-S, with a pronounced dark patch S of center.
19 18 24.9	−01 35 46	NGC 6778	37	13.3	12.3	16.9	PK 34−6.1	Slightly elongated N-S with a PA of 20 degrees; mag 8.47v star E 4′9.
19 22 57.1	+01 30 44	NGC 6790	10	10.2	10.5	11.1	PK 37−6.1	Almost stellar; mag 12.2 star W 0′6.
19 34 33.7	+05 41 01	NGC 6807	8	13.8	12.0	16.3	PK 42−6.1	Stellar; mag 10.13v star NNE 1′5.
18 48 46.5	−05 56 10	PN G27.3−2.1	7	15.9			PK 27−2.1	Mag 10.8 star E 1′5; mag 12.4 star N 1′6.
18 39 21.9	−04 19 51	PN G27.7+0.7	6	17.8		21.0	PK 27+0.1	Mag 10.52v star SW 3′8.
18 57 17.4	−05 59 52	PN G28.2−4.0	6	16.3	15.0		PK 28−4.1	Mag 13.4 star E 1′0; mag 11.1 star WSW 2′6.
18 37 36.9	−03 05 56	PN G28.5+1.6	7	15.0	13.5		PK 28+1.1	Mag 13.4 star WNW 0′9; mag 10.44v star SSE 4′1.
18 34 13.6	−02 27 38	PN G28.7+2.7	8	20.2	15.6		PK 28+2.1	Mag 11.9 star S 1′5; mag 10.49v star SE 2′1.
18 57 49.8	−05 27 36	PN G28.7−3.9	9	16.9			PK 28−3.1	Small triangle of mag 13-14 stars ENE 1′4; pair of mag 11 stars SW 3′6.
18 42 46.9	−03 13 17	PN G29.0+0.4	40	>18.			PK 29+0.1	Strong dark patch in center, in star rich area.
18 44 53.4	−03 20 33	PN G29.2−0.0	6				TDC 1	Mag 12.8 star WNW 3′1.
18 35 22.6	−00 13 50	PN G30.8+3.4	16	>20.9		21.0	PK 30+3.1	= **Abell 47**. Mag 13.5 star ENE 2′2; mag 11.8 star NW 4′5.
18 33 17.6	+00 11 44	PN G31.0+4.1	1	19.7			PK 30+4.1	Close pair of mag 11.3 and 13.4 stars NW 2′6.
18 50 24.6	−01 40 22	PN G31.3−0.5	31				PK 31−0.2	Appears elongated NNE-SSW, matching elongated dark patch down centerline; mag 12.2 star NW 4′5; mag 10.84v star SE 5′7.
18 43 03.4	−00 16 38	PN G31.7+1.7	10				PK 31+1.1	Mag 13 star E 2′1; mag 12.4 star N 3′7.
18 50 40.4	−01 03 12	PN G31.9−0.3	36				PK 31−0.1	Small, bright patches on E and W sides of an elongated N-S dark rift of background sky; mag 10.79v star NE 3′2.
19 00 35.1	−02 12 00	PN G32.0−3.0	4	21.5			PK 32−3.1	Mag 13.3 star E 0′8; mag 10.3 star WSW 5′8.
19 02 10.2	−01 48 46	PN G32.5−3.2		19.1		17.0	PK 32−3.2	Close pair of mag 12 stars NW 1′0; mag 10.7 star ESE 1′5.
18 58 26.3	−01 03 46	PN G32.7−2.0	10	13.2	13.0		PK 32−2.1	Mag 10.58v star E 1′6.
19 01 36.6	−01 19 10	PN G32.9−2.8	1	15.9			PK 32−2.2	Pair of mag 11-12 stars NW 3′1; mag 12.3 star W 3′7.
19 10 25.8	−02 20 25	PN G33.0−5.3	62	15.4		20.5	PK 33−5.1	Numerous stars on disc, in star rich area; mag 14.4 star on SW edge; mag 11.7 star N 2′4.
18 58 51.8	−00 32 53	PN G33.2−1.9				14.0	PK 33−1.1	Mag 11.8 star W 3′0; mag 8.36v star NNE 2′9.
18 45 24.6	+02 01 23	PN G34.0+2.2	4	18.4			PK 34+2.1	Mag 15.4 star W 1′0; mag 8.91v star WNW 4′9.
19 31 07.4	−03 42 33	PN G34.1−10.5	47			16.6	PK 34−10.1	Mag 14.3 star E 1′4; mag 14.4 star SSE 1′2.
18 58 10.6	+01 36 55	PN G35.1−0.7	33			15.1	PK 35−0.1	Mag 12.1 star S 0′6; mag 9.59v star WSW 1′5.
19 14 39.2	+00 13 36	PN G35.7−5.0	10	16.0	14.5		PK 35−5.1	Mag 11.9 star E 1′1.
19 02 00.0	+02 09 23	PN G35.9−1.1	143	12.2	13.2	13.7	PK 36−1.1	Bright portion rectangular, oriented N-S, with large black rift starting at the center and extending SSE; mag 10.70v star W 2′5 from center.
19 02 59.4	+03 02 20	PN G36.9−1.1	15				PK 36−1.2	Has slight N-S rectangular appearance with narrow black line along centerline; mag 12.8 star ESE 1′3.
19 08 02.2	+02 21 21	PN G36.9−2.6	21				PK 36−2.1	Close pair of mag 12 stars S 1′0; mag 9.22v star N 0′6.
19 18 20.8	+01 46 53	PN G37.5−5.1	40	>18.8		19.4	PK 37−5.1	Mag 14.2 star E 1′2; mag 13.9 star NW 1′3, in star rich area.
19 13 06.6	+02 52 49	PN G37.9−3.4	198	>15.5		19.7	PK 37−3.2	= **Abell 56**. Annular, slightly elongated E-W; mag 10.5 star on ESE edge.
19 13 22.6	+03 25 02	PN G38.4−3.3				14.7	PK 38−3.1	Stellar; mag 13.5 star NE 0′9; mag 13.4 star W 1′9.
19 13 54.0	+03 37 39	PN G38.7−3.3	10	13.9	14.0		PK 38−3.2	Mag 10.6 star NE 2′0; mag 11.1 star WSW 1′6.
19 13 34.7	+04 38 02	PN G39.5−2.7	12	14.9	14.5	17.3	PK 39−2.1	Mag 11.4 star WSW 1′3.
19 16 27.8	+05 13 17	PN G40.4−3.1	3	15.7			PK 40−3.1	Close pair of mag 13 stars ENE 2′8; mag 13.7 star NNW 1′3.

106 GALAXIES 106

RA h m s	Dec ° ′ ″	Name	Mag (V)	Dim ′ Maj x min	SB	Type Class	PA	Notes
17 32 59.2	+01 42 49	CGCG 27-1	15.0	0.3 x 0.3	12.2			Almost stellar, in star rich area.
17 50 33.3	+05 40 14	CGCG 55-20	15.6	0.7 x 0.3	13.7	S	33	Mag 9.71v star W 7′2.
17 56 21.9	+05 43 23	CGCG 56-2	14.8	0.4 x 0.3	12.3		141	Mag 12.4 star N 0′8.
17 58 09.3	+03 44 02	CGCG 56-4	14.3	0.5 x 0.5	12.6			Mag 9.65v star W 2′2.
17 58 05.9	+01 19 18	MCG +0-46-2	15.6	0.5 x 0.3	13.4		174	Faint stars superimposed, in star rich area.
17 22 34.4	+00 57 26	UGC 10825	14.5	1.0 x 0.5	13.6	S	89	Stellar nucleus; mag 14.9 star on S edge.
17 42 12.5	+03 11 55	UGC 10943	13.9	1.5 x 0.4	13.2	S?	118	Very knotty; mag 12.2 star on N edge.
17 43 52.8	+04 33 06	UGC 10956	13.8	1.1 x 0.9	13.6	SAB(rs)bc	69	Located 6′2 E of mag 2.77v star "Cebalrai", Beta Ophiuchi.

(Continued from previous page)

GALAXIES

RA h m s	Dec ° ′ ″	Name	Mag (V)	Dim ′ Maj x min	SB	Type Class	PA	Notes
17 54 34.3	+02 52 48	UGC 11030	13.6	1.7 x 0.5	13.3	Sc:	61	A number of superimposed stars, in star rich area.
18 10 22.5	+01 35 32	UGC 11131	14.3	1.6 x 0.5	13.9	Scd?	135	

OPEN CLUSTERS

RA h m s	Dec ° ′ ″	Name	Mag	Diam ′	No. ★	B ★	Type	Notes
17 48 12.0	+01 18 00	Cr 350	6.1	40	20		cl	Few stars; small brightness range; not well detached.
17 46 12.0	+05 43 00	IC 4665	4.2	70	57	6.0	cl	Moderately rich in stars; moderate brightness range; no central concentration; detached.
18 38 54.0	+05 26 00	IC 4756	4.6	40	466	8.0	cl	Rich in stars; large brightness range; slight central concentration; detached. (A) Also known as the **Graff Cluster** or **Graff 1**.
18 01 06.0	+02 54 00	Mel 186	3.0	240			cl	

STAR CLOUDS

RA h m s	Dec ° ′ ″	Name	Mag	Diam ′	No. ★	B ★	Type	Notes
18 39 37.6	−04 45 35	NGC 6682		47			MW	

GLOBULAR CLUSTERS

RA h m s	Dec ° ′ ″	Name	Total V m	B ★ V m	HB V m	Diam ′	Conc. Class Low = 12 High = 1	Notes
17 27 44.3	−05 04 36	NGC 6366	9.5	13.6	15.7	13.0	11	
17 37 36.1	−03 14 45	NGC 6402	7.6	14.0	17.2	11.0		= **M 14**.
17 44 54.7	+03 10 13	NGC 6426	10.9	15.2	18.1	4.2	9	
18 03 50.7	−00 17 49	NGC 6535	9.3	12.8	15.8	3.4	11	
17 58 39.4	−05 04 21	PWM78 2				2.0		

BRIGHT NEBULAE

RA h m s	Dec ° ′ ″	Name	Dim ′ Maj x min	Type	BC	Color	Notes
18 31 36.0	−02 01 00	Sh2-64	20 x 8	E	2-5	3-4	Absorption matter in the central regions divides this object into three main components.
18 30 30.0	+01 11 00	vdB 123	3 x 2	R	3-5	2-4	Nebula involved with a mag 9.1 star; brightest part WNW of the star.

DARK NEBULAE

RA h m s	Dec ° ′ ″	Name	Dim ′ Maj x min	Opacity	Notes
18 38 36.0	−01 47 00	LDN 557		6	Irregular shape.

PLANETARY NEBULAE

RA h m s	Dec ° ′ ″	Name	Diam ″	Mag (P)	Mag (V)	Mag cent ★	Alt Name	Notes
18 14 18.3	−04 59 22	PN G24.2+5.9	44	16.0	14.1	20.3	PK 24+5.1	Mag 14.7 star W 2′.0; mag 10.58v star W 5′.6.
18 26 40.1	−02 42 58	PN G27.6+4.2	15	16.8	15.0		PK 27+4.1	
17 41 41.0	+03 06 57	PN G27.6+16.9	94	16.3		14.9	PK 27+16.1	Mag 10.04v star on S edge; mag 10.28v star N 2′.0.
18 39 21.9	−04 19 51	PN G27.7+0.7	6	17.8		21.0	PK 27+0.1	Mag 10.52v star SW 3′.8.
18 06 00.5	+00 22 42	PN G28.0+10.2	36	16.8		16.8	PK 28+10.1	Mag 12.6 star S 2′.5; pair of N-S mag 12.8 stars W 2′.5.
18 37 36.9	−03 05 56	PN G28.5+1.6	7	15.0	13.5		PK 28+1.1	Mag 13.4 star WNW 0′.9; mag 10.44v star SSE 4′.1.
18 25 00.6	−01 30 54	PN G28.5+5.1	10	19.5	14.5		PK 28+5.1	Located 1′.9 N of mag 9.01v star.
18 34 13.6	−02 27 38	PN G28.7+2.7	8	20.2	15.6		PK 28+2.1	Mag 11.9 star S 1′.5; mag 10.49v star SE 2′.1.
18 25 01.3	+00 52 18	PN G30.6+6.2	561		10.0	16.0	PK 30+6.1	Appears elongated NE-SW with the NE half being more dense than the SW, the SW end being more broken up.
18 35 22.6	−00 13 50	PN G30.8+3.4	16	>20.9		21.0	PK 30+3.1	= **Abell 47**. Mag 13.5 star ENE 2′.2; mag 11.8 star NW 4′.5.
18 33 17.6	+00 11 44	PN G31.0+4.1	1	19.7			PK 30+4.1	Close pair of mag 11.3 and 13.4 stars NW 2′.6.
18 27 09.4	+01 14 24	PN G31.2+5.9	14				PK 31+5.1	Mag 11.2 star W 3′.5.
18 24 44.6	+02 29 27	PN G32.1+7.0	14	12.1	13.0		PK 32+7.2	Mag 15.0 star W 1′.5; mag 12.9 star NW 2′.8.
18 31 00.3	+02 25 23	PN G32.7+5.6	28	15.7	14.6		PK 32+5.1	Slightly elongated NNE-SSW; mag 14.8 star SW 0′.8; mag 10.6 star N 1′.9.
18 23 21.8	+03 36 26	PN G32.9+7.8	8	15.9	14.3		PK 32+7.1	Forms a triangle with a mag 13.0 star NW 2′.9 and a mag 11.6 star SW 2′.6.
18 31 45.9	+04 05 09	PN G34.3+6.2	10	17.1			PK 34+6.1	Mag 12.5 star S 1′.0; mag 14.3 star W 1′.8.

GALAXIES

RA h m s	Dec ° ′ ″	Name	Mag (V)	Dim ′ Maj x min	SB	Type Class	PA	Notes
16 14 42.9	+02 09 38	CGCG 23-28	15.9	0.6 x 0.4	14.2	Sm	54	Mag 12.5 star N 1′.2.
16 15 31.4	+01 32 54	CGCG 23-29	15.0	0.5 x 0.4	13.2	S0	153	Stellar nucleus, faint envelope; mag 9.66v star SW 3′.2.
16 19 47.4	+01 57 32	CGCG 24-2	14.8	0.6 x 0.4	13.1		178	Mag 10.48v star SE 4′.2; UGC 10339 NE 6′.1.
16 23 11.9	+01 45 13	CGCG 24-11	14.4	0.5 x 0.3	12.2	S0	72	Bracketed by a rectangle of very faint stars.
16 41 07.8	+02 38 00	CGCG 24-22	14.4	0.8 x 0.7	13.6	Sc	60	Mag 11.2 star WSW 1′.5.
16 43 04.7	+02 04 59	CGCG 25-2	15.0	0.6 x 0.2	12.8	S0	120	Mag 13.4 star ESE 2′.3; mag 12.1 star S 4′.5.
17 08 22.2	+00 57 47	CGCG 26-2	15.1	0.3 x 0.2	11.9		177	Located 1′.5 NNW of mag 12.7 star.
17 16 34.7	+00 51 42	CGCG 26-3	16.1	0.4 x 0.4	14.0			Low surface brightness; mag 14.2 star N edge.
16 10 35.0	+04 45 13	CGCG 51-59	14.6	0.4 x 0.4	12.5			Mag 12.9 star W 2′.6; mag 10.8 star NE 8′.8.
16 12 59.9	+04 20 40	CGCG 51-66	14.4	0.6 x 0.5	12.9		0	Mag 11.0 star N 4′.6; mag 9.15v star E 8′.8.
16 14 07.1	+02 37 33	CGCG 51-69	15.6	0.5 x 0.4	13.7		6	Located 4′.0 W of mag 6.93v star.
16 19 48.1	+05 09 41	CGCG 52-4	14.1	0.7 x 0.4	12.8	E	24	Mag 13.4 star N 2′.4; mag 12.3 star W 4′.8.
16 27 40.7	+05 05 15	CGCG 52-21	15.4	0.7 x 0.5	14.2	SBbc	33	Mag 13.9 star on E edge and a mag 15.4 star on W edge.
16 30 47.0	+03 30 27	CGCG 52-35	15.2	0.4 x 0.2	12.3		174	**CGCG 52-33** NW 6′.4; mag 9.30v star W 10′.4.
16 31 04.4	+05 09 28	CGCG 52-38	14.1	0.9 x 0.3	12.5		177	**CGCG 52-39** SE 8′.5.
17 04 51.7	+04 28 26	CGCG 53-30	14.5	0.6 x 0.6	13.2			Mag 13.3 star on W edge.
17 08 43.0	+04 02 55	IC 1242	13.8	0.9 x 0.6	13.0	S?	130	Bright E-W oriented center.

(Continued from previous page)

107 GALAXIES 107

RA h m s	Dec ° ′ ″	Name	Mag (V)	Dim ′ Maj x min	SB	Type Class	PA	Notes
16 10 14.6	+01 03 18	MCG +0-41-5	13.7	1.2 x 0.5	13.0	SAB(s)dm: IV-V	15	
16 35 40.9	+00 19 50	MCG +0-42-5	13.9	0.7 x 0.7	13.0	S0		
16 51 42.3	−02 36 52	MCG +0-43-3	13.8	1.2 x 0.8	13.8	E/S0	0	Very bright core or star superimposed; mag 9.72v star S 1′4.
17 06 50.0	+02 14 42	MCG +0-44-1	14.1	0.7 x 0.4	12.5		81	
16 14 17.0	+03 59 37	MCG +1-41-15	14.1	0.4 x 0.2	11.2	S0	21	Anonymous galaxy NE 0′7; mag 7.96v star NW 3′6.
16 30 56.6	+04 04 57	MCG +1-42-8	14.1	0.7 x 0.4	12.6	S?	120	
16 35 59.0	−05 07 05	MCG −1-42-2	14.2	1.5 x 1.4	14.8	SB(rs)b: II	36	Bright N-S bar; mag 12.4 star NE 1′8.
16 38 08.6	−04 49 30	MCG −1-42-3	15.6	1.7 x 0.6	15.5	Sc	45	Mag 10.03v star ESE 7′5.
16 42 03.3	−05 02 00	MCG −1-42-4	14.1	2.4 x 1.3	15.1	SAB(rs)bc:	114	Numerous knots and/or superimposed stars.
16 51 32.5	−03 05 50	MCG −1-43-2	13.4	1.9 x 1.4	14.3	SB(rs)bc I-II	117	Elongated core, branching arms.
16 09 58.7	+00 42 30	NGC 6070	11.8	3.5 x 1.9	13.7	SA(s)cd I	62	Many dark lanes and superimposed stars; **CGCG 23-18** NE 4′2; several small anonymous galaxies in vicinity of CGCG 23-18.
16 12 58.7	+02 10 33	NGC 6080	13.6	1.1 x 1.0	13.5	SA(s)ab? III	48	Multi-galaxy system; small companion on NE edge.
16 16 52.4	+00 50 30	NGC 6100	13.0	1.9 x 1.1	13.7	(R)SAB(r)a:	120	Mag 9.46v star at NW end.
16 21 48.7	−02 17 02	NGC 6118	11.7	4.7 x 2.0	13.9	SA(s)cd I-II	58	Bright nucleus, many superimposed stars and/or knots.
16 22 10.4	−01 30 56	NGC 6172	12.8	0.8 x 0.8	12.4	E+:		
16 47 13.4	−00 16 36	NGC 6220	13.7	1.6 x 0.9	13.9	SA(s)ab?	135	
16 50 46.8	+04 36 14	NGC 6230	14.4	0.9 x 0.8	14.1	E?	36	Stellar anonymous galaxy W 1′0.
16 51 57.5	+04 22 57	NGC 6234	14.4	0.4 x 0.4	12.3			Almost stellar.
16 52 58.8	+02 24 03	NGC 6240	12.9	2.1 x 1.1	13.7	I0: pec	20	Multi-galaxy system, disrupted appearance.
17 08 44.6	+03 53 39	NGC 6296	13.4	0.9 x 0.7	12.7	SAB(s)bc:	130	
16 12 56.8	−00 05 50	UGC 10264	13.7	1.2 x 0.7	13.3	SAB(s)cd: III	35	
16 14 25.1	−00 12 27	UGC 10288	13.3	4.9 x 0.6	14.3	Sc: sp	91	Mag 9.00v star N 4′6.
16 14 33.1	+00 49 18	UGC 10290	13.2	1.8 x 1.5	14.1	SB(s)m III-IV	90	Very knotty.
16 16 43.4	+00 14 46	UGC 10306	13.3	2.2 x 0.7	13.6	(R′)SAB(s)b: III	35	Mag 13.4 star SW end.
16 20 08.6	+02 00 31	UGC 10339	13.9	0.7 x 0.5	12.6	S?	70	CGCG 24-2 SW 6′1.
16 27 50.4	−01 45 49	UGC 10394	14.1	1.5 x 0.5	13.6	SA(s)d: III	170	
16 29 16.5	+03 06 52	UGC 10406	14.3	1.5 x 0.3	13.3	Sbc	120	Very faint dark lane full length along centerline.
16 36 45.1	+01 40 44	UGC 10465	13.9	1.0 x 0.9	13.6	(R′)SB(r)bc: II-III	105	Stellar nucleus; patchy.
16 37 50.4	−01 22 02	UGC 10474	14.2	1.0 x 0.6	13.5	S	10	Mag 14.3 star at N end.
16 39 23.6	−02 24 24	UGC 10492	13.8	1.1 x 0.8	13.5	SA(s)a:	135	
16 48 21.5	−01 37 14	UGC 10554	14.3	1.4 x 0.5	13.8	Scd: sp	155	Almost stellar nucleus, faint envelope.
16 57 51.7	+02 29 43	UGC 10623	14.2	0.9 x 0.6	13.4	SAB(rs)cd:	150	
16 59 46.4	+02 30 33	UGC 10640	13.8	1.1 x 0.7	13.4	SB?	99	EW bar, bright knot or small anonymous galaxy W edge, mag 12.6 star S 1′2.
17 05 00.2	−01 32 16	UGC 10683	14.7	1.0 x 0.6	13.9	Double System	162	Brighter core S end; pair of mag 13 stars N edge; **PGC 59567** NW edge.
17 11 05.7	+05 51 04	UGC 10738	14.3	1.7 x 0.3	13.4	Sbc	49	Mag 9.60v star N 2′6.
17 22 34.4	+00 57 26	UGC 10825	14.5	1.0 x 0.5	13.6	S	89	Stellar nucleus; mag 14.9 star on S edge.

GLOBULAR CLUSTERS

RA h m s	Dec ° ′ ″	Name	Total V m	B ★ V m	HB V m	Diam ′	Conc. Class Low = 12 High = 1	Notes
16 47 14.5	−01 56 52	NGC 6218	6.1	12.0	14.7	16.0	9	= **M 12**.
16 57 08.9	−04 05 58	NGC 6254	6.6	12.0	15.1	20.0	7	= **M 10**.
17 27 44.3	−05 04 36	NGC 6366	9.5	13.6	15.7	13.0	11	
16 59 50.7	−00 31 59	Pal 15	14.2	17.1	19.9	3.0		

PLANETARY NEBULAE

RA h m s	Dec ° ′ ″	Name	Diam ″	Mag (P)	Mag (V)	Mag cent ★	Alt Name	Notes
16 21 04.5	−00 16 12	PN G13.3+32.7	6	12.8	12.8	14.7	PK 13+32.1	Mag 12.3 star W 1′7.
17 12 51.9	−03 16 00	PN G18.0+20.1	15	13.4	13.4	16.6	PK 18+20.1	Mag 14.6 star E 0′8.

108 GALAXIES 108

RA h m s	Dec ° ′ ″	Name	Mag (V)	Dim ′ Maj x min	SB	Type Class	PA	Notes
14 56 33.3	−02 06 30	CGCG 20-34	14.8	1.0 x 0.4	13.6	Irr	48	Mag 13.4 star N 0′9; CGCG 20-35 E 6′4.
14 56 59.1	−02 06 57	CGCG 20-35	14.7	0.6 x 0.4	13.2	E	15	CGCG 20-34 W 6′4.
15 09 40.6	+01 23 06	CGCG 21-20	14.5	0.8 x 0.3	12.9		156	Elongated **CGCG 21-24** E 3′9; mag 10.42v star W 4′8.
15 12 43.8	+01 27 51	CGCG 21-49	14.5	0.7 x 0.4	12.9		141	**CGCG 21-47** NW 3′6; **CGCG 21-46** NW 5′1.
15 16 25.0	+00 42 28	CGCG 21-62	14.7	0.5 x 0.4	12.8	Sbc	27	Mag 6.92v star NW 8′4.
15 17 41.9	+01 21 38	CGCG 21-71	15.4	0.4 x 0.3	13.0		177	Mag 11.7 star WNW 4′4.
15 19 45.3	+00 52 43	CGCG 21-76	15.2	0.6 x 0.3	13.2	Sb	36	Mag 13.6 star SE 1′2.
15 20 40.8	+01 50 46	CGCG 21-78	14.5	0.5 x 0.4	12.6		0	Mag 14.4 star SE 0′8; very faint anonymous galaxy S 2′5.
15 21 38.0	+00 27 55	CGCG 21-80	14.9	0.7 x 0.3	13.0	Sb	168	Mag 12.4 star W 2′7; mag 12.6 star E 4′7.
15 22 59.1	+00 05 52	CGCG 21-84	15.1	0.5 x 0.3	12.9		21	Mag 14.7 star on E edge; mag 10.67v star SE 2′2.
15 26 20.1	−03 32 50	CGCG 21-95	14.5	0.5 x 0.2	11.9		168	Very faint, elongated anonymous galaxy SW 1′7.
15 27 25.5	−02 53 28	CGCG 21-97	15.0	0.5 x 0.5	13.4	Sm		Pair of mag 10-11 stars NE 9′1.
15 28 50.3	+00 46 35	CGCG 21-101	14.6	0.4 x 0.4	12.5			Mag 14 star on S edge.
15 38 35.3	−01 32 57	CGCG 22-20	15.1	0.5 x 0.4	13.2		39	Mag 13.2 star NE 6′0.
15 50 44.7	+01 45 03	CGCG 22-45	15.3	0.7 x 0.4	13.8	Sc	27	Stellar nucleus.
15 55 31.0	−00 49 38	CGCG 23-1	14.4	0.6 x 0.4	12.7	Sb	144	Mag 10.64v star E 7′1.
16 07 52.1	+01 24 11	CGCG 23-14	14.3	0.7 x 0.3	12.5	Sa	27	Located 2′6 SW of mag 12.4 star.
16 14 42.9	+02 09 38	CGCG 23-28	15.9	0.6 x 0.4	14.2	Sm	54	Mag 12.5 star N 1′2.
16 15 31.4	+01 32 54	CGCG 23-29	15.0	0.5 x 0.4	13.2	S0	153	Stellar nucleus, faint envelope; mag 9.66v star SW 3′2.
14 58 34.1	+03 31 03	CGCG 48-83	15.0	0.5 x 0.4	13.1		48	Elongated **CGCG 48-80** SW 8′5.
15 05 56.6	+03 42 23	CGCG 48-115	14.3	1.1 x 0.6	13.7		54	= **Mkn 1392**; **CGCG 48-116** E 2′0.
15 06 34.5	+05 13 24	CGCG 49-1	14.6	0.9 x 0.4	13.3		75	CGCG 49-3 NNE 6′6.
15 06 43.5	+05 19 35	CGCG 49-3	14.4	0.5 x 0.5	12.8			CGCG 49-1 SSW 6′6; mag 10.9 star NNE 3′6.
15 11 56.5	+05 33 34	CGCG 49-39	14.0	0.8 x 0.7	13.2		174	Mag 11.9 star NNE 3′9; many very faint galaxies in the surrounding area.
15 15 10.8	+05 34 36	CGCG 49-75	14.8	0.5 x 0.3	12.5		84	Mag 12.3 star WSW 0′8.
15 18 43.7	+05 53 52	CGCG 49-115	15.2	0.5 x 0.3	12.9		30	Mag 12.2 star E 1′8.
15 22 05.2	+05 51 14	CGCG 49-148	14.1	0.6 x 0.4	12.4	Sab	57	= **Mkn 851**. **CGCG 49-149** SE 1′9; mag 9.99v star NNW 2′5.
15 22 21.9	+05 23 27	CGCG 49-150	14.3	0.6 x 0.4	12.6			Located between two N-S mag 10 stars.
15 24 16.7	+04 51 18	CGCG 49-155	14.5	0.6 x 0.5	13.0		0	Mag 11.7 star NE 5′6.

(Continued from previous page)

GALAXIES

RA h m s	Dec ° ′ ″	Name	Mag (V)	Dim ′ Maj x min	SB	Type Class	PA	Notes
15 29 30.5	+03 28 01	CGCG 49-178	14.2	0.7 x 0.6	13.2	S0	141	Mag 8.48v star W 11′1; MCG +1-39-24 ENE 5′5.
15 33 47.9	+05 17 27	CGCG 50-25	14.8	0.8 x 0.6	13.9	S0	0	Mag 10.9 star N 0′7; **UGC 9894** SE 1′3.
15 36 39.4	+03 21 34	CGCG 50-40	16.1	0.6 x 0.4	14.4	Sm	30	Small triangle of mag 11-14 stars E 4′1; mag 13.4 star SW 1′0.
15 36 53.3	+04 45 27	CGCG 50-43	15.0	0.6 x 0.3	12.9		165	Mag 11.9 star E 5′5.
15 37 51.8	+03 27 10	CGCG 50-49	15.9	0.6 x 0.3	13.8		147	Mag 10.79v star WNW 2′2.
15 38 10.8	+04 46 45	CGCG 50-50	15.0	0.7 x 0.5	13.7		66	Mag 11.2 star NE 4′3.
15 39 02.6	+05 50 24	CGCG 50-59	15.3	0.8 x 0.5	14.2		153	Strong dark lane W.
15 40 54.1	+05 08 30	CGCG 50-68	14.8	0.5 x 0.2	12.1		75	Mag 10.17v star NE 10′0.
15 42 22.0	+02 27 59	CGCG 50-75	15.0	0.4 x 0.3	12.6		39	Mag 10.8 star NE 8′6.
15 43 34.0	+04 52 17	CGCG 50-83	14.3	0.8 x 0.6	13.6	E	129	Mag 13.4 star on NE edge; UGC 9990 S 4′7.
15 44 03.7	+04 46 01	CGCG 50-87	14.3	0.8 x 0.7	13.5	Sb	132	Nucleus offset N; mag 9.97v star SW 3′2.
15 45 00.5	+02 28 02	CGCG 50-91	14.4	0.5 x 0.5	12.8			Mag 14.1 star on E edge; mag 11.4 star ENE 1′8.
15 52 05.7	+04 30 54	CGCG 50-116	15.4	0.5 x 0.3	13.1		12	Mag 9.20v star N 7′3.
15 59 49.8	+02 39 59	CGCG 51-17	14.3	0.6 x 0.3	12.3		177	MCG +1-41-4 N 2′8, with a pair of small, faint anonymous galaxies half way between.
16 07 24.9	+03 35 06	CGCG 51-46	14.6	0.4 x 0.4	12.5			Mag 11.5 star W 3′3; mag 9.80v star SW 6′0.
16 10 35.0	+04 45 13	CGCG 51-59	14.6	0.4 x 0.4	12.5			Mag 12.9 star W 2′6; mag 10.8 star NE 8′8.
16 12 59.9	+04 20 40	CGCG 51-66	14.4	0.6 x 0.5	12.9		0	Mag 11.0 star N 4′6; mag 9.15v star E 8′8.
16 14 07.1	+02 37 33	CGCG 51-69	15.6	0.5 x 0.4	13.7		6	Located 4′0 W of mag 6.93v star.
15 10 56.1	+05 44 42	IC 1101	13.7	1.1 x 0.6	13.2	S0$^-$:	23	IC 1101 at center; surrounding PGC and anonymous GX mostly stellar.
15 11 05.0	+04 17 37	IC 1102	14.0	1.2 x 0.8	13.8	Sb:	25	Short, elongated N-S nucleus.
15 13 14.1	+04 17 14	IC 1105	13.9	0.8 x 0.4	12.5	S?	96	Elongated **MCG +1-39-6** SW 0′6.
15 13 56.3	+04 42 37	IC 1106	14.6	1.0 x 0.4	13.5		30	Anonymous galaxy WNW 1′8.
15 33 05.8	−01 37 44	IC 1125	13.3	1.6 x 1.0	13.7	Sdm:	147	Mag 10.02v star NNW 8′4.
15 37 53.0	−01 44 07	IC 1128	13.6	1.1 x 0.9	13.5	S0:	0	Mag 10.55v star S 5′3.
15 47 34.5	−01 32 46	IC 1136	14.3	0.5 x 0.5	12.6			Mag 10.07v star NNE 2′4.
16 01 34.4	+01 42 28	IC 1158	12.6	2.5 x 1.5	13.9	SAB(r)c: II-III	137	Mag 11.2 star SW 8′3.
14 58 48.8	+02 01 22	MCG +0-38-13	14.1	0.7 x 0.4	12.8	E	0	
15 03 45.2	−03 18 09	MCG +0-38-19	13.4	1.8 x 1.4	14.3	SAB(s)cd III	70	Mag 8.57v star NNW 9′0.
15 07 37.6	+02 01 08	MCG +0-39-3	15.2	0.4 x 0.4	13.0			
15 07 59.4	+01 13 49	MCG +0-39-4	14.4	0.6 x 0.5	12.9	S?	92	UGC 9732 NE 2′8.
15 10 29.6	−02 10 10	MCG +0-39-9	14.4	0.8 x 0.5	13.3	S0	153	**CGCG 21-26** SW 6′4.
15 12 01.2	−03 13 21	MCG +0-39-10	14.7	0.5 x 0.5	13.1	Irr		Mag 12.5 star S 1′1.
15 15 32.1	+01 46 23	MCG +0-39-13	15.5	0.6 x 0.4	13.8		120	
15 15 34.8	+02 14 56	MCG +0-39-14	13.9	1.0 x 0.8	13.5	S0	135	
15 16 42.4	+01 15 44	MCG +0-39-17	14.9	0.3 x 0.3	12.2	S0/a		
15 23 10.5	−03 29 54	MCG +0-39-24	15.4	0.5 x 0.4	13.5	SBd	123	
15 32 18.5	−02 49 23	MCG +0-40-2	14.1	1.2 x 0.7	13.8	SA(s)m? IV-V	97	Stellar nucleus; mag 9.86v star NW 13′4.
15 39 33.3	+01 16 13	MCG +0-40-5	15.3	0.7 x 0.6	14.2		21	
16 10 14.6	+01 03 18	MCG +0-41-5	13.7	1.2 x 0.5	13.0	SAB(s)dm: IV-V	15	
14 56 01.0	+02 27 40	MCG +1-38-20	14.7	1.0 x 0.3	13.3	S0	174	
15 03 27.9	+04 40 59	MCG +1-38-27	14.1	0.7 x 0.3	12.3	Sa	0	
15 05 01.1	+05 47 46	MCG +1-38-28	14.3	1.0 x 0.4	13.2	Sb	21	
15 12 59.8	+03 09 18	MCG +1-39-5	13.8	0.5 x 0.5	12.4	E		
15 18 08.5	+05 18 36	MCG +1-39-16	14.6	0.7 x 0.7	13.7	Sc		
15 18 52.7	+04 51 54	MCG +1-39-17	14.7	0.9 x 0.7	14.0	Sbc	153	
15 19 01.4	+04 31 11	MCG +1-39-18	12.8	1.0 x 0.6	12.3	E?	114	Mag 8.67v star NNW 2′1.
15 29 49.9	+03 30 36	MCG +1-39-24	14.0	1.0 x 0.8	13.7		165	CGCG 49-178 WSW 5′5.
15 39 05.4	+05 34 14	MCG +1-40-10	14.3	0.5 x 0.5	12.8	E		
15 45 46.0	+02 24 32	MCG +1-40-13	15.6	0.9 x 0.2	13.5	S	78	Mag 10.9 star NNE 1′8.
15 59 47.0	+02 42 42	MCG +1-41-4	14.0	0.7 x 0.7	13.1	S0		CGCG 51-17 S 2′8.
16 14 17.0	+03 59 37	MCG +1-41-15	14.1	0.4 x 0.2	11.2	S0	21	Anonymous galaxy NE 0′7; mag 7.96v star NW 3′6.
15 01 18.4	−04 33 21	MCG −1-38-18	14.4	0.5 x 0.4	12.5		60	Mag 9.52v star N 2′1.
15 01 24.9	−04 30 01	MCG −1-38-19	15.6	0.5 x 0.4	13.7	Sbc	171	Mag 9.52v star SW 2′2.
15 04 32.0	−04 36 32	MCG −1-38-21	14.2	0.7 x 0.4	12.6	Sc	150	Bright knot or star S end.
15 23 20.1	−04 09 12	MCG −1-39-5	12.6	1.5 x 0.9	12.8	SB(r)ab pec:	27	Mag 12.0 star N 1′9.
15 31 30.8	−05 09 42	MCG −1-39-6	14.3	0.7 x 0.4	12.8		105	
14 58 23.0	−01 05 28	NGC 5792	11.3	6.9 x 1.7	13.8	SB(rs)b I-II	84	Mag 9.60v star 1′1 NW of nucleus.
15 00 00.5	+01 53 23	NGC 5806	11.7	3.1 x 1.6	13.3	SAB(s)b III:	170	
15 00 27.2	+01 37 23	NGC 5811	13.9	0.9 x 0.7	13.2	SB(s)m: IV-V	98	Mag 9.07v star W 7′2.
15 01 11.3	+01 42 02	NGC 5813	10.5	4.2 x 3.0	13.2	E1-2	145	Located between a pair of mag 12 stars. NGC 5814 SE 4′6.
15 01 21.2	+01 38 12	NGC 5814	13.8	0.9 x 0.5	12.8	(R′)Sab	32	Located 4′6 SE of NGC 5813.
15 04 07.0	+01 13 10	NGC 5831	11.5	2.0 x 1.9	12.9	E3	55	Large bright center, uniform envelope.
15 05 26.4	+02 05 53	NGC 5838	10.9	4.2 x 1.5	12.8	SA0$^-$	43	Slightly brighter middle, uniform envelope.
15 05 27.4	+01 38 05	NGC 5839	12.7	1.3 x 1.2	13.1	SAB(rs)0°:	35	
15 06 00.8	+01 38 00	NGC 5845	12.5	0.8 x 0.5	11.5	E:	150	NGC 5846 E 7′4.
15 06 29.2	+01 36 19	NGC 5846	10.0	4.1 x 3.8	13.0	E0-1	42	NGC 5846A 0′7 S of bright center.
15 06 29.2	+01 35 40	NGC 5846A	12.0	0.5 x 0.3	9.9	cE2-3	42	0′7 S of center of NGC 5846.
15 06 35.2	+02 00 13	NGC 5848	13.8	1.1 x 0.5	13.0	S0$^+$ pec sp	145	
15 07 07.7	+01 32 41	NGC 5850	10.8	4.3 x 3.7	13.6	SB(r)b II	140	Large nucleus in SSE-NNW bar, dark patches N and S of bar.
15 07 47.4	+02 34 04	NGC 5854	11.9	2.8 x 0.8	12.6	SB(s)0$^+$ sp	55	Star or bright knot on W edge; mag 9.45v star NW 5′0.
15 07 49.1	+03 59 01	NGC 5855	14.5	0.4 x 0.4	12.3			Almost stellar; mag 7.67v star ESE 8′0.
15 09 33.5	+03 03 10	NGC 5864	11.8	2.8 x 0.9	12.6	SB(s)0°? sp	68	
15 09 49.2	+00 31 46	NGC 5865	13.4	1.1 x 0.8	13.2	SAB0$^-$	173	
15 09 49.5	+00 28 12	NGC 5869	11.9	2.3 x 1.7	13.3	S0°:	125	Bright center, line of three faint stars extend from center to SE edge.
15 14 44.0	+01 09 15	NGC 5887	13.4	1.3 x 1.1	13.7	(R′)SA(s)0$^+$:	160	**UGC 9782** NE 3′9; mag 8.36v star NNE 7′4.
15 20 18.2	+03 31 06	NGC 5911	13.9	0.8 x 0.8	13.3	S0		**CGCG 49-134** NE 1′6.
15 20 55.4	−02 34 38	NGC 5913	13.2	1.6 x 0.7	13.2	SB(r)a	168	Dark patches N and S of center.
15 21 56.7	+05 04 15	NGC 5921	10.8	4.9 x 4.0	13.9	SB(r)bc I-II	130	Bright nucleus, many branching arms, numerous knots.
15 30 46.1	−02 49 50	NGC 5937	12.3	1.9 x 1.1	13.0	(R′)SAB(rs)b pec II	20	
15 34 56.5	+04 57 29	NGC 5952	15.2	0.6 x 0.4	13.5	S0	126	Stellar.
15 35 12.6	+05 03 43	NGC 5955	14.9	0.9 x 0.6	14.1	Sb-Sbc	6	**MCG +1-40-6A** NNE 2′5.
15 36 18.6	+05 39 52	NGC 5960	14.3	0.7 x 0.6	13.2	Sb	114	
15 37 36.0	+05 58 24	NGC 5964	11.9	4.2 x 3.2	14.6	SB(rs)d	145	Elongated nucleus; many superimposed stars and/or knots.
15 46 16.4	+02 24 57	NGC 5990	12.4	1.5 x 0.9	12.6	(R′)Sa pec?	115	**CGCG 50-98** NNW 2′3.
15 54 19.4	+00 32 31	NGC 6010	12.6	1.9 x 0.5	12.4	S0/a: sp	102	
15 55 57.5	+05 55 55	NGC 6014	12.2	1.7 x 1.6	13.2	S0	171	Pair of stars or bright knots N edge; mag 7.38v star E 10′9.
15 57 15.5	+05 59 52	NGC 6017	13.2	0.8 x 0.7	12.4	S?	140	
16 04 28.0	−02 07 17	NGC 6033	13.7	1.0 x 0.9	13.4	SA(s)bc? II-III	90	
16 04 30.8	+03 52 06	NGC 6036	13.4	1.3 x 0.6	13.0	S0/a	146	Bright bar NE-SW.

(Continued from previous page)

108 GALAXIES 108

RA h m s	Dec ° ′ ″	Name	Mag (V)	Dim ′ Maj x min	SB	Type Class	PA	Notes
16 04 29.9	+03 48 51	NGC 6037	14.1	0.7 x 0.6	13.0	S?	54	Mag 10.97v star SE 3′.0.
16 09 58.7	+00 42 30	NGC 6070	11.8	3.5 x 1.9	13.7	SA(s)cd I	62	Many dark lanes and superimposed stars; **CGCG 23-18** NE 4′.2; several small anonymous galaxies in vicinity of CGCG 23-18.
16 12 58.7	+02 10 33	NGC 6080	13.6	1.1 x 1.0	13.5	SA(s)ab? III	48	Multi-galaxy system; small companion on NE edge.
14 56 01.6	−01 23 19	UGC 9601	13.2	1.4 x 1.3	13.7	SB(s)cd pec? III-IV	162	
15 02 03.2	+01 50 24	UGC 9661	14.0	1.1 x 1.0	14.0	SB(rs)dm III-IV	63	Brighter N-S core slightly offset E.
15 02 19.9	+05 38 46	UGC 9667	13.1	1.8 x 0.9	13.5	S0/a	87	**CGCG 48-93** pair with mag 14.1 star SSE 6′.6.
15 04 30.3	−00 51 06	UGC 9682	13.9	1.7 x 0.6	13.8	SB(s)m IV-V	165	Mag 8.21v star S 3′.6.
15 08 10.0	+01 14 51	UGC 9732	13.4	1.2 x 0.8	13.3	SB(r)b pec II	62	MCG +0-39-4 SW 2′.8; mag 12.1 star WNW 1′.5.
15 09 51.5	+01 46 40	UGC 9744	13.7	1.3 x 1.0	13.8	SB(r)ab	179	Bright core, faint, uniform envelope.
15 10 16.7	+01 56 02	UGC 9746	14.1	1.5 x 0.4	13.4	Sbc: III	144	
15 11 37.2	+05 23 48	UGC 9757	13.8	1.4 x 0.6	13.5	Sbc III	60	Mag 11.1 star SW 2′.9.
15 12 02.2	+01 41 50	UGC 9760	14.1	2.6 x 0.2	13.2	Sd sp	57	Close pair of stars, mags 12.9 and 10.01v, NW 2′.2.
15 14 49.3	+05 32 50	UGC 9781	13.8	1.3 x 1.3	14.2	Sbc		Mag 12.5 star superimposed W of nucleus.
15 15 42.9	+01 27 16	UGC 9787	14.1	0.9 x 0.7	13.4	SABm: V	45	
15 17 49.1	+04 09 44	UGC 9804	14.0	1.3 x 0.8	13.9	Double System	165	Mag 11.9 star S 1′.7.
15 23 01.8	−01 20 48	UGC 9829	13.4	2.1 x 0.9	13.9	Sb pec	167	Distorted; **CGCG 21-81** W 1′.6; **CGCG 21-82** SW 2′.3.
15 29 25.8	+05 05 03	UGC 9864	13.5	1.7 x 0.6	13.4	S0?	25	**CGCG 49-175** SW 2′.9.
15 32 32.1	+04 40 48	UGC 9886	14.1	1.1 x 1.0	14.1	S0⁻:	27	Almost stellar anonymous galaxy W 0′.9.
15 32 44.2	+04 44 50	UGC 9887	14.2	1.2 x 0.4	13.3	S	153	Disrupted. Very faint anonymous galaxies NNW 1′.5; NW 2′.8; and W 2′.1.
15 38 55.0	+04 34 58	UGC 9945	13.2	1.2 x 1.2	13.4	Sc		
15 39 39.3	+03 11 54	UGC 9953	13.8	1.3 x 0.8	13.7	SBcd:	150	Bright knot or star on E edge.
15 40 05.9	+01 56 02	UGC 9960	13.9	1.5 x 0.5	13.4	Sa	42	
15 41 09.9	−01 42 23	UGC 9968	13.7	1.2 x 0.9	13.6	SAB(rs)b II-III	150	
15 41 50.5	+05 51 09	UGC 9976	14.0	1.0 x 0.6	13.3	SB(r)b	144	Stellar nucleus.
15 41 59.7	+00 42 47	UGC 9977	13.2	3.5 x 0.4	13.4	Sc: sp	77	Mag 12.4 star N 1′.7; mag 9.86v star SE 4′.3.
15 42 19.9	+00 28 22	UGC 9979	14.3	1.4 x 0.5	13.8	IB(s)m IV-V	140	Low, uniform surface brightness except for one faint knot NW end.
15 42 34.4	+02 00 51	UGC 9980	13.8	1.0 x 0.6	13.1	SB0	154	Mag 12.5 star S 1′.1.
15 43 29.9	+04 47 45	UGC 9990	14.3	1.3 x 0.3	13.2	Scd:	164	CGCG 50-83 N 4′.7.
15 45 14.6	+00 46 17	UGC 10005	13.9	1.3 x 1.4	14.4	SA(s)d? III-IV	75	Stellar nucleus, pair of faint stars superimposed W, faint envelope.
15 46 54.3	+05 53 27	UGC 10029	13.4	1.3 x 0.7	13.1	S0	100	**CGCG 50-104** ENE 5′.6.
15 47 00.6	−00 59 16	UGC 10030	13.6	1.6 x 0.9	13.8	SB(s)b pec:	0	
15 49 01.2	+05 11 08	UGC 10041	12.9	2.8 x 1.6	14.4	SABdm	165	Several bright knots, mag 12.7 star on NE edge; very faint, elongated anonymous galaxy NW edge.
16 03 05.2	+05 06 26	UGC 10146	14.5	0.7 x 0.6	13.4	Sd:	117	Faint stellar nucleus, uniform envelope.
16 03 16.4	+05 38 24	UGC 10147	13.8	0.9 x 0.6	13.0	Double System	81	Small anonymous galaxy N 0′.7.
16 12 56.8	−00 05 50	UGC 10264	13.7	1.2 x 0.7	13.3	SAB(s)cd: III	35	
16 14 25.1	−00 12 27	UGC 10288	13.3	4.9 x 0.6	14.3	Sc: sp	91	Mag 9.00v star N 4′.6.
16 14 33.1	+00 49 18	UGC 10290	13.2	1.8 x 1.5	14.1	SB(s)m III-IV	90	Very knotty.

GALAXY CLUSTERS

RA h m s	Dec ° ′ ″	Name	Mag 10th brightest	No. Gal	Diam ′	Notes
15 11 00.0	+05 45 00	A 2029	16.0	82	16	
15 15 18.0	+04 22 00	A 2048	16.0	75	16	**MCG +1-39-11** near center; all other members anonymous, stellar/almost stellar.

GLOBULAR CLUSTERS

RA h m s	Dec ° ′ ″	Name	Total V m	B ★ V m	HB V m	Diam ′	Conc. Class Low = 12 High = 1	Notes
15 18 33.8	+02 04 58	NGC 5904	5.7	12.2	15.0	23.0	5	= **M 5**. Slightly oval in shape, NE-SW.
15 16 05.3	−00 07 14	Pal 5	11.8	15.5	17.4	8.0	12	

109 GALAXIES 109

RA h m s	Dec ° ′ ″	Name	Mag (V)	Dim ′ Maj x min	SB	Type Class	PA	Notes
13 51 54.6	+01 50 34	CGCG 17-77	15.0	0.6 x 0.3	13.0		33	CGCG 17-83 NE 10′.3.
13 51 57.2	+02 00 10	CGCG 17-80	14.3	0.4 x 0.4	12.2			NGC 5331 NE 7′.7; CGCG 17-83 ESE 8′.4.
13 52 28.0	+01 56 37	CGCG 17-83	14.5	0.5 x 0.3	12.5	E		CGCG 17-80 WNW 8′.4; NGC 5331 NNW 9′.9; CGCG 17-77 SW 10′.3.
13 53 30.7	−01 03 02	CGCG 17-92	14.6	0.5 x 0.4	12.7	S0		NGC 5334 SW 9′.8; UGC 8801 S 8′.6.
13 53 42.8	+00 03 35	CGCG 17-93	14.5	0.6 x 0.4	12.8		30	Mag 10.31v star NW 2′.8.
14 03 27.9	+00 43 56	CGCG 18-36	14.8	0.6 x 0.5	13.4		90	Pair of mag 9.81v, 10.08v stars W 5′.4; MCG +0-36-15 NE 6′.3.
14 03 35.7	−02 07 58	CGCG 18-37	14.9	0.9 x 0.2	12.9	Sc	78	Mag 9.82v star E 7′.9.
14 06 51.6	−01 32 28	CGCG 18-49	14.5	0.6 x 0.4	12.8		0	CGCG 18-51 S 3′.4.
14 06 56.5	−01 35 42	CGCG 18-51	14.1	0.6 x 0.3	12.3	E	108	CGCG 18-49 N 3′.4.
14 11 10.1	+01 28 26	CGCG 18-69	14.9	0.9 x 0.3	13.3		96	Mag 8.08v star NNE 6′.2; mag 6.41v star SSE 8′.5; **CGCG 18-66** W 7′.1.
14 12 40.1	+01 45 14	CGCG 18-80	15.5	0.5 x 0.4	13.6		123	Mag 14.2 star SSW 1′.8.
14 13 59.5	+01 49 50	CGCG 18-86	15.5	0.5 x 0.2	12.8		30	Much elongated **CGCG 18-85** SW 1′.4.
14 17 01.9	+00 28 32	CGCG 18-104	15.4	0.4 x 0.3	13.0		132	Stellar **CGCG 18-107** E 6′.6.
14 17 34.0	+01 48 42	CGCG 18-109	14.1	1.0 x 0.4	13.0		123	Located 0′.9 N of mag 10.88v star.
14 17 50.0	+00 30 35	CGCG 18-112	15.5	0.4 x 0.2	12.6			Stellar; also stellar **CGCG 18-107** W 5′.8.
14 25 38.3	+01 35 53	CGCG 19-23	16.4	0.6 x 0.4	14.7		6	Mag 9.31v star SW 5′.3.
14 28 37.9	+00 33 08	CGCG 19-41	14.8	0.9 x 0.3	13.3		60	Mag 14.3 star NE 1′.8.
14 31 43.5	−01 59 34	CGCG 19-55	15.5	0.5 x 0.5	13.8	Sc(r)		Mag 12.2 star W 9′.6; mag 10.36v star E 11′.3.
14 34 06.0	+00 37 09	CGCG 19-63	13.8	0.6 x 0.6	12.5			Mag 9.50v star NNE 6′.7.
14 42 29.7	+01 29 59	CGCG 19-84	14.1	0.5 x 0.4	12.3	S0	72	Mag 8.20v star NW 11′.0.
14 49 57.5	+00 33 27	CGCG 20-21	15.0	0.4 x 0.4	12.9			Mag 15.1 star NE 1′.7.
14 56 33.3	−02 06 30	CGCG 20-34	14.8	1.0 x 0.4	13.6	Irr	48	Mag 13.4 star N 0′.9; CGCG 20-35 E 6′.4.
14 56 59.1	−02 06 57	CGCG 20-35	14.7	0.6 x 0.4	13.2	E	15	CGCG 20-34 W 6′.4.
13 44 04.9	+03 54 15	CGCG 45-84	14.8	1.1 x 0.3	13.4	Sa	108	**UGC 8668** W 6′.2.
13 50 20.7	+03 00 49	CGCG 45-113	15.3	0.5 x 0.3	13.1		156	Mag 11.6 star WSW 1′.4.
13 52 12.1	+02 24 15	CGCG 45-123	15.1	0.9 x 0.5	14.1		177	NGC 5329 S 4′.8.
13 56 47.7	+05 31 32	CGCG 46-12	14.5	0.3 x 0.2	11.3		177	Mag 13.3 star NW 1′.9.
13 56 55.7	+05 09 01	CGCG 46-13	15.1	0.5 x 0.4	13.2		126	NGC 5373 NNE 6′.8; NGC 5364 SW 13′.6.
13 58 45.8	+02 50 35	CGCG 46-28	14.2	0.4 x 0.4	12.0			Mag 11.18v star N 1′.6.

RA h m s	Dec ° ′ ″	Name	Mag (V)	Dim ′ Maj x min	SB	Type Class	PA	Notes
14 00 18.6	+05 02 38	CGCG 46-33	14.7	0.6 x 0.3	12.6		36	Mag 12.0 star W 5′.6; mag 9.25v star SW 10′.0.
14 02 12.4	+04 35 06	CGCG 46-38	14.6	0.8 x 0.5	13.4		39	Mag 10.48v star N 3′.5.
14 06 20.9	+04 42 08	CGCG 46-49	16.5	0.5 x 0.3	14.3		27	Mag 11.3 star E 4′.2.
14 08 11.4	+05 48 37	CGCG 46-56	14.7	0.7 x 0.4	13.1		12	Mag 9.61v star E 7′.3.
14 09 54.0	+04 34 32	CGCG 46-62	14.3	0.5 x 0.5	12.6			Mag 11.9 star N 1′.7.
14 14 09.1	+03 13 27	CGCG 46-68	15.0	0.4 x 0.4	12.9			IC 988 ESE 6′.8; mag 6.47v star NE 12′.8.
14 20 39.9	+05 25 35	CGCG 47-14	14.8	0.8 x 0.3	13.1		96	Pair of E-W mag 10.31v, 10.61v stars E 4′.8; CGCG 47-15 N 3′.6.
14 20 42.4	+05 29 06	CGCG 47-15	15.3	0.6 x 0.2	12.9		90	CGCG 47-14 S 3′.6.
14 21 54.0	+03 57 56	CGCG 47-25	14.2	0.5 x 0.3	12.0		96	Mag 11.3 star NE 2′.5.
14 25 54.0	+02 52 24	CGCG 47-35	15.8	0.5 x 0.4	13.9		99	Mag 12.3 star NE 2′.9; mag 12.2 star SE 3′.4.
14 25 53.9	+04 38 30	CGCG 47-36	14.9	0.9 x 0.4	13.6		27	Small triangle of mag 12 stars N 7′.2.
14 27 22.6	+03 34 31	CGCG 47-46	14.6	0.9 x 0.7	14.0		111	Mag 7.46v star ENE 6′.7.
14 27 35.8	+02 57 13	CGCG 47-49	15.0	0.6 x 0.6	13.7			Mag 9.20v star SE 8′.9; very faint anonymous galaxy N 1′.7.
14 28 21.5	+05 02 28	CGCG 47-56	15.0	0.9 x 0.3	13.4		117	Mag 10.66v star NW 6′.7.
14 29 03.2	+04 40 41	CGCG 47-59	15.6	0.8 x 0.3	13.9		6	**CGCG 47-55** WNW 11′.0.
14 33 48.5	+03 57 21	CGCG 47-94	14.9	0.6 x 0.5	13.5	Sb-Sbc	30	Mag 11.8 star NE 2′.1.
14 33 58.1	+03 44 44	CGCG 47-97	15.1	0.7 x 0.5	14.0	E	69	UGC 9371 N 1′.9.
14 34 04.0	+03 44 48	CGCG 47-100	15.3	0.4 x 0.3	13.0	E	111	UGC 9371 NW 2′.2.
14 36 51.3	+05 57 38	CGCG 47-116	15.4	0.4 x 0.3	13.0		9	Mag 11.4 star SE 1′.7.
14 42 26.0	+03 13 51	CGCG 47-147	14.3	0.8 x 0.3	12.6		111	Elongated anonymous galaxy NNW 2′.6.
14 43 25.7	+04 22 43	CGCG 48-8	14.3	0.4 x 0.3	11.8		138	Very faint, elongated anonymous galaxy on NE edge; mag 10.08v star SW 2′.7.
14 49 02.6	+02 26 09	CGCG 48-20	15.5	0.5 x 0.3	13.3	Sc	24	Mag 11.4 star W 7′.9.
14 52 43.2	+03 13 26	CGCG 48-44	14.3	0.5 x 0.4	12.4		22	Mag 11.1 star NW 3′.6; elongated anonymous galaxy SE 3′.7.
14 55 21.3	+05 17 50	CGCG 48-70	15.2	0.5 x 0.3	13.0		90	Mag 10.8 star E 2′.2.
14 58 34.1	+03 31 03	CGCG 48-83	15.0	0.5 x 0.4	13.1		48	Elongated **CGCG 48-80** SW 8′.5.
13 47 43.2	+03 24 39	IC 939	13.8	0.9 x 0.9	13.4			Stellar **CGCG 45-102** S 4′.6; **IC 940** NE 4′.3.
13 50 32.1	+03 11 37	IC 943	14.0	0.7 x 0.7	13.1	Sb		Mag 9.01v star NE 9′.5.
13 52 35.9	+00 49 05	IC 947	12.6	1.5 x 0.9	12.8	S0⁻ pec?	60	8.05v star on NE end.
13 53 42.1	+03 22 37	IC 952	14.0	1.3 x 0.4	13.1	SBbc	93	Mag 9.28v star SE 8′.2.
13 58 14.1	+05 24 29	IC 966	13.1	0.8 x 0.7	12.4	S0	162	Mag 10.34v star NE 9′.0.
14 00 37.4	−02 54 31	IC 968	14.8	1.0 x 0.5	13.9	S0	57	Galaxy pair.
14 08 43.4	−01 09 45	IC 976	13.0	1.4 x 0.6	12.7	S0°	175	Mag 8.03v star NW 11′.3.
14 08 58.2	−02 58 27	IC 978	14.6	0.7 x 0.4	13.1	Sb	33	Mag 8.88v star W 8′.9; **IC 977** WSW 4′.3.
14 11 33.0	−03 13 14	IC 985	13.9	0.5 x 0.3	11.7	Sb	54	Mag 8.72v star SSW 7′.6.
14 11 26.3	+01 17 10	IC 986	14.0	0.6 x 0.6	12.7			**Mkn 1374** SW 2′.2.
14 14 32.1	+03 11 28	IC 988	14.1	0.7 x 0.6	13.2	E?	54	CGCG 46-68 WNW 6′.8; mag 6.47v star NNE 10′.2; IC 989 SE 6′.0.
14 14 51.5	+03 07 51	IC 989	13.1	1.3 x 1.1	13.4	E	63	**UGC 9118** E 4′.3; mag 7.11v star E 7′.1; IC 988 NW 6′.0.
14 18 14.8	+00 53 27	IC 992	13.6	1.1 x 0.9	13.5	SAB(r)c: II	130	Mag 12.8 star NE 1′.2.
14 19 59.3	−04 27 13	IC 997	12.8	1.3 x 0.8	12.7	Sbc	23	Multiple galaxy system; mag 10.98v star WNW 7′.3.
14 24 36.7	+04 33 30	IC 1007	14.1	0.4 x 0.4	12.0			Faint, elongated, anonymous galaxy NE 1′.2.
14 27 20.3	+01 01 30	IC 1010	12.8	1.9 x 1.8	14.0	SB(r)b pec II-III	78	Mag 8.03v star NNW 8′.5.
14 28 04.7	+01 00 17	IC 1011	14.5	0.5 x 0.4	12.6		111	Mag 10.78v star SW 4′.5.
14 31 26.9	+03 00 29	IC 1024	12.9	1.6 x 0.6	12.7	S0?	24	Mag 14.0 star superimposed N edge of core.
14 40 38.0	+03 22 37	IC 1041	14.3	0.8 x 0.5	13.2	Sb	171	Anonymous galaxy E 1′.4, possibly **IC 1043**?
14 40 39.1	+03 28 08	IC 1042	13.3	1.1 x 1.1	13.3	S0?		NGC 5718 E edge.
14 42 57.8	+04 53 31	IC 1048	12.8	2.2 x 0.7	13.1	S	163	Mag 10.2 star NE 7′.7.
14 46 31.2	+01 16 28	IC 1054	13.8	1.0 x 0.6	13.1	S0	13	Mag 11.9 star W 1′.5.
14 52 11.1	+04 40 52	IC 1063	13.4	1.2 x 0.7	13.1	SBb	162	Mag 10.68v star W 3′.9.
14 53 02.8	+03 17 44	IC 1066	13.2	1.2 x 0.7	12.9	S?	70	Mag 13.3 star E 1′.9; IC 1067 N 2′.3.
14 53 05.2	+03 19 51	IC 1067	12.2	2.1 x 1.8	13.5	SB(s)b	122	IC 1066 S 2′.3; mag 8.67v star NE 7′.6.
14 53 32.9	+03 04 36	IC 1068	14.3	0.6 x 0.6	13.1			**CGCG 48-51** S 7′.1.
14 54 12.4	+04 45 01	IC 1071	13.2	1.0 x 0.8	12.8	S0	150	Mag 12.2 star NW 4′.9.
14 19 25.0	−04 29 24	IC 4401	13.4	1.5 x 0.6	13.2	(R′)SAB(r)a?	21	Mag 10.98v star N 4′.8; mag 11.6 star W 4′.5; IC 997 E 8′.8.
14 23 36.8	−05 58 55	IC 4407	14.1	1.5 x 0.9	14.3	SB(s)m IV-V	0	Low surface brightness.
14 27 32.6	+04 49 16	IC 4424	14.1	0.8 x 0.4	12.7	S?	111	UGC 9258 S 2′.6.
13 46 13.3	−03 23 07	MCG +0-35-19	14.5	0.8 x 0.8	13.9	Sa		MCG +0-35-20 SE 3′.0; mag 9.92v star SSW 9′.3.
13 46 20.3	−03 25 41	MCG +0-35-20	14.5	0.9 x 0.3	13.0	Sb	6	Mag 9.92v star SW 8′.5; MCG +0-35-19 NW 3′.0.
13 58 07.1	−00 23 30	MCG +0-36-4	14.2	0.7 x 0.7	13.3	Sb		
13 59 45.0	−03 02 59	MCG +0-36-6	15.6	0.6 x 0.4	13.9	S?	174	
14 01 26.8	−02 42 47	MCG +0-36-10	15.1	0.7 x 0.5	13.9	Sm	135	
14 03 20.6	−00 32 57	MCG +0-36-14	14.3	0.4 x 0.4	12.2	Sa		Mag 7.99v star S 4′.8.
14 03 44.0	+00 48 47	MCG +0-36-15	14.6	0.5 x 0.5	12.9	Sm		Mag 12.0 star SE 3′.0; CGCG 18-36 SW 6′.3.
14 05 11.3	−03 22 12	MCG +0-36-18	12.9	1.7 x 1.0	13.3	(R)Sa pec?	153	Mag 12.4 star NE 1′.9.
14 09 56.5	−01 14 27	MCG +0-36-22	14.7	0.4 x 0.4	12.6	S pec		
14 15 26.5	−03 06 55	MCG +0-36-30	14.7	0.8 x 0.7	13.9	SBc	123	Mag 11.7 star superimposed SE end.
14 28 23.3	−03 36 18	MCG +0-37-9	13.4	1.1 x 0.8	13.1	S?	108	Mag 14.4 star S edge.
14 28 38.2	−03 19 54	MCG +0-37-10	14.4	0.9 x 0.6	13.6	Sbc	156	Small, faint anonymous galaxy W 2′.3.
14 30 42.4	+01 52 48	MCG +0-37-13	14.3	0.7 x 0.5	13.0	Sbc	24	Star or bright knot E edge.
14 31 32.7	−03 17 35	MCG +0-37-14	14.5	0.4 x 0.4	12.4	S?		Mag 12.4 star E 1′.5.
14 35 19.7	+00 28 31	MCG +0-37-17	13.9	0.3 x 0.3	11.4	E		Compact.
14 40 29.3	+00 35 05	MCG +0-37-23	15.3	0.6 x 0.6	14.1	Sb		
14 44 33.7	−03 13 36	MCG +0-38-4	13.6	1.3 x 1.1	13.8	SB(r)bc: III-IV	80	Short SE-NW bar.
14 52 34.5	−03 33 36	MCG +0-38-10	12.9	2.6 x 1.9	14.5	SB(s)dm pec III-IV	105	Several bright knots; mag 8.16v star NW 4′.9.
14 58 48.8	+02 01 22	MCG +0-38-13	14.1	0.7 x 0.4	12.8	E	0	
15 03 45.2	−03 18 09	MCG +0-38-19	13.4	1.8 x 1.4	14.3	SAB(s)cd III	70	Mag 8.57v star NNW 9′.0.
14 17 32.1	+04 33 42	MCG +1-36-31	14.1	0.9 x 0.6	13.3	Sb	177	Mag 12.0 star NW 1′.3.
14 20 51.1	+04 19 52	MCG +1-37-4	14.2	0.5 x 0.4	12.3		129	Mag 9.13v star S 2′.7; note there is a small anonymous galaxy just N of this star.
14 32 45.6	+02 54 54	MCG +1-37-25	13.6	0.9 x 0.6	12.8	SB?	165	Mag 15.2 star W edge; mag 11.8 star E 1′.3.
14 33 09.7	+03 56 46	MCG +1-37-26	14.7	1.0 x 0.5	13.8	Sc	132	UGC 9362 SE 3′.3.
14 39 15.6	+02 43 39	MCG +1-37-41	14.2	0.8 x 0.4	12.8		60	
14 39 37.8	+03 08 46	MCG +1-37-43	14.2	0.9 x 0.8	13.7		27	Mag 11.1 star S 2′.6.
14 42 39.0	+05 21 05	MCG +1-37-50	14.7	0.8 x 0.4	13.3	Sb	33	
14 54 17.9	+04 16 58	MCG +1-38-17	14.1	1.2 x 0.8	14.1	E	144	
14 55 03.3	+02 42 31	MCG +1-38-19	15.7	0.6 x 0.6	14.4	Sb		
14 56 01.0	+02 27 40	MCG +1-38-20	14.7	1.0 x 0.3	13.3	S0	174	
15 03 27.9	+04 40 59	MCG +1-38-27	14.1	0.7 x 0.3	12.3	Sa	0	
13 45 38.3	−05 59 05	MCG −1-35-10	13.8	1.6 x 1.1	14.3	SB(s)m III	57	Several bright knots or superimposed stars; mag 12.4 star SW 2′.2.
13 54 12.1	−04 50 49	MCG −1-35-20	13.3	1.6 x 1.3	14.0	(R′)SB(r)bc: II-III	9	Close pair of stars, mags 14.0 and 13.2, SW 2′.4.
13 55 34.7	−05 58 24	MCG −1-35-21	14.6	0.7 x 0.6	13.5		165	

(Continued from previous page)

GALAXIES

RA h m s	Dec ° ′ ″	Name	Mag (V)	Dim ′ Maj x min	SB	Type Class	PA	Notes
13 55 45.5	−06 00 12	MCG −1-35-22	14.5	1.6 x 0.8	14.6	SB(s)dm pec: IV	147	
13 58 06.6	−04 05 13	MCG −1-36-1	14.0	1.2 x 0.7	13.6	Sc	3	
14 21 22.3	−03 45 37	MCG −1-37-3	14.1	1.2 x 0.5	13.4	Sbc	120	Multi-galaxy system?
14 23 03.9	−04 43 32	MCG −1-37-4	13.5	0.5 x 0.5	11.8			
14 25 45.5	−05 24 16	MCG −1-37-6	13.2	1.2 x 0.6	12.7	I0 pec?	105	Mag 8.71v star SW 12'.7.
14 29 20.3	−05 01 23	MCG −1-37-7	14.4	1.1 x 1.0	14.3	SBbc	21	Mag 8.38v star NNW 5'.9.
14 43 36.6	−05 10 35	MCG −1-38-1	14.1	0.6 x 0.6	13.0	E/S0		
14 48 36.5	−04 44 12	MCG −1-38-2	14.7	1.3 x 1.0	14.8	SA(s)cd: III-IV	12	Mag 13.0 star N edge.
14 48 48.0	−03 43 04	MCG −1-38-3	14.9	0.6 x 0.6	13.6	SB(s)m: IV-V		
14 53 25.5	−04 41 51	MCG −1-38-7	14.7	1.5 x 0.4	14.0	Sc: sp	168	Mag 12.1 star SW 1'.7.
15 01 18.4	−04 33 21	MCG −1-38-18	14.4	0.5 x 0.4	12.5		60	Mag 9.52v star N 2'.1.
15 01 24.9	−04 30 01	MCG −1-38-19	15.6	0.5 x 0.4	13.7	Sbc	171	Mag 9.52v star SW 2'.2.
13 44 25.8	+02 06 36	NGC 5285	13.9	0.8 x 0.8	13.5	E^+ pec		Almost stellar nucleus; mag 9.65v star WNW 7'.0.
13 48 15.9	+03 56 58	NGC 5300	11.4	3.9 x 2.6	13.8	SAB(r)c II	150	Very patchy, many knots and/or superimposed stars.
13 52 04.1	−02 12 22	NGC 5327	12.6	1.9 x 1.6	13.7	SAB(rs)b: II-III	90	Bright nucleus in short bar.
13 52 10.1	+02 19 26	NGC 5329	12.4	1.3 x 1.3	13.0	E:		**CGCG 45-122** NNE 2'.4; **CGCG 45-124** E 3'.4; CGCG 45-123 N 4'.8.
13 52 16.5	+02 06 12	NGC 5331	14.7	1.1 x 0.9	14.5	Sb	0	Double system in contact; CGCG 17-80 SW 7'.7; CGCG 17-83 SSE 9'.9.
13 52 54.8	−01 06 59	NGC 5334	11.3	4.2 x 3.0	13.9	SB(rs)c: III	21	Bright nucleus, filamentary arms with some knots. CGCG 17-92 NE 9'.8; UGC 8801 ESE 10'.5.
13 52 56.6	+02 48 49	NGC 5335	12.8	2.1 x 1.4	13.8	SB(r)b	90	Bright N-S bar, dark patches E and W of bar.
13 53 26.8	+05 12 23	NGC 5338	12.4	2.0 x 0.9	12.9	SB0:	97	Bright nucleus; near center of galaxy cluster A 1809; numerous anonymous galaxies in area.
13 54 14.4	−01 26 10	NGC 5345	12.4	1.6 x 1.5	13.2	SA(r)a pec:	176	Faint star S of nucleus.
13 54 11.2	+05 13 40	NGC 5348	13.1	3.5 x 0.5	13.6	SBbc: sp	177	On E edge of galaxy cluster A 1809.
13 54 58.2	+05 19 56	NGC 5356	12.6	3.1 x 0.9	13.6	SABbc: sp	15	
13 55 38.5	+04 59 02	NGC 5360	13.3	1.5 x 0.5	12.8	I0	70	Faint star on SW end.
13 56 07.0	+05 15 19	NGC 5363	10.1	4.1 x 2.6	12.5	I0?	135	Located 3'.9 SW of mag 8.06v star.
13 56 12.2	+05 00 58	NGC 5364	10.5	6.8 x 4.4	14.1	SA(rs)bc pec I	30	**Leo III**. Bright nucleus, smooth spiral arms.
13 56 25.0	−00 14 51	NGC 5366	13.7	0.9 x 0.8	13.3	(R′)SB(rs)0^+?	57	Bright knot or star NE edge of nucleus; **MCG +0-36-3** S 2'.5.
13 56 37.8	−05 28 12	NGC 5369	13.4	0.9 x 0.8	13.2	E	114	
13 57 07.5	+05 15 05	NGC 5373	14.9	0.5 x 0.4	13.0		174	Mag 9.21v star E 4'.0; CGCG 46-13 SSW 6'.8.
13 59 25.0	−03 12 34	NGC 5392	13.4	1.2 x 0.8	13.3	$S0^+$?	135	Star or knot NE of nucleus.
14 00 37.2	−02 51 29	NGC 5400	12.9	1.5 x 0.9	13.1	(R′)$S0^-$?	80	Anonymous galaxies 0'.9 S and 1'.7 N.
14 06 34.9	−05 27 11	NGC 5468	12.5	2.6 x 2.4	14.3	SAB(rs)cd II-III	105	Almost stellar nucleus, branching, knotty arms; mag 8.34v star S 4'.2; NGC 5472 E 5'.0.
14 06 55.0	−05 27 40	NGC 5472	14.0	1.2 x 0.3	12.7	SA(r)ab? sp	29	Located 5'.0 E of NGC 5468.
14 08 08.5	−01 42 08	NGC 5478	13.6	1.1 x 0.8	13.3	SAB(s)bc II	37	**CGCG 18-54** NW 7'.0.
14 11 29.4	−05 02 37	NGC 5493	11.4	1.6 x 1.3	12.0	S0 pec sp	120	Very bright lens shaped center.
14 11 37.8	−01 09 23	NGC 5496	12.1	4.7 x 0.9	13.6	Sd: sp III:	172	Several superimposed stars.
14 12 20.1	+01 16 20	NGC 5501	13.5	0.8 x 0.6	12.6	(R′)SA(r)0^+?	67	Strong stellar nucleus.
14 13 15.4	−03 12 30	NGC 5506	11.9	2.8 x 0.9	12.8	Sa pec sp	95	Mag 7.04v star SW 10'.0; NGC 5507 N 3'.7.
14 13 20.1	−03 08 55	NGC 5507	12.5	1.7 x 0.9	12.8	SAB(r)0°	55	NGC 5506 S 3'.7.
14 15 23.9	+04 24 27	NGC 5521	13.8	0.6 x 0.4	12.1	S?	96	Star on N side of nucleus.
14 18 54.9	+05 27 04	NGC 5551	14.2	0.6 x 0.4	12.5		111	
14 20 04.8	+03 59 27	NGC 5560	12.4	3.7 x 0.8	13.4	SB(s)b pec	117	Two filamentary arms; mag 8.17v star W 5'.1.
14 20 20.3	+03 56 01	NGC 5566	10.6	6.6 x 2.2	13.3	SB(r)ab II-III	33	Bright center with uniform envelope; NGC 5569 NE end.
14 20 32.2	+03 58 57	NGC 5569	13.2	1.7 x 1.4	14.0	SAB(rs)cd:	72	Located NE end of NGC 5566.
14 20 56.0	+03 14 12	NGC 5574	12.4	1.6 x 1.0	12.8	$SB0:^-$? sp	63	Bright, elongated core.
14 21 03.5	+03 16 14	NGC 5576	11.0	3.8 x 2.7	13.5	E3	95	Bright center, uniform envelope.
14 21 13.0	+03 26 08	NGC 5577	12.3	3.4 x 1.0	13.4	SA(rs)bc: III	56	
14 22 23.9	−00 23 04	NGC 5584	11.4	3.4 x 2.5	13.5	SAB(rs)cd II	140	Stellar nucleus with several knots near N edge.
14 24 42.9	−03 12 44	NGC 5604	12.8	1.8 x 1.0	13.3	Sa pec?	15	
14 27 11.9	−02 15 50	NGC 5618	13.4	1.6 x 1.2	13.9	SB(rs)c II-III	10	Dark lanes E and W of nucleus.
14 27 18.0	+04 48 01	NGC 5619	12.6	2.2 x 1.2	13.5	SAB(rs)b II	8	Strong dark lane NW.
14 28 20.3	−00 12 17	NGC 5632	14.6	0.4 x 0.2	11.7		6	Stellar.
14 29 39.2	+03 15 57	NGC 5636	12.7	1.9 x 1.4	13.6	SAB(r)0^+	40	Stellar nucleus with dark patches N and S; NGC 5638 on S edge.
14 29 40.4	+03 13 57	NGC 5638	11.2	2.7 x 2.4	13.2	E1	150	NGC 5636 N edge.
14 31 01.4	+05 58 40	NGC 5652	12.5	2.0 x 1.4	13.4	SABbc II	117	Diffuse nucleus; mag 10.7 star N 5'.8.
14 21 30.0	+05 04 22	NGC 5665A	14.0	1.2 x 0.5	13.3	Sbc:	40	Almost stellar nucleus in smooth envelope, star or knot on S edge.
14 33 24.3	+04 26 59	NGC 5668	11.5	3.0 x 3.0	13.8	SA(s)d II-III		Bright nucleus, very patchy.
14 33 52.1	+05 27 28	NGC 5674	13.0	1.1 x 1.0	13.0	SABc	30	**CGCG 47-99** N 7'.7.
14 35 06.4	+05 21 21	NGC 5679A	13.7	0.8 x 0.5	12.5	Sc	54	Located W edge NGC 5679B; star or knot N edge.
14 35 08.6	+05 21 31	NGC 5679B	13.6	1.2 x 0.7	13.2	Sb	127	NGC 5679A W edge, NGC 5679C SE edge.
14 35 11.1	+05 21 12	NGC 5679C		0.3 x 0.2			10	Located SE end of NGC 5679B.
14 35 44.7	−00 00 49	NGC 5680	13.7	0.9 x 0.9	13.4	$S0^-$:		Mag 15.5 star NE edge of nucleus, mag 14.0 star on N edge.
14 37 41.7	+02 17 16	NGC 5690	11.8	3.4 x 1.0	13.0	Sc? sp II-III	143	Located 3'.3 E of mag 6.45v star.
14 37 53.4	−00 23 54	NGC 5691	12.3	1.9 x 1.4	13.2	SAB(s)a: pec III-IV	110	
14 38 18.1	+03 24 32	NGC 5692	13.0	0.9 x 0.6	12.2	S?	35	
14 39 11.1	+05 21 48	NGC 5701	10.9	4.3 x 4.1	13.8	(R)SB(rs)0/a	90	Bright N-S bar with strong dark patches E and W.
14 39 49.8	−00 43 07	NGC 5705	12.7	2.9 x 1.7	14.3	SB(rs)d III	75	Elongated nucleus, bright knot or star on S edge.
14 40 11.6	−00 17 25	NGC 5713	11.2	2.8 x 2.5	13.2	SAB(rs)bc pec	10	Large, diffuse core.
14 40 42.9	+03 27 54	NGC 5718	12.9	1.5 x 1.1	13.3	$S0^-$:	117	IC 1042 W edge; mag 10.73v star NE 1'.2.
14 40 56.9	−00 19 07	NGC 5719	12.2	3.2 x 1.2	13.5	SAB(s)ab pec III	107	Strong E-W dark lane south of elongated core; mag 10.48v star N 1'.8.
14 40 58.1	+02 11 08	NGC 5725	13.7	1.1 x 0.7	13.2	SB(s)d: III	40	
14 42 45.9	−00 21 03	NGC 5733	14.0	1.0 x 0.3	12.6	Scd? sp III	25	
14 43 56.5	+01 36 13	NGC 5738	13.9	1.1 x 0.4	12.8	S0°: sp	57	
14 44 24.3	+01 40 50	NGC 5740	11.9	3.0 x 1.5	13.3	SAB(rs)b II	160	Dark lanes N and E; **CGCG 20-10** S 9'.5.
14 44 55.8	+01 57 22	NGC 5746	10.3	7.4 x 1.3	12.6	SAB(rs)b? sp III	170	Very strong dark lane E of center.
14 46 10.9	−00 13 20	NGC 5750	11.6	3.0 x 1.6	13.2	SB(r)0/a	65	Mag 8.35v star SW 12'.0.
14 50 51.0	+05 07 10	NGC 5765A	13.9	0.8 x 0.4	12.5	Sb:	82	Double system with NGC 5765B S.
14 50 51.6	+05 06 51	NGC 5765B	14.6	0.8 x 0.6	13.7	Sa-b	82	Double system with NGC 5765A N.
14 52 08.2	−02 31 48	NGC 5768	12.5	1.9 x 1.4	13.4	SA(rs)c: III-IV	120	
14 53 15.0	+03 57 35	NGC 5770	12.3	1.7 x 1.3	13.0	SB0	108	
14 53 42.5	+03 34 59	NGC 5774	12.1	3.0 x 2.5	14.2	SAB(rs)d	145	Many bright knots and/or superimposed stars.
14 53 57.3	+03 32 40	NGC 5775	11.4	4.2 x 1.0	12.8	SBc? sp	146	**IC 1070** S 4'.0.
14 54 32.8	+02 57 58	NGC 5776	14.1	0.9 x 0.7	13.5		102	Mag 9.14v star S 4'.1.
14 58 23.0	−01 05 28	NGC 5792	11.3	6.9 x 1.7	13.8	SB(rs)b I-II	84	Mag 9.60v star 1'.1 NW of nucleus.
15 00 00.5	+01 53 23	NGC 5806	11.7	3.1 x 1.6	13.3	SAB(s)b III:	170	
15 00 27.2	+01 37 23	NGC 5811	13.9	0.9 x 0.7	13.2	SB(s)m: IV-V	98	Mag 9.07v star W 7'.2.
15 01 11.3	+01 42 02	NGC 5813	10.5	4.2 x 3.0	13.2	E1-2	145	Located between a pair of mag 12 stars. NGC 5814 SE 4'.6.
15 01 21.2	+01 38 12	NGC 5814	13.8	0.9 x 0.5	12.8	(R′)Sab	32	Located 4'.6 SE of NGC 5813.
13 44 23.2	+05 47 30	UGC 8689	13.0	1.6 x 1.1	13.5	S0	155	**CGCG 45-88** NE 5'.1.

(Continued from previous page)

109 GALAXIES 109

RA h m s	Dec ° ′ ″	Name	Mag (V)	Dim ′ Maj x min	SB	Type Class	PA	Notes
13 44 33.7	+04 46 29	UGC 8690	13.9	0.9 x 0.5	12.9	Scd:	156	Bright knot or star E of center; mag 11.22v star SW 1′9.
13 49 38.8	+04 14 17	UGC 8740	12.9	1.5 x 1.4	13.6	SABab		Close pair of mag 12.5 stars NNE 2′3.
13 52 52.6	−01 53 46	UGC 8786	13.9	1.7 x 0.3	13.0	Sab: sp	83	**UGC 8783** S 6′9.
13 52 50.2	+02 15 51	UGC 8787	13.8	1.7 x 0.4	13.2	Sbc: sp	146	Elongated **MCG +1-35-47** E 2′6.
13 53 31.9	−01 11 39	UGC 8801	14.0	1.2 x 0.3	12.8	Sbc? sp	101	Mag 11.9 star NW 1′9; NGC 5334 WNW 10′5; CGCG 17-92 N 8′6.
13 54 04.0	+03 57 01	UGC 8816	14.3	1.1 x 0.5	13.5	Scd:	25	Patchy.
13 54 06.3	+05 21 18	UGC 8818	13.9	0.9 x 0.8	13.4	Sbc	177	Large knot or small galaxy NE edge.
13 55 59.7	−01 15 42	UGC 8844	13.8	1.7 x 0.6	13.7	SB(s)bc: II-III	30	Mag 13.2 star SW 1′9.
13 59 22.6	+05 32 20	UGC 8906	14.2	1.5 x 0.5	13.7	Sd	111	Mag 11.9 star SSE 2′1.
14 00 46.0	+02 01 14	UGC 8924	14.0	1.5 x 0.3	13.0	Scd: sp	122	**CGCG 18-23** N 4′5.
14 02 04.7	−01 21 33	UGC 8939	13.3	1.0 x 0.9	13.0	SAB(r)b III	24	**MCG +0-36-13** SE 2′5; **MCG +0-36-11** S 2′3.
14 04 15.8	+04 06 43	UGC 8986	13.1	1.3 x 1.3	13.6	S0?		Almost stellar nucleus, faint envelope.
14 04 51.6	−00 38 29	UGC 8993	14.0	1.0 x 0.8	13.6	(R)SB(r)a pec	0	
14 04 52.7	−00 36 41	UGC 8994	13.9	1.0 x 0.5	13.0	SAB(s)bc pec: II	132	
14 05 06.2	−00 03 51	UGC 9006	13.9	1.2 x 0.7	13.5	S?	90	
14 10 12.9	−02 34 36	UGC 9057	12.9	2.8 x 0.9	13.8	SB(rs)d? III	173	Many bright patches.
14 15 12.4	+04 49 23	UGC 9120	13.6	1.2 x 0.8	13.4	Scd:	3	Star or very faint anonymous galaxy N edge.
14 23 27.4	+01 43 24	UGC 9215	12.6	2.9 x 1.2	13.8	SB(s)d III	156	Very knotty.
14 24 33.5	+01 10 34	UGC 9229	14.1	0.9 x 0.8	13.6	SAB(rs)bc pec II-III	117	Anonymous galaxy SE 1′1.
14 26 08.5	+05 14 15	UGC 9244	14.0	1.5 x 0.8	14.0	SBbc	6	Very faint, small, elongated anonymous galaxy N edge.
14 26 30.5	+05 58 35	UGC 9246	14.0	1.2 x 0.4	13.1	S?	152	Located 3′8 NW of mag 7.20v star.
14 27 29.6	+04 46 44	UGC 9258	13.9	0.9 x 0.8	13.4	Sbc	3	Three very faint stars in N-S row along E edge.
14 27 45.5	−00 36 08	UGC 9263	14.3	1.0 x 0.8	13.9	Sd:	60	Weak stellar nucleus, star on N edge; mag 9.29v star E 5′4.
14 28 33.2	+03 15 37	UGC 9277	14.2	1.5 x 0.3	13.2	Sb III	48	
14 29 03.7	+03 08 52	UGC 9285	14.4	1.5 x 0.3	13.4	Scd?	88	
14 29 10.6	+02 17 12	UGC 9292	14.0	1.5 x 0.3	13.0	S?	16	
14 29 34.7	−00 01 10	UGC 9299	13.3	1.5 x 0.8	13.3	SAB(s)d pec III	85	
14 30 01.1	+03 13 13	UGC 9310	13.9	2.0 x 0.6	13.9	SBdm	163	Very close pair of stars, mags 12.6 and 14.7, superimposed SW edge.
14 31 11.7	+05 18 24	UGC 9338	14.3	1.0 x 0.4	13.1	Sb II-III	49	Very faint, elongated anonymous galaxy NW 3′1.
14 32 28.5	+00 17 38	UGC 9348	14.2	1.8 x 0.4	13.7	S0°: sp	119	Elongated anonymous galaxy S 1′3.
14 33 17.4	+03 54 10	UGC 9362	13.9	1.0 x 0.9	13.6	Sbc	48	MCG +1-37-26 NW 3′3.
14 33 30.7	+03 41 08	UGC 9365	13.7	1.0 x 0.6	13.0	Sbc	4	**CGCG 47-95** E 5′2; **CGCG 47-101** E 12′1; mag 7.40v star W 8′5.
14 33 59.3	+03 46 37	UGC 9371	14.1	1.2 x 0.7	13.8	Double System	12	Double system, contact; CGCG 47-100 SE 2′1; CGCG 47-97 S 1′8.
14 34 39.0	+04 15 36	UGC 9380	13.6	1.9 x 1.0	14.1	Im	36	Dwarf irregular, very low surface brightness.
14 35 22.7	+05 16 34	UGC 9385	16.0	1.5 x 1.4	16.7	Im:	105	Dwarf, extremely low surface brightness.
14 36 17.1	−01 23 11	UGC 9398	14.8	0.7 x 0.5	13.5	SBb-c	42	
14 36 20.3	+05 19 49	UGC 9400	13.5	1.1 x 0.8	13.2	S0	30	
14 39 04.6	+02 56 59	UGC 9432	14.0	1.2 x 0.9	13.9	Im:	48	Dwarf, very low surface brightness; mag 13.6 star E edge; mag 10.9 star NNW 2′3.
14 41 44.3	−01 48 24	UGC 9469	13.7	1.5 x 1.5	14.4	IAB(s)m: V-VI		Dwarf, low surface brightness.
14 41 48.8	+00 41 14	UGC 9470	13.8	1.1 x 0.6	13.2	SB(rs)dm pec: III-IV	52	Mag 11.7 star SE 2′1.
14 41 43.0	+05 57 08	UGC 9471	13.8	1.5 x 0.5	13.3	SBb	162	CGCG 47-142 NW 10′3.
14 42 32.6	+04 25 50	UGC 9479	14.0	1.6 x 0.4	13.3	Sb III	30	Mag 7.87v star N 3′4.
14 43 02.8	+04 45 55	UGC 9485	14.0	1.5 x 0.4	13.3	S?	148	Mag 9.15v star SW 9′7.
14 44 14.7	+04 13 06	UGC 9491	13.6	1.2 x 1.0	13.6	S?	50	Mag 11.6 star N 2′3.
14 56 01.6	−01 23 19	UGC 9601	13.2	1.4 x 1.3	13.7	SB(s)cd pec? III-IV	162	
15 02 03.2	+01 50 24	UGC 9661	14.0	1.1 x 1.0	14.0	SB(rs)dm III-IV	63	Brighter N-S core slightly offset E.
15 02 19.9	+05 38 46	UGC 9667	13.1	1.8 x 0.9	13.5	S0/a	87	**CGCG 48-93** pair with mag 14.1 star SSE 6′6.

GALAXY CLUSTERS

RA h m s	Dec ° ′ ″	Name	Mag 10th brightest	No. Gal	Diam ′	Notes
13 53 18.0	+05 09 00	A 1809	15.8	78	16	Large concentration of anonymous, stellar GX SW of NGC 5338.

GLOBULAR CLUSTERS

RA h m s	Dec ° ′ ″	Name	Total V m	B ★ V m	HB V m	Diam ′	Conc. Class Low = 12 High = 1	Notes
14 29 37.3	−05 58 35	NGC 5634	9.5		17.8	5.5	4	

110 GALAXIES 110

RA h m s	Dec ° ′ ″	Name	Mag (V)	Dim ′ Maj x min	SB	Type Class	PA	Notes
12 35 50.9	−03 46 00	CGCG 14-74	13.4	0.8 x 0.4	12.1	S0 sp	51	NGC 4546 WSW 5′5.
12 35 54.9	−02 23 10	CGCG 14-76	13.9	0.8 x 0.3	12.3	SA0° sp	132	Located 3′1 SE of mag 9.21v star.
12 37 02.6	+01 49 06	CGCG 14-78	13.9	0.5 x 0.5	12.3	SA0⁻:		Mag 12.1 star N 6′4.
12 40 01.9	−03 36 29	CGCG 14-98	15.4	0.8 x 0.3	13.7		177	Mag 12.4 star N 4′5; mag 10.88v star SE 8′6.
12 44 28.3	−03 00 22	CGCG 15-10	14.2	0.9 x 0.5	13.2	SAB(r)cd?	147	Located 1′4 E of mag 10.60v star.
12 45 52.3	−01 40 49	CGCG 15-17	13.9	0.5 x 0.4	12.1	SA0⁻	57	Mag 12.4 star S 3′0.
12 55 36.8	−00 53 04	CGCG 15-40	15.8	0.6 x 0.6	14.6			Mag 9.13v star NE 6′3; **CGCG 15-41** SSE 6′7.
13 00 54.4	−01 20 51	CGCG 15-54	14.7	0.6 x 0.4	13.0	Sc	39	Mag 10.24v star E 2′7; mag 11.8 star W 1′8.
13 05 58.2	+01 47 29	CGCG 15-61	14.8	0.6 x 0.4	13.1		3	Mag 13.1 star NNW 0′8; mag 10.16v star W 6′7.
13 07 19.2	+02 00 08	CGCG 16-2	15.2	0.6 x 0.6	14.0	Irr		Mag 9.93v star SSW 8′0.
13 12 03.8	+01 38 40	CGCG 16-20	14.9	0.6 x 0.3	12.9		156	Mag 11.07v star W 3′8; mag 10.06v star E 5′2.
13 13 51.2	−03 36 24	CGCG 16-23	14.2	1.0 x 0.3	12.7		93	Located 2′7 NNW of mag 10.60v star.
13 13 52.5	+02 07 55	CGCG 16-24	14.9	0.7 x 0.5	13.6		153	Mag 11.6 star N 1′8.
13 15 21.6	+01 55 36	CGCG 16-31	14.3	1.0 x 0.4	13.1	Sbc	78	Mag 18.0 galaxy **UM 553** E 1′0; mag 16.0 galaxy **UM 550** SW 4′2; mag 12.2 star S 5′3.
13 17 04.9	+01 49 35	CGCG 16-40	14.7	0.8 x 0.5	13.6	Sb	33	Faint, elongated anonymous galaxy SW 1′5.
13 17 11.6	−02 12 04	CGCG 16-41	15.7	0.2 x 0.2	12.3	E		Stellar; mag 11.02v star NW 3′4; IC 4218 SSW 4′1.
13 17 10.8	+00 31 37	CGCG 16-42	14.5	0.5 x 0.4	12.6		30	Mag 13.6 star N 0′6.
13 21 46.4	+00 12 49	CGCG 16-59	15.7	0.6 x 0.5	14.2	Irr	147	Mag 12.4 star NE 1′7; CGCG 16-60 S 4′8.
13 21 51.8	+00 08 18	CGCG 16-60	15.1	0.6 x 0.2	12.9	E/S0	72	CGCG 16-59 N 4′8; mag 15.1 star W 1′0.
13 25 05.4	−00 54 11	CGCG 16-67	15.0	0.6 x 0.3	13.0	Sc	144	Mag 14.9 star NW 0′7; mag 11.8 star E 3′1.
13 32 09.6	−02 44 58	CGCG 17-11	15.2	0.6 x 0.3	13.2	Sm	3	Mag 8.27v star NE 11′3; IC 892 WNW 6′3.
13 34 06.0	+01 45 14	CGCG 17-26	14.7	0.5 x 0.5	13.0	Scd		Faint dark lane S of nucleus.
13 35 38.6	−00 55 14	CGCG 17-30	15.2	0.4 x 0.4	13.0			Stellar; mag 8.81v star SE 4′5.

RA h m s	Dec ° ′ ″	Name	Mag (V)	Dim ′ Maj x min	SB	Type Class	PA	Notes
13 51 54.6	+01 50 34	CGCG 17-77	15.0	0.6 x 0.3	13.0		33	CGCG 17-83 NE 10′.3.
13 51 57.2	+02 00 10	CGCG 17-80	14.3	0.4 x 0.4	12.2			NGC 5331 NE 7′.7; CGCG 17-83 ESE 8′.4.
12 42 56.5	+03 40 32	CGCG 43-3	14.1	0.7 x 0.7	13.4	E?		**CGCG 43-4** SSE 6′.4; **CGCG 43-7** SE 9′.4.
13 01 41.5	+04 40 50	CGCG 43-97	14.5	0.4 x 0.3	12.0			Mag 10.65v star W 4′.2.
13 02 10.8	+03 06 23	CGCG 43-105	14.4	0.5 x 0.4	12.5			Located 2′.8 SSE of mag 11.31v star.
13 03 40.8	+02 34 03	CGCG 43-120	14.7	0.7 x 0.5	13.4		6	Mag 9.52v star ENE 10′.2.
13 19 36.8	+03 01 55	CGCG 44-69	14.2	0.6 x 0.4	12.5		33	Mag 10.60v star E 3′.7.
13 34 06.9	+05 06 07	CGCG 45-26	15.1	0.5 x 0.3	12.9		174	Located 3′.5 SE of mag 10.10v star.
13 36 12.3	+04 44 20	CGCG 45-38	15.0	0.4 x 0.3	12.6			Mag 12.1 star S 4′.3; mag 12.4 star NE 3′.5.
13 36 21.5	+03 19 47	CGCG 45-39	14.2	0.5 x 0.5	12.5			Mag 10.37v star N 2′.3.
13 44 04.9	+03 54 15	CGCG 45-84	14.8	1.1 x 0.3	13.4	Sa	108	**UGC 8668** W 6′.2.
13 50 20.7	+03 00 49	CGCG 45-113	15.3	0.5 x 0.3	13.1		156	Mag 11.6 star WSW 1′.4.
12 44 50.9	−04 26 06	IC 812	13.6	0.8 x 0.5	12.5		120	Almost stellar; mag 9.92v star NW 9′.8.
13 07 38.7	−00 56 34	IC 849	13.0	1.5 x 1.2	13.5	SAB(rs)cd pec? II-III	140	
13 17 58.6	+04 24 12	IC 871	13.4	1.7 x 0.8	13.6	Sb III	70	Mag 11.2 star E 1′.3.
13 18 34.6	+04 29 07	IC 876	14.3	0.8 x 0.7	13.5	Sa	9	Mag 10.73v star NE 7′.7; stellar **IC 873** W 4′.8.
13 30 00.0	+00 18 17	IC 891	14.3	0.5 x 0.5	12.8	E/S0		Mag 8.58v star NE 3′.0.
13 31 45.8	−02 42 47	IC 892	12.9	1.5 x 0.8	13.0	S0 pec	15	CGCG 17-11 ESE 6′.3; **IC 893** N 6′.1; mag 8.89v star SW 12′.8.
13 34 10.1	+04 52 06	IC 896	13.4	1.2 x 0.8	13.3	E	35	Mag 7.92v star ENE 9′.0.
13 38 26.2	−00 13 45	IC 903	13.1	1.9 x 0.8	13.4	(R′)Sab: III	178	
13 38 32.5	+00 32 22	IC 904	13.6	1.0 x 0.6	13.0	S0$^+$?	135	UGC 8631 E 1′.9; **CGCG 17-46** SW 2′.0.
13 41 18.7	−04 20 41	IC 908	14.0	0.9 x 0.6	13.2	SBcd	60	Mag 10.77v star SSW 6′.1.
13 47 43.2	+03 24 39	IC 939	13.8	0.9 x 0.9	13.4			Stellar **CGCG 45-102** S 4′.6; **IC 940** NE 4′.3.
13 50 32.1	+03 11 37	IC 943	14.0	0.7 x 0.7	13.1	Sb		Mag 9.01v star NE 9′.5.
12 32 36.5	+02 39 36	IC 3474	14.2	2.3 x 0.3	13.6	Sd sp	36	Mag 11.1 star S 1′.4.
13 17 03.5	−02 15 41	IC 4218	13.7	1.3 x 0.3	12.5	S?	158	Mag 11.02v star N 5′.4; CGCG 16-41 NNE 4′.1.
13 19 04.7	−02 30 56	IC 4224	14.6	0.9 x 0.5	13.6	Scd	105	mag 9.28v star WSW 2′.6.
13 22 26.1	−02 25 09	IC 4229	13.4	1.0 x 0.7	12.8	(R′)SB(r)b pec:	115	Mag 9.63v star S 3′.8.
12 35 55.3	−02 54 10	MCG +0-32-25	13.8	1.1 x 0.7	13.3	SAcd? IV-V	39	Mag 15.1 star N edge.
12 38 28.9	+01 45 14	MCG +0-32-29	13.7	0.9 x 0.7	13.1	S0	3	Large E-W core; mag 14.1 star N 1′.3.
12 41 22.9	−03 03 31	MCG +0-32-36	13.5	1.2 x 0.5	12.8	SA0$^+$ sp	150	
12 43 03.0	−01 16 59	MCG +0-33-3	13.9	0.6 x 0.6	12.6	Sbc		NGC 4629 SW 8′.7; mag 2.74v star “Porrima”, Gamma Virginis, SW 23′.1.
12 51 56.2	−02 04 39	MCG +0-33-15	14.4	0.7 x 0.5	13.1	Sb	9	Mag 13.1 star N 2′.3.
12 54 19.3	−03 42 54	MCG +0-33-19	13.8	1.2 x 0.5	13.1	SB(s)m? IV	48	Mag 13.6 star W 2′.3.
12 54 41.7	+02 06 14	MCG +0-33-20	14.0	0.8 x 0.6	13.1	S0/a	30	
12 57 26.5	−03 29 28	MCG +0-33-23	13.9	0.9 x 0.9	13.6	S0		Pair of mag 15 stars S edge.
13 04 32.1	−03 34 15	MCG +0-33-28	12.3	3.6 x 2.5	14.5	SAB(s)dm III-IV	60	Many bright knots and/or superimposed stars.
13 06 22.4	−01 40 52	MCG +0-33-29	14.4	0.9 x 0.6	13.6	Sc	69	
13 08 50.3	−00 49 00	MCG +0-34-5	13.9	0.3 x 0.2	10.7		105	Mag 10.33v star SW 6′.2.
13 08 55.3	−00 48 06	MCG +0-34-6	14.7	0.7 x 0.2	12.4	S0	3	
13 08 57.1	−03 42 23	MCG +0-34-7	15.3	0.8 x 0.5	14.1	Sm	174	Located between a pair of stars, mags 10.04v and 10.27v.
13 09 03.7	−00 46 00	MCG +0-34-8	14.2	0.5 x 0.4	12.5	E?	162	
13 10 23.9	−01 00 39	MCG +0-34-12	14.3	0.8 x 0.5	13.2	S0	168	Mag 10.17v star E 1′.7.
13 10 19.5	+02 01 29	MCG +0-34-13	15.2	0.7 x 0.4	13.7		3	
13 14 08.9	−00 26 28	MCG +0-34-18	14.8	0.7 x 0.4	13.3			
13 15 35.8	+00 29 31	MCG +0-34-19	15.3	0.7 x 0.7	14.4	Irr		Mag 13.4 star SW 2′.2.
13 15 47.3	−00 27 46	MCG +0-34-20	15.6	0.7 x 0.6	14.5	LSB	54	
13 17 29.2	−03 10 49	MCG +0-34-24	15.2	0.7 x 0.3	13.3	Sm	129	Very bright core or star superimposed?
13 19 36.7	−02 54 45	MCG +0-34-28	14.6	0.7 x 0.4	13.3	E	42	Triangle of mag 13-14 stars SE 2′.0.
13 20 56.1	+01 30 11	MCG +0-34-30	14.6	0.7 x 0.5	13.3	Sa	108	Mag 9.17v star NE 8′.4.
13 26 38.7	−03 15 45	MCG +0-34-34	14.2	1.4 x 0.6	13.9	SB(s)dm IV-V	157	Mag 12.7 star E 0′.8.
13 28 46.4	−02 02 23	MCG +0-34-35	14.1	0.6 x 0.6	12.8	S?		
13 29 13.9	+00 43 44	MCG +0-34-36	15.3	0.8 x 0.7	14.5		165	
13 29 28.9	−03 40 51	MCG +0-34-37	14.6	1.0 x 0.5	13.7	Sc	126	
13 29 50.3	−01 25 46	MCG +0-34-38	14.6	0.5 x 0.4	12.8	Sbc	45	Mag 8.70v star N 2′.1.
13 32 05.8	+01 35 47	MCG +0-35-4	14.2	0.7 x 0.5	13.2	E	150	Mag 9.20v star N 7′.7.
13 32 20.5	+01 34 01	MCG +0-35-5	13.9	0.9 x 0.8	13.4	Sbc	63	
13 32 50.1	−03 05 00	MCG +0-35-7	14.1	1.5 x 0.3	13.1	Sbc sp	33	Close pair of N-S oriented mag 8.55v, 9.00v stars W 9′.9; **CGCG 17-18** N 3′.6.
13 42 15.1	+01 51 24	MCG +0-35-18	14.0	0.8 x 0.5	12.8	Sb	48	
13 46 13.3	−03 23 07	MCG +0-35-19	14.5	0.8 x 0.8	13.9	Sa		MCG +0-35-20 SE 3′.0; mag 9.92v star SSW 9′.3.
13 46 20.3	−03 25 41	MCG +0-35-20	14.5	0.9 x 0.3	13.0	Sb	6	Mag 9.92v star SW 8′.5; MCG +0-35-19 NW 3′.0.
12 32 50.0	+03 18 01	MCG +1-32-93	14.0	1.0 x 0.6	13.3	SB?	39	
12 33 29.3	+03 47 38	MCG +1-32-96	14.4	0.9 x 0.4	13.2	SB?		
12 36 48.9	+04 06 20	MCG +1-32-113	15.2	0.6 x 0.6	14.0	Sa		
12 39 21.9	+04 56 18	MCG +1-32-126	14.4	0.9 x 0.4	13.1	I?	36	
12 46 49.7	+03 00 27	MCG +1-33-6	14.6	0.7 x 0.7	13.7	LSB		
12 47 28.2	+03 02 16	MCG +1-33-8	14.3	0.5 x 0.5	12.7			Mag 13.5 star N 2′.5.
12 55 26.1	+02 13 55	MCG +1-33-26	16.0	0.7 x 0.7	15.1	Sc		
12 56 20.4	+05 28 35	MCG +1-33-29	14.5	0.5 x 0.4	12.6	Sc		
13 33 54.6	+03 16 58	MCG +1-35-5	14.1	1.0 x 0.8	13.7		159	Very small, faint anonymous galaxy NW edge; **CGCG 45-22** NW 2′.2.
13 36 08.1	+03 30 30	MCG +1-35-13	15.3	0.7 x 0.5	14.0	Sb	0	Mag 10.73v star NE 8′.8.
13 40 00.1	+04 40 36	MCG +1-35-26	16.2	0.7 x 0.6	15.1	Sc	153	Faint, almost stellar anonymous galaxy W 2′.0.
13 40 27.3	+04 46 22	MCG +1-35-28	14.3	1.0 x 0.7	13.8		87	Strong SE-NW bar.
12 33 37.2	−04 53 10	MCG −1-32-23	14.6	2.1 x 0.4	14.2	SBm: sp IV-V	132	Mag 10.48v star S 3′.4.
12 35 00.5	−05 29 26	MCG −1-32-25	14.5	0.9 x 0.7	13.8	Sbc	75	
12 36 22.0	−03 55 17	MCG −1-32-29	14.0	1.1 x 0.4	13.0	SB(s)0$^+$ pec?	129	Stellar nucleus, uniform envelope.
12 42 16.6	−05 47 29	MCG −1-32-38	14.1	0.9 x 0.6	13.3	SAB0$^-$ pec:	6	
12 42 18.8	−05 46 33	MCG −1-32-39	14.3	0.7 x 0.3	12.4	Sab	111	
12 44 03.8	−05 40 38	MCG −1-33-1	12.3	3.9 x 2.8	14.7	SAB(rs)dm V	135	Bright core, knotty branching arms.
12 47 38.1	−05 52 03	MCG −1-33-7	14.0	0.4 x 0.4	11.9	S?		Mag 8.94v star SW 2′.2.
12 48 43.6	−05 15 13	MCG −1-33-11	13.6	1.6 x 1.6	14.5	IB(s)m V-VI		Mag 10.48v star W 5′.2.
12 49 18.3	−04 00 58	MCG −1-33-14	14.5	1.6 x 0.6	14.3	IB(s)m IV-V	135	Mag 14.2 star E 1′.0.
12 51 54.1	−04 34 38	MCG −1-33-30	14.6	0.9 x 0.6	13.8	IB(s)m pec: IV-V	87	Anonymous galaxy SW 1′.0.
12 57 15.8	−05 20 49	MCG −1-33-59	13.2	1.8 x 1.3	14.0	SA(s)dm pec III-IV	102	Numerous bright knots, star on W end.
13 03 46.6	−05 08 01	MCG −1-33-76	13.8	1.6 x 0.4	13.1	Sc? III	36	Mag 7.70v star SE 1′.8.
13 14 09.5	−05 36 01	MCG −1-34-13	14.4	1.4 x 0.3	13.3	Sc: sp	45	
13 30 08.4	−05 24 57	MCG −1-34-16	14.3	0.8 x 0.7	13.5	Sc	96	
13 35 53.3	−04 55 39	MCG −1-35-2	13.8	1.3 x 0.7	13.5	SB?	0	Mag 8.65v star W 7′.1; **MCG −1-35-4** SE 5′.8.
13 45 38.3	−05 59 05	MCG −1-35-10	13.8	1.6 x 1.1	14.3	SB(s)m III	57	Several bright knots or superimposed stars; mag 12.4 star SW 2′.2.

(Continued from previous page)

GALAXIES

RA h m s	Dec ° ′ ″	Name	Mag (V)	Dim ′ Maj x min	SB	Type Class	PA	Notes
12 32 45.8	+00 06 52	NGC 4517	10.4	10.5 x 1.5	13.2	SA(s)cd: sp III-IV	83	Dark lane extends almost full length N of center line; star on N edge near middle.
12 32 27.8	+00 23 19	NGC 4517A	12.5	4.0 x 2.6	14.9	SB(rs)dm: IV-V	30	**Reinmuth 80**. Small nucleus, filamentary arms; mag 10.01v star NW 3′6.
12 34 07.6	+02 39 08	NGC 4527	10.5	6.2 x 2.1	13.2	SAB(s)bc II	67	Bright bulge or bar; two main knotty arms.
12 34 21.8	+02 19 35	NGC 4533	13.8	2.1 x 0.4	13.5	SAd: sp III-IV	161	Mag 8.61v star N 4′0.
12 34 27.8	+02 11 12	NGC 4536	10.6	7.6 x 3.2	13.9	SAB(rs)bc I-II	130	Very small, very bright nucleus; two main filamentary arms with knots.
12 34 41.0	+03 19 22	NGC 4538	14.4	0.9 x 0.3	12.8	S?	81	
12 35 10.6	−00 13 17	NGC 4541	13.0	1.6 x 0.7	13.0	(R′)SAB(r)bc: II	91	Elongated **UGC 7755** and almost stellar **CGCG 14-73** E 7′8.
12 35 36.5	+03 02 04	NGC 4544	13.0	2.0 x 0.6	13.0	SB0/a? sp	161	
12 35 29.4	−03 47 37	NGC 4546	10.3	3.3 x 1.4	11.9	SB(s)0^-:	78	CGCG 14-74 ENE 5′5.
12 37 33.6	+04 22 02	NGC 4576	13.5	1.2 x 0.8	13.3	SAB(rs)bc: II-III	159	Mag 6.99v star SE 6′6.
12 37 48.2	+05 22 04	NGC 4580	11.8	2.1 x 1.6	13.0	SAB(rs)a pec II-III	165	
12 38 05.2	+01 28 34	NGC 4581	12.5	1.9 x 1.1	13.3	E^+	173	**UGC 7807** E 8′4.
12 38 28.4	+04 19 05	NGC 4586	11.7	4.0 x 1.3	13.3	SA(s)a: sp	115	Small bright nucleus with dark lane N.
12 38 35.3	+02 39 23	NGC 4587	13.5	1.1 x 0.6	12.9	SA(r)0°:	48	
12 39 18.1	−00 31 51	NGC 4592	11.6	5.8 x 1.5	13.8	SA(s)dm: III	97	
12 39 39.8	−05 20 39	NGC 4593	10.9	3.9 x 2.9	13.4	(R)SB(rs)b II	114	Bright nucleus in short bar encircled by two asymmetrical arms. **MCG −1-32-33** E 3′8.
12 40 12.6	−05 47 55	NGC 4597	12.0	4.1 x 1.9	14.1	SB(rs)m III	34	Very patchy appearance, many bright knots.
12 40 27.0	+01 11 46	NGC 4599	12.6	1.7 x 0.8	12.8	SA0/a sp	144	Mag 9.39v star NNE 7′2.
12 40 23.0	+03 07 03	NGC 4600	12.7	1.2 x 0.8	12.6	S0	60	Located 2′8 NW of a mag 8.91v star.
12 40 37.0	−05 07 59	NGC 4602	11.5	3.4 x 1.2	12.9	SAB(rs)bc II	99	**PGC 42481** N 2′7; stellar **PGC 42468** S 3′7.
12 40 45.2	−05 18 13	NGC 4604	13.8	1.0 x 0.4	12.6	Im: sp III	114	
12 42 32.8	−01 21 02	NGC 4629	13.2	1.1 x 0.8	12.9	SAB(s)m pec V	101	Mag 2.74v star "Porrima", Gamma Virginis, SW 14′5.
12 42 31.4	+03 57 28	NGC 4630	12.6	1.8 x 1.3	13.3	IB(s)m?	10	Bright, slightly elongated N-S center.
12 42 32.1	−00 04 55	NGC 4632	11.7	3.1 x 1.2	12.9	SAc II-III	63	Stellar **CGCG 14-107** W 5′0.
12 42 49.7	+02 41 16	NGC 4636	9.5	6.0 x 4.7	13.1	E0-1	150	Bright center, several faint stars superimposed; mag 11.27v star N edge.
12 43 17.5	−00 38 45	NGC 4642	12.9	1.9 x 0.6	12.9	SAB(rs)bc pec sp III	37	
12 43 20.3	+01 58 38	NGC 4643	10.8	3.1 x 2.3	12.7	SB(rs)0/a	48	Bright central bulge with short, strong bar SE-NW.
12 43 51.0	−00 33 38	NGC 4653	12.2	3.1 x 2.7	14.4	SAB(rs)cd II-III	30	
12 45 05.8	+03 03 21	NGC 4665	10.5	3.8 x 3.2	13.1	SB(s)0/a	108	Bright N-S bar, bright diffuse nucleus; mag 10.73v star SSW 1′7.
12 45 07.6	−00 27 57	NGC 4666	10.7	4.6 x 1.3	12.5	SABc: II-III	42	
12 45 32.0	−00 32 09	NGC 4668	13.1	1.4 x 0.8	13.1	SB(s)d: III-IV	5	
12 49 42.1	−04 34 46	NGC 4678	14.2	1.0 x 0.5	13.3	Sc	82	Bright knot or superimposed star W of nucleus.
12 47 17.5	−02 43 36	NGC 4684	11.4	2.9 x 1.0	12.4	SB(r)0^+	23	Very faint star N of bright, elongated core.
12 47 47.7	+04 20 09	NGC 4688	11.9	3.2 x 2.8	14.2	SB(s)cd III	123	Very patchy appearance, bright nucleus in short bar, numerous bright knots overall.
12 47 55.6	−01 39 21	NGC 4690	12.9	1.2 x 0.9	12.8	(R′)SA0^-?	150	
12 48 13.9	−03 20 00	NGC 4691	11.1	2.8 x 2.3	13.0	(R)SB(s)0/a pec	85	Very bright, complex bar.
12 48 35.9	−05 48 04	NGC 4697	9.2	7.2 x 4.7	13.0	E6	70	Bright center, many stars superimposed; fairly bright anonymous galaxy WNW 5′9.
12 49 11.4	+03 23 19	NGC 4701	12.4	2.8 x 2.1	14.2	SA(s)cd	45	Bright center; many knotty, branching arms.
12 49 25.0	−05 11 43	NGC 4705	11.7	3.0 x 1.0	12.8	SAB(s)bc: sp II	125	Very patchy arms with several lanes and knots; mag 9.19v star NE 4′9.
12 49 57.3	+05 18 38	NGC 4713	11.7	2.7 x 1.7	13.2	SAB(rs)d III	100	Many bright knots and patches overall.
12 50 33.0	−05 16 58	NGC 4718	13.4	1.8 x 0.6	13.3	SB(rs)b? sp II-III?	98	
12 50 42.9	−04 09 23	NGC 4720	13.3	1.0 x 0.6	12.6	E/S0	123	
12 51 12.8	+04 51 30	NGC 4734	13.5	1.0 x 0.9	13.2	Sc?	145	
12 52 22.2	−01 12 01	NGC 4753	9.9	6.0 x 2.8	12.9	I0	80	Several faint, filamentary dark lanes.
12 53 14.7	+04 27 45	NGC 4765	13.0	1.1 x 0.8	12.7	S0/a? III-IV	80	
12 53 21.0	+01 16 09	NGC 4771	12.3	3.9 x 0.8	13.4	SAd? sp III:	133	Located 2′6 E of mag 10.91v star.
12 53 29.1	+02 10 05	NGC 4772	11.0	3.4 x 1.7	12.8	SA(s)a	147	Bright diffuse nucleus; strong dark lane on E side.
12 55 15.5	+02 53 48	NGC 4799	13.4	1.3 x 0.6	13.0	S?	91	
12 55 49.2	+04 18 11	NGC 4808	11.7	2.8 x 1.1	12.8	SA(s)cd: III	127	
12 54 51.0	+02 39 05	NGC 4809	13.8	1.7 x 0.7	13.8	Im pec	68	Interacting pair with NGC 4810 S.
12 54 51.3	+02 38 24	NGC 4810	14.3	1.3 x 0.5	13.7	Im pec	162	Bright, knotty core; interacting with NGC 4809 N.
12 58 00.7	−03 37 18	NGC 4843	13.0	2.1 x 0.5	12.9	SA(r)0/a?	87	Faint star S of E end.
12 58 01.5	+01 34 35	NGC 4845	11.2	5.0 x 1.3	13.1	SA(s)ab sp III	80	Dark lane across central bulge, partly obscures nucleus.
13 00 39.2	−04 36 14	NGC 4890	13.2	0.9 x 0.7	12.6	Sm	85	
13 00 39.1	+02 30 01	NGC 4900	11.4	2.2 x 2.1	12.9	SB(rs)c III-IV		Very complex; star or bright clump 0′8 SE of nucleus.
13 00 59.1	−00 01 39	NGC 4904	12.0	2.2 x 1.4	13.0	SB(s)cd III	25	
13 01 28.4	−04 32 49	NGC 4915	12.1	1.6 x 1.3	12.8	E0	55	Bright nucleus, several faint stars superimposed on smooth halo; ESO 443-66 NE 2′1.
13 01 50.3	−04 30 05	NGC 4918	14.4	0.9 x 0.4	13.2	SA0^+?	71	Located 1′8 SSE of mag 10.81v star.
13 04 13.1	−05 33 08	NGC 4941	11.1	3.9 x 2.7	13.5	(R)SAB(r)ab: II	15	Large, bright center with moderately dark lanes; faint outer envelope.
13 07 50.2	−05 01 04	NGC 4975	13.9	1.0 x 0.6	13.3	SAB(rs)0° pec?	46	Faint **MCG −1-34-1** WNW 3′9.
13 09 16.1	−05 23 47	NGC 4989	13.3	1.5 x 0.9	13.5	S0°:	171	**MCG −1-34-6** S 1′4.
13 09 17.4	−05 16 23	NGC 4990	13.9	1.0 x 0.8	13.5	S0 pec?	55	Almost stellar nucleus.
13 09 15.1	+02 20 46	NGC 4991	15.2	0.6 x 0.6	13.9	Sb		Double system.
13 09 31.7	+00 51 23	NGC 4996	12.6	1.8 x 1.4	13.5	(R)SB0^+	40	Bright, elongated nucleus, uniform outer halo.
13 09 33.3	+01 40 21	NGC 4999	11.8	2.5 x 1.9	13.4	SB(r)b II	128	Very short NE-SW bar; two main, tightly wound arms.
13 12 07.3	+03 11 58	NGC 5013	15.1	0.8 x 0.5	14.0	Sbc	135	
13 12 22.7	−04 20 13	NGC 5015	12.3	1.8 x 1.5	13.3	(R)SB(r)a:	60	Very faint spiral arms extend beyond published dimensions NW and SE.
13 12 42.4	+04 43 46	NGC 5019	13.6	0.8 x 0.7	12.8	SB?	105	
13 14 42.8	−04 10 44	NGC 5036	14.6	0.6 x 0.6	13.4	$S0^-$: pec?		Mag 9.15v star N 7′9.
13 14 47.5	−04 16 35	NGC 5039	16.1	0.6 x 0.5	14.6	Sa? pec	30	Very faint anonymous galaxy NW 1′2; mag 10.17v star S 2′3.
13 15 41.6	+02 52 40	NGC 5050	13.7	1.1 x 0.5	12.9	S0/a	32	
13 20 37.1	−02 17 24	NGC 5095	13.7	1.2 x 0.5	13.0	(R′)SB(r)0^+:	126	
13 21 23.0	+00 20 32	NGC 5104	13.7	1.2 x 0.4	12.7	Sa:	170	
13 24 52.8	−04 04 57	NGC 5133	13.7	1.0 x 0.7	13.2	E/S0	42	
13 26 19.6	+02 05 56	NGC 5147	11.8	1.9 x 1.5	12.8	SB(s)dm III	120	Stellar galaxy **UM 573** NE 1′2.
13 26 38.8	+02 18 52	NGC 5148	14.2	0.7 x 0.7	13.3	SB(s)d?		Star on NE edge.
13 28 16.1	+02 58 58	NGC 5159	14.2	1.3 x 0.4	13.3	Sd	162	
13 30 06.4	−01 43 15	NGC 5183	12.7	1.9 x 0.8	13.0	Sb pec I-II	122	Bright, elongated nucleus.
13 30 11.3	−01 39 48	NGC 5184	12.6	1.9 x 1.1	13.3	SAB(r)c II	135	star superimposed NW of nucleus.
13 30 51.7	−01 46 43	NGC 5192	14.6	0.6 x 0.3	12.6	Sa?	26	Located near the center of galaxy cluster A 1750.
13 31 19.7	−01 36 56	NGC 5196	14.0	0.7 x 0.5	12.7	$S0^-$	97	
13 31 25.2	−01 41 35	NGC 5197	15.2	0.6 x 0.3	13.2	S0/a	150	Located 2′0 NNW of mag 10.41v star; faint anonymous galaxy NE 0′5.
13 32 00.5	−01 42 00	NGC 5202	14.5	1.2 x 0.3	13.2	Sb: II	0	Anonymous galaxy E 1′8.
13 33 05.2	−01 02 06	NGC 5211	12.3	2.1 x 1.6	13.5	(R′)SAB(rs)ab pec	30	**UGC 8526** S 7′8.
13 34 39.3	+04 07 47	NGC 5213	13.7	0.9 x 0.8	13.2	SBb	132	
13 35 24.5	+01 24 41	NGC 5227	13.1	1.8 x 1.5	14.0	(R′)SB(r)b II	40	Bright bar SE-NW.
13 35 48.3	+02 59 56	NGC 5231	13.4	1.1 x 0.9	13.2	SBa	112	
13 37 23.4	+03 53 47	NGC 5245	17.1	0.8 x 0.4	15.7		90	
13 37 29.4	+04 06 15	NGC 5246	13.6	1.0 x 0.8	13.3	(R′)SB(s)b	63	

(Continued from previous page)

RA h m s	Dec ° ′ ″	Name	Mag (V)	Dim ′ Maj x min	SB	Type Class	PA	Notes
13 38 16.1	+04 32 30	NGC 5252	13.0	1.4 x 0.9	13.2	S0	10	**CGCG 45-60** SE 8′6.
13 39 52.8	+00 50 24	NGC 5257	12.9	1.8 x 0.7	13.0	SAB(s)b pec	121	Interacting with, and one arm linking with, NGC 5258 E.
13 39 57.9	+00 49 55	NGC 5258	12.9	1.4 x 0.9	13.0	SA(s)b: pec	22	Linked with NGC 5257.
13 40 16.2	+05 04 33	NGC 5261	14.4	0.8 x 0.5	13.3		141	Mag 9.80v star W 4′2.
13 42 10.9	+04 15 46	NGC 5270	13.5	1.1 x 0.8	13.2	(R′)SB(s)b	20	
13 44 25.8	+02 06 36	NGC 5285	13.9	0.8 x 0.8	13.5	E+ pec		Almost stellar nucleus; mag 9.65v star WNW 7′0.
13 48 15.9	+03 56 58	NGC 5300	11.4	3.9 x 2.6	13.8	SAB(r)c II	150	Very patchy, many knots and/or superimposed stars.
12 33 55.4	+03 32 38	UGC 7715	14.2	1.1 x 0.9	14.0	Im:		Dwarf, low, uniform surface brightness.
12 34 09.1	−00 21 16	UGC 7720	13.6	1.0 x 0.9	13.3	(R′)SB(r)b:	30	
12 35 30.5	−02 19 37	UGC 7752	14.4	1.5 x 0.4	13.7	SABab? sp	6	Mag 14.4 star SE edge.
12 35 54.7	−01 51 16	UGC 7763	14.3	0.8 x 0.5	13.2	SAB(s)cd III-IV	80	
12 36 42.1	+03 06 24	UGC 7780	15.1	1.5 x 0.3	14.1	Sdm:	48	
12 38 03.1	−02 15 55	UGC 7798	13.8	0.8 x 0.5	12.7	IBm: IV	57	
12 38 05.4	−00 01 36	UGC 7800	14.1	1.0 x 0.3	12.6	SA(r)b? II	133	
12 38 36.1	+01 24 06	UGC 7806	14.3	1.7 x 0.2	13.0	SBcd: sp III	148	Mag 10.97v star E 1′0; **UGC 7807** N 4′5.
12 39 01.2	+00 21 52	UGC 7813	13.9	0.7 x 0.7	13.1	E:		**CGCG 14-88** SW 4′5; **CGCG 14-94** NE 7′8.
12 39 11.6	+00 42 59	UGC 7820	12.9	1.7 x 1.6	13.8	SAB(s)cd: III	108	**CGCG 14-96** ENE 6′7.
12 39 50.6	+01 40 19	UGC 7824	13.8	1.5 x 0.7	13.7	SB(s)m: IV-V	77	Mag 13.1 star W 2′3.
12 41 11.6	+01 24 38	UGC 7841	13.9	0.9 x 0.4	12.6	SA(r)b?	150	
12 42 40.8	+01 20 41	UGC 7873	13.6	1.7 x 0.9	13.9	SA(rs)c: III-IV	150	Mag 8.52v star W 3′1.
12 42 57.4	−01 13 49	UGC 7883	13.1	2.7 x 0.7	13.7	SA(s)cd: sp II-III	71	NGC 4629 SW 9′4; mag 2.74v star "Porrima", Gamma Virginis, SW 23′4.
12 44 29.0	+00 28 09	UGC 7911	12.8	2.6 x 1.9	14.3	(R′)SB(s)m V-VI	20	Dwarf spiral.
12 44 33.6	−02 19 12	UGC 7913	14.1	1.1 x 0.6	13.5	SAB(s)m V-VI	90	Dwarf, very low surface brightness.
12 46 45.7	+05 57 14	UGC 7943	13.1	2.3 x 1.9	14.6	Scd:	30	Mag 6.31v star E 4′1.
12 47 01.0	−01 34 42	UGC 7945	14.0	1.3 x 0.8	13.9	SA(s)m? V-VI	120	Dwarf spiral, very low surface brightness.
12 49 16.0	+04 39 17	UGC 7976	14.3	0.9 x 0.9	13.9	Sdm:		Patchy, several knots.
12 49 50.2	+02 51 03	UGC 7982	13.1	3.0 x 0.8	13.9	S	0	
12 50 39.1	+01 27 51	UGC 7991	14.2	1.7 x 0.2	12.9	Sd: sp	170	
12 55 12.7	+00 06 59	UGC 8041	12.0	3.0 x 1.8	13.7	SB(s)d III-IV	165	Uniform surface brightness.
12 55 49.3	+04 00 48	UGC 8053	13.6	1.5 x 1.2	14.1	SABdm	27	
12 57 00.3	+01 01 42	UGC 8066	16.0	1.7 x 1.3	16.7	IB(s)m V-VI	170	Dwarf spiral, very low surface brightness.
12 57 12.2	−01 42 25	UGC 8067	13.4	1.9 x 0.4	12.9	Sb pec sp	138	Mag 14.1 star NE 0′9.
12 57 44.6	+02 41 30	UGC 8074	14.0	1.1 x 1.0	14.0	Sm V	81	Dwarf spiral, very low surface brightness.
12 58 22.8	+02 47 30	UGC 8084	14.1	1.5 x 1.4	14.8	SB(s)dm IV-V	27	
13 01 03.7	−01 57 16	UGC 8127	14.1	1.5 x 0.4	13.4	IB(s)m: IV-V	15	
13 03 05.6	+03 59 26	UGC 8153	13.5	1.4 x 1.2	13.9	Sd	3	Mag 11.0 star S 2′2.
13 05 58.6	+03 57 22	UGC 8186	14.5	1.7 x 0.3	13.7	Sbc	107	Mag 8.32v star E 4′0.
13 08 43.2	−02 08 07	UGC 8223	13.7	1.2 x 1.0	13.7	SBcd:	123	Mag 10.42v star S 1′3.
13 09 41.0	−01 02 54	UGC 8238	14.1	1.0 x 0.5	13.2	Sd	56	
13 11 26.2	−00 14 59	UGC 8262	13.2	1.2 x 0.7	12.9	SA(r)0+?	160	
13 11 23.2	+03 24 42	UGC 8263	14.7	1.6 x 0.3	13.8	SBc pec:	108	Mag 11.00v star SE 3′5.
13 11 37.5	+00 39 55	UGC 8265	14.1	0.8 x 0.5	13.0	SB?	19	
13 12 08.6	+02 59 16	UGC 8275	14.6	1.5 x 0.2	13.1	Sdm:	36	
13 15 06.0	+03 02 42	UGC 8324	14.8	1.6 x 0.3	13.8	Scd:	51	**CGCG 44-39** S 1′9.
13 16 35.4	−02 05 31	UGC 8340	13.5	1.1 x 0.7	13.0	Scd?	42	Mag 12.6 star NE 1′8.
13 17 58.7	−00 18 42	UGC 8357	13.7	1.5 x 0.5	13.2	SB(s)b pec?	39	
13 18 10.1	+01 14 38	UGC 8360	13.1	1.4 x 1.2	13.5	SAB(r)cd pec: III-IV	60	
13 20 32.3	+05 24 23	UGC 8382	14.2	1.0 x 0.5	13.3	IBm	140	Uniform surface brightness, **CGCG 44-74** NE 5′1; mag 9.74v star S 3′8.
13 32 30.8	+01 50 40	UGC 8521	13.2	1.0 x 0.8	12.8	(R)SB(r)ab pec	51	Bright SE-NW bar, faint envelope.
13 33 14.1	+05 45 53	UGC 8532	14.4	1.1 x 0.9	14.2	S	150	Disrupted? Numerous bright knots SE of core, on SE edge.
13 33 33.2	+01 18 18	UGC 8534	14.1	1.2 x 0.5	13.4	SBcd:	120	
13 34 03.7	+04 45 15	UGC 8543	13.7	2.2 x 0.3	13.1	Sb	141	Mag 15.6 star W edge; mag 9.86v star SW 8′6.
13 35 58.8	+01 43 51	UGC 8581	14.8	1.0 x 0.5	13.9	SAb:	35	Stellar **PGC 47975** on N edge; stellar **CGCG 17-32** just off S edge.
13 37 16.8	−02 09 18	UGC 8607	13.8	0.7 x 0.5	12.6	S0?	80	
13 37 43.1	+05 14 32	UGC 8617	14.1	1.0 x 0.6	13.4	SB	27	Very faint, almost stellar anonymous galaxy E 2′1.
13 38 39.9	+00 32 43	UGC 8631	14.2	0.7 x 0.4	12.7	Sdm:	90	IC 904 W 1′9; mag 12.9 star E 1′4.
13 39 04.7	+02 09 48	UGC 8634	14.8	1.5 x 0.2	13.3	S?	61	Mag 12.3 star SE 1′5.
13 39 13.2	+04 37 20	UGC 8635	13.7	0.9 x 0.8	13.2	Sbc	39	**CGCG 45-62** SSE 7′6.
13 40 12.6	+02 28 48	UGC 8650	13.9	1.7 x 0.3	13.0	Sb pec sp	151	
13 41 05.3	+05 06 20	UGC 8657	13.5	1.5 x 1.1	13.9	Sbc	30	Stellar nucleus; **CGCG 45-71** SE 5′8.
13 41 39.3	+05 01 55	UGC 8663	13.3	1.0 x 0.9	13.1	S0−:	51	Bright nucleus, uniform envelope; mag 9.93v star S 2′0; **CGCG 45-71** W 4′2.
13 43 40.3	+03 53 45	UGC 8686	14.1	1.6 x 0.5	13.7	S	63	**MCG +1-35-33** E 1′3; CGCG 45-84 E 6′2.
13 44 23.2	+05 47 30	UGC 8689	13.0	1.6 x 1.1	13.5	S0	155	**CGCG 45-88** NE 5′1.
13 44 33.7	+04 46 29	UGC 8690	13.9	0.9 x 0.5	12.9	Scd:	156	Bright knot or star E of center; mag 11.22v star SW 1′9.
13 49 38.8	+04 14 17	UGC 8740	12.9	1.5 x 1.4	13.6	SABab		Close pair of mag 12.5 stars NNE 2′3.

GALAXY CLUSTERS

RA h m s	Dec ° ′ ″	Name	Mag 10th brightest	No. Gal	Diam ′	Notes
12 59 24.0	−04 11 00	A 1651	16.0	70	21	All members anonymous, faint, and stellar.
13 30 54.0	−01 50 00	A 1750	15.9	40	27	Other than NGC's, most members anonymous, stellar/almost stellar.
13 42 06.0	+02 14 00	A 1773	15.6	66	16	All members anonymous, stellar/almost stellar.

111 GALAXIES 111

RA h m s	Dec ° ′ ″	Name	Mag (V)	Dim ′ Maj x min	SB	Type Class	PA	Notes
11 20 54.8	+00 10 18	CGCG 11-73	14.8	0.5 x 0.3	12.6		48	Mag 10.02v star E 4′6.
11 21 00.0	+00 31 59	CGCG 11-75	13.9	0.7 x 0.3	12.1	S0 pec?	114	**CGCG 11-68** NW 4′1.
11 26 06.4	−01 50 07	CGCG 11-95	14.2	0.7 x 0.3	12.4	SBb? sp	39	Mag 13.7 star NE 0′9.
11 26 40.7	−01 41 41	CGCG 11-96	13.8	0.3 x 0.3	11.3	cE		Almost stellar; located 2′3 N of mag 9.75v star.
11 32 43.3	−02 12 02	CGCG 12-13	13.9	0.4 x 0.4	11.8	SAB0−:		Mag 12.4 star W 1′2.
11 32 46.0	+00 10 30	CGCG 12-15	14.5	0.3 x 0.2	11.3	SB(s)m IV-V	3	Mag 12.4 star E 4′6.
11 34 39.3	+00 07 27	CGCG 12-27	13.1	0.4 x 0.4	11.0	SA(r)0/a:		CGCG 12-29 N 7′0.
11 34 47.3	+00 14 06	CGCG 12-29	13.8	0.5 x 0.4	12.1	E	18	CGCG 12-27 S 7′0.
11 35 18.5	−02 46 28	CGCG 12-32	13.6	0.3 x 0.3	10.8	SA(s)c? III-IV		Mag 12.0 star E 3′2.

RA h m s	Dec ° ′ ″	Name	Mag (V)	Dim ′ Maj x min	SB	Type Class	PA	Notes
11 36 22.8	−02 49 30	CGCG 12-37	14.1	0.4 x 0.4	12.0	SABa?		**CGCG 12-36** SW 1'.6.
11 37 11.1	+02 07 35	CGCG 12-42	13.2	0.4 x 0.3	10.7	SA(s)ab:	3	Faint star on N edge.
11 39 35.7	+02 02 46	CGCG 12-49	13.5	0.5 x 0.5	11.8	IABm: IV		Mag 11.1 star SW 2'.2.
11 40 06.2	−00 54 06	CGCG 12-53	14.1	0.8 x 0.4	12.7	SA(s)b III	42	Pair of very faint stars on E edge; **CGCG 12-52** NW 3'.7; CGCG 12-54 N 3'.9.
11 40 06.5	−00 50 17	CGCG 12-54	14.0	0.3 x 0.3	11.5	E		Mag 13.4 star W 1'.3; CGCG 12-53 S 3'.9; **CGCG 12-52** SW 4'.0.
11 41 03.7	+01 43 28	CGCG 12-57	13.9	0.5 x 0.3	11.7	SA0°?	30	Mag 13.6 star NW 3'.1.
11 47 00.6	−00 17 41	CGCG 12-81	13.9	0.4 x 0.2	11.1	SA0° pec	123	Mag 12.4 star W 5'.2.
11 48 52.9	+01 03 21	CGCG 12-89	14.3	0.5 x 0.4	12.4	SA(s)c: III-IV	27	Mag 13.0 star NW 5'.5.
11 49 35.4	−03 29 17	CGCG 12-95	14.0	0.5 x 0.3	12.0	E	108	MCG +0-30-27 NW 0'.9; CGCG 12-99 SE 1'.5.
11 49 39.8	−03 31 49	CGCG 12-98	13.8	0.6 x 0.2	11.4	SAB0$^+$:	132	**CGCG 12-97** S 2'.4; CGCG 12-99 N 1'.3.
11 49 38.4	−03 30 31	CGCG 12-99	13.2	0.3 x 0.3	10.5	SB0$^+$?		CGCG 12-95 NNW 1'.5; CGCG 12-98 S 1'.3.
11 49 49.8	−03 31 07	CGCG 12-100	13.9	1.1 x 0.3	12.5	SAB(r)0/a? sp	18	Located 1'.0 S of mag 12.6 star.
11 49 55.3	−03 39 49	CGCG 12-101	13.6	0.4 x 0.4	11.7	E pec:		Mag 15.8 star on SE edge.
11 51 31.0	−00 04 41	CGCG 12-106	13.8	0.2 x 0.2	10.2	SAB(r)0°?		Stellar. **CGCG 12-107** N 1'.6.
11 52 31.0	−03 40 29	CGCG 12-110	13.7	1.3 x 0.9	13.7	IB(s)m IV-V	154	Mag 12.3 star N 6'.8; mag 10.40v star NW 11'.9.
11 56 05.8	+02 06 00	CGCG 13-6	16.3	0.3 x 0.2	13.1			Mag 14.0 star NE 2'.3; pair of faint, anonymous galaxies S 1'.1.
11 57 06.1	+01 07 29	CGCG 13-10	14.6	0.6 x 0.3	12.8	E	144	**CGCG 13-9** WSW 2'.4.
12 01 14.4	−03 40 41	CGCG 13-24	13.5	0.4 x 0.3	11.3	E	168	Mag 9.02v star N 6'.5.
11 59 24.7	+01 25 58	CGCG 13-26	15.4	0.6 x 0.3	13.3	Sc	27	Mag 12.9 star E 3'.4.
11 59 39.2	−03 41 00	CGCG 13-29	14.0	0.7 x 0.3	12.2	S0°	6	Mag 12.4 star W 6'.2.
12 02 47.3	−02 09 22	CGCG 13-44	13.6	0.4 x 0.3	11.1	SAB(rs)bc:	87	Pair of mag 11.8, 12.4 stars SE 3'.3; mag 11.6 star N 4'.3.
12 05 37.8	+01 49 38	CGCG 13-70	14.0	1.0 x 0.4	12.9	SB(r)0/a:	171	Mag 9.84v star SW 8'.9.
12 05 40.9	+01 34 26	CGCG 13-72	13.9	1.2 x 0.3	12.7	SB(s)0$^+$	96	**CGCG 13-73** N 1'.1.
12 08 00.1	+01 23 41	CGCG 13-80	13.4	0.5 x 0.5	11.7	SAb:		Mag 9.69v star W 6'.8; mag 10.89v star E 6'.4.
12 08 31.4	+00 08 07	CGCG 13-84	14.1	0.8 x 0.4	12.7	Sbc?	123	**CGCG 13-83** SW 2'.5.
12 12 39.3	−02 43 20	CGCG 13-102	13.9	0.4 x 0.2	11.1	S0	9	**CGCG 13-100** E 2'.2.
12 14 11.4	−00 49 57	CGCG 13-107	14.6	0.4 x 0.3	12.1	SAB(s)c?	171	Mag 11.7 star SW 1'.4.
12 16 04.6	+01 10 45	CGCG 13-116	14.1	0.5 x 0.3	11.9	(R′)SB(s)b: I-II	21	Mag 10.12v star E 3'.5.
12 16 44.0	−03 25 51	CGCG 13-118	13.6	0.5 x 0.3	11.4	Sb?		Mag 8.83v star NE 13'.0; MCG +0-31-44 SW 12'.2.
12 17 52.9	−00 39 27	CGCG 13-119	14.0	0.3 x 0.2	10.8	SA0°?	141	Mag 13.5 star SW 3'.0; mag 5.91v star SE 14'.1.
12 19 53.1	+01 46 20	CGCG 14-9	15.6	0.4 x 0.2	12.7		39	Mag 13.9 star SE 0'.8.
12 25 36.8	−02 54 05	CGCG 14-30	14.0	0.4 x 0.3	11.6	S0°	132	MCG +0-32-7 S 3'.1.
12 27 07.8	−01 21 18	CGCG 14-41	13.9	0.6 x 0.6	12.7	(R)SA0°		Bright center.
12 27 14.4	−02 02 16	CGCG 14-43	14.4	0.8 x 0.5	13.3	SAdm: IV	33	Close E-W oriented pair of mag 14.7 stars W 1'.2.
12 28 51.0	−02 03 43	CGCG 14-47	14.6	0.6 x 0.4	12.9		63	NGC 4454 N 7'.3.
12 29 41.7	−03 44 59	CGCG 14-52	13.4	0.7 x 0.3	11.6	SAb:	3	Mag 13.5 star W 2'.4.
12 31 04.7	−02 20 41	CGCG 14-53	13.9	0.5 x 0.3	11.8	SA0° sp	36	Forms a triangle with a mag 10.35v star SSW 5'.2 and a mag 11.06v star SSE 5'.6.
12 31 07.6	+00 27 43	CGCG 14-55	13.7	0.5 x 0.2	11.1	SAB(s)c: III-IV	147	Mag 13.8 star W 3'.0; mag 9.73v star S 7'.6.
12 35 50.9	−03 46 00	CGCG 14-74	13.4	0.8 x 0.4	12.1	S0 sp	51	NGC 4546 WSW 5'.5.
12 35 54.9	−02 23 10	CGCG 14-76	13.9	0.8 x 0.3	12.3	SA0° sp	132	Located 3'.1 SE of mag 9.21v star.
12 37 02.6	+01 49 06	CGCG 14-78	13.9	0.5 x 0.5	12.3	SA0$^-$:		Mag 12.1 star N 6'.4.
11 20 17.8	+04 19 12	CGCG 39-125	13.6	0.5 x 0.5	12.2	E		Mag 13.4 star W 1'.7; mag 10.7 star N 6'.2.
11 21 18.3	+03 39 49	CGCG 39-133	13.9	0.4 x 0.4	12.0	E		Forms a triangle with a mag 13.2 star SSE 1'.3 and a mag 13.8 star E 1'.7; **CGCG 39-145** SE 8'.1.
11 22 08.2	+04 41 31	CGCG 39-150	14.3	0.5 x 0.3	12.1		3	Mag 11.2 star NW 7'.9.
11 23 20.3	+05 51 05	CGCG 39-162	14.3	0.4 x 0.4	12.2			Mag 11.8 star WNW 5'.4.
11 23 44.6	+03 00 17	CGCG 39-163	15.6	0.3 x 0.3	12.8			Mag 11.31v star SSE 6'.8.
11 24 55.5	+03 10 03	CGCG 39-173	14.6	0.8 x 0.4	13.2		150	NGC 3664 NW 12'.1; NGC 3664A WNW 8'.2.
11 25 24.7	+04 54 47	CGCG 39-176	15.1	0.5 x 0.4	13.2		159	Mag 12.4 star SE 4'.8.
11 26 09.5	+03 29 50	CGCG 39-180	14.2	0.5 x 0.4	12.3	SAB(s)b:	18	Mag 8.86v star W 5'.1.
11 27 17.4	+03 45 22	CGCG 39-184	14.1	0.7 x 0.5	12.8		3	Very close pair of mag 14-15 stars NW 1'.0; mag 11.8 star N 3'.6.
11 30 23.5	+04 41 43	CGCG 39-197	14.2	0.7 x 0.5	12.9	SB(rs)b III	75	MCG +1-29-47 E 13'.3; mag 14.5 star SW 1'.2.
11 30 37.7	+03 45 21	CGCG 39-198	15.5	0.5 x 0.3	13.3		129	Mag 11.9 star E 3'.8.
11 31 06.2	+03 07 49	CGCG 39-200	14.1	0.8 x 0.3	12.4	SBb sp	81	Mag 10.97v star W 2'.0.
11 34 46.4	+02 55 12	CGCG 40-7	14.0	0.5 x 0.4	12.1	SB(r)b: II	114	Mag 5.76 star 89 Leo NW 10'.3.
11 40 51.6	+04 48 34	CGCG 40-27	14.3	0.8 x 0.2	12.2	Sdm? sp	42	Mag 8.26v star N 8'.8.
11 43 28.7	+03 23 16	CGCG 40-33	15.0	0.8 x 0.5	13.8	Sm	15	Mag 12.1 star W 2'.7; mag 10.57v star NNE 5'.6.
11 47 54.8	+04 00 54	CGCG 40-44	14.0	0.8 x 0.5	13.1	E	36	Mag 12.7 star SE 0'.6.
11 51 33.2	+05 06 04	CGCG 40-53	13.9	0.5 x 0.5	12.5	E		Mag 12.9 star W 2'.1.
11 54 58.1	+04 39 36	CGCG 40-64	13.9	0.8 x 0.5	12.8	SAB(r)0/a?	30	Stellar **CGCG 41-5** SE 5'.3.
11 57 23.5	+04 32 31	CGCG 41-10	15.2	0.4 x 0.3	12.7	Compact	3	= **Mkn 754**. Mag 11.9 star N 1'.6.
12 15 57.8	+03 18 19	CGCG 41-56	14.4	0.8 x 0.5	13.3	S0	156	Mag 10.47v star W 6'.1.
12 19 27.8	+05 02 47	CGCG 42-18	14.5	0.6 x 0.3	12.7	E7	78	= **Mkn 1321**. Mag 10.17v star SE 4'.8.
12 22 42.4	+03 08 27	CGCG 42-55	14.3	0.5 x 0.3	12.3	E6	159	Located 2'.9 NW of MCG +1-32-29.
12 24 14.5	+04 13 30	CGCG 42-81	14.4	0.6 x 0.3	12.6	E$^+$?	21	Mag 10.74v star S 2'.4.
11 26 48.5	−05 00 17	IC 693	13.8	0.9 x 0.4	12.6	Sab		Mag 10.31v star E 10'.3.
11 39 03.3	−00 12 20	IC 716	14.0	1.6 x 0.3	13.1	Sbc pec sp III	132	
11 43 29.4	−01 40 07	IC 725	13.9	0.5 x 0.5	12.5	E:		In galaxy cluster A1364; mag 13.2 star N 1'.1; stellar **CGCG 12-66** NE 2'.6.
11 44 50.4	−01 36 07	IC 728	13.6	1.2 x 0.6	13.1	SB(rs)b I-II	65	Mag 11.09v star E 6'.3; stellar **CGCG 12-74** SE 5'.8; **CGCG 12-76** SE 8'.7.
11 50 31.8	−04 50 09	IC 741	13.9	0.8 x 0.3	12.2	S0/a	135	Mag 8.03v star W 10'.0.
11 54 12.4	+00 08 11	IC 745	13.2	0.7 x 0.7	12.3	S0		Mag 11.5 star ESE 3'.9.
11 59 12.8	−00 31 26	IC 753	14.2	0.4 x 0.3	11.8	SB(r)0°	30	
11 59 23.6	−01 39 21	IC 754	13.1	0.9 x 0.7	12.6	E$^+$	19	Star just off W edge.
12 02 57.7	+04 50 45	IC 756	13.6	1.8 x 0.7	13.7	(R′)SA(s)c: II	99	Mag 9.30v star WNW 11'.5.
12 21 36.9	+05 45 54	IC 782	13.8	1.3 x 0.6	13.4	SB0/SBa(r)	60	Mag 13.2 star on S edge; mag 10.76v star E 3'.9.
12 22 30.3	−04 39 12	IC 784	13.1	1.9 x 0.6	13.1	SAB(rs)bc: II	109	Mag 9.66v star S 2'.9.
11 49 24.8	−05 07 09	IC 2963	13.9	1.4 x 0.3	12.8	S0$^+$: sp	99	Very faint star on N edge.
11 52 31.3	−03 52 22	IC 2969	12.9	1.2 x 0.8	12.7	SB(r)bc? III	100	Mag 13.0 star on W edge; mag 10.65v star W 8'.8.
11 53 48.7	−05 10 05	IC 2974	13.2	2.2 x 0.5	13.1	SA(s)c? sp II-III	99	Mag 9.29v star NNW 6'.9.
12 19 37.0	+05 23 48	IC 3153	14.8	0.5 x 0.5	13.2	Sc(r)I-II		Located in the center of a triangle formed by NGC 4259, NGC 4270 and NGC 4273.
12 19 45.4	+06 00 19	IC 3155	14.1	1.2 x 0.6	13.6	S0 sp	35	On SW edge of NGC 4269.
12 32 36.5	+02 39 36	IC 3474	14.2	2.3 x 0.3	13.6	Sd sp	36	Mag 11.1 star S 1'.4.
11 20 09.5	−03 03 23	MCG +0-29-17	13.6	1.6 x 0.8	13.7	SA(s)c II	147	Mag 13.1 star S 2'.3.
11 20 17.3	−03 42 36	MCG +0-29-18	15.6	0.4 x 0.4	13.5	Sc		Mag 14.2 star NE 1'.4.
11 20 20.7	−02 55 03	MCG +0-29-19	13.9	0.8 x 0.4	12.6	(R)SB(r)0/a	102	Mag 9.10v star ENE 3'.0.
11 20 51.0	−03 38 48	MCG +0-29-21	13.9	1.0 x 0.3	12.5	SA0$^+$:	39	
11 21 12.2	−02 59 05	MCG +0-29-23	14.0	1.4 x 0.7	13.8	SAB(s)b	75	Strong core, two faint arms.
11 23 50.7	+01 44 02	MCG +0-29-26	14.9	0.8 x 0.6	13.9		168	Mag 9.71v star NNE 8'.4.
11 24 08.8	−01 09 31	MCG +0-29-27	14.4	0.5 x 0.3	12.2	SAB(r)0°	57	Stellar nucleus; NGC 3662 NW 6'.4.

(Continued from previous page)

GALAXIES

RA h m s	Dec ° ′ ″	Name	Mag (V)	Dim ′ Maj x min	SB	Type Class	PA	Notes
11 24 09.4	+00 42 01	MCG +0-29-28	13.9	0.9 x 0.4	12.7	S?	81	Mag 13.2 star SE 0′.6.
11 24 18.6	+00 38 35	MCG +0-29-29	13.9	0.6 x 0.6	12.6	SB(s)c II-III		
11 26 56.1	−01 19 17	MCG +0-29-33	14.3	0.6 x 0.4	12.6	SA(rs)bc II-III	30	Mag 10.73v star NE 3′.0.
11 27 15.6	+01 31 59	MCG +0-29-35	15.9	0.4 x 0.4	13.8			
11 28 16.2	+00 53 29	MCG +0-29-36	14.4	0.5 x 0.5	12.7	SAB(s)bc II-III		
11 30 11.8	+01 49 35	MCG +0-29-38	14.8	0.5 x 0.4	12.9	SA(s)c II		
11 31 38.5	−03 38 47	MCG +0-30-2	14.6	0.8 x 0.3	12.9	SA(rs)bc? II-III	141	
11 31 57.8	−02 55 23	MCG +0-30-3	14.3	1.0 x 0.3	12.9	SA0$^+$: sp	177	Mag 14.1 star SW 0′.5.
11 31 58.8	−00 02 58	MCG +0-30-4	14.5	0.8 x 0.4	13.1	SA(s)cd: III-IV		
11 32 52.9	+01 48 48	MCG +0-30-8	14.5	0.5 x 0.5	12.9	SA(rs)c II-III		Mag 11.3 star E 9′.2.
11 33 25.6	−02 22 51	MCG +0-30-9	14.1	0.5 x 0.5	12.4	SA(rs)c: III-IV		
11 33 34.6	−02 16 53	MCG +0-30-10	14.5	0.4 x 0.3	12.0	Sab?	36	
11 34 23.5	−02 31 40	MCG +0-30-13	14.0	0.7 x 0.5	12.7	SB(r)ab	33	
11 36 04.4	−03 06 05	MCG +0-30-15	14.1	1.4 x 0.5	13.6	SA(s)b pec	138	Mag 9.96v star W 2′.8.
11 36 49.1	−02 14 48	MCG +0-30-16	13.9	0.8 x 0.7	13.2	SA(s)bc: III-IV	57	Mag 13.2 star SE 2′.2.
11 44 04.0	−02 33 33	MCG +0-30-20	14.2	0.9 x 0.7	13.5	SA(s)c III-IV	147	Mag 12.3 star N 2′.5.
11 49 26.2	−03 30 45	MCG +0-30-25	14.3	0.3 x 0.3	11.5			Mag 14.0 star N 0′.5; MCG +0-30-27 NNE 2′.6; CGCG 12-95 NE 2′.7; other named galaxies nearby E.
11 49 32.5	−03 28 38	MCG +0-30-27	14.3	0.4 x 0.4	12.2	(R′)S0/a pec?		Mag 10.7 star N 2′.2; CGCG 12-95 SE 0′.9.
11 49 36.5	−01 27 18	MCG +0-30-29	14.4	0.3 x 0.3	11.9	E/S0		
11 50 20.5	−02 48 42	MCG +0-30-30	13.6	1.4 x 0.5	13.1	SAB(s)b pec: II	135	
11 50 33.9	−02 54 36	MCG +0-30-31	13.0	1.3 x 1.1	13.3	(R′)SA(s)0° pec?	126	Bright core, smooth envelope; mag 11.8 star NE 3′.5.
11 55 05.0	+01 43 06	MCG +0-31-1	14.6	0.5 x 0.3	12.5	S0	42	Mag 13,7 star NE 1′.8.
11 58 02.8	−03 44 37	MCG +0-31-5	14.5	0.5 x 0.5	12.8	(R′)SBb I-II		
11 58 33.8	−02 08 08	MCG +0-31-10	13.2	0.9 x 0.7	12.6	SB(r)0$^+$:	33	Mag 10.54v star E 4′.6; NGC 4006 W 7′.1.
11 59 56.3	−00 32 51	MCG +0-31-15	14.2	0.7 x 0.5	12.9	SAbc: III-IV		
12 03 13.2	+01 57 00	MCG +0-31-23	13.9	0.8 x 0.6	13.0	S0/a	42	Mag 10.80v star SW edge; faint almost stellar anonymous galaxy NW 1′.2.
12 03 44.7	+01 41 40	MCG +0-31-25	14.0	1.6 x 0.4	13.3	SAB(r)a:	36	
12 06 09.3	−02 56 57	MCG +0-31-35	13.6	0.9 x 0.8	13.1	(R)SB(r)0$^+$	87	Large N-S core with dark patches E and W.
12 13 41.0	+02 11 17	MCG +0-31-39	14.6	0.5 x 0.4	12.7	SAB(r)a?	174	
12 14 29.2	+00 49 47	MCG +0-31-40	14.3	0.8 x 0.3	12.6	SAB(r)a:	138	
12 14 40.3	−02 26 58	MCG +0-31-41	14.1	0.8 x 0.5	13.0	Sc	135	Mag 12.8 star NE 2′.8.
12 15 11.6	−03 26 27	MCG +0-31-42	14.1	1.2 x 0.3	12.8	SB(s)d pec sp	33	
12 16 08.6	−03 34 16	MCG +0-31-44	13.3	1.2 x 1.0	13.3	SAB(s)cd: II-III	72	Mag 12.7 star NE 1′.5; CGCG 13-118 NE 12′.2.
12 24 11.9	+01 12 47	MCG +0-32-5	14.1	0.8 x 0.8	13.5	SAB(r)bc II		
12 25 41.0	−02 57 05	MCG +0-32-7	13.7	1.5 x 0.5	13.3	(R′)SB(r)ab? II-III	132	CGCG 14-30 N 3′.1; mag 10.03 star S 3′.6.
12 26 00.2	−03 27 52	MCG +0-32-10	14.4	0.5 x 0.3	12.2	SA(r)a?	24	
12 27 04.8	−00 54 23	MCG +0-32-13	13.8	1.0 x 0.7	13.2	SB(r)a pec? I-II	150	Mag 10.04v star W 6′.0.
12 29 21.8	+01 03 22	MCG +0-32-15	13.8	0.8 x 0.6	12.9	SB(rs)bc pec III	9	
12 31 04.1	+01 40 32	MCG +0-32-16	14.2	1.1 x 0.6	13.6	IAB(s)m pec: V	150	
12 31 48.1	−02 58 16	MCG +0-32-18	15.4	0.3 x 0.2	12.4	E	39	Stellar; mag 14.9 star E 0′.7.
12 35 55.3	−02 54 10	MCG +0-32-25	13.8	1.1 x 0.7	13.3	SAcd? IV-V	39	Mag 15.1 star N edge.
12 38 28.9	+01 45 14	MCG +0-32-29	13.7	0.9 x 0.7	13.1	S0	3	Large E-W core; mag 14.1 star N 1′.3.
11 21 10.1	+05 21 47	MCG +1-29-35	13.3	1.0 x 1.0	13.3	E		
11 22 27.5	+04 17 08	MCG +1-29-38	14.1	0.7 x 0.5	12.8	SA(rs)bc pec:	21	
11 23 18.9	+03 57 17	MCG +1-29-39	14.7	0.3 x 0.3	12.2	E		Almost stellar; mag 12.8 star SE 0′.8.
11 28 55.5	+03 24 28	MCG +1-29-46	14.3	0.7 x 0.5	13.0	SAB(s)b II-III	93	Mag 13.7 star W 1′.7.
11 31 16.2	+04 39 08	MCG +1-29-47	15.0	0.5 x 0.5	13.4	Sb		Mag 14.2 star W 1′.4; CGCG 39-197 W 13′.3.
11 38 55.1	+03 34 49	MCG +1-30-5	13.9	0.9 x 0.6	13.0	(R′)SB(r)b:	33	= **Mkn 1302**; mag 12.9 star NW 2′.1.
11 39 26.8	+03 28 13	MCG +1-30-6	14.0	0.7 x 0.5	12.7	SA(s)bc III-IV	3	Mag 12.8 star S 1′.7.
11 40 19.6	+03 00 01	MCG +1-30-8	13.8	0.9 x 0.7	13.2	(R′)SB(r)b: I-II	120	Mag 10.69v star NW 3′.8.
11 41 52.4	+02 34 52	MCG +1-30-10	14.1	0.8 x 0.6	13.2	(R)SA(s)bc III-IV	15	Mag 9.67v star SSE 9′.0.
11 54 51.3	+02 57 30	MCG +1-30-19	14.1	0.7 x 0.6	13.0	SA(s)bc pec:	36	Mag 13.4 star NE edge.
11 56 05.0	+05 31 10	MCG +1-31-2	15.4	0.3 x 0.3	12.6	SBb		
11 58 02.8	+05 54 45	MCG +1-31-7	15.0	0.2 x 0.2	11.3			
12 01 44.3	+05 49 16	MCG +1-31-9	14.2	0.8 x 0.5	13.1	SB?	102	
12 02 54.6	+05 36 44	MCG +1-31-13	15.5	0.5 x 0.3	13.3	SBb		
12 02 55.7	+04 14 01	MCG +1-31-14	13.5	0.6 x 0.6	12.5	E		Mag 9.18v star W 7′.5.
12 05 59.6	+02 29 56	MCG +1-31-21	14.3	1.1 x 0.2	12.5	SA0° sp	66	
12 15 18.4	+05 45 35	MCG +1-31-30	14.5	0.5 x 0.5	13.1	E0 pec?		NGC 4197 W 10′.3.
12 16 00.5	+04 39 00	MCG +1-31-33	14.6	0.7 x 0.3	12.8	IAm? IV-V	123	Double system of superimposed star? mag 11.5 star E 1′.4.
12 19 22.2	+05 54 38	MCG +1-31-53	14.5	0.6 x 0.4	12.8		171	Galaxy **VCC 336** SSW 2′.4.
12 22 49.7	+03 06 12	MCG +1-32-29	14.4	0.6 x 0.6	13.2	Sb		CGCG 42-55 NW 2′.9.
12 24 03.2	+05 10 55	MCG +1-32-41	13.9	0.9 x 0.8	13.4	SB?	138	
12 24 40.1	+03 18 09	MCG +1-32-49	13.1	0.9 x 0.6	12.3	S?	111	
12 25 52.8	+05 48 29	MCG +1-32-56	13.8	0.8 x 0.2	11.7	IB?	30	
12 25 58.6	+04 28 25	MCG +1-32-59	14.2	0.3 x 0.3	11.5	S0 pec?		Almost stellar.
12 32 50.0	+03 18 01	MCG +1-32-93	14.0	1.0 x 0.6	13.3	SB?	39	
12 33 29.3	+03 47 38	MCG +1-32-96	14.4	0.9 x 0.4	13.2	SB?		
12 36 48.9	+04 06 20	MCG +1-32-113	15.2	0.6 x 0.6	14.0	Sa		
12 39 21.9	+04 56 18	MCG +1-32-126	14.4	0.9 x 0.4	13.1	I?	36	
11 20 39.4	−05 40 21	MCG −1-29-10	13.9	1.0 x 0.7	13.4	S0	36	Contains several knots, or superimposed stars?
11 21 48.0	−05 45 32	MCG −1-29-12	13.6	0.5 x 0.5	12.1	E		Mag 7.81v star N 2′.5.
11 24 50.8	−05 21 09	MCG −1-29-17	13.9	0.6 x 0.5	12.5		156	
11 25 02.2	−05 04 09	MCG −1-29-18	13.6	0.7 x 0.5	12.3	Sa	162	
11 25 35.7	−04 59 19	MCG −1-29-20	14.4	0.7 x 0.5	13.1	SBc	78	
11 44 35.7	−03 48 04	MCG −1-30-27A	14.0	1.5 x 0.7	13.9	SB(s)dm IV	75	Mag 12.5 star N 0′.8.
11 46 35.5	−03 51 36	MCG −1-30-32	13.8	1.3 x 0.7	13.5	SAB(rs)b? pec	51	Part of **Wild's Triplet**. On W edge of MCG −1-30-33.
11 46 43.8	−03 50 54	MCG −1-30-33	13.4	3.1 x 0.7	14.1	SB(s)b: pec	79	Part of **Wild's Triplet**. Pair with MCG −1-30-32 W edge, bridge? MCG −1-30-34 N 1′.9.
11 46 49.8	−03 49 26	MCG −1-30-34	15.3	1.0 x 0.6	14.6	SB(s)c pec	24	Part of **Wild's Triplet**. Located 1′.9 N of MCG −1-30-33.
11 52 38.3	−05 12 25	MCG −1-30-41	13.8	1.0 x 0.4	12.6	(R)SB(rs)ab	21	
11 52 59.6	−04 25 36	MCG −1-30-43	13.4	2.2 x 0.7	13.7	SAB(s)dm III-IV	99	Several knots E.
12 02 25.9	−04 18 24	MCG −1-31-2	14.2	1.4 x 0.2	12.7	Sc: sp	57	
12 05 12.6	−03 54 18	MCG −1-31-3	13.7	0.7 x 0.5	12.7	E	120	
12 12 03.4	−04 52 38	MCG −1-31-8	13.8	1.3 x 0.9	13.8	(R′)SB(rs)bc II-III	132	
12 28 14.8	−04 57 34	MCG −1-32-15	13.9	0.5 x 0.5	12.5	E		Mag 12.7 star on NE edge.
12 30 00.9	−05 58 55	MCG −1-32-17	13.4	0.9 x 0.7	12.8	SBab	96	Mag 7.35v star S 2′.5.
12 30 18.3	−05 57 46	MCG −1-32-18	15.1	0.9 x 0.7	14.4	(R′)SA(s)0/a?	93	Mag 8.66v star S 4′.3.
12 33 37.2	−04 53 10	MCG −1-32-23	14.6	2.1 x 0.4	14.2	SBm: sp IV-V	132	Mag 10.48v star S 3′.4.

(Continued from previous page)

GALAXIES

RA h m s	Dec ° ′ ″	Name	Mag (V)	Dim ′ Maj x min	SB	Type Class	PA	Notes
12 35 00.5	−05 29 26	MCG −1-32-25	14.5	0.9 x 0.7	13.8	Sbc	75	
12 36 22.0	−03 55 17	MCG −1-32-29	14.0	1.1 x 0.4	13.0	SB(s)0^+ pec?	129	Stellar nucleus, uniform envelope.
11 20 17.0	+02 57 49	NGC 3630	12.0	2.1 x 0.9	12.6	S0 sp	37	A mag 8.07v star lies SE 9′9; there are several faint, stellar anonymous galaxy midway to this star.
11 20 26.3	+03 35 07	NGC 3633	13.5	1.1 x 0.4	12.5	SAa: sp	72	Mag 9.05v star NE 3′7.
11 21 06.7	+03 14 06	NGC 3640	10.4	4.0 x 3.2	13.2	E3	100	NGC 3641 on S edge.
11 21 08.9	+03 11 41	NGC 3641	13.2	0.9 x 0.9	13.0	E pec		Compact; a pair of stellar anonymous galaxies W 2′1.
11 21 25.0	+03 00 49	NGC 3643	14.1	1.2 x 0.8	13.9	SB(r)0^+: sp	99	Mag 13.2 star on SE edge; mag 11.33v star E 5′4.
11 21 33.0	+02 48 36	NGC 3644	13.7	1.5 x 0.7	13.6	(R′)SBa pec:	48	Mag 12.5 star on S edge; stellar **IC 683** S 3′6; **IC 683E** lies 1′3 W of IC 683.
11 21 40.0	+02 57 51	NGC 3645	14.6	0.6 x 0.6	13.3			Stellar; a small, faint anonymous galaxy S 1′3.
11 21 35.4	+02 53 36	NGC 3647	14.6	0.3 x 0.3	12.1	E		**CGCG 39-141** SW 0′8; stellar anonymous galaxy E 0′9; and another stellar anonymous galaxy NE 0′9.
11 23 46.6	−01 06 15	NGC 3662	12.9	1.4 x 0.9	13.0	SAB(r)bc pec I-II	25	Mag 9.56v star NW 8′6; MCG +0-29-27 SE 6′4.
11 24 25.8	+03 19 33	NGC 3664	12.8	2.0 x 1.9	14.1	SB(s)m pec IV-V	42	Brighter along W edge; NGC 3664A S 6′3; **UGC 6417** S 4′7.
11 24 25.1	+03 13 20	NGC 3664A	14.3	1.1 x 1.0	14.2	(R)SB(s)m: V	15	NGC 3664 N 6′3; CGCG 39-173 ESE 8′2.
11 26 08.6	−05 35 15	NGC 3679	14.2	1.0 x 0.5	13.3	Sbc III:	178	
11 28 16.3	+04 19 38	NGC 3685	14.1	0.6 x 0.3	12.3	E/S0	140	**UGC 6466** SW 1′0.
11 31 41.2	+03 29 15	NGC 3716	13.5	0.7 x 0.6	12.5	(R)SAB0°?	150	
11 32 13.6	+00 49 07	NGC 3719	13.0	1.8 x 1.3	13.8	SA(rs)bc pec: II	15	Very faint outer shell; NGC 3720 E 2′3.
11 32 21.6	+00 48 14	NGC 3720	13.0	1.0 x 0.9	12.7	SAa: I-II	85	NGC 3719 W 2′3.
11 38 18.1	−03 21 15	NGC 3776	15.4	0.4 x 0.3	12.9	Sb	155	Almost stellar.
11 45 35.3	+03 13 48	NGC 3849	13.7	0.8 x 0.6	12.8	S0/a pec?	36	In star poor area; mag 11.7 star SE 7′8.
11 49 30.1	−01 05 12	NGC 3907	13.1	1.2 x 0.7	12.8	SB(s)0^-:	40	Located 1′7 W of NGC 3907.
11 49 23.6	−01 05 05	NGC 3907B	14.0	1.1 x 0.3	12.7	Sb sp III	76	NGC 3907B E 1′7.
11 48 54.8	−04 40 57	NGC 3915	14.2	0.7 x 0.4	12.7	Sm	113	
11 53 40.8	−03 59 50	NGC 3952	13.1	1.6 x 0.7	13.1	IBm: sp III-IV	78	Mag 8.12v star NNW 7′5.
11 56 01.0	−02 43 12	NGC 3979	12.9	1.1 x 0.9	12.8	SB0° pec:	112	Bright knot or star superimposed on E edge of core.
11 58 05.8	−02 07 14	NGC 4006	12.6	1.7 x 1.2	13.3	E pec:	20	MCG +0-31-10 E 7′1.
12 00 23.8	−01 06 04	NGC 4030	10.6	4.2 x 3.0	13.2	SA(s)bc I-II	27	A pair of mag 10 and 13 stars on S end, a mag 12 star N edge.
12 02 23.1	+04 19 46	NGC 4043	13.6	0.7 x 0.6	12.5	(R)SB(r)0°:	135	Mag 9.18v star S 6′7.
12 02 29.7	−00 12 44	NGC 4044	13.0	1.2 x 1.1	13.1	E^+:	51	
12 02 42.3	+01 58 36	NGC 4045	12.0	2.7 x 1.7	13.5	SAB(r)a	95	NGC 4045A S 1′5.
12 02 42.8	+01 57 05	NGC 4045A	14.4	0.7 x 0.4	12.8	SB(r)0° pec sp	153	Located on S edge of NGC 4045.
12 03 48.9	+03 32 52	NGC 4058	13.1	1.2 x 0.6	12.6	SAB(r)0^+:	165	
12 04 05.7	+01 50 46	NGC 4063	13.9	1.2 x 0.4	13.0	SAB0°:	10	
12 04 27.4	+01 53 48	NGC 4073	11.4	3.2 x 2.3	13.4	E^+	105	Stellar galaxy **PGC 38025** lies E edge of nucleus.
12 04 38.0	+02 04 21	NGC 4075	13.5	1.3 x 0.7	13.2	SA(r)0^+	129	**CGCG 13-69** E 10′9.
12 04 38.1	+01 47 16	NGC 4077	13.1	1.3 x 0.9	13.1	SB(r)0°	15	Anonymous stellar galaxy SE 1′2; NGC 4139 NW 1′4.
12 04 50.1	−02 22 59	NGC 4079	12.4	2.2 x 1.6	13.6	(R′)SAB(rs)bc: I	125	Faint outer envelope.
12 07 37.0	+02 41 27	NGC 4116	12.0	3.8 x 2.2	14.1	SB(rs)dm III-IV	155	Bright narrow bar with very small, bright nucleus, several filamentary arms.
12 08 10.8	+02 52 40	NGC 4123	11.4	4.4 x 3.2	14.1	SB(r)c II-III	135	Several bright knots or superimposed stars.
12 04 34.1	+01 48 06	NGC 4139	13.7	1.0 x 0.5	12.8	SB(rs)0°	153	NGC 4077 SE 1′4.
12 12 51.8	+01 18 04	NGC 4179	11.0	4.0 x 1.1	12.4	S0 sp	143	Located 1′9 SW of a mag 10.86v star.
12 14 38.7	+05 48 21	NGC 4197	12.8	3.4 x 0.6	13.5	Sd	36	MCG +1-31-30 E 10′3.
12 18 08.5	−01 03 51	NGC 4202	13.6	1.2 x 0.7	13.2	SAB(rs)bc	127	
12 17 09.0	+03 40 49	NGC 4234	12.7	1.3 x 1.3	13.1	(R′)SB(s)m IV		Stellar **PGC 39445** ESE 8′2.
12 17 59.4	+05 35 52	NGC 4249	14.0	0.6 x 0.6	12.8	$S0_1(0)$		Almost stellar nucleus; **MCG +3-31-43** N 5′4.
12 18 31.0	+05 33 35	NGC 4252	14.0	1.5 x 0.4	13.3	Sb? sp	48	Small, faint galaxy **VCC 286** N 3′5.
12 18 56.0	+04 47 09	NGC 4255	12.8	1.3 x 0.5	12.2	SB(r)0°	115	
12 19 06.4	+05 43 32	NGC 4257	14.0	1.5 x 0.5	13.5	Sab: sp	78	Very faint galaxy **VCC 321** N 0′9.
12 19 22.3	+05 22 36	NGC 4259	13.6	1.1 x 0.4	12.5	S0 sp	143	Mag 14.8 star on NE edge.
12 19 23.2	+05 49 30	NGC 4261	10.4	4.1 x 3.6	13.3	E2-3	160	Galaxy **VCC 344** on S edge; **VCC 336** 3′4 NNW; NGC 4264 ENE 3′5.
12 19 35.8	+05 50 51	NGC 4264	12.8	1.0 x 0.8	12.4	SB(rs)0^+	123	Center of NGC 4261 WSW 3′5.
12 19 41.7	+05 32 16	NGC 4266	13.7	2.0 x 0.4	13.3	SB(s)a? sp	76	Mag 9.21v star on N edge.
12 19 47.2	+05 17 02	NGC 4268	12.8	1.6 x 0.6	12.6	SB0/a: sp	48	Mag 11.6 star S 3′7; located on E edge of galaxy cluster **A 1516**.
12 19 49.8	+05 27 49	NGC 4270	12.2	2.0 x 0.9	12.7	S0	110	Mag 9.12v star and NGC 4266 N 5′4.
12 19 55.6	+05 20 27	NGC 4273	11.9	2.3 x 1.5	13.1	SB(s)c II-III	10	NGC 4277 E 1′9.
12 20 03.7	+05 20 28	NGC 4277	13.4	1.0 x 0.9	13.1	SAB(rs)0/a:	141	Located 1′9 E of NGC 4273.
12 20 21.4	+05 23 15	NGC 4281	11.3	3.0 x 1.5	12.8	$S0^+$: sp	88	NGC 4277 and NGC 4273 WSW 6′0.
12 20 24.2	+05 34 20	NGC 4282	14.0	1.2 x 0.6	13.5	S0?	105	Mag 7.29v star NW 9′2.
12 20 48.5	+05 38 23	NGC 4287	14.3	1.2 x 0.2	12.6	S(on-edge)	72	Mag 14.0 star on N edge; mag 7.29v star W 13′1.
12 21 02.3	+03 43 19	NGC 4289	13.9	3.9 x 0.4	14.2	SA(s)cd: sp	1	Very slender, faint arms. Located 3′1 W of mag 8.37v star.
12 21 16.4	+04 35 46	NGC 4292	12.2	1.7 x 1.1	12.8	(R)SB(r)0°	7	Mag 10.41v star 1′3 NNW; small, round anonymous galaxy N 2′4.
12 21 41.3	+05 23 02	NGC 4300	12.9	1.5 x 0.6	12.7	Sa	42	Mag 6.46v star ESE 13′3.
12 21 33.9	+04 46 46	NGC 4301	13.6	1.4 x 0.5	13.1	SA(r)0/a: sp	132	Bright knot, star or possibly stellar galaxy E of core.
12 21 54.5	+04 28 17	NGC 4303	9.7	6.5 x 5.8	13.4	SAB(rs)bc I-II	162	= **M 61**. Very bright center, branching arms with many small, bright knots.
12 22 27.3	+04 33 57	NGC 4303A	13.0	1.5 x 1.3	13.6	SAB(s)cd III	69	Located 9′8 NE of NGC 4303 (M 61); pair of bright knots, or superimposed stars, on W edge.
12 23 06.2	+05 15 02	NGC 4324	11.6	2.8 x 1.2	12.8	SA(r)0^+	53	Mag 6.46v star NW 8′9.
12 23 54.0	−03 26 32	NGC 4348	12.5	3.2 x 0.7	13.2	SAbc: sp II-III:	38	
12 25 18.1	+05 44 26	NGC 4376	13.3	1.4 x 0.9	13.4	Im	157	Pair of faint stars or bright knots on W side.
12 25 18.2	+04 55 28	NGC 4378	11.7	2.9 x 2.7	13.8	(R)SA(s)a	167	
12 25 42.7	+00 34 23	NGC 4385	12.5	2.2 x 1.4	13.6	SB(rs)0^+: II-III	82	Very strong, irregular bar.
12 26 35.9	+03 57 56	NGC 4412	12.4	1.4 x 1.3	12.9	SB(r)b? pec II	132	Bright bar.
12 26 54.8	−00 52 39	NGC 4418	13.1	1.4 x 0.7	12.9	(R′)SAB(s)a I-II	59	Mag 10.04v star SW 4′4.
12 26 58.5	+02 29 43	NGC 4420	12.1	2.0 x 1.0	12.7	SB(r)bc: III-IV	8	Bright, elongated core.
12 27 12.1	−05 49 51	NGC 4422	13.8	1.3 x 1.3	14.3	$SA0^-$ pec?		
12 27 09.2	+05 52 53	NGC 4423	13.4	2.3 x 0.4	13.2	Sdm:	18	
12 28 50.8	−01 56 20	NGC 4454	11.9	2.3 x 2.1	13.4	(R)SB(r)0/a	118	Bright nucleus in core elongated NNE-SSW; CGCG 14-47 S 7′3.
12 28 59.4	+03 34 15	NGC 4457	10.9	2.7 x 2.3	12.7	(R)SAB(s)0/a II	66	**CGCG 42-133** E 10′4.
12 30 26.8	+04 14 45	NGC 4480	12.4	2.3 x 1.2	13.4	SAB(s)c	175	Several faint knots, irregularly branching arms N-S.
12 31 08.5	+00 36 46	NGC 4493	14.1	1.0 x 0.8	13.9	E^+ pec:	164	Double system, smaller companion SE 0′8.
12 31 39.1	+03 56 17	NGC 4496A	11.4	4.0 x 3.2	14.0	SB(rs)m III-IV	70	With NGC 4496B two overlapping systems. Possibly colliding or strongly interacting.
12 31 41.0	+03 55 33	NGC 4496B	13.9	0.8 x 0.6	13.0	IB(s)m:	102	See notes for NGC 4496A.
12 32 45.8	+00 06 52	NGC 4517	10.4	10.5 x 1.5	13.2	SA(s)cd: sp III-IV	83	Dark lane extends almost full length N of center line; star on N edge near middle.
12 32 27.8	+00 23 19	NGC 4517A	12.5	4.0 x 2.6	14.9	SB(rs)dm: IV-V	30	**Reinmuth 80**. Small nucleus, filamentary arms; mag 10.01v star NW 3′6.
12 34 07.6	+02 39 08	NGC 4527	10.5	6.2 x 2.1	13.2	SAB(s)bc II	67	Bright bulge or bar; two main knotty arms.
12 34 21.8	+02 19 35	NGC 4533	13.8	2.1 x 0.4	13.5	SAd: sp III-IV	161	Mag 8.61v star N 4′0.
12 34 27.8	+02 11 12	NGC 4536	10.6	7.6 x 3.2	13.9	SAB(rs)bc I-II	130	Very small, very bright nucleus; two main filamentary arms with knots.

(Continued from previous page)

GALAXIES

RA h m s	Dec ° ′ ″	Name	Mag (V)	Dim ′ Maj x min	SB	Type Class	PA	Notes
12 34 41.0	+03 19 22	NGC 4538	14.4	0.9 x 0.3	12.8	S?	81	
12 35 10.6	−00 13 17	NGC 4541	13.0	1.6 x 0.7	13.0	(R′)SAB(r)bc: II	91	Elongated **UGC 7755** and almost stellar **CGCG 14-73** E 7′8.
12 35 36.5	+03 02 04	NGC 4544	13.0	2.0 x 0.6	13.0	SB0/a? sp	161	
12 35 29.4	−03 47 37	NGC 4546	10.3	3.3 x 1.4	11.9	SB(s)0$^-$:	78	CGCG 14-74 ENE 5′5.
12 37 33.6	+04 22 02	NGC 4576	13.5	1.2 x 0.8	13.3	SAB(rs)bc: II-III	159	Mag 6.99v star SE 6′6.
12 37 48.2	+05 22 04	NGC 4580	11.8	2.1 x 1.6	13.0	SAB(rs)a pec II-III	165	
12 38 05.2	+01 28 34	NGC 4581	12.5	1.9 x 1.1	13.3	E$^+$	173	**UGC 7807** E 8′4.
12 38 28.4	+04 19 05	NGC 4586	11.7	4.0 x 1.3	13.3	SA(s)a: sp	115	Small bright nucleus with dark lane N.
12 38 35.3	+02 39 23	NGC 4587	13.5	1.1 x 0.6	12.9	SA(r)0°:	48	
12 39 18.1	−00 31 51	NGC 4592	11.6	5.8 x 1.5	13.8	SA(s)dm: III	97	
12 39 39.8	−05 20 39	NGC 4593	10.9	3.9 x 2.9	13.4	(R)SB(rs)b II	114	Bright nucleus in short bar encircled by two asymmetrical arms. **MCG −1-32-33** E 3′8.
11 20 17.2	+02 31 26	UGC 6345	13.4	2.3 x 1.4	14.6	IB(s)m V	75	Irregular, many knots and clumps.
11 20 35.8	−01 29 37	UGC 6359	13.7	0.8 x 0.5	12.6	SAB0 pec	66	Almost stellar anonymous galaxy NE 0′7; mag 12.7 star N 1′0.
11 20 47.8	+00 27 53	UGC 6361	13.4	1.4 x 0.7	13.2	Triple System	147	Triple system; E component much elongated; west component slightly elongated; S component round and compact.
11 23 19.2	−00 55 20	UGC 6402	14.3	0.8 x 0.2	12.2	Sdm? sp	93	
11 23 59.8	+02 41 30	UGC 6413	13.5	1.0 x 1.0	13.3	SAB(rs)bc: II-III		Anonymous galaxy W 2′4.
11 25 17.8	+00 20 59	UGC 6432	14.0	0.7 x 0.6	12.9	SA(rs)bc: II	78	
11 25 34.9	−00 46 05	UGC 6435	12.9	1.1 x 1.1	13.0	S0°:		Mag 9.07v star E 3′2.
11 26 01.2	+01 59 03	UGC 6440	13.6	1.0 x 0.7	13.1	SB(rs)bc II	43	
11 27 12.8	−00 59 39	UGC 6457	13.9	1.2 x 1.1	14.0	Im:		
11 28 17.6	+02 39 13	UGC 6469	13.9	0.9 x 0.4	12.6	SAbc:	123	Mag 7.25v star N 7′9.
11 31 32.2	−02 18 32	UGC 6510	12.5	1.9 x 1.6	13.5	SAB(rs)cd III	10	Mag 10.08v star WNW 2′9.
11 35 37.0	+00 07 37	UGC 6568	14.1	0.9 x 0.4	12.8	SB(s)m IV	2	Mag 8.89v star NNE 5′6.
11 37 09.6	+02 50 15	UGC 6587	14.0	0.8 x 0.5	12.9	SAB(s)b pec:	159	**MCG +1-30-3** NW 0′8.
11 38 33.2	−01 11 05	UGC 6608	13.3	1.0 x 0.6	12.6	SAB(r)ab?	36	
11 42 12.4	+00 20 08	UGC 6665	13.9	0.4 x 0.4	11.8	Sb pec:		
11 45 46.2	+03 01 44	UGC 6736	14.1	1.5 x 0.4	13.3	SBcd? sp	177	Mag 12.4 star S 1′9.
11 46 39.8	−01 59 46	UGC 6750	14.0	1.1 x 0.3	12.7	SB(s)bc: sp	13	Mag 12.9 star N end.
11 47 43.9	+01 49 31	UGC 6769	13.9	1.2 x 0.5	13.1	SB(r)b: I-II	73	
11 48 00.3	+04 29 17	UGC 6771	12.7	1.7 x 1.5	13.6	(R′)SAB(r)ab II-III	63	Almost stellar nucleus in short, bright N-S bar.
11 48 50.4	−02 02 00	UGC 6780	13.0	3.2 x 1.0	14.1	SAB(s)d: III-IV	20	Mag 11.6 star N edge.
11 51 56.2	−02 38 35	UGC 6838	14.2	0.7 x 0.4	12.6	SAab	113	
11 52 37.4	−02 28 09	UGC 6850	13.7	0.6 x 0.5	12.2	Pec	66	
11 52 43.6	+01 44 26	UGC 6854	13.4	1.0 x 0.9	13.1	SB(rs)bc pec	81	Strong dark patch SE of center.
11 54 25.2	−02 19 08	UGC 6879	13.5	1.7 x 0.6	13.4	SAB(r)d? sp IV	168	Mag 8.38v star ESE 8′1.
11 55 37.2	+01 14 12	UGC 6903	12.3	2.3 x 2.0	13.8	SB(s)cd III-IV	150	Knotty, short, bright E-W core.
11 57 31.8	−01 15 13	UGC 6934	14.2	1.7 x 0.2	12.9	SA(r)cd: sp	141	
11 58 23.8	−02 16 37	UGC 6958	13.0	1.2 x 1.2	13.2	SAB(rs)bc II-III		
11 58 45.2	−01 27 43	UGC 6970	14.0	1.5 x 0.7	13.9	SB(s)m IV	75	Mag 11.6 star superimposed W of center.
12 00 47.6	−00 01 25	UGC 6998	15.9	1.7 x 1.3	16.6	SAB(s)m V	60	Dwarf spiral.
12 01 11.4	−01 17 54	UGC 7000	13.8	1.1 x 0.9	13.6	IB(s)m IV	50	Several bright knots or superimposed stars W edge.
12 01 27.7	−00 43 07	UGC 7004	13.5	1.2 x 1.0	13.5	SB(r)c III	60	
12 03 37.6	+02 02 44	UGC 7034	13.8	1.0 x 0.5	12.9	SA(r)0°?	81	Very small, very faint anonymous galaxy NW edge; **PGC 38110** SW 2′0.
12 03 40.3	+02 38 26	UGC 7035	14.0	1.3 x 0.4	13.1	SB(r)a: II-III	149	Very faint, elongated anonymous galaxy SW 1′0.
12 04 20.2	−01 31 55	UGC 7053	14.8	1.7 x 1.4	15.6	IAB(s)m V-VI	3	Dwarf irregular, uniform low surface brightness; mag 11.2 star W 2′1.
12 04 20.3	+01 34 02	UGC 7057	13.6	1.8 x 0.4	13.1	SAb: sp	100	Small, elongated **PGC 38181** S 1′6.
12 04 47.4	−02 43 13	UGC 7065	13.2	1.9 x 1.4	14.1	(R′)SB(r)0/a	102	Bright N-S bar, faint envelope.
12 11 04.1	+00 58 15	UGC 7177	13.5	1.1 x 0.8	13.3	S0$^-$?	90	**CGCG 13-97** NNE 7′1.
12 11 20.0	+01 29 29	UGC 7184	13.7	1.5 x 0.6	13.5	SB(rs)d pec III	148	Mag 13.2 star NE 2′8.
12 11 27.5	+02 55 33	UGC 7185	14.0	1.2 x 1.2	14.2	SA(rs)m: V		Dwarf, very low surface brightness.
12 15 52.6	+00 23 58	UGC 7280	13.5	1.0 x 0.6	12.8	SB(r)a	160	
12 17 55.7	+00 26 11	UGC 7332	13.4	2.1 x 1.4	14.4	IB(s)m V	132	Dwarf irregular, low surface brightness; mag 14.8 star N edge.
12 19 09.9	+03 51 17	UGC 7354	13.5	0.5 x 0.5	12.1	E pec:		Located between a pair of SE-NW mag 13-14 stars.
12 19 40.5	+02 04 46	UGC 7370	14.1	1.3 x 0.4	13.2	SBbc pec sp	151	Stellar **CGCG 14-5** S 4′2.
12 20 17.4	+04 12 03	UGC 7387	14.6	1.9 x 0.3	13.8	Sd: sp	16	Very faint, stellar galaxy **VCC 396** N 4′1.
12 20 27.9	+01 28 05	UGC 7394	14.5	1.7 x 0.3	13.6	SAd sp	146	Mag 9.75v star NE 3′2.
12 20 34.3	+00 47 12	UGC 7396	14.0	1.5 x 0.4	13.3	SB(s)d: sp III-IV	101	
12 23 41.8	+02 57 40	UGC 7464	14.3	1.5 x 0.3	13.3	Sb II-III	147	**Mkn 1323** N 8′1.
12 24 27.3	+00 09 07	UGC 7487	13.9	0.9 x 0.6	13.1	(R)SA(s)b III	7	
12 25 46.5	+04 30 35	UGC 7516	14.1	1.1 x 0.6	13.5	Scd:	163	**CGCG 42-91** W 9′8.
12 25 58.5	+03 25 43	UGC 7522	13.8	2.6 x 0.3	13.4	Scd:	130	**CGCG 42-94** W 6′5.
12 26 11.8	−01 18 17	UGC 7531	14.3	1.0 x 0.7	13.8	IBm pec: IV-V	27	Very patchy.
12 28 51.5	+04 17 26	UGC 7607	14.9	1.9 x 0.2	13.7	Sd	54	Mag 12.6 star S 1′7.
12 29 02.7	+02 43 23	UGC 7612	14.4	2.0 x 1.0	15.0	Sm: V	145	Patchy.
12 30 13.8	+02 37 29	UGC 7642	14.2	1.0 x 0.7	13.7	IB(s)m V-VI	130	Uniform surface brightness; mag 9.03v star W 7′4.
12 30 21.0	+03 44 25	UGC 7644	13.9	2.0 x 0.5	13.8	Sd	50	
12 33 55.4	+03 32 38	UGC 7715	14.2	1.1 x 0.9	14.0	Im:		Dwarf, low, uniform surface brightness.
12 34 09.1	−00 21 16	UGC 7720	13.6	1.0 x 0.9	13.3	(R′)SB(r)b:	30	
12 35 30.5	−02 19 37	UGC 7752	14.4	1.5 x 0.4	13.7	SABab? sp	6	Mag 14.4 star SE edge.
12 35 54.7	−01 51 16	UGC 7763	14.3	0.8 x 0.5	13.2	SAB(s)cd III-IV	80	
12 36 42.1	+03 06 24	UGC 7780	15.1	1.5 x 0.3	14.1	Sdm:	48	
12 38 03.1	−02 15 55	UGC 7798	13.8	0.8 x 0.5	12.7	IBm: IV	57	
12 38 05.4	−00 01 36	UGC 7800	14.1	1.0 x 0.3	12.6	SA(r)b? II	133	
12 38 36.1	+01 24 06	UGC 7806	14.3	1.7 x 0.2	13.0	SBcd: sp III	148	Mag 10.97v star E 1′0; **UGC 7807** N 4′5.
12 39 01.2	+00 21 52	UGC 7813	13.9	0.7 x 0.7	13.1	E:		**CGCG 14-88** SW 4′5; **CGCG 14-94** NE 7′8.
12 39 11.6	+00 42 59	UGC 7820	12.9	1.7 x 1.6	13.8	SAB(s)cd: III	108	**CGCG 14-96** ENE 6′7.
12 39 50.6	+01 40 19	UGC 7824	13.8	1.5 x 0.7	13.7	SB(s)m: IV-V	77	Mag 13.1 star W 2′3.

GALAXY CLUSTERS

RA h m s	Dec ° ′ ″	Name	Mag 10th brightest	No. Gal	Diam ′	Notes
11 23 00.0	+01 05 00	A 1238	16.0	63	16	All members anonymous, faint and stellar.
11 32 54.0	−03 58 00	A 1308	15.7	37	16	**PGC 35654** SE of center 3′9; all other members anonymous and stellar.
11 38 54.0	−04 18 00	A 1334	15.7	39	13	All members anonymous, faint and stellar.
11 43 42.0	−01 45 00	A 1364	16.0	74	20	Except IC 725, all members stellar.
11 51 12.0	−03 05 00	A 1399	16.0	82	20	All members faint, stellar and anonymous.

RA h m s	Dec ° ′ ″	Name	Mag (V)	Dim ′ Maj x min	SB	Type Class	PA	Notes
10 10 41.0	−02 25 58	CGCG 8-56	14.6	0.8 x 0.4	13.2	Sc		Mag 13.6 star 1′4 E.
10 11 05.2	+01 13 17	CGCG 8-60	13.2	1.0 x 0.4	12.1	SB(s)bc pec	9	Dark knot or hole S of center.
10 14 46.3	−02 11 49	CGCG 8-80	15.0	0.4 x 0.4	12.8			Mag 14.0 star S 1′2; mag 13.7 star E 1′2.
10 18 57.1	+01 39 30	CGCG 8-91	14.6	0.5 x 0.3	12.4	Sb	9	Mag 7.94v star SW 8′4.
10 21 53.2	+00 17 41	CGCG 9-8	14.4	0.7 x 0.4	12.8		150	Mag 10.79v star SW 2′2; **CGCG 9-9** N 6′9.
10 22 31.5	+00 41 32	CGCG 9-11	14.4	0.8 x 0.3	12.9	E	24	Mag 9.00v star S 8′1.
10 22 36.2	−03 37 13	CGCG 9-12	14.9	0.4 x 0.4	13.0	E		Mag 6.42v star E 14′2; **CGCG 9-14** E 6′9; mag 8.52v star S 7′5.
10 24 07.7	+00 08 27	CGCG 9-19	15.1	0.5 x 0.2	12.4	S	174	Mag 9.66v star N 4′2.
10 24 38.9	−02 19 36	CGCG 9-22	14.7	0.9 x 0.3	13.1	Sc	144	Mag 9.52v star SE 4′1.
10 25 45.1	+01 26 09	CGCG 9-29	15.0	0.3 x 0.3	12.3	S0/a		Located 2′8 SE of mag 11.8 star.
10 26 54.7	−00 32 30	CGCG 9-41	15.1	0.3 x 0.3	12.6	E		Mag 14.1 star ENE 1′0; mag 10.67v star WSW 3′7.
10 27 10.2	−03 19 08	CGCG 9-42	13.6	1.0 x 0.7	13.1	S0/a	138	Mag 9.67v star SW 3′8; **CGCG 9-45** SE 1′6.
10 27 07.6	+01 29 31	CGCG 9-44	14.5	0.6 x 0.6	13.3	Scd		Located 2′5 SW of mag 8.75v star.
10 29 14.3	−02 30 51	CGCG 9-55	14.5	0.6 x 0.3	12.7	E/S0	18	Mag 12.3 star NW 3′9.
10 29 14.3	−02 01 10	CGCG 9-56	15.1	0.5 x 0.3	12.9		36	Located between a mag 11.58v star N and a mag 11.32v star S.
10 30 15.4	−02 20 17	CGCG 9-64	14.7	0.4 x 0.3	12.3		120	Mag 9.65v star SSE 5′3; mag 7.99v star NNW 8′8.
10 30 50.9	−02 43 18	CGCG 9-70	14.5	0.5 x 0.4	12.8	E	150	Mag 9.95v star NE 3′3; **CGCG 9-69** SSW 2′3.
10 33 03.1	−02 05 11	CGCG 9-78	14.8	0.4 x 0.4	12.7	Sbc		Elongated anonymous galaxy WSW 1′7.
10 37 30.3	−02 44 41	CGCG 9-88	15.0	0.4 x 0.4	12.9			Mag 12.8 star N 4′5.
10 38 08.6	−02 38 28	CGCG 9-90	14.2	1.3 x 0.6	13.8	Interacting	126	UGC 5787 NW 6′1; mag 13.2 star S 2′3.
10 41 06.6	−02 57 51	CGCG 9-98	15.0	0.6 x 0.5	13.5	Scd	39	Mag 14.1 star SW 0′7.
10 45 03.7	+00 25 59	CGCG 10-4	14.0	0.6 x 0.5	12.5	Sa	30	Mag 12.8 star E 0′9.
10 47 54.7	+00 48 41	CGCG 10-16	15.0	0.5 x 0.2	12.4	Sab	60	Mag 14.4 star NE 0′8; UGC 5913 E 6′3.
10 49 17.1	+00 19 40	CGCG 10-26	13.9	0.8 x 0.6	13.0	SB?	12	**CGCG 10-24** NNW 3′4.
10 51 11.1	−03 36 55	CGCG 10-36	14.9	0.6 x 0.3	12.9		102	Mag 13.5 star NW 3′0.
10 58 09.8	−00 46 30	CGCG 10-46	14.3	0.5 x 0.2	11.7	Sc? sp	114	Mag 14.2 star W 3′1; mag 12.9 star NNE 7′9.
10 58 57.4	+01 26 09	CGCG 10-51	14.1	0.6 x 0.5	12.6	SB(rs)b?	75	Faint, elongated anonymous galaxy SE 1′7.
10 59 05.9	+01 10 50	CGCG 10-52	14.0	0.7 x 0.4	12.5	SAab	6	Mag 13.4 star NE 1′1; **CGCG 10-55** SE 3′5.
11 00 01.9	+01 46 33	CGCG 10-58	13.3	0.6 x 0.6	12.1	$SA(r)0^-$		Mag 13.9 star NW 5′1; mag 13.4 star SE 6′0.
11 03 32.7	−01 30 38	CGCG 10-64	13.8	0.4 x 0.3	11.4	$SA(r)0^-$?	48	Located 1′5 SSE of a mag 8.70V star.
11 03 35.6	−01 23 32	CGCG 10-65	14.4	0.4 x 0.4	12.3	Sab		= **Mkn 1277**; mag 8.70v star and CGCG 10-64 S 7′2.
11 05 09.2	−00 47 50	CGCG 10-71	13.9	1.4 x 0.4	13.1	SB(r)0/a: sp	138	Close pair of mag 9.83v, 11.6 stars S 4′2.
11 06 46.4	−02 31 11	CGCG 10-77	13.5	0.5 x 0.3	11.3	SB(r)a?		Close pair of mag 10.44v, 10.73v stars S 7′0; stellar **CGCG 10-78** E 8′6.
11 07 55.5	+02 00 46	CGCG 11-2	13.3	0.7 x 0.4	11.8	SB(s)c: III-IV	45	Mag 13.7 star E 2′0; **CGCG 11-3** N 2′2.
11 12 11.9	−02 23 56	CGCG 11-24	15.3	0.4 x 0.3	12.9		162	Mag 13.8 star E 1′4; MCG +0-29-5 N 3′7.
11 12 32.8	+01 26 58	CGCG 11-25	15.2	0.4 x 0.2	12.3		54	Mag 9.59v star NNE 6′4.
11 13 05.2	+00 00 07	CGCG 11-27	13.7	0.5 x 0.3	11.7	E^+	159	Bright nucleus, faint star at N end.
11 14 46.4	−02 07 26	CGCG 11-30	14.7	0.6 x 0.3	12.7	Sb	135	Mag 12.3 star NE 2′9; mag 11.3 star NW 6′2.
11 15 19.1	−03 46 24	CGCG 11-33	14.0	0.7 x 0.3	12.4	E^+	39	Mag 10.8 star SSE 3′2.
11 16 09.6	+00 34 21	CGCG 11-38	14.5	0.7 x 0.4	13.0	SAB(rs)c: III-IV	123	Mag 12.1 stars SSE 3′2; mag 12.9 star NNW 3′9.
11 17 39.6	−01 18 14	CGCG 11-45	14.1	0.7 x 0.2	11.8	SBbc: sp	78	Mag 10.48v star N 9′1.
11 18 34.8	−02 06 51	CGCG 11-48	13.8	0.4 x 0.4	11.9	E		UGC 6311 W 10′0.
11 20 54.8	+00 10 18	CGCG 11-73	14.8	0.5 x 0.3	12.6		48	Mag 10.02v star E 4′6.
11 21 00.0	+00 31 59	CGCG 11-75	13.9	0.7 x 0.3	12.1	S0 pec?	114	**CGCG 11-68** NW 4′1.
11 26 06.4	−01 50 07	CGCG 11-95	14.2	0.7 x 0.3	12.4	SBb? sp	39	Mag 13.7 star NE 0′9.
11 26 40.7	−01 41 41	CGCG 11-96	13.8	0.3 x 0.3	11.3	cE		Almost stellar; located 2′3 N of mag 9.75v star.
10 08 10.3	+02 27 46	CGCG 36-42	15.0	0.3 x 0.3	12.2			Forms a triangle with mag 11.8 star NE 2′8 and mag 11.8 star SE 3′2.
10 11 19.7	+05 53 45	CGCG 36-52	14.7	0.6 x 0.6	13.5			Mag 15.2 star 0′5 S; mag 11.4 star N 5′0.
10 15 16.8	+05 23 07	CGCG 36-72	14.8	0.6 x 0.3	12.8		27	Pair of mag 10.89v, 10.69v stars N 8′2.
10 19 17.6	+04 45 45	CGCG 36-91	15.8	0.5 x 0.3	13.6		81	Pair of mag 12.5, 13.0 stars ENE 5′0.
10 21 29.5	+05 56 26	CGCG 37-11	14.2	0.6 x 0.5	12.7		30	Triangle of stars, mags 12, 13 and 14, SW 1′2.
10 22 48.4	+03 45 29	CGCG 37-17	14.1	0.9 x 0.5	13.1		126	Mag 8.26v star SE 8′5.
10 33 21.2	+02 36 56	CGCG 37-74	14.2	0.5 x 0.4	12.3	SBb	90	Pair of stars, mags 10.71v and 11.41v, SE 1′7.
10 33 33.8	+04 15 46	CGCG 37-77	15.1	0.7 x 0.6	14.0		6	Mag 9.07v star SE 5′9.
10 34 30.7	+03 44 43	CGCG 37-80	16.0	0.4 x 0.4	13.8			Mag 8.83v star S 7′7; mag 10.24v star NE 8′3.
10 35 42.4	+05 36 55	CGCG 37-87	15.6	0.6 x 0.3	13.6		90	Mag 12.5 star NE 5′0.
10 39 59.9	+03 06 45	CGCG 37-106	15.6	0.7 x 0.4	14.1		162	Double system, stellar companion N.
10 44 21.5	+02 47 32	CGCG 38-6	15.1	0.6 x 0.2	12.6		120	Almost stellar **CGCG 36-4** W 1′2.
10 52 06.3	+03 22 16	CGCG 38-38	14.3	0.6 x 0.4	12.6		0	Mag 12.4 star SSW 2′4.
10 55 12.7	+05 51 41	CGCG 38-48	15.4	0.4 x 0.3	12.9		78	Mag 7.40v star W 6′7.
10 56 46.2	+05 39 35	CGCG 38-58	15.1	0.6 x 0.4	13.4		33	Mag 9.0v star SSE 2′0.
10 58 16.6	+04 34 59	CGCG 38-69	13.9	0.9 x 0.5	12.9	(R)SAB0/a:	63	Mag 10.39v star SE 2′3.
10 59 44.8	+03 59 03	CGCG 38-79	14.1	0.5 x 0.4	12.2	SBa?	147	Mag 11.0 star SW 4′1; mag 10.9 star WSW 5′2.
11 01 21.4	+03 02 22	CGCG 38-89	14.1	0.9 x 0.8	13.6	SAB(r)bc II	51	Bright stellar nucleus.
11 02 43.9	+05 22 49	CGCG 38-96	13.9	0.3 x 0.3	11.4	E		Mag 12.1 star NNW 3′5.
11 04 31.5	+03 43 01	CGCG 38-111	14.2	0.7 x 0.3	12.4		6	A mag 13.4 star W 0′6.
11 06 04.9	+03 19 08	CGCG 38-122	13.7	0.5 x 0.3	11.7	E	9	CGCG 38-124 E 1′3.
11 06 10.3	+03 19 10	CGCG 38-124	13.9	0.6 x 0.4	12.3	SB0° pec	66	An elongated anonymous galaxy S 1′4; CGCG 38-122 W 1′3.
11 07 49.3	+02 21 21	CGCG 39-3	14.1	0.6 x 0.6	12.8	SB(rs)b: I-II		Mag 9.94v star NE 2′4.
11 08 31.6	+05 49 29	CGCG 39-9	14.5	0.5 x 0.4	12.6		30	Mag 14.4 star W 2′7.
11 09 03.0	+05 01 51	CGCG 39-15	16.6	0.4 x 0.4	14.4			A mag 9.32v star NW 2′0; a mag 10.67v star E 2′3.
11 09 04.7	+04 46 30	CGCG 39-17	14.6	0.6 x 0.4	12.9		153	Mag 14.8 star 0′6 NW; NGC 3535 WNW 8′3.
11 09 23.4	+03 57 16	CGCG 39-19	17.3	0.6 x 0.4	15.6		0	Close pair of stars, mags 11.5 and 13, SW 1′2.
11 09 48.5	+03 24 05	CGCG 39-23	14.2	0.6 x 0.4	12.5			Mag 9.15v star ESE 5′2.
11 10 20.1	+04 41 12	CGCG 39-31	14.0	0.9 x 0.3	12.5	SB0°: sp	3	Mag 11.4 star W 3′6; mag 11.10v star ESE 4′8; CGCG 39-33 S 5′4.
11 10 25.5	+04 35 54	CGCG 39-33	13.6	0.7 x 0.3	12.0	E^+:	156	CGCG 39-31 N 5′4; MCG +1-29-9 SSE 7′2.
11 10 23.6	+05 00 17	CGCG 39-35	15.1	0.5 x 0.4	13.2		90	CGCG 39-43 NE 8′9; **CGCG 39-24** W 7′9.
11 10 35.5	+03 15 58	CGCG 39-37	14.0	0.5 x 0.4	12.3	E	0	Mag 9.15v star NW 9′1.
11 10 39.0	+03 45 16	CGCG 39-39	13.9	0.5 x 0.3	11.8	S0	0	Close pair of stars, mags 12.8, 9.97v, WSW 3′7.
11 10 53.1	+05 05 23	CGCG 39-43	16.2	0.5 x 0.4	14.3		150	Mag 10.09v star E 6′9; CGCG 39-95 SW 8′9.
11 11 12.7	+03 04 33	CGCG 39-48	15.5	0.3 x 0.3	12.8			Mag 12.1 star NE 1′7.
11 11 17.8	+03 11 49	CGCG 39-50	13.8	0.6 x 0.4	12.3	E^+	114	Very faint, elongated anonymous galaxy N 0′7; mag 10.98v star NNE 3′6.
11 11 26.7	+04 35 34	CGCG 39-54	14.7	0.6 x 0.3	12.7		126	CGCG 39-61 NNE 5′7; mag 9.98v star W 3′3.
11 11 40.9	+04 40 04	CGCG 39-61	14.3	0.8 x 0.5	13.2		126	CGCG 39-54 SSW 5′7.
11 16 26.3	+04 20 11	CGCG 39-94	15.4	0.9 x 0.3	13.8		84	Mag 10.9 star W 3′7.
11 16 22.3	+04 42 54	CGCG 39-95	15.5	0.6 x 0.5	14.0		12	Mag 11.6 star NNE 1′9.
11 17 20.5	+02 26 48	CGCG 39-101	14.8	0.9 x 0.2	12.8		27	Mag 10.62v star E 3′9.
11 17 31.1	+03 01 54	CGCG 39-102	14.0	0.9 x 0.6	13.2	(R′)SB(r)a:	123	Bright NE-SW bar.
11 20 17.8	+04 19 12	CGCG 39-125	13.6	0.5 x 0.5	12.2	E		Mag 13.4 star W 1′7; mag 10.7 star N 6′2.

(Continued from previous page)

GALAXIES

RA h m s	Dec ° ′ ″	Name	Mag (V)	Dim ′ Maj x min	SB	Type Class	PA	Notes
11 21 18.3	+03 39 49	CGCG 39-133	13.9	0.4 x 0.4	12.0	E		Forms a triangle with a mag 13.2 star SSE 1′.3 and a mag 13.8 star E 1′.7; **CGCG 39-145** SE 8′.1.
11 22 08.2	+04 41 31	CGCG 39-150	14.3	0.5 x 0.3	12.1		3	Mag 11.2 star NW 7′.9.
11 23 20.3	+05 51 05	CGCG 39-162	14.3	0.4 x 0.4	12.2			Mag 11.8 star WNW 5′.4.
11 23 44.6	+03 00 17	CGCG 39-163	15.6	0.3 x 0.3	12.8			Mag 11.31v star SSE 6′.8.
11 24 55.5	+03 10 03	CGCG 39-173	14.6	0.8 x 0.4	13.2		150	NGC 3664 NW 12′.1; NGC 3664A WNW 8′.2.
11 25 24.7	+04 54 47	CGCG 39-176	15.1	0.5 x 0.4	13.2		159	Mag 12.4 star SE 4′.8.
11 26 09.5	+03 29 50	CGCG 39-180	14.2	0.5 x 0.4	12.3	SAB(s)b:	18	Mag 8.86v star W 5′.1.
11 27 17.4	+03 45 22	CGCG 39-184	14.1	0.7 x 0.5	12.8		3	Very close pair of mag 14-15 stars NW 1′.0; mag 11.8 star N 3′.6.
10 07 58.8	−02 29 53	IC 592	13.4	0.8 x 0.7	12.6	Sbc	30	Mag 7.69v star S 1′.5; **CGCG 8-41** W 6′.0; IC 593 E 5′.1.
10 08 18.1	−02 31 35	IC 593	13.6	0.8 x 0.6	12.7	S?	81	Mag 7.69v star W 5′.3.
10 08 32.1	−00 40 04	IC 594	13.8	1.0 x 0.5	12.9	SB(r)bc	127	Mag 9.70v star N 2′.1.
10 13 12.5	−05 37 46	IC 599	14.3	1.1 x 0.3	13.0	Sbc	36	Mag 11.7 star E 5′.4.
10 17 10.8	−03 29 52	IC 600	12.4	2.3 x 1.2	13.4	SB(s)dm III-IV	27	Mag 8.31v star ESE 5′.6.
10 19 25.0	−05 39 22	IC 603	13.6	1.2 x 0.9	13.6	(R)SB(r)a:	157	Mag 9.26v star N 1′.3.
10 22 24.1	+01 11 52	IC 605	13.9	0.6 x 0.5	12.4	S?	12	Mag 9.14v star SW 5′.2; mag 8.44v star SE 6′.0.
10 25 35.2	−02 12 55	IC 609	13.3	1.5 x 0.7	13.2	SAB(rs)bc pec II	28	Mag 9.53v star SE 9′.7.
10 26 51.9	−03 27 54	IC 614	14.3	0.8 x 0.4	13.1	E	27	Mag 8.00v star ESE 11′.3.
10 35 21.1	+03 33 27	IC 623	14.2	1.1 x 0.4	13.2	S	152	Mag 9.74v star N 3′.4.
10 37 19.9	−03 21 29	IC 627	14.2	0.6 x 0.6	13.0	Sab		Mag 9.83v star NE 8′.0.
10 37 36.1	+05 36 11	IC 628	13.7	1.1 x 0.8	13.4	SAB(s)ab	118	Elongated **UGC 5779** N 1′.7.
10 39 11.7	−00 24 35	IC 632	13.8	0.9 x 0.5	12.8	Sa?	30	Mag 12.3 star W 4′.1; IC 633 NE 3′.3.
10 39 24.2	−00 23 21	IC 633	14.3	0.7 x 0.3	12.5	SB(s)a:	102	Mag 12.3 star N 1′.0; IC 632 SW 3′.3.
10 40 55.0	+05 59 27	IC 634	14.2	1.2 x 0.4	13.3	Pec?	116	Mag 10.91v star N 7′.1.
10 41 50.6	+04 19 49	IC 636	14.2	1.0 x 0.4	13.1	SB	48	Mag 8.63v star S 2′.2.
10 50 52.1	+01 09 48	IC 649	14.2	0.7 x 0.5	12.9	Interacting	171	Multi-galaxy system.
10 50 58.3	−02 09 01	IC 651	13.0	0.8 x 0.8	12.4	SB(s)m pec: III		Mag 9.65v star N 5′.7.
10 52 06.6	−00 33 39	IC 653	12.8	1.9 x 0.9	13.2	S0/a:	55	
10 57 53.6	−04 54 18	IC 657	14.7	1.3 x 0.5	14.0	SB(rs)a?	165	Mag 9.84v star NNW 7′.2.
11 07 31.8	+00 46 56	IC 671	13.4	1.3 x 1.0	13.5	SAB(r)bc: III	22	Mag 10.16v star SE 8′.4.
11 09 25.4	−00 05 54	IC 673	13.4	1.7 x 0.7	13.4	(R′)SAB(rs)c I-II	165	
11 17 54.9	−01 56 51	IC 680	14.0	0.7 x 0.4	12.5	(R′)SB(rs)a?	159	Mag 9.68v star W 10′.4.
11 26 48.5	−05 00 17	IC 693	13.8	0.9 x 0.4	12.6	Sab		Mag 10.31v star E 10′.3.
10 08 28.8	+00 22 02	MCG +0-26-22	14.0	0.9 x 0.8	13.7	E	159	Mag 12.5 star SW 1′.9.
10 11 44.8	−02 43 52	MCG +0-26-27	14.4	0.9 x 0.8	13.9	Scd	63	
10 12 02.1	−02 08 48	MCG +0-26-28	14.3	0.5 x 0.5	12.8	E		
10 12 48.5	−02 41 16	MCG +0-26-29	14.1	0.7 x 0.4	12.6	S0	159	
10 13 45.2	−02 48 31	MCG +0-26-31	15.1	0.7 x 0.4	13.6	Sb	18	
10 20 16.7	+01 12 13	MCG +0-27-1	15.0	0.5 x 0.5	13.4	Sc		Mag 12.0 star W 6′.4.
10 21 40.3	−03 27 15	MCG +0-27-2	13.8	0.4 x 0.4	11.9	E		Compact; pair of mag 12-13 stars N 1′.9.
10 23 13.4	−03 14 06	MCG +0-27-4	13.9	0.7 x 0.3	12.3	E?	3	
10 23 56.7	−03 10 59	MCG +0-27-5	12.4	1.9 x 1.2	13.2	SB(r)b I-II	150	Mag 12.1 star NE 3′.2.
10 24 43.3	−00 52 51	MCG +0-27-7	15.1	0.7 x 0.4	13.6	Sdm		
10 25 42.0	+00 45 41	MCG +0-27-10	14.6	0.7 x 0.5	13.3	S0/a	150	Pair mag 12.9 and 13.6 stars NW 1′.7.
10 26 06.2	−03 24 49	MCG +0-27-11	14.3	0.9 x 0.4	13.2	E?	84	Mag 9.67v star E 14′.0.
10 26 34.6	−02 49 52	MCG +0-27-13	14.1	1.4 x 0.4	13.3	Sd pec?	12	
10 27 13.9	−02 34 00	MCG +0-27-16	14.9	0.9 x 0.6	14.1		135	
10 27 29.6	+01 14 43	MCG +0-27-19	13.8	0.8 x 0.3	12.3	E	156	UGC 5667 N edge; **MCG +0-27-17** W 1′.2.
10 29 30.2	−03 45 39	MCG +0-27-22	13.8	0.6 x 0.6	12.8	E?		**CGCG 9-57** NW 1′.6.
10 30 10.7	−03 09 49	MCG +0-27-23	12.7	1.9 x 1.2	13.5	SAB(rs)0^+:	42	Stellar **PGC 30953** 1′.2 SW from center of core.
10 30 30.9	−02 48 46	MCG +0-27-24	15.0	0.7 x 0.7	14.1			Mag 10.28v star S 4′.3.
10 30 42.8	−01 49 37	MCG +0-27-25	14.3	0.9 x 0.4	13.0	Sbc	168	Close pair of mag 12-13 stars W 0′.9.
10 32 05.3	−02 26 27	MCG +0-27-27	14.9	0.8 x 0.4	13.5		3	
10 41 13.1	−02 48 13	MCG +0-27-40	13.5	1.5 x 1.2	14.2	E		Pair of mag 14.3 stars on SE edge.
10 43 58.2	+01 56 03	MCG +0-28-2	15.4	0.9 x 0.7	14.7	SBc	33	Mag 14.6 star SW 0′.9.
10 46 21.8	−02 17 17	MCG +0-28-7	14.3	0.8 x 0.6	13.5	E	93	Mag 13.6 star NE 3′.0.
10 46 28.4	−02 43 09	MCG +0-28-8	14.2	0.3 x 0.3	11.4	Sb		Compact; mag 12.9 star N 2′.2.
10 49 20.9	−00 40 02	MCG +0-28-13	13.4	0.7 x 0.4	11.9	SA0^- pec:	162	
10 49 41.2	+00 21 45	MCG +0-28-15	13.7	0.8 x 0.8	13.3	E?		Mag 10.26v star SW 2′.0.
10 49 53.0	−01 36 54	MCG +0-28-16	16.1	0.8 x 0.8	15.5	SBc		
10 49 59.0	+00 19 18	MCG +0-28-17	14.0	0.7 x 0.5	12.7	S?	66	Elongated **CGCG 10-31** N 1′.9.
10 51 25.0	−03 43 42	MCG +0-28-21	13.6	1.3 x 0.6	13.2	Sb: II	18	
10 52 10.4	−02 24 36	MCG +0-28-23	14.5	0.6 x 0.6	13.4	E		Mag 12.5 star ESE 2′.7.
11 02 29.4	−03 39 17	MCG +0-28-28	13.9	0.8 x 0.7	13.2	SA(s)a:		
11 04 41.5	−01 32 33	MCG +0-28-29	14.4	0.5 x 0.5	13.0	E		Mag 12.4 star NW 2′.5.
11 08 40.5	+00 16 02	MCG +0-29-1	15.0	0.7 x 0.4	13.5	S0	66	Mag 10.37v star SE 2′.6.
11 08 41.1	+00 23 27	MCG +0-29-2	14.8	0.3 x 0.3	12.1	S0		Compact; mag 12.1 star NW 2′.9.
11 11 43.0	−02 27 04	MCG +0-29-4	14.3	0.5 x 0.4	12.4	Sa	111	Mag 11.9 star SW edge; mag 8.38v star W 4′.8.
11 12 09.1	−02 20 19	MCG +0-29-5	13.7	0.3 x 0.3	10.9	Sbc		Mag 13.6 star ENE 1′.1; CGCG 11-24 S 3′.7.
11 12 56.6	+00 09 57	MCG +0-29-6	14.6	0.6 x 0.4	12.9	(R′)SB(s)c: III	130	
11 13 41.0	−02 36 30	MCG +0-29-7	13.7	0.9 x 0.9	13.3	SB(r)b pec? II-III		
11 14 50.5	−02 50 45	MCG +0-29-8	13.1	1.2 x 1.0	13.2	(R)SB(rs)0^+:	153	Slightly elongated N-S core, smooth envelope.
11 17 36.9	−03 10 15	MCG +0-29-10	13.7	1.2 x 0.5	13.0	SAB(r)ab I-II	177	Mag 13.5 star NE 2′.3.
11 18 49.5	+00 37 08	MCG +0-29-13	15.0	0.4 x 0.4	12.8	Sa		Mag 12.4 star SE 2′.2.
11 19 34.8	−02 29 28	MCG +0-29-15	14.2	0.8 x 0.6	13.3	SA(s)c III-IV	81	Forms a triangle with a pair of mag 13.4 and 14.4 stars S.
11 20 09.5	−03 03 23	MCG +0-29-17	13.6	1.6 x 0.8	13.7	SA(s)c II	147	Mag 13.1 star S 2′.3.
11 20 17.3	−03 42 36	MCG +0-29-18	15.6	0.4 x 0.4	13.5	Sc		Mag 14.2 star NE 1′.4.
11 20 20.7	−02 55 03	MCG +0-29-19	13.9	0.8 x 0.4	12.6	(R)SB(r)0/a	102	Mag 9.10v star ENE 3′.0.
11 20 51.0	−03 38 48	MCG +0-29-21	13.9	1.0 x 0.3	12.5	SA0^+:	39	
11 21 12.2	−02 59 05	MCG +0-29-23	14.0	1.4 x 0.7	13.8	SAB(s)b	75	Strong core, two faint arms.
11 23 50.7	+01 44 02	MCG +0-29-26	14.9	0.8 x 0.6	13.9		168	Mag 9.71v star NNE 8′.4.
11 24 08.8	−01 09 31	MCG +0-29-27	14.4	0.5 x 0.3	12.2	SAB(r)0°	57	Stellar nucleus; NGC 3662 NW 6′.4.
11 24 09.4	+00 42 01	MCG +0-29-28	13.9	0.9 x 0.4	12.7	S?	81	Mag 13.2 star SE 0′.6.
11 24 18.6	+00 38 35	MCG +0-29-29	13.9	0.6 x 0.6	12.6	SB(s)c II-III		
11 26 56.1	−01 19 17	MCG +0-29-33	14.3	0.6 x 0.4	12.6	SA(rs)bc II-III	30	Mag 10.73v star NE 3′.0.
11 27 15.6	+01 31 59	MCG +0-29-35	15.9	0.4 x 0.4	13.8			
10 09 46.4	+05 10 19	MCG +1-26-16	14.7	0.9 x 0.7	14.0	Sbc	81	
10 10 37.2	+05 08 58	MCG +1-26-17	16.6	0.7 x 0.5	15.3		15	Mag 9.49v star SE 2′.4.
10 12 41.5	+04 49 16	MCG +1-26-20	15.3	0.9 x 0.9	14.9	Sb		Mag 13.7 star on SE edge; UGC 5506 SE 3′.0.
10 16 00.0	+04 57 15	MCG +1-26-30	15.2	0.4 x 0.2	12.3	Compact		= **Mkn 719**.

(Continued from previous page)

GALAXIES

RA h m s	Dec ° ′ ″	Name	Mag (V)	Dim ′ Maj x min	SB	Type Class	PA	Notes
10 22 31.3	+04 34 54	MCG +1-27-5	13.7	0.7 x 0.4	12.2			
10 22 51.1	+04 46 01	MCG +1-27-6	14.0	1.1 x 0.8	13.7	Sbc	6	
10 24 53.1	+03 50 47	MCG +1-27-7	14.1	0.9 x 0.5	13.1		18	
10 27 33.5	+04 28 00	MCG +1-27-10	15.3	0.5 x 0.5	13.7			Mag 9.47v star E 2′.6; mag 11.2 star SW 1′.1.
10 35 51.2	+03 53 27	MCG +1-27-18	14.1	1.0 x 0.6	13.4		120	
10 36 35.8	+05 54 33	MCG +1-27-20	13.8	0.7 x 0.4	12.3		108	Close pair of mag 14 stars WNW 2′.2.
10 42 27.5	+02 45 38	MCG +1-27-30	14.2	1.1 x 0.5	13.4		153	Mag 8.40v star NW 4′.3.
10 43 39.0	+04 58 12	MCG +1-28-1	15.1	1.0 x 0.5	14.2	Sb	138	Mag 12.8 star NE 2′.3.
10 44 02.3	+04 39 44	MCG +1-28-3	14.2	0.6 x 0.6	12.9	Sb		Pair of mag 13.5 stars E and NE 1′.8.
10 55 36.0	+04 22 37	MCG +1-28-20	14.0	0.8 x 0.5	12.9		144	
11 02 30.7	+02 37 01	MCG +1-28-29	14.3	0.6 x 0.4	12.6	SB(s)bc pec	90	Pair with MCG +1-28-30 SE 1′.1.
11 02 34.0	+02 36 18	MCG +1-28-30	14.5	0.4 x 0.4	12.4	SB(s)bc pec:		Pair with MCG +1-28-29 NW 1′.1.
11 03 35.5	+03 20 11	MCG +1-28-32	17.6	0.7 x 0.5	16.3		6	
11 06 10.9	+04 20 10	MCG +1-28-38	13.4	1.1 x 0.7	13.0	SA0⁻	147	Mag 12.2 star SW 2′.2.
11 08 24.0	+03 29 51	MCG +1-29-3	14.4	0.7 x 0.3	12.6	SB(s)b: I-II	159	Mag 7.82v star NW 3′.4.
11 10 25.9	+03 59 08	MCG +1-29-7	14.0	1.0 x 0.5	13.1	SB(r)a	99	
11 10 45.0	+04 30 35	MCG +1-29-9	13.4	1.1 x 1.0	13.4	(R)SB(r)0/a	117	Strong N-S bar; mag 10.95v star SE 3′.7; CGCG 39-33 NNW 7′.2.
11 11 20.7	+05 49 46	MCG +1-29-12	13.8	0.8 x 0.6	12.8	S0?	132	Almost stellar; NGC 3567 on NW edge.
11 13 08.0	+05 04 23	MCG +1-29-16	14.1	0.9 x 0.6	13.2	SAB(s)c II-III	21	
11 13 12.6	+05 11 48	MCG +1-29-17	15.2	0.9 x 0.9	14.8	SBc		**CGCG 39-70** NE 2′.8.
11 14 59.8	+03 36 05	MCG +1-29-22	14.4	0.6 x 0.6	13.4	E/S0		
11 21 10.1	+05 21 47	MCG +1-29-35	13.3	1.0 x 1.0	13.3	E		
11 22 27.5	+04 17 08	MCG +1-29-38	14.1	0.7 x 0.5	12.8	SA(rs)bc pec:	21	
11 23 18.9	+03 57 17	MCG +1-29-39	14.7	0.3 x 0.3	12.2	E		Almost stellar; mag 12.8 star SE 0′.8.
10 11 00.1	−04 41 50	MCG −1-26-30	11.5	5.5 x 4.5	14.8	IBm V-VI	0	**Sextans A**. Bright star superimposed NE; strong clustering of knots E edge.
10 16 54.5	−05 52 15	MCG −1-26-38	14.4	0.5 x 0.5	12.9	E		Mag 9.12v star NNW 2′.0.
10 19 13.0	−05 37 48	MCG −1-26-40	14.3	0.9 x 0.8	13.8	SBa	55	Mag 9.26v star E 3′.4.
10 24 07.2	−05 37 57	MCG −1-27-7	14.5	1.9 x 0.3	13.8	Sc? sp	75	Mag 10.56v star WSW 6′.1.
10 29 38.8	−03 50 48	MCG −1-27-9	14.4	1.0 x 0.8	14.0	SA(r)c: III	6	Mag 7.72v star SSE 4′.9.
10 29 59.1	−04 59 05	MCG −1-27-10	14.6	0.8 x 0.7	13.9	Sc	87	
10 32 16.7	−05 02 23	MCG −1-27-14	14.1	0.5 x 0.3	12.1	E	18	
10 35 45.2	−03 51 59	MCG −1-27-24	14.6	1.1 x 0.2	12.8	Sbc	153	
10 39 46.5	−05 28 58	MCG −1-27-31	13.6	0.9 x 0.7	13.0	SBc	138	Mag 12.7 star on S edge.
10 54 20.4	−04 20 54	MCG −1-28-5	14.0	1.1 x 0.3	12.7	Sab:	42	
10 58 05.8	−04 45 35	MCG −1-28-11	14.9	1.4 x 0.2	13.4	Sc sp	27	Mag 9.84v star SW 6′.9.
11 06 46.8	−05 54 47	MCG −1-28-25	13.9	0.7 x 0.4	12.6	E/S0	129	
11 06 57.2	−05 36 22	MCG −1-28-26	13.9	0.7 x 0.5	12.6	SBab	123	
11 11 16.1	−04 50 31	MCG −1-29-4	13.4	0.6 x 0.4	11.8	S0/a	54	Mag 9.18v star E 0′.9.
11 12 15.1	−05 45 16	MCG −1-29-5	14.5	1.3 x 0.5	13.9	(R′)SB(s)b: II	81	
11 20 39.4	−05 40 21	MCG −1-29-10	13.9	1.0 x 0.7	13.4	S0	36	Contains several knots, or superimposed stars?
11 21 48.0	−05 45 32	MCG −1-29-12	13.6	0.5 x 0.5	12.1	E		Mag 7.81v star N 2′.5.
11 24 50.8	−05 21 09	MCG −1-29-17	13.9	0.6 x 0.5	12.5		156	
11 25 02.2	−05 04 09	MCG −1-29-18	13.6	0.7 x 0.5	12.3	Sa	162	
11 25 35.7	−04 59 19	MCG −1-29-20	14.4	0.7 x 0.5	13.1	SBc	78	
10 12 41.3	+03 07 42	NGC 3156	12.3	1.9 x 0.9	12.8	S0:	47	A mag 9.39v stars lies 2′.1 SE.
10 13 31.3	+03 22 31	NGC 3165	13.9	1.5 x 0.7	13.8	SA(s)dm:	177	NGC 3166 NE 4′.6; mag 10.51v star W 4′.7.
10 13 45.8	+03 25 32	NGC 3166	10.4	4.8 x 2.3	12.8	SAB(rs)0/a	87	NGC 3165 SW 4′.6.
10 14 14.5	+03 28 03	NGC 3169	10.2	4.4 x 2.8	12.8	SA(s)a pec	45	Mag 10.8 star on E edge.
10 26 21.6	−02 37 21	NGC 3243	12.7	1.4 x 1.1	13.0	S0° pec:	125	
10 26 41.5	+03 51 37	NGC 3246	12.6	2.4 x 1.3	13.7	SABdm	100	Almost stellar nucleus; mag 6.67v star NW 9′.1.
10 39 20.1	−00 12 03	NGC 3325	12.7	1.3 x 1.1	13.0	E:	55	Stellar **PGC 31707** SE 4′.7.
10 39 32.0	+05 06 25	NGC 3326	13.7	0.6 x 0.6	12.4	Sa		Mag 10.23v star N 5′.5; note: **PGC 31711** is 1′.1 S of this star.
10 41 47.7	+04 59 15	NGC 3337	14.3	0.8 x 0.5	13.1		45	
10 42 17.7	−00 22 37	NGC 3340	13.0	1.0 x 0.9	12.7	SB(s)bc I-II	145	Faint, slightly elongated **PGC 31922** ENE 6′.5.
10 42 31.5	+05 02 41	NGC 3341	14.0	1.2 x 0.4	13.1	Pec	24	Faint star on SW edge.
10 46 12.5	+01 48 56	NGC 3365	12.6	4.5 x 0.8	13.8	Scd: sp III	159	
10 48 11.7	+04 55 37	NGC 3385	12.6	1.5 x 0.9	12.8	S0	97	NGC 3386 N 4′.3; **CGCG 38-17** and 38-19 SSE 7′.9.
10 48 12.1	+04 59 53	NGC 3386	13.8	0.5 x 0.5	12.4	E		Compact; NGC 3385 S 4′.3.
10 51 14.6	+05 50 21	NGC 3423	11.1	3.8 x 3.2	13.7	SA(s)cd II-III	10	Many knots and superimposed stars.
10 51 58.0	+03 47 30	NGC 3434	12.0	2.1 x 1.9	13.4	SA(r)b I	41	**CGCG 38-37** 2′.6 N; **CGCG 38-35** 2′.3 SW.
11 01 15.9	+03 37 38	NGC 3495	11.8	4.9 x 1.2	13.6	Sd: III	20	Elongated **CGCG 38-87** lies 3′.9 S of the center.
11 04 24.6	+04 49 42	NGC 3509	12.7	2.1 x 1.0	13.4	SA(s)bc pec	40	Curving extension forms a distorted "U" shape with open end facing SW.
11 05 49.0	−00 02 10	NGC 3521	9.0	11.0 x 5.5	13.3	SAB(rs)bc II-III	163	Mag 7.90v star E 9′.8; **CGCG 10-72** SSW 8′.2 from center.
11 08 34.2	+04 49 49	NGC 3535	13.5	1.7 x 0.8	13.7	SA(s)a pec:	178	CGCG 39-17 ESE 8′.3.
11 11 18.7	+05 50 08	NGC 3567	13.3	0.9 x 0.7	12.6	S0 pec	132	MCG +1-29-12 on SE edge.
11 13 16.0	+03 39 27	NGC 3580	14.0	1.0 x 0.4	12.9	SA0° sp	177	
11 15 33.2	+05 06 53	NGC 3601	13.8	0.5 x 0.5	12.1	SB(r)ab?		
11 17 30.1	+04 33 17	NGC 3611	12.2	2.1 x 1.6	13.3	SA(s)a pec	24	**UGC 6306** and mag 10.40v star N 3′.2, **UGC 6306** is on E edge of star.
11 20 17.0	+02 57 49	NGC 3630	12.0	2.1 x 0.9	12.6	S0 sp	37	A mag 8.07v star lies SE 9′.9; there are several faint, stellar anonymous galaxy midway to this star.
11 20 26.3	+03 35 07	NGC 3633	13.5	1.1 x 0.4	12.5	SAa: sp	72	Mag 9.05v star NE 3′.7.
11 21 06.7	+03 14 06	NGC 3640	10.4	4.0 x 3.2	13.2	E3	100	NGC 3641 on S edge.
11 21 08.9	+03 11 41	NGC 3641	13.2	0.9 x 0.9	13.0	E pec		Compact; a pair of stellar anonymous galaxies W 2′.1.
11 21 25.0	+03 00 49	NGC 3643	14.1	1.2 x 0.8	13.9	SB(r)0⁺: sp	99	Mag 13.2 star on SE edge; mag 11.33v star E 5′.4.
11 21 33.0	+02 48 36	NGC 3644	13.7	1.5 x 0.7	13.6	(R′)SBa pec:	48	Mag 12.5 star on S edge; stellar **IC 683** S 3′.6; **IC 683E** lies 1′.3 W of IC 683.
11 21 40.0	+02 57 51	NGC 3645	14.6	0.6 x 0.6	13.3			Stellar; a small, faint anonymous galaxy S 1′.3.
11 21 35.4	+02 53 36	NGC 3647	14.6	0.3 x 0.3	12.1	E		**CGCG 39-141** SW 0′.8; stellar anonymous galaxy E 0′.9; and another stellar anonymous galaxy NE 0′.9.
11 23 46.6	−01 06 15	NGC 3662	12.9	1.4 x 0.9	13.0	SAB(r)bc pec I-II	25	Mag 9.56v star NW 8′.6; MCG +0-29-27 SE 6′.4.
11 24 25.8	+03 19 33	NGC 3664	12.8	2.0 x 1.9	14.1	SB(s)m pec IV-V	42	Brighter along W edge; NGC 3664A S 6′.3; **UGC 6417** S 4′.7.
11 24 25.1	+03 13 20	NGC 3664A	14.3	1.1 x 1.0	14.2	(R)SB(s)m: V	15	NGC 3664 N 6′.3; CGCG 39-173 ESE 8′.2.
11 26 08.6	−05 35 15	NGC 3679	14.2	1.0 x 0.5	13.3	Sbc III:	178	
10 10 00.0	−02 27 51	UGC 5483	13.9	0.9 x 0.8	13.4	S?	9	
10 10 27.9	+02 13 38	UGC 5487	14.1	1.0 x 0.5	13.2	SBbc	70	
10 11 17.9	+00 26 29	UGC 5493	13.3	1.5 x 1.2	13.8	SAB(rs)c II	15	
10 12 11.7	+04 55 21	UGC 5501	13.9	0.7 x 0.5	12.6	S?	138	Mag 11.6 star on e edge.
10 12 51.4	+04 47 35	UGC 5506	14.0	1.3 x 0.6	13.6	S?	17	Stellar nucleus, uniform envelope; MCG +1-26-20 NW 3′.0.
10 13 38.4	−00 55 34	UGC 5515	12.6	1.4 x 1.1	13.1	E⁺ pec:	90	Mag 9.08v star W 3′.5; located inside W edge of galaxy cluster A 957, numerous PGC galaxies in immediate area.

(Continued from previous page)

112 GALAXIES 112

RA h m s	Dec ° ′ ″	Name	Mag (V)	Dim ′ Maj x min	SB	Type Class	PA	Notes
10 13 52.7	+00 33 00	UGC 5521	13.9	0.9 x 0.8	13.4	SAB(s)c		Pair of stars, mags 10.5v and 10.6v, SW 2′.7.
10 14 39.4	−00 49 56	UGC 5528	13.5	1.0 x 0.8	13.1	SAB(r)a pec:	150	Located on the E edge of galaxy cluster A 957; **PGC 29868** SW 4′.3.
10 15 55.5	+02 41 15	UGC 5539	14.7	2.0 x 1.0	15.3	IBm:	42	Dwarf, low surface brightness.
10 16 20.6	+04 49 16	UGC 5543	13.6	1.3 x 0.7	13.4	Sc	167	
10 20 33.5	−02 28 18	UGC 5586	14.9	1.5 x 0.8	14.9	Double System	63	Double system with envelopes in contract, resembles a double star.
10 22 24.4	+03 59 50	UGC 5607	13.2	1.7 x 0.6	13.1	Sb II-III	62	Stellar **CGCG 37-13** SW 6′.7.
10 26 08.8	+04 22 20	UGC 5648	14.1	1.2 x 0.5	13.4	S	10	**CGCG 37-36** SE 8′.0.
10 27 29.8	+01 15 14	UGC 5667	14.0	0.6 x 0.5	12.5	Double System	57	MCG +0-27-19 S edge; **MCG +0-27-17** SW 1′.4.
10 28 38.3	+03 33 34	UGC 5677	14.6	1.5 x 0.2	13.1	Sdm:	6	Razor thin; mag 14.0 star partially superimposed on E edge.
10 28 42.1	+03 40 37	UGC 5678	13.8	1.0 x 0.7	13.2	Sbc	50	Mag 12.3 star SW 2′.4.
10 30 04.8	+05 02 49	UGC 5702	14.3	0.9 x 0.5	13.3	S	100	
10 31 13.8	+04 27 59	UGC 5708	13.2	3.1 x 0.6	13.7	SBd:	168	Mag 14.1 star N edge.
10 31 35.2	+00 28 29	UGC 5715	13.4	1.0 x 0.6	12.7	Sbc	155	
10 33 51.3	−00 33 40	UGC 5736	13.9	1.0 x 0.8	13.5	Sbc	80	
10 34 44.9	−01 58 11	UGC 5745	12.9	1.3 x 1.0	13.0	SB(rs)0^+	126	Mag 14.0 star superimposed S of center.
10 36 57.4	+00 13 45	UGC 5772	13.7	0.9 x 0.5	12.7	S?	88	
10 38 05.3	+01 44 41	UGC 5783	14.1	1.1 x 0.4	13.0	SB(rs)c? III	54	
10 38 07.0	+04 53 38	UGC 5784	14.1	1.3 x 0.3	13.0	S?	125	
10 38 26.5	−02 34 11	UGC 5787	13.3	1.8 x 0.6	13.2	SBbc: II-III	136	Mag 8.90v star NE 5′.2; CGCG 9-90 SW 6′.1.
10 38 46.9	+05 41 46	UGC 5788	13.6	1.2 x 0.7	13.3	Sb III	10	Mag 7.57v star NNW 2′.5.
10 39 04.8	+04 38 49	UGC 5790	14.7	1.1 x 0.8	14.4	S?	6	
10 39 25.2	+01 43 02	UGC 5797	13.7	0.9 x 0.8	13.2	Im:	156	
10 41 53.5	+00 47 39	UGC 5823	13.9	0.9 x 0.7	13.2	Im:	165	Mag 7.53v star NE 8′.2.
10 43 48.8	−02 31 44	UGC 5847	14.1	1.2 x 0.6	13.6	(R′)SB(s)b:	5	
10 43 52.7	−01 17 35	UGC 5849	13.3	1.0 x 0.6	12.6	I?	9	Strong knot N end.
10 45 10.1	+04 56 37	UGC 5865	14.2	1.0 x 0.6	13.5	Scd?	33	
10 45 04.5	+00 06 14	UGC 5867	13.6	0.8 x 0.7	12.8	SBb	36	**MCG +0-28-5** SE 2′.1.
10 46 51.7	−01 23 33	UGC 5886	13.4	1.0 x 0.6	12.7	SB?	130	
10 47 30.6	−01 29 35	UGC 5896	13.9	1.4 x 0.5	13.3	Sab	132	Mag 12.4 star E 1′.3; **PGC 32240** SSE 2′.9.
10 48 18.8	+00 50 26	UGC 5913	14.0	0.9 x 0.4	12.7	S?	110	Mag 14.0 star ESE 1′.4; CGCG 10-16 W 6′.3.
10 49 01.5	−00 38 18	UGC 5922	13.6	1.0 x 0.6	12.9	SB?	10	Stellar **PGC 32358** SSE 2′.0.
10 49 28.3	+04 47 56	UGC 5929	14.0	1.1 x 0.5	13.2	Sbc	68	
10 50 13.5	−01 17 25	UGC 5943	13.3	1.1 x 0.8	13.0	SAB(r)c II	3	Mag 11.8 star on W edge.
10 51 35.0	+04 35 01	UGC 5974	13.3	1.9 x 0.7	13.5	Scd:	130	Mag 12.8 star SW 1′.2.
10 53 04.1	+04 37 53	UGC 6003	13.7	0.5 x 0.5	12.2	E?		
10 56 02.3	+02 30 37	UGC 6040	14.1	0.9 x 0.5	13.1	S	155	Mag 11.7 star on N edge.
10 57 31.1	+05 41 51	UGC 6049	14.0	1.1 x 0.7	13.5	Sab	67	
10 58 11.0	+01 36 13	UGC 6057	14.0	1.0 x 0.6	13.3	Triple System	144	Two small anonymous galaxies on E edge; a third NE 1′.3.
10 59 14.8	+05 17 28	UGC 6068	13.7	1.3 x 0.6	13.3	SA0^+ pec?	5	
11 00 32.6	+02 06 55	UGC 6087	14.1	0.8 x 0.7	13.3	SBb	0	
11 03 14.6	+03 20 30	UGC 6119	14.3	0.9 x 0.4	13.1	S0°	21	Mag 10.71v star WSW 3′.7.
11 04 39.1	+05 11 57	UGC 6141	13.8	1.1 x 0.6	13.2	SAB(r)b: I-II	13	Bright core, faint envelope; mag 12.1 star on E edge.
11 04 42.0	+04 17 49	UGC 6142	14.0	1.2 x 0.6	13.5	S? I-II	141	**UGC 6141A** E 0′.9.
11 06 02.5	+04 25 43	UGC 6155	13.2	1.4 x 1.1	13.5	SAB(rs)c II-III	153	
11 10 01.6	+05 18 13	UGC 6210	13.5	1.5 x 0.7	13.4	SAB(rs)bc II-III	15	Stellar nucleus; stellar **CGCG 39-21** NW 6′.1.
11 10 24.6	+04 49 47	UGC 6212	13.1	1.1 x 0.8	12.8	Sb	69	Anonymous galaxy S edge; faint anonymous galaxy S 2′.3; Almost stellar anonymous galaxy SW 2′.5.
11 11 59.2	+03 08 07	UGC 6239	13.3	1.1 x 0.9	13.3	E	70	**CGCG 39-65** E 2′.0; mag 10.68v star WSW 2′.1.
11 13 52.8	+04 17 27	UGC 6260	14.0	1.2 x 0.4	13.1	Sbc: sp III	43	Mag 8.02v star E 6′.4; **CGCG 39-86** SE 11′.1.
11 16 17.2	+02 53 20	UGC 6289	13.6	0.8 x 0.7	12.9	SAB(s)bc II	96	Mag 11.8 star W 2′.3.
11 17 55.2	−02 05 33	UGC 6311	13.3	1.4 x 1.1	13.6	SAB(s)c pec I	33	Pair of N-S oriented mag 9.68v, 9.02v stars WNW 13′.3. CGCG 11-48 E 10′.0.
11 18 56.2	+00 10 32	UGC 6329	13.8	1.1 x 1.0	13.8	SAB(rs)c: III	90	Mag 9.50v star SE 2′.7.
11 19 55.3	−00 52 48	UGC 6340	13.0	1.7 x 0.7	13.1	SAB(rs)bc: II-III	155	
11 20 17.2	+02 31 26	UGC 6345	13.4	2.3 x 1.4	14.6	IB(s)m V	75	Irregular, many knots and clumps.
11 20 35.8	−01 29 37	UGC 6359	13.7	0.8 x 0.5	12.6	SAB0 pec	66	Almost stellar anonymous galaxy NE 0′.7; mag 12.7 star N 1′.0.
11 20 47.8	+00 27 53	UGC 6361	13.4	1.4 x 0.7	13.2	Triple System	147	Triple system; E component much elongated; west component slightly elongated; S component round and compact.
11 23 19.2	−00 55 20	UGC 6402	14.3	0.8 x 0.2	12.2	Sdm? sp	93	
11 23 59.8	+02 41 30	UGC 6413	13.5	1.0 x 1.0	13.3	SAB(rs)bc: II-III		Anonymous galaxy W 2′.4.
11 25 17.8	+00 20 59	UGC 6432	14.0	0.7 x 0.6	12.9	SA(rs)bc: II	78	
11 25 34.9	−00 46 05	UGC 6435	12.9	1.1 x 1.1	13.0	S0°:		Mag 9.07v star E 3′.2.
11 26 01.2	+01 59 03	UGC 6440	13.6	1.0 x 0.7	13.1	SB(rs)bc II	43	
11 27 12.8	−00 59 39	UGC 6457	13.9	1.2 x 1.1	14.0	Im:		

GALAXY CLUSTERS

RA h m s	Dec ° ′ ″	Name	Mag 10th brightest	No. Gal	Diam ′	Notes
10 14 00.0	−00 54 00	A 957	15.9	55	20	See notes for UGC 5515, UGC 5528.
10 21 54.0	−04 57 00	A 993	14.9	36	17	**MCG −1-27-4** NW of center 10′.2; other members anonymous and mostly stellar.
10 30 18.0	+04 00 00	A 1032	15.7	31	19	**CGCG 37-50** W of center 3′.4; most other members anonymous and stellar to almost stellar.
10 58 06.0	+01 29 00	A 1139	15.0	36	34	Most members stellar to almost stellar, see notes for UGC 6057.
11 07 30.0	+02 56 00	A 1171	16.2	43	25	**CGCG 38-133** SSW of center 2′.1; all other members anonymous.
11 17 42.0	−04 28 00	A 1216	16.0	57	19	All members anonymous, faint and stellar.
11 23 00.0	+01 05 00	A 1238	16.0	63	16	All members anonymous, faint and stellar.

113 GALAXIES 113

RA h m s	Dec ° ′ ″	Name	Mag (V)	Dim ′ Maj x min	SB	Type Class	PA	Notes
08 57 47.6	−00 12 00	CGCG 5-42	14.6	0.5 x 0.4	12.7	S0	57	Mag 8.27v star W 4′.8.
09 03 30.6	−01 22 51	CGCG 5-51	14.8	0.6 x 0.4	13.1	S0/a	66	Multi-galaxy system?
09 06 39.5	−00 51 56	CGCG 5-56	14.5	0.3 x 0.2	11.5	S0	90	Mag 12.3 star E 2′.3.
09 08 54.4	+01 25 53	CGCG 6-6	15.4	0.6 x 0.3	13.4	Sm	12	Small triangle of mag 13-14 stars SSE 1′.3; mag 11.9 star WNW 3′.8.
09 09 25.2	−01 44 36	CGCG 6-9	14.4	0.5 x 0.3	12.2			Mag 12.2 star E 0′.9.
09 14 11.7	−02 27 40	CGCG 6-19	14.9	0.3 x 0.3	12.1			Mag 10.7 star SSE 4′.4; mag 9.23v star WNW 10′.6.
09 17 45.1	+01 03 15	CGCG 6-27	15.7	0.5 x 0.4	13.8	Sc	9	Mag 10.15v star SSE 3′.1.
09 57 19.5	−02 25 03	CGCG 8-4	13.6	0.6 x 0.4	11.9	Sb	78	Mag 8.41v star NW 11′.5.

(Continued from previous page)

GALAXIES

RA h m s	Dec ° ′ ″	Name	Mag (V)	Dim ′ Maj x min	SB	Type Class	PA	Notes
09 58 52.3	+01 03 34	CGCG 8-6	14.8	0.6 x 0.4	13.1	Sc:	162	Mag 9.61v star SE 2′.4; **UGC 5355** W 3′.5.
10 00 24.5	+00 42 11	CGCG 8-15	14.4	0.7 x 0.3	12.5		3	Mag 9.89v star NNE 8′.1.
10 02 30.9	+00 18 23	CGCG 8-27	14.6	0.7 x 0.4	13.1	S0/a	162	Mag 13.9 star on E edge.
10 03 57.0	+01 30 42	CGCG 8-30	14.8	0.3 x 0.3	12.2	E		Mag 11.8 star NNE 3′.9.
10 06 54.7	−01 14 02	CGCG 8-40	14.7	0.8 x 0.5	13.6	Sc	171	Mag 10.15v star SE 7′.1.
10 10 41.0	−02 25 58	CGCG 8-56	14.6	0.8 x 0.4	13.2	Sc		Mag 13.6 star 1′.4 E.
10 11 05.2	+01 13 17	CGCG 8-60	13.2	1.0 x 0.4	12.1	SB(s)bc pec	9	Dark knot or hole S of center.
10 14 46.3	−02 11 49	CGCG 8-80	15.0	0.4 x 0.4	12.8			Mag 14.0 star S 1′.2; mag 13.7 star E 1′.2.
09 02 15.7	+03 10 07	CGCG 33-49	14.8	0.8 x 0.3	13.1		12	Mag 8.00v star NNW 7′.1.
09 02 29.5	+03 23 02	CGCG 33-50	14.8	0.8 x 0.2	12.7		69	**CGCG 33-47** NW 8′.0; mag 8.00v star SW 8′.8.
09 06 56.9	+03 10 46	CGCG 33-60	14.2	0.6 x 0.4	12.5		141	Mag 11.1 star on S edge; stellar anonymous galaxy SE 2′.9.
09 08 51.2	+03 26 53	CGCG 34-4	14.7	0.6 x 0.4	13.0		3	Mag 9.71v star W 5′.6.
09 23 08.1	+02 29 10	CGCG 34-41	14.7	0.8 x 0.7	13.9		156	UGC 4993 N 7′.1; mag 7.41v star SE 8′.9.
09 44 10.5	+02 24 34	CGCG 35-43	15.2	0.5 x 0.5	13.6			Located 3′.0W of a mag 9.53v star.
09 47 00.0	+02 33 36	CGCG 35-56	16.2	0.5 x 0.4	14.3		21	Patchy.
09 49 00.7	+04 18 10	CGCG 35-63	15.3	0.3 x 0.3	12.5			Mag 10.25v star NW 8′.1.
09 50 35.1	+04 46 22	CGCG 35-66	14.3	0.6 x 0.3	12.3		57	Close pair of mag 10.89v and 12.2 stars E 4′.7.
10 00 57.4	+03 08 30	CGCG 36-19	14.3	0.8 x 0.3	12.6		162	Mag 14.6 star on E edge; mag 9.37v star W 8′.0.
10 05 18.8	+02 29 42	CGCG 36-33	14.7	0.7 x 0.4	13.1		144	Mag 10.64v star WNW 1′.3.
10 07 20.6	+04 04 44	CGCG 36-41	15.0	0.5 x 0.4	13.1		117	Mag 12.4 star SE 2′.8; mag 11.9 star W 5′.7.
10 08 10.3	+02 27 46	CGCG 36-42	15.0	0.3 x 0.3	12.2			Forms a triangle with mag 11.8 star NE 2′.8 and mag 11.8 star SE 3′.2.
10 11 19.7	+05 53 45	CGCG 36-52	14.7	0.6 x 0.6	13.5			Mag 15.2 star 0′.5 S; mag 11.4 star N 5′.0.
10 15 16.8	+05 23 07	CGCG 36-72	14.8	0.6 x 0.3	12.8		27	Pair of mag 10.89v, 10.69v stars N 8′.2.
09 17 51.0	−00 16 41	IC 531	13.5	1.7 x 0.5	13.2	(R′)SB(rs)ab?	60	
09 21 15.4	+03 09 02	IC 534	14.1	1.7 x 0.2	12.8	Sb III	148	Mag 11.3 star SW 10.0.
09 29 08.3	−02 33 00	IC 539	13.4	1.0 x 0.9	13.2	Scd:	156	Mag 10.39v star W 8′.0.
09 40 43.2	+03 57 32	IC 549	14.7	0.6 x 0.3	12.7		3	Mag 10.92v star ESE 4′.4.
09 40 45.2	−05 26 09	IC 553	13.7	1.1 x 0.9	13.5	(R′)SB(rs)a pec:	129	Mag 10.24v star W 11′.4.
09 45 53.6	−00 16 07	IC 560	13.3	1.4 x 0.7	13.2	S0+:	15	
09 45 58.9	+03 08 39	IC 561	14.6	0.4 x 0.4	12.5			Mag 11.9 star SSW 3′.5.
09 46 03.8	−03 58 14	IC 562	14.2	1.4 x 0.3	13.1	Sbc? sp	150	Mag 10.13v star E 5′.9.
09 46 20.2	+03 02 43	IC 563	13.9	0.8 x 0.4	12.5	SB(r)ab: pec	108	Mag 11.2 star S 2′.9; IC 564 N 1′.6; UGC 5224 SW 8′.0.
09 46 21.1	+03 04 14	IC 564	13.4	1.7 x 0.4	12.8	SA(s)cd? pec	68	IC 563 S 1′.6.
09 49 56.5	−00 13 51	IC 566	14.5	0.6 x 0.4	12.9		45	Mag 7.12v star E 3′.9.
10 03 05.3	−02 24 03	IC 587	13.4	1.3 x 0.6	13.0	SAB(r)bc pec?	107	Mag 11.11v star WSW 6′.1.
10 02 07.2	+03 03 25	IC 588	13.7	0.7 x 0.4	12.2	(R)SB(r)a	159	Mag 10.89v star E 5′.1.
10 05 50.2	+00 37 57	IC 590	13.8	1.0 x 0.9	13.7	E	123	Double system.
10 07 58.8	−02 29 53	IC 592	13.4	0.8 x 0.7	12.6	Sbc	30	Mag 7.69v star S 1′.5; **CGCG 8-41** W 6′.0; IC 593 E 5′.1.
10 08 18.1	−02 31 35	IC 593	13.6	0.8 x 0.6	12.7	S?	81	Mag 7.69v star W 5′.3.
10 08 32.1	−00 40 04	IC 594	13.8	1.0 x 0.5	12.9	SB(r)bc	127	Mag 9.70v star N 2′.1.
10 13 12.5	−05 37 46	IC 599	14.3	1.1 x 0.3	13.0	Sbc	36	Mag 11.7 star E 5′.4.
08 58 30.7	+02 55 28	IC 2426	14.7	0.4 x 0.3	12.2		111	Stellar; mag 9.71v star S 4′.4.
09 27 28.7	+03 55 48	IC 2481	13.6	0.9 x 0.6	12.8	S?	160	Mag 11.2 star NNW 5′.5.
09 03 24.0	−02 34 37	MCG +0-23-20	13.8	0.5 x 0.3	11.8	E?	90	
09 14 20.2	−02 47 03	MCG +0-24-3	13.7	0.8 x 0.8	13.1	S?		
09 20 37.5	−00 40 46	MCG +0-24-8	14.4	0.5 x 0.4	12.5	Sb		
09 23 43.2	+01 20 50	MCG +0-24-11	14.1	0.7 x 0.6	13.1	Sa?	36	**MCG +0-24-12** S 1′.1.
09 23 47.8	+02 06 49	MCG +0-24-13	15.0	0.8 x 0.5	13.8	Sc		**CGCG 6-41** E 4′.3; NGC 2861 NW 3′.1.
09 27 57.1	−02 25 56	MCG +0-24-16	15.4	0.7 x 0.6	14.3		114	
09 36 26.2	−00 34 17	MCG +0-25-4	14.4	0.7 x 0.5	13.1	IAB(s)m: V	78	
09 36 35.5	+01 06 56	MCG +0-25-5	14.5	0.7 x 0.4	13.0		111	Mag 13.1 star SW 2′.1.
09 43 02.5	−02 15 12	MCG +0-25-9	14.3	0.4 x 0.4	12.2	S0		Mag 9.07v star W 2′.6.
09 48 18.5	−03 44 03	MCG +0-25-18	14.1	0.8 x 0.7	13.3	Irr	63	Mag 12.7 star SSE 2′.4.
09 50 26.0	+01 34 53	MCG +0-25-23	14.0	0.9 x 0.8	13.5	Sc	18	
09 52 11.6	+01 05 42	MCG +0-25-25	14.4	1.0 x 0.5	13.4	SBb	24	
09 59 26.7	−00 15 12	MCG +0-26-1	14.0	1.0 x 0.9	13.7	Sab		
10 00 10.5	+02 09 17	MCG +0-26-4	14.1	0.6 x 0.4	12.4		105	Mag 11.8 star E edge.
10 00 48.0	+01 14 17	MCG +0-26-9	15.2	0.9 x 0.9	14.8	Sbc		
10 04 24.4	−02 25 32	MCG +0-26-14	13.8	1.0 x 0.5	12.9	S?	66	Mag 13.6 star SW 2′.4.
10 05 24.0	+00 32 37	MCG +0-26-16	14.1	1.1 x 0.3	12.8	S0/a	111	Mag 13.3 star E 1′.2.
10 08 28.8	+00 22 02	MCG +0-26-22	14.0	0.9 x 0.8	13.7	E	159	Mag 12.5 star SW 1′.9.
10 11 44.8	−02 43 52	MCG +0-26-27	14.4	0.9 x 0.8	13.9	Scd	63	
10 12 02.1	−02 08 48	MCG +0-26-28	14.3	0.5 x 0.5	12.8	E		
10 12 48.5	−02 41 16	MCG +0-26-29	14.1	0.7 x 0.4	12.6	S0	159	
10 13 45.2	−02 48 31	MCG +0-26-31	15.1	0.7 x 0.4	13.6	Sb	18	
08 59 49.1	+05 03 36	MCG +1-23-16	14.1	0.6 x 0.4	12.4	S0?	3	Faint star on E edge; mag 12.3 star NE 2′.4.
09 03 40.0	+03 22 18	MCG +1-23-19	14.5	0.9 x 0.4	13.2	S?	24	
09 03 33.4	+03 22 32	MCG +1-23-20	13.8	1.0 x 0.5	12.9	SB0?	120	
09 04 03.9	+03 35 03	MCG +1-23-21	14.1	0.7 x 0.4	12.6	SB?	99	Mag 7.29v star E 9′.5.
09 07 55.0	+03 16 20	MCG +1-24-2	14.8	0.7 x 0.7	13.9	SBbc		Mag 9.81v star NE 3′.3.
09 18 58.1	+05 53 15	MCG +1-24-11	14.0	0.5 x 0.5	12.3			**CGCG 34-26** E 1′.6.
09 22 21.4	+04 03 38	MCG +1-24-15	14.7	0.5 x 0.4	12.8		39	
09 24 01.1	+05 12 33	MCG +1-24-19	14.7	0.9 x 0.7	14.0	Sbc	171	Mag 12.2 star S 1′.5.
09 28 12.2	+03 24 29	MCG +1-24-23	14.1	0.7 x 0.3	12.2	Sb	39	
09 49 08.8	+02 28 48	MCG +1-25-26	14.1	0.8 x 0.8	13.4	SBc		**CGCG 35-62** SW 2′.6.
09 52 00.7	+04 09 00	MCG +1-25-29	14.0	0.9 x 0.7	13.4		60	
09 54 10.5	+02 17 14	MCG +1-25-33	14.4	1.0 x 0.9	14.1	Sb	78	
09 56 58.7	+03 46 38	MCG +1-26-1	14.5	0.4 x 0.3	12.0		108	Mag 8.77v star W 7′.9.
09 56 59.1	+03 48 47	MCG +1-26-2	15.0	0.7 x 0.4	13.4	Sb		
10 02 59.0	+02 20 33	MCG +1-26-11	14.4	0.9 x 0.3	12.9		108	Two very faint extensions N and S.
10 03 57.3	+02 25 02	MCG +1-26-12	14.8	0.9 x 0.3	13.2	Sbc	63	Mag 13.3 star NW 1′.5, with almost stellar anonymous galaxy just S of this (check, text truncated?)
10 06 16.9	+03 49 40	MCG +1-26-15	14.2	0.9 x 0.4	13.0	Sb	159	Pair of stars, mags 10.70v and 9.90v, S 1′.0.
10 09 46.4	+05 10 19	MCG +1-26-16	14.7	0.9 x 0.7	14.0	Sbc	81	
10 10 37.2	+05 08 58	MCG +1-26-17	16.6	0.7 x 0.5	15.3		15	Mag 9.49v star SE 2′.4.
10 12 41.5	+04 49 16	MCG +1-26-20	15.3	0.9 x 0.9	14.9	Sb		Mag 13.7 star on SE edge; UGC 5506 SE 3′.0.
10 16 00.0	+04 57 15	MCG +1-26-30	15.2	0.4 x 0.2	12.3	Compact		= **Mkn 719**.
09 06 05.8	−04 21 24	MCG −1-23-17	13.4	1.5 x 0.7	13.3	SA(s)c? II-III	54	Mag 10.10v star S 4′.5.
09 06 34.3	−05 07 26	MCG −1-23-18	13.9	1.0 x 0.5	13.0		117	Mag 8.81v star E 7′.2.

(Continued from previous page)

GALAXIES

RA h m s	Dec ° ′ ″	Name	Mag (V)	Dim ′ Maj x min	SB	Type Class	PA	Notes
09 32 47.5	−05 43 29	MCG −1-25-2	13.3	1.5 x 0.9	13.5	S	51	Mag 10.29v star NW 3′5; a pair of faint, small anonymous galaxies SW 1′7; another faint, almost stellar anonymous galaxy NE 2′6.
09 38 53.4	−04 51 39	MCG −1-25-8	14.9	0.8 x 0.2	12.7	SB(rs)b pec sp	129	= **Hickson 40C**. **Hickson 40E** (= **MCG −1-25-11**) E 0′5.
09 38 53.5	−04 50 56	MCG −1-25-9	12.8	0.9 x 0.7	12.3	E	0	= **Hickson 40A**. **Hickson 40D** (= **MCG −1-25-12**) NE 1′0.
09 38 54.9	−04 51 58	MCG −1-25-10	14.0	0.5 x 0.4	12.2	SA(r)0⁻: pec	57	= **Hickson 40B**.
09 41 49.0	−04 18 34	MCG −1-25-19	14.2	0.5 x 0.4	12.3		15	Mag 8.19v star WNW 3′6.
09 42 12.2	−04 42 52	MCG −1-25-22	13.4	1.9 x 0.4	13.0	Sbc: II-III	129	**MCG −1-25-23** S 1′5.
09 43 14.0	−04 28 13	MCG −1-25-27	14.2	0.9 x 0.3	12.6	Sbc	171	**MCG −1-25-30** SE 1′3.
09 43 22.6	−05 17 02	MCG −1-25-31	15.4	1.3 x 0.2	13.8	SB(s)dm? sp	90	
09 43 27.6	−05 16 45	MCG −1-25-32	14.3	1.1 x 0.3	12.9	SB(s)m? sp	72	
09 43 28.9	−05 22 03	MCG −1-25-33	13.9	1.3 x 0.7	13.6	SB(s)d pec: III	144	
09 43 36.2	−05 54 51	MCG −1-25-34	13.2	2.1 x 1.4	14.2	SB(s)dm III-IV	9	Mag 9.28v star SSW 8′7.
09 48 48.4	−05 06 01	MCG −1-25-39	14.6	0.7 x 0.7	13.7			Stellar nucleus, uniform envelope; mag 10.10v star SW 7′7.
09 49 22.2	−05 11 29	MCG −1-25-44	14.5	0.9 x 0.7	13.8	S0/a	123	NGC 3022 E 4′6.
09 49 37.2	−04 55 58	MCG −1-25-45	14.0	0.7 x 0.5	12.9	E	99	Mag 9.13v star NW 4′3.
09 50 56.5	−04 59 53	MCG −1-25-49	13.5	2.2 x 0.7	13.8	(R)SB(r)a	21	Mag 10.04v star E 1′5.
09 51 27.3	−05 06 17	MCG −1-25-50	13.2	1.6 x 0.8	13.3	SB(r)ab	174	Mag 9.46v star NE 3′6.
09 51 28.4	−05 01 35	MCG −1-25-51	14.4	0.9 x 0.4	13.1			Mag 8.54v star N 3′2.
10 05 44.8	−04 50 16	MCG −1-26-19	14.9	0.7 x 0.4	13.3	SBcd	21	
10 07 05.3	−05 53 01	MCG −1-26-24	14.1	1.2 x 0.2	12.4	Sb sp	12	
10 11 00.1	−04 41 50	MCG −1-26-30	11.5	5.5 x 4.5	14.8	IBm V-VI	0	**Sextans A**. Bright star superimposed NE; strong clustering of knots E edge.
08 56 12.2	−02 33 43	NGC 2706	13.0	1.8 x 0.6	12.9	Sbc? sp III	167	
08 56 07.9	−03 21 41	NGC 2708	12.0	2.6 x 1.3	13.2	SAB(s)b pec?	25	Several bright knots N end.
08 56 12.8	−03 14 39	NGC 2709	13.7	0.8 x 0.6	12.7	SA0° pec:	81	Mag 9.40v star NE 5′6.
08 57 20.4	+02 55 14	NGC 2713	11.8	3.6 x 1.5	13.4	SB(rs)ab I-II	107	Lies 4′5 E of mag 9.52v star.
08 57 36.0	+03 05 25	NGC 2716	11.8	1.3 x 1.0	12.0	(R)SB(r)0⁺	30	Located on SW edge of galaxy cluster A 732.
08 58 56.6	−04 54 09	NGC 2721	12.2	2.3 x 1.6	13.5	SB(rs)bc pec II	153	Several faint stars just inside W edge.
08 58 46.3	−03 42 38	NGC 2722	12.7	2.0 x 1.3	13.6	SA(rs)bc pec: III	75	
09 00 14.4	+03 10 37	NGC 2723	13.2	0.8 x 0.8	12.6	S0°:		
09 01 28.6	+03 43 12	NGC 2729	13.4	0.8 x 0.5	12.3	S0?	0	Close pair mag 10.06v, 11.03v stars S 3′0.
09 07 37.1	+03 23 31	NGC 2765	12.1	2.1 x 1.0	12.8	S0	107	MCG +1-24-2 SSE 8′4.
09 17 10.6	−04 45 12	NGC 2817	13.0	2.0 x 1.7	14.1	SAB(rs)c II	6	
09 20 57.0	−04 56 25	NGC 2850	14.2	0.9 x 0.7	13.5	S0	30	
09 22 55.0	+03 09 22	NGC 2858	12.6	1.7 x 0.9	12.9	S0/a	117	
09 23 36.5	+02 08 05	NGC 2861	12.7	1.5 x 1.3	13.3	SB(r)bc I-II	6	MCG +0-24-13 SE 3′1; mag 10.73v star W 4′4.
09 24 15.4	+05 56 27	NGC 2864	14.6	0.8 x 0.7	13.8	Sc	12	
09 25 47.0	+02 13 42	NGC 2877	14.3	0.7 x 0.5	13.0	Pec	141	Elongated anonymous galaxy NW 1′4.
09 25 47.5	+02 05 22	NGC 2878	14.2	0.8 x 0.3	12.5	Sa:	174	
09 29 45.8	+02 12 22	NGC 2897	14.5	0.8 x 0.5	13.4	S0	170	Mag 8.38v star E 1′7.
09 29 46.3	+02 03 48	NGC 2898	13.4	1.0 x 0.7	12.9	S0⁺ pec:	125	Bright knot or star on E edge; mag 10.75v star E 8′2.
09 30 15.4	+04 08 31	NGC 2900	13.0	1.3 x 1.1	13.2	SBcd:	150	
09 34 27.1	−02 30 16	NGC 2917	13.6	1.4 x 0.4	12.8	S0⁺	169	Located 2′1 NW of mag 10.18v star.
09 37 43.3	+02 45 12	NGC 2936	13.1	1.6 x 0.9	13.3	I?	33	NE end of galaxy curves sharply due E. Small anonymous galaxy NNW 1′4.
09 37 45.0	+02 44 49	NGC 2937	13.7	0.8 x 0.6	12.9	E	0	Located on E edge of NGC 2936.
09 39 40.8	−00 14 09	NGC 2951	14.3	0.5 x 0.3	12.3	E	90	Star on W edge; mag 9.53v star NE 3′2.
09 40 36.2	+03 34 35	NGC 2960	12.4	1.4 x 1.0	12.6	Sa?	40	
09 40 54.0	+05 09 54	NGC 2962	11.9	2.6 x 1.9	13.5	(R)SAB(rs)0⁺	3	
09 42 11.1	+04 40 20	NGC 2966	12.7	2.2 x 0.9	13.3	SB?	72	Mag 13.2 star appears to intrude into W edge.
09 42 02.9	+00 20 09	NGC 2967	11.6	3.0 x 2.8	13.8	SA(s)c II	65	Numerous bright knots.
09 42 33.2	−03 41 56	NGC 2974	10.9	3.5 x 2.0	13.0	E4	40	Star located 0′7 SW of center, inside envelope.
09 45 41.4	+04 56 31	NGC 2987	12.9	1.5 x 0.7	12.9	Sab	160	
09 46 17.4	+05 42 31	NGC 2990	12.7	1.3 x 0.7	12.5	Sc: II	85	**CGCG 35-52**, and smaller anonymous companion, SE 1′5.
09 49 07.7	−04 44 37	NGC 3014	13.9	1.0 x 0.9	13.7	Sbc II-III	156	Star or bright knot on W edge of core; mag 8.17v star S 8′0.
09 49 22.9	+01 08 42	NGC 3015	13.9	0.6 x 0.4	12.2	SAB0° pec?	95	
09 49 03.1	−02 49 19	NGC 3017	13.1	1.0 x 1.0	13.0	E:		Strong, round core in smooth envelope; mag 8.65v star NW 8′0.
09 49 41.5	+00 37 21	NGC 3018	13.3	1.2 x 0.7	13.0	SB(s)b pec?	27	A mag 9.69v star NW 0′8.
09 49 39.1	−05 09 59	NGC 3022	13.2	1.6 x 1.6	14.1	(R′)SAB0°:		Mag 10.22v star NE 5′2; MCG −1-25-44 W 4′6.
09 49 52.5	+00 37 04	NGC 3023	13.0	2.9 x 1.4	14.3	SAB(s)c pec: II	70	The nucleus appears offset to the E. The small galaxy **Mkn 1236** lies 0′4 E of NGC 3023's nucleus.
09 52 29.7	+02 09 17	NGC 3039	13.4	1.2 x 0.6	12.9	SAB(s)b pec:	12	
09 53 20.3	+00 41 50	NGC 3042	12.9	1.2 x 0.8	12.7	S0°: sp	111	Very pronounced central bulge.
09 53 41.8	+01 34 34	NGC 3044	11.9	4.9 x 0.7	13.1	SB(s)c? sp II-III?	115	
09 54 32.1	−01 17 31	NGC 3047	14.9	0.4 x 0.4	12.8	S?		NGC 3047A W 0′7; mag 8.52v star 1′8 N.
09 54 29.5	−01 17 20	NGC 3047A	13.7	0.3 x 0.3	11.1	E		NGC 3047 E 0′7.
09 55 18.3	+04 16 11	NGC 3055	12.1	2.1 x 1.3	13.1	SAB(s)c II	63	
09 56 35.8	+01 25 39	NGC 3062	14.7	0.6 x 0.3	12.7	Sb	56	Mag 9.86v star SE 5′6.
09 59 49.9	−02 52 37	NGC 3083	13.7	1.0 x 0.4	12.6	Sa?	50	An anonymous galaxy 2′6 ESE.
10 00 11.0	−02 58 34	NGC 3086	13.9	1.1 x 0.4	12.9	Sb II	145	Small, elongated anonymous galaxy SE 1′2.
10 00 30.4	−02 58 13	NGC 3090	12.6	1.7 x 1.4	13.5	E⁺	90	Mag 10.54v star on N edge; **CGCG 8-18** and mag 9.76v star E 4′0.
10 00 47.3	−03 00 46	NGC 3092	13.3	1.4 x 0.7	13.1	SB(s)0⁺?	30	Faint anonymous galaxy S 1′7.
10 00 53.5	−02 58 22	NGC 3093	14.2	0.7 x 0.3	12.5	E	50	Mag 9.76v star NW 2′8; **CGCG 8-18** WNW 3′1.
10 01 35.5	−02 59 42	NGC 3101	14.4	1.0 x 0.3	12.9	Sa? sp	150	Mag 9.30v star N 5′3.
10 06 10.5	+02 54 45	NGC 3117	13.3	1.0 x 1.0	13.3	E?		Small, round core with faint envelope; mag 9.49v star SE 6′1.
10 12 41.3	+03 07 42	NGC 3156	12.3	1.9 x 0.9	12.8	S0:	47	A mag 9.39v stars lies 2′1 SE.
10 13 31.3	+03 22 31	NGC 3165	13.9	1.5 x 0.7	13.8	SA(s)dm:	177	NGC 3166 NE 4′6; mag 10.51v star W 4′7.
10 13 45.8	+03 25 32	NGC 3166	10.4	4.8 x 2.3	12.8	SAB(rs)0/a	87	NGC 3165 SW 4′6.
10 14 14.5	+03 28 03	NGC 3169	10.2	4.4 x 2.8	12.8	SA(s)a pec	45	Mag 10.8 star on E edge.
08 56 40.8	+00 22 28	UGC 4684	13.4	1.5 x 1.2	13.9	SA(rs)dm: III-IV	177	
09 08 10.4	+05 55 32	UGC 4797	13.9	2.0 x 1.5	15.0	Sm: V	0	Mag 12.8 star on W edge.
09 08 23.0	−01 36 44	UGC 4802	14.1	0.8 x 0.6	13.1	SAB(s)c	57	Mag 8.64v star S 4′4; **CGCG 6-1** W 9′9.
09 08 37.7	−01 45 15	UGC 4804	13.8	0.8 x 0.8	13.2	SBbc		Mag 8.64v star NW 6′2; **CGCG 6-2** SW 7′0.
09 13 13.5	+03 13 49	UGC 4857	13.5	1.6 x 0.8	13.6	SBcd:	57	
09 17 29.4	−00 37 14	UGC 4915	14.4	1.0 x 0.3	12.9	Sbc	119	
09 19 33.0	+05 52 56	UGC 4946	13.9	0.9 x 0.5	12.9	SB(r)b	30	
09 20 02.2	+01 02 18	UGC 4956	12.6	1.8 x 1.3	13.5	E:	15	Mag 10.55v star SW 2′1.
09 22 06.2	+03 53 47	UGC 4978	13.8	1.3 x 0.9	13.8	SAd	12	Mag 11.5 star SW 2′1.
09 22 17.6	+04 42 28	UGC 4980	14.0	0.9 x 0.4	12.8	SBb:	172	
09 23 13.3	+02 36 03	UGC 4993	14.2	1.0 x 0.4	13.1	S	120	CGCG 34-41 S 7′1.
09 23 16.2	−00 43 42	UGC 4996	13.4	1.4 x 0.5	12.9	SAB(s)bc: II-III	130	
09 31 41.7	+04 29 58	UGC 5075	13.2	0.8 x 0.8	12.5	S0/a?		

(Continued from previous page)

GALAXIES

RA h m s	Dec ° ′ ″	Name	Mag (V)	Dim ′ Maj x min	SB	Type Class	PA	Notes
09 34 10.8	+00 14 29	UGC 5097	13.0	0.8 x 0.5	11.9	S?	48	Very small, very faint anonymous galaxy N 1′.0.
09 34 34.3	+00 05 20	UGC 5099	13.5	0.9 x 0.4	12.2	SB(s)ab pec:	90	
09 34 38.7	+05 50 30	UGC 5100	13.6	1.5 x 0.9	13.8	SB(s)b	30	Bright NE-SW bar; mag 12.2 star NE end.
09 35 07.2	+05 07 08	UGC 5107	13.9	2.0 x 0.4	13.5	SBd	47	Mag 11.3 star S 1′.4.
09 42 25.0	+04 16 57	UGC 5182	12.9	1.0 x 0.8	12.5	S0	135	**CGCG 35-32** W 2′.9; **CGCG 35-36** N 4′.3; **CGCG 35-38** E 5′.4.
09 44 07.4	−00 39 30	UGC 5205	14.2	1.1 x 0.7	13.8	SBm pec?	152	**MCG +0-25-10** WNW 1′.6.
09 44 38.7	−00 13 19	UGC 5211	14.6	0.9 x 0.9	14.2			Mag 11.25v star on E edge.
09 45 51.9	+02 58 37	UGC 5224	14.7	1.5 x 0.7	14.6	SBdm	70	Mag 11.4 star NW 5′.5.
09 46 03.5	+04 24 12	UGC 5226	12.9	1.0 x 0.9	12.7	S0:	138	Mag 11.4 star E 2′.0.
09 46 04.0	+01 40 04	UGC 5228	13.0	2.4 x 0.8	13.5	SB(s)c: II-III	127	Mag 5.65v star NNE 8′.6.
09 46 53.5	+00 30 24	UGC 5238	13.7	1.8 x 0.7	13.8	SB(s)dm: IV-V	55	
09 47 34.3	−02 01 57	UGC 5245	14.1	2.6 x 0.5	14.2	SBdm? sp IV-V	160	Mag 12.3 star N 2′.2.
09 47 45.2	+02 37 32	UGC 5249	13.2	2.2 x 0.7	13.5	SBd:	15	Knotty, pair of mag 13 stars N.
10 00 00.0	+05 19 42	UGC 5373	11.3	5.5 x 3.7	14.5	IB(s)m IV-V	110	**Sextans B**. Dwarf irregular, low surface brightness.
10 00 27.2	+03 22 27	UGC 5376	13.4	1.9 x 0.6	13.4	Sdm:	151	Mag 12.3 star E edge.
10 00 31.7	+03 12 18	UGC 5377	13.5	1.0 x 0.8	13.1	SABdm:	84	Weak nucleus; mag 9.37v star S 5′.4.
10 00 32.0	+04 24 25	UGC 5378	13.2	1.0 x 0.5	12.3	Sb II-III	103	
10 00 30.8	−02 09 42	UGC 5380	13.3	1.0 x 0.7	12.8	SB(r)ab	155	Faint E-W bar with dark patches N and S.
10 00 45.2	+04 44 04	UGC 5383	13.2	1.1 x 0.7	12.8	S0	155	
10 01 11.5	+00 13 32	UGC 5388	14.1	1.0 x 0.8	13.7	S?	177	**CGCG 8-23** N 6′.1.
10 04 18.6	+03 41 14	UGC 5424	14.3	0.9 x 0.3	12.7	Scd:	79	**CGCG 36-25** W 8′.3.
10 04 53.7	+05 03 44	UGC 5432	13.1	0.8 x 0.7	12.5	E:		Mag 7.35v star S 3′.9.
10 10 00.0	−02 27 51	UGC 5483	13.9	0.9 x 0.8	13.4	S?	9	
10 10 27.9	+02 13 38	UGC 5487	14.1	1.0 x 0.5	13.2	SBbc	70	
10 11 17.9	+00 26 29	UGC 5493	13.3	1.5 x 1.2	13.8	SAB(rs)c II	15	
10 12 11.7	+04 55 21	UGC 5501	13.9	0.7 x 0.5	12.6	S?	138	Mag 11.6 star on e edge.
10 12 51.4	+04 47 35	UGC 5506	14.0	1.3 x 0.6	13.6	S?	17	Stellar nucleus, uniform envelope; MCG +1-26-20 NW 3′.0.
10 13 38.4	−00 55 34	UGC 5515	12.6	1.4 x 1.1	13.1	E+ pec:	90	Mag 9.08v star W 3′.5; located inside W edge of galaxy cluster A 957, numerous PGC galaxies in immediate area.
10 13 52.7	+00 33 00	UGC 5521	13.9	0.9 x 0.8	13.4	SAB(s)c		Pair of stars, mags 10.5v and 10.6v, SW 2′.7.
10 14 39.4	−00 49 56	UGC 5528	13.5	1.0 x 0.8	13.1	SAB(r)a pec:	150	Located on the E edge of galaxy cluster A 957; **PGC 29868** SW 4′.3.
10 15 55.5	+02 41 15	UGC 5539	14.7	2.0 x 1.0	15.3	IBm:	42	Dwarf, low surface brightness.

GALAXY CLUSTERS

RA h m s	Dec ° ′ ″	Name	Mag 10th brightest	No. Gal	Diam ′	Notes
08 57 54.0	+03 10 00	A 732	17.5	65	10	**Hydra II Cluster**. Almost stellar **MCG +1-23-10** at center; **MCG +1-23-12** SE 2′.3; **MCG +1-23-11** N 1′.1.
09 37 06.0	−05 00 00	A 838	15.3	40	17	**MCG −1-25-7** S of center 2′.5; all other members are anonymous, faint and stellar.
10 01 12.0	−00 06 00	A 912	15.9	36	12	Almost stellar **PGC 28996** at center; stellar **PGC 28999** S 2′.1; all others stellar and anonymous.
10 07 42.0	+00 31 00	A 933	15.9	44	22	Members stellar and anonymous.
10 14 00.0	−00 54 00	A 957	15.9	55	20	See notes for UGC 5515, UGC 5528.

GLOBULAR CLUSTERS

RA h m s	Dec ° ′ ″	Name	Total V m	B ★ V m	HB V m	Diam ′	Conc. Class Low = 12 High = 1	Notes
10 05 31.4	+00 04 17	Pal 3	13.9	18.0	20.5	1.6	12	

PLANETARY NEBULAE

RA h m s	Dec ° ′ ″	Name	Diam ″	Mag (P)	Mag (V)	Mag cent ★	Alt Name	Notes
09 39 09.1	−02 48 32	PN G238.0+34.8	270	13.4	12.6	15.5	PK 238+34.1	Mag 7.21v star on SSW edge; mag 12.4 star NNE 4′.0.

GALAXIES

RA h m s	Dec ° ′ ″	Name	Mag (V)	Dim ′ Maj x min	SB	Type Class	PA	Notes
08 22 56.1	−00 22 12	CGCG 4-8	14.5	0.3 x 0.2	11.3		84	Mag 12.8 star N edge; mag 10.15v star S 2′.0.
08 23 57.7	−00 18 19	CGCG 4-11	14.6	0.3 x 0.2	11.6	E	6	Elongated **CGCG 4-13** E 3′.2; stellar **CGCG 4-12** S 4′.3.
08 27 11.3	−01 04 04	CGCG 4-27	15.0	0.4 x 0.3	12.8	E	30	Anonymous galaxy NE 0′.9.
08 28 18.0	+00 17 18	CGCG 4-35	14.7	0.8 x 0.8	14.2	E		Bright N-S bar; mag 7.29v star S 2′.8.
08 35 01.6	−03 36 08	CGCG 4-61	14.7	0.5 x 0.4	12.8		6	Small triangle of mag 13-14 stars W 2′.6.
08 35 32.5	−02 49 33	CGCG 4-67	15.7	0.7 x 0.2	13.4		159	Mag 11.39v N 1′.9.
08 35 41.7	−01 45 24	CGCG 4-71	14.5	0.8 x 0.7	13.7	Sb	156	Mag 8.45v star ENE 6′.7; NGC 2616 S 6′.0.
08 36 59.4	+01 47 30	CGCG 4-86	14.9	0.6 x 0.6	13.7	Sab		Mag 12.7 star W 0′.7; mag 9.87v star WSW 6′.0.
08 37 13.9	−03 21 45	CGCG 4-87	15.7	0.5 x 0.3	13.5		87	Mag 10.64v star on W edge.
08 37 41.4	−03 12 31	CGCG 4-88	15.2	0.4 x 0.3	12.7		15	Mag 12.1 star SE 0′.5.
08 38 14.5	+00 21 03	CGCG 4-95	15.8	0.5 x 0.4	13.9		12	Low, uniform, surface bright ness.
08 50 31.6	−03 05 58	CGCG 5-15	13.8	0.6 x 0.3	12.0	E	0	Mag 8.07v star WSW 11′.1.
08 53 40.8	+01 33 44	CGCG 5-23	15.0	0.6 x 0.4	13.3	S0	93	Mag 9.55v star WNW 8′.6.
08 55 12.5	−02 16 05	CGCG 5-28	14.7	0.5 x 0.5	13.0			UGC 4672 E 7′.8.
08 55 47.6	+00 47 35	CGCG 5-32	14.6	0.7 x 0.3	12.8		87	Mag 8.24v star NNE 6′.0; **CGCG 5-29** WNW 7′.0.
08 57 47.6	−00 12 00	CGCG 5-42	14.6	0.5 x 0.4	12.7	S0	57	Mag 8.27v star W 4′.8.
09 03 30.6	−01 22 51	CGCG 5-51	14.8	0.6 x 0.4	13.1	S0/a	66	Multi-galaxy system?
07 51 45.1	+04 30 22	CGCG 30-21	15.6	0.5 x 0.3	13.4		24	**CGCG 30-22** S 3′.7; mag 6.49v star WSW 13′.7.
07 52 59.0	+05 29 46	CGCG 30-24	14.2	0.7 x 0.5	12.9		51	Mag 12.7 star N 0′.6.
08 01 17.4	+04 17 58	CGCG 31-20	14.5	0.5 x 0.5	12.8			Mag 12.4 star SE 3′.0.
08 07 15.0	+04 28 08	CGCG 31-50	14.3	0.4 x 0.3	11.8		57	Mag 9.78v star S 1′.8.
08 08 32.8	+02 36 59	CGCG 31-57	14.5	0.4 x 0.4	12.4			Mag 13.2 star N 0′.8.
08 10 41.0	+03 10 44	CGCG 31-63	14.6	0.4 x 0.2	11.7		156	Mag 14 star N 0′.7; mag 12.1 star SE 1′.8.
08 19 02.4	+03 38 00	CGCG 31-79	14.5	0.4 x 0.4	12.4			Mag 11.31v star SSE 3′.8.
08 20 17.0	+04 42 09	CGCG 32-2	14.9	0.5 x 0.3	12.6		51	Mag 11.6 star NE 3′.0; mag 9.24v star WNW 6′.5.
08 20 33.4	+04 34 15	CGCG 32-3	14.2	0.6 x 0.5	12.7		159	Mag 13.9 star E 0′.5.
08 22 41.7	+04 19 39	CGCG 32-9	14.3	0.5 x 0.5	12.6			IC 504 S 4′.0.
08 22 55.8	+04 07 17	CGCG 32-11	14.4	0.9 x 0.3	12.8		96	Mag 13.6 star on S edge.

(Continued from previous page)

RA h m s	Dec ° ′ ″	Name	Mag (V)	Dim ′ Maj x min	SB	Type Class	PA	Notes
08 22 57.6	+04 23 08	CGCG 32-13	14.8	0.9 x 0.3	13.2		59	Mag 10.73v star SSE 2'8.
08 25 10.2	+03 22 38	CGCG 32-21	14.3	0.7 x 0.2	12.0		135	Mag 8.43v star NE 10'6; mag 8.41v star SW 11'4.
08 25 17.5	+03 55 59	CGCG 32-22	14.2	0.5 x 0.5	12.5			Mag 12.9 star W 1'4; mag 12.1 star S 2'7.
08 25 35.9	+05 06 56	CGCG 32-24	14.6	0.4 x 0.4	12.5			Very close pair of mag 14 stars E 0'8.
08 26 19.0	+02 55 28	CGCG 32-28	14.3	0.7 x 0.3	12.4		48	Mag 11.1 star SE 2'7.
08 27 56.0	+04 58 39	CGCG 32-32	14.9	0.4 x 0.4	12.7			Mag 13.5 star on W edge.
08 28 20.4	+02 26 41	CGCG 32-33	14.7	0.7 x 0.5	13.4		0	Mag 11.29v star on NE edge.
08 32 07.2	+03 38 03	CGCG 32-36	14.4	0.6 x 0.5	12.9	Sb	75	Partially superimposed mag 11.9 star S of center.
08 39 07.8	+04 34 02	CGCG 32-48	14.4	0.6 x 0.5	12.9	Sc	141	Located between, and just S of, a pair of stars mags 10.11v and 9.53v.
08 46 40.6	+02 24 25	CGCG 33-3	15.5	0.5 x 0.2	12.9		30	Mag 10.90v star 1'2 ESE.
08 47 18.7	+02 45 42	CGCG 33-7	15.0	0.9 x 0.6	14.2		141	**CGCG 33-8** SSE 5'4.
09 02 15.7	+03 10 07	CGCG 33-49	14.8	0.8 x 0.3	13.1		12	Mag 8.00v star NNW 7'1.
09 02 29.5	+03 23 02	CGCG 33-50	14.8	0.8 x 0.2	12.7		69	**CGCG 33-47** NW 8'0; mag 8.00v star SW 8'8.
08 06 24.1	+01 02 07	IC 494	13.1	1.3 x 0.6	12.7	SA0°:	50	Mag 7.30v star N 8'9.
08 09 30.2	+05 16 45	IC 498	13.8	0.9 x 0.8	13.3	S	54	
08 22 10.8	+03 16 05	IC 503	13.0	1.1 x 1.0	12.9	SBa	114	Mag 9.24v star NW 10'3.
08 22 41.3	+04 15 43	IC 504	13.0	1.0 x 0.8	12.7	S0	140	Curving chain of four stars leads to MCG +1-22-7 E 4'6.
08 23 21.6	+04 22 20	IC 505	14.0	1.2 x 0.9	13.9	S	143	IC 506 SSE 4'9.
08 23 30.7	+04 17 59	IC 506	13.7	0.6 x 0.6	12.7	E/S0		IC 505 NNW 4'9.
08 35 31.3	−01 54 04	IC 515	14.7	0.7 x 0.4	13.2		28	NGC 2616 N 5'1.
08 46 44.1	+02 32 14	IC 521	14.1	0.9 x 0.5	13.1		90	Mag 15.5 star superimposed NW of nucleus; mag 10.53v star NW 2'1.
08 11 01.8	+05 05 10	IC 2231	14.0	1.0 x 1.0	14.1	E:		Mag 13.1 star on W edge; mag 9.88v star WNW 7'1.
08 21 28.0	+03 10 10	IC 2327	13.3	1.3 x 0.4	12.4	Sa?	174	Mag 10.38v star SSW 6'2.
08 51 33.7	+03 06 02	IC 2420	14.2	0.6 x 0.6	13.0			Mag 15.3 star on SE edge.
08 58 30.7	+02 55 28	IC 2426	14.7	0.4 x 0.3	12.2		111	Stellar; mag 9.71v star S 4'4.
07 51 50.7	+02 48 56	MCG +0-20-6	13.7	1.0 x 0.8	13.5	E/S0	111	Mag 10.9 star N 1'8.
08 06 11.0	−03 16 58	MCG +0-21-3	14.1	0.5 x 0.4	12.2	Sd	88	
08 20 41.8	−01 24 58	MCG +0-22-3	14.2	0.9 x 0.2	12.2	Sa	42	On SW edge of UGC 4352; mag 9.47v star WSW 1'5.
08 22 33.3	+02 18 26	MCG +0-22-5	15.3	0.5 x 0.4	13.4		36	
08 33 20.7	−03 30 26	MCG +0-22-16	14.4	0.8 x 0.3	12.8	S0/a	171	
08 34 33.7	+01 40 16	MCG +0-22-18	15.1	0.8 x 0.5	14.0	S	177	UGC 4480 SW edge; mag 9.43v star S 2'4.
08 35 06.1	−03 05 48	MCG +0-22-20	14.4	0.5 x 0.5	12.7			Mag 10.99v star SE 3'1.
08 48 18.9	−03 01 06	MCG +0-23-2	13.5	1.6 x 0.7	13.4	SA(s)c: II	54	**CGCG 5-9** N 2'5.
08 50 43.2	+01 21 00	MCG +0-23-4	14.4	0.7 x 0.5	13.2	S0/a	165	Pair of faint, almost stellar anonymous galaxies N 1'8.
09 03 24.0	−02 34 37	MCG +0-23-20	13.8	0.5 x 0.3	11.8	E?	90	
08 03 19.9	+03 03 07	MCG +1-21-8	13.9	0.9 x 0.6	13.1		45	Faint stars superimposed SW edge.
08 07 32.8	+04 30 56	MCG +1-21-14	14.1	0.4 x 0.4	12.0	S0?		Almost stellar; mag 14.4, 13.2 stars S 0'9.
08 18 13.2	+04 46 02	MCG +1-21-21	15.7	0.7 x 0.7	14.7	Sc		
08 21 44.1	+04 32 49	MCG +1-22-3	14.4	0.9 x 0.7	13.8	Sbc	123	Two main arms; mag 9.23v star SW 2'4.
08 22 59.4	+04 16 34	MCG +1-22-7	14.1	0.5 x 0.5	12.5			Mag 11.6 star SW 0'6; IC 504 W 4'6.
08 24 26.0	+04 14 50	MCG +1-22-10	14.0	0.7 x 0.3	12.2	Sab	21	
08 27 32.3	+02 52 58	MCG +1-22-11	14.0	0.7 x 0.4	12.5	Sb	159	Galaxy pair?
08 37 15.0	+04 41 44	MCG +1-22-13	15.2	0.6 x 0.4	13.5		153	Mag 11.7 star NW 2'5.
08 43 51.0	+02 42 48	MCG +1-23-1	14.5	0.9 x 0.3	12.9		144	Mag 9.12v star NE 3'4.
08 59 49.1	+05 03 36	MCG +1-23-16	14.1	0.6 x 0.4	12.4	S0?	3	Faint star on E edge; mag 12.3 star NE 2'4.
09 03 40.0	+03 22 18	MCG +1-23-19	14.5	0.9 x 0.4	13.2	S?	24	
09 03 33.4	+03 22 32	MCG +1-23-20	13.8	1.0 x 0.5	12.9	SB0?	120	
08 24 32.5	−04 53 22	MCG −1-22-12	14.6	1.1 x 0.7	14.2	(R′)Sb?	3	
08 31 03.3	−04 11 47	MCG −1-22-16	13.5	0.6 x 0.6	12.3	SA0⁻:		
08 31 40.9	−04 07 06	MCG −1-22-18	12.7	0.9 x 0.9	12.6	E⁺		Small, round anonymous galaxy E 1'3.
08 32 34.9	−05 39 21	MCG −1-22-19	14.2	0.8 x 0.5	13.0	Sc	78	
08 33 46.6	−04 39 12	MCG −1-22-20	13.4	1.1 x 0.5	12.6	Sb	45	
08 35 03.0	−05 10 32	MCG −1-22-25	15.1	0.8 x 0.4	13.7	Sc	156	
08 35 48.6	−04 05 35	MCG −1-22-26	13.1	1.0 x 0.6	12.4	S0	39	Bright field star on west edge; NGC 2617 W 2'6.
08 38 56.4	−05 44 44	MCG −1-22-30	14.6	0.8 x 0.8	14.0	SBcd		
08 41 18.1	−04 42 55	MCG −1-22-34	13.7	1.4 x 0.9	13.8	(R′)SB(r)a	39	Mag 7.57v star S 5'6.
08 42 41.4	−05 39 28	MCG −1-22-35	14.1	0.8 x 0.5	13.0		126	Two bright field stars, in-line, trailing SE.
07 54 21.0	+04 27 31	NGC 2470	12.7	1.9 x 0.6	12.7	Sab	128	
07 59 07.0	−00 38 17	NGC 2494	13.1	0.9 x 0.7	12.4	(R′)SB(rs)0/a	95	
07 59 52.4	+05 36 25	NGC 2504	14.0	0.5 x 0.4	12.1	S?	96	Mag 8.34v star SW 8'8.
08 11 23.0	+03 37 51	NGC 2538	12.6	1.4 x 1.2	13.0	(R′)SBa	25	
08 17 56.5	+00 44 41	NGC 2555	12.2	1.9 x 1.4	13.2	SB(rs)ab	115	Pair of faint stars superimposed NW of center.
08 19 37.0	+04 39 24	NGC 2561	13.2	1.1 x 0.6	12.6	SB?	138	
08 23 07.9	−05 00 11	NGC 2583	13.4	0.9 x 0.9	13.2	E:		
08 23 15.4	−04 58 17	NGC 2584	13.8	1.0 x 0.6	13.1	SB(s)bc? II	2	
08 23 26.2	−04 54 56	NGC 2585	13.6	1.8 x 0.8	13.8	SB(s)b pec	99	**MCG −1-22-7** NW 2'9; **MCG −1-22-6** NW 4'2.
08 25 01.9	−00 35 30	NGC 2590	13.0	2.2 x 0.7	13.4	SA(s)bc: I-II	77	
08 34 33.4	−02 32 50	NGC 2615	12.5	1.9 x 1.0	13.0	SB(rs)b	40	Three faint stars inline down W edge.
08 35 34.1	−01 51 02	NGC 2616	12.5	1.1 x 0.9	12.4	SAB(rs)0°?	163	CGCG 4-71 N 6'0; **IC 516** ESE 4'3; IC 515 S 5'1.
08 35 38.7	−04 05 16	NGC 2617	13.8	1.1 x 0.8	13.5	S0/a pec:	117	MCG −1-22-26 E 2'6; mag 9.22v star WNW 8'0.
08 35 53.5	+00 42 25	NGC 2618	12.1	2.4 x 1.9	13.6	(R′)SA(rs)ab	140	
08 40 44.2	−04 07 20	NGC 2642	12.6	2.0 x 1.9	13.9	SB(r)bc I-II	90	Located 2'8 NE of mag 9.51v star.
08 41 32.0	+04 58 53	NGC 2644	12.4	2.1 x 0.8	12.8	S?	14	Bright knot or star E of nucleus; mag 9.45v star SW 9'2.
08 52 38.2	−02 36 10	NGC 2690	13.1	1.9 x 0.5	12.9	Sab: sp	19	Mag 9.95v star SE 2'8.
08 54 27.1	−03 04 04	NGC 2695	11.9	1.7 x 1.2	12.5	SAB(s)0°?	174	Mag 13 star on W edge; mag 14 star superimposed just E of bright core.
08 54 59.7	−02 59 17	NGC 2697	12.3	1.8 x 1.0	12.8	SA(s)0⁺:	129	Very small, very faint anonymous galaxy SW 4'4.
08 55 36.5	−03 11 04	NGC 2698	12.6	1.4 x 0.6	12.3	SA0⁺?	96	Mag 7.02v star NW 4'2.
08 55 48.7	−03 07 42	NGC 2699	12.6	1.0 x 1.0	12.6	E:		Mag 7.02v star W 6'7.
08 55 37.2	−03 03 13	NGC 2702	15.2	0.5 x 0.3	13.0		24	Stellar nucleus; mag 7.02v star SSW 6'7.
08 56 12.2	−02 33 43	NGC 2706	13.0	1.8 x 0.6	12.9	Sbc? sp III	167	
08 56 07.9	−03 21 41	NGC 2708	12.0	2.6 x 1.3	13.2	SAB(s)b pec?	25	Several bright knots N end.
08 56 12.8	−03 14 39	NGC 2709	13.7	0.8 x 0.6	12.7	SA0° pec:	81	Mag 9.40v star NE 5'6.
08 57 20.4	+02 55 14	NGC 2713	11.8	3.6 x 1.5	13.4	SB(rs)ab I-II	107	Lies 4'5 E of mag 9.52v star.
08 57 36.0	+03 05 25	NGC 2716	11.8	1.3 x 1.0	12.0	(R)SB(r)0⁺	30	Located on SW edge of galaxy cluster A 732.
08 58 56.6	−04 54 09	NGC 2721	12.2	2.3 x 1.6	13.5	SB(rs)bc pec II	153	Several faint stars just inside W edge.
08 58 46.3	−03 42 38	NGC 2722	12.7	2.0 x 1.3	13.6	SA(rs)bc pec: III	75	
09 00 14.4	+03 10 37	NGC 2723	13.2	0.8 x 0.8	12.6	S0°:		
09 01 28.6	+03 43 12	NGC 2729	13.4	0.8 x 0.5	12.3	S0?	0	Close pair mag 10.06v, 11.03v stars S 3'0.
07 44 52.9	−00 21 30	UGC 4004	14.7	1.2 x 0.2	13.0	S	73	
07 45 44.6	+04 58 53	UGC 4010	13.7	1.4 x 0.4	12.9	Sa?	90	

(Continued from previous page)

114 GALAXIES 114

RA h m s	Dec ° ′ ″	Name	Mag (V)	Dim ′ Maj x min	SB	Type Class	PA	Notes
08 04 05.9	+05 06 47	UGC 4203	13.4	0.6 x 0.6	12.2	S?		Mag 11.9 star SE 2'.4.
08 06 47.8	+05 18 33	UGC 4228	12.4	2.0 x 1.3	13.3	S0	145	
08 07 58.2	+05 12 22	UGC 4239	14.3	0.6 x 0.6	13.0	S		Mag 13.4 star NE 0'.9.
08 08 55.5	+00 18 03	UGC 4248	13.4	1.0 x 0.8	13.0	(R)SAB(rs)a	61	Almost stellar nucleus, smooth envelope.
08 09 14.8	+00 16 57	UGC 4251	13.3	1.1 x 0.7	12.9	S0$^-$:	145	Mag 9.35v star W 1'.5; UGC 4248 W 5'.0.
08 09 24.3	+00 36 35	UGC 4254	13.5	1.5 x 0.7	13.4	SB(rs)m pec? III-IV	92	Mag 11.8 star NE 3'.2.
08 13 51.0	+00 39 39	UGC 4285	13.8	1.4 x 0.8	13.8	Scd:	175	Mag 12.4 star on SE edge; mag 8.44v star S 2'.1.
08 17 15.7	+01 12 18	UGC 4310	13.8	1.1 x 1.1	13.9	SA(s)m V		Stellar nucleus, uniform envelope.
08 17 41.8	+04 36 34	UGC 4316	13.7	1.5 x 1.4	14.4	SABdm		Low surface brightness; mag 10.7 star E 1'.4.
08 20 43.2	−01 24 45	UGC 4352	14.1	1.0 x 0.8	13.7	Triple System	134	Triple system in contact; MCG +0-22-3 on SW edge; **UGC 4349** NW 2'.8.
08 22 28.9	−01 02 49	UGC 4370	13.9	1.2 x 0.7	13.6	SB(rs)d? III	135	**CGCG 4-4** and close star S 3'.3.
08 22 53.2	+03 34 16	UGC 4374	13.3	1.5 x 1.2	13.8	SB(s)cd	55	
08 23 13.5	−00 51 43	UGC 4381	14.1	1.4 x 0.3	13.0	Sc: sp II-III	33	Mag 6.95v star W 6'.4.
08 27 25.7	+01 50 53	UGC 4421	13.9	1.0 x 0.7	13.4	SAB(s)b	177	Mag 11.01v star N 2'.4.
08 28 08.0	−01 32 20	UGC 4430	13.9	1.0 x 0.5	13.0	S0/a	142	**CGCG 4-30** W 9'.1; **CGCG 4-29** SW 10'.5.
08 28 12.8	+00 01 23	UGC 4431	13.6	1.0 x 0.8	13.2	Sbc	145	
08 28 14.9	+01 00 15	UGC 4432	13.6	1.3 x 0.8	13.5	SAbc: III	135	Mag 10.73v star on S edge.
08 29 19.4	+02 38 39	UGC 4439	14.3	1.4 x 0.7	14.1	SB(s)d: III	7	Uniform surface brightness; mag 12.8 star superimposed N of center.
08 31 33.0	−01 11 52	UGC 4455	13.7	0.9 x 0.7	13.0	SB(r)a	20	
08 32 47.0	+00 13 34	UGC 4467	14.2	1.2 x 0.6	13.7	Sd pec? III-IV	170	Mag 13.3 star on N edge.
08 34 31.6	+01 39 55	UGC 4480	13.8	1.0 x 0.4	12.7	SBb	87	In contact with companion MCG +0-22-18 NE edge; mag 9.43v star S 2'.0.
08 35 48.7	+01 43 18	UGC 4491	12.6	2.5 x 0.8	13.2	(R′)SB(rs)a:	60	
08 36 05.4	+01 53 35	UGC 4494	13.8	1.3 x 0.6	13.3	(R′)SAB(s)a:	70	
08 37 56.2	−02 27 44	UGC 4508	14.0	0.4 x 0.4	12.0	cE?		
08 43 26.2	+03 36 51	UGC 4553	13.6	1.2 x 0.5	12.9	SB?	30	
08 48 40.2	+01 02 19	UGC 4613	12.9	1.5 x 1.4	13.6	(R)SB(s)a:	70	Bright N-S core, faint envelope.
08 50 17.9	+03 29 49	UGC 4625	14.7	1.5 x 0.2	13.3	Scd:	124	Mag 11.5 star SE 2'.2.
08 51 38.2	−02 21 58	UGC 4638	13.6	1.3 x 0.6	13.2	S pec:	123	**MCG +0-23-6** N edge.
08 51 43.8	−02 07 53	UGC 4640	13.5	3.1 x 1.1	14.7	SAB(rs)c: II-III	6	Numerous knots.
08 55 43.2	−02 17 24	UGC 4672	13.9	0.8 x 0.7	13.1	S	66	CGCG 5-28 W 7'.8.
08 55 52.5	+02 31 27	UGC 4673	13.4	1.6 x 1.3	14.0	SB(s)d: IV-V	20	
08 56 40.8	+00 22 28	UGC 4684	13.4	1.5 x 1.2	13.9	SA(rs)dm: III-IV	177	

GALAXY CLUSTERS

RA h m s	Dec ° ′ ″	Name	Mag 10th brightest	No. Gal	Diam ′	Notes
08 57 54.0	+03 10 00	A 732	17.5	65	10	**Hydra II Cluster**. Almost stellar **MCG +1-23-10** at center; **MCG +1-23-12** SE 2'.3; **MCG +1-23-11** N 1'.1.

OPEN CLUSTERS

RA h m s	Dec ° ′ ″	Name	Mag	Diam ′	No. ★	B ★	Type	Notes
07 47 02.4	+00 01 05	ADS 6366		4.5	8	8.3	ast?	Discovered by J. Herschel in 1827.
07 46 50.0	−04 40 00	Be 39		8	120	16.0	cl	Rich in stars; moderate brightness range; slight central concentration; detached.
08 13 44.0	−05 45 00	NGC 2548	5.8	30	80	8.0	cl	= **M 48**. Rich in stars; large brightness range; strong central concentration; detached.

PLANETARY NEBULAE

RA h m s	Dec ° ′ ″	Name	Diam ″	Mag (P)	Mag (V)	Mag cent ★	Alt Name	Notes
07 51 39.3	+03 00 29	PN G217.1+14.7	361	13.6	13.5:	17.1	PK 217+14.1	Close pair of mag 12 stars on SSE edge; mag 12.2 star N 3'.7 from center.
08 06 46.0	−02 52 42	PN G224.3+15.3	165	15.4		18.9	PK 224+15.1	Slight brightening along NW and SE edges; mag 12.5 star W 2'.1.

115 GALAXIES 115

RA h m s	Dec ° ′ ″	Name	Mag (V)	Dim ′ Maj x min	SB	Type Class	PA	Notes
07 23 07.8	+02 45 57	CGCG 1-4	14.3	0.3 x 0.3	11.8	E		Mag 10.43v star W 1'.7.
07 15 46.7	+04 46 21	CGCG 29-14	14.5	0.5 x 0.5	12.8			Mag 11.6 star S 1'.2.
07 30 03.1	+05 48 13	CGCG 29-29	14.4	0.5 x 0.3	12.2		27	Close pair of mag 13 stars NW 1'.8; mag 9.48v star ESE 4'.2.
07 38 23.2	+04 58 38	CGCG 30-7	14.8	0.5 x 0.4	12.9		171	Mag 7.97v star SSW 2'.5.
07 40 05.9	+05 34 56	CGCG 30-8	16.0	0.6 x 0.4	14.3		135	Mag 10.8 star E 3'.7; CGCG 30-9 NNE 4'.2.
07 40 13.7	+05 38 41	CGCG 30-9	15.7	0.4 x 0.2	12.8		6	CGCG 30-8 SSW 4'.2.
07 51 45.1	+04 30 22	CGCG 30-21	15.6	0.5 x 0.3	13.4		24	**CGCG 30-22** S 3'.7; mag 6.49v star WSW 13'.7.
07 35 32.7	+00 57 36	MCG +0-20-1	14.4	0.7 x 0.6	13.3		27	Between a very closed pair of mag 13 stars; mag 10.72v star NNE 2'.9.
07 39 42.6	−02 37 18	MCG +0-20-3	16.7	0.9 x 0.9	16.4	S		Between a pair of very close stars, mag 13.3 and 14.3.
07 51 50.7	+02 48 56	MCG +0-20-6	13.7	1.0 x 0.8	13.5	E/S0	111	Mag 10.9 star N 1'.8.
07 22 11.0	−05 55 50	MCG −1-19-1	14.3	1.4 x 0.2	12.8	Sb: sp	99	Mag 5.84v star SE 4'.7.
07 01 03.3	+01 54 41	UGC 3630	14.7	1.5 x 0.7	14.6	SAB(r)b?	0	Numerous stars superimposed, in star rich area.
07 14 56.5	+00 45 28	UGC 3769	14.4	0.9 x 0.4	13.2	S?	115	Mag 12.1 star S 0'.8.
07 22 00.7	+05 08 49	UGC 3819	14.9	1.0 x 0.3	13.4	Im:	130	
07 23 30.8	+02 36 54	UGC 3830	14.3	1.3 x 0.5	13.6	SBd? III-IV	116	Bright knot or star NW end.
07 34 11.6	+04 32 48	UGC 3912	12.7	1.9 x 1.1	13.4	IBm:	114	Very knotty, superimposed stars.
07 37 59.7	+03 18 55	UGC 3946	13.0	1.3 x 0.8	12.9	Im	15	Mag 10.6 star E 1'.6.
07 38 44.1	+02 13 02	UGC 3950	13.5	1.0 x 0.3	12.0	S0/a	176	Pair of stars, mags 11.7 and 12.8, NE 2'.2.
07 40 18.0	−01 34 29	UGC 3964	13.0	1.6 x 1.0	13.4	SAB(s)dm: III-IV	125	Mag 9.57v star NW 3'.6.
07 44 52.9	−00 21 30	UGC 4004	14.7	1.2 x 0.2	13.0	S	73	
07 45 44.6	+04 58 53	UGC 4010	13.7	1.4 x 0.4	12.9	Sa?	90	

(Continued from previous page)

115 OPEN CLUSTERS 115

RA h m s	Dec ° ′ ″	Name	Mag	Diam ′	No. ★	B ★	Type	Notes
07 47 02.4	+00 01 05	ADS 6366		4.5	8	8.3	ast?	Discovered by J. Herschel in 1827.
06 37 47.0	−00 52 42	Be 24		6	25	17.0	cl	
07 09 54.0	+02 43 48	Be 35		5	20	16.0	cl	
07 20 18.0	−00 59 48	Be 37		4	25	15.0	cl	
07 46 50.0	−04 40 00	Be 39		8	120	16.0	cl	Rich in stars; moderate brightness range; slight central concentration; detached.
07 21 28.0	−03 14 00	Be 77		5	30	13.0	cl	(S) Identification uncertain.
07 23 38.0	+05 22 24	Be 78		5	12	16.0	cl	
06 57 46.0	+03 13 42	Bi 9		3	30	15.0	cl	
06 52 07.0	+02 55 00	Bi 10	10.4	3	20	15.0	cl	Few stars similar in brightness; strong central concentration; detached. Probably not a cluster.
06 51 21.0	+05 46 06	Bi 11		2	25	15.0	cl	
06 50 16.0	+05 43 42	Bi 12		4	30	17.0	cl:	
07 00 22.0	−00 14 06	Bi 13		2	20	17.0	cl	
06 48 50.0	+00 22 36	Bo 2	9.7	1.5	10	10.9	cl	Few stars similar in brightness; strong central concentration; detached.
07 03 28.0	−05 00 18	Bo 3	9.9	4	25	11.2	cl	Few stars; small brightness range; not well detached.
06 37 05.0	+03 03 48	ClvdB 1	9.5	5	39		cl	
06 46 52.0	+01 18 54	ClvdB 85		4	15		cl	Nebulosity is NGC 2282.
06 36 30.0	+04 50 00	Cr 104	9.6	20	15		cl	Few stars; moderate brightness range; not well detached; involved in nebulosity. (A) Appears mostly as loose 20′ high "integral sign" chain of stars.
06 37 06.0	+05 58 00	Cr 106	4.6	35	20		cl	Few stars; moderate brightness range; not well detached.
06 37 42.0	+04 45 00	Cr 107	5.1	30	204	7.1	cl	Few stars; moderate brightness range; not well detached. Probably not a cluster.
06 38 39.0	+02 03 00	Cr 110	10.5	18	70	11.0	cl	Moderately rich in stars; small brightness range; not well detached.
06 46 40.0	+01 47 00	Cr 115	9.2	10	50		cl	Moderately rich in stars; moderate brightness range; not well detached.
07 21 48.0	−00 59 00	Cz 28		5	20		cl	(S) Sparse cluster.
06 43 05.7	+00 01 00	Do 23		10	20		cl	Few stars; large brightness range; not well detached. (A) DSS image shows pentagon shaped sparse group of stars.
06 44 44.0	+01 45 00	Do 24		15	40		cl	Moderately rich in stars; moderate brightness range; not well detached. (S) Sparse group.
06 45 00.0	+00 18 00	Do 25	7.6	20	50	8.9	cl:	Moderately rich in stars; small brightness range; not well detached; nebulosity involved. (S) Large V-shaped group.
06 32 19.0	+04 51 00	NGC 2244	4.8	30	100	7.0	cl	Rich in stars; large brightness range; slight central concentration; detached; involved in nebulosity (Rosette Nebula). The brightest star is a non-member.
06 33 47.0	−05 05 00	NGC 2250		10	25	8.7	ast?	Few stars; moderate brightness range; not well detached. (A) Oblong (E-W) 10′ x 5.5′ group.
06 34 19.8	+05 19 00	NGC 2252	7.7	18	30	9.0	cl	Moderately rich in stars; moderate brightness range; no central concentration; detached; involved in nebulosity. (A) DSS image shows obvious N-S stream of bright stars, with large bright nebula involved on SW side.
06 38 04.0	−01 28 00	NGC 2260		20	50	8.0	cl?	
06 39 38.7	+01 08 30	NGC 2262	11.3	4	180		cl	Moderately rich in stars; small brightness range; slight central concentration; detached. (A) Has inner compressed part 2′ in diameter with ~80 stars.
06 43 16.8	+04 37 04	NGC 2269	10.0	3	12	10.0	cl	Few stars; small brightness range; slight central concentration; detached.
06 47 41.0	−03 09 00	NGC 2286	7.5	15	80	9.0	cl	Moderately rich in stars; moderate brightness range; no central concentration; detached. Probably not a cluster.
06 51 46.0	+00 28 00	NGC 2301	6.0	15	80	8.0	cl	Rich in stars; large brightness range; strong central concentration; detached.
06 57 48.0	−04 37 00	NGC 2311	9.6	7	50	12.0	cl	Moderately rich in stars; moderate brightness range; no central concentration; detached.
07 00 37.0	+03 06 00	NGC 2319		16	12	8.9	ast?	(A) Scattered group of brighter stars.
07 04 08.0	+01 03 00	NGC 2324	8.4	8	133	12.0	cl	Rich in stars; moderate brightness range; slight central concentration; detached.
07 07 48.0	−05 43 12	NGC 2338					ast?	(S) Identification uncertain. Nice small cluster.

BRIGHT NEBULAE

RA h m s	Dec ° ′ ″	Name	Dim ′ Maj x min	Type	BC	Color	Notes
07 08 36.0	−04 19 00	IC 466	1 x 1	E+R	1-5	3-4	
06 32 18.0	+05 03 00	NGC 2237-9, 46	80 x 60	E	1-5	3-4	**Rosette Nebula**. A large, bright and mottled nebulosity. This roundish nebula is involved in a sparse cluster of naked eye stars. The NW sector contains many tiny absorption patches.
06 46 54.0	+01 19 00	NGC 2282	3 x 3	R	3-5	2-4	Same as **IC 2172**. A nebulosity with a mag 10 star in the SE sector.
06 44 52.4	+02 57 53	Parsamyan 15	0.6 x 0.4				Mag 12.0 star on NE edge.
06 38 00.0	+01 31 00	Sh2-282	25 x 10	E	3-5	3-4	Nebulosity with several stars liberally sprinkled throughout.

PLANETARY NEBULAE

RA h m s	Dec ° ′ ″	Name	Diam ″	Mag (P)	Mag (V)	Mag cent ★	Alt Name	Notes
07 09 22.4	−00 48 24	NGC 2346	52	11.8	11.6	11.4	PK 215+3.1	
07 00 27.5	+04 20 37	PN G210.0+3.9	230				PK 210+3.1	Intermittent or broken ring shape; mag 10.13v star WSW 3′.3; close pair of mag 11-12 stars SE 2′.8.
06 53 33.8	+03 08 26	PN G210.3+1.9	18	14.2	14.5	21.3	PK 210+1.1	Mag 10.61v star E 1′.5; mag 10.6 star NE 2′.6.
06 35 45.1	−00 05 38	PN G211.2−3.5	5	16.1		15.8	PK 211−3.1	N-S pair of mag 12 stars S 2′.3.
07 05 19.3	+02 46 56	PN G212.0+4.3	12	13.3	13.4	15.6	PK 212+4.1	Mag 10.8 star E 2′.2; mag 9.30v star S 2′.0.
07 22 57.6	+01 45 34	PN G214.9+7.8	67	16.3	14.3	16.5	PK 214+7.1	Mag 9.93v star W 3′.3.
07 36 08.2	+02 42 24	PN G215.6+11.1	122	15.4		19.5	PK 215+11.1	Roughly rectangular in shape, oriented ENE-WSW; mag 12.4 star at ENE end along with small, brighter patch of nebulosity.
06 56 14.5	−02 53 09	PN G216.0−0.2	73	17.5		20.9	PK 216−0.1	
06 41 34.7	−05 02 37	PN G216.3−4.4	15	17.9		19.1	PK 216−4.1	Mag 13.0 star W 2′.6; mag 14.1 star N 2′.9.
07 51 39.3	+03 00 29	PN G217.1+14.7	361	13.6	13.5:	17.1	PK 217+14.1	Close pair of mag 12 stars on SSE edge; mag 12.2 star N 3′.7 from center.
07 06 51.0	−03 05 10	PN G217.4+2.0	19	14.1			PK 217+2.1	Mag 11.7 star W 0′.7.
07 26 34.2	−05 21 52	PN G221.7+5.3	32	14.2	14.8		PK 221+5.1	Slightly elongated NE-SW; mag 11.2 star NE 2′.9; mag 12.6 star SW 3′.

116 GALAXIES 116

RA h m s	Dec ° ′ ″	Name	Mag (V)	Dim ′ Maj x min	SB	Type Class	PA	Notes
05 20 12.5	+05 50 10	CGCG 421-30	14.5	0.8 x 0.4	13.1	Sb	18	Mag 9.75v star SSW 4′.1.
05 21 56.6	+03 29 15	IC 412	13.7	1.3 x 0.6	13.2	S?	3	Multi-galaxy system; broad, faint extension N; pair with IC 413 E edge.
05 21 58.9	+03 28 54	IC 413	13.8	1.0 x 0.9	13.5	S?	159	Multi-galaxy system; mag 12.2 star on N edge.
05 21 55.0	+03 20 31	IC 414	14.0	0.5 x 0.3	11.8		141	Mag 9.34v star SE 2′.2.
05 21 04.2	+02 36 47	MCG +0-14-20	15.3	0.8 x 0.7	14.5		114	Mag 14.4 star NW edge.
05 20 48.4	+03 15 33	MCG +1-14-27	14.2	0.4 x 0.4	12.0	S0?		Pair with MCG +1-14-28 SE 1′.3.
05 20 53.0	+03 15 02	MCG +1-14-28	14.4	0.5 x 0.3	12.2	S?	95	Pair with MCG +1-14-27 WNW 1′.3.
05 23 38.2	+05 34 07	MCG +1-14-36	15.1	0.6 x 0.5	13.7		36	Mag 11.7 star SW 1′.7.
05 28 01.9	-05 18 41	NGC 1924	12.5	1.6 x 1.2	13.1	SB(r)bc II-III	140	Mag 8.04v star NW 6′.7.
05 21 04.1	+04 00 21	UGC 3294	13.0	2.7 x 1.5	14.4	SA(rs)b II-III	130	Mag 6.53v star E 3′.8.
05 21 21.8	+04 53 13	UGC 3296	13.0	1.7 x 1.3	13.8	Sab	140	

(Continued from previous page)

GALAXIES

RA h m s	Dec ° ′ ″	Name	Mag (V)	Dim ′ Maj x min	SB	Type Class	PA	Notes
05 22 59.4	-00 08 28	UGC 3301	14.2	1.3 x 0.9	14.3	SB0/a	44	
05 24 58.9	+04 30 15	UGC 3303	15.4	3.6 x 2.8	17.8	Im?	126	Dwarf? Low surface brightness; mag 11.2 star dead center; numerous other superimposed stars.
05 25 16.7	+00 25 13	UGC 3306	13.6	1.2 x 0.6	13.1	S0⁻:	52	Mag 6.16v star NE 9.′7.
05 37 54.2	+00 07 25	UGC 3331	14.9	1.1 x 0.8	14.6	Sb	132	Short N-S bar; mag 11.7 star E 1.′8.
06 15 06.2	+00 11 30	UGC 3433	15.2	2.0 x 0.6	15.2	SAB(s)bc: II	92	Very knotty, many stars superimposed.
06 21 50.4	+00 21 55	UGC 3457	14.0	1.2 x 0.8	13.8	S0/a	41	

OPEN CLUSTERS

RA h m s	Dec ° ′ ″	Name	Mag	Diam ′	No. ★	B ★	Type	Notes
05 32 37.0	+00 11 18	Be 20		2	20	15.0	cl	
06 37 47.0	−00 52 42	Be 24		6	25	17.0	cl	
06 37 05.0	+03 03 48	ClvdB 1	9.5	5	39		cl	
06 21 33.0	+02 20 00	Cr 91	6.4	14	20		cl:	Few stars; moderate brightness range; not well detached. Probably not a cluster.
06 22 54.0	+05 07 00	Cr 92	8.6	11			cl	Probably not a cluster.
06 30 18.0	+02 52 00	Cr 96	7.3	12	15	8.8	cl	Few stars; moderate brightness range; not well detached.
06 30 58.1	+05 50 00	Cr 97	5.4	25	15		cl	Few stars; moderate brightness range; not well detached.
06 36 30.0	+04 50 00	Cr 104	9.6	20	15		cl	Few stars; moderate brightness range; not well detached; involved in nebulosity. (A) Appears mostly as loose 20′ high "integral sign" chain of stars.
06 37 06.0	+05 58 00	Cr 106	4.6	35	20		cl	Few stars; moderate brightness range; not well detached.
06 37 42.0	+04 45 00	Cr 107	5.1	30	204	7.1	cl	Few stars; moderate brightness range; not well detached. Probably not a cluster.
06 38 39.0	+02 03 00	Cr 110	10.5	18	70	11.0	cl	Moderately rich in stars; small brightness range; not well detached.
06 30 58.0	−04 11 36	Cz 26		5	35		cl	
06 23 18.0	+04 39 00	Do 22		8	10		cl:	Few stars; moderate brightness range; not well detached. Probably not a cluster. **UGC 3459** NE 11.′1.
05 35 16.0	−05 23 00	NGC 1976		47			cl	
05 35 18.0	−04 49 00	NGC 1977					cl	
05 35 26.0	−05 55 00	NGC 1980	2.5	15	30		cl	
05 35 12.0	−04 26 00	NGC 1981	4.2	28	20	10.0	cl	Few stars; large brightness range; no central concentration; detached; involved in nebulosity.
05 53 46.0	+00 25 00	NGC 2112	9.1	18	50	10.0	cl	Moderately rich in stars; moderate brightness range; slight central concentration; detached; involved in nebulosity.
06 03 07.4	+05 44 00	NGC 2143		11	12	10.0	cl?	(A) sparse.
06 09 37.0	+04 42 42	NGC 2180		6	>20	12.0	cl?	
06 11 41.0	−03 36 00	NGC 2184		33	30	9.0	cl?	
06 12 08.0	+05 27 30	NGC 2186	8.7	5	30	12.0	cl	Moderately rich in stars; moderate brightness range; slight central concentration; detached.
06 16 51.0	+06 00 00	NGC 2202		7	12	9.0	cl?	(S) Sparse field.
06 23 42.0	−04 41 00	NGC 2219		10	25	7.3	cl	
06 28 02.0	−04 51 00	NGC 2232	4.2	45	20	5.0	cl	Few stars; moderate brightness range; no central concentration; detached.
06 32 19.0	+04 51 00	NGC 2244	4.8	30	100	7.0	cl	Rich in stars; large brightness range; slight central concentration; detached; involved in nebulosity (Rosette Nebula). The brightest star is a non-member.
06 33 47.0	−05 05 00	NGC 2250		10	25	8.7	ast?	Few stars; moderate brightness range; not well detached. (A) Oblong (E-W) 10′ x 5.5′ group.
06 34 19.8	+05 19 00	NGC 2252	7.7	18	30	9.0	cl	Moderately rich in stars; moderate brightness range; no central concentration; detached; involved in nebulosity. (A) DSS image shows obvious N-S stream of bright stars, with large bright nebula involved on SW side.
06 38 04.0	−01 28 00	NGC 2260		20	50	8.0	cl?	
06 39 38.7	+01 08 30	NGC 2262	11.3	4	180		cl	Moderately rich in stars; small brightness range; slight central concentration; detached. (A) Has inner compressed part 2′ in diameter with ~80 stars.
05 38 45.0	−02 36 00	Sigma Ori		10	4		cl	

BRIGHT NEBULAE

RA h m s	Dec ° ′ ″	Name	Dim ′ Maj x min	Type	BC	Color	Notes
05 33 24.0	−00 37 00	IC 423	6 x 4	R	2-5	1-4	Shaped like a comet-head. It is extended roughly N-S and fans out slightly at the N end. On blue photos the interior is darker, especially in the S part.
05 33 36.0	−00 25 00	IC 424	2 x 2	R	2-5	1-4	Two faint stars (aligned E-W, sep. 20″) involved in the S part.
05 36 48.0	−00 15 00	IC 426	7 x 7	R	2-5	1-4	Very irregular in shape with an extension, about 3′ x 1′, off the SE sector.
05 40 18.0	−01 27 00	IC 431	5 x 3	R	1-5	1-4	
05 40 54.0	−01 29 00	IC 432	8 x 4	R	1-5	1-4	
05 41 00.0	−02 24 00	IC 434	60 x 10	E	1-5	3-4	
05 43 00.0	−02 19 00	IC 435	4 x 3	R	1-5		
05 35 06.0	−04 44 00	NGC 1973	5 x 5	E+R	1-5	1-4	Nebulosity associated with the variable star KX Orionis.
05 35 24.0	−04 41 00	NGC 1975	10 x 5	E+R	1-5	1-4	
05 35 24.0	−05 27 00	NGC 1976	65 x 60	E+R	1-5	3-4	= **M 42**, **Orion Nebula**. Very bright and mottled. This galactic nebula is visible to the naked eye.
05 35 30.0	−04 52 00	NGC 1977	20 x 10	E+R	1-5	1-4	Associated with a sparse group of about a dozen stars; nebulosity brightest in the S part.
05 35 24.0	−05 54 00	NGC 1980	14 x 14	E	3-5	4-4	The southernmost semi-detached part of the **Orion Nebula**.
05 35 36.0	−05 16 00	NGC 1982	20 x 15	E+R	1-5	3-4	= **M 43**. Bright nebulosity with a mag 7 star involved; part of **M 42**, **Orion Nebula** and separated from it by a dark lane.
05 41 36.0	−02 16 00	NGC 2023	10 x 10	E+R	1-5	1-4	
05 41 54.0	−01 51 00	NGC 2024	30 x 30	E	2-5	3-4	**The Flame**. This nebula is approximately bisected by a dark lane 3′ wide, aligned roughly N-S; E section of nebula slightly larger.
05 46 18.0	+00 00 00	NGC 2064	1.5 x 1.0	R	2-5	2-4	A small, relatively bright patch; extended roughly N-S.
05 46 30.0	+00 06 00	NGC 2067	8 x 3	R	2-5	2-4	This nebula is considerably lower in surface brightness than nearby M 78. It is a diffuse patch 3′ across with a narrow extension SSW, nearly to NGC 2064.
05 46 42.0	+00 03 00	NGC 2068	8 x 6	R	1-5	2-4	= **M 78**. Brightest and largest in a group of four nebulae with NGCs 2064, 2067 and 2071. This slightly oval nebula has two 10th mag stars involved.
05 47 12.0	+00 18 00	NGC 2071	7 x 5	R	2-5	2-4	Second brightest in a group of four nebulae with NGCs 2064, 2067 and 2071. There is a 10th mag star involved and another field star off the NW edge.
06 32 18.0	+05 03 00	NGC 2237-9, 46	80 x 60	E	1-5	3-4	**Rosette Nebula**. A large, bright and mottled nebulosity. This roundish nebula is involved in a sparse cluster of naked eye stars. The NW sector contains many tiny absorption patches.
05 48 00.0	+01 00 00	Sh2-276	600 x 30	E	3-5	3-4	**Barnard's Loop**. The NE section is its brightest part.
05 20 00.0	−05 40 00	Sh2-278	25 x 20	E	4-5	3-4	
06 38 00.0	+01 31 00	Sh2-282	25 x 10	E	3-5	3-4	Nebulosity with several stars liberally sprinkled throughout.
05 39 12.0	+04 10 00	vdB 49	8 x 3	R	3-5	1-4	

DARK NEBULAE

RA h m s	Dec ° ′ ″	Name	Dim ′ Maj x min	Opacity	Notes
05 40 52.9	−02 28 00	B 33	6 x 4	4	**Horsehead Nebula**. Dark mass on nebulous strip extending S from Zeta Orionis.

116 (Continued from previous page) PLANETARY NEBULAE 116

RA h m s	Dec ° ′ ″	Name	Diam ″	Mag (P)	Mag (V)	Mag cent ★	Alt Name	Notes
06 04 47.8	+03 56 36	PN G204.0−8.5	153	>16.0	15.3	19.8	PK 204−8.1	Annular with a slightly flattened ring shape, oriented NE-SW; bright ring portion is very thin.
06 23 55.0	+05 30 11	PN G204.8−3.5	11				PK 204−3.1	Mag 0.02v star SW 3′0; mag 10.2 star E 6′1.
06 35 45.1	−00 05 38	PN G211.2−3.5	5	16.1		15.8	PK 211−3.1	N-S pair of mag 12 stars S 2′3.

117 GALAXIES 117

RA h m s	Dec ° ′ ″	Name	Mag (V)	Dim ′ Maj x min	SB	Type Class	PA	Notes
04 20 43.6	+02 19 33	CGCG 393-7	14.1	0.7 x 0.5	12.8		24	Mag 14.0 star NW edge; mag 11.8 star W 1′5.
04 29 32.2	+00 12 40	CGCG 393-24	14.1	0.8 x 0.4	12.7		108	Mag 11.2 star SW 2′8.
04 38 44.3	+00 11 26	CGCG 393-54	13.9	0.3 x 0.3	11.3	E		= **Mkn 1083**.
05 12 49.0	+02 43 07	CGCG 395-16	13.7	0.5 x 0.4	11.8	S?		Mag 11.0 star NW 2′5.
04 53 25.8	+04 03 38	CGCG 420-15	13.8	0.6 x 0.4	12.1	Sa	36	Mag 11.3 star WSW 3′8.
04 58 12.4	+03 59 07	CGCG 420-21	14.2	0.8 x 0.4	12.8		168	Mag 9.13v star W 6′7.
05 03 34.3	+04 40 45	CGCG 420-32	14.0	0.3 x 0.3	11.2			Located 2′2 ESE of a mag 9.14v star.
05 20 12.5	+05 50 10	CGCG 421-30	14.5	0.8 x 0.4	13.1	Sb	18	Mag 9.75v star SSW 4′1.
04 19 14.2	+03 20 48	IC 365	13.9	1.0 x 0.5	13.0	(R′)SB(rs)0$^+$:	30	Close pair of mag 9.93v and 11.3 stars SSW 6′0.
04 30 42.8	−04 52 14	IC 373	13.9	0.8 x 0.8	13.3	(R)SB(rs)0$^+$		Mag 7.83v star E 7′6.
04 46 25.8	+03 30 19	IC 392	12.7	1.0 x 0.8	12.3	S0	167	Mag 10.30v star E 2′3.
04 49 33.9	+00 15 11	IC 395	12.9	0.9 x 0.6	12.1	S0:	130	Mag 10.63v star SW 4′9.
05 19 33.7	+03 19 00	IC 409	13.9	0.9 x 0.7	13.3		30	Mag 11.8 star S 1′3.
05 21 56.6	+03 29 15	IC 412	13.7	1.3 x 0.6	13.2	S?	3	Multi-galaxy system; broad, faint extension N; pair with IC 413 E edge.
05 21 58.9	+03 28 54	IC 413	13.8	1.0 x 0.9	13.5	S?	159	Multi-galaxy system; mag 12.2 star on N edge.
05 21 55.0	+03 20 31	IC 414	14.0	0.5 x 0.3	11.8		141	Mag 9.34v star SE 2′2.
04 46 26.5	−02 40 49	IC 2093	14.3	0.6 x 0.4	12.6		138	Mag 10.06v star E 8′9.
04 48 45.9	−05 07 31	IC 2095	15.2	1.4 x 0.2	13.7	Scd pec sp	54	Mag 7.17v star NE 10′9.
04 50 44.8	−05 25 12	IC 2098	13.7	2.1 x 0.3	13.1	Scd: sp	96	Mag 10.50v star SW 4′5.
04 51 55.2	−04 57 09	IC 2102	13.8	1.2 x 0.9	13.7	SB(rs)cd II-III	144	Mag 12.8 star W 5′6.
04 08 23.6	−01 04 34	MCG +0-11-15	14.9	0.9 x 0.2	12.9	Sb	144	
04 09 13.9	−01 28 17	MCG +0-11-20	14.4	0.4 x 0.4	12.3			**MCG +0-11-19** SW 0′5.
04 12 13.9	+00 02 42	MCG +0-11-25	15.2	0.6 x 0.5	13.7	Sbc	3	Mag 11.1 star SE 1′6.
04 17 23.6	+00 18 21	MCG +0-11-43	14.1	0.7 x 0.7	13.1			Mag 13.3 star NE edge; mag 14.4 star NW edge.
04 17 28.8	+02 13 26	MCG +0-11-45	14.9	0.7 x 0.5	13.6		30	Mag 8.87v star S 4′7.
04 17 35.2	+01 32 23	MCG +0-11-46	14.9	0.7 x 0.3	13.0		120	MCG +0-11-47 NE 2′7.
04 17 44.4	+01 33 49	MCG +0-11-47	14.0	0.9 x 0.6	13.2		126	Star superimposed W of center; MCG +0-11-46 SW 2′7.
04 17 51.8	+00 12 38	MCG +0-11-48	16.3	0.8 x 0.3	14.6	Sbc	18	Mag 15 star N end; mag 11.8 star SSW 1′0.
04 19 07.9	−02 09 20	MCG +0-11-52	14.9	0.9 x 0.4	13.7		144	Mag 13.0 star on NE edge.
04 19 36.7	−00 26 55	MCG +0-12-1	15.1	0.5 x 0.4	13.2	Sc		
04 20 33.0	−02 01 53	MCG +0-12-10	14.0	0.7 x 0.4	12.5	S?	60	Pair of mag 12 stars SE 1′9.
04 20 55.9	+02 37 58	MCG +0-12-11	13.6	1.3 x 1.1	13.9	(R′)SB(rs)0$^+$?	87	Strong E-W core with faint dark patches N and S.
04 21 33.5	−02 20 32	MCG +0-12-14	15.0	0.7 x 0.5	13.7	SBbc	30	
04 21 42.8	−02 12 21	MCG +0-12-15	14.0	0.4 x 0.4	11.9			Mag 14.9 star on NE edge; **MCG +0-12-17** NE 1′9.
04 22 03.9	−02 39 25	MCG +0-12-19	15.0	0.5 x 0.5	13.3	Sbc		
04 27 39.0	−02 32 39	MCG +0-12-30	15.0	1.0 x 0.8	14.6	Sc	114	
04 29 21.1	−00 12 27	MCG +0-12-31	15.5	0.7 x 0.3	13.6	Sb	135	
04 29 54.3	−00 45 03	MCG +0-12-33	14.8	0.9 x 0.7	14.1	Sab		
04 32 00.6	+01 28 40	MCG +0-12-43	14.7	0.8 x 0.3	13.0		90	Mag 10.4 star S 1′9.
04 34 32.0	+01 40 52	MCG +0-12-46	15.5	0.7 x 0.4	14.0	Sb	24	Mag 11.43v star S 2′1.
04 35 53.9	−02 14 01	MCG +0-12-48	14.9	0.9 x 0.2	12.9		12	Located 2′0 W of a mag 7.68v star.
04 35 58.0	−02 02 42	MCG +0-12-49	14.3	0.8 x 0.6	13.4	SBb	45	Strong N-S bar; mag 12.3 star N 0′7.
04 35 58.0	−02 10 15	MCG +0-12-50	14.6	0.8 x 0.3	12.9	S?	9	Mag 7.68v star S 3′7.
04 36 18.6	−02 49 55	MCG +0-12-51	12.7	1.5 x 1.4	13.4	(R′)SB(rs)0$^+$:	135	= **Hickson 30A**. Mag 11.7 star on SE edge; **Hickson 30C** NE 2′2.
04 36 30.5	−02 52 00	MCG +0-12-54	13.2	1.1 x 0.8	13.0	(R′)SAB(rs)0$^+$:	21	= **Hickson 30B**. **Hickson 30D** NE 2′1.
04 37 37.6	−00 17 35	MCG +0-12-58	14.0	0.4 x 0.3	11.5		12	UGC 3109 NW 2′0.
04 40 37.3	+00 30 17	MCG +0-12-65	14.6	0.5 x 0.5	12.9			
04 42 30.2	−01 45 28	MCG +0-12-71	14.3	0.8 x 0.4	13.0	S0	54	Mag 8.91v star NW 7′6.
04 44 19.4	−02 09 58	MCG +0-13-2	14.6	0.5 x 0.5	13.0			Located between a pair of mag 11.8 and 12.8 stars.
04 46 25.8	+00 21 55	MCG +0-13-5	14.0	1.0 x 0.7	13.4	S?	69	Mag 12.5 star WNW 2′2.
04 47 29.5	−01 12 53	MCG +0-13-9	14.9	0.7 x 0.5	13.6		24	
04 48 29.8	−01 32 20	MCG +0-13-13	13.8	0.9 x 0.8	13.3	SB?	177	Strong SE-NW bar; mag 11.05v star NW 2′2.
04 51 09.4	+02 22 01	MCG +0-13-17	14.3	0.6 x 0.5	12.9	S?	153	
04 51 36.4	−02 33 40	MCG +0-13-18	14.4	0.8 x 0.3	12.7	S?	18	Mag 12.8 star S 1′4.
04 53 10.5	+02 20 23	MCG +0-13-23	14.7	0.7 x 0.7	13.7	Sbc		Mag 12.5 star on E edge.
04 53 54.2	+01 40 55	MCG +0-13-24	14.7	0.5 x 0.5	13.1	S0		Mag 6.60v star S 6′8.
04 54 36.0	−01 22 16	MCG +0-13-29	14.5	0.5 x 0.5	12.9	S?		Mag 12.4 star E 2′1.
04 54 40.5	+02 02 04	MCG +0-13-30	15.0	0.7 x 0.2	12.8	Sb	150	Mag 10.48v star NW 2′2.
04 54 48.2	+01 09 53	MCG +0-13-32	14.8	0.7 x 0.3	13.0		81	Mag 9.29v star SW 3′9.
04 55 58.6	+02 14 00	MCG +0-13-35	13.7	0.5 x 0.3	11.5	Sa?	63	
04 56 27.1	+00 42 38	MCG +0-13-40	14.3	0.5 x 0.5	12.6			
04 56 42.8	+01 02 33	MCG +0-13-41	15.4	0.4 x 0.4	13.2			Mag 9.79v star N 2′2.
04 56 52.3	+02 47 51	MCG +0-13-42	14.4	0.6 x 0.6	13.2	S?		Mag 7.81v star SE 1′6.
04 56 53.5	−00 35 16	MCG +0-13-43	14.6	0.8 x 0.3	12.9	Irr	114	
04 57 29.7	−00 51 39	MCG +0-13-45	14.6	0.3 x 0.3	11.8			**CGCG 394-46** WNW 5′6.
04 57 32.8	−01 07 22	MCG +0-13-46	14.2	0.8 x 0.5	13.3	E/S0	123	Mag 6.24v star NW 5′1.
04 57 57.7	+00 47 48	MCG +0-13-47	14.0	0.9 x 0.4	13.0	E/S0	120	Mag 8.26v star NNE 4′9.
04 58 14.7	−00 10 19	MCG +0-13-51	14.3	1.0 x 0.3	12.9		135	UGC 3214 NW 5′3.
04 59 01.8	−01 29 37	MCG +0-13-57	14.9	0.5 x 0.3	12.6	Sd	75	
04 59 13.2	−00 35 23	MCG +0-13-58	14.5	0.8 x 0.2	12.4	Sbc	15	
04 59 23.3	−01 09 37	MCG +0-13-59	14.6	1.0 x 0.2	12.7	SB(s)bc? sp	42	
04 59 46.0	−00 09 24	MCG +0-13-61	14.7	0.7 x 0.2	12.4	Sb	114	Mag 8.09v star E 7′6.
05 01 43.3	−00 30 20	MCG +0-13-62	14.4	0.5 x 0.3	12.2	S?	78	
05 04 11.8	−01 09 20	MCG +0-13-68	14.4	0.9 x 0.4	13.1	SB?	63	Very close pair of mag 11-12 stars NW 2′2.
05 08 50.3	−00 35 13	MCG +0-14-1	14.0	0.7 x 0.3	12.2		21	
05 09 09.8	−00 41 47	MCG +0-14-2	14.4	0.9 x 0.6	13.5	Sc	27	
05 09 40.3	−00 15 08	MCG +0-14-3	14.5	0.7 x 0.4	13.0	S?	123	
05 10 06.3	+00 54 40	MCG +0-14-4	15.3	0.5 x 0.3	13.1	Sb	96	UGC 3256 E 2′2; mag 11.3 star NE 2′8.
05 10 10.7	−00 43 05	MCG +0-14-8	14.3	0.8 x 0.8	13.9	E:		Mag 6.11v star NNW 9′4.
05 10 22.5	−00 43 45	MCG +0-14-9	14.6	0.8 x 0.2	12.4			

(Continued from previous page)

GALAXIES

RA h m s	Dec ° ′ ″	Name	Mag (V)	Dim ′ Maj x min	SB	Type Class	PA	Notes
05 10 48.2	−02 40 57	MCG +0-14-10	13.3	0.7 x 0.5	12.0	I0 pec?	165	= **Mkn 1094**; mag 10.40v star E 4′3; mag 10.58v star SW 5′8.
05 10 56.5	−00 43 03	MCG +0-14-11	15.5	0.7 x 0.5	14.2	Irr		Mag 15.3 star on E edge; close pair of mag 10.61v, 12.2 stars N 3′6.
05 13 04.9	+00 58 23	MCG +0-14-15	14.9	0.6 x 0.5	13.4	SBbc	111	Mag 11.9 star SW 0′7.
05 13 43.9	+01 57 09	MCG +0-14-16	14.3	0.8 x 0.5	13.1		111	Located 3′2 ESE of mag 6.11v star.
05 14 11.7	+01 52 44	MCG +0-14-17	13.8	0.9 x 0.4	12.6		54	Mag 13.0 star NW 1′9.
05 21 04.2	+02 36 47	MCG +0-14-20	15.3	0.8 x 0.7	14.5		114	Mag 14.4 star NW edge.
04 21 19.5	+03 27 14	MCG +1-12-1	15.5	0.8 x 0.7	14.7	Spiral	135	
04 53 56.9	+03 35 35	MCG +1-13-8	14.5	0.9 x 0.9	14.1	Sb		Mag 12.1 star SW 2′2.
05 12 08.6	+05 27 39	MCG +1-14-3	14.3	0.7 x 0.6	13.2	SB?	30	Mag 10.40v star W 1′5.
05 20 48.4	+03 15 33	MCG +1-14-27	14.2	0.4 x 0.4	12.0	S0?		Pair with MCG +1-14-28 SE 1′3.
05 20 53.0	+03 15 02	MCG +1-14-28	14.4	0.5 x 0.3	12.2	S?	95	Pair with MCG +1-14-27 WNW 1′3.
05 23 38.2	+05 34 07	MCG +1-14-36	15.1	0.6 x 0.5	13.7		36	Mag 11.7 star SW 1′7.
04 24 29.9	−03 39 36	MCG −1-12-3	14.1	0.7 x 0.6	13.0	Sb	30	Mag 5.17v star Xi Eridani SW 13′3.
04 31 23.9	−03 55 33	MCG −1-12-15	13.8	1.3 x 0.7	13.5	SB(r)c: II-III	81	Mag 9.41v star SE 13′5.
04 33 01.0	−04 11 21	MCG −1-12-28	14.8	1.6 x 0.2	13.4	Sd: sp	108	NGC 1612 NE 3′2; mag 13.1 star NW 2′9.
04 39 20.0	−04 56 06	MCG −1-12-41	14.2	0.9 x 0.9	13.8			1′8 W mag 8 star.
04 48 22.7	−03 22 00	MCG −1-13-8	13.4	0.9 x 0.9	13.0			Mag 9.09v star NE 5′1.
04 48 34.5	−03 52 09	MCG −1-13-11	13.8	1.2 x 0.3	12.6	S0^{+} pec sp	153	Mag 10.91v star S 3′5.
04 48 43.0	−04 58 40	MCG −1-13-12	13.2	1.3 x 0.6	12.8	SB(s)d III-IV	12	Mag 13.0 star N edge; mag 7.17v star E 7′6.
04 50 29.0	−03 31 21	MCG −1-13-16	13.5	1.2 x 0.5	12.8	SB(rs)0°:	159	Mag 10.21v star NE 2′5.
04 50 38.6	−03 08 31	MCG −1-13-17	14.0	1.0 x 0.4	12.9	SA(s)a pec?	39	Mag 8.31v star and MCG −1-13-22 E 12′2.
04 51 18.5	−05 57 38	MCG −1-13-21	14.5	1.3 x 0.3	13.4	Scd: sp	24	Mag 9.70v star N 8′2; **MCG −1-13-20** W 3′4.
04 51 26.9	−03 07 22	MCG −1-13-22	13.6	1.6 x 0.4	13.0	Sb: sp II-III?	102	Located 1′5 N of mag 8.31v star.
04 51 41.6	−03 48 37	MCG −1-13-25	13.7	0.7 x 0.5	12.5	SAB(s)0^{+} pec:	39	Elongated anonymous galaxy S 2′0.
04 52 25.1	−03 34 00	MCG −1-13-30	13.0	0.8 x 0.8	12.4	SA(s)0° pec?		Mag 9.14v star SSW 9′8.
04 53 18.0	−03 00 42	MCG −1-13-33	13.9	1.3 x 0.9	13.9	SB(s)cd II-III	114	Mag 9.51v star NW 6′7.
04 55 03.4	−04 06 10	MCG −1-13-35	14.1	1.6 x 0.2	12.8	Sb sp	15	Mag 10.19v star E 6′3.
04 56 20.6	−04 26 15	MCG −1-13-37	13.8	0.9 x 0.9	13.4	SBbc		Mag 9.97v star NE 6′9.
05 03 17.3	−02 56 14	MCG −1-13-50	13.4	2.7 x 0.4	13.3	Sb pec sp	96	Faint, narrow dark lane across core; mag 8.29v star S 2′8.
05 15 41.7	−03 26 39	MCG −1-14-7	13.9	0.5 x 0.5	12.3			Mag 7.70v star NW 6′5; note: **MCG −1-14-8** is 2′6 E of this star.
04 17 00.2	+00 50 05	NGC 1541	13.5	1.4 x 0.6	13.2	S0^{+}:	83	Mag 9.01v star NNE 5′9.
04 17 14.1	+04 46 53	NGC 1542	13.9	1.3 x 0.5	13.3	Sab	128	
04 19 38.1	+02 24 36	NGC 1550	12.0	2.2 x 1.9	13.4	SA(s)0^{-}:	30	Small, bright, round nucleus with faint envelope.
04 20 17.6	−00 41 34	NGC 1552	12.9	1.8 x 1.2	13.6	SAB(r)0^{+}?	110	Elongated **MCG +0-12-8** SE 5′3.
04 24 25.4	−00 44 47	NGC 1568	15.1	1.0 x 0.7	14.6	S0/a	135	Stellar UGC 3031 WNW 1′3; stellar **Mkn 615** E 3′5; mag 8.96v star W 6′3.
04 26 19.0	−03 37 17	NGC 1576	13.3	1.3 x 0.8	13.3	S0^{-}:	127	
04 28 18.6	−05 10 42	NGC 1580	13.5	0.9 x 0.8	13.0	Sbc II	95	
04 30 38.2	−00 18 18	NGC 1586	13.2	1.7 x 0.9	13.5	SB(s)bc II-III	155	
04 30 40.0	+00 39 40	NGC 1587	11.7	1.7 x 1.5	12.7	E pec	70	Mag 9.70v star W 5′9; NGC 1588 on E edge.
04 30 44.1	+00 39 42	NGC 1588	12.9	1.1 x 0.6	12.5	E pec?	175	Small, bright core slightly offset to N; NGC 1587 on W edge.
04 30 45.5	+00 51 44	NGC 1589	11.8	3.2 x 1.0	12.9	Sab sp III	160	Low surface brightness **UGC 3072** E 6′0.
04 30 51.7	−05 47 59	NGC 1594	13.0	1.8 x 1.3	13.7	SB(rs)bc I-II	60	Bright knot or star on W edge.
04 31 38.8	−04 35 21	NGC 1599	13.7	0.8 x 0.7	12.9	SB(s)c pec: II-III	174	Mag 9.82 star 1′2 E.
04 31 39.9	−05 05 10	NGC 1600	10.9	2.5 x 1.7	12.5	E3	15	NGC 1601 on N edge; NGC 1603 E 2′6.
04 31 41.9	−05 03 38	NGC 1601	13.8	0.7 x 0.3	12.0	S0: sp	90	Located on N edge of NGC 1600.
04 31 49.9	−05 05 42	NGC 1603	13.8	0.8 x 0.6	12.8	E?	37	Located 2′6 E of NGC 1600.
04 31 58.9	−05 22 11	NGC 1604	13.7	1.2 x 0.8	13.5	S0	71	Mag 10.98v star E 2′3; small, faint anonymous galaxy and mag 14.0 star E 3′5.
04 32 03.3	−05 01 58	NGC 1606	14.9	0.5 x 0.5	13.2	SAB(r)0^{+}:		Mag 7.62v star E 4′5.
04 32 02.9	−04 27 40	NGC 1607	13.2	1.2 x 0.5	12.6	S0/a	50	Mag 7.80v star N 4′8.
04 32 06.1	+00 34 02	NGC 1608	13.4	1.6 x 0.6	13.3	S0°:	130	Small, round core with uniform envelope.
04 32 45.2	−04 22 24	NGC 1609	13.5	1.3 x 0.8	13.5	(R′)S0^{+} pec:	95	Mag 11.8 star W 3′7.
04 34 13.9	−04 42 03	NGC 1610	14.5	1.0 x 0.7	13.9	Sb	159	Mag 9.76v star S 4′2.
04 33 05.9	−04 17 51	NGC 1611	13.4	1.9 x 0.6	13.4	(R′)SB(rs)0^{+} pec?	108	NGC 1613 NE 5′2.
04 33 13.1	−04 10 24	NGC 1612	13.4	1.3 x 1.0	13.5	SB(r)0/a	142	Mag 9.16v star NNE 8′3; MCG −1-12-28 SW 3′2.
04 33 25.5	−04 15 56	NGC 1613	13.7	1.1 x 0.7	13.2	SAB(rs)0^{+}?	45	NGC 1611 SW 5′2.
04 36 06.7	−03 08 54	NGC 1618	12.7	2.3 x 0.8	13.2	SB(r)b: II	26	Mag 9.19v star W 9′1; mag 11.6 star SE 3′8; mag 3.93v star Nu Eridani SSE 12′7.
04 36 37.1	−00 08 39	NGC 1620	12.3	2.9 x 1.0	13.3	SAB(rs)bc II	25	Bright knot or star on N edge; mag 8.77v star E 4′7.
04 36 25.0	−04 59 15	NGC 1621	13.6	1.3 x 0.8	13.6	E^{+}	105	Mag 10.62v star NW 3′9.
04 36 36.6	−03 11 23	NGC 1622	12.5	3.6 x 0.7	13.4	SAB(r)ab:	33	**MCG −1-12-37** NNE 9′3; mag 3.93v star Nu Eridani S 10′8.
04 37 06.1	−03 18 12	NGC 1625	12.3	2.1 x 0.5	12.2	SB(rs)b: II-III:	131	Mag 3.93v star Nu Eridani W 12′2.
04 37 38.2	−04 53 15	NGC 1627	12.8	1.6 x 1.5	13.6	SA(r)c pec II	138	Almost stellar nucleus; mag 11.8 star S 2′8.
04 37 36.3	−04 42 57	NGC 1628	13.4	1.8 x 0.4	12.9	Sb pec sp	6	Small galaxy or star on W edge.
04 40 07.5	−00 32 50	NGC 1635	12.4	1.4 x 1.3	12.9	(R)SB(r)0/a	5	
04 41 28.5	−02 51 17	NGC 1637	10.8	3.3 x 2.7	13.1	SAB(rs)c II	24	Dark lane N and NE of core.
04 41 36.4	−01 48 34	NGC 1638	12.0	1.8 x 1.2	12.8	SAB(rs)0°?	70	Bright knot or star EW of core; mag 9.62v star N 6′2.
04 42 55.3	+00 37 05	NGC 1642	12.6	1.7 x 1.2	13.2	SA(rs)c: I-II	145	Mag 10.37v star W 3′6.
04 43 43.9	−05 19 06	NGC 1643	13.3	1.2 x 1.1	13.4	SB(r)bc pec?	147	Very small, elongated galaxy, or bright knot, extends from S edge.
04 44 06.2	−05 28 00	NGC 1645	13.0	2.1 x 0.9	13.6	(R′)SB(rs)0^{+} pec	85	
04 45 47.3	−02 23 36	NGC 1653	12.0	1.5 x 1.5	12.9	E^{+}:		
04 45 48.4	−02 05 02	NGC 1654	13.4	0.8 x 0.8	12.8	(R′)SB(r)a		Mag 10.18v star 2′2 W.
04 45 53.2	−05 08 12	NGC 1656	12.9	1.5 x 0.8	13.0	S0^{+} pec:	123	Mag 9.65v star N 3′2.
04 46 07.2	−02 04 38	NGC 1657	13.8	1.0 x 0.7	13.3	SAB(rs)bc II-III	150	Mag 9.43v star E 5′1.
04 46 29.9	−04 47 16	NGC 1659	12.5	1.6 x 1.1	12.9	SA(r)bc pec III	50	
04 47 07.7	−02 03 14	NGC 1661	13.2	1.4 x 0.9	13.3	SA(s)bc pec: II-III	35	Short E-W row of four mag 11 - 14 stars S 1′3.
04 48 17.4	−05 25 38	NGC 1665	12.8	1.7 x 1.1	13.4	SA(s)0^{+} pec?	50	Faint, elongated anonymous galaxy N 4′8, possibly **IC 2094**?
04 49 42.7	−02 45 40	NGC 1670	12.7	1.3 x 0.7	12.5	SA0°:	72	
04 50 52.0	−04 53 34	NGC 1677	14.0	1.2 x 0.4	13.1	S0° pec sp	139	
04 51 35.6	−02 37 25	NGC 1678	13.2	1.2 x 0.9	13.1	SA0° pec?	72	Small anonymous galaxy NE 2′5.
04 51 50.4	−05 48 15	NGC 1681	12.9	1.3 x 1.0	13.1	Sb pec:	142	Mag 12 star on W edge.
04 52 19.9	−03 06 22	NGC 1682	13.5	0.9 x 0.9	13.2	E/S0		Mag 7.59v star S 4′1; NGC 1684 E 2′9.
04 52 17.7	−03 01 32	NGC 1683	14.7	0.9 x 0.4	13.4	Sa	162	Mag 10.56v star W 4′7.
04 52 31.5	−03 06 21	NGC 1684	12.0	2.2 x 1.7	13.5	E^{+} pec:	93	Mag 7.59v star SSW 4′6; NGC 1682 W 2′9.
04 52 34.2	−02 56 55	NGC 1685	14.1	1.3 x 0.9	14.1	SB(r)0/a	135	Mag 9.51v star E 6′7.
04 54 16.8	+01 39 58	NGC 1690	14.0	1.2 x 0.4	13.0	SB(s)b	114	**UGC 3198** S 1′7; mag 6.60v star SW 7′9.
04 54 38.2	+03 16 02	NGC 1691	12.0	2.3 x 1.8	13.4	(R)SB(s)0/a:	37	Very faint, broad, outer envelope surrounds brighter, elongated core.
04 55 16.9	−04 39 08	NGC 1694	14.1	1.1 x 0.7	13.7	Sc	34	
04 56 59.8	−04 45 26	NGC 1699	13.9	0.9 x 0.6	13.1	SA(rs)b II-III	155	Mag 7.97v star WSW 4′2.
04 56 56.3	−04 51 57	NGC 1700	11.2	3.3 x 2.1	13.3	E4	120	Mag 7.97v star NNW 5′8.
04 58 44.0	−00 28 43	NGC 1709	14.2	0.9 x 0.7	13.5	SB0°:	49	Mag 13.2 star W 1′0.

(Continued from previous page)

GALAXIES

RA h m s	Dec ° ′ ″	Name	Mag (V)	Dim ′ Maj x min	SB	Type Class	PA	Notes
04 58 54.7	−00 29 24	NGC 1713	12.7	1.4 x 1.2	13.3	E$^+$:	45	Mag 8.90v star SW 10′9; note: **MCG +0-13-52** and **MCG +0-13-53** are 4′7 E of this star.
04 59 34.7	−00 15 37	NGC 1719	13.6	1.1 x 0.3	12.3	Sa: sp	102	Mag 8.09v star NE 10′8; MCG +0-13-61 N 6′9.
05 00 15.6	−03 21 05	NGC 1729	12.9	1.6 x 1.3	13.5	SA(s)c I-II	30	Mag 10.74v star lies 1′0 E.
05 01 54.9	−03 17 49	NGC 1740	12.9	1.5 x 1.2	13.4	SO$^-$:	56	Mag 13 star on S edge.
05 01 38.5	−04 15 31	NGC 1741	12.5	0.9 x 0.3	10.9	Pec	63	= **Hickson 31A**. **Hickson 31C** on SW end of NGC 1741. **Hickson 31B** SW 1′6; **Hickson 31D** W 0′8; and **IC 399** SE 2′3.
05 02 32.2	−03 20 42	NGC 1753	14.5	1.4 x 0.6	14.2	(R′)SBa pec?	165	
05 03 37.1	+01 34 22	NGC 1762	12.6	1.7 x 1.1	13.1	SA(rs)c: II	175	
05 11 46.3	+05 11 59	NGC 1819	12.4	1.3 x 1.0	12.6	SB0	120	NE-SW core in SE-NW bar.
04 09 01.9	−01 09 39	UGC 2969	13.9	1.3 x 0.5	13.3	SB(s)bc: II	127	Mag 9.69v star NE 11′1.
04 12 52.7	+02 22 07	UGC 2983	14.0	2.1 x 0.7	14.2	SB(s)b: II-III	135	
04 14 23.5	+02 43 50	UGC 2994	14.5	1.5 x 0.9	14.7	SAdm: IV-V	35	**MCG +0-11-27** N 8′3.
04 16 34.3	+02 45 31	UGC 2998	13.5	1.5 x 1.3	14.1	SB(rs)b II-III	129	Located 1′9 S of mag 8.30v star.
04 16 54.5	+03 05 48	UGC 3002	14.7	1.1 x 0.7	14.2	SBcd:	33	
04 17 19.0	+02 25 58	UGC 3004	14.0	1.1 x 0.7	13.6	SB? II-III	150	Very slender **UGC 3005** on N edge; pair of stars, mags 11.1 and 9.79v, S 2′2.
04 17 25.3	+02 22 13	UGC 3006	13.5	1.7 x 0.6	13.4	SA0°:	157	Close pair of stars, mags 9.79v and 11.1, NW 2′4.
04 18 40.1	+02 33 36	UGC 3008	13.4	1.2 x 0.7	13.2	(R)SB(rs)0$^+$:	69	Mag 11.03v star S 1′4.
04 18 58.2	+05 26 01	UGC 3010	14.3	1.0 x 0.6	13.6	SBcd:	132	Mag 9.91v star NW 1′7.
04 19 53.6	+02 05 42	UGC 3014	13.5	1.3 x 0.7	13.3	SB?	45	Located between a pair of mag 11 and 12 stars.
04 21 00.2	−00 09 59	UGC 3018	14.4	1.2 x 0.8	14.2	S	170	Mag 5.86v star NE 8′0.
04 21 57.9	+01 50 22	UGC 3023	13.5	1.7 x 1.3	14.2	SB(s)0:	35	Mag 11.6 star W edge of nucleus.
04 23 58.9	−00 49 37	UGC 3029	14.3	1.2 x 0.8	14.1	Sb III	160	Mag 8.96v star N 5′2.
04 24 20.6	−00 44 17	UGC 3031	14.0	0.5 x 0.4	12.1	S?	159	Stellar.
04 28 14.2	+01 03 06	UGC 3054	15.6	1.5 x 0.1	13.4	Sdm:	30	
04 29 42.5	+03 40 53	UGC 3059	14.1	2.2 x 0.6	14.3	SAdm	40	
04 30 58.3	+05 32 28	UGC 3066	13.9	1.4 x 1.0	14.1	SAB(r)d:	110	
04 31 01.1	+05 54 35	UGC 3068	13.2	1.5 x 1.0	13.5	Double System	102	Double or triple system, distorted.
04 30 59.6	−02 00 13	UGC 3070	13.7	1.5 x 0.9	13.9	SAB(s)b pec: II-III	160	Mag 12.2 star superimposed S edge.
04 31 55.2	+01 11 55	UGC 3080	13.4	1.6 x 1.2	13.9	SAB(rs)c II-III	162	Mag 12.2 star W 1′3.
04 32 00.7	+02 23 32	UGC 3081	14.4	1.5 x 0.9	14.5	SB(r)a	10	
04 33 11.2	+05 21 14	UGC 3087	13.3	0.8 x 0.6	12.4	S0:	130	**BW Tauri**. = **3C 120.** Has a variable Seyfert nucleus.
04 33 56.1	+01 06 48	UGC 3091	14.2	1.0 x 0.8	13.8	SAB(s)d: III-IV	24	Mag 10.10 star W 3′4; mag 7.94v star SW 8′3.
04 37 07.7	−02 18 17	UGC 3105	12.9	1.6 x 1.1	13.4	SO$^-$	75	**UGC 3104** W 7′0; stellar **MCG +0-12-53** W 8′0.
04 37 11.3	−01 51 10	UGC 3106	14.4	1.5 x 0.4	13.6	Sa: sp	55	Mag 13.5 star SW end.
04 37 31.6	−00 16 17	UGC 3109	15.5	1.5 x 0.9	15.7	Sdm:	151	Mag 8.60v star NNE 7′7; MCG +0-12-58 SE 2′0.
04 38 50.2	+02 50 41	UGC 3117	13.7	1.6 x 1.2	14.3	SAB(s)c: III	18	
04 40 25.5	−02 01 25	UGC 3127	14.2	1.9 x 0.5	13.9	SB(s)dm: sp IV-V	24	Mag 8.13v star NE 9′0.
04 40 33.5	+04 11 41	UGC 3128	13.9	0.8 x 0.6	13.0	SAB0$^+$	174	Mag 12.9 star on W edge; mag 13.0 star E 0′8.
04 41 48.2	−01 18 09	UGC 3134	13.4	1.2 x 0.9	13.3	SAB(s)c II	18	**MCG +0-12-67** SW 5′0.
04 41 53.6	+01 04 50	UGC 3136	14.2	1.0 x 0.5	13.3	Sb II-III	75	**CGCG 393-70** S 0′9.
04 43 08.6	+00 44 37	UGC 3141	14.4	0.9 x 0.8	13.9	S	42	Strong knot N of nucleus.
04 44 12.7	+00 21 00	UGC 3145	14.1	1.3 x 0.9	14.1	(R′)SB(rs)c: III	80	Stellar nucleus; mag 11.3 star NW 1′4.
04 47 02.4	+00 04 03	UGC 3164	14.3	1.6 x 1.4	15.0	SAB(s)m V		
04 47 32.2	−02 18 42	UGC 3168	14.3	1.0 x 0.7	13.8	Sd	174	
04 47 45.3	+01 49 08	UGC 3171	14.1	0.8 x 0.8	13.4	SBcd:		
04 48 22.8	+00 13 27	UGC 3174	13.9	1.8 x 1.5	14.9	IAB(s)m: V	85	Small, bright center, uniform envelope.
04 49 44.5	+03 19 58	UGC 3179	13.5	0.7 x 0.4	12.0	S0 pec:	9	**CGCG 420-3** WSW 1′7.
04 51 46.3	+03 40 08	UGC 3186	15.3	1.5 x 0.2	13.8	Scd:	48	Small triangle of mag 12-14 stars NE 1′0.
04 52 46.3	+04 23 28	UGC 3191	14.0	1.2 x 0.6	13.5	Sb: III-IV	75	Mag 9.49v star N 1′2.
04 52 49.3	+01 15 29	UGC 3192	13.2	1.0 x 0.8	12.8	SO$^-$:	55	Galaxy **GIN 207** E 2′1.
04 52 52.7	+03 03 22	UGC 3193	13.6	1.5 x 0.7	13.5	SB(rs)ab:	170	Very small, very faint anonymous galaxy and mag 14.6 star E 1′0.
04 52 58.7	+01 19 50	UGC 3194	14.1	1.7 x 0.7	14.1	SB(s)b pec:	6	Mag 10.45v star W 1′1.
04 53 10.9	+04 22 32	UGC 3195	14.4	0.9 x 0.5	13.4	Scd:	110	
04 54 21.4	+02 07 59	UGC 3200	14.0	1.1 x 0.8	13.7	SB(rs)b	107	Short, bright NE-SW bar.
04 55 11.4	−01 11 22	UGC 3202	13.8	1.0 x 0.9	13.5	SBab	48	Mag 10.06v star SW 2′4.
04 55 58.4	+02 56 06	UGC 3206	14.1	1.2 x 0.6	13.6	SAB(s)bc II-III	120	Faint star superimposed on each end.
04 56 09.7	+02 09 24	UGC 3207	13.1	2.6 x 0.6	13.5	SAB(rs)b pec: III	124	Mag 7.68v star S 6′9.
04 56 20.0	+01 36 26	UGC 3208	14.3	0.9 x 0.8	13.8	Sc	0	Mag 6.88v star W 4′9.
04 56 26.4	+03 01 52	UGC 3209	14.1	1.2 x 0.5	13.4	SB(s)bc pec II-III	32	Mag 10.31v star WNW 3′2.
04 57 22.6	+01 53 38	UGC 3211	14.4	1.5 x 0.4	13.7	SBcd: sp III	174	
04 57 56.7	−00 07 36	UGC 3214	13.2	3.0 x 0.6	13.6	Sb sp III	57	MCG +0-13-51 SE 5′3.
04 58 19.4	+00 44 30	UGC 3215	13.3	1.3 x 0.8	13.2	S0° pec:	84	Mag 12.9 star on SE edge; mag 8.26v star NNW 8′0.
04 59 09.4	+04 58 27	UGC 3223	12.9	1.5 x 0.7	12.8	SBa	71	
04 59 21.0	+05 37 03	UGC 3224	13.0	1.5 x 1.2	13.5	Sb II-III	15	Stellar nucleus, uniform envelope.
05 02 34.8	+00 14 42	UGC 3231	13.9	1.3 x 0.6	13.5	SB(s)d: III-IV	138	
05 03 04.3	+00 15 17	UGC 3233	14.2	1.0 x 0.8	13.8	Scd:	34	Very faint, elongated anonymous galaxy NNE 2′3.
05 03 27.6	+00 39 34	UGC 3237	14.3	1.5 x 0.3	13.3	Scd: sp	112	
05 06 02.8	+00 33 26	UGC 3246	13.8	0.9 x 0.8	13.3	S?	30	
05 07 06.6	+03 58 41	UGC 3248	13.8	1.7 x 0.8	13.9	Sb II-III	135	
05 10 14.7	+00 54 17	UGC 3256	13.5	1.1 x 1.0	13.5	SABb	36	MCG +0-14-4 W 2′2.
05 10 43.2	+00 24 27	UGC 3258	13.1	0.7 x 0.6	12.0	SB(r)ab pec	15	Strong dark patch S of nucleus; bright knot of star N of nucleus; mag 10.38v star W 2′5.
05 11 34.8	−00 41 24	UGC 3263	15.1	0.9 x 0.8	14.6	SBc		Mag 13.3 star on S edge; mag 9.26v star E 4′0.
05 11 31.4	−00 35 09	UGC 3264	13.6	1.1 x 0.5	12.8	S0/a	110	**UGC 3262** NW 1′4.
05 16 11.4	−00 08 58	UGC 3271	13.6	1.3 x 0.9	13.7	S?	175	Small bright core, mag 12.2 star on E edge; mag 8.91v star NW 6′2.
05 17 46.1	−01 11 14	UGC 3283	15.1	1.5 x 0.2	13.6	Scd:	105	
05 19 18.9	+01 20 05	UGC 3287	14.0	0.8 x 0.6	13.1	SB(s)cd pec:	24	**PGC 17090** SW 0′8.
05 21 04.1	+04 00 21	UGC 3294	13.0	2.7 x 1.5	14.4	SA(rs)b II-III	130	Mag 6.53v star E 3′8.
05 21 21.8	+04 53 13	UGC 3296	13.0	1.7 x 1.3	13.8	Sab	140	
05 22 59.4	−00 08 28	UGC 3301	14.2	1.3 x 0.9	14.3	SB0/a	44	
05 24 58.9	+04 30 15	UGC 3303	15.4	3.6 x 2.8	17.8	Im?	126	Dwarf? Low surface brightness; mag 11.2 star dead center; numerous other superimposed stars.
05 25 16.7	+00 25 13	UGC 3306	13.6	1.2 x 0.6	13.1	SO$^-$:	52	Mag 6.16v star NE 9′7.

GALAXY CLUSTERS

RA h m s	Dec ° ′ ″	Name	Mag 10th brightest	No. Gal	Diam ′	Notes
04 59 54.0	+05 26 00	A 526	16.4	71	13	Stellar **PGC 16522** NNE of center 3′8; other members anonymous, faint and stellar.

(Continued from previous page)

117 OPEN CLUSTERS 117

RA h m s	Dec ° ′ ″	Name	Mag	Diam ′	No. ★	B ★	Type	Notes
04 46 40.1	−04 40 24	IC 2091		0.6	5		ast	(A) Asterism of 5 stars.

BRIGHT NEBULAE

RA h m s	Dec ° ′ ″	Name	Dim ′ Maj x min	Type	BC	Color	Notes
05 06 54.0	−03 21 00	NGC 1788	5 x 3	R	1-5	1-4	Rather sharply defined on its SW perimeter where it is flanked by the dark nebula LDN 1616. The brightest involved star is 10th mag and lies in the NW sector.
05 20 00.0	−05 40 00	Sh2-278	25 x 20	E	4-5	3-4	

DARK NEBULAE

RA h m s	Dec ° ′ ″	Name	Dim ′ Maj x min	Opacity	Notes
05 06 30.0	−03 26 00	LDN 1616		5	Apparently a part of NGC 1788, involved in the SW part of it. It is most distinct due S of NGC 1788's central region.

118 GALAXIES 118

RA h m s	Dec ° ′ ″	Name	Mag (V)	Dim ′ Maj x min	SB	Type Class	PA	Notes
03 05 48.2	−00 10 39	CGCG 389-80	14.7	0.3 x 0.3	12.0	Compact		Mag 12.1 star SSE 4′.6.
03 32 44.5	−01 04 11	CGCG 391-7	15.2	0.6 x 0.3	13.2	Spiral	132	Forms a triangle with a mag 11.6 star NNE and a mag 11.5 star E.
03 33 04.9	−01 58 43	CGCG 391-8	15.2	0.5 x 0.5	13.6			Mag 7.42v star N 7′.3.
03 34 35.0	−01 56 07	CGCG 391-10	15.4	0.7 x 0.5	14.1	Spiral	33	Mag 10.55v star ESE 5′.5; mag 12.2 star WNW 7′.1.
03 35 50.2	−01 59 06	CGCG 391-13	14.8	0.5 x 0.5	13.2			CGCG 391-15 ENE 3′.9.
03 36 04.3	−01 57 25	CGCG 391-15	15.3	0.7 x 0.3	13.5		51	CGCG 391-13 WSW 3′.9.
03 37 13.7	+01 51 33	CGCG 391-17	14.9	0.4 x 0.4	13.0	E		Mag 12.3 star N 1′.2; mag 10.90v star WSW 3′.5.
03 38 15.5	+01 10 22	CGCG 391-21	14.2	0.5 x 0.3	12.0	Sb pec?	72	Anonymous galaxy WSW 0′.9.
03 40 17.2	+01 05 35	CGCG 391-26	15.0	0.5 x 0.3	12.8		54	Mag 11.8 star N 2′.2; mag 10.40v star S 5′.3.
03 40 36.9	−01 23 41	CGCG 391-27	15.4	0.8 x 0.2	13.2		45	Mag 11.00v star WSW 3′.0; NGC 1409 and NGC 1410 NE 10′.0.
03 52 20.0	−00 58 25	CGCG 391-43	14.7	0.3 x 0.2	11.5	Sb	3	Mag 9.08v star ESE 5′.9.
02 57 33.9	+05 58 33	CGCG 415-40	14.0	0.6 x 0.4	12.4	S0	150	**PGC 11171** W 2′.5; **PGC 11160** W 3′.9.
03 42 07.3	+04 13 53	CGCG 417-5	14.7	0.8 x 0.6	13.7	Sc	39	NE-SW bar, dark lane SE of center.
02 57 10.9	+02 46 29	IC 273	13.4	1.5 x 0.5	12.9	SBa: sp	31	Mag 10.32v star NNE 8′.6.
02 59 50.6	+02 46 16	IC 277	13.1	1.2 x 0.9	13.1	SB(rs)bc II-III	33	Mag 9.76v star NE 11′.7.
03 03 50.2	−00 12 18	IC 283	14.9	0.5 x 0.3	12.7	Spiral	20	Mag 11.3 star S 8′.2.
03 11 19.0	+01 18 49	IC 298	14.6	0.7 x 0.5	13.4	Ring A	126	Galaxy pair; component on W side appears donut shaped with dark hole in center; mag 9.28v star SE 3′.0.
03 12 51.4	+04 42 22	IC 302	12.8	1.9 x 1.5	13.7	SB(rs)bc II	21	Stellar nucleus; mag 11.28v star N 6′.7.
03 13 45.4	−00 14 28	IC 307	13.3	1.7 x 0.8	13.5	(R)SB(r)a pec?	73	
03 19 09.4	+04 02 15	IC 315	14.2	0.8 x 0.4	12.8		36	Mag 11.4 star SW 3′.9.
03 32 01.5	+00 16 45	IC 329	14.3	0.9 x 0.4	13.0	S?	63	Located 1′.9 WSW of mag 8.30v star.
03 32 07.9	+00 21 08	IC 330	14.1	1.2 x 0.3	12.8	Sab	78	Mag 11.7 star N 1′.4.
03 32 19.0	+00 16 58	IC 331	13.9	0.9 x 0.9	13.5	S?		Located 2′.6 E of mag 8.30v star
03 32 37.4	+01 22 56	IC 332	13.7	0.9 x 0.5	12.7	(R)SAB(rs)0/a:	42	Mag 10.66v star W 4′.4.
03 37 38.2	+03 07 04	IC 338	13.8	0.8 x 0.7	13.0	S?	171	Mag 10.46v star WNW 11′.1.
03 41 29.7	−04 39 58	IC 344	14.2	1.0 x 0.5	13.3	SB	45	NGC 1417 ESE 7′.3; mag 7.70v star E 18′.7.
03 42 32.7	−04 17 56	IC 347	13.0	1.3 x 1.0	13.2	SAB(r)0°:	123	Mag 9.79v star SE 5′.0.
02 57 52.8	−02 20 35	IC 1870	13.2	2.8 x 1.6	14.7	SB(s)m III-IV	135	Lies 1′.7 W of mag 7.73v star.
03 07 49.1	+03 08 46	IC 1882	14.0	1.1 x 0.5	13.2	S?	20	Mag 9.16v star N 7′.8.
03 08 03.3	−04 24 00	IC 1886	13.9	1.0 x 0.7	13.6	E	0	Mag 12.2 star W 7′.1.
03 28 57.8	+01 45 00	IC 1931	14.7	0.8 x 0.8	14.1			Mag 11.2 star NW 3′.1.
03 35 33.2	+05 03 57	IC 1956	14.5	1.6 x 0.3	13.5	SBbc:	34	Mag 10.6 star W 5′.0.
03 37 47.8	+03 16 12	IC 1967	15.0	1.0 x 0.5	14.1	Sbc	57	**CGCG 391-19** W 2′.5.
03 00 51.6	+00 48 23	MCG +0-8-66	15.0	0.9 x 0.7	14.4	Sbc	0	Mag 11.8 star N 1′.8.
03 01 44.4	−00 35 31	MCG +0-8-69	15.0	0.7 x 0.7	14.0	Spiral		Mag 9.89v star W 3′.1.
03 03 30.7	−02 14 50	MCG +0-8-75	14.4	0.8 x 0.2	12.3		24	
03 03 52.9	+00 24 54	MCG +0-8-77	14.7	0.7 x 0.7	13.8			Faint, elongated anonymous galaxy NE 1′.6.
03 05 31.0	−00 24 17	MCG +0-8-91	14.3	0.9 x 0.6	13.7	E	147	**UGC 2529** N 1′.5.
03 07 32.8	−00 57 52	MCG +0-9-1	14.1	0.9 x 0.8	13.6			
03 08 11.4	−00 48 35	MCG +0-9-5	14.7	0.7 x 0.4	13.2		3	**CGCG 390-4** W 5′.8.
03 08 28.6	+00 23 12	MCG +0-9-7	15.2	0.7 x 0.5	13.9	Spiral	9	
03 08 46.8	−00 43 58	MCG +0-9-8	14.6	0.8 x 0.4	13.2		45	
03 09 00.9	−00 59 00	MCG +0-9-9	15.3	0.9 x 0.2	13.2	Spiral	123	
03 11 50.0	+00 09 25	MCG +0-9-19	15.1	0.9 x 0.5	14.1		90	Mag 7.03v star SE 4′.3.
03 12 02.6	−00 04 40	MCG +0-9-20	14.7	0.8 x 0.6	13.8		48	Mag 9.34v star ESE 1′.3.
03 12 12.9	−01 03 45	MCG +0-9-22	14.8	0.7 x 0.5	13.5		114	Mag 5.07v star SE 11′.7.
03 13 47.9	+00 41 37	MCG +0-9-26	14.3	0.7 x 0.4	13.0	E/S0	147	Mag 7.47v star NE 4′.6.
03 14 19.5	−01 18 50	MCG +0-9-31	15.0	0.6 x 0.5	13.5	Sm	159	Pair of stars, mags 11.9 and 12.3, N 2′.1.
03 14 45.0	+03 11 14	MCG +0-9-34	14.6	0.8 x 0.7	13.8		90	
03 14 42.3	−00 44 27	MCG +0-9-36	15.2	0.5 x 0.4	13.3	Sbc		
03 15 35.2	−01 01 02	MCG +0-9-37	14.8	0.8 x 0.5	13.7		15	Mag 11.3 star E 2′.6.
03 15 57.1	+01 38 47	MCG +0-9-38	14.9	0.7 x 0.3	13.1		105	
03 17 36.2	−01 43 24	MCG +0-9-47	14.9	0.7 x 0.4	13.4	S0/a	18	**UGC 2650** N 1′.7; MCG +0-9-49 NE 2′.7.
03 17 44.5	−01 41 36	MCG +0-9-49	15.7	0.6 x 0.2	13.2	Sc	15	**UGC 2650** W 2′.4; MCG +0-9-47 SW 2′.7.
03 18 34.0	+00 55 25	MCG +0-9-53	16.2	0.5 x 0.4	14.3	Spiral	18	Mag 12.2 star WNW 2′.6.
03 19 26.0	−02 06 14	MCG +0-9-55	15.2	0.7 x 0.5	13.9		96	Mag 9.07v star NE 5′.0.
03 19 29.8	−02 02 06	MCG +0-9-56	15.4	0.4 x 0.4	13.2	Spiral		Mag 10.72v star N 1′.7 mag 11.8 star NW 1′.3.
03 19 43.2	+00 33 54	MCG +0-9-57	14.6	0.8 x 0.3	12.9	S?	150	= **Mkn 1076**; MCG +0-9-58 NE 1′.5.
03 19 47.1	+00 35 03	MCG +0-9-58	13.8	0.9 x 0.8	13.2	S?	96	MCG +0-9-57 SW 1′.5.
03 20 33.3	−00 20 25	MCG +0-9-64	14.8	0.7 x 0.3	12.9		30	UGC 2692 E 6′.6.
03 20 32.4	−02 00 22	MCG +0-9-65	17.4	1.0 x 0.6	16.7	Sbc	81	
03 22 08.7	−00 50 09	MCG +0-9-70	14.5	0.8 x 0.5	13.4		69	
03 22 41.2	−02 03 39	MCG +0-9-71	14.8	0.5 x 0.3	12.6		93	Mag 9.06v star E 2′.5.
03 22 49.3	−02 01 53	MCG +0-9-73	15.1	0.4 x 0.3	12.6		120	Mag 9.06v star S 1′.7.
03 23 00.5	+01 21 55	MCG +0-9-74	13.7	0.9 x 0.6	12.9	(R)SAB(r)a pec?	123	Mag 12.7 star N 1′.3.
03 23 24.0	+00 26 41	MCG +0-9-76	15.6	0.7 x 0.5	14.3	Sd	66	Mag 13.1 star S 2′.4.

(Continued from previous page)

GALAXIES

RA h m s	Dec ° ′ ″	Name	Mag (V)	Dim ′ Maj x min	SB	Type Class	PA	Notes
03 24 06.6	−01 03 30	MCG +0-9-79	15.3	0.9 x 0.2	13.3	Sb	57	Mag 11.9 star S 1′9.
03 24 27.6	−00 05 08	MCG +0-9-81	15.3	1.0 x 0.3	13.9	Sbc	57	
03 24 40.9	−00 40 53	MCG +0-9-82	15.1	0.7 x 0.4	13.5	Sb	66	
03 29 34.6	−01 32 31	MCG +0-9-88	14.7	0.5 x 0.5	13.3	E		Mag 12.8 star S 1′4.
03 30 17.1	−00 55 07	MCG +0-9-89	14.5	0.5 x 0.5	12.9	Sb		
03 30 28.8	+00 48 39	MCG +0-9-90	15.1	0.7 x 0.3	13.3	Sb	27	**MCG +0-9-91** S 3′6.
03 36 42.6	+00 20 12	MCG +0-10-6	16.1	0.5 x 0.4	14.2	Sc		Pair of mag 11.4 stars W 2′2; mag 4.30v star 10 Tauri NE 4′6.
03 47 51.0	−02 09 16	MCG +0-10-18	14.4	0.6 x 0.5	13.0	Spiral	39	Pair of stars, mags 14.7 and 15.6, W 0′6.
03 48 58.3	−01 08 43	MCG +0-10-20	15.0	0.7 x 0.4	13.4	S0/a	174	Mag 11.9 star W 2′1.
03 49 08.7	+01 09 44	MCG +0-10-21	14.3	0.8 x 0.4	12.9	Sm	78	**UGC 2872** and a mag 10.5 star W 2′7.
03 49 45.5	−01 46 14	MCG +0-10-22	14.4	0.8 x 0.7	13.7	Spiral		
03 51 34.2	−00 28 05	MCG +0-10-23	15.1	0.4 x 0.3	12.7	Sb	117	
03 51 40.7	−00 30 30	MCG +0-10-24	15.2	0.4 x 0.2	12.3	Spiral	54	
03 57 58.9	−00 11 26	MCG +0-11-2	15.2	0.4 x 0.2	12.5	E		
04 07 26.9	+01 45 35	MCG +0-11-12	16.1	0.9 x 0.7	15.5	SBm	6	Mag 9.89v star E 3′0.
04 08 23.6	−01 04 34	MCG +0-11-15	14.9	0.9 x 0.2	12.9	Sb	144	
04 09 13.9	−01 28 17	MCG +0-11-20	14.4	0.4 x 0.4	12.3			**MCG +0-11-19** SW 0′5.
04 12 13.9	+00 02 42	MCG +0-11-25	15.2	0.6 x 0.5	13.7	Sbc	3	Mag 11.1 star SE 1′6.
02 57 45.8	+05 57 05	MCG +1-8-28	15.5	0.9 x 0.4	14.2	Scd	150	**PGC 11195** N 1′4; **PGC 11196** N 2′0; mag 12.4 star N 2′3.
03 00 00.2	+05 58 03	MCG +1-8-34	15.2	0.5 x 0.5	13.5	Sbc		Mag 12.4 star W 1′1.
03 00 08.8	+05 48 13	MCG +1-8-35	14.5	0.5 x 0.5	12.9	Sa/0		
03 00 09.7	+05 42 40	MCG +1-8-36	15.8	0.5 x 0.4	13.9	Sbc	0	UGC 2469 E 2′7.
03 05 13.8	+05 51 10	MCG +1-8-40	14.5	0.6 x 0.5	13.0	Sbc	12	Mag 12.9 star N 2′3.
03 05 55.2	+04 51 27	MCG +1-8-42	14.4	0.6 x 0.6	13.1			
04 05 30.4	+04 24 41	MCG +1-11-10	14.3	0.7 x 0.7	13.3	S?		= **Mkn 1081**; **UGC 2954** N 1′9.
02 58 24.5	−04 17 43	MCG −1-8-24	13.6	2.2 x 1.1	14.4	SB(rs)d III-IV	123	Short, bright E-W bar.
02 59 10.7	−04 28 19	MCG −1-8-25	13.4	1.1 x 0.5	12.6	(R′)SB(r)0/a pec?	93	= **Mkn 1065**.
03 09 31.3	−04 54 43	MCG −1-9-10	14.3	1.6 x 0.7	14.2	SB(rs)d: III-IV	147	
03 11 13.7	−04 14 49	MCG −1-9-13	16.0	1.6 x 1.1	16.5	SA(s)m V	130	Mag 13.0 star N 2′1.
03 12 54.2	−02 54 10	MCG −1-9-15	13.6	1.1 x 0.7	13.2	S0/a	107	Mag 8.63v star W 2′6.
03 14 38.4	−04 46 33	MCG −1-9-21	13.6	1.9 x 1.6	14.7	IB(s)m IV-V	117	Very patchy.
03 15 52.4	−03 37 45	MCG −1-9-22	13.8	0.8 x 0.5	12.6	Sc	66	
03 16 00.6	−05 30 00	MCG −1-9-24	13.1	1.9 x 0.5	12.9	Sab	27	Mag 6.32v star SSE 14′9.
03 24 25.9	−04 49 10	MCG −1-9-34	14.2	0.8 x 0.3	12.6	S0	81	Mag 9.08v star NE 9′3.
03 28 03.4	−04 49 09	MCG −1-9-40	14.0	1.1 x 0.5	13.2	SBb	15	Mag 8.81v star SW 12′1.
03 30 19.2	−04 14 35	MCG −1-9-43	13.4	1.1 x 0.7	13.0	SB(s)c I	117	Mag 7.64v star ESE 5′3.
03 30 45.3	−04 51 03	MCG −1-9-44	14.2	1.0 x 0.5	13.5	E	21	Mag 10.83v star N 4′1.
03 31 23.0	−05 08 32	MCG −1-9-45	13.9	0.9 x 0.7	13.2	Sb	96	Mag 10.12v star N 4′8.
03 32 36.8	−04 45 06	MCG −1-10-1	14.1	0.7 x 0.7	13.1	Spiral		
03 34 32.1	−04 48 39	MCG −1-10-6	13.7	1.1 x 0.7	13.2	Sab	165	Mag 10.22v star SE 1′4.
03 35 55.2	−02 51 59	MCG −1-10-8	14.0	0.5 x 0.5	12.6	E/S0		Pair with round anonymous galaxy 0′5 SE.
03 38 26.5	−04 18 30	MCG −1-10-13	14.1	1.4 x 0.8	14.1	SB(s)m: V	6	
03 38 33.1	−05 28 02	MCG −1-10-14	14.2	1.3 x 0.9	14.2	SAB(r)cd: sp III	174	Mag 10.68v star NE 6′2.
03 38 39.0	−05 20 53	MCG −1-10-15	14.3	1.3 x 0.4	13.4	Sd? III	78	Mag 10.49v star NW 5′7.
03 39 02.8	−03 37 43	MCG −1-10-16	13.7	0.7 x 0.5	12.5	SA(r)0$^+$?	114	
03 44 00.3	−04 01 46	MCG −1-10-27	14.4	1.1 x 1.0	14.4	SA(r)d: III-IV	33	
03 46 36.0	−04 27 13	MCG −1-10-35	14.0	2.5 x 0.3	13.6	Sc sp	24	Mag 9.35v star SE 15′5.
03 46 58.3	−03 27 49	MCG −1-10-36	14.0	1.4 x 0.4	13.2	Sbc: sp	132	
03 47 39.6	−04 21 08	MCG −1-10-37	14.9	1.5 x 1.1	15.2	SB(s)dm: IV-V	21	Mag 11.4 star SSW 8′9.
03 49 05.5	−03 03 44	MCG −1-10-39	14.2	1.5 x 0.4	13.5	Sb sp	30	
03 50 43.5	−02 51 42	MCG −1-10-41	14.1	1.1 x 0.8	13.8	Sd	27	
03 54 33.3	−02 55 06	MCG −1-10-46	13.6	0.5 x 0.5	12.1	E/S0		Mag 4.72v star 32 Eridani SW 4′6.
04 02 43.4	−04 27 45	MCG −1-11-1	14.0	1.1 x 0.4	12.9		0	Mag 8.76v star SW 8′0.
02 57 24.0	−00 18 34	NGC 1149	14.0	0.7 x 0.6	12.9	S0$^-$:	130	A very faint **UGC 2428** NW 6′0.
02 58 10.4	+03 21 40	NGC 1153	12.4	1.2 x 0.7	12.1	S0$^+$?	45	Bright knot SW edge of core; UGC 2446 NE 8′9.
03 03 49.3	−01 06 14	NGC 1194	12.9	1.8 x 0.8	13.1	SA0$^+$:	140	UGC 2517 SE 7′9.
03 06 52.5	−00 47 40	NGC 1211	12.3	2.1 x 1.8	13.6	(R)SB(r)0/a	30	
03 08 26.2	+04 06 40	NGC 1218	12.7	1.3 x 1.0	12.8	S0/a	155	
03 08 28.3	+02 06 26	NGC 1219	13.0	1.0 x 0.8	12.6	SA(s)bc: II	30	
03 08 15.6	−04 15 31	NGC 1221	13.5	1.3 x 0.5	12.9	SB(r)0$^+$? sp	164	Mag 12.7 star S 1′1; mag 11.5 star WNW 4′1.
03 08 56.8	−02 57 15	NGC 1222	12.5	1.1 x 0.9	12.4	S0$^-$ pec:	169	
03 08 20.0	−04 08 19	NGC 1223	13.4	1.2 x 1.0	13.5	cD	4	Faint, elongated anonymous galaxy E 1′9.
03 08 47.3	−04 06 06	NGC 1225	14.0	1.2 x 0.6	13.5	Sbc	142	
03 10 53.8	−02 33 15	NGC 1239	13.6	1.2 x 0.7	13.3	SA0° pec:	44	Elongated **PGC 11876** N 8′0.
03 12 48.9	−05 13 28	NGC 1248	12.5	1.1 x 1.0	12.5	SA(s)0°:	90	Elongated **Mkn 604** S 2′7; mag 8.45v star N 5′5.
03 14 10.2	−02 49 22	NGC 1253	11.7	5.2 x 2.3	14.3	SAB(rs)cd III-IV	81	Several knots, bright knot or star W of core.
03 14 23.7	−02 48 02	NGC 1253A	13.9	1.3 x 0.8	13.8	SB(s)m V	80	Located 4′0 NE of center of NGC 1253 with mag 12.2 star on it's SW edge.
03 14 23.9	+02 40 40	NGC 1254	14.1	0.8 x 0.7	13.3	SA0$^-$:	42	Located between mag 8.61v star NNE 5′5 and mag 9.01v star S 5′2.
03 16 00.9	−02 25 39	NGC 1266	13.0	1.5 x 1.0	13.3	(R′)SB(rs)0° pec:	115	
03 17 56.9	−00 10 11	NGC 1280	13.4	0.9 x 0.8	12.9	SA(rs)c: I-II	55	Mag 8.58v star W 8′4; note: **UGC 2645** and **MCG +0-9-44** lie 4′2 N of this star.
03 18 33.7	−02 43 52	NGC 1287	13.9	0.8 x 0.7	13.1	Sa	130	Mag 7.05v star SE 9′9.
03 18 49.9	−01 58 25	NGC 1289	12.6	1.8 x 1.1	13.2	SB(rs)0°:	100	Stellar galaxy **KUG 0315-021** WNW 7′1.
03 20 13.2	−02 06 52	NGC 1298	14.0	1.1 x 0.8	13.8	E$^+$ pec:	70	Mag 9.07v star WNW 7′5.
03 21 12.6	−04 35 05	NGC 1304	13.5	1.3 x 0.8	13.4	S0$^-$ pec	48	
03 21 23.1	−02 19 02	NGC 1305	13.3	1.4 x 0.9	13.4	(R′)S0$^-$ pec:	130	
03 22 28.6	−02 45 26	NGC 1308	13.7	1.1 x 0.8	13.4	SB(r)0/a	57	Mag 10.05v star E 1′6.
03 22 41.3	−04 11 11	NGC 1314	14.2	1.5 x 1.4	14.8	SA(rs)d IV-V	90	Stellar nucleus; mag 12.9 star on S edge.
03 24 48.8	−03 02 33	NGC 1320	12.5	1.9 x 0.6	12.4	Sa: sp	135	
03 24 48.7	−03 00 58	NGC 1321	13.3	0.9 x 0.6	12.5	Sa	85	
03 24 55.0	−02 55 10	NGC 1322	13.4	1.1 x 0.9	13.3	E/S0	115	
03 24 56.3	−02 49 21	NGC 1323	15.1	0.9 x 0.4	13.8	S0	86	
03 25 01.7	−05 44 44	NGC 1324	12.7	2.1 x 0.8	13.1	Sb: II	135	
03 25 39.1	−04 07 30	NGC 1328	14.0	0.9 x 0.6	13.2	S0	134	Mag 8.71v star NE 4′7.
03 30 13.4	−05 32 36	NGC 1346	13.2	1.1 x 0.7	12.8	Sb pec?	85	**MCG −1-9-41** NW 1′5.
03 31 27.6	+04 22 50	NGC 1349	13.0	0.8 x 0.8	12.4	S0		
03 33 23.6	−04 59 56	NGC 1355	13.3	1.4 x 0.3	12.2	S0: sp	78	NGC 1358 SW 6′8.
03 33 39.8	−05 05 24	NGC 1358	12.1	2.3 x 1.5	13.3	SAB(r)0/a	3	Small elongated core, smooth envelope; mag 9.46v star E 8′5; NGC 1355 NW 6′8.
03 37 05.9	−05 02 36	NGC 1376	12.1	1.7 x 1.7	13.1	SA(s)cd II		
03 39 47.2	−04 40 16	NGC 1397	13.7	1.7 x 1.2	14.3	(R)SB(r)0/a V-VI	105	Very faint arms encircle a bright center.

(Continued from previous page)

118 GALAXIES 118

RA h m s	Dec ° ′ ″	Name	Mag (V)	Dim ′ Maj x min	SB	Type Class	PA	Notes
03 41 10.2	−01 18 08	NGC 1409	13.8	0.9 x 0.7	13.2	SAB0: pec	130	CGCG 391-27 SW 10′0.
03 41 11.4	−01 17 53	NGC 1410	14.3	0.7 x 0.6	13.5	E pec:	120	CGCG 391-27 SW 10′0.
03 41 57.5	−04 42 21	NGC 1417	12.1	2.3 x 1.2	13.1	SAB(rs)b I-II	9	Mag 12.1 star on E edge; mag 7.70v star NE 12′9; IC 344 WNW 7′3.
03 42 16.1	−04 43 57	NGC 1418	13.6	1.0 x 0.6	12.9	SB(s)b: II-III	18	Mag 7.70v star NE 10′2; NGC 1417 W 5′0.
03 43 14.1	−04 43 47	NGC 1424	13.8	1.7 x 0.5	13.5	SAB(rs)b: II	11	Mag 7.70v star NW 10′7.
03 44 40.8	+02 50 05	NGC 1431	14.1	1.0 x 0.8	13.7	S0 pec:	160	Stellar nucleus; mag 9.40v star SW 8′5.
03 45 43.1	−04 05 33	NGC 1441	12.9	1.6 x 0.6	12.7	SB(s)b	90	NGC 1451 E 6′2.
03 46 03.1	−04 08 18	NGC 1449	13.5	0.8 x 0.5	12.3	S0	22	NGC 1451 N 4′2.
03 46 07.2	−04 04 12	NGC 1451	13.3	0.8 x 0.6	12.4	S0$^-$?	33	NGC 1453 NE 7′7; NGC 1449 S 4′2; NGC 1441 W 6′2.
03 46 27.4	−03 58 08	NGC 1453	11.5	2.4 x 1.9	13.2	E2-3	45	NGC 1451 SW 7′7.
04 04 27.3	−02 11 20	NGC 1507	12.3	3.6 x 0.9	13.4	SB(s)m pec? III	11	
02 56 27.2	+00 52 29	UGC 2418	14.1	1.1 x 0.7	13.7	SABbc	109	
02 57 09.3	+05 19 15	UGC 2426	14.1	0.9 x 0.6	13.2	Sb III	65	Mag 10.89v star NE 2′7.
02 58 22.0	+03 51 41	UGC 2441	14.0	2.0 x 0.4	13.6	Sd	13	Faint anonymous galaxy WNW 2′6; mag 9.25v star SW 3′0.
02 58 21.4	−02 02 28	UGC 2443	13.7	1.4 x 0.7	13.5	Scd:	163	
02 58 41.5	+03 26 03	UGC 2446	13.8	0.7 x 0.4	12.3	S?	105	Located on E edge of galaxy cluster **A 403**; **PGC 11298** NE 9′0.
03 00 20.5	+05 42 42	UGC 2469	14.3	1.1 x 0.7	13.9	Pec?	24	MCG +1-8-36 W 2′7.
03 00 40.0	+00 01 09	UGC 2479	15.4	1.3 x 0.4	14.5	Sdm: sp	25	
03 02 33.8	+02 11 39	UGC 2501	13.3	1.3 x 0.6	12.9	SB(s)m pec: III	133	Bright knot SE end; mag 9.52v star W 2′6.
03 03 39.3	+01 56 51	UGC 2513	13.9	1.1 x 0.6	13.3	Sa:	153	
03 04 12.8	−01 11 38	UGC 2517	14.0	1.2 x 0.5	13.3	SB	138	NGC 1194 NW 7′9; b W 4′7.
03 04 57.0	+00 12 58	UGC 2522	15.0	1.3 x 0.4	14.1	Sb	179	Mag 14.4 star SE edge.
03 05 09.0	+01 05 36	UGC 2523	14.4	1.7 x 0.3	13.6	Sd: sp	175	
03 05 20.3	+02 22 42	UGC 2527	14.0	1.1 x 0.8	13.7	SB	65	South half much fainter.
03 11 47.7	−00 24 10	UGC 2585	13.2	1.7 x 1.3	13.9	SB(r)b: II-III	165	Mag 7.07v star S 5′8.
03 12 00.0	−01 09 42	UGC 2587	15.2	1.6 x 0.2	13.8	Sbc: sp III	33	Mag 5.07v star E 11′8.
03 12 45.7	−00 20 02	UGC 2594	13.4	0.9 x 0.8	12.8	SB0/a:		Mag 8.73v star SE 3′9.
03 14 08.8	−01 38 11	UGC 2607	13.5	1.4 x 1.3	14.0	(R′)SB(rs)b: II-III	135	Bright NE-SW bar; mag 12.0 star S 2′0.
03 14 24.6	−01 54 56	UGC 2611	13.4	1.1 x 1.0	13.4	S0	24	Bright nucleus, uniform envelope.
03 16 32.0	−00 28 08	UGC 2628	13.8	1.5 x 0.6	13.6	SB(s)bc: II	127	
03 17 16.1	+03 35 31	UGC 2641	15.7	1.0 x 0.4	14.5	Sdm:	114	
03 17 36.7	−01 49 10	UGC 2649	13.2	1.1 x 1.1	13.2	S0:		
03 19 57.2	+03 35 24	UGC 2677	14.1	1.5 x 0.4	13.4	Sab sp III	122	Mag 11.9 star N 1′3.
03 20 03.9	+01 21 39	UGC 2679	14.1	1.7 x 0.5	13.8	(R′)SB(rs)b:	117	Mag 11.0 star W 1′3.
03 20 07.0	−01 52 55	UGC 2680	14.0	1.3 x 0.6	13.6	Pec?	20	
03 20 20.6	−02 04 54	UGC 2687	14.2	1.2 x 0.8	14.0	SAB(rs)a	81	Bright nucleus in N-S bar, faint envelope; NGC 1298 SW 2′7.
03 20 43.2	−01 06 33	UGC 2690	13.9	1.4 x 0.6	13.5	SAc pec: II	146	= **Hickson 25A**. Mag 12.8 star W 1′2.
03 20 45.5	−01 02 48	UGC 2691	14.1	1.3 x 0.6	13.7	S?	0	= **Hickson 25B**. **Hickson 25F** a "knot" on S end. **Hickson 25A** = UGC 2690 S 3′9; **Hickson 25C** and **E** 2′4 N; **Hickson 25D** E 1′8; **Hickson 25G** ESE 2′0.
03 20 59.1	−00 22 07	UGC 2692	13.1	1.3 x 1.2	13.5	SA(s)c: I-II	35	Mag 8.17v star E 2′5.
03 22 47.5	−01 55 24	UGC 2704	13.2	0.7 x 0.7	12.3	S?		Very small, very faint anonymous galaxy E 3′7.
03 22 47.4	+00 08 58	UGC 2705	13.9	1.0 x 0.7	13.4	SABcd	126	
03 23 29.8	+01 19 26	UGC 2711	14.6	1.2 x 0.3	13.4	(R′)SBb: sp	167	Mag 8.49v star E 11′6.
03 27 17.5	+02 36 32	UGC 2744	14.5	0.9 x 0.4	13.2	Scd:	33	
03 27 54.2	+02 33 40	UGC 2748	13.8	1.5 x 0.9	14.2	E pec?	155	Very faint, small anonymous galaxy E 2′2.
03 37 50.9	+04 56 54	UGC 2802	14.3	1.2 x 0.7	14.0	S?	90	**MCG +1-10-3** W 2′4.
03 39 53.4	−02 06 52	UGC 2812	14.0	0.8 x 0.7	13.2	SB	150	Bright E-W bar with very strong dark patch S; very small anonymous companion galaxy S edge.
03 40 04.6	−02 04 05	UGC 2814	14.3	0.9 x 0.8	13.8	Sbc	150	UGC 2812 SW 5′5.
03 42 14.5	+03 10 38	UGC 2831	16.0	1.5 x 0.9	16.2	SB(s)dm: IV-V	60	Faint, almost stellar anonymous galaxy NW 0′9.
03 43 52.4	+00 36 50	UGC 2842	14.2	0.6 x 0.5	12.8	Scd:	126	
03 44 56.0	+05 54 16	UGC 2852	15.4	1.5 x 0.2	13.9	Sdm:	120	
03 52 14.0	−01 30 31	UGC 2883	14.1	0.9 x 0.8	13.6	Sb	117	
03 59 04.0	+01 21 32	UGC 2913	14.8	1.0 x 0.7	14.3	SB(s)a:	54	Mag 11.2 star NE 2′3.
04 01 02.5	−00 43 06	UGC 2925	15.2	1.1 x 0.9	15.0	SB(rs)dm? V	145	
04 02 48.3	+01 57 52	UGC 2936	14.4	2.4 x 0.7	14.8	SB(s)d III-IV	30	Mag 13.5 star on W edge.
04 07 38.8	+03 58 03	UGC 2963	14.2	2.0 x 0.6	14.2	Sd	145	Located between a close pair of mag 13 stars NW and SE.
04 09 01.9	−01 09 39	UGC 2969	13.9	1.3 x 0.5	13.3	SB(s)bc: II	127	Mag 9.69v star NE 11′1.
04 12 52.7	+02 22 07	UGC 2983	14.0	2.1 x 0.7	14.2	SB(s)b: II-III	135	
04 14 23.5	+02 43 50	UGC 2994	14.5	1.5 x 0.9	14.7	SAdm: IV-V	35	**MCG +0-11-27** N 8′3.

119 GALAXIES 119

RA h m s	Dec ° ′ ″	Name	Mag (V)	Dim ′ Maj x min	SB	Type Class	PA	Notes
01 47 20.4	+02 49 56	CGCG 386-37	14.0	0.6 x 0.3	12.0		81	Mag 15.5 star on N edge.
01 59 11.4	+00 06 32	CGCG 387-20	14.0	0.6 x 0.6	12.7	Sab		Mag 13.7 star WNW 3′2.
02 41 57.2	+01 14 04	CGCG 388-96	14.6	0.7 x 0.4	13.1	S0/a	135	Mag 11.3 star NNE 7′0.
02 47 47.5	+03 09 51	CGCG 389-18	14.2	0.5 x 0.5	12.0	S pec:	171	Located 2′0 W of mag 9.02v star.
02 49 39.9	−00 04 12	CGCG 389-27	14.7	0.4 x 0.2	11.8		171	Mag 9.61v star NW 5′0.
02 44 00.4	+05 25 45	CGCG 415-4	14.1	0.6 x 0.5	12.6	Sc		**CGCG 415-5** NW 0′7.
02 44 43.4	+05 20 01	CGCG 415-6	14.1	0.6 x 0.4	12.4		129	Mag 14.8 star NW 1′0.
02 57 33.9	+05 58 33	CGCG 415-40	14.0	0.6 x 0.4	12.4	S0	150	**PGC 11171** W 2′5; **PGC 11160** W 3′9.
01 49 08.6	−03 54 16	IC 164	13.1	1.4 x 1.2	13.6	E$^+$:	0	Mag 9.95v star S 3′6.
01 54 54.5	+00 48 41	IC 172	14.0	0.5 x 0.3	11.8	S?	90	Mag 9.54v star SW 1′9.
01 55 57.2	+01 17 04	IC 173	14.0	0.8 x 0.7	13.2	SB(rs)bc I	107	Mag 11.4 star N 3′6 stellar **IC 175** NE 6′1.
01 56 16.2	+03 45 43	IC 174	13.3	1.3 x 0.7	13.0	S0?	101	Mag 10.19v star N 8′4.
01 56 53.5	−02 01 09	IC 176	14.0	1.8 x 0.4	13.5	Sc: II-III	94	Mag 11.4 star N 7′5.
01 59 34.1	−05 20 52	IC 183	13.9	1.3 x 0.4	13.1	S0?	96	Mag 10.23v star E 3′8.
02 00 24.7	−01 33 06	IC 186	14.8	0.9 x 0.6	14.0	S0/a	57	Triple system; two almost stellar components E, stacked N-S; larger component W.
02 03 05.2	+02 36 51	IC 194	14.4	1.4 x 0.2	12.9	Sb: sp III	13	Mag 10.62v star E 3′7; stellar **Mkn 585** ESE 6′9.
02 04 05.1	+02 47 13	IC 197	13.5	1.0 x 0.5	12.6	SBbc:	55	Mag 10.82v star W 11′2.
02 07 27.5	−02 05 29	IC 205	13.6	0.9 x 0.9	13.2	SB(s)a		Mag 9.78v star SW 4′0.
02 11 08.1	+03 51 07	IC 211	13.4	2.3 x 1.8	14.8	SAB(s)cd II	45	Mag 9.53v star N 9′3; NGC 851 S 4′5.
02 14 05.8	+05 10 33	IC 214	14.2	0.9 x 0.5	13.2	I?	162	Mag 11.4 star S 5′4.
02 26 28.3	+01 09 37	IC 225	13.5	1.2 x 1.0	13.7	E	168	
02 29 56.3	+01 10 44	IC 231	13.6	1.1 x 0.8	13.4	S0:	160	Mag 9.98v star SE 1′7.
02 31 11.6	+01 15 57	IC 232	12.7	1.3 x 0.7	12.7	E	155	Mag 10.7 star SSE 3′5.

(Continued from previous page)

GALAXIES

RA h m s	Dec ° ′ ″	Name	Mag (V)	Dim ′ Maj x min	SB	Type Class	PA	Notes
02 31 40.8	+02 48 34	IC 233	14.2	0.5 x 0.3	12.0	Sc	105	Almost stellar **CGCG 388-36** S 0′.8.
02 33 31.6	+01 08 20	IC 237	14.7	0.5 x 0.5	13.3	E/S0		Stellar, mag 10.95v star W 2′.4.
02 37 54.5	+02 19 39	IC 241	13.4	1.1 x 0.7	13.0	S?	150	Mag 7.21v star NW 12′.4.
02 39 24.9	+02 43 37	IC 244	14.8	0.7 x 0.4	13.2	Sc	6	Mag 9.95v star N 4′.2.
02 57 10.9	+02 46 29	IC 273	13.4	1.5 x 0.5	12.9	SBa: sp	31	Mag 10.32v star NNE 8′.6.
02 59 50.6	+02 46 16	IC 277	13.1	1.2 x 0.9	13.1	SB(rs)bc II-III	33	Mag 9.76v star NE 11′.7.
03 03 50.2	−00 12 18	IC 283	14.9	0.5 x 0.3	12.7	Spiral	20	Mag 11.3 star S 8′.2.
01 45 19.8	+04 37 04	IC 1726	13.8	0.4 x 0.3	11.3		24	Mag 14.6 star on E edge; mag 10.13v star NW 12′.0; note: **UGC 1222** is 1′.2 SW of this star.
01 54 24.2	+04 48 11	IC 1746	13.8	1.4 x 0.4	13.0	S0	93	Mag 10.3 star SSE 6′.2.
01 56 49.9	+04 01 29	IC 1754	14.5	0.8 x 0.8	13.9			Stellar; mag 10.19v star SSW 9′.8; **IC 1750** WNW 8′.4; **MCG +1-6-16** E 11′.7.
01 57 05.3	−00 28 07	IC 1756	14.6	1.3 x 0.2	13.0	Scd:	155	
02 35 41.8	+04 03 07	IC 1819	14.3	1.6 x 0.6	14.4	S0	36	Pair of stars N end.
02 39 46.5	+01 33 33	IC 1827	13.7	1.1 x 0.2	11.9	Sa	154	Mag 8.19v star S 11′.8; NGC 1038 SE 5′.7.
02 42 48.2	+03 05 00	IC 1834	14.0	0.7 x 0.6	13.0	Sb II-III	156	Located 11′.8 SW of mag 3.57v star "Kaffaljidhma", Gamma Ceti.
02 43 23.5	+03 06 16	IC 1836	15.5	0.5 x 0.5	13.8			Located 8′.4 S of mag 3.57v star "Kaffaljidhma", Gamma Ceti.
02 45 24.7	+02 52 49	IC 1843	13.1	1.1 x 0.7	12.7	SB(s)ab? III	70	Mag 12.1 star ESE 3′.1.
02 45 49.4	+03 13 47	IC 1844	14.4	1.0 x 0.4	13.2	Ia	105	Mag 11.3 star SSW 5′.0.
02 48 50.8	−00 46 03	IC 1856	13.5	1.1 x 0.6	12.9	(R)SB(r)ab:	60	**CGCG 389-26** E 11′.5.
02 57 52.8	−02 20 35	IC 1870	13.2	2.8 x 1.6	14.7	SB(s)m III-IV	135	Lies 1′.7 W of mag 7.73v star.
01 51 13.5	−01 24 36	MCG +0-5-39	15.2	0.4 x 0.4	13.1	Spiral		Compact; mag 14.0 star N 0′.6.
01 51 27.2	−02 15 32	MCG +0-5-40	14.5	0.6 x 0.6	13.5	E/S0		
01 52 34.0	+01 20 48	MCG +0-5-42	15.0	0.6 x 0.4	13.3	Spiral	78	Mag 8.49v star S 5′.5.
01 54 06.2	+00 57 00	MCG +0-5-45	15.2	0.4 x 0.4	13.1	Spiral		
01 54 22.4	−00 37 41	MCG +0-5-46	14.0	0.6 x 0.4	12.3			Mag 9.90v star on S edge.
01 54 26.4	−00 12 41	MCG +0-5-47	14.9	0.7 x 0.2	12.8	E/S0	117	
01 54 55.5	−01 30 56	MCG +0-5-50	15.2	0.7 x 0.5	13.9	Spiral	129	Mag 9.11v star NW 1′.8.
01 58 05.4	+03 04 56	MCG +0-6-10	13.6	0.6 x 0.4	11.9	SBm: pec		UGC 1449 NE edge.
01 58 35.9	−01 27 27	MCG +0-6-15	14.2	0.7 x 0.3	12.4	S0	48	**MCG +0-6-12** SW 1′.8.
02 07 25.4	+02 06 56	MCG +0-6-33	14.9	0.6 x 0.6	13.6	S0?		= **Hickson 15E**. UGC 1617 E 2′.2.
02 07 37.8	+02 11 22	MCG +0-6-36	15.2	0.7 x 0.4	13.6	Sbc	9	= **Hickson 15F**. UGC 1618 = **Hickson 15D** on S edge.
02 08 15.5	+01 11 40	MCG +0-6-42	14.6	0.8 x 0.7	14.0	E	114	MCG +0-6-43 SE 1′.4.
02 08 20.3	+01 11 02	MCG +0-6-43	14.4	0.8 x 0.7	13.6	S?	105	MCG +0-6-42 NW 1′.4.
02 10 09.6	−01 24 04	MCG +0-6-46	16.9	0.9 x 0.4	15.7	Irr	147	
02 11 06.5	−00 39 15	MCG +0-6-47	14.1	1.3 x 0.5	13.5	Sb-c	108	
02 11 08.6	−00 39 11	MCG +0-6-48	14.3	1.0 x 0.3	12.9	S0	0	
02 13 06.7	+01 23 41	MCG +0-6-53	15.6	0.7 x 0.7	14.6			Mag 14.0 star E 1′.9.
02 28 09.7	+00 47 56	MCG +0-7-19	14.6	0.5 x 0.5	13.1	E		Mag 11.9 star WSW 1′.7.
02 31 20.2	+01 50 05	MCG +0-7-29	15.2	0.8 x 0.5	14.0		69	Mag 11.2 star S 2′.5.
02 31 43.3	+00 54 11	MCG +0-7-31	14.6	0.4 x 0.4	12.5	Sc		
02 38 10.4	+02 09 01	MCG +0-7-63	15.5	0.7 x 0.2	13.2		39	Mag 9.56v star NW 5′.5; NGC 1016 SE 3′.1.
02 39 24.2	+01 57 42	MCG +0-7-72	15.7	0.9 x 0.3	14.1	Sd	135	Mag 13.3 star NE 2′.4.
02 42 31.5	−01 00 57	MCG +0-7-84	15.7	0.5 x 0.3	13.5	Sc	114	
02 47 28.4	−00 16 01	MCG +0-8-14	15.0	0.5 x 0.2	12.4	Sb	66	Mag 9.66v star N 3′.7; NGC 1094 on S edge.
02 49 41.0	−00 31 27	MCG +0-8-25	14.0	0.9 x 0.6	13.2	SB(rs)b pec? II-III	51	
02 50 14.6	+00 05 38	MCG +0-8-28	16.5	0.7 x 0.2	14.2	Sc	153	UGC 2324 E edge.
02 51 56.1	−00 50 06	MCG +0-8-37	14.3	0.5 x 0.4	12.6	E		Mag 9.47v star W 6′.0.
02 52 24.3	−01 17 12	MCG +0-8-39	15.4	0.8 x 0.2	13.2	Spiral	161	NGC 1126 WSW 1′.5.
02 54 55.5	+00 06 35	MCG +0-8-45	17.1	0.4 x 0.3	14.6	Spiral	48	Compact; mag 11.3 star N 3′.6.
03 00 51.6	+00 48 23	MCG +0-8-66	15.0	0.9 x 0.7	14.4	Sbc	0	Mag 11.8 star N 1′.8.
03 01 44.4	−00 35 31	MCG +0-8-69	15.0	0.7 x 0.7	14.0	Spiral		Mag 9.89v star W 3′.1.
03 03 30.7	−02 14 50	MCG +0-8-75	14.4	0.8 x 0.2	12.3		24	
03 03 52.9	+00 24 54	MCG +0-8-77	14.7	0.7 x 0.7	13.8			Faint, elongated anonymous galaxy NE 1′.6.
01 44 41.9	+04 53 42	MCG +1-5-30	14.5	0.9 x 0.3	12.9	S?	15	Faint star or bright knot S end; mag 13.3 star SSW 0′.9.
01 47 58.8	+04 59 16	MCG +1-5-33	15.2	0.5 x 0.4	13.3	Sc	6	Mag 9.89v star ENW 4′.1.
01 54 42.8	+05 25 29	MCG +1-5-47	14.5	0.8 x 0.5	13.4	Spiral	135	**UGC 1373** W 4′.6.
01 56 36.9	+05 48 11	MCG +1-6-9	14.8	0.6 x 0.4	13.1		96	
01 57 08.2	+04 33 50	MCG +1-6-13	13.7	0.8 x 0.6	12.8		78	Mag 13.0 star N 1′.9.
01 57 20.7	+04 20 52	MCG +1-6-15	16.4	0.7 x 0.4	14.9	Spiral	24	Mag 12.3 star S 1′.3.
01 58 53.2	+05 25 00	MCG +1-6-23	14.9	0.6 x 0.5	13.4	Sab	51	Mag 11.4 star NW 3′.9.
01 58 50.9	+05 21 33	MCG +1-6-24	14.6	0.6 x 0.3	12.6	Sb	90	Mag 10.9 star S 1′.1.
02 01 32.4	+05 54 56	MCG +1-6-30	15.0	0.7 x 0.5	13.8		156	mag 12.9 star NW 2′.4.
02 22 38.7	+05 32 28	MCG +1-7-3	14.7	0.7 x 0.6	13.6	Sbc	132	Located between a pair of NW-SE mag 14.4 stars.
02 24 12.3	+05 22 20	MCG +1-7-4	14.3	0.9 x 0.3	12.7	S?	48	Mag 14.3 star SW 1′.7.
02 30 51.8	+03 51 06	MCG +1-7-7	15.5	0.7 x 0.4	13.9	Sc	114	
02 51 04.7	+04 27 10	MCG +1-8-8	14.5	0.5 x 0.2	11.9	SBb	132	Located between a pair of stars, mags 8.70v and 10.70v.
02 57 45.8	+05 57 05	MCG +1-8-28	15.5	0.9 x 0.4	14.2	Scd	150	**PGC 11195** N 1′.4; **PGC 11196** N 2′.0; mag 12.4 star N 2′.3.
03 00 00.2	+05 58 03	MCG +1-8-34	15.2	0.5 x 0.5	13.5	Sbc		Mag 12.4 star W 1′.1.
03 00 08.8	+05 48 13	MCG +1-8-35	14.5	0.5 x 0.5	12.9	Sa/0		
03 00 09.7	+05 42 40	MCG +1-8-36	15.8	0.5 x 0.4	13.9	Sbc	0	UGC 2469 E 2′.7.
01 44 41.6	−03 06 41	MCG −1-5-27	14.9	0.9 x 0.4	13.6	SBb	153	Mag 7.40v star E 10′.6.
01 44 45.0	−04 08 06	MCG −1-5-29	13.5	1.0 x 0.8	13.1	SAB(r)0°:	90	
01 45 25.6	−03 49 44	MCG −1-5-31	13.1	1.4 x 0.9	13.2	SB(rs)bc pec:	168	
01 49 15.5	−03 41 02	MCG −1-5-36	14.4	1.1 x 1.1	14.5	SA(rs)cd: II-III		Star on SE edge.
01 50 41.8	−03 56 17	MCG −1-5-40	13.9	1.2 x 0.6	13.4	SB(r)0$^+$ pec?	45	NGC 702 SE 11′.7.
01 51 11.3	−03 29 57	MCG −1-5-42	14.1	1.3 x 0.8	14.0	SB(s)dm IV-V	24	Several bright knots S end.
01 51 50.3	−05 29 48	MCG −1-5-45	14.7	1.6 x 0.5	14.3	SB(s)dm III	66	**MCG −1-5-46** SE edge.
01 52 48.8	−03 26 46	MCG −1-5-47	13.3	3.2 x 0.5	13.7	SA(s)c? sp	162	Mag 12.7 star NE 3′.2.
01 54 21.8	−02 37 14	MCG −1-5-49	14.2	0.9 x 0.3	12.7	Spiral	150	
01 56 10.6	−02 40 41	MCG −1-6-1	14.5	0.5 x 0.4	12.6	S0/a	6	Stellar.
01 56 15.1	−02 42 20	MCG −1-6-2	14.1	0.6 x 0.4	12.7	E	147	Almost stellar.
01 56 16.2	−04 11 12	MCG −1-6-3	14.4	0.8 x 0.3	12.7	Sa	21	0′.8 SW of mag 7 star.
01 56 47.7	−03 44 16	MCG −1-6-5	13.7	1.3 x 0.8	13.6	(R′)S0$^+$ pec:	123	
01 57 41.6	−05 40 33	MCG −1-6-7	13.8	1.6 x 1.1	14.2	SA(rs)cd II-III:	54	Mag 12.0 star SW 1′.7.
01 59 07.1	−03 09 02	MCG −1-6-13	14.1	1.0 x 0.7	13.6	Irr	135	Star or galaxy E edge. **MCG −1-6-14** E 1′.0.
02 00 59.4	−05 30 39	MCG −1-6-24	14.2	1.2 x 0.4	13.3	Sab	165	Mag 10.51v star E 5′.1.
02 02 14.7	−02 50 43	MCG −1-6-29	13.7	0.5 x 0.3	12.2	E		
02 02 26.2	−04 11 02	MCG −1-6-34	14.4	1.0 x 0.7	13.9	Irr	33	
02 04 20.3	−04 54 58	MCG −1-6-38	14.2	1.2 x 0.5	13.5	Irr	138	
02 05 48.5	−05 17 37	MCG −1-6-43	14.0	1.1 x 0.5	13.2	Sbc	144	

RA h m s	Dec ° ′ ″	Name	Mag (V)	Dim ′ Maj x min	SB	Type Class	PA	Notes
02 10 52.9	−02 48 22	MCG −1-6-63	14.2	1.1 x 0.5	13.4	(R′)SB(r)ab	120	1′.0 NE of mag 10 star.
02 11 46.7	−05 54 54	MCG −1-6-68	14.5	0.9 x 0.8	14.0	Sb	21	Star W edge.
02 13 22.6	−03 05 19	MCG −1-6-72	14.3	0.8 x 0.6	13.3	Sc	12	Mag 7.33v star NE 6′.0.
02 13 22.2	−02 20 32	MCG −1-6-73	14.6	0.8 x 0.4	13.3	S0/a	144	Mag 5.65v star 66 Ceti SW 9′.2.
02 17 19.4	−05 28 54	MCG −1-6-80	14.1	1.3 x 0.6	13.7	SB?	141	Low surface brightness.
02 21 28.8	−04 24 50	MCG −1-7-1	15.6	1.4 x 0.4	14.8	SB(s)d: sp IV-V	51	Uniform surface brightness.
02 23 13.4	−04 31 02	MCG −1-7-4	14.1	0.7 x 0.4	12.6	SBc	15	
02 24 02.5	−04 41 38	MCG −1-7-7	14.2	1.1 x 0.6	13.6	SBc	114	Galaxy **PGC 9107** W 0′.8.
02 24 04.7	−02 51 27	MCG −1-7-8	14.8	1.0 x 0.3	13.3	Sc	75	
02 25 56.5	−04 18 59	MCG −1-7-10	14.3	0.8 x 0.5	13.1	Sc	87	
02 28 11.0	−05 50 27	MCG −1-7-15	14.3	1.5 x 0.4	13.6	Sb pec sp	21	Small anonymous galaxy at middle of W edge.
02 29 31.8	−03 12 48	MCG −1-7-16	12.8	1.2 x 0.9	12.7	SB(rs)0/a pec	102	= **Mkn 1043**.
02 30 41.0	−03 48 48	MCG −1-7-20	14.9	2.5 x 1.9	16.4	SB(s)dm: V-VI	12	Low surface brightness; mag 10.98v star E 7′.6.
02 34 24.1	−04 29 28	MCG −1-7-21	14.5	1.1 x 0.7	14.1	SBab	36	
02 36 58.6	−05 20 56	MCG −1-7-24	13.2	1.1 x 0.9	13.1	SA(s)c pec: II	150	
02 40 23.5	−02 43 43	MCG −1-7-28	14.9	2.2 x 0.6	15.0	Sc pec sp	78	**MCG −1-7-29** E 2′.5.
02 42 22.9	−03 06 28	MCG −1-7-37	13.8	0.7 x 0.5	12.7	E	3	Mag 6.06v star SW 10′.8.
02 58 24.5	−04 17 43	MCG −1-8-24	13.6	2.2 x 1.1	14.4	SB(rs)d III-IV	123	Short, bright E-W bar.
02 59 10.7	−04 28 19	MCG −1-8-25	13.4	1.1 x 0.5	12.6	(R′)SB(r)0/a pec?	93	= **Mkn 1065**.
01 48 57.4	+05 54 28	NGC 676	11.9	4.0 x 1.0	13.3	S0/a: sp	172	
01 51 19.2	−04 03 16	NGC 702	13.1	1.2 x 0.8	12.9	SB(s)bc pec	140	Bright knot or faint superimposed star N edge; MCG −1-5-40 NW 11′.7.
01 53 13.3	+04 11 47	NGC 718	11.7	2.3 x 2.0	13.2	SAB(s)a	45	
01 56 20.9	+05 37 39	NGC 741	11.1	3.0 x 2.9	13.5	E0:	90	NGC 742 0′.8 E of nucleus. **IC 1751** NW 1′.5; **MCG +1-6-5** SW 3′.3.
01 56 24.4	+05 37 33	NGC 742	14.3	0.3 x 0.3	11.7	cE0:		Located 0′.8 E of nucleus of NGC 741.
01 56 22.1	−04 28 07	NGC 748	12.6	2.3 x 1.1	13.5	(R′)SA(r)b? II	140	Located 1′.9 SE of mag 10.84v star.
01 55 42.1	−03 04 00	NGC 758	14.4	0.9 x 0.8	13.9	S0	85	Almost stellar nucleus; mag 9.18v star W 8′.2; stellar galaxy **KUG 0153-032** NE 4′.5.
01 56 57.4	−05 24 11	NGC 762	13.0	1.3 x 1.1	13.2	(R′)SB(rs)a	25	
01 58 40.8	+00 31 40	NGC 768	13.2	1.7 x 0.6	13.1	SB(r)bc: I-II	30	**IC 1761** NE 3′.7; mag 8.27v star E 7′.4.
01 59 42.5	−05 57 50	NGC 779	11.2	4.0 x 1.2	12.7	SAB(r)b II	160	MCG −1-6-23 SE 11′.0; mag 10.83v star ESE 11′.6.
02 01 21.7	−05 22 15	NGC 790	13.0	1.3 x 1.3	13.4	SA(r)0°?		Compact core, faint, smooth envelope.
02 02 12.3	−00 06 06	NGC 799	13.0	2.0 x 1.7	14.1	(R′)SB(s)a:	100	Compact core with uniform envelope; star or bright knot on E edge; NGC 800 on S edge.
02 02 11.9	−00 07 52	NGC 800	13.7	1.0 x 0.9	13.4	SA(rs)c: II	10	NGC 799 on N edge.
02 11 13.4	−01 29 08	NGC 850	12.9	1.1 x 1.1	12.9	SAB(s)0⁺		Mag 9.47v star SSE 5′.6; stellar galaxy **UM 406** WNW 4′.0.
02 11 12.2	+03 46 43	NGC 851	13.5	1.2 x 0.6	13.0	SB0⁺:	135	Mag 14.5 star E edge; mag 10.21v star WSW 14′.1; IC 211 N 4′.5.
02 13 38.2	−00 43 01	NGC 856	13.2	1.3 x 0.9	13.3	(R′)SA(rs)0/a:	20	
02 14 33.4	−00 46 01	NGC 863	12.9	1.1 x 1.0	12.9	SA(s)a:	69	
02 15 27.6	+06 00 10	NGC 864	10.9	4.7 x 3.5	13.7	SAB(rs)c II-III	20	A mag 10.74v star lies 0′.8 E of stellar nucleus.
02 15 58.6	−00 42 49	NGC 868	13.9	1.3 x 1.0	14.1	S0⁻:	95	Compact.
02 17 04.8	+01 14 39	NGC 875	12.9	1.1 x 1.1	13.0	S0⁺:		**IC 218**, and small anonymous companion, N 2′.4.
02 18 27.1	−04 12 21	NGC 880	14.6	0.4 x 0.3	12.2	S pec	26	Stellar nucleus; mag 8.46v star N 7′.1..
02 21 36.5	−05 31 16	NGC 895	11.7	3.6 x 2.6	14.0	SA(s)cd I-II	126	Strong core with two prominent arms.
02 26 06.7	−00 19 56	NGC 926	13.3	1.8 x 1.0	13.7	SB(rs)bc: II-III	36	
02 27 32.5	−00 14 38	NGC 934	13.1	1.3 x 0.9	13.2	SAB0⁻	130	Compact core faint envelope.
02 27 37.0	−01 09 19	NGC 936	10.1	4.7 x 4.1	13.2	SB(rs)0⁺	135	Short E-W bar, smooth NW-SE envelope.
02 28 27.7	−01 09 07	NGC 941	12.4	2.6 x 1.9	14.0	SAB(rs)c III-IV	170	
02 30 33.4	−01 06 24	NGC 955	12.0	2.8 x 0.7	12.6	Sab: sp	19	
02 30 42.9	−02 56 18	NGC 958	12.2	2.9 x 1.0	13.2	SB(rs)c: II	10	Moderate dark lane E of core.
02 30 31.0	−04 13 01	NGC 963	13.7	0.7 x 0.7	12.8	Irr		
02 36 46.1	+02 02 59	NGC 993	13.6	1.0 x 0.9	13.3	S0⁻ pec:	110	Faint star on W edge of compact core and N edge of envelope.
02 37 42.0	+01 58 28	NGC 1004	12.7	1.4 x 1.3	13.2	E⁺	115	Two anonymous galaxies NW 2′.1; four more E and NE between 4′ and 6′.
02 37 52.3	+02 09 17	NGC 1007	15.5	0.6 x 0.2	13.1	Pec	48	Mag 9.56v star N 2′.7.
02 37 55.4	+02 04 43	NGC 1008	13.6	0.9 x 0.6	13.0	E	85	Mag 10.6 star SE 2′.9.
02 38 19.0	+02 18 33	NGC 1009	14.4	1.4 x 0.2	12.9	Sb sp III-IV	124	Mag 9.9 star N 5′.9.
02 38 11.6	−01 19 09	NGC 1015	12.1	2.0 x 1.7	13.3	SB(r)a:	16	Mag 8.22v star SE 6′.3.
02 38 19.7	+02 07 09	NGC 1016	11.6	2.0 x 2.0	13.2	E		Mag 9.56v star NW 8′.4; MCG +0-7-63 NW 3′.1.
02 38 27.5	+01 54 25	NGC 1019	13.5	1.0 x 0.9	13.3	SB(rs)bc I	40	**CGCG 388-80** N 3′.7.
02 38 44.4	+02 13 48	NGC 1020	14.1	0.9 x 0.3	12.6	S0 sp	20	Mag 9.56v star W 13′.2; NGC 1021 SE 1′.3.
02 38 48.1	+02 12 59	NGC 1021	14.2	0.8 x 0.6	13.3	SAB(r)bc: II-III	160	NGC 1020 NW 1′.3.
02 39 23.5	+01 05 36	NGC 1032	11.6	3.3 x 1.1	12.9	S0/a sp	68	Mag 12.7 star on NE edge; mag 10.67v star S 4′.2.
02 37 58.8	−01 50 38	NGC 1037	13.4	1.7 x 0.8	13.6	SB(r)b: II	2	**UGC 2106** W 8′.7.
02 40 06.4	+01 30 30	NGC 1038	13.4	1.2 x 0.4	12.4	(R′)S0/a:	64	Mag 8.19v star SW 11′.5; IC 1827 NW 5′.7.
02 40 25.2	−05 26 28	NGC 1041	13.3	1.7 x 1.2	13.9	S0⁻ pec	57	Small, faint galaxy **KUG 0237-055** NNW 7′.0.
02 40 46.7	+01 20 33	NGC 1043	15.5	0.8 x 0.2	13.4	Sb	110	Mag 11.4 star W 5′.0; UGC 2162 SW 9′.0.
02 41 45.8	+00 26 33	NGC 1055	10.6	7.6 x 2.7	13.7	SBb: sp II-III:	105	Irregular dark lane bisects bright central area lengthwise.
02 42 10.1	−05 34 07	NGC 1063	13.8	1.4 x 0.5	13.2	(R′)SA(r)bc: II	111	Mag 10.80v star WSW 3′.9.
02 42 40.9	−00 00 43	NGC 1068	8.9	7.1 x 6.0	12.8	(R)SA(rs)b I-II	70	= **M 77**. Mag 10.81v star lies 1′.2 E of center of bright core.
02 43 22.1	+04 58 01	NGC 1070	11.9	2.3 x 1.9	13.4	Sb III	175	**CGCG 415-3** NE 10′.6.
02 43 31.5	+00 18 22	NGC 1072	13.4	1.5 x 0.5	12.9	SB(rs)b: II	11	Mag 9.23v star SE 6′.8.
02 43 40.6	+01 22 36	NGC 1073	11.0	4.9 x 4.5	14.2	SB(rs)c II-III	15	Numerous bright knots and/or superimposed stars.
02 45 10.0	−04 42 38	NGC 1080	13.5	1.1 x 0.8	13.3	SAB(s)c: I-II	174	
02 46 25.3	+03 36 20	NGC 1085	12.3	1.8 x 1.3	13.0	SA(s)bc: II	15	
02 46 25.3	−00 29 52	NGC 1087	10.9	3.7 x 2.2	13.1	SAB(rs)c III-IV:	5	
02 46 33.7	−00 14 50	NGC 1090	11.8	4.0 x 1.7	13.8	SB(rs)bc II-III	102	
02 47 28.0	−00 17 04	NGC 1094	12.5	1.3 x 1.0	12.7	SAB(s)ab I-II	85	MCG +0-8-14 on N edge; mag 9.66v star N 4′.7.
02 47 38.0	+04 38 13	NGC 1095	13.3	1.3 x 0.7	13.0	SBc	45	Mag 10.51v star NNW 1′.9.
02 48 14.9	+04 34 38	NGC 1101	13.0	1.3 x 1.0	13.2	S0	120	Bright knot or star W edge of compact core.
02 48 39.0	−00 16 19	NGC 1104	13.6	1.2 x 0.9	13.5	(R′)SB(rs)0/a	70	Mag 9.96v star SE 2′.9.
02 50 39.2	−01 44 01	NGC 1121	12.9	0.9 x 0.4	11.7	S0:	10	Mag 10.33v star NNE 1′.5.
02 52 18.7	−01 17 48	NGC 1126	14.6	1.0 x 0.3	13.2	Sb: sp	135	MCG +0-8-39 ENE 1′.5.
02 52 52.0	−01 16 32	NGC 1132	12.3	2.5 x 1.3	13.5	E	140	Mag 9.06v star E 4′.3.
02 54 02.8	+02 57 44	NGC 1137	12.4	2.1 x 1.3	13.4	(R′)SA(rs)b III	20	Compact core, faint, uniform envelope.
02 55 09.4	−00 10 39	NGC 1143	13.7	1.1 x 0.9	13.6	Ring A	110	Mag 6.52v star NW 9′.6; NGC 1144 on SE edge.
02 55 12.1	−00 11 05	NGC 1144	12.9	1.1 x 0.7	12.5	Ring B	130	Located on SE edge of NGC 1144.
02 57 24.0	−00 18 34	NGC 1149	14.0	0.7 x 0.6	12.9	S0⁻:	130	A very faint **UGC 2428** NW 6′.0.
02 58 10.4	+03 21 40	NGC 1153	12.4	1.2 x 0.7	12.1	S0⁺?	45	Bright knot SW edge of core; UGC 2446 NE 8′.9.
03 03 49.3	−01 06 14	NGC 1194	12.9	1.8 x 0.8	13.1	SA0⁺:	140	UGC 2517 SE 7′.9.
01 45 03.4	−00 18 02	UGC 1225	15.6	0.6 x 0.4	13.9	SB?	177	
01 45 35.7	+03 21 27	UGC 1235	14.0	0.9 x 0.4	12.7	Sbc	143	Mag 6.82v star NE 6′.0.
01 46 20.7	+04 15 49	UGC 1240	14.2	1.1 x 0.3	12.8	Sdm:	91	

(Continued from previous page)

GALAXIES

RA h m s	Dec ° ′ ″	Name	Mag (V)	Dim ′ Maj x min	SB	Type Class	PA	Notes
01 49 59.4	+02 05 29	UGC 1293	13.8	0.8 x 0.5	12.7	Sab	143	
01 50 10.6	+02 18 34	UGC 1297	14.2	1.0 x 0.8	13.8	IB(s)m	36	
01 53 59.2	−00 44 58	UGC 1365	13.6	1.0 x 0.4	12.5	S?	118	Low surface brightness dwarf **UGC 1367** E 2′.2.
01 54 41.1	−00 08 35	UGC 1382	13.2	1.4 x 1.4	14.0	E?		
01 56 17.2	+04 38 46	UGC 1410	14.7	1.7 x 0.2	13.3	Sd	93	Close pair of stars, mags 10.6 and 11.6, S 1′.9.
01 56 52.1	+05 46 28	UGC 1425	13.6	1.0 x 0.7	13.1	E/S0	174	**MCG +1-6-11** E 1′.1.
01 57 21.7	+05 36 32	UGC 1435	13.9	0.8 x 0.5	12.7	S?	40	
01 57 33.9	−02 06 06	UGC 1442	14.6	1.1 x 0.8	14.3	Sdm	177	Mag 9.11v star SSE 2′.1.
01 58 00.2	+04 23 39	UGC 1444	14.2	1.4 x 0.5	13.7	S?	140	Stellar nucleus; mag 12.0 star NW 3′.1.
01 58 02.7	+03 22 07	UGC 1446	14.2	1.3 x 0.5	13.5	Im:	51	Stellar **Mkn 1168** NW 9′.5.
01 58 08.6	+02 03 49	UGC 1448	14.2	1.3 x 0.4	13.4	Scd:	110	
01 58 06.7	+03 05 12	UGC 1449	13.3	1.1 x 0.5	12.5	SBm pec:	126	MCG +0-6-10 SW edge; mag 15.2 star on SE end.
01 58 35.2	+03 15 28	UGC 1454	15.9	1.4 x 0.7	15.7	SBm	99	
01 59 07.7	+01 53 23	UGC 1464	16.6	1.3 x 0.8	16.5	IAB(s)m: V-VI	12	Dwarf irregular, low surface brightness.
02 02 17.2	−01 07 41	UGC 1525	14.1	1.2 x 0.8	14.0	E	60	A mag 9.90v star NE 1′.7; a small, faint anonymous galaxy lies NE 1′.1, slightly to the NW of the star.
02 03 10.6	+05 39 21	UGC 1545	14.0	0.9 x 0.3	12.4	S0/a	124	
02 03 35.5	+04 47 07	UGC 1553	14.5	0.9 x 0.5	13.5	Sd	139	Stellar nucleus; mag 9.46v star N 4′.3.
02 06 16.1	−00 17 30	UGC 1597	13.3	1.5 x 1.2	13.8	S0	0	
02 06 42.7	−00 51 37	UGC 1603	14.4	1.8 x 1.5	15.3	SAdm:		Almost stellar nucleus, faint envelope.
02 07 34.1	+02 06 53	UGC 1617	14.0	0.9 x 0.9	13.6	S0$^-$:		= **Hickson 15B**. MCG +0-6-33 = **Hickson 15E** 2′.2 E; UGC 1620 = **Hickson 15C** NE 2′.5.
02 07 37.6	+02 10 50	UGC 1618	14.3	0.6 x 0.5	12.9	S0$^-$:	76	= **Hickson 15D**. MCG +0-6-36 = **Hickson 15F** 0′.6 N.
02 07 39.7	+02 08 56	UGC 1620	13.6	0.8 x 0.7	12.8	S0$^-$:	124	= **Hickson 15C**.
02 07 53.0	+02 10 04	UGC 1624	13.7	1.2 x 0.7	13.4	S0:	130	= **Hickson 15A**.
02 08 04.7	+01 53 39	UGC 1627	14.0	1.0 x 0.3	12.5	S?	70	Very faint anonymous galaxy N 1′.1.
02 09 04.6	+05 06 39	UGC 1646	13.8	2.0 x 0.5	13.7	SBbc	173	Mag 9.52 star W 1′.3.
02 10 37.7	+05 52 10	UGC 1669	13.5	0.7 x 0.6	12.4	S	173	
02 12 20.5	−02 08 30	UGC 1697	13.9	1.4 x 0.5	13.4	Sb III	135	Mag 7.29v star SE 7′.5.
02 12 19.8	−00 48 45	UGC 1698	13.9	1.3 x 1.2	14.2	(R′)SB(r)a:	105	
02 13 28.3	+04 44 51	UGC 1707	14.4	0.8 x 0.5	13.2	SB(s)b	6	Mag 8.44 star 3′.1 W.
02 13 45.2	+04 06 04	UGC 1716	13.6	0.6 x 0.4	11.9	S?	177	Mag 13.0 star N 0′.7.
02 15 53.8	+01 46 51	UGC 1746	14.1	1.1 x 0.9	13.9	SB(rs)d: III	90	Mag 11.5 star SW 2′.8.
02 16 53.7	+02 12 08	UGC 1756	13.9	1.1 x 0.4	12.9	S0°: sp	52	Mag 7.26v star WNW 6′.6.
02 18 26.5	+05 39 12	UGC 1775	12.9	1.3 x 1.2	13.2	S?	141	**UGC 1774** S 6′.0.
02 19 41.0	−00 15 23	UGC 1794	13.5	1.5 x 1.5	14.2	(R′)SAB(s)b pec		Almost stellar nucleus, two main arms; mag 8.56v star E 6′.9; **MCG +0-7-4** ESE 11′.9.
02 19 58.9	+01 55 46	UGC 1797	13.9	0.9 x 0.5	12.9	S0/a	135	**MCG +0-7-2** NW 1′.1.
02 20 53.5	+00 33 23	UGC 1809	13.9	1.0 x 0.7	13.3	S0/a	60	Mag 12.9 star on E edge.
02 22 30.0	−00 36 58	UGC 1839	14.7	3.0 x 0.4	14.7	Sd: sp	45	Mag 7.93v star N 6′.7.
02 24 24.9	−02 09 44	UGC 1862	13.1	1.5 x 1.2	13.5	SAB(rs)d pec: III-IV	13	
02 24 29.1	+01 50 28	UGC 1863	14.3	0.8 x 0.8	13.6	SAB(s)c		
02 26 19.5	−01 33 04	UGC 1905	14.1	0.5 x 0.4	12.3	S0		
02 27 51.4	+00 30 05	UGC 1934	14.5	1.7 x 0.3	13.6	Sbc: sp	112	Faint, stellar galaxy **KUG 0225+002** W 3′.4.
02 28 12.6	−01 20 52	UGC 1945	13.5	2.2 x 0.8	14.0	Sdm: III-IV	12	
02 28 54.5	+00 22 11	UGC 1962	13.6	1.1 x 0.8	13.3	Scd:	105	
02 31 26.2	+01 20 45	UGC 1995	13.6	1.5 x 0.8	13.6	SB(r)c pec: I-II	62	Mag 10.7 star SE 3′.8.
02 31 59.6	+00 54 34	UGC 2004	13.8	1.0 x 0.6	13.1	SAB(s)c	95	Mag 14.6 star superimposed S edge.
02 31 57.9	+01 14 47	UGC 2005	13.1	2.3 x 2.3	14.7	SA(rs)0$^-$:		Small, bright core, uniform envelope; **MCG +0-7-33** WNW 2′.3; **MCG +0-7-34** SW 2′.5.
02 32 07.2	−01 21 46	UGC 2010	13.3	1.2 x 0.8	13.1	SB(rs)bc I	85	**UGC 2009** S 1′.5.
02 32 40.1	+00 15 33	UGC 2018	13.1	1.5 x 0.7	13.1	SB(rs)0$^+$:	75	
02 32 39.2	+00 36 59	UGC 2019	13.6	0.7 x 0.5	12.3	S?	65	Stellar **CGCG 388-45** S 1′.4; stellar **Mkn 1047** N 4′.2.
02 32 52.1	+05 42 00	UGC 2021	13.5	0.7 x 0.6	12.4	S0	45	Mag 10.61v star NW 2′.9.
02 33 01.2	+00 25 11	UGC 2024	13.5	1.0 x 0.5	12.6	Sab	153	Mag 11.3 star NW 1′.2.
02 34 05.0	+01 21 05	UGC 2051	12.9	1.0 x 1.0	12.8	S0		Close pair of mag 13.5, 14.4 stars SW 0′.9; mag 9.65v star SW 3′.8.
02 34 15.6	+02 36 43	UGC 2056	13.9	0.8 x 0.6	13.0	SABbc	117	Mag 11.9 star N 1′.2.
02 34 37.7	−00 58 46	UGC 2061	15.1	1.5 x 0.4	14.4	S	56	
02 34 35.9	+01 20 46	UGC 2062	14.1	0.9 x 0.4	12.8	Sbc	98	Mag 13.0 star on SE edge; mag 7.75v star SE 7′.5.
02 36 01.0	+00 25 08	UGC 2081	13.7	2.1 x 1.4	14.7	SAB(s)cd III	80	
02 36 23.7	+00 42 30	UGC 2091	14.1	1.7 x 0.5	13.8	Sc pec sp	165	
02 38 07.5	+01 23 19	UGC 2120	14.1	1.5 x 0.4	13.4	SBbc:	149	
02 38 03.7	+01 41 11	UGC 2121	13.7	1.3 x 0.9	13.8	SAB(s)c	174	Mag 10.21v star E 2′.7.
02 40 23.6	+01 13 40	UGC 2162	15.7	1.8 x 1.8	16.9	IB(s)m V-VI		Dwarf irregular, very low surface brightness; mag 9.55v stars E 5′.1; NGC 1043 NE 9′.0.
02 41 54.2	+05 56 50	UGC 2176	13.7	1.1 x 0.4	12.6	Sb II-III	14	
02 42 13.7	+02 25 25	UGC 2181	14.0	0.7 x 0.7	13.0	S?		Mag 9.46v star N 5′.4.
02 43 24.0	+01 45 25	UGC 2199	13.5	0.9 x 0.5	12.5	S0/a	153	Located 2′.1 NE of mag 7.82v star.
02 44 21.1	+00 40 35	UGC 2216	15.4	1.3 x 0.4	14.5	Im:	15	**MCG +0-8-2** NNW 7′.4.
02 45 27.6	+00 54 52	UGC 2229	13.8	1.4 x 0.6	13.7	E/S	174	Small, round anonymous galaxy on SW edge.
02 47 44.9	+00 24 53	UGC 2271	14.3	1.5 x 0.9	14.5	SB(rs)b: II-III	10	
02 47 57.6	+03 53 04	UGC 2275	13.1	5.5 x 5.5	16.6	Sm: V		Dwarf spiral, very low surface brightness.
02 48 51.9	+00 59 21	UGC 2292	14.1	1.0 x 0.9	13.8	SAa		Mag 6.82v star ESE 12′.2.
02 48 58.8	+03 10 06	UGC 2295	13.1	1.5 x 0.5	12.7	Sb pec sp	85	
02 49 08.7	+02 07 42	UGC 2302	13.5	5.5 x 5.5	17.1	(R′)SB(rs)m: IV-V		Dwarf spiral, very low surface brightness.
02 49 27.9	−00 52 20	UGC 2311	12.9	1.4 x 1.4	13.5	(R′)SB(r)b		**UGC 2319** SE 10′.2; **MCG +0-8-27** E 10′.5.
02 50 17.0	+00 05 30	UGC 2324	13.9	0.9 x 0.6	13.0	SB(r)b	0	MCG +0-8-28 W edge.
02 51 09.5	+02 35 33	UGC 2338	13.7	1.5 x 0.8	13.7	SB(s)d: III-IV	67	
02 51 35.3	−00 44 03	UGC 2343	13.9	1.0 x 0.7	13.4	Sa:	27	Mag 7.45v star NW 3′.6.
02 51 53.0	−01 10 24	UGC 2345	13.6	4.2 x 3.2	16.3	SB(rs)m: V	168	Dwarf spiral, very low surface brightness; mag 8.53v star W 6′.9.
02 53 58.6	+05 59 22	UGC 2372	14.4	0.9 x 0.8	13.9	SAB(rs)cd:		
02 55 03.2	−01 11 19	UGC 2385	14.4	0.8 x 0.7	13.6	Sc		
02 55 57.3	+00 41 29	UGC 2403	13.7	1.3 x 0.6	13.3	SB(rs)a pec:	149	
02 56 27.2	+00 52 29	UGC 2418	14.1	1.1 x 0.7	13.7	SABbc	109	
02 57 09.3	+05 19 15	UGC 2426	14.1	0.9 x 0.6	13.2	Sb III	65	Mag 10.89v star NE 2′.7.
02 58 22.0	+03 51 41	UGC 2441	14.0	2.0 x 0.4	13.6	Sd	13	Faint anonymous galaxy WNW 2′.6; mag 9.25v star SW 3′.0.
02 58 21.4	−02 02 28	UGC 2443	13.7	1.4 x 0.7	13.5	Scd:	163	
02 58 41.5	+03 26 03	UGC 2446	13.8	0.7 x 0.4	12.3	S?	105	Located on E edge of galaxy cluster **A 403**; **PGC 11298** NE 9′.0.
03 00 20.5	+05 42 42	UGC 2469	14.3	1.1 x 0.7	13.9	Pec?	24	MCG +1-8-36 W 2′.7.
03 00 40.0	+00 01 09	UGC 2479	15.4	1.3 x 0.4	14.5	Sdm: sp	25	
03 02 33.8	+02 11 39	UGC 2501	13.3	1.3 x 0.6	12.9	SB(s)m pec: III	133	Bright knot SE end; mag 9.52v star W 2′.6.
03 03 39.3	+01 56 51	UGC 2513	13.9	1.1 x 0.6	13.3	Sa:	153	

(Continued from previous page)

119 GALAXY CLUSTERS 119

RA h m s	Dec ° ′ ″	Name	Mag 10th brightest	No. Gal	Diam ′	Notes
01 44 42.0	+05 48 00	A 246	16.4	56	25	Difficult to see many GX, even on DSS image.

120 GALAXIES 120

RA h m s	Dec ° ′ ″	Name	Mag (V)	Dim ′ Maj x min	SB	Type Class	PA	Notes
00 41 48.5	−01 44 22	CGCG 383-72	14.6	0.7 x 0.5	13.3		6	Faint, curved dark lane S of center.
01 06 30.1	−02 12 00	CGCG 384-74	15.3	1.5 x 1.2	15.8	SB(rs)cd III-IV	45	Bright core offset to the NE. Faint envelope SW.
01 06 54.2	+01 56 42	CGCG 384-75	14.9	0.4 x 0.4	12.8	Sb		Mag 10.6 star SE 5.′7; mag 11.2 star SW 5.′9.
01 14 10.4	−01 51 17	CGCG 385-36	14.0	0.7 x 0.5	12.7		90	Mag 13.6 star SE 1.′5; mag 12.8 star W 4.′5.
01 22 00.0	−01 02 27	CGCG 385-82	14.9	0.4 x 0.4	12.8	S0		**MCG +0-4-98** N 1.′7.
01 22 35.9	+01 53 22	CGCG 385-87	14.1	0.4 x 0.2	11.2	Pec?	0	= **Mkn 569**. Mag 12.7 star WSW 2.′0.
01 28 59.6	+01 49 37	CGCG 385-161	14.4	0.7 x 0.5	13.2		81	Located on the W edge of galaxy cluster **A 203**; mag 14.2 star S 1.′9.
01 36 23.6	+03 45 16	CGCG 386-17	15.1	0.6 x 0.3	13.1		120	Bright nucleus with faint outer envelope; located 2.′0 S of mag 8.66v star.
01 42 17.5	−02 08 44	CGCG 386-31	14.2	0.7 x 0.4	12.7	Sc	48	Galaxy pair with round component NE and elongated component extending SW; mag 7.94v star ESE 2.′1.
01 47 20.4	+02 49 56	CGCG 386-37	14.0	0.6 x 0.3	12.0		81	Mag 15.5 star on N edge.
00 34 16.7	+05 32 49	CGCG 409-37	14.0	0.4 x 0.4	11.9	Sb		Mag 13.5 star on SE edge.
00 44 48.5	+05 08 05	CGCG 410-2	15.0	0.6 x 0.3	13.0		63	Mag 8.22v star ENE 4.′9.
01 15 53.5	+05 19 28	CGCG 411-14	14.1	0.8 x 0.3	12.4		12	Mag 13.6 star E 1.′2.
01 19 18.2	+04 34 38	CGCG 411-22	13.6	0.5 x 0.5	11.9			= **Mkn 567**; mag 11.7 star NE 0.′7.
00 34 10.8	−02 10 38	IC 29	15.1	0.7 x 0.7	14.1			Mag 10.74v star SW 6.′3; IC 30 N 5.′7.
00 34 14.8	−02 05 05	IC 30	15.9	0.7 x 0.4	14.3		21	IC 29 S 5.′7.
00 35 01.7	−02 08 32	IC 32	14.9	0.4 x 0.4	12.8			Pair with IC 33 E 0.′9.
00 35 05.1	−02 08 18	IC 33	14.5	0.7 x 0.5	13.3	S0		Pair with IC 32 W 0.′9; mag 9.51v star NE 8.′1.
00 39 21.4	+02 27 20	IC 40	14.4	1.1 x 0.5	13.6	S	13	Mag 10.36v star NW 9.′7.
00 43 56.3	+01 51 02	IC 49	13.7	1.5 x 1.3	14.3	SAB(s)c II-III	107	Stellar nucleus; star or bright knot E of center; mag 11.9 star SSE 7.′0.
01 09 22.2	−01 41 45	IC 81	13.8	1.1 x 0.7	13.4	Sbc	126	Mag 13.3 star SE edge.
01 10 30.0	+01 41 18	IC 83	14.6	0.6 x 0.3	12.5		105	Mag 9.80v star NNW 8.′5.
01 11 25.7	+01 38 18	IC 84	14.6	0.7 x 0.3	12.7		12	Mag 11.11v star WSW 7.′1.
01 14 15.9	+00 45 51	IC 87	14.4	0.6 x 0.5	13.0	SB?	6	Mag 10.37v star S 7.′0.
01 16 03.6	+04 17 38	IC 89	12.4	2.0 x 1.6	13.5	(R)SAB0°	54	Small, bright nucleus; mag 12.3 star W 4.′5.
01 22 53.9	−04 38 35	IC 100	13.7	1.0 x 0.7	13.2	SAB0$^-$:	99	Mag 8.30v star WNW 12.′5.
01 24 36.5	+02 02 36	IC 103	14.0	0.9 x 0.6	13.2	S0$^-$:	124	**IC 105** NE 3.′0; galaxy **LEDA 094065** SSW 3.′0.
01 25 13.1	+02 03 58	IC 109	13.5	0.9 x 0.7	12.9	SB0	99	Mag 11.4 star SE 8.′2.
01 26 50.3	−04 58 56	IC 116	13.7	0.9 x 0.5	12.7	Sc	129	Mag 12.6 star E 1.′5; mag 13.3 star N 1.′4.
01 27 55.0	−02 02 28	IC 119	13.7	1.2 x 0.5	13.0	(R′)SB(r)0/a:	77	Lies 1.′5 W of mag 7.04v star.
01 28 21.8	+02 30 47	IC 121	13.5	0.9 x 0.5	12.5	Sbc	108	Mag 9.87v star ESE 6.′5; note: IC 123 is 2.′6 SSE of this star.
01 28 51.5	+02 26 44	IC 123	14.3	0.6 x 0.4	12.6		36	Located 2.′6 SSE of mag 9.87v star.
01 29 47.9	−01 59 04	IC 126	14.1	0.9 x 0.9	13.7	S?		Stellar nucleus, faint envelope; mag 12.6 star S 1.′0.
01 33 02.1	−00 41 22	IC 138	13.9	1.1 x 0.7	13.4	SAB(s)c	30	Mag 10.62v star SE 8.′5; **MCG +0-5-5** and **MCG +0-5-6** NE 4.′3.
01 38 38.4	+00 44 30	IC 145	14.9	0.7 x 0.4	13.4	Sab	177	Mag 7.67v star SE 14.′1.
01 42 57.7	+04 12 00	IC 150	13.9	1.0 x 0.4	12.8	Sb III	143	Mag 11.2 star N 8.′1; **UGC 1204** NNE 5.′9.
01 49 08.6	−03 54 16	IC 164	13.1	1.4 x 1.2	13.6	E$^+$:	0	Mag 9.95v star S 3.′6.
00 40 37.7	−00 19 48	IC 1571	13.7	1.5 x 1.2	14.1	Sb II	42	Mag 9.76v star NE 5.′8.
00 43 33.7	−04 07 06	IC 1575	14.0	1.3 x 1.0	14.1		150	Mag 13.9 star on S edge; MCG −1-3-5 SE 9.′2.
00 53 27.0	+05 46 11	IC 1592	14.1	0.9 x 0.6	13.3	S?	165	Anonymous galaxy W 1.′4.
00 54 41.8	+05 46 27	IC 1598	13.9	1.1 x 0.5	13.1	Sa	2	Mag 7.94v star W 11.′1.
00 58 48.9	+00 35 12	IC 1607	13.4	0.9 x 0.9	13.0	S?		Mag 9.56v star ESE 5.′5.
01 04 49.0	+02 06 49	IC 1613	9.2	16.2 x 14.5	15.0	IB(s)m V-VI	50	Low surface brightness; mag 10.93 star superimposed 3.′0 SW of center; mag 10.68v star on W edge.
01 11 46.6	−00 39 48	IC 1639	13.5	0.9 x 0.8	13.0	Pec?	100	Mag 11.41v star NE 5.′9.
01 12 08.7	−00 24 39	IC 1643	14.2	0.8 x 0.8	13.6			Mag 8.39v star SW 4.′8.
01 21 21.2	+00 05 22	IC 1681	13.8	1.0 x 0.4	12.6	S?	99	Mag 10.75v star S 3.′8.
01 24 47.8	+01 36 26	IC 1694	14.0	0.7 x 0.5	12.7	S	3	Mag 11.8 star SW 3.′0; stellar **PGC 5156** W 6.′3.
01 24 52.5	−01 37 05	IC 1696	13.6	0.7 x 0.7	12.9	E$^+$		NGC 530 NW 3.′1.
01 25 02.9	+00 26 38	IC 1697	13.8	0.9 x 0.5	12.8	S?	110	Mag 10.8 star W 6.′6; **UGC 1001** E 9.′0.
01 26 44.8	−03 30 07	IC 1705	13.4	0.9 x 0.7	12.9	E	3	Mag 10.92v star SW 9.′8.
01 45 19.8	+04 37 04	IC 1726	13.8	0.4 x 0.3	11.3		24	Mag 14.6 star on E edge; mag 10.13v star NW 12.′0; note: **UGC 1222** is 1.′2 SW of this star.
00 32 21.7	−01 45 35	MCG +0-2-65	15.0	0.6 x 0.4	13.3	Sb	84	Mag 7.07v star W 10.′5.
00 32 42.2	−01 49 04	MCG +0-2-68	15.1	0.7 x 0.4	13.6		12	
00 33 14.9	+00 12 01	MCG +0-2-69	15.6	0.7 x 0.4	14.1	Spiral	42	
00 34 42.6	+02 25 28	MCG +0-2-78	15.0	0.7 x 0.5	13.7		24	
00 36 25.0	+02 25 29	MCG +0-2-90	15.2	1.0 x 0.4	14.0	Sbc	15	Mag 14.4 star on NW edge.
00 37 35.8	+00 16 49	MCG +0-2-94	14.4	0.7 x 0.5	13.1	S?	111	= **Mkn 955**.
00 39 20.0	−02 08 16	MCG +0-2-108	14.8	0.9 x 0.6	13.9	Sbc	66	
00 40 00.7	−01 43 05	MCG +0-2-117	14.9	0.6 x 0.6	13.6			
00 40 35.6	+01 40 18	MCG +0-2-120	14.6	0.9 x 0.6	14.0	Compact	54	
00 41 46.1	+01 10 35	MCG +0-2-126	14.4	0.9 x 0.7	13.7	S	18	
00 42 39.4	+02 44 12	MCG +0-2-133	14.9	1.4 x 0.3	13.8	Sb	15	Mag 12.1 star SE 3.′0.
00 42 39.4	+00 16 36	MCG +0-2-134	14.9	0.8 x 0.5	13.7		96	
00 51 59.8	−00 29 12	MCG +0-3-18	14.7	0.9 x 0.4	13.4		170	Mag 8.72v star W 5.′4.
00 52 52.7	+01 12 48	MCG +0-3-21	15.0	1.0 x 0.5	14.1		114	Mag 13 star W end.
00 53 28.8	+02 46 24	MCG +0-3-22	14.7	0.7 x 0.4	13.2	Sbc	84	
00 55 13.7	−01 04 35	MCG +0-3-26	15.8	0.9 x 0.2	13.8	Sbc	9	**UGC 568** NNW 2.′1.
00 55 18.9	−01 16 39	MCG +0-3-28	17.1	0.6 x 0.6	15.8	I		
00 56 22.9	−01 12 36	MCG +0-3-33	14.3	0.6 x 0.3	12.3	Sa	120	**PGC 3350** N 2.′0; UGC 579 SSW 3.′2.
00 56 38.9	−01 17 45	MCG +0-3-36	15.2	0.6 x 0.2	12.8	Sb	108	Mag 11.26v star SW 1.′0.
00 58 20.6	−01 56 15	MCG +0-3-44	15.0	0.8 x 0.5	13.8	Sb	90	
01 00 34.0	−01 41 06	MCG +0-3-51	14.9	0.5 x 0.4	13.0	SAB(s)c	79	Located 2.′2 SW of mag 6.91v star.
01 02 48.3	−01 28 28	MCG +0-3-59	15.2	0.8 x 0.2	13.1	Sb	90	
01 07 09.3	−00 37 16	MCG +0-3-74	14.3	1.1 x 1.0	14.2			
01 08 43.1	−02 08 29	MCG +0-4-10	14.6	0.8 x 0.6	13.7	Sbc	15	Mag 11.6 star SE 2.′4.
01 09 01.5	+01 22 38	MCG +0-4-11	14.9	0.8 x 0.2	12.8		75	Mag 10.77v star N 0.′9.
01 10 01.5	+00 24 40	MCG +0-4-19	14.5	0.7 x 0.5	13.3		18	Mag 12.6 star superimposed NE edge.
01 11 48.9	+02 02 29	MCG +0-4-32	14.9	0.8 x 0.7	14.1	SBbc	165	
01 13 27.4	+00 09 05	MCG +0-4-42	14.9	0.7 x 0.6	13.8	S	144	

(Continued from previous page)

GALAXIES

RA h m s	Dec ° ′ ″	Name	Mag (V)	Dim ′ Maj x min	SB	Type Class	PA	Notes
01 22 41.1	+03 36 03	MCG +0-4-102	14.2	0.8 x 0.7	13.4		15	
01 22 46.5	+01 43 47	MCG +0-4-104	15.1	0.7 x 0.3	13.3	Sd	64	Located 2′.4 E of mag 6.20v star.
01 23 26.5	+01 42 20	MCG +0-4-111	14.4	0.9 x 0.4	13.2		174	Mag 11.8 star WNW 1′.2; mag 10.05v star N 2′.7.
01 23 21.2	−01 58 36	MCG +0-4-112	14.0	1.1 x 0.4	12.9	SB?	162	
01 24 07.5	−01 51 34	MCG +0-4-114	14.3	1.0 x 0.3	12.9	S0	60	
01 24 19.7	−01 44 50	MCG +0-4-116	14.1	0.8 x 0.3	12.4	S?	156	Mag 10.12v star N 1′.0.
01 24 58.9	+01 33 21	MCG +0-4-123	15.2	0.5 x 0.3	12.9	Sab	177	Located between mag 13.6 and 12.2 stars; mag 7.39v star SE 7′.8.
01 24 55.2	+00 12 39	MCG +0-4-126	15.6	0.7 x 0.4	14.3	E?	3	Mag 8.86v star NNE 4′.1.
01 25 39.8	+01 10 38	MCG +0-4-132	15.0	0.4 x 0.3	12.5	Sb	30	
01 25 51.6	−01 19 08	MCG +0-4-140	13.2	0.5 x 0.4	11.3	S?	111	Pair of stars on W edge, mags 13.6 and 14.1.
01 26 24.1	−00 56 17	MCG +0-4-148	14.5	0.3 x 0.3	11.7	S0		**Mkn 570** ESE 6′.8.
01 26 55.8	−00 00 53	MCG +0-4-149	14.4	0.4 x 0.4	12.3			Mag 9.19v star S edge.
01 27 57.0	+02 08 30	MCG +0-4-155	14.5	0.7 x 0.5	13.2		165	Mag 10.77v star W 1′.7.
01 31 12.3	−01 29 20	MCG +0-4-167	14.2	0.9 x 0.5	13.2	SB0:	90	
01 31 48.2	+00 39 43	MCG +0-5-2	15.2	0.6 x 0.5	13.7	Sc	138	Mag 11.52v star S 1′.2.
01 35 29.0	+03 29 46	MCG +0-5-12	15.2	0.8 x 0.5	14.1		117	Mag 9.68v star N 3′.2.
01 35 36.7	+00 00 55	MCG +0-5-13	14.9	0.5 x 0.3	12.7	Irr		Mag 8.03v star W 8′.7.
01 39 15.6	−01 39 59	MCG +0-5-22	15.8	0.5 x 0.4	13.9	Sd	159	Mag 10.83v star E 0′.8.
01 39 27.4	+00 02 17	MCG +0-5-23	15.4	0.5 x 0.4	13.5		30	
01 39 31.9	−00 01 50	MCG +0-5-24	15.3	0.4 x 0.4	13.1			
01 40 14.6	−00 30 55	MCG +0-5-26	15.2	0.6 x 0.3	13.2	Sb	141	
01 51 13.5	−01 24 36	MCG +0-5-39	15.2	0.4 x 0.4	13.1	Spiral		Compact; mag 14.0 star N 0′.6.
01 51 27.2	−02 15 32	MCG +0-5-40	14.5	0.6 x 0.6	13.5	E/S0		
00 32 58.8	+05 17 07	MCG +1-2-23	15.0	0.6 x 0.5	13.6	Sb	12	MCG +1-2-24 S 1′.5.
00 33 00.9	+05 15 45	MCG +1-2-24	15.1	0.7 x 0.5	13.8	Sb	39	MCG +1-2-23 N 1′.1
00 37 55.7	+04 54 40	MCG +1-2-38	14.5	1.1 x 0.2	12.7	S?	174	Mag 12.8 star S 2′.8.
01 13 27.1	+04 14 20	MCG +1-4-3	15.5	0.5 x 0.5	13.8			Faint star on E edge.
01 15 03.7	+05 22 38	MCG +1-4-7	14.1	1.1 x 0.4	13.1	S?	156	Mag 10.7 star E 6′.5.
01 15 53.5	+05 04 54	MCG +1-4-10	15.7	0.7 x 0.3	13.8	Sbc	120	Mag 14.1 star NW 0′.5; mag 10.51v star NE 4′.6.
01 16 13.0	+05 09 58	MCG +1-4-13	15.9	0.4 x 0.4	13.7			Located 3′.8 E of NGC 455.
01 21 05.2	+05 45 41	MCG +1-4-27	15.6	0.9 x 0.3	14.0	Sb	108	
01 32 18.1	+03 51 48	MCG +1-5-1	14.7	0.7 x 0.7	13.8	SBc		Faint, elongated anonymous galaxy NNW 1′.4; mag 11.9 star S 2′.5.
01 32 49.5	+05 44 22	MCG +1-5-4	15.1	0.7 x 0.3	13.3	Sc	60	Mag 10.28v star SE 2′.9.
01 37 07.0	+05 41 34	MCG +1-5-8	15.0	0.7 x 0.4	13.5	Sc	36	Mag 8.18v star WSW 3′.9.
01 41 41.0	+05 51 10	MCG +1-5-20	15.1	0.9 x 0.2	13.1	Sb	96	Mag 12.1 star WNW 1′.1.
01 44 41.9	+04 53 42	MCG +1-5-30	14.5	0.9 x 0.3	12.9	S?	15	Faint star or bright knot S end; mag 13.3 star SSW 0′.9.
01 47 58.8	+04 59 16	MCG +1-5-33	15.2	0.5 x 0.4	13.3	Sc	6	Mag 9.89v star ENW 4′.1.
00 34 01.2	−03 10 00	MCG −1-2-31	13.5	1.4 x 0.5	13.0	S0°:	159	Mag 9.95v star SW 2′.6.
00 35 36.2	−04 44 00	MCG −1-2-38	14.3	1.4 x 0.6	14.0	Sbc pec? III	85	
00 36 09.2	−03 31 18	MCG −1-2-40	14.0	0.9 x 0.7	13.5	E	18	
00 40 52.6	−04 27 10	MCG −1-2-44	14.7	1.0 x 0.7	14.1	Sc		Mag 5.90v star NNW 6′.7.
00 41 16.9	−05 17 06	MCG −1-2-45	14.4	1.1 x 0.8	14.1	Sbc	162	
00 42 47.5	−03 49 46	MCG −1-2-46	14.4	0.9 x 0.6	13.6	S0	141	Mag 7.24v star ESE 3′.9.
00 43 54.3	−04 14 43	MCG −1-3-5	15.0	1.7 x 0.6	14.9	(R′)Sc: II-III	81	IC 1575 NW 9′.2.
00 45 19.5	−03 41 38	MCG −1-3-8	14.7	0.6 x 0.6	13.5	Sb		Mag 9.13v star E 6′.0.
00 45 51.7	−03 50 51	MCG −1-3-9	15.1	0.7 x 0.5	13.9	Sb	153	Mag 10.03v star WSW 10′.3.
00 47 07.1	−02 43 55	MCG −1-3-12	14.4	1.0 x 0.6	13.7	Sc	129	
00 47 30.8	−05 04 47	MCG −1-3-14	14.8	1.0 x 0.3	13.3	Sb	120	Mag 9.83v star W 4′.7.
00 50 03.5	−03 33 32	MCG −1-3-16	14.3	0.8 x 0.7	13.6	SAB(rs)bc III	42	
00 50 27.6	−05 51 33	MCG −1-3-18	13.1	0.8 x 0.8	12.5	(R′)SA(r)0⁻?		
00 51 29.9	−03 08 01	MCG −1-3-23	14.7	1.2 x 0.4	13.8	SB?	3	Mag 7.25v star on SE edge.
00 52 17.2	−03 58 03	MCG −1-3-27	14.5	1.4 x 0.4	13.7	Sd? III-IV	24	
00 53 23.2	−03 27 33	MCG −1-3-28	14.7	0.8 x 0.6	13.7	SBc	90	
00 54 32.1	−04 36 10	MCG −1-3-31	14.6	1.5 x 0.4	13.9	Scd: sp III	168	Mag 6.54v star WNW 13′.5.
00 55 01.3	−03 04 51	MCG −1-3-32	14.2	0.5 x 0.5	12.8	E		
00 55 28.5	−03 26 31	MCG −1-3-35	14.3	0.4 x 0.4	12.2			Compact.
00 57 35.1	−05 00 08	MCG −1-3-41	13.9	1.0 x 0.6	13.2	SB(r)c pec II-III	38	MCG −1-3-42 NNE 3′.7.
00 57 42.5	−04 56 56	MCG −1-3-42	14.3	0.9 x 0.7	13.6	S0/a	114	= **Mkn 966**.
00 57 55.3	−02 25 44	MCG −1-3-44	14.7	0.8 x 0.3	13.0	Sb	18	
00 57 58.2	−03 42 25	MCG −1-3-46	14.6	0.9 x 0.4	13.3	Irr	141	
00 59 23.6	−04 48 16	MCG −1-3-51	13.7	2.1 x 0.6	13.8	Sa pec sp	3	MCG −1-3-52 on NE edge.
00 59 26.2	−04 47 40	MCG −1-3-52	14.1	0.6 x 0.6	12.8	S0		Located on NE edge of MCG −1-3-51.
01 00 03.0	−04 56 14	MCG −1-3-54	14.4	0.6 x 0.5	13.0	Sb	33	Small galaxy or bright knot SE edge; mag 12.3 star SSE 2′.5.
01 01 06.2	−04 36 22	MCG −1-3-61	14.9	0.6 x 0.3	12.9		69	
01 01 13.8	−04 26 18	MCG −1-3-62	14.7	0.8 x 0.5	13.6		99	Star on N edge.
01 01 35.6	−05 58 33	MCG −1-3-66	15.2	0.9 x 0.5	14.1	Sc	21	Mag 8.94v star E 1′.6.
01 02 22.9	−04 30 31	MCG −1-3-72	13.8	0.9 x 0.6	13.0	Im pec V-VI	81	Mottled and disrupted appearance, several bright knots.
01 02 52.3	−04 45 44	MCG −1-3-76	13.7	0.7 x 0.4	12.4	E/S0	90	
01 03 10.7	−03 36 37	MCG −1-3-79	13.7	0.7 x 0.7	13.0	S0?		= **Mkn 970**.
01 03 26.6	−04 52 47	MCG −1-3-82	14.5	0.6 x 0.4	12.8		51	
01 04 36.9	−02 32 38	MCG −1-3-84	14.2	1.1 x 0.8	13.9	Sbc	120	Star NW end.
01 05 06.2	−05 19 54	MCG −1-3-87	14.3	1.0 x 0.7	13.8	SBc	174	
01 05 18.9	−05 25 23	MCG −1-3-89	14.5	1.1 x 0.8	14.2	Scd	108	
01 07 15.1	−03 08 17	MCG −1-3-92	14.3	0.7 x 0.7	13.4	Sbc		
01 07 20.0	−03 09 29	MCG −1-3-93	15.5	0.4 x 0.4	13.4			
01 08 27.1	−05 18 25	MCG −1-4-2	14.5	1.1 x 0.3	13.1	S?	123	
01 09 00.0	−05 31 06	MCG −1-4-3	15.7	1.0 x 0.4	14.5	IB(s)m: V	18	Low uniform surface brightness.
01 10 37.7	−03 13 28	MCG −1-4-6	14.2	1.3 x 0.4	13.4	Sc	15	
01 10 53.8	−05 49 17	MCG −1-4-8	14.0	2.4 x 1.0	14.8	SA(s)c: II-III	162	Mag 11.8 star SSE 2′.5.
01 11 38.5	−03 20 43	MCG −1-4-11	14.1	0.7 x 0.4	12.6	S0/a	168	Mag 9.64v star SSE 5′.5.
01 12 32.0	−04 08 46	MCG −1-4-14	13.7	1.0 x 0.9	13.4	SB0⁻?	171	
01 12 35.6	−04 08 26	MCG −1-4-15	14.0	1.0 x 0.3	12.6	Sa	90	
01 18 08.3	−02 34 05	MCG −1-4-24	14.2	0.8 x 0.5	13.1	S0/a	21	
01 21 45.5	−03 32 24	MCG −1-4-28	14.6	0.7 x 0.4	13.2	E?	9	Mag 8.36v star N 6′.2.
01 24 15.1	−05 58 49	MCG −1-4-35	15.6	1.1 x 0.2	13.8	Sab	147	Mag 12.1 star SE 1′.7; mag 6.79v star E 11′.5.
01 24 16.8	−05 54 51	MCG −1-4-36	15.3	0.9 x 0.3	13.7	Scd	69	Mag 10.59v star NE 1′.9; mag 6.79v star E 11′.2.
01 24 29.5	−04 33 08	MCG −1-4-37	14.9	1.2 x 0.6	14.4	SB(s)cd: II-III	129	Stellar nucleus in faint envelope.
01 25 04.2	−04 39 02	MCG −1-4-38	14.6	0.7 x 0.5	13.3	Sb	27	MCG −1-4-40 S 3′.6.
01 25 00.8	−04 38 18	MCG −1-4-39	14.6	0.5 x 0.5	12.9	S		

RA h m s	Dec ° ′ ″	Name	Mag (V)	Dim ′ Maj x min	SB	Type Class	PA	Notes
01 25 03.9	−04 42 30	MCG −1-4-40	13.3	1.4 x 1.1	13.8	E/S0	123	MCG −1-4-38 and MCG −1-4-39 N 3′.6.
01 25 57.8	−03 58 57	MCG −1-4-42	14.4	1.9 x 1.1	15.0	SB(rs)dm IV-V	81	Mag 6.49v star N 3′.6.
01 27 08.3	−03 13 15	MCG −1-4-50	14.7	0.9 x 0.6	13.8		135	
01 27 33.4	−04 41 00	MCG −1-4-52	14.1	0.7 x 0.4	12.8	E?	18	= **Hickson 12A**. **Hickson 12B** N 1′.9; **Hickson 12C** and **Hickson 12D** NNE 1′.4; **Hickson 12E** W 1′.3.
01 28 25.5	−02 35 21	MCG −1-4-54	14.4	1.5 x 0.2	13.0	Scd? sp	60	Mag 10.90v star E 9′.1.
01 29 46.9	−02 42 15	MCG −1-4-56	13.8	0.6 x 0.6	12.6	SA(s)dm III-IV		Stellar nucleus, faint envelope.
01 29 58.5	−04 42 19	MCG −1-4-58	13.8	0.8 x 0.4	12.5	Sbc	117	
01 30 41.1	−03 56 44	MCG −1-4-59	14.5	1.1 x 0.3	13.2	Sc		
01 33 00.5	−05 00 10	MCG −1-5-6	14.3	1.0 x 0.5	13.3	Sc	150	
01 37 38.5	−05 03 45	MCG −1-5-10	14.2	0.8 x 0.6	13.5	E	30	
01 38 02.5	−05 43 51	MCG −1-5-11	14.6	0.6 x 0.4	12.9	Sb	105	Mag 7.34v star E 10′.5; mag 9.79v star N 4′.0.
01 40 27.9	−05 31 13	MCG −1-5-14	14.3	1.6 x 0.4	13.7	SAB(s)d: III	171	Mag 13.3 star S 1′.5; mag 8.49v star NNE 10′.4.
01 41 06.7	−05 34 09	MCG −1-5-17	13.4	2.1 x 0.8	13.8	SAB(s)m: III-IV	72	Mag 8.49v star N 10′.0.
01 41 44.9	−04 12 01	MCG −1-5-19	14.5	0.9 x 0.7	13.8	Sc	111	
01 43 06.7	−04 02 51	MCG −1-5-23	14.3	0.5 x 0.5	12.8	E/S0		Almost stellar; faint star on W edge of nucleus.
01 43 06.4	−04 42 13	MCG −1-5-24	14.6	0.9 x 0.5	13.6	SBc	132	Mag 6.19v star E 12′.6.
01 43 42.9	−04 00 05	MCG −1-5-25	13.6	1.6 x 0.8	13.7	SA(s)0°?	162	
01 44 41.6	−03 06 41	MCG −1-5-27	14.9	0.9 x 0.4	13.6	SBb	153	Mag 7.40v star E 10′.6.
01 44 45.0	−04 08 06	MCG −1-5-29	13.5	1.0 x 0.8	13.1	SAB(r)0°:	90	
01 45 25.6	−03 49 44	MCG −1-5-31	13.1	1.4 x 0.9	13.2	SB(rs)bc pec:	168	
01 49 15.5	−03 41 02	MCG −1-5-36	14.4	1.1 x 1.1	14.5	SA(rs)cd: II-III		Star on SE edge.
01 50 41.8	−03 56 17	MCG −1-5-40	13.9	1.2 x 0.6	13.4	SB(r)0$^+$ pec?	45	NGC 702 SE 11′.7.
01 51 11.3	−03 29 57	MCG −1-5-42	14.1	1.3 x 0.8	14.0	SB(s)dm IV-V	24	Several bright knots S end.
01 51 50.3	−05 29 48	MCG −1-5-45	14.7	1.6 x 0.5	14.3	SB(s)dm III	66	**MCG −1-5-46** SE edge.
00 35 33.9	−02 50 56	NGC 161	13.4	1.3 x 0.8	13.3	S0°:	27	**MCG −1-2-37** S 1′.6.
00 36 32.9	+02 44 58	NGC 164	15.0	0.4 x 0.4	12.9			Almost stellar; mag 9.78v star W 6′.9.
00 36 46.0	+01 53 08	NGC 170	14.4	0.6 x 0.3	12.4	S0$^-$?	79	Located 1′.9 N of mag 9.28v star.
00 37 12.4	+01 56 30	NGC 173	13.0	3.2 x 2.6	15.2	SA(rs)c II	90	
00 38 12.3	+02 43 39	NGC 182	12.4	2.0 x 1.5	13.5	(R′)SAB(rs)a pec:	75	Very small, faint anonymous galaxies 1′.5 SW and 1′.9 SE.
00 38 25.4	+03 09 56	NGC 186	13.4	1.4 x 0.8	13.4	SB0$^+$ pec:	23	Pair of very faint, very small anonymous galaxies SW 2′.7.
00 39 13.3	+00 51 53	NGC 192	12.6	1.9 x 0.9	13.1	(R′)SB(r)a:	167	Mag 8.78v star NW 8′.0; NGC 197 and NGC 196 N 2′.6.
00 39 18.6	+03 19 51	NGC 193	12.2	1.4 x 1.2	12.6	SAB(s)0$^-$:	55	Mag 13.1 star on SW edge; mag 9.87v star ESE 2′.5.
00 39 18.5	+03 02 17	NGC 194	12.2	1.5 x 1.4	13.0	E	30	Mag 7.35v star N 5′.9.
00 39 17.9	+00 54 42	NGC 196	12.9	1.1 x 0.8	12.6	SB0 pec:	3	NGC 197 S 1′.2.
00 39 18.9	+00 53 29	NGC 197	14.1	0.7 x 0.7	13.3	SB0° pec		Almost stellar nucleus.
00 39 23.0	+02 47 51	NGC 198	13.2	1.2 x 1.2	13.4	SA(r)c II		Mag 9.92v star S 5′.3.
00 39 33.2	+03 08 16	NGC 199	13.6	1.2 x 0.7	13.3	SA0° pec:	160	Mag 7.35v star W 5′.1.
00 39 34.7	+02 53 15	NGC 200	12.6	1.9 x 1.0	13.2	SB(s)bc I	161	Elongated anonymous galaxy SW 2′.7.
00 39 34.7	+00 51 36	NGC 201	12.9	1.8 x 1.4	13.8	SAB(r)c II	155	Stellar nucleus, multi-branching arms.
00 39 40.1	+03 32 05	NGC 202	14.3	1.0 x 0.4	13.2	SB(s)0$^+$?	153	NGC 203 S 5′.5; mag 11.5 star SW 3′.2; mag 7.63v star N 7′.4.
00 39 39.5	+03 26 31	NGC 203	14.0	0.9 x 0.3	12.4	S0° pec:	85	Mag 11.5 star NNW 4′.2; NGC 202 N 5′.5.
00 39 44.3	+03 17 56	NGC 204	12.9	1.2 x 1.1	13.1	SA(s)0° pec:	30	Mag 9.87v star W 4′.2.
00 40 17.7	+02 45 21	NGC 208	14.3	0.7 x 0.7	13.3	SA(r)a?		**MCG +0-2-122** NE 9′.9.
00 42 11.4	+00 54 14	NGC 219	14.3	0.5 x 0.5	12.8	cE?		Almost stellar; mag 7.78v star N 7′.5; NGC 223 S 3′.7.
00 42 15.9	+00 50 43	NGC 223	13.2	1.3 x 0.9	13.2	(R′)SB(r)0/a:	62	NGC 219 N 3′.7.
00 42 36.8	−01 31 44	NGC 227	12.1	1.6 x 1.3	12.8	S0$^-$ pec:	155	
00 43 27.6	+02 57 30	NGC 236	13.5	1.1 x 1.0	13.5	SAB(s)c II	160	
00 43 28.1	−00 07 33	NGC 237	13.0	1.6 x 0.9	13.2	SAB(rs)cd I-II	175	
00 44 37.4	−03 45 35	NGC 239	13.3	1.0 x 0.5	12.4	(R′)Sab	28	Mag 12.3 star E 2′.5.
00 46 05.7	−01 43 24	NGC 245	12.2	1.4 x 1.2	12.6	SA(rs)b pec? II-III	145	
00 48 03.5	−02 46 35	NGC 259	12.8	2.8 x 0.6	13.2	Sbc: sp II	140	
00 50 09.6	−05 11 38	NGC 268	13.1	1.5 x 0.9	13.2	SB(s)bc: I-II	75	Bright knots or stars on E and W edges; mag 8.79v star NE 5′.3.
00 50 41.9	−01 54 38	NGC 271	12.0	2.1 x 1.7	13.2	(R′)SB(rs)ab	130	Mag 8.73v star just E of SE end of galaxy.
00 52 09.0	−02 13 07	NGC 279	12.7	1.6 x 1.2	13.2	(R′)SAB(r)0$^+$ pec:	158	Mag 9.36v star N 7′.2.
00 56 32.8	−01 46 21	NGC 307	12.8	1.6 x 0.7	12.8	S0°:	85	
00 57 39.3	−05 05 11	NGC 321	14.9	0.4 x 0.4	12.8	SB(rs)cd pec II-III		
00 57 47.9	−05 06 44	NGC 325	14.1	1.5 x 0.2	12.6	Sb	91	
00 57 55.3	−05 07 51	NGC 327	13.4	1.6 x 0.7	13.4	SB(s)bc: II	175	
00 58 01.8	−05 04 12	NGC 329	13.3	1.6 x 0.6	13.1	Sb: sp	20	
00 57 04.9	−02 46 14	NGC 331	14.7	0.7 x 0.7	13.7	SBc		Mag 6.98v star NW 5′.2.
01 01 57.6	−01 56 11	NGC 351	13.2	1.4 x 0.8	13.1	(R′)SB(r)0/a:	142	NGC 353 E 6′.9.
01 02 09.0	−04 14 48	NGC 352	12.6	2.4 x 0.9	13.3	(R′)SB(rs)b?	165	
01 02 24.5	−01 57 32	NGC 353	13.7	1.3 x 0.4	12.8	SBa pec:	23	Mag 9.53v star S 6′.0; NGC 351 W 6′.9.
01 04 16.8	−00 45 53	NGC 359	13.3	1.5 x 1.1	13.7	S0$^-$:	135	
01 04 40.7	−00 48 10	NGC 364	13.1	1.4 x 1.3	13.6	(R)SB(s)0°:	30	
01 07 22.6	+00 55 31	NGC 391	13.4	0.9 x 0.7	12.7	(R′)SA0$^-$:	54	Mag 10.71v star 1′.7 NNW.
01 08 08.4	+04 31 51	NGC 396	15.7	0.3 x 0.2	12.5		70	
01 12 31.3	−02 47 36	NGC 413	14.1	1.1 x 0.7	13.7	SB(r)c: I-II	151	Very weak stellar nucleus, uniform envelope.
01 12 48.6	−00 17 24	NGC 426	12.9	1.4 x 1.0	13.2	E$^+$	140	Mag 11.33v star NW 5′.5.
01 12 55.5	+00 58 49	NGC 428	11.5	4.1 x 3.1	14.1	SAB(s)m III-IV	120	Very faint extensions NW of nucleus.
01 12 57.5	−00 20 45	NGC 429	13.4	1.4 x 0.3	12.3	S0°: sp	19	Mag 15.5 star on N edge; mag 11.33v star NW 9′.3.
01 13 00.1	−00 15 08	NGC 430	12.5	1.3 x 1.1	12.9	E:	155	Mag 14.2 star on S edge; mag 11.33v star W 7′.5.
01 13 59.8	+02 04 14	NGC 435	14.2	1.1 x 0.4	13.2	SAB(s)d: III	17	
01 14 22.4	+05 55 37	NGC 437	12.8	1.3 x 1.0	12.9	S0/a	130	
01 14 38.6	−01 01 16	NGC 442	13.4	0.9 x 0.6	12.7	S0/a: sp	157	Lies in the glare of mag 5.70v star 38 Ceti 3′.8 NE.
01 14 52.6	+01 54 59	NGC 445	14.2	0.5 x 0.4	12.3	S0 pec:	153	Mag 12.4 star W 0′.8; UGC 791 SW 6′.4.
01 14 48.1	+04 11 22	NGC 446	14.5	1.0 x 0.2	12.6	S	6	
01 15 16.5	−01 37 34	NGC 448	12.1	1.6 x 0.8	12.3	S0$^-$ sp	116	
01 15 30.6	−00 51 38	NGC 450	11.8	3.1 x 2.3	13.8	SAB(s)cd: III-IV	72	**UGC 807** on NE edge; several stars or bright knots E of core; mag 5.70v star SW 12′.3.
01 15 57.7	+05 10 44	NGC 455	12.7	1.9 x 1.2	13.4	S?	153	MCG +1-4-13 E 3′.8; mag 10.51v star SSE 2′.3.
01 18 11.0	+04 13 33	NGC 462		0.4 x 0.4				Stellar.
01 19 10.1	+03 18 02	NGC 467	11.9	1.7 x 1.7	12.9	SA(s)0° pec?		Mag 7.46v star E 3′.6; **MCG +0-4-83** ESE 7′.4; **MCG +0-4-82** and mag 11.9 star SE 6′.8.
01 19 44.9	+03 24 32	NGC 470	11.8	2.8 x 1.7	13.3	SA(rs)b II-III	155	Located 5′.5 W of core of NGC 474.
01 20 06.7	+03 24 56	NGC 474	11.5	7.1 x 6.3	15.5	SA(s)0°	75	Very faint arms extend out well beyond the bright nucleus and central core to fill out the published dimensions.
01 21 15.8	+03 51 44	NGC 479	13.9	1.1 x 0.9	13.7	SB(rs)bc: III	156	
01 22 10.6	+05 24 41	NGC 486	14.4	0.4 x 0.4	12.3	Sb		NGC 492 NE 1′.0.
01 21 46.8	+05 15 13	NGC 488	10.3	5.2 x 3.9	13.4	SA(r)b I	15	Mag 10.26v star on SSW edge; mag 11.5 star superimposed SE of core.

(Continued from previous page)

GALAXIES

RA h m s	Dec ° ′ ″	Name	Mag (V)	Dim ′ Maj x min	SB	Type Class	PA	Notes
01 22 02.9	+05 21 59	NGC 490	14.4	0.7 x 0.6	13.3	S?	9	Faint, stellar anonymous galaxy NW 8′.0.
01 22 13.6	+05 25 01	NGC 492	16.2	0.7 x 0.6	15.1	SBbc	120	Very small, faint anonymous galaxy NW 4′.9; NGC 486 SW 1′.0.
01 22 09.0	+00 56 43	NGC 493	12.5	3.4 x 1.0	13.6	SAB(s)cd: sp III:	58	
01 22 24.0	−00 52 28	NGC 497	13.0	2.1 x 0.9	13.6	SB(rs)bc: I-II	132	
01 22 39.4	+05 23 12	NGC 500	14.2	0.8 x 0.6	13.5	E/S0	102	Mag 10.9 star N 1′.4.
01 24 28.6	−01 38 29	NGC 519	14.3	0.5 x 0.3	12.2	E$^+$?	140	Centered between a pair of mag 12 stars, 2′.3 E and 2′.3 W.
01 24 34.6	+03 47 38	NGC 520	11.4	3.4 x 1.7	13.2	Pec	130	A broad band of faint material extends S 1′.5 beyond the published width of the galaxy, then curves E.
01 24 34.0	+01 43 48	NGC 521	11.7	2.7 x 2.4	13.6	SB(r)bc I	38	Bright core, uniform envelope; stellar anonymous galaxy WNW 3′.0 from center.
01 24 41.8	−01 35 16	NGC 530	13.0	1.5 x 0.4	12.3	SB0$^+$ sp	134	Star on SE end; faint anonymous galaxy 1′.5 NE.
01 25 31.4	+01 45 33	NGC 533	11.4	3.8 x 2.3	13.7	E3:	50	A number of very small, faint anonymous galaxies E and W up to 4′.0.
01 25 31.2	−01 24 29	NGC 535	13.8	1.0 x 0.3	12.4	S0$^+$ sp	58	Faint, stellar anonymous galaxy NW 2′.4.
01 25 26.1	−01 33 02	NGC 538	13.7	1.3 x 0.6	13.3	SB(s)ab:	40	Star on N edge.
01 25 44.1	−01 22 47	NGC 541	12.1	1.8 x 1.7	13.2	S0$^-$:	54	**Minkowski's Object**. **MCG +0-4-139** NNE 2′.2.
01 25 50.2	−01 17 37	NGC 543	13.1	0.6 x 0.3	11.1	S0$^-$: sp	90	Faint, small anonymous galaxies 2′.1 E and 1′.6 NNE.
01 25 59.2	−01 20 23	NGC 545	12.2	2.4 x 1.6	13.5	SA0$^-$	55	Round, faint, almost stellar anonymous galaxy 2′.4 NE.
01 26 00.9	−01 20 45	NGC 547	12.2	1.3 x 1.3	12.7	E1		Pair with NGC 545 NW.
01 26 02.5	−01 13 33	NGC 548	13.7	0.8 x 0.5	12.6	E$^+$:	135	Mag 12.1 star NW 4′.5.
01 26 42.8	+02 01 17	NGC 550	12.7	1.5 x 0.6	12.4	SB(s)a?	120	Mag 8.96v star NE 8′.9.
01 26 25.0	−01 38 18	NGC 557	13.5	1.8 x 0.6	13.4	SB(rs)0$^+$ pec:	123	Mag 9.00v star SE 4′.8.
01 27 16.2	−01 58 17	NGC 558	14.3	0.5 x 0.2	11.7	E$^+$? sp	110	Mag 12.3 star W 1′.3.
01 27 25.5	−01 54 46	NGC 560	13.0	1.9 x 0.4	12.5	S0° sp	178	Mag 10.49v star N 3′.2.
01 27 48.2	−01 52 45	NGC 564	12.5	1.4 x 1.2	13.1	E	145	Compact core in faint envelope; mag 7.04v star and IC 119 S 9′.8.
01 28 10.4	−01 18 22	NGC 565	13.5	1.3 x 0.4	12.7	Sa:	36	
01 28 58.6	−00 56 56	NGC 570	12.8	1.5 x 1.3	13.3	(R′)SB(rs)a:	175	Short E-W bar; mag 8.91v star SW 7′.9.
01 30 40.5	−01 59 38	NGC 577	12.9	1.8 x 1.4	13.7	(R′)SB(r)a pec	140	Mag 10.48v star E 5′.0.
01 31 41.8	−00 55 59	NGC 585	13.1	2.1 x 0.5	13.0	Sa: sp	86	
01 36 00.3	+00 39 47	NGC 622	12.9	1.6 x 1.0	13.3	SB(rs)b I-II	45	
01 36 47.2	+05 50 05	NGC 631	13.3	1.0 x 0.8	13.1	E	123	Compact nucleus, uniform envelope.
01 37 17.6	+05 52 39	NGC 632	12.4	1.0 x 0.8	12.0	S0?	170	Superimposed star or bright knot on NE edge of core.
01 40 09.0	+05 43 29	NGC 645	12.6	2.6 x 1.2	13.7	SBb:	125	Lies 3′.2 SE of mag 9.40v star.
01 43 45.9	+04 13 22	NGC 664	12.8	1.1 x 0.9	12.6	Sb: II	65	Mag 10.93v star NNW 8′.7; **UGC 1204** NW 9′.2.
01 48 57.4	+05 54 28	NGC 676	11.9	4.0 x 1.0	13.3	S0/a: sp	172	
01 51 19.2	−04 03 16	NGC 702	13.1	1.2 x 0.8	12.9	SB(s)bc pec	140	Bright knot or faint superimposed star N edge; MCG −1-5-40 NW 11′.7.
00 33 22.0	−01 07 20	UGC 328	15.4	1.4 x 0.8	15.4	SB(rs)dm IV-V	144	Close pair of stars, mags 12.9 and 13.9, SW 1′.7.
00 33 36.2	+02 40 53	UGC 329	14.2	1.7 x 0.8	14.4	SAB(s)cd III	170	
00 34 33.0	+03 36 08	UGC 342	14.5	1.1 x 0.9	14.3	SABc	57	Mag 12.3 star WSW 1′.4.
00 35 27.2	+02 56 02	UGC 348	14.6	1.0 x 0.7	14.0	Sdm:	33	Mag 12.7 star E 2′.1.
00 36 13.0	+01 42 43	UGC 358	13.9	0.9 x 0.9	13.5	S?		
00 37 58.0	+05 08 50	UGC 379	15.3	1.5 x 0.2	13.8	Scd:	120	
00 39 18.6	+03 57 10	UGC 402	13.4	1.1 x 0.7	13.0	S0	26	**UGC 416** ESE 3′.7.
00 41 21.6	−01 42 58	UGC 439	13.2	1.1 x 1.1	13.2	(R′)SB(r)a pec:		**UGC 435** NW 7′.4.
00 42 36.3	−01 59 47	UGC 455	14.1	1.4 x 1.3	14.6	Compact	57	Mag 10.80v star superimposed S of nucleus.
00 43 52.0	+00 48 04	UGC 466	14.0	1.1 x 0.5	13.2	SB(s)a pec?	165	
00 48 05.2	−01 33 59	UGC 492	13.3	1.2 x 0.6	12.9	E$^+$: sp	128	Mag 9.17v star SW 3′.5.
00 48 30.6	+01 21 12	UGC 496	13.7	1.1 x 0.8	13.4	Double System	75	Double system, contact.
00 49 25.9	−01 46 17	UGC 505	13.8	1.2 x 0.9	13.8	SB(rs)bc: II	135	
00 49 35.5	+01 06 57	UGC 507	14.3	2.0 x 0.3	13.6	Scd: sp	128	
00 53 32.8	+02 55 29	UGC 544	14.7	1.5 x 0.3	13.7	Sc: sp	42	Mag 11.6 star NW 3′.1.
00 54 59.8	+01 22 50	UGC 563	14.2	1.2 x 0.5	13.5	S	20	
00 56 16.2	−01 15 19	UGC 579	13.3	1.6 x 1.4	14.2	E	40	Several superimposed stars on uniform envelope; UGC 583 E 2′.4.
00 56 25.7	−01 15 48	UGC 583	13.4	0.9 x 0.9	13.2	E:		**CGCG 384-37** S 0′.8.
00 56 57.1	−01 12 45	UGC 587	14.4	0.7 x 0.7	13.5	Sbc		**MCG +0-3-37** S 3′.9.
00 57 02.1	−00 52 30	UGC 588	13.9	1.1 x 1.0	13.8	S?	74	Several stars superimposed on faint envelope.
00 57 35.0	−01 23 28	UGC 595	13.7	1.0 x 1.0	13.5	S?		Very small, very faint anonymous galaxy N 5′.0.
00 57 45.2	−00 25 07	UGC 599	13.9	2.1 x 1.3	14.9	E?	15	Star or bright knot N edge of nucleus.
00 59 42.1	+00 54 57	UGC 618	14.3	1.0 x 1.0	14.1	Scd:		
01 00 37.8	−01 41 53	UGC 626	14.0	0.8 x 0.7	13.2	SAB(s)c	6	Mag 6.91v star N 2′.5; MCG +0-3-51 NW 1′.2.
01 04 16.8	−01 04 33	UGC 661	14.0	1.1 x 0.4	13.0	S0/a	108	**MCG +0-3-68** S 2′.4.
01 06 07.3	+00 46 32	UGC 675	15.8	1.0 x 1.0	15.6	Im:		Dwarf, very low surface brightness.
01 06 14.1	+03 34 25	UGC 678	14.0	0.8 x 0.8	13.4	SA(rs)d: III-IV		
01 07 32.2	+02 10 56	UGC 694	14.6	1.0 x 0.6	13.9	SBb	135	Bright N-S bar, faint arms; very faint, small anonymous galaxy on N edge.
01 07 46.5	+01 03 49	UGC 695	14.0	0.6 x 0.6	12.8	S?		
01 08 11.6	+02 11 48	UGC 701	14.6	1.3 x 1.0	14.7	Double System	156	Double system, contact; **MCG +0-4-6** SSE 1′.5.
01 08 37.5	+01 38 25	UGC 711	13.8	3.6 x 0.4	14.0	SB(s)d? sp	118	Mag 9.71v star E edge.
01 09 57.8	−01 45 01	UGC 726	14.0	1.6 x 0.9	14.2	SB(s)d? III-IV	144	Mag 9.59v star W 5′.3.
01 10 22.3	+03 31 09	UGC 729	14.2	0.6 x 0.3	12.3	S0	131	
01 10 49.3	−01 45 26	UGC 736	14.3	1.0 x 0.6	13.6	Scd:	125	
01 11 30.9	+01 19 17	UGC 749	13.7	1.2 x 0.4	12.8	Sdm:	148	**UGC 747** SW 3′.1.
01 12 34.0	+00 17 34	UGC 757	15.1	0.7 x 0.5	13.8	SB	90	
01 13 10.0	+02 17 19	UGC 768	14.1	1.0 x 0.8	13.7	S0?	45	
01 13 42.3	−00 06 11	UGC 771	13.9	1.1 x 0.6	13.2	Sab	105	
01 13 39.5	+00 52 22	UGC 772	15.6	1.2 x 0.9	15.5	Im:	66	Dwarf irregular, very low surface brightness; mag 10.65v star W 4′.6.
01 14 04.3	−01 44 33	UGC 784	13.8	1.2 x 0.7	13.5	Sb II-III	50	Mag 10.56v star ENE 6′.4; mag 11.7 star WNW 4′.3.
01 14 37.4	+01 10 46	UGC 790	14.0	0.9 x 0.7	13.4	Scd:	162	Mag 11.7 star NNW 2′.2.
01 14 38.5	+01 49 44	UGC 791	13.8	0.9 x 0.7	13.1	SBab	130	Almost stellar **CGCG 385-50** ESE 5′.1.
01 14 48.8	−00 29 50	UGC 793	13.8	1.0 x 0.6	13.1	Interacting	87	
01 14 57.7	+00 25 48	UGC 797	13.9	0.9 x 0.7	13.4	E:	153	**PGC 4513** NE 3′.2; **MCG +0-4-59** SE 2′.2.
01 16 06.6	+01 34 00	UGC 817	14.0	0.8 x 0.4	12.6	SABb:	108	
01 17 51.6	−01 57 20	UGC 830	14.0	2.2 x 1.4	15.1	SAB(r)c? III	120	
01 18 56.8	+05 33 22	UGC 834	14.4	0.7 x 0.7	13.5	SAB(s)bc		Mag 13.5 star SW 0′.9.
01 18 24.1	+04 45 05	UGC 839	15.6	0.8 x 0.7	14.8	S?	60	Stellar nucleus.
01 18 53.7	−01 00 10	UGC 842	13.9	1.2 x 0.8	13.7	S0?	70	
01 19 04.5	−00 08 20	UGC 847	15.7	1.5 x 0.1	13.5	Sdm:	67	Close pair **MCG +0-4-74** and **MCG +0-4-75** SW 9′.2.
01 20 24.2	+05 49 40	UGC 871	15.5	1.8 x 1.6	16.5	Im	162	Dwarf irregular, very low surface brightness.
01 20 46.0	+01 26 39	UGC 875	14.2	0.8 x 0.5	13.0	SAB(s)c	12	
01 20 53.5	+04 48 02	UGC 881	13.7	1.1 x 0.9	13.8	E	43	
01 21 08.6	+05 08 21	UGC 887	14.3	0.7 x 0.6	13.2	Sbc	132	Mag 10.66v star on NW edge.
01 21 08.1	+01 22 22	UGC 890	13.3	1.9 x 0.6	13.3	SB(s)bc: II:	177	**UGC 874** SW 6′.7.

(Continued from previous page)

120 GALAXIES 120

RA h m s	Dec ° ′ ″	Name	Mag (V)	Dim ′ Maj x min	SB	Type Class	PA	Notes
01 21 16.7	−00 32 43	UGC 892	13.1	1.9 x 1.9	14.4	SB(r)ab II		Mag 11.22v star W 1′.6; stellar **MCG +0-4-92** W 2′.6; **MCG +0-4-94** S 3′.2.
01 22 47.3	−01 23 33	UGC 921	13.4	1.8 x 0.5	13.1	S0 pec sp	108	
01 23 12.3	−00 38 29	UGC 928	13.8	1.1 x 0.5	13.0	S0	45	Elongated **UGC 931** S 3′.6; almost stellar **MCG +0-4-101** NW 10′.2.
01 23 14.0	−00 23 08	UGC 929	13.7	1.4 x 1.4	14.3	Scd:		Mag 6.49v star SW 10′.5.
01 24 53.8	−01 30 02	UGC 974	14.0	1.1 x 0.2	12.2	S0/a	160	
01 25 18.0	−01 31 05	UGC 984	13.6	1.1 x 0.4	12.6	S0	136	NGC 538 SE 2′.8.
01 25 32.2	−01 30 12	UGC 996	13.9	0.9 x 0.3	12.3	S0/a	89	NGC 538 SW 3′.1; mag 13.3 star on S edge.
01 25 44.3	−01 27 27	UGC 1003	13.9	0.8 x 0.4	12.5	S0	124	**PGC 5306** ESE 2′.3.
01 25 53.0	+01 30 29	UGC 1006	13.8	0.7 x 0.5	12.5	S0/a	0	Mag 7.39v star WSW 8′.8; stellar **CGCG 385-137** ESE 9′.9.
01 27 07.7	+02 15 57	UGC 1028	14.1	0.9 x 0.3	12.5	S	12	
01 27 16.2	−01 16 22	UGC 1030	13.6	1.0 x 0.5	12.8	E	177	**MCG +0-4-147** WSW 8′.4.
01 27 36.1	−01 06 18	UGC 1040	13.9	1.2 x 0.3	12.6	S0/a	34	**UGC 1043** SE 2′.7; mag 9.31v star S 5′.8.
01 27 43.1	−01 08 24	UGC 1043	13.7	0.7 x 0.4	12.4	E:	27	**UGC 1040** NW 2′.7.
01 28 36.8	−01 43 55	UGC 1055	13.9	1.2 x 0.4	13.0	SBa	157	**UGC 1060** SE 7′.2; **MCG +0-4-156** NW 11′.8.
01 28 59.3	−00 33 42	UGC 1062	12.9	1.5 x 0.9	13.0	(R′)SB(s)0°:	60	
01 29 43.9	−01 14 31	UGC 1072	13.4	1.2 x 0.7	13.0	S0	68	
01 32 29.7	+04 35 59	UGC 1102	13.7	1.3 x 0.9	13.7	S?	0	Dwarf irregular **UGC 1105** NE 3′.7.
01 33 18.8	+03 04 12	UGC 1112	14.0	1.3 x 0.5	13.4	Scd:	108	Mag 7.92v star WSW 9′.0; stellar galaxy **UM 113** WNW 6′.2.
01 33 41.9	+03 32 30	UGC 1118	13.8	2.3 x 0.5	13.8	SA(s)dm: III-IV	66	Faint star on SW edge.
01 34 02.5	−01 04 33	UGC 1120	13.7	1.8 x 0.4	13.1	SBab pec sp	139	UGC 1123 NNE 2′.9.
01 34 08.3	−01 01 58	UGC 1123	13.5	1.2 x 0.3	12.2	Sab: sp	71	UGC 1120 SSW 2′.9.
01 35 03.1	+04 22 50	UGC 1133	13.9	3.6 x 1.8	15.7	Im: V	12	Dwarf irregular, very low surface brightness.
01 35 23.1	+04 52 16	UGC 1138	15.4	1.2 x 0.9	15.3	IABm		Irregular, low surface brightness; mag 8.75v star N 8′.9.
01 36 40.0	−01 23 44	UGC 1151	13.6	0.8 x 0.5	12.5	$S0^-$:	15	
01 37 07.4	+04 53 02	UGC 1155	13.9	0.7 x 0.4	12.3	Scd:	166	
01 37 55.8	+01 00 03	UGC 1163	14.6	1.0 x 0.9	14.3	Compact	156	Small bright nucleus, uniform envelope.
01 38 47.5	+01 04 17	UGC 1169	13.0	1.4 x 0.6	12.7	S0°	74	
01 39 47.2	−02 02 11	UGC 1174	14.0	0.9 x 0.5	13.0	Sbc	103	
01 40 17.8	−00 17 40	UGC 1180	15.0	1.0 x 0.9	14.7	Compact	39	Mag 7.28v star NW 4′.1.
01 42 53.7	+05 50 10	UGC 1203	15.6	1.5 x 0.7	15.5		35	
01 43 58.0	+02 20 58	UGC 1214	12.7	1.3 x 1.0	12.8	(R)SAB(rs)0^+:	108	Bright knot or faint star SW edge.
01 45 03.4	−00 18 02	UGC 1225	15.6	0.6 x 0.4	13.9	SB?	177	
01 45 35.7	+03 21 27	UGC 1235	14.0	0.9 x 0.4	12.7	Sbc	143	Mag 6.82v star NE 6′.0.
01 46 20.7	+04 15 49	UGC 1240	14.2	1.1 x 0.3	12.8	Sdm:	91	
01 49 59.4	+02 05 29	UGC 1293	13.8	0.8 x 0.5	12.7	Sab	143	
01 50 10.6	+02 18 34	UGC 1297	14.2	1.0 x 0.8	13.8	IB(s)m	36	

GALAXY CLUSTERS

RA h m s	Dec ° ′ ″	Name	Mag 10th brightest	No. Gal	Diam ′	Notes
00 48 42.0	+01 22 00	A 102	15.4	39	13	Narrow band of numerous stellar anonymous GX along N-S center line.
00 55 54.0	+00 38 00	A 116	15.7	48	18	Mag 8.05v star on NW edge; **MCG +0-3-30** E of center 3′.0; **PGC 3291** and **PGC 3294** W 2′.1.
00 56 24.0	−01 15 00	A 119	15.0	69	39	Numerous stellar, anonymous GX populate entire cluster area.
01 03 00.0	−02 31 00	A 134	16.0	43	25	Stellar **PGC 3755** and **PGC 3748** at center; mag 9.65v star 8′.9 E of center.
01 08 12.0	+02 09 00	A 147	15.0	32	22	Band of mostly stellar, anonymous GX stretches E-W along center line.
01 15 12.0	+00 14 00	A 168	15.4	89	39	Numerous stellar PGC and anonymous GX grouped near center.
01 23 42.0	+01 38 00	A 189	15.7	50	28	Numerous stellar and almost stellar members aligned N-S inside W boundary.
01 25 36.0	−01 30 00	A 194	13.9	37	56	Many catalog and anonymous GX spread over cluster area, and beyond cluster boundaries.
01 44 42.0	+05 48 00	A 246	16.4	56	25	Difficult to see many GX, even on DSS image.

121 GALAXIES 121

RA h m s	Dec ° ′ ″	Name	Mag (V)	Dim ′ Maj x min	SB	Type Class	PA	Notes
23 35 20.1	−17 25 12	ESO 605-11	14.7	1.4 x 0.5	14.2	SB?	42	Multi-galaxy system; mag 12.3 star NW 4′.8.
23 50 19.6	−17 46 01	ESO 606-13	14.6	1.2 x 0.7	14.3	Im:	81	Multi-galaxy system N 12.4; mag 12.6 star W 6′.0.
00 11 00.9	−12 49 24	IC 2	14.7	0.8 x 0.3	13.0	Sb	141	Mag 10.61v star N 5′.1.
00 17 35.0	−09 32 36	IC 5	13.8	0.9 x 0.7	13.4	E	9	MCG −2-1-45 and mag 12.4 star W 2′.1; MCG −2-1-46 SW 2′.6.
00 19 44.1	−14 07 19	IC 9	14.4	0.6 x 0.4	12.7	Sb(r)	129	Stellar; **MCG −2-2-2** NE 10′.7.
00 28 07.6	−13 05 39	IC 16	14.6	0.7 x 0.5	13.5	E?	51	Mag 10.59v star E 9′.5.
00 28 35.1	−11 35 12	IC 18	14.6	1.1 x 0.6	14.0	Sb	15	Very long, narrow streamer heading NNE beyond published size for this galaxy; mag 7.35v star E 8′.6.
00 28 39.5	−11 38 26	IC 19	14.1	1.0 x 0.7	13.8	E	27	Mag 7.36v star W 9′.6.
00 28 39.7	−13 00 38	IC 20	13.8	0.6 x 0.6	12.6	S0		Almost stellar; mag 10.59v star S 7′.0.
00 29 33.1	−09 04 52	IC 22	13.9	0.7 x 0.3	12.3	E/S0	45	Mag 11.04v star S 3′.4.
00 30 50.9	−12 43 11	IC 23	14.0	0.7 x 0.7	13.3	E		Almost stellar.
00 38 34.2	−15 21 32	IC 37	14.9	0.8 x 0.4	13.5	SBb	171	Mag 15.5 star on E edge; IC 38 S 3′.8.
00 38 38.9	−15 25 13	IC 38	14.0	0.7 x 0.6	12.9	Sbc	84	IC 37 N 3′.8.
23 26 32.2	−12 30 59	IC 1489	14.2	0.4 x 0.3	11.7		30	Located 1′.4 N of mag 9.69v star.
23 29 24.7	−16 19 01	IC 1491	14.1	0.5 x 0.3	11.9	Sbc	159	Mag 10.61v star SSW 8′.4.
23 30 47.7	−13 29 08	IC 1495	13.3	1.3 x 1.0	13.4	SAB(r)b pec: II	170	Mag 9.16v star NE 8′.4.
23 31 53.7	−05 00 29	IC 1498	13.0	1.8 x 0.6	12.9	(R′)SB(r)a:	11	Mag 9.43v star WSW 9′.5.
23 47 16.6	−15 18 24	IC 1509	14.5	1.4 x 0.2	13.0	Sc	12	Mag 13.0 star E 1′.6.
23 57 54.6	−14 02 21	IC 1520	13.8	0.7 x 0.6	12.7	S?	165	Mag 7.23v star SE 11′.0.
00 05 13.3	−11 30 09	IC 1529	13.6	1.2 x 1.1	13.8	(R′)SA(r)0° pec:	0	Mag 13.4 star NE 2′.5.
00 10 23.6	−07 05 05	IC 1533		0.7 x 0.5			138	Mag 9.80v star N 3′.6.
00 39 00.3	−09 00 56	IC 1563	14.0	0.7 x 0.6	12.9	S0 pec sp	143	Located on S edge of NGC 191.
23 26 20.1	−17 57 25	IC 5321	13.0	1.3 x 0.8	12.9	SBa pec?	43	Mag 8.57v star SW 18′.2.
23 55 27.0	−06 14 17	MCG −1-1-3	15.0	0.5 x 0.3	12.8			
23 59 41.5	−06 48 25	MCG −1-1-14	14.3	0.8 x 0.6	13.4	Sbc	84	
00 00 05.1	−05 12 34	MCG −1-1-15	14.7	0.6 x 0.6	13.4	Sbc		
00 00 08.7	−06 22 28	MCG −1-1-16	13.8	1.6 x 0.2	12.4	Sa pec? IV	165	Mag 10.47v star E 6′.9.
00 00 11.4	−05 09 33	MCG −1-1-17	14.2	0.8 x 0.4	12.8	Sbc	33	
00 00 23.4	−06 56 13	MCG −1-1-18	14.8	0.8 x 0.4	13.6	E	144	Close pair of stars just off NE edge.
00 00 48.5	−05 34 49	MCG −1-1-21	15.5	0.7 x 0.4	13.9	S0/a	90	
00 03 35.4	−07 41 27	MCG −1-1-27	15.7	0.6 x 0.5	14.2	SBc	27	
00 05 05.6	−07 05 37	MCG −1-1-28	12.8	2.0 x 1.0	13.4	SB(r)b: II-III	75	Mag 13.3 star NE 2′.2.

(Continued from previous page)

GALAXIES

RA h m s	Dec ° ′ ″	Name	Mag (V)	Dim ′ Maj x min	SB	Type Class	PA	Notes
00 04 40.0	−08 06 00	MCG −1-1-30	13.2	1.2 x 1.2	13.5	SA(r)0$^-$?		Almost stellar **MCG −1-1-29** N 0′.4; mag 8.64v star S 5′.1.
00 05 48.4	−07 37 46	MCG −1-1-32	15.0	0.9 x 0.2	13.0		69	Mag 6.95v star WSW 5′.5.
00 08 35.9	−05 13 01	MCG −1-1-36	14.3	1.1 x 0.5	13.5	(R′)Sa pec:	165	
00 08 43.5	−06 59 51	MCG −1-1-37	13.9	1.1 x 0.9	13.7	S0	132	MCG −1-1-40 NE 8′.1.
00 09 11.0	−06 55 34	MCG −1-1-40	14.4	0.7 x 0.4	12.9	Sc	66	MCG −1-1-37 SW 8′.1.
00 09 56.2	−07 52 22	MCG −1-1-41	14.4	0.8 x 0.4	13.1	Sb	3	
00 10 06.6	−06 19 20	MCG −1-1-42	15.9	0.6 x 0.4	14.2	SAB(s)m IV-V	39	Stellar **PGC 744** ENE 6′.9.
00 10 37.0	−06 38 25	MCG −1-1-45	14.2	0.9 x 0.4	12.9	Sbc	42	
00 12 57.3	−05 31 29	MCG −1-1-49	15.1	0.7 x 0.4	13.6	SBb		
00 13 29.3	−06 47 48	MCG −1-1-51	14.2	0.5 x 0.4	12.5	E	12	
00 13 34.4	−05 05 35	MCG −1-1-52	13.8	1.2 x 0.6	13.3	SA(rs)b pec?	168	
00 14 40.6	−07 17 40	MCG −1-1-57	14.3	1.0 x 0.7	13.8	S0	21	NGC 50 S 3′.1.
00 16 50.9	−05 16 08	MCG −1-1-64	14.0	1.5 x 0.5	13.6	SB(s)a pec:	42	
00 17 05.6	−06 22 23	MCG −1-1-65	14.0	1.1 x 1.1	14.0	SAB(s)cd II-III		Several bright knots or superimposed stars S and E of stellar nucleus.
00 20 00.1	−06 20 04	MCG −1-2-1	13.3	1.5 x 0.8	13.4	SB(s)ab pec? II	141	Mag 10.68v star ESE 14′.9.
00 24 43.3	−06 01 23	MCG −1-2-10	14.1	1.5 x 0.9	14.3	Sbc	9	Mag 10.59v star SW 2′.9.
00 27 14.7	−07 47 17	MCG −1-2-18	14.5	0.8 x 0.4	13.2	S0/a	156	NGC 116 NNW 7′.6.
00 28 02.4	−08 07 17	MCG −1-2-20	14.8	1.2 x 0.7	14.4	S?	162	Bright knot or star E edge.
00 29 48.8	−05 06 57	MCG −1-2-22	14.3	1.0 x 0.8	13.9	Sbc	138	
00 30 59.0	−07 13 02	MCG −1-2-23	14.1	0.9 x 0.5	13.1	S?	24	
00 31 14.4	−07 50 32	MCG −1-2-25	13.7	1.2 x 0.6	13.2	SA(r)b: II-III	117	
00 31 31.6	−06 06 26	MCG −1-2-26	14.0	1.2 x 0.8	13.8	Sc	129	
00 33 04.0	−08 06 54	MCG −1-2-30	14.9	1.0 x 0.6	14.2	SB(s)dm: III-IV	39	
00 34 40.8	−07 54 13	MCG −1-2-34	13.3	1.1 x 0.7	12.9	Sc	57	
00 36 08.9	−07 32 16	MCG −1-2-39	14.4	0.9 x 0.6	13.6	SAB(s)dm? IV-V	144	
00 36 22.5	−06 38 52	MCG −1-2-41	14.7	0.5 x 0.2	12.1	S0/a	120	Mag 11.7 star NE 1′.1; mag 8.15v star NE 9′.7.
23 36 45.2	−05 02 20	MCG −1-60-12	13.7	0.5 x 0.5	12.3	E		Mag 9.35v star ESE 13′.5; MCG −1-60-11 NNW 8′.8.
23 38 48.6	−05 44 52	MCG −1-60-16	14.5	1.7 x 0.7	14.5	SB(s)m: IV-V	54	Mag 7.66v star NE 19′.4.
23 43 52.5	−08 02 02	MCG −1-60-25	14.2	1.0 x 0.7	13.7	SBb	6	
23 44 10.6	−06 10 13	MCG −1-60-26	12.8	1.4 x 0.8	12.8	SB(s)c pec: II-III	45	Located between a close pair of mag 11.5 stars.
23 45 51.2	−06 46 51	MCG −1-60-28	14.1	0.8 x 0.5	12.9	SBc	30	Mag 10.61v star W 1′.9.
23 46 04.1	−07 09 36	MCG −1-60-30	16.0	0.9 x 0.7	15.3	Sbc	15	Elongated **MCG −1-60-29** W 2′.9.
23 47 33.3	−07 48 30	MCG −1-60-37	14.7	1.1 x 0.3	13.3	SBcd	81	
23 49 06.8	−07 02 55	MCG −1-60-44	14.4	0.8 x 0.3	12.7	S?	12	**MCG −4-60-42** WSW 2′.3.
23 55 33.7	−09 39 18	MCG −2-1-1	15.0	0.5 x 0.3	12.8		33	
23 55 52.4	−12 24 34	MCG −2-1-2	15.1	0.8 x 0.3	13.4	Sbc	174	Lies 1′.0 S of mag 11 star.
23 55 55.4	−12 44 01	MCG −2-1-3	15.1	0.8 x 0.3	13.4	Irr	78	
23 56 20.4	−11 07 03	MCG −2-1-4	14.6	0.4 x 0.2	11.7	Sa	0	
23 56 26.6	−11 05 11	MCG −2-1-5	14.6	1.4 x 0.6	14.5	E	165	Close pair of mag 12 star W 1′.6; MCG −2-1-4 SW 2′.4.
23 58 00.8	−11 54 55	MCG −2-1-6	14.5	0.5 x 0.5	12.9			
00 02 06.3	−10 13 07	MCG −2-1-9	15.0	0.9 x 0.3	13.4	Sbc	114	
00 03 07.4	−12 24 12	MCG −2-1-10	14.9	1.0 x 0.5	14.0	Sb	108	
00 03 21.7	−09 59 02	MCG −2-1-11	15.7	0.6 x 0.5	14.3	Sb	3	Stellar nucleus.
00 03 22.2	−10 46 17	MCG −2-1-12	14.0	0.8 x 0.3	12.3	S?	15	0′.4 E of mag 11 star.
00 04 01.6	−11 10 28	MCG −2-1-14	15.4	0.9 x 0.7	14.7	SB(rs)c: II-III	144	MCG −2-1-15 on SW edge.
00 04 00.6	−11 10 45	MCG −2-1-15	14.1	0.3 x 0.2	10.9	Sa?:	9	Stellar; located on SW edge of MCG −2-1-14.
00 04 21.8	−08 02 36	MCG −2-1-17	14.5	0.7 x 0.5	13.2		147	
00 04 34.3	−09 03 12	MCG −2-1-18	14.7	0.6 x 0.2	12.3	Sb	102	
00 05 38.5	−13 36 06	MCG −2-1-21	15.7	0.9 x 0.4	14.4	Sbc	132	Stellar **MCG −2-1-20** NW 1′.1.
00 05 56.2	−13 58 46	MCG −2-1-22	15.1	1.0 x 0.3	13.7	SB?	30	
00 05 56.9	−13 59 46	MCG −2-1-23	14.7	1.1 x 0.2	12.9		57	
00 06 16.9	−13 26 54	MCG −2-1-24	13.9	0.5 x 0.5	12.3	Ring A		Stellar nucleus; NGC 7828 NE 3′.1.
00 07 31.4	−11 47 54	MCG −2-1-26	14.9	0.8 x 0.6	14.0	Sb	168	Mag 13.7 star on NE edge.
00 07 43.9	−11 50 33	MCG −2-1-27	16.4	0.9 x 0.6	15.6	S?	39	Multi-galaxy system.
00 08 34.5	−10 57 00	MCG −2-1-28	14.0	1.2 x 1.2	14.2	SB(r)a		Stellar nucleus.
00 08 41.5	−13 01 24	MCG −2-1-29	14.6	0.7 x 0.4	13.0		27	
00 11 03.3	−09 39 07	MCG −2-1-30	14.9	0.6 x 0.3	12.9		48	Mag 10.44v star NE 1′.1.
00 13 02.1	−13 15 26	MCG −2-1-34	15.0	0.6 x 0.4	13.3	Sbc	36	Mag 10.99v star S 3′.2.
00 13 12.8	−12 33 26	MCG −2-1-35	14.5	1.3 x 0.2	12.8	Sc	105	
00 13 38.7	−08 46 25	MCG −2-1-36	13.7	0.8 x 0.8	13.1	Sc		Bright star on W edge.
00 14 25.2	−13 46 31	MCG −2-1-37	14.9	0.7 x 0.5	13.6		21	
00 16 41.5	−10 33 10	MCG −2-1-41	14.5	0.7 x 0.3	12.7	Sbc	0	= **Mkn 943**.
00 17 02.9	−09 40 46	MCG −2-1-42	15.1	0.9 x 0.4	13.8		6	
00 17 17.1	−12 15 06	MCG −2-1-44	15.0	0.4 x 0.4	12.9	Sb		
00 17 26.6	−09 32 00	MCG −2-1-45	14.4	0.4 x 0.3	11.9		0	Mag 12.4 star on N edge.
00 17 27.5	−09 34 28	MCG −2-1-46	15.3	0.7 x 0.2	13.0	Sb	24	
00 17 59.5	−09 16 22	MCG −2-1-48	14.4	1.0 x 0.6	13.7	Sb	135	
00 18 50.7	−08 27 32	MCG −2-1-50	13.8	1.2 x 0.8	13.6	(R′)SB(r)0/a?	168	
00 18 50.3	−10 22 43	MCG −2-1-51	14.3	0.8 x 0.6	13.4	SB(s)b pec?	51	
00 18 50.4	−10 21 44	MCG −2-1-52	13.6	1.3 x 0.6	13.2	SB(s)c pec	42	Two main, moderately faint arms; MCG −2-1-51 on S edge.
00 19 20.8	−10 56 38	MCG −2-1-54	14.4	1.0 x 0.5	13.5	Sbc	177	Mag 8.16v star W 8′.2.
00 20 35.4	−13 21 09	MCG −2-2-3	15.4	1.0 x 0.7	14.8	SBb	27	
00 21 36.2	−14 16 32	MCG −2-2-4	14.4	0.3 x 0.3	11.7	S0		Stellar.
00 21 51.2	−09 29 37	MCG −2-2-5	14.7	1.3 x 0.5	14.1	Sc	156	
00 22 19.9	−13 16 06	MCG −2-2-6	16.3	0.6 x 0.3	14.3		99	
00 22 34.0	−08 29 12	MCG −2-2-7	13.9	1.6 x 0.4	13.3	(R)S0$^+$? sp	120	Mag 9.24v star SE 12′.7; **MCG −2-2-8** E 10′.4.
00 25 52.4	−09 39 44	MCG −2-2-9	14.3	1.0 x 0.3	12.8	Sa		
00 24 20.5	−14 13 57	MCG −2-2-10	14.2	0.5 x 0.5	12.7	E		Several small, faint galaxies E and SE 1′.2 to 1′.4.
00 24 36.6	−11 29 49	MCG −2-2-12	15.5	1.6 x 0.4	14.8	S?	147	Mag 8.06v star S 13′.5.
00 25 44.7	−08 25 07	MCG −2-2-13	14.3	0.7 x 0.6	13.2	Sbc	36	Mag 7.40v star N 4′.5.
00 27 46.7	−11 31 00	MCG −2-2-15	14.8	0.8 x 0.3	13.1	Sb	147	MCG −2-2-16 and mag 10.07v star S 2′.8.
00 27 48.3	−11 33 47	MCG −2-2-16	14.9	0.4 x 0.4	12.8			Mag 10.07v star S 0′.9; MCG −2-2-15 N 2′.8.
00 28 11.9	−13 53 49	MCG −2-2-18	14.5	0.6 x 0.6	13.3			
00 28 18.7	−13 32 06	MCG −2-2-19	15.4	1.1 x 0.2	13.6	Scd	63	
00 28 18.0	−09 29 35	MCG −2-2-20	14.5	0.9 x 0.4	13.3		78	Mag 8.54v star NE 7′.4.
00 28 44.4	−12 38 40	MCG −2-2-22	16.7	0.6 x 0.4	15.0	Sd	30	
00 29 00.5	−08 19 06	MCG −2-2-25	14.5	0.9 x 0.5	13.7	E	165	
00 29 31.7	−13 59 37	MCG −2-2-26	15.0	0.7 x 0.5	13.7		162	Mag 9.78v star W 3′.4.
00 29 57.5	−11 17 55	MCG −2-2-28	15.0	1.1 x 0.2	13.2	Scd	96	

RA h m s	Dec ° ′ ″	Name	Mag (V)	Dim ′ Maj x min	SB	Type Class	PA	Notes
00 29 55.6	−11 06 59	MCG −2-2-29	15.4	0.7 x 0.2	13.2	Sbc	126	MCG −2-2-30 E 2′9; mag 9.36v star NW 8′9.
00 30 07.2	−11 06 48	MCG −2-2-30	12.8	1.8 x 0.8	13.0	(R)SB(r)a	174	Mag 9.36v star NW 11′2; MCG −2-2-29 W 2′9.
00 30 45.0	−08 24 57	MCG −2-2-31	14.1	0.8 x 0.5	12.9	S0/a	42	Stellar galaxies Kaz 357 and 356 NW 3′5.
00 30 29.9	−08 47 04	MCG −2-2-33	13.8	1.2 x 0.3	12.5	SB(s)b? II-III	150	
00 30 33.0	−11 24 15	MCG −2-2-34	15.3	0.5 x 0.4	13.4	SB(s)cd:	174	Mag 9.97v star ESE 1′2.
00 30 53.8	−09 12 32	MCG −2-2-35	14.4	0.7 x 0.4	12.9	S0	117	
00 30 58.2	−09 24 16	MCG −2-2-36	14.6	0.8 x 0.5	13.4	Sbc	96	Mag 7.43v star W 4′4.
00 31 36.7	−10 53 21	MCG −2-2-37	14.4	0.8 x 0.5	13.3	S0	99	Mag 8.92v star W 6′0.
00 31 13.4	−10 28 51	MCG −2-2-38	13.5	1.2 x 0.6	13.0	SB(r)bc? I-II	3	
00 31 35.8	−10 30 22	MCG −2-2-40	13.5	1.4 x 0.4	12.7	SB(r)bc? sp	48	
00 31 56.8	−11 00 59	MCG −2-2-41	14.8	0.6 x 0.4	13.1		42	
00 31 56.9	−12 58 16	MCG −2-2-42	14.6	0.7 x 0.5	13.3		33	
00 31 57.5	−08 20 41	MCG −2-2-43	15.1	1.0 x 0.6	14.4	SB(s)d: III-IV	177	
00 32 32.9	−12 03 42	MCG −2-2-46	14.4	0.4 x 0.4	12.3			0′3 N mag 12 star.
00 32 45.3	−10 27 06	MCG −2-2-47	15.2	0.6 x 0.4	13.5	Sb	36	
00 32 49.2	−10 00 52	MCG −2-2-48	14.2	0.7 x 0.7	13.3	IB(s)m		Mag 8.55v star E 4′8.
00 32 42.6	−11 19 09	MCG −2-2-49	13.8	1.9 x 0.2	12.6	Sc: sp	33	Mag 10.82v star W 8′3.
00 33 20.5	−09 54 12	MCG −2-2-50	14.4	0.5 x 0.4	12.5		63	Mag 11.0 star on SW end; mag 8.55v star SW 6′1.
00 33 10.3	−13 08 48	MCG −2-2-51	13.9	1.3 x 0.7	13.6	(R′)SB(rs)bc I-II	129	Two main arms for open "S" shape; mag 8.75v star SW 8′3.
00 33 24.3	−14 00 23	MCG −2-2-52	14.8	0.7 x 0.6	13.7		162	
00 34 57.9	−09 20 35	MCG −2-2-57	14.3	0.7 x 0.4	12.8	S?	144	
00 34 54.8	−12 29 55	MCG −2-2-58	14.0	1.0 x 0.7	13.5	SAB(r)c III	9	
00 35 02.9	−09 22 08	MCG −2-2-59	14.4	0.7 x 0.4	12.9	Sbc	33	
00 35 35.0	−12 17 16	MCG −2-2-60	14.0	0.5 x 0.4	12.3	E	51	Almost stellar.
00 35 36.9	−11 27 37	MCG −2-2-61	15.3	0.7 x 0.4	13.8	S0/a	165	
00 35 48.3	−08 35 28	MCG −2-2-62	14.8	0.5 x 0.4	12.9	Sc	141	
00 36 01.9	−09 53 27	MCG −2-2-64	13.9	1.1 x 0.9	13.7	SB(s)c pec: II-III	18	
00 36 23.9	−11 45 34	MCG −2-2-67	14.8	0.9 x 0.3	13.2	Sc	9	Mag 9.43v star NE 8′7.
00 36 12.6	−12 15 32	MCG −2-2-68	13.7	1.2 x 0.6	13.2	SAB(r)a:	141	Compact core in faint, smooth envelope.
00 37 04.3	−12 40 56	MCG −2-2-70	15.0	0.6 x 0.3	13.0	Sb	60	
00 37 42.4	−14 10 49	MCG −2-2-71	14.8	1.0 x 0.4	13.6		165	
00 38 02.5	−13 51 57	MCG −2-2-72	14.7	1.1 x 1.1	14.8	SBbc		Low uniform surface brightness.
00 37 57.4	−09 15 11	MCG −2-2-73	13.0	1.3 x 1.1	13.3	SAB(s)cd II	115	Mag 8.13v star NNW 11′9.
00 38 38.6	−11 02 55	MCG −2-2-74	15.2	0.7 x 0.3	13.3		135	
00 39 20.1	−10 28 59	MCG −2-2-75	15.0	0.6 x 0.3	13.0	Sb	171	
23 22 35.4	−13 05 38	MCG −2-59-13	13.6	1.4 x 0.7	13.5	Sbc	81	
23 26 41.3	−11 37 03	MCG −2-59-17	14.3	1.2 x 0.4	13.3	S?	72	
23 26 53.6	−13 04 21	MCG −2-59-18	14.8	0.7 x 0.4	13.3		135	
23 28 52.6	−08 38 54	MCG −2-59-20	13.9	0.9 x 0.6	13.0	Sc	27	
23 29 56.7	−12 29 06	MCG −2-59-21	14.0	0.7 x 0.3	12.1	Sab	15	
23 30 14.8	−08 52 37	MCG −2-59-22	14.7	0.9 x 0.2	12.7	Sbc		
23 30 23.7	−12 04 49	MCG −2-59-23	14.5	0.9 x 0.3	12.9	Sc	9	
23 34 00.4	−11 47 00	MCG −2-60-1	14.9	0.5 x 0.3	12.7	Sb	135	
23 34 23.4	−11 39 25	MCG −2-60-2	14.3	0.9 x 0.7	13.6	Sc		
23 36 15.0	−11 28 03	MCG −2-60-3	14.3	0.7 x 0.5	13.0	Sc	9	
23 37 05.7	−09 35 55	MCG −2-60-4	14.7	1.0 x 0.6	14.0	IB(s)m V	27	
23 39 49.4	−11 49 04	MCG −2-60-7	14.3	0.8 x 0.5	13.1	Pec	135	
23 41 23.7	−09 53 57	MCG −2-60-9	14.3	0.7 x 0.7	13.4	(R′)SB(s)c pec: II-III		Located on N edge of small galaxy cluster **A 2646**.
23 40 51.6	−13 21 31	MCG −2-60-10	14.8	1.3 x 0.5	14.2	SB(rs)d: III-IV	84	
23 41 02.1	−08 13 21	MCG −2-60-11	14.2	0.7 x 0.5	13.0	S0	99	
23 41 52.7	−08 38 53	MCG −2-60-12	13.9	1.5 x 0.3	12.9	Sb: sp	24	Mag 9.64v star W 7′2.
23 42 13.2	−12 03 00	MCG −2-60-13	15.0	0.7 x 0.2	12.7	Sb	21	
23 44 07.6	−09 03 20	MCG −2-60-14	15.6	1.0 x 0.8	15.2	SB(s)d IV-V	36	
23 45 01.6	−12 54 02	MCG −2-60-15	14.3	0.6 x 0.3	12.3	S0/a	108	
23 48 22.7	−08 30 10	MCG −2-60-16	13.7	1.0 x 0.8	13.3	Sc	12	
23 48 42.1	−14 01 40	MCG −2-60-17	15.0	0.4 x 0.4	12.9	Sc		
23 49 51.7	−12 28 03	MCG −2-60-18	13.7	0.9 x 0.7	13.3	E/S0		
23 51 19.5	−12 45 10	MCG −2-60-19	14.5	0.7 x 0.5	13.4	E?	36	Mag 8.93v star SW 7′2.
23 53 27.4	−11 48 39	MCG −2-60-21	14.1	0.7 x 0.5	12.8	Sc	162	
23 56 47.7	−16 30 37	MCG −3-1-2	14.1	0.9 x 0.7	13.5	(R′)SB(r)a	138	Mag 10.84v star on S edge.
23 56 58.6	−16 45 43	MCG −3-1-3	14.6	0.9 x 0.6	13.7	SAB(s)d: III-IV	0	
23 57 17.0	−17 16 57	MCG −3-1-4	15.0	0.8 x 0.6	14.0	S	45	
23 58 40.5	−17 33 33	MCG −3-1-9	13.7	0.7 x 0.7	12.8	(R′)SB(s)ab		Star on N edge.
23 58 44.6	−16 50 11	MCG −3-1-10	14.9	0.5 x 0.2	12.2	Sc	102	
23 59 29.0	−15 12 42	MCG −3-1-11	14.7	0.6 x 0.5	13.2	SBc	87	
00 01 36.1	−14 44 58	MCG −3-1-14	14.4	0.8 x 0.8	13.7	SBc		
00 01 58.0	−15 26 59	MCG −3-1-15	10.6	9.5 x 3.0	14.1	IB(s)m V	0	Also known as the **WLM** or **Wolf-Lundmark-Mellotte** galaxy.
00 03 45.1	−14 41 40	MCG −3-1-16	14.5	0.8 x 0.6	13.5	Sc	123	
00 04 12.2	−14 31 26	MCG −3-1-17	14.4	1.0 x 0.4	13.2	Sbc	75	Star NE end.
00 04 46.0	−16 01 52	MCG −3-1-18	14.1	1.5 x 1.2	14.6	SB(r)c II	117	Mottled spiral arms with several bright knots.
00 19 13.9	−14 23 49	MCG −3-1-28	14.8	1.0 x 0.5	13.9	Sb	153	
00 21 55.9	−15 59 01	MCG −3-2-1	14.2	0.4 x 0.4	12.0	Sbc		
00 22 15.3	−14 35 36	MCG −3-2-3	14.3	1.0 x 0.3	12.8	Sbc	123	
00 22 32.6	−17 04 29	MCG −3-2-5	14.4	1.1 x 0.9	14.2		165	Stellar nucleus, smooth envelope; mag 10.27v star S 2′5.
00 22 53.0	−14 32 03	MCG −3-2-9	14.1	0.7 x 0.7	13.4	E		
00 22 56.9	−15 59 26	MCG −3-2-10	15.0	1.1 x 0.2	13.2	Scd	153	Mag 6.44v star NNE 3′4.
00 25 18.5	−14 33 28	MCG −3-2-12	14.2	0.8 x 0.5	13.1	S0/a	57	MCG −3-2-13 E 2′7.
00 25 29.6	−14 33 14	MCG −3-2-13	15.7	1.6 x 0.2	14.3	SB(s)d: sp II-III	81	MCG −3-2-12 W 2′7.
00 28 02.4	−14 56 19	MCG −3-2-14	14.9	0.8 x 0.3	13.2	Sbc	6	
00 29 08.0	−15 51 11	MCG −3-2-15	15.8	1.0 x 0.6	15.1	Sc	81	
00 29 33.1	−15 04 52	MCG −3-2-16	14.0	0.5 x 0.3	12.0	E	81	Compact; mag 7.67v star W 4′7; MCG −3-2-17 S 1′1.
00 29 33.2	−15 05 50	MCG −3-2-17	15.0	0.8 x 0.4	13.6	Scd	45	MCG −3-2-16 N 1′1.
00 30 29.4	−16 41 30	MCG −3-2-18	14.7	0.7 x 0.7	13.8	Sc		
00 37 31.5	−17 56 36	MCG −3-2-25	15.2	0.7 x 0.4	13.6	SBbc	174	**MCG −3-2-23** WSW 5′4.
00 38 18.4	−14 50 56	MCG −3-2-27	14.7	0.9 x 0.6	13.9	Sc	51	
23 21 49.3	−16 43 09	MCG −3-59-5	14.2	0.4 x 0.4	12.1	S0		
23 23 51.9	−16 27 25	MCG −3-59-7	14.7	1.0 x 0.3	13.2	Sc	39	
23 37 13.2	−16 37 30	MCG −3-60-6	14.8	1.0 x 0.2	12.9	Sb	72	
23 43 44.6	−14 42 51	MCG −3-60-11	13.7	1.7 x 0.6	13.5	(R′)SB(r)0/a?	129	Mag 11.8 star N 2′7.

(Continued from previous page)

GALAXIES

RA h m s	Dec ° ′ ″	Name	Mag (V)	Dim ′ Maj x min	SB	Type Class	PA	Notes
23 48 57.0	−16 32 26	MCG −3-60-19	13.0	0.5 x 0.2	10.6	E/S0	72	Located on E edge of NGC 7759.
23 54 21.0	−17 18 23	MCG −3-60-23	14.1	0.5 x 0.5	12.5	S		
00 11 07.5	−12 05 58	NGC 34	14.4	2.2 x 0.9	15.0	Pec	30	Very faint, narrow extension NE of bright core accounts for most of published length of this galaxy.
00 11 10.5	−12 01 17	NGC 35	14.1	0.7 x 0.5	12.8	Sb	135	NGC 34 S 5.2′.
00 11 47.0	−05 35 14	NGC 38	13.3	1.1 x 1.1	13.3	(R)SAa:		Strong, round core in smooth envelope.
00 14 30.7	−07 10 06	NGC 47	13.1	2.5 x 2.2	14.8	SB(rs)bc II	81	Close pair of mag 9.54v, 7.32v stars ESE 4.9′.
00 14 44.7	−07 20 44	NGC 50	12.3	2.3 x 1.7	13.6	SO⁻ pec	170	MCG −1-1-57 N 3.1′; **MCG −1-1-59** NNE 6.4′.
00 15 07.7	−07 06 26	NGC 54	13.4	1.2 x 0.4	12.4	SB(r)a?	93	NGC 47 WSW 10.0′.
00 16 24.0	−06 19 03	NGC 61A	13.0	1.1 x 0.6	12.4	S0 pec?	24	Multi-galaxy system with NGC 61B.
00 16 24.4	−06 19 22	NGC 61B	13.2	0.9 x 0.4	11.9	S0	63	NGC 61A on N edge.
00 17 05.4	−13 29 11	NGC 62	13.2	1.0 x 0.7	12.6	(R)SB(r)a:	130	Mag 7.16v star NW 9.0′.
00 17 30.2	−06 49 31	NGC 64	13.2	1.5 x 1.1	13.6	SB(s)bc I-II	39	
00 18 39.2	−15 19 18	NGC 73	13.7	1.6 x 1.1	14.2	SAB(rs)bc: II	145	
00 24 36.6	−13 57 24	NGC 102	13.5	0.9 x 0.6	12.7	S0/a	140	
00 24 43.8	−05 08 56	NGC 106	13.7	0.8 x 0.6	12.8	Sc	100	
00 25 42.1	−08 17 01	NGC 107	14.2	0.8 x 0.6	13.3	Sbc II-III:	142	Mag 7.40v star SSE 4.9′.
00 27 05.2	−07 40 07	NGC 116	13.7	0.8 x 0.4	12.4	S0/a	102	Mag 10.25v star S 2.4′; MCG −1-2-18 SSE 7.6′.
00 33 06.3	−13 22 17	NGC 135	14.8	0.5 x 0.5	13.2			Almost stellar; mag 8.50v star SW 8.8′.
00 31 45.6	−05 09 16	NGC 145	12.7	1.8 x 1.6	13.7	SB(s)dm	57	Mag 8.60v star E 5.6′.
00 34 02.8	−09 42 19	NGC 151	11.6	3.7 x 1.7	13.4	SB(r)bc II	75	A narrow extension to the NE has two bright knots on the end.
00 34 19.5	−12 39 23	NGC 154	14.0	1.1 x 0.9	14.1	E	80	
00 34 40.0	−10 46 02	NGC 155	13.3	1.7 x 1.3	14.0	S0° pec	177	
00 34 47.1	−08 23 57	NGC 157	10.4	3.5 x 2.4	12.6	SAB(rs)bc I-II	36	Mag 8.47v star N 6.3′.
00 35 59.9	−10 07 18	NGC 163	12.7	1.5 x 1.2	13.3	E0	85	Very small, elongated **MCG −2-2-65** S 1.2′.
00 36 29.0	−10 06 25	NGC 165	13.1	1.5 x 1.1	13.5	SB(rs)bc I	77	Almost stellar nucleus; NGC 163 W 7.2′.
00 35 48.8	−13 36 40	NGC 166	14.3	0.9 x 0.3	12.7	Sa	145	
00 39 08.3	−14 10 28	NGC 178	12.6	2.0 x 1.0	13.2	SB(s)m IV-V	178	Two strong knots SW of core.
00 37 46.2	−17 50 58	NGC 179	13.3	0.9 x 0.8	12.8	SAB0⁻	113	Faint star on N edge; mag 10.28v star N 5.1′.
00 39 30.6	−14 39 24	NGC 187	13.2	1.3 x 0.5	12.5	SB(s)cd III	148	
00 38 59.7	−09 00 12	NGC 191	12.4	1.3 x 1.2	12.8	SAB(rs)c: pec	128	Mag 10.49v star NW 5.0′; IC 1563 on S edge.
00 39 35.7	−09 11 42	NGC 195	13.7	1.2 x 0.8	13.5	(R)SB(r)a:	45	Mag 10.35v star S 5.5′.
00 39 40.6	−14 14 13	NGC 207	13.7	0.9 x 0.4	12.4	Sa	85	Mag 9.54v star SSE 4.4′; faint, stellar anonymous galaxy N 3.8′.
23 24 06.9	−11 51 40	NGC 7646	13.3	1.0 x 0.6	12.6	Sc	135	Star on NW end.
23 27 15.1	−09 23 12	NGC 7665	13.3	0.7 x 0.7	12.4	Sm III:		Mag 9.90v star SE 2.7′.
23 32 46.7	−05 35 48	NGC 7692	14.8	0.5 x 0.4	12.9	Irr	100	
23 35 27.5	−16 42 19	NGC 7709	12.6	2.3 x 0.6	12.8	SB0° pec sp	54	
23 37 43.8	−15 07 08	NGC 7717	12.8	1.4 x 1.1	13.1	(R)SB(r)0/a	8	Located 4.8′ SW of mag 6.38v star.
23 38 48.4	−06 31 15	NGC 7721	11.6	2.8 x 1.1	12.7	SA(s)c II	15	Strong dark lane W of center.
23 38 57.1	−12 57 36	NGC 7723	11.2	3.5 x 2.3	13.3	SB(r)b II	40	Tightly wound, branching arms.
23 39 07.3	−12 13 26	NGC 7724	12.9	1.4 x 1.0	13.1	(R′)SB(r)b pec?	37	Strong dark lane N of center.
23 39 53.9	−12 17 33	NGC 7727	10.6	4.7 x 3.5	13.5	SAB(s)a pec	39	Large diffuse core, several superimposed stars SE edge.
23 49 11.3	−16 36 04	NGC 7754	14.5	0.6 x 0.3	12.5	Sa?	128	**PGC 72509** NW 1.7′.
23 48 54.8	−16 32 32	NGC 7759	13.0	1.3 x 0.8	12.9	SAB0° pec:	141	Elongated MCG −3-60-19 on E edge.
23 51 28.8	−13 22 54	NGC 7761	13.1	1.2 x 1.2	13.4	SA(rs)0°:		Mag 10.85v star W 3.5′.
23 50 15.8	−16 35 25	NGC 7763	14.3	0.7 x 0.4	12.8	SB0?	158	Stellar **PGC 72566** on E edge.
23 54 16.5	−13 35 10	NGC 7776	13.9	1.0 x 0.3	12.5	Sa	153	Located 2.3′ NNE of mag 10.63v star.
00 03 32.3	−10 44 39	NGC 7808	13.5	1.3 x 1.3	13.9	(R′)SA0°:		MCG −2-1-12 SW 2.9′.
00 04 09.2	−11 59 04	NGC 7813	14.2	0.8 x 0.3	12.5	Sb	158	
00 05 16.6	−16 28 37	NGC 7821	13.1	1.4 x 0.5	12.6	Scd pec sp	111	
00 06 26.9	−13 24 57	NGC 7828	13.9	0.9 x 0.5	12.9	Ring B	140	Double system with NGC 7829 on SE edge; MCG −2-1-24 SW 3.1′.
00 06 28.9	−13 25 17	NGC 7829	13.9	1.1 x 0.7	13.5	Ring A	130	Bright nucleus, faint envelope, located on SE edge of NGC 7828.

GALAXY CLUSTERS

RA h m s	Dec ° ′ ″	Name	Mag 10th brightest	No. Gal	Diam ′	Notes
23 54 12.0	−10 24 00	A 2670	15.7	142	34	PGC and anonymous galaxies, most stellar/almost stellar.
00 19 24.0	−17 11 00	AS 28	15.8	3	17	All members anonymous, stellar.

BRIGHT NEBULAE

RA h m s	Dec ° ′ ″	Name	Dim ′ Maj x min	Type	BC	Color	Notes
23 43 48.0	−15 17 00	Ced 211	2 x 1	E			

PLANETARY NEBULAE

RA h m s	Dec ° ′ ″	Name	Diam ″	Mag (P)	Mag (V)	Mag cent ★	Alt Name	Notes
00 37 16.0	−13 42 59	PN G108.4−76.1	3	16.3			PK 108−76.1	Mag 14.9 star E 2.2′.

GALAXIES

RA h m s	Dec ° ′ ″	Name	Mag (V)	Dim ′ Maj x min	SB	Type Class	PA	Notes
22 54 35.5	−18 00 00	ESO 603-27	15.1	1.3 x 0.7	14.9	IB(s)m V-VI	36	Mag 8.63v star WSW 9.5′.
22 12 10.6	−12 45 59	IC 1433	14.7	0.7 x 0.5	13.4		111	
22 13 51.5	−10 11 33	IC 1436	14.8	0.6 x 0.4	13.3	E/S0	63	Mag 12.2 star W 2.3′.
22 16 33.3	−16 01 01	IC 1440	13.9	1.2 x 0.3	12.9	E?	78	Mag 10.00v star WNW 7.4′.
22 25 30.3	−17 14 36	IC 1445	12.7	1.5 x 1.1	13.1	SA(s)0⁻:	80	Mag 10.50v star W 10.7′.
22 29 59.7	−05 07 14	IC 1447	12.8	1.4 x 0.8	12.8	SA(r)b? II	108	Mag 9.34v star N 1.5′.
22 46 07.6	−10 22 11	IC 1451	14.2	0.5 x 0.5	12.5	Sc		Mag 10.49v star WSW 4.5′.
22 46 54.4	−13 27 01	IC 1453	14.3	0.7 x 0.6	13.2	SBb	156	Mag 9.46v star SW 10.9′.
22 56 41.5	−07 22 50	IC 1458	13.6	1.3 x 0.8	13.5	SA(rs)cd pec? III	144	Mag 9.50v star W 10.3′.
23 03 11.5	−08 59 24	IC 1464	14.6	0.7 x 0.4	13.1		129	

(Continued from previous page)

GALAXIES

RA h m s	Dec ° ′ ″	Name	Mag (V)	Dim ′ Maj x min	SB	Type Class	PA	Notes
23 08 44.8	−12 38 22	IC 1471	13.6	0.8 x 0.5	12.5	Sb	165	Mag 10.71v star E 3′.8.
23 18 46.5	−10 23 58	IC 1479	13.7	1.1 x 1.0	13.8	E:	18	MCG −2-59-9 W 2′.1.
23 26 32.2	−12 30 59	IC 1489	14.2	0.4 x 0.3	11.7		30	Located 1′.4 N of mag 9.69v star.
23 18 52.6	−10 15 37	IC 5304	13.4	1.4 x 0.9	13.5	SAB0$^-$ pec:	12	Mag 11.4 star SW 1′.7.
23 26 20.1	−17 57 25	IC 5321	13.0	1.3 x 0.8	12.9	SBa pec?	43	Mag 8.57v star SW 18′.2.
22 09 34.7	−06 06 52	MCG −1-56-2	14.2	1.6 x 1.1	14.6	SAB(r)cd: III	129	Several knots around center.
22 19 57.4	−07 40 04	MCG −1-57-1	13.8	1.0 x 1.0	13.6	SBbc		Mag 8.93v star W 3′.5.
22 23 52.8	−05 32 57	MCG −1-57-5	14.3	0.7 x 0.4	12.8	SBd	171	
22 38 13.6	−07 02 08	MCG −1-57-17	13.6	0.9 x 0.8	13.1	Sa pec:	70	
22 38 57.0	−05 51 12	MCG −1-57-18	12.6	2.4 x 0.8	13.1	SAB(s)dm? III-IV	78	Close pair of mag 8.1 stars superimposed W end.
22 55 44.8	−06 27 16	MCG −1-58-5	14.4	0.7 x 0.4	12.9	Sc	129	Mag 8.95v star SE 3′.2.
23 03 43.6	−07 13 33	MCG −1-58-15	13.5	1.2 x 0.7	13.2	I?	27	
23 04 54.8	−06 32 25	MCG −1-58-18	14.8	0.6 x 0.4	13.1	S?	87	
23 07 25.7	−05 09 28	MCG −1-58-22	14.2	1.3 x 0.4	13.4	S?	0	
22 08 47.9	−11 51 01	MCG −2-56-17	15.0	0.7 x 0.3	13.2	Sbc	69	
22 08 52.1	−10 55 45	MCG −2-56-18	15.0	0.8 x 0.5	13.9	Sb	9	
22 09 31.6	−13 36 12	MCG −2-56-19	15.1	0.7 x 0.5	13.8	SBb	150	Mag 9.58v star E 2′.4.
22 10 41.8	−10 27 54	MCG −2-56-20	15.3	0.4 x 0.4	13.4	E/S0		
22 12 50.6	−13 35 14	MCG −2-56-22	15.0	0.8 x 0.4	13.6			
22 13 24.8	−09 16 57	MCG −2-56-23	15.0	0.7 x 0.3	13.1		30	
22 14 20.6	−10 39 50	MCG −2-56-25	14.5	0.6 x 0.3	12.5	Sbc		
22 15 02.6	−11 31 01	MCG −2-56-26	13.8	1.1 x 0.5	13.1	(R′)SB(rs)0$^+$:	87	
22 16 51.3	−09 00 43	MCG −2-56-27	15.3	0.8 x 0.2	13.2		45	Mag 5.79v star N 1′.7.
22 24 00.1	−11 08 33	MCG −2-57-1	14.3	0.5 x 0.4	12.4		174	Mag 8.60v star SE 3′.2.
22 24 43.5	−13 41 13	MCG −2-57-2	14.3	0.7 x 0.5	13.0	Sbc	3	Star on W edge; mag 9.67v star E 5′.4.
22 24 46.7	−13 39 37	MCG −2-57-3	14.5	0.7 x 0.5	12.7		135	Mag 5.77v star NNW 9′.2.
22 25 35.7	−09 57 02	MCG −2-57-4	14.5	1.1 x 0.6	13.9	Sbc	66	Mag 9.61v star E 12′.5.
22 27 12.4	−12 38 51	MCG −2-57-6	14.3	1.0 x 0.6	13.6	Sbc	177	
22 27 41.7	−09 43 39	MCG −2-57-7	14.6	1.1 x 0.2	12.8	Sd	102	
22 29 55.7	−08 16 49	MCG −2-57-8	13.9	0.9 x 0.7	13.2	Sc	156	
22 30 14.0	−10 30 19	MCG −2-57-9	14.2	0.7 x 0.5	12.9		51	
22 31 03.6	−11 42 58	MCG −2-57-12	15.1	0.5 x 0.3	12.9	Sbc	93	
22 32 34.6	−09 31 12	MCG −2-57-14	14.2	0.5 x 0.5	12.6			
22 32 47.0	−14 05 53	MCG −2-57-15	15.2	0.9 x 0.4	13.9	SB(s)d pec: III-IV	129	NGC 7302 WSW 5′.6.
22 34 48.9	−12 54 51	MCG −2-57-18	14.8	1.3 x 0.3	13.6	S?	144	NGC 7308 W 4′.2.
22 35 19.8	−12 49 08	MCG −2-57-19	14.8	0.7 x 0.3	13.0	Sb	135	Mag 8.52v star SE 7′.5.
22 36 06.5	−12 51 18	MCG −2-57-20	14.7	0.5 x 0.4	12.8	Scd	24	Mag 8.52v star W 4′.8.
22 36 14.2	−13 05 23	MCG −2-57-21	14.9	1.3 x 0.2	13.3	S?	63	
22 36 29.7	−12 34 02	MCG −2-57-22	13.8	1.3 x 0.6	13.4	(R′)SB0° pec?	165	Mag 8.74v star SW 4′.4.
22 36 46.5	−12 32 49	MCG −2-57-23	13.9	1.0 x 0.3	12.4	S?	165	= **Mkn 915**.
22 36 54.0	−12 33 44	MCG −2-57-24	15.1	0.7 x 0.5	13.8	Sb	87	
22 37 14.6	−09 52 31	MCG −2-57-25	14.4	0.7 x 0.5	13.1	Sb	3	
22 38 20.5	−13 08 13	MCG −2-57-26	14.7	1.0 x 0.7	14.1	SB(s)m IV-V	144	
22 38 55.9	−12 34 45	MCG −2-57-27	14.9	0.4 x 0.3	12.5		102	
22 42 29.1	−11 28 21	MCG −2-57-28	14.4	0.9 x 0.3	12.8	S0/a	129	Mag 9.00v star E 7′.8.
22 42 51.8	−10 39 33	MCG −2-57-29	14.4	0.7 x 0.4	12.9		3	Very small anonymous galaxy SE edge.
22 46 24.7	−14 10 50	MCG −2-58-3	14.6	1.9 x 0.3	13.8	Sd: sp	111	Mag 10.41v star SE 8′.1.
22 53 12.9	−10 29 52	MCG −2-58-7	14.8	0.7 x 0.5	13.5	Sd	156	
22 54 13.6	−11 11 00	MCG −2-58-8	14.2	0.7 x 0.2	11.9	Sc	153	
22 54 22.4	−10 10 28	MCG −2-58-9	14.9	0.4 x 0.4	12.7	SBm		
22 56 50.9	−08 58 04	MCG −2-58-11	13.7	2.1 x 0.5	13.6	SBc pec?	69	Located between a pair of stars, mags 10.76v and 9.86v; **MCG −2-58-12** perpendicular on W end.
23 00 08.6	−08 39 56	MCG −2-58-14	14.5	0.9 x 0.3	13.0	Sbc	3	
23 00 16.4	−12 57 47	MCG −2-58-17	14.8	1.1 x 0.3	13.4	Sc	120	NGC 7450 ENE 8′.1.
23 00 13.2	−12 38 51	MCG −2-58-18	14.5	0.5 x 0.4	12.6		15	
23 04 43.7	−08 41 12	MCG −2-58-22	14.3	0.7 x 0.5	13.0	Sc	96	Faint anonymous galaxy S 1′.0 .
23 05 22.2	−10 13 24	MCG −2-58-23	13.9	0.7 x 0.5	12.6	Sc	9	
23 06 58.2	−09 17 10	MCG −2-58-24	13.7	0.9 x 0.7	13.3	E/S0	12	
23 08 07.1	−10 32 34	MCG −2-59-1	14.8	1.1 x 0.7	14.3	SB(s)bc pec: II	72	
23 08 15.8	−12 11 42	MCG −2-59-2	14.7	0.9 x 0.4	13.4	Sc	27	
23 10 22.3	−08 41 23	MCG −2-59-4	14.2	1.9 x 0.2	13.0	SB(s)c pec sp	90	MCG −2-59-5 S edge; mag 9.05v star NE 8′.6.
23 10 20.8	−08 41 41	MCG −2-59-5	15.5	0.5 x 0.2	12.8		123	Located on S edge of MCG −2-59-4.
23 13 57.2	−11 30 13	MCG −2-59-6	14.7	0.7 x 0.7	13.8	Sc		
23 14 17.3	−09 50 12	MCG −2-59-7	15.2	1.5 x 0.5	14.7	SB(s)m: sp IV-V	123	Mag 7.53v star NNE 10′.8.
23 17 27.3	−10 01 52	MCG −2-59-8	14.3	1.0 x 0.6	13.6	Sbc	120	
23 18 38.2	−10 24 32	MCG −2-59-9	14.9	1.4 x 0.2	13.4	Sb pec sp	12	IC 1479 E 2′.1.
23 22 35.4	−13 05 38	MCG −2-59-13	13.6	1.4 x 0.7	13.5	Sbc	81	
23 26 41.3	−11 37 03	MCG −2-59-17	14.3	1.2 x 0.4	13.3	S?	72	
23 26 53.6	−13 04 21	MCG −2-59-18	14.8	0.7 x 0.4	13.3		135	
22 12 24.9	−17 43 10	MCG −3-56-10	13.3	0.9 x 0.9	12.9	Sa		
22 13 06.1	−16 38 35	MCG −3-56-11	13.9	0.6 x 0.5	12.4	Sb	6	Mag 9.08v star N 1′.3.
22 20 12.9	−15 56 55	MCG −3-57-1	13.8	1.5 x 0.7	13.7	(R′)SB(r)b?	75	Mag 10.86v star E 8′.4.
22 28 46.9	−15 04 10	MCG −3-57-8	13.7	1.7 x 1.3	14.4	SA(s)d: III-IV	100	Stellar nucleus; mag 12.0 star E 5′.2.
22 30 08.9	−14 18 49	MCG −3-57-12	14.2	1.0 x 0.5	13.3	Sab	153	
22 31 13.3	−17 19 41	MCG −3-57-16	14.1	0.4 x 0.4	11.9			
22 31 55.8	−17 08 49	MCG −3-57-18	15.0	1.2 x 0.6	14.5	SB(s)c II	159	
22 31 58.5	−15 26 26	MCG −3-57-19	14.0	0.7 x 0.7	13.1			
22 34 04.2	−14 19 39	MCG −3-57-22	14.1	1.1 x 1.1	14.2	SB(s)c II-III		
22 35 09.4	−14 52 47	MCG −3-57-23	14.4	0.9 x 0.4	13.2	Sbc	66	
22 35 35.5	−15 12 41	MCG −3-57-24	13.9	0.5 x 0.5	12.2	D		
22 47 42.3	−17 20 58	MCG −3-58-3	14.0	1.1 x 0.5	13.2	S?	160	Star on SW edge.
22 47 48.0	−17 06 03	MCG −3-58-4	14.4	1.3 x 0.2	12.7	S?	108	Mag 9.25v star NW 3′.7.
22 49 05.1	−16 35 36	MCG −3-58-6	13.9	0.6 x 0.3	12.1	E	156	Located at the SE edge of a bright star.
22 52 10.8	−14 52 07	MCG −3-58-8	13.6	1.0 x 0.9	13.6	E/S0	36	
22 54 09.3	−15 14 12	MCG −3-58-11	14.5	0.7 x 0.5	13.2	Sbc	63	**MCG −3-58-10** SSW 2′.9.
22 54 57.0	−16 07 58	MCG −3-58-12	15.0	1.4 x 1.0	15.2	S?	15	
22 56 18.8	−16 34 39	MCG −3-58-13	13.5	0.9 x 0.3	12.0	Sab	57	
22 57 26.5	−17 53 09	MCG −3-58-14	14.1	0.9 x 0.5	13.1	Sa	108	
23 00 19.6	−16 26 52	MCG −3-58-15	15.9	0.5 x 0.5	14.2			Mag 7.82v star WNW 8′.5.

(Continued from previous page)

122 GALAXIES 122

RA h m s	Dec ° ′ ″	Name	Mag (V)	Dim ′ Maj x min	SB	Type Class	PA	Notes
23 00 27.4	−15 22 13	MCG −3-58-16	14.5	0.5 x 0.3	12.3	Sb	108	
23 00 32.5	−14 50 04	MCG −3-58-17	14.6	0.5 x 0.4	12.7		153	
23 11 36.8	−15 28 37	MCG −3-59-1	15.2	1.4 x 0.2	13.7	Sd? sp	174	
23 21 49.3	−16 43 09	MCG −3-59-5	14.2	0.4 x 0.4	12.1	S0		
23 23 51.9	−16 27 25	MCG −3-59-7	14.7	1.0 x 0.3	13.2	Sc	39	
22 10 11.5	−16 39 42	NGC 7218	12.0	2.5 x 1.1	12.9	SB(rs)cd III	24	Very knotty.
22 14 13.2	−17 04 25	NGC 7230	14.1	0.9 x 0.9	13.7	SA(s)bc II-III:		Located inside SW edge of small galaxy cluster **A 3847**; **PGC 68361** NNE 5′.7.
22 15 01.5	−05 03 13	NGC 7239	13.8	1.1 x 0.7	13.4	SAB0$^-$	75	
22 17 42.7	−15 34 17	NGC 7246	12.8	1.6 x 0.8	12.9	(R)SA(r)a:	162	Star on N edge.
22 20 27.3	−15 46 25	NGC 7251	12.6	1.9 x 1.7	13.7	(R′)SA(rs)a?	0	
22 23 08.2	−15 32 32	NGC 7255	14.1	1.3 x 0.4	13.2	S0/a: sp	123	
22 25 46.7	−13 10 00	NGC 7269	13.7	1.0 x 0.6	13.0	Sb II	140	
22 30 50.6	−14 11 14	NGC 7298	13.7	1.3 x 1.0	13.8	SA(s)c: I-II	5	
22 31 00.0	−14 00 11	NGC 7300	12.9	2.0 x 1.0	13.5	SAB(rs)b: I-II	160	
22 30 34.9	−17 34 29	NGC 7301	13.4	1.0 x 0.5	12.4	SB(s)ab pec:	1	
22 32 23.7	−14 07 13	NGC 7302	12.3	1.8 x 1.1	12.9	SA(s)0$^-$:	97	Mag 9.23v star S 3′.0; MCG −2-57-15 ENE 5′.6.
22 34 32.1	−12 56 03	NGC 7308	13.7	1.3 x 0.9	13.7	(R′)SAB(r)0$^-$?	145	Bright, diffuse center; MCG −2-57-18 E 4′.2.
22 34 20.6	−10 21 27	NGC 7309	12.5	1.9 x 1.8	13.6	SAB(rs)c I-II		
22 46 04.1	−11 00 06	NGC 7371	11.5	2.0 x 2.0	12.9	(R)SA(r)0/a: III		Bright nucleus, multi-branching arms.
22 47 47.6	−11 49 00	NGC 7378	12.7	1.4 x 0.9	12.8	(R′)SB(r)ab?	175	Dark patches N and S of nucleus.
22 51 37.7	−05 33 31	NGC 7393	12.9	1.9 x 0.9	13.4	SB(rs)c pec	88	Brightest part offset to E. Dark patch E of nucleus.
22 52 39.3	−09 16 04	NGC 7399	13.7	1.0 x 0.6	13.1	S0/a	150	
22 53 56.4	−06 34 47	NGC 7406	13.6	1.1 x 0.6	13.0	Sbc	75	
22 55 41.3	−05 29 43	NGC 7416	12.4	3.2 x 0.7	13.1	SB(r)b II-III:	110	Mag 8.97v star S 6′.0.
22 57 15.6	−10 57 00	NGC 7425	14.2	0.9 x 0.6	13.4	S0/a	60	
22 59 29.3	−07 03 21	NGC 7441	14.3	1.2 x 1.1	14.5	(R′)SAB(s)0$^+$?	5	
23 00 09.0	−12 48 30	NGC 7443	12.6	1.5 x 0.6	12.3	SB(s)0$^+$: sp	36	Mag 9.69v star WNW 8′.9; NGC 7444 on S edge.
23 00 08.9	−12 50 04	NGC 7444	12.8	1.7 x 0.7	12.9	SB(r)0°? sp	1	
23 00 48.0	−12 55 08	NGC 7450	13.1	1.8 x 1.8	14.3	(R)SB(r)a		MCG −2-58-17 WSW 8′.1.
23 08 05.9	−05 57 59	NGC 7491	13.8	0.9 x 0.6	13.0	Sbc	172	Located 2′.5 N of mag 11.20v star.
23 17 12.0	−06 54 44	NGC 7596	14.2	1.0 x 0.5	13.3	S0	34	
23 18 53.9	−07 34 50	NGC 7600	11.9	2.5 x 1.0	12.8	S0$^-$ sp	70	
23 19 05.4	−08 29 09	NGC 7606	10.8	5.4 x 2.1	13.2	SA(s)b I-II	145	Many dark lanes and knots.
23 24 06.9	−11 51 40	NGC 7646	13.3	1.0 x 0.6	12.6	Sc	135	Star on NW end.
23 27 15.1	−09 23 12	NGC 7665	13.3	0.7 x 0.7	12.4	Sm III:		Mag 9.90v star SE 2′.7.

GALAXY CLUSTERS

RA h m s	Dec ° ′ ″	Name	Mag 10th brightest	No. Gal	Diam ′	Notes
22 31 42.0	−08 26 00	A 2448	16.0	36	16	All members anonymous, faint, stellar.
22 36 36.0	−15 39 00	A 2459	16.0	33	11	All members anonymous, faint, stellar.
22 39 06.0	−17 21 00	A 2462	16.2	40	16	All members anonymous, faint, stellar/almost stellar.
22 58 54.0	−07 36 00	A 2511	16.0	31	16	All members anonymous, faint, stellar/almost stellar.
23 03 36.0	−10 34 00	A 2525	16.0	31	20	All members anonymous, stellar.

GLOBULAR CLUSTERS

RA h m s	Dec ° ′ ″	Name	Total V m	B ★ V m	HB V m	Diam ′	Conc. Class Low = 12 High = 1	Notes
23 08 26.7	−15 36 41	NGC 7492	11.2	15.5	17.6	4.2	12	

123 GALAXIES 123

RA h m s	Dec ° ′ ″	Name	Mag (V)	Dim ′ Maj x min	SB	Type Class	PA	Notes
20 56 52.8	−16 35 10	IC 1337	13.8	0.8 x 0.6	12.9	SAB(r)b pec: I-II	178	Mag 9.73v star S 6′.8.
20 57 55.8	−17 56 35	IC 1339	13.2	1.4 x 0.9	13.3	SAB(rs)b pec: II	55	Mag 9.45v star NW 10′.0.
21 00 16.8	−13 58 38	IC 1341	14.3	0.7 x 0.5	13.0	S0	73	Mag 9.65v star N 2′.8.
21 00 25.5	−14 29 47	IC 1342	13.9	0.8 x 0.3	12.2	S?	78	Mag 10.70v star N 1′.9; mag 8.79v star NW 2′.8.
21 01 16.6	−13 22 50	IC 1344	13.7	1.0 x 0.4	12.5		54	Mag 10.18v star W 6′.5.
21 01 37.4	−13 57 42	IC 1346	14.1	1.1 x 0.6	13.5		102	Round, almost stellar anonymous galaxy N 1′.9.
21 01 44.5	−13 18 48	IC 1347	14.4	0.8 x 0.6	13.5	S0^o pec:	18	Mag 10.41v star W 1′.3.
21 01 52.3	−13 51 12	IC 1350	15.2	0.7 x 0.4	13.7		171	Mag 10.74v star WNW 4′.7.
21 02 53.1	−15 48 43	IC 1356	14.6	1.1 x 0.8	14.3		168	Mag 10.06v star W 3′.0.
21 05 57.2	−10 43 00	IC 1357	14.3	1.2 x 0.5	13.5	(R′)SB(s)a:	39	Mag 8.51v star N 6′.5.
21 53 09.1	−13 20 53	IC 1408	13.9	1.0 x 0.4	12.8	S0	33	Mag 11.4 star E 2′.0; mag 11.7 star NE 2′.3.
21 58 18.4	−17 10 37	IC 1412	13.5	0.9 x 0.4	12.3	(R′)SB0$^+$ pec	99	Mag 12.7 star SSW 3′.2.
22 00 21.9	−13 08 54	IC 1417	13.7	1.4 x 0.4	12.9	Sb? sp II-III?	108	NGC 7171 SE 12′.3.
22 03 04.1	−09 58 42	IC 1421	14.8	0.4 x 0.4	12.7			Stellar; mag 9.98v star NE 12′.5; stellar anonymous galaxy N 3′.6, could be **IC 1419**?
22 07 39.6	−13 30 48	IC 1431	14.2	0.7 x 0.7	13.3	Sa		Mag 8.80v star SE 5′.2.
22 12 10.6	−12 45 59	IC 1433	14.7	0.7 x 0.5	13.4		111	
22 13 51.5	−10 11 33	IC 1436	14.8	0.6 x 0.4	13.3	E/S0	63	Mag 12.2 star W 2′.3.
21 02 31.9	−16 49 04	IC 5078	12.7	4.1 x 0.8	13.9	SA(s)c: III	85	Bright star on N edge; MCG −3-53-24 E 8′.1.
21 02 03.6	−06 17 52	MCG −1-53-20	13.1	1.4 x 1.2	13.6	SB(rs)b II	114	
21 02 14.9	−05 06 47	MCG −1-53-21	14.9	0.7 x 0.5	13.7	SBb	114	
21 05 21.0	−07 50 47	MCG −1-53-22	13.8	0.8 x 0.5	12.7	(R′)SB(s)bc? II-III	135	MCG −1-53-23 and mag 9.85v star N 5′.7.
21 05 23.3	−07 45 07	MCG −1-53-23	14.4	1.6 x 0.2	13.0	SB(r)a pec?	18	Mag 9.85v star N 1′.2.
21 17 07.2	−06 43 30	MCG −1-54-8	14.4	0.6 x 0.4	12.7		30	**MCG −1-54-7** SW 2′.0.
21 19 43.2	−07 33 20	MCG −1-54-11	13.6	1.0 x 0.8	13.2	SB(s)dm pec III	6	
21 32 09.8	−05 43 43	MCG −1-55-1	14.3	0.8 x 0.8	13.6	Sbc		Mag 2.89v star "Sadalsuud", Beta Aquarii NW 13′.0.
21 47 35.4	−06 27 21	MCG −1-55-10	14.4	0.9 x 0.2	12.4	Sb	81	Mag 7.55v star E 6′.3.
21 49 05.4	−07 16 14	MCG −1-55-13	14.4	0.9 x 0.7	13.8	Sb	84	Mag 8.75v star SW 8′.7.
22 09 34.7	−06 06 52	MCG −1-56-2	14.2	1.6 x 1.1	14.6	SAB(r)cd: III	129	Several knots around center.
20 56 23.0	−14 18 20	MCG −2-53-14	14.4	0.6 x 0.4	12.7	Sc	177	
20 58 50.0	−14 02 24	MCG −2-53-15	14.1	1.0 x 1.0	13.9			Almost stellar nucleus; mag 9.19v star SE 4′.7.
21 00 39.6	−13 13 37	MCG −2-53-17	13.8	0.9 x 0.8	13.3	SBc	81	

(Continued from previous page)

GALAXIES

RA h m s	Dec ° ′ ″	Name	Mag (V)	Dim ′ Maj x min	SB	Type Class	PA	Notes
21 02 36.8	−13 47 52	MCG −2-53-22	14.3	1.5 x 0.2	12.9	Sb	156	Mag 11.9 star NW 2′0.
21 03 02.0	−14 15 39	MCG −2-53-23	14.0	1.1 x 1.0	13.9	SAB(s)ab: II-III	39	
21 05 54.9	−11 33 48	MCG −2-53-25	14.2	0.6 x 0.6	12.9			
21 08 06.3	−12 55 07	MCG −2-54-1	14.2	0.9 x 0.4	13.0		6	Mag 7.98v star NE 3′2.
21 11 43.0	−14 04 01	MCG −2-54-2	14.6	0.7 x 0.4	13.1	Scd	126	Mag 8.97v star ENE 3′8.
21 13 08.3	−10 49 40	MCG −2-54-3	14.0	0.8 x 0.6	13.1	Sc	147	
21 20 23.5	−13 10 17	MCG −2-54-5	14.4	1.1 x 0.2	12.6	Sc: sp	147	
21 22 48.2	−11 26 39	MCG −2-54-6	14.4	0.7 x 0.3	12.5		30	
21 23 29.4	−11 07 39	MCG −2-54-7	15.0	0.8 x 0.6	14.1		9	
21 25 57.5	−09 01 57	MCG −2-54-8	14.0	0.8 x 0.5	12.9		39	
21 27 06.0	−09 31 53	MCG −2-54-9	14.0	0.9 x 0.4	12.8		168	
21 41 41.5	−10 38 12	MCG −2-55-1	13.0	0.9 x 0.5	12.0	(R′)SB0°:	135	Mag 10.0 star on NW edge.
21 48 15.0	−13 15 09	MCG −2-55-3	14.7	0.9 x 0.8	14.2	SBbc	69	Mag 10.10v star SW 5′0.
21 51 18.7	−12 17 09	MCG −2-55-4	14.7	0.7 x 0.5	13.4	Sd	177	Mag 9.54v star N 3′9.
21 52 32.5	−10 39 19	MCG −2-55-5	15.0	1.1 x 0.2	13.2	S?	120	Mag 6.88v star NE 8′9.
21 52 50.2	−10 30 41	MCG −2-55-6	15.1	1.5 x 0.2	13.6	S?	48	Mag 11.7 star W 1′8; mag 6.88v star SE 4′0.
21 55 28.4	−12 02 42	MCG −2-55-8	14.3	0.7 x 0.5	13.0	Sbc	156	Mag 8.06v star NW 7′9.
21 59 26.9	−12 14 55	MCG −2-56-1	14.3	0.7 x 0.5	13.0		42	
22 00 02.3	−13 47 44	MCG −2-56-2	14.3	0.4 x 0.4	12.2	S0		Mag 7.55v star ENE 9′2.
22 00 28.3	−13 49 39	MCG −2-56-4	14.4	0.7 x 0.5	13.1	SBc	138	Mag 7.55v star NNE 5′5.
22 01 58.2	−10 16 19	MCG −2-56-6	14.7	0.5 x 0.4	12.8		135	
22 01 59.5	−12 11 27	MCG −2-56-7	14.6	2.2 x 2.0	16.1	SB(s)dm III-IV	135	Mag 9.88v star E 11′6.
22 02 06.6	−09 20 07	MCG −2-56-8	15.6	0.7 x 0.2	13.3	Sbc		Mag 8.23v star N 3′3.
22 03 28.1	−10 43 02	MCG −2-56-10	14.6	0.8 x 0.4	13.2		9	
22 04 09.4	−11 21 41	MCG −2-56-11	14.6	1.0 x 0.4	13.4	Sb	42	
22 04 35.0	−10 54 42	MCG −2-56-12	15.8	0.9 x 0.2	13.8	Sbc	114	Mag 9.08v star NNE 2′0.
22 07 11.3	−12 12 19	MCG −2-56-13	14.8	0.5 x 0.2	12.1	Sb	102	
22 07 25.7	−10 42 44	MCG −2-56-14	14.6	0.7 x 0.2	12.4	Sb	3	
22 07 46.8	−10 44 46	MCG −2-56-16	14.7	1.0 x 0.2	12.8	Sd	81	
22 08 47.9	−11 51 01	MCG −2-56-17	15.0	0.7 x 0.3	13.2	Sbc	69	
22 08 52.1	−10 55 45	MCG −2-56-18	15.0	0.8 x 0.5	13.9	Sb	9	
22 09 31.6	−13 36 12	MCG −2-56-19	15.1	0.7 x 0.5	13.8	SBb	150	Mag 9.58v star E 2′4.
22 10 41.8	−10 27 54	MCG −2-56-20	15.3	0.4 x 0.4	13.4	E/S0		
22 12 50.6	−13 35 14	MCG −2-56-22	15.0	0.8 x 0.4	13.6			
22 13 24.8	−09 16 57	MCG −2-56-23	15.0	0.7 x 0.3	13.1		30	
22 14 20.6	−10 39 50	MCG −2-56-25	14.5	0.6 x 0.3	12.5	Sbc		
22 15 02.6	−11 31 01	MCG −2-56-26	13.8	1.1 x 0.5	13.1	(R′)SB(rs)0^+:	87	
20 59 47.6	−16 38 19	MCG −3-53-15	14.5	0.9 x 0.4	13.3	(R$'_1$)SB(rs)a	138	Bright knot or star SE end.
20 59 56.0	−15 24 08	MCG −3-53-16	14.8	0.4 x 0.4	12.7	Sbc		
21 01 44.0	−14 33 49	MCG −3-53-19	13.8	1.0 x 0.6	13.1	SB(rs)0^+ pec:	51	
21 03 05.8	−14 31 27	MCG −3-53-23	14.8	0.7 x 0.2	12.5		69	
21 03 05.3	−16 48 08	MCG −3-53-24	15.0	0.4 x 0.4	12.8			Stellar; IC 5078 W 8′1.
21 03 09.3	−14 31 29	MCG −3-53-25	13.2	0.7 x 0.4	11.7	SAB(rs)0° pec	177	
21 05 56.5	−14 20 39	MCG −3-53-26	14.3	0.5 x 0.4	12.4		39	
21 06 29.5	−15 26 27	MCG −3-53-27	14.6	0.9 x 0.9	14.2	SA(s)d: III-IV		
21 38 35.3	−17 07 22	MCG −3-55-4	14.2	0.7 x 0.3	12.4	S0/a	123	Mag 7.26v star SW 10′5.
21 42 16.1	−16 56 16	MCG −3-55-6	14.2	0.7 x 0.7	13.2	S		
22 06 23.6	−15 25 04	MCG −3-56-6	14.7	1.3 x 0.3	13.5	S?	120	
22 12 24.9	−17 43 10	MCG −3-56-10	13.3	0.9 x 0.9	12.9	Sa		
22 13 06.1	−16 38 35	MCG −3-56-11	13.9	0.6 x 0.5	12.4	Sb	6	Mag 9.08v star N 1′3.
21 04 39.6	−12 20 18	NGC 7010	13.0	1.9 x 1.0	13.7	E^+ pec:	36	
21 19 51.4	−08 46 55	NGC 7051	13.0	1.3 x 1.0	13.1	SB(r)a pec	162	Pair of stars W 2′2, brightest mag 10.33v.
21 26 42.4	−06 59 45	NGC 7065	13.4	1.0 x 0.8	13.0	SB(r)ab? II-III:	25	
21 26 57.8	−07 01 18	NGC 7065A	13.1	1.0 x 0.9	12.8	SAB(r)c pec?	6	Stellar nucleus; mag 8.93v star N 2′7.
21 29 25.9	−11 29 19	NGC 7073	13.5	0.9 x 0.8	13.0	Sa/b II-III		
21 41 53.7	−06 42 33	NGC 7111	14.1	1.4 x 0.9	14.4	E	35	
21 44 33.2	−06 31 26	NGC 7120	14.0	0.8 x 0.4	12.6	Sb I-II:	141	Faint, elongated anonymous galaxy SW 1′6.
21 47 36.3	−13 10 58	NGC 7131	13.9	1.7 x 1.0	14.4	(R′)SA(rs)0°?	115	Mag 9.03v star NW 4′3.
21 56 56.6	−11 39 35	NGC 7158	14.5	0.6 x 0.3	12.5		90	
21 59 26.3	−16 30 46	NGC 7165	13.2	1.0 x 0.9	13.0	SB(r)ab	67	
22 01 26.4	−05 26 00	NGC 7170	13.8	1.2 x 0.8	13.6	SA0^-?	158	
22 01 01.7	−13 16 09	NGC 7171	12.2	2.6 x 1.5	13.5	SB(rs)b I-II	125	Dark lane S of nucleus; IC 1417 NW 12′3.
22 06 21.9	−08 05 24	NGC 7211	14.3	0.9 x 0.7	13.7	S0	39	
22 10 11.5	−16 39 42	NGC 7218	12.0	2.5 x 1.1	12.9	SB(rs)cd III	24	Very knotty.
22 14 13.2	−17 04 25	NGC 7230	14.1	0.9 x 0.9	13.7	SA(s)bc II-III:		Located inside SW edge of small galaxy cluster **A 3847**; **PGC 68361** NNE 5′7.
22 15 01.5	−05 03 13	NGC 7239	13.8	1.1 x 0.7	13.4	SAB0^-	75	

GALAXY CLUSTERS

RA h m s	Dec ° ′ ″	Name	Mag 10th brightest	No. Gal	Diam ′	Notes
20 58 12.0	−07 45 00	A 2331	16.3	30	13	All members anonymous, faint and stellar.
21 42 48.0	−06 52 00	A 2366	15.9	47	13	**MCG −1-55-4** N of center 2′3; all other members anonymous, stellar.
21 52 00.0	−15 38 00	A 2382	16.0	50	12	All members anonymous, faint and stellar.
21 57 30.0	−07 47 00	A 2399	15.6	52	22	All members anonymous, stellar/almost stellar.
22 02 06.0	−09 53 00	A 2410	16.0	54	22	All members anonymous, stellar.
22 05 24.0	−05 35 00	A 2415	15.9	40	13	All members anonymous; several prominent GX near center.

OPEN CLUSTERS

RA h m s	Dec ° ′ ″	Name	Mag	Diam ′	No. ★	B ★	Type	Notes
20 58 56.0	−12 38 00	NGC 6994	8.9	1.4	4	11.1	ast	= **M 73**. Few stars; small brightness range; not well detached. Not a cluster. Composed of four stars with different proper motions.

123 (Continued from previous page) PLANETARY NEBULAE 123

RA h m s	Dec ° ′ ″	Name	Diam ″	Mag (P)	Mag (V)	Mag cent ★	Alt Name	Notes
21 04 10.9	−11 21 49	NGC 7009	35	8.3	8.0	12.7	PK 37−34.1	**Saturn Nebula**. Bright, oval-shaped. The extensions ("antennae") which give rise to its nickname are faint, and require fairly high power to see.

124 GALAXIES 124

RA h m s	Dec ° ′ ″	Name	Mag (V)	Dim ′ Maj x min	SB	Type Class	PA	Notes
20 03 01.5	−17 13 58	IC 1309	14.2	0.9 x 0.7	13.6	SAB(rs)bc pec II-III	35	Mag 7.58v star S 7′0.
20 18 43.8	−16 56 46	IC 1313	14.1	2.0 x 1.8	15.3	(R′)SB(r)ab	29	Strong dark areas E and W of core.
20 32 12.5	−09 03 23	IC 1324	13.5	1.5 x 1.4	14.2	(R′)SA(rs)0/a:	39	Mag 9.78v star S 3′3.
20 46 15.1	−14 01 27	IC 1330	14.0	1.2 x 0.4	13.0	Sc? sp	120	Mag 9.60v star SSW 6′2.
20 47 48.9	−09 59 48	IC 1331	13.7	1.8 x 0.6	13.6	(R′)SB(r)0/a:	85	Mag 8.49v star SW 4′7; MCG −2-53-4 NNW 3′0.
20 51 51.2	−13 42 46	IC 1332	13.5	0.9 x 0.6	12.7	Sb	66	Mag 8.37v star SW 4′8.
20 52 17.2	−16 17 10	IC 1334	13.6	1.4 x 0.6	13.2	SAa:	69	Mag 9.48v star NE 10′7.
20 56 52.8	−16 35 10	IC 1337	13.8	0.8 x 0.6	12.9	SAB(r)b pec: I-II	178	Mag 9.73v star S 6′8.
20 57 55.8	−17 56 35	IC 1339	13.2	1.4 x 0.9	13.3	SAB(rs)b pec: II	55	Mag 9.45v star NW 10′0.
21 00 16.8	−13 58 38	IC 1341	14.3	0.7 x 0.5	13.0	S0	73	Mag 9.65v star N 2′8.
21 00 25.5	−14 29 47	IC 1342	13.9	0.8 x 0.3	12.2	S?	78	Mag 10.70v star N 1′9; mag 8.79v star NW 2′8.
21 01 16.6	−13 22 50	IC 1344	13.7	1.0 x 0.4	12.5		54	Mag 10.18v star W 6′5.
21 01 37.4	−13 57 42	IC 1346	14.1	1.1 x 0.6	13.5		102	Round, almost stellar anonymous galaxy N 1′9.
21 01 44.5	−13 18 48	IC 1347	14.4	0.8 x 0.6	13.5	S0° pec:	18	Mag 10.41v star W 1′3.
21 01 52.3	−13 51 12	IC 1350	15.2	0.7 x 0.4	13.7		171	Mag 10.74v star WNW 4′7.
21 02 53.1	−15 48 43	IC 1356	14.6	1.1 x 0.8	14.3		168	Mag 10.06v star W 3′0.
20 45 15.0	−05 37 22	IC 5050	13.5	1.3 x 0.9	13.5	SAB(r)b: II-III	75	Mag 10.58v star SE 4′8.
21 02 31.9	−16 49 04	IC 5078	12.7	4.1 x 0.8	13.9	SA(s)c: III	85	Bright star on N edge; MCG −3-53-24 E 8′1.
20 26 26.0	−05 04 11	MCG −1-52-2	14.2	0.5 x 0.4	12.3		3	
20 26 39.8	−05 04 31	MCG −1-52-3	14.3	0.9 x 0.3	12.7	S?	96	
20 29 15.7	−08 00 05	MCG −1-52-4	14.1	0.8 x 0.7	13.3	SB(r)b: II-III	24	Mag 9.19v star S 1′5.
20 30 20.0	−07 53 29	MCG −1-52-5	13.4	1.1 x 0.5	12.6	SB(s)0$^+$?	15	
20 35 23.9	−06 14 41	MCG −1-52-8	14.0	2.2 x 0.2	13.0	Sc: sp	105	Mag 11.4 star S 1′2; mag 8.81v star S 2′6.
20 38 54.4	−05 38 25	MCG −1-52-16	12.9	2.7 x 0.8	13.6	(R′)Sa:	24	Mag 12.8 star SW edge.
20 42 27.0	−07 18 33	MCG −1-52-17	13.3	1.0 x 1.0	13.2	Sbc II		
20 45 14.5	−05 29 05	MCG −1-53-4	13.3	0.8 x 0.6	12.3		120	Mag 10.53v star ESE 7′0.
20 45 40.0	−05 35 26	MCG −1-53-7	14.6	0.6 x 0.5	13.2	SBbc	174	Mag 13.6 star on N edge; mag 10.53v star N 4′2.
20 48 13.6	−05 47 47	MCG −1-53-11	14.9	0.9 x 0.5	13.9	SB?	171	
20 49 52.3	−07 01 18	MCG −1-53-12	14.7	3.2 x 0.3	14.5	Sc sp	0	Tight trio of stars, mags 13-15, NE 1′9.
20 51 19.9	−07 19 00	MCG −1-53-13	14.5	0.7 x 0.5	13.2	Sc	54	Star on SE edge.
21 02 03.6	−06 17 52	MCG −1-53-20	13.1	1.4 x 1.2	13.6	SB(rs)b II	114	
21 02 14.9	−05 06 47	MCG −1-53-21	14.9	0.7 x 0.5	13.7	SBb	114	
19 48 48.1	−10 34 16	MCG −2-50-7	14.6	0.8 x 0.2	12.5	Sb	12	
19 49 23.3	−10 57 59	MCG −2-50-8	14.1	0.9 x 0.7	13.4		9	Mag 8.34v star E 3′5.
20 05 58.9	−10 24 52	MCG −2-51-2	14.8	0.6 x 0.5	13.3		12	
20 06 00.7	−10 26 06	MCG −2-51-3	14.4	0.8 x 0.6	13.4	SBb	6	
20 15 40.0	−13 37 18	MCG −2-51-4	13.1	1.3 x 0.5	12.5		165	
20 16 13.6	−14 21 47	MCG −2-51-5	13.4	0.9 x 0.4	12.1	SBbc	177	
20 17 06.6	−12 05 51	MCG −2-51-8	13.4	0.6 x 0.4	11.7	Sab		Elongated **MCG −2-51-7** SW 1′3.
20 21 47.8	−10 29 22	MCG −2-52-3	14.1	1.7 x 0.4	13.5	Sd: sp	0	Mag 12.4 star N 2′4.
20 24 21.0	−11 02 47	MCG −2-52-4	14.0	0.8 x 0.4	12.6	S?	84	Mag 9.25v star NW 4′8.
20 29 35.5	−12 30 17	MCG −2-52-6	14.1	1.1 x 0.5	13.3	Sb	108	Mag 7.38v star S 5′3.
20 29 40.8	−10 06 32	MCG −2-52-7	14.2	0.7 x 0.3	12.4	Sb	21	
20 30 02.5	−08 43 01	MCG −2-52-9	14.6	1.1 x 0.2	12.8		111	
20 31 42.7	−12 16 01	MCG −2-52-10	14.6	0.9 x 0.5	13.6	SB(s)m: IV-V	3	Double galaxy system.
20 32 13.2	−08 44 07	MCG −2-52-11	15.2	1.2 x 0.4	14.2	IBm pec III-IV	162	
20 32 56.5	−11 52 49	MCG −2-52-13	14.3	0.5 x 0.5	12.6			
20 33 17.9	−11 16 17	MCG −2-52-14	14.3	1.1 x 0.4	13.3	S?	114	
20 33 36.8	−10 41 05	MCG −2-52-15	14.6	1.2 x 0.4	13.6	S?	63	Two stars superimposed SE side.
20 35 16.2	−09 37 21	MCG −2-52-17	14.4	1.3 x 0.2	12.8	Sab? sp	15	Mag 7.38v star NW 4′0.
20 35 37.9	−09 25 35	MCG −2-52-18	14.1	1.1 x 0.4	13.1	Sb	39	
20 41 48.4	−13 50 48	MCG −2-52-19	13.7	0.7 x 0.7	13.0	E:		
20 43 29.6	−10 05 34	MCG −2-52-20	14.6	0.4 x 0.3	12.4	E		
20 46 51.8	−12 50 53	MCG −2-53-3	13.9	2.2 x 1.1	14.7	IB(s)m V-VI	100	**Aquarius Dwarf**. Mag 14.1 star on S edge; mag 8.82v star E 9′1.
20 47 42.4	−09 57 16	MCG −2-53-4	15.2	0.9 x 0.2	13.2	Scd	28	Mag 8.49v star S 6′0; IC 1331 SSE 3′0.
20 48 54.3	−08 27 10	MCG −2-53-7	14.5	0.8 x 0.3	12.8	SBab	48	NGC 6988 NW 7′7.
20 50 03.0	−09 49 31	MCG −2-53-8	13.4	1.0 x 0.7	12.9		36	Mag 8.82v star N 1′4.
20 51 44.5	−09 18 51	MCG −2-53-9	14.1	1.3 x 0.6	13.7	SB(s)m pec: III	108	Mag 8.95v star SE 6′5.
20 51 41.9	−12 12 25	MCG −2-53-10	14.5	1.0 x 0.2	12.6		141	
20 54 48.2	−08 39 00	MCG −2-53-12	14.3	1.0 x 0.7	13.8	SBb	63	
20 55 24.0	−13 26 38	MCG −2-53-13	13.9	1.3 x 0.5	13.2	SAB(rs)d: III-IV	42	Mag 9.13v star E 6′1.
20 56 23.0	−14 18 20	MCG −2-53-14	14.4	0.6 x 0.4	12.7	Sc	177	
20 58 50.0	−14 02 24	MCG −2-53-15	14.1	1.0 x 1.0	13.9			Almost stellar nucleus; mag 9.19v star SE 4′7.
21 00 39.6	−13 13 37	MCG −2-53-17	13.8	0.9 x 0.8	13.3	SBc	81	
21 02 36.8	−13 47 52	MCG −2-53-22	14.3	1.5 x 0.2	12.9	Sb	156	Mag 11.9 star NW 2′0.
21 03 02.0	−14 15 39	MCG −2-53-23	14.0	1.1 x 1.0	13.9	SAB(s)ab: II-III	39	
20 01 02.8	−17 03 12	MCG −3-51-1	14.6	1.0 x 0.5	13.7	SA(rl)b	24	
20 08 16.6	−17 39 36	MCG −3-51-5	14.4	0.7 x 0.7	13.5	SB(s)c		
20 22 11.3	−16 47 34	MCG −3-52-2	12.8	2.1 x 1.5	13.8	SA(rs)cd II-III	129	Mag 8.07v star sits on N edge.
20 25 29.7	−15 51 16	MCG −3-52-5	13.7	1.0 x 0.7	13.1	Sbc	3	
20 30 27.0	−16 44 31	MCG −3-52-14	15.8	1.1 x 1.1	15.9	(R′)SB(rs)c pec?		Brightest portion of galaxy along E edge.
20 37 31.0	−16 32 32	MCG −3-52-20	12.7	0.8 x 0.2	10.6	S?	123	
20 47 37.7	−15 32 04	MCG −3-53-1	14.0	0.9 x 0.5	12.9	SAB(s)m IV-V	150	Mag 8.15v star SSW 5′6.
20 51 27.1	−15 55 00	MCG −3-53-7	14.0	1.1 x 0.2	12.2	Scd? sp	114	
20 59 47.6	−16 38 19	MCG −3-53-15	14.5	0.9 x 0.4	13.3	(R'_1)SB(rs)a	138	Bright knot or star SE end.
20 59 56.0	−15 24 08	MCG −3-53-16	14.8	0.4 x 0.4	12.7	Sbc		
21 01 44.0	−14 33 49	MCG −3-53-19	13.8	1.0 x 0.6	13.1	SB(rs)0$^+$ pec:	51	
21 03 05.8	−14 31 27	MCG −3-53-23	14.8	0.7 x 0.2	12.5		69	
21 03 05.3	−16 48 08	MCG −3-53-24	15.0	0.4 x 0.4	12.8			Stellar; IC 5078 W 8′1.
21 03 09.3	−14 31 29	MCG −3-53-25	13.2	0.7 x 0.4	11.7	SAB(rs)0° pec	177	
19 44 24.5	−06 50 07	NGC 6821	13.1	1.1 x 1.1	13.1	SB(s)d:		Patchy with superimposed stars, in star rich area.

(Continued from previous page)

124 GALAXIES 124

RA h m s	Dec ° ′ ″	Name	Mag (V)	Dim ′ Maj x min	SB	Type Class	PA	Notes
19 44 58.3	−14 48 03	NGC 6822	8.7	15.5 x 13.5	14.4	IB(s)m V	6	**Barnard's Galaxy**. **IC 1308** is an emission nebulosity in object 6′ NNE of center.
19 54 32.8	−12 34 08	NGC 6835	12.5	2.3 x 0.5	12.5	SB(s)a? sp	72	
19 54 40.6	−12 41 17	NGC 6836	12.9	1.5 x 1.3	13.5	SABm III-IV	140	Diffuse center with a number of bright knots on envelope.
20 05 56.7	−09 02 28	NGC 6865	15.0	0.8 x 0.6	14.0	S0	130	
20 21 01.4	−12 15 19	NGC 6897	13.8	0.9 x 0.5	12.7	Sbc II	35	
20 21 08.2	−12 21 34	NGC 6898	13.1	1.1 x 0.7	12.7	(R′)SB(rs)a:	140	Mag 12.3 star on S edge.
20 33 41.7	−11 22 10	NGC 6931	13.5	1.1 x 0.4	12.5	Sb pec?	125	Mag 9.24v star SW 5′.3.
20 48 32.4	−08 21 38	NGC 6968	13.3	1.4 x 1.1	13.7	SA0$^-$ pec?	144	Mag 13.7 star on N edge; MCG −2-53-7 SE 7′.7.
20 52 13.1	−05 47 55	NGC 6975	14.9	0.9 x 0.2	12.9	S?	68	
20 52 26.0	−05 46 23	NGC 6976	14.0	0.8 x 0.7	13.2	SAB(r)bc? II	5	
20 52 29.8	−05 44 47	NGC 6977	13.2	1.2 x 0.9	13.2	SB(r)a pec:	162	
20 52 35.5	−05 42 42	NGC 6978	13.3	1.5 x 0.7	13.2	Sb II	125	
20 45 02.4	−11 06 20	NGC 6985A	13.8	1.3 x 0.9	13.9	SB(s)a pec:	90	Multi-galaxy system; mag 8.66v star E 2′.5.

GALAXY CLUSTERS

RA h m s	Dec ° ′ ″	Name	Mag 10th brightest	No. Gal	Diam ′	Notes
20 48 12.0	−17 48 00	A 2328	16.4	81	16	All members anonymous, faint and stellar.
20 58 12.0	−07 45 00	A 2331	16.3	30	13	All members anonymous, faint and stellar.

OPEN CLUSTERS

RA h m s	Dec ° ′ ″	Name	Mag	Diam ′	No. ★	B ★	Type	Notes
20 58 56.0	−12 38 00	NGC 6994	8.9	1.4	4	11.1	ast	= **M 73**. Few stars; small brightness range; not well detached. Not a cluster. Composed of four stars with different proper motions.

GLOBULAR CLUSTERS

RA h m s	Dec ° ′ ″	Name	Total V m	B ★ V m	HB V m	Diam ′	Conc. Class Low = 12 High = 1	Notes
20 53 27.9	−12 32 13	NGC 6981	9.2	14.2	16.9	6.6	9	= **M 72**.
19 45 14.4	−08 00 26	Pal 11	9.8		17.3	10.0	11	

PLANETARY NEBULAE

RA h m s	Dec ° ′ ″	Name	Diam ″	Mag (P)	Mag (V)	Mag cent ★	Alt Name	Notes
20 31 33.2	−07 05 18	PN G38.1−25.4	42	14.3	14.5	19.1	PK 38−25.1	= **Abell 70**; annular, small jet extends W from N edge; mag 11.02v star.

125 GALAXIES 125

RA h m s	Dec ° ′ ″	Name	Mag (V)	Dim ′ Maj x min	SB	Type Class	PA	Notes
19 29 59.6	−17 40 42	ESO 594-4	13.2	2.9 x 1.5	14.6	IB(s)m: V-VI	81	Mag 8.57v star N 7′.1.
19 37 27.8	−17 27 20	ESO 594-8	14.8	1.5 x 0.3	13.8	Sd? sp	115	Mag 8.53v star NW 13′.2.
19 43 12.6	−06 56 24	MCG −1-50-1	13.5	1.2 x 0.3	12.2	IBm? sp III	99	
19 48 48.1	−10 34 16	MCG −2-50-7	14.6	0.8 x 0.2	12.5	Sb	12	
19 49 23.3	−10 57 59	MCG −2-50-8	14.1	0.9 x 0.7	13.4		9	Mag 8.34v star E 3′.5.
19 34 20.2	−15 32 19	MCG −3-50-1	14.2	0.5 x 0.4	12.3			
19 42 24.2	−15 51 55	MCG −3-50-3	13.8	2.2 x 0.3	13.2	Sc: sp	144	Mag 10.56v star NE 6′.2.
19 42 40.7	−10 19 28	NGC 6814	11.2	3.0 x 2.8	13.4	SAB(rs)bc I-II		Bright, branching spiral arms, numerous superimposed stars and knots.
19 44 24.5	−06 50 07	NGC 6821	13.1	1.1 x 1.1	13.1	SB(s)d:		Patchy with superimposed stars, in star rich area.
19 44 58.3	−14 48 03	NGC 6822	8.7	15.5 x 13.5	14.4	IB(s)m V	6	**Barnard's Galaxy**. **IC 1308** is an emission nebulosity in object 6′ NNE of center.

OPEN CLUSTERS

RA h m s	Dec ° ′ ″	Name	Mag	Diam ′	No. ★	B ★	Type	Notes
18 48 06.0	−05 51 07	Bas 1	8.9	5.5	94	12.6	cl	Few stars; small brightness range; not well detached. Known as **Apriamasvili** cluster by discovery priority.
18 32 54.0	−06 02 00	Do 30		18			ast	Probably not a cluster. (A) Nothing obvious in 24′ field.
18 34 54.0	−06 51 00	Do 31		18			ast	Probably not a cluster. (A) Nothing obvious in 24′ field.
18 32 38.0	−16 53 00	NGC 6645	8.5	15	40	12.0	cl	Moderately rich in stars; small brightness range; not well detached.
18 33 28.3	−10 24 08	NGC 6649	8.9	6	477	13.2	cl	Moderately rich in bright and faint stars with a strong central concentration; detached.
18 36 30.6	−08 11 00	NGC 6664	7.8	12	60	9.0	cl	Moderately rich in stars; moderate brightness range; no central concentration; detached.
18 42 12.7	−06 12 45	NGC 6683	9.4	3	20	11.7	cl	Few stars; small brightness range; slight central concentration; detached; involved in nebulosity.
18 45 15.0	−09 23 00	NGC 6694	8.0	10	120	11.0	cl	= **M 26**. Moderately rich in bright and faint stars; detached.
18 50 45.8	−05 12 18	NGC 6704	9.2	6	71	12.0	cl	Moderately rich in stars; moderate brightness range; strong central concentration; detached.
18 51 04.0	−06 16 00	NGC 6705	5.8	11	682	11.0	cl	= **M 11**.
19 16 36.0	−16 16 00	NGC 6774		20	25	9.0	cl	Few stars, medium brightness range, not detached. (A) Large sparse cluster.
18 32 14.0	−12 14 12	Ru 142		5	15	13.0	cl	(A) Slight condensation in Milky Way star cloud.
18 32 43.0	−12 08 18	Ru 143		6	30	14.0	cl	
18 33 32.7	−11 25 11	Ru 144		2	10	12.0	cl	
18 32 00.7	−16 03 00	Ru 171		12	60	14.0	cl	(A) Rich faint cluster or Milky Way star cloud.
18 39 48.0	−08 28 00	Tr 34	8.6	5	87	11.2	cl	Moderately rich in stars; moderate brightness range; not well detached. Probably not a cluster.

GLOBULAR CLUSTERS

RA h m s	Dec ° ′ ″	Name	Total V m	B ★ V m	HB V m	Diam ′	Conc. Class Low = 12 High = 1	Notes
18 53 04.3	−08 42 22	NGC 6712	8.1	13.3	16.3	9.8	9	
19 45 14.4	−08 00 26	Pal 11	9.8		17.3	10.0	11	

(Continued from previous page)

BRIGHT NEBULAE

RA h m s	Dec ° ′ ″	Name	Dim ′ Maj x min	Type	BC	Color	Notes
18 32 12.0	−11 46 00	Sh2-55	20 x 20	E	5-5	1-4	Very faint diffuse nebulosity in a rich star field. There is a coarse pair of brighter stars (sep. about 30″, NW-SE) on the W edge.

DARK NEBULAE

RA h m s	Dec ° ′ ″	Name	Dim ′ Maj x min	Opacity	Notes
18 32 41.0	−09 09 00	B 100	16 x 16	5	Definite; irregular; curved NW and SE.
18 32 40.0	−08 49 00	B 101	13 x 4	5	Definite; irregular.
18 37 41.8	−13 45 00	B 102	8 x 3	5	Rather definite; elongated NE and SW.
18 39 25.7	−06 40 00	B 103	4 x 4	6	Irregular; one small star in it.
18 47 44.3	−06 55 00	B 105	0.5 x 0.5	4	Very small.
18 48 50.0	−05 04 46	B 106	2 x 2	6	Free of stars.
18 49 30.8	−05 01 00	B 107	5 x 5	6	Irregular; free of stars.
18 49 34.5	−06 18 40	B 108	3 x 3	3	Very small; not black.
18 49 36.1	−07 33 40	B 109	0.7 x 0.7	2	Irregular.
18 51 08.5	−06 40 00	B 112	20 x 20	4	Diffused dark region.
18 53 11.3	−06 57 00	B 114	6 x 6	5	Round.
18 53 19.5	−07 30 00	B 115	7.0 x 1.4	5	Very small and black.
18 53 33.0	−07 11 00	B 116	20 x 3	3	Narrow, semi-vacant region extending S from magnitude 9.29v star.
18 53 43.6	−07 25 00	B 117	1 x 1	6	Round; very black; sharply defined.
18 53 55.7	−07 26 52	B 118	2 x 2	6	Definite; more definite and blacker than B117.
18 54 39.2	−05 11 00	B 119a	30 x 30	3	Irregular dark region; liberally sprinkled with stars in its NW half; several narrow dark lanes in SE part.
19 01 31.9	−05 27 00	B 127	4 x 4	5	Irregular.
19 02 04.5	−05 18 00	B 129	5 x 5	5	Very black; sharply defined.
19 01 56.2	−05 34 00	B 130	7 x 7	5	Dusky; not well defined.
19 06 12.9	−06 54 00	B 133	10 x 3	6	Cometary.
19 06 53.0	−06 15 00	B 134	6 x 6	6	Round; dark, but not sharply defined.
18 32 10.6	−15 35 00	B 312	100 x 30	4	Large dark area.
18 35 56.8	−15 41 00	B 313	15 x 2		Narrow; curved; E and W.
18 37 04.6	−09 43 00	B 314	35 x 25	5	Irregular, dusky marking, NE and SW.
18 45 46.0	−14 12 00	B 317	30 x 5	4	Dusky; NE and SW.
18 49 42.7	−06 24 00	B 318	90 x 2	2	Long, straight, dark line.
18 52 47.1	−05 51 00	B 320	15 x 15	4	Irregular.
18 54 02.0	−11 18 00	B 321	15 x 7	3	Dusky.
19 04 26.9	−05 08 00	B 327	30 x 3	3	Curved, dusky lane; N and S.

PLANETARY NEBULAE

RA h m s	Dec ° ′ ″	Name	Diam ″	Mag (P)	Mag (V)	Mag cent ★	Alt Name	Notes
18 54 37.2	−08 49 33	IC 1295	90	15.0		15.5	PK 25−4.2	
19 16 28.4	−09 02 38	IC 4846	11	12.7	11.9	15.1	PK 27−9.1	Mag 11.0 star S 2′.8.
19 05 55.5	−05 59 33	NGC 6751	26	12.5	11.9	15.4	PK 29−5.1	Mag 11.3 star E 2′.0.
19 43 57.8	−14 09 10	NGC 6818	46	9.9	9.3	16.9	PK 25−17.1	**Little Gem Nebula**. Mag 13.5 stars on E and W edges.
18 35 48.3	−17 36 09	PN G15.4−4.5	6	13.9	13.5	14.3	PK 15−4.1	Forms a small triangle with mag 11.9 star ENE 0′.9 and mag 12.2 star.
18 36 08.3	−16 59 57	PN G16.0−4.3	13	12.5	13.5		PK 16−4.1	Mag 12.5 star NNE 0′.6; mag 11.4 star SE 2′.1.
18 37 46.3	−17 05 47	PN G16.1−4.7	10	13.3	13.5	16.2	PK 16−4.2	Mag 11.8 star SSE 0′.5; mag 11.8 star SW 1′.5.
18 33 58.6	−14 52 24	PN G17.7−2.9	7	15.5	14.7		PK 17−2.1	Mag 11.2 star S 0′.7; mag 11.7 star NNE 1′.4.
18 41 14.8	−15 33 39	PN G17.9−4.8	17	14.6	14.8	17.9	PK 17−4.1	Mag 11.7 star E 0′.3.
18 33 03.8	−13 44 19	PN G18.6−2.2	6	16.6			PK 18−2.1	Mag 11.3 star WSW 1′.2; mag 11.4 star S 2′.3.
18 34 13.8	−13 12 24	PN G19.2−2.2	1	15.0	14.3		PK 19−2.1	Close pair of mag 11-12 stars E 1′.1.
18 45 55.1	−14 27 38	PN G19.4−5.3	2	12.2	12.8	17.1	PK 19−5.1	Stellar, sandwiched between two faint stars on E and W edges; mag 11.3 star SSE 4′.0.
18 43 38.2	−13 44 49	PN G19.7−4.5	10	14.1	14.0		PK 19−4.1	Mag 12.7 star on SW edge.
18 50 44.4	−13 31 02	PN G20.7−5.9	8	14.5		14.1	PK 20−5.1	Mag 6.43v star SW 4′.6; mag 8.40v star NE 6′.2.
18 33 29.0	−11 07 26	PN G20.9−1.1	10	16.7			PK 21−1.1	Appears elongated NE-SW, which could be caused by an adjacent star?.
18 51 31.0	−13 10 38	PN G21.1−5.9	4	14.3	13.8		PK 21−5.1	Forms a triangle with a mag 11.5 star N 1′.3 and a mag 12.4 star W 1′.6.
18 44 06.5	−12 12 50	PN G21.2−3.9	17			21.0	PK 21−3.1	Mag 12.7 star N 2′.0; mag 13.1 star S 1′.6.
18 33 14.9	−10 15 20	PN G21.7−0.6	12	19.9		20.7	PK 21−0.2	Appears to be a small jet extending S; mag 12.6 star on SW edge.
18 32 41.4	−10 05 50	PN G21.8−0.4	24	>14.3		20.1	PK 21−0.1	Appears rectangular, oriented N-S; possibly a star involved on S edge?.
18 42 57.0	−11 06 53	PN G22.0−3.1	6	13.9	13.0		PK 22−3.1	Mag 13.2 star E 2′.4; mag 12.3 star NW 2′.0.
18 40 20.3	−10 39 47	PN G22.1−2.4	8	13.5	13.7	16.3	PK 22−2.1	
19 01 21.8	−11 58 20	PN G23.3−7.6	10	15.3			PK 23−7.1	Mag 12.3 star SW 1′.0; mag 10.20v star W 2′.5.
18 41 07.3	−08 55 59	PN G23.8−1.7	3				PK 23−1.2	Small triangle of mag 112-13 stars NNW 0′.9.
18 43 20.2	−09 04 49	PN G23.9−2.3	5	13.3	12.5		PK 23−2.1	Mag 9.77v star NE 2′.7.
18 54 17.7	−10 05 10	PN G24.2−5.2	21	14.9	13.8	18.7	PK 24−5.1	Forms a triangle with a mag 12.9 star W 0′.6 and a mag 13.3 star NW 0′.7.
18 47 48.8	−09 09 07	PN G24.3−3.3	5	16.7	14.8		PK 24−3.1	Mag 13.0 star NE 1′.1; mag 11.9 star W 3′.0.
18 46 34.6	−08 28 02	PN G24.8−2.7	4	18.3			PK 24−2.1	Mag 9.12v star NNW 2′.2.
19 19 17.8	−12 14 42	PN G25.0−11.6	74	15.7	15.5	18.5	PK 25−11.1	Numerous stars on Disc; mag 11.8 star S 1′.2; mag 12.6 star W 2′.0.
18 54 20.0	−08 47 33	PN G25.3−4.6		14.2	14.0		PK 25−4.1	Mag 12.7 star E 0′.4; planetary nebula IC 1295 SE 4′.7.
18 42 08.1	−06 40 56	PN G25.9−0.9	5	19.2	20.8		PK 25−0.1	Mag 13.0 star E 0′.5; mag 12.0 star N 0′.6.
18 46 24.6	−07 14 34	PN G25.9−2.1	5	16.1		16.2	PK 25−2.1	Mag 13.5 star SE 0′.5; mag 13.3 star NNW 0′.6.
19 18 19.7	−11 06 17	PN G25.9−10.9	16	13.3	13.5	14.0	PK 26−11.1	Appears “egg shaped” E-W, in star rich area.
18 45 27.6	−06 56 58	PN G26.0−1.8	3	17.4			PK 26−1.2	Mag 12.6 star W 0′.6; mag 11.7 star SSE 1′.5.
18 47 32.2	−06 54 06	PN G26.3−2.2	8	14.1	14.3	15.5	PK 26−2.1	Mag 12.2 star NNW 0′.9; mag 9.04v star E 2′.2.
18 49 44.7	−07 01 35	PN G26.5−3.0	4	15.2	14.5	16.4	PK 26−2.3	Mag 12.9 star W 0′.5; mag 11.8 star ENE 1′.6.
18 45 36.3	−06 18 43	PN G26.6−1.5	20	15.7	14.2		PK 26−1.1	Close pair of mag 11-12 stars N 1′.3.
18 48 46.5	−05 56 10	PN G27.3−2.1	7	15.9			PK 27−2.1	Mag 10.8 star E 1′.5; mag 12.4 star N 1′.6.
18 53 28.6	−06 28 35	PN G27.3−3.4	35	16.7		21.0	PK 27−3.1	Numerous stars on disc; mag 13.3 star ESE 1′.2; mag 9.17v star NW 2′.1.
18 54 01.9	−06 26 20	PN G27.4−3.5	15	13.4		15.6	PK 27−3.2	Mag 12.5 star on N edge; small quartet of mag 13-14 stars E 1′.3.
18 57 17.4	−05 59 52	PN G28.2−4.0	6	16.3	15.0		PK 28−4.1	Mag 13.4 star E 1′.0; mag 11.1 star WSW 2′.6.
18 57 49.8	−05 27 36	PN G28.7−3.9	9	16.9			PK 28−3.1	Small triangle of mag 13-14 stars ENE 1′.4; pair of mag 11 stars SW 3′.6.
19 13 55.7	−06 18 53	PN G29.8−7.8	14	15.6			PK 29−7.1	Forms a triangle with an mag 11.5 star E 1′.9 and a mag 12.4 star N 2′.0.
19 27 02.0	-06 35 05	PN G31.0−10.8	11	12.6	12.5	16.3	PK 31−10.1	Mag 13.6 star SW 2′.8; mag 12.4 star NE 3′.0.

RA h m s	Dec ° ′ ″	Name	Mag	Diam ′	No. ★	B ★	Type	Notes
18 22 30.0	−14 32 00	C1819-146		2.8	130		cl	Appears as possible “stellar ring” on DSS.
18 25 30.0	−14 39 00	Do 28		14	16		cl?	Few stars; small brightness range; not well detached. (A) Very sparse group.
18 31 24.0	−06 37 00	Do 29		18			ast	Probably not a cluster. (A) Rich Milky Way field with 4 bright stars.
18 32 54.0	−06 02 00	Do 30		18			ast	Probably not a cluster. (A) Nothing obvious in 24′ field.
18 34 54.0	−06 51 00	Do 31		18			ast	Probably not a cluster. (A) Nothing obvious in 24′ field.
17 59 36.0	−17 23 00	NGC 6507	9.6	15	35	12.0	cl	Moderately rich in stars; large brightness range; not well detached.
18 10 25.0	−16 42 00	NGC 6561		17	100	9.0	cl	(A) Rich cluster of 100 or so faint stars.
18 17 30.0	−16 39 00	NGC 6596		10	30		cl	Moderately rich in stars; moderate brightness range; slight central concentration; detached; involved in nebulosity. (A) Sparse cluster.
18 18 06.0	−12 13 00	NGC 6604	6.5	6	105	7.5	cl	Moderately rich in bright and faint stars with a strong central concentration; detached; involved in nebulosity.
18 16 06.0	−14 59 00	NGC 6605	6.0	29			ast	(A) Elongated E-W group of stars.
18 18 45.0	−13 48 00	NGC 6611	6.0	8	543	11.0	cl	= **M 16**. Moderately rich in bright and faint stars; detached; involved in nebulosity.
18 20 00.0	−17 06 00	NGC 6613	6.9	7	40	8.6	cl	= **M 18**. Few stars; large brightness range; slight central concentration; detached; involved in nebulosity. (A) ~30 bright stars, about 80 stars total.
18 21 09.0	−16 11 00	NGC 6618	6.0	27	660	9.3	cl	Moderately rich in stars; large brightness range; no central concentration; detached; involved in nebulosity (M 17, Swan Nebula). (A) Nebula on west central side.
18 23 12.8	−12 01 00	NGC 6625	9.0	39	30		cl	
18 27 09.5	−12 02 00	NGC 6631	11.7	7	30		cl	Moderately rich in stars; small brightness range; slight central concentration; detached.
18 30 57.8	−13 10 14	NGC 6639		5	25	11.0	cl	
18 32 38.0	−16 53 00	NGC 6645	8.5	15	40	12.0	cl	Moderately rich in stars; small brightness range; not well detached.
18 33 28.3	−10 24 08	NGC 6649	8.9	6	477	13.2	cl	Moderately rich in bright and faint stars with a strong central concentration; detached.
18 36 30.6	−08 11 00	NGC 6664	7.8	12	60	9.0	cl	Moderately rich in stars; moderate brightness range; no central concentration; detached.
17 58 03.3	−11 39 03	Ru 135		4	20		cl	Few stars; moderate brightness range; not well detached.
18 31 18.0	−12 18 00	Ru 141		6	20	12.0	cl	Few stars; small brightness range; not well detached. (A) Slight condensation in Milky Way star cloud.
18 32 14.0	−12 14 12	Ru 142		5	15	13.0	cl	(A) Slight condensation in Milky Way star cloud.
18 32 43.0	−12 08 18	Ru 143		6	30	14.0	cl	
18 33 32.7	−11 25 11	Ru 144		2	10	12.0	cl	
18 25 10.7	−10 02 00	Ru 170		7	30	13.0	cl	
18 32 00.7	−16 03 00	Ru 171		12	60	14.0	cl	(A) Rich faint cluster or Milky Way star cloud.
18 17 10.4	−13 21 00	Tr 32	12.2	12	50	11.8	cl	
18 39 48.0	−08 28 00	Tr 34	8.6	5	87	11.2	cl	Moderately rich in stars; moderate brightness range; not well detached. Probably not a cluster.

GLOBULAR CLUSTERS

RA h m s	Dec ° ′ ″	Name	Total Vm	B ★ Vm	HB Vm	Diam ′	Conc. Class Low = 12 High = 1	Notes
17 27 08.0	−07 05 36	IC 1257	13.1	17.5	19.8	5.0		
18 10 44.2	−07 12 27	IC 1276	10.3	15.7	17.7	8.0	12	Also known as **Palomar 7**.
17 23 35.0	−17 48 47	NGC 6356	8.2	15.1	17.7	10.0	2	
17 27 44.3	−05 04 36	NGC 6366	9.5	13.6	15.7	13.0	11	
18 01 50.6	−08 57 32	NGC 6517	10.1	16.0	18.0	4.0	4	
18 04 49.8	−07 35 09	NGC 6539	8.9	15.9	18.3	7.9	10	
17 58 39.4	−05 04 21	PWM78 2				2.0		

BRIGHT NEBULAE

RA h m s	Dec ° ′ ″	Name	Dim ′ Maj x min	Type	BC	Color	Notes
18 30 24.0	−10 48 00	IC 1287	20 x 10	R	3-5	1-4	
18 18 36.0	−13 58 00	IC 4703	35 x 28	E	1-5	3-4	**Eagle Nebula**. Brightest over an area 10′.0 x 8′.0 off the SE side of the involved open cluster. Approximately centered in this space is a conspicuous absorption patch, which on photos resembles a horse and rider.
18 19 36.0	−16 01 00	IC 4706					
18 20 48.0	−16 11 00	NGC 6618	20 x 15	E	1-5	3-4	= **M 17**, **Swan** or **Omega Nebula**. Very bright and detailed with a conspicuous “2” shape. With the exception of the Orion nebula, it is the brightest galactic nebula visible to northern observers. Cluster is involved.
18 06 06.0	−14 10 00	Sh2-46	30 x 20	E	3-5	3-4	Except to the N, this region is rich in stars, especially to the E.
18 25 12.0	−13 13 00	Sh2-53	15 x 11	E	3-5		Set in a starry field, this object consists of several individual condensations.
18 18 00.0	−11 40 00	Sh2-54	60 x 30	E	1-5	3-4	Mottled and brightest in the NE sector. The brightest single condensation is a patch about 5′ across located 35′ due N of the open cluster NGC 6604.
18 32 12.0	−11 46 00	Sh2-55	20 x 20	E	5-5	1-4	Very faint diffuse nebulosity in a rich star field. There is a coarse pair of brighter stars (sep. about 30″, NW-SE) on the W edge.

DARK NEBULAE

RA h m s	Dec ° ′ ″	Name	Dim ′ Maj x min	Opacity	Notes
17 57 37.4	−17 40 00	B 84a	16 x 16	5	Round.
18 25 53.6	−11 30 00	B 94	15 x 15	4	Roundish; indefinite.
18 25 35.9	−11 45 00	B 95	30 x 30	5	Large; indefinite.
18 26 24.5	−10 18 00	B 96		4	Small; indefinite; elongated E and W.
18 29 05.4	−09 55 00	B 97	50 x 50	4	Irregular; not very definite.
18 32 41.0	−09 09 00	B 100	16 x 16	5	Definite; irregular; curved NW and SE.
18 32 40.0	−08 49 00	B 101	13 x 4	5	Definite; irregular.
18 37 41.8	−13 45 00	B 102	8 x 3	5	Rather definite; elongated NE and SW.
18 39 25.7	−06 40 00	B 103	4 x 4	6	Irregular; one small star in it.
17 50 06.9	−14 22 00	B 284	35 x 5		Curved; NE and SW, outline sharp except at SW end.
17 51 32.3	−12 52 00	B 285	15 x 15		Diffused.
18 18 38.3	−17 57 00	B 307	6 x 1	3	Narrow, dusky mark in star cloud; NE and SW.
18 30 27.2	−17 40 00	B 311	6 x 1	4	Black, elliptical NE and SW.
18 32 10.6	−15 35 00	B 312	100 x 30	4	Large dark area.
18 35 56.8	−15 41 00	B 313	15 x 2		Narrow; curved; E and W.
18 37 04.6	−09 43 00	B 314	35 x 25	5	Irregular, dusky marking, NE and SW.

126 PLANETARY NEBULAE 126

RA h m s	Dec ° ′ ″	Name	Diam ″	Mag (P)	Mag (V)	Mag cent ★	Alt Name	Notes
17 48 20.3	−16 27 35	NGC 6439	14	13.8	12.6	20.2	PK 11+5.1	Mag 12.3 star S 1.′2.
17 26 38.1	−16 48 29	PN G7.9+10.1		17.5		16.4	PK 7+10.1	Mag 12.5 star E 0.′4; mag 14.4 star NE 0.′7.
17 29 02.2	−15 13 07	PN G9.6+10.5	18	17.2		16.5	PK 9+10.1	= **Abell 41**; mag 12.0 star NW 3.′4.
17 52 04.8	−17 36 05	PN G10.4+4.5	6	14.8	14.5		PK 10+4.1	Mag 11.1 star W 0.′5.
17 42 02.0	−15 56 06	PN G10.7+7.4	10	16.2			PK 10+7.1	Mag 11.9 star SW 0.′4.
17 46 54.5	−16 17 25	PN G11.0+6.2	6	14.2	14.6	18.2	PK 11+6.1	Mag 13.0 star S 2.′6; mag 12.4 star NW 5.′8.
17 44 42.1	−15 45 12	PN G11.1+7.0	29			15.5	PK 11+7.1	Appears elongated ENE-WSW; mag 12.3 star ESE 1.′4.
17 28 34.2	−13 26 21	PN G11.1+11.5	7	14.4	14.7	18.8	PK 11+11.1	Mag 13.6 star NW 0.′9.
18 00 08.8	−17 40 43	PN G11.3+2.8	5	18.2			PK 11+2.1	Mag 11.8 star on W edge.
17 56 20.0	−16 29 04	PN G11.9+4.2	8	15.0		17.0	PK 11+4.1	Mag 11.4 star W 5.′1.
17 54 21.1	−15 55 51	PN G12.2+4.9	4			14.6	PM 1−188	Mag 12.8 star ESE 0.′8; mag 10.35v star NE 3.′7.
17 58 58.8	−15 32 15	PN G13.1+4.1	5	13.7	13.3	18.6	PK 13+4.1	
18 01 06.3	−14 30 23	PN G14.2+4.2	6	19.1			PK 14+4.1	Mag 9.93v star E 4.′7; mag 12.2 star W 2.′4.
18 29 11.3	−17 27 12	PN G14.9−3.1	13				SaSt 3−166	Mag 12.7 star NE 0.′4.
17 54 26.4	−12 48 32	PN G14.9+6.4	25	17.6		21.7	PK 14+6.1	Irregular shape; mag 11.6 star ENE 1.′5; mag 14.4 star W 0.′9.
18 35 48.3	−17 36 09	PN G15.4−4.5	6	13.9	13.5	14.3	PK 15−4.1	Forms a small triangle with mag 11.9 star ENE 0.′9 and mag 12.2 star.
18 30 11.3	−16 45 24	PN G15.6−3.0	53	17.4			PK 15−3.1	= **Abell 44**; mag 12.4 star on W edge.
18 07 30.8	−13 28 48	PN G15.9+3.3	5	18.3	15.8		PK 15+3.1	Mag 10.73v star WNW 3.′5; mag 10.52v star WSW 4.′1.
18 36 08.3	−16 59 57	PN G16.0−4.3	13	12.5	13.5		PK 16−4.1	Mag 12.5 star NNE 0.′6; mag 11.4 star SE 2.′1.
17 31 29.2	−08 19 09	PN G16.0+13.5	60	17.8		20.3	PK 16+13.1	Mag 12.5 star 1.′2 E of center.
18 37 46.3	−17 05 47	PN G16.1−4.7	10	13.3	13.5	16.2	PK 16−4.2	Mag 11.8 star SSE 0.′5; mag 11.8 star SW 1.′5.
18 27 56.3	−15 32 55	PN G16.4−1.9	11	14.6	15.4	12.8	PK 16−1.1	Mag 12.4 star SE 0.′5; mag 11.6 star NE 1.′2.
18 29 19.8	−15 07 40	PN G16.9−2.0		16.4			PK 16−2.1	Mag 10.9 star SE 1.′6; mag 10.6 star W 2.′1.
18 33 58.6	−14 52 24	PN G17.7−2.9	7	15.5	14.7		PK 17−2.1	Mag 11.2 star S 0.′7; mag 11.7 star NNE 1.′4.
18 33 03.8	−13 44 19	PN G18.6−2.2	6	16.6			PK 18−2.1	Mag 11.3 star WSW 1.′2; mag 11.4 star S 2.′3.
18 12 09.7	−10 42 59	PN G18.9+3.6	1				PK 18+3.1	Mag 8.48v star ESE 3.′8.
18 10 26.4	−10 29 05	PN G18.9+4.1	12				PK 18+4.1	Mag 12.1 star N 4.′4.
18 34 13.8	−13 12 24	PN G19.2−2.2	1	15.0	14.3		PK 19−2.1	Close pair of mag 11-12 stars E 1.′1.
18 15 17.1	−10 10 12	PN G19.7+3.2	4	15.1	14.3		PK 19+3.1	Mag 13.4 star NW 1.′1.
18 06 59.8	−08 55 34	PN G19.8+5.6	3	16.3			PK 19+5.1	Mag 14.6 star ESE 2.′5.
18 24 07.9	−11 06 41	PN G19.9+0.9	5	19.5			PK 19+0.1	Mag 10.93v star NE 2.′0; mag 13.9 star WSW 2.′6.
18 30 16.1	−11 36 56	PN G20.2−0.6	285			21.1	PK 20−0.1	= **Abell 45**. Numerous stars on the disc, a mag 14.7 star near the center; a mag 11.0 star is 3.′3 SSE of center.
18 33 29.0	−11 07 26	PN G20.9−1.1	10	16.7			PK 21−1.1	Appears elongated NE-SW, which could be caused by an adjacent star?.
18 33 14.9	−10 15 20	PN G21.7−0.6	12	19.9		20.7	PK 21−0.2	Appears to be a small jet extending S; mag 12.6 star on SW edge.
18 32 41.4	−10 05 50	PN G21.8−0.4	24	>14.3		20.1	PK 21−0.1	Appears rectangular, oriented N-S; possibly a star involved on S edge?.
18 28 35.4	−08 43 24	PN G22.5+1.0	17	18.0			MaC 1−13	
18 15 13.4	−06 57 13	PN G22.5+4.8	10			18.0	PK 22+4.1	
18 17 49.5	−06 48 23	PN G23.0+4.3	5	18.0		17.0	PK 23+4.1	Mag 12.9 star SSW 1.′5.
18 30 30.5	−07 27 40	PN G23.9+1.2					PK 23+1.1	Mag 12.8 star S 1.′0; mag 12.5 star N 1.′7.
18 21 23.9	−06 01 57	PN G24.1+3.8	5	16.1			PK 24+3.1	Mag 14.6 star NNW 1.′1.

127 GALAXIES 127

RA h m s	Dec ° ′ ″	Name	Mag (V)	Dim ′ Maj x min	SB	Type Class	PA	Notes
16 28 45.4	−09 08 46	MCG −1-42-1	14.1	1.1 x 0.9	14.0	Sb	177	
16 35 59.0	−05 07 05	MCG −1-42-2	14.2	1.5 x 1.4	14.8	SB(rs)b: II	36	Bright N-S bar; mag 12.4 star NE 1.′8.
16 42 03.3	−05 02 00	MCG −1-42-4	14.1	2.4 x 1.3	15.1	SAB(rs)bc:	114	Numerous knots and/or superimposed stars.
16 17 15.9	−11 43 55	MCG −2-41-1	12.2	3.0 x 2.2	14.1	SAB(r)b?	51	Numerous stars superimposed, many branching arms.

OPEN CLUSTERS

RA h m s	Dec ° ′ ″	Name	Mag	Diam ′	No. ★	B ★	Type	Notes
16 36 30.0	−08 56 00	Do 27		25	15		cl:	Few stars; moderate brightness range; not well detached; involved in nebulosity.

GLOBULAR CLUSTERS

RA h m s	Dec ° ′ ″	Name	Total V m	B ★ V m	HB V m	Diam ′	Conc. Class Low = 12 High = 1	Notes
17 27 08.0	−07 05 36	IC 1257	13.1	17.5	19.8	5.0		
16 32 31.9	−13 03 13	NGC 6171	7.8	13.0	15.6	13.0	10	= **M 107**.
17 23 35.0	−17 48 47	NGC 6356	8.2	15.1	17.7	10.0	2	
17 27 44.3	−05 04 36	NGC 6366	9.5	13.6	15.7	13.0	11	

DARK NEBULAE

RA h m s	Dec ° ′ ″	Name	Dim ′ Maj x min	Opacity	Notes
16 27 09.8	−16 47 00	B 230	60 x 60		Round.

PLANETARY NEBULAE

RA h m s	Dec ° ′ ″	Name	Diam ″	Mag (P)	Mag (V)	Mag cent ★	Alt Name	Notes
17 14 04.5	−12 54 41	NGC 6309	19	10.8	11.5	16.5	PK 9+14.1	**Box Nebula**. Mag 12.6 star on N edge.
17 26 38.1	−16 48 29	PN G7.9+10.1		17.5		16.4	PK 7+10.1	Mag 12.5 star E 0.′4; mag 14.4 star NE 0.′7.
17 05 37.9	−10 08 32	PN G10.8+18.0	59		14.6	15.6	PK 10+18.2	**Butterfly Nebula.** Appears much elongated N-S; mag 14.1 star SW 2.′1.

128 GALAXIES 128

RA h m s	Dec ° ′ ″	Name	Mag (V)	Dim ′ Maj x min	SB	Type Class	PA	Notes
15 11 49.8	−17 56 46	ESO 581-23	16.3	1.9 x 0.3	15.5	SB(s)d pec sp	1	Mag 6.87v star SE 12′2.
14 57 59.8	−06 43 24	IC 1080	13.8	1.4 x 0.9	13.9	$S0^-$ pec:	27	Mag 8.77v star W 4′3.
15 01 14.9	−07 28 33	IC 1084	14.2	0.7 x 0.5	12.9	S?	171	NGC 5812 W 5′0.
15 08 13.6	−11 08 27	IC 1091	13.4	1.1 x 0.8	13.1	SB(s)b? III-IV	120	Mag 9.92v star W 3′3.
14 58 06.7	−06 27 28	MCG −1-38-11	15.1	1.1 x 0.6	14.5	SAB(r)0/a?	36	Almost stellar anonymous galaxy SW 2′0.
14 58 29.7	−06 49 11	MCG −1-38-12	13.8	1.4 x 1.1	14.1	SAB(s)c I-II	36	Mag 13.3 star N edge; mag 8.77v star NW 12′5.
14 59 40.9	−06 59 26	MCG −1-38-14	14.7	1.7 x 0.6	14.6	Sdm: IV-V	81	Almost stellar nucleus, smooth envelope.
15 04 13.5	−07 13 25	MCG −1-38-20	13.3	0.9 x 0.7	12.6	SAB(r)b? III	81	Many knots or superimposed stars.
15 05 40.4	−06 51 41	MCG −1-38-22	16.7	0.5 x 0.4	14.8		90	Mag 9.60v star N 1′2.
15 21 33.3	−07 26 54	MCG −1-39-3	13.8	1.5 x 0.5	13.4	S0 pec	9	Mag 9.93v star W 1′8; NGC 5817 N 4′4.
15 31 30.8	−05 09 42	MCG −1-39-6	14.3	0.7 x 0.4	12.8		105	
15 33 20.7	−08 42 06	MCG −1-40-1	13.6	1.3 x 0.4	12.8	S?	75	
15 37 37.2	−08 43 18	MCG −1-40-4	14.3	1.0 x 0.6	13.6	Sb	72	
15 39 24.6	−06 50 46	MCG −1-40-5	14.2	1.2 x 0.5	13.5	SB?	129	
15 50 43.8	−07 33 39	MCG −1-40-8	13.6	0.6 x 0.6	12.5	E/S0		
14 57 43.6	−13 44 09	MCG −2-38-23	14.1	0.7 x 0.5	12.8	Sc	3	Mag 8.88v star NW 3′0.
15 00 24.3	−13 33 11	MCG −2-38-24	13.6	2.1 x 1.7	14.8	SB(s)m: V	144	Low surface brightness; mag 9.75v star NW 2′0.
15 00 59.8	−15 07 18	MCG −2-38-27	14.6	2.7 x 0.5	14.7	SB(s)d IV-V	129	Mag 8.43v star N 5′8.
15 00 58.6	−15 24 36	MCG −2-38-28	14.7	1.0 x 0.2	12.8	Sb	84	**MCG −2-38-26** at E tip.
15 01 28.0	−14 18 53	MCG −2-38-29	13.5	0.8 x 0.6	12.8	E	36	Stars superimposed E and SW.
15 03 00.3	−13 17 02	MCG −2-38-30	13.6	0.9 x 0.3	12.0	Sa? sp	147	
15 04 02.7	−10 44 20	MCG −2-38-31	14.7	1.3 x 0.2	13.0	S?	24	
15 04 27.3	−14 26 23	MCG −2-38-32	14.4	1.9 x 0.3	13.7	Scd: sp III-IV	168	Mag 10.63v star SSE 4′0.
15 05 48.1	−14 57 55	MCG −2-38-33	14.2	0.5 x 0.4	12.5	E	90	
15 06 24.8	−09 54 27	MCG −2-38-34	14.3	0.8 x 0.4	12.9	Sb	69	
15 07 06.9	−09 57 53	MCG −2-38-36	14.5	0.7 x 0.3	12.6	Irr	135	
15 09 24.6	−10 41 44	MCG −2-39-4	13.6	1.3 x 1.3	14.0	SAB(r)cd pec: III		
15 13 54.2	−13 06 23	MCG −2-39-8	13.9	0.7 x 0.7	13.0	S?		
15 14 11.4	−14 20 59	MCG −2-39-9	14.7	0.8 x 0.5	13.5	SBc	90	Mag 6.76v star N 8′6; NGC 5878 NW 7′8.
15 14 25.6	−14 39 21	MCG −2-39-10	14.5	1.1 x 0.4	13.5	SBb	111	NGC 5883 E 11′0.
15 15 00.3	−14 00 15	MCG −2-39-11	14.8	0.9 x 0.4	13.5	Sb	96	
15 16 52.8	−13 25 12	MCG −2-39-16	14.5	1.1 x 0.9	14.4	SAB(s)dm? IV-V	174	
15 19 07.8	−14 32 10	MCG −2-39-17	14.8	0.7 x 0.5	13.5	S?	93	
15 21 35.9	−12 05 42	MCG −2-39-21	13.4	0.9 x 0.3	11.8	Sbc	15	Mag 8.71v star W 1′0.
15 32 51.3	−15 17 08	MCG −2-40-1	14.6	0.7 x 0.5	13.3	Sm	66	
15 43 10.4	−12 34 30	MCG −2-40-3	14.1	0.9 x 0.3	12.5	SAB(r)a:	75	Mag 9.68v star N 5′1.
14 56 38.6	−16 29 19	MCG −3-38-29	14.2	0.7 x 0.3	12.3	S0/a	45	Mag 8.20v star E 7′0.
15 00 08.9	−16 21 57	MCG −3-38-42	14.3	1.3 x 1.1	14.5	SAm: IV-V	24	
15 19 05.4	−15 59 51	MCG −3-39-5	13.4	0.5 x 0.3	11.2		33	
15 21 51.4	−17 43 49	MCG −3-39-9	14.5	0.6 x 0.4	12.8	(R'_1)SB(l)a	117	Mag 8.84v star N 3′4.
15 37 30.3	−16 35 30	MCG −3-40-3	14.2	0.6 x 0.3	12.2		141	Star on E edge; NGC 5959 W 1′9.
14 56 41.3	−17 14 37	NGC 5781	13.0	1.4 x 0.8	13.0	(R′)SBb pec sp III-IV	25	
14 59 24.9	−16 41 40	NGC 5793	13.2	1.7 x 0.6	13.1	Sb: sp	150	Several faint stars superimposed N of center.
14 59 24.1	−16 37 28	NGC 5796	11.6	2.5 x 1.8	13.3	E0-1	90	
15 00 26.0	−13 54 17	NGC 5801	14.7	0.7 x 0.4	13.2	Sb	51	Faint anonymous galaxy WNW 2′0.
15 00 30.1	−13 55 10	NGC 5802	14.2	0.9 x 0.6	13.4	S0	90	Star E of nucleus.
15 00 34.6	−13 53 41	NGC 5803	14.8	0.6 x 0.5	13.4	E/S0	60	Almost stellar.
15 00 52.4	−14 09 56	NGC 5809	13.4	1.4 x 0.7	13.3	S0/a: sp	155	
15 02 42.7	−17 52 08	NGC 5810	13.3	1.2 x 0.6	12.8	SB(rs)b: III	31	Mag 9.41v star S 5′8.
15 00 55.6	−07 27 28	NGC 5812	11.2	2.1 x 2.1	12.8	E0		Large, bright, diffuse center; IC 1084 E 5′0.
15 00 29.2	−16 50 00	NGC 5815	14.3	0.8 x 0.4	12.9	Sb	25	Located 1′4 SW of mag 12.8 star.
14 59 40.8	−16 10 51	NGC 5817	14.2	0.9 x 0.7	13.5	S0	15	
15 06 50.8	−14 34 18	NGC 5849	13.9	1.0 x 0.8	13.6	S0	30	Bright nucleus with faint star S; possible anonymous companion galaxy on SW edge.
15 08 49.1	−11 12 29	NGC 5858	12.4	1.5 x 0.7	12.4	E6:	137	Mag 12.9 star W 1′8; mag 8.30v star E 17′2.
15 09 15.8	−11 19 17	NGC 5861	11.6	3.0 x 1.7	13.2	SAB(rs)c I-II	155	Patchy, branching arms, several knots; mag 11.4 star SW 2′5; mag 8.30v star ENE 10′9.
15 10 55.8	−11 28 48	NGC 5872	12.6	1.5 x 0.9	12.8	$SAB0^0$:	15	Mag 14 star on E edge; mag 10.46v star WNW 3′0.
15 13 45.8	−14 16 14	NGC 5878	11.5	3.5 x 1.4	13.1	SA(s)b II	0	Strong dark lane W of center; mag 6.76v star NE 8′8.
15 15 01.1	−14 34 45	NGC 5880	14.1	0.7 x 0.6	13.2	E	135	Located 2′1 SE of a mag 10.99v star.
15 15 10.3	−14 37 02	NGC 5883	13.6	1.1 x 0.7	13.2	S0	120	Mag 8.14v star SE 9′5; MCG −2-39-10 W 11′0.
15 15 04.3	−10 05 10	NGC 5885	11.8	3.5 x 3.1	14.2	SAB(r)c II-III	65	Mag 10.16v star on NE edge. Bright nucleus.
15 17 51.3	−17 35 20	NGC 5890	12.7	1.3 x 1.0	12.9	$SA(r)0^+$?	85	Pair of stars E end.
15 16 13.5	−11 29 41	NGC 5891	13.8	0.8 x 0.5	12.6	Sbc II-III:	152	
15 13 48.3	−15 27 48	NGC 5892	11.7	3.5 x 2.8	14.0	SA(s)d III:	105	**Fath 703**. Bright nucleus; multi-branching arms.
15 21 33.4	−13 05 34	NGC 5915	12.3	1.7 x 1.3	13.0	SB(s)ab pec	164	Multi-galaxy system. Extension S with star or knot at end.
15 21 38.0	−13 10 10	NGC 5916	13.1	2.8 x 0.9	13.9	SB(rs)a pec	13	Bright extension N, very faint S.
15 21 13.9	−13 06 03	NGC 5916A	14.0	1.2 x 0.4	13.0	SB(s)c pec	146	Star on W edge.
15 21 32.6	−07 22 38	NGC 5917	13.0	1.5 x 0.8	13.1	Sb pec?	75	MCG −1-39-3 S 4′4.
15 37 22.3	−16 35 45	NGC 5959	13.4	1.9 x 1.6	14.6	E^+:	177	MCG −3-40-3 E 1′9.
15 40 15.6	−08 36 06	NGC 5973	14.4	0.9 x 0.3	12.8	S0/a	140	
15 42 27.3	−13 14 06	NGC 5978	14.0	0.8 x 0.7	13.3	Sa		
15 48 24.9	−13 45 30	NGC 5995	13.6	0.9 x 0.7	13.0	S(B)c	171	

129 GALAXIES 129

RA h m s	Dec ° ′ ″	Name	Mag (V)	Dim ′ Maj x min	SB	Type Class	PA	Notes
14 10 15.2	−17 50 15	ESO 578-32	14.4	1.3 x 0.3	13.2	Sab	106	Mag 11.9 star NNW 7′8.
13 45 24.8	−12 34 29	IC 920	13.5	0.5 x 0.5	12.1	E		Mag 5.50v star 86 Virginis NNE 11′7.
14 03 52.8	−10 08 28	IC 971	12.8	2.2 x 1.2	13.7	SB(s)c I	126	Stellar nucleus; mag 11.9 star SW 7′1; IC 4358 W 4′6.
14 17 48.6	−13 52 25	IC 991	13.1	1.5 x 0.8	13.1	SB(s)c: II-III	109	Mag 8.81v star SW 8′1.
14 47 25.6	−13 42 58	IC 1055	12.6	2.0 x 0.7	12.8	(R′)SA(r)b: II	3	Mag 9.53v star SSW 7′5.
14 57 59.8	−06 43 24	IC 1080	13.8	1.4 x 0.9	13.9	$S0^-$ pec:	27	Mag 8.77v star W 4′3.
15 01 14.9	−07 28 33	IC 1084	14.2	0.7 x 0.5	12.9	S?	171	NGC 5812 W 5′0.
13 58 31.0	−12 36 21	IC 4354	14.5	1.1 x 0.3	13.2	Sb	105	Mag 9.56v star W 3′3.
14 03 33.8	−10 09 03	IC 4358	14.1	1.1 x 0.3	12.8	Sbc pec sp	114	Mag 11.9 star S 3′0; IC 971 E 4′6.
14 04 07.6	−09 46 08	IC 4361	14.8	0.7 x 0.3	13.0	Sb	168	Located 2′8 SE of mag 7.26v star.
14 04 19.7	−09 59 37	IC 4364	14.1	0.7 x 0.5	12.8	Sbc	75	Stellar; mag 12.5 star W 2′9; anonymous galaxy E 6′8, could be **IC 4368**?
14 23 36.8	−05 58 55	IC 4407	14.1	1.5 x 0.9	14.3	SB(s)m IV-V	0	Low surface brightness.

(Continued from previous page)

GALAXIES

RA h m s	Dec ° ′ ″	Name	Mag (V)	Dim ′ Maj x min	SB	Type Class	PA	Notes
13 45 38.3	−05 59 05	MCG −1-35-10	13.8	1.6 x 1.1	14.3	SB(s)m III	57	Several bright knots or superimposed stars; mag 12.4 star SW 2′2.
13 46 40.0	−09 38 47	MCG −1-35-11	14.6	1.1 x 0.5	13.8	Scd	144	Mag 6.05v star ESE 9′1.
13 48 51.3	−07 38 58	MCG −1-35-12	13.3	1.3 x 0.7	13.0	S?	168	
13 48 59.6	−07 11 40	MCG −1-35-13	14.3	2.2 x 0.5	14.3	Sb sp	26	= **Hickson 67B**. NGC 5306 SE 3′4.
13 52 37.1	−07 53 00	MCG −1-35-17	14.2	0.8 x 0.5	13.0	SBc	90	Mag 9.03v star W 4′0.
13 55 34.7	−05 58 24	MCG −1-35-21	14.6	0.7 x 0.6	13.5		165	
13 55 45.5	−06 00 12	MCG −1-35-22	14.5	1.6 x 0.8	14.6	SB(s)dm pec: IV	147	
14 10 14.3	−06 49 18	MCG −1-36-11	14.0	1.0 x 0.7	13.5	SBc	12	
14 10 18.6	−08 09 58	MCG −1-36-12	14.0	0.7 x 0.4	12.5	SBbc	6	
14 25 45.5	−05 24 16	MCG −1-37-6	13.2	1.2 x 0.6	12.7	I0 pec?	105	Mag 8.71v star SW 12′7.
14 29 20.3	−05 01 23	MCG −1-37-7	14.4	1.1 x 1.0	14.3	SBbc	21	Mag 8.38v star NNW 5′9.
14 34 04.7	−07 40 36	MCG −1-37-8	14.6	1.1 x 0.7	14.1	SBc	3	Mag 10.34v star S 6′6.
14 34 57.1	−08 50 27	MCG −1-37-9	14.9	1.5 x 0.2	13.5	S?	0	Located 1′6 SW of mag 9.60v star.
14 38 55.0	−08 37 41	MCG −1-37-10	15.5	2.7 x 0.5	15.7	IB(s)m V-VI	60	Low surface brightness.
14 41 06.5	−08 52 32	MCG −1-37-11	13.2	1.1 x 0.7	12.7	SAB(rs)a pec?	153	
14 43 36.6	−05 10 35	MCG −1-38-1	14.1	0.6 x 0.6	13.0	E/S0		
14 53 06.8	−07 01 17	MCG −1-38-6	15.2	1.0 x 0.7	14.7	SB(rs)b pec: III	84	Mag 10.05v star NW 8′0.
14 55 18.1	−06 02 28	MCG −1-38-8	14.1	1.2 x 1.0	14.1	SB?	111	Small anonymous galaxy E 1′0.
14 58 06.7	−06 27 28	MCG −1-38-11	15.1	1.1 x 0.6	14.5	SAB(r)0/a?	36	Almost stellar anonymous galaxy SW 2′0.
14 58 29.7	−06 49 11	MCG −1-38-12	13.8	1.4 x 1.1	14.1	SAB(s)c I-II	36	Mag 13.3 star N edge; mag 8.77v star NW 12′5.
14 59 40.9	−06 59 26	MCG −1-38-14	14.7	1.7 x 0.6	14.6	Sdm: IV-V	81	Almost stellar nucleus, smooth envelope.
13 48 22.0	−13 33 44	MCG −2-35-16	14.1	0.9 x 0.7	13.4	Sb	162	Lies 1′4 S of mag 11.6 star.
13 52 08.7	−10 37 15	MCG −2-35-18	13.7	0.4 x 0.2	10.8	Sbc	120	**MCG −2-35-17** NNW 0′5; elliptical **MCG −2-35-19** and spherical **MCG −2-35-20** E 1′5.
14 01 42.0	−11 36 27	MCG −2-36-2	13.5	0.7 x 0.7	12.6	SA0$^-$:		Mag 9.06v star N 8′5; located at the center of galaxy cluster A 1836.
14 03 21.3	−14 58 22	MCG −2-36-3	14.0	0.5 x 0.5	12.6	E		Partially covered by glare of mag 12 star. Anonymous galaxy NE 1′5, just N of mag 11 star.
14 04 12.1	−10 45 17	MCG −2-36-8	14.1	0.7 x 0.7	13.1	Sbc		Mag 7.48v star W 10′1.
14 04 20.5	−15 13 23	MCG −2-36-10	14.7	1.4 x 0.2	13.1	S?	78	
14 04 35.4	−10 13 16	MCG −2-36-11	13.2	0.9 x 0.5	12.3	SAB0^0?	0	Mag 10.50v star E 7′7.
14 04 57.5	−14 16 46	MCG −2-36-12	14.3	1.6 x 0.3	13.3	SB(s)b? sp	150	
14 04 59.9	−14 51 27	MCG −2-36-13	14.5	1.0 x 0.9	14.2	Sc	33	Mag 7.25v star E 3′1.
14 05 58.3	−12 04 24	MCG −2-36-15	14.2	0.7 x 0.5	13.0	Sab	138	
14 05 56.8	−15 17 57	MCG −2-36-16	14.2	0.9 x 0.3	12.7		36	
14 12 08.0	−10 04 43	MCG −2-36-17	13.4	0.9 x 0.7	12.8	(R′)SA(r)bc pec:	81	Mag 8.71v star NNW 7′5.
14 14 31.0	−10 42 22	MCG −2-36-18	14.5	0.9 x 0.7	13.8	SBbc	120	
14 23 12.7	−10 51 06	MCG −2-37-1	14.6	0.5 x 0.5	12.9			Mag 7.97v star NW 7′3.
14 23 38.6	−13 20 18	MCG −2-37-2	14.2	1.0 x 0.4	13.1	Sc	18	
14 26 12.3	−11 54 19	MCG −2-37-4	14.8	0.7 x 0.5	13.5	SBb	162	
14 32 52.9	−12 58 33	MCG −2-37-6	14.3	1.6 x 1.1	14.8	SB(rs)bc II-III	20	Mag 9.21v star WSW 5′4.
14 33 09.7	−15 24 22	MCG −2-37-7	14.0	0.6 x 0.5	12.6	S0	162	
14 33 54.1	−13 16 14	MCG −2-37-9	14.5	0.4 x 0.4	12.4			Mag 9.03v star E 3′7.
14 35 24.3	−13 44 09	MCG −2-37-10	14.3	1.1 x 0.7	13.8	SB(s)m IV-V	36	Very amorphous, sits in a triangle of mag 14, 15 and 16 stars.
14 42 36.5	−10 07 46	MCG −2-37-13	14.8	1.0 x 0.3	13.3		3	
14 43 09.7	−10 22 11	MCG −2-37-14	14.6	0.7 x 0.5	13.3	Sc	48	
14 44 00.8	−12 57 07	MCG −2-38-1	14.3	1.5 x 1.1	14.7	SAB(rs)dm: IV-V	0	Mag 9.49v star S 5′2.
14 44 15.7	−11 50 21	MCG −2-38-2	14.7	0.7 x 0.3	12.9	Scd	30	
14 44 31.3	−09 40 16	MCG −2-38-3	14.4	1.0 x 0.6	13.6	Sc	162	Multi-galaxy system.
14 45 19.3	−10 58 14	MCG −2-38-5	14.4	0.5 x 0.5	12.7			
14 45 34.0	−11 39 13	MCG −2-38-6	14.3	0.5 x 0.5	12.6			
14 46 14.6	−11 31 09	MCG −2-38-9	14.8	0.9 x 0.4	13.5	SBbc	60	
14 47 07.3	−13 19 15	MCG −2-38-10	15.2	1.3 x 0.9	15.2	SB(s)d: IV-V	165	
14 47 47.7	−15 08 19	MCG −2-38-13	14.2	0.9 x 0.6	13.3	Sc	114	**MCG −2-38-14** on S edge, equal ion size to MCG −2-38-13 but with bright core.
14 47 56.1	−14 16 58	MCG −2-38-15	13.6	2.2 x 1.7	14.9	SB(s)cd II-III	123	Mag 13.1 star NW 2′2.
14 49 30.4	−10 10 25	MCG −2-38-16	12.8	2.5 x 1.4	14.0	IB(s)m pec III	150	Pair with MCG −2-38-17 overlapping NE edge.
14 49 32.8	−10 09 47	MCG −2-38-17	13.6	2.4 x 1.4	14.7	IB(s)m pec	140	Pair with MCG −2-38-16 overlapping SW edge.
14 50 48.6	−13 31 33	MCG −2-38-18	14.9	0.8 x 0.2	12.7	Sbc	96	Small, round galaxy on N edge.
14 50 47.8	−13 32 15	MCG −2-38-19	14.5	0.7 x 0.5	13.2	SBbc	144	
14 51 37.1	−13 35 13	MCG −2-38-20	14.6	0.7 x 0.4	13.1	SBbc	129	Mag 7.69v star WSW 5′7.
14 52 01.2	−15 42 24	MCG −2-38-21	14.9	0.5 x 0.3	12.7	Sc	90	
14 52 10.6	−10 44 28	MCG −2-38-22	13.5	1.3 x 0.6	13.1	Sb pec? II-III	105	
14 57 43.6	−13 44 09	MCG −2-38-23	14.1	0.7 x 0.5	12.8	Sc	3	Mag 8.88v star NW 3′0.
15 00 24.3	−13 33 11	MCG −2-38-24	13.6	2.1 x 1.7	14.8	SB(s)m: V	144	Low surface brightness; mag 9.75v star NW 2′0.
15 00 59.8	−15 07 18	MCG −2-38-27	14.6	2.7 x 0.5	14.7	SB(s)d IV-V	129	Mag 8.43v star N 5′8.
15 00 58.6	−15 24 36	MCG −2-38-28	14.7	1.0 x 0.2	12.8	Sb	84	**MCG −2-38-26** at E tip.
15 01 28.0	−14 18 53	MCG −2-38-29	13.5	0.8 x 0.6	12.8	E	36	Stars superimposed E and SW.
15 03 00.3	−13 17 02	MCG −2-38-30	13.6	0.9 x 0.3	12.0	Sa? sp	147	
13 45 01.9	−15 52 04	MCG −3-35-17	14.2	0.9 x 0.7	13.5	Sbc	90	Mag 6.19v star NE 10′0.
13 53 08.2	−16 57 40	MCG −3-35-20	14.6	1.1 x 0.2	12.8		60	
13 58 16.9	−15 45 11	MCG −3-36-1	14.5	0.6 x 0.3	12.5	S0	0	
14 05 08.1	−17 30 14	MCG −3-36-4	14.8	1.0 x 0.7	14.3	Sb	171	**MCG −3-36-5** N 1′4.
14 06 18.1	−16 38 29	MCG −3-36-6	14.3	0.8 x 0.5	13.1	Sbc	75	
14 09 27.7	−17 51 58	MCG −3-36-8	13.7	0.9 x 0.7	13.2	E/S0	131	Mag 12.2 star SE 6′2; MCG −3-36-9 WSW 2′1.
14 09 19.7	−17 52 48	MCG −3-36-9	14.1	0.4 x 0.4	12.0	S0		MCG −3-36-8 ENE 2′1.
14 47 24.3	−17 26 48	MCG −3-38-8	13.9	3.2 x 0.5	14.2	SB(s)cd: sp	171	
14 53 01.1	−15 53 00	MCG −3-38-22	15.1	1.1 x 0.5	14.3	SBb	87	Anonymous galaxy SW 0′9.
14 54 40.8	−17 24 24	MCG −3-38-25	13.3	1.9 x 0.8	13.6	SAB(rs)bc: II-III	18	A number of bright knots and/or superimposed stars.
14 56 38.6	−16 29 19	MCG −3-38-29	14.2	0.7 x 0.3	12.3	S0/a	45	Mag 8.20v star E 7′0.
15 00 08.9	−16 21 57	MCG −3-38-42	14.3	1.3 x 1.1	14.5	SAm: IV-V	24	
13 49 11.2	−07 13 28	NGC 5306	12.2	1.4 x 1.0	12.4	S0 pec?	30	= **Hickson 67A**. **Hickson 67C** on N edge; **Hickson 67D** on SW edge.
13 52 05.9	−06 03 32	NGC 5324	11.7	2.3 x 2.1	13.2	SA(rs)c: I-II	0	Bright nucleus, bright branching arms.
13 54 00.3	−07 55 51	NGC 5339	12.0	1.8 x 1.5	12.9	SB(rs)a pec	59	Bright nucleus in E-W bar.
13 54 11.8	−07 35 18	NGC 5343	12.9	1.4 x 1.1	13.2	SA(r)0$^-$?	50	
13 56 37.8	−05 28 12	NGC 5369	13.4	0.9 x 0.8	13.2	E	114	
14 04 00.0	−14 37 02	NGC 5420	13.1	1.7 x 0.7	13.1	Sb: II-III:	138	
14 03 25.0	−06 04 16	NGC 5426	12.1	3.0 x 1.6	13.6	SA(s)c pec I-II	172	Interacting with NGC 5427 N. Many knots W side, strong dark lane E side.
14 03 26.0	−06 01 52	NGC 5427	11.4	2.8 x 2.4	13.3	SA(s)c pec I	66	Interacting with NGC 5426 S. Branching spiral arms.
14 04 43.1	−09 42 50	NGC 5442	13.2	1.2 x 0.5	12.5	SB(s)b pec?	149	Star or stellar galaxy NE edge.
14 06 34.9	−05 27 11	NGC 5468	12.5	2.6 x 2.4	14.3	SAB(rs)cd II-III	105	Almost stellar nucleus, branching, knotty arms; mag 8.34v star S 4′2; NGC 5472 E 5′0.
14 06 55.0	−05 27 40	NGC 5472	14.0	1.2 x 0.3	12.7	SA(r)ab? sp	29	Located 5′0 E of NGC 5468.
14 08 08.7	−06 05 30	NGC 5476	12.8	1.4 x 1.1	13.1	SAB(rs)dm? III-IV	135	Very patchy.

(Continued from previous page)

129 GALAXIES 129

RA h m s	Dec ° ′ ″	Name	Mag (V)	Dim ′ Maj x min	SB	Type Class	PA	Notes
14 11 29.4	−05 02 37	NGC 5493	11.4	1.6 x 1.3	12.0	S0 pec sp	120	Very bright lens shaped center.
14 13 37.4	−17 59 06	NGC 5510	13.9	1.4 x 1.1	14.3	IB(s)m pec: III-IV	40	Bright nucleus, several superimposed stars.
14 17 40.4	−07 25 03	NGC 5534	12.3	1.7 x 1.3	13.1	(R′)SAB(s)ab pec: II	139	Multi-galaxy system; very small, superimposed galaxy **Mkn 1379A** E of center.
14 24 13.7	−16 43 23	NGC 5595	12.0	2.2 x 1.2	12.9	SAB(rs)c II	50	Several bright knots and superimposed stars.
14 24 27.5	−16 45 49	NGC 5597	12.0	2.1 x 1.7	13.3	SAB(s)cd II	95	Several very bright knots.
14 25 07.6	−13 09 47	NGC 5605	12.3	1.6 x 1.3	12.9	(R′)SAB(rs)c pec: II	85	Patchy with several knots or stars S.
14 33 56.4	−16 34 54	NGC 5663	14.2	1.4 x 1.3	14.7	SA0$^-$:	0	
14 33 43.6	−14 37 11	NGC 5664	13.6	0.8 x 0.4	12.3	Sa	30	
14 41 06.0	−17 28 35	NGC 5716	12.6	1.8 x 1.3	13.4	SB(rs)c? II-III	85	Pair of stars on NE edge.
14 42 23.9	−17 15 15	NGC 5728	11.4	3.1 x 1.8	13.1	SAB(r)a:	30	Faint arms extend E and W outside published dimensions.
14 42 06.8	−09 00 35	NGC 5729	12.2	2.5 x 0.6	12.5	Sb pec:	166	
14 45 51.8	−11 54 53	NGC 5741	13.6	1.1 x 1.1	13.9	E		Almost stellar nucleus, faint envelope, star or knot on SE edge.
14 45 37.0	−11 48 35	NGC 5742	13.0	1.3 x 0.7	12.8	SB(r)0°?	73	
14 45 01.9	−13 56 53	NGC 5745	13.4	1.7 x 1.1	13.9	Sa pec sp	77	Strong dark lane N.
14 47 34.0	−14 51 08	NGC 5756	12.3	2.8 x 1.3	13.6	(R′)SB(s)bc pec sp II	58	
14 56 41.3	−17 14 37	NGC 5781	13.0	1.4 x 0.8	13.0	(R′)SBb pec sp III-IV	25	
14 59 24.9	−16 41 40	NGC 5793	13.2	1.7 x 0.6	13.1	Sb: sp	150	Several faint stars superimposed N of center.
14 59 24.1	−16 37 28	NGC 5796	11.6	2.5 x 1.8	13.3	E0-1	90	
15 00 26.0	−13 54 17	NGC 5801	14.7	0.7 x 0.4	13.2	Sb	51	Faint anonymous galaxy WNW 2:0.
15 00 30.1	−13 55 10	NGC 5802	14.2	0.9 x 0.6	13.4	S0	90	Star E of nucleus.
15 00 34.6	−13 53 41	NGC 5803	14.8	0.6 x 0.5	13.4	E/S0	60	Almost stellar.
15 00 52.4	−14 09 56	NGC 5809	13.4	1.4 x 0.7	13.3	S0/a: sp	155	
15 02 42.7	−17 52 08	NGC 5810	13.3	1.2 x 0.6	12.8	SB(rs)b: III	31	Mag 9.41v star S 5:8.
15 00 55.6	−07 27 28	NGC 5812	11.2	2.1 x 2.1	12.8	E0		Large, bright, diffuse center; IC 1084 E 5:0.
15 00 29.2	−16 50 00	NGC 5815	14.3	0.8 x 0.4	12.9	Sb	25	Located 1:4 SW of mag 12.8 star.
14 59 40.8	−16 10 51	NGC 5817	14.2	0.9 x 0.7	13.5	S0	15	

GALAXY CLUSTERS

RA h m s	Dec ° ′ ″	Name	Mag 10th brightest	No. Gal	Diam ′	Notes
14 01 42.0	−11 36 00	A 1836	15.7	41	15	Except MCG −2-36-2, all members anonymous, faint and stellar.
14 01 48.0	−11 09 00	A 1837	15.7	50	18	All members anonymous, faint and stellar.
14 09 30.0	−17 52 00	AS 756	16.2		17	Except for three plotted GX, all others anonymous, stellar.

GLOBULAR CLUSTERS

RA h m s	Dec ° ′ ″	Name	Total V m	B ★ V m	HB V m	Diam ′	Conc. Class Low = 12 High = 1	Notes
14 29 37.3	−05 58 35	NGC 5634	9.5		17.8	5.5	4	

PLANETARY NEBULAE

RA h m s	Dec ° ′ ″	Name	Diam ″	Mag (P)	Mag (V)	Mag cent ★	Alt Name	Notes
14 04 25.9	−17 13 40	IC 972	54	14.9	13.9	17.9	PK 326+42.1	Mag 12.9 stars N 2:5 and NW 3:7.

130 GALAXIES 130

RA h m s	Dec ° ′ ″	Name	Mag (V)	Dim ′ Maj x min	SB	Type Class	PA	Notes
13 15 12.9	−17 58 04	ESO 576-17	14.1	1.2 x 1.2	14.3	SAB(s)d: IV		Mag 11.13v star ESE 4:6.
12 42 08.5	−17 20 58	IC 806	14.4	0.7 x 0.6	13.3	(R′)SB(r)ab pec? II	135	Mag 11.8 star and small anonymous galaxy SW 0:9; mag 10.64v star W 9:2; IC 807 S 3:4.
12 42 12.5	−17 24 13	IC 807	13.3	0.7 x 0.7	12.5	E/S0		IC 806 N 3:4.
12 52 27.5	−15 31 10	IC 829	13.7	0.5 x 0.4	11.9	SB0	126	MCG −2-33-37 on W edge; MCG −2-33-36 E 1:3; spherical anonymous galaxy NE 1:2.
13 17 12.3	−17 15 20	IC 863	13.2	1.2 x 0.9	13.1	SB(rs)0/a? pec	48	Close pair of mag 10.37v, 12.0 stars NE 9:0.
13 45 24.8	−12 34 29	IC 920	13.5	0.5 x 0.5	12.1	E		Mag 5.50v star 86 Virginis NNE 11:7.
12 48 59.8	−14 23 55	IC 3799	13.7	2.3 x 0.3	13.1	Sd: sp	30	Mag 9.00v star W 5:7.
12 49 53.8	−06 43 07	IC 3812	14.1	1.1 x 0.5	13.3	Sc	3	Mag 11.18v star NNW 7:6.
12 50 22.9	−14 19 20	IC 3822	14.5	1.1 x 0.3	13.1	Sc	33	NGC 4724 and NGC 4727 E 8:4.
12 50 40.0	−09 01 49	IC 3826	13.2	1.9 x 0.7	13.4	S0°	177	Mag 10.38v star W 6:9.
12 50 51.8	−14 29 32	IC 3827	13.4	1.0 x 0.7	12.8	Sc	60	IC 3831 SE 8:1.
12 51 18.8	−14 34 28	IC 3831	12.6	1.4 x 0.9	12.8	(R′)SAB(s)0°?	151	Mag 9.30v star S 5:9; IC 3827 NW 8:1.
12 55 13.4	−08 07 14	IC 3883	13.4	0.9 x 0.5	12.4	Sb	12	Mag 10.66v star W 6:4.
12 56 40.7	−07 33 46	IC 3908	12.8	1.7 x 0.6	12.7	SB(s)d? III	171	Mag 10.85v star NW 5:3.
13 02 04.0	−07 36 12	IC 4071	13.8	0.9 x 0.6	13.0		6	Mag 13.8 star on N end; mag 12.0 star E 1:6; NGC 4925 S 6:5.
13 06 29.1	−13 34 17	IC 4177	14.3	0.9 x 0.4	13.0	Sc	87	Mag 11.9 star W 8:8.
13 10 22.6	−07 10 17	IC 4209	13.7	1.3 x 0.5	13.1	SB(rs)bc? I	102	Mag 11.05v star NW 4:0.
13 12 02.9	−06 59 38	IC 4212	15.7	3.1 x 2.3	17.7	SB(s)cd III-IV	54	Low surface brightness; mag 8.95v star SW 5:7.
13 17 02.0	−10 46 17	IC 4216	12.9	1.9 x 0.8	13.2	SAB(rs)cd II-III	51	Mag 10.49v star SW 2:9.
13 17 13.0	−13 09 21	IC 4217	14.6	0.7 x 0.4	13.1	Scd	18	Mag 7.77v star W 3:5.
13 17 54.4	−13 36 23	IC 4220	13.9	0.9 x 0.7	13.3	Sab	111	Mag 9.49v star S 10:9.
13 18 30.4	−14 36 32	IC 4221	12.9	1.3 x 0.8	12.8	SA(r)c pec? II-III:	166	Mag 9.53v star WNW 4:4.
12 35 00.5	−05 29 26	MCG −1-32-25	14.5	0.9 x 0.7	13.8	Sbc	75	
12 35 37.4	−07 52 44	MCG −1-32-28	14.3	4.4 x 2.2	16.6	SAB(s)dm IV-V	56	Faint, stellar anonymous galaxy, or star, SE 5:3.
12 38 05.6	−08 20 02	MCG −1-32-30	14.3	1.7 x 0.6	14.2	(R)SB(r)b:	39	Mag 9.52v star N 4:4.
12 40 16.0	−09 18 00	MCG −1-32-35	13.5	1.1 x 0.7	13.0	SAB(rs)bc: I-II	39	
12 42 16.6	−05 47 29	MCG −1-32-38	14.1	0.9 x 0.6	13.3	SAB0$^-$ pec:	6	
12 42 18.8	−05 46 33	MCG −1-32-39	14.3	0.7 x 0.3	12.4	Sab	111	
12 44 03.8	−05 40 38	MCG −1-33-1	12.3	3.9 x 2.8	14.7	SAB(rs)dm V	135	Bright core, knotty branching arms.
12 45 38.9	−07 45 56	MCG −1-33-2	13.8	1.2 x 0.3	12.6	SAB(s)bc: sp II-III	84	Mag 9.81v star S 2:7.
12 45 41.0	−06 04 15	MCG −1-33-3	12.4	3.3 x 2.4	14.5	SAB(s)m V-VI	144	Mag 8.99v star SW 15:7.
12 47 19.8	−09 29 52	MCG −1-33-6	13.4	0.9 x 0.7	12.9	E/S0	171	
12 47 38.1	−05 52 03	MCG −1-33-7	14.0	0.4 x 0.4	11.9	S?		Mag 8.94v star SW 2:2.
12 48 43.6	−05 15 13	MCG −1-33-11	13.6	1.6 x 1.6	14.5	IB(s)m V-VI		Mag 10.48v star W 5:2.
12 49 33.2	−09 44 25	MCG −1-33-17	13.7	1.3 x 0.4	12.8	Sb?	150	

RA h m s	Dec ° ′ ″	Name	Mag (V)	Dim ′ Maj x min	SB	Type Class	PA	Notes
12 50 34.6	−09 31 12	MCG −1-33-22	15.4	1.4 x 0.3	14.3	Sd pec sp	159	NGC 4717, 4717 N 3′5.
12 51 13.2	−06 33 31	MCG −1-33-27	13.4	1.0 x 0.6	12.7	Im pec	171	Mag 10.53v star N 3′1.
12 51 16.0	−07 46 17	MCG −1-33-28	13.3	1.5 x 0.6	13.1	(R′)SAB(r)0$^+$?	78	
12 52 13.6	−09 29 56	MCG −1-33-31	13.6	0.8 x 0.4	12.2	S0/a	21	
12 52 27.2	−09 45 13	MCG −1-33-32	13.6	2.4 x 0.3	13.1	Scd: sp	69	Mag 9.36v star NW 5′9; mag 13.0 star N 0′8; MCG −1-33-34 SE 2′3.
12 52 34.6	−09 46 37	MCG −1-33-34	13.5	1.1 x 0.9	13.3	SAB(s)ab pec?	129	= **Mkn 1337**. MCG −1-33-32 NW 2′3.
12 52 45.4	−07 25 48	MCG −1-33-35	14.8	1.3 x 0.4	13.9	S?	90	
12 55 09.7	−08 51 34	MCG −1-33-48	14.1	0.7 x 0.4	12.8	E	84	
12 55 30.0	−07 47 45	MCG −1-33-51	13.9	1.0 x 0.4	12.7	Sbc	45	Anonymous galaxy N 1′6.
12 55 37.0	−08 03 07	MCG −1-33-52	13.8	1.3 x 0.8	13.7	SAm: IV-V	66	Mag 10.03v star S 8′5.
12 56 10.4	−08 09 09	MCG −1-33-54	13.7	1.2 x 1.0	13.7	Sdm IV-V	24	Numerous bright knots; mag 10.03v star WSW 6′0.
12 57 15.8	−05 20 49	MCG −1-33-59	13.2	1.8 x 1.3	14.0	SA(s)dm pec III-IV	102	Numerous bright knots, star on W end.
12 57 47.2	−09 38 03	MCG −1-33-60	13.2	3.2 x 0.3	13.0	Sd: sp III	117	Mag 7.26v star S 9′0.
12 58 48.8	−06 06 47	MCG −1-33-61	14.3	1.7 x 0.2	13.0	Sdm: sp	99	Mag 7.26v star NE 3′1.
12 59 50.7	−08 44 44	MCG −1-33-62	14.0	0.8 x 0.3	12.3	Sc	111	
13 00 19.3	−08 05 14	MCG −1-33-63	14.5	1.2 x 0.2	12.8	Sbc	174	
13 00 36.6	−06 28 42	MCG −1-33-68	14.3	1.1 x 0.8	14.0	Sdm pec: IV	168	
13 01 49.3	−08 20 16	MCG −1-33-71	13.5	2.7 x 0.5	13.7	Sc: sp	36	Mag 9.00v star SW 9′5.
13 01 55.6	−06 55 36	MCG −1-33-72	14.3	1.4 x 0.2	12.8	Scd: sp	105	Star just off W end, anonymous galaxy 1′0 off E end.
13 03 46.6	−05 08 01	MCG −1-33-76	13.8	1.6 x 0.4	13.1	Sc? III	36	Mag 7.70v star SE 1′8.
13 10 17.2	−07 27 17	MCG −1-34-8	13.8	0.8 x 0.7	13.0	S	90	
13 14 09.5	−05 36 01	MCG −1-34-13	14.4	1.4 x 0.3	13.3	Sc: sp	45	
13 18 41.3	−08 26 44	MCG −1-34-14	13.5	1.3 x 0.9	13.5	IB(s)m V	33	
13 30 08.4	−05 24 57	MCG −1-34-16	14.3	0.8 x 0.7	13.5	Sc	96	
13 30 33.6	−06 05 34	MCG −1-34-17	14.0	0.6 x 0.4	12.5	E?	81	Mag 10.55v star E 0′9.
13 37 00.5	−08 12 42	MCG −1-35-7	12.9	1.4 x 0.7	12.7	S0°	171	Mag 9.65v star SW 3′6.
13 40 42.3	−07 45 05	MCG −1-35-8	13.2	1.6 x 0.7	13.1	SA(r)c pec? II	66	Mag 12.0 star SW 9′2.
13 45 38.3	−05 59 05	MCG −1-35-10	13.8	1.6 x 1.1	14.3	SB(s)m III	57	Several bright knots or superimposed stars; mag 12.4 star SW 2′2.
13 46 40.0	−09 38 47	MCG −1-35-11	14.6	1.1 x 0.5	13.8	Scd	144	Mag 6.05v star ESE 9′1.
13 48 51.3	−07 38 58	MCG −1-35-12	13.3	1.3 x 0.7	13.0	S?	168	
13 48 59.6	−07 11 40	MCG −1-35-13	14.3	2.2 x 0.5	14.3	Sb sp	26	= **Hickson 67B**. NGC 5306 SE 3′4.
12 33 55.1	−10 40 48	MCG −2-32-15	14.8	1.7 x 0.9	15.1	SAB(s)m: V	90	Mag 14.6 star superimposed S edge.
12 34 08.1	−09 47 55	MCG −2-32-16	14.6	0.8 x 0.4	13.2	Sbc	66	
12 35 50.9	−13 05 23	MCG −2-32-17	14.0	1.5 x 0.4	13.3	SAb: sp III	138	Mag 12.4 star W 7′8.
12 37 23.2	−09 45 12	MCG −2-32-18	14.3	0.9 x 0.9	14.0			mag 8.56v star N 6′0.
12 39 14.4	−10 12 20	MCG −2-32-19	14.9	1.2 x 0.6	14.4	(R′)SAB(s)c pec?	60	
12 40 42.7	−10 50 04	MCG −2-32-21	14.7	0.6 x 0.5	13.2	Sb	114	
12 40 48.1	−14 32 17	MCG −2-32-22	14.7	1.3 x 0.2	13.1	IBm? sp	24	Relatively bright star sitting on N end.
12 41 24.0	−12 36 52	MCG −2-32-23	13.3	1.4 x 0.7	13.1	SAB(rs)bc: III	51	Mag 10.34v star WSW 12′0.
12 41 47.9	−14 31 46	MCG −2-32-24	14.4	0.9 x 0.5	13.4	SBbc	84	
12 42 25.0	−12 40 40	MCG −2-32-25	14.5	1.1 x 0.2	12.7	SB(r)cd? sp III	120	
12 42 49.0	−12 23 29	MCG −2-32-26	13.7	1.3 x 0.6	13.5	E$^+$	39	
12 45 41.2	−10 43 05	MCG −2-33-3	14.0	0.8 x 0.5	12.9	Sa	123	
12 45 44.7	−10 50 42	MCG −2-33-4	14.7	0.9 x 0.6	13.9	(R′)SAc?	126	
12 46 16.2	−11 00 04	MCG −2-33-5	15.1	0.5 x 0.5	13.5	Sd		
12 46 15.1	−13 21 43	MCG −2-33-6	14.2	1.0 x 0.7	13.6	(R′)SB(s)d: III-IV	72	Mag 9.10v star SE 3′3.
12 47 18.9	−14 02 36	MCG −2-33-9	14.2	0.7 x 0.3	12.4	S0/a	75	
12 48 09.8	−10 11 17	MCG −2-33-10	14.4	1.3 x 0.5	13.7	SB(s)m V	135	**PGC 43237** NW 0′8.
12 49 05.2	−12 53 14	MCG −2-33-12	13.8	0.9 x 0.6	13.2	E?	42	
12 49 22.5	−13 21 15	MCG −2-33-14	14.1	1.0 x 0.3	12.6	Sb	9	
12 49 23.9	−10 07 05	MCG −2-33-15	11.5	4.0 x 3.3	14.1	SB(s)m IV	70	Numerous knots and superimposed stars.
12 50 04.7	−14 43 58	MCG −2-33-17	12.0	1.9 x 1.9	13.3	(R)SA(r)0$^+$ pec?		Bright star superimposed S of nucleus; mag 12.7 star W 1′4.
12 50 29.0	−10 51 17	MCG −2-33-20	12.8	3.2 x 1.4	14.3	SB(s)m IV	177	Patchy.
12 50 56.7	−13 27 27	MCG −2-33-24	14.1	1.1 x 0.8	13.8	SB?	105	Mag 11.2 star WNW 2′7; pair of faint stars on N edge; mag 7.99v star ESE 7′4.
12 50 53.3	−09 51 45	MCG −2-33-25	13.1	1.1 x 0.9	13.0	SA(r)0° pec?	90	Star on N edge.
12 51 15.0	−15 12 05	MCG −2-33-28	13.5	1.0 x 0.6	12.8	S0	57	
12 52 05.7	−15 27 33	MCG −2-33-33	13.5	1.3 x 0.6	13.1	Sb	78	Mag 12.6 star NE 1′8.
12 52 33.1	−15 31 02	MCG −2-33-36	14.0	0.5 x 0.5	12.4	Sb		**PGC 43692** NE 0′7.
12 52 25.8	−15 31 04	MCG −2-33-37	13.3	0.7 x 0.5	12.1	SB0	170	IC 829 on E edge.
12 53 36.7	−12 01 09	MCG −2-33-43	14.3	0.9 x 0.7	13.7	SBc	114	Mag 9.12v star NW 8′5.
12 53 47.9	−11 18 49	MCG −2-33-44	14.4	1.1 x 0.3	13.0	SBbc	24	
12 53 47.4	−12 51 09	MCG −2-33-45	14.1	0.6 x 0.6	12.9	Sbc		
12 53 50.7	−13 33 50	MCG −2-33-46	14.6	0.5 x 0.3	12.7	E	120	Mag 8.55v star S 5′3.
12 53 56.8	−12 06 14	MCG −2-33-47	14.1	1.9 x 0.7	14.3	IB(s)m V	138	MCG −2-33-43 NW 7′2.
12 54 37.3	−11 36 22	MCG −2-33-52	15.6	1.9 x 0.2	14.4	S?	168	Mag 6.00v star SW 5′2.
12 54 43.6	−12 45 42	MCG −2-33-54	12.7	0.5 x 0.4	11.1	E	141	Mag 9 stars at N end; anonymous galaxy E 2′1.
12 54 46.8	−13 05 15	MCG −2-33-55	14.1	0.5 x 0.5	12.5	S0		
12 54 50.6	−14 07 29	MCG −2-33-57	13.9	0.6 x 0.5	12.6	E	3	
12 54 51.1	−11 53 32	MCG −2-33-58	14.4	1.5 x 0.2	13.0	Sdm? sp	36	Mag 10.30v star W 3′4.
12 54 55.9	−10 04 05	MCG −2-33-59	13.9	0.7 x 0.5	12.8	E/S0	0	
12 55 51.0	−11 33 44	MCG −2-33-62	14.0	0.9 x 0.8	13.5		132	Anonymous galaxy NW 1′6.
12 55 50.0	−13 19 20	MCG −2-33-63	14.2	0.7 x 0.4	12.6	S0/a	18	
12 56 04.6	−11 35 58	MCG −2-33-64	17.6	1.1 x 0.7	17.1		45	
12 56 11.2	−14 25 13	MCG −2-33-65	14.3	0.7 x 0.5	13.0	Irr	15	
12 56 35.0	−13 23 39	MCG −2-33-66	14.5	1.5 x 0.2	13.0	SB(s)m? sp V	75	Located between a close pair of stars, mags 12.8 and 13.5.
12 57 06.0	−12 16 20	MCG −2-33-68	13.9	2.1 x 0.6	14.0	SB(rs)cd? III-IV?	90	Mag 12.6 star W 2′7.
12 57 13.2	−15 02 13	MCG −2-33-71	13.6	1.4 x 0.9	13.7	SB(rs)bc: III	150	Mag 8.28v star SW 7′5.
12 57 39.5	−10 43 33	MCG −2-33-73	13.9	1.1 x 0.6	13.3	S?	12	Mag 11.29v star W 3′0; NGC 4822 WSW 9′1.
12 58 28.1	−10 34 38	MCG −2-33-75	13.2	0.8 x 0.4	11.9	S0$^-$ pec sp	78	
12 58 30.4	−15 32 00	MCG −2-33-76	14.1	1.2 x 0.4	13.2	Sb		
12 59 39.8	−14 58 03	MCG −2-33-80	13.9	1.6 x 0.4	13.3	S0$^-$: sp	15	Mag 13.6 star N 1′8.
13 00 05.2	−15 22 00	MCG −2-33-82	13.8	1.1 x 0.4	12.7	IB(s)m? sp IV-V	129	
13 00 09.9	−13 00 38	MCG −2-33-83	14.3	0.7 x 0.3	12.5	S0	3	
13 00 15.3	−13 00 29	MCG −2-33-84	13.9	1.1 x 0.2	12.1	SB(r)0°:	81	
13 00 17.1	−12 20 48	MCG −2-33-85	13.8	1.9 x 0.9	14.2	IB(s)m V	111	Mag 10.03v star SW 9′6.
13 00 43.7	−15 42 53	MCG −2-33-88	13.9	1.6 x 1.6	14.8	IB(s)m V-VI		Mag 8.77v star SE 9′2.
13 01 06.0	−14 19 35	MCG −2-33-91	14.3	1.3 x 0.2	12.6	Sb	0	
13 02 03.3	−14 52 57	MCG −2-33-93	13.3	1.2 x 0.6	12.8	SBc	174	
13 02 03.0	−10 24 47	MCG −2-33-95	13.9	1.4 x 0.4	13.1	SB(s)b pec sp	27	Mag 9.18v star S 2′4.

(Continued from previous page)

GALAXIES

RA h m s	Dec ° ′ ″	Name	Mag (V)	Dim ′ Maj x min	SB	Type Class	PA	Notes
13 02 16.4	−11 14 12	MCG −2-33-97	14.1	0.9 x 0.6	13.3	SB(rs)cd: III-IV	51	Weak stellar nucleus; NGC 4920 S 9′.0.
13 02 19.9	−15 46 05	MCG −2-33-98	14.1	1.6 x 0.4	13.4	Sc pec sp	57	Broad streamers off NE end; **MCG −2-33-99** ESE 1′.9.
13 02 23.8	−12 20 35	MCG −2-33-100	13.6	0.9 x 0.6	12.8	S0⁻	111	
13 07 43.8	−12 33 34	MCG −2-34-1	13.6	0.7 x 0.4	12.3	E	144	
13 08 17.9	−14 32 18	MCG −2-34-2	14.5	0.7 x 0.4	13.0	SBbc	96	
13 08 34.4	−10 06 43	MCG −2-34-3	14.6	0.5 x 0.5	13.0			
13 09 40.2	−12 20 19	MCG −2-34-5	14.5	0.7 x 0.7	13.5	Sc		
13 09 47.4	−10 19 10	MCG −2-34-6	13.6	2.9 x 2.4	15.5	SAB(s)d: III-IV	141	Mag 5.96v star 50 Virginis superimposed on SW edge.
13 09 55.1	−10 43 28	MCG −2-34-7	14.6	0.8 x 0.4	13.2	Sc	72	Mag 7.81v star SE 10′.6.
13 11 58.6	−12 03 50	MCG −2-34-8	13.9	1.4 x 0.8	13.9	SB(s)d IV-V	72	mag 8.42v star NW 12′.8.
13 13 34.3	−15 25 57	MCG −2-34-10	13.3	2.6 x 2.6	15.2	SAB(s)dm IV-V		Numerous knotty, branching arms; mag 8.03v star NE 2′.6.
13 16 49.5	−13 36 26	MCG −2-34-12	14.1	0.7 x 0.7	13.1	S?		
13 17 45.0	−15 31 34	MCG −2-34-15	15.1	1.2 x 1.0	15.1	SA(r)c pec:	141	Mag 6.86v star SW 3′.5; MCG −2-34-16 N 1′.1.
13 17 46.2	−15 30 29	MCG −2-34-16	14.5	0.4 x 0.4	12.4	Sbc		MCG −2-34-15 S 1′.1.
13 18 12.3	−15 41 27	MCG −2-34-18	14.0	1.0 x 0.5	13.1	SbII-III	153	
13 18 26.6	−15 46 10	MCG −2-34-19	14.4	0.8 x 0.3	12.7	Sbc	27	
13 19 17.8	−15 09 27	MCG −2-34-24	13.8	1.4 x 0.5	13.2	Sab	102	Anonymous galaxy E 1′.0.
13 19 34.3	−14 46 21	MCG −2-34-28	14.0	0.5 x 0.5	12.6	E0		Mag 9.12v star WNW 5′.2.
13 19 42.8	−11 28 31	MCG −2-34-29	13.9	0.4 x 0.3	11.5	S0/a	162	
13 19 58.0	−12 29 20	MCG −2-34-31	13.5	0.7 x 0.7	12.6	Sc		Mag 9.26v star N 1′.3.
13 20 05.6	−14 45 19	MCG −2-34-32	13.8	0.9 x 0.6	13.0		12	Mag 8.40v star E 8′.3.
13 20 17.2	−14 48 44	MCG −2-34-33	14.3	1.1 x 0.4	13.3	S0	51	Mag 8.40v star ENE 5′.9.
13 20 44.1	−14 06 03	MCG −2-34-36	15.0	0.7 x 0.3	13.2		156	Elongated N-S **MCG −2-34-35** just off N end.
13 21 47.6	−14 06 02	MCG −2-34-38	14.0	0.4 x 0.4	11.9	S0		Anonymous galaxy NW 1′.5.
13 22 25.6	−10 44 39	MCG −2-34-40	14.8	0.8 x 0.5	13.6	SBc	27	Mag 9.14v star ENE 9′.2.
13 24 38.9	−10 29 21	MCG −2-34-45	14.3	0.7 x 0.5	13.0		21	Mag 9.05v star NW 4′.6.
13 25 31.2	−14 21 23	MCG −2-34-46	13.8	0.7 x 0.5	12.6	S0	78	
13 25 57.8	−12 09 31	MCG −2-34-47	13.5	0.9 x 0.8	13.2	E2		
13 26 16.1	−12 36 56	MCG −2-34-48	14.3	1.6 x 1.6	15.2	S0?		Small triangle of mag 13-15 stars SE 1′.9.
13 26 37.4	−10 27 21	MCG −2-34-50	14.1	1.0 x 0.7	13.6	LSB	123	
13 26 44.7	−11 12 46	MCG −2-34-51	14.1	1.5 x 0.4	13.4	Sd? sp	0	Mag 9.20v star SSE 8′.4.
13 27 09.5	−11 47 19	MCG −2-34-52	13.8	0.4 x 0.4	11.8	E		
13 27 24.8	−11 48 34	MCG −2-34-53	15.3	0.9 x 0.9	14.9	Sb		
13 27 55.6	−13 25 24	MCG −2-34-54	12.6	1.9 x 1.2	13.3	SB(rs)bc I	150	Bright SE-NW bar.
13 28 20.3	−11 47 04	MCG −2-34-55	14.4	2.2 x 0.2	13.4	Sc sp	114	Mag 10.58v star SW 4′.9.
13 29 18.6	−10 32 02	MCG −2-34-56	14.7	1.2 x 0.4	13.8	Sbc	135	
13 29 29.9	−12 29 09	MCG −2-34-57	14.3	1.0 x 0.2	12.4	Sab	99	
13 29 49.3	−10 52 59	MCG −2-34-58	13.9	1.3 x 0.2	12.3	S?	69	MCG −2-34-59 and mag 12.4 star N 7′.1.
13 29 54.4	−10 46 43	MCG −2-34-59	13.9	1.1 x 0.6	13.3	SBbc	141	Several bright knots and/or superimposed stars; mag 8.39v star NW 6′.0.
13 30 17.1	−10 03 18	MCG −2-34-60	13.1	1.3 x 0.7	12.8	SB(rs)c II-III	135	
13 31 03.7	−15 06 06	MCG −2-34-61	14.3	1.5 x 0.3	13.3	Sc pec sp	96	Mag 12.3 star NW 3′.4.
13 32 39.2	−10 28 56	MCG −2-35-1	14.1	0.7 x 0.6	13.0	SBb	129	
13 33 50.0	−12 07 52	MCG −2-35-2	13.1	0.8 x 0.7	12.5	E/S0	153	
13 35 15.0	−10 53 34	MCG −2-35-3	14.3	0.7 x 0.5	13.0	SBbc	39	Mag 8.82v star E 3′.4; MCG −2-35-4 on N edge.
13 35 17.1	−10 53 06	MCG −2-35-4	14.4	0.7 x 0.4	12.9	SBb	24	MCG −2-35-3 on S edge.
13 36 05.5	−14 54 06	MCG −2-35-5	13.4	1.1 x 1.0	13.3		117	mag 8.72v star SW 7′.3.
13 37 05.1	−11 49 45	MCG −2-35-6	13.9	1.8 x 1.6	14.9	SB(s)dm IV-V	3	
13 37 10.9	−15 04 17	MCG −2-35-7	14.4	0.5 x 0.2	11.8	Sb	21	
13 37 16.8	−15 04 55	MCG −2-35-8	13.2	1.0 x 0.5	12.3	Sbc	123	
13 37 41.4	−15 06 15	MCG −2-35-9	14.3	0.9 x 0.7	13.7	SBm	9	
13 38 10.3	−09 48 07	MCG −2-35-10	12.2	2.2 x 2.0	13.7	SB(s)m IV	90	Many knots and/or superimposed stars.
13 38 30.7	−10 42 12	MCG −2-35-11	13.1	1.5 x 1.5	13.8	S0 pec		Bright knots N and NE of bright core.
13 41 01.6	−10 46 01	MCG −2-35-13	14.4	1.2 x 1.2	14.6	Sb		
13 48 22.0	−13 33 44	MCG −2-35-16	14.1	0.9 x 0.7	13.4	Sb	162	Lies 1′.4 S of mag 11.6 star.
12 45 11.4	−15 49 00	MCG −3-33-5	14.5	0.8 x 0.5	13.3	SBbc	42	
12 51 38.4	−16 17 49	MCG −3-33-12	14.6	0.5 x 0.5	12.9	Sc		
12 54 28.8	−16 21 06	MCG −3-33-15	14.1	0.6 x 0.6	12.9	S0/a		
12 57 00.5	−17 19 20	MCG −3-33-23	14.0	0.8 x 0.5	12.8	SBc	144	Mag 7.95v star SE 9′.4; in galaxy cluster A 1644, many anonymous and PGC galaxies near.
13 02 26.3	−17 40 50	MCG −3-33-28	13.5	2.8 x 0.3	13.1	Sc: sp II	42	
13 03 16.4	−17 25 23	MCG −3-33-30	13.2	5.0 x 0.8	14.5	IB(s)m sp V	117	Numerous bright knots clustered in central area. Galaxy becomes extremely faint beyond central one third of length.
13 05 15.5	−16 53 17	MCG −3-33-31	14.5	1.5 x 0.9	14.6	SAB(s)m: IV-V	135	Lies between a pair of mag 13.7 stars.
13 07 56.7	−16 41 21	MCG −3-34-2	14.1	0.9 x 0.5	13.1	Sab	132	
13 09 44.0	−16 36 08	MCG −3-34-4	13.2	2.1 x 0.5	13.1	S0⁺:	90	Mag 13.4 star on N edge.
13 10 25.1	−16 55 53	MCG −3-34-8	13.5	1.3 x 0.7	13.2	Ia	171	
13 11 38.1	−16 23 45	MCG −3-34-10	14.4	0.9 x 0.7	13.7	SBb	108	
13 12 34.9	−17 32 27	MCG −3-34-14	12.4	2.5 x 0.8	13.0	SAB(s)c? pec II-III	129	
13 13 05.6	−16 28 40	MCG −3-34-19	13.7	1.2 x 0.5	13.0	SA0°?	177	
13 13 12.4	−16 07 51	MCG −3-34-20	13.9	1.0 x 0.5	13.3	E	69	
13 13 32.2	−17 04 46	MCG −3-34-22	13.2	0.9 x 0.9	12.8	Sa:		
13 14 30.3	−17 32 03	MCG −3-34-25	14.3	0.7 x 0.2	12.0	Sab	153	
13 15 00.0	−17 16 09	MCG −3-34-30	14.9	0.5 x 0.4	13.1	SBb	96	
13 15 17.5	−16 29 10	MCG −3-34-33	13.9	0.7 x 0.5	12.7	S0/a	165	NGC 5044 N 6′.1; faint anonymous galaxy E 3′.4.
13 16 19.8	−16 40 14	MCG −3-34-38	14.1	0.9 x 0.4	12.8	Sb	30	Mag 10.34v star W 1′.0.
13 16 56.1	−16 35 29	MCG −3-34-40	14.5	0.9 x 0.2	12.5	Sm (Interacting)	168	Located on N edge of NGC 5054.
13 17 06.2	−16 15 17	MCG −3-34-41	12.6	2.2 x 0.2	11.6	Sc: sp	171	Mag 10.42v star S 6′.4; MCG −3-34-42 NE 1′.8.
13 17 10.9	−16 13 47	MCG −3-34-42	14.4	0.5 x 0.4	12.7	E⁺	81	MCG −3-34-41 SW 1′.8.
13 18 56.5	−17 38 12	MCG −3-34-44	13.3	0.7 x 0.5	12.1		66	
13 20 04.5	−17 07 21	MCG −3-34-48	13.8	0.4 x 0.4	11.7	S0		
13 20 06.3	−17 07 09	MCG −3-34-49	13.8	0.5 x 0.5	12.2	S0		
13 20 18.5	−16 32 16	MCG −3-34-51	13.6	0.5 x 0.5	12.0	S?		
13 20 37.6	−16 54 53	MCG −3-34-53	14.3	1.0 x 0.2	12.4	Sa	96	**MCG −3-34-52** SW 1′.9.
13 20 40.9	−16 54 01	MCG −3-34-54	14.5	1.0 x 0.3	13.1	Sab	174	
13 21 16.1	−15 54 09	MCG −3-34-55	13.5	0.9 x 0.4	12.5	E/S0	135	
13 21 25.8	−16 18 21	MCG −3-34-56	14.1	0.9 x 0.4	12.8		81	
13 21 38.7	−16 16 42	MCG −3-34-58	13.4	1.0 x 0.7	12.8		102	
13 21 45.0	−16 25 17	MCG −3-34-59	14.9	0.4 x 0.3	12.5		174	
13 21 47.0	−16 25 12	MCG −3-34-60	13.5	0.9 x 0.4	12.2	Sa	93	

(Continued from previous page)

GALAXIES

RA h m s	Dec ° ′ ″	Name	Mag (V)	Dim ′ Maj x min	SB	Type Class	PA	Notes
13 21 51.0	−17 20 12	MCG −3-34-61	13.5	1.2 x 0.8	13.3	(R′)SA(s)0^{-}:	123	Anonymous elongated galaxy SE 1'.5.
13 22 17.9	−16 54 25	MCG −3-34-62	13.7	0.6 x 0.6	12.6	E		Mag 12.1 star SW 0'.5; very faint, spherical galaxy NE 0'.5.
13 22 19.1	−16 42 30	MCG −3-34-63	13.3	1.9 x 0.5	13.1	SB?	45	Mag 10.42v star NW 3'.5; MCG −3-34-64 SE 1'.8.
13 22 24.5	−16 43 43	MCG −3-34-64	13.4	1.1 x 0.9	13.3	SB?	42	MCG −3-34-63 NW 1'.8.
13 22 50.6	−17 01 54	MCG −3-34-65	13.8	0.7 x 0.4	12.3	S0	108	
13 24 16.4	−16 42 17	MCG −3-34-67	14.2	0.9 x 0.5	13.2	Sd	171	
13 26 08.4	−16 31 11	MCG −3-34-76	13.4	1.3 x 1.1	13.7	S?	6	
13 28 00.1	−16 21 23	MCG −3-34-79	13.8	0.8 x 0.5	12.7	S0	6	
13 28 47.0	−17 30 46	MCG −3-34-82	12.9	1.4 x 0.7	12.8	SB(rs)0°	6	Mag 12 star at S end.
13 31 03.3	−16 07 50	MCG −3-34-87	14.2	0.8 x 0.3	12.5	Sa	84	Mag 8.08v star SSW 7'.6.
13 33 12.5	−16 08 24	MCG −3-35-3	14.3	0.6 x 0.4	12.6	Sb	120	
13 33 13.9	−16 07 17	MCG −3-35-4	13.1	2.2 x 1.3	14.1	SAB(s)ab pec:	114	Pair with MCG −3-35-3 S; mag 10.87v star SW 2'.0.
13 35 04.9	−15 52 06	MCG −3-35-5	13.7	0.7 x 0.7	12.7	Sa		Spherical anonymous galaxy NE 0'.7.
13 36 27.9	−17 16 03	MCG −3-35-6	14.1	0.8 x 0.5	12.9	Sb	165	
13 36 41.1	−16 59 15	MCG −3-35-7	14.0	0.7 x 0.3	12.1		33	
13 36 45.4	−16 56 58	MCG −3-35-9	13.4	0.9 x 0.6	12.9	E	171	
13 37 35.3	−16 37 58	MCG −3-35-10	14.2	0.7 x 0.3	12.4		99	
13 45 01.9	−15 52 04	MCG −3-35-17	14.2	0.9 x 0.7	13.5	Sbc	90	Mag 6.19v star NE 10'.0.
12 32 17.5	−07 33 51	NGC 4504	11.2	3.2 x 2.2	13.2	SA(s)cd II-III	153	Bright central part, two main, massive arms.
12 33 49.9	−07 22 34	NGC 4520	14.0	1.1 x 0.5	13.2	SA0^{-} pec sp	104	= **IC 799**.
12 33 54.5	−12 01 38	NGC 4524	13.4	1.2 x 0.7	13.1	SAB(r)bc II-III	80	
12 39 39.8	−05 20 39	NGC 4593	10.9	3.9 x 2.9	13.4	(R)SB(rs)b II	114	Bright nucleus in short bar encircled by two asymmetrical arms. **MCG −1-32-33** E 3'.8.
12 40 00.4	−11 37 35	NGC 4594	8.0	8.7 x 3.5	11.6	SA(s)a sp	87	= **M 104**, **Sombrero Galaxy**. Very bright central bulge, narrow patchy arms in lens, strong dark lane in front.
12 40 12.6	−05 47 55	NGC 4597	12.0	4.1 x 1.9	14.1	SB(rs)m III	34	Very patchy appearance, many bright knots.
12 40 37.0	−05 07 59	NGC 4602	11.5	3.4 x 1.2	12.9	SAB(rs)bc II	99	**PGC 42481** N 2'.7; stellar **PGC 42468** S 3'.7.
12 40 45.2	−05 18 13	NGC 4604	13.8	1.0 x 0.4	12.6	Im: sp III	114	
12 42 25.4	−07 02 41	NGC 4626	13.4	1.4 x 0.4	12.6	SB(s)bc: sp III	39	**PGC 42600** SW 11'.2.
12 42 25.4	−06 58 17	NGC 4628	13.5	1.7 x 0.5	13.2	SA(s)b: sp II-III	46	Pair of mag 12 stars 1'.6 N.
12 44 37.7	−10 05 03	NGC 4658	12.5	2.1 x 0.9	13.0	SB(s)bc II-III	3	Lies 2'.4 E of a mag 8.39v star.
12 44 47.3	−10 11 54	NGC 4663	13.1	1.0 x 0.8	12.8	SB(s)0°?	170	
12 45 47.6	−07 04 12	NGC 4671	12.6	1.3 x 1.0	12.7	E^{+} pec:	141	
12 46 03.7	−08 39 18	NGC 4674	13.1	1.7 x 0.6	13.0	SAB(s)a pec sp III	119	Mag 8.88v star WSW 8'.8.
12 46 54.6	−11 38 12	NGC 4680	12.8	1.4 x 1.2	13.2	Pec	40	Large, bright center; two faint stars on E edge.
12 47 15.5	−10 03 48	NGC 4682	12.2	2.6 x 1.3	13.3	SAB(s)cd III	83	tightly wound, filamentary arms.
12 48 35.9	−05 48 04	NGC 4697	9.2	7.2 x 4.7	13.0	E6	70	Bright center, many stars superimposed; fairly bright anonymous galaxy WNW 5'.9.
12 49 02.4	−08 39 54	NGC 4699	9.5	3.8 x 2.6	11.9	SAB(rs)b	45	Very bright center.
12 49 07.9	−11 24 41	NGC 4700	11.9	3.2 x 0.8	12.8	SB(s)c? sp II-III	48	
12 49 18.9	−09 06 33	NGC 4703	13.2	3.0 x 0.6	13.7	Sb sp	156	Dark lane extends entire length of galaxy.
12 49 25.0	−05 11 43	NGC 4705	11.7	3.0 x 1.0	12.8	SAB(s)bc: sp II	125	Very patchy arms with several lanes and knots; mag 9.19v star NE 4'.9.
12 49 41.4	−11 05 37	NGC 4708	13.1	1.2 x 0.9	13.0	SA(r)ab pec?	55	
12 50 19.4	−13 19 30	NGC 4714	12.7	1.6 x 1.2	13.3	SAB(rs)0^{-}?	150	
12 50 33.2	−09 27 04	NGC 4716	12.9	1.1 x 0.8	12.6	S0° pec?	60	Multi-galaxy system with NGC 4717 S edge.
12 50 34.4	−09 27 50	NGC 4717	13.2	1.5 x 0.7	13.1	SB(rs)a pec?	12	Multi-galaxy system with NGC 4716 N edge.
12 50 33.0	−05 16 58	NGC 4718	13.4	1.8 x 0.6	13.3	SB(rs)b? sp II-III?	98	
12 51 32.3	−13 19 51	NGC 4722	12.6	1.8 x 0.7	12.7	SB(r)0/a	33	Mag 7.99v star S 9'.6.
12 51 02.9	−13 14 13	NGC 4723	14.5	1.1 x 0.6	13.9	SB(s)m pec? IV-V	40	Mag 11.7 star SW 3'.2.
12 50 53.7	−14 19 56	NGC 4724	12.7	1.0 x 0.6	12.0	SB0^{-}:	95	Located on W edge of NGC 4727.
12 51 32.4	−14 13 18	NGC 4726	13.6	0.8 x 0.4	12.3	Sb	85	
12 50 57.5	−14 19 57	NGC 4727	11.8	1.4 x 1.1	12.1	SAB(r)bc pec: II-III?	130	Three faint stars or condensations S edge; NGC 4724 on W edge; IC 3822 W 8'.4.
12 51 00.1	−06 23 22	NGC 4731	11.5	6.6 x 4.2	14.9	SB(s)cd III	95	Short, bright bar SE-NW; two filamentary arms W of bar extending S; many knots and patches E of bar.
12 51 37.2	−08 24 35	NGC 4739	12.5	1.6 x 1.4	13.4	E^{+} pec:	165	Very faint, elongated anonymous galaxy W 8'.0.
12 51 47.8	−10 27 21	NGC 4742	11.3	2.6 x 1.5	12.8	E4:	81	Mag 6.47v star NW 9'.4.
12 52 12.6	−13 24 52	NGC 4748	13.7	1.0 x 0.7	13.2	Sa	45	Mag 10.55v star S 3'.1.
12 52 52.6	−15 24 49	NGC 4756	12.4	1.6 x 1.3	13.1	SAB(s)0°?	50	Located near center of galaxy cluster A 1631, numerous faint anonymous and PGC galaxies N and S.
12 52 50.1	−10 18 39	NGC 4757	13.7	1.5 x 0.4	13.0	S0 sp	60	Very faint stars on edge N and S of center.
12 53 07.2	−10 29 39	NGC 4760	11.4	2.0 x 1.9	12.8	E0?	10	Mag 8.95v star SSW 4'.8.
12 53 09.8	−09 11 51	NGC 4761	13.8	0.9 x 0.4	12.7	E^{+}	173	On E edge of NGC 4778.
12 53 27.2	−17 00 22	NGC 4763	12.6	1.2 x 0.8	12.4	SB(r)a:	132	
12 53 06.8	−09 15 30	NGC 4764	15.1	0.5 x 0.3	12.9	S?	115	Located 2'.0 SSE of mag 9.42v star.
12 53 08.1	−10 22 45	NGC 4766	14.4	1.0 x 0.3	13.0	S0 sp	130	Elongated E-W anonymous galaxy 1'.2 NW.
12 53 32.2	−09 32 31	NGC 4770	12.7	1.4 x 0.7	12.6	S0/a:	90	Mag 4.79v star Psi Virginis E 11'.9.
12 53 36.2	−08 38 21	NGC 4773	12.7	1.3 x 1.0	12.8	E^{+} pec:	88	**MCG −1-33-42** on S edge.
12 53 45.7	−06 37 14	NGC 4775	11.1	2.2 x 2.0	12.5	SA(s)d III	75	Very patchy, many knots on complex arms.
12 53 04.5	−09 11 59	NGC 4776	13.0	0.9 x 0.9	12.7	S0° pec		0'.4 NW of center of NGC 4778.
12 53 58.6	−08 46 32	NGC 4777	13.1	1.9 x 0.9	13.5	(R)SAB(s)a:	15	Bright elongated nucleus, smooth outer envelope.
12 53 05.8	−09 12 15	NGC 4778	12.5	1.8 x 1.8	13.6	S0^{+} sp		The center of NGC 4776 is NW 0'.4; NGC 4761 on E edge; a mag 9.42v star lies 1'.5 S.
12 54 05.0	−08 37 22	NGC 4780	13.4	1.9 x 1.3	14.2	SAB(rs)c III	18	Faint star W of nucleus; elongated anonymous galaxy S 2'.1.
12 54 23.5	−10 32 10	NGC 4781	11.1	3.5 x 1.5	12.7	SB(rs)d III	120	Several knotty, filamentary arms; faint star on W edge.
12 54 36.1	−12 34 11	NGC 4782	11.7	1.8 x 1.7	12.9	E0 pec	2	Interacting with, and connected to, NGC 4783 N.
12 54 36.2	−12 33 29	NGC 4783	11.6	1.3 x 1.3	12.2	E0 pec		Pair of faint stars N and W of large nucleus. See notes for NGC 4782.
12 54 36.9	−10 36 47	NGC 4784	13.6	1.9 x 0.4	13.1	S0 sp	104	Mag 11.06v star W 2'.1.
12 54 32.6	−06 51 30	NGC 4786	11.7	1.6 x 1.2	12.3	E^{+} pec	170	
12 54 52.4	−10 14 52	NGC 4790	12.1	1.7 x 1.1	12.6	SB(rs)c:? III	90	Several small knots; line of four faint stars on N-S line on E end.
12 55 03.6	−12 29 48	NGC 4792	14.2	0.8 x 0.4	12.8	S0 sp	140	Small, faint elongated anonymous galaxy W 2'.4.
12 55 10.6	−12 36 31	NGC 4794	13.3	1.9 x 0.8	13.6	SB(rs)a	153	Bright bar extends NE-SW; faint stars on E and W edges.
12 55 49.9	−12 03 14	NGC 4802	11.4	2.4 x 1.6	12.8	SA(r)0?	20	Mag 10.95v star on E edge of nucleus; mag 9.44v star 1'.9 W.
12 56 36.4	−06 49 01	NGC 4813	13.1	1.2 x 0.5	12.4	S0°	35	**MCG −1-33-58** E 7'.9.
12 56 49.1	−08 31 34	NGC 4818	11.1	4.3 x 1.5	13.0	SAB(rs)ab pec:	0	Bright somewhat elongated core, moderately dark patch N and S of core.
12 57 00.6	−13 43 11	NGC 4820	13.9	1.1 x 0.3	12.6	S0 sp	105	NGC 4825 NNE 4'.3.
12 57 03.8	−10 45 43	NGC 4822	13.3	1.2 x 0.7	13.0	S0^{-}	90	Mag 8.85v star S 7'.5; MCG −2-33-73 ENE 9'.1.
12 57 24.4	−13 44 16	NGC 4823	15.0	0.4 x 0.3	12.8	E	105	Lies half way between NGC 4829 N and a mag 12.7 star S.
12 57 12.2	−13 39 52	NGC 4825	11.7	1.9 x 1.2	12.5	SA0^{-}	138	Mag 14.5 star on N edge; NGC 4829 SE 3'.8; NGC 4820 SSW 4'.3.
12 57 25.7	−13 41 55	NGC 4829	14.7	0.7 x 0.2	12.4	S0/a	175	NGC 4825 NW 3'.8.
12 57 34.4	−12 44 40	NGC 4836	13.4	1.4 x 1.1	13.7	SAB(rs)d pec: III-IV	5	Mag 8.30v star S 4'.1.
12 57 56.4	−13 03 38	NGC 4838	13.0	1.6 x 1.4	13.7	(R′)SB(r)b: II-III	150	Bright nucleus in short bar; mag 9.09v star S 4'.8.
12 58 29.0	−13 08 31	NGC 4847	14.3	0.6 x 0.5	13.1	E	21	Very compact; mag 9.09v star W 9'.6.

(Continued from previous page)

GALAXIES

RA h m s	Dec ° ′ ″	Name	Mag (V)	Dim ′ Maj x min	SB	Type Class	PA	Notes
12 59 18.5	−13 13 49	NGC 4855	12.9	1.8 x 1.3	13.7	(R′)SB(s)0^- pec:	155	Bright core with long, very faint extension S, and shorter extension N.
12 59 20.6	−15 02 40	NGC 4856	10.5	4.3 x 1.2	12.1	SB(s)0/a	35	Triangle of very faint stars on NE edge.
12 59 30.9	−14 07 59	NGC 4862	14.2	1.0 x 0.8	13.8	SB(s)c II-III	153	Stellar nucleus with several tightly wound spiral arms; mag 9.82v star W 4′.6.
12 59 42.6	−14 01 41	NGC 4863	13.7	1.8 x 0.4	13.2	SB(rs)0^+	24	
13 00 26.5	−15 17 01	NGC 4877	12.2	2.2 x 1.0	12.9	SA(s)ab: II-III	7	Located 2′.8 SE of a mag 9.17v star.
13 00 20.2	−06 06 15	NGC 4878	12.9	1.3 x 1.1	13.2	SB(r)0^+	69	Bright, elongated core; uniform envelope.
13 00 33.8	−06 51 13	NGC 4885	14.0	0.5 x 0.4	12.1	Pec	138	
13 00 39.3	−14 39 59	NGC 4887	13.1	1.1 x 0.6	12.5	SA0^+ pec?	155	
13 00 36.4	−06 04 33	NGC 4888	17.4	1.0 x 0.4	16.2	Sab	119	
13 00 52.9	−13 27 02	NGC 4897	11.8	2.6 x 2.3	13.6	SAB(r)bc: II-III	150	Bright nucleus, faint branching arms.
13 00 56.3	−13 56 43	NGC 4899	11.9	2.6 x 1.4	13.1	SAB(rs)c: II-III	20	Many bright knots and patches.
13 01 00.2	−14 30 49	NGC 4902	10.9	3.0 x 2.7	13.0	SB(r)b I-II	70	Pair of mag 10 stars NW and SW.
13 02 04.3	−11 22 42	NGC 4920	13.5	1.0 x 0.7	13.0	IB(s)m III	170	MCG −2-33-97 N 9′.0.
13 02 13.0	−14 58 13	NGC 4924	12.8	1.9 x 1.5	13.8	(R′)SAB(s)0/a pec?	60	Large, bright center, strong dark lane SE.
13 02 07.4	−07 42 40	NGC 4925	13.1	1.0 x 0.7	12.5	S0	135	IC 4071 N 6′.5.
13 03 00.6	−08 05 08	NGC 4928	12.5	1.3 x 1.0	12.6	SA(s)bc pec III-IV	50	
13 03 57.0	−11 29 50	NGC 4933A	11.7	1.8 x 1.1	12.3	S0/a pec	54	Large bright nucleus; interacting with NGC 4933B SW; NGC 4933C located at NE end.
13 03 53.8	−11 30 27	NGC 4933B	14.2	0.8 x 0.4	13.0	E pec	69	Very bright nucleus; faint, asymmetrical extension SW.
13 04 01.4	−11 29 28	NGC 4933C	15.3	0.5 x 0.4	13.4	Sd	135	Extremely faint, low surface brightness.
13 04 14.2	−10 20 26	NGC 4939	11.3	5.5 x 2.8	14.1	SA(s)bc I-II	10	Two main, knotty, filamentary arms.
13 04 13.1	−05 33 08	NGC 4941	11.1	3.9 x 2.7	13.5	(R)SAB(r)ab: II	15	Large, bright center with moderately dark lanes; faint outer envelope.
13 04 19.2	−07 39 00	NGC 4942	13.0	1.8 x 1.2	13.7	SAB(s)d: III	145	Mottled, knotty appearance.
13 04 55.7	−07 56 48	NGC 4948	13.2	2.0 x 0.6	13.3	SB(s)dm: sp III	145	Elongated central core with star or knot on NW end; double star SE 1′.6.
13 05 05.8	−08 09 48	NGC 4948A	13.4	1.5 x 1.2	13.9	SB(s)dm III-IV	20	Mag 8.75v star E 4′.1.
13 05 07.7	−06 29 37	NGC 4951	11.9	3.8 x 1.6	13.7	SAB(rs)cd: II-III	90	Bright core with dark lane NE, several knots and superimposed stars.
13 05 48.9	−08 01 14	NGC 4958	10.7	4.1 x 1.2	12.3	SB(r)0? sp	9	Very bright lens.
13 07 50.2	−05 01 04	NGC 4975	13.9	1.0 x 0.6	13.3	SAB(rs)0° pec?	46	Faint **MCG −1-34-1** WNW 3′.9.
13 08 48.9	−06 46 38	NGC 4981	11.3	2.7 x 1.9	12.9	SAB(r)bc II	159	Many filamentary, branching arms; bright star on S edge.
13 08 57.2	−15 30 59	NGC 4984	11.3	3.6 x 2.5	13.6	(R)SAB(rs)0^+	18	Very bright nucleus; several superimposed stars; very faint material extends beyond published size.
13 09 16.1	−05 23 47	NGC 4989	13.3	1.5 x 0.9	13.5	S0°:	171	**MCG −1-34-6** S 1′.4.
13 09 17.4	−05 16 23	NGC 4990	13.9	1.0 x 0.8	13.5	S0 pec?	55	Almost stellar nucleus.
13 09 40.6	−07 50 00	NGC 4995	11.1	2.4 x 1.5	12.4	SAB(rs)b II	95	Located 3′.4 SSE of mag 8.21v star; bright nucleus, several tightly wound arms.
13 09 51.6	−16 30 58	NGC 4997	12.9	1.3 x 0.8	12.9	$S0^-$	94	Located 2′.2 E of a mag 6.74v star.
13 12 26.2	−15 47 51	NGC 5010	13.3	1.4 x 0.7	13.2	$S0^+$ pec sp	120	
13 12 54.6	−16 45 56	NGC 5017	12.6	1.8 x 1.3	13.4	E^+?	39	Mag 10.30v star E 8′.6.
13 13 46.1	−13 02 34	NGC 5028	12.3	1.8 x 0.9	12.8	E6	130	Mag 12.0 star on W edge.
13 13 54.1	−16 29 23	NGC 5030	12.7	1.8 x 1.3	13.5	SB(r)0^+?	0	Bright, elongated core; mag 13 star on NW edge.
13 14 03.0	−16 07 22	NGC 5031	12.9	1.6 x 0.4	12.2	$S0^+$: sp	114	Large nucleus.
13 14 49.3	−16 29 33	NGC 5035	12.8	1.4 x 1.1	13.1	SAB(r)0^+	156	Located between a mag 10.06v star NNE and a mag 9.76 v star S.
13 14 59.2	−16 35 26	NGC 5037	12.2	2.2 x 0.7	12.5	SA(s)a:	43	Star on NE end. Stellar anonymous galaxy 0′.9 E.
13 15 02.3	−15 57 08	NGC 5038	13.5	1.4 x 0.3	12.4	S0: sp	95	
13 15 24.1	−16 23 04	NGC 5044	10.8	3.0 x 3.0	13.2	E0		Stellar anonymous galaxy W 4′.8; MCG −3-34-33 S 6′.1.
13 15 45.2	−16 19 36	NGC 5046	13.5	0.8 x 0.7	12.9	E?	54	Mag 8.87v star N 3′.2.
13 15 48.5	−16 31 07	NGC 5047	12.6	2.8 x 0.6	13.0	S0 sp	75	Bright central bulge, some faint stars along S edge.
13 15 59.2	−16 23 47	NGC 5049	13.0	1.9 x 0.6	13.0	S0 sp	120	NGC 5044 W 8′.4.
13 16 59.5	−16 38 00	NGC 5054	10.9	5.1 x 3.0	13.7	SA(s)bc I-II	160	Well defined patchy arms wit some knots; MCG −3-34-40 on N edge.
13 18 28.4	−10 14 00	NGC 5066	13.3	0.7 x 0.5	12.0	Sa	0	
13 19 17.8	−12 29 00	NGC 5070	14.4	0.7 x 0.3	12.6	S0/a	160	Mag 7.91v star E 4′.7.
13 19 12.6	−12 32 25	NGC 5072	13.1	1.0 x 1.0	13.0	$SA0^-$:		Star on S edge of bright nucleus; mag 7.91v star NE 7′.0.
13 19 20.5	−14 50 38	NGC 5073	12.6	3.4 x 0.7	13.4	SB(s)c? sp	147	MCG −2-34-28 5′.4 NE, located 1′.0 W of mag 12 star.
13 19 30.5	−12 44 29	NGC 5076	12.7	1.1 x 0.8	12.4	SB(rs)0^+	40	Bright, elongated nucleus.
13 19 31.5	−12 39 25	NGC 5077	11.4	2.2 x 1.6	12.7	E3-4	7	Bright, diffuse, elongated core; round anonymous galaxy on E edge.
13 19 38.1	−12 41 57	NGC 5079	11.7	1.4 x 0.7	11.5	SB(rs)bc pec: II-III	31	Moderately bright nucleus; bright patch NE of nucleus.
13 20 20.3	−12 34 18	NGC 5088	12.4	2.6 x 0.8	13.0	SAB(s)bc: III:	0	Mag 9.26v star and MCG −2-34-31 NNW 7′.8.
13 20 46.8	−14 04 49	NGC 5094	13.0	1.1 x 0.9	13.1	E	105	MCG −2-34-36 SW 1′.4 with **MCG −2-34-35** at N end.
13 20 59.8	−12 28 17	NGC 5097	14.6	0.5 x 0.3	12.7	E	45	Mag 11.8 star NE 2′.4.
13 21 19.5	−13 02 31	NGC 5099	14.3	0.6 x 0.6	13.2	E		Compact.
13 21 49.3	−13 12 27	NGC 5105	11.8	2.0 x 1.5	12.8	SB(rs)c II-III	140	Patchy appearance; pair of faint stars on E edge.
13 22 37.1	−13 03 57	NGC 5110	12.1	0.7 x 0.3	10.5	E	102	Mag 7.81v star S 7′.5.
13 22 56.5	−12 57 53	NGC 5111	12.1	2.0 x 1.7	13.3	S0	120	Very faint, stellar anonymous galaxy ESE 3′.8.
13 24 00.5	−12 16 36	NGC 5119	13.0	1.3 x 0.4	12.1	S0/a	19	
13 24 14.9	−10 39 15	NGC 5122	13.4	0.8 x 0.4	12.0	Sc	115	Very faint extensions NE and SW.
13 24 27.0	−10 12 39	NGC 5130	13.4	0.9 x 0.5	12.4	S0	40	
13 26 37.6	−12 19 25	NGC 5146	12.6	1.9 x 1.2	13.4	E/S0	35	Chain of four faint stars down E side of bright nucleus.
13 29 48.9	−17 57 55	NGC 5170	11.1	8.3 x 1.0	13.3	SA(s)c: sp I	127	Central dark lane; **ESO 576-78** N 3′.0.
13 32 13.4	−08 47 09	NGC 5203	12.6	1.9 x 1.1	13.3	$S0^-$ sp	85	Two faint stars on N edge, two faint stars on S edge form rectangle surrounding galaxy.
13 36 08.2	−08 29 48	NGC 5232	12.1	1.5 x 1.4	12.8	SA(rs)0/a:	110	Large, bright center.
13 36 39.9	−08 24 05	NGC 5241	12.9	1.1 x 0.5	12.1	SB(rs)ab:	47	Mag 10.9v star E 0′.9.
13 38 03.1	−17 53 03	NGC 5247	10.0	5.6 x 4.9	13.4	SA(s)bc I-II	20	Bright nucleus, very knotty arms.
13 39 38.1	−11 29 40	NGC 5254	12.0	3.2 x 1.5	13.5	SA(rs)c: II-III	125	Complex, branching arms NW.
13 49 11.2	−07 13 28	NGC 5306	12.2	1.4 x 1.0	12.4	S0 pec?	30	= **Hickson 67A**. **Hickson 67C** on N edge; **Hickson 67D** on SW edge.

GALAXY CLUSTERS

RA h m s	Dec ° ′ ″	Name	Mag 10th brightest	No. Gal	Diam ′	Notes
12 52 48.0	−15 26 00	A 1631	15.4	34	34	See notes for NGC 4756.
12 57 12.0	−17 21 00	A 1644	15.7	68	30	Prominent mag 14 **PGC 44259** S of center 3′.8; numerous other anonymous/PGC GX, stellar to almost stellar.

OPEN CLUSTERS

RA h m s	Dec ° ′ ″	Name	Mag	Diam ′	No. ★	B ★	Type	Notes
12 35 44.1	−12 02 00	Canali		7.5	6	7.4	ast?	Asterism or possible cluster. Noted 1997 by E. Canali.

RA h m s	Dec ° ′ ″	Name	Mag (V)	Dim ′ Maj x min	SB	Type Class	PA	Notes
11 38 57.5	−17 58 08	ESO 571-12	14.2	1.5 x 0.5	13.7	S?	131	Mag 10.79v star N 2'.3.
11 26 48.5	−05 00 17	IC 693	13.8	0.9 x 0.4	12.6	Sab		Mag 10.31v star E 10'.3.
11 33 12.5	−13 20 18	IC 706	13.9	1.1 x 0.2	12.1	SA(r)0/a? sp	111	Mag 9.98v star S 2'.1.
11 42 57.7	−08 19 58	IC 723	14.1	0.7 x 0.6	13.0	Irr	3	Mag 10.85v star WSW 6'.5; note: **IC 721** is 2'.6 NNW of this star.
11 53 22.4	−13 15 57	IC 743	14.5	1.0 x 0.3	13.0	SBc? sp	147	
12 05 53.7	−12 40 26	IC 761	14.8	0.8 x 0.2	12.7	S0	102	Close pair of mag 12.8, 11.2 stars E 5'.8; MCG −2-31-15 NE 1'.1.
12 10 53.6	−12 39 19	IC 766	14.0	1.5 x 0.3	13.0	SAB(r)0$^+$? sp	165	Mag 12.1 v star NW 10'.4.
12 23 02.2	−13 13 26	IC 785	14.0	0.9 x 0.3	12.4	SB(s)c I-II	75	Mag 10.91v star N 2'.5.
12 23 11.0	−13 12 15	IC 786	13.4	1.0 x 0.9	13.1	(R)SAB0°:	174	Mag 10.91v star NW 1'.9.
11 28 16.2	−12 53 29	IC 2856	13.7	0.9 x 0.5	12.6	Sab	42	Mag 10.53v star SE 6'.3.
11 30 29.1	−13 05 29	IC 2889	13.8	1.2 x 0.7	13.4	SAB(s)c pec: III	156	Mag 7.05v star NE 3'.4.
11 31 54.7	−09 43 32	IC 2910	13.6	1.7 x 1.1	14.1	(R′)SB(rs)0$^+$ pec:	130	Mag 12.2 star NNE 6'.4.
11 49 24.8	−05 07 09	IC 2963	13.9	1.4 x 0.3	12.8	S0$^+$: sp	99	Very faint star on N edge.
11 53 48.7	−05 10 05	IC 2974	13.2	2.2 x 0.5	13.1	SA(s)c? sp II-III	99	Mag 9.29v star NNW 6'.9.
11 20 39.4	−05 40 21	MCG −1-29-10	13.9	1.0 x 0.7	13.4	S0	36	Contains several knots, or superimposed stars?
11 21 48.0	−05 45 32	MCG −1-29-12	13.6	0.5 x 0.5	12.1	E		Mag 7.81v star N 2'.5.
11 22 27.0	−07 39 13	MCG −1-29-13	13.2	1.2 x 0.5	12.6	SB0$^-$: sp	114	
11 22 31.7	−07 03 15	MCG −1-29-14	14.2	0.8 x 0.8	13.5	SBbc		
11 22 44.0	−07 40 39	MCG −1-29-15	13.0	2.1 x 1.0	13.7	SA(s)0° pec:	51	
11 24 50.8	−05 21 09	MCG −1-29-17	13.9	0.6 x 0.5	12.5		156	
11 25 02.2	−05 04 09	MCG −1-29-18	13.6	0.7 x 0.5	12.3	Sa	162	
11 25 22.9	−09 35 14	MCG −1-29-19	13.5	0.7 x 0.7	12.6	S0		Mag 9.34v star W 7'.6.
11 30 28.1	−08 06 09	MCG −1-29-27	14.2	0.7 x 0.5	12.9	Sb	0	
11 34 20.2	−06 23 21	MCG −1-30-2	14.2	1.5 x 0.3	13.1	SAc: sp	18	
11 34 17.0	−09 34 35	MCG −1-30-3	13.0	1.1 x 0.9	12.9	S0/a	18	**MCG −1-30-4** on S edge.
11 34 37.6	−09 38 51	MCG −1-30-8	13.9	0.7 x 0.3	12.1	S0/a	81	NGC 3724 WSW 2'.2.
11 36 49.0	−08 35 14	MCG −1-30-11	13.2	1.2 x 0.3	12.0	SAb: sp	72	
11 36 49.9	−09 34 05	MCG −1-30-12	13.6	0.8 x 0.7	12.9	S0°:	48	
11 37 45.1	−07 16 03	MCG −1-30-13	13.4	1.1 x 1.1	13.5	SB(rs)c II		Very small, faint, field galaxy ENE 0'.5.
11 38 03.4	−06 14 02	MCG −1-30-14	14.3	0.9 x 0.7	13.6	Sc	147	
11 41 22.7	−06 28 56	MCG −1-30-22	13.2	1.3 x 1.1	13.4	SAB(s)m pec: V-VI	54	
11 42 19.2	−08 37 24	MCG −1-30-24	13.8	1.2 x 0.3	12.6	SAB(r)0/a:	51	
11 42 27.7	−07 46 32	MCG −1-30-25	13.4	0.9 x 0.8	12.9	Sc	171	
11 50 36.8	−07 36 32	MCG −1-30-38	14.2	0.7 x 0.6	13.1	Sbc	174	
11 51 15.3	−06 06 25	MCG −1-30-39	14.4	1.3 x 0.2	12.8	Sb: sp	135	Mag 8.97v star S 6'.7.
11 52 38.3	−05 12 25	MCG −1-30-41	13.8	1.0 x 0.4	12.6	(R)SB(rs)ab	21	
11 52 35.2	−08 46 10	MCG −1-30-42	14.3	0.7 x 0.5	13.0	Sbc	168	
11 55 13.7	−06 27 24	MCG −1-30-48	16.3	1.9 x 0.2	15.1	Scd? sp	48	Mag 10.68v star NNW 8'.9.
12 19 51.4	−06 51 21	MCG −1-32-1	13.7	1.2 x 0.4	12.8	SB(s)bc: II-III	78	
12 20 47.6	−06 58 21	MCG −1-32-4	14.8	1.4 x 0.2	13.3	SAc: sp	138	
12 25 31.6	−07 14 14	MCG −1-32-7	14.1	1.0 x 0.3	12.6	Sab	177	
12 27 11.7	−07 42 01	MCG −1-32-11	14.0	0.5 x 0.5	12.4	S0/a		Mag 9.48v star W 5'.6.
12 29 36.8	−08 26 09	MCG −1-32-16	13.9	0.9 x 0.5	12.8	(R′)SA(rs)bc pec?	147	
12 30 00.9	−05 58 55	MCG −1-32-17	13.4	0.9 x 0.7	12.8	SBab	96	Mag 7.35v star S 2'.5.
12 30 18.3	−05 57 46	MCG −1-32-18	15.1	0.9 x 0.7	14.4	(R′)SA(s)0/a?	93	Mag 8.66v star S 4'.3.
12 30 16.1	−08 23 49	MCG −1-32-19	13.2	2.4 x 2.4	15.0	SB(r)bc pec I-II		Strong dark patches E and W of short central bar.
12 30 26.7	−06 00 28	MCG −1-32-20	16.6	0.7 x 0.4	15.2	S0	126	Mag 8.66v star SSW 1'.6.
12 35 00.5	−05 29 26	MCG −1-32-25	14.5	0.9 x 0.7	13.8	Sbc	75	
12 35 37.4	−07 52 44	MCG −1-32-28	14.3	4.4 x 2.2	16.6	SAB(s)dm IV-V	56	Faint, stellar anonymous galaxy, or star, SE 5'.3.
12 38 05.6	−08 20 02	MCG −1-32-30	14.3	1.7 x 0.6	14.2	(R)SB(r)b:	39	Mag 9.52v star N 4'.4.
11 21 54.2	−12 32 20	MCG −2-29-21	14.3	1.2 x 0.8	14.1	SAB(rs)c: III-IV	3	
11 24 04.1	−13 08 37	MCG −2-29-24	14.4	0.7 x 0.7	13.5	Sm		
11 25 37.5	−11 08 24	MCG −2-29-29	13.6	0.9 x 0.7	13.2	E/S0	158	
11 26 17.4	−14 16 51	MCG −2-29-30	14.1	0.8 x 0.5	13.0	S0/a	30	
11 27 22.9	−10 57 09	MCG −2-29-31	14.2	1.1 x 0.8	13.9	SB(s)b II	66	
11 28 43.6	−14 07 12	MCG −2-29-34	14.1	1.2 x 0.9	14.1	(R)SB(r)0$^+$:	144	
11 29 51.2	−12 22 21	MCG −2-29-36	14.4	0.4 x 0.4	12.3	Sb		
11 24 52.8	−13 34 19	MCG −2-29-37	12.9	1.6 x 1.1	13.5	E$^+$	115	Mag 14.8 star N edge; mag 13.2 star NE 1'.9.
11 30 44.7	−15 17 20	MCG −2-29-39	12.8	1.6 x 1.4	13.5	(R′)SB(r)b II-III	80	Mag 9.42v star WSW 5'.0.
11 31 25.5	−13 49 13	MCG −2-29-40	14.1	0.9 x 0.3	12.5	Sbc	45	Mag 11.4 star NE 0'.8.
11 32 26.6	−09 48 58	MCG −2-30-1	14.5	0.9 x 0.6	13.7	Sc	153	Mottled and irregular appearance; mag 11.7 star W 3'.8.
11 33 10.8	−10 13 49	MCG −2-30-3	14.1	1.1 x 0.2	12.3	SABb: sp II-III?	171	
11 35 45.3	−15 42 23	MCG −2-30-7	13.4	1.0 x 0.6	12.7	S0	18	Mag 9.00v star S 5'.7.
11 36 49.0	−12 58 49	MCG −2-30-10	14.6	1.1 x 0.2	12.8	SAcd: sp	96	
11 38 20.1	−13 02 57	MCG −2-30-11	13.9	1.1 x 0.8	13.6	SAB(s)cd pec: III	141	Mag 5.49v star SSE 10'.4.
11 40 34.6	−10 05 09	MCG −2-30-14	13.0	2.2 x 1.7	14.3	(R′)SAB(rs)cd III	153	Mag 10.06v star E 8'.1.
11 41 47.3	−11 02 07	MCG −2-30-15	14.3	0.7 x 0.5	13.0	Sc	171	
11 42 11.8	−10 46 28	MCG −2-30-16	14.2	1.4 x 1.0	14.4	(R′)SA(r)ab I-II	95	
11 43 09.5	−12 51 52	MCG −2-30-18	14.4	1.3 x 0.2	12.9	SB0$^+$? sp	90	NGC 3831 E 2'.4.
11 43 15.1	−11 35 37	MCG −2-30-19	13.5	0.5 x 0.5	11.9	Sab		
11 43 10.4	−12 38 03	MCG −2-30-20	16.8	0.7 x 0.3	15.0	Sb	162	Mag 9.60v star NE 6'.2.
11 43 13.5	−12 45 51	MCG −2-30-21	14.4	1.2 x 0.2	12.7	Sc: sp	39	Mag 9.70v star S 0'.9.
11 43 20.2	−11 40 15	MCG −2-30-22	15.1	0.8 x 0.6	14.1	SBc	105	
11 43 22.6	−12 53 15	MCG −2-30-24	14.5	0.6 x 0.4	12.8	Sab	161	NGC 3831 W 1'.1.
11 43 29.7	−12 24 26	MCG −2-30-25	14.4	1.0 x 0.4	13.2	Sc	99	
11 45 04.5	−13 13 01	MCG −2-30-26	14.4	0.7 x 0.5	13.1	SBb	81	
11 45 26.2	−10 06 11	MCG −2-30-27	13.8	2.8 x 0.7	14.4	SAB(s)d III	141	MCG −2-30-28 N 2'.4.
11 45 25.6	−10 03 47	MCG −2-30-28	12.8	0.7 x 0.4	11.3	S0	15	
11 46 53.7	−14 29 10	MCG −2-30-29	14.2	1.2 x 0.7	13.9	SB(r)cd III	105	
11 48 19.2	−11 38 59	MCG −2-30-31	14.6	0.7 x 0.5	13.4	Sd	108	
11 50 13.1	−12 55 59	MCG −2-30-32	15.8	0.8 x 0.2	13.6	Sb	66	Mag 7.22v star NE 7'.8.
11 50 36.7	−12 28 05	MCG −2-30-33	14.4	0.9 x 0.3	12.8	Sc	102	
11 50 44.8	−12 43 59	MCG −2-30-34	16.5	0.7 x 0.4	15.0	Sc	9	Mag 9.04v star NW 2'.5.
11 52 34.1	−12 39 47	MCG −2-30-36	13.5	1.1 x 0.9	13.4	SAB(rs)c pec: III	39	
11 53 42.7	−15 24 40	MCG −2-30-38	13.6	0.7 x 0.3	11.8	S0/a	90	
11 54 22.3	−12 28 54	MCG −2-30-39	14.6	1.8 x 0.9	15.0	IB(s)m pec IV-V	146	Mag 8.96v star SW 10'.3.
11 56 24.9	−12 10 05	MCG −2-31-2	14.8	1.1 x 0.3	13.5	Sc	30	
11 57 10.5	−14 16 41	MCG −2-31-3	14.4	1.2 x 0.6	13.9	SAB(rs)d: III	141	
11 57 39.0	−12 25 04	MCG −2-31-4	17.2	0.7 x 0.5	15.9	Sb	177	

(Continued from previous page)

GALAXIES

RA h m s	Dec ° ′ ″	Name	Mag (V)	Dim ′ Maj x min	SB	Type Class	PA	Notes
11 57 50.3	−12 50 20	MCG −2-31-5	14.0	1.1 x 0.6	13.4	Sa	138	
11 58 25.2	−14 31 27	MCG −2-31-6	14.2	0.9 x 0.9	13.9	Im pec IV-V		Mottled appearance, star involved e.
11 59 27.0	−15 08 01	MCG −2-31-8	14.9	0.9 x 0.3	13.3	Sbc	3	
11 59 39.2	−12 23 40	MCG −2-31-9	15.7	0.4 x 0.3	13.3	S?		Anonymous elongated galaxy SE 1′.0.
12 00 34.3	−09 47 08	MCG −2-31-10	14.5	0.8 x 0.4	13.2	SBc	177	
12 01 41.2	−15 06 30	MCG −2-31-11	15.1	0.8 x 0.4	13.7	Sc	6	
12 02 36.5	−12 52 35	MCG −2-31-12	14.8	1.1 x 0.2	13.0	Scd?	39	
12 04 20.7	−12 48 43	MCG −2-31-13	13.9	0.7 x 0.5	12.6	Sa	78	Lies 0′.7 SE of mag 11 star.
12 05 57.4	−12 39 49	MCG −2-31-15	14.4	0.6 x 0.2	12.1	E?	33	Close pair of mag 12.8, 11.2 stars E 5′.0; IC 761 SW 1′.1.
12 06 51.1	−11 05 42	MCG −2-31-17	13.8	2.2 x 0.4	13.5	Scd: sp	21	Mag 9.29v star E 10′.7.
12 07 38.4	−14 58 12	MCG −2-31-19	13.1	1.4 x 0.9	13.2	(R′)SB(rs)b pec:	39	Mag 13.1 star SSW 2′.6.
12 08 14.5	−15 43 04	MCG −2-31-19A	13.3	1.0 x 0.8	12.9	SB(rs)c pec: II	9	
12 13 35.4	−13 11 56	MCG −2-31-22	13.9	0.7 x 0.5	12.6	S0	39	
12 15 05.4	−13 14 03	MCG −2-31-25	14.4	1.3 x 0.5	13.8	SABd? IV-V	129	Mag 9.35v star ESE 3′.6.
12 20 13.7	−13 14 44	MCG −2-32-2	15.0	0.5 x 0.5	13.3	Sd		
12 20 44.4	−13 48 19	MCG −2-32-5	13.9	1.6 x 0.9	14.1	SB(s)dm III-IV	123	Mag 8.79v star NE 14′.2.
12 21 44.1	−12 28 25	MCG −2-32-6	14.3	0.9 x 0.2	12.3	SB(r)c pec sp	159	
12 26 24.0	−13 40 50	MCG −2-32-10	14.4	0.7 x 0.4	12.8	Sbc	129	Mag 8.33v star S 7′.0.
12 27 02.4	−11 31 01	MCG −2-32-11	15.0	0.7 x 0.4	13.5	Sbc	87	
12 27 58.8	−13 31 30	MCG −2-32-12	14.3	0.7 x 0.3	12.7	E	93	
12 33 55.1	−10 40 48	MCG −2-32-15	14.8	1.7 x 0.9	15.1	SAB(s)m: V	90	Mag 14.6 star superimposed S edge.
12 34 08.1	−09 47 55	MCG −2-32-16	14.6	0.8 x 0.4	13.2	Sbc	66	
12 35 50.9	−13 05 23	MCG −2-32-17	14.0	1.5 x 0.4	13.3	SAb: sp III	138	Mag 12.4 star W 7′.8.
12 37 23.2	−09 45 12	MCG −2-32-18	14.3	0.9 x 0.9	14.0			mag 8.56v star N 6′.0.
12 39 14.4	−10 12 20	MCG −2-32-19	14.9	1.2 x 0.6	14.4	(R′)SAB(s)c pec?	60	
11 22 33.0	−17 34 22	MCG −3-29-6	14.0	0.6 x 0.4	12.3	Sb	138	
11 32 43.7	−16 44 02	MCG −3-30-2	14.6	0.5 x 0.4	12.7	Sbc	168	Mag 9.66v star E 2′.7.
11 37 59.5	−17 14 03	MCG −3-30-3	14.2	2.4 x 0.3	13.7	Sc: sp II	6	Mag 10.99v star N 4′.7.
11 38 21.7	−16 26 33	MCG −3-30-4	14.0	0.8 x 0.5	12.8	S?	165	
11 42 05.3	−16 27 51	MCG −3-30-7	13.5	0.9 x 0.6	12.7	Sb	51	
11 49 55.3	−15 56 15	MCG −3-30-13	14.8	0.8 x 0.4	13.6	E	69	Mag 6.13v star NE 7′.4.
11 54 49.7	−16 51 49	MCG −3-30-19	13.3	1.7 x 0.8	13.5	SAB(s)cd: III-IV	3	Mag 11.7 star on E edge.
12 03 49.2	−15 48 07	MCG −3-31-18	14.0	0.7 x 0.7	13.0	Sbc		
12 08 10.9	−15 48 16	MCG −3-31-23	13.2	1.4 x 0.6	12.9	SA(s)a?	165	
12 19 59.6	−17 23 36	MCG −3-32-3	13.7	0.7 x 0.5	12.7	E	126	
12 20 13.4	−17 06 51	MCG −3-32-4	14.2	1.5 x 0.3	13.2	SBm: IV-V	33	Uniform surface brightness; mag 11.6 star NE 1′.9.
11 20 30.3	−09 00 50	NGC 3634	14.2	0.5 x 0.5	12.8	E		NGC 3635 on E edge.
11 20 31.6	−09 00 49	NGC 3635	13.4	1.7 x 0.6	13.3	SAB(rs)bc pec: II	18	The nucleus of NGC 3634 sits on the W edge.
11 20 25.1	−10 16 55	NGC 3636	12.4	1.3 x 1.3	13.0	E0		Located 1′.7 NW of a mag 6.54v star.
11 20 39.5	−10 15 25	NGC 3637	12.7	1.9 x 1.6	13.8	(R)SB(r)0°	126	Located 3′.0 NE of a mag 6.54v star.
11 20 10.3	−08 06 22	NGC 3638	13.4	2.2 x 0.7	13.7	SAb: sp	141	
11 23 32.2	−08 39 28	NGC 3660	11.9	2.7 x 2.2	13.7	SB(r)bc II	110	Very mottled spiral surface with many bright knots.
11 23 38.5	−13 49 48	NGC 3661	13.1	1.6 x 0.6	12.9	SA(r)0/a: sp	141	Pair of mag 11 stars S 1′.8.
11 24 00.1	−12 17 46	NGC 3663	12.3	1.9 x 1.3	13.1	SA(rs)bc pec II-III	85	Pair of stars on NE edge; elongated anonymous galaxy N 2′.2.
11 24 17.1	−13 51 25	NGC 3667	13.0	1.5 x 1.0	13.3	(R′)SA(rs)ab: I-II	85	NGC 3667A on E edge.
11 24 21.6	−13 51 18	NGC 3667A	14.0	1.4 x 0.6	13.7	SA(rs)b pec	48	Located on E edge of NGC 3667.
11 25 02.5	−09 47 40	NGC 3672	11.4	4.2 x 1.9	13.5	SA(s)c I-II	12	A mag 10.45v star 3′.9 NW.
11 26 08.6	−05 35 15	NGC 3679	14.2	1.0 x 0.5	13.3	Sbc III:	178	
11 27 44.4	−09 09 56	NGC 3688	14.3	1.2 x 0.9	14.2	(R′)SB(rs)b II	20	Stellar nucleus in smooth envelope.
11 28 12.2	−13 11 40	NGC 3693	12.7	3.2 x 0.7	13.4	(R′)SA(r)b: II	84	
11 28 43.8	−11 16 59	NGC 3696	14.1	1.0 x 0.8	13.7	SAB(s)c pec II	90	Located 3′.2 NW of a mag 10.07v star.
11 30 13.4	−08 51 53	NGC 3702	13.1	1.3 x 0.8	13.0	SAB(r)a pec:	156	Mag 9.35v star ENE 5′.2.
11 30 04.8	−11 32 45	NGC 3704	12.9	1.6 x 1.4	13.7	E$^+$	150	NGC 3707 E 1′.7.
11 30 11.5	−11 32 38	NGC 3707	14.7	0.6 x 0.5	13.4	E	85	Stellar; located 1′.7 E of NGC 3704.
11 29 25.6	−11 04 50	NGC 3711	14.0	0.9 x 0.4	12.7	SBbc	165	
11 31 32.3	−14 13 54	NGC 3715	12.5	1.3 x 0.9	12.5	(R′)SB(rs)bc: I-II	145	
11 34 07.9	−09 28 04	NGC 3721	14.5	0.9 x 0.5	13.5	S0/a	141	Faint, elongated anonymous galaxy SE 7′.2.
11 34 23.3	−09 40 50	NGC 3722	14.1	0.9 x 0.8	13.6	E0	0	Stellar **MCG −1-30-6** on N edge.
11 32 30.6	−09 58 15	NGC 3723	13.3	1.1 x 1.0	13.3	E/S0	85	Compact core, smooth envelope.
11 34 28.9	−09 39 39	NGC 3724	14.1	1.2 x 0.4	13.2	S0$^+$: sp	55	Stellar **MCG −1-30-6** SW 1′.2; MCG −1-30-8 ENE 2′.2.
11 33 41.0	−13 52 45	NGC 3727	14.1	0.8 x 0.6	13.1	S0	78	
11 34 14.0	−09 50 47	NGC 3732	12.5	1.2 x 1.2	12.7	SAB(s)0/a:		
11 34 40.7	−14 04 53	NGC 3734	13.9	1.3 x 1.0	14.0	(R′)SAB(rs)bc: I-II	19	Mag 9.35v star NW 7′.0.
11 36 30.3	−09 50 51	NGC 3763	11.8	1.1 x 1.1	11.9	SAB(rs)c pec II		Located 3′.7 SW of mag 4.69v star Theta, 21 Crateris.
11 39 05.9	−09 20 52	NGC 3771	12.6	1.3 x 1.3	13.1	E:		**MCG −1-30-17** N 1′.0.
11 38 30.3	−08 58 36	NGC 3774	13.8	1.0 x 0.7	13.3	(R′)SB(rs)ab:	56	
11 38 26.8	−10 38 20	NGC 3775	13.8	1.2 x 0.5	13.1	SAB(r)0$^+$:	27	
11 36 07.1	−12 34 09	NGC 3777	13.4	1.1 x 0.6	12.8	SAB(s)ab:	35	
11 38 51.1	−10 35 03	NGC 3779	13.7	1.9 x 1.0	14.3	SB(s)d pec III-IV	92	Faint nucleus; mag 9.46v star NE 5′.9.
11 38 09.1	−09 36 28	NGC 3789	13.4	1.4 x 0.5	12.9	SAB(rs)0$^+$?	179	
11 39 41.9	−09 22 04	NGC 3791	13.4	1.3 x 1.0	13.5	(R)SAB0 pec:	164	Compact core; mag 7.40v star S 6′.0.
11 41 57.2	−06 09 23	NGC 3818	11.7	2.0 x 1.2	12.7	E5	100	Mag 9.26v star N 7′.0.
11 42 14.9	−13 52 00	NGC 3823	12.7	1.5 x 1.2	13.2	E$^+$ pec	88	Mag 13.4 star on E edge.
11 43 18.6	−12 52 45	NGC 3831	12.7	2.7 x 0.6	13.1	SAB(r)0$^+$ pec	23	MCG −2-30-24 E 1′.1; MCG −2-30-18 W 2′.4.
11 43 30.4	−16 47 44	NGC 3836	12.9	1.4 x 1.3	13.4	Sb pec?	33	Multi-galaxy system; mag 9.33v star NE 4′.9.
11 44 52.0	−09 14 00	NGC 3865	12.0	2.0 x 1.5	13.1	SAB(rs)b pec: II	49	Mag 9.21v star N 5′.1.
11 45 11.6	−09 18 51	NGC 3866	13.2	1.4 x 0.8	13.1	(R′)SB(rs)a	56	Mag 12.4 star on W edge.
11 47 04.9	−16 51 16	NGC 3887	10.6	3.3 x 2.5	12.8	SB(r)bc II	20	Number of small knots E; faint star on NNE edge.
11 48 01.2	−10 57 45	NGC 3892	11.5	3.0 x 2.2	13.4	SB(rs)0$^+$	95	
11 49 05.1	−09 43 47	NGC 3905	12.8	1.9 x 1.4	13.7	SB(rs)c I	106	Stellar nucleus, tightly wound arms.
11 51 30.0	−11 25 27	NGC 3942	13.2	1.4 x 0.8	13.1	SAB(rs)c pec: III:	127	Small faint nucleus in smooth envelope; mag 9.09v star NW 9′.1.
11 54 37.7	−07 45 24	NGC 3959	13.7	1.1 x 0.8	13.4	(R′)SB(r)a	35	
11 54 40.0	−13 58 34	NGC 3962	10.7	2.6 x 2.2	12.6	E1	10	
11 55 10.6	−07 50 38	NGC 3967	13.4	1.0 x 0.7	13.0	(R′)S0$^-$:	120	
11 55 28.0	−12 03 39	NGC 3970	13.4	1.3 x 0.6	13.1	SA(r)0°: sp	98	
11 55 40.3	−12 01 41	NGC 3974	13.4	1.1 x 1.1	13.4	(R′)SB(r)0/a:		Almost stellar nucleus in smooth envelope.
12 00 34.6	−17 50 37	NGC 4033	11.7	2.6 x 1.1	12.8	E6	47	**ESO 572-40** NW 4′.2.
12 00 29.4	−15 56 54	NGC 4035	13.4	1.1 x 1.0	13.3	(R′)SAB(rs)bc pec II	0	Mag 8.94v star NNE 5′.8.

(Continued from previous page)

131 GALAXIES 131

RA h m s	Dec ° ′ ″	Name	Mag (V)	Dim ′ Maj x min	SB	Type Class	PA	Notes
12 02 54.1	−16 22 27	NGC 4050	12.2	3.1 x 2.1	14.1	SB(r)ab II	85	Mag 8.25v star SSE 5.′8.
12 05 54.3	−14 31 36	NGC 4094	11.8	4.2 x 1.5	13.7	SAB(rs)cd: III	62	A mag 10.51v star on NE edge.
12 07 12.4	−14 11 07	NGC 4114	13.1	1.9 x 0.9	13.5	(R′)SAB(rs)a: I-II	136	
12 08 53.6	−09 02 12	NGC 4129	12.5	2.3 x 0.6	12.7	SB(s)ab: sp III-IV	94	
12 12 37.0	−09 09 39	NGC 4176	14.5	0.7 x 0.4	13.0	Sb	55	
12 12 41.5	−14 00 54	NGC 4177	12.5	1.7 x 1.1	13.1	SA(rs)c pec: III	69	
12 14 07.6	−12 35 15	NGC 4188	13.7	0.7 x 0.7	12.8	Sb II-III:		Very faint star on N edge.
12 14 41.8	−11 34 59	NGC 4201	13.6	1.0 x 0.8	13.2	S0	80	
12 16 38.3	−12 19 42	NGC 4225	14.0	0.8 x 0.5	12.9	S0/a	60	Located 1.′6 N of mag 9.71v star.
12 17 24.5	−09 57 05	NGC 4240	12.3	1.3 x 1.2	12.7	E$^+$:		Star WSW of nucleus.
12 19 42.1	−12 13 28	NGC 4263	12.6	1.2 x 0.6	12.1	SAB(rs)b pec I-II	125	
12 20 25.1	−11 40 01	NGC 4279	13.6	1.2 x 0.7	13.3	(R)SB(r)0$^+$:	35	
12 20 40.1	−11 38 31	NGC 4285	14.1	0.9 x 0.4	12.8	Sa	50	
12 23 20.9	−12 33 34	NGC 4329	13.5	1.2 x 0.5	13.0	E$^+$:	50	
12 26 12.8	−07 41 07	NGC 4403	12.8	1.6 x 0.5	12.4	SA(r)ab pec sp	27	
12 26 16.4	−07 40 49	NGC 4404	12.7	1.1 x 0.8	12.5	SA(r)0$^-$ pec:	153	Pair with NGC 4403 W.
12 27 12.1	−05 49 51	NGC 4422	13.8	1.3 x 1.3	14.3	SA0$^-$ pec?		
12 27 28.3	−08 10 05	NGC 4428	12.6	1.9 x 0.8	12.9	SAB(rs)c II-III	79	
12 27 38.8	−08 16 44	NGC 4433	12.7	2.2 x 1.0	13.4	SAB(s)ab II-III	5	Bright complex lens seen end-on; stars or bright knots N and S edges.
12 28 52.9	−11 39 07	NGC 4484	13.6	1.5 x 1.5	14.3	SA(s)c pec: II		
12 31 04.5	−08 03 15	NGC 4487	10.9	4.1 x 2.7	13.4	SAB(rs)cd II-III	75	Several strong arms with bright knots and patches overall.
12 32 17.5	−07 33 51	NGC 4504	11.2	3.2 x 2.2	13.2	SA(s)cd II-III	153	Bright central part, two main, massive arms.
12 33 49.9	−07 22 34	NGC 4520	14.0	1.1 x 0.5	13.2	SA0$^-$ pec sp	104	= **IC 799**.
12 33 54.5	−12 01 38	NGC 4524	13.4	1.2 x 0.7	13.1	SAB(r)bc II-III	80	
12 39 39.8	−05 20 39	NGC 4593	10.9	3.9 x 2.9	13.4	(R)SB(rs)b II	114	Bright nucleus in short bar encircled by two asymmetrical arms. **MCG −1-32-33** E 3.′8.
12 40 00.4	−11 37 35	NGC 4594	8.0	8.7 x 3.5	11.6	SA(s)a sp	87	= **M 104**, **Sombrero Galaxy**. Very bright central bulge, narrow patchy arms in lens, strong dark lane in front.

GALAXY CLUSTERS

RA h m s	Dec ° ′ ″	Name	Mag 10th brightest	No. Gal	Diam ′	Notes
11 38 00.0	−09 20 00	A 1332	16.0	46	10	All members anonymous and stellar.

OPEN CLUSTERS

RA h m s	Dec ° ′ ″	Name	Mag	Diam ′	No. ★	B ★	Type	Notes
12 35 44.1	−12 02 00	Canali		7.5	6	7.4	ast?	Asterism or possible cluster. Noted 1997 by E. Canali.

132 GALAXIES 132

RA h m s	Dec ° ′ ″	Name	Mag (V)	Dim ′ Maj x min	SB	Type Class	PA	Notes
10 19 42.0	−17 45 07	ESO 567-48	15.3	1.6 x 0.4	14.7	Sm: sp IV-V	50	Small galaxy, or close pair of stars, on S edge.
10 13 12.5	−05 37 46	IC 599	14.3	1.1 x 0.3	13.0	Sbc	36	Mag 11.7 star E 5.′4.
10 19 25.0	−05 39 22	IC 603	13.6	1.2 x 0.9	13.6	(R)SB(r)a:	157	Mag 9.26v star N 1.′3.
10 24 21.2	−06 02 22	IC 608	13.9	0.8 x 0.4	12.5	Sb	108	Mag 11.5 star SE 6.′7.
10 36 15.4	−08 20 04	IC 624	13.2	2.7 x 0.5	13.4	SABa:	42	Mag 12.1 star E 2.′6.
10 36 57.2	−07 01 29	IC 626	13.9	1.2 x 0.9	13.9	S0$^-$ pec	141	Mag 8.45v star N 8.′4; MCG −1-27-27 SW 7.′5.
10 38 33.8	−07 10 14	IC 630	12.0	1.2 x 1.2	12.3	S0 pec?		Located 0.′7 NE of mag 8.2 star.
10 53 50.4	−11 43 34	IC 654	14.1	1.0 x 0.4	13.0	(R′)SB0$^+$?	126	Mag 11.3 star NW 3.′2.
10 58 03.8	−06 15 41	IC 659	14.3	1.4 x 1.0	14.7	E$^+$	147	Almost stellar nucleus; mag 9.31v star SE 9.′7.
11 08 03.3	−12 29 05	IC 672	18.1	0.6 x 0.2	15.6	Sb	153	Mag 9.42v star ENE 5.′7. MCG −2-29-1 N 2.′3.
11 18 32.0	−12 08 25	IC 681	14.1	0.9 x 0.4	12.8	Sc	18	Mag 10.40v star W 7.′4.
11 26 48.5	−05 00 17	IC 693	13.8	0.9 x 0.4	12.6	Sab		Mag 10.31v star E 10.′3.
11 15 32.3	−14 10 18	IC 2668	13.7	1.3 x 0.6	13.2	SB(rs)a pec	140	Mag 12.0 star ESE 7.′1.
10 10 54.4	−06 54 49	MCG −1-26-29	14.2	1.1 x 0.9	14.0	Scd	129	Stellar nucleus; mag 7.84v star SSE 9.′0.
10 12 46.2	−08 35 21	MCG −1-26-31	14.7	1.0 x 0.6	14.0	Sbc	72	
10 13 21.4	−08 59 25	MCG −1-26-36	14.5	0.7 x 0.4	13.0	Sc	126	
10 16 54.5	−05 52 15	MCG −1-26-38	14.4	0.5 x 0.5	12.9	E		Mag 9.12v star NNW 2.′0.
10 18 27.9	−06 18 45	MCG −1-26-39	14.3	0.8 x 0.5	13.2	SBb	159	
10 19 13.0	−05 37 48	MCG −1-26-40	14.3	0.9 x 0.8	13.8	SBa	55	Mag 9.26v star E 3.′4.
10 20 19.3	−07 53 38	MCG −1-27-1	14.5	0.5 x 0.4	12.8	D	159	Faint, elongated anonymous galaxy WNW 1.′3; mag 8.99v star S 4.′3.
10 20 26.7	−06 31 38	MCG −1-27-2	14.0	0.3 x 0.3	11.3	D		At center of galaxy cluster A 978, surrounded by anonymous galaxies.
10 20 54.5	−07 56 31	MCG −1-27-3	15.7	0.7 x 0.4	14.2	Sbc	63	Mag 9.10v star S 2.′6.
10 24 00.6	−09 14 38	MCG −1-27-6	14.7	0.5 x 0.5	13.1			
10 24 07.2	−05 37 57	MCG −1-27-7	14.5	1.9 x 0.3	13.8	Sc? sp	75	Mag 10.56v star WSW 6.′1.
10 31 19.2	−08 35 30	MCG −1-27-11	13.6	0.9 x 0.7	12.9	(R′)SB(rs)a pec?	12	Mag 10.13v star N 3.′5.
10 32 12.4	−09 14 22	MCG −1-27-13	13.7	1.0 x 0.8	13.4	(R)SA0$^-$:	12	
10 32 16.7	−05 02 23	MCG −1-27-14	14.1	0.5 x 0.3	12.1	E	18	
10 32 54.0	−06 30 30	MCG −1-27-15	13.7	1.1 x 0.6	13.0	S?	153	
10 33 13.2	−07 27 54	MCG −1-27-18	13.7	1.5 x 0.9	13.9	SB(rs)0°	150	Strong NE-SW bar.
10 33 57.1	−08 14 34	MCG −1-27-19	14.5	0.9 x 0.7	13.8	Sc	24	Mag 10.01v star NNE 5.′9.
10 34 55.8	−06 28 18	MCG −1-27-20	14.2	1.1 x 0.5	13.4	SB(rs)c? II	102	Mag 9.02v star S 4.′7.
10 35 17.1	−07 14 54	MCG −1-27-21	13.7	1.0 x 0.5	12.8	SB(s)a pec?	96	
10 35 30.9	−06 11 06	MCG −1-27-22	14.9	0.7 x 0.3	13.1		117	Located 0.′9 W of NGC 3292; mag 9.11v star SSW 4.′8.
10 36 03.3	−06 57 42	MCG −1-27-25	14.7	0.8 x 0.5	13.6	Sb	27	Mag 11.05v star SW 5.′2.
10 36 35.4	−07 06 48	MCG −1-27-27	14.3	1.1 x 0.2	12.5	Sa	42	IC 626 NE 7.′5.
10 39 43.6	−08 41 15	MCG −1-27-30	14.5	0.7 x 0.5	13.5	cD;S0p	80	Located in galaxy cluster A 1069, many anonymous galaxies near by; mag 9.59v star NW 5.′2.
10 39 46.5	−05 28 58	MCG −1-27-31	13.6	0.9 x 0.7	13.0	SBc	138	Mag 12.7 star on S edge.
10 41 52.9	−08 12 31	MCG −1-27-33	14.5	0.6 x 0.5	13.1	S0	36	Mag 7.23v star N 9.′2.
10 51 17.6	−09 46 39	MCG −1-28-3	14.2	1.1 x 0.3	12.9		69	
10 53 12.7	−07 25 44	MCG −1-28-4	14.0	1.2 x 0.9	13.9	SA(rs)c: III	114	
10 55 23.4	−07 39 16	MCG −1-28-6	14.5	0.7 x 0.7	13.6	SBd		Mag 8.70v star N 3.′9.
10 59 51.8	−06 09 02	MCG −1-28-17	14.2	0.5 x 0.5	12.8	E		Elongated galaxies **MCG −1-28-18** and **MCG −1-28-19** W 1.′6.

(Continued from previous page)

GALAXIES

RA h m s	Dec ° ′ ″	Name	Mag (V)	Dim ′ Maj x min	SB	Type Class	PA	Notes
11 00 13.8	−09 42 29	MCG −1-28-20	14.6	0.8 x 0.7	13.9	Sc	162	
11 00 54.0	−06 41 57	MCG −1-28-21	16.9	0.6 x 0.5	15.4		168	
11 02 37.6	−07 30 43	MCG −1-28-22	14.1	1.1 x 0.5	13.3	Sc	153	
11 05 52.0	−09 42 11	MCG −1-28-24	14.0	1.1 x 0.6	13.4	SAB(rs)bc: III	162	Stellar nucleus; mag 10.56v star SE 1′.3.
11 06 46.8	−05 54 47	MCG −1-28-25	13.9	0.7 x 0.4	12.6	E/S0	129	
11 06 57.2	−05 36 22	MCG −1-28-26	13.9	0.7 x 0.5	12.6	SBab	123	
11 07 56.9	−06 50 49	MCG −1-29-1	14.3	0.9 x 0.5	13.3	Sab	90	Mag 8.53v star W 6′.9.
11 09 44.5	−07 35 38	MCG −1-29-3	14.0	0.9 x 0.4	12.8	S0/a	132	
11 12 15.1	−05 45 16	MCG −1-29-5	14.5	1.3 x 0.5	13.9	(R′)SB(s)b: II	81	
11 16 56.0	−06 13 14	MCG −1-29-6	13.5	0.7 x 0.5	12.5	E	18	Very small, very faint, field galaxy E 1′.2.
11 20 39.4	−05 40 21	MCG −1-29-10	13.9	1.0 x 0.7	13.4	S0	36	Contains several knots, or superimposed stars?
11 21 48.0	−05 45 32	MCG −1-29-12	13.6	0.5 x 0.5	12.1	E		Mag 7.81v star N 2′.5.
11 22 27.0	−07 39 13	MCG −1-29-13	13.2	1.2 x 0.5	12.6	SB0$^-$: sp	114	
11 22 31.7	−07 03 15	MCG −1-29-14	14.2	0.8 x 0.8	13.5	SBbc		
11 22 44.0	−07 40 39	MCG −1-29-15	13.0	2.1 x 1.0	13.7	SA(s)0° pec:	51	
11 24 50.8	−05 21 09	MCG −1-29-17	13.9	0.6 x 0.5	12.5		156	
11 25 02.2	−05 04 09	MCG −1-29-18	13.6	0.7 x 0.5	12.3	Sa	162	
11 25 22.9	−09 35 14	MCG −1-29-19	13.5	0.7 x 0.7	12.6	S0		Mag 9.34v star W 7′.6.
10 09 03.3	−11 14 02	MCG −2-26-31	13.3	1.1 x 0.5	12.5	S0°	12	Mag 9.46v star N 2′.9.
10 10 08.8	−10 41 46	MCG −2-26-35	16.0	0.9 x 0.2	14.0	Sd	87	
10 10 23.4	−11 55 36	MCG −2-26-37	14.4	0.5 x 0.4	12.5	S0/a	78	Mag 7.42v star W 11′.3.
10 10 43.3	−15 30 37	MCG −2-26-38	12.6	0.7 x 0.4	11.1		165	Multi-galaxy system?
10 11 08.7	−13 46 55	MCG −2-26-39	14.5	2.2 x 0.8	15.0	SB(s)m IV-V	36	Mag 9.76v star W 3′.4.
10 14 41.3	−09 56 29	MCG −2-26-40	13.8	0.9 x 0.9	13.5	SBc		
10 15 13.3	−15 05 36	MCG −2-26-41	13.3	1.2 x 1.0	13.3	SAB(rs)dm? IV-V	33	Mottled appearance, several knots and/or stars involved.
10 18 23.6	−13 06 17	MCG −2-26-42	13.6	0.8 x 0.8	13.0	(R′)SAB(s)0$^-$:		
10 25 26.6	−15 21 02	MCG −2-27-1	14.3	1.5 x 0.2	12.8	Sd: sp	171	Mag 11.6 star NNE 2′.9.
10 25 50.2	−11 26 33	MCG −2-27-2	14.8	0.4 x 0.4	12.6	Sc		
10 25 49.0	−11 25 34	MCG −2-27-3	12.7	0.6 x 0.5	11.3	S0/a	27	
10 27 34.3	−10 43 16	MCG −2-27-4	13.8	0.7 x 0.2	11.6	S0/a	162	Stars SE edge and S tip.
10 35 27.4	−14 07 49	MCG −2-27-9	13.1	1.9 x 0.5	12.9	SB(rs)0$^+$ pec?	84	Mag 9.95v star N 8′.4.
10 43 36.4	−09 51 28	MCG −2-28-1	13.6	2.0 x 1.6	14.7	SB(s)d pec: III	170	Very patchy, numerous bright knots.
10 43 35.5	−10 09 23	MCG −2-28-2	14.3	1.0 x 0.7	13.7	Sd	147	
10 44 30.8	−14 18 53	MCG −2-28-5	14.3	0.9 x 0.6	13.6	S0	138	
10 45 11.2	−10 03 55	MCG −2-28-6	13.7	1.3 x 1.3	14.1	(R′)SAB(rs)d III-IV		Star or bright knot N edge.
10 45 59.3	−12 22 20	MCG −2-28-7	14.6	1.0 x 0.8	14.2	SBbc	144	
10 49 59.9	−14 45 04	MCG −2-28-9	14.6	0.6 x 0.3	12.6	S0/a	36	
10 50 03.3	−14 45 01	MCG −2-28-10	14.6	0.8 x 0.2	12.4	Sbc	60	
10 51 00.6	−10 25 25	MCG −2-28-14	14.5	0.7 x 0.5	13.2	SBa-b	69	
10 51 23.1	−10 08 02	MCG −2-28-16	14.2	1.7 x 0.2	12.9	S0: sp	150	Lies between a pair of mag 14 stars.
10 53 10.1	−10 31 30	MCG −2-28-17	14.7	0.7 x 0.4	13.2	Sc	39	
10 54 18.2	−11 16 56	MCG −2-28-20	14.4	0.8 x 0.8	13.7	Scd		NGC 3452 S 7′.5.
10 55 49.2	−09 51 36	MCG −2-28-21	14.3	1.1 x 0.8	14.0	SA(s)c? I-II	84	= **Mkn 1270**.
10 56 21.1	−15 52 17	MCG −2-28-22	13.4	0.9 x 0.9	13.0	SBbc		
10 56 44.3	−12 53 17	MCG −2-28-23	14.0	1.0 x 0.3	12.6	Sbc	141	Mag 10.62v star E 1′.3.
10 58 47.0	−09 50 46	MCG −2-28-25	14.4	0.5 x 0.4	12.5	SB0/a	39	= **Mkn 1273**.
10 58 40.6	−15 31 39	MCG −2-28-26	14.6	0.7 x 0.7	13.6	S?		Mag 10.63v star SW 12′.5.
10 59 06.2	−13 01 27	MCG −2-28-28	14.8	0.9 x 0.2	12.8	Sb	39	
10 59 16.3	−09 47 43	MCG −2-28-29	13.7	0.8 x 0.7	13.0	SAB(rs)b pec: II	153	
10 59 15.8	−09 49 03	MCG −2-28-30	14.2	1.1 x 0.2	12.4	Sab pec sp	111	
10 59 38.2	−15 31 36	MCG −2-28-31	14.3	1.9 x 0.2	13.1	Scd: sp III-IV	84	Mag 12.1 star E end.
11 00 23.7	−09 59 07	MCG −2-28-32	13.6	1.5 x 0.6	13.4	SAB(rs)bc pec: II	24	Mag 8.80v star N 2′.8.
11 00 45.2	−14 03 20	MCG −2-28-35	14.4	0.8 x 0.5	13.3		18	Mag 5.85v star W 8′.2.
11 00 44.3	−14 11 17	MCG −2-28-36	13.9	0.8 x 0.8	13.5	E		Mag 5.85v star NW 10′.1.
11 00 59.9	−14 02 23	MCG −2-28-37	13.7	1.1 x 1.0	13.7	(R′)SB(s)0$^+$ pec?	165	Star and bright knot, or small galaxy, N edge; mag 9.75v star N 3′.2.
11 01 16.5	−13 33 38	MCG −2-28-38	13.6	1.0 x 0.6	12.9	Sd	30	
11 01 29.9	−12 26 57	MCG −2-28-39	15.7	0.7 x 0.3	13.9	Sbc	0	
11 01 51.1	−12 29 12	MCG −2-28-40	14.4	0.4 x 0.4	12.5	E		
11 03 13.5	−15 19 52	MCG −2-28-42	14.7	0.9 x 0.6	13.9	SBb	42	Broad band of very faint material extends N and curves NW beyond published size for the galaxy.
11 04 29.0	−09 47 32	MCG −2-28-43	13.8	1.3 x 0.3	12.7	SB(r)0°? sp	105	
11 05 34.1	−09 54 33	MCG −2-28-44	15.1	0.4 x 0.2	12.4	E/S0	45	
11 05 39.1	−09 55 28	MCG −2-28-45	12.3	0.9 x 0.9	12.0	(R′)S0$^-$ pec:		Mag 12 star and small galaxy NE 1′.3; MCG −2-28-44 NW 1′.6.
11 05 34.9	−15 32 21	MCG −2-28-47	14.4	0.9 x 0.3	12.8	Sab	165	
11 05 50.8	−09 55 40	MCG −2-28-48	13.8	1.2 x 0.7	13.5	SB(s)0$^+$ pec:	105	
11 07 15.2	−11 14 25	MCG −2-28-49	14.5	1.1 x 0.5	13.7	SB(s)dm: sp IV-V	150	
11 08 01.4	−12 26 55	MCG −2-29-1	14.5	0.9 x 0.2	12.5	SB(s)bc pec?	177	Mag 9.42v star E 5′.7; IC 672 S 2′.3.
11 08 32.5	−10 29 35	MCG −2-29-3	14.5	1.3 x 1.2	14.8	(R′)SAB(r)d? III	138	
11 08 46.9	−14 35 13	MCG −2-29-5	13.8	0.8 x 0.4	12.5	S0	39	
11 09 29.4	−14 00 03	MCG −2-29-6	14.0	1.0 x 0.3	12.6	SBc	96	
11 10 46.2	−15 44 26	MCG −2-29-8	14.5	1.0 x 0.6	13.8	SB(rs)c: III	153	
11 11 16.2	−09 58 23	MCG −2-29-9	13.4	1.6 x 1.3	14.0	SAB(rs)c I-II	93	Mag 8.81v star NE 6′.3.
11 12 13.9	−13 38 43	MCG −2-29-10	14.1	0.6 x 0.6	12.8	S0/a		
11 13 26.1	−15 46 19	MCG −2-29-11	14.2	0.8 x 0.4	12.9	S0	162	
11 14 07.9	−14 42 20	MCG −2-29-13	14.2	0.9 x 0.4	13.0	S0 pec	168	
11 15 17.0	−13 44 00	MCG −2-29-14	13.5	0.7 x 0.5	12.3	S(r)0° pec?	66	
11 17 17.1	−14 31 27	MCG −2-29-16	14.2	0.8 x 0.4	12.8	Sc	72	Mag 7.36v star W 7′.0.
11 19 20.9	−09 51 47	MCG −2-29-18	14.2	1.1 x 0.9	14.1	SB(s)dm IV-V	123	
11 21 54.2	−12 32 20	MCG −2-29-21	14.3	1.2 x 0.8	14.1	SAB(rs)c: III-IV	3	
11 24 04.1	−13 08 37	MCG −2-29-24	14.4	0.7 x 0.7	13.5	Sm		
11 25 37.5	−11 08 24	MCG −2-29-29	13.6	0.9 x 0.7	13.2	E/S0	158	
11 26 17.4	−14 16 51	MCG −2-29-30	14.1	0.8 x 0.5	13.0	S0/a	30	
11 27 22.9	−10 57 09	MCG −2-29-31	14.2	1.1 x 0.8	13.9	SB(s)b II	66	
11 24 52.8	−13 34 19	MCG −2-29-37	12.9	1.6 x 1.1	13.5	E$^+$	115	Mag 14.8 star N edge; mag 13.2 star NE 1′.9.
10 11 18.8	−17 12 16	MCG −3-26-30	12.9	1.7 x 1.3	13.7	SAB0$^-$ pec:	100	Mag 10.75v star NNE 10′.4.
10 19 08.5	−17 55 53	MCG −3-26-38	14.7	0.8 x 0.6	13.7	SBa-b	3	Double system, NGC 3200 SW 8′.2.
10 20 50.9	−17 19 02	MCG −3-27-5	15.2	0.7 x 0.4	13.6	(R′)SAB:(r)ab	24	Mag 10.60v star on SE edge.
10 23 53.8	−17 17 41	MCG −3-27-8	17.3	0.7 x 0.4	15.8	SA(r)a		Close pair with MCG −3-27-9 with bright star in between.
10 23 51.5	−17 17 21	MCG −3-27-9	16.4	0.4 x 0.2	13.5	(R′)SB(r)a		Anonymous elongated galaxy WSW 1′.0. Pair with MCG −3-27-9 with bright star in between.

(Continued from previous page)

GALAXIES

RA h m s	Dec ° ′ ″	Name	Mag (V)	Dim ′ Maj x min	SB	Type Class	PA	Notes
10 24 57.3	−17 26 06	MCG −3-27-10	14.7	1.0 x 0.8	14.3	SAB(rs)a pec:	174	Small elongated galaxy at NE edge, bright star or galaxy at SW edge.
10 29 20.6	−17 25 16	MCG −3-27-16	14.4	1.1 x 0.2	12.6	Sbc	75	
10 29 40.5	−17 04 41	MCG −3-27-17	14.3	1.0 x 0.4	13.2		30	
10 38 46.1	−16 47 00	MCG −3-27-22	14.0	0.6 x 0.3	12.1	S0	0	Located 6.'3 N of mag 4.91v star Phi 1 Hydrae.
10 39 55.6	−17 39 42	MCG −3-27-23	14.2	0.9 x 0.9	13.8	SA(s)c pec		
10 41 02.9	−17 30 33	MCG −3-27-24	13.6	1.2 x 0.8	13.4	S?	39	
10 42 19.0	−17 38 57	MCG −3-27-26	13.6	1.7 x 0.4	13.1	S0 pec sp	60	Bright, elongated, core with faint, broad filament trailing SW to W.
10 44 07.0	−16 28 12	MCG −3-28-1	13.7	1.0 x 0.9	13.4	SAB(rs)c pec? I	39	Mag 9.08vb star NE 8.'0.
10 46 02.3	−17 41 19	MCG −3-28-2	13.9	0.8 x 0.7	13.2	SA(s)c:	147	
10 46 43.0	−16 08 01	MCG −3-28-3	14.0	2.1 x 1.4	15.1	(R′)SAB(rs)0/a	45	Mag 11.8 star W 2.'3.
10 50 05.3	−17 11 49	MCG −3-28-10	14.8	0.7 x 0.3	13.0	Sm	135	NGC 3420 S 3.'0.
10 50 31.6	−17 03 51	MCG −3-28-13	14.7	1.0 x 0.2	12.8	Sc	156	Mag 9.91v star SE 2.'4; NGC 3409 W 3.'1.
10 52 16.0	−17 07 50	MCG −3-28-16	13.8	0.5 x 0.5	12.2	SA(s)a		Mag 7.78v star NE 8.'2 ; PGC 32573 W 9.'2.
10 53 58.7	−16 07 17	MCG −3-28-17	15.0	0.5 x 0.5	13.3	Sc		NGC 3456 N 5.'7.
10 54 22.1	−17 26 29	MCG −3-28-20	14.1	0.8 x 0.5	13.0	S0/a	132	
10 56 22.5	−15 55 53	MCG −3-28-24	13.4	0.7 x 0.3	11.6	SBbc	39	
11 03 16.6	−16 45 33	MCG −3-28-32	14.3	1.1 x 0.2	12.5	S?	0	Mag 10.18v star S 1.'5.
11 03 35.9	−17 28 45	MCG −3-28-33	13.9	1.5 x 0.2	12.5	S?	150	Mag 9.47v star WNW 4.'6.
11 11 57.8	−16 55 51	MCG −3-29-2	13.6	0.5 x 0.4	11.7	SBa	174	
11 22 33.0	−17 34 22	MCG −3-29-6	14.0	0.6 x 0.4	12.3	Sb	138	
10 09 16.9	−11 57 27	NGC 3138	14.8	1.2 x 0.4	13.8	Sbc	80	Mag 7.42v star NE 6.'4.
10 10 05.4	−11 46 41	NGC 3139	13.5	1.4 x 1.2	13.9	S0° pec:	75	Small core with smooth envelope; mag 7.42v star SW 9.'8.
10 09 27.9	−16 37 44	NGC 3140	14.0	0.9 x 0.8	13.5	Sc I-II	41	Located near E edge of galaxy cluster **A 940**; NGC 3141 SW 2.'6.
10 09 19.9	−16 39 15	NGC 3141	15.4	0.5 x 0.3	13.2	Sbc	26	Almost stellar, NGC 3140 NE 2.'6.
10 10 06.3	−08 28 49	NGC 3142	13.8	1.0 x 0.8	13.4	S0	3	Mag 5.90v star 17 Sextantis N 4.'4.
10 10 04.1	−12 34 54	NGC 3143	14.3	0.8 x 0.7	13.5	SB(s)b II-III	105	**PGC 29578** S 4.'0.
10 10 10.2	−12 26 01	NGC 3145	11.7	3.1 x 1.4	13.2	SB(rs)bc I	18	Located 7.'8 SE of mag 3.81v star Lambda, 41 Hydrae.
10 16 09.0	−15 47 28	NGC 3178	12.7	1.3 x 0.8	12.5	SA(rs)cd pec: III	70	
10 18 36.5	−17 58 57	NGC 3200	12.0	4.2 x 1.3	13.7	SAB(rs)c: I-II	169	MCG −3-26-38 NE 8.'2.
10 32 46.6	−12 38 05	NGC 3280	13.9	0.4 x 0.4	12.0	E/S0		Eastern most of three, NGC 3280A and B on W edge.
10 32 43.9	−12 38 13	NGC 3280A	13.9	0.7 x 0.6	12.9	E/S0	130	Western most of three; NGC 3280B on E edge.
10 32 45.6	−12 38 15	NGC 3280B	13.9	0.5 x 0.3	11.8	E/S0	170	Lies between NGC 3280 E and NGC 3280A W.
10 35 17.5	−17 16 38	NGC 3290	13.5	1.0 x 0.6	12.8	SAB(rs)bc: pec II	60	Mag 8.35v star N 4.'8.
10 35 34.6	−06 10 43	NGC 3292	14.1	1.1 x 0.9	14.0	S0° pec?	175	Mag 9.11v star SSW 5.'3; MCG −1-27-22 W 0.'9.
10 32 45.4	−12 43 03	NGC 3296	13.9	0.7 x 0.7	13.2	E		Mag 9.47v star NW 4.'7.
10 33 11.8	−12 40 19	NGC 3297	14.5	0.8 x 0.5	13.4	S0	160	NGC 3280, NGC 3280A, NGC 3280B group WNW 6.'8.
10 38 50.7	−11 38 56	NGC 3321	13.0	2.5 x 1.2	14.0	SAB(r)c: II-III:	36	Weak stellar nucleus.
10 44 16.4	−11 14 34	NGC 3360	13.7	1.2 x 0.9	13.7	SA(s)c: III	55	Almost stellar nucleus; NGC 3361 NE 3.'8.
10 44 29.3	−11 12 27	NGC 3361	12.8	2.1 x 0.8	13.3	SAB(rs)c? II	159	Mag 15.3 star on W edge; NGC 3360 SW 3.'8.
10 47 00.9	−09 56 30	NGC 3375	12.6	1.5 x 1.1	13.0	(R)SA(rs)0°?	153	
10 50 06.0	−12 40 37	NGC 3402	15.4	0.4 x 0.4	13.4	E⁺		Stellar.
10 50 17.9	−12 06 34	NGC 3404	13.1	2.1 x 0.5	13.0	SBab? sp	84	
10 50 20.3	−17 02 41	NGC 3409	14.1	1.2 x 0.3	12.8	SBc? sp II-III?	10	MCG −3-28-13 E 3.'1.
10 50 26.0	−12 50 38	NGC 3411	11.9	2.1 x 2.1	13.5	E⁺		Anonymous galaxy ESE 2.'1; could be "unknown" **IC 647**?
10 50 09.6	−17 14 34	NGC 3420	13.8	1.3 x 1.1	14.0	(R)SB(rs)a:	30	MCG −3-28-10 N 3.'0.
10 50 57.8	−12 26 57	NGC 3421	13.7	2.0 x 1.5	14.8	(R′)SB(rs)a pec	175	Almost stellar nucleus in smooth envelope; mag 9.53v star NW 8.'8.
10 51 17.6	−12 24 11	NGC 3422	13.6	1.3 x 0.4	12.8	S0⁺? sp	54	10.64v mag star on S edge.
10 51 15.0	−17 00 33	NGC 3431	13.6	1.3 x 0.3	12.4	SABb? I-II	130	PGC 32573 SE 8.'7.
10 54 14.1	−11 24 20	NGC 3452	14.0	1.1 x 0.3	12.6	Sa: sp	65	Mag 9.40v star W 8.'0; MCG −2-28-20 N 7.'5.
10 54 03.5	−16 01 41	NGC 3456	12.6	1.9 x 1.3	13.4	SB(rs)c: II	102	Three faint stars on E end; MCG −3-28-17 S 5.'7.
10 54 44.5	−17 02 34	NGC 3459	13.4	1.6 x 0.5	13.0	(R′)SBab:	156	
10 56 57.4	−14 18 04	NGC 3469	13.1	1.7 x 1.2	13.7	(R′)SB(r)ab	115	
10 58 55.7	−14 57 44	NGC 3479	13.0	1.7 x 1.2	13.7	SAB(r)bc II-III	174	Small elongated nucleus in smooth envelope.
10 59 26.6	−07 32 42	NGC 3481	13.8	0.9 x 0.7	13.1	Sab	136	Lies 2.'2 ENE of mag 10.74v star.
11 02 20.1	−14 08 14	NGC 3502	13.7	1.1 x 1.1	13.8	SB(rs)a pec		Almost stellar nucleus; bright knot or star SE of nucleus.
11 02 59.9	−16 17 18	NGC 3508	12.4	1.1 x 0.8	12.1	SA(r)b pec?	15	
11 08 26.9	−10 15 25	NGC 3537A	12.8	1.4 x 1.3	13.4	S0° pec	54	Small companion galaxy extending S from nucleus?
11 08 45.4	−10 57 51	NGC 3541	14.2	0.8 x 0.5	13.1	S0/a	130	
11 09 46.8	−13 22 53	NGC 3546	13.4	1.6 x 0.9	13.7	SA0⁻ pec:	100	
11 14 03.4	−14 05 14	NGC 3591	13.3	1.3 x 0.9	13.4	(R′)SB(r)0⁺?	150	
11 20 30.3	−09 00 50	NGC 3634	14.2	0.5 x 0.5	12.8	E		NGC 3635 on E edge.
11 20 31.6	−09 00 49	NGC 3635	13.4	1.7 x 0.6	13.3	SAB(rs)bc pec: II	18	The nucleus of NGC 3634 sits on the W edge.
11 20 25.1	−10 16 55	NGC 3636	12.4	1.3 x 1.3	13.0	E0		Located 1.'7 NW of a mag 6.54v star.
11 20 39.5	−10 15 25	NGC 3637	12.7	1.9 x 1.6	13.8	(R)SB(r)0°	126	Located 3.'0 NE of a mag 6.54v star.
11 20 10.3	−08 06 22	NGC 3638	13.4	2.2 x 0.7	13.7	SAb: sp	141	
11 23 32.2	−08 39 28	NGC 3660	11.9	2.7 x 2.2	13.7	SB(r)bc II	110	Very mottled spiral surface with many bright knots.
11 23 38.5	−13 49 48	NGC 3661	13.1	1.6 x 0.6	12.9	SA(r)0/a: sp	141	Pair of mag 11 stars S 1.'8.
11 24 00.1	−12 17 46	NGC 3663	12.3	1.9 x 1.3	13.1	SA(rs)bc pec II-III	85	Pair of stars on NE edge; elongated anonymous galaxy N 2.'2.
11 24 17.1	−13 51 25	NGC 3667	13.0	1.5 x 1.0	13.3	(R′)SA(rs)ab: I-II	85	NGC 3667A on E edge.
11 24 21.6	−13 51 18	NGC 3667A	14.0	1.4 x 0.6	13.7	SA(rs)b pec	48	Located on E edge of NGC 3667.
11 25 02.5	−09 47 40	NGC 3672	11.4	4.2 x 1.9	13.5	SA(s)c I-II	12	A mag 10.45v star 3.'9 NW.
11 26 08.6	−05 35 15	NGC 3679	14.2	1.0 x 0.5	13.3	Sbc III:	178	
11 27 44.4	−09 09 56	NGC 3688	14.3	1.2 x 0.9	14.2	(R′)SB(rs)b II	20	Stellar nucleus in smooth envelope.
10 51 37.8	−17 07 24	PGC 32573	13.2	2.5 x 0.9	13.9	(R)SB(rs)ab pec? III-IV	40	

GALAXY CLUSTERS

RA h m s	Dec ° ′ ″	Name	Mag 10th brightest	No. Gal	Diam ′	Notes
10 20 30.0	−06 31 00	A 978	15.6	55	11	Most members anonymous and stellar.
10 20 24.0	−07 53 00	A 979	15.3	39	17	Most members in N half, anonymous and stellar.
10 39 54.0	−08 36 00	A 1069	15.1	45	13	Most members anonymous and stellar to almost stellar.

RA h m s	Dec ° ′ ″	Name	Mag (V)	Dim ′ Maj x min	SB	Type Class	PA	Notes
09 25 22.7	−12 23 29	IC 537	13.1	1.2 x 1.1	13.3	(R)S0/a?	175	Mag 12.3 star N 1′.4; mag 9.98v star SSW 5′.9.
09 31 06.2	−13 10 55	IC 542	13.7	1.1 x 0.3	12.3	(R′)SB(rs)0/a:	96	Mag 9.71v star WSW 10′.9.
09 34 50.3	−16 23 06	IC 546	13.4	1.2 x 0.7	13.1	SB(rs)0^+	100	Mag 7.83v star W 3′.1; NGC 2924 E 5′.0.
09 40 28.6	−06 56 48	IC 550	13.5	0.9 x 0.7	12.8		30	
09 40 45.2	−05 26 09	IC 553	13.7	1.1 x 0.9	13.5	(R′)SB(rs)a pec:	129	Mag 10.24v star W 11′.4.
09 54 27.2	−06 57 14	IC 574	13.7	1.1 x 1.0	13.7	SA0^-:	3	Mag 14.6 star on E edge; mag 11.10v star E 7′.0.
09 54 33.2	−06 51 30	IC 575	13.2	1.2 x 0.8	13.0	Sa pec sp	126	Faint, SE-NW dark line bisects bright center; mag 12.2 star NW 4′.5; **MCG −1-25-59** NNE 5′.3.
09 56 39.7	−13 46 32	IC 579	13.9	1.2 x 0.4	13.0	SB(rs)ab:	141	
09 59 50.5	−06 55 25	IC 586	13.5	0.6 x 0.6	12.2			Mag 11.27v star NW 2′.7; mag 10.05v star NE 3′.4.
10 13 12.5	−05 37 46	IC 599	14.3	1.1 x 0.3	13.0	Sbc	36	Mag 11.7 star E 5′.4.
09 25 12.2	−06 49 52	IC 2471	13.7	1.3 x 0.7	13.5	S0° pec:	141	Mag 9.66v star W 6′.0.
09 26 59.3	−12 06 35	IC 2482	11.5	2.3 x 1.6	12.9	E^+	145	Mag 9.96v star NW 4′.9.
10 05 47.9	−17 26 04	IC 2541	12.9	1.3 x 0.5	12.2	SB(r)bc? II	5	Mag 10.83v star NNE 7′.8.
08 57 06.4	−08 43 31	MCG −1-23-11	14.5	1.7 x 0.3	13.6	Sbc: II-III	39	Mag 9.26v star ESE 8′.8.
08 58 12.0	−06 11 58	MCG −1-23-13	13.4	1.9 x 0.7	13.6	SB(r)bc: II-III	12	Mag 11.9 star NW edge.
08 59 53.9	−07 25 01	MCG −1-23-16	13.5	1.5 x 1.1	13.9	IB?	108	Several strong knots; mag 13.6 star N 1′.1.
09 06 34.3	−05 07 26	MCG −1-23-18	13.9	1.0 x 0.5	13.0		117	Mag 8.81v star E 7′.2.
09 06 34.1	−07 14 31	MCG −1-23-19	13.0	0.8 x 0.5	11.9	SA(r)0^-?	99	Mag 8.62v star S 0′.9.
09 06 37.3	−06 34 09	MCG −1-23-20	14.6	2.0 x 0.3	13.9	Sd: sp	39	Mag 11.5 star SSE 5′.6.
09 06 59.6	−07 23 41	MCG −1-23-21	12.9	1.6 x 0.5	12.5	SABb	90	Very faint, stellar anonymous galaxy S edge; mag 10.56v star W 2′.0.
09 10 49.4	−08 53 34	MCG −1-24-1	11.3	5.2 x 0.9	12.8	SAB(r)b: sp	30	Bright star on W edge.
09 16 04.9	−09 09 06	MCG −1-24-4	13.7	1.6 x 0.9	13.9	(R′)SB(rs)ab pec:	0	Mag 9.12v star NNE 5′.2.
09 17 39.3	−07 54 30	MCG −1-24-7	14.2	1.4 x 0.2	12.7	S?	162	
09 19 06.1	−06 42 16	MCG −1-24-8	13.8	0.9 x 0.5	12.8	S0	126	
09 20 22.0	−07 52 54	MCG −1-24-10	13.1	1.5 x 0.7	13.0	Sb? III	30	Mag 13.6 star N 1′.6.
09 20 46.3	−08 03 24	MCG −1-24-12	14.0	1.4 x 0.7	13.8	SAB(rs)c: II	33	Star on NE end. Elongated **MCG −1-24-11** N 1′.0.
09 20 47.7	−08 06 08	MCG −1-24-13	14.1	0.9 x 0.7	13.5	SBbc	132	Compact spiral arms.
09 24 27.2	−06 34 51	MCG −1-24-14	14.0	0.8 x 0.7	13.3	Sb	138	
09 31 59.8	−08 43 53	MCG −1-25-1	14.6	1.5 x 0.2	13.2	Sc? sp	159	Mag 9.18v star SE 3′.5.
09 32 47.5	−05 43 29	MCG −1-25-2	13.3	1.5 x 0.9	13.5	S	51	Mag 10.29v star NW 3′.5; a pair of faint, small anonymous galaxies SW 1′.7; another faint, almost stellar anonymous galaxy NE 2′.6.
09 35 20.4	−08 48 40	MCG −1-25-3	14.3	1.0 x 0.9	14.0	S pec	177	Mag 9.66v star NW 4′.6.
09 35 56.9	−07 43 37	MCG −1-25-5	13.1	1.9 x 0.6	13.1	$S0^+$: sp	72	Bright core, smooth envelope.
09 36 12.4	−08 26 06	MCG −1-25-6	13.0	1.6 x 0.7	13.2	E^+?	24	
09 40 31.1	−08 56 38	MCG −1-25-15	12.7	1.8 x 1.4	13.5	SB(s)cd II-III	141	Mag 9.78v star S 4′.5.
09 41 40.2	−09 01 24	MCG −1-25-18	13.9	1.1 x 0.6	13.3		87	
09 41 50.5	−07 19 50	MCG −1-25-20	14.8	1.0 x 0.8	14.4	SAB(s)dm: IV-V	171	Mag 9.80v star NE 3′.8.
09 42 23.4	−06 15 09	MCG −1-25-24	13.5	1.7 x 0.6	13.4	SBd?	153	Mag 9.27v star NW 2′.2.
09 42 49.2	−06 44 27	MCG −1-25-26	14.2	1.9 x 0.4	13.8	Sb: sp	174	Mag 7.54v star NNE 8′.9.
09 43 22.6	−05 17 02	MCG −1-25-31	15.4	1.3 x 0.2	13.8	SB(s)dm? sp	90	
09 43 27.6	−05 16 45	MCG −1-25-32	14.3	1.1 x 0.3	12.9	SB(s)m? sp	72	
09 43 28.9	−05 22 03	MCG −1-25-33	13.9	1.3 x 0.7	13.6	SB(s)d pec: III	144	
09 43 36.2	−05 54 51	MCG −1-25-34	13.2	2.1 x 1.4	14.2	SB(s)dm III-IV	9	Mag 9.28v star SSW 8′.7.
09 48 48.4	−05 06 01	MCG −1-25-39	14.6	0.7 x 0.7	13.7			Stellar nucleus, uniform envelope; mag 10.10v star SW 7′.7.
09 48 45.2	−06 35 54	MCG −1-25-40	13.8	1.1 x 0.3	12.4	Sbc	3	
09 49 02.0	−07 08 18	MCG −1-25-41	13.8	1.4 x 0.5	13.3	SB(rs)0^+ pec	45	
09 49 03.2	−07 09 24	MCG −1-25-42	14.1	0.7 x 0.5	13.0	E:	6	
09 49 22.2	−05 11 29	MCG −1-25-44	14.5	0.9 x 0.7	13.8	S0/a	123	NGC 3022 E 4′.6.
09 50 55.3	−09 19 33	MCG −1-25-48	13.2	1.6 x 1.4	13.9	(R)SB(rs)0/a pec:	63	Bright core, two faint arms.
09 50 56.5	−04 59 53	MCG −1-25-49	13.5	2.2 x 0.7	13.8	(R)SB(r)a	21	Mag 10.04v star E 1′.5.
09 51 27.3	−05 06 17	MCG −1-25-50	13.2	1.6 x 0.8	13.3	SB(r)ab	174	Mag 9.46v star NE 3′.6.
09 51 28.4	−05 01 35	MCG −1-25-51	14.4	0.9 x 0.4	13.1			Mag 8.54v star N 3′.2.
09 52 13.4	−09 30 39	MCG −1-25-53	13.8	1.0 x 0.6	13.1	Sbc	12	
09 52 37.6	−07 37 15	MCG −1-25-54	13.0	0.5 x 0.5	11.6	E/S0		Mag 8.37v star W 4′.4.
09 53 50.2	−09 25 35	MCG −1-25-55	14.8	0.7 x 0.7	13.9			Very irregular shape.
09 56 48.1	−07 10 46	MCG −1-26-2	13.2	1.9 x 0.5	13.0	(R′)SB(s)ab?	0	Mag 13.3 star NE 2′.1.
09 57 04.5	−07 53 05	MCG −1-26-3	14.1	1.0 x 0.7	13.6	SBcd		
10 00 07.2	−06 24 47	MCG −1-26-5	14.0	0.9 x 0.3	12.4	Sc	36	
10 02 36.2	−06 00 51	MCG −1-26-12	14.0	2.3 x 0.2	13.0	Sd: sp IV-V	15	Mag 8.65v star NNE 6′.4.
10 03 57.2	−06 29 50	MCG −1-26-13	14.4	0.7 x 0.2	12.2	$S0^+$ pec	95	NGC 3110 NNE 1′.8.
10 04 50.2	−06 43 34	MCG −1-26-16	13.7	0.7 x 0.4	12.3	E	42	
10 05 12.2	−06 38 01	MCG −1-26-17	14.3	0.4 x 0.3	12.0	E?	72	
10 05 41.6	−07 58 59	MCG −1-26-21	12.5	1.4 x 1.1	12.8	SA(s)0° pec:	171	
10 07 05.3	−05 53 01	MCG −1-26-24	14.1	1.2 x 0.2	12.4	Sb sp	12	
10 10 54.4	−06 54 49	MCG −1-26-29	14.2	1.1 x 0.9	14.0	Scd	129	Stellar nucleus; mag 7.84v star SSE 9′.0.
10 12 46.2	−08 35 21	MCG −1-26-31	14.7	1.0 x 0.6	14.0	Sbc	72	
10 13 21.4	−08 59 25	MCG −1-26-36	14.5	0.7 x 0.4	13.0	Sc	126	
08 59 25.6	−11 50 40	MCG −2-23-7	12.6	1.0 x 0.8	12.2	Sbc	0	Lies between two N-S field stars.
09 01 50.5	−13 31 29	MCG −2-23-8	14.5	0.6 x 0.5	13.0	Sc	33	
09 11 20.3	−15 03 01	MCG −2-24-1	13.8	1.5 x 1.2	14.3	IB(s)m: V	39	Very low surface brightness; mag 12.3 star NW 2′.2.
09 12 06.8	−15 25 57	MCG −2-24-3	14.2	1.3 x 0.4	13.3	Sd: III-IV	91	
09 15 36.7	−13 37 54	MCG −2-24-4	15.9	0.7 x 0.6	14.8		144	Mag 10.08v star E 5′.4.
09 15 55.8	−13 58 56	MCG −2-24-5	13.9	0.9 x 0.7	13.2	SBb	27	Mag 8.88v star W 4′.3.
09 17 41.1	−14 29 57	MCG −2-24-6	14.3	0.9 x 0.6	13.5	SBbc	165	Mag 5.84v star SW 9′.2.
09 18 05.7	−12 05 44	MCG −2-24-7	12.9	0.6 x 0.6	11.7	(R′)SA0^-:		Mag 10.80v star W 1′.1; several faint anonymous galaxies in immediate area.
09 18 31.5	−12 59 35	MCG −2-24-8	14.5	0.9 x 0.3	12.9		120	
09 18 37.8	−11 44 04	MCG −2-24-9	14.2	0.9 x 0.7	13.5		90	
09 19 32.8	−10 29 55	MCG −2-24-10	14.6	0.6 x 0.3	12.6	SB	9	
09 19 48.8	−12 14 52	MCG −2-24-11	14.4	1.0 x 0.5	13.5	IB(s)m pec V	3	
09 20 55.7	−09 43 39	MCG −2-24-13	14.0	0.9 x 0.3	12.4	Sc	69	
09 20 49.2	−12 53 47	MCG −2-24-14	14.5	0.8 x 0.2	12.4		111	
09 22 07.1	−09 45 04	MCG −2-24-17	14.6	1.1 x 0.9	14.5	SA(rs)d: III-IV	114	
09 23 34.6	−15 22 35	MCG −2-24-19	14.4	0.9 x 0.7	13.7	Sb		
09 26 27.4	−15 42 36	MCG −2-24-23	15.1	1.4 x 0.7	15.0	SB(s)m pec: IV	42	
09 28 58.8	−14 48 27	MCG −2-24-27	13.1	1.3 x 0.9	13.1	SB(s)d III-IV		Pair of bright stars superimposed E of center.
09 30 36.3	−12 02 05	MCG −2-24-28	14.0	0.7 x 0.5	12.7		90	
09 30 58.5	−12 41 50	MCG −2-24-29	15.4	0.7 x 0.3	13.5		96	Anonymous elongated galaxy SW 1′.0.
09 31 34.2	−13 02 44	MCG −2-25-1	14.8	0.7 x 0.4	13.3		93	Mag 8.91v star SSE 1′.2.
09 33 43.4	−11 19 19	MCG −2-25-2	12.9	1.6 x 0.5	12.5	SAbc? II-III	141	Mag 12.7 star W 1′.0.

(Continued from previous page)

GALAXIES

RA h m s	Dec ° ′ ″	Name	Mag (V)	Dim ′ Maj x min	SB	Type Class	PA	Notes
09 34 07.0	−15 18 25	MCG −2-25-3	13.9	0.8 x 0.6	12.9	SAB(s)m: V	129	Mag 9.60v star S 5′0.
09 36 13.8	−10 58 30	MCG −2-25-5	13.4	1.3 x 0.6	13.0	SAB(rs)cd: III-IV	95	
09 36 28.4	−11 19 52	MCG −2-25-6	12.1	1.0 x 0.8	11.7		93	Small galaxies located 1′0 S and 0′7 W.
09 36 39.7	−09 53 50	MCG −2-25-7	14.8	0.9 x 0.2	12.8	Sb	12	Mag 1′0.24v star W 1′5.
09 37 46.8	−10 58 28	MCG −2-25-8	13.5	1.4 x 0.4	12.7	Sbc: II-III	174	
09 37 54.6	−11 40 14	MCG −2-25-9	13.9	0.9 x 0.5	12.9	S?	3	
09 39 38.8	−13 24 24	MCG −2-25-10	14.1	0.5 x 0.5	12.4			
09 39 46.4	−11 20 46	MCG −2-25-11	13.1	1.0 x 0.6	12.4	SBbc	21	
09 43 17.8	−09 56 44	MCG −2-25-13	13.3	2.4 x 0.9	14.0	SAB(s)d III-IV	147	Mag 12.2 star W 2′7.
09 46 42.0	−14 51 43	MCG −2-25-17	13.3	0.8 x 0.7	12.5	(R)SB(r)a?	69	
09 47 49.4	−13 44 23	MCG −2-25-18	14.4	0.7 x 0.4	12.9	Sc	120	
09 49 57.3	−12 05 49	MCG −2-25-19	16.7	0.8 x 0.3	15.0	I?	57	Mag 9.99v star NW 4′5.
09 50 13.8	−12 03 26	MCG −2-25-20	13.8	1.7 x 0.3	12.9	Sd: sp	123	Mag 12.2 star NW end.
09 51 22.0	−12 37 28	MCG −2-25-23	14.0	1.0 x 0.7	13.5	SB(s)cd III	51	
09 51 33.9	−12 59 48	MCG −2-25-24	13.9	0.9 x 0.6	13.1	Sb	162	Mag 0.03v star WNW 8′7.
09 51 39.5	−13 42 46	MCG −2-25-25	13.3	1.3 x 0.7	13.1	Sb pec? III	135	Mag 7.67v star W 5′9.
09 55 27.1	−13 47 51	MCG −2-26-1	14.2	0.5 x 0.2	11.8	E	66	
09 55 56.7	−13 45 21	MCG −2-26-2	14.0	1.3 x 0.5	13.4	Sb	9	Small elongated galaxy at S tip.
09 56 17.8	−12 58 19	MCG −2-26-4	14.4	0.7 x 0.5	13.1	Sa	100	**MCG −2-26-3** WNW 1′0.
09 56 38.6	−15 34 11	MCG −2-26-6	14.6	0.7 x 0.7	13.7	Sd		
09 57 34.1	−13 50 59	MCG −2-26-8	13.7	0.9 x 0.3	12.1	Sa	114	**MCG −2-26-7** SW 1′6.
09 57 48.7	−13 35 37	MCG −2-26-9	13.7	0.9 x 0.7	13.3	E	15	
09 59 30.0	−12 20 19	MCG −2-26-10	13.9	0.9 x 0.8	13.6	E/S0	24	
09 59 28.0	−13 29 59	MCG −2-26-11	14.5	0.8 x 0.3	12.8	Scd	60	Mag 9.62v star E 3′5.
10 00 35.0	−14 56 51	MCG −2-26-12	14.0	1.6 x 1.0	14.3	SAB(rs)c II	95	Mag 10.20v star SE 9′7.
10 00 47.3	−13 22 42	MCG −2-26-13	14.8	0.7 x 0.3	13.0	Sb	171	
10 01 02.0	−15 21 50	MCG −2-26-14	14.0	0.7 x 0.5	12.8		3	Mag 7.97v star SE 7′2.
10 01 30.1	−15 10 23	MCG −2-26-15	14.9	0.4 x 0.3	12.5		144	Located on W edge of MCG −2-26-16.
10 01 31.4	−15 10 19	MCG −2-26-16	14.5	1.0 x 0.2	12.6		6	Mag 7.63v star NE 11′8.
10 01 33.6	−13 31 36	MCG −2-26-17	14.3	1.1 x 0.2	12.5		111	
10 01 35.9	−13 40 00	MCG −2-26-18	14.3	0.7 x 0.7	13.4	Sd		
10 01 59.9	−15 14 03	MCG −2-26-19	14.9	0.6 x 0.6	13.7	SBbc		MCG −2-26-20 S 2′9.
10 02 01.2	−15 16 56	MCG −2-26-20	14.2	1.1 x 1.1	14.3	Sb		Mag 8.70v star SSW 10′3; MCG −2-26-21 SE 1′5.
10 02 05.9	−15 17 55	MCG −2-26-21	14.3	0.5 x 0.3	12.1		138	Star, or small galaxy, S edge; MCG −2-26-20 NW 1′5.
10 02 22.6	−10 49 07	MCG −2-26-22	14.6	0.5 x 0.3	12.4	Sb	93	
10 02 57.8	−11 44 02	MCG −2-26-23	15.0	0.7 x 0.5	13.7	Sc	99	
10 03 00.5	−15 21 41	MCG −2-26-24	14.0	1.6 x 0.2	12.6	Sb? sp	102	Mag 9.86v star ENE 6′3.
10 03 07.0	−15 28 34	MCG −2-26-25	14.5	0.5 x 0.5	12.8	Sb		Mag 9.36v star ENE 8′4; **MCG −2-26-26** S 1′2.
10 03 16.8	−14 56 40	MCG −2-26-27	13.1	1.6 x 0.7	13.1	SB(r)cd III	85	Very faint star SE of core; mag 13.8 star N 2′4.
10 03 37.2	−15 06 52	MCG −2-26-28	14.2	0.8 x 0.4	12.8	(R′)SAB(r)0°?	90	
10 04 46.5	−13 38 53	MCG −2-26-29	13.8	0.9 x 0.7	13.2		153	Mag 10.42v star W 3′6.
10 05 06.4	−09 58 25	MCG −2-26-30	14.4	0.7 x 0.6	13.3	Sc	12	
10 09 03.3	−11 14 02	MCG −2-26-31	13.3	1.1 x 0.5	12.5	S0°	12	Mag 9.46v star N 2′9.
10 10 08.8	−10 41 46	MCG −2-26-35	16.0	0.9 x 0.2	14.0	Sd	87	
10 10 23.4	−11 55 36	MCG −2-26-37	14.4	0.5 x 0.4	12.5	S0/a	78	Mag 7.42v star W 11′3.
10 10 43.3	−15 30 37	MCG −2-26-38	12.6	0.7 x 0.4	11.1		165	Multi-galaxy system?
10 11 08.7	−13 46 55	MCG −2-26-39	14.5	2.2 x 0.8	15.0	SB(s)m IV-V	36	Mag 9.76v star W 3′4.
10 14 41.3	−09 56 29	MCG −2-26-40	13.8	0.9 x 0.9	13.5	SBc		
10 15 13.3	−15 05 36	MCG −2-26-41	13.3	1.2 x 1.0	13.3	SAB(rs)dm? IV-V	33	Mottled appearance, several knots and/or stars involved.
09 16 31.1	−17 35 48	MCG −3-24-4	14.1	1.9 x 0.4	13.6	SBd? III	111	Mag 9.50v star E 6′6.
09 19 35.9	−17 07 29	MCG −3-24-6	14.2	1.0 x 0.6	13.5	SBcd	42	
09 28 10.4	−16 54 10	MCG −3-24-11	14.4	1.4 x 0.3	13.3	Sbc	45	
09 30 14.8	−16 38 27	MCG −3-24-13	13.8	0.9 x 0.3	12.2	Sab	36	
09 31 23.4	−16 34 32	MCG −3-25-1	13.5	1.3 x 0.8	13.4	SB?	156	
09 31 41.1	−16 02 28	MCG −3-25-3	13.8	1.9 x 0.2	12.6	Scd? sp	144	Mag 9.23v star NNE 6′1.
09 33 15.4	−16 46 05	MCG −3-25-4	12.3	2.6 x 1.7	13.8	SB(r)b: II	29	Strong, elongated core, patchy envelope.
09 33 50.0	−16 59 25	MCG −3-25-5	14.1	1.1 x 0.5	13.5	E	33	
09 35 38.2	−16 53 13	MCG −3-25-9	14.1	1.1 x 0.3	12.8	Sc	165	
09 35 38.8	−17 23 12	MCG −3-25-10	13.3	0.9 x 0.7	12.6	(L)SB(s)0°	0	Two anonymous, elongated galaxies E 1′7.
09 39 05.4	−17 49 42	MCG −3-25-14	14.1	0.7 x 0.7	13.1	(R′)SA(s)ab		
09 42 29.0	−16 58 36	MCG −3-25-15	13.6	1.1 x 0.9	13.5	SB?	6	Mag 13.7 star E 2′2.
09 42 49.9	−16 06 58	MCG −3-25-16	13.9	1.3 x 0.5	13.3	SAB(s)dm IV-V	120	
09 49 01.0	−16 43 47	MCG −3-25-23	14.3	0.9 x 0.4	13.0	Sd	39	
09 51 08.3	−17 30 02	MCG −3-25-25	13.8	0.8 x 0.3	12.1	Sb	123	
09 53 14.5	−15 58 43	MCG −3-25-27	14.4	0.7 x 0.7	13.4	Sb		Pair of stars S and SW, small anonymous galaxy SSE 0′5.
10 02 49.5	−15 49 13	MCG −3-26-13	14.2	0.5 x 0.5	12.8	E/S0		
10 05 26.2	−17 48 00	MCG −3-26-16	13.9	1.1 x 0.5	13.1	Sb II-III	119	
10 06 15.3	−16 01 29	MCG −3-26-21	14.2	1.7 x 0.2	12.9	SBc? sp	45	Mag 12.5 star NE 1′9.
10 11 18.8	−17 12 16	MCG −3-26-30	12.9	1.7 x 1.3	13.7	SAB0$^-$ pec:	100	Mag 10.75v star NNE 10′4.
09 06 48.8	−15 29 58	NGC 2763	12.0	2.3 x 2.0	13.5	SB(r)cd pec II	120	
09 11 27.5	−14 49 00	NGC 2781	11.6	3.0 x 1.5	13.1	SAB(r)0$^+$	79	Faint outer envelope extends beyond large, bright core.
09 16 11.4	−16 18 41	NGC 2811	11.3	2.5 x 0.9	12.0	SB(rs)a	22	
09 20 09.8	−16 31 33	NGC 2848	11.8	2.7 x 1.7	13.3	SAB(s)c: II-III	40	
09 20 30.3	−16 29 46	NGC 2851	13.3	1.4 x 0.6	13.0	SA0°:	5	Small triangle of stars inside N edge; many knots.
09 21 27.8	−11 54 35	NGC 2855	11.7	2.5 x 2.2	13.4	(R)SA(rs)0/a	132	Several stars superimposed on outer edges; mag 8.97v star N 4′1.
09 23 36.4	−10 25 58	NGC 2863	12.7	1.1 x 0.9	12.5	Sd III	120	Mag 14.1 star on N edge; NGC 2868 W 2′4.
09 23 27.2	−10 25 48	NGC 2868	14.3	0.8 x 0.4	13.1	E/S0	65	Almost stellar nucleus; NGC 2863 E 2′4.
09 25 13.9	−06 43 02	NGC 2876	13.1	1.7 x 1.1	13.6	(R′)S0° pec?	95	Faint, wide jet extends WNW beyond published boundaries; mag 11.16v star NNE 4′9; **MCG −1-24-17** NNE 2′9.
09 25 54.5	−11 59 41	NGC 2881	13.2	1.1 x 0.8	12.9	S?	135	Two bright patches, NW and SE.
09 26 24.6	−11 33 20	NGC 2884	12.6	2.0 x 0.9	13.0	S0/a?	174	Mag 9.91v star SSW 7′4.
09 27 12.7	−11 38 36	NGC 2889	11.7	2.2 x 1.9	13.1	SAB(rs)c II	6	Mag 12.6 star on S edge.
09 26 29.9	−14 31 46	NGC 2890	14.5	0.9 x 0.5	13.5	E/S0	67	Lies 3′9 SW of mag 7.89v star.
09 30 53.0	−14 44 11	NGC 2902	12.2	1.4 x 1.2	12.6	SA(s)0°:	8	Mag 8.01v star SSE 5′7.
09 31 36.7	−16 44 08	NGC 2907	11.6	1.8 x 1.1	12.2	SA(s)a? sp	115	Elongated **PGC 27066** NNE 5′4.
09 35 10.8	−16 23 35	NGC 2924	12.0	1.4 x 1.3	12.6	E$^+$:	150	Mag 9.67v star SE 6′5; IC 546 W 5′0.
09 36 05.8	−12 26 13	NGC 2947	12.1	1.5 x 1.3	12.7	SAB(r)bc II-III	25	A pair of stars, mags 10.6 and 11.8, SE 1′6.
09 37 37.0	−10 11 04	NGC 2952	14.5	0.7 x 0.5	13.2	Sd	9	
09 41 54.5	−08 36 14	NGC 2969	13.1	1.3 x 1.2	13.4	SA(s)c pec: II	145	

(Continued from previous page)

133 GALAXIES 133

RA h m s	Dec ° ′ ″	Name	Mag (V)	Dim ′ Maj x min	SB	Type Class	PA	Notes
09 41 16.1	−16 40 27	NGC 2975	14.8	0.8 x 0.7	14.0	E/S0		Located 1′.8 SSE of mag 10.71v star.
09 43 16.7	−09 44 46	NGC 2978	12.8	1.0 x 0.9	12.5	(R′)SAB(rs)bc? II	85	
09 43 08.6	−10 23 01	NGC 2979	12.3	1.5 x 0.9	12.5	(R′)SA(r)a?	30	
09 43 12.0	−09 36 47	NGC 2980	13.0	1.6 x 0.9	13.2	SAB(s)c? II	164	Mag 11.01v star N 4′.2.
09 45 42.5	−14 18 49	NGC 2992	12.2	3.5 x 1.1	13.5	Sa pec	18	Faint, broad extension from bright nucleus northward, slight brightening at the end.
09 45 48.5	−14 22 09	NGC 2993	12.6	1.3 x 0.9	12.7	Sa pec	95	Very faint extension starts at N edge and curves E and then SE. Some sources indicate this galaxy has a double nucleus.
09 47 45.6	−06 26 17	NGC 3007	13.4	1.4 x 0.6	13.0	S0/a sp	90	
09 49 39.1	−05 09 59	NGC 3022	13.2	1.6 x 1.6	14.1	(R′)SAB0°:		Mag 10.22v star NE 5′.2; MCG −1-25-44 W 4′.6.
09 48 53.9	−08 03 06	NGC 3029	14.0	1.4 x 0.9	14.1	SAB(r)c I-II	46	Stellar nucleus; mag 11.3 star W 4′.5.
09 50 10.6	−12 13 35	NGC 3030	13.3	1.0 x 0.9	13.1	E/S0	120	Mag 8.57v star NW 3′.6.
09 51 55.0	−06 49 25	NGC 3035	12.4	1.6 x 1.4	13.1	SB(rs)bc II	153	
09 53 35.4	−12 28 58	NGC 3058	13.5	1.3 x 0.7	13.3	S0⁺ pec:	30	Double system, small round companion galaxy on W edge.
09 55 41.5	−06 21 53	NGC 3064	14.1	1.1 x 0.3	12.8	Sbc? sp	28	
10 04 02.0	−06 28 31	NGC 3110	12.2	1.5 x 0.7	12.1	SB(rs)b pec	162	MCG −1-26-13 SSW 1′.8.
10 05 14.1	−07 43 07	NGC 3115	8.9	7.2 x 2.4	11.9	S0⁻ sp	40	**Spindle Galaxy**. Galaxy **UGCA 200** ESE 5′.6.
10 06 24.8	−16 07 35	NGC 3127	13.8	1.3 x 0.2	12.2	Sb: sp	55	Located 1′.6 NW of mag 10.21v star; NGC 3128 W 5′.7.
10 06 01.3	−16 07 22	NGC 3128	13.5	1.6 x 0.7	13.5	SB(s)b? I-II	174	Pair of superimposed stars or bright knots N edge; NGC 3127 E 5′.7.
10 07 12.9	−11 57 56	NGC 3133	14.5	0.6 x 0.3	12.5	Sb	25	Mag 10.24v star SW 5′.1.
10 09 16.9	−11 57 27	NGC 3138	14.8	1.2 x 0.4	13.8	Sbc	80	Mag 7.42v star NE 6′.4.
10 10 05.4	−11 46 41	NGC 3139	13.5	1.4 x 1.2	13.9	S0° pec:	75	Small core with smooth envelope; mag 7.42v star SW 9′.8.
10 09 27.9	−16 37 44	NGC 3140	14.0	0.9 x 0.8	13.5	Sc I-II	41	Located near E edge of galaxy cluster **A 940**; NGC 3141 SW 2′.6.
10 09 19.9	−16 39 15	NGC 3141	15.4	0.5 x 0.3	13.2	Sbc	26	Almost stellar, NGC 3140 NE 2′.6.
10 10 06.3	−08 28 49	NGC 3142	13.8	1.0 x 0.8	13.4	S0	3	Mag 5.90v star 17 Sextantis N 4′.4.
10 10 04.1	−12 34 54	NGC 3143	14.3	0.8 x 0.7	13.5	SB(s)b II-III	105	**PGC 29578** S 4′.0.
10 10 10.2	−12 26 01	NGC 3145	11.7	3.1 x 1.4	13.2	SB(rs)bc I	18	Located 7′.8 SE of mag 3.81v star Lambda, 41 Hydrae.

GALAXY CLUSTERS

RA h m s	Dec ° ′ ″	Name	Mag 10th brightest	No. Gal	Diam ′	Notes
09 08 48.0	−09 38 00	A 754	15.2	92	50	Numerous stellar and almost stellar PGC GX extend from SE edge to NW edge.
09 37 06.0	−05 00 00	A 838	15.3	40	17	**MCG −1-25-7** S of center 2′.5; all other members are anonymous, faint and stellar.

PLANETARY NEBULAE

RA h m s	Dec ° ′ ″	Name	Diam ″	Mag (P)	Mag (V)	Mag cent ★	Alt Name	Notes
09 45 35.0	−13 10 10	PN G248.7+29.5	290	14.5	12.9	16.3	PK 248+29.1	Mag 12.9 star N edge; mag 9.71v star SW 5′.6; and elongated anonymous galaxy on W edge, 2′.2 W of center

134 GALAXIES 134

RA h m s	Dec ° ′ ″	Name	Mag (V)	Dim ′ Maj x min	SB	Type Class	PA	Notes
08 33 16.9	−17 57 31	ESO 562-14	13.1	1.5 x 0.3	12.1	SBb pec sp	162	Mag 10.89v star SSE 4′.0.
08 52 18.4	−17 44 42	ESO 563-31	12.6	1.3 x 0.9	12.7	(R′)SB(s)0°:	141	Mag 10.76v star at SE end.
08 12 39.5	−16 03 05	IC 500	12.5	1.3 x 0.5	12.1	E	54	Mag 9.83v star N 5′.7.
08 33 05.2	−12 21 20	IC 513	13.0	1.0 x 0.5	12.2	SB(rs)0°?	40	Mag 9.75v star E 9′.5.
08 26 19.7	−13 18 13	IC 2375	13.1	1.9 x 0.4	12.7	SB(s)b pec:	83	IC 2377 and IC 2379 E 1′.8.
08 26 26.3	−13 18 26	IC 2377	13.4	0.7 x 0.4	11.8	S0/a	36	IC 2379 W; mag 10.73v star E 1′.5.
08 26 27.9	−13 17 36	IC 2379	13.7	0.8 x 0.4	12.3	SB(r)a pec?	144	Pair of stars on N-S line, mags 10.73v and 10.90v, E 2′.8.
08 19 33.9	−07 02 22	MCG −1-22-1	14.9	0.5 x 0.5	13.3	SB(s)d? III		
08 19 58.8	−07 04 36	MCG −1-22-2	13.1	1.0 x 0.7	12.6	SA0⁻	36	
08 23 59.0	−06 53 45	MCG −1-22-11	14.3	1.1 x 0.3	12.9	Sb pec: II	63	Mag 8.27v star NW 6′.5.
08 28 47.3	−06 56 26	MCG −1-22-13	13.4	0.6 x 0.5	12.2	E	69	= **Mkn 1216**.
08 29 54.2	−07 32 15	MCG −1-22-14	14.1	1.2 x 0.2	12.4	S?	168	
08 30 32.2	−07 20 21	MCG −1-22-15	13.5	0.9 x 0.4	12.5	E	18	Mag 11.4 star N edge.
08 32 34.9	−05 39 21	MCG −1-22-19	14.2	0.8 x 0.5	13.0	Sc	78	
08 35 03.0	−05 10 32	MCG −1-22-25	15.1	0.8 x 0.4	13.7	Sc	156	
08 37 38.6	−06 43 07	MCG −1-22-29	13.6	1.1 x 0.6	12.9	Sc	27	Star or bright knot SW end, two field stars on W edge.
08 38 56.4	−05 44 44	MCG −1-22-30	14.6	0.8 x 0.8	14.0	SBcd		
08 39 08.0	−08 51 49	MCG −1-22-31	13.6	1.1 x 0.3	12.3	S0/a:	123	
08 42 41.4	−05 39 28	MCG −1-22-35	14.1	0.8 x 0.5	13.0		126	Two bright field stars, in-line, trailing SE.
08 49 00.4	−07 49 50	MCG −1-23-2	12.9	1.4 x 1.0	13.1	SB(s)bc II	162	Superimposed star or bright knot SW end, two bright stars E edge.
08 51 35.7	−07 15 40	MCG −1-23-5	13.4	1.3 x 0.7	13.1	SA(r)0/a?	147	Mag 5.57v star 15 Hydrae N 5′.1.
08 53 42.4	−07 09 32	MCG −1-23-6	13.6	0.7 x 0.7	12.7	SBab		
08 54 04.8	−07 11 01	MCG −1-23-8	13.3	0.8 x 0.5	12.2	SA0°:	39	Slightly small, fainter, **MCG −1-23-9** NE 2′.2.
08 57 06.4	−08 43 31	MCG −1-23-11	14.5	1.7 x 0.3	13.6	Sbc: II-III	39	Mag 9.26v star ESE 8′.8.
08 58 12.0	−06 11 58	MCG −1-23-13	13.4	1.9 x 0.7	13.6	SB(r)bc: II-III	12	Mag 11.9 star NW edge.
08 59 53.9	−07 25 01	MCG −1-23-16	13.5	1.5 x 1.1	13.9	IB?	108	Several strong knots; mag 13.6 star N 1′.1.
07 57 59.2	−14 17 10	MCG −2-21-1	13.7	1.3 x 0.5	13.1	SB(s)bc II	27	Many bright knots and/or superimposed stars.
08 20 06.0	−10 29 20	MCG −2-22-1	13.7	1.0 x 0.9	13.4	SB(rs)d: III-IV	57	
08 21 33.7	−13 21 06	MCG −2-22-3	14.0	0.7 x 0.6	12.9	Sb	108	Mag 10.21v star W 0′.9.
08 22 10.4	−13 37 19	MCG −2-22-4	15.4	0.7 x 0.2	13.1	Sbc	114	Mag 8.87v star E 1′.0.
08 22 29.7	−11 25 09	MCG −2-22-5	13.7	1.9 x 0.4	13.3	Sbc: sp	111	Mag 9.81v star S 2′.7.
08 23 19.9	−15 01 03	MCG −2-22-6	14.0	0.7 x 0.4	12.4			MCG −2-22-8 E 4′.3.
08 23 36.4	−15 02 13	MCG −2-22-8	12.8	0.8 x 0.8	12.2	SA0⁺:		Several faint stars superimposed on envelope; MCG −2-22-6 W 4′.3.
08 24 16.4	−14 30 09	MCG −2-22-9	15.6	1.2 x 0.1	13.1	S?	54	Mag 7.97v star SE 8′.3.
08 25 52.6	−13 30 41	MCG −2-22-11	13.5	0.7 x 0.5	12.3	S0⁺ pec:	3	**MCG −2-22-10** SW 1′.5.
08 25 58.3	−11 46 46	MCG −2-22-12	14.0	1.5 x 0.2	12.6	Sd? sp	75	Mag 7.78v star E 10′.3.
08 25 59.0	−13 53 10	MCG −2-22-13	13.7	0.9 x 0.8	13.1	SBd	30	
08 27 34.1	−12 45 25	MCG −2-22-17	12.5	2.2 x 1.3	13.5	SA(rs)cd II	25	Numerous stars superimposed; mag 11.06v star N edge.
08 28 21.6	−12 08 28	MCG −2-22-18	14.5	1.0 x 0.9	14.2	SB(s)m IV-V	171	
08 34 05.4	−12 26 50	MCG −2-22-21	14.9	1.1 x 0.3	13.6	Sc	174	Mag 7.87v star N 6′.6.
08 36 30.2	−11 49 52	MCG −2-22-22	14.5	2.2 x 0.2	13.4	Sd: sp	75	Mag 12.1 star NW 2′.5.
08 38 01.5	−09 49 15	MCG −2-22-23	13.5	1.0 x 0.5	12.6	SAB(r)bc II	102	

(Continued from previous page)

134 GALAXIES 134

RA h m s	Dec ° ′ ″	Name	Mag (V)	Dim ′ Maj x min	SB	Type Class	PA	Notes
08 38 46.3	−14 40 54	MCG −2-22-24	14.3	0.6 x 0.3	12.3	SAc	105	Mag 8.21v star E 2.′4.
08 39 04.7	−14 44 24	MCG −2-22-25	13.8	1.1 x 0.6	13.2	SA(s)c: II-III	18	Mag 8.38v star W 5.′2.
08 42 11.4	−13 51 36	MCG −2-22-26	14.4	0.7 x 0.4	12.9	Sb	63	Mag 8.88v star SW 5.′0.
08 43 03.8	−11 28 26	MCG −2-22-27	13.6	1.0 x 0.9	13.4	SAB(r)c: I-II	33	
08 43 56.5	−12 50 51	MCG −2-23-1	12.9	1.6 x 0.8	13.0	SB(r)b: II	90	Mag 10.12v star SE 1.′5.
08 46 28.3	−11 58 30	MCG −2-23-3	13.9	1.0 x 0.3	12.5	Sb	93	
08 47 29.9	−11 21 59	MCG −2-23-4	14.6	1.1 x 0.4	13.6	SBbc	177	Bright knot or star SW tip.
08 51 45.4	−14 54 17	MCG −2-23-6	14.1	0.7 x 0.5	12.8		141	
08 59 25.6	−11 50 40	MCG −2-23-7	12.6	1.0 x 0.8	12.2	Sbc	0	Lies between two N-S field stars.
09 01 50.5	−13 31 29	MCG −2-23-8	14.5	0.6 x 0.5	13.0	Sc	33	
08 13 06.3	−16 06 37	MCG −3-21-8	14.0	0.7 x 0.4	12.4	S	27	Mag 12.6 star N edge; mag 10.81v star SW 0.′9.
08 29 38.4	−17 17 54	MCG −3-22-2	14.4	0.9 x 0.3	12.8		144	
08 37 34.2	−16 56 00	MCG −3-22-10	13.5	0.9 x 0.3	11.9	SBa pec?	75	Mag 9.85v star W 3.′1.
08 45 50.3	−17 31 58	MCG −3-23-3	14.2	0.4 x 0.3	11.7	Sbc	150	
08 50 20.4	−16 34 48	MCG −3-23-9	13.0	1.3 x 1.0	13.1	SB(rs)bc: I-II	135	Star at SE edge.
08 51 06.9	−17 33 53	MCG −3-23-10	13.2	2.6 x 0.7	13.7	SA(rs)d: II-III	75	Several superimposed stars and/or bright knots involved SW part of galaxy.
07 58 30.2	−14 21 18	NGC 2501	13.4	1.3 x 0.9	13.4	SAB(r)0°?	120	
08 02 47.1	−12 19 06	NGC 2517	11.8	1.5 x 1.1	12.2	SAB(rs)0°:	67	Several stars superimposed on outer envelope.
08 05 38.0	−11 25 41	NGC 2525	11.6	2.9 x 1.9	13.3	SB(s)c II	75	Almost stellar nucleus; mag 8.50v star N 6.′2.
08 20 48.1	−08 55 09	NGC 2574	12.3	2.2 x 1.2	13.2	SB(rs)ab:	153	An elongated anonymous galaxy W 1.′4; a mag 10.61v star on S edge.
08 21 24.1	−13 19 05	NGC 2578	12.6	2.0 x 1.1	13.3	SB(r)0/a pec	80	MCG −2-22-3 and a mag 10.21v star SE 2.′4.
08 23 07.9	−05 00 11	NGC 2583	13.4	0.9 x 0.9	13.2	E:		
08 33 50.1	−13 10 29	NGC 2612	12.7	2.7 x 0.6	13.1	$S0^-$ sp	120	Mag 13.4 star on N edge; mag 8.38v star SE 7.′8.
08 45 32.0	−15 07 16	NGC 2662	12.8	1.2 x 1.1	13.0	E	0	
08 49 13.2	−14 17 40	NGC 2674	14.9	0.7 x 0.4	13.4	Sa	65	

GALAXY CLUSTERS

RA h m s	Dec ° ′ ″	Name	Mag 10th brightest	No. Gal	Diam ′	Notes
08 17 24.0	−07 35 00	A 644	16.2	42	13	All members anonymous, faint and stellar.

OPEN CLUSTERS

RA h m s	Dec ° ′ ″	Name	Mag	Diam ′	No. ★	B ★	Type	Notes
07 55 06.0	−17 43 00	NGC 2479	9.6	11	45		cl	Moderately rich in stars; small brightness range; no central concentration; detached.
08 00 02.0	−10 46 00	NGC 2506	7.6	12	807	11.0	cl	Rich in stars; moderate brightness range; strong central concentration; detached.
08 10 38.0	−12 49 00	NGC 2539	6.5	15	59	9.0	cl	Moderately rich in stars; moderate brightness range; no central concentration; detached.
08 13 44.0	−05 45 00	NGC 2548	5.8	30	80	8.0	cl	= **M 48**. Rich in stars; large brightness range; strong central concentration; detached.
07 49 47.0	−17 14 48	Ru 37		4	30	14.0	cl	
07 59 36.0	−16 17 00	Ru 45		9	35	13.0	cl:	Moderately rich in stars; small brightness range; not well detached. Probably not a cluster.

PLANETARY NEBULAE

RA h m s	Dec ° ′ ″	Name	Diam ″	Mag (P)	Mag (V)	Mag cent ★	Alt Name	Notes
08 33 23.4	−16 08 58	NGC 2610	42	13.6	12.7	15.9	PK 239+13.1	Located 3.′4 SW of mag 6.56v star.
07 48 03.8	−14 07 43	PN G232.0+5.7				13.2	PK 232+5.1	Stellar; mag 10.10v star NNW 1.′4; mag 8.25v star NE 3.′2.

135 GALAXIES 135

RA h m s	Dec ° ′ ″	Name	Mag (V)	Dim ′ Maj x min	SB	Type Class	PA	Notes
06 44 27.5	-17 55 58	IC 2171	13.5	1.7 x 0.5	13.2	IB(s)m: IV-V	93	Mag 8.65v star N 5.′6.
07 22 11.0	-05 55 50	MCG −1-19-1	14.3	1.4 x 0.2	12.8	Sb: sp	99	Mag 5.84v star SE 4.′7.
07 24 56.7	-09 39 37	NGC 2377	12.7	1.7 x 1.3	13.4	SA(s)c: II	170	Numerous stars on face of galaxy.

OPEN CLUSTERS

RA h m s	Dec ° ′ ″	Name	Mag	Diam ′	No. ★	B ★	Type	Notes
07 17 08.0	−13 58 24	Bas 11a	8.2	5	89	10.9	cl	Moderately rich in bright and faint stars; detached.
06 41 14.0	−16 29 00	Be 25		7	40	16.0	cl	
07 16 23.0	−13 11 36	Be 36		3	40	17.0	cl	
07 06 40.0	−11 44 00	Be 76		5	40	16.0	cl	
07 03 28.0	−05 00 18	Bo 3	9.9	4	25	11.2	cl	Few stars; small brightness range; not well detached.
07 32 06.0	−16 57 00	Bo 5	7.0	5	50	7.8	cl	Few stars; moderate brightness range; strong central concentration; detached.
07 03 55.0	−11 34 30	ClvdB 92		3	12		cl	
07 06 52.0	−10 32 00	Cr 465	10.1	9			ast	Few stars; moderate brightness range; not well detached. Probably not a cluster.
07 06 56.0	−10 42 54	Cr 466	11.1	4	25	11.1	cl	Few stars; small brightness range; not well detached; involved in nebulosity.
07 28 24.0	−15 23 42	Cz 29	10.3	5	40	12.2	cl:	Moderately rich in stars; moderate brightness range; slight central concentration; detached.
07 31 11.0	−09 57 00	Cz 30		3	20		cl:	
07 04 08.0	−06 07 06	Haf 3		5	20	14.0	cl	(S) Sparse group.
07 06 04.0	−15 00 12	Haf 4		5	30	14.0	cl	
07 20 01.0	−13 10 00	Haf 6	9.2	7	66	16.0	cl	Rich in stars; moderate brightness range; not well detached; involved in nebulosity. (S) Low contrast against sky.
07 23 03.0	−12 17 48	Haf 8	9.1	5	71	12.0	cl	Moderately rich in stars; moderate brightness range; slight central concentration; detached.
07 24 42.0	−16 59 24	Haf 9		4	35	14.0	cl	
07 28 36.0	−15 21 54	Haf 10	11.5	3	40	15.0	cl	Moderately rich in stars; moderate brightness range; no central concentration; detached. Probably not a cluster.
07 09 30.0	−16 56 00	Haf 23		11	40	13.0	cl	Moderately rich in stars; moderate brightness range; not well detached. (A) Several star chains appear on DSS image.
07 37 30.0	−12 03 00	Mel 71	7.1	8	216	10.2	cl	Rich in stars; moderate brightness range; slight central concentration; detached. (A) Also known as "**Wilk's Cluster**."
07 38 29.0	−10 41 36	Mel 72	10.1	5	40		cl	Moderately rich in stars; small brightness range; no central concentration; detached.
06 33 47.0	−05 05 00	NGC 2250		10	25	8.7	ast?	Few stars; moderate brightness range; not well detached. (A) Oblong (E-W) 10′ x 5.5′ group.
06 51 54.0	−07 05 00	NGC 2302	8.9	5	30	12.0	cl	Moderately rich in stars; moderate brightness range; no central concentration; detached.
06 54 36.7	−07 21 00	NGC 2306		20	27	8.0	cl?	

(Continued from previous page)

OPEN CLUSTERS

RA h m s	Dec ° ′ ″	Name	Mag	Diam ′	No. ★	B ★	Type	Notes
06 56 04.0	−07 10 30	NGC 2309	10.5	5	40	13.0	cl	Moderately rich in stars; moderate brightness range; strong central concentration; detached.
06 59 28.0	−13 42 00	NGC 2318				8.0	ast?	(S) Sparse group.
07 02 47.8	−08 23 00	NGC 2323	5.9	15	80	9.0	cl	= **M 50**. Rich in stars; large brightness range; slight central concentration; detached.
07 06 50.0	−10 02 00	NGC 2335	7.2	7	57	10.0	cl	Moderately rich in stars; moderate brightness range; no central concentration; detached; involved in nebulosity.
07 07 48.0	−05 43 12	NGC 2338					ast?	(S) Identification uncertain. Nice small cluster.
07 08 07.0	−10 37 00	NGC 2343	6.7	6	55	8.0	cl	Few stars; moderate brightness range; slight central concentration; detached; involved in nebulosity.
07 08 19.0	−13 12 00	NGC 2345	7.7	12	70	9.0	cl	Rich in stars; large brightness range; slight central concentration; detached.
07 14 31.0	−10 16 00	NGC 2353	7.1	18	106	9.0	cl	Few stars; large brightness range; no central concentration; detached.
07 17 44.0	−15 39 00	NGC 2360	7.2	14	91	10.4	cl	Rich in stars; large brightness range; strong central concentration; detached.
07 20 47.0	−07 33 00	NGC 2364					cl?	
07 21 02.2	−10 22 28	NGC 2368	11.8	3	15		cl	Few stars; small brightness range; not well detached. (A) Appears on the DSS image as small upside-down "Y" group of about 16 stars.
07 23 57.0	−13 16 00	NGC 2374	8.0	12	73	10.7	cl	Few stars; moderate brightness range; not well detached.
07 28 00.0	−11 43 00	NGC 2396	7.4	10	30	11.0	cl	Moderately rich in stars; small brightness range; not well detached. (S) Sparse group.
07 29 25.0	−13 58 00	NGC 2401	12.6	2	20		cl	Few stars similar in brightness; strong central concentration; detached.
07 33 13.0	−15 27 12	NGC 2414	7.9	6	35	8.2	cl	Moderately rich in bright and faint stars with a strong central concentration; detached.
07 36 36.0	−14 29 00	NGC 2422	4.4	25	117	5.0	cl	= **M 47**. Moderately rich in bright and faint stars with a strong central concentration; detached. Includes approximate combined mag for **Struve 1121**, centrally in the cluster.
07 37 07.0	−13 52 00	NGC 2423	6.7	12	86	9.0	cl	Moderately rich in stars; moderate brightness range; slight central concentration; detached.
07 38 18.0	−14 52 42	NGC 2425		5	30	14.0	cl	Moderately rich in stars; small brightness range; slight central concentration; detached.
07 39 18.0	−16 31 00	NGC 2428		13	50		cl?	(A) Has 2 NW-SE streams of stars, extending slightly beyond published diameter.
07 39 30.0	−16 20 00	NGC 2430		7.5	20	8.0	ast?	
07 41 47.0	−14 49 00	NGC 2437	6.1	20	186	10.0	cl	= **M 46**. Rich in stars; moderate brightness range; slight central concentration; detached.
06 36 24.0	−14 09 00	Ru 1		6	15	11.0	cl	Few stars; small brightness range; no central concentration; detached.
06 48 57.0	−10 31 00	Ru 4		6	40	14.0	cl	
06 56 05.0	−13 15 06	Ru 6		1	10		cl	
06 57 50.0	−13 13 12	Ru 7		3	30	14.0	cl	
07 01 46.0	−13 32 42	Ru 8		3	10	12.0	cl:	
07 31 51.0	−12 47 00	Ru 24		9	15	11.0	cl	Few stars; small brightness range; not well detached.
07 37 13.0	−15 40 30	Ru 26		4	20	12.0	cl:	(S) Sparse group.
07 49 47.0	−17 14 48	Ru 37		4	30	14.0	cl	
07 41 18.0	−16 17 00	Ru 151		8	30	12.0	cl	Moderately rich in stars; moderate brightness range; no central concentration; detached. (S) Maybe sparse group.
07 26 08.0	−15 05 39	Wa 7		1.3	13	9.2	cl	

BRIGHT NEBULAE

RA h m s	Dec ° ′ ″	Name	Dim ′ Maj x min	Type	BC	Color	Notes
07 05 12.0	−12 20 00	Ced 90	10 x 10	E+R	1-5	2-4	The W side is sharply defined by a starless region or dark nebula; diameter of dark space about 10′.
07 05 18.0	−10 38 00	IC 2177	120 x 40	E	3-5	3-4	
07 15 42.0	−10 35 00	LBN 1036	60 x 10	E	4-5	3-4	Very faint and diffuse; the S half brightest.
06 48 39.1	−16 54 11	NGC 2296	0.6 x 0.5				This object is likely a galactic nebula, not a galaxy.
06 59 42.0	−07 46 00	NGC 2316	4 x 3	E+R	2-5	2-4	This nebula surrounds a double star; there are three brighter stars close S.
07 04 18.0	−11 18 00	NGC 2327	1.5 x 1.5		1-5	2-4	
07 18 36.0	−13 12 00	NGC 2359	9 x 6	E	2-5	3-4	S half brightest.
07 31 36.0	−16 58 00	Sh2-302	20 x 20	E	4-5	3-4	The E section is split by a narrow dark lane about 12′ in length, aligned roughly N-S. Brightest part is a small area at the NE end.
07 04 18.0	−10 28 00	vdB 93	20 x 20	E+R	1-5	2-4	A prominent, narrow absorption patch projects in from the E edge, ending in the central part.
07 32 30.0	−16 54 00	vdB 97	2 x 2	R	3-5	1-4	Small, round nebulous patch with a mag 9.9 star on the E edge.

PLANETARY NEBULAE

RA h m s	Dec ° ′ ″	Name	Diam ″	Mag (P)	Mag (V)	Mag cent ★	Alt Name	Notes
07 41 50.4	−14 44 06	NGC 2438	76	10.1	10.8	17.7	PK 231+4.2	Mag 10.84v star SW 2′.5; in star rich area.
06 41 34.7	−05 02 37	PN G216.3−4.4	15	17.9		19.1	PK 216−4.1	Mag 13.0 star W 2′.6; mag 14.1 star N 2′.9.
07 26 34.2	−05 21 52	PN G221.7+5.3	32	14.2	14.8		PK 221+5.1	Slightly elongated NE-SW; mag 11.2 star NE 2′.9; mag 12.6 star SW 3′.
07 17 25.5	−10 10 39	PN G224.9+1.0	62			16.8	PK 224+1.1	Close pair of mag 14-15 stars on W edge; mag 11.8 star SSE 1′.4.
07 02 46.9	−13 42 37	PN G226.4−3.7	10	13.3	14.0	16.2	PK 226−3.1	Mag 12.8 star E 1′.7; mag 9.98v star NW 2′.1.
07 37 19.0	−09 38 51	PN G226.7+5.6	14	13.2	13.0	16.9	PK 226+5.1	Mag 13.5 star NNW 0′.6; mag 13.9 star SSE 1′.0.
07 40 22.3	−11 32 32	PN G228.8+5.3	3	13.4	14.0	18.5	PK 228+5.1	Mag 13.0 star W 1′.3; mag 10.46v star SW 2′.1.
07 12 35.9	−16 06 02	PN G229.6−2.7	62			21.0	PK 229−2.1	Contains two E-W bright strips; mag 12.7 star on N edge; mag 13.9 star on E edge; mag 12.4 star SW 2′.7.
07 42 04.2	−14 21 19	PN G231.4+4.3	30	14.0	15.0	20.9	PK 231+4.1	Close pair of N-S mag 12-13 stars SE 0′.8; mag 11.9 star W 2′.0; mag 8.86v star SSW 3′.5.
07 48 03.8	−14 07 43	PN G232.0+5.7				13.2	PK 232+5.1	Stellar; mag 10.10v star NNW 1′.4; mag 8.25v star NE 3′.2.

GALAXIES

RA h m s	Dec ° ′ ″	Name	Mag (V)	Dim ′ Maj x min	SB	Type Class	PA	Notes
06 31 47.2	−17 37 19	ESO 557-2	13.7	1.2 x 0.8	13.5	Sbc	27	Two main, curving arms; mag 10.11v star NW 3′.3.
05 23 56.5	−17 15 37	IC 416	13.3	1.4 x 0.7	13.1	SB(s)c pec: II-III	66	Mag 9.78v star S 6′.7.
05 32 08.7	−07 55 02	IC 421	11.5	3.2 x 2.8	13.7	SAB(rs)bc I-II	39	Bright nucleus with branching arms and numerous bright knots.
05 32 18.6	−17 13 27	IC 422	13.3	0.7 x 0.7	12.4			Mag 9.96v star WNW 8′.0.
05 40 31.4	−11 39 58	IC 433	13.2	0.7 x 0.7	12.5	E/S0		Almost stellar; mag 10.95v star NE 7′.9.
05 53 00.1	−17 52 36	IC 438	11.9	2.8 x 2.1	13.7	SA(rs)c I-II	43	IC 2151 NW 7′.7; mag 8.53v star NNE 6′.8.
06 02 42.6	−12 29 59	IC 441	13.7	1.1 x 1.1	13.7	SB(rs)c II-III		Mag 10.90v star W 2′.3.
05 32 28.6	−13 55 40	IC 2132	13.3	1.5 x 0.7	13.2	Sa pec:	177	Mag 10.21v star SSE 4′.9; NGC 1954 SSE 9′.3.
05 52 36.5	−17 47 15	IC 2151	13.3	1.5 x 0.9	13.5	SB(s)bc: II-III	99	Mag 9.30v star N 3′.3; IC 438 SE 7′.7.
05 23 14.9	−11 25 29	MCG −2-14-15	13.7	1.6 x 0.2	12.3	S0°: sp	51	Located between a close pair of mags 12.3 and 13.0 stars.
05 24 54.6	−12 41 21	MCG −2-14-16	13.5	1.7 x 0.4	13.0	Sab sp	69	
05 31 40.0	−10 23 38	MCG −2-15-1	13.7	1.3 x 0.4	12.8	SB(r)ab? sp	153	Mag 9.72v star SW 8′.2.
05 33 50.2	−13 21 21	MCG −2-15-4	14.6	0.5 x 0.3	12.4		33	MCG −2-15-5 N 1′.4.
05 33 50.4	−13 20 00	MCG −2-15-5	14.2	1.0 x 0.6	13.5	Sbc	33	Mag 9.56v star N 3′.0; MCG −2-15-4 S 1′.4.
05 34 56.5	−10 01 17	MCG −2-15-6	15.5	1.1 x 0.2	13.7	Sc sp	48	Mag 8.52v star N 6′.8.
05 40 53.4	−13 48 12	MCG −2-15-9	14.5	1.0 x 0.8	14.1	SB(s)cd II-III	159	Mag 8.50v star N 7′.6.
05 50 52.5	−14 46 33	MCG −2-15-11	12.4	3.0 x 2.1	14.2	SB(s)d III	126	Very patchy; mag 12.7 star W edge; mag 13.4 star E edge.
05 52 12.1	−14 06 35	MCG −2-15-13	14.1	1.1 x 1.1	14.2	SAB(r)c I-II		Star on N edge.
06 04 34.9	−12 37 29	MCG −2-16-2	14.0	2.4 x 0.5	14.1	Sb? sp	108	Mag 12.4 star NE 8′.5.

136 — GALAXIES — 136

(Continued from previous page)

RA h m s	Dec ° ′ ″	Name	Mag (V)	Dim ′ Maj x min	SB	Type Class	PA	Notes
06 10 35.7	−09 22 29	MCG −2-16-3	15.0	0.5 x 0.5	13.4			Mag 8.48v star WNW 3'.0.
06 16 57.3	−13 05 14	MCG −2-16-8	14.5	0.7 x 0.3	12.7	S	159	Mag 10.81v star SE 1'.7.
05 28 14.2	−16 07 29	MCG −3-14-17	13.0	2.2 x 2.2	14.6	SB(s)cd II-III		Numerous knot in multi-branching arms.
05 35 49.4	−17 46 40	MCG −3-15-4	14.2	1.0 x 0.7	13.6	Irr	130	Mag 9.61v star NNW 8'.2.
05 36 14.3	−16 40 16	MCG −3-15-5	13.9	1.3 x 0.4	13.1	Scd: sp III	75	Mag 8.17v star SE 3'.2.
05 36 53.2	−15 12 15	MCG −3-15-6	13.0	1.1 x 0.8	12.7	(R)SB(r)0/a:	45	Very faint halo extends well beyond published dimensions as seen on Digitized Sky Survey images.
05 39 15.5	−17 01 33	MCG −3-15-7	14.6	1.3 x 0.5	14.0	SB?	144	Mag 7.98v star SW 7'.5.
05 40 17.9	−17 50 13	MCG −3-15-8	14.0	0.5 x 0.4	12.3	E	144	Mag 10.08v star SSE 6'.0; mag 14.3 star E 0'.7.
05 47 12.7	−16 39 36	MCG −3-15-14	14.6	0.6 x 0.4	12.9	Sc	3	Mag 12 star on E edge.
05 54 45.2	−15 08 10	MCG −3-15-27	13.7	1.1 x 0.4	12.6	SB(r)0/a?	168	Mag 10.54v star WNW 3'.8.
05 58 04.3	−16 36 37	MCG −3-16-4	13.8	1.0 x 0.6	13.3	E/S0	126	
06 00 20.6	−16 10 01	MCG −3-16-6	14.3	0.7 x 0.4	12.8	Sbc	171	Mag 10.31v star SE 6'.1.
06 01 55.5	−17 56 31	MCG −3-16-8	13.3	0.7 x 0.4	11.9	S0	108	Mag 9.86v star N 5'.8; **MCG −3-16-9** N 3'.9.
06 11 07.0	−15 29 44	MCG −3-16-18	13.5	0.9 x 0.5	12.4	Sc	84	Mag 8.99v star S 6'.8.
06 19 04.1	−16 56 51	MCG −3-16-23	14.3	1.1 x 0.2	12.5	SBbc? II-III	3	Mag 10.49v star S 1'.8.
06 20 43.0	−16 03 49	MCG −3-17-1	13.5	1.7 x 1.1	14.1	SA(rs)cd: II	130	Several strong knots.
06 23 25.5	−16 09 46	MCG −3-17-4	14.1	1.5 x 0.9	14.3	SAB(s)dm III-IV	10	Mag 6.95v star ESE 13'.9.
06 29 17.5	−17 21 34	MCG −3-17-5	12.6	1.0 x 0.5	11.7	SAB0^{-}: sp	90	Mag 5.76v star SW 11'.6.
05 22 34.7	−11 30 02	NGC 1888	11.9	3.0 x 0.8	12.7	SB(s)c pec	150	NGC 1889 on E edge of core.
05 22 35.4	−11 29 48	NGC 1889	13.1	0.6 x 0.4	11.6	E^{+} pec:	165	Located on the E edge of NGC 1888's core.
05 24 47.4	−15 56 37	NGC 1906	13.6	0.9 x 0.6	12.7	Scd III	145	
05 28 01.9	−05 18 41	NGC 1924	12.5	1.6 x 1.2	13.1	SB(r)bc II-III	140	Mag 8.04v star NW 6'.7.
05 32 48.3	−14 03 47	NGC 1954	11.8	4.2 x 2.2	14.1	SA(rs)bc pec: II-III	150	Numerous superimposed stars and/or bright knots overall; mag 10.21v star NNW 4'.6.
05 32 55.2	−14 07 59	NGC 1957	13.9	1.1 x 0.7	13.5		169	Mag 13.0 star SW 1'.2; NGC 1954 N 4'.6.
05 35 25.6	−17 48 56	NGC 1993	12.4	1.5 x 1.4	13.1	SA(rs)0^{-}:	80	Mag 10.60v star W 4'.3; **MCG −3-15-2** WSW 9'.3.
05 46 47.8	−16 46 55	NGC 2076	13.0	2.2 x 1.3	14.0	S0^{+}: sp	39	Dark, narrow lane runs the length of the bright center.
05 47 51.4	−17 36 11	NGC 2089	12.0	1.9 x 1.1	12.6	SAB0^{-}:	39	Mag 10.46v star 1'.5 SSE.
05 52 11.5	−07 27 21	NGC 2110	12.5	1.7 x 1.2	13.1	SAB0^{-}	173	
06 20 55.6	−08 29 41	UGCA 127	12.2	3.9 x 1.1	13.6	Scd? II	70	

OPEN CLUSTERS

RA h m s	Dec ° ′ ″	Name	Mag	Diam ′	No. ★	B ★	Type	Notes
06 22 05.0	−06 19 06	Be 73		2	15	16.0	cl	(A) Tiny (1′ long), but impressive, N-S star chain at center.
05 35 16.0	−05 23 00	NGC 1976		47			cl	
05 35 26.0	−05 55 00	NGC 1980	2.5	15	30		cl	
05 39 16.0	−17 51 00	NGC 2017		10	8	8.0	ast?	A scattered group of a few stars; not a true cluster. It contains the multiple star h3780.
06 20 50.0	−07 17 00	NGC 2215	8.4	8	40	11.0	cl	Moderately rich in stars; moderate brightness range; slight central concentration; detached.
06 26 37.0	−09 38 30	NGC 2225			40		cl?	(S) NGC 2225/6. Single well-defined object, see notes NGC 2226.
06 26 37.0	−09 38 30	NGC 2226		0.8			ptcl	(A) Core of cluster NGC 2225.
06 33 47.0	−05 05 00	NGC 2250		10	25	8.7	ast?	Few stars; moderate brightness range; not well detached. (A) Oblong (E-W) 10′ x 5.5′ group.
06 36 24.0	−14 09 00	Ru 1		6	15	11.0	cl	Few stars; small brightness range; no central concentration; detached.

BRIGHT NEBULAE

RA h m s	Dec ° ′ ″	Name	Dim ′ Maj x min	Type	BC	Color	Notes
05 38 30.0	−07 05 00	IC 430	11 x 11	R	2-5		Very diffuse, fan-shaped nebulosity centered 10'.0 NW of 49 Orionis. The fan converges to a bright spot at a point 13'.0 NW of the star. This spot, measuring 1′ x 0.3′, is extended roughly E-W and forms an isosceles triangle with two field stars a couple of arcminutes SW.
05 35 24.0	−05 27 00	NGC 1976	65 x 60	E+R	1-5	3-4	= **M 42, Orion Nebula**. Very bright and mottled. This galactic nebula is visible to the naked eye.
05 35 24.0	−05 54 00	NGC 1980	14 x 14	E	3-5	4-4	The southernmost semi-detached part of the **Orion Nebula**.
05 35 36.0	−05 16 00	NGC 1982	20 x 15	E+R	1-5	3-4	= **M 43**. Bright nebulosity with a mag 7 star involved; part of **M 42**, **Orion Nebula** and separated from it by a dark lane.
05 36 30.0	−06 42 00	NGC 1999	2 x 2	E	1-5	1-4	Bright roundish nebula; illuminating star is V380 Ori (V = 10.3).
06 03 30.0	−09 44 00	NGC 2149	3 x 2	R	3-5	1-4	Nebula involved around a mag 9.3 star. There is a small, round absorption patch about 1′ across off the SW sector.
06 07 30.0	−06 24 00	NGC 2170	2 x 2	R	1-5	1-4	A mag 9.5 star involved; connected with vdB 69.
06 09 30.0	−06 20 00	NGC 2182	2.5 x 2.5	R	1-5	1-4	A mag 9.0 star involved in nebulosity.
06 10 48.0	−06 13 00	NGC 2183	1 x 1	R	3-5	1-4	
06 11 06.0	−06 13 00	NGC 2185	1.5 x 1.5	R	3-5	1-4	A small, bright patch. There is also some faint nebulosity involved around a small group of four mag 11-12 stars 3′ to the SW.
06 19 58.1	−10 38 14	Red Rectangle	1.0 x 0.7				
05 20 00.0	−05 40 00	Sh2-278	25 x 20	E	4-5	3-4	
06 08 06.0	−06 14 00	vdB 68	4 x 3	R	3-5	1-4	Nebula with a mag 9.0 star involved. It is similar in structure to the nebulosity involved in the Pleiades star cluster.
06 08 06.0	−06 22 00	vdB 69	2 x 2	R	3-5	2-4	Nebula with a mag 9.0 star involved; connected with NGC 2170.
06 31 48.0	−09 40 00	vdB 80	5 x 3	R	3-5	1-4	Nebulosity around a compact cluster; three brighter stars and two fainter ones involved.

PLANETARY NEBULAE

RA h m s	Dec ° ′ ″	Name	Diam ″	Mag (P)	Mag (V)	Mag cent ★	Alt Name	Notes
05 27 28.2	−12 41 49	IC 418	12	10.7	9.3	10.1	PK 215−24.1	Planetary appears overwhelmed by mag 10.1 central star?.
06 21 42.7	−12 59 14	IC 2165	28	12.9	10.5	17.9	PK 221−12.1	Mag 13.2 star ENE 2'.6.
06 23 37.1	−10 13 24	PN G218.9−10.7	94	15.4		16.2	PK 218−10.1	Brightest along N edge; mag 12.6 star WSW 2'.8.

137 — GALAXIES — 137

RA h m s	Dec ° ′ ″	Name	Mag (V)	Dim ′ Maj x min	SB	Type Class	PA	Notes
04 42 27.0	−17 27 24	ESO 551-30	14.3	1.2 x 1.0	14.4	SB(s)d III-IV	0	Mag 9.99v star E 8'.6.
05 06 33.3	−17 35 12	ESO 553-2	12.4	1.7 x 1.3	13.1	SA(r)0°:	38	Mag 10.59v star S 1'.9; ESO 553-3 N 2'.3; mag 9.25v star NW 11'.1.
05 06 37.7	−17 33 15	ESO 553-3	13.4	1.5 x 1.0	13.7	SAB(rs)c I	138	ESO 553-2 S 2'.3; mag 9.25v star NW 10'.4; **ESO 552-75** NNE 6'.0.
04 16 42.4	−12 12 00	IC 362	13.2	1.7 x 1.1	13.9	E^{+}?	1	Mag 10.38v star NW 5'.0.
04 20 40.9	−14 46 53	IC 367	13.4	1.5 x 0.7	13.3	(R′)SA0^{-}:	141	Mag 9.82v star W 7'.5.
04 22 42.7	−12 36 52	IC 368	13.7	0.9 x 0.7	13.1		174	Mag 9.62v star NE 5'.4.
04 23 28.4	−11 47 27	IC 369	14.3	0.5 x 0.5	12.9	E/S0		Mag 11.9 star SW 8'.5.

(Continued from previous page)

GALAXIES

RA h m s	Dec ° ′ ″	Name	Mag (V)	Dim ′ Maj x min	SB	Type Class	PA	Notes
04 24 01.9	−09 23 46	IC 370	14.0	1.3 x 1.1	14.2	SB(rs)c I-II	150	Stellar nucleus; mag 8.84v star S 8′.9.
04 31 16.5	−12 27 18	IC 376	13.9	0.9 x 0.9	13.5			Anonymous galaxy NNW 1′.5.
04 31 51.0	−07 14 19	IC 379	14.1	1.2 x 0.8	13.9	SBab	9	Mag 12.4 star E 7′.8; **MCG −1-12-24** ESE 8′.6.
04 31 41.2	−12 55 38	IC 380	14.5	0.7 x 0.4	13.0	Sbc	81	Located 2′.5 SW of mag 10.16v star.
04 37 55.3	−09 31 14	IC 382	12.8	2.3 x 1.4	13.9	SAB(rs)c: I	173	Mag 10.47v star NW 6′.8.
04 41 44.4	−07 05 15	IC 387	12.8	1.6 x 1.2	13.4	SAB(rs)c I-II	78	Bright nucleus in short NE-SW bar; several bright knots; mag 11.12v star S 5′.0.
04 41 59.7	−07 18 42	IC 389	13.9	0.9 x 0.7	13.2	$SA0^-$:	51	Close pair of mag 12.5 stars W 1′.6; mag 11.20v star NW 3′.9.
04 42 03.9	−07 12 25	IC 390	14.3	1.0 x 0.3	12.9	$SB(s)0^+$? sp	42	Mag 11.12v star WNW 5′.0.
04 47 52.0	−15 31 31	IC 393	14.0	0.7 x 0.7	13.3	E/S0		Stellar; mag 10.82v star W 12′.6.
04 58 12.7	−07 46 53	IC 398	14.7	1.2 x 0.4	13.8	SB(s)c?	21	Mag 8.72v star SE 9′.8.
05 04 19.9	−10 04 33	IC 401	12.5	1.6 x 0.6	12.3	SB(rs)b? II-III	57	Mag 9.69v star S 4′.4.
05 06 14.9	−09 06 28	IC 402	13.7	2.3 x 1.5	14.9	SAB(rs)cd III	147	Stellar nucleus; mag 10.60v star ENE 6′.1.
05 17 42.8	−15 31 31	IC 407	13.5	2.2 x 0.5	13.4	Sc: II-III	165	Mag 8.56v star S 6′.8.
05 23 56.5	−17 15 37	IC 416	13.3	1.4 x 0.7	13.1	SB(s)c pec: II-III	66	Mag 9.78v star S 6′.7.
04 48 45.9	−05 07 31	IC 2095	15.2	1.4 x 0.2	13.7	Scd pec sp	54	Mag 7.17v star NE 10′.9.
04 50 44.8	−05 25 12	IC 2098	13.7	2.1 x 0.3	13.1	Scd: sp	96	Mag 10.50v star SW 4′.5.
04 51 42.0	−06 13 48	IC 2101	13.7	1.6 x 0.4	13.1	SB(s)c pec? II	147	Mag 10.69v star E 5′.9; MCG −1-13-29 is 4′.3 NE of this star.
04 56 19.1	−15 47 53	IC 2104	13.0	2.7 x 1.1	14.0	(R′)SB(s)bc I-II	105	Mag 9.59v star SSW 7′.0.
04 10 32.8	−07 10 03	MCG −1-11-4	14.1	2.1 x 0.2	13.0	Sd: sp	66	Mag 8.95v star ESE 5′.1.
04 10 55.2	−07 16 53	MCG −1-11-5	13.6	1.0 x 0.7	13.1		132	= **Mkn 614**. Mag 8.95v star N 4′.8.
04 21 54.2	−06 13 53	MCG −1-12-2	14.6	0.8 x 0.3	12.9	Sb	63	Mag 6.44v star SW 5′.3.
04 27 04.8	−06 09 56	MCG −1-12-8	14.2	1.3 x 0.9	14.2	SB(rs)c pec: II-III	105	Mag 11.4 star E 3′.8; double galaxy system **MCG −1-12-9** and **MCG −1-12-10** NNE 3′.1.
04 32 49.7	−06 01 33	MCG −1-12-26	13.9	1.0 x 0.5	13.2	E/S0	12	
04 35 19.4	−07 24 28	MCG −1-12-33	13.8	1.4 x 0.4	13.0	Sb	15	Mag 8.86v star NNW 7′.1.
04 42 57.4	−08 05 27	MCG −1-12-47	14.4	1.6 x 1.2	14.9	SAB(s)dm IV-V	145	Mag 12.3 star N 2′.2.
04 46 32.8	−07 14 09	MCG −1-13-7	14.8	1.2 x 0.3	13.6	Scd: III	114	
04 51 18.5	−05 57 38	MCG −1-13-21	14.5	1.3 x 0.3	13.4	Scd: sp	24	Mag 9.70v star N 8′.2; **MCG −1-13-20** W 3′.4.
04 52 17.0	−06 09 13	MCG −1-13-29	14.4	1.0 x 0.9	14.1	SBbc	114	Mag 10.61v star N 4′.0.
04 55 43.1	−07 20 07	MCG −1-13-36	13.8	0.9 x 0.7	13.1	Sc	48	Mag 10.87v star on S edge.
05 02 37.7	−08 18 06	MCG −1-13-49	14.2	1.3 x 1.1	14.5	SB(rs)d III-IV:	51	Mag 7.81v star NNE 6′.1.
04 08 36.3	−10 11 22	MCG −2-11-19	14.4	0.7 x 0.7	13.5			Mag 9.28v star E 13′.3.
04 08 25.8	−13 59 08	MCG −2-11-20	13.9	0.5 x 0.5	12.4	E?		Mag 10.39v star W 11′.2.
04 09 55.7	−10 16 13	MCG −2-11-21	14.2	0.7 x 0.5	13.2	E	111	Mag 9.34v star NNE 6′.7.
04 10 47.0	−08 59 41	MCG −2-11-22	14.5	0.9 x 0.8	14.0	SBbc	123	Mag 10.50v star SW 5′.9.
04 10 49.6	−09 37 43	MCG −2-11-23	13.9	1.5 x 0.6	13.6	SB(s)cd: III-IV	117	Mag 13.4 star N 2′.3.
04 11 45.2	−12 28 00	MCG −2-11-24	14.6	1.0 x 0.2	12.7	SBc	99	
04 12 29.7	−13 09 06	MCG −2-11-25	14.6	0.7 x 0.6	13.5	Sc	60	Mag 7.91v star SW 7′.3.
04 14 01.5	−13 23 24	MCG −2-11-26	14.4	1.1 x 0.7	14.0	S0?	123	
04 15 45.8	−13 29 18	MCG −2-11-28	13.5	0.7 x 0.7	12.6	S0		
04 16 28.2	−11 16 47	MCG −2-11-29	14.5	0.7 x 0.6	13.4		123	Mag 10.35v star E 7′.3; small, faint anonymous galaxy N 1′.3.
04 16 38.4	−12 23 55	MCG −2-11-30	13.9	1.9 x 0.7	14.0	SB(rs)b II-III	135	Mag 12.4 star E edge; mag 9.43v star SW 2′.8.
04 17 02.4	−11 20 48	MCG −2-11-32	14.0	0.7 x 0.6	12.9	Sbc	30	Mag 10.35v star N 5′.4.
04 17 57.0	−13 59 26	MCG −2-11-33	14.9	0.7 x 0.4	13.4	SBb	81	
04 17 41.4	−12 27 31	MCG −2-11-34	14.2	1.0 x 0.3	12.7		123	
04 17 11.4	−13 51 01	MCG −2-11-35	14.2	0.6 x 0.5	12.8		57	
04 18 41.2	−11 29 42	MCG −2-11-36	14.9	0.9 x 0.6	14.1	Sbc	150	Mag 10.96v star E 8′.4.
04 18 45.6	−14 12 07	MCG −2-11-37	14.3	0.8 x 0.2	12.2	Sab	123	
04 18 54.1	−14 11 47	MCG −2-11-38	14.4	0.5 x 0.5	12.7			
04 19 16.5	−14 06 12	MCG −2-11-39	13.7	0.8 x 0.4	12.4	S0	24	
04 20 50.7	−09 45 52	MCG −2-12-2	14.5	0.4 x 0.4	12.4			MCG −2-12-3 NE 3′.2.
04 21 02.0	−09 44 20	MCG −2-12-3	14.4	0.9 x 0.7	13.8	Sbc	66	MCG −2-12-2 SW 3′.2.
04 21 05.3	−09 47 15	MCG −2-12-4	14.4	0.4 x 0.4	12.3			MCG −2-12-3 N 3′.0.
04 22 01.2	−10 10 20	MCG −2-12-5	13.9	1.5 x 0.4	13.2	Sbc: II-III	69	Mag 8.45v star SSE 6′.2.
04 22 01.8	−12 47 11	MCG −2-12-6	13.7	0.7 x 0.6	12.6	Sc	57	
04 22 12.3	−10 31 27	MCG −2-12-7	14.4	0.7 x 0.5	13.1		81	
04 22 25.6	−10 10 28	MCG −2-12-8	14.4	0.7 x 0.4	12.9	Sc	132	Mag 8.45v star SSW 6′.4.
04 24 11.7	−12 16 22	MCG −2-12-12	14.5	1.0 x 0.3	13.0	Sbc		Mag 8.77v star E 6′.0.
04 25 01.2	−09 54 24	MCG −2-12-13	14.8	0.4 x 0.4	12.7	S0		Mag 6.65v star E 10′.2.
04 26 38.0	−10 33 00	MCG −2-12-15	14.3	0.4 x 0.4	12.1			Mag 7.80v star W 2′.9.
04 26 54.9	−11 59 11	MCG −2-12-16	14.6	0.4 x 0.3	12.2		81	Mag 11.13v star E 1′.2.
04 27 01.5	−12 03 34	MCG −2-12-17	14.5	1.0 x 0.5	13.6	Sb	9	Mag 11.13v star N 3′.9.
04 27 28.8	−10 49 19	MCG −2-12-18	14.1	0.9 x 0.6	13.2		135	
04 27 35.6	−09 24 17	MCG −2-12-19	13.4	1.0 x 1.0	13.3	Sc		
04 27 40.4	−12 35 39	MCG −2-12-20	14.4	0.5 x 0.5	12.8	S0/a		Mag 8.74v star SSW 4′.0.
04 27 44.9	−12 02 03	MCG −2-12-21	13.5	0.7 x 0.4	11.9		123	Mag 8.36v star ENE 3′.2.
04 28 05.6	−14 46 20	MCG −2-12-22	14.2	1.2 x 0.8	14.0	SAB(s)m: V	21	
04 28 16.5	−12 09 18	MCG −2-12-23	13.9	1.0 x 0.3	12.5	Sa	174	Mag 8.01v star S 6′.8.
04 28 45.3	−12 30 44	MCG −2-12-24	14.7	1.3 x 0.9	14.8	SB(s)m IV-V	33	
04 28 59.9	−11 00 55	MCG −2-12-25	14.3	1.3 x 0.2	12.6		27	Mag 7.71v star S 2′.9.
04 28 59.8	−12 13 31	MCG −2-12-26	13.1	0.9 x 0.9	12.8			Lies 1′.6 WSW of mag 10 star.
04 29 08.3	−14 08 20	MCG −2-12-27	14.8	1.0 x 0.2	12.9	Sc	75	Mag 7.80v star NNE 6′.2.
04 29 07.8	−12 34 23	MCG −2-12-28	15.1	0.5 x 0.3	12.9		147	
04 29 25.5	−10 30 25	MCG −2-12-29	14.1	0.7 x 0.4	12.6		114	Mag 8.93vstar NNW 2′.0.
04 29 47.2	−12 30 19	MCG −2-12-30	13.7	1.0 x 0.8	13.3	SBbc	18	
04 31 40.1	−11 42 39	MCG −2-12-33	14.2	0.9 x 0.7	13.5	SBbc	30	Small elongated anonymous galaxy W 0′.5.
04 31 44.5	−12 40 59	MCG −2-12-35	15.1	1.1 x 0.2	13.3	S?	90	Mag 7.16v star NNE 4′.1.
04 32 09.6	−12 41 48	MCG −2-12-36	15.7	1.0 x 0.4	14.6	Sc	6	Mag 7.16v star NW 4′.9.
04 32 56.8	−12 46 43	MCG −2-12-37	14.4	1.1 x 0.4	13.4	SB?	72	Mag 10.70v star E 2′.0.
04 33 01.7	−10 48 50	MCG −2-12-38	14.5	0.7 x 0.6	13.4	Sc	150	Mag 6.06v star ENE 5′.3.
04 33 37.8	−13 15 45	MCG −2-12-39	12.9	1.7 x 1.1	13.6	E^+?	6	Mag 9.55v star N 8′.1; numerous PGC and anonymous galaxies in surrounding area.
04 33 38.8	−12 32 07	MCG −2-12-40	15.1	0.7 x 0.4	13.5	Scd	93	Mag 9.31v star NE 2′.5.
04 33 52.2	−11 42 15	MCG −2-12-41	13.9	2.6 x 0.6	14.2	(R′)SB(rs)ab:	45	Mag 9.26v star NW 2′.8 with an almost stellar anonymous galaxy half way between them.
04 35 11.1	−13 14 42	MCG −2-12-43	14.3	0.9 x 0.4	13.1	S0?	108	**PGC 15577** S 4′.0.
04 35 37.0	−11 42 33	MCG −2-12-44	15.0	0.5 x 0.4	13.1	Sc	141	Mag 9.89v star S 7′.7.
04 36 22.3	−10 22 37	MCG −2-12-45	13.4	0.9 x 0.7	12.7	SB(s)b pec	177	= **Mkn 618**. Mag 10.70v star SE 2′.7.
04 36 24.6	−09 30 54	MCG −2-12-46	15.2	1.4 x 0.3	14.1	IB(s)m: sp IV-V	177	
04 37 06.1	−09 45 19	MCG −2-12-47	14.2	0.9 x 0.5	13.2		6	Mag 8.67v star SE 7′.5.
04 37 49.9	−12 26 34	MCG −2-12-48	14.0	0.8 x 0.5	12.9		24	Mag 10.45v star NW 2′.4.
04 38 14.4	−10 47 46	MCG −2-12-50	14.1	1.0 x 0.7	13.6	Sb	63	Mag 9.63v star E 9′.6.

RA h m s	Dec ° ′ ″	Name	Mag (V)	Dim ′ Maj x min	SB	Type Class	PA	Notes
04 39 33.4	−09 03 29	MCG −2-12-52	13.8	0.7 x 0.5	12.5		42	Mag 9.25v star S 4'.0.
04 39 46.3	−13 16 22	MCG −2-12-53	15.5	1.1 x 0.9	15.3	SA(s)cd: III-IV	138	Mag 8.37v star WSW 13'.1.
04 39 58.5	−08 55 04	MCG −2-12-54	13.9	0.9 x 0.7	13.2	SA(rs)bc pec: II	21	Mag 9.16v star WSW 7'.8.
04 40 11.7	−13 13 43	MCG −2-12-55	14.0	0.5 x 0.4	12.1	Sa	81	Mag 9.88v star SE 10'.3.
04 40 25.4	−13 13 00	MCG −2-12-56	14.6	1.0 x 0.8	14.2	Sc	24	Mag 9.88v star SSE 8'.5.
04 40 58.5	−08 42 43	MCG −2-12-57	16.0	1.5 x 0.2	14.5	Sd: sp III-IV	147	Mag 10.42v star N 2'.2; NGC 1636 NW 7'.7.
04 42 58.7	−12 46 19	MCG −2-12-58	14.3	1.3 x 0.7	14.0	SB(r)b II	39	Mag 10.78v star W 6'.1.
04 46 02.8	−12 26 35	MCG −2-13-1	13.9	1.0 x 0.5	12.9	Sa/b?	66	= **Mkn 1085**; mag 11.06v star W 9'.0.
04 46 21.8	−13 21 02	MCG −2-13-2	14.4	1.2 x 0.2	12.7	Sab sp	105	Mag 9.41v star SSE 4'.9.
04 46 32.8	−11 55 47	MCG −2-13-3	15.0	0.6 x 0.6	13.7	Sb		Mag 10.90v star W 1'.9.
04 46 54.6	−11 28 19	MCG −2-13-4	14.5	0.7 x 0.5	13.2		39	Mag 10.89v star ESE 3'.8.
04 46 56.7	−14 51 39	MCG −2-13-5	14.2	0.5 x 0.3	12.0			Mag 9.58v star NNW 7'.7.
04 47 20.7	−10 14 21	MCG −2-13-6	13.4	1.1 x 0.8	13.2	E$^+$ pec	108	Mag 10.19v star N 6'.4.
04 47 41.0	−10 14 16	MCG −2-13-7	14.1	0.9 x 0.6	13.3	SBbc	27	Mag 14.8 star on N edge; mag 10.19v star NW 8'.3; MCG −2-13-8 E 1'.8.
04 47 47.8	−10 14 06	MCG −2-13-8	14.4	0.5 x 0.4	12.5			MCG −2-13-7 W 1'.8.
04 48 12.6	−13 40 02	MCG −2-13-9	12.5	1.3 x 1.0	12.9	E$^+$?	153	Small, faint, elongated anonymous galaxy W 1'.7.
04 48 47.8	−14 22 46	MCG −2-13-10	14.0	0.6 x 0.4	12.3		90	
04 49 24.2	−12 44 47	MCG −2-13-11	16.1	1.4 x 0.8	16.0	SB(s)m: V	21	Mag 11.7 star NNE 4'.9.
04 49 42.7	−10 42 25	MCG −2-13-12	13.6	0.9 x 0.8	13.1	SB(s)a pec?	90	Mag 9.10v star NW 7'.9.
04 49 53.9	−10 45 18	MCG −2-13-13	14.4	1.1 x 0.6	13.7	SB(s)bc II-III	30	Mag 10.54v star S 6'.4.
04 50 33.5	−11 25 38	MCG −2-13-14	14.4	1.1 x 0.7	13.9	Sbc	27	Mag 10.87v star NNE 6'.5.
04 51 18.5	−13 50 17	MCG −2-13-15	14.7	1.3 x 0.3	13.5	S?	159	Mag 9.50v star N 7'.4.
04 51 32.2	−14 36 17	MCG −2-13-16	13.9	0.4 x 0.3	11.5		99	Lies 1'.2 SW of mag 8 star.
04 52 34.5	−13 53 01	MCG −2-13-17	14.5	0.7 x 0.4	13.0		171	Mag 10.30v star NE 12'.6.
04 54 15.8	−11 46 28	MCG −2-13-18	15.1	1.3 x 0.4	14.2	Sd? III-IV	87	Mag 10.76v star E 3'.8.
04 54 39.4	−12 12 13	MCG −2-13-19	14.4	2.2 x 0.7	14.7	SAB(s)c: II	120	MCG −2-13-23 ENE 9'.0.
04 54 46.5	−10 16 43	MCG −2-13-20	15.5	1.1 x 0.7	15.1	SBc	141	Superimposed star just NW of center.
04 55 05.3	−10 41 35	MCG −2-13-21	13.0	1.9 x 0.5	12.8	SA(s)b: II	63	Mag 12.7 star W end; mag 12.6 star N edge.
04 55 06.8	−12 05 40	MCG −2-13-22	13.8	0.3 x 0.3	11.1	SA(r)0°:		Almost stellar; mag 9.89v star NE 10'.2.
04 55 15.6	−12 09 27	MCG −2-13-23	13.8	1.3 x 1.3	14.2	SAB(rs)c: II-III		Mag 10.48v star E 5'.5.
04 56 37.7	−10 13 08	MCG −2-13-24	13.5	1.0 x 0.5	12.6	SAB(s)0°	24	Mag 8.95v star NW 8'.8.
04 56 45.5	−10 35 44	MCG −2-13-25	13.3	2.1 x 0.9	13.8	SB(rs)bc II	21	Mag 10.99v star SW 5'.6.
04 58 16.1	−09 47 48	MCG −2-13-26	14.1	1.5 x 0.4	13.4	Scd sp	156	Mag 9.01v star NW 3'.4.
04 59 41.5	−11 16 18	MCG −2-13-31	14.0	1.7 x 0.3	13.1	SB(s)c? sp	159	Mag 8.90v star E 8'.7.
04 59 53.0	−12 02 13	MCG −2-13-32	13.2	1.2 x 1.0	13.3	SAB0$^-$?	6	Mag 10.95v star S 1'.6.
04 59 49.1	−12 27 01	MCG −2-13-33	14.0	0.7 x 0.6	12.9	Sc	174	Mag 4.78v star 64 Eridani S 5'.5.
05 00 18.5	−12 41 30	MCG −2-13-34	14.6	1.6 x 0.4	14.0	SAd: sp	57	Located between a pair of mag 13.6 stars.
05 00 11.7	−13 46 08	MCG −2-13-35	14.2	0.9 x 0.9	13.8	SBbc		Mag 10.37v star NE 7'.0.
05 01 01.8	−08 57 27	MCG −2-13-36	14.2	1.2 x 1.0	14.3	SB(rs)dm: III	60	Mag 10.93v star NNW 2'.1.
05 00 55.2	−13 25 20	MCG −2-13-37	13.4	1.5 x 0.8	13.7	E/S0	27	Mag 7.48v star SSW 6'.0.
05 02 19.6	−10 21 26	MCG −2-13-38	12.9	2.4 x 1.1	13.8	(R′)SA(rs)a:	117	Bright, elongated core, smooth envelope.
05 02 32.3	−12 22 52	MCG −2-13-39	14.5	0.8 x 0.5	13.4		33	Mag 8.56v star ESE 7'.1.
05 04 53.2	−10 14 53	MCG −2-13-40A	14.2	0.9 x 0.4	12.9	Radio galaxy	159	Mag 9.69v star NW 9'.7.
05 09 09.3	−09 15 37	MCG −2-14-1	16.6	1.2 x 0.7	16.2	IB(s)m V	153	Mag 10.91v star E 1'.1.
05 10 04.6	−14 56 49	MCG −2-14-2	13.5	1.3 x 0.5	12.9	SB(s)dm: III-IV	162	Mag 9.73v star N 8'.1.
05 11 08.3	−09 23 22	MCG −2-14-3	13.7	2.6 x 0.6	14.0	SB(s)c: sp II	90	
05 11 40.8	−14 47 18	MCG −2-14-4	12.4	2.6 x 2.2	14.2	SAB(rs)cd III	90	
05 12 22.9	−14 21 37	MCG −2-14-5	14.1	1.1 x 0.3	12.7	IBm pec sp III-IV	90	Mag 9.29v star WSW 4'.1.
05 13 02.6	−09 12 44	MCG −2-14-6	14.0	1.3 x 0.2	12.4	S?	18	Mag 9.39v star SW 7'.6.
05 13 43.6	−12 45 44	MCG −2-14-7	14.0	1.4 x 1.0	14.2	SB(rs)d III-IV	24	Mottled surface, several bright knots.
05 16 21.2	−10 33 44	MCG −2-14-9	13.6	0.8 x 0.7	12.9	SBb		Star superimposed inside S edge.
05 16 21.0	−13 28 18	MCG −2-14-10	14.0	1.2 x 0.3	12.7	S?	171	Mag 10.63v star SE 5'.0.
05 16 46.2	−12 20 59	MCG −2-14-11	13.9	0.9 x 0.4	12.9	E/S0	165	Mag 8.35v star SW 6'.7.
05 18 38.0	−14 09 02	MCG −2-14-12	14.5	0.8 x 0.5	13.3		39	Mag 9.06v star N 8'.1.
05 23 14.9	−11 25 29	MCG −2-14-15	13.7	1.6 x 0.2	12.3	S0°: sp	51	Located between a close pair of mags 12.3 and 13.0 stars.
05 24 54.6	−12 41 21	MCG −2-14-16	13.5	1.7 x 0.4	13.0	Sab sp	69	
04 08 33.0	−16 45 29	MCG −3-11-14	14.4	1.0 x 0.7	13.9	Sbc	6	Mag 9.74v star NW 6'.2.
04 11 07.2	−14 52 11	MCG −3-11-17	13.8	0.9 x 0.4	12.6	Sbc	60	
04 11 19.9	−16 13 48	MCG −3-11-18	14.4	1.5 x 0.4	13.7	SB(s)m: sp IV-V	108	Mag 9.29v star WNW 5'.3.
04 16 12.9	−16 45 08	MCG −3-11-19	13.8	1.5 x 0.6	13.5	SB(s)dm pec: IV	165	
04 19 22.1	−17 36 50	MCG −3-12-1	13.8	1.1 x 0.7	13.4	I?	58	
04 21 37.2	−15 59 40	MCG −3-12-3	13.7	0.9 x 0.8	13.2	SBa	117	
04 25 29.8	−16 32 36	MCG −3-12-8	14.1	0.7 x 0.5	12.8	Sc	129	Mag 13.1 star on S edge.
04 30 52.3	−16 18 24	MCG −3-12-11	13.1	1.1 x 0.9	12.9	SAB(r)a pec:	156	Mag 8.79v star NW 3'.8.
04 31 40.7	−15 46 30	MCG −3-12-13	14.1	0.5 x 0.5	12.5			Mag 10.32v star NW 2'.3.
04 45 13.2	−15 06 54	MCG −3-13-2	14.4	0.4 x 0.4	12.2			Mag 10.79v star SW 11'.5.
04 45 36.7	−16 13 30	MCG −3-13-3	14.9	0.9 x 0.2	12.9		39	0'.2 E of mag 11 star.
04 45 53.9	−17 16 45	MCG −3-13-4	12.8	1.6 x 0.9	13.0	SB(rs)bc pec: II	114	Several bright knots or superimposed stars; mag 8.93v star SW 5'.1.
04 46 08.0	−17 03 55	MCG −3-13-5	14.1	1.2 x 0.4	13.2	Sb	0	Star at southern tip.
04 46 07.2	−15 49 18	MCG −3-13-6	15.1	0.9 x 0.4	13.8		147	Mag 11.9 star W 8'.2.
04 46 58.2	−16 28 24	MCG −3-13-8	14.0	0.9 x 0.5	13.0	SBb	168	Broad, faint streamers extending E and W.
04 47 30.6	−17 35 50	MCG −3-13-9	13.5	1.1 x 1.0	13.5	SA(r)0°	22	Star on NE edge.
04 47 34.3	−17 20 50	MCG −3-13-10	14.2	0.9 x 0.7	13.5	S	99	Mag 8.41v star NW 7'.6.
04 47 38.5	−17 26 05	MCG −3-13-11	14.0	0.7 x 0.5	12.7		41	Mag 10.34v star E 6'.3.
04 47 58.2	−16 04 50	MCG −3-13-13	14.9	0.8 x 0.7	14.1		54	mag 9.43v star N 6'.2.
04 50 13.6	−17 15 58	MCG −3-13-16	13.4	2.3 x 0.5	13.4	SB(r)bc? sp I-II	96	Mag 12.2 star on S edge; mag 10.51v star ESE 9'.1.
04 51 20.6	−17 30 12	MCG −3-13-17	13.4	1.0 x 0.7	12.9	SA(s)0$^-$ pec:	3	Mag 10.90v star N 2'.8.
04 51 27.0	−14 57 59	MCG −3-13-18	14.3	0.9 x 0.7	13.7	SBc	21	Mag 11.26v star N 2'.8.
04 53 34.4	−17 55 22	MCG −3-13-20	14.3	1.0 x 0.7	13.7	S?	63	Mag 9.05v star N 5'.7.
04 54 43.2	−17 27 09	MCG −3-13-24	14.0	0.5 x 0.5	12.3			Mag 10.87v star S 6'.0.
04 55 09.5	−16 08 57	MCG −3-13-27	13.4	1.0 x 0.6	12.7	Sb	156	Mag 9.06v star NW 11'.3.
04 55 31.4	−15 06 03	MCG −3-13-28	14.9	0.8 x 0.3	13.2	Sb	84	Mag 10.30v star SE 4'.3; MCG −3-13-30 E 1'.5.
04 55 37.5	−15 05 50	MCG −3-13-30	13.5	0.9 x 0.9	13.1	SBb		Mag 10.30v star SSE 3'.3.
04 55 43.4	−14 59 54	MCG −3-13-31	14.0	1.0 x 1.0	13.9	SAB(rs)d III-IV:		Mag 10.97v star N 3'.9.
04 56 15.5	−17 25 32	MCG −3-13-33	14.8	1.1 x 0.7	14.4	Sb	0	Mag 9.31v star NW 6'.1.
04 56 30.0	−17 40 06	MCG −3-13-35	13.8	1.2 x 0.9	13.7	S0?	1	Mag 7.39v star SW 5'.7.
04 59 27.1	−16 33 35	MCG −3-13-42	12.9	1.3 x 1.2	13.2	SB(s)c II-III	48	Mag 10.64v star N 6'.2.
04 59 33.4	−17 27 28	MCG −3-13-44	13.9	0.9 x 0.5	12.8	S0/a	95	MCG −3-13-46 E 6'.2; mag 11.35v star WNW 8'.5.
04 59 59.5	−17 27 35	MCG −3-13-46	13.3	1.5 x 0.7	13.2	(R)SB(r)a	82	Mag 12.6 star E end; mag 8.62v star SE 12'.9; very faint, stellar anonymous galaxy SE 2'.0.

(Continued from previous page)

GALAXIES

RA h m s	Dec ° ′ ″	Name	Mag (V)	Dim ′ Maj x min	SB	Type Class	PA	Notes
05 00 29.3	−17 26 04	MCG −3-13-48	14.3	0.8 x 0.7	13.5	Sc		Mag 8.62v star SSE 9′.6.
05 01 32.5	−16 09 59	MCG −3-13-51	13.3	1.7 x 0.5	13.0	(R′)SB(s)b:	87	Mag 8.58v star E 11′.9.
05 01 42.7	−15 09 47	MCG −3-13-52	13.5	1.1 x 0.9	13.3	Sb	3	Mag 10.16v star SE 5′.8.
05 01 45.1	−15 26 54	MCG −3-13-53	13.5	0.8 x 0.6	12.6	S0?	159	= **Hickson 32A**. **Hickson 32C** and **Hickson 32D** 1′.2 N. **Hickson 32B** N 3′.2.
05 02 29.2	−15 32 44	MCG −3-13-58	14.1	1.5 x 0.7	14.0		75	Mag 9.85v star NE 4′.7; center of large planetary nebula **Abell 7** (**PK 215-30.1**) ESE 9′.9.
05 04 38.2	−15 46 11	MCG −3-13-62	14.5	0.7 x 0.4	13.0		9	Broad streamer extending N.
05 04 31.9	−16 35 04	MCG −3-13-63	13.7	1.3 x 1.3	14.1	SB?		Mag 11.30v star N edge.
05 04 34.3	−17 05 31	MCG −3-13-64	14.2	1.5 x 0.2	12.7	Sc: sp	66	Mag 13.2 star N 1′.5.
05 05 16.7	−17 21 05	MCG −3-13-67	14.1	0.7 x 0.5	12.8	Irr	165	Mag 9.25v star SE 12′.8.
05 06 05.4	−16 44 39	MCG −3-13-69	14.3	1.3 x 0.7	14.1	S?	96	Close pair of mag 9.90v, 9.93v stars SE 6′.6.
05 07 47.0	−16 17 34	MCG −3-14-1	13.1	2.4 x 1.8	14.5	SB(s)m IV-V	150	Short E-W bar; mag 14.1 star on E edge; mag 11.13v star S 1′.9.
05 08 35.4	−17 10 53	MCG −3-14-3	14.9	0.5 x 0.5	13.2			Mag 8.81v star WSW 10′.8.
05 08 42.7	−16 57 45	MCG −3-14-4	14.2	1.3 x 1.1	14.4	S?	168	
05 15 42.9	−16 06 32	MCG −3-14-11	13.1	1.1 x 0.9	12.9	Sc	162	Mag 10.17v star NNE 8′.3.
04 08 07.5	−08 49 46	NGC 1516A	13.6	0.5 x 0.5	11.9	SBbc		
04 08 08.8	−08 50 14	NGC 1516B	13.8	0.7 x 0.5	12.5	Sc	126	
04 08 07.7	−17 11 34	NGC 1519	12.9	2.1 x 0.5	12.8	SB(r)b? II-III?	108	Mag 8.54v star SSE 4′.6.
04 14 36.1	−13 10 30	NGC 1538	14.4	0.9 x 0.7	13.8	E	125	Almost stellar nucleus; mag 9.04v star SW 5′.5.
04 17 12.6	−17 51 30	NGC 1547	13.4	1.3 x 0.6	13.0	SB(rs)bc pec: II-III	133	
04 23 01.1	−15 50 47	NGC 1561	13.8	1.1 x 0.8	13.5	E/S0	175	Mag 8.73v star SW 2′.2.
04 21 47.7	−15 45 20	NGC 1562	14.3	0.6 x 0.6	13.0	S0		Very faint anonymous galaxy W 3′.8.
04 22 54.0	−15 43 58	NGC 1563	15.0	0.5 x 0.5	13.4	E1?		Anonymous galaxy NW 5′.5; NGC 1564 E 1′.7.
04 23 01.1	−15 44 20	NGC 1564	14.6	0.6 x 0.5	13.1	S0	20	NGC 1563 W 1′.7.
04 23 23.5	−15 44 44	NGC 1565	14.0	0.9 x 0.8	13.5	Sc	170	Stellar anonymous galaxy N 3′.7, possibly **IC 2064**?
04 26 20.6	−10 05 53	NGC 1577	12.6	1.5 x 1.3	13.2	SBb	130	
04 28 18.6	−05 10 42	NGC 1580	13.5	0.9 x 0.8	13.0	Sbc II	95	
04 28 20.8	−17 35 44	NGC 1583	13.7	0.7 x 0.7	12.8	SA0$^-$		**ESO 551-7** SW 3′.1; NGC 1584 NNW 5′.0.
04 28 10.2	−17 31 24	NGC 1584	14.0	0.6 x 0.6	12.8	SA(s)0$^-$ pec:		Almost stellar; NGC 1583 SSE 5′.0.
04 30 51.7	−05 47 59	NGC 1594	13.0	1.8 x 1.3	13.7	SB(rs)bc I-II	60	Bright knot or star on W edge.
04 31 13.5	−11 17 26	NGC 1597	13.9	1.0 x 0.8	13.6	E/S0	95	Small, elongated anonymous galaxy 2′.0 WSW.
04 31 39.9	−05 05 10	NGC 1600	10.9	2.5 x 1.7	12.5	E3	15	NGC 1601 on N edge; NGC 1603 E 2′.6.
04 31 41.9	−05 03 38	NGC 1601	13.8	0.7 x 0.3	12.0	S0: sp	90	Located on N edge of NGC 1600.
04 31 49.9	−05 05 42	NGC 1603	13.8	0.8 x 0.6	12.8	E?	37	Located 2′.6 E of NGC 1600.
04 31 58.9	−05 22 11	NGC 1604	13.7	1.2 x 0.8	13.5	S0	71	Mag 10.98v star E 2′.3; small, faint anonymous galaxy and mag 14.0 star E 3′.5.
04 32 03.3	−05 01 58	NGC 1606	14.9	0.5 x 0.5	13.2	SAB(r)0$^+$:		Mag 7.62v star E 4′.5.
04 34 00.6	−08 34 54	NGC 1614	12.9	1.3 x 0.9	13.0	SB(s)c pec	142	Very faint jet extends 0′.7 SW from SW edge.
04 35 32.4	−13 33 21	NGC 1623	15.6	0.8 x 0.5	14.5	SBaa	15	
04 39 58.4	−09 27 24	NGC 1632	14.4	0.8 x 0.5	13.5	E	40	
04 40 40.2	−08 36 32	NGC 1636	13.0	1.2 x 0.8	12.8	(R′)SB(rs)ab:	0	MCG −2-12-57 SE 7′.7.
04 43 43.9	−05 19 06	NGC 1643	13.3	1.2 x 1.1	13.4	SB(r)bc pec?	147	Very small, elongated galaxy, or bright knot, extends from S edge.
04 44 06.2	−05 28 00	NGC 1645	13.0	2.1 x 0.9	13.6	(R′)SB(rs)0$^+$ pec	85	
04 44 23.3	−08 31 49	NGC 1646	12.4	2.2 x 1.3	13.7	E/S0	159	Small, elongated anonymous galaxy on E edge; mag 5.78v star W 4′.9.
04 44 34.9	−08 28 44	NGC 1648	14.6	0.7 x 0.5	13.3	S0	4	Stellar nucleus; NGC 1646 SW 4′.3; mag 5.78v star W 7′.6.
04 45 11.6	−15 52 12	NGC 1650	12.7	2.3 x 1.1	13.6	E$^+$	170	Pair of stars, or a star and small, round galaxy S edge.
04 45 53.2	−05 08 12	NGC 1656	12.9	1.5 x 0.8	13.0	S0$^+$ pec:	123	Mag 9.65v star N 3′.2.
04 48 17.4	−05 25 38	NGC 1665	12.8	1.7 x 1.1	13.4	SA(s)0$^+$ pec?	50	Faint, elongated anonymous galaxy N 4′.8, possibly **IC 2094**?
04 48 32.9	−06 34 14	NGC 1666	12.6	1.3 x 1.0	12.8	SB(r)0$^+$	143	
04 48 36.9	−06 19 15	NGC 1667	12.1	1.4 x 1.0	12.3	SAB(r)c II	168	
04 51 50.4	−05 48 15	NGC 1681	12.9	1.3 x 1.0	13.1	Sb pec:	142	Mag 12 star on W edge.
04 52 54.7	−15 20 49	NGC 1686	13.7	1.7 x 0.3	12.8	Sbc sp	30	
04 57 17.0	−15 17 18	NGC 1710	12.7	1.5 x 1.2	13.2	SA0$^-$:	15	Lies 1′.2 NW of a mag 10.63v star.
04 59 20.8	−07 51 35	NGC 1720	12.4	2.0 x 1.2	13.2	SB(s)ab	78	Mag 8.09v star N 4′.8.
04 59 17.4	−11 07 08	NGC 1721	12.8	1.9 x 1.0	13.4	(R′)SAB(s)0° pec	132	NGC 1725 on E edge; mag 12.2 star SSW 3′.4.
04 59 26.0	−10 58 52	NGC 1723	11.7	2.6 x 1.8	13.2	SB(r)a pec	144	Lies between a mag 9.97v star N and an 11.05v star S.
04 59 22.9	−11 07 57	NGC 1725	12.8	1.2 x 1.0	12.9	E/S0	20	NGC 1721 on W edge; NGC 1728 on E edge.
04 59 42.0	−07 45 20	NGC 1726	11.7	1.7 x 1.4	12.5	SA(s)0°:	170	Star on S edge; mag 8.09v star W 6′.2.
04 59 27.8	−11 07 24	NGC 1728	13.9	2.0 x 0.7	14.1	Sa pec sp	177	Mag 10.41v star ENE 6′.8; NGC 1725 on W edge.
04 59 32.0	−15 49 26	NGC 1730	12.3	2.2 x 1.0	13.0	SB(r)a	100	Short N-S bar; mag 8.63v star NW 5′.3; mag 8.38v star E 7′.2.
05 02 09.6	−08 14 27	NGC 1752	12.4	2.6 x 0.8	13.0	SB(r)c: II	70	Mag 7.81v star E 9′.9.
05 05 18.1	−09 08 52	NGC 1779	12.1	2.3 x 1.3	13.1	(R′)SAB(r)0/a?	108	
05 05 27.2	−11 52 17	NGC 1784	11.7	4.0 x 2.5	14.0	SB(r)c II	105	Numerous bright knots and/or superimposed stars.
05 07 44.8	−08 01 08	NGC 1797	13.5	1.6 x 1.0	13.9	(R′)SB(rs)a pec	90	Short N-S bar with faint E-W envelope; NGC 1799 N 3′.0.
05 07 44.7	−07 58 11	NGC 1799	13.7	1.1 x 0.6	13.1	SB0° pec?	116	NGC 1797 S 3′.0.
05 11 46.0	−15 08 06	NGC 1821	13.3	1.2 x 0.8	13.1	IB(s)m III	125	Lies 2′.6 NE of a mag 10.34v star.
05 12 03.0	−15 41 25	NGC 1832	11.3	2.6 x 1.7	12.8	SB(r)bc I-II	12	Star on E edge; mag 9.09v star NE 5′.4.
05 14 06.2	−10 37 36	NGC 1843	12.7	2.0 x 1.6	13.8	SAB(s)cd: II-III	122	
05 22 34.7	−11 30 02	NGC 1888	11.9	3.0 x 0.8	12.7	SB(s)c pec	150	NGC 1889 on E edge of core.
05 22 35.4	−11 29 48	NGC 1889	13.1	0.6 x 0.4	11.6	E$^+$ pec:	165	Located on the E edge of NGC 1888's core.
05 24 47.4	−15 56 37	NGC 1906	13.6	0.9 x 0.6	12.7	Scd III	145	

GALAXY CLUSTERS

RA h m s	Dec ° ′ ″	Name	Mag 10th brightest	No. Gal	Diam ′	Notes
04 33 36.0	−13 14 00	A 496	15.3	50	112	MCG −2-12-39 at center; numerous stellar PGC and anonymous GX along W half.

BRIGHT NEBULAE

RA h m s	Dec ° ′ ″	Name	Dim ′ Maj x min	Type	BC	Color	Notes
05 06 54.0	−07 13 00	IC 2118	180 x 60	SNR-R	3-5	1-4	**Witch Head Nebula**. Brightest part 70′ ENE of 65 Eridani.
05 20 00.0	−05 40 00	Sh2-278	25 x 20	E	4-5	3-4	

(Continued from previous page)

137 PLANETARY NEBULAE 137

RA h m s	Dec ° ′ ″	Name	Diam ″	Mag (P)	Mag (V)	Mag cent ★	Alt Name	Notes
05 27 28.2	−12 41 49	IC 418	12	10.7	9.3	10.1	PK 215−24.1	Planetary appears overwhelmed by mag 10.1 central star?.
04 14 15.8	−12 44 21	NGC 1535	51	9.6	9.6	12.1	PK 206−40.1	Close pair of mag 13.5, 13.9 stars W 2′1.
05 03 41.9	−06 10 04	PN G205.8−26.7		14.8		16.7	PK 205−26.1	Stellar; mag 11.06v star NW 1′2; mag 11.17v star ESE 2′4.
05 03 07.5	−15 36 23	PN G215.5−30.8	760	13.2		15.4	PK 215−30.1	Mag 8.15v star SE 10′5; mag 9.02v star SW 10′6; galaxy MCG −3-13-58 WNW 9′9.

138 GALAXIES 138

RA h m s	Dec ° ′ ″	Name	Mag (V)	Dim ′ Maj x min	SB	Type Class	PA	Notes
03 09 29.9	−17 24 58	ESO 547-11	14.8	1.2 x 0.8	14.6	SAB(s)m V	123	Mag 15.0 star N edge; mag 10.33v star NE 10′2.
03 09 36.9	−17 49 53	ESO 547-12	15.8	1.5 x 0.2	14.3	Sc:	111	Mag 9.24v star E 5′4.
03 12 57.5	−17 55 44	ESO 547-20	14.6	1.3 x 1.1	14.8	IB(s)m V	75	Mag 10.54v star WNW 11′6; double galaxy system **MCG −3-9-8** E 2′9.
03 26 00.2	−17 39 30	ESO 548-14	13.8	1.5 x 0.3	12.8	Sa:	77	Mag 8.96v star WSW 6′6; note: **MCG −3-9-39** located 3′5 WNW of this star; NGC 1329 N 4′0.
03 28 14.5	−17 25 13	ESO 548-23	14.4	1.2 x 0.5	13.8	E^+?	21	Mag 10.72v star ESE 3′0.
03 32 18.8	−17 43 07	ESO 548-32	14.2	2.2 x 0.9	14.8	SB(s)m V	144	Mag 12.2 star NE 9′7; mag 12.7 star SW 6′0.
03 34 38.3	−17 28 30	ESO 548-46	14.1	1.0 x 0.8	13.7	Sab	164	Mag 11.4 star SW 12′1.
03 41 23.0	−17 45 27	ESO 548-75	13.2	1.7 x 1.3	13.9	SB(rs)c II	23	Mag 9.53v star SSW 8′7.
03 55 09.3	−17 28 11	ESO 549-36	13.4	1.8 x 1.0	13.9	SB(rs)c: III	6	
04 04 29.7	−17 54 21	ESO 550-2	14.3	1.5 x 0.3	13.3	S?	44	Mag 9.40v star ESE 6′3.
04 06 00.1	−17 46 26	ESO 550-5	14.0	2.4 x 0.7	14.4	SB(s)m sp V	58	Close pair of mag 10.62v, 9.59v stars S 3′8.
02 55 59.5	−12 00 29	IC 271	13.8	1.0 x 0.8	13.4		120	Several knots and/or superimposed stars.
02 56 06.3	−14 11 14	IC 272	14.1	1.2 x 0.7	13.8	Sc	171	Very faint anonymous galaxies N 1′6 and S 2′5; IC 270 W 5′5.
02 58 41.1	−15 42 12	IC 276	13.1	1.9 x 0.5	12.9	S0° pec sp	60	Mag 8.03v star SW 12′4; mag 9.77v star SSW 6′0.
03 04 06.2	−12 00 55	IC 285	14.2	1.2 x 0.3	12.9	S?	117	Located 3′2 SE of NGC 1200.
03 07 26.5	−12 35 17	IC 291	13.7	1.4 x 0.7	13.5	(R)SB(r)0/a?	90	Mag 12.2 star NE 2′6.
03 13 00.0	−11 43 00	IC 306	14.6	0.5 x 0.3	12.4	SBc		Small, round companion galaxy on NW edge.
03 18 55.3	−12 44 26	IC 317	14.0	0.9 x 0.7	13.3	SBc	39	Mag 10.68v star S 5′9.
03 20 43.8	−14 34 06	IC 318	13.9	0.9 x 0.5	12.9	Sb	138	Mag 11.6 star SSE 7′3.
03 24 29.6	−14 59 08	IC 321	15.2	0.6 x 0.6	13.9			Mag 7.99v star ESE 11′4.
03 30 36.5	−14 25 33	IC 326	14.2	1.1 x 0.7	13.8	$SA0^-$?	114	Mag 9.57v star NE 9′2.
03 31 10.1	−14 41 34	IC 327	14.5	0.7 x 0.4	13.0	Sc	60	Mag 12.2 star SE 4′9; IC 328 N 3′2.
03 31 11.1	−14 38 20	IC 328	14.1	0.6 x 0.4	12.4	Sc	15	Mag 10.57v star N 7′7; IC 327 S 3′2.
03 39 28.7	−13 06 54	IC 340	13.5	1.5 x 0.6	13.3	S0°:	90	Mag 10.60v star SW 8′5.
03 44 36.7	−11 48 03	IC 350	13.8	1.2 x 0.8	13.6	Sab	171	Mag 9.79v star E 2′2.
03 06 28.5	−09 43 50	IC 1880	13.0	1.6 x 1.1	13.5	$SA0^-$ pec:	30	Mag 10.41v star E 1′9.
03 10 46.0	−10 47 46	IC 1897	14.1	0.5 x 0.5	12.4	S0/a		NGC 1238 NNE 3′4.
04 03 50.8	−11 10 54	IC 2026	14.6	0.6 x 0.3	12.6	Sb	114	NGC 1509 E 1′2.
03 08 48.4	−07 02 30	MCG −1-9-6	14.0	1.3 x 0.4	13.1	SB(rs)cd? III	12	Mag 10.37v star E 4′0.
03 13 24.0	−08 12 48	MCG −1-9-17	15.0	0.8 x 0.3	13.6	E	78	Mag 10.37v star NW 9′0.
03 14 14.8	−06 05 13	MCG −1-9-20	14.6	0.8 x 0.5	13.5	SBcd	105	
03 16 00.6	−05 30 00	MCG −1-9-24	13.1	1.9 x 0.5	12.9	Sab	27	Mag 6.32v star SSE 14′9.
03 19 26.1	−06 07 17	MCG −1-9-27	13.8	0.9 x 0.3	12.3	Spiral	168	= **Mkn 1075**. Mag 11.72v star N 4′4.
03 22 17.8	−07 05 28	MCG −1-9-31	13.1	1.1 x 0.8	12.9	SA(r)ab pec:	108	Mag 9.53v star SE 4′7.
03 25 11.3	−06 10 59	MCG −1-9-39	13.5	0.9 x 0.6	12.7	S	24	Mag 10.32v star WNW 6′2; stellar galaxies **Mkn 609**, **Mkn 610** NE 5′0.
03 31 23.0	−05 08 32	MCG −1-9-45	13.9	0.9 x 0.7	13.2	Sb	96	Mag 10.12v star N 4′8.
03 33 58.8	−07 08 27	MCG −1-10-4	17.3	1.9 x 0.2	16.1	SAB(r)cd: sp	120	Mag 14.0 star SW 1′1.
03 36 39.5	−07 46 39	MCG −1-10-10	14.4	0.9 x 0.5	13.4	Scd	18	
03 37 58.8	−06 16 19	MCG −1-10-12	14.1	1.3 x 0.8	14.0	SAB(r)cd III	60	Bright knot or star NW end.
03 38 33.1	−05 28 02	MCG −1-10-14	14.2	1.3 x 0.9	14.2	SAB(r)cd: sp III	174	Mag 10.68v star NE 6′2.
03 38 39.0	−05 20 53	MCG −1-10-15	14.3	1.3 x 0.4	13.4	Sd? III	78	Mag 10.49v star NW 5′7.
03 40 43.0	−06 24 58	MCG −1-10-19	14.2	1.7 x 1.3	15.0	SAB(r)cd II-III	15	Mag 10.69v star N 1′2.
03 42 10.4	−06 45 57	MCG −1-10-23	14.3	0.8 x 0.5	13.1	(R′)SAB(rs)0/a:	117	Mag 10.19v star SE 6′1.
03 45 41.4	−07 30 47	MCG −1-10-30	14.5	1.1 x 0.4	13.5	SBbc	15	Small anonymous elongated galaxy 0′5 N.
03 48 25.0	−06 37 38	MCG −1-10-38	14.2	0.9 x 0.5	13.2	(R′)SB(rs)b pec:	39	Mag 8.54v star NW 3′2.
03 51 25.1	−07 40 48	MCG −1-10-44	14.5	1.9 x 0.3	13.7	SBm: sp III	18	Mag 9.28v star NNE 3′1.
03 55 04.7	−06 13 19	MCG −1-10-47	14.2	0.8 x 0.5	13.1	SAB(rs)dm: IV-V	126	
04 06 15.8	−08 38 12	MCG −1-11-2	14.8	1.1 x 0.4	13.7	IB(s)m pec IV	80	Mag 9.55v star S 4′2.
04 10 32.8	−07 10 03	MCG −1-11-4	14.1	2.1 x 0.2	13.0	Sd: sp	66	Mag 8.95v star ESE 5′1.
04 10 55.2	−07 16 53	MCG −1-11-5	13.6	1.0 x 0.7	13.1		132	= **Mkn 614**. Mag 8.95v star N 4′8.
02 56 09.9	−13 41 17	MCG −2-8-31	14.5	0.7 x 0.5	13.3	Sab	48	Mag 8.06v star W 3′9.
02 57 11.4	−11 14 10	MCG −2-8-32	14.5	0.4 x 0.3	12.0	Sc	171	
02 57 49.6	−10 10 02	MCG −2-8-33	12.9	2.7 x 0.7	13.5	(R)SB(r)0/a?	147	Mag 9.97v star SSW 8′6.
02 59 31.6	−13 37 17	MCG −2-8-37	15.9	0.7 x 0.2	13.7	Sc	138	Mag 13 star NW end.
03 00 04.3	−10 49 27	MCG −2-8-38	14.4	0.9 x 0.4	13.1	SBb	39	Small, elongated anonymous galaxy NW 2′0.
03 00 31.0	−11 24 56	MCG −2-8-39	13.6	1.4 x 0.9	13.7	SAB(rs)a pec:	9	Mag 7.36v star SW 3′9.
03 01 51.3	−10 37 12	MCG −2-8-40	14.2	1.0 x 0.7	13.9	E	45	Mag 10.69v star S 4′8.
03 06 55.4	−09 37 45	MCG −2-8-50	14.8	0.6 x 0.5	13.3	Sd	24	= **Hickson 23D**.
03 06 55.8	−11 59 02	MCG −2-8-52	13.8	0.9 x 0.4	12.7	cD	114	Mag 10.10v star SSE 2′3.
03 08 18.9	−13 54 16	MCG −2-9-2	13.7	1.0 x 0.3	12.2	Sa	54	= **Mkn 1069**. Mag 10.17v star SW 8′2.
03 09 09.1	−10 17 40	MCG −2-9-3	13.1	1.4 x 0.6	12.8	(R′)SB(r)a pec?	3	
03 09 33.2	−09 49 00	MCG −2-9-5	14.4	0.6 x 0.5	13.2	E/S0	159	
03 09 52.7	−10 03 02	MCG −2-9-6	13.7	1.5 x 0.4	12.9	(R′)SB(r)a?	132	
03 09 50.6	−11 58 30	MCG −2-9-7	14.3	1.0 x 0.3	12.8	Sbc	144	Mag 10.54v star E 5′5.
03 10 16.1	−10 44 36	MCG −2-9-8	14.3	0.7 x 0.3	12.5	S-	81	Mag 12.3 star N 3′7.
03 11 33.9	−10 36 37	MCG −2-9-13	13.9	1.1 x 0.4	12.9	S?	126	Mag 11.7 star NNW 4′5.
03 14 24.6	−10 45 17	MCG −2-9-16	14.1	1.2 x 0.6	13.7	S0/a	84	
03 14 21.7	−09 26 16	MCG −2-9-17	14.8	0.8 x 0.7	14.3	E	162	
03 14 27.6	−08 56 24	MCG −2-9-18	14.6	0.8 x 0.5	13.5	Sbc	9	
03 15 46.2	−12 01 27	MCG −2-9-19	13.6	2.2 x 0.5	13.6	$S0^+$: sp	24	Mag 10.57v star W 1′5.
03 16 08.8	−10 40 32	MCG −2-9-20	14.5	1.1 x 0.5	13.7	SB(r)0/a?	135	Mag 11.7 star N 4′2.
03 16 05.5	−13 02 09	MCG −2-9-21	14.5	1.3 x 0.2	12.9	Sc? sp	18	
03 17 42.3	−13 26 13	MCG −2-9-23	14.0	1.0 x 0.8	13.6	Sc	147	Mag 10.58v star SW 6′1.
03 18 42.5	−13 37 36	MCG −2-9-24	14.6	0.7 x 0.4	13.0	Sbc	24	Mag 10.35v star WSW 3′1.
03 18 59.5	−11 56 53	MCG −2-9-27	14.7	0.9 x 0.4	13.4		135	
03 19 07.0	−11 20 14	MCG −2-9-28	14.9	1.4 x 0.9	15.0	SB(s)dm IV-V	39	Mag 7.90v star ESE 8′2.
03 19 17.9	−12 06 14	MCG −2-9-29	13.5	1.4 x 0.8	13.5	SA(r)c: II	168	
03 20 15.2	−10 51 49	MCG −2-9-31	14.8	0.5 x 0.5	13.2	S?		**Hickson 24C** W 0′5.
03 20 22.6	−10 52 04	MCG −2-9-32	14.6	0.4 x 0.4	12.4	S?		Star N edge. **Hickson 24D** NW 1′0.

(Continued from previous page)

GALAXIES

RA h m s	Dec ° ′ ″	Name	Mag (V)	Dim ′ Maj x min	SB	Type Class	PA	Notes
03 21 24.1	−11 08 45	MCG −2-9-33	14.6	0.5 x 0.4	12.7		6	Mag 9.14v star N 4′3.
03 21 54.2	−12 22 42	MCG −2-9-34	14.5	1.1 x 0.5	13.7		57	Mag 8.36v star N 3′1.
03 21 55.3	−13 39 02	MCG −2-9-35	15.1	1.1 x 0.2	13.3	SBd pec sp	66	Part of the **Hickson 26** group of galaxies.
03 22 55.2	−11 12 11	MCG −2-9-36	13.9	2.5 x 1.7	15.4	SB(s)d III-IV	159	Mag 12.4 star N 2′1; mag 9.60v star ESE 9′1.
03 23 33.7	−10 32 40	MCG −2-9-37	15.0	0.8 x 0.7	14.2		18	Mag 12.5 star W 2′4.
03 23 38.5	−11 11 48	MCG −2-9-38	14.4	1.7 x 0.5	14.0	(R′)S0/a:	57	Mag 9.60v star S 4′3.
03 24 26.3	−11 40 13	MCG −2-9-39	14.6	0.9 x 0.5	13.6	SBc	87	Mag 11.3 star S 2′9.
03 25 05.0	−12 18 32	MCG −2-9-40	13.9	1.0 x 0.3	12.4	S0/a	135	Mag 8.84v star S 2′9.
03 27 20.9	−13 44 57	MCG −2-9-41	14.3	1.3 x 0.4	13.4	S?	120	Mag 11.7 star E 5′5.
03 28 50.4	−13 48 06	MCG −2-9-43	14.2	0.7 x 0.5	13.0	S0	21	
03 29 50.9	−13 23 01	MCG −2-9-45	14.5	0.8 x 0.2	12.4	Sab	174	Mag 12.6 star NNE 1′2.
03 30 17.0	−14 01 22	MCG −2-9-46	14.4	1.0 x 0.2	12.5		78	Mag 10.09v star S 5′7.
03 33 49.1	−10 15 05	MCG −2-10-2	14.2	1.2 x 0.3	12.9	Sc	153	Mag 10.37v star N 3′1.
03 38 51.2	−09 44 14	MCG −2-10-3	14.7	0.8 x 0.7	13.9	Sc	120	
03 39 21.3	−09 37 07	MCG −2-10-4	14.4	0.9 x 0.6	13.6	Sbc	6	Mag 12.0 star N 2′8.
03 39 39.7	−12 42 19	MCG −2-10-6	14.7	0.7 x 0.6	13.6	Sc	0	
03 41 59.9	−09 36 35	MCG −2-10-7	14.3	0.8 x 0.7	13.5	Sc	3	Mag 10.23v star SSE 8′3.
03 42 56.7	−12 54 50	MCG −2-10-9	13.4	3.2 x 0.5	13.8	Sc: sp	33	Mag 10.87v star E 2′1.
03 47 16.8	−11 42 17	MCG −2-10-11	14.0	1.0 x 0.7	13.4	Sb	162	
03 48 13.4	−12 27 10	MCG −2-10-12	14.7	0.9 x 0.6	13.9	SBbc	81	Mag 11.2 star S 1′3.
03 49 31.7	−13 53 46	MCG −2-10-13	14.9	0.9 x 0.3	13.3	Sm	165	Mag 8.85v star E 2′9.
03 50 57.7	−08 30 52	MCG −2-10-14	14.1	0.7 x 0.5	12.8	Sb	15	Mag 9.22v star S 4′9.
03 52 11.4	−09 07 11	MCG −2-10-17	14.0	1.0 x 0.4	12.9	SBc	6	Stellar anonymous galaxy S end; NGC 1470 N 7′2.
03 53 18.3	−10 26 51	MCG −2-10-18	14.4	1.0 x 0.5	13.5	SBc	141	Mag 7.61v star S 6′1.
03 53 36.9	−09 27 26	MCG −2-10-19	14.0	1.0 x 0.4	12.9	SA0°:	171	
03 53 37.8	−12 20 46	MCG −2-10-20	14.9	0.9 x 0.4	13.7		114	Mag 7.33v star SE 8′2.
03 54 13.5	−08 44 34	MCG −2-10-21	14.1	0.9 x 0.7	13.5	SAB(s)d III-IV	114	Mag 10.13v star ENE 7′5.
03 55 31.5	−09 35 05	MCG −2-11-1	14.3	0.8 x 0.4	13.0	Spiral	39	Mag 10.36v star W 4′2.
03 58 29.2	−09 56 34	MCG −2-11-2	14.6	1.0 x 0.3	13.2	Sbc	51	Mag 10.95v star E 6′1.
03 58 38.1	−10 19 26	MCG −2-11-3	14.9	0.6 x 0.6	13.7			Mag 10.68v star NW 8′1.
04 00 26.0	−09 09 11	MCG −2-11-4	15.1	0.7 x 0.4	13.6	Sb-c	120	Mag 9.73v star NE 11′6.
04 00 57.2	−10 16 21	MCG −2-11-5	14.2	0.9 x 0.8	13.7	SB? I-II	84	Close pair of mag 10.95v, 11.6 stars W 3′1.
04 01 32.8	−10 26 22	MCG −2-11-6	14.4	0.7 x 0.5	13.1	Sb	135	Mag 10.56v star WSW 2′2.
04 02 03.2	−09 27 24	MCG −2-11-7	14.4	1.3 x 0.3	13.3	Sb	153	Mag 10.59v star SW 9′7.
04 02 57.0	−09 01 30	MCG −2-11-10	14.8	1.1 x 0.8	14.6		105	Mag 10.41v star NE 11′8.
04 03 11.9	−14 17 13	MCG −2-11-11	14.5	0.7 x 0.7	13.5	Sc		Mag 11.8 star N 6′2.
04 04 27.5	−10 10 42	MCG −2-11-14	14.8	0.8 x 0.6	13.8	Scd	18	Mag 8.12v star S 3′4.
04 04 32.7	−10 18 06	MCG −2-11-15	15.0	0.9 x 0.5	14.0	Sc	162	Mag 8.12v star N 4′3.
04 07 12.0	−10 13 49	MCG −2-11-16	14.3	0.7 x 0.5	13.0	Sa	123	Mag 11.2 star and **Mkn 1193** NNW 4′5.
04 08 36.3	−10 11 22	MCG −2-11-19	14.4	0.7 x 0.7	13.5			Mag 9.28v star E 13′3.
04 08 25.8	−13 59 08	MCG −2-11-20	13.9	0.5 x 0.5	12.4	E?		Mag 10.39v star W 11′2.
04 09 55.7	−10 16 13	MCG −2-11-21	14.2	0.7 x 0.5	13.2	E	111	Mag 9.34v star NNE 6′7.
04 10 47.0	−08 59 41	MCG −2-11-22	14.5	0.9 x 0.8	14.0	SBbc	123	Mag 10.50v star SW 5′9.
04 10 49.6	−09 37 43	MCG −2-11-23	13.9	1.5 x 0.6	13.6	SB(s)cd: III-IV	117	Mag 13.4 star N 2′3.
04 11 45.2	−12 28 00	MCG −2-11-24	14.6	1.0 x 0.2	12.7	SBc	99	
04 12 29.7	−13 09 06	MCG −2-11-25	14.6	0.7 x 0.6	13.5	Sc	60	Mag 7.91v star SW 7′3.
04 14 01.5	−13 23 24	MCG −2-11-26	14.4	1.1 x 0.7	14.0	S0?	123	
04 15 45.8	−13 29 18	MCG −2-11-28	13.5	0.7 x 0.7	12.6	S0		
02 56 11.0	−15 23 52	MCG −3-8-46	14.0	1.4 x 0.9	14.1	SB(s)b II-III	171	Bright knot or superimposed star SW of center.
02 57 10.7	−16 39 44	MCG −3-8-50	14.8	0.7 x 0.4	13.2		48	Mag 9.60v star N 2′4.
02 57 43.8	−16 34 20	MCG −3-8-51	14.1	1.0 x 0.7	13.6	Scd	78	Mag 9.60v star WSW 8′8.
02 57 48.0	−14 53 16	MCG −3-8-52	14.7	0.9 x 0.4	13.5	Sbc	33	NGC 1151 and NGC 1150 SW 13′9.
02 59 32.5	−15 16 16	MCG −3-8-55	15.2	0.7 x 0.4	13.6		129	Mag 10.90v star NE 2′5.
03 00 31.8	−15 44 13	MCG −3-8-57	13.0	1.4 x 1.4	13.5	SA(s)d III		Numerous bright knots.
03 01 04.0	−15 10 45	MCG −3-8-58	15.0	0.9 x 0.7	14.3	SBb	144	Elongated anonymous galaxy on E edge.
03 03 54.8	−14 17 49	MCG −3-8-69	14.3	1.5 x 0.3	13.3	Sa? sp	108	Mag 9.58v star N 4′6.
03 05 41.0	−16 43 39	MCG −3-8-72	14.6	0.8 x 0.3	12.9	Sab	114	Mag 9.98v star S 6′9.
03 09 39.1	−15 51 58	MCG −3-9-3	15.4	1.1 x 0.5	14.6		3	
03 11 44.7	−16 55 38	MCG −3-9-4	15.1	0.5 x 0.5	13.4			Stellar; mag 9.56v star SW 11′6.
03 12 38.3	−16 41 26	MCG −3-9-5	14.4	0.6 x 0.5	13.0		168	Mag 9.02v star S 5′3.
03 12 42.4	−17 06 48	MCG −3-9-6	14.3	1.1 x 0.3	13.0	Sa	9	Mag 10.49v star NE 11′5.
03 13 31.5	−15 15 27	MCG −3-9-10	14.1	0.7 x 0.7	13.2	SBc		
03 18 09.5	−16 16 31	MCG −3-9-16	14.3	0.4 x 0.4	12.1			Mag 11.6 star N 5′0.
03 20 19.6	−15 25 17	MCG −3-9-20	13.9	0.7 x 0.3	12.0		165	Mag 10.62v star NE 10′8.
03 20 29.5	−15 12 53	MCG −3-9-21	14.5	0.9 x 0.4	13.3	SBb	72	Mag 10.62v star SE 7′0.
03 20 52.7	−17 53 14	MCG −3-9-24	14.7	0.9 x 0.6	13.9	Sd	42	Numerous bright knots.
03 21 35.5	−16 43 19	MCG −3-9-25	14.2	0.2 x 0.2	10.5			Stellar; mag 10.13v star WSW 9′2; MCG −3-9-26 ESE 2′3.
03 21 43.4	−16 44 37	MCG −3-9-26	15.0	0.3 x 0.2	11.8			MCG −3-9-25 WNW 2′3.
03 21 51.8	−15 42 39	MCG −3-9-27	14.7	1.8 x 1.1	15.3	SB(s)m V	85	Mag 8.88v star SW 8′0.
03 22 18.0	−17 10 52	MCG −3-9-29	14.7	0.7 x 0.4	13.1		132	MCG −3-9-29 W 2′5.
03 22 28.8	−17 10 53	MCG −3-9-30	15.1	0.5 x 0.3	12.9		21	MCG −3-9-30 E 2′5.
03 23 24.2	−15 38 53	MCG −3-9-31	14.9	0.9 x 0.4	13.7	Sbc	108	Mag 12.1 star SW 1′4.
03 23 34.3	−16 46 49	MCG −3-9-32	13.9	1.1 x 0.6	13.3	S?	144	
03 24 05.3	−16 18 39	MCG −3-9-34	14.3	0.9 x 0.7	13.6	SBbc	138	Mag 8.95v star WNW 8′9.
03 25 07.7	−15 04 56	MCG −3-9-36	14.6	0.9 x 0.3	13.0	Sb	21	Mag 7.99v star NE 3′1.
03 25 24.9	−16 14 11	MCG −3-9-41	14.1	3.0 x 0.3	13.8	Sdm: sp	9	Mag 10.54v star W 11′4; **MCG −3-9-37** NW 5′3.
03 27 42.7	−17 00 25	MCG −3-9-44	13.9	0.9 x 0.4	12.6	Sb	174	
03 29 33.4	−15 14 33	MCG −3-9-47	14.3	2.2 x 2.0	15.7	SB(s)m V	90	Mag 10.21v star NE 13′6.
03 30 35.9	−17 56 29	MCG −3-9-48	13.4	1.3 x 0.9	13.4	SB(rs)0⁺ pec?	169	Mag 9.71v star N 4′4.
03 34 00.1	−16 20 51	MCG −3-10-9	14.4	1.0 x 0.4	13.2			Mag 8.92v star NE 12′1.
03 37 51.1	−14 52 56	MCG −3-10-16	13.9	1.1 x 0.6	13.3	SBb	90	Mag 11.2 star NE 3′5.
03 39 42.3	−14 34 10	MCG −3-10-24	14.6	1.5 x 0.2	13.1	Sc: sp	150	Mag 10.40v star WNW 12′4.
03 42 31.3	−16 00 10	MCG −3-10-38	14.2	1.1 x 0.5	13.4	Sb	6	Mag 7.93v star WNW 5′0.
03 43 35.5	−16 00 52	MCG −3-10-41	13.8	2.2 x 1.1	14.6	SB(s)dm III-IV	171	Mag 12.7 star NNE 2′9.
03 44 00.8	−14 21 28	MCG −3-10-42	12.6	2.1 x 0.6	12.7	SAB(s)bc: II-III	138	Mag 10.35v star NE 12′4.
03 46 38.5	−16 32 56	MCG −3-10-45	13.0	1.1 x 0.5	12.2	IBm pec: III	30	Close pair of mag 11.9, 12.4 stars NW 3′3.
03 53 29.0	−17 07 24	MCG −3-10-50	14.8	0.9 x 0.4	13.6	Sb	168	Mag 10.05v star W 9′2.
03 53 53.6	−17 35 36	MCG −3-10-51	13.9	1.0 x 0.4	12.7	Sb	90	Mag 8.19v star NE 14′7; ESO 549-36 located E 5′4 of this star.
04 02 03.3	−16 26 52	MCG −3-11-7	14.6	1.2 x 0.2	12.9	S?	144	

RA h m s	Dec ° ′ ″	Name	Mag (V)	Dim ′ Maj x min	SB	Type Class	PA	Notes
04 05 56.0	−15 09 02	MCG −3-11-10	14.4	0.7 x 0.4	12.9	Sm	3	Mag 11.6 star SE 9.′7.
04 07 12.6	−17 12 13	MCG −3-11-12	14.3	2.0 x 2.0	15.7	SB(s)dm V		Mag 10.25v star S 6.′3.
04 08 33.0	−16 45 29	MCG −3-11-14	14.4	1.0 x 0.7	13.9	Sbc	6	Mag 9.74v star NW 6.′2.
04 11 07.2	−14 52 11	MCG −3-11-17	13.8	0.9 x 0.4	12.6	Sbc	60	
04 11 19.9	−16 13 48	MCG −3-11-18	14.4	1.5 x 0.4	13.7	SB(s)m: sp IV-V	108	Mag 9.29v star WNW 5.′3.
02 57 04.3	−07 41 09	NGC 1148	15.3	1.3 x 0.9	15.3	SB(rs)d pec? III-IV	84	Stellar nucleus; mag 9.03v star W 9.′4.
02 57 01.3	−15 02 55	NGC 1150	14.1	1.0 x 0.7	13.6	S0	73	Mag 7.90v star W 8.′1.
02 57 04.7	−15 00 49	NGC 1151	15.0	0.5 x 0.4	13.4	E	10	Mag 7.90v star W 8.′7.
02 57 33.6	−07 45 31	NGC 1152	13.5	1.1 x 0.7	13.1	S0	29	Mag 12.0 star S 2.′4; NGC 1148 NW 8.′5.
02 58 07.8	−10 21 48	NGC 1154	13.2	1.0 x 0.6	12.5	SB	124	Mag 9.75v star S 5.′0; NGC 1155 NE 1.′5.
02 58 13.0	−10 21 01	NGC 1155	13.4	1.0 x 0.9	13.1	S0	165	Compact core, faint, uniform envelope; NGC 1154 SW 1.′5.
02 58 06.7	−15 07 08	NGC 1157	14.8	0.5 x 0.2	12.1	Sb	169	
02 57 11.4	−14 23 45	NGC 1158	14.8	0.9 x 0.6	14.0	S0	147	Compact core, uniform envelope; mag 12.3 star E 3.′8.
02 58 56.2	−12 23 53	NGC 1162	12.5	1.4 x 1.4	13.2	E:		Bright, round core, uniform envelope.
03 00 22.2	−17 09 09	NGC 1163	13.8	2.2 x 0.3	13.2	SBbc? sp	141	
03 01 36.1	−14 50 11	NGC 1172	11.9	2.3 x 1.8	13.3	E$^+$:	27	Round core, uniform envelope; mag 10.45v star E 2.′0.
03 01 51.1	−15 01 49	NGC 1180	14.9	0.6 x 0.4	13.4	E/S0	5	
03 01 42.9	−15 03 09	NGC 1181	15.4	0.8 x 0.2	13.3	Sb	100	
03 03 28.5	−09 40 14	NGC 1182	13.9	0.8 x 0.4	12.5	Sa	115	Stellar **PGC 11535** SSE 7.′9.
03 02 59.4	−09 07 55	NGC 1185	13.7	1.2 x 0.4	12.8	SB(r)b? II-III?	30	
03 03 43.3	−15 29 08	NGC 1188	13.8	1.1 x 0.6	13.2	SAB(r)0°:	170	Faint, elongated anonymous galaxy E 9.′3.
03 03 24.3	−15 37 22	NGC 1189	13.8	1.7 x 1.5	14.7	SB(s)dm: IV-V	96	Almost stellar nucleus, smooth envelope.
03 03 26.2	−15 39 43	NGC 1190	14.2	0.7 x 0.3	12.4	S0°: sp	95	
03 03 31.0	−15 41 08	NGC 1191	14.3	0.4 x 0.3	11.9	SB0$^-$:	60	
03 03 34.6	−15 40 45	NGC 1192	14.4	0.5 x 0.2	11.9	E?	102	
03 03 32.8	−12 02 24	NGC 1195	14.5	0.6 x 0.5	13.1	E/S0		Almost stellar nucleus; mag 13.8 star SE 0.′7; NGC 1196 S 2.′3.
03 03 35.5	−12 04 34	NGC 1196	13.2	1.4 x 1.4	13.8	SB(rs)0°		NGC 1195 N 2.′3.
03 03 38.2	−15 36 50	NGC 1199	11.4	2.4 x 1.9	13.0	E3:	63	Faint anonymous galaxy on S edge.
03 03 54.5	−11 59 29	NGC 1200	12.0	2.9 x 2.3	13.9	SA(s)0$^-$	93	IC 285 SE 3.′2.
03 05 02.4	−06 29 31	NGC 1202	14.2	0.6 x 0.5	12.7	Sb	94	Mag 10.33v star N 4.′8.
03 05 14.1	−14 22 53	NGC 1203A	14.6	0.3 x 0.3	11.8	E1?		Pair with NGC 1203B on N edge. Located between mag 8.10v star SW 2.′5 and mag 9.38v star NE 3.′1.
03 05 14.3	−14 22 39	NGC 1203B	15.1	0.1 x 0.1	9.9	E3?		NGC 1203A on S edge; see notes for NGC 1203A.
03 04 39.9	−12 20 29	NGC 1204	13.3	1.2 x 0.4	12.3	S0/a:	69	Star on S edge.
03 06 09.8	−08 49 59	NGC 1206	14.5	0.5 x 0.5	12.9	E		Almost stellar nucleus; mag 10.72v star S 6.′3.
03 06 11.9	−09 32 28	NGC 1208	12.3	1.8 x 0.9	12.7	SA(r)0/a?	80	Mag 12.3 star NW 3.′0; faint anonymous galaxy SW 3.′3.
03 06 03.1	−15 36 40	NGC 1209	11.4	2.4 x 1.1	12.5	E6:	85	NGC 1231 ENE 6.′8.
03 06 55.9	−09 32 41	NGC 1214	14.0	1.3 x 0.4	13.2	SB(rs)0$^+$?	40	= **Hickson 23A**.
03 07 09.6	−09 35 33	NGC 1215	14.1	1.8 x 1.2	14.8	(R)SAB(r)a pec	65	= **Hickson 23B**. **Hickson 23E** on N edge.
03 07 18.6	−09 36 44	NGC 1216	14.8	0.9 x 0.3	13.3	S0$^+$? sp	65	= **Hickson 23C**.
03 06 29.4	−15 34 09	NGC 1231	14.2	0.8 x 0.8	13.6	Sc		Mag 10.22v star SE 6.′2; NGC 1209 WSW 6.′8.
03 09 39.4	−07 50 48	NGC 1234	14.2	1.8 x 0.9	14.6	SB(r)cd pec III	141	Stellar nucleus; mag 9.48v star NE 4.′3.
03 10 52.6	−10 44 52	NGC 1238	13.3	1.6 x 1.2	14.0	E$^+$	110	IC 1897 SSW 3.′4.
03 11 14.5	−08 55 19	NGC 1241	12.0	3.4 x 1.6	13.7	SB(rs)b I-II	140	Mag 9.32v star 3.′1 N of center; NGC 1242 on NE edge.
03 11 19.4	−08 54 11	NGC 1242	13.7	1.2 x 0.5	13.0	SB(rs)c: III	130	Located on NE edge of NGC 1241.
03 12 14.1	−10 28 51	NGC 1247	12.5	3.4 x 0.5	12.9	Sbc sp	70	Mag 9.92v star WNW 6.′2; stellar **Mkn 1071** S 2.′3.
03 12 48.9	−05 13 28	NGC 1248	12.5	1.1 x 1.0	12.5	SA(s)0°:	90	Elongated **Mkn 604** S 2.′7; mag 8.45v star N 5.′5.
03 15 33.8	−15 52 48	NGC 1262	14.2	0.9 x 0.7	13.6	SAB(s)c: II	135	
03 15 39.5	−15 05 50	NGC 1263	14.1	0.8 x 0.7	13.3	S0/a	0	Stellar nucleus, mag 10.23v star N 5.′1.
03 17 45.4	−10 17 23	NGC 1284	13.8	1.3 x 0.9	13.9	SA0°:	30	Bright nucleus, uniform envelope; mag 12.1 star E 3.′8.
03 17 53.5	−07 17 56	NGC 1285	12.8	1.6 x 1.0	13.2	(R′)SB(r)b pec	34	
03 17 48.6	−07 37 02	NGC 1286	13.9	0.9 x 0.6	13.1	S0	150	Mag 9.66v star W 4.′9.
03 19 25.1	−13 59 23	NGC 1290	14.8	0.5 x 0.4	12.9	E2?	85	Almost stellar.
03 20 03.4	−13 59 54	NGC 1295	14.3	0.9 x 0.5	13.2	S0	175	Mag 10.32v star E 3.′2.
03 18 49.7	−13 03 48	NGC 1296	14.2	1.1 x 0.8	14.0	SB(rs)ab pec II	0	Strong E-W oriented core.
03 20 09.5	−06 15 49	NGC 1299	13.0	1.2 x 0.6	12.5	SB(rs)b? II	39	Mag 9.95v star W 9.′0.
03 20 40.8	−07 23 42	NGC 1303	13.9	0.8 x 0.6	13.0	S0	28	Mag 9.67v star SSE 6.′0.
03 22 06.1	−15 23 59	NGC 1309	11.5	2.2 x 2.0	13.0	SA(s)bc: II	45	Mag 7.26v star S 4.′0.
03 25 01.7	−05 44 44	NGC 1324	12.7	2.1 x 0.8	13.1	Sb: II	135	
03 26 02.7	−17 35 32	NGC 1329	12.7	1.4 x 1.1	13.0	SA(rs)a?	35	ESO 548-14 S 4.′0; mag 8.96v star SW 9.′3.
03 28 06.0	−08 23 21	NGC 1337	11.9	5.8 x 1.5	14.1	SA(s)cd II-III	144	Mag 9.70v star W 6.′7.
03 28 54.1	−12 09 15	NGC 1338	12.9	1.4 x 1.3	13.4	(R′)SAB(rs)b pec II-III	55	Mag 9.99v star E 2.′0; mag 8.81v star NE 5.′8.
03 29 31.6	−17 46 44	NGC 1345	13.8	1.5 x 1.1	14.1	SB(s)c pec: III-IV	108	Mag 9.45v star S 7.′6.
03 30 13.4	−05 32 36	NGC 1346	13.2	1.1 x 0.7	12.8	Sb pec?	85	**MCG −1-9-41** NW 1.′5.
03 32 29.2	−15 13 15	NGC 1354	12.3	2.2 x 0.8	12.7	SAB(rs)0/a:	150	
03 33 23.6	−04 59 56	NGC 1355	13.3	1.4 x 0.3	12.2	S0: sp	78	NGC 1358 SW 6.′8.
03 33 17.0	−13 39 48	NGC 1357	11.5	3.2 x 2.4	13.6	SA(s)ab	73	Mag 8.92v star N 3.′7.
03 33 39.8	−05 05 24	NGC 1358	12.1	2.3 x 1.5	13.3	SAB(r)0/a	3	Small elongated core, smooth envelope; mag 9.46v star E 8.′5; NGC 1355 NW 6.′8.
03 34 17.7	−06 15 54	NGC 1361	13.9	1.6 x 1.2	14.6	E$^+$ pec:	39	
03 34 49.5	−09 50 33	NGC 1363	13.1	0.8 x 0.7	12.3	Sbc I-II:	69	Mag 6.25v star SW 3.′4; NGC 1364 E 2.′3.
03 34 59.0	−09 50 19	NGC 1364	14.7	0.5 x 0.5	13.1	Scd III:		Stellar nucleus; NGC 1363 W 2.′3.
03 34 58.9	−15 39 23	NGC 1368	14.2	1.3 x 0.6	13.9	S0/a	108	Mag 10.04v star NW 8.′7.
03 36 59.8	−15 52 55	NGC 1372	14.3	0.6 x 0.6	13.2	E		
03 37 05.9	−05 02 36	NGC 1376	12.1	1.7 x 1.7	13.1	SA(s)cd II		
03 38 12.1	−15 53 56	NGC 1388	13.8	0.7 x 0.7	13.1	E		Mag 9.69v star N 6.′7.
03 40 18.7	−15 31 42	NGC 1405	14.1	1.5 x 0.5	13.6	S0: sp	153	Mag 9.42v star NW 8.′6; NGC 1413 S 5.′1.
03 40 11.6	−15 36 38	NGC 1413	14.3	1.0 x 0.8	14.1	E	57	Stellar nucleus; NGC 1405 N 5.′1.
03 42 29.3	−13 29 25	NGC 1421	11.4	3.5 x 0.9	12.5	SAB(rs)bc: II-III	0	
03 42 40.2	−06 22 55	NGC 1423	14.1	0.8 x 0.5	12.9	SBbc	22	Mag 8.91v star W 10.′9.
03 46 12.9	−09 40 58	NGC 1434	14.3	0.8 x 0.4	13.0	S0/a	165	Almost stellar nucleus; mag 8.57v star SE 5.′9.
03 44 56.2	−09 51 22	NGC 1445	14.0	0.8 x 0.5	13.1	E	14	
03 45 47.1	−09 01 08	NGC 1447	14.7	1.0 x 0.6	14.0	S0	105	Stellar nucleus; mag 7.94v star E 2.′7.
03 45 36.7	−09 14 06	NGC 1450	14.1	0.8 x 0.7	13.4	E/S0	40	
03 48 27.0	−16 23 30	NGC 1461	11.8	3.0 x 0.9	12.7	SA(r)0°	156	Mag 10.18v star NNW 3.′3.
03 51 24.3	−15 24 09	NGC 1464	13.8	0.7 x 0.5	12.5	Pec	44	
03 51 52.8	−08 50 19	NGC 1467	14.1	1.2 x 0.7	13.8	(R′)SA(rs)0$^+$?	55	
03 52 12.6	−06 20 56	NGC 1468	14.2	1.2 x 0.8	14.0	S0$^-$	141	
03 52 09.9	−09 00 00	NGC 1470	13.7	1.3 x 0.3	12.5	Sab sp	168	MCG −2-10-17 S 7.′2.
03 53 47.3	−08 34 07	NGC 1472	14.4	0.6 x 0.6	13.2	S0		

(Continued from previous page)

138 GALAXIES 138

RA h m s	Dec ° ′ ″	Name	Mag (V)	Dim ′ Maj x min	SB	Type Class	PA	Notes
03 54 02.9	−08 34 31	NGC 1477	14.2	0.6 x 0.6	13.0	E		
03 54 07.3	−08 33 21	NGC 1478	14.8	0.6 x 0.4	13.1	S0	52	
04 02 29.8	−09 20 10	NGC 1504	14.4	0.7 x 0.6	13.3	E/S0	20	Mag 11.7 star W 1′.7; NGC 1505 NE 1′.8.
04 02 36.5	−09 19 20	NGC 1505	13.7	1.0 x 0.8	13.3	S0	80	NGC 1504 SW 1′.8.
04 03 55.3	−11 10 46	NGC 1509	13.7	0.8 x 0.6	12.8	Sa	50	IC 2026 W 1′.2; mag 10.11v star NNW 10′.3.
04 08 07.5	−08 49 46	NGC 1516A	13.6	0.5 x 0.5	11.9	SBbc		
04 08 08.8	−08 50 14	NGC 1516B	13.8	0.7 x 0.5	12.5	Sc	126	
04 08 07.7	−17 11 34	NGC 1519	12.9	2.1 x 0.5	12.8	SB(r)b? II-III?	108	Mag 8.54v star SSE 4′.6.
04 14 36.1	−13 10 30	NGC 1538	14.4	0.9 x 0.7	13.8	E	125	Almost stellar nucleus; mag 9.04v star SW 5′.5.

GALAXY CLUSTERS

RA h m s	Dec ° ′ ″	Name	Mag 10th brightest	No. Gal	Diam ′	Notes
03 06 48.0	−12 02 00	A 415	16.3	67	24	

PLANETARY NEBULAE

RA h m s	Dec ° ′ ″	Name	Diam ″	Mag (P)	Mag (V)	Mag cent ★	Alt Name	Notes
04 14 15.8	−12 44 21	NGC 1535	51	9.6	9.6	12.1	PK 206−40.1	Close pair of mag 13.5, 13.9 stars W 2′.1.

139 GALAXIES 139

RA h m s	Dec ° ′ ″	Name	Mag (V)	Dim ′ Maj x min	SB	Type Class	PA	Notes
02 41 23.9	−17 56 08	ESO 546-5	14.5	1.0 x 0.8	14.1	SB?	156	
01 46 24.7	−08 38 19	IC 159	13.2	1.4 x 0.7	13.0	SB(rs)b pec:	36	Mag 10.73v star E 4′.8.
01 46 29.5	−13 14 54	IC 160	13.9	1.4 x 0.9	14.1	SAB0⁻:	85	Mag 10.11v star SSE 6′.5.
01 50 27.6	−08 31 23	IC 168	14.2	0.9 x 0.3	12.6	Sa	105	Mag 12.0 star E 3′.8.
01 51 57.4	−08 31 04	IC 170	14.4	0.8 x 0.5	13.4	E		NGC 707 W 7′.5.
01 59 34.1	−05 20 52	IC 183	13.9	1.3 x 0.4	13.1	S0?	96	Mag 10.23v star E 3′.8.
01 59 51.2	−06 50 24	IC 184	13.8	1.1 x 0.6	13.2	SB(r)a:	177	Mag 10.51v star SSE 2′.7.
02 09 30.6	−06 58 08	IC 206	14.2	0.9 x 0.4	12.9	Spiral	138	Mag 11.4 star N 3′.9; IC 207 NE 3′.6.
02 09 39.9	−06 55 22	IC 207	13.9	2.2 x 0.4	13.6	S0⁺ pec sp	99	IC 206 SW 3′.6; mag 11.4 star W 3′.4; IC 206 SW 3′.6.
02 08 58.8	−07 03 33	IC 209	13.1	1.5 x 1.1	13.5	SB(r)bc: II-III	60	Mag 11.9 star W 8′.3.
02 09 28.9	−09 40 47	IC 210	13.1	2.3 x 0.6	13.3	SBbc? II-III	66	Mag 9.26v star NE 3′.4.
02 14 09.5	−06 48 24	IC 215	14.8	1.0 x 0.3	13.3	Sab	81	Located 1′.5 N of mag 9.69v star.
02 16 10.7	−11 55 31	IC 217	13.6	2.2 x 0.4	13.3	Scd? sp	39	Mag 9.75v star NE 5′.9.
02 18 38.9	−06 54 12	IC 219	13.4	1.0 x 0.6	12.9	E?	18	MCG −1-6-67 on NW edge.
02 19 11.8	−12 46 55	IC 220	14.5	0.9 x 0.4	13.3	Sbc	36	Mag 8.86v star E 6′.8; MCG −2-6-55 W 10′.4.
02 24 45.2	−12 33 54	IC 224	14.3	0.7 x 0.4	12.8	Sbc	132	Mag 10.62v star E 11′.9.
02 28 47.3	−10 49 53	IC 230	14.8	0.6 x 0.4	13.4	E/S0	90	Mag 10.54v star WNW 7′.0.
02 38 32.0	−06 54 08	IC 243	14.2	1.0 x 0.6	13.5	(R′)SB(rs)0/a	30	Mag 12.0 star N 1′.4.
02 38 54.6	−14 18 25	IC 245	14.3	1.1 x 0.3	12.9	Sbc	21	MCG −3-7-47 SE 1′.2.
02 40 08.8	−11 44 02	IC 247	14.0	1.3 x 0.9	14.0	SB(rs)bc pec?	45	Mag 4.83v star Epsilon Ceti SW 11′.9.
02 40 54.4	−13 18 50	IC 250	13.9	0.9 x 0.7	13.2	SBab	126	Pair of N-S oriented mag 10.36v, 10.60v stars SE 4′.1.
02 41 13.8	−14 57 31	IC 251	14.0	0.6 x 0.6	12.7			Located 0′.8 NW of mag 10.63v star.
02 42 05.8	−15 02 50	IC 253	13.9	0.9 x 0.6	13.3	E/S0	155	Mag 7.28v star WSW 9′.1; NGC 1065 S 2′.8.
02 55 27.1	−14 06 12	IC 268	14.6	1.0 x 0.6	13.9	SBbc	72	Mag 11.4 star NE 2′.2; IC 269 N 2′.3; anonymous galaxy S 2′.1.
02 55 26.5	−14 04 01	IC 269	14.4	1.1 x 0.4	13.4		126	IC 268 S 2′.3; mag 11.4 star SE 2′.6.
02 55 44.3	−14 12 28	IC 270	12.9	1.3 x 1.3	13.4	SA0⁻:		Mag 12.0 star W 1′.5; IC 272 E 5′.5.
02 55 59.5	−12 00 29	IC 271	13.8	1.0 x 0.8	13.4		120	Several knots and/or superimposed stars.
02 56 06.3	−14 11 14	IC 272	14.1	1.2 x 0.7	13.8	Sc	171	Very faint anonymous galaxies N 1′.6 and S 2′.5; IC 270 W 5′.5.
02 58 41.1	−15 42 12	IC 276	13.1	1.9 x 0.5	12.9	S0° pec sp	60	Mag 8.03v star SW 12′.4; mag 9.77v star SSW 6′.0.
01 51 07.8	−09 47 32	IC 1738	13.9	0.9 x 0.7	13.2	(R′)SAB(rs)b II	80	Mag 9.83v star E 9′.2; NGC 701 N 5′.4.
01 51 56.7	−16 47 17	IC 1741	14.2	0.8 x 0.6	13.3	S0/a	171	Mag 8.49v star WNW 7′.3.
01 59 59.1	−11 04 45	IC 1767	13.6	1.7 x 0.7	13.6	(R′)SA(r)0/a:	75	Bright knot or star W end; mag 12.7 star ENE 4′.5.
02 43 42.1	−15 42 22	IC 1840	14.3	0.7 x 0.7	13.4	Sb		Three faint stars superimposed.
02 48 04.4	−13 59 37	IC 1853	14.2	1.0 x 0.4	13.1	SB(rs)b? III	91	NGC 1103 N 1′.9; mag 10.79v star SSE 7′.9.
02 54 53.1	−15 39 08	IC 1866	14.0	1.1 x 0.7	13.8	E/S0	90	Mag 11.8 star N 8′.4.
01 44 38.8	−07 37 13	MCG −1-5-28	13.8	1.3 x 1.0	13.9	(R′)SA(r)0⁺:	0	Compact core, faint, smooth envelope.
01 50 00.0	−06 29 19	MCG −1-5-39	14.5	1.3 x 0.4	13.6	S?	117	Stellar **MCG −1-5-38** N 2′.8.
01 51 28.4	−06 30 59	MCG −1-5-44	14.7	1.5 x 0.4	14.0	SAd pec sp III	90	Mag 12.3 star ENE 7′.0.
01 51 50.3	−05 29 48	MCG −1-5-45	14.7	1.6 x 0.5	14.3	SB(s)dm III	66	**MCG −1-5-46** SE edge.
01 57 41.6	−05 40 33	MCG −1-6-7	13.8	1.6 x 1.1	14.2	SA(s)cd II-III:	54	Mag 12.0 star SW 1′.7.
01 58 34.4	−07 42 42	MCG −1-6-11	14.5	0.9 x 0.4	13.3	Compact	6	Mag 8.22v star N 5′.7.
01 58 52.2	−08 09 46	MCG −1-6-12	13.1	1.3 x 1.0	13.2	SB(r)c pec I-II	63	Pair of superimposed stars or bright knots SE of core.
01 59 44.5	−07 50 22	MCG −1-6-18	13.4	1.2 x 0.8	13.2	Sbc	30	Located on W edge of small galaxy cluster **A 287**.
01 59 49.8	−07 03 31	MCG −1-6-20	14.5	1.6 x 0.6	14.4	SA0° pec	153	= **Hickson 14A**. MCG −1-6-22 = **Hickson 14B** S 1′.8; **Hickson 14C** N 1′.8; mag 8.22v star NNE 2′.9.
01 59 52.4	−07 05 17	MCG −1-6-22	14.1	1.4 x 0.4	13.3	SAb pec?	177	= **Hickson 14B**; mag 9.17v star S 2′.0.
02 00 15.5	−06 05 27	MCG −1-6-23	14.2	0.7 x 0.7	13.3	Spiral		Mag 10.83v star NE 4′.6; NGC 779 NW 11′.0.
02 00 59.4	−05 30 39	MCG −1-6-24	14.2	1.2 x 0.4	13.3	Sab	165	Mag 10.51v star E 5′.1.
02 01 30.1	−07 17 28	MCG −1-6-27	14.2	1.0 x 0.4	13.0	Spiral	126	Mag 8.93v star SW 12′.6.
02 01 59.3	−08 14 42	MCG −1-6-30	14.4	1.0 x 0.7	13.8	Sb	90	Mag 10.47v star NW 9′.1.
02 02 13.3	−06 04 42	MCG −1-6-31	14.4	1.5 x 0.3	13.4	Sc pec sp	173	Mag 9.26v star SW 4′.4.
02 03 47.8	−07 25 48	MCG −1-6-36	14.6	0.7 x 0.5	13.3	Sbc	90	Close pair of mag 9.81v, 9.60v stars W 10′.0.
02 04 31.1	−06 12 01	MCG −1-6-39	13.4	3.2 x 1.3	14.8	SB(s)m III-IV	72	Patchy, mag 13.1 star superimposed E end.
02 05 05.3	−06 30 24	MCG −1-6-41	14.0	1.1 x 1.1	14.1	SA(r)cd III		Stellar nucleus, MCG −1-6-42 NNE 5′.0.
02 05 19.5	−06 27 04	MCG −1-6-42	13.7	1.2 x 0.4	12.7	(R′)SB(r)ab?	96	MCG −1-6-41 SSW 5′.0.
02 05 48.5	−05 17 37	MCG −1-6-43	14.0	1.1 x 0.5	13.2	Sbc	144	
02 07 51.1	−06 32 14	MCG −1-6-47	14.3	0.9 x 0.6	13.4	SBbc	21	Star W edge.
02 09 28.8	−06 13 00	MCG −1-6-52	13.9	1.1 x 1.0	13.9	SB(s)c: II-III	3	Star at N end.
02 10 02.0	−07 37 28	MCG −1-6-56	14.2	0.5 x 0.5	12.5	Compact		= **Mkn 1024**. Mag 9.99v star E 12′.3.
02 11 16.4	−06 26 16	MCG −1-6-66	16.4	0.9 x 0.7	15.7	SBbc	12	MCG −1-6-67 ESE 7′.6.
02 11 43.9	−06 29 26	MCG −1-6-67	14.7	1.5 x 0.2	13.2	Sbc: sp	36	MCG −1-6-66 WNW 7′.6.

RA h m s	Dec ° ′ ″	Name	Mag (V)	Dim ′ Maj x min	SB	Type Class	PA	Notes
02 11 46.7	−05 54 54	MCG −1-6-68	14.5	0.9 x 0.8	14.0	Sb	21	Star W edge.
02 13 15.9	−07 39 42	MCG −1-6-70	13.5	1.5 x 1.2	14.0	(R′)SB(rs)a	135	Mag 9.31v star W 4′.8.
02 13 10.9	−06 20 28	MCG −1-6-71	14.1	0.5 x 0.5	12.4			
02 13 27.7	−07 41 54	MCG −1-6-75	13.0	0.7 x 0.4	11.5	Sb	21	**MCG −1-6-74** SSW 2′.2.
02 14 25.9	−07 22 07	MCG −1-6-77	13.7	2.4 x 0.4	13.5	Sb pec sp	66	Mag 10.51v star S 2′.9; elongated **MCG −1-6-78** NNE 4′.4.
02 17 19.4	−05 28 54	MCG −1-6-80	14.1	1.3 x 0.6	13.7	SB?	141	Low surface brightness.
02 18 16.1	−06 42 24	MCG −1-6-85	14.7	0.5 x 0.5	13.0	Sbc		Mag 8.39v star N 6′.6.
02 23 53.7	−07 21 12	MCG −1-7-6	13.9	0.4 x 0.4	11.7			Mag 8.06v star SE 5′.1; **MCG −1-7-5** NNW 7′.0.
02 24 42.1	−07 56 47	MCG −1-7-9	14.8	1.0 x 0.7	14.2	Sbc	135	Mag 9.70-v star NW 3′.6.
02 28 11.0	−05 50 27	MCG −1-7-15	14.3	1.5 x 0.4	13.6	Sb pec sp	21	Small anonymous galaxy at middle of W edge.
02 34 48.5	−07 40 59	MCG −1-7-22	13.5	1.1 x 1.1	13.5	SB(s)c I-II		Mag 8.98v star SE 4′.3.
02 36 58.6	−05 20 56	MCG −1-7-24	13.2	1.1 x 0.9	13.1	SA(s)c pec: II	150	
02 49 18.2	−08 10 33	MCG −1-8-11	13.8	0.8 x 0.5	12.7	Sbc	72	Mag 13.5 star on N edge.
02 50 44.1	−06 44 43	MCG −1-8-14	14.2	1.3 x 0.4	13.4	Sb pec?	9	Mag 7.24v star SW 7′.0.
02 54 54.0	−06 22 24	MCG −1-8-15	13.3	1.6 x 1.3	13.9	(R′)SB(r)a	141	Strong NE-SW bar.
01 44 34.1	−08 46 24	MCG −2-5-39	15.0	0.7 x 0.5	13.7		3	
01 44 53.6	−10 56 14	MCG −2-5-40	14.3	0.7 x 0.7	13.4			
01 45 28.4	−10 05 37	MCG −2-5-41	14.1	1.0 x 0.8	13.7	S0⁻	165	
01 46 39.6	−13 24 30	MCG −2-5-46	14.0	0.9 x 0.4	12.7	Sa	39	
01 47 18.3	−14 02 44	MCG −2-5-48	14.3	1.0 x 0.8	13.9	Sbc	156	
01 47 47.8	−13 35 36	MCG −2-5-49	14.1	0.7 x 0.4	12.8	E/S0	33	
01 48 26.1	−12 22 59	MCG −2-5-50	13.2	3.0 x 2.5	15.3	IAB(rs)m IV-V	20	Mag 10.74v star N 6′.8; MCG −2-5-50A on E edge.
01 48 32.2	−12 23 19	MCG −2-5-50A	14.4	0.5 x 0.4	12.5	Sc	171	Located on E edge of MCG −2-5-50.
01 48 36.3	−10 19 39	MCG −2-5-51	14.2	1.2 x 0.3	13.0	S0°: sp	162	NGC 681 SE 10′.4.
01 49 10.3	−10 03 45	MCG −2-5-53	13.2	2.7 x 2.2	15.0	SB(s)d III	120	Mag 13.5 star 2′.4 E.
01 49 23.3	−08 26 34	MCG −2-5-54	14.7	0.7 x 0.7	13.8			
01 49 22.1	−14 07 25	MCG −2-5-55	14.4	1.1 x 0.2	12.6	Sbc	63	
01 49 38.0	−13 34 17	MCG −2-5-56	14.0	1.7 x 0.6	13.9	SB(r)b? II-III	63	Anonymous galaxy NE 1′.0 and NW 2′.7.
01 49 40.1	−12 49 29	MCG −2-5-57	14.4	1.5 x 1.0	14.7	SB(s)m V	15	Located between a pair of mag 12 stars.
01 51 10.8	−12 30 53	MCG −2-5-62	13.8	1.3 x 0.6	13.3	Spiral	3	
01 51 26.0	−08 24 57	MCG −2-5-64	13.6	0.6 x 0.4	11.9	Sb	24	NGC 707 S 5′.6.
01 51 33.9	−08 23 57	MCG −2-5-65	14.3	0.9 x 0.5	13.3	Sb	105	Multi-galaxy system.
01 52 49.2	−08 16 01	MCG −2-5-67	14.8	0.7 x 0.6	13.7	Irr	126	Stellar galaxy **KUG 0149-084** NW 8′.0.
01 53 18.2	−12 22 09	MCG −2-5-70	14.7	1.0 x 0.6	14.2	E?	33	Mag 9.35v star W 3′.9.
01 54 18.3	−09 42 51	MCG −2-5-71	14.3	1.1 x 0.7	13.9	SB0⁻:	90	Stellar galaxy **KUG 0152-100** SE 7′.4.
01 54 03.4	−14 15 11	MCG −2-5-72	13.8	1.5 x 0.4	13.1	(R)S0/a? sp	78	
01 54 52.7	−13 39 14	MCG −2-5-74	13.9	1.0 x 0.7	13.3	Sm pec: III-IV:	27	
01 55 18.1	−12 54 30	MCG −2-5-76	14.8	1.1 x 0.8	14.5	SB(s)dm pec IV-V	6	
01 55 17.5	−14 09 22	MCG −2-6-1	14.8	0.7 x 0.5	13.5	Sd	150	
01 55 15.9	−10 29 34	MCG −2-6-2	14.6	0.7 x 0.4	13.1	Sa	108	
01 55 51.3	−09 58 00	MCG −2-6-4	14.0	1.7 x 1.3	14.7	SB(r)c pec II	147	Compact core, faint envelope.
01 56 59.2	−11 46 50	MCG −2-6-6	13.9	2.7 x 0.6	14.3	SB(s)dm: III-IV	84	Mag 11.02v star S 1′.2.
01 58 13.4	−10 37 47	MCG −2-6-9	15.0	0.7 x 0.7	14.1			Stellar nucleus, faint envelope; mag 8.78v star S 3′.9.
02 00 38.4	−10 42 38	MCG −2-6-13	14.8	1.2 x 0.3	13.5	Sbc		
02 00 37.3	−13 30 44	MCG −2-6-14	14.8	0.5 x 0.5	13.2	Sc		Mag 7.99v star WSW 8′.0.
02 00 55.2	−08 50 28	MCG −2-6-16	14.5	1.1 x 0.9	14.4	IB(s)m: IV-V	25	Very close pair of mag 14-15 stars W 2′.2.
02 01 51.2	−10 28 00	MCG −2-6-17	13.5	1.3 x 1.1	13.7	SB(rs)c: II	27	Mag 9.81v star NW 7′.3.
02 03 01.2	−12 33 23	MCG −2-6-18	15.0	1.1 x 0.3	13.6	Sbc	36	Mag 10.16v star SW 5′.7.
02 03 02.2	−09 39 19	MCG −2-6-19	13.9	3.2 x 0.3	13.7	Sd? sp	39	Mag 10.38v star E 6′.3.
02 03 17.1	−12 38 16	MCG −2-6-20	14.3	0.7 x 0.4	12.8	S0	147	Mag 11.36v star W 6′.1.
02 04 12.2	−10 04 51	MCG −2-6-22	15.4	0.9 x 0.7	14.7		29	Mag 11.16v star SW 3′.8; NGC 811 ESE 5′.7.
02 05 03.1	−12 54 08	MCG −2-6-25	14.5	0.8 x 0.4	13.1	Sab	75	Mag 10.28v star NE 7′.9.
02 05 49.5	−09 49 22	MCG −2-6-26	14.5	1.9 x 0.9	15.0	SB(s)m pec: III	114	**PGC 7990** S 1′.4.
02 06 21.4	−08 52 18	MCG −2-6-27	15.0	1.0 x 0.6	14.3		18	Mag 9.64v star SW 8′.8.
02 08 01.3	−10 21 18	MCG −2-6-28	14.2	1.0 x 0.8	13.8	Spiral	3	Mag 9.18v star SSW 9′.4.
02 08 35.8	−08 37 56	MCG −2-6-29	14.1	1.0 x 0.5	13.2	S0/a	96	Mag 11.8 star SW 2′.1.
02 10 14.1	−09 42 47	MCG −2-6-35	13.4	2.5 x 0.8	14.0	SBc pec?	36	Mag 9.26v star WNW 9′.8; IC 210 is 3′.4 SW of this star.
02 11 22.2	−10 07 14	MCG −2-6-37	15.7	0.9 x 0.4	14.4	Sc	66	Mag 6.00v star N 4′.1.
02 12 04.9	−13 42 20	MCG −2-6-39	14.9	0.4 x 0.3	12.4		129	Mag 10.70v star N 1′.6.
02 13 14.2	−14 00 29	MCG −2-6-40	14.8	0.4 x 0.3	12.4		81	Mag 10.48v star N 3′.2.
02 14 06.0	−12 49 13	MCG −2-6-41	14.7	0.7 x 0.5	13.7	E/S0	168	Mag 9.42v star SE 9′.1.
02 14 30.5	−13 15 50	MCG −2-6-42	14.8	0.5 x 0.3	12.6	Scd		
02 15 26.4	−13 02 53	MCG −2-6-43	14.5	0.7 x 0.5	13.2	SBbc	165	Mag 9.68v star W 8′.3.
02 15 37.3	−13 43 09	MCG −2-6-44	14.4	1.0 x 0.3	12.9	Sc	33	Mag 10.93v star N 5′.0.
02 15 47.6	−13 49 27	MCG −2-6-45	14.3	0.4 x 0.4	12.2			MCG −2-6-44 N 6′.8.
02 16 20.3	−10 59 58	MCG −2-6-47	14.6	0.9 x 0.6	13.8	Sc	54	Mag 10.26v star SSE 3′.0.
02 17 26.8	−11 42 08	MCG −2-6-49	15.4	0.7 x 0.6	14.3	SB(r)d pec: III-IV	36	
02 17 32.6	−11 31 09	MCG −2-6-50	14.9	1.4 x 0.2	13.4	Sdm	87	Mag 12.2 star E 7′.2.
02 17 35.3	−09 11 24	MCG −2-6-51	15.0	0.7 x 0.4	13.4		3	Mag 10.30v star NNW 1′.9.
02 18 14.2	−12 13 53	MCG −2-6-52	13.6	1.0 x 0.7	13.1	S0	129	Mag 10.79v star NNW 7′.5.
02 18 21.9	−09 17 27	MCG −2-6-53	15.0	0.9 x 0.6	14.2	Irr	69	MCG −2-5-51 and mag 10.30v star NW 13′.0.
02 18 24.0	−12 11 23	MCG −2-6-54	15.1	0.7 x 0.4	13.5	Spiral	114	Star SW edge; MCG −2-6-54 NNE 3′.6.
02 18 29.9	−12 48 55	MCG −2-6-55	14.8	1.1 x 0.3	13.4	Sbc	45	Mag 9.33v star NNW 11′.7; IC 220 E 10′.4.
02 18 45.1	−11 29 18	MCG −2-6-56	14.0	1.1 x 0.4	12.9	Sb	141	
02 19 18.8	−11 48 03	MCG −2-7-1	14.6	0.9 x 0.3	13.0	Sb	153	Mag 10.89v star S 6′.3.
02 21 47.3	−10 01 26	MCG −2-7-2	13.5	0.9 x 0.4	12.3	S0⁻? sp	39	Mag 12.1 star SE 2′.6.
02 23 47.7	−10 22 36	MCG −2-7-3	15.2	0.7 x 0.7	14.3	Sd		Mag 9.13v star NE 1′.1.
02 24 41.7	−11 46 34	MCG −2-7-4	14.2	1.0 x 0.5	13.3	Spiral	81	Mag 9.65v star S 3′.3.
02 25 50.9	−11 11 26	MCG −2-7-6	13.4	0.9 x 0.6	12.7	(R′)SB(rs)0°?	45	Mag 9.59v star NE 7′.2.
02 26 22.5	−09 50 34	MCG −2-7-7	14.1	2.8 x 1.0	15.0	SAB(s)dm: IV	108	Mag 8.25v star SW 7′.9.
02 26 57.7	−09 16 03	MCG −2-7-8	14.5	0.9 x 0.6	13.7	Spiral	174	
02 27 32.3	−10 10 00	MCG −2-7-10	13.9	1.3 x 0.4	13.0	Sc? sp III	88	Mag 11.4 star NW 2′.0.
02 27 40.4	−09 54 47	MCG −2-7-11	15.6	0.5 x 0.5	14.0			Mag 8.60v star N 6′.6.
02 28 32.4	−11 17 16	MCG −2-7-12	13.7	1.0 x 0.3	12.3	Sa	48	Mag 7.06v star SE 4′.9.
02 28 41.4	−10 41 19	MCG −2-7-14	15.0	0.7 x 0.5	13.7		177	Mag 8.66v star N 4′.1.
02 28 59.1	−08 30 57	MCG −2-7-17	14.8	0.9 x 0.7	14.1		174	
02 29 02.9	−11 01 22	MCG −2-7-20	14.7	0.7 x 0.3	12.9	Sab	36	Located 2′.1 W of NGC 950.
02 29 54.6	−11 07 01	MCG −2-7-22	14.3	1.0 x 0.3	12.8	Sbc	36	Mag 8.06v star W 8′.2.

RA h m s	Dec ° ′ ″	Name	Mag (V)	Dim ′ Maj x min	SB	Type Class	PA	Notes
02 29 58.8	−13 15 53	MCG −2-7-23	14.1	0.6 x 0.5	12.7		18	Mag 8.57v star WNW 2′9; SW of the center of galaxy cluster A 358, many anonymous galaxies in the area.
02 30 05.4	−08 59 54	MCG −2-7-24	14.4	1.0 x 0.9	14.2	S?	171	= **Mkn 1044**. Mag 10.37v star E 5′4.
02 30 07.0	−10 27 22	MCG −2-7-25	14.3	1.0 x 0.8	13.9	Spiral	168	
02 30 26.0	−10 44 57	MCG −2-7-26	13.9	1.7 x 1.3	14.6	IB(s)m IV-V	84	Mag 9.86v star SE 10′9.
02 32 01.6	−13 20 05	MCG −2-7-29	15.0	0.8 x 0.6	14.1	SB(s)dm IV-V	126	
02 33 15.9	−11 44 43	MCG −2-7-32	15.0	1.6 x 1.1	15.5	SB(s)cd III	85	Small, N-S elongated nucleus; mag 12.9 star N 8′2.
02 34 24.5	−10 50 38	MCG −2-7-33	13.2	3.2 x 1.1	14.5	SBbc? II-III	69	Mag 9.81v star ESE 12′0.
02 34 25.9	−14 16 18	MCG −2-7-34	14.0	0.7 x 0.7	13.1	S0		Mag 7.96v star NW 8′5.
02 34 54.1	−13 39 33	MCG −2-7-36	13.1	1.9 x 0.8	13.4	SAB(rs)c: II	138	Mag 8.60v star SW 11′1.
02 35 36.0	−12 16 33	MCG −2-7-38	13.8	1.4 x 0.3	12.8	S0?	120	Mag 10.17v star E 11′5.
02 35 42.5	−10 32 17	MCG −2-7-39	13.4	0.7 x 0.4	11.9	Sb	135	
02 35 43.3	−13 40 54	MCG −2-7-40	13.9	1.0 x 0.3	12.7	E/S0	123	MCG −2-7-41 N 1′5.
02 35 44.8	−13 39 21	MCG −2-7-41	12.8	1.8 x 1.5	13.7	(R′)SB(rs)0/a	117	Mag 10.63v star NE 9′7; MCG −2-7-40 S 1′5.
02 36 13.3	−09 18 36	MCG −2-7-42	14.5	0.9 x 0.6	13.7	Sbc	171	Mag 9.94v star E 3′5.
02 36 34.9	−09 21 16	MCG −2-7-43	14.9	0.4 x 0.4	12.7			Mag 9.94v star NW 2′7.
02 39 48.9	−10 36 22	MCG −2-7-49	14.7	0.7 x 0.5	13.4		135	
02 39 56.9	−14 12 07	MCG −2-7-50	14.4	0.7 x 0.6	13.3	Sb	3	Mag 9.14v star S 4′8.
02 40 10.5	−12 30 02	MCG −2-7-51	14.3	0.4 x 0.4	12.3	E		Mag 9.88v star S 3′3.
02 40 20.4	−12 29 19	MCG −2-7-56	14.6	0.7 x 0.4	13.2	E/S0	168	**MCG −2-7-55** N edge; mag 9.88v star SSW 4′6.
02 40 27.4	−12 52 35	MCG −2-7-57	14.3	0.3 x 0.2	11.1	Sa	72	Located on W edge of MCG −2-7-60.
02 40 28.9	−12 52 39	MCG −2-7-60	14.5	0.4 x 0.3	12.0		177	Mag 10.40v star WSW 1′9; MCG −2-7-57 on W edge.
02 40 24.9	−12 51 49	MCG −2-7-61	14.1	0.3 x 0.3	11.3			Mag 10.40v star S 1′9.
02 40 47.1	−13 23 34	MCG −2-7-63	14.1	1.2 x 0.3	12.9	Sbc	6	Pair of N-S oriented mag 10.36v, 10.60v stars ENE 5′5.
02 41 04.6	−13 26 47	MCG −2-7-65	14.3	0.9 x 0.4	13.1	S0?	117	Pair of N-S oriented mag 10.36v, 10.60v stars N 5′4.
02 41 17.9	−09 55 59	MCG −2-7-66	13.8	1.2 x 0.7	13.5	Sbc	156	
02 41 20.3	−13 26 16	MCG −2-7-67	14.6	1.1 x 0.2	12.8	Sbc	102	Pair of N-S oriented mag 10.36v, 10.60v stars NNW 5′6.
02 41 26.4	−13 07 43	MCG −2-7-68	14.1	1.0 x 0.6	13.4	S0$^-$ pec	147	
02 41 39.1	−12 21 37	MCG −2-7-69	13.8	0.9 x 0.6	13.0	Sa	111	Mag 9.00v star ENE 11′7.
02 42 13.2	−09 32 20	MCG −2-7-70	14.7	0.8 x 0.3	13.0	SBc	141	MCG −2-7-72 NE 6′5.
02 42 33.4	−09 28 07	MCG −2-7-72	14.1	0.9 x 0.7	13.7	E?	6	MCG −2-7-70 SW 6′5.
02 42 38.2	−12 25 17	MCG −2-7-73	13.2	0.8 x 0.5	12.1	S0$^-$ pec:	33	= **Hickson 19A**.
02 42 41.9	−12 25 40	MCG −2-7-74	14.8	0.9 x 0.2	12.7	(R)SB(r)a pec?	99	= **Hickson 19B**. **Hickson 19D** S 1′2.
02 42 46.8	−12 23 54	MCG −2-7-75	14.8	1.1 x 0.5	14.0	SBm pec? IV-V	108	= **Hickson 19C**. Multi-galaxy system; mag 9.00v star NW 7′3.
02 42 59.4	−11 40 18	MCG −2-7-76	14.3	0.8 x 0.4	12.9	Sab	96	
02 44 22.5	−09 44 17	MCG −2-8-2	14.8	1.0 x 0.3	13.3	Sb		Mag 12.0 star N 1′9; MCG −2-8-3 on E edge.
02 44 25.5	−09 44 09	MCG −2-8-3	14.9	1.0 x 0.2	13.1	Sb	48	MCG −2-8-2 on W edge.
02 45 08.0	−13 14 13	MCG −2-8-4	14.4	1.5 x 0.3	13.3	Sc	162	Pair of mag 13 stars W and NW 1′9.
02 48 15.7	−10 15 11	MCG −2-8-7	14.5	1.2 x 0.4	13.5	S?	72	Mag 11.27v star NNW 2′7.
02 48 16.4	−08 55 51	MCG −2-8-8	14.5	0.6 x 0.4	12.8	Sc	18	Mag 8.16v star SW 12′4; stellar **Mkn 1055** S 1′9.
02 49 10.4	−10 40 51	MCG −2-8-9	14.5	1.1 x 0.6	13.9	Sc	177	Mag 7.91v star N 8′2.
02 49 40.7	−12 15 14	MCG −2-8-10	14.8	0.5 x 0.5	13.2	Sm		Mag 12.0 star NNE 1′6; NGC 1118 NE 7′0.
02 50 17.6	−08 35 53	MCG −2-8-12	14.2	2.4 x 0.3	13.7	Sc sp	93	Mag 10.46v star SW 11′0.
02 51 29.2	−13 18 22	MCG −2-8-13	14.2	1.1 x 0.8	14.1	E/S0	135	
02 52 23.4	−08 30 39	MCG −2-8-14	14.1	1.5 x 0.2	12.7	Sa? sp	99	Mag 9.57v star NW 20′5.
02 53 51.4	−13 52 17	MCG −2-8-16	14.7	1.3 x 0.7	14.4	SBbc	138	Mag 11.9 star WNW 3′2.
02 54 04.3	−12 52 05	MCG −2-8-17	14.7	1.3 x 0.5	14.1	SBbc	60	Mag 11.7 star NE end.
02 54 14.9	−08 28 50	MCG −2-8-18	13.8	0.8 x 0.8	13.4	E		= **Mkn 1062**.
02 54 46.5	−13 27 26	MCG −2-8-20	14.7	1.0 x 0.4	13.6		30	Mag 9.83v star ENE 8′3.
02 55 06.8	−10 45 12	MCG −2-8-21	14.4	0.7 x 0.7	13.5	S-		Mag 9.93v star E 8′6.
02 55 14.4	−13 32 13	MCG −2-8-22	14.2	1.2 x 0.4	13.3	S0	27	Star S end. Anonymous galaxy 1′0 E; mag 9.832v star N 7′6.
02 55 36.6	−09 59 36	MCG −2-8-25	14.4	0.9 x 0.6	13.6	S+	6	
02 55 36.2	−10 08 26	MCG −2-8-26	14.1	1.1 x 0.7	13.8	E	60	
02 55 40.9	−10 29 26	MCG −2-8-27	14.1	0.5 x 0.5	12.6	E		Mag 11.6 star W 4′5.
02 56 09.9	−13 41 17	MCG −2-8-31	14.5	0.7 x 0.5	13.3	Sab	48	Mag 8.06v star W 3′9.
02 57 11.4	−11 14 10	MCG −2-8-32	14.5	0.4 x 0.3	12.0	Sc	171	
02 57 49.6	−10 10 02	MCG −2-8-33	12.9	2.7 x 0.7	13.5	(R)SB(r)0/a?	147	Mag 9.97v star SSW 8′6.
02 59 31.6	−13 37 17	MCG −2-8-37	15.9	0.7 x 0.2	13.7	Sc	138	Mag 13 star NW end.
03 00 04.3	−10 49 27	MCG −2-8-38	14.4	0.9 x 0.4	13.1	SBb	39	Small, elongated anonymous galaxy NW 2′0.
03 00 31.0	−11 24 56	MCG −2-8-39	13.6	1.4 x 0.9	13.7	SAB(rs)a pec:	9	Mag 7.36v star SW 3′9.
03 01 51.3	−10 37 12	MCG −2-8-40	14.2	1.0 x 0.7	13.9	E	45	Mag 10.69v star S 4′8.
01 46 29.1	−15 07 18	MCG −3-5-19	14.5	0.4 x 0.4	12.6	E/S0		
01 46 55.7	−14 27 09	MCG −3-5-20	14.3	0.7 x 0.3	12.4	Sbc	63	
01 57 23.0	−15 16 07	MCG −3-6-2	14.2	0.5 x 0.4	12.3		30	
01 57 30.0	−15 00 34	MCG −3-6-3	14.6	0.7 x 0.4	13.1	Sb	96	
02 02 56.8	−14 36 31	MCG −3-6-5	15.3	1.4 x 0.8	15.3	SB(s)d II-III	170	Stellar nucleus, faint arms; mag 10.33v star N 4′7; NGC 815 S 3′8.
02 05 43.6	−17 48 52	MCG −3-6-6	14.8	0.7 x 0.7	13.8	Sc		
02 07 55.5	−16 40 35	MCG −3-6-7	14.9	0.9 x 0.6	14.1	Sb	21	
02 08 22.9	−17 24 07	MCG −3-6-8	14.9	0.6 x 0.6	13.7	(R′)SAB(s)0$^+$		Mag 7.90v star E 5′3.
02 09 32.0	−16 47 13	MCG −3-6-9	14.3	0.5 x 0.3	12.3	E	63	Mag 11.7 star NE 1′5.
02 12 56.3	−15 08 18	MCG −3-6-15	14.4	0.9 x 0.6	13.6	SB(s)cd II-III	108	
02 15 50.3	−16 41 47	MCG −3-6-20	14.3	0.8 x 0.5	13.1	Sb	129	Mag 8.85v star NNE 5′7.
02 17 42.0	−16 35 38	MCG −3-6-22	14.7	0.7 x 0.4	13.1	Sbc	105	
02 18 05.1	−15 28 01	MCG −3-6-23	14.7	0.9 x 0.4	13.4		147	Mag 10.20v star S 3′5.
02 22 44.5	−15 23 25	MCG −3-7-6	14.9	0.4 x 0.3	12.5	Sb	30	Small anonymous companion galaxy E edge.
02 23 15.2	−15 09 00	MCG −3-7-7	14.6	0.7 x 0.3	12.8	SBc	150	Mag 10.69v star E 2′8.
02 24 33.3	−14 29 55	MCG −3-7-10	14.8	1.0 x 0.2	12.9	Sc	51	Mag 10.64v star NW 1′9.
02 25 36.5	−16 06 37	MCG −3-7-13	14.2	0.7 x 0.4	12.7	Sbc	6	Mag 11.03v star E 4′6.
02 26 25.0	−15 13 34	MCG −3-7-14	14.6	0.4 x 0.4	12.6	E		Mag 5.86v star SW 9′2.
02 27 16.2	−14 22 16	MCG −3-7-18	14.6	0.9 x 0.3	13.0	Irr	57	Small, elongated anonymous galaxy E 1′1.
02 27 21.0	−15 25 19	MCG −3-7-19	14.5	1.2 x 0.3	13.2	SBdm: sp III-IV:	117	Mag 10.43v star SW 5′0.
02 29 38.6	−17 02 41	MCG −3-7-23	14.3	0.6 x 0.4	12.6	Sb	30	Multi-galaxy system or star involved? **MCG −3-7-25** SSE 18′7.
02 29 50.7	−15 39 50	MCG −3-7-24	14.6	0.7 x 0.7	13.7			Mag 10.82v star N 5′5.
02 31 26.1	−17 39 13	MCG −3-7-27	13.6	0.6 x 0.6	12.4	S?		Mag 10.47v star SE 6′3.
02 31 41.5	−15 27 33	MCG −3-7-28	14.9	0.3 x 0.3	12.2			Mag 10.43v star N 3′1.
02 33 26.8	−15 20 52	MCG −3-7-32	13.5	0.3 x 0.2	10.3	Sc	12	0′25 S of mag 9.87v star.
02 34 00.9	−15 21 58	MCG −3-7-35	14.5	0.4 x 0.4	12.4	Sb		Mag 9.87v star and MCG −3-7-32 W 8′4.
02 36 02.0	−17 14 41	MCG −3-7-38	14.1	0.5 x 0.3	11.9	Sbc		Mag 7.789v star N 6′1; MCG −3-7-40 S 1′2.
02 36 03.7	−17 22 03	MCG −3-7-39	14.6	0.6 x 0.4	12.9	Sb	114	Mag 9.21v star E 11′7.

(Continued from previous page)

RA h m s	Dec ° ′ ″	Name	Mag (V)	Dim ′ Maj x min	SB	Type Class	PA	Notes
02 36 04.2	−17 15 51	MCG −3-7-40	14.0	1.0 x 0.5	13.1	S/Irr?	171	MCG −3-7-38 N 1′.2.
02 37 16.7	−15 49 20	MCG −3-7-41	14.6	0.9 x 0.7	13.9	SBc	129	MCG −3-7-42 ENE 6′.2.
02 37 40.8	−15 47 19	MCG −3-7-42	14.3	1.3 x 0.9	14.3	SB(s)d: III	30	Mag 11.12v star N 7′.3; NGC 1034 E 8′.1.
02 38 58.2	−14 19 11	MCG −3-7-47	14.6	1.5 x 0.2	13.1	S?	45	Pair with IC 245 NW.
02 39 42.7	−16 00 10	MCG −3-7-50	14.0	0.8 x 0.6	13.1	SBc	0	Mag 7.97v star SSE 7′.0.
02 40 36.7	−15 46 08	MCG −3-7-51	14.3	0.6 x 0.6	13.0			Mag 10.89v star SSW 8′.4.
02 40 46.5	−15 08 31	MCG −3-7-52	13.3	1.3 x 1.1	13.6	SAB(s)c I-II	141	Mag 7.28v star E 10′.9.
02 40 45.9	−17 07 14	MCG −3-7-53	14.1	0.5 x 0.5	12.7	E/S0		Mag 10.70v star NW 7′.2.
02 41 35.3	−16 40 52	MCG −3-7-57	14.1	0.8 x 0.4	12.8	SBc	168	Mag 9.02v star S 6′.4.
02 43 16.3	−15 15 42	MCG −3-7-61	14.4	0.9 x 0.6	13.5		135	Mag 10.65v star N 8′.7.
02 43 52.4	−14 48 42	MCG −3-8-5	13.7	0.7 x 0.5	12.4	Sb	78	Mag 10.39v star N 2′.8; NGC 1076 NW 6′.6.
02 44 39.9	−17 43 38	MCG −3-8-6	15.4	0.7 x 0.2	13.1	Sbc	54	Mag 9.18v star SE 5′.5; MCG −3-8-7 N 1′.0.
02 44 41.5	−17 42 49	MCG −3-8-7	14.7	0.5 x 0.2	12.1	Sb	132	NGC 1098 NNE 4′.6, MCG −3-8-6 S 1′.0.
02 45 12.1	−17 19 10	MCG −3-8-12	15.0	0.5 x 0.4	13.1	S?	36	Mag 11.7 star on S edge; **MCG −3-8-9** W 4′.4.
02 45 35.7	−15 12 25	MCG −3-8-17	14.8	0.7 x 0.3	12.9	Sc	90	Mag 9.87v star E 6′.2.
02 45 47.1	−15 30 39	MCG −3-8-18	14.4	0.9 x 0.4	13.1	Sbc	54	Mag 7.61v star NW 9′.7.
02 46 15.8	−17 06 44	MCG −3-8-19	14.5	0.9 x 0.2	12.5	Sb	159	Mag 10.63v star W 5′.0.
02 46 22.2	−14 47 02	MCG −3-8-21	15.3	1.0 x 0.4	14.2	Sb		Mag 11.7 star N 6′.9.
02 47 19.8	−14 48 07	MCG −3-8-25	14.2	0.9 x 0.6	13.4	Sa	87	MCG −3-8-26 N edge.
02 47 22.0	−14 47 34	MCG −3-8-26	14.6	0.7 x 0.5	13.3	Sb	90	MCG −3-8-25 S edge.
02 49 35.6	−14 28 00	MCG −3-8-30	14.6	0.4 x 0.4	12.5			NGC 1120 W 7′.6.
02 51 06.6	−14 43 24	MCG −3-8-32	14.3	1.1 x 0.7	13.9	S?	150	Bright knot or star N end of elongated nucleus; mag 11.6 star W 3′.7.
02 51 11.7	−14 41 03	MCG −3-8-33	14.4	0.7 x 0.3	12.6	Sab	27	Mag 11.31v star E 1′.4.
02 51 37.7	−16 39 35	MCG −3-8-34	15.1	0.4 x 0.2	12.2	Sa	177	Located on SW edge of NGC 1125.
02 52 22.3	−14 33 15	MCG −3-8-37	14.2	1.1 x 0.8	13.9	SB?	114	Mag 10.15v star SW 9′.8; NGC 1139 E 6′.1.
02 53 25.0	−17 36 11	MCG −3-8-41	14.3	0.7 x 0.4	12.7	S?	111	
02 54 57.4	−17 03 06	MCG −3-8-45	14.9	1.9 x 1.6	15.9	SB(rs)cd: III-IV	168	Mag 12.2 star on W edge; mag 11.2 stars E 7′.1.
02 56 11.0	−15 23 52	MCG −3-8-46	14.0	1.4 x 0.9	14.1	SB(s)b II-III	171	Bright knot or superimposed star SW of center.
02 57 10.7	−16 39 44	MCG −3-8-50	14.8	0.7 x 0.4	13.2		48	Mag 9.60v star N 2′.4.
02 57 43.8	−16 34 20	MCG −3-8-51	14.1	1.0 x 0.7	13.6	Scd	78	Mag 9.60v star WSW 8′.8.
02 57 48.0	−14 53 16	MCG −3-8-52	14.7	0.9 x 0.4	13.5	Sbc	33	NGC 1151 and NGC 1150 SW 13′.9.
02 59 32.5	−15 16 16	MCG −3-8-55	15.2	0.7 x 0.4	13.6		129	Mag 10.90v star NE 2′.5.
03 00 31.8	−15 44 13	MCG −3-8-57	13.0	1.4 x 1.4	13.5	SA(s)d III		Numerous bright knots.
03 01 04.0	−15 10 45	MCG −3-8-58	15.0	0.9 x 0.7	14.3	SBb	144	Elongated anonymous galaxy on E edge.
03 03 54.8	−14 17 49	MCG −3-8-69	14.3	1.5 x 0.3	13.3	Sa? sp	108	Mag 9.58v star N 4′.6.
01 49 11.1	−10 25 32	NGC 681	12.0	2.6 x 1.6	13.4	SAB(s)ab sp	60	Moderately dark lane extends across nucleus NE-SW.
01 49 04.2	−14 58 31	NGC 682	13.4	1.4 x 1.1	13.8	SA0$^-$	95	
01 47 48.1	−16 43 19	NGC 690	14.2	1.2 x 0.8	14.0	SAB(s)c: I-II	154	Mag 10.22v star SSW 7′.2.
01 50 43.5	−12 02 08	NGC 699	14.1	1.5 x 0.3	13.1	SBbc? sp	135	Mag 10.60v star W 5′.8; mag 8.78v star SW 7′.3.
01 51 04.1	−09 42 08	NGC 701	12.2	2.5 x 1.2	13.2	SB(rs)c III-IV	40	Mag 12.4 star N 5′.5; IC 1738 S 5′.4.
01 51 27.2	−08 30 21	NGC 707	13.6	1.4 x 0.9	13.7	(R′)SAB(s)0$^-$:	95	Star or knot on SE edge of nucleus; MCG −2-5-64 and MCG −2-5-65 N 5′.6; IC 170 E 7′.5.
01 55 21.4	−09 05 01	NGC 713	14.4	1.0 x 0.3	12.9	Sb II-III?	95	Stellar galaxy **KUG 0152-092** N 3′.4.
01 53 12.6	−12 52 25	NGC 715	14.0	0.8 x 0.4	12.6	Sb	175	
01 53 00.5	−13 44 15	NGC 720	10.2	4.7 x 2.4	12.8	E5	140	
01 52 35.4	−16 31 03	NGC 725	14.3	0.8 x 0.5	13.1	Sc II-III:	78	Mag 8.33v star E 6′.4.
01 55 32.2	−10 47 54	NGC 726	14.3	1.2 x 0.6	13.8	SB(s)dm pec III-IV	100	
01 54 56.2	−09 00 39	NGC 731	12.1	1.7 x 1.7	13.1	E$^+$:		Large circular core, faint, smooth envelope.
01 54 57.4	−17 04 46	NGC 734	15.9	0.7 x 0.4	14.4	SAB(rs)a?	24	
01 57 30.6	−09 27 49	NGC 747	13.7	1.0 x 0.5	12.8	Sb: II	175	
01 56 22.8	−09 03 41	NGC 755	12.6	3.4 x 0.9	13.6	SB(rs)b? II-III:	45	
01 54 29.3	−16 42 30	NGC 756	14.5	0.9 x 0.6	13.7	S0: pec	32	
01 56 57.4	−05 24 13	NGC 762	13.0	1.3 x 1.1	13.2	(R′)SB(rs)a	25	
01 58 50.8	−09 35 18	NGC 767	14.5	1.1 x 0.4	13.4	SBb pec?	165	Mag 8.64v star E 6′.8.
01 58 52.0	−11 30 51	NGC 773	13.1	1.3 x 0.7	12.8	SAB(r)a pec	179	
01 59 42.5	−05 57 50	NGC 779	11.2	4.0 x 1.2	12.7	SAB(r)b II	160	MCG −1-6-23 SE 11′.0; mag 10.83v star ESE 11′.6.
02 00 48.5	−09 00 10	NGC 787	12.9	2.5 x 1.9	14.4	(R)SA(rs)b: II	90	Mag 10.28v star SE 4′.1; stellar galaxy **KUG 0158-092** SW 4′.7.
02 01 06.5	−06 48 58	NGC 788	12.1	1.6 x 1.4	12.8	SA(s)0/a:	111	
02 01 21.7	−05 22 15	NGC 790	13.0	1.3 x 1.3	13.4	SA(r)0°?		Compact core, faint, smooth envelope.
02 03 31.3	−09 55 58	NGC 806	13.9	1.2 x 0.4	13.0	Scd pec?	60	Stellar galaxy **KUG 0200-101** NW 4′.4; mag 10.26v star W 6′.4.
02 04 19.1	−08 44 07	NGC 809	13.8	1.2 x 0.9	13.8	(R)S0$^+$:	179	
02 04 35.0	−10 06 31	NGC 811	14.6	1.1 x 0.6	14.0	SB(s)d: III	35	Stellar nucleus; mag 12.4 star W 3′.6; MCG −2-6-22 WNW 5′.7.
02 10 37.6	−15 46 25	NGC 814	13.8	1.3 x 0.4	12.9	SBbc	4	
02 02 54.3	−14 40 24	NGC 815	15.3	1.0 x 0.5	14.4	SB(s)m: III-IV	50	MCG −3-6-5 N 3′.8.
02 08 42.3	−07 47 23	NGC 829	13.5	1.2 x 0.8	13.3	SB(s)c pec?	60	**Mkn 1019** 2′.5 S.
02 08 58.8	−07 46 00	NGC 830	13.3	1.2 x 0.7	13.0	SB0$^-$?	72	Mag 9.92v star SSE 6′.9.
02 09 20.8	−10 07 58	NGC 833	12.7	1.5 x 0.7	12.6	(R′)Sa: pec	75	NGC 835 on E edge.
02 09 24.9	−10 08 10	NGC 835	12.1	1.3 x 1.0	12.2	SAB(r)ab: pec	15	Pair with NGC 833 on W edge; mag 9.58 star 2′.6 S.
02 09 38.7	−10 08 46	NGC 838	13.0	1.1 x 0.9	12.8	SA(rs)0° pec:	85	Located 3′.4 E of NGC 835.
02 09 42.9	−10 11 02	NGC 839	13.1	1.3 x 0.5	12.5	S0: pec sp	84	Mag 9.58v star W 4′.2.
02 09 50.9	−07 45 46	NGC 842	12.7	1.6 x 0.8	12.8	SAB(r)0?	150	Mag 10.73v star W 5′.5.
02 10 17.9	−10 19 18	NGC 848	13.0	1.5 x 1.0	13.3	(R′)SB(s)ab pec?	141	Mag 12 star NNE 1′.8.
02 11 41.2	−09 18 21	NGC 853	12.8	1.5 x 1.3	13.4	Sm pec? III	69	
02 15 25.2	−17 46 56	NGC 872	13.7	1.5 x 0.5	13.2	SB(s)c? II-III	174	
02 16 32.3	−11 20 57	NGC 873	12.6	1.6 x 1.1	13.0	Sc pec:	145	
02 16 51.0	−08 57 48	NGC 879	14.7	0.7 x 0.6	13.6	Irr IV:	5	Uniform surface brightness.
02 18 45.3	−06 38 25	NGC 881	12.4	2.4 x 1.3	13.5	SAB(r)c I	135	Mag 8.39v star WNW 5′.3.
02 19 05.1	−06 47 26	NGC 883	12.6	1.2 x 1.1	12.8	SA(s)0$^-$:	100	Star on S edge; mag 9.09v star S 3′.3.
02 19 32.7	−16 04 08	NGC 887	12.7	1.9 x 1.5	13.7	SAB(rs)c I	5	
02 21 36.5	−05 31 16	NGC 895	11.7	3.6 x 2.6	14.0	SA(s)cd I-II	126	Strong core with two prominent arms.
02 22 21.7	−16 40 46	NGC 902	14.0	0.6 x 0.5	12.6	SB(rs)bc?	15	
02 22 43.7	−08 43 09	NGC 905	15.1	0.5 x 0.3	12.9	S0/a	115	Stellar; mag 9.81v star N 4′.4.
02 26 33.5	−15 50 52	NGC 921	14.3	1.3 x 0.6	13.9	SB(r)c? II-III:	84	
02 27 18.3	−12 05 12	NGC 929	14.2	0.9 x 0.4	12.9	Sa pec?	170	Mag 9.73v star NNE 3′.4.
02 29 09.6	−10 49 40	NGC 942	13.4	1.7 x 0.9	13.7	S0$^+$ pec:	14	In collision, or strong reaction, with NGC 943 on S edge.
02 29 10.6	−10 50 08	NGC 943	13.7	1.3 x 0.9	13.7	I0? pec	30	Separation from NGC 942 0′.6.
02 26 41.5	−14 30 59	NGC 944	14.0	1.1 x 0.3	12.7	S0$^+$? sp	15	Mag 9.01v star NE 4′.3.
02 28 37.3	−10 32 19	NGC 945	12.1	2.4 x 2.0	13.6	SB(rs)c I-II	10	Mag 8.66v star S 5′.1; NGC 948 NE 2′.6.
02 28 45.6	−10 30 49	NGC 948	13.7	1.3 x 1.2	14.0	SB(s)c: II-III	12	Located 2′.6 NE of NGC 945.
02 29 11.7	−11 01 31	NGC 950	13.8	1.2 x 0.7	13.4	SB(rs)b: II-III:	40	Mag 8.06v star SSE 5′.2; MCG −2-7-20 W 2′.1.
02 31 41.3	−09 18 02	NGC 960	14.2	1.1 x 0.3	12.9	Sb:	125	

(Continued from previous page)

GALAXIES

RA h m s	Dec ° ′ ″	Name	Mag (V)	Dim ′ Maj x min	SB	Type Class	PA	Notes
02 32 12.8	−17 13 00	NGC 967	12.5	1.6 x 1.0	12.8	SAB0^{-}	33	Very faint, slightly elongated anonymous galaxy SW 8'.8.
02 33 03.5	−10 45 37	NGC 977	13.5	1.9 x 1.6	14.5	(R′)SAB(r)a:	26	Mag 8.75v star NNW 7'.7; faint, stellar galaxy **KUG 0231-110** ESE 10'.2.
02 32 59.8	−10 58 29	NGC 981	13.9	0.9 x 0.5	12.9	SC III:	5	
02 34 37.4	−08 47 09	NGC 985	13.4	1.0 x 0.9	13.1	Ring pec	69	Stellar nucleus, smooth envelope.
02 35 28.3	−09 21 28	NGC 988	11.0	4.1 x 1.6	12.9	SB(s)cd: II	114	Partially hidden behind the glare of mag 7.11v star 79 Ceti.
02 33 46.1	−16 30 43	NGC 989	14.0	0.8 x 0.7	13.2	S0	80	Compact; mag 9.66v star SSW 6'.0.
02 35 32.4	−07 09 18	NGC 991	11.7	3.0 x 2.7	13.8	SAB(rs)c II-III	60	
02 37 35.0	−11 01 32	NGC 1010	14.1	0.8 x 0.8	13.5	Sc		NGC 1011 NNE 1'.6.
02 37 38.8	−11 00 22	NGC 1011	14.3	0.6 x 0.6	13.0	Sa		NGC 1010 SSW 1'.6; NGC 1017 E 2'.7.
02 37 50.4	−11 30 26	NGC 1013	14.0	0.9 x 0.7	13.4	S0/a	75	Mag 8.15v star NNE 8'.8.
02 37 49.8	−11 00 36	NGC 1017	14.4	0.7 x 0.6	13.3	Sm	50	NGC 1011 W 2'.7.
02 38 10.4	−09 32 40	NGC 1018	14.2	1.0 x 0.6	13.5	(R′)SB(r)0/a	170	Stellar nucleus.
02 38 33.0	−06 40 38	NGC 1022	11.3	2.4 x 2.2	13.0	(R′)SB(s)a	12	Mag 10.28v star ENE 9'.9.
02 40 16.0	−08 46 36	NGC 1033	13.8	1.3 x 1.1	14.0	SA(s)c: III	0	Almost stellar nucleus; mag 8.58v star SW 7'.1.
02 38 14.0	−15 48 35	NGC 1034	14.0	0.9 x 0.7	13.3	Irr	142	Mag 11.12v star NW 11'.5; MCG −3-7-42 W 8'.1.
02 39 29.1	−08 07 56	NGC 1035	12.2	2.2 x 0.7	12.5	SA(s)c? III:	150	Mag 9.23v star S 7'.4.
02 40 25.2	−05 26 28	NGC 1041	13.3	1.7 x 1.2	13.9	S0^{-} pec	57	Small, faint galaxy **KUG 0237-055** NNW 7'.0.
02 40 23.6	−08 26 01	NGC 1042	11.0	4.7 x 3.6	13.9	SAB(rs)cd I-II	18	Almost stellar nucleus; NGC 1048 and NGC 1048A SSE 7'.1.
02 40 29.0	−11 16 41	NGC 1045	12.4	1.6 x 1.2	13.0	SA0^{-} pec?	40	Located on W edge of galaxy cluster **A 371**; mag 11.6 star N 6'.7.
02 40 32.9	−08 08 52	NGC 1047	13.5	1.3 x 0.6	13.1	S0^{+}: sp	95	NGC 1052 SE 10'.1.
02 40 38.0	−08 32 03	NGC 1048	14.5	1.0 x 0.4	13.4	S0^{+}? sp	101	
02 40 35.6	−08 32 50	NGC 1048A	14.5	0.8 x 0.4	13.1	SBb pec sp	0	
02 41 02.5	−06 56 11	NGC 1051	12.8	1.9 x 1.4	13.8	SB(rs)m pec III-IV	42	Mag 13.2 star NE edge of nucleus.
02 41 04.8	−08 15 21	NGC 1052	10.5	3.0 x 2.1	12.5	E4	126	NGC 1047 NW 10'.1.
02 42 10.1	−05 34 07	NGC 1063	13.8	1.4 x 0.5	13.2	(R′)SA(r)bc: II	111	Mag 10.80v star WSW 3'.9.
02 42 23.5	−09 21 44	NGC 1064	14.3	1.0 x 0.9	14.0	SB(s)c: II	30	Mag 12.0 star N 4'.3; MCG −2-7-72 SSE 6'.8.
02 42 06.3	−15 05 31	NGC 1065	14.1	0.8 x 0.7	13.3	E/S0	170	Mag 7.28v star W 8'.7; IC 253 N 2'.8.
02 42 59.8	−08 17 28	NGC 1069	13.5	1.4 x 0.9	13.6	SAB(s)c I-II	144	Almost stellar nucleus; mag 9.06v star E 5'.0.
02 43 07.8	−08 46 29	NGC 1071	14.4	1.1 x 0.5	13.6	SB(rs)a	160	Mag 10.06v star SW 7'.2.
02 43 36.2	−16 17 49	NGC 1074	14.3	1.9 x 1.2	15.1	SAB(r)ab pec:	167	Mag 11.03v star SE 5'.8; NGC 1075 N 5'.8.
02 43 33.6	−16 12 04	NGC 1075	14.3	0.8 x 0.6	13.4	S0/a	138	NGC 1074 S 5'.8.
02 43 29.4	−14 45 18	NGC 1076	12.7	1.9 x 1.1	13.3	SB0/a pec?	99	Mag 10.39v star E 5'.5; MCG −3-8-5 SE 6'.6.
02 44 08.1	−09 27 10	NGC 1078	14.3	0.8 x 0.7	13.6	E/S0	18	Almost stellar nucleus; mag 10.26v star NNW 5'.8.
02 45 05.6	−15 35 15	NGC 1081	13.3	1.7 x 0.6	13.2	SB(s)b? II	27	Mag 7.61v star N 9'.8.
02 45 41.3	−08 10 48	NGC 1082	14.0	0.9 x 0.6	13.2	S0	120	Small knot SE of core; stellar galaxy **Mkn 1053** ESE 7'.9.
02 45 40.7	−15 21 26	NGC 1083	14.4	1.6 x 0.3	13.4	Sb pec sp	21	Mag 7.61v star SW 7'.9.
02 45 59.9	−07 34 47	NGC 1084	10.7	2.8 x 1.4	12.1	SA(s)c II	30	Knotty; small, faint anonymous galaxy SSE 3'.7.
02 46 10.0	−15 04 24	NGC 1089	13.5	1.1 x 0.8	13.4	E	15	Mag 10.70v star SW 8'.2.
02 45 22.4	−17 32 01	NGC 1091	14.1	0.8 x 0.7	13.3	(R′)SAB(rs)a pec?	77	= **Hickson 21E**.
02 45 29.5	−17 32 30	NGC 1092	13.4	0.8 x 0.8	13.0	E?		= **Hickson 21D**.
02 44 53.6	−17 39 31	NGC 1098	12.6	1.8 x 1.3	13.4	SA(s)0^{-}:	102	= **Hickson 21C** . Mag 8.72v star N 5'.3; MCG −3-8-7 and MCG −3-8-6 SSW 4'.6.
02 45 17.8	−17 42 34	NGC 1099	13.1	1.8 x 0.6	13.0	SB(rs)b II	10	= **Hickson 21A**. Mag 9.18v star SW 6'.3.
02 45 36.0	−17 41 20	NGC 1100	13.0	1.7 x 0.7	13.1	SAB(r)a:	58	= **Hickson 21A.** Mag 8.72v star NW 10'.2.
02 48 05.6	−13 57 42	NGC 1103	12.9	2.1 x 0.5	12.8	SB(s)b: II-III	45	IC 1853 S 1'.9; mag 10.79v star SSE 9'.1.
02 52 17.8	−15 42 37	NGC 1105	14.3	0.8 x 0.6	13.4	S0/a	116	
02 48 38.5	−07 57 04	NGC 1108	13.9	0.8 x 0.4	12.5	E/S0	85	Almost stellar nucleus; NGC 1110 NE 10'.3.
02 49 09.6	−07 50 12	NGC 1110	13.9	3.0 x 0.5	14.2	SB(s)m: sp III-IV?	18	Mag 10.82v star S 4'.8; NGC 1108 SE 10'.3.
02 49 07.3	−16 59 42	NGC 1114	12.5	1.7 x 0.8	12.6	SA(r)c I-II	9	
02 49 58.9	−12 09 50	NGC 1118	13.2	2.1 x 0.6	13.3	(R′)SB(r)0°?	90	Mag 10.34v star W 9'.0; MCG −2-8-10 SW 7'.0.
02 48 17.0	−17 59 15	NGC 1119	14.2	0.5 x 0.4	12.3	(R′)SB0°?	0	Mag 10.15v star E 3'.0.
02 49 04.2	−14 28 15	NGC 1120	13.6	1.5 x 0.9	13.8	S0^{-}	40	MCG −3-8-30 E 7'.6.
02 51 40.4	−16 39 01	NGC 1125	12.6	1.8 x 0.9	13.0	(R′)SB(r)0/a?	51	Mag 10.40v star W 13'.5; MCG −3-8-34 on SW edge.
02 52 42.2	−08 48 17	NGC 1133	14.0	1.1 x 0.8	13.7	Sa	157	Mag 7.90v star E 11'.5.
02 52 47.0	−14 31 46	NGC 1139	14.6	1.0 x 0.9	14.4	SB(r)0/a pec:	36	MCG −3-8-37 W 6'.1.
02 54 33.5	−10 01 44	NGC 1140	12.5	1.9 x 0.8	12.8	IBm pec:	169	Small companion galaxy, or small streamer, SW edge, protruding W.
02 57 04.3	−07 41 09	NGC 1148	15.3	1.3 x 0.9	15.3	SB(rs)d pec? III-IV	84	Stellar nucleus; mag 9.03v star W 9'.4.
02 57 01.3	−15 02 55	NGC 1150	14.1	1.0 x 0.7	13.6	S0	73	Mag 7.90v star W 8'.1.
02 57 04.7	−15 00 49	NGC 1151	15.0	0.5 x 0.4	13.4	E	10	Mag 7.90v star W 8'.7.
02 57 33.6	−07 45 31	NGC 1152	13.5	1.1 x 0.7	13.1	S0	29	Mag 12.0 star S 2'.4; NGC 1148 NW 8'.5.
02 58 07.8	−10 21 48	NGC 1154	13.2	1.0 x 0.6	12.5	SB	124	Mag 9.75v star S 5'.0; NGC 1155 NE 1'.5.
02 58 13.0	−10 21 01	NGC 1155	13.4	1.0 x 0.9	13.1	S0	165	Compact core, faint, uniform envelope; NGC 1154 SW 1'.5.
02 58 06.7	−15 07 08	NGC 1157	14.8	0.5 x 0.2	12.1	Sb	169	
02 57 11.4	−14 23 45	NGC 1158	14.8	0.9 x 0.6	14.0	S0	147	Compact core, uniform envelope; mag 12.3 star E 3'.8.
02 58 56.2	−12 23 53	NGC 1162	12.5	1.4 x 1.4	13.2	E:		Bright, round core, uniform envelope.
03 00 22.2	−17 09 09	NGC 1163	13.8	2.2 x 0.3	13.2	SBbc? sp	141	
03 01 36.1	−14 50 11	NGC 1172	11.9	2.3 x 1.8	13.3	E^{+}:	27	Round core, uniform envelope; mag 10.45v star E 2'.0.
03 01 51.1	−15 01 49	NGC 1180	14.9	0.6 x 0.4	13.4	E/S0	5	
03 01 42.9	−15 03 09	NGC 1181	15.4	0.8 x 0.2	13.3	Sb	100	
03 03 28.5	−09 40 14	NGC 1182	13.9	0.8 x 0.4	12.5	Sa	115	Stellar **PGC 11535** SSE 7'.9.
03 02 59.4	−09 07 55	NGC 1185	13.7	1.2 x 0.4	12.8	SB(r)b? II-III?	30	
03 03 43.3	−15 29 08	NGC 1188	13.8	1.1 x 0.6	13.2	SAB(r)0°:	170	Faint, elongated anonymous galaxy E 9'.3.
03 03 24.3	−15 37 22	NGC 1189	13.8	1.7 x 1.5	14.7	SB(s)dm: IV-V	96	Almost stellar nucleus, smooth envelope.
03 03 26.2	−15 39 43	NGC 1190	14.2	0.7 x 0.3	12.4	S0°: sp	95	
03 03 31.0	−15 41 08	NGC 1191	14.3	0.4 x 0.3	11.9	SB0^{-}:	60	
03 03 34.6	−15 40 45	NGC 1192	14.4	0.5 x 0.2	11.9	E?	102	
03 03 32.8	−12 02 24	NGC 1195	14.5	0.6 x 0.5	13.1	E/S0		Almost stellar nucleus; mag 13.8 star SE 0'.7; NGC 1196 S 2'.3.
03 03 35.5	−12 04 34	NGC 1196	13.2	1.4 x 1.4	13.8	SB(rs)0°		NGC 1195 N 2'.3.
03 03 38.2	−15 36 50	NGC 1199	11.4	2.4 x 1.9	13.0	E3:	63	Faint anonymous galaxy on S edge.
03 03 54.5	−11 59 29	NGC 1200	12.0	2.9 x 2.3	13.9	SA(s)0^{-}	93	IC 285 SE 3'.2.

GALAXY CLUSTERS

RA h m s	Dec ° ′ ″	Name	Mag 10th brightest	No. Gal	Diam ′	Notes
01 54 42.0	−06 16 00	A 274	16.3	140	11	All members very faint and stellar.
01 55 48.0	−07 22 00	A 277	15.6	50	15	Numerous stellar anonymous GX,, spread out mostly S and SW of center.
02 30 12.0	−13 11 00	A 358	15.6	34	39	Numerous faint, stellar anonymous GX.
02 29 42.0	−17 02 00	AS 268	16.1	15	17	Except MCG −3-7-23, all members anonymous, stellar.
02 31 24.0	−17 38 00	AS 273	16.0		17	Except MCG −3-7-27, all members anonymous, stellar.

RA h m s	Dec ° ′ ″	Name	Mag (V)	Dim ′ Maj x min	SB	Type Class	PA	Notes
01 20 36.0	−17 20 27	ESO 542-6	13.4	1.8 x 1.0	13.9	SAB(s)0⁻:	177	ESO 542-7, ESO 542-8 S 2′8.
01 20 35.8	−17 23 14	ESO 542-7	13.9	1.1 x 0.4	12.8	SBb pec?	88	Pair with ESO 542-8 on S edge; ESO 542-6 N 2′8.
01 20 35.5	−17 23 48	ESO 542-8	13.6	1.5 x 0.6	13.3	Sb pec sp	147	Pair with ESO 542-7 on N edge.
01 34 53.8	−17 51 40	ESO 542-30	14.0	0.6 x 0.5	12.5		129	Mag 10.32v star NNW 10′1.
00 38 34.2	−15 21 32	IC 37	14.9	0.8 x 0.4	13.5	SBb	171	Mag 15.5 star on E edge; IC 38 S 3′8.
00 38 38.9	−15 25 13	IC 38	14.0	0.7 x 0.6	12.9	Sbc	84	IC 37 N 3′8.
00 41 05.9	−15 25 41	IC 42	14.8	0.8 x 0.5	13.6	SBc	45	Mag 9.28v star NW 8′6.
00 43 34.4	−08 11 11	IC 48	13.1	1.0 x 0.8	12.8	SAB(rs)0° pec?	25	Almost stellar; **MCG −1-3-4A** NE 2′8.
00 46 05.7	−09 30 11	IC 50	14.1	0.7 x 0.7	13.3	E		Stellar; mag 8.50v star SW 7′5.
00 46 24.3	−13 26 29	IC 51	12.8	1.5 x 1.1	13.2	S0 pec?	30	Mag 11.7 star S 9′5.
00 51 30.1	−12 50 42	IC 56	14.3	0.7 x 0.6	13.2	SAB(r)cd: III	9	Mag 8.04v star SW 7′6; **IC 56A** NE 6′9.
00 55 02.5	−13 40 43	IC 58	14.4	0.7 x 0.4	13.1	E	111	Mag 10.32v star N 8′4.
00 56 04.3	−13 21 30	IC 60	14.5	0.7 x 0.6	13.4		60	MCG −2-3-42 SW 9′8.
01 08 51.4	−15 24 24	IC 77	12.9	0.5 x 0.3	11.0	D	90	Very small, faint anonymous galaxy contact SW edge and 0′5 NW.
01 08 47.7	−15 50 44	IC 78	13.5	1.7 x 0.7	13.6	SA(rs)a pec:	124	Stellar anonymous galaxy W 1′9; **MCG −3-4-6** W 8′7.
01 08 49.8	−15 56 57	IC 79	14.2	0.7 x 0.7	13.3	SA0⁻		Mag 15.1 star on N edge; stellar **PGC 4065** W 2′2.
01 08 52.5	−15 25 21	IC 80A		0.5 x 0.3		S?	57	**MCG −3-4-8** W 2′1; numerous small, faint PGC and anonymous galaxies in surrounding area.
01 08 53.3	−15 20 15	IC 80B	15.0	0.4 x 0.4	12.9	SA0⁻ pec:		Almost stellar, numerous small, faint PGC and anonymous galaxies SW.
01 09 05.9	−16 00 02	IC 82	13.8	0.8 x 0.7	13.1	S	111	Mag 10.35v star SE 6′0; **PGC 4100** SW 0′9.
01 16 30.4	−07 58 40	IC 90	13.6	0.9 x 0.7	13.2	E⁺	144	Mag 12.3 star W 6′5.
01 19 02.5	−17 03 41	IC 93	13.2	1.3 x 0.5	12.6	Sb pec:	170	Mag 8.44v star E 11′5; MCG −3-4-44 N 2′2.
01 20 54.9	−12 36 17	IC 98	14.8	0.7 x 0.7	13.8			Stellar; mag 8.78v star S 8′4; MCG −2-4-28 NNE 2′4.
01 22 27.5	−12 57 05	IC 99	14.7	0.5 x 0.5	13.0	Sc		Mag 7.83v star E 8′6.
01 24 39.1	−12 38 11	IC 108	14.2	0.9 x 0.3	12.7	S0/a	3	Mag 9.37v star NNW 4′6.
01 29 18.3	−13 16 48	IC 125	14.6	0.5 x 0.5	13.0			Mag 7.16v star NE 6′5.
01 29 47.8	−06 58 52	IC 127	13.7	1.6 x 0.4	13.1	Sb: sp	120	Mag 12.9 star on S edge; mag 9.70v star W 4′2.
01 31 23.9	−12 37 26	IC 128	14.6	0.8 x 0.6	13.6	SBc	165	Mag 11.4 star NE 9′3; IC 129 SE 2′4.
01 31 31.2	−12 39 17	IC 129	14.1	1.3 x 0.7	13.9	(R′)SAB(rs)0⁺?	60	IC 128 NW 2′4; mag 11.4 star NE 9′4.
01 31 28.9	−15 35 30	IC 130	14.2	0.5 x 0.5	12.6	S0		Mag 8.41v star NE 17′3.
01 32 51.8	−14 48 54	IC 141	13.6	1.3 x 1.0	13.7	SB(rs)bc: II		Mag 11.25v star SW 8′9.
01 37 40.8	−13 18 55	IC 144	14.9	0.7 x 0.7	13.9			Mag 11.7 star SE 5′9.
01 39 59.9	−14 51 50	IC 147	14.5	0.8 x 0.6	13.5	Sbc	69	Mag 7.83v star E 11′9.
01 42 25.3	−16 18 02	IC 149	14.2	1.1 x 0.4	13.2		81	Mag 12.2 star 0′9 NE.
01 46 24.7	−08 38 19	IC 159	13.2	1.4 x 0.7	13.0	SB(rs)b pec:	36	Mag 10.73v star E 4′8.
01 46 29.5	−13 14 54	IC 160	13.9	1.4 x 0.9	14.1	SAB0⁻:	85	Mag 10.11v star SSE 6′5.
01 50 27.6	−08 31 23	IC 168	14.2	0.9 x 0.3	12.6	Sa	105	Mag 12.0 star E 3′8.
01 51 57.4	−08 31 04	IC 170	14.4	0.8 x 0.5	13.4	E		NGC 707 W 7′5.
00 39 00.3	−09 00 56	IC 1563	14.0	0.7 x 0.6	12.9	S0 pec sp	143	Located on S edge of NGC 191.
00 55 51.9	−09 59 08	IC 1602	14.3	0.7 x 0.6	13.4	E	174	NGC 309 ENE 13′3.
01 01 42.6	−15 34 06	IC 1610	12.9	1.2 x 1.1	13.1	(R′)SAB0⁺:	110	Mag 12.6 star on N edge; mag 11.9 star S 3′9.
01 07 36.8	−17 32 20	IC 1622	13.7	0.7 x 0.5	12.4	S?	165	Mag 9.89v star SSE 8′0; IC 1623 and IC 1623A NE 3′0; **ESO 541-21** NW 4′5.
01 07 46.8	−17 30 27	IC 1623	13.2	0.8 x 0.6	12.2	Irr	12	Pair with NGC 1623A E edge; IC 1622 SW 3′0.
01 07 48.3	−17 30 24	IC 1623A	13.4	0.4 x 0.4	11.3	S pec:		Pair with IC 1623 W edge.
01 18 42.4	−17 03 01	IC 1667	13.6	1.0 x 0.9	13.3		69	Mag 11.8 star W 7′4; IC 93 E 4′9.
01 18 48.9	−16 48 15	IC 1670A	13.6	1.9 x 0.4	13.1	Sbc? sp	123	Pair with IC 1670B E edge; mag 11.9 star N 3′1.
01 18 52.9	−16 48 12	IC 1670B	13.5	1.4 x 0.4	12.8	S0⁺ pec:	94	Pair with IC 1670A W edge; mag 11.9 star N 3′1.
01 51 07.8	−09 47 32	IC 1738	13.9	0.9 x 0.7	13.2	(R′)SAB(rs)b II	80	Mag 9.83v star E 9′2; NGC 701 N 5′4.
01 51 56.7	−16 47 17	IC 1741	14.2	0.8 x 0.6	13.3	S0/a	171	Mag 8.49v star WNW 7′3.
00 33 04.0	−08 06 54	MCG −1-2-30	14.9	1.0 x 0.6	14.2	SB(s)dm: III-IV	39	
00 34 40.8	−07 54 13	MCG −1-2-34	13.3	1.1 x 0.7	12.9	Sc	57	
00 36 08.9	−07 32 16	MCG −1-2-39	14.4	0.9 x 0.6	13.6	SAB(s)dm? IV-V	144	
00 36 22.5	−06 38 52	MCG −1-2-41	14.7	0.5 x 0.2	12.1	S0/a	120	Mag 11.7 star NE 1′1; mag 8.15v star NE 9′7.
00 41 16.9	−05 17 06	MCG −1-2-45	14.4	1.1 x 0.8	14.1	Sbc	162	
00 43 18.0	−06 38 36	MCG −1-2-47	14.0	0.7 x 0.7	13.1	Sbc		
00 43 19.3	−06 39 57	MCG −1-2-48	14.8	0.7 x 0.3	13.0	Sc	177	
00 43 28.7	−06 20 58	MCG −1-2-49	13.9	2.2 x 0.4	13.6	Sbc sp	99	Mag 7.40v star W 2′5.
00 45 58.7	−07 17 00	MCG −1-3-10	13.3	1.4 x 0.5	12.7	Sbc	99	
00 46 49.0	−06 31 53	MCG −1-3-11	14.1	1.1 x 0.3	12.8	S0	36	Mag 8.56v star W 8′4.
00 47 30.8	−05 04 47	MCG −1-3-14	14.8	1.0 x 0.3	13.3	Sb	120	Mag 9.83v star W 4′7.
00 50 27.6	−05 51 33	MCG −1-3-18	13.1	0.8 x 0.8	12.5	(R′)SA(r)0⁻?		
00 56 14.9	−07 40 41	MCG −1-3-37	15.0	0.8 x 0.5	13.8	Sc	63	Stellar nucleus; mag 7.73v star N 6′1.
00 57 35.1	−05 00 08	MCG −1-3-41	13.9	1.0 x 0.6	13.2	SB(r)c pec II-III	38	MCG −1-3-42 NNE 3′7.
00 58 10.8	−08 13 06	MCG −1-3-49	13.9	1.4 x 1.0	14.1	SAB(s)0⁻?	57	
00 58 08.7	−06 14 59	MCG −1-3-50	15.0	1.2 x 0.3	13.7	S?	36	
01 00 38.0	−07 58 46	MCG −1-3-56	13.6	1.1 x 1.1	13.6	SAB(rs)c II		
01 00 48.8	−06 55 14	MCG −1-3-57	16.0	0.8 x 0.6	15.1	SBd	30	Located 4′7 SE of NGC 340.
01 00 53.4	−07 02 29	MCG −1-3-59	14.9	0.9 x 0.8	14.4	SBc	51	
01 01 35.6	−05 58 33	MCG −1-3-66	15.2	0.9 x 0.5	14.1	Sc	21	Mag 8.94v star E 1′6.
01 01 59.1	−06 23 11	MCG −1-3-70	15.2	1.3 x 0.3	14.0	Sc	177	Mag 10.63v star ESE 2′2.
01 02 18.2	−06 52 27	MCG −1-3-73	14.9	0.6 x 0.4	13.2		126	Almost stellar.
01 02 36.2	−06 23 35	MCG −1-3-74	14.7	0.5 x 0.4	12.8		90	Almost stellar; MCG −1-3-75 E 3′2.
01 02 48.5	−06 24 47	MCG −1-3-75	14.4	1.3 x 0.3	13.2		144	Anonymous galaxies E and SE 1′9; MCG −1-3-74 W 3′2.
01 05 04.9	−06 12 46	MCG −1-3-85	11.6	4.4 x 3.3	14.4	SAB(rs)d IV	112	MCG −1-3-88 S 4′1; mag 13.0 star E 3′7.
01 05 01.9	−06 44 11	MCG −1-3-86	14.6	1.0 x 0.4	13.5	Sbc	165	
01 05 06.2	−05 19 54	MCG −1-3-87	14.3	1.0 x 0.7	13.8	SBc	174	
01 05 08.8	−06 16 47	MCG −1-3-88	15.1	1.6 x 0.3	14.1	SBm? sp	75	MCG −1-3-85 N 4′1.
01 05 18.9	−05 25 23	MCG −1-3-89	14.5	1.1 x 0.8	14.2	Scd	108	
01 06 20.4	−06 11 51	MCG −1-3-91	14.4	0.5 x 0.5	12.9	E/S0		Compact; stellar anonymous galaxy E 1′4.
01 08 27.1	−05 18 25	MCG −1-4-2	14.5	1.1 x 0.3	13.1	S?	123	
01 09 00.0	−05 31 06	MCG −1-4-3	15.7	1.0 x 0.4	14.5	IB(s)m: V	18	Low uniform surface brightness.
01 10 37.0	−07 37 14	MCG −1-4-7	13.5	1.3 x 1.1	13.8	SB(s)0°:	168	Two stars superimposed, one just S of the nucleus, the other just E.
01 10 53.8	−05 49 17	MCG −1-4-8	14.0	2.4 x 1.0	14.8	SA(s)c: II-III	162	Mag 11.8 star SSE 2′5.
01 11 12.2	−07 32 49	MCG −1-4-9	13.9	1.0 x 0.7	13.3	SAc: III	102	Star at SE edge.
01 11 22.7	−06 49 46	MCG −1-4-10	14.6	0.8 x 0.6	13.7	Sbc	129	
01 15 37.0	−06 35 58	MCG −1-4-18	14.5	0.3 x 0.3	11.8	S0		Almost stellar, on S edge of MCG −1-4-19.
01 15 36.9	−06 35 14	MCG −1-4-19	14.3	1.1 x 0.3	13.0	SB(s)c pec?	141	Weak nucleus with smooth envelope.
01 15 57.1	−06 23 01	MCG −1-4-20	14.4	1.0 x 0.7	14.0	E	9	
01 16 07.4	−06 37 20	MCG −1-4-21	14.5	0.5 x 0.5	12.8	SBc		Mag 9.60v star N 4′5.
01 16 07.1	−06 43 24	MCG −1-4-22	14.0	1.6 x 0.5	13.6	Sb: II	66	Mag 11.25v star S 2′6.

(Continued from previous page)

GALAXIES

RA h m s	Dec ° ′ ″	Name	Mag (V)	Dim ′ Maj x min	SB	Type Class	PA	Notes
01 18 42.6	−07 27 01	MCG −1-4-25	14.2	1.0 x 1.0	14.1	S?		Weak stellar nucleus in uniform envelope.
01 21 51.5	−07 18 50	MCG −1-4-29	14.0	0.5 x 0.5	12.4	S0/a		
01 23 12.1	−06 38 12	MCG −1-4-32	13.3	0.6 x 0.5	11.8		30	**MCG −1-4-31** W 2′5.
01 24 15.1	−05 58 49	MCG −1-4-35	15.6	1.1 x 0.2	13.8	Sab	147	Mag 12.1 star SE 1′7; mag 6.79v star E 11′5.
01 24 16.8	−05 54 51	MCG −1-4-36	15.3	0.9 x 0.3	13.7	Scd	69	Mag 10.59v star NE 1′9; mag 6.79v star E 11′2.
01 25 36.5	−07 44 26	MCG −1-4-41	14.3	0.9 x 0.4	13.0	S0/a	63	
01 26 14.3	−06 05 42	MCG −1-4-44	13.9	3.2 x 0.5	14.3	SBc: II	24	Mag 13.4 star NE edge.
01 26 20.3	−08 05 23	MCG −1-4-47	15.6	0.8 x 0.6	14.6	S?	150	Star N end.
01 27 24.3	−06 08 37	MCG −1-4-51	13.6	0.9 x 0.5	12.6	S0	60	
01 32 20.6	−07 51 38	MCG −1-5-2	14.8	1.1 x 0.3	13.4	SA(s)b:	105	= **Hickson 13A**. **Hickson 13B** and 13C SSE 1′.1. **Hickson 13D** S 2′.2.
01 33 00.5	−05 00 10	MCG −1-5-6	14.3	1.0 x 0.5	13.3	Sc	150	
01 37 38.5	−05 03 45	MCG −1-5-10	14.2	0.8 x 0.6	13.5	E	30	
01 38 02.5	−05 43 51	MCG −1-5-11	14.6	0.6 x 0.4	12.9	Sb	105	Mag 7.34v star E 10′5; mag 9.79v star N 4′0.
01 38 00.0	−07 34 04	MCG −1-5-12	14.4	1.0 x 0.5	13.5	Sbc	54	
01 40 27.9	−05 31 13	MCG −1-5-14	14.3	1.6 x 0.4	13.7	SAB(s)d: III	171	Mag 13.3 star S 1′5; mag 8.49v star NNE 10′4.
01 40 25.6	−07 54 09	MCG −1-5-16	14.0	1.2 x 0.3	12.7	S pec:	57	**MCG −1-15-15** SW 1′2.
01 41 06.7	−05 34 09	MCG −1-5-17	13.4	2.1 x 0.8	13.8	SAB(s)m: III-IV	72	Mag 8.49v star N 10′0.
01 42 51.0	−06 08 39	MCG −1-5-22	14.7	1.1 x 0.6	14.1	SB?	144	Star E edge.
01 44 38.8	−07 37 13	MCG −1-5-28	13.8	1.3 x 1.0	13.9	(R′)SA(r)0⁺:	0	Compact core, faint, smooth envelope.
01 50 00.0	−06 29 19	MCG −1-5-39	14.5	1.3 x 0.4	13.6	S?	117	Stellar **MCG −1-5-38** N 2′8.
01 51 28.4	−06 30 59	MCG −1-5-44	14.7	1.5 x 0.4	14.0	SAd pec sp III	90	Mag 12.3 star ENE 7′0.
01 51 50.3	−05 29 48	MCG −1-5-45	14.7	1.6 x 0.5	14.3	SB(s)dm III	66	**MCG −1-5-46** SE edge.
00 32 32.9	−12 03 42	MCG −2-2-46	14.4	0.4 x 0.4	12.3			0′3 N mag 12 star.
00 32 45.3	−10 27 06	MCG −2-2-47	15.2	0.6 x 0.4	13.5	Sb	36	
00 32 49.2	−10 00 52	MCG −2-2-48	14.2	0.7 x 0.7	13.3	IB(s)m		Mag 8.55v star E 4′8.
00 32 42.6	−11 19 09	MCG −2-2-49	13.8	1.9 x 0.2	12.6	Sc: sp	33	Mag 10.82v star W 8′3.
00 33 20.5	−09 54 12	MCG −2-2-50	14.4	0.5 x 0.4	12.5		63	Mag 11.0 star on SW end; mag 8.55v star SW 6′1.
00 33 10.3	−13 08 48	MCG −2-2-51	13.9	1.3 x 0.7	13.6	(R′)SB(rs)bc I-II	129	Two main arms for open "S" shape; mag 8.75v star SW 8′3.
00 33 24.3	−14 00 23	MCG −2-2-52	14.8	0.7 x 0.6	13.7		162	
00 34 57.9	−09 20 35	MCG −2-2-57	14.3	0.7 x 0.4	12.8	S?	144	
00 34 54.8	−12 29 55	MCG −2-2-58	14.0	1.0 x 0.7	13.5	SAB(r)c III	9	
00 35 02.9	−09 22 08	MCG −2-2-59	14.4	0.7 x 0.4	12.9	Sbc	33	
00 35 35.0	−12 17 16	MCG −2-2-60	14.0	0.5 x 0.4	12.3	E	51	Almost stellar.
00 35 36.9	−11 27 37	MCG −2-2-61	15.3	0.7 x 0.4	13.8	S0/a	165	
00 35 48.3	−08 35 28	MCG −2-2-62	14.8	0.5 x 0.4	12.9	Sc	141	
00 36 01.9	−09 53 27	MCG −2-2-64	13.9	1.1 x 0.9	13.7	SB(s)c pec: II-III	18	
00 36 23.9	−11 45 34	MCG −2-2-67	14.8	0.9 x 0.3	13.2	Sc	9	Mag 9.43v star NE 8′7.
00 36 12.6	−12 15 32	MCG −2-2-68	13.7	1.2 x 0.6	13.2	SAB(r)a:	141	Compact core in faint, smooth envelope.
00 37 04.3	−12 40 56	MCG −2-2-70	15.0	0.6 x 0.3	13.0	Sb	60	
00 37 42.4	−14 10 49	MCG −2-2-71	14.8	1.0 x 0.4	13.6		165	
00 38 02.5	−13 51 57	MCG −2-2-72	14.7	1.1 x 1.1	14.8	SBbc		Low uniform surface brightness.
00 37 57.4	−09 15 11	MCG −2-2-73	13.0	1.3 x 1.1	13.3	SAB(s)cd II	115	Mag 8.13v star NNW 11′9.
00 38 38.6	−11 02 55	MCG −2-2-74	15.2	0.7 x 0.3	13.3		135	
00 39 20.1	−10 28 59	MCG −2-2-75	15.0	0.6 x 0.3	13.0	Sb	171	
00 40 29.4	−10 18 23	MCG −2-2-80	13.9	0.6 x 0.4	12.2	Sc	75	Mag 10.85v star S 2′0.
00 40 55.1	−13 46 32	MCG −2-2-82	13.4	1.1 x 0.9	13.2	SBa	9	NGC 210 SSW 7′6.
00 41 12.9	−13 32 44	MCG −2-2-83	14.1	1.3 x 1.3	14.5	IB(s)m IV-V		
00 41 50.5	−09 18 13	MCG −2-2-86	13.7	0.8 x 0.6	12.8	SB(r)0° pec?	144	Near center of galaxy cluster A 85; **MCG −2-2-87** and **MCG −2-2-88** S 11′5.
00 42 07.4	−14 07 22	MCG −2-2-89	14.3	1.3 x 0.7	14.0	Sc	63	
00 42 14.9	−11 28 21	MCG −2-2-92	13.9	1.1 x 0.6	13.3	SA(rs)0⁻:	75	
00 43 58.5	−13 46 05	MCG −2-3-2	14.1	0.5 x 0.5	12.7	E/S0		Almost stellar nucleus, faint envelope.
00 43 59.6	−11 18 25	MCG −2-3-3	14.3	0.5 x 0.4	12.4		6	Compact.
00 44 12.9	−12 35 31	MCG −2-3-4	14.4	1.0 x 0.2	12.6	S?	0	
00 44 34.5	−08 20 56	MCG −2-3-5	14.2	1.0 x 0.4	13.0	Sc	162	Star at S tip.
00 45 09.1	−09 37 54	MCG −2-3-6	14.3	0.7 x 0.5	13.0	Sc	12	Star at S tip.
00 45 24.0	−12 32 14	MCG −2-3-7	15.0	0.5 x 0.3	12.8	S?	63	
00 45 52.0	−09 19 43	MCG −2-3-8	14.3	0.8 x 0.8	13.7	Sc		
00 46 03.2	−11 30 17	MCG −2-3-9	13.7	2.2 x 1.7	15.0	SAB(s)m V	103	Very low surface brightness; mag 7.69v star NE 4′2.
00 46 16.3	−09 17 23	MCG −2-3-12	14.5	1.1 x 0.3	13.2	S0/a	0	
00 47 24.8	−11 17 23	MCG −2-3-13	14.8	0.7 x 0.5	13.5	SBc	147	Mag 8.00v star NW 10′0.
00 47 25.6	−09 51 13	MCG −2-3-14	15.9	0.7 x 0.2	13.6	Sbc	93	Mag 9.18v star SW 7′7.
00 47 46.4	−09 50 07	MCG −2-3-15	13.6	1.7 x 0.5	13.3	(R′)SBb: II	21	MCG −2-3-16 S 3′9.
00 47 47.1	−09 54 01	MCG −2-3-16	13.6	2.6 x 0.4	13.5	SBd pec sp	84	Mag 9.18v star WSW 11′1; MCG −2-3-15 N 3′9.
00 47 48.4	−09 22 47	MCG −2-3-18	13.6	1.0 x 0.6	12.9	Sc	24	Several bright knots or stars at N end.
00 48 35.5	−12 43 03	MCG −2-3-19	13.8	0.7 x 0.5	12.5	(R)SB0° pec?	18	Stars or bright knots E and NE edges.
00 48 45.7	−11 45 59	MCG −2-3-20	15.9	1.0 x 0.3	14.5	SB(rs)b? III	57	Star SW end.
00 49 07.5	−12 29 18	MCG −2-3-22	13.5	0.5 x 0.3	11.3		63	
00 49 07.5	−12 27 56	MCG −2-3-23	14.9	0.6 x 0.4	13.2	Sc	138	
00 49 56.5	−12 13 07	MCG −2-3-25	14.4	0.4 x 0.4	12.5	E?		Mag 9.52v star W 4′8; **MCG −2-3-24** S 5′2.
00 50 05.8	−12 11 04	MCG −2-3-26	15.6	0.5 x 0.3	13.6	E	39	Mag 9.52v star W 7′2.
00 51 23.0	−08 31 09	MCG −2-3-29	15.4	0.8 x 0.3	13.7	SB(rs)d III-IV	66	NGC 277 S 4′9.
00 53 45.1	−09 06 38	MCG −2-3-36	14.4	0.8 x 0.5	13.2	Sc	63	
00 53 48.6	−13 51 18	MCG −2-3-37	14.3	0.8 x 0.4	13.1	E?	3	Almost stellar galaxies **Arp 251B** SSW 0′4, and **Arp 251C** SSE 0′4.
00 54 07.2	−10 50 31	MCG −2-3-38	14.6	0.5 x 0.5	12.9	Sc		Mag 10.63v star W 3′6.
00 54 46.9	−10 58 23	MCG −2-3-39	15.4	0.9 x 0.2	13.4	Sbc	150	
00 54 56.9	−11 03 08	MCG −2-3-40	15.2	0.7 x 0.6	14.1		42	Stellar nucleus; mag 8.61v star E 10′5; MCG −2-3-39 NW 5′3.
00 55 32.8	−13 27 42	MCG −2-3-42	15.1	0.9 x 0.4	13.8		12	Mag 10.32v star SW 9′9; IC 60 NE 9′8.
00 55 28.9	−10 56 14	MCG −2-3-44	14.8	1.0 x 0.7	14.3			Mag 8.61v star SSE 7′7.
00 55 35.5	−10 05 25	MCG −2-3-45	15.0	0.5 x 0.4	13.1		177	Almost stellar nucleus; mag 12.4 star NW 6′1.
00 55 45.2	−14 08 59	MCG −2-3-46	14.5	0.5 x 0.4	12.6	Sc	129	Stellar anonymous galaxy SE 1′8; **MCG −2-3-48** NE 7′0.
00 58 08.1	−10 07 13	MCG −2-3-51	14.5	0.7 x 0.7	13.5	S0/a		
00 58 23.9	−08 24 29	MCG −2-3-52	13.0	1.5 x 1.1	13.4	(R′)SB(r)a pec?	111	Bright diffuse center, smooth envelope.
00 58 30.8	−10 09 29	MCG −2-3-53	14.6	1.2 x 0.3	13.4	SB(s)m? sp	39	
00 58 38.8	−10 07 09	MCG −2-3-54	16.1	0.8 x 0.3	14.4	SBbc	102	
00 58 56.6	−10 55 48	MCG −2-3-55	15.0	0.8 x 0.7	14.3	S?	105	Mag 13.7 star on SW edge; mag 9.35v star N 5′7.
00 59 09.2	−13 45 52	MCG −2-3-56	15.7	1.3 x 0.2	14.0	Scd	117	
00 59 17.5	−13 41 31	MCG −2-3-57	15.2	0.5 x 0.5	13.6			Mag 8.96v star W 7′2.
00 59 37.3	−10 19 58	MCG −2-3-58	14.5	0.8 x 0.4	13.2	S0	0	
00 59 35.3	−12 26 22	MCG −2-3-59	14.4	1.1 x 0.3	13.1	S?	30	

(Continued from previous page)

GALAXIES

RA h m s	Dec ° ′ ″	Name	Mag (V)	Dim ′ Maj x min	SB	Type Class	PA	Notes
00 59 59.3	−13 59 48	MCG −2-3-60	14.6	0.6 x 0.4	13.0	S0	105	
01 00 04.1	−11 04 58	MCG −2-3-61	13.4	1.5 x 1.3	14.0	(R′)SAB(r)bc I	138	**MCG −2-3-62** NE 3′.4.
01 00 47.7	−09 11 18	MCG −2-3-64	14.4	0.3 x 0.2	11.2		9	On E edge of NGC 341.
01 01 19.5	−09 50 43	MCG −2-3-66	14.3	0.7 x 0.6	13.2	Sc	78	Located on NE edge of small galaxy cluster **A 129**.
01 02 40.4	−12 51 01	MCG −2-3-67	14.4	0.5 x 0.5	12.9	Compact		= **Mkn 969**.
01 03 23.7	−10 51 16	MCG −2-3-68	13.9	1.0 x 0.4	12.7		165	
01 04 02.8	−11 38 34	MCG −2-3-69	15.3	0.2 x 0.2	11.6			
01 04 53.7	−11 11 35	MCG −2-3-70	14.7	0.9 x 0.6	13.9	Sbc		
01 05 19.1	−11 21 31	MCG −2-3-71	14.3	0.8 x 0.5	13.1	S0/a		
01 07 32.9	−13 16 19	MCG −2-3-74	15.1	0.3 x 0.3	12.4	I:		Stellar; located on outside SW edge of small galaxy cluster **A 148**.
01 08 45.4	−11 48 58	MCG −2-4-1	14.7	0.5 x 0.4	12.8		33	Stellar.
01 08 46.9	−12 36 35	MCG −2-4-3	14.4	0.4 x 0.4	12.3	S?		
01 10 00.3	−14 13 34	MCG −2-4-4	14.8	0.7 x 0.3	12.9		69	
01 10 20.9	−09 34 12	MCG −2-4-5	14.5	0.7 x 0.4	12.9	Sc	147	
01 10 31.8	−13 58 50	MCG −2-4-6	14.8	0.5 x 0.5	13.1			
01 10 38.0	−12 02 10	MCG −2-4-7	14.7	0.4 x 0.4	12.6			Almost stellar; mag 8.23v star E 2′.0.
01 10 40.3	−13 38 30	MCG −2-4-8	15.0	0.7 x 0.4	13.4	Sbc	162	
01 10 58.0	−13 39 13	MCG −2-4-9	14.5	0.7 x 0.7	13.6			
01 11 09.2	−13 57 42	MCG −2-4-10	14.3	0.6 x 0.6	13.1	Sc		
01 11 38.1	−14 02 44	MCG −2-4-11	15.1	0.6 x 0.3	13.1	Sc	33	Star S edge. Elongated anonymous galaxy W 1′.4.
01 12 26.8	−12 46 34	MCG −2-4-12	14.8	0.6 x 0.4	13.1	Sb	126	
01 13 36.7	−10 34 49	MCG −2-4-13	14.3	0.8 x 0.4	12.9		126	
01 14 04.8	−13 02 56	MCG −2-4-14	15.1	0.7 x 0.3	13.2	S?	159	
01 14 23.4	−10 06 25	MCG −2-4-15	14.9	0.7 x 0.4	13.4		141	
01 17 18.3	−10 07 59	MCG −2-4-17	15.1	0.7 x 0.4	13.5		30	Mag 8.77v star S 3′.3.
01 17 06.6	−13 59 53	MCG −2-4-18	14.8	0.6 x 0.6	13.6	S0		
01 17 33.0	−08 15 14	MCG −2-4-19	15.0	0.6 x 0.6	13.8			
01 17 48.3	−08 36 30	MCG −2-4-20	14.9	1.7 x 1.3	15.6	SB(s)m V	108	Low surface brightness.
01 17 45.6	−12 00 22	MCG −2-4-21	14.6	0.8 x 0.5	13.5		57	
01 19 21.0	−11 52 08	MCG −2-4-22	13.8	1.0 x 0.3	12.4	S?	48	
01 19 54.0	−09 56 53	MCG −2-4-23	15.4	0.6 x 0.3	13.4		84	Mag 6.66v star NNE 4′.5; mag 8.97v star S 1′.8.
01 20 33.3	−08 33 17	MCG −2-4-24	15.1	1.1 x 0.9	14.9	Sb	147	Very faint spiral arms extending from bright, elongated core. Star NW end.
01 21 00.0	−12 34 16	MCG −2-4-28	15.3	0.3 x 0.3	12.5			IC 98 SSW 2′.4; mag 8.78v star S 10′.4.
01 20 58.6	−13 51 03	MCG −2-4-29	15.5	0.6 x 0.6	14.3	S		Anonymous galaxy NW 1′.2.
01 21 27.6	−10 37 13	MCG −2-4-31	15.0	1.1 x 0.4	14.0	SB?	24	
01 21 43.0	−11 46 11	MCG −2-4-32	13.3	1.0 x 1.0	13.2	SAB(r)c II		
01 23 15.9	−09 01 45	MCG −2-4-35	14.6	0.6 x 0.4	12.9	Sb	156	Mag 10.07v star S 1′.3.
01 24 03.3	−11 15 46	MCG −2-4-38	14.7	0.9 x 0.4	13.4	SBc	87	Star on S edge.
01 24 31.2	−08 41 24	MCG −2-4-39	14.7	0.7 x 0.5	13.4	Sb	168	
01 24 53.3	−08 49 14	MCG −2-4-42	14.0	0.6 x 0.3	12.0	Sc	6	**MCG −2-4-40** SW 2′.7.
01 24 50.3	−12 38 44	MCG −2-4-43	15.2	0.7 x 0.5	13.9	Irr	42	Mag 11.31v star SE 2′.9.
01 24 52.9	−11 35 21	MCG −2-4-44	13.9	1.0 x 0.6	13.1	Sb	177	Mag 8.88v star E 7′.1.
01 25 16.5	−08 52 27	MCG −2-4-45	16.7	0.5 x 0.3	14.5	Sc	30	
01 25 16.3	−11 54 52	MCG −2-4-46	14.4	0.4 x 0.4	12.2			Almost stellar.
01 25 48.0	−12 30 37	MCG −2-4-47	15.2	0.3 x 0.3	12.7	E		Stellar.
01 25 53.8	−09 38 47	MCG −2-4-48	15.3	0.6 x 0.4	13.6	Sbc	39	
01 26 06.7	−13 25 14	MCG −2-4-49	14.8	0.4 x 0.4	12.9	E		
01 26 25.9	−08 33 52	MCG −2-4-50	14.7	0.4 x 0.4	12.6			Almost stellar; mag 6.57v star NW 8′.5; stellar **Mkn 994** NE 7′.4.
01 26 15.6	−13 31 05	MCG −2-4-51	15.1	0.3 x 0.3	12.3	S0/a		
01 26 26.6	−13 29 04	MCG −2-4-52	14.7	0.4 x 0.4	12.5			
01 27 29.9	−13 44 27	MCG −2-4-54	14.2	0.7 x 0.7	13.5	E		
01 28 01.1	−09 28 30	MCG −2-4-56	14.5	0.8 x 0.6	13.5	SBbc	138	
01 28 24.1	−13 34 21	MCG −2-4-57	14.7	0.8 x 0.2	12.6	Sbc	57	
01 29 14.3	−09 13 08	MCG −2-4-59	15.3	0.9 x 0.2	13.3	Sc	87	
01 30 20.4	−12 14 34	MCG −2-4-61	14.7	0.8 x 0.4	13.3	S?	6	Mag 10.06v star W 3′.1.
01 30 24.9	−11 58 48	MCG −2-4-62	15.0	0.7 x 0.7	14.1			Stellar nucleus, faint envelope.
01 31 33.4	−10 51 15	MCG −2-4-64	14.7	0.7 x 0.6	13.6		24	Uniform surface brightness; mag 8.02v star SW 5′.5.
01 32 17.0	−09 29 57	MCG −2-5-2	15.0	0.6 x 0.3	13.0		51	Mag 9.41v star E 2′.0.
01 32 59.0	−10 27 10	MCG −2-5-6	14.3	0.9 x 0.7	13.6		129	
01 35 34.3	−13 59 49	MCG −2-5-8	14.6	0.5 x 0.5	13.0	Sbc		
01 35 43.8	−09 59 20	MCG −2-5-9	15.5	0.7 x 0.5	14.2	Sd	171	NGC 624 E 2′.0.
01 35 52.0	−09 56 44	MCG −2-5-11	15.4	0.8 x 0.2	13.3	Sc	33	NGC 624 S 3′.5.
01 36 06.7	−11 01 38	MCG −2-5-12	14.5	1.1 x 0.8	14.2	SBbc	111	
01 36 18.5	−13 41 50	MCG −2-5-13	14.3	1.5 x 0.7	14.2	SA(rs)c pec: II	15	Mag 13.6 star NW 1′.4.
01 36 38.4	−13 52 56	MCG −2-5-14	15.3	0.7 x 0.3	13.5	Sb	33	
01 36 46.1	−14 03 11	MCG −2-5-15	14.5	0.8 x 0.5	13.4	SBbc	99	
01 36 40.7	−13 53 25	MCG −2-5-16	15.2	0.9 x 0.4	13.9	Sbc	171	
01 36 53.6	−14 00 00	MCG −2-5-17	13.9	1.1 x 0.9	13.8	S0	36	
01 36 56.1	−13 56 39	MCG −2-5-18	14.7	0.4 x 0.4	12.8	E		
01 37 01.7	−12 18 48	MCG −2-5-19	14.6	0.7 x 0.4	13.0		114	
01 37 15.4	−09 11 52	MCG −2-5-20	13.3	0.8 x 0.6	12.4	$SA0^-$	165	Mag 12.2 star on W edge.
01 37 13.7	−13 38 54	MCG −2-5-21	15.4	0.5 x 0.2	12.8	Sb	111	
01 37 23.5	−09 10 09	MCG −2-5-22	15.4	0.4 x 0.4	13.4	E/S0		Compact; stellar MCG −2-5-22 W 4′.3.
01 37 23.1	−09 16 16	MCG −2-5-23	14.3	0.5 x 0.4	12.4		171	Pair with MCG −2-5-24 on W edge; mag 6.25v star SSE 8′.7.
01 37 21.7	−09 16 15	MCG −2-5-24	15.4	0.5 x 0.2	13.0	E/S0	168	Pair with MCG −2-5-23 on E edge.
01 37 18.9	−13 33 11	MCG −2-5-25	15.4	0.7 x 0.3	13.5	Sbc	99	
01 37 30.8	−08 53 09	MCG −2-5-26	14.7	1.0 x 0.5	13.8	Sbc	51	
01 37 37.3	−12 31 11	MCG −2-5-27	14.3	0.8 x 0.5	13.1		108	Uniform surface brightness; mag 8.38v star NNW 5′.3.
01 38 20.6	−09 51 31	MCG −2-5-29	14.6	0.8 x 0.5	13.4	Sbc	117	
01 39 08.9	−10 30 15	MCG −2-5-30	16.0	1.1 x 0.2	14.2	Sd	81	Mag 7.94v star W 2′.3.
01 39 34.7	−12 04 37	MCG −2-5-32	13.8	1.2 x 1.0	13.8	(R′)SAB(r)b: II	9	Stellar nucleus; small faint patches N and S of nucleus; mag 8.74v star WNW 4′.2.
01 40 53.5	−10 16 43	MCG −2-5-35	14.4	0.9 x 0.5	13.4	Sbc	45	
01 43 12.6	−13 38 12	MCG −2-5-38	14.4	0.9 x 0.5	13.3	Sc	84	
01 44 34.1	−08 46 24	MCG −2-5-39	15.0	0.7 x 0.5	13.7		3	
01 44 53.6	−10 56 14	MCG −2-5-40	14.3	0.7 x 0.7	13.4			
01 45 28.4	−10 05 37	MCG −2-5-41	14.1	1.0 x 0.8	13.7	$S0^-$	165	
01 46 39.6	−13 24 30	MCG −2-5-46	14.0	0.9 x 0.4	12.7	Sa	39	
01 47 18.3	−14 02 44	MCG −2-5-48	14.3	1.0 x 0.8	13.9	Sbc	156	
01 47 47.8	−13 35 36	MCG −2-5-49	14.1	0.7 x 0.4	12.8	E/S0	33	

RA h m s	Dec ° ′ ″	Name	Mag (V)	Dim ′ Maj x min	SB	Type Class	PA	Notes
01 48 26.1	−12 22 59	MCG −2-5-50	13.2	3.0 x 2.5	15.3	IAB(rs)m IV-V	20	Mag 10.74v star N 6'.8; MCG −2-5-50A on E edge.
01 48 32.2	−12 23 19	MCG −2-5-50A	14.4	0.5 x 0.4	12.5	Sc	171	Located on E edge of MCG −2-5-50.
01 48 36.3	−10 19 39	MCG −2-5-51	14.2	1.2 x 0.3	13.0	S0°: sp	162	NGC 681 SE 10'.4.
01 49 10.3	−10 03 45	MCG −2-5-53	13.2	2.7 x 2.2	15.0	SB(s)d III	120	Mag 13.5 star 2'.4 E.
01 49 23.3	−08 26 34	MCG −2-5-54	14.7	0.7 x 0.7	13.8			
01 49 22.1	−14 07 25	MCG −2-5-55	14.4	1.1 x 0.2	12.6	Sbc	63	
01 49 38.0	−13 34 17	MCG −2-5-56	14.0	1.7 x 0.6	13.9	SB(r)b? II-III	63	Anonymous galaxy NE 1'.0 and NW 2'.7.
01 49 40.1	−12 49 29	MCG −2-5-57	14.4	1.5 x 1.0	14.7	SB(s)m V	15	Located between a pair of mag 12 stars.
01 51 10.8	−12 30 53	MCG −2-5-62	13.8	1.3 x 0.6	13.3	Spiral	3	
01 51 26.0	−08 24 57	MCG −2-5-64	13.6	0.6 x 0.4	11.9	Sb	24	NGC 707 S 5'.6.
01 51 33.9	−08 23 57	MCG −2-5-65	14.3	0.9 x 0.5	13.3	Sb	105	Multi-galaxy system.
00 37 31.5	−17 56 36	MCG −3-2-25	15.2	0.7 x 0.4	13.6	SBbc	174	**MCG −3-2-23** WSW 5'.4.
00 38 18.4	−14 50 56	MCG −3-2-27	14.7	0.9 x 0.6	13.9	Sc	51	
00 41 11.1	−14 55 42	MCG −3-2-37	14.3	0.8 x 0.7	13.5		90	
00 41 25.9	−15 34 48	MCG −3-2-38	14.8	0.7 x 0.5	13.5	Sbc	102	
00 41 40.8	−16 51 10	MCG −3-2-39	13.9	0.5 x 0.3	11.7	Sbc	84	Pair with MCG −3-2-40 E 1'.1.
00 41 45.5	−16 51 42	MCG −3-2-40	13.5	1.7 x 1.2	14.1	SB(s)d III	171	
00 43 19.8	−17 36 32	MCG −3-2-42	15.1	0.5 x 0.5	13.4			
00 43 49.3	−17 44 23	MCG −3-3-1	14.8	0.8 x 0.3	13.1	Sc	78	Located 15'.3 N of mag 2.05v star Diphda, Beta Ceti.
00 50 00.9	−14 40 58	MCG −3-3-4	14.4	0.9 x 0.4	13.3	E	171	
00 50 40.2	−15 16 45	MCG −3-3-5	14.8	1.3 x 0.5	14.2	SAB(s)dm? IV-V	69	Mag 12.6 star on NE end.
00 52 00.6	−16 24 25	MCG −3-3-6	13.5	0.9 x 0.9	13.2	SAB(r)bc: II		
00 56 12.5	−14 16 36	MCG −3-3-8	14.2	0.8 x 0.5	13.1	SB	9	= **Mkn 1149**.
00 57 35.1	−16 29 40	MCG −3-3-9	14.1	0.5 x 0.5	12.4	S0/a		
00 58 20.0	−15 23 32	MCG −3-3-10	13.6	1.0 x 0.5	12.7	(R)SB(r)ab:	60	Star E edge.
00 58 56.2	−16 52 11	MCG −3-3-14	14.7	0.5 x 0.3	12.5		15	
01 00 15.9	−15 17 59	MCG −3-3-17	14.4	0.7 x 0.4	12.8	Sa	171	Mag 10.86v star N 4'.1.
01 00 24.9	−15 10 34	MCG −3-3-18	14.5	0.8 x 0.5	13.4		156	Mag 10.86v star S 3'.8.
01 01 46.5	−15 16 12	MCG −3-3-19	14.9	0.5 x 0.5	13.3			Stellar; mag 10.95v star S 2'.2.
01 05 25.0	−17 42 56	MCG −3-3-21	15.4	0.6 x 0.3	13.4	S	78	NGC 369 SW 4'.7.
01 07 39.6	−17 14 01	MCG −3-4-2	14.7	0.6 x 0.5	13.2	(R′$_1$)SB(s)a:	51	
01 10 09.1	−16 51 12	MCG −3-4-14	13.9	0.4 x 0.4	12.0	E?		
01 10 54.6	−17 11 51	MCG −3-4-18	14.6	0.5 x 0.3	12.4		33	
01 11 14.1	−17 04 18	MCG −3-4-22	14.8	0.7 x 0.6	13.7	BCG	51	
01 11 22.1	−17 04 19	MCG −3-4-23	14.4	0.6 x 0.4	12.7	SBb	129	At center of galaxy cluster A 2881.
01 11 33.8	−17 04 04	MCG −3-4-26	15.6	0.4 x 0.4	13.4			
01 14 15.9	−16 40 06	MCG −3-4-32	14.6	0.8 x 0.4	13.2		30	
01 15 16.0	−14 26 19	MCG −3-4-34	14.8	1.1 x 0.2	13.0		138	
01 17 23.8	−16 03 44	MCG −3-4-38	14.4	0.9 x 0.6	13.6		51	
01 19 03.0	−17 01 26	MCG −3-4-44	14.9	0.9 x 0.3	13.3	Sb	78	Mag 8.44v star E 11'.5; IC 93 S 2'.2.
01 19 24.9	−15 42 00	MCG −3-4-46	14.5	0.9 x 0.7	13.8	SBb	90	
01 19 31.3	−17 25 24	MCG −3-4-47	14.4	0.5 x 0.5	12.7	Sbc		Star S edge.
01 19 33.4	−17 41 38	MCG −3-4-48	14.5	0.5 x 0.5	12.9			
01 19 48.8	−16 15 38	MCG −3-4-49	14.8	0.7 x 0.4	13.3	SBb	30	Star or bright knot S end.
01 20 43.0	−14 43 57	MCG −3-4-54	14.7	0.8 x 0.4	13.3		3	
01 21 58.3	−16 50 44	MCG −3-4-57	15.0	0.4 x 0.4	12.9	S0		0.30′ NW of mag 11 star.
01 23 10.9	−17 31 23	MCG −3-4-58	15.0	0.5 x 0.4	13.1		153	Mag 10.64v star W 3'.5.
01 24 09.6	−16 47 56	MCG −3-4-59	14.5	0.7 x 0.4	13.0		111	Star just off N end.
01 24 20.0	−16 30 29	MCG −3-4-60	14.4	0.8 x 0.3	12.7	Sc	141	
01 24 30.8	−15 32 19	MCG −3-4-61	13.5	1.1 x 1.1	13.5	(R)SB(r)0/a		
01 25 19.9	−16 27 31	MCG −3-4-62	14.6	0.9 x 0.2	12.6	S0/a	63	
01 26 49.2	−17 02 43	MCG −3-4-66	14.4	1.0 x 0.6	13.7	Sc	135	Star NW edge.
01 27 03.7	−15 11 33	MCG −3-4-67	15.0	1.2 x 0.5	14.3	SB(s)dm IV-V	147	
01 27 01.9	−17 21 45	MCG −3-4-68	14.9	0.5 x 0.5	13.2			
01 28 01.4	−16 51 24	MCG −3-4-71	14.7	0.8 x 0.4	13.3		30	
01 28 53.4	−15 46 22	MCG −3-4-74	14.9	0.7 x 0.2	12.7		165	
01 28 56.8	−16 31 00	MCG −3-4-75	14.5	1.0 x 0.2	12.6	Sc	96	
01 29 19.3	−15 31 28	MCG −3-4-76	14.5	0.9 x 0.4	13.2		168	
01 30 42.5	−16 27 52	MCG −3-4-78	14.2	0.6 x 0.6	13.1	E		
01 31 15.8	−17 41 58	MCG −3-4-79	15.2	0.7 x 0.4	13.6	Sb-c	51	1'.0 SW of mag 9 star.
01 32 18.1	−16 08 43	MCG −3-5-2	14.8	0.9 x 0.3	13.2	Sb	6	
01 32 28.9	−15 05 52	MCG −3-5-3	15.0	0.7 x 0.5	13.7		0	
01 33 41.1	−17 22 21	MCG −3-5-6	15.2	0.6 x 0.5	13.8		141	
01 34 25.1	−15 49 08	MCG −3-5-7	13.4	1.0 x 0.9	13.2	Sbrst	159	
01 34 35.0	−15 30 15	MCG −3-5-8	15.1	1.6 x 0.2	13.7	S?	54	Mag 5.62v star 49 Ceti S 10'.4.
01 37 08.1	−15 03 42	MCG −3-5-9	14.5	0.6 x 0.4	12.8	S0	30	
01 37 51.4	−17 59 33	MCG −3-5-10	14.7	0.8 x 0.6	13.8	S(r)	159	
01 39 14.1	−14 16 43	MCG −3-5-12	14.4	0.7 x 0.5	13.2	S0	27	
01 41 36.0	−16 08 50	MCG −3-5-14	14.6	1.5 x 0.7	14.5	IB(s)m V	70	IC 149 and mag 12.2 star SE 15'.0.
01 43 46.8	−15 57 26	MCG −3-5-18	13.0	0.8 x 0.5	11.8	Sc	162	
01 46 29.1	−15 07 18	MCG −3-5-19	14.5	0.4 x 0.4	12.6	E/S0		
01 46 55.7	−14 27 09	MCG −3-5-20	14.3	0.7 x 0.3	12.4	Sbc	63	
00 33 06.3	−13 22 17	NGC 135	14.8	0.5 x 0.5	13.2			Almost stellar; mag 8.50v star SW 8'.8.
00 34 02.8	−09 42 19	NGC 151	11.6	3.7 x 1.7	13.4	SB(r)bc II	75	A narrow extension to the NE has two bright knots on the end.
00 34 19.5	−12 39 23	NGC 154	14.0	1.1 x 0.9	14.1	E	80	
00 34 40.0	−10 46 02	NGC 155	13.3	1.7 x 1.3	14.0	S0° pec	177	
00 34 47.1	−08 23 57	NGC 157	10.4	3.5 x 2.4	12.6	SAB(rs)bc I-II	36	Mag 8.47v star N 6'.3.
00 35 59.9	−10 07 18	NGC 163	12.7	1.5 x 1.2	13.3	E0	85	Very small, elongated **MCG −2-2-65** S 1'.2.
00 36 29.0	−10 06 25	NGC 165	13.1	1.5 x 1.1	13.5	SB(rs)bc I	77	Almost stellar nucleus; NGC 163 W 7'.2.
00 35 48.8	−13 36 40	NGC 166	14.3	0.9 x 0.3	12.7	Sa	145	
00 39 08.3	−14 10 28	NGC 178	12.6	2.0 x 1.0	13.2	SB(s)m IV-V	178	Two strong knots SW of core.
00 37 46.2	−17 50 58	NGC 179	13.3	0.9 x 0.8	12.8	SAB0$^-$	113	Faint star on N edge; mag 10.28v star N 5'.1.
00 39 30.6	−14 39 24	NGC 187	13.2	1.3 x 0.5	12.5	SB(s)cd III	148	
00 38 59.7	−09 00 12	NGC 191	12.4	1.3 x 1.2	12.8	SAB(rs)c: pec	128	Mag 10.49v star NW 5'.0; IC 1563 on S edge.
00 39 35.7	−09 11 42	NGC 195	13.7	1.2 x 0.8	13.5	(R)SB(r)a:	45	Mag 10.35v star S 5'.5.
00 39 40.6	−14 14 13	NGC 207	13.7	0.9 x 0.4	12.4	Sa	85	Mag 9.54v star SSE 4'.4; faint, stellar anonymous galaxy N 3'.8.
00 40 35.1	−13 52 26	NGC 210	10.9	5.0 x 3.3	13.8	SAB(s)b I	165	Mag 8.40v star W 7'.6; MCG −2-2-82 NNE 7'.6.
00 41 34.2	−10 01 21	NGC 217	12.4	2.6 x 0.6	12.8	S0/a: sp	115	Elongated N-S anonymous galaxy, or knot?, NW end.

(Continued from previous page)

GALAXIES

RA h m s	Dec ° ′ ″	Name	Mag (V)	Dim ′ Maj x min	SB	Type Class	PA	Notes
00 45 46.9	−15 35 51	NGC 244	12.9	1.2 x 1.0	13.0	S0 pec?	50	Mag 10.10v star S 3.′5.
00 47 47.1	−11 28 08	NGC 255	11.9	3.0 x 2.5	13.9	SAB(rs)bc II-III	15	
00 48 48.5	−13 06 27	NGC 263	14.3	0.7 x 0.4	12.7	Sbc I-II:	45	
00 50 09.6	−05 11 38	NGC 268	13.1	1.5 x 0.9	13.2	SB(s)bc: I-II	75	Bright knots or stars on E and W edges; mag 8.79v star NE 5.′3.
00 50 32.5	−08 39 08	NGC 270	12.9	1.7 x 1.2	13.5	S0^{+}	33	Bright core in faint, smooth envelope.
00 50 48.5	−06 53 12	NGC 273	12.9	1.9 x 0.5	12.7	S0 sp	102	
00 51 01.9	−07 03 30	NGC 274	11.8	1.4 x 1.2	12.3	SAB(r)0^{-} pec	39	On NW edge of NGC 275.
00 51 04.5	−07 03 57	NGC 275	12.5	1.5 x 1.0	12.8	SB(rs)cd pec	126	On SE edge of NGC 274.
00 51 17.2	−08 35 49	NGC 277	13.7	1.4 x 1.2	14.2	S0^{-}	50	Almost stellar nucleus; mag 10.87v star on E edge; MCG −2-3-29 N 4.′9.
00 53 13.2	−13 09 51	NGC 283	14.1	1.6 x 1.0	14.5	Sc I-II	154	Two main arms, short NE-SW bar.
00 53 24.3	−13 09 34	NGC 284	14.4	0.6 x 0.5	13.2	E	4	**MCG −2-3-32** N 0.′8.
00 53 30.0	−13 09 41	NGC 285	14.7	0.7 x 0.5	13.4	E/S0	10	Faint, elongated, anonymous galaxy SE 1.′2.
00 53 30.4	−13 06 49	NGC 286	14.1	1.2 x 0.7	13.8	SAB(s)0° pec?	10	
00 53 29.8	−08 46 05	NGC 291	13.9	1.1 x 0.5	13.1	(R′)SB(r)a:	45	Mag 6.16v star 16 Ceti E 11.′9.
00 54 16.0	−07 14 10	NGC 293	14.2	1.1 x 0.8	13.9	(R′)SB(rs)b II-III	19	Two main arms; mag 8.98v star E 6.′7.
00 55 02.1	−07 20 00	NGC 298	13.8	1.6 x 0.4	13.2	Scd: sp II-III	90	Mag 8.98v star NW 7.′4.
00 56 18.4	−10 40 26	NGC 301	14.8	0.7 x 0.5	13.5		70	Almost stellar core; mag 9.22v star SE 7.′6.
00 54 54.7	−16 39 21	NGC 303	14.3	0.7 x 0.3	12.4	S0: pec	158	
00 56 42.6	−09 54 49	NGC 309	11.9	3.0 x 2.5	14.0	SAB(r)c I	175	Pronounced spiral arms with dark lanes; IC 1603 WSW 13.′3.
00 57 39.3	−05 05 11	NGC 321	14.9	0.4 x 0.4	12.8	SB(rs)cd pec II-III		
00 57 47.9	−05 06 44	NGC 325	14.1	1.5 x 0.2	12.6	Sb	91	
00 57 55.3	−05 07 51	NGC 327	13.4	1.6 x 0.7	13.4	SB(s)bc: II	175	
00 58 01.8	−05 04 12	NGC 329	13.3	1.6 x 0.6	13.1	Sb: sp	20	
00 58 51.2	−16 28 08	NGC 333	13.9	1.6 x 0.9	14.4	E/S0	119	Almost stellar NGC 333B S 0.′9.
00 58 51.0	−16 29 03	NGC 333B	14.7	0.3 x 0.2	11.5		68	**MCG −3-3-12** S 0.′3.
00 59 50.1	−07 34 43	NGC 337	11.6	2.8 x 2.0	13.3	SB(s)d III	141	
01 01 33.6	−07 35 18	NGC 337A	12.2	6.6 x 5.0	15.8	SAB(s)dm	10	Very weak nucleus with uniform envelope, mag 12.9 star superimposed 1.′2 SE of center.
01 00 34.9	−06 51 59	NGC 340	13.7	0.8 x 0.4	12.3	Sb	65	Small, very faint anonymous galaxy SE 1.′7; MCG −1-3-57 SE 4.′7.
01 00 45.7	−09 11 08	NGC 341	13.0	1.1 x 1.0	13.0	SAB(r)bc II	55	MCG −2-3-64 on E edge.
01 00 49.8	−06 46 23	NGC 342	14.4	0.9 x 0.4	13.4	E	107	Mag 7.21v star E 11.′3, see notes for NGC 347.
01 01 22.1	−06 53 05	NGC 345	13.9	1.4 x 0.8	13.8	SA(s)a:	138	Mag 7.21v star NNE 6.′0 (see notes for NGC 347).
01 01 35.2	−06 44 02	NGC 347	14.8	0.6 x 0.5	13.3		118	Mag 7.21v star S 3.′9; note: an anonymous galaxy lies 1.′5 WSW of this star.
01 01 50.8	−06 48 01	NGC 349	13.1	1.3 x 1.2	13.5	SA0^{-}	49	A mag 10.74 star lies 2.′6 E with NGC 350 in the middle; mag 7.21 star 4.′2 W.
01 01 56.6	−06 47 47	NGC 350	14.4	0.6 x 0.4	12.8	E/S0	82	Located between mag 10.74v star E and NGC 349 W.
01 03 06.9	−06 19 27	NGC 355	15.3	0.9 x 0.4	14.1	SB0° pec:	62	Stellar nucleus in faint envelope.
01 03 07.1	−06 59 19	NGC 356	13.1	1.8 x 1.2	13.8	SAB(s)bc pec:	53	Faint, stellar anonymous galaxy N 4.′2; mag 10.47v star N 7.′0.
01 03 21.9	−06 20 22	NGC 357	12.0	2.4 x 1.7	13.4	SB(r)0/a:	33	Strong nucleus in short E-W bar; bright knot or star E edge of nucleus.
01 06 15.8	−16 32 33	NGC 363	14.7	0.7 x 0.6	13.6	S0	49	
01 05 49.0	−12 07 47	NGC 367	14.7	0.7 x 0.3	12.9	Sc	15	
01 05 08.9	−17 45 37	NGC 369	13.7	1.0 x 0.8	13.3	SAB(r)b: I-II	52	MCG −3-3-21 NE 4.′7.
01 20 34.5	−09 52 50	NGC 480	15.2	0.5 x 0.2	12.5	Sa	65	Almost stellar; mag 6.66v star W 7.′7.
01 21 12.5	−09 12 42	NGC 481	13.3	1.3 x 0.9	13.4	SA(r)0^{-} pec?	37	
01 21 55.1	−16 22 12	NGC 487	13.4	1.1 x 0.7	13.0	SB(r)a:	112	
01 27 02.4	−10 15 56	NGC 567	14.2	1.2 x 0.5	13.4	S0	134	
01 31 20.9	−06 52 06	NGC 584	10.5	4.1 x 2.0	12.8	E4	72	Mag 7.45v star NE 13.′4; NGC 586 ESE 4.′1.
01 31 36.9	−06 53 40	NGC 586	13.2	1.5 x 0.7	13.1	SA(s)a:?	0	NGC 584 NW 4.′1.
01 32 39.9	−12 02 34	NGC 589	14.1	1.1 x 0.9	13.9	(R)SB(r)0/a	90	Almost stellar nucleus in short E-W bar.
01 32 20.9	−12 21 19	NGC 593	14.6	1.3 x 0.3	13.5	SB(r)0°? sp	12	
01 32 56.8	−16 32 10	NGC 594	13.5	1.3 x 0.6	13.1	Sbc: I-II	32	
01 32 52.1	−07 01 57	NGC 596	10.9	3.2 x 2.0	12.9	E^{+} pec:	30	**PGC 5754** W 5.′6; Mag 5.76v star E 12.′7.
01 32 54.0	−12 11 26	NGC 599	13.4	1.4 x 1.3	13.9	SAB0^{-} pec:	135	Compact core, smooth envelope; mag 8.59v star E 11.′9.
01 33 05.4	−07 18 43	NGC 600	12.4	2.8 x 2.1	14.1	(R′)SB(rs)d II-III	18	Strong, thin, elongated nucleus in uniform envelope.
01 35 05.7	−07 20 29	NGC 615	11.6	2.7 x 0.9	12.4	SA(rs)b II	155	Mag 9.23v star SW 5.′3.
01 34 02.6	−09 46 30	NGC 617	14.5	0.5 x 0.5	12.9	Sab		Almost stellar nucleus.
01 35 51.0	−10 00 10	NGC 624	13.3	1.4 x 0.8	13.3	(R′)SB(r)b pec II	100	Mag 14.4 star on S edge; MCG −2-5-9 W 2.′0; MCG −2-5-11 N 3.′5.
01 39 06.6	−07 30 47	NGC 636	11.5	2.1 x 2.1	13.1	E3		
01 39 24.8	−09 24 03	NGC 640	14.4	0.7 x 0.4	12.8	S	145	Compact.
01 39 56.0	−09 14 37	NGC 647	13.4	1.5 x 1.1	13.8	SB(r)0°?	3	Mag 8.03v star E 5.′0.
01 38 39.7	−17 49 54	NGC 648	14.4	0.7 x 0.4	13.0	SA0^{-} pec:	114	
01 40 07.6	−09 16 18	NGC 649	14.3	0.9 x 0.3	12.7	Sa	20	2.′9 SW of a mag 8.03v star.
01 41 55.5	−13 04 53	NGC 655	14.0	1.2 x 0.8	13.8	S0 pec	78	Very faint, narrow extension curving SW from W end, beyond published dimensions.
01 49 11.1	−10 25 32	NGC 681	12.0	2.6 x 1.6	13.4	SAB(s)ab sp	60	Moderately dark lane extends across nucleus NE-SW.
01 49 04.2	−14 58 31	NGC 682	13.4	1.4 x 1.1	13.8	SA0^{-}	95	
01 47 48.1	−16 43 19	NGC 690	14.2	1.2 x 0.8	14.0	SAB(s)c: I-II	154	Mag 10.22v star SSW 7.′2.
01 50 43.5	−12 02 08	NGC 699	14.1	1.5 x 0.3	13.1	SBbc? sp	135	Mag 10.60v star W 5.′8; mag 8.78v star SW 7.′3.
01 51 04.1	−09 42 08	NGC 701	12.2	2.5 x 1.2	13.2	SB(rs)c III-IV	40	Mag 12.4 star N 5.′5; IC 1738 S 5.′4.
01 51 27.2	−08 30 21	NGC 707	13.6	1.4 x 0.9	13.7	(R′)SAB(s)0^{-}:	95	Star or knot on SE edge of nucleus; MCG −2-5-64 and MCG −2-5-65 N 5.′6; IC 170 E 7.′5.
00 55 55.9	−10 39 11	PGC 3311	15.3	0.5 x 0.1	11.9		12	Located 2.′4 SE of a mag 9.21v star.

GALAXY CLUSTERS

RA h m s	Dec ° ′ ″	Name	Mag 10th brightest	No. Gal	Diam ′	Notes
00 41 36.0	−09 20 00	A 85	15.7	59	34	MCG −2-2-86 4.′1 ENE of cluster center.
00 56 00.0	−10 01 00	A 117	16.0	40	17	IC 1602 at center.
00 57 30.0	−07 00 00	A 121	16.0	67	18	All cluster members very faint and stellar.
01 08 54.0	−15 25 00	A 151	15.0	72	39	**Haufen A**. IC 77 and 80A at center with numerous stellar and almost stellar GX in surrounding area.
01 11 18.0	−17 05 00	A 2881	15.6	36	22	
01 02 54.0	−17 06 00	AS 118	16.3		17	All members anonymous, faint, stellar.

PLANETARY NEBULAE

RA h m s	Dec ° ′ ″	Name	Diam ″	Mag (P)	Mag (V)	Mag cent ★	Alt Name	Notes
00 47 03.6	−11 52 20	NGC 246	245	8.0	10.9	11.9	PK 118−74.1	Several stars superimposed, brightest along W edge; mag 11.9 star on NW edge.
00 37 16.0	−13 42 59	PN G108.4−76.1	3	16.3			PK 108−76.1	Mag 14.9 star E 2.′2.

RA h m s	Dec ° ′ ″	Name	Mag (V)	Dim ′ Maj x min	SB	Type Class	PA	Notes
00 01 19.1	−27 25 13	ESO 409-1	13.8	1.0 x 0.6	13.1	SAB0⁻	117	Bright knot or star on SW edge; mag 9.72v star S 7′.4.
00 01 55.8	−27 37 40	ESO 409-3	13.7	1.3 x 1.0	13.9	(R′)SA(s)b: I-II	177	Bright core, smooth envelope; mag 9.72v star NW 9′.5.
00 07 21.3	−28 07 14	ESO 409-21	14.0	1.8 x 0.7	14.2		153	Double system, bridge.
00 11 22.0	−28 51 16	ESO 409-25	12.9	1.5 x 0.8	13.1	E⁺4	30	
00 36 51.1	−28 22 06	ESO 410-27	13.8	1.7 x 0.8	14.0	SB(s)c: II	42	MCG −5-2-25 NW 6′.9; mag 10.28v star WSW 10′.7.
00 39 18.3	−27 20 57	ESO 411-3	13.7	1.3 x 0.6	13.3	SA(rs)b: II-III	57	Mag 10.43v star NNE 8′.2.
23 32 03.4	−27 43 36	ESO 470-18	13.7	1.3 x 1.0	13.8	(R′)SB(rs)0⁺	117	MCG −5-55-24 SE 5′.6.
23 40 54.5	−28 20 19	ESO 471-2	13.8	1.7 x 0.7	13.8	SA(rs)bc: II-III	144	Mag 12.2 star W 2′.7, mag 7.64v star W 9′.0.
23 49 51.4	−27 57 07	ESO 471-24	13.7	1.4 x 0.3	12.7	S0?	115	Located 16′.3 NE of mag 4.58v star Delta Sculptoris; and a mag 6.95v star NW 9′.1.
23 56 24.5	−29 01 24	ESO 471-47	14.1	1.3 x 0.9	14.1		101	Double system, contact;, = **IC 5364**?
23 57 28.3	−29 02 48	ESO 471-49	13.0	1.7 x 0.9	13.3	SB(r)b: II	168	Mag 12.9 star W 1′.3; mag 9.40v star WSW 13′.1.
23 52 13.8	−25 24 21	ESO 472-4	15.3	1.6 x 0.7	15.2	SB(r)b: II	135	Mag 12.4 star W 4′.0.
23 55 14.5	−23 50 00	ESO 472-10	13.7	1.5 x 0.7	13.6	SB(s)c II-III	175	Stellar nucleus; mag 15.4 star on NE edge; mag 10.78v star N 8′.1.
00 15 10.9	−23 52 55	ESO 473-5	13.6	1.4 x 0.6	13.3	S	49	Mag 10.02v star superimposed SW end.
00 20 41.8	−23 16 59	ESO 473-16	13.5	1.5 x 0.6	13.3	(R′)SB(r)ab	78	Mag 15.0 star on E edge; mag 10.23v star SSW 7′.4.
00 22 38.9	−24 07 40	ESO 473-18	14.0	1.2 x 0.4	13.0	Sb pec	17	Star on W edge; mag 10.56v star W 4′.4.
00 31 22.5	−22 46 01	ESO 473-24	15.7	1.1 x 0.6	15.1	IB(s)m V-VI	27	Mag 13.4 star ESE 1′.4; NGC 144 N 7′.2.
00 31 49.7	−26 43 14	ESO 473-25	13.5	2.5 x 0.3	13.0	Sc sp	81	Star superimposed E end; mag 9.68v star SE 7′.0.
23 19 02.7	−22 42 59	ESO 535-15	13.5	1.3 x 0.9	13.5	SAB(s)bc pec I	99	Mag 10.80v star W 2′.5; **MCG −4-55-4** NNE 5′.0.
23 19 10.8	−23 26 44	ESO 535-16	14.5	1.5 x 0.6	14.2	Sa: pec	135	Star SE end.
23 22 01.0	−23 30 31	ESO 536-2	13.3	1.2 x 0.8	13.1	SA(s)0⁻: pec	0	Small galaxy on E edge.
23 39 02.5	−25 40 11	ESO 536-14	13.8	1.0 x 0.6	13.1	(R)SB(r)b: I-II	39	Mag 10.48v star S 7′.1.
23 57 32.2	−22 00 03	ESO 538-8	15.5	1.5 x 0.4	14.8	Sb pec sp	65	Mag 8.79v star NE 10′.3; MCG −4-1-8 N 1′.4.
23 57 40.1	−21 34 49	ESO 538-10	13.9	0.8 x 0.8	13.3	S0?		Mag 8.50v star SSW 4′.9.
00 07 10.2	−21 47 12	ESO 538-22	14.3	1.7 x 0.4	13.7	(R′)SB(s)ab	160	Mag 8.88v star SSW 3′.4.
00 10 17.8	−18 15 53	ESO 538-24	14.5	1.5 x 1.2	15.0	IB(s)m V-VI	75	Mag 12.2 star ESE 6′.2.
00 17 10.1	−19 18 04	ESO 539-5	12.8	1.6 x 1.0	13.1	SAB(rs)c? II-III	39	Triangle of mag 10-12 stars E 3′.1.
00 18 49.3	−19 00 23	ESO 539-7	14.3	1.8 x 1.3	15.1	SAB(s)m V	165	Mag 6.46v star W 18′.3.
00 31 28.3	−19 45 44	ESO 539-14	14.9	1.7 x 0.2	13.6	Sd? sp	179	Mag 12.4 star N 10′.0.
00 34 13.8	−21 26 18	ESO 540-1	12.8	1.4 x 1.0	13.0	(R′)SB(r)bc	9	= **Hickson 4A**. **Hickson 40E** W 1′.3, **Hickson 40C**]N 2′.0
00 34 14.0	−21 28 12	ESO 540-2	14.9	1.1 x 0.4	13.8	(R′)SB(r)b: pec	132	= **Hickson 4B**. **Hickson 40D** on SE end.
00 35 39.7	−20 07 40	ESO 540-3	13.1	2.2 x 1.0	13.8	(R′)SAB(r)b II	165	Star on W edge.
00 39 34.4	−20 23 00	ESO 540-10	14.8	1.4 x 0.7	14.6	SB(s)cd III-IV	144	Mag 12.2 star SW 1′.6; mag 10.94v star W 9′.8.
00 41 11.8	−21 07 57	ESO 540-14	14.5	0.6 x 0.4	12.9	E⁺ pec?	153	Located in between a NW-SE pair of mag 13.5 stars; NGC 216 NNE 6′.4.
00 42 14.8	−18 09 40	ESO 540-16	13.6	3.2 x 0.3	13.4	SB(s)cd? sp III-IV	75	Located 21′.8 SW of mag 2.05v star Diphda, Beta Ceti.
23 21 14.2	−21 14 11	ESO 605-3	14.0	1.7 x 0.4	13.4	S?	33	Mag 10.98v star N 5′.7.
23 23 38.5	−19 00 37	ESO 605-4	13.8	1.5 x 1.0	14.1	(R′)SB(rs)c: II-III	138	Mag 7.91v star S 6′.1.
23 35 20.1	−17 25 12	ESO 605-11	14.7	1.4 x 0.5	14.2	SB?	42	Multi-galaxy system; mag 12.3 star NW 4′.8.
23 35 32.2	−19 23 51	ESO 605-12	14.0	0.9 x 0.5	13.0	S0? pec	42	Mag 10.49v star SE 4′.3.
23 37 06.8	−20 27 49	ESO 605-16	12.5	1.7 x 1.5	13.4	SB(r)c I	95	mag 10.86v star SE 2′.8.
23 46 09.2	−19 21 36	ESO 606-7	14.8	1.1 x 0.5	14.0	IB(s)m IV	36	Pair of N-S oriented mag 10.66v, 10.64v stars E 7′.7.
23 49 00.8	−19 41 35	ESO 606-11	13.9	1.3 x 1.0	14.0	SB(rs)c I	85	Mag 9.52v star S 2′.8.
23 50 19.6	−17 46 01	ESO 606-13	14.6	1.2 x 0.7	14.3	Im:	81	Multi-galaxy system N 12.4; mag 12.6 star W 6′.0.
00 32 40.2	−25 36 32	IC 1553	13.6	1.2 x 0.3	12.4	SA0	15	Mag 9.52v star SW 12′.3.
00 35 47.2	−25 22 32	IC 1558	12.2	3.4 x 2.3	14.3	SAB(s)m V	150	Mag 8.23v star W 3′.6.
00 38 32.7	−24 20 26	IC 1561	14.1	1.4 x 0.5	13.6	SB?	97	IC 1562 N 3′.9.
00 38 33.9	−24 16 29	IC 1562	12.8	1.6 x 1.5	13.6	SB(s)c I-II	170	Mag 13.7 star on N edge; mag 10.78v star NW 9′.3; IC 1561 S 3′.9; MCG −4-2-31 NE 4′.7.
00 43 03.8	−22 14 47	IC 1574	13.7	2.0 x 0.7	13.9	IB(s)m V-VI	175	Mag 10.06v star WSW 13′.3.
23 26 20.1	−17 57 25	IC 5321	13.0	1.3 x 0.8	12.9	SBa pec?	43	Mag 8.57v star SW 18′.2.
23 29 35.1	−28 49 54	IC 5326	13.9	1.1 x 0.4	12.8	Sb	117	Mag 9.17v star SW 6′.0.
23 39 22.3	−22 29 49	IC 5343	13.6	0.8 x 0.6	12.7	Irr		Mag 7.31v star ESE 5′.0; **IC 5345** N 5′.4.
23 47 14.6	−27 57 30	IC 5350	13.4	0.9 x 0.9	13.2	E?		**PGC 72393** N 1′.8.
23 47 28.6	−28 06 35	IC 5353	13.0	1.4 x 1.0	13.2	S0⁻	137	**PGC 72423** on E edge.
23 47 28.9	−28 08 08	IC 5354	14.0	0.8 x 0.5	13.0	E?	71	Dual system?
23 47 45.0	−28 08 27	IC 5358	12.6	2.5 x 1.0	13.5	E⁺4 pec	131	**PGC 72437** on W edge. Several other faint PGC galaxies in surrounding area within 2′.0 to 3′.0.
23 51 37.0	−28 21 56	IC 5362	12.8	1.3 x 1.3	13.3	SA0° pec:		Mag 8.18v star NE 4′.5.
23 56 29.2	−20 06 42	MCG −3-1-1	14.4	0.7 x 0.3	12.6	SB:(r)0⁺	138	
23 57 17.0	−17 16 57	MCG −3-1-4	15.0	0.8 x 0.6	14.0	S	45	
23 57 12.8	−18 37 37	MCG −3-1-5	16.1	0.8 x 0.6	15.1	(R′)SA(s)a	48	
23 57 50.7	−18 46 50	MCG −3-1-6	14.6	0.8 x 0.6	13.7	Sb	126	
23 58 06.2	−19 47 44	MCG −3-1-7	14.6	0.7 x 0.3	12.7	S	30	
23 58 40.5	−17 33 33	MCG −3-1-9	13.7	0.7 x 0.7	12.8	(R′)SB(s)ab		Star on N edge.
00 00 03.3	−18 00 34	MCG −3-1-12	14.5	0.6 x 0.3	12.4	Sc	144	0′.8 E of mag 8 star.
00 00 55.4	−18 57 30	MCG −3-1-13	14.9	0.9 x 0.5	13.9	S	15	Mag 10.39v star N 6′.2.
00 06 32.3	−18 01 54	MCG −3-1-20	15.1	0.7 x 0.3	13.3	S	12	Mag 7.32v star W 1′.8; located on W edge of small galaxy cluster **A 2712**.
00 14 05.0	−19 39 42	MCG −3-1-22	14.1	1.0 x 0.9	13.9	Sb	75	Elongated **MCG −3-1-23** E 5′.3.
00 17 46.0	−18 00 53	MCG −3-1-25	14.0	1.2 x 0.9	13.9	Sb	103	
00 21 56.2	−18 58 04	MCG −3-2-2	14.5	0.7 x 0.5	13.2	S	60	Mag 10.26v star W 2′.2.
00 22 32.6	−17 04 29	MCG −3-2-5	14.4	1.1 x 0.9	14.2		165	Stellar nucleus, smooth envelope; mag 10.27v star S 2′.5.
00 22 30.3	−19 39 44	MCG −3-2-7	15.4	0.8 x 0.4	14.0	S	27	MCG −3-2-7 SE 0′.7.
00 22 55.7	−19 02 01	MCG −3-2-8	13.9	0.9 x 0.5	12.9	S0	6	
00 24 46.2	−19 58 34	MCG −3-2-11	14.3	0.4 x 0.4	12.4	E		
00 32 48.4	−18 35 14	MCG −3-2-21	15.1	0.8 x 0.5	13.9		30	
00 37 31.5	−17 56 36	MCG −3-2-25	15.2	0.7 x 0.4	13.6	SBbc	174	**MCG −3-2-23** WSW 5′.4.
00 39 20.9	−18 55 07	MCG −3-2-32	13.7	1.0 x 0.7	13.2	S?	174	Mag 8.31v star WNW 7′.8.
00 43 19.8	−17 36 32	MCG −3-2-42	15.1	0.5 x 0.5	13.4			
00 43 49.3	−17 44 23	MCG −3-3-1	14.8	0.8 x 0.3	13.1	Sc	78	Located 15′.3 N of mag 2.05v star Diphda, Beta Ceti.
23 35 59.8	−19 28 31	MCG −3-60-4	15.1	0.6 x 0.4	13.4	(R'_2)SB(s)ab	15	Mag 10.49v star NW 3′.6.
23 36 05.0	−19 42 37	MCG −3-60-5	14.7	0.7 x 0.3	12.9	S	78	
23 37 31.2	−19 37 17	MCG −3-60-7	13.7	0.8 x 0.8	13.1	S0		Mag 9.60v star E 4′.2.
23 37 51.8	−19 34 34	MCG −3-60-9	14.7	0.7 x 0.5	13.4	Sb	150	Mag 9.60v star S 2′.7.
23 47 08.8	−18 14 49	MCG −3-60-13	14.6	0.7 x 0.4	13.1	Sc	153	
23 47 13.2	−20 01 13	MCG −3-60-15	14.4	0.5 x 0.5	12.7	S?		Pair of faint stars E edge.
23 48 02.5	−18 35 55	MCG −3-60-17	13.8	0.7 x 0.7	12.9	(R′)SB(s)ab		**MCG −3-60-16** WNW 7′.5.
23 54 21.0	−17 18 23	MCG −3-60-23	14.1	0.5 x 0.5	12.5	S		
23 54 04.7	−25 27 18	MCG −4-1-1	13.9	0.6 x 0.4	12.2	S?	91	
23 55 04.6	−20 52 50	MCG −4-1-3	14.3	0.8 x 0.5	13.2	SAB(r)0⁺:	18	
23 55 50.6	−20 17 23	MCG −4-1-6	14.5	0.7 x 0.4	13.0	Sb	96	
23 57 34.5	−21 58 47	MCG −4-1-8	14.7	0.9 x 0.4	13.5	SBbc	57	Mag 8.79v star NE 9′.0; ESO 538-8 S 1′.4.

RA h m s	Dec ° ′ ″	Name	Mag (V)	Dim ′ Maj x min	SB	Type Class	PA	Notes
23 59 05.4	−20 45 15	MCG −4-1-9	14.3	0.7 x 0.4	12.8	S0	102	Multi-galaxy system?
00 14 55.5	−24 05 25	MCG −4-1-22	14.3	1.1 x 0.8	14.0	SB(rs)ab	125	
00 15 05.8	−24 41 31	MCG −4-1-23	14.8	0.9 x 0.7	14.1	Sbc	129	
00 15 07.2	−24 05 44	MCG −4-1-24	14.7	0.8 x 0.4	13.3	Sbc	114	
00 15 08.1	−24 02 40	MCG −4-1-25	14.1	1.1 x 0.5	13.3	Sab	150	
00 15 22.6	−24 03 38	MCG −4-1-27	14.3	0.7 x 0.3	12.5	S0	17	Bright knots or superimposed stars E and W edges.
00 17 04.8	−24 20 52	MCG −4-1-28	14.9	0.7 x 0.6	13.8	Sbc	93	
00 19 11.3	−22 40 07	MCG −4-2-3	15.2	0.4 x 0.4	13.1	Im: pec		Very compact; located on SW edge of galaxy cluster **A 20**.
00 19 09.3	−24 31 21	MCG −4-2-5	14.2	1.2 x 0.5	13.5	SB?	147	Broad, faint, streamer extending 0.′7 SW.
00 22 33.2	−22 18 11	MCG −4-2-7	14.9	0.6 x 0.5	13.4	Sc	21	
00 24 56.0	−20 43 53	MCG −4-2-13	13.5	0.5 x 0.5	11.8			**MCG −4-2-12** NW 1.′2; **MCG −4-2-11** W 2.′1.
00 35 06.7	−26 06 12	MCG −4-2-21	14.8	0.8 x 0.5	13.7	Sc	27	
00 36 23.0	−23 01 05	MCG −4-2-25	14.7	0.8 x 0.6	13.8	Sbc	3	
00 38 49.0	−24 13 25	MCG −4-2-31	16.1	0.7 x 0.5	15.1	E?	90	IC 1562 SW 4.′7.
00 42 40.0	−21 48 03	MCG −4-2-38	14.7	0.3 x 0.3	11.9	Sb		
00 42 42.4	−21 46 28	MCG −4-2-39	13.9	0.4 x 0.2	11.2	E	150	Immediately NE of mag 11 star.
00 43 44.9	−24 25 17	MCG −4-2-44	15.0	1.3 x 0.8	14.9	SB?	54	Mag 12.3 star N 3.′7.
23 27 21.2	−23 30 01	MCG −4-55-8	14.4	0.7 x 0.3	12.5	Sb	144	
23 29 41.8	−21 17 46	MCG −4-55-9	14.9	0.4 x 0.4	12.9	E		
23 37 34.7	−23 29 41	MCG −4-55-13	14.7	0.7 x 0.2	12.4	Sbc	63	
23 38 13.3	−20 46 59	MCG −4-55-14	14.0	1.0 x 0.7	13.5	SBb	142	Mag 7.70v star SE 8.′3.
23 39 25.5	−21 18 11	MCG −4-55-18	15.1	0.7 x 0.3	13.3	Sbc	6	
23 39 58.4	−22 47 19	MCG −4-55-21	14.5	0.5 x 0.2	11.8	Sbc	102	
23 44 38.7	−23 48 41	MCG −4-56-1	14.5	0.9 x 0.4	13.5	E	99	Pair of spherical galaxies. **MCG −4-56-2** E 0.′5.
23 54 36.1	−27 34 05	MCG −5-1-2	14.1	0.4 x 0.4	12.2	E		
23 56 09.7	−27 28 04	MCG −5-1-5	14.4	1.2 x 0.2	12.7	Sa-b	61	
23 56 24.3	−29 07 55	MCG −5-1-9	14.7	0.5 x 0.4	12.8	Sc	0	Mag 9.40v star NE 1.′9.
23 57 59.3	−29 52 00	MCG −5-1-14	13.8	0.8 x 0.4	12.5	S0?	125	Mag 10.28v star WNW 4.′8.
23 58 05.8	−29 45 42	MCG −5-1-16	13.6	0.4 x 0.3	11.2		87	Pair of mag 10-11 stars on W edge.
23 58 39.6	−29 50 45	MCG −5-1-18	14.3	0.3 x 0.3	11.7	E		**ESO 471-53** N 9.′0.
23 59 11.3	−28 35 56	MCG −5-1-19	14.6	0.7 x 0.4	13.0	S	18	
23 59 55.4	−27 14 05	MCG −5-1-21	14.5	0.6 x 0.4	12.8	Sc	15	
00 05 02.9	−27 42 55	MCG −5-1-29	14.6	0.4 x 0.4	12.5	Sab		
00 10 39.5	−29 15 12	MCG −5-1-39	14.4	0.7 x 0.5	13.1	S	42	
00 12 14.6	−27 11 24	MCG −5-1-42	14.0	1.1 x 0.6	13.4	S0(r)	6	
00 15 07.2	−29 12 41	MCG −5-1-46	14.1	1.0 x 0.3	12.6	Sb	12	
00 19 19.2	−26 42 40	MCG −5-2-1	14.4	0.7 x 0.5	13.3	E	9	
00 19 21.3	−26 41 36	MCG −5-2-2	14.8	0.5 x 0.4	13.2	E	30	
00 28 17.7	−27 58 57	MCG −5-2-5	15.3	1.5 x 0.4	14.6	S	123	Mag 11.10v star S 1.′1.
00 29 55.9	−27 29 52	MCG −5-2-9	14.3	1.0 x 0.4	13.2	Sa-b	82	
00 30 45.8	−29 20 38	MCG −5-2-12	14.3	0.9 x 0.6	13.5	S?	165	
00 31 13.6	−28 24 47	MCG −5-2-13	14.6	0.8 x 0.3	13.0	S0	3	
00 35 01.9	−26 27 11	MCG −5-2-20	14.5	0.7 x 0.4	13.2	E	0	
00 35 46.5	−28 29 05	MCG −5-2-22	14.8	0.7 x 0.3	13.0	S	0	
00 36 27.5	−27 47 08	MCG −5-2-23	13.6	1.0 x 0.5	12.7	S0?	102	
00 36 37.5	−27 47 20	MCG −5-2-24	14.6	0.6 x 0.4	12.9	S0?	27	
00 36 26.4	−28 17 55	MCG −5-2-25	15.0	0.7 x 0.3	13.2	SA(s)a	6	Star at N end; mag 9.77v star WNW 6.′0; ESO 410-27 SE 6.′9.
00 36 59.2	−29 33 01	MCG −5-2-29	14.9	0.4 x 0.2	12.2	E	9	
00 37 49.2	−26 38 58	MCG −5-2-30	13.8	1.0 x 1.0	13.7	SAB(r)0$^+$		
00 39 18.8	−29 56 48	MCG −5-2-32	14.4	0.5 x 0.5	12.8	S0?		Star on NE edge.
23 18 43.5	−27 07 52	MCG −5-55-1	13.9	0.8 x 0.5	12.8	S	147	
23 19 24.7	−28 01 24	MCG −5-55-2	14.5	0.7 x 0.7	13.6	S		
23 19 19.7	−28 56 34	MCG −5-55-3	13.6	1.2 x 0.8	13.5	(R)SB(s)0°	33	
23 21 43.3	−26 17 32	MCG −5-55-4	13.7	0.7 x 0.3	11.9	(R)SB(l)0$^+$	71	
23 23 08.2	−28 45 45	MCG −5-55-6	15.1	0.5 x 0.3	12.8	S	90	Mag 8.27v star NNE 6.′9.
23 24 55.1	−29 28 55	MCG −5-55-8	14.8	0.5 x 0.4	12.9	S	24	
23 26 27.4	−27 41 12	MCG −5-55-11	14.5	0.8 x 0.8	13.8	(R′)SB(rs)d:		
23 27 55.1	−27 58 55	MCG −5-55-13	14.5	0.9 x 0.7	13.9	SB(rs)bc	159	
23 28 45.7	−29 32 46	MCG −5-55-14	14.2	0.7 x 0.5	12.9	S0	99	
23 30 32.9	−28 48 53	MCG −5-55-18	13.9	1.0 x 0.6	13.2	(R)SB(r)ab:	43	Lies 1.′8 N of a mag 10.86v star.
23 30 55.8	−27 40 16	MCG −5-55-19	14.1	0.5 x 0.5	12.5	S0?		
23 31 00.0	−27 31 33	MCG −5-55-20	14.4	0.7 x 0.4	12.9	S0	90	
23 32 25.0	−27 46 17	MCG −5-55-24	13.8	0.8 x 0.8	13.2	S?		ESO 470-18 NW 5.′6.
23 35 35.9	−26 40 33	MCG −5-55-26	14.2	0.7 x 0.4	12.6	S0	27	
23 36 49.8	−26 59 32	MCG −5-55-27	14.2	0.7 x 0.7	13.3	Sc		Lies 1.′0 NE of mag 11 star.
23 36 37.8	−29 04 14	MCG −5-55-28	14.9	0.4 x 0.4	12.7	Sb?		
23 41 29.8	−29 19 17	MCG −5-55-32	14.2	0.4 x 0.4	12.3	E		faint star on E edge.
23 41 35.5	−29 14 11	MCG −5-55-33	14.3	0.5 x 0.5	12.9	E		
23 41 50.9	−28 01 30	MCG −5-55-34	14.4	0.9 x 0.4	13.1	S	156	
23 46 15.4	−28 05 59	MCG −5-56-4	13.9	1.1 x 0.4	12.9	S0?	118	**IC 5349** N 5.′9.
23 46 36.9	−29 04 14	MCG −5-56-7	13.5	0.9 x 0.6	12.7	S0?	170	
23 46 59.9	−29 10 20	MCG −5-56-8	14.8	0.7 x 0.3	13.0	S	96	
23 47 44.8	−27 29 13	MCG −5-56-12	14.1	1.0 x 0.4	12.9	Sa	19	
23 47 55.9	−28 19 15	MCG −5-56-15	14.4	0.8 x 0.5	13.2	S?	135	
23 49 15.8	−29 01 52	MCG −5-56-18	14.0	0.9 x 0.6	13.2	Sa:	22	
23 50 13.6	−29 00 34	MCG −5-56-21	14.0	0.6 x 0.6	12.9	E		**MCG −5-56-20** S 7.′6.
23 50 38.0	−28 26 09	MCG −5-56-22	14.0	0.7 x 0.5	12.9	E	24	Very slender anonymous galaxy perpendicular to E edge; a second anonymous galaxy ENE 1.′8; mag 12.0 star NE 5.′0.
23 51 54.3	−27 55 48	MCG −5-56-25	14.0	0.8 x 0.6	13.1	S0	141	**MCG −5-56-24** 2.′3 SSW.
23 52 10.3	−29 04 46	MCG −5-56-26	13.9	0.8 x 0.8	13.5	E		
23 52 24.3	−29 01 24	MCG −5-56-29	14.4	0.4 x 0.4	12.5	E		
23 53 25.4	−29 32 19	MCG −5-56-33	13.9	0.5 x 0.5	12.5	E		
00 08 20.6	−29 55 02	NGC 7	13.9	2.2 x 0.5	13.8	SB(s)c? sp III-IV	29	
00 09 56.6	−24 57 40	NGC 24	11.6	5.8 x 1.3	13.7	SA(s)c III	46	
00 14 05.2	−23 10 55	NGC 45	10.6	8.5 x 5.9	14.7	SA(s)dm IV	142	Numerous bright knots, stars and clumps. A mag 9.88v star lies 1.′7 S of the center; a mag 6.88 star lies outside, 4.′8 SW.
00 15 25.5	−21 26 43	NGC 59	12.4	2.6 x 1.3	13.6	SA(rs)0$^-$:	127	
00 18 58.7	−22 52 48	NGC 65	13.9	0.8 x 0.6	12.9	SAB(rs)0$^-$:	5	Located 2.′7 NW of mag 9.55v star.
00 19 04.8	−22 56 12	NGC 66	13.5	1.2 x 0.7	13.1	SB(r)b pec I-II	32	Mag 9.55v star 1.′4 N.

(Continued from previous page)

141 GALAXIES 141

RA h m s	Dec ° ′ ″	Name	Mag (V)	Dim ′ Maj x min	SB	Type Class	PA	Notes
00 20 01.7	−22 31 57	NGC 77	14.8	0.4 x 0.4	12.7	SA0$^-$:		Stellar nucleus; mag 9.67v star NNW 8'.7; located on NE edge of galaxy cluster **A 20**.
00 31 07.6	−22 37 09	NGC 142	13.8	1.1 x 0.6	13.2	SB(s)b? pec I	101	NGC 144 ESE 3'.4; close pair of N-S oriented mag 14-15 stars N 0'.9; NGC 143 N 3'.9.
00 31 15.5	−22 33 39	NGC 143	14.4	1.0 x 0.3	13.0	SB(r)b? sp	20	Mag 13.5 star E 3'.3; NGC 142 S 3'.9.
00 31 20.9	−22 38 48	NGC 144	13.8	0.8 x 0.8	13.1	Sc: pec I-II		
00 34 15.7	−27 48 15	NGC 150	11.4	3.9 x 1.9	13.4	SB(rs)b: II	118	**ESO 410-17** W 8'.3.
00 35 23.0	−23 22 30	NGC 167	13.7	1.0 x 0.7	13.2	SB(r)c? II	171	
00 36 38.8	−22 35 34	NGC 168	14.0	1.2 x 0.3	12.8	S0$^-$? sp	26	Mag 10.39v star N 5'.5.
00 37 13.6	−22 35 14	NGC 172	13.4	2.0 x 0.3	12.7	SB(r)bc? sp	12	NGC 177 E 5'.3.
00 36 59.1	−29 28 38	NGC 174	12.9	1.4 x 0.6	12.6	SB(rs)0/a:	152	Lies in a small triangle of mag 14 stars.
00 37 21.7	−19 56 03	NGC 175	12.2	2.1 x 1.9	13.5	SB(r)ab I-II		Strong SE-NW bar with prominent dark patches NE and SW of bar.
00 37 34.4	−22 32 59	NGC 177	13.3	2.2 x 0.5	13.3	SA(r)ab	9	Mag 10.73v star E 6'.4; NGC 172 W 5'.3.
00 37 46.2	−17 50 58	NGC 179	13.3	0.9 x 0.8	12.8	SAB0$^-$	113	Faint star on N edge; mag 10.28v star N 5'.1.
00 39 03.7	−18 36 30	NGC 209	12.9	1.1 x 1.1	13.0	SA0$^-$ pec:		Almost stellar nucleus, faint envelope; mag 9.57v star NNW 8'.1.
00 41 27.2	−21 02 43	NGC 216	13.2	2.0 x 0.7	13.5	S0°? sp	27	ESO 540-14 SSW 6'.4.
00 42 27.0	−23 37 47	NGC 230	14.5	1.2 x 0.2	12.8	Sa? pec sp	44	Mag 11.9 star NW 0'.8; **IC 1573** NW 4'.6.
00 42 45.9	−23 33 41	NGC 232	13.3	0.9 x 0.7	12.6	SB(r)a? pec	42	NGC 235 and NGC 235A NE 2'.0.
00 42 52.8	−23 32 32	NGC 235	13.2	1.3 x 0.7	12.9	S0 pec	117	NGC 235A on SE edge of core; NGC 232 SW 2'.0.
00 42 53.8	−23 32 48	NGC 235A	13.0	0.4 x 0.4	11.0	E0? pec		Located SE of core of NGC 235.
23 16 26.6	−22 09 18	NGC 7573	13.7	1.2 x 1.0	13.7	SB(s)c I	42	Located on W edge of small galaxy cluster **A 2568**.
23 22 33.0	−29 16 49	NGC 7636	13.6	0.8 x 0.5	12.5	S0°	30	
23 23 47.5	−29 23 18	NGC 7645	12.9	1.4 x 1.2	13.3	SB(r)c I-II	165	
23 24 32.1	−19 03 26	NGC 7656	13.5	1.3 x 0.9	13.5	S0 pec:	50	Multi-galaxy system; broad, faint extensions NE and NW from bright core forming a "V."
23 38 02.6	−22 58 30	NGC 7719	14.2	0.8 x 0.6	13.3	SA(rs)ab? pec	82	Faint star S edge.
23 40 46.1	−20 30 41	NGC 7730	14.0	0.9 x 0.6	13.2	S	129	
23 42 25.3	−19 27 07	NGC 7736	12.9	1.7 x 1.5	13.8	(R′)SA(r)0°?		
23 45 47.4	−29 31 07	NGC 7749	12.8	1.6 x 1.1	13.3	SA0°	28	
23 48 55.1	−22 01 28	NGC 7758	14.1	0.7 x 0.5	12.9	SAB(s)0$^-$:	118	Almost stellar nucleus.
00 00 26.7	−18 50 31	NGC 7807	14.6	0.7 x 0.5	13.3	S0	27	Mag 10.39v star E 6'.6.

GALAXY CLUSTERS

RA h m s	Dec ° ′ ″	Name	Mag 10th brightest	No. Gal	Diam ′	Notes
00 15 12.0	−23 53 00	A 14	15.2	35	39	ESO 473-5 and mag 10.02v star at center of cluster.
00 38 54.0	−22 18 00	A 74	15.9	49	13	Mag 9.00v star on W edge; all members faint and stellar.
00 42 30.0	−21 47 00	A 86	15.9	31	13	Pair of mag 8.58v, 8.74v stars 8'.9 NE of cluster center.
23 45 18.0	−25 58 00	A 2660	16.4	45	34	**ESO 537-5** near N edge, other members anonymous, stellar.
00 02 54.0	−27 10 00	A 2716	16.3	44	17	All members anonymous.
00 11 18.0	−28 52 00	A 2734	16.3	58	17	ESO 409-25 and **MCG −5-1-40** at center; **ESO 409-24** W 11'.6; other members anonymous.
00 38 00.0	−25 05 00	A 2800	15.8	59	17	All members anonymous, stellar/almost stellar.
23 47 42.0	−28 08 00	A 4038	14.2	117	28	Numerous PGC galaxies within 10'.0 radius from center.
23 51 36.0	−28 22 00	A 4049	15.0	39	22	IC 5362 at center; **MCG −5-56-30** E 11'.8.
00 02 48.0	−29 55 00	AS 2	16.1	11	17	All members anonymous, stellar.
00 19 24.0	−17 11 00	AS 28	15.8	3	17	All members anonymous, stellar.
23 46 00.0	−23 15 00	AS 1146	16.1		17	All members anonymous, stellar.
23 50 12.0	−29 01 00	AS 1155	15.6	3	22	Most members anonymous, stellar.
23 56 00.0	−18 10 00	AS 1163	16.1	8	17	All members anonymous, stellar.
23 58 42.0	−29 51 00	AS 1165	15.7	3	17	Most members anonymous, stellar.
00 01 24.0	−27 32 00	AS 1171	16.1	8	17	Except ESO 409-1, 409-3, all are anonymous, stellar.

OPEN CLUSTERS

RA h m s	Dec ° ′ ″	Name	Mag	Diam ′	No. ★	B ★	Type	Notes
00 04 12.0	−29 56 00	Bl 1	4.5	70	30	8.0	cl	Zeta Scl cluster, large and bright.

142 GALAXIES 142

RA h m s	Dec ° ′ ″	Name	Mag (V)	Dim ′ Maj x min	SB	Type Class	PA	Notes
21 57 55.2	−28 48 18	ESO 466-21	12.4	1.1 x 0.8	12.2	SA0$^-$:	178	Mag 10.90v star on E edge.
21 58 10.6	−27 23 33	ESO 466-24	13.0	1.1 x 0.6	12.4	Sb	124	Mag 10.41v star N 3'.6.
21 58 43.4	−28 28 01	ESO 466-26	12.6	1.4 x 1.1	13.0	SAB0$^-$	49	Mag 12.4 star S 4'.6.
21 59 13.4	−28 41 26	ESO 466-28	14.0	1.9 x 0.5	13.8	SAB(s)cd: II-III	21	Mag 13;0 star N edge; mag 7.40v star NE 8'.8.
22 06 14.6	−27 57 32	ESO 467-3	14.3	2.2 x 1.3	15.3	(R′)SAB(r)bc pec:	19	Double system, small companion galaxy NW edge; mag 8.53v star S 4'.0.
22 09 14.2	−27 46 58	ESO 467-13	14.6	0.9 x 0.9	14.3	S?		= **Hickson 91C**. Almost stellar MCG −5-52-35 NNE of center; **ESO 457-13** NE 2'.1.
22 09 16.3	−27 43 51	ESO 467-15	14.8	1.1 x 0.2	13.0	S?	167	= **Hickson 91B**.
22 09 50.4	−27 32 10	ESO 467-16	13.8	1.2 x 1.1	14.0	(R′)SB(r)a	164	Slightly elongated E-W core with star superimposed E side; faint envelope.
22 13 31.1	−27 33 38	ESO 467-23	12.9	1.5 x 0.4	12.2	Sab pec:	39	Mag 8.88v star N 6'.2.
22 14 35.6	−29 09 36	ESO 467-26	12.9	1.0 x 0.9	12.6	Sa	37	Mag 8.88v star NW 4'.2.
22 14 39.3	−27 27 52	ESO 467-27	12.6	1.7 x 0.7	12.7	(R′)SAB(rs)bc: II	60	**MCG −5-52-53** WSW 5'.1; mag 8.88v star W 13'.4.
22 14 55.0	−27 56 01	ESO 467-30	13.6	1.8 x 0.5	13.3	(R′)SA(rs)b	102	Mag 12.3 star N 3'.1; **ESO 467-31** S 7'.1; mag 5.44v star Lambda PsA NW 12'.8.
22 16 14.9	−27 24 11	ESO 467-37	13.0	1.3 x 0.7	12.8	S0$^-$: sp	66	Pair of N-S oriented mag 11.7, 11.9 stars W 9'.6; **MCG −5-52-58** SW 8'.7.
22 23 17.4	−28 58 54	ESO 467-51	13.9	3.0 x 0.3	13.7	Scd: sp	99	NGC 7259 NW 3'.0.
22 24 35.3	−27 49 34	ESO 467-53	14.0	0.9 x 0.8	13.5	(R′)SBa	27	Mag 9.46v star W 10'.2.
22 31 19.2	−28 24 07	ESO 468-6	14.5	1.5 x 1.2	14.9	SB(s)d pec	165	Mag 10.35v star E 4'.1; **MCG −5-53-12** S 8'.3.
23 03 50.6	−28 44 31	ESO 469-8	15.3	1.6 x 0.5	14.9	SB(s)m: V	54	Mag 12.9 star W 3'.0.
23 07 26.7	−27 48 57	ESO 469-14	13.7	0.5 x 0.4	11.8		122	Compact, star superimposed or very bright core; mag 12.4 star S 1'.1.
23 10 25.3	−29 43 53	ESO 469-17	13.4	1.6 x 0.6	13.2	Sbc?	120	Mag 12.4 star S 4'.7.
22 01 45.8	−23 42 25	ESO 532-11	14.1	1.0 x 0.7	13.5	SB(s)cd II	129	Mag 10.39v star E 2'.4; mag 7.50v star NE 8'.3.
22 02 57.3	−22 28 23	ESO 532-14	14.3	1.3 x 0.6	13.9	SAB(s)cd pec: III	109	Mag 10.64v star SSW 6'.3; **ESO 532-15** NNE 5'.4.
22 06 04.0	−26 11 04	ESO 532-19	14.8	1.5 x 0.1	12.6	Sc:	142	Lies 0'.8 SW of 9.51v mag star.
22 08 13.8	−25 03 39	ESO 532-21	13.8	1.0 x 0.5	12.8	SB(r)a?	156	Mag 12.0 star E 4'.4.
22 09 11.3	−27 09 59	ESO 532-22	13.8	1.5 x 0.7	13.7	SB(rs)cd: III-IV	127	Mag 11.9 star E 2'.6; mag 9.34v stars SW 3'.5; MCG −5-52-33A SSW 4'.9.
22 10 53.3	−25 04 24	ESO 532-26	13.6	1.5 x 0.6	13.4	S0$^+$ sp	105	Strong dark lanes N of center; mag 10.04v star W 10'.1.
22 14 03.1	−26 56 17	ESO 533-4	13.3	2.4 x 0.4	13.1	Sc: sp I	151	Mag 11.28v star WSW 7'.8.
22 15 44.2	−25 58 36	ESO 533-5	14.5	1.4 x 0.4	13.7	Sab? sp	69	Mag 7.91v star W 7'.4.

(Continued from previous page)

GALAXIES

RA h m s	Dec ° ′ ″	Name	Mag (V)	Dim ′ Maj x min	SB	Type Class	PA	Notes
22 17 48.8	−23 53 30	ESO 533-9	13.5	1.0 x 0.9	13.3	S0	146	Mag 10.22v star W 2′.8.
22 18 06.8	−25 41 22	ESO 533-10	14.9	1.5 x 1.0	15.2	SA(s)m V	42	Mag 10.52v star E 3′.2.
22 19 50.6	−26 20 30	ESO 533-14	13.1	0.9 x 0.6	12.3	SA?(r)	136	Mag 9.74v star NNE 7′.2.
22 22 20.8	−26 52 53	ESO 533-18	13.9	0.9 x 0.7	13.3	S0/a	19	Mag 12.3 star WNW 7′.8.
22 22 32.2	−23 31 27	ESO 533-20	13.8	1.4 x 0.8	13.8	S0 pec	6	Pair with ESO 533-21 overlapping; mag 11.11v star NNW 5′.1; **MCG −4-52-39** SSW 6′.9.
22 22 33.1	−23 30 54	ESO 533-21	13.4	1.4 x 0.8	13.3	S0 pec	0	Pair with ESO 533-20 overlapping; mag 11.11v star NNW 5′.1.
22 25 30.8	−25 38 44	ESO 533-25	12.6	1.7 x 1.5	13.5	(R)SB(rs)0$^+$	73	Mag 9.37v star NE 9′.9.
22 26 39.2	−24 52 17	ESO 533-28	13.1	1.2 x 0.6	12.6	Sc? III-IV	110	Mag 9.52v star NW 15′.9; note: MCG −4-53-1 located 3′.5 E of this star.
22 30 07.6	−26 46 26	ESO 533-35	13.7	1.5 x 1.4	14.3	(R)SB(rs)0$^+$:	24	Mag 10.70v star W 4′.6.
22 30 38.7	−26 55 58	ESO 533-37	14.7	1.1 x 0.5	13.9	SB(s)d III-IV	99	Mag 9.49v star WSW 6′.8.
22 32 29.8	−25 39 43	ESO 533-45	12.6	2.0 x 1.7	13.8	SB(r)b II	38	Pair of mag 9.45v, 8.96v stars SSE 11′.7.
22 34 50.0	−25 40 39	ESO 533-50	13.2	1.5 x 1.3	13.8	(R)SB(r)0/a	60	Mag 10.31v star E 4′.0.
22 36 06.8	−26 18 56	ESO 534-1	13.7	1.6 x 0.5	13.3	SB(s)m: pec IV-V	8	Located 1′.2 N of mag 9.59v star; ESO 534-3 NE 7′.5.
22 36 27.8	−24 20 32	ESO 534-2	12.5	1.1 x 0.8	12.5	E?	24	MCG −4-53-19 W 1′.4.
22 36 36.6	−26 15 34	ESO 534-3	13.9	1.5 x 0.4	13.2	IBm? pec sp	68	Mag 10.68v star N 7′.1; ESO 534-1 SW 7′.5.
22 37 03.1	−25 14 14	ESO 534-4	13.2	1.9 x 1.7	14.4	SB(s)dm: IV-V	76	Mag 9.62v star NW 13′.2.
22 38 41.4	−25 51 06	ESO 534-9	12.3	3.1 x 0.9	13.2	SA(s)ab? sp	28	Mag 13.0 star on E edge; mag 10.79v star NW 1′.8.
22 38 51.5	−25 42 33	ESO 534-10	12.7	1.0 x 1.0	12.6	S0		Mag 7.82v star WNW 9′.3.
22 39 17.0	−26 30 37	ESO 534-13	13.0	1.3 x 0.8	12.9	SAB(s)0$^-$ pec	39	**MCG −5-53-24** N 6′.9.
22 45 14.8	−22 43 49	ESO 534-24	12.9	1.8 x 1.2	13.6	SAB(rs)d: III	111	Mag 10.67v star E 2′.9; mag 9.39v star N 6′.8.
22 56 45.2	−24 57 14	ESO 534-32	13.9	1.6 x 0.6	13.7	SAB(s)c: I	171	Mag 10.49v star W 7′.6.
22 59 01.3	−25 31 42	ESO 535-1	13.3	1.7 x 1.4	14.1	(R)SA(rs)b: I-II	45	Very small galaxy SW edge.
23 19 02.7	−22 42 59	ESO 535-15	13.5	1.3 x 0.9	13.5	SAB(s)bc pec I	99	Mag 10.80v star W 2′.5; **MCG −4-55-4** NNE 5′.0.
23 19 10.8	−23 26 44	ESO 535-16	14.5	1.5 x 0.6	14.2	Sa: pec	135	Star SE end.
23 22 01.0	−23 30 31	ESO 536-2	13.3	1.2 x 0.8	13.1	SA(s)0$^-$: pec	0	Small galaxy on E edge.
22 01 30.5	−22 04 15	ESO 601-4	13.8	1.8 x 0.5	13.5	SB(s)c: II-III	57	Mag 9.41v star E 10′.0.
22 01 37.8	−21 30 43	ESO 601-5	14.0	1.6 x 0.3	13.1	Sb:	65	Mag 10.33v star N 4′.3.
22 02 21.5	−21 02 23	ESO 601-7	14.3	1.7 x 0.4	13.7	SB(s)m: IV-V	148	Mag 10.70v star W 6′.1.
22 03 51.4	−19 52 51	ESO 601-12	13.5	1.8 x 0.5	13.2	(R′)SAB(r)ab: II	178	Mag 10.33v star W 9′.1.
22 06 21.8	−21 04 17	ESO 601-18	13.3	1.5 x 0.6	13.2	E/S0	160	
22 07 27.1	−20 50 34	ESO 601-19	14.1	1.5 x 0.5	13.7	SB(s)cd: III-IV	67	Mag 7.40v star NNE 5′.7.
22 10 12.4	−18 51 54	ESO 601-25	14.2	2.0 x 0.7	14.5	SAB(s)m: IV-V	70	Mag 10.16v star S 3′.6; **ESO 601-26** E 2′.6.
22 13 39.9	−21 44 03	ESO 601-31	14.6	1.2 x 0.7	14.3	IB(s)m V	111	Mag 11.51v star W 6′.0.
22 14 54.5	−21 15 51	ESO 601-34	16.1	0.8 x 0.5	14.9	dE	12	Mag 9.78v star N 4′.7; mag 5.33v star 41 Aquarii NW 14′.2.
22 16 51.2	−21 15 02	ESO 602-3	14.3	2.3 x 0.7	14.7	IB(s)m: IV	90	Double system; mag 9.37v star WSW 11′.7.
22 22 51.1	−20 20 59	ESO 602-15	14.5	1.7 x 0.3	13.7	SB(s)d III-IV	82	Mag 8.23v star E 9′.6.
22 31 25.6	−19 02 04	ESO 602-25	13.3	1.7 x 0.6	13.1	(R′)SB(rs)bc pec? II-III	171	Mag 9.66v star SW 4′.9.
22 36 51.3	−19 48 24	ESO 602-30	13.4	1.1 x 0.9	13.3	SB?	66	Mag 10.85v star NNW 9′.4.
22 36 56.0	−22 13 17	ESO 602-31	13.6	1.6 x 0.6	13.4	(R)SB(r)b: II-III	55	Mag 10.40v star NW 11′.5.
22 39 05.9	−19 39 45	ESO 603-1	13.8	1.2 x 0.5	13.1	SB(r)b I	147	Close pair of N-S oriented mag 10.35v, 9.14v stars E 9′.3.
22 44 26.9	−20 02 07	ESO 603-8	15.5	1.6 x 1.5	16.3	IB(s)m V-VI	111	Mag 10.44v star SE 7′.5.
22 46 05.2	−19 24 54	ESO 603-12	13.9	2.2 x 0.4	13.6	SB(s)dm: III-IV	96	Pair of E-W oriented mag 10.39v, 10.49v stars SSW 5′.8.
22 51 01.6	−20 16 06	ESO 603-20	14.1	1.9 x 0.3	13.3	S pec sp	56	**ESO 603-21** E 5′.0.
22 54 35.5	−18 00 00	ESO 603-27	15.1	1.3 x 0.7	14.9	IB(s)m V-VI	36	Mag 8.63v star WSW 9′.5.
22 56 24.1	−21 22 28	ESO 603-29	13.8	1.0 x 0.5	12.9	S0	83	Mag 11.4 star N 6′.5.
23 14 54.3	−20 59 46	ESO 604-6	14.2	1.8 x 0.3	13.4	Sbc sp	113	Mag 10.44v star SW 12′.1.
23 21 14.2	−21 14 11	ESO 605-3	14.0	1.7 x 0.4	13.4	S?	33	Mag 10.98v star N 5′.7.
23 23 38.5	−19 00 37	ESO 605-4	13.8	1.5 x 1.0	14.1	(R′)SB(rs)c: II-III	138	Mag 7.91v star S 6′.1.
21 58 18.4	−17 10 37	IC 1412	13.5	0.9 x 0.4	12.3	(R′)SB0$^+$ pec	99	Mag 12.7 star SSW 3′.2.
22 13 26.1	−22 05 50	IC 1435	13.1	1.2 x 0.7	12.8	SA(rs:)b	8	Mag 12.0 star NE 6′.0.
22 16 29.1	−21 25 52	IC 1438	11.9	2.4 x 2.0	13.5	(R′)SAB(rs)a:	2	Mag 9.37v star NW 10′.4; IC 1439 SE 4′.2.
22 16 40.2	−21 29 12	IC 1439	13.7	1.3 x 0.7	13.5	(R)SB(rs)a:	33	IC 1438 NW 4′.2.
22 19 03.6	−20 56 25	IC 1443	12.5	0.8 x 0.7	11.9	E2:	36	Mag 8.62v star SW 9′.3.
22 25 30.3	−17 14 36	IC 1445	12.7	1.5 x 1.1	13.1	SA(s)0$^-$:	80	Mag 10.50v star W 10′.7.
21 58 59.2	−27 24 50	IC 5149	13.6	1.3 x 0.6	13.2	(R′)Sa	32	Mag 11.7 star NW 2′.2.
22 12 33.4	−22 57 18	IC 5178	13.8	1.1 x 0.9	13.7	SB(s)c II-III	88	Star superimposed S of nucleus.
22 22 31.3	−18 52 12	IC 5210	13.0	1.2 x 1.1	13.1	SA0$^-$:		IC 5211 E 2′.3; mag 12.0 star S 2′.4.
22 22 43.1	−18 52 48	IC 5211	13.5	1.2 x 0.5	12.8	S?	160	IC 5210 W 2′.3.
22 31 02.6	−25 20 37	IC 5225	14.6	1.4 x 0.5	14.1	Sa-b	117	Mag 10.21v star N 7′.6; note: **ESO 533-40** is 1′.4 SW of this star.
22 54 25.2	−20 21 46	IC 5261	13.2	1.5 x 1.3	13.7	SB(s)cd II-III	59	Mag 9.39v star W 5′.5.
23 12 53.4	−23 28 12	IC 5290	13.2	1.3 x 0.8	13.1	(R)Sa:	54	Mag 10.90v star NW 10′.8.
22 08 22.6	−19 04 22	MCG −3-56-7	13.9	0.8 x 0.7	13.1	SB(r)bc	60	Mag 9.29v star on S edge.
22 12 24.9	−17 43 10	MCG −3-56-10	13.3	0.9 x 0.9	12.9	Sa		
22 22 02.8	−19 34 26	MCG −3-57-3	13.7	1.0 x 0.4	12.5	Sb-c	99	Mag 7.86v star N 8′.3.
22 29 33.5	−19 49 24	MCG −3-57-10	13.5	1.1 x 0.8	13.3	SB?	117	Mag 9.86v star and MCG −3-57-11 N 5′.3.
22 29 39.1	−19 45 02	MCG −3-57-11	14.0	1.1 x 0.8	14.0	E	30	Mag 9.86v star NE 1′.3.
22 31 13.3	−17 19 41	MCG −3-57-16	14.1	0.4 x 0.4	11.9			
22 31 55.8	−17 08 49	MCG −3-57-18	15.0	1.2 x 0.6	14.5	SB(s)c II	159	
22 33 00.0	−20 05 37	MCG −3-57-20	13.8	1.2 x 1.1	14.0	S?	38	Mag 9.58v star NE 4′.4.
22 47 42.3	−17 20 58	MCG −3-58-3	14.0	1.1 x 0.5	13.2	S?	160	Star on SW edge.
22 47 48.0	−17 06 03	MCG −3-58-4	14.4	1.3 x 0.2	12.7	S?	108	Mag 9.25v star NW 3′.7.
22 49 37.0	−19 16 28	MCG −3-58-7	14.0	0.5 x 0.5	12.4	(R′)SAB(s)0/a		Located on the W edge of galaxy cluster A 2492.
22 57 26.5	−17 53 09	MCG −3-58-14	14.1	0.9 x 0.5	13.1	Sa	108	
21 58 15.4	−24 53 22	MCG −4-51-16	13.9	0.9 x 0.5	12.9	(R′)SBa	14	
21 59 21.2	−23 15 18	MCG −4-51-17	14.5	0.7 x 0.5	13.2	Sc	108	**MCG −4-51-18** at E end.
22 01 51.4	−22 29 10	MCG −4-52-6	13.8	1.2 x 0.6	13.3	(R′)SAB(r)0/a	174	mag 9.66v star S 2′.4.
22 05 32.8	−22 17 01	MCG −4-52-13	14.7	0.9 x 0.7	14.1	SBbc	141	
22 09 11.6	−20 28 42	MCG −4-52-17	14.5	0.9 x 0.4	13.2	SBb	9	
22 10 34.0	−22 39 23	MCG −4-52-18	13.3	1.6 x 1.5	14.1	SB(r)bc II	102	Short N-S bar; knot or star SW edge.
22 13 10.4	−22 26 43	MCG −4-52-24	13.4	0.9 x 0.5	12.3	Sbc	42	
22 15 09.3	−20 59 03	MCG −4-52-27	14.1	1.0 x 0.8	13.7	(R′)IB(rs)m V	8	Mag 5.33v star 41 Aquarii SW 13′.2.
22 19 06.4	−24 10 58	MCG −4-52-34	14.4	0.5 x 0.5	12.7	Sa		Mag 9.73v star SW 2′.8; numerous stellar anonymous galaxies W and SW.
22 20 18.2	−22 01 55	MCG −4-52-35	14.5	0.7 x 0.4	12.9	SBc	129	Mag 7.70v star W 4′.1.
22 21 52.8	−25 34 35	MCG −4-52-37	14.5	0.4 x 0.4	12.6	E		
22 21 53.5	−25 35 26	MCG −4-52-38	14.3	0.4 x 0.4	12.2	S0		
22 23 19.3	−22 31 36	MCG −4-52-44	14.0	1.0 x 0.9	13.7	SB(r)b:	148	Mag 10.67v star ESE 3′.8.
22 25 34.0	−24 14 32	MCG −4-52-46	14.0	1.2 x 0.4	13.1	Sbc	141	
22 25 56.2	−24 43 49	MCG −4-53-1	14.1	0.8 x 0.5	13.0	Sc	69	Mag 9.52v star W 3′.5.
22 28 17.7	−22 26 15	MCG −4-53-3	13.5	1.0 x 0.9	13.3	(R)SB0$^+$	153	

GALAXIES

RA h m s	Dec ° ′ ″	Name	Mag (V)	Dim ′ Maj x min	SB	Type Class	PA	Notes
22 31 58.8	−25 21 40	MCG −4-53-8	15.1	0.7 x 0.2	12.8	Sc	147	NGC 7294 SE 3′.0.
22 33 38.5	−24 45 52	MCG −4-53-12	14.3	0.6 x 0.5	12.9	S0	42	**MCG −4-53-11** W 7′.9.
22 34 29.8	−24 39 54	MCG −4-53-13	14.1	0.5 x 0.4	12.4	E?	165	
22 34 31.8	−22 41 34	MCG −4-53-14	14.0	1.6 x 0.5	13.6	S?	36	Mag 8.47v star N 5′.9.
22 35 26.2	−25 04 33	MCG −4-53-17	13.8	0.9 x 0.7	13.1	S0	34	
22 36 22.4	−24 20 50	MCG −4-53-19	14.5	0.5 x 0.3	12.5	E		ESO 534-2 E 1′.4.
22 37 21.0	−24 14 46	MCG −4-53-25	13.2	1.1 x 0.7	13.0	E/S0	58	Mag 9.97v star N 3′.0.
22 39 13.9	−25 50 42	MCG −4-53-28	14.1	1.3 x 0.5	13.5	Sbc	12	Multi-galaxy system; mag 10.63v star SW 3′.1.
22 42 47.5	−21 09 58	MCG −4-53-30	13.4	1.3 x 0.9	13.4	Pec	38	Mag 8.74v star on S edge; **ESO 603-7** S 3′.7.
22 43 01.0	−24 19 51	MCG −4-53-31	14.3	0.7 x 0.4	12.7	Sb	12	Mag 8.68v star ESE 6′.8.
22 44 34.8	−22 59 32	MCG −4-53-32	13.9	1.0 x 0.3	12.5	Sc	86	Galaxy **ESO 534-21A** on N edge.
22 45 52.1	−20 50 35	MCG −4-53-37	14.1	0.6 x 0.5	12.6	S?	23	Star on E edge.
22 55 40.2	−23 57 17	MCG −4-54-2	13.7	1.2 x 0.8	13.5	S?	116	Mag 10.32v star N 1′.1.
23 02 33.0	−24 19 24	MCG −4-54-5	14.5	0.6 x 0.6	13.2	Sc		
23 09 58.8	−25 28 43	MCG −4-54-9	15.0	0.7 x 0.3	13.1	SBbc	60	
23 10 07.7	−25 29 21	MCG −4-54-10	14.7	0.4 x 0.3	12.2	Sbc	3	Lies 0′.8 W of mag 10 star.
23 12 06.9	−23 33 18	MCG −4-54-12	15.2	0.4 x 0.3	12.7	Sb		MCG −4-54-13 S 3′.6; mag '0.77v star S 5′.5.
23 12 13.5	−23 36 31	MCG −4-54-13	14.8	0.7 x 0.5	13.5	S0	60	Mag 10.77v star S 2′.2; MCG −4-54-12 N 3′.6.
21 56 44.0	−28 24 54	MCG −5-51-24	13.9	1.1 x 0.4	12.9	S0	68	
21 57 01.9	−28 37 12	MCG −5-51-25	13.4	1.0 x 0.8	13.1	SB0	76	Multi-galaxy system.
21 57 27.7	−28 44 17	MCG −5-51-26	13.5	0.9 x 0.6	12.7	(R′)SB(r)a	151	Mag 7.92v star SW 10′.1.
21 57 29.6	−29 23 41	MCG −5-51-27	14.5	0.6 x 0.4	12.9	S0	54	
22 00 29.9	−26 54 30	MCG −5-52-2	14.7	0.7 x 0.7	13.8	Sb-c		
22 00 39.5	−28 12 28	MCG −5-52-3	14.2	0.7 x 0.5	12.9	S	84	Star on S edge.
22 00 57.2	−28 21 48	MCG −5-52-4	13.5	1.1 x 0.5	12.8	S0	148	Faint anonymous galaxy NE 1′.6; mag 5.63v star Eta Piscis Austrini N 5′.8.
22 01 18.5	−28 29 43	MCG −5-52-5	14.1	0.9 x 0.9	13.7	Sc		Mag 5.63v star Eta Piscis Austrini SW 6′.7.
22 02 39.6	−28 59 20	MCG −5-52-13	14.1	1.0 x 0.7	13.6	Sb	145	Mag 9.51v star ESE 6′.2.
22 02 42.4	−28 05 41	MCG −5-52-15	14.3	0.5 x 0.5	12.7	S0		
22 03 12.9	−26 24 27	MCG −5-52-16	14.0	0.7 x 0.4	12.5	S?	99	
22 03 22.9	−28 46 16	MCG −5-52-17	13.6	1.3 x 0.3	12.4	S0/a	145	
22 03 44.5	−27 47 56	MCG −5-52-18	13.3	1.2 x 1.0	13.4	S?	66	**MCG −5-52-19** S 8′.1.
22 06 34.1	−27 59 33	MCG −5-52-25	14.6	1.1 x 0.5	13.8	Sa	102	Star or bright knot E end; mag 8.53v star WSW 5′.6.
22 07 12.5	−29 50 20	MCG −5-52-30	14.4	0.8 x 0.7	13.6	S	96	
22 08 55.8	−27 13 25	MCG −5-52-33A	14.0	0.6 x 0.4	12.3	S	100	Mag 9.34v star N 1′.8; ESO 532-22 NNE 4′.9.
22 09 08.9	−27 48 05	MCG −5-52-35	14.7	0.5 x 0.3	12.5	S0?	86	= **Hickson 91D**; on N edge of NGC 7214.
22 09 14.5	−27 24 15	MCG −5-52-37	14.7	0.9 x 0.5	13.6	Sc/Irr	144	
22 09 37.1	−27 14 51	MCG −5-52-40	14.2	0.8 x 0.5	13.0	Sb-c	11	Mag 10.94v star N 5′.6; ESO 532-22 NW 7′.5.
22 10 32.3	−28 58 30	MCG −5-52-41	14.5	0.8 x 0.4	13.3	E	102	Bright knot or galaxy W end.
22 10 41.1	−27 53 49	MCG −5-52-42	13.7	1.3 x 0.8	13.6	SA(rs)bc	38	Bright star SW end.
22 11 43.2	−28 48 09	MCG −5-52-44	14.5	0.5 x 0.5	12.9	S		Mag 10.40v star E 5′.0.
22 11 51.1	−27 57 45	MCG −5-52-45	14.7	0.6 x 0.5	13.3	Sb?	150	Star on E edge.
22 12 10.5	−27 49 42	MCG −5-52-46	14.9	0.6 x 0.3	13.0	S	156	**MCG −5-52-47** S 2′.0.
22 13 23.6	−27 13 14	MCG −5-52-48	14.0	1.0 x 0.6	13.3	Sb	62	Mag 7.42v star NW 13′.1.
22 14 24.2	−29 58 53	MCG −5-52-52	13.4	1.3 x 0.4	12.5	S?	142	Mag 9.40v star S 5′.2.
22 14 39.5	−27 39 47	MCG −5-52-55	13.6	1.0 x 0.5	12.8	S0	94	Mag 5.44v star Lambda Piscis Austrini SW 7′.8.
22 16 04.9	−26 40 49	MCG −5-52-59	13.6	1.3 x 0.5	13.0	S?	173	
22 16 34.3	−26 49 16	MCG −5-52-62	13.5	0.8 x 0.7	12.8	Sa-b	165	
22 17 41.2	−27 21 57	MCG −5-52-63	15.9	0.6 x 0.4	14.2	Sbc	30	Mag 7.03v star WSW 5′.5.
22 19 14.6	−28 24 16	MCG −5-52-64	13.5	1.1 x 0.8	13.2	(R′)SB(s)ab	121	Anonymous elongated galaxy 0′.5 S.
22 22 43.8	−27 21 18	MCG −5-52-67	14.4	1.0 x 0.2	12.5	Sb?	57	Mag 9.11v star S 4′.4.
22 23 42.6	−26 36 47	MCG −5-52-71	13.4	1.3 x 0.6	13.0	Sb-c	67	
22 28 09.2	−29 54 05	MCG −5-53-8	13.3	1.0 x 0.4	12.2	S0	47	
22 33 44.2	−27 14 46	MCG −5-53-15	13.8	0.7 x 0.6	12.9	E	97	Almost stellar; NGC 7306 W 6′.1.
22 40 52.2	−27 10 59	MCG −5-53-25	14.4	0.6 x 0.4	12.7	S0/a	135	Mag 4.19v star Epsilon Piscis Austrini NNW 8′.8.
22 40 55.8	−27 11 45	MCG −5-53-26	14.6	0.6 x 0.5	13.1	Sc	165	Mag 4′19v star Epsilon Piscis Austrini NNW 9′.8.
22 42 28.7	−28 35 03	MCG −5-53-28	14.9	0.7 x 0.4	13.4	S0	30	
22 47 07.8	−29 28 21	MCG −5-53-32	14.2	0.9 x 0.8	13.6	Sb	13	
22 52 07.9	−28 36 31	MCG −5-54-2	14.6	0.8 x 0.4	13.2	S0	27	
22 52 38.6	−29 03 24	MCG −5-54-3	14.1	0.8 x 0.7	13.4	Sa?	48	Almost stellar nucleus; mag 7.69v star S 7′.6.
22 54 45.2	−26 53 28	MCG −5-54-4	14.5	0.7 x 0.3	12.7	S0		Mag 8.31v star S 3′.3.
23 04 14.7	−28 24 33	MCG −5-54-10	14.9	0.6 x 0.4	13.2	S	21	
23 07 29.0	−27 20 32	MCG −5-54-15	14.6	0.4 x 0.3	12.3	E	174	
23 09 26.5	−28 48 12	MCG −5-54-17	15.0	0.5 x 0.4	13.1	(R_2)SAB(r)0/a	96	
23 13 50.8	−29 35 07	MCG −5-54-24	14.8	0.4 x 0.4	12.9	E		Stellar; mag 8.55v star SW 4′.7.
23 18 43.5	−27 07 52	MCG −5-55-1	13.9	0.8 x 0.5	12.8	S	147	
23 19 24.7	−28 01 24	MCG −5-55-2	14.5	0.7 x 0.7	13.6	S		
23 19 19.7	−28 56 34	MCG −5-55-3	13.6	1.2 x 0.8	13.5	(R)SB(s)0°	33	
23 21 43.3	−26 17 32	MCG −5-55-4	13.7	0.7 x 0.3	11.9	(R)SB(l)0^+	71	
23 23 08.2	−28 45 45	MCG −5-55-6	15.1	0.5 x 0.3	12.8	S	90	Mag 8.27v star NNE 6′.9.
21 56 56.8	−25 21 03	NGC 7157	14.0	1.1 x 0.5	13.2	(R)SB(r)a:	5	Located 2′.8 ENE of mag 10.26v star.
22 00 31.0	−24 38 02	NGC 7167	12.5	1.8 x 1.4	13.4	SB(s)c: II	145	Strong dark lane E of center; mag 10.19v star E 1′.3.
22 02 18.5	−20 32 54	NGC 7180	12.6	1.6 x 0.7	12.6	S0°?	68	
22 02 21.8	−18 54 59	NGC 7183	11.9	3.8 x 1.1	13.3	$S0^+$ pec sp	77	Prominent dark lane along centerline.
22 02 40.8	−20 48 38	NGC 7184	10.9	6.0 x 1.5	13.1	SB(r)c II	62	
22 02 56.7	−20 28 13	NGC 7185	12.6	2.3 x 1.5	13.8	SAB(rs)0^- pec?	15	
22 03 29.0	−20 19 02	NGC 7188	13.2	1.6 x 0.7	13.2	(R′)SB(rs)bc III:	44	
22 08 24.6	−29 03 04	NGC 7208	13.1	0.9 x 0.5	12.1	SAB0°?	142	
22 09 07.8	−27 48 33	NGC 7214	12.7	2.2 x 1.4	13.8	SB(s)bc pec:	115	= **Hickson 91A**; IC 5188 SW 5′.5.
22 11 31.1	−22 57 12	NGC 7220	13.5	0.8 x 0.6	12.6	SAB(s)0°:	3	
22 13 08.1	−26 08 53	NGC 7225	12.3	2.0 x 1.0	12.9	SA(s)0/a: pec	147	Prominent dark lane through center NW-SE.
22 14 03.1	−29 23 01	NGC 7229	12.5	1.8 x 1.5	13.4	SB(s)c I-II	157	Knotty, multi-branching arms; mag 8.77v star E 6′.9.
22 14 13.2	−17 04 25	NGC 7230	14.1	0.9 x 0.9	13.7	SA(s)bc II-III:		Located inside SW edge of small galaxy cluster **A 3847**; **PGC 68361** NNE 5′.7.
22 17 41.2	−23 43 49	NGC 7247	12.6	1.4 x 0.9	12.7	SB(s)b: II	2	Mag 8.92v star NW 2′.5.
22 20 44.8	−24 40 45	NGC 7252	12.1	1.9 x 1.6	13.1	(R)SA(r)0°:	119	
22 22 36.0	−21 44 12	NGC 7256	13.2	1.4 x 0.6	12.9	SBb? III-IV	122	
22 22 58.5	−28 20 49	NGC 7258	13.2	1.4 x 0.6	12.8	Sb:	141	
22 23 05.5	−28 57 17	NGC 7259	13.1	1.1 x 0.9	13.0	Sb	49	**ESO 457-51** SE 3′.0.
22 28 35.5	−24 50 43	NGC 7284	12.1	2.1 x 1.5	13.2	SB(s)0° pec	133	Interacting with NGC 7285; centers separated by 0′.5.

(Continued from previous page)

GALAXIES

RA h m s	Dec ° ′ ″	Name	Mag (V)	Dim ′ Maj x min	SB	Type Class	PA	Notes
22 28 38.3	−24 50 26	NGC 7285	12.0	2.1 x 1.5	13.1	SB(rs)a pec	100	
22 28 48.5	−22 12 14	NGC 7287	14.5	0.6 x 0.3	12.6	S0	6	Multi-galaxy system with NGC 7287A.
22 28 49.0	−22 12 09	NGC 7287A	14.4	0.4 x 0.3	12.0	S0°: sp	162	
22 32 07.9	−25 23 55	NGC 7294	12.5	1.9 x 1.2	13.3	(R′)SAB(rs)0° pec	47	MCG −4-53-8 NW 3′.0.
22 30 34.9	−17 34 29	NGC 7301	13.4	1.0 x 0.5	12.4	SB(s)ab pec:	1	
22 33 16.5	−27 14 49	NGC 7306	12.9	1.7 x 0.7	12.9	(R′)SBb? II-III	60	Almost stellar MCG −5-53-15 E 6′.1.
22 34 36.9	−22 29 09	NGC 7310	13.8	0.9 x 0.7	13.2	SB(rs)bc: II	36	Mag 8.47v star SSW 7′.8.
22 35 32.5	−26 06 07	NGC 7313	14.4	0.6 x 0.4	12.7	SB(s)b pec:	170	Located 4′.3 SW of NGC 7314.
22 35 46.2	−26 03 05	NGC 7314	11.0	4.6 x 2.1	13.3	SAB(rs)bc II-III	3	NGC 7313 SW 4′.3.
22 39 05.5	−22 39 59	NGC 7341	12.3	2.4 x 1.0	13.1	SAB(r)ab:	94	Located 1′.7 S of mag 8.26v star.
22 41 14.8	−21 47 46	NGC 7349	14.3	1.1 x 0.4	13.2	SBb pec? III-IV	166	
22 44 48.1	−23 41 18	NGC 7359	12.5	2.3 x 0.6	12.7	S0°: sp	55	Anonymous galaxy SE 2′.5.
22 45 10.1	−19 57 06	NGC 7365	12.9	1.5 x 1.0	13.2	SA(rs)0⁻:	34	Mag 10.44v star SSW 10′.1; ESO 603-8 SW 11′.4.
22 47 47.7	−22 18 39	NGC 7377	11.1	3.0 x 2.5	13.2	SA(s)0⁺	101	
22 50 08.2	−19 43 29	NGC 7381	14.2	0.7 x 0.5	12.9	(R)Scd pec?	123	
22 51 48.8	−20 36 29	NGC 7392	11.9	2.1 x 1.3	12.8	SA(s)bc I-II	123	
23 06 13.8	−19 49 11	NGC 7481	14.9	0.6 x 0.5	13.4		135	Star or companion galaxy on E edge.
23 08 58.8	−24 22 11	NGC 7494	14.6	0.6 x 0.5	13.4	E1	81	
23 09 56.1	−24 25 32	NGC 7498	14.0	0.9 x 0.4	12.8	(R′)Sab:	177	Mag 11.3 star N 4′.8; located in NW quadrant of small galaxy cluster **A 2542**.
23 12 07.5	−28 32 19	NGC 7507	10.4	2.8 x 2.7	12.6	E0		Very large, bright center.
23 13 13.8	−28 21 27	NGC 7513	11.4	3.2 x 2.1	13.3	(R′)SB(s)b pec	108	Bright ENE-WSW bar.
23 16 26.6	−22 09 18	NGC 7573	13.7	1.2 x 1.0	13.7	SB(s)c I	42	Located on W edge of small galaxy cluster **A 2568**.
23 22 33.0	−29 16 49	NGC 7636	13.6	0.8 x 0.5	12.5	S0°	30	
23 23 47.5	−29 23 18	NGC 7645	12.9	1.4 x 1.2	13.3	SB(r)c I-II	165	

GALAXY CLUSTERS

RA h m s	Dec ° ′ ″	Name	Mag 10th brightest	No. Gal	Diam ′	Notes
22 04 12.0	−21 26 00	A 2412	15.9	43	21	All members anonymous, stellar/almost stellar.
22 39 06.0	−17 21 00	A 2462	16.2	40	16	All members anonymous, faint, stellar/almost stellar.
22 00 00.0	−19 12 00	AS 983	16.3	7	17	All members anonymous, stellar.
22 01 54.0	−22 25 00	AS 987	16.0	20	17	Most members anonymous, stellar.
22 10 30.0	−20 54 00	AS 999	16.3	5	17	**ESO 601-27** N of center 6′.1; all others anonymous, stellar.

PLANETARY NEBULAE

RA h m s	Dec ° ′ ″	Name	Diam ″	Mag (P)	Mag (V)	Mag cent ★	Alt Name	Notes
22 29 38.4	−20 50 12	NGC 7293	1054	7.5	7.3	13.5	PK 36−57.1	**Helix Nebula**. Overall shape is rectangular, oriented SE-NW, with a prominent central ring approximately 720″ in diameter.

GALAXIES

RA h m s	Dec ° ′ ″	Name	Mag (V)	Dim ′ Maj x min	SB	Type Class	PA	Notes
20 56 57.8	−27 57 59	ESO 464-1	13.6	1.6 x 0.6	13.4	(R′)SA(s)0⁺?	167	Mag 11.7 star E 2′.9.
21 02 23.8	−28 10 30	ESO 464-16	14.4	1.8 x 0.4	13.9	SB(r)b? pec	27	Mag 10.09v star ESE 5′.7.
21 04 52.8	−29 07 07	ESO 464-21	14.0	0.9 x 0.8	13.5	(R)SB0⁺	34	Pair of stars, mags 11.1 and 10.53v, S and SE 2′.6.
21 42 32.8	−29 22 05	ESO 466-1	13.6	1.5 x 0.4	12.9	Sab sp	105	Mag 13.1 star N 0′.5.
21 44 23.5	−29 54 26	ESO 466-4	14.5	1.7 x 1.3	15.2	SB(s)d IV	15	Mag 12.0 star NNE 1′.9; mag 9.73v star N 6′.8; **ESO 466-3** SSW 6′.4.
21 57 55.2	−28 48 18	ESO 466-21	12.4	1.1 x 0.8	12.2	SA0⁻:	178	Mag 10.90v star on E edge.
21 58 10.6	−27 23 33	ESO 466-24	13.0	1.1 x 0.6	12.4	Sb	124	Mag 10.41v star N 3′.6.
21 58 43.4	−28 28 01	ESO 466-26	12.6	1.4 x 1.1	13.0	SAB0⁻	49	Mag 12.4 star S 4′.6.
21 59 13.4	−28 41 26	ESO 466-28	14.0	1.9 x 0.5	13.8	SAB(s)cd: II-III	21	Mag 13;0 star N edge; mag 7.40v star NE 8′.8.
20 36 04.8	−24 10 41	ESO 528-23	13.6	1.5 x 0.7	13.5	(R′)SB(s)b:	134	Mag 9.41v star E 6′.6.
20 42 08.8	−24 12 50	ESO 528-33	13.8	1.1 x 0.6	13.2		65	Double system, two bright cores with thin extensions ENE and W; mag 10.00v star WNW 7′.0.
20 42 29.1	−24 09 01	ESO 528-34	13.8	1.7 x 0.8	14.0	SB(s)c II-III	79	Mag 10.00v star W 11′.1.
20 43 45.8	−26 33 03	ESO 528-36	12.9	0.9 x 0.9	12.7	E0 pec:		Mag 12.5 star NW edge; **ESO 528-35** SW 3′.3.
21 09 27.0	−22 52 45	ESO 530-10	13.2	1.0 x 1.0	13.1	SA(r)0⁺:		Mag 8.92v star WSW 9′.2; note: **ESO 530-4** located 1′.7 NW of this star.
21 17 06.5	−23 07 18	ESO 530-29	13.9	0.8 x 0.4	12.6	S0	104	Mag 14.1 star SW 3′.0.
21 17 28.4	−22 48 57	ESO 530-30	13.6	1.4 x 0.9	13.7	SB(rs)a	177	Mag 8.75v star W 10′.7; MCG −4-50-14 NE 4′.2; ESO 530-34 E 7′.1; ESO 530-32 E 5′.5.
21 17 51.5	−22 50 20	ESO 530-32	14.5	1.1 x 0.3	13.1	SB(s)bc: I-II	113	ESO 530-34 E 1′.7; ESO 530-30 W 5′.5.
21 17 58.8	−22 50 05	ESO 530-34	13.6	1.6 x 0.4	13.0	S?	122	ESO 530-32 W 1′.7; ESO 530-30 W 7′.1.
21 27 10.5	−23 01 08	ESO 530-48	14.1	0.5 x 0.5	12.5	S0:		Compact; **MCG −4-50-25** NW 2′.3.
21 32 27.5	−23 56 55	ESO 531-2	14.1	1.5 x 1.1	14.5	SA(s)0°:	168	**ESO 531-1** W 8′.2; mag 10.40v star SW 12′.1.
21 40 29.5	−26 31 44	ESO 531-22	13.1	2.6 x 0.3	12.7	Sbc: sp	8	star on N end.
22 01 45.8	−23 42 25	ESO 532-11	14.1	1.0 x 0.7	13.5	SB(s)cd II	129	Mag 10.39v star E 2′.4; mag 7.50v star NE 8′.3.
22 02 57.3	−22 28 23	ESO 532-14	14.3	1.3 x 0.6	13.9	SAB(s)cd pec: III	109	Mag 10.64v star SSW 6′.3; **ESO 532-15** NNE 5′.4.
20 40 56.5	−20 28 15	ESO 597-26	12.5	1.0 x 1.0	12.4	S0		Mag 11.07v star NW 7′.0; **ESO 597-23** NW 4′.4.
20 48 15.0	−19 50 59	ESO 597-36	14.3	1.5 x 0.2	12.9	S0° pec sp II	58	= **Hickson 87A**. **Hickson 87D** 0′.5 W., **Hickson 87C** 1′.2 N.
20 50 15.1	−19 25 34	ESO 597-41	13.6	2.4 x 1.7	15.0	SAB(rs)b pec: II-III	62	Mag 9.55v star SSE 9′.2.
20 55 27.6	−19 33 52	ESO 598-2	14.0	0.8 x 0.3	12.4	S0	167	Mag 9.80v star E 11′.9.
20 56 11.5	−18 18 38	ESO 598-6	13.8	0.9 x 0.6	13.0	S	32	Faint anonymous galaxies 1′.6 SW and 2′.7 NW.
21 04 24.9	−21 46 51	ESO 598-20	13.4	1.5 x 1.1	13.8	(R′)SB(r)bc II	32	Mag 10.30v star NW 8′.5; mag 9.33v star and **ESO 598-23**, MCG −4-49-11 E 8′.6.
21 06 24.0	−21 40 13	ESO 598-25	13.7	1.0 x 0.5	12.8	Sa	48	Mag 11.5 star W 7′.9.
21 11 58.6	−19 47 38	ESO 598-31	13.2	0.9 x 0.8	12.7	SAB0⁻:	3	**ESO 598-30** NW 3′.6.
21 16 14.7	−22 14 07	ESO 599-4	13.8	1.1 x 0.6	13.2	SAB(s)0°:	108	Mag 9.13v star N 3′.3; ESO 599-5 E 4′.6.
21 16 33.4	−22 12 58	ESO 599-5	14.5	1.7 x 0.2	13.2	Sab?	116	ESO 599-4 W 4′.6; mag 9.13v star WNW 5′.4.
21 34 08.3	−20 28 14	ESO 599-20	14.1	1.5 x 0.4	13.4	SAB(s)ab pec sp II-III	130	Mag 10.06v star N 6′.8.
21 38 51.6	−22 13 10	ESO 600-4	14.4	0.8 x 0.7	13.7	S0?	84	Pair of E-W oriented mag 10.00v, 10.75v stars SE 6′.4.
22 01 30.5	−22 04 15	ESO 601-4	13.8	1.8 x 0.5	13.5	SB(s)c: II-III	57	Mag 9.41v star E 10′.0.
22 01 37.8	−21 30 43	ESO 601-5	14.0	1.6 x 0.3	13.1	Sb:	65	Mag 10.33v star N 4′.3.
22 02 21.5	−21 02 23	ESO 601-7	14.3	1.7 x 0.4	13.7	SB(s)m: IV-V	148	Mag 10.70v star W 6′.1.
22 03 51.4	−19 52 51	ESO 601-12	13.5	1.8 x 0.5	13.2	(R′)SAB(r)ab: II	178	Mag 10.33v star W 9′.1.
20 55 05.1	−18 02 19	IC 1336	14.3	0.7 x 0.4	12.8	Sa?	36	Mag 12.1 star WNW 2′.6.
20 57 55.8	−17 56 35	IC 1339	13.2	1.4 x 0.9	13.3	SAB(rs)b pec: II	55	Mag 9.45v star NW 10′.0.

RA h m s	Dec ° ′ ″	Name	Mag (V)	Dim ′ Maj x min	SB	Type Class	PA	Notes
21 29 37.4	−21 11 43	IC 1386	13.1	1.5 x 1.1	13.5	SA0$^-$:	130	Mag 10.77v star SW 7′.3.
21 32 08.0	−18 01 11	IC 1389	14.5	0.7 x 0.4	12.9	S	120	
21 40 14.4	−22 24 43	IC 1393	14.6	0.8 x 0.6	13.7	S0	174	MCG −4-51-9 S 1′.3; NGC 7104 WSW 2′.7.
21 58 18.4	−17 10 37	IC 1412	13.5	0.9 x 0.4	12.3	(R′)SB0$^+$ pec	99	Mag 12.7 star SSW 3′.2.
20 43 14.2	−29 51 10	IC 5039	12.7	2.1 x 0.6	12.8	Sbc: sp III	156	Mag 10.76v star ENE 7′.0; mag 10.30v star NW 10′.8.
20 43 34.5	−29 42 14	IC 5041	12.6	1.7 x 1.2	13.2	SAB(s)d IV	31	Mag 10.30v star W 10′.0.
20 51 46.1	−29 50 54	IC 5065	13.7	1.0 x 0.7	13.2	SB0: pec	148	Mag 7.24v star N 4′.2.
21 08 32.1	−29 46 10	IC 5086	12.9	1.3 x 1.3	13.3	SA0$^-$		Bright round core, faint envelope; MCG −5-50-1 NNW 8′.4.
21 39 45.8	−22 24 27	IC 5122	15.4	0.8 x 0.4	14.0	S0:	45	**ESO 531-13** WSW 4′.9.
21 58 59.2	−27 24 50	IC 5149	13.6	1.3 x 0.6	13.2	(R′)Sa	32	Mag 11.7 star NW 2′.2.
20 47 45.3	−19 53 38	MCG −3-53-2	13.9	0.9 x 0.5	12.9	SAB(r)cd: II-III	84	Mag 10.68v star W 4′.0.
20 48 10.7	−19 51 22	MCG −3-53-3	14.4	0.5 x 0.4	12.6	SA(r)0$^+$ pec?	84	= **Hickson 87B**.
20 58 28.5	−19 58 52	MCG −3-53-14	13.3	1.3 x 0.5	12.7	Sb pec II-III	145	Star on NE edge.
21 02 30.9	−19 27 54	MCG −3-53-20	15.0	0.9 x 0.7	14.3	S	69	
21 28 55.2	−19 40 23	MCG −3-54-2	14.5	0.5 x 0.4	12.6	S	117	Mag 8.97v star N 8′.2.
21 29 09.9	−19 41 17	MCG −3-54-3	14.6	0.4 x 0.4	12.5	S		Mag 8.84v star SE 7′.2.
21 38 35.3	−17 07 22	MCG −3-55-4	14.2	0.7 x 0.3	12.4	S0/a	123	Mag 7.26v star SW 10′.5.
20 38 19.6	−25 07 03	MCG −4-48-25	15.7	0.4 x 0.4	13.6			Mag 6.38v star W 6′.1.
20 38 39.3	−25 07 33	MCG −4-48-26	14.0	0.5 x 0.5	12.6	E/S0		Mag 8.48v star SSE 4′.9.
20 40 40.2	−25 53 09	MCG −4-48-27	14.0	0.8 x 0.4	12.6	Sbc	150	
20 46 37.1	−23 37 53	MCG −4-49-1	13.1	0.8 x 0.7	12.4	S0$^-$:	168	Mag 7.54v star SSW 7′.4.
20 47 21.0	−22 32 27	MCG −4-49-2	13.2	0.5 x 0.5	11.6	S0		**MCG −4-49-3** NE 3′.6.
20 49 47.4	−25 41 57	MCG −4-49-4	13.1	1.0 x 0.9	12.9	(R)SB0$^+$	18	Mag 5.86v star SW 8′.3.
20 51 50.4	−26 18 01	MCG −4-49-5	13.9	0.7 x 0.4	12.4	Sbc	150	
20 57 47.2	−24 37 31	MCG −4-49-9	14.2	1.3 x 1.0	14.3	S?	9	Stellar nucleus; mag 8.72v star NNE 3′.9.
21 04 56.0	−21 43 52	MCG −4-49-11	13.4	0.8 x 0.8	12.8	S0		Mag 9.33v star SE 1′.9; **ESO 598-23** S 2′.5.
21 10 18.2	−24 13 38	MCG −4-50-2	13.9	1.2 x 0.5	13.2	S?	44	
21 11 26.0	−23 12 50	MCG −4-50-4	15.1	0.5 x 0.3	12.9	SBc		
21 11 39.4	−23 10 21	MCG −4-50-6	14.3	0.4 x 0.4	12.4	E/S0		**MCG −4-50-5** W 4′.1.
21 11 40.1	−23 09 43	MCG −4-50-7	13.8	0.5 x 0.4	12.1	E	176	Mag 8.61v star N 3′.4.
21 11 37.1	−23 10 54	MCG −4-50-8	14.1	0.3 x 0.2	10.9	S0/a	27	
21 12 58.1	−20 32 23	MCG −4-50-9	14.2	0.7 x 0.4	12.7	(R′)SB(s)a	81	
21 17 42.2	−22 46 17	MCG −4-50-14	13.9	0.9 x 0.5	12.8	SB(rs)ab:	167	Mag 11.6 star NE 1′.9, small, faint elongated anonymous galaxy E 2′.9.
21 17 54.7	−23 02 12	MCG −4-50-16	13.8	0.6 x 0.5	12.4	Sb	21	
21 19 47.8	−24 17 15	MCG −4-50-18	14.1	0.6 x 0.6	12.8	Sbc		
21 25 10.4	−24 56 00	MCG −4-50-21	14.6	0.6 x 0.4	12.9	Sc	15	Mag 9.25v star NE 3′.6.
21 25 55.8	−22 45 41	MCG −4-50-22	13.8	1.3 x 1.1	14.0	(R′)SB(r)cd I-II	103	
21 25 58.5	−22 43 47	MCG −4-50-23	13.8	1.0 x 1.0	13.7	(R′)SB(r)cd I-II		
21 26 43.5	−21 50 01	MCG −4-50-24	13.4	0.7 x 0.5	12.2	S0?	31	Mag 11.2 star N edge; mag 7.85v star W 6′.9.
21 27 37.0	−25 57 01	MCG −4-50-26	14.5	0.9 x 0.4	13.3	Sbc	153	
21 27 44.2	−23 40 08	MCG −4-50-27	14.6	0.9 x 0.7	14.0	Sd	105	Mag 7.72v star NW 5′.4.
21 37 38.9	−24 56 23	MCG −4-51-2	15.4	0.8 x 0.2	13.3	Sc	105	Mag 9.25v star S 4′.5.
21 39 02.4	−22 47 37	MCG −4-51-4	13.9	0.8 x 0.7	13.1	Sb	102	
21 39 59.4	−22 41 36	MCG −4-51-7	14.7	0.4 x 0.4	12.5	S?		
21 40 14.6	−22 25 59	MCG −4-51-9	14.6	0.4 x 0.3	12.2	S0	30	IC 1393 N 1′.3.
21 58 15.4	−24 53 22	MCG −4-51-16	13.9	0.9 x 0.5	12.9	(R′)SBa	14	
21 59 21.2	−23 15 18	MCG −4-51-17	14.5	0.7 x 0.5	13.2	Sc	108	**MCG −4-51-18** at E end.
22 01 51.4	−22 29 10	MCG −4-52-6	13.8	1.2 x 0.6	13.3	(R′)SAB(r)0/a	174	mag 9.66v star S 2′.4.
20 37 32.3	−27 34 37	MCG −5-48-24	13.4	1.1 x 0.6	12.8	SB(r)b:	3	**MCG −5-48-25** NE 1′.2.
20 39 34.7	−28 29 22	MCG −5-48-27	13.7	1.0 x 0.6	13.0	Sb-c	90	Mag 7.96v star NW 8′.0.
20 47 44.5	−26 43 42	MCG −5-49-3	14.0	0.9 x 0.7	13.4	Sc	156	Numerous bright knots.
21 03 01.9	−28 20 25	MCG −5-49-14	13.5	0.5 x 0.5	12.1	E0:		Mag 10.09v star NNW 8′.4.
21 03 24.2	−27 51 21	MCG −5-49-15	14.1	0.7 x 0.5	12.8	Sbc II	99	Mag 6.25v star NNW 8′.0.
21 08 10.9	−29 39 07	MCG −5-50-1	14.6	0.7 x 0.4	13.1	S	60	IC 5086 SSE 8′.4; mag 8.72v star NE 10′.1.
21 09 05.0	−26 54 30	MCG −5-50-3	13.8	0.7 x 0.5	12.5	S(r)	126	Mag 9.01v star S 6′.0.
21 10 19.2	−29 31 38	MCG −5-50-4	13.9	1.5 x 0.4	13.2	(R′)SB(s)ab?	144	Mag 12.8 star N 0′.8.
21 18 21.8	−27 20 57	MCG −5-50-9	14.0	1.2 x 0.5	13.3	Sa? pec	142	Multi-galaxy system with **ESO 464-31A** E of core.
21 31 20.8	−29 15 45	MCG −5-50-16	14.7	0.7 x 0.4	13.2	Sb	63	
21 33 29.8	−29 30 35	MCG −5-50-17	14.7	0.7 x 0.5	13.4	S(r:)	0	
21 34 56.4	−29 43 50	MCG −5-51-1	14.3	0.6 x 0.4	12.6	Sc	84	Mag 6.44v star S 2′.1.
21 35 24.8	−26 39 34	MCG −5-51-2	14.4	0.9 x 0.6	13.6	(R′$_2$)SB(s)c?	111	Mag 9.37v star SE 3′.0.
21 36 37.6	−27 43 32	MCG −5-51-3	14.3	0.6 x 0.5	13.1	E	21	Mag 10.78v star NW 5′.6.
21 37 31.3	−29 26 44	MCG −5-51-4	14.5	0.5 x 0.3	12.6	E		
21 38 52.4	−27 59 14	MCG −5-51-5	14.4	0.7 x 0.4	12.9	S	84	Mag 7.65v star NNW 7′.2.
21 39 10.8	−29 50 07	MCG −5-51-6	14.6	0.4 x 0.4	12.7	E		
21 39 15.4	−29 52 18	MCG −5-51-7	14.1	0.7 x 0.4	12.7	S	75	
21 46 31.2	−29 31 34	MCG −5-51-14	14.8	0.5 x 0.4	12.9	S	33	
21 47 37.0	−29 41 42	MCG −5-51-15	13.9	0.6 x 0.6	12.9	E		
21 50 39.0	−29 11 03	MCG −5-51-18	13.6	0.7 x 0.4	12.1	SB0	169	
21 54 29.1	−28 21 48	MCG −5-51-23	14.1	0.8 x 0.4	12.7	S0/a	73	
21 56 44.0	−28 24 54	MCG −5-51-24	13.9	1.1 x 0.4	12.9	S0	68	
21 57 01.9	−28 37 12	MCG −5-51-25	13.4	1.0 x 0.8	13.1	SB0	76	Multi-galaxy system.
21 57 27.7	−28 44 17	MCG −5-51-26	13.5	0.9 x 0.6	12.7	(R′)SB(r)a	151	Mag 7.92v star SW 10′.1.
21 57 29.6	−29 23 41	MCG −5-51-27	14.5	0.6 x 0.4	12.9	S0	54	
22 00 29.9	−26 54 30	MCG −5-52-2	14.7	0.7 x 0.7	13.8	Sb-c		
22 00 39.5	−28 12 28	MCG −5-52-3	14.2	0.7 x 0.5	12.9	S	84	Star on S edge.
22 00 57.2	−28 21 48	MCG −5-52-4	13.5	1.1 x 0.5	12.8	S0	148	Faint anonymous galaxy NE 1′.6; mag 5.63v star Eta Piscis Austrini N 5′.8.
22 01 18.5	−28 29 43	MCG −5-52-5	14.1	0.9 x 0.9	13.7	Sc		Mag 5.63v star Eta Piscis Austrini SW 6′.7.
22 02 39.6	−28 59 20	MCG −5-52-13	14.1	1.0 x 0.7	13.6	Sb	145	Mag 9.51v star ESE 6′.2.
22 02 42.4	−28 05 41	MCG −5-52-15	14.3	0.5 x 0.5	12.7	S0		
22 03 12.9	−26 24 27	MCG −5-52-16	14.0	0.7 x 0.4	12.5	S?	99	
22 03 22.9	−28 46 16	MCG −5-52-17	13.6	1.3 x 0.3	12.4	S0/a	145	
22 03 44.5	−27 47 56	MCG −5-52-18	13.3	1.2 x 1.0	13.4	S?	66	**MCG −5-52-19** S 8′.1.
20 56 30.7	−18 34 01	NGC 6986	13.5	1.0 x 0.6	12.8	SB0$^-$?	8	
21 01 37.8	−28 01 56	NGC 6998	14.0	0.6 x 0.6	13.0	E0		Mag 13.1 star S 1′.5; NGC 6999 ESE 5′.1.
21 01 59.6	−28 03 33	NGC 6999	13.8	0.8 x 0.6	12.9	SA(s)0°	179	Located 2′.1 N of mag 10.28v star; NGC 6998 WNW 5′.1.
21 07 16.4	−25 28 10	NGC 7016	13.7	0.8 x 0.8	13.3	E0		Faint, stellar anonymous galaxy NE 1′.5.
21 07 20.8	−25 29 18	NGC 7017	13.8	0.7 x 0.5	12.5	S0	90	Multi-galaxy system.
21 07 25.5	−25 25 45	NGC 7018	13.4	1.4 x 0.9	13.7	E	86	Multi-galaxy system; faint anonymous galaxy and close star N 0′.9.

(Continued from previous page)

143 GALAXIES 143

RA h m s	Dec ° ′ ″	Name	Mag (V)	Dim ′ Maj x min	SB	Type Class	PA	Notes
21 06 25.9	−24 24 46	NGC 7019	14.2	0.6 x 0.3	12.1	SB(s)b: I-II	137	**ESO 529-23** N 2′.8.
21 11 13.3	−20 29 13	NGC 7030	13.7	0.9 x 0.7	13.0	(R′)SB(r)ab pec? I-II	70	
21 10 46.6	−23 08 14	NGC 7035	14.2	1.0 x 0.7	13.7	SAB(s)0:	105	Is a pair of galaxies with a separation of 0′.45; common corona but no sign of interaction.
21 39 51.6	−22 28 26	NGC 7103	14.2	1.3 x 1.2	14.7	$E^{+}1$	63	Stellar **ESO 531-17** ENE 2′.0.
21 40 03.3	−22 25 30	NGC 7104	14.2	0.8 x 0.7	13.7	E0 pec:	51	IC 1393 ENE 2′.6; MCG −4-51-9 E 2′.7; **IC 5124** W 1′.9.
21 43 39.4	−25 21 05	NGC 7115	13.4	1.9 x 0.5	13.2	Sb pec	66	
21 53 59.0	−29 17 23	NGC 7152	13.6	1.0 x 0.6	12.9	SB(rs)b? pec II	17	
21 54 35.4	−29 03 53	NGC 7153	13.2	1.9 x 0.3	12.5	Sb? sp	68	
21 56 56.8	−25 21 03	NGC 7157	14.0	1.1 x 0.5	13.2	(R)SB(r)a:	5	Located 2′.8 ENE of mag 10.26v star.
22 00 31.0	−24 38 02	NGC 7167	12.5	1.8 x 1.4	13.4	SB(s)c: II	145	Strong dark lane E of center; mag 10.19v star E 1′.3.
22 02 18.5	−20 32 54	NGC 7180	12.6	1.6 x 0.7	12.6	S0°?	68	
22 02 21.8	−18 54 59	NGC 7183	11.9	3.8 x 1.1	13.3	$S0^{+}$ pec sp	77	Prominent dark lane along centerline.
22 02 40.8	−20 48 38	NGC 7184	10.9	6.0 x 1.5	13.1	SB(r)c II	62	
22 02 56.7	−20 28 13	NGC 7185	12.6	2.3 x 1.5	13.8	$SAB(rs)0^{-}$ pec?	15	
22 03 29.0	−20 19 02	NGC 7188	13.2	1.6 x 0.7	13.2	(R′)SB(rs)bc III:	44	

GALAXY CLUSTERS

RA h m s	Dec ° ′ ″	Name	Mag 10th brightest	No. Gal	Diam ′	Notes
20 48 12.0	−17 48 00	A 2328	16.4	81	16	All members anonymous, faint and stellar.
21 29 30.0	−22 12 00	A 2347	16.4	79	10	All members anonymous, faint and stellar.
21 52 18.0	−19 32 00	A 2384	15.9	61	11	**ESO 600-14** on S edge; all other members anonymous, stellar.
20 36 00.0	−25 16 00	A 3698	15.0	71	22	Except NGC and ESO members, all others anonymous, stellar.
21 02 00.0	−28 03 00	A 3733	15.6	59	22	Except plotted catalog members, all other anonymous, stellar.
21 07 12.0	−25 28 00	A 3744	14.7	70	28	Heavy concentration of GX around center, and NNE of center.
20 40 42.0	−20 44 00	AS 891	15.5	2	22	**ESO 597-24** at center; **ESO 597-22** SW 9′.1; **ESO 597-25** SSE 5′.3.
20 43 36.0	−26 25 00	AS 894	16.0	5	17	See notes ESO 528-36; all others anonymous, stellar.
21 17 48.0	−22 46 00	AS 929	16.3		17	All but Uranometria plotted galaxies anonymous, stellar.
21 23 12.0	−21 05 00	AS 940	16.1		17	All members anonymous, stellar.
21 31 48.0	−19 38 00	AS 951	16.3		17	All members anonymous, stellar.
21 39 54.0	−22 28 00	AS 963	15.2	10	22	Line of galaxies extend from center to S edge.
22 00 00.0	−19 12 00	AS 983	16.3	7	17	All members anonymous, stellar.
22 01 54.0	−22 25 00	AS 987	16.0	20	17	Most members anonymous, stellar.

GLOBULAR CLUSTERS

RA h m s	Dec ° ′ ″	Name	Total V m	B ★ V m	HB V m	Diam ′	Conc. Class Low = 12 High = 1	Notes
21 40 22.0	−23 10 45	NGC 7099	6.9	12.1	15.1	12.0	5	= **M 30**.
21 46 38.8	−21 15 03	Pal 12	11.7	14.6	17.1	2.9	12	

144 GALAXIES 144

RA h m s	Dec ° ′ ″	Name	Mag (V)	Dim ′ Maj x min	SB	Type Class	PA	Notes
19 17 30.2	−28 15 11	ESO 459-10	14.1	1.7 x 0.2	12.8	Sbc: sp	160	Mag 10.06v star SE 7′.1.
19 28 17.1	−29 31 47	ESO 460-4	12.5	1.4 x 1.0	12.9	$E^{+}4$	118	Mag 8.57v star W 5′.8.
19 38 43.1	−29 48 34	ESO 460-18	12.7	2.1 x 0.8	13.1	SA(s)bc? II-III	132	Mag 9.93v star superimposed on center.
19 41 15.3	−28 04 12	ESO 460-23	13.5	1.2 x 1.0	13.5	SB(r)bc: II	32	Several stars superimposed N edge; mag 12.6 star SW 2′.3.
19 42 52.8	−28 30 08	ESO 460-25	13.5	1.5 x 0.8	13.5	SB(r)ab III-IV	80	Mag 11.7 star SW 1′.4; mag 9.19v star S 9′.5.
19 42 59.8	−27 25 11	ESO 460-26	13.4	1.2 x 0.7	13.1	$SA0^{+}$ pec:	162	Close N-S oriented pair of mag 10.06v, 10.26v stars SW 13′.2.
19 44 02.3	−28 24 05	ESO 460-30	12.9	1.2 x 0.8	12.8	$SA(rs)0^{-}$:	80	Mag 12.4 star on SW edge; NGC 6816 S 5′.3.
19 44 21.8	−27 24 27	ESO 460-31	14.4	2.8 x 0.3	14.1	Sc sp	92	Mag 11.04v star SW 3′.2; mag 8.29v star E 11′.9.
19 44 39.0	−27 58 46	ESO 460-32	13.7	1.6 x 0.8	13.8	SA(rs)bc: II-III	54	Mag 10.13v star NNW 8′.8.
19 44 44.9	−28 23 11	ESO 460-33	13.4	1.1 x 0.7	13.0	$SB0^{-}$? pec	126	Mag 10.82v star S 3′.9.
19 44 53.1	−28 11 25	ESO 460-34	15.0	2.0 x 1.0	15.6	IB(s)m V	111	Mag 10.39v star SW 7′.8.
19 46 18.7	−28 19 43	ESO 460-35	13.1	1.2 x 0.4	12.2	S	130	Mag 10.80v star S 7′.0.
19 50 38.8	−28 15 45	ESO 461-3	13.7	1.8 x 0.6	13.7	SA(rs)bc? II-III	77	Mag 9.47v star E 10′.4.
20 04 03.2	−29 35 45	ESO 461-37	13.9	1.3 x 0.3	12.8	S0	86	Mag 9.14v star NNE 8′.2.
20 08 05.1	−29 19 10	ESO 461-44	13.9	1.5 x 0.3	12.8	Sbc: sp	9	Mag 13.6 star W 1′.3; mag 12.4 star SE 5′.5.
20 21 24.1	−27 20 49	ESO 462-8	13.8	1.2 x 0.5	13.1	$(R)SB0^{+}$	148	Mag 11.18v star W 8′.4; MCG −5-48-2 WSW 4′.7.
20 22 14.9	−29 51 26	ESO 462-10	14.5	1.5 x 0.6	14.2	SAB(s)c: III-IV	135	Mag 11.9 star SSE 2′.5; MCG −5-48-5 ENE 4′.8.
20 23 14.1	−27 42 54	ESO 462-15	11.9	1.6 x 1.2	12.7	E3	170	Mag 9.39v star W 11′.0.
20 23 39.0	−28 16 43	ESO 462-16	13.0	1.8 x 1.2	13.6	SA(s)cd III-IV	117	Knotty; mag 8.56v star SE 9′.2.
20 26 56.7	−29 07 09	ESO 462-20	14.2	0.6 x 0.3	12.2		97	Mag 10.70v star NE 1′.1.
19 49 35.8	−26 25 57	ESO 526-7	13.4	2.5 x 1.8	14.9	SA(r)bc: II	130	Mag 8.26v star NE 6′.3.
19 52 55.9	−25 09 04	ESO 526-11	15.6	1.5 x 0.2	14.1	Sc:	49	Mag 8.89v star S 3′.4.
20 08 31.2	−25 27 32	ESO 527-11	14.2	1.6 x 0.6	14.0	SA(r)c: I-II	176	Mag 10.01v star SSW; mag 9.52v star ESE 8′.5.
20 20 45.8	−24 07 44	ESO 527-19	12.8	1.7 x 1.2	13.4	SB(rs)bc: II-III	113	Mag 9.83v star N 5′.5.
20 29 12.8	−22 40 19	ESO 528-8	12.9	0.9 x 0.9	12.6	$S0^{-}$		Mag 10.7 star on W edge; mag 9.87v star E 8′.2.
20 33 20.7	−27 05 53	ESO 528-17	15.2	1.7 x 0.2	13.9	SBcd? sp	102	Mag 9.89v star ENE 4′.2.
20 35 06.1	−25 20 28	ESO 528-21	13.7	1.5 x 0.8	13.7	SB?	102	**ESO 528-20** SW 1′.8; mag 9.67v star NNW 7′.9.
20 36 04.8	−24 10 41	ESO 528-23	13.6	1.5 x 0.7	13.5	(R′)SB(s)b:	134	Mag 9.41v star E 6′.6.
20 42 08.8	−24 12 50	ESO 528-33	13.8	1.1 x 0.6	13.2		65	Double system, two bright cores with thin extensions ENE and W; mag 10.00v star WNW 7′.0.
20 42 29.1	−24 09 01	ESO 528-34	13.8	1.7 x 0.8	14.0	SB(s)c II-III	79	Mag 10.00v star W 11′.1.
20 43 45.8	−26 33 03	ESO 528-36	12.9	0.9 x 0.9	12.7	E0 pec:		Mag 12.5 star NW edge; **ESO 528-35** SW 3′.3.
19 29 22.9	−20 45 45	ESO 594-2	12.6	0.8 x 0.5	11.5	$SAB(s)0^{-}$	48	Mag 10.11v star N 7′.5.
19 29 59.6	−17 40 42	ESO 594-4	13.2	2.9 x 1.5	14.6	IB(s)m: V-VI	81	Mag 8.57v star N 7′.1.
19 37 27.8	−17 27 20	ESO 594-8	14.8	1.5 x 0.3	13.8	Sd? sp	115	Mag 8.53v star NW 13′.2.
19 48 21.2	−18 04 41	ESO 594-17	14.2	2.5 x 1.5	15.4	(R′)SB(s)m pec: V	18	Mag 9.68v star WNW 7′.8.
20 02 21.3	−20 13 33	ESO 595-8	14.0	1.4 x 0.9	14.1	SB(r)0/a	141	Strong elongated N-S core in faint envelope.
20 04 18.6	−19 03 46	ESO 595-10	13.7	1.4 x 0.7	13.6	SB(s)dm pec: III	165	Mag 9.50v star NW 11′.6.
20 05 37.7	−20 04 18	ESO 595-12	14.3	0.9 x 0.9	14.0	Sc		Mag 8.68v star ESE 12′.7.
20 07 37.4	−21 07 28	ESO 595-14	12.6	1.2 x 1.0	12.7	SB(s)cd II	172	Mag 10.98v star S 7′.6.
20 15 43.6	−21 31 02	ESO 596-12	14.2	1.7 x 1.3	15.0	(R′)SA(rs)0°	70	Mag 8.63v star NW 4′.3.
20 23 16.7	−21 11 16	ESO 596-27	14.1	0.6 x 0.5	12.6		125	Very slender **ESO 596-26** E 1′.9.

GALAXIES

RA h m s	Dec ° ′ ″	Name	Mag (V)	Dim ′ Maj x min	SB	Type Class	PA	Notes
20 24 06.5	−20 09 50	ESO 596-30	13.7	1.9 x 1.6	14.7	SB(r)d: III-IV	145	Mag 9.28v star SSW 5.′4.
20 30 28.7	−19 11 20	ESO 596-49	13.6	0.4 x 0.3	11.2		12	Compact; mag 14.9 star on W edge.
20 32 18.9	−19 44 02	ESO 597-6	13.1	0.9 x 0.6	12.3	SA(rs)0$^-$:	138	Small anonymous galaxy on SW edge.
20 40 56.5	−20 28 15	ESO 597-26	12.5	1.0 x 1.0	12.4	S0		Mag 11.07v star NW 7.′0; **ESO 597-23** NW 4.′4.
20 03 01.5	−17 13 58	IC 1309	14.2	0.9 x 0.7	13.6	SAB(rs)bc pec II-III	35	Mag 7.58v star S 7.′0.
20 26 01.2	−18 30 15	IC 1319	13.8	0.8 x 0.5	12.7	Sb	9	MCG −3-52-6 SW 2.′9; mag 8.24v star NW 2.′2.
20 23 56.0	−26 00 53	IC 4999	12.5	1.8 x 1.0	13.0	SB(rs)c? III	95	Several strong knots; mag 9.26v star NW 12.′2.
20 25 20.2	−25 49 45	IC 5005	12.7	1.9 x 1.4	13.7	SB(s)cd: III	67	Central core surrounded by five stars.
20 43 14.2	−29 51 10	IC 5039	12.7	2.1 x 0.6	12.8	Sbc: sp III	156	Mag 10.76v star ENE 7.′0; mag 10.30v star NW 10.′8.
20 43 34.5	−29 42 14	IC 5041	12.6	1.7 x 1.2	13.2	SAB(s)d IV	31	Mag 10.30v star W 10.′0.
20 01 02.8	−17 03 12	MCG −3-51-1	14.6	1.0 x 0.5	13.7	SA(rl)b	24	
20 08 16.6	−17 39 36	MCG −3-51-5	14.4	0.7 x 0.7	13.5	SB(s)c		
20 15 03.1	−19 59 54	MCG −3-51-7	14.0	0.8 x 0.4	12.6	Sb:	38	Mag 10.09v star on N edge.
20 19 33.7	−18 09 28	MCG −3-51-9	14.5	0.8 x 0.7	13.7	Sd	156	Stars on E and W edge.
20 21 54.4	−18 22 47	MCG −3-52-1	13.6	0.5 x 0.5	12.0	S0		
20 25 50.3	−18 31 29	MCG −3-52-6	14.4	0.8 x 0.6	13.5	Sc	90	Mag 8.24v star N 2.′6; IC 1319 NE 2.′9.
20 28 01.1	−18 46 49	MCG −3-52-10	14.5	0.8 x 0.6	13.6	SB(r)0/a:	9	**MCG −3-52-9** SW 7.′1.
20 28 30.2	−19 00 06	MCG −3-52-12	14.6	0.5 x 0.4	12.7	SB(s)c	18	Mag 10.66v star S 2.′7.
20 30 14.5	−19 06 59	MCG −3-52-13	12.9	0.8 x 0.4	11.7	E?	153	Mag 8.56v star W 7.′4.
20 30 42.0	−19 08 41	MCG −3-52-15	14.3	0.8 x 0.5	13.1	(R$'_1$)SB(rs)bc	12	Very faint, elongated galaxy at N edge; mag 9.07v star E 4.′5.
20 32 17.9	−19 40 03	MCG −3-52-17	14.1	0.5 x 0.5	12.4	Sb		**ESO 597-2** NW 7.′7.
19 50 16.3	−22 32 48	MCG −4-46-3	13.9	0.8 x 0.7	13.1	Scd	162	Almost uniform faint surface brightness, star on E edge.
19 58 31.1	−20 50 44	MCG −4-47-4	14.1	0.6 x 0.6	12.8	Sbc		
20 05 37.9	−25 08 45	MCG −4-47-8	14.3	0.7 x 0.4	12.7	SBc	33	
20 09 03.7	−21 28 29	MCG −4-47-11	14.3	0.8 x 0.4	13.0	Sd	84	
20 19 00.8	−22 09 26	MCG −4-48-1	13.9	0.8 x 0.7	13.1	SB(s)b: pec	51	
20 26 20.4	−23 07 03	MCG −4-48-8	13.6	0.9 x 0.6	12.8	S0	116	
20 29 33.0	−22 05 36	MCG −4-48-10	14.5	0.6 x 0.3	12.5	SBcd	22	
20 33 31.2	−24 37 29	MCG −4-48-15	14.0	1.1 x 0.4	13.0	(R)S(r)b?	138	Mag 8.26v star SE 4.′8.
20 34 21.5	−26 33 49	MCG −4-48-17	13.8	0.4 x 0.4	11.7			MCG −4-48-15 E 0.′4, **MCG −4-48-16** NW 1.′3.
20 34 56.0	−23 06 19	MCG −4-48-19	13.2	0.9 x 0.6	12.4	S0	45	
20 38 19.6	−25 07 03	MCG −4-48-25	15.7	0.4 x 0.4	13.6			Mag 6.38v star W 6.′1.
20 38 39.3	−25 07 33	MCG −4-48-26	14.0	0.5 x 0.5	12.6	E/S0		Mag 8.48v star SSE 4.′9.
20 40 40.2	−25 53 09	MCG −4-48-27	14.0	0.8 x 0.4	12.6	Sbc	150	
19 32 20.4	−29 23 13	MCG −5-46-2	13.3	1.1 x 0.5	12.5	Sc	164	Star or bright knot S of center.
19 41 57.5	−29 03 17	MCG −5-46-4	13.2	1.3 x 1.1	13.4	SA(r)0$^-$	93	Mag 10.43v star N 4.′4.
19 56 02.0	−27 27 12	MCG −5-47-5	13.5	0.5 x 0.5	11.9	S0		
20 02 04.1	−27 33 44	MCG −5-47-20	13.9	0.8 x 0.6	12.9	SA(r)cd	128	Mag 4.49v star 62 Sagittarii SE 11.′9.
20 03 36.9	−29 52 25	MCG −5-47-21	13.5	1.0 x 0.7	13.0	S0	19	
20 07 47.5	−29 48 45	MCG −5-47-23	13.8	1.1 x 0.9	13.6	SAB(rs)c: I-II	43	Mag 6.71v star NW 7.′7.
20 11 18.0	−28 59 47	MCG −5-47-25	13.9	1.1 x 0.8	13.6	(R′)SB(rs)ab	21	
20 21 05.0	−27 22 34	MCG −5-48-2	14.3	0.5 x 0.2	11.7	(R)SB0$^+$:	81	Mag 9.44v star WNW 4.′8; ESO 462-8 ENE 4.′7.
20 22 35.0	−29 49 26	MCG −5-48-5	14.3	0.4 x 0.4	12.1			ESO 462-10 WSW 4.′8.
20 37 32.3	−27 34 37	MCG −5-48-24	13.4	1.1 x 0.6	12.8	SB(r)b:	3	**MCG −5-48-25** NE 1.′2.
20 39 34.7	−28 29 22	MCG −5-48-27	13.7	1.0 x 0.6	13.0	Sb-c	90	Mag 7.96v star NW 8.′0.
19 43 59.0	−28 29 09	NGC 6816	12.9	2.2 x 1.0	13.6	SB(rs)bc pec	105	Companion galaxy or bright knot E of bright nucleus; ESO 460-30 N 5.′3.
20 23 44.8	−19 19 35	NGC 6903	11.9	1.5 x 1.5	12.6	SAB0$^-$ pec:		Star NE of nucleus.
20 25 06.7	−24 48 26	NGC 6907	11.2	3.3 x 2.7	13.4	SB(s)bc II	46	Bright center forms an open "S" shape; dark patches N and E.
20 26 52.2	−18 37 03	NGC 6912	13.2	1.2 x 1.1	13.3	SB(s)c I	55	
20 33 19.3	−25 28 28	NGC 6924	12.8	2.0 x 1.7	14.0	SA(s)0$^-$:	154	Faint star S of nucleus.
20 35 56.4	−25 16 50	NGC 6936	12.7	1.9 x 0.8	13.0	SA0$^-$ sp:	4	Mag 10.02v star NW 11.′7; **ESO 528-24** SSE 9.′8; **ESO 528-26** ENE 13.′8.

GALAXY CLUSTERS

RA h m s	Dec ° ′ ″	Name	Mag 10th brightest	No. Gal	Diam ′	Notes
20 36 00.0	−25 16 00	A 3698	15.0	71	22	Except NGC and ESO members, all others anonymous, stellar.
20 40 42.0	−20 44 00	AS 891	15.5	2	22	**ESO 597-24** at center; **ESO 597-22** SW 9.′1; **ESO 597-25** SSE 5.′3.
20 43 36.0	−26 25 00	AS 894	16.0	5	17	See notes ESO 528-36; all others anonymous, stellar.

OPEN CLUSTERS

RA h m s	Dec ° ′ ″	Name	Mag	Diam ′	No. ★	B ★	Type	Notes
19 27 18.0	−23 35 00	ESO 525-8		5			cl	

GLOBULAR CLUSTERS

RA h m s	Dec ° ′ ″	Name	Total V m	B ★ V m	HB V m	Diam ′	Conc. Class Low = 12 High = 1	Notes
20 06 04.8	−21 55 17	NGC 6864	8.6	14.6	17.5	6.8	1	= **M 75**.

PLANETARY NEBULAE

RA h m s	Dec ° ′ ″	Name	Diam ″	Mag (P)	Mag (V)	Mag cent ★	Alt Name	Notes
19 58 14.1	−26 28 26	PN G14.8−25.6	46			17.7	PK 14−25.1	Two pairs of very faint stars form an E-W line on S edge.
19 46 34.3	−23 08 14	PN G17.3−21.9	136	15.2	13.2	15.9	PK 17−21.1	Elongated SE-NW; mag 13.3 star on SE edge; mag 13.2 star NW 1.′8 from center, just beyond NW edge.
19 17 04.2	−18 01 38	PN G19.4−13.6	32	14.0			PK 19−13.1	Mag 12.3 star 0.′9 NE; mag 12.7 star WSW 1.′9.
19 40 29.1	−20 27 05	PN G19.4−19.6	140			19.3	PK 19−19.1	Mag 13.6 star at center; mag 12.7 star on NE edge.
19 57 31.8	−21 36 37	PN G19.8−23.7	267	14.9		17.3	PK 19−23.1	= **Abell 66**. Two stars inside planetary, a mag 13.1 star 1.′2 ENE of the center and a mag 14.6 star 1.′4 WSW of the center.

GALAXIES

RA h m s	Dec ° ′ ″	Name	Mag (V)	Dim ′ Maj x min	SB	Type Class	PA	Notes
19 17 30.2	−28 15 11	ESO 459-10	14.1	1.7 x 0.2	12.8	Sbc: sp	160	Mag 10.06v star SE 7′.1.
19 06 18.7	−19 56 47	ESO 592-10	13.0	1.0 x 0.4	11.8	SB(r)bc? II-III	118	Mag 8.79v star ESE 10′.0.
19 11 47.2	−21 09 59	ESO 593-3	15.1	1.3 x 0.2	13.5	Scd	73	Mag 8.16v star NNE 5′.7.
19 12 27.4	−29 02 37	MCG −5-45-4	14.0	0.5 x 0.4	12.1	S(r:)a:	57	

OPEN CLUSTERS

RA h m s	Dec ° ′ ″	Name	Mag	Diam ′	No. ★	B ★	Type	Notes
18 14 05.7	−28 38 08	AL 3		1.3	30		cl	Sparse, circular.
18 02 00.0	−23 42 00	Bo 14	9.3	2	11	10.3	cl	Few stars; small brightness range; no central concentration; detached; involved in nebulosity.
18 08 36.0	−21 26 00	ClvdB 113		15	20		cl	
18 09 51.0	−23 50 00	Cr 367	6.4	40	30		cl	(A) Nebulous cluster of brighter stars.
18 17 16.0	−19 45 00	Cr 371	7.0	20	30		cl	
18 16 33.7	−18 18 42	Cr 469	9.1	2.6	51		cl	
18 09 24.0	−24 32 00	ESO 521-38		3			cl	
18 12 54.0	−24 22 00	ESO 522-5		4			cl	
18 56 38.9	−26 58 11	ESO 524-1		6			cl	
18 02 12.0	−21 55 00	ESO 589-26		2.5			cl	
18 31 45.0	−19 07 00	IC 4725	4.6	26	601	8.0	cl	= **M 25**. Moderately rich in bright and faint stars; strong central concentration; detached.
18 15 16.6	−18 59 33	Mrk 38		2			cl	
17 56 56.0	−19 01 00	NGC 6494	5.5	25	150	10.0	cl	= **M 23**. Rich in stars; moderate brightness range; slight central concentration; detached.
17 59 36.0	−17 23 00	NGC 6507	9.6	15	35	12.0	cl	Moderately rich in stars; large brightness range; not well detached.
18 02 31.0	−23 01 00	NGC 6514	6.3	30	70	6.0	cl	Involved in nebulosity (Trifid Nebula).
18 03 25.0	−27 53 30	NGC 6520	7.6	5	60	9.0	cl	Rich in stars; moderate brightness range; strong central concentration; detached; involved in nebulosity.
18 04 25.0	−24 23 00	NGC 6530	4.6	15	113	6.0	cl	Moderately rich in stars; moderate brightness range; slight central concentration; detached; involved in nebulosity. (A) Lagoon Nebula and cluster.
18 04 13.0	−22 30 00	NGC 6531	5.9	16	100	8.0	cl	= **M 21**. Rich in stars; large brightness range; strong central concentration; detached.
18 07 12.0	−23 18 00	NGC 6546	8.0	15	150	10.6	cl	Moderately rich in stars; small brightness range; slight central concentration; detached.
18 12 45.0	−21 35 00	NGC 6568	8.6	12	50		cl	Moderately rich in stars; small brightness range; not well detached.
18 13 50.0	−22 08 30	NGC 6573		6	40		cl?	(A) Three "mini-groups" of stars in a 6′ group, much as J. Herschel described it (his h 1999).
18 15 50.0	−22 08 12	NGC 6583	10.0	5	35		cl	Moderately rich in stars; moderate brightness range; strong central concentration; detached.
18 18 28.0	−18 24 24	NGC 6603	11.1	4	100	14.0	cl	Rich in stars; moderate brightness range; strong central concentration; detached; involved in nebulosity. (A) Cluster in IC 4715 = **M 24** (this is not M 24). Over 100 stars.
18 20 00.0	−17 06 00	NGC 6613	6.9	7	40	8.6	cl	= **M 18**. Few stars; large brightness range; slight central concentration; detached; involved in nebulosity. (A) ~30 bright stars, about 80 stars total.
18 54 34.0	−19 55 00	NGC 6716	7.5	10	38	8.3	cl	Few stars; small brightness range; not well detached.
17 59 17.8	−24 41 30	Ru 136		3	40	13.0	cl	
17 59 54.0	−25 10 00	Ru 137		5	30	13.0	cl	
17 59 54.0	−24 40 00	Ru 138		6	20	13.0	cl	
18 01 12.0	−23 32 00	Ru 139		12	80	12.0	cl:	
18 50 19.4	−18 12 00	Ru 145		35		10.0	cl	Moderately rich in stars; small brightness range; no central concentration; detached.
18 52 30.0	−21 04 54	Ru 146		4	20	12.0	cl	
17 59 30.0	−24 46 00	Ru 169		3	10	14.0	cl	
17 59 48.0	−28 10 00	Tr 31	9.8	5	25		cl	Few stars; moderate brightness range; no central concentration; detached.
18 24 38.8	−19 43 57	Tr 33	7.8	6	74	9.7	cl	Few stars; moderate brightness range; slight central concentration; detached.

STAR CLOUDS

RA h m s	Dec ° ′ ″	Name	Mag	Diam ′	No. ★	B ★	Type	Notes
18 17 00.0	−18 36 00	IC 4715		120			MW	(A) Small Sagittarius star cloud. Includes NGC 6603, B 92, B 93, B 304, B 307.

GLOBULAR CLUSTERS

RA h m s	Dec ° ′ ″	Name	Total V m	B ★ V m	HB V m	Diam ′	Conc. Class Low = 12 High = 1	Notes
18 08 21.8	−19 49 47	2MASS-GC 1				3.3		
18 08 36.5	−20 46 44	2MASS-GC 2				1.9		
18 01 49.1	−27 49 33	ESO 456-38	9.9			9.9		
18 06 08.6	−27 45 55	NGC 6540	14.6			1.5		Few stars similar in brightness; strong central concentration; detached; involved in nebulosity.
18 07 20.6	−24 59 51	NGC 6544	7.5	12.8	14.9	9.2	5	
18 09 17.3	−25 54 28	NGC 6553	8.3	15.3	16.9	9.2	11	
18 24 32.9	−24 52 12	NGC 6626	6.9	12.0	15.7	13.8	4	= **M 28**.
18 30 56.2	−25 29 47	NGC 6638	9.2	14.2	16.5	7.3	6	
18 31 54.3	−23 28 35	NGC 6642	8.9		16.3	5.8		
18 36 24.2	−23 54 12	NGC 6656	5.2	10.7	14.2	32.0	7	= **M 22**.
18 55 06.2	−22 42 03	NGC 6717	8.4	14.0	15.6	5.4	8	Also known as **Palomar 9**.
18 41 29.9	−19 49 33	Pal 8	10.9	15.4	17.3	5.2	10	
18 01 38.8	−26 50 23	Terzan 9	16.0	17.2	20.3	0.2		
18 02 57.0	−26 04 00	Terzan 10	14.9	19.7	22:	1.5		
18 12 15.8	−22 44 31	Terzan 11	16.4	18.5	20.5	1.0		

BRIGHT NEBULAE

RA h m s	Dec ° ′ ″	Name	Dim ′ Maj x min	Type	BC	Color	Notes
18 09 30.0	−23 44 00	IC 1274	7 x 7	E	1-5	3-4	This is the northernmost and second brightest in a group of nebulae including ICs 1275, 4685 and NGC 6559. The nebulosity is involved around a N-S chain of three equally spaced stars, two 9th and one 10th mag.
18 10 00.0	−23 50 00	IC 1275	10 x 6	E	1-5	3-4	Roundish nebula around two stars, mags 8 and 9. The N border is defined by the dark nebula Barnard 91.
18 17 48.0	−19 40 00	IC 1283/84	17 x 15	E+R	2-5	3-4	Irregular in shape; many fainter stars involved.
18 09 06.0	−23 25 00	IC 4684	3 x 2	R	3-5	1-4	A tiny patch of nebulosity surrounding a 9th mag star.
18 09 18.0	−23 59 00	IC 4685	15 x 10	R	3-5	3-4	Roughly centered around a 7th mag star, this is the largest and faintest in a group of nebulae including ICs 1274, 1275 and NGC 6559.
18 02 18.0	−23 02 00	NGC 6514	20 x 20	E+R	1-5	3-4	= **M 20. Trifid Nebula**. Approximately trisected by three prominent dark lanes aligned NE, S and W. Cluster involved.

(Continued from previous page)

BRIGHT NEBULAE

145

RA h m s	Dec ° ′ ″	Name	Dim ′ Maj x min	Type	BC	Color	Notes
18 03 48.0	−24 23 00	NGC 6523	45 x 30	E	1-5	2-4	= **M 8**, **Lagoon Nebula**. The densest section of the nebulosity is known as the **Hourglass**. One of the brightest nebulae, it can be seen with the unaided eye. A prominent dark lane splits the southern one-third of the object.
18 04 48.0	−23 35 00	NGC 6526	40 x 30	E	3-5	3-4	
18 10 00.0	−24 06 00	NGC 6559	8 x 5	E	1-5	3-4	The brightest condensation in a large nebular complex including ICs 1274, 1275 and 4685.
18 16 54.0	−19 47 00	NGC 6589	5 x 3	R	1-5	1-4	A pair of 10th mag stars (h2827, sep. 20″, ENE-WSW) involved in nebulosity.
18 17 06.0	−19 52 00	NGC 6590	4 x 3	R	1-5	1-4	Small, roundish nebula surrounding a pair of 11th mag stars.
18 15 54.0	−20 15 00	Sh2-35	10 x 7	E	3-5	3-4	Very irregular in shape and fairly diffuse. This nebulosity is set on a background extremely rich in very faint stars.

DARK NEBULAE

RA h m s	Dec ° ′ ″	Name	Dim ′ Maj x min	Opacity	Notes
17 57 37.4	−17 40 00	B 84a	16 x 16	5	Round.
18 02 25.2	−23 01 00	B 85			Trifid nebula.
18 03 01.5	−27 52 00	B 86	5 x 5	5	Several small stars in it; cluster NGC 6520 close E.
18 04 35.0	−24 06 52	B 88	2.0 x 0.5		Extended N and S; in M8.
18 04 59.8	−24 21 50	B 89	0.5 x 0.5		Very small; in M 8.
18 10 14.1	−28 16 50	B 90	3 x 1	5	Irregular; elongated N and S.
18 10 07.5	−23 42 00	B 91	5 x 2	5	Edge of diffused nebulosity.
18 15 35.2	−18 14 00	B 92	15 x 9	6	Black spot.
18 16 53.6	−18 04 00	B 93	15 x 2	4	Cometary; a sharply defined black head, with a diffused tail.
18 33 16.4	−26 03 40	B 98	3 x 3	6	Very small; black; sharply defined.
18 33 17.5	−21 29 00	B 99	11 x 3	3	Definite.
17 56 36.8	−29 01 00	B 289	35 x 7	4	Dark space between star clouds.
18 01 30.3	−28 36 30	B 294	3 x 3	3	Definite; star in SW part.
18 04 04.4	−24 32 00	B 296	6 x 1		Narrow; black; on S edge of M8.
18 04 20.9	−18 45 00	B 297	90 x 60	3	Region of irregular, dark markings, 2° E of M23.
18 06 20.4	−27 18 00	B 299		3	Several small spots in star cloud, each about 3′ in diameter.
18 08 50.8	−18 42 00	B 301	45 x 2	3	Dusky lane; NE and SW.
18 09 13.5	−23 58 00	B 302	0.5 x 0.5	5	Dark spot.
18 09 28.6	−24 00 00	B 303	1 x 1	5	Very black; crescent shaped, convex to E.
18 13 20.8	−18 43 00	B 304		3	Dusky lanes in star cloud extending toward the SW from B92 for 1one half degrees.
18 15 44.6	−25 43 00	B 306	4.0 x 0.5	3	Small, narrow, black; elongated NE and SW.
18 18 38.3	−17 57 00	B 307	6 x 1	3	Narrow, dusky mark in star cloud; NE and SW.
18 19 08.4	−22 14 00	B 308	7 x 5	3	Curved; dusky; E and W.
18 23 08.5	−24 02 00	B 309	5 x 5		Irregular; dusky.
18 30 13.1	−18 35 11	B 310	2 x 2	3	Small star near center.
18 30 27.2	−17 40 00	B 311	6 x 1	4	Black, elliptical NE and SW.
18 42 18.4	−20 02 00	B 315	5 x 5	3	Round; dark.

PLANETARY NEBULAE

RA h m s	Dec ° ′ ″	Name	Diam ″	Mag (P)	Mag (V)	Mag cent ★	Alt Name	Notes
18 03 18.5	−27 06 22	IC 4673	15	12.9	13.0	17.6	PK 3−2.3	Mag 11.9 star NE 0′.5; close pair of mag 10.53v and 11.9 stars W 1′.9.
18 33 54.6	−22 38 41	IC 4732	13	13.3	12.1	16.2	PK 10−6.1	Mag 10.14v star W 2′.5.
18 05 13.1	−19 50 35	NGC 6537	10	12.5	11.6	18.8	PK 10+0.1	Almost stellar; mag 11.6 star W 1′.5.
18 11 52.6	−28 10 42	NGC 6565	14	13.2	11.6	18.5	PK 3−4.5	Almost stellar; mag 10.9 star W 2′.5, in star rich area.
18 13 45.2	−19 04 33	NGC 6567	12	11.7	11.0	14.3	PK 11−0.2	Very faint stars on N and E edges; mag 10.79v star N 1′.9; mag 9.63v star SE 3′.4.
18 16 16.6	−20 27 03	NGC 6578	11	13.1	12.9	15.8	PK 10−1.1	Mag 10.9 star W 0′.4; small, tight grouping of three mag 9-12 stars ESE 1′.1.
18 22 54.3	−26 49 18	NGC 6620	8	15.0	12.7	19.6	PK 5−6.1	Stellar; mag 10.1 star ESE 1′.0; mag 11.0 star W 2′.1.
18 25 42.4	−23 12 10	NGC 6629	16	11.6	11.3	12.9	PK 9−5.1	Mag 12.8 star E 0′.9; mag 9.52v star SE 2′.0.
18 32 34.7	−25 07 44	NGC 6644	12	12.2	10.7	15.6	PK 8−7.2	Mag 12.2 star N 0′.9.
17 56 39.9	−29 37 23	PN G0.6−2.3					PK 0−2.3	Stellar; small triangle of mag 10-11 stars W 2′.6.
17 58 09.6	−29 44 20	PN G0.7−2.7	7	13.5	13.6	14.7	PK 0−2.4	Mag 12.6 star on SE edge; mag 10.1 star NW 2′.0.
17 56 02.8	−29 11 16	PN G0.9−2.0	5	18.0		16.2	PK 0−2.1	Mag 10.6 star WSW 3′.2; mag 12.4 star E 3′.3.
17 58 26.1	−29 20 46	PN G1.0−2.6	5	17.9			PK 1−2.2	Mag 12.2 star E 0′.8; mag 11.3 star NE 2′.4.
18 00 37.7	−29 21 50	PN G1.2−3.0	5			16.2	PK 1−3.1	Mag 12.0 star E 0′.9; mag 11.2 star S 1′.2.
18 03 53.6	−29 51 21	PN G1.2−3.9	4	15.8		16.5	PK 1−3.7	Mag 12.4 star E 0′.9; mag 8.04v star W 3′.7.
18 02 27.0	−29 25 13	PN G1.4−3.4	13	18.2		15.8	PK 1−3.3	Mag 12.5 star E 0′.8; mag 12.4 star S 1′.1.
17 56 02.4	−28 14 11	PN G1.7−1.6					PK 1−1.3	Mag 12.2 star E 3′.3; mag 9.86v star NW 2′.7.
18 07 14.6	−29 41 24	PN G1.7−4.4	5	17.7		16.5	PK 1−4.1	Mag 11.9 star NW 0′.9; mag 11.3 star WSW 1′.0.
18 07 54.0	−29 44 36	PN G1.7−4.6	3	14.4	14.5	17.0	PK 1−4.2	Located 0′.8 S of mag 8.15v star.
18 05 05.6	−29 20 13	PN G1.8−3.8	12	16.2		14.7	PK 1−3.9	Close pair of mag 11.5, 12.5 stars NW 0′.6.
17 58 21.9	−28 14 53	PN G2.0−2.0	6	15.5		16.8	PK 2−2.1	Mag 13.0 star E 1′.1; mag 12.4 star SW 1′.3.
17 59 19.4	−28 13 49	PN G2.1−2.2	7	14.3	14.3		PK 2−2.2	Triangle of mag 11-12 stars W 3′.4; mag 12.8 star SSE 0′.7.
18 07 07.3	−29 13 06	PN G2.1−4.2	7	14.1	14.0	15.4	PK 2−4.1	Close pair of mag 10.5, 12.8 stars E 1′.0.
18 01 00.0	−28 16 14	PN G2.2−2.5	5	19.6			PK 2−2.5	Line of three mag 11-12 stars S and SW.
18 01 42.7	−28 25 44	PN G2.2−2.7	9	12.4	12.4	16.7	PK 2−2.4	Close pair of mag 11.4, 11.6 stars SW 1′.5.
18 04 28.9	−28 37 34	PN G2.3−3.4	4	15.8			PK 2−3.2	Mag 11.4 star SE 0′.4; mag 11.2 star WSW 0′.8.
18 04 05.0	−28 27 53	PN G2.4−3.2	12	17.2			PK 2−3.7	Located between a pair of E-W mag 11-12 stars.
18 06 05.8	−28 40 30	PN G2.4−3.7	3			14.4	PK 2−3.5	Mag 10.9 star NE 2′.1.
17 58 30.8	−27 37 07	PN G2.5−1.7	5	19.5		21.0	PK 2−1.1	Mag 12.7 star NE 0′.4; mag 11.6 star W 0′.5.
18 13 00.5	−29 25 12	PN G2.5−5.4	17	14.6		17.7	PK 2−5.2	Mag 11.0 star N 0′.5; mag 9.4 star E 2′.4.
18 05 25.8	−28 22 04	PN G2.6−3.4	3			14.9	PK 2−3.3	Mag 12.4 star W 1′.1; mag 10.5 star ESE 1′.8.
18 11 05.0	−28 59 01	PN G2.7−4.8	9	13.1	13.8	17.4	PK 2−4.2	Mag 12.7 star S 1′.3; mag 11.9 star ESE 1′.7 and mag 12.8 star NE 1′.4.
18 01 10.3	−27 38 20	PN G2.8−2.2	5				PK 2−2.3	Mag 12.1 star E 0′.5; mag 11.3 star N 1′.2.
18 08 05.8	−28 26 11	PN G2.9−3.9	4	15.4			PK 2−3.6	Mag 12.5 star E 0′.4; mag 4.56v star S 1′.2.
18 02 51.7	−27 41 01	PN G3.0−2.6	3	17.6		21.8	PK 3−2.4	Mag 10.4 star on N edge; mag 10.6 star WSW 1′.0.
18 10 30.8	−28 19 24	PN G3.2−4.4	3	17.2		20.0	PK 3−4.1	Mag 12.1 star SW 1′.8; mag 12.1 star NW 2′.5.
18 17 41.4	−29 08 20	PN G3.2−6.2	8	13.0		15.7	PK 3−6.1	Mag 11.6 star S 1′.6; mag 10.00v star W 2′.7.
18 11 35.0	−28 22 37	PN G3.3−4.6	12			13.2	PK 3−4.7	Close pair of mag 12.2 stars SW 2′.2; mag 10.8 star N 2′.7.
18 23 08.9	−29 43 25	PN G3.3−7.5	8	16.7		17.0	PK 3−7.1	Mag 11.9 star NW 3′.4; mag 11.6 star SW 4′.6.
18 12 48.0	−28 20 00	PN G3.4−4.8	9	15.5			PK 3−4.9	Mag 9.21v star ENE 2′.6.
18 03 11.9	−26 58 31	PN G3.6−2.3	9	16.3	14.3		PK 3−2.2	Mag 10.2 star E 1′.6, mag 10.7 star S 0′.6.
18 12 34.6	−27 58 10	PN G3.7−4.6	9	13.1	13.2	17.1	PK 3−4.8	Mag 11.9 star WNW 0′.8; mag 11.5 star E 1′.1.
18 11 29.3	−27 46 17	PN G3.8−4.3	6	14.8	14.8	16.2	PK 3−4.3	Mag 11.1 star S 0′.9; mag 11.2 star NW 3′.9.
18 12 23.8	−27 52 14	PN G3.8−4.5	8	16.0			PK 3−4.4	Mag 13.3 star SE 0′.4; mag 13.3 star NW 0′.7.

(Continued from previous page)

145 PLANETARY NEBULAE 145

RA h m s	Dec ° ′ ″	Name	Diam ″	Mag (P)	Mag (V)	Mag cent ★	Alt Name	Notes
18 03 39.3	−26 43 34	PN G3.9−2.3	5	14.6	14.6		PK 3−2.1	Mag 9.88v star ESE 1′.9; mag 10.3 star WNW 1′.9.
18 06 49.8	−27 06 16	PN G3.9−3.1	5	18.2			PK 3−3.2	Mag 12.8 star W 0′.5; mag 9.09v star E 2′.6.
18 06 40.9	−26 54 57	PN G4.0−3.0	4	14.7		15.5	PK 4−3.1	Mag 12.3 star NNE 1′.4; mag 12.2 star SW 1′.4.
18 17 42.4	−28 17 17	PN G4.0−5.8	10	16.0		17.2	PK 4−5.3	Mag 12.6 star E 0′.4; mag 6.36v star W 4′.1.
18 10 12.2	−27 16 36	PN G4.1−3.8	3	17.5		20.4	PK 4−3.3	Close pair of mag 11-12 stars NW 1′.5; mag 9.69v star S 3′.4.
18 08 01.4	−26 54 02	PN G4.2−3.2	6	16.6		19.8	PK 4−3.2	Mag 12.2 star on SW edge; mag 11.0 star S 1′.5.
18 12 25.2	−27 29 14	PN G4.2−4.3	6	14.7		16.4	PK 4−4.1	Mag 12.4 star ESE 0′.7; mag 12.4 star NW 0′.9.
18 18 38.5	−28 08 04	PN G4.2−5.9	7	17.8		16.8	PK 4−5.5	Mag 12.3 star W 1′.2; mag 12.4 star E 1′.8.
18 05 57.5	−26 29 42	PN G4.3−2.6		15.7	15.3		PK 4−2.1	Small triangle of mag 11-12 stars SSW 1′.7; mag 11.5 star NE 2′.3.
18 16 11.5	−27 14 58	PN G4.8−5.0	9	13.8	14.2	17.2	PK 4−5.1	Mag 12.1 star ENE 0′.8; mag 12.7 star W 1′.1.
18 16 17.4	−27 04 31	PN G4.9−4.9	4	16.1	16.5		PK 4−4.2	Very close pair of mag 13 stars S 1′.0; mag 11.4star NNE 2′.9.
18 12 22.4	−26 32 53	PN G5.0−3.9	12	17.1		20.6	PK 5−3.2	Mag 13.7 star WNW 1′.0; mag 10.8 star WSW 2′.8.
18 09 13.9	−26 02 29	PN G5.1−3.0	6	16.4		17.4	PK 5−3.1	Mag 11.6 star E 0′.8; mag 13.3 star N 1′.5.
18 32 30.9	−28 43 20	PN G5.1−8.9	19	15.7	16.7	18.0	PK 5−8.1	Mag 12.2 star NE 1′.9; mag 12.4 star WSW 1′.8.
18 05 25.3	−25 13 33	PN G5.4−1.9					PBOZ 34	Mag 12.5 star NW 0′.8.
18 07 54.0	−25 24 03	PN G5.5−2.5	10	14.2			PK 5−2.1	Mag 12.6 star E 0′.8; mag 11.4 star NW 2′.2.
18 13 40.5	−26 08 40	PN G5.5−4.0	7	16.5		17.4	PK 5−4.1	Mag 12.5 star ESE 1′.1; mag 13.1 star E 1′.6.
18 16 53.9	−26 23 21	PN G5.6−4.7	12	16.7		18.9	PK 5−4.2	Mag 12.4 star S 0′.5; mag 12.5 star W 0′.6.
18 12 45.2	−25 44 06	PN G5.7−3.6	14	16.1		18.1	PK 5−3.3	Mag 12.8 star ENE 0′.9; mag 13.1 star WNW 1′.5.
18 19 25.3	−26 35 22	PN G5.7−5.3	9	15.1		17.4	PK 5−5.1	Mag 13.0 star WSW 0′.4; mag 12.2 star N 1′.7.
18 09 12.7	−25 04 29	PN G5.9−2.6	10				PK 5−2.2	Mag 10.8 star NNW 1′.0; mag 13.4 star E 1′.4.
18 13 16.1	−25 30 06	PN G6.0−3.6	5	13.6			PK 6−3.3	Mag 13.4 star E 1′.1; mag 11.2 star NW 3′.1.
18 14 19.4	−25 20 51	PN G6.2−3.7	9	17.8		20.7	PK 6−3.4	Mag 9.9 star SE 4′.4; mag 8.43v star NW 3′.3.
18 18 13.4	−25 38 08	PN G6.4−4.6	7	13.4	15.0	17.3	PK 6−4.1	Mag 13.1 star SE 1′.2; mag 9.48v star N 5′.0.
18 12 34.0	−24 50 00	PN G6.5−3.1	25	18.1			PK 6−3.1	Mag 13.3 star E1′.1.
18 09 30.3	−24 12 26	PN G6.7−2.2	8	17.0			PK 6−2.1	Mag 13.2 S 1′.4; mag 12.3 NE 1′.6.
18 14 28.8	−24 43 38	PN G6.8−3.4	5	15.7		17.4	PK 6−3.2	Mag 13.2 star SE 0′.6; mag 9.52v star ENE 1′.9.
18 34 55.4	−27 06 32	PN G6.8−8.6	13	15.2		18.0	PK 6−8.1	Mag 13.6 star E 1′.8; mag 12.8 star NW 1′.5.
18 24 57.5	−25 41 55	PN G7.0−6.0	8	14.4	13.2		PK 7−6.1	Mag 11.9 star SE 0′.6; mag 12.7 star N 0′.7.
18 27 59.7	−26 06 48	PN G7.0−6.8	7	13.3	13.1	16.6	PK 7−6.2	Mag 12.5 star on N edge; mag 10.86v star NE 2′.2.
18 17 15.9	−23 58 54	PN G7.8−3.7	8	16.2			PK 7−3.1	Mag 13.5 star S 0′.3; mag 9.65v star W 2′.1.
18 20 08.8	−24 15 04	PN G7.8−4.4	8			14.6	PK 7−4.1	Mag 10.5 star SW 2′.5; mag 10.5 star E 4′.5.
18 22 01.2	−24 10 40	PN G8.1−4.7	3	14.5		15.8	PK 8−4.1	
18 22 32.1	−24 09 29	PN G8.2−4.8	4	14.0		18.2	PK 8−4.2	Mag 12.4 star S 1′.9; mag 9.91v star WNW 2′.9.
18 08 26.0	−22 16 53	PN G8.3−1.1	5	14.7	14.8	18.3	PK 8−1.1	Mag 10.9 star SE 3′.0; mag 8.08v star W 3′.8.
18 18 23.9	−23 24 55	PN G8.4−3.6	8	18.8			PK 8−3.1	Mag 12.0 star N 0′.6; mag 11.5 star W 1′.4.
18 14 50.9	−22 43 55	PN G8.6−2.6		17.2			PK 8−2.1	Small triangle of mag 10-12 stars S 1′.6; mag 9.78v star W 4′.1.
18 31 52.9	−24 46 16	PN G8.6−7.0	3	16.8	15.6		PK 8−7.1	Mag 12.9 star SW 1′.0; mag 12.3 star NE 2′.3.
17 56 00.6	−19 29 27	PN G9.3+2.8					PK 9+2.1	Mag 12.0 star E 4′.0.
18 44 43.2	−25 21 33	PN G9.4−9.8	6	13.6	13.3	17.1	PK 9−9.1	Located 0′.8 NE of a mag 8.47v star.
18 48 12.2	−25 28 53	PN G9.6−10.6	5	13.5	13.7	15.9	PK 9−10.1	Mag 11.2 star E 3′.6; mag 11.6 star W 3′.8.
18 25 04.9	−22 34 50	PN G9.8−4.6	6	14.0	13.7		PK 9−4.1	Mag 12.3 star W 0′.8; mag 10.46v star SE 2′.6.
18 36 23.4	−23 55 20	PN G9.8−7.5	9	15.::		14.3	PK 9−7.1	Located approximately 1′.2 S to SSE of the center of M 22. Note that specialty finder charts will probably be needed to identify this elusive planetary nebula.
17 57 06.6	−18 06 43	PN G10.6+3.2	25	16.8		18.3	PK 10+3.1	Mag 11.3 star W 1′.3; mag 9.67v star NE 1′.6.
18 34 51.6	−22 43 17	PN G10.7−6.7	8	15.1	14.7	17.3	PK 10−6.2	Mag 14.8 star on NE edge; mag 12.7 star SSW 1′.3.
18 29 11.0	−21 46 55	PN G11.0−5.1	5	13.1	13.5	15.3	PK 11−5.1	Mag 12.3 star SE 2′.0; mag 12.0 star N 3′.7.
18 00 08.8	−17 40 43	PN G11.3+2.8	5	18.2			PK 11+2.1	Mag 11.8 star on W edge.
18 46 35.1	−23 26 47	PN G11.3−9.4	2	14.1	14.0	13.4	PK 11−9.1	Mag 11.4 star E 3′.1; mag 11.6 star WNW 3′.9.
18 11 48.9	−18 46 21	PN G11.7−0.0	5	17.5		16.0	PK 11−0.1	Mag 11.1 star ENE 1′.7; mag 11.7 star W 1′.6.
18 36 33.8	−21 49 03	PN G11.7−6.6	25			13.9	PK 11−6.1	Mag 13.3 star E 2′.0; mag 11.4 star W 2′.1.
18 50 26.0	−22 34 23	PN G12.5−9.8	4	13.2	13.8	17.4	PK 12−9.1	Mag 13.7 star W 1′.8; mag 11.9 star ESE 3′.4.
18 23 08.0	−19 17 05	PN G12.6−2.7	3			16.8	PK 12−2.1	Forms a triangle with a mag 10.9 star W 1′7 and a mag 11.5 star NW 2′.1.
18 29 59.6	−19 40 38	PN G13.0−4.3	5	16.0	15.5		PK 13−4.1	Mag 10.53v star E 2′.1; mag 11.6 star NW 3′.8.
18 29 30.0	−19 05 46	PN G13.4−3.9	5	15.7	14.8		PK 13−3.1	Mag 11.1 star NE 2′.6; mag 12.2 star W 3′.0.
18 55 30.7	−21 49 40	PN G13.7−10.6	15	13.2		16.7	PK 13−10.1	Mag 12.4 star ESE 1′.1; mag 12.1 star NW 2′.3.
18 26 03.9	−18 12 37	PN G13.8−2.8	110	15.2			PK 13−2.1	Mag 10.46v star NE 6′.3; mag 10.7 star N 3′.9.
18 45 35.2	−20 34 59	PN G13.8−7.9	13	13.9	14.0	18.4	PK 13−7.1	Mag 12.1 star SW 0′.3; mag 12.9 star NNE 0′.8.
18 36 32.3	−19 19 29	PN G14.0−5.5		15.1		13.2	PK 14−5.1	Stellar; pair with mag 12.2 star on W edge.
18 44 01.8	−19 54 53	PN G14.2−7.3	7	14.3	14.2	17.4	PK 14−7.1	Mag 9.96v star ENE 0′.7; mag 8.80v star NW 1′.9.
18 37 11.2	−19 02 22	PN G14.3−5.5	1	14.4			PK 14−5.2	Forms triangle with mag 11.7 star SE 4′.0 and mag 11.5 star NE 4′.4; mag 8.00v star SSW 6′.7.
18 33 20.9	−18 16 37	PN G14.6−4.3	6	13.0	13.3		PK 14−4.1	Mag 11.9 star W 2′.3.
19 02 16.9	−21 26 53	PN G14.7−11.8	85			20.0	PK 14−11.1	Mag 13.1 star on E edge; mag 13.0 star ~ 0′.5 W of W edge.
18 29 11.3	−17 27 12	PN G14.9−3.1	13				SaSt 3-166	Mag 12.7 star NE 0′.4.
18 35 48.3	−17 36 09	PN G15.4−4.5	6	13.9	13.5	14.3	PK 15−4.1	Forms a small triangle with mag 11.9 star ENE 0′.9 and mag 12.2 star.
18 36 08.3	−16 59 57	PN G16.0−4.3	13	12.5	13.5		PK 16−4.1	Mag 12.5 star NNE 0′.6; mag 11.4 star SE 2′.1.
18 37 46.3	−17 05 47	PN G16.1−4.7	10	13.3	13.5	16.2	PK 16−4.2	Mag 11.8 star SSE 0′.5; mag 11.8 star SW 1′.5.
19 01 01.4	−18 12 14	PN G17.6−10.2	67	15.4	15.0	15.4	PK 17−10.1	Mag 13.1 star SSE 1′.2 from center; mag 10.32v star SW 2′.0 from center.
19 17 04.2	−18 01 38	PN G19.4−13.6	32	14.0			PK 19−13.1	Mag 12.3 star 0′.9 NE; mag 12.7 star WSW 1′.9.

146 GALAXIES 146

RA h m s	Dec ° ′ ″	Name	Mag (V)	Dim ′ Maj x min	SB	Type Class	PA	Notes
16 47 02.7	−29 05 29	ESO 453-4	15.6	1.5 x 0.2	14.1	S	20	Mag 10.36v star SW 2′.1.
16 58 06.0	−21 16 27	ESO 586-6	13.1	1.5 x 1.4	13.8	SAB(r)b? II	93	Close pair of mag 10.24v, 9.22v stars NNW 4′.4.

OPEN CLUSTERS

RA h m s	Dec ° ′ ″	Name	Mag	Diam ′	No. ★	B ★	Type	Notes
18 02 00.0	−23 42 00	Bo 14	9.3	2	11	10.3	cl	Few stars; small brightness range; no central concentration; detached; involved in nebulosity.
17 46 18.9	−29 20 00	Cr 347	8.8	10	40	10.6	cl	Moderately rich in stars; moderate brightness range; slight central concentration; detached; involved in nebulosity.
17 49 51.0	−28 46 00	Cr 351	9.3	8	30		cl	Moderately rich in stars; moderate brightness range; not well detached. (A) As Collinder says, a slight condensation in a Milky Way star cloud.
17 53 16.4	−27 22 19	Cz 37		3	35		cl	
16 47 06.0	−25 49 00	ESO 518-3		5			cl	
17 04 22.3	−19 27 25	ESO 587-4		2.5			cl	

(Continued from previous page)

OPEN CLUSTERS

RA h m s	Dec ° ′ ″	Name	Mag	Diam ′	No. ★	B ★	Type	Notes
18 02 12.0	−21 55 00	ESO 589-26		2.5			cl	
17 52 56.5	−22 19 00	NGC 6469	8.2	8	50		cl	Moderately rich in stars; moderate brightness range; not well detached.
17 56 56.0	−19 01 00	NGC 6494	5.5	25	150	10.0	cl	= **M 23**. Rich in stars; moderate brightness range; slight central concentration; detached.
17 59 36.0	−17 23 00	NGC 6507	9.6	15	35	12.0	cl	Moderately rich in stars; large brightness range; not well detached.
18 02 31.0	−23 01 00	NGC 6514	6.3	30	70	6.0	cl	Involved in nebulosity (Trifid Nebula).
18 03 25.0	−27 53 30	NGC 6520	7.6	5	60	9.0	cl	Rich in stars; moderate brightness range; strong central concentration; detached; involved in nebulosity.
17 47 08.2	−29 36 05	Ru 129		4	10	12.0	cl	
17 49 06.9	−29 14 00	Ru 131		8	15	11.0	cl	Few stars; small brightness range; not well detached.
17 52 34.1	−28 40 38	Ru 133		5	20	12.0	cl:	
17 52 48.0	−29 33 00	Ru 134		5	30	12.0	cl	
17 59 17.8	−24 41 30	Ru 136		3	40	13.0	cl	
17 59 54.0	−25 10 00	Ru 137		5	30	13.0	cl	
17 59 54.0	−24 40 00	Ru 138		6	20	13.0	cl	
18 01 12.0	−23 32 00	Ru 139		12	80	12.0	cl:	
17 52 48.0	−28 27 00	Ru 168		4	20	12.0	cl	
17 59 30.0	−24 46 00	Ru 169		3	10	14.0	cl	
17 28 29.9	−29 30 00	Tr 26	9.5	7	40		cl	Moderately rich in stars; small brightness range; slight central concentration; detached.
17 59 48.0	−28 10 00	Tr 31	9.8	5	25		cl	Few stars; moderate brightness range; no central concentration; detached.
17 46 14.4	−29 42 05	vdB-Ha 245		1	12		cl	

STAR CLOUDS

RA h m s	Dec ° ′ ″	Name	Mag	Diam ′	No. ★	B ★	Type	Notes
17 25 23.9	−30 00 17	NGC 6360					MW	
17 54 12.0	−29 07 12	NGC 6476		20			MW	(A) A small Milky Way star cloud as described by J. Herschel (his h 3712). Dark nebula on E side.

GLOBULAR CLUSTERS

RA h m s	Dec ° ′ ″	Name	Total V m	B ★ V m	HB V m	Diam ′	Conc. Class Low = 12 High = 1	Notes
16 39 25.0	−28 23 54	ESO 452-11	12.0			1.2		
18 01 49.1	−27 49 33	ESO 456-38	9.9			9.9		
17 31 05.2	−29 58 54	HP 1	12.5	16.0	18.6	1.2		
16 53 25.4	−22 10 38	NGC 6235	8.9	14.0	16.7	5.0	10	
17 02 37.7	−26 16 05	NGC 6273	6.8	14.0	17.0	17.0	8	= **M 19**.
17 04 28.8	−24 45 53	NGC 6284	8.9		16.6	6.2	9	
17 05 09.4	−22 42 29	NGC 6287	9.3	14.5	17.1	4.8	7	
17 10 10.4	−26 34 54	NGC 6293	8.3	14.3	16.5	8.2	4	
17 14 32.5	−29 27 44	NGC 6304	8.3	14.5	16.2	8.0	6	
17 16 37.4	−28 08 24	NGC 6316	8.1	15.0	17.8	5.4	3	
17 17 59.2	−23 45 57	NGC 6325	10.2	14.7	17.3	4.1	4	
17 19 11.8	−18 30 59	NGC 6333	7.8	13.5	16.2	12.0	8	= **M 9**.
17 21 10.2	−19 35 14	NGC 6342	9.5	15.0	16.9	4.4	4	
17 23 58.6	−26 21 13	NGC 6355	8.6		17.2	4.2		
17 23 35.0	−17 48 47	NGC 6356	8.2	15.1	17.7	10.0	2	
17 38 36.9	−23 54 32	NGC 6401	7.4	15.5	18.0	4.8	8	
17 48 52.6	−20 21 34	NGC 6440	9.3	16.7	18.7	4.4	5	
17 43 42.2	−26 13 21	Pal 6	11.6		19.1	1.2	11	
17 48 04.9	−24 46 47	Terzan 5	13.9	20.5	22.5	2.4		
18 01 38.8	−26 50 23	Terzan 9	16.0	17.2	20.3	0.2		
18 02 57.0	−26 04 00	Terzan 10	14.9	19.7	22:	1.5		
17 54 27.2	−24 08 43	UKS 1	17.3			2.0		

BRIGHT NEBULAE

RA h m s	Dec ° ′ ″	Name	Dim ′ Maj x min	Type	BC	Color	Notes
18 02 18.0	−23 02 00	NGC 6514	20 x 20	E+R	1-5	3-4	= **M 20**. **Trifid Nebula**. Approximately trisected by three prominent dark lanes aligned NE, S and W. Cluster involved.
18 03 48.0	−24 23 00	NGC 6523	45 x 30	E	1-5	2-4	= **M 8**, **Lagoon Nebula**. The densest section of the nebulosity is known as the **Hourglass**. One of the brightest nebulae, it can be seen with the unaided eye. A prominent dark lane splits the southern one-third of the object.
17 46 36.0	−29 18 00	Sh2-16	12 x 12	E	3-5	4-4	In a rich star field. There is a small patch of nebulosity 3′ across, centered about 5′ off the NE edge of Sh 2-16.

DARK NEBULAE

RA h m s	Dec ° ′ ″	Name	Dim ′ Maj x min	Opacity	Notes
16 40 33.6	−24 04 00	B 44		5	Dark lane E from Rho Ophiuchi region.
16 46 26.2	−21 36 00	B 45		5	Rather definite.
16 57 12.9	−22 44 00	B 46	12 x 12	6	Irregular; definite; a string of small stars in the middle.
16 59 42.9	−22 39 00	B 47	15 x 15	5	Irregular; definite; connected with B51 by two sharp lanes.
17 04 44.0	−22 16 00	B 51	20 x 20	6	Definite.
17 08 22.2	−23 40 00	B 57	5 x 5	6	Elongated NE and SW.
17 11 23.6	−27 30 00	B 59	60 x 60	5	Sink hole; at the W end of a wide broken lane from B78.
17 11 51.2	−22 26 00	B 60	13 x 13	3	Curved; extended NE and SW.
17 14 59.9	−20 29 00	B 61	10 x 4	6	Small; elongated E and W .
17 16 12.3	−20 53 00	B 62	19 x 19	6	Very black in NW part.
17 16 28.3	−21 29 00	B 63	100 x 20	3	Large, definite, curved figure, convex to N.
17 17 18.7	−18 29 00	B 64	20 x 20	6	Cometary; W of M 9.
17 19 36.4	−26 42 00	B 65	12 x 12	6	Definite; elongated E and W.
17 19 57.1	−26 54 00	B 66	8 x 8	6	Definite; extended E and W.
17 20 57.1	−26 52 00	B 67		6	Definite; extended N and S; connects B65 and 66.
17 22 30.0	−21 53 00	B 67a	13 x 13	6	Irregular; definite.
17 22 36.4	−23 47 00	B 68	4 x 4	6	Small; irregular; sharply defined; about 20′ SW of B72.
17 22 55.9	−23 55 00	B 69	4 x 4	6	Very small; irregular; sharply defined.
17 23 32.3	−24 02 00	B 70	4 x 4	4	Sharply defined on W side.
17 23 02.2	−24 00 00	B 71			Very small.

(Continued from previous page)

DARK NEBULAE

RA h m s	Dec ° ′ ″	Name	Dim ′ Maj x min	Opacity	Notes
17 23 35.9	−23 37 00	B 72	4 x 4	6	**Snake Nebula**. S-shaped.
17 24 08.2	−24 17 10	B 73	1 x 1		Very small; extended N and S.
17 25 08.0	−24 12 00	B 74	15 x 10	5	Slightly curved, sharpest on W and S sides.
17 25 17.7	−22 02 00	B 75		5	Curved and scalloped marking.
17 25 38.8	−24 25 00	B 76	30 x 30	4	Irregular, narrow, black strip E and W.
17 28 37.0	−23 51 00	B 77	60 x 60	3	Indefinite; connected by a vacant strip with B78.
17 32 43.3	−25 36 00	B 78	200 x 140	5	Very large.
17 37 23.2	−19 37 00	B 79	30 x 30	6	Narrow; sharply defined SE and NW.
17 37 42.7	−21 16 40	B 80	3 x 1	3	Small; definite; elongated SW and NE.
17 38 32.7	−23 56 00	B 81			Very small; diffused; close SW of NGC 6401.
17 38 37.2	−23 47 00	B 82		3	Small; angular; well defined; 8′ N of NGC 6401.
17 38 58.6	−24 11 00	B 83	7 x 7	7	Irregular; several small stars in it; narrow extension to the S.
17 45 18.7	−20 00 00	B 83a	4 x 4	6	Small; definite.
17 46 25.6	−20 15 00	B 84		6	Irregular looped figure; sharply defined.
17 57 37.4	−17 40 00	B 84a	16 x 16	5	Round.
18 02 25.2	−23 01 00	B 85			Trifid nebula.
18 03 01.5	−27 52 00	B 86	5 x 5	5	Several small stars in it; cluster NGC 6520 close E.
16 48 54.4	−29 48 00	B 236			Center of system of indistinct dark lanes.
16 48 55.0	−29 58 00	B 237	30 x 6	1	Dark lane N of magnitude 8 star.
16 52 21.8	−23 08 00	B 238	13 x 13	6	Irregular.
17 09 46.1	−29 35 00	B 243	25 x 25	3	Very irregular.
17 10 51.8	−28 25 00	B 244	30 x 20	5	Irregular; sharpest on W side.
17 11 55.6	−29 24 00	B 245	8 x 8	3	Round; diffused; several bright stars in it
17 12 01.9	−22 39 00	B 246	20 x 20	3	Irregular.
17 13 04.2	−28 59 00	B 248	10 x 3	5	Irregular; elongated E and W.
17 13 04.8	−29 09 00	B 249	5.0 x 0.3	5	Very thin, short, black line.
17 13 02.0	−28 24 00	B 250	15 x 15	4	Diffused.
17 13 48.8	−20 09 00	B 251	20 x 5	3	Dusky; elongated NE and SW.
17 15 31.8	−22 34 00	B 253	60 x 60		Irregular dark region.
17 20 35.2	−23 28 00	B 255	5 x 5	5	Dusky.
17 22 12.4	−28 50 00	B 256	50 x 10	5	Very irregular and diffused; longest E and W.
17 22 01.5	−19 18 00	B 259	30 x 30	4	Irregular dark area.
17 24 48.9	−25 37 00	B 260	12 x 7		Diffused; elongated N and S.
17 25 04.0	−23 02 00	B 261	14 x 5	4	Elongated E and W; diffused on S side.
17 25 32.6	−22 37 00	B 262	30 x 30	4	Large dark region; diffused except on N border.
17 27 12.8	−25 32 00	B 264	10 x 1	4	Narrow; black; E and W; bends N at E end.
17 27 36.6	−25 12 00	B 265	18 x 7	4	Irregular; N and S; diffused on E side.
17 28 07.2	−20 56 00	B 266	30 x 5	3	Curved; NE and SW.
17 28 31.8	−25 13 00	B 267	5 x 3	5	Small, very black; elongated NW and SE.
17 31 25.9	−20 31 00	B 268	100 x 50	4	Irregular dark area.
17 32 13.5	−22 46 00	B 269	60 x 28		Dusky region, slightly elongated NW and SE.
17 32 43.0	−19 36 00	B 270	11 x 11	5	Round; in NE part of B268.
17 37 35.9	−23 25 00	B 272	45 x 45	3	Region of irregular dark markings.
17 38 03.6	−22 44 00	B 274	18 x 3.5	4	Triangular; definite; elongated N and S.
17 39 39.0	−19 49 00	B 276	40 x 40	6	Broken region of dark markings.
17 40 44.9	−23 04 00	B 277	18 x 18	4	Consists of two narrow arms in V shape; E and W.
17 44 33.2	−22 34 00	B 279	60 x 60	3	Irregular dark area.
17 44 57.1	−20 43 00	B 280	60 x 60	3	Irregular dusky area.
17 46 47.3	−23 43 00	B 281		2	Dark loop.
17 48 36.5	−23 28 00	B 282	18 x 4		Elongated NE and SW; sharp on NW side.
17 56 36.8	−29 01 00	B 289	35 x 7	4	Dark space between star clouds.
18 01 30.3	−28 36 30	B 294	3 x 3	3	Definite; star in SW part.
17 21 06.0	−26 47 00	LDN 1773	60 x 10	6	**Pipe Nebula**.

PLANETARY NEBULAE

RA h m s	Dec ° ′ ″	Name	Diam ″	Mag (P)	Mag (V)	Mag cent ★	Alt Name	Notes
17 01 33.8	−21 49 34	IC 4634	24	10.7	10.9	13.9	PK 0+12.1	Small triangle of mag 11.9-13.5 stars E 1′.5.
17 55 07.0	−21 44 40	IC 4670	5	13.1	12.0	14.7	PK 7+1.1	= **PN G7.2+1.8**; mag 10.60v star NW 2í.0.
18 03 18.5	−27 06 22	IC 4673	15	12.9	13.0	17.6	PK 3−2.3	Mag 11.9 star NE 0′.5; close pair of mag 10.53v and 11.9 stars W 1′.9.
17 29 20.5	−23 45 34	NGC 6369	38	12.9	11.4	15.9	PK 2+5.1	Ring shape with only slight darkening in center.
17 49 14.9	−20 00 36	NGC 6445	44	13.2	11.2	19.0	PK 8+3.1	Rectangular with slight "dumbbell" shape, PA 150°, slight darkening in center; mag 7.60v star E 4′.8. Some sources call this "Little Gem," a nickname given to NGC 6818 by J. Mallas.
17 50 24.2	−29 25 18	PN G0.1−1.1	4			21.0	PK 0−1.1	Small triangle of mag 12.3, 12.9 and 13.0 stars ENE 2′.2.
17 55 18.1	−29 57 40	PN G0.1−2.3	9	14.9			PK 0−2.2	Four stars for a small square 2′.4 N, their mags are 9.84v, 11.8, 11.4 and 11.2.
17 35 35.5	−27 24 49	PN G0.1+2.6	9				PK 0+2.1	Mag 10.20v star NW 3′.6.
17 29 23.4	−26 26 05	PN G0.1+4.3	5	15.5			PK 0+4.2	
16 43 53.9	−18 57 14	PN G0.1+17.2	12	14.4	13.9	15.2	PK 0+17.1	Mag 15.2 star E 1′.4; mag 13.5 star SSE 3′.6.
17 53 45.8	−29 43 46	PN G0.2−1.9	8	16.2	15.5	17.3	PK 0−1.5	Mag 12.8 star NNW 0′.5; mag 11.37v star S 1′.4.
17 20 22.0	−24 51 51	PN G0.3+6.9	12				Trz 41	Mag 10.40v star WNW 5′.9.
17 54 25.4	−29 36 09	PN G0.4−1.9	7	14.8	15.1	16.1	PK 0−1.6	Mag 12.8 star W 1′.1; close pair of mag 11 stars SW 2′.0.
17 53 24.0	−29 16 57	PN G0.5−1.6					PK 0−1.7	Stellar; small triangle of mag 12 stars W 1′.7; mag 11.1 star E 1′.5.
17 52 35.9	−29 06 38	PN G0.6−1.3	3				PK 0−1.2	Mag 9.84v star WNW 2′.9.
17 56 39.9	−29 37 23	PN G0.6−2.3					PK 0−2.3	Stellar; small triangle of mag 10-11 stars W 2′.6.
17 58 09.6	−29 44 20	PN G0.7−2.7	7	13.5	13.6	14.7	PK 0−2.4	Mag 12.6 star on SE edge; mag 10.1 star NW 2′.0.
17 34 54.8	−26 35 58	PN G0.7+3.2	5	17.1			PK 0+3.1	
17 29 25.9	−25 49 07	PN G0.7+4.7	3	19.1		21.0	PK 0+4.1	
17 53 49.7	−28 59 12	PN G0.8−1.5				17.6	PK 0−1.3	Stellar; mag 11.9 star E edge; mag 8.45v star SW 2′.2.
17 56 02.8	−29 11 16	PN G0.9−2.0	5	18.0		16.2	PK 0−2.1	Mag 10.6 star WSW 3′.2; mag 12.4 star E 3′.3.
17 40 26.8	−27 01 03	PN G1.0+1.9	37			20.3	PK 1+1.1	Mag 12.2 star on NE edge.
17 58 26.1	−29 20 46	PN G1.0−2.6	5	17.9			PK 1−2.2	Mag 12.2 star E 0′.8; mag 11.3 star NE 2′.4.
17 54 52.1	−28 48 55	PN G1.1−1.6	6	18.4			PK 1−1.4	Mag 12.1 star N 0′.5; mag 9.53v star NNE 3′.9.
17 40 12.9	−26 44 23	PN G1.2+2.1	4				PK 1+2.1	Mag 11.8 star E 3′.9.
18 00 37.7	−29 21 50	PN G1.2−3.0	5			16.2	PK 1−3.1	Mag 12.0 star E 0′.9; mag 11.2 star S 1′.2.
18 03 53.6	−29 51 21	PN G1.2−3.9	4	15.8		16.5	PK 1−3.7	Mag 12.4 star E 0′.9; mag 8.04v star W 3′.7.
17 53 47.3	−28 27 15	PN G1.3−1.2	4	20.1			PK 1−1.1	Mag 11.0 star SSE 1′.5; mag 10.76v star NE 2′.2.
18 02 27.0	−29 25 13	PN G1.4−3.4	13	18.2		15.8	PK 1−3.3	Mag 12.5 star E 0′.8; mag 12.4 star S 1′.1.
17 28 37.7	−24 51 07	PN G1.4+5.3	5	16.3		16.3	PK 1+5.2	Located between two NNW-SSE mag 12 stars.

(Continued from previous page)

PLANETARY NEBULAE

RA h m s	Dec ° ′ ″	Name	Diam ″	Mag (P)	Mag (V)	Mag cent ★	Alt Name	Notes
17 54 34.9	−28 12 44	PN G1.6−1.3	5	18.0			PK 1−1.2	Mag 10.6 star W 1′.4.
17 56 02.4	−28 14 11	PN G1.7−1.6					PK 1−1.3	Mag 12.2 star E 3′.3; mag 9.86v star NW 2′.7.
17 28 01.8	−24 25 23	PN G1.7+5.7	7	16.2			PK 1+5.1	Mag 12.1 star SW 1′.9.
17 58 21.9	−28 14 53	PN G2.0−2.0	6	15.5		16.8	PK 2−2.1	Mag 13.0 star E 1′.1; mag 12.4 star SW 1′.3.
17 59 19.4	−28 13 49	PN G2.1−2.2	7	14.3	14.3		PK 2−2.2	Triangle of mag 11-12 stars W 3′.4; mag 12.8 star SSE 0′.7.
17 37 51.2	−25 20 44	PN G2.1+3.3					PBOZ 24	Stellar.
17 48 44.7	−26 43 29	PN G2.2+0.5	28				Te 2337	Faint star on SE edge; forms a triangle with mag 11.4 star WNW 4′.8 and mag 11.4 star WSW 4′.6.
18 01 00.0	−28 16 14	PN G2.2−2.5	5	19.6			PK 2−2.5	Line of three mag 11-12 stars S and SW.
18 01 42.7	−28 25 44	PN G2.2−2.7	9	12.4	12.4	16.7	PK 2−2.4	Close pair of mag 11.4, 11.6 stars SW 1′.5.
17 42 30.1	−25 45 27	PN G2.3+2.2	8	18.3			Te 5	Mag 9.44v star S 1′.2.
17 58 30.8	−27 37 07	PN G2.5−1.7	5	19.5		21.0	PK 2−1.1	Mag 12.7 star NE 0′.4; mag 11.6 star W 0′.5.
17 43 39.5	−25 36 43	PN G2.6+2.1	11	17.3			Te 1580	Mag 11.3 star N 3′.9.
17 35 58.6	−24 25 28	PN G2.6+4.2	10	15.8			PK 2+4.1	
17 21 17.7	−22 18 35	PN G2.6+8.1	6	14.4			PK 2+8.1	
17 45 39.9	−25 40 00	PN G2.8+1.7	4	19.8			PK 2+1.1	
17 45 29.7	−25 38 14	PN G2.8+1.8	10	17.8			Te 1567	Mag 12.0 star N 3′.3.
18 01 10.3	−27 38 20	PN G2.8−2.2	5				PK 2−2.3	Mag 12.1 star E 0′.5; mag 11.3 star N 1′.2.
17 27 53.7	−22 57 18	PN G2.9+6.5					PM 1−149	Located 0′.5 SW of a mag 11.33v star.
18 02 51.7	−27 41 01	PN G3.0−2.6	3	17.6		21.8	PK 3−2.4	Mag 10.4 star on N edge; mag 10.6 star WSW 1′.0.
17 41 52.8	−24 42 08	PN G3.1+2.9	6	13.1	13.0	17.0	PK 3+2.1	Mag 11.9 star N 3′.3; mag 12.3 star ESE 4′.0.
17 40 07.5	−24 25 42	PN G3.1+3.4	4			21.0	PK 3+3.1	Mag 9.76v star SW 5′.5.
18 03 11.9	−26 58 31	PN G3.6−2.3	9	16.3	14.3		PK 3−2.2	Mag 10.2 star E 1′.6, mag 10.7 star S 0′.6.
17 41 57.3	−24 11 16	PN G3.6+3.1		18.0			PK 3+3.2	Mag 11.7 star WNW 2′.4; mag 10.8 star SW 4′.2.
17 24 46.0	−21 33 38	PN G3.7+7.9	4			21.0	PK 3+7.1	
17 34 26.9	−22 53 19	PN G3.8+5.3	4	17.3			PK 3+5.1	Mag 13.1 star NW 1′.1; mag 10.67v star SW 5′.0.
17 48 28.5	−24 41 21	PN G3.9+1.6	8				Te 2111	Mag 12.2 star NW 0′.7; mag 12.1 star E 1′.3.
18 03 39.3	−26 43 34	PN G3.9−2.3	5	14.6	14.6		PK 3−2.1	Mag 9.88v star ESE 1′.9; mag 10.3 star WNW 1′.9.
17 48 36.6	−24 16 34	PN G4.3+1.8	5	18.3			PK 4+1.1	Close pair of mag 11.3 stars NW 4′.4; mag 10.82v star SE 5′.2.
17 30 35.9	−21 28 51	PN G4.5+6.8	5				PK 4+6.1	Close pair of mag 12 stars SE 2′.9; mag 12.6 star NW 2′.7.
17 33 37.6	−21 46 25	PN G4.6+6.0	9	16.2			PK 4+6.2	Mag 12.3 star SW 3′.4; mag 10.5 star NE 4′.2.
17 49 00.6	−23 42 54	PN G4.8+2.0	4	19.2			PK 4+2.1	Mag 12.3 star NW 3′.9.
17 38 30.4	−22 08 39	PN G4.9+4.9	5	14.0	13.3	17.9	PK 4+4.1	Mag 12.4 star NW 1′.4; mag 12.4 star SW 2′.6.
17 45 36.8	−23 02 25	PN G5.0+3.0	12	17.9		21.0	PK 5+3.1	Mag 11.9 star E 3′.0; mag 12.0 star NNE 3′.0.
17 40 18.0	−22 19 18	PN G5.0+4.4	5	15.6	15.5		PK 5+4.1	Close pair of mag 10.7, 11.7 stars N 1′.8; mag12.5 star W 2′.3.
17 41 36.6	−22 13 03	PN G5.2+4.2		18.5			PK 5+4.2	Pair of N-S mag 13 stars N 1′.3; mag 11.2 star W 1′.6.
17 36 22.6	−21 31 13	PN G5.2+5.6	6	14.8	15.3		PK 5+5.1	Mag 11.6 star N 1′.5; mag 12.1 star ENE 1′.4.
17 48 07.5	−22 46 50	PN G5.5+2.7	2	18.3			PK 5+2.1	Mag 13.0 star SE 2′.7; mag 11.2 star NW 2′.5.
17 35 21.4	−20 57 23	PN G5.5+6.1	7	17.7		17.2	PK 5+6.1	Mag 9.36v star SE 4′.1.
17 39 54.8	−21 14 14	PN G5.8+5.1	17	15.3			PK 5+5.2	Mag 12.8 star N 1′.1.
17 48 37.4	−22 16 49	PN G6.0+2.8					PK 6+2.1	Stellar; mag 12.4 star W 1′.0; mag 12.1 star NNW 1′.3.
17 47 38.4	−22 06 20	PN G6.0+3.1	15	16.8	17.:	20.8	PK 6+3.2	Mag 12.2 star E 1′.7; mag 12.1 star S 1′.9.
17 28 57.7	−19 15 53	PN G6.1+8.3	7	13.4	13.:	17.1	PK 6+8.1	Close pair of mag 12.1, 12.9 stars WSW 2′.5.
17 55 56.2	−22 59 02	PN G6.2+1.0	12				PK 6+1.1	Forms a triangle with a mag 12.4 star E 1′.9 and mag 12.3 star NNE 1′.6.
17 47 33.9	−21 47 23	PN G6.3+3.3	6				PK 6+3.1	Mag 13.0 star SSW 1′.0; mag 12.9 star E 1′.2.
17 43 28.8	−21 09 52	PN G6.3+4.4	4	16.1	15.0		PK 6+4.1	Mag 11.9 star NNE 2′.7.
17 52 41.5	−22 21 57	PN G6.4+2.0		13.6	14.5	19.0	PK 6+2.5	Mag 12.4 star N 2′.6; mag 11.7 star SW 2′.2.
17 53 37.2	−21 58 42	PN G6.8+2.0	4	18.3			PK 6+2.4	Mag 12.1 star W 2′.5; mag 11.8 star NE 6′.9.
17 52 22.6	−21 51 14	PN G6.8+2.3	6				PK 6+2.3	Mag 13.3 star E 1′.0; mag 11.9 star SSW 1′.7.
17 45 31.7	−20 58 02	PN G6.8+4.1	4	14.6	14.5		PK 6+4.2	Mag 12.9 star NNW 1′.7; mag 11.3 star SE 5′.0.
17 38 11.6	−19 37 38	PN G7.0+6.3	6	14.2	14.:	16.0	PK 7+6.2	Mag 12.5 star NE 2′.5; mag 12.5 star SW 2′.7.
17 46 20.8	−20 13 49	PN G7.5+4.3		17.3		18.2	PK 7+4.1	Mag 11.7 star SE 4′.2; mag 11.1 star SW 4′.4.
17 35 10.1	−18 34 20	PN G7.5+7.4	9	13.3			PK 7+7.1	Mag 11.7 star ENE 5′.3; mag 11.6 star NW 4′.2.
17 37 22.0	−18 46 42	PN G7.6+6.9	7	14.2	14.5		PK 7+6.1	Mag 11.1 star W 5′.0; mag 11.1 star NW 3′.9.
17 38 57.4	−18 17 35	PN G8.2+6.8	10		11.0	14.2	PK 8+6.1	Mag 12.5 star WSW 0′.8; mag 11.7 star SW 1′.6.
17 46 09.4	−18 39 28	PN G8.8+5.2	19	16.5			PK 8+5.1	Mag 12.0 star SW 0′.4.
17 50 28.3	−19 03 08	PN G9.0+4.1	6	16.2	14.7		PK 9+4.1	Mag 12.9 star E 1′.0; mag 10.76v star W 2′.2.
17 56 00.6	−19 29 27	PN G9.3+2.8					PK 9+2.1	Mag 12.0 star E 4′.0.
17 50 57.2	−18 46 49	PN G9.3+4.1		15.6			PK 9+4.2	Mag 11.4 star S 1′.1; close pair of mag 11-12 stars W 2′.8.
17 52 04.8	−17 36 05	PN G10.4+4.5	6	14.8	14.5		PK 10+4.1	Mag 11.1 star W 0′.5.
17 57 06.6	−18 06 43	PN G10.6+3.2	25	16.8		18.3	PK 10+3.1	Mag 11.3 star W 1′.3; mag 9.67v star NE 1′.6.
18 00 08.8	−17 40 43	PN G11.3+2.8	5	18.2			PK 11+2.1	Mag 11.8 star on W edge.
16 44 49.1	−28 04 06	PN G352.9+11.4	23			12.7	PK 352+11.2	Forms a triangle with a mag 13.9 star NNW 1′.4 and a mag 13.7 star WNW 1′.6.
17 17 20.3	−28 59 27	PN G356.5+5.1	17	16.4			PK 356+5.1	Mag 12.0 star E 1′.1; mag 12.4 star ESE 2′.7.
17 25 06.0	−29 45 15	PN G356.8+3.3	1	20.3			PK 356+3.1	Mag 13.1 star SE 1′.8.
17 21 04.4	−29 02 59	PN G356.9+4.4	2	16.5	15.5		PK 356+4.2	
17 20 33.3	−29 00 39	PN G356.9+4.5	15	14.4	14.:		PK 356+4.1	Mag 11.7 star N 2′.1.
17 24 34.5	−29 24 19	PN G357.1+3.6	6	14.9	15.7	16.4	PK 357+3.1	Mag 12.0 star WNE 5′.0; mag 11.3 star SE 4′.8.
17 21 37.9	−28 55 07	PN G357.1+4.4	8	18.4			TeJu 18	Mag 11.1 star SE 6′.6.
17 10 41.7	−27 08 45	PN G357.2+7.4	8	14.1	13.5	17.9	PK 357+7.1	Mag 12.7 star NE 7′.2.
17 25 59.8	−29 21 50	PN G357.3+3.3	4			16.3	PK 357+3.2	Mag 11.7 star S 1′.4; mag 11.0 star ENE 1′.4.
17 23 24.9	−28 59 06	PN G357.3+4.0	5	16.9		20.5	PK 357+4.1	N-S pair of mag 11.1, 11.7 stars S 1′.2.
17 27 24.3	−29 21 14	PN G357.5+3.1	6				PK 357+3.5	Mag 11.4 star ENE 0′.6; mag 12.2 star W 1′.6.
17 26 59.8	−29 15 32	PN G357.5+3.2	5	16.9			PK 357+3.4	Mag 10.64v star N 1′.7.
17 32 46.9	−30 00 15	PN G357.6+1.7	3	15.6			PK 357+1.1	Close pair of mag 11-12 star S 1′.8.
17 29 42.7	−29 32 50	PN G357.6+2.6	2	16.4			PK 357+2.4	Mag 12.1 star on W edge; mag 10.15v star ENE 0′.5.
17 30 21.4	−29 10 12	PN G358.0+2.6					PK 358+2.2	Mag 9.72v star S 2′.2.
17 12 33.6	−26 25 29	PN G358.0+7.5	17				TeJu 8	Mag 12.0 star SSW 1′.5.
17 05 44.6	−25 25 02	PN G358.0+9.3	10	14.9		17.8	PK 358+9.1	Mag 11.4 star W 1′.4; mag 11.7 star E 1′.4.
17 27 32.9	−28 31 07	PN G358.2+3.5	5	16.0			PK 358+3.2	
17 27 20.1	−28 27 52	PN G358.2+3.6	3	14.2	13.6		PK 358+3.1	Mag 11.6 star E 1′.3.
17 24 52.1	−28 05 55	PN G358.2+4.2	5	16.7		17.5	PK 358+4.1	Mag 12.2 star N 3′.8.
17 36 59.8	−29 40 08	PN G358.3+1.2		19.3			PK 358+1.4	Mag 12.0 star N 2′.7; mag 12.6 star WNW 2′.3.
17 29 40.6	−28 40 22	PN G358.3+3.0	5	15.3	15.5		PK 358+3.7	Mag 9.04v star N 2′.3; mag 10.9 star S 2′.3.
17 28 41.8	−28 27 19	PN G358.4+3.3	1				PK 358+3.3	Mag 11.7 star SE 2′.1; mag 10.8 star NW 2′.6.
17 31 47.3	−28 42 04	PN G358.5+2.6	37	14.9			PK 358+2.5	Mag 12.6 star E 1′.1.
17 30 30.4	−28 35 55	PN G358.5+2.9		18.4			PK 358+2.4	Mag 11.9 star NW 2′.0; mag 10.72v star WSW 2′.4.
17 27 47.0	−28 11 02	PN G358.5+3.7					PK 358+3.9	Mag 9.03v star NNE 0′.9; mag 8.71v star W 1′.5.
17 21 11.7	−27 11 37	PN G358.5+5.4	18	15.2	15.2		PK 358+5.1	Mag 12.0 star W 10′.2.

146

(Continued from previous page)

PLANETARY NEBULAE

RA h m s	Dec ° ′ ″	Name	Diam ″	Mag (P)	Mag (V)	Mag cent ★	Alt Name	Notes
17 35 14.0	−29 03 10	PN G358.6+1.8	2	18.3		21.0	PK 358+1.1	Mag 12.9 star W 1′.1.
17 12 39.3	−25 43 37	PN G358.6+7.8	4	14.8	15.3	17.1	PK 358+7.1	Mag 10.8 star WSW 2′.5.
17 22 28.3	−27 08 43	PN G358.7+5.2	3			17.2	PK 358+5.2	Stellar.
17 42 42.5	−29 51 35	PN G358.8−0.0	12			14.9	Te 2022	Mag 11.7 star NW 2′.9; mag 11.3 star SW 5′.1.
17 31 09.4	−28 14 49	PN G358.8+3.0	6	16.9		16.5	PK 358+3.8	Mag 10.75v star NNW 4′.7.
17 27 10.5	−27 43 57	PN G358.8+4.0	6				PK 358+4.2	
17 27 00.2	−27 40 38	PN G358.8+4.1	70				PK 358+4.3	Mag 12.0 star E 2′.6; mag 13.4 star NNW 2′.3.
17 30 43.8	−28 04 07	PN G358.9+3.2	4	15.9			PK 358+3.6	Mag 12.2 star SW 1′.1.
17 30 02.5	−27 59 18	PN G358.9+3.4	1	17.5			PK 358+3.4	Close pair of mag 12.0, 13.1 stars ENE 2′.9.
17 32 22.7	−28 14 26	PN G359.0+2.8					PK 359+2.5	Mag 10.5 star E 1′.9.
16 48 34.5	−21 00 51	PN G359.1+15.1	34	16.8		19.7	PK 359+15.1	Mag 8.69v star NW 6′.5; mag 14.5 star E 3′.1.
17 39 03.0	−28 56 36	PN G359.2+1.2	24			17.2	PM 1-166	Mag 9.91v star E 1′.4.
17 25 44.0	−26 57 47	PN G359.2+4.7	3			15.8	PK 359+4.1	Mag 9.31v star S 1′.7.
17 47 56.2	−29 59 40	PN G359.3−0.9	15	13.6	11.8	18.6	PK 359−0.1	Faint star on SW edge; mag 11.5 star NW 2′.2.
17 38 42.4	−28 42 48	PN G359.3+1.4					PK 359+1.1	Mag 12.3 star ESE 1′.1; mag 13.1 star NW 2′.0.
17 30 14.5	−27 30 19	PN G359.3+3.6		19.7			PK 359+3.4	Mag 112.8 star SW 1′.3; mag 11.3 star S 3′.1.
17 35 15.7	−28 07 01	PN G359.4+2.3					PK 359+2.4	Mag 11.2 star N 2′.4.
17 34 13.8	−27 56 00	PN G359.5+2.6					PK 359+2.7	Mag 10.07v star NW 2′.6; mag 12.5 star E 2′.5.
17 36 14.2	−28 00 45	PN G359.6+2.2					PK 359+2.6	Mag 13.3 star SSW 3′.6; mag 11.1 star SW 6′.8.
17 35 48.0	−27 43 22	PN G359.8+2.4	6				PK 359+2.3	Mag 12.9 star W 1′.3.
17 30 46.8	−27 06 00	PN G359.8+3.7	7	19.6			PK 359+3.2	Mag 12.9 star SSE 2′.5.
17 25 23.6	−26 11 53	PN G359.8+5.2	8				TeJu 19	
17 24 01.3	−25 59 21	PN G359.8+5.6	5			14.7	PK 359+5.1	Mag 12.3 star SW 2′.4; mag 12.8 star ENE 2′.7.
17 19 13.3	−25 17 17	PN G359.8+6.9	10	16.4			PK 359+6.1	Mag 10.94v star SW 4′.7.
17 25 43.4	−26 11 56	PN G359.9+5.1	17		15.0	18.8	PK 359+5.2	Mag 12.3 star N 2′.6; mag 10.01v star S 2′.7.

147

GALAXIES

RA h m s	Dec ° ′ ″	Name	Mag (V)	Dim ′ Maj x min	SB	Type Class	PA	Notes
15 44 17.7	−28 18 30	ESO 450-11	15.1	1.3 x 0.2	13.5	Sb:	47	Staggered N-S line of four mag 11-13 stars SSE 6′.0.
15 49 19.6	−28 54 06	ESO 450-18	14.5	1.9 x 0.4	14.1	I0:	151	Lies 1′.9 W of mag 8.69v star.
15 57 48.4	−28 22 46	ESO 451-1	13.7	1.3 x 1.3	14.1	Sc:		Mag 11.7 star WSW 1′.4; mag 9.04v star ENE 11′.0.
16 06 16.4	−29 16 56	ESO 451-8	14.0	2.0 x 0.3	13.3	(R′)SB(s)b:	83	Mag 10.69v star SE 3′.0.
16 31 40.3	−28 06 10	ESO 452-5	13.5	1.0 x 1.0	13.3	SA(r)cd III		Stellar nucleus; ESO 452-7 E 7′.1.
16 32 11.6	−28 05 27	ESO 452-7	13.8	2.8 x 1.1	14.9	(R′)SAB(s)b pec:	121	ESO 452-5 W 7′.1; mag 11.9 star on E edge; mag 9.69v star N 4′.1.
16 32 37.2	−29 22 50	ESO 452-8	15.1	1.8 x 0.3	14.3	Sd? sp	35	Mag 11.41v star SE 2′.7.
15 19 03.6	−23 49 17	ESO 514-5	12.9	1.8 x 0.6	12.9	Sa: sp	21	Mag 10.72v star E 0′.7; mag 9.35v star W 4′.5.
15 19 11.1	−24 25 52	ESO 514-6	13.6	1.2 x 0.9	13.6	SB(s)c II-III	120	Mag 10.35v star W 9′.4.
15 23 44.6	−26 16 38	ESO 514-15	14.6	1.5 x 1.1	15.0	Sc	62	Mag 8.80v star ENE 11′.7.
15 24 29.0	−23 06 33	ESO 514-16	14.0	0.4 x 0.4	11.9	Sa		Compact; star SE edge.
15 26 55.6	−23 45 32	ESO 514-17	13.8	1.2 x 0.3	12.6	Sa:	154	Mag 12.2 star W 5′.1.
15 32 03.2	−27 37 58	ESO 514-23	13.8	1.8 x 0.3	12.9	SBcd? sp	142	**ESO 514-24** S 1′.3.
15 37 14.9	−26 25 52	ESO 514-27	13.7	1.1 x 0.5	12.9	Sb	131	Mag 14.3 star S edge; mag 7.38v star SW 4′.1.
15 38 57.0	−22 39 53	ESO 515-3	13.3	1.8 x 1.2	14.0	SB(r)c II-III	17	Mag 12.8 star superimposed E edge; mag 10.02v star N 4′.3.
15 53 00.8	−27 17 29	ESO 515-13	13.3	2.1 x 0.8	13.7	SA(rs)b II	55	Mag 10.09v star W 5′.0; mag 8.25v star N 12′.8.
16 08 34.0	−25 01 11	ESO 516-8	15.8	1.3 x 0.2	14.2	Sb:	31	Mag 8.60v star S 7′.1.
15 17 18.2	−22 17 51	ESO 582-1	13.5	1.7 x 0.7	13.6	SA(rs)b: II-III	36	Mag 5.51v star SW 14′.1; elongated **ESO 582-3** NE 8′.4.
15 20 46.0	−18 20 56	ESO 582-4	14.4	1.2 x 0.6	13.9	Sc:	163	Mag 14.6 star n edge; mag 9.40v star S 7′.7.
15 26 07.9	−22 16 53	ESO 582-12	12.0	2.5 x 1.5	13.3	SAB(rs)c II	48	Mag 9.51v star W 2′.9.
15 29 55.5	−18 37 11	ESO 582-13	13.7	1.6 x 0.9	14.0	SAB(r)c I	60	Mag 9.09v star N 10′.9.
15 49 54.8	−21 03 01	ESO 583-7	13.3	1.5 x 0.7	13.2	SAB(rs)c: II-III	108	Mag 10.95v star SSE 5′.2; mag 8.17v star NE 14′.8.
15 57 50.2	−22 29 39	ESO 583-8	15.5	1.5 x 0.2	14.0	S?	109	Mag 9.03v star SSE 3′.8.
16 06 43.8	−18 12 44	ESO 584-5	14.1	0.8 x 0.6	13.2	SB(s)0°: pec	6	Mag 10.62v star SSW 7′.6; mag 8.13v star SE 12′.9.
16 15 02.6	−20 47 17	ESO 584-8	14.4	0.8 x 0.5	13.3	Sab	33	Stellar nucleus; mag 12.8 star SE 2′.3.
15 21 11.6	−23 39 33	IC 4538	12.1	2.6 x 2.0	13.7	SAB(rs)c: I-II	32	Stellar nucleus; mag 10.88v star N 7′.6.
16 16 03.7	−22 37 31	IC 4596	14.0	1.5 x 0.4	13.3	(R′)SB(rs)ab? II	54	Small triangle of mag 12-13 star N 3′.6.
15 21 51.4	−17 43 49	MCG −3-39-9	14.5	0.6 x 0.4	12.8	(R'_1)SB(l)a	117	Mag 8.84v star N 3′.4.
15 24 50.0	−21 21 38	MCG −3-39-10	14.6	0.9 x 0.2	12.6	Sb:	45	
16 00 32.8	−20 24 11	MCG −3-41-1	14.3	0.7 x 0.4	12.7	Sb	120	
15 18 36.1	−24 07 15	MCG −4-36-7	13.5	1.0 x 0.3	12.3	E^+	119	NGC 5903 N 3′.2.
15 19 59.8	−23 55 31	MCG −4-36-11	14.2	0.9 x 0.5	13.1	Sd	4	Mag 10.61v star E 7′.0; ESO 514-5 WNW 14′.3.
15 20 30.8	−24 28 11	MCG −4-36-12	13.6	1.4 x 0.5	13.1	S0	34	Bright knot of star NE edge.
15 48 14.5	−25 22 51	MCG −4-37-4	14.3	0.5 x 0.4	12.4	Sb	42	
15 48 13.5	−24 53 10	MCG −4-37-5	13.8	0.7 x 0.4	12.3	SBbc	39	
15 55 41.2	−24 40 26	MCG −4-37-6	14.0	0.5 x 0.4	12.1		177	Lies 1′.1 NW of mag 10 star.
16 02 56.0	−22 31 43	MCG −4-38-2	14.1	0.7 x 0.4	12.6	S0/a	27	
15 46 43.2	−28 43 56	MCG −5-37-2	13.7	1.2 x 1.0	13.8	SA(s)d pec IV-V	1	Numerous bright knots.
15 17 51.3	−17 35 20	NGC 5890	12.7	1.3 x 1.0	12.9	$SA(r)0^+$?	85	Pair of stars E end.
15 18 13.7	−24 05 52	NGC 5898	11.4	2.7 x 2.6	13.5	E0	42	Very faint, very small anonymous galaxy 0′.9 N of center.
15 18 36.4	−24 04 06	NGC 5903	11.2	2.7 x 2.1	13.1	E2	168	MCG −4-36-7 S 3′.2.
15 49 49.4	−29 23 16	NGC 6000	11.9	1.9 x 1.6	12.9	SB(s)bc: II-III	154	Bright knots N; mag 9.111v star SW 3′.2.

OPEN CLUSTERS

RA h m s	Dec ° ′ ″	Name	Mag	Diam ′	No. ★	B ★	Type	Notes
15 49 28.0	−28 35 48	NGC 5998		3.5	10		ast	(A) Almost certainly asterism of 10 stars here.

GLOBULAR CLUSTERS

RA h m s	Dec ° ′ ″	Name	Total V m	B ★ V m	HB V m	Diam ′	Conc. Class Low = 12 High = 1	Notes
16 39 25.0	−28 23 54	ESO 452-11	12.0			1.2		
15 17 24.5	−21 00 37	NGC 5897	8.4	13.3	16.3	11.0	11	
16 17 02.5	−22 58 30	NGC 6093	7.3	12.5	16.2	10.0	2	= **M 80**.

147 GLOBULAR CLUSTERS 147

(Continued from previous page)

RA h m s	Dec ° ′ ″	Name	Total V m	B ★ V m	HB V m	Diam ′	Conc. Class Low = 12 High = 1	Notes
16 23 35.5	−26 31 31	NGC 6121	5.4	10.8	13.4	36.0	9	= **M 4**. A band of slightly more luminous stars which bisects the cluster in a roughly N-S direction, may be visible to visual observers.
16 27 14.1	−26 01 29	NGC 6144	9.0	13.4	16.5	7.4	11	

BRIGHT NEBULAE

RA h m s	Dec ° ′ ″	Name	Dim ′ Maj x min	Type	BC	Color	Notes
16 12 18.0	−27 56 00	IC 4591	12 x 10	R	5-5	1-4	
16 12 00.0	−19 28 00	IC 4592	150 x 60	R	3-5	1-4	Brightest in the region surrounding 14 (ν) Scorpii.
16 20 00.0	−20 02 00	IC 4601	20 x 10	R	3-5	1-4	One 6th mag and a very wide pair of 7th mag stars involved; nebulosity most prominent around the 7th mag stars.
16 25 36.0	−24 28 00	IC 4603	35 x 20	R	3-5	1-4	
16 25 36.0	−23 26 00	IC 4604	60 x 50	R	3-5	1-4	
16 30 12.0	−25 06 00	IC 4605	30 x 30	R	3-5	1-4	
16 31 36.0	−26 03 00	IC 4606	60 x 40	R	1-5	3-4	**Antares Nebula**. Nebulosity surrounding Antares (α Scorpii).
15 58 54.0	−26 09 00	Sh2-1	90 x 10	E+R	4-5	1-4	
16 21 06.0	−25 35 00	Sh2-9	60 x 15	E+R	2-5	4-4	For the most part very diffuse. The brightest and most prominent feature is a very narrow nebulous strip E of σ Scorpii. It begins at a point 17′ W of the star extending NNE for a distance of 10′ where it breaks for a distance of 5′ and then fading slightly, it continues on in the same direction for nearly 15′.

DARK NEBULAE

RA h m s	Dec ° ′ ″	Name	Dim ′ Maj x min	Opacity	Notes
16 14 38.8	−18 58 00	B 40	15 x 15	3	Diffused dark spot; small star in it.
16 22 17.6	−19 38 00	B 41	40 x 40	3	Diffused dark spot.
16 25 29.6	−23 26 00	B 42	20 x 6	6	Great nebula of Rho Ophiuchi.
16 30 18.9	−19 47 00	B 43		4	Large dark region.
16 40 33.6	−24 04 00	B 44		5	Dark lane E from Rho Ophiuchi region.
16 17 40.9	−27 19 00	B 229	45 x 45		Indefinite, partially vacant region.

PLANETARY NEBULAE

RA h m s	Dec ° ′ ″	Name	Diam ″	Mag (P)	Mag (V)	Mag cent ★	Alt Name	Notes
16 43 53.9	−18 57 14	PN G0.1+17.2	12	14.4	13.9	15.2	PK 0+17.1	Mag 15.2 star E 1′.4; mag 13.5 star SSE 3′.6.
15 22 19.4	−23 37 33	PN G342.1+27.5	16	11.5	11.6	18.8	PK 342+27.1	Mag 9.86v star W 0′.8.

148 GALAXIES 148

RA h m s	Dec ° ′ ″	Name	Mag (V)	Dim ′ Maj x min	SB	Type Class	PA	Notes
13 57 18.4	−28 11 43	ESO 445-83	14.8	1.5 x 0.2	13.4	Sc	51	Mag 12.1 star E 3′.3.
14 08 38.4	−29 34 12	ESO 446-18	13.8	2.5 x 0.3	13.4	Sb sp	5	Mag 9.18v star E 5′.3.
14 09 16.6	−29 11 45	ESO 446-19	14.0	0.9 x 0.6	13.1	(R′)SB(s)b pec:	142	Mag 10.58v star NW 11′.9.
14 12 56.5	−28 46 26	ESO 446-29	13.6	1.0 x 0.7	13.1	Sab: pec	43	Very sinuous **ESO 446-28** SW edge.
14 13 30.5	−29 35 40	ESO 446-31	12.9	2.0 x 1.7	14.1	SB(s)cd IV	169	A number of superimposed stars on envelope; mag 12.8 star NNE 2′.1.
14 14 37.2	−29 33 27	ESO 446-35	14.1	1.5 x 1.1	14.4	SAB(r)c pec	51	Very faint outer arms surround bright core.
14 15 03.2	−29 21 39	ESO 446-36	13.9	0.9 x 0.5	12.8	S0/a	66	Mag 8.91v star N 1′.7.
14 20 53.8	−29 01 19	ESO 446-51	15.2	1.3 x 0.2	13.6	Sc	9	Mag 12.7 star W 3′.3; mag 11.2 star NW 6′.8.
14 21 17.1	−29 15 48	ESO 446-53	14.0	1.5 x 0.7	13.9	IB(s)m V	170	Small triangle of mag 12-13 stars centered WNW 2′.6; NGC 5556 W 9′.6.
14 30 40.8	−28 40 25	ESO 447-10	14.3	0.8 x 0.3	12.6	S0:	63	Mag 13.8 star SW 2′.2; mag 9.79v star ESE 4′.5.
14 31 22.8	−28 45 00	ESO 447-17	12.9	1.9 x 0.7	13.1	SA(s)bc	153	Mag 14.1 star on E edge; mag 12.4 star W 1′.1; mag 9.79v star WNW 5′.6.
14 32 07.6	−29 56 53	ESO 447-18	14.2	1.0 x 0.7	13.7	Sb	107	Stellar nucleus, smooth envelope; mag 12.7 star WSW 2′.9.
14 32 22.8	−28 22 45	ESO 447-19	13.6	1.9 x 0.3	12.8	(R′)SB(s)ab: sp	98	Mag 11.9 star S 0′.8.
14 32 28.7	−29 55 29	ESO 447-21	13.8	1.3 x 0.6	13.4	(R′)SB(r)b	30	ESO 447-18 W 4′.8; mag 10.64v star SW 9′.7.
14 34 35.8	−27 59 45	ESO 447-23	13.0	2.1 x 1.1	13.8	SAB(rs)cd: II-III	117	Mag 12.3 star S 1′.5; mag 10.04v star SSW 8′.2.
14 46 53.1	−27 42 56	ESO 447-36	13.4	2.7 x 0.4	13.3	SB(s)bc: sp	148	Mag 11.8 star E 3′.1; mag 7.54v star E 11′.1.
13 57 35.8	−25 47 30	ESO 510-26	14.9	1.1 x 0.6	14.3	SB(rs)m pec V-VI	24	Mag 10.91v star E 10′.7; **ESO 510-21** W 8′.7.
14 00 24.0	−23 18 33	ESO 510-36	14.0	0.4 x 0.4	11.9	S?		
14 00 25.8	−23 18 17	ESO 510-36A	13.8	1.1 x 0.6	13.3		175	
14 01 34.6	−23 08 24	ESO 510-43	13.9	1.2 x 0.7	13.6	SB(s)c pec? III-IV	141	**ESO 510-45** 2′.2 NNE.
14 01 37.6	−26 25 59	ESO 510-44	13.6	0.7 x 0.7	12.7	S0		Mag 10.93v star E 4′.6.
14 03 34.7	−27 16 50	ESO 510-52	14.5	0.6 x 0.5	13.1		34	Mag 9.79v star SW 11′.2.
14 04 03.2	−26 12 57	ESO 510-54	13.1	1.1 x 0.6	12.5	SAB0°: pec	21	Mag 9.36v star E 2′.5.
14 04 08.6	−25 37 54	ESO 510-55	14.9	0.8 x 0.4	13.5	(R′)SAB(r)b II	144	Mag 9.50v star N 7′.0; ESO 510-56 E 1′.7.
14 04 15.9	−25 38 23	ESO 510-56	13.1	1.6 x 1.3	13.8	SA(rs)ab	130	Mag 9.50v star N 7′.0; ESO 510-55 W 1′.7.
14 04 37.3	−24 49 52	ESO 510-58	13.2	1.2 x 0.7	12.9	SB(s)cd? II-III	3	ESO 510-59 E 2′.3.
14 04 47.3	−24 49 54	ESO 510-59	13.0	2.0 x 1.7	14.2	SB(s)cd	41	Mag 9.93v star E 3′.5; ESO 510-58 W 2′.3.
14 05 31.0	−25 21 49	ESO 510-61	14.1	0.9 x 0.8	13.6	SA?(r)ab	175	Mag 12.4 star W 7′.5.
14 07 12.2	−26 50 10	ESO 510-65	13.8	1.2 x 0.9	13.7	SB(rs)bc II-III	24	Mag 11.18v star S 2′.1; mag 3.26v star Pi Hydrae NW 14′.5.
14 07 40.2	−26 33 43	ESO 510-70	13.4	1.3 x 0.9	13.4	SAB(rs)0⁻?	174	Mag 3.26v Pi Hydrae WSW 18′.8.
14 07 41.0	−25 07 02	ESO 510-71	13.0	1.3 x 0.9	13.1	SAB(s)0⁻?	12	Mag 10.63v star SE edge; mag 9.95v star NE 2′.2.
14 08 54.0	−26 56 47	ESO 510-74	13.0	1.9 x 0.4	12.6	Sa: sp	128	MCG −4-33-51 SE 8′.2.
14 09 48.9	−27 13 45	ESO 511-1	14.2	0.7 x 0.4	12.6	Sa	24	Mag 13.2 star NE edge; **ESO 511-1A** NW 1′.2.
14 10 00.0	−27 41 22	ESO 511-3	14.2	0.8 x 0.7	13.4	Sc	90	Mag 13.9 star SW 2′.7.
14 11 08.5	−27 01 39	ESO 511-6	14.2	0.7 x 0.4	12.7	S	108	Mag 13.2 star W 1′.8.
14 11 28.6	−26 12 16	ESO 511-8	13.8	0.7 x 0.5	12.5	S?	140	Mag 13.5 star SE 1′.0.
14 16 49.5	−27 13 59	ESO 511-17	13.8	1.1 x 0.3	12.5	S0	139	Mag 10.14v star E 7′.2.
14 17 39.6	−24 11 05	ESO 511-18	13.7	0.8 x 0.6	12.8	(R)SAB(r)ab pec II	36	**ESO 511-20** E 2′.3.
14 18 14.8	−27 24 52	ESO 511-21	13.2	1.0 x 0.6	12.7	E?	163	Mag 11.8 star W 1′.2; mag 8.86v star S 6′.3.
14 18 26.8	−27 22 44	ESO 511-23	12.7	1.0 x 0.8	12.3	SAB(s)0⁻	72	**ESO 511-24** S 1′.0; **ESO 511-25** NNE 3′.3.
14 18 50.7	−27 24 37	ESO 511-26	13.2	1.1 x 0.8	13.0	S0	98	Mag 9.41v star S 1′.7; MCG −4-34-8 NE 2′.8.
14 19 22.6	−26 38 41	ESO 511-30	12.6	3.0 x 1.9	14.3	SA(rs)c pec I-II	75	Very faint arms extend well beyond bright central area.
14 19 23.4	−27 22 16	ESO 511-31	13.2	1.2 x 0.8	13.1	SA0 pec	167	Mag 9.31v star NW 7′.2; **ESO 511-28** NNW 10′.8.
14 19 30.4	−27 22 35	ESO 511-32	13.3	0.9 x 0.6	12.5	S0?	166	ESO 511-31 W 1′.6.

RA h m s	Dec ° ′ ″	Name	Mag (V)	Dim ′ Maj x min	SB	Type Class	PA	Notes
14 20 42.1	−27 23 08	ESO 511-33	13.8	2.0 x 0.4	13.4	SAb? sp	176	Mag 12.0 star SW 2′.1; mag 7.49v star E 13′.5.
14 21 13.2	−27 15 41	ESO 511-34	13.1	1.3 x 1.1	13.3	(R′)SB(rs)b? pec	159	Strong knot or star S edge; mag 12.8 star E edge; mag 7.49v star ESE 7′.1.
14 22 07.0	−26 51 28	ESO 511-35	13.2	1.5 x 0.8	13.2	SB(s)0°? pec	57	Mag 7.87v star ENE 8′.8.
14 28 14.3	−23 52 24	ESO 511-42	15.1	1.9 x 0.3	14.3	Sc sp	53	Mag 9.94v star N 4′.9.
14 29 22.2	−22 55 42	ESO 511-44	14.0	1.2 x 0.5	13.3	SBb	39	Pair of mag 7.82v, 7.92v stars WSW 13′.2.
14 34 43.2	−25 26 39	ESO 512-5	15.7	1.7 x 0.2	14.4	Sc:	118	Mag 9.82v star N 2′.1.
14 40 10.9	−25 46 34	ESO 512-12	13.0	3.2 x 0.4	13.1	Sb sp I	113	Mag 12.4 star N edge; mag 9.85v star E 3′.6.
14 42 46.9	−24 12 37	ESO 512-16	14.7	1.5 x 0.2	13.3	S	168	Mag 10.56v star SE 7′.4.
14 43 34.0	−24 27 39	ESO 512-18	11.9	1.9 x 0.8	12.3	SAB0°: pec	31	Pair with ESO 512-19 on E edge; Mag 8.55v star 0.7 W.
14 43 36.9	−24 27 58	ESO 512-19	12.1	2.3 x 0.7	12.5	SAa? pec	34	Pair with ESO 512-18 on W edge.
14 51 14.1	−26 37 57	ESO 512-23	13.1	1.4 x 1.1	13.4	(R)SB(r)a:	15	Mag 8.52v star E 13′.3.
14 55 19.6	−25 28 27	ESO 513-4	14.0	1.4 x 1.0	14.2	SB(s)dm IV-V	174	Mag 12.2 star E 1′.2; mag 8.34v star S 8′.6.
15 00 27.2	−26 27 12	ESO 513-11A	14.4	0.6 x 0.2	11.9		85	Pair with ESO 513-11B NE.
15 00 29.1	−26 26 54	ESO 513-11B	14.1	0.5 x 0.2	11.4		158	Pair with ESO 513-11A SW.
15 03 46.7	−25 52 24	ESO 513-15	13.2	1.1 x 1.1	13.3	(R′)SAB(s)0/a:		Mag 12.0 star N 1′.1; mag 9.51v star NW 7′.4.
15 15 08.7	−22 48 19	ESO 513-30	13.6	1.1 x 0.6	13.0	SB(r)0/a:	54	Mag 8.38v star S 6′.8.
15 19 03.6	−23 49 17	ESO 514-5	12.9	1.8 x 0.6	12.9	Sa: sp	21	Mag 10.72v star E 0′.7; mag 9.35v star W 4′.5.
15 19 11.1	−24 25 52	ESO 514-6	13.6	1.2 x 0.9	13.6	SB(s)c II-III	120	Mag 10.35v star W 9′.4.
15 23 44.6	−26 16 38	ESO 514-15	14.6	1.5 x 1.1	15.0	Sc	62	Mag 8.80v star ENE 11′.7.
13 56 49.8	−21 48 19	ESO 578-10	14.1	0.8 x 0.6	13.2		26	Mag 13.0 star NW 2′.4.
14 03 21.9	−22 33 28	ESO 578-19	13.3	1.3 x 1.1	13.5	(R′)SB(r)bc pec II	69	Mag 6.30v star NNE 10′.9.
14 08 15.6	−20 00 20	ESO 578-25	14.0	1.3 x 0.4	13.1	Sa:	64	Mag 11.6 star ESE 5′.1; **ESO 578-27** E 10′.1.
14 08 42.0	−21 35 53	ESO 578-26	13.6	1.8 x 1.2	14.3	SB(r)c pec: II	57	Anonymous galaxy W 2′.1.
14 10 15.2	−17 50 15	ESO 578-32	14.4	1.3 x 0.3	13.2	Sab	106	Mag 11.9 star NNW 7′.8.
14 11 49.1	−19 51 15	ESO 578-33	14.0	1.2 x 0.7	13.7	SA?(r)0	36	Mag 13.4 star W edge.
14 12 16.8	−21 11 14	ESO 578-34	14.3	1.2 x 0.7	13.9	SA0$^-$	162	Mag 10.02v star E 10′.6; mag 12.5 star NNW 4′.0.
14 17 31.4	−19 51 51	ESO 579-11	14.5	0.9 x 0.8	14.0	SA(rs)d: IV	42	Mag 6.95v star S 6′.0.
14 19 21.0	−20 57 08	ESO 579-18	13.9	1.0 x 0.6	13.2	(R′)SB(r)0/a pec	34	Mag 8.12v star NE 7′.9.
14 23 40.9	−19 26 40	ESO 579-22	14.5	1.0 x 0.8	14.1	SA?(r:)a:	28	Mag 12.2 star N 2′.9; mag 10.51v star SW 3′.6.
14 31 29.7	−22 22 03	ESO 579-25	13.8	1.0 x 0.5	12.8	Sb	173	Mag 6.96v star SW 12′.8.
14 37 14.2	−18 40 47	ESO 580-4	16.6	1.5 x 1.1	17.0	IB(s)m V-VI	24	Multi-galaxy system; mag 12.0 star NW 8′.8.
14 45 26.1	−20 41 12	ESO 580-18	14.3	1.7 x 1.0	14.7	SB(rs)m pec IV-V	176	Mag 11.9 star SE 4′.4; mag 11.1 star NW 9′.2.
14 45 45.6	−20 47 05	ESO 580-20	14.8	1.9 x 1.3	15.6	SB(s)m V-VI	15	Pair with ESO 580-21 S edge; mag 11.9 star N 3′.5.
14 45 49.5	−20 48 40	ESO 580-21	15.3	1.8 x 1.4	16.2	IB(s)m V-VI	85	Pair with ESO 580-20 N edge.
14 46 04.9	−18 01 24	ESO 580-22	13.2	2.0 x 1.3	14.1	SAB(s)dm IV	157	Mag 8.20v star W 11′.2.
14 47 28.1	−22 09 29	ESO 580-26	13.9	1.1 x 0.8	13.6	S0?	168	Mag 12.8 star N 1′.7; mag 8.82v star W 9′.7; ESO 580-27 S 7′.3.
14 47 28.6	−22 16 43	ESO 580-27	12.6	1.7 x 1.1	13.1	(R′)SB(r)a	158	ESO 580-26 N 7′.3; mag 8.82v star NW 11′.8; IC 4501 S 7′.7.
14 47 33.5	−19 45 53	ESO 580-29	13.2	2.2 x 0.3	12.6	Sc: sp II	74	Mag 10.01v star WSW 10′.0.
14 47 35.7	−18 04 19	ESO 580-30	13.3	1.7 x 1.3	14.0	SB(s)cd: II-III	2	Stellar nucleus; mag 9.72v star E 13′.0.
14 47 54.0	−19 07 59	ESO 580-34	14.8	1.6 x 0.3	13.9	SBc? sp III	36	Mag 10.12v star E 5′.3.
14 49 01.0	−20 50 54	ESO 580-37	13.8	1.5 x 0.8	13.8	SA(r)d: III-IV	17	Mag 13.6 star NW 1′.2; mag 12.3 star SSE 6′.6.
14 50 36.3	−18 09 05	ESO 580-41	13.5	2.0 x 0.4	13.1	Sbc? sp	78	Mag 9.23v star ESE 11′.6.
14 50 58.2	−18 28 20	ESO 580-43	12.9	1.7 x 1.6	13.8	S0° pec:	6	Close pair of mag 8.30v, 9.04v stars ESE 8′.5.
14 51 12.4	−20 26 33	ESO 580-45	13.3	1.7 x 0.9	13.6	SB(s)dm pec: III	137	Mag 10.63v star ENE 8′.3.
14 53 07.9	−19 44 14	ESO 580-49	14.3	1.6 x 0.3	13.3	Sbc: sp	126	Mag 11.8 star W 3′.2.
14 55 13.0	−19 39 53	ESO 580-52	13.5	1.5 x 1.0	13.8	SB(s)d pec? III-IV	174	Mag 10.49v star N 3′.4.
14 57 29.9	−18 27 17	ESO 581-4	14.4	2.1 x 0.3	13.8	SB(s)cd: sp III-IV	12	Mag 10.62v star ESE 4′.5.
14 58 03.2	−19 23 26	ESO 581-6	14.7	1.7 x 0.2	13.3	SB(s)d pec sp	145	Mag 10.39v star SW 7′.3; MCG −3-38-33 NNW 7′.3.
14 58 50.2	−19 03 37	ESO 581-8	13.8	1.0 x 0.3	12.4	S0	11	Mag 10.87v star W 10′.2.
14 59 22.4	−20 01 10	ESO 581-10	14.5	1.2 x 0.7	14.1	SAB(rs)d pec: IV	129	Mag 8.96v star NNW 6′.0.
14 59 38.5	−18 43 58	ESO 581-11	14.4	1.5 x 0.4	13.7	Sc: sp III-IV	118	Mag 7.82v star NE 12′.2; mag 12.0 star W 2′.4.
15 00 16.7	−18 32 44	ESO 581-13	14.0	0.8 x 0.6	13.1	SB(s)m III-IV	40	Mag 7.82v star S 5′.2.
15 01 41.7	−19 17 23	ESO 581-16	12.8	1.4 x 0.6	12.6	E6:	26	Mag 8.58v star NW 3′.1.
15 01 55.0	−18 36 32	ESO 581-17	15.5	0.7 x 0.5	14.2	IB(s)m V-VI	177	Mag 11.4 star ESE 7′.9.
15 11 49.8	−17 56 46	ESO 581-23	16.3	1.9 x 0.3	15.5	SB(s)d pec sp	1	Mag 6.87v star SE 12′.2.
15 13 30.4	−20 40 35	ESO 581-25	12.7	3.5 x 0.7	13.5	SBd pec sp III	57	Mag 11.29v star NW 8′.8; mag 9.57v star E 13′.2.
15 17 18.2	−22 17 51	ESO 582-1	13.5	1.7 x 0.7	13.6	SA(rs)b: II-III	36	Mag 5.51v star SW 14′.1; elongated **ESO 582-3** NE 8′.4.
15 20 46.0	−18 20 56	ESO 582-4	14.4	1.2 x 0.6	13.9	Sc:	163	Mag 14.6 star n edge; mag 9.40v star S 7′.7.
14 57 21.7	−19 12 51	IC 1077	12.6	1.4 x 1.1	12.9	(R′)SB(s)bc pec? II	33	MCG −3-38-33 ESE 7′.8.
14 58 55.1	−19 14 25	IC 1081	13.7	1.4 x 0.5	13.1	SAB(s)0/a:	147	NGC 5791 SW 2′.7.
13 57 13.8	−25 14 47	IC 4350	12.6	1.6 x 0.8	12.7	(R′)SA(r)0$^+$ pec:	136	Mag 10.05v star N 4′.2.
13 57 54.1	−29 18 52	IC 4351	11.7	6.0 x 0.9	13.4	SA(s)b: sp III	17	Mag 9.22v star W 7′.3.
14 07 29.7	−27 01 06	IC 4374	12.4	1.5 x 1.2	12.9	SA(rs)0$^-$:	113	Mag 9.35v star N 2′.3.
14 34 28.9	−27 31 08	IC 4453	12.2	2.2 x 1.0	12.9	SAB(rs)0°: sp	160	Mag 9.91v star SW 3′.7.
14 38 26.7	−22 22 01	IC 4468	12.9	2.2 x 0.6	13.1	SB(s)c? I-II	155	Mag 7.41v star E 6′.9; **ESO 580-5** NNW 3′.4.
14 47 25.4	−22 24 25	IC 4501	13.6	1.2 x 0.9	13.5	SB(s)dm	63	ESO 580-27 N 7′.7; mag 7.34v star W 13′.3.
15 13 17.0	−18 08 14	IC 4536	13.0	2.2 x 1.8	14.3	SB(s)dm III-IV	10	Mag 6.87v star W 12′.4.
15 21 11.6	−23 39 33	IC 4538	12.1	2.6 x 2.0	13.7	SAB(s)c: I-II	32	Stellar nucleus; mag 10.88v star N 7′.6.
13 59 03.2	−19 03 31	MCG −3-36-2	13.7	1.0 x 0.5	12.8	SB?	20	
14 05 08.1	−17 30 14	MCG −3-36-4	14.8	1.0 x 0.7	14.3	Sb	171	**MCG −3-36-5** N 1′.4.
14 09 27.7	−17 51 58	MCG −3-36-8	13.7	0.9 x 0.7	13.2	E/S0	131	Mag 12.2 star SE 6′.2; MCG −3-36-9 WSW 2′.1.
14 09 19.7	−17 52 48	MCG −3-36-9	14.1	0.4 x 0.4	12.0	S0		MCG −3-36-8 ENE 2′.1.
14 43 55.9	−18 29 02	MCG −3-38-1	13.9	1.0 x 0.9	13.6	SB(s)d III-IV	122	
14 44 09.7	−18 09 17	MCG −3-38-2	13.8	1.0 x 1.0	13.6	SA(s)cd		
14 47 24.3	−17 26 48	MCG −3-38-8	13.9	3.2 x 0.5	14.2	SB(s)cd: sp	171	
14 47 33.5	−20 28 04	MCG −3-38-10	13.8	0.7 x 0.7	12.9			**MCG −3-38-11** N 5′.0.
14 54 40.8	−17 24 24	MCG −3-38-25	13.3	1.9 x 0.8	13.6	SAB(rs)bc: II-III	18	A number of bright knots and/or superimposed stars.
14 57 27.1	−21 39 55	MCG −3-38-32	13.7	1.0 x 0.8	13.3	(R_1)SB0$^+$	120	Star just off SE edge.
14 57 50.8	−19 16 42	MCG −3-38-33	14.1	0.8 x 0.5	12.9	SB(r)b:	57	ESO 581-6 SSE 7′.3; IC 1077 WNW 7′.8.
15 21 51.4	−17 43 49	MCG −3-39-9	14.5	0.6 x 0.4	12.8	(R'_1)SB(l)a	117	Mag 8.84v star N 3′.4.
13 56 13.4	−26 09 36	MCG −4-33-15	14.1	0.6 x 0.3	12.0		114	
13 56 20.9	−22 38 20	MCG −4-33-17	14.0	0.7 x 0.5	12.7	Sbc	45	
13 57 04.9	−24 48 29	MCG −4-33-18	14.1	0.5 x 0.4	12.4	E/S0	15	Mag 9.64v star E 4′.1.
13 57 22.1	−27 10 25	MCG −4-33-20	14.1	0.8 x 0.7	13.3	Sb	8	Pair of bright knots or stars W edge.
13 59 36.4	−24 22 06	MCG −4-33-22	13.4	1.0 x 0.7	12.9	S0	155	Mag 8.98v star NW 7′.9.
14 01 21.0	−26 11 50	MCG −4-33-28	14.2	1.3 x 0.9	14.2	SB(s)cd II-III	83	Close pair of stars, mags 13.1 and 13.8, SW 2′.3.
14 01 32.4	−25 15 05	MCG −4-33-30	14.4	1.0 x 0.7	13.9	S?	15	
14 01 59.2	−22 21 51	MCG −4-33-31	14.1	1.0 x 0.7	13.6	SAB(r)bc	36	
14 01 58.4	−25 32 23	MCG −4-33-32	13.4	1.2 x 1.1	13.6	(R′)SB(rs)b II	21	Lies between a pair of stars, mags 10.80v and 11.24v.

(Continued from previous page)

148 GALAXIES 148

RA h m s	Dec ° ′ ″	Name	Mag (V)	Dim ′ Maj x min	SB	Type Class	PA	Notes
14 03 35.8	−25 25 44	MCG −4-33-35	13.3	1.1 x 0.5	12.6	SB(s)0$^-$?	174	
14 04 11.4	−22 34 37	MCG −4-33-37	13.8	0.6 x 0.5	12.6	E	75	Mag 10.35v star N 2′3; mag 6.30v star NNW 10′2.
14 05 22.2	−26 36 01	MCG −4-33-41	13.4	0.8 x 0.7	12.6	SAB(rs)ab:	19	Mag 8.68v star E 3′9.
14 06 16.1	−25 47 59	MCG −4-33-42	13.6	0.8 x 0.4	12.3	SAB(s)0$^-$	159	
14 07 15.7	−27 09 31	MCG −4-33-44	13.1	0.9 x 0.9	12.7	SA(rs)0$^-$		Star on S edge; IC 4374 N 9′0.
14 09 23.9	−27 01 41	MCG −4-33-51	13.8	0.7 x 0.4	12.3		30	ESO 510-74 NW 8′2.
14 11 17.3	−25 02 13	MCG −4-33-52	13.9	0.9 x 0.3	12.3	Sbc	74	
14 19 02.2	−27 23 27	MCG −4-34-8	13.9	0.4 x 0.4	11.7			ESO 511-26 SW 2′8.
14 30 43.3	−27 27 20	MCG −4-34-17	13.6	1.0 x 0.6	12.9	Sbc	76	
14 31 32.2	−25 23 14	MCG −4-34-19	13.4	0.7 x 0.5	12.1	Sc	77	Mag 7.78v star WNW 8′3.
14 39 22.3	−25 28 12	MCG −4-35-2	13.2	0.8 x 0.8	12.6	S0		
14 43 27.2	−23 26 38	MCG −4-35-5	13.8	0.9 x 0.5	12.7	Scd	155	
14 44 22.8	−22 52 10	MCG −4-35-7	14.5	0.5 x 0.5	12.8			Mag 8.87v star E 3′8.
14 44 56.8	−23 47 39	MCG −4-35-8	13.4	1.1 x 1.0	13.4	(R)SB0$^+$	113	
14 50 43.0	−26 19 03	MCG −4-35-11	14.9	0.4 x 0.3	12.4	Sb	138	
14 51 44.9	−24 34 20	MCG −4-35-13	14.2	0.6 x 0.5	12.7	SA(r)b:	19	Mag 9.77v star E 2′4; on SW edge of galaxy cluster **A 1977**.
14 52 41.3	−24 48 30	MCG −4-35-14	13.8	0.5 x 0.5	12.1			
14 55 10.0	−23 58 13	MCG −4-35-15	14.5	0.9 x 0.9	14.1	Sb		Very faint outer arms surrounding stellar nucleus.
14 56 33.7	−24 30 10	MCG −4-35-17	14.2	1.1 x 0.6	13.6	Sc	109	
15 10 41.1	−22 01 45	MCG −4-36-1	13.9	1.1 x 0.8	13.6	SBc	126	
15 12 00.7	−24 02 39	MCG −4-36-2	14.2	1.0 x 0.5	13.3	SBbc	102	
15 18 36.1	−24 07 15	MCG −4-36-7	13.5	1.0 x 0.3	12.3	E$^+$	119	NGC 5903 N 3′2.
15 19 59.8	−23 55 31	MCG −4-36-11	14.2	0.9 x 0.5	13.1	Sd	4	Mag 10.61v star E 7′0; ESO 514-5 WNW 14′3.
15 20 30.8	−24 28 11	MCG −4-36-12	13.6	1.4 x 0.5	13.1	S0	34	Bright knot of star NE edge.
14 07 20.4	−28 58 14	MCG −5-33-46	13.5	1.3 x 0.3	12.4	Sa-b	55	
14 20 14.6	−29 44 52	MCG −5-34-7	13.0	1.1 x 0.9	12.9	(R′)SA(r)0$^+$:	21	
14 00 32.0	−28 52 28	NGC 5393	13.0	0.9 x 0.7	12.3	(R′)SB(r)a pec:	153	
14 12 23.5	−27 06 32	NGC 5495	12.6	1.6 x 1.4	13.3	(R′)SAB(r)c I-II	38	Short SE-NW bar; mag 12.1 star on NE edge.
14 13 37.4	−17 59 06	NGC 5510	13.9	1.4 x 1.1	14.3	IB(s)m pec: III-IV	40	Bright nucleus, several superimposed stars.
14 18 48.2	−19 08 22	NGC 5555	14.5	0.9 x 0.4	13.2	Sb	115	Mag 9.05v star NW 6′9; note: **ESO 579-14** lies 3′0 NNE of this star.
14 20 34.0	−29 14 25	NGC 5556	11.8	4.0 x 3.2	14.4	SAB(rs)d III-IV	148	Patchy with many superimposed stars; ESO 446-53 E 9′6.
14 23 55.5	−28 41 19	NGC 5592	12.8	1.5 x 0.8	12.8	SB(s)bc? II	88	Mag 11.28v star E 2′1.
14 29 49.0	−29 44 56	NGC 5626	12.9	1.1 x 0.9	12.7	SA(s)0$^+$	127	Mag 8.74v star NW 4′8; note: **ESO 447-7** lies 1′5 N of this star.
14 41 06.0	−17 28 35	NGC 5716	12.6	1.8 x 1.3	13.4	SB(rs)c? II-III	85	Pair of stars on NE edge.
14 42 55.9	−18 26 43	NGC 5726	12.7	1.3 x 1.0	12.9	S0$^-$ pec sp	133	Pair of mag 9.11v, 8.75v stars NNW 8′7.
14 42 23.9	−17 15 15	NGC 5728	11.4	3.1 x 1.8	13.1	SAB(r)a:	30	Faint arms extend E and W outside published dimensions.
14 45 09.1	−20 52 18	NGC 5734	12.6	1.5 x 1.0	12.9	S0° pec sp	38	Star N end, strong dark lane W; NGC 5743 S 2′5.
14 45 10.8	−20 54 49	NGC 5743	12.9	1.3 x 0.5	12.3	Sb: II	95	NGC 5734 N 2′5.
14 46 38.6	−18 30 50	NGC 5744	13.5	1.0 x 0.7	12.9	(R′)SB(rs)d:	108	
14 47 46.2	−19 04 44	NGC 5757	11.9	2.0 x 1.6	13.0	(R)SB(r)b II	160	Strong bar N-S, dark patch W of center.
14 49 08.6	−20 22 35	NGC 5761	12.4	1.4 x 1.3	12.9	S0° pec:	102	**PGC 52918** on N edge; **ESO 580-38** SW 1′9.
14 53 09.5	−21 23 41	NGC 5766	13.3	1.0 x 0.7	12.7	SAB(s)bc: II-III	149	
14 56 41.3	−17 14 37	NGC 5781	13.0	1.4 x 0.8	13.0	(R′)SBb pec sp III-IV	25	
14 58 46.4	−19 16 05	NGC 5791	11.7	2.6 x 1.3	13.0	E6:	163	IC 1081 NE 2′7; mag 9.87v star SSE 4′7.
15 02 42.7	−17 52 08	NGC 5810	13.3	1.2 x 0.6	12.8	SB(rs)b: III	31	Mag 9.41v star S 5′8.
15 10 48.3	−18 25 54	NGC 5863	12.9	1.2 x 1.0	12.9	(R′)SAB(rs)a	30	Bright nucleus.
15 17 51.3	−17 35 20	NGC 5890	12.7	1.3 x 1.0	12.9	SA(r)0$^+$?	85	Pair of stars E end.
15 18 13.7	−24 05 52	NGC 5898	11.4	2.7 x 2.6	13.5	E0	42	Very faint, very small anonymous galaxy 0′9 N of center.
15 18 36.4	−24 04 06	NGC 5903	11.2	2.7 x 2.1	13.1	E2	168	MCG −4-36-7 S 3′2.

GALAXY CLUSTERS

RA h m s	Dec ° ′ ″	Name	Mag 10th brightest	No. Gal	Diam ′	Notes
13 57 30.0	−24 43 00	A 3578	15.1	52	22	**ESO 510-25** at center; **ESO 510-28** NE 7′4; **ESO 510-27** N 8′3; most others anonymous.
14 07 30.0	−27 01 00	A 3581	15.6	42	22	
14 19 12.0	−19 28 00	A 3593	16.4	32	17	**ESO 579-16** SW of center 7′7; **PGC 51213** E 9′0; all others anonymous, stellar.
14 09 30.0	−17 52 00	AS 756	16.2		17	Except for three plotted GX, all others anonymous, stellar.
14 18 48.0	−27 25 00	AS 761	15.8	7	17	**ESO 511-22** SSW of center 15′9 and 3′1 NE of mag 9.32v star.

OPEN CLUSTERS

RA h m s	Dec ° ′ ″	Name	Mag	Diam ′	No. ★	B ★	Type	Notes
14 37 24.0	−29 25 00	ESO 447-29		12			cl	

GLOBULAR CLUSTERS

RA h m s	Dec ° ′ ″	Name	Total V m	B ★ V m	HB V m	Diam ′	Conc. Class Low = 12 High = 1	Notes
13 56 21.0	−27 09 42	AM 4	15.9	20.5		3.0		
14 39 36.5	−26 32 18	NGC 5694	10.2	15.5	18.5	4.3	7	
15 17 24.5	−21 00 37	NGC 5897	8.4	13.3	16.3	11.0	11	

PLANETARY NEBULAE

RA h m s	Dec ° ′ ″	Name	Diam ″	Mag (P)	Mag (V)	Mag cent ★	Alt Name	Notes
14 04 25.9	−17 13 40	IC 972	54	14.9	13.9	17.9	PK 326+42.1	Mag 12.9 stars N 2′5 and NW 3′7.
15 22 19.4	−23 37 33	PN G342.1+27.5	16	11.5	11.6	18.8	PK 342+27.1	Mag 9.86v star W 0′8.

149 GALAXIES 149

RA h m s	Dec ° ′ ″	Name	Mag (V)	Dim ′ Maj x min	SB	Type Class	PA	Notes
12 37 13.6	−28 29 40	ESO 442-13	12.8	2.0 x 0.8	13.1	SB(rs)cd III-IV	174	Mag 8.44v star NE 6′2.
12 42 42.5	−29 37 14	ESO 442-14	13.9	0.7 x 0.5	12.7	Sb	20	Mag 9.34v star SW 6′0.

(Continued from previous page)

GALAXIES

RA h m s	Dec ° ′ ″	Name	Mag (V)	Dim ′ Maj x min	SB	Type Class	PA	Notes
12 51 33.5	−28 14 42	ESO 442-25	13.8	1.1 x 0.9	13.6	SB(r)c II	75	Mag 8.51v star WSW 11'.3.
12 52 13.2	−29 50 29	ESO 442-26	11.6	2.6 x 0.9	12.4	SB0?	9	Mag 11.7 star E 6'.4.
12 54 22.2	−29 00 52	ESO 443-4	13.4	1.1 x 0.7	13.0	S0	174	Mag 14.9 star SE of core; stellar **PGC 43914** E 0'.9; **PGC 43905** N 2'.5; almost stellar **PGC 43881** WSW 2'.5.
12 57 44.8	−29 46 00	ESO 443-17	12.9	1.5 x 1.0	13.2	(R)SB0/a	18	Mag 10.51v star N 8'.0.
12 59 46.0	−29 35 59	ESO 443-21	13.6	2.2 x 0.3	13.0	Scd: sp	162	Mag 8.73v star NNW 8'.7.
13 03 29.8	−29 49 40	ESO 443-42	13.0	3.2 x 0.5	13.3	Sb: sp	128	Mag 8.59v star S 1'.9.
13 04 22.6	−28 42 05	ESO 443-50	13.4	0.9 x 0.7	12.8	S0	140	**ESO 443-51** S 2'.6.
13 05 35.4	−28 27 35	ESO 443-59	13.8	2.0 x 0.5	13.7	SA(s)d IV	100	Mag 7.68v star SW 6'.6; MCG −5-31-33 NE 3'.3; **ESO 443-57** NW 7'.1.
13 06 11.0	−29 43 38	ESO 443-66	13.7	0.9 x 0.6	12.9	S0?	24	NGC 4955 SW 2'.1.
13 06 53.1	−28 33 29	ESO 443-69	12.5	2.3 x 1.8	13.9	SB(rs)d III-IV	161	Elongated N-S core, few superimposed stars; mag 10.10v star NNE 7'.1.
13 10 22.8	−27 58 20	ESO 443-79	14.7	1.9 x 0.4	14.2	IB(s)m V	177	N-S line of four mag 12-13 stars W 5'.3.
13 11 09.3	−28 00 32	ESO 443-80	13.6	1.6 x 0.9	13.9	SB(s)m IV-V	146	Mag 10.04v star N 10'.1.
13 16 45.2	−27 53 11	ESO 444-2	14.7	0.9 x 0.6	13.9	SAB(s)dm V	86	Mag 11.9 star SE 3'.2.
13 29 50.9	−29 30 53	ESO 444-55	13.4	1.4 x 0.8	13.3	S0?	30	= **Hickson 65A**. **Hickson 65B** NE 1'.3; **Hickson 65C** NNE 1'.4; **Hickson 65D** NE edge of core of ESO 444-55; **Hickson 65E** NE 1'.7.
13 37 20.1	−28 02 46	ESO 444-84	14.8	1.3 x 0.9	14.8	Im	125	Low surface brightness; mag 13.9 star N 1'.8.
13 40 31.5	−28 32 51	ESO 445-8	13.8	0.9 x 0.5	12.7	S0?	72	Mag 13.4 star SW 2'.3.
13 43 49.0	−29 44 43	ESO 445-20	14.8	1.5 x 0.2	13.4	Sc	15	IC 4319 SW 6'.1; mag 11.09v star NNE 6'.7.
13 49 21.1	−28 12 06	ESO 445-51	13.6	1.2 x 0.7	13.3	(R)SAB(r)a:	110	Close pair of E-W oriented mag 7.77v, 8.97v stars E 7'.4.
13 52 21.1	−27 53 39	ESO 445-64	13.5	1.8 x 1.5	14.4	(R)SB(rs)0/a	104	Very faint outer arms extend well beyond published dimensions of this galaxy.
13 52 46.0	−29 55 47	ESO 445-65	13.2	1.5 x 0.2	11.8	S0?	13	mag 9.45v star S 5'.4.
13 53 58.4	−29 51 35	ESO 445-73	13.4	1.2 x 0.7	13.1	S?	114	Mag 14.0 star, with faint anonymous galaxy on it's W, edge N 1'.7.
13 54 54.7	−28 22 06	ESO 445-75	12.9	1.4 x 0.8	13.0	(R′)SAB(s)0°? pec	44	Mag 11.5 star NW 8'.5.
13 54 55.6	−29 08 15	ESO 445-76	14.2	1.1 x 1.1	14.3	S?		Mag 9.60v star NE 3'.3.
13 57 18.4	−28 11 43	ESO 445-83	14.8	1.5 x 0.2	13.4	Sc	51	Mag 12.1 star E 3'.3.
12 38 54.6	−27 18 28	ESO 506-27	13.4	1.2 x 0.5	12.7	S0$^+$: pec sp	72	Mag 9.92v star WNW 5'.8; several faint anonymous galaxies in the area E and W.
12 39 50.9	−23 04 40	ESO 506-32	13.9	0.7 x 0.5	12.6	S?	127	Mag 13.1 star NNE 2'.5.
12 40 13.9	−25 19 39	ESO 506-33	12.0	1.9 x 0.6	12.0	S0°: sp	177	Mag 9.49v star N 5'.7.
12 44 34.5	−24 16 40	ESO 507-6	14.4	0.9 x 0.4	13.1	Sa:	12	
12 45 42.7	−26 14 41	ESO 507-7	13.1	2.8 x 0.4	13.0	Sbc: sp	130	Mag 11.07v star NW 9'.0.
12 45 41.8	−26 38 04	ESO 507-8	13.3	1.4 x 1.2	13.7	SB(rs)c III-IV	61	Mag 11.5 star W 5'.9.
12 48 05.5	−27 34 42	ESO 507-13	13.8	1.7 x 0.4	13.2	SB(s)bc? sp II-III	63	Located between a mag 5.66v star E and a mag 7.95v star W.
12 48 20.6	−26 27 54	ESO 507-14	12.7	1.7 x 0.9	13.0	(R′)SB(r)0°:	115	Mag 9.35v star ESE 13'.0.
12 49 45.9	−23 51 38	ESO 507-17	13.8	0.8 x 0.6	12.8	Sb	168	Mag 12.0 star NE 5'.5.
12 50 28.8	−26 50 32	ESO 507-21	12.4	2.0 x 0.8	12.7	SA0$^-$:	28	ESO 507-24 E 13'.1; mag 6.14v star ENE 20'.9.
12 51 26.9	−26 48 27	ESO 507-24	12.8	1.4 x 1.0	13.0	(R)SB(r)0°	162	Mag 6.14v star NE 8'.1; ESO 507-21 W 13'.1.
12 51 31.9	−26 27 09	ESO 507-25	11.6	2.0 x 1.9	13.0	SA0$^-$	97	Mag 9.05v star W 6'.1; mag 8.33v star E 8'.0.
12 51 35.9	−26 05 26	ESO 507-26	16.8	1.0 x 0.8	16.5	IB(s)m pec V-VI	91	Dwarf irregular, very faint; ESO 507-27 S 1'.7.
12 51 37.6	−26 07 00	ESO 507-27	12.6	1.7 x 0.4	12.1	S0 sp	167	Mag 10.24v star SW 10'.0; ESO 507-28 NE 2'.4.
12 51 45.9	−26 05 24	ESO 507-28	12.6	1.5 x 0.8	12.7	SB(r)0/a:	28	ESO 507-27 SW 2'.4;.
12 52 00.0	−26 37 19	ESO 507-29	13.7	1.5 x 1.1	14.1	SAB(s)d pec IV	176	Mag 6.14v star S 7'.1.
12 52 14.9	−26 18 16	ESO 507-32	12.5	1.5 x 0.9	12.7	(R′)SAB0$^-$:	63	Mag 10.03v star N 2'.6.
12 52 56.3	−26 41 50	ESO 507-35	13.4	0.7 x 0.6	12.3	S?	83	Mag 9.66v star E 5'.4; mag 6.14v star W 13'.3.
12 52 59.4	−24 03 28	ESO 507-36	15.3	1.5 x 0.2	13.8	Scd? sp pec	106	Mag 9.72v star S 11'.9.
12 53 11.4	−27 27 48	ESO 507-37	13.4	1.2 x 0.4	12.5	S?	29	Mag 7.71v star NE 7'.9.
12 53 40.1	−26 39 17	ESO 507-43	13.5	1.0 x 0.9	13.3	S0	159	Elongated **ESO 507-39** WSW 1'.6.
12 55 35.6	−26 49 29	ESO 507-45	11.8	2.0 x 1.4	12.8	(R′)SA(s)0°: pec	164	Pair with ESO 507-46 E 2'.1.
12 55 44.2	−26 48 34	ESO 507-46	13.0	1.2 x 0.9	13.1	E2:	38	Pair with ESO 507-45 W 2'.1.
12 59 23.1	−27 25 37	ESO 507-62	13.2	1.6 x 0.5	12.8	SB(s)b? II	139	Mag 9.60v star N 2'.1.
13 01 48.6	−27 07 27	ESO 507-67	13.4	2.1 x 1.6	14.5	SB(rs)c II	154	Mag 10.76v star W 6'.4.
13 07 07.6	−24 06 39	ESO 508-7	14.4	1.5 x 1.0	14.7	SB(s)d: IV-V	73	Mag 11.9 star S 3'.2; NGC 4970 NE 8'.5.
13 07 29.0	−27 23 20	ESO 508-8	13.0	1.4 x 1.1	13.4	(R)SB(s)0$^+$	153	Mag 10.50v star SW 3'.4.
13 07 45.0	−22 51 29	ESO 508-11	13.5	3.2 x 0.5	13.9	SB(s)d: sp IV-V	94	Mag 10.48v star W 7'.6; **MCG −4-31-28** W 12'.2.
13 09 51.1	−24 14 31	ESO 508-19	13.4	1.7 x 0.7	13.4	SB(s)m IV	45	Star at N end; mag 10.55v star W 5'.2.
13 10 46.3	−23 51 57	ESO 508-24	13.1	2.4 x 2.1	14.7	SB(rs)c: II-III	60	Mag 8.66v star W 5'.8.
13 11 09.1	−25 54 04	ESO 508-25	14.0	0.8 x 0.6	13.0	S?	116	Mag 8.39v star W 8'.6; note **ESO 508-22** is 3'.0 WNW of this star.
13 14 55.1	−23 08 42	ESO 508-30	14.9	1.6 x 0.5	14.5	IB(s)m V-VI	135	**ESO 508-29** N 1'.7.
13 16 23.3	−26 33 41	ESO 508-33	13.3	0.7 x 0.4	11.7	S0?	83	Mag 9.27v star SE 6'.1.
13 16 57.3	−25 20 14	ESO 508-34	14.0	1.1 x 0.6	13.4	SAB(s)m: IV-V	132	Mag 9.65v star SE 13'.5.
13 19 06.3	−24 25 09	ESO 508-43	13.7	1.4 x 0.3	12.6	Sb:	85	Pair of small, faint anonymous galaxies on E end; another anonymous galaxy 0'.5 W.
13 19 28.0	−26 59 52	ESO 508-44	13.4	1.5 x 0.6	13.2	SAB(rs)b: II	37	Mag 9.65v star E 19'.3; mag 11.8 star SE 8'.3; note: **ESO 508-39** located 1'.0 E of this star.
13 20 29.8	−26 05 06	ESO 508-51	13.9	1.5 x 0.9	14.1	SAB(s)dm pec V	50	Mag 11.7 star NW 2'.9; mag 11.8 star S 6'.7.
13 21 12.9	−27 33 27	ESO 508-56	14.1	0.8 x 0.5	13.0	S0	142	Mag 10.43v star W 3'.3; **ESO 508-54** and mag 12.7 star SW 6'.6.
13 23 28.1	−24 03 31	ESO 508-62	13.6	1.0 x 0.5	12.7	S0/a	114	Mag 12.0 star SW 7'.7.
13 23 31.5	−23 53 37	ESO 508-63	13.8	0.6 x 0.6	12.6	Sb		Mag 11.9 star WNW 4'.6; **ESO 508-67** E 8'.9.
13 24 03.4	−24 39 45	ESO 508-66	13.9	1.6 x 1.4	14.6	IB(s)m V	119	Uniform surface brightness. Mag 9.26v star on NNW edge.
13 25 40.7	−26 27 52	ESO 508-78	14.0	1.0 x 0.9	13.7	(R′)SB(r)a	66	Elongated N-S core; mag 14.6 star S edge.
13 26 10.5	−27 25 38	ESO 509-3	13.8	1.3 x 0.6	13.4	S0?	157	Mag 9.96v star NW 1'.3.
13 26 44.2	−27 26 22	ESO 509-8	13.0	1.5 x 0.8	13.2	E?	18	Several very small, very faint galaxies within 2'.5 N, S and E; mag 9.96v star and ESO 509-3 W 8'.5.
13 26 48.8	−27 08 39	ESO 509-9	13.8	1.0 x 0.9	13.6	S0?	149	Mag 11.05v star N 1'.3 with **PGC 47075** NE edge of star.
13 27 04.8	−26 59 39	ESO 509-12	13.8	0.8 x 0.8	13.3	S0?		Star NW edge.
13 27 56.4	−25 51 24	ESO 509-19	13.9	2.8 x 0.3	13.5	Sbc: sp	52	Mag 11.9 star N 5'.1; **ESO 509-24** ENE 5'.7.
13 28 00.8	−27 06 18	ESO 509-21	13.8	0.8 x 0.6	12.9		5	Galaxy pair, mag 13.4 star NW edge; mag 9.75v star NNW 2'.3.
13 28 10.4	−24 58 12	ESO 509-23	13.8	1.8 x 0.5	13.5	SAB(rs)c I-II	113	Mag 14.0 star on N edge; pair of stellar anonymous galaxies NE 4'.2 and NE 5'.6; **ESO 509-22** N 8'.4.
13 28 25.1	−27 34 16	ESO 509-26	14.6	1.3 x 1.0	14.7	SAB(s)m V-VI	132	Pair with **ESO 509-25** at NW end.
13 32 08.9	−24 51 39	ESO 509-44	14.3	1.6 x 0.3	13.3	Sb:	145	Pair of stars N end.
13 32 08.6	−22 57 07	ESO 509-45	14.0	1.5 x 0.8	14.0	Sc	52	Pair of mag 9.5v stars S 1'.0.
13 32 49.6	−23 38 45	ESO 509-53	13.8	0.9 x 0.7	13.2	S0	173	Mag 10.01v star ESE 6'.1; **MCG −4-32-34** S 3'.1.
13 32 55.8	−25 10 45	ESO 509-55	13.9	1.0 x 0.8	13.5	Sbc	57	Mag 10.59v star SE 2'.2.
13 35 41.3	−24 04 26	ESO 509-74	12.8	2.6 x 0.5	13.0	S?	140	Mag 10.53v star NW 13'.7.
13 42 43.7	−24 20 11	ESO 509-95	14.1	0.8 x 0.6	13.1	Sa	11	Mag 9.39v star SW 4'.9.
13 43 18.0	−25 15 59	ESO 509-97	13.9	0.9 x 0.6	13.1	Sb	19	Mag 10.53v star W 6'.4.
13 43 21.9	−25 55 56	ESO 509-98	13.8	1.1 x 0.8	13.5	(R′)SB(s)a	32	Elongated core in smooth envelope.
13 44 30.9	−24 20 43	ESO 509-103	13.7	1.0 x 0.9	13.5	S0	116	Mag 13.1 star E edge.
13 51 10.4	−26 18 31	ESO 510-3	14.7	1.5 x 0.2	13.2	Sc:	145	Mag 13.0 star E 0'.6; mag 11.30v star E 5'.4.
13 53 58.7	−27 37 28	ESO 510-7	13.2	2.1 x 0.6	13.3	SB(rs)a:	29	Mag 10.66v star NW 11'.8.

(Continued from previous page)

GALAXIES

RA h m s	Dec ° ′ ″	Name	Mag (V)	Dim ′ Maj x min	SB	Type Class	PA	Notes
13 54 19.1	−26 52 26	ESO 510-9	12.6	1.3 x 0.4	11.8	S0 sp	28	Pair with ESO 510-10 S edge; mag 7.63v star NW 13.′0.
13 54 18.6	−26 53 37	ESO 510-10	12.7	1.4 x 1.0	12.9	(R′)SB(s)0/a pec:	56	Pair with ESO 510-9 N edge; mag 7.63v star NW 13.′0.
13 55 04.6	−26 46 48	ESO 510-13	12.3	2.3 x 1.3	13.4	Sa: pec sp	119	Strong dark lane along centerline; mag 8.96v star NE 8.′4.
13 55 22.7	−27 41 13	ESO 510-17	14.6	1.5 x 0.3	13.6	SB(r)0$^+$: pec	33	Mag 10.11v star NNE 2.′7.
13 57 35.8	−25 47 30	ESO 510-26	14.9	1.1 x 0.6	14.3	SB(rs)m pec V-VI	24	Mag 10.91v star E 10.′7; **ESO 510-21** W 8.′7.
14 00 24.0	−23 18 33	ESO 510-36	14.0	0.4 x 0.4	11.9	S?		
14 00 25.8	−23 18 17	ESO 510-36A	13.8	1.1 x 0.6	13.3		175	
14 01 34.6	−23 08 24	ESO 510-43	13.9	1.2 x 0.7	13.6	SB(s)c pec? III-IV	141	**ESO 510-45** 2.′2 NNE.
14 01 37.6	−26 25 59	ESO 510-44	13.6	0.7 x 0.7	12.7	S0		Mag 10.93v star E 4.′6.
14 03 34.7	−27 16 50	ESO 510-52	14.5	0.6 x 0.5	13.1		34	Mag 9.79v star SW 11.′2.
12 43 34.0	−20 50 39	ESO 574-24	13.6	1.7 x 0.8	13.8	(R′)SB(rs)ab: II	141	**ESO 574-23** N 6.′0; mag 11.9 star NNE 5.′9.
12 44 48.8	−20 47 07	ESO 574-28	14.5	1.5 x 0.3	13.5	SA(s)c: sp III	9	Mag 10.70v star NW 9.′1.
12 44 51.6	−20 25 35	ESO 574-29	13.0	1.1 x 1.0	12.9	SAB(s)c pec: I	63	Close pair of mag 11.7, 11.9 stars NW 7.′8.
12 46 46.2	−21 23 34	ESO 574-31	13.4	1.7 x 0.8	13.5	(R′)SAB(rs)0/a:	129	Mag 8.02v star ESE 10.′4.
12 47 54.7	−21 55 40	ESO 574-32	14.4	1.6 x 0.4	13.7	Sb	140	Mag 12.4 star SW 1.′1.
12 47 58.5	−22 16 07	ESO 574-33	13.0	2.4 x 1.4	14.1	SB(rs)bc II	102	Mag 7.42v star NW 15.′1; mag 12.3 star N 6.′2.
12 49 13.0	−21 35 21	ESO 575-1	13.9	1.1 x 0.3	12.6	S0	103	Almost stellar anonymous galaxy E 1.′1.
12 51 34.9	−20 42 29	ESO 575-10	13.9	1.3 x 0.6	13.5	Sa	24	Mag 12.1 star WSW 13.′4.
12 52 08.4	−21 51 54	ESO 575-13	12.8	1.2 x 1.2	13.2	E?		Sits in a triangle of faint stars.
12 54 05.5	−20 11 00	ESO 575-21	13.0	1.5 x 1.2	13.5	SB(r)a: II-III	118	Mag 9.61v star S 2.′7; mag 6.72v star NE 7.′7.
12 55 59.7	−19 16 11	ESO 575-29	13.1	2.5 x 1.7	14.5	SAB(rs)dm: IV	14	Star just inside W edge.
12 56 56.4	−20 28 21	ESO 575-32	13.0	0.8 x 0.8	12.4	S0		Mag 9.28v star NW 5.′9; ESO 575-33 on SE edge.
12 57 00.6	−20 28 53	ESO 575-33	12.8	1.0 x 1.0	12.7	S0		**ESO 575-22** on NW edge; MCG −3-33-20 S 1.′5.
12 59 34.1	−21 05 52	ESO 575-42	13.7	1.5 x 0.5	13.3	SB(r)0$^+$?	10	Mag 0.54v star W 11.′0; **ESO 575-41** NW 9.′8.
12 59 45.5	−22 10 35	ESO 575-43	12.5	1.9 x 1.7	13.6	SAB(s)0$^-$	164	Mag 7.94v star NW 5.′3.
13 00 20.0	−22 41 35	ESO 575-44	12.2	1.1 x 0.9	12.3	S0	15	ESO 575-44A on E edge; mag 9.96v star WSW 12.′6.
13 00 22.2	−22 41 43	ESO 575-44A	14.3	0.3 x 0.3	11.7			Located on E edge of ESO 575-44.
13 01 09.3	−18 11 51	ESO 575-47	12.7	1.8 x 1.3	13.4	(R′)SB(r)ab	119	Mag 7.57v star N 4.′5.
13 05 05.0	−22 23 05	ESO 575-53	13.7	1.1 x 0.4	12.7	Sb? sp	96	Mag 10.71v star N 3.′9.
13 07 43.9	−19 23 46	ESO 575-59	13.2	2.2 x 0.5	13.2	SA0$^+$? sp	178	Mag 8.81v star SW 7.′3.
13 08 15.4	−21 00 05	ESO 575-61	15.0	2.0 x 0.2	13.9	SBd: sp	174	Mag 9.26 star W 7.′0; note: **ESO 575-60** located 2.′1 NE of this star.
13 10 34.9	−21 44 52	ESO 576-3	13.4	2.2 x 0.9	14.0	SB(s)d pec: III-IV	93	MCG −3-34-6 NNW 4.′7.
13 11 17.9	−19 49 44	ESO 576-5	14.7	1.4 x 0.4	13.9	SBdm? sp	120	Mag 10.85v star W 4.′2.
13 12 18.9	−19 26 48	ESO 576-8	14.1	1.3 x 0.3	12.9	S0? sp	150	Mag 9.43v star E 5.′6; NGC 5018 ESE 10.′9.
13 13 05.1	−19 58 42	ESO 576-11	13.2	3.2 x 0.5	13.6	SB(s)cd? sp III	0	Mag 5.32v star 55 Virginis E 15.′6.
13 14 09.6	−21 39 42	ESO 576-15	16.7	1.2 x 1.0	16.8	SB(s)m VI	108	Dwarf.
13 15 12.9	−17 58 04	ESO 576-17	14.1	1.2 x 1.2	14.3	SAB(s)d: IV		Mag 11.13v star ESE 4.′6.
13 15 41.3	−22 44 19	ESO 576-18	13.5	0.8 x 0.8	12.9	S0		Mag 10.14v star SW 9.′0.
13 17 57.1	−21 51 55	ESO 576-24	15.1	1.5 x 0.4	14.4	dE?	7	Mag 9.30v star SW 16.′2.
13 18 30.1	−20 41 19	ESO 576-25	15.4	0.9 x 0.8	14.9	dE1? pec	90	Mag 9.12v star NW 6.′3.
13 18 34.5	−21 18 01	ESO 576-26	14.7	1.9 x 0.3	14.0	SBd? sp	61	Located at the top edge of galaxy cluster A 1709; mag 9.63v star SW 14.′9.
13 18 59.0	−18 35 16	ESO 576-30	13.0	1.6 x 0.5	12.6	SA0$^-$?	17	Close pair of mag 9.77v, 9.33v, stars ENE 8.′9.
13 19 47.4	−21 54 01	ESO 576-31	13.1	1.6 x 0.9	13.4	SA(rs)0°	170	Located 8.′2 SE of the center of NGC 5084.
13 19 51.7	−22 16 43	ESO 576-32	12.4	2.3 x 1.7	13.8	SB(s)cd pec III	139	Several strong dark patches; mag 10.70v star S 2.′1; mag 8.20v star NE 12.′5.
13 20 28.8	−19 50 35	ESO 576-37	15.2	1.5 x 0.3	14.2	SB(s)m? sp IV-V	15	Mag 10.26v star W 7.′2.
13 20 42.0	−18 02 07	ESO 576-39	14.8	1.9 x 0.6	14.8	SB(s)m IV-V	104	Mag 9.93v star N 3.′1.
13 20 43.7	−22 03 02	ESO 576-40	13.9	2.4 x 0.5	13.9	SBd? pec sp	157	Mag 8.20v star SSW 5.′9.
13 23 18.3	−19 37 13	ESO 576-44	13.6	1.4 x 0.7	13.4	SA0°: pec	129	Mag 11.9 star NE 5.′9.
13 24 42.0	−19 42 07	ESO 576-50	12.4	3.1 x 2.0	14.2	SB(rs)c pec II	0	Mag 10.9 star SE 2.′2; **ESO 567-49** S 7.′9; ESO 576-51 SE 7.′6.
13 25 08.9	−19 46 05	ESO 576-51	12.9	1.3 x 0.8	12.8	SB(s)bc pec I-II	78	Multi-galaxy system; mag 8.04v star ESE 11.′3; ESO 576-50 NW 7.′6.
13 26 07.8	−19 46 37	ESO 576-54	13.7	1.1 x 0.6	13.1	(R′)SAB(s)0$^-$?	45	Mag 8.04v star SW 3.′9.
13 26 17.1	−19 38 09	ESO 576-56	13.2	1.8 x 0.5	13.0	SAB(rs)0$^+$:	129	MCG −3-34-78 SE 3.′4; mag 15.2 star SE end.
13 26 35.5	−22 14 22	ESO 576-59	14.8	1.5 x 0.8	14.8	IB(s)m V-VI	90	Mag 8.26v star S 3.′9.
13 26 47.0	−21 02 37	ESO 576-60	13.5	1.3 x 0.4	12.7	S0/a	92	Mag 12.1 star NE edge.
13 29 55.8	−20 41 07	ESO 576-66	12.9	1.1 x 1.0	12.9	SAB(rs)0$^-$:	12	ESO 576-67 N 8.′2.
13 30 02.3	−20 33 06	ESO 576-67	13.3	1.4 x 0.6	12.9	SAB0$^-$	63	ESO 576-66 S 8.′2.
13 30 05.4	−20 56 01	ESO 576-69	13.8	1.0 x 0.6	13.1	Sa? pec	173	Located between a pair of NE-SW oriented mag 12.1 and 13.4 stars.
13 30 11.7	−21 00 45	ESO 576-70	13.9	1.1 x 0.3	12.5	S0	151	Mag 12.3 star W 5.′0.
13 30 14.8	−21 13 12	ESO 576-71	14.0	0.9 x 0.3	12.4	Sa	79	Mag 10.74v star S 0.′5.
13 30 42.8	−22 25 15	ESO 576-76	13.2	1.6 x 0.7	13.2	SA0$^-$: pec	166	
13 30 43.1	−20 57 08	ESO 576-77	13.5	1.0 x 0.9	13.2	S0	171	Mag 13.2 star WNW 2.′0; note: **ESO 576-74** is 0.′3 N of this star.
13 30 58.4	−21 50 52	ESO 577-1	14.3	1.5 x 0.5	13.8	SB(r)cd? II-III	166	Mag 10.64v star N 3.′8.
13 31 02.4	−20 31 55	ESO 577-2	15.4	1.0 x 1.0	15.3	S0		Mag 10.56v star W 1.′2.
13 31 23.6	−21 15 03	ESO 577-3	13.9	1.0 x 1.0	13.7	Sa		Triangle of mag 11.10v, 12.1, 12.3 stars centered SE 2.′6.
13 36 10.1	−22 31 15	ESO 577-11	14.1	0.5 x 0.5	12.5	S0:		Compact.
13 45 59.9	−21 22 33	ESO 577-35	14.3	0.7 x 0.7	13.4			Stellar nucleus; mag 9.55v star NW 3.′4.
13 48 25.6	−18 52 22	ESO 577-38	16.0	1.0 x 0.2	14.1	Sd: sp	6	Mag 10.57v star N 11.′3.
13 52 24.2	−20 22 59	ESO 578-1	13.7	0.9 x 0.8	13.2	S0	58	Mag 14.8 star NW 0.′9.
13 53 23.1	−20 07 03	ESO 578-3	13.8	1.5 x 0.8	13.8	(R′)SAB(r)a:	1	Almost stellar nucleus, smooth envelope; mag 9.76v star N 6.′5.
13 56 49.8	−21 48 19	ESO 578-10	14.1	0.8 x 0.6	13.2		26	Mag 13.0 star NW 2.′4.
14 03 21.9	−22 33 28	ESO 578-19	13.3	1.3 x 1.1	13.5	(R′)SB(r)bc pec II	69	Mag 6.30v star NNE 10.′9.
12 42 08.5	−17 20 58	IC 806	14.4	0.7 x 0.6	13.3	(R′)SB(r)ab pec? II	135	Mag 11.8 star and small anonymous galaxy SW 0.′9; mag 10.64v star W 9.′2; IC 807 S 3.′4.
12 42 12.5	−17 24 13	IC 807	13.3	0.7 x 0.7	12.5	E/S0		IC 806 N 3.′4.
13 17 12.3	−17 15 20	IC 863	13.2	1.2 x 0.9	13.1	SB(rs)0/a? pec	48	Close pair of mag 10.37v, 12.0 stars NE 9.′0.
13 19 00.7	−27 37 43	IC 874	12.7	1.1 x 1.0	12.7	SB(rs)0°	17	Close pair of mag 10.60v, 10.79v stars N 2.′9.
13 19 40.5	−27 25 44	IC 879	13.2	1.4 x 1.2	13.6	SB(s)ab pec	64	Located 2.′3 SW of NGC 5078.
12 50 02.6	−25 55 12	IC 3813	12.6	1.3 x 1.0	12.7	SAB(r)0°	171	Mag 8.44v star N 4.′9.
12 51 32.6	−27 46 59	IC 3829	12.9	2.3 x 1.2	13.9	SAB0°? sp II-III	113	Mag 9.66v star NNE 2.′6.
12 58 10.5	−22 52 34	IC 3927	12.5	1.2 x 1.0	12.6	SAB(rs)0$^-$:	162	Mag 9.41v star E 6.′4.
13 06 56.5	−23 55 02	IC 4180	12.7	0.9 x 0.7	12.1	(R)SB(r)0$^+$:	165	NGC 4970 SE 10.′2.
13 08 04.3	−23 47 49	IC 4197	12.5	1.5 x 0.9	12.7	SAB0$^-$	163	Mag 7.53v star E 8.′3.
13 23 13.3	−26 18 05	IC 4231	13.3	1.9 x 0.5	13.1	Sbc sp II-III	32	Mag 9.36v star SE 14.′2.
13 23 22.5	−26 06 38	IC 4232	13.6	1.2 x 0.4	12.7	Sbc: pec I	2	Mag 10.9 star E 3.′2.
13 24 32.8	−21 08 17	IC 4237	12.3	2.5 x 1.5	13.6	SB(r)b? II-III	140	Mag 9.94v star S 8.′5; NGC 5134 E 10.′7.
13 25 59.9	−26 40 43	IC 4245	14.1	1.0 x 0.5	13.2	Sa	107	Galaxy pair.
13 26 47.2	−29 52 54	IC 4248	13.2	1.1 x 0.8	12.9	S?	103	Mag 11.6 star SE 1.′1; mag 10.09v star WSW 8.′9.
13 27 06.3	−27 57 24	IC 4249	13.7	1.3 x 0.4	12.9	S?	107	Mag 10.29v star on E end.
13 27 24.3	−29 26 41	IC 4251	14.0	0.9 x 0.8	13.5	S0?	142	Mag 11.9 star SW 1.′3.
13 27 28.2	−27 19 28	IC 4252	13.1	1.3 x 0.7	12.9	SA(rs)0$^-$:	138	Mag 12.3 star W 2.′2; stellar **PGC 47157** E 0.′9.

RA h m s	Dec ° ′ ″	Name	Mag (V)	Dim ′ Maj x min	SB	Type Class	PA	Notes
13 27 32.3	−27 52 19	IC 4253	14.0	1.4 x 0.6	13.7	(R′)SB(rs)bc II	133	Mag 10.29v star SW 7′5; note: this star is on E end of IC 4249.
13 28 00.1	−27 21 17	IC 4255	13.2	0.9 x 0.6	12.5	E?	35	Mag 7.19v star ESE 7′1; stellar **PGC 47217** SE 0′7; many PGC and anonymous galaxies in surrounding area.
13 29 47.7	−28 00 24	IC 4261	13.7	1.1 x 0.8	13.4	S0	131	IC 4264 NE 8′1.
13 30 17.6	−27 55 42	IC 4264	14.5	1.7 x 0.4	14.0	SA(s)d: sp	103	IC 4261 SW 8′1.
13 30 36.2	−26 15 22	IC 4267	14.4	1.3 x 0.5	13.8	S0?	137	Mag 10.59v star SSE 5′2.
13 30 49.2	−25 20 03	IC 4270	13.9	0.8 x 0.8	13.2	Sb-c		Mag 11.20v star NW 6′9.
13 31 29.8	−28 53 39	IC 4273	14.5	1.0 x 0.6	13.8	S0?	33	Mag 11.9 star E 1′3.
13 31 51.4	−29 43 55	IC 4275	13.6	0.9 x 0.5	12.6	SB?	22	Mag 11.02v star NW 1′7; faint anonymous galaxy S 4′7, many others in area.
13 32 06.1	−28 09 25	IC 4276	14.5	1.4 x 0.2	13.0	S?	62	Mag 6.49v star WNW 7′8.
13 32 31.0	−27 07 43	IC 4279	14.3	0.7 x 0.4	12.8	S pec	51	Mag 8.45v star NE 10′4; note: **ESO 509-52** is 3′9 W of this star.
13 32 53.3	−24 12 27	IC 4280	12.6	1.2 x 1.0	12.6	S?	54	Mag 10.71v star ENE 3′8; note: **ESO 509-56** is 4′0 N of this star.
13 32 38.4	−27 10 13	IC 4281	13.4	1.2 x 0.7	13.1	S?	94	See notes for IC 4279 located NNW 3′0.
13 34 30.6	−27 18 16	IC 4288	13.6	1.0 x 0.5	12.7	S?	112	Mag 9.24v star E 4′1.
13 34 47.8	−27 07 40	IC 4289	13.2	0.9 x 0.7	12.8	E?	6	Mag 10.73v star NW 7′7.
13 35 19.6	−28 01 18	IC 4290	13.3	1.5 x 1.3	13.8	(R′)SB(r)b	98	E-W bar with strong dark patches N and S; mag 10.92v star ESE 7′2.
13 35 46.7	−27 40 33	IC 4292	14.3	0.5 x 0.4	12.4	S	29	Mag 11.9 star N 3′7.
13 36 02.4	−25 52 55	IC 4293	12.5	1.4 x 1.0	12.7	SA0$^-$	177	Mag 9.99v star S 6′5.
13 36 34.8	−26 33 20	IC 4298	13.2	1.5 x 0.7	13.1	(R′)SA(s)bc	177	Mag 5.72v star NE 4′6.
13 38 57.3	−25 50 44	IC 4310	12.2	2.4 x 0.9	12.9	S0 pec	73	Mag 9.24v star W 8′2.
13 40 03.2	−25 28 29	IC 4315	14.1	1.7 x 0.3	13.3	SB(s)cd: sp	132	Mag 9.09v star NNW 3′8.
13 40 18.6	−28 53 33	IC 4316	14.4	1.0 x 0.7	13.8	IBm? pec	45	Mag 9.96v star E 7′3.
13 43 22.6	−28 58 05	IC 4318	13.7	1.1 x 1.0	13.7	S?	44	Double nucleus or superimposed star. Dark patch NE.
13 43 26.5	−29 48 14	IC 4319	13.2	1.5 x 0.5	12.8	SA(s)bc?	70	ESO 445-20 NE 6′1; faint, round anonymous galaxy S 1′3.
13 44 03.6	−27 13 55	IC 4320	13.2	0.9 x 0.9	12.8	S0?		Mag 10.55v star WNW 4′9; **MCG −4-32-52** S 8′2.
13 47 39.8	−29 26 05	IC 4325	13.8	1.2 x 0.4	12.9	SB?	102	Mag 10.19v star ESE 10′7.
13 48 21.6	−29 37 37	IC 4326	13.5	0.8 x 0.7	12.7	S?	168	Mag 8.11v star E 7′1.
13 49 02.8	−29 56 14	IC 4328	13.9	0.9 x 0.6	13.1	S?	172	Mag 9.58v star SW 10′3; **PGC 49009** SSW 5′6.
13 47 14.8	−28 19 55	IC 4330	13.8	1.3 x 0.6	13.4	SB(rs)c I-II	94	Mag 10.84v star NNE 6′1.
13 57 13.8	−25 14 47	IC 4350	12.6	1.6 x 0.8	12.7	(R′)SA(r)0$^+$ pec:	136	Mag 10.05v star N 4′2.
13 57 54.1	−29 18 52	IC 4351	11.7	6.0 x 0.9	13.4	SA(s)b: sp III	17	Mag 9.22v star W 7′3.
12 36 24.4	−19 23 44	MCG −3-32-15	13.0	0.8 x 0.8	12.6	E		
12 37 13.5	−20 03 06	MCG −3-32-16	13.4	0.7 x 0.4	12.0	E$^+$:	75	Lies 1′6 N of mag 9 star.
12 38 00.5	−20 07 53	MCG −3-32-17	14.4	0.7 x 0.4	13.0	E/S0	105	
12 40 35.0	−20 33 47	MCG −3-32-18	13.7	0.7 x 0.5	12.6	E$^+$	21	
12 50 40.7	−20 20 06	MCG −3-33-9	14.2	0.9 x 0.5	13.1	S	119	Mag 8.19v star W 7′4.
12 50 52.9	−20 22 21	MCG −3-33-10	14.0	0.7 x 0.6	12.9	S?	15	Mag 9.40v star E 4′9.
12 51 38.3	−19 52 44	MCG −3-33-11	13.2	1.0 x 0.7	12.7	S0	135	
12 54 37.3	−18 18 32	MCG −3-33-16	13.5	1.2 x 1.1	13.7	(R′)SB(rs)a pec:	109	Star on NE edge, and star or bright knot N of bright core.
12 56 43.5	−21 01 15	MCG −3-33-18	13.8	0.6 x 0.3	11.8	Sa	90	
12 57 02.1	−20 30 19	MCG −3-33-20	13.4	0.7 x 0.5	12.1	S0	116	ESO 575-33 N 1′5.
12 57 02.7	−19 31 07	MCG −3-33-22	14.2	1.0 x 1.0	14.0	SAB(s)m: IV-V		
12 57 00.5	−17 19 20	MCG −3-33-23	14.0	0.8 x 0.5	12.8	SBc	144	Mag 7.95v star SE 9′4; in galaxy cluster A 1644, many anonymous and PGC galaxies near.
13 02 26.3	−17 40 50	MCG −3-33-28	13.5	2.8 x 0.3	13.1	Sc: sp II	42	
13 03 16.4	−17 25 23	MCG −3-33-30	13.2	5.0 x 0.8	14.5	IB(s)m sp V	117	Numerous bright knots clustered in central area. Galaxy becomes extremely faint beyond central one third of length.
13 09 36.1	−18 58 25	MCG −3-34-3	14.0	0.8 x 0.5	12.8	SA(s)b	5	Mag 8.68v star NE 8′9.
13 10 24.0	−21 41 06	MCG −3-34-6	13.9	1.0 x 0.5	13.0	(R′$_2$:)SB(r)bc	158	**ESO 576-2** SSE 4′7; multi-galaxy system **MCG −3-34-7** N 6′0.
13 12 34.9	−17 32 27	MCG −3-34-14	12.4	2.5 x 0.8	13.0	SAB(s)c? pec II-III	129	
13 13 32.2	−17 04 46	MCG −3-34-22	13.2	0.9 x 0.9	12.8	Sa:		
13 14 30.3	−17 32 03	MCG −3-34-25	14.3	0.7 x 0.2	12.0	Sab	153	
13 14 41.9	−18 46 45	MCG −3-34-26	14.2	0.7 x 0.7	13.3	Sc		
13 15 00.0	−17 16 09	MCG −3-34-30	14.9	0.5 x 0.4	13.1	SBb	96	
13 18 56.5	−17 38 12	MCG −3-34-44	13.3	0.7 x 0.5	12.1		66	
13 18 41.5	−19 04 52	MCG −3-34-45	13.8	0.9 x 0.8	13.3	Sa	90	Mag 10.51v star W 6′0.
13 20 04.5	−17 07 21	MCG −3-34-48	13.8	0.4 x 0.4	11.7	S0		
13 20 06.3	−17 07 09	MCG −3-34-49	13.8	0.5 x 0.5	12.2	S0		
13 21 51.0	−17 20 12	MCG −3-34-61	13.5	1.2 x 0.8	13.3	(R′)SA(s)0$^-$:	123	Anonymous elongated galaxy SE 1′5.
13 22 50.6	−17 01 54	MCG −3-34-65	13.8	0.7 x 0.4	12.3	S0	108	
13 25 16.4	−20 39 01	MCG −3-34-72	13.9	0.7 x 0.4	12.4	S(r)0$^+$?	27	
13 25 52.2	−20 34 10	MCG −3-34-74	14.2	1.1 x 0.8	13.9	SB(rs)b	31	Mag 10.03v star E 6′3.
13 26 08.3	−21 02 06	MCG −3-34-75	13.8	1.0 x 0.4	12.7	Sa-b	70	Mag 9.67v star SW 4′0.
13 26 28.8	−19 40 11	MCG −3-34-78	14.1	0.9 x 0.5	13.1	SA0$^-$	81	ESO 576-56 NW 3′4; mag 14.5 star on S edge.
13 28 07.3	−20 58 00	MCG −3-34-80	13.6	0.6 x 0.6	12.6	E$^+$2 pec		Very faint, elongated galaxy extending NE from bright core.
13 28 29.8	−20 56 11	MCG −3-34-81	13.7	0.6 x 0.6	12.6	E		Mag 10.78v star ENE 4′4; **ESO 576-62** E 8′3.
13 28 47.0	−17 30 46	MCG −3-34-82	12.9	1.4 x 0.7	12.8	SB(rs)0°	6	Mag 12 star at S end.
13 29 44.7	−21 10 58	MCG −3-34-83	13.8	0.7 x 0.6	12.7	Sa:	50	ESO 576-71 ESE 7′3.
13 31 51.4	−20 10 26	MCG −3-35-1	14.1	1.0 x 0.5	13.3	S0	12	
13 32 12.9	−21 12 08	MCG −3-35-2	14.0	1.2 x 0.3	12.8	Sb	5	
13 36 27.9	−17 16 03	MCG −3-35-6	14.1	0.8 x 0.5	12.9	Sb	165	
13 36 47.9	−18 18 52	MCG −3-35-8	14.1	0.5 x 0.4	12.2		42	
13 38 11.3	−20 40 50	MCG −3-35-12	14.2	0.9 x 0.9	13.8	SAB(rs)d: III-IV		Mag 9.16v star S 5′0.
13 42 49.9	−18 49 17	MCG −3-35-15	14.0	1.0 x 0.7	13.4	Sa?	150	
13 43 55.0	−19 45 34	MCG −3-35-16	13.7	1.0 x 0.2	11.9	S0	148	Mag 11.6 star E 1′3; **PGC 48637, 40, 41** S 5′2.
13 49 29.5	−18 05 41	MCG −3-35-18	14.2	0.4 x 0.3	11.8	Sab	147	Mag 4.96v star 89 Virginis SE 6′0.
13 50 11.6	−19 47 28	MCG −3-35-19	13.8	0.5 x 0.5	12.3	E/S0		Lies 1′2 N of mag 11 star.
13 59 03.2	−19 03 31	MCG −3-36-2	13.7	1.0 x 0.5	12.8	SB?	20	
12 39 15.7	−26 54 37	MCG −4-30-9	13.5	1.1 x 0.7	13.0	SB(s)cd IV	14	Located 10′3 S of the center of M 68 (NGC 4590).
12 41 02.9	−25 57 34	MCG −4-30-11	14.6	1.0 x 0.6	13.9	Sc	69	Anonymous galaxies lie E 1′4 and SSW 1′4.
12 42 40.3	−23 46 37	MCG −4-30-12	14.3	0.8 x 0.5	13.2	Sa	92	Mag 8.34v star SE 6′9.
12 47 40.3	−26 12 03	MCG −4-30-16	13.9	1.1 x 0.5	13.1	SB(rs)bc pec	9	Spiral galaxy **MCG −4-30-17** E 1′8.
12 49 31.1	−23 00 00	MCG −4-30-22	13.3	0.8 x 0.8	12.9	E/S0		
12 49 40.3	−25 09 38	MCG −4-30-23	14.1	1.1 x 0.9	13.9	(R′)SAB(r)0/a	17	Very faint outer shell.
12 52 36.1	−21 54 49	MCG −4-30-35	14.7	0.9 x 0.7	14.0	S0	21	Double system, contact.
12 52 40.3	−22 11 40	MCG −4-30-36	14.1	0.8 x 0.4	12.8	SBb	24	
12 53 22.6	−22 41 19	MCG −4-31-2	14.2	0.5 x 0.5	12.6	SB(r)0/a		
12 53 36.3	−26 17 40	MCG −4-31-3	13.5	1.2 x 0.6	13.0	Sb II-III	79	
12 53 34.7	−25 23 05	MCG −4-31-4	14.2	0.9 x 0.7	13.5	SBcd	167	
12 56 03.7	−25 00 25	MCG −4-31-7	13.8	0.8 x 0.4	12.4	S0	27	

(Continued from previous page)

GALAXIES

RA h m s	Dec ° ′ ″	Name	Mag (V)	Dim ′ Maj x min	SB	Type Class	PA	Notes
12 56 19.4	−25 45 43	MCG −4-31-8	14.0	0.9 x 0.5	13.0	S0	92	
12 56 58.0	−22 49 55	MCG −4-31-9	13.7	1.2 x 0.9	13.7	(R′)SB(r)0+	26	Very faint outer shell.
12 58 14.0	−26 20 26	MCG −4-31-13	13.7	0.4 x 0.3	11.3	Sc	24	
12 58 18.4	−25 11 17	MCG −4-31-14	14.3	0.9 x 0.4	13.1	SBc	71	Lies 1′.3 S of mag 8 star.
13 01 39.7	−23 59 17	MCG −4-31-20	14.4	0.5 x 0.5	13.0	E?		
13 02 44.8	−22 04 23	MCG −4-31-22	14.0	1.0 x 0.5	13.1	Sbc	9	
13 02 52.5	−23 55 20	MCG −4-31-23	13.6	1.2 x 0.4	12.7	Sm	75	
13 04 53.8	−24 50 33	MCG −4-31-24	13.8	0.5 x 0.3	11.6	Sa	123	
13 06 24.5	−24 09 54	MCG −4-31-27	13.9	0.8 x 0.4	12.5	Sab? pec	99	Mag 12.3 star SE 5′.3 ESO 508-7 ENE 10′.4.
13 07 37.8	−23 34 46	MCG −4-31-34	13.8	0.6 x 0.4	12.2	SAB0−	85	Mag 8.39v star E 6′.8.
13 09 19.2	−24 23 07	MCG −4-31-37	14.2	1.1 x 0.7	13.7	IB(s)m V-VI	128	
13 09 36.6	−27 08 31	MCG −4-31-38	14.2	0.7 x 0.4	12.7		135	Uniform surface brightness.
13 14 17.7	−26 35 00	MCG −4-31-42	13.7	0.6 x 0.3	11.9	E	0	
13 15 48.5	−24 15 22	MCG −4-31-44	14.5	0.7 x 0.4	12.9	SBc	18	Mag 9.73v star N 5′.0.
13 17 45.4	−24 19 12	MCG −4-31-46	16.4	0.5 x 0.5	15.0	E/S0		Lies due W 1′.8 from a mag 10 star.
13 18 55.1	−24 41 51	MCG −4-31-49	14.3	0.8 x 0.4	12.9	SBbc	78	Mag 8.27v star NE 7′.6.
13 25 14.0	−25 21 12	MCG −4-32-11	14.0	1.3 x 0.4	13.1	(R′)SA(s)ab	24	Lies 2′.0 N of a mag 10 star.
13 26 20.7	−22 37 55	MCG −4-32-14	13.8	1.3 x 1.1	14.1	(R′)SA(r)ab pec: III-IV	16	
13 26 35.3	−26 44 33	MCG −4-32-15	14.0	0.7 x 0.6	12.9	SAB(rs)bc II-III	124	
13 30 10.7	−24 08 43	MCG −4-32-22	14.1	0.5 x 0.5	12.5	SBbc		
13 30 36.4	−22 30 32	MCG −4-32-25	14.5	0.7 x 0.7	13.5	Sc		
13 31 13.9	−25 24 12	MCG −4-32-29	13.7	0.7 x 0.4	12.2	Sab	35	Mag 12.2 star SW 3′.8; mag 10.85v star SE 7′.6.
13 33 06.9	−22 36 00	MCG −4-32-37	12.7	0.7 x 0.5	11.6	E?	175	
13 33 48.3	−24 45 28	MCG −4-32-38	14.0	1.0 x 0.5	13.1	Sb	80	
13 34 39.3	−23 40 50	MCG −4-32-40	13.8	0.7 x 0.4	12.3	S?	10	**ESO 509-65** S 6′.6.
13 35 46.4	−23 15 41	MCG −4-32-43	15.0	0.7 x 0.7	14.1			
13 38 37.8	−24 08 53	MCG −4-32-46	13.6	0.9 x 0.5	12.6	Sc	164	**ESO 509-89** E 6′.1.
13 39 34.2	−22 29 54	MCG −4-32-48	14.0	0.9 x 0.7	13.4	Sc	66	
13 40 34.2	−21 55 49	MCG −4-32-51	13.8	1.0 x 0.8	13.4	S?	176	
13 46 23.6	−24 05 09	MCG −4-33-1	13.8	0.6 x 0.4	12.2	S0	18	
13 47 52.7	−25 55 34	MCG −4-33-3	14.5	0.5 x 0.5	12.9	SA(r)cd:		
13 49 56.0	−22 16 36	MCG −4-33-4	14.4	0.6 x 0.5	12.9	Sb	135	
13 51 14.5	−23 03 14	MCG −4-33-5	14.3	0.7 x 0.4	13.0	E/S0	45	
13 53 04.1	−24 42 18	MCG −4-33-6	14.4	0.5 x 0.5	12.7	SAB(r)d: III-IV		Mag 9.16v star W 1′.8.
13 54 26.5	−26 34 36	MCG −4-33-10	13.4	1.4 x 0.7	13.2	SAB(r)a:	68	Mag 7.63v star SW 12′.6; MCG −4-33-11 NE 2′.2.
13 54 34.7	−26 33 23	MCG −4-33-11	13.8	1.1 x 0.9	13.7	(R′)SB(s)0°:	170	MCG −4-33-10 SW 2′.2; MCG −4-33-12 N 2′.6.
13 54 35.3	−26 30 46	MCG −4-33-12	14.9	0.4 x 0.4	12.8			Mag 7.88v star NE 11′.0; MCG −4-33-11 S 2′.2.
13 56 13.4	−26 09 36	MCG −4-33-15	14.1	0.6 x 0.3	12.0		114	
13 56 20.9	−22 38 20	MCG −4-33-17	14.0	0.7 x 0.5	12.7	Sbc	45	
13 57 04.9	−24 48 29	MCG −4-33-18	14.1	0.5 x 0.4	12.4	E/S0	15	Mag 9.64v star E 4′.1.
13 57 22.1	−27 10 25	MCG −4-33-20	14.1	0.8 x 0.7	13.3	Sb	8	Pair of bright knots or stars W edge.
13 59 36.4	−24 22 06	MCG −4-33-22	13.4	1.0 x 0.7	12.9	S0	155	Mag 8.98v star NW 7′.9.
14 01 21.0	−26 11 50	MCG −4-33-28	14.2	1.3 x 0.9	14.2	SB(s)cd II-III	83	Close pair of stars, mags 13.1 and 13.8, SW 2′.3.
14 01 32.4	−25 15 05	MCG −4-33-30	14.4	1.0 x 0.7	13.9	S?	15	
14 01 59.2	−22 21 51	MCG −4-33-31	14.1	1.0 x 0.7	13.6	SAB(r)bc	36	
14 01 58.4	−25 32 23	MCG −4-33-32	13.4	1.2 x 1.1	13.6	(R′)SB(rs)b II	21	Lies between a pair of stars, mags 10.80v and 11.24v.
14 03 35.8	−25 25 44	MCG −4-33-35	13.3	1.1 x 0.5	12.6	SB(s)0−?	174	
12 54 41.2	−29 13 43	MCG −5-31-1	13.1	0.7 x 0.7	12.2	S?		
12 55 32.6	−28 27 34	MCG −5-31-2	13.9	1.0 x 0.2	12.0	Sa	80	Mag 9.63v star W 4′.0.
12 57 28.9	−28 12 19	MCG −5-31-6	13.5	0.9 x 0.4	12.2	SB(rs)cd:	6	
12 59 37.0	−29 15 33	MCG −5-31-8	14.2	0.7 x 0.7	13.5	E/S0		
13 05 46.9	−28 25 20	MCG −5-31-33	14.1	1.0 x 0.5	13.2	SB	99	ESO 443-59 SW 3′.3.
13 20 50.0	−29 28 49	MCG −5-32-1	13.5	1.2 x 0.9	13.4	S?	138	Star on NW edge.
13 22 15.5	−29 33 25	MCG −5-32-2	14.3	0.7 x 0.5	13.0		33	
13 25 42.5	−29 32 43	MCG −5-32-14	14.0	1.0 x 0.7	13.5	Sc	40	
13 27 00.8	−29 11 33	MCG −5-32-19	13.7	0.9 x 0.6	12.9	S0/a	53	
13 28 15.0	−27 58 37	MCG −5-32-27	13.8	0.6 x 0.6	12.6	S0/a		Mag 10.64v star W 5′.6; small, faint anonymous spiral galaxy NNW 4′.6.
13 28 36.3	−28 11 48	MCG −5-32-30	14.8	0.7 x 0.5	13.5	Sb	18	**MCG −5-32-29** WNW 1′.7.
13 44 09.6	−28 01 27	MCG −5-32-72	14.4	0.7 x 0.4	12.9		21	
13 47 18.7	−29 48 33	MCG −5-33-4	13.1	0.8 x 0.6	12.2	S0?	153	
13 53 15.0	−28 25 43	MCG −5-33-29	13.8	1.1 x 0.4	12.8	SB(r)0/a	134	Mag 11.8 star W 4′.0.
12 53 27.2	−17 00 22	NGC 4763	12.6	1.2 x 0.8	12.4	SB(r)a:	132	
12 56 12.4	−29 30 10	NGC 4806	12.8	1.2 x 1.0	12.8	SB(s)c? III	50	Several branching, patchy arms.
12 57 27.9	−19 41 30	NGC 4830	12.1	2.2 x 1.4	13.2	SAB0−	157	Round anonymous galaxy 1′.3 SE; mag 10.45v star S 2′.0.
12 57 36.5	−27 17 32	NGC 4831	12.5	1.7 x 0.9	12.8	SAB0−	178	
13 06 04.9	−29 45 17	NGC 4955	12.2	1.8 x 1.3	13.1	E2	23	
13 07 09.5	−28 13 43	NGC 4965	12.2	2.6 x 2.0	13.9	SAB(s)d III	136	Moderately bright nucleus, patchy, broad arms.
13 07 06.2	−23 40 38	NGC 4968	12.8	1.9 x 0.9	13.3	(R′)SAB0°	56	Bright center; mag 8.39v star ENE 15′.0; MCG −4-31-34 NE 9′.4.
13 07 33.8	−24 00 34	NGC 4970	12.2	2.0 x 1.1	13.0	S0°:	140	Bright, elongated core; very small, very faint anonymous galaxy 1′.2 SE; ESO 508-7 SW 8′.5.
13 09 10.5	−28 38 30	NGC 4980	12.9	1.7 x 0.8	13.1	SAB(rs)a pec?	168	Nucleus slight offset to the W.
13 09 47.8	−23 23 03	NGC 4993	12.4	1.3 x 1.1	12.6	(R′)SAB(rs)0−:	168	Mag 8.08v star ESE 5′.4.
13 09 25.8	−22 28 51	NGC 4994	12.4	0.8 x 0.2	10.3	(R′)SAB(rs)0-:	130	
13 11 45.7	−19 15 45	NGC 5006	12.4	2.0 x 1.7	13.5	(R)SB(r)0+	170	Star 0′.6 NW of nucleus.
13 13 00.6	−19 31 06	NGC 5018	10.8	3.3 x 2.5	13.1	E3:	112	Bright, diffuse center; mag 9.43v star NW 6′.2.
13 13 30.8	−19 32 47	NGC 5022	12.8	2.4 x 0.4	12.6	SBb pec sp	21	Mag 12 star 2′.4 N of center.
13 15 30.3	−23 59 01	NGC 5042	11.7	4.2 x 2.2	13.9	SAB(rs)c II	22	Bright nucleus with patchy arms. Mag 8.33v star 2′.6 W.
13 16 08.6	−28 24 37	NGC 5048	12.8	1.5 x 0.8	12.9	SA0−:	36	**ESO 444-3** SE 11′.4.
13 16 20.2	−28 17 08	NGC 5051	13.3	1.5 x 0.6	13.0	SAB(rs)bc: II-III	50	Very faint star on NE edge; mag 11.6 star and **MCG −5-31-40** W 11′.8.
13 18 05.2	−26 50 09	NGC 5061	10.4	3.5 x 3.0	12.9	E0		Located 2′.2 W of mag 8.96v star. **ESO 508-36** NW 3′.9; **ESO 508-35** SW 5′.5.
13 18 54.6	−21 02 18	NGC 5068	10.0	7.2 x 6.3	14.0	SAB(rs)cd III	110	Short, bright bar, many bright knots and patches.
13 19 49.6	−27 24 38	NGC 5078	11.0	4.0 x 1.9	13.0	SA(s)a: sp	148	Strong, dark lane runs whole length of galaxy cutting through middle of bright central lens; IC 879 SW 2′.3; mag 7.79v star E 9′.0.
13 20 17.6	−21 49 31	NGC 5084	10.5	9.3 x 1.7	13.4	S0 sp	80	Bright nucleus, very faint anonymous galaxy at E end; mag 8.17v star E 11′.5.
13 20 18.2	−24 26 22	NGC 5085	11.1	3.4 x 3.0	13.5	SA(s)c II	38	Bright nucleus; mag 8.44v star 4′.1 S; double stars NE.
13 20 24.9	−20 36 35	NGC 5087	11.4	2.7 x 2.2	13.2	SA0:	10	Elongated, diffuse nucleus.
13 21 46.2	−27 25 50	NGC 5101	10.7	5.4 x 4.6	14.0	(R)SB(rs)0/a	123	Bright, diffuse nucleus in short bar; several small, faint anonymous galaxies on SW edge; mag 10.12v star NNW 3′.3.
13 25 18.5	−21 08 01	NGC 5134	11.3	2.8 x 1.7	12.9	SA(s)b? III	155	Mag 9.67v star ENE 9′.3; IC 4237 W 10′.7.
13 25 44.1	−29 50 01	NGC 5135	12.1	2.6 x 1.8	13.6	SB(s)ab II	29	Bright nucleus in bright SE-NW bar, two main arms.

(Continued from previous page)

149 GALAXIES 149

RA h m s	Dec ° ′ ″	Name	Mag (V)	Dim ′ Maj x min	SB	Type Class	PA	Notes
13 27 36.5	−29 33 44	NGC 5150	12.6	1.3 x 1.0	12.7	SBbc: II	115	Located 1.′7 SW of mag 9.90v star.
13 27 50.0	−29 37 03	NGC 5152	12.4	2.0 x 0.6	12.4	SB(s)b	117	Interacting with NGC 5153; nucleus offset E; star N of nucleus.
13 27 54.5	−29 37 03	NGC 5153	11.8	2.1 x 1.4	13.0	E1 pec	175	Interacting with NGC 5152 W.
13 29 48.9	−17 57 55	NGC 5170	11.1	8.3 x 1.0	13.3	SA(s)c: sp I	127	Central dark lane; **ESO 576-78** N 3.′0.
13 30 41.1	−28 09 02	NGC 5182	12.4	1.9 x 1.3	13.3	(R′)SB(r)bc II	11	Bright nucleus; row of faint stars oriented N-S on E side; mag 6.49v star E 11.′7.
13 37 01.0	−29 51 44	NGC 5236	7.5	12.9 x 11.5	12.8	SAB(s)c II	44	= **M 83**. Galaxy **ESO 444-85** lies on E edge.
13 38 03.1	−17 53 03	NGC 5247	10.0	5.6 x 4.9	13.4	SA(s)bc I-II	20	Bright nucleus, very knotty arms.
13 40 20.0	−23 51 26	NGC 5260	12.8	1.6 x 1.4	13.5	SB(s)c I		Short bar with tightly wound arms.
13 41 36.7	−29 54 46	NGC 5264	12.0	2.5 x 1.5	13.3	IB(s)m IV-V	54	Very faint envelope, many stars superimposed.
13 52 53.3	−28 29 23	NGC 5328	11.6	1.7 x 1.3	12.5	E1:	87	Anonymous galaxy 1.′3 SW.
13 52 59.4	−28 28 17	NGC 5330	13.7	0.7 x 0.5	12.6	E	6	Anonymous galaxy N 1.′2.
14 00 32.0	−28 52 28	NGC 5393	13.0	0.9 x 0.7	12.3	(R′)SB(r)a pec:	153	

GALAXY CLUSTERS

RA h m s	Dec ° ′ ″	Name	Mag 10th brightest	No. Gal	Diam ′	Notes
12 57 12.0	−17 21 00	A 1644	15.7	68	30	Prominent mag 14 **PGC 44259** S of center 3.′8; numerous other anonymous/PGC GX, stellar to almost stellar.
13 18 42.0	−21 27 00	A 1709	16.4	43	16	**ESO 576-28** SE of center 4.′6; several anonymous, almost stellar GX near center.
13 26 54.0	−27 06 00	A 1736	14.8	41	39	Members appear mostly spread along NW-SE line with a concentration on SE edge.
12 54 18.0	−29 01 00	A 3528	16.3	70	17	**ESO 443-2** near N edge; includes a number of PGC galaxies.
13 20 48.0	−28 58 00	A 3555	16.4	61	17	All members anonymous, faint, stellar.
13 29 54.0	−29 31 00	A 3559	15.7	141	17	See notes for ESO 444-55; numerous other anonymous, stellar/almost stellar GX
13 54 18.0	−27 50 00	A 3577	15.8	103	17	**PGC 49355** WSW of center 12.′1.
13 57 30.0	−24 43 00	A 3578	15.1	52	22	**ESO 510-25** at center; **ESO 510-28** NE 7.′4; **ESO 510-27** N 8.′3; most others anonymous.
12 51 12.0	−22 32 00	AS 713	16.4	3	17	**MCG −4-30-27** W of center 5.′9; **ESO 575-12** NNE of center 17.′3 ; most members anonymous, stellar.
12 51 30.0	−26 27 00	AS 714	13.8		56	
13 03 36.0	−19 31 00	AS 719	16.3	3	17	**ESO 575-51** SW of center 8.′8; all others anonymous, stellar.
13 28 30.0	−20 56 00	AS 735	15.7	21	17	See notes MCG −3-34-81; most others anonymous, stellar.
13 44 00.0	−19 48 00	AS 741	16.4	10	17	See notes for MCG −3-35-16; all others anonymous, stellar.

GLOBULAR CLUSTERS

RA h m s	Dec ° ′ ″	Name	Total V m	B ★ V m	HB V m	Diam ′	Conc. Class Low = 12 High = 1	Notes
13 56 21.0	−27 09 42	AM 4	15.9	20.5		3.0		
12 39 28.0	−26 44 34	NGC 4590	7.3	12.6	15.6	11.0	10	= **M 68**.

PLANETARY NEBULAE

RA h m s	Dec ° ′ ″	Name	Diam ″	Mag (P)	Mag (V)	Mag cent ★	Alt Name	Notes
12 53 41.4	−22 51 42	PN G303.6+40.0	770	12.0	12.7	9.6	PK 303+40.1	= **Abell 35**; Mag 11.3 star near center with mag 9.69v star w of center 2.′6; galaxy **ESO 507-38** located 3.′0 N of the mag 9.69v star.
13 40 41.3	−19 52 57	PN G318.4+41.4	370	13.0		11.5	PK 318+41.1	Mag 13.1 star W 4.′0; mag 10.37v star NE 9.′7.

150 GALAXIES 150

RA h m s	Dec ° ′ ″	Name	Mag (V)	Dim ′ Maj x min	SB	Type Class	PA	Notes
11 17 18.2	−27 49 23	ESO 438-17	13.3	1.8 x 0.6	13.2	SB(s)c III-IV	15	Mag 11.03v star on NW edge; mag 9.82v star W 2.′5.
11 17 33.3	−27 53 43	ESO 438-18	14.2	1.2 x 0.5	13.5	SB?	10	ESO 438-17 NW 5.′6.
11 20 53.2	−29 24 10	ESO 438-23	13.1	1.5 x 0.9	13.3	S0?	38	Mag 11.7 star N 3.′9.
11 27 23.8	−29 15 29	ESO 439-9	13.6	1.9 x 0.5	13.4	(R)SB(r)a pec:	99	ESO 439-10 N 5.′3.
11 27 32.3	−29 10 38	ESO 439-10	14.7	1.6 x 0.5	14.3	Pec sp	145	Mag 12.3 star E 2.′4; mag 10.55v star NW 8.′6; ESO 439-9 S 5.′3.
11 38 15.8	−29 43 44	ESO 439-20	13.4	1.7 x 0.7	13.4	SAB(r)bc II-III	99	Mag 14.7 star N edge; mag 12.6 star S 1.′1.
11 45 41.8	−28 21 59	ESO 440-4	13.8	2.8 x 1.3	15.0	SB(s)dm IV-V	69	Mag 8.46v star W 3.′4.
11 48 45.5	−28 17 40	ESO 440-11	12.3	2.8 x 2.1	14.1	SB(s)d: III-IV	176	Mag 13.1 star E edge; mag 13.4 star SE edge; mag 7.58v star W 11.′5.
11 53 23.8	−28 33 12	ESO 440-27	12.4	4.6 x 0.6	13.4	Sd: sp II-III	77	Mag 9.56v star S 5.′9.
11 59 17.0	−28 54 19	ESO 440-37	13.1	1.5 x 1.5	13.9	(R′)SA(rs)0°		Mag 9.45v star SE 3.′7.
12 02 17.9	−29 30 54	ESO 440-43	14.1	0.8 x 0.5	13.0	Sc	91	
12 02 46.0	−29 05 36	ESO 440-44	13.3	1.7 x 0.8	13.5	SB(s)m pec IV	0	Faint star superimposed W of center; mag 12.7 star NE 3.′0.
12 04 41.0	−28 07 01	ESO 440-46	12.9	2.3 x 1.7	14.3	SB(s)m: V	38	Numerous stars along W half.
12 06 42.5	−28 20 02	ESO 440-55	14.5	1.5 x 0.3	13.5	Sc	75	Mag 12.4 star N 3.′0; **ESO 440-57** NNE 4.′9.
12 07 12.7	−29 56 18	ESO 441-1	14.2	1.0 x 0.2	12.3	S0	46	Located between a pair of mag 14.9 and 15.6 stars.
12 09 25.1	−28 48 05	ESO 441-11	14.8	1.6 x 0.2	13.4	Sc	136	Mag 13.4 star N 0.′6; mag 13.3 star W 1.′7.
12 33 22.2	−28 43 39	ESO 442-2	15.6	1.7 x 0.2	14.2	Sc:	8	Mag 8.36v star NNE 9.′1.
12 37 13.6	−28 29 40	ESO 442-13	12.8	2.0 x 0.8	13.1	SB(rs)cd III-IV	174	Mag 8.44v star NE 6.′2.
12 42 42.5	−29 37 14	ESO 442-14	13.9	0.7 x 0.5	12.7	Sb	20	Mag 9.34v star SW 6.′0.
11 17 44.5	−26 02 25	ESO 503-11	14.9	0.9 x 0.6	14.0	SAB(s)cd IV	99	Mag 8.87v star SW 10.′8; NGC 3617 S 5.′8.
11 33 29.6	−26 56 51	ESO 503-22	14.0	1.7 x 0.7	14.0	IB(s)m: pec III-IV	117	Mag 8.38v star NNE 9.′2.
11 35 30.9	−25 08 42	ESO 504-1	13.9	1.0 x 0.8	13.6	(R)SA0/a pec	43	Very close pair of mag 11.00v, 12.4 stars ESE 8.′1.
11 36 22.9	−24 00 45	ESO 504-3	13.9	0.7 x 0.5	12.6	Sab	127	Mag 11.7 star SW 2.′1.
11 36 54.3	−24 35 41	ESO 504-4	14.2	0.9 x 0.4	12.9	Sa	15	Mag 14.2 star S 1.′8.
11 38 10.1	−23 25 44	ESO 504-5	14.2	0.9 x 0.4	12.9	Sb	94	Mag 13.1 star E 1.′9.
11 43 02.8	−23 26 04	ESO 504-10	15.1	1.1 x 0.5	14.3	SB(s)m: V	21	Mag 10.67v star NNE 4.′4.
11 48 46.4	−27 22 44	ESO 504-17	13.4	0.8 x 0.6	12.5	S	104	Mag 8.36v star S 8.′2.
11 51 42.4	−27 44 26	ESO 504-19	14.0	0.9 x 0.4	12.8		62	Mag 10.53v star NE 5.′5.
11 53 37.8	−26 59 48	ESO 504-24	14.6	1.3 x 1.0	14.7	SB(s)m VI	153	Mag 11.04v star SW 2.′7; mag 9.53v star SE 9.′7.
11 53 50.1	−27 21 02	ESO 504-25	13.9	1.5 x 1.3	14.5	SB(s)m V-VI	14	Mag 8.82v star W 10.′2.
11 54 01.4	−23 49 07	ESO 504-27	14.0	1.0 x 0.4	12.9	Sa?	113	Mag 9.57v star SW 7.′9.
11 54 54.7	−27 15 00	ESO 504-28	13.2	1.4 x 1.3	13.7	SAB(rs)d: III-IV	90	Mag 9.53v star NW 13.′7.
11 57 15.0	−27 42 00	ESO 504-30	14.1	1.1 x 1.0	14.0	SB(r)dm:	130	Mag 12.0 star W 2.′7.
12 00 31.6	−24 43 22	ESO 505-2	13.4	1.3 x 1.2	13.7	IAB(s)m: IV	90	Very close pair of mag 10.66v, 11.58v stars NW 2.′7.
12 01 07.7	−24 34 12	ESO 505-3	13.3	2.6 x 0.5	13.4	Sdm? sp	128	Mag 8.73v star NW 11.′2.
12 03 30.5	−25 28 32	ESO 505-7	16.9	2.2 x 1.7	18.2	IB(s)m VI	158	Mag 10.72v star SW 3.′3; ESO 505-8 N 5.′9.

(Continued from previous page)

GALAXIES

RA h m s	Dec ° ′ ″	Name	Mag (V)	Dim ′ Maj x min	SB	Type Class	PA	Notes
12 03 35.0	−25 22 54	ESO 505-8	14.3	1.6 x 0.4	13.7	S?	11	ESO 505-7 S 5′.9.
12 03 51.1	−27 36 07	ESO 505-9	14.5	1.6 x 0.5	14.1	SA(s)c: II-III	77	Mag 14.7 star on N edge; mag 10.23v star ESE 13′.3.
12 05 59.3	−27 00 55	ESO 505-12	14.2	0.7 x 0.4	12.7	S0:	157	Mag 10.68v star E 3′.8.
12 06 07.3	−22 51 00	ESO 505-13	12.6	2.4 x 2.1	14.2	SB(s)m pec: IV	0	Mag 10.91v star on W edge.
12 07 05.4	−27 41 47	ESO 505-14	12.9	1.1 x 0.9	12.8	SA0$^-$	80	Pair of E-W oriented mag 9.35v, 9.01v stars N 9′.3.
12 07 08.0	−25 41 34	ESO 505-15	13.0	1.0 x 0.7	12.7	E$^+$3:	108	Mag 8.87v star NW 13′.9.
12 09 36.0	−23 24 49	ESO 505-23	14.6	1.7 x 0.5	14.3	SB(s)m IV-V	47	Mag 9.19v star NW 4′.2.
12 20 10.1	−26 04 03	ESO 506-2	13.5	2.6 x 0.3	13.1	Sbc: sp	3	Mag 8.41v star WSW; note: IC 3152 located 3′.1 SE of this star.
12 21 49.5	−25 04 35	ESO 506-3	14.2	1.6 x 0.5	13.8	Sab: sp	80	Mag 10.63v star E 6′.0, part of a small asterism of six mag 10-12 stars.
12 21 49.3	−24 10 08	ESO 506-4	12.4	2.1 x 1.0	13.1	SAB(r)ab: II	88	Mag 10.03v star SW 4′.9; mag 7.02v star SE 9′.5.
12 38 54.6	−27 18 28	ESO 506-27	13.4	1.2 x 0.5	12.7	S0$^+$: pec sp	72	Mag 9.92v star WNW 5′.8; several faint anonymous galaxies in the area E and W.
12 39 50.9	−23 04 40	ESO 506-32	13.9	0.7 x 0.5	12.6	S?	127	Mag 13.1 star NNE 2′.5.
12 40 13.9	−25 19 39	ESO 506-33	12.0	1.9 x 0.6	12.0	S0°: sp	177	Mag 9.49v star N 5′.7.
11 20 12.1	−21 28 15	ESO 570-19	13.7	1.3 x 0.9	13.7	(R′)SAc: pec III-IV	72	Mag 8.18v star WNW 10′.9.
11 38 57.5	−17 58 08	ESO 571-12	14.2	1.5 x 0.5	13.7	S?	131	Mag 10.79v star N 2′.3.
11 40 58.9	−22 28 41	ESO 571-15	14.2	1.7 x 0.2	12.9	Sb sp	37	Mag 11.17v star E 3′.2; mag 9.22v star NW 10′.6.
11 42 09.7	−18 10 10	ESO 571-16	13.4	1.8 x 0.4	12.9	SB(s)bc: sp I-II	96	Mag 10.13v star E 6′.5; note: **ESO 571-17** located 3′.9 N of this star.
11 52 27.9	−20 06 14	ESO 572-7	13.8	1.0 x 0.5	12.9	S0	59	Mag 8.85v star N 8′.1.
11 53 05.6	−18 22 42	ESO 572-8	13.9	0.8 x 0.5	12.8	S?	105	Mag 11.25v star E 2′.0.
11 53 22.9	−18 10 00	ESO 572-9	16.5	1.8 x 1.4	17.3	IB(s)m VI	124	Mag 8.05v star NW 8′.7.
11 55 51.0	−18 11 45	ESO 572-18	13.2	1.6 x 0.8	13.3	SB(r)c pec: III-IV	36	Mag 9.45v star NW 10′.0.
11 56 22.5	−19 33 09	ESO 572-22	14.1	1.5 x 0.4	13.4	(R)SBd pec? IV-V	43	Mag 15.0 star on NE end; mag 11.00v star E 7′.7.
11 56 58.0	−19 51 17	ESO 572-23	13.0	1.4 x 1.2	13.4	SAB(s)0°:	148	Stellar galaxy **1SZ57** SSW 1′.1; mag 8.88v star NW 5′.8.
11 56 59.6	−19 59 15	ESO 572-24	14.9	1.4 x 0.4	14.2	SBd? IV-V	114	NGC 3981 WNW 13′.5; mag 10.05v star S 6′.9.
11 57 28.0	−19 37 29	ESO 572-25	14.3	1.1 x 0.4	13.2	S0	4	Mag 10.28v star SW 2′.0.
11 58 25.4	−22 26 27	ESO 572-30	13.3	2.8 x 1.4	14.6	SB(s)m: V	111	Superimposed star or bright knot W of center; close pair of mag 10.71v, 11.9 stars W 4′.1.
11 58 45.5	−19 50 43	ESO 572-32	16.3	1.9 x 1.2	17.1	SB(s)d pec IV-V	87	Double system, diffuse, extended envelope.
11 59 27.7	−20 26 50	ESO 572-35	13.9	1.0 x 0.3	12.4	Sa	100	Mag 12.9 star NE 1′.3.
12 01 08.5	−20 29 19	ESO 572-44	13.8	1.8 x 0.3	13.0	Sb sp	141	MCG −3-31-13 SE 2′.4; mag 8.49v star SSW 9′.3.
12 03 24.1	−19 31 20	ESO 572-49	14.0	2.0 x 0.4	13.6	SB(s)cd: sp III-IV	70	Mag 9.41v star SW 12′.5; very faint, stellar anonymous galaxy W 4′.0.
12 07 40.8	−18 04 30	ESO 573-2	14.3	0.9 x 0.8	13.8	SA(s)c I-II	8	Stellar nucleus; mag 7.07v star SSW 7′.7.
12 12 55.0	−20 25 13	ESO 573-3	14.4	1.7 x 0.8	14.6	IB(s)m V	135	Star on W edge.
12 17 44.9	−22 10 27	ESO 573-6	15.6	1.5 x 0.6	15.3	Sc:	137	Mag 12.4 star N 3′.6.
12 19 06.5	−22 06 56	ESO 573-11	14.9	1.3 x 0.9	14.9	SAB(s)cd III-IV	162	Mag 12.9 star W 1′.7; mag 5.95v star E 15′.2.
12 20 37.2	−18 39 59	ESO 573-12	13.5	2.0 x 0.6	13.5	(R′)SAB(rs)bc? I-II	153	Mag 12.5 star SE 1′.9.
12 21 15.9	−21 59 46	ESO 573-14	14.4	1.0 x 0.5	13.5	S?	98	Mag 15.2 star NE of center; mag 9.74v star SE 8′.1; mag 5.21v star Seta Corvi SSW 16′.4.
12 22 24.0	−21 11 39	ESO 573-16	14.1	1.6 x 0.3	13.1	SAB(r)a? sp	63	Mag 10.03v star SSE 5′.6.
12 22 29.3	−22 20 44	ESO 573-17	14.4	1.7 x 0.9	14.7	SAB(r)a?	70	Mag 9.95v star S 6′.3.
12 26 45.0	−22 00 51	ESO 573-22	16.3	1.6 x 0.3	15.3	S	8	Very small, faint, elongated anonymous galaxy on E edge; mag 9.22v star E 7′.4.
12 43 34.0	−20 50 39	ESO 574-24	13.6	1.7 x 0.8	13.8	(R′)SB(rs)ab: II	141	**ESO 574-23** N 6′.0; mag 11.9 star NNE 5′.9.
12 05 53.4	−29 17 28	IC 760	12.5	1.7 x 0.5	12.2	S0°: sp	148	Mag 10.06v star NE 8′.8.
12 10 14.2	−29 44 09	IC 764	12.0	4.8 x 1.6	14.1	SA(s)c? II-III	177	Mag 8.94v star N 7′.8.
12 42 08.5	−17 20 58	IC 806	14.4	0.7 x 0.6	13.3	(R′)SB(r)ab pec? II	135	Mag 11.8 star and small anonymous galaxy SW 0′.9; mag 10.64v star W 9′.2; IC 807 S 3′.4.
12 42 12.5	−17 24 13	IC 807	13.3	0.7 x 0.7	12.5	E/S0		IC 806 N 3′.4.
11 27 05.1	−28 58 50	IC 2764	12.2	1.6 x 1.4	13.0	(R)SA(rs)0/a	0	Mag 13.4 star N edge; mag 8.98v star E 5′.7.
12 05 47.1	−27 56 25	IC 2995	12.2	3.2 x 1.0	13.3	SB(s)c: III	117	Mag 9.55v star E 5′.3.
12 05 48.5	−29 58 22	IC 2996	13.5	1.5 x 0.4	12.8	S?	21	Star on S end.
12 19 35.8	−26 08 44	IC 3152	12.5	1.8 x 1.2	13.2	SA0$^-$	52	Mag 8.41v star NW 3′.1; ESO 506-2 NE 9′.0.
12 24 57.6	−26 01 50	IC 3289	13.0	1.0 x 1.0	12.8	S0?		Mag 6.85v star NE 4′.6.
11 22 33.0	−17 34 22	MCG −3-29-6	14.0	0.6 x 0.4	12.3	Sb	138	
11 31 01.9	−18 26 14	MCG −3-29-9	14.3	0.7 x 0.6	13.2	Sc	30	
11 37 59.5	−17 14 03	MCG −3-30-3	14.2	2.4 x 0.3	13.7	Sc: sp II	6	Mag 10.99v star N 4′.7.
11 38 22.1	−21 05 49	MCG −3-30-5	14.2	0.7 x 0.4	12.7	SAB(rs)a	30	Round, anonymous galaxy S 1′.8.
11 42 50.9	−19 04 09	MCG −3-30-9	14.2	0.7 x 0.3	12.3	Sb:	30	
11 44 18.2	−18 12 08	MCG −3-30-11	14.4	0.7 x 0.4	12.8	SB(rs)b	117	Located 11′.1 NW of mag 4.71v star Zeta Crateris.
11 53 57.2	−20 08 20	MCG −3-30-15	14.8	1.4 x 0.4	14.1	S	141	**ESO 572-10** W 2′.4.
11 54 19.8	−18 57 03	MCG −3-30-18	14.0	0.5 x 0.5	12.5	E		
11 58 46.5	−20 19 59	MCG −3-31-5	14.4	0.7 x 0.5	13.1	Sb	30	Very faint, slightly elongated, anonymous galaxy W 0′.5.
11 58 58.4	−19 01 46	MCG −3-31-6	13.6	0.9 x 0.5	12.5	IBm III-IV	26	
12 00 31.4	−21 19 20	MCG −3-31-9	14.0	0.7 x 0.7	13.1	SB(r)0/a		Mag 8.16v star NE 7′.4.
12 01 17.5	−20 30 36	MCG −3-31-13	14.0	0.7 x 0.4	12.5	SB(rs)ab	120	ESO 572-44 NW 2′.4; mag 8.49v star SSW 9′.1.
12 06 38.8	−19 37 22	MCG −3-31-21	14.1	0.9 x 0.6	13.3	SA(s)bc:	81	
12 18 02.1	−20 51 13	MCG −3-31-25	14.2	0.9 x 0.7	13.6	Sb	6	Lies between a pair of mag 7 and 9 stars.
12 18 59.0	−20 04 54	MCG −3-31-26	13.6	0.7 x 0.5	12.3	SA(s)bc	12	
12 19 59.6	−17 23 36	MCG −3-32-3	13.7	0.7 x 0.5	12.7	E	126	
12 20 13.4	−17 06 51	MCG −3-32-4	14.2	1.5 x 0.3	13.2	SBm: IV-V	33	Uniform surface brightness; mag 11.6 star NE 1′.9.
12 25 47.8	−21 44 48	MCG −3-32-8	13.9	1.1 x 0.9	13.7	SB(r)a: II-III	39	
12 29 36.0	−18 23 13	MCG −3-32-9	14.0	1.0 x 0.6	13.3	SB(rs)a	135	
12 32 03.5	−20 34 39	MCG −3-32-11	14.4	0.8 x 0.4	13.0	Sb	75	
12 36 24.4	−19 23 44	MCG −3-32-15	13.0	0.8 x 0.8	12.6	E		
12 37 13.5	−20 03 06	MCG −3-32-16	13.4	0.7 x 0.4	12.0	E$^+$:	75	Lies 1′.6 N of mag 9 star.
12 38 00.5	−20 07 53	MCG −3-32-17	14.4	0.7 x 0.4	13.0	E/S0	105	
12 40 35.0	−20 33 47	MCG −3-32-18	13.7	0.7 x 0.5	12.6	E$^+$	21	
11 16 51.5	−23 59 25	MCG −4-27-7	14.9	0.4 x 0.4	12.8			Stellar, appears with three stars forming a dipper like asterism.
11 23 41.0	−22 16 15	MCG −4-27-9	14.3	0.7 x 0.4	12.8	S?	60	
11 34 29.1	−26 52 11	MCG −4-27-13	14.3	0.7 x 0.4	12.7	Sb	174	Mag 13.6 star on E edge; mag 8.38v star WNW 9′.2.
11 34 53.4	−27 24 27	MCG −4-28-1	14.3	0.6 x 0.6	13.0			
11 39 08.5	−23 18 11	MCG −4-28-2	13.8	0.9 x 0.7	13.1	S?	13	
11 58 11.1	−21 48 27	MCG −4-28-7	14.0	1.5 x 0.2	12.6	Sbc	18	Mag 10.51v star NNE 7′.0.
12 01 29.8	−23 19 00	MCG −4-29-2	14.5	0.8 x 0.6	13.6	SBc	134	
12 09 26.5	−24 17 09	MCG −4-29-9	14.4	0.7 x 0.5	13.1	SB(rs)b	174	
12 09 24.5	−26 13 25	MCG −4-29-10	14.7	0.9 x 0.4	13.4	Sc	123	
12 12 13.6	−27 06 46	MCG −4-29-12	14.2	0.7 x 0.7	13.3	S0/a		Mag 7.63v star W 9′.1.
12 15 12.8	−27 39 39	MCG −4-29-13	14.6	0.8 x 0.4	13.2	SBd	36	Mag 8.96v star NE 4′.0.
12 16 17.9	−26 39 28	MCG −4-29-14	13.6	1.0 x 0.5	12.8	S0	54	Mag 8.27v star NNW 6′.9.
12 16 57.3	−26 12 35	MCG −4-29-15	14.1	0.6 x 0.4	12.4	Sa	148	**ESO 505-31** E 0′.7.
12 18 36.2	−27 27 26	MCG −4-29-17	13.5	0.9 x 0.3	11.9	Sc	86	
12 31 17.0	−26 17 20	MCG −4-30-3	13.4	1.1 x 0.7	13.0	SBb	176	
12 34 07.4	−26 17 20	MCG −4-30-5	13.7	1.3 x 0.5	13.1	Sb	125	

(Continued from previous page)

150 GALAXIES 150

RA h m s	Dec ° ′ ″	Name	Mag (V)	Dim ′ Maj x min	SB	Type Class	PA	Notes
12 39 15.7	−26 54 37	MCG −4-30-9	13.5	1.1 x 0.7	13.0	SB(s)cd IV	14	Located 10′.3 S of the center of M 68 (NGC 4590).
12 41 02.9	−25 57 34	MCG −4-30-11	14.6	1.0 x 0.6	13.9	Sc	69	Anonymous galaxies lie E 1′.4 and SSW 1′.4.
12 42 40.3	−23 46 37	MCG −4-30-12	14.3	0.8 x 0.5	13.2	Sa	92	Mag 8.34v star SE 6′.9.
11 18 53.6	−29 25 29	MCG −5-27-9	13.7	0.9 x 0.5	12.7	S?	152	**ESO 438-21** on E edge.
12 03 54.9	−29 37 33	MCG −5-29-4	14.0	0.7 x 0.5	12.7		105	**ESO 440-45** NE 5′.1.
12 06 18.8	−28 11 51	MCG −5-29-11	14.3	0.7 x 0.5	13.0	(R′$_2$)SB(s)b	171	MCG −5-29-12 NE 2′.5.
12 06 28.0	−28 10 29	MCG −5-29-12	14.3	0.6 x 0.4	12.6	SA:(rs)bc	177	Elongated anonymous galaxy E 2′.3; MCG −5-29-11 SW 2′.5.
11 17 50.9	−26 08 04	NGC 3617	12.8	1.8 x 1.3	13.7	E^+	147	Mag 9.51v star SSW 4′.6; ESO 503-11 N 5′.8.
11 25 12.7	−26 44 09	NGC 3673	11.5	3.6 x 2.4	13.7	SB(rs)b II	73	A few very faint stars sprinkled on surface; a mag 11.43v star on E edge.
11 46 46.7	−27 55 21	NGC 3885	11.9	2.4 x 1.0	12.7	SA(s)0/a III	123	Star on SE end.
11 49 13.1	−29 16 29	NGC 3904	10.9	2.7 x 1.9	12.6	E2-3:	8	
11 51 01.4	−28 48 22	NGC 3923	9.8	5.9 x 3.9	13.2	E4-5	50	Numerous stars superimposed around the bright nucleus.
11 52 20.1	−26 54 22	NGC 3936	12.1	3.9 x 0.6	12.9	SB(s)bc: sp II	63	Mag 9.83v star N 4′.7.
11 53 57.2	−23 09 42	NGC 3955	12.0	2.9 x 0.9	12.8	S0/a pec	165	Mag 9.75v star NE 4′.8.
11 54 00.8	−20 33 59	NGC 3956	12.0	3.4 x 1.0	13.2	SA(s)c: II-III	58	
11 54 01.4	−19 34 00	NGC 3957	11.8	3.1 x 0.7	12.5	SA0^+: sp	173	
11 55 09.3	−18 55 42	NGC 3969	12.9	1.4 x 0.9	13.0	SA(r)0^+:	64	Mag 10.67v star N 3′.6; stellar galaxy **1SZ 18** N 5′.7.
11 56 07.3	−19 53 53	NGC 3981	11.2	5.2 x 2.3	13.7	SA(rs)bc I-II	15	Very patchy, very bright inner arms, faint distorted outer arms; mag 9.79v star NW 4′.3.
11 58 31.2	−18 20 48	NGC 4024	11.7	1.9 x 1.5	12.7	SAB0^-	70	Small, faint anonymous spiral galaxy NE 1′.5.
11 59 30.7	−19 15 56	NGC 4027	11.1	3.2 x 2.4	13.2	SB(s)dm III-IV	167	One main, asymmetrical knotty arm; NGC 4027A S 4′.2.
11 59 29.5	−19 19 57	NGC 4027A	14.5	0.9 x 0.6	13.7	IB(s)m: V-VI	159	Located 4′.2 S of NGC 4027; faint, stellar anonymous galaxy E 2′.1.
12 00 34.6	−17 50 37	NGC 4033	11.7	2.6 x 1.1	12.8	E6	47	**ESO 572-40** NW 4′.2.
12 01 52.9	−18 52 05	NGC 4038	10.3	3.4 x 1.7	12.0	SB(s)m pec II-III:	94	**Antennae.** Bright complex center with dark markings and very bright, knotty arms. Colliding system with NGC 4039.
12 01 54.3	−18 53 31	NGC 4039	10.3	3.2 x 2.1	12.2	SA(s)m pec III	62	Very long, very faint extensions go N and S from E end where it joins with NGC 4038.
12 05 35.5	−26 31 23	NGC 4087	12.1	2.1 x 1.7	13.4	SA0^-:	39	Faint stars superimposed on envelope S and E edges.
12 06 40.9	−29 45 39	NGC 4105	10.7	2.7 x 2.0	12.5	E3	151	Interacting with NGC 4106 E; Mag 11.03v star S 2′.3.
12 06 45.8	−29 46 07	NGC 4106	11.4	1.6 x 1.3	12.0	SB(s)0^+	77	Interacting with NGC 4105 W.
12 29 21.4	−23 10 01	NGC 4462	11.8	3.2 x 1.2	13.1	SB(r)ab II	124	Short bar SE-NW, very diffuse arms.

GALAXY CLUSTERS

RA h m s	Dec ° ′ ″	Name	Mag 10th brightest	No. Gal	Diam ′	Notes
12 06 42.0	−28 16 00	AS 687	16.3		17	**ESO 440-57** at center; most members anonymous, stellar.

OPEN CLUSTERS

RA h m s	Dec ° ′ ″	Name	Mag	Diam ′	No. ★	B ★	Type	Notes
12 34 06.0	−29 25 00	ESO 442-4		10			cl	

GLOBULAR CLUSTERS

RA h m s	Dec ° ′ ″	Name	Total V m	B ★ V m	HB V m	Diam ′	Conc. Class Low = 12 High = 1	Notes
12 39 28.0	−26 44 34	NGC 4590	7.3	12.6	15.6	11.0	10	= **M 68**.

PLANETARY NEBULAE

RA h m s	Dec ° ′ ″	Name	Diam ″	Mag (P)	Mag (V)	Mag cent ★	Alt Name	Notes
12 24 30.8	−18 47 02	NGC 4361	126	10.3	10.9	13.2	PK 294+43.1	Bright center of about 65″ diameter with fainter outer halo to full dimensions.
12 33 13.1	−27 48 58	PN G298.0+34.8	7	17.1		16.8	PK 298+34.1	Located 0′.4 W of mag 11.9 star.

151 GALAXIES 151

RA h m s	Dec ° ′ ″	Name	Mag (V)	Dim ′ Maj x min	SB	Type Class	PA	Notes
10 04 03.9	−27 20 01	Antlia Dwarf	14.0	2.0 x 1.5	15.0		153	Mag 8.41v star E 9′.5.
09 57 48.2	−28 30 26	ESO 435-14	13.4	2.5 x 0.3	13.0	Sc: sp	54	
09 58 46.5	−28 37 18	ESO 435-16	12.6	1.9 x 1.3	13.4	I0? pec	118	Mag 14.4 star on N edge; mag 11.14v star NW 8′.2.
09 59 21.2	−28 08 00	ESO 435-20	13.6	0.9 x 0.5	12.6		92	Mag 10.83v star S 6′.8.
10 07 03.3	−28 32 32	ESO 435-43	13.0	0.5 x 0.4	11.2	S0:	54	Mag 11.5 star superimposed center.
10 12 47.6	−27 50 26	ESO 436-1	13.4	2.5 x 0.3	12.9	Sbc: sp	137	Mag 10.88v star superimposed near center.
10 25 49.0	−29 36 08	ESO 436-19	13.8	0.9 x 0.9	13.4	Sa:		Slightly elongated N-S core, smooth envelope.
10 31 25.3	−29 57 14	ESO 436-31	14.1	1.0 x 0.6	13.4	SB(s)d IV	21	Mag 10.39v star E 2′.3.
10 32 44.5	−28 36 40	ESO 436-34	13.3	2.5 x 0.4	13.1	Sb sp	60	Mag 9.81v star NW 12′.5; **PGC 31091** W 10′.9.
10 34 47.6	−28 29 50	ESO 436-44	12.9	1.2 x 0.8	12.7	S0?	101	**ESO 436-45** SE 1′.2; mag 10.38v star W 3′.2.
10 34 50.5	−28 35 02	ESO 436-46	12.6	2.4 x 1.7	14.0	SB(rs)bc I	138	Star superimposed just N of center; **PGC 31288** W 3′.3; MCG −5-25-17 E 2′.7.
10 35 23.4	−28 18 52	ESO 437-4	13.1	1.5 x 0.9	13.2	SAB(r)bc pec: II	137	Mag 10.47v star SW 5′.0; stellar **PGC 31401** E 6′.8.
10 36 34.8	−28 12 56	ESO 437-9	13.8	1.1 x 0.4	12.9	E?	15	Mag 11.06v star SW 3′.1.
10 36 57.9	−28 10 42	ESO 437-15	12.6	2.5 x 0.5	12.7	SB(s)0°: sp	31	Mag 11.06v star SW 3′.0.
10 38 10.7	−28 47 02	ESO 437-21	12.9	1.3 x 1.3	13.3	S0?		**ESO 437-27** E 5′.2.
10 38 17.9	−28 53 14	ESO 437-22	14.1	1.7 x 0.5	13.8	SBbc: sp	164	MCG −5-25-26 W 5′.0.
10 39 22.1	−29 35 06	ESO 437-31	13.9	1.5 x 0.9	14.1	SAB(rs)d	123	Mag 10.39v star NE 3′.4.
10 41 42.4	−28 46 45	ESO 437-44	13.2	2.2 x 2.0	14.6	SA(s)c	21	ESO 437-45 E 3′.9; note: stellar **PGC 31868** located between them; **ESO 437-39** W 9′.4.
10 41 59.8	−28 46 34	ESO 437-45	13.3	1.6 x 1.2	13.9	SA0° pec	82	Mag 9.39v star SE 6′.2; note: very faint, elongated anonymous galaxy on N edge of this star.
11 08 58.6	−28 22 18	ESO 438-5	14.1	3.2 x 0.6	14.6	SB(s)m: sp V	61	Mag 8.48v star NE 7′.9.
11 10 48.1	−28 30 07	ESO 438-9	13.4	0.9 x 0.7	12.8	(R′)SB(r)ab pec	44	Mag 11.12v star NNW 5′.3.
11 10 52.0	−27 53 51	ESO 438-10	14.2	1.6 x 0.7	14.1	SAB(s)m IV-V	12	Mag 9.98v star W 5′.5.
11 14 01.6	−29 57 16	ESO 438-14	14.0	0.3 x 0.3	11.3			Compact, almost stellar; mag 14.6 star on N edge.
11 15 03.4	−28 23 31	ESO 438-15	12.9	1.9 x 0.9	13.3	SB(rs)bc II	35	Mag 8.67v star NW 6′.4.
11 17 18.2	−27 49 23	ESO 438-17	13.3	1.8 x 0.6	13.2	SB(s)c III-IV	15	Mag 11.03v star on NW edge; mag 9.82v star W 2′.5.
11 17 33.3	−27 53 43	ESO 438-18	14.2	1.2 x 0.5	13.5	SB?	10	ESO 438-17 NW 5′.6.
11 20 53.2	−29 24 10	ESO 438-23	13.1	1.5 x 0.9	13.3	S0?	38	Mag 11.7 star N 3′.9.

(Continued from previous page)

GALAXIES

RA h m s	Dec ° ′ ″	Name	Mag (V)	Dim ′ Maj x min	SB	Type Class	PA	Notes
09 56 04.5	−27 06 22	ESO 499-22	14.8	1.6 x 0.3	13.8	Sc:	18	Mag 13.0 star N end; mag 8.93v star N 3′7.
09 56 25.8	−26 05 41	ESO 499-23	11.9	2.0 x 1.3	12.8	SA(rs)0^-	104	Mag 9.89v star on E edge.
09 58 20.8	−25 10 32	ESO 499-26	13.4	3.1 x 0.5	13.7	SB(s)dm: IV	22	Located 3′0 SSE of mag 6.77v star.
09 59 07.3	−25 59 42	ESO 499-28	13.4	1.0 x 0.7	12.9	S0	24	Faint anonymous galaxy N 1′0.
09 59 24.4	−24 58 17	ESO 499-30	13.7	1.0 x 0.7	13.2	S0	41	Several faint stars on NE edge.
09 59 28.4	−26 51 49	ESO 499-32	12.5	2.2 x 1.4	13.6	SB(r)a	147	Mag 15.6 star superimposed NW of center; mag 10.30v star W 2′0.
10 00 19.5	−24 48 16	ESO 499-34	14.4	1.2 x 0.9	14.3	SB(s)m pec V	153	Mag 13.1 star W 0′9.
10 03 41.9	−27 01 40	ESO 499-37	12.9	3.7 x 1.8	14.8	SAB(s)d: III-IV	51	Several bright knots or stars NE end; mag 7.40v star N 7′0.
10 09 38.3	−25 58 54	ESO 500-4	14.0	0.8 x 0.4	12.7	Sb	66	Mag 12.9 star N 0′5.
10 10 35.0	−25 49 24	ESO 500-6	13.5	2.3 x 1.7	14.8	SB(s)dm: IV-V	122	Several stars superimposed; mag 13.2 star W 3′8.
10 14 38.6	−27 24 37	ESO 500-17	14.1	1.2 x 0.4	13.1	(R′)SB(s)b	31	Mag 8.57v star NNE 7′0.
10 14 53.7	−23 03 02	ESO 500-18	13.0	1.5 x 0.7	13.0	SAB(rs)0°?	178	Mag 12.7 star on E edge; mag 9.78v star SSW 5′9.
10 20 53.3	−26 56 08	ESO 500-31	14.3	1.5 x 0.4	13.6	Sab	121	Mag 9.64v star N 4′0.
10 22 51.2	−24 20 17	ESO 500-32	14.1	1.5 x 1.3	14.6	SAB(s)dm: IV-V	67	Uniform surface brightness. Lies 1′8 W of mag 6.82v star.
10 29 37.1	−24 06 47	ESO 501-1	13.2	1.4 x 0.7	13.0	SAB(s)d III-IV	141	Pair of bright knots or superimposed stars SE of center; mag 10.97v star E 5′7.
10 31 48.2	−26 33 56	ESO 501-3	12.8	1.4 x 1.0	13.0	SAB0^-	64	Mag 6.51v star WNW 13′6.
10 33 07.5	−27 05 47	ESO 501-10	13.7	1.5 x 0.2	12.3	S?	171	Mag 10.89v star E 5′8; ESO 501-12 SSE 7′1.
10 33 10.5	−24 32 37	ESO 501-11	14.4	1.5 x 0.2	12.9	Sbc	95	Mag 8.33v star W 11′0.
10 50 58.6	−23 39 32	ESO 501-12	15.1	1.0 x 0.8	14.7	Irr	114	Mag 9.97v star E 5′6.
10 33 30.0	−26 53 52	ESO 501-13	12.9	1.3 x 0.5	12.4	S0°:	57	**ESO 501-14** S 7′7; pair of mag 8.51v, 9.04v stars N 5′5.
10 35 23.5	−24 45 21	ESO 501-23	12.6	3.4 x 2.7	14.8	SB(s)dm V	12	Mag 14.1 star superimposed n of center; mag 9.92v star WNW 5′5.
10 35 27.4	−24 23 03	ESO 501-24	13.5	1.3 x 0.7	13.3	(R)SAB(r)0^+:	101	Mag 13.4 star E edge; mag 10.77v star N 7′9.
10 35 25.3	−26 39 24	ESO 501-25	13.2	1.7 x 0.7	13.2	SAB0°	155	Mag 6.33v star E 8′9.
10 36 24.5	−26 59 58	ESO 501-35	13.2	1.5 x 0.5	12.7	SB(r)0°:	119	Mag 9.90v star S 6′0; MCG −4-25-37 ESE 7′1.
10 37 28.9	−26 19 00	ESO 501-51	12.0	2.6 x 1.5	13.3	(R′)SA(rs)a:	116	Star or bright knot E of center; mag 9.19v star W 4′1; **PGC 31510** W 6′4.
10 37 45.1	−26 37 53	ESO 501-56	12.9	1.8 x 0.5	12.6	S0 sp	74	Mag 9.10v star NNW 3′9; **MCG −4-25-40** WSW 7′6.
10 38 33.3	−27 44 14	ESO 501-65	13.2	1.5 x 0.7	13.1	SB(s)d: pec	94	Mag 14.6 star on SW edge; mag 10.62v star SE 3′6.
10 38 55.3	−26 38 24	ESO 501-66	13.8	1.2 x 0.4	12.9	S0?	53	Mag 8.85v star SE 4′0.
10 39 03.6	−26 52 34	ESO 501-67	14.3	0.7 x 0.3	12.5	SB(rs)b II	138	Mag 10.86v star on NE edge.
10 39 18.1	−26 50 23	ESO 501-68	13.4	2.0 x 0.7	13.6	S?	14	Located 2′0 SSW of a mag star.
10 40 59.0	−27 05 00	ESO 501-75	12.7	2.3 x 1.3	13.7	SA(s)c: II	24	Mag 9.04v star E 7′1.
10 41 25.8	−23 23 03	ESO 501-79	13.5	1.5 x 1.0	13.8	SAB(s)m: V-VI	41	Mag 8.80v star SE 2′2.
10 42 37.7	−23 56 08	ESO 501-80	13.2	2.4 x 0.5	13.3	Sc? sp	105	Mag 10.26v star N 7′3.
10 43 11.6	−26 15 02	ESO 501-82	13.3	1.9 x 0.8	13.6	SB(rs)bc: II	65	Pair of mag 10.33v, 10.92v stars on W edge; mag 8.75v star W 8′9.
10 43 28.3	−25 51 59	ESO 501-84	13.1	1.8 x 1.2	13.8	SA(s)0°	156	Mag 8.27v star W 4′2.
10 43 47.7	−24 22 05	ESO 501-86	13.1	2.0 x 1.0	13.7	SAB(s)bc II	19	Close pair of mag 9.46v, 11.09v stars NNE 4′5.
10 44 18.5	−22 49 35	ESO 501-88	13.2	1.4 x 1.0	13.4	(R′)SAB(rs)ab: II-III	77	MCG −4-26-5 W 7′1; mag 8.05v star S 9′0.
10 57 37.0	−25 25 40	ESO 502-5	12.8	1.3 x 0.8	12.7	S0?	114	Mag 9.16v star N 8′6; MCG −4-26-16 S 4′7.
10 58 13.2	−26 18 27	ESO 502-7	13.9	0.6 x 0.3	11.9	(R′)SB(s)0/a	156	Faint anonymous galaxy NE 1′7.
11 02 31.4	−26 10 04	ESO 502-11	14.0	1.1 x 0.5	13.2	Pec	88	Mag 12.5 star WNW 8′4.
11 02 50.3	−23 35 28	ESO 502-12	13.3	1.5 x 0.8	13.4	(R′)SAB(rs)ab	30	Mag 8.67v star SW 2′5.
11 05 14.9	−23 50 46	ESO 502-15	14.3	1.2 x 0.6	13.7		161	Mag 11.8 star ESE 3′8.
11 05 13.6	−26 37 31	ESO 502-16	13.7	1.6 x 0.5	13.3	SB(s)m IV-V	81	Small triangle of mag 12.6, 12.1 and 10.49v stars S 2′9.
11 05 54.8	−25 35 23	ESO 502-17	14.5	0.7 x 0.3	12.8	E?	93	Mag 11.9 star NNW 3′3.
11 07 18.3	−25 34 26	ESO 502-18	16.0	1.4 x 0.8	16.0	IB(s)m VI	27	Star on E edge.
11 09 22.0	−22 52 36	ESO 502-20	13.6	1.2 x 0.9	13.5	SAB(s)c: III-IV	21	Mag 8.69v star E 10′0.
11 12 13.5	−24 14 02	ESO 502-23	15.2	1.0 x 0.5	14.3	IB(s)m IV-V	84	Mag 10.66v star N 2′8; mag 10.85v star ENE 5′0.
11 15 15.6	−27 39 40	ESO 503-5	13.5	0.7 x 0.5	12.2	S?	34	Faint star NE edge.
11 15 48.8	−27 06 40	ESO 503-7	13.9	1.3 x 0.5	13.3	SB0^-? sp	33	Mag 9.63v star SW 10′8.
11 17 44.5	−26 02 25	ESO 503-11	14.9	0.9 x 0.6	14.0	SAB(s)cd IV	99	Mag 8.87v star SW 10′8; NGC 3617 S 5′8.
09 56 08.3	−21 59 20	ESO 566-30	13.9	1.9 x 1.0	14.4	SAB(rs)bc pec III-IV	16	Mag 10.08v star W 5′3.
10 01 31.5	−19 32 25	ESO 567-5	14.3	1.4 x 0.4	13.5	Sb? sp	72	Mag 8.93v star S 8′5; **MCG −3-26-10** NNW 7′9.
10 01 32.4	−20 23 08	ESO 567-6	15.1	1.9 x 0.3	14.3	Sd: sp	7	Mag 9.78v star N 7′1.
10 11 37.4	−21 15 15	ESO 567-25	13.6	0.6 x 0.6	12.3	S?		Compact; mag 12.4 star S 1′7.
10 14 03.5	−21 58 37	ESO 567-26	13.3	2.4 x 0.4	13.1	Sbc sp II	167	Mag 8.97v star S 4′0.
10 14 52.4	−20 00 47	ESO 567-29	14.2	0.5 x 0.5	12.5			A close pair of SE-NW oriented mag 10.27v, 9.14v stars NE 11′1.
10 15 44.5	−20 17 45	ESO 567-32	13.2	1.5 x 0.7	13.1	SB(r)a pec:	130	Mag 9.78v star SSW 7′1.
10 15 47.2	−21 44 12	ESO 567-33	13.8	0.8 x 0.3	12.2		76	Star superimposed center?
10 16 54.1	−22 13 53	ESO 567-37	14.0	0.4 x 0.4	12.1	E/S0:		Compact; mag 14.3 star on NW edge.
10 17 13.2	−21 04 05	ESO 567-39	13.8	0.7 x 0.6	12.7	S pec?	142	Mag 7.233v star E 9′7.
10 19 42.0	−17 45 07	ESO 567-48	15.3	1.6 x 0.4	14.7	Sm: sp IV-V	50	Small galaxy, or close pair of stars, on S edge.
10 20 08.1	−21 41 42	ESO 567-52	13.2	1.3 x 1.0	13.4	S0	43	MCG −3-27-3 S 1′6; mag 12.3 star NE 2′4.
10 26 29.4	−21 19 01	ESO 568-10	15.7	1.5 x 0.1	13.5	Sd:	66	Mag 8.51v star SE 15′1; mag 10.44v star S 7′6.
10 26 40.5	−19 03 08	ESO 568-11	13.9	2.2 x 1.2	14.8	SAB(s)bc: pec II	116	Two main arms; small, faint, elongated anonymous galaxy SE 2′5.
10 26 44.2	−21 51 48	ESO 568-12	14.9	1.5 x 0.2	13.5	Sc:	25	Mag 10.04v star SE 2′9; mag 9.33v star N 6′8.
10 36 49.4	−18 06 30	ESO 568-19	13.4	1.5 x 0.9	13.5	(R′)SB(r)ab pec	144	Mag 8.90v star SW 4′4.
10 44 09.0	−20 48 10	ESO 569-1	14.2	0.6 x 0.4	12.7	E?	83	Mag 12.4 star ESE 2′1.
10 48 29.4	−21 38 04	ESO 569-9	13.9	1.5 x 0.9	14.1	SAB(s)d? III-IV	107	Mag 10.59v star NNE 7′5.
10 49 16.4	−19 38 12	ESO 569-12	12.7	1.7 x 1.0	13.1	SAB0° pec:	102	**MCG −3-28-7** WSW 6′9; mag 9.37v star W 8′0.
10 51 23.8	−19 53 22	ESO 569-14	13.0	3.9 x 0.7	13.9	SB(s)cd: II	152	Mag 8.68 star SW 12′8.
10 53 16.6	−22 19 11	ESO 569-16	13.6	1.3 x 0.8	13.5	SB(rs)0^+:	0	Mag 10.41v star NE 8′5.
10 56 59.8	−20 10 09	ESO 569-24	12.2	1.4 x 1.3	12.7	Sbc?	50	Mag 9.74v star superimposed on center.
11 05 49.5	−20 47 31	ESO 570-2	13.4	1.2 x 0.6	12.9	SB(rs)bc II-III	60	Mag 8.20v star NE 12′3.
11 07 08.7	−18 01 32	ESO 570-4	13.8	1.1 x 0.8	13.8	E^+4 pec	165	Double system, common envelope.
11 20 12.1	−21 28 15	ESO 570-19	13.7	1.3 x 0.9	13.7	(R′)SAc: pec III-IV	72	Mag 8.18v star WNW 10′9.
10 39 26.6	−23 45 21	IC 625	14.1	1.6 x 0.3	13.2	Sb?	111	Mag 9.38v star NNE 5′7; mag 7.80v star NNE 11′0.
09 59 55.5	−29 37 03	IC 2531	12.0	6.9 x 0.6	13.4	Sc: sp	75	Single, narrow dust lane center one third of galaxy.
10 03 52.3	−27 34 13	IC 2537	12.1	2.6 x 1.7	13.6	SAB(rs)c II	26	Numerous bright knots or stars over face of galaxy.
10 05 47.9	−17 26 04	IC 2541	12.9	1.3 x 0.5	12.2	SB(r)bc? II	5	Mag 10.83v star NNE 7′8.
10 31 02.4	−28 43 03	IC 2586	12.5	0.9 x 0.5	11.7	E4	79	Mag 10.63v star SW 4′6.
10 32 21.0	−24 02 18	IC 2589	13.4	0.9 x 0.6	12.6	S?	12	Mag 10.69v star W 0′9.
10 36 04.4	−24 19 24	IC 2594	12.4	1.4 x 1.4	13.1	SA0^-		Mag 10.77v star NW 10′0; mag 11.5 star E 4′1.
10 37 47.4	−27 04 54	IC 2597	11.8	1.5 x 1.2	12.5	E^+4	4	= **Hickson 48A**. **Hickson 48C**, **Hickson 48D** NW 2′3.
11 03 51.1	−20 05 36	IC 2623	13.5	0.9 x 0.6	12.8	E?	70	Mag 12.9 star SE 7′5.
11 09 53.5	−23 43 32	IC 2627	12.0	2.4 x 2.3	13.7	SA(s)bc: I-II	66	Mag 9.86v star W 8′3.
09 59 40.2	−20 47 17	MCG −3-26-5	13.4	1.0 x 0.4	12.3	S0	103	**MCG −3-26-4** W 2′8.
10 00 10.3	−19 37 20	MCG −3-26-6	13.2	0.3 x 0.3	10.7	E?		= **Hickson 42C**; located on NW edge of NGC 3091.
10 00 43.5	−20 22 09	MCG −3-26-9	13.2	1.0 x 0.7	12.7	(R)SB(r)0^+	135	Several stars on E edge.
10 03 18.9	−21 25 51	MCG −3-26-15	13.8	1.4 x 1.2	14.2	SB(s)cd III-IV	176	Mottled, with bright star at southern tip.

(Continued from previous page)

GALAXIES

RA h m s	Dec ° ′ ″	Name	Mag (V)	Dim ′ Maj x min	SB	Type Class	PA	Notes
10 05 26.2	−17 48 00	MCG −3-26-16	13.9	1.1 x 0.5	13.1	Sb II-III	119	
10 06 25.2	−18 16 30	MCG −3-26-23	14.1	1.0 x 0.5	13.2	S	12	
10 08 18.6	−20 10 11	MCG −3-26-26	13.7	1.3 x 0.4	12.9	S0	125	
10 09 19.3	−20 00 13	MCG −3-26-27	13.7	0.7 x 0.6	12.6	(RL)SB(s)0^+	54	Mag 9.28v star ENE 6.′5.
10 11 18.8	−17 12 16	MCG −3-26-30	12.9	1.7 x 1.3	13.7	SAB0^- pec:	100	Mag 10.75v star NNE 10.′4.
10 19 08.5	−17 55 53	MCG −3-26-38	14.7	0.8 x 0.6	13.7	SBa-b	3	Double system, NGC 3200 SW 8.′2.
10 19 49.4	−19 57 20	MCG −3-27-1	13.6	0.8 x 0.7	12.9	S0	78	**ESO 567-49A** N 1.′0.
10 20 02.5	−21 31 08	MCG −3-27-2	13.2	0.7 x 0.7	12.5	E?		
10 20 08.9	−21 43 20	MCG −3-27-3	14.0	1.2 x 0.5	13.3	S0 pec	58	Bright core with long, faint, wide filament extending SW; ESO 567-52 N 1.′6.
10 20 50.9	−17 19 02	MCG −3-27-5	15.2	0.7 x 0.4	13.6	(R′)SAB:(r)ab	24	Mag 10.60v star on SE edge.
10 21 48.3	−19 29 44	MCG −3-27-6	14.1	0.9 x 0.3	12.5	SB(rl)0/a	92	Mag 9.61v star N 3.′5.
10 23 49.6	−20 57 32	MCG −3-27-7	14.4	0.5 x 0.4	12.5		36	
10 23 53.8	−17 17 41	MCG −3-27-8	17.3	0.7 x 0.4	15.8	SA(r)a		Close pair with MCG −3-27-9 with bright star in between.
10 23 51.5	−17 17 21	MCG −3-27-9	16.4	0.4 x 0.2	13.5	(R′)SB(r)a		Anonymous elongated galaxy WSW 1.′0. Pair with MCG −3-27-9 with bright star in between.
10 24 57.3	−17 26 06	MCG −3-27-10	14.7	1.0 x 0.8	14.3	SAB(rs)a pec:	174	Small elongated galaxy at NE edge, bright star or galaxy at SW edge.
10 26 08.2	−20 14 09	MCG −3-27-11	13.5	1.0 x 0.6	12.9	S0	44	Faint stars SE and W edges.
10 26 28.7	−20 02 31	MCG −3-27-12	13.6	1.1 x 0.8	13.3	S?	161	
10 27 30.1	−18 48 37	MCG −3-27-14	13.3	0.7 x 0.5	12.0	SA(s)bc:	15	
10 28 34.8	−20 55 52	MCG −3-27-15	14.3	0.5 x 0.3	12.1	SBb	105	
10 29 20.6	−17 25 16	MCG −3-27-16	14.4	1.1 x 0.2	12.6	Sbc	75	
10 29 40.5	−17 04 41	MCG −3-27-17	14.3	1.0 x 0.4	13.2		30	
10 34 54.6	−20 32 56	MCG −3-27-19	13.7	1.0 x 0.6	13.1	S0:	48	
10 39 55.6	−17 39 42	MCG −3-27-23	14.2	0.9 x 0.9	13.8	SA(s)c pec		
10 41 02.9	−17 30 33	MCG −3-27-24	13.6	1.2 x 0.8	13.4	S?	39	
10 41 15.2	−21 01 26	MCG −3-27-25	13.6	0.9 x 0.6	12.8	(R′)SB(rs)b	155	
10 42 19.0	−17 38 57	MCG −3-27-26	13.6	1.7 x 0.4	13.1	S0 pec sp	60	Bright, elongated, core with faint, broad filament trailing SW to W.
10 46 02.3	−17 41 19	MCG −3-28-2	13.9	0.8 x 0.7	13.2	SA(s)c:	147	
10 50 00.9	−19 15 46	MCG −3-28-9	14.4	0.5 x 0.3	12.2	Sbc	126	
10 50 05.3	−17 11 49	MCG −3-28-10	14.8	0.7 x 0.3	13.0	Sm	135	NGC 3420 S 3.′0.
10 50 31.6	−17 03 51	MCG −3-28-13	14.7	1.0 x 0.2	12.8	Sc	156	Mag 9.91v star SE 2.′4; NGC 3409 W 3.′1.
10 52 16.0	−17 07 50	MCG −3-28-16	13.8	0.5 x 0.5	12.2	SA(s)a		Mag 7.78v star NE 8.′2 ; PGC 32573 W 9.′2.
10 54 22.1	−17 26 29	MCG −3-28-20	14.1	0.8 x 0.5	13.0	S0/a	132	
10 57 50.7	−20 00 09	MCG −3-28-27	13.0	1.6 x 1.0	13.4	SB0° pec:	164	Star superimposed N of bright center.
10 58 00.4	−18 25 41	MCG −3-28-28	14.0	1.1 x 0.9	13.8	SA(rs)b:	73	Star on N edge.
10 58 24.4	−19 09 15	MCG −3-28-29	14.3	1.0 x 0.6	13.6	SB	10	Double system.
11 03 35.9	−17 28 45	MCG −3-28-33	13.9	1.5 x 0.2	12.5	S?	150	Mag 9.47v star WNW 4.′6.
11 22 33.0	−17 34 22	MCG −3-29-6	14.0	0.6 x 0.4	12.3	Sb	138	
10 05 26.3	−21 46 10	MCG −4-24-16	14.4	0.7 x 0.6	13.5	E?	150	
10 05 55.6	−23 03 26	MCG −4-24-17	13.7	1.1 x 0.5	12.9	(R)SB(r)0^+:	90	
10 10 02.1	−24 19 43	MCG −4-24-18	13.7	1.2 x 0.8	13.5	SAB(r)b	103	
10 11 27.3	−25 18 02	MCG −4-24-20	14.0	1.1 x 0.4	12.9	Sab	117	Mag 9.05v star S 4.′6.
10 19 07.6	−26 59 05	MCG −4-25-1	14.0	1.0 x 0.4	12.8	Sb	99	
10 22 44.7	−22 36 05	MCG −4-25-5	14.0	1.0 x 0.3	12.5	Sa	164	
10 24 31.5	−23 33 13	MCG −4-25-6	13.3	1.1 x 0.6	12.8	(R)SB(s)0/a	153	Elongated **ESO 500-36** NE 4.′7.
10 26 56.3	−24 05 22	MCG −4-25-8	13.3	0.9 x 0.8	12.8	(R)SA(r)ab	150	Bright knot or superimposed star E edge.
10 28 02.4	−23 15 24	MCG −4-25-10	13.6	1.1 x 0.8	13.3	SBc	56	
10 33 59.9	−27 20 12	MCG −4-25-20	13.9	0.7 x 0.4	12.3	SB(rs)ab	151	Mag 7.41v star NE 2.′0.
10 34 23.8	−26 29 31	MCG −4-25-21	14.0	1.0 x 0.3	12.6	S?	15	
10 34 47.8	−27 12 52	MCG −4-25-23	13.5	1.0 x 0.7	13.0	S0?	32	
10 35 20.8	−27 21 46	MCG −4-25-25	13.5	1.4 x 0.3	12.4	S0?	164	**PGC 31371** ESE 2.′6.
10 36 22.2	−25 22 37	MCG −4-25-30	13.6	0.7 x 0.8	12.8	SB?	163	
10 36 53.2	−27 03 12	MCG −4-25-37	14.5	0.6 x 0.4	12.8	SB?	39	ESO 501-35 WNW 7.′1; mag 13.1 star WNW 2.′8.
10 36 57.0	−26 11 41	MCG −4-25-38	13.3	1.1 x 0.5	12.5	S?	124	Mag 15.1 star on W edge; mag 11.5 star NW 2.′7.
10 37 17.0	−27 28 11	MCG −4-25-43	13.8	0.9 x 0.4	12.6	SB0°:	71	Mag 4.88v star N 3.′5.
10 37 38.0	−26 16 38	MCG −4-25-48	13.9	1.0 x 0.3	12.5	S?	129	Mag 10.23 star S 1.′0; ESO 501-51 SW 2.′9.
10 37 49.7	−27 07 18	MCG −4-25-50	14.2	0.9 x 0.6	13.4	S?	57	= **Hickson 48B**. IC 2597 N 2.′4.
10 43 48.2	−22 49 03	MCG −4-26-5	13.9	0.7 x 0.6	12.8	Sc	151	Mag 9.41v star W 6.′4; ESO 501-88 E 7.′1.
10 44 26.1	−25 22 39	MCG −4-26-7	12.8	0.8 x 0.5	11.9	E?	174	Mag 0.05v star W 4.′6.
10 46 13.7	−21 48 00	MCG −4-26-8	14.8	0.6 x 0.5	13.3	SBc	162	Multi-galaxy system?
10 57 38.7	−25 30 17	MCG −4-26-16	13.5	1.1 x 0.6	12.9	Sc	49	Mag 11.14v star SSW 5.′4; ESO 502-5 N 4.′7.
10 59 51.1	−25 29 44	MCG −4-26-17	13.5	0.8 x 0.5	12.4	S?	96	
11 16 51.5	−23 59 25	MCG −4-27-7	14.9	0.4 x 0.4	12.8			Stellar, appears with three stars forming a dipper like asterism.
11 23 41.0	−22 16 15	MCG −4-27-9	14.3	0.7 x 0.4	12.8	S?	60	
10 10 47.2	−28 54 08	MCG −5-24-25	13.5	0.7 x 0.7	12.8	E?		
10 14 48.3	−28 57 32	MCG −5-24-29	13.9	1.4 x 0.3	12.8	Sa:	39	NGC 3175 N 5.′4.
10 16 13.5	−29 12 16	MCG −5-24-30	13.9	1.1 x 0.5	13.1	Sb	145	Star on N edge.
10 34 38.9	−28 35 02	MCG −5-25-17	12.6	0.5 x 0.3	10.4	SB(rs)bc I	15	Located 2.′7 W of ESO 436-46.
10 34 59.8	−28 04 45	MCG −5-25-18	13.8	1.0 x 0.4	12.7	S?	154	
10 36 32.9	−28 03 53	MCG −5-25-21	13.6	1.3 x 0.3	12.5	S0?	154	Mag 9.81v star S 1.′7.
10 36 50.7	−27 55 11	MCG −5-25-22	13.3	1.2 x 0.6	12.8	S0?	141	Pair with MCG −5-25-24 E edge.
10 36 54.1	−27 55 05	MCG −5-25-24	13.6	1.0 x 0.4	12.5	S0?	12	Pair with MCG −5-25-22 W edge; **PGC 31523** NE 3.′9.
10 37 56.3	−28 54 29	MCG −5-25-26	13.2	1.2 x 1.0	13.3	SB?	59	ESO 437-22 E 5.′0.
10 38 40.3	−28 34 08	MCG −5-25-29	13.6	1.1 x 0.5	12.8	S?	31	
10 39 25.1	−27 54 46	MCG −5-25-32	13.9	0.8 x 0.7	13.1	SB(rs)bc pec: II	20	Mag 8.69v star S 2.′3.
10 40 50.4	−27 58 01	MCG −5-25-37	13.3	1.1 x 0.6	12.8	S0?	83	
10 47 12.9	−28 55 19	MCG −5-26-6	14.8	0.6 x 0.6	13.6	SBb		Star superimposed E edge.
10 53 33.8	−29 23 41	MCG −5-26-11	13.4	1.4 x 0.4	12.6	S	86	
11 12 17.7	−28 00 03	MCG −5-27-1	13.3	1.1 x 0.7	12.9	S?	36	
11 18 53.6	−29 25 29	MCG −5-27-9	13.7	0.9 x 0.5	12.7	S?	152	**ESO 438-21** on E edge.
09 57 23.9	−19 21 20	NGC 3072	12.8	1.9 x 0.6	12.8	S0/a? sp	71	**ESO 566-32** N 7.′9.
09 57 37.7	−18 10 42	NGC 3076	13.3	1.0 x 0.9	13.1	Sab pec:		
09 58 24.6	−26 55 38	NGC 3078	11.1	2.5 x 2.1	12.9	E2-3	177	Mag 14.4 star on W edge.
09 59 29.5	−22 49 36	NGC 3081	12.0	2.1 x 1.6	13.1	(R)SAB(r)0/a	74	Very faint galactic material extends beyond the published size to the N and S.
09 59 06.9	−27 07 46	NGC 3084	12.3	1.8 x 1.6	13.3	(R′)SB(s)ab pec	20	Several stars superimposed S of nucleus.
09 59 29.3	−19 29 33	NGC 3085	13.2	1.2 x 0.4	12.3	S0°: sp	119	Anonymous galaxy NW 2.′7.
09 59 36.5	−28 19 51	NGC 3089	12.2	1.8 x 1.0	12.7	SAB(rs)b II	139	Lies 2.′2 W of mag 8.20v star.
10 00 14.2	−19 38 10	NGC 3091	11.1	3.0 x 1.9	13.0	E3:	149	= **Hickson 42A**; **Hickson 42D** lies S 2.′2.

(Continued from previous page)

GALAXIES

RA h m s	Dec ° ′ ″	Name	Mag (V)	Dim ′ Maj x min	SB	Type Class	PA	Notes
10 00 33.2	−19 39 39	NGC 3096	13.4	1.0 x 0.8	13.0	SB(rs)0°	170	= **Hickson 42B.** A mag 12 star 1′.2 NE; a mag 11.1 star 1′.8 NW; and an anonymous galaxy 1′.3 SW.
10 03 05.9	−26 09 24	NGC 3109	9.9	19.1 x 3.7	14.3	SB(s)m sp IV-V	93	Numerous knots and superimposed stars.
10 03 59.2	−20 46 56	NGC 3112	14.9	0.9 x 0.2	12.9	Sb?	47	
10 04 26.1	−28 26 41	NGC 3113	12.5	3.3 x 1.2	13.9	SAB(s)d:	87	Several faint knots; mag 7.29v star N 4′.3.
10 06 39.9	−19 13 23	NGC 3124	12.1	3.0 x 2.5	14.2	SAB(rs)bc I	9	Tightly wound branching arms; mag 8.67v star S 5′.3.
10 06 33.0	−29 56 08	NGC 3125	13.0	1.1 x 0.7	12.7	E?	114	Lies 2′.4 SE of mag 11.24v star.
10 09 07.5	−29 03 46	NGC 3137	11.5	6.3 x 2.2	14.2	SA(s)cd III-IV	1	Many stars superimposed overall.
10 11 09.9	−20 52 15	NGC 3146	13.0	1.0 x 0.9	12.8	SB(r)b? II	100	Mag 8.87v star N 3′.8.
10 15 37.0	−20 38 52	NGC 3171	12.8	1.8 x 1.2	13.5	S0$^-$ pec	176	
10 14 34.8	−27 41 36	NGC 3173	12.8	2.1 x 1.7	14.0	SA(s)c II-III	7	A mag 10.05v star lies 1′.2 E and a mag 10.43v star lies SE 2′.3.
10 14 42.5	−28 52 16	NGC 3175	11.2	5.0 x 1.3	13.1	SAB(s)a? III	56	Mag 9.36v stars 1′.7 N of NE end of galaxy; MCG −5-24-29 S 5′.4.
10 18 36.5	−17 58 57	NGC 3200	12.0	4.2 x 1.3	13.7	SAB(rs)c: I-II	169	MCG −3-26-38 NE 8′.2.
10 19 34.2	−26 41 53	NGC 3203	12.1	2.9 x 0.6	12.6	SA(r)0$^+$? sp	58	Almost stellar **ESO 500-23** W 3′.1.
10 19 41.3	−25 48 57	NGC 3208	12.7	1.8 x 1.5	13.7	SAB(rs)bc II	20	Mag 10.79v star W 1′.6.
10 21 57.6	−22 16 04	NGC 3233	12.5	1.7 x 0.9	12.8	(R′)SB(r)0/a	140	
10 24 30.8	−21 47 27	NGC 3240	13.1	1.1 x 0.9	12.9	SA(rs)cd: II-III	78	
10 32 22.0	−22 18 10	NGC 3282	12.9	1.9 x 0.6	12.9	(R)SB(r)0°?	82	
10 33 35.9	−27 27 19	NGC 3285	12.0	2.6 x 1.3	13.2	SB(s)a pec	108	Mag 7.41v star NNE 7′.3.
10 32 48.8	−27 31 20	NGC 3285A	13.7	1.2 x 0.8	13.5	SB(rs)cd: II-III	171	Elongated **MCG −4-25-14** W 1′.7.
10 34 36.9	−27 39 08	NGC 3285B	13.1	1.5 x 1.1	13.5	SAB(r)b: III	43	Stellar nucleus; **PGC 31268** W 5′.0.
10 35 17.5	−17 16 38	NGC 3290	13.5	1.0 x 0.6	12.8	SAB(rs)bc: pec II	60	Mag 8.35v star N 4′.8.
10 36 11.9	−27 09 43	NGC 3305	12.8	1.1 x 1.1	12.9	E0		**PGC 31454** and **PGC 31453** SE 5′.6.
10 36 17.2	−27 31 49	NGC 3307	14.5	0.9 x 0.3	12.9	SB(r)0/a pec?	28	
10 36 22.2	−27 26 19	NGC 3308	11.9	1.7 x 1.3	12.6	SAB(s)0$^-$:	32	**PGC 31419** SW 2′.7; **PGC 31450** SSE 3′.1.
10 36 35.9	−27 31 07	NGC 3309	11.6	1.9 x 1.6	12.8	E3	31	Galaxies **PGC 31464** and **PGC 31450** NNW 2′.3.
10 36 42.8	−27 31 42	NGC 3311	11.7	2.3 x 2.1	13.4	E$^+$2	19	Galaxy **Anon 1034-27A** 2′.0 S.
10 37 02.6	−27 33 50	NGC 3312	11.9	3.3 x 1.3	13.3	SA(s)b pec?	175	Galaxy **ESO 501-49** E 3′.9.
10 37 25.4	−25 19 06	NGC 3313	11.4	3.9 x 3.2	14.0	(R′)SB(rs)ab II	55	Elongated galaxy **ESO 501-44** WSW 5′.0.
10 37 12.8	−27 41 01	NGC 3314A	13.1	1.5 x 0.7	13.0	Sab: sp	143	Almost stellar NGC 3314B NNW 1′.6.
10 37 09.7	−27 39 32	NGC 3314B	12.8	0.3 x 0.2	9.6	SA(s)c:	98	Very faint star on E edge.
10 37 19.4	−27 11 33	NGC 3315	13.4	1.1 x 1.0	13.3	S0$^-$?	80	Mag 9.80v star on W edge.
10 37 37.4	−27 35 40	NGC 3316	12.7	1.3 x 1.1	12.9	SB(rs)0°:	36	Small anonymous galaxy S edge of core.
10 40 09.1	−23 49 14	NGC 3331	13.2	1.2 x 0.9	13.1	SB(s)c: II	27	Mag 7.80v star N 12′.0.
10 39 34.1	−23 55 21	NGC 3335	13.1	1.1 x 0.9	13.0	SAB(r)0$^+$:	130	
10 40 16.9	−27 46 35	NGC 3336	12.2	1.9 x 1.5	13.2	SAB(rs)c pec II	123	**PGC 31766** and **PGC 31776** S 6′.1.
10 46 44.7	−25 14 39	NGC 3369	13.6	1.4 x 0.8	13.6	SA0$^-$?	114	
10 47 19.3	−24 26 19	NGC 3383	12.8	1.5 x 1.1	13.2	SB(rs)bc? II	24	Mag 8.12v star S 7′.3; note: **ESO 501-98** lies 2′.4 E of this star.
10 48 23.7	−25 09 42	NGC 3393	12.2	1.8 x 1.5	13.1	(R′)SB(rs)a:	48	**ESO 501-99** W 4′.2; mag 9.26v star E 2′.9.
10 50 20.3	−17 02 41	NGC 3409	14.1	1.2 x 0.3	12.8	SBc? sp II-III?	10	MCG −3-28-13 E 3′.1.
10 50 09.6	−17 14 34	NGC 3420	13.8	1.3 x 1.1	14.0	(R)SB(rs)a:	30	MCG −3-28-10 N 3′.0.
10 51 15.0	−17 00 33	NGC 3431	13.6	1.3 x 0.3	12.4	SABb? I-II	130	PGC 32573 SE 8′.7.
10 48 03.6	−20 51 03	NGC 3450	11.8	2.5 x 2.2	13.5	SB(r)b II	128	Strong nucleus in short bar, uniform envelope; **ESO 569-7** N 3′.7.
10 53 40.5	−21 47 36	NGC 3453	12.9	1.1 x 0.6	12.3	SB(s)b: II-III	4	
10 54 44.5	−17 02 34	NGC 3459	13.4	1.6 x 0.5	13.0	(R′)SBab:	156	
10 55 13.4	−26 08 28	NGC 3463	12.8	1.5 x 0.7	12.7	Sb II-III	77	
10 54 40.0	−21 04 00	NGC 3464	12.6	2.6 x 1.7	14.0	SB(rs)c II	112	small, weak nucleus, multi-branching arms.
10 59 00.5	−28 28 37	NGC 3483	12.0	1.8 x 1.3	12.8	(R′)SA(r)0$^+$:	105	Located 1′.8 SW of mag 9.13v star.
11 03 23.6	−23 05 04	NGC 3511	11.0	5.8 x 2.0	13.5	SA(s)c II-III	76	Very patchy, several knots and dark rifts.
11 03 46.7	−23 14 38	NGC 3513	11.5	2.8 x 2.2	13.3	SB(rs)c II-III	75	Short, moderately bright SE-NW bar; dark patch S of center.
11 04 00.0	−18 46 53	NGC 3514	12.8	1.1 x 0.9	12.7	SAB(s)c? II	115	
11 06 16.0	−18 15 33	NGC 3520	14.9	1.1 x 0.2	13.2	S?	141	
11 07 18.2	−19 28 20	NGC 3528	11.9	2.6 x 1.4	13.2	SA(s)0°:	59	
11 07 19.2	−19 33 20	NGC 3529	13.1	1.0 x 0.8	12.7	SB(rs)b pec:	55	
11 07 47.9	−20 01 21	NGC 3565	14.3	0.6 x 0.5	12.9	SB0$^-$ pec?	129	Compact.
11 11 30.2	−18 17 25	NGC 3571	12.1	3.0 x 1.0	13.1	(R′)SAB(rs)a:	94	
11 13 16.3	−26 45 15	NGC 3585	9.9	4.6 x 2.5	12.6	E6	107	
11 14 40.8	−23 43 48	NGC 3597	12.8	1.9 x 1.5	13.8	S0$^+$: pec	64	Nucleus offset NE with faint material extending SW.
11 17 50.9	−26 08 04	NGC 3617	12.8	1.8 x 1.3	13.7	E$^+$	147	Mag 9.51v star SSW 4′.6; ESO 503-11 N 5′.8.
10 51 37.8	−17 07 24	PGC 32573	13.2	2.5 x 0.9	13.9	(R)SB(rs)ab pec? III-IV	40	

GALAXY CLUSTERS

RA h m s	Dec ° ′ ″	Name	Mag 10th brightest	No. Gal	Diam ′	Notes
10 36 54.0	−27 31 00	A 1060	12.7	50	168	**Hydra I Cluster**. Very large population of MCG, PGC and anonymous galaxies surround brighter NGC members.

OPEN CLUSTERS

RA h m s	Dec ° ′ ″	Name	Mag	Diam ′	No. ★	B ★	Type	Notes
10 14 00.0	−29 11 00	ESO 436-2		5			cl	
11 12 12.0	−21 19 00	ESO 570-12		13			cl	

PLANETARY NEBULAE

RA h m s	Dec ° ′ ″	Name	Diam ″	Mag (P)	Mag (V)	Mag cent ★	Alt Name	Notes
10 24 46.2	−18 38 34	NGC 3242	64	8.6	7.7	13.3	PK 261+32.1	**Ghost of Jupiter**. Slightly elongated SE-NW; mag 11.16v star S 2′.6.
10 34 30.7	−29 11 16	PN G270.1+24.8	54			16.7	PK 270+24.1	Mag 11.9 star on E edge.

RA h m s	Dec ° ′ ″	Name	Mag (V)	Dim ′ Maj x min	SB	Type Class	PA	Notes
09 06 27.7	−28 01 32	ESO 433-2	14.7	2.0 x 1.7	15.9	SB(s)dm V	27	Mag 8.52v star N 3′2; **ESO 433-3** E 8′7.
09 15 04.9	−28 15 51	ESO 433-10	12.5	1.7 x 1.2	13.2	SAB(rs)c III	98	Several knots or superimposed stars; mag 7.97v star ENE 11′1.
09 41 20.6	−28 57 42	ESO 434-21	14.0	1.0 x 0.6	13.3	(R′)SAB(s)0/a:	168	Faint star W edge; mag 13.3 star N 1′0.
09 41 43.2	−28 11 31	ESO 434-23	13.5	1.8 x 0.9	13.9	SAB(s)d III-IV	8	Almost stellar nucleus, smooth envelope; mag 8.45v star ESE 8′4.
09 44 13.2	−28 50 55	ESO 434-28	12.2	1.0 x 0.7	11.7	SA0$^-$:	101	Mag 10.08v star S 4′4.
09 44 37.9	−29 19 22	ESO 434-32	13.9	1.1 x 1.1	13.9	Sm:		Stellar nucleus, smooth envelope; faint star S edge.
09 50 05.3	−29 25 36	ESO 435-1	14.3	0.7 x 0.6	13.2	S0:	33	Almost stellar nucleus; mag 10.48v star NW 2′9.
09 52 19.8	−29 26 16	ESO 435-3	13.9	1.7 x 0.7	13.9	SA(s)0$^+$ pec:	102	Interacting with spiral companion 0′3 S.
09 57 48.2	−28 30 26	ESO 435-14	13.4	2.5 x 0.3	13.0	Sc: sp	54	
09 58 46.5	−28 37 18	ESO 435-16	12.6	1.9 x 1.3	13.4	I0? pec	118	Mag 14.4 star on N edge; mag 11.14v star NW 8′2.
09 59 21.2	−28 08 00	ESO 435-20	13.6	0.9 x 0.5	12.6		92	Mag 10.83v star S 6′8.
08 36 14.9	−26 24 33	ESO 495-21	11.6	1.7 x 1.3	12.3	I0? pec	169	Two bright knots or superimposed stars W of center; mag 12.1 star W 1′6.
08 52 57.1	−25 18 07	ESO 496-13	13.5	1.3 x 1.2	13.8	SAB(rs)c II	175	Mag 8.55v star NE 6′1.
08 55 21.1	−25 05 31	ESO 496-19	14.4	1.5 x 0.7	14.3	SB(s)cd II	177	Mag 11.00v star N 5′1; note: this star is on the W edge of **MCG −4-21-12**.
08 57 06.8	−24 45 04	ESO 496-22	13.2	1.6 x 1.5	14.0	SB(rs)d II-III	60	NGC 2717 N 4′9; mag 7.11v star SE 9′4; note: **ESO 496-23** lies 4′1 E of this star.
09 00 22.2	−25 52 10	ESO 497-1	15.9	1.6 x 0.4	15.2	IB(s)m V	123	Mag 9.21v star W 6′7.
09 01 27.4	−26 18 02	ESO 497-2	14.0	1.9 x 1.1	14.7	SB(rs)dm IV	176	Mag 14.9 star superimposed NW of core; mag 12.4 star NW 2′9.
09 09 45.8	−23 00 36	ESO 497-17	14.5	1.3 x 0.9	14.5	IAB(s)m V-VI	60	Mag 10.76v star N 1′9; **ESO 497-16** S 4′7.
09 10 25.5	−23 29 30	ESO 497-18	14.2	1.8 x 0.3	13.4	S0/a: sp	87	Mag 10.30v star E 5′6.
09 12 00.6	−26 49 37	ESO 497-22	13.5	1.2 x 0.5	12.8	Sb	174	Mag 11.01v star W 3′1.
09 15 22.8	−25 23 25	ESO 497-26	13.9	1.1 x 0.9	13.7	S0	67	Mag 10.43v star SE 2′8.
09 15 28.0	−25 34 03	ESO 497-27	13.8	1.1 x 1.0	13.7	SA?(r)	79	Mag 10.89v star S 8′0.
09 15 34.3	−23 50 56	ESO 497-28	14.5	0.8 x 0.6	13.5	Sa	135	Small triangle of mag 13-14 stars W edge.
09 15 43.6	−23 42 05	ESO 497-29	14.7	1.5 x 0.2	13.2	Sbc	63	NGC 2815 NE 9′2; mag 7.58v star N 12′5.
09 19 50.4	−24 13 42	ESO 497-39	14.6	0.8 x 0.3	12.9		81	Pair of stars, mags 10.93v and 12.8, SE 2′7.
09 23 11.9	−26 56 27	ESO 497-42	14.7	1.7 x 0.3	13.8	Sc	2	Star on SE edge; mag 10.14v star W 2′4.
09 23 36.5	−26 52 52	ESO 498-3	13.3	1.5 x 0.4	12.6	Sb? sp	33	ESO 497-42 SW 6′7.
09 23 47.5	−25 38 14	ESO 498-4	12.3	2.2 x 1.4	13.4	SA(rs)0°	142	Mag 13.7 star on E edge; mag 10.32v star NNW 4′6.
09 24 40.7	−25 05 38	ESO 498-5	13.1	1.7 x 1.1	13.6	SAB(s)bc pec I-II	126	Almost stellar nucleus; mag 10.29v star N 4′2.
09 24 52.4	−25 47 17	ESO 498-6	12.8	1.2 x 0.5	12.1	S0	79	Pair of E-W oriented mag 13.2 stars E edge.
09 33 34.0	−25 47 21	ESO 498-13	14.1	1.0 x 0.4	13.0	(R′)SB(r)ab	17	Strong anonymous galaxy SW edge; mag 8.56v star NE 3′2.
09 45 21.8	−27 05 28	ESO 499-2	13.9	1.7 x 1.3	14.6	SAB(s)d III-IV	120	Stellar nucleus; mag 15.4 star on W edge; mag 13.1 star W 4′7.
09 46 34.2	−23 55 08	ESO 499-4	14.1	1.5 x 0.5	13.6	SB(rs)cd II-III	82	Mag 10.19v star ESE 2′8; mag 10.39v star E 4′0.
09 47 13.3	−24 50 22	ESO 499-5	12.9	2.6 x 0.6	13.2	SAB(s)c: II	144	Several bright knots; mag 12.6 star N 1′2.
09 48 55.1	−25 58 30	ESO 499-7	13.7	1.0 x 0.7	13.2	S0:	19	Mag 10.7 star on S edge.
09 49 59.2	−25 00 34	ESO 499-8	13.6	2.0 x 1.0	14.2	SB(rs)cd: III-IV	86	Mag 10.05v star W 4′7.
09 50 18.3	−23 01 28	ESO 499-9	12.7	2.2 x 1.0	13.4	SB(r)b: II-III	11	Mag 6.67v star E 10′6.
09 52 00.3	−25 18 46	ESO 499-11	14.2	1.5 x 1.0	14.5	SB(rs)d IV-V	168	Pair of mag 12.6, 13.2 stars W 1′0; mag 9.87v star E 4′8.
09 53 16.6	−25 55 45	ESO 499-13	12.9	1.4 x 1.3	13.4	SB(s)0°	2	Mag 13.5 star on E edge; mag 10.24v star WNW 3′3; mag 4.87v star E 12′4.
09 55 53.5	−23 03 22	ESO 499-21	14.3	0.9 x 0.3	12.7	Sa? pec sp	45	Mag 9.90v star S 6′4.
09 56 04.5	−27 06 22	ESO 499-22	14.8	1.6 x 0.3	13.8	Sc:	18	Mag 13.0 star N end; mag 8.93v star N 3′7.
09 56 25.8	−26 05 41	ESO 499-23	11.9	2.0 x 1.3	12.8	SA(rs)0$^-$	104	Mag 9.89v star on E edge.
09 58 20.8	−25 10 32	ESO 499-26	13.4	3.1 x 0.5	13.7	SB(s)dm: IV	22	Located 3′0 SSE of mag 6.77v star.
09 59 07.3	−25 59 42	ESO 499-28	13.4	1.0 x 0.7	12.9	S0	24	Faint anonymous galaxy N 1′0.
09 59 24.4	−24 58 17	ESO 499-30	13.7	1.0 x 0.7	13.2	S0	41	Several faint stars on NE edge.
09 59 28.4	−26 51 49	ESO 499-32	12.5	2.2 x 1.4	13.6	SB(r)a	147	Mag 15.6 star superimposed NW of center; mag 10.30v star W 2′0.
10 00 19.5	−24 48 16	ESO 499-34	14.4	1.2 x 0.9	14.3	SB(s)m pec V	153	Mag 13.1 star W 0′9.
10 03 41.9	−27 01 40	ESO 499-37	12.9	3.7 x 1.8	14.8	SAB(s)d: III-IV	51	Several bright knots or stars NE end; mag 7.40v star N 7′0.
08 36 34.9	−20 28 16	ESO 562-23	12.4	2.3 x 0.5	12.4	SB(r)0$^+$? sp	170	Mag 10.78v star N 2′0; **ESO 562-20** SW 9′3.
08 37 11.3	−22 14 59	ESO 563-2	14.3	1.7 x 0.4	13.7	SB(s)c? sp IV-V	95	Mag 10.80v star N 6′2.
08 37 18.8	−20 56 27	ESO 563-3	13.5	1.5 x 0.3	12.5	SB(r)0/a pec sp	88	Mag 8.17v star SW 16′5.
08 41 35.3	−20 18 55	ESO 563-11	13.1	1.7 x 0.9	13.4	SB(rs)bc: III	82	Several stars superimposed; mag 10.6 star E 2′3.
08 41 38.9	−20 44 39	ESO 563-12	13.6	1.7 x 0.8	13.8	SB(s)d pec: III-IV	69	Mag 10.65v star WNW 3′6.
08 42 32.3	−19 52 20	ESO 563-13	14.5	1.5 x 0.3	13.5	SB(rs)c: IV	131	mag 10.42v star W 3′0.
08 42 57.1	−20 03 06	ESO 563-14	13.0	2.3 x 0.5	13.0	SBd? sp	83	Mag 9.76v star E 5′9.
08 43 15.6	−20 39 45	ESO 563-16	13.2	2.5 x 1.0	14.0	SAB(s)m IV	92	Mag 9.40v star W 4′4.
08 44 30.8	−20 21 02	ESO 563-17	12.1	1.8 x 1.0	12.6	SA(rs)a?	29	Stars superimposed on E and W edges; mag 7.40v star NNW 5′8.
08 47 16.8	−20 02 09	ESO 563-21	13.2	3.3 x 0.5	13.6	SAbc: sp	166	**ESO 563-22** NNE 5′9; mag 9.51v star NW 10′3.
08 50 25.3	−19 31 48	ESO 563-26	15.4	1.5 x 0.2	14.0	Scd	149	Mag 10.74v star SW 2′6; **ESO 568-27** N 5′9, located 1′8 E of mag 11.21v star.
08 50 45.6	−21 57 47	ESO 563-28	14.8	1.8 x 0.5	14.6	SBa: pec	15	Mag 9.80v star NW 6′7.
08 52 18.4	−17 44 42	ESO 563-31	12.6	1.3 x 0.9	12.7	(R′)SB(s)0°:	141	Mag 10.76v star at SE end.
08 57 11.4	−20 34 37	ESO 563-36	13.0	1.5 x 0.9	13.2	Sa	135	Triangle of mag 10.09v, 10.59v and 11.08v stars W 7′5; note: this triangle includes **MCG −3-23-12**; **ESO 563-34** NW 4′6.
09 02 46.2	−20 43 32	ESO 564-11	12.8	0.8 x 0.5	11.7	Sa?	156	Pair with MCG −3-23-15 N edge; mag 10.84v star W 9′0.
09 11 54.8	−20 07 08	ESO 564-27	13.4	5.0 x 0.4	14.0	Scd: sp	169	Mag 10.25v star NW 5′3.
09 13 12.0	−19 24 31	ESO 564-30	14.2	1.9 x 0.9	14.7	IB(s)m V-VI	125	Mag 8.84v star NW 5′4.
09 14 36.9	−19 33 08	ESO 564-31	13.6	2.3 x 0.6	13.8	SAB(s)c II	114	Mag 7.46v star E 9′2.
09 14 36.6	−21 58 07	ESO 564-32	14.5	2.0 x 0.4	14.1	SB(s)bc? II-III	15	Mag 9.14v star SW 11′4.
09 18 12.4	−18 26 38	ESO 564-36	13.6	1.7 x 1.2	14.2	SB(r)c II-III	66	mag 9.61v star NNW 7′7.
09 21 27.6	−22 29 59	ESO 565-1	14.1	2.4 x 0.5	14.1	IB(s)m sp V	28	Mag 10.03v star E 2′3.
09 27 37.4	−19 41 42	ESO 565-7	12.4	1.2 x 1.0	12.5	S0	117	Mag 9.61v star W 9′0.
09 28 08.1	−19 54 47	ESO 565-8	13.0	0.7 x 0.4	11.5	SA?(r)0	146	Mag 10.42v star NNW 7′9.
09 29 16.5	−20 22 47	ESO 565-11	12.9	1.7 x 1.3	13.6	(R)SB(r)0/a pec	153	Mag 8.35v star SW 5′2.
09 34 43.7	−21 55 42	ESO 565-19	12.8	1.2 x 0.7	12.6	E?	170	Pair with MCG −4-23-7 SW edge; mag 9.84v star NW 9′1.
09 36 09.8	−21 43 40	ESO 565-21	13.8	0.9 x 0.4	12.6	S0	1	Mag 10.44v star N 1′4.
09 36 34.8	−21 13 33	ESO 565-22	13.9	0.8 x 0.3	12.2	Sa:	105	Mag 11.6 star NE 2′8; NGC 2935 N 6′3.
09 37 06.5	−18 11 34	ESO 565-24	13.8	0.8 x 0.6	12.9	Sab	156	Mag 13.6 star NE 1′5.
09 37 07.1	−22 10 21	ESO 565-25	13.8	0.8 x 0.4	12.4	S	111	Mag 10.59v star NW 10′5.
09 37 22.8	−21 01 32	ESO 565-27	14.1	0.7 x 0.5	12.8	SBa	128	Small, faint anonymous galaxy SW 5′6; NGC 2935 SW 10′8.
09 37 49.0	−22 24 13	ESO 565-29	13.6	2.0 x 1.2	14.4	SB(s)d: IV	60	Mag 12.3 star NE edge; mag 9.69v star SW 10′9.
09 38 01.4	−20 20 38	ESO 565-30	12.3	1.7 x 1.1	12.8	SA(rs)0$^-$:	15	Mag 121.8 star superimposed N of center; mag 10.12v star NW 1′5; **ESO 565-26** S 20′3.
09 39 12.8	−21 03 54	ESO 565-33	14.6	1.7 x 0.6	14.5	SB(s)m V	103	Mag 9.77v star W 2′6; mag 9.34v star NE 6′4.
09 43 16.9	−19 52 20	ESO 566-2	14.3	1.5 x 0.4	13.7	SB0/a? sp	81	Mag 10.62v star NW 9′9.
09 44 58.8	−19 43 37	ESO 566-7	14.7	1.7 x 0.9	15.0	SBb pec?	9	Thin arms form a open "S" shape blending into ESO 566-8 N.
09 45 00.3	−19 42 37	ESO 566-8	14.7	0.9 x 0.9	14.4			Single arm extends E then NW from bright core.
09 46 07.7	−21 08 57	ESO 566-10	14.0	1.4 x 0.5	13.5	S0°: pec	69	Mag 9.91v star NNW 9′0.
09 49 10.8	−20 21 54	ESO 566-14	14.1	1.5 x 0.3	13.1	Sbc	9	Mag 9.03v star NE 12′9.
09 51 13.2	−18 28 32	ESO 566-19	13.3	1.7 x 1.6	14.2	SB(rs)cd II-III	164	Mag 10.74v star SE 5′4.
09 53 33.4	−19 35 02	ESO 566-24	12.9	1.5 x 1.3	13.5	SB(r)bc I-II	75	Mag 7.69v star N 5′9.

GALAXIES

RA h m s	Dec ° ′ ″	Name	Mag (V)	Dim ′ Maj x min	SB	Type Class	PA	Notes
09 56 08.3	−21 59 20	ESO 566-30	13.9	1.9 x 1.0	14.4	SAB(rs)bc pec III-IV	16	Mag 10.08v star W 5′3.
10 01 31.5	−19 32 25	ESO 567-5	14.3	1.4 x 0.4	13.5	Sb? sp	72	Mag 8.93v star S 8′5; **MCG −3-26-10** NNW 7′9.
10 01 32.4	−20 23 08	ESO 567-6	15.1	1.9 x 0.3	14.3	Sd: sp	7	Mag 9.78v star N 7′1.
08 58 12.7	−19 11 33	IC 524	13.9	1.1 x 0.6	13.3	S0	17	Mag 10.05v star NNW 2′3.
09 05 33.1	−19 12 27	IC 2437	12.9	1.8 x 1.1	13.5	SAB(rs)0$^-$:	123	Mag 13.9 star on S edge; mag 9.11v star SW 10′7.
09 59 55.5	−29 37 03	IC 2531	12.0	6.9 x 0.6	13.4	Sc: sp	75	Single, narrow dust lane center one third of galaxy.
10 03 52.3	−27 34 13	IC 2537	12.1	2.6 x 1.7	13.6	SAB(rs)c II	26	Numerous bright knots or stars over face of galaxy.
08 36 19.8	−20 17 04	MCG −3-22-7	13.5	1.1 x 0.7	13.1	Sc	171	
08 41 22.5	−20 15 52	MCG −3-22-11	14.5	1.5 x 0.2	13.0	Sc:	45	Mag 13.5 start SW edge; mag 10.6 star and ESO 563-11 SE 6′0.
08 45 50.3	−17 31 58	MCG −3-23-3	14.2	0.4 x 0.3	11.7	Sbc	150	
08 49 19.0	−19 00 17	MCG −3-23-7	13.4	0.6 x 0.4	11.8	S0	12	
08 50 01.8	−19 44 26	MCG −3-23-8	14.2	1.0 x 0.8	13.8	SB(rs)ab:	147	
08 51 06.9	−17 33 53	MCG −3-23-10	13.2	2.6 x 0.7	13.7	SA(rs)d: II-III	75	Several superimposed stars and/or bright knots involved SW part of galaxy.
09 02 45.1	−20 42 54	MCG −3-23-15	12.9	0.5 x 0.5	11.3	IBm pec:		Pair with ESO 564-11 S edge; mag 10.84v star W 9′0.
09 05 08.1	−18 31 13	MCG −3-23-17	13.1	1.0 x 0.3	11.7	S	85	Mag 9.81v star ESE 2′3.
09 05 27.6	−18 31 30	MCG −3-23-18	13.7	0.8 x 0.7	13.0	S	70	Mag 9.81v star WSW 3′0.
09 16 31.1	−17 35 48	MCG −3-24-4	14.1	1.9 x 0.4	13.6	SBd? III	111	Mag 9.50v star E 6′6.
09 19 35.9	−17 07 29	MCG −3-24-6	14.2	1.0 x 0.6	13.5	SBcd	42	
09 27 14.1	−19 31 59	MCG −3-24-9	13.1	1.1 x 0.5	12.3	(R_1)SB(s)a	59	
09 35 38.8	−17 23 12	MCG −3-25-10	13.3	0.9 x 0.7	12.6	(L)SB(s)0°	0	Two anonymous, elongated galaxies E 1′7.
09 38 01.2	−20 51 37	MCG −3-25-12	13.9	1.0 x 0.5	13.0	(R'_2)SB(s)b	63	Mag 9.00v star ESE 8′2.
09 39 05.4	−17 49 42	MCG −3-25-14	14.1	0.7 x 0.7	13.1	(R′)SA(s)ab		
09 44 06.4	−21 17 11	MCG −3-25-18	13.3	1.3 x 0.4	12.4	Sa	11	Anonymous elongated galaxy SW 1′2.
09 51 08.3	−17 30 02	MCG −3-25-25	13.8	0.8 x 0.3	12.1	Sb	123	
09 52 11.9	−20 48 26	MCG −3-25-26	13.3	0.9 x 0.8	12.8	S(rs)ab?	32	
09 54 35.4	−19 53 34	MCG −3-25-31	14.1	1.1 x 0.9	13.9	SB(r)c	168	
09 59 40.2	−20 47 17	MCG −3-26-5	13.4	1.0 x 0.4	12.3	S0	103	**MCG −3-26-4** W 2′8.
10 00 10.3	−19 37 20	MCG −3-26-6	13.2	0.3 x 0.3	10.7	E?		= **Hickson 42C**; located on NW edge of NGC 3091.
10 00 43.5	−20 22 09	MCG −3-26-9	13.2	1.0 x 0.7	12.7	(R)SB(r)0$^+$	135	Several stars on E edge.
10 03 18.9	−21 25 51	MCG −3-26-15	13.8	1.4 x 1.2	14.2	SB(s)cd III-IV	176	Mottled, with bright star at southern tip.
08 39 45.3	−23 27 33	MCG −4-21-6	13.0	1.0 x 0.8	12.7	S0	53	
08 40 57.1	−23 54 50	MCG −4-21-7	13.7	1.2 x 0.4	12.8	Sb	69	Pair of bright stars SW end.
08 49 06.1	−26 19 19	MCG −4-21-9	14.0	0.6 x 0.5	12.5	Sab	142	
09 27 41.2	−24 14 06	MCG −4-23-4	13.7	0.7 x 0.6	12.6		9	
09 33 35.2	−25 08 13	MCG −4-23-6	14.0	0.6 x 0.5	12.8	E/S0	177	
09 34 40.6	−21 56 14	MCG −4-23-7	14.0	0.5 x 0.4	12.1		67	Pair with ESO 565-19 NE edge; mag 9.84v star NW 9′1.
09 35 57.8	−24 58 51	MCG −4-23-9	14.1	1.1 x 0.5	13.3	Sbc	151	Small anonymous galaxy NW 1′7.
09 40 06.8	−25 03 29	MCG −4-23-12	14.1	1.1 x 1.0	14.0	SAB(r)b pec	49	
09 45 16.9	−27 24 23	MCG −4-23-14	13.8	1.3 x 0.4	13.0	SB(rs)0/a	85	Mag 8.62v star WNW 6′0.
09 50 27.4	−21 48 10	MCG −4-24-1	13.3	1.4 x 1.0	13.6	SB(s)bc I-II	168	Bright spiral arms, many knots.
09 55 14.1	−22 29 01	MCG −4-24-6	14.1	1.2 x 0.3	12.9	Sbc	159	
09 46 18.4	−29 07 56	MCG −5-23-13	13.9	1.2 x 0.5	13.2	SB(r)a	0	
08 46 00.8	−19 18 12	NGC 2665	12.2	2.0 x 1.5	13.3	(R)SB(rs)a	144	Very faint, elliptical outer shell NW and SE.
08 57 01.3	−24 40 25	NGC 2717	12.3	2.1 x 1.5	13.4	SAB(s)0$^-$	11	Several faint stars on E edge; a mag 10.21v star lies 2′0 NE; ESO 496-22 S 4′9.
09 05 11.2	−19 05 07	NGC 2754	14.1	0.6 x 0.5	12.7	S0° pec	163	Almost stellar; mag 11.17v star WSW 11′3; NGC 2758 NE 5′4.
09 05 31.3	−19 02 35	NGC 2758	13.4	1.9 x 0.5	13.2	(R′)SBbc pec? III-IV	19	NGC 2754 SW 5′4; mag 10.84v star ESE 9′8.
09 07 41.9	−23 37 11	NGC 2772	13.0	1.5 x 0.4	12.3	Sb: pec sp	170	Very small galaxy **ESO 497-14A** on SW tip of galaxy.
09 12 19.6	−24 10 22	NGC 2784	10.2	5.5 x 2.2	12.7	SA(s)0°:	73	Numerous stars scattered over the surface.
09 16 19.8	−23 38 00	NGC 2815	11.9	3.5 x 1.1	13.2	SB(r)b: II	10	Several superimposed stars or bright knots; mag 7.58v star NW 12′8.
09 16 48.0	−26 48 58	NGC 2821	13.0	2.0 x 0.4	12.6	Sbc: sp	100	
09 17 53.5	−22 21 10	NGC 2835	10.5	6.6 x 4.4	14.0	SB(rs)c I-II	8	Numerous superimposed stars and bright knots. Located between a pair of mag 10 stars.
09 23 30.5	−23 09 43	NGC 2865	11.7	2.5 x 1.8	13.3	E3-4	162	Large core in smooth envelope; mag 9.64v star N 4′8.
09 26 19.7	−28 02 06	NGC 2888	12.6	1.4 x 1.0	13.0	E$^+$:	158	
09 26 56.7	−24 46 58	NGC 2891	12.6	1.5 x 1.4	13.3	SA0$^-$:	145	Large bright core, faint smooth envelope.
09 34 12.1	−20 51 33	NGC 2920	13.1	0.9 x 0.6	12.3	Sa pec:	129	
09 34 31.2	−20 55 14	NGC 2921	12.0	2.8 x 1.0	13.0	(R′)SAB(rs)a pec	83	
09 36 45.0	−21 07 39	NGC 2935	11.3	3.6 x 2.8	13.7	(R′)SAB(s)b I-II	0	Small, faint anonymous galaxy NE 5′3.
09 37 41.1	−22 02 03	NGC 2945	12.1	1.6 x 1.2	12.7	SA0$^-$:	168	Mag 12.1 star N 4′9.
09 39 17.1	−19 06 04	NGC 2956	14.2	0.9 x 0.3	12.6	SB(s)b? II	55	Mag 9.96v star ENE 3′3.
09 43 41.1	−20 28 36	NGC 2983	11.8	2.5 x 1.5	13.1	SB(rs)0$^+$	95	Strong NNE-SSW core in uniform E-W envelope; mag 6.57v star N 7′2.
09 44 16.2	−21 16 41	NGC 2986	10.8	3.2 x 2.6	13.0	E2	30	Chain of 10 faint stars cascade from just inside SE edge southward.
09 45 25.2	−18 22 25	NGC 2989	13.0	1.7 x 0.9	13.4	SAB(s)bc: I-II	38	Located 2′8 NW of mag 10.77v star.
09 46 30.2	−21 34 17	NGC 2996	12.7	1.5 x 1.3	13.3	S0$^+$ pec:	115	Anonymous galaxy NW 3′0; mag 10.64v star ENE 1′1.
09 49 29.1	−21 44 32	NGC 3025	12.9	1.5 x 1.2	13.4	SA0° pec:	110	Located 2′6 NW of a mag 9.34v star.
09 49 54.1	−19 11 06	NGC 3028	12.9	0.9 x 0.8	12.4	Sb pec: II	108	
09 51 24.1	−27 00 37	NGC 3037	13.0	1.2 x 1.1	13.2	IB(s)m III-IV	43	Mag 9.74v star N 5′5.
09 53 17.7	−18 38 41	NGC 3045	13.0	1.4 x 0.6	12.7	SA(r)b? II-III	20	**ESO 566-25** E 6′1.
09 53 58.8	−27 17 13	NGC 3051	11.8	1.9 x 1.9	13.1	(R′)SB(s)0$^-$ pec:		
09 54 28.1	−18 38 22	NGC 3052	12.2	2.0 x 1.3	13.1	SAB(r)c: I-II	102	
09 54 28.5	−25 42 13	NGC 3054	11.8	3.8 x 2.3	14.0	SAB(r)b I-II	118	Tightly wound multi-branching arms.
09 54 32.8	−28 17 53	NGC 3056	11.7	1.8 x 1.1	12.3	(R)SA(s)0$^+$:	16	Faint star just N of nucleus.
09 57 23.9	−19 21 20	NGC 3072	12.8	1.9 x 0.6	12.8	S0/a? sp	71	**ESO 566-32** N 7′9.
09 57 37.7	−18 10 42	NGC 3076	13.3	1.0 x 0.9	13.1	Sab pec:		
09 58 24.6	-26 55 38	NGC 3078	11.1	2.5 x 2.1	12.9	E2-3	177	Mag 14.4 star on W edge.
09 59 29.5	-22 49 36	NGC 3081	12.0	2.1 x 1.6	13.1	(R)SAB(r)0/a	74	Very faint galactic material extends beyond the published size to the N and S.
09 59 06.9	-27 07 46	NGC 3084	12.3	1.8 x 1.6	13.3	(R′)SB(s)ab pec	20	Several stars superimposed S of nucleus.
09 59 29.3	-19 29 33	NGC 3085	13.2	1.2 x 0.4	12.3	S0°: sp	119	Anonymous galaxy NW 2′7.
09 59 36.5	-28 19 51	NGC 3089	12.2	1.8 x 1.0	12.7	SAB(rs)b II	139	Lies 2′2 W of mag 8.20v star.
10 00 14.2	-19 38 10	NGC 3091	11.1	3.0 x 1.9	13.0	E3:	149	= **Hickson 42A**; **Hickson 42D** lies S 2′2.
10 00 33.2	-19 39 39	NGC 3096	13.4	1.0 x 0.8	13.0	SB(rs)0°	170	= **Hickson 42B.** A mag 12 star 1′2 NE; a mag 11.1 star 1′8 NW; and an anonymous galaxy 1′3 SW.
10 03 05.9	-26 09 24	NGC 3109	9.9	19.1 x 3.7	14.3	SB(s)m sp IV-V	93	Numerous knots and superimposed stars.
10 03 59.2	-20 46 56	NGC 3112	14.9	0.9 x 0.2	12.9	Sb?	47	

152

(Continued from previous page)

GALAXY CLUSTERS

152

RA h m s	Dec ° ′ ″	Name	Mag 10th brightest	No. Gal	Diam ′	Notes
09 32 12.0	−24 53 00	A 3420	16.3	50	17	**ESO 498-12** on SSW edge; all others anonymous, stellar.
09 37 54.0	−20 20 00	AS 617	15.6	23	22	See notes for ESO 565-30; all others anonymous, stellar.

OPEN CLUSTERS

RA h m s	Dec ° ′ ″	Name	Mag	Diam ′	No. ★	B ★	Type	Notes
08 42 54.0	−27 52 00	ESO 432-3		8			cl	
09 55 30.0	−28 59 00	ESO 435-9		15			cl	
08 37 15.0	−29 57 00	NGC 2627	8.4	9	60	11.0	cl	Rich in stars; moderate brightness range; slight central concentration; detached.

PLANETARY NEBULAE

RA h m s	Dec ° ′ ″	Name	Diam ″	Mag (P)	Mag (V)	Mag cent ★	Alt Name	Notes
08 40 17.0	−20 54 14	PN G244.5+12.5	400	14.3		18.3	PK 244+12.1	Brightest on NE and SW edges; mag 12.7 star 1′2 N of center; mag 10.51v star on N edge; mag 7.98v star NE 7′9.
08 57 45.9	−28 57 36	PN G253.5+10.7	58	15.3		16.6	PK 253+10.1	Mag 14.2 star on SW edge; mag 13.5 star S 1′3; mag 11.9 star N 2′6.

153

GALAXIES

153

RA h m s	Dec ° ′ ″	Name	Mag (V)	Dim ′ Maj x min	SB	Type Class	PA	Notes
07 16 26.3	−29 37 15	ESO 428-13	14.1	1.2 x 0.4	13.1	S?	126	Several stars superimposed.
07 16 31.1	−29 19 29	ESO 428-14	12.2	1.8 x 1.1	12.8	SAB(r)0° pec	136	Group of four stars, the brightest mag 9.48v, SW 2′7.
07 17 13.7	−28 19 16	ESO 428-16	14.2	1.0 x 0.6	13.5	Sbc	108	Several stars superimposed; pair of stars, mags 11.14v and 10.93v, NE 2′3.
07 18 37.3	−29 22 22	ESO 428-20	14.4	1.5 x 0.2	13.0	Sc	154	Mag 8.74v star N 3′6.
07 20 21.0	−28 59 16	ESO 428-22	13.1	1.4 x 1.2	13.5		37	Numerous stars superimposed overall; mag 10.12v star SE edge.
07 22 09.7	−29 14 06	ESO 428-23	12.6	2.3 x 1.4	13.7	(R′)SB(rs)ab: III-IV	39	Numerous stars superimposed; mag 10.36v star W 2′2.
07 23 42.2	−29 39 07	ESO 428-29	12.3	1.6 x 1.3	12.9	SAB(rs)b: III-IV	148	Mag 10.00v star E 3′3.
07 49 26.8	−29 21 17	ESO 429-19	15.9	1.5 x 0.3	14.8	S	164	Mag 10.10v star N 1′2.
07 55 12.3	−28 09 58	ESO 430-1	12.4	1.9 x 1.9	13.6	SB(r)b?		Mag 7.19v star S 7′2.
08 07 08.5	−28 03 14	ESO 430-20	13.6	1.7 x 0.7	13.6	SAB(s)d? III-IV	110	Many knots; mag 10.13v star WNW 7′3.
08 17 38.5	−29 43 52	ESO 431-1	12.9	1.7 x 1.5	13.8	SA(rs)c: III-IV	45	Mag 7.14v star N 2′4.
07 55 29.7	−22 43 51	ESO 494-4	13.9	1.5 x 0.3	12.9	Sb:	87	Mag 11.3 star SSE 4′7.
07 56 54.4	−24 54 24	ESO 494-7	13.3	3.0 x 0.4	13.3	Sbc: sp	51	mag 9.06v star NW 4′5.
07 59 57.3	−24 34 17	ESO 494-10	13.9	1.1 x 0.4	12.8	Sa:	105	Mag 14.8 star superimposed W end.
08 02 11.1	−22 56 32	ESO 494-12	13.8	1.3 x 0.4	13.0	Sc:	38	Mag 12.8 star on W edge.
08 03 48.8	−25 37 01	ESO 494-16	13.8	1.2 x 0.8	13.6	SBab	144	Mag 8.98v star ESE 11′0.
08 04 57.5	−22 58 31	ESO 494-19	13.3	1.3 x 1.0	13.4	S0/a	106	Numerous stars superimposed; mag 9.30v star E 4′8.
08 05 29.3	−22 55 31	ESO 494-21	15.2	1.0 x 0.2	13.3	Sc:	148	Mag 10.44v star SW 3′8.
08 05 32.2	−24 48 59	ESO 494-22	12.8	2.1 x 0.5	12.7	Sa sp	46	Mag 13.5 star superimposed E of center; mag 10.16v star N 4′4.
08 05 50.7	−25 58 47	ESO 494-24	13.7	0.9 x 0.5	12.7	Sa:	164	Pair of stars, mags 11.4 and 12.4, S 2′4.
08 05 59.2	−27 23 47	ESO 494-25	12.4	2.2 x 1.1	13.3	SA(rs)0/a?	148	Mag 8.84v star SW 5′1; ESO 494-26 S 8′4.
08 06 11.0	−27 31 41	ESO 494-26	11.5	4.6 x 2.9	14.1	SAB(s)b pec: II	152	Numerous superimposed stars; mag 8.84v star NW 8′6.
08 06 42.7	−22 54 18	ESO 494-27	13.5	1.3 x 0.4	12.6	Sab	172	Mag 11.0 star superimposed S of center.
08 07 49.1	−25 57 18	ESO 494-29	14.3	1.7 x 0.3	13.4	Sc	87	Mag 8.99v star E 8′2.
08 10 34.8	−22 42 33	ESO 494-31	13.1	1.2 x 0.6	12.7	S0	110	Mag 12.8 star E 2′4.
08 12 50.4	−27 33 16	ESO 494-35	12.2	1.6 x 0.7	12.2	S0°: sp	23	Mag 13.9 star on E edge; mag 9.30v star E 5′4.
08 14 53.4	−22 52 42	ESO 494-36	13.1	1.2 x 1.1	13.3	S0	172	Close pair of stars, mags 12.3 and 13.3, W 2′1.
08 17 02.6	−23 25 40	ESO 494-39	13.7	1.0 x 0.6	13.0	Sb	118	Mag 10.22v star SSW 2′8.
08 17 26.8	−24 40 55	ESO 494-42	13.1	1.8 x 0.8	13.4	S0°:	55	Mag 14.4 star superimposed SW 0f center; mag 10.96v star E 1′7.
08 19 14.8	−25 11 20	ESO 495-5	12.7	1.0 x 1.0	12.5	SAB(r)a:		Mag 10.68v star W 1′1.
08 19 24.4	−24 47 11	ESO 495-6	13.4	1.1 x 0.5	12.6	SB0°? pec	102	Mag 9.41v star W 4′8.
08 21 17.5	−25 46 27	ESO 495-9	14.1	2.0 x 0.3	13.4	Sb: sp	34	Mag 9.61v star S 3′2; mag 10.14v star NE 3′9.
08 23 19.0	−26 11 48	ESO 495-11	12.8	1.3 x 1.2	13.1	SB(s)m III-IV	7	Mag 9.57v star NE 2′6; mag 5.89v star SSW 11′2.
08 23 51.9	−25 50 19	ESO 495-12	13.8	1.9 x 0.4	13.3	SB(s)b? II-III	86	Mag 11.9 star on N edge.
08 26 26.4	−27 21 39	ESO 495-13	13.5	0.9 x 0.9	13.4	E/S0:		Numerous stars superimposed; mag 8.67v star W 5′9.
08 32 57.4	−22 54 02	ESO 495-17	13.7	1.0 x 0.3	12.3	S	110	Mag 9.04v star NNW 1′1; NGC 2613 SE 8′2.
08 36 14.9	−26 24 33	ESO 495-21	11.6	1.7 x 1.3	12.3	I0? pec	169	Two bright knots or superimposed stars W of center; mag 12.1 star W 1′6.
07 46 25.3	−18 32 52	ESO 560-12	14.4	2.2 x 0.5	14.4	SB(s)cd? sp IV	145	Mag 8.87v star N 2′8.
07 47 51.9	−18 44 56	ESO 560-13	13.0	3.9 x 0.7	14.0	SB(s)bc? II-III	141	Mag 9.76v star NE 1′3.
07 48 13.9	−18 38 35	ESO 560-14	15.3	1.6 x 0.5	14.9	IBm V-VI	3	Mag 10.66v star E 5′7.
07 55 25.8	−21 20 30	ESO 561-2	12.5	2.2 x 1.7	13.8	SB(s)c: V-VI	141	Mag 10.20v star on W edge.
07 57 27.4	−19 14 37	ESO 561-3	14.0	1.9 x 0.3	13.3	SB(s)cd pec sp	105	Mag 7.77v star N 6′6.
08 03 02.5	−18 42 11	ESO 561-9	14.2	1.5 x 0.6	13.9	(R′)SB(r)a	31	Mag 10.74v star W 2′0; mag 7.88v star NNE 8′0.
08 11 52.0	−18 18 04	ESO 561-23	14.2	2.4 x 0.3	13.7	Sbc: sp	91	Mag 11.11v star W 4′7; mag 8.11v star WSW 12′3.
08 12 14.1	−19 19 26	ESO 561-24	15.6	1.0 x 0.2	13.7	S	118	Mag 8.66v star SE 3′9.
08 12 22.7	−21 31 57	ESO 561-25	15.5	1.7 x 0.2	14.2	S	158	Mag 9.61v star E 1′9.
08 13 37.1	−18 58 06	ESO 561-27	16.5	1.7 x 0.4	15.9	Sc:	13	Mag 8.65v star N 6′7.
08 14 18.3	−18 17 22	ESO 561-30	15.0	1.8 x 0.6	15.0	IB(s)m V-VI	98	Several stars superimposed; mag 9.07v star NNW 9′5.
08 14 56.2	−22 01 03	ESO 561-32	15.6	1.7 x 0.2	14.3	Sc:	138	Close pair of mag 11.5, 10.78v stars SW 4′7.
08 15 33.1	−20 52 35	ESO 561-33	14.4	1.5 x 0.8	14.4	SAB(rs)c: III	84	Two bright knots or stars at center; mag 10.59v star WNW 8′3.
08 26 19.3	−20 50 54	ESO 562-6	14.2	1.1 x 0.6	13.6	Sa	144	Mag 10.58v star WNW 7′2.
08 28 41.2	−21 44 13	ESO 562-7	12.5	2.4 x 0.7	12.9	(R′)SB(rs)ab?	104	Pair of mag 10.79v, 10.74v stars SW 4′1.
08 33 10.8	−20 53 57	ESO 562-13	14.4	1.5 x 0.9	14.6	SAB(s)c II	52	Mag 10.03v star N 6′5.
08 33 16.9	−17 57 31	ESO 562-14	13.1	1.5 x 0.3	12.1	SBb pec sp	162	Mag 10.89v star SSE 4′0.
08 33 56.5	−21 52 57	ESO 562-19	13.2	2.2 x 0.5	13.2	SB(s)dm: sp IV	68	Mag 8.14v star S 9′2.
08 36 34.9	−20 28 16	ESO 562-23	12.4	2.3 x 0.5	12.4	SB(r)0$^+$? sp	170	Mag 10.78v star N 2′0; **ESO 562-20** SW 9′3.
08 37 11.3	−22 14 59	ESO 563-2	14.3	1.7 x 0.4	13.7	SB(s)c? sp IV-V	95	Mag 10.80v star N 6′2.
08 37 18.8	−20 56 27	ESO 563-3	13.5	1.5 x 0.3	12.5	SB(r)0/a pec sp	88	Mag 8.17v star SW 16′5.
08 41 35.3	−20 18 55	ESO 563-11	13.1	1.7 x 0.9	13.4	SB(rs)bc: III	82	Several stars superimposed; mag 10.6 star E 2′3.
08 41 38.9	−20 44 39	ESO 563-12	13.6	1.7 x 0.8	13.8	SB(s)d pec: III-IV	69	Mag 10.65v star WNW 3′6.
08 42 32.3	−19 52 20	ESO 563-13	14.5	1.5 x 0.3	13.5	SB(rs)c: IV	131	mag 10.42v star W 3′0.
08 42 57.1	−20 03 06	ESO 563-14	13.0	2.3 x 0.5	13.0	SBd? sp	83	Mag 9.76v star E 5′9.
08 43 15.6	−20 39 45	ESO 563-16	13.2	2.5 x 1.0	14.0	SAB(s)m IV	92	Mag 9.40v star W 4′4.
08 18 46.0	−25 22 15	IC 2311	11.5	1.3 x 1.3	12.1	E0:		**MCG −4-20-6** SW 1′7.

(Continued from previous page)

GALAXIES

RA h m s	Dec ° ′ ″	Name	Mag (V)	Dim ′ Maj x min	SB	Type Class	PA	Notes
08 24 10.2	−18 46 33	IC 2367	11.9	2.4 x 1.7	13.3	SB(r)b II-III	55	Mag 10.57v star NW 7'.1.
08 02 05.2	−18 46 04	MCG −3-21-1	14.0	0.7 x 0.7	13.1	S		
08 29 38.4	−17 17 54	MCG −3-22-2	14.4	0.9 x 0.3	12.8		144	
08 36 19.8	−20 17 04	MCG −3-22-7	13.5	1.1 x 0.7	13.1	Sc	171	
08 41 22.5	−20 15 52	MCG −3-22-11	14.5	1.5 x 0.2	13.0	Sc:	45	Mag 13.5 start SW edge; mag 10.6 star and ESO 563-11 SE 6'.0.
08 06 38.7	−25 59 12	MCG −4-20-1	13.4	1.0 x 0.4	12.2	Sa	78	Mag 10.40v star on E edge.
08 18 30.5	−25 24 42	MCG −4-20-5	13.6	1.1 x 0.3	12.5	E/S0	108	Mag 8.26v star NW 8'.7; IC 2311 NE 4'.3.
08 32 41.1	−24 02 33	MCG −4-21-1	13.5	0.7 x 0.4	12.0	Sab	177	Anonymous galaxy S 0'.4.
08 39 45.3	−23 27 33	MCG −4-21-6	13.0	1.0 x 0.8	12.7	S0	53	
08 40 57.1	−23 54 50	MCG −4-21-7	13.7	1.2 x 0.4	12.8	Sb	69	Pair of bright stars SW end.
07 23 54.6	−27 31 41	NGC 2380	11.2	2.0 x 1.9	12.5	SAB0°:	168	Numerous superimposed stars.
08 17 06.2	−27 27 31	NGC 2559	10.9	3.0 x 1.4	12.3	SB(s)bc pec: II	6	Mag 9.36v star on E edge, many stars superimposed on surface.
08 18 30.1	−21 49 01	NGC 2564	12.4	1.0 x 0.6	11.8	S0⁻	81	Mag 8.52v star W 9'.3.
08 18 45.4	−25 30 01	NGC 2566	11.0	2.9 x 1.7	12.6	(R′)SB(rs)ab pec: II	62	A stellar nucleus with numerous stars superimposed; mag 9.29v star SSW 6'.3.
08 33 23.0	−22 58 26	NGC 2613	10.3	6.5 x 1.4	12.5	SA(s)b II	113	Numerous stars superimposed; mag 9.04v star and ESO 495-17 NW 8'.2.

GALAXY CLUSTERS

RA h m s	Dec ° ′ ″	Name	Mag 10th brightest	No. Gal	Diam ′	Notes
08 26 24.0	−27 21 00	AS 610	16.2		17	Except ESO 495-13, all members anonymous, stellar.

OPEN CLUSTERS

RA h m s	Dec ° ′ ″	Name	Mag	Diam ′	No. ★	B ★	Type	Notes
07 31 45.1	−19 27 00	Bo 6	9.9	10	40	10.6	cl	Moderately rich in bright and faint stars with a strong central concentration; detached; involved in nebulosity.
08 24 12.0	−29 10 12	Cr 187	9.6	5	20		cl	Few stars; small brightness range; not well detached. (S) Sparse.
07 36 58.0	−20 31 42	Cz 31		5	40		cl:	(S) Sparse group.
07 50 28.0	−29 50 54	Cz 32		3	30		cl	
07 33 24.0	−28 11 00	ESO 429-2		5			cl	
08 42 54.0	−27 52 00	ESO 432-3		8			cl	
07 39 42.0	−27 18 00	ESO 493-3					cl	
07 59 18.0	−27 35 00	ESO 494-9		3			cl	
07 18 06.0	−18 36 00	ESO 559-2		1.2			cl	
07 28 42.0	−20 49 00	ESO 559-13		10			cl	
07 59 18.0	−22 41 00	ESO 561-5		7			cl	
07 18 06.0	−22 39 00	Haf 5		7	50	15.0	cl	
07 22 54.0	−29 29 48	Haf 7		3	40	14.0	cl	
07 35 22.0	−27 42 18	Haf 11		5	35	16.0	cl	
07 44 48.0	−28 23 00	Haf 14		10	50	14.0	cl	(S) Sparse group.
07 50 21.0	−25 27 06	Haf 16	10.0	5	30	12.0	cl	Moderately rich in stars; moderate brightness range; strong central concentration; detached.
07 52 44.0	−26 22 42	Haf 18	9.3	5	25	11.0	cl	
07 52 46.0	−26 16 42	Haf 19	9.4	2	39	14.0	cl	
08 01 11.0	−27 12 30	Haf 21	10.3	3	51	15.0	cl	Few stars similar in brightness; strong central concentration; detached.
08 12 36.0	−27 52 00	Haf 22		7	30	15.0	cl	(A) Rich field of stars, but no obvious cluster.
07 48 40.0	−26 56 48	Haf 25		2	20	14.0	cl:	(S) Sparse group.
07 30 04.0	−18 32 10	Mayer 3					cl	
07 18 42.0	−24 57 18	NGC 2362	3.8	6	60	8.0	cl	Rich in stars; large brightness range; strong central concentration; detached. (S) Tau CMa Cluster.
07 20 05.0	−21 53 00	NGC 2367	7.9	5	30	9.4	cl	Moderately rich in bright and faint stars; detached.
07 24 40.0	−20 56 54	NGC 2383	8.4	5	40	9.8	cl	Moderately rich in bright and faint stars; detached.
07 25 12.0	−21 01 24	NGC 2384	7.4	5	15	8.6	cl	Few stars; large brightness range; not well detached.
07 36 12.0	−20 37 00	NGC 2421	8.3	8	70	11.0	cl	Rich in stars; moderate brightness range; strong central concentration; detached.
07 40 54.0	−19 05 00	NGC 2432	10.2	7	50		cl	Moderately rich in stars; moderate brightness range; slight central concentration; detached.
07 44 30.0	−23 51 00	NGC 2447	6.2	10	80	9.0	cl	= **M 93**. Rich in stars; large brightness range; strong central concentration; detached.
07 47 35.0	−27 11 42	NGC 2453	8.3	4	76	9.5	cl	Moderately rich in bright and faint stars with a strong central concentration; detached.
07 48 59.0	−21 18 00	NGC 2455	10.2	15	50	12.0	cl	Moderately rich in stars; moderate brightness range; no central concentration; detached.
07 52 30.0	−26 26 00	NGC 2467	7.1	15	50		cl	Moderately rich in stars; large brightness range; strong central concentration; detached; involved in nebulosity. (A)Involved in Sh2-311.
07 55 06.0	−17 43 00	NGC 2479	9.6	11	45		cl	Moderately rich in stars; small brightness range; no central concentration; detached.
07 55 10.0	−24 16 00	NGC 2482	7.3	10	40	10.0	cl	Moderately rich in stars; small brightness range; not well detached.
07 55 39.0	−27 53 00	NGC 2483	7.6	9	45	9.3	cl	Moderately rich in stars; moderate brightness range; no central concentration; detached. Probably not a cluster.
08 00 48.0	−19 03 00	NGC 2509	9.3	12	70		cl	Moderately rich in stars; small brightness range; strong central concentration; detached.
08 04 53.0	−28 08 00	NGC 2527	6.5	10	45	8.6	cl	Moderately rich in stars; moderate brightness range; slight central concentration; detached.
08 07 00.0	−29 52 00	NGC 2533	7.6	6	124	9.0	cl	Rich in stars; moderate brightness range; slight central concentration; detached.
08 18 57.0	−29 45 00	NGC 2571	7.0	7	49	8.8	cl	Moderately rich in bright and faint stars; detached.
08 23 25.0	−29 30 00	NGC 2587	9.2	10	40		cl	Moderately rich in stars; moderate brightness range; no central concentration; detached. (S) Small group.
08 37 15.0	−29 57 00	NGC 2627	8.4	9	60	11.0	cl	Rich in stars; moderate brightness range; slight central concentration; detached.
07 19 34.0	−19 37 36	Ru 15		2	20	12.0	cl	(S) Small group.
07 23 09.0	−19 28 42	Ru 16		4	15	13.0	cl	Few stars; small brightness range; not well detached. Probably not a cluster. (S) Sparse group.
07 23 35.0	−23 11 18	Ru 17		3	15	12.0	cl	(S) Sparse group.
07 24 39.0	−26 11 00	Ru 18	9.4	8	40	12.0	cl	Moderately rich in stars; small brightness range; no central concentration; detached.
07 25 36.0	−21 30 00	Ru 19		5	20	13.0	cl:	
07 26 41.0	−28 49 54	Ru 20	9.5	6	30	11.0	cl	Moderately rich in stars; moderate brightness range; no central concentration; detached. Doubtful cluster. (S) Sparse group.
07 29 17.0	−29 10 30	Ru 22		3	15	12.0	cl:	(S) Sparse group.
07 30 40.0	−23 22 24	Ru 23		4	20	13.0	cl	(S) Sparse group.
07 36 47.0	−23 23 06	Ru 25		2	10	14.0	cl:	
07 37 36.0	−26 30 00	Ru 27		6	30	12.0	cl	Moderately rich in stars; moderate brightness range; not well detached. (S) Sparse group, identification uncertain. (A) DSS image shows N-S elongated possible cluster.
07 41 06.0	−24 19 00	Ru 29		5	20	13.0	cl:	
07 45 06.0	−25 31 36	Ru 32	8.4	6	30	10.0	cl:	Moderately rich in bright and faint stars; detached; involved in nebulosity.
07 45 48.0	−21 57 00	Ru 33		9	20	13.0	cl	(S) Sparse group.
07 45 54.0	−20 23 00	Ru 34	9.5	7	35	12.0	cl	Moderately rich in bright and faint stars; detached. Probably not a cluster.
07 48 24.0	−26 17 54	Ru 36	9.6	5	30	12.0	cl	Moderately rich in stars; small brightness range; not well detached.
07 49 47.0	−17 14 48	Ru 37		4	30	14.0	cl	
07 50 31.0	−20 11 18	Ru 38		2	15	13.0	cl:	(S) Small group.
07 52 20.0	−22 26 30	Ru 39		2	15	14.0	cl	

(Continued from previous page)

153 OPEN CLUSTERS 153

RA h m s	Dec ° ′ ″	Name	Mag	Diam ′	No. ★	B ★	Type	Notes
07 33 27.0	−20 31 06	Ru 40		2	10	13.0	cl:	(S) Sparse group.
07 53 48.0	−26 58 12	Ru 41		2	20	14.0	cl	(S) Sparse group.
07 57 37.0	−25 55 00	Ru 42		4	20	14.0	cl	
07 58 48.0	−28 54 00	Ru 43		7	25	12.0	cl	Few stars; small brightness range; not well detached.
07 58 50.0	−28 34 42	Ru 44	7.2	5	84	12.0	cl?	Moderately rich in stars; moderate brightness range; not well detached.
08 02 10.0	−19 27 54	Ru 46	9.1	3	15	12.0	cl	Few stars; moderate brightness range; strong central concentration; detached.
08 03 17.0	−26 46 12	Ru 49	9.6	2	10	13.0	cl	Few stars; small brightness range; slight central concentration; detached.
08 10 50.0	−27 00 00	Ru 53		6	40	10.0	cl	Moderately rich in stars; large brightness range; not well detached. (S) Sparse.
08 14 48.0	−26 58 00	Ru 57		5	30	12.0	cl	
08 32 30.0	−19 40 00	Ru 62		7	20	11.0	cl:	Few stars; moderate brightness range; not well detached. Probably not a cluster. (S) Sparse group.
08 29 52.0	−19 06 00	Ru 157		7	30	11.0	cl	Moderately rich in stars; small brightness range; not well detached. (S) Sparse group.
07 26 23.0	−24 12 42	Tr 6	10.0	6			cl	Few stars; moderate brightness range; no central concentration; detached.
07 27 21.0	−23 58 00	Tr 7	7.9	5	30	10.0	cl	Moderately rich in bright and faint stars; detached; involved in nebulosity. (S) Irregular structure.
07 55 40.0	−25 53 12	Tr 9	8.7	6	20	10.1	cl	Few stars; moderate brightness range; slight central concentration; detached.
07 55 06.4	−25 21 54	Wa 3		1.9	20	12.9	cl	

BRIGHT NEBULAE

RA h m s	Dec ° ′ ″	Name	Dim ′ Maj x min	Type	BC	Color	Notes
07 35 30.0	−18 46 00	Sh2-307	4 x 4	E	3-5	4-4	
07 52 30.0	−26 24 00	Sh2-311	16 x 12	E	1-5	3-4	This bright, circular nebula is involved in the open cluster NGC 2467. There is a prominent E-W lane, a little N of center,
07 19 36.0	−24 02 00	vdB 96	10 x 5	R	3-5	1-4	Three 9th mag stars in a chain with nebulosity; brightest around the NW star.
07 36 24.0	−25 20 00	vdB 98	10 x 10	R	3-5	1-4	Nebula with a mag 7.3 star involved.

PLANETARY NEBULAE

RA h m s	Dec ° ′ ″	Name	Diam ″	Mag (P)	Mag (V)	Mag cent ★	Alt Name	Notes
07 41 55.4	−18 12 31	NGC 2440	79	10.8	9.4	17.6	PK 234+2.1	Elongated E-W; mag 8.43v star E 2′.9.
07 47 26.5	−27 20 09	NGC 2452	29	12.6	12.0	17.7	PK 243−1.1	Appears slightly elongated E-W; mag 11.12v star ESE 3′.2.
07 21 14.9	−18 08 37	PN G232.4−1.8	36	12.9	12.6		PK 232−1.1	Slightly elongated SSE-NNW; mag 11.2 star S 0′.8; mag 11.4 star N 3′; mag 9.65v star E 5′.1.
07 27 56.6	−20 13 25	PN G234.9−1.4	13	13.4	13.5	14.1	PK 235−1.1	Mag 9.00v star WNW 2′.0; mag 8.43v star NW 2′.8.
07 19 21.6	−21 43 58	PN G235.3−3.9	3	15.2		14.0	PK 235−3.1	Small triangle of mag 12 stars SW 2′.1.
07 50 11.6	−19 18 16	PN G236.7+3.5	37	15.9		21.0	PK 236+3.1	Mag 11.9 star on N edge; mag 10.92v star SW 2′.5.
08 08 44.1	−19 14 01	PN G238.9+7.3	40				PK 238+7.2	Mag 10.57v star S 1′.8; mag 4.40v star 16 Puppis E 4′.2.
08 10 41.7	−20 31 33	PN G240.3+7.0	8	14.4		15.6	PK 240+7.1	Mag 12.4 star SW 2′.1; mag 11.9 star NE 3′.3.
07 55 11.4	−23 38 13	PN G241.0+2.3	34	11.8		15.7	PK 241+2.1	Rectangular shape, oriented SE-NW; mag 10.96v star NE 3′.8; mag 10.86v star SSE 4′.0.
08 40 17.0	−20 54 14	PN G244.5+12.5	400	14.3		18.3	PK 244+12.1	Brightest on NE and SW edges; mag 12.7 star 1′.2 N of center; mag 10.51v star on N edge; mag 7.98v star NE 7′.9.
08 02 29.1	−27 41 57	PN G245.4+1.6	7	13.8	14.2	17.9	PK 245+1.1	Mag 10.021v star NW 1′.2; mag 10.4 star SE 2′.8.
08 31 42.8	−27 45 34	PN G249.0+6.9	11			13.0	PK 249+6.1	Mag 12.2 star SW 0′.5; mag 9.42v star W 1′.1.

154 GALAXIES 154

RA h m s	Dec ° ′ ″	Name	Mag (V)	Dim ′ Maj x min	SB	Type Class	PA	Notes
06 00 34.7	−28 59 31	ESO 425-2	13.7	2.5 x 1.3	14.8	SB(s)d: IV	154	Pair of mag 10.74v, 10.09v stars NW 9′.0.
06 08 57.6	−27 48 16	ESO 425-10	13.2	1.4 x 0.8	13.2	SB(s)dm III-IV	178	Many bright knots and/or superimposed stars.
06 08 56.4	−29 09 23	ESO 425-11	14.6	1.8 x 0.2	13.4	Scd	154	Mag 12.2 star E 1′.5; mag 10.33v star E 6′.7.
06 11 06.8	−28 42 26	ESO 425-12	15.1	1.5 x 0.2	13.6	Sc	144	Mag 8.40v star N 6′.5.
06 13 02.9	−27 43 46	ESO 425-14	12.9	1.9 x 0.6	12.9	$SA0^-$:	105	Lies 1′.8 E of mag 8.02v star.
06 20 47.2	−27 54 56	ESO 425-18	14.2	1.3 x 1.1	14.4	SAB(s)d IV	80	Stellar nucleus, faint arms; mag 15.3 star N edge.
06 21 26.5	−28 06 44	ESO 425-19	12.3	1.0 x 0.7	11.8	$SA0^-$:	102	Mag 10.96v star NW 7′.6.
06 21 42.9	−27 33 36	ESO 426-1	14.5	1.2 x 0.5	13.8	SAB(s)m: V	44	Mag 8.10v star N 3′.0.
06 25 52.4	−27 59 14	ESO 426-8	14.8	1.5 x 0.2	13.3	Sbc: sp	84	Mag 10.81v star S 3′.1.
06 28 49.5	−28 44 58	ESO 426-14	13.3	1.1 x 0.7	12.9	S0	22	Strong almost stellar nucleus, smooth envelope.
07 02 06.8	−28 27 30	ESO 427-26	13.6	1.6 x 0.4	12.9	$S0^+$ sp	63	Mag 9.03v star NNE 6′.8.
07 02 45.4	−29 25 45	ESO 427-29	12.3	1.9 x 1.3	13.1	S0°: pec	150	Mag 13.3 star on W edge; mag 12.1 star N 4′.2.
07 07 25.4	−28 13 35	ESO 427-34	13.7	2.2 x 1.1	14.5	SB(s)dm IV	19	Mag 8.54v star (part of open cluster Ru 12) NW 4′.0; pair of mag 8.46v, 9.04v stars SE 2′.8.
07 10 55.5	−28 04 45	ESO 428-2	13.9	0.6 x 0.6	12.6	S0:		Compact nucleus; mag 10.87v star SW 0′.7.
07 12 36.0	−28 39 15	ESO 428-4	12.7	1.4 x 1.3	13.2	SA?(r)0:	116	Mag 8.45v star NE 11′.4.
07 15 15.4	−29 16 54	ESO 428-9	14.1	1.7 x 0.4	13.6	Sc	74	ESO 428-11 SE 5′.7; mag 9.78v star NNE 5′.0.
07 15 31.1	−29 21 33	ESO 428-11	11.9	1.7 x 1.2	12.5	$SA0^-$:	16	Mag 14.4 star on S edge; mag 10.56v star WNW 5′.1; ESO 428-9 NW 5′.7.
07 16 26.3	−29 37 15	ESO 428-13	14.1	1.2 x 0.4	13.1	S?	126	Several stars superimposed.
07 16 31.1	−29 19 29	ESO 428-14	12.2	1.8 x 1.1	12.8	SAB(r)0° pec	136	Group of four stars, the brightest mag 9.48v, SW 2′.7.
07 17 13.7	−28 19 16	ESO 428-16	14.2	1.0 x 0.6	13.5	Sbc	108	Several stars superimposed; pair of stars, mags 11.14v and 10.93v, NE 2′.3.
07 18 37.3	−29 22 22	ESO 428-20	14.4	1.5 x 0.2	13.0	Sc	154	Mag 8.74v star N 3′.6.
07 20 21.0	−28 59 16	ESO 428-22	13.1	1.4 x 1.2	13.5		37	Numerous stars superimposed overall; mag 10.12v star SE edge.
07 22 09.7	−29 14 06	ESO 428-23	12.6	2.3 x 1.4	13.7	(R′)SB(rs)ab: III-IV	39	Numerous stars superimposed; mag 10.36v star W 2′.2.
07 23 42.2	−29 39 07	ESO 428-29	12.3	1.6 x 1.3	12.9	SAB(rs)b: III-IV	148	Mag 10.00v star E 3′.3.
05 58 39.8	−25 24 57	ESO 488-49	14.4	1.9 x 0.6	14.4	SB(s)dm IV	136	Mag 9.02v star NE 3′.6.
05 58 52.5	−23 20 22	ESO 488-51	13.3	2.2 x 0.9	13.9	SAB(r)a	129	Pair of mag 9.31v, 9.33v stars NW 9′.2.
06 03 39.2	−26 39 55	ESO 488-59	13.4	0.9 x 0.7	12.8	$SA0^-$:	102	Mag 8.57v star W 12′.4.
06 04 43.1	−26 07 26	ESO 488-60	12.9	2.8 x 1.2	14.1	SB(rs)cd III	152	Star W of center.
06 06 45.7	−26 20 25	ESO 489-4	14.3	1.0 x 0.4	13.1	Sb	137	Several stars superimposed; mag 13.2 star NE 2′.9.
06 07 20.2	−23 22 04	ESO 489-5	13.9	1.0 x 0.5	13.0	S0	27	Mag 12.1 double star S 0′.8.
06 07 44.8	−23 29 14	ESO 489-6	12.7	1.9 x 1.2	13.5	SB(s)d: III-IV	57	Several bright knots or superimposed stars; mag 8.61v star W 8′.4.
06 08 05.7	−23 59 32	ESO 489-7	14.1	0.7 x 0.7	13.2	S		Almost stellar core, smooth envelope; mag 11.31v star NE 2′.4.
06 09 48.2	−24 41 27	ESO 489-10	13.8	1.1 x 0.3	12.4	S0	97	Mag 10.87v star NW 7′.2; mag 9.02v star E 10′.6.
06 10 01.9	−22 37 32	ESO 489-11	14.2	1.5 x 1.1	14.6	SAB(s)dm IV-V	123	Mag 5.71v star SSW 9′.5.
06 12 08.3	−23 32 20	ESO 489-15	14.0	0.5 x 0.5	12.3			Mag 11.7 star W 2′.1.
06 14 33.4	−26 03 54	ESO 489-19	14.3	1.1 x 0.9	14.1	Sc	89	Mag 14.1 star superimposed SW of nucleus.
06 15 19.0	−26 34 40	ESO 489-22	14.7	1.7 x 1.3	15.4	IAB(s)m V-VI	39	Pair of mag 11.04v, 11.18v stars SE 2′.1.
06 15 28.9	−22 36 11	ESO 489-23	14.2	1.5 x 1.0	14.5	SB(rs)d IV	153	Elongated **ESO 489-21** WSW 11′.7; mag 8.72v star SSE 10′.9.
06 16 06.6	−26 48 23	ESO 489-28	13.9	1.1 x 0.7	13.5	S0	56	Located 2′.8 SSE of NGC 2206.
06 17 05.0	−27 23 10	ESO 489-29	12.5	2.9 x 0.4	12.5	Sbc sp II-III	1	Mag 8.93v star SE 1′.7.

(Continued from previous page)

RA h m s	Dec ° ′ ″	Name	Mag (V)	Dim ′ Maj x min	SB	Type Class	PA	Notes
06 17 31.3	−23 04 24	ESO 489-31	15.4	1.0 x 0.4	14.3	SB(s)m V	54	Mag 9.30v star WSW 8′9.
06 18 59.4	−24 37 48	ESO 489-35	12.7	1.5 x 0.5	12.3	SAB0⁻	136	Mag 9.68v star NE 4′1; note: **ESO 489-36** lies 1′6 NW of this star; **ESO 489-34** N 4′7.
06 19 17.4	−24 27 56	ESO 489-37	12.7	1.5 x 0.5	12.3	SB(r)0°	54	Mag 12.3 star W 3′4; mag 9.68v star S 8′4.
06 20 41.7	−23 09 49	ESO 489-40	14.0	0.9 x 0.6	13.2	Sab	33	Mag 11.24v star S edge.
06 23 51.1	−23 11 40	ESO 489-47	13.4	1.5 x 1.0	13.7	SA(r)bc: II	92	Lies 2′1 NE of mag 8.42v star.
06 24 38.9	−22 35 50	ESO 489-50	14.9	1.5 x 0.2	13.4	S0: sp	89	Mag 8.77v star E 7′5.
06 25 35.6	−22 56 28	ESO 489-53	13.5	0.9 x 0.7	12.9	Sbc II-III	33	Mag 8.87v star NW 1′0.
06 26 00.4	−27 00 01	ESO 489-54	14.1	1.2 x 1.0	14.1	SAB(rs)bc	9	Star superimposed N edge of core.
06 26 52.8	−24 37 04	ESO 489-57	13.1	0.9 x 0.8	12.9	E/S0	107	Located in between a pair of mag 10.33v and 12.2 stars.
06 29 04.4	−27 19 40	ESO 490-5	14.0	1.5 x 0.2	12.6	Sab	48	Mag 10.27v star NW 4′0.
06 29 13.8	−26 29 47	ESO 490-6	12.4	1.2 x 1.1	12.6	SA(r)0/a	58	1′0 W of mag 8.88v star.
06 30 57.6	−23 43 43	ESO 490-7	13.7	1.9 x 1.2	14.5	SB(s)m V	84	Located between a close pair of E-W mag 11.9 stars; mag 13.9 star on W edge.
06 31 56.6	−26 46 14	ESO 490-10	12.7	2.0 x 1.1	13.4	SB(rs)d? pec IV	34	Mag 13.7 star N edge; mag 10.75v star SW 3′1.
06 32 52.3	−25 43 23	ESO 490-12	13.1	1.1 x 0.7	12.7	SB(r)0⁺:	117	Close pair of mag 10.53v, 11.00v stars NNE 4′3.
06 33 43.0	−24 59 06	ESO 490-14	12.8	1.7 x 1.0	13.2	SB(s)bc: II-III	14	Star on N end.
06 37 24.5	−24 11 22	ESO 490-16	13.9	1.4 x 0.3	12.8	S0/a	118	Mag 9.97v star E 5′0.
06 37 57.2	−26 00 09	ESO 490-17	13.0	1.5 x 1.3	13.6	IAB(s)m? V-VI	151	Mag 10.69v star on W edge.
06 38 00.0	−24 12 06	ESO 490-18	13.9	1.0 x 0.6	13.2	SB(r)a	66	Mag 9.97v star WNW 4′0; **ESO 490-18A** N 1′0.
06 38 52.2	−25 09 54	ESO 490-20	14.2	1.9 x 1.8	15.4	IB(s)m V-VI	48	Multi-galaxy system; mag 11.1 star N 4′1.
06 39 44.3	−27 14 42	ESO 490-22	11.0	2.6 x 2.6	12.9			Mag 10.03v star lies almost on center. Some sources question that ESO 490-22 might be a nebula.
06 40 11.7	−25 53 39	ESO 490-26	13.0	1.5 x 1.0	13.4	S? pec	15	Mag 13.9 star S of core; mag 10.06v star S 3′6; mag 8.48v star E 11′6.
06 40 20.7	−27 05 48	ESO 490-28	13.9	1.5 x 0.2	12.4	S pec?	177	ESO 490-22 SW 11′9; very small, very faint anonymous galaxy E 0′7.
06 42 22.1	−26 53 36	ESO 490-31	14.2	1.5 x 0.6	13.9	SB(rs)c: III-IV	54	Mag 13.3 star SW end; mag 9.72v star S 3′6.
06 44 07.1	−27 10 29	ESO 490-36	13.7	1.8 x 0.4	13.2	Scd: sp	27	Mag 6.44v star SE 14′1; mag 11.4 star N 4′5.
06 44 22.5	−26 06 32	ESO 490-37	12.0	2.2 x 1.4	13.1	(R′)SA(r)0/a:	167	Mag 14.2 star S of center; mag 8.19v star WSW 11′8.
06 45 25.9	−26 39 38	ESO 490-38	13.3	1.3 x 0.7	13.0	SAB(rs)0/a:	18	Mag 9.25v star ESE 8′7.
06 46 22.0	−26 06 32	ESO 490-41	12.5	3.0 x 1.0	13.6	(R′)SAB(s)0/a pec:	164	Mag 13.2 star NE of center; mag 9.18v star N 6′7.
06 46 47.2	−26 28 29	ESO 490-45	13.4	2.6 x 0.8	14.1	SB(s)d IV	88	Numerous stars superimposed, in star rich area.
06 49 20.0	−26 26 30	ESO 491-1	14.0	1.1 x 0.7	13.6	Sa?	138	Mag 11.11v star NE 2′4.
06 53 02.0	−26 31 27	ESO 491-6	12.5	1.7 x 1.3	13.4	E/S0	168	Mag 11.05v, 11,7 double star SE 2′1.
06 55 25.2	−26 36 26	ESO 491-9	13.9	1.7 x 0.5	13.5	SAB(s)dm IV-V	97	Mag 9.61v star E 3′4.
06 55 26.0	−26 45 32	ESO 491-10	14.9	1.2 x 0.8	14.7	IB(s)m V-VI	36	Mag 11.05v star N 1′1.
06 57 24.6	−24 43 33	ESO 491-12	13.7	2.3 x 0.5	13.7	Sc? sp	164	Located 6′2 SSW of mag 5.45v star.
06 59 21.2	−25 54 01	ESO 491-13	13.3	1.9 x 0.8	13.6	(R)SAB(r)0/a:	7	Bright knot or star S of elongated center; mag 11.3 star N 6′7.
07 00 42.8	−27 22 10	ESO 491-15	13.2	2.5 x 0.4	13.1	SABc: sp II-III	62	Mag 10.01v star NE 7′1; mag 6.95v star NE 14′4.
07 09 47.1	−27 34 11	ESO 491-20	12.9	1.3 x 0.9	12.9	SB(rs)b: pec	71	Pair with ESO 491-21 E edge; mag 5.45v star NE 8′6.
07 09 49.8	−27 34 33	ESO 491-21	12.8	1.6 x 0.5	12.4	SB(r)ab? pec	21	Pair with ESO 491-20 W edge; mag 9.15v star E 2′2.
07 10 25.9	−23 32 18	ESO 491-22	13.7	1.2 x 1.1	13.8	Sbc	175	Numerous stars superimposed, in star rich area.
07 11 40.9	−26 42 26	ESO 492-2	12.2	1.9 x 1.0	12.7	SAB(rs)b pec	134	**ESO 492-3** N 3′1; mag 10.11v star W 3′8.
05 56 25.0	−19 41 30	ESO 555-13	13.2	0.9 x 0.7	12.6	Sa	147	Mag 9.81v star NW 2′6.
05 57 38.4	−18 35 40	ESO 555-14	14.2	1.0 x 0.6	13.6	Sb pec:	21	Mag 9.91v star NW 5′5.
06 00 21.7	−21 39 57	ESO 555-19	14.7	0.8 x 0.6	13.8	Im pec III-IV	42	Pair with ESO 555-20 E edge.
06 00 25.3	−21 40 22	ESO 555-20	14.3	0.9 x 0.6	13.6		96	Pair with ESO 555-19 W edge.
06 01 08.8	−21 44 06	ESO 555-22	12.7	2.5 x 0.7	13.2	SB(s)bc? II	61	Mag 10.64v star W 3′1.
06 01 36.0	−20 20 32	ESO 555-23	13.9	1.1 x 0.2	12.1	S0/a	148	Mag 12.6 star N 1′5.
06 03 36.5	−20 39 17	ESO 555-27	12.8	2.1 x 1.9	14.1	SB(s)d pec III	39	Mag 8.67v star E 7′4.
06 04 27.9	−19 37 21	ESO 555-28	16.4	1.6 x 0.5	16.0	IB(s)m V-VI	147	Mag 9.82v star S 1′3; mag 11.11v star W 1′8.
06 04 27.4	−20 21 15	ESO 555-29	14.2	2.0 x 0.3	13.5	Sd? sp	14	Mag 9.61v star NNW 10′1.
06 07 41.9	−19 54 46	ESO 555-36	14.4	2.1 x 0.2	13.3	SB(s)c? sp	145	Star on E edge.
06 09 01.1	−21 43 23	ESO 555-39	15.1	1.4 x 0.8	15.0	IB(s)m V-VI	30	Located between a mag 10.00 star 3′8 W and a mag 9.42v star E 3′6.
06 09 37.0	−20 06 09	ESO 555-40	14.6	1.2 x 0.3	13.3	Scd? pec sp	39	Mag 10.39v star NW 2′3.
06 09 48.3	−19 43 41	ESO 556-1	13.4	1.0 x 0.7	12.9	(R′)SAB0⁻:	135	Mag 9.14v star NNW 8′8.
06 11 16.4	−21 35 56	ESO 556-2	13.7	1.2 x 0.9	13.6	IA(s)m V-VI	4	Mag 9.37v star W 2′0.
06 15 07.1	−19 25 07	ESO 556-5	13.6	1.2 x 0.7	13.3	SB(s)bc? II	10	Many bright knots; MCG −3-16-20 SSE 1′8; mag 7.95v star SE 8′1.
06 17 49.0	−21 03 38	ESO 556-12	14.1	1.5 x 0.6	13.8	SB(s)m IV-V	124	Small triangle of mag 10.72v, 10.65v and 10.85v stars S 4′8.
06 21 05.8	−20 02 48	ESO 556-15	11.7	3.2 x 2.2	13.7	SB(s)a pec	138	Strong dark lane W of center; mag 9.69v star E 6′6; **ESO 556-18** SE 10′4, which is 2′9 SE of mag 10.49 star.
06 21 50.8	−20 13 38	ESO 556-19	14.1	1.1 x 0.8	13.9	SAB(s)m V	38	Mag 10.49v star NNW 8′2; note: **ESO 555-18** located 2′9 SE of this star.
06 25 36.8	−19 48 47	ESO 556-22	14.2	0.9 x 0.5	13.2	Sb	94	Several stars superimposed; mag 12.8 star on NW edge.
06 31 47.2	−17 37 19	ESO 557-2	13.7	1.2 x 0.8	13.5	Sbc	27	Two main, curving arms; mag 10.11v star NW 3′3.
06 31 53.3	−20 10 03	ESO 557-3	13.6	2.0 x 0.7	13.8	S0° pec sp	85	Mag 9.80vstar N 2′4; mag 9.14v star S 7′4.
06 35 02.4	−19 27 48	ESO 557-5	14.3	1.4 x 0.3	13.2	Sc:	156	Mag 9.91v star NE 6′9.
06 35 00.8	−21 46 15	ESO 557-6	12.7	1.5 x 1.4	13.3	SB(s)m IV-V	169	Mag 10.92v star S 10′1.
06 38 42.5	−20 17 09	ESO 557-9	13.0	1.9 x 1.0	13.5	SB(s)c II-III	97	Several stars superimposed; mag 10.12v star E 9′7.
07 00 43.1	−21 14 52	ESO 558-5	14.4	1.2 x 0.8	14.2	SA(r)ab?	81	Mag 9.48v star W 7′6.
05 57 53.2	−23 10 51	IC 2152	12.6	1.5 x 1.2	13.1	(R)SB(r)a:	54	Mag 9.31v star E 6′5; **ESO 488-48** E 3′5.
06 05 17.9	−27 51 24	IC 2158	12.1	1.7 x 1.3	12.9	SB(r)ab: II-III	107	Mag 9.31v star SW 9′8.
06 16 30.0	−21 22 35	IC 2163	11.7	2.2 x 1.0	12.4	SB(rs)c pec	86	Overlaps E edge of NGC 2207.
06 44 27.5	−17 55 58	IC 2171	13.5	1.7 x 0.5	13.2	IB(s)m: IV-V	93	Mag 8.65v star N 5′6.
05 56 27.7	−18 13 06	MCG −3-16-1	13.4	0.5 x 0.4	11.6	SA(r)0⁺	18	Mag 8.11v star SE 6′2.
06 00 23.8	−19 42 33	MCG −3-16-5	13.3	0.9 x 0.5	12.3	SABb? II-III	174	Mag 8.79v star SSW 10′1.
06 01 56.9	−20 08 16	MCG −3-16-7	14.0	0.5 x 0.3	11.8		6	Mag 8.51v star NNE 8′2.
06 01 55.5	−17 56 31	MCG −3-16-8	13.3	0.7 x 0.4	11.9	S0	108	Mag 9.86v star N 5′8; **MCG −3-16-9** N 3′9.
06 07 25.0	−18 54 38	MCG −3-16-16	14.1	0.5 x 0.3	11.9	Sa	36	Mag 11.02v star WSW 2′9.
06 15 10.8	−19 26 40	MCG −3-16-20	13.9	0.8 x 0.6	12.9	Sc	75	Mag 7.95v star ESE 6′5; ESO 556-5 NNW 1′8.
06 29 17.5	−17 21 34	MCG −3-17-5	12.6	1.0 x 0.5	11.7	SAB0⁻: sp	90	Mag 5.76v star SW 11′6.
06 51 48.2	−20 14 26	MCG −3-18-4	14.4	0.7 x 0.4	12.9	Sb	99	Mag 9.55v star SW 4′1.
06 02 51.0	−23 41 29	MCG −4-15-6	13.9	0.4 x 0.4	12.0	E/S0		Stellar; mag 10.57v star SSW 3′4.
06 04 42.7	−21 55 48	MCG −4-15-7	14.3	0.6 x 0.4	12.6	SBc	126	Mag 9.56v star NW 1′5.
06 06 30.0	−24 41 12	MCG −4-15-9	13.5	1.1 x 0.6	12.9	SBab	78	Mag 12.4 star on N edge; mag 10.36v star W 3′5.
06 08 53.4	−25 08 40	MCG −4-15-12	13.9	1.2 x 0.5	13.1	SB(s)d III-IV	144	Mag 10.13v star NW 7′1.
06 14 35.2	−25 47 53	MCG −4-15-15	13.4	0.6 x 0.4	11.7	Sa? pec	165	Mag 9.37v star N 7′0, MCG −4-15-16 E 1′3.
06 14 40.6	−25 47 29	MCG −4-15-16	13.6	0.4 x 0.4	11.5			Mag 9.37v star N 6′8; MCG −4-15-15 W 1′3.
06 15 02.7	−25 49 04	MCG −4-15-17	14.3	0.6 x 0.3	12.3		141	Mag 8.86v star SSW 4′3.
06 18 25.9	−24 51 08	MCG −4-15-22	13.8	0.6 x 0.6	12.7	E0:		Mag 9.90v star S 2′4; **ESO 489-33** SSW 5′5.
06 29 34.4	−26 37 57	MCG −4-16-6	13.1	0.5 x 0.5	11.5	SAB0°:		Mag 8.66v star and ESO 490-6 NNW 8′7.
06 34 07.4	−25 40 56	MCG −4-16-11	13.1	0.7 x 0.6	12.0	Sbc	66	Close pair of mag 11.20v, 11.9 stars N 6′4.
06 40 16.8	−26 16 49	MCG −4-16-16	14.2	0.6 x 0.4	12.5	Sbc	97	Mag 8.42v star SE 3′2.

(Continued from previous page)

GALAXIES

RA h m s	Dec ° ′ ″	Name	Mag (V)	Dim ′ Maj x min	SB	Type Class	PA	Notes
05 57 28.8	−29 57 38	MCG −5-15-1	14.2	1.0 x 0.9	13.9	Sc:	156	Mag 8.74v star SSW 9.′5.
06 09 07.4	−27 44 09	MCG −5-15-6	13.2	0.4 x 0.3	10.8	SB(s)dm	171	Situated between two faint E-W stars.
06 22 25.1	−27 58 15	MCG −5-16-1	14.0	0.9 x 0.4	12.8	Sb	111	Forms a triangle with mag 9.55v star NNW 8.′4 and mag 9.28v star NNE 8.′1.
06 26 39.3	−28 39 56	MCG −5-16-3	13.4	1.1 x 0.3	12.0	Sb:	177	Mag 9.70v star WSW 6.′6.
06 26 37.7	−29 56 08	MCG −5-16-4	13.6	1.0 x 0.7	13.1	S(r)b	84	Several stars involved in fainter outer envelope.
06 33 13.4	−28 01 37	MCG −5-16-11	13.8	1.0 x 0.8	13.4	(R′)SB(r)ab	22	Mag 10.51v star NW 1.′9.
06 40 30.2	−27 12 25	MCG −5-16-14	14.2	0.6 x 0.6	12.9	Sbc		Anonymous elongated galaxy W 1.′8.
06 43 44.4	−27 16 00	MCG −5-16-18	13.4	0.9 x 0.7	12.8	Sb:	15	Mag 6.44v star ESE 15.′6.
06 46 44.2	−27 16 45	MCG −5-16-22	13.5	1.2 x 0.6	13.0	Sb?	149	Mag 9.58v star WSW 8.′6.
07 09 03.2	−28 29 15	MCG −5-17-8	14.4	0.6 x 0.4	12.7	S	147	Mag 8.25v star ENE 7.′1.
05 57 52.3	−20 05 08	NGC 2124	12.2	2.7 x 0.8	12.9	SA(s)b? II-III	5	Several knots or superimposed stars; mag 8.57v star SE 9.′7.
05 58 47.6	−26 39 13	NGC 2131	14.0	1.1 x 0.4	12.9	IB(s)m: III-IV	118	Star on N edge.
06 01 08.4	−23 40 32	NGC 2139	11.6	2.4 x 1.9	13.1	SAB(rs)cd II-III	140	Numerous bright knots.
06 08 02.3	−21 44 49	NGC 2179	12.3	1.1 x 0.9	12.2	SA(s)0/a	170	Star at N and S ends.
06 12 09.7	−21 48 22	NGC 2196	11.0	2.7 x 2.1	12.7	(R′)SA(s)a I-II	35	**PGC 18617** SE 8.′1.
06 16 00.0	−26 46 00	NGC 2206	12.2	2.4 x 1.3	13.3	SAB(rs)bc: II	138	Several stars superimposed; mag 12 star E 2.′7; ESO 489-28 SE 2.′7.
06 16 22.1	−21 22 28	NGC 2207	10.9	3.9 x 2.2	13.1	SAB(rs)bc pec I-II	119	IC 2163 overlaps E edge.
06 18 30.3	−18 32 15	NGC 2211	12.7	1.4 x 0.7	12.5	SB(r)0°:	22	
06 18 35.7	−18 31 12	NGC 2212	13.4	1.5 x 0.8	13.5	SB(rs)0⁺ pec:	136	Star superimposed SE of nucleus.
06 21 30.6	−22 05 10	NGC 2216	12.8	1.4 x 1.1	13.1	SAB(r)ab:	20	Bracketed by a triangle of mag 12 stars.
06 21 39.6	−27 13 58	NGC 2217	10.7	4.5 x 4.2	13.8	(R)SB(rs)0⁺	21	A pair of mag 12 stars lie halfway to the W edge, with an inline trio of mag 15 stars on the E edge.
06 24 35.7	−22 50 19	NGC 2223	11.5	3.2 x 2.8	13.8	SAB(r)b II	175	Strong branching arms; mag 11.1 star N 3.′1; **ESO 489-48** NNW 3.′5.
06 25 57.9	−22 00 18	NGC 2227	12.5	2.1 x 1.1	13.2	SB(rs)c II-III	19	Several knots, moderate dark lane E of core.
06 38 29.0	−24 50 55	NGC 2263	11.9	2.6 x 2.0	13.6	(R′)SB(r)ab	143	Much fainter outer envelope surrounds bright, elongated core. Numerous stars on outer envelope, especially N.
06 42 52.9	−23 28 34	NGC 2271	12.1	2.1 x 1.4	13.2	SAB0⁻	83	Several superimposed stars on envelope; mag 9.26v star N 4.′8.
06 42 41.3	−27 27 32	NGC 2272	11.5	2.3 x 1.6	12.8	SAB(s)0⁻	123	Several faint stars very close to bright core.
06 44 49.2	−27 38 23	NGC 2280	10.3	6.3 x 3.0	13.3	SA(s)cd I-II	163	Numerous stars on the face of the galaxy; several dark lanes surround the bright, elongated core.
06 45 52.8	−18 12 38	NGC 2283	11.5	3.4 x 2.7	13.8	SB(s)cd II	2	Numerous superimposed stars and knots in multi-branching arms.
06 47 41.3	−26 45 02	NGC 2292	11.0	4.1 x 3.6	13.8	SAB0° pec	124	Shares common envelope with NGC 2293; respective bright centers have 0.′9 separation.
06 47 43.2	−26 45 18	NGC 2293	11.2	4.2 x 3.3	13.9	SAB(s)0⁺ pec	125	Shares common envelope with NGC 2292. Respective bright centers have 0.′9 separation.
06 47 23.6	−26 44 08	NGC 2295	12.5	2.1 x 0.6	12.6	Sab: sp	46	Located 4.′0 W of NGC 2292 and NGC 2293; **ESO 490-42** W 10.′5.
07 02 40.4	−28 41 55	NGC 2325	11.3	3.0 x 1.5	13.0	E4	6	Numerous superimposed stars; faint, elongated anonymous galaxy W 3.′1.
07 23 54.6	−27 31 41	NGC 2380	11.2	2.0 x 1.9	12.5	SAB0°:	168	Numerous superimposed stars.

GALAXY CLUSTERS

RA h m s	Dec ° ′ ″	Name	Mag 10th brightest	No. Gal	Diam ′	Notes
05 56 54.0	−21 15 00	A 3374	16.1	34	17	**ESO 555-15** E of center 11.′8, all others anonymous, stellar.
06 02 36.0	−27 46 00	AS 562	16.1		17	All members anonymous, faint, stellar.
06 09 12.0	−27 30 00	AS 568	16.4		17	Most members anonymous, stellar.
06 15 30.0	−29 16 00	AS 578	16.1		17	
06 17 48.0	−23 41 00	AS 581	16.0	18	17	**ESO 489-30** SW of center 9.′7; all others anonymous, stellar.

OPEN CLUSTERS

RA h m s	Dec ° ′ ″	Name	Mag	Diam ′	No. ★	B ★	Type	Notes
07 04 15.3	−19 44 44	Auner 1		3	30		cl	
06 40 00.0	−27 18 00	ClvdB 83		25	10		cl	
06 56 20.0	−24 44 00	Cr 121	2.6	90	33	3.8	cl	Few stars; large brightness range; not well detached.
06 04 48.0	−29 11 00	ESO 425-6		5			cl	
06 14 36.0	−29 22 00	ESO 425-15		5			cl	
06 05 00.0	−26 44 00	ESO 489-1		10			cl	
07 18 06.0	−18 36 00	ESO 559-2		1.2			cl	
07 18 06.0	−22 39 00	Haf 5		7	50	15.0	cl	
07 22 54.0	−29 29 48	Haf 7		3	40	14.0	cl	
06 15 33.0	−18 40 00	NGC 2204	8.6	10	353	13.0	cl	Rich in stars; moderate brightness range; slight central concentration; detached.
06 46 00.0	−20 45 00	NGC 2287	4.5	39	80	8.0	cl	= **M 41**. Rich in stars; large brightness range; strong central concentration; detached.
07 14 16.0	−25 42 00	NGC 2354	6.5	18	297	9.1	cl	Rich in stars; moderate brightness range; no central concentration; detached.
07 18 42.0	−24 57 18	NGC 2362	3.8	6	60	8.0	cl	Rich in stars; large brightness range; strong central concentration; detached. (S) Tau CMa Cluster.
07 20 05.0	−21 53 00	NGC 2367	7.9	5	30	9.4	cl	Moderately rich in bright and faint stars; detached.
06 41 01.0	−29 33 00	Ru 2		3	10	12.0	cl:	Few stars; moderate brightness range; not well detached.
06 42 06.0	−29 27 12	Ru 3		3	15	11.0	cl	Few stars; small brightness range; no central concentration; detached.
06 55 24.0	−18 43 48	Ru 5		2	10	13.0	cl:	
07 06 27.6	−20 07 00	Ru 10		7	40	12.0	cl:	
07 07 23.0	−20 47 06	Ru 11		5	25	11.0	cl	Few stars; moderate brightness range; no central concentration; detached.
07 07 10.0	−28 09 06	Ru 12		5	20	14.0	cl:	
07 08 00.0	−25 52 00	Ru 13		6	15	13.0	cl	
07 19 34.0	−19 37 36	Ru 15		2	20	12.0	cl	(S) Small group.
07 23 09.0	−19 28 42	Ru 16		4	15	13.0	cl	Few stars; small brightness range; not well detached. Probably not a cluster. (S) Sparse group.
07 23 35.0	−23 11 18	Ru 17		3	15	12.0	cl	(S) Sparse group.
06 52 12.0	−23 36 00	Ru 149		5	25	13.0	cl:	
07 05 55.0	−28 24 30	Ru 150		2	12	13.0	cl	(S) Small group.
07 00 27.0	−20 34 12	Tom 1		6	45	14.0	cl	
07 03 06.0	−20 49 06	Tom 2		3	50	16.0	cl	

BRIGHT NEBULAE

RA h m s	Dec ° ′ ″	Name	Dim ′ Maj x min	Type	BC	Color	Notes
07 09 48.0	−18 29 00	Sh2-301	8 x 7	E	1-5	3-4	
07 19 36.0	−24 02 00	vdB 96	10 x 5	R	3-5	1-4	Three 9th mag stars in a chain with nebulosity; brightest around the NW star.

(Continued from previous page)

154 PLANETARY NEBULAE 154

RA h m s	Dec ° ′ ″	Name	Diam ″	Mag (P)	Mag (V)	Mag cent ★	Alt Name	Notes
07 21 14.9	−18 08 37	PN G232.4−1.8	36	12.9	12.6		PK 232−1.1	Slightly elongated SSE-NNW; mag 11.2 star S 0′.8; mag 11.4 star N 3′; mag 9.65v star E 5′.1.
07 11 16.8	−19 51 05	PN G232.8−4.7	10	15.9	15.5	13.9	PK 232−4.1	Mag 10.26v star SSE 1′.4; mag 12.9 star N 1′.5.
06 50 40.7	−22 26 15	PN G233.0−10.1	55	16.3			PK 233−10.1	Mag 14.8 star on SE edge; mag 10.92v star NW 4′.0.
06 27 02.0	−25 22 50	PN G233.5−16.3	34	16.3	15.6	16.1	PK 233−16.1	Mag 12.0 star N 1′.1; mag 10.45v star NE 2′.6.
07 06 57.7	−22 02 21	PN G234.3−6.6	120		14.5	21.0	PK 234−6.1	Elongated NE-SW; mag 10.49v star on SE edge.
07 19 21.6	−21 43 58	PN G235.3−3.9	3	15.2		14.0	PK 235−3.1	Small triangle of mag 12 stars SW 2′.1.
06 54 20.8	−25 24 33	PN G236.0−10.6	148			19.8	PK 236−10.1	Mag 14.4 star S of center 0′.6; mag 10.05v star SW 5′.1; mag 9.68v star W 6′.1.
06 55 12.3	−29 07 28	PN G239.6−12.0	24	15.1		19.0	PK 239−12.1	Mag 14.8 star on N edge; mag 13.3 star W 1′.4.
07 14 49.8	−27 50 24	PN G240.3−7.6	8	15.7	14.7	21.1	PK 240−7.1	Mag 12.2 star W 0′.7; mag 10.89v star NNE 1′.9.

155 GALAXIES 155

RA h m s	Dec ° ′ ″	Name	Mag (V)	Dim ′ Maj x min	SB	Type Class	PA	Notes
04 49 12.5	−29 12 26	ESO 421-19	12.5	3.2 x 2.6	14.7	SAB(s)m: V	56	Numerous knots and superimposed stars; small triangle of mag 12-13 stars E 3′.7.
04 58 42.6	−28 14 10	ESO 422-16	13.4	0.9 x 0.8	12.9	S0	55	Located between a pair of mag 12.8 and 12.0 stars.
05 06 02.3	−27 48 08	ESO 422-29	14.1	0.8 x 0.6	13.1		15	Mag 12.2 star ESE 5′.1.
05 06 49.7	−27 39 39	ESO 422-33	13.9	1.5 x 0.8	13.9	IB(s)m pec: IV	123	Mag 9.21v star NW 12′.2.
05 10 13.9	−29 24 15	ESO 422-40	13.7	1.5 x 0.4	12.9	Sbc	23	Star on W edge.
05 34 07.4	−28 27 57	ESO 423-20	12.8	1.4 x 0.9	12.9	SAB(s)c: II	93	Lies 1′.1 N of mag 9.5v star; **ESO 423-25** E 8′.0.
05 34 41.4	−29 13 55	ESO 423-24	12.2	1.2 x 1.2	12.5	SA0°? pec		Mag 10.59v star WNW 4′.8.
05 35 34.7	−29 09 01	ESO 424-1	13.9	0.9 x 0.5	13.1	E^{+}4 pec:	147	Strong core, faint envelope; mag 9.84v star NE 4′.5.
05 52 58.1	−28 21 01	ESO 424-36	13.9	0.8 x 0.6	13.0	S0/a	133	Mag 13.0 star NE 2′.4.
06 00 34.7	−28 59 31	ESO 425-2	13.7	2.5 x 1.3	14.8	SB(s)d: IV	154	Pair of mag 10.74v, 10.09v stars NW 9′.0.
04 39 11.8	−24 10 54	ESO 485-4	14.9	1.7 x 0.2	13.6	Sc:	143	Pair with MCG −4-12-3 W 1′.2; mag 11.01v star E 4′.4.
04 48 03.4	−25 13 50	ESO 485-12	13.9	1.8 x 0.3	13.1	Sc: sp	102	Mag 10.34v star N 1′.3.
04 48 55.9	−23 43 48	ESO 485-16	14.2	1.1 x 0.7	13.8	SAB(rs)ab:	69	Mag 9.14v star NW 3′.3.
04 52 52.9	−25 14 48	ESO 485-21	12.7	3.8 x 3.3	15.3	SB(s)m IV-V	129	Mag 8.45v star S 6′.3.
04 59 33.9	−22 41 49	ESO 486-3	15.0	1.8 x 0.8	15.2	SAB(s)m V-VI	40	Mag 9.43v star WNW 8′.8; note: **ESO 485-29** is located 1′.9 SW from this star.
04 59 49.8	−23 55 37	ESO 486-4	14.2	1.8 x 0.9	14.6	SAB(rs)a	9	Mag 11.01v star WNW 8′.6.
05 00 41.8	−25 04 34	ESO 486-7	13.8	1.7 x 0.8	14.0	SA(r)b:	167	Mag 7.42v star W 10′.0.
05 03 13.2	−22 49 58	ESO 486-19	12.4	1.5 x 0.5	12.0	S0^{-}: sp	155	Pair with MCG −4-12-38 N edge; mag 5.74v star W 6′.9.
05 03 43.2	−27 14 47	ESO 486-22	13.7	1.2 x 1.0	13.7	S?	51	Mag 10.37v star S 4′.9.
05 04 01.2	−23 59 48	ESO 486-23	12.8	1.5 x 1.5	13.5	SA(r)0^{-}:		Mag 9.66v star NNE 5′.0; **ESO 486-18** W 11′.7.
05 07 35.0	−23 04 02	ESO 486-29	13.4	1.0 x 0.7	12.9	S0	138	**ESO 486-30** S 1′.7.
05 09 49.4	−25 36 20	ESO 486-32	13.5	1.6 x 1.1	13.9	SB(rs)bc I-II	136	Mag 10.39v star SW 4′.0; Double galaxy system **ESO 486-31** and **ESO 486-31A** NW 9′.6.
05 12 20.6	−26 03 30	ESO 486-34	13.8	1.1 x 0.8	13.5	Sb	122	Mag 13.2 star S edge.
05 14 29.6	−23 53 55	ESO 486-37	13.4	1.2 x 0.9	13.4	SA0^{-}	24	Mag 9.94v star WSW 8′.9.
05 15 13.8	−22 32 59	ESO 486-38	13.5	1.0 x 0.9	13.3	S0	87	Mag 9.82v star W 9′.8.
05 15 21.4	−26 28 17	ESO 486-39	13.6	0.9 x 0.4	12.3	Pec?	169	Irregular shape; multiple galaxy system?
05 15 48.4	−23 28 32	ESO 486-41	13.8	1.5 x 0.5	13.3	SA(r)b pec II	78	Mag 9.62v star W 4′.4.
05 17 26.8	−23 44 44	ESO 486-49	13.8	1.5 x 0.5	13.4	S?	105	Mag 9.49v star S 5′.6; **ESO 486-44** WSW 11′.2.
05 19 38.2	−24 53 36	ESO 486-52	13.8	0.9 x 0.8	13.3	SA(r)a	49	Pair of stars, mags 12.9 and 13.9, NNE 2′.5.
05 19 49.6	−25 06 08	ESO 486-53A/B	15.6	0.9 x 0.9	15.3			Double system; IC 2121 N 2′.5; mag 7.02v star WSW 4′.4.
05 20 35.2	−26 47 04	ESO 486-57	13.9	0.9 x 0.7	13.3	S0	102	Mag 11.9 star W 10′.5.
05 22 22.0	−22 47 12	ESO 487-5	14.1	1.0 x 0.5	13.2	Sdm?	103	Close pair of stars S edge.
05 30 29.3	−24 52 37	ESO 487-17	15.7	2.1 x 1.4	16.7	IB(s)m VI	39	Mag 9.13v star SE 7′.5.
05 33 26.8	−26 59 11	ESO 487-22	13.2	0.8 x 0.7	12.4		61	Star on SW edge.
05 37 19.0	−26 25 49	ESO 487-30	15.2	2.1 x 0.3	14.6	Sd? sp IV	154	Mag 11.20v star NNW 2′.6.
05 42 00.9	−22 56 45	ESO 487-35	13.0	2.8 x 0.7	13.6	SB(s)dm: sp III	104	Mag 11.23v star W 2′.0.
05 42 20.9	−25 32 30	ESO 487-36	13.4	1.2 x 0.7	13.1	SA(s)0° pec:	27	Pair with MCG −4-14-17 W 1′.4.
05 42 31.1	−26 47 21	ESO 487-37	14.3	1.4 x 0.4	13.5	Sb	6	**ESO 487-37A** E 0′.9.
05 43 08.6	−24 57 29	ESO 488-1	13.4	0.6 x 0.5	12.0		82	Very bright center or star? Mag 9.98v star SE 1′.9.
05 45 27.6	−25 55 52	ESO 488-7	12.9	0.6 x 0.6	11.7	S0		Pair with ESO 488-9 E edge; mag 8.94v star ENE 9′.5; numerous faint galaxies in surrounding area.
05 45 29.8	−25 55 58	ESO 488-9	13.1	0.6 x 0.6	12.0	E?		Pair with ESO 488-7 W edge; mag 8.94v star ENE 9′.5; numerous faint galaxies in surrounding area.
05 46 20.7	−23 28 25	ESO 488-12	13.9	1.6 x 0.4	13.2	S0°: pec sp	34	Pair of N-S oriented mag 10.49v, 9.34v star E 4′.9.
05 47 24.6	−25 15 24	ESO 488-15	13.8	0.8 x 0.6	13.0	E?	39	Pair with ESO 488-19 NE edge; mag 14.5 star S edge; mag 9.58v star W 5′.6.
05 47 26.5	−25 14 50	ESO 488-19	13.7	0.9 x 0.7	13.2	E?	42	Pair with ESO 488-15 SW; star N of nucleus; mag 12.0 star NW edge.
05 48 38.4	−25 28 42	ESO 488-27	13.0	0.8 x 0.8	12.6	E^{+}1		Pair with MCG −4-14-35; located near E edge of galaxy cluster A 548, innumerable cataloged and anonymous galaxies in area.
05 49 39.4	−24 25 27	ESO 488-35	15.0	1.7 x 0.2	13.6	S	105	Mag 8.68v star SW 4′.4; **ESO 488-37** NE 3′.9.
05 58 39.8	−25 24 57	ESO 488-49	14.4	1.9 x 0.6	14.4	SB(s)dm IV	136	Mag 9.02v star NE 3′.6.
05 58 52.5	−23 20 22	ESO 488-51	13.3	2.2 x 0.9	13.9	SAB(r)a	129	Pair of mag 9.31v, 9.33v stars NW 9′.2.
06 03 39.2	−26 39 55	ESO 488-59	13.4	0.9 x 0.7	12.8	SA0^{-}:	102	Mag 8.57v star W 12′.4.
04 42 27.0	−17 27 24	ESO 551-30	14.3	1.2 x 1.0	14.4	SB(s)d III-IV	0	Mag 9.99v star E 8′.6.
04 42 29.1	−21 41 31	ESO 551-31	14.3	1.5 x 0.6	14.0	SB(s)d: sp III-IV	177	Mag 9.75v star NNW 7′.2.
04 46 04.6	−22 21 54	ESO 552-3	13.7	1.3 x 1.0	13.9	SA(rs)0/a	93	Mag 10.09v star SW 6′.9.
04 49 58.1	−18 05 04	ESO 552-11	13.9	1.4 x 0.4	13.2	S0?	161	Close pair of mag 13.3, 13.4 star S 5′.9.
04 50 22.7	−18 14 56	ESO 552-12	14.0	0.9 x 0.4	12.7	Sab:	118	Mag 12.9 star N 0′.9.
04 54 52.0	−18 06 54	ESO 552-20	12.2	2.6 x 1.3	13.5	E^{+}	146	Mag 10.71v star E 6′.3.
04 58 47.2	−21 34 07	ESO 552-40	13.4	1.2 x 0.8	13.2	SB(s)ab: pec	45	MCG −4-12-25 W 4′.3.
04 59 35.2	−19 11 58	ESO 552-43	14.0	1.5 x 0.8	14.1	SAB(rs)cd: II-III	139	Close pair of mag 11.20v, 10.61v stars S 5′.4.
05 03 10.3	−20 16 24	ESO 552-58	13.9	1.0 x 0.3	12.5	S0	107	Mag 11.6 star S 2′.6.
05 05 14.1	−18 23 28	ESO 552-66	15.5	1.6 x 0.9	15.7	IB(s)m V-VI	135	Mag 10.36v star ENE 7′.3; **ESO 552-68** ENE 4′.2.
05 06 33.3	−17 35 12	ESO 553-2	12.4	1.7 x 1.3	13.1	SA(r)0°:	38	Mag 10.59v star S 1′.9; ESO 553-3 N 2′.3; mag 9.25v star NW 11′.1.
05 06 37.7	−17 33 15	ESO 553-3	13.4	1.5 x 1.0	13.7	SAB(rs)c I	138	ESO 553-2 S 2′.3; mag 9.25v star NW 10′.4; **ESO 552-75** NNE 6′.0.
05 08 08.9	−22 07 32	ESO 553-9	13.8	1.0 x 0.6	13.1	S0	129	Faint, elongated anonymous galaxy NE 1′.6; almost stellar anonymous galaxy SW 2′.4.
05 08 18.4	−19 50 04	ESO 553-10	13.7	1.3 x 0.6	13.3	S0?	107	Located between a pair of mag 12.2 and 13.0 stars.
05 09 17.0	−18 42 57	ESO 553-14	13.6	1.0 x 0.9	13.4	S0	164	Mag 12.9 star E 2′.0.
05 11 05.9	−18 25 41	ESO 553-16	14.3	2.0 x 0.5	14.1	SB(s)m IV-V	117	Mag 11.00v star WSW 7′.6.
05 11 38.2	−20 25 36	ESO 553-20	12.4	2.5 x 1.2	13.4	(R′)SB(rs)ab pec: II-III	113	Mag 12.6 star S 1′.6.
05 12 21.7	−19 09 47	ESO 553-23	14.0	0.9 x 0.4	12.7	SA?(r)a	115	Mag 9.52v star NW 10′.8.
05 13 18.2	−22 13 33	ESO 553-26	14.0	1.7 x 0.5	13.7	S?	63	Mag 10.14v star SE 7′.7.
05 19 01.7	−21 32 40	ESO 553-33	14.7	1.8 x 0.7	14.8	IB(s)m V-VI	15	Mag 11.41v star NNW 6′.9.
05 25 02.4	−20 27 04	ESO 553-42	13.8	1.1 x 0.6	13.1	S0/a	39	Mag 9.32v star NNW 9′.1.
05 26 27.4	−21 17 11	ESO 553-43	13.5	0.7 x 0.7	12.6	S0^{+} pec		Close pair of E-W oriented mag 9.55v, 9.61v stars W 10′.0.

(Continued from previous page)

GALAXIES

RA h m s	Dec ° ′ ″	Name	Mag (V)	Dim ′ Maj x min	SB	Type Class	PA	Notes
05 26 44.4	−19 12 37	ESO 553-44	13.4	1.7 x 0.3	12.5	Sc: sp I-II	61	Mag 10.24v star W 5′6; mag 7.98v star NE 13′5.
05 27 05.8	−20 40 38	ESO 553-46	13.4	1.0 x 0.6	12.7	S	162	Mag 7.66v star WSW 10′0.
05 29 07.3	−19 56 05	ESO 554-2	13.5	1.5 x 1.2	14.0	SB(s)cd III-IV	9	Mag 9.02v star W 11′4.
05 29 38.7	−19 33 42	ESO 554-4	13.7	0.9 x 0.6	12.8	Sa	26	Mag 10.07v star NE 1′2; elongated anonymous galaxy S 1′5; almost stellar anonymous galaxy NW 1′8.
05 36 47.2	−18 36 21	ESO 554-18	14.7	1.6 x 0.2	13.3	Sc:	153	Mag 10.42v star E 4′4.
05 36 46.8	−22 24 23	ESO 554-19	13.6	1.3 x 0.7	13.4	SA(s)b II	9	Mag 10.35v star W 5′6.
05 41 19.8	−18 16 41	ESO 554-24	13.6	1.7 x 0.9	13.9	SB(rs)cd pec: II-III	54	Mag 9.17v star SW 9′9.
05 43 06.1	−20 31 16	ESO 554-27	13.5	0.8 x 0.4	12.1		101	Mag 9.90v star W 5′9.
05 43 13.7	−18 38 22	ESO 554-28	14.3	1.6 x 0.2	13.0	S	129	Mag 5.74v star N 5′1.
05 43 52.2	−19 17 49	ESO 554-29	13.3	2.0 x 0.6	13.3	SB(s)cd pec: IV	9	**ESO 554-30** NE 7′5; mag 9.62v star NNW 8′7.
05 48 36.1	−18 40 17	ESO 554-38	12.7	1.9 x 0.7	12.9	(R′)SB(r)0/a?	134	Mag 8.92v star SW 3′2.
05 49 31.8	−19 33 02	ESO 555-1	13.5	1.2 x 0.7	13.1	(R′)SAB(r)ab:	31	Mag 10.09v star N 3′6; mag 7.31v star NNW 10′2.
05 50 26.8	−19 43 37	ESO 555-2	13.4	2.6 x 0.4	13.3	Sbc: sp	105	Mag 9.82v star SW 7′2.
05 51 04.6	−18 54 09	ESO 555-4	14.0	0.7 x 0.3	12.2	S0:	155	Stellar core, faint envelope; close pair of mag 12.5 stars NW 2′5.
05 51 39.9	−18 01 25	ESO 555-5	14.6	1.2 x 0.8	14.4	SB(s)dm IV-V	96	Mag 8.17v star NNW 7′9; ESO 555-6 E 7′0.
05 52 08.9	−18 01 27	ESO 555-6	12.9	1.5 x 0.4	12.2	Sab	85	Mag 8.17v star NW 12′3; ESO 555-5 W 7′0.
05 52 57.4	−20 42 43	ESO 555-10	13.9	1.0 x 0.3	12.5	Im?	137	Mag 9.49v star on NW end.
05 56 25.0	−19 41 30	ESO 555-13	13.2	0.9 x 0.7	12.6	Sa	147	Mag 9.81v star NW 2′6.
05 57 38.4	−18 35 40	ESO 555-14	14.2	1.0 x 0.6	13.5	Sb pec:	21	Mag 9.91v star NW 5′5.
06 00 21.7	−21 39 57	ESO 555-19	14.7	0.8 x 0.6	13.8	Im pec III-IV	42	Pair with ESO 555-20 E edge.
06 00 25.3	−21 40 22	ESO 555-20	14.3	0.9 x 0.6	13.6		96	Pair with ESO 555-19 W edge.
06 01 08.8	−21 44 06	ESO 555-22	12.7	2.5 x 0.7	13.2	SB(s)bc? II	61	Mag 10.64v star W 3′1.
06 01 36.0	−20 20 32	ESO 555-23	13.9	1.1 x 0.2	12.1	S0/a	148	Mag 12.6 star N 1′5.
06 03 36.5	−20 39 17	ESO 555-27	12.8	2.1 x 1.9	14.1	SB(s)d pec III	39	Mag 8.67v star E 7′4.
05 20 18.6	−25 19 29	IC 411	13.0	1.2 x 0.7	12.7	S0?	129	Mag 8.91v star NW 9′1; numerous small, faint anonymous galaxies in surrounding area.
05 23 56.5	−17 15 37	IC 416	13.3	1.4 x 0.7	13.1	SB(s)c pec: II-III	66	Mag 9.78v star S 6′7.
05 32 18.6	−17 13 27	IC 422	13.3	0.7 x 0.7	12.4			Mag 9.96v star WNW 8′0.
05 53 00.1	−17 52 36	IC 438	11.9	2.8 x 2.1	13.7	SA(rs)c I-II	43	IC 2151 NW 7′7; mag 8.53v star NNE 6′8.
04 56 34.0	−28 30 20	IC 2106	13.0	1.7 x 0.9	13.3	(R′)SB(rs)b III-IV	157	Mag 8.13v star WSW 12′0; **ESO 422-13** E 10′0.
05 06 51.1	−20 20 42	IC 2119	13.6	1.3 x 0.7	13.3	(R)SB(r)b	56	Mag 10.99v star SSW 6′6.
05 19 44.9	−25 03 52	IC 2121	12.3	1.9 x 1.1	13.0	SAB(s)0°: pec	160	Double system ESO 486-53A/B S 2′5; mag 7.02v star SW 4′7.
05 24 28.1	−27 00 58	IC 2125	13.3	1.0 x 0.7	12.8	S0	125	Mag 9.79v star S 4′3.
05 31 50.8	−23 08 46	IC 2130	13.1	1.8 x 0.9	13.4	SB(s)dm IV	103	Mag 9.94v star ESE 5′1.
05 34 21.8	−23 32 00	IC 2138	13.0	1.2 x 0.8	12.8	SAB(r)ab	79	Mag 8.47v star E 2′0; **ESO 487-25** N 2′7; **MCG −4-14-5** N 6′5.
05 46 52.3	−18 43 36	IC 2143	12.5	1.9 x 0.8	12.8	SB(rs)b? II	98	Mag 8.63v star W 11′3.
05 52 36.5	−17 47 15	IC 2151	13.3	1.5 x 0.9	13.5	SB(s)bc: II-III	99	Mag 9.30v star N 3′3; IC 438 SE 7′7.
05 57 53.2	−23 10 51	IC 2152	12.6	1.5 x 1.2	13.1	(R)SB(r)a:	54	Mag 9.31v star E 6′5; **ESO 488-48** E 3′5.
04 36 26.8	−20 50 45	MCG −3-12-16	14.4	0.7 x 0.5	13.1	Sb	24	Mag 9.90v star N 4′5.
04 45 53.9	−17 16 45	MCG −3-13-4	12.8	1.6 x 0.9	13.0	SB(rs)bc pec: II	114	Several bright knots or superimposed stars; mag 8.93v star SW 5′1.
04 46 08.0	−17 03 55	MCG −3-13-5	14.1	1.2 x 0.4	13.2	Sb	0	Star at southern tip.
04 46 53.6	−20 18 33	MCG −3-13-7	14.1	0.7 x 0.7	13.2	Sb		Located on E side of galaxy cluster A 514, many anonymous galaxies in area; mag 8.85v star ESE 12′4.
04 47 30.6	−17 35 50	MCG −3-13-9	13.5	1.1 x 1.0	13.5	SA(r)0°	22	Star on NE edge.
04 47 34.3	−17 20 50	MCG −3-13-10	14.2	0.9 x 0.7	13.5	S	99	Mag 8.41v star NW 7′6.
04 47 38.5	−17 26 05	MCG −3-13-11	14.0	0.7 x 0.5	12.7		41	Mag 10.34v star E 6′3.
04 48 43.0	−20 20 47	MCG −3-13-14	14.3	0.8 x 0.7	13.5	Sa:	17	Mag 12.2 star on N edge; mag 8.62v star S 5′4.
04 50 13.6	−17 15 58	MCG −3-13-16	13.4	2.3 x 0.5	13.4	SB(r)bc? sp I-II	96	Mag 12.2 star on S edge; mag 10.51v star ESE 9′1.
04 51 20.6	−17 30 12	MCG −3-13-17	13.4	1.0 x 0.7	12.9	SA(s)0^- pec:	3	Mag 10.90v star N 2′8.
04 53 34.4	−17 55 22	MCG −3-13-20	14.3	1.0 x 0.7	13.7	S?	63	Mag 9.05v star N 5′7.
04 54 18.6	−18 55 16	MCG −3-13-22	14.3	0.4 x 0.3	11.8			MCG −3-13-23 S edge.
04 54 18.5	−18 55 47	MCG −3-13-23	13.6	0.5 x 0.4	11.7		171	MCG −3-13-22 N edge.
04 54 43.2	−17 27 09	MCG −3-13-24	14.0	0.5 x 0.5	12.3			Mag 10.87v star S 6′0.
04 55 48.0	−18 56 56	MCG −3-13-32	13.7	1.0 x 0.9	13.4	Sa-b	131	**MCG −3-13-36** NE 12′3.
04 56 15.5	−17 25 32	MCG −3-13-33	14.8	1.1 x 0.7	14.4	Sb	0	Mag 9.31v star NW 6′1.
04 56 30.0	−17 40 06	MCG −3-13-35	13.8	1.2 x 0.9	13.7	S0?	1	Mag 7.39v star SW 5′7.
04 58 41.2	−19 35 31	MCG −3-13-41	13.9	0.5 x 0.5	12.3	S0		Mag 10.22v star NE 6′3.
04 59 33.4	−17 27 28	MCG −3-13-44	13.9	0.9 x 0.5	12.8	S0/a	95	MCG −3-13-46 E 6′2; mag 11.35v star WNW 8′5.
04 59 59.5	−17 27 35	MCG −3-13-46	13.3	1.5 x 0.7	13.2	(R)SB(r)a	82	Mag 12.6 star E end; mag 8.62v star SE 12′9; very faint, stellar anonymous galaxy SE 2′0.
05 00 29.3	−17 26 04	MCG −3-13-48	14.3	0.8 x 0.7	13.5	Sc		Mag 8.62v star SSE 9′6.
05 01 14.6	−20 16 55	MCG −3-13-50	14.7	0.7 x 0.6	13.6	S	174	Mag 4.90v star Upsilon 1 Eridani N 14′0.
05 01 41.4	−20 02 47	MCG −3-13-56	13.6	0.7 x 0.4	12.1	Sb	114	Mag 4.90v star Upsilon 1 Eridani W 3′8.
05 02 18.2	−20 00 04	MCG −3-13-57	13.3	1.2 x 1.1	13.5	SB(r)b	4	Mag 4.90v star Upsilon 1 Eridani W 12′6.
05 02 54.2	−20 22 01	MCG −3-13-60	13.3	0.9 x 0.9	13.0	S0		Mag 10.48v star SE 11′5.
05 03 26.3	−20 02 36	MCG −3-13-61	13.8	0.9 x 0.4	12.6	S0	88	Mag 10.47v star N 3′5.
05 04 34.3	−17 05 31	MCG −3-13-64	14.2	1.5 x 0.2	12.7	Sc: sp	66	Mag 13.2 star N 1′5.
05 05 16.7	−17 21 05	MCG −3-13-67	14.1	0.7 x 0.5	12.8	Irr	165	Mag 9.25v star SE 12′8.
05 05 48.0	−19 48 59	MCG −3-13-68	14.3	0.9 x 0.6	13.5	Scd	131	Mag 10.64v star E 1′3.
05 07 04.2	−18 25 23	MCG −3-13-74	14.1	0.4 x 0.4	12.0	Sd		Very close pair of mag 9.87v, 10.60v stars N 7′0.
05 08 35.4	−17 10 53	MCG −3-14-3	14.9	0.5 x 0.5	13.2			Mag 8.81v star WSW 10′8.
05 09 02.1	−20 15 21	MCG −3-14-5	12.9	1.1 x 0.5	12.2	S?	93	Mag 8.73v star NE 2′6.
05 17 43.2	−20 52 39	MCG −3-14-12	14.3	0.6 x 0.4	12.6		90	Mag 9.27v star WNW 7′0.
05 29 53.9	−18 42 54	MCG −3-14-19	13.8	0.8 x 0.5	12.6		30	Mag 8.74v star NE 10′0.
05 30 09.1	−20 20 24	MCG −3-14-20	13.2	1.3 x 0.4	12.3	S0/a	110	Mag 10.92v star ENE 12′0.
05 35 49.4	−17 46 40	MCG −3-15-4	14.2	1.0 x 0.7	13.6	Irr	130	Mag 9.61v star NNW 8′2.
05 39 15.5	−17 01 33	MCG −3-15-7	14.6	1.3 x 0.5	14.0	SB?	144	Mag 7.98v star SW 7′5.
05 40 17.9	−17 50 13	MCG −3-15-8	14.0	0.5 x 0.4	12.3	E	144	Mag 10.08v star SSE 6′0; mag 14.3 star E 0′7.
05 47 47.6	−19 52 02	MCG −3-15-15	13.1	0.8 x 0.6	12.2	S0	174	Mag 10.04v star E 3′8.
05 48 16.4	−19 34 14	MCG −3-15-17	13.5	1.1 x 0.3	12.1	S?	31	Mag 9.96v star SSE 3′2.
05 50 46.8	−18 10 15	MCG −3-15-21	14.0	0.6 x 0.6	12.8	S0° pec:		
05 52 25.8	−19 26 31	MCG −3-15-23	13.7	0.8 x 0.4	12.3	Sa	128	Mag 7.56v star WNW 9′5.
05 53 33.1	−20 15 58	MCG −3-15-26	14.0	0.7 x 0.5	12.7	Sb	126	Mag 8.15v star N 7′4.
05 56 27.7	−18 13 06	MCG −3-16-1	13.4	0.5 x 0.4	11.6	SA(r)0^+	18	Mag 8.11v star SE 6′2.
06 00 23.8	−19 42 33	MCG −3-16-5	13.3	0.9 x 0.5	12.3	SABb? II-III	174	Mag 8.79v star SSW 10′1.
06 01 56.9	−20 08 16	MCG −3-16-7	14.0	0.5 x 0.3	11.8		6	Mag 8.51v star NNE 8′2.
06 01 55.5	−17 56 31	MCG −3-16-8	13.3	0.7 x 0.4	11.9	S0	108	Mag 9.86v star N 5′8; **MCG −3-16-9** N 3′9.
04 36 03.5	−23 43 49	MCG −4-11-24	14.5	0.4 x 0.4	12.4	Sc		Mag 10.84v star ESE 11′3.
04 37 11.4	−24 05 43	MCG −4-11-25	14.3	0.7 x 0.7	13.4	Sbc		Mag 9.80v star NNE 3′6.
04 39 11.9	−21 20 53	MCG −4-12-2	13.2	1.1 x 0.4	12.2	Sb	51	Mag 9.83v star N 3′0.

(Continued from previous page)

GALAXIES

RA h m s	Dec ° ′ ″	Name	Mag (V)	Dim ′ Maj x min	SB	Type Class	PA	Notes
04 39 06.4	−24 11 06	MCG −4-12-3	13.4	1.1 x 0.4	12.4	Sab	165	Pair with ESO 485-4 E 1′.2; mag 12.0 star W 4′.0.
04 39 18.9	−22 12 45	MCG −4-12-5	14.6	0.9 x 0.6	13.8	Sbc	99	Mag 9.82v star NW 5′.9.
04 40 14.7	−24 19 10	MCG −4-12-6	13.7	1.1 x 1.0	13.6	SB?	50	Mag 10.05v star E 4′.6.
04 41 26.2	−21 50 43	MCG −4-12-9	14.2	0.6 x 0.4	12.7	E?	15	Mag 10.72v star NNE 6′.0.
04 50 07.9	−23 53 34	MCG −4-12-17	14.0	0.9 x 0.5	13.0	Sbc	3	Mag 9.50v star W 5′.0.
04 52 21.0	−22 59 56	MCG −4-12-18	14.3	0.7 x 0.5	13.0	Sc	60	Mag 10.07v star NNW 9′.6.
04 54 42.8	−24 31 59	MCG −4-12-20	14.4	0.7 x 0.5	13.1	SBc	138	Mag 10.59v star SW 8′.8.
04 55 00.1	−26 02 08	MCG −4-12-21	14.0	1.0 x 0.8	13.6	Sbc	2	
04 55 16.8	−22 43 40	MCG −4-12-22	14.3	1.2 x 0.8	14.1	S?	165	
04 58 09.0	−21 37 12	MCG −4-12-23	13.5	1.1 x 0.9	13.4	S?	84	Mag 10.77v star SSW 5′.3.
04 58 28.8	−21 34 05	MCG −4-12-25	14.4	0.5 x 0.5	12.7	Sb		**MCG −4-12-24** NW 1′.5.
05 00 14.5	−22 11 18	MCG −4-12-30	14.7	0.9 x 0.4	13.4	SBab	99	Mag 9.72v star SE 5′.2.
05 01 13.6	−21 39 45	MCG −4-12-32	14.4	0.4 x 0.3	12.0	Sb		Mag 8.04v star E 3′.9.
05 02 00.7	−21 08 15	MCG −4-12-34	13.0	1.1 x 0.9	13.1	E?	73	Mag 8.17v star E 10′.5.
05 02 44.8	−25 32 12	MCG −4-12-37	13.6	1.1 x 0.9	13.5	S?	4	
05 03 11.5	−22 48 50	MCG −4-12-38	13.8	1.2 x 0.3	12.6	S0°: sp	138	Pair with ESO 486-19 S edge; mag 5.74v star W 6′.9.
05 03 18.7	−25 41 51	MCG −4-12-40	13.8	1.1 x 0.4	12.8	S?	106	
05 03 19.9	−25 25 26	MCG −4-12-41	13.5	1.2 x 0.8	13.3	S?	107	
05 03 44.2	−23 19 25	MCG −4-12-42	14.5	0.5 x 0.5	12.8			Mag 10.13v star E 4′.0.
05 09 56.9	−22 17 59	MCG −4-13-2	14.0	1.3 x 0.7	13.7	S?	62	1′.2 ENE mag 9 star.
05 10 30.9	−25 42 37	MCG −4-13-3	13.8	1.1 x 0.6	13.2	Sbc	38	Mag 10.18v star ESE 4′.9; mag 12.4 star N 3′.8.
05 11 23.0	−22 14 48	MCG −4-13-4	13.6	1.0 x 0.8	13.2	S?	159	Mag 8.48v star N 4′.7.
05 15 38.8	−22 42 31	MCG −4-13-5	14.0	1.2 x 0.6	13.4	SB?	24	Small, faint anonymous galaxy ESE 3′.7.
05 18 14.0	−24 06 09	MCG −4-13-8	14.4	0.5 x 0.5	12.7	Sb		Mag 12 star sitting on N edge.
05 20 15.3	−25 42 51	MCG −4-13-10	14.1	0.5 x 0.5	12.5	Sbc		
05 22 21.1	−22 20 57	MCG −4-13-14	13.7	1.1 x 0.5	12.9	SBc	140	Mag 8.77v star SSW 6′.1.
05 27 01.0	−21 33 25	MCG −4-13-15	13.9	1.2 x 0.4	12.9	Sc	92	Mag 10.53v star N 5′.9.
05 29 40.6	−24 42 53	MCG −4-13-16	14.0	0.4 x 0.4	11.8	Sc		Mag 9.86v star SW 5′.6.
05 30 32.0	−23 06 25	MCG −4-14-1	14.8	0.9 x 0.4	13.5		150	Mag 9.38v star NE 3′.5.
05 40 11.8	−22 00 11	MCG −4-14-8	14.0	1.0 x 0.6	13.2	S?	124	Star N edge, bright knot or star NW edge.
05 42 04.7	−26 07 22	MCG −4-14-12	14.4	0.5 x 0.5	12.7	Sb		Mag 10.44v star WNW 0′.9.
05 42 04.6	−26 08 43	MCG −4-14-13	14.2	0.4 x 0.3	11.8	S0	36	Located 1′.4 S of MCG −4-14-12.
05 42 07.3	−26 11 53	MCG −4-14-14	12.4	0.5 x 0.5	11.0	E/S0		Mag 9.77v star WNW 6′.3.
05 42 18.8	−26 05 54	MCG −4-14-16	14.1	0.5 x 0.5	12.7	E		Mag 10.19v star N 3′.7.
05 42 14.6	−25 32 25	MCG −4-14-17	14.6	0.4 x 0.4	12.5	Sa		Pair with ESO 487-36 E 1′.4; mag 9.46v star W 3′.2.
05 42 36.4	−20 59 54	MCG −4-14-18	14.1	0.6 x 0.6	12.8	S0/a		Mag 9.22v star ESE 4′.6.
05 43 52.4	−26 28 01	MCG −4-14-19	14.5	0.8 x 0.5	13.3	SBc	114	Mag 9.75v star ESE 4′.3.
05 44 51.2	−25 05 50	MCG −4-14-20	13.6	1.2 x 0.6	13.3	S?	139	
05 45 22.0	−25 47 32	MCG −4-14-21	13.7	1.0 x 0.8	13.3	SA(rs)0^+	116	Bright knots or stars E and W edges; numerous faint galaxies in surrounding area.
05 45 41.2	−25 32 10	MCG −4-14-23	14.0	1.0 x 0.8	13.6	(R′)SAB(r)ab	34	Galaxies **PGC 17753** 1′.9 NW, and **PGC 17765** SE 2′.0.
05 46 55.5	−25 38 10	MCG −4-14-26	14.0	1.0 x 0.6	13.4	SA(r)0^+	64	Spiral galaxy **PGC 17809** WSW 1′.3.
05 47 28.9	−23 34 37	MCG −4-14-27	14.1	1.1 x 0.3	12.8	Sc	6	Mag 9.34v star WNW 10′.8.
05 47 25.2	−25 34 22	MCG −4-14-28	14.1	1.0 x 0.7	13.6	S?	42	**PGC 17845** NE 1′.7 and **PGC 17847** NE 2′.5.
05 47 30.4	−25 07 25	MCG −4-14-30	14.1	0.9 x 0.8	13.6	SBbc	142	**MCG −4-14-29** NW 1′.4; mag 11.13v star NW 2′.4.
05 47 43.4	−25 54 59	MCG −4-14-31	14.2	0.6 x 0.3	12.2	S0	123	Mag 10.57v star N 6′.2; stellar **PGC 17857** NE 1′.6.
05 47 52.8	−25 09 03	MCG −4-14-32	14.2	0.8 x 0.6	13.2	S?	153	
05 48 48.4	−24 22 46	MCG −4-14-34	13.8	1.0 x 0.7	13.2	Sc	42	Several stars N edge; mag 7.82v star SW 8′.2.
05 48 43.2	−25 28 39	MCG −4-14-35	13.0	0.5 x 0.4	11.1	S0	12	Pair with ESO 488-27 W 1′.1; located near E edge of galaxy cluster A 548, innumerable cataloged and anonymous galaxies in area.
05 48 53.3	−25 38 05	MCG −4-14-37	13.8	0.6 x 0.6	12.6	S?		Bright knot or superimposed star N.
05 49 21.8	−25 20 53	MCG −4-14-38	13.5	0.8 x 0.6	12.8	E?	72	Close pair of N-S mag 11.25v, 11.7 stars WSW 2′.7; numerous cataloged and anonymous galaxies in surrounding area.
05 50 03.8	−24 38 15	MCG −4-14-39	13.9	1.0 x 0.8	13.6	SB(r)0/a	169	
05 50 44.5	−22 58 47	MCG −4-14-41	13.9	1.0 x 0.4	12.8	Sab	11	Mag 5.87v star W 11′.8.
06 02 51.0	−23 41 29	MCG −4-15-6	13.9	0.4 x 0.4	12.0	E/S0		Stellar; mag 10.57v star SSW 3′.4.
04 38 03.4	−29 37 17	MCG −5-12-1	14.5	0.7 x 0.3	12.6	S	45	
04 52 05.5	−28 35 38	MCG −5-12-5	14.4	0.9 x 0.7	13.8	Im:	33	
04 58 38.3	−27 50 43	MCG −5-12-12	13.9	1.1 x 0.6	13.3	Sb	72	
05 00 45.5	−29 51 29	MCG −5-12-13	14.0	1.0 x 0.9	13.7	Sc	84	Mag 9.41v star E 6′.9.
05 04 33.1	−29 04 16	MCG −5-13-1	14.2	0.5 x 0.5	12.5	S0/a		Mag 10.12v star ENE 6′.5.
05 05 49.5	−28 35 20	MCG −5-13-3	13.5	0.8 x 0.6	12.6	S0?	85	Mag 11.04v star S 2′.9.
05 17 57.9	−29 42 03	MCG −5-13-14	14.2	1.4 x 0.2	12.7	Sc	149	Mag 10.47v star S 1′.1.
05 31 16.3	−29 19 08	MCG −5-14-1	13.4	0.5 x 0.4	11.7	E	60	Mag 11.16v star S 4′.2.
05 32 39.0	−29 37 38	MCG −5-14-2	13.8	1.1 x 0.4	12.8	Sb	83	Mag 9.29v star E 9′.9.
05 39 21.9	−29 22 20	MCG −5-14-9	14.1	1.5 x 0.5	13.7	Sb?	21	**ESO 424-5A** SE edge.
05 42 56.6	−29 43 45	MCG −5-14-10	13.4	1.2 x 1.0	13.5	S/Irr	142	Star NE edge.
05 43 33.0	−27 39 07	MCG −5-14-12	13.4	1.0 x 0.8	13.1	.S0	37	Star SW end, star or small galaxy NW edge.
05 44 39.4	−28 33 19	MCG −5-14-14	13.5	1.2 x 0.3	12.3	S0	50	Mag 9.61v star SW 7′.7.
05 57 28.8	−29 57 38	MCG −5-15-1	14.2	1.0 x 0.9	13.9	Sc:	156	Mag 8.74v star SSW 9′.5.
04 37 15.5	−18 54 04	NGC 1630	14.0	0.7 x 0.5	12.7	SB0^+ pec?	140	Mag 10.23v star WSW 6′.0.
04 38 24.1	−20 38 59	NGC 1631	13.3	1.4 x 0.9	13.4	(R′)SAB(r)0/a:	44	Mag 7.26v star WSW 6′.2.
04 42 14.3	−20 26 05	NGC 1640	11.7	2.8 x 2.5	13.6	SB(r)b II	45	
04 55 23.6	−20 34 15	NGC 1692	13.0	1.3 x 1.2	13.3	SA(s)0°:	5	Galaxy **ESO 552-22** SE 1′.1.
04 55 51.3	−29 53 00	NGC 1701	12.8	1.2 x 0.9	12.7	(R)SA(r)b: II-III	137	Mag 10.68v star 1′.3 SE.
04 58 13.5	−20 21 51	NGC 1716	13.1	1.4 x 1.1	13.4	SAB(s)bc pec I-II	20	
05 01 46.3	−18 09 30	NGC 1738	12.9	1.3 x 0.7	12.7	SB(s)bc pec: I-II	44	NGC 1739 on NS edge.
05 01 47.6	−18 10 03	NGC 1739	13.5	1.4 x 0.7	13.3	SB(s)bc pec: II	105	NGC 1738 on N edge.
04 59 58.0	−26 01 22	NGC 1744	11.1	7.4 x 3.5	14.5	SB(s)d III-IV	168	Elongated core, numerous knots.
05 06 20.7	−19 28 02	NGC 1780	13.7	0.9 x 0.5	12.7	(R′)SB(r)0°:	84	
05 07 55.0	−18 11 24	NGC 1794	12.6	1.3 x 1.1	12.8	(R)SB(s)0° pec:	45	Mag 7.53v star NE 9′.6.
05 08 42.8	−29 16 36	NGC 1811	13.4	1.7 x 0.4	12.8	Sa: pec sp	60	
05 08 52.8	−29 15 08	NGC 1812	12.6	1.2 x 0.9	12.5	Sa pec	8	
05 21 48.2	−23 48 37	NGC 1886	12.7	3.1 x 0.4	12.8	Sbc sp	60	Mag 9.97v star SW 2′.6.
05 33 22.0	−21 56 39	NGC 1964	10.8	5.6 x 1.8	13.2	SAB(s)b I-II	32	Mag 10.20v star W 1′.4.
05 34 01.1	−23 18 35	NGC 1979	11.9	1.8 x 1.8	13.1	SA0:		
05 35 25.6	−17 48 56	NGC 1993	12.4	1.5 x 1.4	13.1	SA(rs)0^-:	80	Mag 10.60v star W 4′.3; **MCG −3-15-2** WSW 9′.3.
05 45 53.9	−21 59 57	NGC 2073	12.5	1.5 x 1.4	13.2	SA(rs)0^-:		
05 47 51.4	−17 36 11	NGC 2089	12.0	1.9 x 1.1	12.6	SAB0^-:	39	Mag 10.46v star 1′.5 SSE.
05 50 46.6	−21 34 05	NGC 2106	12.1	2.7 x 1.3	13.4	SB(s)0°	103	

(Continued from previous page)

155 GALAXIES 155

RA h m s	Dec ° ′ ″	Name	Mag (V)	Dim ′ Maj x min	SB	Type Class	PA	Notes
05 57 52.3	−20 05 08	NGC 2124	12.2	2.7 x 0.8	12.9	SA(s)b? II-III	5	Several knots or superimposed stars; mag 8.57v star SE 9′7.
05 58 47.6	−26 39 13	NGC 2131	14.0	1.1 x 0.4	12.9	IB(s)m: III-IV	118	Star on N edge.
06 01 08.4	−23 40 32	NGC 2139	11.6	2.4 x 1.9	13.1	SAB(rs)cd II-III	140	Numerous bright knots.

GALAXY CLUSTERS

RA h m s	Dec ° ′ ″	Name	Mag 10th brightest	No. Gal	Diam ′	Notes
04 38 54.0	−22 06 00	A 500	15.8	53	15	Most members faint and stellar.
04 47 42.0	−20 25 00	A 514	15.2	78	39	Numerous anonymous GX, mostly in southern half.
05 01 30.0	−22 36 00	A 533	15.8	31	12	Several relatively bright, almost stellar, anonymous GX seen on DSS image.
05 47 00.0	−25 36 00	A 548	13.7	79	56	Majority of PGC and anonymous cluster members stretch from SW to NE edges.
05 11 24.0	−28 59 00	A 3323	16.2	42	17	**ESO 422-43**, **43A**, **43B** at center, all others anonymous, stellar.
05 34 42.0	−28 30 00	A 3354	15.3	54	22	See notes for ESO 423-20; all others anonymous.
05 49 24.0	−24 28 00	A 3367	15.6	33	22	Majority of members anonymous, stellar.
05 50 30.0	−22 32 00	A 3368	16.1	32	17	All members anonymous, faint, stellar.
05 56 54.0	−21 15 00	A 3374	16.1	34	17	**ESO 555-15** E of center 11′8, all others anonymous, stellar.
06 02 36.0	−27 46 00	AS 562	16.1		17	All members anonymous, faint, stellar.

OPEN CLUSTERS

RA h m s	Dec ° ′ ″	Name	Mag	Diam ′	No. ★	B ★	Type	Notes
05 16 42.0	−24 02 00	ESO 486-45		3			cl	
05 39 16.0	−17 51 00	NGC 2017		10	8	8.0	ast?	A scattered group of a few stars; not a true cluster. It contains the multiple star h3780.

GLOBULAR CLUSTERS

RA h m s	Dec ° ′ ″	Name	Total V m	B ★ V m	HB V m	Diam ′	Conc. Class Low = 12 High = 1	Notes
05 24 10.6	−24 31 27	NGC 1904	7.7	13.1	16.2	9.6	5	= **M 79**.

PLANETARY NEBULAE

RA h m s	Dec ° ′ ″	Name	Diam ″	Mag (P)	Mag (V)	Mag cent ★	Alt Name	Notes
05 55 06.8	−22 54 02	PN G228.2−22.1	132				PK 11+17.1	Small triangle of stars in center; the brightest of the triangle, mag 12.9 is the center of the planetary, a mag 14.4 star is W and a mag 15.5 star is S.

156 GALAXIES 156

RA h m s	Dec ° ′ ″	Name	Mag (V)	Dim ′ Maj x min	SB	Type Class	PA	Notes
03 29 56.9	−28 46 33	ESO 418-7A	12.9	0.8 x 0.7	12.2	SAB(s)0° pec	117	Pair with ESO 418-7 overlapping N; mag 11.22v star S 2′1.
03 29 55.9	−28 46 03	ESO 418-7	13.9	1.2 x 1.1	14.0	SB(s)b pec	30	Pair with ESO 418-7A overlapping S; mag 11.22v star S 2′1.
03 42 11.1	−27 51 53	ESO 419-3	13.2	1.7 x 1.0	13.6	(R′)SAB(s)c pec: II	133	Several bright knots E edge; mag 10.29v star W 7′4.
03 57 10.4	−28 52 37	ESO 419-12	13.1	1.9 x 0.3	12.3	Sa: sp	10	Mag 10.79v star W 10′5.
04 07 45.7	−29 51 34	ESO 420-3	12.8	1.9 x 1.3	13.7	SA(rs)c I-II	144	Mag 10.27v star NW 4′7; mag 10.02v star SE 4′0.
04 12 28.1	−29 03 36	ESO 420-10	14.5	0.9 x 0.8	14.1		51	Almost stellar core, faint envelope; mag 9.06v star S 2′5.
04 24 18.2	−27 56 39	ESO 420-18	14.7	1.6 x 0.2	13.3	Sb: sp	127	Mag 8.01v star NW 10′7.
03 16 09.0	−24 12 03	ESO 481-16	14.7	1.1 x 0.8	14.4	IB(s)m V-VI	96	Mag 10.39v star E 5′2.
03 17 04.6	−22 51 59	ESO 481-17	12.3	1.9 x 1.7	13.4	SB(rs)a pec:	163	Bright core, uniform envelope; mag 10.54v star E 5′7.
03 18 33.0	−25 50 13	ESO 481-18	12.7	2.8 x 0.9	13.6	SB(rs)cd IV	41	Mag 10.46v star NE 7′2.
03 18 43.1	−23 46 57	ESO 481-19	14.9	2.1 x 1.1	15.7	IB(s)m V-VI	174	Mag 12.4 star N 3′0.
03 20 18.1	−26 27 53	ESO 481-21	14.7	1.8 x 0.3	13.9	SBd? sp	108	Mag 10.83v star SW 4′7 mag 6.41v star SSE 10′3.
03 29 02.6	−26 26 45	ESO 481-29	13.9	0.6 x 0.5	12.4	S0?	79	Mag 10.05v star W 3′0.
03 29 37.9	−23 21 08	ESO 481-30	14.9	1.5 x 0.2	13.5	S	50	Mag 6.87v star S 7′5.
03 31 18.9	−26 33 47	ESO 482-1	13.9	1.5 x 0.4	13.2	S?	44	Mag 8.19v star SW 5′1.
03 33 02.0	−24 08 01	ESO 482-5	14.7	2.1 x 0.3	14.0	SB(s)dm: sp IV	78	Mag 10.04v star W 4′0.
03 40 41.5	−26 47 13	ESO 482-32	14.6	1.1 x 0.4	13.6	IB(s)m pec: IV	66	Mag 13.9 star S 1′3; NGC 1412 SW 5′2.
03 41 14.6	−23 50 16	ESO 482-35	12.8	1.9 x 1.3	13.7	SB(rs)ab	2	Mag 11.9 star SE 2′1; mag 11.6 star N 5′2.
03 42 19.2	−22 45 24	ESO 482-36	14.1	1.8 x 1.1	14.7	SAB(s)m V-VI	3	Mag 11.7 star WNW 10′2.
03 42 17.6	−22 44 54	ESO 482-36A	14.0	0.6 x 0.3	12.0	SAB(s)m V	107	On NW edge of ESO 482-36.
03 49 42.4	−26 59 37	ESO 482-46	13.3	3.8 x 0.5	13.8	Sc: sp	69	Mag 8.43v star S 7′0.
03 50 39.5	−25 16 45	ESO 482-47	13.1	1.5 x 1.4	13.7	SAB(rs)c I	31	Mag 10.48v star N 6′2.
03 56 46.0	−24 29 52	ESO 483-1	14.3	0.4 x 0.2	11.5		36	**ESO 483-1A** E 0′9; mag 12.2 star NNW 2′4.
04 00 26.3	−25 10 58	ESO 483-6	13.2	2.8 x 0.5	13.4	(R′)SB(s)b: sp	132	Mag 9.19v star E 7′7.
04 07 42.3	−22 42 54	ESO 483-9	13.7	1.5 x 0.6	13.4	SAB0° pec	141	Mag 14.5 star E 1′3.
04 10 22.7	−23 37 02	ESO 483-12	13.4	2.1 x 0.5	13.3	S0/a: sp	18	Mag 12.2 star E 3′6; mag 12.2 star S 2′9.
04 12 41.1	−23 09 36	ESO 483-13	13.3	1.6 x 0.9	13.5	SA0$^-$:	135	Mag 6.78v star ENE 5′5.
04 19 22.7	−26 47 48	ESO 484-5	14.8	1.9 x 0.2	13.6	Sc sp	20	Mag 10.38v star W 2′5; **MCG −4-11-8** N 4′1.
04 39 11.8	−24 10 54	ESO 485-4	14.9	1.7 x 0.2	13.6	Sc:	143	Pair with MCG −4-12-3 W 1′2; mag 11.01v star E 4′4.
03 21 52.4	−19 28 55	ESO 548-1	14.3	1.0 x 0.3	12.8	Sa:	38	Anonymous galaxy E 1′9; mag 12.4 star S 1′9.
03 23 47.1	−19 45 13	ESO 548-5	13.2	1.8 x 1.5	14.2	SAB(s)m IV-V	58	Stellar nucleus; mag 11.4 star S 4′3.
03 26 00.2	−17 39 30	ESO 548-14	13.8	1.5 x 0.3	12.8	Sa:	77	Mag 8.96v star WSW 6′6; note: **MCG −3-9-39** located 3′5 WNW of this star; NGC 1329 N 4′0.
03 27 35.2	−21 13 43	ESO 548-21	13.8	2.3 x 0.4	13.6	SB(s)dm? IV-V	68	Mag 9.85v star S 10′4.
03 28 14.5	−17 25 13	ESO 548-23	14.4	1.2 x 0.5	13.8	E$^+$?	21	Mag 10.72v star ESE 3′0.
03 29 00.7	−22 08 48	ESO 548-25	14.1	1.5 x 0.8	14.1	(R′)SB(s)a pec	81	Mag 11.6 star E 9′3; NGC 1347 SE 12′5.
03 32 18.8	−17 43 07	ESO 548-32	14.2	2.2 x 0.9	14.8	SB(s)m V	144	Mag 12.2 star NE 9′7; mag 12.7 star SW 6′0.
03 32 28.5	−18 56 56	ESO 548-33	13.3	1.5 x 0.8	13.3	SB(rs)0° pec?	109	Mag 13.1 star on E edge; mag 12.0 star NNW 6′3.
03 33 29.2	−18 08 47	ESO 548-35	12.9	1.6 x 1.3	13.5	SAB(rs)c: I-II	39	Star superimposed NE of center.
03 33 27.6	−21 33 53	ESO 548-36	13.3	1.0 x 0.8	12.9	S pec	168	Located 6′3 NW of mag 4.27v star 19 Eridani.
03 33 51.2	−21 27 19	ESO 548-40	13.8	1.6 x 1.2	14.3	S?	129	Located 10′5 N of mag 4.278v star 19 Eridani; IC 1953 SW 2′6.
03 34 38.3	−17 28 30	ESO 548-46	14.1	1.0 x 0.8	13.7	Sab	164	Mag 11.4 star SW 12′1.
03 34 43.2	−19 01 46	ESO 548-47	12.7	3.0 x 0.7	13.4	(R′)SB(r)0$^+$? sp	72	Close pair of mag 10.45v, 9.10v stars W 15′8.
03 40 02.4	−19 22 03	ESO 548-65	14.7	1.5 x 0.3	13.7	SBa? sp	38	Mag 9.00v star SE 7′6, there is a mag 6.82v star 3′6 S of this star; MCG −3-10-26 S 3′6.

GALAXIES

RA h m s	Dec ° ′ ″	Name	Mag (V)	Dim ′ Maj x min	SB	Type Class	PA	Notes
03 40 18.9	−18 55 52	ESO 548-68	13.1	1.7 x 0.9	13.5	SB(s)0⁻: sp	135	Mag 10.70v star SW 2′.1; mag 7.78v star WSW 12′.1.
03 40 41.1	−22 17 11	ESO 548-70	15.0	1.7 x 0.3	14.1	SBd? sp IV	64	Mag 8.57v star NW 11′.6.
03 41 23.0	−17 45 27	ESO 548-75	13.2	1.7 x 1.3	13.9	SB(rs)c II	23	Mag 9.53v star SSW 8′.7.
03 42 04.1	−21 14 42	ESO 548-81	12.0	0.8 x 0.8	11.4	SB(rs)a pec?		Mag 8.34v star immediately NW.
03 42 57.4	−19 01 17	ESO 549-2	13.9	1.2 x 0.8	13.7	IB(s)m pec V	33	Mag 8.84v star W 10′.9.
03 43 38.3	−21 14 12	ESO 549-6	14.5	1.6 x 0.6	14.3	IB(s)m V	28	
03 44 37.3	−20 10 14	ESO 549-8	14.2	0.3 x 0.3	11.5			Compact.
03 48 14.2	−21 28 27	ESO 549-18	12.6	2.4 x 1.4	13.7	SAB(rs)c II-III	18	Mag 8.40v star S 6′.3; **MCG −4-10-3** NNE 3′.6.
03 48 28.6	−18 45 10	ESO 549-21	14.1	1.5 x 0.2	12.7	Sa:	125	Bright knot or superimposed star SE end.
03 48 51.7	−18 58 46	ESO 549-22	13.8	1.5 x 0.7	13.7	SA(rs)b: I-II	38	Mag 12.8 star on W edge; **ESO 549-25** E 3′.5; mag 10.14v star SE 2′.2.
03 48 57.8	−22 07 55	ESO 549-23	13.1	1.2 x 0.7	12.8	(R′)SBa	145	Mag 8.63vb star WSW 9′.2; **MCG −4-10-5** S 6′.6.
03 55 09.3	−17 28 11	ESO 549-36	13.4	1.8 x 1.0	13.9	SB(rs)c: III	6	
03 57 10.3	−18 46 41	ESO 549-40	13.0	1.5 x 0.6	12.8	S?	143	
04 04 29.7	−17 54 21	ESO 550-2	14.3	1.5 x 0.3	13.3	S?	44	Mag 9.40v star ESE 6′.3.
04 06 00.1	−17 46 26	ESO 550-5	14.0	2.4 x 0.7	14.4	SB(s)m sp V	58	Close pair of mag 10.62v, 9.59v stars S 3′.8.
04 09 32.0	−21 46 23	ESO 550-14	14.2	1.8 x 1.1	14.8	SB(s)d III-IV	103	Mag 8.93v star WSW 9′.6.
04 21 14.1	−21 50 51	ESO 550-24	12.1	5.6 x 1.9	14.5	SB(s)d: IV	131	Mag 7.53v star NE 8′.9.
04 21 20.0	−18 48 36	ESO 550-25	14.0	0.6 x 0.5	12.5		150	
04 35 40.5	−21 59 20	ESO 551-16	14.1	1.6 x 0.4	13.5	SBbc? pec sp	4	Mag 11.8 star S edge; mag 10.00v star S 8′.4.
04 42 27.0	−17 27 24	ESO 551-30	14.3	1.2 x 1.0	14.4	SB(s)d III-IV	0	Mag 9.99v star E 8′.6.
04 42 29.1	−21 41 31	ESO 551-31	14.3	1.5 x 0.6	14.0	SB(s)d: sp III-IV	177	Mag 9.75v star NNW 7′.2.
03 40 07.2	−18 26 37	IC 343	13.2	1.6 x 0.8	13.3	SB(rs)0⁺:	118	Mag 12.7 star E 1′.6; NGC 1407 S 8′.1.
03 41 09.1	−18 18 53	IC 345	13.8	0.8 x 0.7	13.0	S0/a	42	Mag 10.19v star SSW 6′.0.
03 41 44.5	−18 16 00	IC 346	12.6	2.0 x 1.3	13.5	SB(rs)0⁺	78	Faint, elongated anonymous galaxy W 3′.5; mag 10.40v star SSW 10′.1.
03 27 29.2	−21 33 33	IC 1928	13.1	1.6 x 0.4	12.5	Sab? sp	30	Mag 10.58v star W 9′.8.
03 33 26.8	−23 42 47	IC 1952	12.7	2.6 x 0.6	13.0	SB(s)bc? II	141	Mag 9.86v star on E edge.
03 33 42.0	−21 28 41	IC 1953	11.7	2.8 x 2.1	13.5	SB(rs)d II-III	121	Located 9′.2 N of mag 4.27v star 19 Eridani; ESO 548-40 NE 2′.6.
03 35 37.3	−21 18 01	IC 1962	14.1	2.7 x 0.5	14.2	SB(s)dm IV-V	2	Mag 12.0 star SSW 3′.0; **MCG −4-9-30** NNW 5′.2.
03 55 22.9	−28 09 31	IC 2007	13.0	1.3 x 0.8	12.9	SB(rs)bc?	52	Mag 9.76v star SE 12′.3.
03 20 52.7	−17 53 14	MCG −3-9-24	14.7	0.9 x 0.6	13.9	Sd	42	Numerous bright knots.
03 22 18.0	−17 10 52	MCG −3-9-29	14.7	0.7 x 0.4	13.1		132	MCG −3-9-29 W 2′.5.
03 22 28.8	−17 10 53	MCG −3-9-30	15.1	0.5 x 0.3	12.9		21	MCG −3-9-30 E 2′.5.
03 27 42.7	−17 00 25	MCG −3-9-44	13.9	0.9 x 0.4	12.6	Sb	174	
03 30 35.9	−17 56 29	MCG −3-9-48	13.4	1.3 x 0.9	13.4	SB(rs)0⁺ pec?	169	Mag 9.71v star N 4′.4.
03 34 19.4	−19 25 28	MCG −3-10-10	13.3	1.3 x 0.5	12.8	SA(r)0⁺:	59	NGC 1359 SW 8′.6.
03 38 10.3	−19 44 11	MCG −3-10-18	14.6	1.3 x 0.5	14.0	SB?	48	Mag 13.5 star SW end.
03 40 00.0	−19 25 36	MCG −3-10-26	13.7	1.0 x 0.7	13.2	S0	132	Mag 9.00v star E 7′.6, there is a mag 6.82v star 3′.6 S of this star; ESO 548-65 N 3′.6.
03 41 31.8	−19 54 21	MCG −3-10-34	13.8	1.3 x 0.9	13.9	S0?	16	
03 41 56.1	−18 53 45	MCG −3-10-36	13.3	1.4 x 1.1	13.7	S?	163	Star on E edge.
03 42 03.0	−18 29 03	MCG −3-10-37	13.8	1.3 x 0.6	13.4	S?	75	Mag 9.87v star S 1′.5.
03 47 49.0	−20 02 34	MCG −3-10-46	14.1	0.8 x 0.6	13.4	E	129	Mag 10.30v star SW 3′.6.
03 53 29.0	−17 07 24	MCG −3-10-50	14.8	0.9 x 0.4	13.6	Sb	168	Mag 10.05v star W 9′.2.
03 53 53.6	−17 35 36	MCG −3-10-51	13.9	1.0 x 0.4	12.7	Sb	90	Mag 8.19v star NE 14′.7; ESO 549-36 located E 5′.4 of this star.
03 54 24.5	−19 11 29	MCG −3-10-52	13.7	1.0 x 0.4	12.6	Sb	19	Anonymous galaxy NE 1′.7.
03 56 37.5	−18 03 44	MCG −3-11-1	13.9	1.1 x 0.7	13.5	SA(r)bc?	40	
04 02 25.7	−18 02 56	MCG −3-11-8	13.2	0.7 x 0.5	11.9	Sb	16	Mag 9.27v star SW 10′.7.
04 07 12.6	−17 12 13	MCG −3-11-12	14.3	2.0 x 2.0	15.7	SB(s)dm V		Mag 10.25v star S 6′.3.
04 19 22.1	−17 36 50	MCG −3-12-1	13.8	1.1 x 0.7	13.4	I?	58	
04 26 21.8	−20 43 05	MCG −3-12-9	14.3	0.7 x 0.5	13.0	S	126	Mag 10.24v star N 6′.4.
04 33 22.4	−18 40 42	MCG −3-12-15	13.7	1.4 x 0.7	13.6	SB?	15	Mag 10.70v star N 6′.6.
04 36 26.8	−20 50 45	MCG −3-12-16	14.4	0.7 x 0.5	13.1	Sb	24	Mag 9.90v star N 4′.5.
03 16 23.0	−23 48 40	MCG −4-8-54	14.5	0.7 x 0.4	13.0	SBc	159	Mag 10.87v star NNE 3′.4.
03 16 21.4	−25 51 20	MCG −4-8-55	13.7	0.6 x 0.6	12.5	S0/a		
03 20 40.4	−22 55 53	MCG −4-9-1	13.6	0.5 x 0.5	11.9	S?		
03 25 44.6	−26 23 20	MCG −4-9-9	13.5	0.7 x 0.7	12.8	E/S0		Mag 9.53v star N 4′.4.
03 26 01.4	−25 44 33	MCG −4-9-10	14.6	0.5 x 0.3	12.4	Sbc	132	NGC 1327 WNW 9′.3; mag 8.34v star W 15′.7.
03 30 47.0	−21 03 32	MCG −4-9-18	13.4	1.2 x 0.8	13.2	SB?	98	
03 31 38.2	−25 00 32	MCG −4-9-21	14.0	1.3 x 0.8	13.8	S?	89	
03 32 57.7	−21 05 24	MCG −4-9-23	13.5	1.2 x 1.1	13.7	SB?	133	Mag 12.1 star N 1′.1.
03 33 51.2	−21 03 17	MCG −4-9-27	14.4	0.7 x 0.4	12.9	SBcd	48	Mag 10.38v star N 4′.6.
03 36 17.5	−25 36 17	MCG −4-9-32	14.2	1.5 x 0.5	13.7	Sbc	63	Mag 12.7 star SW 1′.6.
03 38 17.6	−23 25 12	MCG −4-9-38	13.7	0.9 x 0.9	13.4	S?		
03 39 21.5	−21 24 57	MCG −4-9-43	14.3	0.9 x 0.4	13.0	SBbc	48	Mag 9.63v star NW 10′.9.
03 41 02.0	−24 36 20	MCG −4-9-49	15.5	0.4 x 0.2	12.6	Sc	135	Mag 9.49v star S 1′.3.
03 44 56.9	−23 42 01	MCG −4-9-57	14.8	0.5 x 0.4	12.9	Sb	9	Star or galaxy on W edge.
03 57 13.2	−25 23 56	MCG −4-10-9	14.1	1.5 x 0.5	13.6	SB(s)c II	156	Mag 12.1 star N 2′.5.
04 02 46.2	−21 07 09	MCG −4-10-12	13.3	1.0 x 0.6	12.6	Sbc	7	
04 07 36.3	−21 25 55	MCG −4-10-14	14.2	0.8 x 0.6	13.5	E?	76	
04 08 50.9	−21 43 18	MCG −4-10-16	13.8	0.7 x 0.7	12.9	SBa		
04 13 40.7	−24 02 00	MCG −4-11-2	14.8	0.3 x 0.3	12.0			
04 13 44.7	−24 01 19	MCG −4-11-3	14.8	0.6 x 0.5	13.4	SBc	54	
04 15 52.0	−24 47 42	MCG −4-11-5	14.4	1.0 x 0.5	13.5	S0/a	135	mag 9.36v star E 5′.3.
04 20 59.6	−23 44 37	MCG −4-11-9	13.9	0.4 x 0.4	11.8	S0		Lies 1′.2 WSW of mag 11 star.
04 23 40.2	−24 06 41	MCG −4-11-11	14.9	0.6 x 0.4	13.2	Sbc	27	Lies 1′.0 N of mag 11 star.
04 23 45.2	−23 05 00	MCG −4-11-12	14.2	0.5 x 0.5	12.6	S0		Elongated anonymous galaxy, next to star, SE 1′.6.
04 29 42.1	−26 46 24	MCG −4-11-16	13.9	1.3 x 0.3	12.7	Sa	126	NGC 1591 NNW 4′.4.
04 29 59.7	−26 50 17	MCG −4-11-18	13.5	0.8 x 0.5	12.5	E?	27	Almost stellar nucleus; mag 10.72v star SW 4′.4.
04 33 15.0	−24 40 25	MCG −4-11-19	14.7	0.8 x 0.5	13.6	SBc	45	Mag 10.75v star NNE 8′.8.
04 34 29.4	−24 55 21	MCG −4-11-20	13.8	1.1 x 0.6	13.2	Sa	80	Mag 7.38v star SSE 9′.0.
04 35 39.5	−25 08 00	MCG −4-11-23	13.8	1.3 x 0.3	12.6	Sbc	50	Mag 7.38v star NW 11′.6.
04 36 03.5	−23 43 49	MCG −4-11-24	14.5	0.4 x 0.4	12.4	Sc		Mag 10.84v star ESE 11′.3.
04 37 11.4	−24 05 43	MCG −4-11-25	14.3	0.7 x 0.7	13.4	Sbc		Mag 9.80v star NNE 3′.6.
04 39 11.9	−21 20 53	MCG −4-12-2	13.2	1.1 x 0.4	12.2	Sb	51	Mag 9.83v star N 3′.0.
04 39 06.4	−24 11 06	MCG −4-12-3	13.4	1.1 x 0.4	12.4	Sab	165	Pair with ESO 485-4 E 1′.2; mag 12.0 star W 4′.0.
04 39 18.9	−22 12 45	MCG −4-12-5	14.6	0.9 x 0.6	13.8	Sbc	99	Mag 9.82v star NW 5′.9.
04 40 14.7	−24 19 10	MCG −4-12-6	13.7	1.1 x 1.0	13.6	SB?	50	Mag 10.05v star E 4′.6.
04 41 26.2	−21 50 43	MCG −4-12-9	14.2	0.6 x 0.4	12.7	E?	15	Mag 10.72v star NNE 6′.0.
03 29 16.6	−27 39 24	MCG −5-9-6	14.2	0.7 x 0.4	12.7	S	156	Mag 9.93v star W 6′.9.
03 29 21.2	−28 08 03	MCG −5-9-7	13.9	0.9 x 0.5	12.9	Sb	11	

(Continued from previous page)

GALAXIES

RA h m s	Dec ° ′ ″	Name	Mag (V)	Dim ′ Maj x min	SB	Type Class	PA	Notes
03 46 48.2	−29 56 23	MCG −5-10-1	14.0	1.0 x 0.8	13.6	Sb	175	Mag 6.55v star NE 7′.2.
03 49 24.5	−26 43 49	MCG −5-10-3	15.0	0.8 x 0.2	12.9	S	24	
04 10 50.5	−29 55 35	MCG −5-10-14	13.5	1.3 x 0.9	13.6	SB(s)bc II	128	Star on W edge.
04 29 09.6	−29 47 06	MCG −5-11-10	14.3	0.5 x 0.4	12.4	Sa	27	Mag 8.01v star W 7′.7.
04 31 30.8	−29 45 03	MCG −5-11-12	14.2	0.7 x 0.5	13.2	E	120	Pair with MCG −5-11-13; located 1′.5 N of mag 8.19v star.
04 31 35.3	−29 45 17	MCG −5-11-13	12.7	0.5 x 0.4	11.0	E	45	
04 31 42.1	−29 40 31	MCG −5-11-15	13.9	0.5 x 0.5	12.2	S		
04 38 03.4	−29 37 17	MCG −5-12-1	14.5	0.7 x 0.3	12.6	S	45	
03 18 15.1	−27 36 40	NGC 1292	12.2	3.0 x 1.3	13.5	SA(s)c II-III	7	Close pair of mag 10.29v, 11.6 stars NNE 3′.3.
03 19 14.1	−19 06 01	NGC 1297	11.8	2.2 x 1.9	13.2	SAB(s)0° pec:	3	
03 19 41.3	−19 24 36	NGC 1300	10.4	6.2 x 4.1	13.8	SB(rs)bc I	106	Small bright core in E-W bar with two main arms.
03 20 35.3	−18 42 56	NGC 1301	13.4	2.2 x 0.4	13.1	SB(rs)b? II	140	Mag 7.05v star SW 13′.9.
03 19 51.4	−26 03 36	NGC 1302	10.7	3.9 x 3.7	13.5	(R)SB(r)0/a		Large bright core with smooth envelope.
03 21 03.2	−25 30 46	NGC 1306	12.9	0.8 x 0.8	12.3	Sb?		**ESO 481-24** N 4′.9.
03 23 06.6	−21 22 32	NGC 1315	12.6	1.6 x 1.4	13.4	SB(rs)0$^+$?	55	
03 23 56.5	−21 31 37	NGC 1319	12.9	1.3 x 0.7	12.7	S0 pec sp	27	Mag 10.36v star W 6′.5.
03 24 24.7	−21 32 49	NGC 1325	11.5	4.7 x 1.5	13.5	SA(s)bc II	56	Star superimposed E of center; mag 8.87v star S 7′.4.
03 24 48.6	−21 20 09	NGC 1325A	12.7	2.1 x 2.0	14.1	SAB(rs)d: IV	144	Stellar nucleus.
03 25 23.3	−25 40 48	NGC 1327	14.5	1.1 x 0.3	13.2	(R′)SB(s)b I-II	176	Mag 10.75v star W 2′.5.
03 26 02.7	−17 35 32	NGC 1329	12.7	1.4 x 1.1	13.0	SA(rs)a?	35	ESO 548-14 S 4′.0; mag 8.96v star SW 9′.3.
03 26 28.4	−21 21 19	NGC 1331	13.4	0.9 x 0.7	12.9	E2:	3	Located on E edge of NGC 1332.
03 26 17.0	−21 20 06	NGC 1332	10.3	4.5 x 1.4	12.2	S(s)0$^-$: sp	112	**ESO 548-16** W 3′.5; NGC 1331 on E edge.
03 29 31.6	−17 46 44	NGC 1345	13.8	1.5 x 1.1	14.1	SB(s)c pec: III-IV	108	Mag 9.45v star S 7′.6.
03 29 41.8	−22 16 57	NGC 1347	13.2	1.7 x 1.3	13.9	SB(s)c: pec II	12	Several bright knots S end; mag 11.6 star N 7′.2; ESO 548-25 NW 12′.5.
03 31 33.0	−19 16 42	NGC 1352	13.3	1.0 x 0.7	12.8	SB(s)0°:	116	Mag 8.45v star SE 4′.4.
03 32 03.1	−20 49 05	NGC 1353	11.5	3.4 x 1.4	13.0	SB(rs)b: II-III	138	
03 33 46.4	−19 29 12	NGC 1359	12.2	2.4 x 1.7	13.6	SB(s)m? pec III-IV	139	Mag 10.25v star WSW 7′.7; MCG −3-10-10 NE 8′.6; **ESO 548-43** SE 6′.8.
03 33 53.1	−20 16 56	NGC 1362	12.8	1.2 x 1.1	12.9	S0° pec:	7	Mag 8.85v star S 5′.5.
03 35 14.4	−20 22 26	NGC 1370	12.6	1.5 x 1.0	12.9	E$^+$:	50	
03 35 01.4	−24 55 59	NGC 1371	10.7	5.6 x 3.9	13.9	SAB(rs)a	135	Mag 8.74v star NE 4′.4.
03 36 39.0	−20 54 07	NGC 1377	12.5	1.8 x 0.9	12.9	S0°	92	
03 37 39.1	−18 20 22	NGC 1383	12.5	2.0 x 0.9	13.0	SAB(s)0°	91	
03 37 28.7	−24 30 08	NGC 1385	10.9	3.4 x 2.0	12.9	SB(s)cd II-III	171	
03 37 52.1	−19 00 29	NGC 1390	13.7	1.4 x 0.5	13.1	SBa pec:	19	Mag 9.76v star S 6′.3.
03 38 52.9	−18 21 16	NGC 1391	13.3	1.1 x 0.5	12.6	SB(s)0°	65	
03 38 38.5	−18 25 41	NGC 1393	12.0	1.7 x 1.3	12.8	SA(r)0°:	170	
03 39 06.9	−18 17 34	NGC 1394	12.8	1.3 x 0.4	12.0	S0°: sp	5	
03 38 30.0	−23 01 40	NGC 1395	9.6	5.0 x 4.5	13.0	E2	120	
03 38 51.7	−26 20 11	NGC 1398	9.7	7.1 x 5.4	13.5	(R′)SB(r)ab I	100	
03 39 30.6	−18 41 14	NGC 1400	11.0	2.3 x 2.0	12.5	SA0$^-$	40	Mag 9.44v star W 15′.0; NGC 1407 NE 11′.7.
03 39 22.0	−22 43 32	NGC 1401	12.3	2.4 x 0.6	12.5	SB(s)0°: sp	130	**ESO 482-24** NNW 9′.8.
03 39 30.5	−18 31 37	NGC 1402	13.6	0.8 x 0.6	12.6	SB0° pec:	88	NGC 1407 E 10′.2.
03 39 10.8	−22 23 19	NGC 1403	12.7	0.9 x 0.7	12.1	SAB(s)0°	163	Mag 8.38v star N 4′.1.
03 40 12.0	−18 34 45	NGC 1407	9.7	4.6 x 4.3	12.9	E0	35	Large, bright core, smooth envelope; faint, elongated anonymous galaxy ESE 8′.5 from center.
03 40 29.5	−26 51 44	NGC 1412	12.5	1.9 x 0.8	12.8	SAB0°: sp	131	Mag 10.34v star SW 4′.1; ESO 482-32 NE 5′.2.
03 40 57.0	−21 42 48	NGC 1414	14.0	1.4 x 0.3	12.9	SB(s)bc? sp III	172	
03 40 56.9	−22 33 49	NGC 1415	11.9	3.5 x 1.8	13.7	(R)SAB(s)0/a	148	Numerous stars superimposed SE half.
03 41 02.9	−22 43 09	NGC 1416	12.9	1.2 x 1.2	13.4	E1:		Lies 1′.6 NNE of mag 9.75v star.
03 41 31.4	−21 40 52	NGC 1422	13.2	2.5 x 0.6	13.5	SBab pec sp	65	
03 42 11.2	−29 53 27	NGC 1425	10.6	5.8 x 2.6	13.4	SA(s)b II	129	Mag 10.53v star N edge; mag 9.09v star W 13′.3.
03 42 49.3	−22 06 26	NGC 1426	11.4	2.8 x 1.8	13.1	E4	111	Mag 8.04v star N 8′.3.
03 45 17.5	−23 00 06	NGC 1438	12.4	2.0 x 0.9	12.9	SB(r)0/a:	69	Mag 11.21v star on E edge.
03 44 49.9	−21 55 15	NGC 1439	11.4	2.5 x 2.3	13.3	E1	27	Brighter round core with faint envelope.
03 45 03.3	−18 15 57	NGC 1440	11.5	2.1 x 1.6	12.7	(R′)SB(rs)0°:	28	Mag 9.99v star W 7′.2.
03 45 22.1	−18 38 01	NGC 1452	11.8	2.8 x 1.5	13.2	(R′)SB(r)0/a	113	Almost stellar, NGC 1452 WNW 7′.6.
03 45 52.0	−18 40 48	NGC 1455	11.8	0.6 x 0.4	10.1	Sa:	166	Strong N-S bar; NGC 1455 ESE 7′.6.
03 46 57.8	−25 31 18	NGC 1459	12.9	1.7 x 1.1	13.4	SB(s)bc? I-II	167	Mag 9.77v star W 6′.0.
03 54 29.0	−20 25 38	NGC 1481	13.4	0.9 x 0.6	12.6	SA0$^-$:	133	Lies 2′.5 NW of mag 8.56v star.
03 54 39.2	−20 30 10	NGC 1482	12.2	2.5 x 1.4	13.4	SA0$^+$ pec sp	103	Lies 2′ SW and 3′ S of a pair of mag 8.6v stars.
03 56 18.7	−21 49 18	NGC 1486	14.2	0.9 x 0.6	13.4	Sbc: pec	2	
03 57 38.3	−19 12 58	NGC 1489	13.8	1.5 x 0.5	13.3	SB(r)b II	12	Mag 10.80v star W 2′.7.
04 06 48.6	−21 10 44	NGC 1518	11.8	3.0 x 1.3	13.1	SB(s)dm III-IV	35	Mag 10.11v star 1′.5 E of S end of galaxy.
04 08 07.7	−17 11 34	NGC 1519	12.9	2.1 x 0.5	12.8	SB(r)b? II-III?	108	Mag 8.54v star SSE 4′.6.
04 08 18.8	−21 03 04	NGC 1521	11.4	2.5 x 1.8	13.1	E3:	10	Mag 8.31v star S 4′.7.
04 15 10.4	−28 29 20	NGC 1540	14.3	1.0 x 0.5	13.3	Sa? pec	144	NGC 1540A on N edge; mag 9.35v star NE 7′.9.
04 15 10.4	−28 28 45	NGC 1540A	13.5	0.9 x 0.6	12.7	Sa? pec	72	Located on N edge of NGC 1540.
04 17 12.6	−17 51 30	NGC 1547	13.4	1.3 x 0.6	13.0	SB(rs)bc pec: II-III	133	
04 28 20.8	−17 35 44	NGC 1583	13.7	0.7 x 0.7	12.8	SA0$^-$		**ESO 551-7** SW 3′.1; NGC 1584 NNW 5′.0.
04 28 10.2	−17 31 24	NGC 1584	14.0	0.6 x 0.6	12.8	SA(s)0$^-$ pec:		Almost stellar; NGC 1583 SSE 5′.0.
04 29 30.6	−26 42 48	NGC 1591	12.9	1.2 x 0.8	12.7	SB(r)ab pec	30	MCG −4-11-16 SSE 4′.4.
04 29 40.0	−27 24 33	NGC 1592	14.0	1.7 x 1.0	14.4	Pec	96	Double system, distorted.
04 37 15.5	−18 54 04	NGC 1630	14.0	0.7 x 0.5	12.7	SB0$^+$ pec?	140	Mag 10.23v star WSW 6′.0.
04 38 24.1	−20 38 59	NGC 1631	13.3	1.4 x 0.9	13.4	(R′)SAB(r)0/a:	44	Mag 7.26v star WSW 6′.2.
04 42 14.3	−20 26 05	NGC 1640	11.7	2.8 x 2.5	13.6	SB(r)b II	45	

GALAXY CLUSTERS

RA h m s	Dec ° ′ ″	Name	Mag 10th brightest	No. Gal	Diam ′	Notes
04 38 54.0	−22 06 00	A 500	15.8	53	15	Most members faint and stellar.
03 40 30.0	−28 42 00	A 3151	16.0	52	17	All members anonymous, faint, stellar.
03 18 06.0	−29 38 00	AS 337	16.3		17	All members anonymous, stellar.
04 19 48.0	−27 52 00	AS 440	16.4		17	All members anonymous, stellar.
04 24 18.0	−27 44 00	AS 449	15.9		17	**ESO 420-19**, **19A** at center.
04 28 36.0	−28 08 00	AS 459	16.3	0	17	All members anonymous, stellar.
04 31 30.0	−29 45 00	AS 465	15.3	4	22	MCG −5-11-12 at center; **ESO 421-5** W 8′.6.

(Continued from previous page)

GLOBULAR CLUSTERS

RA h m s	Dec ° ′ ″	Name	Total V m	B ★ V m	HB V m	Diam ′	Conc. Class Low = 12 High = 1	Notes
04 24 44.5	−21 11 13	Eridanus Cluster	14.7	17.6	20.4	1.0		

PLANETARY NEBULAE

RA h m s	Dec ° ′ ″	Name	Diam ″	Mag (P)	Mag (V)	Mag cent ★	Alt Name	Notes
03 33 15.4	−25 52 13	NGC 1360	385	9.6	9.4	11.3	PK 220−53.1	Several faint stars on disc; mag 10.13v star 8'.7 SE of center.

157 GALAXIES 157

RA h m s	Dec ° ′ ″	Name	Mag (V)	Dim ′ Maj x min	SB	Type Class	PA	Notes
02 02 12.1	−28 39 28	ESO 414-25	13.2	1.7 x 0.8	13.4	SB(s)c II	161	Mag 12.9 star W 7'.4; mag 11.9 star ENE 9'.6.
02 06 12.3	−28 35 50	ESO 414-28	13.4	1.1 x 0.7	13.0	(R′)SA(r)0^-?	108	Mag 7.32v star W 12'.6; **ESO 414-30** and **ESO 414-30A** E 4'.6.
02 08 07.6	−28 38 13	ESO 414-32	13.8	1.7 x 0.5	13.5	(R′)SAB(r)a pec?	70	Mag 8.70v star S 6'.0.
02 13 16.4	−29 43 28	ESO 415-7	13.9	1.0 x 0.7	13.4	S0	25	Mag 10.07v star SW 8'.7.
02 21 41.1	−27 16 52	ESO 415-22	13.4	1.7 x 1.2	14.0	(R′)SB(r)b	146	**MCG −5-6-15** NW 2'.2; mag 12.4 star ENE 4'.1.
02 32 34.3	−29 41 45	ESO 416-1	13.8	0.5 x 0.4	11.9		156	Mag 10.92v star SW 5'.3.
02 41 43.4	−27 18 23	ESO 416-8	14.6	1.4 x 0.7	14.4	SB(s)cd III-IV	141	Mag 8.74v star SE 6'.6.
02 51 18.2	−28 12 06	ESO 416-37	13.3	1.7 x 0.5	12.9	Sb:	19	Mag 8.77v star NW 5'.9.
02 55 40.5	−27 25 21	ESO 417-3	12.7	1.9 x 1.6	13.8	SAB(rs)c: I-II	28	Tightly wound arms, several strong lanes.
03 01 15.4	−28 28 07	ESO 417-11	13.3	1.7 x 1.1	13.8	(R)SB(r)0^+	0	Mag 7.83v star NE 3'.3.
01 56 16.0	−22 54 04	ESO 477-16	13.8	2.2 x 0.2	12.8	Sbc: sp	155	Mag 12.2 star S 1'.4; mag 10.37v star E 6'.9.
02 01 31.0	−24 55 30	ESO 477-22	13.4	1.6 x 0.7	13.4	SB(s)b pec: I-II	96	Mag 13.8 star S 0'.9; mag 9.54v star S 2'.9.
02 09 18.1	−23 24 55	ESO 478-6	12.5	1.8 x 1.0	13.0	Sbc I	102	Mag 11.16v star ENE 6'.3; **MCG −4-6-10** NNE 6'.7.
02 25 28.3	−25 38 17	ESO 479-1	13.5	1.3 x 1.1	13.7	SB(s)cd: II	14	Mag 12.6 star NW 4'.7.
02 26 21.9	−24 17 23	ESO 479-4	12.2	2.8 x 1.3	13.5	SB(s)dm III-IV	55	Many bright knots and/or superimposed stars.
02 39 07.9	−22 39 41	ESO 479-20	14.2	1.9 x 0.5	13.9	SB(s)c? II-III	153	Pair of N-S oriented mag 10.6, 12.6 stars N 7'.4.
02 42 07.0	−24 07 55	ESO 479-25	13.4	1.6 x 0.3	12.5	Sdm?	53	Mag 12.8 star E 5'.8.
02 44 47.6	−24 30 53	ESO 479-31	13.6	1.3 x 0.9	13.7	SB(s)0°	138	Mag 8.38v star NNW 7'.3.
02 45 39.4	−24 48 55	ESO 479-33	13.6	1.7 x 0.4	13.1	S0° sp	157	MCG −4-7-35 S 4'.3; mag 10.54v star N 6'.2.
02 45 49.2	−23 14 04	ESO 479-34	14.0	1.0 x 0.3	12.6	S0	144	Mag 11.9 star NW 2'.6.
02 46 30.1	−26 14 58	ESO 479-37	13.2	1.5 x 1.0	13.5	(R′)SAB(rs)bc pec: II	24	N-S oriented pair of mag 8.75v, 10.50v stars NW 5'.2; MCG −5-7-23 SW 5'.2.
02 46 33.5	−24 51 59	ESO 479-38	13.1	1.3 x 0.9	13.2	SAB(s)0°? pec	6	Mag 10.22v star SE 8'.2; ESO 479-33 and MCG −4-7-35 W 11'.9.
02 46 41.6	−25 20 53	ESO 479-40	13.7	1.7 x 0.6	13.6	SB(s)bc: II	51	Mag 7.38v star W 7'.6; elongated **ESO 479-41** SSE 3'.0.
02 47 35.5	−25 08 48	ESO 479-43	14.2	1.2 x 0.5	13.5	(R′)SB(r)a:	132	Mag 10.52v star SE 8'.4.
02 52 59.0	−24 51 46	ESO 480-9	15.2	1.5 x 0.2	13.7	S	116	MCG −4-7-49 N 4'.7.
03 02 34.5	−23 07 46	ESO 480-22	14.3	0.9 x 0.3	12.8	S0	175	Mag 13.0 star on E edge; mag 10.79v star W 7'.9.
03 03 50.5	−25 16 23	ESO 480-25	14.4	1.7 x 1.1	14.9	SB(s)m: V	109	Mag 11.9 star NW 6'.2.
03 12 08.1	−25 07 54	ESO 481-7	13.3	1.3 x 1.3	13.7	(R′)SA0°:		Mag 10.14v star ESE 8'.6; MCG −4-8-45 E 2'.3.
03 12 24.8	−24 32 46	ESO 481-11	14.4	1.5 x 0.2	13.0	Sb	12	Mag 13.8 star N 1'.5; mag 10.70v star NW 8'.4; MCG −4-8-44 S 4'.7.
03 13 40.5	−25 11 25	ESO 481-14	13.1	3.8 x 1.0	14.4	SB(s)m: sp IV-V	163	Several bright knots or superimposed stars S end; mag 10.14v star W 12'.9.
03 16 09.0	−24 12 03	ESO 481-16	14.7	1.1 x 0.8	14.4	IB(s)m V-VI	96	Mag 10.39v star E 5'.2.
03 17 04.6	−22 51 59	ESO 481-17	12.3	1.9 x 1.7	13.4	SB(rs)a pec:	163	Bright core, uniform envelope; mag 10.54v star E 5'.7.
03 18 33.0	−25 50 13	ESO 481-18	12.7	2.8 x 0.9	13.6	SB(rs)cd IV	41	Mag 10.46v star NE 7'.2.
03 18 43.1	−23 46 57	ESO 481-19	14.9	2.1 x 1.1	15.7	IB(s)m V-VI	174	Mag 12.4 star N 3'.0.
03 20 18.1	−26 27 53	ESO 481-21	14.7	1.8 x 0.3	13.9	SBd? sp	108	Mag 10.83v star SW 4'.7 mag 6.41v star SSE 10'.3.
01 58 00.9	−19 48 25	ESO 544-2	14.3	0.6 x 0.3	12.3		42	**ESO 544-1** NW edge.
01 58 47.8	−22 09 27	ESO 544-4	14.2	0.7 x 0.5	13.0		166	Pair of stars, mags 9.79v and 9.41v, S 1'.8.
02 11 46.8	−18 16 01	ESO 544-20	14.7	1.6 x 0.4	14.0	SB(r)bc? III	121	N-S line of four mag 12.9.12.8, 12.7 and 11.07v stars W 12'.6.
02 14 57.9	−20 12 46	ESO 544-30	12.8	2.0 x 1.4	13.8	SB(s)dm pec: III-IV	101	Mag 10.69v star W 6'.2.
02 19 15.0	−18 55 55	ESO 545-2	14.2	1.9 x 0.8	14.5	SB(s)m V	53	Mag 9.94v star NNW 4'.2.
02 20 06.2	−19 45 02	ESO 545-5	13.1	2.3 x 0.7	13.5	Im pec sp	60	MCG −3-7-4 E 10'.0.
02 24 40.6	−19 08 29	ESO 545-13	13.2	1.2 x 0.9	13.1	SAB(s)c I-II	6	MCG −3-7-12 E 2'.6; mag 10.70v star SSW 7'.5.
02 26 00.0	−21 25 12	ESO 545-16	13.7	2.1 x 1.4	14.7	SAB(s)m: V-VI	31	Mag 9.14v star NNW 7'.5.
02 33 32.2	−20 10 54	ESO 545-34	13.6	0.8 x 0.4	12.2	S0	45	MCG −3-7-36 E 5'.2.
02 38 11.4	−20 10 00	ESO 545-40	13.0	1.5 x 1.0	13.3	SA(rs)0°?	36	**ESO 540-41** NE 1'.6.
02 39 29.2	−19 50 31	ESO 545-42	14.0	1.7 x 0.7	14.0	SA(rs)0°: IV-V	12	Almost stellar nucleus, mag 12.1 star E 4'.2.
02 41 23.9	−17 56 08	ESO 546-5	14.5	1.0 x 0.8	14.1	SB?	156	
02 41 47.9	−20 47 56	ESO 546-6	15.3	1.1 x 0.6	14.6	dE?	48	Pair with ESO 546-7 S 1'.0.
02 41 50.3	−20 48 48	ESO 546-7	14.2	1.1 x 0.4	13.2		65	Pair with ESO 546-6 N 1'.0.
02 54 56.2	−21 53 56	ESO 546-31	14.2	1.0 x 0.5	13.2	SB(s)cd: III	33	Mag 10.34v star E 3'.0; faint, round anonymous galaxy W 2'.3.
02 58 37.4	−18 42 00	ESO 546-34	15.0	1.5 x 0.5	14.5	SB(s)m pec: V	117	Mag 10.82v star E 11'.2; mag 10.77v star WSW 12'.5.
03 03 31.0	−20 10 36	ESO 547-4	14.3	1.5 x 0.6	14.0	SAB(rs)c: pec III	35	Mag 8.63v star SW 3'.6.
03 03 34.6	−18 22 07	ESO 547-5	14.7	1.2 x 0.9	14.6	IB(s)m V	30	Mag 10.37v star E 1'.3.
03 06 00.1	−19 23 18	ESO 547-9	15.1	2.4 x 0.5	15.1	IB(s)m V-VI	57	Mag 9.18v star N 6'.3.
03 09 29.9	−17 24 58	ESO 547-11	14.8	1.2 x 0.8	14.6	SAB(s)m V	123	Mag 15.0 star N edge; mag 10.33v star NE 10'.2.
03 09 36.9	−17 49 53	ESO 547-12	15.8	1.5 x 0.2	14.3	Sc:	111	Mag 9.24v star E 5'.4.
03 09 55.2	−20 05 21	ESO 547-15	13.9	0.7 x 0.5	12.6	Sb?	140	Mag 8.29v star NW 6'.3.
03 12 57.5	−17 55 44	ESO 547-20	14.6	1.3 x 1.1	14.8	IB(s)m V	75	Mag 10.54v star WNW 11'.6; double galaxy system **MCG −3-9-8** E 2'.9.
03 21 52.4	−19 28 55	ESO 548-1	14.3	1.0 x 0.3	12.8	Sa:	38	Anonymous galaxy E 1'.9; mag 12.4 star S 1'.9.
03 23 47.1	−19 45 13	ESO 548-5	13.2	1.8 x 1.5	14.2	SAB(s)m IV-V	58	Stellar nucleus; mag 11.4 star S 4'.3.
02 22 00.9	−20 44 40	IC 223	13.1	1.2 x 0.7	12.7	IB(s)m pec? IV	152	Mag 11.34v star NW 6'.0; NGC 899 S 4'.9.
01 59 11.6	−27 48 36	IC 1763	13.8	1.1 x 0.9	13.6	Sb	32	Mag 8.34v star SSE 8'.7.
02 00 49.8	−25 01 38	IC 1768	13.0	0.7 x 0.7	12.1	SA0		Equilateral triangle of mag 9.52v, 9.70v and 10.36v stars centered SE 6'.8; MCG −4-5-24 SW 3'.4.
02 39 03.6	−27 26 34	IC 1830	12.8	1.7 x 1.4	13.6	SAB(rs)0^+:	137	Lies 1'.3 E of mag 10.42v star.
02 41 38.7	−28 10 16	IC 1833	13.1	1.5 x 0.8	13.2	SAB0°	61	Mag 10.74v star W 5'.7.
02 45 37.2	−27 57 42	IC 1845	13.0	1.4 x 0.8	13.0	Sb	123	Anonymous galaxy NNE 2'.2.
03 04 32.3	−27 27 38	IC 1876	14.1	1.1 x 1.0	14.1	SA0^+? pec	3	Mag 10.93v star WNW 10'.5; close pair MCG −5-8-14 and 15 ESE 6'.0.
03 08 27.2	−23 03 21	IC 1892	13.3	1.9 x 1.0	13.8	SB(s)d pec IV	10	**MCG −4-8-29** on N edge.
03 09 36.4	−25 15 16	IC 1895	13.3	1.5 x 1.1	13.7	(R)SAB(rs)0/a pec:	153	Mag 12.5 star E edge; mag 12.3 star N 1'.7.
03 10 20.0	−22 24 17	IC 1898	12.7	3.6 x 0.6	13.3	SB(s)c: sp II-III	73	Mag 10.08v star WSW 12'.8.
03 12 13.1	−25 18 20	IC 1899	13.4	1.3 x 0.4	12.6	S0/a	165	MCG −4-8-46 NE 1'.7.
02 05 43.6	−17 48 52	MCG −3-6-6	14.8	0.7 x 0.7	13.8	Sc		
02 08 22.9	−17 24 07	MCG −3-6-8	14.9	0.6 x 0.6	13.7	(R′)SAB(s)0^+		Mag 7.90v star E 5'.3.
02 10 44.2	−19 46 01	MCG −3-6-11	14.1	0.7 x 0.4	12.8	E	177	Mag 11.9 star SE 4'.3.
02 11 32.4	−20 02 24	MCG −3-6-12	14.1	0.8 x 0.5	12.9	S0/a	146	**MCG −3-6-14** N 7'.8.

(Continued from previous page)

GALAXIES

RA h m s	Dec ° ′ ″	Name	Mag (V)	Dim ′ Maj x min	SB	Type Class	PA	Notes
02 15 51.6	−18 54 51	MCG −3-6-21	14.4	0.9 x 0.3	12.8	S	165	Star E edge.
02 20 45.6	−18 38 30	MCG −3-7-3	14.5	0.9 x 0.6	13.7	(R′$_2$)SB(s)b	63	Mag 12.3 star on E edge.
02 20 48.2	−19 47 06	MCG −3-7-4	14.8	0.6 x 0.5	13.3	S	132	Mag 6.53v star SE 9′.9.
02 23 18.8	−18 50 26	MCG −3-7-8	14.2	0.9 x 0.6	13.6	E	168	
02 24 14.9	−19 49 33	MCG −3-7-9	14.8	0.7 x 0.4	13.3	S	87	
02 24 51.3	−19 08 08	MCG −3-7-12	14.8	0.7 x 0.4	13.3	S	177	ESO 545-13 W 2′.6.
02 27 02.2	−19 15 17	MCG −3-7-17	13.6	1.0 x 0.7	13.1	S0	129	Mag 9.94v star SSE 2′.5.
02 27 17.9	−19 11 26	MCG −3-7-20	14.2	0.5 x 0.5	12.7	E		Small, faint anonymous galaxies SW 1′.4 and S 2′.1.
02 29 38.6	−17 02 41	MCG −3-7-23	14.3	0.6 x 0.4	12.6	Sb	30	Multi-galaxy system or star involved? **MCG −3-7-25** SSE 18′.7.
02 31 04.4	−19 20 04	MCG −3-7-26	13.6	1.2 x 1.0	13.7	SAB(r)bc: II	119	
02 31 26.1	−17 39 13	MCG −3-7-27	13.6	0.6 x 0.6	12.4	S?		Mag 10.47v star SE 6′.3.
02 33 53.6	−20 12 18	MCG −3-7-36	14.7	0.7 x 0.7	13.8	Sc		ESO 545-34 W 5′.2.
02 36 02.0	−17 14 41	MCG −3-7-38	14.1	0.5 x 0.3	11.9	Sbc		Mag 7.789v star N 6′.1; MCG −3-7-40 S 1′.2.
02 36 03.7	−17 22 03	MCG −3-7-39	14.6	0.6 x 0.4	12.9	Sb	114	Mag 9.21v star E 11′.7.
02 36 04.2	−17 15 51	MCG −3-7-40	14.0	1.0 x 0.5	13.1	S/Irr?	171	MCG −3-7-38 N 1′.2.
02 39 00.5	−19 19 36	MCG −3-7-48	14.4	0.5 x 0.4	12.5	S0	144	Stellar; mag 10.56v star NW 10′.1.
02 40 45.9	−17 07 14	MCG −3-7-53	14.1	0.5 x 0.5	12.7	E/S0		Mag 10.70v star NW 7′.2.
02 42 12.4	−19 49 48	MCG −3-7-60	14.4	0.9 x 0.6	13.6	Sb-c	141	
02 44 39.9	−17 43 38	MCG −3-8-6	15.4	0.7 x 0.2	13.1	Sbc	54	Mag 9.18v star SE 5′.5; MCG −3-8-7 N 1′.0.
02 44 41.5	−17 42 49	MCG −3-8-7	14.7	0.5 x 0.2	12.1	Sb	132	NGC 1098 NNE 4′.6, MCG −3-8-6 S 1′.0.
02 45 12.1	−17 19 10	MCG −3-8-12	15.0	0.5 x 0.4	13.1	S?	36	Mag 11.7 star on S edge; **MCG −3-8-9** W 4′.4.
02 46 15.8	−17 06 44	MCG −3-8-19	14.5	0.9 x 0.2	12.5	Sb	159	Mag 10.63v star W 5′.0.
02 46 20.9	−18 46 53	MCG −3-8-22	14.9	0.7 x 0.4	13.4	Sbc	33	Mag 8.55v star S 7′.2.
02 46 30.6	−18 46 21	MCG −3-8-23	15.1	0.6 x 0.4	13.4	Sbc	30	Mag 8.55v star S 7′.3.
02 48 48.5	−19 58 18	MCG −3-8-27	13.4	1.4 x 0.3	12.3	Sbc	33	Mag 8.85v star W 8′.9.
02 53 25.0	−17 36 11	MCG −3-8-41	14.3	0.7 x 0.4	12.7	S?	111	
02 54 57.4	−17 03 06	MCG −3-8-45	14.9	1.9 x 1.6	15.9	SB(rs)cd: III-IV	168	Mag 12.2 star on W edge; mag 11.2 stars E 7′.1.
02 56 12.3	−19 39 02	MCG −3-8-47	13.8	1.2 x 0.5	13.1	S?	165	Mag 7.62v star SW 9′.5.
03 12 42.4	−17 06 48	MCG −3-9-6	14.3	1.1 x 0.3	13.0	Sa	9	Mag 10.49v star NE 11′.5.
03 15 22.3	−18 59 20	MCG −3-9-13	14.9	0.7 x 0.7	14.0	Sc		
03 20 52.7	−17 53 14	MCG −3-9-24	14.7	0.9 x 0.6	13.9	Sd	42	Numerous bright knots.
03 22 18.0	−17 10 52	MCG −3-9-29	14.7	0.7 x 0.4	13.1		132	MCG −3-9-29 W 2′.5.
03 22 28.8	−17 10 53	MCG −3-9-30	15.1	0.5 x 0.3	12.9		21	MCG −3-9-30 E 2′.5.
02 00 36.1	−25 03 07	MCG −4-5-24	13.6	0.9 x 0.3	12.1	SA0°	71	Equilateral triangle of mag 9.52v, 9.70v and 10.36v stars centered ESE 8′.3; IC 1768 NE 3′.4.
02 02 17.0	−21 45 49	MCG −4-6-1	13.8	1.2 x 0.5	13.1	S?	106	Mag 9.08v star WNW 4′.1.
02 02 35.4	−24 27 56	MCG −4-6-2	14.7	0.6 x 0.4	13.0	SBc	39	Mag 9.90v star S 2′.5.
02 08 20.5	−24 32 54	MCG −4-6-7	14.1	0.9 x 0.6	13.2	SBc	45	Mag 8.96v star W 6′.7.
02 09 15.2	−22 40 40	MCG −4-6-8	14.5	0.8 x 0.4	13.2	Sbc	6	
02 10 53.3	−21 12 47	MCG −4-6-14	14.6	0.7 x 0.4	13.1	Sbc	21	Mag 11.20v star WNW 6′.1.
02 11 56.5	−20 22 37	MCG −4-6-15	13.8	1.1 x 0.5	13.0	Sbc	179	Mag 10.12v star SE 2′.0.
02 14 49.1	−24 51 04	MCG −4-6-17	13.8	0.8 x 0.6	12.8	(R′)SB(s)b pec	144	Mag 8.34v star ENE 14′.0.
02 15 45.1	−23 31 14	MCG −4-6-18	14.7	0.8 x 0.4	13.3	SBc	6	Mag 10.23v star NNE 3′.0.
02 17 31.0	−22 53 47	MCG −4-6-20	13.9	1.6 x 0.6	13.7	S?	15	Mag 9.32v star SE 3′.1.
02 17 52.5	−23 44 41	MCG −4-6-22	14.5	0.7 x 0.5	13.3	SBb	66	Lies 1′.3 S of mag 10 star.
02 18 35.0	−25 45 28	MCG −4-6-23	13.9	1.1 x 0.7	13.5	SBbc	149	Mag 11.06v star W 10′.6.
02 18 39.3	−22 29 49	MCG −4-6-24	14.1	0.9 x 0.5	13.1	Sab	94	MCG −4-6-24 SSE 1′.2.
02 19 29.5	−21 26 09	MCG −4-6-26	14.6	1.5 x 0.2	13.1	Sbc	117	**ESO 545-4** NE 6′.5.
02 19 31.7	−22 52 01	MCG −4-6-27	15.0	0.4 x 0.4	12.8	Sb		
02 21 00.9	−20 49 25	MCG −4-6-28	14.6	0.7 x 0.3	12.7	Sbc	3	Mag 8.50v star W 12′.4.
02 21 34.0	−20 34 38	MCG −4-6-29	14.3	0.5 x 0.4	12.4	Sc	0	
02 22 10.3	−20 23 21	MCG −4-6-33	14.0	0.9 x 0.5	13.0	Sbc		Mag 9.94v star WNW 4′.9.
02 23 51.8	−21 01 42	MCG −4-6-36	14.4	1.0 x 0.9	14.2	S?	144	
02 25 58.3	−24 42 36	MCG −4-6-39	14.1	1.0 x 0.5	13.2	(R′)SB(r)a:	9	Mag 10.33v star W 6′.4.
02 26 35.3	−22 59 14	MCG −4-6-42	14.6	0.9 x 0.3	13.0	Sc	102	Mag 8.69v star NE 2′.7.
02 26 36.3	−23 25 02	MCG −4-6-43	13.9	0.5 x 0.5	12.2	S0/a		Mag 10.48v star W 3′.1.
02 27 00.2	−22 54 44	MCG −4-6-44	14.3	0.7 x 0.3	12.5	Sab	102	Mag 8.69v star SW 4′.9.
02 29 01.7	−23 45 22	MCG −4-7-3	14.2	0.7 x 0.7	13.3	Sb		Mag 10.55v star SE 6′.7.
02 30 37.1	−24 04 56	MCG −4-7-6	13.7	0.9 x 0.3	12.2	S0/a	146	Mag 12.1 star NE 7′.7.
02 31 18.6	−20 28 10	MCG −4-7-7	14.7	0.7 x 0.5	13.4	SBa	6	Mag 10.95v star WNW 3′.5.
02 31 28.4	−22 56 19	MCG −4-7-8	14.5	0.4 x 0.3	12.1	Sbc	106	Mag 7.92v star SE 11′.5.
02 35 21.8	−25 31 43	MCG −4-7-11	14.3	1.0 x 0.6	13.6	SB(rs)bc	93	Pair of N-S oriented mag 10.81v, 10.24v stars E 6′.8.
02 37 31.0	−20 20 18	MCG −4-7-12	14.5	0.9 x 0.6	13.7		18	Star S end.
02 39 42.4	−21 14 24	MCG −4-7-14	14.3	0.8 x 0.4	12.9	Sc	141	Mag 10.36v star W 1′.3.
02 40 07.2	−25 12 57	MCG −4-7-15	14.2	0.8 x 0.8	13.6	SBc		Several bright knots E of center.
02 40 31.4	−20 22 22	MCG −4-7-16	14.5	0.5 x 0.3	12.3	Sa	149	MCG −4-7-17 S 1′.2.
02 40 30.8	−20 23 30	MCG −4-7-17	14.6	0.3 x 0.3	11.8	Sb		MCG −4-7-16 N 1′.2.
02 40 35.7	−20 23 51	MCG −4-7-18	14.1	0.9 x 0.6	13.3	S0	143	Mag 7.34v star W 13′.4; MCG −4-7-17 W 1′.3; MCG −4-7-16 NNW 1′.8.
02 40 39.2	−21 00 11	MCG −4-7-19	13.8	1.0 x 0.9	13.6	SBc	127	
02 41 55.1	−20 57 50	MCG −4-7-22	13.9	1.2 x 1.0	13.9	(R′)SB(r)b	74	
02 41 56.8	−21 17 16	MCG −4-7-23	14.5	1.1 x 0.9	14.3	(R′)SB(s)m:	138	Mag 8.67v star NNW 7′.6.
02 42 30.7	−21 35 05	MCG −4-7-25	14.6	0.9 x 0.8	14.1	Sd		Mag 8.11v star N 7′.3.
02 42 45.8	−21 28 02	MCG −4-7-26	13.9	1.3 x 1.1	14.2	SB?	76	Mag 8.11v star W 4′.1.
02 42 45.4	−25 34 41	MCG −4-7-27	14.8	0.4 x 0.4	12.7	S0		MCG −4-7-29 E 1′.5.
02 42 51.8	−25 27 20	MCG −4-7-28	13.7	1.1 x 0.6	13.1	(R′)SB(rs)a	87	Mag 8.39v star N 8′.4.
02 42 51.5	−25 34 56	MCG −4-7-29	15.2	0.7 x 0.4	13.7	SBbc	144	Mag 10.39v star SE 6′.6; MCG −4-7-27 W 1′.5.
02 45 25.3	−20 49 17	MCG −4-7-33	14.1	0.7 x 0.4	12.6	Sab	153	Mag 8.93v star WSW 5′.8.
02 45 42.4	−24 53 13	MCG −4-7-35	14.5	0.7 x 0.5	13.2	Sbc	30	ESO 479-33 N 4′.3; mag 14.7 star W 1′.0.
02 46 23.1	−21 11 00	MCG −4-7-36	14.8	0.7 x 0.4	13.2	Sbc		Mag 9.42v star NE 8′.1.
02 46 53.7	−22 38 46	MCG −4-7-39	14.0	1.3 x 0.4	13.1	SB(rs)m: pec III-IV	24	Mag 9.07v star W 2′.6.
02 47 24.5	−22 16 06	MCG −4-7-41	14.5	0.4 x 0.4	12.4	Sa		NGC 1102 NNW 4′.4.
02 47 53.8	−26 02 39	MCG −4-7-44	13.5	1.3 x 0.3	12.4	S0/a	114	Mag 7.71v star S 3′.1.
02 48 00.0	−22 00 07	MCG −4-7-45	15.3	0.7 x 0.7	14.3	Sc		Mag 9.96v star W 7′.0.
02 52 54.1	−24 47 14	MCG −4-7-49	14.3	0.6 x 0.4	12.6		73	ESO 480-9 S 4′.7.
02 54 45.1	−25 06 42	MCG −4-8-1	14.4	0.8 x 0.3	12.7	S?	24	Mag 11.7 star S 3′.8.
02 55 19.4	−24 45 56	MCG −4-8-3	15.1	0.7 x 0.2	12.8	Sbc	60	Mag 9.42v star SSW 8′.3.
02 55 23.5	−24 34 53	MCG −4-8-4	14.7	0.7 x 0.5	13.4	SBbc	42	Mag 13.2 star E 2′.1.
02 56 04.5	−22 15 14	MCG −4-8-6	15.4	1.1 x 0.2	13.6	Sab	121	Mag 8.58v star ESE 8′.4.
02 58 21.6	−23 03 06	MCG −4-8-7	15.1	0.3 x 0.3	12.3	Sbc		Mag 10.51v star E 3′.5; stellar **MCG −4-8-8** ESE 1′.0.
02 59 40.1	−24 17 40	MCG −4-8-10	14.7	0.4 x 0.4	12.5	Sc		Mag 11.9 star N 8′.1.

RA h m s	Dec ° ′ ″	Name	Mag (V)	Dim ′ Maj x min	SB	Type Class	PA	Notes
03 00 42.4	−22 08 28	MCG −4-8-13	14.4	0.9 x 0.5	13.4	SB?	32	Mag 11.40v star E 2′.0.
03 01 44.7	−25 18 50	MCG −4-8-14	14.4	0.3 x 0.2	11.2	S0/a	90	Mag 9.60v star W 2′.0.
03 02 28.6	−22 48 04	MCG −4-8-15	14.1	0.9 x 0.4	12.9	Sbc	16	0′.9 NW of mag 8 star; NGC 1187 SSE 4′.5.
03 03 07.2	−22 12 24	MCG −4-8-18	14.3	0.8 x 0.6	13.4	Sc	16	**MCG −4-8-17** SSW 0′.7.
03 03 09.5	−23 11 57	MCG −4-8-19	14.1	0.8 x 0.8	13.5	SB(rs)b		Stellar nucleus; mag 12.0 star NE 1′.5.
03 03 45.5	−22 15 17	MCG −4-8-20	15.0	0.9 x 0.2	13.0	Scd	153	**ESO 547-6** WNW 1′.5.
03 03 54.0	−24 33 57	MCG −4-8-22	14.7	0.4 x 0.4	12.5	Sbc		Mag 9.16v star SW 12′.6.
03 08 15.9	−24 40 22	MCG −4-8-28	14.5	0.7 x 0.4	13.0	Sc	90	Mag 10.20v star S 5′.7.
03 09 51.4	−24 28 48	MCG −4-8-34	14.5	0.8 x 0.5	13.4	Sc	147	
03 10 08.6	−21 35 21	MCG −4-8-35	14.7	0.8 x 0.5	13.6	S0/a		Mag 8.71v star S 6′.7.
03 12 11.4	−23 37 47	MCG −4-8-39	14.4	0.4 x 0.4	12.3	S0		Mag 11.08v star E 6′.2.
03 12 16.4	−21 02 57	MCG −4-8-40	14.2	0.6 x 0.6	13.0	Sc		Mag 10.62v star S 7′.2; MCG −4-8-47 NE 4′.2.
03 12 14.0	−25 12 14	MCG −4-8-43	14.5	0.4 x 0.4	12.4	Sa		ESO 481-7 N 4′.5; mag 10.14v star ENE 7′.1.
03 12 18.2	−24 37 16	MCG −4-8-44	13.5	1.0 x 0.7	13.0	Sab	121	Mag 10.32v star S 5′.2; ESO 481-11 N 4′.7.
03 12 17.7	−25 08 50	MCG −4-8-45	14.6	0.4 x 0.3	12.2	S0/a	63	ESO 481-7 W 2′.3.
03 12 17.8	−25 17 00	MCG −4-8-46	14.8	0.5 x 0.4	12.9	Sc	150	IC 1899 SW 1′.7.
03 12 30.2	−21 00 16	MCG −4-8-47	14.9	1.2 x 1.0	15.0	Sc	3	MCG −4-8-40 SW 4′.2.
03 13 23.7	−24 52 19	MCG −4-8-49	14.8	0.7 x 0.4	13.3	Sc	12	Mag 8.36v star E 2′.3.
03 16 23.0	−23 48 40	MCG −4-8-54	14.5	0.7 x 0.4	13.0	SBc	159	Mag 10.87v star NNE 3′.4.
03 16 21.4	−25 51 20	MCG −4-8-55	13.7	0.6 x 0.6	12.5	S0/a		
03 20 40.4	−22 55 53	MCG −4-9-1	13.6	0.5 x 0.5	11.9	S?		
02 04 43.1	−29 18 08	MCG −5-6-3	13.4	0.7 x 0.3	11.6	S0	14	Lies 3′ E of mag 5 star Nu Fornacis.
02 15 16.6	−26 22 47	MCG −5-6-10	14.0	0.9 x 0.6	13.1	Sa	176	Mag 7.90v star SSE 9′.9.
02 16 39.5	−29 57 05	MCG −5-6-12	14.6	1.1 x 0.4	13.6	S	141	Mag 12.3 star ENE 8′.1.
02 29 28.0	−26 32 08	MCG −5-7-4	13.7	0.9 x 0.6	12.9	Sb?	155	Elongated **ESO 479-10** NE 9′.8.
02 31 26.1	−29 33 55	MCG −5-7-6	13.7	0.8 x 0.6	12.9	E	165	Mag 10.46v star E 6′.7.
02 35 13.6	−29 36 17	MCG −5-7-7	13.9	0.9 x 0.5	12.8	Sa:	169	Small, anonymous spiral galaxy WSW 1′.6.
02 38 25.1	−27 13 51	MCG −5-7-11	14.7	0.6 x 0.5	13.3	S0/a	129	Small, very faint galaxy with pair of stars SE 0′.7.
02 45 58.5	−28 16 11	MCG −5-7-20	13.9	1.2 x 0.3	12.6	Sb	113	Mag 10.25v star W 9′.1.
02 46 12.8	−26 18 27	MCG −5-7-23	13.8	0.9 x 0.7	13.1	(R)SA(r)a	24	N-S oriented pair of mag 8.75v, 10.50v stars N 7′.3; ESO 479-37 NE 5′.2.
02 46 37.8	−26 58 21	MCG −5-7-25	13.8	1.1 x 0.6	13.2	Sb	76	Anonymous elongated galaxies at 1′.8 NNE and 1′.8 SSW.
02 48 26.3	−29 35 39	MCG −5-7-27	13.8	1.3 x 0.4	13.0	S?	67	Mag 10.55v star W 6′.8; **ESO 416-22** SSW 7′.8.
02 48 42.2	−27 27 39	MCG −5-7-28	14.2	0.7 x 0.4	12.9	E	110	Mag 9.99v star SSW 6′.1.
02 48 57.1	−28 32 31	MCG −5-7-31	13.6	1.3 x 0.4	12.8	SB(r)0/a	146	Mag 9.37v star W 4′.8.
02 51 08.8	−26 56 51	MCG −5-7-41	14.3	0.6 x 0.3	12.4	S0:	134	Mag 10.30v star NNW 6′.7; MCG −5-7-43 SE 2′.6.
02 51 19.0	−26 58 08	MCG −5-7-43	14.6	0.5 x 0.4	12.7	S	60	MCG −5-7-41 NW 2′.6.
02 55 15.9	−29 50 02	MCG −5-8-3	14.3	0.9 x 0.3	12.9	E	111	Mag 7.79v star W 8′.5.
02 56 13.2	−28 02 32	MCG −5-8-5	13.6	0.7 x 0.7	12.7	S		Mag 11.9 star SE 5′.6.
02 58 31.4	−26 49 31	MCG −5-8-8	14.7	0.8 x 0.7	13.9	SB(rs)cd	99	Mag 8.72v star SSE 6′.6.
03 04 02.6	−26 35 47	MCG −5-8-12	14.3	0.8 x 0.6	13.4	SB	140	Mag 9.49v star S 8′.4.
03 04 55.6	−27 30 30	MCG −5-8-14	15.3	0.4 x 0.2	12.4	Sdm	38	Pair with MCG −5-8-15 E edge.
03 04 56.8	−27 30 23	MCG −5-8-15	15.3	0.4 x 0.3	12.8	Sdm	123	Pair with MCG −5-8-14 W edge.
03 05 08.1	−27 20 15	MCG −5-8-16	14.1	1.1 x 0.3	12.7	Sa	25	Mag 10.98v star NE 1′.8.
03 05 53.0	−27 13 54	MCG −5-8-17	14.2	0.8 x 0.5	13.1	Sb	36	Mag 11.25v star N 3′.0.
03 07 46.0	−27 04 16	MCG −5-8-19	14.9	0.5 x 0.3	12.6	Sc	81	
01 58 32.5	−26 17 34	NGC 775	12.6	1.7 x 1.2	13.2	SA(rs)c: I-II	167	Mag 6.78v star SE 13′.4.
02 03 56.6	−23 18 46	NGC 808	13.5	1.2 x 0.6	13.0	(R′)SB(r)bc: II	7	
02 07 20.2	−25 26 29	NGC 823	12.8	1.8 x 1.3	13.6	SA(r)0⁻?	102	
02 10 24.9	−22 03 20	NGC 836	13.4	1.0 x 0.8	13.0	(R′)SA(s)0⁺:	110	
02 10 16.2	−22 25 54	NGC 837	14.1	1.0 x 0.4	12.9	Sb? II	12	Pair of faint stars on E edge; very faint, elongated anonymous galaxy E 1′.5.
02 10 11.2	−22 19 23	NGC 849	14.3	0.6 x 0.4	12.7	(R)S0?	117	Anonymous galaxy SW 3′.1.
02 12 30.2	−22 28 17	NGC 858	13.7	1.3 x 1.1	14.0	SB(rs)c? I		**ESO 478-14** E 8′.1, 1′.8 S of mag 10.81v star.
02 15 25.2	−17 46 56	NGC 872	13.7	1.5 x 0.5	13.2	SB(s)c? II-III	174	
02 16 02.0	−23 18 20	NGC 874	14.2	1.0 x 0.6	13.5	Sab? pec	173	Located 2′.9 SW of a mag 10.98v star.
02 17 54.2	−23 23 03	NGC 878	13.8	0.8 x 0.5	12.6	Sa?	112	
02 20 52.1	−23 06 49	NGC 892	14.7	0.7 x 0.4	13.2	Sab?	6	
02 21 53.8	−20 49 25	NGC 899	12.5	1.9 x 1.3	13.3	IB(s)m III-IV	116	IC 223 N 4′.9.
02 23 01.8	−20 42 44	NGC 907	12.6	1.8 x 0.6	12.6	SBdm? sp III-IV	87	
02 23 06.4	−21 13 55	NGC 908	10.2	6.0 x 2.6	13.0	SA(s)c I-II	75	Many dark lanes.
02 25 03.7	−24 47 24	NGC 922	12.1	1.9 x 1.6	13.2	SB(s)cd pec III	172	Brightest portion to the E side; mag 10.96v star WSW 9′.3.
02 28 33.0	−19 02 32	NGC 947	12.6	2.0 x 1.1	13.3	SA(r)c I-II	50	Small, anonymous spiral galaxy SE 2′.4.
02 28 57.0	−22 21 00	NGC 951	14.6	1.0 x 0.6	13.9	SB(r)ab:	48	**MCG −4-7-2** S 1′.9; 13.3 mag star SE 2′.9.
02 32 25.1	−18 38 29	NGC 965	14.1	1.0 x 0.8	13.7	SB(s)cd pec? II-III	10	
02 31 46.9	−19 52 53	NGC 966	13.2	1.2 x 0.8	13.0	SA0° pec:	112	Star on S edge.
02 32 12.8	−17 13 00	NGC 967	12.5	1.6 x 1.0	12.8	SAB0⁻	33	Very faint, slightly elongated anonymous galaxy SW 8′.8.
02 43 44.8	−29 00 10	NGC 1079	11.5	5.5 x 3.1	14.4	(R)SAB(rs)0/a pec	87	Large core with faint envelope; mag 10.34v star S 5′.9.
02 45 22.4	−17 32 01	NGC 1091	14.1	0.8 x 0.7	13.3	(R′)SAB(rs)a pec?	77	= **Hickson 21E**.
02 45 29.5	−17 32 30	NGC 1092	13.4	0.8 x 0.8	13.0	E?		= **Hickson 21D**.
02 44 53.6	−17 39 31	NGC 1098	12.6	1.8 x 1.3	13.4	SA(s)0⁻:	102	= **Hickson 21C** . Mag 8.72v star N 5′.3; MCG −3-8-7 and MCG −3-8-6 SSW 4′.6.
02 45 17.8	−17 42 34	NGC 1099	13.1	1.8 x 0.6	13.0	SB(rs)b II	10	= **Hickson 21A**. Mag 9.18v star SW 6′.3.
02 45 36.0	−17 41 20	NGC 1100	13.0	1.7 x 0.7	13.1	SAB(r)a:	58	= **Hickson 21A**. Mag 8.72v star NW 10′.2.
02 47 12.9	−22 12 32	NGC 1102	14.5	0.8 x 0.6	13.6	SAB(rl?)ab?	90	Mag 10.63v star WSW 10′.7; MCG −4-7-41 SSE 4′.4.
02 49 07.3	−16 59 42	NGC 1114	12.5	1.7 x 0.8	12.6	SA(r)c I-II	9	
02 48 17.0	−17 59 15	NGC 1119	14.2	0.5 x 0.4	12.3	(R′)SB0°?	0	Mag 10.15v star E 3′.0.
02 51 35.9	−25 42 06	NGC 1124	14.0	1.0 x 0.8	13.6	(R)SB0⁺:	164	Compact core, faint envelope; mag 10.26v star NE 1′.8.
02 54 33.3	−18 38 07	NGC 1145	12.5	3.2 x 0.5	12.8	Sc: sp II	60	Mag 10.89v star E 3′.3.
03 00 22.2	−17 09 09	NGC 1163	13.8	2.2 x 0.3	13.2	SBbc? sp	141	
03 02 38.5	−18 53 53	NGC 1179	12.0	4.9 x 3.8	15.0	SAB(r)cd II:	35	Almost stellar nucleus; mag 10.71v star SW 8′.3.
03 02 37.4	−22 52 02	NGC 1187	10.8	4.2 x 3.2	13.5	SB(r)c I-II	130	MCG −4-8-15 NNW 4′.5.
03 04 07.9	−26 04 03	NGC 1201	10.7	3.6 x 2.1	12.8	SA(r)0°:	7	Mag 10.66v star NNE 3′.8.
03 06 45.4	−25 43 02	NGC 1210	12.7	1.0 x 0.9	12.4	(R′)SB(rs)0⁺ pec	156	
03 08 11.7	−22 55 25	NGC 1228	13.3	1.5 x 0.9	13.5	(R′)SB(rs)0⁺ pec:	77	Mag 11.20v star W 6′.7.
03 08 11.0	−22 57 39	NGC 1229	14.1	1.4 x 0.9	14.2	SBb: pec	74	NGC 1230 SSE 1′.9.
03 08 16.5	−22 59 05	NGC 1230	14.4	0.7 x 0.3	12.5	SB0°? pec	107	Located 1′.9 SSE of NGC 1229.
03 09 45.1	−20 34 47	NGC 1232	9.9	7.4 x 6.5	13.9	SAB(rs)c I-II	108	Many branching, knotty arms; NGC 1232A on E edge; mag 9.16v star 7′.5 E of center.
03 10 02.1	−20 36 01	NGC 1232A	14.6	0.9 x 0.7	14.0	SB(s)m IV	5	Located on E edge of NGC 1232.
03 13 31.8	−25 43 48	NGC 1255	10.9	4.2 x 2.6	13.4	SAB(rs)bc II	123	
03 13 58.1	−21 59 09	NGC 1256	13.6	1.1 x 0.4	12.6	SAB0⁻?	108	Mag 9.31v star W 6′.5.
03 14 05.7	−21 46 26	NGC 1258	13.2	1.3 x 0.9	13.2	SAB(s)cd: II-III	17	Mag 10.29v star SE 3′.0.

(Continued from previous page)

157 GALAXIES 157

RA h m s	Dec ° ′ ″	Name	Mag (V)	Dim ′ Maj x min	SB	Type Class	PA	Notes
03 18 15.1	−27 36 40	NGC 1292	12.2	3.0 x 1.3	13.5	SA(s)c II-III	7	Close pair of mag 10.29v, 11.6 stars NNE 3′.3.
03 19 14.1	−19 06 01	NGC 1297	11.8	2.2 x 1.9	13.2	SAB(s)0° pec:	3	
03 19 41.3	−19 24 36	NGC 1300	10.4	6.2 x 4.1	13.8	SB(rs)bc I	106	Small bright core in E-W bar with two main arms.
03 20 35.3	−18 42 56	NGC 1301	13.4	2.2 x 0.4	13.1	SB(rs)b? II	140	Mag 7.05v star SW 13′.9.
03 19 51.4	−26 03 36	NGC 1302	10.7	3.9 x 3.7	13.5	(R)SB(r)0/a		Large bright core with smooth envelope.
03 21 03.2	−25 30 46	NGC 1306	12.9	0.8 x 0.8	12.3	Sb?		**ESO 481-24** N 4′.9.
03 23 06.6	−21 22 32	NGC 1315	12.6	1.6 x 1.4	13.4	SB(rs)0^+?	55	
03 23 56.5	−21 31 37	NGC 1319	12.9	1.3 x 0.7	12.7	S0 pec sp	27	Mag 10.36v star W 6′.5.

GALAXY CLUSTERS

RA h m s	Dec ° ′ ″	Name	Mag 10th brightest	No. Gal	Diam ′	Notes
02 51 18.0	−24 54 00	A 389	15.9	97	20	All members anonymous, faint, and stellar.
03 08 30.0	−23 38 00	A 419	15.7	32	12	Almost stellar **PGC 11753** at center of cluster, other members very faint and stellar.
03 11 24.0	−26 55 00	A 3094	16.3	80	17	**ESO 481-6** at center, all other members anonymous, stellar.
03 12 24.0	−27 08 00	A 3095	16.3	49	17	**ESO 481-12**, **12A** at center, all other members anonymous, stellar.
02 20 06.0	−25 57 00	AS 244	16.0		17	All members anonymous, stellar.
02 25 48.0	−29 36 00	AS 258	16.3	8	17	All members anonymous, stellar/almost stellar.
02 26 36.0	−23 25 00	AS 263	15.5	16	22	Except MCG −4-6-42, 43, all members anonymous, stellar.
02 29 42.0	−17 02 00	AS 268	16.1	15	17	Except MCG −3-7-23, all members anonymous, stellar.
02 31 24.0	−17 38 00	AS 273	16.0		17	Except MCG −3-7-27, all members anonymous, stellar.
03 15 12.0	−29 14 00	AS 333	15.8	24	17	All members anonymous, stellar.
03 18 06.0	−29 38 00	AS 337	16.3		17	All members anonymous, stellar.

158 GALAXIES 158

RA h m s	Dec ° ′ ″	Name	Mag (V)	Dim ′ Maj x min	SB	Type Class	PA	Notes
00 36 51.1	−28 22 06	ESO 410-27	13.8	1.7 x 0.8	14.0	SB(s)c: II	42	MCG −5-2-25 NW 6′.9; mag 10.28v star WSW 10′.7.
00 39 18.3	−27 20 57	ESO 411-3	13.7	1.3 x 0.6	13.3	SA(rs)b: II-III	57	Mag 10.43v star NNE 8′.2.
00 57 46.7	−27 30 09	ESO 411-34	12.7	1.8 x 1.8	13.8	SAB(r)c: II		Many tightly wound branching arms.
01 03 46.5	−27 45 14	ESO 412-3	14.3	0.8 x 0.4	12.9		162	Mag 13.3 star NW 1′.6.
01 22 32.8	−29 58 57	ESO 412-27	14.5	1.8 x 0.2	13.2	S?	54	Mag 11.14v star W 1′.6.
01 39 57.2	−28 41 49	ESO 413-16	13.0	2.0 x 0.5	12.9	Sbc sp	167	Mag 12.4 star SSW 5′.1; **MCG −5-5-4** NW 14′.7.
01 40 00.6	−28 02 06	ESO 413-18	13.4	2.2 x 0.5	13.4	(R′)SB(r)a pec	161	Mag 12.5 star and **MCG −5-5-5** N 4′.0; mag 9.65v star E 10′.1.
02 02 12.1	−28 39 28	ESO 414-25	13.2	1.7 x 0.8	13.4	SB(s)c II	161	Mag 12.9 star W 7′.4; mag 11.9 star ENE 9′.6.
00 47 07.7	−22 35 44	ESO 474-25	15.3	1.0 x 0.8	14.9	SAB(s)m: V-VI	39	Stellar nucleus; mag 13.6 star SE 3′.0; mag 5.50v star WNW 13′.7.
00 47 07.5	−24 22 18	ESO 474-26	13.7	0.8 x 0.6	12.8	S	104	Mag 7.52v star N 9′.5; note: elongated **IC 1582** is 10′.5 WSW of this star.
01 15 33.2	−26 26 57	ESO 475-14	13.7	2.2 x 0.4	13.4	SB(s)m pec III-IV	129	Mag 11.36v star NNW 5′.6.
01 15 45.9	−26 50 36	ESO 475-15	13.1	1.6 x 1.1	13.5	(R′)SB(s)b? II-III	102	Mag 12.3 star SW 2′.5.
01 21 07.3	−26 43 34	ESO 476-4	12.8	1.9 x 0.7	13.0	$SAB0^-$? sp	112	Mag 9.46v star W 8′.0; mag 9.03v star ENE 8′.6.
01 21 17.8	−22 48 05	ESO 476-5	13.0	2.1 x 0.6	13.1	SB(r)bc II	10	Mag 14.1 star on S edge; close pair of mag 11.3, 11.5 stars E 5′.0.
01 26 34.2	−23 13 32	ESO 476-8	12.7	2.2 x 2.0	14.1	SAB(rs)c II	47	= **Hickson 11A**. **Hickson 11B** 2′.4 S; **Hickson 11C** NNW 2′.4; **Hickson 11D** E 2′.7.
01 26 42.3	−25 18 52	ESO 476-10	14.4	1.4 x 0.6	14.1	SB(s)m IV	21	Mag 8.60v star E 12′.2.
01 30 25.9	−26 46 51	ESO 476-16	13.6	1.6 x 0.8	13.7	SB(r)bc II-III	141	Mag 15.3 star on NW edge; mag 8.63v star SSW 12′.6.
01 34 13.0	−25 34 03	ESO 476-25	13.6	1.5 x 0.3	12.6	S?	145	**ESO 476-24** W 4′.7; mag 8.36v star SW 14′.2.
01 49 24.4	−26 44 47	ESO 477-7	13.7	1.4 x 0.8	13.7	S0°	60	Mag 11.9 star E 9′.3.
01 54 59.8	−26 01 09	ESO 477-14	13.3	1.7 x 0.8	13.5	SA0°	63	Mag 12.6 star NNE 7′.3, note: **MCG −4-5-18** ESE of this star 1′.9.
01 56 16.0	−22 54 04	ESO 477-16	13.8	2.2 x 0.2	12.8	Sbc: sp	155	Mag 12.2 star S 1′.4; mag 10.37v star E 6′.9.
02 01 31.0	−24 55 30	ESO 477-22	13.4	1.6 x 0.7	13.4	SB(s)b pec: I-II	96	Mag 13.8 star S 0′.9; mag 9.54v star S 2′.9.
00 39 34.4	−20 23 00	ESO 540-10	14.8	1.4 x 0.7	14.6	SB(s)cd III-IV	144	Mag 12.2 star SW 1′.6; mag 10.94v star W 9′.8.
00 41 11.8	−21 07 57	ESO 540-14	14.5	0.6 x 0.4	12.9	E^+ pec?	153	Located in between a NW-SE pair of mag 13.5 stars; NGC 216 NNE 6′.4.
00 42 14.8	−18 09 40	ESO 540-16	13.6	3.2 x 0.3	13.4	SB(s)cd? sp III-IV	75	Located 21′.8 SW of mag 2.05v star Diphda, Beta Ceti.
00 45 57.2	−20 36 30	ESO 540-19	14.1	1.0 x 0.7	13.6	S?	103	Mag 9.43v star SE 4′.5.
00 49 49.1	−21 00 58	ESO 540-31	14.8	1.9 x 0.8	15.1	IB(s)m: V-VI	37	Mag 7.05v star SW 12′.9.
00 50 24.5	−19 54 23	ESO 540-32	15.7	2.1 x 0.9	16.2	IAB(s)m pec: VI	0	Mag 10.19v star E 3′.0.
00 55 03.4	−19 00 19	ESO 541-1	13.1	1.9 x 1.1	13.8	SB(rs)bc I-II	1	Strong dark lane W of center.
00 58 58.1	−18 44 37	ESO 541-4	12.5	3.0 x 1.4	13.9	SAB(rs)bc II	22	
00 59 18.3	−20 34 43	ESO 541-5	15.0	1.2 x 0.5	14.3	SB(rs)dm? IV-V	87	Mag 9.68v star NE 11′.1.
01 02 03.6	−19 27 03	ESO 541-11	14.0	1.2 x 0.7	13.7	SA(s)0° pec?	150	Mag 8.24v star WNW 4′.4.
01 02 41.5	−21 52 53	ESO 541-13	13.6	1.8 x 1.0	14.2	E^+3 pec	21	Bright, slightly elongated core with faint envelope.
01 03 51.8	−20 36 28	ESO 541-15	13.5	0.7 x 0.7	12.6	S		Mag 7.51v star SE 15′.4.
01 12 58.9	−19 00 21	ESO 541-26	14.0	1.0 x 0.7	13.5		33	Double system, interaction.
01 20 36.0	−17 20 27	ESO 542-6	13.4	1.8 x 1.0	13.9	$SAB(s)0^-$:	177	ESO 542-7, ESO 542-8 S 2′.8.
01 20 35.8	−17 23 14	ESO 542-7	13.9	1.1 x 0.4	12.8	SBb pec?	88	Pair with ESO 542-8 on S edge; ESO 542-6 N 2′.8.
01 20 35.5	−17 23 48	ESO 542-8	13.6	1.5 x 0.6	13.3	Sb pec sp	147	Pair with ESO 542-7 on N edge.
01 34 53.8	−17 51 40	ESO 542-30	14.0	0.6 x 0.5	12.5		129	Mag 10.32v star NNW 10′.1.
01 42 42.3	−18 13 41	ESO 543-12	13.3	1.6 x 0.8	13.5	(R)SB(r)a:	123	Mag 8.92v star NW 10′.2.
01 52 03.7	−20 09 59	ESO 543-20	13.4	2.4 x 1.1	14.3	SAB(rs)b: III	100	Bright core with uniform envelope.
01 58 00.9	−19 48 25	ESO 544-2	14.3	0.6 x 0.3	12.3		42	**ESO 544-1** NW edge.
01 58 47.8	−22 09 27	ESO 544-4	14.2	0.7 x 0.5	13.0		166	Pair of stars, mags 9.79v and 9.41v, S 1′.8.
01 19 02.5	−17 03 41	IC 93	13.2	1.3 x 0.5	12.6	Sb pec:	170	Mag 8.44v star E 11′.5; MCG −3-4-44 N 2′.2.
00 38 32.7	−24 20 26	IC 1561	14.1	1.4 x 0.5	13.6	SB?	97	IC 1562 N 3′.9.
00 38 33.9	−24 16 29	IC 1562	12.8	1.6 x 1.5	13.6	SB(s)c I-II	170	Mag 13.7 star on N edge; mag 10.78v star NW 9′.3; IC 1561 S 3′.9; MCG −4-2-31 NE 4′.7.
00 43 03.8	−22 14 47	IC 1574	13.7	2.0 x 0.7	13.9	IB(s)m V-VI	175	Mag 10.06v star WSW 13′.3.
00 44 14.1	−25 06 35	IC 1576	14.7	0.7 x 0.4	13.2	Sb	144	Located between a close pair of mag 14 and 15 stars.
00 45 32.6	−26 33 57	IC 1579	14.1	1.0 x 0.5	13.2	SB(r)bc:	5	Mag 11.22v star ESE 4′.1.
00 48 43.4	−23 33 43	IC 1587	14.3	1.4 x 0.4	13.6	S0	36	Mag 11.9 star N 4′.3.
00 50 57.9	−23 33 31	IC 1588	14.9	0.7 x 0.3	13.0	S0	159	Mag 7.94v star E 7′.7.
00 55 35.0	−24 09 15	IC 1601	13.7	1.0 x 0.5	12.8	S?	117	**MCG −4-3-33** SW 1′.0.
01 04 56.2	−27 25 48	IC 1616	12.6	1.6 x 1.4	13.3	SB(rs)bc I	4	E-W line of three mag 11.36v, 12.6, 12.5 stars along S edge.
01 07 36.8	−17 32 20	IC 1622	13.7	0.7 x 0.5	12.4	S?	165	Mag 9.89v star SSE 8′.0; IC 1623 and IC 1623A NE 3′.0; **ESO 541-21** NW 4′.5.
01 07 46.8	−17 30 27	IC 1623	13.2	0.8 x 0.6	12.2	Irr	12	Pair with NGC 1623A E edge; IC 1622 SW 3′.0.
01 07 48.3	−17 30 24	IC 1623A	13.4	0.4 x 0.4	11.3	S pec:		Pair with IC 1623 W edge.
01 08 47.6	−28 34 56	IC 1628	12.5	1.1 x 1.1	12.7	$SA(r)0^-$: pec		Mag 10.07v star N 7′.6; MCG −5-3-28 SSE 1′.1.
01 18 42.4	−17 03 01	IC 1667	13.6	1.0 x 0.9	13.3		69	Mag 11.8 star W 7′.4; IC 93 E 4′.9.
01 40 21.4	−28 54 49	IC 1720	12.9	1.5 x 1.1	13.3	Sbc II-III	162	Mag 7.43v star SW 15′.5.

RA h m s	Dec ° ′ ″	Name	Mag (V)	Dim ′ Maj x min	SB	Type Class	PA	Notes
01 47 55.2	−26 53 32	IC 1729	12.6	1.7 x 0.9	13.1	SAB(s)0$^-$:	150	Mag 8.95v star N 9′.4.
01 59 11.6	−27 48 36	IC 1763	13.8	1.1 x 0.9	13.6	Sb	32	Mag 8.34v star SSE 8′.7.
02 00 49.8	−25 01 38	IC 1768	13.0	0.7 x 0.7	12.1	SA0		Equilateral triangle of mag 9.52v, 9.70v and 10.36v stars centered SE 6′.8; MCG −4-5-24 SW 3′.4.
00 37 31.5	−17 56 36	MCG −3-2-25	15.2	0.7 x 0.4	13.6	SBbc	174	**MCG −3-2-23** WSW 5′.4.
00 39 20.9	−18 55 07	MCG −3-2-32	13.7	1.0 x 0.7	13.2	S?	174	Mag 8.31v star WNW 7′.8.
00 43 19.8	−17 36 32	MCG −3-2-42	15.1	0.5 x 0.5	13.4			
00 43 49.3	−17 44 23	MCG −3-3-1	14.8	0.8 x 0.3	13.1	Sc	78	Located 15′.3 N of mag 2.05v star Diphda, Beta Ceti.
01 00 09.3	−18 37 37	MCG −3-3-16	13.4	1.2 x 0.8	13.2	Sc	24	
01 05 25.0	−17 42 56	MCG −3-3-21	15.4	0.6 x 0.3	13.4	S	78	NGC 369 SW 4′.7.
01 07 39.6	−17 14 01	MCG −3-4-2	14.7	0.6 x 0.5	13.2	(R$'_1$)SB(s)a:	51	
01 10 00.3	−19 44 19	MCG −3-4-15	14.7	0.6 x 0.4	13.0	S	150	
01 10 02.1	−19 44 05	MCG −3-4-16	15.1	0.7 x 0.4	13.5	Sb	33	
01 10 15.1	−19 45 43	MCG −3-4-17	15.2	0.6 x 0.5	13.8	Sc	132	
01 10 54.6	−17 11 51	MCG −3-4-18	14.6	0.5 x 0.3	12.4		33	
01 11 21.6	−18 16 38	MCG −3-4-20	14.4	0.7 x 0.5	13.1		174	
01 11 03.0	−19 31 26	MCG −3-4-21	14.7	0.7 x 0.5	13.4	S	120	
01 11 14.1	−17 04 18	MCG −3-4-22	14.8	0.7 x 0.6	13.7	BCG	51	
01 11 22.1	−17 04 19	MCG −3-4-23	14.4	0.6 x 0.4	12.7	SBb	129	At center of galaxy cluster A 2881.
01 11 33.8	−17 04 04	MCG −3-4-26	15.6	0.4 x 0.4	13.4			
01 13 08.2	−19 00 06	MCG −3-4-28	14.0	0.4 x 0.3	11.8	E/S0	78	**MCG −3-4-29** and **MCG −3-4-31** S 1′.2.
01 13 04.8	−18 59 53	MCG −3-4-30	15.6	0.4 x 0.4	13.5	SA0$^-$: pec		Bridge to ESO 541-26 W.
01 18 44.6	−19 37 35	MCG −3-4-42	13.3	0.9 x 0.6	12.5	SA(r)0$^+$	72	
01 19 03.0	−17 01 26	MCG −3-4-44	14.9	0.9 x 0.3	13.3	Sb	78	Mag 8.44v star E 11′.5; IC 93 S 2′.2.
01 19 02.8	−19 14 15	MCG −3-4-45	15.0	0.4 x 0.4	12.8			
01 19 31.3	−17 25 24	MCG −3-4-47	14.4	0.5 x 0.5	12.7	Sbc		Star S edge.
01 19 33.4	−17 41 38	MCG −3-4-48	14.5	0.5 x 0.5	12.9			
01 20 15.9	−18 11 07	MCG −3-4-50	14.8	0.6 x 0.5	13.4	Sc	36	
01 23 10.9	−17 31 23	MCG −3-4-58	15.0	0.5 x 0.4	13.1		153	Mag 10.64v star W 3′.5.
01 25 42.1	−18 12 39	MCG −3-4-64	15.1	0.6 x 0.5	13.6	Sb	87	NGC 539 NW 5′.6.
01 26 49.2	−17 02 43	MCG −3-4-66	14.4	1.0 x 0.6	13.7	Sc	135	Star NW edge.
01 27 01.9	−17 21 45	MCG −3-4-68	14.9	0.5 x 0.5	13.2			
01 27 44.0	−19 02 57	MCG −3-4-70	13.8	1.0 x 0.8	13.4	S?	118	
01 28 06.7	−18 48 34	MCG −3-4-72	14.9	0.6 x 0.4	13.2		33	**MCG −3-4-73** SE 2′.2.
01 31 15.8	−17 41 58	MCG −3-4-79	15.2	0.7 x 0.4	13.6	Sb-c	51	1′.0 SW of mag 9 star.
01 33 41.1	−17 22 21	MCG −3-5-6	15.2	0.6 x 0.5	13.8		141	
01 37 51.4	−17 59 33	MCG −3-5-10	14.7	0.8 x 0.6	13.8	S(r)	159	
00 36 23.0	−23 01 05	MCG −4-2-25	14.7	0.8 x 0.6	13.8	Sbc	3	
00 38 49.0	−24 13 25	MCG −4-2-31	16.1	0.7 x 0.5	15.1	E?	90	IC 1562 SW 4′.7.
00 42 40.0	−21 48 03	MCG −4-2-38	14.7	0.3 x 0.3	11.9	Sb		
00 42 42.4	−21 46 28	MCG −4-2-39	13.9	0.4 x 0.2	11.2	E	150	Immediately NE of mag 11 star.
00 43 44.9	−24 25 17	MCG −4-2-44	15.0	1.3 x 0.8	14.9	SB?	54	Mag 12.3 star N 3′.7.
00 45 58.7	−21 38 20	MCG −4-3-2	14.7	0.8 x 0.6	13.7	Sbc	165	
00 46 56.9	−21 50 53	MCG −4-3-4	14.2	0.7 x 0.4	12.7	S?	102	
00 47 28.2	−23 01 25	MCG −4-3-7	14.8	0.6 x 0.5	13.4	Sc	105	
00 47 34.8	−22 48 03	MCG −4-3-8	14.4	0.6 x 0.4	12.7	Sbc	36	
00 47 35.2	−20 25 44	MCG −4-3-10	13.7	1.0 x 0.4	12.5	SB(s)b pec?	24	Brightest member of the **Burbidge Chain**.
00 47 36.8	−20 29 10	MCG −4-3-11	15.0	0.9 x 0.5	14.0	SBbc	88	Member of the **Burbidge Chain**.
00 47 37.6	−20 27 12	MCG −4-3-12	16.7	0.6 x 0.2	14.2	Sb	43	Member of the **Burbidge Chain**.
00 47 37.8	−20 31 20	MCG −4-3-13	14.0	1.0 x 0.8	13.6	SB(s)c pec:	130	Member of the **Burbidge Chain**. Strong dark area E of center.
00 47 41.4	−21 29 26	MCG −4-3-14	13.7	0.8 x 0.7	13.0	S?	16	
00 49 01.6	−23 48 42	MCG −4-3-16	15.0	0.3 x 0.3	12.3	Sbc		
00 49 14.8	−23 51 32	MCG −4-3-17	15.1	0.4 x 0.4	12.9	SBc		
00 49 41.4	−21 32 09	MCG −4-3-18	15.4	0.7 x 0.4	13.8	Sc	96	
00 50 29.9	−21 15 28	MCG −4-3-20	15.1	1.2 x 0.2	13.4	Sbc	162	
00 52 46.8	−25 40 19	MCG −4-3-23	15.0	0.4 x 0.4	12.9	Sc		
00 53 59.9	−21 40 40	MCG −4-3-25	15.2	0.7 x 0.2	12.9	Sc	108	
00 54 07.5	−21 42 52	MCG −4-3-26	14.2	0.4 x 0.4	12.3	E		
00 54 20.6	−23 33 11	MCG −4-3-28	14.4	0.7 x 0.4	12.8	S?	63	= **Hickson 9A** Multi-galaxy system. **Hickson 9D** W 1′.0; **Hickson 9B** NW 1′.5; **Hickson 9C** SSW 1′.0.
00 56 51.1	−22 06 11	MCG −4-3-34	15.1	0.4 x 0.4	12.9	SBc		
00 57 08.5	−21 55 13	MCG −4-3-35	15.1	0.4 x 0.4	13.2	E		
00 58 46.2	−22 07 36	MCG −4-3-36	14.8	0.5 x 0.4	12.9	Sc	129	
00 58 54.9	−25 46 13	MCG −4-3-38	14.3	0.7 x 0.5	13.1	S0?	84	
01 00 02.0	−21 29 20	MCG −4-3-40	13.7	1.3 x 1.1	14.2	E/S0	179	
01 01 57.1	−20 54 05	MCG −4-3-43	14.2	0.9 x 0.8	13.7	SA(r)0/a pec:	174	
01 02 39.9	−21 20 10	MCG −4-3-45	14.8	0.8 x 0.5	13.7	SBbc	30	
01 02 50.5	−24 13 10	MCG −4-3-46	14.4	1.1 x 0.8	14.1	S?	135	Many bright knots and/or superimposed stars.
01 03 53.4	−23 29 19	MCG −4-3-48	14.6	0.8 x 0.5	13.4	Sbc	171	
01 06 32.5	−23 40 46	MCG −4-3-52	14.2	1.3 x 0.9	14.2	SBc	153	
01 08 25.2	−23 30 24	MCG −4-3-54	14.2	1.0 x 0.7	13.6	SBc	114	
01 17 02.9	−25 37 32	MCG −4-4-2	14.8	0.7 x 0.5	13.5	S?	21	
01 18 45.8	−23 56 44	MCG −4-4-3	13.2	1.4 x 0.7	13.0	(R)SB(r)ab pec	57	
01 19 29.6	−20 46 40	MCG −4-4-4	14.1	1.1 x 0.9	13.9	SB?	163	Star on E and W edges.
01 24 24.7	−24 06 47	MCG −4-4-7	14.6	0.8 x 0.5	13.5	SBb	120	
01 24 41.3	−22 49 46	MCG −4-4-8	14.9	0.7 x 0.7	14.0	Sbc		
01 27 13.4	−21 46 26	MCG −4-4-12	13.4	0.9 x 0.9	13.1	S0?		
01 27 09.8	−20 24 16	MCG −4-4-15	13.8	1.0 x 0.7	13.2	Sab	57	Mag 6.92v star NE 4′.8.
01 28 44.4	−23 24 38	MCG −4-4-17	14.5	0.7 x 0.5	13.3	Sa	90	
01 29 38.1	−22 17 37	MCG −4-4-19	15.0	0.9 x 0.2	13.0	Sb	165	Much elongated MCG −4-4-19 S 1′.8.
01 30 47.1	−23 35 17	MCG −4-4-21	14.1	1.2 x 0.7	13.7	SBbc	101	
01 31 16.3	−23 36 46	MCG −4-4-22	14.2	0.7 x 0.7	13.3	Sa		
01 37 29.7	−21 14 22	MCG −4-5-1	14.4	0.8 x 0.7	13.6	S?	81	
01 38 17.7	−22 55 44	MCG −4-5-2	14.6	0.5 x 0.5	12.9	Sbc		
01 41 05.2	−26 01 21	MCG −4-5-5	14.2	1.2 x 0.8	14.0		15	Double system, common envelope.
01 47 45.6	−21 50 05	MCG −4-5-7	13.9	0.8 x 0.6	13.2	E	129	
01 52 48.8	−22 44 39	MCG −4-5-14	14.5	0.7 x 0.4	13.0	Sab	111	
02 00 36.1	−25 03 07	MCG −4-5-24	13.6	0.9 x 0.3	12.1	SA0°	71	Equilateral triangle of mag 9.52v, 9.70v and 10.36v stars centered ESE 8′.3; IC 1768 NE 3′.4.
02 02 17.0	−21 45 49	MCG −4-6-1	13.8	1.2 x 0.5	13.1	S?	106	Mag 9.08v star WNW 4′.1.
02 02 35.4	−24 27 56	MCG −4-6-2	14.7	0.6 x 0.4	13.0	SBc	39	Mag 9.90v star S 2′.5.
00 36 27.5	−27 47 08	MCG −5-2-23	13.6	1.0 x 0.5	12.7	S0?	102	

(Continued from previous page)

GALAXIES

RA h m s	Dec ° ′ ″	Name	Mag (V)	Dim ′ Maj x min	SB	Type Class	PA	Notes
00 36 37.5	−27 47 20	MCG −5-2-24	14.6	0.6 x 0.4	12.9	S0?	27	
00 36 26.4	−28 17 55	MCG −5-2-25	15.0	0.7 x 0.3	13.2	SA(s)a	6	Star at N end; mag 9.77v star WNW 6′.0; ESO 410-27 SE 6′.9.
00 36 59.2	−29 33 01	MCG −5-2-29	14.9	0.4 x 0.2	12.2	E	9	
00 37 49.2	−26 38 58	MCG −5-2-30	13.8	1.0 x 1.0	13.7	SAB(r)0$^+$		
00 39 18.8	−29 56 48	MCG −5-2-32	14.4	0.5 x 0.5	12.8	S0?		Star on NE edge.
00 47 11.2	−26 22 59	MCG −5-3-4	13.9	1.0 x 0.7	13.3	Sa-b	61	
00 53 43.4	−27 03 07	MCG −5-3-11	14.0	1.2 x 0.2	12.3	S?	17	
01 05 13.5	−28 47 52	MCG −5-3-23	13.9	1.0 x 0.6	13.2	Sbc	72	
01 08 06.3	−27 37 45	MCG −5-3-26	14.4	0.9 x 0.7	13.8	Sa	57	Star on E edge.
01 08 50.8	−28 35 58	MCG −5-3-28	14.0	0.6 x 0.4	12.4	SB(rs)0	147	Mag 11.11v star SSW 3′.6; IC 1628 NNW 1′.1.
01 11 50.3	−27 41 10	MCG −5-4-5	14.4	0.9 x 0.9	14.0	Sb-c		
01 23 51.2	−27 47 22	MCG −5-4-35	13.8	1.3 x 0.9	13.9	(R)SB(r)0$^+$	168	
01 28 20.3	−28 43 13	MCG −5-4-39	14.2	1.0 x 0.5	13.3	S	66	
01 31 13.9	−26 51 32	MCG −5-4-42	13.4	1.3 x 1.0	13.6	S0	174	Mag 12.8 star NW 1′.4; mag 10.04v star ESE 7′.4.
01 41 12.4	−26 16 44	MCG −5-5-10	14.9	1.5 x 0.2	13.4	Scd	105	Mag 11.8 star S 0′.9.
01 45 54.9	−29 02 19	MCG −5-5-13	13.6	0.8 x 0.8	13.0	S0		Elongated anonymous galaxy WNW 1′.9.
01 49 31.0	−27 04 54	MCG −5-5-17	13.3	1.2 x 0.9	13.2	S?	44	Elongated anonymous galaxy SSW 1′.0.
01 49 51.0	−26 17 05	MCG −5-5-18	14.0	1.0 x 0.8	13.6	Sb	178	
01 49 57.2	−27 46 47	MCG −5-5-20	14.2	0.5 x 0.5	12.6	S pec		Pair of anonymous galaxies S 2′.6.
01 50 14.3	−28 52 19	MCG −5-5-21	14.2	0.7 x 0.5	12.9	S	171	
00 36 38.8	−22 35 34	NGC 168	14.0	1.2 x 0.3	12.8	S0$^-$? sp	26	Mag 10.39v star N 5′.5.
00 37 13.6	−22 35 14	NGC 172	13.4	2.0 x 0.3	12.7	SB(r)bc? sp	12	NGC 177 E 5′.3.
00 36 59.1	−29 28 38	NGC 174	12.9	1.4 x 0.6	12.6	SB(rs)0/a:	152	Lies in a small triangle of mag 14 stars.
00 37 21.7	−19 56 03	NGC 175	12.2	2.1 x 1.9	13.5	SB(r)ab I-II		Strong SE-NW bar with prominent dark patches NE and SW of bar.
00 37 34.4	−22 32 59	NGC 177	13.3	2.2 x 0.5	13.3	SA(r)ab	9	Mag 10.73v star E 6′.4; NGC 172 W 5′.3.
00 37 46.2	−17 50 58	NGC 179	13.3	0.9 x 0.8	12.8	SAB0$^-$	113	Faint star on N edge; mag 10.28v star N 5′.1.
00 39 03.7	−18 36 30	NGC 209	12.9	1.1 x 1.1	13.0	SA0$^-$ pec:		Almost stellar nucleus, faint envelope; mag 9.57v star NNW 8′.1.
00 41 27.2	−21 02 43	NGC 216	13.2	2.0 x 0.7	13.5	S0°? sp	27	ESO 540-14 SSW 6′.4.
00 42 27.0	−23 37 47	NGC 230	14.5	1.2 x 0.2	12.8	Sa? pec sp	44	Mag 11.9 star NW 0′.8; **IC 1573** NW 4′.6.
00 42 45.9	−23 33 41	NGC 232	13.3	0.9 x 0.7	12.6	SB(r)a? pec	42	NGC 235 and NGC 235A NE 2′.0.
00 42 52.8	−23 32 32	NGC 235	13.2	1.3 x 0.7	12.9	S0 pec	117	NGC 235A on SE edge of core; NGC 232 SW 2′.0.
00 42 53.8	−23 32 48	NGC 235A	13.0	0.4 x 0.4	11.0	E0? pec		Located SE of core of NGC 235.
00 47 07.0	−20 44 36	NGC 247	9.1	19.2 x 5.5	14.0	SAB(s)d IV	172	Numerous knots and HII regions; mag 9.47v star inside S edge.
00 47 33.5	−25 17 28	NGC 253	7.2	29.0 x 6.8	12.8	SAB(s)c II	52	**Sculptor Galaxy**. Numerous dark patches along NW edge.
00 52 06.5	−22 40 52	NGC 276	14.7	1.1 x 0.4	13.6	SAB(r)b? pec	90	Mag 7.31v star N 4′.1.
00 58 46.6	−20 50 25	NGC 320	13.8	0.9 x 0.5	12.8	(R)SB(rs)0/a:	159	Mag 12.1 star N 1′.5.
00 59 19.6	−18 14 03	NGC 335	14.0	1.2 x 0.3	12.8	Sbc? sp	137	PGC 3562 SE 6′.1.
01 01 35.7	−23 15 56	NGC 344	14.9	0.7 x 0.5	13.6	Sc III	128	Mag 12.4 star 0′.8 N.
01 05 08.9	−17 45 37	NGC 369	13.7	1.0 x 0.8	13.3	SAB(r)b: I-II	52	MCG −3-3-21 NE 4′.7.
01 06 35.0	−20 19 57	NGC 377	15.1	1.0 x 0.3	13.6	Sbc	30	
01 11 05.4	−18 08 56	NGC 417	14.0	0.8 x 0.5	12.9	SAB0$^-$:	55	MCG −3-4-20 on N edge.
01 11 22.5	−29 14 05	NGC 423	13.4	1.1 x 0.4	12.4	S0/a? pec sp	114	**ESO 412-12** SE 5′.9.
01 20 09.4	−22 22 41	NGC 478	13.9	0.8 x 0.7	13.2	S pec	83	W half brighter, faint, curving extension occupies E half.
01 25 21.7	−18 09 50	NGC 539	13.5	1.5 x 1.3	14.1	SB(rs)c I	145	Stellar nucleus; mag 9.68v star SW 11′; 5. MCG −3-4-64 SE 5′.6.
01 27 08.9	−20 02 09	NGC 540	14.5	0.9 x 0.4	13.3	SB(s)0:	178	
01 27 09.7	−22 43 30	NGC 554	13.6	0.7 x 0.5	12.3	S0 pec	177	Mag 12.6 star E 1′.8.
01 27 11.9	−22 45 44	NGC 555	14.1	0.7 x 0.6	13.0	(R′)SB(r)0/a:	13	Two very small, very faint anonymous galaxies on N edge.
01 27 12.6	−22 41 52	NGC 556	14.5	0.4 x 0.3	12.1	S0$^-$: pec	136	Almost stellar.
01 27 10.0	−18 39 12	NGC 563	13.3	1.0 x 0.8	12.9	SA(s)0°:	20	2′.4 SE of mag 10.36v star.
01 30 28.7	−22 40 09	NGC 578	10.9	4.9 x 3.1	13.7	SAB(rs)c I-II	110	Two small anonymous galaxies on SW and W edges.
01 29 43.9	−18 20 22	NGC 583	13.9	0.7 x 0.6	12.9	SB(rs)0°	40	
01 34 17.8	−29 25 02	NGC 613	10.1	5.5 x 4.2	13.3	SB(rs)bc II	120	A mag 9.55v star on NNE edge.
01 38 59.0	−29 55 32	NGC 639	14.1	1.0 x 0.2	12.2	Sa?	31	
01 39 06.4	−29 54 57	NGC 642	12.9	2.0 x 1.1	13.6	SB(s)c II	31	Loose extensions NE; NGC 639 W 1′.7.
01 38 39.7	−17 49 54	NGC 648	14.4	0.7 x 0.4	13.0	SA0$^-$ pec:	114	
01 44 56.8	−22 55 10	NGC 667	14.3	0.6 x 0.5	12.8	SA0°:	86	Lies 1′.5 SE of mag 11.38v star.
01 48 56.2	−23 47 54	NGC 686	12.2	1.8 x 1.4	13.1	SA0$^-$:	14	Mag 7.91v star SSW 5′.3.
01 49 51.7	−27 27 58	NGC 689	13.5	1.0 x 0.6	12.8	(R′)SAB(r)ab pec:	68	
01 53 45.7	−23 45 28	NGC 723	12.5	1.5 x 1.3	13.0	SA(r)bc: II	166	
01 54 57.4	−17 04 46	NGC 734	15.9	0.7 x 0.4	14.4	SAB(rs)a?	24	
01 55 41.0	−29 55 20	NGC 749	12.2	1.9 x 1.4	13.1	(R′)SB(s)0/a:	111	Strong, elongated core in smooth envelope.
01 58 32.5	−26 17 34	NGC 775	12.6	1.7 x 1.2	13.2	SA(rs)c: I-II	167	Mag 6.78v star SE 13′.4.
02 03 56.6	−23 18 46	NGC 808	13.5	1.2 x 0.6	13.0	(R′)SB(r)bc: II	7	
00 59 39.3	−18 18 08	PGC 3562	14.8	0.9 x 0.6	14.0	SB(s)a pec:	27	

GALAXY CLUSTERS

RA h m s	Dec ° ′ ″	Name	Mag 10th brightest	No. Gal	Diam ′	Notes
00 38 54.0	−22 18 00	A 74	15.9	49	13	Mag 9.00v star on W edge; all members faint and stellar.
00 42 30.0	−21 47 00	A 86	15.9	31	13	Pair of mag 8.58v, 8.74v stars 8′.9 NE of cluster center.
00 53 42.0	−21 40 00	A 114	15.9	43	15	Numerous faint, stellar PGC and anonymous GX extend W well beyond A114 boundaries.
01 02 36.0	−21 47 00	A 133	15.9	47	19	Close pair of magnitude 7.98v, 8.10v stars on N edge; all members, except ESO 541-13, faint and stellar.
00 38 00.0	−25 05 00	A 2800	15.8	59	17	All members anonymous, stellar/almost stellar.
00 48 36.0	−21 20 00	A 2824	15.5	46	22	Small grouping of anonymous GX SW of center.
01 11 18.0	−17 05 00	A 2881	15.6	36	22	
01 02 54.0	−17 06 00	AS 118	16.3		17	All members anonymous, faint, stellar.
01 12 54.0	−19 00 00	AS 138	16.0		17	Four MCG galaxies at center, all others anonymous, stellar.

GLOBULAR CLUSTERS

RA h m s	Dec ° ′ ″	Name	Total V m	B ★ V m	HB V m	Diam ′	Conc. Class Low = 12 High = 1	Notes
00 52 45.5	−26 34 51	NGC 288	8.1	12.6	15.3	13.0	10	

RA h m s	Dec ° ′ ″	Name	Mag (V)	Dim ′ Maj x min	SB	Type Class	PA	Notes
23 53 47.5	−40 01 30	ESO 293-10	13.8	0.9 x 0.7	13.4	E	71	Galaxy pair, in contact.
23 56 03.1	−40 53 30	ESO 293-14	14.3	1.0 x 1.0	14.2	S? pec		Double system **ESO 293-16** E 3.′5.
23 56 27.5	−39 10 09	ESO 293-17	14.9	2.0 x 0.8	15.2		42	Triple system, interaction.
00 00 29.4	−40 28 59	ESO 293-27	13.6	2.2 x 0.5	13.6	SB(r)bc: III	154	Mag 8.56v star SSW 8.′3; double system **ESO 293-28** S 6.′0.
00 00 55.8	−40 42 42	ESO 293-29	14.2	1.4 x 0.6	13.8	SB(rs)c: I-II	19	Mag 7.31v star W 4.′6.
00 09 07.2	−37 21 27	ESO 293-43	13.9	1.4 x 1.0	14.2	S?	168	**MCG −6-1-25** NW 2.′5; mag 11.26v star SE 2.′5.
00 14 41.1	−37 39 28	ESO 293-49	13.8	1.2 x 0.8	13.6	SB(s)b II	59	Mag 11.9 star superimposed N edge.
00 23 28.7	−39 59 25	ESO 294-7	13.9	0.9 x 0.4	12.7	Sa: pec	135	Mag 11.27v star NE 9.′5.
00 27 09.2	−40 59 07	ESO 294-16	14.0	1.3 x 0.7	13.7	(R′)SB(rs)ab:	137	Mag 13.2 star N 1.′4.
00 32 10.0	−40 16 01	ESO 294-20	13.6	1.9 x 1.4	14.6	SB(s)dm V	8	Close pair of N-S oriented mag 10.54v, 10.10v stars N 5.′6.
00 35 13.1	−37 33 29	ESO 294-23	13.7	1.3 x 1.0	13.8	(R)SB(r)a	49	Strong SE-NW bar.
00 43 25.9	−38 45 24	ESO 295-2	14.1	1.0 x 0.5	13.3	SAB(s)0° pec	42	Mag 10.79v star E 4.′7.
23 26 56.3	−37 20 47	ESO 347-17	15.4	1.8 x 0.4	14.9	SB(s)m: sp IV	96	Double system; mag 11.7 star W 7.′6.
23 31 49.1	−37 51 57	ESO 347-21	14.8	1.7 x 0.3	13.9	S?	172	Galaxy **AM 2329-380** NW 2.′0.
23 36 09.9	−39 46 30	ESO 347-27	13.8	0.9 x 0.5	12.8	Sa	108	Mag 11.5 star W 1.′1.
23 36 27.3	−38 47 04	ESO 347-29	13.2	3.6 x 1.5	14.9	SB(s)dm IV	151	Mag 10.20v star on S edge.
23 49 23.6	−37 46 23	ESO 348-9	16.1	2.0 x 0.5	15.9	IB(s)m V-VI	83	Mag 14.8 star on E end; mag 11.9 star S 1.′3.
23 51 30.4	−34 27 10	ESO 349-1	13.1	1.1 x 0.9	13.1	E$^+$1	66	Mag 9.44v star S 2.′2.
23 52 02.3	−34 54 45	ESO 349-2	14.0	0.8 x 0.7	13.2	Sbc	158	Small, faint, elongated anonymous galaxy W 1.′9; mag 12.8 star SSE 1.′7.
23 52 11.6	−34 35 24	ESO 349-3	14.0	1.0 x 0.8	13.7	(R′)SA(rs)cd	148	Mag 9.78v star SE 4.′4; **MCG −6-1-1** E 7.′2.
23 54 58.0	−34 36 06	ESO 349-5	13.7	0.8 x 0.7	13.0	Sc	42	Mag 11.7 star W 7.′5.
23 57 00.7	−34 40 51	ESO 349-9	13.6	1.4 x 0.9	13.7	(R′)SB(rs)b	23	Mag 12.4 star S 1.′0; **PGC 72986** SW 2.′3.
23 57 00.9	−34 45 34	ESO 349-10	12.7	1.9 x 1.0	13.5	E$^+$4	147	Mag 10.35v star W 6.′3; stellar **PGC 73007** S 2.′7.
23 58 01.5	−34 17 26	ESO 349-13	13.7	1.2 x 0.8	13.5	S0?	150	Mag 9.94v star S 2.′7.
00 00 58.3	−33 36 45	ESO 349-17	13.6	1.3 x 1.1	13.8	S?	105	Mag 12.6 star SE edge.
00 01 33.9	−36 19 02	ESO 349-18	14.0	1.2 x 0.7	13.6	S0/a	28	Mag 9.50v star NNE 9.′7.
00 01 53.1	−36 51 00	ESO 349-19	14.3	1.1 x 0.8	14.0	(R′)SAB(rs)ab	15	Mag 10.49v star SSE 5.′9.
00 02 03.9	−33 28 04	ESO 349-20	13.8	1.1 x 0.7	13.4	(R′)SB(s)a	14	Mag 15.2 star E 0.′7.
00 03 12.9	−35 56 16	ESO 349-22	13.5	1.1 x 1.1	13.6	S0?		Almost stellar core, smooth envelope; almost stellar anonymous galaxies E 1.′6 and W 1.′2; mag 9.14v star SE 2.′9.
00 05 41.6	−35 57 05	ESO 349-25	14.1	0.9 x 0.7	13.4	Sb	1	Mag 7.76v star SW 8.′8; elongated **MCG −6-1-20** W 2.′8.
00 05 59.1	−36 06 55	ESO 349-26	13.5	1.5 x 0.9	13.8	S0 pec sp	110	Quadruple system, connected; ESO 349-27 S 1.′2.
00 05 58.7	−36 08 12	ESO 349-27	13.6	1.2 x 0.8	13.4	SAB(rs)ab pec	17	ESO 349-26 N 1.′2; Faint, round anonymous galaxy S 1.′8.
00 09 03.3	−32 54 52	ESO 349-33	13.8	1.5 x 0.3	12.8	S?	20	Mag 10.77v star W 3.′8.
00 10 57.0	−35 12 33	ESO 349-37	14.0	0.9 x 0.8	13.5	(R′)SAB0/a	85	Mag 5.24v star Theta Scuptoris NE 10.′5.
00 11 12.6	−33 34 44	ESO 349-38	13.1	1.4 x 0.6	12.8	(R′)SAB(rs)b: II	56	Mag 11.6 star NNW 3.′1.
00 11 37.1	−36 56 16	ESO 349-39	15.1	1.7 x 0.7	15.1	SB(s)d pec IV	163	Very small, very faint **ESO 349-39B** on E edge.
00 15 44.0	−33 01 49	ESO 350-4	13.5	1.2 x 0.9	13.4	SAB(rs)0/a	24	Very small, very faint anonymous galaxies W 1.′4 and 4.′6.
00 18 04.8	−33 52 33	ESO 350-7	13.2	1.7 x 1.3	13.9	(R)SB(rs)0/a	80	Small, faint, elongated anonymous galaxy SW 2.′1.
00 19 47.7	−34 28 23	ESO 350-9	13.6	1.5 x 1.3	14.2	(R)SB(r)a	129	Mag 11.6 start W 6.′3.
00 25 31.1	−33 02 47	ESO 350-15	13.1	1.7 x 1.0	13.7	E$^+$3	23	Mag 12.8 star on N edge; mag 9.42v star S 4.′0; **ESO 350-16** S 5.′5 **MCG −6-2-4** SE 3.′9.
00 27 34.2	−34 11 52	ESO 350-19	14.7	1.5 x 0.2	13.2	Sab: sp	61	Mag 8.79v star N 3.′6; **MCG −6-2-8** N 7.′4.
00 28 45.7	−32 24 23	ESO 350-20	13.9	1.1 x 1.1	14.0	SB(rs)bc I		Mag 10.62v star ESE 6.′2.
00 30 08.1	−36 52 48	ESO 350-22	14.0	1.0 x 0.7	13.5	SAB(r)c: II-III	41	Pair of stars, mags 13.4 and 14.4, N 2.′5.
00 31 02.1	−36 53 20	ESO 350-27	14.0	1.5 x 0.9	14.2	(R′)SA(rs)0°:	33	Mag 15.2 star on E edge; mag 10.87v star S 3.′9.
00 31 14.1	−36 51 14	ESO 350-28	13.7	1.1 x 0.6	13.1	SB(r)0°?	39	Mag 15.1 star E 0.′8; ESO 350-27 SW 3.′1.
00 33 59.8	−34 16 40	ESO 350-34	14.9	2.1 x 1.2	15.8		69	
00 36 06.3	−32 36 15	ESO 350-37	13.8	1.0 x 0.6	13.1	Sa	58	**MCG −6-2-22** NE 2.′0.
00 36 52.6	−33 33 20	ESO 350-38	13.6	0.6 x 0.4	11.9		103	Mag 15.3 star E 1.′3.
00 42 50.3	−36 52 47	ESO 351-3	13.7	0.5 x 0.5	12.1	Sb		Mag 10.00v star N edge.
23 17 39.5	−34 47 29	ESO 407-14	12.8	1.9 x 1.2	13.5	SB(s)c? III-IV	40	Mag 10.69v star NW 7.′7.
23 26 27.0	−32 23 20	ESO 407-18	13.3	1.5 x 1.3	13.9	IB(s)m pec: V-VI	8	Pair of stars inside E edge on N-S line.
23 31 50.5	−34 03 20	ESO 408-8	13.5	2.3 x 1.0	14.3	SAB(rs)0°	165	Mag 12.1 star E 1.′1.
23 37 34.4	−36 59 48	ESO 408-12	14.5	1.3 x 0.7	14.3	SAB(s)d: III-IV	84	Small triangle of mag 14 stars SE 4.′6.
23 41 49.0	−36 37 33	ESO 408-17	14.2	0.5 x 0.4	12.4		75	Mag 11.5 star E 3.′8.
23 43 47.5	−36 42 51	ESO 408-21	14.0	1.1 x 0.6	13.4	SAB(r)bc	44	Mag 9.52v star NNW 3.′7.
23 43 47.6	−34 54 25	ESO 408-22	13.6	1.0 x 0.7	13.0	S?	94	Mag 13.3 star NW 1.′4.
23 46 13.1	−36 45 51	ESO 408-28	15.1	1.3 x 0.3	13.9	Sb: pec sp	75	Mag 11.9 star WSW 7.′1.
23 49 57.8	−35 28 40	ESO 408-37	13.8	1.1 x 1.1	13.9	E?		Mag 9.55v star SE 10.′3; stellar **PGC 72563** SE 5.′6.
00 15 31.7	−32 10 55	ESO 410-5	14.2	1.3 x 1.0	14.4	E3:	54	Mag 7.89v star N 8.′4; mag 8.04v star WNW 8.′3.
00 34 10.9	−30 46 28	ESO 410-18	13.5	1.5 x 1.5	14.2	SAB(s)m: V		Mag 12.0 star SW 1.′9.
00 40 04.0	−30 35 50	ESO 411-6	14.3	1.5 x 1.0	14.6	SB(rs)bc pec I-II	140	Stellar nucleus; mag 9.78v star NW 6.′0.
23 30 03.5	−31 09 26	ESO 470-13	14.2	1.1 x 0.3	12.8	Sb pec	109	MCG −5-55-17 N 1.′9.
23 43 45.9	−31 57 33	ESO 471-6	13.2	4.0 x 0.8	14.3	SB(s)m: sp V	48	Mag 10.92v star N 2.′7; mag 9.18v star NW 7.′7; note: multi-galaxy system **ESO 471-5** NNE 2.′2 from this star.
23 49 15.6	−30 14 56	ESO 471-22	13.7	1.5 x 0.8	13.8	IB0? pec	120	Mag 10.54v star SW 8.′4.
23 51 55.8	−30 35 57	ESO 471-28	14.1	0.7 x 0.5	12.8	Sb	105	Very small, faint anonymous galaxy N edge; mag 13.5 star S 1.′6.
23 56 24.5	−29 01 24	ESO 471-47	14.1	1.3 x 0.9	14.1		101	Double system, contact;, = **IC 5364**?
23 57 28.3	−29 02 48	ESO 471-49	13.0	1.7 x 0.9	13.3	SB(r)b: II	168	Mag 12.9 star W 1.′3; mag 9.40v star WSW 13.′1.
23 59 19.4	−31 43 51	ESO 471-54	14.0	0.8 x 0.5	12.8	SB pec	0	Very close pair of mag 15.3 stars W 1.′4.
00 09 35.6	−32 16 38	IC 1531	12.4	1.8 x 1.4	13.3	SAB0$^-$ pec:	138	Mag 10.60v star NE 11.′0.
00 33 07.3	−32 15 30	IC 1554	12.9	1.4 x 0.8	12.9	SB(rs)0$^+$ pec:	24	Mag 10.86v star NW 5.′6.
00 34 33.1	−30 01 10	IC 1555	13.7	1.4 x 0.9	13.8	SA(s)d IV	126	Mag 11.9 star W 1.′7.
23 34 27.4	−36 06 11	IC 5332	10.5	8.9 x 8.2	15.0	SA(s)d II-III	159	Numerous bright knots and superimposed stars.
23 57 10.9	−37 00 08	IC 5365	14.7	0.8 x 0.6	13.7	SB(r)a	123	Mag 8.48v star SE 10.′5.
23 56 23.8	−31 21 15	MCG −5-1-6	13.2	1.0 x 0.7	12.9	E?	2	
23 56 24.3	−29 07 55	MCG −5-1-9	14.7	0.5 x 0.4	12.8	Sc	0	Mag 9.40v star NE 1.′9.
23 56 56.9	−32 07 59	MCG −5-1-10	14.3	1.0 x 0.6	13.6	Sbc	140	
23 57 28.0	−30 27 43	MCG −5-1-13	14.7	0.5 x 0.3	12.5	Sc	18	MCG −5-1-13 on W edge.
23 57 59.3	−29 52 00	MCG −5-1-14	13.8	0.8 x 0.4	12.5	S0?	125	Mag 10.28v star WNW 4.′8.
23 57 54.3	−31 48 56	MCG −5-1-15	14.0	0.6 x 0.5	12.5	S	9	
23 58 05.8	−29 45 42	MCG −5-1-16	13.6	0.4 x 0.3	11.2		87	Pair of mag 10-11 stars on W edge.
23 58 17.6	−30 06 46	MCG −5-1-17	14.0	0.8 x 0.5	12.9	SB?	98	
23 58 39.6	−29 50 45	MCG −5-1-18	14.3	0.3 x 0.3	11.7	E		**ESO 471-53** N 9.′0.
23 59 37.9	−30 39 25	MCG −5-1-20	14.3	0.5 x 0.3	12.3	E	105	
00 02 32.4	−30 37 20	MCG −5-1-26	13.5	0.7 x 0.6	12.4	Sa	160	
00 04 42.3	−30 29 03	MCG −5-1-28	13.0	0.9 x 0.6	12.4	E?	155	
00 04 59.7	−30 30 22	MCG −5-1-30	14.0	0.8 x 0.5	12.9	(R)SB0°	6	b SE 8.′3; **MCG −5-1-34** SE 15.′7.
00 10 39.5	−29 15 12	MCG −5-1-39	14.4	0.7 x 0.5	13.1	S	42	

(Continued from previous page)

159 GALAXIES 159

RA h m s	Dec ° ′ ″	Name	Mag (V)	Dim ′ Maj x min	SB	Type Class	PA	Notes
00 13 59.0	−30 42 37	MCG −5-1-44	14.5	0.8 x 0.7	13.7	Sb-c	177	
00 15 07.2	−29 12 41	MCG −5-1-46	14.1	1.0 x 0.3	12.6	Sb	12	
00 26 39.0	−30 33 04	MCG −5-2-4	14.6	0.5 x 0.5	13.0	Pec		
00 29 04.0	−30 50 28	MCG −5-2-7	13.8	0.8 x 0.6	13.1	E	162	MCG −5-2-7 N 0′.5.
00 30 45.8	−29 20 38	MCG −5-2-12	14.3	0.9 x 0.6	13.5	S?	165	
00 36 59.2	−29 33 01	MCG −5-2-29	14.9	0.4 x 0.2	12.2	E	9	
00 39 18.8	−29 56 48	MCG −5-2-32	14.4	0.5 x 0.5	12.8	S0?		Star on NE edge.
00 40 53.7	−30 27 39	MCG −5-2-35	15.1	0.5 x 0.4	13.2	SB(r)d	27	Mag 14.0 star NW 1′.7; mag 13.1 star E 5′.8.
23 24 55.1	−29 28 55	MCG −5-55-8	14.8	0.5 x 0.4	12.9	S	24	
23 28 45.7	−29 32 46	MCG −5-55-14	14.2	0.7 x 0.5	12.9	S0	99	
23 30 07.0	−31 07 36	MCG −5-55-17	14.3	0.8 x 0.6	13.4	S?	30	Small anonymous elongated galaxy SW edge; **ESO 470-12** N 9′.2.
23 34 19.5	−32 05 01	MCG −5-55-25	14.7	0.4 x 0.4	12.5	S		
23 36 37.8	−29 04 14	MCG −5-55-28	14.9	0.4 x 0.4	12.7	Sb?		
23 39 10.2	−31 34 15	MCG −5-55-30	15.0	0.6 x 0.4	13.3	S	168	Star or small galaxy S edge.
23 41 29.8	−29 19 17	MCG −5-55-32	14.2	0.4 x 0.4	12.3	E		faint star on E edge.
23 41 35.5	−29 14 11	MCG −5-55-33	14.3	0.5 x 0.5	12.9	E		
23 46 36.9	−29 04 14	MCG −5-56-7	13.5	0.9 x 0.6	12.7	S0?	170	
23 46 59.9	−29 10 20	MCG −5-56-8	14.8	0.7 x 0.3	13.0	S	96	
23 49 15.8	−29 01 52	MCG −5-56-18	14.0	0.9 x 0.6	13.2	Sa:	22	
23 50 13.6	−29 00 34	MCG −5-56-21	14.0	0.6 x 0.6	12.9	E		**MCG −5-56-20** S 7′.6.
23 52 10.3	−29 04 46	MCG −5-56-26	13.9	0.8 x 0.8	13.5	E		
23 52 22.1	−30 13 51	MCG −5-56-28	13.8	0.9 x 0.6	13.1	(R)SAB(r)0/a:	80	
23 52 24.3	−29 01 24	MCG −5-56-29	14.4	0.4 x 0.4	12.5	E		
23 52 31.9	−30 10 53	MCG −5-56-31	13.5	1.1 x 0.8	13.2	(R′)SB(s)a pec	19	
23 53 25.4	−29 32 19	MCG −5-56-33	13.9	0.5 x 0.5	12.5	E		
23 48 10.1	−38 24 46	MCG −7-48-23	14.1	0.9 x 0.9	13.8	SA0$^-$:		Mag 10.70v star N 5′.5.
00 08 20.6	−29 55 02	NGC 7	13.9	2.2 x 0.5	13.8	SB(s)c? sp III-IV	29	
00 08 34.2	−33 51 30	NGC 10	12.5	2.5 x 1.2	13.6	SAB(rs)bc I-II	26	Strong N-S dark lane E of center; mag 10.58v star NE 7′.4.
00 15 05.9	−39 13 01	NGC 55	7.9	32.4 x 5.6	13.4	SB(s)m: sp III-IV	108	
00 23 54.7	−32 32 08	NGC 101	12.8	2.2 x 2.0	14.3	SAB(rs)cd: III-IV	89	Prominent round knot E of center; mag 12.0 star NE 6′.2.
00 26 46.5	−33 40 45	NGC 115	13.1	1.9 x 0.9	13.5	SB(s)bc: II-III	124	Mag 10.47v star E 5′.8.
00 29 38.6	−33 15 35	NGC 131	13.2	1.9 x 0.6	13.2	SB(s)b: sp	63	Located 8′.6 W of NGC 134; mag 9.95v star S 5′.0.
00 30 20.8	−33 14 48	NGC 134	10.4	8.2 x 2.0	13.3	SAB(s)bc II-III	50	**ESO 350-26** E 9′.0; NGC 131 W 8′.6.
00 34 15.6	−31 47 12	NGC 148	12.2	2.0 x 0.8	12.6	S0°: sp	90	
00 36 59.1	−29 28 38	NGC 174	12.9	1.4 x 0.6	12.6	SB(rs)0/a:	152	Lies in a small triangle of mag 14 stars.
23 22 33.0	−29 16 49	NGC 7636	13.6	0.8 x 0.5	12.5	S0°	30	
23 23 47.5	−29 23 18	NGC 7645	12.9	1.4 x 1.2	13.3	SB(r)c I-II	165	
23 26 24.5	−39 12 56	NGC 7658A	14.2	0.6 x 0.2	11.7	S0 pec	137	
23 26 25.1	−39 13 37	NGC 7658B	13.9	0.9 x 0.3	12.3	S0 pec	128	NGC 7658A N 0′.8; an anonymous galaxy SW 1′.8; **ESO 347-13** SW 7′.3.
23 36 15.0	−37 56 17	NGC 7713	11.2	4.5 x 1.8	13.3	SB(r)d: III-IV	168	
23 37 08.9	−37 42 52	NGC 7713A	12.5	1.8 x 1.4	13.4	SB(r)c: II	85	Patchy arms.
23 45 47.4	−29 31 07	NGC 7749	12.8	1.6 x 1.1	13.3	SA0°	28	
23 47 51.8	−30 31 18	NGC 7755	11.9	3.8 x 2.9	14.3	SB(rs)c: II	20	Bright center, knotty, branching arms; almost stellar anonymous galaxy on SE edge.
23 50 54.5	−40 43 55	NGC 7764	12.2	1.9 x 1.3	13.0	IB(s)m III-IV	148	
23 57 49.8	−32 35 26	NGC 7793	9.1	9.3 x 6.3	13.4	SA(s)d IV	98	Very bright center, numerous knots.
00 02 54.5	−34 14 07	NGC 7812	13.2	1.2 x 0.9	13.1	SAB(rs)b:	146	

GALAXY CLUSTERS

RA h m s	Dec ° ′ ″	Name	Mag 10th brightest	No. Gal	Diam ′	Notes
00 03 18.0	−35 57 00	A 2717	15.6	52	22	ESO 349-22 at center; other members anonymous.
00 37 30.0	−39 07 00	A 2799	16.2	63	17	**ESO 294-25** at center; other members anonymous, stellar/almost stellar.
23 56 42.0	−34 40 00	A 4059	15.5	66	22	Numerous PGC, ESO galaxies stretch N and S from center.
00 02 48.0	−29 55 00	AS 2	16.1	11	17	All members anonymous, stellar.
00 04 48.0	−30 29 00	AS 6	15.6	7	22	Contains four MCG galaxies, all others anonymous, stellar.
00 09 18.0	−35 21 00	AS 12	15.3	3	22	**PGC 710**, **MCG −6-1-31** NE of center 8′.9.
00 25 30.0	−33 01 00	AS 41	15.5	26	22	**PGC 1624** NNE of center 13′.8.
23 29 54.0	−31 06 00	AS 1129	16.3	3	17	See notes MCG −5-55-17; most others anonymous.
23 41 18.0	−30 13 00	AS 1142	15.7	20	17	Elongated **ESO 471-1** W of center 6′.7; all others anonymous, stellar.
23 50 12.0	−29 01 00	AS 1155	15.6	3	22	Most members anonymous, stellar.
23 51 54.0	−34 25 00	AS 1157	14.9	19	22	
23 58 42.0	−29 51 00	AS 1165	15.7	3	17	Most members anonymous, stellar.
00 01 24.0	−38 46 00	AS 1172	16.3	8	17	All members anonymous, stellar.

OPEN CLUSTERS

RA h m s	Dec ° ′ ″	Name	Mag	Diam ′	No. ★	B ★	Type	Notes
00 04 12.0	−29 56 00	Bl 1	4.5	70	30	8.0	cl	Zeta Scl cluster, large and bright.

160 GALAXIES 160

RA h m s	Dec ° ′ ″	Name	Mag (V)	Dim ′ Maj x min	SB	Type Class	PA	Notes
22 22 17.4	−40 05 22	ESO 345-2	12.8	1.5 x 0.5	12.3	Sb	76	Mag 9.26v star N 7′.4.
22 26 16.1	−37 22 27	ESO 345-11	13.1	1.1 x 0.5	12.3	S0$^+$: pec sp	172	Mag 9.42v star N 6′.7.
22 31 07.3	−38 01 40	ESO 345-17	13.9	1.0 x 0.6	13.2	E$^+$4	55	Mag 10.49v star ESE 7′.5.
22 32 37.6	−38 02 56	ESO 345-21	14.0	1.0 x 0.4	12.9	S	88	Mag 9.37v star E 6′.1.
22 34 15.2	−38 09 50	ESO 345-28	13.5	0.4 x 0.4	11.4			Compact; mag 13.0 star NW 1′.7.
22 35 26.2	−37 23 54	ESO 345-32	13.9	0.9 x 0′.6	13.1	Sb	153	ESO 345-33 NE 2′.6.
22 35 35.2	−37 22 02	ESO 345-33	13.8	0.9 x 0.6	13.0	Sb	127	ESO 345-32 SW 2′.6; mag 13.1 star WNW 1′.4.
22 42 31.5	−37 18 49	ESO 345-42	14.4	0.8 x 0.5	13.3	SAB(s)0°:	161	Mag 7.93v star SE 3′.4.
22 43 14.3	−40 02 57	ESO 345-45A	13.4	0.6 x 0.5	12.2	E/S0	140	Stellar **ESO 345-45** S 1′.0; mag 7.77v star N 2′.5.
22 43 15.8	−39 52 08	ESO 345-46	12.8	2.4 x 2.2	14.4	SA(s)d: III-IV	45	Mag 7.77v star S 8′.5; **ESO 345-41** SW 11′.6; **ESO 345-40** SW 22′.5.
22 45 36.8	−40 55 00	ESO 345-50	14.8	1.5 x 0.5	14.3	IBm: sp V-VI	66	Mag 7.06v star NW 12′.7.
22 48 33.8	−39 38 58	ESO 346-1	13.8	1.9 x 0.3	13.0	Sc sp	64	Mag 9.92v star NE 8′.2; **ESO 346-2** S 2′.6.

RA h m s	Dec ° ′ ″	Name	Mag (V)	Dim ′ Maj x min	SB	Type Class	PA	Notes
22 49 22.9	−37 28 20	ESO 346-3	12.9	1.8 x 1.2	13.7	E^+2 pec:	128	An anonymous galaxy on NE edge. **MCG −6-50-4** and anonymous galaxy SE 1′.6.
22 52 39.1	−40 19 48	ESO 346-6	13.9	1.0 x 0.9	13.7	Sc: pec I-II	125	Mag 12.3 star WNW 3′.1.
22 53 24.0	−38 48 01	ESO 346-7	14.4	0.9 x 0.6	13.5	Im:	85	Mag 12.8 star on NW edge.
22 54 14.5	−40 08 57	ESO 346-9	14.5	0.4 x 0.3	12.0	Pec	83	Almost stellar; mag 12.1 star S 5′.1.
22 55 06.3	−38 35 07	ESO 346-14	14.3	1.9 x 0.2	13.1	SB(s)d? sp	58	Mag 10.93v star SW 2′.0.
22 57 11.6	−40 05 55	ESO 346-18	13.9	2.6 x 1.0	14.7	SA(s)dm V-VI	3	Mag 12.3 star SW 7′.7.
23 01 56.1	−40 39 28	ESO 346-24	14.0	1.2 x 0.8	13.8	SBab	117	Large NE-SW core.
23 08 17.0	−37 46 16	ESO 346-32	14.0	0.9 x 0.9	13.6	(R)SB0°		Mag 9.66v star SE 5′.6.
23 08 46.4	−39 48 44	ESO 346-33	15.9	1.2 x 0.4	15.0	IBm VI	176	Mag 10.79v star W 2′.5.
23 14 14.7	−40 37 53	ESO 347-2	14.0	2.1 x 0.3	13.3	SBd? sp	86	
23 14 52.2	−37 51 21	ESO 347-3	13.7	1.0 x 0.7	13.2	S?	158	Mag 10.50v star NNE 1′.6.
21 56 45.4	−36 29 36	ESO 404-11	14.0	1.1 x 0.6	13.4	SB(rs)ab? III-IV	132	Close pair of mag 12.1, 12.2 stars NW 5′.8.
21 57 06.8	−34 35 00	ESO 404-12	12.2	2.4 x 1.9	13.7	SAB(r)c: I-II	135	Mag 11.4 star E 4′.0; mag 9.95v star WNW 7′.7.
22 00 29.0	−33 22 16	ESO 404-15	13.7	1.3 x 0.4	12.8	S?	147	Mag 8.66v star S 8′.7.
22 00 55.5	−35 17 14	ESO 404-17	14.4	1.9 x 0.6	14.4	SB(s)dm IV-V	127	N-S oriented pair of mag 12.9, 12.7 stars S 3′.5.
22 01 10.1	−32 34 46	ESO 404-18	14.8	2.6 x 0.2	13.9	SB(s)d? sp III-IV	34	Mag 10.11v star W 5′.5.
22 01 23.9	−33 54 45	ESO 404-19	14.0	1.0 x 0.4	12.9	Sb	57	Anonymous galaxy NW 2′.6; mag 11.8 star W 1′.6.
22 02 06.4	−32 53 12	ESO 404-21	14.0	1.2 x 0.8	13.8	S?	0	Mag 10.59v star NNW 5′.4.
22 02 41.7	−33 48 19	ESO 404-23	13.2	1.8 x 1.1	13.8	SB(rs)bc II	14	Mag 10.77v star W 7′.2; **IC 5166** E 7′.2.
22 03 48.0	−32 17 07	ESO 404-27	12.7	2.2 x 0.8	13.1	SB(s)c I-II	128	Mag 8.90v star S 4′.3.
22 04 15.3	−32 36 17	ESO 404-28	14.1	1.2 x 0.4	13.2	Sab	131	Mag 8.94v star NE 7′.5.
22 07 55.1	−35 30 21	ESO 404-30	13.8	1.0 x 0.9	13.6	S?	34	Mag 12.9 star SSW 3′.1.
22 08 13.9	−32 27 41	ESO 404-31	14.1	1.5 x 0.5	13.6	SA(s)c: I-II	77	Mag 8.82v star NNW 5′.6; stellar **ESO 404-29** SW 9′.5.
22 11 41.5	−33 53 11	ESO 404-39	13.4	1.2 x 0.3	12.2	S?	70	Compact **ESO 404-39A** E 0′.9.
22 12 34.0	−35 56 15	ESO 404-40	13.5	1.3 x 0.6	13.1	S0	119	Low, uniform surface brightness **ESO 404-41** S 1′.4.
22 13 49.8	−33 24 01	ESO 404-43	14.8	1.5 x 0.2	13.4	Sc	1	Mag 10.20v star SW 6′.1.
22 14 09.7	−33 14 08	ESO 404-45	14.5	2.0 x 0.3	13.8	Sc: sp	110	Mag 9.22v star NE 9′.9; **ESO 405-1** E 12′.5.
22 16 48.5	−36 24 00	ESO 405-6	14.3	1.3 x 0.8	14.2	Sc	55	Bright knot NE edge.
22 20 25.8	−32 41 16	ESO 405-9	14.0	1.1 x 0.6	13.4	Sa	139	Mag 9.86v star W 11′.5.
22 21 31.9	−37 01 57	ESO 405-11	14.0	1.1 x 0.5	13.2	(R′)SB(s)a	72	Mag 12.9 star N 7′.1.
22 21 53.7	−35 12 21	ESO 405-13	14.2	1.0 x 0.6	13.5	S?	96	Mag 10.55v star W 5′.3.
22 22 52.3	−36 57 04	ESO 405-15	13.9	0.6 x 0.6	12.7			Mag 13.2 star SE 2′.3.
22 34 01.8	−32 23 53	ESO 405-29	14.2	1.1 x 0.9	14.0	SB(s)c: III-IV	43	Mag 10.79v star E 5′.4; stellar **ESO 405-30** E 7′.4.
22 35 29.0	−35 07 21	ESO 405-31	14.0	1.0 x 0.5	13.1	Sb	94	Mag 12.6 star SW 3′.6.
22 42 33.5	−37 11 09	ESO 406-4	13.5	0.9 x 0.5	12.5		50	Mag 9.35v star WNW 5′.4.
22 44 53.8	−34 12 46	ESO 406-10	14.6	0.8 x 0.5	13.5		3	Double system, strongly interacting.
22 45 47.9	−35 12 22	ESO 406-11	13.7	1.0 x 0.3	12.2	S0/a	157	Mag 11.00v star S 2′.1.
22 54 03.6	−34 03 33	ESO 406-17	13.2	1.6 x 1.4	13.9	(R)SB(rs)bc: I-II	43	Mag 13.4 star superimposed on elongated anonymous galaxy NW 2′.4; mag 12.2 star NW 3′.5.
22 54 15.7	−37 05 01	ESO 406-18	14.2	0.6 x 0.5	12.7	S pec?	97	Two nuclei, or superimposed star; mag 10.51v star SE 0′.8.
22 55 52.0	−34 11 33	ESO 406-21	14.2	0.6 x 0.4	12.5	Sa?	137	Mag 13.4 star E 2′.1.
22 55 52.6	−34 33 18	ESO 406-22	13.7	0.7 x 0.5	12.4	S0:	173	Mag 12.6v star SW 3′.8.
22 57 40.8	−35 23 50	ESO 406-31	13.9	1.4 x 0.3	12.8	Sb	11	Mag 13.5 star NNE 2′.8.
22 59 08.1	−34 04 25	ESO 406-37	14.1	0.9 x 0.4	12.9	Sa	121	Mag 10.33v star W 5′.8.
23 02 14.6	−37 05 03	ESO 406-42	13.9	1.7 x 1.2	14.5	SAB(s)m: V	66	Mag 8.19v star E 9′.2.
23 04 41.4	−34 03 30	ESO 407-2	13.2	1.0 x 0.6	12.5	S0/a	121	Mag 12.8 star W 1′.4.
23 09 39.3	−36 25 07	ESO 407-7	12.8	3.0 x 0.5	13.1	SAb? sp	168	Mag 13.4 star on W edge; mag 12.0 star N 2′.1.
23 12 44.8	−37 12 29	ESO 407-9	13.2	2.0 x 0.8	13.6	SB(s)d: IV	30	Stellar nucleus; mag 12.0 star N 2′.2.
23 17 39.5	−34 47 29	ESO 407-14	12.8	1.9 x 1.2	13.5	SB(s)c? III-IV	40	Mag 10.69v star NW 7′.7.
22 01 20.3	−31 31 48	ESO 466-36	13.4	1.1 x 0.4	12.3	S?	123	Mag 9.28v star WNW 3′.7; Mag 9.31v star NE 7′.3; **ESO 466-37** NNE 7′.3.
22 02 14.9	−31 13 16	ESO 466-46	14.1	1.5 x 0.2	12.7	Sb:	53	Pair of N-S oriented mag 10.84v, 11.08v stars N 3′.5.
22 03 48.2	−31 57 22	ESO 466-51	13.3	1.5 x 0.4	12.6	S0?	65	Mag 10.06v star NW 1′.8.
22 14 35.6	−29 09 36	ESO 467-26	12.9	1.0 x 0.9	12.6	Sa	37	Mag 8.88v star NW 4′.2.
22 16 06.1	−30 22 08	ESO 467-36	13.8	1.2 x 0.7	13.4	SA(s)bc	165	Mag 9.51v star SSE 3′.3; **ESO 467-32** NW 7′.3.
22 19 46.5	−31 14 30	ESO 467-43	13.9	0.7 x 0.4	12.5	S0:	148	ESO 467-43, with a mag 12.4 stars on it's SW edge, NW 1′.8.
22 22 42.1	−32 12 00	ESO 467-46	13.5	0.9 x 0.8	13.1	S0	8	Galaxy **APMBGC 467-068+126** NE 3′.0.
22 26 05.9	−30 52 07	ESO 467-58	13.8	1.6 x 0.5	13.4	SA0/a: pec	72	Mag 13.8 star N 2′.3; MCG −5-53-5 E 10′.3.
22 40 44.2	−30 48 02	ESO 468-20	15.2	1.7 x 1.1	15.9	E3: pec	43	Mag 5.88v star NNW 9′.7.
23 05 50.1	−30 36 48	ESO 469-11	13.5	1.9 x 1.0	14.0	Pec	103	Bright core offset to W.
23 08 55.6	−30 51 34	ESO 469-15	13.9	1.9 x 0.3	13.2	Sb: sp	149	Mag 9.19v star W 9′.1.
23 10 25.3	−29 43 53	ESO 469-17	13.4	1.6 x 0.6	13.2	Sbc?	120	Mag 12.4 star S 4′.7.
22 57 10.4	−36 27 37	IC 1459	10.0	5.2 x 3.8	13.2	E3-4	40	Mag 10.87v star W 2′.8; IC 5264 SSW 6′.6.
22 03 14.9	−33 50 18	IC 5156	12.2	2.2 x 0.8	12.6	SB(s)ab pec: III	175	Mag 12.2 star E 7′.4; ESO 404-23 W 7′.2.
22 03 27.1	−34 56 32	IC 5157	12.7	1.3 x 1.3	13.3	E:		Mag 11.08v star ENE 2′.7.
22 10 10.1	−36 05 21	IC 5169	12.9	1.9 x 0.5	12.7	(R′)SAB(r)0^+ pec:	22	Mag 8.75v star N 6′.7.
22 12 45.1	−38 10 18	IC 5174	14.0	2.1 x 1.1	14.7	SB(s)b pec	1	Very faint arms extend N and S from bright center; mag 7.60v star SW 11′.4; IC 5175 N 2′.8.
22 12 48.3	−38 07 40	IC 5175	13.6	1.4 x 0.5	13.0	S?	94	Mag 10.37v star on S edge; IC 5174 S 2′.8.
22 16 09.8	−36 50 37	IC 5179	11.8	2.3 x 1.1	12.7	SA(rs)bc II-III	57	Mag 9.09v star SSW 4′.6.
22 18 46.5	−36 48 06	IC 5186	11.9	1.7 x 1.2	12.5	(R′)SAB(rs)b: II-III	106	Mag 12.1 star W 1′.7; mag 10.81v star NNE 8′.1.
22 19 33.1	−37 32 03	IC 5199	13.9	1.5 x 0.2	12.4	SAb: sp	155	Mag 10.58v star NW 11′.9.
22 23 30.5	−38 02 18	IC 5212	13.5	1.0 x 0.7	13.0	Sbc: II	42	Mag 11.8 star 1′.0 NW.
22 55 21.5	−33 53 29	IC 5262	13.4	1.3 x 0.8	13.2	S?	130	Multiple-galaxy system.
22 56 52.7	−36 33 18	IC 5264	12.5	2.5 x 0.5	12.6	Sab pec sp	82	IC 1459 NNE 6′.6; mag 10.87v star N 6′.3.
22 57 43.5	−36 01 35	IC 5269	12.2	1.8 x 0.8	12.5	SAB(rs)0:	51	Mag 11.8 star SW 7′.5.
22 55 55.5	−36 20 50	IC 5269A	13.4	1.3 x 1.0	13.5	SB(s)m: V	35	Mag 6.41v star WSW 8′.5.
22 56 37.1	−36 15 02	IC 5269B	12.4	4.1 x 0.8	13.6	SB(rs)cd: IV	96	Close pair of N-S oriented mag 10.52v, 10.83v stars WNW 10′.7.
23 00 49.1	−35 22 08	IC 5269C	13.6	2.1 x 0.7	13.8	SB(s)d pec:	66	Mag 9.72v star S 6′.0.
22 57 54.9	−35 51 28	IC 5270	12.2	3.2 x 0.6	12.8	SB(r)cd: sp IV	103	Mag 9.86v star N 6′.7.
22 58 01.8	−33 44 32	IC 5271	11.6	2.6 x 0.9	12.4	Sb? II	138	Mag 9.35v star N 9′.3.
22 59 26.6	−37 42 23	IC 5273	11.4	2.7 x 1.8	13.0	SB(rs)cd: II-III	56	Mag 12.8 star NW 2′.3.
23 11 17.2	−32 27 11	IC 5289	13.4	0.9 x 0.7	12.7	Sb	23	Multi-galaxy system.
21 57 29.6	−29 23 41	MCG −5-51-27	14.5	0.6 x 0.4	12.9	S0	54	
21 58 11.5	−30 19 12	MCG −5-51-30	14.8	0.5 x 0.5	13.1	Sb:		
21 58 24.5	−32 14 16	MCG −5-51-31	13.8	1.0 x 0.5	12.9	Sa:	30	Mag 7.92v star SSE 7′.7.
21 59 56.6	−31 27 47	MCG −5-52-1	14.4	0.4 x 0.4	12.3	S0		Mag 7.06v star S 3′.9.
22 02 09.6	−30 27 33	MCG −5-52-12	14.0	0.7 x 0.6	12.9	S0/a	122	
22 02 44.0	−31 59 26	MCG −5-52-14	13.8	1.2 x 0.5	13.2	S?	18	**ESO 466-47** N 2′.1.
22 07 12.5	−29 50 20	MCG −5-52-30	14.4	0.8 x 0.7	13.6	S	96	
22 14 24.2	−29 58 53	MCG −5-52-52	13.4	1.3 x 0.4	12.5	S?	142	Mag 9.40v star S 5′.2.
22 24 50.5	−31 19 38	MCG −5-52-72	13.4	0.5 x 0.4	11.6	$S0^-$	56	Double galaxy system **ESO 467-55** ESE 6′.8.

(Continued from previous page)

160 GALAXIES 160

RA h m s	Dec ° ′ ″	Name	Mag (V)	Dim ′ Maj x min	SB	Type Class	PA	Notes
22 26 53.7	−30 53 22	MCG −5-53-5	14.4	0.9 x 0.7	13.8	Irr	60	ESO 467-58 W 10′.3; **ESO 467-63** SE 9′.5.
22 27 43.8	−30 02 20	MCG −5-53-7	14.0	0.9 x 0.7	13.4	SA(r)cd	140	
22 28 09.2	−29 54 05	MCG −5-53-8	13.3	1.0 x 0.4	12.2	S0	47	
22 36 10.5	−31 43 44	MCG −5-53-18	13.9	1.0 x 0.5	13.0	S0	12	Mag 5.82v star NE 6′.6.
22 35 48.0	−31 42 05	MCG −5-53-20	14.0	0.7 x 0.4	12.5	S0	135	Mag 10.40v star NNW 3′.5.
22 43 17.8	−30 05 44	MCG −5-53-29	14.2	0.8 x 0.6	13.3	S	12	
22 47 07.8	−29 28 21	MCG −5-53-32	14.2	0.9 x 0.8	13.6	Sb	13	
22 52 38.6	−29 03 24	MCG −5-54-3	14.1	0.8 x 0.7	13.4	Sa?	48	Almost stellar nucleus; mag 7.69v star S 7′.6.
22 57 19.5	−31 27 22	MCG −5-54-5	14.7	0.7 x 0.3	12.8	S	54	
22 58 28.2	−32 14 25	MCG −5-54-6	13.4	1.2 x 0.8	13.2	S?	2	
22 58 58.8	−30 29 41	MCG −5-54-7	14.0	1.2 x 0.6	13.5	SAB(rs)c II	4	Mag 9.67v star N 2′.1.
23 02 05.0	−32 00 44	MCG −5-54-8	14.6	0.5 x 0.5	12.9	S		
23 02 39.9	−31 03 50	MCG −5-54-9	14.7	0.7 x 0.5	13.4	S	12	Mag 7.91v star SE 8′.2.
23 04 53.3	−31 17 19	MCG −5-54-12	14.7	0.6 x 0.5	13.2	S	6	
23 13 50.8	−29 35 07	MCG −5-54-24	14.8	0.4 x 0.4	12.9	E		Stellar; mag 8.55v star SW 4′.7.
21 59 20.4	−31 53 02	NGC 7163	13.4	1.9 x 1.1	14.0	(R′)SB(rs)ab pec: II	101	
22 02 01.5	−31 52 14	NGC 7172	11.9	2.5 x 1.4	13.1	Sa pec sp	100	Strong dark patches E and W of center; mag 10.55v star SE 2′.7.
22 02 03.0	−31 58 25	NGC 7173	12.0	1.2 x 0.9	12.1	E^+ pec:	143	
22 02 05.8	−31 59 33	NGC 7174	13.3	2.3 x 1.2	14.3	Sab pec sp	88	NGC 7176 on E edge of bright center.
22 02 08.7	−31 59 24	NGC 7176	11.4	0.8 x 0.8	11.0	E pec:		On E edge of NGC 7174's bright center.
22 02 25.2	−35 47 23	NGC 7178	14.1	1.2 x 0.5	13.4	Sb: II	172	Located 2′.6 N of mag 8.04v star.
22 02 44.6	−32 48 15	NGC 7187	12.5	1.4 x 1.3	13.1	$(R)SA(r)0^+$	171	Moderately dark ring around bright center; pair of mag 11.34v, 12.2 stars W 5′.5.
22 06 32.2	−31 15 50	NGC 7201	12.9	1.6 x 0.5	12.5	SAa:	128	
22 06 43.9	−31 09 46	NGC 7203	12.7	1.6 x 0.9	13.0	(R′)SB(r)0/a:	72	**ESO 467-2** NW 8′.5.
22 06 53.8	−31 03 07	NGC 7204	14.0	1.6 x 0.8	14.1	S0/a pec:	91	Multi-galaxy system.
22 08 24.6	−29 03 04	NGC 7208	13.1	0.9 x 0.5	12.1	SAB0°?	142	
22 11 15.6	−30 33 52	NGC 7221	12.1	2.0 x 1.6	13.2	SB(rs)bc pec:	10	
22 14 03.1	−29 23 01	NGC 7229	12.5	1.8 x 1.5	13.4	SB(s)c I-II	157	Knotty, multi-branching arms; mag 8.77v star E 6′.9.
22 23 28.4	−32 21 55	NGC 7262	14.1	0.8 x 0.7	13.3	SA0°:	96	
22 24 22.2	−33 41 34	NGC 7267	12.2	1.6 x 1.3	12.9	(R′)SB(rs)a pec	6	Strong E-W bar; mag 8.40v star SW 3′.6.
22 25 41.2	−31 12 02	NGC 7268	13.2	1.1 x 0.6	12.6	E0 pec?	83	Double galaxy system; NGC 7277 NE 7′.1.
22 26 11.1	−31 08 49	NGC 7277	13.3	1.3 x 0.6	12.9	(R)Sb: I-II	125	NGC 7268 SW 7′.1.
22 27 12.4	−35 08 28	NGC 7279	13.8	1.2 x 0.8	13.6	SB(rs)c II-III	68	
22 29 20.2	−35 28 22	NGC 7289	13.1	1.4 x 1.1	13.5	SA(r)0°:	165	
22 31 10.4	−37 49 36	NGC 7297	13.2	0.8 x 0.6	12.3	SA(s)bc: I	130	
22 31 33.2	−37 48 36	NGC 7299	13.9	0.7 x 0.7	13.0	SB(rs)c? II		
22 33 52.8	−40 56 05	NGC 7307	12.6	3.5 x 0.8	13.5	SB(s)cd pec: II-III	9	
22 37 51.7	−37 13 52	NGC 7322	13.6	1.0 x 0.6	12.9	$S0^+$?	114	
22 43 30.5	−36 51 57	NGC 7355	14.3	1.0 x 0.5	13.3	SAB(rs)b:	43	
22 42 18.0	−30 03 26	NGC 7361	12.3	3.8 x 1.0	13.6	S(r)c:? III-IV	4	Located 5′.4 E of a mag 7.78v star.
22 45 31.8	−39 20 34	NGC 7368	12.0	3.0 x 0.6	12.5	SB(s)b? sp II-III	130	
22 50 23.9	−36 51 27	NGC 7382	13.3	1.2 x 0.3	12.0	(R′)SAa? sp	109	Star on NW end.
22 54 18.4	−39 18 54	NGC 7404	12.8	1.5 x 0.9	12.9	$SAB(s)0^-$?	2	Mag 12.1 star 1′.7 S.
22 55 00.9	−39 39 38	NGC 7410	10.3	5.2 x 1.6	12.5	SB(s)a	45	
22 56 35.1	−37 01 38	NGC 7418	11.1	3.5 x 2.6	13.3	SAB(rs)cd II-III	139	Very knotty multi-branching arms.
22 56 41.4	−36 46 21	NGC 7418A	13.2	3.7 x 1.8	15.1	SA(rs)d: IV-V	83	Bright nucleus with faint envelope.
22 56 55.0	−37 20 48	NGC 7421	11.9	2.0 x 1.8	13.2	SB(rs)bc III	78	
23 02 10.1	−39 34 17	NGC 7456	11.8	5.5 x 1.6	14.1	SA(s)cd: III	23	Many bright knots.
23 02 47.1	−40 50 07	NGC 7462	12.0	4.2 x 0.7	13.1	SB(s)bc? sp II-III	75	Star on W end; pair of stars mags 10.38v and 11.04v W 3′.9; mag 9.96v star SE 3′.7.
23 07 05.1	−36 16 29	NGC 7484	11.8	1.8 x 1.7	13.0	E	56	Star on S edge.
23 15 32.3	−38 32 06	NGC 7545	13.1	1.1 x 0.7	12.7	(R′)SB(s)b pec?	80	Star on W edge; mag 10.92v star SW 2′.1.
23 22 33.0	−29 16 49	NGC 7636	13.6	0.8 x 0.5	12.5	S0°	30	
23 23 47.5	−29 23 18	NGC 7645	12.9	1.4 x 1.2	13.3	SB(r)c I-II	165	

GALAXY CLUSTERS

RA h m s	Dec ° ′ ″	Name	Mag 10th brightest	No. Gal	Diam ′	Notes
22 27 48.0	−30 34 00	A 3880	15.7	31	17	**PGC 68924** at center, all others anonymous, stellar.
22 38 54.0	−36 41 00	A 3895	15.7	47	17	**ESO 406-1** N of center 6′.1; all others anonymous.
22 46 24.0	−36 05 00	A 3912	15.8	41	17	All members anonymous, stellar.
22 09 00.0	−35 07 00	AS 997	15.6		22	**MCG −6-48-23** NW of center 2′.9; **MCG −6-48-24** W 1′.9; **ESO 404-35** S 4′.9.
22 14 06.0	−36 42 00	AS 1005	16.3		17	**MCG −6-48-30** at center; **ESO 405-4, 4A** E of center 17′.3, most all members anonymous, stellar.
22 39 18.0	−36 09 00	AS 1050	15.7	26	17	All members anonymous, stellar.
22 43 18.0	−39 52 00	AS 1055	15.6	0	22	See notes ESO 345-46; most others anonymous, stellar.
22 49 18.0	−37 29 00	AS 1065	15.8		17	See notes ESO 346-3, all others anonymous, stellar.

PLANETARY NEBULAE

RA h m s	Dec ° ′ ″	Name	Diam ″	Mag (P)	Mag (V)	Mag cent ★	Alt Name	Notes
21 59 35.2	−39 23 09	IC 5148-50	132	12.9	11.::	16.5	PK 2−52.1	Mag 10.35v star 1′.8 S of center.

161 GALAXIES 161

RA h m s	Dec ° ′ ″	Name	Mag (V)	Dim ′ Maj x min	SB	Type Class	PA	Notes
20 41 13.0	−38 11 46	ESO 341-4	12.7	1.5 x 1.0	13.0	$S0^+$: pec	66	Brightest portion at NE end.
20 41 36.6	−37 59 06	ESO 341-6	14.0	0.6 x 0.4	12.3		95	Mag 9.35v star SW 2′.2.
20 45 29.9	−38 10 29	ESO 341-11	13.3	1.2 x 1.0	13.4	SB(s)b pec	156	Mag 13.9 star on N edge; mag 12.1 star NW 3′.8; note: faint anonymous galaxy 0′.5 NE of this star.
20 45 37.0	−38 13 53	ESO 341-12	13.3	0.6 x 0.6	12.1	S0:		Almost stellar; mag 10.80v star ESE 5′.1.
20 47 09.0	−38 05 20	ESO 341-13	12.9	1.1 x 1.0	12.9	S0?	5	Mag 9.97v star ESE 6′.1; note: there is a small anonymous galaxy on the E edge of this star.
20 49 21.2	−38 49 56	ESO 341-17	13.8	0.8 x 0.3	12.1	Sa:	76	Contact with companion SW edge.
20 54 31.8	−40 48 40	ESO 341-21	13.9	1.0 x 0.5	13.0	(R′)SB(r)ab	75	Bright star superimposed.
20 59 00.4	−39 27 43	ESO 341-23	13.7	1.7 x 0.5	13.4	(R′)SB(s)c: sp	154	Mag 10.95v star W 13′.1.
20 59 31.1	−38 20 17	ESO 341-26	13.8	1.0 x 0.6	13.2	S0	138	Pair of stars, mags 9.54v and 10.07v, WNW 2′.7.

(Continued from previous page)

GALAXIES

RA h m s	Dec ° ′ ″	Name	Mag (V)	Dim ′ Maj x min	SB	Type Class	PA	Notes
21 00 59.6	−38 34 37	ESO 341-29	13.8	0.9 x 0.6	13.0	(R′)SAB(rs)ab	160	Faint, almost stellar anonymous galaxy NE 2′9.
21 03 34.6	−39 26 54	ESO 341-32	13.2	1.5 x 1.2	13.7	SB(rs)m IV-V	99	Mag 10.22v star SE 6′8.
21 04 26.8	−39 53 35	ESO 342-1	14.2	0.7 x 0.4	12.7	S?	151	Close pair of mag 14.4 stars NW 1′5.
21 06 44.1	−40 29 32	ESO 342-5	14.1	1.0 x 0.3	12.7	Sbc	108	Mag 9.65v star SSE 6′5.
21 08 23.6	−37 56 53	ESO 342-6	13.9	0.8 x 0.8	13.3	S0?		Mag 6.87v star E 10′8.
21 08 40.6	−39 52 10	ESO 342-8	13.2	1.0 x 0.8	12.8	S0	48	Mag 10.28v star E 3′7; ESO 342-10 NE.
21 08 47.2	−39 51 08	ESO 342-10	13.6	1.0 x 0.7	13.2	S0 pec	42	Mag 13.5 star on E edge; mag 10.28v star E 2′7; ESO 342-8 SW.
21 08 59.5	−40 01 19	ESO 342-12	14.0	0.8 x 0.4	12.6	S	26	Bright star superimposed N end; mag 12.3 star NE 1′3.
21 10 00.8	−37 30 16	ESO 342-13	13.1	1.3 x 0.5	12.5	Pec	162	Mag 10.39v star E 9′6; note: **ESO 342-14** is 3′8 N of this star.
21 12 30.9	−39 19 30	ESO 342-19	13.3	1.0 x 0.8	12.9	S0	143	Mag 5.26v star SE 8′8.
21 12 56.3	−37 52 12	ESO 342-22	13.9	0.7 x 0.6	12.8	Sb	134	**MCG −6-46-12** SE 1′6.
21 16 22.2	−38 58 54	ESO 342-25	13.8	1.0 x 0.4	12.7	Sbc	71	Mag 7.56v star ESE 11′4.
21 18 19.4	−38 12 12	ESO 342-34	13.9	1.0 x 0.9	13.6	Sc	80	Mag 14.3 star SW edge.
21 20 07.7	−39 46 09	ESO 342-35	12.8	1.3 x 1.2	13.2	(R′)SB(rs)a	39	Mag 9.61v star SW 7′7.
21 23 29.5	−37 30 37	ESO 342-38	13.6	1.2 x 0.6	13.1	S0?	43	Mag 9.39v star SSE 3′7; **ESO 342-37** SW 5′2.
21 25 21.6	−40 07 15	ESO 342-42	13.8	0.9 x 0.5	12.8	S pec	115	Mag 12.3 star NE 1′7; **ESO 342-44** E 4′4.
21 25 54.3	−40 00 39	ESO 342-45	13.4	0.9 x 0.8	12.9	SB?	23	Mag 11.6 star ENE 6′0.
21 26 21.5	−39 53 18	ESO 342-48	13.6	1.1 x 0.6	13.0	S0?	32	Mag 10.76v star NNE 5′9.
21 28 15.0	−37 51 43	ESO 342-50	12.3	2.4 x 1.4	13.4	SA(rs)c pec? II	21	Pair of N-S oriented mag 11.4 stars NNW 3′8.
21 30 22.0	−40 26 19	ESO 343-1	13.6	1.3 x 0.6	13.2	(R)SB(s)ab:	57	Mag 14.0 star S edge.
21 31 15.0	−38 37 00	ESO 343-3	13.1	1.5 x 0.5	12.7	S0	95	0′6 W of mag 10.32 star.
21 34 19.7	−40 49 41	ESO 343-7	13.9	1.1 x 0.4	12.9	Sa	79	NGC 7087 E 2′7.
21 34 48.5	−40 43 15	ESO 343-9	14.1	0.8 x 0.4	12.7	S?	3	Mag 13.4 star E edge; mag 11.8 star N 2′5.
21 35 20.9	−37 41 29	ESO 343-11	13.8	0.9 x 0.7	13.2	S0	30	Mag 12.3 star NE 2′6.
21 36 10.9	−38 32 39	ESO 343-13	13.6	1.8 x 1.0	14.1	S?	127	Double system, strongly interacting; mag 11.9 star SW 1′4.
21 37 45.6	−38 29 43	ESO 343-14	15.1	1.5 x 0.2	13.6	Sc: pec	158	Mag 10.91v star ENE 6′5.
21 40 19.4	−39 26 12	ESO 343-15	14.2	1.1 x 0.3	12.8	Im:	154	Mag 11.41v star SSW 6′2.
21 41 28.8	−39 46 03	ESO 343-18	13.9	1.5 x 0.3	12.8	Sc: sp	137	Mag 10.16v star W 2′1.
21 41 59.6	−38 50 02	ESO 343-20	14.0	1.1 x 0.3	12.6	Sb:	172	Mag 10.48v star W 3′9.
21 43 10.2	−38 37 50	ESO 343-21	13.9	0.8 x 0.7	13.2	SAa	72	Mag 11.44v star NW 10′5; note: **MCG −7-44-33** is 2′6 S of this star.
21 43 18.0	−39 11 15	ESO 343-23	13.8	0.9 x 0.6	13.2	E$^+$3:	48	Mag 10.78v star N 2′1; pair of mag 10.10v, 10.21v stars S 5′6.
21 43 26.7	−40 07 33	ESO 343-24	13.6	1.0 x 0.3	12.2	S	32	Pair of mag 13.6 and 13.9 stars E 0′9.
21 44 41.7	−39 11 06	ESO 343-26	13.8	1.0 x 0.6	13.1	Sb	43	Mag 11.11v star NW 3′7.
21 47 24.5	−38 42 49	ESO 343-28	14.9	1.5 x 0.4	14.2	S?	115	Mag 8.30v star W 10′5.
21 49 48.8	−37 47 08	ESO 343-31	13.8	0.9 x 0.7	13.1	SB(r)ab pec:	31	Pair of E-W oriented mag 9.82v, 10.56v stars S 3′1.
21 51 38.9	−37 30 19	ESO 343-34	13.2	1.1 x 0.3	11.8	S0?	28	Mag 8.89v star N 6′3.
21 54 55.8	−40 50 08	ESO 343-36	14.1	0.5 x 0.5	12.4			Compact; mag 13.6 star W edge; mag 10.30v star NE 2′1.
20 37 42.1	−35 29 08	ESO 400-43	13.9	0.4 x 0.3	11.4		156	Compact.
20 45 18.5	−35 14 27	ESO 401-7	13.9	1.7 x 0.6	13.7	SAB(r)a pec:	91	Stellar nucleus; mag 9.21v star S 1′9; mag 6.82v star N 4′7.
20 59 16.7	−32 41 37	ESO 401-25	14.1	1.2 x 0.8	14.0	SAB0$^-$ pec:	120	Mag 10.55v star W 6′0.
20 59 51.2	−37 17 32	ESO 401-26	15.8	1.4 x 0.4	15.1	SB(s)dm V	24	Mag 10.15v star W 4′3.
21 03 28.1	−35 12 20	ESO 402-3	14.0	1.4 x 0.6	13.7	Sab	78	**MCG −6-46-3** SE 2′7.
21 08 07.8	−33 14 34	ESO 402-10	13.4	1.9 x 0.7	13.5	SB0? pec	147	**MCG −6-46-8** NW 1′4.
21 17 03.4	−33 59 19	ESO 402-21	14.1	1.5 x 0.5	13.6	SB(s)a? pec	150	Mag 8.27v star E 7′9.
21 22 31.5	−36 40 51	ESO 402-26	12.9	3.1 x 1.0	14.0	(R′)SB(rs)ab:	106	**ESO 402-25** NW 4′3; mag 9.38v star ESE 6′5.
21 30 34.1	−33 38 57	ESO 403-3	14.5	1.6 x 0.2	13.1	Sc	155	Mag 10.85v star SW 8′2.
21 31 28.3	−36 10 14	ESO 403-4	14.2	1.0 x 0.7	13.6	(R′)SB(s)a	154	Mag 10.04v star NW 3′2.
21 32 25.1	−34 22 23	ESO 403-6	13.7	0.9 x 0.4	12.4	Sb	141	Mag 8.48v star NE 3′1.
21 34 27.6	−33 23 11	ESO 403-9	13.6	1.4 x 0.5	13.1	Sa pec sp	97	Mag 11.6 star S 2′0.
21 39 52.0	−36 02 12	ESO 403-12	12.8	1.4 x 1.3	13.3	SAB(r)a pec	168	**MCG −6-47-10** ESE 7′9; mag 10.08v star W 3′7.
21 44 06.9	−36 10 47	ESO 403-17	14.1	1.5 x 0.6	13.8	S?	138	Mag 11.9 star on S edge.
21 46 39.6	−36 12 21	ESO 403-24	14.1	1.5 x 0.4	13.3	SABcd:	48	Star or bright knot E of nucleus; mag 11.8 star NW 8′5.
21 47 09.5	−35 47 47	ESO 403-26	13.5	0.8 x 0.5	12.4	S	122	Mag 8.37v star NW 6′7; **ESO 403-28** ESE 6′0.
21 52 39.3	−36 30 08	ESO 404-3	13.4	1.5 x 0.7	13.3	SB(rs)bc pec II-III	13	Mag 10.43v star W 3′2.
21 56 45.4	−36 29 36	ESO 404-11	14.0	1.1 x 0.6	13.4	SB(rs)ab? III-IV	132	Close pair of mag 12.1, 12.2 stars NW 5′8.
21 57 06.8	−34 35 00	ESO 404-12	12.2	2.4 x 1.9	13.7	SAB(r)c: I-II	135	Mag 11.4 star E 4′0; mag 9.95v star WNW 7′7.
22 00 29.0	−33 22 16	ESO 404-15	13.7	1.3 x 0.4	12.8	S?	147	Mag 8.66v star S 8′7.
22 00 55.5	−35 17 14	ESO 404-17	14.4	1.9 x 0.6	14.4	SB(s)dm IV-V	127	N-S oriented pair of mag 12.9, 12.7 stars S 3′5.
22 01 10.1	−32 34 46	ESO 404-18	14.8	2.6 x 0.2	13.9	SB(s)d? sp III-IV	34	Mag 10.11v star W 5′5.
22 01 23.9	−33 54 45	ESO 404-19	14.0	1.0 x 0.4	12.9	Sb	57	Anonymous galaxy NW 2′6; mag 11.8 star W 1′6.
22 02 06.4	−32 53 12	ESO 404-21	14.0	1.2 x 0.8	13.8	S?	0	Mag 10.59v star NNW 5′4.
22 02 41.7	−33 48 19	ESO 404-23	13.2	1.8 x 1.1	13.8	SB(rs)bc II	14	Mag 10.77v star W 7′2; **IC 5166** E 7′2.
22 03 48.0	−32 17 07	ESO 404-27	12.7	2.2 x 0.8	13.1	SB(s)c I-II	128	Mag 8.90v star S 4′3.
20 38 23.0	−31 55 21	ESO 463-10	14.3	1.2 x 0.5	13.6	SB(s)c I-II	51	Mag 9.85v star W 8′4.
20 48 06.3	−30 13 37	ESO 463-25	14.4	1.6 x 0.2	13.1	SBd: sp	76	Mag 9.84v star W 2′3; **PGC 65403** N 7′1.
20 48 38.3	−30 52 33	ESO 463-26	13.9	1.1 x 0.4	12.8	Sa	80	Mag 11.8 star superimposed W of nucleus.
20 58 02.5	−32 17 25	ESO 464-3	13.9	1.1 x 0.6	13.3	Sb	54	Mag 13.9 star, small anonymous galaxy, NE 2′3.
20 58 25.1	−31 28 15	ESO 464-5	14.0	1.2 x 0.9	13.9	S?	168	
21 03 00.3	−30 58 46	ESO 464-17	13.6	1.0 x 0.6	12.9	SA(s)0$^-$:	22	Mag 11.9 star E 2′4; mag 10.34v star NNW 4′6.
21 04 52.8	−29 07 07	ESO 464-21	14.0	0.9 x 0.8	13.5	(R)SB0$^+$	34	Pair of stars, mags 11.1 and 10.53v, S and SE 2′6.
21 42 32.8	−29 22 05	ESO 466-1	13.6	1.5 x 0.4	12.9	Sab sp	105	Mag 13.1 star N 0′5.
21 44 23.5	−29 54 26	ESO 466-4	14.5	1.7 x 1.3	15.2	SB(s)d IV	15	Mag 12.0 star NNE 1′9; mag 9.73v star N 6′8; **ESO 466-3** SSW 6′4.
22 01 20.3	−31 31 48	ESO 466-36	13.4	1.1 x 0.4	12.3	S?	123	Mag 9.28v star WNW 3′7; Mag 9.31v star NE 7′3; **ESO 466-37** NNE 7′3.
22 02 14.9	−31 13 16	ESO 466-43	14.1	1.5 x 0.2	12.7	Sb:	53	Pair of N-S oriented mag 10.84v, 11.08v stars N 3′5.
22 03 48.2	−31 57 22	ESO 466-51	13.3	1.5 x 0.4	12.6	S0?	65	Mag 10.06v star NW 1′8.
20 43 14.2	−29 51 10	IC 5039	12.7	2.1 x 0.6	12.8	Sbc: sp III	156	Mag 10.76v star ENE 7′0; mag 10.30v star NW 10′8.
20 43 34.5	−29 42 14	IC 5041	12.6	1.7 x 1.2	13.2	SAB(s)d IV	31	Mag 10.30v star W 10′0.
20 47 23.6	−38 25 00	IC 5049	13.3	1.0 x 0.6	12.8	E pec	168	Galaxy pair.
20 51 46.1	−29 50 54	IC 5065	13.7	1.0 x 0.7	13.2	SB0: pec	148	Mag 7.24v star N 4′2.
21 08 32.1	−29 46 10	IC 5086	12.9	1.3 x 1.3	13.3	SA0$^-$		Bright round core, faint envelope; MCG −5-50-1 NNW 8′4.
21 24 22.1	−40 32 16	IC 5105	11.6	1.9 x 1.3	12.7	E$^+$	40	Mag 10.02v star NW 5′2.
21 25 31.2	−40 16 27	IC 5105A	12.8	2.1 x 1.1	13.6	SB(rs)c I	103	Mag 9.47v star SW 4′4.
21 26 00.1	−40 50 08	IC 5105B	13.0	1.3 x 0.5	12.4	SB(rs)ab:	65	Mag 11.5 star on N edge.
21 43 11.9	−38 58 07	IC 5128	13.0	1.1 x 0.6	12.5	(R)SB0$^+$	135	Mag 11.13v star N 7′6.
21 47 25.2	−34 53 01	IC 5131	12.3	1.0 x 1.0	12.2	SB0$^-$		Mag 10.97v star SE 3′9; NGC 7130 ESE 11′8.
21 50 25.5	−30 59 42	IC 5139	12.3	2.1 x 0.9	12.9	SB0$^-$?	27	Elongated **ESO 466-9** NW 6′1.
22 03 14.9	−33 50 18	IC 5156	12.2	2.2 x 0.8	12.6	SB(s)ab pec: III	175	Mag 12.2 star E 7′4; ESO 404-23 W 7′2.
22 03 27.1	−34 56 32	IC 5157	12.7	1.3 x 1.3	13.3	E:		Mag 11.08v star ENE 2′7.
20 51 55.7	−30 53 51	MCG −5-49-5	13.5	1.1 x 0.7	13.1	Sc	72	

(Continued from previous page)

161 GALAXIES 161

RA h m s	Dec ° ′ ″	Name	Mag (V)	Dim ′ Maj x min	SB	Type Class	PA	Notes
21 08 10.9	−29 39 07	MCG −5-50-1	14.6	0.7 x 0.4	13.1	S	60	IC 5086 SSE 8′.4; mag 8.72v star NE 10′.1.
21 10 19.2	−29 31 38	MCG −5-50-4	13.9	1.5 x 0.4	13.2	(R′)SB(s)ab?	144	Mag 12.8 star N 0′.8.
21 14 28.6	−30 03 20	MCG −5-50-5	14.6	0.7 x 0.4	13.1	S	18	Mag 9.29v star ESE 6′.7.
21 14 42.1	−31 35 00	MCG −5-50-6	14.4	0.6 x 0.3	12.4	Sb-c	168	
21 16 07.0	−30 40 42	MCG −5-50-7	14.3	0.7 x 0.6	13.2	S	105	
21 20 36.1	−32 23 10	MCG −5-50-11	14.4	0.9 x 0.3	12.8	(R′?)SB(r)0/a:	72	Mag 7.48v star S 8′.8.
21 20 46.8	−30 34 44	MCG −5-50-12	14.9	0.5 x 0.3	12.9	E/S0	159	
21 20 49.9	−30 31 44	MCG −5-50-13	14.4	0.7 x 0.4	13.1	E	147	Mag 7.34v star NW 6′.5; MCG −5-50-12, MCG −5-50-14 S 2′.8.
21 20 52.4	−30 34 31	MCG −5-50-14	14.3	0.5 x 0.5	12.6	S0		
21 31 20.8	−29 15 45	MCG −5-50-16	14.7	0.7 x 0.4	13.2	Sb	63	
21 33 29.8	−29 30 35	MCG −5-50-17	14.7	0.7 x 0.5	13.4	S(r:)	0	
21 34 56.4	−29 43 50	MCG −5-51-1	14.3	0.6 x 0.4	12.6	Sc	84	Mag 6.44v star S 2′.1.
21 37 31.3	−29 26 44	MCG −5-51-4	14.5	0.5 x 0.3	12.6	E		
21 39 10.8	−29 50 07	MCG −5-51-6	14.6	0.4 x 0.4	12.7	E		
21 39 15.4	−29 52 18	MCG −5-51-7	14.1	0.7 x 0.4	12.7	S	75	
21 46 30.8	−30 01 41	MCG −5-51-13	14.7	0.4 x 0.4	12.6	S		Mag 11.04v star S 1′.3.
21 46 31.2	−29 31 34	MCG −5-51-14	14.8	0.5 x 0.4	12.9	S	33	
21 47 37.0	−29 41 42	MCG −5-51-15	13.9	0.6 x 0.6	12.9	E		
21 48 33.8	−32 10 38	MCG −5-51-16	13.7	1.1 x 1.0	13.6	Sc	85	
21 50 39.0	−29 11 03	MCG −5-51-18	13.6	0.7 x 0.4	12.1	SB0	169	
21 57 29.6	−29 23 41	MCG −5-51-27	14.5	0.6 x 0.4	12.9	S0	54	
21 58 11.5	−30 19 12	MCG −5-51-30	14.8	0.5 x 0.5	13.1	Sb:		
21 58 24.5	−32 14 16	MCG −5-51-31	13.8	1.0 x 0.5	12.9	Sa:	30	Mag 7.92v star SSE 7′.7.
21 59 56.6	−31 27 47	MCG −5-52-1	14.4	0.4 x 0.4	12.3	S0		Mag 7.06v star S 3′.9.
22 02 09.6	−30 27 33	MCG −5-52-12	14.0	0.7 x 0.6	12.9	S0/a	122	
22 02 44.0	−31 59 26	MCG −5-52-14	13.8	1.2 x 0.5	13.2	S?	18	**ESO 466-47** N 2′.1.
20 41 15.0	−32 29 12	NGC 6947	13.5	1.7 x 1.0	13.9	(R′)SB(rs)b II-III	51	Faint dark patches E and W of nucleus.
20 48 42.4	−37 59 51	NGC 6958	11.4	2.5 x 1.9	13.0	E^+	107	Mag 9.47v star NW 2′.8.
21 31 32.9	−38 37 06	NGC 7075	12.7	1.2 x 0.9	12.7	E^+?	116	ESO 343-3 W 3′.5.
21 34 33.5	−40 49 06	NGC 7087	13.0	1.1 x 0.6	12.4	SB(s)ab:	39	ESO 343-7 W 2′.7.
21 34 08.2	−36 39 13	NGC 7091	12.6	2.1 x 1.7	13.9	SB(s)dm: IV-V	78	Very patchy.
21 41 58.4	−34 26 45	NGC 7109	13.3	0.7 x 0.7	12.6	E:		
21 42 12.1	−34 09 48	NGC 7110	13.2	1.3 x 0.6	12.7	SB(r)b: III-IV	76	
21 48 19.5	−34 57 08	NGC 7130	12.1	1.5 x 1.4	12.8	Sa pec	105	Mag 10.97v star W 8′.5; **ESO 403-33** E 6′.9.
21 49 46.2	−34 52 31	NGC 7135	11.7	2.1 x 1.8	13.1	$SA0^-$ pec	47	Multi-galaxy system; mag 10.65v star N 2′.3.
21 53 59.0	−29 17 23	NGC 7152	13.6	1.0 x 0.6	12.9	SB(rs)b? pec II	17	
21 54 35.4	−29 03 53	NGC 7153	13.2	1.9 x 0.3	12.5	Sb? sp	68	
21 55 21.8	−34 48 43	NGC 7154	12.4	2.1 x 1.6	13.6	SB(s)m pec: II-III	102	Nucleus and bright knot, or two bright knots, offset S.
21 59 20.4	−31 53 02	NGC 7163	13.4	1.9 x 1.1	14.0	(R′)SB(rs)ab pec: II	101	
22 02 01.5	−31 52 14	NGC 7172	11.9	2.5 x 1.4	13.1	Sa pec sp	100	Strong dark patches E and W of center; mag 10.55v star SE 2′.7.
22 02 03.0	−31 58 25	NGC 7173	12.0	1.2 x 0.9	12.1	E^+ pec:	143	
22 02 05.8	−31 59 33	NGC 7174	13.3	2.3 x 1.2	14.3	Sab pec sp	88	NGC 7176 on E edge of bright center.
22 02 08.7	−31 59 24	NGC 7176	11.4	0.8 x 0.8	11.0	E pec:		On E edge of NGC 7174's bright center.
22 02 25.2	−35 47 23	NGC 7178	14.1	1.2 x 0.5	13.4	Sb: II	172	Located 2′.6 N of mag 8.04v star.
22 02 44.6	−32 48 15	NGC 7187	12.5	1.4 x 1.3	13.1	$(R)SA(r)0^+$	171	Moderately dark ring around bright center; pair of mag 11.34v, 12.2 stars W 5′.5.

GALAXY CLUSTERS

RA h m s	Dec ° ′ ″	Name	Mag 10th brightest	No. Gal	Diam ′	Notes
20 42 12.0	−38 20 00	A 3706	16.2	41	17	Except ESO 341-4, all members anonymous, stellar.
20 40 30.0	−40 04 00	AS 890	15.6	14	22	**MCG −7-42-17** at center; **ESO 341-3**, **3A** NE 11′.4; all others anonymous.
20 41 42.0	−37 39 00	AS 892	16.1	16	17	**ESO 341-7** ESE of center 6′.2; all others anonymous, stellar.
20 47 24.0	−38 24 00	AS 897	15.7	25	17	
20 48 42.0	−37 59 00	AS 900	14.6	12	28	Most members anonymous, stellar/almost stellar.
21 01 42.0	−38 30 00	AS 917	15.6	18	22	**MCG −7-43-11** at center.
21 02 54.0	−37 59 00	AS 918	15.5	7	22	**ESO 341-31** at center; **ESO 341-30** WSW 15′.9; **ESO 341-35/37** SE 16′.1.
21 06 54.0	−39 38 00	AS 922	15.3	24	22	Majority of members anonymous, stellar.
21 28 54.0	−39 31 00	AS 947	16.3	9	17	**ESO 342-54** NE of center 7′.0; **MCG −7-44-13** E 12′.7; **MCG −7-44-11**, **12** SE 13′.0.
21 32 06.0	−35 11 00	AS 952	16.2	23	17	**MCG −6-47-5** at center; all others anonymous, stellar.
21 43 24.0	−38 50 00	AS 964	16.2		17	Except Uranometria plotted GX, all are anonymous, stellar.

PLANETARY NEBULAE

RA h m s	Dec ° ′ ″	Name	Diam ″	Mag (P)	Mag (V)	Mag cent ★	Alt Name	Notes
21 59 35.2	−39 23 09	IC 5148−50	132	12.9	11.::	16.5	PK 2−52.1	Mag 10.35v star 1′.8 S of center.
21 05 53.7	−37 08 42	PN G6.0−41.9	8	16.7		17.5	PK 6−41.1	Located 1′.3 SW of mag 12.9 star.

162 GALAXIES 162

RA h m s	Dec ° ′ ″	Name	Mag (V)	Dim ′ Maj x min	SB	Type Class	PA	Notes
19 35 32.3	−40 01 27	ESO 338-12	14.0	0.8 x 0.6	13.1	Sb	8	Mag 10.11v star E 7′.1.
19 38 16.6	−38 53 44	ESO 338-17	13.8	1.1 x 0.8	13.6	SAB(s)ab	63	Bright core, faint envelope.
19 41 44.5	−39 04 51	ESO 338-21	13.8	0.8 x 0.5	12.7	SAB(r)0/a:	154	Mag 10.92v star N 1′.5.
19 46 44.8	−40 21 19	ESO 339-1	13.4	1.1 x 0.6	12.8	Sb	57	Mag 12.8 star W edge.
19 50 56.6	−40 12 43	ESO 339-3	13.4	1.1 x 0.8	13.1	S0	6	Pair of mag 13 stars on SW edge.
19 54 00.3	−37 59 05	ESO 339-4	14.4	1.1 x 0.3	13.0	Sbc: pec sp	168	Mag 10.47v star NNW 7′.7; stellar **ESO 339-4A** S 1′.5.
19 55 06.5	−39 22 23	ESO 339-6	13.5	1.5 x 0.8	13.6	(R′)SB(s)c II	31	Mag 8.49v star SW 4′.0.
19 56 34.9	−38 06 03	ESO 339-8	13.7	1.1 x 0.3	12.3	Sa:	82	Mag 11.8 star NE 1′.3.
19 56 37.3	−38 36 47	ESO 339-9	13.2	1.3 x 0.6	12.7	(R′)SB(r)a	4	Star or knot S edge; mag 10.27v star SW 3′.7.
19 57 37.7	−37 56 09	ESO 339-11	13.7	1.1 x 0.4	12.7	Sb	74	Mag 12.8 star superimposed E end, mag 14.4 star W end.
19 58 30.8	−40 48 58	ESO 339-12	13.0	1.3 x 1.0	13.1	SAB(rs)cd	7	Mag 12.9 star on SW edge; mag 9.56v star SW 6′.9.
20 00 10.2	−38 30 58	ESO 339-17	13.2	1.4 x 0.9	13.3	$SB(r)0^+$	70	**MCG −6-44-4**, which is involved with a star, WNW 2′.1.
20 00 35.9	−39 39 03	ESO 339-20	13.0	1.2 x 1.0	13.1	SA(rs)0/a	94	Mag 15.5 star on E edge; mag 9.72v star W 3′.7.
20 01 34.0	−38 24 46	ESO 339-25	12.7	0.7 x 0.5	11.4	SA(rs)ab:	131	Lies 0′.7 WNW of mag 8.28v star; **ESO 339-24** NW 8′.5.

(Continued from previous page)

RA h m s	Dec ° ′ ″	Name	Mag (V)	Dim ′ Maj x min	SB	Type Class	PA	Notes
20 02 03.2	−37 34 54	ESO 339-26	13.4	1.3 x 1.2	13.7	SAB(rs)ab pec II	150	Mag 9.59v star NW 4'5.
20 02 04.5	−39 00 50	ESO 339-27	14.0	1.0 x 0.6	13.3	(R′)SB(r)a	126	Mag 10.07v star W 5'0.
20 02 21.8	−38 36 54	ESO 339-28	14.3	0.9 x 0.7	13.7	Sc	125	Pair of N-S oriented mag 7.81v, 7.73v stars W 13'3.
20 04 14.6	−38 10 02	ESO 339-31	13.5	1.5 x 0.4	12.8	(R′)SB(s)a?	173	Mag 11.6 star WNW 8'0.
20 07 42.2	−39 07 40	ESO 339-34	13.8	0.9 x 0.7	13.1	SB(r)a	17	Mag 13.1 star NE 2'3.
20 11 27.2	−38 11 56	ESO 339-36	14.0	1.3 x 0.5	13.4	S?	72	Mag 13.1 star W 1'9.
20 12 39.6	−40 07 59	ESO 340-1	13.8	1.4 x 0.3	12.7	Sbc	114	Mag 11.65v star NW 2'6.
20 13 46.0	−40 03 24	ESO 340-2	13.6	1.3 x 1.0	13.7	Sc	70	Mag 11.02v star N 5'9.
20 14 15.7	−38 10 00	ESO 340-3	13.2	1.2 x 0.8	13.1	SB(s)0^-:	156	Galaxy **Tololo 2010-383** NW 2'8; mag 9.01v star S 2'1.
20 15 05.8	−40 02 49	ESO 340-5	14.1	0.8 x 0.5	13.0	Sa	163	
20 15 23.6	−39 59 54	ESO 340-6	13.7	1.0 x 0.4	12.6	(R)SB(r)0^+	40	Mag 8.79v star SE 2'3.
20 15 25.3	−37 30 43	ESO 340-7	13.3	1.2 x 1.0	13.4	S?	120	
20 17 11.5	−40 55 30	ESO 340-8	14.6	1.8 x 0.3	13.8	Scd sp	32	Mag 7.02v star W 4'1.
20 17 20.8	−38 40 27	ESO 340-9	13.9	2.5 x 0.3	13.4	Sd sp	99	Mag 11.5 star ENE 3'8.
20 18 33.4	−39 20 11	ESO 340-12	13.5	1.2 x 1.0	13.6	SAB(rs)cd: II	88	Mag 10.71v star SSW 3'9.
20 19 32.3	−40 43 33	ESO 340-16	13.1	1.2 x 0.7	12.8	(R′)SB(s)a	148	Mag 11.7 star SW 3'0.
20 19 41.1	−39 17 15	ESO 340-17	12.8	1.7 x 0.8	12.9	SB(s)dm: pec	115	Mag 9.58v star WNW 3'2.
20 24 24.5	−40 21 58	ESO 340-25	13.4	1.1 x 0.9	13.3	S0?	163	Small, faint, stellar anonymous galaxy W 0'9.
20 25 24.4	−40 09 44	ESO 340-28	14.2	1.0 x 0.4	13.1	S	135	Mag 10.67v star NW 4'2.
20 26 06.1	−40 55 54	ESO 340-29	13.1	0.9 x 0.9	12.7	SB?		Mag 10.15v star E 3'0; mag 6.09v star NNW 8'7.
20 26 56.0	−39 36 55	ESO 340-32	13.7	0.7 x 0.3	11.9		146	Mag 12.8 star SSW 2'0.
20 29 51.3	−38 58 45	ESO 340-36	13.3	1.2 x 1.0	13.4	SB(r)b	72	Mag 10.65v star S 1'1.
20 30 46.9	−39 14 57	ESO 340-40	13.9	0.9 x 0.4	12.6	S	161	Mag 8.89v star SW 1'9.
20 34 55.9	−40 37 29	ESO 340-42	13.6	1.5 x 0.9	13.7	SB(s)dm IV-V	68	Mag 9.52v star SE 1'8.
20 41 13.0	−38 11 46	ESO 341-4	12.7	1.5 x 1.0	13.0	$S0^+$: pec	66	Brightest portion at NE end.
20 41 36.6	−37 59 06	ESO 341-6	14.0	0.6 x 0.4	12.3		95	Mag 9.35v star SW 2'2.
19 24 10.8	−35 10 55	ESO 397-18	15.1	2.2 x 0.2	14.1	Sc sp	179	Mag 11.21v star NW 2'1; mag 9.89v star N 3'2.
19 36 12.2	−33 59 44	ESO 398-12	13.4	1.2 x 1.2	13.7	Sa		Mag 12.0 star S 1'9; mag 10.30v star W 9'7.
19 41 31.1	−36 20 21	ESO 398-20	14.9	1.7 x 0.3	14.0	Scd: sp	17	Mag 8.14v star W 4'2.
19 44 46.0	−33 32 01	ESO 398-27	14.5	1.4 x 0.5	13.9	SAB(r)ab	149	Mag 11.8 star S 5'2.
19 45 16.7	−34 44 57	ESO 398-29	15.3	1.5 x 0.2	13.9	Sb	118	Mag 10.43v star WSW 12'6.
19 58 00.0	−32 39 50	ESO 399-10	13.0	1.0 x 1.0	12.9	SA(s)0^- pec:		Mag 7.77v star E 7'4.
19 58 42.7	−34 24 08	ESO 399-13	14.0	1.1 x 1.0	14.0	Sc	21	Numerous stars superimposed.
20 00 35.3	−34 46 46	ESO 399-14	13.0	1.7 x 1.3	13.7	SA(s)c pec I	119	Mag 5.31v star WNW 10'3.
20 00 46.7	−34 38 01	ESO 399-15	13.8	0.9 x 0.5	12.8	S0	90	Mag 13.5 star E 2'1; mag 5.31v star WSW 11'9.
20 02 26.7	−34 54 02	ESO 399-18	13.6	0.8 x 0.5	12.4	(R′)SABa:	154	Mag 11.5 star WNW 10'3.
20 04 25.2	−36 01 01	ESO 399-19	13.6	1.5 x 0.2	12.1	S0/a	148	Mag 8.40v star E 3'1.
20 09 48.0	−37 19 50	ESO 399-23	13.2	1.3 x 0.8	13.1	SB(r)bc I	59	Mag 12.1 star WSW 6'8.
20 13 27.8	−37 11 19	ESO 399-25	13.0	2.2 x 1.0	13.7	(R′)SA(s)0/a:	154	Bright core, uniform envelope; mag 10.92v star W 2'3.
20 15 06.6	−37 14 13	ESO 399-26	13.8	1.2 x 0.6	13.3	S?	15	Mag 9.28v star SW 9'8.
20 16 23.5	−35 54 42	ESO 400-4	13.5	0.8 x 0.6	12.6	SAB(r)ab	78	Mag 9.81v star NNE 6'9.
20 16 51.5	−36 59 13	ESO 400-5	13.3	1.6 x 0.8	13.4	SA(s)c I-II	149	Small triangle of mag 10-12 stars SE 5'8; **ESO 400-2** WSW 6'2.
20 18 26.0	−35 21 54	ESO 400-6	13.7	0.8 x 0.6	12.8	Sa	66	Mag 10.16v star S 1'5.
20 19 35.8	−32 50 51	ESO 400-7	14.0	1.2 x 0.5	13.3	S?	154	Double system, interaction; mag 12.5 star W edge; mag 8.75v star SE 2'5.
20 24 02.1	−36 56 18	ESO 400-13	13.9	0.9 x 0.2	11.9	S0	144	Mag 12.5 star NE 3'2.
20 24 43.1	−34 12 19	ESO 400-15	14.3	0.5 x 0.5	12.6			Almost stellar; mag 12.9 star N 1'8; ESO 400-20 E 13'3.
20 25 40.3	−33 32 46	ESO 400-17	14.0	1.5 x 0.5	13.6	SB(s)bc: pec	109	Mag 8.00v star SSE 6'3; **MCG −6-44-34** E 2'0.
20 25 42.9	−32 32 45	ESO 400-19	13.5	1.6 x 1.2	14.0	SB(rs)c: II-III	90	Center lies 0'9 SE of mag 8.37v star.
20 25 47.6	−34 11 35	ESO 400-20	13.2	2.2 x 0.7	13.5	SA(s)b: II	160	Mag 10.18v star S edge; mag 9.30v star W 2'7.
20 27 12.4	−37 09 51	ESO 400-23	14.2	1.4 x 0.4	13.4	S	75	Mag 14.1 star superimposed E end; mag 9.76v star N 5'8.
20 27 11.1	−35 14 37	ESO 400-24	14.0	1.0 x 0.8	13.6	(R)SB(r)cd	30	Mag 12.5 star NW 3'1.
20 27 38.1	−32 57 27	ESO 400-25	14.5	1.5 x 0.8	14.5	SAB(s)dm: V	136	Mag 13.1 star S edge, mag 10.03v star N 2'0.
20 28 25.4	−33 04 28	ESO 400-28	13.4	0.8 x 0.7	12.7	SA(rs)bc	76	Mag 10.90v star W 2'0.
20 28 35.0	−36 02 40	ESO 400-30	13.0	0.6 x 0.4	11.6	E	172	Located on S edge of IC 5013.
20 30 15.5	−34 48 43	ESO 400-33	13.8	1.0 x 0.3	12.3	Sb	34	Mag 9.16v star N 7'6.
20 31 13.2	−33 28 41	ESO 400-37	13.8	1.5 x 0.7	13.7	SB(s)cd pec III-IV	66	Mag 12.7 star S 1'1; IC 5020 W 7'4.
20 32 31.5	−34 14 32	ESO 400-38	13.4	0.9 x 0.4	12.1	S? pec	6	Located between a close pair of mag 12.8 and 12.6 stars.
20 34 45.3	−35 49 16	ESO 400-40	14.5	1.1 x 0.5	13.9		165	Double system, connected.
20 35 28.1	−33 51 59	ESO 400-41	14.5	1.2 x 0.3	13.3	Im:	23	Mag 14.3 star S end; mag 14.7 star NE edge.
20 37 42.1	−35 29 08	ESO 400-43	13.9	0.4 x 0.3	11.4		156	Compact.
19 28 17.1	−29 31 47	ESO 460-4	12.5	1.4 x 1.0	12.9	E^+4	118	Mag 8.57v star W 5'8.
19 29 20.4	−32 11 35	ESO 460-8	13.5	1.1 x 0.5	12.7	SAb:	177	Elongated **ESO 460-5** WNW 12'5; mag 8.81v star W 8'9.
19 30 26.7	−31 52 41	ESO 460-9	13.6	0.9 x 0.5	12.6	(R′)SB(s)0/a	143	Mag 9.04v star N 3'3.
19 34 49.7	−31 31 45	ESO 460-13	14.7	1.2 x 0.7	14.4	SAB(s)m V	156	Mag 7.60v star SSE 5'9.
19 38 43.1	−29 48 34	ESO 460-18	12.7	2.1 x 0.8	13.1	SA(s)bc? II-III	132	Mag 9.93v star superimposed on center.
19 50 33.8	−30 52 21	ESO 461-2	13.0	1.2 x 0.7	12.8	E^+3:	145	Lies 1'2 NE of mag 8.15v star.
19 51 04.8	−30 45 26	ESO 461-5	13.3	0.9 x 0.7	12.7	S0	74	Mag 8.15v star and **MCG −5-46-8** SW 10'7; small, elongated anonymous galaxy E 5'2.
19 51 55.5	−31 58 51	ESO 461-6	14.9	1.7 x 0.2	13.6	Sc sp	15	Mag 7.11v star E 10'4.
19 52 08.9	−30 49 30	ESO 461-7	13.1	1.3 x 0.8	13.2	E2	49	= **Hickson 86A**.
19 58 07.1	−31 52 19	ESO 461-24	13.9	1.2 x 0.3	12.6	Sa: pec sp	24	NGC 6841 NW 5'2.
19 58 16.9	−31 46 19	ESO 461-25	13.8	1.6 x 0.8	13.9	SB(rs)c pec? II	27	Short chain of three N-S orients mag 13-14 stars NE 2'6; NGC 6841 WSW 6'3.
20 00 03.7	−32 05 07	ESO 461-29	13.9	1.7 x 0.4	13.3	SAB(r)ab?	28	**ESO 461-28** SSW 8'2; mag 7.44v star SE 9'2.
20 04 03.2	−29 35 45	ESO 461-37	13.9	1.3 x 0.3	12.8	S0	86	Mag 9.14v star NNE 8'2.
20 08 05.1	−29 19 10	ESO 461-44	13.9	1.5 x 0.3	12.8	Sbc: sp	9	Mag 13.6 star W 1'3; mag 12.4 star SE 5'5.
20 22 14.9	−29 51 26	ESO 462-10	14.5	1.5 x 0.6	14.2	SAB(s)c: III-IV	135	Mag 11.9 star SSE 2'5; MCG −5-48-5 ENE 4'8.
20 22 49.7	−31 33 46	ESO 462-13	14.2	1.4 x 0.8	14.2	SB(s)cd III-IV	87	Strong N-S bar; **MCG −5-48-7** NE 4'2; MCG −5-48-8 NE 8'8.
20 26 56.7	−29 07 09	ESO 462-20	14.2	0.6 x 0.3	12.2		97	Mag 10.70v star NE 1'1.
20 38 23.0	−31 55 21	ESO 463-10	14.3	1.2 x 0.5	13.6	SB(s)c I-II	51	Mag 9.85v star W 8'4.
19 56 47.7	−37 19 42	IC 4913	12.9	1.1 x 0.7	12.6	$SA0^-$	130	Mag 9.72v star S 8'6.
20 00 12.2	−38 34 43	IC 4926	12.7	1.0 x 0.9	12.7	E?	40	IC 4931 E 7'4; ESO 339-17 N 3'8.
20 00 50.4	−38 34 30	IC 4931	11.9	2.4 x 1.9	13.5	E^+:	127	Mag 7.81v star E 4'4.
20 22 10.4	−38 18 33	IC 4998	13.2	1.5 x 1.0	13.5	SB(s)c: II	105	Mag 10.03v star E 7'9.
20 28 33.7	−36 01 41	IC 5013	11.7	1.8 x 0.6	11.7	SB(s)0^0: pec	13	ESO 400-30 on S edge; mag 10.72v star E 4'6.
20 30 47.1	−36 04 40	IC 5019	14.5	0.8 x 0.3	12.8	(R′)SAB(r)bc:	91	Mag 10.19v star S 8'4.
20 30 38.4	−33 29 11	IC 5020	12.3	3.0 x 2.1	14.2	SA(s)bc II	153	Mag 11.7 star S 4'0; ESO 400-37 E 7'4.
20 43 14.2	−29 51 10	IC 5039	12.7	2.1 x 0.6	12.8	Sbc: sp III	156	Mag 10.76v star ENE 7'0; mag 10.30v star NW 10'8.
20 43 34.5	−29 42 14	IC 5041	12.6	1.7 x 1.2	13.2	SAB(s)d IV	31	Mag 10.30v star W 10'0.
19 22 07.4	−30 23 02	MCG −5-45-6	14.4	0.3 x 0.3	11.6	Sb		**ESO 459-15** ESE 1'6.
19 32 20.4	−29 23 13	MCG −5-46-2	13.3	1.1 x 0.5	12.5	Sc	164	Star or bright knot S of center.

(Continued from previous page)

162 GALAXIES 162

RA h m s	Dec ° ′ ″	Name	Mag (V)	Dim ′ Maj x min	SB	Type Class	PA	Notes
19 41 57.5	−29 03 17	MCG −5-46-4	13.2	1.3 x 1.1	13.4	SA(r)0$^-$	93	Mag 10.43v star N 4'.4.
19 51 51.8	−30 48 30	MCG −5-47-1	14.8	0.5 x 0.2	12.1	S?	159	= **Hickson 86D**.
19 51 57.5	−30 51 25	MCG −5-47-2	15.0	0.4 x 0.4	12.9	S?		= **Hickson 86C**.
19 51 59.0	−30 48 57	MCG −5-47-3	13.6	0.5 x 0.5	12.1	E2		= **Hickson 86B**.
19 56 43.9	−31 28 31	MCG −5-47-8	13.7	0.8 x 0.4	12.3	S(r)0	76	
19 57 52.1	−32 21 29	MCG −5-47-12	14.6	0.5 x 0.4	12.8	S0/a	147	Mag 7.93v star E 8'.5.
19 57 53.2	−32 27 58	MCG −5-47-13	13.6	1.1 x 0.3	12.3	S0	53	**MCG −5-47-9** W 7'.5.
20 00 19.5	−31 54 03	MCG −5-47-19	13.5	0.7 x 0.7	12.6	Sa		
20 03 36.9	−29 52 25	MCG −5-47-21	13.5	1.0 x 0.7	13.0	S0	19	
20 07 07.3	−30 23 54	MCG −5-47-22	13.4	1.2 x 0.9	13.4	(R′)SB(rs)b	176	
20 07 47.5	−29 48 45	MCG −5-47-23	13.8	1.1 x 0.9	13.6	SAB(rs)c: I-II	43	Mag 6.71v star NW 7'.7.
20 11 18.0	−28 59 47	MCG −5-47-25	13.9	1.1 x 0.8	13.6	(R′)SB(rs)ab	21	
20 19 39.7	−31 35 29	MCG −5-48-1	14.2	0.7 x 0.3	12.4		120	Mag 7.34v star S 8'.4.
20 21 51.4	−31 17 25	MCG −5-48-3	13.7	1.1 x 0.8	13.4	(R′)SB(r)a	113	Mag 9.81v star S 1'.5.
20 22 35.0	−29 49 26	MCG −5-48-5	14.3	0.4 x 0.4	12.1			ESO 462-10 WSW 4'.8.
20 23 26.5	−31 29 42	MCG −5-48-8	14.6	0.5 x 0.4	12.7	Sc	102	Pair of N-S oriented mag 13.1, 13.3 stars N 1'.8; **MCG −5-48-7** W 5'.1.
20 27 02.7	−30 31 13	MCG −5-48-13	14.6	0.5 x 0.4	12.7	Sm	165	Mag 9.86v star W 1'.1.
20 30 08.0	−32 12 40	MCG −5-48-15	14.2	0.6 x 0.3	12.2	(R_1)SB(r)0$^+$	144	
20 31 52.4	−30 47 55	MCG −5-48-18	13.8	1.2 x 0.3	12.6	S?	3	NGC 6923 SW 3'.6; mag 9.59v star E 6'.3.
20 33 09.3	−31 17 54	MCG −5-48-19	14.2	0.5 x 0.5	12.6	S0		**MCG −5-48-20** NE 1'.9.
20 33 46.5	−31 01 31	MCG −5-48-21	14.4	0.8 x 0.7	13.6	Sa	0	Bright knot or superimposed star S of center.
19 16 32.6	−40 12 37	NGC 6768	12.2	1.2 x 1.1	12.5	E4	36	**MCG −7-39-9** SW 0'.9.
19 28 03.9	−38 55 08	NGC 6794	12.9	1.7 x 1.5	13.7	SA(rs)ab:	80	Bright center, faint envelope.
19 36 45.7	−37 33 18	NGC 6805	12.8	1.1 x 0.9	12.8	E1	163	
19 57 49.1	−31 48 42	NGC 6841	12.6	1.5 x 1.4	13.4	E0:	149	Bright nucleus, uniform envelope; ESO 461-24 SE 5'.2; ESO 461-25 ENE 6'.3.
20 06 15.7	−40 11 52	NGC 6849	12.1	1.9 x 1.1	12.8	SB0$^-$:	18	
20 31 39.2	−30 49 56	NGC 6923	11.9	2.6 x 1.3	13.1	SB(rs)b: II	78	MCG −5-48-18 NE 3'.6; mag 9.59v star NE 9'.5.
20 34 20.7	−31 58 50	NGC 6925	11.3	4.5 x 1.2	13.0	SA(s)bc I-II	5	
20 41 15.0	−32 29 12	NGC 6947	13.5	1.7 x 1.0	13.9	(R′)SB(rs)b II-III	51	Faint dark patches E and W of nucleus.

GALAXY CLUSTERS

RA h m s	Dec ° ′ ″	Name	Mag 10th brightest	No. Gal	Diam ′	Notes
20 00 30.0	−38 31 00	A 3656	13.6	35	56	
20 29 12.0	−36 57 00	A 3682	16.3	66	17	**ESO 400-31**, **31A** at center; all others anonymous, stellar.
20 34 48.0	−35 49 00	A 3695	16.2	123	17	MCG −6-45-7 at center; all others anonymous, stellar.
20 42 12.0	−38 20 00	A 3706	16.2	41	17	Except ESO 341-4, all members anonymous, stellar.
19 58 00.0	−32 39 00	AS 836	14.9		22	Most members anonymous, stellar.
20 40 30.0	−40 04 00	AS 890	15.6	14	22	**MCG −7-42-17** at center; **ESO 341-3**, **3A** NE 11'.4; all others anonymous.
20 41 42.0	−37 39 00	AS 892	16.1	16	17	**ESO 341-7** ESE of center 6'.2; all others anonymous, stellar.

GLOBULAR CLUSTERS

RA h m s	Dec ° ′ ″	Name	Total V m	B ★ V m	HB V m	Diam ′	Conc. Class Low = 12 High = 1	Notes
19 28 44.1	−30 21 14	Arp 2	13.0	15.5	18.2	2.3		
19 39 59.4	−30 57 44	NGC 6809	6.3	11.2	14.4	19.0	11	= **M 55**.
19 17 43.7	−34 39 27	Terzan 7	12.0	15.0	17.9	1.2		
19 41 45.0	−34 00 01	Terzan 8	12.4	15.0	18.0	3.5		

PLANETARY NEBULAE

RA h m s	Dec ° ′ ″	Name	Diam ″	Mag (P)	Mag (V)	Mag cent ★	Alt Name	Notes
19 17 23.5	−39 36 48	IC 1297	24	10.6	10.7	14.2	PK 358−21.1	Central star, RU, a variable, brightens to mag 9.8. Pair of stars, mag 13.5 and 13.9, SSW 2'.2.
19 32 06.8	−34 12 58	PN G4.8−22.7	10	14.1	14.5		PK 4−22.1	Mag 15.1 star E 0'.8; mag 10.79v star SSW 2'.7.
19 22 10.5	−31 30 39	PN G6.8−19.8		13.4			PK 6−19.1	Stellar; mag 13.3 star E 2'.2; mag 13.6 star WNW 2'.0.

163 GALAXIES 163

RA h m s	Dec ° ′ ″	Name	Mag (V)	Dim ′ Maj x min	SB	Type Class	PA	Notes
18 25 03.1	−38 48 48	ESO 335-11	13.2	1.7 x 1.1	13.7	Sc	51	Mag 9.20v star ENE 8'.6.
18 36 07.2	−37 56 45	ESO 336-6	12.4	1.7 x 0.5	12.0	SB(r)bc II-III	135	Mag 8.14v star WNW 9'.0.
18 45 29.2	−39 10 50	ESO 336-13	13.2	2.2 x 0.9	13.8	SAb:	138	Mag 10.57v star W 6'.8.
18 22 26.4	−35 40 42	ESO 395-2	14.2	2.1 x 0.5	14.1	Sb sp	90	Close pair of mag 9.17v, 10.67v stars NW 9'.1.
18 48 40.5	−34 56 15	ESO 396-3	14.6	1.6 x 0.2	13.2	S0? sp	162	Mag 11.2 star N 3'.5.
18 54 22.2	−34 36 41	ESO 396-7	13.1	2.1 x 0.8	13.5	Sc: II	114	Mag 10.40v star N 5'.8.
19 02 35.2	−36 22 18	ESO 396-16	14.0	2.5 x 2.3	15.8	SAB(s)c? I-II	84	Small bright core, many superimposed stars; mag 8.37v star W 5'.8.
18 31 55.8	−31 47 42	ESO 457-15	13.7	1.5 x 0.7	13.6	SA(r)bc?	75	Mag 10.4 star N 1'.7.
18 59 07.6	−30 05 38	ESO 458-10	14.6	1.0 x 0.3	13.1	SBb?	0	Mag 10.65v star E 2'.5.
19 12 08.7	−32 07 33	ESO 459-6	13.1	1.7 x 0.6	13.0	SAB(s)bc?	41	Mag 13.0 star S edge; mag 9.01v star W 5'.9.
19 08 25.5	−32 13 23	MCG −5-45-2	13.7	0.5 x 0.5	12.0	S		
19 12 27.4	−29 02 37	MCG −5-45-4	14.0	0.5 x 0.4	12.1	S(r:)a:	57	
19 14 54.5	−30 59 08	MCG −5-45-5	13.6	0.6 x 0.4	11.9		15	
19 22 07.4	−30 23 02	MCG −5-45-6	14.4	0.3 x 0.3	11.6	Sb		**ESO 459-15** ESE 1'.6.
19 16 32.6	−40 12 37	NGC 6768	12.2	1.2 x 1.1	12.5	E4	36	**MCG −7-39-9** SW 0'.9.

OPEN CLUSTERS

RA h m s	Dec ° ′ ″	Name	Mag	Diam ′	No. ★	B ★	Type	Notes
19 04 06.0	−33 23 00	ESO 397-1		6			cl	
18 21 49.8	−33 11 51	Ru 140		5	15	11.0	cl	Few stars; small brightness range; not well detached.
17 56 24.0	−35 19 00	Tr 30	8.8	20	20		cl	Few stars; small brightness range; not well detached.

(Continued from previous page)

GLOBULAR CLUSTERS

RA h m s	Dec ° ′ ″	Name	Total Vm	B ★ Vm	HB Vm	Diam ′	Conc. Class Low = 12 High = 1	Notes
18 03 35.0	−30 02 02	NGC 6522	9.9	14.1	16.9	9.4	6	
18 04 49.6	−30 03 21	NGC 6528	9.6	15.5	17.1	5.0	5	
18 10 17.6	−31 45 47	NGC 6558	8.6		16.7	4.2		
18 13 38.9	−31 49 35	NGC 6569	8.4		17.1	6.4	8	
18 23 40.5	−30 21 40	NGC 6624	7.6	14.0	16.1	8.8	6	
18 31 23.2	−32 20 53	NGC 6637	7.7	13.7	15.9	9.8	5	= **M 69**.
18 35 45.7	−32 59 25	NGC 6652	8.5	13.3	15.9	6.0	6	
18 43 12.7	−32 17 31	NGC 6681	7.8	13.0	15.6	8.0	5	= **M 70**.
18 55 03.3	−30 28 42	NGC 6715	7.7	15.2	18.2	12.0	3	= **M 54**.
18 59 33.2	−36 37 54	NGC 6723	6.8	12.8	15.5	13.0	7	
19 17 43.7	−34 39 27	Terzan 7	12.0	15.0	17.9	1.2		

BRIGHT NEBULAE

RA h m s	Dec ° ′ ″	Name	Dim ′ Maj x min	Type	BC	Color	Notes
19 01 06.0	−37 04 00	IC 4812	10 x 7	R	2-5	1-4	Bright on blue but very faint on red photos. Immersed in the nebulosity is a bright pair of 6.5 mag stars separated by 13″ rough E-W.
19 01 42.0	−36 53 00	NGC 6726/27	9 x 7	R	1-5		High surface brightness nebula surrounding two fairly bright stars; illuminated by the variable star TY Coronae Australis.
19 01 54.0	−36 57 00	NGC 6729	1 x 1	E+R	2-5		Comet-like ellipse about 2′ long. The variable star R Coronae Australis (9.7–13.5) is near the NW apex.

DARK NEBULAE

RA h m s	Dec ° ′ ″	Name	Dim ′ Maj x min	Opacity	Notes
18 04 10.6	−33 20 00	B 87	12 x 12	4	**Parrot's Head**. Several smaller stars in it.
17 57 06.4	−37 05 19	B 288	2 x 2		Black; diffuses toward NE.
17 56 36.8	−29 01 00	B 289	35 x 7	4	Dark space between star clouds.
17 59 20.7	−37 09 00	B 290	3 x 3		Very small.
17 59 43.4	−33 54 00	B 291	5 x 5		Small, round, black.
18 00 34.1	−33 21 00	B 292	60 x 60		Irregular, broken, dark region.
18 01 12.6	−35 21 00	B 293	18 x 18		Dusky; curved; like an inverted U.
18 04 05.2	−32 00 00	B 295	50 x 50	4	Irregular, dark region.
18 05 11.0	−30 06 00	B 298	4 x 4		Small; 5′ SE of NGC 6528.
18 07 01.3	−32 39 00	B 300		4	Broken dark region about 30′ E of B87.
18 14 37.7	−31 48 00	B 305	13 x 13		Dark; irregular.
19 02 54.0	−37 08 00	Be 157	110 x 28	6	
19 10 20.0	−37 08 00	SL 42	12 x 7		

PLANETARY NEBULAE

RA h m s	Dec ° ′ ″	Name	Diam ″	Mag (P)	Mag (V)	Mag cent ★	Alt Name	Notes
19 17 23.5	−39 36 48	IC 1297	24	10.6	10.7	14.2	PK 358−21.1	Central star, RU, a variable, brightens to mag 9.8. Pair of stars, mag 13.5 and 13.9, SSW 2′.2.
18 45 50.9	−33 20 36	IC 4776	18	11.7	10.8	14.1	PK 2−13.1	Mag 13.1 star S 0′.9.
18 12 02.5	−33 52 06	NGC 6563	48	13.8	11.0	17.3	PK 358−7.1	Mag 12.4 star N 0′.9; mag 11.9 star NW 1′.3.
18 13 18.0	−32 19 42	PN G0.0−6.8	25			14.9	PK 359−6.1	Mag 12.9 star on NW edge; mag 10.16v star N 1′.3.
18 08 28.0	−31 36 31	PN G0.1−5.6	12	17.4		21.0	PK 0−5.1	Mag 10.07v star NE 2′.5; mag 11.6 star SSW 1′.3.
18 04 44.0	−31 02 48	PN G0.2−4.6	6				PK 0−4.3	
17 57 43.8	−30 02 27	PN G0.3−2.8		18.7			PK 0−2.5	Mag 11.8 star SW 0′.4.
18 05 02.7	−30 58 17	PN G0.3−4.6	5	15.4			PK 0−4.1	Close pair of mag 12 stars S 2′.8.
17 58 19.3	−30 00 39	PN G0.4−2.9	6	16.4		16.9	PK 0−2.6	Mag 9.78v star SE 2′.5.
17 59 15.7	−30 02 48	PN G0.5−3.1	8	19.0		17.3	PK 0−3.2	Small triangle of mag 12 stars S 1′.2.
17 56 39.9	−29 37 23	PN G0.6−2.3					PK 0−2.3	Stellar; small triangle of mag 10-11 stars W 2′.6.
17 58 09.6	−29 44 20	PN G0.7−2.7	7	13.5	13.6	14.7	PK 0−2.4	Mag 12.6 star on SE edge; mag 10.1 star NW 2′.0.
18 02 19.2	−30 14 26	PN G0.7−3.7	7	15.1	15.1	17.8	PK 0−3.1	Mag 12.1 star NNE 2′.0; mag 10.3 star SE 2′.8.
18 17 37.2	−31 56 46	PN G0.7−7.4	5	15.3	14.9		PK 0−7.1	Mag 10.01v star NNE 2′.4.
18 18 37.4	−31 54 45	PN G0.8−7.6	4	15.7	15.5		PK 0−7.2	Short line of three stars, mags 10.91v, 12.9 and 13.1, W 1′.6.
17 56 02.8	−29 11 16	PN G0.9−2.0	5	18.0		16.2	PK 0−2.1	Mag 10.6 star WSW 3′.2; mag 12.4 star E 3′.3.
18 07 06.1	−30 34 17	PN G0.9−4.8	11	13.8	14.3		PK 0−4.2	Close pair of mag 11-12 stars N 1′.4; mag 12.2 star W 1′.1.
17 58 26.1	−29 20 46	PN G1.0−2.6	5	17.9			PK 1−2.2	Mag 12.2 star E 0′.8; mag 11.3 star NE 2′.4.
18 00 37.7	−29 21 50	PN G1.2−3.0	5			16.2	PK 1−3.1	Mag 12.0 star E 0′.9; mag 11.2 star S 1′.2.
18 03 53.6	−29 51 21	PN G1.2−3.9	4	15.8		16.5	PK 1−3.7	Mag 12.4 star E 0′.9; mag 8.04v star W 3′.7.
18 02 27.0	−29 25 13	PN G1.4−3.4	13	18.2		15.8	PK 1−3.3	Mag 12.5 star E 0′.8; mag 12.4 star S 1′.1.
18 16 12.3	−30 52 09	PN G1.5−6.7	14	12.0	13.0	11.7	PK 1−6.2	Mag 12.7 star on SE edge; mag 11.6 star N 0′.6.
18 07 14.6	−29 41 24	PN G1.7−4.4	5	17.7		16.5	PK 1−4.1	Mag 11.9 star NW 0′.9; mag 11.3 star WSW 1′.0.
18 07 54.0	−29 44 36	PN G1.7−4.6	3	14.4	14.5	17.0	PK 1−4.2	Located 0′.8 S of mag 8.15v star.
18 05 05.6	−29 20 13	PN G1.8−3.8	12	16.2		14.7	PK 1−3.9	Close pair of mag 11.5, 12.5 stars NW 0′.6.
18 15 06.5	−30 15 33	PN G2.0−6.2	6	13.1	14.0	14.4	PK 2−6.1	Located in a small triangle of stars with a mag 10.28v star NW 0′.7.
18 07 07.3	−29 13 06	PN G2.1−4.2	7	14.1	14.0	15.4	PK 2−4.1	Close pair of mag 10.5, 12.8 stars E 1′.0.
18 16 19.4	−30 07 37	PN G2.2−6.3	7	12.8		14.7	PK 2−6.2	Mag 9.10v star on NW edge.
18 29 11.6	−31 29 59	PN G2.2−9.4	12	12.6	11.9	15.2	PK 2−9.1	Mag 9.22v star SSW 1′.4; mag 8.18v star SSE 2′.9.
18 22 34.6	−30 43 30	PN G2.3−7.8	14	14.7	13.8		PK 2−7.1	Located mid-point between a pair of N-S mag 11.4 stars.
18 13 00.5	−29 25 12	PN G2.5−5.4	17	14.6		17.7	PK 2−5.2	Mag 11.0 star N 0′.5; mag 9.4 star E 2′.4.
18 17 41.4	−29 08 20	PN G3.2−6.2	8	13.0		15.7	PK 3−6.1	Mag 11.6 star S 1′.6; mag 10.00v star W 2′.7.
18 23 08.9	−29 43 25	PN G3.3−7.5	8	16.7		17.0	PK 3−7.1	Mag 11.9 star NW 3′.4; mag 11.6 star SW 4′.6.
19 05 35.9	−33 11 37	PN G3.8−17.1	5	13.4	12.5		PK 3−17.1	Mag 12.8 star ESE 1′.8; mag 10.7 star W 5′.5.
18 55 37.8	−32 15 48	PN G3.9−14.9	4	10.9		13.9	PK 3−14.1	Mag 13.1 star N 0′.6; mag 10.5 star WNW 2′.3.
18 39 25.8	−30 40 36	PN G4.0−11.1	8	13.1	13.8	15.5	PK 4−11.1	Located in the center of a small triangle of mag 11-12 stars.
18 44 14.6	−30 19 36	PN G4.7−11.8	13	15.1	14.0	15.6	PK 4−11.2	Close pair of mag 12 stars NE 2′.0; mag 10.35v star W 2′.6.
19 14 23.4	−32 34 18	PN G5.2−18.6	5	16.1			PK 5−18.1	Mag 13.0 star W 0′.5; mag 13.2 star E 1′.8.
19 22 10.5	−31 30 39	PN G6.8−19.8		13.4			PK 6−19.1	Stellar; mag 13.3 star E 2′.2; mag 13.6 star WNW 2′.0.
18 00 11.8	−38 49 53	PN G352.9−7.5	2	11.4		14.3	PK 352−7.1	Forms a triangle with a mag 10.40v star WSW 0′.8 and a mag 12.3 star SSE 0′.5.
18 26 41.4	−40 29 51	PN G353.7−12.8	30	15.5			PK 353−12.1	Close pair of mag 11.3, 12.9 stars N 1′.3.
18 04 57.5	−37 38 08	PN G354.4−7.8	12	14.7			PK 354−7.1	Mag 13.0 star E 1′.2; mag 12.6 star WSW 1′.1.
18 02 32.2	−36 39 11	PN G355.1−6.9	5	11.7	12.7	15.3	PK 355−6.1	Mag 11.3 star W 1′.1; mag 8.94v star S 1′.5.
18 02 31.5	−35 13 12	PN G356.3−6.2	10	16.5			PK 356−6.1	Mag 11.2 star S 2′.0.

(Continued from previous page)

PLANETARY NEBULAE

RA h m s	Dec ° ′ ″	Name	Diam ″	Mag (P)	Mag (V)	Mag cent ★	Alt Name	Notes
8 09 48.9	−35 44 11	PN G356.6−7.8	13	15.9			PK 356−7.2	Mag 9.78v star E 2.′4; mag 11.0 star S 2.′0.
7 57 19.1	−34 09 49	PN G356.7−4.8	10	13.0		16.2	PK 356−4.2	Mag 11.1 ESE 0.′9; mag 10.67v star NNW 1.′0.
8 04 29.1	−34 58 00	PN G356.7−6.4	13	16.7			PK 356−6.2	Mag 11.6 star NW 2.′2; mag 10.1 star NE 3.′0.
8 00 18.6	−34 27 40	PN G356.8−5.4	11	16.0			PK 356−5.1	Mag 11.2 star E 2.′2.
8 27 50.6	−37 15 55	PN G356.8−11.7	104			16.1	PK 356−11.1	Mag 12.1 star W 2.′1; mag 10.62v star E 2.′0.
8 02 02.8	−34 27 46	PN G356.9−5.8	7	14.6	14.6		PK 356−5.2	Mag 12.1 star E 1.′7; mag 11.2 star ESE 2.′0.
7 58 14.5	−33 47 37	PN G357.1−4.7	2			15.4	PK 357−4.3	Mag 12.8 star WSW 2.′1; mag 12.2 star ENE 2.′5.
8 04 05.0	−34 28 38	PN G357.1−6.1	4	15.9			PK 357−6.1	Mag 10.9 star E 2.′5.
7 57 25.2	−33 35 43	PN G357.2−4.5	6	14.3	13.3	17.3	PK 357−4.1	Line of three mag 13.1-13.7 stars W 1.′2.
7 58 32.6	−33 28 37	PN G357.4−4.6	5	15.2		16.5	PK 357−4.2	Mag 12.0 star W 0.′4; mag 12.6 star SW 0.′6.
7 56 13.4	−32 37 27	PN G357.9−3.8	11	17.5		21.0	PK 358−3.2	Mag 9.18v star N 2.′3.
8 01 22.2	−33 17 42	PN G357.9−5.1	11	14.9	13.3		PK 357−5.1	
8 01 42.8	−33 15 27	PN G358.0−5.1	9	16.5			PK 358−5.1	Mag 12.0 star NW 2.′6.
7 59 02.4	−32 21 43	PN G358.5−4.2	1	14.2	14.2	16.2	PK 358−4.1	Mag 10.0 star NE 2.′4.
8 04 56.6	−32 54 03	PN G358.6−5.5	9	17.1			PK 358−5.4	Mag 12.5 star E 0.′4; mag 11.3 star ENE 2.′2.
8 03 53.4	−32 41 42	PN G358.7−5.2	10	14.5	13.5	17.1	PK 358−5.3	Mag 10.41v star SW 1.′4; mag11.5 star ENE 2.′4.
7 58 10.7	−31 42 57	PN G358.9−3.7	4	18.6			PK 358−3.1	Mag 11.1 star N 0.′6; mag 12.3 star 2.′1.
7 59 56.9	−31 54 27	PN G359.0−4.1	5	17.8		21.0	PK 359−4.1	Mag 12.4 star N 1.′6; mag 12.5 star WSW 1.′4.
8 02 46.5	−32 09 28	PN G359.0−4.8	14	15.0	14.5		PK 359−4.3	Mag 10.9 star SSW 0.′7; mag 11.1 star S 2.′6.
7 56 25.5	−31 04 16	PN G359.3−3.1	5	18.1		17.1	PK 359−3.1	Mag 11.8 star NW 2.′5; mag 12.4 star SE 2.′4.
7 58 12.5	−31 08 06	PN G359.4−3.4	6	18.2		21.0	PK 359−3.2	Mag 10.03v star E 0.′9; mag 11.1 star NW 1.′9.
8 04 07.8	−31 39 11	PN G359.6−4.8	10	16.7			PK 359−4.4	Mag 8.08v star E 1.′7.
8 02 54.1	−31 23 47	PN G359.7−4.4	14	16.6		18.7	PK 359−4.5	Mag 12.3 star N 3.′2.
8 14 50.7	−32 36 55	PN G359.8−7.2	10	12.4	13.7	15.7	PK 359−7.1	Close pair of mag 12.4, 13.2 stars WSW 2.′4.
8 03 52.5	−31 17 46	PN G359.9−4.5	5	14.0	14.2		PK 359−4.2	Mag 12.4 star SE 0.′5; mag 13.0 star N 1.′0.
8 07 19.3	−31 42 54	PN G359.9−5.4	12	16.4			PK 359−5.1	Mag 10.4 star SE 3.′2; mag 9.84v star N 3.′2.

GALAXIES

RA h m s	Dec ° ′ ″	Name	Mag (V)	Dim ′ Maj x min	SB	Type Class	PA	Notes
6 47 02.7	-29 05 29	ESO 453-4	15.6	1.5 x 0.2	14.1	S	20	Mag 10.36v star SW 2.′1.

OPEN CLUSTERS

RA h m s	Dec ° ′ ″	Name	Mag	Diam ′	No. ★	B ★	Type	Notes
7 28 54.0	−31 33 00	Ant 1		35	24	7.0	cl	(A) In 35′ scattered group. 10′ long straggling group just to SW of center.
7 29 42.0	−32 29 00	Ant 2	8.8	3	8	9.7	cl	Probably not a cluster.
7 30 34.0	−32 13 00	Ant 3		21	14		cl	(A) Involved with RCW 132 = **Gum 67**.
7 32 39.0	−32 57 24	Ant 4		3.5	9	10.2	cl	(A) Originally called **NGC 6374** by Antalov, named Ant 4 here.
7 52 28.4	−30 06 23	Bas 5		5	30		cl	
7 17 20.7	−35 33 00	Bo 13	7.2	15	35	8.0	cl	Moderately rich in stars; large brightness range; no central concentration; detached. (A) Mostly two brighter stars, with a few fainter background stars.
6 55 30.0	−40 49 00	Cr 316	3.4	105		14.0	cl	Moderately rich in stars; moderate brightness range; strong central concentration; detached. (A) Large group of bright stars, superposed on Tr 24. May just be brighter stars of Tr 24?
7 30 48.0	−37 04 00	Cr 332	8.9	2	12		cl	Few stars; small brightness range; not well detached.
7 31 31.3	−34 01 00	Cr 333	9.8	8	30		cl	Moderately rich in stars; moderate brightness range; slight central concentration; detached.
7 38 12.0	−37 33 00	Cr 338	8.0	20	40		cl	Moderately rich in stars; moderate brightness range; no central concentration; detached. (A) Scattering of bright stars in rich Milky Way.
7 44 35.0	−33 52 00	Cr 345	10.9	4.8	30	9.4	cl?	Probably not a cluster. (A) An elongated N-S group in Milky Way star cloud.
7 46 18.9	−29 20 00	Cr 347	8.8	10	40	10.6	cl	Moderately rich in stars; moderate brightness range; slight central concentration; detached; involved in nebulosity.
7 04 54.0	−38 19 00	ESO 332-20		7			cl	
7 07 30.0	−40 49 00	ESO 332-22		7			cl	
7 26 52.0	−34 41 00	ESO 392-13		10	25		ast	(A) Sparse group.
7 53 54.0	−32 29 00	ESO 456-9		6			cl	
7 31 03.0	−36 48 00	Ha 16		15	70	8.5	cl	Rich in stars; moderate brightness range; no central concentration; detached. (A) Scattered group standing out some from Milky Way.
7 19 00.5	−38 48 52	HM 1		6	20		cl	
6 55 30.6	−39 28 00	NGC 6242	6.4	9	23	7.3	cl	Moderately rich in bright and faint stars with a strong central concentration; detached. (A) Dark nebula on E side.
7 02 04.1	−39 43 16	NGC 6268	9.5	6			cl	Few stars; moderate brightness range; slight central concentration; detached.
7 04 47.2	−37 53 00	NGC 6281	5.4	8	70	9.0	cl	Few stars; moderate brightness range; weak central concentration; detached.
7 16 13.4	−39 25 26	NGC 6318	11.8	5	35	12.0	cl	Few stars; moderate brightness range; no central concentration; detached.
7 34 42.2	−32 35 00	NGC 6383	5.5	20	40	10.2	cl	Moderately rich in stars; large brightness range; weak central concentration, involved in nebulosity; detached.
7 37 38.0	−35 01 42	NGC 6396	8.5	3	30	9.8	cl	Moderately rich in bright and faint stars; detached. (A) Compact cluster, with faint extensions well off to N and NE.
7 40 10.0	−36 58 00	NGC 6400	8.8	12	60	9.0	cl	Moderately rich in stars; moderate brightness range; slight central concentration; detached.
7 39 36.0	−33 14 00	NGC 6404	10.6	6	50		cl	Moderately rich in stars; moderate brightness range; no central concentration; detached.
7 40 17.0	−32 16 00	NGC 6405	4.2	33	331	5.5	cl	= **M 6**, **Butterfly Cluster**. Rich in stars; large brightness range; slight central concentration; detached. (A) Variable in brightness. Brightest star is BM Scorpii, V = 5.5 to 7.
7 44 18.0	−32 21 00	NGC 6416	5.7	15	304	8.4	cl	Moderately rich in stars; moderate brightness range; no central concentration; detached. (A) Large scattered cluster, not obvious on DSS image.
7 46 53.0	−31 37 00	NGC 6425	7.2	10	73	10.1	cl	Moderately rich in stars; small brightness range; slight central concentration; detached.
7 49 34.1	−34 49 00	NGC 6444		12	50	11.0	cl	Few stars; small brightness range; not well detached.
7 50 41.0	−30 13 00	NGC 6451	8.2	8	80	12.0	cl	Rich in stars; moderate brightness range; strong central concentration; detached; involved in nebulosity.
7 53 46.0	−34 47 00	NGC 6475	3.3	75	80	7.0	cl	= **M 7**. Rich in stars; large brightness range; strong central concentration; detached. (A) 50′ diameter, 75′ diameter with outliers.
7 24 43.4	−34 12 23	Pi 24	9.6	2	15	10.4	cl	Few stars; moderate brightness range; no central concentration; detached; involved in nebulosity.
6 55 08.3	−40 56 35	Ru 122		3			cl	
7 23 24.0	−37 55 00	Ru 123		10	50	10.0	cl	Moderately rich in stars; large brightness range; no central concentration; detached. (A) Sparse group, brightest star on S side.
7 27 55.9	−40 43 12	Ru 124		2	15	12.0	cl	
7 35 04.4	−34 18 33	Ru 126		6	20	13.0	cl	(A) Sparse group.
7 37 50.4	−36 18 25	Ru 127	8.8	5	20	11.0	cl	Few stars; moderate brightness range; slight central concentration; detached.
7 44 19.3	−34 53 12	Ru 128		5	45	13.0	cl	
7 47 08.2	−29 36 05	Ru 129		4	10	12.0	cl	
7 47 32.0	−30 05 48	Ru 130		3	20	14.0	cl	
7 49 06.9	−29 14 00	Ru 131		8	15	11.0	cl	Few stars; small brightness range; not well detached.
7 52 48.0	−29 33 00	Ru 134		5	30	12.0	cl	

(Continued from previous page)

OPEN CLUSTERS

RA h m s	Dec ° ′ ″	Name	Mag	Diam ′	No. ★	B ★	Type	Notes
16 57 00.0	−40 38 00	Tr 24	8.6	60			cl	Few stars; moderate brightness range; not well detached; involved in nebulosity. (A) Large group of stars, with IC 4628 nebulosity involved on N side.
17 24 30.3	−39 01 16	Tr 25	11.7	4	40		cl	
17 28 29.9	−29 30 00	Tr 26	9.5	7	40		cl	Moderately rich in stars; small brightness range; slight central concentration; detached.
17 36 12.7	−33 29 00	Tr 27	6.7	7	82	8.4	cl	Moderately rich in stars; large brightness range; no central concentration; detached.
17 36 59.0	−32 28 24	Tr 28	7.7	6	85	9.8	cl	Moderately rich in stars; moderate brightness range; no central concentration; detached; involved in nebulosity.
17 41 36.4	−40 07 00	Tr 29	7.5	12	30		cl	Moderately rich in stars; moderate brightness range; no central concentration; detached.
17 56 24.0	−35 19 00	Tr 30	8.8	20	20		cl	Few stars; small brightness range; not well detached.
16 56 11.3	−40 40 06	vdB-Ha 205		4	14	6.7	cl	
17 05 50.3	−36 33 21	vdB-Ha 214		4	6		ast?	(A) Appears to be asterism of just 6 stars.
17 16 20.4	−40 49 50	vdB-Ha 217		5	30		cl	
17 18 34.0	−32 22 00	vdB-Ha 221		10	20		cl?	(A) Not a real cluster?
17 18 47.1	−38 17 24	vdB-Ha 222		2	20		cl	
17 20 42.0	−35 53 00	vdB-Ha 223		7	16		cl?	(A) Sparse group, elongated NE-SW, with stream of faint stars continuing to SW.
17 31 55.7	−31 54 36	vdB-Ha 231		2	15		cl	
17 46 14.4	−29 42 05	vdB-Ha 245		1	12		cl	

STAR CLOUDS

RA h m s	Dec ° ′ ″	Name	Mag	Diam ′	No. ★	B ★	Type	Notes
17 25 23.9	−30 00 17	NGC 6360					MW	
17 42 54.0	−35 08 18	NGC 6415		23			MW	(A) Identity (J. Herschel's h 3701) apparently refers to patch ("projection") of Milky Way here.
17 45 14.0	−33 41 12	NGC 6421		45			MW	(A) Confirmed against J. Herschel's drawing.
17 49 00.0	−35 51 06	NGC 6437		40			MW	(A) Star cloud with dark nebulae on E, W, and SW sides.
17 51 48.0	−35 10 30	NGC 6455		58			MW	(A) Milky Way star cloud, with dark nebula on NE, SE, and SW.
17 54 12.0	−29 07 12	NGC 6476		20			MW	(A) A small Milky Way star cloud as described by J. Herschel (his h 3712). Dark nebula on E side.
17 54 34.7	−30 26 29	NGC 6480		17		12.0	MW	(A) "V" shaped eastward pointing star cloud, **LDN 1788** dark nebula on eastern side.

GLOBULAR CLUSTERS

RA h m s	Dec ° ′ ″	Name	Total V m	B ★ V m	HB V m	Diam ′	Conc. Class Low = 12 High = 1	Notes
17 47 28.3	−33 03 56	Djorg 1	13.6					
17 31 05.2	−29 58 54	HP 1	12.5	16.0	18.6	1.2		
17 33 24.5	−33 23 20	Liller 1	15.8					
16 59 32.6	−37 07 17	NGC 6256	11.3	15.3	18.2	4.1		
17 01 12.6	−30 06 44	NGC 6266	6.4	13.2	16.3	15.0	4	= **M 62**.
17 14 32.5	−29 27 44	NGC 6304	8.3	14.5	16.2	8.0	6	
17 34 28.0	−39 04 09	NGC 6380	11.5	17.0	19.5	3.6		A mag 8.5 star is projected on southern edge.
17 50 12.9	−37 03 04	NGC 6441	7.2	15.4	17.1	9.6	3	
17 50 51.8	−34 35 55	NGC 6453	10.2	14.3	17.5	7.6	4	
18 03 35.0	−30 02 02	NGC 6522	9.9	14.1	16.9	9.4	6	
17 35 47.0	−30 28 46	Terzan 1	15.9	18.5	21.4	2.4		
17 27 33.4	−30 48 08	Terzan 2	14.3		19.8	0.6	9	
17 30 38.9	−31 35 44	Terzan 4	16.0		21.6	0.7		
17 50 46.4	−31 16 31	Terzan 6	13.9	20.5	22.3	1.4		
17 36 10.5	−38 33 12	Ton 2	12.2		18.2	2.2		

BRIGHT NEBULAE

RA h m s	Dec ° ′ ″	Name	Dim Maj x min	Type	BC	Color	Notes
16 57 00.0	−40 20 00	IC 4628	90 x 60	E	2-5	2-4	= **Gum 56**. Elongated and very irregular in shape.
17 20 24.0	−35 51 00	NGC 6334	35 x 20	E	2-5	3-4	= **RCW 127**. A series of dark lanes divides this nebular complex into four larger (7′–10′) sections roughly centered around a smaller (2′–3′) cometary shaped condensation.
17 24 42.0	−34 12 00	NGC 6357	25 x 25	E	2-5	3-4	Considerably mottled and very irregular in shape. On red photos it is vaguely reminiscent of a fainter version of the Orion nebula.
17 16 54.0	−36 21 00	RCW 126	16 x 4	E	4-5	3-4	Very faint and diffuse, extended ENE-WSW; irregular surface brightness.
17 34 42.0	−32 35 00	RCW 132	80 x 30	E	3-5	3-4	In an area extremely rich in faint stars. The NW section contains several small patches of dark nebulosities.
17 12 18.0	−38 29 00	Sh2-3	12 x 8	E	2-5		Egg-shaped object extended NNW-SSE with an 8th mag star in the N part. The northern two-thirds of the object is for the most part obscured by absorbing matter, especially on the E side. Brightest part is a 6′ x 3′ crescent on S end, centered roughly 6′ SSE of the involved star.
17 29 06.0	−31 33 00	Sh2-13	40 x 35	E	3-5	3-4	The entire region surrounding Sh 2-13 is littered with faint nebulosities and is extremely rich in faint stars.
17 46 36.0	−29 18 00	Sh2-16	12 x 12	E	3-5	4-4	In a rich star field. There is a small patch of nebulosity 3′ across, centered about 5′ off the NE edge of Sh 2-16.

DARK NEBULAE

RA h m s	Dec ° ′ ″	Name	Dim ′ Maj x min	Opacity	Notes
16 44 45.0	−40 20 00	B 44a	5 x 5	5	Irregular; sharpest on SE side.
17 01 42.0	−40 41 00	B 48	40 x 15	5	Fairly well defined NE and SW.
17 02 39.5	−33 16 00	B 49		3	Small.
17 02 54.1	−34 23 00	B 50	15 x 15	6	Large, irregular dark space.
17 06 11.3	−33 35 00	B 53	30 x 10	4	Large; diffuse; curved toward the E at N end.
17 06 34.0	−34 15 00	B 54	5 x 5	5	Small, round; close S of 11 mag star.
17 07 33.1	−32 00 00	B 55	16 x 16	5	Irregular.
17 08 48.6	−32 06 00	B 56	3 x 3	5	Small.
17 11 12.2	−40 25 00	B 58	30 x 20	6	Slightly extended N and S; a darker core at NE side.
16 38 24.3	−35 25 00	B 231	50 x 40	6	Dusky; sharpest on W side, diffuses to the E.
16 43 44.4	−39 49 00	B 232	10 x 10		Dusky; N and S; two small stars in it.
16 44 45.4	−35 24 00	B 233	55 x 20	5	Diffused; N and S.
16 46 26.5	−30 29 00	B 234	30 x 6	1	Narrow.
16 48 54.4	−29 48 00	B 236			Center of system of indistinct dark lanes.
16 48 55.0	−29 58 00	B 237	30 x 6	1	Dark lane N of magnitude 8 star.
16 55 00.1	−31 07 00	B 239	15 x 2.5	1	Curved; very black and narrow.
16 59 17.6	−35 22 00	B 240	20 x 20		Black.
16 59 27.2	−30 12 00	B 241	18 x 6	2	Dusky; diffused; about 20′ W of M62.

(Continued from previous page)

RA h m s	Dec ° ′ ″	Name	Dim ′ Maj x min	Opacity	Notes
17 05 06.5	−32 26 00	B 242	30 x 8		Dusky; elongated E and W.
17 09 46.1	−29 35 00	B 243	25 x 25	3	Very irregular.
17 11 55.6	−29 24 00	B 245	8 x 8	3	Round; diffused; several bright stars in it
17 13 06.9	−30 15 00	B 247	4 x 4	3	Black.
17 13 04.8	−29 09 00	B 249	5.0 x 0.3	5	Very thin, short, black line.
17 15 18.5	−32 09 00	B 252	20 x 5	5	Triangular; eastern side diffuse.
17 20 29.2	−30 08 00	B 254	60 x 29	5	Irregular; diffused; elongated E and W.
17 22 46.7	−35 38 00	B 257	10 x 7	5	Dusky spot.
17 22 57.8	−34 43 00	B 258	40 x 40		Irregular area of dark markings.
17 34 16.8	−34 15 00	B 271	120 x 10		Dusky; curved; elongated N and S.
17 38 29.3	−33 21 00	B 273	15 x 15		Dark spot at N end of B271.
17 38 59.1	−32 20 00	B 275	13 x 13	4	Round; about 20′ W of M6.
17 42 39.3	−32 19 00	B 278	15 x 15	4	Round; about 30′ E of M6 .
17 51 16.1	−33 52 00	B 283	90 x 60	5	Irregular, dusky area, long axis E and W.
17 53 03.6	−35 37 00	B 286	15 x 15		Diffused.
17 54 26.8	−35 12 00	B 287	25 x 15	5	Irregular, semi-vacancy; SE of M7.
17 57 06.4	−37 05 19	B 288	2 x 2		Black; diffuses toward NE.
17 56 36.8	−29 01 00	B 289	35 x 7	4	Dark space between star clouds.
17 59 20.7	−37 09 00	B 290	3 x 3		Very small.
17 59 43.4	−33 54 00	B 291	5 x 5		Small, round, black.
18 00 34.1	−33 21 00	B 292	60 x 60		Irregular, broken, dark region.
18 01 12.6	−35 21 00	B 293	18 x 18		Dusky; curved; like an inverted U.
17 20 42.0	−31 57 00	LDN 1710	10 x 8	5	

PLANETARY NEBULAE

RA h m s	Dec ° ′ ″	Name	Diam ″	Mag (P)	Mag (V)	Mag cent ★	Alt Name	Notes
17 05 10.5	−40 53 09	IC 4637	22	13.6	12.5	12.5	PK 345+0.1	Forms a triangle with a pair of mag 11.37v and 11.8 stars S 2′4.
17 13 44.3	−37 06 13	NGC 6302	89	12.8	9.6	21.1	PK 349+1.1	**Bug Nebula**. Very much elongated with a PA of 90 degrees; brightest portion, with central star, actually slightly E of center.
17 22 15.6	−38 29 02	NGC 6337	51	11.9	12.3	14.9	PK 349−1.1	Ring shape with dark center, central star offset to NE; mag 9.89v star SE 3′5.
17 50 24.2	−29 25 18	PN G0.1−1.1	4			21.0	PK 0−1.1	Small triangle of mag 12.3, 12.9 and 13.0 stars ENE 2′2.
17 55 18.1	−29 57 40	PN G0.1−2.3	9	14.9			PK 0−2.2	Four stars for a small square 2′4 N, their mags are 9.84v, 11.8, 11.4 and 11.2.
17 53 45.8	−29 43 46	PN G0.2−1.9	8	16.2	15.5	17.3	PK 0−1.5	Mag 12.8 star NNW 0′5; mag 11.37v star S 1′4.
17 57 43.8	−30 02 27	PN G0.3−2.8		18.7			PK 0−2.5	Mag 11.8 star SW 0′4.
17 54 25.4	−29 36 09	PN G0.4−1.9	7	14.8	15.1	16.1	PK 0−1.6	Mag 12.8 star W 1′1; close pair of mag 11 stars SW 2′0.
17 58 19.3	−30 00 39	PN G0.4−2.9	6	16.4		16.9	PK 0−2.6	Mag 9.78v star SE 2′5.
17 53 24.0	−29 16 57	PN G0.5−1.6					PK 0−1.7	Stellar; small triangle of mag 12 stars W 1′7; mag 11.1 star E 1′5.
17 59 15.7	−30 02 48	PN G0.5−3.1	8	19.0		17.3	PK 0−3.2	Small triangle of mag 12 stars S 1′2.
17 52 35.9	−29 06 38	PN G0.6−1.3	3				PK 0−1.2	Mag 9.84v star WNW 2′9.
17 56 39.9	−29 37 23	PN G0.6−2.3					PK 0−2.3	Stellar; small triangle of mag 10-11 stars W 2′6.
17 58 09.6	−29 44 20	PN G0.7−2.7	7	13.5	13.6	14.7	PK 0−2.4	Mag 12.6 star on SE edge; mag 10.1 star NW 2′0.
18 02 19.2	−30 14 26	PN G0.7−3.7	7	15.1	15.1	17.8	PK 0−3.1	Mag 12.1 star NNE 2′0; mag 10.3 star SE 2′8.
17 56 02.8	−29 11 16	PN G0.9−2.0	5	18.0		16.2	PK 0−2.1	Mag 10.6 star WSW 3′2; mag 12.4 star E 3′3.
17 58 26.1	−29 20 46	PN G1.0−2.6	5	17.9			PK 1−2.2	Mag 12.2 star E 0′8; mag 11.3 star NE 2′4.
18 00 37.7	−29 21 50	PN G1.2−3.0	5			16.2	PK 1−3.1	Mag 12.0 star E 0′9; mag 11.2 star S 1′2.
18 03 53.6	−29 51 21	PN G1.2−3.9	4	15.8		16.5	PK 1−3.7	Mag 12.4 star E 0′9; mag 8.04v star W 3′7.
18 02 27.0	−29 25 13	PN G1.4−3.4	13	18.2		15.8	PK 1−3.3	Mag 12.5 star E 0′8; mag 12.4 star S 1′1.
16 44 23.6	−40 03 19	PN G343.6+3.7	20	18.4		20.5	PK 343+3.1	Very close pair of mag 13 stars N 1′1; mag 13.1 star S 1′2.
16 42 33.3	−38 54 32	PN G344.2+4.7	10	14.9	12.:	16.7	PK 344+4.1	Mag 9.20v star S 1′4.
16 50 33.1	−40 03 00	PN G344.4+2.8		16.8	14.8		PK 344+2.1	Mag 12.4 star SSE 2′9.
16 49 32.7	−39 21 09	PN G344.8+3.4		19.0			PK 344+3.1	Mag 11.9 star NW 0′5.
16 50 25.4	−39 08 21	PN G345.0+3.4	5	15.6			PK 345+3.1	Mag 11.6 star S 0′6.
16 46 45.1	−38 36 58	PN G345.0+4.3					PK 345+4.1	Mag 11.8 star N 3′0.
16 39 28.3	−36 34 18	PN G345.6+6.7	21	14.9	15.0		PK 345+6.1	Elongated E-W; mag 12.4 star NW 4′7.
16 54 27.3	−38 44 10	PN G345.9+3.0	20	15.3			PK 345+3.2	Mag 11.5 star WNW 1′5.
16 48 54.0	−35 47 09	PN G347.4+5.8	1	13.0			PK 347+5.1	Mag 11.5 star NW 4′2.
17 04 33.7	−37 53 15	PN G347.7+2.0		17.1			PK 347+1.1	On the W edge of NGC 6281; mag 9.00v star S 1′3.
16 48 48.5	−35 00 58	PN G348.0+6.3	9		12.7	18.2	PK 348+6.1	Mag 13.0 star S 1′1.
17 32 48.2	−40 58 24	PN G348.4−4.1		15.1			PK 348−4.1	Mag 9.51v star NNE 3′6.
17 32 20.0	−39 51 23	PN G349.2−3.5	20	15.8			PK 349−3.1	Mag 13.6 star NW 0′8.
17 35 39.8	−40 11 59	PN G349.3−4.2	83	13.3		16.8	PK 349−4.1	Mag 12.0 star NNE 1′3; mag 11.0 star E 1′9.
17 01 06.2	−34 49 38	PN G349.8+4.4	5	13.4	14.0	17.0	PK 349+4.1	Mag 10.8 star W 1′1; mag 11.7 star E 1′7.
17 36 29.8	−39 21 57	PN G350.1−3.9	18	14.4	14.0		PK 350−3.1	Mag 9.86v star E 1′5.
17 42 54.1	−39 36 24	PN G350.5−5.0	8	17.2			PK 350−5.1	Mag 11.8 star NW 2′1; mag 11.4 star SW 2′7.
17 32 22.0	−37 57 23	PN G350.8−2.4	25	16.7			PK 350−2.1	Mag 12.1 star N 2′6.
17 04 36.2	−33 59 18	PN G350.9+4.4	6	14.4	13.6	13.2	PK 350+4.1	Mag 12.1 star ENE 2′4.
17 03 46.7	−33 29 43	PN G351.1+4.8	8	14.2	14.5	16.9	PK 351+4.1	Mag 9.68v star W 3′2.
17 02 19.1	−33 10 05	PN G351.2+5.2	5	16.4		16.2	PK 351+5.1	Close pair of mag 13 stars S 0′3.
16 53 37.0	−31 40 33	PN G351.3+7.6		14.6		16.3	PK 351+7.1	Mag 12.4 star S 3′0.
17 50 44.6	−39 17 28	PN G351.6−6.2	9	13.8	15.:		PK 351−6.1	Mag 12.0 star N 3′4.
17 33 00.6	−36 43 54	PN G351.9−1.9		16.0			PK 351−1.1	Mag 11.3 star SE 0′9; mag 12.2 star WNW 1′6.
16 50 17.0	−30 19 56	PN G351.9+9.0	7	14.1	14.0	16.0	PK 351+9.1	Mag 13.1 star NE 0′8; a second mag 13.1 star NE 1′7.
17 45 06.8	−38 08 50	PN G352.0−4.6	5	16.1			PK 352−4.1	Mag 11.5 star SE 2′0; mag 9.85v star NE 4′5.
17 05 30.7	−32 32 09	PN G352.1+5.1	4	14.9	14.7		PK 352+5.1	Mag 11.1 star WSW 1′9; close N-S pair of mag 11.6 stars N 1′3.
17 26 24.2	−35 01 41	PN G352.6+0.1	7	18.4			PK 352+0.1	Mag 12.2 star N 1′8.
17 14 42.9	−33 24 47	PN G352.6+3.0	3	16.9			PK 352+3.2	Mag 9.36v star N 0′6; mag 8.9 star SE 2′1.
17 28 27.5	−35 07 31	PN G352.8−0.2	10	16.1			PK 352−0.1	Mag 10.27v star N 0′6.
18 00 11.8	−38 49 53	PN G352.9−7.5	2	11.4		14.3	PK 352−7.1	Forms a triangle with a mag 10.40v star WSW 0′8 and a mag 12.3 star SSE 0′5.
17 50 45.3	−37 23 51	PN G353.2−5.2	7	18.3			PK 353−5.1	Mag 11.6 star NNE 0′6; mag 10.4 star W 1′8.
17 04 18.2	−30 53 28	PN G353.3+6.3	8	13.7	14.5	16.4	PK 353+6.2	Very faint star on E edge; mag 11.6 star NE 2′7.
17 49 48.2	−37 01 28	PN G353.5−4.9	10	12.0	12.1	16.8	PK 353−4.1	Located 1′2 NNW of mag 3.19v star.
17 05 13.5	−30 32 17	PN G353.7+6.3	8	16.5		13.9	PK 353+6.1	Mag 10.6 star NNE 1′6; mag 11.24v star SSW 2′5.
17 14 06.9	−31 19 42	PN G354.2+4.3	4	16.8			PK 354+4.1	Mag 10.96v star N 3′4.
17 18 51.8	−31 39 05	PN G354.5+3.3		16.4			PK 354+3.1	Close pair of mag 10.6, 11.3 stars NNW 2′1.
17 19 20.1	−31 12 39	PN G354.9+3.5		19.0			PK 355+3.3	Mag 12.9 star N 0′6; mag 9.94v star ESE 2′8.
17 24 26.1	−31 43 21	PN G355.1+2.3					PK 355+2.3	Mag 12.1 star WNW 2′3.
17 45 32.1	−34 33 55	PN G355.1−2.9	5	14.5			PK 355−2.4	Mag 9.33v star E 1′2.

(Continued from previous page)

164 PLANETARY NEBULAE 164

RA h m s	Dec ° ′ ″	Name	Diam ″	Mag (P)	Mag (V)	Mag cent ★	Alt Name	Notes
18 02 32.2	−36 39 11	PN G355.1−6.9	5	11.7	12.7	15.3	PK 355−6.1	Mag 11.3 star W 1′.1; mag 8.94v star S 1′.5.
17 44 13.8	−34 17 34	PN G355.2−2.5	10	15.0	14.3		PK 355−2.2	Close pair of mag 11.2, 12.4 stars S 1′.8.
17 44 20.6	−34 06 41	PN G355.4−2.4	7	14.5	15.0		PK 355−2.1	Mag 11.8 star N 3′.3; another mag 11.8 star W 3′.8.
17 51 12.2	−34 55 26	PN G355.4−4.0	9	13.7	14.0		PK 355−4.1	Small triangle of mag 11-12 stars centered W 1′.1.
17 46 07.0	−34 03 42	PN G355.6−2.7	5	16.9	14.3		PK 355−2.3	Close pair of mag 11.0, 11.9 stars E 1′.7.
17 47 49.4	−34 08 05	PN G355.7−3.0	3	15.4	15.5		PK 355−3.1	Mag 12.4 star S 2′.7; mag 10.63v star W 5′.6.
17 48 58.0	−34 21 53	PN G355.7−3.4		15.2	15.5		PK 355−3.2	
17 49 13.9	−34 22 52	PN G355.7−3.5	2	12.7	13.1	15.4	PK 355−3.3	Mag 12.4 star W 1′.4; mag 12.2 star N 1′.4.
17 24 40.7	−30 51 58	PN G355.9+2.7	2	19.5			PK 355+2.2	Mag 11.6 star on NE edge.
17 21 31.8	−30 20 48	PN G355.9+3.6	7	15.6		15.8	PK 355+3.2	Mag 11.8 star N 0′.5.
17 52 58.9	−34 38 25	PN G355.9−4.2	5	14.7	14.7	16.4	PK 355−4.2	Mag 12.7 star WSW 0′.8; mag 10.0 star W 1′.5.
17 25 19.2	−30 40 41	PN G356.1+2.7	2	18.3			PK 356+2.1	Mag 11.04 star E 2′.1; mag 10.57v star W 2′.7.
17 49 50.6	−34 00 32	PN G356.1−3.3	5	18.5			PK 356−3.1	Mag 11.1 star NE 1′.8; mag 12.0 star N 1′.6.
17 54 32.9	−34 22 22	PN G356.2−4.4	2	13.9	12.2		PK 356−4.1	Mag 12.1 star NW 1′.5; mag 11.4 star SSE 2′.4.
18 02 31.5	−35 13 12	PN G356.3−6.2	10	16.5			PK 356−6.1	Mag 11.2 star S 2′.0.
17 30 58.8	−31 01 06	PN G356.5+1.5	9	19.7			PK 356+1.2	Mag 12.4 star WSW 2′.4; mag 12.2 star WW 2′.9.
17 46 45.5	−33 08 35	PN G356.5−2.3	5			14.5	PK 356−2.2	Mag 12.2 star W 2′.4; mag 10.57v star E 2′.2.
17 51 50.5	−33 47 36	PN G356.5−3.6		19.0			PK 356−3.2	Mag 11.0 star E 2′.4.
17 53 21.0	−33 55 58	PN G356.5−3.9	5	17.7		16.3	PK 356−3.3	Mag 12.0 star E 1′.5; mag 11.6 star WSW 1′.4.
17 57 19.1	−34 09 49	PN G356.7−4.8	10	13.0		16.2	PK 356−4.2	Mag 11.1 ESE 0′.9; mag 10.67v star NNW 1′.0.
17 25 06.0	−29 45 15	PN G356.8+3.3	1	20.3			PK 356+3.1	Mag 13.1 star SE 1′.8.
18 00 18.6	−34 27 40	PN G356.8−5.4	11	16.0			PK 356−5.1	Mag 11.2 star E 2′.2.
17 21 04.4	−29 02 59	PN G356.9+4.4	2	16.5	15.5		PK 356+4.2	
17 20 33.3	−29 00 39	PN G356.9+4.5	15	14.4	14.:		PK 356+4.1	Mag 11.7 star N 2′.1.
18 02 02.8	−34 27 46	PN G356.9−5.8	7	14.6	14.6		PK 356−5.2	Mag 12.1 star E 1′.7; mag 11.2 star ESE 2′.0.
17 28 50.3	−30 07 45	PN G357.0+2.4	7	18.2		21.0	PK 357+2.5	Close pair of mag 12.4, 12.8 stars W 1′.6.
17 30 51.4	−30 17 12	PN G357.1+1.9		19.8			PK 357+2.7	Mag 12.3 star WSW 1′.4; mag 12.1 star NE 2′.4.
17 24 34.5	−29 24 19	PN G357.1+3.6	6	14.9	15.7	16.4	PK 357+3.1	Mag 12.0 star WNE 5′.0; mag 11.3 star SE 4′.8.
17 58 14.5	−33 47 37	PN G357.1−4.7	2			15.4	PK 357−4.3	Mag 12.8 star WSW 2′.1; mag 12.2 star ENE 2′.5.
17 33 17.0	−30 26 28	PN G357.2+1.4					PK 357+1.2	Stellar; mag 10.83v star NE 0′.7.
17 31 08.1	−30 10 28	PN G357.2+2.0		16.6			PK 357+2.6	Mag 10.6 star SW 2′.0.
17 57 25.2	−33 35 43	PN G357.2−4.5	6	14.3	13.3	17.3	PK 357−4.1	Line of three mag 13.1-13.7 stars W 1′.2.
17 25 59.8	−29 21 50	PN G357.3+3.3	4			16.3	PK 357+3.2	Mag 11.7 star S 1′.4; mag 11.0 star ENE 1′.4.
17 52 34.4	−32 45 50	PN G357.4−3.2	5	14.8	14.3		PK 357−3.2	Small triangle of mag 11-12 star SW 2′.5; mag 12.1 star NE 1′.8.
17 53 37.9	−32 58 48	PN G357.4−3.5	5	15.5	15.0	17.3	PK 357−3.4	Mag 10.8 star NE 0′.7; mag 11.6 star W 1′.7.
17 58 32.6	−33 28 37	PN G357.4−4.6	5	15.2		16.5	PK 357−4.2	Mag 12.0 star W 0′.4; mag 12.6 star SW 0′.6.
17 27 24.3	−29 21 14	PN G357.5+3.1	6				PK 357+3.5	Mag 11.4 star ENE 0′.6; mag 12.2 star W 1′.6.
17 26 59.8	−29 15 32	PN G357.5+3.2	5	16.9			PK 357+3.4	Mag 10.64v star N 1′.7.
17 35 43.2	−30 21 27	PN G357.6+1.0	30				PK 357+1.3	Mag 8.98v star NNW 1′.2.
17 32 46.9	−30 00 15	PN G357.6+1.7	3	15.6			PK 357+1.1	Close pair of mag 11-12 star S 1′.8.
17 29 42.7	−29 32 50	PN G357.6+2.6	2	16.4			PK 357+2.4	Mag 12.1 star on W edge; mag 10.15v star ENE 0′.5.
17 53 16.8	−32 40 38	PN G357.6−3.3	5	20.0		21.0	PK 357−3.3	Mag 12.7 star E 0′.4.
17 56 13.4	−32 37 27	PN G357.9−3.8	11	17.5		21.0	PK 358−3.2	Mag 9.18v star N 2′.3.
18 01 22.2	−33 17 42	PN G357.9−5.1	11	14.9	13.3		PK 357−5.1	
17 30 21.4	−29 10 12	PN G358.0+2.6					PK 358+2.2	Mag 9.72v star S 2′.2.
18 01 42.8	−33 15 27	PN G358.0−5.1	9	16.5			PK 358−5.1	Mag 12.0 star NW 2′.6.
17 46 02.5	−31 03 36	PN G358.2−1.1	13				PK 358−1.1	Mag 11.1 star NNE 3′.0.
17 36 59.8	−29 40 08	PN G358.3+1.2		19.3			PK 358+1.4	Mag 12.0 star N 2′.7; mag 12.6 star WNW 2′.3.
17 51 44.7	−31 36 00	PN G358.3−2.5	5				PK 358−2.4	
17 51 44.5	−31 36 00	PN G358.5−2.5	7				PK 358−2.1	N-S pair of mag 10.55v, 12.1 stars N 1′.3.
17 59 02.4	−32 21 43	PN G358.5−4.2	1	14.2	14.2	16.2	PK 358−4.1	Mag 10.0 star NE 2′.4.
17 35 14.0	−29 03 10	PN G358.6+1.8	2	18.3		21.0	PK 358+1.1	Mag 12.9 star W 1′.1.
17 53 37.5	−31 25 32	PN G358.7−2.7		19.5			PK 358−2.5	Mag 11.8 star NW 2′.1; mag 13.2 star NE 2′.6.
18 03 53.4	−32 41 42	PN G358.7−5.2	10	14.5	13.5	17.1	PK 358−5.3	Mag 10.41v star SW 1′.4; mag11.5 star ENE 2′.4.
17 42 42.5	−29 51 35	PN G358.8−0.0	12			14.9	Te 2022	Mag 11.7 star NW 2′.9; mag 11.3 star SW 5′.1.
17 45 57.7	−30 12 01	PN G358.9−0.7	9	13.7	13.8	12.7	PK 358−0.2	Mag 10.7 star WSW 1′.1.
17 58 10.7	−31 42 57	PN G358.9−3.7	4	18.6			PK 358−3.1	Mag 11.1 star N 0′.6; mag 12.3 star 2′.1.
17 59 56.9	−31 54 27	PN G359.0−4.1	5	17.8		21.0	PK 359−4.1	Mag 12.4 star N 1′.6; mag 12.5 star WSW 1′.4.
18 02 46.5	−32 09 28	PN G359.0−4.8	14	15.0	14.5		PK 359−4.3	Mag 10.9 star SSW 0′.7; mag 11.1 star S 2′.6.
17 50 18.0	−30 34 55	PN G359.1−1.7	8	13.6	14.7		PK 359−1.1	Mag 12.4 star N 1′.8; mag 12.0 star W 1′.8.
17 52 45.9	−30 49 35	PN G359.1−2.3	8	15.0	15.5	17.2	PK 359−2.2	Mag 12.1 star N 1′.8; mag 11.3 star WSW 2′.1.
17 55 06.0	−31 12 14	PN G359.1−2.9	4	19.3		17.9	PK 359−2.4	Mag 1.8 star NW 1′.9; mag 12.9 star E 2′.1.
17 47 56.2	−29 59 40	PN G359.3−0.9	15	13.6	11.8	18.6	PK 359−0.1	Faint star on SW edge; mag 11.5 star NW 2′.2.
17 51 18.9	−30 23 53	PN G359.3−1.8	4		16.0		PK 359−1.2	Mag 11.9 star N 0′.8; mag 11.0 star SSW 1′.2.
17 56 25.5	−31 04 16	PN G359.3−3.1	5	18.1		17.1	PK 359−3.1	Mag 11.8 star NW 2′.5; mag 12.4 star SE 2′.4.
17 58 12.5	−31 08 06	PN G359.4−3.4	6	18.2		21.0	PK 359−3.2	Mag 10.03v star E 0′.9; mag 11.1 star NW 1′.9.
17 52 06.0	−30 05 14	PN G359.7−1.8	6	16.5	16.:		PK 359−1.3	Mag 12.5 star E 0′.8; mag 11.9 star WNW 1′.4.
17 55 36.0	−30 33 32	PN G359.7−2.6	4	14.9			PK 359−2.3	Mag 8.68v star NW 0′.8; mag 7.59v star ESE 2′.1.
18 02 54.1	−31 23 47	PN G359.7−4.4	14	16.6		18.7	PK 359−4.5	Mag 12.3 star N 3′.2.
18 03 52.5	−31 17 46	PN G359.9−4.5	5	14.0	14.2		PK 359−4.2	Mag 12.4 star SE 0′.5; mag 13.0 star N 1′.0.

165 GALAXIES 165

RA h m s	Dec ° ′ ″	Name	Mag (V)	Dim ′ Maj x min	SB	Type Class	PA	Notes
15 18 23.8	-38 30 26	ESO 328-41	12.9	2.6 x 0.5	13.1	(R′)SB(r)b: II-III	0	Mag 10.92v star NW 8′.3.
15 22 24.4	-38 11 59	ESO 328-46	13.2	1.4 x 0.7	13.1	SB(rs)cd? II-III	98	Mag 9.35v star S 1′.7; mag 6.55v star W 10′.9.
15 30 08.6	-38 39 04	ESO 329-7	13.2	2.1 x 0.5	13.1	SA(rs)b: II	175	Mag 6.44v star W 9′.6.
15 39 29.8	-38 32 18	ESO 329-12	13.1	1.5 x 1.5	13.8	SAB(rs)c I-II		Mag 9.72v star S 2′.7.
15 42 37.0	-38 42 04	ESO 329-15	13.7	2.1 x 0.4	13.4	SBab? sp	16	Mag 11.17v star SW 1′.4.
15 49 37.0	-40 19 36	ESO 329-22	15.3	1.2 x 0.8	15.1	SB(s)cd: II	48	Mag 9.14v star N 5′.6.
15 16 42.1	-36 48 09	ESO 387-16	13.2	1.4 x 0.7	13.2	E5	94	Mag 9.95v star W 7′.3; mag 9.54v star NE 8′.6; note: **ESO 387-18** is 1′.8 SW of this star.
15 17 34.2	-36 24 28	ESO 387-19	13.8	0.3 x 0.3	11.0	S0:		Stellar.
15 18 06.9	-33 25 09	ESO 387-21	13.3	1.8 x 1.5	14.2	SB(r)bc II-III	85	Stellar nucleus, numerous superimposed stars.
15 20 46.9	-36 55 59	ESO 387-26	12.4	2.3 x 1.0	13.2	SAB(s)c: I-II	7	Several strong dark bands; mag 10.96v star W 1′.8; elongated **ESO 387-27** SE 8′.2.
15 23 24.8	-37 22 41	ESO 387-29	13.3	1.4 x 0.3	12.2	Sa	70	Mag 11.8 star NE 2′.8.
15 25 57.6	-37 09 21	ESO 387-33	13.1	1.3 x 0.5	12.5	SB(rs)a:	124	Mag 9.10v star W 4′.7.
16 00 56.1	-35 41 59	ESO 389-6	14.5	1.3 x 0.4	13.6	SA(r)c:	93	Mag 10.87v star W 11′.0; mag 7.65v star SW 16′.5.

(Continued from previous page)

165 GALAXIES 165

RA h m s	Dec ° ′ ″	Name	Mag (V)	Dim ′ Maj x min	SB	Type Class	PA	Notes
15 19 48.6	-31 24 39	ESO 449-4	14.7	0.8 x 0.6	13.8	SAa? sp	10	Mag 14.8 star N edge; ESO 449-5 E 6′.4.
15 20 17.9	-31 26 00	ESO 449-5	13.3	1.1 x 0.7	12.9	Sa: pec	76	ESO 449-4 W 6′.4; mag 10.71v star E 3′.2.
15 42 10.6	-31 33 02	ESO 450-9	13.5	1.0 x 0.5	12.6	S0/a	176	Mag 12.2 star E 1′.3.
16 06 16.4	-29 16 56	ESO 451-8	14.0	2.0 x 0.3	13.3	(R′)SB(s)b:	83	Mag 10.69v star SE 3′.0.
16 32 37.2	-29 22 50	ESO 452-8	15.1	1.8 x 0.3	14.3	Sd? sp	35	Mag 11.41v star SE 2′.7.
16 17 39.7	-34 21 59	IC 4597	13.7	1.1 x 1.0	13.7	S	120	Several stars superimposed on envelope.
16 18 13.4	-31 26 35	IC 4598	13.8	1.5 x 0.4	13.1	Sb:	175	Mag 11.09v star W 5′.7.
15 39 56.9	-30 33 07	NGC 5968	12.2	2.1 x 1.9	13.6	SAB(r)ab I-II		Bright nucleus, strong dark lanes.
15 49 49.4	-29 23 16	NGC 6000	11.9	1.9 x 1.6	12.9	SB(s)bc: II-III	154	Bright knots N; mag 9.111v star SW 3′.2.
15 25 56.2	-30 16 21	PGC 55070	14.6	0.8 x 0.6	13.6	(R′)SB(s)b	57	Mag 12.3 star NE 1′.2; mag 13.0 star SW 1′.1.

OPEN CLUSTERS

RA h m s	Dec ° ′ ″	Name	Mag	Diam ′	No. ★	B ★	Type	Notes
15 57 50.2	−36 12 23	ESO 389-5		1.8	7		ast	
16 25 17.2	−40 40 00	NGC 6124	5.8	40	100	9.0	cl	Rich in stars; large brightness range; strong central concentration; detached.

GLOBULAR CLUSTERS

RA h m s	Dec ° ′ ″	Name	Total V m	B ★ V m	HB V m	Diam ′	Conc. Class Low = 12 High = 1	Notes
15 46 03.5	−37 47 10	NGC 5986	7.6	13.2	16.5	9.6	7	
16 27 40.4	−38 50 56	NGC 6139	9.1	15.0	17.9	8.2	2	
16 28 40.1	−35 21 13	Terzan 3	12.0	15.0	17.3	3.0		

DARK NEBULAE

RA h m s	Dec ° ′ ″	Name	Dim ′ Maj x min	Opacity	Notes
15 44 57.9	−34 31 00	B 228	240 x 20	6	Large vacant region NW and SE.
16 38 24.3	−35 25 00	B 231	50 x 40	6	Dusky; sharpest on W side, diffuses to the E.
16 43 44.4	−39 49 00	B 232	10 x 10		Dusky; N and S; two small stars in it.
16 09 24.0	−39 08 00	Be 149	60 x 12	6	
15 57 00.0	−37 48 00	SL 11	150 x 40	6	

PLANETARY NEBULAE

RA h m s	Dec ° ′ ″	Name	Diam ″	Mag (P)	Mag (V)	Mag cent ★	Alt Name	Notes
16 01 21.0	−34 32 36	NGC 6026	40	13.2	12.9	13.2	PK 341+13.1	Several faint stars superimposed on NE edge.
16 12 58.2	−36 13 48	NGC 6072	98	14.1	11.7	19.3	PK 342+10.1	Rectangular shape with PA of 65 degrees; center one third brightest portion.
16 31 30.6	−40 15 12	NGC 6153	24	11.5	10.9	16.1	PK 341+5.1	Mag 8.27v star NE 2′.5; mag 9.84v star N 2′.6 and mag 11.32v star NW 2′.6.
15 43 05.0	−39 18 15	PN G335.5+12.4	180			12.4	PK 335+12.1	Mag 12.2 star S 2′.1; mag 12.5 star E 1′.9.
16 08 26.4	−37 08 47	PN G340.8+10.8	79			18.5	PK 340+10.1	Mag 13.3 star NE 2′.9.
16 03 22.0	−36 00 54	PN G340.8+12.3	60			19.0	PK 340+12.1	Mag 13.9 star N 1′.3; mag 13.8 star SE 1′.3.
16 13 28.1	−34 35 40	PN G343.4+11.9	2	14.0	14.5	17.3	PK 343+11.1	Mag 9.65v star W 1′.2; mag 10.02v star SW 1′.1.
16 42 33.3	−38 54 32	PN G344.2+4.7	10	14.9	12.:	16.7	PK 344+4.1	Mag 9.20v star S 1′.4.
16 09 45.8	−30 55 06	PN G345.5+15.1	71				PK 345+15.1	Mag 14.8 star on W edge; mag 13.8 star ESE 2′.1.
16 39 28.3	−36 34 18	PN G345.6+6.7	21	14.9	15.0		PK 345+6.1	Elongated E-W; mag 12.4 star NW 4′.7.
16 34 04.3	−35 05 27	PN G346.0+8.5	10	15.0			PK 346+8.1	Mag 13.0 star E 2′.4.
16 23 18.7	−31 44 58	PN G346.9+12.4	151	>15.7		20.5	PK 346+12.1	Annular, elongated E-W; mag 10.84v star N 1′.5.

166 GALAXIES 166

RA h m s	Dec ° ′ ″	Name	Mag (V)	Dim ′ Maj x min	SB	Type Class	PA	Notes
13 58 02.5	-40 49 41	ESO 325-38	14.2	1.0 x 0.7	13.7	Sc	13	Stellar nucleus, few stars superimposed.
13 58 39.5	-38 24 34	ESO 325-40	13.4	0.8 x 0.5	12.5	E	167	Located between a close pair of stars, mags 12.4 and 13.6.
13 59 20.4	-40 04 08	ESO 325-42	13.2	1.5 x 0.4	12.5	Sb:	144	Mag 7.92v star S 3′.4.
13 59 57.7	-37 51 48	ESO 325-43	13.5	1.2 x 1.0	13.6	SAB(s)c pec: II-III	115	Stellar nucleus; mag 9.95v star SW 10′.9.
14 02 18.0	-38 50 23	ESO 325-44	13.8	0.9 x 0.7	13.1	Sa:	49	Mag 8.97v star SSW 9′.1.
14 03 08.5	-38 28 36	ESO 325-45	13.7	1.2 x 1.0	13.8	SB(s)d IV	57	Stellar nucleus, filamentary.
14 05 41.5	-39 36 31	ESO 325-52	13.7	1.7 x 0.7	13.8	(R′)SB(r)ab	167	Mag 9.48v star NNW 6′.9.
14 06 14.1	-39 16 49	ESO 325-53	14.0	0.9 x 0.8	13.5	Sc	65	Mag 9.50v star W 7′.0.
14 07 42.1	-38 10 01	ESO 325-55	15.0	1.5 x 0.2	13.6	Sb:	65	Mag 7.89v star E 12′.9.
14 11 08.5	-40 06 21	ESO 326-6	14.3	0.7 x 0.6	13.3	S0: pec	174	Mag 10.81v star WNW 11′.1.
14 41 06.5	-39 59 07	ESO 327-7	13.4	1.1 x 0.8	13.1	SAB(rs)c II-III	65	Three faint stars in N-S line along E edge; mag 8.72v star NE 10′.8.
14 48 11.0	-37 49 30	ESO 327-20	12.9	1.1 x 1.0	12.9	SBc	63	Mag 9.18v star ESE 6′.2.
14 51 23.2	-37 59 11	ESO 327-23	12.9	0.9 x 0.9	12.5	$SA0^-$ pec:		Mag 9.39v star E 3′.5.
14 53 01.6	-39 55 20	ESO 327-27	14.7	1.5 x 0.3	13.7	Sbc	5	Mag 9.45v star SSW 6′.4.
14 55 34.6	-38 16 42	ESO 327-31	13.0	2.2 x 0.4	12.7	SB(s)c? sp II	48	Mag 8.96v start WNW 8′.7.
14 55 53.4	-39 12 13	ESO 327-32	14.0	1.0 x 0.7	13.5	(R′)SB(rs)bc III-IV	4	Mag 10.96v star N 1′.3.
14 58 19.3	-38 53 06	ESO 327-36	14.1	1.5 x 0.4	13.4	S	41	Mag 7.81v star SE 4′.4.
15 01 30.8	-37 59 37	ESO 328-5	13.7	2.4 x 0.5	13.7	(R′)SA(s)a	5	Mag 5.87v star SSW 5′.2; **ESO 328-6** NE 3′.8; **ESO 328-7** E 6′.2.
15 18 23.8	-38 30 26	ESO 328-41	12.9	2.6 x 0.5	13.1	(R′)SB(r)b: II-III	0	Mag 10.92v star NW 8′.3.
15 22 24.4	-38 11 59	ESO 328-46	13.2	1.4 x 0.7	13.1	SB(rs)cd? II-III	98	Mag 9.35v star S 1′.7; mag 6.55v star W 10′.9.
13 56 17.0	-34 32 26	ESO 384-14	13.7	1.9 x 0.4	13.3	S0° sp	143	Mag 12.1 star W 3′.2.
13 57 01.1	-35 20 00	ESO 384-16	14.2	1.0 x 0.6	13.7	E3?	67	Mag 12.9 star WNW 1′.6.
13 57 22.3	-34 46 36	ESO 384-18	14.5	1.5 x 0.3	13.5	S?	119	Mag 7.40v star SSW 8′.2.
13 57 39.3	-34 13 09	ESO 384-19	12.8	1.5 x 1.2	13.3	(R′)SA(r)0°:	92	Mag 8.31v star ENE 4′.5.
13 57 57.5	-34 00 32	ESO 384-21	13.2	1.3 x 1.0	13.3	SB?	177	Located 2′.2 SW of a mag 7.87v star.
13 58 30.2	-34 14 30	ESO 384-23	13.3	1.1 x 0.5	12.6	S0:	61	Small triangle of mag 12-14 stars SW edge; mag 8.31v star WNW 7′.1.
13 59 48.0	-33 40 48	ESO 384-25	14.4	1.5 x 0.8	14.4	Sb: pec	135	Double system, connected.
14 00 14.6	-34 02 16	ESO 384-26	13.1	1.6 x 0.4	12.5	S0 sp	110	Mag 8.75v star SE 9′.6.
14 00 28.7	-34 37 19	ESO 384-27	14.2	1.3 x 0.4	13.3	Sb:	20	Star superimposed E edge.

RA h m s	Dec ° ′ ″	Name	Mag (V)	Dim ′ Maj x min	SB	Type Class	PA	Notes
14 00 47.2	-34 13 29	ESO 384-29	12.9	1.4 x 1.2	13.4	(R′)SAB(s)0⁺	121	Mag 8.75v star N 6′.0.
14 01 29.8	-34 14 33	ESO 384-33	13.7	0.9 x 0.7	13.1	SB(r)0⁺	91	Mag 8.75v star NW 10′.3.
14 02 32.7	-33 23 57	ESO 384-35	12.7	1.6 x 1.3	13.3	I?	57	
14 03 34.9	-34 04 25	ESO 384-37	13.9	1.1 x 0.5	13.1	S?	48	Mag 14.9 star S edge; **ESO 384-42** SE 15′.1.
14 04 14.6	-33 01 42	ESO 384-41	14.0	0.8 x 0.7	13.2	Sb	20	Large core offset N.
14 04 27.8	-33 31 04	ESO 384-43	15.0	1.5 x 0.2	13.6	Sd: sp	63	Mag 8.95v star W 10′.7.
14 05 27.4	-32 49 52	ESO 384-47	13.8	1.2 x 0.6	13.3	S0?	21	**ESO 384-48** NE 2′.1; **ESO 384-46** SW 2′.1.
14 06 06.9	-33 55 24	ESO 384-49	13.1	1.0 x 0.6	12.6	E?	75	Star superimposed NE of core?
14 06 12.2	-33 04 24	ESO 384-51	13.5	1.6 x 1.0	13.9	SAB(rs)0	63	Bright core, faint envelope, mag 12.8 star on SE edge.
14 06 35.5	-34 18 41	ESO 384-53	12.1	2.6 x 2.2	13.8	SAB(rs)bc I	66	Mag 9.61v star on NW edge; **MCG −6-31-22** W 6′.3.
14 07 40.9	-37 17 11	ESO 384-55	13.1	1.5 x 0.7	13.1	E⁺3 pec	176	Mag 10.80v star NNE 3′.3.
14 07 56.9	-33 28 55	ESO 384-57	14.0	1.4 x 0.8	14.0	(R′)SAB(r)b	71	Elongated core with dark patches N and S.
14 15 32.9	-35 31 58	ESO 385-2	14.1	0.6 x 0.4	12.4	S?	119	Mag 9.85v star S 3′.4.
14 15 36.3	-37 11 48	ESO 385-3	14.4	0.9 x 0.7	13.7	Sc	16	Stellar nucleus, smooth envelope; pair of mag 13 stars E 0′.9.
14 16 05.2	-35 03 57	ESO 385-5	13.7	0.8 x 0.3	12.0	Sa	125	Superimposed star or large knot N edge.
14 19 04.0	-34 51 12	ESO 385-8	14.3	1.9 x 0.2	13.1	Scd	46	Line of three N-S oriented mag 9-10 stars W 10′.6.
14 21 15.6	-36 13 36	ESO 385-12	13.7	1.3 x 1.0	13.9	Sdm:	142	Mag 12.2 star WNW 3′.0; mag 8.71v star W 13′.3.
14 22 24.4	-36 01 21	ESO 385-14	14.1	1.1 x 0.5	13.3	Sb?	128	Mag 10.36v star SE 2′.9.
14 22 25.6	-34 21 52	ESO 385-15	12.8	1.9 x 0.7	13.0	(R′)SA(r)ab	10	ESO 385-17 NE 8′.9; mag 10.03v star ESE 10′.1.
14 23 02.4	-34 17 32	ESO 385-17	13.0	1.0 x 0.7	12.7	E4	100	Mag 14.4 star on N edge; mag 10.72v star N 5′.7.
14 27 31.3	-33 56 39	ESO 385-25	13.3	1.5 x 1.2	13.8	SAB(r)c: I-II	60	Mag 9.96v star NW 7′.0.
14 28 30.9	-34 42 39	ESO 385-28	13.2	1.4 x 1.1	13.5	SA?(r)a	101	N-S oriented bar, dark lanes E and W sides of bar; mag 12.4 star N 2′.5.
14 29 19.3	-33 27 20	ESO 385-30	11.9	2.4 x 1.2	12.9	SB(s)0°	18	Mag 8.64v star NNE 6′.4; mag 8.51v star NW 8′.6.
14 30 12.0	-36 57 52	ESO 385-32	12.5	2.1 x 1.4	13.5	SAB(s)c: II	125	Mag 7.84v star SW 8′.9.
14 30 15.3	-34 14 13	ESO 385-33	12.8	1.7 x 1.5	13.6	SAB(s)d pec: II-III	29	Strong dark lane E of center; mag 10.83v star W 5′.3.
14 34 45.5	-32 58 42	ESO 385-47	14.3	1.3 x 0.8	14.2	SBdm:	71	Mag 13.0 star on E edge; mag 10.44v star W 8′.0.
14 37 19.6	-36 07 41	ESO 385-49	14.0	1.0 x 0.3	12.6	S0	115	Mag 10.49v star W edge.
14 40 03.8	-36 37 40	ESO 386-2	13.0	1.1 x 1.0	13.0	S0	175	Mag 8.99v star S 6′.8.
14 40 29.2	-35 07 27	ESO 386-4	12.6	1.2 x 1.0	12.7	SB(r)0⁺	19	Star on SW edge; ESO 386-6 ESE 3′.8.
14 40 45.8	-35 08 57	ESO 386-6	13.8	1.7 x 0.4	13.2	SAab pec:	112	ESO 386-4 WNW 3′.8.
14 41 38.1	-33 18 44	ESO 386-9	12.5	1.3 x 1.0	12.6	Sb	67	Mag 8.60v star N 11′.9.
14 42 24.5	-36 34 21	ESO 386-11	13.2	1.0 x 1.0	13.2	E?		Mag 11.9 star S 2′.7.
14 42 38.7	-35 04 34	ESO 386-12	13.6	0.9 x 0.5	12.6	Sb	68	Mag 4.05v star ESE 13′.7.
14 43 08.0	-36 29 24	ESO 386-14	13.0	1.0 x 0.6	12.4	S0	144	Several stars superimposed.
14 46 10.4	-37 41 09	ESO 386-19	12.9	0.7 x 0.7	12.0	SA(r)0/a		Mag 10.19v star NE 7′.1.
14 46 54.5	-37 17 26	ESO 386-21	13.8	1.0 x 0.4	12.7	S0	9	Mag 11.7 star N 1′.0.
14 55 19.2	-35 43 32	ESO 386-31	13.3	1.3 x 0.6	12.9	S0	67	Mag 10.42v star N 6′.0.
14 56 06.5	-37 41 49	ESO 386-33	12.6	1.1 x 1.0	12.6	SB?	154	Mag 10.8 star SE 3′.0; **ESO -6-33-5** SE 2′.9; **ESO 327-30** SW 9′.4.
14 56 08.6	-37 38 55	ESO 386-34	12.7	1.2 x 0.8	12.5	SB(r)a: pec	70	Mag 10.86v star W 5′.3; ESO 386-33 S 3′.0.
14 56 19.5	-37 28 47	ESO 386-38	13.1	1.3 x 1.0	13.3	S0	168	Mag 10.48v star SSE 2′.5; mag 10.57v star S 3′.1.
14 56 25.3	-37 36 05	ESO 386-39	12.4	1.6 x 0.9	12.7	SB(s)c: pec	19	Multi-galaxy system; mag 9.89v star E 5′.6; **ESO 386-35** NW 3′.2.
14 56 54.9	-35 56 42	ESO 386-40	12.6	0.8 x 0.5	11.5	S0:	129	Mag 9.13v star WSW 5′.2; ESO 386-41 N 3′.9.
14 57 00.3	-35 53 07	ESO 386-41	12.9	1.2 x 0.8	12.7	S?	173	ESO 386-40 S 3′.9.
14 58 19.9	-37 33 25	ESO 386-43	13.1	1.4 x 0.7	12.9	SA(rs)bc:	115	Mag 12.0 star N 1′.7.
15 00 23.8	-37 21 45	ESO 386-44	13.9	1.1 x 0.7	13.4	Sc	141	Mag 11.19v star NE 3′.7, in star rich area.
15 02 38.2	-35 41 48	ESO 386-47	13.9	0.8 x 0.6	12.9	Sb	13	Mag 10.36v star ENE 4′.5.
15 08 41.3	-37 40 04	ESO 387-5	13.5	1.1 x 0.5	12.7	Sa:	115	Mag 7.66v star N 4′.7.
15 15 55.8	-36 45 15	ESO 387-12	13.2	0.8 x 0.6	12.3	S0	77	Mag 9.95v star SE 2′.9.
15 16 42.1	-36 48 09	ESO 387-16	13.2	1.4 x 0.7	13.2	E5	94	Mag 9.95v star W 7′.3; mag 9.54v star NE 8′.6; note: **ESO 387-18** is 1′.8 SW of this star.
15 17 34.2	-36 24 28	ESO 387-19	13.8	0.3 x 0.3	11.0	S0:		Stellar.
15 18 06.9	-33 25 09	ESO 387-21	13.3	1.8 x 1.5	14.2	SB(r)bc II-III	85	Stellar nucleus, numerous superimposed stars.
15 20 46.9	-36 55 59	ESO 387-26	12.4	2.3 x 1.0	13.2	SAB(s)c: I-II	7	Several strong dark bands; mag 10.96v star W 1′.8; elongated **ESO 387-27** SE 8′.2.
15 23 24.8	-37 22 41	ESO 387-29	13.3	1.4 x 0.3	12.2	Sa	70	Mag 11.8 star NE 2′.8.
13 56 34.9	-32 37 56	ESO 445-81	13.3	1.9 x 0.3	12.5	Sbc: sp	4	Mag 14.3 star on E edge; pair of mag 8.04v, 8.21v, stars SW 11′.8.
13 58 28.5	-31 46 32	ESO 445-85	15.1	1.6 x 0.2	13.7	Sc	64	Mag 8.12v star E 11′.6.
13 59 37.2	-32 33 07	ESO 445-86	13.6	1.0 x 0.2	11.7	S0?	24	Mag 9.82v star NNE 2′.7.
14 01 04.2	-30 19 41	ESO 445-89	12.6	2.5 x 2.2	14.3	SB(s)d pec: IV-V	116	Branching spiral arms; mag 10.57v star SSW 8′.8.
14 03 16.6	-31 20 55	ESO 446-1	13.7	1.5 x 0.9	13.9	SA(s)bc: II-III	138	Sits in a tight triangle of stars.
14 03 43.0	-32 43 01	ESO 446-2	14.0	1.7 x 0.4	13.5	Sab: sp	156	Small, anonymous galaxy SW 0′.6.
14 03 51.2	-31 57 00	ESO 446-3	13.7	1.1 x 0.7	13.3	S0?	27	Mag 9.05v star E 11′.8.
14 06 16.6	-30 15 27	ESO 446-7	14.3	0.6 x 0.4	12.6	Sb pec:	33	Mag 15.4 star SW edge.
14 06 35.3	-32 34 39	ESO 446-8	14.6	1.2 x 0.8	14.4	SB(rs)c pec: III-IV	53	Mag 11.8 star E 2′.6; mag 9.02v star ESE 4′.6; **ESO 466-13** E 6′.9.
14 07 07.4	-32 35 50	ESO 446-13	16.4	1.5 x 0.2	14.9	Sd	35	Mag 9.02v star W 2′.7; ESO 446-8 W 6′.9.
14 07 59.1	-32 03 06	ESO 446-17	12.6	2.0 x 1.3	13.5	(R)SB(s)b I-II	154	Mag 10.02v star W 5′.1.
14 08 38.4	-29 34 12	ESO 446-18	13.8	2.5 x 0.3	13.4	Sb sp	5	Mag 9.18v star E 5′.3.
14 09 16.6	-29 11 45	ESO 446-19	14.0	0.9 x 0.6	13.1	(R′)SB(s)b pec:	142	Mag 10.58v star NW 11′.9.
14 11 36.1	-30 24 05	ESO 446-23	14.2	1.7 x 0.2	12.9	Sc	178	Located in the center of a triangle of mag 10.46v, 10.98v and 11.9 stars.
14 12 25.9	-31 34 48	ESO 446-26	13.7	0.9 x 0.5	12.8	S0	142	Compact core; mag 15.0 star W edge.
14 12 44.8	-32 07 24	ESO 446-27	13.5	1.2 x 0.5	12.9	SB0⁻? pec	132	Star or bright knot N edge; mag 9.54v star W 5′.3.
14 13 30.5	-29 35 40	ESO 446-31	12.9	2.0 x 1.7	14.1	SB(s)cd IV	169	A number of superimposed stars on envelope; mag 12.8 star NNE 2′.1.
14 14 37.2	-29 33 27	ESO 446-35	14.1	1.5 x 1.1	14.4	SAB(r)c pec	51	Very faint outer arms surround bright core.
14 15 03.2	-29 21 39	ESO 446-36	13.9	0.9 x 0.5	12.8	S0/a	66	Mag 8.91v star N 1′.7.
14 17 51.7	-32 18 15	ESO 446-45	14.3	1.2 x 0.6	13.8	SAB(s)d: IV-V	92	Star superimposed E end; mag 12.0 star ENE 2′.7.
14 20 53.8	-29 01 19	ESO 446-51	15.2	1.3 x 0.2	13.6	Sc	9	Mag 12.7 star W 3′.3; mag 11.2 star NW 6′.8.
14 21 17.1	-29 15 48	ESO 446-53	14.0	1.5 x 0.7	13.9	IB(s)m V	170	Small triangle of mag 12-13 stars centered WNW 2′.6; NGC 5556 W 9′.6.
14 32 07.6	-29 56 53	ESO 447-18	14.2	1.0 x 0.7	13.7	Sb	107	Stellar nucleus, smooth envelope; mag 12.7 star WSW 2′.9.
14 32 28.7	-29 55 29	ESO 447-21	13.8	1.3 x 0.6	13.4	(R′)SB(r)b	30	ESO 447-18 W 4′.8; mag 10.64v star SW 9′.7.
14 32 38.7	-30 08 12	ESO 447-22	13.6	1.0 x 0.4	12.5	Sa	1	Mag 10.93v star N 3′.0.
14 39 46.4	-32 40 09	ESO 447-30	12.4	1.7 x 1.2	13.1	(R)SAB(r)0°	25	Bright center with stars superimposed on envelope; mag 11.9 star S 5′.1.
14 40 55.6	-32 22 33	ESO 447-31	12.8	1.8 x 0.9	13.2	(R′)SB(rs)0⁺	154	Mag 8.09v star N 3′.1.
15 19 48.6	-31 24 39	ESO 449-4	14.7	0.8 x 0.6	13.8	SAa? sp	10	Mag 14.8 star N edge; ESO 449-5 E 6′.4.
15 20 17.9	-31 26 00	ESO 449-5	13.3	1.1 x 0.7	12.9	Sa: pec	76	ESO 449-4 W 6′.4; mag 10.71v star E 3′.2.
13 57 54.1	-29 18 52	IC 4351	11.7	6.0 x 0.9	13.4	SA(s)b: sp III	17	Mag 9.22v star W 7′.3.
13 58 24.9	-34 31 02	IC 4352	12.4	1.8 x 0.6	12.3	Sab:	88	Pair of E-W oriented 7.55v, 7.80v stars N 7′.3.
14 05 11.9	-33 45 44	IC 4366	12.6	1.5 x 1.3	13.2	SAB(rs)c pec I-II	140	Mag 10.17v star S 1′.4; mag 8.26v star W 7′.4.
14 05 36.8	-39 12 14	IC 4367	12.2	1.8 x 1.5	13.1	(R)SAB(r)c II-III	14	Mag 9.50v star S 4′.0.
14 12 09.5	-34 15 53	IC 4378	13.5	0.8 x 0.5	12.4	S0	155	**IC 4379**, elongated E-W, on S edge; **MCG −6-31-29** ESE 4′.4.
14 16 03.6	-31 45 13	IC 4388	13.6	0.9 x 0.8	13.1	Sb	66	Mag 9.18v star N 3′.6.
14 16 27.2	-31 41 07	IC 4391	13.7	0.8 x 0.6	12.7	Sa	57	Mag 9.18v star W 5′.4.

166

(Continued from previous page)

GALAXIES

166

RA h m s	Dec ° ′ ″	Name	Mag (V)	Dim ′ Maj x min	SB	Type Class	PA	Notes
14 17 48.8	-31 20 57	IC 4393	13.6	2.4 x 0.3	13.1	Scd? sp	77	Mag 9.71v star W 4′.0.
14 28 31.4	-37 35 02	IC 4421	12.3	1.1 x 0.7	12.1	E?	164	Mag 11.6 star E 1′.6; mag 10.44v star W 6′.1.
14 34 36.8	-36 17 12	IC 4451	12.0	1.8 x 1.1	12.7	E$^+$:	89	Mag 11.4 star on N edge; mag 9.15v star W 11′.0.
14 37 49.1	-36 52 42	IC 4464	12.6	1.6 x 0.6	12.4	SAB(r)0°:	61	
14 03 12.9	-33 21 30	MCG −5-33-39	13.1	1.3 x 0.8	13.0	SA0$^-$	79	
14 03 37.8	-33 36 12	MCG −5-33-41	13.6	0.9 x 0.5	12.6	Sb?	164	Bright knot or superimposed star N end; mag 8.95v star N 7′.0.
14 06 39.8	-31 12 52	MCG −5-33-43	13.9	1.1 x 0.8	13.6	Sc	171	Several bright knots or stars on edges.
14 16 05.9	-30 16 51	MCG −5-34-4	13.8	0.8 x 0.4	12.6	S?	56	Lies 1′.2 E of mag 11 star.
14 20 14.6	-29 44 52	MCG −5-34-7	13.0	1.1 x 0.9	12.9	(R′)SA(r)0$^+$:	21	
14 20 33.6	-33 20 47	MCG −5-34-8	13.8	0.8 x 0.5	12.6	Sb	106	
14 28 15.9	-33 18 33	MCG −5-34-13	13.7	0.9 x 0.6	12.9	SB(r)bc	135	
14 44 27.0	-31 08 39	MCG −5-35-4	13.6	1.2 x 1.0	13.6	Sb	31	
14 47 26.8	-30 38 43	MCG −5-35-6	13.6	1.0 x 0.7	13.0	Sa:	16	
14 47 51.1	-30 19 20	MCG −5-35-7	14.4	0.6 x 0.5	12.9		21	Mag 8.12v star WSW 5′.9.
13 55 59.7	-30 20 30	NGC 5357	12.0	1.5 x 1.3	12.7	E	23	Located between a pair of 11 and 12 mag stars.
14 01 10.4	-33 56 46	NGC 5397	12.7	1.4 x 1.0	13.0	SAB(s)0$^-$:	60	Mag 12.1 star W 4′.8.
14 01 22.0	-33 03 43	NGC 5398	12.2	2.8 x 1.7	13.7	(R′)SB(s)dm pec: IV	172	Patchy arms, star or bright knot SW of nucleus.
14 03 38.6	-33 58 42	NGC 5419	10.9	4.2 x 3.3	13.7	E	77	Bright center, faint envelope, many stars superimposed; **MCG −6-31-20** E 11′.6.
14 07 04.5	-30 01 01	NGC 5464	13.0	1.3 x 0.7	12.7	IB(s)m? III-IV	85	Mag 13.8 star on E edge; mag 10.87v star W 3′.9.
14 08 03.5	-33 18 50	NGC 5488	11.7	3.4 x 1.0	12.9	SABbc? II	22	Mag 9.04v star sits on S end.
14 12 24.0	-30 38 45	NGC 5494	11.8	2.2 x 1.9	13.2	SA(s)c I-II	32	Nucleus and pair of stars center; several spiral arms.
14 20 34.0	-29 14 25	NGC 5556	11.8	4.0 x 3.2	14.4	SAB(rs)d III-IV	148	Patchy with many superimposed stars; ESO 446-53 E 9′.6.
14 29 49.0	-29 44 56	NGC 5626	12.9	1.1 x 0.9	12.7	SA(s)0$^+$	127	Mag 8.74v star NW 4′.8; note: **ESO 447-7** lies 1′.5 N of this star.
15 07 27.9	-36 19 42	NGC 5843	12.2	1.9 x 1.1	12.8	SB(s)b: II	70	Bright nucleus in bar.

GALAXY CLUSTERS

RA h m s	Dec ° ′ ″	Name	Mag 10th brightest	No. Gal	Diam ′	Notes
14 33 18.0	−31 48 00	A 3603	16.2	39	17	All members anonymous, faint, stellar.
14 03 36.0	−33 58 00	AS 753	13.9	18	56	
14 06 18.0	−39 49 00	AS 754	16.4		17	All members anonymous, stellar; **ESO 325-49** W of center 18′.3.
14 12 18.0	−33 08 00	AS 757	15.9	10	17	All members anonymous, stellar.
14 12 24.0	−34 19 00	AS 758	15.8	16	17	See notes IC 4378, all others anonymous, stellar.
14 41 42.0	−37 57 00	AS 770	16.3		17	**ESO 327-9** N of center 2′.6; **ESO 327-11** S 7′.3; **ESO 327-14** ESE 14′.8.
14 51 24.0	−37 59 00	AS 775	15.6	6	22	Except ESO 327-23, all members anonymous, stellar.
14 56 30.0	−37 36 00	AS 778	14.8	4	28	

OPEN CLUSTERS

RA h m s	Dec ° ′ ″	Name	Mag	Diam ′	No. ★	B ★	Type	Notes
14 37 24.0	−29 25 00	ESO 447-29		12			cl	

GLOBULAR CLUSTERS

RA h m s	Dec ° ′ ″	Name	Total V m	B ★ V m	HB V m	Diam ′	Conc. Class Low = 12 High = 1	Notes
15 03 58.5	−33 04 04	NGC 5824	9.1	15.5	18.5	7.4	1	

BRIGHT NEBULAE

RA h m s	Dec ° ′ ″	Name	Dim ′ Maj x min	Type	BC	Color	Notes
13 57 42.0	−39 59 00	NGC 5367	2.5 x 2.5	R	2-5	1-4	A round and fairly bright nebula surrounding a pair of 10th and 11th mag stars separated by 4″ in PA 33°.

PLANETARY NEBULAE

RA h m s	Dec ° ′ ″	Name	Diam ″	Mag (P)	Mag (V)	Mag cent ★	Alt Name	Notes
15 12 51.1	−38 07 33	NGC 5873	13	13.3	11.0	15.5	PK 331+16.1	Mag 14.1 star on SW edge; mag 12.4 star W 1′.9.

167

GALAXIES

167

RA h m s	Dec ° ′ ″	Name	Mag (V)	Dim ′ Maj x min	SB	Type Class	PA	Notes
12 37 11.8	−38 33 32	ESO 322-32	14.0	0.8 x 0.3	12.6		102	Mag 10.62v star WNW 5′.4.
12 38 28.3	−40 22 05	ESO 322-40	13.3	1.3 x 0.7	13.0	Sbc II	71	Mag 8.58v star E 8′.2.
12 42 00.1	−39 49 07	ESO 322-54	14.0	1.1 x 0.6	13.4		128	Double system, contact; superimposed star involved?
12 46 26.0	−40 45 12	ESO 322-75	12.8	1.7 x 0.9	13.2	(R)SA0°:	101	Mag 11.8 star SW 5′.2; note: very small, faint **PGC 43070** is 2′.1 E of this star.
12 46 41.6	−40 01 34	ESO 322-76	13.5	1.1 x 0.7	13.0	SA(r)bc: II-III	76	Mag 15.4 star E edge; mag 13.8 star NNW 2′.0.
12 48 00.5	−39 37 55	ESO 322-84	13.5	1.1 x 0.4	12.4	S0°: sp	136	Mag 10.32v star SW 7′.1; NGC 4679 NW 6′.8.
12 48 02.2	−40 35 40	ESO 322-85	14.5	1.5 x 0.4	13.8	Sc? pec sp	134	Mag 10.65v star S 2′.7.
12 48 55.3	−40 34 59	ESO 322-92	13.6	1.1 x 0.3	12.2	S0?	10	Mag 9.56v star SE 4′.8; note: **PGC 43339** is 1′.1 S of this star and **PGC 43349** is 1′.8 E of this star.
12 50 03.8	−40 07 36	ESO 323-2	13.7	1.7 x 0.3	12.8	S0$^-$: sp	26	Mag 10.80v star WSW 3′.4.
12 50 10.5	−38 41 52	ESO 323-4	14.1	0.9 x 0.5	13.1	S?	12	Mag 12.9 star NW 2′.3.
12 50 24.1	−40 50 45	ESO 323-7	13.1	1.4 x 0.9	13.2	SAB(s)0$^+$ pec	176	Bright star superimposed S edge of core; mag 10.70v star E 2′.5.
12 51 17.5	−40 49 09	ESO 323-12	13.5	0.9 x 0.6	12.7	S?	114	Mag 10.45v star ENE 2′.5; **PGC 43520** NW 2′.5.
12 51 43.7	−38 49 49	ESO 323-15	12.8	2.0 x 1.1	13.6	SA0°	152	Mag 9.90v star WSW 6′.9.
12 52 25.9	−40 42 30	ESO 323-23	13.1	1.3 x 0.9	13.2	S?	19	Moderately strong core, smooth envelope.
12 52 29.4	−39 08 47	ESO 323-24	13.9	1.2 x 0.4	12.9	S0?	127	Mag 9.92v star E 2′.9.
12 52 38.9	−39 01 46	ESO 323-25	12.6	1.5 x 0.9	12.7	(R′)SB(s)bc: I-II	97	Line of three N-S mag 8.69v, 9.35, 10.30v stars centered W 5′.8.
12 52 50.3	−40 27 09	ESO 323-27	13.0	1.8 x 1.0	13.5	SA(r)c I-II	99	Mag 13.5 stars on N and E edges; mag 9.22v star SE 4′.0.
12 55 01.2	−40 58 19	ESO 323-42	13.2	2.2 x 0.6	13.3	S?	77	Mag 10.33v star W 2′.3.

(Continued from previous page)

RA h m s	Dec ° ′ ″	Name	Mag (V)	Dim ′ Maj x min	SB	Type Class	PA	Notes
12 55 38.6	−39 32 26	ESO 323-44	13.3	1.0 x 0.9	13.1	S?	86	Bright knot or superimposed star NW of nucleus.
12 57 04.2	−39 46 04	ESO 323-49	14.0	0.9 x 0.5	12.9	I?	26	Located 1′.4 NW of mag 7.76v star.
12 59 18.2	−40 57 42	ESO 323-54	14.1	1.0 x 0.8	13.7	S?	169	Stellar nucleus, filamentary envelope; mag 8.50v star SW 2′.7.
13 01 03.3	−40 25 37	ESO 323-63	13.4	1.1 x 0.6	12.9	S0?	150	Almost stellar **PGC 44845** W 0′.8; slightly larger **PGC 44827** W 2′.3.
13 03 09.9	−39 14 58	ESO 323-71	13.5	1.4 x 0.6	13.1	(R′)SB(s)a	65	Mag 12.2 star SW 3′.0.
13 04 01.7	−38 11 58	ESO 323-73	13.5	1.4 x 0.8	13.5	S?	177	Mag 9.30v star SW 7′.3.
13 06 25.9	−40 24 52	ESO 323-77	12.5	2.0 x 1.2	13.3	(R)SAB(rs)0°	155	Mag 11.38v star W edge.
13 06 40.5	−38 16 32	ESO 323-79	13.0	1.2 x 0.9	13.0	S0?	39	Mag 14.9 star on E edge.
13 07 12.3	−40 24 29	ESO 323-81	14.2	0.5 x 0.3	12.0		47	Compact.
13 10 35.6	−39 36 20	ESO 323-89	12.4	2.0 x 1.3	13.3	(R)SA(s)0°	68	Mag 8.94v star NE 3′.0.
13 12 15.5	−39 56 16	ESO 323-92	12.9	1.1 x 0.8	12.8	E?	7	Close pair of stars, mags 13.4 and 12.8, NW 1′.8.
13 13 57.6	−39 15 40	ESO 323-97	13.9	1.3 x 0.4	13.1	S	25	Mag 12.9 star SSE 2′.2.
13 18 03.6	−39 20 47	ESO 324-3	13.8	1.1 x 0.7	13.4	SA?(r)0	165	Mag 9.43v star W 11′.0; **ESO 324-2** N 3′.4.
13 19 25.6	−39 09 14	ESO 324-5	13.9	1.0 x 0.4	12.8	Sb	155	Pair of stars, mag 13.6 and 15.3, N 1′.0.
13 19 40.1	−40 47 47	ESO 324-7	14.8	1.5 x 0.2	13.3	Sc:	115	Mag 9.20v star ENE 7′.1.
13 21 32.9	−38 47 43	ESO 324-9	14.3	0.9 x 0.4	13.0	Sa	60	Mag 7.13v star SE 5′.9.
13 22 16.1	−38 17 56	ESO 324-11	14.1	0.7 x 0.5	12.8	S0?	156	Mag 9.62v star NNW 5′.6.
13 24 48.7	−38 06 50	ESO 324-18	14.2	0.6 x 0.5	12.7	S0?	87	Mag 7.13v star NW 6′.9.
13 26 52.1	−40 06 45	ESO 324-21	13.7	1.4 x 1.3	14.2	Sc	141	Mag 6.41v star SE 6′.5; mag 12.8 star N 1′.4.
13 27 29.2	−38 10 41	ESO 324-23	12.6	3.2 x 0.6	13.2	SB(s)d: sp III-IV	127	Pair of NE-SW oriented mag 10.17v, 10.19v stars WSW 10′.2.
13 28 22.0	−38 10 12	ESO 324-26	14.3	1.2 x 0.8	14.1	IB(s)m VI	45	Line of three N-S mag 11-12 stars centered W 4′.0; mag 9.78v star ESE7′.5.
13 28 40.4	−40 11 41	ESO 324-27	13.8	1.0 x 0.9	13.5	Sa	33	Several knots or superimposed stars NE of nucleus.
13 32 27.3	−38 10 08	ESO 324-33	12.9	1.1 x 0.9	12.8	SB(rs)ab	98	Mag 10.81v star W 2′.3; **ESO 324-32** SSW 5′.4.
13 33 17.9	−40 29 47	ESO 324-35	14.3	1.0 x 0.7	13.8	SB	171	Stars superimposed on edges; mag 9.28v star W 2′.8.
13 33 35.6	−38 52 53	ESO 324-36	14.4	1.1 x 1.0	14.4	Scd	75	Faint stellar nucleus, uniform envelope; located between a mag 10.35v star SE and a mag 7.60v star W.
13 35 23.0	−40 04 19	ESO 324-40	13.8	0.9 x 0.5	12.8	S0	141	Small group of three mag 12-13 stars N 1′.4; stellar **ESO 324-41** NE 2′.4.
13 38 06.3	−39 50 26	ESO 324-44	12.7	1.6 x 0.8	12.8	SAB(rs)cd III-IV	49	Mag 9.08v star NW 13′.4.
13 41 56.8	−38 18 32	ESO 325-1	13.6	0.9 x 0.6	12.8	S0	20	Mag 10.49v star W 2′.2.
13 43 33.2	−38 10 36	ESO 325-4	12.9	1.0 x 1.0	12.8	SA(r)0⁻:		**ESO 325-5** SE 2′.0; **ESO 325-8** NE 8′.7.
13 44 01.9	−38 27 19	ESO 325-6	13.9	1.0 x 0.7	13.3	Sa:	108	Almost stellar nucleus in short E-W bar; **MCG −6-30-19** W 10′.0.
13 46 24.1	−37 58 18	ESO 325-16	13.8	0.5 x 0.5	12.2			Almost stellar.
13 46 47.1	−37 54 36	ESO 325-19	13.8	0.6 x 0.4	12.1	S?	17	Pair of compacts.
13 50 36.9	−38 26 34	ESO 325-28	13.4	1.3 x 1.1	13.6	(R′)SAB(r)b	81	Mag 9.27v star WNW 10′.3.
13 51 26.0	−38 13 03	ESO 325-30	13.8	0.8 x 0.5	12.6	S0/a	49	Mag 10.87v star NW 1′.8.
13 58 02.5	−40 49 41	ESO 325-38	14.2	1.0 x 0.7	13.7	Sc	13	Stellar nucleus, few stars superimposed.
13 58 39.5	−38 24 34	ESO 325-40	13.4	0.8 x 0.5	12.5	E	167	Located between a close pair of stars, mags 12.4 and 13.6.
13 59 20.4	−40 04 08	ESO 325-42	13.2	1.5 x 0.4	12.5	Sb:	144	Mag 7.92v star S 3′.4.
13 59 57.7	−37 51 48	ESO 325-43	13.5	1.2 x 1.0	13.6	SAB(s)c pec: II-III	115	Stellar nucleus; mag 9.95v star SW 10′.9.
14 02 18.0	−38 50 23	ESO 325-44	13.8	0.9 x 0.7	13.1	Sa:	49	Mag 8.97v star SSW 9′.1.
14 03 08.5	−38 28 36	ESO 325-45	13.7	1.2 x 1.0	13.8	SB(s)d IV	57	Stellar nucleus, filamentary.
12 39 09.5	−34 46 53	ESO 381-4	13.5	1.0 x 0.5	12.6	S?	132	Located between a pair of mag 12.4 stars.
12 40 32.9	−36 58 08	ESO 381-5	14.1	1.1 x 0.8	13.8	S?	12	Mag 13.8 star E edge; mag 12.1 star S 0′.9.
12 40 58.5	−36 43 57	ESO 381-9	13.0	1.4 x 1.2	13.4	SB(s)bc pec III-IV	150	Mag 10.03v star N 3′.9, IC 3639 SW 1′.8.
12 44 05.4	−34 12 10	ESO 381-12	12.6	0.9 x 0.9	12.3	S?		Mag 9.99v star SE 4′.8; **ESO 381-13** NNE 1′.6; **ESO 381-15** NNE 3′.4.
12 44 09.7	−36 30 37	ESO 381-14	14.3	1.8 x 0.3	13.5	Sbc: sp	6	Mag 6.38v star N 10′.0.
12 44 34.7	−33 54 24	ESO 381-17	13.6	1.3 x 1.0	13.8	SB(rs)d IV	43	Mag 10.01v star SW 12′.1.
12 46 00.7	−33 50 21	ESO 381-20	13.6	2.2 x 1.0	14.3	IB(s)m V-VI	136	Mag 9.87v star E 2′.8.
12 47 14.5	−33 38 39	ESO 381-23	13.7	0.8 x 0.4	12.3		74	Located between very close mag 14.7 and 15.3 stars.
12 56 28.3	−36 22 19	ESO 381-29	12.7	1.1 x 0.6	12.3	S?	43	Mag 10.21v star S 1′.5.
12 58 33.4	−37 33 28	ESO 381-32	13.6	1.1 x 0.7	13.1	SB(r)0⁺	88	Mag 10.16v star S 5′.2.
13 00 05.3	−34 26 04	ESO 381-38	13.8	0.9 x 0.6	13.0	Sb	22	Mag 10.20v star and ESO 381-41 NE 5′.9.
13 00 32.0	−34 22 30	ESO 381-41	14.3	1.5 x 0.5	13.8	IB(s)m: V	77	Mag 10.20v star S edge.
13 00 38.7	−32 47 59	ESO 381-42	13.3	0.9 x 0.5	12.3	Sb:	12	Mag 10.46v star NW 10′.8.
13 00 49.3	−33 21 15	ESO 381-44	14.3	0.5 x 0.4	12.5		108	Almost stellar.
13 01 02.1	−35 55 55	ESO 381-46	13.7	1.3 x 0.5	13.1	Sc	37	Mag 12.3 star W 2′.4.
13 01 05.4	−35 37 02	ESO 381-47	13.1	1.0 x 1.0	13.0	S0?		Mag 9.94v star NNW 9′.1.
13 01 05.6	−36 36 07	ESO 381-48	14.3	0.7 x 0.4	12.7		110	Mag 8.21v star NW 2′.7.
13 02 07.8	−32 47 13	ESO 381-50	13.6	1.3 x 1.0	13.7	SB(r)bc pec	169	= **Hickson 63B**; stellar **Hickson 63D** NW edge.
13 02 07.8	−33 07 14	ESO 381-51	13.5	1.5 x 0.3	12.4	Sb	58	Mag 9.37v star S 4′.3.
13 03 01.2	−32 50 30	ESO 382-2	13.3	1.0 x 0.6	12.6	S0	101	Mag 10.33v star NNW 4′.3.
13 05 43.0	−37 41 09	ESO 382-7	13.9	0.8 x 0.5	12.8	S0	33	Mag 9.50v star WNW 5′.5.
13 07 07.8	−33 51 53	ESO 382-10	13.9	0.9 x 0.6	13.1	S?	33	Core offset SE; small companion galaxy SE edge; double system?
13 08 15.8	−37 08 52	ESO 382-12	17.3	1.1 x 0.6	16.7	SB(s)m VI	153	Mag 11.03v star WSW 6′.5.
13 13 12.4	−36 43 23	ESO 382-16	12.7	0.9 x 0.7	12.1	SA0° pec:	159	Mag 9.67v star N 5′.5.
13 13 23.0	−36 41 19	ESO 382-17	13.9	0.9 x 0.2	11.9	S0/a	149	Bright star on W edge; mag 12.4 star NW 1′.6; mag 9.67v star NW 5′.1.
13 15 06.9	−37 08 36	ESO 382-23	13.3	0.8 x 0.7	12.5	SAB(r)b	75	Mag 10.56v star SE 3′.0.
13 15 21.9	−34 56 05	ESO 382-24	13.3	1.1 x 0.9	13.1	S0	84	Faint, elongated anonymous galaxy WNW 2′.8; mag 12.8 star W 1′.9.
13 15 22.4	−34 47 40	ESO 382-25	13.6	0.8 x 0.6	12.7		155	Double? system, contact; mag 10.08v star NE 3′.4.
13 16 03.5	−37 20 49	ESO 382-27	13.6	0.9 x 0.5	12.6	Sbc	148	**MCG −6-29-20** N 0′.9.
13 17 40.1	−37 16 48	ESO 382-31	13.8	2.5 x 0.4	13.7	SB(s)d: sp	162	Mag 10.13v star W 4′.2; numerous anonymous galaxies in surrounding area.
13 17 42.2	−34 21 13	ESO 382-32	14.4	1.7 x 0.3	13.5	Sc?	11	Mag 11.7 star W 2′.8; mag 9.52v star ESE 9′.0.
13 17 47.5	−37 39 41	ESO 382-33	13.6	1.7 x 0.6	13.5	(R′)SA(rs)0/a pec	59	Mag 9.97v star N 9′.3; **ESO 382-30** W 8′.8.
13 18 01.8	−36 57 03	ESO 382-34	12.6	1.2 x 1.1	12.8	SAB(r)0°	171	Mag 7.56v star SW 6′.2, several anonymous galaxies in the area.
13 19 29.0	−35 06 10	ESO 382-41	14.7	2.2 x 0.2	13.6	Sc: sp	12	Pair of N-S oriented mag 11.4, 12.2 stars NW 9′.7.
13 19 31.5	−33 29 15	ESO 382-43	13.9	1.3 x 0.6	13.5	S pec	143	Star superimposed NW of nucleus; mag 9.66v star SE 1′.7.
13 20 07.7	−37 11 26	ESO 382-44	14.0	1.0 x 0.3	12.5	S	13	Mag 10.76v star SW 6′.7; numerous small, faint anonymous galaxies in surrounding area.
13 20 14.9	−36 02 44	ESO 382-45	13.4	3.2 x 1.1	14.6	SB(s)m V-VI	96	Several bright knots E of center; mag 10.84v star E 3′.5; **MCG −6-29-29** NE 9′.2.
13 22 08.1	−37 23 10	ESO 382-51	13.7	1.1 x 0.6	13.1	SAB(rs)cd	70	Mag 10.81v star WSW 4′.3.
13 23 01.0	−35 46 37	ESO 382-54	13.4	0.9 x 0.7	12.7	Sb	103	Several stars superimposed; mag 12.4 star NW 1′.4.
13 25 12.4	−33 39 23	ESO 382-58	13.3	2.8 x 0.4	13.3	SB(r)bc: sp II	157	Mag 11.32v star NE 6′.0; ESO 382-60 S 8′.8.
13 25 27.0	−33 47 49	ESO 382-60	13.5	1.0 x 0.6	12.8	S0/a	143	Mag 11.04v star S 3′.1. ESO 382-58 N 8′.8.
13 26 49.5	−36 45 25	ESO 382-67	13.5	0.9 x 0.6	12.7	Sa	128	Mag 13.4 star NE 3′.1.
13 28 37.9	−37 34 25	ESO 383-2	13.7	1.6 x 0.8	13.8	SAB(rs)c II	34	Mag 10.58v star W 6′.0.
13 29 23.8	−34 16 20	ESO 383-5	13.2	3.7 x 0.6	13.9	Sbc sp	133	Narrow dust lane runs length of bright center.
13 31 19.7	−35 05 24	ESO 383-8	13.6	1.3 x 0.5	13.0	S?	36	E-W line of four mag 11.5 to 13 stars N 2′.4.
13 31 43.5	−33 22 49	ESO 383-12	14.0	1.0 x 0.5	13.2	SA?(r)0	137	Mag 10.98v star W 5′.9; note: small, faint anonymous galaxy on S edge of this star.
13 33 47.9	−37 43 57	ESO 383-19	13.3	1.8 x 1.6	14.3	SAB(r)bc II-III	102	Triangle of mag 10.55v, 10.68v, 10.94v stars centered NE 3′.8; **ESO 383-23** E 7′.0.
13 34 04.7	−36 50 41	ESO 383-20	13.9	0.5 x 0.4	12.0	SA?(r)	68	Compact; mag 10.83v star SW 0′.9.

(Continued from previous page)

GALAXIES

RA h m s	Dec ° ′ ″	Name	Mag (V)	Dim ′ Maj x min	SB	Type Class	PA	Notes
13 34 46.6	−34 18 42	ESO 383-25	13.0	1.3 x 1.0	13.2	SB(rs)0$^+$ pec	29	Mag 7.55v star SW 7'.9.
13 34 51.4	−36 31 19	ESO 383-26	13.8	0.6 x 0.5	12.4	Sa	157	Mag 11.7 star E 2'.0.
13 35 04.5	−35 16 13	ESO 383-27	13.5	1.7 x 0.4	13.0	S0 sp	79	Mag 9.92v star NE 6'.4.
13 35 18.1	−33 53 55	ESO 383-30	13.2	1.3 x 0.8	13.1	SAB(rs)0$^+$:	10	Mag 13.9 star superimposed S of center; mag 7.77v star S 7'.5.
13 35 21.7	−34 12 25	ESO 383-31	13.6	1.2 x 0.6	13.1	SA(rs)bc:	82	**ESO 383-32** N 2'.7; **MCG −6-30-14** NE 1'.3.
13 35 53.8	−34 17 46	ESO 383-35	12.9	1.0 x 0.6	12.2	S?	115	Mag 10.79v star S 2'.0.
13 36 18.6	−33 13 36	ESO 383-38	13.4	0.8 x 0.5	12.4	S0	109	Mag 10.98v star NNW 8'.2.
13 37 27.9	−33 00 23	ESO 383-44	13.4	1.7 x 0.8	13.6	SA(s)d:	34	Mag 10.75v star NW 7'.6; note: stellar **ESO 383-43** is 1'.1, ESE of this star.
13 37 39.2	−33 48 44	ESO 383-45	13.0	1.2 x 0.5	12.3	S0?	62	Mag 10.17v star E 3'.1.
13 37 50.5	−36 03 03	ESO 383-47	13.5	0.9 x 0.2	11.6	S0	39	Mag 14.0 star W 1'.7.
13 38 01.7	−33 31 09	ESO 383-48	13.3	2.5 x 0.4	13.2	S0 sp	0	Mag 11.13v star NE 4'.6.
13 38 21.2	−35 21 46	ESO 383-50	13.7	0.6 x 0.2	11.3		59	Exceptionally bright center, or superimposed star? Mag 9.66v star WNW 2'.6.
13 38 29.3	−36 23 43	ESO 383-52	14.5	0.9 x 0.2	12.5	S	120	Mag 10.95v star NNW 6'.3.
13 38 54.1	−36 26 49	ESO 383-53	15.1	1.5 x 0.2	13.7	S	127	Mag 10.29v star NE 5'.1.
13 39 27.3	−34 11 14	ESO 383-55	14.1	0.9 x 0.3	12.6	S	105	Mag 13.7 star NNW 2'.5.
13 40 31.0	−33 39 19	ESO 383-60	12.9	1.9 x 1.3	13.8	SB(rs)d: IV	15	Mag 9.13v star E 8'.0.
13 42 07.9	−36 21 02	ESO 383-64	13.6	0.8 x 0.5	12.5	Sb	11	Large knot or companion galaxy NE edge; mag 12.8 star NW 2'.6.
13 43 38.5	−32 55 09	ESO 383-67	14.6	1.6 x 0.2	13.2	Sc	41	Mag 9.98v star SW 2'.8.
13 44 59.9	−36 09 02	ESO 383-71	13.8	1.2 x 0.9	13.7	SB(r)bc	148	Two strong arms W side.
13 45 13.4	−33 40 48	ESO 383-73	14.0	1.0 x 0.4	12.9	Sa	33	Mag 10.63v star SE 2'.5.
13 47 28.9	−32 51 57	ESO 383-76	12.1	2.7 x 1.3	13.4	E$^+$5:	5	Mag 10.91v star W 5'.7; ESO 383-81 SE 4'.1.
13 47 43.7	−32 54 34	ESO 383-81	13.8	0.4 x 0.3	11.4	S	95	Compact, almost stellar; very bright center or star? ESO 383-76 NW 4'.1.
13 49 18.5	−36 03 42	ESO 383-87	10.5	4.4 x 3.6	13.4	SB(s)dm IV	93	Numerous stars superimposed; mag 9.67v star SSE 8'.4.
13 49 33.3	−33 04 11	ESO 383-88	13.3	1.8 x 0.7	13.4	SAB(r)bc? II	94	Mag 8.48v star SE 13'.0; **ESO 383-85** WSW 5'.2; **ESO 383-90** E 8'.8.
13 50 32.3	−37 17 23	ESO 383-91	13.7	2.7 x 0.4	13.6	SB(s)d: sp	103	Mag 11.7 star S 1'.6; mag 7.59v star SE 12'.3.
13 50 42.2	−35 55 01	ESO 383-92	14.3	0.6 x 0.5	12.8	S	44	Mag 10.07v star N 8'.6.
13 50 44.5	−36 45 14	ESO 383-93	13.9	0.9 x 0.5	12.9	S pec?	138	Mag 10.11v star SSW 3'.9.
13 51 19.2	−33 48 29	ESO 384-2	11.9	4.8 x 2.6	14.5	SB(s)dm V	135	Mag 8.73v star NW edge.
13 51 22.3	−37 37 43	ESO 384-3	13.6	1.8 x 0.4	13.0	Sb: sp	84	**ESO 384-4** S 2'.2; mag 10.97v star S 4'.8.
13 52 16.4	−37 27 25	ESO 384-5	14.9	1.6 x 0.3	14.0	SB(s)dm: sp	73	Mag 9.86v star SSE 4'.6.
13 52 24.5	−34 08 40	ESO 384-6	13.6	1.0 x 0.7	13.1	S0	178	Mag 10.24v star E 4'.4.
13 52 25.2	−34 56 00	ESO 384-7	13.6	1.3 x 0.8	13.5	S0?	11	Located between a pair of mag 13.0 and 13.6 stars.
13 53 40.7	−33 57 03	ESO 384-9	13.9	1.5 x 1.1	14.3	SB(s)c II	138	Mag 9.63v star NW 12'.3.
13 55 12.4	−36 19 06	ESO 384-11	14.1	1.1 x 0.7	13.7	S?	161	Several knots or superimposed stars; mag 12.9 star E 2'.4.
13 55 33.9	−33 54 03	ESO 384-12	13.2	1.0 x 0.8	12.9	S0 pec	121	Mag 10.36v star W 2'.1.
13 55 41.8	−33 43 33	ESO 384-13	13.0	1.3 x 1.0	13.1	SA(r)0°?	88	Mag 12.4 star N 4'.0.
13 56 17.0	−34 32 26	ESO 384-14	13.7	1.9 x 0.4	13.3	S0° sp	143	Mag 12.1 star W 3'.2.
13 57 01.1	−35 20 00	ESO 384-16	14.2	1.0 x 0.6	13.7	E3?	67	Mag 12.9 star WNW 1'.6.
13 57 22.3	−34 46 36	ESO 384-18	14.5	1.5 x 0.3	13.5	S?	119	Mag 7.40v star SSW 8'.2.
13 57 39.3	−34 13 09	ESO 384-19	12.8	1.5 x 1.2	13.3	(R′)SA(r)0°:	92	Mag 8.31v star ENE 4'.5.
13 57 57.5	−34 00 25	ESO 384-21	13.2	1.3 x 1.0	13.3	SB?	177	Located 2'.2 SW of a mag 7.87v star.
13 58 30.2	−34 14 30	ESO 384-23	13.3	1.1 x 0.5	12.6	S0:	61	Small triangle of mag 12-14 stars SW edge; mag 8.31v star WNW 7'.1.
13 59 48.0	−33 40 48	ESO 384-25	14.4	1.5 x 0.8	14.4	Sb: pec	135	Double system, connected.
14 00 14.6	−34 02 16	ESO 384-26	13.1	1.6 x 0.4	12.5	S0 sp	110	Mag 8.75v star SE 9'.6.
14 00 28.7	−34 37 19	ESO 384-27	14.2	1.3 x 0.4	13.3	Sb:	20	Star superimposed E edge.
14 00 47.2	−34 13 29	ESO 384-29	12.9	1.4 x 1.2	13.4	(R′)SAB(s)0$^+$	121	Mag 8.75v star N 6'.0.
14 01 29.8	−34 14 33	ESO 384-33	13.7	0.9 x 0.7	13.1	SB(r)0$^+$	91	Mag 8.75v star NW 10'.3.
14 02 32.7	−33 23 57	ESO 384-35	12.7	1.6 x 1.3	13.3	I?	57	
14 03 34.9	−34 04 25	ESO 384-37	13.9	1.1 x 0.5	13.1	S?	48	Mag 14.9 star S edge; **ESO 384-42** SE 15'.1.
12 42 42.5	−29 37 14	ESO 442-14	13.9	0.7 x 0.5	12.7	Sb	20	Mag 9.34v star SW 6'.0.
12 42 50.9	−30 24 35	ESO 442-15	13.1	0.8 x 0.7	12.3	(R′)SB(r)a	27	Mag 10.49v star SE 0'.9.
12 52 13.2	−29 50 29	ESO 442-26	11.6	2.6 x 0.9	12.4	SB0?	9	Mag 11.7 star E 6'.4.
12 52 34.7	−31 53 19	ESO 442-28	14.4	1.7 x 1.0	14.9	SB(s)m V	168	Mag 10.29v star SE 1'.8; mag 8.60v star E 5'.6; **ESO 442-27** S 3'.4.
12 54 22.2	−29 00 52	ESO 443-4	13.4	1.1 x 0.7	13.0	S0	174	Mag 14.9 star SE of core; stellar **PGC 43914** E 0'.9; **PGC 43905** N 2'.5; almost stellar **PGC 43881** WSW 2'.5.
12 54 36.2	−31 52 53	ESO 443-6	15.5	1.4 x 1.2	15.9	IAB(s)m V-VI	96	Small triangle of mag 10.06v, 10.87v, 12.8 stars N 10'.2; **ESO 443-3** NNW 7'.0.
12 55 36.0	−30 20 52	ESO 443-11	13.1	1.6 x 0.7	13.2	SAB0° pec	133	Disturbed, NW of two.
12 56 58.4	−31 19 46	ESO 443-14	13.7	1.0 x 0.7	13.1	S	74	Close pair of mag 14.6 stars SE; mag 11.04v star N 3'.9.
12 57 08.4	−32 30 45	ESO 443-15	12.5	1.2 x 0.9	12.5	Sa	144	MCG −5-31-5 on SE edge.
12 57 44.8	−29 46 00	ESO 443-17	12.9	1.5 x 1.0	13.2	(R)SB0/a	18	Mag 10.51v star N 8'.0.
12 59 46.0	−29 35 59	ESO 443-21	13.6	2.2 x 0.3	13.0	Scd: sp	162	Mag 8.73v star NNW 8'.7.
13 01 00.8	−32 26 30	ESO 443-24	11.8	1.9 x 1.5	12.8	SA(s)0$^-$	165	Mag 12.3 star on NW edge; mag 10.19v star NW 6'.4.
13 01 08.4	−30 48 52	ESO 443-28	13.4	1.0 x 0.8	13.0	Sa	35	Mag 13.6 star on E edge; mag 7.39v star E 8'.6.
13 01 41.9	−30 54 44	ESO 443-34	12.7	2.3 x 1.9	14.1	(R′)SAB(rs)0/a: pec	23	Bright N-S bar, very extensive faint arms; mag 9.89v star E 2'.3.
13 02 12.0	−30 26 19	ESO 443-36	14.0	1.3 x 1.0	14.1	SBb	155	Narrow, elongated core; mag 12.6 star NE 1'.8.
13 02 17.0	−32 45 37	ESO 443-37	14.1	1.3 x 0.5	13.4	(R′)SB(s)a	177	= **Hickson 63A**; **Hickson 63C** W edge.
13 03 03.1	−30 47 30	ESO 443-39	12.8	1.4 x 0.8	12.7	S0?	16	Star just inside W edge.
13 03 23.7	−32 14 16	ESO 443-41	13.0	1.6 x 0.7	13.0	(R′)SAB(rs)b II	42	Mag 10.06v star SW 6'.0.
13 03 29.8	−29 49 40	ESO 443-42	13.0	3.2 x 0.5	13.3	Sb: sp	128	Mag 8.59v star S 1'.9.
13 04 00.7	−30 22 08	ESO 443-43	13.0	1.3 x 1.2	13.4	S0?	44	Mag 11.44v star SSW 3'.5.
13 04 20.1	−30 09 49	ESO 443-48	13.8	1.1 x 0.3	12.5	S0	142	ESO 443-53 E 2'.5.
13 04 31.6	−30 10 17	ESO 443-53	12.9	1.1 x 0.9	12.9	E?	62	ESO 443-48 W 2'.5; mag 9.51v star E 7'.4.
13 04 49.9	−30 14 41	ESO 443-54	12.8	1.2 x 0.8	12.6	S?	8	Mag 9.51v star NE 4'.4.
13 04 50.7	−30 29 27	ESO 443-55	13.3	1.3 x 1.1	13.6	SAB(r)0/a	28	Mags 15.3 star S edge; mag 13.7 star E 1'.8.
13 05 08.5	−32 11 31	ESO 443-56	13.8	1.0 x 0.6	13.1	S?	140	Mag 14.4 star N edge.
13 06 11.0	−29 43 38	ESO 443-66	13.7	0.9 x 0.6	12.9	S0?	24	NGC 4955 SW 2'.1.
13 06 13.2	−30 09 39	ESO 443-67	13.7	1.0 x 0.6	13.0	SAB(r)0/a:	156	Bright knots at both ends.
13 08 35.3	−32 08 54	ESO 443-72	13.8	0.9 x 0.5	12.8	Sa	44	Mag 12.2 star N 2'.8; mag 8.28v star NE 12'.1.
13 09 00.5	−32 07 32	ESO 443-73	14.1	0.7 x 0.4	12.6	Sa	8	Star superimposed W edge; mag 8.28v star NNE 7'.4.
13 09 50.8	−30 30 54	ESO 443-77	13.8	1.2 x 0.3	12.6	S0?	134	Mag 13.0 star NE 2'.7.
13 15 14.1	−32 15 15	ESO 443-85	14.0	1.3 x 0.7	13.7	SB(rs)dm	51	Faint, elongated anonymous galaxy W 1'.9; mag 13.9 star NW 1'.6.
13 18 29.2	−32 13 45	ESO 444-7	13.3	1.0 x 0.9	13.0	(R′)SAB(r)a	16	Mag 15.0 star S edge.
13 20 27.7	−30 54 37	ESO 444-10	14.4	1.4 x 0.2	12.9	S?	2	**ESO 444-10A** NE edge.
13 22 53.9	−32 40 15	ESO 444-16	13.9	0.8 x 0.6	13.0	(R′)SB(s)cd pec:	90	ESO 444-18 S 3'.5.
13 22 55.7	−31 44 17	ESO 444-17	13.9	0.9 x 0.7	13.3	S0?	75	Stars superimposed N and S of core.
13 22 56.8	−32 43 42	ESO 444-18	13.6	1.2 x 0.9	13.5	S0?	153	Almost stellar anonymous galaxy NE 0'.9.
13 23 30.4	−30 06 54	ESO 444-21	14.9	1.8 x 0.2	13.6	S?	65	**PGC 46803** NE 4'.5; mag 9.48v star SW 11'.3.
13 26 04.7	−32 07 48	ESO 444-33	14.3	1.7 x 0.4	13.7	SBm? sp IV	85	Mag 10.52v star ENE 3'.8.
13 26 59.7	−30 04 26	ESO 444-37	14.3	1.8 x 1.1	14.9	SB(s)dm: IV-V	115	Bright knot or core offset SE; mag 11.6 star S 3'.6.

RA h m s	Dec ° ′ ″	Name	Mag (V)	Dim ′ Maj x min	SB	Type Class	PA	Notes
13 27 56.8	−31 29 45	ESO 444-46	12.8	1.8 x 1.0	13.4	E^{+}4	157	Mag 8.11v star SE 4′8; stellar **PGC 47197** S 2′6; numerous faint galaxies in the area.
13 28 25.0	−31 51 31	ESO 444-47	14.0	1.7 x 0.5	13.7	SB(r)cd? sp	21	Mag 10.74v star SSW 4′8; **PGC 47177** WNW 9′2.
13 29 50.9	−29 30 53	ESO 444-55	13.4	1.4 x 0.8	13.3	S0?	30	= **Hickson 65A**. **Hickson 65B** NE 1′3; **Hickson 65C** NNE 1′4; **Hickson 65D** NE edge of core of ESO 444-55; **Hickson 65E** NE 1′7.
13 33 30.3	−32 43 11	ESO 444-71	13.7	1.5 x 0.5	13.3	S0/a: sp	24	Mag 10.29v star E 6′5.
13 33 34.7	−31 40 22	ESO 444-72	13.3	1.7 x 0.9	13.6	SAB(rs)0°	75	Mag 11.10v star WNW 3′5; at center of A 3562.
13 37 57.2	−30 55 56	ESO 444-86	13.6	1.6 x 0.4	12.9	S?	70	Mag 10.01v star S 4′3.
13 39 22.5	−32 13 28	ESO 445-1	13.0	1.5 x 0.9	13.2	SA(s)0°:	96	Mag 14.4 star on S edge; mag 11.8 star S 6′2; **ESO 445-3** N 7′5.
13 39 23.5	−30 46 27	ESO 445-2	11.4	2.1 x 1.9	12.8	SA0^{-}	94	Star 0′5 ESE of bright nucleus.
13 42 08.6	−30 45 51	ESO 445-14	12.6	1.7 x 0.8	12.8	S?	108	Mag 9.25v star NNE 6′9.
13 42 14.6	−31 02 16	ESO 445-15	13.6	1.5 x 0.6	13.3	S?	57	Mag 15.3 star on SW end; mag 12.2 star E 3′5.
13 42 37.1	−32 07 34	ESO 445-16	13.8	0.4 x 0.4	11.7	S0?		Compact; mag 9.58v star ESE 2′0.
13 43 49.0	−29 44 43	ESO 445-20	14.8	1.5 x 0.2	13.4	Sc	15	IC 4319 SW 6′1; mag 11.09v star NNE 6′7.
13 46 32.9	−30 52 45	ESO 445-26	14.0	1.9 x 0.2	12.8	S?	102	Mag 13.6 star on N edge; small, round, faint anonymous galaxies 2′3 ENE and 4′0 N.
13 47 53.0	−32 27 48	ESO 445-33	13.7	1.6 x 0.3	12.8	S0 sp	171	Mag 8.98v star NW 3′0.
13 48 39.5	−30 48 41	ESO 445-40	13.1	1.8 x 0.5	12.9	SAB(r)0°	86	Mag 8.19v star NW 5′0; group of four faint anonymous galaxies centered W 8′1, with mag 15.1 star on E edge of one galaxy.
13 49 00.6	−30 37 35	ESO 445-44	14.0	0.7 x 0.7	13.1	Sb		Star NE edge; stellar **PGC 48979** W 6′7.
13 49 09.8	−31 09 57	ESO 445-49	12.5	1.8 x 0.6	12.4	SA0/a sp	170	Dark dust lane runs the length of the galaxy.
13 50 25.8	−31 06 57	ESO 445-53	13.8	0.6 x 0.2	11.3	S	112	**ESO 445-55** SE 2′7; mag 7.35v star NNW 3′0.
13 50 32.6	−30 02 59	ESO 445-54	13.8	1.1 x 0.9	13.6	SB0?	101	Strong N-S core; mag 12.4 star E 2′4.
13 51 09.0	−30 46 51	ESO 445-57	13.2	1.4 x 0.9	13.3	SB?	127	Mag 9.97v star SE 1′8; mag 9.69v star E 8′5; note: **ESO 445-63** is located 4′4 ESE of this star.
13 51 19.9	−31 08 56	ESO 445-58	13.0	1.9 x 0.8	13.3	SB(rs)bc I-II	156	Bright, elongated core; mag 7.02v star WSW 9′1.
13 51 39.8	−30 29 23	ESO 445-59	12.4	1.2 x 0.5	11.8	E?	115	Pair with ESO 445-59A S; mag 10.87v star NE 2′9.
13 51 41.2	−30 30 09	ESO 445-59A	14.0	0.6 x 0.6	12.7			Pair with ESO 445-59 N.
13 52 46.0	−29 55 47	ESO 445-65	13.2	1.5 x 0.2	11.8	S0?	13	mag 9.45v star S 5′4.
13 52 50.4	−30 42 41	ESO 445-66	13.6	1.2 x 1.0	13.7	SB(r)ab	119	Mag 7.74v star SW 8′3; note: **ESO 445-63** is 3′4 SSW of this star.
13 53 09.7	−30 42 49	ESO 445-69	14.0	1.5 x 0.8	14.0	(R′)SB(r)0/a:	89	Close pair of mag 13 stars S 1′6; ESO 445-66 W 4′2.
13 53 58.4	−29 51 35	ESO 445-73	13.4	1.2 x 0.7	13.1	S?	114	Mag 14.0 star, with faint anonymous galaxy on it's W, edge N 1′7.
13 54 55.6	−29 08 15	ESO 445-76	14.2	1.1 x 1.1	14.3	S?		Mag 9.60v star NE 3′3.
13 56 34.9	−32 37 56	ESO 445-81	13.3	1.9 x 0.3	12.5	Sbc: sp	4	Mag 14.3 star on E edge; pair of mag 8.04v, 8.21v, stars SW 11′8.
13 58 28.5	−31 46 32	ESO 445-85	15.1	1.6 x 0.2	13.7	Sc	64	Mag 8.12v star E 11′6.
13 59 37.2	−32 33 07	ESO 445-86	13.6	1.0 x 0.2	11.7	S0?	24	Mag 9.82v star NNE 2′7.
14 01 04.2	−30 19 41	ESO 445-89	12.6	2.5 x 2.2	14.3	SB(s)d pec: IV-V	116	Branching spiral arms; mag 10.57v star SSW 8′8.
14 03 16.6	−31 20 55	ESO 446-1	13.7	1.5 x 0.9	13.9	SA(s)bc: II-III	138	Sits in a tight triangle of stars.
14 03 43.0	−32 43 01	ESO 446-2	14.0	1.7 x 0.4	13.5	Sab: sp	156	Small, anonymous galaxy SW 0′6.
14 03 51.2	−31 57 00	ESO 446-3	13.7	1.1 x 0.7	13.3	S0?	27	Mag 9.05v star E 11′8.
13 03 18.2	−30 31 18	IC 844	12.8	1.6 x 0.4	12.1	S0 sp	100	Mag 9.24v star SE 6′2.
12 40 52.9	−36 45 25	IC 3639	12.3	1.2 x 1.2	12.5	SB(rs)bc: II		Mag 10.03v star N 5′2, mag 14.3 star on SW edge; ESO 381-9 NE 1′8.
13 01 32.3	−32 17 28	IC 3986	12.3	1.7 x 1.0	12.7	SAB0^{-}:	132	Small, faint, elongated anonymous galaxies S 2′6 and WNW 3′2.
13 17 42.9	−32 06 05	IC 4214	11.4	2.6 x 1.7	12.9	(R′)SB(r)ab	0	Mag 11 star on S edge.
13 18 29.8	−31 37 53	IC 4219	13.0	1.1 x 1.0	13.0	SB(rs)b pec: II		Mag 11.7 star SW 2′1.
13 26 44.2	−30 21 46	IC 4247	13.8	1.3 x 0.6	13.4	S?	153	Mag 10.57v star S 3′5; **PGC 47047** N 8′8.
13 26 47.2	−29 52 54	IC 4248	13.2	1.1 x 0.8	12.9	S?	103	Mag 11.6 star SE 1′1; mag 10.09v star WSW 8′9.
13 27 24.3	−29 26 41	IC 4251	14.0	0.9 x 0.8	13.5	S0?	142	Mag 11.9 star SW 1′3.
13 29 28.3	−30 08 05	IC 4259	14.2	1.0 x 0.6	13.5	S?	162	Mag 9.98v star NE 6′4.
13 31 51.4	−29 43 55	IC 4275	13.6	0.9 x 0.5	12.6	SB?	22	Mag 11.02v star NW 1′7; faint anonymous galaxy S 4′7, many others in area.
13 36 39.2	−33 57 58	IC 4296	10.6	2.8 x 2.8	12.9	E		Large, bright core with faint envelope; mag 12.3 star NW 2′9; IC 4299 S 6′4.
13 36 47.3	−34 04 01	IC 4299	12.6	1.9 x 0.9	13.1	SAB(s)a	58	IC 4296 N 6′4; mag 11.6 star W 3′0.
13 43 26.5	−29 48 14	IC 4319	13.2	1.5 x 0.5	12.8	SA(s)bc?	70	ESO 445-20 NE 6′1; faint, round anonymous galaxy S 1′3.
13 44 31.1	−30 08 24	IC 4321	14.2	0.8 x 0.6	13.2	Sa-b	24	Mag 10.10v star NW 1′2.
13 45 27.4	−30 13 39	IC 4324	13.9	0.7 x 0.3	12.0	Sa:	133	Mag 9.83v star ESE 7′4; note: a faint, elongated anonymous galaxy is 2′4 N of this star.
13 47 39.8	−29 26 05	IC 4325	13.8	1.2 x 0.4	12.9	SB?	102	Mag 10.19v star ESE 10′7.
13 48 21.6	−29 37 37	IC 4326	13.5	0.8 x 0.7	12.7	S?	168	Mag 8.11v star E 7′1.
13 48 43.9	−30 13 05	IC 4327	13.9	1.1 x 0.5	13.1	SB(s)c? II-III	56	Mag 9.53v star WSW 2′4.
13 49 02.8	−29 56 14	IC 4328	13.9	0.9 x 0.6	13.1	S?	172	Mag 9.58v star SW 10′3; **PGC 49009** SSW 5′6.
13 49 05.2	−30 17 45	IC 4329	11.3	3.4 x 1.9	13.2	SAB(s)0^{-}	63	Mag 10.31v star S 4′5; IC 4329A E 3′0.
13 49 19.3	−30 18 37	IC 4329A	13.0	1.4 x 0.4	12.3	SA0^{+}: sp	45	Located 3′0 E of IC 4329.
13 57 54.1	−29 18 52	IC 4351	11.7	6.0 x 0.9	13.4	SA(s)b: sp III	17	Mag 9.22v star W 7′3.
13 58 24.9	−34 31 02	IC 4352	12.4	1.8 x 0.6	12.3	Sab:	88	Pair of E-W oriented 7.55v, 7.80v stars N 7′3.
12 54 41.2	−29 13 43	MCG −5-31-1	13.1	0.7 x 0.7	12.2	S?		
12 57 12.4	−32 31 04	MCG −5-31-5	12.5	0.4 x 0.3	10.1	S0/a: pec	132	Located on SE edge of ESO 443-15.
12 59 37.0	−29 15 33	MCG −5-31-8	14.2	0.7 x 0.7	13.5	E/S0		
13 00 23.8	−31 26 21	MCG −5-31-10	13.5	1.3 x 0.4	12.7	S?	96	
13 01 20.5	−32 24 16	MCG −5-31-14	13.8	0.9 x 0.7	13.1	SA(r)cd	12	Several faint knots, star on SW edge; ESO 443-24 SW 4′7.
13 01 36.7	−30 03 56	MCG −5-31-17	13.2	1.3 x 0.6	12.7	S?	37	Faint anonymous galaxy E 1′9.
13 02 45.5	−32 40 46	MCG −5-31-22	14.5	0.6 x 0.4	12.8	Sbc	33	Mag 11 star on E edge.
13 03 32.4	−32 52 08	MCG −5-31-27	13.7	1.0 x 0.7	13.2	S?	164	Almost stellar anonymous galaxy N 5′2.
13 05 23.1	−30 06 11	MCG −5-31-31	14.0	0.9 x 0.7	13.3	Sb-c	14	Mag 9.48v star W 1′8.
13 12 55.7	−32 41 30	MCG −5-31-39	12.7	3.2 x 0.6	13.3	SB(s)d: sp II-III	152	**Hardcastle Nebula**.
13 20 50.0	−29 28 49	MCG −5-32-1	13.5	1.2 x 0.9	13.4	S?	138	Star on NW edge.
13 22 15.5	−29 33 25	MCG −5-32-2	14.3	0.7 x 0.5	13.0		33	
13 22 23.7	−31 04 36	MCG −5-32-3	13.8	0.9 x 0.9	13.4	S?		
13 23 06.4	−32 14 39	MCG −5-32-4	13.5	1.2 x 0.9	13.5	SAB(rs)0°	20	**ESO 444-23** ENE 10′8.
13 24 06.8	−31 40 12	MCG −5-32-7	13.1	1.3 x 0.8	13.1	S0?	158	Many bright knots and/or superimposed stars.
13 24 28.9	−30 25 56	MCG −5-32-8	12.9	1.3 x 0.7	12.7	SB	81	Mag 9 star superimposed on S edge.
13 25 14.8	−30 32 42	MCG −5-32-12	13.6	1.0 x 0.3	12.1	Sa	27	
13 25 42.5	−29 32 43	MCG −5-32-14	14.0	1.0 x 0.7	13.5	Sc	40	
13 27 00.8	−29 11 33	MCG −5-32-19	13.7	0.9 x 0.6	12.9	S0/a	53	
13 35 08.1	−30 07 04	MCG −5-32-42	14.4	0.5 x 0.4	12.5	Sc	144	
13 35 47.6	−30 52 36	MCG −5-32-45	13.7	1.3 x 0.6	13.3	I?	93	
13 35 49.7	−30 22 57	MCG −5-32-47	13.6	0.8 x 0.8	13.0	(R)SB0^{+}		**ESO 444-77** S 2′9.
13 36 05.3	−33 00 43	MCG −5-32-48	13.1	1.1 x 0.9	13.0	SB0?	68	
13 38 03.8	−33 52 28	MCG −5-32-55	13.4	1.1 x 0.4	12.6	E^{+}:	122	Mag 10.17v star NNW 4′0.
13 38 12.4	−31 25 02	MCG −5-32-57	13.3	1.3 x 1.0	13.4	(R)SB(r)a	107	Bright knots or superimposed stars E of center and W edge.
13 39 59.4	−32 35 31	MCG −5-32-61	13.2	1.1 x 0.6	12.6	S?	76	Mag 9.82v star NNW 7′2.
13 40 49.8	−32 39 30	MCG −5-32-64	12.8	0.8 x 0.7	12.1	SA(rs)c	89	
13 41 27.4	−32 00 13	MCG −5-32-65	14.0	0.8 x 0.5	12.8	S	109	

(Continued from previous page)

GALAXIES

RA h m s	Dec ° ′ ″	Name	Mag (V)	Dim ′ Maj x min	SB	Type Class	PA	Notes
13 43 24.6	−30 22 12	MCG −5-32-70	14.0	0.9 x 0.7	13.3	Sa	105	
13 45 21.9	−30 01 07	MCG −5-32-74	13.8	0.5 x 0.2	11.2	S0	78	
13 47 18.7	−29 48 33	MCG −5-33-4	13.1	0.8 x 0.6	12.2	S0?	153	
13 47 23.2	−30 25 00	MCG −5-33-5	15.1	0.4 x 0.3	12.6	S0	50	**Seashell Galaxy**.
13 48 11.3	−30 27 07	MCG −5-33-10	13.0	1.3 x 0.8	12.9	SB(rs)b	169	NGC 5298 E 5.′5.
13 48 16.6	−32 14 27	MCG −5-33-13	13.3	0.9 x 0.5	12.3	S?	130	Lies 1.′3 E of mag 9 star.
13 48 48.9	−31 09 18	MCG −5-33-17	13.5	1.3 x 0.2	11.9	S0?	135	
13 50 53.6	−30 17 20	MCG −5-33-23	13.6	1.2 x 0.2	11.9	S0	133	Mag 9.26v star NW 3.′6.
14 03 12.9	−33 21 30	MCG −5-33-39	13.1	1.3 x 0.8	13.0	SA0$^-$	79	
14 03 37.8	−33 36 12	MCG −5-33-41	13.6	0.9 x 0.5	12.6	Sb?	164	Bright knot or superimposed star N end; mag 8.95v star N 7.′0.
12 36 07.7	−39 26 19	NGC 4553	12.2	2.1 x 1.0	12.9	SA(r)0$^+$	176	Bright, elongated nucleus; small anonymous galaxy SW 2.′6.
12 37 43.6	−35 31 06	NGC 4574	13.0	1.7 x 1.1	13.5	SAB(s)c II	113	Short, bright bar; patchy arms; several stars superimposed E edge.
12 37 51.2	−40 32 14	NGC 4575	12.6	2.0 x 1.3	13.5	SB(s)bc pec:	106	Very patchy inner arms; faint outer arms; star on W edge; mag 10.64v star SW 1.′6.
12 40 46.7	−40 53 39	NGC 4601	13.4	1.8 x 0.5	13.2	SAB(r)0$^+$	16	Located 2.′8 SW of a mag 11.22v star; **ESO 322-53** E 3.′9.
12 40 55.6	−40 58 34	NGC 4603	11.4	3.4 x 2.5	13.5	SA(s)c: I-II	27	Faint stellar nucleus, many knots and/or superimposed stars; mag 9.12v star S 4.′3.
12 39 37.0	−40 44 27	NGC 4603A	13.3	1.9 x 0.6	13.3	SB(s)c? III-IV	90	Mag 8.74v star E 4.′6.
12 40 43.1	−40 45 50	NGC 4603C	13.1	1.8 x 0.4	12.6	S0 sp	160	**MCG −7-26-23** W 4.′7; mag 8.74v star W 7.′8.
12 42 07.9	−40 49 19	NGC 4603D	13.2	1.5 x 1.1	13.6	SAB(s)d: III-IV	74	Almost stellar nucleus, uniform envelope.
12 42 16.7	−40 38 34	NGC 4616	13.4	0.9 x 0.9	13.1	E$^+$:		Mag 11.9 star on NE edge; **PGC 42667** S 1.′0.
12 42 37.5	−40 44 42	NGC 4622	12.4	1.7 x 1.6	13.4	(R′)SA(r)a pec	173	Very thin, tightly wound arms.
12 43 49.0	−40 42 53	NGC 4622A	13.6	0.6 x 0.6	12.4	SAB0$^-$ pec:		Interacting pair with NGC 4622B on SE edge.
12 43 50.2	−40 43 07	NGC 4622B	13.8	1.0 x 0.5	13.0	S0 pec	63	Bright nucleus offset NE, fainter extension SW; pair with NGC 4622A on NW edge.
12 44 19.2	−40 43 59	NGC 4650	11.6	3.2 x 2.8	13.9	SB(s)0/a pec	164	Bright, diffuse nucleus with short, bright bar E-W; **PGC 42911** E 1.′9.
12 44 49.2	−40 42 53	NGC 4650A	13.3	1.6 x 0.8	13.4	S0/a pec:	158	Multi-galaxy system. The much elongated NGC 4650A has a short, stout companion laying at a right angle at its center forming an elongated plus sign.
12 45 14.8	−40 49 29	NGC 4661	13.5	1.1 x 0.4	12.5	E$^+$: sp	121	A pair of mag 13 stars N 1.′1.
12 47 30.1	−39 34 15	NGC 4679	12.4	2.4 x 1.0	13.2	SB(s)c pec: II	4	Faint, patchy outer arms; mag 8.81v star N 5.′6; ESO 322-84 SE 6.′8.
12 48 02.6	−40 49 06	NGC 4696C	13.7	1.9 x 0.3	12.9	Sb: sp	139	Mag 11.7 star SSE 3.′3.
12 48 26.0	−40 56 11	NGC 4696E	13.6	1.8 x 0.7	13.7	SB(s)0° sp	0	Stellar nucleus, faint envelope; mag 11.7 star NW 4.′9.
12 53 53.2	−39 42 55	NGC 4767	11.5	2.6 x 1.2	12.7	E	123	Stellar galaxy **PGC 43861** E 1.′8.
12 53 01.5	−39 50 08	NGC 4767A	15.2	0.8 x 0.3	13.5	Scd pec sp	114	Mag 12.6 star W 1.′4.
12 54 44.8	−39 51 09	NGC 4767B	12.9	1.3 x 1.0	13.0	SB(s)cd: II-III	99	Mag 10.26v star NNW 7.′2; note: **ESO 323-40** lies 1.′0 S of this star.
12 56 12.4	−29 30 10	NGC 4806	12.8	1.2 x 1.0	12.8	SB(s)c? III	50	Several branching, patchy arms.
12 57 47.4	−39 45 44	NGC 4832	12.2	1.9 x 1.2	13.0	S0°	25	Bright star on E edge.
13 01 22.2	−30 56 07	NGC 4903	12.9	1.8 x 1.4	13.7	SB(rs)c II	73	E-W bar, many branching arms; mag 7.39v star NE 8.′2.
13 01 30.8	−30 52 09	NGC 4905	13.3	1.8 x 1.1	13.9	SB(s)0$^+$ pec	26	Mag 7.39v star NE 4.′3.
13 04 17.2	−30 31 33	NGC 4936	10.7	2.7 x 2.3	12.7	E0	168	Mag 12 star on E edge; mag 9.24v star SW 8.′2.
13 05 20.5	−35 20 12	NGC 4947	11.9	2.4 x 1.3	13.0	SAB(r)b pec II	10	Small, bright nucleus, patchy arms.
13 04 20.8	−35 13 47	NGC 4947A	14.5	1.7 x 0.7	14.5	SB(s)m: IV-V	11	
13 06 10.4	−37 35 10	NGC 4953	12.9	1.1 x 0.9	12.7	S0$^+$: pec	46	In center of a group; six objects in a common envelope. **ESO 382-8A** at SW edge; mag 10.67v star NW 1.′6; **ESO 382-9** E 8.′0.
13 06 04.9	−29 45 17	NGC 4955	12.2	1.8 x 1.3	13.1	E2	23	
13 18 23.6	−35 27 33	NGC 5062	12.2	2.2 x 0.7	12.5	S0° pec sp	130	Mag 11.01v star SW 1.′7; very faint anonymous spiral galaxy with stellar nucleus at NW edge.
13 18 25.6	−35 21 08	NGC 5063	12.3	2.3 x 1.8	13.7	(R′)SA(rs)a	143	Bright nucleus with several stars superimposed.
13 21 57.4	−36 37 51	NGC 5102	9.6	8.7 x 2.8	13.0	SA0$^-$	48	17.′1 ENE of mag 2.76v star Iota Centauri.
13 23 18.8	−32 20 36	NGC 5108	13.9	1.2 x 0.3	12.7	SB(s)bc? sp II	2	Mag 10.78v star E 3.′5.
13 24 01.6	−32 20 38	NGC 5114	12.4	1.7 x 1.0	12.9	SAB0$^-$	80	Mag 10.78v star W 5.′9.
13 24 45.5	−37 40 58	NGC 5121	11.5	1.9 x 1.5	12.5	(R′)SA(s)a	36	
13 25 32.8	−37 22 43	NGC 5121A	14.9	1.7 x 0.5	14.6	SAB(s)dm IV-V	116	Located 2.′2 NW of mag 10.36v star. Very low surface brightness, faint nucleus and arms.
13 24 50.2	−30 18 31	NGC 5124	12.1	2.2 x 0.7	12.5	E6:	9	NGC 5126 SSE 1.′6.
13 24 53.7	−30 20 01	NGC 5126	13.1	1.4 x 0.4	12.3	S0/a pec sp	57	**PGC 46903** on S edge.
13 25 44.1	−29 50 01	NGC 5135	12.1	2.6 x 1.8	13.6	SB(s)ab II	29	Bright nucleus in bright SE-NW bar, two main arms.
13 26 21.6	−33 52 08	NGC 5140	11.8	2.0 x 1.7	13.0	SAB(s)0$^-$:	33	Faint stars superimposed on envelope; mag 10.02v star E 5.′6.
13 27 36.5	−29 33 44	NGC 5150	12.6	1.3 x 1.0	12.7	SBbc: II	115	Located 1.′7 SW of mag 9.90v star.
13 27 50.0	−29 37 03	NGC 5152	12.4	2.0 x 0.6	12.4	SB(s)b	117	Interacting with NGC 5153; nucleus offset E; star N of nucleus.
13 27 54.5	−29 37 03	NGC 5153	11.8	2.1 x 1.4	13.0	E1 pec	175	Interacting with NGC 5152 W.
13 29 13.5	−33 10 25	NGC 5161	11.2	5.6 x 2.2	13.8	SA(s)c: I-II	77	Bright thin arms with many knots and HII regions; mag 10.27v star 3.′9 W.
13 31 28.9	−34 47 41	NGC 5188	12.1	3.0 x 1.1	13.3	(R′)SAB(s)b pec: II	104	Very faint stars E and W of bright center.
13 31 53.8	−33 14 05	NGC 5193	11.6	1.5 x 1.5	12.5	E pec:		May be interacting with NGC 5193A on W edge; mag 8.09v star E 4.′6.
13 31 49.2	−33 14 24	NGC 5193A	13.0	0.9 x 0.3	11.4	S0°: sp	47	Faint anonymous galaxy W 2.′0.
13 35 06.3	−33 28 51	NGC 5215A	13.1	0.6 x 0.4	11.4	S0 pec	105	Connected with NGC 5215B E; bright nucleus with faint extension W.
13 35 09.7	−33 29 03	NGC 5215B	12.9	1.1 x 0.5	12.1	S0 pec	80	Connected with NGC 5215A W.
13 35 56.9	−33 27 12	NGC 5220	12.0	2.3 x 0.7	12.4	SAa pec sp	97	Bright nucleus with E-W dark lane S; star E of nucleus; mag 9.06v star SE 2.′8.
13 37 01.0	−29 51 44	NGC 5236	7.5	12.9 x 11.5	12.8	SAB(s)c II	44	= **M 83**. Galaxy **ESO 444-85** lies on E edge.
13 39 56.1	−31 38 39	NGC 5253	10.4	5.0 x 1.9	12.7	Pec	42	**ESO 445-7** SE 6.′4.
13 41 36.7	−29 54 46	NGC 5264	12.0	2.5 x 1.5	13.3	IB(s)m IV-V	54	Very faint envelope, many stars superimposed.
13 47 24.5	−30 24 29	NGC 5291	14.1	1.2 x 0.8	14.0	E pec:	168	Pair of mag 10.4v and 10.7v stars NW 2.′1; MCG −5-33-5 **Seashell Galaxy** on SW edge.
13 47 40.2	−30 56 21	NGC 5292	11.8	1.8 x 1.5	12.8	(R′)SA(rs)ab	55	Pair of stars on NE edge.
13 48 36.3	−30 25 42	NGC 5298	13.1	1.4 x 0.7	12.9	SB(r)b	69	Mag 10.31v star NE 5.′9; faint anonymous galaxy N 1.′0; MCG −5-33-10 W 5.′5.
13 48 50.1	−30 30 45	NGC 5302	12.1	1.8 x 1.1	12.7	SB(s)0$^+$:	153	**MCG −5-33-9** and **MCG −5-33-11** WSW 8.′0.
13 50 01.6	−30 34 41	NGC 5304	12.5	1.5 x 1.0	13.0	E$^+$ pec:	146	Two faint stars S and E of nucleus.
13 55 59.7	−30 20 30	NGC 5357	12.0	1.5 x 1.3	12.7	E	23	Located between a pair of 11 and 12 mag stars.
14 01 10.4	−33 56 46	NGC 5397	12.7	1.4 x 1.0	13.0	SAB(s)0$^-$:	60	Mag 12.1 star W 4.′8.
14 01 22.0	−33 03 43	NGC 5398	12.2	2.8 x 1.7	13.7	(R′)SB(s)dm pec: IV	172	Patchy arms, star or bright knot SW of nucleus.
14 03 38.6	−33 58 42	NGC 5419	10.9	4.2 x 3.3	13.7	E	77	Bright center, faint envelope, many stars superimposed; **MCG −6-31-20** E 11.′6.
13 32 25.3	−33 08 16	PGC 47626	14.6	2.0 x 0.8	15.2	E$^+$5	63	Mag 8.09v star SSW 7.′1; note: NGC 5139 W 4.′6 from this star.

GALAXY CLUSTERS

RA h m s	Dec ° ′ ″	Name	Mag 10th brightest	No. Gal	Diam ′	Notes
12 54 18.0	−29 01 00	A 3528	16.3	70	17	**ESO 443-2** near N edge; includes a number of PGC galaxies.
12 55 36.0	−30 21 00	A 3530	16.0	34	17	Overlaps A 3532; ESO 443-11 at center; remaining GX also included in A 3532.
12 57 18.0	−30 22 00	A 3532	16.2	36	17	**ESO 443-18** on SE edge; numerous PGC galaxies.
13 01 00.0	−32 26 00	A 3537	14.3	35	28	Large concentration of galaxies in SE quadrant.
13 08 42.0	−34 33 00	A 3542	16.4	45	17	**MCG −6-29-13** near center; all other members anonymous, stellar.
13 19 12.0	−37 10 00	A 3553	16.2	36	17	Numerous anonymous GX stretch from NE to SW edges.
13 19 30.0	−33 28 00	A 3554	16.4	59	17	ESO 382-43 at center; **ESO 382-40** N 12.′6; **ESO 382-37** NW 16.′4; **ESO 382-39** SSW 8.′5.
13 24 06.0	−31 39 00	A 3556	16.4	49	17	**PGC 46821** W of center 1.′8; **PGC 46860** SE 5.′0; **PGC 46863** WSW 4.′8.

(Continued from previous page)

167 GALAXY CLUSTERS 167

RA h m s	Dec ° ′ ″	Name	Mag 10th brightest	No. Gal	Diam ′	Notes
13 27 54.0	−31 29 00	A 3558	15.1	226	22	ESO 444-46 at center; four PGC galaxies towards E edge.
13 29 54.0	−29 31 00	A 3559	15.7	141	17	See notes for ESO 444-55; numerous other anonymous, stellar/almost stellar GX
13 31 48.0	−33 13 00	A 3560	15.1	184	22	**ESO 383-11** N of center 17′0; most members anonymous, stellar.
13 33 30.0	−31 40 00	A 3562	15.5	129	22	Most members anonymous, stellar/almost/stellar.
13 34 24.0	−35 13 00	A 3564	15.6	53	22	Most members anonymous, faint, stellar.
13 36 42.0	−33 58 00	A 3565	14.0	64	56	
13 39 00.0	−35 33 00	A 3566	15.8	100	17	Most members anonymous, stellar/almost stellar.
13 46 48.0	−37 54 00	A 3570	15.8	31	17	
13 47 30.0	−32 51 00	A 3571	15.8	126	17	Most members anonymous, stellar/almost stellar.
13 48 12.0	−33 22 00	A 3572	15.7	49	17	**ESO 383-83** S of center 5′9.
13 49 12.0	−30 17 00	A 3574	13.4	31	56	
13 52 36.0	−32 52 00	A 3575	15.8	49	17	**ESO 384-10** on E edge; all other members anonymous, stellar.
12 59 42.0	−33 40 00	AS 718	16.4	16	17	All members anonymous, stellar.
13 06 06.0	−37 35 00	AS 721	16.3	26	17	See notes NGC 4953; most members anonymous, stellar.
13 23 00.0	−34 52 00	AS 731	16.4	29	17	**ESO 382-55** at center; all others anonymous, stellar.
13 43 30.0	−38 11 00	AS 740	15.6	2	22	Many members anonymous, stellar.
13 44 36.0	−34 18 00	AS 742	15.8	11	17	**ESO 383-72** E of center 5′0; all others anonymous, stellar.
13 46 18.0	−39 53 00	AS 743	16.3		17	**ESO 325-18** E of center 7′4; **ESO 325-12** W 12′5; **ESO 325-15** WSW 8′1; others anonymous.
13 49 48.0	−34 58 00	AS 746	16.1	8	17	All members anonymous, stellar.
14 03 36.0	−33 58 00	AS 753	13.9	18	56	

OPEN CLUSTERS

RA h m s	Dec ° ′ ″	Name	Mag	Diam ′	No. ★	B ★	Type	Notes
13 31 30.0	−35 04 00	ESO 383-10		5			cl	(A) Galaxy ESO 383-8 on SW edge.
13 54 38.1	−31 57 13	ESO 445-74		2.5			cl	

BRIGHT NEBULAE

RA h m s	Dec ° ′ ″	Name	Dim ′ Maj x min	Type	BC	Color	Notes
13 57 42.0	−39 59 00	NGC 5367	2.5 x 2.5	R	2-5	1-4	A round and fairly bright nebula surrounding a pair of 10th and 11th mag stars separated by 4″ in PA 33°.

PLANETARY NEBULAE

RA h m s	Dec ° ′ ″	Name	Diam ″	Mag (P)	Mag (V)	Mag cent ★	Alt Name	Notes
13 25 36.3	−37 36 26	PN G310.3+24.7	115			12.8	PK 310+24.1	Mag 11.7 star near center; N-S line of three mag 14-15 stars extend S from center; mag 9.77v star S 2′9.

168 GALAXIES 168

RA h m s	Dec ° ′ ″	Name	Mag (V)	Dim ′ Maj x min	SB	Type Class	PA	Notes
11 17 52.9	−40 35 37	ESO 319-11	14.2	2.5 x 1.5	15.4	SB(r)cd: I-II	115	Numerous stars scattered over faint outer bands.
11 22 08.3	−38 04 07	ESO 319-16	13.7	1.7 x 0.9	14.0	SB(s)c pec: III-IV	27	Mag 13.7 star on N edge; mag 11.04v star N 5′5.
11 34 43.8	−38 15 01	ESO 320-4	14.0	1.5 x 0.4	13.3	SB(s)c: IV	51	Mag 9.73v star NE 11′8.
11 34 55.7	−38 31 06	ESO 320-5	13.2	0.9 x 0.6	12.4	SAB(r)0/a	106	Mag 9.73v star E edge.
11 35 45.2	−38 21 53	ESO 320-7	13.2	1.1 x 0.6	12.6	S0/a:	3	Mag 12.4 star N 2′4.
11 44 45.8	−39 36 52	ESO 320-19	13.7	1.1 x 0.3	12.3	S	2	Located between a pair of mag 14 stars.
11 46 00.1	−39 20 04	ESO 320-20	13.9	1.4 x 0.3	12.8	Sb:	1	Mag 12.4 star W 0′7; mag 10.49v star W 2′5.
11 49 26.0	−38 49 36	ESO 320-24	13.4	1.0 x 0.8	13.0	Sc?	5	Mag 12.8 star S edge; mag 11.04v star NW 5′6.
11 49 50.0	−38 47 06	ESO 320-26	12.0	2.6 x 1.0	12.9	Sb I-II	161	Mag 11.04v star W 8′2.
11 50 25.0	−38 38 52	ESO 320-27	13.7	1.4 x 0.6	13.3	IB? pec	41	Mag 9.69v star N 8′6.
11 53 11.8	−39 07 48	ESO 320-30	12.3	2.4 x 1.3	13.4	(R′)SAB(r)a:	126	Mag 10.31v star S 5′2.
11 54 05.9	−39 51 50	ESO 320-31	13.0	2.9 x 0.3	12.7	SB(s)c? sp II	143	0′6 W of mag 8.6v star.
11 54 39.1	−40 55 54	ESO 320-32	13.7	0.9 x 0.5	12.7	(R′)SB(r)b	5	Mag 13.1 star SW 2′3.
11 56 46.5	−38 11 31	ESO 320-35	12.8	2.5 x 1.8	14.3	SAB(s)c pec: II	8	Stellar nucleus; mag 8.85v star S 8′2.
11 58 19.0	−39 33 01	ESO 321-1	12.6	2.1 x 1.3	13.5	SA(rs)cd III	139	Several knots; mag 8.43v star S 5′1.
12 03 03.1	−38 40 29	ESO 321-4	14.6	0.5 x 0.2	11.9	S	41	Almost stellar, mag 12.6 star N 2′0.
12 05 47.3	−38 51 17	ESO 321-5	12.6	1.2 x 1.0	12.7	SB(rs)cd	1	Mag 9.66v star E 6′6.
12 07 22.0	−40 13 06	ESO 321-7	13.4	0.8 x 0.5	12.3	Sa	113	Located 2′5 E of NGC 4112.
12 11 42.0	−38 32 55	ESO 321-10	13.0	2.0 x 0.3	12.3	S0/a: sp	72	Mag 9.44v star WSW 2′3.
12 12 36.5	−40 02 11	ESO 321-12	14.2	1.2 x 0.3	12.9	S	99	Mag 9.53v star ESE 9′5.
12 13 49.3	−38 13 51	ESO 321-14	14.3	1.6 x 0.5	13.9	IBm V-VI	21	Mag 10.00v star E 7′6.
12 15 26.8	−38 08 42	ESO 321-16	12.8	2.6 x 0.7	13.3	SB(s)cd pec: III-IV	125	Mag 10.53v star SW 5′5; ESO 321-18 NE 6′3.
12 15 54.5	−38 05 34	ESO 321-18	13.2	1.2 x 0.8	13.0	IB(s)m IV-V	39	ESO 321-16 SW 6′3.
12 17 04.3	−39 03 04	ESO 321-19	12.1	0.9 x 0.7	11.5	SAB(r)0°	134	
12 19 57.2	−39 39 57	ESO 321-20	11.3	1.3 x 0.8	11.2		143	Bright star superimposed on SE quadrant of galaxy; mag 12.1 star NW 1′8.
12 20 16.4	−40 23 26	ESO 321-21	12.5	1.0 x 0.8	12.1	Sc	109	Mag 9.45v star W 1′5.
12 21 43.3	−39 46 14	ESO 321-25	11.9	2.4 x 1.2	12.8	SB(rs)cd pec: III-IV	14	Mag 11.05v star S edge.
12 21 52.5	−38 47 24	ESO 321-26	11.8	1.0 x 0.4	10.7		61	Star superimposed center.
12 25 15.5	−39 05 44	ESO 322-5	13.5	0.8 x 0.7	12.8	S0	160	Mag 15.0 star superimposed S edge; mag 10.42v star NW 8′6.
12 25 36.5	−40 43 32	ESO 322-7	14.5	1.4 x 1.1	14.9	SB(s)m V-VI	1	Stellar nucleus, few stars superimposed, faint envelope.
12 26 09.6	−39 07 30	ESO 322-9	13.2	1.5 x 0.9	13.4	SB(s)b pec	131	Short line of three E-W oriented mag 10.48v, 10.91v, 10.02v stars S 5′4.
12 27 16.1	−40 28 30	ESO 322-11	13.8	1.1 x 0.7	13.3	S?	24	Mag 12.6 star N 1′3.
12 28 29.5	−38 14 36	ESO 322-18	14.1	1.1 x 0.8	13.8	Sa?	58	Mag 14.3 star SW edge.
12 29 07.5	−40 40 28	ESO 322-19	13.9	1.5 x 0.3	12.9	SB(s)cd? sp	117	Pair with ESO 322-20 SE edge; mag 10.50v star N 7′1.
12 29 12.7	−40 41 35	ESO 322-20	13.4	1.2 x 1.1	13.6	SB(rs)dm IV	105	Mag 15.5 star on S edge; pair with ESO 322-19 NW edge.
12 30 16.0	−40 52 20	ESO 322-21	13.9	0.6 x 0.4	12.3	Pec	40	Mag 10.84v star WNW 5′3.
12 32 52.0	−40 29 24	ESO 322-25	13.5	0.9 x 0.4	12.3	S0?	113	Mag 12.3 star E 2′0.
12 34 42.8	−40 18 02	ESO 322-27	12.6	1.3 x 1.1	12.8	(R)SAB(r)0/a	146	Mag 9.76v star S 4′6.
12 37 11.8	−38 33 32	ESO 322-32	14.0	0.8 x 0.3	12.6		102	Mag 10.62v star WNW 5′4.
12 38 28.3	−40 22 05	ESO 322-40	13.3	1.3 x 0.7	13.0	Sbc II	71	Mag 8.58v star E 8′2.
12 42 00.1	−39 49 07	ESO 322-54	14.0	1.1 x 0.6	13.4		128	Double system, contact; superimposed star involved?
11 16 00.8	−33 58 04	ESO 377-31	13.3	1.8 x 0.9	13.7	SAB(s)bc: I-II	146	Mag 12.6 star E 2′0.

(Continued from previous page)

GALAXIES

RA h m s	Dec ° ′ ″	Name	Mag (V)	Dim ′ Maj x min	SB	Type Class	PA	Notes
11 17 04.6	−34 57 19	ESO 377-34	13.2	2.4 x 0.6	13.5	SB(s)bc: II	160	Mag 6.88v star NNW 2′.1.
11 19 00.1	−36 14 45	ESO 377-38	14.0	0.9 x 0.7	13.3	(R′)SB(s)b	100	Mag 11.8 star NNW 5′.6.
11 24 20.9	−35 01 00	ESO 377-41	13.3	1.2 x 1.1	13.5	S0	18	Mag 10.20v star NE 3′.3.
11 25 53.0	−35 23 41	ESO 377-46	13.1	1.1 x 1.0	13.1	(R)SAB(rs)0°	8	Mag 12.8 star NE 2′.2; **ESO 377-43** W 8′.1; **MCG −6-25-18** SSW 10′.8.
11 28 04.3	−36 32 34	ESO 378-3	13.1	2.1 x 1.3	14.1	SAB(s)c: II-III	176	Mag 10.29v star E 8′.6.
11 28 45.6	−35 02 11	ESO 378-5	13.9	1.5 x 1.0	14.2	SA(s)cd IV	80	Mag 10.37v star W 4′.4.
11 30 04.7	−35 32 22	ESO 378-8	13.8	0.8 x 0.4	12.4	Sc	2	Faint, stellar anonymous galaxy WNW 1′.0.
11 34 43.7	−37 12 59	ESO 378-11	14.2	1.6 x 0.3	13.2	Sc: sp III-IV	160	Mag 10.31v star NNE 8′.7.
11 37 02.5	−36 48 55	ESO 378-12	13.4	1.4 x 0.6	13.0	SAab: II-III	43	Triangle of mag 13-14 stars NE 2′.1.
11 47 16.4	−37 33 01	ESO 378-20	12.8	1.4 x 0.7	12.6	SB(rs)0°?	35	Mag 10.70v star NE 3′.9.
11 50 29.0	−35 30 30	ESO 378-27	14.0	0.6 x 0.6	12.7	Sa		Mag 11.7 star on SW edge.
11 51 05.6	−34 33 47	ESO 379-1	15.0	1.7 x 0.7	15.0	SB(r)d: IV	95	Mag 12.1 star WSW 1′.1; mag 10.82v star SW 7′.2.
11 53 03.4	−36 38 22	ESO 379-6	13.4	2.6 x 0.4	13.3	Sc: sp II	69	Mag 6.63v star NNE 4′.6.
11 56 43.8	−34 25 37	ESO 379-13	14.9	1.7 x 1.1	15.5	Sm? IV-V?	149	Mag 9.85v star NE 10′.2; **ESO 379-14** E 5′.3.
11 59 31.8	−36 43 09	ESO 379-19	13.9	0.7 x 0.6	12.8	(R′)SAB(r)ab	47	Mag 9.61v star E 9′.0.
12 00 57.8	−35 11 39	ESO 379-20	14.5	1.6 x 0.8	14.7	SB(s)a pec	166	Very faint arms extend N and S from bright center; mag 10.51v star WNW 4′.4; ESO 379-21 S edge.
12 01 01.3	−35 12 59	ESO 379-21	14.0	1.7 x 0.3	13.1	SB(s)b? pec	4	Mag 11.00v star SW 1′.3; ESO 379-20 N edge.
12 01 20.7	−33 52 43	ESO 379-22	13.2	1.1 x 0.8	13.0	S?	8	Mag 11.18vb star W 4′.6.
12 06 17.0	−35 58 52	ESO 379-26	13.4	1.0 x 0.9	13.2	S0	15	Small, faint, elongated anonymous galaxy S 1′.5; mag 10.87v star NE 2′.2.
12 06 53.3	−36 16 25	ESO 379-27	14.3	1.7 x 0.6	14.2	Sc	140	Mag 11.12v star NE 6′.4.
12 08 59.7	−36 42 13	ESO 379-30	13.9	1.0 x 0.3	12.5	Sb	5	Mag 8.81v star E 2′.8.
12 10 39.4	−34 03 31	ESO 379-31	13.3	1.2 x 1.0	13.4	S?	164	Strong core, smooth envelope.
12 13 39.4	−34 29 40	ESO 379-35	13.3	0.5 x 0.5	11.7			Mag 10.09v star SW 0′.7; **ESO 379-33** and double system **ESO 379-32** W 6′.8.
12 14 45.2	−35 30 37	ESO 380-1	12.0	3.0 x 1.4	13.4	(R′)SB(s)b:	9	Mag 10.18v star W 8′.4; **ESO 380-2** SSE 5′.7.
12 15 33.6	−35 37 48	ESO 380-6	11.6	4.0 x 1.7	13.6	(R′)SA(s)b:	78	**ESO 380-2** W 8′.1.
12 15 44.2	−34 54 00	ESO 380-7	13.9	0.8 x 0.7	13.1	SAB(s)c I-II	146	Mag 9.11v star SSW 8′.0.
12 16 58.5	−37 28 16	ESO 380-8	14.1	1.6 x 0.8	14.2	SB(s)dm IV-V	34	Mag 11.8 star SSW 6′.6.
12 19 50.9	−36 27 36	ESO 380-14	13.3	1.7 x 0.7	13.3	SB(rs)c: II	75	Mag 9.93v star SW 5′.6.
12 22 02.0	−35 47 32	ESO 380-19	12.3	3.2 x 0.5	12.6	Scd: sp II	121	Mag 9.99v star SW 3′.6.
12 24 34.3	−35 24 31	ESO 380-25	13.8	1.7 x 0.3	12.9	SB(s)d: sp III-IV	11	Mag 5.32v star W 12′.0.
12 26 54.8	−36 25 32	ESO 380-29	14.1	1.8 x 0.5	13.8	SB(s)m: IV-V	75	Pair of E-W oriented mag 12.3, 11.33v stars SW 7′.6.
12 27 23.7	−33 31 36	ESO 380-30	13.6	0.8 x 0.7	12.9	SA(r)a	76	
12 27 44.0	−34 25 24	ESO 380-33	13.8	0.9 x 0.5	12.8	S	144	Bright knot NW end; mag 13.5 star N 1′.3.
12 28 00.1	−37 19 46	ESO 380-34	14.6	1.7 x 0.3	13.7	SB(s)m: sp V-VI	156	Mag 10.33v star S edge.
12 29 44.3	−33 08 39	ESO 380-35	14.1	1.0 x 1.0	14.0	SB(s)c II		Stellar nucleus.
12 32 28.0	−36 22 14	ESO 380-40	13.7	0.9 x 0.7	13.1	Sb	174	Mag 10.59v star SSE 2′.9.
12 33 04.1	−33 34 56	ESO 380-41	14.1	0.8 x 0.5	13.0	Sa	174	Mag 10.35v star N 2′.9.
12 33 48.3	−35 06 56	ESO 380-42	13.6	1.1 x 0.9	13.5	(R′)SB(r)a pec	86	Mag 10.67v star SE 3′.1.
12 33 49.1	−37 45 22	ESO 380-43	14.1	0.8 x 0.5	13.0	Sb	48	Mag 9.14v star NW 10′.8.
12 34 58.4	−36 52 05	ESO 380-46	14.0	0.6 x 0.5	12.5	Sb	4	Mag 11.08v star W 6′.5.
12 39 09.5	−34 46 53	ESO 381-4	13.5	1.0 x 0.5	12.6	S?	132	Located between a pair of mag 12.4 stars.
12 40 32.9	−36 58 08	ESO 381-5	14.1	1.1 x 0.8	13.8	S?	12	Mag 13.8 star E edge; mag 12.1 star S 0′.9.
12 40 58.5	−36 43 57	ESO 381-9	13.0	1.4 x 1.2	13.4	SB(s)bc pec III-IV	150	Mag 10.03v star N 3′.9, IC 3639 SW 1′.8.
11 20 53.2	−29 24 10	ESO 438-23	13.1	1.5 x 0.9	13.3	S0?	38	Mag 11.7 star N 3′.9.
11 27 23.8	−29 15 29	ESO 439-9	13.6	1.9 x 0.5	13.4	(R)SB(r)a pec:	99	ESO 439-10 N 5′.3.
11 27 32.3	−29 10 38	ESO 439-10	14.7	1.6 x 0.5	14.3	Pec sp	145	Mag 12.3 star E 2′.4; mag 10.55v star NW 8′.6; ESO 439-9 S 5′.3.
11 27 41.0	−30 50 14	ESO 439-11	15.2	1.5 x 0.4	14.5	S	38	Mag 10.92v star SE 3′.2.
11 38 15.8	−29 43 44	ESO 439-20	13.4	1.7 x 0.7	13.4	SAB(r)bc II-III	99	Mag 14.7 star N edge; mag 12.6 star S 1′.1.
11 46 32.5	−30 05 58	ESO 440-6	13.3	1.7 x 0.5	13.0	S0⁺? pec	43	Mag 9.77v star N 4′.4.
11 51 52.3	−31 28 43	ESO 440-19	14.0	0.9 x 0.6	13.2	SB(r)bc	108	Mag 9.78v star ENE 6′.8; note: **ESO 440-22** is 3′.4 N of this star.
11 53 20.0	−32 39 02	ESO 440-26	13.9	1.1 x 0.8	13.6	SB(r)a	97	Strong N-S core; mag 10.51v star SE 1′.8; MCG −5-28-14 NNW 5′.6.
11 56 41.4	−32 23 11	ESO 440-32	13.1	1.0 x 0.6	12.4	SB(r)0/a	84	Mag 12.1 star S edge; bright anonymous galaxy W edge; mag 11.9 star W 1′.4.
11 56 59.9	−31 45 28	ESO 440-34	13.8	0.9 x 0.3	12.2	Sa:	121	Mag 10.21v star SW 11′.2.
12 01 42.8	−31 42 15	ESO 440-38	12.8	1.8 x 1.1	13.4	SAB(r)0⁻:	111	Close pair of mag 12.1, 12.2 stars NW 11′.0.
12 01 57.8	−30 14 11	ESO 440-39	14.2	1.5 x 0.8	14.3	IB(s)m V-VI	117	Double system; mag 9.65v star SE 9′.8.
12 02 06.4	−32 00 29	ESO 440-41	14.0	1.4 x 0.3	12.9	Sa:	18	Mag 10.50v star W 7′.8; anonymous galaxy N 3′.4.
12 02 17.9	−29 30 54	ESO 440-43	14.1	0.8 x 0.5	13.0	Sc	91	
12 02 46.0	−29 05 36	ESO 440-44	13.3	1.7 x 0.8	13.5	SB(s)m pec IV	0	Faint star superimposed W of center; mag 12.7 star NE 3′.0.
12 05 34.1	−31 25 26	ESO 440-49	13.0	1.7 x 1.4	13.8	SAB(s)d III-IV	76	Stellar nucleus; pair of E-W oriented mag 9.89v, 11.16v stars E 9′.1.
12 07 12.7	−29 56 18	ESO 441-1	14.2	1.0 x 0.2	12.3	S0	46	Located between a pair of mag 14.9 and 15.6 stars.
12 08 15.8	−32 00 36	ESO 441-7	14.2	2.2 x 0.3	13.6	Sbc: sp	161	Triangle of mag 10.66v, 10.72v, 9.78v stars centered E 3′.5; MCG −5-29-19 W 6′.8.
12 09 27.2	−32 30 54	ESO 441-12	13.5	1.8 x 0.4	13.0	Sab: sp	155	Mag 10.62v star E 5′.5.
12 10 14.2	−30 04 47	ESO 441-14	14.1	1.6 x 0.5	13.7	S pec?	150	MCG −5-29-27 E 2′.9; mag 8.35v star SE 10′.2.
12 11 08.3	−31 07 38	ESO 441-17	12.8	1.2 x 0.7	12.5	S?	85	**ESO 441-18** S 1′.5; mag 10.15v star SW 6′.9.
12 15 08.1	−30 46 00	ESO 441-22	13.0	2.4 x 0.6	13.2	SAB(rs)bc? I	179	Mag 10.78v star at N end.
12 34 06.2	−31 13 02	ESO 442-6	13.9	0.7 x 0.7	13.1	S0?		Mag 11.20v star W 7′.2.
12 42 42.5	−29 37 14	ESO 442-14	13.9	0.7 x 0.5	12.7	Sb	20	Mag 9.34v star SW 6′.0.
12 42 50.9	−30 24 35	ESO 442-15	13.1	0.8 x 0.7	12.3	(R′)SB(r)a	27	Mag 10.49v star SE 0′.9.
12 05 53.4	−29 17 28	IC 760	12.5	1.7 x 0.5	12.2	S0°: sp	148	Mag 10.06v star NE 8′.8.
12 10 14.2	−29 44 09	IC 764	12.0	4.8 x 1.6	14.1	SA(s)c? II-III	177	Mag 8.94v star N 7′.8.
11 31 51.4	−30 24 40	IC 2913	13.0	0.8 x 0.8	12.4	SA(r)0⁺:		Mag 9.28v star E 5′.7.
11 55 14.8	−37 41 50	IC 2977	12.3	1.6 x 0.7	12.2	I0?	137	Mag 6.46v star WSW 10′.1.
12 05 48.5	−29 58 22	IC 2996	13.5	1.5 x 0.4	12.8	S?	21	Star on S end.
12 07 14.2	−30 01 27	IC 3005	13.0	2.3 x 0.4	12.8	SBcd? sp II-III	160	Mag 10.35v star SSE 8′.0.
12 07 57.3	−30 20 25	IC 3010	12.2	1.9 x 1.8	13.4	(R′)SB(rs)0⁺:	84	Mag 9.47v star NE 12′.9.
12 09 00.4	−31 31 14	IC 3015	12.2	2.9 x 0.7	12.8	Sbc: sp II	166	Mag 9.55v star E 3′.7.
12 23 45.2	−34 37 20	IC 3253	11.6	2.5 x 1.1	12.6	SA(s)c: II-III	23	Many knots; strong dark lane E side; mag 10.10v star E 11′.2.
12 25 09.0	−39 46 35	IC 3290	12.0	2.0 x 1.4	12.9	(R′)SB(s)0/a	43	Located on SW edge of NGC 4373.
12 27 37.2	−39 20 20	IC 3370	11.0	2.1 x 1.6	12.4	E2-3	45	Mag 8.99v star SW 3′.2.
12 40 52.9	−36 45 25	IC 3639	12.3	1.2 x 1.2	12.5	SB(rs)bc: II		Mag 10.03v star N 5′.2, mag 14.3 star on SW edge; ESO 381-9 NE 1′.8.
11 17 32.8	−30 31 18	MCG −5-27-6	14.0	0.7 x 0.4	12.5	(R′)SA(r)0/a pec?	138	
11 18 53.6	−29 25 29	MCG −5-27-9	13.7	0.9 x 0.5	12.7	S?	152	**ESO 438-21** on E edge.
11 24 07.2	−30 27 25	MCG −5-27-11	14.4	0.5 x 0.3	12.2		3	
11 29 43.3	−31 51 08	MCG −5-27-14	14.2	0.7 x 0.4	12.7	S?	18	**ESO 439-14** NNE 5′.8.
11 33 22.3	−33 09 25	MCG −5-27-17	14.0	0.7 x 0.4	12.5	Sbc		
11 38 00.5	−32 19 31	MCG −5-28-1	13.5	1.3 x 0.8	13.4	SAB(r)bc I-II	106	Star superimposed E tip of bright core.
11 38 05.9	−32 33 38	MCG −5-28-2	13.9	1.0 x 0.4	12.8	S(r?)0	107	
11 48 54.8	−30 10 27	MCG −5-28-8	13.5	0.6 x 0.5	12.1	Sb	69	

(Continued from previous page)

GALAXIES

RA h m s	Dec ° ′ ″	Name	Mag (V)	Dim ′ Maj x min	SB	Type Class	PA	Notes
11 52 32.9	−32 50 03	MCG −5-28-13	13.8	0.8 x 0.5	12.7		48	
11 53 08.2	−32 34 00	MCG −5-28-14	13.1	1.1 x 0.8	12.8	S0?	179	ESO 440-26 and mag 10.51v star SSE 5′.6.
11 54 51.5	−31 34 28	MCG −5-28-16	13.4	0.4 x 0.3	11.0	Interacting	0	Small elongated galaxy E edge.
12 03 54.9	−29 37 33	MCG −5-29-4	14.0	0.7 x 0.5	12.7		105	**ESO 440-45** NE 5′.1.
12 05 33.1	−30 09 41	MCG −5-29-6	14.1	0.9 x 0.4	12.9	Sb	104	
12 06 52.1	−31 56 51	MCG −5-29-17	13.8	1.2 x 0.8	13.6		74	Double system, interaction.
12 07 43.9	−32 01 12	MCG −5-29-19	13.8	1.1 x 1.0	13.8	SB(r)b	74	ESO 441-7 E 6′.8; mag 10.46v star SSE 4′.2.
12 08 32.4	−31 21 19	MCG −5-29-22	13.8	0.7 x 0.4	12.3	Sa	105	
12 10 27.0	−30 05 21	MCG −5-29-27	14.5	0.6 x 0.3	12.5	Interacting	123	ESO 441-14 W 2′.9; mag 8.35v star SE 7′.6.
12 25 49.5	−30 41 18	MCG −5-29-35	14.6	0.7 x 0.4	13.0	Sbc	12	
12 33 25.1	−31 21 49	MCG −5-30-4	13.2	0.6 x 0.6	12.2	E/S0		Mag 8.80v star WNW 11′.8.
11 16 15.7	−33 49 41	NGC 3606	12.3	1.5 x 1.3	13.0	E0:	168	Mag 15.3 star on W edge; mag 10.97v star NNE 4′.5.
11 18 16.4	−32 48 36	NGC 3621	9.7	12.3 x 6.8	14.3	SA(s)d III-IV	159	Many stars superimposed overall; a pair of mag 10 stars S of the bright central core; faint outer envelope.
11 29 44.6	−36 23 27	NGC 3706	11.3	3.0 x 1.8	13.0	SA(rs)0^{-}	78	
11 31 32.1	−30 18 23	NGC 3717	11.2	6.0 x 1.1	13.1	SAb: sp III	33	Strong dark lane entire W side; mag 9.11v star W 6′.9.
11 35 32.1	−37 57 20	NGC 3742	12.1	2.4 x 1.7	13.4	(R′)SAB(rs)ab pec	116	Star or bright knot W edge of large core.
11 35 54.0	−37 59 57	NGC 3749	12.1	4.2 x 1.2	13.7	SA(s)a pec sp	107	
11 39 01.9	−37 44 20	NGC 3783	11.9	1.9 x 1.7	13.0	(R′)SB(r)ab I-II	160	Mag 9.20v star on SE edge.
11 49 03.6	−37 31 02	NGC 3903	12.8	1.1 x 1.0	12.7	SAB(rs)c pec: II	116	**ESO 378-23** W 2′.3.
11 49 13.1	−29 16 29	NGC 3904	10.9	2.7 x 1.9	12.6	E2-3:	8	
12 06 40.9	−29 45 39	NGC 4105	10.7	2.7 x 2.0	12.5	E3	151	Interacting with NGC 4106 E; Mag 11.03v star S 2′.3.
12 06 45.8	−29 46 07	NGC 4106	11.4	1.6 x 1.3	12.0	SB(s)0^{+}	77	Interacting with NGC 4105 W.
12 07 09.4	−40 12 27	NGC 4112	11.5	1.6 x 0.9	11.8	SAB(rs)b pec	5	Galaxy ESO 321-7 E 2′.5; mag 10.41 star S 1′.5; mag 9.25v star S 2′.1.
12 22 12.7	−33 29 03	NGC 4304	11.7	3.0 x 2.4	13.6	(R′)SB(s)bc pec: II	12	Pair of arms close to bright nucleus form a flattened "S" shape. Faint outer envelope N and S of bright center.
12 25 18.1	−39 45 36	NGC 4373	10.9	3.4 x 2.5	13.1	SAB(rs)0^{-}:	43	
12 25 37.2	−39 19 06	NGC 4373A	11.6	2.6 x 0.8	12.3	SA0^{+}: sp	149	Located 1′.5 N of mag 10.39v star.
12 26 43.8	−39 08 08	NGC 4373B	14.1	1.5 x 0.9	14.3	SB(s)dm IV-V	173	ESO 322-9 W 6′.7.
12 27 52.6	−30 05 52	NGC 4456	13.3	1.2 x 0.6	12.7	Sbc II-III	150	Star on E edge.
12 32 05.1	−39 58 59	NGC 4499	13.2	1.8 x 1.3	13.9	SB(rs)bc: II-III	93	Almost stellar anonymous galaxy S 1′.0.
12 35 37.0	−39 54 35	NGC 4507	12.1	1.7 x 1.3	12.8	(R′)SAB(rs)b I-II	56	Faint star on NE edge.
12 36 07.7	−39 26 19	NGC 4553	12.2	2.1 x 1.0	12.9	SA(r)0^{+}	176	Bright, elongated nucleus; small anonymous galaxy SW 2′.6.
12 37 43.6	−35 31 06	NGC 4574	13.0	1.7 x 1.1	13.5	SAB(s)c II	113	Short, bright bar; patchy arms; several stars superimposed E edge.
12 37 51.2	−40 32 14	NGC 4575	12.6	2.0 x 1.3	13.5	SB(s)bc pec:	106	Very patchy inner arms; faint outer arms; star on W edge; mag 10.64v star SW 1′.6.
12 40 46.7	−40 53 39	NGC 4601	13.4	1.8 x 0.5	13.2	SAB(r)0^{+}	16	Located 2′.8 SW of a mag 11.22v star; **ESO 322-53** E 3′.9.
12 40 55.6	−40 58 34	NGC 4603	11.4	3.4 x 2.5	13.5	SA(s)c: I-II	27	Faint stellar nucleus, many knots and/or superimposed stars; mag 9.12v star S 4′.3.
12 39 37.0	−40 44 27	NGC 4603A	13.3	1.9 x 0.6	13.3	SB(s)c? III-IV	90	Mag 8.74v star E 4′.6.
12 40 43.1	−40 45 50	NGC 4603C	13.1	1.8 x 0.4	12.6	S0 sp	160	**MCG −7-26-23** W 4′.7; mag 8.74v star W 7′.8.
12 42 07.9	−40 49 19	NGC 4603D	13.2	1.5 x 1.1	13.6	SAB(s)d: III-IV	74	Almost stellar nucleus, uniform envelope.
12 42 16.7	−40 38 34	NGC 4616	13.4	0.9 x 0.9	13.1	E^{+}:		Mag 11.9 star on NE edge; **PGC 42667** S 1′.0.
12 42 37.5	−40 44 44	NGC 4622	12.4	1.7 x 1.6	13.4	(R′)SA(r)a pec	173	Very thin, tightly wound arms.
12 43 49.0	−40 42 53	NGC 4622A	13.6	0.6 x 0.6	12.4	SAB0^{-} pec:		Interacting pair with **NGC 4622B** on SE edge.
12 43 50.2	−40 43 07	NGC 4622B	13.8	1.0 x 0.5	13.0	S0 pec	63	Bright nucleus offset NE, fainter extension SW; pair with NGC 4622A on NW edge.

GALAXY CLUSTERS

RA h m s	Dec ° ′ ″	Name	Mag 10th brightest	No. Gal	Diam ′	Notes
12 00 00.0	−31 23 00	A 3497	16.4	40	17	All members anonymous, faint, stellar.
12 08 42.0	−34 26 00	A 3505	16.3	53	17	All members anonymous, faint, stellar.
11 25 48.0	−35 23 00	AS 665	16.0	12	17	See notes for ESO 377-46; all others anonymous, stellar.
11 50 00.0	−32 31 00	AS 677	16.4		17	All members anonymous, stellar.

OPEN CLUSTERS

RA h m s	Dec ° ′ ″	Name	Mag	Diam ′	No. ★	B ★	Type	Notes
12 34 06.0	−29 25 00	ESO 442-4		10			cl	

PLANETARY NEBULAE

RA h m s	Dec ° ′ ″	Name	Diam ″	Mag (P)	Mag (V)	Mag cent ★	Alt Name	Notes
11 26 43.7	−34 22 18	PN G283.6+25.3	180	12.6	12.1:	17.4	PK 283+25.1	Mag 12.3 star SW 4′.4; mag 10.53v star SE 6′.4.

GALAXIES

RA h m s	Dec ° ′ ″	Name	Mag (V)	Dim ′ Maj x min	SB	Type Class	PA	Notes
09 56 05.7	−37 46 15	ESO 316-4	13.8	1.4 x 1.1	14.1	SAB(rs)bc	4	Bright knot or superimposed star N of stellar nucleus; mag 10.83v star W 2′.6.
09 57 49.2	−39 21 45	ESO 316-8	13.6	0.6 x 0.4	11.9		86	Almost stellar; mag 11.16v star S 1′.5.
09 59 55.9	−39 09 31	ESO 316-13	13.4	1.3 x 0.4	12.6	S0	41	Mag 13.5 star W 0′.7.
10 03 17.2	−38 50 07	ESO 316-20	13.7	0.9 x 0.7	13.1	S?	6	Mag 10.54v star S edge.
10 07 15.5	−39 56 47	ESO 316-28	13.7	0.8 x 0.5	12.5	S0/a	175	Mag 12.0 star SW 4′.3.
10 08 40.1	−40 05 38	ESO 316-30	13.8	1.0 x 0.6	13.1	(R)SA0^{+}	170	Mag 15 star E edge.
10 09 01.1	−40 19 04	ESO 316-31	13.4	1.0 x 0.7	12.9	S0	113	Mag 14.4 star W edge; pair of mag 13 stars E 2′.3.
10 09 05.4	−38 24 36	ESO 316-32	12.4	1.4 x 0.9	12.5	(R)SAB(r)0/a pec	161	Mag 9.03v star on E edge; ESO 316-33 NE edge.
10 09 08.3	−38 23 48	ESO 316-33	12.5	1.1 x 0.7	12.3	E4: pec	134	ESO 316-32 SW edge; mag 9.03 star S 1′.2.
10 09 38.6	−39 56 16	ESO 316-34	12.8	1.3 x 0.8	12.8	SA0^{-} pec	58	Mag 8.07v star NNW 7′.3; **ESO 316-41** ESE 10′.7.
10 10 08.5	−38 08 03	ESO 316-38	13.1	1.2 x 0.9	13.1	(R)SB(rs)0^{+} pec:	63	Mag 9.30v star S 4′.6.
10 10 22.3	−38 01 57	ESO 316-40	13.7	0.6 x 0.5	12.2	(R′)Sa	156	Mag 10.46v star NW 6′.3.
10 10 35.6	−39 08 36	ESO 316-42	13.6	1.0 x 0.7	13.1	(R)SB(r)0/a	167	Mag 10.48v star WSW 2′.7.
10 11 02.2	−39 41 38	ESO 316-44	12.9	1.1 x 0.6	12.3	(R)S0/a: pec	62	Small anonymous galaxy on SE edge.
10 11 41.6	−37 55 43	ESO 316-46	13.1	1.3 x 0.8	13.0	(R)SAB(rs)0°:	123	Mag 8.96v star WNW 6′.5.
10 12 08.8	−38 53 07	ESO 316-47	13.5	1.0 x 0.7	13.0	SAB(r)b	121	Pair of stars, mags 12.8 and 13.3, W 1′.4.

(Continued from previous page)

RA h m s	Dec ° ′ ″	Name	Mag (V)	Dim ′ Maj x min	SB	Type Class	PA	Notes
10 13 27.5	−38 11 48	ESO 317-3	12.9	1.3 x 1.0	13.3	E1	149	Mag 10.12v star W 4'.8.
10 14 47.4	−39 48 27	ESO 317-5	13.0	1.0 x 0.9	12.8	SAB(r)0^{+}	152	Pair of faint stars S of nucleus; mag 10.44v star ESE 1'.5.
10 14 49.4	−38 11 27	ESO 317-6	13.5	1.8 x 0.5	13.2	SB(rs)b pec?	123	Mag 11.19v star WSW 3'.7.
10 17 03.8	−38 27 47	ESO 317-9	13.9	0.7 x 0.5	12.6	S	119	Mag 12.1 star W 1'.1.
10 20 24.4	−40 24 10	ESO 317-14	13.5	0.8 x 0.6	12.6	S0	94	Mag 9.60v star N 2'.3.
10 20 28.4	−39 21 10	ESO 317-15	13.4	1.0 x 0.5	12.5	Sa	35	Pair of mag 13 stars NE 0'.8.
10 20 43.9	−37 46 44	ESO 317-16	14.0	1.1 x 0.8	13.7	SAB(s)c pec: II	4	Mag 14.1 star S edge.
10 21 18.0	−39 47 58	ESO 317-17	13.3	1.4 x 1.0	13.5	(R)SAB(r)a	174	Several stars superimposed on edges.
10 23 02.2	−39 10 01	ESO 317-19	13.7	1.2 x 0.9	13.7	(R′)SAB(r)a	63	Mag 12.9 star W 1'.2.
10 23 08.3	−39 37 28	ESO 317-21	12.8	1.3 x 1.0	13.0	SA0^{-}	115	Mag 11.23v star ESE 6'.9.
10 23 35.2	−38 39 23	ESO 317-22	14.1	1.3 x 1.1	14.4	S?	155	Mag 14.4 star N edge; mag 10.62v star SW 1'.6.
10 24 42.4	−39 18 22	ESO 317-23	12.8	1.7 x 0.7	12.8	(R′)SB(rs)a	13	Two main arms; mag 10.41v star NNE 5'.6.
10 27 00.4	−40 01 31	ESO 317-27	13.6	1.3 x 0.3	12.4	S0/a	148	Mag 10.31 star W 2'.3.
10 29 33.7.	−39 50 36	ESO 317-36	12.7	1.1 x 0.8	12.4	SB(s)a pec	78	Most of faint envelope offset E of core; mag 10.02v star E 6'.3.
10 29 45.4	−38 20 57	ESO 317-38	13.5	1.3 x 0.4	12.6	(R)SBa pec?	72	Mag 10.49v star SE 7'.2.
10 31 32.7	−39 33 34	ESO 317-42	13.1	1.1 x 0.6	12.5	S0^{-}	60	Mag 9.62v star SW 9'.6.
10 31 47.7	−39 45 36	ESO 317-45	14.1	1.3 x 0.3	12.9	Sa:	139	Mag 12.4 star N 2'.3.
10 33 02.2	−38 58 57	ESO 317-46	14.2	2.2 x 0.4	13.9	SB(s)d: sp	12	Mag 10.49v star W 4'.1.
10 38 14.6	−38 05 38	ESO 317-54	13.8	2.2 x 2.1	15.3	SB(s)cd III	138	Mag 8.61v star WNW 8'.5.
10 42 06.5	−40 34 37	ESO 318-2	13.6	1.5 x 0.4	12.9	SA(s)a:	166	Mag 10.01v star W 3'.1.
10 43 50.3	−38 15 53	ESO 318-4	12.5	2.8 x 0.6	12.9	SA(s)c? sp II-III	62	Mag 10.97v star N edge.
10 43 58.9	−40 07 11	ESO 318-6	13.6	0.6 x 0.5	12.1	SA?(r:) pec	52	Mag 9.37v star NE 2'.3.
10 44 43.3	−40 35 11	ESO 318-8	13.8	0.9 x 0.8	13.3	SBa	116	Mag 9.62v star E 6'.6; **ESO 318-9** S 3'.5; **ESO 318-7** SW 4'.8.
10 47 41.7	−38 51 19	ESO 318-13	14.2	1.5 x 0.4	13.5	SB(s)d: sp	74	Star on N edge, center.
10 49 45.4	−39 51 14	ESO 318-17	13.6	1.0 x 0.7	13.1	SB0	23	Mag 12.8 star N 1'.6.
10 50 21.4	−38 51 14	ESO 318-19	13.8	1.1 x 0.5	13.0	(R′)SAB(s)ab	69	Strong E-W elongated core, smooth envelope.
10 53 06.0	−40 19 48	ESO 318-21	12.6	1.1 x 0.7	12.4	E?	131	Mag 7.99v star S 5'.2.
10 57 51.8	−39 26 21	ESO 318-24	13.3	1.7 x 0.5	12.9	SB(s)m pec	162	Numerous knots along length; mag 10.28v star SW 8'.9.
11 17 52.9	−40 35 37	ESO 319-11	14.2	2.5 x 1.5	15.4	SB(r)cd: I-II	115	Numerous stars scattered over faint outer bands.
11 22 08.3	−38 04 07	ESO 319-16	13.7	1.7 x 0.9	14.0	SB(s)c pec: III-IV	27	Mag 13.7 star on N edge; mag 11.04v star N 5'.5.
09 57 49.2	−32 49 09	ESO 374-13	14.1	0.9 x 0.6	13.3	S0	55	Mag 7.19v star WSW 7'.3.
09 58 00.7	−32 52 09	ESO 374-14	14.2	1.1 x 0.9	14.0	SB:b:	43	Stellar nucleus; mag 10.33v star NE edge; mag 7.19v star W 9'.4.
10 04 33.9	−37 20 10	ESO 374-28	13.7	1.2 x 0.5	13.0	S?	115	Mag 11.02v star N 5'.4.
10 08 53.2	−32 47 01	ESO 374-38	13.8	0.5 x 0.3	11.8	E/S0	30	Mag 8.72v star S 5'.4.
10 13 19.8	−35 58 59	ESO 374-44	14.2	1.1 x 0.8	13.9	(R′)SB(rs)a:	111	Located in a rectangle of mag 12-14 stars.
10 13 42.7	−34 51 35	ESO 374-45	14.5	1.2 x 0.8	14.3	(R)SAB0°:	153	Pair of galaxies, interaction.
10 14 01.6	−35 08 26	ESO 374-46	13.4	1.1 x 0.7	13.2	E0	12	Two ellipticals; connected?
10 15 53.6	−34 06 52	ESO 375-3	15.4	1.3 x 0.4	14.5	SB(s)m V-VI	72	Mag 7.40v star E 12'.0.
10 19 01.1	−37 40 21	ESO 375-7	14.0	1.1 x 0.3	12.7	S0?	53	Mag 7.91v star NE 3'.5.
10 24 32.5	−36 16 23	ESO 375-20	13.9	0.8 x 0.7	13.1	S?	21	Mag 9.68v star SSE 4'.7.
10 24 33.3	−36 55 55	ESO 375-22	13.9	1.0 x 0.8	13.6	(R′)SA0°: pec	17	Mag 11.8 star S 2'.9.
10 27 02.5	−36 13 39	ESO 375-26	13.5	2.0 x 0.3	12.8	Sbc: sp	29	Mag 9.33v star E 4'.1.
10 29 30.8	−35 15 34	ESO 375-41	13.6	1.7 x 0.6	13.5	S0^{-} sp	150	Mag 10.36v star W 5'.4.
10 30 19.6	−34 24 16	ESO 375-47	13.7	1.3 x 0.4	12.9	S0^{+}: sp pec	10	Mag 10.07v star ENE 5'.4.
10 31 00.3	−36 58 31	ESO 375-52	13.4	1.2 x 0.8	13.2	SB0/a: pec	2	Mag 9.67v star E edge.
10 31 56.2	−34 59 30	ESO 375-57	13.9	1.0 x 0.6	13.3	S0	69	Mag 9.34v star WNW 3'.3; NGC 3258E E 5'.8.
10 32 48.6	−34 23 59	ESO 375-62	13.6	1.4 x 0.2	12.1	S0: sp	172	Mag 10.46v star W 2'.7.
10 34 00.8	−35 16 58	ESO 375-64	13.7	1.7 x 1.1	14.2	SAB(rs)a pec	8	NGC 3289 S 2'.8.
10 35 18.7	−36 52 46	ESO 375-69	13.7	0.7 x 0.6	12.6	S?	152	Mag 12.7 star SE 1'.8.
10 35 51.1	−34 16 12	ESO 375-70	14.3	1.0 x 0.8	13.9	S?	40	Located in a small triangle of mag 13-14 stars.
10 36 09.4	−37 14 17	ESO 375-71	12.5	3.2 x 2.6	14.6	IB(s)m V-VI	79	Mag 10.87v star E edge; mag 7.14v star SW 11'.2.
10 36 39.3	−34 45 26	ESO 375-72	14.9	1.5 x 0.3	13.9	SB(s)m: sp IV-V	119	Mag 8.69v star NE 5'.8.
10 41 08.7	−33 28 54	ESO 376-6	13.8	0.8 x 0.4	12.4	S0/a	177	Mag 12.5 star S 1'.6.
10 41 11.5	−37 08 39	ESO 376-7	13.0	1.7 x 0.9	13.3	(R′)SAB0° pec:	98	Mag 10.23v star E 7'.5; note: **ESO 376-8** is 1'.0 W of this star.
10 42 02.0	−33 14 42	ESO 376-9	12.8	1.7 x 0.4	12.3	SB0°? sp	126	Mag 10.69v star W 2'.6.
10 42 10.5	−37 10 32	ESO 376-11	14.1	1.6 x 1.4	14.8	SB(s)m IV-V	8	Mag 10.23v star NW 4'.8; note: **ESO 376-8** is 1'.0 W of this star.
10 42 22.7	−36 10 37	ESO 376-12	13.9	0.9 x 0.5	12.9	SB?	16	Mag 8.51v star S 1'.6.
10 46 38.5	−36 21 14	ESO 376-20	14.2	1.2 x 0.3	12.9	Sb: pec	21	Mag 14.7 star NE edge; mag 14.0 star SW edge.
10 51 21.0	−34 25 52	ESO 376-22	13.5	1.9 x 0.5	13.3	IBm? sp	49	Mag 7.92v star SW 8'.4.
10 51 33.7	−35 28 33	ESO 376-23	14.5	1.9 x 0.2	13.3	Sc: sp	106	Mag 9.13v star N 4'.7.
10 54 08.2	−33 07 13	ESO 376-26	12.4	2.1 x 1.2	13.3	(R′)SAB(rs)0^{+}:	101	Star just N of bright, elongated core.
11 03 55.1	−34 21 32	ESO 377-3	14.1	0.7 x 0.4	12.6	S pec	69	Mag 14.3 star W edge; mag 10.51v star NW 1'.0.
11 06 26.4	−36 38 24	ESO 377-6	14.0	1.1 x 0.2	12.2	Sab	103	Mag 13.0 star SSW 1'.3; mag 10.82v star W 10'.4.
11 06 26.5	−36 41 47	ESO 377-7	14.8	1.6 x 0.3	13.6	Sc	133	ESO 377-6 N 3'.5; mag 10.82v star WNW 10'.8.
11 06 31.7	−37 39 08	ESO 377-10	12.5	2.8 x 0.7	13.1	SB(rs)b: II	150	Mag 12.2 star ENE 2'.6.
11 08 20.1	−37 37 29	ESO 377-12	13.0	1.4 x 1.0	13.2	SB(rs)ab III-IV	138	Bright knots or superimposed stars N and S of nucleus.
11 10 45.4	−35 21 00	ESO 377-19	13.6	1.3 x 0.5	12.9	Sc	9	Mag 12.2 star W 1'.3.
11 10 56.5	−35 59 00	ESO 377-21	12.9	1.9 x 0.8	13.2	SAB(rs)b pec?	129	Mag 10.06v star SW 2'.8.
11 12 33.3	−36 25 30	ESO 377-24	12.5	1.4 x 1.2	12.9	SA(rs)c: pec II-III	100	Mag 11.41v star SSW 5'.2.
11 14 38.8	−33 54 22	ESO 377-29	12.7	1.5 x 0.4	12.0	S0 sp	126	Mag 9.33v star N 5'.3.
11 16 00.8	−33 58 04	ESO 377-31	13.3	1.8 x 0.9	13.7	SAB(s)bc: I-II	146	Mag 12.6 star E 2'.0.
11 17 04.6	−34 57 19	ESO 377-34	13.2	2.4 x 0.6	13.5	SB(s)bc: II	160	Mag 6.88v star NNW 2'.1.
11 19 00.1	−36 14 45	ESO 377-38	14.0	0.9 x 0.7	13.3	(R′)SB(s)b	100	Mag 11.8 star NNW 5'.6.
09 56 21.5	−31 18 00	ESO 435-10	13.8	1.7 x 0.6	13.6	SB(s)cd IV	4	Mag 9.01v star W 3'.2.
09 59 06.0	−30 15 01	ESO 435-19	13.1	3.6 x 0.4	13.3	SB(s)c? sp	152	Mag 8.97v star W 8'.4.
10 00 38.5	−32 22 03	ESO 435-29	14.0	1.0 x 1.0	13.9	SB(rs)cd		Several stars superimposed W of center.
10 10 50.5	−30 25 25	ESO 435-50	15.0	2.1 x 0.2	13.9	Sc	71	Mag 9.78v star SE 6'.1.
10 25 49.0	−29 36 08	ESO 436-19	13.8	0.9 x 0.9	13.4	Sa:		Slightly elongated N-S core, smooth envelope.
10 28 43.0	−31 02 18	ESO 436-26	13.9	0.8 x 0.5	12.8	SBa	114	Mag 13.1 star on E edge.
10 28 53.6	−31 36 39	ESO 436-27	11.7	2.8 x 1.4	13.1	(R′)SA(s)0° pec	4	Mag 10.39v star W 2'.2; IC 2580 NW 9'.4.
10 30 23.6	−30 23 39	ESO 436-29	12.7	1.6 x 1.4	13.5	SAB(rs)c: I-II	134	Mag 5.58v star Delta Antliae SW 16'.6.
10 31 25.3	−29 57 14	ESO 436-31	14.1	1.0 x 0.6	13.4	SB(s)d IV	21	Mag 10.39v star E 2'.3.
10 31 30.0	−32 42 50	ESO 436-32	14.2	1.1 x 0.4	13.2	Sa:	162	Mag 12.7 star W 2'.0.
10 32 50.0	−30 16 00	ESO 436-35	13.7	1.2 x 1.0	13.7	SAB(r)dm IV	109	Star superimposed S of nucleus; mag 12.0 star NW 6'.6.
10 33 59.2	−30 10 06	ESO 436-39	13.9	1.8 x 0.3	13.0	Sbc: sp	82	N-S oriented pair of mag 11.07, 11.9 stars N 6'.4.
10 36 52.7	−32 20 53	ESO 437-14	12.4	2.6 x 0.9	13.1	SA(s)ab: sp	85	Mag 10.78v star E 8'.2; note: **ESO 437-16** located 4'.8 S of this star.
10 37 51.9	−30 40 08	ESO 437-18	14.1	1.9 x 0.4	13.6	Sb:	172	Mag 11.7 star E 7'.3.
10 39 15.4	−30 17 57	ESO 437-30	12.7	3.4 x 0.6	13.3	SB(rs)bc: sp	124	Mag 10.68v star WNW 5'.7.
10 39 22.1	−29 35 06	ESO 437-31	13.9	1.5 x 0.9	14.1	SAB(rs)d	123	Mag 10.39v star NE 3'.4.

(Continued from previous page)

GALAXIES

RA h m s	Dec ° ′ ″	Name	Mag (V)	Dim ′ Maj x min	SB	Type Class	PA	Notes
10 39 58.5	−30 11 36	ESO 437-33	12.9	1.6 x 1.1	13.4	(R′)SAB(r)a:	9	Mag 10.59v star WNW 7′8; ESO 437-35 S 4′8.
10 40 04.4	−30 16 08	ESO 437-35	13.3	1.5 x 0.6	13.0	SB(s)bc:	14	Mag 9.62v star S 5′2; ESO 437-33 N 4′8.
10 41 27.6	−31 46 48	ESO 437-42	13.5	1.5 x 1.5	14.2	SB(rs)c		Galaxy **AM 1039-313** SW 2′6.
10 43 09.8	−30 02 58	ESO 437-49	13.8	1.8 x 1.4	14.7	SAB(s)c II	132	Stellar nucleus; mag 10.40v star S 5′3.
10 44 24.6	−32 12 37	ESO 437-56	13.0	1.6 x 1.0	13.4	SB(rs)bc II-III	169	Mag 7.18v star WNW 13′7.
10 49 13.6	−31 18 21	ESO 437-65	13.5	1.9 x 0.9	13.9	SB(rs)bc pec	52	Mag 10.19v star SSW 6′5.
10 52 16.0	−32 40 15	ESO 437-67	12.8	2.2 x 1.9	14.2	(R′)SB(r)ab	117	Very faint, wide outer arms extending out and encircling the bright, elongated core.
10 56 13.6	−31 56 20	ESO 437-72	13.9	2.0 x 0.3	13.2	Sc:	163	Mag 9.63v star SE 2′8; mag 7.34v star NNW 10′3.
11 10 44.5	−30 20 46	ESO 438-8	13.4	1.0 x 0.6	12.7	S?	119	Two faint arms SE and NW.
11 14 01.6	−29 57 16	ESO 438-14	14.0	0.3 x 0.3	11.3			Compact, almost stellar; mag 14.6 star on N edge.
11 20 53.2	−29 24 10	ESO 438-23	13.1	1.5 x 0.9	13.3	S0?	38	Mag 11.7 star N 3′9.
09 57 03.0	−32 15 23	IC 2526	12.6	2.1 x 0.7	12.8	SAB(s)0°:	55	Mag 8.49v star W 8′0.
09 59 55.5	−29 37 03	IC 2531	12.0	6.9 x 0.6	13.4	Sc: sp	75	Single, narrow dust lane center one third of galaxy.
10 00 05.4	−34 13 43	IC 2532	13.1	1.4 x 1.0	13.3	(R′)SB(rs)a	38	Mag 10.57v star at NE edge.
10 00 31.5	−31 14 40	IC 2533	12.0	1.8 x 1.3	12.8	SA0⁻:	1	Mag 7.20v star SW 9′8.
10 01 29.8	−34 06 46	IC 2534	12.5	1.0 x 0.6	11.8	(R′)SA(rs)0⁺ pec:	88	Mag 7.09v star SE 7′3.
10 03 29.9	−33 57 06	IC 2536	13.6	2.0 x 0.4	13.2	SBc pec sp	45	Mag 11.02v star NE 5′1; **ESO 374-23** W 8′2.
10 03 56.6	−34 48 30	IC 2538	13.7	1.5 x 0.8	13.7	SA(r)c pec I-II	1	Mag 8.33v star S 3′8.
10 04 16.4	−31 21 48	IC 2539	13.1	1.9 x 0.5	12.9	SA(s)bc: III-IV	25	Mag 12.2 star SSW 1′2.
10 07 55.0	−35 13 47	IC 2548	13.1	1.9 x 1.7	14.2	(R′)SB(r)b I-II	65	Very faint arms encircle galaxy N and S edges.
10 10 46.2	−34 50 41	IC 2552	12.5	1.6 x 1.5	13.3	SAB(s)0⁻:	96	Mag 8.48v start NNE 2′5.
10 12 37.8	−34 43 45	IC 2556	13.7	2.0 x 1.0	14.3	(R′)SB(s)d III-IV	108	Mag 12.2 star S 2′6; **ESO 374-43** SSE 7′8.
10 14 44.4	−34 20 16	IC 2558	13.7	1.2 x 0.7	13.4	Pec:	13	Mag 9.39v star ESE 4′2.
10 14 45.3	−34 03 32	IC 2559	13.4	1.7 x 0.7	13.4	SB(s)b? III-IV	18	Mag 12.3 star on W edge; mag 9.68v star NNE 9′4.
10 16 18.7	−33 33 48	IC 2560	11.7	3.2 x 2.0	13.6	(R′)SB(r)b: II	45	Pair of mag 11.7, 11.8 stars NE edge.
10 18 51.7	−32 35 49	IC 2563	15.0	1.0 x 0.4	13.9	(R)SBd? pec	106	Mag 7.74v star NW 4′6.
10 21 34.4	−33 37 25	IC 2570	14.9	1.0 x 0.4	13.7	SB(rs)bc:	165	Mag 10.59v star E 1′0.
10 23 30.1	−35 27 11	IC 2573	14.5	1.5 x 0.3	13.5	SB(s)d: sp III-IV	2	Mag 11.9 star NE 4′7.
10 25 59.0	−32 54 13	IC 2576	14.0	0.7 x 0.6	12.9	S?	48	Strong dark lane E of center.
10 27 23.0	−33 52 43	IC 2578	14.2	1.5 x 0.3	13.2	Sc: sp	141	Mag 10.48v star SW 4′4.
10 28 18.5	−31 31 05	IC 2580	12.5	1.9 x 1.7	13.7	SB(rs)c II	154	Small, faint anonymous galaxy E 6′7; ESO 436-27 SE 9′4.
10 29 10.9	−30 20 39	IC 2582	12.9	1.4 x 1.1	13.2	SAB(r)c I-II	19	Mag 12.0 star W 3′0.
10 29 51.7	−34 54 44	IC 2584	12.7	1.4 x 0.4	11.9	S0: sp	133	Mag 8.55v star SW 9′4.
10 30 59.7	−34 33 51	IC 2587	12.3	2.0 x 1.5	13.4	(R′)SB(s)0⁻	26	Mag 11.9 star E 2′2.
10 31 50.2	−30 23 04	IC 2588	12.7	1.4 x 1.2	13.1	(R′)SB(r)a:	141	Strong dark lanes E and W of center; mag 11.7 star W 7′2.
09 56 45.0	−31 52 03	MCG −5-24-7	13.4	1.0 x 0.4	12.3	a:	21	
09 59 20.0	−30 44 15	MCG −5-24-13	14.2	1.2 x 1.0	14.3	Sc	8	Mag 8.37v star N 2′9.
10 16 13.5	−29 12 16	MCG −5-24-30	13.9	1.1 x 0.5	13.1	Sb	145	Star on N edge.
10 34 21.2	−32 11 10	MCG −5-25-15	13.8	1.2 x 0.4	12.8	(R′)SB(rs)ab: pec	47	
10 38 46.0	−31 25 53	MCG −5-25-30	13.9	1.2 x 1.1	14.1	Sc	145	Several stars superimposed W side.
10 43 31.0	−30 46 21	MCG −5-26-2	13.4	1.0 x 0.7	12.9	SB?	172	**MCG −5-26-3** S 2′1; **ESO 437-52** N 8′0; **ESO 437-47** WSW 14′8.
10 43 56.4	−31 06 34	MCG −5-26-4	14.1	0.9 x 0.6	13.3	Sa-b	33	Mag 8.14v star W 4′0.
10 53 33.8	−29 23 41	MCG −5-26-11	13.4	1.4 x 0.4	12.6	S	86	
10 55 19.2	−30 28 11	MCG −5-26-13	13.6	1.4 x 1.1	13.9	Sb	151	
10 56 58.4	−33 09 51	MCG −5-26-15	14.5	0.9 x 0.6	13.7	S	144	Multi-galaxy system, or very bright knot N end. Galaxy **ESO 376-28** NE 1′4.
11 17 32.8	−30 31 18	MCG −5-27-6	14.0	0.7 x 0.4	12.5	(R′)SA(r)0/a pec?	138	
11 18 53.6	−29 25 29	MCG −5-27-9	13.7	0.9 x 0.5	12.7	S?	152	**ESO 438-21** on E edge.
09 58 53.0	−30 21 29	NGC 3082	12.4	1.8 x 0.7	12.6	SA(s)0⁻:	26	A pair of stars on the N end; ESO 435-19 N 7′2.
09 59 08.6	−34 13 34	NGC 3087	11.6	2.0 x 2.0	13.2	E⁺:		Mag 11.3 star N 1′6; mag 12.7 star on W edge; mag 13.2 star E edge of core.
10 00 05.7	−31 33 14	NGC 3095	11.7	3.5 x 2.0	13.7	SAB(rs)c II	126	Mag 10.32v star 3′0 NNE; numerous stars superimposed on the surface.
10 00 40.9	−31 39 53	NGC 3100	11.1	3.2 x 1.6	12.8	SAB(s)0° pec	154	Mag 10.50v star E edge.
10 02 29.0	−31 40 35	NGC 3108	11.8	2.5 x 1.8	13.3	SA(s)0⁺	110	Several stars superimposed; mag 10.19v star S 1′4.
10 05 23.1	−34 13 10	NGC 3120	12.8	1.8 x 1.3	13.6	SAB(s)bc: II-III	1	Mag 9.52v star ENE 5′6.
10 06 33.0	−29 56 08	NGC 3125	13.0	1.1 x 0.7	12.7	E?	114	Lies 2′4 SE of mag 11.24v star.
10 09 07.5	−29 03 46	NGC 3137	11.5	6.3 x 2.2	14.2	SA(s)cd III-IV	1	Many stars superimposed overall.
10 11 42.4	−31 38 31	NGC 3157	13.2	2.5 x 0.5	13.3	SB(s)bc: sp II-III	38	
10 21 35.5	−34 15 58	NGC 3223	11.0	4.1 x 2.5	13.3	SA(s)b I-II	135	Mag 11.6 star on E edge.
10 21 41.2	−34 41 46	NGC 3224	12.0	1.9 x 1.5	13.1	E⁺	133	
10 24 17.3	−32 29 01	NGC 3241	12.2	2.2 x 1.5	13.3	SA(r)ab: II-III	123	Mag 11.4 star just off NW end.
10 25 29.1	−39 49 44	NGC 3244	12.3	2.2 x 1.6	13.5	SA(rs)cd II	170	Mag 10.18v star N 1′4.
10 26 22.0	−34 57 47	NGC 3249	12.9	1.6 x 1.3	13.6	SAB(rs)cd: II	139	Mag 10.49v star and small, faint anonymous galaxy NW 4′3.
10 26 32.5	−39 56 33	NGC 3250	11.1	2.8 x 2.0	13.0	E4	148	A number of stars superimposed on envelope.
10 27 53.9	−40 04 55	NGC 3250A	14.3	1.1 x 0.2	12.5	Sb: sp	89	A faint star at E end.
10 27 44.8	−40 26 09	NGC 3250B	12.7	2.3 x 0.7	13.0	SBa pec?	6	
10 27 42.6	−40 00 12	NGC 3250C	13.1	1.8 x 0.7	13.2	(R′)SA(rs)ab:	56	NGC 3250A S 5′1.
10 27 58.0	−39 48 56	NGC 3250D	13.2	1.7 x 0.3	12.3	S0: sp	29	
10 29 00.6	−40 04 57	NGC 3250E	12.5	2.1 x 1.4	13.6	SAB(s)cd: II-III	142	Mag 9.40v star NW 4′4.
10 28 47.0	−35 39 30	NGC 3257	13.1	1.0 x 0.9	12.9	SAB(s)0⁻:	0	A mag 10.82v star 2′5 W; a chain of four faint stars on E-W line above the center.
10 28 53.3	−35 36 20	NGC 3258	11.5	2.9 x 2.5	13.6	E1	68	NGC 3258 W 1′3.
10 28 19.1	−35 27 17	NGC 3258A	13.2	1.1 x 0.5	12.4	SAB0⁺:	169	Pair of mag 10 stars 1′0 and 1′3 N.
10 31 23.9	−35 13 16	NGC 3258C	13.6	1.1 x 0.8	13.3	SB(r)a	48	Mag 7.24v star N 5′8.
10 31 55.6	−35 24 38	NGC 3258D	13.0	1.6 x 0.9	13.2	SB(s)b:	5	Faint extensions go beyond published length N and S.
10 32 24.8	−34 59 54	NGC 3258E	14.3	1.6 x 0.3	13.4	Sb? sp	27	Mag 11.07v star NE 3′0; ESO 375-57 W 5′8.
10 29 06.4	−35 35 41	NGC 3260	12.7	1.2 x 1.0	12.8	E pec:	2	Faint star S of nucleus.
10 29 48.5	−35 19 21	NGC 3267	12.4	1.5 x 1.0	12.7	SAB(r)0°	148	NGC 3268 E 2′4; three small anonymous galaxies in N-S line S ~4′2.
10 30 00.5	−35 19 32	NGC 3268	11.4	3.0 x 2.5	13.7	E2	46	**MCG +6-23-0** SW 3′3; NGC 3267 W 2′4.
10 29 57.0	−35 13 28	NGC 3269	12.2	2.5 x 1.0	13.0	SA(r)0⁺	8	Located 2′4 SSE of mag 10.05v star; **PGC 30985** E 5′6.
10 30 26.5	−35 21 36	NGC 3271	11.8	3.3 x 1.8	13.6	SB(r)0°	94	Small, almost stellar anonymous galaxy SSE 2′1.
10 30 29.3	−35 36 41	NGC 3273	12.5	1.7 x 0.8	12.7	SA(r)0°	97	Faint elongated anonymous galaxy N 3′0.
10 30 51.8	−36 44 09	NGC 3275	11.8	2.8 x 2.1	13.6	SB(r)ab I	121	Mag 9.08v star S 2′9; mag 8.56v star WNW 8′6; note: **ESO 375-46** is 1′4 W of this star.
10 31 09.1	−39 56 43	NGC 3276	13.4	1.1 x 0.6	12.8	S0:	74	lies 2′3 S of mag 9.25v star.
10 31 35.5	−39 57 22	NGC 3278	12.2	1.3 x 0.9	12.2	SA(s)c?	62	Close pair of mag 10 stars 1′2 NE.
10 31 51.8	−34 51 17	NGC 3281	11.7	3.3 x 1.6	13.4	SA(s)ab pec:	140	Mag 9.09v star WNW 7′1.
10 31 58.5	−35 11 58	NGC 3281A	13.3	1.2 x 1.0	13.4	SAB(rs)0°:	121	**MCG −6-23-49**, also known as **NGC 3281B** in some references, lies ESE 1′3, with a much fainter anonymous galaxy on it's west edge.
10 32 59.4	−34 53 07	NGC 3281C	13.4	1.4 x 0.3	12.3	S0 sp	160	A mag 10.6v star 2′0 NNW.
10 34 18.8	−34 24 08	NGC 3281D	13.6	2.0 x 0.4	13.2	SB(s)d: sp III-IV	160	
10 34 07.5	−35 19 25	NGC 3289	12.5	2.2 x 0.6	12.7	SB(rs)0⁺: sp	150	Close pair of E-W oriented mag 10.23v, 10.98v stars SSW 3′5; ESO 375-64 N 2′8.
10 35 47.5	−32 21 33	NGC 3302	12.3	1.7 x 1.2	12.9	SA0°	118	Mag 9.43v star S 8′9; **ESO 437-5** SSW 7′7.

(Continued from previous page)

169 GALAXIES 169

RA h m s	Dec ° ′ ″	Name	Mag (V)	Dim ′ Maj x min	SB	Type Class	PA	Notes
10 39 49.6	−36 02 06	NGC 3333	13.2	2.0 x 0.4	12.8	SABbc pec sp	160	
10 42 46.9	−36 21 06	NGC 3347	11.3	3.6 x 2.1	13.3	SB(rs)b I-II	173	Forms very open "S" shape.
10 40 20.5	−36 24 41	NGC 3347A	12.6	2.0 x 0.7	12.8	SB(s)cd: sp III-IV	5	NGC 3347C NE 9′.9.
10 42 00.1	−36 56 09	NGC 3347B	12.8	3.2 x 0.8	13.7	SA(s)dm: sp III-IV	95	Stellar nucleus, several strong knots.
10 40 53.7	−36 17 21	NGC 3347C	14.2	1.5 x 1.2	14.7	SB(s)d III-IV	20	Weak nucleus; NGC 3347A SW 9′.9.
10 43 02.9	−36 21 49	NGC 3354	13.2	0.7 x 0.7	12.3	S: pec		Faint star on N edge.
10 43 32.7	−36 24 38	NGC 3358	11.4	3.3 x 1.9	13.3	(R)SAB(s)0/a I	141	Several faint stars superimposed around nucleus.
10 46 43.6	−40 00 58	NGC 3378	12.5	1.5 x 1.4	13.1	Sbc: II	11	Mag 10.04v star NE 7′.2.
10 48 04.1	−31 32 00	NGC 3390	11.9	3.5 x 0.6	12.5	Sb sp	177	Mag 5.88v star S 9′.4.
10 52 53.8	−32 55 40	NGC 3449	12.2	3.3 x 1.0	13.3	SA(s)ab: II	148	Several knots; mag 8.40v star SE 5′.7.
11 07 07.9	−37 10 23	NGC 3533	12.6	2.8 x 0.6	13.0	(R′)SAB(r)ab: sp	65	
11 09 58.0	−37 32 19	NGC 3557	10.4	4.1 x 3.0	13.1	E3	21	Located 3′.3 NW of a mag 10.46v star.
11 09 32.2	−37 21 00	NGC 3557B	12.6	2.0 x 0.6	12.8	E5:	110	Slightly elongated, compact core in faint envelope.
11 10 36.3	−37 32 54	NGC 3564	12.1	1.8 x 0.8	12.4	S0: sp	15	NGC 3557 W 7′.6.
11 10 48.3	−37 26 56	NGC 3568	12.0	2.5 x 0.8	12.6	SB(s)c:	7	An elongated N-S triangle of mag 12 stars on E side; mag 9.31v star E 5′.4.
11 11 18.8	−36 52 33	NGC 3573	12.3	3.6 x 1.0	13.6	SA0/a pec sp	4	Dust lane bisects nucleus N-S; very faint distorted arms.
11 16 15.7	−33 49 41	NGC 3606	12.3	1.5 x 1.3	13.0	E0:	168	Mag 15.3 star on W edge; mag 10.97v star NNE 4′.5.
11 18 16.4	−32 48 36	NGC 3621	9.7	12.3 x 6.8	14.3	SA(s)d III-IV	159	Many stars superimposed overall; a pair of mag 10 stars S of the bright central core; faint outer envelope.

GALAXY CLUSTERS

RA h m s	Dec ° ′ ″	Name	Mag 10th brightest	No. Gal	Diam ′	Notes
10 00 42.0	−38 11 00	AS 622	16.1		17	All members anonymous, stellar; overlaps AS 624.
10 01 12.0	−38 17 00	AS 624	15.8		17	All members anonymous, stellar; overlaps AS 622.
10 06 06.0	−39 44 00	AS 628	15.1		22	Except ESO 316-28, all members anonymous, stellar.
10 09 36.0	−39 56 00	AS 631	15.6		22	Except ESO galaxies, most members anonymous, stellar.
10 30 06.0	−35 19 00	AS 636	13.4	1	56	Numerous galaxies stretch from NE through center to SW.
10 42 42.0	−30 48 00	AS 640	16.1		17	See notes for MCG −5-26-2; all others anonymous, stellar.
10 44 24.0	−40 37 00	AS 643	16.2		17	See notes ESO 318-8, all others anonymous, stellar.

OPEN CLUSTERS

RA h m s	Dec ° ′ ″	Name	Mag	Diam ′	No. ★	B ★	Type	Notes
09 58 48.0	−30 55 00	ESO 435-17		4			cl	
10 03 54.0	−32 03 00	ESO 435-33		3			cl	
10 14 00.0	−29 11 00	ESO 436-2		5			cl	

PLANETARY NEBULAE

RA h m s	Dec ° ′ ″	Name	Diam ″	Mag (P)	Mag (V)	Mag cent ★	Alt Name	Notes
10 07 01.8	−40 26 09	NGC 3132	88	8.2	9.2	10.0	PK 272+12.1	**Eight-Burst Nebula**. Elongated N-S; pair of mag 11 stars SE 2′.4.
10 34 30.7	−29 11 16	PN G270.1+24.8	54			16.7	PK 270+24.1	Mag 11.9 star on E edge.

170 GALAXIES 170

RA h m s	Dec ° ′ ″	Name	Mag (V)	Dim ′ Maj x min	SB	Type Class	PA	Notes
08 57 28.7	−39 16 07	ESO 314-2	12.5	1.5 x 1.5	13.3	S?		Numerous stars superimposed; mag 10.94v star SW edge.
09 23 22.3	−37 59 22	ESO 315-3	13.4	1.1 x 0.8	13.2	S0	164	Several stars superimposed around core; mag 11.10v star ENE 2′.1.
09 23 23.4	−38 43 24	ESO 315-4	13.6	1.2 x 0.4	12.7	S	130	Pair of stars superimposed S of nucleus; mag 9.82v star SW 1′.8.
09 24 29.6	−37 45 13	ESO 315-6	13.7	1.1 x 1.1	13.8	SA(rs)0/a:		Mag 6.47v star W 8′.8.
09 29 01.7	−37 50 30	ESO 315-12	14.7	2.2 x 0.2	13.7	Sc sp	158	Lies 0′.5 E of a pair of mag 8.4v and 10.4v stars which lie in a N-S line.
09 38 19.4	−39 00 28	ESO 315-17	13.5	1.8 x 1.7	14.6	SA(r)cd II-III	134	Mag 11.22v star SSE 4′.6.
09 42 17.0	−38 53 23	ESO 315-19	14.5	1.0 x 0.5	13.6	S0	27	Mag 10.24v star NE 5′.4.
09 49 45.6	−39 01 14	ESO 316-2	13.8	1.0 x 0.6	13.1	S	48	Bright knot or star SW of nucleus.
09 56 05.7	−37 46 15	ESO 316-4	13.8	1.4 x 1.1	14.1	SAB(rs)bc	4	Bright knot or superimposed star N of stellar nucleus; mag 10.83v star W 2′.6.
09 57 49.2	−39 21 45	ESO 316-8	13.6	0.6 x 0.4	11.9		86	Almost stellar; mag 11.16v star S 1′.5.
09 59 55.9	−39 09 31	ESO 316-13	13.4	1.3 x 0.4	12.6	S0	41	Mag 13.5 star W 0′.7.
10 03 17.2	−38 50 07	ESO 316-20	13.7	0.9 x 0.7	13.1	S?	6	Mag 10.54v star S edge.
08 40 04.4	−34 24 31	ESO 371-3	13.6	1.2 x 0.8	13.4	SB(s)m: IV-V	163	Mag 10.65v star ENE 4′.3.
08 47 05.6	−33 45 51	ESO 371-16	11.6	3.9 x 3.2	14.2	(R′)SB(rs)0/a:	148	Numerous stars superimposed; mag 7.17v star SW 12′.3.
08 48 16.5	−33 49 15	ESO 371-17	14.6	1.3 x 0.5	14.0	S	33	Mag 8.98v star SSE 7′.9.
08 50 55.9	−34 32 14	ESO 371-20	12.5	3.7 x 0.9	13.6	SA(s)c: II-III	60	Mag 10.78v star superimposed NE of center.
08 52 39.3	−33 27 53	ESO 371-24	13.8	1.8 x 0.9	14.1	SAB(s)m V-VI	6	Mag 10.37v star W 6′.0.
08 54 32.5	−32 56 18	ESO 371-26	12.5	3.3 x 0.7	13.3	SAB(rs)0°: sp	69	Mag 10.34v star NNW 4′.9.
08 56 01.2	−37 11 54	ESO 371-28	14.6	1.7 x 0.5	14.3	S	85	Mag 6.87v star NNE 6′.0.
09 00 19.0	−34 04 48	ESO 371-30	14.0	2.1 x 1.5	15.1	IAB(s)m V-VI	152	Many stars superimposed; mag 8.93v star W 6′.7.
09 10 04.1	−33 09 19	ESO 372-8	14.4	2.2 x 1.4	15.5	SB(s)m: IV-V	156	Mag 8.54v star ENE 7′.9.
09 13 54.0	−35 04 11	ESO 372-9	14.7	1.3 x 0.5	14.0	SB(rs)d: IV	130	Mag 11.39v star E 2′.0.
09 15 35.2	−35 38 50	ESO 372-12	13.0	1.7 x 0.9	13.3	(R′)SAB(r)0/a:	40	Mag 10.04v star S 3′.0.
09 21 00.5	−33 11 33	ESO 372-16	13.9	2.6 x 0.4	13.8	SAb? sp	72	Mag 9.63v star NNE 4′.4.
09 21 20.8	−34 25 44	ESO 372-18	15.0	1.5 x 0.2	13.5	Sc?	38	Mag 9.10v star NE 1′.1.
09 25 07.0	−37 10 05	ESO 372-23	14.7	2.3 x 0.2	13.7	Sc: sp	51	Mag 8.84v star SE 8′.6.
09 30 54.1	−35 41 19	ESO 373-5	12.3	2.8 x 2.6	14.3	SA(rs)c II	18	Stellar nucleus, several stars superimposed; mag 5.85v star E 8′.1.
09 33 21.8	−33 01 59	ESO 373-8	12.0	5.9 x 0.9	13.7	Scd sp	89	Mag 8.26v star 3′.8 SW of center.
09 36 27.9	−37 20 29	ESO 373-10	13.6	1.2 x 0.5	12.9	SAB0⁻:	172	
09 38 07.4	−33 55 40	ESO 373-12	14.5	1.9 x 1.0	15.1	SAB(s)dm IV-V	151	Mag 10.43v star SW 5′.3; **ESO 373-11** W 6′.0.
09 38 20.7	−33 51 41	ESO 373-13	13.5	1.7 x 0.2	12.2	S0/a? sp	24	
09 40 28.7	−32 50 38	ESO 373-19	13.8	1.5 x 0.3	12.7	Sa: pec sp	114	Mag 8.05v star SW 9′.1; note: **ESO 373-17** is 3′.0 S of this star.
09 43 38.3	−32 44 31	ESO 373-20	14.5	1.0 x 0.9	14.2	IAB(s)m: pec V	142	Mag 10.87v star inside W edge.
09 43 53.7	−34 33 31	ESO 373-21	15.0	1.9 x 0.4	14.5	SB(s)dm: IV	27	Situated between two 9.9v mag stars on NE-SW line.
09 46 49.2	−33 36 17	ESO 373-26	13.3	1.7 x 1.6	14.2	SB(s)d IV	102	Mag 6.93v star NW 8′.9.

(Continued from previous page)

GALAXIES

RA h m s	Dec ° ′ ″	Name	Mag (V)	Dim ′ Maj x min	SB	Type Class	PA	Notes
09 48 21.3	−34 13 31	ESO 373-30	15.8	1.0 x 0.9	15.5	SA?(r)	84	Mag 9.23v star NW 10′5.
09 51 57.5	−33 04 32	ESO 374-3	13.4	2.4 x 0.6	13.7	SAB(rs)cd: II-III	153	Mag 7.19v star N 5′6.
09 54 23.2	−33 07 01	ESO 374-8	14.9	1.5 x 0.3	13.9	Sb? pec sp	132	Mag 7.72v star SSW 9′3; mag 9.67v star E 9′8.
09 57 49.2	−32 49 09	ESO 374-13	14.1	0.9 x 0.6	13.3	S0	55	Mag 7.19v star WSW 7′3.
09 58 00.7	−32 52 09	ESO 374-14	14.2	1.1 x 0.9	14.0	SB:b:	43	Stellar nucleus; mag 10.33v star NE edge; mag 7.19v star W 9′4.
08 40 50.6	−32 02 43	ESO 432-2	14.7	2.5 x 0.2	13.8	SB(s)c? sp II	129	Mag 9.94v star W 6′8; mag 8.38v star SW 10′1.
08 55 15.9	−32 02 49	ESO 432-12	13.6	1.7 x 0.4	13.1	SA(s)0°?	98	Mag 9.31v star SE 3′7.
08 56 27.6	−31 59 04	ESO 432-13	16.4	1.2 x 0.5	15.7	SB(s)m: IV-V	174	Mag 9.63v star SSE 7′1.
09 07 19.9	−32 10 33	ESO 433-4	15.2	1.6 x 0.2	13.8	Sc:	8	Mag 8.47v star SE 8′2.
09 11 20.3	−30 52 20	ESO 433-7	13.2	1.1 x 0.5	12.4	IB(s)m IV	64	Mag 15.0 star on NE end; mag 10.13v star NW 8′7.
09 12 12.8	−30 54 41	ESO 433-8	12.6	2.1 x 0.3	11.9	S0°: sp	166	Much elongated **ESO 433-9** E 0′8; mag 10.15v star NW 4′0.
09 18 13.3	−32 28 59	ESO 433-12	14.3	1.4 x 0.3	13.2	Sbc: pec sp	148	mag 8.20v star WSW 10′0; E-W oriented pair of mag 9.82v, 10.35v stars E 1′9.
09 18 50.4	−32 26 33	ESO 433-13	16.3	1.5 x 0.4	15.6	Sm? IV-V?	115	E-W oriented pair of mag 9.82v, 10.35v stars WSW 5′1.
09 21 08.7	−31 53 31	ESO 433-15	14.3	1.4 x 0.8	14.3	SB(s)c: III-IV	9	Mag 7.81v star NW 9′4.
09 27 25.6	−32 00 44	ESO 434-5	14.4	1.7 x 1.3	15.1	IB(s)m V-VI	27	Mag 10.33v star N 3′9.
09 30 16.5	−30 36 40	ESO 434-7	14.2	1.7 x 0.4	13.6	SB0: pec	79	Mag 12.2 star W 2′5; mag 10.22v star W 8′9.
09 31 07.0	−30 21 27	ESO 434-9	12.8	1.1 x 1.1	13.1	E/S0		Mag 10.43v star ESE 3′9; mag 10.59v star NW 2′9.
09 36 57.1	−32 12 19	ESO 434-15	13.7	0.6 x 0.2	11.3		15	Stellar nucleus; mag 10.60v star NW 1′2; mag 5.62v star NE 3′1.
09 40 01.6	−31 02 34	ESO 434-18	13.0	1.0 x 0.7	12.6	S0	74	Located between a pair of mag 14 stars NE-SW.
09 44 04.8	−32 09 58	ESO 434-27	14.7	1.0 x 0.7	14.2	IAB(s)m: V-VI	29	Mag 10.05v star NE 2′6.
09 44 37.9	−29 19 22	ESO 434-32	13.9	1.1 x 1.1	13.9	Sm:		Stellar nucleus, smooth envelope; faint star S edge.
09 44 47.7	−31 49 31	ESO 434-33	12.6	2.2 x 1.9	14.0	SB(s)m pec: IV-V	169	Several strong knots or superimposed stars S of center; IC 2507 NW 3′6.
09 45 29.9	−30 20 34	ESO 434-34	14.0	2.8 x 0.5	14.2	IBm: sp IV-V	57	Mag 6.45v star N 8′6.
09 47 45.3	−31 30 27	ESO 434-41	14.1	1.8 x 0.7	14.2	IB(s)m V-VI	99	Mag 13.6 star superimposed N of center; mag 9.04v star S 7′5.
09 50 05.3	−29 25 36	ESO 435-1	14.3	0.7 x 0.6	13.2	S0:	33	Almost stellar nucleus; mag 10.48v star NW 2′9.
09 52 19.8	−29 26 16	ESO 435-3	13.9	1.7 x 0.7	13.9	SA(s)0$^+$ pec:	102	Interacting with spiral companion 0′3 S.
09 53 58.7	−31 18 09	ESO 435-5	14.2	1.8 x 0.3	13.4	SB(rs)c sp I-II	83	Mag 14.4 star on S edge; mag 10.43v star E 3′2.
09 56 21.5	−31 18 00	ESO 435-10	13.8	1.7 x 0.6	13.6	SB(s)cd IV	4	Mag 9.01v star W 3′2.
09 59 06.0	−30 15 01	ESO 435-19	13.1	3.6 x 0.4	13.3	SB(s)c? sp	152	Mag 8.97v star W 8′4.
10 00 38.5	−32 22 03	ESO 435-29	14.0	1.0 x 1.0	13.9	SB(rs)cd		Several stars superimposed W of center.
09 23 01.3	−32 27 00	IC 2469	11.1	4.7 x 1.0	12.6	SB(rs)ab II-III	36	Mag 10.29v star W 5′8.
09 44 33.5	−31 47 32	IC 2507	12.7	1.7 x 0.8	12.9	IB(s)m pec: III-IV	49	Mag 10.32v star N 2′8; ESO 434-33 SE 3′6.
09 47 43.6	−32 50 15	IC 2510	12.5	1.3 x 0.7	12.2	SB(rs)ab: III-IV	148	Mag 9.84v star W 3′3.
09 49 24.6	−32 50 24	IC 2511	12.1	2.9 x 0.6	12.6	(R′)SAB(s)a: sp	38	Mag 8.48v star W 5′7.
09 50 00.8	−32 53 01	IC 2514	12.5	3.0 x 0.6	13.0	SA(s)ab: sp	61	Mag 12.3 star on E edge; mag 10.67v star E 4′0.
09 55 08.9	−33 08 29	IC 2522	12.2	2.8 x 2.0	13.9	SB(s)c pec II	171	Mag 9.67v star N 2′1; IC 2523 S 4′3.
09 55 09.5	−33 12 40	IC 2523	13.0	1.3 x 0.8	12.9	SB(s)bc pec? II-III	25	IC 2522 and mag 9.67v star N 4′3.
09 57 03.0	−32 15 23	IC 2526	12.6	2.1 x 0.7	12.8	SAB(s)0°:	55	Mag 8.49v star W 8′0.
09 59 55.5	−29 37 03	IC 2531	12.0	6.9 x 0.6	13.4	Sc: sp	75	Single, narrow dust lane center one third of galaxy.
10 00 05.4	−34 13 43	IC 2532	13.1	1.4 x 1.0	13.3	(R′)SB(rs)a	38	Mag 10.57v star at NE edge.
10 00 31.5	−31 14 40	IC 2533	12.0	1.8 x 1.3	12.8	SA0$^-$:	1	Mag 7.20v star SW 9′8.
10 01 29.8	−34 06 46	IC 2534	12.5	1.0 x 0.6	11.8	(R′)SA(rs)0$^+$ pec:	88	Mag 7.09v star SE 7′3.
10 03 29.9	−33 57 06	IC 2536	13.6	2.0 x 0.4	13.2	SBc pec sp	45	Mag 11.02v star NE 5′1; **ESO 374-23** W 8′2.
10 03 56.6	−34 48 30	IC 2538	13.7	1.5 x 0.8	13.7	SA(r)c pec I-II	1	Mag 8.33v star S 3′8.
09 31 37.1	−30 18 26	MCG −5-23-5	13.4	1.2 x 0.4	12.4	S	171	N-S oriented pair of mag 10.83v, 10.86v stars N 4′9.
09 46 18.4	−29 07 56	MCG −5-23-13	13.9	1.2 x 0.5	13.2	SB(r)a	0	
09 47 40.2	−30 56 56	MCG −5-23-16	13.1	1.2 x 0.5	12.4	S0?	50	Star SW of bright core.
09 56 45.0	−31 52 03	MCG −5-24-7	13.4	1.0 x 0.4	12.3	a:	21	
09 59 20.0	−30 44 15	MCG −5-24-13	14.2	1.2 x 1.0	14.3	Sc	8	Mag 8.37v star N 2′9.
08 45 08.1	−33 47 40	NGC 2663	10.6	3.5 x 2.4	12.9	E	110	Galaxy **ESO 371-13** W 4′0.
09 18 36.6	−38 00 40	NGC 2845	11.7	2.3 x 1.2	12.6	SA(r)0/a:	70	Many superimposed stars.
09 25 18.5	−34 06 15	NGC 2883	13.1	2.2 x 0.8	13.6	IB(s)m pec:	176	Numerous stars superimposed; faint lane N-S along centerline.
09 30 17.0	−30 23 05	NGC 2904	12.4	1.2 x 0.8	12.2	SAB(s)0$^-$?	84	Mag 8.74v star E 4′1; **ESO 434-8** E 2′9.
09 37 59.7	−30 08 57	NGC 2973	13.1	1.2 x 0.8	12.9	Sa	36	
09 45 39.5	−31 11 26	NGC 2997	9.5	8.9 x 6.8	13.8	SAB(rs)c I-II	110	Numerous superimposed stars and knots scattered over entire surface.
09 46 18.7	−30 26 14	NGC 3001	11.9	2.9 x 1.9	13.6	SAB(rs)bc I-II	6	Pair of stars superimposed N of nucleus; N-S oriented pair of mag 11.67v, 11.02v stars W 4′2.
09 51 15.6	−32 45 13	NGC 3038	11.6	2.5 x 1.3	12.7	SA(rs)b II	130	Several bright knots or superimposed stars SE of core.
09 58 53.0	−30 21 29	NGC 3082	12.4	1.8 x 0.7	12.6	SA(s)0$^-$:	26	A pair of stars on the N end; ESO 435-19 N 7′2.
09 59 08.6	−34 13 34	NGC 3087	11.6	2.0 x 2.0	13.2	E$^+$:		Mag 11.3 star N 1′6; mag 12.7 star on W edge; mag 13.2 star E edge of core.
10 00 05.7	−31 33 14	NGC 3095	11.7	3.5 x 2.0	13.7	SAB(rs)c II	126	Mag 10.32v star 3′0 NNE; numerous stars superimposed on the surface.
10 00 40.9	−31 39 53	NGC 3100	11.1	3.2 x 1.6	12.8	SAB(s)0° pec	154	Mag 10.50v star E edge.
10 02 29.0	−31 40 35	NGC 3108	11.8	2.5 x 1.8	13.3	SA(s)0$^+$	110	Several stars superimposed; mag 10.19v star S 1′4.

GALAXY CLUSTERS

RA h m s	Dec ° ′ ″	Name	Mag 10th brightest	No. Gal	Diam ′	Notes
10 00 42.0	−38 11 00	AS 622	16.1		17	All members anonymous, stellar; overlaps AS 624.
10 01 12.0	−38 17 00	AS 624	15.8		17	All members anonymous, stellar; overlaps AS 622.

OPEN CLUSTERS

RA h m s	Dec ° ′ ″	Name	Mag	Diam ′	No. ★	B ★	Type	Notes
08 45 06.0	−31 37 00	Cr 196	10.5	3	12		cl:	Few stars; moderate brightness range; not well detached. Probably not a cluster.
08 45 24.0	−31 45 00	Cr 198	11.2	5			cl	Probably not a cluster.
08 43 54.0	−40 14 00	ESO 313-12		10			cl	
09 19 42.0	−38 07 00	ESO 314-14		10			cl	
09 35 24.0	−39 32 00	ESO 315-14		3			cl	
08 53 00.0	−35 29 00	ESO 371-25		3			cl	
09 58 48.0	−30 55 00	ESO 435-17		4			cl	
10 03 54.0	−32 03 00	ESO 435-33		3			cl	
08 37 15.0	−29 57 00	NGC 2627	8.4	9	60	11.0	cl	Rich in stars; moderate brightness range; slight central concentration; detached.
08 38 27.0	−34 46 18	NGC 2635	11.2	3	15	13.0	cl	Few stars; small brightness range; slight central concentration; detached. Probably not a cluster.
08 43 28.0	−32 39 00	NGC 2658	9.2	10	80	12.0	cl	Rich in stars; moderate brightness range; strong central concentration; detached. (S) Diffuse.
09 16 10.0	−36 37 00	NGC 2818	8.2	8	298	11.3	clpn	Moderately rich in stars; small brightness range; no central concentration; detached; involved in nebulosity. (A) The NGC 2818 designation applies to the cluster and planetary nebula.
09 19 23.0	−40 31 12	NGC 2849	12.5	3	40		cl	Moderately rich in stars; small brightness range; strong central concentration; detached.

(Continued from previous page)

170 OPEN CLUSTERS 170

RA h m s	Dec ° ′ ″	Name	Mag	Diam ′	No. ★	B ★	Type	Notes
08 37 39.0	−39 34 48	Pi 5	9.9	2	10	11.0	cl	Few stars; moderate brightness range; no central concentration; detached; involved in nebulosity. Probably not a cluster.
08 41 09.0	−38 42 06	Pi 7		3	35	13.0	cl	
08 37 24.0	−40 05 00	Ru 64		70	80	9.0	cl	Rich in stars; large brightness range; slight central concentration; detached.
08 40 32.0	−38 04 36	Ru 66		2	20	15.0	cl	
08 44 30.0	−35 55 00	Ru 68		10	80	14.0	cl	
08 52 05.0	−37 36 06	Ru 72		3	12	13.0	cl	
09 21 00.0	−37 07 00	Ru 74		2	20	13.0	cl	(S) Sparse group.
08 52 23.0	−37 33 36	Ru 158		2	10	12.0	cl	
08 56 10.9	−39 30 40	vdB-Ha 55		2	30		cl	

GLOBULAR CLUSTERS

RA h m s	Dec ° ′ ″	Name	Total V m	B ★ V m	HB V m	Diam ′	Conc. Class Low = 12 High = 1	Notes
09 07 57.8	−37 13 17	Pyxis Cluster	12.9	15.2	18.7	4.0		

PLANETARY NEBULAE

RA h m s	Dec ° ′ ″	Name	Diam ″	Mag (P)	Mag (V)	Mag cent ★	Alt Name	Notes
09 16 01.5	−36 37 37	NGC 2818	93	13.0	11.6	19.5	PK 261+8.1	Elongated E-W with dark patches on either side of bright center, numerous faint stars superimposed on edges. (A) The NGC 2818 designation applies to the cluster and planetary nebula.
08 40 40.3	−32 22 34	PN G253.9+5.7	19	10.7	10.9:	13.9	PK 254+5.1	Close pair of mag 13-14 stars W 1′3; mag 12.1 star NW 2′3.
08 36 16.4	−35 15 04	PN G255.7+3.3	20				PK 255+3.1	Mag 10.72v star E 2′2.
08 37 07.7	−39 25 06	PN G259.1+0.9	75	13.8	14.1	18.9	PK 259+0.1	Mag 12.1 star on W edge.
08 42 16.6	−40 44 12	PN G260.7+0.9	36				Vo 3	Mag 8.39v star NW 2′5.
08 53 30.8	−40 03 44	PN G261.6+3.0	49	14.1	12.9:	17.8	PK 261+2.1	Elongated NE-SW; mag 8.10v star S 4′8.

171 GALAXIES 171

RA h m s	Dec ° ′ ″	Name	Mag (V)	Dim ′ Maj x min	SB	Type Class	PA	Notes
07 16 35.2	−38 29 20	ESO 310-6	14.0	1.2 x 1.0	14.0	SAB(r)ab	126	Stellar nucleus.
07 25 06.6	−38 25 01	ESO 310-14	13.9	0.7 x 0.4	12.4	S0?	74	Few faint stars superimposed.
07 16 22.0	−35 30 12	ESO 367-7	13.2	1.5 x 0.9	13.4	SB(r)0⁺	87	N-S bar, several stars superimposed on envelope; mag 12.3 star S 2′7.
07 16 39.2	−35 22 27	ESO 367-8	12.1	2.4 x 1.4	13.3	SB(r)0°	69	Mag 9.98v star WNW 4′9.
07 19 26.6	−35 39 26	ESO 367-17	12.7	2.3 x 0.7	13.1	(R′)SB(rs)ab:	71	Mag 9.33v star S 8′0.
07 21 08.9	−34 23 41	ESO 367-18	14.1	1.0 x 0.5	13.2	SAB(s)bc? II	36	Several superimposed stars; mag 7.94v star N 8′4.
07 23 39.2	−36 40 12	ESO 367-22	14.1	1.8 x 1.0	14.5	(R′)SB(s)b pec	55	Mag 9.13v star E 4′9.
08 18 39.5	−33 39 52	ESO 370-6	13.4	0.9 x 0.3	11.8		7	Mag 10.80v star W 4′2.
08 40 04.4	−34 24 31	ESO 371-3	13.6	1.2 x 0.8	13.4	SB(s)m: IV-V	163	Mag 10.65v star ENE 4′3.
07 16 26.3	−29 37 15	ESO 428-13	14.1	1.2 x 0.4	13.1	S?	126	Several stars superimposed.
07 16 31.1	−29 19 29	ESO 428-14	12.2	1.8 x 1.1	12.8	SAB(r)0° pec	136	Group of four stars, the brightest mag 9.48v, SW 2′7.
07 18 37.3	−29 22 22	ESO 428-20	14.4	1.5 x 0.2	13.0	Sc	154	Mag 8.74v star N 3′6.
07 22 09.7	−29 14 06	ESO 428-23	12.6	2.3 x 1.4	13.7	(R′)SB(rs)ab: III-IV	39	Numerous stars superimposed; mag 10.36v star W 2′2.
07 23 38.8	−30 03 05	ESO 428-28	13.3	2.5 x 0.4	13.2	Sc: sp	58	Mag 9.03v star SW 4′5.
07 23 42.2	−29 39 07	ESO 428-29	12.3	1.6 x 1.3	12.9	SAB(rs)b: III-IV	148	Mag 10.00v star E 3′3.
07 25 01.6	−32 30 00	ESO 428-31	13.6	2.0 x 0.4	13.2	Scd: sp II-III	79	Mag 8.58v star E 7′5.
07 25 40.7	−30 24 17	ESO 428-32	12.4	1.7 x 1.6	13.3	SB(s)c pec II-III	23	Mag 8.85v star SW 6′3.
07 30 17.8	−31 35 56	ESO 428-37	13.4	1.9 x 0.7	13.5	SA(s)b: II	76	Star or bright knot E of center; mag 10.65v star N 3′3.
07 40 20.6	−30 57 12	ESO 429-9	15.5	1.5 x 0.2	14.0	S	174	Mag 10.27v star W 4′0.
07 49 26.8	−29 21 17	ESO 429-19	15.9	1.5 x 0.3	14.8	S	164	Mag 10.10v star N 1′2.
08 14 39.2	−31 20 51	ESO 430-26	14.7	1.5 x 0.3	13.7	S	8	Mag 10.40v star N 1′8.
08 14 59.5	−30 52 00	ESO 430-28	13.5	1.5 x 1.1	14.1	E	168	Mag 7.71v star NNE 0′5.
08 17 38.5	−29 43 52	ESO 431-1	12.9	1.7 x 1.5	13.8	SA(rs)c: III-IV	45	Mag 7.14v star N 2′4.
08 17 43.3	−30 07 51	ESO 431-2	12.4	3.1 x 2.2	14.4	SB(rs)c: II-III	50	Numerous stars superimposed; mag 9.74v star E 3′2; mag 6.42v star NNE 8′1.
08 35 35.7	−32 08 50	ESO 431-18	14.1	2.8 x 0.6	14.5	SB(s)m: IV	166	Mag 8.15v star S 4′3.
08 40 50.6	−32 02 43	ESO 432-2	14.7	2.5 x 0.2	13.8	SB(s)c? sp II	129	Mag 9.94v star W 6′8; mag 8.38v star SW 10′1.
08 34 49.0	−31 58 22	MCG −5-21-1	13.0	2.6 x 0.7	13.5	SB(s)m:		

OPEN CLUSTERS

RA h m s	Dec ° ′ ″	Name	Mag	Diam ′	No. ★	B ★	Type	Notes
07 38 44.0	−33 50 24	AM 2		3		16.3	cl	An old open cluster (rather than a globular cluster).
07 40 15.0	−33 32 36	Bo 15	6.3	3	33	7.7	cl	Few stars; moderate brightness range; not well detached; involved in nebulosity. (S) Knot of three stars plus nebula.
07 17 24.5	−36 50 00	Cr 135	2.1	49	77	2.7	cl	Few stars; moderate brightness range; not well detached. Without brightest star, visual mag = 3.0.
07 24 26.7	−31 51 00	Cr 140	3.5	30	76	5.4	cl	Moderately rich in stars; large brightness range; no central concentration; detached.
08 23 19.0	−36 20 00	Cr 185	7.8	10	30	10.1	cl	Moderately rich in stars; moderate brightness range; no central concentration; detached.
08 24 12.0	−29 10 12	Cr 187	9.6	5	20		cl	Few stars; small brightness range; not well detached. (S) Sparse.
07 50 28.0	−29 50 54	Cz 32		3	30		cl	
08 43 54.0	−40 14 00	ESO 313-12		10			cl	
07 17 06.0	−35 17 00	ESO 367-10		3			cl	
07 47 00.0	−32 58 00	ESO 368-14		4			cl	
07 41 06.0	−30 44 00	ESO 429-13					cl	
08 03 36.0	−31 26 00	ESO 430-14		5			cl	
08 06 54.0	−30 50 00	ESO 430-18		12			cl	
07 22 54.0	−29 29 48	Haf 7		3	40	14.0	cl	
07 40 26.0	−30 05 00	Haf 13		15	15	8.0	cl	Few stars; large brightness range; no central concentration; detached. (S) Sparse group.
07 45 30.0	−32 50 30	Haf 15	9.4	3	35	12.0	cl	Moderately rich in stars; moderate brightness range; slight central concentration; detached.
07 51 36.0	−31 48 54	Haf 17		2.5	20	15.0	cl	
07 56 15.0	−30 22 24	Haf 20	11.0	3	33	14.0	cl	Few stars; small brightness range; slight central concentration; detached.
08 15 24.0	−30 49 00	Haf 26		6	20	14.0	cl:	(A) DSS image shows scattering of stars.
07 40 46.0	−31 42 00	NGC 2439	6.9	9	181	9.0	cl	Rich in stars; large brightness range; slight central concentration; detached. (A) The DSS image shows this to be a "ring" shaped cluster.

(Continued from previous page)

OPEN CLUSTERS

RA h m s	Dec ° ′ ″	Name	Mag	Diam ′	No. ★	B ★	Type	Notes
07 45 24.0	−37 58 00	NGC 2451	2.8	50	153	6.0	ast	Moderately rich in stars; moderate brightness range; slight central concentration; detached.
07 52 10.0	−38 32 00	NGC 2477	5.8	20	1911	12.0	cl	Rich in stars; moderate brightness range; strong central concentration; detached.
07 56 16.0	−30 03 54	NGC 2489	7.9	5	112	11.0	cl	Moderately rich in stars; moderate brightness range; strong central concentration; detached.
08 07 00.0	−29 52 00	NGC 2533	7.6	6	124	9.0	cl	Rich in stars; moderate brightness range; slight central concentration; detached.
08 12 24.0	−37 37 00	NGC 2546	6.3	70	40	7.0	cl	Moderately rich in stars; moderate brightness range; no central concentration; detached. (A) DSS image shows 90′ group of bright stars, with 30′ group of fainter stars at center, all in nebulosity.
08 18 30.0	−30 39 00	NGC 2567	7.4	11	117	11.0	cl	Moderately rich in stars; moderate brightness range; slight central concentration; detached.
08 18 19.0	−37 06 18	NGC 2568	10.7	3	30	11.0	cl	= **Pi 1**. Moderately rich in stars; moderate brightness range; slight central concentration; detached.
08 18 57.0	−29 45 00	NGC 2571	7.0	7	49	8.8	cl	Moderately rich in bright and faint stars; detached.
08 20 53.0	−36 13 00	NGC 2579	7.5	19	10	9.5	ast?	Few stars; moderate brightness range; not well detached. Probably an asterism, not a cluster.
08 21 28.0	−30 18 00	NGC 2580	9.7	8	50		cl	Moderately rich in stars; moderate brightness range; slight central concentration; detached.
08 23 25.0	−29 30 00	NGC 2587	9.2	10	40		cl	Moderately rich in stars; moderate brightness range; no central concentration; detached. (S) Small group.
08 23 10.0	−32 58 30	NGC 2588	11.8	2	20		cl	Few stars; small brightness range; slight central concentration; detached.
08 37 15.0	−29 57 00	NGC 2627	8.4	9	60	11.0	cl	Rich in stars; moderate brightness range; slight central concentration; detached.
08 38 27.0	−34 46 18	NGC 2635	11.2	3	15	13.0	cl	Few stars; small brightness range; slight central concentration; detached. Probably not a cluster.
08 43 28.0	−32 39 00	NGC 2658	9.2	10	80	12.0	cl	Rich in stars; moderate brightness range; strong central concentration; detached. (S) Diffuse.
08 31 21.0	−38 39 00	Pi 3		6	50	13.0	cl	
08 37 39.0	−39 34 48	Pi 5	9.9	2	10	11.0	cl	Few stars; moderate brightness range; no central concentration; detached; involved in nebulosity. Probably not a cluster.
08 41 09.0	−38 42 06	Pi 7		3	35	13.0	cl	
07 27 06.0	−31 10 00	Ru 21		11		11.0	cl	Moderately rich in stars; small brightness range; no central concentration; detached. (S) DSS shows a slight concentration of stars in a rich field.
07 29 17.0	−29 10 30	Ru 22		3	15	12.0	cl:	(S) Sparse group.
07 39 36.0	−30 55 42	Ru 28		5	20	14.0	cl:	(S) Sparse group, identification uncertain. (A) DSS image barely shows rise in faint stars of Milky Way background.
07 42 22.0	−31 28 00	Ru 30		6.5	30	11.0	cl	Moderately rich in stars; moderate brightness range; not well detached. (A) Sparse 6.′5 diameter group of brighter stars, on 24′ DSS image.
07 43 00.0	−35 35 24	Ru 31		5	15	11.0	cl	Few stars; moderate brightness range; no central concentration; detached.
07 46 14.0	−31 16 48	Ru 35		1.5	10	14.0	cl	
08 02 18.0	−31 04 42	Ru 47	9.6	4	20	13.0	cl	Few stars; small brightness range; slight central concentration; detached.
08 02 42.0	−32 02 30	Ru 48		2	20	12.0	cl	
08 03 24.0	−30 52 00	Ru 50		3	30	12.0	cl	
08 03 36.0	−30 37 00	Ru 51		4	30	14.0	cl	
08 05 00.0	−31 56 00	Ru 52		3	15	13.0	cl	
08 11 23.0	−31 57 00	Ru 54		3	20	15.0	cl	
08 12 19.0	−32 34 00	Ru 55	7.8	6	36	11.0	cl	Few stars; small brightness range; no central concentration; detached. Probably not a cluster. (S) Sparse group.
08 12 36.0	−40 27 00	Ru 56		40	40	9.0	cl:	Moderately rich in stars; moderate brightness range; no central concentration; detached. Probably not a cluster.
08 14 49.0	−31 56 12	Ru 58		6	30	12.0	cl	(S) Sparse group.
08 19 22.0	−34 28 51	Ru 59	9.0	3	21	12.0	cl	Few stars; small brightness range; no central concentration; detached. Probably not a cluster.
08 25 15.0	−34 08 42	Ru 61		2	25	14.0	cl	
08 37 24.0	−40 05 00	Ru 64		70	80	9.0	cl	Rich in stars; large brightness range; slight central concentration; detached.
08 40 32.0	−38 04 36	Ru 66		2	20	15.0	cl	
07 54 29.0	−38 14 18	Ru 152		3	35	16.0	cl	
08 00 18.0	−30 13 54	Ru 153		2.8		14.0	cl	
08 04 54.0	−31 46 00	Ru 155		3	10	13.0	cl:	
08 25 25.2	−39 37 57	SSWZ94 2		3			cl	
07 37 45.0	−36 03 48	vdB-Ha 4		2	20		cl	
08 07 04.0	−32 21 30	vdB-Ha 19		3	12		cl	
08 14 24.0	−36 23 00	vdB-Ha 23		40	15		cl	Few stars; moderate brightness range; no central concentration; detached.

BRIGHT NEBULAE

RA h m s	Dec ° ′ ″	Name	Dim ′ Maj x min	Type	BC	Color	Notes
08 35 36.0	−40 40 00	NGC 2626	5 x 5	E+R	1-5	1-4	A 10th mag star involved in the N part of an irregularly round nebula; an extensive absorption region some 3′ by 4′ in size off N side of NGC 2626.

PLANETARY NEBULAE

RA h m s	Dec ° ′ ″	Name	Diam ″	Mag (P)	Mag (V)	Mag cent ★	Alt Name	Notes
07 28 53.8	−35 45 15	PN G248.8−8.5	6	12.9	13.2	17.0	PK 248−8.1	Mag 14.1 star N 1.′7.
07 43 18.0	−34 45 12	PN G249.3−5.4	54		13.1		PK 249−5.1	Mag 11.05v star W 1.′5; mag 11.12v star NNE 1.′7.
08 09 01.6	−32 40 25	PN G250.3+0.1	40	18.1			PK 250+0.1	Mag 11.03v star E 2.′3; mag 10.04v star WSW 3.′3.
08 04 14.3	−34 16 06	PN G251.1−1.5	35	16.0			PK 251−1.1	Mag 12.0 star E 1.′7; mag 10.72v star N 2.′4.
08 31 52.5	−32 06 10	PN G252.6+4.4	63	>17.9		13.9	PK 252+4.1	Elongated NE-SW; mag 10.20v star NE 4.′9.
08 40 40.3	−32 22 34	PN G253.9+5.7	19	10.7	10.9:	13.9	PK 254+5.1	Close pair of mag 13-14 stars W 1.′3; mag 12.1 star NW 2.′3.
08 20 54.8	−36 12 58	PN G254.6+0.2	162	11.5		10.5	ESO 370-09	Complex system, oriented SE-NW, consisting of two bright areas, the northwestern area being the larger; mag 10.05v star W 2.′3.
08 36 16.4	−35 15 04	PN G255.7+3.3	20				PK 255+3.1	Mag 10.72v star E 2.′2.
08 16 10.0	−39 51 51	PN G257.1−2.6	8	18.2			Vo 2	Close pair of mag 8.61v, 8.99v stars E 2.′1; mag 12.5 star N 1.′3.
08 30 53.9	−38 18 03	PN G257.5+0.6	101	14.0			PK 257+0.1	Rectangular, oriented ESE-WNW; mottled appearance; brighter on SE edge; mag 11.00v star S 1.′5.
08 28 28.1	−39 23 42	PN G258.1−0.3	10	14.4	14.8	16.2	PK 258−0.1	Mag 13.7 star W 1.′7.
08 37 07.7	−39 25 06	PN G259.1+0.9	75	13.8	14.1	18.9	PK 259+0.1	Mag 12.1 star on W edge.
08 42 16.6	−40 44 12	PN G260.7+0.9	36				Vo 3	Mag 8.39v star NW 2.′5.

RA h m s	Dec ° ′ ″	Name	Mag (V)	Dim ′ Maj x min	SB	Type Class	PA	Notes
05 56 07.5	−38 09 19	ESO 307-5	13.9	1.4 x 0.6	13.6	SB(r)a pec:	101	Mag 7.47v star NW 5.′7; faint, elongated anonymous galaxy N 1.′3.
06 00 41.1	−40 02 42	ESO 307-13	13.4	1.1 x 0.6	13.0	E^+3	62	Almost stellar nucleus, smooth envelope; **PGC 18233** SW 1.′7; other small galaxies in surrounding area.
06 06 27.3	−39 51 55	ESO 307-17	14.1	1.4 x 0.3	13.1	Sa? pec sp	106	Mag 10.47v star W 9.′7.
06 13 46.8	−37 41 07	ESO 307-25	13.5	2.2 x 0.7	13.8		12	Very faint, broad, band of material extends S 1.′2 from small, bright core.
06 35 38.7	−39 15 32	ESO 308-16	13.3	1.2 x 0.9	13.2	SAB(s)c I	168	Mag 10.26v star E 8.′0; **ESO 308-17** E 3.′2.
06 39 49.9	−38 14 43	ESO 308-23	13.7	1.1 x 0.4	12.6	SB(r)b: II-III	94	Mag 6.59v star N 5.′3.
06 40 04.8	−37 45 57	ESO 308-24	13.3	1.3 x 0.7	13.0	Sa	14	Mag 9.30v star WNW 5.′0; **MCG −6-15-15** SW 6.′8.
06 41 03.2	−38 02 10	ESO 308-25	13.8	1.3 x 0.4	13.0	Sa	75	Mag 9.23v star NNE 7.′3.
06 53 03.0	−39 16 15	ESO 309-5	13.2	1.2 x 0.7	12.9	SB(rs)b?	156	Mag 11.6 star SW 1.′9.
06 54 35.4	−38 05 16	ESO 309-8	13.5	1.0 x 0.7	12.9	Sa:	67	**ESO 309-10** E 6.′1.
06 59 08.4	−39 40 24	ESO 309-13	13.4	0.8 x 0.6	12.5	SA?(r:)	178	Mag 8.78v star NNE 5.′3; **ESO 309-12** N 3.′3.
07 00 45.7	−37 44 49	ESO 309-15	13.5	1.1 x 0.8	13.2	Sb:	21	Mag 10.91v star NW 6.′0.
07 06 50.9	−37 59 16	ESO 309-19	14.9	1.3 x 0.3	13.7	Sbc: III-IV	132	Mag 9.08v star SE 2.′5.
07 10 48.5	−38 13 11	ESO 310-2	14.7	1.8 x 0.3	13.9	Sc:	44	Mag 10.60v star SSW 1.′7; note: **ESO 310-1** on W edge of this star.
07 16 35.2	−38 29 20	ESO 310-6	14.0	1.2 x 1.0	14.0	SAB(r)ab	126	Stellar nucleus.
05 56 27.1	−33 25 52	ESO 364-16	14.3	1.3 x 0.5	13.6	Pec	78	Very transparent; mag 7.30v star E edge.
05 56 26.4	−34 06 08	ESO 364-17	13.5	1.1 x 0.9	13.3	SAB(r)cd	158	Mag 10.31v star NW 1.′4.
05 57 12.5	−37 28 38	ESO 364-18	13.8	1.0 x 0.5	13.1	E^+2	138	Mag 11.20v star NW 1.′6; **ESO 307-6** SW 14.′2.
05 58 18.1	−34 53 19	ESO 364-19	13.9	0.9 x 0.5	12.8	SB	50	Almost stellar, faint anonymous galaxy S 0.′6.
06 00 08.1	−33 00 53	ESO 364-23	14.1	1.0 x 0.4	12.9	S	84	Pair of mag 14.7 stars N edge; mag 13.1 star SW 1.′9.
06 04 41.8	−33 04 31	ESO 364-28	14.0	1.0 x 0.9	13.8	Sc	23	Very knotty; mag 10.45v star partially superimposed S edge.
06 05 44.5	−33 04 54	ESO 364-29	13.3	2.6 x 1.6	14.7	IB(s)m V-VI	53	Mag 12 star superimposed inside NE edge.
06 08 46.0	−33 55 00	ESO 364-33	13.7	0.9 x 0.8	13.2	S?	62	Mag 10.19v star S 2.′5.
06 09 55.2	−33 39 00	ESO 364-35	13.9	1.5 x 0.9	14.0	S?	105	Galaxy **CSRO 0481** S 2.′0.
06 10 00.9	−33 38 23	ESO 364-36	13.8	0.8 x 0.7	13.0	S?	13	Pair with ESO 364-35 SW edge; galaxy **CSRO 0481** SSW 2.′9, lying between two stars, mags 11.8 and 13.0.
06 10 51.3	−33 52 56	ESO 364-39	12.9	1.7 x 1.3	13.7	SA(rs)ab: pec	10	Mag 7.36v star E 5.′1.
06 10 59.4	−32 34 07	ESO 364-40	13.9	1.0 x 0.9	13.6	S0/a	32	Mag 15.0 star N edge; mag 12.2 star SSE 2.′9.
06 10 54.6	−36 13 35	ESO 364-41	13.8	1.2 x 0.4	12.9	S	94	Very faint, stellar anonymous galaxy E edge; mag 9.75v star E 2.′1.
06 12 54.0	−33 26 04	ESO 364-43	14.4	1.0 x 0.4	13.3	S?	75	Mag 14.6 star S edge; mag 12.5 star SE 0.′6.
06 14 08.6	−33 30 00	ESO 365-1	14.9	1.7 x 0.3	14.0	Sbc	105	Mag 10.34v star NNW 4.′7.
06 18 45.2	−35 18 17	ESO 365-5	13.7	0.9 x 0.7	13.1	SBc	173	E-W core, two main arms.
06 20 43.8	−33 25 33	ESO 365-7	14.1	1.0 x 0.4	12.9	Sc	1	Mag 10.46v star NNE 3.′8.
06 21 57.6	−34 41 38	ESO 365-9	13.3	0.8 x 0.7	12.6	Sa	63	Mag 7.93v star S 4.′4.
06 21 57.0	−36 33 24	ESO 365-10	13.7	2.5 x 0.8	14.3	SA(s)b: pec	104	Faint, narrow filaments extent E and W of center.
06 23 38.7	−35 38 36	ESO 365-11	13.4	1.1 x 0.5	12.6	Sc:	8	Several small companion galaxies on edges; mag 10.92v star NW 1.′2.
06 23 55.1	−35 01 31	ESO 365-13	14.0	1.0 x 0.6	13.3		135	Double system; mag 6.53v star W 8.′8.
06 25 10.5	−37 20 27	ESO 365-16	14.1	1.4 x 0.6	13.8	(R)SB(r)ab	146	Mag 8.64v star ENE 8.′9.
06 26 29.8	−34 50 23	ESO 365-20	13.8	1.1 x 0.6	13.2	Sa	151	Located between a pair of mag 12.3 and 12.8 stars.
06 29 05.0	−36 18 57	ESO 365-21	14.2	1.8 x 0.3	13.3	Sbc pec	111	Close pair of stars on N-S line at E end.
06 31 11.1	−36 49 16	ESO 365-25	13.4	1.3 x 0.4	12.6	S0	159	Mag 6.29v star SSE 8.′7.
06 32 05.6	−35 37 27	ESO 365-26B	13.7	0.6 x 0.4	12.0		100	Pair with ESO 365-26A E.
06 32 07.2	−35 37 40	ESO 365-26A	13.6	0.7 x 0.3	11.8		32	Pair with ESO 365-26B W; mag 12.4 star NE 2.′3.
06 32 48.9	−34 07 43	ESO 365-27	13.7	1.5 x 1.0	14.0	SB(r)c I-II	126	Mag 10.59v star N 2.′1; mag 8.83v star SW 7.′3.
06 33 35.6	−34 15 39	ESO 365-28	13.7	1.5 x 0.8	13.7	SAB(s)c: pec	51	Mag 9.97v star WSW 7.′8; ESO 365-29 SE 1.′4.
06 33 41.6	−34 16 20	ESO 365-29	13.6	1.0 x 0.5	12.7	S0	139	Pair with ESO 365-28 NW 1.′4.
06 36 45.4	−35 04 17	ESO 365-35	13.8	1.4 x 1.1	14.1	(R′)SB0/a	134	Strong core with faint dark patches N and S, faint envelope.
06 38 44.8	−35 15 30	ESO 366-4	15.0	1.7 x 0.3	14.1	SB(r)d sp	78	Mag 9.82v star SE 1.′8; mag 7.98v star NNE 8.′1.
06 38 57.1	−36 41 12	ESO 366-5	14.0	0.8 x 0.6	13.1	Sc:	85	Mag 10.47v star N 2.′4.
06 41 37.4	−34 44 32	ESO 366-8	13.6	1.0 x 0.5	12.7	S0/a	148	Mag 13.3 star E edge; mag 11.8 star SW 1.′9.
06 42 50.0	−35 34 20	ESO 366-9	13.5	1.6 x 0.9	13.7	(R′)SAB(s)c: II-III	40	Mag 8.53v star SE 6.′9.
06 43 19.7	−35 09 29	ESO 366-11	14.1	1.0 x 0.4	13.0	SB(r)0/a:	159	Mag 13.6 star W 1.′9.
06 45 29.2	−37 00 37	ESO 366-16	13.1	1.3 x 0.8	13.1	S0	168	Mag 12.2 star E edge; faint, almost stellar anonymous galaxy N 0.′9.
06 55 57.5	−36 56 19	ESO 366-28	13.5	1.5 x 1.1	13.9	Sc	84	Located in a triangle of mag 12 stars.
07 00 04.5	−32 37 10	ESO 366-30	15.0	1.1 x 0.3	13.6	IB(s)m? pec V	51	Mag 9.83v star E 7.′2.
07 14 43.4	−36 48 27	ESO 367-6	13.9	1.0 x 0.8	13.5	(R′)SB(r)a	72	Close pair of stars, mags 12.6 and 13.0, ENE 2.′1.
07 16 22.0	−35 30 12	ESO 367-7	13.2	1.5 x 0.9	13.4	SB(r)0^+	87	N-S bar, several stars superimposed on envelope; mag 12.3 star S 2.′7.
07 16 39.2	−35 22 27	ESO 367-8	12.1	2.4 x 1.4	13.3	SB(r)0°	69	Mag 9.98v star WNW 4.′9.
07 19 26.6	−35 39 26	ESO 367-17	12.7	2.3 x 0.7	13.1	(R′)SB(rs)ab:	71	Mag 9.33v star S 8.′0.
07 21 08.9	−34 23 41	ESO 367-18	14.1	1.0 x 0.5	13.2	SAB(s)bc? II	36	Several superimposed stars; mag 7.94v star N 8.′4.
07 23 39.2	−36 40 12	ESO 367-22	14.1	1.8 x 1.0	14.5	(R′)SB(s)b pec	55	Mag 9.13v star E 4.′9.
06 00 11.0	−31 47 14	ESO 425-1	14.5	0.9 x 0.8	14.0	S	20	Mag 12.9 star SW edge.
06 08 56.4	−29 09 23	ESO 425-11	14.6	1.8 x 0.2	13.4	Scd	154	Mag 12.2 star E 1.′5; mag 10.33v star E 6.′7.
06 14 36.8	−30 25 34	ESO 425-16	13.8	1.0 x 0.8	13.4	S	150	Several faint stars superimposed; mag 11.16v star W 2.′6.
06 23 46.3	−32 13 00	ESO 426-2	13.4	1.3 x 1.0	13.6	(R′)SB(r)0/a	42	Mag 14.1 star on N edge; mag 9.17v star NW 8.′2.
06 25 29.7	−32 17 12	ESO 426-7	13.3	1.1 x 0.4	12.2	Sa?	149	Pair of stars on E edge.
06 30 21.0	−32 16 50	ESO 426-18	14.0	0.8 x 0.7	13.2	(R′)SB(s)b	139	Pair of stars, mags 12.1 and 12.5, S 1.′7.
06 48 44.1	−32 05 39	ESO 427-8	15.7	1.5 x 0.2	14.2	Sc	171	Close pair of mag 12.7, 12.9 stars SE 1.′1; mag 14.7 star superimposed S of center.
06 51 11.0	−30 25 05	ESO 427-13	13.9	1.0 x 0.6	13.2	SA(rs)b	35	Mag 12.6 star NE 2.′9.
06 57 28.5	−31 04 41	ESO 427-21	13.9	0.9 x 0.7	13.3	S0	54	Mag 8.66v star E 8.′7.
07 02 45.4	−29 25 45	ESO 427-29	12.3	1.9 x 1.3	13.1	S0°: pec	150	Mag 13.3 star on W edge; mag 12.1 star N 4.′2.
07 15 15.4	−29 16 54	ESO 428-9	14.1	1.7 x 0.4	13.6	Sc	74	ESO 428-11 SE 5.′7; mag 9.78v star NNE 5.′0.
07 15 31.1	−29 21 33	ESO 428-11	11.9	1.7 x 1.2	12.5	SA0^-:	16	Mag 14.4 star on S edge; mag 10.56v star WNW 5.′1; ESO 428-9 NW 5.′7.
07 16 26.3	−29 37 15	ESO 428-13	14.1	1.2 x 0.4	13.1	S?	126	Several stars superimposed.
07 16 31.1	−29 19 29	ESO 428-14	12.2	1.8 x 1.1	12.8	SAB(r)0° pec	136	Group of four stars, the brightest mag 9.48v, SW 2.′7.
07 18 37.3	−29 22 22	ESO 428-20	14.4	1.5 x 0.2	13.0	Sc	154	Mag 8.74v star N 3.′6.
07 22 09.7	−29 14 06	ESO 428-23	12.6	2.3 x 1.4	13.7	(R′)SB(rs)ab: III-IV	39	Numerous stars superimposed; mag 10.36v star W 2.′2.
07 23 38.8	−30 03 05	ESO 428-28	13.3	2.5 x 0.4	13.2	Sc: sp	58	Mag 9.03v star SW 4.′5.
07 23 42.2	−29 39 07	ESO 428-29	12.3	1.6 x 1.3	12.9	SAB(rs)b: III-IV	148	Mag 10.00v star E 3.′3.
07 00 17.4	−30 09 46	IC 456	11.9	2.1 x 1.3	12.8	SB(rs)0°	110	Mag 9.51v star on E edge.
06 00 04.9	−33 55 11	IC 2153	13.5	1.0 x 0.8	13.1	Pec	72	Double system, strongly interacting; mag 8.03v star NW 12.′8.
05 57 28.8	−29 57 38	MCG −5-15-1	14.2	1.0 x 0.9	13.9	Sc:	156	Mag 8.74v star SSW 9.′5.
06 03 39.8	−32 08 51	MCG −5-15-3	12.9	1.3 x 0.7	12.7	SB(r)a	14	Several small galaxies at S edge.
06 05 59.8	−32 50 39	MCG −5-15-5	14.4	0.8 x 0.6	13.4	(RL)SB(r)0^+	165	Star NW edge.
06 26 23.0	−31 47 39	MCG −5-16-2	13.8	0.6 x 0.4	12.1	SB(r)bc	47	Mag 7.90v star SE 7.′4.
06 26 37.7	−29 56 08	MCG −5-16-4	13.6	1.0 x 0.7	13.1	S(r)b	84	Several stars involved in fainter outer envelope.
06 29 13.1	−30 20 22	MCG −5-16-5	14.1	0.9 x 0.5	13.1	Sb	70	Mag 8.88v star SW 6.′2; **MCG −5-16-6** SE 2.′4.
06 30 00.5	−32 54 20	MCG −5-16-7	14.2	0.5 x 0.5	12.5			Mag 9.77v star E 5.′7.

(Continued from previous page)

172 GALAXIES 172

RA h m s	Dec ° ′ ″	Name	Mag (V)	Dim ′ Maj x min	SB	Type Class	PA	Notes
06 45 46.8	−31 13 51	MCG −5-16-21	13.2	0.9 x 0.3	11.6	S	2	Mag 9.86v star WSW 5.′3.
05 59 51.7	−39 08 06	MCG −7-13-1	13.2	0.9 x 0.6	12.5	SAB(rs)0°: pec	105	Mag 8.36v star S 6.′2.
06 10 10.1	−34 06 15	NGC 2188	11.7	4.2 x 1.1	13.2	SB(s)m sp III-IV	175	Mag 8.47v star SW 8.′0.
06 33 58.9	−34 48 42	NGC 2255	13.5	1.5 x 0.7	13.4	SAB(r)c: II	152	A pair of bright knots or superimposed stars S edge of core.
06 40 51.8	−32 28 58	NGC 2267	12.5	1.7 x 1.3	13.2	SB(r)0°	120	Bright knot or superimposed star on W edge of core.
06 53 55.0	−40 51 33	NGC 2310	11.8	4.4 x 0.8	13.0	S0 sp	47	Mag 8.54v star E 10.′2.

GALAXY CLUSTERS

RA h m s	Dec ° ′ ″	Name	Mag 10th brightest	No. Gal	Diam ′	Notes
06 00 42.0	−40 02 00	A 3376	15.4	42	22	Staggered band of numerous PGC GX stretches ENE edge to WSW edge.
06 09 54.0	−33 35 00	A 3381	14.7	69	28	Numerous PGC galaxies concentrated around center.
06 25 00.0	−37 20 00	A 3390	14.7	63	28	Most members anonymous, stellar/almost stellar.
06 27 06.0	−35 28 00	A 3392	15.5	77	22	**ESO 365-17** on WSW edge, 8.′9 N of mag 7.89v star; all others anonymous.
05 57 12.0	−37 28 00	AS 555	16.1		17	See notes ESO 364-18; all others anonymous, stellar.
06 01 18.0	−39 00 00	AS 559	15.1		22	Most members anonymous, stellar.
06 03 24.0	−32 44 00	AS 563	16.1	17	17	**ESO 364-26** at center, all other members anonymous, stellar.
06 11 00.0	−32 33 00	AS 570	16.0	0	17	Except ESO 364-40, all members anonymous, stellar.
06 11 36.0	−33 07 00	AS 571	16.2		17	Overlaps AS 573; **PGC 18592** at center; **PGC 18593** N 3.′5; **ESO 364-42**, **42A** S 12.′9.
06 12 00.0	−32 57 00	AS 573	16.0	9	17	Overlaps AS 571; **PGC 18598** at center.
06 15 30.0	−29 16 00	AS 578	16.1		17	
06 28 36.0	−32 29 00	AS 589	16.4	1	17	All members anonymous, stellar.
06 34 18.0	−34 56 00	AS 591	16.1		17	Except NGC 2255, all members anonymous, stellar.
06 40 06.0	−37 45 00	AS 593	15.0	0	22	See notes ESO 308-24; except ESO GX, all anonymous, stellar.
06 45 24.0	−37 00 00	AS 595	15.9	1	17	ESO 366-16 at center; **ESO 366-18** NE 13.′7; all others anonymous, stellar.
06 49 24.0	−33 01 00	AS 597	15.8	5	17	**ESO 366-24** SSE of center 6.′6; most others anonymous, stellar.

OPEN CLUSTERS

RA h m s	Dec ° ′ ″	Name	Mag	Diam ′	No. ★	B ★	Type	Notes
07 15 21.1	−30 41 00	Cr 132	3.8	80	25		cl	
07 17 24.5	−36 50 00	Cr 135	2.1	49	77	2.7	cl	Few stars; moderate brightness range; not well detached. Without brightest star, visual mag = 3.0.
07 17 06.0	−35 17 00	ESO 367-10		3			cl	
06 04 48.0	−29 11 00	ESO 425-6		5			cl	
06 14 36.0	−29 22 00	ESO 425-15		5			cl	
06 36 18.0	−30 51 00	ESO 426-26		4			cl	
07 06 12.0	−30 10 00	ESO 427-32		10			cl	
07 22 54.0	−29 29 48	Haf 7		3	40	14.0	cl	
06 29 35.0	−31 17 00	NGC 2243	9.4	8.3	368	11.8	cl	Rich in stars; moderate brightness range; strong central concentration; detached.
06 41 01.0	−29 33 00	Ru 2		3	10	12.0	cl:	Few stars; moderate brightness range; not well detached.
06 42 06.0	−29 27 12	Ru 3		3	15	11.0	cl	Few stars; small brightness range; no central concentration; detached.
07 14 55.0	−31 21 36	Ru 14		3	10	13.0	cl	(S) Small group.

GLOBULAR CLUSTERS

RA h m s	Dec ° ′ ″	Name	Total V m	B ★ V m	HB V m	Diam ′	Conc. Class Low = 12 High = 1	Notes
06 48 59.2	−36 00 19	NGC 2298	9.3	13.4	16.2	5.0	6	

PLANETARY NEBULAE

RA h m s	Dec ° ′ ″	Name	Diam ″	Mag (P)	Mag (V)	Mag cent ★	Alt Name	Notes
06 55 12.3	−29 07 28	PN G239.6−12.0	24	15.1		19.0	PK 239−12.1	Mag 14.8 star on N edge; mag 13.3 star W 1.′4.
07 02 49.8	−31 35 30	PN G242.6−11.6	28	11.9	12.3:	15.5	PK 242−11.1	Slightly elongated SE-NW; mag 9.27v star N 1.′9.

173 GALAXIES 173

RA h m s	Dec ° ′ ″	Name	Mag (V)	Dim ′ Maj x min	SB	Type Class	PA	Notes
04 36 49.4	−39 38 48	ESO 304-6	14.2	0.9 x 0.7	13.6	Sc	101	Mag 13.1 star W 3.′1.
05 08 07.7	−38 18 33	ESO 305-9	12.5	3.8 x 3.1	15.0	SB(s)dm V	47	Several bright knots; mag 9.95v star SSW 9.′1.
05 12 34.1	−39 51 37	ESO 305-14	13.5	1.7 x 0.8	13.6	SB(s)c I-II	107	Mag 10.63v star W 1.′2.
05 22 42.5	−39 03 50	ESO 305-21	13.7	1.1 x 0.8	13.4	SA?(r) pec	48	A pair of very small, very faint anonymous galaxies NE end? Mag 13.4 star E 1.′4.
05 23 00.2	−38 52 15	ESO 305-22	14.0	0.8 x 0.5	12.8	S	9	Mag 9.68v star E 5.′9.
05 25 46.9	−39 54 26	ESO 305-25	13.4	1.1 x 0.7	13.0	SA(s)bc: II-III	85	Mag 12.1 star S 3.′9.
05 29 07.7	−39 25 17	ESO 306-3	13.6	1.2 x 0.6	13.1	(R)SB(s)bc? II	15	Mag 9.23v star WNW 8.′2.
05 31 16.1	−40 06 58	ESO 306-7	13.5	0.7 x 0.7	12.6			Mag 12.3 star N 1.′7.
05 39 53.5	−40 30 49	ESO 306-16	13.8	1.0 x 0.6	13.0	S?	161	Mag 12.6 star SW 2.′2; **ESO 306-20** ENE 9.′3.
05 40 06.6	−40 50 11	ESO 306-17	12.3	2.5 x 1.6	13.8	E^+3	177	Mag 5.81v star NW 17.′5; mag 9.81v star NNW 6.′4.
05 41 57.9	−38 27 22	ESO 306-22	13.9	1.1 x 0.5	13.2	S pec	77	Disturbed, strong knot W end; stellar companion galaxy 0.′4 W.
05 45 50.9	−39 29 42	ESO 306-25	13.9	1.0 x 0.7	13.3	(R)SA(r)0/a	21	Mag 13.8 star S edge; mag 12.0 star WNW 2.′1.
05 50 56.3	−38 55 33	ESO 306-30	13.9	1.1 x 0.4	12.8	S0?	111	Mag 12.5 star S 0.′9.
05 56 07.5	−38 09 19	ESO 307-5	13.9	1.4 x 0.6	13.6	SB(r)a pec:	101	Mag 7.47v star NW 5.′7; faint, elongated anonymous galaxy N 1.′3.
06 00 41.1	−40 02 42	ESO 307-13	13.4	1.1 x 0.6	13.0	E^+3	62	Almost stellar nucleus, smooth envelope; **PGC 18233** SW 1.′7; other small galaxies in surrounding area.
04 46 58.8	−35 54 58	ESO 361-9	14.5	1.7 x 1.0	14.9	IB(s)m V-VI	49	Mag 112.0 star on W edge; mag 7.75v star NW 9.′4.
04 49 51.6	−36 06 12	ESO 361-12	14.4	1.7 x 0.2	13.0	Sbc? sp	139	Pair of E-W oriented mag 10.85v, 12.4 stars SE 10.′3.
04 51 57.5	−33 10 47	ESO 361-15	13.0	3.2 x 0.6	13.5	SB(s)d? sp III-IV	94	Mag 7.75v star S 5.′8.
04 54 53.7	−37 19 22	ESO 361-19	13.6	1.1 x 0.4	12.6	SBa? pec	102	Mag 10.19v star SW 9.′6; stellar **PGC 16320** N 4.′0.
04 58 17.4	−33 48 31	ESO 361-23	15.1	1.5 x 0.9	15.3	SB(s)cd III-IV	150	Mag 9.07v star N 7.′4.
05 01 40.1	−34 01 36	ESO 361-25	13.6	2.3 x 1.1	14.5	S pec	168	Long, faint extension N; mag 11.00v star W 5.′5; ESO 362-1 E 3.′2.
05 01 55.8	−34 01 43	ESO 362-1	14.1	0.8 x 0.4	12.7	S0/a	39	Mag 14.8 star S end; ESO 361-25 W 3.′2.

(Continued from previous page)

GALAXIES

RA h m s	Dec ° ′ ″	Name	Mag (V)	Dim ′ Maj x min	SB	Type Class	PA	Notes
05 04 02.6	−36 01 53	ESO 362-3	14.8	1.9 x 0.2	13.5	Sc	135	Mag 9.40v star SW 4′.1.
05 11 08.9	−34 23 35	ESO 362-8	12.8	1.2 x 0.5	12.1	S0?	166	Mag 9.56v star W 5′.6.
05 11 59.3	−32 58 21	ESO 362-9	12.2	3.7 x 3.0	14.7	SAB(s)m V	18	Mag 10.78v star W 3′.0; mag 11.10v star N 3′.3.
05 16 39.4	−37 06 08	ESO 362-11	12.2	5.2 x 0.8	13.6	Sbc: sp	76	Mag 9.08v star S 1′.8; ESO 362-12 NE 3′.9.
05 17 04.1	−37 04 08	ESO 362-12	13.5	1.1 x 0.7	13.1	SA?(r)ab	161	ESO 362-11 SW 3′.9; mag 8.75v star NE 5′.6.
05 18 00.1	−33 54 45	ESO 362-13	14.0	1.3 x 0.5	13.4	SA0$^-$:	108	Mag 8.37v star S 7′.1.
05 19 17.3	−37 08 40	ESO 362-15	13.3	0.8 x 0.7	12.5	S	86	Mag 8.36v star SE 12′.1; **ESO 362-16** N 2′.4; IC 2122 NW 4′.6.
05 19 35.7	−32 39 28	ESO 362-18	13.0	1.0 x 0.8	12.7	SB(s)0/a: pec	141	Mag 8.55v star E 9′.9; **ESO 362-17** S 6′.4.
05 21 04.0	−36 57 25	ESO 362-19	13.5	2.4 x 0.7	13.9	SB(s)m IV-V	178	Numerous strong knots; mag 9.48v star SW 12′.4.
05 30 40.3	−33 23 17	ESO 363-3	14.1	0.9 x 0.7	13.4	SAB(rs)b pec	48	Mag 12.0 star SE 1′.8.
05 41 00.7	−35 42 27	ESO 363-15	13.2	2.8 x 2.0	14.9	SA(s)d III-IV	0	Pair of mag 12 stars NE edge.
05 43 11.4	−34 36 53	ESO 363-17	13.8	1.1 x 1.0	13.8	Sc II	100	Mag 9.40v star S 2′.8.
05 43 54.6	−33 16 34	ESO 363-18	13.5	0.7 x 0.5	12.2	Sb	4	Mag 12.7 star SE 2′.8.
05 50 28.6	−33 44 30	ESO 364-2	14.2	1.0 x 0.6	13.5	(R′)SAB(r)ab	58	Mag 10.61v star SSW 6′.1.
05 51 00.8	−34 46 42	ESO 364-4	14.0	1.1 x 0.4	13.0	S0	10	**MCG −6-13-12** W 0′.9; **ESO 364-3** SSW 8′.1; **MCG −6-13-13, 14, 15** S 7′.7.
05 52 19.0	−34 20 56	ESO 364-7	13.3	1.5 x 1.2	13.8	SB(r)c: pec I-II	78	Bright knot, or small galaxy, E edge.
05 52 32.5	−36 57 54	ESO 364-8	13.9	1.0 x 0.6	13.2	Sc	41	Mag 12.9 star E edge.
05 53 51.7	−32 44 35	ESO 364-12	13.3	1.7 x 1.6	14.2	Sb	85	Mag 7.74v star SW 8′.8; MCG −5-14-23 S 1′.8.
05 56 27.1	−33 25 52	ESO 364-16	14.3	1.3 x 0.5	13.6	Pec	78	Very transparent; mag 7.30v star E edge.
05 56 26.4	−34 06 08	ESO 364-17	13.5	1.1 x 0.9	13.3	SAB(r)cd	158	Mag 10.31v star NW 1′.4.
05 57 12.5	−37 28 38	ESO 364-18	13.8	1.0 x 0.5	13.1	E$^+$2	138	Mag 11.20v star NW 1′.6; **ESO 307-6** SW 14′.2.
05 58 18.1	−34 53 19	ESO 364-19	13.9	0.9 x 0.5	12.8	SB	50	Almost stellar, faint anonymous galaxy S 0′.6.
06 00 08.1	−33 00 53	ESO 364-23	14.1	1.0 x 0.4	12.9	S	84	Pair of mag 14.7 stars N edge; mag 13.1 star SW 1′.9.
04 49 12.5	−29 12 26	ESO 421-19	12.5	3.2 x 2.6	14.7	SAB(s)m: V	56	Numerous knots and superimposed stars; small triangle of mag 12-13 stars E 3′.7.
04 52 14.1	−30 04 14	ESO 422-6	14.5	0.9 x 0.3	12.9	Pec?	76	Close pair of N-S oriented mag 10.80v, 10.81v stars NNE 5′.4.
04 55 30.2	−31 48 38	ESO 422-10	13.5	1.5 x 0.6	13.2	SAB(rs)b:	109	Mag 8.90v star ENE 14′.7.
04 58 48.6	−30 32 50	ESO 422-18	13.8	0.6 x 0.4	12.1		157	Mag 12.8 star NW 3′.6.
05 02 17.9	−30 27 20	ESO 422-23	13.4	1.6 x 0.8	13.5	SB(rs)b pec	17	Mag 9.97v star NNW 5′.5.
05 05 07.1	−31 47 04	ESO 422-27	13.6	1.5 x 0.9	13.8	SAB(r)c II-III	9	Mag 8.87v star WNW 5′.4.
05 10 13.9	−29 24 15	ESO 422-40	13.7	1.5 x 0.4	12.9	Sbc	23	Star on W edge.
05 10 46.6	−31 35 52	ESO 422-41	12.7	2.7 x 2.0	14.4	SAB(rs)dm: IV-V	79	Mag 12.3 star ESE 5′.1; mag 10.69v star WSW 10′.9.
05 15 16.1	−30 31 42	ESO 423-2	12.4	3.0 x 0.9	13.3	SB(s)d: sp III-IV	19	Mag 10.80v star WNW 10′.6.
05 26 34.6	−31 50 38	ESO 423-16	13.7	0.9 x 0.5	13.2	(R)SB(s)0/a	33	Mag 9.14v star NE 11′.9.
05 34 41.4	−29 13 55	ESO 423-24	12.2	1.2 x 1.2	12.5	SA0°? pec		Mag 10.59v star WNW 4′.8.
05 35 34.7	−29 09 01	ESO 424-1	13.9	0.9 x 0.5	13.1	E$^+$4 pec:	147	Strong core, faint envelope; mag 9.84v star NE 4′.5.
05 50 50.0	−31 44 27	ESO 424-27	13.9	1.0 x 0.5	13.0	SA0$^-$: pec	178	Mag 8.35v star E 9′.6; note: **ESO 424-32** is located 3′.5 E of this star; **ESO 424-29** E 5′.4.
05 51 56.0	−31 12 32	ESO 424-33	13.6	1.3 x 0.4	12.7	S0/a	36	Mag 14.2 star superimposed NW edge; mag 10.97v star SW 1′.3.
06 00 11.0	−31 47 14	ESO 425-1	14.5	0.9 x 0.8	14.0	S	20	Mag 12.9 star SW edge.
05 19 01.3	−37 05 22	IC 2122	12.7	1.5 x 1.3	13.3	SAB(s)0$^-$:	67	Mag 9.73v star N 7′.8; **ESO 362-16** E 3′.5; ESO 362-15 SE 4′.6.
05 33 13.1	−36 23 59	IC 2135	12.5	2.8 x 0.6	12.9	Scd: sp	109	Mag 9.69v star SW 5′.5; **MCG −6-13-5** E 3′.4.
05 43 28.3	−30 29 42	IC 2147	12.7	1.8 x 1.4	13.5	SB(rs)dm	90	Mag 8.49v star SE 10′.2.
05 51 17.9	−38 19 16	IC 2150	12.8	2.6 x 0.8	13.4	SB(r)c: I	87	Mag 11.02v star W 7′.6; **ESO 306-31** S 2′.2.
06 00 04.9	−33 55 11	IC 2153	13.5	1.0 x 0.8	13.1	Pec	72	Double system, strongly interacting; mag 8.03v star NW 12′.8.
04 38 03.4	−29 37 17	MCG −5-12-1	14.5	0.7 x 0.3	12.6	S	45	
04 53 49.0	−32 26 23	MCG −5-12-7	13.6	1.2 x 1.0	13.7	(R′)SB(r)ab:	98	
05 00 45.5	−29 51 29	MCG −5-12-13	14.0	1.0 x 0.9	13.7	Sc	84	Mag 9.41v star E 6′.9.
05 04 33.1	−29 04 16	MCG −5-13-1	14.2	0.5 x 0.5	12.5	S0/a		Mag 10.12v star ENE 6′.5.
05 17 57.9	−29 42 03	MCG −5-13-14	14.2	1.4 x 0.2	12.7	Sc	149	Mag 10.47v star S 1′.1.
05 25 35.4	−31 36 06	MCG −5-13-19	14.2	0.5 x 0.5	12.5	SA0$^-$ pec		Mag 10.04v star SE 3′.8.
05 25 38.7	−31 33 14	MCG −5-13-20	14.4	0.5 x 0.3	12.2		156	Mag 8.84v star NW 8′.4; **MCG −5-13-18** W 3′.6.
05 25 41.8	−31 32 12	MCG −5-13-21	13.9	0.4 x 0.4	11.7			Mag 8.84v star WNW 8′.6.
05 31 16.3	−29 19 08	MCG −5-14-1	13.4	0.5 x 0.4	11.7	E	60	Mag 11.16v star S 4′.2.
05 32 39.0	−29 37 38	MCG −5-14-2	13.8	1.1 x 0.4	12.8	Sb	83	Mag 9.29v star E 9′.9.
05 34 29.3	−30 41 07	MCG −5-14-5	13.9	0.8 x 0.4	12.6	S0	166	Mag 9.44v star S 2′.0.
05 36 05.1	−32 54 54	MCG −5-14-8	15.9	0.4 x 0.4	13.8	SBbc		
05 39 21.9	−29 22 20	MCG −5-14-9	14.1	1.5 x 0.5	13.7	Sb?	21	**ESO 424-5A** SE edge.
05 42 56.6	−29 43 45	MCG −5-14-10	13.4	1.2 x 1.0	13.5	S/Irr	142	Star NE edge.
05 48 27.6	−32 58 40	MCG −5-14-17	13.1	1.2 x 0.9	13.3	E$^+$1:	176	Anonymous galaxy SE 2′.2.
05 49 02.3	−31 29 28	MCG −5-14-19	13.9	0.7 x 0.4	12.4		15	Mag 9.81v star N 8′.6.
05 51 40.1	−31 31 25	MCG −5-14-22	15.5	0.6 x 0.4	13.8		18	Mag 9.18v star NW 2′.4.
05 53 49.2	−32 46 27	MCG −5-14-23	13.8	1.0 x 0.4	12.7	SA0$^+$	112	Mag 7.74v star WSW 7′.5; ESO 364-12 N 1′.8.
05 55 04.7	−31 37 38	MCG −5-14-24	14.2	0.6 x 0.4	12.5	Sd	141	Mag 9.67v star NW 1′.2.
05 57 28.8	−29 57 38	MCG −5-15-1	14.2	1.0 x 0.9	13.9	Sc:	156	Mag 8.74v star SSW 9′.5.
06 03 39.8	−32 08 51	MCG −5-15-3	12.9	1.3 x 0.7	12.7	SB(r)a	14	Several small galaxies at S edge.
05 59 51.7	−39 08 06	MCG −7-13-1	13.2	0.9 x 0.6	12.5	SAB(rs)0°: pec	105	Mag 8.36v star S 6′.2.
04 49 55.7	−31 58 03	NGC 1679	11.6	2.7 x 2.0	13.3	SB(s)m IV	150	Star inside N edge; several bright knots and/or superimposed stars.
04 51 21.0	−33 56 26	NGC 1687	13.9	1.4 x 0.5	13.4	SAB(r)ab:	40	Mag 8.79v star SW 6′.8.
04 55 51.3	−29 53 00	NGC 1701	12.8	1.2 x 0.9	12.7	(R)SA(r)b: II-III	137	Mag 10.68v star 1′.3 SE.
05 00 49.2	−38 40 28	NGC 1759	12.8	1.2 x 1.2	13.1	E$^+$:		Star or bright knot S of round nucleus; smooth, faint envelope.
05 05 14.8	−37 58 48	NGC 1792	10.2	5.2 x 2.6	12.9	SA(rs)bc II-III	137	**ESO 305-5** S 8′.9.
05 06 25.7	−31 57 15	NGC 1800	12.6	2.0 x 1.1	13.3	IB(s)m IV-V	113	
05 07 42.8	−37 30 51	NGC 1808	9.9	6.5 x 3.9	13.3	(R)SAB(s)a	139	Strong dark lane along SW side of large core; close galaxy pair **ESO 305-10, 10A** SE 12′.6.
05 08 42.8	−29 16 36	NGC 1811	13.4	1.7 x 0.4	12.8	Sa: pec sp	60	
05 08 52.8	−29 15 08	NGC 1812	12.6	1.2 x 0.9	12.5	Sa pec	8	
05 10 04.9	−36 57 38	NGC 1827	12.4	3.0 x 0.7	13.1	SAB(s)cd: sp IV	120	
05 19 48.5	−32 08 31	NGC 1879	12.8	2.5 x 1.7	14.2	SB(s)m IV-V	60	Mag 9.62v star NW 3′.6.
05 34 23.4	−30 48 04	NGC 1989	12.9	1.4 x 0.9	13.1	SA(s)0$^-$:	106	Mag 9.66v star N 3′.1.
05 34 31.8	−30 53 50	NGC 1992	13.5	1.1 x 0.7	13.1	SA(rs)0/a?	45	Mag 9.75v star W 2′.3.
05 43 15.0	−30 04 42	NGC 2049	12.7	2.0 x 1.0	13.3	SA(s)a?	168	
05 47 01.7	−34 15 02	NGC 2090	11.2	4.9 x 2.4	13.7	SA(rs)c II-III	13	Numerous stars on faint outer envelope.

GALAXY CLUSTERS

RA h m s	Dec ° ′ ″	Name	Mag 10th brightest	No. Gal	Diam ′	Notes
05 00 48.0	−38 40 00	A 3301	15.5	172	22	NGC 1759 at center, **MCG −6-12-1** S 1′.8, most other members anonymous, stellar/almost stellar.
05 25 36.0	−31 35 00	A 3341	14.7	87	28	String of MCG GX at center, most other members anonymous.
05 31 48.0	−38 21 00	A 3351	16.2	114	17	All members anonymous, faint, stellar.
05 35 00.0	−38 08 00	A 3356	16.0	30	17	All members anonymous, faint, stellar.

(Continued from previous page)

GALAXY CLUSTERS

RA h m s	Dec ° ′ ″	Name	Mag 10th brightest	No. Gal	Diam ′	Notes
05 55 48.0	−34 47 00	A 3372	16.3	35	17	All members anonymous, faint, stellar.
06 00 42.0	−40 02 00	A 3376	15.4	42	22	Staggered band of numerous PGC GX stretches ENE edge to WSW edge.
04 36 36.0	−39 38 00	AS 476	16.1		17	Except ESO 304-6, all members anonymous, stellar.
04 38 06.0	−31 40 00	AS 478	16.3	2	17	All members anonymous, stellar.
04 40 06.0	−35 37 00	AS 484	16.1	5	17	**ESO 361-3**, **3A** at center, all others anonymous, stellar.
04 41 48.0	−36 56 00	AS 489	16.0	3	17	**MCG −6-11-4** at center, all others anonymous, stellar.
04 45 30.0	−35 55 00	AS 496	15.8		17	All members anonymous, stellar.
05 09 00.0	−37 49 00	AS 512	15.9	21	17	**ESO 305-12**, **12A** ESE of center 4′.5; **ESO 305-11** NW 6′.0.
05 19 00.0	−37 05 00	AS 521	14.6	0	28	
05 33 18.0	−36 20 00	AS 535	15.5		22	See notes for IC 2135; all others anonymous, stellar.
05 34 24.0	−30 48 00	AS 536	14.8	11	28	Except three catalog GX at center, all others anonymous, stellar.
05 40 06.0	−40 50 00	AS 540	15.4	22	22	See notes ESO 306-16; except three ESO GX, all GX anonymous, stellar
05 46 48.0	−32 34 00	AS 545	15.6	28	22	**ESO 363-22** at center; most others anonymous, stellar.
05 50 54.0	−34 47 00	AS 550	15.6	3	22	See notes ESO 364-4; most others anonymous, stellar.
05 57 12.0	−37 28 00	AS 555	16.1		17	See notes ESO 364-18; all others anonymous, stellar.
06 01 18.0	−39 00 00	AS 559	15.1		22	Most members anonymous, stellar.
06 03 24.0	−32 44 00	AS 563	16.1	17	17	**ESO 364-26** at center, all other members anonymous, stellar.

OPEN CLUSTERS

RA h m s	Dec ° ′ ″	Name	Mag	Diam ′	No. ★	B ★	Type	Notes
05 49 48.0	−32 28 00	ESO 424-25		8			cl	
05 21 17.2	−35 44 00	NGC 1891		27		9.7	cl?	
05 32 10.8	−36 23 00	NGC 1963		14	32	11.0	cl	(A)In form of a figure "3."
05 44 06.0	−33 56 00	NGC 2061		10	11	13.0	ast?	(A) Includes galaxy **ESO 363-19** on NE edge.

GLOBULAR CLUSTERS

RA h m s	Dec ° ′ ″	Name	Total V m	B ★ V m	HB V m	Diam ′	Conc. Class Low = 12 High = 1	Notes
05 14 06.3	−40 02 50	NGC 1851	7.1	13.2	16.1	12.0	2	

PLANETARY NEBULAE

RA h m s	Dec ° ′ ″	Name	Diam ″	Mag (P)	Mag (V)	Mag cent ★	Alt Name	Notes
05 03 01.8	−39 45 45	PN G243.8−37.1	23			15.6	PK 242−37.1	Mag 14.8 star SW 0′.6.

GALAXIES

RA h m s	Dec ° ′ ″	Name	Mag (V)	Dim ′ Maj x min	SB	Type Class	PA	Notes
03 23 54.3	−37 30 35	ESO 301-11	13.2	0.9 x 0.8	12.7	Ring: pec	46	Mag 12.0 star W 6′.4.
03 25 15.1	−39 53 48	ESO 301-14	14.2	1.5 x 1.2	14.7	SA(rs)d IV	2	Mag 11.8 star NNE 4′.7.
03 40 05.4	−37 47 35	ESO 301-22	14.7	1.4 x 1.1	15.0	Sa? pec	139	
03 41 59.8	−38 01 59	ESO 301-25	15.0	0.9 x 0.4	13.8	S pec	133	Mag 12.4 star E 4′.0.
03 46 35.4	−40 38 55	ESO 302-6	13.3	1.1 x 0.6	12.7	SA(rs)0^{-}:	135	Mag 9.97v star SW 1′.3.
03 47 06.4	−37 22 43	ESO 302-7	13.4	1.3 x 0.8	13.3	SAB(s)c: II-III	10	Located between a pair of stars, mags 11.7 and 11.33v.
03 47 13.5	−38 58 10	ESO 302-8	14.4	1.0 x 0.5	13.5	SAB0°	72	Mag 9.93v star NNE 5′.8.
03 47 33.5	−38 34 43	ESO 302-9	13.7	2.5 x 0.5	13.8	SB(s)dm? IV-V	48	Mag 11.4 star on S edge.
03 50 30.5	−39 22 36	ESO 302-12	14.6	1.3 x 0.9	14.6	SB(s)d IV	93	Mag 12.8 star W 4′.9.
03 50 59.7	−38 16 16	ESO 302-13	15.2	1.2 x 0.4	14.2		100	Triple system, strongly interacting; mag 9.28v star NE 5′.0.
03 51 40.6	−38 27 03	ESO 302-14	14.4	1.7 x 1.4	15.2	IB(s)m V-VI	120	Uniform surface brightness; mag 10.51v star N 7′.8.
03 58 28.4	−40 13 38	ESO 302-23	13.4	1.2 x 0.7	13.1		60	Double system, contact; faint companion NE 0′.5 from center.
04 13 58.8	−38 05 53	ESO 303-5	13.9	2.3 x 1.6	15.3	E^{+}1 pec		Mag 13.7 star on S edge; mag 10.54v star S 5′.5.
04 28 17.7	−40 28 13	ESO 303-20	14.1	1.2 x 1.0	14.1	S?	171	Mag 7.14v star W 8′.7.
04 29 06.0	−37 28 46	ESO 303-21	13.8	0.8 x 0.5	12.6	S?	168	Mag 8.02v star SW 11′.1.
04 36 49.4	−39 38 48	ESO 304-6	14.2	0.9 x 0.7	13.6	Sc	101	Mag 13.1 star W 3′.1.
03 16 53.4	−35 32 32	ESO 357-12	13.0	3.0 x 1.8	14.7	SB(s)d IV-V	122	Mag 7.05v star S 9′.9.
03 23 37.4	−35 46 43	ESO 357-25	14.2	1.2 x 0.5	13.6	SAB0°:	30	Mag 10.61v star NE 9′.2.
03 27 16.3	−33 29 11	ESO 358-5	13.9	1.3 x 1.0	14.0	SAB(s)m pec: V	111	Uniform surface brightness; mag 10.03v star W 5′.5.
03 27 17.9	−34 31 40	ESO 358-6	13.1	1.1 x 0.6	12.5	SB0^{-}?	32	Mag 9.22v star NE 1′.7.
03 29 43.4	−33 33 27	ESO 358-10	13.7	0.9 x 0.5	12.7	SA(s)0°	59	Mag 7.15v star SW 11′.7.
03 34 57.2	−32 38 22	ESO 358-20	13.8	1.2 x 0.8	13.6	IB(s)m? pec IV	167	Mag 10.60v star W 9′.6; ESO 358-22 ENE 5′.3.
03 35 20.3	−32 36 11	ESO 358-22	12.9	1.4 x 1.1	13.3	(R′)SA(r)0°	99	ESO 358-20 WSW 5′.3.
03 35 33.2	−32 27 54	ESO 358-25	12.8	1.5 x 0.8	12.9	SA0^{-}:	59	Mag 13.3 star WNW 3′.3; ESO 358-22 SSW 8′.7.
03 38 09.4	−34 31 12	ESO 358-42	14.4	1.0 x 0.6	13.8	SB0°:	141	Mag 8.37v star SW 12′.7.
03 39 11.8	−33 31 57	ESO 358-47	13.4	1.3 x 0.5	12.8		30	Multiple galaxy system, compact, interaction.
03 41 03.7	−33 46 49	ESO 358-50	12.9	1.5 x 0.6	12.7	SA0°?	172	Star N end.
03 41 32.5	−34 53 21	ESO 358-51	13.2	1.4 x 0.6	12.9	S0/a:	3	Mag 8.59v star W 4′.4.
03 45 03.6	−35 58 23	ESO 358-59	13.1	0.9 x 0.7	12.5	SAB0^{-}	153	Close pair of stars, mags 14.4 and 15.6, WNW 1′.5.
03 45 12.2	−35 34 14	ESO 358-60	15.3	1.7 x 0.4	14.8	IB(s)m: sp V-VI	103	Mag 12.4 star N 2′.1.
03 45 54.7	−36 21 32	ESO 358-61	13.1	2.4 x 0.7	13.5	I0:	2	Mag 11.8 star NNW 2′.5.
03 46 18.8	−34 56 31	ESO 358-63	11.8	4.8 x 1.1	13.4	I0?	132	Pair of E-W oriented mag 10.94v, 12.1 stars NE 6′.0.
03 47 52.8	−36 28 20	ESO 358-66	14.0	0.9 x 0.5	13.1	SB(s)0^{-}:	39	Mag 9.52v star SW 5′.5.
03 48 07.1	−36 42 09	ESO 358-67	14.1	0.7 x 0.5	12.9	S?	97	Mag 8.60v star SW 7′.0.
03 50 36.8	−35 54 36	ESO 359-2	13.4	1.0 x 0.6	12.8	SB0^{-}?	45	Mag 14.0 star N edge; mag 11.18v star ESE 9′.3.
03 52 01.1	−33 28 10	ESO 359-3	13.6	1.7 x 0.6	13.5	Sab: sp	131	Mag 12.5 star on N edge; mag 10.63v star NW 9′.3.
03 59 42.0	−36 42 21	ESO 359-13	14.1	1.3 x 0.6	13.7	S?	151	Mag 10.09v star N 6′.2.
04 04 25.0	−36 10 53	ESO 359-16	14.5	1.5 x 0.3	13.5	IBm? sp III-IV	52	Mag 10.02v star NW 6′.7.
04 05 04.6	−35 00 25	ESO 359-18	15.1	1.5 x 0.4	14.4	SB0°? sp	9	Mag 8.12v star ESE 10′.9.
04 12 50.4	−33 00 13	ESO 359-29	14.2	1.4 x 0.7	14.0	IB(s)m V	25	NGC 1532 NW 12′.6; mag 7.05v star N 12′.8.
04 13 09.6	−34 25 35	ESO 359-31	14.2	1.1 x 1.0	14.2	SB(s)dm IV	45	Mag 9.57v star SSE 8′.3.
04 15 32.2	−35 20 35	ESO 360-2	14.1	1.1 x 0.7	13.7	SB?	79	Diffuse, broad arms.
04 16 27.0	−35 44 00	ESO 360-4	14.0	0.9 x 0.5	13.0	S?	127	Mag 10.78v star NE 11′.6.

(Continued from previous page)

GALAXIES

RA h m s	Dec ° ′ ″	Name	Mag (V)	Dim ′ Maj x min	SB	Type Class	PA	Notes
04 20 07.8	−36 41 07	ESO 360-7	15.4	1.0 x 0.4	14.2	SB(s)m V-VI	114	Mag 8.69v star NNW 3′.4.
04 27 24.0	−33 31 16	ESO 360-10	15.6	1.0 x 0.5	14.7	Sm: IV?	9	Mag 9.74v star E 5′.2; **MCG −6-10-9** SSE 7′.2.
04 32 38.0	−33 24 52	ESO 360-14	13.9	0.7 x 0.7	13.0	SAB(r)bc pec: II		Mag 13.1 star SE 1′.6.
03 31 30.9	−30 12 46	ESO 418-8	13.3	1.3 x 0.9	13.3	SAB(rs)dm IV	133	Mag 8.24v star SW 14′.0.
03 31 55.6	−31 20 18	ESO 418-9	13.4	1.5 x 1.2	13.9	IB(s)m	114	Mag 9.95v star NW 6′.0.
03 34 16.6	−30 43 59	ESO 418-11	13.5	1.6 x 1.1	13.9	(R′)SA(r)b pec II	98	Stellar nucleus; mag 7.16v star NW 8′.0.
03 37 12.7	−32 00 01	ESO 418-14	14.2	0.8 x 0.4	12.9	S?	27	Mag 10.39v star N 6′.8.
04 00 41.7	−30 49 52	ESO 419-13	13.8	1.2 x 0.5	13.1	S0⁺? pec	3	Mag 9.69v star NE 8′.5.
04 07 45.7	−29 51 34	ESO 420-3	12.8	1.9 x 1.3	13.7	SA(rs)c I-II	144	Mag 10.27v star NW 4′.7; mag 10.02v star SE 4′.0.
04 09 13.8	−30 24 57	ESO 420-5	13.1	1.2 x 0.9	13.1	S?	99	Mag 9.99v star covers SE end.
04 10 01.7	−31 15 29	ESO 420-6	14.5	1.7 x 0.7	14.5	IB(s)m V-VI	133	Mag 11.32v star SE 8′.5.
04 11 00.5	−31 24 27	ESO 420-9	13.0	1.7 x 1.3	13.8	SB(s)c II	36	Mag 12.3 star NE 3′.9; mag 11.32v star W 8′.4.
04 12 28.1	−29 03 36	ESO 420-10	14.5	0.9 x 0.8	14.1		51	Almost stellar core, faint envelope; mag 9.06v star S 2′.5.
04 13 49.8	−32 00 27	ESO 420-13	12.5	0.8 x 0.8	11.9	SA(r)0⁺ pec?		Mag 6.97v star SE 2′.8.
03 35 31.2	−34 26 49	IC 335	12.1	2.6 x 0.7	12.6	S0 sp	84	Mag 11.8 star W 7′.5.
03 16 05.7	−34 21 38	IC 1906	13.5	1.1 x 0.5	12.7	(R′)SB(s)bc? II	63	Mag 11.8 star NE 4′.4.
03 17 20.0	−33 41 28	IC 1909	13.5	1.1 x 0.6	12.9	SB(r)b II-III	60	String of four mag 14 and 15 stars extending N from N edge.
03 19 34.8	−32 27 58	IC 1913	13.7	1.9 x 0.3	12.9	SBb? sp	149	Mag 10.59v star N 7′.8.
03 26 02.3	−32 53 43	IC 1919	13.0	1.3 x 1.0	13.2	SA(rs)0⁻?	84	Mag 9.85v star N 5′.1.
03 47 05.1	−33 42 38	IC 1993	11.7	2.5 x 2.1	13.3	(R′)SAB(rs)b II-III	56	Mag 9.27v star on W edge.
03 54 28.8	−35 58 03	IC 2006	11.3	2.1 x 1.8	12.7	(R)SA0⁻	36	Mag 9.90v star SSW 5′.3; **ESO 359-5** SSE 7′.7.
04 09 55.5	−39 41 20	IC 2036	13.5	0.9 x 0.7	12.8	(R′)SB(s)bc	82	E-W bar with dark patches N and S.
04 12 59.5	−32 33 14	IC 2040	13.1	1.4 x 0.8	13.0	S0°: pec	65	
04 12 34.9	−32 49 07	IC 2041	14.0	1.0 x 0.5	13.1	SAB(s)0°:	136	N end of NGC 1532 W 4′.4; mag 7.05v star E 6′.0.
04 20 26.2	−31 43 28	IC 2059	12.9	1.3 x 0.4	12.0	SAB(r)0°:	172	Mag 12.2 star W 8′.2.
03 20 52.0	−30 47 22	MCG −5-9-2	14.7	0.9 x 0.3	13.1	S	60	
03 21 12.4	−30 58 54	MCG −5-9-3	14.6	0.5 x 0.4	12.7	S	0	Mag 9.84v star SE 1′.1.
03 36 42.5	−30 44 10	MCG −5-9-18	13.9	1.1 x 0.4	12.9	Sb:	76	Mag 8.47v star W 12′.1.
03 46 48.2	−29 56 23	MCG −5-10-1	14.0	1.0 x 0.8	13.6	Sb	175	Mag 6.55v star NE 7′.2.
04 03 17.5	−30 48 13	MCG −5-10-8	14.6	0.5 x 0.4	12.7	S	153	
04 10 50.5	−29 55 35	MCG −5-10-14	13.5	1.3 x 0.9	13.6	SB(s)bc II	128	Star on W edge.
04 11 56.1	−30 52 22	MCG −5-11-3	14.9	0.5 x 0.5	13.2	Sc		
04 28 20.6	−32 25 45	MCG −5-11-8	14.2	0.7 x 0.5	12.9	Sb	15	Mag 7.89v star SE 3′.9; **MCG −5-11-9** E 0′.8.
04 29 09.6	−29 47 06	MCG −5-11-10	14.3	0.5 x 0.4	12.4	Sa	27	Mag 8.01v star W 7′.7.
04 31 30.8	−29 45 03	MCG −5-11-12	14.2	0.7 x 0.5	13.2	E	120	Pair with MCG −5-11-13; located 1′.5 N of mag 8.19v star.
04 31 35.3	−29 45 17	MCG −5-11-13	12.7	0.5 x 0.4	11.0	E	45	
04 31 37.6	−31 28 24	MCG −5-11-14	14.6	0.5 x 0.3	12.4	S	153	Mag 7.55v star S 2′.7.
04 31 42.1	−29 40 31	MCG −5-11-15	13.9	0.5 x 0.5	12.2	S		
04 38 03.4	−29 37 17	MCG −5-12-1	14.5	0.7 x 0.3	12.6	S	45	
03 17 13.2	−32 34 35	NGC 1288	12.1	2.3 x 1.9	13.6	SAB(rs)c I-II	166	Tightly wound multi-branching arms.
03 21 03.8	−36 06 06	NGC 1310	12.1	2.0 x 1.5	13.1	SA(s)c: III	95	Mag 9.31v star NE 7′.7.
03 22 41.0	−37 12 28	NGC 1316	8.5	11.0 x 7.2	13.2	SAB(s)0° pec	50	**Fornax A**.
03 24 58.4	−37 00 35	NGC 1316C	13.5	1.6 x 0.7	13.5	(R′)SA0°:	85	Mag 7.16v star S 9′.0; faint, elongated anonymous galaxy N 5′.2.
03 22 44.6	−37 06 11	NGC 1317	11.0	2.8 x 2.4	12.9	SAB(r)a	78	
03 23 56.4	−36 27 50	NGC 1326	10.5	3.9 x 2.9	13.0	(R)SB(r)0⁺	86	
03 25 08.8	−36 21 50	NGC 1326A	13.1	1.9 x 1.7	14.3	SB(s)m: IV-V	50	
03 25 18.9	−36 22 59	NGC 1326B	13.3	3.7 x 1.1	14.7	SB(s)m: sp IV	130	
03 26 32.3	−35 42 49	NGC 1336	12.3	2.1 x 1.5	13.4	SA0⁻	22	Mag 5.70v star E 12′.5.
03 28 06.7	−32 17 13	NGC 1339	11.6	1.5 x 1.1	12.2	E⁺ pec:	172	Mag 9.09v star NW 5′.9.
03 27 58.4	−37 09 00	NGC 1341	12.3	1.2 x 1.1	12.4	SAB(s)ab III	89	
03 28 19.8	−31 04 07	NGC 1344	10.4	4.8 x 3.1	13.4	E5	165	Mag 9.64v star E 5′.9.
03 31 08.4	−33 37 42	NGC 1350	10.3	5.9 x 3.1	13.3	(R′)SB(r)ab II	0	Mag 7.24v star 5′.9 NNE of center.
03 30 35.2	−34 51 13	NGC 1351	11.6	2.8 x 1.7	13.1	SA0⁻ pec:	140	Mag 9.44v star NW 9′.0.
03 28 49.0	−35 10 46	NGC 1351A	13.3	2.7 x 0.5	13.5	SB(rs)bc: sp	132	
03 33 36.7	−36 08 20	NGC 1365	9.6	11.0 x 6.2	14.1	SB(s)b I	32	Two main arms with many knots.
03 33 53.7	−31 11 33	NGC 1366	12.0	2.1 x 0.9	12.5	S0° sp	2	Mag 6.21v star N 6′.9.
03 36 45.3	−36 15 23	NGC 1369	12.8	1.5 x 1.4	13.4	SB(rs)0/a	12	Mag 7.24v star SE 4′.3.
03 34 59.3	−35 10 20	NGC 1373	13.3	0.7 x 0.7	12.6	E⁺:		NGC 1374 SE 4′.9.
03 35 16.9	−35 13 36	NGC 1374	11.1	2.5 x 2.3	13.0	E	124	NGC 1373 NW 4′.9; NGC 1375 S 2′.3.
03 35 17.0	−35 15 58	NGC 1375	12.4	2.2 x 0.9	13.0	SAB0°: sp	91	NGC 1374 N 2′.3.
03 36 04.1	−35 26 31	NGC 1379	10.9	2.4 x 2.3	12.8	E	17	Very faint **PGC 13287** NW 7′.6.
03 36 27.6	−34 58 35	NGC 1380	9.9	4.0 x 2.4	12.2	SA0	7	**ESO 358-31** N 4′.5.
03 36 47.6	−34 44 16	NGC 1380A	12.4	2.4 x 0.7	12.8	S0°: sp	179	Mag 8.86v star SW 4′.6; mag 8.37v star NE 8′.8.
03 36 31.7	−35 17 45	NGC 1381	11.5	2.7 x 0.7	12.0	SA0: sp	139	**MCG −6-9-8** SE 6′.6.
03 37 09.0	−35 11 43	NGC 1382	12.9	1.5 x 1.3	13.5	SAB(s)0⁻:	26	Small, faint elongated anonymous galaxy and a mag 14.3 star NNE 3′.2.
03 36 46.3	−35 59 56	NGC 1386	11.2	3.4 x 1.1	12.5	SB(s)0⁺	25	Mag 9.39v star S 5′.3.
03 36 57.2	−35 30 25	NGC 1387	10.7	2.8 x 2.6	12.7	SAB(s)0⁻	119	**MCG −6-9-8** N 8′.1.
03 37 12.0	−35 44 50	NGC 1389	11.5	2.3 x 1.2	12.5	SAB(s)0⁻:	30	Mag 9.78v star N 3′.1.
03 38 06.6	−35 26 27	NGC 1396	13.8	0.7 x 0.4	12.3	SAB0⁻:	90	Located 4′.7 W of center of NGC 1399.
03 38 29.7	−35 26 53	NGC 1399	9.6	6.9 x 6.5	13.7	E1 pec	76	NGC 1396 4′.7 W.
03 38 51.7	−35 35 40	NGC 1404	10.0	3.3 x 3.0	12.5	E1	163	Mag 8.09v star SSE 2′.8.
03 39 23.4	−31 19 12	NGC 1406	11.8	3.8 x 0.8	12.9	SB(s)bc: sp II	15	Mag 9.67v star E 6′.7.
03 40 42.3	−37 30 41	NGC 1419	12.6	0.9 x 0.9	12.4	E pec:		**ESO 301-24** N 6′.1.
03 42 11.2	−29 53 27	NGC 1425	10.6	5.8 x 2.6	13.4	SA(s)b II	129	Mag 10.53v star N edge; mag 9.09v star W 13′.3.
03 42 19.8	−35 23 29	NGC 1427	10.9	3.6 x 2.5	13.2	E⁺	76	Mag 13.1 star on W edge; mag 9.11v star W 12′.8.
03 40 10.5	−35 37 17	NGC 1427A	12.9	2.3 x 1.5	14.1	IB(s)m IV-V	70	Faint envelope extends NE; mag 11.04v star W 4′.0.
03 42 22.9	−35 09 15	NGC 1428	12.9	1.1 x 0.6	12.3	SAB0⁻ pec:	118	Faint, elongated anonymous galaxy E 5′.7; mag 10.70v star SE 6′.0.
03 43 37.0	−35 51 12	NGC 1437	11.7	3.0 x 2.0	13.5	(R′)SAB(rs)ab II-III	150	Mag 9.65v star NE 10′.9.
03 43 02.2	−36 16 21	NGC 1437A	13.3	1.9 x 1.4	14.2	SAB(s)dm	83	= **MCG −6-9-24**.
03 46 13.7	−36 41 50	NGC 1460	12.6	1.4 x 1.0	12.8	SB(rs)0°	63	Mag 12.8 star on S edge; mag 10.57v star N 2′.7.
03 54 17.3	−36 58 18	NGC 1484	13.1	2.5 x 0.6	13.4	SB(s)b? II-III	80	
03 58 13.1	−35 26 49	NGC 1492	13.5	0.8 x 0.7	12.8	Sa?	10	
04 11 59.0	−32 51 04	NGC 1531	12.5	1.3 x 0.7	12.3	S0⁻ pec:	122	On W edge of NGC 1532.
04 12 02.5	−32 53 01	NGC 1532	9.8	12.6 x 3.0	13.6	SB(s)b pec sp I-II	33	Strong dark lane along E side; NGC 1531 on W edge; mag 7.05v star ENE 13′.5.
04 13 41.3	−31 38 44	NGC 1537	10.6	3.9 x 2.6	13.0	SAB0⁻ pec?	110	
04 22 42.8	−40 36 03	NGC 1572	12.4	2.5 x 1.2	13.5	(R′)SB(s)a:	0	Mag 12.0 star on E edge.
04 30 20.9	−36 38 37	PGC 15321	14.6	0.8 x 0.8	14.0	(R′)SA0°?		Mag 9.26v star SE 6′.7; **ESO 360-13** W 11′.2.

(Continued from previous page)

174 GALAXY CLUSTERS 174

RA h m s	Dec ° ′ ″	Name	Mag 10th brightest	No. Gal	Diam ′	Notes
03 34 00.0	−39 00 00	A 3135	15.5	111	22	All members anonymous, faint, stellar.
03 42 00.0	−32 03 00	A 3154	16.3	48	17	All members anonymous, faint, stellar.
03 44 54.0	−35 38 00	A 3161	16.4	36	17	All members besides ESO 358-60 are anonymous, stellar.
04 08 36.0	−30 49 00	A 3223	15.6	100	22	ESO 420-5 on N edge, most other members anonymous, stellar.
03 18 06.0	−29 38 00	AS 337	16.3		17	All members anonymous, stellar.
03 38 30.0	−35 27 00	AS 373	10.3		180	**Fornax Galaxy Cluster**.
04 06 00.0	−38 51 00	AS 418	15.4	12	22	**MCG −7-9-8** ENE of center, all others anonymous, stellar.
04 30 18.0	−36 38 00	AS 464	16.1	27	17	See notes for PGC 15321, all others anonymous, stellar.
04 31 30.0	−29 45 00	AS 465	15.3	4	22	MCG −5-11-12 at center; **ESO 421-5** W 8′.6.
04 36 36.0	−39 38 00	AS 476	16.1		17	Except ESO 304-6, all members anonymous, stellar.
04 38 06.0	−31 40 00	AS 478	16.3	2	17	All members anonymous, stellar.
04 40 06.0	−35 37 00	AS 484	16.1	5	17	**ESO 361-3**, **3A** at center, all others anonymous, stellar.
04 41 48.0	−36 56 00	AS 489	16.0	3	17	**MCG −6-11-4** at center, all others anonymous, stellar.

175 GALAXIES 175

RA h m s	Dec ° ′ ″	Name	Mag (V)	Dim ′ Maj x min	SB	Type Class	PA	Notes
01 56 36.7	−39 14 43	ESO 297-34	13.7	1.5 x 0.8	13.7	SAB0° pec	90	Mag 11.45v star NW 3′.0.
01 57 58.8	−38 01 10	ESO 297-36	14.4	1.2 x 0.6	13.8	Sc: II	82	Mag 9.98v star NW 14′.2.
01 58 12.7	−39 32 44	ESO 297-37	13.9	1.9 x 0.4	13.4	SB(rs)bc: sp	63	Mag 11.26v star S 8′.0.
02 06 14.2	−37 20 00	ESO 298-7	14.2	1.2 x 0.9	14.2	S?	94	Mag 10.36v star SE 2′.5.
02 08 43.1	−40 07 52	ESO 298-11	14.4	1.1 x 0.9	14.2	Sb	120	**ESO 298-12** NE 3′.0.
02 10 38.5	−40 55 06	ESO 298-15	13.6	2.0 x 0.7	13.8	SB(rs)cd pec: IV	58	Mag 6.50v star WNW 10′.0; note: **ESO 298-13** is 2′.5 WSW of this star.
02 10 54.9	−39 21 59	ESO 298-16	12.8	2.1 x 0.5	12.7	Sa:	53	Mag 9.65v star N 2′.8.
02 11 56.3	−39 12 22	ESO 298-19	13.5	1.7 x 1.0	13.9	SAB(rs)bc II-III	57	Mag 8.86v star WNW 13′.3.
02 13 38.2	−39 44 31	ESO 298-21	13.9	1.5 x 0.3	12.9	(R)SB0^{+}?	169	Mag 10.96v star ESE 10′.2.
02 14 27.7	−39 11 09	ESO 298-23	14.0	1.3 x 1.0	14.1	S	45	Mag 9.51v star NE 8′.5.
02 19 37.5	−37 49 10	ESO 298-28	12.8	2.6 x 1.3	14.0	SA(rs)bc II	21	Strong nucleus, smooth envelope; mag 8.95v star SW 14′.6.
02 20 09.0	−37 19 17	ESO 298-30	14.6	1.5 x 0.6	14.3	SA0^{-}:	45	Mag 12.2 star WNW 9′.9.
02 21 59.0	−37 27 22	ESO 298-36	14.1	1.5 x 0.4	13.4	SAB(rs)bc: II	144	Mag 11.16v star N 1′.9.
02 22 03.8	−38 47 17	ESO 298-37	14.3	0.7 x 0.6	13.2	Pec	168	Mag 8.37v star N 5′.3.
02 25 14.2	−40 25 37	ESO 299-1	13.1	1.2 x 0.8	12.9		164	Double system, interaction.
02 43 33.8	−38 37 38	ESO 299-13	13.4	0.7 x 0.4	12.1	E?	2	Mag 14.3 star SE 0′.6.
02 48 27.8	−40 33 22	ESO 299-18	14.4	2.1 x 0.3	13.8	Scd: sp	55	Mag 9.50v star W 7′.8.
02 49 33.7	−38 46 14	ESO 299-20	13.0	1.7 x 0.7	13.1	(R)SB(r)a?	32	Mag 9.07v star E 12′.1.
03 02 02.3	−39 59 33	ESO 300-5	13.2	1.0 x 0.7	12.7	Sa	118	Mag 10.77v star E 10′.0.
03 06 05.8	−39 01 12	ESO 300-10A	13.3	0.7 x 0.5	12.1		40	NGC 1217 on S edge.
03 07 40.4	−39 36 20	ESO 300-12	14.0	1.2 x 0.4	13.1	SB(s)c: III-IV	10	Mag 10.43v star NNE 10′.1.
03 23 54.3	−37 30 35	ESO 301-11	13.2	0.9 x 0.8	12.7	Ring: pec	46	Mag 12.0 star W 6′.4.
01 58 00.3	−35 34 25	ESO 354-19	13.7	1.3 x 0.8	13.6	Sb	56	Mag 14.1 star SW 3′.2.
01 58 33.2	−34 51 48	ESO 354-21	14.0	1.2 x 0.4	13.1	S pec?	36	Mag 12.9 star S 2′.4.
02 00 17.7	−34 19 02	ESO 354-25	12.7	1.5 x 0.9	12.9	S0 pec?	110	Mag 10.83v star S 1′.1; ESO 354-26 NNE 4′.3.
02 00 29.3	−34 15 21	ESO 354-26	12.9	1.3 x 1.0	13.1	E4	149	ESO 354-25 SSW 4′.3.
02 01 10.2	−33 12 31	ESO 354-29	13.6	1.1 x 1.0	13.6	Sc	49	Mag 10.52v star N 2′.1.
02 05 46.3	−32 40 38	ESO 354-34	13.6	1.0 x 1.0	13.5	(R)SAB0^{+}		Mag 9.24v star SE 5′.2.
02 06 22.4	−36 18 03	ESO 354-36	15.2	1.0 x 0.2	13.3	Sc	140	Mag 10.78v star SSE 6′.8.
02 07 49.8	−35 12 05	ESO 354-41	13.5	1.3 x 0.7	13.2	(R′)S(r)a	156	Mag 9.47v star SW 5′.9; **MCG −6-5-29** NW 7′.1.
02 17 55.1	−34 48 13	ESO 355-4	13.8	1.4 x 1.0	14.0	SB(s)b pec	128	Mag 11.11v star ENE 6′.8.
02 21 18.4	−34 19 07	ESO 355-8	13.8	1.3 x 0.9	13.9	S0° pec	165	Mag 12.3 star E 6′.9; note: faint, stellar anonymous galaxy 0′.7 SE of this star.
02 21 36.9	−33 48 33	ESO 355-10	14.6	1.1 x 0.4	13.6	SBc? III-IV	66	Mag 9.23v star SW 8′.7; **ESO 355.11** S 4′.3.
02 32 17.8	−35 01 49	ESO 355-26	13.0	1.7 x 0.9	13.4	SB(s)bc: II-III	154	Located between a close pair of E-W oriented mag 10.30v, 10.64v stars.
02 37 36.2	−32 55 33	ESO 355-30	12.9	1.9 x 0.7	13.0	SB(rs)bc: II	82	Mag 9.04v star W 5′.6.
02 38 02.2	−33 42 19	ESO 355-31	13.7	0.8 x 0.4	12.4		42	Stellar galaxy **ESO 355-31A** E edge; mag 7.33v star NW 2′.5.
02 38 56.3	−35 29 26	ESO 356-2	14.5	1.2 x 0.5	13.8	SB(s)b: pec	75	Mag 8.54v star E 10′.6.
02 39 59.0	−34 27 00	ESO 356-4	8.0	35 x 30	15.6	E4	55	**Fornax Dwarf**. Dwarf galaxy; globular clusters NGC 1049 N 11′.7 and Fornax 4 S 5′.9 from center of galaxy.
02 43 08.6	−32 48 12	ESO 356-9	13.4	1.0 x 0.6	12.7	Sc:	7	Mag 11.24v star E 3′.7.
02 48 47.4	−36 42 57	ESO 356-13	13.7	1.2 x 0.9	13.7	SA(s)cd: pec II-III	44	Mag 12.4 star SW 2′.3.
02 50 30.1	−34 51 58	ESO 356-14A	13.5	0.8 x 0.6	12.6		127	**MCG −6-7-6** NW 2′.4; **MCG −6-7-8** SE 2′.0; **MCG −6-7-9** SE 3′.2; mag 10.76v star S 2′.0.
02 55 12.6	−36 18 00	ESO 356-18	14.4	1.5 x 0.3	13.4	S	75	Mag 11.33v star N 5′.0.
02 57 25.0	−35 34 13	ESO 356-20	13.4	1.8 x 0.7	13.5	SA(r)ab:	40	Mag 14.4 star on E edge; mag 10.83v star W 5′.1.
02 57 49.3	−36 43 08	ESO 356-22	12.4	2.1 x 1.3	13.4	SB(r)c II	151	Mag 7.39v star W 13′.1.
02 58 55.3	−36 36 53	ESO 356-24	14.0	0.8 x 0.5	12.9		31	Mag 14.2 star W edge.
03 00 21.7	−36 55 15	ESO 356-26	14.4	1.2 x 0.3	13.2	S	142	PGC 11343 W 4′.8; mag 9.17v star WNW 14′.5.
03 02 09.6	−35 41 11	ESO 357-1	14.4	1.0 x 0.8	14.0	SA(s)d: III-IV	18	Mag 10.95v star S 1′.7.
03 10 24.9	−33 09 28	ESO 357-7	14.0	2.4 x 0.4	13.8	SB(s)m sp IV-V	130	Mag 7.84v star N 6′.4.
03 15 42.3	−33 32 34	ESO 357-10	13.0	1.1 x 0.5	12.2	SAB(s)dm? V	18	Mag 9.84v star SSW 6′.9.
03 16 53.4	−35 32 32	ESO 357-12	13.0	3.0 x 1.8	14.7	SB(s)d IV-V	122	Mag 7.05v star S 9′.9.
03 23 37.4	−35 46 43	ESO 357-25	14.2	1.2 x 0.5	13.6	SAB0°:	30	Mag 10.61v star NE 9′.2.
02 10 56.1	−31 35 59	ESO 415-3	14.1	0.7 x 0.6	13.0	S	142	Mag 9.61v star SW 11′.1; note **ESO 415-2** is 4′.8 WNW of this star.
02 13 16.4	−29 43 28	ESO 415-7	13.9	1.0 x 0.7	13.4	S0	25	Mag 10.07v star SW 8′.7.
02 14 12.9	−31 09 02	ESO 415-10	14.5	1.5 x 0.3	13.5	SB(s)d III-IV	30	Mag 10.18v star NW 11′.3.
02 21 01.4	−31 56 30	ESO 415-19	14.0	1.3 x 0.5	13.4	SB(s)ab: pec	144	Size indicates bright core only. Very long, faint arms extend NE and S.
02 21 07.1	−31 54 54	ESO 415-20	14.5	2.6 x 1.5	15.9	S	156	Small, bright core with very faint arms.
02 28 20.2	−31 52 55	ESO 415-26	13.7	1.6 x 1.1	14.2	S0°: pec	15	Pair of N-S oriented mag 9.61v, 10.20v stars ENE 10′.0.
02 29 07.4	−30 59 52	ESO 415-28	14.5	1.7 x 0.9	14.8	SAB(s)bc II	18	Mag 6.11v star SW 9′.3.
02 30 36.0	−31 35 52	ESO 415-31	13.9	1.7 x 0.4	13.4	SB(s)cd? sp III-IV	77	Mag 10.98v star SE 6′.8.
02 32 34.3	−29 41 45	ESO 416-1	13.8	0.5 x 0.4	11.9		156	Mag 10.92v star SW 5′.3.
02 38 45.4	−30 48 27	ESO 416-5	14.3	0.9 x 0.6	13.4	SA?(r)	15	Mag 14.9 star SSE edge; mag 13.7 star NE 2′.6.
02 42 22.7	−30 19 24	ESO 416-9	13.3	1.2 x 0.9	13.3	S0? pec	72	Mag 7.65v star SW 8′.3.
02 43 37.8	−31 56 47	ESO 416-12	13.7	1.7 x 0.7	13.7	SAB(rs)c: III	48	Mag 9.36v star E 8′.3.
02 46 00.5	−32 09 19	ESO 416-18	13.3	1.9 x 0.9	13.7	(R′)SAB(rs)0^{+}?	99	Mag 12.4 star W 1′.8; **ESO 416-17** N 6′.5.
02 48 41.0	−31 32 08	ESO 416-25	13.7	2.2 x 0.4	13.4	Sb: sp	29	Mag 9.12v star SE 3′.4.
02 49 36.0	−30 34 47	ESO 416-32	13.1	0.8 x 0.8	12.5	S0?		Mag 9.96v star N 6′.6.
02 52 22.4	−31 49 03	ESO 416-39	13.8	0.8 x 0.7	13.1	S0?	8	Almost stellar nucleus; mag 12.6 star N 4′.8.
02 53 29.5	−30 51 40	ESO 416-41	13.5	1.7 x 0.6	13.3	SAB(rs)bc: II	93	Mag 10.31v star S 6′.7.
02 54 41.5	−30 25 42	ESO 417-1	13.2	1.8 x 0.9	13.6	SA0° pec	179	Close pair of N-S oriented mag 10.49v, 10.59v stars W 8′.2.

(Continued from previous page)

GALAXIES

RA h m s	Dec ° ′ ″	Name	Mag (V)	Dim ′ Maj x min	SB	Type Class	PA	Notes
03 07 13.1	−31 24 00	ESO 417-18	13.3	2.2 x 1.2	14.2	SAB(r)cd I-II	177	Numerous knots; mag 10.71v star ESE 3′.6.
03 12 48.3	−31 29 12	ESO 417-20	14.0	1.5 x 0.9	14.2	SB?	64	Mag 9.97v star N 5′.8.
01 57 55.4	−32 59 18	IC 1759	12.9	1.5 x 1.4	13.6	SA(rs)bc:	169	Mag 6.37v star SE 8′.1.
01 57 48.7	−33 14 25	IC 1762	13.5	2.0 x 0.6	13.5	SB(s)bc I-II	43	Mag 10.45v star SSE 5′.4.
02 10 06.3	−32 56 28	IC 1783	12.5	2.0 x 0.8	12.9	SA(rs)b? II	3	Mag 11.36v star SSE 11′.5.
02 15 49.9	−31 12 10	IC 1788	12.3	2.6 x 1.1	13.3	SB(s)bc? II	27	Mag 10.26v star SE 6′.8.
02 30 38.1	−34 15 53	IC 1811	13.4	1.3 x 0.7	13.1	(R′)SB(r)ab	7	Mag 10.58v star SW 11′.9; IC 1813 NE 3′.6.
02 30 49.6	−34 13 17	IC 1813	13.2	1.2 x 0.9	13.1	SA(rs)0$^+$ pec:	102	IC 1811 SW 3′.6; mag 12.9 star N 1′.7.
02 31 51.1	−36 40 20	IC 1816	12.8	1.4 x 1.2	13.2	SB(r)ab pec?	167	Mag 10.16v star NW 5′.2.
02 49 08.3	−31 17 23	IC 1858	13.1	1.8 x 0.6	13.1	SA0$^+$:	176	Mag 7.21v star SW 14′.9.
02 49 03.7	−31 10 29	IC 1859	13.3	1.3 x 0.9	13.3	S?	21	Located 6′.4 W of IC 1860; mag 13.4 star W 1′.3.
02 49 33.8	−31 11 25	IC 1860	12.7	2.2 x 1.7	14.1	E4	6	MCG −5-7-36 on S edge of core; mag 11.7 star NE 2′.8; several PGC galaxies N and NE.
02 51 58.9	−33 20 23	IC 1862	13.3	3.0 x 0.3	13.0	SAbc: sp	3	Mag 8.11v star SSE 7′.9.
02 53 39.4	−34 11 52	IC 1864	12.6	1.2 x 0.7	12.4	E?	63	Mag 11.35v star W 8′.8.
03 03 56.6	−39 26 28	IC 1875	12.5	1.4 x 1.2	12.9	SAB(rs)0$^-$:	3	Elongated **ESO 300-8** and **ESO 300-7** NNE 5′.1.
03 06 40.5	−32 51 51	IC 1885	14.1	1.4 x 0.5	13.6	Sc? II-III	138	Mag 8.81v star N 9′.5.
03 15 00.7	−30 42 28	IC 1904	13.4	1.4 x 0.6	13.0	(R′)SB(rs)ab:	108	Mag 10.54v star SW 2′.4.
03 16 05.7	−34 21 38	IC 1906	13.5	1.1 x 0.5	12.7	(R′)SB(s)bc? II	63	Mag 11.8 star NE 4′.4.
03 17 20.0	−33 41 28	IC 1909	13.5	1.1 x 0.6	12.9	SB(r)b II-III	60	String of four mag 14 and 15 stars extending N from N edge.
03 19 34.8	−32 27 58	IC 1913	13.7	1.9 x 0.3	12.9	SBb? sp	149	Mag 10.59v star N 7′.8.
02 01 14.5	−31 43 48	MCG −5-5-26	13.6	1.2 x 0.9	13.6	(R)SB(rs)a:	126	Mag 8.68v star SW 11′.7.
02 02 10.4	−31 01 30	MCG −5-5-28	13.3	0.8 x 0.8	12.7	S0?		Mag 9.82v star WNW 4′.0.
02 03 56.4	−31 47 12	MCG −5-6-2	13.7	0.8 x 0.8	13.3	E		Mag 12.0 star N 4′.4.
02 04 43.1	−29 18 08	MCG −5-6-3	13.4	0.7 x 0.3	11.6	S0	14	Lies 3′ E of mag 5 star Nu Fornacis.
02 05 03.9	−32 23 10	MCG −5-6-4	13.7	0.9 x 0.7	13.0	Sa:	68	Mag 10.94v star E 3′.7.
02 16 39.5	−29 57 05	MCG −5-6-12	14.6	1.1 x 0.4	13.6	S	141	Mag 12.3 star ENE 8′.1.
02 31 26.1	−29 33 55	MCG −5-7-6	13.7	0.8 x 0.6	12.9	E	165	Mag 10.46v star E 6′.7.
02 35 13.6	−29 36 17	MCG −5-7-7	13.9	0.9 x 0.5	12.8	Sa:	169	Small, anonymous spiral galaxy WSW 1′.6.
02 47 07.2	−31 29 07	MCG −5-7-26	13.8	0.9 x 0.8	13.3	(R′)SBa	72	Relatively bright star N edge.
02 48 26.3	−29 35 39	MCG −5-7-27	13.8	1.3 x 0.4	13.0	S?	67	Mag 10.55v star W 6′.8; **ESO 416-22** SSW 7′.8.
02 48 45.2	−32 20 54	MCG −5-7-30	13.2	1.0 x 0.6	12.6	SAB0$^+$	98	Mag 4.46v star Beta Fornaeis SE 5′.6.
02 49 33.1	−31 12 01	MCG −5-7-36	14.6	0.5 x 0.2	12.1	E	75	Located on S edge of IC 1860's core.
02 50 26.0	−31 33 33	MCG −5-7-38	13.6	1.2 x 0.9	13.6	SA(r)0/a	118	Mag 9.99v star W 3′.5.
02 50 35.6	−31 31 15	MCG −5-7-39	13.6	1.0 x 0.7	13.1	S?	19	Stellar **PGC 10793** NE 1′.7.
02 50 31.7	−31 23 45	MCG −5-7-40	13.6	1.1 x 0.9	13.4	(R)SA(rs)0/a	91	Mag 10.21v star E 11′.0.
02 52 26.9	−30 46 40	MCG −5-7-44	13.2	1.1 x 0.7	12.7	S?	60	Mag 6.39v star WSW 7′.1.
02 55 15.9	−29 50 02	MCG −5-8-3	14.3	0.9 x 0.3	12.9	E	111	Mag 7.79v star W 8′.5.
02 56 21.5	−32 11 13	MCG −5-8-6	13.4	1.1 x 1.1	13.5	(R)SA0/a?		Bright knot or star SE edge.
02 59 51.0	−31 44 06	MCG −5-8-10	14.5	0.7 x 0.4	13.0	S	24	Mag 9.11v star NE 7′.3.
03 09 41.4	−31 41 38	MCG −5-8-20	15.7	0.7 x 0.4	14.4	E/S0	57	Mag 9.16v star SSE 7′.2.
03 09 57.9	−31 08 29	MCG −5-8-21	13.9	1.2 x 0.7	13.5	SB:	139	Bright knot or superimposed star W of center.
03 13 15.9	−31 39 15	MCG −5-8-23	13.2	1.0 x 0.6	12.5	SAB0$^-$	151	Star at N edge.
03 20 52.0	−30 47 22	MCG −5-9-2	14.7	0.9 x 0.3	13.1	S	60	
03 21 12.4	−30 58 54	MCG −5-9-3	14.6	0.5 x 0.4	12.7	S	0	Mag 9.84v star SE 1′.1.
02 06 53.4	−36 27 16	NGC 824	13.1	1.4 x 1.2	13.5	SB(r)bc I	26	Short N-S bar with small dark patches E and W.
02 11 30.8	−35 50 07	NGC 854	12.9	1.8 x 0.6	12.8	SB(rs)c: II	0	
02 12 36.9	−31 56 44	NGC 857	12.7	1.5 x 1.3	13.2	SA(rs)0°:	92	
02 21 06.5	−33 43 13	NGC 897	11.8	2.1 x 1.3	12.8	SA(rs)a	17	Mag 9.23v star S 10′.0; mag 11.54v star on E edge.
02 31 06.2	−36 02 02	NGC 964	12.4	2.0 x 0.5	12.2	Sab	31	**IC 1814** NW 3′.5.
02 33 34.3	−39 02 35	NGC 986	10.9	3.9 x 3.0	13.4	SB(rs)ab II	150	with faint extensions forms an open "S" shape.
02 32 41.9	−39 17 45	NGC 986A	14.0	1.8 x 0.7	14.1	IBm: V	72	Mag 8.91v star SE 5′.1.
02 43 44.8	−29 00 10	NGC 1079	11.5	5.5 x 3.1	14.4	(R)SAB(rs)0/a pec	87	Large core with faint envelope; mag 10.34v star S 5′.9.
02 46 18.5	−30 16 09	NGC 1097	9.5	9.3 x 6.3	13.7	SB(s)b I-II	130	Many knots; NGC 1097A located 3′.4 NNW of center.
02 46 10.1	−30 13 43	NGC 1097A	13.5	0.9 x 0.5	12.6	E pec:	105	Located 3′.4 NNW of center of NGC 1097.
02 58 47.8	−32 05 58	NGC 1165	12.7	2.5 x 1.0	13.6	(R′)SB(r)a:	115	**ESO 417-9** E 5′.2.
03 06 06.1	−39 02 11	NGC 1217	12.1	1.8 x 1.3	12.9	(R)SA(r)a:	50	ESO 300-10A on N edge; mag 9.13v star SE 7′.1.
03 17 13.2	−32 34 35	NGC 1288	12.1	2.3 x 1.9	13.6	SAB(rs)c I-II	166	Tightly wound multi-branching arms.
03 21 03.8	−37 06 06	NGC 1310	12.1	2.0 x 1.5	13.1	SA(s)c: III	95	Mag 9.31v star NE 7′.7.
03 22 41.0	−37 12 28	NGC 1316	8.5	11.0 x 7.2	13.2	SAB(s)0° pec	50	**Fornax A**.
03 22 44.6	−37 06 11	NGC 1317	11.0	2.8 x 2.4	12.9	SAB(r)a	78	
03 23 56.4	−36 27 50	NGC 1326	10.5	3.9 x 2.9	13.0	(R)SB(r)0$^+$	86	
02 34 26.1	−32 34 53	PGC 9803	14.1	1.2 x 0.7	13.8	(R)S0°	90	
02 59 58.5	−36 56 30	PGC 11343	14.7	1.2 x 0.8	14.5	S0	117	

GALAXY CLUSTERS

RA h m s	Dec ° ′ ″	Name	Mag 10th brightest	No. Gal	Diam ′	Notes
02 30 30.0	−33 05 00	A 3027	16.3	44	17	**ESO 355-19** at SW edge; all other members anonymous, stellar.
03 08 12.0	−36 42 00	A 3089	15.8	30	17	All members anonymous, stellar.
02 01 06.0	−40 35 00	AS 215	16.3		17	All members anonymous, stellar.
02 25 48.0	−29 36 00	AS 258	16.3	8	17	All members anonymous, stellar/almost stellar.
02 49 36.0	−31 11 00	AS 301	14.9	5	22	See notes for IC 1860.
03 00 18.0	−37 02 00	AS 316	15.7	6	17	Except ESO 356-26, PGC 11343, all members anonymous, stellar.
03 13 54.0	−39 05 00	AS 332	15.8		17	All members anonymous, stellar.
03 15 12.0	−29 14 00	AS 333	15.8	24	17	All members anonymous, stellar.
03 18 06.0	−29 38 00	AS 337	16.3		17	All members anonymous, stellar.

GLOBULAR CLUSTERS

RA h m s	Dec ° ′ ″	Name	Total V m	B ★ V m	HB V m	Diam ′	Conc. Class Low = 12 High = 1	Notes
02 37 02.0	−34 11 00	Fornax 1	15.6	18.3	21.3	0.9		
02 38 44.0	−34 48 36	Fornax 2	13.5	19.0	21.3	0.8		
02 40 07.7	−34 32 10	Fornax 4	13.6	18.6	21.3	0.8	4	
02 42 21.0	−34 06 12	Fornax 5	13.4	18.6	21.3	1.7	3	
02 40 06.8	−34 25 14	Fornax 6				0.6		
02 39 48.0	−34 15 24	NGC 1049	12.6	18.4	21.3	0.8		Cluster in Fornax Dwarf Galaxy.

RA h m s	Dec ° ′ ″	Name	Mag (V)	Dim ′ Maj x min	SB	Type Class	PA	Notes
00 43 25.9	−38 45 24	ESO 295-2	14.1	1.0 x 0.5	13.3	SAB(s)0° pec	42	Mag 10.79v star E 4'.7.
00 48 32.1	−40 10 20	ESO 295-7	14.9	1.5 x 0.2	13.4	Sc	148	Mag 10.98v star NE 9'.7.
00 49 09.3	−39 32 21	ESO 295-9	14.4	1.5 x 0.2	13.0	Sc	168	Mag 11.04v star NE 7'.0.
00 49 54.0	−39 38 27	ESO 295-10	13.2	1.5 x 0.5	12.8	(R)SB(r)a	8	Mag 9.92v star NNE 6'.5.
00 51 00.0	−38 45 39	ESO 295-13	13.8	0.9 x 0.5	12.7	Sab	150	Mag 10.55v star NW 9'.2.
00 55 46.5	−37 24 29	ESO 295-22	13.8	0.9 x 0.6	13.0		166	**MCG −6-3-10** N 1'.2.
01 04 18.7	−40 08 53	ESO 295-31	13.8	1.0 x 0.8	13.4	(R')SB(rs)ab	121	Mag 10.80v star NW 4'.4.
01 04 42.2	−37 38 16	ESO 295-32	14.0	1.3 x 0.4	13.2	SB(s)cd? sp II-III	17	Mag 12.5 star WSW 1'.8.
01 09 02.8	−37 17 25	ESO 296-2	13.6	1.0 x 0.6	12.9	(R)SB(r)a pec:	166	Mag 9.67v star ESE 3'.3.
01 12 33.8	−40 56 46	ESO 296-5	14.2	0.6 x 0.6	13.0	S?		Mag 10.01v star W 5'.4.
01 12 37.5	−37 53 58	ESO 296-6	14.3	1.5 x 0.5	13.9	SB(s)b: II-III	24	Mag 5.94v star N 2'.8.
01 24 17.1	−37 20 07	ESO 296-19	14.0	1.2 x 0.7	13.6	SB(rs)c pec I-II	94	
01 25 49.4	−39 06 21	ESO 296-28	14.0	1.5 x 0.6	13.8	(R')SAB(s)0/a:	15	Mag 10.67v star NNE 6'.9.
01 25 51.3	−37 20 01	ESO 296-29	13.4	1.0 x 0.7	12.9	S0 pec	100	Triple system.
01 30 41.9	−37 43 15	ESO 296-35	14.1	0.9 x 0.7	13.4	SB(s)b pec I-II	31	Mag 8.05vc star S 4'.2.
01 32 27.7	−38 40 48	ESO 296-38	13.3	1.2 x 1.1	13.4	SB(s)bc? II	115	Mag 10.6 star N 5'.1.
01 34 13.6	−38 37 01	ESO 297-3	13.7	1.3 x 0.9	13.7	SAB0$^-$	173	Group of three small, faint anonymous galaxies NW 2'.6.
01 35 30.8	−39 23 04	ESO 297-8	12.9	1.4 x 1.1	13.2	(R')SAB(rs)a pec	64	Mag 10.42v star WNW 10'.6; NGC 630 NE 1'.8.
01 36 24.2	−37 20 29	ESO 297-12	14.1	0.9 x 0.5	13.3	E	1	NGC 633 N edge.
01 37 52.2	−40 32 22	ESO 297-15	13.9	1.0 x 0.4	12.8		112	Mag 13.9 star SW 2'.4.
01 37 59.5	−40 04 09	ESO 297-16	14.0	1.7 x 0.4	13.4	SB(s)cd? sp II	39	Mag 14 star N end.
01 38 20.0	−40 44 02	ESO 297-17	14.4	1.1 x 0.9	14.2	Sc	108	Stellar nucleus.
01 38 37.3	−40 00 42	ESO 297-18	13.3	1.5 x 0.4	12.6	Sa: sp	143	
01 44 28.6	−40 39 52	ESO 297-23	13.6	1.3 x 0.4	12.8	SAB(r)ab pec:	4	Mag 13.3 star N 2'.1.
01 44 55.7	−38 43 59	ESO 297-25	14.6	0.8 x 0.4	13.2	Pec	162	Mag 9.44v star NW 3'.1.
01 54 16.0	−37 47 11	ESO 297-32	13.7	1.3 x 1.0	13.8	(R')SB(s)a:	73	Mag 14.1 star NE 1'.7.
01 56 36.7	−39 14 43	ESO 297-34	13.7	1.5 x 0.8	13.7	SAB0° pec	90	Mag 11.45v star NW 3'.0.
01 57 58.8	−38 01 10	ESO 297-36	14.4	1.2 x 0.6	13.8	Sc: II	82	Mag 9.98v star NW 14'.2.
01 58 12.7	−39 32 44	ESO 297-37	13.9	1.9 x 0.4	13.4	SB(rs)bc: sp	63	Mag 11.26v star S 8'.0.
00 36 06.3	−32 36 15	ESO 350-37	13.8	1.0 x 0.6	13.1	Sa	58	**MCG −6-2-22** NE 2'.0.
00 36 52.6	−33 33 20	ESO 350-38	13.6	0.6 x 0.4	11.9		103	Mag 15.3 star E 1'.3.
00 42 50.3	−36 52 47	ESO 351-3	13.7	0.5 x 0.5	12.1	Sb		Mag 10.00v star N edge.
00 52 22.9	−35 00 05	ESO 351-16	13.6	1.0 x 0.6	12.9	Sa	105	Mag 11.8 star SW 1'.6.
00 54 59.7	−35 19 17	ESO 351-21	13.4	1.0 x 1.0	13.3	SA0$^-$		Mag 14.2 star NW 1'.9.
00 59 29.6	−36 11 14	ESO 351-28	13.9	1.7 x 0.6	13.8	SB(s)c: pec II-III	79	Mag 6.97v star E 13'.7.
01 00 09.3	−33 42 37	ESO 351-30	10.0	30 x 18	16.9	E?	110	**Sculptor Dwarf**. Dwarf elliptical.
01 04 30.4	−33 39 17	ESO 352-2	13.5	0.9 x 0.5	12.5	S?	115	Mag 6.46v star N 7'.4.
01 07 20.8	−36 45 20	ESO 352-6	13.8	0.8 x 0.5	12.6	(R')SB(s)ab	56	Faint, small, elongated anonymous galaxy SW 1'.1.
01 07 35.6	−33 38 18	ESO 352-7	13.8	1.6 x 0.9	14.1	(R')SB(rs)a	52	Mag 12.1 star NNE 5'.1.
01 08 46.8	−36 20 44	ESO 352-8	13.5	1.5 x 0.8	13.6	SAB0$^-$ pec	42	Mag 10.87v star NW 7'.1.
01 10 19.3	−35 44 09	ESO 352-15	13.6	1.5 x 1.1	13.9	SAB(s)d III-IV	126	Mag 13.3 star WNW 2'.5.
01 12 09.3	−32 14 33	ESO 352-18	13.6	1.5 x 1.4	14.3	(R')SA(r)a:	156	Mag 8.13v star W 12'.7.
01 12 38.9	−33 40 06	ESO 352-20	14.2	1.8 x 1.1	14.9	S0$^-$ pec	0	Mag 10.27v star SE 9'.4.
01 15 00.8	−32 14 40	ESO 352-28	14.0	1.5 x 1.3	14.6	(R)SB(rs)0/a	138	Mag 13.4 star on S edge; mag 12.5 star WNW 2'.1; MCG −5-4-20 W 6'.4.
01 18 18.8	−37 06 15	ESO 352-38	13.6	1.0 x 0.6	12.9	SA(s)0°: pec	113	Mag 11.56v star NW 5'.7.
01 19 07.7	−34 06 16	ESO 352-41	12.8	1.9 x 1.1	13.4	SAB(rs)0°?	17	Mag 10.43v star N 2'.5; mag 10.27v star E 3'.2.
01 20 33.9	−34 07 20	ESO 352-47	14.8	1.2 x 0.8	14.6	IB(s)m IV-V	84	Mag 9.76v star SW 3'.8.
01 21 04.8	−36 07 04	ESO 352-49	13.3	1.6 x 0.7	13.3	Sab: pec	149	Mag 7.46v star SW 12'.4.
01 21 07.2	−35 12 07	ESO 352-50	14.6	1.5 x 0.2	13.1	Sbc? sp	145	**ESO 352-52** ESE 1'.6.
01 21 13.4	−36 28 21	ESO 352-51	13.7	1.0 x 0.7	13.3	E?	55	ESO 352-54 SSE 5'.2; **ESO 352-48** N 9'.7.
01 21 29.5	−36 32 37	ESO 352-54	14.1	1.5 x 0.2	12.7	Sb: sp	127	Mag 6.71v star ESE 15'.0; note: **ESO 352-60** lies on N edge of this star and **ESO 253-58** is 4'.2 W of this star.
01 21 32.9	−33 09 21	ESO 352-55	13.3	1.2 x 0.9	13.3	SA(r)0$^-$:	153	Mag 13.4 star SE 2'.4.
01 22 02.3	−34 11 49	ESO 352-57	13.7	1.1 x 0.4	12.7	SB(s)0° pec	15	Mag 10.18v star E 2'.5; **ESO 352-56** N 3'.2.
01 23 04.7	−34 59 08	ESO 352-61	14.2	0.8 x 0.6	13.2	S?	7	Mag 8.12v star W 5'.6.
01 23 06.4	−34 44 09	ESO 352-62	14.2	1.7 x 0.2	12.9	Sc: sp	54	Mag 7.54v star NW 9'.0; note: **ESO 352-59** is 3'.1 E of this star.
01 23 15.0	−32 50 29	ESO 352-63	13.5	1.3 x 0.6	13.0	Sbc II	127	Mag 15.2 star superimposed on NW end.
01 23 34.9	−34 56 10	ESO 352-64	13.8	1.2 x 0.3	12.6	S0$^-$: sp	67	Mag 11.9 star E 8'.3; ESO 352-61 SW 6'.8.
01 24 13.9	−34 43 37	ESO 352-69	13.2	1.4 x 0.9	13.3	SB(s)ab? pec	95	Mag 9.83v star NW 10'.6.
01 24 34.5	−33 10 28	ESO 352-71	13.4	1.5 x 0.7	13.3	SB(s)bc? II	168	Close pair of mag 13.1 stars N 7'.5.
01 25 15.3	−33 24 31	ESO 352-73	13.6	0.9 x 0.5	12.6	Sa	159	Mag 10.76v star S 2'.8.
01 25 51.0	−34 29 29	ESO 352-76	14.1	0.9 x 0.9	13.8	SA(rs)0/a:		Almost stellar nucleus, faint envelope; **ESO 352-74** SW 11'.7.
01 28 22.8	−35 59 31	ESO 353-5	13.1	1.4 x 1.0	13.4	SA(r)0°:	30	Mag 9.06v star S 5'.1.
01 30 24.8	−33 02 13	ESO 353-7	12.6	1.0 x 0.8	12.2	Sa pec	142	Mag 9.23v star W 5'.8.
01 31 50.5	−33 07 11	ESO 353-9	12.7	1.6 x 1.5	13.5	SB(r)bc: II	179	Mag 11.07v star NW 7'.8.
01 34 00.4	−34 23 22	ESO 353-14	13.6	1.7 x 0.3	12.7	Sab: sp	164	Mag 9.16v star NE 5'.1; **ESO 353-17** S 3'.6.
01 34 51.5	−36 08 14	ESO 353-20	13.1	1.5 x 0.5	12.7	S0$^+$? sp	76	Mag 10.40v star NNE 2'.6; **MCG −6-4-54** E 9'.8.
01 36 09.4	−36 18 11	ESO 353-25	13.9	0.8 x 0.6	13.0	S?	10	Mag 13.3 star N edge.
01 36 27.7	−36 22 41	ESO 353-26	13.4	1.7 x 0.6	13.3	SB(s)b: pec II	168	Mag 7.65v star SW 7'.4.
01 38 28.3	−33 36 29	ESO 353-29	13.8	1.1 x 0.6	13.2	SBbc? pec	35	Mag 12.6 star N 1'.0.
01 40 35.6	−33 37 19	ESO 353-31	13.8	0.9 x 0.3	12.3	S0	144	Mag 12.2 star NNW 3'.7.
01 42 14.9	−33 15 35	ESO 353-33	14.0	0.8 x 0.5	12.9	S?	150	Mag 13.1 star N 3'.1.
01 43 37.3	−33 42 23	ESO 353-38	13.3	1.1 x 0.5	12.6	SB0°: pec	114	Mag 11.48v star N 4'.7.
01 43 44.5	−36 05 21	ESO 353-40	12.4	1.9 x 1.2	13.2	(R)SB0$^+$ pec:	168	Mag 12.4 star N 3'.1.
01 45 02.8	−36 07 08	ESO 353-41	14.0	1.0 x 0.4	12.9	(R')S0/a:	64	Mag 8.59v star E 4'.8.
01 46 53.2	−33 55 20	ESO 353-45	13.9	1.0 x 0.8	13.5	SB(rs)ab	31	Mag 10.95v star W 7'.6.
01 49 29.2	−33 50 20	ESO 353-49	14.4	1.1 x 0.8	14.1	SAB(s)d III-IV	48	Mag 12.7 star NNE 5'.5.
01 50 52.7	−36 00 47	ESO 354-3	13.4	1.6 x 1.2	14.0	SA0$^-$:	168	Bright, slightly elongated N-S core with faint envelope.
01 51 41.9	−36 11 18	ESO 354-4	13.6	1.2 x 0.9	13.6	(R')SA(rs)b pec	53	Mag 14.3 star ENE 1'.4.
01 58 00.3	−35 34 25	ESO 354-19	13.7	1.3 x 0.8	13.6	Sb	56	Mag 14.1 star SW 3'.2.
01 58 33.2	−34 51 48	ESO 354-21	14.0	1.2 x 0.4	13.1	S pec?	36	Mag 12.9 star S 2'.4.
02 00 17.7	−34 19 02	ESO 354-25	12.7	1.5 x 0.9	12.9	S0 pec?	110	Mag 10.83v star S 1'.1; ESO 354-26 NNE 4'.3.
02 00 29.3	−34 15 21	ESO 354-26	12.9	1.3 x 1.0	13.1	E4	149	ESO 354-25 SSW 4'.3.
02 01 10.2	−33 12 31	ESO 354-29	13.6	1.1 x 1.0	13.6	Sc	49	Mag 10.52v star N 2'.1.
00 40 04.0	−30 35 50	ESO 411-6	14.3	1.5 x 1.0	14.6	SB(rs)bc pec I-II	140	Stellar nucleus; mag 9.78v star NW 6'.0.
00 45 51.8	−31 11 57	ESO 411-10	14.0	1.7 x 1.1	14.5	SB(r)bc I	75	Size includes faint outer arms.
00 53 54.0	−31 05 48	ESO 411-28	12.9	1.3 x 1.1	13.2	SAB0$^-$:	37	Mag 11.25v star NNE 6'.9.
00 54 54.4	−32 01 56	ESO 411-29	13.9	1.1 x 0.6	13.3	S0$^-$: pec sp	178	Pair with ESO 411-30 NE edge; mag 10.19v star NW 4'.0.
00 54 57.2	−32 01 18	ESO 411-30	14.0	0.8 x 0.6	13.1	SBab? pec	103	Pair with ESO 411-29 SW edge; mag 10.19v star NW 4'.0.

(Continued from previous page)

GALAXIES

RA h m s	Dec ° ′ ″	Name	Mag (V)	Dim ′ Maj x min	SB	Type Class	PA	Notes
00 57 23.8	−30 58 02	ESO 411-33	14.5	1.2 x 0.5	13.8	SAB(s)0°	6	Mag 12.5 star S 1′.9; mag 9.62v star SW 3′.3; MCG −5-3-17 and 18 N 1′.8.
01 12 49.8	−31 12 00	ESO 412-15	14.4	1.8 x 0.2	13.2	Sbc	5	Mag 7.78v star ENE 12′.8; note: MCG −5-4-14 located 3′.2 N of this star.
01 14 30.0	−31 10 50	ESO 412-21	13.9	1.5 x 0.4	13.1	S?	38	Mag 7.78v star WNW 10′.5; note: MCG −5-4-14 located 3′.2 N of this star.
01 22 32.8	−29 58 57	ESO 412-27	14.5	1.8 x 0.2	13.2	S?	54	Mag 11.14v star W 1′.6.
01 24 50.5	−31 45 27	ESO 413-4	13.4	1.2 x 1.1	13.6	SAB(rs)c I	20	Stellar nucleus; mag 6.86v star WSW 8′.9.
01 41 58.7	−31 00 37	ESO 413-24	12.9	1.9 x 0.4	12.4	S?	177	Mag 10.6v star superimposed on center.
01 51 56.8	−31 44 11	ESO 414-8	15.0	1.5 x 0.7	14.9	SB(s)dm: IV	120	Mag 14.2 star on E edge; mag 11.17v star W 3′.1.
00 59 24.1	−34 19 43	IC 1608	12.4	2.0 x 0.9	12.9	(R)SA(r)0^+ pec?	170	Mag 9.78v star SW 11′.3.
00 59 47.0	−40 20 05	IC 1609	12.6	1.4 x 1.4	13.2	SA0°:		Close pair of E-W oriented mag 10.77v, 11.06v stars N 6′.6.
01 11 01.3	−30 26 20	IC 1637	12.8	1.7 x 1.3	13.5	SB(rs)c: I-II	90	Tightly wound, multi-branching arms; mag 9.12v star SE 10′.4.
01 14 06.9	−32 39 04	IC 1657	12.4	2.3 x 0.5	12.4	(R)SB(s)bc: I-II	170	**ESO 352-24A** W 1′.8; **ESO 352-24B** NE 2′.0.
01 37 35.8	−33 55 31	IC 1719	12.6	1.6 x 1.2	13.2	SA(s)0°:	174	Mag 9.88v star NE 7′.1.
01 43 02.6	−34 11 19	IC 1722	13.9	1.5 x 0.5	13.4	SAB(s)bc pec	50	Mag 9.85v star ENE 10′.5; **MCG −6-4-69** E 3′.4.
01 43 09.7	−34 14 33	IC 1724	12.9	1.3 x 0.5	12.3	$S0^+$:	126	Mag 12.2 star W 6′.6; IC 1722 N 3′.6 **ESO 353-39** ESE 5′.9.
01 47 44.5	−33 36 09	IC 1728	13.3	1.3 x 0.9	13.4	(R′)SAB(s)bc II	3	Mag 9.00v star E 12′.6.
01 49 17.2	−32 44 39	IC 1734	12.8	1.6 x 1.4	13.5	SB(rs)c II	28	Mottled; mag 9.59v star NW 12′.5.
01 50 29.2	−34 03 21	IC 1739	14.1	1.0 x 0.6	13.4	Sb	94	Mag 13.7 star on N edge; **MCG −6-5-8** SE 1′.7.
01 57 55.4	−32 59 18	IC 1759	12.9	1.5 x 1.4	13.6	SA(rs)bc:	169	Mag 6.37v star SE 8′.1.
01 57 48.7	−33 14 25	IC 1762	13.5	2.0 x 0.6	13.5	SB(s)bc I-II	43	Mag 10.45v star SSE 5′.4.
00 36 59.2	−29 33 01	MCG −5-2-29	14.9	0.4 x 0.2	12.2	E	9	
00 39 18.8	−29 56 48	MCG −5-2-32	14.4	0.5 x 0.5	12.8	S0?		Star on NE edge.
00 40 53.7	−30 27 39	MCG −5-2-35	15.1	0.5 x 0.4	13.2	SB(r)d	27	Mag 14.0 star NW 1′.7; mag 13.1 star E 5′.8.
00 50 42.7	−31 23 02	MCG −5-3-8	14.2	1.2 x 0.5	13.5	Sa-b	167	
00 51 04.4	−32 25 27	MCG −5-3-9	13.5	1.1 x 0.3	12.2	S(r)0/a	160	
00 56 58.9	−30 47 00	MCG −5-3-16	14.1	0.9 x 0.4	12.8	S	63	
00 57 19.0	−30 56 17	MCG −5-3-17	14.0	0.6 x 0.5	12.8	E	30	MCG −5-3-18 E 1′.5; ESO 411-33 S 1′.8.
00 57 25.9	−30 56 03	MCG −5-3-18	14.9	0.4 x 0.3	12.7	E	15	Mag 15.2 star SW 0′.4; ESO 411-33 S 1′.8; MCG −5-3-17 W 1′.5.
01 00 13.5	−30 48 39	MCG −5-3-21	13.9	1.0 x 0.8	13.5	Sb-c	36	Mag 10.38v star NE 8′.0.
01 12 46.7	−30 58 42	MCG −5-4-8	14.3	1.0 x 0.7	13.7	S	9	Mag 10.43v star N 6′.9.
01 12 57.0	−31 27 00	MCG −5-4-10	13.6	1.1 x 1.0	13.6	(R)SB(s)0/a	128	Mag 9.94v star SE 10′.1.
01 13 27.9	−31 49 03	MCG −5-4-12	15.3	0.5 x 0.3	13.1	S	144	Mag 8.24v star S 5′.4.
01 13 32.4	−31 48 29	MCG −5-4-13	14.9	0.4 x 0.4	12.9	E		Mag 8.24v star S 6′.4.
01 13 45.5	−31 03 51	MCG −5-4-14	13.7	1.1 x 1.1	13.8	(R)SBa		Mag 7.78v star S 3′.2.
01 13 53.5	−32 13 02	MCG −5-4-17	13.8	1.0 x 0.7	13.3	S0	31	Mag 10.90v star SE 4′.0; **MCG −5-4-19** ESE 5′.2.
01 14 11.0	−31 49 41	MCG −5-4-18	14.1	0.7 x 0.5	13.1	E	102	Mag 8.24v star SW 11′.4; NGC 441 NW 4′.8.
01 14 30.9	−32 15 58	MCG −5-4-20	13.5	1.5 x 1.1	13.9	SB(s)bc pec: I-II	96	Mag 10.90v star W 5′.1; **MCG −5-4-19** W 3′.4.
01 15 03.9	−31 47 04	MCG −5-4-23	14.1	0.7 x 0.4	12.8	E	42	Mag 12.0 star E 6′.5.
01 15 51.5	−31 35 27	MCG −5-4-26	14.5	0.8 x 0.4	13.1	S	72	Mag 9.92v star WSW 4′.2.
01 15 53.9	−32 28 46	MCG −5-4-27	13.5	1.4 x 1.3	14.0	SAB(s)cd: II	173	Stellar nucleus, branching spiral arms.
01 16 40.9	−31 26 04	MCG −5-4-29	14.1	0.7 x 0.5	13.0	E	39	Mag 8.70v star NE 2′.0.
01 23 00.9	−30 37 03	MCG −5-4-34	14.6	0.7 x 0.4	13.1	S	51	Mag 8.88v star W 3′.2.
01 24 55.2	−30 37 18	MCG −5-4-37	14.5	0.7 x 0.7	13.5	S		Mag 9.19v star SSE 3′.2.
01 25 37.2	−30 21 56	MCG −5-4-38	13.2	0.8 x 0.5	12.3	E?	162	Mag 11.5 star W 1′.2; mag 7.47v star NE 12′.4.
01 28 55.6	−32 00 47	MCG −5-4-40	14.3	0.6 x 0.5	12.9	S	60	Mag 8.86v star S 1′.7.
01 38 47.6	−31 49 20	MCG −5-5-1	14.0	1.0 x 0.5	13.1	SA(r)a:	168	Mag 10.98v star S 7′.6.
01 40 39.6	−31 49 22	MCG −5-5-9	13.4	1.0 x 0.2	11.6	S0 sp	127	Mag 8.57v star N 6′.4.
01 45 54.9	−29 02 19	MCG −5-5-13	13.6	0.8 x 0.8	13.0	S0		Elongated anonymous galaxy WNW 1′.9.
02 01 14.5	−31 43 48	MCG −5-5-26	13.6	1.2 x 0.9	13.6	(R)SB(rs)a:	126	Mag 8.68v star SW 11′.7.
02 02 10.4	−31 01 30	MCG −5-5-28	13.3	0.8 x 0.8	12.7	S0?		Mag 9.82v star WNW 4′.0.
02 03 56.4	−31 47 12	MCG −5-6-2	13.7	0.8 x 0.8	13.3	E		Mag 12.0 star N 4′.4.
00 36 59.1	−29 28 38	NGC 174	12.9	1.4 x 0.6	12.6	SB(rs)0/a:	152	Lies in a small triangle of mag 14 stars.
00 47 27.7	−31 25 22	NGC 254	11.7	2.5 x 1.5	13.0	(R)SAB(r)0^+:	137	Mag 7.09v star NNE 4′.9; stellar **ESO 411-12** SW 10′.7.
00 48 21.2	−38 14 07	NGC 264	13.5	1.1 x 0.4	12.4	SB0: sp	113	Mag 8.97v star ESE 8′.7.
00 52 42.3	−31 12 22	NGC 289	11.0	5.1 x 3.6	14.0	SB(rs)bc I-II	130	Very faint arms encircle galaxy beyond published size.
00 54 53.3	−37 41 02	NGC 300	8.1	19.0 x 12.9	14.0	SA(s)d III-IV	120	Many superimposed stars, some bright; also, many knots and clumps.
00 56 52.4	−31 57 48	NGC 314	13.4	1.0 x 0.8	13.0	(R)SB(rs)0^+:	168	Lies 2′.1 NW of mag 12 star.
00 57 14.8	−40 57 35	NGC 324	12.9	1.4 x 0.5	12.3	S?	95	
00 58 49.7	−35 07 03	NGC 334	13.6	1.0 x 0.6	12.9	(R′)SB(s)b pec: II	1	
01 04 18.5	−35 07 17	NGC 365	13.4	1.0 x 0.6	12.7	SB(r)bc pec: II	5	
01 06 12.1	−30 10 43	NGC 378	13.0	1.5 x 1.2	13.5	SB(r)c: I	105	**MCG −5-3-25** NE 1′.3.
01 09 33.2	−35 48 20	NGC 409	13.0	1.3 x 1.1	13.4	E:	174	Mag 13.5 star on NW edge.
01 10 05.5	−35 29 28	NGC 415	13.4	1.5 x 0.9	13.5	SB(rs)b II	55	
01 10 35.7	−30 13 19	NGC 418	12.5	2.0 x 1.7	13.7	SB(s)c I	12	Prominent multi-branching arms; mag 10.55v star N 7′.3.
01 11 22.5	−29 14 05	NGC 423	13.4	1.1 x 0.4	12.4	S0/a? pec sp	114	**ESO 412-12** SE 5′.9.
01 11 27.8	−38 05 05	NGC 424	12.8	2.3 x 0.8	13.3	(R)SB(r)0/a:	60	
01 12 19.2	−32 03 44	NGC 427	14.0	1.0 x 0.6	13.3	(R)SB(r)a:	0	
01 13 34.0	−37 54 07	NGC 438	12.8	1.4 x 1.1	13.1	(R′)SAB(s)b: II	36	Mag 5.94v star WNW 10′.0.
01 13 47.2	−31 44 51	NGC 439	11.5	2.5 x 1.5	12.8	SAB(rs)0^-?	156	NGC 441 S 2′.6.
01 13 51.2	−31 47 19	NGC 441	12.7	1.4 x 1.1	13.0	(R′)SB(rs)0/a:	135	Mag 8.24v star SW 9′.4; NGC 439 N 2′.6.
01 17 20.3	−33 50 31	NGC 461	13.3	1.2 x 0.9	13.3	SAB(s)c: I-II	23	
01 20 20.6	−40 58 01	NGC 482	13.3	2.2 x 0.5	13.2	SAab: sp	84	Faint lane runs full length just N of center.
01 21 20.1	−34 03 49	NGC 491	12.5	1.4 x 1.0	12.8	SB(rs)b: II	93	Mag 13.9 star on SW edge; mag 11.9 star W 4′.7.
01 20 05.0	−33 54 02	NGC 491A	13.8	2.0 x 0.9	14.3	SB(s)dm: IV	102	Patchy, bright knot or clump on W end.
01 23 54.2	−35 03 55	NGC 526	13.8	1.2 x 0.6	13.3	S0?	72	Pair with NGC 526B E edge.
01 23 57.0	−35 04 10	NGC 526B	13.0	0.8 x 0.5	11.9	SB0: pec	162	Pair with NGC 526 W edge.
01 23 58.0	−35 06 58	NGC 527	13.0	1.7 x 0.4	12.4	SB(r)0/a?	14	Pair with NGC 527B S edge; mag 10.55v star S 6′.9.
01 23 59.2	−35 07 42	NGC 527B	14.7	0.7 x 0.3	12.9	S	39	Pair with NGC 527 N edge.
01 24 44.7	−38 07 46	NGC 534	13.3	0.9 x 0.8	12.8	SAB(rs)0°	142	Mag 10.98v star S 4′.2.
01 25 12.0	−38 05 42	NGC 544	13.3	1.5 x 1.1	13.7	$SAB0^-$:	2	Mag 8.87v star SE 7′.3; NGC 546 N edge.
01 25 13.1	−38 04 03	NGC 546	13.5	1.4 x 0.5	13.0	SB(s)b: pec	35	Narrow plume extends NE; **ESO 296-22** N 3′.9.
01 25 28.0	−38 16 06	NGC 549	13.2	2.2 x 0.5	13.2	SB(s)0^+: sp	68	Galaxy **ESO 296-26A** SSE 1′.2; mag 8.87v star N 5′.3.
01 27 56.8	−35 43 02	NGC 568	12.6	2.2 x 1.2	13.5	$SA0^-$ pec:	140	**IC 1709** NE 4′.5.
01 28 36.5	−39 18 28	NGC 572	14.0	0.9 x 0.7	13.4	SAB(s)0°:	124	
01 29 02.9	−35 35 57	NGC 574	13.3	1.3 x 0.7	13.1	(R′)SB(rs)b	2	
01 32 14.9	−33 29 50	NGC 597	13.2	1.4 x 1.3	13.7	SB(s)bc? II	23	
01 33 57.6	−36 29 36	NGC 612	12.9	1.4 x 0.9	13.0	$SA0^+$ pec sp	172	A weak N-S dark lane W of core; mag 10.62v star lies 1′.0 W.
01 34 17.8	−29 25 02	NGC 613	10.1	5.5 x 4.2	13.3	SB(rs)bc II	120	A mag 9.55v star on NNE edge.
01 34 51.6	−36 29 22	NGC 619	13.3	1.7 x 1.1	13.9	(R′)SB(r)b	130	**ESO 353-19** WSW 3′.9; NGC 623 E 2′.9.
01 35 06.5	−36 29 25	NGC 623	12.5	2.0 x 1.5	13.6	E^+:	89	NGC 619 W 2′.9.

(Continued from previous page)

176 GALAXIES 176

RA h m s	Dec ° ′ ″	Name	Mag (V)	Dim ′ Maj x min	SB	Type Class	PA	Notes
01 35 12.2	−39 08 40	NGC 626	12.7	2.2 x 1.7	14.0	SB(rs)c: II-III	4	Large core, faint envelope; mag 9.23v star SE 8′.7.
01 35 36.3	−39 21 29	NGC 630	12.5	1.6 x 1.4	13.3	SA(rs)0^-:	60	Mag 9.23v star N 7′.0; ESO 297-8 SW 1′.8.
01 36 23.5	−37 19 19	NGC 633	12.8	1.6 x 1.4	13.5	SB(r)b: II	148	Very faint envelope NW and SE; a pair of mag 9 stars NW 3′.1.
01 38 59.0	−29 55 32	NGC 639	14.1	1.0 x 0.2	12.2	Sa?	31	
01 39 06.4	−29 54 57	NGC 642	12.9	2.0 x 1.1	13.6	SB(s)c II	31	Loose extensions NE; NGC 639 W 1′.7.
01 49 31.1	−34 54 19	NGC 696	13.2	1.7 x 0.6	13.0	SAB(s)0^+ pec?	25	**MCG −5-5-6** ESE 10′.9.
01 49 43.7	−34 49 52	NGC 698	13.9	0.9 x 0.7	13.3	(R′)SAB(rs)ab:	164	Pair of very faint, stellar anonymous galaxies 2′.5 NE.
01 53 49.5	−35 51 22	NGC 727	13.9	1.0 x 0.6	13.2	(R′)SAB(s)ab II-III	76	
01 55 41.0	−29 55 20	NGC 749	12.2	1.9 x 1.4	13.1	(R′)SB(s)0/a:	111	Strong, elongated core in smooth envelope.

GALAXY CLUSTERS

RA h m s	Dec ° ′ ″	Name	Mag 10th brightest	No. Gal	Diam ′	Notes
00 37 30.0	−39 07 00	A 2799	16.2	63	17	**ESO 294-25** at center; other members anonymous, stellar/almost stellar.
01 04 06.0	−39 46 00	A 2860	16.2	41	17	All members anonymous and stellar.
01 26 00.0	−37 58 00	A 2911	16.3	72	17	Mostly anonymous and stellar members.
00 55 48.0	−37 24 00	AS 102	16.4	29	17	ESO 295-22 at center; most others anonymous, stellar.
00 57 18.0	−30 55 00	AS 109	16.0		17	
01 00 36.0	−40 13 00	AS 113	16.3	12	17	**MCG −7-3-5** N of center7′.0.
01 13 42.0	−31 45 00	AS 141	15.6	12	22	
01 44 12.0	−35 17 00	AS 186	16.2	9	17	All members anonymous, stellar.
02 01 06.0	−40 35 00	AS 215	16.3		17	All members anonymous, stellar.

177 GALAXIES 177

RA h m s	Dec ° ′ ″	Name	Mag (V)	Dim ′ Maj x min	SB	Type Class	PA	Notes
23 56 36.4	−48 25 26	ESO 193-6	13.9	1.5 x 1.2	14.4	SB(s)c II	7	Mag 14.8 star, or possible stellar anonymous galaxy, N 2′.2.
00 03 24.5	−49 47 30	ESO 193-11	13.3	1.4 x 0.7	13.2	S0°: pec	147	Mag 10.62v star E 5′.0.
00 04 28.5	−50 37 12	ESO 193-15	14.4	0.9 x 0.8	13.8	Sc	57	**ESO 193-14** SW 2′.7.
00 05 07.9	−50 49 16	ESO 193-17	13.4	1.1 x 0.9	13.2	S0?	172	**PGC 346** SW 0′.9; anonymous galaxy W 1′.3.
00 05 29.4	−49 37 14	ESO 193-18	14.0	1.1 x 0.6	13.4	(R)SA(r)b: II-III	84	Mag 11.9 star W 7′.2.
00 05 29.2	−50 16 13	ESO 193-19	13.6	1.8 x 1.1	14.1	SAB(rs)c? pec I	63	Stellar **PGC 391** S 2′.2; stellar **PGC 390** S 3′.2; stellar **PGC 362** SSW 5′.5.
00 06 52.7	−50 24 54	ESO 193-26	13.3	1.2 x 0.8	13.1	$SA0^-$	89	**PGC 528** S 1′.3.
00 07 32.3	−48 49 36	ESO 193-29	13.7	1.5 x 0.9	13.9	S0	0	Faint, stellar **ESO 193-28** WNW 6′.1.
00 13 15.6	−49 20 31	ESO 193-36	14.0	1.1 x 0.6	13.4	S0?	2	Pair with ESO 193-37 SSE 1′.6.
00 13 21.6	−49 21 43	ESO 193-37	13.8	1.3 x 0.9	13.9	S0	88	Mag 12.4 star E 2′.8; ESO 193-36 NW 1′.6.
00 16 35.6	−48 15 56	ESO 193-41	13.6	1.2 x 0.9	13.6	S0	150	Mag 11.58v star W 3′.6.
00 19 39.8	−51 16 02	ESO 194-4	13.5	1.5 x 0.5	13.1	SB(r)0^+ pec:	27	**ESO 194-4A** NE 1′.4.
00 22 38.5	−48 34 52	ESO 194-13	13.6	1.1 x 1.0	13.6	SAB(r)cd III-IV	46	NE/SW oriented line of three mag 12-13 stars E 1′.0.
00 24 28.2	−50 41 09	ESO 194-15	14.0	1.0 x 0.7	13.5	S?	30	Trio of faint stars E edge; mag 14.2 star NE 1′.2.
00 29 41.6	−51 31 16	ESO 194-21	12.2	1.3 x 1.1	12.4	$SA0^-$	97	Mag 12.8 star W 1′.9.
00 30 23.9	−48 41 06	ESO 194-22	13.8	1.7 x 0.4	13.2	Sab: sp	112	Located 12′.2 NW of mag 4.76v star Lambda 1 Phoenicis.
00 31 25.2	−49 35 26	ESO 194-24	13.7	1.1 x 0.7	13.3	S	54	Mag 10.49v star SE 3′.9; note: **ESO 194-25** is 4′.2 SSE of this star.
00 44 02.3	−50 37 41	ESO 194-32	14.3	0.6 x 0.4	12.6	Sbc	101	Mag 9.31v star W 8′.4.
00 46 19.5	−47 25 54	ESO 194-36	14.4	0.9 x 0.6	13.6	S0/a	51	Double system; mag 5.80v star SSW 9′.3.
00 47 03.9	−48 43 07	ESO 194-38	13.8	1.0 x 0.7	13.3	Scd: III-IV	99	Mag 15 stars N and S edges.
00 47 27.4	−48 14 16	ESO 195-1	13.9	0.7 x 0.7	13.0	S0		Mag 12.5 star NE 1′.3.
00 49 18.8	−49 28 42	ESO 195-3	13.8	0.9 x 0.7	13.2	(R′)SA(rs)b	70	Mag 11.05v star SW 6′.0.
23 15 12.2	−49 23 53	ESO 239-17	14.8	1.7 x 1.4	15.6	SAB(s)c III-IV	118	Mag 9.99v star S 6′.8.
23 23 59.5	−50 31 30	ESO 240-2	14.4	0.5 x 0.3	12.2		86	Mag 10.11v star S 2′.3.
23 27 49.1	−47 22 54	ESO 240-3	13.9	0.8 x 0.4	12.5	(R)SB(rs)c pec	174	Pair of knots N end; mag 10.48v star SSW 5′.6; mag 12.4 star N 1′.8.
23 27 52.7	−51 07 52	ESO 240-4	15.0	1.5 x 0.3	14.0	Sd: sp	142	Dual galaxy system; mag 10.61v star S 2′.8.
23 37 44.4	−47 30 24	ESO 240-10	11.7	2.6 x 1.4	13.0	SAB0°: pec	133	Mag 9.49v star S 7′.4.
23 37 49.7	−47 43 39	ESO 240-11	12.4	5.6 x 0.6	13.5	Sc sp	129	Mag 9.49v star N 6′.4.
23 38 49.8	−51 51 40	ESO 240-12	13.8	1.0 x 0.6	13.0	S?	102	Mag 11.8 star SSW 6′.5.
23 39 27.3	−47 46 35	ESO 240-13	13.3	1.3 x 0.6	12.9	(R′)SB(s)b II-III	175	Mag 9.64v star SE 5′.8.
23 56 15.0	−43 25 39	ESO 241-6	13.8	1.2 x 0.7	13.5	SB(s)m IV-V	152	Mag 8.34v star NE 8′.6.
00 01 02.7	−43 19 48	ESO 241-10	12.9	1.7 x 1.3	13.7	E$^+$2	177	Small, elongated anonymous galaxy on W edge; a second anonymous galaxy lies W 1′.9.
00 09 35.7	−46 11 30	ESO 241-20	13.8	1.0 x 0.9	13.6	Sbc	164	Mag 13.9 star NW 3′.1.
00 10 20.0	−46 25 08	ESO 241-21	13.1	2.2 x 0.5	13.1	(R′)SB(s)b: I-II	54	Galaxy **CSRG 20** at NE end.
00 10 24.9	−46 29 30	ESO 241-22	13.5	3.2 x 1.3	14.9	SB(s)bc II	68	ESO 241-21 N 4′.5.
00 10 44.8	−47 04 30	ESO 241-23	13.1	1.8 x 0.8	13.3	(R′)SAB(r)ab	164	Mag 10.52v star S 2′.2.
00 24 22.0	−45 30 30	ESO 242-7	14.1	1.7 x 0.3	13.2	S	146	Star W edge.
00 26 55.3	−44 38 18	ESO 242-9	14.1	0.9 x 0.7	13.5	S	137	Mag 11.9 star SW 4′.2.
00 32 38.0	−45 13 58	ESO 242-12	14.4	0.8 x 0.7	13.6	Sab pec	161	Small companion galaxy S edge.
00 34 32.6	−43 39 22	ESO 242-14	13.2	1.8 x 1.5	14.2	(R)SB0°	119	Large core, smooth envelope.
00 35 11.2	−43 59 07	ESO 242-16	13.9	1.2 x 0.5	13.2	S0 pec:	122	Bright star E edge.
00 35 18.8	−44 04 52	ESO 242-17	14.0	1.0 x 1.0	13.9	(R)SAB(rs)ab:		Mag 8.70v star W 13′.5.
00 37 06.4	−46 38 40	ESO 242-18	13.3	1.8 x 1.0	13.8	SB(rs)cd: III	83	Mag 11.9 star NW 5′.7; mag 11.8 star S 4′.4.
00 38 02.2	−46 31 06	ESO 242-20	13.6	1.6 x 1.0	13.9	SB(s)dm IV	140	Mag 10.61v star N 8′.9.
00 39 15.3	−43 04 31	ESO 242-23	13.2	1.6 x 0.6	13.0	Sc pec	127	Mag 12 star on N edge.
00 40 31.6	−45 59 06	ESO 242-24	13.5	1.5 x 0.3	12.5	(R′)SB(s)0/a: sp	39	Mag 12 star at NE end.
00 49 34.7	−46 52 34	ESO 243-2	14.0	0.9 x 0.8	13.5	(R′)SA(rs)b	9	Mag 13.8 star N edge; mag 9.80v star ENE 2′.4.
00 51 56.9	−43 28 38	ESO 243-6	14.1	1.3 x 0.3	12.9	Sb	50	Located between a pair of mag 13.6 and 14.2 stars.
23 11 11.3	−42 50 51	ESO 291-3	15.0	1.8 x 0.2	13.7	SBdm? sp	70	Mag 5.83v star W 13′.5.
23 11 12.7	−45 41 44	ESO 291-4	13.8	1.3 x 0.5	13.2	(R′)SB(rs)0/a:	125	Mag 9.43v star NW 3′.4.
23 12 21.9	−43 51 49	ESO 291-6	14.2	1.1 x 0.5	13.4	$SA0^-$:	162	Mag 7.17v star SSW 8′.1.
23 13 58.8	−42 43 42	ESO 291-9	13.6	1.5 x 0.5	13.3	E$^+$4	61	Mag 8.82v star SW 7′.5; **ESO 291-8** NW 13′.5.
23 23 41.0	−42 24 10	ESO 291-24	14.1	1.6 x 0.4	13.5	SB(s)c: pec	91	Mag 11.15v star W 7′.2.
23 34 43.2	−46 12 21	ESO 291-32	14.4	0.8 x 0.5	13.3	S?	26	Mag 12.0 star NW 4′.5.
23 39 26.4	−45 58 40	ESO 292-7	13.6	1.2 x 0.9	13.6	S0	123	Mag 15.3 star superimposed E edge.
23 40 24.1	−44 30 52	ESO 292-9	13.7	1.9 x 0.7	13.8	SB(r)ab pec	98	Elongated **ESO 292-8** NW 4′.5.
23 42 31.7	−42 56 02	ESO 292-13	14.0	0.9 x 0.7	13.3	S	153	Star superimposed E of center; mag 12.6 star W 1′.5.
23 42 35.2	−44 54 17	ESO 292-14	13.1	2.6 x 0.4	13.0	S?	83	Mag 12.0 star E 5′.2.
23 46 27.4	−46 58 44	ESO 292-22	13.8	1.3 x 0.8	13.8	$SA0^-$ pec:	121	Double system **ESO 292-19** W 10′.9; **ESO 292-23** SSE 10′.1.
23 47 00.7	−44 45 47	ESO 292-24	13.9	0.5 x 0.5	12.3			Located 1′.5 N of mag 9.85v star.

(Continued from previous page)

177 GALAXIES 177

RA h m s	Dec ° ′ ″	Name	Mag (V)	Dim ′ Maj x min	SB	Type Class	PA	Notes
23 47 02.7	−43 19 56	ESO 292-25	14.0	0.8 x 0.5	12.9	SA(r)0°	27	Mag 9.13v star S 7′.1.
23 53 47.5	−40 01 30	ESO 293-10	13.8	0.9 x 0.7	13.4	E	71	Galaxy pair, in contact.
23 56 03.1	−40 53 30	ESO 293-14	14.3	1.0 x 1.0	14.2	S? pec		Double system **ESO 293-16** E 3′.5.
23 58 01.0	−41 31 57	ESO 293-22	13.7	0.9 x 0.6	12.9	S0?	25	**ESO 293-23** S 4′.7.
00 00 29.4	−40 28 59	ESO 293-27	13.6	2.2 x 0.5	13.6	SB(r)bc: III	154	Mag 8.56v star SSW 8′.3; double system **ESO 293-28** S 6′.0.
00 00 55.8	−40 42 42	ESO 293-29	14.2	1.4 x 0.6	13.8	SB(rs)c: I-II	19	Mag 7.31v star W 4′.6.
00 06 20.3	−41 29 36	ESO 293-34	13.0	3.1 x 0.8	13.8	SB(s)cd pec sp	13	**MCG −7-1-10** SE 1′.2.
00 07 06.2	−41 21 24	ESO 293-37	13.4	1.2 x 1.0	13.5	S?	50	Mag 10.61v star W 9′.4.
00 11 25.2	−41 23 59	ESO 293-45	14.6	1.4 x 0.4	13.8	SB(s)dm?	136	Mag 10.49v star N 7′.3.
00 26 33.7	−41 51 20	ESO 294-10	14.7	1.5 x 0.6	14.5	E5 pec:	6	Mag 11 &13 stars, on N-S line, W 1′.6.
00 27 09.2	−40 59 07	ESO 294-16	14.0	1.3 x 0.7	13.7	(R′)SB(rs)ab:	137	Mag 13.2 star N 1′.4.
00 30 12.1	−41 06 06	ESO 294-17	14.8	2.0 x 0.2	13.7	Sc: sp	107	Mag 6.18v star N 10′.1.
00 32 10.0	−40 16 01	ESO 294-20	13.6	1.9 x 1.4	14.6	SB(s)dm V	8	Close pair of N-S oriented mag 10.54v, 10.10v stars N 5′.6.
00 32 14.8	−41 23 11	ESO 294-21	13.3	1.5 x 0.5	12.8	SAB(rs)b II	176	Mag 9.81v star N 4′.2.
00 32 23.6	−42 10 25	ESO 294-22	14.0	1.2 x 1.0	14.1	(R′)SA(rs)b	110	Mag 12.6 star NW 3′.2.
00 48 32.1	−40 10 20	ESO 295-7	14.9	1.5 x 0.2	13.4	Sc	148	Mag 10.98v star NE 9′.7.
00 50 20.7	−41 14 40	ESO 295-12	13.7	1.0 x 0.6	13.0	S?	160	Faint star on W edge.
23 14 14.7	−40 37 53	ESO 347-2	14.0	2.1 x 0.3	13.3	SBd? sp	86	
23 20 48.8	−41 43 57	ESO 347-8	14.5	2.1 x 1.5	15.6	SB(s)m V	46	Mag 9.92v star SW 7′.3; note: almost stellar **MCG −7-47-34** is located half way to this star.
23 31 31.8	−42 08 49	ESO 347-20	14.3	1.0 x 0.4	13.1	S	17	Located 1′.3 NE of mag 8.28v star.
23 28 43.5	−41 19 57	IC 5325	11.3	2.8 x 2.5	13.2	SAB(rs)bc II-III	8	Star on S edge.
23 33 16.8	−45 00 56	IC 5328	11.4	2.5 x 1.5	12.8	E4	40	Mag 6.90v star SW 9′.0. IC 5328A 0′.7 SW of center.
23 33 13.9	−45 01 30	IC 5328A	13.7	0.5 x 0.2	11.1	SB(s)0/a? sp	151	Located 0′.7 SW of the center of IC 5328.
23 33 58.9	−45 12 35	IC 5328B	14.3	1.5 x 0.5	13.9	SB(s)c: IV	169	Mag 10.83v star S 1′.9.
00 02 10.6	−43 58 39	MCG −7-1-8	13.8	1.1 x 0.8	13.5	SAB0⁻	168	Mag 8.54v star E 4′.8; note: stellar galaxy **Fair 626** is 2′.1 SE of this star; **ESO 241-11** W 10′.3.
00 21 14.4	−48 37 43	NGC 87	14.3	0.9 x 0.7	13.7	IBm pec	171	NGC 88 SE 1′.4.
00 21 21.9	−48 38 23	NGC 88	14.4	0.8 x 0.5	13.3	SB(rs)0/a: pec	145	Nucleus slightly offset towards NE; NGC 89 S 1′.6; NGC 87 NW 1′.4.
00 21 24.4	−48 39 57	NGC 89	13.5	1.2 x 0.6	13.0	SB(s)0/a? pec	148	Two main arms; NGC 88 N 1′.6.
00 21 31.7	−48 37 33	NGC 92	13.1	1.9 x 0.9	13.5	SAa: pec	148	Faint extension SE; NGC 88 W 1′.8.
00 22 49.5	−45 16 09	NGC 98	12.7	1.7 x 1.3	13.4	SB(rs)bc I-II	0	
00 43 26.2	−50 11 04	NGC 238	12.5	1.9 x 1.6	13.6	(R′)SB(r)b pec	96	Short SE-NW bar.
23 09 47.1	−43 25 37	NGC 7496	11.4	3.3 x 3.0	13.7	SB(s)b II-III	2	Bright nucleus in short bar with bright knot N.
23 12 23.7	−43 46 43	NGC 7496A	13.9	1.4 x 0.7	13.7	SB(s)m V	82	Uniform surface brightness, faint star superimposed W; **ESO 291-5** NW 5′.9.
23 14 48.5	−43 36 05	NGC 7531	11.3	4.5 x 1.8	13.5	SAB(r)bc II	15	Bright elongated center, faint outer envelope.
23 16 10.9	−42 35 02	NGC 7552	10.6	3.4 x 2.7	12.8	(R′)SB(s)ab II	1	Member of **Grus Quartet**. Bright E-W bar, two main arms; mag 9.27v star W 4′.4.
23 18 23.7	−42 22 08	NGC 7582	10.6	5.0 x 2.1	13.0	(R′)SB(s)ab	157	Member of **Grus Quartet**. Strong dark lane N of center.
23 18 55.0	−42 14 17	NGC 7590	11.5	2.7 x 1.0	12.5	SA(rs)bc: III	36	Member of **Grus Quartet**. Star on NE edge.
23 19 21.5	−42 15 25	NGC 7599	11.5	4.4 x 1.3	13.2	SA(s)c II-III	57	Member of **Grus Quartet**. Strong dark lane NW of center.
23 22 01.0	−42 28 51	NGC 7632	12.1	2.2 x 1.1	12.9	(R′)SB(s)0°:	92	Close pair **ESO 291-22** and 291-23 S 7′.2.
23 33 02.6	−51 41 55	NGC 7690	12.4	2.2 x 0.9	13.0	SA(r)b:? II	132	
23 44 59.1	−42 54 38	NGC 7744	11.9	2.2 x 1.7	13.2	SAB(s)0⁻	105	Bright slightly elongated N-S center, uniform envelope.
23 50 54.5	−40 43 55	NGC 7764	12.2	1.9 x 1.3	13.0	IB(s)m III-IV	148	

GALAXY CLUSTERS

RA h m s	Dec ° ′ ″	Name	Mag 10th brightest	No. Gal	Diam ′	Notes
00 04 36.0	−45 37 00	AS 5	16.2		17	**ESO 241-12** at center; **ESO 241-13** N 8′.4; all others anonymous.
00 07 06.0	−44 22 00	AS 9	16.2		17	**PGC 546** at center; **ESO 241-16** SW 9′.9; most others anonymous.
00 23 42.0	−42 15 00	AS 37	16.2		17	**ESO 294-8** N of center 8′.1; all others anonymous, stellar.
00 45 42.0	−49 56 00	AS 74	15.9	21	17	**ESO 194-35, 35A** SW of center 8′.2; others anonymous, stellar.
00 47 30.0	−48 14 00	AS 78	16.1		17	**ESO 95-1** at center; all other members anonymous, stellar.
00 49 54.0	−47 22 00	AS 85	16.2		17	**ESO 195-4, 4A** at center, most others anonymous, stellar.
23 14 00.0	−42 43 00	AS 1101	16.4	28	17	See notes ESO 291-9, all others anonymous, stellar.
23 19 06.0	−42 05 00	AS 1111	15.7	25	17	**MCG −7-47-31, 32** at center; except NGC's, all others anonymous.
23 46 24.0	−46 58 00	AS 1147	16.4	4	17	See notes ESO 292-22, all others anonymous, stellar.
00 01 48.0	−43 57 00	AS 1173	15.5	27	22	See notes MCG −7-1-8; all others anonymous, stellar.

178 GALAXIES 178

RA h m s	Dec ° ′ ″	Name	Mag (V)	Dim ′ Maj x min	SB	Type Class	PA	Notes
21 32 13.9	−48 00 13	ESO 236-18	13.5	1.2 x 0.2	11.9	S0 pec?	48	Close pair of stars, mags 14.4 and 13.1, NE 2′.9.
21 32 18.4	−48 12 21	ESO 236-19	14.2	1.0 x 0.6	13.5		25	Double system, interaction, bright star E.
21 37 34.2	−47 19 43	ESO 236-25	13.6	0.9 x 0.7	12.9	S?	37	Mag 11.7 star ENE 1′.3.
21 38 13.7	−47 46 24	ESO 236-26	14.3	0.9 x 0.7	13.7	SAB(r)b:	36	Mag 10.26v star S 8′.2.
21 42 46.5	−51 17 18	ESO 236-34	14.4	1.6 x 1.0	14.7	IB(s)m V	16	Multi-galaxy system.
21 42 43.9	−47 59 29	ESO 236-35	13.3	2.4 x 1.2	14.3	SAB(rs)cd IV	48	Mag 11.7 star on S edge; ESO 236-36 N 8′.1.
21 42 53.3	−47 51 28	ESO 236-36	14.9	1.3 x 0.4	14.0	IB(s)m: pec	150	Mag 10.61v star W 12′.1; ESO 236-35 S 8′.1.
21 43 21.6	−48 19 12	ESO 236-37	13.4	1.6 x 0.9	13.7	SA(rs)bc: I-II	35	Mag 10.60v star SE 2′.2.
21 57 20.0	−51 13 04	ESO 237-19	14.9	1.3 x 0.6	14.5	IBm: V-VI	85	Mag 10.10v star N 8′.9.
22 03 04.2	−50 26 33	ESO 237-30	13.9	1.3 x 0.7	13.7	SAB(rs)ab pec:	33	Mag 7.99v star N 7′.5; stellar **ESO 237-34** E 9′.8.
22 04 20.5	−50 02 53	ESO 237-35	13.4	0.8 x 0.8	12.9	E0:		Mag 10.47v star SE 8′.7; **ESO 237-29** SW 16′.6.
22 09 45.6	−49 47 53	ESO 237-42	13.9	0.9 x 0.5	12.9	Pec	40	Mag 9.54v star SW 8′.9.
22 13 12.0	−48 50 39	ESO 237-47	13.8	0.8 x 0.7	13.0	Sc	83	Mag 7.92v star NW 7′.6.
22 14 33.2	−49 17 30	ESO 237-48	14.0	0.8 x 0.8	13.3	SB?		Mag 13.7 star SW edge; mag 11.16v star NW 2′.0.
22 16 02.9	−47 39 47	ESO 237-49	14.3	2.1 x 0.4	14.0	SBcd? sp	34	Mag 12.2 star NE 10′.6.
22 17 15.9	−48 18 06	ESO 237-51	13.8	1.2 x 0.9	14.0	E⁺3 pec	171	Mag 9.76v star N 8′.1.
22 17 46.8	−49 15 41	ESO 237-52	13.8	1.6 x 0.8	13.9	SB(s)dm pec: IV	62	Mag 10.03v star SE 10′.0.
22 21 15.9	−48 21 06	ESO 238-4	13.6	1.0 x 1.0	13.5	SB(r)bc: II-III		Mag 10.45v star SSW 4′.1; **ESO 238-3** WSW 3′.3.
22 22 30.3	−48 24 15	ESO 238-5	13.5	1.5 x 1.0	13.8	IABm V-VI	177	Uniform surface brightness; pair of N-S oriented mag 11.8, 11.7 stars E 8′.2.
23 15 12.2	−49 23 53	ESO 239-17	14.8	1.7 x 1.4	15.6	SAB(s)c III-IV	118	Mag 9.99v star S 6′.8.
21 34 31.7	−44 18 55	ESO 287-37	13.0	1.7 x 1.5	13.9	SB(s)dm: IV-V	38	Mag 12.0 star W 2′.6; mag 10.03v star E 9′.4.
21 36 34.2	−45 04 03	ESO 287-39	14.3	1.1 x 0.5	13.5	Pec:	34	Distorted; small companion 0′.9 NW.
21 37 28.1	−47 02 06	ESO 287-40	13.3	1.3 x 1.1	13.6	(R)SAB(r)bc pec: I	124	Mag 12.8 star E 2′.3.
21 38 07.5	−42 45 39	ESO 287-41	13.9	0.9 x 0.5	12.9	Pec:	100	Faint, almost stellar anonymous galaxy SW 0′.9.
21 38 08.1	−42 36 19	ESO 287-42	14.0	0.8 x 0.4	12.7	S0?	120	Mag 9.30v star SE 8′.7.
21 38 11.5	−43 56 00	ESO 287-43	13.8	2.0 x 0.3	13.1	Scd? sp	110	Pair of mag 9.65v, 9.26v stars WSW 8′.9.

(Continued from previous page)

GALAXIES

RA h m s	Dec ° ′ ″	Name	Mag (V)	Dim ′ Maj x min	SB	Type Class	PA	Notes
21 39 54.4	−42 52 51	ESO 287-45	13.7	1.1 x 0.7	13.2	(R′)SB(rs)b	9	
21 40 00.7	−44 05 45	ESO 287-46	13.6	1.5 x 0.9	13.8	SB(rs)c II	3	Mag 13.1 star N 1′3.
21 44 03.3	−43 12 37	ESO 287-55	13.7	1.4 x 1.1	14.0	SAB(rs)c pec II	112	Located between a pair of mag 13 stars SE-NW.
21 51 46.3	−43 07 28	ESO 288-13	14.0	1.2 x 1.2	14.2	SAB(s)m V-VI		Pair of superimposed stars or bright knots N and S of center.
21 55 47.8	−43 13 15	ESO 288-21	13.4	1.5 x 0.4	12.7	SB(s)c: I	177	Mag 9.70v star S 4′9; **ESO 288-20** SW 4′0.
21 59 17.5	−43 52 04	ESO 288-25	13.0	2.4 x 0.3	12.5	Sbc sp	54	Mag 10.08v star WNW 13′3.
22 01 32.3	−42 27 15	ESO 288-30	13.9	0.9 x 0.7	13.2	Sb: pec	42	Mag 13.6 star N 1′3; ESO 288-32 NE 1′2.
22 01 37.3	−42 26 24	ESO 288-32	14.1	0.9 x 0.6	13.3	(R′)SB(s)ab	93	Forms a triangle with a mag 13.6 star W and ESO 288-30 SW.
22 06 34.3	−42 51 36	ESO 288-40	15.6	1.5 x 0.9	15.8	IB(s)m V-VI	165	Mag 8.31v star S 7′1.
22 10 38.1	−43 15 13	ESO 288-45	14.0	1.3 x 0.6	13.6	SB(rs)c: III-IV	178	Pair of mag 13-14 stars close NW.
22 11 49.2	−45 35 28	ESO 288-49	13.4	2.2 x 1.3	14.4	SB(rs)dm: IV-V	57	Mag 10.74v star N 8′2.
22 13 57.7	−42 35 30	ESO 289-4	13.1	1.3 x 0.7	12.8	SAB0°?	70	Mag 12.1 star NW 5′8.
22 15 07.2	−46 16 58	ESO 289-5	14.1	1.1 x 0.5	13.3	S	166	Mag 15.1 star SE edge.
22 16 43.7	−47 07 11	ESO 289-10	14.1	2.0 x 0.4	13.7	SB(s)d? sp	121	Mag 8.94v star S 7′1.
22 17 17.2	−45 04 03	ESO 289-11	14.1	0.9 x 0.7	13.4	IB(s)m V-VI	69	Mag 11.6 star on E edge.
22 18 07.7	−42 41 23	ESO 289-15	14.2	0.5 x 0.4	12.3	S0?	107	Compact.
22 23 33.1	−42 16 25	ESO 289-26	13.8	2.2 x 1.0	14.5	SB(s)dm IV	146	Mag 12.3 star NW 4′1.
22 31 16.5	−46 28 57	ESO 289-32	14.2	0.9 x 0.5	13.1	(R′)SB(r)a	13	Faint star S edge; mag 12.6 star SW 2′5.
22 33 19.9	−44 32 59	ESO 289-37	13.9	0.9 x 0.5	13.0	S0 + S0?	119	Mag 6.79v star SSE 10′4.
22 35 30.8	−44 24 36	ESO 289-42	14.0	0.9 x 0.4	12.8	Sa? pec	93	Mag 13.1 star E 2′3.
22 37 59.5	−45 50 13	ESO 289-44	13.6	1.4 x 0.8	13.6	(R′)SB(rs)ab:	23	Mag 10.17v star W 6′5.
22 39 51.2	−43 51 39	ESO 289-47	13.3	0.9 x 0.8	12.8	Sa	155	Mag 11.25v star WSW 12′0.
22 40 57.6	−45 39 44	ESO 289-48	13.8	1.9 x 0.4	13.3	SB(s)cd: sp II-III	141	Mag 9.36v star N 5′9.
22 41 52.7	−46 05 35	ESO 290-1	13.8	0.5 x 0.5	12.2			Compact.
22 42 38.2	−42 54 32	ESO 290-4	13.3	1.0 x 0.8	13.0	SA0° pec:	78	Mag 10.43v star NW 1′5.
22 43 33.2	−45 19 37	ESO 290-6	13.3	1.2 x 1.0	13.3	(R′)SAB(rs)cd?	102	Mag 10.31v star S 5′5.
22 43 35.1	−43 53 49	ESO 290-7	14.2	1.3 x 0.3	13.1	(R′)SB(s)b:	107	Mag 13.7 star S 1′0.
22 44 16.3	−45 07 19	ESO 290-9	15.4	1.8 x 0.5	15.1		24	Quadruple system, linear, bridges.
22 45 26.9	−44 23 19	ESO 290-10	13.8	1.2 x 0.7	13.4	(R′)SAB(r)ab pec:	151	Mag 12.3 star WSW 8′1.
22 50 43.5	−45 19 53	ESO 290-20	13.2	1.1 x 0.8	13.2	E$^+$2	54	Mag 12.6 star E 5′5; **ESO 290-19, 19A** N 9′2.
22 52 30.8	−46 41 45	ESO 290-21	14.1	0.7 x 0.6	13.1		90	Mag 9.23v star SW 8′3.
22 55 53.4	−42 16 56	ESO 290-25	13.7	1.0 x 0.5	12.9	SA0° pec	65	Mag 10.84v star WSW 4′0.
23 01 32.2	−46 38 46	ESO 290-35	14.1	2.0 x 0.3	13.4	Sc sp	160	Mag 8.55v star ESE 6′7.
23 03 29.7	−46 01 56	ESO 290-39	14.6	1.0 x 0.7	14.0	SB(s)m IV-V	147	Mag 8.13v star ENE 6′6.
23 03 43.0	−46 31 58	ESO 290-40	15.2	0.4 x 0.4	13.1			Almost stellar.
23 04 05.6	−47 08 25	ESO 290-42	13.8	1.3 x 0.4	13.0	SB(r)0$^+$?	59	Mag 7.79v star SW 11′5.
23 04 42.6	−43 35 00	ESO 290-44	13.8	1.0 x 0.6	13.1	SAB(rs)cd: II-III	34	Mag 12.2 star E 5′8.
23 06 55.9	−42 53 46	ESO 290-51	13.8	1.0 x 0.6	13.1	SAB(rs)0$^-$: pec	174	Galaxy pair, in common envelope; **MCG −7-47-18** SE 1′8.
23 07 02.4	−42 43 49	ESO 290-52	14.1	1.5 x 0.3	13.1	Sbc: sp	97	Mag 10.83v star NW 7′4.
23 11 11.3	−42 50 51	ESO 291-3	15.0	1.8 x 0.5	13.7	SBdm? sp	70	Mag 5.83v star W 13′5.
23 11 12.7	−45 41 44	ESO 291-4	13.8	1.3 x 0.5	13.2	(R′)SB(rs)0/a:	125	Mag 9.43v star NW 3′4.
23 12 21.9	−43 51 49	ESO 291-6	14.2	1.1 x 0.5	13.4	SA0$^-$:	162	Mag 7.17v star SSW 8′1.
23 13 58.8	−42 43 42	ESO 291-9	13.6	1.5 x 0.5	13.3	E$^+$4	61	Mag 8.82v star SW 7′5; **ESO 291-8** NW 13′5.
21 34 19.7	−40 49 41	ESO 343-7	13.9	1.1 x 0.4	12.9	Sa	79	NGC 7087 E 2′7.
21 34 48.5	−40 43 15	ESO 343-9	14.1	0.8 x 0.4	12.7	S?	3	Mag 13.4 star E edge; mag 11.8 star N 2′5.
21 43 26.7	−40 07 33	ESO 343-24	13.6	1.0 x 0.3	12.2	S	32	Pair of mag 13.6 and 13.9 stars E 0′9.
21 54 55.8	−40 50 08	ESO 343-36	14.1	0.5 x 0.5	12.4			Compact; mag 13.6 star W edge; mag 10.30v star NE 2′1.
22 22 17.4	−40 05 22	ESO 345-2	12.8	1.5 x 0.5	12.3	Sb	76	Mag 9.26v star N 7′4.
22 42 54.9	−42 02 52	ESO 345-44	13.7	0.5 x 0.5	12.1			Mag 11.3 star W 11′3; **ESO 345-43** N 4′0.
22 43 14.3	−40 02 57	ESO 345-45A	13.4	0.6 x 0.5	12.2	E/S0	140	Stellar **ESO 345-45** S 1′0; mag 7.77v star N 2′5.
22 45 36.8	−40 55 00	ESO 345-50	14.8	1.5 x 0.5	14.3	IBm: sp V-VI	66	Mag 7.06v star NW 12′7.
22 52 39.1	−40 19 48	ESO 346-6	13.9	1.0 x 0.9	13.7	Sc: pec I-II	125	Mag 12.3 star WNW 3′1.
22 54 14.5	−40 08 57	ESO 346-9	14.5	0.4 x 0.3	12.0	Pec	83	Almost stellar; mag 12.1 star S 5′1.
22 57 11.6	−40 05 55	ESO 346-18	13.9	2.6 x 1.0	14.7	SA(s)dm V-VI	3	Mag 12.3 star SW 7′7.
23 01 56.1	−40 39 28	ESO 346-24	14.0	1.2 x 0.8	13.8	SBab	117	Large NE-SW core.
23 02 06.7	−41 09 54	ESO 346-25	14.1	1.3 x 0.8	14.0	(R′)SB(r)0°? pec	6	Mag 12.6 star W 8′1.
23 14 14.7	−40 37 53	ESO 347-2	14.0	2.1 x 0.3	13.3	SBd? sp	86	
22 02 41.6	−51 17 45	IC 5152	10.6	5.0 x 3.2	13.4	IA(s)m IV-V	100	Mag 7.76v star inside N edge.
22 12 29.6	−47 13 22	IC 5170	12.5	1.8 x 0.8	12.7	SABa: sp	25	Mag 10.28v star E 8′0.
22 10 56.5	−46 04 50	IC 5171	12.6	3.0 x 0.4	12.6	SAB(rs)bc: sp II	158	Mag 11.37v star N 5′8.
22 13 22.3	−46 01 05	IC 5181	11.5	2.6 x 0.8	12.2	SA0 sp	74	Mag 10.46v star E 5′6.
22 20 57.6	−46 02 07	IC 5201	10.8	8.5 x 3.9	14.5	SB(rs)cd III	33	Slender, elongated core, many knots; mag 9.39v star SE 7′7.
22 30 30.0	−45 59 42	IC 5224	13.6	1.4 x 0.4	12.8	Sab? sp	167	Mag 7.48v star SW 8′2.
22 41 53.0	−44 46 03	IC 5240	11.9	2.8 x 1.9	13.6	SB(r)a	100	Strong E-W bar; mag 8.91v star NW 19′3.
22 57 13.3	−43 23 44	IC 5267	10.5	5.2 x 3.9	13.7	SA(s)0/a	140	Mag 9.66v star NW 9′7.
22 55 56.2	−43 26 08	IC 5267A	13.4	2.4 x 0.8	14.0	SAB(s)bc? III-IV	172	Mag 10.35v star W 6′8.
22 56 57.2	−43 45 39	IC 5267B	12.7	1.9 x 0.8	13.0	S0°:	106	Mag 10.32v star NW 12′9.
21 32 35.4	−44 04 05	NGC 7079	11.6	2.1 x 1.3	12.6	SB(s)0°	82	
21 34 33.5	−40 49 06	NGC 7087	13.0	1.1 x 0.6	12.4	SB(s)ab:	39	ESO 343-7 W 2′7.
21 40 12.9	−42 32 20	NGC 7097	11.7	1.9 x 1.3	12.7	E5	20	Mag 6.64v star NW 8′1.
21 40 38.0	−42 28 52	NGC 7097A	14.3	0.7 x 0.4	13.0	E:	134	Mag 6.64v star W 10′5.
21 42 26.7	−44 47 39	NGC 7107	12.8	2.0 x 1.5	13.8	SB(s)m IV	128	
21 45 47.2	−48 25 11	NGC 7117	12.8	1.4 x 0.9	12.9	SA(s)0$^-$:	27	Mag 9.92v star SW 4′0.
21 46 09.7	−48 21 14	NGC 7118	12.5	1.5 x 1.2	13.1	SA0$^-$ pec:	50	**ESO 236-47** SE 2′7.
21 46 15.9	−46 30 59	NGC 7119A	12.8	1.1 x 0.8	12.5	SB(rs)bc pec: II	130	NGC 7119B on SW edge.
21 46 14.8	−46 31 16	NGC 7119B	12.9	0.6 x 0.3	10.8	Sc? pec	22	
21 48 05.7	−50 33 51	NGC 7124	12.3	3.0 x 1.3	13.7	SB(rs)bc I-II	143	Strong dark lane E of center.
21 52 42.3	−48 15 11	NGC 7144	10.8	3.7 x 3.6	13.6	E0		Mag 11.08v star NE 3′0.
21 53 20.7	−47 52 57	NGC 7145	11.1	2.5 x 2.4	13.1	E0	173	Star SE of center; mag 11.22v star SE 2′3.
21 55 04.6	−50 39 28	NGC 7151	12.7	3.0 x 1.2	13.9	SAB(rs)cd III-IV	75	Patchy, several superimposed stars.
21 56 10.0	−49 31 20	NGC 7155	12.2	2.2 x 1.8	13.5	SB(r)0°	4	Bright center with uniform envelope.
21 59 39.3	−43 18 20	NGC 7162	12.7	2.8 x 1.0	13.7	SA(s)c II-III	10	Bright, elongated center.
22 00 36.0	−43 08 15	NGC 7162A	12.5	2.6 x 2.3	14.3	SAB(s)m V	70	Elongated center, filamentary arms.
22 00 32.9	−43 23 23	NGC 7166	11.9	2.5 x 0.9	12.6	SA0$^-$	14	
22 02 07.6	−51 44 37	NGC 7168	11.9	2.0 x 1.5	13.1	E3	68	
22 02 48.7	−47 41 54	NGC 7169	13.6	1.0 x 0.5	12.7	SAB0$^-$ sp:	72	Located 3′2 ESE of mag 8.78v star.
22 05 54.9	−50 07 13	NGC 7196	11.5	2.5 x 1.8	13.1	E:	53	Star or stellar galaxy NE of center 1′1.
22 07 09.7	−49 59 42	NGC 7200	12.9	1.2 x 0.8	12.7	E$^+$:	33	
22 09 16.7	−47 09 57	NGC 7213	10.1	3.1 x 2.8	12.3	SA(s)a:	124	Very bright center.

(Continued from previous page)

178 GALAXIES 178

RA h m s	Dec ° ′ ″	Name	Mag (V)	Dim ′ Maj x min	SB	Type Class	PA	Notes
22 15 38.6	−45 51 00	NGC 7232	12.0	2.6 x 0.8	12.6	SB(rs)a:	99	NGC 7233 on E end.
22 13 41.8	−45 53 38	NGC 7232A	13.0	2.2 x 0.4	12.7	SB(rs)ab sp	111	Mag 8.53v star SE 2′.2.
22 15 53.0	−45 46 53	NGC 7232B	13.6	1.7 x 1.5	14.5	SB(s)m IV-V	0	Located 2′.1 N of mag 8.64v star.
22 15 49.0	−45 50 48	NGC 7233	12.5	1.7 x 1.3	13.2	SAB(s)0/a	133	NGC 7232 W; mag 8.64v star 2′.0 N.
22 33 52.8	−40 56 05	NGC 7307	12.6	3.5 x 0.8	13.5	SB(s)cd pec: II-III	9	
22 54 21.0	−45 20 48	NGC 7400	12.9	2.6 x 0.5	13.0	Sbc	2	
22 55 46.4	−42 38 24	NGC 7412	11.4	3.9 x 2.9	13.8	SB(s)b I-II	75	Bright nucleus; mag 7.26v star NNE 5′.6.
22 57 08.3	−42 48 19	NGC 7412A	13.9	3.5 x 0.5	14.4	SBdm: sp V	91	Low surface brightness.
22 57 18.2	−41 04 12	NGC 7424	10.5	9.5 x 8.1	15.0	SAB(rs)cd II-III	101	Bright center, branching arms with many knots and superimposed stars.
23 02 47.1	−40 50 07	NGC 7462	12.0	4.2 x 0.7	13.1	SB(s)bc? sp II-III	75	Star on W end; pair of stars mags 10.38v and 11.04v W 3′.9; mag 9.96v star SE 3′.7.
23 05 14.2	−50 06 43	NGC 7470	13.9	1.4 x 0.9	14.0	SA(s)bc pec II	84	Anonymous galaxy S 1′.9.
23 05 11.8	−43 05 54	NGC 7476	12.8	1.4 x 1.0	13.0	(R′)SB(r)ab:	175	Mag 7.37v star E 5′.8; mag 7.73v star N 4′.5; **ESO 290-46** NNE 6′.5.
23 09 47.1	−43 25 37	NGC 7496	11.4	3.3 x 3.0	13.7	SB(s)b II-III	2	Bright nucleus in short bar with bright knot N.
23 12 23.7	−43 46 43	NGC 7496A	13.9	1.4 x 0.7	13.7	SB(s)m V	82	Uniform surface brightness, faint star superimposed W; **ESO 291-5** NW 5′.9.
23 14 48.5	−43 36 05	NGC 7531	11.3	4.5 x 1.8	13.5	SAB(r)bc II	15	Bright elongated center, faint outer envelope.
21 45 30.0	−51 36 28	PGC 67298	14.3	1.2 x 0.6	14.0	E^+5 pec:	126	Anonymous elliptical galaxy NE 2′.0; **ESO 236-38** S 5′.9; **ESO 237-3** on E edge.
21 51 06.8	−46 16 32	PGC 67483	14.2	0.7 x 0.5	13.0	(R′)SA(s)0°: pec	138	Mag 11.8 star W 1′.5.

GALAXY CLUSTERS

RA h m s	Dec ° ′ ″	Name	Mag 10th brightest	No. Gal	Diam ′	Notes
21 47 00.0	−43 54 00	A 3809	16.0	73	17	All members anonymous, stellar.
22 15 54.0	−51 33 00	A 3849	16.0	42	17	**PGC 68430** NW of center 4′.2; all others anonymous, stellar.
22 51 48.0	−46 35 00	A 3925	16.4	38	17	Except ESO 290-21, all anonymous, stellar.
21 37 30.0	−47 02 00	AS 959	15.7	7	17	Except ESO 287-40, 236-25, all members anonymous, stellar.
21 45 36.0	−51 36 00	AS 968	15.4	20	22	See notes for PGC 67298; all others anonymous, stellar.
21 46 12.0	−46 31 00	AS 971	16.1	0	17	Most members anonymous, stellar.
21 48 12.0	−46 01 00	AS 974	15.3	18	22	**ESO 288-6**, **6A** at center; **ESO 288-7**, **288-9** S 3′.5; **ESO 288-8**, **8A** N 10′.8.
22 04 24.0	−50 04 00	AS 989	15.5	25	22	Most members anonymous, stellar.
22 46 36.0	−43 45 00	AS 1059	16.3		17	**ESO 290-11**, **11A** at center, all others anonymous, stellar.
22 46 42.0	−46 01 00	AS 1060	15.8	22	17	All members anonymous, stellar.
22 50 42.0	−45 20 00	AS 1067	16.1	20	17	See notes ESO 290-20, all others anonymous, stellar.
23 14 00.0	−42 43 00	AS 1101	16.4	28	17	See notes ESO 291-9, all others anonymous, stellar.

179 GALAXIES 179

RA h m s	Dec ° ′ ″	Name	Mag (V)	Dim ′ Maj x min	SB	Type Class	PA	Notes
20 04 37.3	−50 18 37	ESO 233-22	13.9	0.9 x 0.5	12.8	S0?	152	Almost stellar nucleus, smooth envelope; mag 12.1 star E 2′.4.
20 06 21.7	−48 04 55	ESO 233-27	13.6	0.9 x 0.3	12.0	S0	16	Mag 10.04v star NE 6′.2.
20 09 25.5	−48 17 04	ESO 233-35	13.3	1.0 x 0.4	12.2	S0	149	Mag 10.65v star W 6′.1.
20 09 34.1	−49 13 21	ESO 233-36	13.8	1.4 x 0.3	12.7	Sab? sp	55	Mag 10.5 star E 5′.9.
20 09 39.2	−49 19 55	ESO 233-37	13.6	1.3 x 0.3	12.4	Sb sp	13	Mag 11.8 star W 4′.7.
20 11 48.5	−51 23 20	ESO 233-44	13.9	0.7 x 0.3	12.1		145	Mag 10.19v star NE 4′.0.
20 16 09.9	−49 18 49	ESO 233-49	13.1	0.9 x 0.6	12.3	SA(rs)0^-?	126	Large core, small, faint envelope.
20 17 02.9	−47 46 18	ESO 233-50	14.1	1.1 x 0.7	13.7	S?	97	Mag 12.16v star E 2′.3; mag 10.32v star WSW 8′.7.
20 20 29.3	−51 37 24	ESO 234-3	13.9	0.9 x 0.4	12.7	S0:	12	
20 20 45.0	−49 07 35	ESO 234-4	14.4	1.5 x 0.6	14.1	SB(s)d: pec III-IV	112	Mag 10.06v star N 7′.7.
20 21 49.4	−48 00 18	ESO 234-9	14.3	0.7 x 0.7	13.4	S?		ESO 234-11 NE 2′.5; mag 12.0 star W 1′.9.
20 22 10.4	−51 42 47	ESO 234-10	14.1	1.2 x 0.4	13.2	Sb: II-III	124	Mag 10.77v star S 2′.6.
20 22 03.1	−47 59 14	ESO 234-11	13.5	1.6 x 0.7	13.5	(R)SA0°:	60	Mag 9.83v star S 6′.8.
20 22 32.6	−50 45 12	ESO 234-13	13.7	1.0 x 0.5	12.8	Sbc II-III	174	Mag 7.78v star W 8′.3.
20 22 54.9	−49 30 27	ESO 234-14	13.5	0.8 x 0.4	12.1	SAB0^-	178	Mag 8.03v star NW 7′.1.
20 23 25.6	−50 32 44	ESO 234-16	13.7	1.2 x 1.0	13.7	SA(rs)c II-III	53	Stellar nucleus, smooth envelope.
20 23 51.2	−48 21 34	ESO 234-19	14.3	1.6 x 0.3	13.3	Sbc: sp	69	Mag 8.52v star NE 7′.5; **ESO 234-18** S 5′.1.
20 24 20.1	−49 41 06	ESO 234-21	12.7	1.3 x 1.0	12.9	SA0°: pec	164	Close pair of mag 9.62v, 10.24v stars SW 8′.4.
20 25 27.8	−51 31 56	ESO 234-24	13.9	1.9 x 0.4	13.4	Sbc sp II	148	Mag 12.4 star on E edge; mag 9.53v star NW 10′.9.
20 27 31.9	−51 39 17	ESO 234-28	13.5	1.2 x 0.5	12.8	SB(s)b pec	175	Mag 8.99v star E 4′.4.
20 28 06.3	−51 41 32	ESO 234-32	13.6	1.6 x 0.6	13.4	(R′)SB(s)bc	73	Two main, open arms; mag 8.99v star N 1′.7.
20 32 13.9	−48 31 50	ESO 234-43	13.6	1.1 x 0.8	13.3	SB(s)m IV-V	53	Mag 9.45v star SSW 8′.7; **ESO 234-41** NW 6′.8, just W of a mag 12.8 star.
20 33 29.9	−49 34 34	ESO 234-44	14.1	0.9 x 0.5	13.1	S?	73	Mag 12.6 star N 1′.7.
20 34 25.4	−49 26 11	ESO 234-47	13.8	1.5 x 0.8	13.8	(R)SAab:	20	Mag 12.1 star W 1′.4; mag 11.18v star W 3′.0.
20 35 18.2	−49 51 52	ESO 234-49	12.8	1.7 x 1.5	13.6	Sbc: pec	38	Star on NE edge, **PGC 65018** NE 0′.7.
20 35 58.0	−50 11 35	ESO 234-50	13.5	0.6 x 0.4	11.9		21	Mag 11.6 star S 2′.0.
20 35 57.9	−49 36 15	ESO 234-51	12.7	1.5 x 1.0	13.1	SA0^- pec?	141	Mag 10.24v star N 6′.9; small, faint, anonymous galaxy S 3′.1.
20 36 24.8	−49 15 34	ESO 234-53	13.2	1.9 x 0.4	12.8	S0 sp	79	Mag 12.0 star W 4′.6.
20 37 00.6	−51 09 56	ESO 234-55	14.0	0.9 x 0.7	13.3	S?	66	Bright bar near S edge, smooth envelope.
20 45 52.4	−51 06 26	ESO 234-68	13.2	1.6 x 0.8	13.3	SAB(s)0°:	168	Mag 9.10v star W 3′.5.
20 45 54.4	−51 23 29	ESO 234-69	13.4	2.4 x 0.6	13.6	SB(s)c: II-III	110	Mag 9.08v star NW 3′.7.
20 46 03.6	−51 36 58	ESO 234-70	13.5	0.9 x 0.8	13.0	S0	11	Strong elongated core. Smooth envelope.
20 49 55.8	−50 08 04	ESO 235-4	13.7	1.5 x 0.7	13.6	SB(s)dm IV	78	Multiple galaxy system; star on N edge.
20 55 14.1	−49 11 15	ESO 235-16	13.3	1.5 x 1.0	13.6	(R′)SB(r)bc II	31	Mag 11.9 star NNE 3′.7; **ESO 235-15** NNW 8′.8.
20 58 18.3	−49 00 46	ESO 235-22	13.3	0.9 x 0.4	12.1	SA(s)0°: pec	162	Mag 9.96v star NE 2′.8.
20 58 32.8	−49 17 05	ESO 235-23	13.2	2.0 x 1.5	14.3	SBa: pec	159	Two galaxies, common irregular envelope.
21 00 51.0	−48 35 56	ESO 235-33	12.9	1.8 x 1.4	13.8	SA(s)c I-II	115	Mag 9.89v star NE 6′.4.
21 02 43.6	−48 21 29	ESO 235-39	13.5	1.1 x 0.8	13.2	SA(r)0°:	57	Mag 10.40v star W 8′.6; **ESO 235-37** NW 8′.9.
21 03 26.7	−48 12 16	ESO 235-42	13.1	1.0 x 0.8	12.7	(R′)SAB(r)0/a	6	Mag 12.2 star NW 6′.0.
21 03 49.5	−50 21 57	ESO 235-45	14.0	1.2 x 0.5	13.3	SB(s)b: pec	54	Mag 6.58v star SE 5′.8.
21 04 27.5	−48 14 25	ESO 235-47	14.0	0.9 x 0.5	13.0	S?	69	ESO 235-49 NE 3′.7.
21 04 41.0	−48 11 26	ESO 235-49	12.6	1.4 x 0.9	12.8	E:	96	Mag 10.38v star E 2′.7; ESO 235-47 SW 3′.7.
21 04 51.4	−51 49 24	ESO 235-50	13.1	1.6 x 1.2	13.7	SA(r)0^-	102	Mag 9.05v star N 5′.4.
21 05 01.0	−51 56 56	ESO 235-51	13.4	1.2 x 0.5	12.7	SB(s)0^+	141	Mag 8.19v star NW 6′.7.
21 05 10.4	−47 47 24	ESO 235-53	13.2	1.9 x 0.4	12.8	Sb sp	48	Mag 8.59v star SE 4′.8; mag 12.3 star superimposed NE edge; **ESO 235-52** N 4′.6.
21 05 44.7	−51 42 38	ESO 235-54	13.7	1.0 x 0.5	12.8	(R′)SA(r)a:	33	Mag 9.05v star W 8′.3.
21 05 55.3	−48 12 29	ESO 235-55	12.0	2.6 x 2.5	13.9	(R′)SAB(rs)bc I-II	1	Mag 10.44v star on N edge; ESO 235-57 NE 4′.9.
21 06 21.8	−48 10 13	ESO 235-57	13.5	2.2 x 0.5	13.4	Sbc: sp	133	Located between ESO 235-58 N and ESO 235-55 SW.
21 06 28.5	−48 07 16	ESO 235-58	14.3	1.3 x 0.8	14.2	SB(rs)d IV	120	Mag 12.2 star N 1′.3.

RA h m s	Dec ° ′ ″	Name	Mag (V)	Dim ′ Maj x min	SB	Type Class	PA	Notes
21 06 33.8	−49 13 10	ESO 235-59	13.6	0.9 x 0.4	12.4	S0	62	Mag 8.47v star NE 7'.8.
21 06 56.8	−48 51 49	ESO 235-60	13.7	1.0 x 0.3	12.3	S0?	63	Mag 14.6 star on SE edge; galaxy **Fair 948** SE 2'.1.
21 07 14.0	−47 33 26	ESO 235-61	14.1	1.3 x 0.8	14.0	SB(s)d: pec III-IV	9	Double system, contact.
21 17 52.7	−48 18 48	ESO 235-83	13.8	0.9 x 0.7	13.1	SB(rs)0° pec:	44	Mag 12.4 star S 2'.0; ESO 235-84 S 5'.4.
21 17 57.6	−48 24 08	ESO 235-84	13.9	1.4 x 0.8	13.9	SB(rs)dm? IV-V	23	Mag 10.63v star S 3'.8; ESO 235-83 N 5'.4.
21 18 16.2	−48 32 19	ESO 235-85	13.1	0.8 x 0.4	11.7	Sa?	149	NGC 7049 E 7'.4.
21 20 00.6	−51 03 29	ESO 236-3	14.3	0.9 x 0.8	13.8	Sb	156	Mag 14.0 star S edge.
21 21 08.9	−47 30 57	ESO 236-6	14.2	1.3 x 0.7	13.9	SB(s)dm IV	53	Mag 9.63v star W 7'.2.
21 24 26.1	−49 16 47	ESO 236-8	13.5	1.3 x 1.1	13.8	(R')SB(s)b II-III	166	Strong core, dark patch SW; mag 12.9 star NE 1'.3.
21 26 02.6	−48 50 54	ESO 236-11	14.0	1.4 x 0.8	14.0	(R')SAB(rs)c II-III	27	Mag 9.73v star SE 10'.0.
21 27 29.6	−49 06 16	ESO 236-14	14.0	0.9 x 0.4	12.8	S0?	24	NGC 7061 N 2'.6.
21 32 13.9	−48 00 13	ESO 236-18	13.5	1.2 x 0.2	11.9	S0 pec?	48	Close pair of stars, mags 14.4 and 13.1, NE 2'.9.
21 32 18.4	−48 12 21	ESO 236-19	14.2	1.0 x 0.6	13.5		25	Double system, interaction, bright star E.
21 37 34.2	−47 19 43	ESO 236-25	13.6	0.9 x 0.7	12.9	S?	37	Mag 11.7 star ENE 1'.3.
21 38 13.7	−47 46 24	ESO 236-26	14.3	0.9 x 0.7	13.7	SAB(r)b:	36	Mag 10.26v star S 8'.2.
19 58 17.1	−46 06 49	ESO 284-4	13.4	1.6 x 0.8	13.5	SB(rs)bc II	11	Mag 7.49v star E 7'.4.
20 00 08.8	−46 21 27	ESO 284-7	13.9	1.5 x 0.3	12.8	Sa:	94	**ESO 284-6** W 1'.6.
20 01 21.0	−46 40 04	ESO 284-9	12.8	1.0 x 0.5	12.1	E5:	150	Mag 9.92v star NW 8'.4.
20 02 14.1	−45 27 00	ESO 284-11	13.8	1.2 x 1.0	13.8	S?	9	Mag 15.1 star E edge.
20 05 36.0	−42 42 13	ESO 284-13	14.2	1.5 x 0.4	13.5	Sb: sp	142	Mag 12.2 star SW 5'.4.
20 06 20.7	−42 45 49	ESO 284-16	13.4	1.9 x 0.9	13.8	SAB(rs)c II	130	Mag 12.0 star S 5'.4.
20 07 06.5	−42 48 50	ESO 284-17	13.2	1.3 x 0.6	12.8	SAB(s)0$^-$	5	Mag 12.0 star W 6'.9; mag 10.69v star S 7'.1.
20 09 14.7	−46 21 02	ESO 284-20	14.8	1.6 x 0.3	13.8	Sc: pec sp	31	Close pair of E-W oriented mag 8.92v, 8.23v stars NW 6'.0.
20 09 43.9	−44 09 15	ESO 284-21	13.3	1.3 x 0.8	13.2	SB(s)cd III-IV	14	Mag 12.1 star E 0'.8.
20 12 11.7	−44 08 48	ESO 284-26	13.1	1.1 x 0.9	12.9	S0$^-$:	111	Mag 14.4 star S edge; mag 12.6 star E 2'.5.
20 13 56.3	−44 21 09	ESO 284-32	12.8	1.3 x 0.9	12.8	SB(r)0/a:	145	Star inside W edge, several faint stars superimposed.
20 14 12.4	−44 37 24	ESO 284-33	13.2	1.1 x 0.7	12.8	SAB0°	134	Mag 11.5 star NNW 3'.0.
20 15 55.6	−46 09 03	ESO 284-37	13.4	1.0 x 0.7	12.9	SAB(rs)0°	52	Mag 9.98v star NW 9'.4.
20 15 54.6	−44 29 40	ESO 284-38	13.1	1.2 x 0.7	12.8	SA0°	123	Mag 8.42v star NW 7'.0.
20 16 06.6	−44 25 18	ESO 284-39	13.7	1.0 x 0.5	12.8	S?	105	Mag 13.4 star NW 2'.7; mag 8.42v star W 6'.8.
20 16 13.7	−46 35 51	ESO 284-40	14.2	1.1 x 0.9	14.0	S?	55	Mag 9.31v star N 2'.1.
20 16 18.5	−44 18 02	ESO 284-41	12.1	1.2 x 0.6	11.6	Pec	39	Double system, strongly interacting.
20 16 30.9	−44 42 32	ESO 284-43	13.7	1.2 x 0.5	13.0	SBd: IV	88	ESO 284-46 E 2'.9; mag 10.69v star SE 3'.5.
20 16 36.0	−46 58 00	ESO 284-44	12.6	1.5 x 1.0	12.9	SA0$^+$ pec	37	Mag 10.48v star E 2'.7.
20 16 48.4	−46 31 54	ESO 284-45	13.3	1.6 x 1.0	13.7	SB(rs)b: pec	175	Mag 9.31v star W 6'.1.
20 16 47.0	−44 41 48	ESO 284-46	13.6	1.4 x 1.0	13.8	SB(s)b? pec III-IV	134	Strong dark patch SE center; ESO 284-43 W 2'.9.
20 16 53.1	−47 19 43	ESO 284-47	13.7	1.0 x 0.4	12.6	S?	130	Mag 11.4 star NE 8'.4.
20 17 15.3	−45 47 08	ESO 284-48	13.5	1.1 x 0.8	13.2	Sb: pec	60	Strong dark patch W of bright core.
20 18 03.3	−45 31 10	ESO 284-51	13.6	0.9 x 0.7	13.0	S0:	152	**ESO 284-50** N 1'.4.
20 18 14.9	−45 50 32	ESO 284-53	13.4	0.9 x 0.5	12.4	Sa	63	Close pair of mag 12 star W 1'.8.
20 21 29.2	−44 36 37	ESO 285-1	13.9	1.8 x 0.4	13.5	(R')SB0$^+$? sp	142	Mag 10.33v star E 1'.0.
20 27 42.3	−43 43 53	ESO 285-13	13.6	1.1 x 0.9	13.5	SA(s)0/a:	79	Mag 9.58v star SE 3'.5.
20 27 55.9	−44 36 53	ESO 285-14	14.2	1.3 x 0.5	13.6	Pec	40	Core offset NE; mag 10.16v star NE 4'.9.
20 29 32.6	−42 30 26	ESO 285-19	13.6	1.0 x 0.7	13.1	SAB(r) pec?	175	Soft almost stellar nucleus; mag 10.60v star E 2'.4.
20 29 37.5	−43 50 30	ESO 285-20	13.9	1.7 x 0.7	13.9	SA(s)bc: II-III	90	Mag 9.88v star N 3'.6.
20 30 46.9	−44 22 59	ESO 285-23	13.7	0.7 x 0.5	12.4	SB?	83	Star on E edge; ESO 285-23 S 1'.3.
20 30 47.6	−44 24 17	ESO 285-24	13.5	0.9 x 0.8	13.1	SA0$^-$?	172	ESO 285-23 N.
20 31 40.6	−47 06 15	ESO 285-26	14.1	0.6 x 0.6	12.8	Sa		Pair of mag 14.9, 14.6 stars on N and S edges.
20 32 02.8	−44 25 01	ESO 285-28	13.9	1.1 x 0.3	12.6	Sa:	169	Mag 14.0 star E edge; mag 12.5 star SE 0'.7.
20 33 20.2	−42 28 45	ESO 285-30	15.0	1.0 x 0.5	14.1	dE?	176	Mag 10.40v star SW 7'.1.
20 34 16.6	−44 18 18	ESO 285-32	13.6	1.2 x 0.5	12.9	SB0/a? pec	73	Star superimposed E of center.
20 34 38.2	−43 50 28	ESO 285-35	13.9	1.3 x 0.8	13.8	Pec	12	Mag 12.5 star on N edge.
20 35 07.3	−43 19 14	ESO 285-36	14.0	0.7 x 0.4	12.5		86	Double system, contact? **MCG −7-42-12** SW 8'.0.
20 36 47.2	−45 18 32	ESO 285-40	14.1	1.6 x 0.3	13.2	Sab: sp	52	Mag 9.56v star WNW 8'.4.
20 39 18.1	−44 36 25	ESO 285-42	13.4	1.2 x 0.7	13.1	(R')SAB(rs)a:	87	Mag 14.9 star on N edge.
20 40 53.9	−44 48 54	ESO 285-44	15.3	1.0 x 0.7	14.8	Sdm:	108	Close pair of mag 12.0, 12.8 stars W 7'.4.
20 44 40.2	−45 58 43	ESO 285-48	12.9	2.3 x 1.0	13.6	SB(s)cd II-III	84	Pair of E-W oriented mag 10.88v, 10.00v stars N 5'.4.
20 45 20.5	−45 36 58	ESO 285-49	13.2	1.3 x 0.6	12.8	(R')S0°:	112	Located between a pair of stars, mags 12.3 and 13.2.
20 46 46.5	−44 38 05	ESO 285-52	13.6	0.7 x 0.7	12.7	S0$^-$ pec		Mag 10.98v star NW 10'.9.
20 48 36.7	−43 17 02	ESO 285-55	13.9	1.0 x 0.5	13.0		37	Double system; mag 9.15v star S 5'.4.
20 54 21.0	−44 03 47	ESO 286-10	12.4	2.1 x 1.6	13.5	(R')SA(rs)a	138	Very close pair of mag 11.1v, 13.3 stars SW 3'.2.
20 57 38.6	−46 36 19	ESO 286-16	13.6	1.5 x 0.3	12.6	Sab? sp	166	Mag 9.67v star NNE 8'.0; stellar **ESO 286-15** N 2'.2.
20 57 49.2	−43 21 05	ESO 286-17	13.5	1.1 x 0.5	12.7	S0°?	120	Pair with elongated ESO 286-18 S 1'.4.
20 57 50.3	−43 22 33	ESO 286-18	13.2	2.7 x 0.3	12.8	SB(s)bc? sp	56	Pair with ESO 286-17 N 1'.4.
20 58 26.9	−42 39 03	ESO 286-19	14.2	0.7 x 0.4	12.7		132	Very faint, narrow streamer S.
21 00 28.8	−42 25 25	ESO 286-22	14.1	0.5 x 0.4	12.3	cE?	177	Compact, almost stellar.
21 01 31.8	−46 02 17	ESO 286-26	13.8	0.9 x 0.6	13.0	S0/a	3	**ESO 286-25** N 7'.6.
21 02 20.1	−44 26 00	ESO 286-27	14.5	1.3 x 0.2	12.9	Sb? sp	119	Mag 9.86v star E 7'.5.
21 03 04.4	−47 08 46	ESO 286-29	13.7	0.8 x 0.5	12.6	S0$^-$ pec?	90	Mag 12.9 star N 1'.7.
21 03 36.7	−43 44 41	ESO 286-31	14.1	0.9 x 0.5	13.1	S pec?	16	Mag 10.41v star SSE 8'.0.
21 03 48.4	−47 11 17	ESO 286-32	14.7	1.1 x 0.7	14.3	SB(s)dm V	3	Mag 13.0 star S 1'.0.
21 04 08.5	−43 32 11	ESO 286-33	14.1	0.8 x 0.4	12.7		17	Double system, contact; mag 12.8 star ESE 1'.3.
21 04 11.2	−43 35 33	ESO 286-35	13.7	1.1 x 0.3	12.3	S pec	23	Mag 12.8 star WSW 2'.5.
21 04 13.3	−44 54 09	ESO 286-36	14.8	1.5 x 0.2	13.4	Sb:	91	Mag 12.5 star N 1'.3.
21 04 52.7	−47 07 24	ESO 286-37	13.5	1.5 x 0.5	13.0	(R')SA(s)a:	162	Mag 11.29v star S 1'.4.
21 05 04.5	−43 25 13	ESO 286-41	13.1	1.1 x 0.6	12.6	SA(r)0$^+$	20	Mag 10.06v star N 1'.8; mag 7.22v star SE 7'.9.
21 05 31.2	−47 02 48	ESO 286-42	13.6	0.7 x 0.5	12.3	SA0°:	80	Almost stellar; mag 11.29v star and ESO 286-37 SW 8'.8.
21 05 39.0	−42 46 54	ESO 286-44	13.7	0.9 x 0.5	12.7	S0	122	Mag 12.1 star NW 5'.0.
21 06 04.9	−45 25 58	ESO 286-46	13.2	1.2 x 0.6	12.7	SAB(r)0$^+$	119	Mag 7.94v star NE 9'.5.
21 06 23.4	−43 30 29	ESO 286-47	13.5	1.1 x 0.6	12.9	SA0° pec	135	Mag 7.22v star W 9'.3; **ESO 286-45** SSW 10'.9.
21 06 47.8	−47 11 14	ESO 286-49	13.1	1.3 x 0.6	12.8	E$^+$:	3	Mag 10.76v star NE 3'.0; **ESO 235-56** SW 11'.8.
21 06 41.2	−42 33 27	ESO 286-50	12.9	1.4 x 0.7	12.7	SA(rs)0$^-$?	134	Mag 10.62v star NW 8'.9.
21 06 51.4	−44 49 35	ESO 286-52	13.1	0.8 x 0.4	11.7		91	**ESO 286-54** S 4'.3; **ESO 286-48** SW 4'.6.
21 08 18.0	−43 54 57	ESO 286-58	14.1	1.3 x 0.7	13.9	SB(s)c I-II	157	Mag 9.85v star SE 4'.5.
21 08 39.3	−43 29 08	ESO 286-59	13.3	1.1 x 0.6	12.7	SA0$^-$ pec	38	**MCG −7-43-23** S 1'.6. Possible anonymous galaxy W 1'.7.
21 08 57.0	−43 41 12	ESO 286-60	13.6	1.0 x 0.7	13.1	SAB(s)0°	131	Mag 12.1 star NW 3'.2.
21 09 52.2	−45 31 40	ESO 286-63	13.2	1.1 x 0.6	12.6	SAB(rs)bc:	42	Mag 10.10v star WNW 8'.0; double system **ESO 286-62** N 5'.8.
21 10 54.6	−42 38 13	ESO 286-69	13.9	1.0 x 0.7	13.4		43	Triple system, interaction; mag 9.56v star E 3'.3.
21 11 15.0	−46 48 49	ESO 286-71	14.2	1.1 x 0.7	13.8	SB(s)dm: IV	113	Mag 12.4 star WSW 5'.7.

(Continued from previous page)

GALAXIES

RA h m s	Dec ° ′ ″	Name	Mag (V)	Dim ′ Maj x min	SB	Type Class	PA	Notes
21 15 33.2	−46 46 50	ESO 286-80	13.8	1.0 x 0.6	13.1	SB(s)c II-III	35	Stellar galaxy **Fair 961** SW 4′6.
21 15 45.7	−42 25 41	ESO 286-82	13.8	1.1 x 0.8	13.5	SB(s)c II	129	Mag 10.87v star NE 1′6.
21 17 28.0	−42 20 25	ESO 287-2	13.8	1.0 x 0.7	13.3	S0	9	Mag 13.7 star NE 2′7; mag 9.61v star S 6′0.
21 18 06.2	−46 18 01	ESO 287-4	13.5	1.7 x 0.5	13.2	SB(rs)b: II	170	Mag 7.73v star NE 4′4.
21 19 41.2	−45 07 04	ESO 287-7	13.6	0.7 x 0.6	12.5	S0?	138	**ESO 287-7A** SE 0′9; **ESO 287-7B** E 1′2; mag 6.11v star NE 7′6.
21 21 16.5	−46 09 11	ESO 287-9	13.8	1.8 x 0.3	13.0	Sbc: sp	104	Mag 9.12v star NE 4′4.
21 21 22.4	−46 35 59	ESO 287-10	13.9	1.1 x 0.7	13.4		140	Double system, contact.
21 23 14.1	−45 46 30	ESO 287-13	12.6	3.4 x 0.8	13.5	SAbc I-II	62	Small, elongated **ESO 287-12** W 3′0.
21 24 35.5	−42 35 29	ESO 287-16	13.5	0.9 x 0.9	13.2	S0:		**MCG −7-44-3** ESE 2′1.
21 26 21.2	−43 13 17	ESO 287-26	13.7	1.0 x 0.4	12.6	Sa	27	Very slender **ESO 287-24** SW 1′9.
21 30 35.2	−44 25 50	ESO 287-30	14.1	0.6 x 0.5	12.7	S0: pec	78	Mag 11.10v star NW 11′6.
21 31 24.6	−43 15 36	ESO 287-32	14.0	0.5 x 0.4	12.1	S0? pec	27	Compact; mag 12.7 star W 1′9; located near center of galaxy cluster **A 3775**.
21 31 35.9	−46 53 17	ESO 287-33	13.5	1.2 x 0.9	13.5	SB(r)c I	178	Several faint stars superimposed on edges.
21 31 52.8	−46 03 33	ESO 287-35	13.4	1.2 x 0.6	12.8	SAB(r)a:	18	Mag 13.5 star superimposed SE of center; galaxy pair **AM 2128-461** NE 2′0.
21 34 31.7	−44 18 55	ESO 287-37	13.0	1.7 x 1.5	13.9	SB(s)dm: IV-V	38	Mag 12.0 star W 2′6; mag 10.03v star E 9′4.
21 36 34.2	−45 04 03	ESO 287-39	14.3	1.1 x 0.5	13.5	Pec:	34	Distorted; small companion 0′9 NW.
21 37 28.1	−47 02 06	ESO 287-40	13.3	1.3 x 1.1	13.6	(R)SAB(r)bc pec: I	124	Mag 12.8 star E 2′3.
21 38 07.5	−42 45 39	ESO 287-41	13.9	0.9 x 0.5	12.9	Pec:	100	Faint, almost stellar anonymous galaxy SW 0′9.
21 38 08.1	−42 36 19	ESO 287-42	14.0	0.8 x 0.4	12.7	S0?	120	Mag 9.30v star SE 8′7.
21 38 11.5	−43 56 00	ESO 287-43	13.8	2.0 x 0.3	13.1	Scd? sp	110	Pair of mag 9.65v, 9.26v stars WSW 8′9.
21 39 54.4	−42 52 51	ESO 287-45	13.7	1.1 x 0.7	13.2	(R′)SB(rs)b	9	
21 40 00.7	−44 05 45	ESO 287-46	13.6	1.5 x 0.9	13.8	SB(rs)c II	3	Mag 13.1 star N 1′3.
19 58 30.8	−40 48 58	ESO 339-12	13.0	1.3 x 1.0	13.1	SAB(rs)cd	7	Mag 12.9 star on SW edge; mag 9.56v star SW 6′9.
20 00 44.9	−41 57 40	ESO 339-21	13.8	0.6 x 0.5	12.4	S0?	81	Mag 11.04v star E 2′3.
20 12 39.6	−40 07 59	ESO 340-1	13.8	1.4 x 0.3	12.7	Sbc	114	Mag 11.65v star NW 2′6.
20 13 46.0	−40 03 24	ESO 340-2	13.6	1.3 x 1.0	13.7	Sc	70	Mag 11.02v star N 5′9.
20 15 05.8	−40 02 49	ESO 340-5	14.1	0.8 x 0.5	13.0	Sa	163	
20 15 23.6	−39 59 54	ESO 340-6	13.7	1.0 x 0.4	12.6	(R)SB(r)0^+	40	Mag 8.79v star SE 2′3.
20 17 11.5	−40 55 30	ESO 340-8	14.6	1.8 x 0.3	13.8	Scd sp	32	Mag 7.02v star W 4′1.
20 17 25.9	−41 08 02	ESO 340-10	13.8	1.0 x 0.6	13.1	S0	6	Located between a pair of stars, mags 12.9 and 13.7.
20 18 37.2	−41 03 21	ESO 340-13	13.6	1.3 x 1.0	13.7	SB(r)0^+	58	Small star superimposed NE end; IC 4991 W.
20 19 14.9	−41 20 36	ESO 340-14	14.4	0.6 x 0.6	13.1	SA(rs)a		Mag 10.31v star W 2′6; ESO 340-15 E 2′4..
20 19 26.6	−41 19 59	ESO 340-15	12.7	1.0 x 0.9	12.4	SA(rs)a	90	ESO 340-14 W 2′4; **MCG −7-41-27** NW 6′7.
20 19 32.3	−40 43 33	ESO 340-16	13.1	1.2 x 0.7	12.8	(R′)SB(s)a	148	Mag 11.7 star SW 3′0.
20 22 29.0	−41 54 04	ESO 340-21	12.9	1.3 x 1.1	13.1	SAB(r)ab	4	Mag 5.58v star Kappa 1 Sagittarii S 9′0.
20 24 24.5	−40 21 58	ESO 340-25	13.4	1.1 x 0.9	13.3	S0?	163	Small, faint, stellar anonymous galaxy W 0′9.
20 24 43.4	−41 30 46	ESO 340-26	13.9	2.2 x 0.3	13.3	Sc: sp	45	Mag 10.90v star NW 10′0.
20 25 24.4	−40 09 44	ESO 340-28	14.2	1.0 x 0.4	13.1	S	135	Mag 10.67v star NW 4′2.
20 26 06.1	−40 55 54	ESO 340-29	13.1	0.9 x 0.9	12.7	SB?		Mag 10.15v star E 3′0; mag 6.09v star NNW 8′7.
20 30 38.8	−42 18 33	ESO 340-38	14.6	0.8 x 0.5	13.4	S pec	139	Stellar nucleus; mag 11.6 star W 2′6.
20 34 55.9	−40 37 29	ESO 340-42	13.6	1.5 x 0.9	13.7	SB(s)dm IV-V	68	Mag 9.52v star SE 1′8.
20 35 04.7	−41 41 05	ESO 340-43	12.7	1.3 x 1.2	13.1	(R′)SA0^+:	76	Mag 11.20v star WNW 8′8.
20 54 31.8	−40 48 40	ESO 341-21	13.9	1.0 x 0.5	13.0	(R′)SB(r)ab	75	Bright star superimposed.
21 06 44.1	−40 29 32	ESO 342-5	14.1	1.0 x 0.3	12.7	Sbc	108	Mag 9.65v star SSE 6′5.
21 08 59.5	−40 01 19	ESO 342-12	14.0	0.8 x 0.4	12.6	S	26	Bright star superimposed N end; mag 12.3 star NE 1′3.
21 16 46.4	−42 15 45	ESO 342-26	13.0	1.2 x 0.7	12.7	SB(s)0° pec	107	Mag 10.27v star SW 9′5; ESO 342-27 E 1′7.
21 16 55.4	−42 15 37	ESO 342-27	12.3	1.4 x 1.0	12.6	SA0^-: pec	63	ESO 342-26 W 1′7.
21 17 16.6	−41 50 09	ESO 342-30	13.9	1.1 x 0.2	12.2	Sa:	59	Mag 10.94v star E 2′8.
21 17 34.0	−41 14 59	ESO 342-32	14.1	0.7 x 0.2	11.8		1	Mag 10.90v star SW 5′2.
21 25 21.6	−40 07 15	ESO 342-42	13.8	0.9 x 0.5	12.8	S pec	115	Mag 12.3 star NE 1′7; **ESO 342-44** E 4′4.
21 25 54.3	−40 00 39	ESO 342-45	13.4	0.9 x 0.8	12.9	SB?	23	Mag 11.6 star ENE 6′0.
21 29 05.7	−41 46 40	ESO 342-52	13.5	1.1 x 0.9	13.3	SB(r)ab	135	Mag 14.7 star N edge.
21 30 22.0	−40 26 19	ESO 343-1	13.6	1.3 x 0.6	13.2	(R)SB(s)ab:	57	Mag 14.0 star S edge.
21 34 19.7	−40 49 41	ESO 343-7	13.9	1.1 x 0.4	12.9	Sa	79	NGC 7087 E 2′7.
21 34 48.5	−40 43 15	ESO 343-9	14.1	0.8 x 0.4	12.7	S?	3	Mag 13.4 star E edge; mag 11.8 star N 2′5.
19 56 45.9	−50 03 22	IC 4909	13.8	0.9 x 0.4	12.5	SA(s)bc? II-III	83	Mag 8.82v star NW 8′5.
19 57 42.1	−51 59 11	IC 4911	14.7	0.5 x 0.4	12.8	S0:	135	Mag 9.93v star E 2′3; pair of mag 7.64v, 8.21v stars NE 6′3.
19 58 19.2	−50 16 21	IC 4916	13.7	0.7 x 0.5	12.4	SA(s)b? pec II	7	Mag 9.31v star SW 9′9.
20 06 28.4	−48 22 33	IC 4943	12.7	0.6 x 0.6	11.6	E pec:		Mag 10.32v star NE 3′9; NGC 6861 E 8′5.
20 23 58.3	−43 59 44	IC 4946	11.8	2.5 x 1.0	12.7	SAB(rs)0/a:	68	Mag 9.98v star E 3′0.
20 11 31.5	−45 35 34	IC 4956	12.4	1.7 x 1.6	13.5	E0:	14	Mag 11.01v star W 4′1.
20 18 23.7	−41 03 05	IC 4991	11.6	2.6 x 1.8	13.1	SA(r)0°? pec	135	Numerous stars or bright knots surround core.
21 24 22.1	−40 32 16	IC 5105	11.6	1.9 x 1.3	12.7	E^+	40	Mag 10.02v star NW 5′2.
21 25 31.2	−40 16 27	IC 5105A	12.8	2.1 x 1.1	13.6	SB(rs)c I	103	Mag 9.47v star SW 4′4.
21 26 00.1	−40 50 08	IC 5105B	13.0	1.3 x 0.5	12.4	SB(rs)ab:	65	Mag 11.5 star on N edge.
20 01 00.8	−47 04 08	NGC 6845A	13.0	2.2 x 1.0	13.7	SB(s)b: pec	72	Multi-galaxy system; long, bright extension NE with NGC 6845B at it's end.
20 01 05.5	−47 03 33	NGC 6845B	13.9	0.9 x 0.7	13.2	SBb? pec	6	Multi-galaxy system; narrow extension NE.
20 00 57.0	−47 05 07	NGC 6845C	15.4	0.8 x 0.3	13.7	$S0^+$: pec sp	138	Located on S edge of NGC 6845A.
20 00 53.6	−47 05 44	NGC 6845D	14.8	0.8 x 0.4	13.4	S0: pec sp	156	Located 0′9 SW of NGC 6845C.
20 06 15.7	−40 11 52	NGC 6849	12.1	1.9 x 1.1	12.8	$SB0^-$:	18	
20 03 34.5	−48 17 05	NGC 6851	11.8	2.0 x 1.5	13.0	E:	160	
20 05 39.9	−47 58 47	NGC 6851A	16.1	1.2 x 0.2	14.4	SB(rs)d pec IV	70	Faint stellar nucleus; pair with NGC 6851B 1′5 E.
20 05 48.8	−47 58 43	NGC 6851B	13.8	1.3 x 0.6	13.4	Sbc: sp	54	Uniform brightness, very diffuse; mag 10.19v star NE 11′0.
20 07 19.4	−48 22 14	NGC 6861	11.1	2.8 x 1.8	12.7	SA(s)0^-:	140	Mag 10.32v star WNW 5′3; IC 4943 W 8′5.
20 06 05.6	−48 28 30	NGC 6861B	14.1	1.2 x 0.3	12.8	SB0° pec? sp	97	Mag 9.88v star SW 10′4; **PGC 64075** SW 6′0.
20 06 41.6	−48 38 57	NGC 6861C	14.1	1.2 x 0.3	12.9	S0 pec? sp	35	
20 08 19.7	−48 12 42	NGC 6861D	12.4	2.3 x 0.7	12.8	SA(s)0^-:	154	Mag 10.44v star W 1′3.
20 11 01.3	−48 41 29	NGC 6861E	14.1	1.3 x 0.3	13.0	Sab:	39	
20 11 11.7	−48 16 33	NGC 6861F	14.5	1.6 x 0.3	13.6	SBdm? sp	86	Mag 9.02v star S 6′5.
20 09 53.6	−48 22 44	NGC 6868	10.7	3.5 x 2.8	13.1	E2	86	Small, faint anonymous galaxy on NE edge 1′6 from center.
20 10 11.1	−48 17 14	NGC 6870	12.3	2.6 x 1.2	13.4	SAB(r)ab	85	Faint, elongated anonymous galaxy NW 7′6.
20 13 12.4	−46 09 44	NGC 6875	12.1	2.3 x 1.3	13.1	SAB(s)0^- pec:	22	Small, anonymous galaxy on N edge; mag 8.26v star SE 3′4.
20 11 55.6	−46 08 38	NGC 6875A	13.0	2.7 x 0.5	13.2	SB(r)bc? sp II-III	75	Mag 9.07v star E 4′8.
20 13 53.1	−44 31 30	NGC 6878	12.7	1.6 x 1.2	13.3	SA(s)b I-II	111	Two main arms.
20 13 36.4	−44 49 00	NGC 6878A	13.2	1.9 x 0.8	13.5	SAB(rs)b II-III	68	
20 18 18.4	−44 48 26	NGC 6890	12.3	1.5 x 1.2	12.7	SA(rs)b III	152	
20 20 49.5	−48 14 21	NGC 6893	11.8	2.6 x 1.7	13.3	SAB(s)0°	10	Bright, slightly elongated center, many stars superimposed.
20 24 22.7	−50 26 02	NGC 6899	12.8	1.7 x 1.0	13.3	SAB(r)bc: I-II	112	Located 2′3 ESE of mag 10v star.
20 24 28.0	−43 39 07	NGC 6902	10.9	5.6 x 3.9	14.1	SA(r)b II	153	Very bright center, branching arms, a number of superimposed stars; a mag 11.9 star SE 4′4.

179

(Continued from previous page)

GALAXIES

RA h m s	Dec ° ′ ″	Name	Mag (V)	Dim ′ Maj x min	SB	Type Class	PA	Notes
20 22 59.6	−44 16 19	NGC 6902A	13.4	1.3 x 1.0	13.6	SB(s)m pec IV-V	48	Multi-galaxy system.
20 23 06.7	−43 52 07	NGC 6902B	13.6	1.5 x 1.3	14.2	SAB(s)cd: II-III	99	Bright center, patchy; mag 9.95v star NW 9′.5.
20 27 38.8	−47 01 37	NGC 6909	11.7	2.2 x 1.1	12.6	E$^+$:	68	Stellar galaxies **Fair 898** and **899** SSE 3′.4.
20 30 47.2	−47 28 28	NGC 6918	13.4	0.9 x 0.7	12.8	(R′)SB(r)0/a:	178	N-S bar with strong dark line E.
20 31 38.2	−44 13 05	NGC 6919	13.0	1.7 x 1.2	13.7	SAB(s)c: I-II	145	Strong dark patch S of nucleus.
20 52 09.4	−48 46 44	NGC 6970	12.6	1.0 x 0.6	11.9	SB(rs)ab pec: II	123	Almost stellar anonymous galaxy N 1′.8.
20 57 18.5	−51 51 46	NGC 6982	13.7	1.1 x 0.6	13.1	(R)SB(rs)a pec?	68	Mag 10.61v star NNE 4′.9.
20 56 43.5	−43 59 11	NGC 6983	13.4	0.8 x 0.6	12.5	Sab pec?	147	
20 57 54.1	−51 52 18	NGC 6984	12.7	1.8 x 1.2	13.4	SB(r)c I-II	110	Faint star S of nucleus; mag 10.61v star N 4′.4; galaxy **Fair 927** NNE 5′.7.
20 58 10.5	−48 37 51	NGC 6987	12.4	1.4 x 1.2	13.0	E0:	92	Star on W edge; mag 10.91v star SW 3′.1.
21 03 44.9	−49 01 48	NGC 7002	12.4	1.5 x 1.2	13.0	E1 pec?	3	
21 04 02.1	−49 06 54	NGC 7004	13.8	1.3 x 0.6	13.4	(R′)SAB(s)0/a:	70	
21 06 45.4	−44 48 54	NGC 7012	12.6	2.5 x 1.3	13.8	E$^+$4 pec	115	ESO 286-52 on SE edge. Several very small, faint anonymous galaxies also on SE edge.
21 07 52.3	−47 10 45	NGC 7014	12.4	1.9 x 1.5	13.4	E$^+$:	130	Mag 10.21v star SSW 8′.6.
21 09 35.3	−49 18 13	NGC 7022	13.0	1.5 x 1.1	13.4	SB(rs)0°	16	
21 11 52.2	−49 17 00	NGC 7029	11.5	2.6 x 1.4	12.9	E6:	71	
21 15 07.6	−47 13 09	NGC 7038	11.6	3.2 x 1.6	13.3	SAB(s)c: I-II	127	Bright nucleus, branching arms.
21 16 32.3	−48 21 47	NGC 7041	11.2	3.6 x 1.5	12.9	SA(rs)0$^-$:	85	Mag 10.73v star ESE 2′.0.
21 19 00.6	−48 33 42	NGC 7049	10.7	4.3 x 3.0	13.3	SA(s)0°	57	Mag 10.16v star NNW 3′.7; ESO 235-85 W 7′.4.
21 24 58.6	−42 27 38	NGC 7057	12.6	1.4 x 1.0	12.8	SA0$^-$?	151	
21 25 53.6	−42 24 44	NGC 7060	12.9	2.0 x 1.2	13.7	(R′)SAB(r)a	120	Strong N-S bar, dark patches E and W.
21 27 27.0	−49 03 48	NGC 7061	13.3	1.2 x 0.7	13.1	E$^+$4	137	ESO 236-14 S 2′.5.
21 30 25.4	−43 05 13	NGC 7070	12.3	2.3 x 1.8	13.6	SA(s)cd III	22	Moderately dark patch S of center.
21 31 47.4	−42 50 48	NGC 7070A	12.3	1.7 x 1.6	13.3	I0	2	
21 30 37.2	−43 09 08	NGC 7072	13.5	0.8 x 0.7	12.8	SAB(s)d? IV	90	
21 30 25.9	−43 12 11	NGC 7072A	13.9	0.8 x 0.6	12.9	SAB(s)c: II	125	
21 32 35.4	−44 04 05	NGC 7079	11.6	2.1 x 1.3	12.6	SB(s)0°	82	
21 34 33.5	−40 49 06	NGC 7087	13.0	1.1 x 0.6	12.4	SB(s)ab:	39	ESO 343-7 W 2′.7.
20 21 50.9	−49 42 09	PGC 64536	12.6	1.1 x 0.6	12.1	SAB0°: pec	162	= **Fair 892**.

GALAXY CLUSTERS

RA h m s	Dec ° ′ ″	Name	Mag 10th brightest	No. Gal	Diam ′	Notes
21 06 42.0	−47 08 00	A 3742	15.3	35	22	Except NGC, ESO galaxies, all members anonymous, stellar.
21 08 42.0	−43 29 00	A 3747	15.2	44	22	ESO 286-59 at center; **ESO 286-61** ESE 10′.0.
21 15 42.0	−42 36 00	A 3756	16.4	39	17	**ESO 286-84** ESE 7′.5 from center; most members anonymous, faint, stellar.
21 29 30.0	−50 48 00	A 3771	16.3	42	17	**PGC 66845** S of center 5′.4; most members anonymous, stellar.
20 09 54.0	−48 23 00	AS 851	13.5	6	56	
20 21 48.0	−49 42 00	AS 866	15.2		22	Except Uranometria plotted GX, all others anonymous, stellar.
20 35 30.0	−43 26 00	AS 882	16.2		17	**MCG −7-42-14** at center; **ESO 285-39** S 10′.6; **ESO 285-34** SW 12′.5.
20 40 30.0	−40 04 00	AS 890	15.6	14	22	**MCG −7-42-17** at center; **ESO 341-3**, **3A** NE 11′.4; all others anonymous.
20 52 24.0	−51 57 00	AS 906	15.9	29	17	Several PGC GX S of center; **ESO 235-10** NE 7′.6.
20 54 18.0	−43 03 00	AS 907	16.0	3	17	**MCG −7-43-1** near center; **ESO 286-9** W 7′.1; most others anonymous.
20 55 30.0	−43 31 00	AS 909	15.5	25	22	**ESO 286-11** S of center 4′.4; all others anonymous, stellar.
21 01 42.0	−41 25 00	AS 916	16.0	3	17	**MCG −7-43-9** WNW of center 10′.4; all others anonymous, stellar.
21 05 06.0	−43 26 00	AS 919	15.8	6	17	Except ESO 286-41, all members anonymous, stellar.
21 06 48.0	−44 49 00	AS 921	15.4	27	22	
21 07 54.0	−47 10 00	AS 924	15.9	13	17	Superimposed on A 3742.
21 37 30.0	−47 02 00	AS 959	15.7	7	17	Except ESO 287-40, 236-25, all members anonymous, stellar.

OPEN CLUSTERS

RA h m s	Dec ° ′ ″	Name	Mag	Diam ′	No. ★	B ★	Type	Notes
21 21 30.0	−51 49 00	ESO 236-7		30			cl	

PLANETARY NEBULAE

RA h m s	Dec ° ′ ″	Name	Diam ″	Mag (P)	Mag (V)	Mag cent ★	Alt Name	Notes
20 19 27.8	−41 31 30	PN G359.2−33.5	8			10.5	He 3-1863	Very bright central star.

180

GALAXIES

RA h m s	Dec ° ′ ″	Name	Mag (V)	Dim ′ Maj x min	SB	Type Class	PA	Notes
18 49 02.3	−51 15 58	ESO 231-1	12.7	1.2 x 1.1	12.9	S?	1	Mag 8.01v star N 8′.8.
18 49 00.9	−48 31 33	ESO 231-2	14.1	0.4 x 0.4	12.0			Compact; mag 10.13v star N 3′.4.
18 49 24.3	−48 46 11	ESO 231-3	14.3	1.5 x 0.3	13.3	Sb sp	115	Mag 9.98v star S 1′.2.
18 52 37.5	−51 28 04	ESO 231-6	14.8	1.5 x 0.2	13.3	Sc	57	Mag 8.28v star W 8′.3; mag 7.44v star WSW 11′.8.
18 56 02.1	−49 57 06	ESO 231-9	13.6	1.0 x 0.9	13.4	Sb	34	Mag 8.02v star SW 9′.5.
18 57 40.3	−47 33 03	ESO 231-11	13.2	1.7 x 0.6	13.0	SB(rs)c? III-IV	1	Mag 8.12v star WSW 5′.9.
19 00 37.8	−49 06 42	ESO 231-14A	13.7	0.7 x 0.4	12.2		85	Pair with ESO 231-14 N; **Fair 501** NW 14′.2.
19 00 39.7	−49 05 57	ESO 231-14	13.8	0.9 x 0.7	13.4	E$^+$3	135	Pair with ESO 231-14A S; mag 10.28v star NNE 7′.0.
19 04 46.5	−47 50 58	ESO 231-17	12.7	1.1 x 0.4	11.9	E6?	17	Mag 10.42v star NW 3′.1.
19 05 06.2	−51 13 44	ESO 231-18	13.5	1.3 x 0.6	13.1	SA(r)b: III-IV	140	Mag 10.42v star W 3′.4.
19 08 44.1	−51 02 48	ESO 231-23	13.3	1.7 x 0.4	12.7	Sb sp	102	Mag 9.96v star SE 6′.1.
19 09 41.8	−50 46 23	ESO 231-24	12.8	1.0 x 0.7	12.3	SA(r)0$^+$	139	Lies 0′.5 NE of mag 8.72v star.
19 15 37.4	−47 53 16	ESO 231-29	14.3	1.5 x 0.3	13.3	Scd: sp	56	Mag 7.42v star W 4′.0.
19 19 13.5	−47 45 44	ESO 232-1	14.0	1.0 x 0.4	12.8	S?	151	Mag 10.43v star NNE 2′.1.
19 22 46.8	−51 00 09	ESO 232-4	13.6	1.3 x 0.7	13.4	S?	22	Close pair of mag 12 stars S edge.
19 23 44.3	−47 54 28	ESO 232-5	14.0	1.1 x 0.5	13.2	Sa:	119	Stellar nucleus.
19 24 28.5	−51 57 44	ESO 232-6	13.1	1.6 x 0.9	13.3	(R′)SB(r)b: II	130	Pair of stars on SW edge.
19 30 52.7	−47 44 41	ESO 232-9	13.2	1.0 x 0.7	12.7	S pec	7	Bright star superimposed SE edge.
19 43 24.6	−48 32 08	ESO 232-19	13.8	1.0 x 0.5	13.0	S0	100	Mag 13.6 star N 0′.7.

(Continued from previous page)

GALAXIES

RA h m s	Dec ° ′ ″	Name	Mag (V)	Dim ′ Maj x min	SB	Type Class	PA	Notes
19 44 36.4	−51 36 09	ESO 232-21	13.0	1.8 x 0.3	12.2	S0$^+$: sp	61	Mag 9.45v star W 4'.8; stellar galaxy **Fair 873** S 3'.3; **ESO 232-22** SE 5'.7.
19 45 52.4	−50 17 42	ESO 232-23	14.3	1.2 x 1.0	14.3	SA(rs)cd III-IV	15	Stellar nucleus; mag 9.85v star NW 5'.9.
18 23 18.1	−43 08 37	ESO 280-13	13.1	3.0 x 0.4	13.1	SA(s)bc? sp	10	Mag 9.05v star W 2'.5.
18 29 44.9	−42 45 16	ESO 281-1	14.1	1.7 x 1.0	14.5	(R')SB(rs)ab	48	Numerous stars superimposed.
18 33 40.4	−46 45 12	ESO 281-8	12.8	1.1 x 0.5	12.1	S0 pec:	6	Mag 10.37v star N edge; **ESO 281-11** E 6'.0.
18 49 20.0	−43 59 50	ESO 281-28	13.5	1.5 x 0.6	13.3	S0$^+$: pec	129	Mag 8.73v star W 6'.5.
18 52 58.4	−42 32 18	ESO 281-33	12.9	1.7 x 0.5	12.5	Sb pec? sp	98	Mag 10.88v star N 4'.0.
18 56 20.9	−43 08 55	ESO 281-38	12.9	1.2 x 0.4	12.0	SAa:	20	Mag 9.85v star W 7'.6.
18 57 18.8	−47 03 36	ESO 281-40	15.5	1.1 x 0.2	13.7	Sbc	163	Mag 11.3 star SSE 3'.7.
19 05 55.3	−43 43 32	ESO 282-10	13.1	0.9 x 0.6	12.5	E/S0	123	Mag 12.5 star N edge.
19 06 35.6	−46 35 14	ESO 282-12	13.6	1.0 x 0.8	13.2	SBb:	49	Mag 13.8 star E edge.
19 09 06.0	−44 43 02	ESO 282-14	13.5	1.4 x 0.4	12.7	(R')SB(s)a	97	Located between a close pair of mag 12.8 and 13.2 s tars.
19 10 04.6	−44 40 19	ESO 282-17	13.9	0.7 x 0.6	12.8	Sa:	39	Mag 12.1 star SW 2'.9.
19 10 40.2	−44 10 48	ESO 282-18	13.4	1.2 x 0.5	12.7	(R')SB(s)cd	78	Strong dark patch W of center; mag 13.2 star E 2'.4.
19 11 58.2	−47 22 58	ESO 282-20	13.6	1.6 x 0.5	13.2	SAB(rs)c: IV-V	6	Mag 9.9v star NE edge.
19 12 47.9	−46 00 43	ESO 282-21	12.9	1.7 x 1.1	13.4	(R')SB(s)b pec I	118	Mag 9.86v star SSW 1'.8.
19 13 31.3	−47 03 43	ESO 282-24	12.3	1.6 x 1.2	13.0	E4	27	Mag 8.04v star S 4'.5; mag 11.6 star and **ESO 282-23** SW 4'.3.
19 14 14.5	−46 15 25	ESO 282-27	13.4	0.8 x 0.7	12.6	Sa:	41	Mag 12.4 star SW 3'.0.
19 14 34.7	−46 35 49	ESO 282-28	12.9	1.4 x 0.5	12.4	S0?	147	Mag 11.82v star on E edge; mag 9.77v star N 5'.1.
19 32 52.8	−45 13 42	ESO 283-4	13.7	1.5 x 0.4	13.1	SA(r)0/a	153	Mag 5.59v star ESE 5'.6.
19 42 15.6	−46 09 33	ESO 283-15	13.8	0.9 x 0.4	12.6	S0	122	Mag 14.7 star SE end.
19 51 20.2	−44 52 40	ESO 283-19	13.9	1.1 x 0.8	13.7	(R')SA(rs)0$^+$: pec	26	Almost stellar nucleus, smooth envelope; mag 15.1 star superimposed S edge; ESO 283-20 NNE 2'.4.
19 51 26.6	−44 50 36	ESO 283-20	13.3	1.5 x 0.7	13.5	SAB(s)0$^-$: pec	20	ESO 283-19 SSW 2'.4; mag 10.15v star SW 8'.9.
19 53 56.8	−45 54 31	ESO 284-2	14.9	1.6 x 0.2	13.6	Sc	28	Mag 10.66v star W 4'.4.
19 58 17.1	−46 06 49	ESO 284-4	13.4	1.6 x 0.8	13.5	SB(rs)bc II	11	Mag 7.49v star E 7'.4.
20 00 08.8	−46 21 27	ESO 284-7	13.9	1.5 x 0.3	12.8	Sa:	94	**ESO 284-6** W 1'.6.
20 01 21.0	−46 40 04	ESO 284-9	12.8	1.0 x 0.5	12.1	E5:	150	Mag 9.92v star NW 8'.4.
20 02 14.1	−45 27 00	ESO 284-11	13.8	1.2 x 1.0	13.8	S?	9	Mag 15.1 star E edge.
18 32 46.1	−41 30 41	ESO 336-3	12.4	2.3 x 1.2	13.4	SAB(rs)c: II	45	Mag 12.3 star superimposed SW of center; mag 10.54v star W 5'.7; ESO 336-4 NE 3'.1.
18 33 00.8	−41 29 09	ESO 336-4	13.2	1.5 x 0.6	13.0	SAB(s)a	13	Mag 11.9 star on N edge; mag 10.71v star on W edge; ESO 336-3 SW 3'.1.
18 40 41.1	−41 34 26	ESO 336-8	12.9	1.7 x 1.0	13.3	SA(r)0$^+$:	155	Mag 9.46v star WSW 8'.3.
18 43 14.7	−41 24 36	ESO 336-12	13.9	1.7 x 1.4	14.7	SAB(s)b II		Mag 11.1 star S 3'.4.
18 47 23.1	−41 43 06	ESO 336-16	13.4	2.3 x 1.0	14.1	SA(s)c:	161	Mag 10.06v star on S edge.
18 59 39.2	−41 45 47	ESO 337-6	12.9	1.5 x 0.8	13.0	SAB(rs)bc: I-II	29	Mag 9.99v star S 5'.3.
19 27 58.0	−41 34 35	ESO 338-4	13.0	0.5 x 0.3	10.9		73	Stellar with star on SW edge.
19 35 32.3	−40 01 27	ESO 338-12	14.0	0.8 x 0.6	13.1	Sb	8	Mag 10.11v star E 7'.1.
19 46 44.8	−40 21 19	ESO 339-1	13.4	1.1 x 0.6	12.8	Sb	57	Mag 12.8 star W edge.
19 50 56.6	−40 12 43	ESO 339-3	13.4	1.1 x 0.8	13.1	S0	6	Pair of mag 13 stars on SW edge.
19 58 30.8	−40 48 58	ESO 339-12	13.0	1.3 x 1.0	13.1	SAB(rs)cd	7	Mag 12.9 star on SW edge; mag 9.56v star SW 6'.9.
20 00 44.9	−41 57 40	ESO 339-21	13.8	0.6 x 0.5	12.4	S0?	81	Mag 11.04v star E 2'.3.
19 01 07.5	−45 18 53	IC 4808	12.3	1.9 x 0.8	12.6	SA(s)c II-III	45	Mag 10.43v star SE 5'.0.
19 36 21.5	−47 16 01	IC 4874	13.9	1.2 x 0.7	13.6	SAB(rs)bc: III-IV	60	Mag 14.8 star on SW end.
19 37 56.2	−51 59 33	IC 4877	13.7	1.1 x 0.3	12.4	SB0^0?	82	Mag 10.57v star ENE 8'.9.
19 43 14.6	−51 48 27	IC 4886	13.2	1.4 x 0.7	13.0	Sb: pec	113	Mag 10.00v star E 6'.9; **ESO 232-20**, **ESO 232-20A** and **ESO 232-20B** SE 8'.6.
19 46 59.2	−51 50 48	IC 4894	13.9	0.9 x 0.6	13.1	SB(rs)b pec	165	Faint stellar nucleus, SE-NW bar.
19 49 19.8	−51 52 08	IC 4897	14.6	0.7 x 0.4	13.1	SB(rs)b:	117	Mag 9.28v star NE 8'.6.
19 56 45.9	−50 03 22	IC 4909	13.8	0.9 x 0.4	12.5	SA(s)bc? II-III	83	Mag 8.82v star NW 8'.5.
19 57 42.1	−51 59 11	IC 4911	14.7	0.5 x 0.4	12.8	S0:	135	Mag 9.93v star E 2'.3; pair of mag 7.64v, 8.21v stars NE 6'.3.
19 58 19.2	−50 16 21	IC 4916	13.7	0.7 x 0.5	12.4	SA(s)b? pec II	7	Mag 9.31v star SW 9'.9.
19 05 59.5	−42 21 59	MCG −7-39-5	12.5	1.3 x 1.1	12.7	S0?	42	
19 11 25.7	−50 38 31	NGC 6754	12.1	1.9 x 0.9	12.5	SB(rs)b II	80	Bright anonymous galaxy NW 1'.1.
19 15 05.0	−50 39 24	NGC 6761	13.4	1.6 x 1.2	14.0	(R)SB(r)ab III	19	Bright nucleus in N-S bar.
19 16 32.6	−40 12 37	NGC 6768	12.2	1.2 x 1.1	12.5	E4	36	**MCG −7-39-9** SW 0'.9.
19 37 05.2	−42 17 50	NGC 6806	13.2	1.2 x 0.8	13.0	SAB(rs)c pec I-II	24	Stellar nucleus, patchy.
20 01 00.8	−47 04 08	NGC 6845A	13.0	2.2 x 1.0	13.7	SB(s)b: pec	72	Multi-galaxy system; long, bright extension NE with NGC 6845B at it's end.
20 01 05.5	−47 03 33	NGC 6845B	13.9	0.9 x 0.7	13.2	SBb? pec	6	Multi-galaxy system; narrow extension NE.
20 00 57.0	−47 05 07	NGC 6845C	15.4	0.8 x 0.3	13.7	S0$^+$: pec sp	138	Located on S edge of NGC 6845A.
20 00 53.6	−47 05 44	NGC 6845D	14.8	0.8 x 0.4	13.4	S0: pec sp	156	Located 0'.9 SW of NGC 6845C.
20 03 34.5	−48 17 05	NGC 6851	11.8	2.0 x 1.5	13.0	E:	160	

GALAXY CLUSTERS

RA h m s	Dec ° ′ ″	Name	Mag 10th brightest	No. Gal	Diam ′	Notes
18 27 30.0	−51 32 00	AS 801	15.4		22	**ESO 230-4** N of center 4'.5; all others anonymous, stellar.
19 00 42.0	−49 05 00	AS 808	16.1	17	17	See notes for ESO 231-14A; other members anonymous, stellar.

OPEN CLUSTERS

RA h m s	Dec ° ′ ″	Name	Mag	Diam ′	No. ★	B ★	Type	Notes
19 16 54.0	−51 29 00	ESO 231-30		20			cl	
18 40 00.0	−44 12 00	ESO 281-24		6			cl	
19 13 54.0	−42 37 00	ESO 282-26		20			cl	

PLANETARY NEBULAE

RA h m s	Dec ° ′ ″	Name	Diam ″	Mag (P)	Mag (V)	Mag cent ★	Alt Name	Notes
18 26 41.4	−40 29 51	PN G353.7−12.8	30	15.5			PK 353−12.1	Close pair of mag 11.3, 12.9 stars N 1'.3.

GALAXIES

RA h m s	Dec ° ′ ″	Name	Mag (V)	Dim ′ Maj x min	SB	Type Class	PA	Notes
18 17 58.9	−42 46 15	ESO 280-7	13.2	1.5 x 0.7	13.1	SA0⁻:	126	Mag 12.4 star on E edge; mag 10.73v star SW 2′.8.
18 23 18.1	−43 08 37	ESO 280-13	13.1	3.0 x 0.4	13.1	SA(s)bc? sp	10	Mag 9.05v star W 2′.5.

GALAXY CLUSTERS

RA h m s	Dec ° ′ ″	Name	Mag 10th brightest	No. Gal	Diam ′	Notes
18 27 30.0	−51 32 00	AS 801	15.4		22	**ESO 230-4** N of center 4′.5; all others anonymous, stellar.

OPEN CLUSTERS

RA h m s	Dec ° ′ ″	Name	Mag	Diam ′	No. ★	B ★	Type	Notes
16 55 30.0	−40 49 00	Cr 316	3.4	105		14.0	cl	Moderately rich in stars; moderate brightness range; strong central concentration; detached. (A) Large group of bright stars, superposed on Tr 24. May just be brighter stars of Tr 24?
18 09 06.0	−46 25 00	ESO 280-6		1.5			cl	
17 07 30.0	−40 49 00	ESO 332-22		7			cl	
18 16 12.0	−41 02 00	ESO 335-5		9.3	14		ast?	(A) Sparse 9–10′ diameter group.
17 03 57.0	−48 07 00	Ha 13		15	70	12.0	cl	Few stars; small brightness range; not well detached. (A) Irregular in shape, stars faint and difficult to identify from Milky Way background on DSS image.
16 44 24.0	−47 33 00	Ho 20		4			cl	
16 45 36.9	−47 44 08	Ho 21		4	12		cl?	(A) E-W elongated group of 12 or so stars.
16 46 35.0	−47 04 57	Ho 22	6.7	1.2	10	7.3	cl	Few stars; large brightness range; not well detached.
17 24 42.0	−49 55 00	IC 4651	6.9	10	102	10.0	cl	Rich in stars; moderate brightness range; slight central concentration; detached.
16 46 03.0	−50 46 00	Lynga 12		6	35		cl	
16 48 54.0	−43 26 00	Lynga 13		9	30		cl	
16 55 05.0	−45 14 48	Lynga 14	9.7	3	15	11.2	cl	Few stars; small brightness range; slight central concentration; detached.
16 44 08.6	−47 28 00	NGC 6200	7.4	15	40	9.2	cl	Moderately rich in stars; moderate brightness range; slight central concentration; detached.
16 46 07.9	−47 00 44	NGC 6204	8.2	6	45	9.6	cl	Moderately rich in bright and faint stars with a strong central concentration; detached.
16 49 23.6	−44 43 39	NGC 6216	10.1	4	40	12.0	cl	Moderately rich in stars; moderate brightness range; strong central concentration; detached.
16 54 09.8	−41 50 00	NGC 6231	2.6	14	93	6.0	cl	Few stars; large brightness range; strong central concentration; detached.
16 57 41.7	−44 48 17	NGC 6249	8.2	6	30	9.8	cl	Moderately rich in stars; moderate brightness range; slight central concentration; detached.
16 57 55.0	−45 56 00	NGC 6250	5.9	16	35	7.6	cl	Rich in stars; large brightness range; slight central concentration; detached. (A) 6 brighter stars and ~30 fainter stars, standing out on NE to SW trending dark nebulosity.
17 00 42.0	−44 39 00	NGC 6259	8.0	15	162	11.0	cl	Rich in stars; moderate brightness range; slight central concentration; detached.
17 18 30.0	−42 56 00	NGC 6322	6.0	5	38	7.5	cl	Moderately rich in bright and faint stars with a strong central concentration; detached.
16 55 08.3	−40 56 35	Ru 122		3			cl	
17 27 55.9	−40 43 12	Ru 124		2	15	12.0	cl	
16 57 00.0	−40 38 00	Tr 24	8.6	60			cl	Few stars; moderate brightness range; not well detached; involved in nebulosity. (A) Large group of stars, with IC 4628 nebulosity involved on N side.
17 41 36.4	−40 07 00	Tr 29	7.5	12	30		cl	Moderately rich in stars; moderate brightness range; no central concentration; detached.
16 46 30.3	−45 51 36	vdB-Ha 197		4	15		cl	(A) Slight condensation in Milky Way star cloud, which contains We 1 to SE. Strong dark nebula to W of cloud.
16 50 00.0	−44 12 00	vdB-Ha 200		4	20		cl	
16 56 11.3	−40 40 06	vdB-Ha 205		4	14	6.7	cl	
17 02 07.9	−41 06 35	vdB-Ha 211		4			cl	(A) Brighter stars of long E-W chain.
17 16 20.4	−40 49 50	vdB-Ha 217		5	30		cl	
16 46 43.0	−45 57 00	We 1		9	20	14.5	cl	(A) Faint heavily reddened cluster.

GLOBULAR CLUSTERS

RA h m s	Dec ° ′ ″	Name	Total V m	B ★ V m	HB V m	Diam ′	Conc. Class Low = 12 High = 1	Notes
17 25 29.2	−48 25 22	NGC 6352	7.8	13.4	15.2	9.0	11	
17 36 17.0	−44 44 06	NGC 6388	6.8	14.8	17.2	10.4	3	
17 59 02.0	−44 15 54	NGC 6496	8.6	14.3	16.5	5.6	12	
18 08 02.2	−43 42 20	NGC 6541	6.3	12.1	15.3	15.0	3	

BRIGHT NEBULAE

RA h m s	Dec ° ′ ″	Name	Dim ′ Maj x min	Type	BC	Color	Notes
16 57 00.0	−40 20 00	IC 4628	90 x 60	E	2-5	2-4	= **Gum 56**. Elongated and very irregular in shape.
17 04 00.0	−51 05 00	vdBH 81	6 x 4	R	2-5	1-4	Shaped like a slight gibbous moon, convex to the N. The involved star is 9th mag and is situated near the S edge.

DARK NEBULAE

RA h m s	Dec ° ′ ″	Name	Dim ′ Maj x min	Opacity	Notes
16 44 45.0	−40 20 00	B 44a	5 x 5	5	Irregular; sharpest on SE side.
17 01 42.0	−40 41 00	B 48	40 x 15	5	Fairly well defined NE and SW.
17 11 12.2	−40 25 00	B 58	30 x 20	6	Slightly extended N and S; a darker core at NE side.
16 47 06.4	−44 29 00	B 235	7 x 7	6	Elongated NE and SW; darkest part at SW end.
17 26 56.4	−42 47 00	B 263	30 x 30	5	Very slightly elongated.
16 53 00.0	−43 35 00	SL 17	40 x 5	5	

PLANETARY NEBULAE

RA h m s	Dec ° ¢ ¢¢	Name	Diam ¢¢	Mag (P)	Mag (V)	Mag cent	Alt Name	Notes
17 45 35.5	−46 05 24	IC 1266	10	12.1		11.3	PK 345−8.1	= **PN G345.2-8.8**; mag 11.18v star S 2i.5.
17 05 10.5	−40 53 09	IC 4637	22	13.6	12.5	12.5	PK 345+0.1	Forms a triangle with a pair of mag 11.37v and 11.8 stars S 2′.4.
17 45 28.8	−44 54 18	IC 4663	20	13.1	12.5	15.2	PK 346−8.1	Mag 11.2 star SE 1′.4.
18 18 32.1	−45 59 02	IC 4699	14	11.9	13.0	15.1	PK 348−13.1	Located between a close pair of mag 12.3 stars S and an 11.8 star N.
17 20 46.7	−51 45 18	NGC 6326	19	12.2		16.7	PK 338−8.1	Mag 12.5 star E 0′.5; in star rich area.
16 45 00.0	−51 12 18	PN G335.2−3.6	23	16.6		17.0	PK 335−3.1	Mag 13.4 star E 0′.5; mag 11.16v star NW 2′.6.

(Continued from previous page)

181 PLANETARY NEBULAE 181

RA h m s	Dec ° ′ ″	Name	Diam ″	Mag (P)	Mag (V)	Mag cent	Alt Name	Notes
16 48 40.1	−51 09 13	PN G335.6−4.0	24			20.3	PK 335−4.1	Mag 13.5 star on S edge.
16 47 57.3	−50 42 32	PN G335.9−3.6	9	18.4		21.1	PK 335−3.2	Close pair of mag 12.9, 13.6 stars S 0′.6.
16 59 36.1	−51 42 04	PN G336.3−5.6	3	13.4		16.6	PK 336−5.1	Mag 12.9 star NNW 1′.2; mag 11.29v star SE 2′.6.
17 01 37.4	−50 22 56	PN G337.5−5.1	6	15.3		12.7	PK 337−5.1	Mag 14.4 star SE 0′.7.
16 57 29.0	−49 46 54	PN G337.6−4.2	17	16.6		21.9	PK 337−4.1	Mag 10.23v star SW 2′.0.
17 11 27.3	−47 25 00	PN G340.9−4.6		16.7			PK 340−4.1	Mag 11.9 star SE 1′.9; mag 12.0 star N 2′.5.
17 36 07.5	−49 25 35	PN G341.5−9.1	5	13.9	15.5	15.4	PK 341−9.1	Mag 12.9 star W 1′.8; mag 10.33v star E 1′.9.
18 07 15.9	−51 01 10	PN G342.5−14.3	68	11.9		12.5	PK 342−14.1	Elongated ESE-WNW; mag 12.4 star SSE 0′.8.
16 53 31.3	−42 39 24	PN G342.7+0.7	16	16.2			PK 342+0.1	Mag 9.96v star on NW edge.
17 27 48.3	−46 55 34	PN G342.8−6.6	8	12.8		16.2	PK 342−6.1	Mag 12.8 star on SE edge; pair of mag 8.15v, 8.88v stars S 1′.5.
17 06 22.7	−44 13 12	PN G342.9−2.0	19	15.0	14.0		PK 342−2.1	Mag 10.44v star WSW 4′.8.
17 19 32.4	−45 53 09	PN G342.9−4.9	35	12.6	12.:		PK 342−4.1	Mag 12.1 star W 2′.6.
17 05 39.1	−43 57 27	PN G343.0−1.7		18.5			PK 343−1.1	Mag 10.4 star WSW 1′.9; mag 11.2 star NE 2′.4.
17 01 28.0	−43 05 55	PN G343.3−0.6	120			16.4	PK 343−0.1	Mag 11.7 star NE 2′.8.
17 35 41.1	−46 59 51	PN G343.5−7.8	5	13.4	14.1	14.6	PK 343−7.1	Mag 12.7 star S 2′.1.
16 44 23.6	−40 03 19	PN G343.6+3.7	20	18.4		20.5	PK 343+3.1	Very close pair of mag 13 stars N 1′.1; mag 13.1 star S 1′.2.
16 57 23.7	−41 37 55	PN G343.9+0.8	5	17.9			PK 344+0.1	Mag 12.9 star NNW 0′.5; mag 12.1 star NNE 1′.2.
17 06 59.5	−42 41 15	PN G344.2−1.2	12	16.7			PK 344−1.1	Mag 12.9 star NNE 1′.1.
16 50 33.1	−40 03 00	PN G344.4+2.8		16.8	14.8		PK 344+2.1	Mag 12.4 star SSE 2′.9.
17 30 03.9	−45 22 49	PN G344.4−6.1		17.6			PK 344−6.1	Close pair of mag 10.6, 11.8 stars WNW 1′.9.
17 26 11.8	−44 11 29	PN G345.0−4.9	5	11.9		14.3	PK 345−4.1	Mag 11.3 star E 1′.9.
17 10 27.4	−41 52 48	PN G345.2−1.2	9	13.4	13.5		PK 345−1.1	Mag 12.1 star SSE 1′.4; mag 12.5 star ESE 1′.7.
17 52 47.1	−46 41 53	PN G345.3−10.2	55			17.3	PK 345−10.1	Mag 11.9 star on NE edge; mag 10.17v star N 2′.9.
17 59 36.5	−46 38 51	PN G345.9−11.2	24				PK 345−11.1	Mag 12.4 star E 0′.5.
17 39 19.9	−44 09 37	PN G346.3−6.8	5	12.6			PK 346−6.1	Mag 10.7 star SW 1′.8; mag 10.71v star SSE 1′.7.
17 32 48.2	−40 58 24	PN G348.4−4.1		15.1			PK 348−4.1	Mag 9.51v star NNE 3′.6.
17 56 33.1	−43 03 16	PN G348.8−9.0	3		14.0	16.8	PK 348−9.1	Mag 13.2 star ENE 1′.9.
17 35 39.8	−40 11 59	PN G349.3−4.2	83	13.3		16.8	PK 349−4.1	Mag 12.0 star NNE 1′.3; mag 11.0 star E 1′.9.
18 08 57.8	−41 48 24	PN G351.0−10.4	162				PK 351−10.2	Mag 12.6 star 0′.8 E of center; mag 12.8 star 1′.0 W of center.
18 12 53.0	−41 30 28	PN G351.7−10.9	8	14.8		16.3	PK 351−10.1	Mag 11.9 start SSW 1′.3; mag 10.8 star NNE 2′.5.
18 26 41.4	−40 29 51	PN G353.7−12.8	30	15.5			PK 353−12.1	Close pair of mag 11.3, 12.9 stars N 1′.3.

182 GALAXIES 182

RA h m s	Dec ° ′ ″	Name	Mag (V)	Dim ′ Maj x min	SB	Type Class	PA	Notes
15 14 13.2	−46 48 45	ESO 274-1	10.8	10.5 x 1.5	13.6	SAd: sp	36	Numerous stars superimposed, in star rich area.
15 16 09.9	−44 00 56	ESO 274-6	13.2	1.0 x 0.6	12.7	E3	173	Several stars superimposed, in star rich area.
15 28 55.4	−42 47 02	ESO 274-16	14.5	1.3 x 0.5	13.9	IB(s)m V-VI	43	Mag 10.05v star SW 10′.4.
15 38 15.5	−44 24 41	ESO 274-19	14.2	1.1 x 0.8	14.0	SA(s)cd III-IV	108	Mag 7.59v star WNW 11′.9.
15 12 28.2	−42 01 03	ESO 328-31	15.7	1.2 x 0.2	14.0	Sbc	146	Mag 9.26v star N 7′.4.
15 19 18.2	−41 13 50	ESO 328-43	13.3	2.5 x 1.9	14.8	(R′)SB(s)dm IV-V	7	Pair of mag 8.19v, 8.72v stars S 5′.1.
15 44 04.9	−41 14 50	ESO 329-16	13.6	1.7 x 0.8	13.8	SB(rs)bc:	115	Mag 10.23v star NNW 4′.8.
15 49 37.0	−40 19 36	ESO 329-22	15.3	1.2 x 0.8	15.1	SB(s)cd: II	48	Mag 9.14v star N 5′.6.

OPEN CLUSTERS

RA h m s	Dec ° ′ ″	Name	Mag	Diam ′	No. ★	B ★	Type	Notes
16 35 18.0	−50 57 00	Cr 307	9.2	5	12	11.1	cl	Few stars; moderate brightness range; no central concentration; detached. Doubtful cluster.
16 24 12.0	−51 09 00	ESO 226-6		4			cl	
15 27 30.0	−41 03 00	ESO 329-2		8			cl	
16 28 51.7	−49 07 25	Ho 19		4			cl	
16 44 24.0	−47 33 00	Ho 20		4			cl	
16 45 36.9	−47 44 08	Ho 21		4	12		cl?	(A) E-W elongated group of 12 or so stars.
16 46 35.0	−47 04 57	Ho 22	6.7	1.2	10	7.3	cl	Few stars; large brightness range; not well detached.
16 04 51.0	−51 57 18	Lynga 6	9.5	5	60	10.7	cl	(A) Includes arc of 7 stars.
16 22 57.0	−50 10 55	Lynga 8		1			cl	
16 20 39.0	−48 31 24	Lynga 9		5			cl	
16 38 10.0	−46 19 06	Lynga 11		4	15		cl	
16 46 03.0	−50 46 00	Lynga 12		6	35		cl	
16 48 54.0	−43 26 00	Lynga 13		9	30		cl	
16 24 26.8	−51 57 12	NGC 6115	9.8	3.4	20	11.0	cl?	
16 25 17.2	−40 40 00	NGC 6124	5.8	40	100	9.0	cl	Rich in stars; large brightness range; strong central concentration; detached.
16 27 43.6	−49 09 51	NGC 6134	7.2	6	179	11.0	cl	Moderately rich in bright and faint stars; detached.
16 34 35.0	−49 46 00	NGC 6167	6.7	7	218	7.4	cl	Moderately rich in bright and faint stars; detached. Probably not a cluster.
16 34 05.0	−44 03 00	NGC 6169	6.6	12	40		cl	Moderately rich in stars; small brightness range; no central concentration; detached.
16 35 50.8	−45 38 48	NGC 6178	7.2	5	19	8.4	cl	Few stars; large brightness range; no central concentration; detached.
16 40 20.8	−43 22 00	NGC 6192	8.5	9	60	11.0	cl	Rich in stars; moderate brightness range; strong central concentration; detached.
16 41 20.3	−48 46 00	NGC 6193	5.2	14	14	5.7	cl	Few stars; large brightness range; slight central concentration; detached. (A) Substantial bright and dark nebula on W side.
16 44 08.6	−47 28 00	NGC 6200	7.4	15	40	9.2	cl	Moderately rich in stars; moderate brightness range; slight central concentration; detached.
16 46 07.9	−47 00 44	NGC 6204	8.2	6	45	9.6	cl	Moderately rich in bright and faint stars with a strong central concentration; detached.
16 49 23.6	−44 43 39	NGC 6216	10.1	4	40	12.0	cl	Moderately rich in stars; moderate brightness range; strong central concentration; detached.
16 13 11.2	−51 54 10	Pi 22		4	30	7.8	cl	
16 23 58.0	−48 53 30	Pi 23		1		15.0	cl	
16 23 24.3	−51 59 48	Ru 116		5		9.0	cl	Few stars; moderate brightness range; slight central concentration; detached.
16 23 30.7	−51 52 06	Ru 117		1.7		12.0	cl	(A) Very weak condensation of stars on DSS image.
16 28 09.3	−51 31 00	Ru 119	8.8	8	13	10.0	cl	Few stars; small brightness range; slight central concentration; detached. Probably not a cluster. (A) Weak condensation of stars on DSS image.
16 35 12.2	−48 17 13	Ru 120		3.4	11	12.0	cl?	(A) "Y"-shaped group, with top to W.
16 41 47.6	−46 09 13	Ru 121		6	35	13.0	cl	(A) Weak condensation of stars, with similar, but larger condensation of Milky Way to NW.
16 46 30.3	−45 51 36	vdB-Ha 197		4	15		cl	(A) Slight condensation in Milky Way star cloud, which contains We 1 to SE. Strong dark nebula to W of cloud.
16 50 00.0	−44 12 00	vdB-Ha 200		4	20		cl	
16 46 43.0	−45 57 00	We 1		9	20	14.5	cl	(A) Faint heavily reddened cluster.

(Continued from previous page)

182 GLOBULAR CLUSTERS 182

RA h m s	Dec ° ′ ″	Name	Total V m	B ★ V m	HB V m	Diam ′	Conc. Class Low = 12 High = 1	Notes
15 28 00.5	−50 40 22	NGC 5927	8.0	14.5	16.6	6.0	8	
15 35 28.5	−50 39 34	NGC 5946	8.4		17.2	3.0	9	
15 39 07.3	−50 03 02	vdB-Ha 176	14.0			3.0		

BRIGHT NEBULAE

RA h m s	Dec ° ′ ″	Name	Dim ′ Maj x min	Type	BC	Color	Notes
16 33 54.0	−48 07 00	NGC 6164/65	6 x 3	E	2-5		A shell surrounding an 8th mag Wolf-Rayet star; NGC 6164 is the NW condensation, NGC 6165 the SE one.
16 40 30.0	−48 47 00	NGC 6188	20 x 12	E+R	2-5	1-4	Very irregular and mottled. The whole region is littered with dark and bright nebulosities, and it is brightest in the region around and to the W of the open cluster NGC 6193.
16 17 48.0	−51 55 00	RCW 102	12 x 8	E	3-5		Faint irregular nebula, somewhat extended NE-SW; in an extremely rich star field.
16 17 06.0	−51 07 00	RCW 103	7 x 7	SNR	2-5		This supernova remnant lies in a field rich in faint stars and consists of two segments of a broken ring. The larger and brightest segment about 5′ long and 2′ broad forms the SE part while the NW portion is about half as large.

DARK NEBULAE

RA h m s	Dec ° ′ ″	Name	Dim ′ Maj x min	Opacity	Notes
16 44 45.0	−40 20 00	B 44a	5 x 5	5	Irregular; sharpest on SE side.
16 47 06.4	−44 29 00	B 235	7 x 7	6	Elongated NE and SW; darkest part at SW end.
16 01 48.0	−41 52 00	SL 7	60 x 10	6	
16 14 12.0	−44 04 00	SL 8	25 x 5	6	

PLANETARY NEBULAE

RA h m s	Dec ° ′ ″	Name	Diam ″	Mag (P)	Mag (V)	Mag cent ★	Alt Name	Notes
16 19 23.0	−42 15 35	IC 4599	15	12.4		16.3	PK 338+5.1	= **PN G338.8+5.6**; mag 10.35v star ESE 2′.1.
15 16 50.0	−45 39 00	NGC 5882	20	10.5	9.4	13.4	PK 327+10.1	Mag 11.02v star NE 4′.4.
16 31 30.6	−40 15 12	NGC 6153	24	11.5	10.9	16.1	PK 341+5.1	Mag 8.27v star NE 2′.5; mag 9.84v star N 2′.6 and mag 11.32v star NW 2′.6.
15 25 08.1	−51 19 42	PN G325.8+4.5	5	14.4		15.9	PK 325+4.1	Close pair of mag 11-12 stars NNW 0′.7; mag 8.49v star NE 2′.5; mag 6.96v star E 4′.4.
15 51 41.2	−51 31 23	PN G329.0+1.9	72	13.6	12.6	14.0	PK 329+2.1	Annular; located on the edge of a small triangle of mag 11-12 stars.
15 55 08.7	−51 23 46	PN G329.5+1.7	100			13.5	Wray 16−191	Mag 9.56v star S 4′.0.
15 42 13.3	−47 40 45	PN G330.2+5.9	107			19.5	PK 330+5.1	Mag 12.9 star S 1′.6; mag 12.2 star NNE 3′.0.
15 51 16.2	−48 45 00	PN G330.7+4.1	1	12.9		11.1	PK 330+4.1	Mag 9.77v star N 2′.6.
15 51 19.8	−48 26 03	PN G330.9+4.3	15	14.9		17.7	PK 330+4.2	Mag 13.0 star ENE 1′.2.
16 08 59.3	−51 02 01	PN G331.4+0.5	11	14.9			PK 331+0.1	Mag 11.5 star N 3′.2.
16 17 13.4	−51 59 15	PN G331.7−1.0	25			17.6	PK 331−1.1	**Ant Nebula**. Mag 12.4 star S 0′.7.
16 00 22.1	−48 15 37	PN G332.2+3.5	8				PK 332+3.1	Mag 13.1 star W 1′.6; mag 12.5 star N 2′.0.
16 15 20.2	−49 13 25	PN G333.4+1.1	11	13.5			PK 333+1.1	Mag 10.79v star SW 0′.6; mag 11.1 star N 1′.1.
16 31 06.7	−50 27 12	PN G334.3−1.4	17				PK 334−1.1	Mag 10.01v star NE 1′.4.
16 45 00.0	−51 12 18	PN G335.2−3.6	23	16.6		17.0	PK 335−3.1	Mag 13.4 star E 0′.5; mag 11.16v star NW 2′.6.
16 34 14.8	−49 21 19	PN G335.4−1.1	22	17.3	17.4		PK 335−1.1	Mag 11.7 star NW 2′.4.
15 53 12.7	−41 50 25	PN G335.4+9.2	29	17.1		21.0	PK 335+9.1	Mag 9.34v star SW 3′.3.
16 48 40.1	−51 09 13	PN G335.6−4.0	24			20.3	PK 335−4.1	Mag 13.5 star on S edge.
16 47 57.3	−50 42 32	PN G335.9−3.6	9	18.4		21.1	PK 335−3.2	Close pair of mag 12.9, 13.6 stars S 0′.6.
16 23 54.5	−46 42 16	PN G336.2+1.9	12	16.3	16.1		PK 336+1.1	Mag 12.5 star SSE 0′.6.
16 02 13.1	−41 33 38	PN G336.9+8.3		15.2		15.4	PK 336+8.1	Stellar; mag 12.7 star W 1′.5.
16 30 25.8	−46 02 54	PN G337.4+1.6	5	17.6		16.7	PK 337+1.1	Mag 12.9 star NW 3′.3.
16 44 23.6	−40 03 19	PN G343.6+3.7	20	18.4		20.5	PK 343+3.1	Very close pair of mag 13 stars N 1′.1; mag 13.1 star S 1′.2.
16 50 33.1	−40 03 00	PN G344.4+2.8		16.8	14.8		PK 344+2.1	Mag 12.4 star SSE 2′.9.

183 GALAXIES 183

RA h m s	Dec ° ′ ″	Name	Mag (V)	Dim ′ Maj x min	SB	Type Class	PA	Notes
13 33 57.4	−50 29 34	ESO 220-19	13.6	1.5 x 0.5	13.1	SBd: pec III-IV	149	Mag 12.2 star N 0′.8; mag 10.93v star N 3′.2.
13 36 50.0	−49 59 03	ESO 220-20	13.9	1.3 x 0.3	12.7	Pec	59	Mag 12.6 star NW 0′.9; ESO 220-21 E 2′.7.
13 37 05.8	−49 59 47	ESO 220-21	14.1	1.3 x 0.4	13.3	S	91	ESO 220-20 W 2′.7; mag 9.92v star SE 2′.3.
13 37 14.1	−49 47 24	ESO 220-22	14.2	1.2 x 0.6	13.7	SB(s)m? V	37	Located between mag 10.44v and 10.38v stars; star N edge of nucleus; ESO 220-23 N 2′.6.
13 37 18.7	−49 44 57	ESO 220-23	13.4	1.6 x 0.3	12.5	S0 pec sp	29	ESO 220-22 S 2′.5; mag 10.44v star SSE 1′.9.
13 39 46.1	−48 17 58	ESO 220-26	13.3	2.0 x 2.0	14.7	SB(s)d IV		Many stars superimposed; mag 7.55v star S 8′.1.
13 40 13.1	−51 08 34	ESO 220-28	13.7	2.0 x 0.3	13.0	Sbc sp	81	Mag 8.37v star SW 2′.8.
13 42 00.4	−48 18 26	ESO 220-32	13.7	1.4 x 0.7	13.5	SB?	170	Many superimposed stars; mag 10.33v star E 1′.9.
13 45 42.1	−50 27 43	ESO 220-37	13.9	1.1 x 0.9	13.7	Sc	87	Several stars superimposed, in star rich area.
13 48 18.9	−50 25 38	ESO 221-1	13.9	0.7 x 0.5	12.6	S	91	Mag 12.4 star NE 2′.6.
13 49 08.8	−48 37 49	ESO 221-2	13.8	2.0 x 0.6	13.9	SB(s)dm IV-V	135	Bright knot E of nucleus; mag 10.14v star S 3′.2.
13 49 46.7	−48 45 11	ESO 221-3	13.4	0.9 x 0.5	12.4	S	66	Mag 12.7 star SW 2′.5; mag 10.60v star SW 4′.2.
13 49 52.0	−48 34 40	ESO 221-4	13.4	1.4 x 0.9	13.5	Sdm:	161	Mag 10.97v star E edge; mag 12.1 star S edge; mag 9.13v star NNE 4′.8.
13 50 18.5	−48 54 16	ESO 221-5	13.6	1.1 x 0.2	11.8	Pec:	179	ESO 221-9 SE 5′.0; mag 9.34v star SW 5′.4.
13 50 22.0	−48 22 36	ESO 221-6	12.4	2.0 x 2.0	13.7	SA(r)cd III-IV		Almost stellar nucleus; numerous stars superimposed; mag 8.83v star W 2′.9.
13 50 27.4	−48 16 41	ESO 221-8	13.1	1.1 x 0.6	12.5	Pec	132	Double system; mag 8.09v star W 1′.2.
13 50 44.7	−48 56 52	ESO 221-9	13.4	1.0 x 0.9	13.1	Sb	164	Mag 13.4 star on NE edge; ESO 221-5 NW 5′.0; mag 9.34v star W 7′.8.
13 50 57.0	−49 03 19	ESO 221-10	12.4	1.3 x 1.1	12.6		14	
13 51 32.3	−48 04 57	ESO 221-12	13.4	1.5 x 0.5	12.9	SBm? pec	164	ESO 221-14 SE 7′.8; ESO 221-13 N 3′.4.
13 51 35.1	−48 01 35	ESO 221-13	12.7	1.6 x 1.1	13.2	(R)SB(r)0/a	103	
13 52 07.1	−48 10 13	ESO 221-14	12.5	1.7 x 1.3	13.2	SB(s)c II-III	42	Close double star, mag 7.34v, 7.43v SW 9′.3.
13 56 44.9	−48 29 50	ESO 221-18	14.0	1.0 x 0.8	13.6	Sa? pec	133	Numerous stars superimposed; mag 12.2 star SW 1′.0.
13 58 23.0	−48 28 34	ESO 221-20	12.1	1.2 x 0.7	11.8	SA0$^-$: pec	106	Mag 7.15v star NE 1′.1.
13 58 34.1	−48 31 37	ESO 221-21	13.8	1.3 x 0.4	12.9	Sdm:	171	Mag 9.67v star SW 2′.8.
14 00 11.5	−48 16 18	ESO 221-22	13.3	2.7 x 0.4	13.2	SBd? sp	33	Mag 10.26v star NNW 4′.3.
14 07 34.9	−48 23 39	ESO 221-25	13.7	1.3 x 0.6	13.3	Sbc	41	Mag 9.30 star N edge; mag 9.45v star SW 2′.3.
14 08 23.5	−47 58 13	ESO 221-26	11.0	2.8 x 1.8	12.8	E5 pec:	0	Numerous superimposed stars; mag 10.37v star SSW 8′.3.
14 08 35.0	−49 14 06	ESO 221-27	12.9	1.3 x 0.4	12.1	Sa:	35	Mag 9.45v star N 5′.1.
14 09 02.3	−51 10 12	ESO 221-28	14.5	0.7 x 0.7	13.6	S		Mag 9.68v star NW 7′.7.

(Continued from previous page)

GALAXIES

RA h m s	Dec ° ′ ″	Name	Mag (V)	Dim ′ Maj x min	SB	Type Class	PA	Notes
14 12 09.6	−49 23 23	ESO 221-32	11.9	1.8 x 1.3	12.7	SA(r)c pec II	179	Mag 10.4v star SE edge.
14 18 10.7	−48 00 38	ESO 221-37	12.2	1.4 x 1.3	12.8	(R′)SA(s)0$^+$ pec	21	Mag 11.3 star NE 4′.2.
14 18 48.9	−47 44 18	ESO 222-1	13.2	1.7 x 0.6	13.1	SB(s)m VI	155	Mag 11.00v star N 7′.3.
14 23 40.6	−49 39 09	ESO 222-4	13.5	2.4 x 1.7	14.9	SB(s)m V-VI	114	
14 44 26.7	−49 24 09	ESO 222-15	13.4	2.1 x 0.5	13.3	Scd?	35	Mag 10.37v star superimposed center.
14 52 24.2	−49 41 25	ESO 223-2	15.3	1.0 x 0.2	13.4	S	30	Mag 10.52v star S 5′.2.
14 59 19.4	−49 04 32	ESO 223-7	13.7	1.6 x 1.0	14.1	SAB(s)dm: sp IV-V		Mag 10.35v star S 6′.5.
15 01 08.3	−48 17 33	ESO 223-9	13.2	2.5 x 2.0	14.8	IAB(s)m V-VI	135	Numerous superimposed stars; mag 10.42v star NW 6′.4; planetary nebula **ESO 223-10** SE 6′.4.
13 34 48.2	−45 33 01	ESO 270-17	10.7	12.3 x 1.5	13.7	SB(s)m: IV-V	110	**Fourcade-Figueroa Galaxy**.
13 37 18.9	−46 12 57	ESO 270-21	14.5	1.1 x 0.4	13.5	Im? pec	123	Mag 8.40v star ENE 4′.8.
13 43 12.7	−44 51 37	ESO 270-26	14.3	1.2 x 1.1	14.5	SB(s)cd: pec II-III	14	Bright knot, or galaxy, NE of center is designated **ESO 270-26A**.
13 44 34.6	−47 10 03	ESO 270-28	13.7	1.3 x 0.5	13.1	Sb? pec sp	178	Mag 7.15v star S 3′.1.
13 55 22.2	−44 28 42	ESO 271-4	13.9	1.3 x 0.3	12.7	SB(s)dm sp IV	163	Mag 12.6 star on S edge; mag 9.14vb star W 2′.6.
13 56 09.0	−45 39 31	ESO 271-5	14.7	0.9 x 0.6	13.9	Sm? IV-V?	165	
14 00 46.2	−45 25 06	ESO 271-10	12.2	2.3 x 1.9	13.6	SAB(s)cd III-IV	43	Numerous stars superimposed; mag 9.25v star N 4′.2.
14 09 27.4	−42 44 19	ESO 271-17	13.5	0.9 x 0.8	13.0		178	Mag 13.7 star NW edge; pair of mag 13 stars N 1′.9.
14 10 03.8	−46 13 25	ESO 271-18	14.3	1.1 x 0.4	13.3	SBm? IV-V?	174	Mag 8.19v star S 2′.8.
14 10 29.1	−42 46 59	ESO 271-20	14.0	1.0 x 0.7	13.5	S	61	Several stars superimposed NE half.
14 13 29.6	−45 24 47	ESO 271-22	13.1	2.7 x 0.5	13.2	SB(s)cd: sp II-III	135	Mag 9.49v star SW 13′.5.
14 15 46.0	−47 38 29	ESO 271-26	13.4	1.3 x 0.7	13.2	(R′)SB(rs)0$^+$ pec?	106	Three mag 14 stars superimposed; mag 11.4 star NNW 2′.1.
14 16 41.5	−42 50 01	ESO 271-27	13.6	1.7 x 0.4	13.0	SAB(s)b? II-III	150	Mag 8.55v star ESE 4′.5.
14 17 20.6	−45 12 08	ESO 272-1	13.9	1.3 x 0.3	12.7	S	17	Mag 11.5 star SE 0′.7.
14 19 11.8	−45 19 04	ESO 272-4	14.2	1.0 x 0.4	13.0	SB(s)0/a? pec	111	Mag 12 star E end.
14 26 55.4	−45 52 39	ESO 272-9	14.6	1.2 x 1.0	14.6	SB(s)dm V-VI	13	Low surface brightness; located between two stars, mag 12.2 N and mag.
14 41 55.6	−43 37 11	ESO 272-24	13.7	1.1 x 0.5	12.9	SBa: pec	35	Mag 12.0 star NE 1′.9.
14 43 25.5	−44 42 21	ESO 272-25	13.9	1.2 x 0.6	13.4		61	Double system; mag 8.97v star E 11′.1.
14 48 42.4	−43 55 53	ESO 273-4	13.1	1.6 x 0.8	13.2	S pec	56	Several stars superimposed; mag 10.85v star W 11′.0.
14 58 25.5	−47 41 58	ESO 273-14	12.3	3.8 x 2.8	14.7	IB(s)m pec: IV-V	114	This object has two entries in the ESO catalog. There seems to be some doubt if it is an irregular galaxy; or, a galaxy plus nebula?
15 14 13.2	−46 48 45	ESO 274-1	10.8	10.5 x 1.5	13.6	SAd: sp	36	Numerous stars superimposed, in star rich area.
13 33 17.9	−40 29 47	ESO 324-35	14.3	1.0 x 0.7	13.8	SB	171	Stars superimposed on edges; mag 9.28v star W 2′.8.
13 35 23.0	−40 04 19	ESO 324-40	13.8	0.9 x 0.5	12.8	S0	141	Small group of three mag 12-13 stars N 1′.4; stellar **ESO 324-41** NE 2′.4.
13 35 40.3	−41 37 25	ESO 324-42	13.9	1.0 x 0.5	13.0	Sab	127	Mag 14.9 star S edge, in star rich area.
13 45 00.8	−41 51 33	ESO 325-11	13.6	2.5 x 1.2	14.6	IB(s)m	128	Mag 10.13v star WSW 6′.4.
13 49 01.8	−41 56 05	ESO 325-25	13.7	0.7 x 0.5	12.4	S0	13	Mag 8.92v star WSW 11′.6; mag 3.40v star Nu Centauri NNE 15′.7.
13 58 02.5	−40 49 41	ESO 325-38	14.2	1.0 x 0.7	13.7	Sc	13	Stellar nucleus, few stars superimposed.
13 59 20.4	−40 04 08	ESO 325-42	13.2	1.5 x 0.4	12.5	Sb:	144	Mag 7.92v star S 3′.4.
14 11 08.5	−40 06 21	ESO 326-6	14.3	0.7 x 0.6	13.3	S0: pec	174	Mag 10.81v star WNW 11′.1.
14 59 01.4	−42 34 51	ESO 327-39	12.9	2.1 x 0.5	12.8	Sb: sp	176	Mag 8.15v star on NE edge.
15 12 28.2	−42 01 03	ESO 328-31	15.7	1.2 x 0.2	14.0	Sbc	146	Mag 9.26v star N 7′.4.
13 40 08.2	−51 02 14	IC 4311	13.4	1.0 x 0.7	12.8	S?	148	IC 4312 SE 4′.2; mag 9.79v star NNE 3′.4.
13 40 30.9	−51 04 17	IC 4312	12.3	0.9 x 0.7	11.7	SAB(r)0°:	30	Mag 12.0 star N 1′.4; IC 4311 NW 4′.2.
14 05 23.2	−45 16 16	IC 4359	13.0	1.2 x 0.7	12.7	SA(rs)bc	17	Mag 8.35v star W 7′.9.
14 05 22.4	−41 49 11	IC 4362	12.8	1.7 x 0.5	12.4	SBc	146	Mag 7.00v star W 13′.1.
14 15 01.8	−43 57 47	IC 4386	12.5	3.0 x 1.5	14.0	SAB(s)dm pec: IV-V	147	Very knotty, numerous superimposed stars; elongated extension S curving W. IC 4387 appears as a large "knot" at the end of this extension.
14 15 01.6	−43 59 29	IC 4387	14.4	0.9 x 0.6	13.6	SAB(s)m pec: V	70	S of IC 4386; see notes for IC 4386.
14 16 59.3	−44 58 40	IC 4390	13.0	1.8 x 0.6	12.9	SAB(s)bc: II-III	11	Mag 10.54v star SSW 3′.9.
14 21 13.8	−46 18 01	IC 4402	11.6	4.2 x 0.9	12.9	SA(s)b? sp	127	Mag 9.41v star SSE 5′.5.
14 30 17.2	−43 33 55	IC 4441	14.0	1.1 x 0.6	13.4	SA(s)b pec	44	Very faint, narrow extension SW from bright, nearly round core. Mag 8.88v star NW 1′.6.
14 31 39.0	−43 25 09	IC 4444	11.4	1.7 x 1.4	12.2	SAB(rs)bc: II	90	Mag 10.82v star N 8′.2.
14 31 54.5	−46 02 07	IC 4445	14.0	1.1 x 0.4	13.0	Sc?	153	Mag 8.58v star S 4′.3.
14 40 10.8	−44 18 57	IC 4472	12.9	2.1 x 0.5	12.8	Sc? sp	0	Mag 13.9 star on S edge; mag 11.8 star NE 2′.6.
14 57 43.7	−43 08 01	IC 4518	14.4	2.2 x 0.6	14.5	Sc pec	108	Multi-galaxy system.
15 05 10.6	−43 30 36	IC 4523	13.2	1.9 x 1.4	14.1	SAB(s)c II-III		Mag 9.38v star SSW 5′.9.
15 05 40.6	−42 27 01	IC 4527	13.3	1.3 x 0.7	13.0	SB(r)bc: II-III	46	Mag 9.39v star NW 10′.4.
13 33 43.8	−48 09 04	NGC 5206	10.9	3.7 x 3.2	13.5	(R′)SB(rs)0° pec:	37	Bright stellar nucleus, very faint envelope, many stars superimposed.
13 37 30.1	−49 50 12	NGC 5234	13.0	1.3 x 0.8	12.9	(R′)SAB(s)0$^+$	39	Many stars superimposed, star rich area; ESO 220-22 NW 3′.9; mag 9.01v star ENE 9′.7.
13 37 38.8	−42 50 52	NGC 5237	12.5	1.9 x 1.6	13.5	I0?	115	
13 38 42.3	−45 51 18	NGC 5244	12.6	1.1 x 0.7	12.2	Sb:	17	Superimposed stars and/or knots.
13 43 01.7	−48 10 10	NGC 5266	11.1	3.2 x 2.1	13.0	SA0$^-$:	103	Bright center, many stars superimposed.
13 40 37.1	−48 20 33	NGC 5266A	12.6	3.1 x 2.4	14.6	SAcd: III	44	Small NNE-SSW bar, faint filamentary arms, many stars superimposed; mag 7.70v star E 6′.0.
13 54 24.2	−48 30 43	NGC 5333	11.7	1.9 x 1.0	12.2	SB(r)0°:	52	Stars superimposed; mag 8.43v star E 2′.7.
13 57 50.6	−43 55 54	NGC 5365	11.4	3.8 x 2.1	13.5	(R)SB(s)0$^-$	7	Many superimposed stars; NGC 5365B E 9′.1.
13 56 39.7	−44 00 38	NGC 5365A	12.4	2.6 x 0.5	12.6	SAb: sp II-III	92	Star E end.
13 58 39.5	−43 57 51	NGC 5365B	13.1	1.6 x 0.3	12.1	Sab: sp	51	NGC 5365 W 9′.1.
14 03 21.7	−41 22 45	NGC 5408	11.6	2.0 x 1.2	12.4	IB(s)m	62	Very filamentary appearance, many superimposed stars; between two mag 10 stars, a mag 6.09v star 2′.9 SSE.
14 10 25.3	−43 19 29	NGC 5483	11.2	3.7 x 3.4	13.8	SA(s)c II	25	Bright, elongated nucleus; many superimposed stars; in star rich area.
14 12 01.0	−46 05 19	NGC 5489	12.2	1.5 x 1.0	12.5	SA(s)a: III-IV	129	
14 15 54.6	−48 06 55	NGC 5516	12.0	1.8 x 1.2	12.7	SA(s)0$^-$:	169	**ESO 221-34A** SE 1′.8.
14 18 27.5	−43 23 12	NGC 5530	11.3	4.2 x 1.9	13.4	SA(rs)bc III	127	Star superimposed on nucleus, many superimposed stars overall.
14 32 40.7	−44 10 29	NGC 5643	10.0	4.6 x 4.0	13.0	SAB(rs)c III	153	Bright nucleus, numerous superimposed stars, in star rich area.
14 35 36.3	−45 58 04	NGC 5670	12.0	1.5 x 0.7	11.9	SA(rs)0°	74	Mag 8.20v star SW 2′.1.
14 39 35.4	−45 01 13	NGC 5688	11.9	3.1 x 1.9	13.6	SAB(rs)c III-IV	85	Bright nucleus, many superimposed stars.
14 58 56.0	−42 00 45	NGC 5786	11.5	2.3 x 1.1	12.3	(R′)SB(s)bc I-II	63	Bright nucleus, numerous superimposed stars; mag 3.13v star Kappa Centauri SE 6′.0.

GALAXY CLUSTERS

RA h m s	Dec ° ′ ″	Name	Mag 10th brightest	No. Gal	Diam ′	Notes
13 33 18.0	−42 51 00	A 3561	16.3	36	17	All members anonymous, faint, stellar.

183

(Continued from previous page)

OPEN CLUSTERS

183

RA h m s	Dec ° ′ ″	Name	Mag	Diam ′	No. ★	B ★	Type	Notes
14 07 36.0	−48 18 00	NGC 5460	5.6	35	61	9.0	cl	Moderately rich in bright and faint stars; strong central concentration; detached.

GLOBULAR CLUSTERS

RA h m s	Dec ° ′ ″	Name	Total V m	B ★ V m	HB V m	Diam ′	Conc. Class Low = 12 High = 1	Notes
13 46 26.5	−51 22 24	NGC 5286	7.4	13.5	16.5	11.0	5	

PLANETARY NEBULAE

RA h m s	Dec ° ′ ″	Name	Diam ″	Mag (P)	Mag (V)	Mag cent ★	Alt Name	Notes
14 22 26.5	−44 09 05	IC 4406	106	10.6	10.2	17.4	PK 319+15.1	Pair of mag 13.0 and 13.7 stars NW 2'.0.
13 51 03.4	−51 12 24	NGC 5307	18	12.1	11.2	14.6	PK 312+10.1	Pair of mag 13-14 stars SE.
14 11 52.1	−51 26 23	PN G315.4+9.4	5	14.0		17.9	PK 315+9.1	Mag 13.3 star S 0'.8.
14 46 34.9	−50 23 26	PN G321.0+8.3	27	17.3		20.6	PK 321+8.1	Mag 13.7 star on S edge; mag 8.13v star SW 3'.8.
15 01 40.8	−48 21 04	PN G324.1+9.0	18	15.9		17.1	PK 324+9.1	E-W line of four mag 10-13 stars S 0'.8.
15 06 13.7	−42 59 58	PN G327.5+13.3	5	12.7	12.6	18.7	PK 327+13.1	Galaxy **ESO 273-17** N 1'.0.

184

GALAXIES

184

RA h m s	Dec ° ′ ″	Name	Mag (V)	Dim ′ Maj x min	SB	Type Class	PA	Notes
11 58 03.0	−51 53 07	ESO 217-15	13.0	1.1 x 0.8	12.7	SB0?	119	Mag 14.5 star on N edge; mag 10.01v star E 10'.1.
11 58 17.6	−51 08 06	ESO 217-16	13.3	1.3 x 0.7	13.1	SAB(r)b pec: II	155	Mag 12.4 star SE edge; mag 10.24v star NW edge.
11 58 20.6	−50 58 23	ESO 217-17	13.5	1.6 x 0.5	13.1	Sc: II-III	165	Mag 12 star superimposed N end.
12 01 48.2	−50 35 01	ESO 217-20	13.6	1.6 x 0.4	13.0	SBbc? sp	172	Mag 10.92v star NW 3'.2.
12 04 47.8	−50 15 07	ESO 217-22	13.3	1.6 x 0.6	13.1	Sc I-II	44	Several bright knots; mag 6.96v star E 5'.3.
12 31 53.8	−51 44 53	ESO 218-8	15.0	2.5 x 0.2	14.1	SB?	31	Stellar nucleus, extremely faint arms.
12 55 21.9	−49 32 58	ESO 219-7	12.8	0.9 x 0.6	12.1	S0	31	Mag 10.06v star N 2'.4.
12 57 35.7	−49 45 00	ESO 219-14	12.4	1.5 x 0.6	12.1	SAB(rs)a:	134	Mag 8.34v star S 1'.9.
12 59 23.2	−49 37 17	ESO 219-16	15.5	2.2 x 1.4	16.5	IB(s)m V-VI	0	Mag 12.3 star on S edge; mag 10.03v star N 8'.1.
13 00 07.0	−50 56 17	ESO 219-18	13.6	0.9 x 0.7	13.1	E/S0	17	**ESO 219-19** N 1'.5; mag 9.22v star NW 2'.4.
13 02 20.9	−50 20 01	ESO 219-21	12.4	5.2 x 1.3	14.4	SAB(s)d III-IV	32	Numerous stars superimposed; mag 9.55v star W 11'.0.
13 02 23.9	−49 28 26	ESO 219-22	13.4	2.1 x 0.4	13.1	SAB(s)dm: IV-V	106	Multi-galaxy system; mag 4.83v star Xi 1 Centauri ESE 11'.7.
13 06 29.7	−49 49 41	ESO 219-27	16.9	1.7 x 0.4	16.3		151	Located 6'.2 NW of mag 4.28v star Xi 2 Centauri.
13 09 12.5	−51 00 47	ESO 219-30	14.4	1.7 x 0.6	14.3	IB(s)m V-VI	49	Mag 9.30v star E 9'.4.
13 09 33.5	−48 14 40	ESO 219-34	14.2	1.0 x 0.7	13.7	SB0/a	6	Mag 12.3 star SW 1'.9; close pair of mag 14 stars N 1'.4.
13 10 26.0	−51 37 22	ESO 219-37	14.0	1.5 x 1.4	14.7	SAB(r)dm IV-V	80	Mag 7.71v star WSW 10'.8.
13 13 58.2	−49 28 44	ESO 219-41	12.0	3.2 x 0.8	12.8	(R')SAB(s)ab pec: II	77	Mag 12.7 star on N edge; mag 10.92v star NE 6'.0; **ESO 219-39** NNW 2'.5.
13 16 01.2	−50 16 20	ESO 219-43	14.3	1.9 x 0.5	14.1	SAB(s)d pec IV	162	Mag 10.15v star W 1'.4.
13 22 57.4	−50 00 28	ESO 220-7	14.8	1.7 x 0.3	13.9	Sc	113	Mag 13.6 star on N edge; mag 10.19v star ESE 4'.8.
13 26 05.8	−48 14 11	ESO 220-8	13.2	1.9 x 0.5	13.0	SA(s)b: sp II	170	Mag 9.79v star W 3'.1; mag 9.43v star E 3'.5.
13 26 53.9	−48 09 45	ESO 220-9	13.3	0.7 x 0.5	12.0	Sa:	175	Mag 10.37v star W 0'.9.
13 26 58.7	−49 19 31	ESO 220-10	14.2	0.7 x 0.5	13.0		173	**ESO 220-10A** SE 0'.9; mag 6.27v star SE 4'.9.
13 28 00.5	−49 01 42	ESO 220-12	14.1	0.9 x 0.7	13.4	S	53	Mag 7.69v star NE 6'.3.
13 33 57.4	−50 29 34	ESO 220-19	13.6	1.5 x 0.5	13.1	SBd: pec III-IV	149	Mag 12.2 star N 0'.8; mag 10.93v star N 3'.2.
13 36 50.0	−49 59 03	ESO 220-20	13.9	1.3 x 0.3	12.7	Pec	59	Mag 12.6 star NW 0'.9; ESO 220-21 E 2'.7.
13 37 05.8	−49 59 47	ESO 220-21	14.1	1.3 x 0.4	13.3	S	91	ESO 220-20 W 2'.7; mag 9.92v star SE 2'.3.
13 37 14.1	−49 47 24	ESO 220-22	14.2	1.2 x 0.6	13.7	SB(s)m? V	37	Located between mag 10.44v and 10.38v stars; star N edge of nucleus; ESO 220-23 N 2'.6
13 37 18.7	−49 44 57	ESO 220-23	13.4	1.6 x 0.3	12.5	S0 pec sp	29	ESO 220-22 S 2'.5; mag 10.44v star SSE 1'.9.
13 39 46.1	−48 17 58	ESO 220-26	13.3	2.0 x 2.0	14.7	SB(s)d IV		Many stars superimposed; mag 7.55v star S 8'.1.
11 59 28.0	−47 09 45	ESO 267-5	13.9	1.1 x 0.6	13.3	Sc	66	Several knots and/or superimposed stars.
12 03 31.6	−43 39 16	ESO 267-11	12.6	1.0 x 0.8	12.3	S0: pec	48	Mag 10.75v star N 0'.8.
12 04 52.8	−43 43 57	ESO 267-13	13.3	1.3 x 0.7	13.0	(R')SAB(rs)ab V-VI	68	Mag 11.09v star S 2'.8.
12 06 23.2	−44 26 54	ESO 267-16	13.4	0.9 x 0.8	12.9	Sa:	10	Mag 12.2 star NW 3'.0.
12 06 38.3	−47 20 47	ESO 267-17	14.1	1.0 x 0.7	13.6	(R')SB(r)ab	95	Numerous stars superimposed.
12 13 28.2	−44 22 59	ESO 267-27	14.1	0.9 x 0.4	12.8	Pec	14	Star or knot S end.
12 13 52.5	−47 16 29	ESO 267-29	13.3	1.6 x 1.0	13.6	SB(rs)ab pec	140	Mag 9.42v star S 2'.6; ESO 267-30 NE 4'.4.
12 14 13.0	−47 13 46	ESO 267-30	13.1	1.2 x 0.6	12.6	SA(rs)b pec	123	ESO 267-29 SW 4'.4.
12 15 16.5	−43 01 28	ESO 267-34	13.5	1.0 x 0.7	13.0	S0⁻:	101	Mag 15.0 star N edge; **ESO 267-35** ESE 2'.8.
12 15 36.3	−44 27 21	ESO 267-36	15.0	1.0 x 0.5	14.1	SB(s)m V	38	Mag 10.41v star WSW 8'.4.
12 21 57.3	−43 20 09	ESO 267-41	13.5	0.8 x 0.6	12.5	S?	30	Streamer extending S along E edge; **MCG −7-26-3** E 1'.2.
12 22 02.1	−46 12 57	ESO 267-43	13.4	1.2 x 0.7	13.1	(R)SB(rs)0° pec:	162	Mag 12.2 star E 0'.8.
12 25 17.8	−43 26 45	ESO 268-3	13.1	1.0 x 0.7	12.6	SB(s)0°	124	Mag 9.79v star E 4'.3.
12 25 21.2	−47 28 29	ESO 268-4	12.5	1.0 x 0.6	11.8	SA0°: pec	122	Mag 11.8 star SE edge; mag 6.62v star NW 11'.7.
12 25 34.0	−45 15 50	ESO 268-5	13.5	0.7 x 0.5	12.2	S0	153	Mag 11.8 star W 5'.6.
12 26 41.6	−45 41 51	ESO 268-8	13.3	0.8 x 0.8	12.6	S?		Several bright knots or stars on S edge.
12 32 01.8	−44 46 55	ESO 268-15	13.7	1.0 x 0.7	13.2	SB(rs)bc	117	Stellar nucleus; mag 13.2 star E 0'.7.
12 34 48.9	−44 44 11	ESO 268-22	13.5	1.1 x 0.5	12.7	S0	167	Mag 5.77v star N 4'.0.
12 35 49.2	−43 03 29	ESO 268-23	13.8	0.6 x 0.4	12.2	S0?	76	**ESO 268-24** NE 2'.9.
12 38 30.0	−42 51 53	ESO 268-27	13.4	1.0 x 0.8	13.0	SA(rs)b: III-IV	135	Mag 9.69v star N 1'.6.
12 41 02.0	−44 13 57	ESO 268-30	14.8	1.0 x 0.4	13.6	SB(s)d pec III-IV	125	Mag 6.74v star N 8'.0.
12 42 29.7	−47 33 33	ESO 268-33	13.6	1.9 x 0.4	13.2	SB(s)bc? sp II	167	Mag 7.83v star WSW 6'.4.
12 43 03.7	−42 54 43	ESO 268-34	13.7	1.9 x 0.5	13.5	Sb: pec sp II-III	173	Mag 10.24v star W 3'.7; stellar galaxy **Fair 457** E 4'.2.
12 43 58.3	−45 35 43	ESO 268-35	14.2	0.6 x 0.4	12.5	Sc?	7	Mag 10.76v star S 2'.4.
12 44 10.7	−42 47 19	ESO 268-36	13.6	0.8 x 0.5	12.4	Sab	87	Bright star superimposed E end; mag 8.73v star E 1'.3.
12 45 05.0	−44 00 22	ESO 268-37	13.3	1.1 x 0.7	12.8	Sc I-II	131	Mag 8.33v star E 12'.4.
12 46 00.6	−45 24 58	ESO 268-38	13.9	0.7 x 0.7	12.9			Mag 8.31v star S 5'.2.
12 48 42.2	−45 00 33	ESO 268-44	13.5	1.0 x 0.4	12.4	(R')Sb: III-IV	66	Mag 10.95v star SE 2'.7; this star is part of a larger group of 6-8 brighter stars S and SE.
12 50 19.5	−44 25 39	ESO 268-46	13.8	2.6 x 0.8	14.5	IB(s)m IV-V	165	Mag 9.81v star NW 8'.0.
12 51 06.4	−47 38 58	ESO 268-47	13.6	1.0 x 0.9	13.3	S0?	158	Mag 9.80v star SSW 6'.1.
12 53 56.3	−45 52 40	ESO 269-2	13.9	1.3 x 0.8	13.8	SB(s)dm IV	95	Mag 12.6 star SE 1'.7.
12 55 00.6	−45 54 21	ESO 269-6	14.0	1.0 x 0.8	13.6	SAB(s)bc: III-IV	82	Stellar nucleus, several stars superimposed.
12 55 23.2	−44 48 51	ESO 269-8	13.2	1.0 x 0.9	12.9	S0?	175	**MCG −7-27-17** E 1'.2.
12 55 31.3	−45 13 59	ESO 269-9	14.2	1.0 x 0.5	13.3	IB(s)m pec: V	48	Double system? Mag 9.81v star NW 10'.3; note: **ESO 269-5** 0'.6 W of this star.
12 56 40.6	−46 55 35	ESO 269-12	13.7	1.0 x 0.9	13.4	S0?	15	Mag 10.02v star SW 2'.8.

(Continued from previous page)

RA h m s	Dec ° ′ ″	Name	Mag (V)	Dim ′ Maj x min	SB	Type Class	PA	Notes
12 56 51.3	−43 07 38	ESO 269-13	12.4	1.6 x 1.2	13.0	(R′)SB(r)0/a	140	Mag 10.08v star NW 5′.8.
12 57 06.9	−46 52 19	ESO 269-14	13.8	1.0 x 0.3	12.4	S0?	81	Pair of mag 14.5-15 stars on N edge.
12 58 08.1	−43 19 48	ESO 269-20	13.6	1.8 x 0.9	14.0	Pec	88	Mag 10.77v star WNW 3′.0.
12 58 51.0	−43 52 34	ESO 269-22	13.5	0.8 x 0.7	12.8		131	Star or knot on S edge.
12 59 31.1	−44 27 16	ESO 269-25	13.4	1.3 x 0.7	13.1	(R)SA(r)a pec:	171	Mag 9.45v star SSE 2′.2.
13 00 26.9	−45 21 43	ESO 269-30	14.7	1.5 x 0.6	14.5	SB(s)m pec V	10	Mag 10.65v star W 2′.8; ESO 269-31 S 3′.3.
13 00 32.0	−45 24 54	ESO 269-31	13.4	0.7 x 0.7	12.5	SA0$^-$: pec		ESO 269-30 N 3′.3.
13 03 48.9	−42 57 55	ESO 269-38	13.6	1.0 x 0.7	13.0		20	Star superimposed S edge; mag 11.04v star N 2′.3.
13 05 04.7	−46 47 26	ESO 269-43	13.8	0.7 x 0.4	12.3	S0	50	N-S row of three mag 13-14 stars E 1′.7.
13 06 29.8	−44 10 01	ESO 269-48	14.6	1.5 x 0.5	14.1	SB(s)dm: IV-V	91	Mag 9.57v star NW 7′.1.
13 07 32.6	−45 00 49	ESO 269-49	13.3	1.3 x 0.8	13.2	SAB(r)c pec III-IV	136	Mag 10.96v star N 2′.9.
13 08 47.2	−43 40 35	ESO 269-52	15.0	1.9 x 0.2	13.8	S	140	Mag 10.08v star S 4′.3.
13 08 50.6	−42 56 33	ESO 269-53	15.1	1.5 x 0.6	14.8	IB(s)m pec V-VI	161	Mag 10.73v star S 3′.0.
13 10 01.1	−43 12 45	ESO 269-56	13.5	2.9 x 0.9	14.4	SB(s)m pec: IV-V	84	Mag 10.74v star S 2′.6.
13 10 04.3	−46 26 11	ESO 269-57	11.7	3.6 x 2.7	14.0	(R′)SAB(r)ab	50	Very faint outer arms, many stars on face.
13 10 32.9	−46 59 36	ESO 269-58	12.5	2.1 x 1.4	13.5	I0 pec	62	Mag 8.82v star N 8′.7.
13 10 58.8	−46 13 25	ESO 269-60	13.2	0.9 x 0.7	12.6	SB(s)0°	97	Mag 9.84v star SSW 3′.0.
13 11 01.9	−44 35 33	ESO 269-61	12.9	2.3 x 0.4	12.7	Sb: sp II-III	78	Mag 9.14v star W 8′.6.
13 13 09.0	−44 53 25	ESO 269-66	13.1	1.5 x 1.2	13.7	E2? pec	57	Uniform surface brightness, stars involved on face.
13 13 17.3	−43 35 03	ESO 269-69	14.0	1.1 x 0.6	13.4	(R′)SB(r)a	135	Mag 12.2 star SSW 1′.5.
13 13 27.4	−43 22 59	ESO 269-70	13.5	1.1 x 0.4	12.5	SB(r)0°	6	Mag 12.1 star NW 2′.4.
13 13 33.8	−43 31 35	ESO 269-72	12.8	1.4 x 0.6	12.6	E4:	19	Mag 9.81v star WNW 3′.5.
13 14 23.5	−46 06 49	ESO 269-74A	12.3	1.8 x 1.4	13.2	(R)SAB(rs)0/a pec	60	Mag 10.40v star NE edge; pair with ESO 269-74 SW edge.
13 14 20.1	−46 07 28	ESO 269-74	14.7	1.9 x 1.4	15.6	SA?(r)0	0	Double system interacting with companion **ESO 264-74A**.
13 15 05.0	−47 08 00	ESO 269-75	13.9	1.6 x 0.6	13.7	SAB(s)dm IV	150	Mag 13.4 star on E edge; mag 9.71v star W 8′.2.
13 16 38.3	−45 53 43	ESO 269-78	14.0	0.8 x 0.4	12.7	SB(s)d IV-V	101	Pair of mag 9.59v, 9.49v stars N 5′.6.
13 19 01.0	−47 15 27	ESO 269-80	12.7	1.3 x 1.1	13.0	SA0$^-$	9	ESO 269-82 S 4′.1.
13 19 01.9	−47 02 03	ESO 269-81	13.7	1.1 x 0.5	12.9	Sa:	176	Mag 10.70v star SE edge; mag 10.57v star E 1′.0.
13 19 02.5	−47 19 33	ESO 269-82	14.3	0.9 x 0.5	13.3	SB(s)c: III-IV	16	Mag 10.18v star W 3′.4; ESO 269-80 N 4′.1.
13 19 58.5	−47 16 53	ESO 269-85	12.0	2.7 x 1.7	13.5	SA(rs)c I-II	53	Numerous bright knots an/or stars; mag 9.56v star N 2′.9.
13 20 50.8	−47 13 08	ESO 269-90	12.7	1.0 x 0.5	11.9	SAB0$^-$	120	Mag 9.70v star NW 1′.8.
13 21 38.2	−45 55 54	ESO 270-5	13.5	1.4 x 0.7	13.3	(R′)SB(s)b	133	Mag 9.50v star NE 6′.0; ESO 270-6 E 3′.4.
13 21 57.6	−45 56 15	ESO 270-6	13.3	1.9 x 0.5	13.1	Sb: sp	4	Mag 9.50v star N 4′.0; ESO 270-5 W 3′.4.
13 23 15.9	−43 32 42	ESO 270-7	13.4	1.1 x 0.5	12.6	SB(s)0$^+$ pec:	177	Mag 13.1 star W 0′.9.
13 28 26.2	−44 28 46	ESO 270-13	14.1	0.9 x 0.6	13.3	(R′)Sa? pec	146	Mag 9.71v star E 5′.6.
13 28 27.3	−44 10 22	ESO 270-14	12.7	1.1 x 0.8	12.5	S0$^-$ pec	34	Mag 9.03v star W 1′.2.
13 29 22.5	−47 05 42	ESO 270-15	13.6	1.9 x 0.5	13.4	SB(r)d? III-IV	57	Mag 9.70v star NE 8′.6.
13 34 48.2	−45 33 01	ESO 270-17	10.7	12.3 x 1.5	13.7	SB(s)m: IV-V	110	**Fourcade-Figueroa Galaxy**.
13 37 18.9	−46 12 57	ESO 270-21	14.5	1.1 x 0.4	13.5	Im? pec	123	Mag 8.40v star ENE 4′.8.
12 01 08.7	−42 44 50	ESO 321-2	13.2	1.1 x 1.0	13.1	Sb	160	Mag 11.01v star N 2′.9.
12 01 20.4	−42 40 15	ESO 321-3	13.9	0.9 x 0.4	12.6	Sb	121	Mag 11.01v star S 2′.1.
12 07 22.0	−40 13 06	ESO 321-7	13.4	0.8 x 0.5	12.3	Sa	113	Located 2′.5 E of NGC 4112.
12 12 36.5	−40 02 11	ESO 321-12	14.2	1.2 x 0.3	12.9	S	99	Mag 9.53v star ESE 9′.5.
12 15 48.0	−42 44 35	ESO 321-17	14.8	1.7 x 0.3	13.9	SB(s)cd? sp IV	132	Mag 9.93v star NNE 9′.6.
12 20 16.4	−40 23 26	ESO 321-21	12.5	1.0 x 0.8	12.1	Sc	109	Mag 9.45v star W 1′.5.
12 25 36.5	−40 43 32	ESO 322-7	14.5	1.4 x 1.1	14.9	SB(s)m V-VI	1	Stellar nucleus, few stars superimposed, faint envelope.
12 27 16.1	−40 28 30	ESO 322-11	13.8	1.1 x 0.7	13.3	S?	24	Mag 12.6 star N 1′.3.
12 29 07.5	−40 40 28	ESO 322-19	13.9	1.5 x 0.3	12.9	SB(s)cd? sp	117	Pair with ESO 322-20 SE edge; mag 10.50v star N 7′.1.
12 29 12.7	−40 41 35	ESO 322-20	13.4	1.2 x 1.1	13.6	SB(rs)dm IV	105	Mag 15.5 star on S edge; pair with ESO 322-19 NW edge.
12 30 16.0	−40 52 20	ESO 322-21	13.9	0.6 x 0.4	12.3	Pec	40	Mag 10.84v star WNW 5′.3.
12 32 52.0	−40 29 24	ESO 322-25	13.5	0.9 x 0.4	12.3	S0?	113	Mag 12.3 star E 2′.0.
12 34 42.8	−40 18 02	ESO 322-27	12.6	1.3 x 1.1	12.8	(R)SAB(r)0/a	146	Mag 9.76v star S 4′.6.
12 34 46.0	−42 32 03	ESO 322-28	13.5	1.4 x 0.7	13.3	(R′)Sab: III-IV	119	Mag 15 star superimposed W edge.
12 36 48.9	−42 08 29	ESO 322-31	12.9	1.7 x 1.1	13.4	SB(r)0$^+$	0	Mag 9.78v star SSW 7′.4.
12 37 33.4	−41 18 31	ESO 322-34	13.3	0.9 x 0.3	11.8	S0?	37	Mag 13.4 SE 1′.3.
12 37 52.6	−42 40 38	ESO 322-37	13.7	0.7 x 0.5	12.6	E	163	Mag 11.9 star W 0′.6.
12 38 18.4	−41 30 13	ESO 322-38	12.6	1.7 x 0.7	12.6	SAB0$^-$	85	Mag 10.64v star N 5′.7.
12 38 28.3	−40 22 05	ESO 322-40	13.3	1.3 x 0.7	13.0	Sbc II	71	Mag 8.58v star E 8′.2.
12 38 40.6	−42 12 57	ESO 322-42	13.6	1.6 x 0.5	13.2	SB(rs)c III	43	Mag 12.8 star N 1′.5.
12 40 02.0	−42 02 36	ESO 322-45	13.3	1.5 x 0.6	13.0	SB(s)c: II-III	131	Mag 9.25v star N 6′.0.
12 40 54.2	−41 36 27	ESO 322-51	12.7	1.9 x 0.5	12.5	SB0$^-$? sp	51	Mag 9.81v star E 11′.2.
12 46 16.6	−41 16 34	ESO 322-74	13.9	0.9 x 0.4	12.6	S?	93	Mag 12.4 star NE 2′.7.
12 46 26.0	−40 45 12	ESO 322-75	12.8	1.7 x 0.9	13.2	(R)SA0°:	101	Mag 11.8 star SW 5′.2; note: very small, faint **PGC 43070** is 2′.1 E of this star.
12 46 41.6	−40 01 34	ESO 322-76	13.5	1.1 x 0.7	13.0	SA(r)bc: II-III	76	Mag 15.4 star E edge; mag 13.8 star NNW 2′.0.
12 48 02.2	−40 35 40	ESO 322-85	14.5	1.5 x 0.4	13.8	Sc? pec sp	134	Mag 10.65v star S 2′.7.
12 48 22.6	−41 07 26	ESO 322-89	14.0	0.5 x 0.5	12.4			Mag 14.8 star N edge; mag 11.8 star S 2′.3.
12 48 55.3	−40 34 59	ESO 322-92	13.6	1.1 x 0.3	12.2	S0?	10	Mag 9.56v star SE 4′.8; note: **PGC 43339** is 1′.1 S of this star and **PGC 43349** is 1′.8 E of this star.
12 49 06.6	−41 50 02	ESO 322-94	13.6	1.0 x 0.6	12.9	SA0$^-$	71	Mag 13.4 star NW 1′.6.
12 49 12.1	−41 32 42	ESO 322-96	14.0	1.1 x 0.6	13.3	S0?	33	Mag 10.28v star NE 1′.0.
12 49 26.3	−41 29 23	ESO 322-99	13.3	1.1 x 0.7	12.9	S0°: sp	169	Mag 10.28v star and ESO 322-96 SW 3′.2; ESO 322-100 N 1′.0.
12 49 26.7	−41 27 49	ESO 322-100	13.7	1.0 x 0.4	12.5	S0?	89	ESO 322-99 S 1′.6; **PGC 4353** N 2′.0.
12 49 34.6	−41 03 19	ESO 322-101	13.3	0.8 x 0.4	12.1	E?	78	Mag 9.11v star W 8′.2.
12 49 38.0	−41 23 20	ESO 322-102	13.8	1.4 x 0.3	12.7	SB(s)0°: sp	45	NGC 4709 E 4′.9; almost stellar **PGC 43379** N 1′.3.
12 50 03.8	−40 07 36	ESO 323-2	13.7	1.7 x 0.3	12.8	S0$^-$: sp	26	Mag 10.80v star WSW 3′.4.
12 50 12.3	−41 30 57	ESO 323-5	13.1	1.3 x 0.6	12.7	S0: sp	9	Mag 10.38v star SE 5′.6.
12 50 24.1	−40 50 45	ESO 323-7	13.1	1.4 x 0.9	13.2	SAB(s)0$^+$ pec	176	Bright star superimposed S edge of core; mag 10.70v star E 2′.5.
12 50 34.4	−41 28 15	ESO 323-8	13.6	1.0 x 0.4	12.4	S0 sp	142	Mag 10.83v star and ESO 323-9 NE 2′.3.
12 50 43.3	−41 25 49	ESO 323-9	13.5	1.1 x 0.5	12.7	S0 sp	81	Mag 10.83v star S 0′.9; almost stellar **PGC 43500** NE 3′.2.
12 50 44.8	−42 32 55	ESO 323-10	12.6	1.6 x 1.6	13.5	SAB(s)bc? I-II		Mag 9.33v star WNW 3′.9.
12 50 46.7	−41 53 42	ESO 323-11	13.7	0.7 x 0.4	12.3	E?	144	Mag 8.93v star W 3′.1.
12 51 17.5	−40 49 09	ESO 323-12	13.5	0.9 x 0.6	12.7	S?	114	Mag 10.45v star ENE 2′.5; **PGC 43520** NW 2′.5.
12 51 41.3	−42 36 22	ESO 323-14	13.5	1.1 x 0.8	13.2	S0°	87	Mag 12.2 star WNW 5′.1.
12 52 01.5	−41 16 48	ESO 323-18	13.8	1.4 x 0.7	13.6	SAB(s)bc: pec	68	Close pair of stars, mags 12.5 and 12.2, N 2′.2; **PGC 43652** NE 2′.9; mag 9.52v star WSW 3′.4.
12 52 03.4	−41 27 39	ESO 323-19	12.8	1.5 x 0.7	12.7	E$^+$	109	**PGC 43651** SE 2′.5.
12 52 13.0	−41 20 25	ESO 323-20	14.1	1.2 x 1.0	14.1	S?	132	Stellar nucleus, smooth envelope; pair of stars, mags 12.2 and 9.95v, E 1′.2.
12 52 25.9	−40 42 30	ESO 323-23	13.1	1.3 x 0.9	13.2	S?	19	Moderately strong core, smooth envelope.
12 52 50.3	−40 27 09	ESO 323-27	13.0	1.8 x 1.0	13.5	SA(r)c I-II	99	Mag 13.5 stars on N and E edges; mag 9.22v star SE 4′.0.

RA h m s	Dec ° ′ ″	Name	Mag (V)	Dim ′ Maj x min	SB	Type Class	PA	Notes
12 52 50.2	−41 20 16	ESO 323-28	13.0	1.7 x 1.0	13.4	SA(s)0/a sp	15	**PGC 43721** on N edge; mag 9.95v star and ESO 323-20 W 5′8.
12 53 20.4	−41 38 09	ESO 323-32	12.8	1.4 x 1.3	13.3	(R)SAB0$^+$	120	Mag 12.0 star SE 3′5.
12 53 22.9	−42 08 28	ESO 323-33	14.4	1.6 x 0.3	13.5	Sc: sp	16	Mag 10.42v star SE 2′0.
12 53 25.9	−41 12 16	ESO 323-34	11.9	2.1 x 0.9	12.6	E5:	162	Mag 8.49v star W 7′8.
12 54 18.4	−41 49 20	ESO 323-38	13.3	1.3 x 0.5	12.7	S?	0	Star superimposed E of center; mag 12.6 star N edge.
12 55 01.2	−40 58 19	ESO 323-42	13.2	2.2 x 0.6	13.3	S?	77	Mag 10.33v star W 2′3.
12 59 18.2	−40 57 42	ESO 323-54	14.1	1.0 x 0.8	13.7	S?	169	Stellar nucleus, filamentary envelope; mag 8.50v star SW 2′7.
13 01 03.3	−40 25 37	ESO 323-63	13.4	1.1 x 0.6	12.9	S0?	150	Almost stellar **PGC 44845** W 0′8; slightly larger **PGC 44827** W 2′3.
13 01 51.5	−41 24 20	ESO 323-66	13.6	0.9 x 0.9	13.3	S?		Mag 10.9 star E 7′1.
13 01 51.3	−41 52 32	ESO 323-67	13.3	1.4 x 1.3	13.8	S?	48	Pair of stars, mags 10.67v and 12.0, E and SE.
13 01 57.2	−41 04 16	ESO 323-68	13.7	1.3 x 0.5	13.1	S?	169	Mag 9.13v star WSW 9′9.
13 06 25.9	−40 24 52	ESO 323-77	12.5	2.0 x 1.2	13.3	(R)SAB(rs)0°	155	Mag 11.38v star W edge.
13 07 12.3	−40 24 29	ESO 323-81	14.2	0.5 x 0.3	12.0		47	Compact.
13 07 12.9	−42 05 46	ESO 323-82	14.1	1.2 x 0.7	13.7	Sab	100	Mag 10.59v star W 1′2.
13 08 38.8	−41 27 47	ESO 323-85	13.5	1.3 x 0.9	13.5	SB(rs)bc	167	Mag 8.43v star NW 6′4.
13 10 38.7	−41 33 39	ESO 323-90	13.0	1.3 x 1.0	13.1	SB(r)0/a	34	Located between close pair of E-W mag 10.55v, 11.20v stars; mag 12.3 star on E edge.
13 13 13.9	−42 17 16	ESO 323-93	12.3	1.5 x 0.9	12.5	SAB0$^-$ pec	105	Mag 8.96v star N 2′5.
13 14 40.0	−42 40 34	ESO 323-99	12.6	2.3 x 2.0	14.1	SAB(s)d: III-IV	166	Numerous bright knots and/or superimposed stars.
13 15 52.8	−42 28 18	ESO 324-1	14.0	1.3 x 0.7	13.8	S?	14	Mag 13.3 star NW 2′5., in star rich area.
13 19 40.1	−40 47 47	ESO 324-7	14.8	1.5 x 0.2	13.3	Sc:	115	Mag 9.20v star ENE 7′1.
13 26 52.1	−40 06 45	ESO 324-21	13.7	1.4 x 1.3	14.2	Sc	141	Mag 6.41v star SE 6′5; mag 12.8 star N 1′4.
13 27 38.2	−41 29 00	ESO 324-24	12.4	3.1 x 2.2	14.3	IABm:	54	Numerous superimposed stars; mag 5.66v star W 7′7.
13 28 40.4	−40 11 41	ESO 324-27	13.8	1.0 x 0.9	13.5	Sa	33	Several knots or superimposed stars NE of nucleus.
13 29 03.9	−41 59 46	ESO 324-29	12.2	2.6 x 0.9	13.0	SAB(r)0$^+$	9	Mag 8.01v star W 7′1.
13 33 17.9	−40 29 47	ESO 324-35	14.3	1.0 x 0.7	13.8	SB	171	Stars superimposed on edges; mag 9.28v star W 2′8.
13 35 23.0	−40 04 19	ESO 324-40	13.8	0.9 x 0.5	12.8	S0	141	Small group of three mag 12-13 stars N 1′4; stellar **ESO 324-41** NE 2′4.
13 35 40.3	−41 37 25	ESO 324-42	13.9	1.0 x 0.5	13.0	Sab	127	Mag 14.9 star S edge, in star rich area.
12 56 43.0	−50 20 50	IC 3896	10.9	2.5 x 1.9	12.6	E	10	Mag 10.23v star WSW 3′7.
12 55 31.7	−50 04 19	IC 3896A	12.1	2.2 x 1.7	13.4	SB(rs)cd pec: IV-V	27	Mag 9.81v star N 4′2.
13 09 34.5	−51 58 06	IC 4200	12.7	1.4 x 0.9	12.8	(R′)SA(r)0°:	152	Mag 9.12v star SE 3′9.
12 07 09.4	−40 12 27	NGC 4112	11.5	1.6 x 0.9	11.8	SAB(rs)b pec	5	Galaxy ESO 321-7 E 2′5; mag 10.41 star S 1′5; mag 9.25v star S 2′1.
12 16 27.3	−43 19 32	NGC 4219	11.9	3.7 x 1.1	13.2	SA(s)bc III	30	Very patchy appearance.
12 17 59.7	−43 32 25	NGC 4219A	13.4	1.3 x 0.6	13.0	SB(s)c? I-II	31	Pair of mag 13 stars NE 2′2.
12 28 36.4	−43 15 45	NGC 4444	12.2	2.5 x 2.3	13.9	SAB(rs)bc II	119	Stellar nucleus with numerous stars and/or knots overall.
12 37 43.8	−43 37 13	NGC 4573	13.0	2.6 x 2.0	14.6	(R)SB(rs)0/a	138	Small, bright inner ring 0′7 X 0′5; much fainter outer ring with mag 10.53v star superimposed.
12 37 51.2	−40 32 14	NGC 4575	12.6	2.0 x 1.3	13.5	SB(s)bc pec:	106	Very patchy inner arms; faint outer arms; star on W edge; mag 10.64v star SW 1′6.
12 40 46.7	−40 53 39	NGC 4601	13.4	1.8 x 0.5	13.2	SAB(r)0$^+$	16	Located 2′8 SW of a mag 11.22v star; **ESO 322-53** E 3′9.
12 40 55.6	−40 58 34	NGC 4603	11.4	3.4 x 2.5	13.5	SA(s)c: I-II	27	Faint stellar nucleus, many knots and/or superimposed stars; mag 9.12v star S 4′3.
12 39 37.0	−40 44 27	NGC 4603A	13.3	1.9 x 0.6	13.3	SB(s)c? III-IV	90	Mag 8.74v star E 4′6.
12 40 29.6	−41 04 14	NGC 4603B	14.3	1.4 x 0.3	13.2	Sb pec: III-IV	37	**MCG −7-26-21** WSW 7′8; mag 9.12v star ENE 3′9.
12 40 43.1	−40 45 50	NGC 4603C	13.1	1.8 x 0.4	12.6	S0 sp	160	**MCG −7-26-23** W 4′7; mag 8.74v star W 7′8.
12 42 07.9	−40 49 19	NGC 4603D	13.2	1.5 x 1.1	13.6	SAB(s)d: III-IV	74	Almost stellar nucleus, uniform envelope.
12 42 16.7	−40 38 34	NGC 4616	13.4	0.9 x 0.9	13.1	E$^+$:		Mag 11.9 star on NE edge; **PGC 42667** S 1′0.
12 42 37.5	−40 44 44	NGC 4622	12.4	1.7 x 1.6	13.4	(R′)SA(r)a pec	173	Very thin, tightly wound arms.
12 43 49.0	−40 42 53	NGC 4622A	13.6	0.6 x 0.6	12.4	SAB0$^-$ pec:		Interacting pair with NGC 4622B on SE edge.
12 43 50.2	−40 43 07	NGC 4622B	13.8	1.0 x 0.5	13.0	S0 pec	63	Bright nucleus offset NE, fainter extension SW; pair with NGC 4622A on NW edge.
12 44 09.8	−41 45 00	NGC 4645	11.8	2.2 x 1.4	12.9	E$^+$	52	Several faint stars superimposed N and E of bright core.
12 43 05.6	−41 21 32	NGC 4645A	12.2	3.0 x 0.8	13.0	SB0° pec sp	33	Mag 11.5 star 0′9 SE; very small, faint, elongated anonymous galaxy at N end; it extends NW to a mag 15 star.
12 43 31.1	−41 21 42	NGC 4645B	12.1	1.9 x 0.7	12.3	SAB0:	154	Located 2′4 S of mag 7.17v star. Faint star on N edge, pair of fainter stars on W edge.
12 44 19.2	−40 43 59	NGC 4650	11.6	3.2 x 2.8	13.9	SB(s)0/a pec	164	Bright, diffuse nucleus with short, bright bar E-W; **PGC 42911** E 1′9.
12 44 49.2	−40 42 53	NGC 4650A	13.3	1.6 x 0.8	13.4	S0/a pec:	158	Multi-galaxy system. The much elongated NGC 4650A has a short, stout companion laying at a right angle at its center forming an elongated plus sign.
12 45 14.8	−40 49 29	NGC 4661	13.5	1.1 x 0.4	12.5	E$^+$: sp	121	A pair of mag 13 stars N 1′1.
12 46 15.0	−41 42 28	NGC 4672	13.1	2.0 x 0.6	13.1	SA(s)a pec sp	47	Mottled appearance, several stars superimposed; bright patch E side.
12 46 57.1	−41 34 54	NGC 4677	12.8	1.7 x 0.7	12.9	SB(s)0$^+$:	167	Bright nucleus, two faint stars along W edge.
12 47 28.8	−43 20 08	NGC 4681	12.5	1.2 x 1.0	12.5	Sab II	166	Mag 7.83v star NW 6′2.
12 47 42.4	−41 31 44	NGC 4683	12.7	1.4 x 0.8	12.7	SB(s)0$^-$	130	Faint star on SE end; mag 12.8 star SW 1′1.
12 48 49.3	−41 18 43	NGC 4696	10.4	4.5 x 3.2	13.3	E$^+$1 pec	107	Numerous stars superimposed. **PGC 43301** E 0′9 of center; **ESO 322-93** ESE 3′2; **PGC 43294** N 2′6; **PGC 43280** NNW 3′1; and **PGC 43269** W 3′5.
12 46 55.6	−41 29 54	NGC 4696A	13.7	1.4 x 0.5	13.2	SA(s)b? II-III	174	Lies between a mag 14 star E and a mag 13 star W.
12 47 21.7	−41 14 18	NGC 4696B	12.7	1.3 x 0.7	12.4	SA0$^-$:	40	Large center with a mag 13.9 star on SW end.
12 48 02.6	−40 49 06	NGC 4696C	13.7	1.9 x 0.3	12.9	Sb: sp	139	Mag 11.7 star SSE 3′3.
12 48 21.3	−41 42 52	NGC 4696D	12.8	1.9 x 0.4	12.4	SB(r)0° pec sp	132	Mag 13 star on SW edge; mag 14.6 star SE of center.
12 48 26.0	−40 56 11	NGC 4696E	13.6	1.8 x 0.7	13.7	SB(s)0° sp	0	Stellar nucleus, faint envelope; mag 11.7 star NW 4′9.
12 49 54.0	−41 16 45	NGC 4706	12.9	1.4 x 0.6	12.6	SAB(s)0° sp	24	Very faint **PGC 43412** 1′2 N.
12 50 03.7	−41 22 54	NGC 4709	10.9	2.4 x 2.0	12.6	E1	112	Bright nucleus, several stars superimposed on smooth envelope; **MCG −7-26-57** SE 1′2.
12 51 46.3	−41 07 59	NGC 4729	12.3	1.6 x 1.6	13.4	E$^+$:		Numerous faint stars superimposed on envelope surrounding nucleus.
12 52 00.4	−41 08 51	NGC 4730	12.8	1.0 x 1.0	12.7	SA(r)0$^-$:		Very faint star S of nucleus; **PGC 43596** SW 2′9.
12 52 16.0	−41 23 24	NGC 4743	12.9	1.3 x 0.5	12.3	SA0$^+$:	176	Two faint stars N end; ESO 323-20 N 3′2; ESO 323-19 S 4′8.
12 52 19.5	−41 03 38	NGC 4744	12.6	2.1 x 1.0	13.3	SB(s)0/a	125	Bright nucleus in a bright bar.
12 52 50.5	−42 39 30	NGC 4751	11.8	3.0 x 1.1	13.0	SA0$^-$:	175	Number of superimposed stars S of bright center; **ESO 323-30** S 4′7.
12 53 26.9	−48 44 57	NGC 4785	12.4	1.9 x 1.0	13.0	(R′)SB(r)b: II	81	Numerous stars superimposed on outer envelope, in star rich area.
12 56 52.5	−41 47 51	NGC 4811	13.0	1.3 x 0.9	13.1	SAB(rs)0$^+$ pec	35	Distorted, interacting with NGC 4812 S.
12 56 52.6	−41 48 50	NGC 4812	12.9	1.1 x 0.4	11.9	S0 sp	36	Interacting with NGC 4811 N.
12 58 07.7	−46 15 49	NGC 4835	11.6	4.0 x 0.9	12.9	SAB(rs)bc: II	150	Several superimposed stars or faint knots. Mag 9.56v star SE 5′3; mag 7.48v star S 9′8.
12 57 13.1	−46 22 37	NGC 4835A	13.3	2.6 x 0.4	13.2	Scd: sp	0	Mag 7.48v star E 9′3; **ESO 269-18** SSE 8′3.
13 02 01.8	−42 46 20	NGC 4909	12.7	1.9 x 1.7	13.8	(R′)SA(rs)a	28	Bright, almost stellar nucleus in bright lens, faint outer ring; stars superimposed.
13 04 05.1	−41 24 42	NGC 4930	11.1	4.5 x 3.7	14.0	SB(rs)b II	40	bright nucleus in NE-SW bar, many superimposed stars; mag 8.37v and 10.12v stars just off E edge.
13 05 00.1	−47 14 14	NGC 4940	13.1	1.0 x 1.0	12.9	Sa		Large nucleus.
13 05 26.4	−49 28 34	NGC 4945	8.4	20.0 x 3.8	12.9	SB(s)cd: sp IV	43	Very patchy; prominent dark lanes; many stars superimposed.
13 06 34.0	−49 41 31	NGC 4945A	12.1	2.5 x 1.4	13.3	SB(s)m? sp V	55	Mag 8.53v star superimposed NE end; mag 4.28v star Xi 2 Centauri S 13′3.
13 05 29.4	−43 35 30	NGC 4946	12.4	1.5 x 1.1	12.9	E$^+$?	131	Large nucleus; faint star S edge.
13 05 36.4	−43 30 02	NGC 4950	13.7	0.9 x 0.7	13.0	SB(s)0°:		Compact.
13 08 37.4	−49 30 17	NGC 4976	10.0	5.6 x 3.0	13.1	E4 pec:	161	Very bright core; numerous superimposed stars; mag 10.89v star on NE edge; mag 7.80v star E 3′6.
13 09 54.3	−43 06 24	NGC 4988	13.1	1.7 x 0.5	12.8	S0$^+$:	26	Pair of faint stars S of elongated center; mag 9.42v star N 2′9.

(Continued from previous page)

184 GALAXIES 184

RA h m s	Dec ° ′ ″	Name	Mag (V)	Dim ′ Maj x min	SB	Type Class	PA	Notes
13 12 52.0	−43 05 46	NGC 5011	11.4	2.4 x 2.0	13.1	E1-2	154	Very bright nucleus with numerous superimposed stars.
13 12 09.6	−43 18 30	NGC 5011A	13.4	1.7 x 0.8	13.5	SAB(s)c: II	90	Stellar nucleus.
13 13 12.1	−43 14 49	NGC 5011B	13.3	0.8 x 0.4	12.0	S0: sp	78	NGC 5011C S 1′2.
13 13 11.8	−43 15 56	NGC 5011C	13.2	1.2 x 0.8	13.1	S0: pec	132	Low mostly uniform surface brightness.
13 14 13.6	−42 57 37	NGC 5026	11.3	3.2 x 2.0	13.2	(R′)SB(rs)0/a	52	Bright nucleus between two very faint stars; short, open "S" shaped bar.
13 19 00.2	−47 54 35	NGC 5064	11.8	1.9 x 0.9	12.2	(R′)SAab: II-III	32	
13 20 39.9	−43 42 04	NGC 5082	12.6	1.7 x 1.0	13.0	SB(rs)0°	31	Located 6′0 W of NGC 5090.
13 20 55.5	−43 44 22	NGC 5086	15.9	0.4 x 0.2	13.2	E	105	Almost stellar.
13 21 12.9	−43 42 17	NGC 5090	11.4	2.9 x 2.4	13.5	E2	109	Mag 6.67v star NNE 5′0; NGC 5091 SE edge of nucleus.
13 19 21.2	−43 38 59	NGC 5090A	12.8	1.7 x 0.6	12.7	S0°	148	Faint star on W edge.
13 20 17.6	−43 51 56	NGC 5090B	13.0	1.9 x 0.9	13.4	SB(r)b	125	Bright nucleus, two main arms, several stars superimposed.
13 21 18.8	−43 43 20	NGC 5091	12.9	1.8 x 0.4	12.4	Sb pec sp	130	Located SE of NGC 5090's bright nucleus; **ESO 270-3** SSW 2′5.
13 25 29.0	−43 01 00	NGC 5128	6.8	25.7 x 20.0	13.5	S0 pec	35	**Centaurus A**. Wide dark lane.
13 28 44.1	−48 55 00	NGC 5156	11.7	2.3 x 2.0	13.2	SB(r)b I-II	110	Large, bright, diffused center; mag 7.69v star SW 3′8.
13 33 43.8	−48 09 04	NGC 5206	10.9	3.7 x 3.2	13.5	(R′)SB(rs)0° pec:	37	Bright stellar nucleus, very faint envelope, many stars superimposed.
13 37 30.1	−49 50 12	NGC 5234	13.0	1.3 x 0.8	12.9	(R′)SAB(s)0⁺	39	Many stars superimposed, star rich area; ESO 220-22 NW 3′9; mag 9.01v star ENE 9′7.
13 37 38.8	−42 50 52	NGC 5237	12.5	1.9 x 1.6	13.5	I0?	115	
13 38 42.3	−45 51 18	NGC 5244	12.6	1.1 x 0.7	12.2	Sb:	17	Superimposed stars and/or knots.

GALAXY CLUSTERS

RA h m s	Dec ° ′ ″	Name	Mag 10th brightest	No. Gal	Diam ′	Notes
12 48 54.0	−41 18 00	A 3526	13.2	33	180	**Centaurus Cluster**. Numerous PGC galaxies mixed in with NGC and ESO galaxies.
13 33 18.0	−42 51 00	A 3561	16.3	36	17	All members anonymous, faint, stellar.
12 11 06.0	−46 40 00	AS 689	15.9	0	17	All members anonymous, stellar; **ESO 267-24** S of center 12′2 ; **ESO 267-25** S of center 16′1.
12 43 30.0	−42 56 00	AS 707	15.3		22	Stellar galaxy **Fair 457** at center; most members anonymous, stellar.
12 49 36.0	−44 04 00	AS 712	15.7		17	**PGC 43366** at center; **ESO 268-45** S 12′3; others anonymous, stellar.

OPEN CLUSTERS

RA h m s	Dec ° ′ ″	Name	Mag	Diam ′	No. ★	B ★	Type	Notes
13 23 42.0	−41 53 00	ESO 324-15		4			cl	

GLOBULAR CLUSTERS

RA h m s	Dec ° ′ ″	Name	Total V m	B ★ V m	HB V m	Diam ′	Conc. Class Low = 12 High = 1	Notes
13 26 45.9	−47 28 37	NGC 5139	3.9	11.5	14.5	55.0	8	**Omega Centauri**. Largest and brightest globular cluster in the sky; visible to the naked eye; impressive even in binoculars, and spectacular in telescopes.
12 38 40.2	−51 09 01	Ru 106	10.9	14.8	17.8	2.0		

PLANETARY NEBULAE

RA h m s	Dec ° ′ ″	Name	Diam ″	Mag (P)	Mag (V)	Mag cent ★	Alt Name	Notes
12 00 43.5	−47 33 12	PN G294.1+14.4	70	15.2			PK 294+14.1	Annular; mag 10.29v star SE 2′6.
12 30 52.5	−44 14 20	PN G299.0+18.4	65	12.6		18.7	PK 299+18.1	Slightly elongated N-S; mag 13.1 star E 1′9.

185 GALAXIES 185

RA h m s	Dec ° ′ ″	Name	Mag (V)	Dim ′ Maj x min	SB	Type Class	PA	Notes
10 23 11.7	−49 28 18	ESO 214-2	13.7	1.1 x 0.6	13.0	(R′)SB(r)a:	150	Mag 9.12v star E 4′3.
10 31 01.4	−49 02 54	ESO 214-13	13.8	1.1 x 0.5	13.0	SB(rs)d? III-IV	126	Mag 10.48v star NNW 7′7.
10 38 09.4	−50 09 31	ESO 214-16	15.1	2.1 x 0.2	14.0	Sc: sp	160	Mag 10.16v star WSW 8′2.
10 40 18.7	−48 34 09	ESO 214-17	11.3	4.1 x 3.3	14.0	SB(rs)d II-III	60	Many superimposed stars, mag 11.85star superimposed near NE edge; mag 10.15v star W 8′3.
10 55 34.8	−48 14 51	ESO 215-5	14.1	0.8 x 0.7	13.3	SB	115	Mag 11.6 star WSW 2′2.
10 56 21.2	−50 33 38	ESO 215-7	12.7	1.5 x 0.7	12.7	(R′)SAB(r)0⁺:	83	Mag 7.28v star SE 4′2.
10 57 06.0	−48 39 45	ESO 215-8	13.1	1.1 x 0.8	12.9	S0	14	Mag 14.7 star N edge; mag 14.3 star W of core; mag 10.94v star in star rich area.
10 58 12.6	−50 26 38	ESO 215-10	13.4	1.2 x 0.9	13.3	S0	144	Mag 14.4 star S end; mag 11.9 star W edge; mag 9.71v star E 3′0.
10 58 43.0	−50 19 29	ESO 215-12	14.3	1.8 x 0.3	13.5	Sbc sp	0	Mag 8.18v star N 3′3.
10 58 53.4	−49 58 35	ESO 215-13	13.5	1.5 x 1.0	13.8	IB(s)m IV	173	Elongated N-S core; a number of stars superimposed W side; mag 10.31v star SSW 8′4.
11 01 14.3	−49 54 28	ESO 215-15	14.3	1.5 x 0.3	13.3	SB(rs)0/a: sp	81	Mag 9.49v star ESE 9′9.
11 05 09.2	−48 26 39	ESO 215-19	14.3	1.1 x 0.7	13.9	Sc	127	Star knot E of core; mag 13.4 star W 1′4.
11 06 32.7	−48 02 29	ESO 215-21	13.7	1.2 x 0.3	12.4	(R′)SAB(r)a:	167	Mag 12.1 star N 1′2; in star rich area.
11 06 44.1	−48 29 02	ESO 215-23	13.7	1.0 x 0.6	12.9	S0/a	6	Large core; mag 13.4 star SW 1′3.
11 08 22.3	−47 55 53	ESO 215-27	14.2	1.7 x 0.3	13.3	Sb sp	98	Mag 9.65v star SW 11′3.
11 10 34.9	−49 06 13	ESO 215-31	12.8	2.4 x 1.5	14.0	(R′)SB(rs)b II-III	133	Mag 9.72v star N 4′3.
11 11 22.7	−48 01 14	ESO 215-32	13.3	1.1 x 0.7	12.8	SAB0°?	43	Mag 12.4 star NW 1′5.
11 11 54.7	−47 55 21	ESO 215-33	14.5	0.8 x 0.8	13.9			**ESO 215-34** E 1′6.
11 15 35.8	−48 45 36	ESO 215-37	12.8	2.6 x 2.3	14.6	SA(r)c II-III	175	Numerous superimposed stars; mag 11.05v star on S edge.
11 17 04.2	−49 12 08	ESO 215-39	13.0	1.5 x 1.0	13.3	SAB(s)c II	19	Several superimposed stars; mag 11.17v star W 3′7.
11 17 18.0	−48 26 35	ESO 215-40	13.5	1.1 x 0.6	12.9	SAB(r)0°?	90	Mag 10.08v star NNE 6′3.
11 18 15.5	−48 44 49	ESO 216-3	12.9	1.5 x 1.3	13.5	(R′)SB(rs)0⁺	94	Mag 7.31v star S 2′6.
11 25 45.6	−48 08 32	ESO 216-8	13.6	1.6 x 0.4	13.0	SAB(s)c: sp II	88	Mag 11.3 star N 4′4.
11 28 53.0	−50 32 07	ESO 216-10	13.7	0.4 x 0.4	11.6			Compact.
11 35 47.3	−49 01 49	ESO 216-21	13.5	1.6 x 1.0	13.9	SAB(rs)c: II	52	Mag 5.50v star SW 10′5.
11 37 44.8	−49 10 39	ESO 216-24	12.4	1.0 x 1.0	12.3	(R)SAB(rs)c I-II		Mag 11 star on W edge.
11 39 09.9	−50 40 08	ESO 216-27	13.8	1.0 x 0.7	13.3	(R′)SB(rs)b? II	20	Mag 11.6 star E 2′5.
11 39 11.7	−49 24 57	ESO 216-28	13.0	1.1 x 0.9	12.9	(R′)SAB(r)bc II	7	E-W bar with small dark patches N and S.
11 40 01.0	−48 57 43	ESO 216-31	13.9	2.2 x 0.5	13.9	SAab: sp	152	Mag 9.64v star N 3′2.
11 44 53.0	−50 32 41	ESO 216-37	13.5	1.2 x 0.7	13.2	SB(r)bc pec II-III	174	**ESO 216-39** ESE 6′2; mag 9.58v star SE 5′9.
11 44 55.6	−51 54 22	ESO 216-38	14.1	1.4 x 0.7	13.9	SB(r)ab	161	Mag 10.04v star N 7′2.
11 50 08.9	−49 37 05	ESO 217-9	14.2	1.0 x 0.4	13.0	S?	105	Pair of mag 10.25v and 10.9 stars E 1′7.

RA h m s	Dec ° ′ ″	Name	Mag (V)	Dim ′ Maj x min	SB	Type Class	PA	Notes
11 53 19.6	−49 35 11	ESO 217-12	13.0	1.3 x 1.2	13.4	SAB(s)c: I-II	136	Mag 10.46v star E 2′.1.
11 55 24.7	−50 18 01	ESO 217-14	13.3	1.5 x 0.5	12.8	IB(s)m: III-IV	15	Mag 9.81v star W 3′.4.
11 58 03.0	−51 53 07	ESO 217-15	13.0	1.1 x 0.8	12.7	SB0?	119	Mag 14.5 star on N edge; mag 10.01v star E 10′.1.
11 58 17.6	−51 08 06	ESO 217-16	13.3	1.3 x 0.7	13.1	SAB(r)b pec: II	155	Mag 12.4 star SE edge; mag 10.24v star NW edge.
11 58 20.6	−50 58 23	ESO 217-17	13.5	1.6 x 0.5	13.1	Sc: II-III	165	Mag 12 star superimposed N end.
12 01 48.2	−50 35 01	ESO 217-20	13.6	1.6 x 0.4	13.0	SBbc? sp	172	Mag 10.92v star NW 3′.2.
10 21 08.4	−42 54 02	ESO 263-28	14.1	0.4 x 0.4	12.0			Stellar.
10 21 10.4	−46 00 50	ESO 263-29	12.8	1.8 x 0.4	12.4	(R)SB(rs)0^+:	1	Mag 10.14v star ESE 3′.9.
10 22 59.6	−42 49 41	ESO 263-30	13.8	0.7 x 0.6	12.7	SB pec	63	Mag 11.4 star E edge.
10 23 13.5	−47 20 24	ESO 263-31	14.1	1.9 x 0.7	14.3	SB(rs)c: III-IV	160	Mag 11.07v star E 3′.8.
10 24 47.4	−43 57 54	ESO 263-33	13.0	0.8 x 0.8	12.4	(R′)SA(s)0^-:		Mag 11.6 star NE 1′.4; mag 7.56v star SW 7′.8.
10 26 14.2	−45 44 13	ESO 263-35	13.0	1.9 x 1.0	13.6	SA(rs)b: I-II	110	Mag 13.1 star on S edge; mag 11.6 star W 5′.9; **ESO 263-36** ENE 9′.8.
10 27 24.1	−43 08 14	ESO 263-37	13.5	0.7 x 0.7	12.6	SA(rs)b		Mag 9.96v star NNW 6′.0.
10 30 15.3	−44 18 34	ESO 263-46	14.4	0.8 x 0.3	12.7	SB(s)d: IV	41	Mag 6.85v star ESE 8′.3.
10 30 32.0	−46 16 13	ESO 263-47	14.0	1.3 x 1.0	14.1	SAB(s)m: V-VI	105	Mag 10.60v star NW 3′.8.
10 31 11.9	−46 15 08	ESO 263-48	11.5	1.9 x 1.2	12.3	SA(s)0°?	168	**ESO 263-49** S 2′.0.
10 31 40.2	−45 29 02	ESO 263-51	13.3	1.3 x 0.5	12.6	SB(rs)cd pec III-IV	162	Mag 9.20v star N 1′.4.
10 34 24.4	−46 33 36	ESO 264-4	13.7	0.9 x 0.4	12.5		30	Mag 11.4 star N 2′.7.
10 36 33.5	−44 31 55	ESO 264-11	14.2	1.1 x 0.9	14.0	SB(s)dm pec V	57	Numerous stars superimposed.
10 38 07.1	−44 27 14	ESO 264-18	13.9	0.8 x 0.6	13.0	S?	141	Mag 8.49v star NE 2′.1.
10 38 41.6	−45 54 32	ESO 264-21	13.5	1.2 x 0.6	13.0	Sab	119	Mag 10.56v star NNE 1′.6.
10 39 08.7	−46 30 21	ESO 264-23	13.7	1.1 x 0.5	12.9	S0	21	Mag 11.2 star SW 3′.7.
10 39 11.3	−46 08 30	ESO 264-24	13.2	1.9 x 0.6	13.2	$S0^-$: sp	147	Mag 9.72v star SW 6′.5; **ESO 264-20** SW 8′.4; **ESO 264-22** S 8′.2.
10 39 19.0	−44 59 57	ESO 264-25	13.9	1.7 x 0.5	13.5	SAB(rs)c II	160	Mag 7.61v star ESE 10′.3; **ESO 264-27** SE 9′.0.
10 40 06.1	−46 06 59	ESO 264-28	13.6	1.2 x 0.8	13.4	SB(r)0^+	173	Mag 8.08v star NE 4′.4.
10 40 15.4	−46 19 42	ESO 264-30	13.3	2.1 x 0.8	13.9	E	87	**ESO 264-26** W 5′.6.
10 40 33.3	−46 11 29	ESO 264-31	12.8	1.2 x 0.8	12.8	E?	25	Mag 9.58 star W 2′.4.
10 40 38.8	−46 18 59	ESO 264-32	13.6	1.5 x 1.1	14.0	(R′)SB(r)a	70	Knotty, several stars superimposed; mag 13.1 stars on E and W edges.
10 42 25.5	−44 53 11	ESO 264-34	13.7	0.9 x 0.4	12.5	SAB(s)0°:	108	Mag 10.03v star ESE 4′.6; stellar galaxy **Fair 435** W 5′.1.
10 42 51.2	−47 37 02	ESO 264-35	13.3	1.1 x 0.7	12.8	SB(s)d pec: III-IV	58	Very small, very faint anonymous galaxy S edge.
10 43 07.7	−46 12 47	ESO 264-36	13.1	1.0 x 0.7	12.6	SB?	107	Mag 12.9 star W 0′.9.
10 46 04.9	−45 19 38	ESO 264-39	13.0	1.7 x 1.2	13.7	(R′)SB(rs)a	152	Mag 9.08v star NE 14′.5; mag 10.59v star S 7′.9.
10 48 25.4	−45 41 29	ESO 264-41	13.7	1.5 x 1.2	14.1	SA(rs)c II-III	48	Mag 10.33v star N 4′.7.
10 48 38.5	−47 18 08	ESO 264-42	14.4	2.1 x 0.5	14.3	Pec:	54	
10 48 43.7	−45 25 11	ESO 264-43	13.4	2.1 x 0.5	13.3	Sb sp	1	Mag 8.31v star N 7′.1.
10 52 25.0	−46 22 08	ESO 264-46	12.8	1.5 x 1.1	13.2	(R′)SB(s)bc pec	162	Mag 9.07v star NE 1′.0.
10 52 37.8	−45 09 48	ESO 264-47	13.6	1.2 x 1.0	13.7	(R′)SAB(r)a	62	Mag 10.62v star W 2′.2.
10 52 47.9	−45 40 43	ESO 264-48	13.5	2.0 x 0.5	13.3	SAB(s)c: II-III	69	Mag 9.57v star W 10′.4.
10 54 05.1	−45 49 01	ESO 264-49	13.4	1.3 x 0.7	13.2	SAB(r)0^+:	164	Mag 11.6 star WSW 6′.6.
10 54 28.8	−46 12 42	ESO 264-50	13.9	1.0 x 0.5	13.0	SB(r)ab	169	Mag 9.93v star NNW 9′.6.
10 59 02.1	−43 26 31	ESO 264-57	13.7	0.9 x 0.5	12.9	SA(rs)cd:	124	Mag 7.29v star ESE 6′.8 **MCG −7-23-3** N 4′.0.
11 01 09.0	−44 02 26	ESO 265-3	12.9	2.8 x 1.3	14.1	SAB(s)cd III-IV	108	Mag 10.01v star SSW 2′.5.
11 07 49.5	−46 31 43	ESO 265-7	11.4	4.1 x 1.3	13.0	SB(s)cd II-III	141	Several strong dark patches; mag 10.21v star N 4′.0.
11 09 34.5	−46 45 00	ESO 265-9	13.3	0.9 x 0.9	13.0	SA(r)0/a:		Mag 9.89v star W 6′.4.
11 10 01.1	−47 01 13	ESO 265-11	14.5	0.9 x 0.2	12.4	S	88	**ESO 265-13** SE 2′.7; pair of stars, mags 9.90v and 10.69v, E 2′.1.
11 13 36.7	−44 01 07	ESO 265-16	14.2	1.9 x 0.4	13.7	SAc: sp II	49	Mag 10.65v star NE 3′.2.
11 18 42.6	−46 40 42	ESO 265-22	13.7	1.5 x 1.2	14.2	(R′)SB(rs)a	90	Mag 10.75v star on N edge.
11 23 00.1	−43 20 59	ESO 265-29	13.5	0.8 x 0.4	12.2	Sa:	11	Mag 9.65v star W 7′.5.
11 24 39.7	−46 15 08	ESO 265-31	13.6	0.8 x 0.6	12.9	E/S0	14	Mag 13.0 star E 1′.3.
11 26 34.8	−43 03 33	ESO 265-33	13.5	1.1 x 0.4	12.5	S0:	127	Close pair of stars, mags 12.9 and 13.6, NNE 1′.9.
11 28 11.0	−43 58 15	ESO 265-35	13.8	0.8 x 0.5	12.6	Sa?	166	Mag 8.77v star S 7′.7.
11 33 08.3	−45 52 29	ESO 266-2	14.0	1.1 x 0.7	13.5	Sa	118	Bright knot or star S edge.
11 33 40.6	−44 50 03	ESO 266-3	13.5	0.5 x 0.5	11.9	$SA0^-$		Mag 7.82v star NW 7′.4.
11 36 28.8	−45 03 45	ESO 266-8	14.2	1.5 x 0.3	13.2	Sb: pec sp	95	Mag 9.81v star NE 7′.7.
11 39 49.6	−44 33 38	ESO 266-12	13.8	1.0 x 0.6	13.1	(R′)SAB(s)bc	137	Mag 9.70v star ENE 7′.1; note: **ESO 266-13** is 1′.9 E of this star.
11 40 56.0	−44 28 55	ESO 266-15	12.4	1.5 x 1.0	12.7	Sbc I-II	139	Mag 7.79v star N 4′.8; mag 9.70v star SW 5′.7; note: **ESO 266-13** is 1′.9 E of this star.
11 43 35.7	−44 23 57	ESO 266-19	13.6	1.0 x 0.9	13.3	S	89	Bright star E of nucleus; mag 10.68v star NE 1′.7.
11 44 54.1	−44 05 58	ESO 266-20	13.9	0.9 x 0.6	13.0	(R′)SB(r)a	119	Mag 12.0 star NE edge.
11 45 24.2	−44 25 28	ESO 266-22	13.3	0.8 x 0.7	12.7	E?	138	**ESO 266-22A** SW 1′.6.
11 45 30.4	−43 34 28	ESO 266-23	13.7	1.0 x 0.8	13.3	SB(r)b II	9	Stellar nucleus; mag 12.3 star SE 2′.3.
11 59 28.0	−47 09 45	ESO 267-5	13.9	1.1 x 0.6	13.3	Sc	66	Several knots and/or superimposed stars.
12 03 31.6	−43 39 16	ESO 267-11	12.6	1.0 x 0.8	12.3	S0: pec	48	Mag 10.75v star N 0′.8.
10 20 24.4	−40 24 10	ESO 317-14	13.5	0.8 x 0.6	12.6	S0	94	Mag 9.60v star N 2′.3.
10 23 07.7	−42 14 18	ESO 317-20	12.8	2.0 x 1.8	14.1	SA(rs)c: II-III	128	Mag 10.37v star ENE 7′.3; **ESO 317-18** NNW 3′.4.
10 27 00.4	−40 01 31	ESO 317-27	13.6	1.3 x 0.3	12.4	S0/a	148	Mag 10.31 star W 2′.3.
10 28 01.8	−42 06 40	ESO 317-32	13.6	1.5 x 0.6	13.3	(R′)SB(r)b? II	72	Mag 9.79v star NNW 4′.8.
10 31 22.7	−42 03 39	ESO 317-41	13.5	1.7 x 0.5	13.2	SB(r)bc: pec II	105	Mag 12.7 star superimposed E of center; mag 10.27v star NNW 4′.3.
10 42 06.5	−40 34 37	ESO 318-2	13.6	1.5 x 0.4	12.9	SA(s)a:	166	Mag 10.01v star W 3′.1.
10 43 58.9	−40 07 11	ESO 318-6	13.6	0.6 x 0.5	12.1	SA?(r:) pec	52	Mag 9.37v star NE 2′.3.
10 44 43.3	−40 35 11	ESO 318-8	13.8	0.9 x 0.8	13.3	SBa	116	Mag 9.62v star E 6′.6; **ESO 318-9** S 3′.5; **ESO 318-7** SW 4′.8.
10 49 07.0	−41 33 18	ESO 318-14	13.7	1.2 x 0.3	12.5	Sa	54	Mag 9.85v star ESE 4′.9.
10 53 06.0	−40 19 48	ESO 318-21	12.6	1.1 x 0.7	12.4	E?	131	Mag 7.99v star S 5′.2.
11 17 52.9	−40 35 37	ESO 319-11	14.2	2.5 x 1.5	15.4	SB(r)cd: I-II	115	Numerous stars scattered over faint outer bands.
11 29 29.2	−41 04 39	ESO 319-24	15.0	1.3 x 0.6	14.6	Sc:	176	Mag 9.11v star N 6′.0.
11 30 20.8	−41 04 01	ESO 319-26	15.0	1.6 x 0.2	13.6	Sc sp	69	Mag 10.76v star E 3′.5.
11 32 07.9	−41 25 38	ESO 320-2	14.5	1.7 x 0.3	13.6	Sbc: sp	29	Mag 8.30v star SW 10′.8.
11 38 34.8	−42 00 55	ESO 320-15	14.0	1.0 x 0.7	13.5	SA?(r)a	10	Pair of stars, mags 11.8 and 11.08v, SE 2′.3.
11 54 39.1	−40 55 54	ESO 320-32	13.7	0.9 x 0.5	12.7	(R′)SB(r)b	5	Mag 13.1 star SW 2′.3.
12 01 08.7	−42 44 50	ESO 321-2	13.2	1.1 x 1.0	13.1	Sb	160	Mag 11.01v star N 2′.9.
12 01 20.4	−42 40 15	ESO 321-3	13.9	0.9 x 0.4	12.6	Sb	121	Mag 11.01v star S 2′.1.
10 27 53.9	−40 04 55	NGC 3250A	14.3	1.1 x 0.2	12.5	Sb: sp	89	A faint star at E end.
10 27 44.8	−40 26 09	NGC 3250B	12.7	2.3 x 0.7	13.0	SBa pec?	6	
10 27 42.6	−40 00 12	NGC 3250C	13.1	1.8 x 0.7	13.2	(R′)SA(rs)ab:	56	NGC 3250A S 5′.1.
10 29 00.6	−40 04 57	NGC 3250E	12.5	2.1 x 1.4	13.6	SAB(s)cd: II-III	142	Mag 9.40v star NW 4′.4.
10 27 56.5	−43 54 04	NGC 3256	11.5	3.8 x 2.1	13.6	Pec	85	Bright core with very faint, broad arm extending E.
10 25 51.5	−43 44 53	NGC 3256A	14.2	1.3 x 0.6	13.7	SB(s)m pec: V	80	
10 29 00.5	−44 24 07	NGC 3256B	12.8	1.8 x 0.5	12.5	SB(s)bc: III-IV	135	Located 1′.5 E of mag 10.09v star.
10 29 05.7	−43 51 00	NGC 3256C	12.5	1.5 x 1.1	12.9	SB(rs)d III-IV	159	Several faint stars superimposed.
10 29 01.5	−44 39 27	NGC 3261	11.2	3.7 x 2.8	13.6	SB(rs)b I-II	85	Many stars superimposed; mag 12.5 star NE; mag 11.7 star SW.
10 29 06.1	−44 09 39	NGC 3262	13.2	1.1 x 0.7	12.7	SAB(rs)0^+ pec	108	NGC 3263 N 2′.5; mag 11.17v star E 4′.3.

(Continued from previous page)

185 GALAXIES 185

RA h m s	Dec ° ′ ″	Name	Mag (V)	Dim ′ Maj x min	SB	Type Class	PA	Notes
10 29 13.4	−44 07 22	NGC 3263	11.9	3.0 x 0.7	12.5	SB(rs)cd: sp	103	Close pair of mag 10.28v, 10.21v stars NE 7′3; NGC 3262 S 2′5.
10 37 16.4	−41 37 39	NGC 3318	11.6	2.4 x 1.3	12.7	SAB(rs)b II-III	78	
10 35 31.4	−41 44 26	NGC 3318A	15.0	1.4 x 0.4	14.2	SAB(rs)c:	3	Located between a pair of mag 12 stars 2′6 NE and 2′4 SW.
10 37 33.9	−41 27 56	NGC 3318B	13.6	1.5 x 1.1	14.0	SB(s)c III-IV	110	Mag 7.01v star NW 10′1.
10 35 08.0	−43 41 36	NGC 3366	11.3	1.8 x 0.8	11.5	(R′)SB(r)b: II	37	Lies 1′7 S of mag 6.11v star.
10 46 43.6	−40 00 58	NGC 3378	12.5	1.5 x 1.4	13.1	Sbc: II	11	Mag 10.04v star NE 7′2.
10 58 34.3	−46 35 04	NGC 3482	12.4	1.9 x 1.4	13.3	(R′)SB(rs)a	14	Several superimposed stars; mag 8.20v star E 8′8.
11 38 21.8	−50 42 57	NGC 3778	12.9	1.2 x 0.9	12.8	SAB0⁻:	24	Two faint stars on NW edge of nucleus.

GALAXY CLUSTERS

RA h m s	Dec ° ′ ″	Name	Mag 10th brightest	No. Gal	Diam ′	Notes
10 40 36.0	−46 11 00	AS 639	14.7	14	28	
10 44 24.0	−40 37 00	AS 643	16.2		17	See notes ESO 318-8, all others anonymous, stellar.
11 10 48.0	−46 55 00	AS 655	15.4		22	**ESO 265-14** S of center 7′8; **ESO 265-15** SSE 12′2.
11 32 06.0	−50 44 00	AS 669	15.9	60	17	**ESO 216-14** E of center 10′2; **ESO 216-13**, **13A** SE 11′0; all others anonymous.
11 40 18.0	−46 41 00	AS 673	16.1	12	17	**IC 2949** NNE of center 9′9 with mag 8.39v star W 5′1.

OPEN CLUSTERS

RA h m s	Dec ° ′ ″	Name	Mag	Diam ′	No. ★	B ★	Type	Notes
10 21 23.0	−51 43 24	NGC 3228	6.0	5	23	7.9	cl	Few stars; large brightness range; slight central concentration; detached.
10 52 17.0	−45 09 36	NGC 3446		6.5	26		cl?	(A) Galaxy ESO 264-47 on east side.
11 25 38.0	−43 15 00	NGC 3680	7.6	7	66	10.0	cl	Moderately rich in stars; moderate brightness range; strong central concentration; detached.
11 49 30.0	−48 16 00	NGC 3909		35	34	9.0	cl?	(S) Widespread group.

PLANETARY NEBULAE

RA h m s	Dec ° ′ ″	Name	Diam ″	Mag (P)	Mag (V)	Mag cent ★	Alt Name	Notes
10 54 40.7	−48 47 06	PN G283.9+9.7	300			12.2	PK 283+9.1	Mag 10.10v star E 3′6; mag 12.3 star WSW 3′4; very small, faint, elongated anonymous galaxy 0′7 N of central star.
11 13 50.5	−48 05 32	PN G286.5+11.6	215	11.6			PK 286+11.1	A triangle of mag 8.87v, 11.05v and 11.8 stars occupy the southern half of the nebula.
11 52 29.4	−42 17 39	PN G291.4+19.2	32	16.2			PK 291+19.1	Mag 13.2 star ENE 1′9; mag 14.2 star S 2′1.
11 53 06.9	−50 50 57	PN G293.6+10.9	82	12.2		18.0	PK 293+10.1	Mag 9.59v star on SE edge.
12 00 43.5	−47 33 12	PN G294.1+14.4	70	15.2			PK 294+14.1	Annular; mag 10.29v star SE 2′6.

186 GALAXIES 186

RA h m s	Dec ° ′ ″	Name	Mag (V)	Dim ′ Maj x min	SB	Type Class	PA	Notes
10 16 55.4	−48 52 52	ESO 213-11	11.2	2.6 x 1.8	12.7	SA(s)c I-II	10	Numerous stars cover the face of the galaxy.
10 23 11.7	−49 28 18	ESO 214-2	13.7	1.1 x 0.6	13.0	(R′)SB(r)a:	150	Mag 9.12v star E 4′3.
09 46 22.5	−46 39 06	ESO 262-4	14.5	2.3 x 0.3	13.9	Scd? sp	3	Mag 11v star on side of S end.
10 05 51.8	−44 13 32	ESO 263-3	13.6	1.4 x 0.7	13.4	Pec	117	Double system, strongly interacting.
10 06 46.3	−47 41 48	ESO 263-4	13.4	1.0 x 0.8	13.1	SB(r)a	115	Mag 13.0 star W edge.
10 06 55.8	−45 02 53	ESO 263-5	13.0	2.0 x 0.5	12.9	SB(s)b pec? II-III	66	Mag 11.27v star on N edge.
10 07 00.9	−43 03 30	ESO 263-6	14.0	0.6 x 0.3	12.1	S0?	116	Pair with ESO 263-7 S 1′1.
10 07 01.1	−43 04 36	ESO 263-7	13.6	1.5 x 0.3	12.6	Sb: sp	105	Mag 8.41v star ESE 5′4.
10 07 45.5	−43 29 55	ESO 263-8	14.5	1.5 x 0.3	13.5	S	32	Mag 9.46v star E 4′4.
10 09 17.1	−43 19 55	ESO 263-12	13.7	1.1 x 1.0	13.7	Sbc	42	Mag 9.98v star NW 10′0.
10 09 48.2	−42 48 42	ESO 263-13	13.7	1.3 x 1.0	13.9	SAB(rs)b	13	Almost stellar nucleus, uniform envelope.
10 10 59.7	−45 09 07	ESO 263-14	12.9	1.1 x 0.6	12.2	SB(s)b? II-III	106	Mag 8.53v star SW 2′5.
10 12 19.8	−47 17 41	ESO 263-15	13.3	3.2 x 0.3	13.1	Scd: sp	108	Mag 10.19v star S 4′4.
10 12 32.0	−45 14 17	ESO 263-16	13.0	1.4 x 0.7	12.8	SA(s)c pec	165	Mag 8.72v star NE 6′3.
10 13 30.0	−43 42 57	ESO 263-18	13.7	2.4 x 0.3	13.2	Sbc sp	128	Mag 8.64v star SW 6′4; ESO 263-19 S 6′1.
10 13 30.9	−43 49 06	ESO 263-19	13.0	1.3 x 0.8	12.9	SAB(r)bc III-IV	38	Mag 8.64v star W 4′9; ESO 263-18 N 6′1.
10 14 42.0	−44 51 05	ESO 263-21	12.5	1.2 x 0.8	12.3	IB(s)m IV	27	Four stars in an E-W row across S end; mag 8.73v star S 1′8.
10 14 57.4	−43 37 10	ESO 263-23	12.8	2.4 x 0.4	12.6	SA0/a? pec	81	Mag 9.11v star W 3′4.
10 15 33.4	−45 08 00	ESO 263-24	13.4	1.5 x 0.8	13.5	SB(r)0°	0	Mag 11.28v star SE 8′3.
10 21 08.4	−42 54 02	ESO 263-28	14.1	0.4 x 0.4	12.0			Stellar.
10 21 10.4	−46 00 50	ESO 263-29	12.8	1.8 x 0.4	12.4	(R)SB(rs)0⁺:	1	Mag 10.14v star ESE 3′9.
10 22 59.6	−42 49 41	ESO 263-30	13.8	0.7 x 0.6	12.7	SB pec	63	Mag 11.4 star E edge.
10 23 13.5	−47 20 24	ESO 263-31	14.1	1.9 x 0.7	14.3	SB(rs)c: III-IV	160	Mag 11.07v star E 3′8.
10 24 47.4	−43 57 54	ESO 263-33	13.0	0.8 x 0.8	12.4	(R′)SA(s)0⁻:		Mag 11.6 star NE 1′4; mag 7.56v star SW 7′8.
10 26 14.2	−45 44 13	ESO 263-35	13.0	1.9 x 1.0	13.6	SA(rs)b: I-II	110	Mag 13.1 star on S edge; mag 11.6 star W 5′9; **ESO 263-36** ENE 9′8.
10 27 24.1	−43 08 14	ESO 263-37	13.5	0.7 x 0.7	12.6	SA(rs)b		Mag 9.96v star NNW 6′0.
09 42 21.9	−41 49 00	ESO 315-20	15.2	1.7 x 0.3	14.3	S	56	Mag 8.71v star NE 9′6.
10 02 44.2	−42 05 34	ESO 316-18	13.6	2.6 x 0.3	13.1	Sc sp	116	Mag 10.39v star S 6′6.
10 04 26.2	−41 25 00	ESO 316-21	13.8	2.7 x 0.3	13.5	Sb? sp	18	Mag 10.79v star E 1′4.
10 06 21.7	−41 57 31	ESO 316-25	13.9	0.7 x 0.3	12.1	S0	113	Mag 11.9 star SE 0′6.
10 06 41.5	−41 07 08	ESO 316-26	13.2	0.9 x 0.5	12.2	Sa:	14	Mag 11.8 star N 7′7.
10 07 45.3	−41 19 56	ESO 316-29	13.5	1.7 x 0.7	13.5	(R′)SB(rs)ab?	77	Mag 10.05v star N 4′1.
10 08 40.1	−40 05 38	ESO 316-30	13.8	1.0 x 0.6	13.1	(R)SA0⁺	170	Mag 15 star E edge.
10 09 01.1	−40 19 04	ESO 316-31	13.4	1.0 x 0.7	12.9	S0	113	Mag 14.4 star W edge; pair of mag 13 stars E 2′3.
10 15 59.8	−41 33 22	ESO 317-7	13.6	0.9 x 0.6	12.8	S pec	21	Mag 13.1 star SE 1′7.
10 16 26.1	−42 14 48	ESO 317-8	13.8	0.9 x 0.7	13.2	(R′)SAB(rs)b II-III	88	Mag 10.22v star N 1′6.
10 20 24.4	−40 24 10	ESO 317-14	13.5	0.8 x 0.6	12.6	S0	94	Mag 9.60v star N 2′3.
10 23 07.7	−42 14 18	ESO 317-20	12.8	2.0 x 1.8	14.1	SA(rs)c: II-III	128	Mag 10.37v star ENE 7′3; **ESO 317-18** NNW 3′4.
10 27 00.4	−40 01 31	ESO 317-27	13.6	1.3 x 0.3	12.4	S0/a	148	Mag 10.31 star W 2′3.
10 27 53.9	−40 04 55	NGC 3250A	14.3	1.1 x 0.2	12.5	Sb: sp	89	A faint star at E end.
10 27 44.8	−40 26 09	NGC 3250B	12.7	2.3 x 0.7	13.0	SBa pec?	6	
10 27 42.6	−40 00 12	NGC 3250C	13.1	1.8 x 0.7	13.2	(R′)SA(rs)ab:	56	NGC 3250A S 5′1.
10 27 56.5	−43 54 04	NGC 3256	11.5	3.8 x 2.1	13.6	Pec	85	Bright core with very faint, broad arm extending E.
10 25 51.5	−43 44 53	NGC 3256A	14.2	1.3 x 0.6	13.7	SB(s)m pec: V	80	

(Continued from previous page)

OPEN CLUSTERS

RA h m s	Dec ° ′ ″	Name	Mag	Diam ′	No. ★	B ★	Type	Notes
08 44 48.0	−45 58 00	Bo 7	6.8	20	12	7.6	cl	Few stars; large brightness range; not well detached. (A) Several bright stars running N-S on E side of field, otherwise only rich field.
08 44 40.0	−41 17 00	Cr 197	6.7	25	40	7.3	cl	Moderately rich in stars; large brightness range; no central concentration; detached; involved in nebulosity.
09 55 00.0	−50 57 00	Cr 213	8.7	17	21	8.7	cl	Few stars; moderate brightness range; not well detached. (A)Irregularly shaped.
09 04 48.0	−46 02 00	ESO 260-17		7			cl	
09 11 24.0	−47 02 00	ESO 261-3		10			cl	
09 24 18.0	−44 42 00	ESO 261-7		33			cl	
09 40 33.0	−50 35 00	Ho 1		7			ast?	(A) Scattering of 5 bright stars and some fainter ones.
09 00 31.0	−48 59 06	Mrk 18	7.8	5	30	9.3	cl	Moderately rich in bright and faint stars with a strong central concentration; detached.
08 45 30.0	−48 48 00	NGC 2670	7.8	7	30	13.0	cl	Moderately rich in stars; moderate brightness range; no central concentration; detached.
08 46 12.0	−41 52 36	NGC 2671	11.6	5	40		cl	Moderately rich in stars; moderate brightness range; strong central concentration; detached.
09 19 23.0	−40 31 12	NGC 2849	12.5	3	40		cl	Moderately rich in stars; small brightness range; strong central concentration; detached.
09 22 06.0	−51 06 00	NGC 2866		20			cl?	(S) Compact group.
09 40 12.0	−50 19 18	NGC 2972	9.9	5	25	11.4	cl	Few stars; small brightness range; slight central concentration; detached.
09 42 01.0	−44 02 00	NGC 2982		12		10.0	cl	Few stars; small range in brightness; no central concentration; detached. (S) Sparse group elongated E-W.
10 21 23.0	−51 43 24	NGC 3228	6.0	5	23	7.9	cl	Few stars; large brightness range; slight central concentration; detached.
09 15 53.0	−50 00 42	Pi 11		2.5	20	12.0	cl	
09 20 00.0	−45 07 00	Pi 12	9.7	5	20	12.0	cl	Few stars; small brightness range; no central concentration; detached. Doubtful cluster.
09 22 06.5	−51 06 08	Pi 13	10.2	2	30	12.0	cl	Moderately rich in stars; moderate brightness range; strong central concentration; detached.
09 34 45.0	−48 02 06	Pi 15		5	35	14.0	cl	
08 44 49.1	−47 36 04	Ru 69		2	10	14.0	cl	
08 49 06.0	−46 49 00	Ru 70		5		13.0	cl	
08 49 30.0	−46 47 00	Ru 71		7	30	11.0	cl	Moderately rich in stars; moderate brightness range; no central concentration; detached.
09 01 12.0	−50 54 30	Ru 73		3	25	14.0	cl	
09 24 18.0	−51 39 54	Ru 76	10.8	5	20	13.0	cl	Few stars; moderate brightness range; not well detached. (S) Sparse group.
09 45 13.0	−44 06 18	Ru 81		5		12.0	cl:	(S) Sparse group.
10 15 30.0	−50 42 18	Ru 87		4	20	11.0	cl	Few stars; moderate brightness range; no central concentration; detached. (S) Small group.
09 55 30.0	−47 01 00	Ru 160		1.7		14.0	cl	(S) Small group.
08 47 45.0	−42 30 00	Tr 10	5.0	30	40	6.4	cl	Moderately rich in bright and faint stars; detached. (S) Small group of bright stars.
08 49 41.0	−44 21 12	vdB-Ha 54		4	20		cl	
08 57 08.0	−43 15 00	vdB-Ha 56		20	35		cl	Moderately rich in stars; large brightness range; not well detached; involved in nebulosity. (S) Large circlet.
09 20 38.0	−49 13 18	vdB-Ha 63		1.5			cl	
09 26 43.0	−51 16 30	vdB-Ha 67		4	40		cl	(S) Rich cluster.
09 31 56.0	−50 13 12	vdB-Ha 73		1.5	20		cl	(S) Rich cluster.
10 02 00.0	−49 35 30	vdB-Ha 85		5	50		cl	(A) Group of 4 brighter stars with two included clumps (clusters?) NW and SE.
10 06 15.0	−51 36 48	vdB-Ha 88		4	20		cl	

GLOBULAR CLUSTERS

RA h m s	Dec ° ′ ″	Name	Total Vm	B ★ Vm	HB Vm	Diam ′	Conc. Class Low = 12 High = 1	Notes
10 17 36.8	−46 24 40	NGC 3201	6.9	11.7	14.8	20.0	10	

BRIGHT NEBULAE

RA h m s	Dec ° ′ ″	Name	Dim ′ Maj x min	Type	BC	Color	Notes
08 44 48.0	−41 20 00	Gum 15	20 x 15	E	3-5	2-4	= **RCW 32**. Irregularly round and somewhat mottled in appearance. On photos a narrow rift of obscuring matter 1′ wide and 7′ long (E-W) is visible very close S of the brightest involved star.
08 51 00.0	−42 05 00	Gum 17	100 x 65	E	3-5	4-4	= **RCW 33**. Very faint and diffuse nebulosity; fairly even surface brightness.
08 59 30.0	−47 27 00	Gum 23	25 x 20	E	2-5	2-4	= **RCW 38**. A nebular complex which appears largely obscured, effectively breaking it up into one small and three larger nebulosities.
09 02 24.0	−48 42 00	Gum 25	7 x 6	E	2-5		= **RCW 40**. Roundish and of rather even surface brightness. The smaller **RCW 39** (not plotted) lies 20′ NNE. It is of similar surface brightness and about 2′.5 across.
09 00 18.0	−45 57 00	NGC 2736	20 x 3	SNR-E	2-5	2-4	Part of Gum Nebula. A long, narrow streak less than 1′ wide, aligned NNE-SSW. It is brightest along the northern two-thirds of its length and best defined along the E edge. Off the W edge are many short, faint, nebulous strands of nebulosity, filamentary in nature.

DARK NEBULAE

RA h m s	Dec ° ′ ″	Name	Dim ′ Maj x min	Opacity	Notes
08 53 36.0	−42 13 00	SL 4	15 x 7	5	Involved in emission nebula Gum 17.

PLANETARY NEBULAE

RA h m s	Dec ° ′ ″	Name	Diam ″	Mag (P)	Mag (V)	Mag cent ★	Alt Name	Notes
09 12 26.8	−42 25 42	NGC 2792	21	13.5	11.6	17.2	PK 265+4.1	Close pair of mag 11 stars SE 3′.6.
10 07 01.8	−40 26 09	NGC 3132	88	8.2	9.2	10.0	PK 272+12.1	**Eight-Burst Nebula**. Elongated N-S; pair of mag 11 stars SE 2′.4.
08 53 30.8	−40 03 44	PN G261.6+3.0	49	14.1	12.9:	17.8	PK 261+2.1	Elongated NE-SW; mag 8.10v star S 4′.8.
08 48 40.7	−42 54 05	PN G263.2+0.4	185	13.4		11.3	PK 263+0.1	A line of four mag 12 stars stretch E-W across the center; there is a close pair of mag 11-12 stars on the N edge.
09 16 09.7	−45 28 45	PN G268.4+2.4	5	14.2		18.7	PK 268+2.1	Close N-S pair of mag 13-14 stars S 1′.8; mag 11.24v star NW 3′.2; mag 6.28v star SW 10′.7.
08 54 18.4	−50 32 24	PN G269.7−3.6	14	13.4	13.6	13.6	PK 269−3.1	Mag 12.8 star NW 2′.9; mag 12.9 star E 1′.8.
09 52 43.6	−46 16 56	PN G273.6+6.1	90			12.4	PK 273+6.1	Mag 12.5 star near center; mag 11.5 star on NW edge; mag 8.77v star WNW 2′.6.
09 41 09.8	−49 22 49	PN G274.1+2.5	10	17.5		16.5	PK 274+2.1	Close pair of mag 13.6, 14.0 stars WNW 1′.5; mag 13.4 star S 2′.4.
10 05 45.8	−44 21 39	PN G274.3+9.1	48	16.0		20.5	PK 274+9.1	Mag 10.41v star N 3′.9; mag 11.8 star SW 4′.7.
09 41 36.3	−49 57 47	PN G274.6+2.1	5	13.7	14.3	16.2	PK 274+2.2	Mag 13.8 star SE 0′.3; mag 11.6 star E 0′.7.
09 47 25.0	−48 58 17	PN G274.6+3.5	23	13.9	14.:		PK 274+3.1	Mag 9.95v star SW 2′.1; mag 13.6 star N 1′.5.
10 13 16.1	−50 20 01	PN G278.8+4.9	18		12.8	17.6	PK 278+5.1	Mag 14.8 star on N edge; mag 14.9 star S 0′.5.

GALAXIES

RA h m s	Dec ° ′ ″	Name	Mag (V)	Dim ′ Maj x min	SB	Type Class	PA	Notes
07 09 10.3	−51 28 03	ESO 207-22	14.7	1.1 x 1.0	14.6	IA(s)m V-VI	96	Uniform surface brightness, several stars involved.
07 10 52.5	−51 48 24	ESO 207-25	13.7	1.3 x 0.6	13.3	(R)SB(s)0⁺:	170	Mag 10.81v star S 9′.4.
07 21 20.2	−50 45 31	ESO 208-3	14.0	0.7 x 0.7	13.1	SA(rs)d: pec V-VI		Mag 7.31v star SSW 6′.5.
07 27 12.1	−51 22 16	ESO 208-15	13.5	1.1 x 1.0	13.4	SB(r)a:	40	Bright core, uniform envelope; mag 6.72v star SE 2′.8.
07 33 56.3	−50 26 37	ESO 208-21	11.2	2.1 x 1.4	12.3	SAB0⁻	108	Mag 6.69v star SSW 9′.6.
07 35 19.8	−50 15 10	ESO 208-25	13.2	1.0 x 0.8	12.8	Sc	100	Mag 11.7 star SW 5′.3.
07 35 21.4	−50 02 37	ESO 208-26	13.8	1.2 x 0.5	13.1	(R)SB0⁺	107	Pair of mag 14.4 and 14.9 stars E edge.
07 38 22.3	−50 45 26	ESO 208-33	13.5	1.5 x 0.7	13.4	SB(s)m V-VI	164	Mag 10.32v star NNW 9′.0; **ESO 208-29** SW 9′.2.
07 58 15.9	−49 51 20	ESO 209-9	12.0	6.3 x 1.0	13.8	SB(s)cd: sp	152	Numerous stars superimposed whole length of galaxy.
07 35 07.1	−46 55 34	ESO 257-19	13.6	2.7 x 0.4	13.6	SB(s)cd? sp	131	Mag 13.7 star on W edge; mag 9.91v star SSW 11′.1.
07 44 55.7	−41 46 57	ESO 311-7	13.7	1.3 x 1.3	14.1	SA(s)d		**ESO 311-16** S 1′.0.
07 47 34.1	−41 27 10	ESO 311-12	11.8	3.4 x 0.5	12.2	S0/a? sp	14	Mag 6.92v star SW 5′.9.
07 36 30.5	−47 38 20	NGC 2427	11.5	5.2 x 2.2	14.0	SAB(s)dm IV	122	Numerous superimposed stars; **ESO 257-21** ENE 11′.4.

GALAXY CLUSTERS

RA h m s	Dec ° ′ ″	Name	Mag 10th brightest	No. Gal	Diam ′	Notes
07 08 30.0	−49 12 00	A 3408	15.5	41	22	**PGC 20233** at center; **ESO 207-24** E 19′.6; **ESO 207-21** W 14′.0; all others anonymous.
07 09 30.0	−50 09 00	AS 602	16.3	3	17	**ESO 207-23** S of center 5′.9; all others anonymous, stellar.

OPEN CLUSTERS

RA h m s	Dec ° ′ ″	Name	Mag	Diam ′	No. ★	B ★	Type	Notes
08 44 48.0	−45 58 00	Bo 7	6.8	20	12	7.6	cl	Few stars; large brightness range; not well detached. (A) Several bright stars running N-S on E side of field, otherwise only rich field.
08 44 40.0	−41 17 00	Cr 197	6.7	25	40	7.3	cl	Moderately rich in stars; large brightness range; no central concentration; detached; involved in nebulosity.
07 45 48.0	−46 43 00	ESO 258-1		10			cl	
07 49 24.0	−42 42 00	ESO 311-14		6			cl	
08 26 54.0	−41 15 00	ESO 312-4		1			cl	
08 31 42.0	−41 47 00	ESO 313-3		2.5			cl	
08 43 54.0	−40 14 00	ESO 313-12		10			cl	
08 42 31.0	−48 07 00	IC 2395	4.6	13	40	5.5	cl	Moderately rich in bright and faint stars; detached.
07 26 19.0	−47 41 00	Mel 66	7.8	15	821	11.4	cl	Moderately rich in stars; small brightness range; slight central concentration; detached.
08 10 11.0	−49 14 00	NGC 2547	4.7	25	112	7.0	cl	Rich in stars; large brightness range; strong central concentration; detached; involved in nebulosity.
08 39 04.0	−46 13 36	NGC 2645	7.0	3	12	9.0	cl	
08 42 34.0	−45 00 00	NGC 2659	8.6	15	80	10.0	cl	Rich in stars; moderate brightness range; no central concentration; detached. (S) Group extends to NE.
08 42 39.0	−47 12 00	NGC 2660	8.8	3	381	13.0	cl	Moderately rich in stars; small brightness range; strong central concentration; detached.
08 45 30.0	−48 48 00	NGC 2670	7.8	7	30	13.0	cl	Moderately rich in stars; moderate brightness range; no central concentration; detached.
08 46 12.0	−41 52 36	NGC 2671	11.6	5	40		cl	Moderately rich in stars; moderate brightness range; strong central concentration; detached.
08 17 54.0	−41 40 24	Pi 2		4	100	15.0	cl	
08 34 36.0	−44 22 00	Pi 4	5.9	25	45	8.0	cl	Moderately rich in stars; large brightness range; no central concentration; detached; involved in nebulosity. (S) Large group in Gum 12 nebula.
08 41 35.0	−46 16 18	Pi 8	9.5	3	25	10.0	cl	Few stars; moderate brightness range; slight central concentration; detached.
08 12 36.0	−40 27 00	Ru 56		40	40	9.0	cl:	Moderately rich in stars; moderate brightness range; no central concentration; detached. Probably not a cluster.
08 24 26.0	−47 12 42	Ru 60		3	30	13.0	cl:	
08 32 40.0	−48 18 00	Ru 63		3	20	13.0	cl	
08 37 24.0	−40 05 00	Ru 64		70	80	9.0	cl	Rich in stars; large brightness range; slight central concentration; detached.
08 39 22.0	−44 03 18	Ru 65		5	20	13.0	cl:	Few stars; moderate brightness range; not well detached. Probably not a cluster.
08 41 45.0	−43 23 00	Ru 67	9.1	7	35	12.0	cl:	Moderately rich in stars; moderate brightness range; no central concentration; detached. Probably not a cluster.
08 44 49.1	−47 36 04	Ru 69		2	10	14.0	cl	
08 49 06.0	−46 49 00	Ru 70		5		13.0	cl	
08 49 30.0	−46 47 00	Ru 71		7	30	11.0	cl	Moderately rich in stars; moderate brightness range; no central concentration; detached.
08 01 42.0	−44 25 00	Ru 154		3	10	13.0	cl	
08 47 45.0	−42 30 00	Tr 10	5.0	30	40	6.4	cl	Moderately rich in bright and faint stars; detached. (S) Small group of bright stars.
08 31 12.0	−44 30 00	vdB-Ha 34		25	20		cl	Few stars; moderate brightness range; not well detached; involved in nebulosity. (S) Sparse group of brighter stars in Gum nebula region.
08 35 48.0	−43 36 06	vdB-Ha 37		3	40		cl	
08 49 41.0	−44 21 12	vdB-Ha 54		4	20		cl	
08 40 20.7	−46 07 26	Wa 6	8.4	2	30	9.2	cl	Few stars; large brightness range; slight central concentration; detached. (S) Small group.

BRIGHT NEBULAE

RA h m s	Dec ° ′ ″	Name	Dim ′ Maj x min	Type	BC	Color	Notes
08 30 00.0	−45 00 00	Gum 12	1200 x 720	SNR-E	4-5		**Vela Supernova Remnant** or **Gum Nebula**. Very filamentary structure.
08 44 48.0	−41 20 00	Gum 15	20 x 15	E	3-5	2-4	= **RCW 32**. Irregularly round and somewhat mottled in appearance. On photos a narrow rift of obscuring matter 1′ wide and 7′ long (E-W) is visible very close S of the brightest involved star.
08 51 00.0	−42 05 00	Gum 17	100 x 65	E	3-5	4-4	= **RCW 33**. Very faint and diffuse nebulosity; fairly even surface brightness.
08 35 36.0	−40 40 00	NGC 2626	5 x 5	E+R	1-5	1-4	A 10th mag star involved in the N part of an irregularly round nebula; an extensive absorption region some 3′ by 4′ in size off N side of NGC 2626.

DARK NEBULAE

RA h m s	Dec ° ′ ″	Name	Dim ′ Maj x min	Opacity	Notes
07 19 00.0	−44 35 00	Be 135	13 x 5	6	

PLANETARY NEBULAE

RA h m s	Dec ° ′ ″	Name	Diam ″	Mag (P)	Mag (V)	Mag cent ★	Alt Name	Notes
07 14 49.4	−46 57 40	PN G258.0−15.7	82	11.8			PK 258−15.1	Annular; mag 10.7 star on S edge; mag 11.3 star on N edge.
08 42 16.6	−40 44 12	PN G260.7+0.9	36				Vo 3	Mag 8.39v star NW 2′.5.
08 23 40.4	−43 12 43	PN G260.7−3.3	75	16.9			PK 260−3.1	Elongated E-W, very irregular shape; mag 12.9 star NE edge; mag 12.8 star S 0′.9.

(Continued from previous page)

187 PLANETARY NEBULAE 187

RA h m s	Dec ° ′ ″	Name	Diam ″	Mag (P)	Mag (V)	Mag cent ★	Alt Name	Notes
08 23 53.9	−45 31 11	PN G262.6−4.6	20	15.3			PK 262−4.1	Mag 11.15v star SE 1'.2.
08 20 39.9	−46 20 15	PN G263.0−5.5	3	13.2			PK 263−5.1	Mag 12.3 star WSW 2'.1; mag 13.4 star E 1'.9.
08 48 40.7	−42 54 05	PN G263.2+0.4	185	13.4		11.3	PK 263+0.1	A line of four mag 12 stars stretch E-W across the center; there is a close pair of mag 11-12 stars on the [illegible] edge.
08 05 10.9	−48 23 33	PN G263.3−8.8	24	15.3			PK 263−8.1	Mag 10.3 star on NE edge.
08 11 31.9	−48 43 15	PN G264.1−8.1	45	12.4	12.4	16.9	PK 264−8.1	Mag 11.8 star W 2'.4; mag 12.6 star E 1'.9.
08 34 28.9	−46 25 38	PN G264.4−3.6	60				PK 264−3.1	Mag 10.73v star E 2'.6.
07 47 20.1	−51 15 05	PN G264.4−12.7	3	12.3		14.7	PK 264−12.1	Mag 15.3 star on N edge; mag 12.1 star WSW 2'.5.
08 43 24.1	−46 08 28	PN G265.1−2.2	5	17.5			PK 265−2.1	Mag 11.9 star E 2'.5; mag 11.7 star N 0'.7.
08 34 06.9	−47 16 41	PN G265.1−4.2	28	14.2			PK 265−4.1	Mag 14.2 star on N edge; mag 11.7 star SW 2'.1.

188 GALAXIES 188

RA h m s	Dec ° ′ ″	Name	Mag (V)	Dim ′ Maj x min	SB	Type Class	PA	Notes
05 32 16.2	−50 34 20	ESO 204-13	14.3	1.4 x 0.6	14.0	S0° pec sp	9	
05 32 32.3	−49 54 14	ESO 204-14	14.1	1.2 x 0.5	13.4	Sb? III-IV	74	Mag 10.61v star SE 2'.2.
05 44 06.6	−49 53 20	ESO 204-32	14.2	0.8 x 0.7	13.5	S?	2	Mag 13.8 star N edge.
05 44 41.5	−51 57 51	ESO 204-34	14.3	1.4 x 1.1	14.6	IB(s)m V-VI	68	Uniform surface brightness; mag 10.52v star superimposed E of center.
05 45 00.7	−48 05 13	ESO 204-35	14.2	0.9 x 0.6	13.4	SAB(rs)0^-:	9	**ESO 204-35A** NE 0'.5; **ESO 204-33** SW 10'.9; **ESO 204-31** SW 15'.0.
05 57 41.4	−51 58 13	ESO 205-11	13.9	1.2 x 0.9	13.8	(R)SB(r)b: II	131	Strong N-S bar.
05 58 58.7	−48 19 23	ESO 205-12	13.8	0.8 x 0.4	12.5	SAB(rs)0° pec	9	Star or knot N edge.
05 59 09.8	−51 28 13	ESO 205-13	13.0	1.6 x 0.8	13.1	Sa:	66	Mag 8.13v star E 13'.3.
06 05 10.1	−49 18 02	ESO 205-18	13.5	1.2 x 0.9	13.4	Sbc	158	Three bright knots.
06 08 35.5	−48 49 52	ESO 205-23	14.6	1.6 x 0.3	13.7	Sa? sp	18	Mag 10.84v star on N end.
06 09 40.3	−47 36 55	ESO 205-27A	13.8	0.5 x 0.3	11.6		120	ESO 205-27 S edge; mag 9.57v star N 5'.7.
06 09 39.3	−47 37 22	ESO 205-27	13.5	0.6 x 0.5	12.1	SA0^- pec	27	ESO 205-27A on N edge; small, faint almost stellar anonymous galaxy SE 1'.3; a pair of almost stella[r] anonymous galaxies NW 2'.5.
06 10 23.0	−49 49 14	ESO 205-28	13.6	1.2 x 1.0	13.7	Pec	52	Located between a pair of stars, mags 13.1 and 13.5.
06 10 51.2	−50 41 39	ESO 205-31	13.7	0.9 x 0.7	13.1	Sb	153	Mag 10.08v star W 2'.4.
06 13 18.4	−51 19 02	ESO 205-34	13.0	1.3 x 1.2	13.4	SAB(s)m IV	39	Numerous knots; mag 10.48v star W 7'.3.
06 16 12.7	−51 50 25	ESO 206-1	13.8	1.1 x 0.5	13.2	E$^+$5	2	Mag 11.90v star W 5'.1; **ESO 206-3** N 15'.1.
06 28 28.4	−48 45 50	ESO 206-14	14.0	1.6 x 1.3	14.6	SAB(s)c III-IV	173	Several bright knots and/or superimposed stars S half.
06 38 19.6	−51 57 02	ESO 206-17	14.5	1.7 x 0.3	13.6	Scd: sp	178	Mag 9.809v star NE 3'.2.
06 41 36.7	−50 57 58	ESO 206-A20	6.8	22 x 15	13.2	E3	70	**Carina Dwarf**. Mag 9.14 star 2'.1 S of galaxy center.
07 01 24.2	−47 34 57	ESO 207-13	14.1	0.8 x 0.6	13.2	Pec	67	Mag 12.6 star N 1'.8.
07 02 22.6	−49 43 31	ESO 207-15	13.7	0.9 x 0.8	13.2	S0	116	Mag 9.99v star N 1'.6; **ESO 207-16** NNE 8'.8; most others anonymous, stellar.
07 03 33.5	−49 11 04	ESO 207-18	14.0	1.0 x 0.2	12.1	S	7	Pair of stars E 0'.5, mags 12.5 and 13.3.
07 04 59.2	−49 05 00	ESO 207-19	13.2	1.2 x 1.0	13.3	S0	108	Mag 8.53v star E 1'.7; **ESO 207-20** S 8'.7.
07 09 10.3	−51 28 03	ESO 207-22	14.7	1.1 x 1.0	14.6	IA(s)m V-VI	96	Uniform surface brightness, several stars involved.
07 10 52.5	−51 48 24	ESO 207-25	13.7	1.3 x 0.6	13.3	(R)SB(s)0^+:	170	Mag 10.81v star S 9'.4.
05 32 21.0	−45 55 55	ESO 253-12	14.3	1.5 x 0.4	13.6	S?	75	Mag 10.43v star partially covering E end.
05 37 06.3	−46 37 47	ESO 253-15	13.7	1.0 x 0.6	13.0	S0	166	Mag 9.84v star NW 10'.7.
05 47 14.4	−45 28 49	ESO 253-26	14.4	1.1 x 0.8	14.1		160	Double system, bridge.
05 48 08.1	−47 24 19	ESO 253-27	13.6	1.1 x 0.6	13.1	E$^+$3	161	Mag 9.87v star SE 7'.5.
05 49 04.1	−47 10 02	ESO 253-29	13.9	1.1 x 0.8	13.6	(R′)SB(r)ab	161	Stellar nucleus in E-W bar; mag 13.9 star on E edge.
05 49 52.8	−43 23 58	ESO 253-30	13.4	1.3 x 1.0	13.5	S0?	149	Mag 7.57v star N 2'.1.
05 57 51.2	−45 51 28	ESO 254-6	13.5	0.9 x 0.7	12.9	(R)SB(r)0^+	18	Mag 15.2 star on SE edge.
06 06 35.8	−47 29 56	ESO 254-17	13.3	0.9 x 0.7	12.9	E2? pec	123	Mag 14.5 star E of nucleus.
06 07 45.1	−47 25 01	ESO 254-22	13.5	1.3 x 0.7	13.3	SA(rs)c II	133	Mag 12.5 star SW 1'.6.
06 10 57.0	−45 00 09	ESO 254-31	14.1	1.1 x 0.3	12.8	S pec	59	Mag 7.21v star NNE 6'.2.
06 12 34.7	−43 02 06	ESO 254-35	14.0	0.8 x 0.6	13.1	S0/a	40	Located 1'.4 SE of a close pair of stars, mags 9.54v and 7.78v.
06 12 25.8	−44 39 39	ESO 254-36	13.3	1.0 x 1.0	13.2	S0		**ESO 254-34** W 6'.5; **ESO 254-33** SW 11'.8.
06 12 32.4	−45 04 27	ESO 254-37	13.6	1.3 x 0.9	13.7		116	Double system, connected; **ESO 254-42, 42A, 42B** NE 16'.2.
06 13 43.7	−42 47 42	ESO 254-41	13.6	0.6 x 0.6	12.4	S0:		Close pair of mag 14.9 stars N 1'.5.
06 14 03.3	−42 51 30	ESO 254-43	14.1	1.1 x 0.6	13.5	(R′)SA(r)b	55	Stellar nucleus, smooth envelope; mag 12.5 star N 2'.1.
06 21 34.9	−44 42 31	ESO 255-5	13.8	1.1 x 0.6	13.2	SA(r)ab	99	Mag 12.4 star SE 2'.6.
06 27 22.7	−47 10 56	ESO 255-7	14.0	0.9 x 0.5	13.0		151	Triple system, contact, linear.
06 28 29.8	−46 03 22	ESO 255-10	13.3	1.1 x 0.7	12.9	S0:	96	**ESO 255-9** W 6'.1.
06 28 52.0	−44 40 00	ESO 255-11	13.8	1.2 x 0.4	12.8	SB(r)a	88	Located between a mag 11.6 star 1'.3 S and a pair of stars, mags 11.39v and 8.99v, 2'.3 N.
06 36 17.2	−44 52 07	ESO 255-13	13.8	1.3 x 0.4	12.9	Sab	54	Mag 12.9 star E 1'.2.
06 45 04.2	−46 26 50	ESO 255-18	13.8	1.2 x 0.9	13.7	Ring	21	Strong dark patch N of bright core; mag 14.9 star on W edge.
06 45 48.2	−47 31 53	ESO 255-19	13.8	1.9 x 1.5	14.8	SA(s)m V	37	Mag 9.76v star SSW 1'.8.
06 55 20.4	−47 06 47	ESO 256-7	13.8	1.1 x 0.9	13.6	(R′)SB(r)b II	101	E-W bar in diffuse envelope; mag 13.5 star W 0'.5.
06 57 35.3	−45 48 42	ESO 256-11	13.0	2.1 x 0.9	13.5	SA(s)0°:	140	Mag 6.21v star E 11'.9.
07 00 15.2	−44 52 59	ESO 256-12	13.5	1.5 x 0.7	13.4	Sab	29	Mag 9.98v star SW 7'.5.
05 37 19.5	−42 24 39	ESO 306-12	14.1	1.1 x 0.6	13.5	S?	155	Located between a pair of stars, mags 12.9 and 13.6.
05 38 58.2	−41 44 14	ESO 306-13	13.3	1.0 x 0.8	12.9	SB?	136	Mag 8.85v star E 2'.7.
05 39 53.5	−40 30 49	ESO 306-16	13.8	1.0 x 0.6	13.0	S?	161	Mag 12.6 star SW 2'.2; **ESO 306-20** ENE 9'.3.
05 40 06.6	−40 50 11	ESO 306-17	12.3	2.5 x 1.6	13.8	E$^+$3	177	Mag 5.81v star NW 17'.5; mag 9.81v star NNW 6'.4.
05 46 11.2	−41 48 25	ESO 306-26	14.3	0.8 x 0.5	13.2	Sc	97	Mag 10.71v star S 4'.8.
06 00 18.6	−42 06 39	ESO 307-12	13.9	0.8 x 0.4	12.5	Sbc	5	Mag 10.05v star NE 3'.0.
06 00 41.1	−40 02 42	ESO 307-13	13.4	1.1 x 0.6	13.0	E$^+$3	62	Almost stellar nucleus, smooth envelope; **PGC 18233** SW 1'.7; other small galaxies in surrounding area.
06 03 31.6	−41 17 14	ESO 307-15	13.8	1.0 x 0.5	12.9	S0	100	Mag 11.8 star WNW 1'.5.
07 04 04.8	−41 52 49	ESO 309-17	13.8	1.8 x 0.3	13.0	Sb? sp	18	Mag 9.64v star S 2'.6.
05 33 15.7	−48 41 44	NGC 1998	14.2	1.0 x 0.6	13.5	(R)SA(r)0^+	20	
05 34 59.3	−50 55 20	NGC 2007	13.9	1.7 x 0.6	13.8	SB(rs)c: II	78	NGC 2008 S 2'.8; mag 10.37v star NNE 7'.2.
05 35 03.5	−50 58 02	NGC 2008	13.8	1.5 x 0.7	13.7	Sc I	105	Mag 11.19v star E 3'.0; NGC 2007 N 2'.8.
05 47 04.5	−51 33 07	NGC 2104	12.7	2.0 x 0.9	13.2	SB(s)m pec III-IV	160	
05 51 20.0	−50 35 01	NGC 2115	13.0	0.8 x 0.7	12.3	(S0?) + (I0?)	50	E-W row of three mag 8 - 10 stars S 3'.0; NGC 2115A on S edge.
05 51 21.5	−50 35 32	NGC 2115A	15.0	0.6 x 0.4	13.4	S0+? pec	56	On S edge of NGC 2115.
06 00 55.5	−50 44 28	NGC 2152	13.5	1.1 x 0.8	13.3	(R′)SB(r)a: pec	69	Star on N edge, two more faint stars superimposed on surface.
06 13 17.5	−43 39 49	NGC 2200	14.0	0.9 x 0.6	13.2	SB(r)c II-III	161	Pair of mag 8.31v, 9.24v stars NW 7'.0; NGC 2201 SE 3'.6.
06 13 31.6	−43 42 20	NGC 2201	13.3	1.1 x 0.5	12.5	SAB(r)b:	113	NGC 2200 NW 3'.6.
06 53 55.0	−40 51 33	NGC 2310	11.8	4.4 x 0.8	13.0	S0 sp	47	Mag 8.54v star E 10'.2.
07 02 36.1	−42 04 06	NGC 2328	12.8	1.3 x 0.8	12.7	(R′)SAB0^-?	115	

(Continued from previous page)

188 GALAXY CLUSTERS 188

RA h m s	Dec ° ′ ″	Name	Mag 10th brightest	No. Gal	Diam ′	Notes
06 00 42.0	−40 02 00	A 3376	15.4	42	22	Staggered band of numerous PGC GX stretches ENE edge to WSW edge.
06 06 06.0	−42 17 00	A 3379	16.4	115	17	All members anonymous, faint, stellar.
06 07 00.0	−49 29 00	A 3380	15.7	40	17	**ESO 205-20** E of center 2′6; **ESO 205-22** NE 7′7; all others anonymous.
07 05 00.0	−49 04 00	A 3407	15.3	57	22	Four ESO galaxies, other members anonymous, stellar.
07 08 30.0	−49 12 00	A 3408	15.5	41	22	**PGC 20233** at center; **ESO 207-24** E 19′6; **ESO 207-21** W 14′0; all others anonymous.
05 35 42.0	−42 48 00	AS 538	15.6		22	All members anonymous, stellar.
05 40 06.0	−40 50 00	AS 540	15.4	22	22	See notes ESO 306-16; except three ESO GX, all GX anonymous, stellar
05 45 00.0	−48 05 00	AS 542	15.7	24	17	See notes ESO 204-35; all others anonymous, stellar.
06 05 48.0	−42 45 00	AS 564	16.4		17	**ESO 254-14** NE of center 9′5; all others anonymous, stellar.
06 07 06.0	−45 11 00	AS 566	15.7	7	17	**ESO 254-19** at center; **ESO 254-18**, **18A** N 9′1; **ESO 254-21** E 4′6; **ESO 254-24** E 11′4.
06 11 30.0	−43 59 00	AS 572	16.0		17	
06 12 30.0	−45 04 00	AS 574	15.4	19	22	Other than ESO galaxies, all members anonymous, stellar.
06 16 12.0	−51 51 00	AS 580	16.1		17	See notes ESO 206-1; all others anonymous, stellar.
07 01 06.0	−49 44 00	AS 601	16.0	16	17	
07 09 30.0	−50 09 00	AS 602	16.3	3	17	**ESO 207-23** S of center 5′9; all others anonymous, stellar.

OPEN CLUSTERS

RA h m s	Dec ° ′ ″	Name	Mag	Diam ′	No. ★	B ★	Type	Notes
06 50 42.0	−42 23 00	ESO 309-3		6			cl	
06 20 35.0	−44 41 00	NGC 2220		22	20	7.5	cl?	(A) Sparse group of bright stars, with 3 brightest on SE side. Galaxy ESO 255-5 on E side.

PLANETARY NEBULAE

RA h m s	Dec ° ′ ″	Name	Diam ″	Mag (P)	Mag (V)	Mag cent ★	Alt Name	Notes
07 14 49.4	−46 57 40	PN G258.0−15.7	82	11.8			PK 258−15.1	Annular; mag 10.7 star on S edge; mag 11.3 star on N edge.

189 GALAXIES 189

RA h m s	Dec ° ′ ″	Name	Mag (V)	Dim ′ Maj x min	SB	Type Class	PA	Notes
04 00 29.2	−49 01 45	ESO 201-14	13.0	1.5 x 0.6	12.8	S0°:	162	Mag 9.22v star NE 3′4.
04 03 26.5	−51 49 55	ESO 201-16	14.3	0.9 x 0.7	13.7	SB(s)c? I	23	Mag 8.20v star S 6′2.
04 04 21.8	−47 56 59	ESO 201-17	13.2	1.2 x 0.6	12.7	(R′)SAB(r)a pec:	4	A filamentary **ESO 201-18** SE 2′3.
04 08 40.7	−49 35 00	ESO 201-21	13.7	1.6 x 0.5	13.3	Sab	60	Mag 9.48v star ESE 9′4.
04 09 00.2	−48 43 38	ESO 201-22	14.0	2.5 x 0.3	13.5	Sc: sp	59	Star on NE end; mag 12.4 star E 8′1.
04 16 29.9	−47 50 56	ESO 202-1	13.6	1.2 x 0.7	13.3	SAB(r)0°: pec	65	Mag 12.6 star on E end.
04 19 25.5	−50 53 55	ESO 202-7	13.3	1.3 x 1.1	13.5	S0?	64	Large, strong core, uniform envelope.
04 24 20.7	−47 31 33	ESO 202-15	12.7	1.3 x 1.2	13.1	(R′)SA(rs)0$^+$ pec:	22	Mag 9.03v star N 7′3.
04 27 59.8	−47 54 41	ESO 202-23	12.5	1.5 x 1.2	13.0	Pec	164	**Carafe Galaxy**. Mag 13.3 star E 0′9; mag 11.8 star WSW 5′2.
04 32 15.7	−49 40 29	ESO 202-35	12.6	2.8 x 0.4	12.6	Sb: sp	133	Mag 10.14v star E 2′0.
04 32 49.7	−50 46 27	ESO 202-37	13.1	0.5 x 0.4	11.2		49	Mag 9.39v star W 5′0; mag 10.09v star SE 3′2.
04 36 52.4	−48 32 54	ESO 202-40	15.0	0.8 x 0.8	14.4	S?		Mag 13.1 star NE 2′5.
04 37 20.6	−48 14 14	ESO 202-42	14.4	1.4 x 1.3	14.9	S?	48	Many knots and/or superimposed stars.
04 37 47.5	−51 25 23	ESO 202-43	13.5	1.6 x 0.8	13.7	E$^+$4	132	Mag 8.86v star E 10′4; **ESO 202-44** S 15′1.
04 39 49.1	−47 31 41	ESO 202-48	14.1	1.0 x 0.5	13.2		130	Mag 13.7 star SW 1′3.
04 44 49.4	−49 36 42	ESO 202-55	14.0	0.5 x 0.4	12.3	E	174	Mag 11.25v star NE 3′9.
04 46 29.0	−50 38 26	ESO 202-56	14.5	0.4 x 0.2	11.6		78	Mag 8.69v star W 5′0.
04 47 05.9	−50 03 02	ESO 203-2	13.9	1.2 x 0.8	13.7	SB(r)bc	5	Mag 9.23v star E 11′7.
04 50 52.6	−47 46 05	ESO 203-7	13.4	0.9 x 0.8	13.0	S0	174	Mag 15.4 star S edge; mag 12.6 star NW 2′6.
04 51 49.8	−48 17 07	ESO 203-9	14.3	0.8 x 0.5	13.1	SABm: V-VI	41	Mag 10.07v star NW 10′2.
05 00 40.9	−49 44 48	ESO 203-15	13.8	1.3 x 0.6	13.4	SAB(rs)bc II	135	Mag 11.47v star NE 4′9.
05 02 00.6	−48 38 48	ESO 203-16	13.3	1.2 x 0.9	13.2	(R)SB(r)0°	179	Mag 9.39v star NNE 6′4.
05 03 30.9	−49 27 51	ESO 203-17	13.9	0.9 x 0.4	12.7	Sc:	104	Located 3′0 NE of mag 7.16v star.
05 05 33.8	−49 35 48	ESO 203-19	13.2	0.7 x 0.6	12.1	SAB(r)0$^+$	168	NGC 1803 NW 2′1; mag 5.04v star Eta 2 Pictoris W 5′9.
05 13 24.8	−49 02 10	ESO 203-24	14.9	0.3 x 0.2	11.6	Pec	48	Faint star E edge.
05 22 16.8	−50 50 25	ESO 204-3	14.0	1.0 x 0.5	13.1	Sa	109	Small galaxy, **ESO 204-3A**, on N edge.
05 25 26.6	−51 38 44	ESO 204-8	14.8	1.2 x 0.6	14.3	SB(s)b	55	Mag 11.17v star S 7′3.
05 31 27.9	−49 24 04	ESO 204-11	13.5	1.0 x 0.3	12.1	S0	135	Mag 9.83v star N 4′8.
05 32 16.2	−50 34 20	ESO 204-13	14.3	1.4 x 0.6	14.0	S0° pec sp	9	
05 32 32.3	−49 54 14	ESO 204-14	14.1	1.2 x 0.5	13.4	Sb? III-IV	74	Mag 10.61v star SE 2′2.
03 58 56.6	−45 51 35	ESO 249-35	15.7	0.9 x 0.2	13.7	SBcd: sp	98	ESO 249-36 E 3′3; mag 11.26v star N 3′2; stellar galaxy **Fair 392** S 2′5.
03 59 15.3	−45 52 14	ESO 249-36	14.9	1.7 x 1.4	15.7	IB(s)m V-VI	158	Mag 11.26v star NW 4′2; ESO 249-35 W 3′3; stellar galaxy **Fair 392** SW 3′7.
04 04 35.5	−46 02 32	ESO 250-5	13.5	1.4 x 1.0	13.7	SB(s)0°?	79	Mag 8.18v star N 8′2.
04 12 48.2	−46 37 12	ESO 250-12	14.1	0.5 x 0.5	12.5			Mag 10.8 star superimposed E half; = **Fair 403**.
04 26 31.2	−42 26 31	ESO 250-21	14.2	1.0 x 0.8	13.8	(R)SAB(s)bc: II-III	86	Mag 9.57v star W 6′7.
04 27 40.5	−42 46 50	ESO 251-1	13.3	1.1 x 0.8	13.2	E/S0	179	Double system, connected; mag 8.45v star NE 4′3.
04 27 59.7	−42 50 19	ESO 251-2	13.0	1.2 x 0.6	12.6	SB(s)0°: pec	8	Mag 8.45v star N 5′1.
04 28 49.1	−44 51 44	ESO 251-4	13.3	1.2 x 0.6	13.0	E5	24	Mag 10.72v star ENE 4′0.
04 28 50.0	−46 57 53	ESO 251-5	13.8	0.7 x 0.6	12.7	S?	27	Mag 9.48v star N 6′8.
04 30 09.1	−42 40 54	ESO 251-6	15.4	1.5 x 0.2	13.9	Sb? pec sp	179	Mag 8.73v star NW 11′1.
04 30 47.3	−43 39 29	ESO 251-7	14.1	1.2 x 0.3	12.8	Sb? sp	171	Mag 13.4 star SE 2′8.
04 33 08.9	−43 36 33	ESO 251-11	14.2	1.0 x 0.6	13.5	S0	10	Mag 8.81v star NE 10′7.
04 38 12.5	−44 14 19	ESO 251-16	14.2	1.1 x 0.6	13.5	SBa:	62	Pair of stars superimposed E end; mag 11.6 star NW 2′8.
04 40 26.6	−44 37 57	ESO 251-21	13.0	1.4 x 0.8	13.1	SAB0$^-$ pec	24	Mag 9.46v star S 6′2; **ESO 251-20** S 15′2.
04 42 02.5	−44 34 49	ESO 251-23	13.3	1.2 x 0.7	13.0	S?	77	Pair of stars 1′9 E, mags 15.0 and 14.4.
04 42 23.3	−45 46 09	ESO 251-26	13.5	1.0 x 0.6	12.9	S0	130	Mag 10.44v star NW 5′8.
04 43 36.9	−44 58 33	ESO 251-28	13.7	1.2 x 0.4	12.7	S?	58	Mag 11.7 star S 2′6.
04 47 18.9	−44 40 54	ESO 251-32	13.8	0.9 x 0.8	13.3	Sc	23	Mag 13.1 star W 1′9.
04 52 33.6	−43 52 11	ESO 251-37	14.3	0.8 x 0.6	13.4	Sc	85	Located between a pair of mag 12.6 and 13.4 stars.
04 53 05.3	−45 29 01	ESO 251-39	13.7	1.3 x 0.4	12.9	Sb: pec	112	**ESO 251-40** N 0′5; mag 13.1 star NNE 1′4.
04 54 13.9	−43 52 30	ESO 251-41	13.7	1.4 x 0.3	12.7	SA0$^+$? sp	107	Mag 10.63v star W 4′5.
04 59 40.3	−45 58 28	ESO 252-4	13.6	1.5 x 0.7	13.5	SA0°:	57	Mag 112.7 star W 4′6; mag 12.5 star S 2′9.
05 03 01.5	−43 17 56	ESO 252-7	13.7	1.2 x 0.6	13.2	SB?	45	Narrow, elongated core, smooth envelope.
05 06 09.5	−45 02 51	ESO 252-10	13.2	1.3 x 0.9	13.3	SAB(r)ab	170	Mag 9.28v star NW 9′8.

(Continued from previous page)

189 GALAXIES 189

RA h m s	Dec ° ′ ″	Name	Mag (V)	Dim ′ Maj x min	SB	Type Class	PA	Notes
05 08 21.4	−46 20 49	ESO 252-12	14.3	1.2 x 0.4	13.4	(R′)SB(s)bc	172	Star S end of elongated core.
05 19 08.6	−43 09 32	ESO 252-18	14.9	1.0 x 0.5	14.1	S0:	131	Mag 12.1 star NE 2′9.
05 23 18.8	−42 42 39	ESO 252-22	13.9	1.1 x 0.4	12.9	S pec?	54	Mag 13.3 star W 2′3.
05 24 20.3	−46 44 50	ESO 253-1	13.9	1.2 x 0.8	13.7	SB(rs)cd IV	38	Numerous bright knots.
05 24 35.9	−46 02 32	ESO 253-2	14.6	0.8 x 0.7	13.8	SB(s)m V-VI	42	Mag 9.09v star N 9′2; **ESO 253-3** E 7′6.
05 28 56.7	−47 23 30	ESO 253-7	13.7	1.0 x 0.6	13.0	Sbc	137	Mag 13.0 star W 0′9.
05 32 21.0	−45 55 55	ESO 253-12	14.3	1.5 x 0.4	13.6	S?	75	Mag 10.43v star partially covering E end.
05 37 06.3	−46 37 47	ESO 253-15	13.7	1.0 x 0.6	13.0	S0	166	Mag 9.84v star NW 10′7.
03 58 28.4	−40 13 38	ESO 302-23	13.4	1.2 x 0.7	13.1		60	Double system, contact; faint companion NE 0′5 from center.
03 59 14.7	−42 16 27	ESO 302-24	14.1	1.1 x 0.9	14.0	IB(s)m pec: IV	76	Mag 8.12v star W 2′3.
04 13 10.0	−42 19 21	ESO 303-2	14.0	1.0 x 0.5	13.1	Sa	89	Mag 12.6 star SW 2′2.
04 13 24.0	−41 23 45	ESO 303-3	13.9	1.0 x 0.6	13.2	Sa	75	Mag 8.52v star E 2′7.
04 28 17.7	−40 28 13	ESO 303-20	14.1	1.2 x 1.0	14.1	S?	171	Mag 7.14v star W 8′7.
04 44 07.8	−41 11 03	ESO 304-17	13.8	0.6 x 0.5	12.3	Sb	3	Mag 9.94v star NE 1′4.
04 45 07.6	−41 34 44	ESO 304-19	13.8	1.0 x 0.3	12.3	SAB(s)0°:	147	Mag 9.81v star SE 6′8.
04 49 40.2	−42 03 42	ESO 304-21	14.2	1.1 x 0.6	13.5	SAB(rs)bc	32	Mag 14.6 star S edge; mag 15.4 star N edge.
05 12 50.8	−41 38 01	ESO 305-15	13.7	1.6 x 0.4	13.0	Sc I-II	156	Mag 7.43v star W 9′1; stellar galaxy **Fair 1133** SSE 7′2.
05 15 01.0	−41 23 28	ESO 305-17	14.1	1.9 x 0.6	14.1	IB(s)m IV-V	68	Double system; mag 10.29v star NNW 6′1.
05 29 13.6	−42 12 36	ESO 306-4	13.6	1.4 x 0.7	13.6	E$^+$4 pec	17	Two strong knots S of core; mag 10.95v star ESE 2′9.
05 31 16.1	−40 06 58	ESO 306-7	13.5	0.7 x 0.7	12.6			Mag 12.3 star N 1′7.
05 31 40.9	−42 09 53	ESO 306-9	13.2	1.1 x 1.1	13.2	(R′)SA(r)ab pec:		Mag 9.81v star E 1′4.
05 37 19.5	−42 24 39	ESO 306-12	14.1	1.1 x 0.6	13.5	S?	155	Located between a pair of stars, mags 12.9 and 13.6.
05 38 58.2	−41 44 14	ESO 306-13	13.3	1.0 x 0.8	12.9	SB?	136	Mag 8.85v star E 2′7.
05 39 53.5	−40 30 49	ESO 306-16	13.8	1.0 x 0.6	13.0	S?	161	Mag 12.6 star SW 2′2; **ESO 306-20** ENE 9′3.
04 09 01.8	−45 31 04	IC 2035	11.8	1.2 x 0.9	11.7	S0 pec?	86	Mag 10.82v star S 7′3.
04 26 36.8	−42 05 37	IC 2068	13.3	1.2 x 0.8	13.1	(R′)SA(rs)0/a:	153	Mag 9.98v star SE 7′8.
03 57 27.6	−46 12 39	NGC 1493	11.3	2.8 x 2.8	13.4	SB(r)cd III		Many multi-branching arms with many knots.
03 57 42.5	−48 54 26	NGC 1494	11.7	3.2 x 1.9	13.5	SAB(s)d: II-III	179	Numerous knots on branching arms; mag 10.35v star NNW 7′2.
03 58 21.8	−44 27 59	NGC 1495	12.6	3.0 x 0.5	12.9	Sc? sp	105	Mag 9.39v star E 5′8.
04 03 32.8	−43 24 02	NGC 1510	13.0	0.9 x 0.9	12.7	SA0° pec?		Located 5′0 SW of NGC 1512.
04 03 54.3	−43 20 58	NGC 1512	10.3	8.9 x 5.6	14.4	SB(r)a I	90	Bright NE-SW core with faint E-W envelope; NGC 1510 SW 5′0.
04 08 24.1	−47 53 48	NGC 1527	10.8	3.7 x 1.4	12.4	SAB(r)0$^-$:	78	
04 17 44.8	−50 09 50	NGC 1556	13.1	1.7 x 0.5	12.7	Sab pec?	167	
04 20 16.9	−45 01 52	NGC 1558	12.4	2.5 x 0.9	13.1	(R′)SAB(r)bc: II	72	**ESO 250-18** E 8′1; mag 7.71v star E 10′4.
04 21 08.6	−48 15 16	NGC 1567	12.2	1.3 x 1.3	12.7	E0 pec:		**ESO 202-9** S 3′2; mag 10.53v star SSW 5′3.
04 22 09.0	−43 37 47	NGC 1571	12.3	1.5 x 1.2	13.0	E$^+$ pec	172	
04 22 42.8	−40 36 03	NGC 1572	12.4	2.5 x 1.2	13.5	(R′)SB(s)a:	0	Mag 12.0 star on E edge.
04 23 46.5	−51 35 58	NGC 1578	13.1	1.2 x 1.1	13.3	SA(s)a pec:	177	
04 27 32.8	−42 09 58	NGC 1585	13.5	1.1 x 0.7	13.0	SAc: II	175	Mag 12.6 star on W edge; mag 9.68v star SE 5′0.
04 28 21.8	−47 48 56	NGC 1595	12.7	1.3 x 0.9	12.9	E3	15	NGC 1598 NE 2′8.
04 28 33.6	−47 46 59	NGC 1598	13.3	1.4 x 0.8	13.3	(R′)SB(s)c pec: I-II	123	NGC 1595 SW 2′8.
04 32 42.2	−43 42 56	NGC 1616	12.6	1.8 x 0.9	13.0	SAB(rs)bc pec:	36	Mag 10.88v star W 4′4; **ESO 251-12** SE 6′0.
04 44 01.2	−41 27 47	NGC 1658	13.5	1.4 x 0.5	12.9	Sbc: I-II	124	Mag 9.91v star SW 4′9.
04 44 11.4	−41 29 52	NGC 1660	14.0	1.1 x 0.5	13.2	Sa? pec	32	Mag 9.91v star W 5′7.
04 46 05.9	−44 44 02	NGC 1668	12.8	1.6 x 0.9	13.0	SAB0$^-$:	107	Faint star on N edge.
04 48 34.0	−47 48 58	NGC 1680	13.6	1.2 x 0.6	13.1	(R)SB(s)b? II	102	
05 05 26.7	−49 34 04	NGC 1803	12.9	1.3 x 0.8	12.7	SB(s)bc: I-II	62	Mag 5.04v star Eta 2 Pictoris W 4′6; ESO 203-19 SE 2′1.
05 25 56.8	−46 43 43	NGC 1930	12.4	1.9 x 1.2	13.1	SAB(s)0$^+$:	32	
05 33 15.7	−48 41 44	NGC 1998	14.2	1.0 x 0.6	13.5	(R)SA(r)0$^+$	20	
05 34 59.3	−50 55 20	NGC 2007	13.9	1.7 x 0.6	13.8	SB(rs)c: II	78	NGC 2008 S 2′8; mag 10.37v star NNE 7′2.
05 35 03.5	−50 58 02	NGC 2008	13.8	1.5 x 0.7	13.7	Sc I	105	Mag 11.19v star E 3′0; NGC 2007 N 2′8.
04 59 56.8	−45 49 36	PGC 16516	14.7	0.7 x 0.4	13.4	E	57	

GALAXY CLUSTERS

RA h m s	Dec ° ′ ″	Name	Mag 10th brightest	No. Gal	Diam ′	Notes
04 37 48.0	−51 26 00	AS 479	16.2		17	See notes for ESO 202-43, all others anonymous.
04 40 24.0	−44 38 00	AS 487	15.2		22	Most members anonymous, stellar
04 46 06.0	−44 43 00	AS 497	15.8		17	Most members anonymous, stellar.
04 58 06.0	−46 36 00	AS 505	15.7		17	All members anonymous, stellar.
05 12 54.0	−41 46 00	AS 515	16.0	16	17	See notes ESO 305-15; all others anonymous, stellar.
05 30 54.0	−49 14 00	AS 534	16.0		17	Except ESO 204-11, all members anonymous, stellar.
05 35 42.0	−42 48 00	AS 538	15.6		22	All members anonymous, stellar.

GLOBULAR CLUSTERS

RA h m s	Dec ° ′ ″	Name	Total V m	B ★ V m	HB V m	Diam ′	Conc. Class Low = 12 High = 1	Notes
05 14 06.3	−40 02 50	NGC 1851	7.1	13.2	16.1	12.0	2	

190 GALAXIES 190

RA h m s	Dec ° ′ ″	Name	Mag (V)	Dim ′ Maj x min	SB	Type Class	PA	Notes
02 22 15.7	−51 06 32	ESO 198-8	13.1	0.9 x 0.9	12.8	S0		Mag 5.90v star E 6′3.
02 27 52.7	−50 42 43	ESO 198-11	13.6	1.1 x 0.4	12.5	(R′)SB(s)b	177	Mag 10.91v star E 0′7; with faint anonymous galaxy in between.
02 29 16.2	−48 29 29	ESO 198-13	12.6	1.6 x 1.0	12.9	(R′)SA(r)ab:	130	Strong dark ring around bright core; mag 10.85v star SW 3′2.
02 29 35.9	−48 37 54	ESO 198-14	13.7	0.9 x 0.4	12.5	(R′)SB(rs)ab?	50	Close pair of faint, almost stellar anonymous galaxies S 2′0.
02 35 27.6	−48 12 43	ESO 198-19	13.5	1.0 x 0.7	13.0	(R)SA(r)0$^+$	141	Mag 13.9 star S 2′2.
02 35 33.2	−50 19 16	ESO 198-20	14.0	0.9 x 0.4	12.8	S	116	Almost stellar **ESO 198-22** E 5′6.
02 36 04.1	−51 20 47	ESO 198-21	13.9	1.1 x 0.4	12.9	S	99	Mag 12.1 star NNW 3′7.
02 42 10.1	−51 54 15	ESO 198-27	12.9	1.1 x 1.0	12.9	S0	12	Close pair of mag 13.5, 13.8 stars SE 2′5.
02 49 03.3	−49 56 24	ESO 199-1	13.7	1.0 x 0.8	13.3	Sa pec	119	Mag 10.48v star NW 8′4.
03 30 04.8	−47 51 42	ESO 200-29	13.5	0.7 x 0.5	12.2	S0?	101	Mag 12.0 star SSE 5′3.
03 37 30.0	−51 15 02	ESO 200-53	13.6	1.3 x 0.6	13.1	S0/a:	156	Mag 8.63v star W 12′7.
03 37 34.1	−51 11 20	ESO 200-54	13.7	1.0 x 0.8	13.3	SA(s)c? II	150	Mag 10.77v star N 2′0.

(Continued from previous page)

GALAXIES

RA h m s	Dec ° ′ ″	Name	Mag (V)	Dim ′ Maj x min	SB	Type Class	PA	Notes
03 50 23.5	−50 18 20	ESO 201-4	13.7	1.2 x 0.6	13.2	SB(rs)ab: pec	156	Double system, interacting with companion S.
04 00 29.2	−49 01 45	ESO 201-14	13.0	1.5 x 0.6	12.8	S0°:	162	Mag 9.22v star NE 3.′4.
04 03 26.5	−51 49 55	ESO 201-16	14.3	0.9 x 0.7	13.7	SB(s)c? I	23	Mag 8.20v star S 6.′2.
02 28 20.4	−42 45 52	ESO 246-13	13.1	0.5 x 0.5	11.6	E?		Mag 10.99v star SW 6.′7.
02 28 47.0	−43 04 27	ESO 246-15	13.7	1.3 x 0.5	13.1	(R)SA(s)ab:	147	Mag 10.20v star SSE 5.′4.
02 30 28.3	−43 01 38	ESO 246-21	12.7	2.3 x 1.5	13.8	(R′)SB(r)b I-II	141	Mag star on SE edge.
02 31 00.3	−44 25 38	ESO 246-22	14.0	1.1 x 0.6	13.4	SBc III-IV	134	Mag 9.57v star SW 1.′5.
02 33 25.4	−43 31 09	ESO 246-25	13.7	0.8 x 0.7	12.9	(R)SB(r)a:	151	Mag 12.6 star S 3.′7.
02 51 42.5	−42 45 03	ESO 247-8	13.4	1.2 x 1.0	13.4	SAB0^{-}	19	Mag 12.3 star NW edge.
03 05 06.2	−45 57 45	ESO 248-2	13.3	3.3 x 0.7	14.0	SB(s)d IV	14	Mag 11.28v star W 2.′7; mag 11.33v star NE 1.′3.
03 17 57.9	−44 14 19	ESO 248-6	13.9	1.2 x 0.5	13.2	S0?	6	ESO 248-6 at center, numerous PGC GX overall.
03 21 28.8	−43 35 06	ESO 248-12	13.8	1.2 x 0.4	12.9	S?	139	Almost stellar **ESO 248-11** SW 1.′4.
03 32 53.2	−42 23 46	ESO 249-1	14.0	1.1 x 0.3	12.6	Sa:	3	Mag 9.45v star N 3.′2.
03 37 19.0	−43 35 13	ESO 249-8	13.5	1.0 x 0.6	12.9	Pec	175	Mag 14.0 star NNE 2.′7.
03 52 49.0	−43 23 58	ESO 249-25	14.3	0.4 x 0.3	11.9		47	Stellar.
03 54 29.0	−44 45 13	ESO 249-27	14.8	1.4 x 0.4	14.0	IB(s)m: sp IV-V	93	Double galaxy system; mag 10.11v star W 7.′5.
03 58 56.6	−45 51 35	ESO 249-35	15.7	0.9 x 0.2	13.7	SBcd: sp	98	ESO 249-36 E 3.′3; mag 11.26v star N 3.′2; stellar galaxy **Fair 392** S 2.′5.
03 59 15.3	−45 52 14	ESO 249-36	14.9	1.7 x 1.4	15.7	IB(s)m V-VI	158	Mag 11.26v star NW 4.′2; ESO 249-35 W 3.′3; stellar galaxy **Fair 392** SW 3.′7.
02 21 11.8	−42 00 08	ESO 298-31	14.0	1.1 x 0.4	13.0	(R′)SB(r)a:	109	Mag 12.4 star W 2.′6.
02 25 14.2	−40 25 37	ESO 299-1	13.1	1.2 x 0.8	12.9		164	Double system, interaction.
02 48 27.8	−40 33 22	ESO 299-18	14.4	2.1 x 0.3	13.8	Scd: sp	55	Mag 9.50v star W 7.′8.
03 09 38.9	−41 02 05	ESO 300-14	12.4	4.5 x 1.9	14.5	SAB(s)m V-VI	166	Numerous bright knots and/or superimposed stars.
03 21 24.4	−41 59 52	ESO 301-7	15.6	0.8 x 0.7	14.8	dE?	18	Mag 10.96v star NW 11.′1.
03 22 55.0	−42 11 23	ESO 301-9	13.4	2.4 x 0.4	13.3	S0: sp	80	Mag 10.27v star SSW 6.′4.
03 26 55.3	−41 41 15	ESO 301-15	14.1	0.8 x 0.7	13.3	Sa	25	Mag 6.33v star WNW 8.′6.
03 46 35.4	−40 38 55	ESO 302-6	13.3	1.1 x 0.6	12.7	SA(rs)0^{-}:	135	Mag 9.97v star SW 1.′3.
03 52 21.6	−42 08 25	ESO 302-16	14.9	1.6 x 0.5	14.7	E6:	94	Mag 10.05v star N 8.′4.
03 58 28.4	−40 13 38	ESO 302-23	13.4	1.2 x 0.7	13.1		60	Double system, contact; faint companion NE 0.′5 from center.
03 59 14.7	−42 16 27	ESO 302-24	14.1	1.1 x 0.9	14.0	IB(s)m pec: IV	76	Mag 8.12v star W 2.′3.
02 22 47.3	−41 22 17	IC 1796	13.0	1.1 x 0.9	12.9	SAB(s)0^{-}?	86	Mag 10.18v star SE 3.′4.
02 29 26.9	−43 04 36	IC 1810	13.2	1.3 x 1.0	13.3	(R)SAB(rs)ab	122	**MCG −7-6-7** N 5.′5.
02 29 31.9	−42 48 40	IC 1812	12.2	2.0 x 1.6	13.5	E^{+}:	24	Mag 10.82v star ESE 4.′5.
03 03 09.7	−50 30 42	IC 1877	15.4	0.8 x 0.2	13.3	Sb: pec	155	Elongated N-S **ESO 199-2** ENE 2.′5.
03 13 12.7	−50 34 42	IC 1903	15.7	0.7 x 0.3	14.1	E/S0	6	Mag 10.98v star NW 7.′4; **ESO 199-24A** NNW 1.′1.
03 19 25.2	−49 35 56	IC 1914	12.9	3.4 x 1.6	14.6	SAB(s)d IV	99	Mag 10.02v star NE 9.′1.
03 25 25.6	−51 16 02	IC 1929	14.1	0.8 x 0.5	12.9	S?	143	**IC 1932** SE 6.′4; **IC 1936** ESE 10.′4.
03 26 13.6	−50 00 44	IC 1935	14.3	1.2 x 0.8	14.1	SA(s)cd? pec II	65	Brightest on N edge.
03 26 47.1	−48 42 15	IC 1937	14.1	1.0 x 0.6	13.4	S	52	Mag 9.81v star SW 6.′5.
03 30 52.8	−47 58 46	IC 1949	13.1	1.1 x 0.9	12.9	S?	27	**IC 1948** NNW 0.′8.
03 31 04.1	−50 25 54	IC 1950	14.5	1.4 x 0.4	13.7	Sc: sp II	153	Mag 10.48v star WSW 4.′1.
03 31 30.6	−51 54 24	IC 1954	11.6	3.2 x 1.5	13.2	SB(s)b: II-III	66	Patchy; mag 10.63v star N 8.′2.
03 33 12.8	−50 24 52	IC 1959	12.8	2.8 x 0.7	13.4	SB(s)m: sp III-IV	147	Mag 5.66v star W 6.′2.
03 36 14.0	−45 10 47	IC 1969	14.4	1.1 x 0.3	13.1	Sa	42	Mag 12.4 star W 4.′0; note: stellar **PGC 13294** is 1.′5 SE of this star.
03 36 31.5	−43 57 28	IC 1970	12.0	3.1 x 0.6	12.5	SAb: sp	78	Mag 10.19v star NW 10.′8.
03 40 35.2	−45 21 22	IC 1986	14.3	1.0 x 0.7	13.7	Irr	120	Several knots; mag 10.65v star SSW 2.′7.
03 41 54.8	−50 57 29	IC 1989	13.5	0.9 x 0.6	12.7	SA0^{-}	131	Mag 12.6 star SE 1.′6; mag 10.14v star SE 2.′6.
03 49 07.3	−48 51 31	IC 2000	11.9	4.1 x 0.8	13.0	SB(s)cd: sp II-III	83	Mag 12.2 star W 12.′0.
03 51 45.6	−49 25 11	IC 2004	14.2	0.8 x 0.5	13.0	Sa	46	Mag 10.31v star N 6.′8.
03 53 34.4	−48 59 39	IC 2009	13.8	1.6 x 1.1	14.3	IAB(s)m V	51	Mag 11.41v star SW 1.′6.
02 26 21.8	−44 26 47	NGC 939	13.0	1.2 x 1.0	13.2	E^{+}:	110	
02 28 51.6	−41 24 13	NGC 954	12.8	1.8 x 0.9	13.2	SB(rs)c: II	19	Faint knots in tightly wound arms; mag 9.43v star NE 7.′5.
02 31 38.9	−44 31 28	NGC 979	12.7	0.9 x 0.7	12.1	(R)SB(r)0°	0	Mag 11.9 star W 1.′8.
03 17 18.7	−41 06 00	NGC 1291	8.5	11.0 x 9.5	13.4	(R)SB(s)0/a	72	Bright N-S oriented core with faint shell oriented more E-W.
03 30 40.8	−50 18 32	NGC 1356	13.0	1.7 x 0.9	13.3	SAB(r)bc pec:	173	**IC 1947** SW 2.′2.
03 38 44.7	−44 06 02	NGC 1411	11.3	1.8 x 1.4	12.2	SA(r)0^{-}:	6	Stellar **PGC 13390** W 8.′6.
03 42 01.6	−47 13 15	NGC 1433	9.9	6.5 x 5.9	13.7	(R′)SB(r)ab II	99	Strong core in short bar, moderately dark patches N and S of bar with faint halo beyond.
03 44 32.4	−44 38 32	NGC 1448	10.7	7.6 x 1.7	13.3	SAcd: sp II-III	41	Numerous knots; located between a pair of N-S oriented mag 9.21v, 9.67v stars.
03 52 08.9	−44 31 59	NGC 1476	13.3	1.4 x 0.4	12.5	Sa? pec	86	Mag 10.75v star W 6.′0.
03 52 47.7	−47 28 39	NGC 1483	12.7	1.6 x 1.3	13.3	SB(s)bc: III-IV	125	
03 55 44.8	−42 22 04	NGC 1487	11.9	2.4 x 1.7	13.3	Pec	55	Multi-galaxy system.
03 57 27.6	−46 12 39	NGC 1493	11.3	2.8 x 2.8	13.4	SB(r)cd III		Many multi-branching arms with many knots.
03 57 42.5	−48 54 26	NGC 1494	11.7	3.2 x 1.9	13.5	SAB(s)d: II-III	179	Numerous knots on branching arms; mag 10.35v star NNW 7.′2.
03 58 21.8	−44 27 59	NGC 1495	12.6	3.0 x 0.5	12.9	Sc? sp	105	Mag 9.39v star E 5.′8.
04 03 32.8	−43 24 02	NGC 1510	13.0	0.9 x 0.9	12.7	SA0° pec?		Located 5.′0 SW of NGC 1512.
04 03 54.3	−43 20 58	NGC 1512	10.3	8.9 x 5.6	14.4	SB(r)a I	90	Bright NE-SW core with faint E-W envelope; NGC 1510 SW 5.′0.
02 22 07.1	−48 33 54	PGC 9006	13.9	1.2 x 0.9	14.1	E^{+}2	108	Mag 12.6 star S 2.′1; stellar galaxy **Fair 379** N 6.′9.

GALAXY CLUSTERS

RA h m s	Dec ° ′ ″	Name	Mag 10th brightest	No. Gal	Diam ′	Notes
02 22 06.0	−48 34 00	A 3009	16.3	54	17	PGC 9006 at center, most other members anonymous.
03 10 54.0	−47 23 00	A 3093	16.4	93	17	**ESO 199-19** on W edge, all other members anonymous, stellar.
03 13 48.0	−47 47 00	A 3100	16.1	46	17	**Fair 744** SE of center 6.′3, all others anonymous, stellar.
03 14 18.0	−45 24 00	A 3104	16.2	37	17	All members anonymous, stellar.
03 15 12.0	−47 37 00	A 3108	16.2	73	17	**ESO 199-26** near N edge; all other members anonymous, stellar.
03 16 30.0	−50 54 00	A 3110	16.2	37	17	Three PGC GX near center; **IC 1912** near N edge, 2.′6 W of mag 8.63v star.
03 17 48.0	−45 44 00	A 3111	16.3	54	17	**Fair 1103** E of center 9.′3; all others anonymous, stellar.
03 17 54.0	−44 14 00	A 3112	16.1	116	17	
03 22 18.0	−41 20 00	A 3122	15.8	100	17	All members anonymous, stellar/almost stellar.
03 32 42.0	−45 56 00	A 3133	16.2	33	17	All members anonymous, faint, stellar.
02 22 18.0	−51 06 00	AS 250	16.3		17	Except ESO 198-8, all members anonymous, stellar.
02 33 24.0	−49 34 00	AS 277	16.2		17	**ESO 198-18** N of center 3.′5; all others anonymous, stellar.
02 47 30.0	−41 49 00	AS 297	16.4	6	17	**ESO 299-17** NE of center 13.′1; all others anonymous, stellar.
02 54 00.0	−50 58 00	AS 307	16.1	5	17	All members anonymous, stellar.
02 59 12.0	−51 01 00	AS 315	16.3		17	**Fair 733** at center, all other members anonymous, stellar.
03 16 18.0	−45 07 00	AS 334	16.2	5	17	All members anonymous, stellar.
03 17 06.0	−46 31 00	AS 335	16.3		17	All members anonymous, stellar.
03 17 30.0	−44 42 00	AS 336	16.1	5	17	**PGC 12228** at center, most others anonymous, stellar.

(Continued from previous page)

190 GALAXY CLUSTERS 190

RA h m s	Dec ° ′ ″	Name	Mag 10th brightest	No. Gal	Diam ′	Notes
03 21 48.0	−45 32 00	AS 345	16.3	9	17	All members anonymous, stellar.
03 22 24.0	−49 19 00	AS 346	16.2	8	17	**PGC 12638** at center; most other members anonymous, stellar.
03 26 24.0	−51 24 00	AS 353	16.0		17	See notes for IC 1929; other members anonymous, stellar.
03 29 36.0	−45 58 00	AS 356	16.2	10	17	All members anonymous, stellar.
03 33 18.0	−49 07 00	AS 360	16.3		17	All members anonymous, stellar; **IC 1961** N of center 10′6 and 4′6 SE of mag 8.18v star.
03 35 30.0	−45 10 00	AS 367	16.2		17	**PGC 13276** at center; **PGC 13302** NE 10′9.
03 45 48.0	−41 10 00	AS 384	15.7	6	17	**MCG −7-8-6** at center, all others anonymous, stellar.
03 48 18.0	−45 32 00	AS 393	16.3	11	17	**PGC 13876** at center, several PGC GX E, most others anonymous.

GLOBULAR CLUSTERS

RA h m s	Dec ° ′ ″	Name	Total V m	B ★ V m	HB V m	Diam ′	Conc. Class Low = 12 High = 1	Notes
03 55 02.7	−49 36 52	AM 1	15.8	18.2	20.9	0.5		

PLANETARY NEBULAE

RA h m s	Dec ° ′ ″	Name	Diam ″	Mag (P)	Mag (V)	Mag cent ★	Alt Name	Notes
02 56 58.2	−44 10 19	PN G255.3−59.6	373			15.4	PK 255−59.1	Mag 13.7 star NW of center 5′5; mag 10.53v star ESE of center 13′2.

191 GALAXIES 191

RA h m s	Dec ° ′ ″	Name	Mag (V)	Dim ′ Maj x min	SB	Type Class	PA	Notes
00 44 02.3	−50 37 41	ESO 194-32	14.3	0.6 x 0.4	12.6	Sbc	101	Mag 9.31v star W 8′4.
00 46 19.5	−47 25 54	ESO 194-36	14.4	0.9 x 0.6	13.6	S0/a	51	Double system; mag 5.80v star SSW 9′3.
00 47 03.9	−48 43 07	ESO 194-38	13.8	1.0 x 0.7	13.3	Scd: III-IV	99	Mag 15 stars N and S edges.
00 47 27.4	−48 14 16	ESO 195-1	13.9	0.7 x 0.7	13.0	S0		Mag 12.5 star NE 1′3.
00 49 18.8	−49 28 42	ESO 195-3	13.8	0.9 x 0.7	13.2	(R′)SA(rs)b	70	Mag 11.05v star SW 6′0.
00 55 45.0	−50 27 53	ESO 195-17	13.8	0.9 x 0.6	13.0	Sa	94	Faint, elongated anonymous galaxy N 1′5.
01 01 44.4	−51 52 10	ESO 195-24	13.1	0.9 x 0.8	12.6	(R′)SAB(rs)a:	96	Mag 9.61v star WNW 5′8.
01 04 30.0	−51 27 39	ESO 195-29	14.2	1.2 x 0.6	13.7	SAB(s)c II-III	117	Mag 13.4 star SW 2′1.
01 15 55.3	−50 11 23	ESO 195-35	13.5	1.5 x 1.1	13.9	(R)SAB(rs)0^+	140	Strong core with dark patches N and S.
01 19 28.5	−51 46 48	ESO 196-3	13.7	1.2 x 1.0	13.8	(R)SB0/a	66	Located between a pair of stars, NE-SW, mags 10.99v and 13.1.
01 30 43.7	−51 08 33	ESO 196-11	13.6	1.6 x 0.6	13.4	SB(r)cd? II-III	12	Mag 7.77v star SE 5′2; **ESO 196-9** SW 7′0.
01 42 17.0	−47 31 49	ESO 196-19	13.7	1.3 x 0.8	13.6	(R′)SB(rs)a	61	Mag 13.1 star S 1′1.
01 46 23.9	−47 27 52	ESO 196-21	13.9	1.1 x 0.8	13.6	SB	102	Close pair of mag 15 stars NE 1′8.
01 48 37.0	−48 49 15	ESO 197-2	14.1	1.5 x 0.3	13.1	SB(s)cd? sp	151	Mag 12.4 star E 6′5.
01 52 46.2	−48 53 30	ESO 197-9	13.9	1.3 x 0.6	13.5	SB(s)bc? II	119	Mag 10.92v star ENE 3′7.
01 53 12.7	−49 33 41	ESO 197-10	12.4	1.6 x 1.3	13.1	S0 pec	178	Close pair of bright stars W 7′7.
01 59 17.6	−50 43 13	ESO 197-16	13.4	1.0 x 0.4	12.3	S0?	48	Stellar **PGC 7543** NE 4′0.
02 02 31.0	−50 55 55	ESO 197-18	12.5	1.4 x 0.8	12.6	SA(r)0^-?	172	Mag 10.56v star NE 10′3.
02 16 45.2	−47 49 11	ESO 198-1	13.4	1.0 x 0.7	13.1	E^+4	114	Mag 11.22v star E 7′8.
02 18 02.8	−47 32 15	ESO 198-2	13.3	1.1 x 0.5	12.5	S0?	3	Mag 13.8 star SW 1′8.
02 19 27.4	−50 51 15	ESO 198-6	13.6	0.9 x 0.5	12.6	S?	123	Strong nucleus, smooth envelope.
02 22 15.7	−51 06 32	ESO 198-8	13.1	0.9 x 0.9	12.8	S0		Mag 5.90v star E 6′3.
02 27 52.7	−50 42 43	ESO 198-11	13.6	1.1 x 0.4	12.5	(R′)SB(s)b	177	Mag 10.91v star E 0′7; with faint anonymous galaxy in between.
00 49 34.7	−46 52 34	ESO 243-2	14.0	0.9 x 0.8	13.5	(R′)SA(rs)b	9	Mag 13.8 star N edge; mag 9.80v star ENE 2′4.
00 51 56.9	−43 28 38	ESO 243-6	14.1	1.3 x 0.3	12.9	Sb	50	Located between a pair of mag 13.6 and 14.2 stars.
00 55 13.7	−43 55 03	ESO 243-11	13.7	1.3 x 0.8	13.6	(R′)SAB(rs)a	49	Mag 12.8 star 1′0 SW.
00 56 15.2	−47 03 55	ESO 243-12	13.9	1.1 x 0.6	13.3	(R)SA(r)0^+?	57	Mag 12.3 star W 7′7.
01 05 11.1	−46 21 18	ESO 243-24	13.7	1.1 x 0.5	13.0	S0	74	Almost stellar anonymous galaxy S 0′9; larger anonymous galaxy SW 2′3.
01 06 33.1	−46 38 40	ESO 243-29	13.4	0.7 x 0.6	12.3	(R′)SAB(r)0^+:	144	Located 6′6 NE of mag 3.32v star Beta Phoenicis.
01 07 09.7	−42 23 39	ESO 243-30	14.0	0.9 x 0.6	13.2	SA(s)c? II	145	Mag 10.11v star NW 7′6.
01 07 14.8	−46 37 20	ESO 243-31	13.9	1.0 x 0.3	12.5	Sa	137	Mag 14.7 star SW edge; mag 3.32v star Beta Phoenicis SW 6′5.
01 08 11.5	−47 08 13	ESO 243-35	13.5	1.3 x 0.9	13.5	(R)SB(rs)0^0	2	Small, faint, elongated anonymous galaxy 1′5 NW; **ESO 243-38** NE 3′0.
01 08 18.9	−47 04 21	ESO 243-37	13.8	1.3 x 0.9	13.9	(R)SB(r)0^+	116	**ESO 243-38** ESE 1′8; **ESO 243-42** SE 7′3.
01 08 51.6	−45 50 00	ESO 243-41	13.4	1.2 x 0.4	12.5	$S0^0$: pec	89	ESO 243-45 NNE 4′2; IC 1633 SE 12′5.
01 08 54.9	−45 18 19	ESO 243-43	13.7	1.0 x 0.8	13.3	SBa	166	**ESO 243-44** S 2′7.
01 09 04.5	−45 46 24	ESO 243-45	12.4	1.3 x 1.3	12.9	$S0^-$		Mag 12.1 star NNW 4′5; ESO 243-41 SSW 4′2.
01 10 27.8	−46 04 28	ESO 243-49	13.8	1.2 x 0.3	12.6	S0/a	62	Faint anonymous galaxy WSW 5′8.
01 11 19.2	−45 55 58	ESO 243-51	13.3	1.2 x 0.7	12.9	SAa?	50	Pair with ESO 243-52 E.
01 11 27.8	−45 56 20	ESO 243-52	13.8	1.3 x 0.4	12.9	$S0^0$: sp	106	Pair with ESO 243-51 W.
01 15 09.8	−44 28 57	ESO 244-6	13.5	1.5 x 1.1	13.9	$SA0^-$	14	Mag 8.38v star S 8′4.
01 15 18.3	−46 30 49	ESO 244-8	13.4	0.8 x 0.7	12.6	Sa:	64	**ESO 244-5** W 3′4.
01 17 03.5	−44 10 54	ESO 244-10	12.8	1.3 x 1.2	13.2	SAB(rs)ab	117	Star superimposed E of core.
01 18 08.9	−44 27 49	ESO 244-12	14.7	1.6 x 0.9	14.9	Sb: pec	177	Multi-galaxy system.
01 19 45.2	−43 34 16	ESO 244-16	13.6	1.1 x 0.8	13.3	S0	165	**ESO 244-15** N 2′0.
01 20 19.6	−44 07 41	ESO 244-17	13.5	1.3 x 1.2	13.8	(R)SB(r)a pec:	21	Small anonymous galaxy, or double star, W 1′4.
01 21 50.2	−42 25 08	ESO 244-20	13.0	0.8 x 0.8	12.4	Sc?		Mag 13.1 star W 1′1.
01 21 49.1	−44 03 28	ESO 244-21	13.4	1.4 x 1.0	13.6	SB(r)bc I-II	170	Mag 12.6 star S 2′3.
01 22 52.7	−42 33 58	ESO 244-22	13.8	0.9 x 0.6	13.0	S0	0	Mag 11.13v star W 10′2.
01 23 02.7	−43 07 39	ESO 244-23	14.0	1.7 x 0.3	13.2	Sa: pec sp	149	Mag 9.06v star ESE 2′2.
01 29 51.8	−42 19 38	ESO 244-30	13.9	1.1 x 0.6	13.3	Pec	46	Mag 11.9 star W 6′6.
01 30 05.1	−42 41 08	ESO 244-31	13.0	1.7 x 1.3	13.7	SA(r)c I	87	Stellar nucleus, multi-branching arms; mag 10.00v star E 4′8.
01 30 47.1	−42 27 46	ESO 244-32	14.3	0.9 x 0.3	12.7	Pec	9	**MCG −7-4-13** NE 11′1.
01 36 25.7	−42 35 39	ESO 244-36	13.9	1.7 x 1.1	14.4	(R)SAB(rs)bc pec: II-III	139	Mag 10.17v star W 10′1.
01 36 46.1	−46 49 34	ESO 244-37	13.9	1.0 x 0.8	13.5	Sa	49	Galaxy **Fair 703** NNW 6′5.
01 37 33.7	−42 40 30	ESO 244-39	13.7	1.5 x 0.5	13.2	(R)SB0/a: sp	18	**ESO 244-38** WSW 7′1.
01 38 54.0	−46 50 11	ESO 244-44	13.2	1.7 x 0.8	13.4	(R′)SB(rs)a:	77	Mag 9.66v star W 2′9.
01 38 58.5	−46 34 24	ESO 244-45	12.7	1.2 x 1.1	12.8	(R′)$SA0^-$?	20	Mag 9.89v star NE 8′4.
01 39 06.7	−43 21 39	ESO 244-47	13.4	1.3 x 0.5	12.8	S0 pec	27	**ESO 244-46** 0′7 SW of center.
01 39 09.3	−47 07 41	ESO 244-48	14.5	1.5 x 0.2	13.0	Sb sp	150	Mag 9.23v star NE 3′8; galaxy **Fair 709** E 5′4.
01 39 30.9	−46 59 11	ESO 244-49	14.2	1.7 x 0.2	12.9	SB(r)0^+: sp	67	Mag 9.23v star S 5′7.
01 43 06.2	−44 12 55	ESO 245-3	13.7	1.1 x 0.4	12.6	Sb	80	**ESO 245-4** NE 5′5.
01 45 03.8	−43 35 44	ESO 245-5	12.3	3.9 x 3.0	14.8	IB(s)m V-VI	122	Mag 11.34v star on W edge; mag 8.49v star N 3′1.

(Continued from previous page)

GALAXIES

RA h m s	Dec ° ′ ″	Name	Mag (V)	Dim ′ Maj x min	SB	Type Class	PA	Notes
01 50 28.8	−47 09 58	ESO 245-6	14.4	1.7 x 1.5	15.2	(R)SA(r)ab? pec	177	Almost stellar nucleus, faint envelope; mag 10.47v star N 9'.4.
01 51 06.3	−44 26 51	ESO 245-7	12.4	3.4 x 3.0	14.7	IAm VI	93	**Phoenix Dwarf**, has the appearance of an open star cluster.
01 56 44.7	−43 58 26	ESO 245-10	13.4	1.8 x 0.5	13.1	Sb pec	24	Mag 9.51v star NW 13'.6.
02 02 37.7	−42 47 47	ESO 245-12	13.6	1.7 x 0.7	13.7	(R′)SB(rs)ab pec:	99	
02 19 44.4	−43 15 08	ESO 246-8	14.5	1.2 x 0.4	13.6	Sc: II	13	Mag 11.09v star ESE 9'.0.
00 48 32.1	−40 10 20	ESO 295-7	14.9	1.5 x 0.2	13.4	Sc	148	Mag 10.98v star NE 9'.7.
00 50 20.7	−41 14 40	ESO 295-12	13.7	1.0 x 0.6	13.0	S?	160	Faint star on W edge.
01 04 18.7	−40 08 53	ESO 295-31	13.8	1.0 x 0.8	13.4	(R′)SB(rs)ab	121	Mag 10.80v star NW 4'.4.
01 07 10.7	−41 55 07	ESO 295-38	13.1	1.2 x 0.7	12.7	SAB(s)0^-	42	**MCG −7-3-10** NE 2'.0.
01 12 33.8	−40 56 46	ESO 296-5	14.2	0.6 x 0.6	13.0	S?		Mag 10.01v star W 5'.4.
01 19 56.8	−41 14 12	ESO 296-11	14.2	1.3 x 0.9	14.3	S?	162	Double system, each elongated, at right angles, interacting.
01 30 28.3	−41 17 45	ESO 296-34	13.7	1.4 x 1.0	13.9	(R′)SB(rs)0^+:	69	Strong, elongated core, smooth envelope; mag 12.8 star NW 2'.8.
01 37 52.2	−40 32 22	ESO 297-15	13.9	1.0 x 0.4	12.8		112	Mag 13.9 star SW 2'.4.
01 37 59.5	−40 04 09	ESO 297-16	14.0	1.7 x 0.4	13.4	SB(s)cd? sp II	39	Mag 14 star N end.
01 38 20.0	−40 44 02	ESO 297-17	14.4	1.1 x 0.9	14.2	Sc	108	Stellar nucleus.
01 38 37.3	−40 00 42	ESO 297-18	13.3	1.5 x 0.4	12.6	Sa: sp	143	
01 40 47.9	−41 12 47	ESO 297-20	14.8	1.2 x 0.4	13.9	SB(s)d? III-IV	75	Mag 9.41v star N 4'.3.
01 44 28.6	−40 39 52	ESO 297-23	13.6	1.3 x 0.4	12.8	SAB(r)ab pec:	4	Mag 13.3 star N 2'.1.
01 45 10.6	−41 52 49	ESO 297-27	14.1	1.4 x 0.9	14.2	SA(rs)b: II-III	78	Mag 13.4 star ESE 1'.6.
02 02 40.7	−41 24 53	ESO 298-3	14.0	1.0 x 0.8	13.6	S?	128	Mag 10.96v star W 11'.7.
02 06 16.4	−41 31 28	ESO 298-8	13.8	1.4 x 0.5	13.3	SAB(rs)cd III-IV	128	Mag 12.5 star N 1'.6.
02 08 43.1	−40 07 52	ESO 298-11	14.4	1.1 x 0.9	14.2	Sb	120	**ESO 298-12** NE 3'.0.
02 10 38.5	−40 55 06	ESO 298-15	13.6	2.0 x 0.7	13.8	SB(rs)cd pec: IV	58	Mag 6.50v star WNW 10'.0; note: **ESO 298-13** is 2'.5 WSW of this star.
02 21 11.8	−42 00 08	ESO 298-31	14.0	1.1 x 0.4	13.0	(R′)SB(r)a:	109	Mag 12.4 star W 2'.6.
02 25 14.2	−40 25 37	ESO 299-1	13.1	1.2 x 0.8	12.9		164	Double system, interaction.
00 53 47.0	−45 11 17	IC 1595	14.3	1.4 x 0.2	12.7	Sb? sp	12	Mag 7.44v star E 12'.8.
00 56 59.4	−45 24 49	IC 1603	13.8	1.5 x 0.8	13.9	SA(s)c: II	111	Mag 12.5 star NW 8'.3.
00 57 37.7	−48 54 11	IC 1605	12.9	1.5 x 1.3	13.5	(R′)SB(r)bc I-II	138	Mag 12.8 star E 4'.0.
00 59 47.0	−40 20 05	IC 1609	12.6	1.4 x 1.4	13.2	SA0°:		Close pair of E-W oriented mag 10.77v, 11.06v stars N 6'.6.
01 04 06.9	−51 07 58	IC 1615	13.4	1.3 x 0.6	12.9	SB(r)bc I-II	147	Mag 13.6 star E 3'.7; IC 1617 N 6'.2.
01 04 16.9	−51 02 00	IC 1617	13.6	1.4 x 0.6	13.3	(R′)S(r)0/a?	125	Mag 11.28v star E 5'.6.
01 06 23.0	−46 43 31	IC 1621	13.2	1.3 x 0.4	12.3	Sb-c	4	Located 3'.0 E of mag 3.32v star Beta Phoenicis.
01 07 42.8	−46 54 27	IC 1625	12.0	1.5 x 1.2	12.5	SA(rs)0^- pec:	7	Mag 3.32v star Beta Phoenicis NW 20'.2; **ESO 243-32** NW 6'.2.
01 08 10.9	−46 05 40	IC 1627	12.9	2.5 x 0.6	13.2	SB(s)b: sp II	139	Mag 10.76v star NW 2'.9.
01 08 17.1	−46 45 15	IC 1630	14.3	1.5 x 0.3	13.3	Sb pec sp	65	Mag 6.98v star NNE 5'.9.
01 08 45.0	−46 28 33	IC 1631	13.3	0.9 x 0.6	12.5	Sab pec	82	Mag 12.0 star WNW 10'.1.
01 09 55.2	−45 55 50	IC 1633	11.5	2.6 x 2.4	13.6	E^+1	93	Large, bright core with faint envelope; several small, faint anonymous galaxies near by.
01 12 19.0	−50 24 09	IC 1650	13.6	1.3 x 0.7	13.4	SB(s)b pec II	61	Mag 10.28v star NE 9'.1.
01 19 18.5	−50 57 50	IC 1674	14.1	1.1 x 0.8	13.8	SA:(r)c	28	Mag 10.34v star ENE 4'.6.
02 22 47.3	−41 22 17	IC 1796	13.0	1.1 x 0.9	12.9	SAB(s)0^-?	86	Mag 10.18v star SE 3'.4.
00 56 57.4	−43 50 18	NGC 319	13.4	1.0 x 0.8	13.0	(R′)SAB(s)a:	150	
00 57 10.1	−43 43 39	NGC 322	13.4	1.0 x 0.5	12.4	S0? pec sp	153	
00 57 14.8	−40 57 35	NGC 324	12.9	1.4 x 0.5	12.3	S?	95	
01 04 22.1	−43 16 38	NGC 368	13.8	0.7 x 0.6	12.7	(R′)SAB(rs)0^+?	169	Located 3'.1 NE of mag 8.76v star.
01 20 20.6	−40 58 01	NGC 482	13.3	2.2 x 0.5	13.2	SAab: sp	84	Faint lane runs full length just N of center.
01 28 57.5	−51 35 54	NGC 576	13.4	0.9 x 0.7	12.8	(R′)SAB(rs)0^+	18	
01 35 05.0	−41 26 11	NGC 625	11.2	5.8 x 1.9	13.6	SB(s)m? sp III-IV	92	Several faint stars or bright knots superimposed around large core; **ESO 297-10** NE 11'.1.
01 38 39.2	−42 31 38	NGC 641	12.1	1.4 x 1.3	12.6	SA(s)0^-:	4	Mag 9.88v star WSW 4'.2; **ESO 244-40** W 11'.7.
01 38 53.0	−42 35 08	NGC 644	14.0	1.3 x 0.6	13.6	SB(r)bc: I-II	155	Mag 9.88v star W 6'.8.
01 48 41.5	−48 38 54	NGC 692	12.3	2.1 x 1.8	13.6	(R′)SB(r)bc: I	34	Almost stellar **ESO 197-4** N 7'.4; **ESO 197-7** E 11'.9.
02 06 39.1	−41 09 27	NGC 822	13.1	1.1 x 0.6	12.7	E:	77	
02 13 03.0	−42 02 04	NGC 862	12.8	0.8 x 0.8	12.4	E:		
02 19 07.0	−41 44 59	NGC 889	13.3	0.9 x 0.8	13.0	E:	103	Mag 6.35v star SSE 6'.8.
02 19 58.3	−41 24 11	NGC 893	12.7	1.3 x 1.0	12.9	SA(s)c pec:	119	Located 3'.2 WSW of mag 8.64v star.
02 26 21.8	−44 26 47	NGC 939	13.0	1.2 x 1.0	13.2	E^+:	110	
01 13 51.0	−45 24 38	PGC 4427	14.9	0.7 x 0.7	14.2	E?		Mag 8.92v star NE 12'.2; **PGC 4409** NNW 5'.2.
02 22 07.1	−48 33 54	PGC 9006	13.9	1.2 x 0.9	14.1	E^+2	108	Mag 12.6 star S 2'.1; stellar galaxy **Fair 379** N 6'.9.

GALAXY CLUSTERS

RA h m s	Dec ° ′ ″	Name	Mag 10th brightest	No. Gal	Diam ′	Notes
00 53 42.0	−47 36 00	A 2836	16.3	41	17	**IC 1594** S of center 2'.9; **ESO 195-11** near N edge; all others anonymous.
01 00 48.0	−50 31 00	A 2854	15.8	64	17	**ESO 195-22** and **22A** at center; all others anonymous and stellar.
01 07 42.0	−46 54 00	A 2870	15.2	33	22	Numerous anonymous stellar and almost stellar GX near center.
01 09 48.0	−45 54 00	A 2877	14.3	30	28	
02 22 06.0	−48 34 00	A 3009	16.3	54	17	PGC 9006 at center, most other members anonymous.
00 45 42.0	−49 56 00	AS 74	15.9	21	17	**ESO 194-35**, **35A** SW of center 8'.2; others anonymous, stellar.
00 47 30.0	−48 14 00	AS 78	16.1		17	**ESO 95-1** at center; all other members anonymous, stellar.
00 49 54.0	−47 22 00	AS 85	16.2		17	**ESO 195-4**, **4A** at center, most others anonymous, stellar.
00 53 24.0	−48 09 00	AS 92	16.4	27	17	**ESO 195-13** NE 13'.0; **ESO 195-10**, **10A**, **10B** WSW 14'.2.
00 55 30.0	−49 55 00	AS 101	15.8	26	17	**ESO 195-15** WSW of center 7'.4; all others anonymous, stellar/almost stellar.
01 00 36.0	−40 13 00	AS 113	16.3	12	17	**MCG −7-3-5** N of center7'.0.
01 01 42.0	−51 51 00	AS 116	15.8	25	17	ESO 195-24 at center, all others anonymous, stellar.
01 03 48.0	−42 52 00	AS 120	15.6		22	**ESO 243-22** E of center 5'.2; **ESO 243-25** E 16'.5; **ESO 243-20** SW 10'.6.
01 31 42.0	−51 20 00	AS 162	15.8	22	17	Elongated **ESO 196-12** N of center 5'.2; most others anonymous, stellar.
02 01 06.0	−40 35 00	AS 215	16.3		17	All members anonymous, stellar.
02 01 36.0	−44 39 00	AS 217	16.2	13	17	Mag 5.14v star near center; **ESO 245-11**, **11A** SE of star 7'.5; all others anonymous.
02 11 54.0	−47 50 00	AS 230	16.2	10	17	Faint **ESO 197-28**, **28A** at center; most other members anonymous.
02 22 18.0	−51 06 00	AS 250	16.3		17	Except ESO 198-8, all members anonymous, stellar.

OPEN CLUSTERS

RA h m s	Dec ° ′ ″	Name	Mag	Diam ′	No. ★	B ★	Type	Notes
01 53 42.0	−45 57 00	ESO 245-9		12			cl	(A) Sparse group of stars on DSS. Faint, anonymous galaxy on N edge.

RA h m s	Dec ° ′ ″	Name	Mag (V)	Dim ′ Maj x min	SB	Type Class	PA	Notes
00 23 54.1	−62 16 21	ESO 78-23	14.1	0.7 x 0.5	12.8		122	Mag 10.72v star NE 2.′2.
00 25 49.2	−62 19 49	ESO 79-1	13.9	1.1 x 0.6	13.3	SAB(rs)c: I-II	126	Mag 8.14v star NW 9.′6.
00 54 14.0	−62 27 20	ESO 79-8	13.8	1.3 x 0.8	13.7	SB(s)bc I	153	Mag 9.35v star SW 9.′9.
23 03 46.1	−62 31 19	ESO 109-32	14.2	1.7 x 0.3	13.3	SAB(s)c? sp III	128	Mag 7.78v star NNW 9.′3.
23 36 42.8	−62 48 54	ESO 110-16	13.9	1.0 x 0.5	13.0	SAB(r)0/a	66	Mag 8.13v star N 3.′8.
23 52 10.6	−62 34 20	ESO 110-29	13.7	1.3 x 0.2	12.0	Sa	96	Mag 15 star on N edge; **ESO 78-9** ENE 2.′0.
23 52 45.4	−60 38 12	ESO 111-6	13.3	1.5 x 1.1	13.7	SAB(rs)c I-II	111	Stellar nucleus; mag 8.58v star NW 8.′4.
23 53 26.8	−59 41 44	ESO 111-7	14.2	1.2 x 0.5	13.5	Sb? pec	147	Mag 7.06v star NNE 10.′9.
23 55 55.5	−60 40 55	ESO 111-9	13.9	1.9 x 0.6	13.8	SAc:	89	Pair with ESO 111-10 NE edge; mag 9.81v star N 5.′0.
23 56 07.3	−60 40 18	ESO 111-10	13.4	1.6 x 1.0	13.7	(R′)SAB(rs)bc	127	Pair with ESO 111-9 SW edge; mag 9.81v star N 4.′9.
00 08 18.7	−59 30 55	ESO 111-14	13.6	1.9 x 1.1	14.3	SAB(rs)cd II	161	Mag 10.31v star NW 4.′4.
00 15 16.5	−57 14 38	ESO 111-22	14.0	1.9 x 0.9	14.4	(R)SA(r)b:	9	Mag 9.91v star N 2.′4.
00 18 01.8	−59 05 05	ESO 111-24	14.4	1.5 x 0.4	13.7	Sab: sp	156	Strong dark lane through center; mag 9.47v star ESE 7.′5.
00 23 11.5	−59 37 06	ESO 112-2	14.0	1.1 x 0.6	13.4	Sc	58	Stellar nucleus, smooth envelope.
00 28 42.7	−59 56 41	ESO 112-5	13.8	1.1 x 0.8	13.6	SB?	168	Mag 9.35v star SW 11.′2.
00 35 50.8	−59 41 41	ESO 112-8	15.2	1.3 x 0.5	14.6	S0: pec:	36	Lies 1.′9 NW of mag 7 star.
00 45 02.3	−60 45 37	ESO 112-9	13.9	1.1 x 0.5	13.2	SA0°: sp	24	Mag 10.12v star E 6.′6.
00 55 57.1	−59 40 02	ESO 112-11	14.4	1.5 x 0.3	13.4	Sb: sp	129	Mag 9.35v star E 4.′1.
01 00 34.3	−57 44 57	ESO 113-4	14.4	0.4 x 0.3	11.9		107	Mag 8.97v star NE 2.′4.
01 01 34.7	−59 44 32	ESO 113-6	15.1	1.5 x 0.2	13.6	Sc: sp	64	Lies 1.′2 N of a pair of mag 12 stars.
22 56 05.5	−58 23 46	ESO 147-14	13.6	0.9 x 0.6	12.8	SB(r)c? III-IV	108	Mag 13.4 star NE 2.′9.
23 01 44.7	−59 10 26	ESO 147-20	13.9	1.2 x 1.2	14.2	SAB(r)bc II-III		Three stars on W edge in N-S line.
23 14 57.0	−61 28 04	ESO 148-1	13.8	1.0 x 0.5	12.8	SB(r)a	171	Mag 11.8 star E 5.′0.
23 15 46.8	−59 03 21	ESO 148-2	14.7	1.2 x 0.7	14.4		173	Peculiar, plumes; mag 14.0 star S 1.′5.
23 19 24.8	−61 12 30	ESO 148-4	13.9	1.5 x 0.2	12.5	Im:	104	Mag 11.25v star SW 5.′4.
23 21 15.6	−58 04 56	ESO 148-7	13.1	0.9 x 0.7	12.5		150	Mag 10.34v star SW 7.′7.
23 30 07.1	−59 27 57	ESO 148-17	12.4	2.2 x 0.8	13.0	E6	27	Mag 7.14v star W 9.′9.
23 31 10.2	−60 55 55	ESO 148-18	13.3	1.7 x 1.1	13.8	SB(s)cd II-III	22	Mag 11.19v star N 4.′9.
23 33 58.7	−60 13 55	ESO 148-19	14.5	1.5 x 0.6	14.2	SAB(s)d IV-V	7	Mag 13.5 star NW 2.′0.
23 36 20.1	−57 37 48	ESO 148-20	14.9	2.2 x 0.2	13.9	SBd? sp	66	Mag 7.75v star NW 11.′5.
23 38 32.4	−57 59 59	ESO 148-21	13.8	1.3 x 0.4	12.9	SB(s)c? III-IV	148	Mag 10.34v star S 2.′9.
23 47 48.9	−57 04 19	ESO 149-1	13.0	3.8 x 0.7	13.9	SB(s)dm IV	34	Mag 11 star superimposed N of center.
23 52 02.8	−52 34 40	ESO 149-3	14.4	1.3 x 0.5	13.8	IB(s)m: sp IV	152	Mag 7.93v star N 14.′8.
23 59 23.1	−56 20 00	ESO 149-11	13.8	1.3 x 1.3	14.2	SAB(rs)d IV		Almost stellar nucleus; mag 12.3 star E 3.′2.
00 02 46.2	−52 46 18	ESO 149-13	14.7	1.6 x 0.5	14.3	IB(s)m V	100	Mag 10.12v star SW 8.′9.
00 04 51.8	−54 29 13	ESO 149-16	13.7	1.1 x 0.6	13.2	(R′)SAB(s)0^{+}?	10	Mag 11.7 star N 5.′2.
00 07 14.5	−52 37 19	ESO 149-18	15.2	1.0 x 0.8	14.8	SB(s)m VI	27	Mag 10.52v star N 5.′9.
00 10 39.1	−56 48 13	ESO 149-21	14.1	1.1 x 0.6	13.5	(R′)SB(rs)a	96	Mag 14.4 star NW 2.′0.
00 20 06.4	−54 31 26	ESO 150-1	14.1	1.3 x 0.7	13.9	SAB(s)c I-II	15	Mag 9.13v star ESE 11.′3.
00 22 26.0	−53 38 46	ESO 150-5	13.5	3.1 x 2.0	15.3	SAB(s)dm IV-V	27	Mag 10.48v star SSW 6.′0.
00 35 26.7	−55 31 18	ESO 150-13	13.7	1.1 x 0.8	13.4	(R)SAB(rs)a	127	Star or bright knot S edge.
00 36 37.7	−56 54 24	ESO 150-14	14.1	1.9 x 0.2	12.9	S0^{+}: sp	107	Mag 10.91v star N 2.′8.
00 37 16.4	−53 15 41	ESO 150-15	13.9	1.2 x 0.9	13.8	SB(s)c: pec I-II	154	Mag 12.0 star SW 7.′9.
00 43 21.1	−55 19 39	ESO 150-20	13.5	1.5 x 0.6	13.3	(R′)SB(r)bc: I-II	3	Mag 12.4 star NNW 6.′3.
00 44 59.5	−56 22 48	ESO 150-21	12.9	1.6 x 1.3	13.6	SAB(rs)0^{+}	120	Very small, very faint elongated anonymous galaxy NE of nucleus; mag 9.89v star NE 6.′4.
00 46 09.6	−55 34 07	ESO 150-22	13.9	1.3 x 0.4	13.0	(R′)SAB(rs)bc? II-III	88	Mag 8.97v star SE 6.′2.
00 50 47.1	−55 36 28	ESO 150-24	14.4	1.6 x 0.2	13.0	Sab? pec sp	154	**ESO 150-25** N 7.′1.
00 56 07.4	−53 11 22	ESO 151-4	13.6	1.3 x 0.3	12.4	S0° sp	160	Mag 6.62v star W 2.′2.
00 56 08.0	−52 49 49	ESO 151-5	14.0	1.0 x 0.6	13.3	(R′)SAB(s)0/a	171	NGC 312 and a mag 11.6 star N 2.′5.
00 56 47.2	−53 05 53	ESO 151-12	12.9	1.4 x 0.8	12.9	SA(s)0°:	118	Mag 6.62v star and ESO 151-4 SW 9.′9.
01 01 35.5	−53 11 57	ESO 151-18	13.7	1.5 x 0.3	12.7	Sa: sp	168	Mag 8.38v star NE 3.′2.
01 02 20.2	−54 19 36	ESO 151-19	14.1	1.5 x 1.1	14.5	SA(s)m: V	49	Mag 10.74v star NNE 8.′5.
23 38 45.7	−56 28 47	ESO 192-11	14.5	1.2 x 0.9	14.5	SA(r)bc	79	
23 39 08.5	−56 25 04	ESO 192-12	14.2	0.9 x 0.7	13.6	S?	7	Mag 12.4 star N 2.′9.
00 06 03.5	−52 13 05	ESO 193-22	13.5	1.7 x 0.9	13.8	SA0^{-}:	12	Mag 7.54v star WNW 9.′5; **ESO 193-25** E 7.′1.
00 19 39.8	−51 16 02	ESO 194-4	13.5	1.5 x 0.5	13.1	SB(r)0^{+} pec:	27	**ESO 194-4A** NE 1.′4.
00 29 41.6	−51 31 16	ESO 194-21	12.2	1.3 x 1.1	12.4	SA0^{-}	97	Mag 12.8 star W 1.′9.
00 47 03.3	−52 03 08	ESO 194-39	14.1	1.6 x 0.6	13.9	SAa? pec	80	Two spirals, interaction; trio of mag 9-10 stars NW 3.′6.
01 01 44.4	−51 52 10	ESO 195-24	13.1	0.9 x 0.8	12.6	(R′)SAB(rs)a:	96	Mag 9.61v star WNW 5.′8.
23 27 52.7	−51 07 52	ESO 240-4	15.0	1.5 x 0.3	14.0	Sd: sp	142	Dual galaxy system; mag 10.61v star S 2.′8.
23 38 49.8	−51 51 40	ESO 240-12	13.8	1.0 x 0.6	13.0	S?	102	Mag 11.8 star SSW 6.′5.
23 39 31.1	−52 02 53	ESO 240-14	14.2	1.1 x 0.5	13.4	Sb	152	Mag 9.79v star ENE 2.′8.
00 53 32.2	−58 06 30	IC 1597	14.2	1.7 x 0.3	13.3	(R′)SB(rs)b pec:	151	Mag 11.5 star E 8.′6.
00 09 59.8	−57 01 13	NGC 25	13.0	1.4 x 0.8	13.0	SB0^{-}? pec	88	Mag 10.46v star NW 2.′8; galaxy **Fair 2** SSE 6.′0.
00 10 25.2	−56 59 20	NGC 28	13.8	0.8 x 0.6	12.9	E1	119	
00 10 38.4	−56 59 14	NGC 31	13.7	1.2 x 0.6	13.2	SB(rs)cd II	5	Very small, faint anonymous galaxy SW 1.′7.
00 11 22.9	−56 57 26	NGC 37	13.7	1.0 x 0.6	13.0	(R′)SAB(s)0^{+}:	35	Mag 13.6 star on W edge; small, faint, elongated anonymous galaxy NE 3.′0.
00 14 42.8	−60 19 45	NGC 53	12.6	2.0 x 1.4	13.6	(R′)SB(r)b	172	Very faint outer shell. Mag 10.81 star W 1.′6.
00 26 58.0	−56 58 41	NGC 119	13.0	1.0 x 1.0	12.9	SA0^{-} pec		Bright nucleus.
00 34 35.9	−55 47 29	NGC 159	13.7	1.4 x 0.5	13.1	(R)SB(r)0/a	95	
00 40 13.4	−56 09 12	NGC 212	13.3	1.0 x 0.9	13.1	SA0^{-}	131	Anonymous galaxy SW 2.′0; mag 9.65v star S 2.′5.
00 40 49.1	−56 12 54	NGC 215	13.0	1.1 x 0.9	12.8	SA0^{-}	120	Anonymous galaxy NW 2.′9.
00 56 16.2	−52 46 59	NGC 312	12.4	1.4 x 1.1	12.9	E2:	62	Mag 11.6 star W 2.′5.
00 56 41.8	−52 58 35	NGC 323	12.6	1.3 x 1.1	12.9	E0:	175	Small faint anonymous galaxy on S edge; NGC 328 NNE 4.′0.
00 56 58.0	−52 55 28	NGC 328	13.2	2.7 x 0.5	13.3	SB(rs)a pec:	100	Galaxy **ESO 151-10** NW 2.′7.
01 00 52.1	−53 14 43	NGC 348	13.7	0.8 x 0.7	13.0	Sb II	94	Mag 14.4 star on N edge; mag 8.40v star SE 8.′8.
23 21 38.6	−62 07 07	NGC 7622	13.4	1.3 x 0.3	12.3	SB0^{-}?	60	On the side of a triangle of two mag 11 stars and one mag 13 star.
23 25 21.0	−57 47 30	NGC 7650	12.7	1.4 x 1.1	13.0	SB(s)cd pec: II-III	126	Multi-galaxy system.
23 25 37.7	−57 53 15	NGC 7652	13.5	1.0 x 0.7	12.9	(R′)SA(r)a pec	106	Bright nucleus in smooth envelope; mag 12.3 star 2.′1 W.
23 26 47.0	−57 48 20	NGC 7657	14.1	1.4 x 0.4	13.4	SB(s)dm pec: IV	106	
23 29 01.4	−59 43 01	NGC 7676	12.5	1.7 x 0.9	12.8	SA0^{-}: pec	85	
23 33 16.9	−54 05 39	NGC 7689	11.7	2.9 x 1.9	13.4	SAB(rs)cd II-III	135	Bright center, pair of dark patches NW.
23 33 02.6	−51 41 55	NGC 7690	12.4	2.2 x 0.9	13.0	SA(r)b:? II	132	
23 35 29.2	−56 00 47	NGC 7702	12.2	2.2 x 1.2	13.1	(R)SA(r)0^{+}	117	Mag 8.21v star W 3.′4.
23 59 00.0	−55 27 29	NGC 7796	11.5	2.2 x 1.9	13.0	E^{+}	168	Bright center, smooth envelope.
00 04 46.0	−62 03 42	NGC 7823	12.6	1.1 x 1.0	12.6	SB(s)bc? II-III	21	Small triangle of faint stars S.

(Continued from previous page)

192 GALAXY CLUSTERS 192

RA h m s	Dec ° ′ ″	Name	Mag 10th brightest	No. Gal	Diam ′	Notes
00 10 12.0	−56 59 00	A 2731	15.3	39	22	
00 40 12.0	−56 09 00	A 2806	15.3	37	22	NGC 212 and 215 near center, all other members anonymous.
00 25 42.0	−56 58 00	AS 45	16.4	20	17	Most members anonymous, stellar.
01 01 42.0	−51 51 00	AS 116	15.8	25	17	ESO 195-24 at center, all others anonymous, stellar.

193 GALAXIES 193

RA h m s	Dec ° ′ ″	Name	Mag (V)	Dim ′ Maj x min	SB	Type Class	PA	Notes
21 35 47.3	−62 46 40	ESO 107-37	13.8	1.2 x 1.0	13.8	Sa	157	Bright knots E and W of center; mag 9.68 star SW edge.
22 12 17.1	−62 43 31	ESO 108-18	14.1	0.6 x 0.4	12.5		30	Stellar **ESO 108-18A** NE 0′.7.
23 03 46.1	−62 31 19	ESO 109-32	14.2	1.7 x 0.3	13.3	SAB(s)c? sp III	128	Mag 7.78v star NNW 9′.3.
21 17 00.0	−59 19 54	ESO 145-1	13.8	0.8 x 0.5	12.7		127	Mag 9.93v star SW 3′.3; small group **Fair 89, 90, 91, 92, 93, 94, 95** S 10′.9.
21 18 51.3	−57 38 31	ESO 145-4	13.6	2.1 x 0.4	13.2	SA(s)c: III	121	Mag 8.33v star NW 12′.2.
21 27 32.3	−60 49 30	ESO 145-6	13.5	1.4 x 0.3	12.4	S0?	99	Bright star W edge; mag 9.21v star S 2′.3.
21 49 10.9	−59 02 10	ESO 145-16	13.9	1.0 x 0.7	13.4	(R′)SAB(r)ab	58	Small, faint anonymous galaxy SW 4′.9.
21 52 59.8	−59 23 33	ESO 145-21	14.7	1.5 x 0.3	13.7	S	84	Mag 9.71v star W 11′.0; IC 5141 S 6′.5.
21 53 39.8	−61 04 38	ESO 145-24	14.3	1.7 x 0.2	13.0	SBc: sp	90	Mag 10 star on N edge.
21 54 05.7	−57 36 33	ESO 145-25	13.3	1.9 x 1.3	14.2	SB(s)m V	137	Uniform surface brightness; stellar **PGC 67632** ESE 7′.9.
21 56 20.5	−59 17 32	ESO 146-1	14.4	1.0 x 0.6	13.7	SB(s)d IV	168	Mag 8.41v star S 3′.5.
21 57 55.2	−60 18 35	ESO 146-2A	15.6	1.6 x 1.3	16.2	IB(s)m V-VI	24	
22 05 15.5	−60 40 42	ESO 146-6	14.7	1.5 x 0.3	13.7	Sc: sp	44	Mag 12.3 star NW 8′.5.
22 09 12.9	−57 52 57	ESO 146-11	15.5	1.5 x 0.2	14.0	Sc: sp	78	Mag 10.92v star W 12′.6.
22 13 00.0	−62 04 09	ESO 146-14	14.7	2.8 x 0.2	13.9	SBd? sp	51	Mag 9.82v star N 5′.3.
22 28 51.2	−60 52 54	ESO 146-28	12.8	1.7 x 1.0	13.4	E$^+$3	156	Mag 15.3 star on NW edge.
22 05 30.9	−60 30 21	ESO 146-206	13.6	1.0 x 1.0	13.5			Mag 9.08v star WNW 6′.3.
22 41 36.7	−57 36 22	ESO 147-5	13.3	1.6 x 1.2	13.9	SB(r)c: I-II	179	Mag 12.7 star W 1′.6.
22 43 18.6	−59 24 52	ESO 147-6	13.6	1.1 x 0.3	12.3	Sb:	73	Mag 11.01v star S 3′.0.
22 48 57.9	−57 53 46	ESO 147-10	14.9	1.5 x 0.3	13.8	Sbc? pec sp	66	Mag 9.34v star WSW 4′.8; stellar galaxies **Fair 144** E 5′.3 and **Fair 143** SE 3′.9.
22 53 09.6	−60 41 56	ESO 147-13	13.5	1.4 x 1.2	14.0	S0°: pec	42	Mag 9.87v star NW 11′.0.
22 56 05.5	−58 23 46	ESO 147-14	13.6	0.9 x 0.6	12.8	SB(r)c? III-IV	108	Mag 13.4 star NE 2′.9.
23 01 44.7	−59 10 26	ESO 147-20	13.9	1.2 x 1.2	14.2	SAB(r)bc II-III		Three stars on W edge in N-S line.
20 56 43.6	−56 37 50	ESO 187-34	13.8	0.7 x 0.5	12.5	Sa:	25	Star or knot E edge; mag 13.1 star W edge.
20 56 56.3	−55 43 15	ESO 187-35	14.4	1.4 x 0.6	14.1	SB(s)m V	123	Mag 10.71v star E 6′.0.
20 57 08.5	−53 15 44	ESO 187-36	13.7	0.8 x 0.4	12.4	S0:	121	Mag 9.67v star W 13′.2.
20 58 42.9	−54 11 18	ESO 187-38	14.0	0.8 x 0.7	13.3	S?	178	Stellar nucleus.
20 58 46.9	−52 48 30	ESO 187-39	14.2	1.0 x 0.7	13.7	SAB(s)dm? V	39	Mag 10.69v star W 11′.7; ESO 187-42 NNE 11′.7.
20 59 31.1	−52 39 53	ESO 187-42	14.0	0.9 x 0.4	12.8	(R′)SB(s)a	126	Mag 10.27v star WNW 13′.2; ESO 187-39 SSW 11′.7.
21 07 33.0	−54 57 11	ESO 187-51	14.1	2.1 x 1.1	14.8	SB(s)m V	6	Mag 7.41v star W 9′.1.
21 12 10.6	−57 16 43	ESO 187-58	13.8	2.1 x 0.3	13.2	SBc? sp	14	Mag 8.88v star N 4′.1.
21 13 00.2	−55 02 50	ESO 187-59	13.7	1.1 x 0.6	13.1	S0	172	Mag 11.01v star on E edge.
21 37 51.2	−56 13 44	ESO 188-12A	14.1	1.0 x 0.8	13.7	SBbc	109	
21 43 03.4	−52 41 14	ESO 188-18	13.8	2.1 x 0.7	14.1	S0° pec	128	Double system, bridge and streamers; NGC 7106 W 4′.2.
21 50 58.1	−54 33 07	ESO 189-5	14.1	0.9 x 0.8	13.6	Sa	102	Mag 14.7 star S edge.
21 55 17.0	−55 34 22	ESO 189-10	14.0	1.1 x 0.9	13.8	(R′)SB(s)0/a	170	Small core, uniform envelope; mag 14.6 star E edge; **ESO 189-9** W 9′.8.
21 55 38.9	−54 52 37	ESO 189-12	14.6	1.7 x 0.2	13.2	SAc: sp	40	Mag 10.88v star NW 10′.3.
21 58 07.3	−57 08 34	ESO 189-16	13.6	1.3 x 0.7	13.3	(R)SB(rs)0/a:	0	Mag 13.1 star, involved with faint anonymous galaxy, ENE 1′.2.
22 02 36.9	−54 04 44	ESO 189-21	14.6	1.7 x 0.9	14.9	SB(s)m V-VI	134	Mag 7.90v star SW 9′.6.
22 14 16.1	−54 07 15	ESO 189-31	13.4	1.2 x 0.8	13.2	S?	101	Mag 12.1 star W 2′.3.
22 15 02.4	−55 58 31	ESO 189-32	14.5	1.7 x 0.4	14.0	SA0/a: sp	85	Mag 10.40v star SW 5′.4.
22 26 11.8	−52 55 34	ESO 190-10	14.0	1.3 x 0.3	12.8	(R′)SB(s)ab:	37	Mag 13.1 star S 1′.7.
22 27 03.7	−52 15 43	ESO 190-11	13.6	1.6 x 1.6	14.5	SB(s)m V		Mag 11.9 star NE 8′.0.
22 27 47.6	−54 44 30	ESO 190-12	13.7	1.4 x 1.2	14.1	(R′)SAB(rs)0$^+$ pec	42	Two very faint, elongated N-S anonymous galaxies on N edge; mag 9.18v star NW 10′.9.
20 59 13.1	−52 00 23	ESO 235-26	13.9	0.9 x 0.4	12.6	Pec	82	Mag 9.65v star NE 10′.1; compact anonymous galaxy ENE 3′.7.
21 00 55.1	−52 13 26	ESO 235-32	13.5	1.5 x 0.8	13.6	SA(r)0$^+$	99	Mag 8.35v star WNW 12′.0.
21 01 30.3	−52 00 53	ESO 235-35	13.9	1.3 x 0.3	12.7	Sb sp	12	Mag 13.5 star E 0′.9.
21 04 51.4	−51 49 24	ESO 235-50	13.1	1.6 x 1.2	13.7	SA(r)0$^-$	102	Mag 9.05v star N 5′.4.
21 05 01.0	−51 56 56	ESO 235-51	13.4	1.2 x 0.5	12.7	SB(s)0$^+$	141	Mag 8.19v star NW 6′.7.
21 05 44.7	−51 42 38	ESO 235-54	13.7	1.0 x 0.5	12.8	(R′)SA(r)a:	33	Mag 9.05v star W 8′.3.
21 20 00.6	−51 03 29	ESO 236-3	14.3	0.9 x 0.8	13.8	Sb	156	Mag 14.0 star S edge.
21 42 46.5	−51 17 18	ESO 236-34	14.4	1.6 x 1.0	14.7	IB(s)m V	16	Multi-galaxy system.
21 57 20.0	−51 13 04	ESO 237-19	14.9	1.3 x 0.6	14.5	IBm: V-VI	85	Mag 10.10v star N 8′.9.
21 06 06.5	−56 14 58	IC 5079	18.8	2.0 x 2.0	20.2			Mag 8.94v star NW 11′.5.
21 30 43.3	−60 00 06	IC 5110	13.1	1.3 x 1.0	13.3		49	Strong N-S bar.
21 41 50.5	−52 46 26	IC 5125	13.4	0.7 x 0.7	12.5	S(r)bc pec: II		Mag 12.2 star SE 3′.1.
21 53 17.6	−59 29 32	IC 5141	12.7	1.5 x 1.3	13.2	SA(rs)bc: II	30	Mag 8.97v star S 5′.5; ESO 145-21 N 6′.5.
22 02 41.6	−51 17 45	IC 5152	10.6	5.0 x 3.2	13.4	IA(s)m IV-V	100	Mag 7.76v star inside N edge.
22 08 03.1	−52 42 51	IC 5162	14.4	1.0 x 0.4	13.2	S	114	Mag 9.90v star ESE 9′.2; stellar galaxy **ESO 189-25B** SE 0′.8.
22 18 18.5	−59 36 29	IC 5187	14.1	0.5 x 0.5	12.5	S		Mag 9.15v star N 7′.0; IC 5188 SSE 2′.3.
22 18 26.9	−59 38 30	IC 5188	13.3	1.3 x 1.1	13.5	SAB(s)c: I	65	IC 5187 NNW 2′.3.
22 19 01.3	−59 53 07	IC 5190	14.2	1.2 x 0.8	14.0	SB(s)bc pec II	138	Mag 10.79v star NW 5′.5.
22 19 50.0	−60 08 13	IC 5197	14.1	1.0 x 0.4	12.9	SB(s)b?	30	Located 12′.3 NE of 2.86v star Alpha Tucanae.
22 22 34.8	−59 46 26	IC 5203	14.2	1.5 x 0.8	14.3	SB(s)c pec II	42	Galaxy **IC 5205** SE 1′.9.
22 23 29.5	−60 33 58	IC 5207	14.5	1.5 x 0.9	14.7	SAB(s)cd II	0	Mag 9.01v star WNW 9′.7.
22 28 02.4	−59 43 22	IC 5220	14.3	1.0 x 0.3	12.9	Sc	111	Mag 12.4 star N 8′.2.
20 57 18.5	−51 51 46	NGC 6982	13.7	1.1 x 0.6	13.1	(R)SB(rs)a pec?	68	Mag 10.61v star NNE 4′.9.
20 57 54.1	−51 52 18	NGC 6984	12.7	1.8 x 1.2	13.4	SB(r)c I-II	110	Faint star S of nucleus; mag 10.61v star N 4′.4; galaxy **Fair 927** NNE 5′.7.
20 59 57.1	−55 33 43	NGC 6990	13.2	1.1 x 0.5	12.4	Sa:	0	Star at S end.
21 05 27.7	−52 33 08	NGC 7007	12.0	1.9 x 1.1	12.7	SA0$^-$:	2	
21 27 21.5	−60 00 51	NGC 7059	11.9	3.5 x 1.7	13.6	SAB(rs)c III	98	Numerous knot and/or superimposed stars.
21 29 02.3	−52 46 04	NGC 7064	12.5	3.8 x 0.6	13.2	SB(s)c? sp II-III	91	Mag 10.09v star S 1′.3.
21 36 28.4	−54 33 18	NGC 7090	10.7	7.4 x 1.3	13.0	SBc? sp	127	Broken dark lane along centerline.
21 42 37.0	−52 41 59	NGC 7106	13.4	1.7 x 1.0	13.8	SAB(rs)c: II	75	Mag 12.7 star N 1′.7; ESO 188-18 E 4′.2.
21 49 16.4	−60 42 48	NGC 7125	12.4	3.1 x 2.1	14.3	SAB(rs)c I-II	110	Knotty arms with superimposed stars.
21 49 18.1	−60 36 39	NGC 7126	12.2	2.8 x 1.3	13.5	SA(rs)c II-III	71	
21 52 15.4	−55 34 13	NGC 7140	11.5	4.2 x 3.0	14.1	SAB(rs)bc II	18	Bright nucleus in elongated center with faint outer halo.
22 02 07.6	−51 44 37	NGC 7168	11.9	2.0 x 1.5	13.1	E3	68	
22 08 33.7	−57 26 38	NGC 7205	11.0	4.1 x 2.0	13.1	SA(s)bc II	73	Bracketed by a mag 10.65v star NE and a mag 8.81v star SW.

(Continued from previous page)

193 GALAXIES 193

RA h m s	Dec ° ′ ″	Name	Mag (V)	Dim ′ Maj x min	SB	Type Class	PA	Notes
22 07 32.3	−57 27 53	NGC 7205A	14.0	1.1 x 0.9	13.8	SAB(s)cd: II	65	**PGC 68123** SE 10′9.
22 20 31.1	−55 07 29	NGC 7249	13.4	1.0 x 0.6	12.7	SA0°:	136	Anonymous galaxy, or galaxy pair, NE 2′4.
22 28 22.6	−60 10 13	NGC 7278	14.4	0.7 x 0.6	13.3	SB(rs)c pec? II	4	
21 45 30.0	−51 36 28	PGC 67298	14.3	1.2 x 0.6	14.0	E⁺5 pec:	126	Anonymous elliptical galaxy NE 2′0; **ESO 236-38** S 5′9; **ESO 237-3** on E edge.
21 49 49.0	−59 19 24	PGC 67429	14.0	0.8 x 0.8	13.4	SA0⁻		Mag 9.24v star WSW 11′8.

GALAXY CLUSTERS

RA h m s	Dec ° ′ ″	Name	Mag 10th brightest	No. Gal	Diam ′	Notes
21 34 30.0	−62 01 00	A 3782	16.4	40	17	Most members anonymous, stellar.
21 34 30.0	−53 37 00	A 3785	16.2	45	17	**PGC 66976** SW of center 2′2; **PGC 66975** NW 4′4.
21 46 36.0	−57 17 00	A 3806	16.2	115	17	**Fair 116** at center; all other members anonymous, stellar.
21 50 24.0	−55 18 00	A 3816	15.3	39	22	**PGC 67496** E of center 9′6; most others anonymous, stellar.
21 54 06.0	−57 50 00	A 3822	16.4	113	17	
21 58 24.0	−60 23 00	A 3825	16.0	77	17	Except ESO 146-2A, all members anonymous, stellar.
21 59 54.0	−56 09 00	A 3826	15.1	62	22	**ESO 189-19** SE of center 6′6; most others anonymous.
22 15 54.0	−51 33 00	A 3849	16.0	42	17	**PGC 68430** NW of center 4′2; all others anonymous, stellar.
22 16 42.0	−52 35 00	A 3851	16.0	33	17	
22 21 24.0	−55 07 00	A 3869	15.3	49	22	**ESO 190-2**, **2A** NNW of center 14′8; most others anonymous.
22 31 42.0	−54 44 00	A 3886	16.2	31	17	Most members anonymous, faint, stellar.
21 17 00.0	−59 24 00	AS 927	16.2	29	17	
21 35 36.0	−52 32 00	AS 957	16.2	12	17	All members anonymous, stellar.
21 45 36.0	−51 36 00	AS 968	15.4	20	22	See notes for PGC 67298; all others anonymous, stellar.
21 55 42.0	−55 34 00	AS 981	14.7	21	28	Small trio **Fair 595**, **596**, **597** E of center 7′0, also N of mag 7.49v star 8′0.
22 04 30.0	−58 07 00	AS 988	16.0	16	17	**Fair 126** E of center3′0; **PGC 68001** S of center 11′0.
22 13 24.0	−61 55 00	AS 1003	16.3		17	Except ESO 146-14, all members anonymous, stellar.
22 24 18.0	−56 28 00	AS 1020	15.9	25	17	**ESO 190-5** near center, all others anonymous, stellar.
22 25 36.0	−55 49 00	AS 1023	16.1	27	17	All members anonymous, stellar.
22 34 54.0	−52 27 00	AS 1039	16.4	24	17	All members anonymous, stellar.

OPEN CLUSTERS

RA h m s	Dec ° ′ ″	Name	Mag	Diam ′	No. ★	B ★	Type	Notes
21 21 30.0	−51 49 00	ESO 236-7		30			cl	

194 GALAXIES 194

RA h m s	Dec ° ′ ″	Name	Mag (V)	Dim ′ Maj x min	SB	Type Class	PA	Notes
19 25 01.2	−62 58 07	ESO 104-51	13.6	1.1 x 0.8	13.3	S0	35	Pair of galaxies, bridge; mag 6.59v star WSW 5′7.
19 00 55.3	−62 03 41	ESO 141-17B	14.1	1.9 x 0.8	14.4		135	Pair with ESO 141-17A W edge.
19 00 51.0	−62 03 40	ESO 141-17A	13.4	1.5 x 0.6	13.1		15	Pair with ESO 141-17B E edge.
19 03 07.3	−60 25 22	ESO 141-21	13.2	1.0 x 0.6	12.5	S?	144	Mag 15 star E edge; mag 13.7 star W 1′4.
19 05 01.2	−57 59 08	ESO 141-24	13.8	0.9 x 0.4	12.6	(R)SA(r)0⁺?	42	Two stars at center.
19 12 52.4	−60 23 10	ESO 141-32	13.3	0.9 x 0.7	12.7		23	Mag 12.2 star S edge.
19 16 12.6	−62 21 40	ESO 141-42	12.4	3.0 x 0.5	12.7	SB(s)m sp	111	Mag 8.12v star S 2′1; IC 4833 WNW 4′0.
19 21 14.2	−58 40 15	ESO 141-55	13.5	0.8 x 0.5	12.4	S?	70	Mag 9.74v star SSW 11′0.
19 28 56.4	−61 18 46	ESO 142-13	13.8	1.1 x 0.3	12.4	SBb? pec	68	Mag 8.87v star SW 7′8.
19 33 18.4	−58 06 47	ESO 142-19	11.8	4.1 x 1.1	13.3	Sb: sp	161	Strong dark lane through center; mag 9.38v star NNW 6′6.
19 35 42.7	−59 47 25	ESO 142-23	13.0	1.0 x 1.0	12.9	SA(r)0°		Mag 10.73v star W 4′7.
19 40 39.1	−60 02 56	ESO 142-30	13.5	1.8 x 0.3	12.7	Sc sp	81	Mag 10.87v star S 3′5.
19 43 00.8	−59 56 49	ESO 142-34	14.8	1.6 x 0.2	13.4	Sc:	43	Mag 11.7 star NW 2′7.
19 50 43.8	−61 16 35	ESO 142-48	14.2	1.1 x 0.7	13.7	SB(s)dm IV-V	165	Mag 9.22v star SW 8′7.
19 52 55.4	−60 59 03	ESO 142-49	12.8	1.3 x 0.9	13.0	E4	85	Mag 9.96v star NE 5′6.
19 54 57.9	−58 43 54	ESO 142-51	13.7	0.9 x 0.5	12.7	S?	30	Located 4′5 E of IC 4901.
19 57 17.6	−61 51 30	ESO 142-58	14.4	1.0 x 0.4	13.2	(R′)SB(s)dm: IV-V	24	Mag 12.2 star on W edge.
19 58 50.0	−57 50 02	ESO 142-59	13.7	0.6 x 0.5	12.3		89	Mag 12.4 star E 0′9; mag 10.67v star ESE 2′7.
20 05 08.2	−60 09 53	ESO 143-5	14.5	0.6 x 0.4	12.8	S0	3	Mag 9.98v star S 5′1.
20 21 52.7	−61 58 24	ESO 143-21	14.3	1.0 x 0.4	13.1	SB(s)m V	30	Mag 12.2 star on E edge.
20 27 44.3	−58 39 12	ESO 143-27	15.0	0.8 x 0.3	13.3	SB(s)m? sp	144	Star on N edge.
20 36 34.7	−57 37 25	ESO 143-35	13.7	0.8 x 0.6	12.8	S0	91	Mag 10.32v star W 3′3; elongated **ESO 143-36** E 3′5; **ESO 143-32, 32A** SW 9′7.
20 37 56.1	−61 54 39	ESO 143-38	13.0	1.0 x 1.0	12.9	SB(rs)0⁺ pec		Almost stellar nucleus, smooth envelope; mag 12.5 star S 0′9.
18 56 56.0	−54 32 42	ESO 183-30	11.7	1.9 x 1.5	12.7	SA0⁻ pec?	22	Mag 7.78v star W 3′4.
18 59 31.1	−54 21 38	ESO 183-34	13.1	1.3 x 0.5	12.5	S0/a?	102	Mag 13.0 star W 1′7.
19 03 35.9	−53 10 22	ESO 184-4	13.7	1.1 x 0.8	13.4	S?	71	Mag 10.66v star W edge; galaxy **Fair 505** SW 3′3.
19 03 39.5	−53 56 05	ESO 184-5	12.7	1.0 x 0.7	12.3	SA0⁻: pec	157	Mag 10 star S 0′7.
19 04 31.0	−53 59 12	ESO 184-6	13.5	1.6 x 0.5	13.1	(R′)SB(rs)a:	179	Mag 11.4 star E 3′8.
19 06 23.0	−56 09 53	ESO 184-11	13.0	1.5 x 1.0	13.3	IB(s)m pec: III-IV	171	Several bright stars superimposed; IC 4817 W 1′6.
19 07 43.5	−55 18 35	ESO 184-15	14.1	0.8 x 0.4	12.7	Pec	52	Very faint anonymous galaxy with stellar core NE 2′0.
19 08 45.6	−55 40 18	ESO 184-16	13.7	0.7 x 0.4	12.2	S0	158	Mag 9.46v star SW 1′8; mag 6.45v star S 3′1.
19 11 54.8	−56 16 33	ESO 184-26	13.1	1.6 x 0.8	13.2	SAB0⁻:	105	Mag 7.38v star WSW 5′0.
19 12 42.4	−53 52 18	ESO 184-32	13.6	1.3 x 0.4	12.8	I0 pec	11	Double system; mag 11.8 star W 4′6.
19 12 54.7	−56 16 48	ESO 184-33	13.3	1.1 x 1.0	13.3	SA(s)0°: pec	17	Stars superimposed along E and W edges.
19 12 59.8	−52 30 52	ESO 184-34	14.0	0.8 x 0.3	12.3	S0?	91	Mag 9.81v star NE 1′1.
19 14 18.0	−55 53 07	ESO 184-41	14.4	1.5 x 0.2	12.9	Sc? sp	102	Mag 10.12v star NE 8′3.
19 14 22.1	−54 33 58	ESO 184-42	13.0	0.7 x 0.7	12.1	S0 pec		Mag 10.56v star SW 2′7.
19 14 37.5	−54 13 43	ESO 184-43	13.4	0.9 x 0.5	12.4	SAB(rs)0°	0	**ESO 184-44** SSE 7′4; mag 12.4 star SE 1′8; mag 11.45v star N 3′6.
19 16 56.2	−53 42 06	ESO 184-51	13.2	2.2 x 0.7	13.5	SAB(s)bc: II-III	42	Mag 12.1 star SE 3′8.
19 20 23.6	−54 20 47	ESO 184-57	14.1	0.6 x 0.3	12.1	S0	121	Mag 12.5 star E 1′4.
19 21 06.0	−55 37 24	ESO 184-58	13.9	1.2 x 0.5	13.2	SAB(r)d? IV	105	Star on S edge.
19 21 38.7	−52 51 00	ESO 184-59	13.9	0.6 x 0.4	12.3		45	Small triangle of mag 15 stars W 0′9.
19 22 41.5	−54 35 08	ESO 184-60	14.2	1.7 x 0.7	14.2	(R′)SAB(rs)d: IV	163	Mag 7.04v star NE 3′8.
19 23 40.5	−55 03 52	ESO 184-63	13.6	2.4 x 0.3	13.0	Sb: pec sp	1	Mag 7.63v star E 11′5.
19 28 28.1	−53 22 22	ESO 184-71	13.4	0.7 x 0.5	12.1	SAB(s)0° pec	9	Mag 10.23v star N 5′4.
19 29 03.7	−52 32 03	ESO 184-72	13.8	1.0 x 0.3	12.3	Sa:	86	Mag 10.81v star S edge.

(Continued from previous page)

GALAXIES

RA h m s	Dec ° ′ ″	Name	Mag (V)	Dim ′ Maj x min	SB	Type Class	PA	Notes
19 30 30.4	−57 16 55	ESO 184-74	13.7	1.6 x 0.4	13.1	(R′)SB(s)a: sp	161	Mag 9.32v star SE 8'.5.
19 30 58.6	−53 32 41	ESO 184-75	13.8	1.0 x 0.7	13.3	SB(rs)c I-II	7	Mag 10.71v star S 2'.7.
19 31 36.4	−53 33 10	ESO 184-77	14.6	1.5 x 0.3	13.6	Sb: sp	73	Mag 10.71v star SE 4'.4.
19 34 56.9	−52 55 31	ESO 184-81	13.8	0.7 x 0.5	12.6	S0	155	Mag 9.69v star E 2'.2.
19 45 00.7	−54 15 06	ESO 185-13	14.8	0.4 x 0.3	12.4		53	Almost stellar; mag 13.3 star WNW 1'.6.
19 46 20.4	−55 43 32	ESO 185-19	14.5	1.5 x 0.3	13.5	SB(s)b? sp	12	Mag 9.56v star E 1'.8.
19 49 20.4	−54 12 04	ESO 185-21	13.8	1.6 x 0.4	13.1	Sc? sp	159	Small galaxy or star NE 1'.5; mag 10.78v star NE 7'.8.
19 54 51.7	−55 06 37	ESO 185-25	13.7	1.0 x 0.7	13.2	SA(rs)0°:	132	**ESO 185-26** NNE 8'.1.
19 56 13.4	−56 54 40	ESO 185-27	12.9	0.7 x 0.5	11.6	SAB0^{-}	81	Mag 8.67v star WNW 10'.3; stellar **ESO 185-29** ESE 3'.9.
19 57 33.8	−55 34 42	ESO 185-32	13.4	1.1 x 0.5	12.6	SAB(s)0^{-}	97	Mag 10.42v star SW 1'.7.
19 57 53.4	−55 39 11	ESO 185-34	13.7	1.0 x 0.3	12.3	S0°: sp	24	Mag 13.1 star N edge; **ESO 185-36** SE 3'.1; mag 10.42v star NW 4'.9; galaxy **Fair 523** NE 6'.2.
19 58 10.1	−57 20 18	ESO 185-35	14.3	1.1 x 0.4	13.3	SB(s)c: III-IV	131	Mag 14.2 star N edge; mag 12.7 star W 1'.7.
19 58 14.5	−55 26 45	ESO 185-37	13.4	1.1 x 0.5	12.6	SB(s)0^{-}:	72	Pair of very faint anonymous galaxies SW 1'.8.
20 03 01.3	−55 56 56	ESO 185-53	13.2	1.1 x 0.4	12.2	SB0° pec: sp	160	ESO 185-54 E 3'.6; mag 9.65v star W 4'.9.
20 03 27.2	−55 56 51	ESO 185-54	11.3	3.0 x 2.1	13.4	E2	124	Mag 13.8 star S edge; **ESO 185-58** SE 5'.2.
20 06 52.6	−54 18 15	ESO 185-64	14.0	1.1 x 0.5	13.2	(R′)SAB(r)0^{+}:	78	Mag 8.11v star N 7'.0.
20 07 14.0	−53 48 12	ESO 185-68	15.1	1.5 x 0.3	14.1	SB(s)d: sp	171	Mag 9.52v star NW 6'.7.
20 09 50.7	−53 43 43	ESO 186-4	13.7	0.4 x 0.3	11.2		165	Almost stellar.
20 12 23.8	−54 21 37	ESO 186-12	13.6	0.9 x 0.6	12.8	(R)SB(r)0/a	100	Mag 12.6 star W 2'.3.
20 15 08.9	−52 43 06	ESO 186-21	13.3	1.7 x 0.9	13.6	SAB(rs)c I	3	IC 4975 W 10'.0; **IC 4984** E 10'.4 with mag 11.8 star on N edge.
20 19 32.7	−53 30 13	ESO 186-32	14.0	1.5 x 0.4	13.3	S0^{+}: sp	32	Mag 11.00v star SSW 2'.3; IC 4994 NNE 3'.9.
20 21 19.3	−53 45 53	ESO 186-36	12.6	2.0 x 0.6	12.6	SBa? sp	19	Mag 10.89v star WSW 8'.7.
20 23 12.8	−53 55 44	ESO 186-39	13.6	0.6 x 0.5	12.1	Pec:	11	Mag 10.65v star E 8'.1.
20 26 17.6	−52 24 20	ESO 186-43	13.6	1.2 x 0.6	13.1	(R)SAB(rs)a	98	Mag 11.8 star S 2'.8; mag 9.95v star WNW 10'.0.
20 27 47.8	−56 21 58	ESO 186-46	14.4	0.9 x 0.8	13.9	Sc	95	Low, uniform surface brightness; mag 9.38v star NW 0'.9.
20 27 45.7	−52 23 05	ESO 186-47	13.6	1.5 x 0.5	13.2	SB(rs)bc? I-II	14	Mag 9.54v star ESE 2'.5.
20 28 22.7	−52 25 34	ESO 186-52	14.0	0.5 x 0.5	12.4			Mag 10.75v star S 1'.7; mag 8.91v star NE 2'.8.
20 31 05.9	−52 36 52	ESO 186-55	13.7	0.9 x 0.4	12.5	S(r)ab?	102	Mag 10.63v star ESE 4'.8.
20 31 52.8	−53 46 32	ESO 186-57	13.8	0.9 x 0.4	12.5	SAB(s)dm pec	102	Mag 9.99v star W 12'.4; ESO 186-59 NNE 2'.4.
20 32 03.1	−53 44 38	ESO 186-59	12.9	1.7 x 1.1	13.5		24	Pair of bright knots on W edge; ESO 186-57 SSW 2'.4; **ESO 186-58** N 6'.7.
20 33 25.1	−54 31 48	ESO 186-60	14.1	1.3 x 1.1	14.3	SB(s)m V	130	Very filamentary, low surface brightness; IC 5021 ENE 1'.5.
20 34 02.7	−52 58 53	ESO 186-62	12.2	2.1 x 1.8	13.5	SB(s)d III-IV	151	Mag 8.26v star N 8'.6.
20 35 57.5	−54 17 59	ESO 186-65	13.1	1.8 x 0.4	12.6	S0^{-}: sp	46	Mag 8.97v star N 8'.8.
20 37 12.2	−54 39 30	ESO 186-69	14.0	1.7 x 0.6	13.9		2	Double system, interaction.
20 41 07.5	−53 29 50	ESO 186-75	14.9	1.5 x 0.2	13.4	Sc: sp	145	Mag 8.32v star NW 12'.4; **ESO 186-72** SW 7'.6; stellar galaxy **Fair 555** ESE 5'.0.
20 43 02.9	−52 30 59	ESO 187-6	14.1	1.7 x 0.6	14.0	(R′)SB(rs)c: III-IV	118	Several bright knots; mag 9.22v star W 5'.6.
20 43 25.0	−56 12 14	ESO 187-8	14.8	1.5 x 0.2	13.3	Scd sp	81	Mag 10.71v star W 4'.0.
20 51 19.9	−52 38 10	ESO 187-20	13.4	1.0 x 1.0	13.3	S0^{-} pec		Strong core, smooth envelope; mag 8.94v star NW 0'.9.
20 51 57.5	−52 48 07	ESO 187-25	13.3	0.8 x 0.8	12.9	E/S0		
20 51 57.0	−52 37 48	ESO 187-26	13.2	1.2 x 0.7	12.9	SA0^{-}:	52	Mag 8.94v star and ESO 187-20 W 6'.3; **ESO 187-22** NW 3'.1; numerous anonymous galaxies in area.
20 52 01.5	−53 28 21	ESO 187-27	13.4	0.9 x 0.7	12.8	SA0^{-}:	74	Mag 11.6 star S 3'.8.
20 56 43.6	−56 37 50	ESO 187-34	13.8	0.7 x 0.5	12.5	Sa:	25	Star or knot E edge; mag 13.1 star W edge.
20 56 56.3	−55 43 15	ESO 187-35	14.4	1.4 x 0.6	14.1	SB(s)m V	123	Mag 10.71v star E 6'.0.
20 57 08.5	−53 15 44	ESO 187-36	13.7	0.8 x 0.4	12.4	S0:	121	Mag 9.67v star W 13'.2.
20 58 42.9	−54 11 18	ESO 187-38	14.0	0.8 x 0.7	13.3	S?	178	Stellar nucleus.
20 58 46.9	−52 48 30	ESO 187-39	14.2	1.0 x 0.7	13.7	SAB(s)dm? V	39	Mag 10.69v star W 11'.7; ESO 187-42 NNE 11'.7.
20 59 31.1	−52 39 53	ESO 187-42	14.0	0.9 x 0.4	12.8	(R′)SB(s)a	126	Mag 10.27v star WNW 13'.2; ESO 187-39 SSW 11'.7.
19 05 06.2	−51 13 44	ESO 231-18	13.5	1.3 x 0.6	13.1	SA(r)b: III-IV	140	Mag 10.42v star W 3'.4.
19 08 44.1	−51 02 48	ESO 231-23	13.3	1.7 x 0.4	12.7	Sb sp	102	Mag 9.96v star SE 6'.1.
19 22 46.8	−51 00 09	ESO 232-4	13.6	1.3 x 0.7	13.4	S?	22	Close pair of mag 12 stars S edge.
19 24 28.5	−51 57 44	ESO 232-6	13.1	1.6 x 0.9	13.3	(R′)SB(r)b: II	130	Pair of stars on SW edge.
19 44 36.4	−51 36 09	ESO 232-21	13.0	1.8 x 0.3	12.2	S0^{+}: sp	61	Mag 9.45v star W 4'.8; stellar galaxy **Fair 873** S 3'.3; **ESO 232-22** SE 5'.7.
19 58 41.3	−52 21 26	ESO 233-14	14.0	0.5 x 0.4	12.1	S0? pec	145	Almost stellar.
20 11 48.5	−51 23 20	ESO 233-44	13.9	0.7 x 0.3	12.1		145	Mag 10.19v star NE 4'.0.
20 20 29.3	−51 37 24	ESO 234-3	13.9	0.9 x 0.4	12.7	S0:	12	
20 22 10.4	−51 42 47	ESO 234-10	14.1	1.2 x 0.4	13.2	Sb: II-III	124	Mag 10.77v star S 2'.6.
20 23 22.7	−52 05 02	ESO 234-15	13.7	1.0 x 0.6	13.0	(R′)SB(rs)a:	167	Mag 9.96v star W 5'.5.
20 25 27.8	−51 31 56	ESO 234-24	13.9	1.9 x 0.4	13.4	Sbc sp II	148	Mag 12.4 star on E edge; mag 9.53v star NW 10'.9.
20 27 31.9	−51 39 17	ESO 234-28	13.5	1.2 x 0.5	12.8	SB(s)b pec	175	Mag 8.99v star E 4'.4.
20 28 06.3	−51 41 32	ESO 234-32	13.6	1.6 x 0.6	13.4	(R′)SB(s)bc	73	Two main, open arms; mag 8.99v star N 1'.7.
20 36 27.5	−52 17 06	ESO 234-52	13.8	0.8 x 0.4	12.4	S0^{-}: pec	111	Close pair of stars, mags 10.09v and 11.08v, S 2'.0.
20 37 00.6	−51 09 56	ESO 234-55	14.0	0.9 x 0.7	13.3	S?	66	Bright bar near S edge, smooth envelope.
20 45 52.4	−51 06 26	ESO 234-68	13.2	1.6 x 0.8	13.3	SAB(s)0°:	168	Mag 9.10v star W 3'.5.
20 45 54.4	−51 23 29	ESO 234-69	13.4	2.4 x 0.6	13.6	SB(s)c: II-III	110	Mag 9.08v star NW 3'.7.
20 46 03.6	−51 36 58	ESO 234-70	13.5	0.9 x 0.8	13.0	S0	11	Strong elongated core. Smooth envelope.
20 52 45.5	−52 09 59	ESO 235-9	14.0	1.0 x 0.5	13.1	SB(rs)a:	21	Mag 15.2 star E edge; mag 12.3 star W 2'.0.
20 59 13.1	−52 00 23	ESO 235-26	13.9	0.9 x 0.4	12.6	Pec	82	Mag 9.65v star NE 10'.1; compact anonymous galaxy ENE 3'.7.
21 00 55.1	−52 13 26	ESO 235-32	13.5	1.5 x 0.8	13.6	SA(r)0^{+}	99	Mag 8.35v star WNW 12'.0.
21 01 30.3	−52 00 53	ESO 235-35	13.9	1.3 x 0.3	12.7	Sb sp	12	Mag 13.5 star E 0'.9.
18 56 56.0	−61 24 03	IC 4793	13.6	1.1 x 0.5	12.8	Sb	120	Mag 10.8 star WSW 10'.2.
18 57 09.9	−62 05 29	IC 4794	13.0	0.9 x 0.5	12.1	S0	25	Mag 9.60v star W 11'.0.
18 56 28.3	−54 12 53	IC 4796	12.3	1.6 x 0.9	12.5	SB(s)0°:	140	Mag 12.3 star on N edge; mag 10.08v star ESE 9'.0.
18 56 29.8	−54 18 22	IC 4797	11.3	2.1 x 1.1	12.3	E^{+} pec:	146	Mag 10.08v star W 8'.5.
18 58 20.9	−62 07 05	IC 4798	12.2	1.5 x 0.8	12.2	SA(s)0°: pec	108	Mag 10.59v star SW 10'.0, part of a "V" shaped asterism.
19 01 07.5	−61 50 00	IC 4804	13.8	1.2 x 0.8	13.6	SA(s)c I-II	147	Stellar nucleus, faint envelope.
19 01 30.9	−57 31 55	IC 4806	12.2	2.3 x 0.6	12.4	(R′)SA(s)b: sp II	9	Mag 9.44v star NNE 6'.0.
19 02 17.8	−56 55 49	IC 4807	13.4	1.0 x 0.6	12.7	SB(s)c? I	7	Mag 8.31v star SW 3'.6.
19 04 05.4	−62 11 39	IC 4809	14.1	0.9 x 0.5	13.1	Dwarf	14	Uniform surface brightness.
19 02 59.9	−56 09 37	IC 4810	12.6	3.5 x 0.4	12.8	SAB(s)d? sp III-IV	136	Mag 10.90v star NNE 5'.8.
19 04 59.2	−58 34 47	IC 4814	13.4	1.1 x 0.4	12.3	S	93	Mag 9.25v star WNW 10'.2.
19 06 51.1	−61 42 07	IC 4815	13.3	0.8 x 0.6	12.4	S0	162	Mag 8.29v star W 10'.7.
19 06 12.5	−56 09 36	IC 4817	13.6	1.6 x 0.5	13.2	SAB(r)bc pec:	12	Mag 10.37v star NW 6'.2; ESO 184-11 E 1'.6.
19 06 03.3	−55 08 13	IC 4818	14.3	1.5 x 0.3	13.3	(R′)SB(r)0^{+}:	78	Mag 9.51v star W 8'.9.
19 07 05.8	−59 27 53	IC 4819	13.4	2.9 x 0.3	13.1	Scd: sp	125	Mag 10.96v star N 4'.4.
19 09 32.0	−55 01 03	IC 4821	12.9	1.8 x 0.7	13.0	SA(s)c? II	4	Numerous bright knots and/or superimposed stars.
19 13 16.8	−62 05 23	IC 4824	16.2	1.7 x 1.0	16.6	IAB(s)m V	103	IC 4828 E 2'.8; mag 10.43v star NW 9'.9.
19 12 21.2	−57 12 07	IC 4826	13.5	1.4 x 0.8	13.5	SA(s)ab pec?	45	Mag 10.70v star N 6'.4.
19 13 21.3	−60 51 37	IC 4827	12.2	2.8 x 0.5	12.4	SA(s)ab: sp II	166	Mag 8.80v star W 1'.6; mag 9.14v star NE 2'.9.

(Continued from previous page)

RA h m s	Dec ° ′ ″	Name	Mag (V)	Dim ′ Maj x min	SB	Type Class	PA	Notes
19 13 40.6	−62 04 58	IC 4828	14.0	1.0 x 0.5	13.1	S?	68	Uniform surface brightness; IC 4824 W 2′8; mag 10.65v star SSE 4′3.
19 12 33.8	−56 32 27	IC 4829	13.8	1.9 x 0.4	13.4	SA(s)b? sp II-III	23	Mag 12 star on E edge; mag 9.18v star SSW 5′3.
19 13 48.8	−59 17 37	IC 4830	12.3	1.7 x 1.3	13.0	SB(r)bc II	28	Mag 12.0 star SW 4′0.
19 14 45.0	−62 16 22	IC 4831	12.3	3.6 x 0.9	13.4	(R′)SA(s)ab II-III	111	Mag 8.44v star N 2′2.
19 14 03.9	−56 36 37	IC 4832	12.7	2.2 x 0.5	12.7	SA(s)a: sp	144	Mag 9.22v star N 2′1.
19 15 41.9	−62 19 48	IC 4833	14.0	0.6 x 0.4	12.3		99	ESO 141-42 ESE 4′0; mag 8.12v star ESE 4′7.
19 15 27.5	−58 14 17	IC 4835	13.7	0.8 x 0.5	12.6	S	166	**ESO 141-39** S 8′2.
19 16 18.6	−60 12 02	IC 4836	12.7	1.5 x 1.3	13.2	SB(s)bc I-II	12	Mag 10.48v star SW 5′3.
19 15 14.2	−54 39 54	IC 4837	12.5	2.1 x 1.0	13.2	SB(s)cd pec II	8	IC 4839 NE 3′7; mag 10.58v star SW 7′6.
19 15 16.1	−54 07 55	IC 4837A	11.5	4.1 x 0.6	12.4	SA(s)b: sp I	165	Mag 9.31v star N 8′0.
19 16 46.3	−61 36 52	IC 4838	13.7	1.3 x 0.5	13.0		64	Mag 10.66v star W 5′4.
19 15 34.1	−54 37 33	IC 4839	12.3	2.3 x 1.5	13.5	(R)SAB(s)bc pec I-II	147	Mag 10.26v star ESE 7′0; IC 4837 SW 3′7.
19 15 51.7	−56 12 32	IC 4840	13.7	1.1 x 0.7	13.2	SB0$^+$?	131	Mag 8.44v star SE 11′4.
19 19 24.6	−60 38 40	IC 4842	12.4	1.5 x 0.8	12.7	E:	20	NGC 6771 NW 8′1; mag 11.23v star W 9′6.
19 19 21.3	−59 18 34	IC 4843	14.2	1.3 x 0.2	12.6	SB	89	Mag 8.65v star SE 4′6.
19 19 02.9	−56 01 37	IC 4844	13.1	1.4 x 1.0	13.3	SB(r)b: III-IV	164	Mag 6.81v star S 7′7.
19 20 22.8	−60 23 22	IC 4845	11.6	1.5 x 1.0	11.9	SA(rs)b: II	87	NGC 6670 SW 14′5; mag 12.2 star NE 10′1.
19 22 54.5	−56 46 52	IC 4848	14.2	0.6 x 0.5	12.7	S	32	Mag 9.74v star NE 9′8.
19 25 36.2	−62 55 58	IC 4849	13.6	1.0 x 0.4	12.4	SA:(s)c	116	Mag 10.66v star SE 3′2.
19 25 29.7	−57 40 15	IC 4851	13.0	1.6 x 0.6	12.8	SB(r)b? II	13	Mag 9.03v star WNW 9′6.
19 26 25.6	−60 20 11	IC 4852	13.0	1.6 x 1.3	13.6	(R′)SB(s)bc pec II	178	Mag 9.27v star SW 6′5.
19 27 21.4	−59 18 55	IC 4854	13.0	1.2 x 1.1	13.2	SAB(r)cd	28	Mag 9.93v star S 2′6.
19 27 30.5	−54 54 31	IC 4856	13.7	1.3 x 0.5	13.0	IB(s)m pec: IV	33	Mag 13.3 star NE 1′1.
19 28 39.5	−58 46 02	IC 4857	13.0	1.8 x 1.2	13.7	SB(s)c pec II	35	**ESO 142-10** SW 1′6.
19 34 35.0	−61 08 45	IC 4866	13.0	1.5 x 1.1	13.4	(R)SAB(s)b?	157	Mag 10.06v star N 7′5; stellar galaxies **Fair 60** and **61** SE 8′3.
19 36 03.2	−61 01 36	IC 4869	13.0	1.1 x 1.0	12.9	SA(s)m V	76	Mag 12.1 star on S edge; mag 8.82v star NNE 6′2.
19 35 42.2	−57 31 18	IC 4872	13.1	3.5 x 0.4	13.3	SAB(s)d? sp	6	Mag 10.47v star W 3′2.
19 37 38.7	−52 04 35	IC 4875	13.8	0.7 x 0.4	12.2	Im? pec	73	Mag 11.7 star SW 6′1.
19 37 42.9	−52 50 37	IC 4876	13.6	1.3 x 0.9	13.7	SAB(rs)cd? II	120	Mag 10.82v star ESE 4′8; **ESO 185-5** E 5′1.
19 37 56.2	−51 59 33	IC 4877	13.7	1.1 x 0.3	12.4	SB0°?	82	Mag 10.57v star ENE 8′9.
19 38 49.6	−58 13 44	IC 4878	14.2	1.5 x 0.3	13.1	S	41	**ESO 142-28** NW 1′1.
19 39 36.7	−52 22 09	IC 4879	13.5	0.8 x 0.4	12.2	Sa?	80	Mag 10.13v star W 4′6.
19 40 30.7	−56 24 34	IC 4880	13.3	1.7 x 0.4	12.7	SB(s)0$^+$:	116	Mag 10.54v star SW 2′3.
19 40 26.1	−55 51 29	IC 4881	13.8	0.9 x 0.5	12.8	SB(s)bc? I-II	156	faint stellar nucleus, faint envelope.
19 40 23.4	−55 11 50	IC 4882	13.4	0.8 x 0.6	12.5	SB(r)b: II	165	Bright stellar nucleus, faint envelope.
19 42 00.8	−55 32 42	IC 4883	13.6	1.4 x 0.4	12.9	SB(s)b? I	163	Mag 8.95v star NE 7′7.
19 42 41.4	−58 07 42	IC 4884	13.6	1.5 x 0.4	12.9	(R)SA(rl)0$^+$	166	Mag 9.10v star NNW 8′9.
19 43 51.9	−60 39 04	IC 4885	13.6	1.9 x 0.4	13.1	SA(s)c:	113	Pair of N-S oriented mag 10.26v, 10.48v stars SW 8′6.
19 43 14.6	−51 48 27	IC 4886	13.2	1.4 x 0.7	13.0	Sb: pec	113	Mag 10.00v star E 6′9; **ESO 232-20**, **ESO 232-20A** and **ESO 232-20B** SE 8′6.
19 44 52.4	−54 27 25	IC 4888	13.8	0.9 x 0.6	13.0	SA0$^-$	93	Mag 9.84v star NW 2′7.
19 45 15.5	−54 20 41	IC 4889	11.1	2.2 x 1.5	12.5	E5-6	0	Mag 12.7 star NW edge; mag 9.85v star W 11′1.
19 45 35.8	−56 32 46	IC 4890	13.9	0.9 x 0.7	13.3	SB(r)b pec II-III	75	N-S pair of mag 12 stars S 1′0.
19 46 59.2	−51 50 48	IC 4894	13.9	0.9 x 0.6	13.1	SB(rs)b pec	165	Faint stellar nucleus, SE-NW bar.
19 49 19.8	−51 52 08	IC 4897	14.6	0.7 x 0.4	13.1	SB(rs)b:	117	Mag 9.28v star NE 8′6.
19 54 23.5	−58 42 47	IC 4901	11.5	3.8 x 2.7	13.9	SAB(rs)c: II-III	126	Mag 9.64v star W 6′5; ESO 142-51 E 4′5.
19 56 06.4	−61 13 17	IC 4905	14.7	1.4 x 0.3	13.6	Sa?	117	Mag 6.21v star W 10′8.
19 56 47.8	−60 28 09	IC 4906	12.5	1.6 x 0.9	12.8	SAB0$^-$?	63	Mag 10.32v star SSW 4′0.
19 56 56.8	−55 47 32	IC 4908	14.1	0.6 x 0.6	12.9	S?		Mag 9.44v star ENE 4′7.
19 57 42.1	−51 59 11	IC 4911	14.7	0.5 x 0.4	12.8	S0:	135	Mag 9.93v star E 2′3; pair of mag 7.64v, 8.21v stars NE 6′3.
19 58 32.0	−52 38 35	IC 4915	13.5	0.9 x 0.6	12.8	SA0$^-$	103	Mag 9.40v star N 12′1; galaxy **Fair 877** N 3′4.
20 00 09.3	−55 22 24	IC 4919	14.3	1.5 x 0.7	14.2	SB(s)dm pec III-IV	31	Lies 0′8 W of mag 9.31v star.
20 00 57.4	−52 37 57	IC 4923	13.4	0.9 x 0.8	13.0	(R)SB(r)0° pec:	87	Faint star superimposed N of nucleus.
20 03 29.4	−54 58 46	IC 4933	12.3	1.9 x 1.7	13.4	(R′)SB(r)bc: I	0	**ESO 185-57** E 2′6; mag 9.43v star ESE 10′1.
20 04 34.3	−57 35 57	IC 4935	12.9	1.6 x 0.4	12.3	(R′)SB(r)a: sp	177	Star at S end.
20 05 52.1	−61 25 37	IC 4936	13.6	1.3 x 0.8	13.5	SB(s)dm IV	10	Mag 9.76v star SW 6′8.
20 05 17.9	−56 15 22	IC 4937	13.8	1.9 x 0.4	13.3	Sb sp	0	Mag 10.64v star E 3′0; **ESO 185-62** E 4′6.
20 06 12.2	−60 12 42	IC 4938	12.7	1.3 x 1.2	13.0	(R)SB(s)ab:	29	Mag 9.98v star W 9′1.
20 07 11.3	−60 44 20	IC 4939	14.3	1.3 x 0.6	13.9	Sc?	144	Mag 11.08v star S 2′5.
20 06 58.6	−53 39 08	IC 4941	15.1	0.8 x 0.7	14.3	SB(s)dm	0	Mag 9.52v star SSW 5′7.
20 06 49.5	−52 36 40	IC 4942	14.7	0.6 x 0.4	13.0	(R′)SB(s)ab? pec	130	Mag 7.76v star NNW 6′7.
20 07 09.0	−54 26 49	IC 4944	13.7	1.0 x 0.4	12.6	SA(rs)0$^+$:	6	Mag 11.16v star SE 5′2.
20 09 32.3	−61 51 12	IC 4951	13.3	2.8 x 0.5	13.5	SB(s)dm: sp IV	176	Mag 10.88v star NW 6′1.
20 08 37.9	−55 27 16	IC 4952	13.1	1.3 x 0.6	12.7	Sa:	8	Mag 9.65v star E 6′6.
20 11 28.6	−53 07 33	IC 4961	14.2	1.5 x 0.4	13.5	SAB(s)dm: III-IV	90	Mag 10.05v star E 1′4.
20 12 05.7	−55 14 46	IC 4963	13.3	1.0 x 0.5	12.4	(R′)SAB(rs)a pec:	160	Mag 9.40v star NW 12′9.
20 12 27.5	−56 49 37	IC 4965	13.7	1.0 x 1.0	13.6	SA0°		Bright nucleus, faint envelope.
20 14 03.0	−52 43 22	IC 4975	13.3	1.1 x 1.0	13.3	(R′)SB(rs)0°:		Several faint stars on W edge; ESO 186-21 E 10′0.
20 14 41.8	−53 27 33	IC 4979	14.1	1.1 x 0.8	13.8	SAB(r)cd III-IV	4	Faint nucleus; uniform surface brightness; mag 15.6 star on S edge.
20 15 29.5	−57 54 48	IC 4980	13.1	1.5 x 0.7	13.0	SB(s)0/a: pec	122	Mag 10.00v star SW 10′8.
20 16 06.2	−52 05 14	IC 4983	13.6	1.0 x 0.9	13.3	SB(r)c II-III	42	Mag 9.86v star W 8′8.
20 17 11.5	−55 02 07	IC 4986	13.1	1.8 x 0.9	13.5	SAB(s)dm IV-V	20	Mag 13.4 star on S edge; mag 10.39v star NW 11′8.
20 19 44.4	−53 26 53	IC 4994	12.9	0.7 x 0.7	12.0	SA0°: pec		Mag 12.1 star E 1′6; ESO 186-32 SSW 3′9.
20 19 59.1	−52 37 21	IC 4995	13.3	0.9 x 0.6	12.5	SA0° pec?	153	Mag 10.07v star E 10′8; double system **ESO 186-35** ESE 8′9.
20 26 40.2	−54 47 59	IC 5002	13.1	1.3 x 0.7	12.9	S0°	92	Mag 9.78v star E 4′8 **IC 5001** WNW 3′2.
20 33 34.6	−54 31 16	IC 5021	14.3	0.7 x 0.5	13.0		2	ESO 186-60 SW 1′6.
20 43 41.8	−57 01 50	IC 5034	13.6	1.5 x 0.7	13.5	SA(s)bc II	22	Mag 11.5 star and **IC 5036** SE 8′3.
20 52 02.1	−57 04 08	IC 5063	11.9	2.1 x 1.4	12.9	SA(s)0$^+$:	116	Mag 12.0 star N 3′3.
20 52 38.6	−57 13 56	IC 5064	13.2	1.0 x 0.7	12.7	(R)SB(s)a?	57	Mag 11.8 star W 4′1.
19 00 51.0	−57 45 33	NGC 6721	12.0	1.7 x 1.4	13.0	E$^+$:	155	
19 01 56.6	−53 51 48	NGC 6725	12.2	2.2 x 0.5	12.2	S0 sp	40	Double system **ESO 183-35** SW 10′4.
19 06 10.6	−62 11 49	NGC 6733	12.3	1.8 x 1.2	13.0	SAB0° pec:	110	Line of four stars W of bright center.
19 07 48.9	−61 22 03	NGC 6739	12.2	2.4 x 0.9	12.9	SA(r)0°	171	Bright, elongated center, uniform envelope.
19 10 22.2	−61 58 09	NGC 6746	12.6	1.4 x 0.9	12.7	SA0$^-$:	173	Mag 9.41v star SE 3′0.
19 11 23.4	−57 02 56	NGC 6753	11.1	2.5 x 2.1	12.8	(R)SA(r)b I-II	30	Bright center, wide dark lane NE; mag 10.28v star N 3′1.
19 13 52.3	−56 18 34	NGC 6758	11.6	2.2 x 1.7	13.0	E$^+$:	121	**ESO 184-36** N 3′9.
19 18 22.9	−60 30 03	NGC 6769	11.8	2.3 x 1.5	12.9	SAB(r)b pec II	123	Bright center, dark lane S; star on E edge with NGC 6770.
19 18 37.3	−60 29 46	NGC 6770	11.9	2.3 x 1.7	13.3	SAB(rs)b pec	20	Bright center, short E-W bar; NGC 6769 on W edge.
19 18 39.8	−60 32 47	NGC 6771	12.5	2.3 x 0.5	12.5	SB(r)0$^+$? sp	118	IC 4842 SE 8′1.
19 22 51.0	−55 46 29	NGC 6780	12.6	1.9 x 1.6	13.7	SAB(rs)c II	168	Stellar nucleus, many superimposed stars.

(Continued from previous page)

194 GALAXIES 194

RA h m s	Dec ° ′ ″	Name	Mag (V)	Dim ′ Maj x min	SB	Type Class	PA	Notes
19 23 58.2	−59 55 18	NGC 6782	11.8	2.4 x 2.0	13.4	(R)SAB(r)a	73	Bright center N-S, faint envelope E-W.
19 26 49.7	−54 57 03	NGC 6788	11.9	2.9 x 0.9	12.8	SA(s)ab II	71	Star on W end.
19 32 17.0	−55 54 27	NGC 6799	12.4	1.6 x 1.2	13.0	SA0$^-$:	98	Stars superimposed on W, SW and SE edges.
19 43 34.1	−58 39 16	NGC 6810	11.4	3.2 x 0.9	12.4	SA(s)ab: sp	176	
19 45 24.3	−55 20 53	NGC 6812	12.5	1.1 x 0.6	11.9	S0: pec	97	Small, faint anonymous galaxy on SW edge; another faint anonymous galaxy WNW 1′.8; a mag 11.11v star NW 2′.8.
20 02 47.7	−56 05 18	NGC 6848	12.1	2.5 x 1.0	12.9	Sa: sp	157	Strong dark lane W of centerline.
20 03 30.2	−54 50 45	NGC 6850	12.5	2.1 x 0.9	13.0	SB(s)0$^+$ pec:	153	Mag 10.50v star NNE 8′.4.
20 05 38.8	−54 22 30	NGC 6854	12.2	2.0 x 1.3	13.1	SAB(s)0$^-$ pec:	166	Faint star N edge of nucleus; anonymous galaxy NE 1′.9; mag 11.33v star SE 3′.6.
20 06 50.3	−56 23 23	NGC 6855	12.9	1.5 x 1.2	13.4	(R′)SB(rs)0$^+$:	113	Star SE edge of nucleus; mag 10.33v star NE 2′.4.
20 08 47.1	−61 06 03	NGC 6860	12.6	1.3 x 0.8	12.5	(R′)SB(r)b	34	
20 08 54.7	−56 23 30	NGC 6862	12.7	1.4 x 1.1	13.0	SB(rs)b: II	149	Stellar nucleus.
20 10 29.9	−54 46 58	NGC 6867	13.1	2.0 x 0.7	13.3	SB(rs)bc: I-II	156	Galaxy **Fair 533** N 8′.4.
20 17 17.1	−52 47 46	NGC 6887	12.1	3.4 x 1.3	13.5	SAbc: II	102	Faint stellar nucleus; mag 9.64v star SE 5′.6.
20 18 53.0	−53 57 29	NGC 6889	12.9	0.9 x 0.7	12.3	SBbc? pec	63	Several stars superimposed NE.
20 38 20.3	−52 06 40	NGC 6935	12.0	2.0 x 1.7	13.2	(R′)SA(r)a	8	Very bright nucleus.
20 38 46.2	−52 08 31	NGC 6937	12.9	2.5 x 2.1	14.6	(R′)SB(r)c: II-III	51	Bright nucleus, faint, slender, branching arms.
20 40 38.0	−54 18 12	NGC 6942	11.7	2.0 x 1.5	12.8	(R′)SB(rs)0/a:	150	**ESO 186-70** WSW 6′.5.
20 43 29.0	−53 21 26	NGC 6948	12.8	2.2 x 1.0	13.5	SA(s)a: sp	115	Dark lane N of center.
20 57 18.5	−51 51 46	NGC 6982	13.7	1.1 x 0.6	13.1	(R)SB(rs)a pec?	68	Mag 10.61v star NNE 4′.9.
20 57 54.1	−51 52 18	NGC 6984	12.7	1.8 x 1.2	13.4	SB(r)c I-II	110	Faint star S of nucleus; mag 10.61v star N 4′.4; galaxy **Fair 927** NNE 5′.7.
20 59 57.1	−55 33 43	NGC 6990	13.2	1.1 x 0.5	12.4	Sa:	0	Star at S end.
19 21 09.5	−58 48 39	PGC 63102	14.6	0.7 x 0.5	13.3	SA0/a pec	33	

GALAXY CLUSTERS

RA h m s	Dec ° ′ ″	Name	Mag 10th brightest	No. Gal	Diam ′	Notes
19 52 12.0	−55 05 00	A 3651	15.4	75	22	**ESO 185-23, 23A** E of center 6′.3, all others anonymous, stellar.
20 12 30.0	−56 48 00	A 3667	15.4	85	22	Many PGC galaxies stretch from NW edge, below center, to SSE edge.
20 51 30.0	−52 42 00	A 3716	15.0	66	22	Heavy concentration of PGC GX center one third diameter.
19 39 30.0	−53 10 00	AS 823	15.5	21	22	**ESO 185-7, 7A, 7B** N of center 4′.0; **ESO 185-6** N 12′.2; most others anonymous.
19 58 30.0	−52 37 00	AS 835	15.7	15	17	Most members anonymous, stellar.
20 01 18.0	−52 55 00	AS 839	15.3	17	22	**ESO 185-47, 47A** below center 3′.0; **IC 4925** N 3′.5; **IC 4932** NE 9′.8.
20 03 24.0	−55 57 00	AS 840	13.9	15	56	
20 07 24.0	−53 08 00	AS 844	16.3	14	17	**IC 4947** at center; all other members anonymous, stellar.
20 11 24.0	−56 44 00	AS 854	15.7	24	17	Superimposed on A 3667, see notes A 3667.
20 36 30.0	−57 37 00	AS 885	16.1	22	17	See notes ESO 143-35; all others anonymous, stellar.
20 40 00.0	−53 03 00	AS 889	16.0	7	17	**ESO 186-71, 71A** at center; all others anonymous, stellar.
20 52 24.0	−51 57 00	AS 906	15.9	29	17	Several PGC GX S of center; **ESO 235-10** NE 7′.6.

OPEN CLUSTERS

RA h m s	Dec ° ′ ″	Name	Mag	Diam ′	No. ★	B ★	Type	Notes
19 18 00.0	−57 54 00	ESO 141-47		7			cl	
19 16 54.0	−51 29 00	ESO 231-30		20			cl	

GLOBULAR CLUSTERS

RA h m s	Dec ° ′ ″	Name	Total V m	B ★ V m	HB V m	Diam ′	Conc. Class Low = 12 High = 1	Notes
19 10 51.8	−59 58 55	NGC 6752	5.3	10.5	13.8	29.0	6	

PLANETARY NEBULAE

RA h m s	Dec ° ′ ″	Name	Diam ″	Mag (P)	Mag (V)	Mag cent ★	Alt Name	Notes
19 09 48.2	−55 35 10	PN G341.2−24.6	60				PK 341−24.1	Annular; mag 12.8 star W 0′.9; mag 9.59v star NW 4′.3.

195 GALAXIES 195

RA h m s	Dec ° ′ ″	Name	Mag (V)	Dim ′ Maj x min	SB	Type Class	PA	Notes
16 57 58.4	−62 43 33	ESO 101-20	12.6	1.9 x 0.8	12.9	SB(rs)bc pec: II	117	Several stars superimposed; mag 10.96v star W 5′.8.
17 34 16.3	−62 44 24	ESO 102-9	13.9	1.2 x 0.4	13.0	(R′)SA(s)ab:	58	Mag 10.31v star NW 5′.7.
16 56 45.2	−62 24 13	ESO 138-7	12.9	2.3 x 0.7	13.2	SB(rs)bc II-III	49	Mag 8.62v star S 8′.6.
16 58 02.8	−60 53 13	ESO 138-9	13.9	1.5 x 0.6	13.6	IB(s)m V-VI	71	Mag 8.70v star WSW 9′.6.
16 59 02.5	−60 12 55	ESO 138-10	10.8	3.0 x 2.1	12.6	SA(s)dm IV-V	53	Numerous stars superimposed; mag 9.00v star NNW 10′.6.
16 59 36.3	−62 32 43	ESO 138-11	13.6	1.0 x 0.6	12.9	Sdm:	1	Numerous superimposed stars; mag 12.9 star S edge, mag 9.71v star SE 1′.6.
16 59 50.3	−58 41 22	ESO 138-12	13.5	1.4 x 0.7	13.4	IB(s)m IV	100	Filamentary, numerous stars superimposed; mag 12.4 star ENE 2′.4.
17 03 03.7	−61 46 04	ESO 138-13	13.3	1.0 x 0.7	12.7	Sc	23	Mag 8.80v star ENE 3′.4.
17 07 00.0	−62 05 05	ESO 138-14	12.6	3.1 x 0.4	12.7	SB(s)d: sp III-IV	135	Mag 9.89v star N 4′.2.
17 07 47.5	−59 51 03	ESO 138-15	14.1	1.9 x 0.2	12.9	Sdm:	94	Mag 10.95v star NW 1′.3.
17 10 15.6	−61 55 04	ESO 138-16	13.1	1.3 x 0.4	12.3	SB(s)bc? pec	120	Mag 11.4 star N 1′.8; mag 9.77v star W 10′.5.
17 13 38.9	−59 06 25	ESO 138-17	12.4	2.0 x 1.6	13.5	SAB(s)c: II	178	Numerous stars superimposed; mag 9.89v star SE 4′.1.
17 13 54.3	−60 48 40	ESO 138-18	13.8	0.9 x 0.3	12.2	S0$^+$?	11	Mag 12.0 star SE 1′.0.
17 20 08.7	−60 09 27	ESO 138-21	13.1	0.9 x 0.5	12.1	S	84	Mag 7.55v star NE 11′.8.
17 22 39.4	−59 30 28	ESO 138-22	13.6	1.3 x 0.3	12.4	Sa? sp	111	Mag 11.0 star superimposed; in star rich area.
17 24 07.0	−59 22 55	ESO 138-24	13.1	2.1 x 0.3	12.4	SB(s)c: III	53	Mag 8.75v star W 4′.6.
17 26 35.2	−60 16 30	ESO 138-26	12.9	1.2 x 0.6	12.4	(R′)SAB(rs)a?	178	Mag 9.05v star S 2′.1.
17 26 43.5	−59 55 59	ESO 138-27	12.9	0.7 x 0.6	11.8	Sbc	46	Mag 9.67v star E 4′.5.
17 29 10.8	−62 26 53	ESO 138-29	11.7	2.5 x 1.4	12.9	(R)SAB(s)0$^+$ pec	39	Mag 10.48v star S 1′.9.
17 29 25.3	−62 28 52	ESO 138-30	14.0	1.2 x 0.5	13.3	SB(s)ab? pec	154	Numerous stars superimposed on faint envelope.
17 30 28.2	−62 01 43	ESO 139-1	13.5	0.9 x 0.6	12.7	S0?	45	Mag 14.1 star NE edge; mag 10.41v star E 2′.5.
17 33 21.1	−60 46 35	ESO 139-4	13.8	0.9 x 0.4	12.6	SAB(s)0°	129	Mag 11.8 star NE 1′.9.

(Continued from previous page)

RA h m s	Dec ° ′ ″	Name	Mag (V)	Dim ′ Maj x min	SB	Type Class	PA	Notes
17 34 19.7	−60 52 20	ESO 139-5	13.2	0.9 x 0.6	12.5	SAB(s)0°	111	Faint stars superimposed along S edge.
17 35 00.3	−61 00 10	ESO 139-8	14.0	0.9 x 0.5	13.0	SAB(rs)c	17	Mag 12.3 star NW 1′3.
17 36 06.1	−61 53 57	ESO 139-9	13.6	0.9 x 0.6	12.8	(R)SAB(rs)0°	105	Mag 8.62v star N 8′4.
17 36 13.0	−58 15 15	ESO 139-11	13.4	1.2 x 0.4	12.5	Sab? sp	95	Mag 8.00v star S 6′9.
17 37 39.3	−59 56 29	ESO 139-12	13.0	1.1 x 0.7	12.6	(R′)SA(rs)bc pec:	31	Mag 10.20v star E 6′3.
17 40 01.0	−59 35 31	ESO 139-14	14.1	1.0 x 0.4	13.0	S	153	Mag 10.83v star W 3′4.
17 44 09.7	−60 58 44	ESO 139-21	13.5	2.1 x 0.4	13.2	Sb sp	107	Mag 10.02v star W 7′0.
17 45 38.8	−59 18 30	ESO 139-24	13.5	1.1 x 0.8	13.3	(R′)SB(r)ab	175	Mag 8.56v star ESE 4′5.
17 46 51.0	−59 15 31	ESO 139-26	13.1	1.0 x 0.7	12.6	SB(s)0°: pec	113	Mag 9.29v star N 6′2.
17 48 51.2	−60 42 25	ESO 139-28	13.8	1.1 x 0.5	13.0	S0?	73	Mag 8.68v star NNW 2′0.
17 49 55.8	−60 45 12	ESO 139-32	14.0	1.1 x 0.8	13.7	(R′)SB(rs)ab	167	Three superimposed stars form a triangle; mag 12.0 star NE 3′0.
17 50 22.2	−59 33 40	ESO 139-34	14.1	1.1 x 0.7	13.6	S	138	Mag 11.5 star W 8′1.
17 51 24.0	−60 07 42	ESO 139-36	13.5	1.3 x 0.5	12.9	Sb:	60	Located 2′6 NW of mag 5.78v star.
17 52 14.6	−60 44 57	ESO 139-37	14.2	1.1 x 0.4	13.1	S0?	11	Mag 12.3 star NW 1′8.
17 53 41.6	−59 50 37	ESO 139-38	14.1	1.3 x 0.3	13.0	S0	45	Mag 8.66v star N 8′4.
17 56 11.7	−60 16 33	ESO 139-41	13.5	1.3 x 0.5	12.9	SB(s)b:	179	Mag 7.77v star SE 6′6.
17 58 39.7	−58 22 20	ESO 139-45	13.0	0.8 x 0.4	11.7	SB(s)0°	54	Located between a N-S pair of mag 10.15v, 10.18v stars.
17 59 17.1	−59 30 31	ESO 139-46	14.3	1.8 x 0.3	13.5	SB(s)cd: sp III-IV	173	Mag 10.33v star N 2′3.
17 59 23.3	−58 06 15	ESO 139-47	14.3	1.0 x 0.8	13.9	SB(s)cd: IV	149	Stellar nucleus, smooth envelope, few superimposed stars on edges.
17 59 47.6	−59 49 05	ESO 139-48	14.3	1.1 x 0.4	13.3		82	Mag 13 star E edge.
18 00 35.0	−59 08 17	ESO 139-49	13.2	2.6 x 0.6	13.5	SB(s)cd: pec II-III	19	Mag 10.21v star S 2′4.
18 02 59.1	−58 01 52	ESO 139-51	14.3	1.2 x 0.6	13.8	SAB(s)c II-III	8	Mag 12.7 star NE 1′1; mag 11.9 star W 2′7.
18 04 21.7	−59 37 24	ESO 139-53	14.0	1.0 x 0.4	12.8	S0?	80	Mag 10.85v star WSW 2′5.
18 07 00.6	−57 43 54	ESO 139-55	12.8	1.0 x 0.7	12.3	(R′)SA(rs)0^+:	7	Star on NE edge of bright nucleus.
18 07 56.6	−58 43 08	ESO 140-2	13.8	1.1 x 0.6	13.2	S?	111	Mag 10.40v star NW 3′5.
18 12 00.8	−58 58 35	ESO 140-6	13.8	1.4 x 0.5	13.2	SB(s)cd pec III-IV	50	Mag 13.0 star NE edge; mag 7.81v star WSW 1′5.
18 14 16.5	−60 05 30	ESO 140-12	12.5	1.9 x 1.4	13.4	(R)SAB(s)a pec:	135	Mag 12 star on SE edge.
18 22 16.5	−61 52 12	ESO 140-18	13.9	1.3 x 0.8	13.8	S?	6	Large knot or star E edge; mag 12.2 star SW edge.
18 32 40.5	−57 40 17	ESO 140-23	12.7	1.3 x 1.1	12.9	SB(s)d IV-V	48	Mag 11.6 star N 5′7.
18 37 53.8	−57 36 42	ESO 140-31	12.6	1.2 x 0.8	12.6	E1	61	Mag 11.8 star E 2′5.
18 42 45.5	−61 26 46	ESO 140-38	13.3	1.2 x 1.0	13.3	S0?	173	Mag 15 stars superimposed NE and NW edges.
18 44 53.8	−62 21 55	ESO 140-43	14.0	1.6 x 0.9	14.2	(R′)SB(s)b: III-IV	167	Mag 9.04v star SSE 4′3; NGC 6673 N 4′3.
18 46 56.1	−57 41 44	ESO 141-3	12.8	1.0 x 0.5	12.0	$SA0^-$	29	Mag 9.18v star NE 7′4.
19 00 55.3	−62 03 41	ESO 141-17B	14.1	1.9 x 0.8	14.4		135	Pair with ESO 141-17A W edge.
19 00 51.0	−62 03 40	ESO 141-17A	13.4	1.5 x 0.6	13.1		15	Pair with ESO 141-17B E edge.
19 03 07.3	−60 25 22	ESO 141-21	13.2	1.0 x 0.6	12.5	S?	144	Mag 15 star E edge; mag 13.7 star W 1′4.
17 37 54.8	−55 31 05	ESO 181-2	13.8	1.5 x 0.7	13.7	SAB(s)bc: II-III	36	Mag 12.8, 13.1 stars on E edge; mag 13.5 star on S edge.
17 58 42.4	−53 48 02	ESO 182-1	12.8	1.2 x 0.7	12.5	(R)SAB(rs)0^+:	19	Mag 10.7 star E 1′1; mag 9.64v star WNW 6′6.
18 09 31.0	−52 34 00	ESO 182-4	13.3	1.4 x 1.2	13.7	SB(r)0/a	65	Strong core, faint envelope, stars superimposed, in star rich area.
18 16 33.4	−52 54 54	ESO 182-6	13.8	0.8 x 0.3	12.1	S0?	132	1′0 NE of mag 10 star.
18 16 49.5	−57 13 56	ESO 182-7	12.9	1.3 x 0.5	12.5	E?	172	Possible planetary nebula Sa 2-323 ENE 11′0.
18 17 10.9	−54 34 06	ESO 182-8	13.8	0.8 x 0.5	12.7	S pec	55	Mag 10.72v star SE 1′1.
18 18 30.7	−54 41 42	ESO 182-10	13.0	1.9 x 1.3	13.9	SAB(rs)c pec III-IV	35	Mag 8.09v star E 12′9.
18 21 59.9	−52 56 21	ESO 182-12	14.0	1.5 x 0.2	12.5	Sa:	17	**ESO 182-11** W 12′6.
18 22 53.9	−56 29 10	ESO 182-13	13.1	0.7 x 0.7	12.2	S		Mag 9.26v star S 9′0.
18 27 52.1	−53 12 38	ESO 182-17	13.4	1.5 x 0.4	12.7	S0: sp	54	Mag 5.97v star NW 9′4.
18 33 19.9	−55 46 24	ESO 183-3	13.9	0.6 x 0.4	12.2	S0	154	Mag 10.52v star NW 4′2.
18 39 03.3	−55 37 08	ESO 183-5	14.9	1.5 x 0.2	13.5	SB(s)d sp IV	141	Mag 9.16v star E 8′8.
18 39 09.8	−57 26 42	ESO 183-6	13.8	1.0 x 0.5	12.9	Sa:	7	Mag 10.18v star NW 2′6; mag 8.88v star W 4′7.
18 40 33.9	−55 04 26	ESO 183-7	14.0	1.1 x 0.4	13.0	Pec	84	Mag 12.8 star N 1′5.
18 41 33.0	−54 46 24	ESO 183-8	14.0	1.1 x 0.4	13.0	S0?	114	Mag 9.90v star NE 1′2.
18 44 21.4	−56 44 49	ESO 183-11	14.3	1.3 x 0.2	12.7	Sb? sp III-IV	7	Mag 11.05v star WNW 3′7.
18 51 22.6	−52 35 27	ESO 183-19	13.9	1.1 x 0.5	13.1	SB(s)c II	103	Stellar nucleus, smooth envelope.
18 56 56.0	−54 32 42	ESO 183-30	11.7	1.9 x 1.5	12.7	$SA0^-$ pec?	22	Mag 7.78v star W 3′4.
18 59 31.1	−54 21 38	ESO 183-34	13.1	1.3 x 0.5	12.5	S0/a?	102	Mag 13.0 star W 1′7.
19 03 35.9	−53 10 22	ESO 184-4	13.7	1.1 x 0.8	13.4	S?	71	Mag 10.66v star W edge; galaxy **Fair 505** SW 3′3.
19 03 39.5	−53 56 05	ESO 184-5	12.7	1.0 x 0.7	12.3	$SA0^-$: pec	157	Mag 10 star S 0′7.
18 49 02.3	−51 15 58	ESO 231-1	12.7	1.2 x 1.1	12.9	S?	1	Mag 8.01v star N 8′8.
18 52 37.5	−51 28 04	ESO 231-6	14.8	1.5 x 0.2	13.3	Sc	57	Mag 8.28v star W 8′3; mag 7.44v star WSW 11′8.
18 53 23.2	−52 23 28	ESO 231-7	14.3	1.6 x 1.1	14.7	SAB(s)dm? V	51	Mag 7.64v star N 8′2.
17 23 53.8	−59 59 56	IC 4646	11.9	2.8 x 1.7	13.4	SA(rs)c: I-II	0	Numerous stars superimposed; mag 10.12v star WNW 6′2.
17 26 26.7	−59 43 41	IC 4652	13.3	0.8 x 0.4	11.9	Sb: II-III	10	Mag 11.21v star N 2′4.
17 27 07.4	−60 52 45	IC 4653	12.5	1.1 x 0.6	12.0	SB(r)0/a pec:	52	Mag 9.95v star ESE 6′2.
18 08 13.2	−62 23 43	IC 4674	14.3	1.1 x 0.4	13.2	SAB(s)d: IV	89	Mag 7.99v star N 1′6.
18 11 24.2	−56 15 17	IC 4679	13.1	2.2 x 0.9	13.7	SB(s)cd II	99	Pair of mag 7.14v, 7.56v stars S 9′6.
18 13 38.8	−57 43 59	IC 4686	14.2	0.5 x 0.4	12.3		125	Pair with IC 4687 N.
18 13 40.2	−57 43 31	IC 4687	13.2	0.9 x 0.5	12.2	Sb: pec	54	Pair with IC 4686 S.
18 13 40.5	−57 44 58	IC 4689	13.9	0.9 x 0.3	12.3	Sa: pec	139	IC 4686 and IC 4687 N 1′2.
18 14 50.0	−58 41 35	IC 4692	13.3	1.8 x 0.9	13.7	SB(r)cd II-III	0	Mag 7.82v star W 3′9.
18 15 27.1	−58 12 28	IC 4694	13.0	2.2 x 0.8	13.5	SB(s)c: III-IV	18	Mag 8.98v star N 4′9.
18 23 04.3	−59 14 22	IC 4702	12.8	1.5 x 1.0	13.1	(R)SB(rs)0/a	151	Mag 11.14v star NE 4′5.
18 24 19.8	−56 22 13	IC 4709	13.4	1.1 x 0.4	12.4	$(R_1)SB(r)0^+$	4	Mag 10.98v star N 4′6.
18 33 17.3	−57 58 34	IC 4717	13.3	1.5 x 0.3	12.3	Sb? sp	93	Mag 10.04v star W 4′0.
18 33 50.6	−60 07 48	IC 4718	12.9	1.3 x 0.6	12.5	Sb: III-IV	116	Mag 9.05v star W 5′1.
18 33 12.2	−56 44 01	IC 4719	13.0	1.4 x 1.0	13.2	SA(s)dm pec: IV	116	Mag 8.33v star N 5′4.
18 33 32.8	−58 24 14	IC 4720	12.8	2.5 x 0.9	13.6	SB(rs)cd: IV	163	Mag 9.58v star ENE 8′3.
18 34 25.3	−58 29 48	IC 4721	11.6	5.2 x 1.5	13.7	SB(s)cd: III	146	Almost stellar **IC 4721A** on SW edge.
18 34 31.7	−57 47 33	IC 4722	12.8	1.6 x 1.2	13.4	SA(s)c pec: II	46	Mag 10.40v star ENE 8′6.
18 36 59.4	−62 51 15	IC 4726	13.4	1.1 x 0.8	13.1	S0?	23	Mag 14.8 star superimposed N of bright center; mag 9.63v star SW 3′5.
18 37 56.2	−62 42 02	IC 4727	12.8	1.1 x 1.1	12.9	$S0^-$:		Mag 10.32v star S 3′3.
18 37 57.3	−62 31 54	IC 4728	12.9	2.0 x 0.7	13.1	SB(rs)ab III-IV	173	Group of three stars N 3′5, brightest is mag 10.81v.
18 38 43.2	−62 56 37	IC 4731	12.5	1.5 x 0.7	12.4	SB(r)0^+:	82	Mag 10.17v star SW 10′2; **ESO 103-36** S 5′2; **IC 4735** E 7′7.
18 38 25.5	−57 29 28	IC 4734	13.2	1.3 x 0.8	13.0	(R′)SB(s)ab pec: II-III	94	Mag 8.88v star N 2′9.
18 38 40.1	−57 53 36	IC 4736	13.9	1.0 x 0.9	13.6	SB(s)dm IV-V	117	Mag 10.22v star N 7′6.
18 39 58.7	−62 35 54	IC 4737	13.7	0.9 x 0.6	12.9	(R′)SB(s)a	108	Bright N-S bar; mag 8.98v star SE 1′6.
18 41 29.5	−61 46 21	IC 4743	13.7	0.8 x 0.4	12.4	S0	98	Mag 10.86v star S 4′1; **IC 4739** and **IC 4738** SSW 9′8.
18 43 19.5	−62 06 47	IC 4751	13.1	1.0 x 0.6	12.4	(R)SB0°	3	Mag 9.89v star SSW 3′8; IC 4753 E 1′6.
18 43 32.6	−62 06 30	IC 4753	13.5	0.7 x 0.6	12.5	E?	5	Mag 10.73v star N 3′; IC 4751 W 1′6.
18 44 00.3	−61 59 27	IC 4754	13.3	1.4 x 1.3	13.8	(R′)SB(r)b: II-III	122	Strong N-S bar, mag 14.3 star on S edge ; mag 10.73v star SSW 5′0.

(Continued from previous page)

195 GALAXIES 195

RA h m s	Dec ° ′ ″	Name	Mag (V)	Dim ′ Maj x min	SB	Type Class	PA	Notes
18 43 56.2	−57 10 07	IC 4757	13.4	0.9 x 0.4	12.2	SA(s)0/a: sp	57	Central dark lane runs length of galaxy.
18 43 55.8	−52 51 15	IC 4761	13.1	1.6 x 1.3	13.8	(R′)SB(r)bc: II-III	115	Brighter core with very faint outer regions; numerous stars superimposed.
18 48 10.4	−57 56 10	IC 4774	13.4	1.0 x 0.9	13.2	SB(r)bc II	114	Star or bright knot N of nucleus; mag 11.3 star N 1′1; mag 10.39v star E 2′5.
18 48 26.5	−57 11 05	IC 4775	13.9	1.5 x 0.3	12.9	Sb sp	17	Mag 9.34v star WSW 5′9.
18 48 11.4	−53 08 53	IC 4777	13.7	0.8 x 0.6	12.8	SB0?	128	Bright, elongated center, faint envelope.
18 50 00.6	−61 43 09	IC 4778	13.0	1.2 x 0.9	13.0	(R′)S0/a?	52	Mag 10.82v star SSW 5′9.
18 49 56.6	−59 15 10	IC 4780	13.9	0.7 x 0.5	12.6	S	99	Mag 9.83v star ESE 3′1.
18 50 54.8	−55 29 30	IC 4782	14.3	1.4 x 0.2	12.7	Scd: III-IV	99	Mag 10.62v star E 2′6; elongated N-S **ESO 183-17** S 1′1.
18 51 33.6	−58 48 49	IC 4783	13.6	1.1 x 1.0	13.6	Sb:		Mag 9.52v star SW 4′6.
18 52 55.2	−59 15 17	IC 4785	12.2	3.1 x 1.4	13.7	(R′)SB(rs)b II	140	Pair of N-S oriented mag 11.24v, 10.45v star W 8′0.
18 55 42.1	−56 24 15	IC 4792	14.1	1.1 x 0.3	12.7	Sa:	156	Located 1′2 NW of mag 6.77v star.
18 56 56.0	−61 24 03	IC 4793	13.6	1.1 x 0.5	12.8	Sb	120	Mag 10.8 star WSW 10′2.
18 57 09.9	−62 05 29	IC 4794	13.0	0.9 x 0.5	12.1	S0	25	Mag 9.60v star W 11′0.
18 56 28.3	−54 12 53	IC 4796	12.3	1.6 x 0.9	12.5	SB(s)0°:	140	Mag 12.3 star on N edge; mag 10.08v star ESE 9′0.
18 56 29.8	−54 18 22	IC 4797	11.3	2.1 x 1.1	12.3	E$^+$ pec:	146	Mag 10.08v star W 8′5.
18 58 20.9	−62 07 05	IC 4798	12.2	1.5 x 0.8	12.2	SA(s)0°: pec	108	Mag 10.59v star SW 10′0, part of a "V" shaped asterism.
19 01 07.5	−61 50 00	IC 4804	13.8	1.2 x 0.8	13.6	SA(s)c I-II	147	Stellar nucleus, faint envelope.
19 01 30.9	−57 31 55	IC 4806	12.2	2.3 x 0.6	12.4	(R′)SA(s)b: sp II	9	Mag 9.44v star NNE 6′0.
19 02 17.8	−56 55 49	IC 4807	13.4	1.0 x 0.6	12.7	SB(s)c? I	7	Mag 8.31v star SW 3′6.
19 02 59.9	−56 09 37	IC 4810	12.6	3.5 x 0.4	12.8	SAB(s)d? sp III-IV	136	Mag 10.90v star NNE 5′8.
17 17 00.4	−62 49 12	NGC 6300	10.2	4.3 x 2.8	12.8	SB(rs)b II	118	Bright nucleus with strong dark patches NW and SE; many superimposed stars.
17 18 01.0	−59 10 19	NGC 6305	12.2	1.4 x 0.8	12.2	SB(s)0$^-$ pec?	133	Many superimposed stars in star rich area.
17 42 43.8	−61 41 39	NGC 6398	12.7	2.0 x 1.7	13.9	(R)SB(r)a	6	Bright nucleus, very faint outer envelope, few superimposed stars.
17 43 23.8	−61 40 58	NGC 6403	13.5	0.5 x 0.5	11.9	SA(rs)0°:		Almost stellar,; mag 11v star 1′1 S.
17 44 57.5	−60 44 28	NGC 6407	12.3	2.1 x 1.6	13.5	SA0°? pec	60	Bright nucleus, superimposed stars, mag 9.37v star SE 1′7.
18 45 06.4	−62 17 47	NGC 6673	11.6	2.2 x 0.9	12.3	SAB(s)0$^-$:	26	ESO 140-43 S 4′3.
18 52 02.6	−57 19 17	NGC 6699	12.0	1.5 x 1.5	12.7	SAB(rs)bc: I-II		Stellar nucleus, tightly wound arms.
18 55 22.0	−53 49 05	NGC 6707	12.6	2.1 x 0.9	13.2	SB(s)c? II-III	143	Mag 10.64v star NE 2′0.
18 55 35.6	−53 43 28	NGC 6708	12.7	1.1 x 0.9	12.5	Sb: III-IV	167	
19 00 51.0	−57 45 33	NGC 6721	12.0	1.7 x 1.4	13.0	E$^+$:	155	
19 01 56.6	−53 51 48	NGC 6725	12.2	2.2 x 0.5	12.2	S0 sp	40	Double system **ESO 183-35** SW 10′4.

GALAXY CLUSTERS

RA h m s	Dec ° ′ ″	Name	Mag 10th brightest	No. Gal	Diam ′	Notes
18 27 30.0	−51 32 00	AS 801	15.4		22	**ESO 230-4** N of center 4′5; all others anonymous, stellar.

OPEN CLUSTERS

RA h m s	Dec ° ′ ″	Name	Mag	Diam ′	No. ★	B ★	Type	Notes
17 37 42.0	−58 08 00	ESO 139-13		5			cl	
18 04 42.0	−58 31 00	ESO 139-54		6			cl	
16 59 06.1	−52 42 57	NGC 6253	10.2	4	30	13.0	cl	Moderately rich in stars; small brightness range; slight central concentration; detached.

GLOBULAR CLUSTERS

RA h m s	Dec ° ′ ″	Name	Total V m	B ★ V m	HB V m	Diam ′	Conc. Class Low = 12 High = 1	Notes
17 40 41.3	−53 40 25	NGC 6397	5.3	10.0	12.9	31.0	9	
18 18 37.7	−52 12 54	NGC 6584	7.9	13.5	16.5	6.6	8	

BRIGHT NEBULAE

RA h m s	Dec ° ′ ″	Name	Dim ′ Maj x min	Type	BC	Color	Notes
17 04 00.0	−51 05 00	vdBH 81	6 x 4	R	2-5	1-4	Shaped like a slight gibbous moon, convex to the N. The involved star is 9th mag and is situated near the S edge.

PLANETARY NEBULAE

RA h m s	Dec ° ′ ″	Name	Diam ″	Mag (P)	Mag (V)	Mag cent ★	Alt Name	Notes
17 11 45.3	−55 24 02	IC 4642	24	13.4		16.0	PK 334−9.1	Mag 12.0 star W 0′8.
17 20 46.7	−51 45 18	NGC 6326	19	12.2		16.7	PK 338−8.1	Mag 12.5 star E 0′5; in star rich area.
17 51 52.9	−60 23 20	PN G332.8−16.4	36	16.7			PK 332−16.1	Located 1′4 W of mag 7.44v star.
17 09 00.5	−56 54 50	PN G332.9−9.9				11.5	He 3-1333	Stellar; mag 11.5 star W 1′6.
17 03 03.0	−53 55 49	PN G334.8−7.4		14.0		11.5	PK 334−7.1	Stellar; mag 11.4 star WNW 4′9.
17 06 14.6	−52 29 59	PN G336.2−6.9	7	12.7		16.5	PK 336−6.1	Located 1′1 S of mag 8.27v star.
16 59 36.1	−51 42 04	PN G336.3−5.6	3	13.4		16.6	PK 336−5.1	Mag 12.9 star NNW 1′2; mag 11.29v star SE 2′6.
17 09 35.9	−52 13 05	PN G336.8−7.2	38			18.0	PK 336−7.1	Mag 11.9 star SE 1′1; mag 10.7 star WNW 2′9.
17 34 28.1	−54 28 55	PN G336.9−11.5	60			19.2	PK 336−11.1	Mag 10.34v star NE 3′3.
17 22 37.1	−52 46 37	PN G337.4−9.1	18	16.4			PK 337−9.1	mag 10.5 star WSW 2′1; mag 10.12v star SW 1′6.
18 00 59.5	−52 44 23	PN G340.4−14.1	11	15.3		16.2	PK 340−14.1	Mag 10.9 star SW 3′2.
18 07 15.9	−51 01 10	PN G342.5−14.3	68	11.9		12.5	PK 342−14.1	Elongated ESE-WNW; mag 12.4 star SSE 0′8.

196 GALAXIES 196

RA h m s	Dec ° ′ ″	Name	Mag (V)	Dim ′ Maj x min	SB	Type Class	PA	Notes
16 25 24.1	−62 38 16	ESO 100-22	12.6	1.5 x 0.7	12.5	SA(s)0°	30	Mag 11 start N end.
16 48 41.3	−62 36 22	ESO 101-14	12.0	1.2 x 1.2	12.3	SA0$^-$		Numerous stars involved.
16 57 58.4	−62 43 33	ESO 101-20	12.6	1.9 x 0.8	12.9	SB(rs)bc pec: II	117	Several stars superimposed; mag 10.96v star W 5′8.
15 30 42.5	−58 08 51	ESO 135-10	13.5	2.9 x 1.0	14.5		32	Mag 8.52v star N 2′7.
16 02 36.3	−61 46 32	ESO 136-12	13.1	2.7 x 1.5	14.5	SB(s)c: II	130	Mag 10.90v star S 1′0.

GALAXIES

RA h m s	Dec ° ′ ″	Name	Mag (V)	Dim ′ Maj x min	SB	Type Class	PA	Notes
16 03 28.5	−59 39 46	ESO 136-15	14.7	1.7 x 0.5	14.4	SBc? sp	117	Mag 11.45v star N 1′.2.
16 03 49.0	−60 58 47	ESO 136-16	13.6	3.3 x 0.5	14.0	SB(s)c: sp	132	
16 13 35.8	−60 51 52	ESO 137-2	12.8	0.9 x 0.2	10.9	S0? sp	168	Mag 10.07v star S 1′.6.
16 13 47.8	−61 00 11	ESO 137-3	12.8	3.3 x 1.7	14.5		160	Mag 10.67v star E 1′.3; ESO 137-4 N 3′.9.
16 13 57.5	−60 56 24	ESO 137-4	13.8	0.8 x 0.2	11.7	S0/a	21	ESO 137-3 S 3′.9.
16 15 03.9	−60 54 26	ESO 137-6	12.1	0.8 x 0.8	11.7	E1		Mag 9.23v star E 3′.1; ESO 137-8 E 5′.1.
16 15 32.9	−60 39 55	ESO 137-7	12.8	0.5 x 0.5	11.1	S0?		Mag 10.69v star N 1′.6.
16 15 46.3	−60 55 07	ESO 137-8	11.7	0.5 x 0.5	10.1	S0⁻:		Mag 9.23v star W 2′.2.
16 15 50.4	−60 48 11	ESO 137-10	11.6	2.0 x 1.1	12.4	(R′)SAB(s)0⁻:	167	Mag 9.26v star W 4′.3; ESO 137-11 S 2′.9.
16 15 53.6	−60 50 58	ESO 137-11	13.6	1.3 x 0.4	12.8	S0⁺ pec	168	ESO 137-10 N 2′.9.
16 16 19.6	−61 17 46	ESO 137-12	12.2	1.0 x 0.5	11.4	SB(s)0°	14	Mag 8.63v star WNW 8′.9.
16 19 17.1	−61 01 15	ESO 137-16	13.4	1.2 x 0.3	12.2	SB(s)0⁺: sp	9	Mag 11.31v star WNW 5′.2.
16 20 59.4	−60 29 15	ESO 137-18	11.5	3.2 x 1.1	12.7	SA(s)c: III-IV	30	Mag 9.02v star E 1′.0.
16 21 55.1	−60 40 09	ESO 137-19	12.5	1.6 x 0.9	12.8	SAB(s)0°:	158	Mag 10.66v star N 1′.2.
16 23 37.3	−61 42 28	ESO 137-21	13.2	1.4 x 0.3	12.1	S0°: sp	89	Located between a pair of stars, mags 10.92v and 12, in star rich area.
16 25 50.4	−60 45 08	ESO 137-24	12.6	1.1 x 1.0	12.6	S0°:	30	Mag 10.44v star N 4′.5.
16 26 05.1	−60 31 36	ESO 137-25	13.4	1.4 x 0.4	12.6	SA0°	168	Mag 9.69v star N 3′.7.
16 26 10.5	−61 39 43	ESO 137-26	12.6	1.6 x 0.6	12.4	S0/a? sp	75	Mag 10.17v star N 2′.9.
16 26 42.5	−59 57 23	ESO 137-27	13.4	3.2 x 0.6	13.9	SB(s)dm: IV-V	115	Mag 8.39v star E 11′.8.
16 29 07.1	−61 38 37	ESO 137-29	12.6	2.0 x 1.1	13.4	SA(r)0°	5	Mag 5.18v star E 12′.2.
16 34 35.3	−62 02 38	ESO 137-32	14.8	2.5 x 0.2	13.9	Sc	103	Mag 6.95v star SW 8′.5.
16 35 14.0	−58 04 48	ESO 137-34	11.4	1.0 x 1.0	11.2	SAB(s)0/a?		Mag 9.32v star E 1′.5.
16 35 51.2	−61 28 00	ESO 137-35	13.1	1.0 x 0.5	12.2	SB(s)0⁺	54	Mag 10.64v star E 3′.7.
16 36 54.0	−61 01 41	ESO 137-36	13.0	1.5 x 0.5	12.5	S0?	36	Mag 9.02v star N 1′.3.
16 39 18.3	−59 53 10	ESO 137-37	14.0	1.9 x 0.3	13.3	S?	73	Mag 10.54v star N 5′.0.
16 40 52.7	−60 23 38	ESO 137-38	12.0	3.4 x 0.9	13.1	SAB(r)bc: II	105	Mag 6.22v star S 3′.1.
16 41 07.6	−60 59 00	ESO 137-39	14.3	1.6 x 0.2	12.9	Sc? sp	129	Mag 9.70v star N 5′.9.
16 47 40.7	−60 08 57	ESO 137-42	12.8	1.4 x 0.7	12.6	SB(s)bc II-III	25	Several stars superimposed on envelope; mag 9.29v star SW 7′.7.
16 50 55.9	−61 48 53	ESO 137-44	12.4	1.1 x 0.9	12.3	SAB(s)0°?	74	Multi-galaxy group, connected.
16 51 03.2	−60 48 32	ESO 137-45	12.0	0.9 x 0.9	11.8	E:		Mag 9.83v star SW 8′.0.
16 51 20.4	−59 14 05	ESO 138-1	13.4	1.0 x 0.5	12.5	E?	157	Strong core with several stars superimposed on envelope; star rich area.
16 53 53.9	−58 46 42	ESO 138-5	11.7	1.3 x 0.8	11.7	SB0°? pec	140	Mag 11.1 star on N edge; mag 9.70v star E 6′.5.
16 54 20.1	−61 53 05	ESO 138-6	14.1	1.0 x 0.6	13.3	Sm? IV-V?	104	Low surface brightness; mag 13.8 star SW edge.
16 56 45.2	−62 24 13	ESO 138-7	12.9	2.3 x 0.7	13.2	SB(rs)bc II-III	49	Mag 8.62v star S 8′.6.
16 58 02.8	−60 53 13	ESO 138-9	13.9	1.5 x 0.6	13.6	IB(s)m V-VI	71	Mag 8.70v star WSW 9′.6.
16 59 02.5	−60 12 55	ESO 138-10	10.8	3.0 x 2.1	12.6	SA(s)dm IV-V	53	Numerous stars superimposed; mag 9.00v star NNW 10′.6.
16 59 36.3	−62 32 43	ESO 138-11	13.6	1.0 x 0.6	12.9	Sdm:	1	Numerous superimposed stars; mag 12.9 star S edge, mag 9.71v star SE 1′.6.
16 59 50.3	−58 41 22	ESO 138-12	13.5	1.4 x 0.7	13.4	IB(s)m IV	100	Filamentary, numerous stars superimposed; mag 12.4 star ENE 2′.4.
17 03 03.7	−61 46 04	ESO 138-13	13.3	1.0 x 0.7	12.7	Sc	23	Mag 8.80v star ENE 3′.4.
16 47 20.4	−57 26 26	ESO 179-13	13.1	2.8 x 1.6	14.6	SB(s)m IV-V	36	Double system, interaction.
15 09 29.2	−52 33 23	ESO 223-12	15.0	2.5 x 0.4	14.8	SA(s)bc? sp II-III	50	Mag 9.33v star N 6′.9.
16 34 52.7	−60 37 09	NGC 6156	11.5	1.6 x 1.4	12.2	SAB(rs)c pec: II	0	Bright center, faint, patchy envelope, many stars superimposed.
16 51 07.2	−58 59 34	NGC 6215	11.5	2.1 x 1.8	12.8	SA(s)c II-III	78	Patchy, branching arms, numerous superimposed stars, in star rich area.
16 52 48.8	−58 56 49	NGC 6215A	13.4	1.5 x 0.5	12.9	SAb? III	8	Many superimposed stars.
16 52 46.5	−59 12 57	NGC 6221	9.9	3.5 x 2.5	12.1	SB(s)c II-III	5	Numerous superimposed stars; prominent dark patch N of center.

GALAXY CLUSTERS

RA h m s	Dec ° ′ ″	Name	Mag 10th brightest	No. Gal	Diam ′	Notes
16 15 30.0	−60 54 00	A 3627	13.5	59	56	

OPEN CLUSTERS

RA h m s	Dec ° ′ ″	Name	Mag	Diam ′	No. ★	B ★	Type	Notes
15 50 48.0	−57 39 00	Cr 292	7.9	15	50		cl	Moderately rich in stars; moderate brightness range; not well detached.
16 24 24.0	−61 44 00	ESO 137-23		5			cl	
16 48 24.0	−61 46 00	ESO 137-43		6			cl	
16 24 12.0	−51 09 00	ESO 226-6		4			cl	
16 19 54.0	−54 58 00	Ha 10	6.9	25	30		cl	Moderately rich in stars; large brightness range; no central concentration; detached. (A) Scattering of stars, elongated E-W, with one non-member bright star to south. Also known as **Cr 299**.
15 08 23.4	−58 42 00	Lo 2045		39		8.8	cl	
15 33 05.9	−58 12 00	Lo 2158		49		8.6	cl	
15 40 44.5	−52 32 00	Lo 2313		21		10.1	cl	
15 16 47.0	−58 22 42	Lynga 3	12.6	5			cl	
15 33 20.0	−55 14 12	Lynga 4	11.4	3	30	12.5	cl	Moderately rich in stars; moderate brightness range; slight central concentration; detached. Probably not a cluster.
15 41 57.1	−56 39 30	Lynga 5		3	20		cl	
16 04 51.0	−51 57 18	Lynga 6	9.5	5	60	10.7	cl	(A) Includes arc of 7 stars.
16 01 36.1	−54 09 00	Moffat 1		8	20		cl	
15 04 12.6	−54 25 00	NGC 5822	6.5	35	150	10.0	cl	Rich in stars; moderate brightness range; slight central concentration; detached.
15 05 26.0	−55 37 00	NGC 5823	7.9	12	103	13.0	cl	Rich in stars; moderate brightness range; slight central concentration; detached.
15 27 42.0	−54 31 00	NGC 5925	8.4	20	120		cl	Rich in stars; moderate brightness range; no central concentration; detached.
15 52 14.0	−56 28 12	NGC 5999	9.0	3	40	12.0	cl	Moderately rich in stars; moderate brightness range; strong central concentration; detached.
15 55 47.4	−57 26 37	NGC 6005	10.7	5	35	11.8	cl	Moderately rich in stars; moderate brightness range; strong central concentration; detached. Probably not a cluster.
16 03 18.0	−60 25 00	NGC 6025	5.1	15	139	7.0	cl	Rich in stars; large brightness range; slight central concentration; detached. (A) Scattering of stars just to W of a dark nebula
16 07 34.4	−54 01 00	NGC 6031	8.5	3	121	10.9	cl	Few stars; large brightness range; strong central concentration; detached.
16 13 09.0	−54 13 00	NGC 6067	5.6	15	315	10.0	cl	Rich in stars; large brightness range; strong central concentration; detached.
16 18 51.8	−57 54 00	NGC 6087	5.4	15	349	8.0	cl	Moderately rich in stars; moderate brightness range; slight central concentration; detached. (A) Irregularly shaped.
16 24 26.8	−51 57 12	NGC 6115	9.8	3.4	20	11.0	cl?	
16 32 42.0	−52 37 00	NGC 6152	8.1	25	70	11.0	cl	Rich in stars; large brightness range; no central concentration; detached.
16 49 25.8	−53 42 00	NGC 6208	7.2	18	303	10.0	cl	Rich in stars; moderate brightness range; no central concentration; detached.
16 59 06.1	−52 42 57	NGC 6253	10.2	4	30	13.0	cl	Moderately rich in stars; small brightness range; slight central concentration; detached.
15 15 23.1	−59 04 24	Pi 20	7.8	4.5	12	8.2	cl	Few stars; large brightness range; strong central concentration; detached.
15 16 47.7	−59 39 41	Pi 21		2		13.0	cl	
16 13 11.2	−51 54 10	Pi 22		4	30	7.8	cl	
14 57 00.0	−62 32 00	Ru 112		10	80	14.0	cl	(A) Rich group (Milky Way patch?) of faint stars.
15 57 12.0	−59 27 00	Ru 113		25	20	9.0	cl	Few stars; moderate brightness range; not well detached.

(Continued from previous page)

196

OPEN CLUSTERS

196

RA h m s	Dec ° ′ ″	Name	Mag	Diam ′	No. ★	B ★	Type	Notes
16 06 31.0	−56 53 00	Ru 114		15	50	12.0	cl	(A) Large, loose group of faint stars.
16 12 51.1	−52 23 15	Ru 115		5		13.0	cl	
16 23 24.3	−51 59 48	Ru 116		5		9.0	cl	Few stars; moderate brightness range; slight central concentration; detached.
16 23 30.7	−51 52 06	Ru 117		1.7		12.0	cl	(A) Very weak condensation of stars on DSS image.
16 28 09.3	−51 31 00	Ru 119	8.8	8	13	10.0	cl	Few stars; small brightness range; slight central concentration; detached. Probably not a cluster. (A) Weak condensation of stars on DSS image.
16 00 50.7	−53 32 00	Tr 23	11.2	9	40		cl	Moderately rich in stars; small brightness range; no central concentration; detached.

GLOBULAR CLUSTERS

RA h m s	Dec ° ′ ″	Name	Total V m	B ★ V m	HB V m	Diam ′	Conc. Class Low = 12 High = 1	Notes
16 11 03.0	−55 18 52	Lynga 7				2.2		

BRIGHT NEBULAE

RA h m s	Dec ° ′ ″	Name	Dim ′ Maj x min	Type	BC	Color	Notes
16 17 48.0	−51 55 00	RCW 102	12 x 8	E	3-5		Faint irregular nebula, somewhat extended NE-SW; in an extremely rich star field.
16 17 06.0	−51 07 00	RCW 103	7 x 7	SNR	2-5		This supernova remnant lies in a field rich in faint stars and consists of two segments of a broken ring. The larger and brightest segment about 5′ long and 2′ broad forms the SE part while the NW portion is about half as large.
17 04 00.0	−51 05 00	vdBH 81	6 x 4	R	2-5	1-4	Shaped like a slight gibbous moon, convex to the N. The involved star is 9th mag and is situated near the S edge.

PLANETARY NEBULAE

RA h m s	Dec ° ′ ″	Name	Diam ″	Mag (P)	Mag (V)	Mag cent ★	Alt Name	Notes
15 47 41.3	−61 13 06	NGC 5979	18	11.8	11.5	15.3	PK 322−5.1	Mag 10.10v star S 2′.6.
15 04 08.7	−60 53 20	PN G318.3−2.0	37	14.4		16.9	PK 318−2.1	Mag 8.43v star N 4′.3.
15 06 00.7	−61 21 24	PN G318.3−2.5	51	15.0			PK 318−2.2	Mag 11.9 star N 2′.0.
15 08 42.7	−61 44 03	PN G318.4−3.0	20	17.4			PK 318−3.1	Mag 9.60v star S 1′.2; mag 11.8 star N 1′.1.
15 05 59.6	−55 59 19	PN G320.9+2.0	5	14.9		17.9	PK 321+2.2	Mag 10.95v star N 1′.5.
15 30 18.5	−61 01 39	PN G321.0−3.8	12	18.1			PK 321−3.1	Mag 11.1 star S 0′.7.
14 59 53.7	−54 18 06	PN G321.0+3.9		15.8		12.2	PK 321+3.1	Stellar, appears to have several very faint stars on surrounding edges.
15 05 17.0	−55 11 13	PN G321.3+2.8	3	14.9		16.2	PK 321+2.1	Mag 13.2 star N 1′.3; mag 12.8 star W 1′.5.
15 11 56.5	−55 39 50	PN G321.8+1.9	41	14.5		21.2	PK 321+1.1	Elongated N-S; mag 9.53v star NE 3′.1.
15 52 09.6	−62 30 52	PN G322.1−6.6	10	14.5	12.5	17.2	PK 322−6.1	Mag 11.19v star NE 1′.6; mag 11.3 star SW 1′.4.
15 23 42.1	−57 09 17	PN G322.4−0.1	2	17.8			PK 322−0.1	Mag 10.8 star E 2′.3.
15 34 16.7	−59 09 09	PN G322.4−2.6	26	12.5	12.1		PK 322−2.1	Mag 12.4 star E 1′.7.
15 38 01.2	−58 44 43	PN G323.1−2.5	18	14.8		17.0	PK 323−2.1	Mag 11.0 star N 0′.9; mag 10.06v star E 2′.4.
15 22 19.6	−54 08 16	PN G323.9+2.4	5	15.0		16.8	PK 323+2.1	Close pair of mag 13 stars N 1′.0.
15 19 08.8	−53 09 50	PN G324.0+3.5	27		16.0	16.2	PK 324+3.1	Mag 9.03v star S 1′.8.
15 23 35.1	−53 51 23	PN G324.2+2.5	3			17.0	PK 324+2.1	Mag 11.4 star S 1′.2; mag 11.18v star E 1′.8.
15 41 57.4	−56 36 45	PN G324.8−1.1	4	16.5			PK 324−1.1	Mag 13.2 star SSW 0′.5.
15 25 32.7	−52 50 41	PN G325.0+3.2	2	14.9			PK 325+3.1	Small triangle of mag 12 stars centered SW 1′.6.
15 59 09.1	−58 23 51	PN G325.4−4.0	14	12.3	12.2	13.9	PK 325−4.1	Close pair of mag 11.5, 11.9 stars E 0′.9.
15 25 08.1	−51 19 42	PN G325.8+4.5	5	14.4		15.9	PK 325+4.1	Close pair of mag 11-12 stars NNW 0′.7; mag 8.49v star NE 2′.5; mag 6.96v star E 4′.4.
16 15 42.4	−59 54 01	PN G326.0−6.5	3			13.0	PK 326−6.1	mag 11.1 star N 1′.1.
15 52 58.5	−56 23 54	PN G326.1−1.9	12	16.9			PK 326−1.2	Mag 12.3 star SW 0′.6; mag 12.0 star E 2′.6.
15 58 08.3	−55 41 51	PN G327.1−1.8	3	17.1		17.2	PK 327−1.2	Mag 11.9 star ESE 0′.6.
15 59 58.1	−55 55 25	PN G327.1−2.2	4	15.9		15.1	PK 327−2.1	Faint star on S edge; mag 11.3 star NNE 1′.6.
16 19 15.7	−57 58 24	PN G327.7−5.4	14			19.3	PK 327−5.1	Mag 12.2 star N 0′.3.
16 00 59.5	−55 05 41	PN G327.8−1.6	5	15.2			PK 327−1.1	Mag 11.4 star WSW 4′.1.
16 23 30.6	−58 19 23	PN G327.8−6.1	2	14.5		15.4	PK 327−6.1	Mag 13.7 star E 0′.3; mag 10.38v star E 2′.3.
16 29 31.4	−59 09 27	PN G327.8−7.2	20	13.4			PK 327−7.1	Mag 12.6 star NE 0′.7.
16 14 00.7	−56 59 31	PN G327.9−4.3		16.8		16.9	PK 327−4.1	Mag 11.6 star ENE 1′.6; mag 11.7 star SE 1′.8.
15 49 32.5	−52 30 31	PN G328.2+1.3	12				PK 328+1.1	Mag 10.60v star NNW 0′.9.
16 10 40.8	−54 57 31	PN G328.9−2.4	22	16.7			PK 328−2.1	Mag 8.72v star SW 2′.7.
15 51 41.2	−51 31 23	PN G329.0+1.9	72	13.6	12.6	14.0	PK 329+2.1	Annular; located on the edge of a small triangle of mag 11-12 stars.
16 14 32.0	−54 57 03	PN G329.3−2.8	23	12.6			PK 329−2.2	Mag 13.7 star on E edge.
16 14 24.5	−54 47 39	PN G329.4−2.7	3	15.1		16.1	PK 329−2.1	Mag 12.8 star SW 0′.8.
15 55 08.7	−51 23 46	PN G329.5+1.7	100			13.5	Wray 16-191	Mag 9.56v star S 4′.0.
16 12 34.4	−54 23 35	PN G329.5−2.2	24	16.6		15.7	PK 330−2.3	Located between a close pair of mag 11-12 stars.
16 17 14.7	−53 32 06	PN G330.6−2.1	13	14.3			PK 330−2.1	Mag 10.77v star on NE edge.
16 24 21.3	−54 36 03	PN G330.6−3.6	10	13.8			PK 330−3.1	Located 0′.4 S of mag 10.9 star.
16 22 13.8	−53 40 58	PN G331.0−2.7	3	16.1		17.6	PK 331−2.1	Mag 12.3 star on N edge.
16 37 41.4	−55 42 23	PN G331.1−5.7	5	11.9		12.6	PK 331−5.1	Mag 11.6 star on E edge.
16 08 59.3	−51 02 01	PN G331.4+0.5	11	14.9			PK 331+0.1	Mag 11.5 star N 3′.2.
16 27 50.9	−54 01 30	PN G331.4−3.5	5	16.5		13.4	PK 331−3.1	Mag 11.3 star NW 0′.9; mag 11.9 star WSW 0′.9.
16 24 37.8	−53 22 37	PN G331.5−2.7	10	14.8			PK 331−2.2	Mag 11.16v star NW 1′.7.
16 29 59.5	−54 09 36	PN G331.5−3.9	50	13.8			PK 331−3.2	Mag 11.8 star S 2′.2; mag 11.8 star SE 1′.7.
16 17 13.4	−51 59 15	PN G331.7−1.0	25			17.6	PK 331−1.1	**Ant Nebula**. Mag 12.4 star S 0′.7.
16 29 53.3	−53 23 15	PN G332.0−3.3	16	13.9			PK 332−3.1	Mag 10.24v star SW 2′.4; mag 12.4 star N 2′.4.
16 35 21.2	−53 50 10	PN G332.3−4.2	5	13.5		16.4	PK 332−4.1	Mag 12.4 star S 2′.1; mag 11.4 star N 2′.8.
16 39 37.4	−52 49 13	PN G333.4−4.0	42	16.1			PK 333−4.1	Pair of mag 13.0, 12.8 stars SE 0′.8; mag 10.67v star NW 3′.3.
17 03 03.0	−53 55 49	PN G334.8−7.4		14.0		11.5	PK 334−7.1	Stellar; mag 11.4 star WNW 4′.9.
16 45 00.0	−51 12 18	PN G335.2−3.6	23	16.6		17.0	PK 335−3.1	Mag 13.4 star E 0′.5; mag 11.16v star NW 2′.6.
16 48 40.1	−51 09 13	PN G335.6−4.0	24			20.3	PK 335−4.1	Mag 13.5 star on S edge.
16 59 36.1	−51 42 04	PN G336.3−5.6	3	13.4		16.6	PK 336−5.1	Mag 12.9 star NNW 1′.2; mag 11.29v star SE 2′.6.

97

GALAXIES

197

RA h m s	Dec ° ′ ″	Name	Mag (V)	Dim ′ Maj x min	SB	Type Class	PA	Notes
2 57 40.6	−53 23 34	ESO 172-12	14.7	1.3 x 0.7	14.5	SAB(rs)cd III-IV	3	Mag 11.2 star N 3′.0.
3 54 33.8	−53 18 41	ESO 174-2	15.6	1.5 x 0.2	14.1	SBm? sp V	62	Mag 5.89v star SW 8′.5.
3 57 34.8	−52 46 21	ESO 174-3	14.0	1.5 x 1.1	14.4	SB(s)cd III	153	Close pair of mag 12.0, 10.37v stars on N edge.

(Continued from previous page)

GALAXIES

RA h m s	Dec ° ′ ″	Name	Mag (V)	Dim ′ Maj x min	SB	Type Class	PA	Notes
14 06 40.7	−55 21 30	ESO 175-1	12.6	1.9 x 1.3	13.4	(R′)SAB(rs)a:	169	Numerous stars superimposed.
14 17 47.0	−52 49 49	ESO 175-5	15.1	1.5 x 0.2	13.6	S	83	Mag 10.31v star N 3′.6.
13 09 12.5	−51 00 47	ESO 219-30	14.4	1.7 x 0.6	14.3	IB(s)m V-VI	49	Mag 9.30v star E 9′.4.
13 10 26.0	−51 37 22	ESO 219-37	14.0	1.5 x 1.4	14.7	SAB(r)dm IV-V	80	Mag 7.71v star WSW 10′.8.
13 22 51.3	−52 44 43	ESO 220-6	12.7	0.9 x 0.9	12.4			Almost stellar nucleus; mag 11.0 star E 1′.7.
13 40 13.1	−51 08 34	ESO 220-28	13.7	2.0 x 0.3	13.0	Sbc sp	81	Mag 8.37v star SW 2′.8.
13 44 23.8	−52 22 14	ESO 220-35	13.7	1.6 x 0.4	13.1	SB(s)m V	170	Mag 10.10v star N 1′.1.
13 50 28.6	−52 39 38	ESO 221-7	13.3	1.6 x 1.0	13.6	SA(rs)c II-III	120	Mag 12.2 star N edge; mag 5.26v star SE 17′.2.
14 09 02.3	−51 10 12	ESO 221-28	14.5	0.7 x 0.7	13.6	S		Mag 9.68v star NW 7′.7.
14 16 04.4	−52 36 33	ESO 221-35	13.5	1.5 x 0.3	12.5	SA0/a: sp	29	Mag 12.3 star on S edge; mag 10.43v star W 1′.7.
13 09 34.5	−51 58 06	IC 4200	12.7	1.4 x 0.9	12.8	(R′)SA(r)0°:	152	Mag 9.12v star SE 3′.9.
13 40 08.2	−51 02 14	IC 4311	13.4	1.0 x 0.7	12.8	S?	148	IC 4312 SE 4′.2; mag 9.79v star NNE 3′.4.
13 40 30.9	−51 04 17	IC 4312	12.3	0.9 x 0.7	11.7	SAB(r)0°:	30	Mag 12.0 star N 1′.4; IC 4311 NW 4′.2.

OPEN CLUSTERS

RA h m s	Dec ° ′ ″	Name	Mag	Diam ′	No. ★	B ★	Type	Notes
13 27 49.0	−62 18 42	Bas 18	8.2	3	154	11.0	cl	Few stars; small brightness range; no central concentration; detached. (S) Outline of cluster uncertain.
13 30 30.0	−61 16 00	Cr 272	7.7	10	102	10.5	cl	Moderately rich in stars; moderate brightness range; no central concentration; detached. (S) Sparse group.
13 35 24.0	−60 11 00	Cr 275	10.2	11	20		cl	Few stars; large brightness range; no central concentration; detached.
13 12 26.8	−62 42 01	Danks 1		1	25		cl	
13 12 53.8	−62 40 48	Danks 2		0.6	80		cl	
14 44 48.0	−59 10 00	ESO 134-12		4			cl	
14 18 30.0	−56 55 00	ESO 175-6		18			cl	
13 29 11.0	−61 12 00	Ho 16	8.4	7	20	10.1	cl	Few stars; moderate brightness range; slight central concentration; detached.
14 33 50.4	−61 21 51	Ho 17	8.3	4	41	9.6	cl	Few stars; moderate brightness range; not well detached. Without brightest star, visual mag = 8.7.
14 50 41.4	−52 15 51	Ho 18	8.0	5	34	8.8	cl	Few stars; moderate brightness range; no central concentration; detached.
13 36 55.0	−62 05 36	IC 4291	9.7	4	35	10.0	cl	
13 24 54.0	−62 25 00	Lo 807	7.9	20	35		cl	
13 23 39.0	−59 44 00	Lo 821		26	26	8.7	cl	
13 33 44.0	−59 16 00	Lo 915		30	36	9.1	cl	(A) Nice NE-SW group of faint stars.
13 45 24.0	−62 00 00	Lo 991		5	20		cl	
13 45 24.0	−60 15 00	Lo 1010		30	35		cl	(A) Large (30′) group of bright stars on rich Milky Way background.
13 53 30.0	−59 44 00	Lo 1095		20	10		cl	
13 58 40.6	−61 44 00	Lo 1101		16	30	10.1	cl	(A) Seven brighter stars with more in background.
13 59 30.0	−59 22 00	Lo 1152		9	10		cl	
13 59 31.4	−58 22 47	Lo 1171		6.8	35		cl	
14 05 48.0	−59 42 00	Lo 1194		14	10		cl	(A) DSS image shows mostly scattering of stars, with two bright stars on E side.
14 04 24.0	−58 41 00	Lo 1202		20	30		cl	
14 09 12.0	−59 42 00	Lo 1225		20	30		cl	(A) DSS image shows triangle of stars 3′ E of brighter one.
14 18 12.0	−61 25 00	Lo 1256		12	30		cl	(A) NE to SW chain of stars.
14 15 54.0	−59 08 00	Lo 1282		15	30		cl	
14 16 56.9	−57 51 00	Lo 1289		14	15	10.4	cl	
14 33 30.0	−61 59 00	Lo 1339		30	30	10.9	cl	(A) DSS image shows scattering of stars with small triangle at given position.
14 27 25.0	−58 06 00	Lo 1378		30	25	8.9	cl	
14 44 42.0	−61 41 00	Lo 1409		30	35	9.6	cl	(A) E-W chain of stars with one bright star to N and S.
14 00 02.0	−62 09 30	Lynga 1		3	15		cl	
14 24 24.0	−61 21 00	Lynga 2	6.4	10	104	7.7	cl	Moderately rich in bright and faint stars; detached.
13 00 05.0	−59 37 00	NGC 4852	8.9	12	60		cl	Rich in stars; large brightness range; strong central concentration; detached.
13 16 40.0	−60 03 00	NGC 5043		10	18	7.0	ast?	
13 27 16.0	−59 03 00	NGC 5138	7.6	8	92	10.3	cl	Moderately rich in stars; moderate brightness range; slight central concentration; detached. (S) Very rich background.
13 31 08.0	−60 56 24	NGC 5168	9.1	4	50	10.4	cl	Moderately rich in stars; moderate brightness range; strong central concentration; detached.
13 44 45.0	−62 55 00	NGC 5269		3		12.0	MWcl	(S) Sparse group at this position.
13 46 36.7	−62 55 00	NGC 5281	5.9	8	40	8.0	cl	Moderately rich in bright and faint stars with a strong central concentration; detached. Without brightest star, visual mag = 6.8.
13 54 00.0	−61 51 00	NGC 5316	6.0	15	129	11.0	cl	Rich in stars; moderate brightness range; slight central concentration; detached.
14 00 36.0	−59 33 00	NGC 5381		11	50	12.0	cl:	Moderately rich in stars; moderate brightness range; slight central concentration; detached. Probably not a cluster.
14 25 54.0	−54 47 00	NGC 5593		6	20		cl	Few stars; moderate brightness range; no central concentration; detached.
14 27 48.0	−59 38 00	NGC 5606	7.7	3	15	7.9	cl	Few stars; large brightness range; strong central concentration; detached.
14 29 42.0	−60 42 00	NGC 5617	6.3	10	292	10.0	cl	Rich in stars; large brightness range; strong central concentration; detached.
14 35 30.8	−56 40 00	NGC 5662	5.5	30	280	10.0	cl	Rich in stars; large brightness range; slight central concentration; detached.
14 43 39.0	−57 34 00	NGC 5715	9.8	7	30	11.0	cl	Moderately rich in stars; moderate brightness range; no central concentration; detached.
14 48 48.8	−54 30 00	NGC 5749	8.8	10	30		cl	Moderately rich in stars; moderate brightness range; slight central concentration; detached.
14 53 31.8	−52 40 16	NGC 5764	12.6	3	12		cl	Few stars; small brightness range; no central concentration; detached.
14 30 43.0	−60 53 06	Pi 19		3	60	12.0	cl	Moderately rich in stars; small brightness range; strong central concentration; detached.
13 32 12.0	−58 28 00	Ru 108	7.5	10	15	9.0	cl	Few stars; large brightness range; no central concentration; detached.
14 35 48.0	−59 56 00	Ru 111		8	40	13.0	cl:	(A) E-S scattering of stars, barely rising above Milky Way background.
14 57 00.0	−62 32 00	Ru 112		10	80	14.0	cl	(A) Rich group (Milky Way patch?) of faint stars.
14 18 12.0	−58 57 00	Ru 167		15	40	11.0	cl	Moderately rich in stars; moderate brightness range; no central concentration; detached.
13 18 38.4	−62 32 00	St 16	6.9	20	83	6.9	cl	Few stars; large brightness range; no central concentration; detached; involved in nebulosity.
13 32 14.0	−62 47 18	Tr 21	7.7	5	20	8.9	cl	Few stars; moderate brightness range; strong central concentration; detached.
14 31 12.0	−61 09 00	Tr 22	7.9	10	68	12.0	cl:	Moderately rich in stars; moderate brightness range; no central concentration; detached. Probably not a cluster. (A) Includes westward opening semi-circle of 7 stars (e.g. a "stellar-ring"?).
13 40 12.0	−61 43 48	vdB-Ha 151		5	30		cl	

STAR CLOUDS

RA h m s	Dec ° ′ ″	Name	Mag	Diam ′	No. ★	B ★	Type	Notes
13 47 06.0	−59 13 39	NGC 5284		20		10.6	MW	

(Continued from previous page)

197 GLOBULAR CLUSTERS 197

RA h m s	Dec ° ′ ″	Name	Total V m	B ★ V m	HB V m	Diam ′	Conc. Class Low = 12 High = 1	Notes
13 46 26.5	−51 22 24	NGC 5286	7.4	13.5	16.5	11.0	5	

PLANETARY NEBULAE

RA h m s	Dec ° ′ ″	Name	Diam ″	Mag (P)	Mag (V)	Mag cent ★	Alt Name	Notes
13 51 03.4	−51 12 24	NGC 5307	18	12.1	11.2	14.6	PK 312+10.1	Pair of mag 13-14 stars SE.
13 00 41.2	−56 53 42	PN G304.2+5.9	36	13.4			PK 304+5.2	Close pair of mag 11.7, 12.0 stars N 2'.1.
13 05 48.2	−57 39 28	PN G304.8+5.1	10	14.3		15.3	PK 304+5.1	Mag 8.57v star E 1'.4.
13 09 36.4	−61 19 38	PN G305.1+1.4	10	13.7		15.6	PK 305+1.1	Mag 11.8 star W 2'.6; mag 12.1 star N 1'.6.
13 24 22.1	−57 31 24	PN G307.3+5.0	18	14.3			PK 307+5.1	Tight group of three mag 10-11 stars W 2'.3.
13 28 05.1	−54 42 00	PN G308.2+7.7	18	16.2		17.0	PK 308+7.1	Mag 10.80v star SW 2'.2.
13 42 36.3	−61 22 30	PN G309.0+0.8	4	14.4			PK 309+0.1	Close chain of four N-S mag 13-14 stars N 1'.1.
13 43 59.8	−60 49 47	PN G309.2+1.3	96	16.6			PK 309+1.1	Elongated NNE-SSW, brighter on E edge; mag 11.21v star on S edge.
13 53 23.2	−60 33 51	PN G310.4+1.3		19.0			Vo 4	Mag 13.0 star E 0'.4.
13 55 43.7	−59 22 43	PN G311.0+2.4	112			12.2	PK 311+2.2	Elongated SE-NW; mag 12.0 star at center; mag 7.85v star on E edge.
13 54 56.0	−58 27 18	PN G311.1+3.4		18.3			PK 311+3.1	Mag 13.0 star WNW 1'.8; mag 12.2 star NE 2'.5.
13 58 13.9	−58 54 33	PN G311.4+2.8	9	13.5		17.2	PK 311+2.1	Mag 11.7 star SSW 1'.0.
14 33 18.3	−60 49 38	PN G315.0−0.3	32	13.2	14.::	16.7	PK 315−0.1	Mag 12.7 star on W edge; mag 11.6 star S 1'.2; mag 10.37v star SE 3'.0; mag 7.76v star NE 4'.6.
14 20 49.0	−55 28 02	PN G315.4+5.2	7	15.0		18.9	PK 315+5.1	Mag 12.2 star NE 1'.7; mag 10.43v star NW 1'.3.
14 11 52.1	−51 26 23	PN G315.4+9.4	5	14.0		17.9	PK 315+9.1	Mag 13.3 star S 0'.8.
14 41 17.6	−61 19 55	PN G315.7−1.2	28				PK 315−1.1	Mag 13.2 star W 1'.2; mag 11.4 star ESE 2'.7.
14 21 59.9	−55 02 17	PN G315.7+5.5	19	16.8			PK 315+5.2	Mag 10.20v star N 2'.1.
14 18 09.0	−52 10 40	PN G316.1+8.4	16	13.7	13.9	12.7	PK 316+8.1	Mag 12.1 star S 0'.4.
14 46 21.3	−61 13 47	PN G316.3−1.3	54				PK 316−1.1	Pair of mag 9.96v, 9.25v stars NW 1'.7.
14 41 35.9	−56 15 15	PN G317.8+3.3	50	16.8			PK 317+3.1	Annular; mag 9.36v star NW 2'.9.
14 40 31.0	−52 34 59	PN G319.2+6.8	15	13.2		18.1	PK 319+6.1	Mag 12.5 star S 2'.1; mag 11.2 star WSW 4'.6.
14 59 53.7	−54 18 06	PN G321.0+3.9		15.8		12.2	PK 321+3.1	Stellar, appears to have several very faint stars on surrounding edges.

198 GALAXIES 198

RA h m s	Dec ° ′ ″	Name	Mag (V)	Dim ′ Maj x min	SB	Type Class	PA	Notes
12 22 38.7	−58 37 02	ESO 130-12	14.8	1.2 x 0.3	13.6	SB(r)0+:	147	
11 26 08.5	−54 14 07	ESO 170-3	13.7	1.5 x 0.4	13.0	(R)SAB(r)ab?	127	ESO 170-4 E 4'.8.
11 26 41.2	−54 15 24	ESO 170-4	15.1	1.3 x 0.3	13.9	Sc? pec sp	89	ESO 170-3 W 4'.8.
11 59 09.4	−53 24 37	ESO 171-4	12.5	1.6 x 1.0	12.9	SAB(r)0°?	60	Mag 8.71v star SW 5'.1.
12 01 36.1	−53 21 30	ESO 171-5	13.5	1.1 x 0.6	12.9	SAB(rs)a: pec	27	Numerous stars superimposed, in star rich area; mag 10.55v star NE 2'.7.
12 04 46.1	−53 10 30	ESO 171-8	13.2	1.3 x 0.9	13.2	SA0− pec:	13	Diffuse core, stars superimposed on envelope.
12 10 48.3	−52 51 25	ESO 171-13	14.8	1.5 x 0.2	13.4	Sd: sp	120	Mag 12 star on S edge.
12 47 33.1	−53 56 55	ESO 172-9	13.5	1.2 x 0.9	13.4	IB(s)m pec	103	Mag 10.60v star SW 5'.1.
12 57 40.6	−53 23 34	ESO 172-12	14.7	1.3 x 0.7	14.5	SAB(rs)cd III-IV	3	Mag 11.2 star N 3'.0.
11 19 28.5	−52 22 19	ESO 216-5	13.4	1.5 x 0.7	13.3	(R′)SAB(rs)a:	100	Mag 7.33v star S 7'.5.
11 44 55.6	−51 54 22	ESO 216-38	14.1	1.4 x 0.7	13.9	SB(r)ab	161	Mag 10.04v star N 7'.2.
11 58 03.0	−51 53 07	ESO 217-15	13.0	1.1 x 0.8	12.7	SB0?	119	Mag 14.5 star on N edge; mag 10.01v star E 10'.1.
11 58 17.6	−51 08 06	ESO 217-16	13.3	1.3 x 0.7	13.1	SAB(r)b pec: II	155	Mag 12.4 star SE edge; mag 10.24v star NW edge.
12 12 23.6	−52 27 44	ESO 217-30	13.5	1.2 x 1.0	13.6	SA(r)bc: III	69	Numerous stars superimposed, in star rich area.
12 13 50.1	−52 18 53	ESO 217-31	14.3	1.3 x 1.1	14.5	IB(s)m V-VI	39	Numerous stars involved.
12 21 08.8	−52 35 10	ESO 218-2	12.7	1.1 x 1.1	12.9	E2:		Several stars superimposed E of center; mag 11.07v star N 6'.1.
12 31 53.8	−51 44 53	ESO 218-8	15.0	2.5 x 0.2	14.1	SB?	31	Stellar nucleus, extremely faint arms.
11 46 05.8	−56 23 17	NGC 3882	12.0	2.3 x 1.3	13.0	SB(s)bc II-III	126	Numerous stars superimposed overall. Mag 8.82v star E 2'.4.
12 02 47.7	−53 50 11	PGC 38038	13.4	0.8 x 0.5	12.3	SB0°? pec	126	Mag 9.87v star NE 4'.6.

OPEN CLUSTERS

RA h m s	Dec ° ′ ″	Name	Mag	Diam ′	No. ★	B ★	Type	Notes
11 10 32.9	−59 02 00	Bas 17	7.7	10	20	9.6	cl	(A) Also known as **Lo 282**. (S) Sparse group elongated SE-NW.
10 57 18.0	−61 44 00	Bo 12	9.7	10	20	11.4	cl	Few stars; large brightness range; no central concentration; detached.
10 59 38.1	−60 59 00	Cr 236	7.7	7	20		cl	Few stars; moderate brightness range; no central concentration; detached.
11 11 40.2	−60 19 00	Cr 240	3.9	32	42	4.6	ast	Moderately rich in stars; moderate brightness range; no central concentration; detached; involved in nebulosity. Collinder's 32′ by 23′ scattering of stars. Includes NGC 3572, Hogg 10, Hogg 11.
11 21 12.0	−58 14 00	ESO 129-19		5			cl	
11 44 12.0	−61 05 00	ESO 129-32		4			cl	
12 22 49.6	−59 38 00	ESO 130-13		10			cl	
12 29 36.0	−57 53 00	ESO 131-9		5			cl	
11 06 00.0	−59 48 00	Fei 1	4.7	25	40	6.6	cl:	Moderately rich in stars; large brightness range; not well detached. (A) A large scattering of stars.
12 27 16.8	−60 46 44	Ha 5	7.1	5	25	9.0	cl	Few stars; large brightness range; slight central concentration; detached. (A) Ha 5 is not Cr 257 which lies 13'.1 E.
11 10 42.1	−60 23 04	Ho 10	6.9	3	23	7.1	cl	(A) Part of Cr 240.
11 11 32.6	−60 22 38	Ho 11	8.1	2	10	8.3	cl	(A) Part of Cr 240.
11 12 15.0	−60 46 12	Ho 12	8.8	2	11		cl	(S) Small group.
12 28 39.0	−59 48 00	Ho 14	9.5	3	11	10.5	cl	Few stars; large brightness range; slight central concentration; detached. Probably not a cluster. (S) Small group.
11 17 18.0	−62 43 00	IC 2714	8.2	15	100	10.0	cl	Rich in stars; moderate brightness range; slight central concentration; detached. (S) Weak concentration.
11 06 00.7	−61 11 00	Lo 306		23	27	7.6	cl	(A) A scattering of bright and faint stars.
11 30 52.0	−58 29 00	Lo 372		6	15		cl	(S) Sparse group.
11 32 49.0	−60 43 00	Lo 402		25	35	7.2	cl	(A) Mostly rich Milky Way field.
11 51 06.6	−61 15 00	Lo 481		28	29	9.3	cl	(A) Includes star semi-circle (opening N).
12 08 24.4	−60 51 00	Lo 565		11	10	10.0	cl	
12 28 25.7	−61 48 00	Lo 624		40	45	6.2	cl	(A) Center has 5 stars in NNW-SSE line, with chains of fainter stars to SW and NE, over 40′ field.
12 47 19.0	−60 38 00	Lo 682		28	31	8.3	cl	(A) Mostly a NW-SE group of faint stars.
12 53 31.0	−60 50 00	Lo 694		16	31	8.7	cl	
10 59 34.0	−60 20 00	NGC 3496	8.2	7	110	11.8	cl	Moderately rich in stars; small brightness range; slight central concentration; detached.
11 01 18.0	−59 50 48	NGC 3503	9.4	0.6	9	9.0	cl	Few stars; moderate brightness range; no central concentration; detached.
11 04 03.0	−61 22 00	NGC 3519	7.7	8	93	11.0	cl	
11 05 30.0	−58 44 00	NGC 3532	3.0	50	677	8.0	cl	Rich in stars; large brightness range; slight central concentration; detached.
11 10 26.6	−60 15 00	NGC 3572	6.6	7	35	7.0	cl	Moderately rich in bright and faint stars; detached; involved in nebulosity.
11 13 00.0	−60 47 18	NGC 3590	8.2	2	30	10.3	cl	Few stars; moderate brightness range; strong central concentration; detached.

(Continued from previous page)

OPEN CLUSTERS

RA h m s	Dec ° ′ ″	Name	Mag	Diam ′	No. ★	B ★	Type	Notes
11 15 07.0	−61 15 42	NGC 3603	9.1	4	44	11.3	cl	Moderately rich in bright and faint stars; detached; involved in nebulosity.
11 36 15.0	−61 37 00	NGC 3766	5.3	15	137	8.0	cl	Rich in stars; large brightness range; strong central concentration; detached.
11 50 34.0	−55 41 00	NGC 3960	8.3	7	317	11.5	cl	Moderately rich in stars; moderate brightness range; strong central concentration; detached.
12 06 40.0	−61 15 00	NGC 4103	7.4	6	45	10.0	cl	Moderately rich in stars; moderate brightness range; strong central concentration; detached.
12 13 38.5	−62 42 20	NGC 4184		2	20	13.0	cl	(S) = **Ru 102**.
12 17 16.0	−55 05 12	NGC 4230	9.4	5	15		cl:	Few stars; moderate brightness range; not well detached. (S) Sparse group around very bright star.
12 24 04.0	−58 07 24	NGC 4337	8.9	3.5	16	10.5	cl	Few stars; large brightness range; slight central concentration; detached. Probably not a cluster.
12 24 07.0	−61 52 12	NGC 4349	7.4	4	204	11.0	cl	Moderately rich in stars; moderate brightness range; slight central concentration; detached.
12 28 27.0	−60 06 12	NGC 4439	8.4	4	20	10.3	cl	Few stars; small brightness range; slight central concentration; detached.
12 42 20.0	−62 59 36	NGC 4609	6.9	6	52	10.0	cl	Moderately rich in stars; moderate brightness range; slight central concentration; detached.
12 53 38.0	−60 21 00	NGC 4755	4.2	10	218	7.0	cl	**Jewel Box**. Rich in stars; large brightness range; strong central concentration; detached.
13 00 05.0	−59 37 00	NGC 4852	8.9	12	60		cl	Rich in stars; large brightness range; strong central concentration; detached.
11 43 40.0	−61 08 00	Ru 95		4	30	12.0	cl	
11 50 53.0	−62 08 18	Ru 96		5	15	13.0	cl	
11 57 23.0	−62 42 18	Ru 97	9.1	5	142	12.0	cl	Few stars; small brightness range; not well detached.
12 05 42.0	−62 31 00	Ru 100		7	30	12.0	cl	
12 09 40.0	−63 00 00	Ru 101		15	60	13.0	cl	(A) Appears as obvious increase in fainter stars in rich field.
12 15 54.0	−58 23 00	Ru 103		3	20	13.0	cl	
12 24 51.0	−60 25 54	Ru 104		3.6		13.0	cl	
12 34 00.0	−61 32 00	Ru 105		12		12.0	cl	Few stars; moderate brightness range; no central concentration; detached.
11 31 06.0	−60 47 00	Ru 164		6	40	14.0	cl	(S) Identification uncertain. Some small (~1′) clumps nearby.
12 28 36.0	−56 26 00	Ru 165		20	35	7.0	cl:	Moderately rich in bright and faint stars; detached. Probably not a cluster. (S) Large group of bright stars.
11 01 03.0	−60 14 00	Sh 1	8.8	1	11		cl	
11 13 06.0	−58 54 00	St 13	7.0	5	23	10.0	cl	Few stars; large brightness range; strong central concentration; detached; involved in nebulosity.
11 43 51.0	−62 32 00	St 14	6.3	8	20	10.0	cl	Few stars; large brightness range; no central concentration; detached. (S) Sparse group.
12 10 30.0	−59 29 00	St 15		12		10.0	cl	Few stars; small brightness range; not well detached.
10 56 24.0	−59 12 18	Tr 17	8.4	5	44	10.3	cl	Moderately rich in stars; moderate brightness range; slight central concentration; detached.
11 11 24.0	−60 39 00	Tr 18	6.9	6	30	8.1	cl	Moderately rich in bright and faint stars; detached.
11 14 18.0	−57 34 00	Tr 19	9.6	10	40		cl	Moderately rich in stars; moderate brightness range; not well detached.
12 39 32.0	−60 38 00	Tr 20	10.1	7			cl	Moderately rich in stars; moderate brightness range; slight central concentration; detached.
11 07 18.0	−61 27 00	vdB-Ha 110		2	20		cl	
11 22 54.0	−58 31 00	vdB-Ha 118		2	20		cl	

GLOBULAR CLUSTERS

RA h m s	Dec ° ′ ″	Name	Total V m	B ★ V m	HB V m	Diam ′	Conc. Class Low = 12 High = 1	Notes
12 38 40.2	-51 09 01	Ru 106	10.9	14.8	17.8	2.0		

BRIGHT NEBULAE

RA h m s	Dec ° ′ ″	Name	Dim ′ Maj x min	Type	BC	Color	Notes
12 44 46.1	−54 31 10	Boomerang Nebula	1.4 x 0.6				
11 28 54.0	−62 41 00	Gum 39	20 x 15	E	3-5	3-4	= **RCW 60** North. Irregular in shape and slightly extended E-W. The nebula is rather evenly illuminated except for a slightly brighter narrow strip along the NW perimeter where it is best defined.
11 28 54.0	−62 56 00	IC 2872	15 x 6	E	3-5	3-4	= **RCW 60** South. Extended NNE-SSW and somewhat mottled being brighter in the S part; several stars involved.
11 01 18.0	−59 51 00	NGC 3503	3 x 3	E+R	3-5	2-4	Roundish nebula with three 10th mag stars involved and several others of comparable brightness in the nearby surrounding field. The field is very rich in faint stars, except a region to the W and NW.
11 11 30.0	−61 22 00	NGC 3576	3 x 3	E	2-5		A bright knot; westernmost of six; part of the RCW 57 complex.
11 11 54.0	−61 14 00	NGC 3579	20 x 15	E	2-5		A knot; second in a group of six; part of the RCW 57 complex.
11 12 06.0	−61 18 00	NGC 3581		E	2-5		A fan-shaped nebulosity with a 12th mag star involved; third in a group of six; part of the RCW 57 complex.
11 12 06.0	−61 17 00	NGC 3582		E	2-5		This is the brightest in a group of six small nebulae. The group is somewhat wedge-shaped and about 8′ x 5′ in extent, being essentially a single nebula divided by dark lanes; part of the RCW 57 complex.
11 12 18.0	−61 13 00	NGC 3584		E	2-5		Fifth nebula in a group of six; part of the RCW 57 complex.
11 12 36.0	−61 21 00	NGC 3586		E	2-5		Elongated roughly E-W; easternmost of six; part of the RCW 57 complex.
11 12 00.0	−61 12 00	RCW 57	50 x 25	E	2-5		A large and very irregular nebular complex consisting of two main parts catalogued as RCW 57 West and RCW 57 East, each part with many individual condensations.

DARK NEBULAE

RA h m s	Dec ° ′ ″	Name	Dim ′ Maj x min	Opacity	Notes
12 53 00.0	−63 00 00	Coalsack	55 x 55		A dark patch on the Milky Way. It appears completely black to the naked eye, but is not uniformly dark.

PLANETARY NEBULAE

RA h m s	Dec ° ′ ″	Name	Diam ″	Mag (P)	Mag (V)	Mag cent ★	Alt Name	Notes
11 27 57.1	−59 57 28	NGC 3699	45	11.0	11.3		PK 292+1.1	Appears to be in two parts with dark, narrow band NE-SW through center. .
11 50 18.1	−57 10 58	NGC 3918	23	8.4	8.1	15.6	PK 294+4.1	**Blue Planetary**. Bright central star; mag 10.52v star E 4′.2.
11 18 09.7	−52 10 06	PN G288.7+8.1	36	15.5		15.6	PK 288+8.1	Mag 13.8 star SW 0′.7; mag 12.2 star SW 1′.2.
10 56 02.5	−61 28 02	PN G289.6−1.6	20	16.5			PK 289−1.1	Mag 13.4 star on W edge; mag 12.4 star SW 1′.2.
11 24 01.2	−52 51 20	PN G289.8+7.7	3	14.6			PK 289+7.1	Mag 8.24v star WNW 1′.7.
11 03 55.6	−60 36 05	PN G290.1−0.4	19	15.0			PK 290−0.1	Mag 11.6 star E 1′.6; mag 8.66v star NW 1′.8.
11 28 36.4	−52 56 04	PN G290.5+7.9	49	11.6	11.4		PK 290+7.1	Triangle of mag 10-12 stars W 4′.6.
11 27 24.2	−57 18 02	PN G291.7+3.7	8	17.4			PK 291+3.1	Mag 11.6 star WNW 1′.2; mag 10.21v star SW 2′.1.
11 33 18.0	−57 06 16	PN G292.4+4.1	5	13.0	12.8	13.6	PK 292+4.1	Small triangle of mag 11-12 stars centered E 1′.6.
11 30 48.6	−59 17 03	PN G292.7+1.9	12	16.0			PK 292+1.3	Forms a triangle with mag 11.3 star NE 3′.0 and mag 11.07v star SE 3′.0.
11 28 47.9	−60 06 32	PN G292.8+1.1	5	13.2	13.8		PK 292+1.2	Mag 12.4 star W 1′.8; mag 9.93v star SE 1′.3.
11 35 12.5	−60 16 53	PN G293.6+1.2	35	13.8			PK 293+1.1	Close pair of mag 10.9, 12.0 stars W 1′.4.
11 41 37.8	−62 28 56	PN G294.9−0.6	68	14.4			PK 294−0.1	Chain of five mag 11-12 stars extend E from E edge.
12 09 10.2	−58 42 39	PN G297.4+3.7	3	17.5			PK 297+3.1	Located 0′.6 W of mag 9.39v star.
12 23 53.1	−60 13 13	PN G299.5+2.4	24	15.3			PK 299+2.1	Mag 11.8 star NW 1′.5; mag 11.2 star SE 3′.5.
12 28 45.6	−62 05 34	PN G300.2+0.6	15	19.6			PK 300+0.1	Pair of mag 8.23v, 10.99v stars W 2′.5.
12 45 55.0	−60 20 22	PN G302.2+2.5	12	16.3			PK 302+2.1	Close pair of mag 12 stars with mag 10.16v star N 2′.6; mag 13.1 star ESE 1′.1.
13 00 41.2	−56 53 42	PN G304.2+5.9	36	13.4			PK 304+5.2	Close pair of mag 11.7, 12.0 stars N 2′.1.

GALAXIES

RA h m s	Dec ° ′ ″	Name	Mag (V)	Dim ′ Maj x min	SB	Type Class	PA	Notes
09 17 30.9	−62 53 04	ESO 91-7	13.2	2.1 x 1.1	13.9	SB(s)cd III-IV	176	Mag 8.43v star SW 2′.4.
09 12 37.7	−58 50 35	ESO 126-1	13.5	0.6 x 0.5	12.1		20	Mag 12.1 star N edge, in star rich area.
09 13 30.5	−60 47 22	ESO 126-2	12.8	1.0 x 1.0	12.7	(R′)SB(r)b:		Mag 6.36v star SW 9′.1.
09 14 39.2	−60 26 04	ESO 126-3	12.6	1.5 x 0.8	12.6	SB(s)bc II-III	144	Mag 9.40v star N 3′.0; mag 7.45v star SSW 7′.2.
09 14 43.3	−60 44 46	ESO 126-4	14.1	1.6 x 1.3	14.7	SB(s)d IV-V	13	Mag 9.85v star WNW 6′.2.
09 22 25.1	−61 02 58	ESO 126-10	12.5	1.5 x 0.5	12.1	SB(r)b: II	150	Mag 8.60v star SE 4′.1.
09 26 23.1	−60 36 59	ESO 126-11	13.6	1.3 x 0.6	13.2	SB(s)m V	57	Mag 9.31v star SSE 4′.1.
09 27 31.8	−60 46 13	ESO 126-13	13.2	2.0 x 0.6	13.3	S	103	Numerous stars superimposed; mag 9.31v star NW 8′.6; ESO 126-14 E 6′.9.
09 28 26.4	−60 48 09	ESO 126-14	12.4	1.4 x 0.7	12.3	SA0°:	61	Mag 9.76v star NE 7′.5; note: **ESO 126-16** is 1′.8 NNE of this star; **ESO 125-15** SSE 3′.9.
09 29 58.4	−62 10 59	ESO 126-17	12.3	1.7 x 1.1	12.9	S0 pec: sp	124	Bright central core with very faint outer surface, many stars superimposed.
09 34 14.2	−61 16 49	ESO 126-19	13.1	1.8 x 1.6	14.1	IB(s)m pec V-VI	57	Numerous stars superimposed; mag 12.8 star on SE edge.
09 36 30.4	−62 05 56	ESO 126-21	13.3	1.0 x 1.0	13.2	S0		Numerous stars superimposed, in star rich area; mag 10.17v star W 7′.5.
09 37 00.8	−60 56 01	ESO 126-22	14.7	1.2 x 0.4	13.8	SB(s)m IV-V	129	Mag 9.93v star SW 4′.6.
09 37 51.4	−62 09 09	ESO 126-23	13.1	1.6 x 0.4	12.4	SA(s)c: sp III	10	Mag 10.9 star N 2′.0.
09 38 29.5	−61 49 47	ESO 126-24	13.2	1.5 x 0.8	13.2	SAB(s)bc II	150	Mag 10.08v star N 3′.7; ESO 126-25 N 5′.5.
09 38 33.8	−61 44 18	ESO 126-25	13.3	1.0 x 0.7	12.8	S0	67	Mag 10.08v star S 2′.0; ESO 126-24 S 5′.5.
10 24 44.3	−54 47 50	ESO 168-2	15.0	1.7 x 0.5	14.7	S	158	Mag 8.45v star NNW 6′.9.

OPEN CLUSTERS

RA h m s	Dec ° ′ ″	Name	Mag	Diam ′	No. ★	B ★	Type	Notes
10 35 42.0	−60 07 00	Bo 9	6.3	16	30	7.9	cl	Moderately rich in stars; large brightness range; no central concentration; detached; involved in nebulosity. Doubtful cluster.
10 42 08.0	−59 09 00	Bo 10	6.2	20	40	8.6	cl	Moderately rich in bright and faint stars; detached; involved in nebulosity.
10 47 12.0	−60 06 00	Bo 11	7.9	22	20	8.4	cl	Few stars; large brightness range; not well detached; involved in nebulosity.
10 57 18.0	−61 44 00	Bo 12	9.7	10	20	11.4	cl	Few stars; large brightness range; no central concentration; detached.
10 25 52.0	−57 56 00	Cr 220		9.5	50		cl	(A) This is not NGC 3247 as Collinder and most later catalogues claim.
10 30 24.0	−60 05 00	Cr 223	9.4	8	35		cl	Moderately rich in stars; moderate brightness range; slight central concentration; detached.
10 44 00.0	−60 05 00	Cr 228	4.4	14	98	6.3	cl	
10 44 39.0	−59 33 36	Cr 232	6.8	4			cl	
10 45 12.9	−59 44 47	Cr 234		3	18	9.5	ptcl	Southern part of Trumpler 16.
10 59 38.1	−60 59 00	Cr 236	7.7	7	20		cl	Few stars; moderate brightness range; no central concentration; detached.
10 53 42.0	−58 14 00	ESO 128-16		10			cl	
09 05 12.0	−55 57 00	ESO 165-9		15	20		cl	(A) Nice cluster on DSS, about 20 stars.
09 10 30.0	−53 53 00	ESO 166-4		4			cl	
09 51 30.0	−56 17 00	Ho 2		7	20		cl:	(A) Faint.
09 57 40.0	−54 40 36	Ho 3		2	10		cl?	(A) Scattered group with 3 brighter stars. May not be a cluster? Hogg 4 to NE.
09 57 53.0	−54 36 36	Ho 4		4			cl?	(A) Scattered group. May not be a cluster? Hogg 3 to SW.
10 06 09.0	−60 23 06	Ho 5		3	20		cl	(A) This is not Tr 12, but Hogg suggests it may be an extension of it.
10 06 28.0	−60 29 54	Ho 6		3			cl	(A) Appears as asterism on DSS image. Hogg suggests this may be an extension of Tr 12.
10 29 06.0	−60 43 00	Ho 7		4			ast?	(A) Very slight condensation.
09 27 27.0	−56 57 00	IC 2488	7.4	18	70	10.0	cl	Rich in stars; large brightness range; slight central concentration; detached. (S) Sparse.
10 27 27.0	−57 37 30	IC 2581	4.3	5	398	4.6	cl	Few stars; moderate brightness range; slight central concentration; detached; involved in nebulosity. Without brightest star, visual mag = 5.9.
10 04 25.0	−55 47 00	Lo 1		16	29	8.0	cl	
10 11 06.0	−56 35 00	Lo 27		21	30	10.7	cl	(A) Three brighter stars and a scattering of fainter ones.
10 03 08.6	−58 09 00	Lo 28		15			cl?	
10 23 16.0	−54 43 00	Lo 46		27	50	7.8	cl	
10 30 22.0	−54 07 06	Lo 59		5.5	25	12.6	cl	
10 28 47.6	−56 45 00	Lo 89		30	26	8.9	cl?	
10 32 40.0	−56 42 00	Lo 112		22	8	9.3	cl	(A) Mostly eight 9-10th magnitude stars.
10 34 46.7	−58 09 00	Lo 153		7	10	9.4	ast?	(A) Northwest of NGC 3293.
10 35 56.0	−58 44 00	Lo 165		11	30	11.2	cl	(A) A scattering of about 30 stars, 12th magnitude and somewhat fainter.
10 27 51.0	−60 35 00	Lo 172		22	17	9.0	cl	(A) Just NE (and associated with?) NGC 3255.
10 36 40.0	−61 53 00	Lo 213		24	24	8.5	cl	
09 22 06.0	−51 06 00	NGC 2866		20			cl?	(S) Compact group.
09 30 30.0	−52 54 48	NGC 2910	7.2	6	59	9.3	cl	Moderately rich in stars; large brightness range; no central concentration; detached.
09 33 11.0	−53 24 00	NGC 2925	8.3	15	40		cl	Moderately rich in bright and faint stars; detached. (S) Sparse.
09 48 40.0	−56 25 00	NGC 3033	8.8	12	50	10.6	cl	Moderately rich in stars; moderate brightness range; slight central concentration; detached.
09 49 16.0	−62 40 30	NGC 3036		20			cl	(S) asterism 4′ long elongated NE-SW.
10 00 40.0	−54 47 18	NGC 3105	9.7	2	97	12.4	cl	Few stars; moderate brightness range; strong central concentration; detached.
10 02 42.0	−60 06 00	NGC 3114	4.2	35	171	9.0	cl	Rich in stars; large brightness range; slight central concentration; detached.
10 21 23.0	−51 43 24	NGC 3228	6.0	5	23	7.9	cl	Few stars; large brightness range; slight central concentration; detached.
10 24 01.0	−57 45 30	NGC 3247	10.5	1.1	25	11.4	cl	Few stars; small brightness range; not well detached; involved in nebulosity. (A) Includes nebula **GUM 29** = **RCW 49** = **Bran 300B**.
10 26 32.0	−60 40 42	NGC 3255	11.0	2	30	12.4	cl	Moderately rich in stars; moderate brightness range; strong central concentration; detached. Probably not a cluster.
10 35 49.0	−58 13 30	NGC 3293	4.7	5	93	8.0	cl	Rich in stars; large brightness range; strong central concentration; detached. (S) Beautiful cluster.
10 37 19.0	−58 39 36	NGC 3324	6.7	5	44	8.2	cl	Rich in stars; large brightness range; strong central concentration; detached; involved in nebulosity.
10 38 48.0	−54 06 54	NGC 3330	7.4	6	39	8.8	cl	Moderately rich in stars; moderate brightness range; no central concentration; detached.
10 59 34.0	−60 20 00	NGC 3496	8.2	7	110	11.8	cl	Moderately rich in stars; small brightness range; slight central concentration; detached.
11 01 18.0	−59 50 48	NGC 3503	9.4	0.6	9	9.0	cl	Few stars; moderate brightness range; no central concentration; detached.
09 22 06.5	−51 06 08	Pi 13	10.2	2	30	12.0	cl	Moderately rich in stars; moderate brightness range; strong central concentration; detached.
09 29 18.0	−52 42 00	Pi 14		1.5	12	14.0	cl	
09 51 14.0	−53 10 36	Pi 16	8.0	2	12	9.0	cl	Few stars; large brightness range; strong central concentration; detached.
09 21 54.0	−56 18 42	Ru 75		2	20	12.0	cl	
09 24 18.0	−51 39 54	Ru 76	10.8	5	20	13.0	cl	Few stars; moderate brightness range; not well detached. (S) Sparse group.
09 27 02.0	−55 07 18	Ru 77	10.4	5	58	14.0	cl	Moderately rich in stars; small brightness range; slight central concentration; detached.
09 29 07.9	−53 41 57	Ru 78		3	30	15.0	cl	
09 41 03.0	−53 50 00	Ru 79	9.2	6	109	11.0	cl	Few stars; moderate brightness range; no central concentration; detached.
09 45 49.0	−53 59 42	Ru 82	8.1	6	143	12.0	cl	Few stars; moderate brightness range; not well detached.
09 49 15.0	−54 36 00	Ru 83	9.8	3	44	12.0	cl	Moderately rich in stars; moderate brightness range; strong central concentration; detached.
10 01 26.0	−55 06 36	Ru 85		3	40	15.0	cl:	
10 01 27.7	−59 31 00	Ru 86		12		12.0	cl	Moderately rich in stars; moderate brightness range; no central concentration; detached. (A) DSS image shows 3′ x 15′ E-W grouping of stars standing out slightly from Milky Way background.
10 28 30.0	−58 10 00	Ru 89		4	25	13.0	cl	
10 31 00.0	−58 27 00	Ru 90		5	15	12.0	cl	Few stars; moderate brightness range; not well detached.
10 47 42.0	−57 28 00	Ru 91		2	15	6.3	cl	Few stars; moderate brightness range; no central concentration; detached.
10 53 49.0	−61 45 00	Ru 92	8.6	8	58	12.0	cl	Few stars; small brightness range; not well detached.

(Continued from previous page)

OPEN CLUSTERS

RA h m s	Dec ° ′ ″	Name	Mag	Diam ′	No. ★	B ★	Type	Notes
09 20 08.0	−60 24 12	Ru 159		2	50	15.0	cl	(A) Faint and small.
10 08 48.0	−61 11 00	Ru 161		32		11.0	cl	Few stars; moderate brightness range; no central concentration; detached.
10 53 00.0	−62 16 00	Ru 162		6	60	12.0	cl:	
10 04 39.0	−55 51 29	Sch 1		1.3	50		cl	(A) Very faint cluster. Possibly a globular cluster?
11 01 03.0	−60 14 00	Sh 1	8.8	1	11		cl	
10 04 54.0	−61 36 00	Tr 11	8.1	5	6		cl	Moderately rich in bright and faint stars; detached.
10 06 30.0	−60 18 00	Tr 12	8.8	4	11		cl	Few stars; large brightness range; strong central concentration; detached.
10 23 48.0	−60 08 30	Tr 13	11.3	5	40		cl	Moderately rich in stars; moderate brightness range; slight central concentration; detached.
10 43 57.0	−59 32 54	Tr 14	5.5	5	44	7.0	cl	
10 44 44.0	−59 21 00	Tr 15	7.0	15	39	8.4	cl	Few stars; moderate brightness range; no central concentration; detached; involved in nebulosity.
10 45 10.0	−59 43 00	Tr 16	5.0	10	90	6.2	cl	Involved in Eta Carina Nebula.
10 56 24.0	−59 12 18	Tr 17	8.4	5	44	10.3	cl	Moderately rich in stars; moderate brightness range; slight central concentration; detached.
09 10 28.0	−56 19 00	vdB-Ha 58		2	25		cl	(S) Identification uncertain.
09 25 16.0	−54 43 06	vdB-Ha 66		1.5	20		cl	(A) Many references incorrectly identify this as a globular cluster.
09 26 43.0	−51 16 30	vdB-Ha 67		4	40		cl	(S) Rich cluster.
09 31 23.0	−53 02 12	vdB-Ha 72		2	25		cl	
09 34 42.0	−54 40 54	vdB-Ha 75		3	30		cl	
09 43 48.0	−56 34 24	vdB-Ha 78		1.5	15		cl	
10 01 19.0	−58 13 06	vdB-Ha 84		4.5			cl	
10 04 34.0	−55 23 00	vdB-Ha 87		3	30		cl	
10 06 15.0	−51 36 48	vdB-Ha 88		4	20		cl	
10 12 10.0	−58 04 12	vdB-Ha 90	10.3	4	8	12.1	cl	
10 17 18.0	−58 42 00	vdB-Ha 91		5			cl	(S) Sparse group.
10 19 04.0	−56 25 00	vdB-Ha 92		2	15		cl	
10 38 45.0	−59 11 00	vdB-Ha 99		20	40		cl	Moderately rich in stars; large brightness range; no central concentration; detached; involved in nebulosity.
10 52 38.6	−54 16 00	vdB-Ha 106		3	20		cl	(A) ~2.5′ group that barely stands out from background.

BRIGHT NEBULAE

RA h m s	Dec ° ′ ″	Name	Dim ′ Maj x min	Type	BC	Color	Notes
10 37 30.0	−60 58 00	Anonymous					
10 46 18.0	−58 39 00	Gum 32	7 x 7	E	3-5		The brightest part is shaped like a broad irregular crescent, 7′ x 2′ and convex to the ENE; several stars involved.
10 16 48.0	−57 57 00	NGC 3199	20 x 15	E	3-5		Ring-shaped. The brightest part forms a broad crescent roughly 10′ x 0′.3 which is convex to the WSW. The opposite edge is well defined and there are many stars involved in the nebulosity, including a double.
10 23 54.0	−57 45 00	NGC 3247	7 x 7	E	2-5		Irregularly roundish and somewhat mottled. There are several smaller, detached nebulosities in the outlying areas. (A) Includes nebula **GUM 29 = RCW 49 = Bran 300B**.
10 37 42.0	−58 40 00	NGC 3324	16 x 14	E+R	2-5	2-4	A rather bright nebula, very irregular in shape and best defined along the NW perimeter. It lies in a rich star field and contains the double star h4338 (mag 8.5, 9.5, sep. 6″, E-W).
10 43 48.0	−59 52 00	NGC 3372	120 x 120	E	1-5	3-4	**Eta Carinae Nebula**. Very bright and beyond description, surpassing even the Great Orion Nebula. It is segmented by prominent dark lanes of absorbing matter.
11 01 18.0	−59 51 00	NGC 3503	3 x 3	E+R	3-5	2-4	Roundish nebula with three 10th mag stars involved and several others of comparable brightness in the nearby surrounding field. The field is very rich in faint stars, except a region to the W and NW.
10 05 12.0	−58 57 00	RCW 47	25 x 20	E	3-5		Irregular in shape and mottled; in a field very rich in faint stars. At the W border of this nebula is a nova shell involved with a 9th mag star.
10 27 24.0	−57 10 00	RCW 50	10 x 7	E	4-5		Slightly elongated, irregularly oval nebulosity of low surface brightness. The nebula and surrounding field is very rich in very faint stars.

PLANETARY NEBULAE

RA h m s	Dec ° ′ ″	Name	Diam ″	Mag (P)	Mag (V)	Mag cent ★	Alt Name	Notes
09 38 47.5	−60 05 28	IC 2501	2	11.3	10.4	14.4	PK 281−5.1	
10 09 21.7	−62 36 40	IC 2553	9	13.0	10.3	15.5	PK 285−5.1	
09 21 25.4	−58 18 43	NGC 2867	24	9.7	9.7	16.6	PK 278−5.1	Mag 10.32v star E 2′.6.
09 27 03.5	−56 06 18	NGC 2899	117	12.2	11.8	15.9	PK 277−3.1	Slightly curved, elongated, kidney bean shape, oriented ESE-WNW with strong dark patch W of center.
10 17 50.7	−62 40 17	NGC 3211	19	11.8	10.7	18.0	PK 286−4.1	Pair of mag 11.3, 11.4 stars NW 3′.4.
09 08 40.0	−53 19 15	PN G273.2−3.7	11	14.8			PK 273−3.1	Mag 11.2 star N 3′.4; mag 9.75v star W 4′.7.
09 15 07.8	−54 52 46	PN G275.0−4.1	10	12.7	12.8	16.1	PK 275−4.1	Mag 10.51v star W 1′.5; mag 10.72v star N 2′.8.
09 22 06.9	−54 09 40	PN G275.2−2.9	10	14.8		17.0	PK 275−2.1	Mag 10.04v star W 1′.2.
09 18 01.3	−54 39 28	PN G275.2−3.7	5	15.6		16.9	PK 275−3.1	Mag 13.1 star SW 0′.5; mag 13.1 star ENE 1′.3.
09 13 53.0	−55 28 18	PN G275.3−4.7	2	13.5			PK 275−4.2	Mag 14.3 star NW 0′.3; small triangle of mag 11-12 stars S 1′.2.
09 30 48.4	−53 10 01	PN G275.5−1.3	7	15.5			PK 275−1.1	Mag 10.63v star N 2′.7; mag 10.1 star S 2′.7.
09 24 45.8	−54 36 18	PN G275.8−2.9	14	13.3	13.3		PK 275−2.2	Close pair of mag 12.4, 13.4 stars N 0′.5. .
09 31 19.7	−56 17 39	PN G277.7−3.5	159	14.6		18.1	PK 277−3.2	Very slightly elongated E-W; three stars form triangle inside boundaries, mag 12.4 N, 12.6 E and 13.7 S.
09 30 55.9	−57 36 57	PN G278.5−4.5	56	16.1			PK 278−4.1	Elongated N-S, several small bright patches; mag 11.01v star SSE 2′.7.
09 19 27.5	−59 12 02	PN G278.6−6.7	10	11.9		16.7	PK 278−6.1	Mag 13.3 star ESE 0′.9; mag 12.7 star SSW 0′.6.
09 43 26.0	−57 16 59	PN G279.6−3.1	22	13.3	11.8	11.3	PK 279−3.1	Mag 12.7 star S 0′.9; mag11.1 star N 2′.0.
10 11 57.8	−52 38 19	PN G280.0+2.9	20	13.4			PK 280+2.1	Very close pair of mag 11.2, 11.7 stars W 0′.9; mag 12.1 star E 1′.1.
10 31 32.0	−53 33 30	PN G282.9+3.8	14	15.9			PK 282+3.1	Mag 13.2 star NE 1′.2; mag 10.19v star W 3′.4.
10 34 19.0	−53 41 04	PN G283.3+3.9	12	14.2			PK 283+3.1	Forms a triangle with a mag 11.7 star E 1′.5 and mag 11.2star NE 1′.2.
10 31 33.5	−55 20 51	PN G283.8+2.2	8	12.7	13.3	17.2	PK 283+2.1	Forms a small triangle with a mag 12.0 star N 0′.7 and mag 11.8 star NW 1′.0.
10 03 49.3	−60 43 50	PN G283.8−4.2	10	14.2		16.4	PK 283−4.1	Mag 11.2 star WSW 1′.6; mag 12.4 star W 0′.9.
10 15 33.9	−58 51 10	PN G283.9−1.8	21	16.6			PK 283−1.1	Mag 11.9 star NE 1′.8; mag 12.4 star SW 1′.8.
10 28 43.3	−59 04 10	PN G285.4−1.1	90	16.1			PK 285−1.1	Mag 9.83v star N 2′.8; mag 11.4 star W 3′.5.
10 38 32.7	−56 46 34	PN G285.4+1.5	3	14.0			PK 285+1.1	Mag 12.1 star ENE 1′.8; mag 12.9 star SE 0′.5.
10 41 19.9	−56 09 18	PN G285.4+2.2		15.4			PK 285+2.1	Mag 12.4 star N 1′.4; mag 11.9 star E 2′.4.
10 23 09.1	−60 32 42	PN G285.6−2.7	5	14.1		12.9	PK 285−2.1	Mag 12.0 stars E 0′.7 and W 0′.7; mag 10.74v star N 0′.8.
10 39 32.2	−57 06 08	PN G285.7+1.2	5	15.0			PK 285+1.2	Very faint star on SE edge.
10 48 43.3	−56 03 09	PN G286.3+2.8	18	15.1		17.4	PK 286+2.1	Mag 12.2 star on SE edge.
10 54 35.0	−59 09 47	PN G288.4+0.3	25	13.7			PK 288+0.1	Appears brighter on eastern half; mag 13.0 star E 1′.0.
10 44 31.5	−61 39 40	PN G288.4−2.4	8	14.0			PK 288−2.1	Mag 13.0 star N 0′.3; mag 12.8 star SE 0′.4.
10 53 58.9	−60 26 41	PN G288.9−0.8	72	10.4		11.1	PK 288−0.1	Mag 11.7 star near center with mag 11.5 star 0′.4 S of center; small triangle of mag 11 stars W 1′.3.
10 56 02.5	−61 28 02	PN G289.6−1.6	20	16.5			PK 289−1.1	Mag 13.4 star on W edge; mag 12.4 star SW 1′.2.
11 03 55.6	−60 36 05	PN G290.1−0.4	19	15.0			PK 290−0.1	Mag 11.6 star E 1′.6; mag 8.66v star NW 1′.8.

200 GALAXIES 200

RA h m s	Dec ° ′ ″	Name	Mag (V)	Dim ′ Maj x min	SB	Type Class	PA	Notes
07 28 18.9	−62 53 38	ESO 88-15	15.1	1.6 x 0.2	13.8	Scd: sp III-IV	71	Mag 10.66v star NNW 6′.3.
06 59 05.0	−59 07 42	ESO 122-11	13.4	0.9 x 0.6	12.6		159	Double system, contact.
07 12 03.6	−60 30 29	ESO 122-16	13.3	0.9 x 0.8	12.9	SB0°? pec	53	Located between a pair of mag 13 stars.
07 20 31.2	−58 03 49	ESO 123-4	14.5	1.3 x 0.3	13.3	Sa? sp	22	Mag 6.72v star W 9′.6.
07 22 50.8	−62 01 45	ESO 123-9	13.5	2.1 x 0.9	14.0	SA(rs)c III	144	Mag 10.40v star N 1′.4; mag 6.93v star NW 9′.5.
07 30 25.9	−61 47 26	ESO 123-16	13.7	2.0 x 0.5	13.6	SB(s)dm IV	108	Mag 8.77v star NE 11′.9.
07 30 41.2	−62 01 37	ESO 123-17	13.4	1.3 x 0.5	12.8	SAB0°:	60	Pair of stars superimposed NE end; mag 12.6 star SE 2′.7.
07 42 03.0	−59 04 54	ESO 123-19	14.0	0.9 x 0.6	13.2	SB	173	Star on E edge.
07 44 38.1	−58 09 15	ESO 123-23	14.0	2.3 x 0.3	13.5	SB(s)cd? sp II-III	105	Mag 10.00v star NNE 4′.8.
08 07 20.1	−61 43 16	ESO 124-11	14.0	1.4 x 0.3	12.9	S0 sp	111	Mag 14.2 star S edge; close pair of stars, mags 12.9 and 15, N 1′.7; mag 9.05v star N 7′.7.
08 09 12.6	−61 39 38	ESO 124-14	12.8	1.4 x 0.6	12.5	$S0^-$	135	Mag 9.33v star N 4′.6; **ESO 124-13** W 3′.7.
08 23 40.8	−60 52 35	ESO 124-15	13.5	1.7 x 0.8	13.7	SB(s)bc II-III	42	Numerous bright knots and/or stars on surface.
08 31 39.2	−59 47 07	ESO 124-18	13.4	1.7 x 1.2	14.0	SB(r)bc: pec I-II	32	Pair with ESO 124-19 E.
08 31 53.6	−59 46 58	ESO 124-19	13.9	0.6 x 0.6	12.6	S?		Several stars superimposed; pair with ESO 124-18 W.
07 15 23.9	−55 04 34	ESO 162-15	13.9	0.9 x 0.7	13.3	SB(s)c III-IV	123	Mag 14.3 star on E edge; mag 7.37v star NNE 8′.6.
07 15 54.5	−57 20 39	ESO 162-17	12.6	1.9 x 0.6	12.6	Sb? pec sp	62	Bright knot or star SW end; mag 11.9 star W 4′.6.
07 38 05.3	−55 11 23	ESO 163-11	12.9	2.5 x 0.6	13.2	SB(s)b? sp	3	**ESO 163-10** 1′.8 W.
07 43 43.9	−56 46 08	ESO 163-14	13.2	1.8 x 0.4	12.7	SA(s)ab: pec	73	Mag 12.4 star on N edge; mag 9.38v star ENE 3′.5.
07 49 17.3	−54 27 28	ESO 163-19	13.8	1.5 x 0.7	13.7	IB(s)m pec IV-V	12	Mag 13 star S end.
08 06 51.1	−53 30 37	ESO 164-4	15.9	0.5 x 0.5	14.3	S0		Mag 8.93v star E 7′.4.
08 26 13.7	−54 02 06	ESO 164-10	16.3	1.5 x 1.5	17.0	SABm: V-VI		Very amorphous, many superimposed stars.
08 34 47.5	−57 38 49	ESO 165-1	13.9	1.5 x 0.3	12.9	(R′)SB(r)ab: sp	106	Mag 10 star S edge.
07 09 10.3	−51 28 03	ESO 207-22	14.7	1.1 x 1.0	14.6	IA(s)m V-VI	96	Uniform surface brightness, several stars involved.
07 10 52.5	−51 48 24	ESO 207-25	13.7	1.3 x 0.6	13.3	(R)SB(s)0^+:	170	Mag 10.81v star S 9′.4.
07 27 12.1	−51 22 16	ESO 208-15	13.5	1.1 x 1.0	13.4	SB(r)a:	40	Bright core, uniform envelope; mag 6.72v star SE 2′.8.
07 37 36.9	−52 18 17	ESO 208-31	14.1	1.8 x 0.2	12.9	Sc sp I-II	166	**ESO 208-30** N 1′.4.
07 28 17.6	−62 21 12	IC 2200	13.2	1.3 x 0.7	13.0	(R)SABb pec II	58	Pair with IC 2200A W edge; mag 13.4 star on NE edge; mag 10.59v star NW 13′.7.
07 28 06.6	−62 21 47	IC 2200A	12.7	1.3 x 0.9	12.8	SB(s)0^- pec?	44	Pair with IC 2200 E edge.
07 16 38.1	−62 20 38	NGC 2369	12.3	3.5 x 1.1	13.6	SB(s)a I-II	177	
07 18 43.5	−62 56 11	NGC 2369A	12.9	1.9 x 1.2	13.7	SAB(rs)bc II-III	33	Surrounded by a quartet of faint stars.
07 20 30.2	−62 03 14	NGC 2369B	13.3	1.5 x 1.4	13.9	(R′)SB(r)bc: III	14	Mag 6.93v star Delta Volantis NE 10′.3.
07 30 11.8	−62 15 05	NGC 2417	12.0	2.8 x 2.0	13.8	SA(rs)bc I-II	81	Numerous stars on outer envelope; mag 9.73v star Delta Volantis ENE 6′.9.
07 55 51.8	−52 18 26	NGC 2502	12.0	2.0 x 1.1	12.7	SAB(s)0°	126	
08 37 24.8	−55 07 27	NGC 2640	11.1	2.2 x 1.9	12.6	$SAB0^-$	104	Numerous stars superimposed on, and surround, bright center.
08 53 29.7	−59 13 03	NGC 2714	12.9	1.1 x 1.1	13.2	E0		Several faint stars superimposed on envelope, in star rich area of the sky.

OPEN CLUSTERS

RA h m s	Dec ° ′ ″	Name	Mag	Diam ′	No. ★	B ★	Type	Notes
07 52 30.0	−60 20 00	ESO 123-26		16			cl	
08 47 12.0	−52 04 00	ESO 211-1		4			cl	
08 40 18.0	−52 55 00	IC 2391	2.6	60	30	4.0	cl	Moderately rich in bright and faint stars; detached.
07 58 00.0	−60 45 00	NGC 2516	3.8	22	103	7.0	cl	Rich in stars; large brightness range; strong central concentration; detached.
08 29 30.0	−61 06 36	NGC 2609		6	10		cl?	(S) Sparse group.
08 46 19.0	−52 56 00	NGC 2669	6.1	14	90	7.7	cl	Moderately rich in stars; large brightness range; no central concentration; detached. (A) Includes **Harvard 3**.

BRIGHT NEBULAE

RA h m s	Dec ° ′ ″	Name	Dim ′ Maj x min	Type	BC	Color	Notes
07 56 49.0	−59 08 00	IC 2220	5 x 5				**Toby Jug Nebula.**

PLANETARY NEBULAE

RA h m s	Dec ° ′ ″	Name	Diam ″	Mag (P)	Mag (V)	Mag cent ★	Alt Name	Notes
07 47 20.1	−51 15 05	PN G264.4−12.7	3	12.3		14.7	PK 264−12.1	Mag 15.3 star on N edge; mag 12.1 star WSW 2′.5.
08 53 36.7	−54 05 08	PN G272.4−5.9	110	14.9		19.4	PK 271−5.1	Close pair of mag 13.1, 14.1 stars on N edge; mag 12.1 star WNW 2′.4.

201 GALAXIES 201

RA h m s	Dec ° ′ ″	Name	Mag (V)	Dim ′ Maj x min	SB	Type Class	PA	Notes
05 07 43.8	−62 59 25	ESO 85-47	13.8	1.8 x 1.1	14.4	SB(s)m V-VI	22	Mag 7.95v star SW 12′.0.
06 33 19.0	−62 59 39	ESO 87-28	13.5	1.7 x 1.0	14.0	S0° pec	125	Mag 10.02v star W 5′.1.
04 57 16.1	−59 07 16	ESO 119-23	17.7	1.5 x 0.9	17.9	IBm? VI	60	Group of three mag 9.25v, 9.47v and 9.83v stars W 7′.8.
04 59 08.5	−58 39 17	ESO 119-25	13.4	1.3 x 0.6	13.0	(R)SB(r)a	112	Mag 10.76v star S 0′.9.
05 08 27.4	−61 22 08	ESO 119-38	14.1	1.0 x 1.0	13.9	SAB(s)c pec II		Mag 13.3 star on E edge; mag 9.85v star W 12′.6.
05 10 37.2	−61 31 18	ESO 119-43	13.8	1.4 x 0.4	13.1	S0° sp	12	Mag 9.63v star SSW 6′.3.
05 13 28.8	−61 33 13	ESO 119-44	14.3	1.2 x 1.0	14.3	SA(s)c: II-III	144	Multiple branching arms; mag 13.4 star SW 1′.5.
05 14 21.7	−60 47 27	ESO 119-45	14.1	1.0 x 0.8	13.7	SB(s)bc II	20	Stellar nucleus; mag 10.63v star SSE 5′.9.
05 14 29.5	−62 10 16	ESO 119-46	14.0	1.1 x 0.3	12.6	Sb? pec sp	56	Mag 9.07v star E 5′.5.
05 14 31.1	−62 13 42	ESO 119-47	13.3	1.2 x 0.6	12.8	SB(s)b pec	51	Mag 9.07v star NE 7′.0.
05 14 36.0	−61 28 53	ESO 119-48	12.8	1.5 x 0.9	13.0	(R′)SA(s)0^+ pec	57	Pair of superimposed stars or bright knots SW end; mag 11.15v star N 9′.2.
05 14 43.3	−61 11 25	ESO 119-49	13.8	1.2 x 0.5	13.1	SA0°: sp	93	Mag 10.80v star N 7′.4.
05 19 18.4	−61 39 40	ESO 119-52	13.9	1.3 x 0.9	13.9	SB(s)bc: II-III	84	Mag 10.72v star NNE 5′.0; ESO 119-53 SSE 5′.3.
05 19 39.2	−61 44 21	ESO 119-53	13.5	1.5 x 1.2	14.0	(R′)SB(rs)a:	3	ESO 119-52 NNW 5′.3; mag 10.98v star W 7′.9.
05 20 19.6	−61 15 36	ESO 119-54	13.5	1.3 x 0.7	13.2	(R′)SB(r)0/a pec	178	Mag 7.90v star N 5′.7; pair with ESO 119-55 S 2′.3.
05 20 21.2	−61 17 48	ESO 119-55	12.9	1.9 x 1.0	13.4	SA(s)a: pec	176	Pair with ESO 119-54 N 2′.3.
05 20 53.8	−61 24 15	ESO 119-57	14.0	1.2 x 0.7	13.7		115	Double system, contact.
05 21 15.4	−61 03 34	ESO 119-58	13.5	1.7 x 0.7	13.6	SB(r)0/a pec:	123	Mag 11.18v star NW 3′.5.
05 39 58.6	−58 35 08	ESO 120-12	13.5	2.2 x 1.5	14.6	(R′)SB(rs)d IV-V	87	Mag 7.94v star NE 11′.0.
05 51 35.3	−59 02 47	ESO 120-16	13.6	2.2 x 0.4	13.3	Sb sp	2	Mag 10.82v star W 6′.1.
05 53 14.2	−59 03 59	ESO 120-21	15.3	1.5 x 0.8	15.3	IBm pec	117	Double galaxy system; mag 11.11v star N 5′.2.
05 55 53.0	−61 24 11	ESO 120-23	14.3	1.1 x 0.5	13.5	$SA0^-$	9	Mag 8.05v star SE 3′.1.

(Continued from previous page)

GALAXIES

RA h m s	Dec ° ′ ″	Name	Mag (V)	Dim ′ Maj x min	SB	Type Class	PA	Notes
05 59 37.6	−60 04 21	ESO 120-26	13.9	1.7 x 1.3	14.6	SAB(rs)c pec II	0	**ESO 126-26A** SE 0′5.
06 07 30.3	−61 48 27	ESO 121-6	12.6	4.2 x 0.6	13.5	Sc pec: sp III	40	Mag 9.79v star W 1′3.
06 21 39.1	−59 44 26	ESO 121-26	11.9	3.0 x 2.0	13.7	SB(rs)bc II	112	Mag 11.15v star NW 7′7.
06 37 29.7	−59 39 05	ESO 121-41	15.3	1.2 x 0.4	14.4	IB(s)m V-VI	8	Mag 8.78v stars SSE 8′4.
06 40 43.5	−58 31 33	ESO 122-1	12.3	2.0 x 1.0	12.9	SB(s)b pec I-II	170	Mag 10.73v star S 2′9; ESO 122-2 N 3′3.
06 40 46.6	−58 28 14	ESO 122-2	13.9	0.7 x 0.6	12.8		66	ESO 122-1 S 3′3.
06 41 38.8	−60 15 48	ESO 122-3	13.5	0.8 x 0.6	12.6	S0	153	Double system, in contact with small companion S.
06 47 59.5	−57 48 56	ESO 122-6	13.7	0.8 x 0.7	12.9	SBab	70	Mag 12.5 star NE 2′4.
06 59 05.0	−59 07 42	ESO 122-11	13.4	0.9 x 0.6	12.6		159	Double system, contact.
05 04 35.2	−55 31 28	ESO 158-17	13.6	1.0 x 0.7	13.0	Sab	70	Mag 8.40v star NNE 1′6.
05 06 09.4	−55 36 50	ESO 158-18	13.9	1.1 x 0.9	13.7	SAB(s)d: III-IV	146	Mag 8.40v star NW 14′2.
05 15 07.1	−53 44 15	ESO 159-2	13.0	1.9 x 1.4	13.9	SB(s)bc: II	59	Mag 10.23v star NNE 6′8.
05 16 08.9	−54 06 17	ESO 159-3	13.2	1.9 x 1.1	13.9	SB0°: pec	49	Mag 15.7 star on E edge; ESO 159-4 W 9′0.
05 17 07.8	−54 04 02	ESO 159-4	13.4	1.5 x 0.9	13.6	(R)SB(r)0$^+$:	85	Large core, smooth envelope; small, elongated N-S galaxy at E edge; mag 10.83v star S 4′0.
05 28 41.8	−56 56 10	ESO 159-12	13.4	1.1 x 0.8	13.1	(R)SAB(r)0°:	177	Mag 15.6 star N edge; mag 14.3 star W 1′1.
05 33 11.7	−52 38 31	ESO 159-19	13.3	1.8 x 1.2	14.0	S0/a pec sp	9	Mag 7.00v star E 11′0.
05 34 18.0	−55 52 55	ESO 159-20	14.0	1.6 x 1.3	14.6	SAB(rs)d IV-V	0	Mag 9.37v star SE 8′8.
05 40 09.9	−55 32 26	ESO 159-23	13.3	1.5 x 0.8	13.3	Sb II	119	Pair of E-W oriented mag 10.02v, 9.80v stars SW 5′8.
05 43 06.5	−52 42 21	ESO 159-25	14.5	1.5 x 1.1	14.9	IB(s)m V-VI	55	Lies between a 14 mag star E and a 12 mag star W.
05 51 15.1	−53 34 31	ESO 160-2	12.9	1.9 x 0.4	12.5	Sb? sp	84	Mag 9.45v star on S edge; mag 7.91v star S 5′3.
06 04 31.2	−53 05 50	ESO 160-11	13.9	1.2 x 0.7	13.6	(R′)SA(rs)a pec:	72	Mag 10.82v star E 5′1.
06 09 10.5	−54 24 26	ESO 160-16	14.1	1.1 x 0.9	14.0	SB0?	88	Stellar nucleus in bright core with faint envelope.
06 13 35.5	−53 14 04	ESO 160-19	14.2	1.0 x 0.7	13.7	SAB(s)bc III-IV	15	Mag 9.15v star S 2′9.
06 14 30.5	−53 56 31	ESO 160-20	13.9	1.3 x 0.7	13.7	(R)SB(r)0/a	11	Close pair of stars, mags 13.3 and 14.8, SE 1′4.
06 17 16.6	−55 33 01	ESO 160-22	13.3	1.3 x 0.9	13.3	(R′)SB(r)b: I-II	27	Knot or star on W edge; mag 12.3 star W 2′0.
06 17 42.2	−56 36 39	ESO 160-23	14.1	0.9 x 0.8	13.6	SB0?	160	Mag 11.2 star NW 2′6.
06 20 30.4	−57 29 48	ESO 161-1	14.4	1.1 x 0.5	13.6	SB(s)dm pec: IV	147	NGC 2222 SSW 2′9.
06 27 36.3	−54 27 01	ESO 161-8	15.7	1.8 x 0.4	15.3	E$^+$5	126	Numerous very small, very faint galaxies in surrounding area.
06 37 17.9	−55 21 44	ESO 161-19	13.6	1.4 x 0.9	13.7	(R′)SB(r)b	176	Located 2′8 SW of mag 6.89v star.
06 44 03.6	−55 54 32	ESO 161-23	14.0	1.0 x 0.5	13.1	Sab	172	Pair of stars, mags 13.9 and 15.0, on NW edge.
06 44 08.8	−56 41 15	ESO 161-24	13.6	1.0 x 0.6	12.9		53	Strong knot or star E edge; mag 10.09v star W 1′6.
06 45 45.7	−55 32 07	ESO 161-25	13.7	0.7 x 0.6	12.6		153	Double system? Mag 12.6 star on NE edge; pair of mag 12 stars NE 1′3.
05 09 40.0	−52 11 37	ESO 203-22	15.2	1.5 x 0.1	13.0	Sc sp	164	Mag 9.34v star NNE 7′6.
05 25 26.6	−51 38 44	ESO 204-8	14.8	1.2 x 0.6	14.3	SB(s)b	55	Mag 11.17v star S 7′3.
05 36 26.0	−52 11 02	ESO 204-22	14.5	1.4 x 0.9	14.6	SB(s)m: pec V-VI	18	Mag 12 star SE edge.
05 44 41.5	−51 57 51	ESO 204-34	14.3	1.4 x 1.1	14.6	IB(s)m V-VI	68	Uniform surface brightness; mag 10.52v star superimposed E of center.
05 45 00.2	−52 21 46	ESO 204-36	16.4	1.1 x 0.7	15.9	SB(s)m	120	Mag 11.6 star N 1′6; mag 10.27v star WNW 9′0.
05 57 17.8	−52 22 18	ESO 205-9	14.2	1.8 x 0.3	13.4	Sbc sp	113	Mag 11.4 star S 4′8.
05 57 41.4	−51 58 13	ESO 205-11	13.9	1.2 x 0.9	13.8	(R)SB(r)b: II	131	Strong N-S bar.
05 59 09.8	−51 28 13	ESO 205-13	13.0	1.6 x 0.8	13.1	Sa:	66	Mag 8.13v star E 13′3.
05 59 49.2	−52 24 31	ESO 205-14	13.7	1.6 x 0.4	13.0	SB(rs)ab?	114	Mag 10.93v star E 3′4; **PGC 18182** WSW 7′7 with galaxy **Fair 801** 1′5 S of PGC 18182.
06 13 18.4	−51 19 02	ESO 205-34	13.0	1.3 x 1.2	13.4	SAB(s)m IV	39	Numerous knots; mag 10.48v star W 7′3.
06 16 12.7	−51 50 25	ESO 206-1	13.8	1.1 x 0.5	13.2	E$^+$5	2	Mag 11.90v star W 5′1; **ESO 206-3** N 15′1.
06 38 19.6	−51 57 02	ESO 206-17	14.5	1.7 x 0.3	13.6	Scd: sp	178	Mag 9.809v star NE 3′2.
06 50 39.8	−52 08 25	ESO 207-7	13.3	2.5 x 1.6	14.7	SB(s)m V	9	Several stars superimposed N of center; mag 11.28v star NE 9′3.
04 58 24.4	−62 01 44	NGC 1765	13.0	1.2 x 1.0	13.0	SB(rs)0$^-$:	150	Stellar nucleus.
05 02 43.0	−61 08 21	NGC 1796	12.3	1.9 x 1.0	12.9	SB(rs)c pec: III	102	
05 05 03.2	−61 29 02	NGC 1796A	13.9	1.7 x 0.5	13.6	(R′)SA(rs)ab:	150	
05 07 53.7	−61 11 25	NGC 1796B	14.5	1.1 x 0.4	13.4	SAB(rs)bc: II	21	**ESO 119-37A** on E edge.
05 06 55.8	−59 43 22	NGC 1824	12.6	3.0 x 0.9	13.5	SB(s)m IV	160	
05 12 16.3	−57 23 59	NGC 1853	13.0	2.0 x 0.7	13.2	SB(s)d? III-IV	43	Mag 12 star 1′0 NE.
05 44 16.3	−55 31 58	NGC 2087	13.6	0.8 x 0.6	12.7	SB(r)a: pec	136	**ESO 159-27** S 8′0; mag 7.55v star E 10′8.
05 46 23.3	−52 05 24	NGC 2101	13.7	1.9 x 1.1	14.3	IB(s)m pec III-IV	85	Brighter core slightly offset to W of center.
05 47 04.5	−51 33 07	NGC 2104	12.7	2.0 x 0.9	13.2	SB(s)m pec III-IV	160	
05 58 46.1	−59 07 35	NGC 2148	13.6	1.1 x 0.8	13.3	SA(rs)b pec: III-IV	150	Mag 12.4 star on E edge.
06 08 23.8	−52 30 43	NGC 2191	12.3	1.7 x 0.9	12.7	SB(r)0°:	118	Lies NE of mag 8.92v star.
06 10 33.1	−62 32 18	NGC 2205	12.7	1.3 x 0.9	12.7	(R′)SAB(rs)0$^-$:	80	Faint star or knot E of core.
06 20 16.0	−57 34 40	NGC 2221	13.2	1.9 x 0.4	12.8	Sa? pec sp	0	
06 20 16.7	−57 31 58	NGC 2222	13.8	1.2 x 0.3	12.5	SBa? pec sp	150	ESO 161-1 NNE 2′9.
06 00 49.6	−58 35 28	PGC 18246	13.0	1.3 x 1.1	13.3	SB(s)0° pec	3	**ESO 18246** E 10′2; **ESO 120-25** SW 19′0; **ESO 121-4** SE 19′3.
06 24 33.3	−55 43 30	PGC 18976	13.6	1.5 x 0.6	13.3	SAB(s)cd	36	Mag 7.53v star E 4′6.

GALAXY CLUSTERS

RA h m s	Dec ° ′ ″	Name	Mag 10th brightest	No. Gal	Diam ′	Notes
06 26 18.0	−53 40 00	A 3391	16.1	40	17	**ESO 161-7**, **7A** at center; **Fair 814** ESE 10′4; all others anonymous, stellar/almost stellar.
06 27 30.0	−54 23 00	A 3395	15.9	54	17	Numerous PGC galaxies, mostly in SW quadrant.
05 20 24.0	−61 17 00	AS 524	15.2		22	**ESO 119-60** E of center 16′6; other non-ESO members anonymous, stellar.
05 34 12.0	−59 24 00	AS 537	15.8		17	All members anonymous, stellar.
05 40 12.0	−59 42 00	AS 541	16.3		17	All members anonymous, stellar.
05 51 06.0	−57 06 00	AS 552	16.2		17	All members anonymous, stellar.
05 58 00.0	−59 51 00	AS 558	14.9	17	22	**PGC 18154** at center.
06 00 42.0	−58 36 00	AS 560	15.4		22	See notes PGC 18246; all others anonymous, stellar.
06 00 54.0	−60 35 00	AS 561	15.8		17	All members anonymous, stellar; **ESO 121-1** N of center 17′2.
06 16 12.0	−51 51 00	AS 580	16.1		17	See notes ESO 206-1; all others anonymous, stellar.
06 23 00.0	−53 35 00	AS 584	16.0	1	17	All members anonymous, stellar.
06 55 18.0	−55 32 00	AS 599	15.7		17	All members anonymous, stellar.

OPEN CLUSTERS

RA h m s	Dec ° ′ ″	Name	Mag	Diam ′	No. ★	B ★	Type	Notes
05 36 21.8	−61 47 14	ESO 120-8		1.4			cl	
05 44 15.5	−62 47 05	NGC 2097		1.8			cl	
05 55 09.1	−59 55 00	NGC 2132		45	15	8.0	ast	(A) Large group of 6 bright stars (and several just fainter).

RA h m s	Dec ° ′ ″	Name	Mag (V)	Dim ′ Maj x min	SB	Type Class	PA	Notes
02 57 21.3	−62 52 36	ESO 82-6	14.4	1.0 x 0.5	13.5	SBc pec	128	Mag 9.50v star SW 10′.1.
03 32 00.1	−62 34 11	ESO 83-6	14.1	1.7 x 0.7	14.1	(R′)SB(rs)c III	101	Mag 7.41v star ESE 11′.6.
04 14 23.0	−62 48 45	ESO 84-9	13.5	1.0 x 0.6	12.8	SB(s)c I-II	52	Mag 15.3 star N edge.
04 27 46.8	−62 35 33	ESO 84-18	13.2	1.5 x 1.3	13.7	SA(rs)a:	108	Almost stellar nucleus; mag 5.74v star N 4′.3.
04 29 23.1	−62 53 21	ESO 84-20	13.9	1.0 x 0.3	12.5	Sb: sp III-IV	170	Mag 12.0 star ESE 1′.2.
04 37 17.6	−62 35 03	ESO 84-28	13.3	1.7 x 0.4	12.8	$S0^-$ sp	122	Mag 11.10v star SW 1′.9.
04 42 12.1	−62 54 49	ESO 84-35	14.2	1.2 x 0.4	13.3	SAb: II	1	Star or bright knot N end.
04 44 59.8	−62 42 21	ESO 84-40	14.4	0.8 x 0.5	13.3	Im:	39	Mag 10.76v star NW 8′.3; note: anonymous galaxy 2′.2 NE of this star.
04 46 44.3	−62 27 52	ESO 85-1	13.3	1.1 x 0.9	13.1	SAB(s)cd III	153	Mag 9.14v star NE 8′.9.
04 47 50.4	−62 36 45	ESO 85-4	14.0	1.5 x 0.3	13.0	SB(rs)b pec: II-III	26	
04 54 42.7	−62 47 58	ESO 85-14	12.8	3.2 x 1.1	14.0	SB(s)m IV	78	Double system; mag 7.97v star E 2′.6.
03 02 33.0	−57 51 39	ESO 116-5	13.3	1.0 x 0.6	12.6	(R)SB(r)0/a	33	Lies 1′.4 WSW of mag 10 star. Mag 13 star on E edge.
03 13 03.9	−57 21 28	ESO 116-12	12.4	3.5 x 1.0	13.6	SB(s)d: IV	25	Lies 4′.6 SE of mag 6 star.
03 15 50.6	−58 34 28	ESO 116-14	14.2	1.7 x 0.3	13.3	SBcd: sp	56	Mag 10.43v star N 7′.3.
03 17 43.7	−57 26 52	ESO 116-15	13.9	1.3 x 0.5	13.3		113	Double? system, interaction.
03 24 53.6	−60 44 19	ESO 116-18	14.3	1.3 x 0.3	13.1	(R)SAB(r)0^+: sp	90	Mag 9.85v star E 8′.2.
03 56 48.0	−60 25 38	ESO 117-16	14.3	1.1 x 0.6	13.6	S?	0	Mag 10.90v star ESE 10′.7.
04 00 07.9	−61 17 02	ESO 117-18	13.2	1.3 x 0.8	13.1	SB(s)bc II	115	Mag 11.8 star SW 3′.1.
04 02 32.9	−62 18 58	ESO 117-19	13.6	1.9 x 0.3	12.8	SBbc? sp II	77	Mag 9.54v star SSE 4′.7.
04 08 30.7	−57 35 08	ESO 118-3	15.1	0.5 x 0.3	12.9		138	Double system; located between a pair of mag 12.6 and 12.9 stars.
04 13 30.4	−61 50 45	ESO 118-12	13.5	1.1 x 0.5	12.7	S0°:	56	Mag 12.1 star SW 3′.0.
04 15 56.6	−57 38 52	ESO 118-15	14.1	1.3 x 1.0	14.3	(R′)SB(r)b: II	36	Strong SE-NW bar; mag 15.9 star, or knot?, N edge.
04 18 59.5	−58 15 26	ESO 118-19	14.0	0.9 x 0.7	13.4	S0? pec	34	Superimposed star or double nucleus?
04 21 35.0	−59 20 10	ESO 118-21	13.9	0.6 x 0.4	12.3	S0:	81	Close pair of mag 13 stars N 1′.9.
04 21 52.2	−61 46 56	ESO 118-22	14.0	1.1 x 0.6	13.4	(R)SB(r)0^+:	95	Mag 8.71v star SE 5′.1.
04 30 37.7	−58 27 42	ESO 118-28	14.3	0.8 x 0.7	13.5	(R′)SB(s)b	98	Strong dark patches N and S of center.
04 31 13.1	−61 27 13	ESO 118-30	13.2	2.1 x 1.0	14.0	E^+4	69	Several bright knots and/or superimposed stars.
04 36 11.3	−58 51 48	ESO 118-31	13.5	3.7 x 3.2	16.1	Sm? IV-V?		
04 40 17.3	−58 44 43	ESO 118-34	13.1	0.9 x 0.8	12.6	S0° pec:	1	Mag 9.39v star ESE 12′.8; mag 11.5 star SE 2′.8.
04 48 21.0	−59 25 02	ESO 119-4	14.2	0.7 x 0.4	12.6	Sa?	114	Extremely faint **ESO 119-3** W 1′.7.
04 48 57.1	−57 39 34	ESO 119-8	13.9	1.0 x 1.0	13.8	(R)SB(r)a:		Mag 11.8 star E 1′.5.
04 49 11.8	−62 21 38	ESO 119-11	13.9	0.5 x 0.4	12.0		107	Mag 9.61v star NNE 4′.8.
04 50 18.9	−61 15 07	ESO 119-12	14.1	1.1 x 0.7	13.7	Sc: pec III-IV	163	Mag 12.4 star E 2′.8.
04 50 27.0	−61 20 42	ESO 119-13	12.9	1.5 x 1.2	13.4	SB(s)c III-IV	45	**ESO 119-14** 1′.7 ESE.
04 51 23.3	−59 14 03	ESO 119-15	13.9	1.0 x 0.6	13.3	S0°:	54	Mag 10.00v star NE 9′.5.
04 51 29.1	−61 39 04	ESO 119-16	14.3	2.3 x 1.0	15.1	IB(s)m V	26	Mottled surface, numerous bright knots and/or superimposed stars.
04 55 45.2	−60 12 28	ESO 119-21	13.7	1.5 x 0.8	13.7	(R)SB(s)0°	72	Mag 12.8 star W 1′.4; mag 11.3 star NE 3′.9.
04 57 16.1	−59 07 16	ESO 119-23	17.7	1.5 x 0.9	17.9	IBm? VI	60	Group of three mag 9.25v, 9.47v and 9.83v stars W 7′.8.
04 59 08.8	−58 39 17	ESO 119-25	13.4	1.3 x 0.6	13.0	(R)SB(r)a	112	Mag 10.76v star S 0′.9.
02 56 51.7	−54 34 18	ESO 154-23	12.4	7.4 x 1.3	14.7	SB(s)m V-VI	40	Mag 9.39v star on S edge.
03 02 08.0	−54 24 42	ESO 154-28	14.9	1.2 x 0.8	14.7	SAB(s)d IV	9	Mag 9.56v star NNE 6′.4.
03 17 32.2	−54 21 29	ESO 155-14	13.1	1.0 x 1.0	13.0	S0?		ESO 155-16 SE 2′.3.
03 17 43.7	−54 22 59	ESO 155-16	13.9	1.1 x 1.0	13.9		16	Double or triple system, interaction; Mag 10.36v star NE 2′.8; ESO 155-14 NW 2′.3.
03 20 38.3	−54 17 46	ESO 155-20	13.8	1.4 x 0.7	13.6	S?	78	**PGC 12503** W 2′.1; **PGC 12493** NW 2′.8; **PGC 12500** NW 3′.5.
03 27 59.7	−55 25 18	ESO 155-35	14.3	0.6 x 0.5	12.8	Pec	123	Mag 12.8 star NE 0′.9.
03 30 25.7	−53 16 19	ESO 155-41	14.1	0.9 x 0.7	13.4	SB?	9	Strong dark patches N and S of center; mag 12.6 star NW 1′.5.
03 35 06.4	−55 05 35	ESO 155-54	14.3	1.0 x 0.5	13.4	S0?	176	Mag 14.2 star N edge.
03 42 56.5	−53 38 07	ESO 156-8	14.1	2.2 x 0.8	14.7		111	Pair of galaxies. Surrounding area contains many small, faint galaxies.
03 45 34.6	−54 31 39	ESO 156-11	14.5	1.5 x 0.5	14.0	SB(s)b? pec	9	Mag 7.32v star W 11′.7.
03 52 13.9	−54 53 17	ESO 156-18	13.3	1.7 x 1.0	13.8	SAB(s)0°: pec	22	Mag 11.8 star S 1′.1; mag 7.75v star E 10′.1.
04 00 41.5	−52 44 04	ESO 156-29	14.5	1.5 x 1.0	14.8	SA(rs)cd pec: III-IV	165	Mag 9.91v star W 4′.1; IC 2028 E 5′.9.
04 22 24.5	−56 13 35	ESO 157-23	14.0	1.2 x 0.7	13.7	(R′)SB(s)bc	138	Mag 14.2 star N edge; very faint, elongated anonymous galaxy S edge; mag 13.0 star N 2′.4.
04 27 32.5	−54 11 48	ESO 157-30	14.0	1.4 x 0.5	13.6	E4:	41	Mag 13.5 star N edge.
04 29 49.6	−53 48 55	ESO 157-36	14.1	1.0 x 0.3	12.6	S?	115	**PGC 15299** NW end; **PGC 15287** SW 2′.0; mag 8.11v star S 7′.0.
04 35 12.2	−54 12 19	ESO 157-42	14.1	1.3 x 0.7	13.8	(R′)SAB(r)ab	13	Mag 12.0 star S 2′.6.
04 35 15.3	−54 18 59	ESO 157-43	14.3	1.2 x 0.4	13.4	S pec	25	Knot or small galaxy S end; mag 12.9 star E 1′.2.
04 37 18.3	−55 55 26	ESO 157-44	14.3	1.0 x 0.6	13.6	IB(s)m IV	73	Mag 11.7 star S 5′.8.
04 39 37.1	−53 00 44	ESO 157-49	13.6	1.9 x 0.4	13.1	S?	27	Mag 9.72v star N 3′.1; small, elongated **ESO 157-48** SSW 2′.6.
04 40 43.9	−52 45 28	ESO 157-50	14.3	1.6 x 0.3	13.4	Sbc: pec sp	63	Mag 9.94v star S 1′.8.
04 43 33.7	−54 09 29	ESO 158-2	13.9	1.7 x 1.0	14.5	E^+3	96	Mag 10.38v star N 3′.8.
04 46 16.7	−57 20 35	ESO 158-3	13.3	1.5 x 1.3	13.8	SB(s)m pec III-IV	172	Mag 10.33v star E 4′.5.
04 49 37.0	−53 54 39	ESO 158-7	13.8	1.6 x 0.9	14.0	SAB(r)b: II	50	Lies 1′.7 S mag 8.6 star.
04 55 41.9	−56 14 11	ESO 158-15	14.8	1.2 x 0.8	14.6	Pec	65	Knotty.
03 37 30.0	−51 15 02	ESO 200-53	13.6	1.3 x 0.6	13.1	S0/a:	156	Mag 8.63v star W 12′.7.
03 37 34.1	−51 11 20	ESO 200-54	13.7	1.0 x 0.8	13.3	SA(s)c? II	150	Mag 10.77v star N 2′.0.
04 03 26.5	−51 49 55	ESO 201-16	14.3	0.9 x 0.7	13.7	SB(s)c? I	23	Mag 8.20v star S 6′.2.
04 31 31.1	−52 02 08	ESO 202-31	14.0	1.0 x 0.4	12.8	S0 sp	44	Mag 12.4 star SSE 3′.2.
04 36 57.2	−52 10 32	ESO 202-41	14.3	1.4 x 1.0	14.5	SB(s)m V	165	Filamentary, many knots E edge.
04 37 47.5	−51 25 23	ESO 202-43	13.5	1.6 x 0.8	13.7	E^+4	132	Mag 8.86v star E 10′.4; **ESO 202-44** S 15′.1.
03 15 05.1	−54 49 12	IC 1908	14.0	1.3 x 1.0	14.1	SB(rs)b pec	45	Double system; the very small galaxy **ESO 155-13A** extends S from the center of IC 1908.
03 25 25.6	−51 16 02	IC 1929	14.1	0.8 x 0.5	12.9	S?	143	**IC 1932** SE 6′.4; **IC 1936** ESE 10′.4.
03 25 40.4	−52 47 09	IC 1933	12.5	2.2 x 1.2	13.4	SAB(s)d: III-IV	55	Many knots; mag 12.2 star superimposed on S edge.
03 27 42.4	−52 08 26	IC 1940	13.7	0.9 x 0.9	13.3	(R′)SB(r)ab		Strong dark patch N edge.
03 31 30.6	−51 54 24	IC 1954	11.6	3.2 x 1.5	13.2	SB(s)b: II-III	66	Patchy; mag 10.63v star N 8′.2.
03 36 58.9	−57 58 24	IC 1980	14.2	1.0 x 0.4	13.0	Sb? sp pec	20	Mag 10.45v star NNE 3′.3.
03 37 42.5	−57 46 36	IC 1982	14.8	0.5 x 0.5	13.2	(R′)Sab pec:		Mag 8.66v star NE 9′.5.
03 44 51.8	−59 08 16	IC 1997	13.4	1.2 x 0.6	12.9	Sa? pec	73	Mag 9.76v star S 4′.7.
03 51 57.5	−59 55 48	IC 2010	13.7	1.1 x 0.4	12.6	Sa?	70	Pair of E-W oriented mag 12.0, 10.79v stars N 2′.8.
03 56 39.6	−59 23 42	IC 2017	13.8	0.9 x 0.5	12.8	SA0°: pec	19	Mag 12.2 star N 2′.7.
04 01 18.4	−52 42 29	IC 2028	14.2	0.7 x 0.5	12.9	Scd: III-IV	53	ESO 156-29 W 5′.9; mag 10.80v star and IC 2029 S 5′.3.
04 01 17.9	−52 48 04	IC 2029	14.6	0.9 x 0.4	13.3	SB(s)c pec	177	Mag 10.80 star W 1′.1; **PGC 14278** W 4′.1.
04 07 04.7	−55 19 26	IC 2032	14.1	1.3 x 0.8	14.0	IAB(s)m pec: IV-V	78	Several knots or stars W half.
04 07 14.5	−53 40 54	IC 2033	13.9	1.2 x 0.7	13.6	SAB(rs)a?	127	Mag 9.84v star SE 9′.6.
04 08 19.2	−58 45 05	IC 2037	13.8	1.6 x 0.3	12.8	Sb sp	92	Mag 10.14v star SW 7′.8.
04 08 54.4	−55 59 37	IC 2038	14.8	1.7 x 0.4	14.2	Sd pec: IV	152	IC 2039 SE 1′.6.
04 09 02.6	−56 00 43	IC 2039	14.2	0.8 x 0.5	13.1	S0°: pec	131	IC 2038 NW 1′.6; mag 9.73v star SE 2′.1.
04 11 09.6	−53 41 10	IC 2043	14.6	1.4 x 0.2	13.0	Sb: sp	9	Mag 9.49v star NE 8′.3.
04 12 04.4	−58 33 31	IC 2049	14.2	1.0 x 0.9	14.0	SAB(s)d? III-IV	168	Mag 9.84v star SE 5′.8.
04 13 55.9	−53 28 36	IC 2050	13.7	1.0 x 0.8	13.3	SA(rs)b pec	23	Mag 11.5 star NW 2′.8.

(Continued from previous page)

GALAXIES

RA h m s	Dec ° ′ ″	Name	Mag (V)	Dim ′ Maj x min	SB	Type Class	PA	Notes
04 16 24.7	−60 12 26	IC 2056	11.9	1.9 x 1.5	12.9	(R)SAB(r)bc: II	8	Mag 10.75v star NW 10′5.
04 17 54.5	−55 56 00	IC 2058	13.2	3.0 x 0.4	13.2	Sd? sp	18	Mag 11.25v star NW 8′7.
04 17 53.8	−56 37 01	IC 2060	14.0	1.5 x 1.1	14.4	S0⁻: pec	148	Mag 10.59v star N 6′8.
04 21 28.0	−55 56 04	IC 2065	13.7	1.1 x 0.5	12.9	Sa?	37	Mag 9.65v star N 5′7.
04 24 35.9	−57 58 53	IC 2070	13.8	1.4 x 0.7	13.6	SAB(s)c: I	85	Mag 11.31v star E 2′9.
04 26 33.7	−53 11 14	IC 2073	13.8	1.4 x 0.6	13.4	SB(s)cd? pec	49	Very patchy.
04 28 30.9	−53 44 17	IC 2079	13.9	1.2 x 0.4	13.0	SB(rs)ab	123	Galaxies **ESO-LV 1570332** N 1′7; **ESO-LV 1570331** WNW 1′1.
04 29 01.1	−53 36 49	IC 2081	13.3	1.0 x 0.8	13.0	SB0⁻?	92	Numerous small, faint PGC galaxies close in surrounding area.
04 29 07.6	−53 49 40	IC 2082	12.7	1.3 x 0.8	12.6	S0 pec	136	Mag 8.11v star SE 9′2; entire surrounding area filled with stellar PGC galaxies.
04 30 44.5	−53 58 54	IC 2083	14.1	1.0 x 0.6	13.4	SB(s)0/a pec	128	Bright E-W bar.
04 31 24.2	−54 25 01	IC 2085	13.1	2.4 x 0.5	13.2	S0° pec sp	119	Mag 11.08v star N 0′9.
04 31 32.2	−53 38 53	IC 2086	14.6	0.4 x 0.4	12.7	E		Mag 10.21v star SE 5′4; many stellar PGC galaxies in surrounding area.
03 10 05.2	−53 19 59	NGC 1249	11.8	4.9 x 2.3	14.2	SB(s)cd III	86	Several strong knots; mag 10.04v star S 7′1.
03 20 06.7	−52 11 13	NGC 1311	13.0	3.0 x 0.8	13.8	SB(s)m? sp IV	40	**ESO 200-2** SW 11′9.
03 46 15.6	−59 48 36	NGC 1463	13.5	1.4 x 1.2	13.9	(R)SA(rs)a?	45	Mag 10.31v star NNW 4′5.
03 58 14.1	−52 19 44	NGC 1500	13.8	1.0 x 0.9	13.8	SA(r)0⁻:	101	Round core with faint outer shell; mag 10.15v star SE 4′6; **PGC 14188** S 6′2.
04 00 21.7	−52 34 25	NGC 1506	13.4	1.0 x 0.7	12.9	SA0⁻	80	Mag 12.4 star N 5′0; mag 9.91v star S 8′2; **IC 2021**, **IC 2023** SW 9′6.
04 04 02.2	−54 06 10	NGC 1515	11.2	5.4 x 1.3	13.2	SAB(s)bc II	17	NGC 1515A on W edge.
04 03 50.2	−54 06 47	NGC 1515A	14.3	1.2 x 0.9	14.2	SB(r)b? II	146	Located on W edge of NGC 1515.
04 06 08.4	−52 40 15	NGC 1522	13.6	1.2 x 0.8	13.4	(R′)S0°: pec	42	
04 07 20.4	−62 53 59	NGC 1529	13.3	1.2 x 0.3	12.1	S0°: sp	164	Mag 8.91v star N 9′0.
04 09 52.5	−56 07 09	NGC 1533	10.7	2.8 x 2.3	12.6	SB0⁻	151	Mag 112.2 star NE 2′1; mag 9.73v star NW 7′5.
04 08 46.5	−62 47 54	NGC 1534	12.7	1.7 x 0.8	12.9	SA(rs)0/a: sp	76	Mag 13.5 star on S edge; mag 8.91v star WNW 6′6.
04 11 00.2	−56 29 00	NGC 1536	12.5	2.0 x 1.4	13.5	SB(s)c pec: III-IV	155	Mag 10.54v star E 2′9; **PGC 14585** W 9′3.
04 12 42.2	−57 44 11	NGC 1543	10.5	3.8 x 2.8	13.0	(R)SB(s)0°	93	Published size does not include large, extremely faint envelope that has a diameter of 7′0.
04 14 36.8	−56 03 36	NGC 1546	10.9	3.0 x 1.7	12.6	SA0⁺?	147	Mag 10.84v star 1′8 W.
04 15 45.0	−55 35 31	NGC 1549	9.8	4.9 x 4.1	13.0	E0-1	135	Mag 8.71v star 3′4 S; mag 12.7 star 1′9 E of large, bright center.
04 16 11.5	−55 46 51	NGC 1553	9.4	4.5 x 2.8	12.0	SA(r)0°	150	Star superimposed S of core; mag 11.25v star E 9′5.
04 17 37.5	−62 46 59	NGC 1559	10.7	3.5 x 2.0	12.6	SB(s)cd III	64	Many bright knots.
04 20 00.8	−54 56 14	NGC 1566	9.7	8.3 x 6.6	13.9	SAB(s)bc I-II	60	Multi-branching arms with knots; mag 12.2 star superimposed E of core; mag 8.14v star W 5′3.
04 21 58.9	−56 58 26	NGC 1574	10.4	4.0 x 3.8	13.3	SA(s)0⁻:	134	Mag 9.71v star in SE quadrant.
04 23 46.5	−51 35 58	NGC 1578	13.1	1.2 x 1.1	13.3	SA(s)a pec:	177	
04 24 44.9	−54 56 33	NGC 1581	12.9	1.8 x 0.7	13.0	S0⁻	80	Close pair of mag 9.84v, 10.13v stars E 6′5.
04 27 38.5	−55 01 34	NGC 1596	11.2	3.7 x 1.0	12.4	SA0: sp	20	Mag 15.6 star on W edge.
04 27 55.1	−55 03 25	NGC 1602	13.0	1.9 x 1.1	13.6	IB(s)m pec: IV-V	83	Large, bright knot W end; several faint stars superimposed.
04 31 40.3	−54 36 12	NGC 1617	10.4	4.3 x 2.1	12.7	SB(s)a	113	**PGC 15473** ENE 9′1.
04 45 42.2	−59 14 41	NGC 1672	9.7	6.6 x 5.5	13.4	SB(s)b II	170	Numerous stars superimposed, especially along the E side.
04 48 23.9	−59 47 44	NGC 1688	12.1	2.6 x 1.9	13.7	SB(rs)d III	177	Several superimposed stars and/or bright knots.
04 52 52.9	−59 44 40	NGC 1703	11.3	2.8 x 2.6	13.3	SB(r)b I-II	6	Numerous small stars and bright knots; one brighter star in SE quadrant.
04 54 14.4	−53 21 36	NGC 1705	12.4	1.9 x 1.4	13.3	SA0⁻ pec:	50	
04 52 31.0	−62 59 08	NGC 1706	12.6	1.4 x 1.0	12.8	SA(rs)ab I	124	
04 58 24.4	−62 01 44	NGC 1765	13.0	1.2 x 1.0	13.0	SB(rs)0⁻:	150	Stellar nucleus.
05 02 43.0	−61 08 21	NGC 1796	12.3	1.9 x 1.0	12.9	SB(rs)c pec: III	102	

GALAXY CLUSTERS

RA h m s	Dec ° ′ ″	Name	Mag 10th brightest	No. Gal	Diam ′	Notes
03 23 00.0	−52 01 00	A 3123	16.2	37	17	**PGC 12668** at center; **PGC 12632** at NNW edge, all others anonymous.
03 27 24.0	−53 30 00	A 3125	15.8	46	17	**ESO 155-31**, **31A** N of center 8′8; **ESO 155-37** SW 10′0.
03 30 12.0	−52 33 00	A 3128	15.3	140	22	Numerous IC and PGC GX included; **IC 1945**, **46** SW of center 8′7.
03 37 06.0	−55 01 00	A 3144	15.8	54	17	**PGC 13351** at center; most members anonymous, stellar.
03 43 00.0	−53 38 00	A 3158	15.8	85	17	Tight concentration of PGC GX in center one third diameter.
03 45 48.0	−57 02 00	A 3164	15.7	33	17	**IC 1999** ENE of center 16′4; most other members anonymous.
03 58 12.0	−52 20 00	A 3193	15.6	41	22	NGC 1500 at center, most other members anonymous.
04 00 12.0	−53 39 00	A 3202	15.8	65	17	**IC 2024** on N edge, most other members anonymous, stellar.
04 09 18.0	−59 36 00	A 3225	15.9	37	17	**PGC 14575** at center; **ESO 118-8**, **8A**, **8B** E of center 12′6.
04 31 12.0	−61 28 00	A 3266	15.5	91	22	Staggered line of PGC GX stretch N and S of ESO 118-30 at center.
02 59 12.0	−51 01 00	AS 315	16.3		17	**Fair 733** at center, all other members anonymous, stellar.
03 19 18.0	−53 52 00	AS 339	15.5	15	22	Line of 7 PGC galaxies extend N-S through center.
03 26 24.0	−51 24 00	AS 353	16.0		17	See notes for IC 1929; other members anonymous, stellar.
03 34 54.0	−53 35 00	AS 366	16.3		17	**PGC 13251** at center; **PGC 13293** SE 13′7; all others anonymous.
03 37 12.0	−55 25 00	AS 372	16.1	7	17	**PGC 13358** at center; **ESO 156-2** NW 9′3; all others anonymous.
03 40 42.0	−55 12 00	AS 377	15.6	19	22	**PGC 13533** at center; **IC 1987** NNW 9′5.
03 56 24.0	−53 52 00	AS 404	15.7	23	17	**PGC 14135** at center; **Fair 760** N 9′5; **Fair 761** S 8′1.
03 58 36.0	−59 35 00	AS 410	15.9		17	**Fair 18** at center; **Fair 17** N 11′9; **Fair 20** E 12′6.
04 01 42.0	−56 53 00	AS 412	16.3		17	**PGC 14306** S of center 5′1; all others anonymous, stellar.
04 03 42.0	−60 48 00	AS 415	15.6	1	22	**ESO 117-19A** W of center 6′8; **ESO 117-20**, **20A** NE 8′9; all others anonymous.
04 08 30.0	−60 49 00	AS 426	15.6		22	**ESO 118-6** at center; **ESO 118-5** N 15′7, all others anonymous.
04 08 48.0	−61 26 00	AS 429	16.1		17	**ESO 118-4**, **4A** N of center 10′6, all others anonymous.
04 29 12.0	−53 49 00	AS 463	15.3	26	22	Numerous PGC galaxies, mostly on N-S line through center.
04 37 48.0	−51 26 00	AS 479	16.2		17	See notes for ESO 202-43, all others anonymous.
04 46 42.0	−62 27 00	AS 500	14.9		22	

OPEN CLUSTERS

RA h m s	Dec ° ′ ″	Name	Mag	Diam ′	No. ★	B ★	Type	Notes
03 10 56.0	−57 46 00	NGC 1252		9	18	6.7	ast	

GLOBULAR CLUSTERS

RA h m s	Dec ° ′ ″	Name	Total V m	B ★ V m	HB V m	Diam ′	Conc. Class Low = 12 High = 1	Notes
03 12 15.3	−55 13 01	NGC 1261	8.3	13.5	16.8	6.8	2	
04 36 11.3	−58 51 48	Reticulum		14	2	5.9		

RA h m s	Dec ° ′ ″	Name	Mag (V)	Dim ′ Maj x min	SB	Type Class	PA	Notes
01 04 13.1	−62 31 17	ESO 79-15	14.0	1.1 x 1.0	13.9	SB(rs)cd	161	Mag 11.2 star E 9′.0.
01 13 58.4	−62 24 02	ESO 80-1	14.3	1.3 x 0.9	14.3	SB(s)cd IV	93	Mag 10.18v star SW 2′.5.
01 43 08.0	−62 28 24	ESO 80-4	13.9	0.9 x 0.5	12.9	S pec	32	Mag 13.0 star NW 1′.0.
01 47 16.4	−62 58 19	ESO 80-6	13.9	1.6 x 0.9	14.1	SB(s)m IV	66	Strong knot of superimposed star N of center; mag 10.81v star NE 2′.1.
01 53 12.6	−62 25 32	ESO 81-1	13.7	1.1 x 0.5	12.9	S0	100	Galaxy **AM 0151-624** SSW 2′.6.
02 42 17.8	−62 35 05	ESO 82-2	13.9	1.0 x 0.3	12.4	Sb	15	Mag 10.31v star SW 3′.0.
02 57 21.3	−62 52 36	ESO 82-6	14.4	1.0 x 0.5	13.5	SBc pec	128	Mag 9.50v star SW 10′.1.
01 00 34.3	−57 44 57	ESO 113-4	14.4	0.4 x 0.3	11.9		107	Mag 8.97v star NE 2′.4.
01 01 34.7	−59 44 32	ESO 113-6	15.1	1.5 x 0.2	13.6	Sc: sp	64	Lies 1′.2 N of a pair of mag 12 stars.
01 05 17.5	−58 26 15	ESO 113-10	13.7	1.2 x 0.9	13.6	(R)SB(r)a?	114	Strong dark patches N and S of core.
01 06 24.5	−58 47 10	ESO 113-11	14.2	1.1 x 0.9	14.0	S?	104	Mag 11.9 star N 1′.0.
01 08 09.1	−58 27 18	ESO 113-14	13.5	1.0 x 0.5	12.6	S?	46	Faint star NW edge.
01 14 02.0	−57 56 09	ESO 113-27	13.6	1.9 x 0.9	14.0	(R′)SB(s)c II-III	81	Several bright knots along S spiral arm.
01 14 50.8	−58 40 27	ESO 113-29	13.9	0.9 x 0.5	12.9	S	15	Mag 11.7 star ESE 1′.6.
01 15 35.2	−61 15 37	ESO 113-30	13.9	1.1 x 0.8	13.6	(R′)SAB(s)cd	46	Mag 10.22v star NNE 6′.7.
01 15 39.7	−61 18 53	ESO 113-31	13.9	1.2 x 0.3	12.6	Sa	27	ESO 113-30 N 3′.3.
01 16 04.7	−61 37 23	ESO 113-32	13.4	1.6 x 0.6	13.2	Sb II	60	Mag 8.91v star E 6′.7.
01 18 41.5	−58 43 56	ESO 113-35	13.6	1.0 x 0.9	13.4	SB(r)0$^+$	87	Strong N-S bar, smooth envelope; mag 12.2 star N 2′.7.
01 21 28.8	−61 22 35	ESO 113-41	13.7	1.5 x 0.6	13.5	SB(s)b pec	172	ESO 113-42 N 1′.5; mag 7.24v star ENE 6′.8.
01 21 30.7	−61 21 13	ESO 113-42	14.1	0.6 x 0.6	12.9	SAB0$^-$:		ESO 113-41 S 1′.5; mag 7.24v star E 6′.3.
01 21 29.7	−61 45 50	ESO 113-43	13.8	1.1 x 0.5	13.0	SB	35	Mag 14.1 star NE 1′.8.
01 23 46.0	−58 48 23	ESO 113-45	13.2	0.7 x 0.6	12.1		109	Stellar **PGC 5109** SE edge.
01 27 08.2	−58 21 22	ESO 113-48	14.3	1.2 x 1.1	14.4	S?	80	Faint star or knot N edge.
01 29 27.1	−61 27 49	ESO 113-50	13.4	1.4 x 1.1	13.7	S0$^-$ pec	44	Mag 10.55v star W 5′.1.
01 37 40.3	−60 51 52	ESO 114-1	14.0	1.3 x 0.4	13.1	SAB(r)ab	65	Small triangle of mag 13–14 stars centered NNE 4′.9.
01 44 07.1	−61 07 25	ESO 114-4	13.2	1.1 x 0.7	12.8	S0	32	Mag 6.79v star NNE 7′.1.
01 46 30.7	−58 40 20	ESO 114-7	13.8	1.8 x 1.4	14.7	SB(s)m IV-V	60	
01 47 19.8	−61 11 22	ESO 114-8	13.8	1.5 x 0.4	13.1	(R)SB0/a sp	67	Mag 11.8 star E 5′.8.
01 54 40.6	−62 06 21	ESO 114-14	13.3	1.3 x 1.3	13.8	(R′)SA(rs)bc: II-III		Almost stellar nucleus; mag 11.6 star NE 3′.5.
01 57 51.8	−61 12 52	ESO 114-16	12.9	1.7 x 1.2	13.5	(R′)SB(r)b II	68	Mag 12.0 star WSW 1′.6; mag 10.69v star NNE 7′.3.
02 01 45.7	−58 22 39	ESO 114-19	13.7	1.2 x 0.6	13.2	(R)SBa	110	Strong dark patches N and S of core.
02 01 59.3	−58 34 15	ESO 114-21	13.4	1.9 x 1.2	14.1	Sc I-II	15	Mag 9.70v star NW 9′.8.
02 03 32.4	−60 51 11	ESO 114-23	13.9	1.4 x 0.6	13.5	(R′)SAB(r)ab: III-IV	28	Mag 13.6 star E 2′.1.
02 04 54.3	−57 29 00	ESO 114-24	14.0	1.0 x 0.8	13.6	(R′)SB(r)b:	54	SE-NW bar; mag 145.7 star E 0′.9.
02 06 21.6	−61 57 47	ESO 114-27	13.7	0.8 x 0.6	12.8	S0	79	Mag 12.3 star SW 2′.4.
02 24 22.8	−58 23 49	ESO 115-8	13.2	0.8 x 0.6	12.4	E1:	136	ESO 115-9 S 2′.4.
02 24 29.5	−58 26 04	ESO 115-9	13.6	1.8 x 0.3	12.8	Sab pec sp	123	Mag 9.58v star SSW 6′.9; ESO 115-8 N 2′.4.
02 30 16.5	−58 10 13	ESO 115-13	13.6	1.1 x 1.1	13.7	SAB(r)bc		Stellar nucleus, smooth envelope.
02 31 09.4	−57 55 07	ESO 115-15	13.2	1.0 x 0.7	12.7	SB(r)0$^+$?	52	Stellar galaxy **Fair 223** S 0′.8; **ESO 115-14** N 7′.2; **ESO 115-16** N 9′.9.
02 32 23.6	−58 01 17	ESO 115-17	13.4	1.1 x 0.6	12.8	S0°: sp	109	Mag 10.70v star SSE 4′.7.
02 37 44.9	−61 20 28	ESO 115-21	12.4	4.9 x 0.7	13.6	SBdm: sp IV-V	42	Mag 7.99v star E 6′.2.
02 38 23.7	−58 14 24	ESO 115-22	15.8	1.5 x 1.5	16.5	SB(s)m: V-VI		Mag 9.56v star NE 5′.9.
02 42 15.1	−59 53 51	ESO 115-25	14.1	1.0 x 0.5	13.2		123	Double system, contact, tails; mag 9.35v star SW 7′.4.
02 52 20.9	−57 56 52	ESO 116-1	13.2	1.3 x 1.0	13.4	(R′)SB(r)bc I	147	Mag 10.06v star W 9′.1.
03 02 33.0	−57 51 39	ESO 116-5	13.3	1.0 x 0.6	12.6	(R)SB(r)0/a	33	Lies 1′.4 WSW of mag 10 star. Mag 13 star on E edge.
00 56 07.4	−53 11 22	ESO 151-4	13.6	1.3 x 0.3	12.4	S0° sp	160	Mag 6.62v star W 2′.2.
00 56 08.0	−52 49 49	ESO 151-5	14.0	1.0 x 0.6	13.3	(R′)SAB(s)0/a	171	NGC 312 and a mag 11.6 star N 2′.5.
00 56 47.2	−53 05 53	ESO 151-12	12.9	1.4 x 0.8	12.9	SA(s)0°:	118	Mag 6.62v star and ESO 151-4 SW 9′.9.
01 01 35.5	−53 11 57	ESO 151-18	13.7	1.5 x 0.3	12.7	Sa: sp	168	Mag 8.38v star NE 3′.2.
01 02 20.2	−54 19 36	ESO 151-19	14.1	1.5 x 1.1	14.5	SA(s)m: V	49	Mag 10.74v star NNE 8′.5.
01 10 34.2	−52 33 26	ESO 151-26	13.3	1.1 x 0.6	12.8	E4:	122	Pair of mag 15 stars NW edge.
01 11 00.8	−55 52 15	ESO 151-27	13.0	1.2 x 1.0	13.1	SA0$^-$	49	Mag 10.60v star W 7′.8.
01 12 16.3	−54 27 34	ESO 151-32	14.6	1.5 x 0.5	14.2	SB(rs)b? II-III	54	Mag 12.2 star NE 7′.4.
01 18 48.8	−53 17 33	ESO 151-40	13.3	1.9 x 1.3	14.1	SAB(rs)c pec I	18	Bright knot or star N of core; mag 12.2 star N 2′.0.
01 23 29.7	−56 41 16	ESO 151-43	14.3	1.4 x 0.7	14.1	SB(rs)cd: III-IV	62	Mag 8.89v star NNE 5′.7.
01 48 02.5	−56 00 08	ESO 152-25	13.9	1.1 x 0.3	12.6	S0/a	26	Mag 13.0 star N 3′.6.
01 49 28.0	−56 03 12	ESO 152-26	13.1	2.0 x 1.3	14.0	(R)SB(r)a	25	Faint anonymous galaxy E 5′.6.
01 54 40.9	−55 39 48	ESO 152-34	13.4	0.9 x 0.6	12.6	SB0	102	Bright knot or star N edge.
01 57 25.8	−55 08 24	ESO 153-1	13.9	1.0 x 0.3	12.5	S0/a:	84	Mag 9.20v star W 2′.3.
01 58 18.1	−54 13 01	ESO 153-3	12.7	1.5 x 0.7	12.7	S0$^-$: sp	22	Mag 9.57v star WNW 13′.1.
01 58 28.3	−56 15 20	ESO 153-4	13.5	1.0 x 0.7	13.0	S0$^-$ pec	9	ESO 153-4A NE 1′.3.
01 58 34.0	−56 14 26	ESO 153-4A	13.9	0.7 x 0.6	12.8	S0/a	42	ESO 153-4 SW 1′.3.
02 04 19.7	−55 13 54	ESO 153-13	13.8	1.1 x 0.4	12.8	Sa:	55	Mag 11.24v star W 2′.6.
02 04 35.0	−55 07 10	ESO 153-15	13.8	1.1 x 0.5	13.1	S0	154	Star or very small galaxy S edge.
02 04 51.0	−55 12 59	ESO 153-16	14.5	1.6 x 0.6	14.3	SB(s)cd pec	6	Very open arms.
02 05 05.5	−55 06 46	ESO 153-17	13.1	2.3 x 1.6	14.3	SAB(r)c II-III	115	Mag 12 star SE edge.
02 05 41.9	−52 48 10	ESO 153-18	13.2	1.2 x 0.8	13.1	E3:	138	Mag 12.4 star SE 1′.3.
02 05 39.5	−56 04 12	ESO 153-19	14.3	1.6 x 0.7	14.3	SB(s)cd? IV-V	52	Mag 9.55v star SW 8′.8.
02 06 03.6	−55 11 41	ESO 153-20	13.0	1.6 x 1.0	13.4	(R′)SB(rs)b pec	2	Mag 7.02v star NNE 7′.6; almost stellar **ESO 153-21** E 4′.3.
02 06 43.9	−56 56 09	ESO 153-23	13.7	1.2 x 0.8	13.5	(R′)SAB(r)0/a	1	Stellar nucleus with smooth envelope; mag 11.7 star SSW 1′.7.
02 06 53.5	−52 34 43	ESO 153-24	14.0	0.9 x 0.8	13.5	SAB(rs)ab:	170	Mag 12.6 star WNW 8′.4.
02 10 31.7	−53 50 12	ESO 153-27	13.5	1.6 x 0.6	13.3	(R)SB(r)0/a:	149	Mag 11.06v star W 7′.3.
02 14 13.2	−54 41 09	ESO 153-29	13.8	0.9 x 0.5	12.8	(R′)SAB(r)bc:	72	ESO 153-30 SE 3′.6.
02 14 28.0	−54 44 02	ESO 153-30	14.1	1.1 x 0.3	12.7	S	4	Mag 12.2 star E 2′.5.
02 25 49.9	−53 05 20	ESO 153-34	13.0	1.2 x 0.9	13.0	SA(rs)0$^+$?	15	Bright nucleus, smooth envelope; mag 12.4 star E 0′.8.
02 32 22.0	−52 30 19	ESO 154-2	14.3	0.8 x 0.6	13.3		18	Faint anonymous galaxy NE 1′.4.
02 42 55.8	−54 34 35	ESO 154-9	13.9	1.7 x 0.6	13.8	SA0$^-$:	82	Mag 10.68v star N 2′.9.
02 45 08.8	−55 44 25	ESO 154-10	12.3	2.6 x 1.1	13.3	(R′)SB(r)a:	93	Mag 9.92v star S 7′.3.
02 46 17.6	−55 27 29	ESO 154-13	12.7	1.4 x 0.8	12.7	Sb: II	178	Star on E edge.
02 54 12.4	−52 54 49	ESO 154-22	14.2	1.4 x 0.8	14.2	SA0$^-$	140	Mag 12.0 star SW 9′.6; stellar galaxy **Fair 380** is 2′.9 SW of this star; **ESO 154-21** N 8′.9.
02 56 51.7	−54 34 18	ESO 154-23	12.4	7.4 x 1.3	14.7	SB(s)m V-VI	40	Mag 9.39v star on S edge.
03 02 08.0	−54 24 42	ESO 154-28	14.9	1.2 x 0.8	14.7	SAB(s)d IV	9	Mag 9.56v star NNE 6′.4.
01 01 44.4	−51 52 10	ESO 195-24	13.1	0.9 x 0.8	12.6	(R′)SAB(rs)a:	96	Mag 9.61v star WNW 5′.8.
01 04 30.0	−51 27 39	ESO 195-29	14.2	1.2 x 0.6	13.7	SAB(s)c II-III	117	Mag 13.4 star SW 2′.1.
01 19 28.5	−51 46 48	ESO 196-3	13.7	1.2 x 1.0	13.8	(R)SB0/a	66	Located between a pair of stars, NE-SW, mags 10.99v and 13.1.
01 30 43.7	−51 08 33	ESO 196-11	13.6	1.6 x 0.6	13.4	SB(r)cd? II-III	12	Mag 7.77v star SE 5′.2; **ESO 196-9** SW 7′.0.
01 59 03.5	−52 10 01	ESO 197-15	15.6	1.1 x 0.9	15.4		156	Double system, contact.
02 04 32.4	−52 10 23	ESO 197-21	12.3	1.4 x 1.4	13.0	S0$^-$:		Mag 11.24v star W 11′.7; small, faint anonymous galaxy NE 4′.9.
02 06 20.8	−52 01 44	ESO 197-24	14.4	1.6 x 0.2	13.0	Sc: sp	60	Mag 7.49v star W 7′.4.

(Continued from previous page)

203 GALAXIES 203

RA h m s	Dec ° ′ ″	Name	Mag (V)	Dim ′ Maj x min	SB	Type Class	PA	Notes
02 22 15.7	−51 06 32	ESO 198-8	13.1	0.9 x 0.9	12.8	S0		Mag 5.90v star E 6′.3.
02 36 04.1	−51 20 47	ESO 198-21	13.9	1.1 x 0.4	12.9	S	99	Mag 12.1 star NNW 3′.7.
02 42 10.1	−51 54 15	ESO 198-27	12.9	1.1 x 1.0	12.9	S0	12	Close pair of mag 13.5, 13.8 stars SE 2′.5.
01 04 06.9	−51 07 58	IC 1615	13.4	1.3 x 0.6	12.9	SB(r)bc I-II	147	Mag 13.6 star E 3′.7; IC 1617 N 6′.2.
01 04 16.9	−51 02 00	IC 1617	13.6	1.4 x 0.6	13.3	(R′)S(r)0/a?	125	Mag 11.28v star E 5′.6.
01 11 51.7	−55 51 31	IC 1649	13.7	1.7 x 0.3	12.8	Sbc sp	136	Mag 10.24v star ENE 7′.6.
00 56 16.2	−52 46 59	NGC 312	12.4	1.4 x 1.1	12.9	E2:	62	Mag 11.6 star W 2′.5.
00 56 41.8	−52 58 35	NGC 323	12.6	1.3 x 1.1	12.9	E0:	175	Small faint anonymous galaxy on S edge; NGC 328 NNE 4′.0.
00 56 58.0	−52 55 28	NGC 328	13.2	2.7 x 0.5	13.3	SB(rs)a pec:	100	Galaxy **ESO 151-10** NW 2′.7.
01 00 52.1	−53 14 43	NGC 348	13.7	0.8 x 0.7	13.0	Sb II	94	Mag 14.4 star on N edge; mag 8.40v star SE 8′.8.
01 11 46.7	−61 31 38	NGC 432	12.9	1.3 x 1.2	13.3	S0^{-}	117	
01 12 14.3	−58 14 55	NGC 434	12.0	2.1 x 1.2	12.8	SAB(s)ab I-II	6	Superimposed star or bright knot NE edge of core; NGC 434A NNE 3′.2.
01 12 29.8	−58 12 31	NGC 434A	15.0	1.2 x 0.3	13.8	SB(s)0/a pec sp	51	Very thin, curving arms; similar to the "Integral Sign" galaxy UGC 3697.
01 12 48.7	−58 17 00	NGC 440	13.2	1.1 x 0.7	12.7	SA(s)bc pec: III-IV	45	Located 2′.7 NW of a mag 10.45v star.
01 14 23.2	−55 24 15	NGC 454	12.3	2.0 x 2.0	13.7	Pec		Double system, strongly interacting.
01 17 13.6	−58 54 35	NGC 466	12.7	1.8 x 1.5	13.7	SA(rs)0^{+}:	103	
01 19 34.6	−58 31 32	NGC 484	12.1	1.9 x 1.4	13.0	SA0^{-}	94	
01 28 57.5	−51 35 54	NGC 576	13.4	0.9 x 0.7	12.8	(R′)SAB(rs)0^{+}	18	
01 47 42.5	−52 45 48	NGC 685	11.0	3.7 x 3.3	13.6	SAB(r)c II-III	74	Several knots and superimposed stars.
01 54 09.6	−56 41 26	NGC 745	12.6	1.3 x 0.8	12.5	S0^{+}: pec	50	Multi-galaxy system; mag 10.15v star E 4′.4.
01 54 21.1	−56 45 39	NGC 754	14.2	0.6 x 0.5	12.9	E0 pec:	93	Compact.
01 57 41.1	−57 47 29	NGC 782	11.9	2.3 x 2.0	13.4	SB(r)b II	23	Large core with faint outer envelope.
01 59 49.7	−55 49 28	NGC 795	13.2	1.2 x 0.7	13.0	S0^{-}:	141	
02 08 55.5	−56 44 14	NGC 852	13.4	1.3 x 1.0	13.5	SB(rs)bc: III-IV	74	
02 17 27.3	−59 51 43	NGC 888	13.0	0.8 x 0.8	12.6	E1: pec		Star or knot E of core; **ESO 115-3** NNE 9′.2.
02 36 20.1	−54 51 50	NGC 1025	13.8	0.9 x 0.5	12.8	Sb II	6	
02 36 38.7	−54 51 35	NGC 1031	12.5	1.9 x 1.1	13.2	SB(r)a:	23	
02 43 49.4	−59 54 49	NGC 1096	12.8	1.9 x 1.8	13.9	SB(rs)bc II	50	Two main arms; mag 9.83v star S 9′.0.
02 50 47.7	−54 55 47	NGC 1135	14.8	0.7 x 0.4	13.2	Sd? pec	72	NGC 1136 S 3′.0.
02 50 53.8	−54 58 33	NGC 1136	13.0	1.4 x 1.2	13.4	SB(r)a? II	80	Strong dark area S of core; mag 8.33v star SSE 7′.2.
01 36 59.6	−62 32 23	PGC 5993	14.2	1.0 x 0.8	13.8	SA0°	39	Mag 10.63v star W 10′.6; faint anonymous galaxy NNE 7′.1.

GALAXY CLUSTERS

RA h m s	Dec ° ′ ″	Name	Mag 10th brightest	No. Gal	Diam ′	Notes
01 01 42.0	−51 51 00	AS 116	15.8	25	17	ESO 195-24 at center, all others anonymous, stellar.
01 11 48.0	−61 32 00	AS 137	15.6		22	NGC 432 at center; **ESO 113-19** W 10′.2; most others anonymous.
01 31 42.0	−51 20 00	AS 162	15.8	22	17	Elongated **ESO 196-12** N of center 5′.2; most others anonymous, stellar.
02 07 06.0	−61 12 00	AS 225	16.3		17	All members anonymous, stellar; Mag 7.50v star N of center 13′.0, **ESO 114-28** W of this star 4′.4.
02 22 18.0	−51 06 00	AS 250	16.3		17	Except ESO 198-8, all members anonymous, stellar.
02 31 12.0	−57 55 00	AS 274	15.6		22	
02 53 48.0	−61 53 00	AS 309	16.4		17	All members anonymous, stellar
02 54 48.0	−61 42 00	AS 312	16.4		17	All members anonymous, stellar.
02 59 12.0	−51 01 00	AS 315	16.3		17	**Fair 733** at center, all other members anonymous, stellar.

204 GALAXIES 204

RA h m s	Dec ° ′ ″	Name	Mag (V)	Dim ′ Maj x min	SB	Type Class	PA	Notes
23 43 17.1	−72 34 20	ESO 28-1	13.5	1.1 x 1.0	13.4	Sc	33	Mag 12.9 star W edge.
23 58 00.9	−73 27 12	ESO 28-7	14.8	1.6 x 0.9	15.0	SAB(s)m: IV-V	171	Mag 10.24v star SW 6′.8.
00 18 19.8	−73 09 12	ESO 28-12	13.7	1.1 x 0.6	13.1	(R)S0^{+}?	116	Mag 11.3 star NNW 3′.3.
23 44 27.2	−69 51 26	ESO 50-2	14.0	1.3 x 1.1	14.2	Scd III	144	
00 13 58.9	−70 01 26	ESO 50-6	13.6	1.6 x 1.4	14.3	SB(s)cd II-III	81	Stellar nucleus; mag 12.7 star SSE 1′.9.
01 00 00.3	−68 07 42	ESO 51-11	13.4	1.1 x 0.9	13.2	(R)SA(r)ab pec? I-II	108	Mag 8.86v star NE 4′.1.
22 42 12.7	−67 47 05	ESO 76-21	13.7	1.0 x 0.4	12.6	Sa	163	Mag 8.84v star S 2′.9.
22 46 23.7	−71 11 14	ESO 76-22	14.1	1.0 x 0.9	13.8	SB(rs)d IV	20	Star E of nucleus.
22 55 25.4	−70 37 53	ESO 76-30	13.2	1.3 x 0.6	12.8	SB(rs)0/a:	179	Mag 11.9 star SW 1′.0; mag 10.39v star W 7′.8.
22 55 57.0	−70 34 27	ESO 76-31	12.9	1.1 x 0.9	12.9	S0^{-} pec:	176	Mag 8.33v star E 9′.2.
23 04 39.7	−72 06 59	ESO 77-2	13.7	0.9 x 0.6	12.9	S0	56	Mag 8.53v star NW 12′.5.
23 09 39.4	−72 01 31	ESO 77-3	14.1	1.1 x 0.7	13.6	S?	71	Mag 10.99v star S 5′.5; ESO 77-4 N 1′.6.
23 09 44.3	−72 00 01	ESO 77-4	14.5	0.7 x 0.2	12.2	Pec	122	ESO 77-3 S 1′.6.
23 34 28.1	−70 37 54	ESO 77-23	13.1	1.2 x 0.9	13.1	(R)SB(r)a:	10	Mag 9.55v star NE 11′.2.
23 43 50.1	−67 42 57	ESO 77-28	14.0	1.6 x 0.8	14.1	(R′)SB(r)0/a pec	33	Star on E and W edges.
23 57 11.4	−64 40 56	ESO 78-10	14.2	0.7 x 0.4	12.7		25	Faint anonymous galaxy SE 1′.2; mag 10.93v star SW 2′.0.
00 20 56.9	−63 51 25	ESO 78-22	13.8	2.1 x 0.4	13.5	Sb: sp III-IV	170	Mag 9.11v star NE 15′.0.
00 23 54.1	−62 16 21	ESO 78-23	14.1	0.7 x 0.5	12.8		122	Mag 10.72v star NE 2′.2.
00 25 49.2	−62 19 49	ESO 79-1	13.9	1.1 x 0.6	13.3	SAB(rs)c: I-II	126	Mag 8.14v star NW 9′.6.
00 32 00.8	−64 23 26	ESO 79-2	13.7	1.9 x 0.4	13.2	SB(s)d: pec	115	Mag 10.54v star NW 8′.1; ESO 79-3 N 8′.2.
00 32 01.9	−64 15 09	ESO 79-3	12.6	2.8 x 0.5	12.8	SBb? sp	132	Mag 11.5 star SE end; mag 10.54v star SW 6′.2; ESO 79-2 S 8′.2.
00 40 43.4	−63 26 38	ESO 79-5	13.0	2.1 x 1.0	13.7	SB(rs)d pec: IV	1	Mag 12.3 star NW 7′.9.
00 50 02.0	−66 33 11	ESO 79-7	12.8	1.6 x 1.2	13.3	SA0° pec	179	Mag 11.43v star W 6′.7.
00 54 14.0	−62 27 20	ESO 79-8	13.8	1.3 x 0.8	13.7	SB(s)bc I	153	Mag 9.35v star SW 9′.9.
00 54 39.7	−63 17 09	ESO 79-10	13.8	1.3 x 0.5	13.2	Sbc	171	Mag 12.8 star NE 2′.2.
00 56 51.2	−63 28 54	ESO 79-13	14.1	1.0 x 0.6	13.4	SA(r)c pec	145	Pair with ESO 79-13A SW edge.
00 56 45.9	−63 29 17	ESO 79-13A	14.0	0.7 x 0.6	13.1	E0-1	143	Pair with ESO 79-13 NE edge.
01 04 13.1	−62 31 17	ESO 79-15	14.0	1.1 x 1.0	13.9	SB(rs)cd	161	Mag 11.2 star E 9′.0.
01 04 27.0	−64 07 34	ESO 79-16	13.5	1.0 x 0.3	12.0	S?	177	Bright star superimposed N end; mag 8.73v star SW 3′.0.
01 13 58.4	−62 24 02	ESO 80-1	14.3	1.3 x 0.9	14.3	SB(s)cd IV	93	Mag 10.18v star SW 2′.5.
22 48 33.8	−66 43 18	ESO 109-23	14.3	0.4 x 0.4	12.2	Pec		Between a pair of stars on N-S line, mags 11.16 and 10.29.
23 03 46.1	−62 31 19	ESO 109-32	14.2	1.7 x 0.3	13.3	SAB(s)c? sp III	128	Mag 7.78v star NNW 9′.3.
23 04 18.3	−63 26 00	ESO 109-34	14.1	1.4 x 1.1	14.4	SB(s)d II-III	59	Strong SE-NW bar, patchy envelope.
23 19 09.4	−66 40 35	ESO 110-8	13.6	0.8 x 0.5	12.5	SAB(rs)ab	43	Mag 13.6 star SW 1′.1.
23 34 58.7	−66 40 48	ESO 110-13	13.8	1.0 x 0.5	12.9	Sb II	143	Mag 12.3 star W edge.
23 35 11.4	−65 35 13	ESO 110-14	14.5	1.3 x 0.7	14.3	SB(rs)cd IV-V	39	Mag 11.01v star NNW 9′.6.
23 36 42.8	−62 48 54	ESO 110-16	13.9	1.0 x 0.5	13.0	SAB(r)0/a	66	Mag 8.13v star N 3′.8.
23 45 45.2	−63 18 15	ESO 110-25	14.3	1.4 x 1.0	14.5	Sc	172	Double system? Open arms.

(Continued from previous page)

GALAXIES

RA h m s	Dec ° ′ ″	Name	Mag (V)	Dim ′ Maj x min	SB	Type Class	PA	Notes
23 47 08.7	−63 21 22	ESO 110-27	13.4	1.1 x 0.5	12.6	S0 pec	131	Mag 11.7 star W 2′.9.
23 47 22.0	−63 46 26	ESO 110-28	13.5	1.1 x 0.9	13.3	(R′)SB(r)ab:	37	Mag 10.05v star N 6′.0.
23 52 10.6	−62 34 20	ESO 110-29	13.7	1.3 x 0.2	12.0	Sa	96	Mag 15 star on N edge; **ESO 78-9** ENE 2′.0.
00 09 54.4	−64 22 17	IC 1532	14.0	1.6 x 0.4	13.4	Sb sp III-IV	74	Mag 9.85v star N 5′.9.
22 44 13.8	−64 02 31	IC 5244	12.6	2.9 x 0.4	12.6	Sb sp	178	Mag 6.93v star N 5′.9.
22 46 39.6	−64 53 53	IC 5246	13.8	1.0 x 0.6	13.1	S	151	Mag 10.98v star W 6′.8; IC 5249 NNE 4′.8.
22 46 50.1	−65 16 26	IC 5247	13.9	1.4 x 0.4	13.1	S	124	Mag 9.51v star SE 10′.6.
22 47 06.5	−64 49 52	IC 5249	13.6	3.6 x 0.4	13.8	SBd sp	14	Mag 12.4 star 1′.4 W of south end; IC 5246 SSW 4′.8.
22 47 18.5	−65 03 23	IC 5250	11.1	3.2 x 3.2	13.5	S0 pec?		Multiple galaxy system.
22 48 09.5	−68 54 11	IC 5252	12.9	1.2 x 0.8	12.7	(R′)SB(rs)bc II	172	Mag 8.66v star NE 7′.1.
22 49 45.3	−68 41 26	IC 5256	13.9	1.0 x 0.5	13.0	SB(s)dm pec IV	19	Mag 8.66v star SSW 8′.2.
22 58 13.9	−69 03 09	IC 5263	13.2	1.4 x 0.7	13.1	SB(r)0$^+$	146	Mag 10.67v star NE 6′.6.
22 58 21.0	−65 07 47	IC 5266	13.7	1.5 x 0.4	13.0	Sb: sp	33	Mag 10.06v star SSW 4′.7.
22 59 30.9	−65 11 39	IC 5272	14.1	1.2 x 0.8	13.9	I?	33	Mag 9.64v star N 9′.8; stellar galaxy **Fair 201** NNE 4′.7.
23 03 02.6	−69 12 36	IC 5279	13.5	1.3 x 0.7	13.2	Sa: pec	27	Mag 11.7 star E 6′.4.
23 03 50.2	−65 12 29	IC 5280	14.0	1.4 x 0.5	13.5	(R′)SA(r)ab:	4	Almost stellar **IC 5277** E 11′.7.
23 11 44.5	−68 05 38	IC 5288	13.7	0.8 x 0.8	13.1	S0°: pec		Mag 9.80v star S 5′.1.
23 18 59.4	−69 33 44	IC 5301	14.6	1.4 x 0.6	14.2	S	87	Mag 14.7 star on E edge; mag 10.69v star S 4′.3.
23 19 36.4	−64 34 10	IC 5302	13.9	1.1 x 0.3	12.6	SB0°?	23	Located 2′.2 N of mag 9.63v star.
23 27 36.3	−67 48 56	IC 5323	13.0	1.6 x 1.1	13.4	SB(rs)0/a	160	Mag 10.90v star WNW 9′.0.
23 28 17.8	−67 49 17	IC 5324	13.0	1.1 x 1.1	13.2	SA0$^-$:		**IC 5320** and **IC 5322** N 3′.7.
23 38 06.0	−68 26 34	IC 5339	13.5	1.0 x 0.9	13.3	SA0$^-$ pec:	114	Mag 9.95v star N 7′.5.
00 57 58.6	−72 23 23	NGC 292	2.3	320 x 205	14.1	SB(s)m pec IV	45	**Small Magellanic Cloud.**
01 02 51.4	−65 36 35	NGC 360	12.5	3.5 x 0.5	12.9	Sbc sp	144	Mag 6.29v star N 9′.2.
01 07 24.4	−69 52 36	NGC 406	12.5	2.6 x 1.0	13.4	SA(s)c: II-III	160	
22 40 23.6	−66 28 44	NGC 7329	11.3	3.9 x 2.6	13.7	SB(r)b II	107	Short, bright bar E-W.
22 45 36.4	−65 07 18	NGC 7358	12.8	1.9 x 0.5	12.6	SA0°	176	
22 55 57.3	−63 41 40	NGC 7408	12.6	1.5 x 1.2	13.1	SB(s)cd: II-III	167	Very patchy, numerous knots.
22 57 49.7	−65 02 21	NGC 7417	12.3	2.4 x 1.6	13.6	(R′)SB(r)ab	2	Bright center with dark lanes E and W; mag 9.64v star E 10′.3.
23 21 38.6	−62 07 07	NGC 7622	13.4	1.3 x 0.3	12.3	SB0$^-$?	60	On the side of a triangle of two mag 11 stars and one mag 13 star.
23 23 03.3	−67 39 10	NGC 7633	12.4	2.3 x 2.0	13.9	(R)SB(r)0/a	142	Bright center in E-W bar.
23 26 45.9	−68 01 40	NGC 7655	13.2	0.7 x 0.6	12.2	(R′)SA0°?		Mag 9.05v star W 2′.0.
23 27 14.4	−65 16 18	NGC 7661	13.4	1.8 x 1.2	14.1	SB(s)cd II-III	25	Patchy; dark lanes E and W of bright center.
23 34 52.5	−65 23 46	NGC 7697	13.5	1.6 x 0.3	12.5	Sb pec sp	87	Mag 11.01v star SSW 3′.9.
23 42 33.2	−65 57 27	NGC 7733	13.6	1.3 x 0.6	13.2	(R′)SB(rs)b pec	107	Strong dark lanes N and S of center bar.
23 42 43.0	−65 56 38	NGC 7734	13.1	1.4 x 1.1	13.4	(R′)SB(r)b pec: II	119	Strong dark patches E and W of center.
00 04 46.0	−62 03 42	NGC 7823	12.6	1.1 x 1.0	12.6	SB(s)bc? II-III	21	Small triangle of faint stars S.
22 41 49.6	−64 25 11	Tucana Dwarf	11.8	2.9 x 1.2	13.2	E4		= **PGC 69519**; mag 9.63v star W 8′.5.

GALAXY CLUSTERS

RA h m s	Dec ° ′ ″	Name	Mag 10th brightest	No. Gal	Diam ′	Notes
00 45 48.0	−63 35 00	A 2819	16.0	90	17	All members anonymous, faint and stellar.
00 58 24.0	−66 48 00	AS 112	16.3	16	17	**Fair 216** at center, all others anonymous, stellar.
01 09 12.0	−67 43 00	AS 131	16.1		17	All members anonymous, stellar.
22 46 06.0	−71 19 00	AS 1057	16.3		17	**PGC 69679** at center, except ESO 76-22, all others anonymous.

OPEN CLUSTERS

RA h m s	Dec ° ′ ″	Name	Mag	Diam ′	No. ★	B ★	Type	Notes
00 59 48.1	−72 20 02	IC 1611	12.0	1	336	14.7	cl	
01 00 01.4	−72 22 09	IC 1612	12.5	0.7	104	14.4	cl	
01 05 21.5	−72 02 35	IC 1624	12.9	0.7	130	15.7	cl	
01 11 54.4	−71 19 48	IC 1655		1			cl	
00 03 53.2	−73 28 22	Lind 1		4			cl	
00 32 55.5	−73 06 59	NGC 152		1.7			cl	
00 35 58.6	−73 09 57	NGC 176	12.7	1			cl	
00 40 30.6	−73 24 10	NGC 220	12.4	0.8	256	14.8	cl	
00 40 44.1	−73 23 00	NGC 222	12.2	0.6	128	15.3	cl	
00 41 06.2	−73 21 07	NGC 231	12.7	0.8	131	15.0	cl	
00 43 28.5	−73 36 00	NGC 241		7.5	60		ass	(A) This is not NGC 242 or part of NGC 242, which is about 10′ to the north.
00 43 33.9	−73 26 28	NGC 242		0.9			2cl	(A) Two clusters. This is not, and does not contain, NGC 241.
00 45 24.1	−73 22 44	NGC 248	13.6	0.8	20	14.8	cl	
00 45 54.4	−73 30 24	NGC 256	12.7	0.6	142	14.7	cl	
00 46 33.0	−73 05 55	NGC 261	13.4	0.4	16	13.9	cl	
00 47 11.6	−73 28 38	NGC 265	12.2	1	301	15.0	cl	
00 48 21.3	−73 31 49	NGC 269	13.3	0.6	218	15.2	cl	
00 51 14.2	−73 09 41	NGC 290	12.0	0.8	228	13.8	cl	
00 53 05.6	−73 22 49	NGC 294	12.7	0.8	410	16.1	cl	
00 53 24.8	−72 11 47	NGC 299		0.7			cl	
00 54 14.8	−72 14 28	NGC 306		0.8			cl	
00 56 18.7	−72 27 50	NGC 330	9.6	1.4	874	12.6	cl	
00 59 06.0	−72 11 00	NGC 346	10.3	5.2			cl	Involved in nebulosity which also carries the same designation.
01 02 11.3	−71 36 25	NGC 361		1.6			cl	
01 03 18.0	−72 05 00	NGC 371		7.5			cl	Involved in nebulosity.
01 03 53.5	−72 49 34	NGC 376	10.9	1	470	12.8	cl	
01 05 08.0	−71 59 50	NGC 395					cl	
01 07 55.7	−71 46 09	NGC 411		1.3			cl	
01 07 59.0	−72 21 19	NGC 416	12.6	1.2	985	15.3	cl	
01 08 19.5	−72 53 02	NGC 419	11.2	2.4	1762	15.4	cl	
01 09 21.5	−72 46 00	NGC 422	13.4				cl	
01 13 43.9	−73 17 35	NGC 456		2			cl	Open cluster and emission nebula; westernmost in a chain of small clusters; in the SMC. Involved in nebulosity which also carries the same designation.
01 14 53.8	−71 32 58	NGC 458		1.5			cl	
01 14 37.9	−73 18 28	NGC 460	12.5	1			cl	(A) Involved nebula is **Henize N 84A**.
01 15 42.7	−73 19 27	NGC 465	11.5	2			cl	

(Continued from previous page)

204 GLOBULAR CLUSTERS 204

RA h m s	Dec ° ′ ″	Name	Total V m	B ★ V m	HB V m	Diam ′	Conc. Class Low = 12 High = 1	Notes
00 24 05.2	−72 04 51	NGC 104	4.0	11.7	14.1	50.0	3	**47 Tucanae**.
00 26 41.3	−71 31 24	NGC 121				1.5		
01 03 14.3	−70 50 54	NGC 362	6.8	12.7	15.4	14.0	3	

BRIGHT NEBULAE

RA h m s	Dec ° ′ ″	Name	Dim ′ Maj x min	Type	BC	Color	Notes
00 45 24.0	−73 23 00	NGC 248		E			Emission nebula in the SMC.
00 45 24.0	−73 05 00	NGC 249		E			A small, round emission nebula in a rich star field; in the SMC.
00 45 54.0	−73 30 00	NGC 256					
00 46 30.0	−73 07 00	NGC 261		E?			Larger and more diffuse than NGC 249; includes a central star. This nebula lies in a rich star field in the SMC.
00 48 00.0	−73 17 00	NGC 267					Nebula involved in open cluster; in SMC.
00 53 06.0	−73 23 00	NGC 294					
00 54 00.0	−72 10 00	NGC 299					
00 54 54.0	−72 13 00	NGC 306					
00 59 06.0	−72 11 00	NGC 346	14 x 11	E			
01 14 00.0	−73 17 00	NGC 456	15 x 15	E			

205 GALAXIES 205

RA h m s	Dec ° ′ ″	Name	Mag (V)	Dim ′ Maj x min	SB	Type Class	PA	Notes
21 52 15.3	−72 29 13	ESO 48-10A	14.0	0.8 x 0.6	13.0		111	Very slender **ESO 48-10** N 1′.2.
21 52 35.0	−73 17 48	ESO 48-11	13.9	1.0 x 0.4	12.7	Sbc pec?	102	Mag 14.6 star S edge; **ESO 48-12** E 0′.9.
21 58 28.0	−73 51 37	ESO 48-13	14.1	1.1 x 0.7	13.7	(R′)SA0°: pec	123	Mag 8.99v star E 6′.2.
22 02 21.8	−73 42 08	ESO 48-14	14.1	1.0 x 0.4	13.0	Sdm:	1	Mag 13.5 star W 1′.7.
22 08 54.6	−73 22 53	ESO 48-17	14.0	1.5 x 1.0	14.3	SB(rs)d IV-V	54	Mag 14.3 star NE end; mag 9.65v star W 7′.7.
22 18 31.8	−73 36 58	ESO 48-21	13.5	1.0 x 0.5	12.6	S0/a	30	Mag 12.7 star W 2′.5.
22 25 06.8	−72 35 17	ESO 48-22	13.8	0.9 x 0.8	13.3	S0	163	Mag 13.5 star W 1′.1.
20 40 42.7	−70 41 48	ESO 74-4	14.6	1.0 x 0.3	13.1	SB	70	Mag 9.87v star SSE 7′.7.
20 45 18.3	−71 23 53	ESO 74-7	13.4	1.3 x 0.3	12.2	Sa	53	Mag 10.01v star NE 6′.9.
20 48 57.1	−69 05 41	ESO 74-9A	13.4	0.9 x 0.5	12.4		8	Located between close mag 12.7 and 12.1 stars.
20 54 10.8	−69 02 19	ESO 74-20	14.9	1.1 x 0.8	14.7		139	Double system, nuclei form strong NE-SW bar; mag 9.58v star S 2′.0.
21 08 39.0	−67 19 05	ESO 74-25	14.2	0.9 x 0.8	13.7	S?	101	Mag 11.9 star WNW 2′.4.
21 22 51.1	−67 39 31	ESO 75-5	13.9	1.0 x 0.5	13.1	S0	38	Mag 8.75v star E 12′.7.
21 23 29.6	−69 41 09	ESO 75-6	14.1	1.2 x 0.8	13.9	(R′)SB(r)c pec: III-IV	162	Mag 6.39v star SE 5′.0.
21 50 58.5	−71 24 53	ESO 75-28	12.3	1.6 x 1.1	13.0	E2	15	Mag 8.95v star N 4′.0; mag 9.11v star NNW 6′.7.
21 52 39.3	−68 45 35	ESO 75-32	13.6	1.2 x 0.6	13.1	S?	135	Mag 14.3 star N edge.
21 55 32.7	−69 47 44	ESO 75-39	14.1	1.5 x 0.3	13.1	S	80	Mag 10.38v star W 6′.3.
21 57 06.8	−69 41 25	ESO 75-41	13.3	1.5 x 0.7	13.3	SA0⁻	129	Mag 9.04v star NE 6′.6.
21 57 36.2	−70 10 23	ESO 75-43	13.5	1.0 x 0.9	13.3	S0	111	**ESO 75-42** W 1′.6.
22 06 52.0	−69 56 09	ESO 75-56	14.4	1.2 x 0.8	14.2		51	
22 12 47.0	−71 06 12	ESO 76-4	14.1	1.0 x 0.3	12.6	S	66	Mag 10.4 star SSE 2′.7.
22 13 10.7	−71 44 46	ESO 76-5	14.4	0.8 x 0.6	13.5		21	Triple system, interaction.
22 16 17.3	−71 30 15	ESO 76-11	14.1	1.0 x 0.6	13.4	Sc	21	Mag 13.8 star WNW 1′.8.
22 35 26.2	−70 16 55	ESO 76-19	13.2	1.3 x 0.9	13.2	SAB(rs)bc I-II	144	Mag 13.1 star W 1′.7.
22 42 12.7	−67 47 05	ESO 76-21	13.7	1.0 x 0.4	12.6	Sa	163	Mag 8.84v star S 2′.9.
22 46 23.7	−71 11 14	ESO 76-22	14.1	1.0 x 0.9	13.8	SB(rs)d IV	20	Star E of nucleus.
22 55 25.4	−70 37 53	ESO 76-30	13.2	1.3 x 0.6	12.8	SB(rs)0/a:	179	Mag 11.9 star SW 1′.0; mag 10.39v star W 7′.8.
22 55 57.0	−70 34 27	ESO 76-31	12.9	1.1 x 0.9	12.9	S0⁻ pec:	176	Mag 8.33v star E 9′.2.
20 20 59.6	−65 55 21	ESO 106-2	14.1	1.5 x 0.3	13.1	SB(s)c pec	27	Mag 10.28v star E 8′.5.
20 40 31.5	−66 35 27	ESO 106-9	14.9	1.9 x 0.1	12.9	Sc	93	Located 13′.3 NW of mag 5.15 star Upsilon Pavonis.
21 01 17.5	−64 57 53	ESO 107-2	14.0	0.9 x 0.8	13.5	S pec	158	Mag 12.4 star NE 3′.2.
21 03 29.6	−67 10 53	ESO 107-4	12.0	1.9 x 1.5	13.1	E1:	136	Mag 10.23v star SW 10′.2.
21 11 54.7	−64 13 41	ESO 107-14	15.0	1.2 x 0.4	14.1	SB(s)d V	177	Mag 8.42v star NE 9′.0.
21 13 43.3	−63 20 11	ESO 107-15	13.8	2.0 x 0.3	13.1	SB(s)cd: sp	42	Mag 12.1 stars S 1′.9; mag 11.9 star N 4′.5.
21 16 07.2	−64 48 59	ESO 107-16	14.5	2.6 x 0.3	14.0	SB(s)dm: sp	85	Mag 6.33v star NE 14′.6; mag 12.6 star W 4′.5.
21 33 11.3	−64 38 24	ESO 107-32	13.9	1.1 x 0.6	13.3	(R′)SB(rs)a	167	Mag 9.65v star SSE 4′.7.
21 33 43.7	−63 34 40	ESO 107-35	14.1	0.5 x 0.5	12.5	S0?		Mag 13.8 star NW edge.
21 35 47.3	−62 46 40	ESO 107-37	13.8	1.2 x 1.0	13.8	Sa	157	Bright knots E and W of center; mag 9.68 star SW edge.
21 36 23.9	−64 25 32	ESO 107-39	13.6	0.4 x 0.4	11.5			Mag 12.8 star NW 1′.8.
21 40 03.4	−63 54 35	ESO 107-44	13.0	1.2 x 0.8	12.8	SAB(s)0°?	1	Mag 14 star on E edge.
22 10 21.7	−65 30 08	ESO 108-16	13.6	1.1 x 0.7	13.2	(R)SA0°	21	Strong core, smooth envelope; mag 9.89v star SE 2′.1.
22 10 47.5	−66 52 08	ESO 108-17	13.4	1.0 x 0.6	12.7	I0? pec?	76	Mag 12.5 star W 5′.5.
22 12 17.1	−62 43 31	ESO 108-18	14.1	0.6 x 0.4	12.5		30	Stellar **ESO 108-18A** NE 0′.7.
22 15 16.4	−65 33 05	ESO 108-21	12.6	1.5 x 0.7	12.5		60	Mag 14.5 star on E edge; mag 10.49v star NE 4′.6; **IC 5182** NE 7′.7.
22 16 32.0	−64 23 24	ESO 108-23	12.9	2.0 x 1.9	14.2	SAB(s)cd? III-IV	128	Numerous faint knots or superimposed stars.
22 39 48.8	−66 43 35	ESO 109-10	14.5	1.5 x 0.5	14.2	E2 pec	24	Multiple galaxy system; mag 10.26v star SW 9′.7; note: stellar PCG 69399 is 1′.5 N of this star.
22 48 33.8	−66 43 18	ESO 109-23	14.3	0.4 x 0.4	12.2	Pec		Between a pair of stars on N-S line, mags 11.16 and 10.29.
22 13 00.0	−62 04 09	ESO 146-14	14.7	2.8 x 0.2	13.9	SBd? sp	51	Mag 9.82v star N 5′.3.
20 17 24.1	−73 53 07	IC 4964	13.8	1.3 x 0.5	13.2	SA(s)d: III-IV	162	Mag 10.80v star W 7′.5.
20 16 23.4	−70 33 54	IC 4967	13.8	0.7 x 0.5	12.6	E:	90	Mag 12.4 star W 1′.5.
20 16 57.6	−70 44 59	IC 4970	13.9	0.7 x 0.2	11.6	SA0⁻ pec:	6	Located on N edge of NGC 6872.
20 19 39.3	−70 50 54	IC 4981	13.1	0.9 x 0.3	11.5	I pec sp	135	Located on N edge of NGC 6880.
20 20 44.1	−70 59 15	IC 4985	13.9	0.9 x 0.6	13.1	S0?	66	**IC 4982** SW 2′.8; mag 12.4 star SE 2′.3.
20 23 26.9	−71 33 58	IC 4992	13.8	2.2 x 0.2	12.8	SB(s)c? sp	58	Mag 7.72v star SE 8′.3.
20 32 34.4	−72 10 05	IC 5009	13.5	1.2 x 0.6	13.0	SB(r)0⁺	90	Mag 9.71v star N 5′.7.
20 35 15.9	−73 27 11	IC 5014	13.6	0.5 x 0.5	11.9			Almost stellar; mag 10.05v star SSE 11′.1.
20 38 10.7	−67 11 09	IC 5023	13.3	1.3 x 0.4	12.4	S(r)a:	118	Mag 10.93v star E 1′.6.
20 40 09.4	−71 06 30	IC 5024	14.3	0.9 x 0.8	13.7	SB(s)c:	101	Bright N-S bar.
20 45 20.0	−67 32 26	IC 5031	12.9	0.6 x 0.4	11.2		31	Connected to IC 5032 S by faint bridge.
20 45 22.3	−67 33 10	IC 5032	12.9	0.7 x 0.4	11.4		156	Galaxy pair with IC 5031 N.
20 46 51.5	−65 01 02	IC 5038	13.4	1.4 x 0.7	13.2	(R′)SB(s)cd:	75	Mag 6.70v star N 6′.5.
20 47 46.1	−65 05 07	IC 5042	13.2	1.2 x 0.8	13.0	S?	43	Close pair or E-W oriented mag 11.4, 10.52v stars N 6′.0.
20 52 06.0	−69 12 00	IC 5052	11.2	5.9 x 0.8	12.7	SBd: sp	143	Mag 10.05v star N 4′.8.
20 53 36.0	−71 08 29	IC 5053	13.7	1.6 x 0.6	13.5	Sa: sp	55	Mag 12.2 star N 1′.7; mag 10.47v star W 7′.5.
20 53 45.7	−71 01 28	IC 5054	13.9	2.0 x 0.4	13.5	(R′)SB(r)a:	5	Mag 9.89v star on W edge.

(Continued from previous page)

205 GALAXIES 205

RA h m s	Dec ° ′ ″	Name	Mag (V)	Dim ′ Maj x min	SB	Type Class	PA	Notes
21 00 10.7	−71 48 38	IC 5069	13.6	0.9 x 0.9	13.2	(R′)SB(rs)b		Mag 10.83v star W 2'.9.
21 01 20.0	−72 38 36	IC 5071	12.5	3.4 x 0.6	13.1	SAB(s)c? II	15	Mag 11.9 star E 6'.0.
21 03 19.9	−72 41 18	IC 5073	14.0	1.1 x 0.9	13.9	SB(rs)c I-II	20	Stellar nucleus; mag 11.9 star NW 4'.3.
21 01 00.5	−63 09 12	IC 5074	14.0	0.7 x 0.6	12.9	S	166	Mag 11.6 star SE 0'.9.
21 04 38.5	−71 52 06	IC 5075	13.7	1.0 x 0.4	12.6	(R′)SA:(s:)a	149	Mag 8.75v star W 12'.4.
21 09 14.4	−63 17 25	IC 5084	12.6	1.6 x 0.8	12.7	SB(s)b? pec	137	Mag 10.16v star SW 5'.7.
21 14 22.4	−73 46 27	IC 5087	14.0	0.7 x 0.4	12.5	SB(s)b: pec	104	Mag 8.64v star NE 5'.6.
21 16 15.7	−64 27 57	IC 5092	12.0	2.9 x 2.4	13.9	SB(rs)c: III-IV	8	Strong E-W bar; mag 10.44v star WSW 12'.5.
21 17 49.6	−66 25 38	IC 5094	13.7	1.0 x 0.6	13.0	SB(r)c II	148	Mag 14.4 star on NW edge.
21 18 21.6	−63 45 39	IC 5096	12.3	3.2 x 0.5	12.6	Sbc sp	148	Mag 10.73v star W 1'.5; mag 10.26v star N 4'.2.
21 21 43.9	−65 56 00	IC 5100	13.9	1.9 x 0.4	13.5	SBc? II	114	Mag 10.89v star S 4'.4.
21 21 56.0	−65 50 11	IC 5101	13.3	1.3 x 0.9	13.3	SB(s)bc pec I-II	0	Mag 10.41v star E 2'.7.
21 26 13.7	−73 18 39	IC 5102	13.7	0.8 x 0.5	12.6	S0	124	Stellar nucleus with several stars superimposed.
21 28 38.2	−70 50 05	IC 5106	12.9	1.4 x 1.0	13.1	SB0$^-$	143	Mag 10.75v star WSW 6'.5.
21 37 05.8	−70 58 56	IC 5116	13.4	1.6 x 1.2	14.0	S?	27	Mag 9.03v star NNW 8'.7.
21 38 48.6	−64 21 03	IC 5120	13.4	1.9 x 0.6	13.4	SAB(s)bc pec? III-IV	102	Mag 11.8 star W 8'.5.
21 44 49.5	−72 25 17	IC 5123	13.6	1.5 x 0.5	13.1	SAB(s)bc? II-III	18	Mag 9.71v star N 8'.8.
21 50 24.8	−73 59 53	IC 5130	13.6	1.7 x 0.5	13.3	SA(s)bc: II	109	Mag 10.24v star ENE 9'.0.
21 53 22.2	−68 57 15	IC 5138	13.8	0.9 x 0.4	12.5	(R′)SB(r)b:	128	Mag 10.45v star W 8'.6.
21 54 16.6	−67 19 55	IC 5140	14.6	1.9 x 0.3	13.9	SBd? sp	136	Mag 10.04v star NW 8'.1.
21 55 20.2	−65 30 35	IC 5142	13.7	1.6 x 0.8	13.8	SAB(r)cd: IV	72	Mag 9.94v star SSW 4'.8.
22 04 30.1	−66 06 53	IC 5154	13.6	0.7 x 0.4	12.0	Irr	175	Mag 12.2 star NNE 8'.7.
22 10 07.4	−64 34 41	IC 5165	13.9	1.4 x 0.4	13.2	SAB(rs)ab:	45	Mag 9.80v star N 8'.5.
22 14 45.0	−69 21 56	IC 5173	14.5	1.4 x 0.3	13.4	SBc:	73	Multi-galaxy system.
22 14 55.2	−66 51 04	IC 5176	12.5	4.5 x 0.5	13.2	SAB(s)bc? sp	28	Mag 7.59v star NW 8'.6.
22 17 43.8	−65 51 29	IC 5185	14.1	0.5 x 0.4	12.2	S	68	Mag 13.3 star E 5'.0.
22 22 56.6	−65 48 15	IC 5202	13.8	1.5 x 0.6	13.5	SB(rs)c II	127	Mag 10.47v star W 7'.9; **IC 5200** and mag 12.4 star WNW 5'.4.
22 24 34.5	−65 13 39	IC 5208	13.8	1.5 x 0.3	12.8	S0$^+$: sp	64	Mag 7.67v star S 8'.5.
22 29 55.4	−65 39 43	IC 5222	12.8	1.9 x 1.1	13.4	(R)SB(s)bc?	93	Mag 11.8 star S 4'.0.
22 34 03.9	−64 41 53	IC 5227	13.1	1.4 x 1.0	13.3	(R)SB(rs)ab	177	Close pair of mag 9.97v, 10.42v stars N 7'.3.
22 44 13.8	−64 02 31	IC 5244	12.6	2.9 x 0.4	12.6	Sb sp	178	Mag 6.93v star N 5'.9.
22 46 39.6	−64 53 53	IC 5246	13.8	1.0 x 0.6	13.1	S	151	Mag 10.98v star W 6'.8; IC 5249 NNE 4'.8.
22 46 50.1	−65 16 26	IC 5247	13.9	1.4 x 0.4	13.1	S	124	Mag 9.51v star SE 10'.6.
22 47 06.5	−64 49 52	IC 5249	13.6	3.6 x 0.4	13.8	SBd sp	14	Mag 12.4 star 1'.4 W of south end; IC 5246 SSW 4'.8.
22 47 18.5	−65 03 23	IC 5250	11.1	3.2 x 3.2	13.5	S0 pec?		Multiple galaxy system.
22 48 09.5	−68 54 11	IC 5252	12.9	1.2 x 0.8	12.7	(R′)SB(rs)bc II	172	Mag 8.66v star NE 7'.1.
22 49 45.3	−68 41 26	IC 5256	13.9	1.0 x 0.5	13.0	SB(s)dm pec IV	19	Mag 8.66v star SSW 8'.2.
20 16 58.0	−70 46 06	NGC 6872	11.8	6.0 x 1.5	14.0	SB(s)b pec	66	Mag 10.41v star SE of bright center; single, long, faint, narrow arms NE and SW. IC 4970 on N edge; **PGC 64439** SE 4'.0.
20 18 18.8	−70 51 31	NGC 6876	11.3	2.8 x 2.2	13.2	E3	80	NGC 6877 on E edge; **IC 4972** SSW 4'.5.
20 18 36.0	−70 51 14	NGC 6877	12.2	1.1 x 0.6	11.8	E6	169	Located on E edge of NGC 6876.
20 19 30.0	−70 51 35	NGC 6880	12.1	2.0 x 0.9	12.6	SAB(s)0$^+$:	21	IC 4981 on N edge.
20 42 08.7	−73 37 08	NGC 6932	12.3	2.1 x 1.5	13.4	(R)SAB(r)0$^+$	115	Mag 9.40v star W 2'.8.
20 44 35.2	−68 44 54	NGC 6943	11.4	4.0 x 2.0	13.5	SAB(r)cd: I-II	130	Located 4'.9 SW of mag 9.29v star.
21 11 20.1	−64 01 29	NGC 7020	11.8	3.7 x 1.8	13.8	(R)SA(r)0$^+$	165	Bright, elongated center with dark patches E and W.
21 15 23.0	−68 17 16	NGC 7032	13.1	1.1 x 1.0	13.0	SA(r)c I-II	85	
21 35 45.0	−63 54 10	NGC 7083	11.2	3.9 x 2.3	13.5	SA(s)bc I-II	5	Numerous superimposed stars/knots S of center.
21 41 19.2	−63 54 31	NGC 7096	11.9	1.9 x 1.6	13.0	SA(s)a I	130	**ESO 107-45** NW 8'.3.
21 50 46.9	−70 20 02	NGC 7123	12.2	3.0 x 1.1	13.4	Sa sp	146	Prominent dark lane extends full length of galaxy.
22 04 49.4	−64 02 48	NGC 7179	12.8	2.0 x 0.8	13.1	SB(rs)bc	48	
22 06 51.3	−64 38 00	NGC 7191	12.9	1.6 x 0.6	12.7	SAB(s)c: I-II	136	
22 06 50.0	−64 18 59	NGC 7192	11.2	2.0 x 1.9	12.7	E$^+$:	27	
22 08 29.8	−64 42 25	NGC 7199	13.1	1.1 x 0.9	12.9	SB(r)a	30	
22 12 36.2	−68 39 42	NGC 7216	12.6	1.7 x 1.0	13.0	SA(rs)0$^-$:	133	Mag 8.75v star SE 4'.7; **ESO 76-7** NE 11'.1.
22 13 09.2	−64 50 56	NGC 7219	12.5	1.7 x 1.0	12.9	(R)SA(r)0/a pec	27	
22 40 23.6	−66 28 44	NGC 7329	11.3	3.9 x 2.6	13.7	SB(r)b II	107	Short, bright bar E-W.
22 45 36.4	−65 07 18	NGC 7358	12.8	1.9 x 0.5	12.6	SA0°	176	
22 55 57.3	−63 41 40	NGC 7408	12.6	1.5 x 1.2	13.1	SB(s)cd: II-III	167	Very patchy, numerous knots.
22 41 49.6	−64 25 11	Tucana Dwarf	11.8	2.9 x 1.2	13.2	E4		= **PGC 69519**; mag 9.63v star W 8'.5.

GALAXY CLUSTERS

RA h m s	Dec ° ′ ″	Name	Mag 10th brightest	No. Gal	Diam ′	Notes
20 33 24.0	−63 01 00	A 3687	16.4	46	17	**PGC 64937** W of center 2'.8; all others anonymous, stellar.
21 34 30.0	−62 01 00	A 3782	16.4	40	17	Most members anonymous, stellar.
21 41 36.0	−72 43 00	A 3799	15.5	50	22	
22 27 48.0	−69 01 00	A 3879	16.2	114	17	All members anonymous, stellar.
21 56 06.0	−72 02 00	AS 980	16.2		17	All members anonymous, stellar.
22 46 06.0	−71 19 00	AS 1057	16.3		17	**PGC 69679** at center, except ESO 76-22, all others anonymous.

206 GALAXIES 206

RA h m s	Dec ° ′ ″	Name	Mag (V)	Dim ′ Maj x min	SB	Type Class	PA	Notes
18 06 35.4	−73 53 01	ESO 44-25	14.0	1.1 x 0.5	13.2	SAB(rs)b? III-IV	52	Mag 7.42v star S 7'.6.
18 23 22.6	−72 44 22	ESO 45-4	14.1	1.4 x 0.8	14.1	SB(rs)d: III-IV	122	Mag 12.4 star SSE 3'.6; mag 12.0 star E 5'.3.
18 11 48.6	−67 38 19	ESO 71-4	14.4	1.6 x 0.4	13.7	SB(s)d? sp III-IV	18	Mag 10.78v star SE 3'.7.
18 27 09.8	−69 46 13	ESO 71-10	15.2	1.7 x 0.6	15.1	SAB(s)dm IV-V	105	Mag 11.00v star NW 12'.6.
18 28 51.9	−72 05 33	ESO 71-13	12.7	1.3 x 0.9	12.8	SA(r)0°:	144	Mag 10.45v star NW 3'.7.
18 38 51.6	−71 47 24	ESO 71-18	14.2	1.3 x 0.5	13.5	(R′)SB(rs)c: I-II	84	Mag 10.90v star ESE 10'.2.
18 57 49.1	−69 47 59	ESO 72-6A	14.1	1.6 x 0.9	14.3		3	
19 18 50.4	−67 26 28	ESO 72-13	13.5	0.9 x 0.8	13.0	S?	23	Mag 9.53v star S 5'.8.
19 33 52.1	−68 13 45	ESO 72-17	13.8	1.3 x 0.3	12.7	S0?	137	Mag 10.82v star W 7'.1.
19 40 39.7	−70 26 52	ESO 73-1	13.3	1.3 x 0.8	13.3		4	**ESO 73-1A** N edge; mag 10.86v star S edge.
19 44 54.9	−69 53 04	ESO 73-5	14.7	1.2 x 0.4	13.8	SAa?	13	Mag 7.82v star S 8'.0; ESO 73-6 SE 4'.3.
19 45 36.4	−69 55 15	ESO 73-6	13.8	0.6 x 0.2	11.4		58	Pair of mag 13.6 and 14.9 stars E 1'.0; mag 7.82v star SSW 6'.2.
19 52 28.4	−69 40 48	ESO 73-13	13.4	1.2 x 0.8	13.2	S0	87	Mag 11.8 star E edge.

(Continued from previous page)

GALAXIES

RA h m s	Dec ° ′ ″	Name	Mag (V)	Dim ′ Maj x min	SB	Type Class	PA	Notes
20 01 04.6	−71 12 04	ESO 73-18	13.9	0.6 x 0.4	12.2	S0?	177	Mag 9.49v star E 1′.6.
20 01 04.4	−70 20 13	ESO 73-19	13.8	0.9 x 0.9	13.4	S0?		Almost stellar nucleus; mag 11.9 star NW 2′.1.
17 54 27.7	−63 46 32	ESO 102-18	14.3	1.0 x 0.4	13.2	S?	132	Stellar companion galaxy SE end; mag 12.5 star S 1′.4.
18 02 34.1	−63 19 01	ESO 102-23	14.1	1.5 x 0.2	12.6	Sbc: sp	74	Pair of E-W oriented mag 8.81v, 10.44v stars NE 6′.0.
18 09 41.4	−63 12 59	ESO 103-5	13.8	1.0 x 0.5	12.9	S0?	51	Mag 10.30v star SE 8′.5.
18 25 52.0	−63 21 11	ESO 103-19	14.0	1.3 x 0.3	12.9	Sb sp	6	Mag 10.86v star E 3′.0.
18 38 20.6	−65 25 46	ESO 103-35	13.6	1.0 x 0.4	12.5	S0?	34	Very small, very faint anonymous galaxy N edge.
18 39 22.4	−66 42 00	ESO 103-39	14.2	1.4 x 0.9	14.3	Sc?	22	Numerous faint stars superimposed on envelope and core, in star rich area.
18 41 24.8	−64 00 51	ESO 103-46	13.2	1.3 x 1.0	13.3	S?	85	Mag 11.23v star WNW 4′.0.
18 42 01.8	−65 05 46	ESO 103-49	13.7	1.0 x 0.4	12.6	S0?	139	Mag 11.6 star N 4′.5.
18 43 33.7	−64 06 27	ESO 103-56	13.9	0.8 x 0.5	12.7	S0?	164	**IC 4752** 2′.1 NE; mag 11.5 star S 1′.1; mag 9.39v star N 8′.7.
18 45 01.7	−63 31 01	ESO 103-59	13.8	1.4 x 0.2	12.3	S0	155	Mag 14.1 star W edge; mag 12.8 star N 2′.4.
18 47 18.3	−63 21 35	ESO 104-7	12.9	0.8 x 0.6	12.2	E?	93	Mag 10.29v star on S edge; IC 4765 N edge.
18 55 42.2	−64 48 40	ESO 104-22	14.8	1.9 x 1.5	15.8	IB(s)m V-VI	129	Uniform surface brightness; mag 11.4 star NW edge.
19 11 23.7	−64 13 20	ESO 104-44	14.1	1.5 x 1.2	14.6	SABm V-VI	150	Two mag 15 stars on W edge.
19 25 01.2	−62 58 07	ESO 104-51	13.6	1.1 x 0.8	13.3	S0	35	Pair of galaxies, bridge; mag 6.59v star WSW 5′.7.
19 32 24.8	−64 24 48	ESO 105-4	12.7	1.5 x 1.3	13.3	SA(s)0$^-$?	151	Mag 11.35v star S 5′.7; close pair **ESO 105-6** and **ESO 105-6A** E 10′.2.
19 34 55.0	−64 44 28	ESO 105-7	13.3	1.2 x 0.9	13.2	S0?	46	Mag 13.9 star S edge.
19 36 24.0	−64 35 02	ESO 105-9	13.9	1.1 x 0.3	12.6	Sa:	87	Mag 12.2 star W 1′.4.
19 46 44.2	−65 14 51	ESO 105-12	13.8	1.2 x 0.2	12.2	S0:	117	Mag 9.03v star SE 2′.6.
20 02 41.3	−64 04 42	ESO 105-20	15.1	1.7 x 0.2	13.8	S?	117	Mag 9.28v star W 4′.8.
20 03 39.1	−65 04 33	ESO 105-22	15.4	1.0 x 0.2	13.5		164	Pair of E-W oriented mag 10.97v, 10.53v stars W 8′.7.
20 09 28.1	−66 13 00	ESO 105-26	15.1	1.0 x 0.3	13.7	S pec	140	**McLeish's Object.**
20 20 59.6	−65 55 21	ESO 106-2	14.1	1.5 x 0.3	13.1	SB(s)c pec	27	Mag 10.28v star E 8′.5.
18 44 53.8	−62 21 55	ESO 140-43	14.0	1.6 x 0.9	14.2	(R′)SB(s)b: III-IV	167	Mag 9.04v star SSE 4′.3; NGC 6673 N 4′.3.
19 00 55.3	−62 03 41	ESO 141-17B	14.1	1.9 x 0.8	14.4		135	Pair with ESO 141-17A W edge.
19 00 51.0	−62 03 40	ESO 141-17A	13.4	1.5 x 0.6	13.1		15	Pair with ESO 141-17B E edge.
19 16 12.6	−62 21 40	ESO 141-42	12.4	3.0 x 0.5	12.7	SB(s)m sp	111	Mag 8.12v star S 2′.1; IC 4833 WNW 4′.0.
18 08 13.2	−62 23 43	IC 4674	14.3	1.1 x 0.4	13.2	SAB(s)d: IV	89	Mag 7.99v star N 1′.6.
18 13 29.6	−64 28 38	IC 4680	12.8	1.7 x 0.8	13.0	(R′)SB(r)ab: II-III	75	Mag 10.46v star S 3′.3.
18 16 25.8	−71 34 52	IC 4682	12.2	2.3 x 1.6	13.5	SB(rs)bc II	144	Mag 10.87v star W 8′.0.
18 20 17.5	−64 44 05	IC 4696	12.4	2.3 x 0.9	13.0	SAB(rs)bc: II	77	Mag 10.74v star NNE 5′.0.
18 21 00.5	−63 20 53	IC 4698	14.2	1.3 x 0.3	13.0	Sb:	39	Mag 8.48v star N 5′.9.
18 27 53.9	−71 36 38	IC 4704	12.1	1.5 x 1.2	12.6	SA0$^-$	165	Mag 8.35v star NE 10.7; IC 4705 S 5′.2.
18 28 10.6	−71 41 41	IC 4705	12.7	1.2 x 0.8	12.5	SB(s)a? pec	48	Mag 7.70v star S 5′.8; IC 4704 N 5′.2.
18 28 38.9	−66 59 05	IC 4710	11.9	3.6 x 2.8	14.3	SB(s)m IV-V	5	Mottled, numerous stars superimposed; mag 7.57v star E 10′.8.
18 31 07.1	−71 41 38	IC 4712	12.2	2.3 x 1.2	13.1	SA(rs)bc: II-III	60	Mag 9.87v star S 6′.4.
18 29 58.9	−67 13 29	IC 4713	13.5	0.9 x 0.7	12.8	SBm? sp	105	Small group of four stars NE 3′.6, brightest is 10.12v.
18 35 56.8	−63 22 41	IC 4723	13.5	0.6 x 0.6	12.3	S		A close pair of mag 12 stars E 2′.0.
18 36 59.4	−62 51 15	IC 4726	13.4	1.1 x 0.8	13.1	S0?	23	Mag 14.8 star superimposed N of bright center; mag 9.63v star SW 3′.5.
18 37 56.2	−62 42 02	IC 4727	12.8	1.1 x 1.1	12.9	S0$^-$:		Mag 10.32v star S 3′.3.
18 37 57.3	−62 31 54	IC 4728	12.9	2.0 x 0.7	13.1	SB(rs)ab III-IV	173	Group of three stars N 3′.5, brightest is mag 10.81v.
18 39 56.3	−67 25 33	IC 4729	12.6	1.6 x 1.3	13.2	SAB(r)c II-III	150	Mag 9.35v star SSE 6′.2.
18 38 50.7	−63 21 02	IC 4730	13.2	1.4 x 0.7	13.0	SB(rs)0$^+$:	152	Mag 9.10v star S 1′.9.
18 38 43.2	−62 56 37	IC 4731	12.5	1.5 x 0.7	12.4	SB(r)0$^+$:	82	Mag 10.17v star SW 10′.2; **ESO 103-36** S 5′.2; **IC 4735** E 7′.7.
18 39 58.7	−62 35 54	IC 4737	13.7	0.9 x 0.6	12.9	(R′)SB(s)a	108	Bright N-S bar; mag 8.98v star SE 1′.6.
18 41 44.0	−63 56 54	IC 4741	12.6	1.4 x 0.9	12.7	SA(r)ab	35	Mag 11.04v star N 1′.5.
18 41 52.6	−63 51 43	IC 4742	11.9	1.7 x 1.3	12.7	E1	20	Mag 9.73v star just N of bright center.
18 42 36.3	−64 56 38	IC 4745	12.7	2.1 x 0.9	13.3	Sab pec: sp	179	Strong dark lane E of center; mag 10.12v star W 14′.1; mag 4.78v star ENE 18′.7.
18 42 46.6	−64 04 24	IC 4748	13.2	0.8 x 0.8	12.6	S0?		Mag 13.3 star on E edge; mag 9.39v star NE 9′.5; **ESO 103-51** S 5′.5.
18 42 49.6	−63 12 31	IC 4749	13.1	0.7 x 0.7	12.2	S0?		Mag 10.72v star N 2′.3; **IC 4744** W 6′.3.
18 43 19.5	−62 06 47	IC 4751	13.1	1.0 x 0.6	12.4	(R)SB0°	3	Mag 9.89v star SSW 3′.8; IC 4753 E 1′.6.
18 43 32.6	−62 06 30	IC 4753	13.5	0.7 x 0.6	12.5	E?	5	Mag 10.73v star N 3′; IC 4751 W 1′.6.
18 46 18.8	−65 45 30	IC 4758	13.1	1.1 x 0.9	13.0	Sc	140	Mag 10.60v star W 3′.8.
18 47 07.8	−63 29 06	IC 4764	13.6	1.2 x 0.3	12.3	S?	127	Bright knot of stars NW end; mag 11.10v star S 5′.1.
18 47 17.8	−63 19 57	IC 4765	11.2	3.5 x 1.9	13.3	E$^+$4	115	**ESO 104-8** N 1′.5; **ESO 104-2** SW 3′.3; **PGC 62391** WNW 3′.1; ESO 104-7 S edge; mag 7.41v star E 12′.3.
18 47 35.7	−63 17 32	IC 4766	13.8	1.1 x 0.4	12.8	SA(r)0$^+$	110	**ESO 104-8** SW 1′.8.
18 47 41.8	−63 24 21	IC 4767	13.4	1.5 x 0.5	12.9	S0$^+$: pec sp	30	Mag 10.29v star and ESO 104-7 NNW.
18 47 44.4	−63 09 26	IC 4769	13.1	1.9 x 1.2	13.9	(R′)SB(s)b pec	130	**ESO 104-12** SE 2′.8.
18 51 22.0	−69 55 36	IC 4773	13.6	1.6 x 0.7	13.6	SAB(r)dm: IV	137	Mag 8.49v star E 8′.0.
18 50 30.9	−63 00 46	IC 4779	13.9	0.9 x 0.7	13.2	SA(rs)ab:	121	Mag 13.3 star on SW edge; row of three mag 12-13 stars S 2′.0.
18 52 48.3	−63 15 37	IC 4784	12.8	1.1 x 1.1	12.9	S0?		Anonymous galaxy SE 2′.2.
18 56 04.6	−68 40 57	IC 4787	14.0	1.2 x 1.0	14.0	SB(s)m V	167	Low, uniform surface brightness; mag 12.9 star W 2′.1; mag 9.54v star S 1′.9.
18 54 41.2	−63 27 08	IC 4788	14.9	1.6 x 0.2	13.5	Sc	12	Mag 10.14v star N 9′.3.
18 56 18.9	−68 34 03	IC 4789	13.4	1.5 x 0.6	13.1	SA(s)c: III-IV	174	IC 4787 and mag 9.54v star S 7′.9.
18 56 32.4	−64 55 44	IC 4790	13.0	1.2 x 0.7	12.6	(R′)SB(rs)bc?	57	Mag 11.43v star E 7′.3.
18 57 09.9	−62 05 29	IC 4794	13.0	0.9 x 0.5	12.1	S0	25	Mag 9.60v star W 11′.0.
18 58 20.9	−62 07 05	IC 4798	12.2	1.5 x 0.8	12.2	SA(s)0°: pec	108	Mag 10.59v star SW 10′.0, part of a "V" shaped asterism.
18 58 56.7	−63 55 52	IC 4799	13.1	1.5 x 1.1	13.5	SB(rs)ab	30	Mag 6.94v star SSW 5′.3.
18 58 43.8	−63 08 20	IC 4800	12.8	1.8 x 1.0	13.3	(R′)SB(s)a:	164	Mag 12.8 star on N edge, mag 10.79v star W 4′.3.
18 59 38.7	−64 40 32	IC 4801	12.5	1.7 x 1.0	13.0	SAB(r)0$^+$	90	Mag 9.48v star NW 3′.8.
19 02 01.7	−63 02 52	IC 4805	13.6	1.5 x 0.4	12.9	Sa:	21	Mag 7.40v star NNE 11′.2.
19 04 05.4	−62 11 39	IC 4809	14.1	0.9 x 0.5	13.1	Dwarf	14	Uniform surface brightness.
19 05 42.0	−66 31 21	IC 4813	13.6	1.3 x 1.0	13.7	Dwarf Spiral	35	Mag 12.1 star N 4′.5.
19 09 14.0	−63 28 00	IC 4820	14.6	1.3 x 0.9	14.6	SAB(s)d: IV-V	110	Multi-galaxy system, knotty and disrupted.
19 12 15.1	−63 58 48	IC 4823	13.5	1.2 x 0.9	13.5		20	Dual galaxy system; smaller, almost stellar component S.
19 13 16.8	−62 05 23	IC 4824	16.2	1.7 x 1.0	16.6	IAB(s)m V	103	IC 4828 E 2′.8; mag 10.43v star NW 9′.9.
19 13 40.6	−62 04 58	IC 4828	14.0	1.0 x 0.5	13.1	S?	68	Uniform surface brightness; IC 4824 W 2′.8; mag 10.65v star SSE 4′.3.
19 14 45.0	−62 16 22	IC 4831	12.3	3.6 x 0.9	13.4	(R′)SA(s)ab II-III	111	Mag 8.44v star N 2′.2.
19 15 41.9	−62 19 48	IC 4833	14.0	0.6 x 0.4	12.3		99	ESO 141-42 ESE 4′.0; mag 8.12v star ESE 4′.7.
19 16 30.9	−64 00 23	IC 4834	13.6	1.0 x 0.5	12.7	Sa?	125	Mag 13.6 star on S edge.
19 20 42.8	−72 13 38	IC 4841	13.9	0.9 x 0.9	13.6	SAB(rs)bc:		Stellar nucleus.
19 23 32.6	−65 30 25	IC 4847	14.0	0.9 x 0.6	13.2	S0?	152	Mag 7.27v star S 4′.4; **ESO 104-50** SE 6′.6.
19 25 36.2	−62 55 58	IC 4849	13.6	1.0 x 0.4	12.4	SA:(s)c	116	Mag 10.66v star SE 3′.2.
19 30 47.6	−71 04 14	IC 4853	14.0	1.0 x 0.3	12.5	S	169	Mag 9.91v star NNE 9′.9.
19 30 46.7	−66 18 54	IC 4859	13.7	1.1 x 0.7	13.2	SA(s)bc	35	Close pair of mag 13 stars E 1′.7.
19 31 28.5	−67 22 14	IC 4860	13.9	1.1 x 0.7	13.5	SB(s)c? pec III-IV	141	Mag 7.69v star W 11′.1. IC 4862 N 3′.1.
19 31 40.4	−67 19 21	IC 4862	13.7	1.4 x 0.4	12.9	SAB(rs)bc: II-III	0	Mag 7.69v star W 11′.1; IC 4860 S 3′.1.

(Continued from previous page)

206 GALAXIES 206

RA h m s	Dec ° ′ ″	Name	Mag (V)	Dim ′ Maj x min	SB	Type Class	PA	Notes
19 37 37.9	−65 48 43	IC 4870	13.1	1.6 x 0.9	13.3	IBm? pec	136	Mag 8.02v star NW 5.′6.
19 48 21.1	−69 35 14	IC 4887	13.4	1.0 x 0.8	13.0	SB(r)ab	95	Mag 12.1 star SSE 2.′3.
19 49 31.5	−70 13 42	IC 4892	14.0	1.7 x 0.4	13.5	Sbc? sp	3	Mag 8.78v star NW 3.′2.
19 54 26.7	−70 35 28	IC 4899	13.7	1.0 x 0.7	13.2	S?	9	Mag 9.29v star SW 12.′6.
19 58 13.3	−70 27 13	IC 4903	14.3	1.4 x 0.4	13.6	Sa-b	177	Mag 9.45v star SW 3.′9.
19 58 39.0	−70 11 03	IC 4904	14.1	1.0 x 0.8	13.7	Sa?	62	Very filamentary.
20 06 42.0	−71 40 56	IC 4929	13.5	1.4 x 0.4	12.7	Sb: pec	19	Mag 8.74v star E 8.′9.
20 07 15.1	−69 28 47	IC 4934	14.2	1.6 x 0.2	12.8	Scd	30	Mag 7.46v star S 4.′5.
20 11 16.9	−71 00 46	IC 4945	13.9	1.1 x 0.4	12.9	Sb sp	178	Mag 12.6 star N 1.′6; mag 10.43v star NW 11.′8.
20 15 23.6	−70 32 15	IC 4960	13.4	1.1 x 0.4	12.4	SAB0°: sp	166	Mag 10.20v star W 1.′5.
20 17 24.1	−73 53 07	IC 4964	13.8	1.3 x 0.5	13.2	SA(s)d: III-IV	162	Mag 10.80v star W 7.′5.
20 16 23.4	−70 33 54	IC 4967	13.8	0.7 x 0.5	12.6	E:	90	Mag 12.4 star W 1.′5.
20 14 50.3	−64 47 59	IC 4968	14.2	0.6 x 0.5	12.8	(R:)SAB(rs)a:	44	Mag 7.44v star ESE 3.′5.
20 16 57.6	−70 44 59	IC 4970	13.9	0.7 x 0.2	11.6	SA0^{-} pec:	6	Located on N edge of NGC 6872.
20 19 39.3	−70 50 54	IC 4981	13.1	0.9 x 0.3	11.5	I pec sp	135	Located on N edge of NGC 6880.
20 20 44.1	−70 59 15	IC 4985	13.9	0.9 x 0.6	13.1	S0?	66	**IC 4982** SW 2.′8; mag 12.4 star SE 2.′3.
20 23 26.9	−71 33 58	IC 4992	13.8	2.2 x 0.2	12.8	SB(s)c? sp	58	Mag 7.72v star SE 8.′3.
17 59 30.9	−63 40 09	NGC 6483	11.9	1.5 x 0.9	12.2	E4	122	Located between two close mag 10 stars; **ESO 102-20A** SE 2.′0.
18 02 48.1	−66 25 53	NGC 6492	11.5	2.5 x 1.2	12.6	SA(rs)bc: I-II	75	Mag 11.8 star on E edge.
18 04 13.5	−65 24 36	NGC 6502	12.5	1.3 x 1.1	12.7	SA0^{-}:	42	Star or bright knot W of nucleus.
18 12 15.1	−63 46 34	NGC 6545	13.0	1.0 x 0.9	12.9	E1	168	
18 25 07.6	−63 14 50	NGC 6614	12.5	1.4 x 1.1	12.9	SB(s)0^{-}:	65	Mag 11.6 star S 1.′5; mag 9.39v star NNW 3.′1.
18 32 34.9	−63 17 33	NGC 6630	13.9	0.8 x 0.7	13.2	SA0^{-}?		
18 44 39.2	−73 15 48	NGC 6653	12.2	1.7 x 1.5	13.2	E1:		Bright center, uniform envelope, few stars superimposed.
18 45 06.4	−62 17 47	NGC 6673	11.6	2.2 x 0.9	12.3	SAB(s)0^{-}:	26	ESO 140-43 S 4.′3.
18 48 57.9	−65 10 31	NGC 6684	10.4	4.6 x 2.9	13.1	(R′)SB(s)0°	35	Bright, large center with short bar.
18 52 22.7	−64 49 54	NGC 6684A	12.4	3.0 x 1.9	14.2	IB(s)m V-VI	140	Very diffuse; mag 9.74v star NW 2.′5.
18 56 51.4	−63 10 01	NGC 6706	12.9	1.6 x 0.5	12.5	S0^{-}: sp	123	Mag 9.84v star WSW 13.′1.
19 01 29.0	−66 06 36	NGC 6718	13.2	1.4 x 0.8	13.2	(R)SB(r)b III-IV	172	
19 03 07.5	−68 35 16	NGC 6719	12.8	1.7 x 0.8	13.0	SA(r)cd pec III-IV	107	Very patchy, a few faint knots.
19 03 40.3	−64 53 44	NGC 6722	12.5	2.9 x 0.4	12.5	Sb sp	166	Pair of stars superimposed N end.
19 07 33.9	−68 54 48	NGC 6730	12.2	1.7 x 1.5	13.2	E1	25	Mag 7.13v star NE 2.′1.
19 06 10.6	−62 11 49	NGC 6733	12.3	1.8 x 1.2	13.0	SAB0° pec:	110	Line of four stars W of bright center.
19 07 14.6	−65 27 45	NGC 6734	12.7	1.3 x 1.1	12.9	S0^{-}	8	Galaxy **Fair 507** SE 1.′2.
19 07 29.5	−65 25 43	NGC 6736	13.3	1.1 x 0.8	13.1	E2	73	
19 09 46.4	−63 51 28	NGC 6744	8.5	20.1 x 12.9	14.4	SAB(r)bc II	15	Bright center with numerous superimposed stars.
19 08 44.0	−63 43 47	NGC 6744A	14.4	1.8 x 0.7	14.5	IB(s)m	110	Uniform surface brightness, star on W end.
19 25 19.3	−63 51 38	NGC 6776	12.1	1.7 x 1.4	13.0	E^{+} pec:	15	
19 25 06.3	−63 41 02	NGC 6776A	13.9	1.4 x 0.3	12.8	SB(s)cd: sp	83	Star on N edge.
19 26 36.3	−65 37 05	NGC 6784	13.8	0.9 x 0.5	12.8	2 S0 galaxies	30	
19 26 31.3	−65 37 35	NGC 6784A	13.8	0.9 x 0.5	12.8	SA0^{-} pec:	162	
19 43 54.3	−70 37 59	NGC 6808	12.5	1.5 x 0.8	12.5	SA(r)ab pec: II	40	
20 02 50.1	−65 13 47	NGC 6844	12.7	1.4 x 1.1	13.0	(R)SAB(rs)b	0	Mag 8.15v star SE 4.′6.
20 16 58.0	−70 46 06	NGC 6872	11.8	6.0 x 1.5	14.0	SB(s)b pec	66	Mag 10.41v star SE of bright center; single, long, faint, narrow arms NE and SW. IC 4970 on N edge; **PGC 64439** SE 4.′0.
20 18 18.8	−70 51 31	NGC 6876	11.3	2.8 x 2.2	13.2	E3	80	NGC 6877 on E edge; **IC 4972** SSW 4.′5.
20 18 36.0	−70 51 14	NGC 6877	12.2	1.1 x 0.6	11.8	E6	169	Located on E edge of NGC 6876.
20 19 30.0	−70 51 35	NGC 6880	12.1	2.0 x 0.9	12.6	SAB(s)0^{+}:	21	IC 4981 on N edge.

GALAXY CLUSTERS

RA h m s	Dec ° ′ ″	Name	Mag 10th brightest	No. Gal	Diam ′	Notes
7 52 24.0	−65 29 00	AS 797	15.7		17	Faint triple **ESO 102-17**, **17A**, **17B** near center; all others anonymous, stellar.
8 47 12.0	−63 19 00	AS 805	14.7	8	28	Large number of catalog galaxies spread NNE from center.

207 GALAXIES 207

RA h m s	Dec ° ′ ″	Name	Mag (V)	Dim ′ Maj x min	SB	Type Class	PA	Notes
7 09 22.7	−73 19 16	ESO 44-1	13.4	1.6 x 0.8	13.5	SAB(s)c II-III	53	Mag 8.99v star ENE 11.′0.
7 16 37.6	−73 08 53	ESO 44-9	13.5	1.5 x 0.6	13.3	SB(s)0/a pec:	78	Mag 9.20v star N 4.′7.
8 06 35.4	−73 53 01	ESO 44-25	14.0	1.1 x 0.5	13.2	SAB(rs)b? III-IV	52	Mag 7.42v star S 7.′6.
5 46 31.9	−68 01 29	ESO 68-11	14.2	0.9 x 0.2	12.2	SB(s)dm IV-V	108	Mag 8.79v star SW 13.′1.
5 54 45.8	−68 43 02	ESO 68-12	12.4	1.1 x 0.6	11.8	S?	88	
5 55 48.2	−68 52 11	ESO 68-13	14.9	1.3 x 0.6	14.5	IB(s)m V	0	Mag 9.77v star S 1.′9.
6 06 26.0	−67 43 15	ESO 68-16	13.7	2.1 x 1.0	14.3	S0?	63	Pair of galaxies, connected.
6 38 13.2	−68 27 25	ESO 69-6	12.9	1.4 x 0.5	12.4	Sbc	3	Mag 10.92v star S 2.′7.
6 48 15.6	−69 08 27	ESO 69-9	12.9	1.8 x 1.3	13.7	SB(rs)c I-II	72	Mag 1.91v star Alpha Trianguli Australis NNE 7.′1; mag 8.25v star W 5.′0.
7 24 34.2	−70 28 15	ESO 70-9	13.9	1.1 x 0.4	12.9	SAab?	159	**ESO 70-10** N 1.′3.
7 46 31.8	−68 15 41	ESO 70-13	14.0	1.5 x 1.1	14.4	SB(s)dm IV	171	Mag 9.42v star SE 8.′3.
5 45 11.4	−66 17 37	ESO 99-8	15.0	2.1 x 0.2	13.9	Sd sp	11	Mag 10.40v star W 4.′8.
6 16 32.5	−65 39 09	ESO 100-14	13.2	1.5 x 0.8	13.2	(R′)SB(r)a:	138	Numerous knots and/or superimposed stars.
6 25 24.1	−62 38 16	ESO 100-22	12.6	1.5 x 0.7	12.5	SA(s)0°	30	Mag 11 start N end.
6 26 35.8	−63 11 44	ESO 100-23	14.1	1.9 x 0.9	14.5	SAB(s)cd: pec III	32	Mag 6.14v star NW 9.′3.
6 27 52.3	−66 47 13	ESO 100-25	15.6	1.0 x 0.4	14.4	Sbc	84	Mag 9.11v star WNW 13.′9.
6 36 35.4	−66 00 24	ESO 101-1	13.9	1.0 x 0.5	13.0	S	178	Elongated core; mag 12.0 star S 2.′3, in star rich area.
6 37 52.8	−64 48 49	ESO 101-3	13.9	0.8 x 0.3	12.2	S	51	Mag 7.95v star W 6.′9.
6 38 18.0	−64 21 37	ESO 101-4	13.2	1.1 x 0.9	13.1	S	144	Stellar nucleus, pair of mag 13 stars on S edge, in star rich area.
6 39 06.3	−66 54 34	ESO 101-5	14.5	1.6 x 0.3	13.5	SB(s)c: sp II-III	164	Mag 8.97v star W 2.′5.
6 40 34.1	−67 27 11	ESO 101-7	13.9	1.1 x 0.2	12.1	Sa:	56	Mag 8.00v star E 9.′5.
6 41 05.6	−66 35 51	ESO 101-8	14.0	1.4 x 0.2	12.5	S?	109	Mag 8.95v star SW 3.′3.
6 45 08.2	−66 37 04	ESO 101-11	13.4	1.0 x 0.8	13.0	SAB(rs)b:	5	Almost stellar nucleus, stars superimposed, in star rich area; mag 10.24v star E 2.′8; ESO 101-13 NE.
6 45 31.5	−66 34 02	ESO 101-13	13.6	1.2 x 0.9	13.6	SAB(s)cd IV	27	ESO 101-11 SW 3.′8; mag 10.24v star S 2.′6.
6 48 41.3	−62 36 22	ESO 101-14	12.0	1.2 x 1.2	12.3	SA0^{-}		Numerous stars involved.
6 54 46.9	−63 06 32	ESO 101-17	14.0	1.4 x 0.4	13.3	SABbc? III-IV	63	Patchy, faint star NW edge.
6 57 00.0	−65 57 12	ESO 101-18	12.3	2.0 x 1.0	12.9	SB(rs)a II	75	Mag 11.9 star E 1.′4; mag 10.32v star S 5.′3.

207

(Continued from previous page)

GALAXIES

RA h m s	Dec ° ′ ″	Name	Mag (V)	Dim ′ Maj x min	SB	Type Class	PA	Notes
16 58 01.7	−66 26 02	ESO 101-19	14.5	1.5 x 0.2	13.0	S	71	Mag 9.18v star SSE 4'.5.
16 57 58.4	−62 43 33	ESO 101-20	12.6	1.9 x 0.8	12.9	SB(rs)bc pec: II	117	Several stars superimposed; mag 10.96v star W 5'.8.
17 05 51.4	−63 27 16	ESO 101-21	13.4	1.1 x 0.7	13.0		127	Double system, bridge; mag 10.05v star NE 3'.0.
17 14 32.0	−65 24 53	ESO 101-23	14.0	1.0 x 0.5	13.1	Sa:	1	Mag 12.2 star NW 1'.5.
17 15 15.9	−66 15 23	ESO 101-24	13.9	1.0 x 0.7	13.4		44	Several stars superimposed, in star rich area; mag 9.40v star NE 3'.1.
17 21 38.7	−65 10 29	ESO 102-2	13.2	1.6 x 1.3	13.9	SB(s)cd II-III	168	Mag 13 star N end.
17 26 54.3	−64 53 37	ESO 102-4	13.8	1.5 x 0.3	12.8	Sc	16	Mag 8.03v star SSE 7'.1.
17 27 46.5	−64 33 39	ESO 102-5	13.4	1.2 x 0.4	12.5	Sbc	23	Pair of close mag 10.17v, 8.97v stars WNW 6'.9.
17 28 55.9	−66 09 06	ESO 102-6	13.9	1.1 x 0.3	12.5	Im: IV?	73	Bright star W edge.
17 29 31.2	−64 01 22	ESO 102-7	12.6	1.6 x 1.0	12.9	SAB(rs)c: II-III	111	Many bright knots and/or superimposed stars.
17 34 16.3	−62 44 24	ESO 102-9	13.9	1.2 x 0.4	13.0	(R′)SA(s)ab:	58	Mag 10.31v star NW 5'.7.
17 42 37.7	−63 47 04	ESO 102-12	14.6	1.5 x 0.4	13.9	Sdm: IV-V:	77	Mag 9.08v star W 5'.4.
17 54 27.7	−63 46 32	ESO 102-18	14.3	1.0 x 0.4	13.2	S?	132	Stellar companion galaxy SE end; mag 12.5 star S 1'.4.
18 02 34.1	−63 19 01	ESO 102-23	14.1	1.5 x 0.2	12.6	Sbc: sp	74	Pair of E-W oriented mag 8.81v, 10.44v stars NE 6'.0.
16 34 35.3	−62 02 38	ESO 137-32	14.8	2.5 x 0.2	13.9	Sc	103	Mag 6.95v star SW 8'.5.
16 56 45.2	−62 24 13	ESO 138-7	12.9	2.3 x 0.7	13.2	SB(rs)bc II-III	49	Mag 8.62v star S 8'.6.
16 59 36.3	−62 32 43	ESO 138-11	13.6	1.0 x 0.6	12.9	Sdm:	1	Numerous superimposed stars; mag 12.9 star S edge, mag 9.71v star SE 1'.6.
17 07 00.0	−62 05 05	ESO 138-14	12.6	3.1 x 0.4	12.7	SB(s)d: sp III-IV	135	Mag 9.89v star N 4'.2.
17 29 10.8	−62 26 53	ESO 138-29	11.7	2.5 x 1.4	12.9	(R)SAB(s)0$^+$ pec	39	Mag 10.48v star S 1'.9.
17 29 25.3	−62 28 52	ESO 138-30	14.0	1.2 x 0.5	13.3	SB(s)ab? pec	154	Numerous stars superimposed on faint envelope.
17 30 28.2	−62 01 43	ESO 139-1	13.5	0.9 x 0.6	12.7	S0?	45	Mag 14.1 star NE edge; mag 10.41v star E 2'.5.
15 29 55.6	−70 35 08	IC 4541	13.3	2.3 x 0.6	13.5	Sb: III-IV	151	Mag 10.37v star NW 9'.2.
15 48 51.6	−67 19 23	IC 4571	13.1	1.4 x 0.5	12.5	SB(rs)b pec: II	142	Mag 9.97v star W 2'.7.
16 00 12.9	−66 22 59	IC 4584	15.1	1.7 x 1.5	15.9	SA(rs)cd II	96	IC 4585 N 3'.8.
16 00 17.6	−66 19 22	IC 4585	12.3	2.7 x 0.6	12.7	SB(s)b I-II	54	Mag 11.24v star on N edge; IC 4584 S 3'.8.
16 20 43.8	−70 08 38	IC 4595	12.7	2.7 x 0.5	12.9	Sc: sp	63	Mag 11.8 star E 4'.2.
17 24 37.1	−73 56 24	IC 4644	13.9	1.9 x 0.3	13.1	Sb? pec	134	Mag 12.3 star N 1'.0; mag 10.23v star NNW 7'.9.
17 37 44.1	−63 43 50	IC 4656	12.7	2.3 x 0.5	12.7	SA(s)c: I-II	90	Mag 9.37v star SW 3'.5.
17 47 09.1	−64 38 34	IC 4662	11.3	3.2 x 1.9	13.1	IBm III-IV	105	Multi-galaxy system.
17 51 36.4	−64 57 34	IC 4662A	13.4	1.4 x 0.9	13.5	SB(rs)c: II	44	Mag 11 star NE edge.
17 48 58.9	−63 15 14	IC 4664	12.9	1.8 x 1.0	13.4	(R′)SB(r)b II-III	105	Mag 9.75v star W 12'.2.
15 36 25.9	−66 51 32	NGC 5938	11.8	2.8 x 2.5	13.8	SB(rs)bc I-II	177	Numerous stars superimposed, in very star rich area.
16 39 08.0	−73 14 51	NGC 6151	15.9	0.3 x 0.2	12.7		88	Almost stellar; located 0'.9 ESE of mag 10.89v star.
16 41 42.4	−69 22 21	NGC 6183	12.2	1.7 x 0.5	11.8	SAa	36	Mag 8.52v star SW 3'.9.
16 54 58.2	−72 35 14	NGC 6209	12.2	2.0 x 1.6	13.3	(R′)SA(rs)bc II	10	Bright nucleus, many superimposed stars.
17 17 00.4	−62 49 12	NGC 6300	10.2	4.3 x 2.8	12.8	SB(rs)b II	118	Bright nucleus with strong dark patches NW and SE; many superimposed stars.
17 23 41.1	−65 00 37	NGC 6328	12.1	2.4 x 1.4	13.2	SAB(s)ab III	157	Numerous superimposed stars.
17 43 30.9	−69 47 07	NGC 6392	11.6	1.3 x 1.3	12.0	(R′)SAB(rs)ab:		Several superimposed stars S of nucleus.
17 59 30.9	−63 40 09	NGC 6483	11.9	1.5 x 0.9	12.2	E4	122	Located between two close mag 10 stars; **ESO 102-20A** SE 2'.0.
18 02 48.1	−66 25 53	NGC 6492	11.5	2.5 x 1.2	12.6	SA(rs)bc: I-II	75	Mag 11.8 star on E edge.
18 04 13.5	−65 24 36	NGC 6502	12.5	1.3 x 1.1	12.7	SA0$^-$:	42	Star or bright knot W of nucleus.
17 41 11.1	−63 23 27	PGC 60692	12.6	1.0 x 0.9	12.4	(R)SA(s)0°:	165	

GALAXY CLUSTERS

RA h m s	Dec ° ′ ″	Name	Mag 10th brightest	No. Gal	Diam ′	Notes
17 28 36.0	−66 41 00	AS 794	15.6		22	All members anonymous, stellar.
17 52 24.0	−65 29 00	AS 797	15.7		17	Faint triple **ESO 102-17**, **17A**, **17B** near center; all others anonymous, stellar.

OPEN CLUSTERS

RA h m s	Dec ° ′ ″	Name	Mag	Diam ′	No. ★	B ★	Type	Notes
15 29 48.0	−64 52 00	ESO 99-6		8			cl	

GLOBULAR CLUSTERS

RA h m s	Dec ° ′ ″	Name	Total V m	B ★ V m	HB V m	Diam ′	Conc. Class Low = 12 High = 1	Notes
16 25 48.6	−72 12 06	NGC 6101	9.2	13.5	16.6	5.0	10	
17 31 54.8	−67 02 53	NGC 6362	8.1	12.7	15.3	15.0	10	

PLANETARY NEBULAE

RA h m s	Dec ° ′ ″	Name	Diam ″	Mag (P)	Mag (V)	Mag cent ★	Alt Name	Notes
15 37 11.6	−71 54 54	PN G315.1−13.0	6	11.7	11.6	11.0	PK 315−13.1	Bright central star; mag 13.3 star NW 1'.9.
15 56 01.2	−66 09 10	PN G320.1−9.6	7		13.2	10.9	PK 320−9.1	Bright central star; mag 12.7 star NW 0'.7.
17 01 16.7	−70 06 05	PN G321.3−16.7	5	12.3		16.4	PK 321−16.1	Mag 14.0 star W 0'.5; mag 11.80v star ESE 1'.4.
15 52 09.6	−62 30 52	PN G322.1−6.6	10	14.5	12.5	17.2	PK 322−6.1	Mag 11.19v star NE 1'.6; mag 11.3 star SW 1'.4.
16 54 35.2	−64 14 31	PN G325.8−12.8	3	12.0		13.4	PK 325−12.1	Mag 13.2 star on E edge; mag 11.15v star W 1'.0.

208

GALAXIES

RA h m s	Dec ° ′ ″	Name	Mag (V)	Dim ′ Maj x min	SB	Type Class	PA	Notes
15 27 39.7	−73 56 40	ESO 42-7	13.7	2.3 x 2.0	15.2	SB(s)m IV-V	30	Numerous stars superimposed; mag 9.04v star N 6'.1.
15 19 42.2	−70 54 39	ESO 68-2	17.1	1.2 x 0.4	16.1	SB(s)m V	5	Pair of N-S oriented mag 10.44v, 10.71v stars WNW 3'.5.
14 10 36.7	−65 34 56	ESO 97-12	12.8	0.8 x 0.3	11.1	S	107	Mag 10.42v star NNW 2'.6.
14 13 09.9	−65 20 18	ESO 97-13	10.6	6.9 x 3.0	13.8	SA(s)b: II	38	**Circinus Galaxy**. Stellar nucleus, numerous stars superimposed; mag 10.10v star on SE edge.
15 14 52.4	−63 48 18	ESO 99-2	15.7	1.7 x 0.6	15.6		18	Many superimposed stars, mag 12.1 star on SE edge; mag 7.96v star SW 10'.0.
15 27 35.1	−64 10 14	ESO 99-5	12.4	1.5 x 1.3	13.0	SB(r)ab	59	Mag 8.37v star WSW 7'.5.
14 47 45.3	−73 18 26	IC 4484	14.0	2.6 x 0.3	13.6	Scd: sp	132	Mag 9.57v star N 6'.7.
15 29 55.6	−70 35 08	IC 4541	13.3	2.3 x 0.6	13.5	Sb: III-IV	151	Mag 10.37v star NW 9'.2.
15 05 35.6	−72 26 02	NGC 5799	12.9	0.9 x 0.7	12.3	S0$^+$: pec	130	Bright center, in star rich area, few stars superimposed.

(Continued from previous page)

208 GALAXIES 208

RA h m s	Dec ° ′ ″	Name	Mag (V)	Dim ′ Maj x min	SB	Type Class	PA	Notes
15 11 54.5	−72 51 36	NGC 5833	12.0	2.1 x 1.0	12.7	SA(s)bc pec: I-II	128	Bright nucleus, numerous stars superimposed, in star rich area .
15 36 25.9	−66 51 32	NGC 5938	11.8	2.8 x 2.5	13.8	SB(rs)bc I-II	177	Numerous stars superimposed, in very star rich area.

OPEN CLUSTERS

RA h m s	Dec ° ′ ″	Name	Mag	Diam ′	No. ★	B ★	Type	Notes
13 15 16.0	−65 55 18	AL 1		1.3			cl	Faint, circular. = **vdB-Ha 144** and **ESO 96-SC4**.
13 27 49.0	−62 18 42	Bas 18	8.2	3	154	11.0	cl	Few stars; small brightness range; no central concentration; detached. (S) Outline of cluster uncertain.
13 22 36.0	−66 07 00	Cr 269	9.2	15			ast	Few stars; moderate brightness range; not well detached. Probably not a cluster.
13 30 00.0	−64 11 00	Cr 271	8.7	5	20	9.5	cl	Few stars; moderate brightness range; not well detached.
13 48 42.0	−66 04 00	Cr 277	9.2	12	30		cl	Moderately rich in stars; small brightness range; not well detached.
13 12 26.8	−62 42 01	Danks 1		1	25		cl	
13 12 53.8	−62 40 48	Danks 2		0.6	80		cl	
13 29 18.0	−71 16 00	ESO 65-7		4			cl	
15 29 48.0	−64 52 00	ESO 99-6		8			cl	
13 18 12.0	−67 05 00	Ha 8	9.5	5	30	11.6	cl	Moderately rich in stars; moderate brightness range; no central concentration; detached.
13 36 55.0	−62 05 36	IC 4291	9.7	4	35	10.0	cl	
13 12 24.0	−65 17 00	Lo 757		20	20		cl	
13 24 54.0	−62 25 00	Lo 807	7.9	20	35		cl	
13 39 44.3	−64 20 00	Lo 894		8	6	9.7	cl	
13 45 24.0	−62 00 00	Lo 991		5	20		cl	
13 52 21.0	−64 54 00	Lo 995		22	21	9.1	cl	(A) Not obvious on DSS.
13 54 18.0	−65 18 00	Lo 1002		15	20		cl	(A) DSS image shows only rich Milky Way.
14 00 02.0	−62 09 30	Lynga 1		3	15		cl	
13 25 39.0	−63 27 30	NGC 5120	10.8	4	30	12.0	MWcl	
13 44 45.0	−62 55 00	NGC 5269		3		12.0	MWcl	(S) Sparse group at this position.
13 46 36.7	−62 55 00	NGC 5281	5.9	8	40	8.0	cl	Moderately rich in bright and faint stars with a strong central concentration; detached. Without brightest star, visual mag = 6.8.
13 48 45.6	−64 41 11	NGC 5288	11.8	3	25		cl	Few stars; small brightness range; slight central concentration; detached.
13 59 45.2	−70 24 00	NGC 5359		8	12		cl?	(A) Scattered group of stars here fits W. Herschel's description.
13 19 46.0	−64 56 48	Ru 107	9.7	3	20	12.0	cl	Few stars; moderate brightness range; no central concentration; detached.
14 06 30.0	−67 34 00	Ru 110		18	35	10.0	cl	Moderately rich in stars; moderate brightness range; no central concentration; detached.
14 57 00.0	−62 32 00	Ru 112		10	80	14.0	cl	(A) Rich group (Milky Way patch?) of faint stars.
13 18 38.4	−62 32 00	St 16	6.9	20	83	6.9	cl	Few stars; large brightness range; no central concentration; detached; involved in nebulosity.
13 32 14.0	−62 47 18	Tr 21	7.7	5	20	8.9	cl	Few stars; moderate brightness range; strong central concentration; detached.
13 38 02.0	−63 20 48	vdB-Ha 150		3	15		cl	(A) Small group of faint stars at given position. There is also a N-S group of brighter stars on W side.
14 44 06.0	−66 22 00	vdB-Ha 164		29			cl	(A) Nothing obvious on 45′ DSS image.

STAR CLOUDS

RA h m s	Dec ° ′ ″	Name	Mag	Diam ′	No. ★	B ★	Type	Notes
13 17 02.6	−63 24 54	NGC 5045		60	>180		MW	(A) 1 degree Milky Way star cloud at this position, not well separated from surrounding area.
13 29 36.0	−63 24 44	NGC 5155		60	>200		MW	(A) Milky Way star cloud, broken by dark nebula at several places (e.g. the center).

BRIGHT NEBULAE

RA h m s	Dec ° ′ ″	Name	Dim ′ Maj x min	Type	BC	Color	Notes
13 25 24.0	−64 01 00	Ced 122	150 x 150	E	1-5		Visible to the unaided eye! Brightest part lies NE of 4th mag star m Centauri.
14 49 24.0	−65 14 00	vdBH 63	1.5 x 1.5	R	2-5	1-4	A small, round and well-defined nebulosity; the involved star is on the W side.
15 01 06.0	−63 17 00	vdBH 65a	0.3 x 0.3	R	1-5	3-4	Cometary in shape. It forms an approximate equilateral triangle with a brighter star 2′ SW and a fainter one 2′ NW.

DARK NEBULAE

RA h m s	Dec ° ′ ″	Name	Dim ′ Maj x min	Opacity	Notes
14 48 36.0	−65 15 00	Be 145		4	

PLANETARY NEBULAE

RA h m s	Dec ° ′ ″	Name	Diam ″	Mag (P)	Mag (V)	Mag cent ★	Alt Name	Notes
13 08 48.3	−67 38 32	IC 4191	5	12.0	10.6	16.4	PK 304−4.1	
13 33 30.8	−65 58 29	NGC 5189	140	10.3		14.9	PK 307−3.1	Bright, narrow "bar" NE-SW across center, several superimposed stars, very filamentary.
13 53 57.2	−66 30 52	NGC 5315	14	13.0	9.8	14.4	PK 309−4.2	Located 3′.9 E of mag 7.12v star.
15 10 40.6	−64 40 26	NGC 5844	73	13.2			PK 317−5.1	Elongated E-W; mag 10.29v star NE 2′.9.
13 19 30.2	−66 09 07	PN G305.7−3.4	24	14.5			PK 305−3.1	Mag 13.3 star N 0′.8; mag 12.5 star W 1′.5.
13 22 34.0	−63 21 02	PN G306.4−0.6	23	14.4		14.4	PK 306−0.1	Mag 9.61v star on W edge.
13 45 24.2	−71 28 52	PN G307.2−9.0	5	12.6		15.3	PK 307−9.1	Mag 12.6 star on SW edge; mag 7.15v star S 3′.3.
13 39 35.4	−67 22 58	PN G307.5−4.9	25	12.2	12.9	14.4	PK 307−4.1	**Hourglass Nebula**; Elongated SSE-NNW, close pair of mag 12 stars W 1′.0.
13 52 30.9	−66 23 27	PN G309.0−4.2	17			14.0	PK 309−4.1	Located between mag 9.55v star ENE 1′.8 and mag 8.81v star WSW 1′.7.
13 54 27.0	−64 59 27	PN G309.5−2.9		16.4			MaC 1-2	Stellar; mag 9.11v star NW 3′.4.
14 05 36.9	−64 41 01	PN G310.7−2.9	20	13.9	14.0		PK 310−2.1	Mag 10.69v star E 3′.4.
14 15 24.2	−67 31 57	PN G310.8−5.9	16	14.2			PK 310−5.1	Mag 10.57v star SE 2′.7.
14 18 43.5	−63 07 09	PN G312.6−1.8	10	16.3		15.1	PK 312−1.1	Mag 10.59v star E 3′.5.
15 21 09.7	−72 14 06	PN G313.8−12.6	112			19.0	PK 313−12.1	Mag 11.9 star E 2′.6; mag 10.20v star SW 3′.8.
15 37 11.6	−71 54 54	PN G315.1−13.0	6	11.7	11.6	11.0	PK 315−13.1	Bright central star; mag 13.3 star NW 1′.9.

209 GALAXIES 209

RA h m s	Dec ° ′ ″	Name	Mag (V)	Dim ′ Maj x min	SB	Type Class	PA	Notes
11 13 08.4	−69 15 57	ESO 63-11	14.1	3.3 x 0.5	14.5	SB(s)dm: sp	173	Mag 10.70v star W 1′.0; **ESO 63-10** W 7′.5.
11 53 09.8	−68 36 33	ESO 64-3	15.0	1.0 x 0.3	13.5	S	31	Mag 9.28v star NNW 7′.4.

(Continued from previous page)

GALAXIES

RA h m s	Dec ° ′ ″	Name	Mag (V)	Dim ′ Maj x min	SB	Type Class	PA	Notes
10 59 26.5	−66 20 01	ESO 93-3	13.8	1.5 x 0.9	14.0	SAB(r)0/a?	144	Mag 10.8 star S 4′.4.
11 57 31.1	−73 41 04	IC 2980	12.3	1.2 x 1.0	12.5	E3	42	Mag 12.9 star N edge; mag 10.80v star WSW 5′.5.

OPEN CLUSTERS

RA h m s	Dec ° ′ ″	Name	Mag	Diam ′	No. ★	B ★	Type	Notes
13 15 16.0	−65 55 18	AL 1		1.3			cl	Faint, circular. = **vdB-Ha 144** and **ESO 96-SC4**.
11 38 20.4	−63 22 00	Cr 249	2.9	65	61		cl	The associated nebula (near Lambda Cen) is **IC 2944**.
13 12 26.8	−62 42 01	Danks 1		1	25		cl	
13 12 53.8	−62 40 48	Danks 2		0.6	80		cl	
12 24 30.0	−68 28 00	ESO 64-5		10	10		ast?	(A) Irregular, likely asterism of about 10 stars.
12 51 36.0	−69 44 00	ESO 65-3		6			cl	
11 19 42.0	−65 13 00	ESO 93-8					ast	
10 56 32.0	−63 00 48	Graham 1		5	20		cl	(S) Small group.
12 37 54.0	−68 23 00	Ha 6	10.7	9	100		cl	Moderately rich in stars; small brightness range; slight central concentration; detached.
13 18 12.0	−67 05 00	Ha 8	9.5	5	30	11.6	cl	Moderately rich in stars; moderate brightness range; no central concentration; detached.
12 43 36.0	−63 06 06	Ho 15	10.3	2	21	12.1	cl	Few stars; small brightness range; slight central concentration; detached.
10 42 57.0	−64 24 00	IC 2602	1.6	100	60	3.0	cl	**Southern Pleiades**. Rich in stars; large brightness range; strong central concentration; detached.
11 17 18.0	−62 43 00	IC 2714	8.2	15	100	10.0	cl	Rich in stars; moderate brightness range; slight central concentration; detached. (S) Weak concentration.
12 42 48.6	−63 03 00	Lo 670		33	27	5.3	cl	
13 12 24.0	−65 17 00	Lo 757		20	20		cl	
10 42 00.0	−65 06 00	Mel 101	8.0	16	50	9.7	cl	Moderately rich in stars; small brightness range; no central concentration; detached.
11 19 41.0	−63 29 06	Mel 105	8.5	5	73	11.1	cl	Rich in stars; moderate brightness range; strong central concentration; detached.
12 02 01.0	−63 14 00	NGC 4052	8.8	10	80		cl	Rich in stars; moderate brightness range; no central concentration; detached.
12 13 38.5	−62 42 20	NGC 4184		2	20	13.0	cl	(S) = **Ru 102**.
12 29 56.0	−64 47 24	NGC 4463	7.2	6	30	8.3	cl	Moderately rich in bright and faint stars with a strong central concentration; detached.
12 42 20.0	−62 59 36	NGC 4609	6.9	6	52	10.0	cl	Moderately rich in stars; moderate brightness range; slight central concentration; detached.
12 57 59.0	−64 57 42	NGC 4815	8.6	5	100	9.6	cl	Rich in stars; moderate brightness range; strong central concentration; detached. Probably not a cluster.
11 30 24.0	−63 25 00	Ru 94		20	60	10.0	cl	Rich in stars; moderate brightness range; no central concentration; detached.
11 50 53.0	−62 08 18	Ru 96		5	15	13.0	cl	
11 57 23.0	−62 42 18	Ru 97	9.1	5	142	12.0	cl	Few stars; small brightness range; not well detached.
11 58 48.0	−64 34 00	Ru 98	7.0	15	50	9.0	cl	Moderately rich in stars; moderate brightness range; no central concentration; detached.
12 03 20.0	−63 52 00	Ru 99		5	15	12.0	cl	(A) 6′ long N-S group barely standing out from background. Has obvious star chain.
12 05 42.0	−62 31 00	Ru 100		7	30	12.0	cl	
12 09 40.0	−63 00 00	Ru 101		15	60	13.0	cl	(A) Appears as obvious increase in fainter stars in rich field.
13 19 46.0	−64 56 48	Ru 107	9.7	3	20	12.0	cl	Few stars; moderate brightness range; no central concentration; detached.
10 53 00.0	−62 16 00	Ru 162		6	60	12.0	cl:	
11 04 47.0	−67 56 42	Ru 163		2	10	12.0	cl	
11 43 51.0	−62 32 00	St 14	6.3	8	20	10.0	cl	Few stars; large brightness range; no central concentration; detached. (S) Sparse group.
13 18 38.4	−62 32 00	St 16	6.9	20	83	6.9	cl	Few stars; large brightness range; no central concentration; detached; involved in nebulosity.
11 09 30.0	−63 50 00	vdB-Ha 111		2	8		cl	
12 26 25.7	−63 24 51	vdB-Ha 131		6	20		cl	
12 26 58.0	−64 03 06	vdB-Ha 132		4	20		cl	
12 53 30.0	−67 10 00	vdB-Ha 140		5	50		cl	(S) Identification uncertain.

STAR CLOUDS

RA h m s	Dec ° ′ ″	Name	Mag	Dia m ′	No. ★	B ★	Type	Notes
13 17 02.6	−63 24 54	NGC 5045		60	>180		MW	(A) 1 degree Milky Way star cloud at this position, not well separated from surrounding area.

GLOBULAR CLUSTERS

RA h m s	Dec ° ′ ″	Name	Total V m	B ★ V m	HB V m	Diam ′	Conc. Class Low = 12 High = 1	Notes
12 25 45.4	−72 39 33	NGC 4372	7.2	12.2	15.6	5.0	12	
12 59 35.0	−70 52 29	NGC 4833	8.4	12.4	15.5	14.0	8	
12 14 04.6	−63 35 37	SSWZ94 4				1.0		

BRIGHT NEBULAE

RA h m s	Dec ° ′ ″	Name	Dim ′ Maj x min	Type	BC	Color	Notes
11 28 54.0	−62 41 00	Gum 39	20 x 15	E	3-5	3-4	= **RCW 60** North. Irregular in shape and slightly extended E-W. The nebula is rather evenly illuminated except for a slightly brighter narrow strip along the NW perimeter where it is best defined.
11 30 24.0	−63 50 00	Gum 41	15 x 15	E	3-5	3-4	= **RCW 61**. Irregularly round and slightly mottled nebula.
11 28 54.0	−62 56 00	IC 2872	15 x 6	E	3-5	3-4	= **RCW 60** South. Extended NNE-SSW and somewhat mottled being brighter in the S part; several stars involved.
11 37 18.0	−63 11 00	IC 2948	75 x 50	E			**Running Chicken Nebula**. Somewhat mottled, especially the NW part; S part brighter.
11 50 12.0	−64 52 00	IC 2966	3 x 2	E+R	2-5	2-4	= **vdBH 56**. Even surface brightness and slightly extended E-W; fairly well-defined outline.
11 06 18.0	−65 32 00	RCW 58	8 x 5		3-5		Wolf-Rayet shell in the shape of a fragmented oval ring in which is approximately centered a star of 8th mag. The two brightest segments are in the SW part; the NW perimeter is completely missing.

DARK NEBULAE

RA h m s	Dec ° ′ ″	Name	Dim ′ Maj x min	Opacity	Notes
12 53 00.0	−63 00 00	Coalsack	55 x 55		A dark patch on the Milky Way. It appears completely black to the naked eye, but is not uniformly dark.
12 25 00.0	−71 30 00	Dark Doodad	150 x 15		

(Continued from previous page)

209 PLANETARY NEBULAE 209

RA h m s	Dec ° ′ ″	Name	Diam ″	Mag (P)	Mag (V)	Mag cent ★	Alt Name	Notes
11 00 19.5	−65 14 54	IC 2621	5	11.3	11.2	15.4	PK 291−4.1	
13 08 48.3	−67 38 32	IC 4191	5	12.0	10.6	16.4	PK 304−4.1	
12 04 15.2	−67 18 35	NGC 4071	76	12.9	13.0	19.2	PK 298−4.1	Slightly elongated NE-SW; brightest along NW edge.
11 41 37.8	−62 28 56	PN G294.9−0.6	68	14.4			PK 294−0.1	Chain of five mag 11-12 stars extend E from E edge.
11 31 45.6	−65 58 14	PN G294.9−4.3	13	14.9			PK 294−4.1	Mag 12.5 star NNE 1′.0; mag 12.9 star W 1′.7.
11 17 43.4	−70 49 37	PN G295.3−9.3	10	13.8			PK 295−9.1	Forms a very small triangle with two faint stars; mag 10.87v star SW 1′.7.
11 48 38.3	−65 08 37	PN G296.3−3.0	4	12.8		16.1	PK 296−3.1	Mag 11.8 star NNE 2′.0; mag 12.0 star S 1′.7.
11 39 11.6	−68 52 11	PN G296.4−6.9	5	14.7			PK 296−6.1	Mag 12.2 star S 1′.4; mag 11.07v star SE 1′.8.
12 09 01.1	−63 07 05	PN G298.1−0.7	22	17.3			PK 298−0.1	Very close pair of mag 11.6, 12.9 stars NE 0′.8; mag 9.68v star SW 2′.7.
12 08 25.2	−64 12 11	PN G298.2−1.7	18	14.8			PK 298−1.2	Mag 12.4 star 0′.5 N.
12 16 33.2	−66 45 37	PN G299.4−4.1	72	17.4			PK 299−4.1	Mag 12.4 star on SW edge; mag 11.4 star W 1′.4.
12 23 02.6	−64 02 07	PN G299.8−1.3	6	17.0			PK 299−1.1	Mag 12.0 star S 1′.9; mag 12.6 star NE 1′.3.
12 28 45.6	−62 05 34	PN G300.2+0.6	15	19.6			PK 300+0.1	Pair of mag 8.23v, 10.99v stars W 2′.5.
12 28 46.7	−63 44 34	PN G300.4−0.9	28	14.1			PK 300−0.1	Mag 11.5 star SW 2′.5; mag 11.2 star ENE 4′.5.
12 30 07.9	−63 53 02	PN G300.5−1.1	10	14.4		16.5	PK 300−1.1	Mag 11.9 star SSE 3′.0.
12 30 31.8	−64 51 11	PN G300.7−2.0	4	14.2		19.0	PK 300−2.1	Mag 9.80v star W 2′.9; mag 11.9 star SW 3′.2.
12 30 24.5	−66 14 25	PN G300.8−3.4	18	15.2			PK 300−3.1	Mag 11.4 star W 1′.2.
12 45 51.2	−64 09 37	PN G302.3−1.3	75			15.3	We 21	Mag 11.1 star WNW 4′.5.
12 48 31.0	−63 49 56	PN G302.6−0.9	48	14.9			PK 302−0.3	Mag 13.2 star SW 0′.9.
13 19 30.2	−66 09 07	PN G305.7−3.4	24	14.5			PK 305−3.1	Mag 13.3 star N 0′.8; mag 12.5 star W 1′.5.

210 GALAXIES 210

RA h m s	Dec ° ′ ″	Name	Mag (V)	Dim ′ Maj x min	SB	Type Class	PA	Notes
08 38 19.5	−73 32 39	ESO 36-5	12.8	1.0 x 1.0	12.7	SA(s)0°:		Star E side of nucleus.
08 52 47.2	−73 37 21	ESO 36-10	13.5	0.9 x 0.5	12.6	SA0: pec	100	Mag 11.9star NW 7′.7; note: **ESO 36-9** is 1′.0 SE of this star.
08 58 40.1	−73 19 38	ESO 36-15	14.1	1.0 x 0.9	13.8	SB0?	120	Almost stellar nucleus, smooth, low surface brightness envelope.
10 17 59.9	−73 49 52	ESO 37-13	13.8	0.7 x 0.3	12.0	Sa?	126	Mag 15.1 star S edge.
08 22 40.9	−69 46 22	ESO 60-4	13.6	1.2 x 0.4	12.7	SB(rs)b pec II-III	83	Pair of E-W oriented mag 10.76v, 9.27v stars NE 7′.6.
08 38 37.1	−67 56 14	ESO 60-10	14.2	0.8 x 0.7	13.4	Sc	2	Mag 12.0 star E 1′.3.
08 42 43.5	−67 48 56	ESO 60-11	13.4	1.6 x 1.2	13.9	SB:(r)b:	30	Mag 9.02v star ESE 5′.0; mag 8.77v star W 7′.9.
08 42 56.8	−68 05 09	ESO 60-12	13.5	1.1 x 0.5	12.8	S0	128	Mag 6.31v star SE 9′.3.
08 50 04.2	−70 07 36	ESO 60-15	14.0	1.5 x 0.3	13.0	Sc:	25	Lies 0′.8 E of mag 9.51 star.
08 56 40.1	−67 52 14	ESO 60-18	13.0	2.2 x 1.5	14.1	(R′)SB(s)b III-IV	136	Mag 9.71v star W 10′.4.
08 57 27.3	−69 03 40	ESO 60-19	12.2	3.5 x 1.4	13.8	SB(s)d III-IV	159	Numerous bright knots and superimposed stars; mag 9.01v star N 2′.5.
09 02 40.3	−68 13 37	ESO 60-24	12.8	3.2 x 0.5	13.2	Sb: sp	54	Mag 10.32v star W 7′.6. **ESO 60-23** S 4′.5.
09 04 01.6	−72 03 18	ESO 60-26	13.6	0.9 x 0.6	12.8	Ring pec	135	Pair with ESO 60-27 E edge.
09 04 08.3	−72 03 01	ESO 60-27	13.6	0.9 x 0.7	13.0	Ring pec	120	Pair with ESO 60-26 W edge.
09 16 00.0	−72 34 06	ESO 61-5	14.1	0.9 x 0.5	13.0	Sc:	43	Mag 14.8 star W edge.
09 19 34.4	−68 37 58	ESO 61-7	14.1	1.3 x 0.7	13.8	SB?	23	Numerous stars superimposed; mag 11.9 star NE 3′.4.
09 19 56.5	−68 54 41	ESO 61-8	13.0	1.9 x 0.6	13.0	SA(rs)c: II-III	105	Mag 9.83v star N 9′.8.
09 27 28.8	−70 23 58	ESO 61-9	15.2	1.8 x 0.2	13.9	S	55	Mag 9.19v star SW 3′.3.
09 38 20.1	−70 05 31	ESO 61-11	13.7	1.8 x 0.3	12.9	S	168	Mag 10.39v star SW 4′.5.
09 39 46.0	−68 49 21	ESO 61-13	13.8	1.3 x 0.8	13.7	S	153	Numerous stars superimposed; mag 14.4 star on SW edge.
09 46 17.4	−68 54 49	ESO 61-15	12.8	1.7 x 1.0	13.2	SA(s)c II	21	Mag 11.31v star at N end.
09 52 33.7	−69 03 58	ESO 61-17	14.1	1.2 x 0.6	13.6		120	Mag 8.82v star WNW 10′.3.
09 58 36.6	−68 18 44	ESO 61-20	14.1	0.8 x 0.3	12.4	S	105	Mag 11.9 star NW edge.
09 58 32.4	−69 00 33	ESO 61-21	14.0	1.2 x 0.7	13.7	SB	35	Mag 13.5 star E edge.
08 38 37.6	−64 20 35	ESO 90-4	13.7	1.5 x 1.1	14.1	IB(s)m V	92	Pair of E-W oriented mag 9.53v, 9.01v stars WNW 12′.6.
08 55 36.4	−67 27 10	ESO 90-9	12.7	1.6 x 0.5	12.3	SAB(s)c: III	29	Mag 10.60v star NE 9′.3.
08 58 23.0	−66 43 48	ESO 90-12	12.7	2.0 x 0.4	12.3	S0 sp	50	Mag 8.89v star N 3′.7.
09 01 36.7	−64 16 22	ESO 90-14	14.6	1.5 x 0.3	13.6	Sbc	150	Double system, interaction.
09 02 06.0	−64 54 18	ESO 90-15	13.2	1.8 x 0.3	12.4	SAB(r)b: III-IV	144	Mag 7.18v star W 5′.2.
09 13 32.2	−63 37 40	ESO 91-3	12.1	2.1 x 1.4	13.1	SA(rs)ab III	78	Numerous stars superimposed on face, right to the core.
09 15 56.9	−65 13 47	ESO 91-6	14.6	1.5 x 0.3	13.6	SAB(rs)d? IV-V	178	Mag 10.27v star W 2′.0.
09 17 30.9	−62 53 04	ESO 91-7	13.2	2.1 x 1.1	13.9	SB(s)cd III-IV	176	Mag 8.43v star SW 2′.4.
09 23 26.7	−63 40 48	ESO 91-11	14.4	1.5 x 0.3	13.4	S	48	Close pair of mag 10.13v, 10.15v stars NW 1′.9; mag 8.53v star ESE 6′.6.
09 38 13.7	−63 29 14	ESO 91-16	13.3	1.4 x 0.5	12.8	Sb? sp	123	Mag 10.06v star SSW 3′.7.
09 46 08.5	−63 16 21	ESO 91-18	15.1	1.9 x 0.3	14.3	Sc sp	62	Mag 10.57v star NE 3′.6.
09 56 47.6	−65 37 40	ESO 92-1	13.3	1.1 x 1.0	13.3	S0?	82	Mag 11.45v star NNW 3′.2.
09 59 28.6	−67 18 20	ESO 92-3	14.4	1.5 x 0.5	13.9	SBdm: sp IV-V	95	Mag 9.53v star S 5′.9.
09 59 45.2	−67 38 33	ESO 92-4	13.9	1.6 x 0.5	13.5	SAB(s)m: V	62	Mag 8.69v star SSE 1′.9.
10 03 18.9	−64 58 03	ESO 92-6	12.8	1.7 x 0.8	12.9	SB(r)b II-III	38	Mag 11.8 star on NE edge.
10 10 52.7	−66 38 56	ESO 92-14	12.3	1.4 x 0.6	12.0	SB(rs)0°?	116	Mag 7.35v star NW 6′.2.
10 11 00.9	−67 08 43	ESO 92-15	13.5	1.0 x 0.9	13.2	Sb:	45	Numerous stars superimposed; in star rich area.
10 21 05.8	−66 29 36	ESO 92-21	12.9	1.2 x 0.7	12.6	SA(s)0^-:	49	Mag 10.6 star S 6′.2.
10 31 57.6	−63 42 31	ESO 92-22	13.7	0.8 x 0.5	12.6	S	67	Mag 12.4 star E 1′.1; mag 10.41v star S 3′.0.
09 29 58.4	−62 10 59	ESO 126-17	12.3	1.7 x 1.1	12.9	S0 pec: sp	124	Bright central core with very faint outer surface, many stars superimposed.
09 36 30.4	−62 05 56	ESO 126-21	13.3	1.0 x 1.0	13.2	S0		Numerous stars superimposed, in star rich area; mag 10.17v star W 7′.5.
09 37 51.4	−62 09 09	ESO 126-23	13.1	1.6 x 0.4	12.4	SA(s)c: sp III	10	Mag 10.9 star N 2′.0.
10 08 50.5	−67 01 53	IC 2554	11.8	3.1 x 1.3	13.2	SB(s)bc pec:	175	Multiple galaxy system.
08 25 30.3	−68 07 05	NGC 2601	12.5	1.6 x 1.1	13.0	SAB(r)a pec?	120	
09 09 03.8	−67 56 01	NGC 2788	12.7	1.8 x 0.4	12.2	Sab? sp	114	
09 03 35.1	−67 58 01	NGC 2788B	13.7	1.5 x 0.4	13.0	Sb pec?	171	
09 13 49.3	−69 38 36	NGC 2822	10.9	3.3 x 2.2	13.1	E?	90	Almost lost in the glare of mag 1.67v star "Miaplacidus", Beta Carinae.
09 13 45.0	−69 20 02	NGC 2836	12.2	2.6 x 1.9	13.8	SA(rs)bc:	112	Numerous stars superimposed on bright core.
09 15 36.3	−63 04 10	NGC 2842	12.5	1.5 x 1.3	13.1	(R′)SB(rs)0/a:	120	Numerous stars superimposed.
09 23 24.2	−63 48 44	NGC 2887	11.7	2.1 x 1.5	12.8	SA(s)0^-?	78	Many stars superimposed from edges to core; mag 8.53v star NE 8′.6.
09 50 09.0	−73 55 21	NGC 3059	11.0	3.6 x 3.2	13.5	SB(rs)bc II-III	70	Numerous superimposed stars; mag 8.94v star W 5′.2.
10 05 48.2	−67 22 39	NGC 3136	10.7	3.1 x 2.1	12.7	E:	30	Numerous stars superimposed on envelope.
10 03 32.8	−67 26 56	NGC 3136A	14.9	2.0 x 0.3	14.2	IB(s)m? sp V	83	Mag 9.26v star W 6′.8.
10 10 13.0	−67 00 18	NGC 3136B	11.7	1.5 x 0.9	12.0	E^+	30	Stellar nucleus; mag 9.10v star E 3′.1.

(Continued from previous page)

210 OPEN CLUSTERS 210

RA h m s	Dec ° ′ ″	Name	Mag	Diam ′	No. ★	B ★	Type	Notes
10 21 42.0	−69 20 00	ESO 62-8		5			cl	
10 38 54.0	−69 02 00	ESO 62-11		6			cl	
10 14 58.0	−64 36 48	ESO 92-18		5	1100	15.7	cl	
10 42 57.0	−64 24 00	IC 2602	1.6	100	60	3.0	cl	**Southern Pleiades**. Rich in stars; large brightness range; strong central concentration; detached.
10 42 00.0	−65 06 00	Mel 101	8.0	16	50	9.7	cl	Moderately rich in stars; small brightness range; no central concentration; detached.
09 49 16.0	−62 40 30	NGC 3036			20		cl	(S) asterism 4′ long elongated NE-SW.
09 49 04.0	−65 15 06	Ru 84		4	20	11.0	cl	Few stars; moderate brightness range; no central concentration; detached.
10 18 48.0	−63 08 12	Ru 88		5	30	12.0	cl	(S) Sparse group.
10 53 00.0	−62 16 00	Ru 162		6	60	12.0	cl:	

GLOBULAR CLUSTERS

RA h m s	Dec ° ′ ″	Name	Total V m	B ★ V m	HB V m	Diam ′	Conc. Class Low = 12 High = 1	Notes
09 12 02.6	−64 51 47	NGC 2808	6.2	13.5	16.2	14.0	1	

PLANETARY NEBULAE

RA h m s	Dec ° ′ ″	Name	Diam ″	Mag (P)	Mag (V)	Mag cent ★	Alt Name	Notes
09 07 06.4	−69 56 35	IC 2448	27	11.5	10.4	14.2	PK 285−14.1	Mag 11.4 star N 3'.4.
10 09 21.7	−62 36 40	IC 2553	9	13.0	10.3	15.5	PK 285−5.1	
10 17 50.7	−62 40 17	NGC 3211	19	11.8	10.7	18.0	PK 286−4.1	Pair of mag 11.3, 11.4 stars NW 3'.4.
10 07 23.8	−63 54 32	PN G286.0−6.5	10	13.5			PK 286−6.1	Mag 12.0 star ESE 2'.3; mag 11.03v star NW 3'.9.
10 06 59.9	−64 21 55	PN G286.2−6.9	60	14.9		16.9	PK 286−6.2	Mag 11.0 star on SW edge.
10 35 45.9	−64 19 12	PN G288.8−5.2	14	14.2	13.8	15.6	PK 288−5.1	Mag 12.3 star on NW edge; mag 8.96v star NW 1'.7.

211 GALAXIES 211

RA h m s	Dec ° ′ ″	Name	Mag (V)	Dim ′ Maj x min	SB	Type Class	PA	Notes
06 43 31.5	−72 35 41	ESO 34-12	13.2	1.7 x 0.9	13.5	SB(r)c II-III	136	Mag 10.79v star W 4'.1.
06 43 48.2	−73 40 31	ESO 34-13	13.9	1.3 x 0.5	13.3	SA(rs)bc pec	45	Mag 12.0 star ENE 2'.7.
07 10 43.0	−73 30 40	ESO 35-1	13.7	0.7 x 0.5	12.4		104	Mag 10.22v star SW 5'.1.
07 50 19.9	−72 52 30	ESO 35-17	14.2	1.0 x 0.7	13.7	SB(s)bc II	123	Mag 11.08v star WSW 11'.3.
06 11 30.7	−70 03 00	ESO 57-73	13.8	0.6 x 0.4	12.4		103	Pair with ESO 57-73A NW.
06 11 23.8	−70 02 33	ESO 57-73A	13.6	0.5 x 0.3	11.4		118	Pair with ESO 57-73 SE.
06 23 09.4	−68 44 01	ESO 57-80	13.9	1.6 x 0.3	13.0	Sb? sp	114	Mag 12.8 star on W edge; mag 8.94v star ESE 10'.2.
06 31 01.2	−71 30 05	ESO 58-3	15.0	1.6 x 0.2	13.6	Scd: sp	5	Mag 8.19v star W 7'.7.
06 32 20.0	−67 38 51	ESO 58-4	13.8	0.9 x 0.8	13.3	SB(rs)bc pec	22	Mag 13.2 star NE 2'.4.
06 32 21.0	−67 53 42	ESO 58-5	13.4	1.0 x 0.8	13.0	(R)SAB0°	124	Line of four NE-SW oriented mag 10.61v, 9.95v, 10.80v, 10.57v stars centered W NW 4'.8.
06 40 15.0	−72 27 21	ESO 58-9	14.4	1.5 x 0.3	13.4	SA(s)d:	96	Mag 11.04v star W 10'.2.
06 44 40.9	−71 27 29	ESO 58-13	13.4	2.2 x 0.3	12.8	Sb: sp	3	Mag 8.68v star NE 6'.4.
06 46 36.1	−70 36 54	ESO 58-14	13.7	1.1 x 0.6	13.1	S0?	97	Mag 13.0 star W 2'.5.
06 52 56.6	−71 45 45	ESO 58-19	12.5	1.6 x 1.1	13.0	SAB0⁻	99	**ESO 58-20** S 6'.2.
07 13 24.3	−68 21 33	ESO 58-25	13.9	1.7 x 0.4	13.3	SB(rs)c?	114	Mag 10.52v star SW 7'.4.
07 31 18.3	−68 11 17	ESO 59-1	13.1	1.8 x 1.5	14.0	IB(s)m V-VI	158	Mag 13.5 and 12.6 stars on S edge; mag 9.34v star WNW 9'.8.
07 34 51.6	−69 46 56	ESO 59-6	14.3	1.5 x 0.7	14.2	IAB(s)m? V	114	Mag 8.57v star SW 8'.7; ESO 59-7 E 7'.1.
07 36 12.3	−69 47 48	ESO 59-7	13.9	1.1 x 0.5	13.1	(R′)SB(rs)0/a:	103	**ESO 59-10** ENE 7'.6; mag 10.35v star S 4'.9; ESO 59-6 W 7'.1.
07 36 12.8	−70 42 42	ESO 59-9	13.3	1.5 x 1.5	14.0	(R′)SA(s)dm? IV-V		Mag 10.23v star WSW 6'.3.
07 38 11.6	−69 28 34	ESO 59-11	12.4	2.0 x 1.1	13.2	SB(s)0/a pec:	160	NGC 2442 SW 10'.8.
07 38 31.8	−68 46 16	ESO 59-12	13.1	2.0 x 0.9	13.6	(R′)SAB(s)a: IV-V	22	Mag 13.1 star E 1'.6.
07 42 54.8	−71 13 14	ESO 59-16	14.0	1.0 x 0.4	12.9	Sab	80	Mag 9.00v star W 9'.0.
07 45 29.0	−67 48 08	ESO 59-17	13.8	1.1 x 1.0	13.8	SA(r)c II-III	167	Mag 9.66v star E 7'.8.
07 47 28.9	−69 42 36	ESO 59-19	13.6	1.3 x 0.4	12.8	SB(s)b? II	63	Mag 9.14v star WNW 1'.0.
07 56 06.2	−68 16 40	ESO 59-23	13.8	1.6 x 0.7	13.8	SB(rs)c I-II	6	Mag 10.52v star W 7'.0; mag 9.30v star NW 12'.2.
08 03 27.9	−70 22 00	ESO 59-25	14.4	1.9 x 0.3	13.7	SB(s)c: sp II	57	Mag 7.22v star NE 2'.7.
08 11 40.0	−69 39 11	ESO 59-27	13.9	1.3 x 0.6	13.5	SAB(s)bc III-IV	114	Mag 9.27v star S 3'.4.
08 22 40.9	−69 46 22	ESO 60-4	13.6	1.2 x 0.4	12.7	SB(rs)b pec II-III	83	Pair of E-W oriented mag 10.76v, 9.27v stars NE 7'.6.
06 02 35.9	−64 50 32	ESO 86-52	14.2	1.1 x 0.9	14.0	Sc	167	Mag 7.02v star W 4'.6.
06 03 59.6	−63 41 59	ESO 86-56	13.4	1.3 x 0.5	12.8	SAB(r)a	164	Mag 8.49v star SW 6'.3.
06 08 52.4	−65 43 52	ESO 86-62	12.6	1.5 x 1.0	13.2	E⁺4:	173	Mag 9.54v star NNE 4'.3.
06 18 20.1	−66 01 02	ESO 87-3	13.5	1.7 x 0.5	13.1	(R′)SB(s)bc: II	111	Mag 8.38v star SE 6'.8.
06 21 30.8	−65 24 55	ESO 87-10	13.5	0.9 x 0.8	13.0	S?	44	Mag 9.18v star NW 10'.5.
06 26 12.3	−63 45 17	ESO 87-20	13.7	1.2 x 1.0	13.8	S0?	17	Mag 6.26v star SW 9'.6.
06 33 19.0	−62 59 39	ESO 87-28	13.5	1.7 x 1.0	14.0	S0° pec	125	Mag 10.02v star W 5'.1.
06 34 57.2	−66 05 28	ESO 87-29	13.7	0.8 x 0.6	12.8	S0?	158	Mag 10.54v star NE 2'.0.
06 36 41.5	−66 31 52	ESO 87-32	13.4	1.5 x 1.0	13.7	(R′)SB(s)bc I-II	7	Mag 10.80v star NE 5'.1; **ESO 87-33** NNE 5'.2.
06 42 42.1	−64 35 24	ESO 87-34	13.7	1.0 x 0.5	12.8	Sa	134	Mag 8.16v star NW 5'.4.
06 43 51.9	−64 14 28	ESO 87-38	13.9	1.0 x 0.5	13.0	Sa:	150	Mag 10.85v star S 0'.9.
06 47 36.0	−65 42 01	ESO 87-42	13.9	0.9 x 0.8	13.4	S?	86	Mag 9.97v star S 2'.2.
06 49 47.1	−64 07 25	ESO 87-46	14.7	1.7 x 0.3	13.8	SBm: pec sp	48	Mag 12.2 star W 6'.5; **ESO 87-48** E 8'.9.
06 53 02.9	−64 55 08	ESO 87-49	13.8	0.8 x 0.7	13.0	S?	111	Mag 11.18v star W 9'.3.
06 53 58.4	−63 13 09	ESO 87-50	14.9	1.6 x 0.3	13.9	Sc? sp	122	Mag 8.77v star ENE 5'.5; note: **ESO 87-53** located 2'.9 SE of this star.
06 55 28.3	−65 29 49	ESO 87-54	13.6	1.2 x 0.8	13.4	(R′)SA(rs)b II	151	Mag 9.28v star NNW 9'.2; **ESO 87-52** W 5'.2.
07 10 06.9	−63 15 46	ESO 88-4	12.9	0.7 x 0.5	11.7		98	Mag 10.91v star W 3'.4; mag 6.04v star ENE 13'.8.
07 18 33.0	−65 47 57	ESO 88-9	14.1	1.3 x 0.3	12.9	Sb:	118	Mag 10.25v star SE 3'.0.
07 28 18.9	−62 53 38	ESO 88-15	15.1	1.6 x 0.2	13.8	Scd: sp III-IV	71	Mag 10.66v star NNW 6'.3.
07 28 59.5	−66 53 59	ESO 88-17	12.7	1.8 x 0.7	12.8	SAB(r)bc pec:	63	ESO 88-18 S 1'.1; mag 8.26v star SW 8'.2.
07 29 06.0	−66 54 44	ESO 88-18	14.5	0.7 x 0.2	12.4	E5 pec	116	ESO 88-17 N 1'.1; mag 8.26v star SW 8'.2.
07 33 30.6	−65 23 42	ESO 88-21	14.0	0.9 x 0.4	12.8	S	119	Mag 12.5 star SW 0'.7.
07 34 12.8	−67 34 50	ESO 88-22	13.8	1.0 x 0.9	13.5	SAcd? III-IV	155	Mag 9.51v star NE 11'.1.
07 35 23.9	−66 21 15	ESO 88-23	14.1	1.1 x 0.9	14.0	SB(r)c II-III	132	Mag 10.84v star E 0'.9.
07 37 36.6	−66 07 12	ESO 88-26	13.7	1.0 x 0.2	11.8	Sa	110	Mag 10.76v star NNW 8'.5.
07 42 44.4	−65 19 02	ESO 89-1	16.5	0.9 x 0.9	16.1	Sb?		Mag 10.75v star E 4'.1.
08 05 09.4	−67 35 14	ESO 89-9	13.4	1.0 x 0.6	12.7	SAB(s)0°	123	Mag 10.09v star WNW 12'.5.
08 10 00.7	−64 56 15	ESO 89-12	13.3	2.2 x 0.3	12.7	SB(r)bc? sp II	100	Mag 8.58v star NW 14'.0.

(Continued from previous page)

GALAXIES

RA h m s	Dec ° ′ ″	Name	Mag (V)	Dim ′ Maj x min	SB	Type Class	PA	Notes
08 14 42.5	−66 30 23	ESO 89-13	14.2	1.3 x 0.6	13.8	SB(s)d? IV-V	105	Mag 9.18v star SW 7′9.
07 22 50.8	−62 01 45	ESO 123-9	13.5	2.1 x 0.9	14.0	SA(rs)c III	144	Mag 10.40v star N 1′4; mag 6.93v star NW 9′5.
07 30 41.2	−62 01 37	ESO 123-17	13.4	1.3 x 0.5	12.8	SAB0°:	60	Pair of stars superimposed NE end; mag 12.6 star SE 2′7.
07 28 17.6	−62 21 12	IC 2200	13.2	1.3 x 0.7	13.0	(R)SABb pec II	58	Pair with IC 2200A W edge; mag 13.4 star on NE edge; mag 10.59v star NW 13′7.
07 28 06.6	−62 21 47	IC 2200A	12.7	1.3 x 0.9	12.8	SB(s)0$^-$ pec?	44	Pair with IC 2200 E edge.
07 27 55.2	−67 34 31	IC 2202	12.9	2.0 x 0.7	13.1	SA(s)bc: II	165	Mag 10.32v star SE 2′3.
05 55 46.3	−69 33 42	NGC 2150	13.0	1.1 x 0.9	12.8	(R)SAB(r)ab:	143	Stellar nucleus.
06 02 47.9	−63 45 51	NGC 2178	12.9	0.9 x 0.9	12.8	E1		Lies 2′6 WSW of mag 8.49v star.
06 03 53.0	−69 34 39	NGC 2187	12.2	1.3 x 0.9	12.5	SA(s)a?	99	
06 03 44.2	−69 35 19	NGC 2187A	12.2	1.7 x 0.9	12.5	SA(s)a?	99	
06 04 45.0	−73 24 03	NGC 2199	12.8	1.9 x 0.8	13.1	(R′)SA(r)a:	37	
06 10 33.1	−62 32 18	NGC 2205	12.7	1.3 x 0.9	12.7	(R′)SAB(rs)0$^-$:	80	Faint star or knot E of core.
06 21 15.4	−64 27 33	NGC 2228	13.3	0.8 x 0.7	12.6	SA0°:		Star or bright knot E of core.
06 21 23.6	−64 57 25	NGC 2229	13.2	1.4 x 0.4	12.5	SAB(s)0$^-$? sp	133	An elongated anonymous galaxy N 2′3; **ESO 87-4** NW 10′8.
06 21 27.4	−64 59 35	NGC 2230	12.8	1.1 x 0.9	12.8	(R′)SA0$^-$?	77	Mag 10.33v star SW 3′4; NGC 2229 N 2′2.
06 21 40.1	−65 02 01	NGC 2233	13.7	0.9 x 0.3	12.1	S0$^-$: sp	39	Mag 10.33v star W 4′0.
06 22 22.2	−64 56 05	NGC 2235	12.7	1.3 x 1.0	13.0	E2:	68	Mag 10.84v star on N edge; **ESO 87-12** N 12′4.
06 44 24.7	−63 43 04	NGC 2297	12.7	1.9 x 1.5	13.7	SAB(rs)bc	101	Very faint envelope extends E and W beyond bright, elongated center.
06 48 37.6	−64 16 23	NGC 2305	11.7	2.0 x 1.5	12.9	E2: pec	142	Mag 13 star S edge and a fainter star E of central core; NGC 2307 S 4′0.
06 48 51.1	−64 20 09	NGC 2307	12.0	1.7 x 1.6	13.0	SB(rs)b II-III	145	Mag 9.74v star W 10′1; NGC 2305 N 4′0.
07 16 38.1	−62 20 38	NGC 2369	12.3	3.5 x 1.1	13.6	SB(s)a I-II	177	
07 18 43.5	−62 56 11	NGC 2369A	12.9	1.9 x 1.2	13.7	SAB(rs)bc II-III	33	Surrounded by a quartet of faint stars.
07 20 30.2	−62 03 14	NGC 2369B	13.3	1.5 x 1.4	13.9	(R′)SB(r)bc: III	14	Mag 6.93v star Delta Volantis NE 10′3.
07 19 57.6	−63 04 04	NGC 2381	12.8	1.4 x 1.4	13.4	(R′)SB(rs)a:		
07 21 20.9	−69 00 07	NGC 2397	11.8	2.5 x 1.2	12.9	SB(s)b: II-III	123	Small, anonymous spiral galaxy with stellar nucleus NNE 2′1; NGC 2397A S 7′0.
07 21 08.4	−69 06 57	NGC 2397A	14.0	1.1 x 0.9	13.9	SA(rs)cd:	36	NGC 2397 N 7′0.
07 21 55.8	−68 50 48	NGC 2397B	14.1	0.9 x 0.5	13.1	IB(s)m pec IV-V	97	Located 2′5 N of mag 9.93v star.
07 30 11.8	−62 15 05	NGC 2417	12.0	2.8 x 2.0	13.8	SA(rs)bc I-II	81	Numerous stars on outer envelope; mag 9.73v star Delta Volantis ENE 6′9.
07 34 51.3	−69 17 04	NGC 2434	11.3	2.5 x 2.3	13.1	E0-1	133	Mag 12.5 star on NNW edge; pair of E-W oriented mag 12 stars N 2′0.
07 36 16.5	−69 32 24	NGC 2442	10.4	5.5 x 4.9	13.8	SAB(s)bc pec II	27	Arms form prominent "S" shape, several strong knots and dust lanes.
07 45 16.0	−71 24 41	NGC 2466	13.0	1.5 x 1.4	13.6	SA(s)c: I-II	7	Numerous knots and/or superimposed stars.
08 25 30.3	−68 07 05	NGC 2601	12.5	1.6 x 1.1	13.0	SAB(r)a pec?	120	
06 15 46.2	−70 53 40	PGC 18727	13.7	2.0 x 0.8	14.1	SB(s)dm V-VI	87	

GALAXY CLUSTERS

RA h m s	Dec ° ′ ″	Name	Mag 10th brightest	No. Gal	Diam ′	Notes
06 21 48.0	−64 57 00	A 3389	14.6	35	28	Majority of members anonymous, stellar/almost stellar.
06 22 24.0	−64 55 00	AS 585	14.8	1	28	

OPEN CLUSTERS

RA h m s	Dec ° ′ ″	Name	Mag	Diam ′	No. ★	B ★	Type	Notes
05 52 23.7	−67 20 03	NGC 2130		1.2			cl	Relatively faint; relatively few stars.
05 53 35.1	−67 25 43	NGC 2135		1			cl	Relatively faint; relatively few stars.
05 52 58.9	−69 29 30	NGC 2136	10.5	1.9		14.5	cl	Bright and moderately rich in stars; detached.
05 53 11.9	−69 28 58	NGC 2137		0.8			cl	Faint; relatively few stars.
05 54 49.7	−65 50 17	NGC 2138		1			cl	Extremely faint; nearly stellar.
05 54 16.0	−68 35 58	NGC 2140		1.7			cl	Relatively faint; few stars; extended NW-SE.
05 54 22.6	−70 54 03	NGC 2145		1.7			cl	Relatively faint.
05 55 46.4	−68 12 14	NGC 2147		1.1			cl	A small knot of stars in the N central part of a relatively large, faint nebulosity.
05 56 20.8	−69 01 05	NGC 2151		1			cl	
05 57 52.3	−66 24 04	NGC 2153		1.3			cl	
05 57 38.5	−67 15 44	NGC 2154		2.4			cl	Quite bright; contains a prominent core surrounded by a faint outer halo of stars.
05 58 32.3	−65 28 40	NGC 2155		2.1			cl	
05 57 50.0	−68 27 40	NGC 2156	11.4	1.1			cl	Relatively bright; moderately rich; slight central concentration; detached.
05 57 34.8	−69 11 47	NGC 2157	10.2	2.7		14.5	cl	Very rich in stars; strong central concentration; detached.
05 58 02.7	−68 37 26	NGC 2159	11.4	0.9			cl	Relatively bright; moderately rich; detached.
05 58 13.0	−68 17 27	NGC 2160		1.2			cl	Relatively faint; irregular in shape.
06 00 31.4	−63 43 18	NGC 2162		2.1			cl	Very faint; slightly elongated E-W.
05 58 55.6	−68 30 56	NGC 2164	10.3	2.5		14.7	cl	Rich in stars; strong central concentration; detached.
05 59 33.0	−67 56 33	NGC 2166		1.2			cl	Relatively faint; compact group.
05 58 52.9	−67 53 52	NGC 2171		0.7			cl	Almost stellar appearance in star rich area.
06 00 05.6	−68 38 14	NGC 2172	11.8	1.7			cl	Relatively bright; moderately rich; detached.
05 57 58.0	−72 58 44	NGC 2173		2.4			cl	A faint globular cluster with a small condensed core.
06 01 19.4	−66 51 19	NGC 2176		1.3			cl	Very faint and diffuse.
06 01 15.7	−67 44 06	NGC 2177		1.2			cl	
06 02 43.1	−65 15 54	NGC 2181		1.6			cl	Very faint; very compact.
06 06 17.4	−65 05 53	NGC 2193		1.9			cl	Very faint.
06 06 07.4	−67 05 54	NGC 2197		1.7			cl	Faint and compact.
06 08 34.8	−73 50 22	NGC 2209		2.8			cl	Very faint and diffuse.
06 10 42.2	−71 31 47	NGC 2213		2.1			cl	Relatively faint.
06 12 56.1	−68 15 38	NGC 2214	10.9	3.6		14.5	cl	Rich in stars; somewhat extended E-W; detached.
06 20 43.7	−67 31 04	NGC 2231		2			cl	Very faint; contains a tiny condensed core.
06 22 52.0	−68 55 32	NGC 2241		1.3		19.0	cl	Faint; very compact group.
06 25 49.5	−68 55 13	NGC 2249	12.2	1.7			cl	Relatively bright.
07 03 02.7	−67 24 00	NGC 2348		11	30	9.9	cl?	(A) Very small anonymous galaxy located 7′ to north of center.

GLOBULAR CLUSTERS

RA h m s	Dec ° ′ ″	Name	Total V m	B ★ V m	HB V m	Diam ′	Conc. Class Low = 12 High = 1	Notes
06 14 22.8	−69 50 50	H60b 11				2.5		
06 11 31.3	−69 07 22	NGC 2210	10.2			2.1		
06 30 12.4	−64 19 34	NGC 2257	13.5			2.2		

(Continued from previous page)

211 BRIGHT NEBULAE 211

RA h m s	Dec ° ′ ″	Name	Dim ′ Maj x min	Type	BC	Color	Notes
05 56 00.0	−68 12 00	NGC 2147					

212 GALAXIES 212

RA h m s	Dec ° ′ ″	Name	Mag (V)	Dim ′ Maj x min	SB	Type Class	PA	Notes
03 37 43.6	−72 23 31	ESO 31-18	14.4	1.1 x 0.6	13.8	SA(s)c II	99	Mag 13.4 star on N edge, mag 1.6 star N 2′.4.
04 57 47.7	−73 13 55	ESO 33-3	12.7	1.2 x 0.7	12.5	E4?	20	Pair of faint stars or bright knots S of core.
05 05 06.9	−73 39 06	ESO 33-11	13.2	1.0 x 0.9	12.9	(R)SB(rs)a	65	Mag 8.26v star NE 4′.9.
05 07 18.7	−73 33 20	ESO 33-14	14.3	0.9 x 0.6	13.5	S	167	Several stars superimposed, in star rich area.
05 31 41.5	−73 44 59	ESO 33-22	14.8	2.1 x 0.2	13.7	Sd sp	170	Mag 9.44v star N 7′.2; mag 5.84v star E 12′.9.
05 49 03.2	−72 33 20	ESO 33-30	14.0	1.1 x 0.5	13.2	SBa	179	Mag 12.6 star NE 1′.5.
03 49 48.1	−71 38 29	ESO 54-21	12.1	4.5 x 2.1	14.4	SAB(s)dm IV-V	91	Many bright knots and superimposed stars.
04 17 15.5	−70 01 28	ESO 55-18	14.0	1.0 x 0.5	13.2	(R′)SAB(r)0/a:	148	Mag 10.14v star N 8′.0.
04 25 48.4	−67 48 58	ESO 55-23	13.3	1.6 x 0.4	12.6	SA(s)b: sp II	151	Mag 10.26v star ESE 7′.7.
04 37 20.7	−69 12 13	ESO 55-29	13.3	1.5 x 0.9	13.5	SAB(s)d: III-IV	81	**ESO 55-28** N 3′.7.
04 38 50.6	−69 30 25	ESO 55-33	13.4	1.0 x 0.7	13.0	S0:	93	Mag 10.60v star N 1′.5.
04 43 03.7	−67 55 37	ESO 55-35	13.1	1.3 x 1.0	13.2	Sm?	55	Mag 8.04v star SE 12′.9.
04 52 13.8	−69 42 18	ESO 56-15	15.0	0.9 x 0.2	12.9	S	47	Globular cluster NGC 1751 SE 11′.9; mag 7.23v star SW 10′.7.
04 53 18.3	−70 35 55	ESO 56-19	13.0	2.1 x 1.1	13.8		50	Globular cluster NGC 1754 NE 10′.6.
05 20 26.6	−69 51 27	ESO 56-104	13.0	1.0 x 0.5	12.1		131	Open clusters NGC 1938 and NGC 1939 SE 7′.5.
05 23 17.8	−69 45 21	ESO 56-115	0.4	650 x 550	14.1	SB(s)m III-IV	170	**Large Magellanic Cloud** (**LMC**).
05 33 59.8	−71 45 26	ESO 56-154	13.3	1.2 x 1.1	13.5	SA?(r)0	100	Mag 8.39v star E 4′.1; open cluster NGC 2025 W 4′.4.
03 32 00.1	−62 34 11	ESO 83-6	14.1	1.7 x 0.7	14.1	(R′)SB(rs)c III	101	Mag 7.41v star ESE 11′.6.
03 33 41.1	−66 07 45	ESO 83-7	16.1	1.5 x 0.2	14.7	Im? V?	9	Mag 10.56v star NW 10′.2.
03 45 40.6	−64 18 04	ESO 83-10	14.3	1.0 x 0.7	13.7	S?	34	Mag 8.82v star E 10′.5.
03 54 57.0	−65 56 45	ESO 83-12	13.7	1.8 x 0.2	12.5	Sb?	41	Mag 12.9 star S 1′.5; NGC 1490 SW 9′.3.
04 14 23.0	−62 48 45	ESO 84-9	13.5	1.0 x 0.6	12.8	SB(s)c I-II	52	Mag 15.3 star N edge.
04 18 57.6	−63 29 23	ESO 84-11	13.9	1.1 x 0.3	12.5	S	146	Mag 11.9 star NE 3′.0.
04 21 44.9	−64 13 25	ESO 84-14	13.4	1.0 x 0.9	13.2	SAB(r)bc	120	Mag 11.9 star W 1′.4.
04 22 11.8	−63 36 41	ESO 84-15	14.7	1.7 x 0.5	14.3	SB(s)m: V	84	Mag 9.35v star W 4′.4.
04 27 46.8	−62 35 33	ESO 84-18	13.2	1.5 x 1.3	13.7	SA(rs)a:	108	Almost stellar nucleus; mag 5.74v star N 4′.3.
04 29 23.1	−62 53 21	ESO 84-20	13.9	1.0 x 0.3	12.5	Sb: sp III-IV	170	Mag 12.0 star ESE 1′.2.
04 31 08.1	−63 19 25	ESO 84-21	13.6	1.2 x 0.6	13.1	SB(r)b II	121	Mag 12.3 star W edge.
04 35 01.0	−65 18 02	ESO 84-22	13.5	1.1 x 0.5	13.0	E	57	Mag 13.9 star NE edge; mag 15.0 star S edge.
04 36 20.7	−65 08 32	ESO 84-26	13.2	1.8 x 0.5	13.0	SAB(r)0^{+}: sp	100	Mag 10.74v star E 5′.6.
04 36 53.4	−64 34 31	ESO 84-27	14.7	0.6 x 0.4	13.0		168	Double (triple?) system, connected.
04 37 17.6	−62 35 03	ESO 84-28	13.3	1.7 x 0.4	12.8	S0^{-} sp	122	Mag 11.10v star SW 1′.9.
04 39 00.3	−63 02 24	ESO 84-32	13.9	1.1 x 1.1	14.0	SAB(s)c II-III		Mag 9.96v star S 6′.6; mag 9.41v star E 10.9.
04 38 53.7	−64 12 30	ESO 84-33	13.3	1.6 x 0.5	12.9	Sb: II	13	Mag 9.43v star NW 10.2.
04 40 26.6	−63 06 31	ESO 84-34	14.4	1.7 x 0.2	13.0	Sb sp I-II	10	Mag 9.41v star N 6′.3.
04 42 12.1	−62 54 49	ESO 84-35	14.2	1.2 x 0.4	13.3	SAb: II	1	Star or bright knot N end.
04 44 59.8	−62 42 21	ESO 84-40	14.4	0.8 x 0.5	13.3	Im:	39	Mag 10.76v star NW 8′.3; note: anonymous galaxy 2′.2 NE of this star.
04 20 13.7	−63 57 58	ESO 84-212	13.6	1.2 x 0.4	12.7		69	Mag 9.99v star NE 7′.2.
04 46 44.3	−62 27 52	ESO 85-1	13.3	1.1 x 0.9	13.1	SAB(s)cd III	153	Mag 9.14v star NE 8′.9.
04 46 46.2	−63 01 55	ESO 85-2	14.1	1.4 x 0.3	13.0	SB(s)d pec: III-IV	87	Mag 10.42v star SW 7′.0.
04 47 50.4	−62 36 45	ESO 85-4	14.0	1.5 x 0.3	13.0	SB(rs)b pec: II-III	26	
04 48 15.2	−63 27 50	ESO 85-5	13.8	1.0 x 0.6	13.1	SBc	153	Mag 7.15v star SW 1′.4.
04 54 42.7	−62 47 58	ESO 85-14	12.8	3.2 x 1.1	14.0	SB(s)m IV	78	Double system; mag 7.97v star E 2′.6.
04 58 04.0	−63 55 08	ESO 85-24	13.5	1.3 x 0.3	12.3	Im:	133	Mag 9.29v star SE 4′.3.
05 01 29.9	−63 17 35	ESO 85-30	12.8	1.5 x 0.7	12.7	S0^{+}? pec	147	Mag 8.55v star NE 13′.5; triple system **ESO 85-33**, **ESO 85-33A**, **ESO 85-33B** NE 9′.4.
05 03 18.0	−63 45 05	ESO 85-34	12.8	1.2 x 0.8	12.6	Sa?	85	Mag 11.4 star N 3′.9.
05 04 18.9	−63 34 57	ESO 85-38	12.7	1.4 x 0.9	12.8	(R′)SB(rs)b: I-II	89	Close pair of N-S oriented mag 11.2, 9.40v stars E 2′.1.
05 07 43.8	−62 59 25	ESO 85-47	13.8	1.8 x 1.1	14.4	SB(s)m V-VI	22	Mag 7.95v star SW 12′.0.
05 27 09.2	−63 14 30	ESO 85-88	16.2	1.5 x 0.7	16.1	IB(s)m V-VI	132	Mag 9.19v star NW 5′.8.
05 33 13.8	−65 46 44	ESO 86-10	13.5	2.6 x 1.0	14.4		37	Obscured by LMC.
05 38 31.2	−64 29 26	ESO 86-18	13.7	1.4 x 0.4	13.0	S	168	Numerous stars superimposed; mag 12.1 star NE 1′.5.
05 42 41.0	−65 48 41	ESO 86-23	13.9	1.1 x 0.8	13.6	S	56	Located between a pair of N-S mag 11.4 and 11.9 stars.
06 02 35.9	−64 50 32	ESO 86-52	14.2	1.1 x 0.9	14.0	Sc	167	Mag 7.02v star W 4′.6.
06 03 59.6	−63 41 59	ESO 86-56	13.4	1.3 x 0.5	12.8	SAB(r)a	164	Mag 8.49v star SW 6′.3.
04 02 32.9	−62 18 58	ESO 117-19	13.6	1.9 x 0.3	12.8	SBbc? sp II	77	Mag 9.54v star SSE 4′.7.
04 49 11.8	−62 21 38	ESO 119-11	13.9	0.5 x 0.4	12.0		107	Mag 9.61v star NNE 4′.8.
05 14 29.5	−62 10 16	ESO 119-46	14.0	1.1 x 0.3	12.6	Sb? pec sp	56	Mag 9.07v star E 5′.5.
05 14 31.1	−62 13 42	ESO 119-47	13.3	1.2 x 0.6	12.8	SB(s)b pec	51	Mag 9.07v star NE 7′.0.
03 47 26.3	−68 13 17	NGC 1473	12.9	1.5 x 0.8	12.9	IB(s)m: IV	36	
03 53 34.1	−66 01 06	NGC 1490	12.4	1.3 x 1.1	12.8	E1	142	Mag 9.32v star S 3′.9; ESO 83-12 NW 9′.3.
03 56 33.2	−66 02 27	NGC 1503	13.6	0.9 x 0.8	13.1	SB(rs)0^{+}?	140	Mag 12.1 star NE 4′.8; ESO 83-12 NW 11′.4.
03 59 37.0	−67 38 09	NGC 1511	11.3	3.5 x 1.2	12.7	SAa pec:	125	NGC 1511A 7′.5 E.
04 00 19.6	−67 48 28	NGC 1511A	13.3	1.7 x 0.4	12.8	SBa? sp	110	
04 00 55.4	−67 36 45	NGC 1511B	14.5	1.7 x 0.3	13.6	SBd? sp IV-V	92	Located 7′.5 E of NGC 1511.
04 05 12.4	−65 50 25	NGC 1526	13.7	0.8 x 0.6	12.7	Sbc I-II	36	Mag 9.76v star N 3′.2.
04 07 20.4	−62 53 59	NGC 1529	13.3	1.2 x 0.3	12.1	S0°: sp	164	Mag 8.91v star N 9′.0.
04 08 46.5	−62 47 54	NGC 1534	12.7	1.7 x 0.8	12.9	SA(rs)0/a: sp	76	Mag 13.5 star on S edge; mag 8.91v star WNW 6′.6.
04 17 37.5	−62 46 59	NGC 1559	10.7	3.5 x 2.0	12.6	SB(s)cd III	64	Many bright knots.
04 43 00.5	−65 48 55	NGC 1669	13.8	0.7 x 0.4	12.3	Sa?	97	
04 52 31.0	−62 59 08	NGC 1706	12.6	1.4 x 1.0	12.8	SA(rs)ab I	124	
04 58 24.4	−62 01 44	NGC 1765	13.0	1.2 x 1.0	13.0	SB(rs)0^{-}:	150	Stellar nucleus.
04 58 56.2	−63 17 56	NGC 1771	13.4	1.9 x 0.5	13.2	SAB(r)c: sp I-II	136	
05 02 04.7	−69 34 01	NGC 1809	12.1	3.2 x 0.8	13.0	Sc:	143	Open cluster NGC 1801 SW 8′.3.
05 17 09.0	−64 57 35	NGC 1892	12.2	2.9 x 0.8	13.0	Scd:	74	
05 26 47.3	−63 45 38	NGC 1947	10.6	3.0 x 2.6	12.7	S0^{-} pec	119	Numerous stars superimposed on face, in star rich area.
05 41 51.3	−64 18 07	NGC 2082	12.1	1.3 x 1.2	12.4	SB(r)b III	37	
05 55 46.3	−69 33 42	NGC 2150	13.0	1.1 x 0.9	12.8	(R)SAB(r)ab:	143	Stellar nucleus.
06 02 47.9	−63 45 51	NGC 2178	12.9	0.9 x 0.9	12.8	E1		Lies 2′.6 WSW of mag 8.49v star.
06 03 53.0	−69 34 39	NGC 2187	12.2	1.3 x 0.9	12.5	SA(s)a?	99	
06 03 44.2	−69 35 19	NGC 2187A	12.2	1.7 x 0.9	12.5	SA(s)a?	99	
06 04 45.0	−73 24 03	NGC 2199	12.8	1.9 x 0.8	13.1	(R′)SA(r)a:	37	

(Continued from previous page)

GALAXY CLUSTERS

RA h m s	Dec ° ′ ″	Name	Mag 10th brightest	No. Gal	Diam ′	Notes
03 32 48.0	−64 14 00	AS 362	16.2		17	All members anonymous, stellar.
04 00 48.0	−64 58 00	AS 411	16.1		17	**Fair 394** N of center 5.′3; all others anonymous, stellar.
04 46 42.0	−62 27 00	AS 500	14.9		22	

OPEN CLUSTERS

RA h m s	Dec ° ′ ″	Name	Mag	Diam ′	No. ★	B ★	Type	Notes
05 02 29.1	−70 17 32	HS66 94		0.9			cl	
04 57 14.5	−68 26 30	IC 2117		1.8			cl	(A) Not a planetary nebula as the IC indicates.
05 40 23.9	−69 40 11	IC 2145		0.5			cl	
04 53 30.4	−66 55 31	KMHK 211		0.8			ptcl	
04 12 55.0	−70 30 00	NGC 1557		22	20		cl?	(A) Irregular sparse cluster.
04 29 36.3	−71 50 17	NGC 1629		1			cl	Extremely faint.
04 36 06.0	−65 46 54	NGC 1641		5	10		cl?	(A) Includes background galaxy **SL 6** = **ESO 84-G25**.
04 37 39.4	−66 11 58	NGC 1644		1.6			cl	
04 38 06.9	−68 46 41	NGC 1649	11.2	0.7			cl	
04 37 31.8	−70 35 07	NGC 1651		2.5			cl	Very faint.
04 38 22.6	−68 40 21	NGC 1652	9.3	1.2			cl	Relatively faint.
04 42 39.2	−69 49 09	NGC 1673		0.7			cl	Very faint; brighter star on N border.
04 43 52.6	−68 49 34	NGC 1676		0.8			cl	Extremely faint; very few stars.
04 47 39.9	−69 20 35	NGC 1693		0.7			cl	Faint.
04 47 44.8	−69 22 28	NGC 1695		1.5			cl	Relatively faint; brighter than nearby NGC 1693.
04 48 29.8	−68 14 33	NGC 1696		0.9			cl	
04 48 37.0	−68 33 38	NGC 1697		2.6			cl	
04 49 04.0	−69 06 54	NGC 1698		1.6			cl	
04 49 25.9	−69 51 11	NGC 1702		1			cl	
04 49 55.5	−69 45 16	NGC 1704	11.5	1.7			cl	
04 50 37.4	−69 59 02	NGC 1711	10.1	2.4		13.5	cl	
04 50 57.1	−69 24 15	NGC 1712		2.5			cl	Southwestern most of three star clouds in a chain including NGCs 1722 and 1727; involved in nebulosity.
04 52 06.6	−66 55 25	NGC 1714		1.2			cl	
04 52 25.6	−67 03 09	NGC 1718	12.3	2			cl	Compact and moderately bright.
04 51 51.5	−69 23 55	NGC 1722		1			cl	
04 52 12.3	−69 20 16	NGC 1727		2.5			cl	Northwestern most in a chain of three star clouds including NGCs 1712 and 1722; involved in nebulosity.
04 53 29.5	−66 55 00	NGC 1731		8			cl	
04 53 11.5	−68 39 03	NGC 1732	12.3	0.9			cl	Relatively faint.
04 54 04.9	−66 40 56	NGC 1733		1.2			cl	Very faint.
04 53 33.9	−68 46 03	NGC 1734		1.3			cl	Very faint.
04 54 20.2	−67 06 04	NGC 1735		1.6			cl	
04 53 02.3	−68 03 12	NGC 1736		1.8			cl	
04 54 03.4	−69 11 57	NGC 1743		0.5			cl?	
04 54 56.5	−68 11 19	NGC 1749		1.2			cl	
04 54 11.9	−69 48 30	NGC 1751		1.7			cl	Relatively faint with a brighter core surrounded by a halo of faint stars. There is a prominent star chain extending to the SW off the central core.
04 55 14.7	−68 12 20	NGC 1755	9.9	2.6			cl	
04 54 49.9	−69 14 14	NGC 1756	12.2	1.1			cl	
04 56 34.4	−66 28 24	NGC 1761		1.2			cl	A rich scattered star cloud or cluster involved in a faint nebulosity.
04 56 27.4	−67 41 46	NGC 1764		1			cl	
04 55 57.7	−70 13 36	NGC 1766	12.2	0.7			cl	Moderately bright; relatively few stars.
04 56 27.4	−69 24 10	NGC 1767		1.6			cl	Small group of stars involved in nebulosity.
04 57 02.7	−68 14 54	NGC 1768		0.9			cl	
04 57 16.8	−68 24 41	NGC 1770		1.6			cl	
04 56 53.9	−69 33 23	NGC 1772		1.5			cl	Small group of stars involved in nebulosity.
04 58 06.2	−67 14 32	NGC 1774	10.8	1.8			cl	
04 56 53.6	−70 25 48	NGC 1775	12.6	0.7			cl	Few stars.
04 58 39.2	−66 25 46	NGC 1776		1.1			cl	Moderately bright; slight central concentration; detached.
04 57 51.5	−69 23 37	NGC 1782		1.2			cl	Small group of stars involved in nebulosity.
04 59 08.7	−65 59 16	NGC 1783		3			cl	
04 58 45.0	−68 49 24	NGC 1785		4	12	12.4	ast?	(A) Has NE-SW line of 9 stars.
05 00 00.0	−65 50 00	NGC 1787	10.9	23	150		ass	
04 57 51.5	−71 54 08	NGC 1789		1.5			cl	
04 59 06.0	−70 10 03	NGC 1791		1.3			cl	
04 59 38.3	−69 33 27	NGC 1793	12.4	1.3			cl	Relatively faint; detached.
04 59 46.8	−69 48 03	NGC 1795		1.6			cl	Very faint with a slightly brighter core.
05 00 35.1	−69 36 46	NGC 1801	12.2	2.2			cl	Moderately bright; slight central concentration; detached.
05 01 05.1	−69 05 03	NGC 1804	11.9	1			cl	
05 02 21.2	−66 06 41	NGC 1805	10.6	2.2			cl	Rich in stars; strong central concentration; detached.
05 02 11.3	−67 59 11	NGC 1806		2.2			cl	Large bright core in a small halo of faint stars.
05 03 22.8	−66 22 55	NGC 1810	11.9	1.2		12.8	cl	Moderately bright; detached.
05 03 45.4	−67 18 09	NGC 1814	9.0	1			cl	Coarse group of relatively bright stars involved in nebulosity; not well detached.
05 02 26.9	−70 37 13	NGC 1815		1.2			cl	Relatively bright; slight central concentration.
05 03 51.2	−67 15 38	NGC 1816	9.0	1			cl	Coarse group of relatively bright stars; not well detached.
05 04 13.0	−66 25 58	NGC 1818	9.7	3.4		13.3	cl	Very rich; strong central concentration; detached.
05 04 06.0	−67 16 00	NGC 1820	9.0				cl	Coarse group of relatively bright stars; not well detached.
05 05 08.1	−66 12 43	NGC 1822		0.8			cl	Faint.
05 03 23.8	−70 20 12	NGC 1823		0.9			cl	Relatively few stars; slight central concentration; not well detached.
05 04 06.0	−68 55 00	NGC 1825	12.0	0.5			cl	
05 05 33.7	−66 13 51	NGC 1826		0.9			cl	Faint.
05 04 21.0	−69 23 19	NGC 1828	12.5	0.7			cl	Relatively bright; detached.
05 04 58.5	−68 03 39	NGC 1829		2.1			cl	
05 04 39.5	−69 20 30	NGC 1830	12.6	0.7			cl	Relatively bright; detached.
05 06 16.2	−64 55 07	NGC 1831	11.2	3.9		14.3	cl	Very rich; strong central concentration; detached.
05 04 22.3	−70 43 52	NGC 1833		2.1	50		cl	
05 05 35.6	−68 37 42	NGC 1836	12.2	0.8			cl	
05 04 56.0	−70 43 18	NGC 1837		1.3	20		cl	Part of large and beautiful star cloud with many faint stars; in a large complex in the LMC. (A) Appears as N-S string of stars, with 20″ clump of stars at N end.
05 06 08.3	−68 26 42	NGC 1838	12.9	0.7			cl	Faint; relatively few stars; not well detached.
05 05 19.9	−71 45 50	NGC 1840		0.6			cl	
05 07 18.4	−67 16 23	NGC 1842		0.8			cl	Extremely faint.

(Continued from previous page)

OPEN CLUSTERS

RA h m s	Dec ° ′ ″	Name	Mag	Diam ′	No. ★	B ★	Type	Notes
05 07 30.4	−67 19 25	NGC 1844	12.1	1.3		14.3	cl	Relatively faint; relatively few stars.
05 05 12.0	−70 35 00	NGC 1845		20			ass	(A) Association with NGC 1833 and NGC 1837, plus others not plotted on map.
05 07 33.9	−67 27 39	NGC 1846		2.8			cl	Only moderately rich; relatively small central core.
05 07 08.1	−68 58 17	NGC 1847	11.1	1		14.3	cl	Bright; detached.
05 09 34.5	−66 19 00	NGC 1849	12.8	1.3			cl	Bright; round; detached.
05 08 46.7	−68 45 42	NGC 1850	9.0	3.4		14.3	cl	Rich in stars; involved in nebulosity. (A) May physically contain one to three star clusters.
05 09 24.1	−67 46 47	NGC 1852	12.0	1.8			cl	Compact cluster. The faint emission nebula Henize 25 lies just off the SSW edge.
05 09 20.6	−68 50 51	NGC 1855	10.4	3		14.2	cl	Rich in stars; detached. (A) Includes **NGC 1854** as its central portion.
05 09 30.6	−69 07 42	NGC 1856	10.1	1.8		15.0	cl	Very rich; strong central concentration; detached.
05 11 32.6	−65 14 55	NGC 1859		2			cl	Relatively bright; slight central concentration; detached.
05 10 39.5	−68 45 12	NGC 1860	11.0	0.8			cl	Faint; detached.
05 10 21.1	−70 46 40	NGC 1861		1.2			cl	Relatively faint; detached.
05 12 33.6	−66 09 19	NGC 1862		0.3			cl	
05 11 39.8	−68 43 38	NGC 1863	11.0	0.5		16.0	cl	Relatively bright; detached.
05 12 41.1	−67 37 25	NGC 1864		0.9			cl	
05 12 25.4	−68 46 19	NGC 1865	12.9	0.8			cl	
05 13 38.6	−65 27 51	NGC 1866	9.7	4.5		15.0	cl	Very rich; strong central concentration; well detached. Only a few stars brighter than magnitude 15.8.
05 13 41.6	−66 17 36	NGC 1867		1.3			cl	(A) The DSS image shows an unusual short "arc" of some sort, just NE. Is it a real LMC object or plate defect?
05 14 36.1	−63 57 17	NGC 1868		3.9			cl	
05 13 53.9	−67 23 00	NGC 1869		14			ass	
05 13 11.0	−69 07 02	NGC 1870	11.3	0.5		15.0	cl	Moderately bright; detached.
05 13 52.4	−67 27 09	NGC 1871		2			cl	
05 13 11.5	−69 18 45	NGC 1872	11.0	1			cl	Bright; very rich; strong central concentration; detached.
05 14 01.7	−67 19 54	NGC 1873		3.5			cl	
05 12 49.7	−70 28 19	NGC 1878		1.1			cl	Relatively faint; detached.
05 15 32.8	−66 07 44	NGC 1882		1.2			cl	Relatively bright; slight central concentration; detached.
05 15 06.9	−68 58 43	NGC 1885	12.0	0.8			cl	Moderately bright; detached.
05 16 05.3	−66 19 12	NGC 1887		1.1			cl	Relatively bright.
05 13 47.2	−72 04 41	NGC 1890		1.2			cl	
05 15 51.4	−69 28 13	NGC 1894	12.2	0.8			cl	Relatively bright; moderately rich in stars; slight central concentration; detached.
05 17 31.6	−67 26 56	NGC 1897		1			cl	Very faint.
05 19 08.8	−63 01 25	NGC 1900		1.7			cl	
05 17 48.0	−68 26 00	NGC 1901		40	40		cl	Extremely faint.
05 18 18.3	−66 37 35	NGC 1902		1.6			cl	Relatively bright; detached.
05 17 22.8	−69 20 13	NGC 1903	11.9	1			cl	Rich in stars; strong central concentration; detached.
05 18 23.0	−67 16 42	NGC 1905		1			cl	
05 18 23.8	−69 14 00	NGC 1910	11.2	8.5			ass	
05 19 25.2	−66 40 55	NGC 1911		0.4			cl	
05 18 19.2	−69 32 15	NGC 1913	11.1	0.6			cl	
05 17 39.4	−71 15 19	NGC 1914		2.2			cl	
05 19 28.1	−66 44 21	NGC 1915		0.4			cl	
05 20 17.1	−66 52 53	NGC 1919		1.7			cl	
05 20 32.1	−66 46 45	NGC 1920		1.8			cl	
05 19 50.0	−69 30 01	NGC 1922	11.5	0.5			cl	
05 21 34.5	−65 29 13	NGC 1923		0.9			cl	
05 21 31.0	−65 49 00	NGC 1925		11			ass	(A) Large association of scattered stars.
05 20 35.2	−69 31 33	NGC 1926	11.8	0.7			cl	
05 20 57.2	−69 28 40	NGC 1928	11.9	0.5			cl	
05 22 27.4	−66 09 06	NGC 1932	11.8	1.2			cl	Relatively bright; detached.
05 21 27.1	−69 57 00	NGC 1938	11.8	0.6			cl	Relatively faint; nearly in contact with NGC 1939 to S.
05 21 24.3	−69 56 20	NGC 1939	12.9	0.5			cl	Relatively bright; nearly in contact with NGC 1938 to N.
05 22 44.2	−67 11 10	NGC 1940		1.0			cl	Relatively bright; detached.
05 23 07.1	−66 22 41	NGC 1941		0.9			cl	
05 24 42.7	−63 56 19	NGC 1942		1.1			cl	
05 21 58.1	−72 29 39	NGC 1944		3.3			cl	
05 25 16.0	−66 23 34	NGC 1946		1			cl	
05 25 05.3	−68 28 20	NGC 1949		1.3			cl	
05 24 32.1	−69 54 07	NGC 1950	13.2	1			cl	
05 26 06.4	−66 35 50	NGC 1951	10.6	1.9			cl	Moderately rich; slight central concentration; extended N-S.
05 26 04.2	−67 29 52	NGC 1955	9.0	1.8		11.3	cl	Relatively few stars; no central concentration; large brightness range; extended E-W; involved in nebulosity.
05 25 30.3	−69 50 10	NGC 1958	13.0	0.7			cl	In a very rich star field.
05 25 36.1	−69 55 31	NGC 1959	12.2	0.5			cl	
05 26 19.4	−68 50 15	NGC 1962		0.5			cl	In a large complex in the LMC.
05 26 44.6	−69 06 05	NGC 1967	10.8	0.4			cl	Relatively bright.
05 27 39.7	−67 27 20	NGC 1968	9.0	1.1		12.6	cl	Relatively few stars; no central concentration; large brightness range; extended E-W; involved in nebulosity.
05 26 32.1	−69 50 31	NGC 1969	12.5	0.8			cl	In a very rich star field.
05 26 45.0	−69 51 06	NGC 1971	11.9	0.8			cl	Relatively bright; in a very rich star field.
05 26 48.1	−69 50 17	NGC 1972	12.6	0.9			cl	Relatively faint and compact group; in a very rich star field.
05 28 00.6	−67 25 22	NGC 1974	9.0	1.7		12.7	cl	Relatively few stars; no central concentration; moderate brightness range; involved in nebulosity. (A) This is also **NGC 1991**, due to a positional error by J. Herschel.
05 28 44.6	−66 14 08	NGC 1978		3.9			cl	Conspicuously oval in shape NNW-SSE; the W side appears somewhat flattened.
05 27 37.9	−69 58 19	NGC 1986	11.1	0.7			cl	Rich in stars; fairly strong central concentration; detached.
05 27 16.4	−70 44 14	NGC 1987		1.7			cl	The slightly brighter core appears off-centered to NNE.
05 28 23.3	−69 08 29	NGC 1994	9.8	0.6			cl	Relatively bright.
05 30 34.9	−63 12 10	NGC 1997		1.3			cl	
05 27 30.5	−71 52 50	NGC 2000	11.9	1.7			cl	Relatively bright; detached.
05 29 08.2	−68 45 17	NGC 2001		1.7			cl	Stellar association with a moderate brightness range; extended NNE-SSW about 3:1; fairly even distribution; not well detached.
05 30 21.4	−66 53 02	NGC 2002	10.1	2			cl	Relatively bright; moderately rich; detached.
05 30 54.3	−66 27 53	NGC 2003	11.3	2.1			cl	
05 30 40.9	−67 17 17	NGC 2004	9.6	2.7		13.0	cl	Rich in stars; strong central concentration; detached.
05 31 19.9	−66 58 22	NGC 2006	11.5	1			cl	Relatively bright; detached. Open cluster SL 538, similar in size and brightness, lies 1′ N.
05 30 59.6	−69 10 56	NGC 2009	11.0	0.8			cl	Relatively bright; relatively few stars; detached.
05 30 34.6	−70 49 10	NGC 2010		2.2			cl	Irregular in shape.
05 32 19.8	−67 31 16	NGC 2011		1			cl	A knot of stars involved in nebulosity.
05 32 21.7	−67 41 58	NGC 2014		1.8			cl	
05 31 48.0	−69 14 54	NGC 2015		5.6			ass	
05 31 30.0	−69 56 00	NGC 2016		0.3			cl	Moderately rich; fairly even distribution; slightly extended E-W.
05 31 11.6	−71 04 11	NGC 2018		2.8			cl	
05 33 30.3	−67 27 12	NGC 2021		0.9			cl	A knot of stars involved in faint nebulosity.

(Continued from previous page)

OPEN CLUSTERS

RA h m s	Dec ° ′ ″	Name	Mag	Diam ′	No. ★	B ★	Type	Notes
05 32 33.6	−71 43 02	NGC 2025	10.9	1.9			cl	Relatively bright; detached.
05 35 00.9	−66 54 59	NGC 2027	11.9	0.7		13.6	cl	Relatively few stars; moderate brightness range; not well detached.
05 33 48.7	−69 57 03	NGC 2028	12.9	0.5			cl	Relatively faint; somewhat extended NE-SW; greatest concentration of stars in SW end.
05 35 41.1	−66 02 04	NGC 2030		1.3			cl	
05 33 41.3	−70 59 11	NGC 2031	10.8	3.4			cl	Bright; rich in stars; irregular shape; detached.
05 34 29.8	−69 46 49	NGC 2033	11.6	0.5			cl	SW part of a large stellar association including NGCs 2037 and 2048; involved in nebulosity.
05 34 32.1	−70 03 56	NGC 2036	12.8	0.6			cl	Relatively faint.
05 34 40.1	−69 44 48	NGC 2037	10.3	0.5			cl	Part of a large stellar association including NGCs 2033 and 2048; involved in nebulosity.
05 34 42.4	−70 33 40	NGC 2038	11.9	1.7			cl	Relatively bright.
05 36 06.1	−67 33 58	NGC 2040		1			cl	
05 36 28.0	−66 59 29	NGC 2041	10.4	0.7			cl	Moderately rich; detached.
05 36 09.6	−68 56 00	NGC 2042		9	60		ass	Rich in stars; moderate brightness range; slight central concentration.
05 36 06.1	−69 11 46	NGC 2044		2			3cl	Moderately rich; large brightness range; irregular in shape; involved in nebulosity. (A) Three clusters.
05 35 37.4	−70 14 30	NGC 2046	12.6	1.3			cl	Relatively faint.
05 35 54.2	−70 11 30	NGC 2047	13.2	0.8			cl	Very faint.
05 36 40.0	−69 22 49	NGC 2050		1			cl	The most concentrated part near the N central edge of a star cloud, very rich in small groups and star knots; in a large complex in the LMC.
05 36 07.5	−71 00 43	NGC 2051		1.7			cl	
05 37 40.5	−67 24 50	NGC 2053		1.2			cl	Faint; relatively few stars; no central concentration.
05 36 41.8	−69 29 49	NGC 2055		0.6			cl	The most concentrated part near the S boundary of a star cloud, very rich in small groups and star knots; in a large complex in the LMC.
05 36 34.6	−70 40 15	NGC 2056	12.3	1.7			cl	Relatively bright.
05 36 55.1	−70 16 14	NGC 2057	12.2	1.8			cl	Relatively faint.
05 36 54.7	−70 09 47	NGC 2058	11.9	1.8		16.0	cl	Rich in stars; strong central concentration; detached.
05 37 00.8	−70 07 43	NGC 2059	12.9	1			cl	
05 37 48.0	−69 10 18	NGC 2060		2			cl	Moderately rich; no central concentration; involved in nebulosity. (A) Large sparse group in nebulosity.
05 40 03.3	−66 52 36	NGC 2062		0.9			cl	Faint; few stars.
05 37 35.9	−70 14 07	NGC 2065	11.2	2.6		16.0	cl	Rich in stars; strong central concentration; detached.
05 37 41.2	−70 09 58	NGC 2066	13.1	0.7			cl	Very faint.
05 38 42.4	−69 06 03	NGC 2070	8.3	5			cl	**30 Doradus Cluster**. Several stars associated with very bright nebulosity (Tarantula Nebula).
05 38 23.6	−70 13 56	NGC 2072	13.2	1.3			cl	Faint and very compact group.
05 39 02.7	−69 29 41	NGC 2074		1			cl	
05 38 20.5	−70 41 09	NGC 2075		2			cl	
05 39 34.9	−69 39 19	NGC 2077		0.6			cl	
05 39 44.6	−69 38 42	NGC 2080		0.7			cl	
05 40 09.2	−69 40 21	NGC 2085		0.4			cl	
05 40 59.7	−68 27 51	NGC 2088		1.7			cl	Few stars; relatively faint.
05 40 58.0	−69 26 11	NGC 2091		1.7			cl	Relatively faint; somewhat extended E-W.
05 41 22.2	−69 13 25	NGC 2092		1.2			cl	Relatively faint; few stars.
05 41 49.1	−68 55 14	NGC 2093		1.7			cl	
05 42 07.2	−68 21 47	NGC 2094		0.4			cl	Relatively faint; slightly extended N-S.
05 42 51.7	−67 19 16	NGC 2095		0.8			cl	Few stars; not well detached.
05 42 16.5	−68 27 29	NGC 2096		1.2			cl	Few stars; relatively faint.
05 44 15.5	−62 47 05	NGC 2097		1.8			cl	
05 42 29.8	−68 16 26	NGC 2098		1.6			cl	
05 42 08.6	−69 12 42	NGC 2100	9.6	2.8		11.8	cl	Rich in stars; moderate central concentration; detached.
05 42 19.7	−69 29 14	NGC 2102		1			cl	Relatively faint.
05 44 19.8	−66 55 02	NGC 2105		1.7			cl	Faint; relatively few stars.
05 43 11.9	−70 38 27	NGC 2107	11.5	2.1			cl	Rich in stars; detached.
05 43 56.9	−69 10 50	NGC 2108		2			cl	Faint; a large core surrounded by a small faint halo of stars.
05 44 22.7	−68 32 52	NGC 2109		2			cl	
05 44 32.5	−70 59 37	NGC 2111		1.6			cl	Relatively faint.
05 45 24.5	−69 46 27	NGC 2113		1.8			cl	
05 46 12.3	−68 02 56	NGC 2114		1			cl	Faint; few stars; extended NW-SE; not well detached.
05 47 14.5	−68 30 24	NGC 2116		1			cl	Faint; few stars.
05 47 47.2	−67 27 07	NGC 2117		1.3			cl	Relatively bright; relatively few stars; slight central concentration.
05 47 39.6	−69 07 54	NGC 2118	12.0	1.3			cl	Relatively faint.
05 50 34.8	−63 40 33	NGC 2120		2			cl	
05 48 12.8	−71 28 47	NGC 2121		2.7			cl	Not round; no central concentration of stars although there is a small sprinkling of brighter stars in the central area.
05 48 55.0	−70 04 00	NGC 2122		4.5			cl	
05 51 43.5	−65 19 17	NGC 2123		1.2			cl	Relatively bright and compact group.
05 50 53.0	−69 28 42	NGC 2125		1			cl	
05 51 21.7	−69 21 38	NGC 2127		1.2			cl	Relatively faint.
05 52 23.7	−67 20 03	NGC 2130		1.2			cl	Relatively faint; relatively few stars.
05 51 29.3	−71 10 24	NGC 2133		1.7			cl	Relatively bright.
05 51 56.5	−71 05 52	NGC 2134	11.1	2.5		16.2	cl	Rich in stars; strong central concentration; detached.
05 53 35.1	−67 25 43	NGC 2135		1			cl	Relatively faint; relatively few stars.
05 52 58.9	−69 29 30	NGC 2136	10.5	1.9		14.5	cl	Bright and moderately rich in stars; detached.
05 53 11.9	−69 28 58	NGC 2137		0.8			cl	Faint; relatively few stars.
05 54 49.7	−65 50 17	NGC 2138		1			cl	Extremely faint; nearly stellar.
05 54 16.0	−68 35 58	NGC 2140		1.7			cl	Relatively faint; few stars; extended NW-SE.
05 54 22.6	−70 54 03	NGC 2145		1.7			cl	Relatively faint.
05 55 46.4	−68 12 14	NGC 2147		1.1			cl	A small knot of stars in the N central part of a relatively large, faint nebulosity.
05 56 20.8	−69 01 05	NGC 2151		1			cl	
05 57 52.3	−66 24 04	NGC 2153		1.3			cl	
05 57 38.5	−67 15 44	NGC 2154		2.4			cl	Quite bright; contains a prominent core surrounded by a faint outer halo of stars.
05 58 32.3	−65 28 40	NGC 2155		2.1			cl	
05 57 50.0	−68 27 40	NGC 2156	11.4	1.1			cl	Relatively bright; moderately rich; slight central concentration; detached.
05 57 34.8	−69 11 47	NGC 2157	10.2	2.7		14.5	cl	Very rich in stars; strong central concentration; detached.
05 58 02.7	−68 37 26	NGC 2159	11.4	0.9			cl	Relatively bright; moderately rich; detached.
05 58 13.0	−68 17 27	NGC 2160		1.2			cl	Relatively faint; irregular in shape.
06 00 31.4	−63 43 18	NGC 2162		2.1			cl	Very faint; slightly elongated E-W.
05 58 55.6	−68 30 56	NGC 2164	10.3	2.5		14.7	cl	Rich in stars; strong central concentration; detached.
05 59 33.0	−67 56 33	NGC 2166		1.2			cl	Relatively faint; compact group.
05 58 52.9	−67 53 52	NGC 2171		0.7			cl	Almost stellar appearance in star rich area.
06 00 05.6	−68 38 14	NGC 2172	11.8	1.7			cl	Relatively bright; moderately rich; detached.
05 57 58.0	−72 58 44	NGC 2173		2.4			cl	A faint cluster with a small condensed core.
06 01 19.4	−66 51 19	NGC 2176		1.3			cl	Very faint and diffuse.
06 01 15.7	−67 44 06	NGC 2177		1.2			cl	

(Continued from previous page)

OPEN CLUSTERS

RA h m s	Dec ° ′ ″	Name	Mag	Diam ′	No. ★	B ★	Type	Notes
06 02 43.1	−65 15 54	NGC 2181		1.6			cl	Very faint; very compact.
06 06 17.4	−65 05 53	NGC 2193		1.9			cl	Very faint.
06 06 07.4	−67 05 54	NGC 2197		1.7			cl	Faint and compact.

GLOBULAR CLUSTERS

RA h m s	Dec ° ′ ″	Name	Total V m	B ★ V m	HB V m	Diam ′	Conc. Class Low = 12 High = 1	Notes
03 44 32.4	−71 40 16	NGC 1466	11.4			1.9		
04 54 18.4	−70 26 31	NGC 1754	12.0			1.6		Quite bright; a fairly bright field star very near to SE.
04 59 07.8	−67 44 40	NGC 1786	10.9			1.6	2	There is a relatively bright field star on the NW edge and a fainter one on the SW edge.
05 16 42.0	−69 39 24	NGC 1898	11.9					
05 17 26.7	−69 22 34	NGC 1916	10.4			0.5		Rich in stars; slight central concentration; detached.
05 30 10.3	−69 45 09	NGC 2005	11.6			1.8		Relatively bright and compact; detached.
05 31 56.7	−70 09 35	NGC 2019	10.9			1.0		Irregularly roundish in shape.

BRIGHT NEBULAE

RA h m s	Dec ° ′ ″	Name	Dim ′ Maj x min	Type	BC	Color	Notes
05 43 12.0	−67 53 00	LH 120-N070	7 x 7	ESNR?			Emission nebula; a faint ring with several stars involved.
04 49 24.0	−69 10 00	NGC 1698					
04 51 00.0	−69 26 00	NGC 1712					
04 52 06.0	−66 56 00	NGC 1714	1 x 1	E			A bright, irregularly round emission nebula; brightest on the NW side.
04 52 06.0	−66 55 00	NGC 1715	0.5 x 0.5	E			Rather faint, round emission nebula; <2′ NE of NGC 1714; two stars nearby.
04 51 54.0	−69 24 00	NGC 1722					A very small, round nebula (IC 2111) involved in the NE part of a coarse cluster of faint stars, which itself is involved in a larger region of nebulosity including NGCs 1712 and 1727.
04 52 12.0	−69 20 00	NGC 1727					
04 53 30.0	−66 56 00	NGC 1731		E			Emission nebula and open cluster.
04 54 18.0	−67 06 00	NGC 1735	0.8 x 0.5				Stars involved; extended roughly NNW-SSE.
04 53 00.0	−68 03 00	NGC 1736		E			Emission nebula in the LMC.
04 54 00.0	−69 10 00	NGC 1737		E			Emission nebula; in a large complex in the LMC.
04 54 00.0	−69 12 00	NGC 1743	2 x 2	E			Irregular in form; the largest and brightest in a group of nebulae; in a large complex in the LMC. (A) Nebulous cluster or group of stars.
04 54 24.0	−69 10 00	NGC 1745		E			A large emission nebulosity surrounding a group of smaller nebulae including NGCs 1737, 1743 and 1748; in a large complex in the LMC.
04 55 12.0	−67 10 00	NGC 1747		E			Emission nebula and open cluster.
04 54 24.0	−69 12 00	NGC 1748	0.5 x 0.5	E			= **IC 2114**. Emission nebula; in a large complex in the LMC.
04 56 36.0	−66 32 00	NGC 1760		E			Emission nebula.
04 56 30.0	−66 29 00	NGC 1761					
04 56 48.0	−66 24 00	NGC 1763	5 x 3	E			Bright and well-defined emission nebula with stars involved. This bean-shaped nebula is extended ENE-WSW and set in a very rich star field. The W and E components are also catalogued as **IC 2115** and **IC 2116**, respectively.
04 56 24.0	−69 24 00	NGC 1767					
04 57 42.0	−66 28 00	NGC 1769		E			Bright and well-defined; a single bright star involved near the center.
04 57 12.0	−68 25 00	NGC 1770	3 x 2				Nebula and open cluster. The nebula is also catalogued as IC 2117 = **Henize 91A**.
04 56 54.0	−69 33 00	NGC 1772					
04 58 12.0	−66 22 00	NGC 1773		E			Irregularly round, moderately bright nebula; brightest on SW side.
04 57 48.0	−69 24 00	NGC 1782					
04 59 00.0	−70 10 00	NGC 1791	2 x 2				
05 04 54.0	−68 03 00	NGC 1829					Nebula and open cluster; in the LMC.
05 04 24.0	−70 44 00	NGC 1833					Nebula and open cluster; in a large complex in the LMC. (A) Elongated NE to SW.
05 04 54.0	−70 43 00	NGC 1837					
05 06 00.0	−68 38 00	NGC 1839					Relatively small group of stars involved in nebulosity.
05 07 24.0	−71 11 00	NGC 1848					A scattered group of relatively bright stars involved in nebulosity.
05 09 24.0	−67 46 00	NGC 1852					
05 09 48.0	−68 54 00	NGC 1858					Nebula and open cluster; in the LMC.
05 13 48.0	−67 23 00	NGC 1869		E			Emission nebula and open cluster. (A) Includes NGC 1871, and NGC 1873.
05 13 48.0	−67 27 00	NGC 1871					Nebula and open cluster; in the LMC. (A) Part of NGC 1869 association.
05 13 54.0	−67 20 00	NGC 1873		E			Emission nebula and open cluster. (A) Part of NGC 1869 association.
05 13 18.0	−69 22 00	NGC 1874/76/77	1 x 1	E			Regarded as a single object. This emission nebula and open cluster lies in a large complex in the LMC.
05 13 36.0	−69 23 00	NGC 1880		E			Emission nebula involved in a large complex in the LMC.
05 13 12.0	−69 21 00	NGC 1881					In a large complex in the LMC.
05 16 54.0	−67 20 00	NGC 1895		E			
05 17 48.0	−67 53 00	NGC 1899		E			Emission nebula in the LMC; three stars about 10th mag close W.
05 17 42.0	−71 15 00	NGC 1914					Nebula and open cluster; in the LMC. (A) Center of NW-SE elongated nebulous cluster.
05 19 06.0	−69 39 00	NGC 1918		SNR			In the LMC.
05 20 12.0	−66 53 00	NGC 1919		E			Emission nebula and open cluster.
05 20 36.0	−66 48 00	NGC 1920		E			
05 19 18.0	−69 48 00	NGC 1921					
05 21 30.0	−65 29 00	NGC 1923		E			Emission nebula and open cluster.
05 21 48.0	−65 48 00	NGC 1925					
05 21 36.0	−67 56 00	NGC 1929		E			Compact nebulous object; westernmost in a group of four objects; in a large complex in the LMC.
05 21 48.0	−67 56 00	NGC 1934		E			Nebulous object, larger and more diffuse than NGC 1929; a bright star on W side.
05 22 00.0	−67 57 00	NGC 1935		E			A round and diffuse nebulous knot with several stars involved; in a large complex in the LMC.
05 22 06.0	−67 59 00	NGC 1936		E			A bright round nebulous object with a faint extensive wisp in the direction of NGC 1935; in a large complex in the LMC.
05 23 06.0	−66 23 00	NGC 1941		E			Emission nebula with several stars involved.
05 22 30.0	−70 09 00	NGC 1943					Several stars involved in nebulosity.
05 25 24.0	−66 22 00	NGC 1945		E			Emission nebula.
05 25 42.0	−66 16 00	NGC 1948		E			Emission nebula and open cluster.
05 25 06.0	−68 29 00	NGC 1949		E			Emission nebula in the LMC.
05 26 24.0	−68 50 00	NGC 1962					Regarded as one object with NGCs 1965,66,70. In a large complex in the LMC.
05 26 36.0	−68 48 00	NGC 1965					Regarded as one object with NGCs 1962,66,70. In a large complex in the LMC.
05 26 48.0	−68 49 00	NGC 1966	13 x 13				Regarded as one object with NGCs 1962,65,70. In a large complex in the LMC.
05 27 24.0	−67 28 00	NGC 1968	13 x 12				Nebula and open cluster; in the LMC.
05 26 54.0	−68 50 00	NGC 1970					Regarded as one object with NGCs 1962,65,66. In a large complex in the LMC.
05 28 00.0	−67 25 00	NGC 1974		E			Emission nebula and open cluster. (A) This is also **NGC 1991**, due to a positional error by J. Herschel.
05 27 42.0	−68 59 00	NGC 1983					In a complex region of scattered, irregular star clouds.

(Continued from previous page)

212 BRIGHT NEBULAE 212

RA h m s	Dec ° ′ ″	Name	Dim ′ Maj x min	Type	BC	Color	Notes
05 27 42.0	−69 08 00	NGC 1984					In a complex region of scattered, irregular star clouds.
05 32 18.0	−67 31 00	NGC 2011					
05 32 18.0	−67 41 00	NGC 2014					Nebula and open cluster; in the LMC.
05 32 06.0	−69 14 00	NGC 2015					In a field of a coarse, scattered group of stars.
05 30 36.0	−71 04 00	NGC 2018	30 x 20	E			Nebula and open cluster; 10th mag star involved in W side; in the LMC.
05 33 12.0	−67 43 00	NGC 2020	25 x 18	SNR?			In the LMC.
05 33 30.0	−67 27 00	NGC 2021					
05 35 00.0	−67 33 00	NGC 2029		E			Faintest part of the **Seagull Nebula** with NGCs 2032 and 2035.
05 35 36.0	−66 02 00	NGC 2030	3 x 3				Nebula and open cluster.
05 35 18.0	−67 34 00	NGC 2032		E			The brightest part of the **Seagull Nebula** with NGCs 2029 and 2035; fan-shaped and extended N-S.
05 35 36.0	−66 54 00	NGC 2034					In a concentration of stars.
05 35 30.0	−67 35 00	NGC 2035	3 x 3	E			Irregular outline; part of the **Seagull Nebula** with NGCs 2029 and 2032.
05 35 00.0	−69 44 00	NGC 2037					
05 36 06.0	−67 34 00	NGC 2040	3 x 3	E			Compact nebulous object with several stars involved; in the LMC.
05 35 54.0	−69 39 00	NGC 2048	18 x 18				Several stars involved in nebulosity; in the LMC.
05 36 42.0	−69 23 00	NGC 2050					
05 37 12.0	−69 46 00	NGC 2052	18 x 12	E			Emission nebula; in the LMC.
05 36 42.0	−69 29 00	NGC 2055					
05 38 30.0	−69 05 00	NGC 2069	18 x 11	E			Elongated emission nebula; part of NGC 2070.
05 38 36.0	−69 05 00	NGC 2070	30 x 20	E	1-5	3-4	**Tarantula Nebula**. A high surface brightness nebula; a very complex ribbon-like or looped structure in the central regions. Involved with 30 Doradus Cluster.
05 39 00.0	−69 30 00	NGC 2074		E			A conspicuous S-shaped nebula with many stars involved.
05 38 18.0	−70 41 00	NGC 2075		E			Emission nebula and open cluster.
05 39 36.0	−69 40 00	NGC 2077	15 x 15	E			Southwestern most of two; forms a dumbbell shape with NGC 2080. (A) NGC 2080 to NE.
05 39 36.0	−69 45 00	NGC 2078		E			Emission nebula; northwestern most in a group of seven; in the NGC 2079 group.
05 39 36.0	−69 47 00	NGC 2079		E			Emission nebula; brightest in a group of seven.
05 39 42.0	−69 39 00	NGC 2080		E			Northeastern most of two; forms a dumbbell shape with NGC 2077. (A) NGC 2077 to SW.
05 40 06.0	−69 24 00	NGC 2081		E			Emission nebula and open cluster.
05 40 00.0	−69 44 00	NGC 2083					Emission nebula; northeastern most in a group of seven; in the NGC 2079 group.
05 40 06.0	−69 46 00	NGC 2084					Emission nebula; southeastern most in a group of seven; in the NGC 2079 group.
05 40 12.0	−69 39 00	NGC 2085					Very small emission nebula; southwestern most of two; a 10th mag star very near. (A) NGC 2086 just to E.
05 40 12.0	−69 40 00	NGC 2086					Very small emission nebula; northeastern most of two; a 10th mag star close W.
05 41 48.0	−68 55 00	NGC 2093					
05 42 30.0	−68 17 00	NGC 2098					
05 41 42.0	−71 20 00	NGC 2103		E			Emission nebula; stars involved.
05 44 24.0	−68 33 00	NGC 2109					
05 45 24.0	−69 46 00	NGC 2113		E			Emission nebula and open cluster.
05 48 48.0	−70 04 00	NGC 2122		E			Emission nebula and open cluster. (A) Stars mostly hidden by very bright nebula on DSS image.
05 56 00.0	−68 12 00	NGC 2147					

213 GALAXIES 213

RA h m s	Dec ° ′ ″	Name	Mag (V)	Dim ′ Maj x min	SB	Type Class	PA	Notes
02 11 32.8	−73 09 47	ESO 30-11	14.3	0.5 x 0.5	12.6	(R′)SA(s)ab?		Mag 11.11v star S 3′.3.
02 53 54.6	−72 45 30	ESO 31-4	13.7	0.9 x 0.8	13.2	(R′)SAB(r)b: III-IV	144	Stellar nucleus, smooth envelope; mag 15.1 star on NW edge; mag 9.98v star W 1′.9.
03 37 43.6	−72 23 31	ESO 31-18	14.4	1.1 x 0.6	13.8	SA(s)c II	99	Mag 13.4 star on N edge, mag 1.6 star N 2′.4.
01 24 48.1	−68 37 21	ESO 52-1	13.7	0.9 x 0.8	13.3	SB(r)0/a	40	Numerous faint anonymous as well as named galaxies in surrounding area.
01 29 15.1	−67 57 55	ESO 52-4	14.3	1.4 x 1.0	14.5	Sc	123	Mag 10.93v star N 4′.8.
01 36 47.9	−67 57 27	ESO 52-8	14.1	1.0 x 0.9	13.8	Sc	38	Mag 12.1 star W 3′.6.
01 59 25.3	−67 47 20	ESO 52-14	13.8	1.2 x 0.7	13.4	SAB(s)0/a?	45	Mag 11.7 star W 5′.1.
02 02 12.6	−69 26 53	ESO 52-17	13.8	1.5 x 1.0	14.3	E	96	Galaxy pair, in contact.
02 05 43.4	−71 06 57	ESO 52-20	13.7	1.4 x 0.7	13.6	SB(s)bc pec I-II	10	Mag 13.7 star E edge; ESO 52-21 ESE 1′.4.
02 05 57.4	−71 07 49	ESO 52-21	14.1	0.6 x 0.4	12.4	S	178	ESO 52-20 WNW 1′.4.
02 13 12.6	−70 54 51	ESO 53-2	15.3	1.5 x 0.2	13.8	Sc sp	109	Mag 7.09v star W 6′.8.
02 56 16.4	−71 51 14	ESO 53-23	14.1	1.8 x 0.5	13.8	(R′)SB(rs)a: II	71	Mag 8.01v star N 2′.6.
01 04 13.1	−62 31 17	ESO 79-15	14.0	1.1 x 1.0	13.9	SB(rs)cd	161	Mag 11.2 star E 9′.0.
01 04 27.0	−64 07 34	ESO 79-16	13.5	1.0 x 0.3	12.0	S?	177	Bright star superimposed N end; mag 8.73v star SW 3′.0.
01 13 58.4	−62 24 02	ESO 80-1	14.3	1.3 x 0.9	14.3	SB(s)cd IV	93	Mag 10.18v star SW 2′.5.
01 42 22.1	−65 38 04	ESO 80-3	14.3	1.3 x 1.0	14.4	Sc	0	Mag 10.16v star ESE 12′.1; **PGC 6240** W 5′.3.
01 43 08.0	−62 28 24	ESO 80-4	13.9	0.9 x 0.5	12.9	S pec	32	Mag 13.0 star NW 1′.0.
01 47 16.4	−62 58 19	ESO 80-6	13.9	1.6 x 0.9	14.1	SB(s)m IV	66	Strong knot of superimposed star N of center; mag 10.81v star NE 2′.1.
01 53 12.6	−62 25 32	ESO 81-1	13.7	1.1 x 0.5	12.9	S0	100	Galaxy **AM 0151-624** SSW 2′.6.
02 18 56.2	−66 03 39	ESO 81-6	14.2	1.1 x 0.6	13.6	(R′)SB(rs)b	47	Mag 12.8 star 1′.4 N.
02 21 03.8	−63 37 31	ESO 81-7	14.8	1.5 x 0.2	13.3	Sc	76	Mag 8.78v star NW 9′.5.
02 21 37.8	−64 36 38	ESO 81-8	13.6	1.0 x 0.4	12.4	Pec?	79	Mag 14.0 star N 1′.2.
02 42 17.8	−62 35 05	ESO 82-2	13.9	1.0 x 0.3	12.4	Sb	15	Mag 10.31v star SW 3′.0.
02 57 21.3	−62 52 36	ESO 82-6	14.4	1.0 x 0.5	13.5	SBc pec	128	Mag 9.50v star SW 10′.1.
03 07 40.2	−66 40 10	ESO 82-10	13.4	1.2 x 1.0	13.4	SAB(r)c II	87	Mag 13.1 star SW 1′.7; NGC 1244 SSW 9′.3.
03 32 00.1	−62 34 11	ESO 83-6	14.1	1.7 x 0.7	14.1	(R′)SB(rs)c III	101	Mag 7.41v star ESE 11′.6.
03 33 41.1	−66 07 45	ESO 83-7	16.1	1.5 x 0.2	14.7	Im? V?	9	Mag 10.56v star NW 10′.2.
01 54 40.6	−62 06 21	ESO 114-14	13.3	1.3 x 1.3	13.8	(R′)SA(rs)bc: II-III		Almost stellar nucleus; mag 11.6 star NE 3′.5.
01 07 24.4	−69 52 36	NGC 406	12.5	2.6 x 1.0	13.4	SA(s)c: II-III	160	
01 37 23.5	−64 53 53	NGC 646	13.4	2.0 x 1.5	14.5	SAB(s)c pec:	78	Looks like a snake. Bright, round nucleus E with narrow arm or extension W, then curving sharply S.
01 59 06.2	−67 52 16	NGC 802	13.7	0.9 x 0.6	12.9	SAB(s)0$^+$ pec:	152	Mag 12.1 star SW 10′.1; ESO 52-14 N 5′.4; mag 11.7 star NNW 5′.2.
02 01 36.3	−68 26 19	NGC 813	12.8	1.3 x 0.9	12.9	SAB(r)0/a: pec	99	
03 06 30.5	−66 46 32	NGC 1244	12.8	1.9 x 0.4	12.4	SA(r)ab pec:	2	ESO 82-10 NNE 9′.3; mag 10.15v star S 6′.0.
03 07 01.2	−66 56 22	NGC 1246	12.9	1.3 x 0.8	12.9	E5:	31	Mag 10.15v star NW 6′.8.
03 18 13.5	−66 30 26	NGC 1313	8.7	9.2 x 6.9	13.1	SB(s)d IV	38	Numerous bright knots.
03 20 05.5	−66 42 08	NGC 1313A	13.8	1.2 x 0.3	12.6	Sb:	30	
01 36 59.6	−62 32 23	PGC 5993	14.2	1.0 x 0.8	13.8	SA0°	39	Mag 10.63v star W 10′.6; faint anonymous galaxy NNE 7′.1.

GALAXY CLUSTERS

RA h m s	Dec ° ′ ″	Name	Mag 10th brightest	No. Gal	Diam ′	Notes
01 54 30.0	−71 28 00	A 2954	16.3	121	17	All members anonymous, stellar.
01 09 12.0	−67 43 00	AS 131	16.1		17	All members anonymous, stellar.

(Continued from previous page)

GALAXY CLUSTERS

RA h m s	Dec ° ′ ″	Name	Mag 10th brightest	No. Gal	Diam ′	Notes
01 32 54.0	−67 46 00	AS 164	16.3		17	**Fair 702** ESE of center 6.′6; all others anonymous, stellar.
01 36 36.0	−73 34 00	AS 176	16.4		17	**ESO 29-47** W of center 10.′1; all others anonymous, stellar.
01 45 00.0	−72 57 00	AS 191	16.4	4	17	Very faint **ESO 30-2**, **2A**, **2B** S of center 4.′9; all others anonymous, stellar.
01 57 36.0	−64 23 00	AS 210	16.3		17	All members anonymous, stellar/almost stellar.
03 32 48.0	−64 14 00	AS 362	16.2		17	All members anonymous, stellar.

OPEN CLUSTERS

RA h m s	Dec ° ′ ″	Name	Mag	Diam ′	No. ★	B ★	Type	Notes
01 05 21.5	−72 02 35	IC 1624	12.9	0.7	130	15.7	cl	
01 11 54.4	−71 19 48	IC 1655		1			cl	
01 05 08.0	−71 59 50	NGC 395					cl	
01 07 55.7	−71 46 09	NGC 411		1.3			cl	
01 07 59.0	−72 21 19	NGC 416	12.6	1.2	985	15.3	cl	
01 08 19.5	−72 53 02	NGC 419	11.2	2.4	1762	15.4	cl	
01 09 21.5	−72 46 00	NGC 422	13.4				cl	
01 13 43.9	−73 17 35	NGC 456		2			cl	Open cluster and emission nebula; westernmost in a chain of small clusters; in the SMC. Involved in nebulosity which also carries the same designation.
01 14 53.8	−71 32 58	NGC 458		1.5			cl	
01 14 37.9	−73 18 28	NGC 460	12.5	1			cl	(A) Involved nebula is **Henize N 84A**.
01 15 42.7	−73 19 27	NGC 465	11.5	2			cl	
01 42 52.9	−73 20 11	WG71 1		0.8			cl	

BRIGHT NEBULAE

RA h m s	Dec ° ′ ″	Name	Dim ′ Maj x min	Type	BC	Color	Notes
01 14 00.0	−73 17 00	NGC 456	15 x 15	E			
01 29 36.0	−73 33 00	NGC 602					

GALAXIES

RA h m s	Dec ° ′ ″	Name	Mag (V)	Dim ′ Maj x min	SB	Type Class	PA	Notes
00 52 41.0	−83 51 20	ESO 2-10	13.2	1.1 x 0.6	12.7	$S0^-$ sp:	114	Mag 7.96 v star NW 7.′9.
01 38 47.7	−83 21 58	ESO 3-3	13.7	1.6 x 0.6	13.5	SB(s)cd pec III-IV	128	Mag 7.33v star N 7.′0.
01 41 32.6	−83 12 44	ESO 3-4	13.9	2.0 x 0.3	13.2	Sbc: sp II-III	136	Mag 7.33v star SW 6.′0.
02 00 15.3	−83 59 13	ESO 3-7	12.4	1.6 x 1.4	13.2	SAB(rs)c: pec II	99	Mag 11.24v star SSE 1.′5.
21 40 59.8	−82 47 57	ESO 11-5	15.2	1.5 x 0.3	14.1	SB(s)cd: sp IV	47	Mag 6.36v star NW 15.′1.
23 44 14.8	−80 10 32	ESO 12-4	14.1	2.3 x 0.3	13.5	Sc: sp	160	Mag 9.90v star WSW 6.′6.
23 56 27.0	−81 33 59	ESO 12-10	13.6	3.1 x 1.4	15.0	SB(s)dm	152	Mag 10.97v star NNE 4.′9.
00 00 21.8	−80 47 31	ESO 12-12	13.0	1.3 x 1.2	13.3	(R′)SAB(rs)bc pec I-II	22	Mag 9.30v star N 7.′1.
00 02 44.2	−80 20 48	ESO 12-14	14.5	2.1 x 0.8	14.9	SB(s)m V-VI	31	Mag 7.93v star SE 4.′4.
00 07 08.0	−80 18 35	ESO 12-15	14.8	1.5 x 0.2	13.4	Sc: sp	119	Mag 7.93v star SW 9.′4.
00 40 45.9	−79 14 26	ESO 12-21	13.7	0.5 x 0.3	11.5		147	Mag 12.3 star SW 4.′0.
00 47 19.6	−80 48 03	ESO 13-1	14.0	0.9 x 0.4	12.7	S pec	109	Mag 13.4 star W 1.′8.
01 02 44.1	−80 14 04	ESO 13-9	14.1	1.7 x 0.5	13.8	SAB(s)d? IV	121	ESO 13-10 NE 8.′5; ESO 13-12 ESE 11.′6.
01 05 07.3	−80 08 06	ESO 13-10	14.6	1.1 x 0.5	13.8	SAB(s)d: pec IV	156	Mag 10.11v star N 4.′3.
01 07 00.9	−80 18 25	ESO 13-12	12.6	2.9 x 0.7	13.2	S0/a: sp	156	Strong dark lane entire length along centerline; mag 12.7 star SSE 2.′9.
01 32 48.5	−79 28 27	ESO 13-16	12.8	2.2 x 1.3	13.8	SB(s)dm IV	166	Mag 9.26v star SW 12.′0.
01 36 33.5	−80 20 51	ESO 13-18	15.3	1.7 x 0.2	14.0	Scd sp	33	Mag 9.64v star S 7.′8; **ESO 13-17** WNW 4.′3.
01 45 42.3	−77 56 47	ESO 13-21	13.8	1.1 x 0.5	13.0	Sa	4	Mag 10.94v star S 4.′2; **ESO 13-20** W 5.′6.
01 45 58.4	−78 54 16	ESO 13-22	14.0	0.8 x 0.6	13.1	SA(s)c pec I-II	45	Double system, stellar companion SW of nucleus; mag 10.74v star N 2.′5.
01 53 18.8	−78 14 23	ESO 13-24	13.8	1.2 x 0.4	12.9	Sab	139	Mag 6.15v star SE 10.′1.
01 58 46.8	−78 23 54	ESO 13-26	13.5	1.3 x 0.9	13.5	SB(s)bc pec	117	Mag 6.15v star W 9.′4.
02 02 32.7	−79 40 23	ESO 13-28	14.0	1.9 x 0.3	13.2	Sbc sp	7	Mag 10.91v star W 3.′5; **ESO 13-27** S 2.′4.
21 52 26.5	−81 31 54	ESO 27-1	11.5	2.8 x 2.7	13.6	SB(s)c III-IV	121	Short bar NE-SW with dark patch S.
21 52 19.2	−80 34 27	ESO 27-2	14.8	1.2 x 0.7	14.5	IAB(s)m VI	21	
21 53 44.1	−81 39 11	ESO 27-3	14.0	0.7 x 0.6	12.9	IB(s)m IV	51	Lies 1.′5 NW of 9.03v star.
22 23 02.3	−79 59 44	ESO 27-8	12.0	2.8 x 1.0	13.0	SB(s)c: II-III	144	Mag 11.00v star NE 2.′7; small triangle of faint stars on S edge.
22 28 32.7	−80 59 59	ESO 27-9	13.7	1.3 x 1.0	13.8	SAB(s)d III-IV	168	Mag 9.79v star N 2.′7.
22 30 25.6	−79 41 39	ESO 27-12A	14.6	1.1 x 0.6	14.0		74	**ESO 27-12** SW edge.
22 36 11.2	−81 03 56	ESO 27-14	13.0	0.9 x 0.6	12.1	SB(s)b? pec V	114	Mag 10.07v star NNE 9.′6.
23 04 19.3	−79 28 00	ESO 27-21	13.1	0.9 x 0.7	12.5	(R)SAa: pec	98	
23 13 32.7	−81 37 34	ESO 27-24	13.2	1.0 x 0.6	12.5	SA(rs)b: II-III	45	Mag 10.18v star W 4.′6.
23 46 12.6	−75 16 32	ESO 28-2	12.5	1.5 x 1.3	13.1	SAB(rs)bc II	179	Mag 13 star SE edge.
23 46 33.9	−76 46 57	ESO 28-3	13.9	1.2 x 0.8	13.7	SAB(s)d: IV	165	Mag 12.4 star S edge.
23 58 00.9	−73 27 12	ESO 28-7	14.8	1.6 x 0.9	15.0	SAB(s)m: IV-V	171	Mag 10.24v star SW 6.′8.
00 05 43.5	−75 42 24	ESO 28-9	13.5	1.0 x 0.4	12.3	Sb	149	
00 18 19.8	−73 09 12	ESO 28-12	13.7	1.1 x 0.6	13.1	(R)$S0^+$?	116	Mag 11.3 star NNW 3.′3.
00 19 58.4	−77 05 31	ESO 28-14	15.0	1.7 x 1.3	15.7	SB(s)d V		**ESO 28-14A** SW 1.′5.
02 09 01.1	−75 56 22	ESO 30-8	14.1	2.2 x 0.4	13.8	SB(s)d: sp III-IV	18	Mag 6.89v star E 6.′9.
02 10 40.9	−75 02 19	ESO 30-9	13.3	2.5 x 0.6	13.6	SA(rs)c II	134	Mag 10.82v star S 4.′0.
02 11 32.8	−73 09 47	ESO 30-11	14.3	0.5 x 0.5	12.6	(R′)SA(s)ab?		Mag 11.11v star S 3.′3.
02 16 48.2	−74 24 16	ESO 30-13	14.1	0.8 x 0.6	13.2	Sab	117	Mag 12.5 star NW 3.′1.
02 17 56.3	−76 04 55	ESO 30-14	13.2	1.6 x 1.4	13.9	SA(s)c I-II	80	Mag 9.30v star W 12.′9; anonymous galaxy WNW 5.′0.
21 45 45.3	−74 49 53	ESO 48-7	14.1	1.3 x 0.4	13.2	(R′)SB(s)b: II-III	147	Close pair of stars, mags 12.5 and 14.0, N 2.′2.
21 52 35.0	−73 17 48	ESO 48-11	13.9	1.0 x 0.4	12.7	Sbc pec?	102	Mag 14.6 star S edge; **ESO 48-12** E 0.′9.
21 58 28.0	−73 51 37	ESO 48-13	14.1	1.1 x 0.7	13.7	(R′)SA0°: pec	123	Mag 8.99v star E 6.′2.
22 02 21.8	−73 42 08	ESO 48-14	14.1	1.0 x 0.4	13.0	Sdm:	1	Mag 13.5 star W 1.′7.
22 08 54.6	−73 22 53	ESO 48-17	14.0	1.5 x 1.0	14.3	SB(rs)d IV-V	54	Mag 14.3 star NE end; mag 9.65v star W 7.′7.
22 14 23.9	−76 03 33	ESO 48-20	14.6	1.1 x 0.4	13.5	Sa: pec sp	66	Mag 6.13v star WSW 9.′5.
22 18 31.8	−73 36 58	ESO 48-21	13.5	1.0 x 0.5	12.6	S0/a	30	Mag 12.7 star W 2.′5.
21 50 24.8	−73 59 53	IC 5130	13.6	1.7 x 0.5	13.3	SA(s)bc: II	109	Mag 10.24v star ENE 9.′0.
01 39 13.7	−75 00 43	NGC 643B	13.4	1.5 x 0.3	12.4	SB0?	113	Mag 8.09v star NE 3.′0.
01 41 49.4	−75 16 07	NGC 643C	14.6	1.3 x 0.2	13.0	Scd: sp	150	

(Continued from previous page)

214 GALAXIES 214

RA h m s	Dec ° ′ ″	Name	Mag (V)	Dim ′ Maj x min	SB	Type Class	PA	Notes
21 44 16.4	−75 06 43	NGC 7098	11.3	4.1 x 2.6	13.7	(R)SAB(rs)a	74	Bright center, strong dark areas SE and NW.
23 26 27.0	−81 54 41	NGC 7637	12.4	2.1 x 1.9	13.8	SA(r)c I-II	36	Bright nucleus, several superimposed stars.

GALAXY CLUSTERS

RA h m s	Dec ° ′ ″	Name	Mag 10th brightest	No. Gal	Diam ′	Notes
00 15 54.0	−84 04 00	AS 21	16.1		17	**ESO 2-4** NW of center 12.′6; all others anonymous, stellar.
00 54 24.0	−83 52 00	AS 107	16.2		17	Except ESO 2-10; other members anonymous, stellar.
01 36 36.0	−73 34 00	AS 176	16.4		17	**ESO 29-47** W of center 10.′1; all others anonymous, stellar.
22 24 12.0	−80 10 00	AS 1014	15.5	16	22	**PGC 68764** at center; **ESO 27-7**, **7A**, **7B** SW 6.′6.

OPEN CLUSTERS

RA h m s	Dec ° ′ ″	Name	Mag	Diam ′	No. ★	B ★	Type	Notes
00 03 53.2	−73 28 22	Lind 1		4			cl	
00 32 55.5	−73 06 59	NGC 152		1.7			cl	
00 35 58.6	−73 09 57	NGC 176	12.7	1			cl	
00 40 30.6	−73 24 10	NGC 220	12.4	0.8	256	14.8	cl	
00 40 44.1	−73 23 00	NGC 222	12.2	0.6	128	15.3	cl	
00 41 06.2	−73 21 07	NGC 231	12.7	0.8	131	15.0	cl	
00 43 28.5	−73 36 00	NGC 241		7.5	60		ass	(A) This is not NGC 242 or part of NGC 242, which is about 10′ to the north.
00 43 33.9	−73 26 28	NGC 242		0.9			2cl	(A) Two clusters. This is not, and does not contain, NGC 241.
00 45 24.1	−73 22 44	NGC 248	13.6	0.8	20	14.8	cl	
00 45 54.4	−73 30 24	NGC 256	12.7	0.6	142	14.7	cl	
00 46 33.0	−73 05 55	NGC 261	13.4	0.4	16	13.9	cl	
00 47 11.6	−73 28 38	NGC 265	12.2	1	301	15.0	cl	
00 48 21.3	−73 31 49	NGC 269	13.3	0.6	218	15.2	cl	
00 51 14.2	−73 09 41	NGC 290	12.0	0.8	228	13.8	cl	
00 53 05.6	−73 22 49	NGC 294	12.7	0.8	410	16.1	cl	
00 57 45.4	−74 28 20	NGC 339		2.2			cl	Bright foreground star in field.
01 13 43.9	−73 17 35	NGC 456		2			cl	Open cluster and emission nebula; westernmost in a chain of small clusters; in the SMC. Involved in nebulosity which also carries the same designation.
01 14 37.9	−73 18 28	NGC 460	12.5	1			cl	(A) Involved nebula is **Henize N 84A**.
01 15 42.7	−73 19 27	NGC 465	11.5	2			cl	
01 35 00.8	−75 33 26	NGC 643		1.5			cl	(A) Do not confuse with the galaxy "NGC 643B."
01 56 44.5	−74 13 09	NGC 796		0.8			cl	
01 42 52.9	−73 20 11	WG71 1		0.8			cl	

BRIGHT NEBULAE

RA h m s	Dec ° ′ ″	Name	Dim ′ Maj x min	Type	BC	Color	Notes
00 45 24.0	−73 23 00	NGC 248		E			Emission nebula in the SMC.
00 45 24.0	−73 05 00	NGC 249		E			A small, round emission nebula in a rich star field; in the SMC.
00 45 54.0	−73 30 00	NGC 256					
00 46 30.0	−73 07 00	NGC 261		E?			Larger and more diffuse than NGC 249; includes a central star. This nebula lies in a rich star field in the SMC.
00 48 00.0	−73 17 00	NGC 267					Nebula involved in open cluster; in SMC.
00 53 06.0	−73 23 00	NGC 294					
01 14 00.0	−73 17 00	NGC 456	15 x 15	E			
01 29 36.0	−73 33 00	NGC 602					
01 56 42.0	−74 13 00	NGC 796					

215 GALAXIES 215

RA h m s	Dec ° ′ ″	Name	Mag (V)	Dim ′ Maj x min	SB	Type Class	PA	Notes
19 20 16.1	−84 47 38	ESO 10-3	13.8	0.9 x 0.7	13.1	S?	160	Mag 13.9 star E edge; mag 8.99v star W 2.′9.
19 28 49.5	−82 41 22	ESO 10-5	14.0	1.2 x 0.8	13.8	SB(r)bc pec III	23	Stellar nucleus, smooth envelope; mag 12.4 star NE 2.′1.
19 40 13.6	−84 16 26	ESO 10-6	14.9	0.9 x 0.5	13.9	SB(s)m V	162	Mag 11.64v star NE 2.′0.
21 35 32.8	−82 54 11	ESO 11-3	13.3	1.3 x 0.7	13.1		174	Pair of mag 14-15 stars on W edge; mag 6.36v star NNW 13.′5.
21 39 29.3	−83 20 10	ESO 11-4	14.0	1.8 x 0.9	14.5	E^{+}4: pec	105	Mag 9.91v star S 2.′2.
21 40 59.8	−82 47 57	ESO 11-5	15.2	1.5 x 0.3	14.1	SB(s)cd: sp IV	47	Mag 6.36v star NW 15.′1.
17 48 58.1	−81 43 13	ESO 24-8	14.3	1.0 x 0.5	13.4	SB(s)c:	15	Mag 10.21v star E 5.′9.
18 23 06.8	−77 42 30	ESO 24-15	13.0	2.1 x 0.8	13.4		0	Double system, possible bridge. Galaxy at N end and S end with star at center on E edge.
18 47 12.4	−82 07 39	ESO 24-19	14.9	1.7 x 0.2	13.6	Scd sp	170	Mag 10.96v star N 3.′7.
18 54 40.9	−78 53 50	ESO 25-2	13.5	1.0 x 1.0	13.4	(R)SB(r)b pec:		Mag 10.60v star NNE 2.′8.
20 24 57.1	−81 34 33	ESO 26-1	13.6	1.7 x 1.4	14.4	(R′)SB(rs)cd: IV	47	Mag 9.08v star E 6.′8.
20 46 02.2	−80 04 59	ESO 26-5	14.9	1.1 x 0.9	14.7	SAB(s)m V-VI	54	Stellar nucleus; mag 11.21v star SE 4.′5.
21 52 26.5	−81 31 54	ESO 27-1	11.5	2.8 x 2.7	13.6	SB(s)c III-IV	121	Short bar NE-SW with dark patch S.
21 52 19.2	−80 34 27	ESO 27-2	14.8	1.2 x 0.7	14.5	IAB(s)m VI	21	
21 53 44.1	−81 39 11	ESO 27-3	14.0	0.7 x 0.6	12.9	IB(s)m IV	51	Lies 1.′5 NW of 9.03v star.
17 52 59.9	−76 51 47	ESO 44-22	14.4	1.1 x 0.6	13.8	SB(s)d IV-V	168	Mag 9.70v star W 2.′2.
18 06 35.4	−73 53 01	ESO 44-25	14.0	1.1 x 0.5	13.2	SAB(rs)b? III-IV	52	Mag 7.42v star S 7.′6.
18 21 17.7	−74 01 53	ESO 45-2	14.6	1.6 x 1.2	15.2	IB(s)m V-VI	63	Mag 10.05v star W 3.′6; mag 9.42v star SSE 4.′2.
18 22 57.5	−76 41 46	ESO 45-3	14.3	1.6 x 0.7	14.3	SB(s)d IV	165	Close pair of mag 12.3, 9.94v stars NNE 6.′9; NGC 6557 NW 8.′6.
18 28 37.4	−76 56 23	ESO 45-5	14.2	1.1 x 0.5	13.5	(R′)SAB(r)0^{+}:	45	Mag 9.12v star NW 7.′5.
21 31 43.6	−76 28 50	ESO 47-34	13.1	1.8 x 0.3	12.3	S0 sp	159	Pair of E-W oriented mag 10.41v, 10.11v stars SW 4.′6.
21 36 00.5	−76 20 48	ESO 48-2	13.5	2.3 x 0.6	13.7	Sb sp	90	Mag 12.0 star NNW 1.′5; mag 8.95v star S 11.′1.
21 39 34.7	−76 24 27	ESO 48-3	13.9	1.2 x 0.4	13.0	SB(s)bc: II-III	114	Mag 13.6 star ESE 1.′2; mag 8.95v star SW 10.′4.
21 45 45.3	−74 49 53	ESO 48-7	14.1	1.3 x 0.4	13.2	(R′)SB(s)b: II-III	147	Close pair of stars, mags 12.5 and 14.0, N 2.′2.
21 52 35.0	−73 17 48	ESO 48-11	13.9	1.0 x 0.4	12.7	Sbc pec?	102	Mag 14.6 star S edge; **ESO 48-12** E 0.′9.
21 58 28.0	−73 51 37	ESO 48-13	14.1	1.1 x 0.7	13.7	(R′)SA0°: pec	123	Mag 8.99v star E 6.′2.
22 02 21.8	−73 42 08	ESO 48-14	14.1	1.0 x 0.4	13.0	Sdm:	1	Mag 13.5 star W 1.′7.
22 08 54.6	−73 22 53	ESO 48-17	14.0	1.5 x 1.0	14.3	SB(rs)d IV-V	54	Mag 14.3 star NE end; mag 9.65v star W 7.′7.

(Continued from previous page)

215 GALAXIES 215

RA h m s	Dec ° ′ ″	Name	Mag (V)	Dim ′ Maj x min	SB	Type Class	PA	Notes
22 14 23.9	−76 03 33	ESO 48-20	14.6	1.1 x 0.4	13.5	Sa: pec sp	66	Mag 6.13v star WSW 9′5.
22 18 31.8	−73 36 58	ESO 48-21	13.5	1.0 x 0.5	12.6	S0/a	30	Mag 12.7 star W 2′5.
17 51 03.5	−74 02 00	IC 4661	12.7	1.6 x 1.2	13.2	SA(s)c II	31	Mag 8.69v star WNW 14′3; **ESO 44-23** SSE 9′6.
19 40 06.0	−77 33 28	IC 4864	13.5	1.5 x 0.4	12.8	(R′)SAB(s)a: sp	176	Mag 7.58v star ESE 5′4; **ESO 25-12** WNW 5′4.
20 17 24.1	−73 53 07	IC 4964	13.8	1.3 x 0.5	13.2	SA(s)d: III-IV	162	Mag 10.80v star W 7′5.
20 35 15.9	−73 27 11	IC 5014	13.6	0.5 x 0.5	11.9			Almost stellar; mag 10.05v star SSE 11′1.
20 44 59.8	−76 59 07	IC 5025	14.4	1.5 x 0.3	13.4	Sbc: sp	108	Mag 9.82v star NNW 8′8.
20 48 28.8	−78 04 11	IC 5026	14.4	2.3 x 0.3	13.8	Sd sp	76	Mag 8.19v star ESE 9′2.
21 14 22.4	−73 46 27	IC 5087	14.0	0.7 x 0.4	12.5	SB(s)b: pec	104	Mag 8.64v star NE 5′6.
21 26 13.7	−73 18 39	IC 5102	13.7	0.8 x 0.5	12.6	S0	124	Stellar nucleus with several stars superimposed.
21 29 13.1	−74 04 14	IC 5103	14.1	0.8 x 0.5	13.0	(R′)SB(s)c: II	33	Mag 9.10v star SE 9′1.
21 33 42.7	−74 06 43	IC 5109	13.6	0.8 x 0.6	12.9	E:	174	Almost stellar.
21 50 24.8	−73 59 53	IC 5130	13.6	1.7 x 0.5	13.3	SA(s)bc: II	109	Mag 10.24v star ENE 9′0.
18 21 25.2	−76 35 00	NGC 6557	12.9	1.6 x 1.1	13.3	SA(r)0°	80	Close pair of mag 12.3, 9.94v stars E 8′6; ESO 45-3 SE 8′6.
18 44 39.2	−73 15 48	NGC 6653	12.2	1.7 x 1.5	13.2	E1:		Bright center, uniform envelope, few stars superimposed.
20 43 57.7	−80 00 08	NGC 6920	12.4	1.8 x 1.5	13.3	SA(rs)0°:	142	Mag 12.2 star W 2′7; ESO 26-5 SE 7′2.
20 42 08.7	−73 37 08	NGC 6932	12.3	2.1 x 1.5	13.4	(R)SAB(r)0^+	115	Mag 9.40v star W 2′8.
21 44 16.4	−75 06 43	NGC 7098	11.3	4.1 x 2.6	13.7	(R)SAB(rs)a	74	Bright center, strong dark areas SE and NW.

GALAXY CLUSTERS

RA h m s	Dec ° ′ ″	Name	Mag 10th brightest	No. Gal	Diam ′	Notes
18 28 12.0	−77 10 00	AS 800	16.2		17	Except ESO 45-5, all members anonymous, stellar.

OPEN CLUSTERS

RA h m s	Dec ° ′ ″	Name	Mag	Diam ′	No. ★	B ★	Type	Notes
20 26 36.0	−80 00 00	ESO 26-2		8			cl	
20 15 18.0	−79 17 00	Mel 227	5.3	70	40		cl	Moderately rich in stars; large brightness range; no central concentration; detached.

PLANETARY NEBULAE

RA h m s	Dec ° ′ ″	Name	Diam ″	Mag (P)	Mag (V)	Mag cent ★	Alt Name	Notes
19 33 50.0	−74 32 59	PN G320.3−28.8	10	12.1		14.9	PK 320−28.1	Mag 14.4 star and small, faint, elongated anonymous galaxy ESE 2′2.

216 GALAXIES 216

RA h m s	Dec ° ′ ″	Name	Mag (V)	Dim ′ Maj x min	SB	Type Class	PA	Notes
13 57 48.6	−83 04 50	ESO 8-4	13.5	2.1 x 1.9	14.8	SB(rs)d III-IV	9	Mag 12.9 star on E edge; mag 10.75v star E 2′0; mag 10.64v star NE 3′5.
15 03 31.1	−82 47 46	ESO 8-7	13.7	2.0 x 1.0	14.3	SB(s)m IV	163	Mag 11.9 star on SE edge; mag 11.8 star S 3′6.
15 59 35.8	−82 54 53	ESO 9-1	13.9	1.8 x 0.7	14.0	SAB(s)d: III-IV	91	Close pair of mag 12.1, 11.30v stars N 6′6.
13 42 27.1	−79 00 44	ESO 21-5	14.1	1.3 x 0.3	13.0	Sb? sp	115	Mag 13.8 star N edge; mag 10.87v star SSE 6′5.
14 45 50.6	−77 55 27	ESO 22-3	13.5	1.8 x 1.2	14.1	SA(rs)c pec: III	159	Mag 9.61v star W 6′5.
15 33 35.1	−78 07 26	ESO 22-10	12.7	1.2 x 0.9	12.7	S0	175	Bright core, faint envelope; mag 10.34v star ENE 3′9.
17 48 58.1	−81 43 13	ESO 24-8	14.3	1.0 x 0.5	13.4	SB(s)c:	15	Mag 10.21v star E 5′9.
14 19 49.0	−76 19 30	ESO 41-5	16.1	0.9 x 0.5	15.0	Im? IV-V?	27	Mag 12.4 star NW 6′9.
14 23 50.6	−75 19 18	ESO 41-6	14.7	1.7 x 1.2	15.3	SB(s)m V-VI	11	Mag 11.9 star NE 2′7; mag 8.84v star W 13′1.
15 10 17.3	−76 00 32	ESO 42-1	14.5	1.5 x 1.3	15.1	IAB(s)m V-VI	174	Mag 12.2 star NE 3′6; IC 4522 NNE 10′0.
15 27 39.7	−73 56 40	ESO 42-7	13.7	2.3 x 2.0	15.2	SB(s)m IV-V	30	Numerous stars superimposed; mag 9.04v star N 6′1.
17 09 22.7	−73 19 16	ESO 44-1	13.4	1.6 x 0.8	13.5	SAB(s)c II-III	53	Mag 8.99v star ENE 11′0.
17 11 38.0	−74 08 26	ESO 44-2	13.2	1.5 x 1.1	13.6	SA(rs)ab pec:	6	Mag 10.20v star W 4′3.
17 16 37.6	−73 08 53	ESO 44-9	13.5	1.5 x 0.6	13.3	SB(s)0/a pec:	78	Mag 9.20v star N 4′7.
17 19 54.3	−77 21 55	ESO 44-10	14.4	1.8 x 0.5	14.1	SB(s)d pec? IV	123	Close pair of E-W oriented mag 12.0, 11.28v stars ESE 3′7.
17 52 59.9	−76 51 47	ESO 44-22	14.4	1.1 x 0.6	13.8	SB(s)d IV-V	168	Mag 9.70v star W 2′2.
18 06 35.4	−73 53 01	ESO 44-25	14.0	1.1 x 0.5	13.2	SAB(rs)b? III-IV	52	Mag 7.42v star S 7′6.
14 05 20.9	−84 16 23	IC 4333	13.4	1.6 x 0.4	12.8	S0? sp	62	Mag 10.46v star S 4′8.
14 16 59.7	−75 38 51	IC 4377	12.9	1.7 x 0.6	12.9	S0^+:	94	Mag 7.86v star SW 11′6.
14 40 29.5	−78 48 37	IC 4448	13.5	1.0 x 0.8	13.1	(R)SB(s)d pec? IV-V	168	Strong dark patch S of center; mag 10.43v star W 2′5.
14 47 45.3	−73 18 26	IC 4484	14.0	2.6 x 0.3	13.6	Scd: sp	132	Mag 9.57v star N 6′7.
15 11 30.2	−75 51 37	IC 4522	12.3	2.0 x 0.5	12.2	Sb III-IV	115	Several faint stars on N edge; mag 10.76v star NW 4′7.
15 41 28.4	−81 37 35	IC 4545	12.9	1.9 x 0.9	13.3	SAB(s)cd III	158	Mag 10.09v star ESE 9′2.
15 48 15.2	−78 10 45	IC 4555	12.8	1.9 x 0.5	12.6	SB(s)cd: sp III-IV	64	Mag 10.67v star ESE 3′4.
15 53 11.6	−74 49 34	IC 4578	13.1	1.2 x 0.5	12.4	(R′)SB(s)b II-III	133	Mag 11.12v star N 4′3; **ESO 42-14** NNE 6′5.
16 46 54.4	−77 29 23	IC 4608	13.7	0.9 x 0.8	13.2	SB(s)m pec	80	Mag 11.0 star N 2′5; mag 4.23v star Beta Apodis WSW 12′6.
16 57 50.0	−76 59 34	IC 4618	12.0	1.7 x 1.3	12.8	SB(rs)bc pec II-III	118	Several strong knots; mag 10.32v star NW 7′3.
17 13 47.1	−77 32 11	IC 4633	13.0	4.0 x 3.0	15.5	SA(r)cd II	145	Almost stellar nucleus, uniform envelope; mag 10.28v star W 11′0.
17 15 40.7	−77 29 26	IC 4635	14.0	3.0 x 0.7	14.6	SAB(s)b: sp	165	Mag 10.09v star E 4′3.
17 23 57.1	−80 03 51	IC 4640	13.6	0.9 x 0.8	13.1	SAB(r)d: III-IV	90	Mag 9.78v star WSW 5′7.
17 24 09.2	−80 08 50	IC 4641	13.6	1.3 x 1.3	14.0	SA(rs)c: III-IV		Stellar nucleus; mag 9.78v star WNW 6′2.
17 24 37.1	−73 56 24	IC 4644	13.9	1.9 x 0.3	13.1	Sb? pec	134	Mag 12.3 star N 1′0; mag 10.23v star NNW 7′9.
17 26 03.0	−80 11 42	IC 4647	13.3	1.6 x 0.8	13.4	SAB(s)0°	0	Mag 10.18v star N 3′2.
17 37 08.7	−74 22 52	IC 4654	12.4	1.4 x 1.1	12.8	SAB(rs)bc: II	102	Mag 13.3 star on E edge; mag 10.11v star SE 6′6.
17 51 03.5	−74 02 00	IC 4661	12.7	1.6 x 1.2	13.2	SA(s)c II	31	Mag 8.69v star WNW 14′3; **ESO 44-23** SSE 9′6.
14 34 02.2	−78 23 16	NGC 5612	12.1	1.9 x 1.0	12.7	SAab: II	63	
15 48 16.1	−75 40 23	NGC 5967	12.0	2.9 x 1.7	13.6	SAB(rs)c: II-III	90	Weak nucleus, many superimposed stars.
15 46 59.2	−75 47 14	NGC 5967A	13.5	2.0 x 1.3	14.3	SB(rs)cd: III-IV	43	Bright nucleus, very patchy, numerous superimposed stars.
16 39 08.0	−73 14 51	NGC 6151	15.9	0.3 x 0.2	12.7		88	Almost stellar; located 0′9 ESE of mag 10.89v star.

(Continued from previous page)

216 OPEN CLUSTERS 216

RA h m s	Dec ° ′ ″	Name	Mag	Diam ′	No. ★	B ★	Type	Notes
14 55 12.0	−83 25 00	ESO 8-6		10			cl	
14 15 54.0	−78 30 00	ESO 21-6		9			cl	(A) Large "Z" shaped chain of bright stars.
17 04 06.0	−74 20 00	ESO 43-13		8			cl	

GLOBULAR CLUSTERS

RA h m s	Dec ° ′ ″	Name	Total V m	B ★ V m	HB V m	Diam ′	Conc. Class Low = 12 High = 1	Notes
15 00 18.5	−82 12 49	IC 4499	10.1	14.6	17.7	8.0		

PLANETARY NEBULAE

RA h m s	Dec ° ′ ″	Name	Diam ″	Mag (P)	Mag (V)	Mag cent ★	Alt Name	Notes
14 15 25.7	−74 12 49	PN G308.6−12.2	31	13.4		14.7	PK 308−12.1	Mag 12.2 star on SW edge.

217 GALAXIES 217

RA h m s	Dec ° ′ ″	Name	Mag (V)	Dim ′ Maj x min	SB	Type Class	PA	Notes
12 42 07.1	−84 27 53	ESO 7-6	14.2	1.5 x 1.1	14.6	SA(s)dm IV-V	60	Mag 9.21v star SW 1′.3.
13 34 32.3	−83 07 48	ESO 8-1	15.1	1.7 x 0.3	14.2	SB(s)d? sp	173	Mag 14.8 star on W edge; mag 9.99v star NW 9′.1.
13 57 48.6	−83 04 50	ESO 8-4	13.5	2.1 x 1.9	14.8	SB(rs)d III-IV	9	Mag 12.9 star on E edge; mag 10.75v star E 2′.0; mag 10.64v star NE 3′.5.
10 38 01.1	−81 05 50	ESO 19-3	12.3	1.6 x 0.8	12.5	SB(rs)c: pec III-IV	137	Mag 7.77v star WSW 12′.5.
11 16 10.6	−79 23 54	ESO 19-6	14.7	2.0 x 0.3	14.0		75	Mag 9.06v star NE 4′.5.
13 23 39.7	−78 40 59	ESO 21-2	13.4	1.2 x 0.5	12.7	SAB(s)d: IV-V	111	Mag 13.8 star SE of center.
13 32 26.7	−80 25 57	ESO 21-3	12.8	1.7 x 1.1	13.3	SB(s)bc III-IV	54	Bright knot or star N of center; mag 12.2 star N 4′.0.
13 32 41.8	−77 50 41	ESO 21-4	12.1	2.3 x 0.9	12.7	SA(s)0/a:	100	Dark lane runs the length of the galaxy S of the center.
13 42 27.1	−79 00 44	ESO 21-5	14.1	1.3 x 0.3	13.0	Sb? sp	115	Mag 13.8 star N edge; mag 10.87v star SSE 6′.5.
09 46 27.8	−74 35 45	ESO 37-5	13.2	1.7 x 0.6	13.1	SAB(s)b pec	89	Mag 8.92v star W 6′.3.
10 04 18.5	−74 55 49	ESO 37-9	13.5	1.2 x 0.6	13.1	SAB(s)0	48	Mag 10.02v star S 7′.5.
10 04 16.9	−75 28 43	ESO 37-10	12.5	2.5 x 2.3	14.2	SAB(s)c II-III	163	Many stars superimposed on surface.
10 17 59.9	−73 49 52	ESO 37-13	13.8	0.7 x 0.3	12.0	Sa?	126	Mag 15.1 star S edge.
10 38 57.0	−76 35 22	ESO 38-4	16.3	1.7 x 0.5	15.9	SB(s)d IV-V	63	Lies 1′.2 S of mag 9.59v star.
10 58 11.2	−77 38 02	ESO 38-6	16.5	1.6 x 0.3	15.6	Sb? sp	114	Mag 9.21v star E 10′.3.
11 50 21.7	−75 22 22	ESO 39-2	13.5	2.5 x 1.4	14.7	IAB(s)m V	37	Mag 10.02v star E 8′.1.
13 21 39.2	−77 32 03	ESO 40-7	13.6	3.0 x 0.5	13.9	SB(s)c: sp III	28	Mag 10.61v star W 10′.1.
13 35 18.5	−77 38 29	ESO 40-12	13.8	2.5 x 0.4	13.6	Sbc sp II	113	Mag 6.51v star WNW 7′.9.
13 35 30.5	−74 39 11	ESO 40-13	13.3	1.7 x 0.9	13.6	(R′)SA(rs)c: III-IV	81	Mag 9.37v star N 1′.1.
14 19 49.0	−76 19 30	ESO 41-5	16.1	0.9 x 0.5	15.0	Im? IV-V?	27	Mag 12.4 star NW 6′.9.
11 57 31.1	−73 41 04	IC 2980	12.3	1.2 x 1.0	12.5	E3	42	Mag 12.9 star N edge; mag 10.80v star WSW 5′.5.
12 18 48.0	−79 43 38	IC 3104	12.6	3.8 x 1.6	14.4	IB(s)m: V-VI	45	Numerous stars superimposed; mag 10.61v star NW 2′.7.
14 05 20.9	−84 16 23	IC 4333	13.4	1.6 x 0.4	12.8	S0? sp	62	Mag 10.46v star S 4′.8.
14 16 59.7	−75 38 51	IC 4377	12.9	1.7 x 0.6	12.9	S0⁺:	94	Mag 7.86v star SW 11′.6.
09 50 09.0	−73 55 21	NGC 3059	11.0	3.6 x 3.2	13.5	SB(rs)bc II-III	70	Numerous superimposed stars; mag 8.94v star W 5′.2.
10 03 45.6	−80 25 18	NGC 3149	12.6	2.0 x 1.9	13.9	(R)SA(rs)b: III		Numerous faint stars superimposed.
11 16 05.0	−76 13 01	NGC 3620	13.3	2.8 x 1.1	14.4	(R′)SB(s)ab pec:	78	Numerous faint stars superimposed on envelope.

OPEN CLUSTERS

RA h m s	Dec ° ′ ″	Name	Mag	Diam ′	No. ★	B ★	Type	Notes
14 15 54.0	−78 30 00	ESO 21-6		9			cl	(A) Large "Z" shaped chain of bright stars.

PLANETARY NEBULAE

RA h m s	Dec ° ′ ″	Name	Diam ″	Mag (P)	Mag (V)	Mag cent ★	Alt Name	Notes
10 09 21.0	−80 51 34	NGC 3195	42	11.5	11.6	15.3	PK 296−20.1	Small triangle of mag 12.5-14 stars W 0′.9; mag 12.0 star SE 2′.1.
13 34 14.8	−75 46 28	PN G305.6−13.1	72	14.6		16.4	PK 305−13.1	Mag 12.5 star NW 2′.1; mag 13.4 star SW 1′.7.
14 15 25.7	−74 12 49	PN G308.6−12.2	31	13.4		14.7	PK 308−12.1	Mag 12.2 star on SW edge.

218 GALAXIES 218

RA h m s	Dec ° ′ ″	Name	Mag (V)	Dim ′ Maj x min	SB	Type Class	PA	Notes
07 26 19.4	−84 31 18	ESO 5-6	13.6	0.9 x 0.4	12.3	Sb pec?	57	Mag 13.1 star NE 2′.1.
07 32 32.6	−84 04 23	ESO 5-10	12.7	1.4 x 0.8	12.9	E4	33	Pair of N-S oriented mag 10.34v, 10.20v star W 4′.0.
08 20 31.7	−84 58 39	ESO 6-2	13.3	1.6 x 0.7	13.3	SAB(r)b: II	117	Mag 8.26v star S 6′.6.
08 36 37.2	−83 27 47	ESO 6-3	14.4	1.7 x 0.3	13.5	SAB(s)cd? III-IV	8	Mag 11.02v star NE 4′.8.
08 19 13.6	−78 41 49	ESO 18-2	12.7	2.7 x 1.3	13.9	SA(rs)bc II	150	Very small galaxy inside N edge, pair of superimposed stars W of center.
08 43 48.7	−78 56 54	ESO 18-13	12.9	2.1 x 0.7	13.1	SB(s)c II	121	Mag 5.46v star Eta Chamaeleonis W 7′.2.
09 10 12.7	−79 14 11	ESO 18-15	13.6	1.4 x 0.8	13.6	SB(r)cd: IV	80	Mag 13 star NE edge.
06 34 50.5	−74 53 34	ESO 34-9	13.8	1.2 x 1.1	13.9	SAB(r)c II-III	69	Mag 14.0 star NW edge.
06 43 25.5	−74 15 27	ESO 34-11B	14.2	1.1 x 0.5	13.4	S0⁻	175	ESO 34-11 NW.
06 43 05.4	−74 14 31	ESO 34-11	12.9	1.8 x 1.3	13.7	Ring	6	"Lindsay-Shapely Ring"
06 43 48.2	−73 40 31	ESO 34-13	13.9	1.3 x 0.5	13.3	SA(rs)bc pec	45	Mag 12.0 star ENE 2′.7.
06 57 33.4	−75 55 43	ESO 34-14	13.9	1.1 x 0.8	13.6	Sa?	159	Mag 10.28v star S 3′.7.
07 10 43.0	−73 30 40	ESO 35-1	13.7	0.7 x 0.5	12.4		104	Mag 10.22v star SW 5′.1.
07 16 38.1	−75 33 59	ESO 35-4	13.8	0.7 x 0.6	12.7	SA?(r)0/a	99	Slightly elongated core, smooth envelope; mag 12.3 star NW 2′.3.
07 21 00.7	−75 23 12	ESO 35-5	13.9	0.9 x 0.7	13.3	SAB(r)c: III-IV	63	Mag 8.79v star ENE 3′.5.
07 25 36.4	−75 10 51	ESO 35-7	13.9	0.7 x 0.4	12.4		51	Mag 12.3 star N 1′.7.
07 28 40.6	−75 03 22	ESO 35-9	15.1	1.8 x 0.6	15.1	IB(s)m V-VI	19	Mag 10.34v star E 2′.9; stellar **ESO 35-9A** on W edge.
07 35 17.5	−76 55 09	ESO 35-11	13.2	1.0 x 0.8	12.8	SB(r)ab pec: II-III	48	Mag 10.02v star SW 7′.5.

(Continued from previous page)

218 GALAXIES 218

RA h m s	Dec ° ′ ″	Name	Mag (V)	Dim ′ Maj x min	SB	Type Class	PA	Notes
07 55 04.3	−76 24 41	ESO 35-18	12.9	2.8 x 0.7	13.5	SA(s)c: sp III	135	Mag 8.61v star S 5′.2.
08 38 19.5	−73 32 39	ESO 36-5	12.8	1.0 x 1.0	12.7	SA(s)0°:		Star E side of nucleus.
08 38 47.9	−75 09 21	ESO 36-6	13.7	2.4 x 1.1	14.6	IAB(rs)m V	164	Mag 7.90v star NNW 6′.5.
08 52 47.2	−73 37 21	ESO 36-10	13.5	0.9 x 0.5	12.6	SA0: pec	100	Mag 11.9star NW 7′.7; note: **ESO 36-9** is 1′.0 SE of this star.
08 58 40.1	−73 19 38	ESO 36-15	14.1	1.0 x 0.9	13.8	SB0?	120	Almost stellar nucleus, smooth, low surface brightness envelope.
09 06 36.0	−75 49 35	ESO 36-19	13.0	1.9 x 0.6	13.0	(R′)SBbc pec: II-III	153	Mag 11.5 star NNW 2′.7; mag 10.81v star E 3′.6.
09 23 15.2	−75 00 45	ESO 37-2	13.9	1.2 x 0.8	13.7	Sc	53	Mag 10.29v star NNE 5′.8.
09 46 27.8	−74 35 45	ESO 37-5	13.2	1.7 x 0.6	13.1	SAB(s)b pec	89	Mag 8.92v star W 6′.3.
10 04 18.5	−74 55 49	ESO 37-9	13.5	1.2 x 0.6	13.1	SAB(s)0	48	Mag 10.02v star S 7′.5.
10 04 16.9	−75 28 43	ESO 37-10	12.5	2.5 x 2.3	14.2	SAB(s)c II-III	163	Many stars superimposed on surface.
10 17 59.9	−73 49 52	ESO 37-13	13.8	0.7 x 0.3	12.0	Sa?	126	Mag 15.1 star S edge.
05 55 27.9	−76 55 13	IC 2160	13.1	2.1 x 1.0	13.8	(R′)SB(s)c pec: I-II	107	Mag 8.47v star N 7′.2.
06 06 52.6	−75 21 55	IC 2164	13.4	1.0 x 0.9	13.1	(R′)SA(rs)ab: II	131	Stars superimposed N; strong dark lane S.
05 40 57.5	−82 07 10	NGC 2144	13.0	1.4 x 1.1	13.3	(R′)SA(rs)a:	93	Mag 9.90v star E 6′.1.
06 04 45.0	−73 24 03	NGC 2199	12.8	1.9 x 0.8	13.1	(R′)SA(r)a:	37	
09 26 12.2	−76 37 37	NGC 2915	12.7	1.9 x 1.0	13.2	I0	129	Mag 7.79v star SW 8′.4.
09 50 09.0	−73 55 21	NGC 3059	11.0	3.6 x 3.2	13.5	SB(rs)bc II-III	70	Numerous superimposed stars; mag 8.94v star W 5′.2.
10 03 45.6	−80 25 18	NGC 3149	12.6	2.0 x 1.9	13.9	(R)SA(rs)b: III		Numerous faint stars superimposed.

OPEN CLUSTERS

RA h m s	Dec ° ′ ″	Name	Mag	Diam ′	No. ★	B ★	Type	Notes
05 57 25.5	−75 08 24	IC 2161		1.5			cl	
05 55 42.3	−74 21 14	NGC 2161	13.0	2.2			cl	
06 01 02.3	−74 43 34	NGC 2190		2			cl	
06 04 41.6	−75 26 13	NGC 2203		3.3		19.0	cl	Compact cluster.
06 08 34.8	−73 50 22	NGC 2209		2.8			cl	Very faint and diffuse.

GLOBULAR CLUSTERS

RA h m s	Dec ° ′ ″	Name	Total V m	B ★ V m	HB V m	Diam ′	Conc. Class Low = 12 High = 1	Notes
09 20 59.3	−77 16 57	E 3	11.4					

PLANETARY NEBULAE

RA h m s	Dec ° ′ ″	Name	Diam ″	Mag (P)	Mag (V)	Mag cent ★	Alt Name	Notes
10 09 21.0	−80 51 34	NGC 3195	42	11.5	11.6	15.3	PK 296−20.1	Small triangle of mag 12.5-14 stars W 0′.9; mag 12.0 star SE 2′.1.
05 57 02.1	−75 40 21	PN G286.8−29.5	46	16.5		16.7	PK 286−29.1	Mag 11.15v star 1′.2 N; mag 9.43v star S 2′.2.
06 59 27.2	−79 38 50	PN G291.3−26.2				15.8	Vo 1	Mag 15.2 star SE 0′.4; mag 13.2 star SSE 2′.1.

219 GALAXIES 219

RA h m s	Dec ° ′ ″	Name	Mag (V)	Dim ′ Maj x min	SB	Type Class	PA	Notes
01 41 32.6	−83 12 44	ESO 3-4	13.9	2.0 x 0.3	13.2	Sbc: sp II-III	136	Mag 7.33v star SW 6′.0.
02 00 15.3	−83 59 13	ESO 3-7	12.4	1.6 x 1.4	13.2	SAB(rs)c: pec II	99	Mag 11.24v star SSE 1′.5.
02 53 38.1	−83 08 34	ESO 3-13	12.8	1.1 x 1.0	12.8	SA(r)0°:	82	Pair of E-W oriented mag 10.35v, 10.38v stars E 7′.1.
03 12 24.4	−83 07 30	ESO 3-14	14.4	1.1 x 0.6	13.8	SB(s)dm IV	100	Mag 9.99v star W 13′.1.
03 24 06.5	−84 09 40	ESO 3-15	14.0	1.3 x 0.4	13.2	SB(s)dm: IV-V	79	Several strong knots; mag 13.9 star S edge.
03 49 36.9	−83 57 42	ESO 4-6	14.6	1.5 x 1.2	15.1	SA(s)m: V	150	Mag 11.32v star WSW 9′.4; IC 2051 NNE 8′.7.
03 59 48.1	−84 05 39	ESO 4-10	13.1	1.4 x 0.7	12.9	SA(r)0⁻ pec?	133	Very large, bright core; mag 12.7 star W 2′.6.
04 00 37.5	−84 22 01	ESO 4-11	13.9	1.4 x 0.7	13.7	SA(s)dm: IV	68	Mag 10.21v star E 2′.3.
04 27 30.2	−83 33 04	ESO 4-13	13.9	1.2 x 0.5	13.2	SB(s)c: pec II-III	108	Mag 12.3 star N 3′.0.
01 45 42.3	−77 56 47	ESO 13-21	13.8	1.1 x 0.5	13.0	Sa	4	Mag 10.94v star S 4′.2; **ESO 13-20** W 5′.6.
01 45 58.4	−78 54 16	ESO 13-22	14.0	0.8 x 0.6	13.1	SA(s)c pec I-II	45	Double system, stellar companion SW of nucleus; mag 10.74v star N 2′.5.
01 53 18.8	−78 14 23	ESO 13-24	13.8	1.2 x 0.4	12.9	Sab	139	Mag 6.15v star SE 10′.1.
01 58 46.8	−78 23 54	ESO 13-26	13.5	1.3 x 0.9	13.5	SB(s)bc pec	117	Mag 6.15v star W 9′.4.
02 02 32.7	−79 40 23	ESO 13-28	14.0	1.9 x 0.3	13.2	Sbc sp	7	Mag 10.91v star W 3′.5; **ESO 13-27** S 2′.4.
02 24 18.9	−79 55 31	ESO 14-1	13.8	0.9 x 0.7	13.1	SB(s)m: pec III-IV	151	Bright knot or star superimposed W edge.
02 37 59.0	−79 11 32	ESO 14-3	15.2	1.1 x 0.7	14.7	SB(s)dm:	84	Mag 9.90v star NE 5′.2.
03 43 45.4	−80 06 19	ESO 15-1	13.9	1.9 x 0.8	14.2	IB(s)m V	105	Located between mag 8.21v, 8.45v stars.
03 58 31.2	−81 03 47	ESO 15-5	12.4	1.8 x 1.5	13.3	(R′)SB(r)b: II	63	Mag 10.00v star NNE 4′.4; anonymous galaxy WNW 4′.5.
04 04 10.2	−82 20 31	ESO 15-6	13.8	1.7 x 0.8	14.0	(R′)SA(r)a:	74	Mag 10.89v star S 4′.2.
04 07 12.5	−82 17 05	ESO 15-8	12.1	1.0 x 0.4	10.9		2	Bright star superimposed; mag 10.57v star N 5′.5.
04 48 19.1	−81 41 01	ESO 15-15	13.7	1.1 x 1.0	13.6	S?	96	Mag 10.94v star W 5′.5.
05 04 58.8	−81 18 40	ESO 15-18	13.2	1.7 x 0.9	13.7	SA0⁻:	9	Mag 9.34v star WSW 10′.1.
02 09 01.1	−75 56 22	ESO 30-8	14.1	2.2 x 0.4	13.8	SB(s)d: sp III-IV	18	Mag 6.89v star E 6′.9.
02 10 40.9	−75 02 19	ESO 30-9	13.3	2.5 x 0.6	13.6	SA(rs)c II	134	Mag 10.82v star S 4′.0.
02 11 32.8	−73 09 47	ESO 30-11	14.3	0.5 x 0.5	12.6	(R′)SA(s)ab?		Mag 11.11v star S 3′.3.
02 16 48.2	−74 24 16	ESO 30-13	14.1	0.8 x 0.6	13.2	Sab	117	Mag 12.5 star NW 3′.1.
02 17 56.3	−76 04 55	ESO 30-14	13.2	1.6 x 1.4	13.9	SA(s)c I-II	80	Mag 9.30v star W 12′.9; anonymous galaxy WNW 5′.0.
02 35 47.4	−74 14 33	ESO 30-17	16.7	1.3 x 0.3	15.5	S	102	Mag 14.4 star superimposed W end; mag 12.7 star W 1′.6.
02 43 47.6	−74 43 20	ESO 30-19	14.0	0.4 x 0.4	11.9			Compact.
02 58 06.8	−74 27 26	ESO 31-5	13.5	1.5 x 0.5	13.0	SAB(s)b: sp II	10	Mag 10.18v star NNE 6′.9.
03 41 11.3	−77 08 11	ESO 31-20	14.4	1.1 x 0.5	13.6	(R′)SB(r)bc pec III	174	Mag 10.52v star S 5′.6.
04 55 58.5	−75 32 33	ESO 33-2	13.7	1.0 x 0.9	13.5	SB0	11	Mag 12.9 star W 2′.0.
04 57 47.7	−73 13 55	ESO 33-3	12.7	1.2 x 0.7	12.5	E4?	20	Pair of faint stars or bright knots S of core.
04 57 18.1	−75 25 08	ESO 33-4	13.2	1.5 x 0.8	13.3	SB(rs)bc II	9	Pair with ESO 33-5 S 1′.5.
04 57 22.1	−75 26 44	ESO 33-5	13.4	1.0 x 0.5	12.5	Sa	40	Pair with ESO 33-4 N 1′.5; almost stellar anonymous galaxy SE 1′.5.
04 58 32.6	−74 45 01	ESO 33-6	13.9	0.5 x 0.5	12.3	S		Compact; mag 12.4 star NNE 3′.0.
04 58 35.1	−75 04 44	ESO 33-7	14.4	1.0 x 0.7	13.9	Sb	19	Mag 13.1 star N 1′.0.
05 05 06.9	−73 39 06	ESO 33-11	13.2	1.0 x 0.9	12.9	(R)SB(rs)a	65	Mag 8.26v star NE 4′.9.
05 07 18.7	−73 33 20	ESO 33-14	14.3	0.9 x 0.6	13.5	S	167	Several stars superimposed, in star rich area.

(Continued from previous page)

219 GALAXIES 219

RA h m s	Dec ° ′ ″	Name	Mag (V)	Dim ′ Maj x min	SB	Type Class	PA	Notes
05 31 41.5	−73 44 59	ESO 33-22	14.8	2.1 x 0.2	13.7	Sd sp	170	Mag 9.44v star N 7′.2; mag 5.84v star E 12′.9.
03 52 01.3	−83 49 57	IC 2051	11.4	2.8 x 1.7	13.0	SB(r)bc: II	67	Numerous knots or superimposed stars along edges; mag 10.95v star W 2′.0.
04 39 47.8	−76 50 16	IC 2103	13.6	1.8 x 0.3	12.8	Sc: sp III-IV	89	Mag 8.67v star W 11′.2.
05 55 27.9	−76 55 13	IC 2160	13.1	2.1 x 1.0	13.8	(R′)SB(s)c pec: I-II	107	Mag 8.47v star N 7′.2.
06 06 52.6	−75 21 55	IC 2164	13.4	1.0 x 0.9	13.1	(R′)SA(rs)ab: II	131	Stars superimposed N; strong dark lane S.
01 41 49.4	−75 16 07	NGC 643C	14.6	1.3 x 0.2	13.0	Scd: sp	150	
05 19 35.9	−77 43 48	NGC 1956	12.9	1.9 x 0.8	13.2	SA(s)a	68	Mag 9.41v star NE 6′.1.
05 22 35.4	−79 51 09	NGC 2012	12.9	1.1 x 0.6	12.3	SA0$^-$:	117	
05 40 57.5	−82 07 10	NGC 2144	13.0	1.4 x 1.1	13.3	(R′)SA(rs)a:	93	Mag 9.90v star E 6′.1.
06 04 45.0	−73 24 03	NGC 2199	12.8	1.9 x 0.8	13.1	(R′)SA(r)a:	37	

OPEN CLUSTERS

RA h m s	Dec ° ′ ″	Name	Mag	Diam ′	No. ★	B ★	Type	Notes
05 23 06.6	−75 26 49	IC 2134		1			cl	
05 33 22.2	−75 22 35	IC 2140		2.1			cl	
05 37 47.4	−74 46 58	IC 2146		2.6			cl	
05 39 12.1	−75 33 48	IC 2148		1.1			cl	
05 57 25.5	−75 08 24	IC 2161		1.5			cl	
01 56 44.5	−74 13 09	NGC 796		0.8			cl	
03 57 51.1	−76 48 20	NGC 1520		5	11	9.0	ast	
04 55 48.5	−74 17 00	NGC 1777		2.1		16.2	cl	
05 55 42.3	−74 21 14	NGC 2161	13.0	2.2			cl	
06 01 02.3	−74 43 34	NGC 2190		2			cl	
06 04 41.6	−75 26 13	NGC 2203		3.3		19.0	cl	Compact cluster.
06 08 34.8	−73 50 22	NGC 2209		2.8			cl	Very faint and diffuse.
01 42 52.9	−73 20 11	WG71 1		0.8			cl	

GLOBULAR CLUSTERS

RA h m s	Dec ° ′ ″	Name	Total V m	B ★ V m	HB V m	Diam ′	Conc. Class Low = 12 High = 1	Notes
04 45 24.2	−83 59 53	NGC 1841	14.1			2.4		

BRIGHT NEBULAE

RA h m s	Dec ° ′ ″	Name	Dim ′ Maj x min	Type	BC	Color	Notes
01 56 42.0	−74 13 00	NGC 796					

PLANETARY NEBULAE

RA h m s	Dec ° ′ ″	Name	Diam ″	Mag (P)	Mag (V)	Mag cent ★	Alt Name	Notes
05 57 02.1	−75 40 21	PN G286.8−29.5	46	16.5		16.7	PK 286−29.1	Mag 11.15v star 1′.2 N; mag 9.43v star S 2′.2.

220 GALAXIES 220

RA h m s	Dec ° ′ ″	Name	Mag (V)	Dim ′ Maj x min	SB	Type Class	PA	Notes
14 27 22.5	−87 46 21	ESO 1-6	13.4	2.4 x 0.8	13.9	SB(rs)cd: sp III-IV	31	Mag 11.17v star E 4′.2.
00 59 03.3	−86 17 10	ESO 2-11	14.4	1.2 x 0.8	14.2	SB(s)d IV-V	45	Mag 10.43v star E 5′.1.
01 00 50.0	−85 31 22	ESO 2-12	14.3	1.8 x 0.2	13.0	Scd? sp	49	Mag 10.42v star N 4′.7.
01 32 00.3	−85 10 41	ESO 3-1	12.8	1.3 x 1.2	13.1	(R)SB(rs)0/a	0	Mag 9.68v star NE 9′.5.
03 24 06.5	−84 09 40	ESO 3-15	14.0	1.3 x 0.4	13.2	SB(s)dm: IV-V	79	Several strong knots; mag 13.9 star S edge.
03 59 48.1	−84 05 39	ESO 4-10	13.1	1.4 x 0.7	12.9	SA(r)0$^-$ pec?	133	Very large, bright core; mag 12.7 star W 2′.6.
04 00 37.5	−84 22 01	ESO 4-11	13.9	1.4 x 0.7	13.7	SA(s)dm: IV	68	Mag 10.21v star E 2′.3.
05 04 31.5	−87 01 48	ESO 4-17	15.4	1.2 x 0.8	15.2	IB(s)m: V-VI	60	Mag 9.76v star E 4′.2.
05 18 49.8	−86 23 09	ESO 4-19	14.3	1.5 x 1.3	14.9	SA(rs)cd IV	168	Mag 12.3 star WSW 7′.7.
05 29 24.8	−85 55 46	ESO 4-20	13.9	1.0 x 0.5	13.0	Sb	76	Mag 6.78 star E 11′.3.
06 05 39.7	−86 37 54	ESO 5-4	12.4	3.9 x 0.6	13.2	Sb: sp	97	Mag 10.75v star N 4′.8.
07 26 19.4	−84 31 18	ESO 5-6	13.6	0.9 x 0.4	12.3	Sb pec?	57	Mag 13.1 star NE 2′.1.
07 32 32.6	−84 04 23	ESO 5-10	12.7	1.4 x 0.8	12.9	E4	33	Pair of N-S oriented mag 10.34v, 10.20v star W 4′.0.
07 42 03.5	−85 25 21	ESO 5-11	13.7	1.2 x 0.4	12.7	S0$^-$ sp	110	Mag 8.98v star W 2′.4.
08 20 31.7	−84 58 39	ESO 6-2	13.3	1.6 x 0.7	13.3	SAB(r)b: II	117	Mag 8.26v star S 6′.6.
09 15 42.8	−86 46 02	ESO 6-6	13.4	0.9 x 0.7	13.0	E2:	179	Mag 12.4 star E 0′.9.
10 46 48.7	−86 17 18	ESO 7-1	15.2	1.5 x 0.4	14.4	SBm pec? sp	36	Mag 8.04v star WSW 13′.1.
12 42 07.1	−84 27 53	ESO 7-6	14.2	1.5 x 1.1	14.6	SA(s)dm IV-V	60	Mag 9.21v star SW 1′.3.
15 31 18.5	−87 26 05	ESO 8-8	13.3	2.3 x 0.5	13.3	SB(s)dm sp IV	175	Mag 9.97v star E 4′.2.
17 39 32.1	−85 18 34	ESO 9-10	11.9	2.1 x 1.5	13.0	SA(s)bc: II	171	Mag 10.79v star N 7′.8.
19 20 16.1	−84 47 38	ESO 10-3	13.8	0.9 x 0.7	13.1	S?	160	Mag 13.9 star E edge; mag 8.99v star W 2′.9.
19 31 00.2	−86 00 55	ESO 10-4	13.3	2.4 x 1.1	14.2	SB(s)dm IV-V	108	Several bright knots and/or superimposed stars.
19 40 13.6	−84 16 26	ESO 10-6	14.9	0.9 x 0.5	13.9	SB(s)m V	162	Mag 11.64v star NE 2′.0.
14 05 20.9	−84 16 23	IC 4333	13.4	1.6 x 0.4	12.8	S0? sp	62	Mag 10.46v star S 4′.8.
01 42 09.1	−89 20 05	NGC 2573	13.5	2.0 x 0.8	13.9	SAB(s)cd: III-IV	85	**Polarissima Australis.**
23 12 04.5	−89 07 35	NGC 2573A	14.3	2.1 x 0.6	14.4	SBb? pec	18	Pair with NGC 2573B W edge.
23 07 32.6	−89 07 00	NGC 2573B	14.6	1.7 x 0.6	14.5	IBm? pec	120	Brightest portion offset to the NE.
18 22 16.9	−85 24 08	NGC 6438	11.1	1.6 x 1.3	11.8	Ring A	156	Multi-galaxy system with NGC 6438A E.
18 22 44.1	−85 24 36	NGC 6438A	11.8	2.7 x 1.0	12.7	Ring B	32	Multi-galaxy system, long extension S; NGC 6438 W.

(Continued from previous page)

GALAXY CLUSTERS

RA h m s	Dec ° ′ ″	Name	Mag 10th brightest	No. Gal	Diam ′	Notes
00 15 54.0	−84 04 00	AS 21	16.1		17	**ESO 2-4** NW of center 12.′6; all others anonymous, stellar.

GLOBULAR CLUSTERS

RA h m s	Dec ° ′ ″	Name	Total V m	B ★ V m	HB V m	Diam ′	Conc. Class Low = 12 High = 1	Notes
04 45 24.2	−83 59 53	NGC 1841	14.1			2.4		

Messier Object Index

Messier	Designation	Type	Pages
M 1	**NGC 1952**	BN	77, VI, XIII, XIV
M 2	**NGC 7089**	GC	103, VIII
M 3	**NGC 5272**	GC	71, IV, XI
M 4	**NGC 6121**	GC	147, X, XVII
M 5	**NGC 5904**	GC	108, X, XI
M 6	**NGC 6405**	OC	164, A20, IX, X, XVII
M 7	**NGC 6475**	OC	164, A20, IX, X, XVII
M 8	**NGC 6523**	BN	145, 146, A17, IX, X, XVII
M 9	**NGC 6333**	GC	146, X, XVII
M 10	**NGC 6254**	GC	107, X
M 11	**NGC 6705**	OC	125, A14, IX
M 12	**NGC 6218**	GC	107, X
M 13	**NGC 6205**	GC	50, 51, III, X
M 14	**NGC 6402**	GC	106, IX, X
M 15	**NGC 7078**	GC	83, II, VIII, IX
M 16	**NGC 6611**	OC	126, IX, X, XVII
M 17	**NGC 6618**	BN	126, IX, X, XVII
M 18	**NGC 6613**	OC	126, 145, IX, X, XVII
M 19	**NGC 6273**	GC	146, X, XVII
M 20	**NGC 6514**	BN	145, 146, A17, IX, X, XVII
M 21	**NGC 6531**	OC	145, A17, IX, X, XVII
M 22	**NGC 6656**	GC	145, IX, X, XVII
M 23	**NGC 6494**	OC	145, 146, IX, X, XVII
M 24	**IC 4715**	Star Cloud	145, IX, X, XVII
M 25	**IC 4725**	OC	145, IX, X, XVII
M 26	**NGC 6694**	OC	125, IX, XVII
M 27	**NGC 6853**	PN	66, II, III, IX
M 28	**NGC 6626**	GC	145, IX, X, XVII
M 29	**NGC 6913**	OC	48, A2, II, III
M 30	**NGC 7099**	GC	143, VIII, XVI
M 31	**NGC 224**	GX	30, II, VII
M 32	**NGC 221**	GX	30, 45, 62, II, VII
M 33	**NGC 598**	GX	62, VII, XV
M 34	**NGC 1039**	OC	43, VII
M 35	**NGC 2168**	OC	76, VI, XIII, XIV
M 36	**NGC 1960**	OC	59, VI, XIII, XIV
M 37	**NGC 2099**	OC	59, VI, XIII, XIV
M 38	**NGC 1912**	OC	59, VI, XIV
M 39	**NGC 7092**	OC	32, I, II
M 40	**Double Star**		24, I, IV, V
M 41	**NGC 2287**	OC	154, XIII, XX
M 42	**NGC 1976**	BN	116, 136, XIII, XIV
M 43	**NGC 1982**	BN	116, 136, XIII, XIV
M 44	**NGC 2632**	OC	74, 75, V, XII, XIII
M 45	**Pleiades**	OC	78, A12, VII, XIV
M 46	**NGC 2437**	OC	135, XIII, XX
M 47	**NGC 2422**	OC	135, XIII, XX
M 48	**NGC 2548**	OC	114, 134, XIII, XIX
M 49	**NGC 4472**	GX	91, A15, IV, XI, XII
M 50	**NGC 2323**	OC	135, XIII, XX
M 51	**NGC 5194**	GX	37, IV
M 52	**NGC 7654**	OC	18, I, II, VII
M 53	**NGC 5024**	GC	71, IV, XI
M 54	**NGC 6715**	GC	163, IX, XVII
M 55	**NGC 6809**	GC	162, IX, XVI, XVII
M 56	**NGC 6779**	GC	48, 49, III, IX
M 57	**NGC 6720**	PN	49, III, IX
M 58	**NGC 4579**	GX	90, 91, A13, IV, XI
M 59	**NGC 4621**	GX	90, IV, XI
M 60	**NGC 4649**	GX	90, IV, XI
M 61	**NGC 4303**	GX	111, A15, XI, XII
M 62	**NGC 6266**	GC	164, X, XVII
M 63	**NGC 5055**	GX	37, IV
M 64	**NGC 4826**	GX	71, IV, XI
M 65	**NGC 3623**	GX	92, V, XII
M 66	**NGC 3627**	GX	91, 92, V, XII
M 67	**NGC 2682**	OC	94, V, XII, XIII
M 68	**NGC 4590**	GC	149, 150, XI, XVIII
M 69	**NGC 6637**	GC	163, IX, X, XVII
M 70	**NGC 6681**	GC	163, IX, XVII
M 71	**NGC 6838**	GC	66, III, IX
M 72	**NGC 6981**	GC	124, VIII, IX, XVI
M 73	**NGC 6994**	OC	123, 124, VIII, IX, XVI
M 74	**NGC 628**	GX	100, VII, XV
M 75	**NGC 6864**	GC	144, IX, XVI, XVII
M 76	**NGC 650-51**	PN	29, 44, I, VII
M 77	**NGC 1068**	GX	119, XIV, XV
M 78	**NGC 2068**	BN	116, XIII, XIV
M 79	**NGC 1904**	GC	155, XIII, XIV, XX
M 80	**NGC 6093**	GC	147, X, XVII
M 81	**NGC 3031**	GX	14, I, IV, V, VI
M 82	**NGC 3034**	GX	14, I, IV, V, VI
M 83	**NGC 5236**	GX	149, 167, XI, XVIII
M 84	**NGC 4374**	GX	91, A13, IV, XI, XII
M 85	**NGC 4382**	GX	72, IV, XI, XII
M 86	**NGC 4406**	GX	91, A13, IV, XI, XII
M 87	**NGC 4486**	GX	91, A13, IV, XI, XII
M 88	**NGC 4501**	GX	90, 91, A13, IV, XI, XII
M 89	**NGC 4552**	GX	90, 91, A13, IV, XI
M 90	**NGC 4569**	GX	90, 91, A13, IV, XI
M 91	**NGC 4548**	GX	90, 91, A13, IV, XI
M 92	**NGC 6341**	GC	34, III
M 93	**NGC 2447**	OC	153, XIII, XX
M 94	**NGC 4736**	GX	37, IV
M 95	**NGC 3351**	GX	92, V, XII
M 96	**NGC 3368**	GX	92, V, XII
M 97	**NGC 3587**	PN	24, I, IV, V
M 98	**NGC 4192**	GX	91, IV, XI, XII
M 99	**NGC 4254**	GX	91, A13, IV, XI, XII
M 100	**NGC 4321**	GX	91, A13, IV, XI, XII
M 101	**NGC 5457**	GX	23, I, IV
M 102	**= M 101**		
M 103	**NGC 581**	OC	29, I, II, VI, VII
M 104	**NGC 4594**	GX	130, 131, XI, XVIII
M 105	**NGC 3379**	GX	92, V, XII
M 106	**NGC 4258**	GX	37, IV, V
M 107	**NGC 6171**	GC	127, X, XVII
M 108	**NGC 3556**	GX	24, I, IV, V
M 109	**NGC 3992**	GX	24, I, IV, V
M 110	**NGC 205**	GX	30, II VII

Key

GX = Galaxy	GC = Globular Cluster	DN = Dark Nebula
GXCL = Galaxy Cluster	Star Cld = Star Cloud	PN = Planetary Nebula
OC = Open Cluster	BN = Bright Nebula	

Common Name Index

Key

GX = Galaxy	GC = Globular Cluster	DN = Dark Nebula	R + X = Radio + X-ray
GXCL = Galaxy Cluster	Star Cld = Star Cloud	PN = Planetary Nebula	QSO = Quasi-stellar Object
OC = Open Cluster	BN = Bright Nebula	RAD = Radio	SNR = Supernova Remnant

Combined Non-Stellar Index

2MASS-GC 1 GC 145
2MASS-GC 2 GC 145, A17
47 Tucanae GC 204
A 14 GXCL 141
A 21 GXCL 63
A 43 GXCL 63, 81
A 63 GXCL 30
A 71 GXCL 45, 62, 63, 80
A 72 GXCL 30
A 74 GXCL 141, 158
A 75 GXCL 63, 80
A 76 GXCL 81, 100
A 85 GXCL 140
A 86 GXCL 141, 158
A 102 GXCL 120
A 104 GXCL 80
A 114 GXCL 158
A 116 GXCL 120
A 117 GXCL 140
A 119 GXCL 120
A 121 GXCL 140
A 133 GXCL 158
A 134 GXCL 120
A 147 GXCL 120
A 151 GXCL 140
A 154 GXCL 80, 100
A 158 GXCL 100
A 160 GXCL 100
A 161 GXCL 62
A 165 GXCL 62, A7
A 168 GXCL 120
A 171 GXCL 100
A 174 GXCL 62
A 179 GXCL 80
A 189 GXCL 120
A 193 GXCL 100
A 194 GXCL 120, A16
A 195 GXCL 80
A 225 GXCL 80
A 240 GXCL 100
A 245 GXCL 99, 100
A 246 GXCL 99, 100, 119, 120
A 260 GXCL 62
A 262 GXCL 62, A6
A 274 GXCL 139
A 276 GXCL 44
A 277 GXCL 139
A 278 GXCL 61, 62
A 347 GXCL 43, 44
A 358 GXCL 139
A 372 GXCL 43
A 376 GXCL 61
A 389 GXCL 157
A 397 GXCL 98, 99
A 399 GXCL 98, 99
A 400 GXCL 98, 99
A 401 GXCL 98, 99
A 407 GXCL 61
A 415 GXCL 138
A 419 GXCL 157
A 426 GXCL 43, A4
A 449 GXCL 7
A 450 GXCL 78, A12
A 496 GXCL 137
A 500 GXCL 155, 156
A 505 GXCL 7
A 514 GXCL 155
A 526 GXCL 97, 117
A 527 GXCL 7, 16
A 533 GXCL 155
A 539 GXCL 97
A 548 GXCL 155
A 553 GXCL 41
A 559 GXCL 15
A 564 GXCL 15
A 566 GXCL 15
A 568 GXCL 58
A 569 GXCL 40, 41
A 576 GXCL 26
A 582 GXCL 40
A 592 GXCL 95
A 595 GXCL 26
A 602 GXCL 57, 75
A 610 GXCL 75
A 628 GXCL 57
A 634 GXCL 26
A 644 GXCL 134
A 671 GXCL 57
A 692 GXCL 74, 75
A 732 GXCL 113, 114
A 754 GXCL 133
A 757 GXCL 39
A 762 GXCL 6
A 779 GXCL 56, A5
A 834 GXCL 14
A 838 GXCL 113, 133
A 912 GXCL 113
A 933 GXCL 113
A 957 GXCL 112, 113
A 978 GXCL 132
A 979 GXCL 132
A 993 GXCL 112
A 999 GXCL 92
A 1016 GXCL 92
A 1020 GXCL 92
A 1032 GXCL 112
A 1035 GXCL 38, 55
A 1060 GXCL 151, A19
A 1069 GXCL 132
A 1097 GXCL 55
A 1100 GXCL 73
A 1126 GXCL 92
A 1132 GXCL 24, 25
A 1139 GXCL 112
A 1142 GXCL 92
A 1145 GXCL 92
A 1149 GXCL 92
A 1171 GXCL 112
A 1177 GXCL 73
A 1185 GXCL 73
A 1187 GXCL 55
A 1213 GXCL 54, 55, 72, 73
A 1216 GXCL 112
A 1218 GXCL 24, 38
A 1225 GXCL 24
A 1228 GXCL 54, 55
A 1238 GXCL 111, 112
A 1257 GXCL 54
A 1267 GXCL 72
A 1270 GXCL 24
A 1275 GXCL 54
A 1291 GXCL 24
A 1297 GXCL 5
A 1308 GXCL 111
A 1314 GXCL 38
A 1318 GXCL 24
A 1332 GXCL 131
A 1334 GXCL 111
A 1336 GXCL 54
A 1362 GXCL 91
A 1364 GXCL 111
A 1365 GXCL 54
A 1367 GXCL 72, A11
A 1377 GXCL 24
A 1382 GXCL 13
A 1383 GXCL 24
A 1390 GXCL 91
A 1399 GXCL 111
A 1412 GXCL 5, 13
A 1436 GXCL 24
A 1452 GXCL 24, 37, 38
A 1468 GXCL 24, 37
A 1474 GXCL 91
A 1496 GXCL 24
A 1500 GXCL 5
A 1507 GXCL 24
A 1541 GXCL 91, A15
A 1616 GXCL 24
A 1631 GXCL 130
A 1638 GXCL 71
A 1644 GXCL 130, 149
A 1651 GXCL 110
A 1656 GXCL 71, A8
A 1691 GXCL 53
A 1709 GXCL 149
A 1736 GXCL 149
A 1749 GXCL 53
A 1750 GXCL 110
A 1767 GXCL 23
A 1773 GXCL 110
A 1775 GXCL 71
A 1781 GXCL 53, 71
A 1783 GXCL 23
A 1793 GXCL 53
A 1795 GXCL 71
A 1800 GXCL 71
A 1809 GXCL 89, 109
A 1813 GXCL 53
A 1825 GXCL 70, 71
A 1831 GXCL 70, 71
A 1836 GXCL 129
A 1837 GXCL 129
A 1873 GXCL 70
A 1890 GXCL 89
A 1899 GXCL 70, 89
A 1904 GXCL 36
A 1913 GXCL 89
A 1927 GXCL 70
A 1983 GXCL 89
A 1991 GXCL 70
A 1999 GXCL 23
A 2005 GXCL 70
A 2019 GXCL 70
A 2020 GXCL 88, 89
A 2022 GXCL 70
A 2028 GXCL 88
A 2029 GXCL 88, 108
A 2033 GXCL 88
A 2036 GXCL 70
A 2040 GXCL 88
A 2048 GXCL 108
A 2052 GXCL 88
A 2055 GXCL 88
A 2061 GXCL 51, 52
A 2063 GXCL 88
A 2065 GXCL 69, 70
A 2067 GXCL 51, 52
A 2079 GXCL 69
A 2089 GXCL 69
A 2092 GXCL 51
A 2107 GXCL 69
A 2108 GXCL 69, 88
A 2124 GXCL 51
A 2142 GXCL 69
A 2147 GXCL 88, A9
A 2148 GXCL 69
A 2149 GXCL 22
A 2151 GXCL 69, 88, A9
A 2152 GXCL 88, A9
A 2159 GXCL 87, 88, A9
A 2162 GXCL 51, 69
A 2169 GXCL 35
A 2170 GXCL 69
A 2175 GXCL 51, 69
A 2184 GXCL 35
A 2197 GXCL 35, 51, A3
A 2199 GXCL 51, A3
A 2241 GXCL 50
A 2247 GXCL 4
A 2248 GXCL 4
A 2249 GXCL 50
A 2255 GXCL 11
A 2256 GXCL 4
A 2271 GXCL 4
A 2293 GXCL 21
A 2295 GXCL 10, 11
A 2296 GXCL 3, 4
A 2301 GXCL 10
A 2308 GXCL 10
A 2309 GXCL 3
A 2311 GXCL 10
A 2312 GXCL 10
A 2315 GXCL 10
A 2319 GXCL 33
A 2328 GXCL 124, 143
A 2331 GXCL 123, 124
A 2347 GXCL 143
A 2366 GXCL 123
A 2382 GXCL 123
A 2384 GXCL 143
A 2399 GXCL 123
A 2410 GXCL 123
A 2412 GXCL 142
A 2415 GXCL 103, 123
A 2440 GXCL 102
A 2448 GXCL 122
A 2457 GXCL 102
A 2459 GXCL 122
A 2462 GXCL 122, 142
A 2503 GXCL 46, 64
A 2511 GXCL 122
A 2525 GXCL 122
A 2572 GXCL 63, 64
A 2589 GXCL 81, 82
A 2593 GXCL 81, 82
A 2618 GXCL 63
A 2622 GXCL 63
A 2625 GXCL 63
A 2626 GXCL 63
A 2630 GXCL 81
A 2634 GXCL 63
A 2656 GXCL 101
A 2657 GXCL 81
A 2660 GXCL 141
A 2665 GXCL 81
A 2666 GXCL 63
A 2670 GXCL 121
A 2675 GXCL 81
A 2700 GXCL 101
A 2716 GXCL 141
A 2717 GXCL 159
A 2731 GXCL 192
A 2734 GXCL 141
A 2799 GXCL 159, 176
A 2800 GXCL 141, 158
A 2806 GXCL 192
A 2819 GXCL 204
A 2824 GXCL 158
A 2836 GXCL 191
A 2854 GXCL 191
A 2860 GXCL 176
A 2870 GXCL 191
A 2877 GXCL 191
A 2881 GXCL 140, 158
A 2911 GXCL 176
A 2954 GXCL 213
A 3009 GXCL 190, 191
A 3027 GXCL 175
A 3089 GXCL 175
A 3093 GXCL 190
A 3094 GXCL 157
A 3095 GXCL 157
A 3100 GXCL 190
A 3104 GXCL 190
A 3108 GXCL 190
A 3110 GXCL 190
A 3111 GXCL 190
A 3112 GXCL 190
A 3122 GXCL 190
A 3123 GXCL 202
A 3125 GXCL 202
A 3128 GXCL 202
A 3133 GXCL 190
A 3135 GXCL 174
A 3144 GXCL 202
A 3151 GXCL 156
A 3154 GXCL 174
A 3158 GXCL 202
A 3161 GXCL 174
A 3164 GXCL 202
A 3193 GXCL 202
A 3202 GXCL 202
A 3223 GXCL 174
A 3225 GXCL 202
A 3266 GXCL 202
A 3301 GXCL 173
A 3323 GXCL 155
A 3341 GXCL 173
A 3351 GXCL 173
A 3354 GXCL 155
A 3356 GXCL 173
A 3367 GXCL 155
A 3368 GXCL 155
A 3372 GXCL 173
A 3374 GXCL 154, 155
A 3376 GXCL 172, 173, 188
A 3379 GXCL 188
A 3380 GXCL 188
A 3381 GXCL 172
A 3389 GXCL 211
A 3390 GXCL 172
A 3391 GXCL 201
A 3392 GXCL 172
A 3395 GXCL 201
A 3407 GXCL 188
A 3408 GXCL 187, 188
A 3420 GXCL 152
A 3497 GXCL 168
A 3505 GXCL 168
A 3526 GXCL 184, A23
A 3528 GXCL 149, 167, A21
A 3530 GXCL 167, A21
A 3532 GXCL 167, A21
A 3537 GXCL 167, A21
A 3542 GXCL 167
A 3553 GXCL 167
A 3554 GXCL 167
A 3555 GXCL 149
A 3556 GXCL 167
A 3558 GXCL 167
A 3559 GXCL 149, 167
A 3560 GXCL 167
A 3561 GXCL 183, 184
A 3562 GXCL 167
A 3564 GXCL 167
A 3565 GXCL 167
A 3566 GXCL 167
A 3570 GXCL 167
A 3571 GXCL 167
A 3572 GXCL 167
A 3574 GXCL 167, A18
A 3575 GXCL 167
A 3577 GXCL 149
A 3578 GXCL 148, 149
A 3581 GXCL 148
A 3593 GXCL 148
A 3603 GXCL 166
A 3627 GXCL 196
A 3651 GXCL 194
A 3656 GXCL 162
A 3667 GXCL 194
A 3682 GXCL 162
A 3687 GXCL 205
A 3695 GXCL 162
A 3698 GXCL 143, 144
A 3706 GXCL 161, 162
A 3716 GXCL 194
A 3733 GXCL 143
A 3742 GXCL 179
A 3744 GXCL 143
A 3747 GXCL 179
A 3756 GXCL 179
A 3771 GXCL 179
A 3782 GXCL 193, 205
A 3785 GXCL 193
A 3799 GXCL 205
A 3806 GXCL 193
A 3809 GXCL 178
A 3816 GXCL 193
A 3822 GXCL 193
A 3825 GXCL 193
A 3826 GXCL 193
A 3849 GXCL 178, 193
A 3851 GXCL 193
A 3869 GXCL 193
A 3879 GXCL 205
A 3880 GXCL 160
A 3886 GXCL 193
A 3895 GXCL 160
A 3912 GXCL 160
A 3925 GXCL 178
A 4038 GXCL 141
A 4049 GXCL 141
A 4059 GXCL 159
ADS 6366 OC 114, 115
ADS 13292 OC 48
AL 1 OC 208, 209
AL 3 OC 145
AL 5 OC 105, A14
AM 1 GC 190
AM 2 OC 171
AM 4 GC 148, 149
Ant 1 OC 164
Ant 2 OC 164
Ant 3 OC 164
Ant 4 OC 164
Archinal 1 OC 85, 105
Arp 2 GC 162
AS 2 GXCL 141, 159
AS 5 GXCL 177
AS 6 GXCL 159
AS 9 GXCL 177
AS 12 GXCL 159
AS 21 GXCL 214, 220
AS 28 GXCL 121, 141
AS 37 GXCL 177
AS 41 GXCL 159
AS 45 GXCL 192
AS 74 GXCL 177, 191
AS 78 GXCL 177, 191
AS 85 GXCL 177, 191
AS 92 GXCL 191
AS 101 GXCL 191
AS 102 GXCL 176

Key

GX = Galaxy	GC = Globular Cluster	DN = Dark Nebula
GXCL = Galaxy Cluster	Star Cld = Star Cloud	PN = Planetary Nebula
OC = Open Cluster	BN = Bright Nebula	

AS 107 GXCL 214
AS 109 GXCL 176
AS 112 GXCL 204
AS 113 GXCL 176, 191
AS 116 GXCL 191, 192, 203
AS 118 GXCL 140, 158
AS 120 GXCL 191
AS 131 GXCL 204, 213
AS 137 GXCL 203
AS 138 GXCL 158
AS 141 GXCL 176
AS 162 GXCL 191, 203
AS 164 GXCL 213
AS 176 GXCL 213, 214
AS 186 GXCL 176
AS 191 GXCL 213
AS 210 GXCL 213
AS 215 GXCL 175, 176, 191
AS 217 GXCL 191
AS 225 GXCL 203
AS 230 GXCL 191
AS 244 GXCL 157
AS 250 GXCL 190, 191, 203
AS 258 GXCL 157, 175
AS 263 GXCL 157
AS 268 GXCL 139, 157
AS 273 GXCL 139, 157
AS 274 GXCL 203
AS 277 GXCL 190
AS 297 GXCL 190
AS 301 GXCL 175
AS 307 GXCL 190
AS 309 GXCL 203
AS 312 GXCL 203
AS 315 GXCL 190, 202, 203
AS 316 GXCL 175
AS 332 GXCL 175
AS 333 GXCL 157, 175
AS 334 GXCL 190
AS 335 GXCL 190
AS 336 GXCL 190
AS 337 GXCL 156, 157, 174, 175
AS 339 GXCL 202
AS 345 GXCL 190
AS 346 GXCL 190
AS 353 GXCL 190, 202
AS 356 GXCL 190
AS 360 GXCL 190
AS 362 GXCL 212, 213
AS 366 GXCL 202
AS 367 GXCL 190
AS 372 GXCL 202
AS 373 GXCL 174
AS 377 GXCL 202
AS 384 GXCL 190
AS 393 GXCL 190
AS 404 GXCL 202
AS 410 GXCL 202
AS 411 GXCL 212
AS 412 GXCL 202
AS 415 GXCL 202
AS 418 GXCL 174
AS 426 GXCL 202
AS 429 GXCL 202
AS 440 GXCL 156
AS 449 GXCL 156
AS 459 GXCL 156
AS 463 GXCL 202
AS 464 GXCL 174
AS 465 GXCL 156, 174
AS 476 GXCL 173, 174
AS 478 GXCL 173, 174
AS 479 GXCL 189, 202
AS 484 GXCL 173, 174
AS 487 GXCL 189
AS 489 GXCL 173, 174
AS 496 GXCL 173
AS 497 GXCL 189
AS 500 GXCL 202, 212
AS 505 GXCL 189
AS 512 GXCL 173
AS 515 GXCL 189
AS 521 GXCL 173
AS 524 GXCL 201
AS 534 GXCL 189
AS 535 GXCL 173
AS 536 GXCL 173
AS 537 GXCL 201
AS 538 GXCL 188, 189
AS 540 GXCL 173, 188
AS 541 GXCL 201
AS 542 GXCL 188
AS 545 GXCL 173
AS 550 GXCL 173
AS 552 GXCL 201
AS 555 GXCL 172, 173
AS 558 GXCL 201
AS 559 GXCL 172, 173
AS 560 GXCL 201
AS 561 GXCL 201
AS 562 GXCL 154, 155
AS 563 GXCL 172, 173
AS 564 GXCL 188
AS 566 GXCL 188
AS 568 GXCL 154
AS 570 GXCL 172
AS 571 GXCL 172
AS 572 GXCL 188
AS 573 GXCL 172
AS 574 GXCL 188
AS 578 GXCL 154, 172
AS 580 GXCL 188, 201
AS 581 GXCL 154
AS 584 GXCL 201
AS 585 GXCL 211
AS 589 GXCL 172
AS 591 GXCL 172
AS 593 GXCL 172
AS 595 GXCL 172
AS 597 GXCL 172
AS 599 GXCL 201
AS 601 GXCL 188
AS 602 GXCL 187, 188
AS 610 GXCL 153
AS 617 GXCL 152
AS 622 GXCL 169, 170
AS 624 GXCL 169, 170
AS 628 GXCL 169
AS 631 GXCL 169
AS 636 GXCL 169
AS 639 GXCL 185
AS 640 GXCL 169
AS 643 GXCL 169, 185
AS 655 GXCL 185
AS 665 GXCL 168
AS 669 GXCL 185
AS 673 GXCL 185
AS 677 GXCL 168
AS 687 GXCL 150
AS 689 GXCL 184
AS 707 GXCL 184, A23
AS 712 GXCL 184
AS 713 GXCL 149
AS 714 GXCL 149
AS 718 GXCL 167, A21
AS 719 GXCL 149
AS 721 GXCL 167
AS 731 GXCL 167
AS 735 GXCL 149
AS 740 GXCL 167
AS 741 GXCL 149
AS 742 GXCL 167
AS 743 GXCL 167
AS 746 GXCL 167
AS 753 GXCL 166, 167
AS 754 GXCL 166
AS 756 GXCL 129, 148
AS 757 GXCL 166
AS 758 GXCL 166
AS 761 GXCL 148
AS 770 GXCL 166
AS 775 GXCL 166
AS 778 GXCL 166
AS 794 GXCL 207
AS 797 GXCL 206, 207
AS 800 GXCL 215
AS 801 GXCL 180, 181, 195
AS 805 GXCL 206
AS 808 GXCL 180
AS 823 GXCL 194
AS 835 GXCL 194
AS 836 GXCL 162
AS 839 GXCL 194
AS 840 GXCL 194
AS 844 GXCL 194
AS 851 GXCL 179
AS 854 GXCL 194
AS 866 GXCL 179
AS 882 GXCL 179
AS 885 GXCL 194
AS 889 GXCL 194
AS 890 GXCL 161, 162, 179
AS 891 GXCL 143, 144
AS 892 GXCL 161, 162
AS 894 GXCL 143, 144
AS 897 GXCL 161
AS 900 GXCL 161
AS 906 GXCL 179, 194
AS 907 GXCL 179
AS 909 GXCL 179
AS 916 GXCL 179
AS 917 GXCL 161
AS 918 GXCL 161
AS 919 GXCL 179
AS 921 GXCL 179
AS 922 GXCL 161
AS 924 GXCL 179
AS 927 GXCL 193
AS 929 GXCL 143
AS 940 GXCL 143
AS 947 GXCL 161
AS 951 GXCL 143
AS 952 GXCL 161
AS 957 GXCL 193
AS 959 GXCL 178, 179
AS 963 GXCL 143
AS 964 GXCL 161
AS 968 GXCL 178, 193
AS 971 GXCL 178
AS 974 GXCL 178
AS 980 GXCL 205
AS 981 GXCL 193
AS 983 GXCL 142, 143
AS 987 GXCL 142, 143
AS 988 GXCL 193
AS 989 GXCL 178
AS 997 GXCL 160
AS 999 GXCL 142
AS 1003 GXCL 193
AS 1005 GXCL 160
AS 1014 GXCL 214
AS 1020 GXCL 193
AS 1023 GXCL 193
AS 1039 GXCL 193
AS 1050 GXCL 160
AS 1055 GXCL 160
AS 1057 GXCL 204, 205
AS 1059 GXCL 178
AS 1060 GXCL 178
AS 1065 GXCL 160
AS 1067 GXCL 178
AS 1101 GXCL 177, 178
AS 1111 GXCL 177
AS 1129 GXCL 159
AS 1142 GXCL 159
AS 1146 GXCL 141
AS 1147 GXCL 177
AS 1155 GXCL 141, 159
AS 1157 GXCL 159
AS 1163 GXCL 141
AS 1165 GXCL 141, 159
AS 1171 GXCL 141
AS 1172 GXCL 159
AS 1173 GXCL 177
Auner 1 OC 154
B 1 DN 60
B 2 DN 60
B 3 DN 60
B 4 DN 60
B 5 DN 60
B 6 DN 28
B 7 DN 78
B 9 DN 28
B 10 DN 78
B 11 DN 28
B 12 DN 28
B 13 DN 28
B 14 DN 77, 78
B 15 DN 42
B 16 DN 42
B 17 DN 42
B 18 DN 78
B 19 DN 78
B 20 DN 42
B 21 DN 28
B 22 DN 77, 78
B 23 DN 59, 60, 77, 78
B 24 DN 59, 60, 77, 78
B 25 DN 42
B 26 DN 59
B 27 DN 59
B 28 DN 59
B 29 DN 59
B 30 DN 96
B 31 DN 96
B 32 DN 96
B 33 DN 116
B 34 DN 59
B 35 DN 96
B 36 DN 96
B 37 DN 95, 96
B 38 DN 95, 96
B 39 DN 95, 96
B 40 DN 147
B 41 DN 147
B 42 DN 147
B 43 DN 147
B 44 DN 146, 147
B 44a DN 164, 181, 182, A22
B 45 DN 146
B 46 DN 146
B 47 DN 146
B 48 DN 164, 181, A22
B 49 DN 164
B 50 DN 164
B 51 DN 146
B 53 DN 164
B 54 DN 164
B 55 DN 164
B 56 DN 164
B 57 DN 146
B 58 DN 164, 181
B 59 DN 146
B 60 DN 146
B 61 DN 146
B 62 DN 146
B 63 DN 146
B 64 DN 146
B 65 DN 146
B 66 DN 146
B 67 DN 146
B 67a DN 146
B 68 DN 146
B 69 DN 146
B 70 DN 146
B 71 DN 146
B 72 DN 146
B 73 DN 146
B 74 DN 146
B 75 DN 146
B 76 DN 146
B 77 DN 146
B 78 DN 146
B 79 DN 146
B 80 DN 146
B 81 DN 146
B 82 DN 146
B 83 DN 146
B 83a DN 146
B 84 DN 146
B 84a DN 126, 145, 146
B 85 DN 145, 146, A17
B 86 DN 145, 146
B 87 DN 163
B 88 DN 145, A17
B 89 DN 145, A17
B 90 DN 145
B 91 DN 145, A17
B 92 DN 145
B 93 DN 145
B 94 DN 126
B 95 DN 126
B 96 DN 126
B 97 DN 126
B 98 DN 145
B 99 DN 145
B 100 DN 125, 126
B 101 DN 125, 126
B 102 DN 125, 126
B 103 DN 125, 126
B 104 DN 105, A14
B 105 DN 125, A14
B 106 DN 105, 125, A14
B 107 DN 105, 125, A14
B 108 DN 125, A14
B 109 DN 125, A14
B 110 DN 105, A14
B 111 DN 105, A14
B 112 DN 125, A14
B 113 DN 105, A14
B 114 DN 125, A14
B 115 DN 125, A14
B 116 DN 125, A14
B 117 DN 125, A14
B 117a DN 105, A14
B 118 DN 125, A14
B 119 DN 105, A14
B 119a DN 105, 125, A14
B 120 DN 105, A14
B 121 DN 105, A14
B 122 DN 105, A14
B 123 DN 105, A14
B 124 DN 105, A14
B 125 DN 105, A14
B 126 DN 105, A14
B 127 DN 105, 125, A14
B 128 DN 105, A14
B 129 DN 105, 125
B 130 DN 105, 125, A14
B 131 DN 105
B 132 DN 105
B 133 DN 125
B 134 DN 125
B 135 DN 105
B 136 DN 105
B 137 DN 105
B 138 DN 105
B 139 DN 105
B 140 DN 85, 105
B 141 DN 105
B 142 DN 85
B 143 DN 85
B 144 DN 48
B 145 DN 48
B 146 DN 48
B 147 DN 48
B 148 DN 20
B 149 DN 20
B 150 DN 20
B 151 DN 19
B 152 DN 19
B 153 DN 19
B 154 DN 19
B 155 DN 31, 32
B 156 DN 31, 32
B 157 DN 19
B 158 DN 31, 32
B 159 DN 31, 32
B 160 DN 19
B 161 DN 19
B 162 DN 19
B 163 DN 19
B 164 DN 19, 31
B 165 DN 19
B 166 DN 19
B 167 DN 19
B 168 DN 31
B 169 DN 19
B 170 DN 19
B 171 DN 19
B 173 DN 19
B 174 DN 19
B 201 DN 29
B 202 DN 60
B 203 DN 60
B 204 DN 60
B 205 DN 60
B 206 DN 60
B 207 DN 78
B 208 DN 78
B 209 DN 78
B 210 DN 78
B 211 DN 78
B 212 DN 78
B 213 DN 78
B 214 DN 78
B 215 DN 78
B 216 DN 78
B 217 DN 78
B 218 DN 78
B 219 DN 60, 78
B 220 DN 77, 78
B 221 DN 59, 60
B 222 DN 59
B 223 DN 96, 97
B 224 DN 96, 97
B 225 DN 96
B 226 DN 59
B 227 DN 76
B 228 DN 165
B 229 DN 147
B 230 DN 127
B 231 DN 164, 165
B 232 DN 164, 165
B 233 DN 164
B 234 DN 164
B 235 DN 181, 182
B 236 DN 146, 164
B 237 DN 146, 164
B 238 DN 146
B 239 DN 164
B 240 DN 164
B 241 DN 164
B 242 DN 164
B 243 DN 146, 164
B 244 DN 146
B 245 DN 146, 164
B 246 DN 146
B 247 DN 164
B 248 DN 146
B 249 DN 146, 164
B 250 DN 146
B 251 DN 146
B 252 DN 164
B 253 DN 146
B 254 DN 164
B 255 DN 146
B 256 DN 146
B 257 DN 164
B 258 DN 164
B 259 DN 146
B 260 DN 146
B 261 DN 146
B 262 DN 146
B 263 DN 181
B 264 DN 146
B 265 DN 146
B 266 DN 146
B 267 DN 146
B 268 DN 146
B 269 DN 146
B 270 DN 146
B 271 DN 164
B 272 DN 146
B 273 DN 164, A20
B 274 DN 146
B 275 DN 164, A20
B 276 DN 146
B 277 DN 146
B 278 DN 164, A20
B 279 DN 146
B 280 DN 146
B 281 DN 146
B 282 DN 146
B 283 DN 164, A20
B 284 DN 126
B 285 DN 126
B 286 DN 164, A20
B 287 DN 164, A20
B 288 DN 163, 164
B 289 DN 145, 146, 163, 164
B 290 DN 163, 164
B 291 DN 163, 164, A20
B 292 DN 163, 164
B 293 DN 163, 164
B 294 DN 145, 146
B 295 DN 163
B 296 DN 145, A17
B 297 DN 145
B 298 DN 163
B 299 DN 145
B 300 DN 163
B 301 DN 145
B 302 DN 145, A17
B 303 DN 145, A17
B 304 DN 145
B 305 DN 163
B 306 DN 145
B 307 DN 126, 145
B 308 DN 145
B 309 DN 145
B 310 DN 145
B 311 DN 126, 145
B 312 DN 125, 126
B 313 DN 125, 126
B 314 DN 125, 126
B 315 DN 145
B 316 DN 105
B 317 DN 125
B 318 DN 125, A14
B 319 DN 105
B 320 DN 105, 125, A14
B 321 DN 125
B 322 DN 105, A14
B 323 DN 105, A14
B 324 DN 105, A14
B 325 DN 105, A14
B 326 DN 105
B 327 DN 105, 125
B 328 DN 105
B 329 DN 105
B 330 DN 85
B 331 DN 85
B 332 DN 85
B 333 DN 85
B 334 DN 85
B 335 DN 85
B 336 DN 85
B 337 DN 85
B 338 DN 85
B 339 DN 84, 85
B 340 DN 84, 85
B 341 DN 48
B 342 DN 32

B 343 DN 32, 48, A2
B 344 DN 32, 48, A2
B 345 DN 32
B 346 DN 32
B 347 DN 48, A2
B 348 DN 32, A2
B 349 DN 32, A1
B 350 DN 32, A1
B 351 DN 32
B 352 DN 32, A1
B 353 DN 32, A1
B 354 DN 19, 20
B 355 DN 32, A1
B 356 DN 32, A1
B 357 DN 19, 20
B 358 DN 32, A1
B 359 DN 19
B 360 DN 19
B 361 DN 32
B 362 DN 32
B 363 DN 32
B 364 DN 19
B 365 DN 19
B 366 DN 19
B 367 DN 19
B 368 DN 19
B 369 DN 19
B 370 DN 19
Bar 1 OC 32, A1
Bar 2 OC 19, 31
Bas 1 OC 105, 125, A14
Bas 4 OC 59
Bas 5 OC 164
Bas 7 OC 95, 96
Bas 8 OC 95, 96
Bas 10 OC 29
Bas 11a OC 135
Bas 11b OC 76, 77
Bas 12 OC 32
Bas 13 OC 32
Bas 14 OC 32
Bas 17 OC 198
Bas 18 OC 197, 208
Be 1 OC 18
Be 2 OC 18
Be 4 OC 8
Be 5 OC 17, 29
Be 6 OC 29
Be 7 OC 17, 29
Be 8 OC 2, 7
Be 9 OC 28
Be 10 OC 16, 17
Be 11 OC 42
Be 12 OC 42
Be 13 OC 28
Be 14 OC 42
Be 15 OC 42
Be 17 OC 59
Be 18 OC 42
Be 19 OC 59, 77
Be 20 OC 116
Be 21 OC 77
Be 22 OC 96
Be 23 OC 76
Be 24 OC 115, 116
Be 25 OC 135
Be 29 OC 95
Be 35 OC 115
Be 36 OC 135
Be 37 OC 115
Be 39 OC 114, 115
Be 43 OC 85
Be 44 OC 66, 67
Be 45 OC 85
Be 47 OC 66, 85
Be 49 OC 48
Be 51 OC 48
Be 52 OC 66
Be 53 OC 19, 20, 32
Be 54 OC 32, 47
Be 55 OC 19, 32
Be 56 OC 32
Be 58 OC 18
Be 59 OC 8
Be 60 OC 18
Be 61 OC 8
Be 62 OC 8
Be 63 OC 17
Be 64 OC 17
Be 65 OC 29
Be 66 OC 28
Be 67 OC 42
Be 68 OC 42
Be 69 OC 59
Be 70 OC 42
Be 71 OC 59
Be 72 OC 77
Be 73 OC 136
Be 76 OC 135
Be 77 OC 115
Be 78 OC 95, 115
Be 79 OC 105
Be 80 OC 105
Be 82 OC 85
Be 83 OC 66
Be 84 OC 48
Be 85 OC 48, A2
Be 86 OC 48, A2
Be 87 OC 48, A2
Be 89 OC 32
Be 90 OC 32
Be 92 OC 19
Be 93 OC 9
Be 94 OC 19
Be 95 OC 19
Be 96 OC 19
Be 97 OC 19
Be 98 OC 19
Be 99 OC 8
Be 100 OC 8
Be 101 OC 8
Be 102 OC 18
Be 103 OC 18
Be 104 OC 8
Be 135 DN 187
Be 145 DN 208
Be 149 DN 165
Be 157 DN 163
Bergeron 1 OC 18
Bi 1 OC 48
Bi 2 OC 48
Bi 7 OC 95
Bi 8 OC 95
Bi 9 OC 115
Bi 10 OC 115
Bi 11 OC 95, 115
Bi 12 OC 95, 115
Bi 13 OC 115
Bl 1 OC 141, 159
Bo 1 OC 76
Bo 2 OC 115
Bo 3 OC 115, 135
Bo 5 OC 135
Bo 6 OC 153
Bo 7 OC 186, 187
Bo 9 OC 199
Bo 10 OC 199
Bo 11 OC 199
Bo 12 OC 198, 199
Bo 13 OC 164
Bo 14 OC 145, 146, A17
Bo 15 OC 171
C1819-146 OC 126
Canali OC 130, 131
Ced 33 BN 78
Ced 34 BN 78
Ced 59 BN 96
Ced 90 BN 135
Ced 122 BN 208
Ced 174 BN 48
Ced 211 BN 121
Ced 214 BN 8
CGCG 1-4 GX 115
CGCG 4-8 GX 114
CGCG 4-11 GX 114
CGCG 4-27 GX 114
CGCG 4-35 GX 114
CGCG 4-61 GX 114
CGCG 4-67 GX 114
CGCG 4-71 GX 114
CGCG 4-86 GX 114
CGCG 4-87 GX 114
CGCG 4-88 GX 114
CGCG 4-95 GX 114
CGCG 5-15 GX 114
CGCG 5-23 GX 114
CGCG 5-28 GX 114
CGCG 5-32 GX 114
CGCG 5-42 GX 113, 114
CGCG 5-51 GX 113, 114
CGCG 5-56 GX 113
CGCG 6-6 GX 113
CGCG 6-9 GX 113
CGCG 6-19 GX 113
CGCG 6-27 GX 113
CGCG 8-4 GX 113
CGCG 8-6 GX 113
CGCG 8-15 GX 113
CGCG 8-27 GX 113
CGCG 8-30 GX 113
CGCG 8-40 GX 113
CGCG 8-56 GX 112, 113
CGCG 8-60 GX 112, 113
CGCG 8-80 GX 112, 113
CGCG 8-91 GX 112
CGCG 9-8 GX 112
CGCG 9-11 GX 112
CGCG 9-12 GX 112
CGCG 9-19 GX 112
CGCG 9-22 GX 112
CGCG 9-29 GX 112
CGCG 9-41 GX 112
CGCG 9-42 GX 112
CGCG 9-44 GX 112
CGCG 9-55 GX 112
CGCG 9-56 GX 112
CGCG 9-64 GX 112
CGCG 9-70 GX 112
CGCG 9-78 GX 112
CGCG 9-88 GX 112
CGCG 9-90 GX 112
CGCG 9-98 GX 112
CGCG 10-4 GX 112
CGCG 10-16 GX 112
CGCG 10-26 GX 112
CGCG 10-36 GX 112
CGCG 10-46 GX 112
CGCG 10-51 GX 112
CGCG 10-52 GX 112
CGCG 10-58 GX 112
CGCG 10-64 GX 112
CGCG 10-65 GX 112
CGCG 10-71 GX 112
CGCG 10-77 GX 112
CGCG 11-2 GX 112
CGCG 11-24 GX 112
CGCG 11-25 GX 112
CGCG 11-27 GX 112
CGCG 11-30 GX 112
CGCG 11-33 GX 112
CGCG 11-38 GX 112
CGCG 11-45 GX 112
CGCG 11-48 GX 112
CGCG 11-73 GX 111, 112
CGCG 11-75 GX 111, 112
CGCG 11-95 GX 111, 112
CGCG 11-96 GX 111, 112
CGCG 12-13 GX 111
CGCG 12-15 GX 111
CGCG 12-27 GX 111
CGCG 12-29 GX 111
CGCG 12-32 GX 111
CGCG 12-37 GX 111
CGCG 12-42 GX 111
CGCG 12-49 GX 111
CGCG 12-53 GX 111
CGCG 12-54 GX 111
CGCG 12-57 GX 111
CGCG 12-81 GX 111
CGCG 12-89 GX 111
CGCG 12-95 GX 111
CGCG 12-98 GX 111
CGCG 12-99 GX 111
CGCG 12-100 GX 111
CGCG 12-101 GX 111
CGCG 12-106 GX 111
CGCG 12-110 GX 111
CGCG 13-6 GX 111
CGCG 13-10 GX 111
CGCG 13-24 GX 111
CGCG 13-26 GX 111
CGCG 13-29 GX 111
CGCG 13-44 GX 111
CGCG 13-70 GX 111
CGCG 13-72 GX 111
CGCG 13-80 GX 111
CGCG 13-84 GX 111
CGCG 13-102 GX 111
CGCG 13-107 GX 111
CGCG 13-116 GX 111
CGCG 13-118 GX 111
CGCG 13-119 GX 111
CGCG 14-9 GX 111
CGCG 14-30 GX 111
CGCG 14-41 GX 111
CGCG 14-43 GX 111
CGCG 14-47 GX 111
CGCG 14-52 GX 111
CGCG 14-53 GX 111
CGCG 14-55 GX 111
CGCG 14-74 GX 110, 111
CGCG 14-76 GX 110, 111
CGCG 14-78 GX 110, 111
CGCG 14-98 GX 110
CGCG 15-10 GX 110
CGCG 15-17 GX 110
CGCG 15-40 GX 110
CGCG 15-54 GX 110
CGCG 15-61 GX 110
CGCG 16-2 GX 110
CGCG 16-20 GX 110
CGCG 16-23 GX 110
CGCG 16-24 GX 110
CGCG 16-31 GX 110
CGCG 16-40 GX 110
CGCG 16-41 GX 110
CGCG 16-42 GX 110
CGCG 16-59 GX 110
CGCG 16-60 GX 110
CGCG 16-67 GX 110
CGCG 17-11 GX 110
CGCG 17-26 GX 110
CGCG 17-30 GX 110
CGCG 17-77 GX 109, 110
CGCG 17-80 GX 109, 110
CGCG 17-83 GX 109
CGCG 17-92 GX 109
CGCG 17-93 GX 109
CGCG 18-36 GX 109
CGCG 18-37 GX 109
CGCG 18-49 GX 109
CGCG 18-51 GX 109
CGCG 18-69 GX 109
CGCG 18-80 GX 109
CGCG 18-86 GX 109
CGCG 18-104 GX 109
CGCG 18-109 GX 109
CGCG 18-112 GX 109
CGCG 19-23 GX 109
CGCG 19-41 GX 109
CGCG 19-55 GX 109
CGCG 19-63 GX 109
CGCG 19-84 GX 109
CGCG 20-21 GX 109
CGCG 20-34 GX 108, 109
CGCG 20-35 GX 108, 109
CGCG 21-20 GX 108
CGCG 21-49 GX 108
CGCG 21-62 GX 108
CGCG 21-71 GX 108
CGCG 21-76 GX 108
CGCG 21-78 GX 108
CGCG 21-80 GX 108
CGCG 21-84 GX 108
CGCG 21-95 GX 108
CGCG 21-97 GX 108
CGCG 21-101 GX 108
CGCG 22-20 GX 108
CGCG 22-45 GX 108
CGCG 23-1 GX 108
CGCG 23-14 GX 108
CGCG 23-28 GX 107, 108
CGCG 23-29 GX 107, 108
CGCG 24-2 GX 107
CGCG 24-11 GX 107
CGCG 24-22 GX 107
CGCG 25-2 GX 107
CGCG 26-2 GX 107
CGCG 26-3 GX 107
CGCG 27-1 GX 106
CGCG 29-3 GX 95
CGCG 29-14 GX 115
CGCG 29-25 GX 95
CGCG 29-29 GX 95, 115
CGCG 30-5 GX 95
CGCG 30-7 GX 115
CGCG 30-8 GX 95, 115
CGCG 30-9 GX 95, 115
CGCG 30-11 GX 95
CGCG 30-16 GX 94, 95
CGCG 30-20 GX 94, 95
CGCG 30-21 GX 114, 115
CGCG 30-24 GX 94, 114
CGCG 31-16 GX 94
CGCG 31-18 GX 94
CGCG 31-20 GX 114
CGCG 31-24 GX 94
CGCG 31-37 GX 94
CGCG 31-50 GX 114
CGCG 31-57 GX 114
CGCG 31-63 GX 114
CGCG 31-65 GX 94
CGCG 31-79 GX 114
CGCG 32-2 GX 114
CGCG 32-3 GX 114
CGCG 32-9 GX 114
CGCG 32-11 GX 114
CGCG 32-13 GX 114
CGCG 32-14 GX 94
CGCG 32-21 GX 114
CGCG 32-22 GX 114
CGCG 32-23 GX 94
CGCG 32-24 GX 94, 114
CGCG 32-28 GX 114
CGCG 32-32 GX 114
CGCG 32-33 GX 114
CGCG 32-36 GX 114
CGCG 32-46 GX 94
CGCG 32-48 GX 114
CGCG 33-3 GX 114
CGCG 33-7 GX 114
CGCG 33-43 GX 93, 94
CGCG 33-49 GX 113, 114
CGCG 33-50 GX 113, 114
CGCG 33-60 GX 113
CGCG 34-4 GX 113
CGCG 34-20 GX 93
CGCG 34-41 GX 113
CGCG 35-3 GX 93
CGCG 35-23 GX 93
CGCG 35-43 GX 113
CGCG 35-44 GX 93
CGCG 35-55 GX 93
CGCG 35-56 GX 113
CGCG 35-57 GX 93
CGCG 35-63 GX 113
CGCG 35-66 GX 113
CGCG 35-75 GX 93
CGCG 35-82 GX 93
CGCG 36-19 GX 113
CGCG 36-27 GX 93
CGCG 36-33 GX 113
CGCG 36-41 GX 113
CGCG 36-42 GX 112, 113
CGCG 36-49 GX 92, 93
CGCG 36-51 GX 92, 93
CGCG 36-52 GX 92, 93, 112, 113
CGCG 36-67 GX 92, 93
CGCG 36-72 GX 92, 93, 112, 113
CGCG 36-84 GX 92
CGCG 36-91 GX 112
CGCG 37-5 GX 92
CGCG 37-11 GX 92, 112
CGCG 37-17 GX 112
CGCG 37-24 GX 92
CGCG 37-28 GX 92
CGCG 37-38 GX 92
CGCG 37-39 GX 92
CGCG 37-43 GX 92
CGCG 37-55 GX 92
CGCG 37-67 GX 92
CGCG 37-74 GX 112
CGCG 37-77 GX 112
CGCG 37-78 GX 92
CGCG 37-80 GX 112
CGCG 37-81 GX 92
CGCG 37-82 GX 92
CGCG 37-87 GX 92, 112
CGCG 37-106 GX 112
CGCG 38-6 GX 112
CGCG 38-38 GX 112
CGCG 38-48 GX 92, 112
CGCG 38-58 GX 92, 112
CGCG 38-64 GX 92
CGCG 38-68 GX 92
CGCG 38-69 GX 112
CGCG 38-78 GX 92
CGCG 38-79 GX 112
CGCG 38-89 GX 112
CGCG 38-96 GX 92, 112
CGCG 38-103 GX 92
CGCG 38-104 GX 92
CGCG 38-105 GX 92
CGCG 38-107 GX 92
CGCG 38-111 GX 112
CGCG 38-118 GX 92
CGCG 38-122 GX 112
CGCG 38-124 GX 112
CGCG 39-3 GX 112
CGCG 39-9 GX 92, 112
CGCG 39-14 GX 92
CGCG 39-15 GX 92, 112
CGCG 39-16 GX 92
CGCG 39-17 GX 112
CGCG 39-19 GX 112
CGCG 39-22 GX 92
CGCG 39-23 GX 112
CGCG 39-31 GX 112
CGCG 39-33 GX 112
CGCG 39-35 GX 92, 112
CGCG 39-37 GX 112
CGCG 39-39 GX 112
CGCG 39-43 GX 92, 112
CGCG 39-48 GX 112
CGCG 39-50 GX 112
CGCG 39-54 GX 112
CGCG 39-59 GX 92
CGCG 39-61 GX 112
CGCG 39-80 GX 92
CGCG 39-94 GX 112
CGCG 39-95 GX 112
CGCG 39-101 GX 112
CGCG 39-102 GX 112
CGCG 39-104 GX 92
CGCG 39-113 GX 92
CGCG 39-125 GX 111, 112
CGCG 39-133 GX 111, 112
CGCG 39-150 GX 111, 112
CGCG 39-162 GX 91, 92, 111, 112
CGCG 39-163 GX 111, 112
CGCG 39-167 GX 91, 92
CGCG 39-173 GX 111, 112
CGCG 39-176 GX 111, 112
CGCG 39-180 GX 111, 112
CGCG 39-184 GX 111, 112
CGCG 39-197 GX 111
CGCG 39-198 GX 111
CGCG 39-200 GX 111
CGCG 40-7 GX 111
CGCG 40-27 GX 111
CGCG 40-33 GX 111
CGCG 40-42 GX 91
CGCG 40-44 GX 111
CGCG 40-51 GX 91
CGCG 40-53 GX 91, 111
CGCG 40-54 GX 91
CGCG 40-64 GX 111
CGCG 41-10 GX 111
CGCG 41-50 GX 91, A15
CGCG 41-56 GX 111, A15
CGCG 42-9 GX 91, A15
CGCG 42-18 GX 91, 111, A15
CGCG 42-55 GX 111, A15
CGCG 42-81 GX 111, A15
CGCG 42-120 GX 91, A15
CGCG 42-153 GX 90, 91
CGCG 43-3 GX 110
CGCG 43-48 GX 90
CGCG 43-97 GX 110
CGCG 43-105 GX 110
CGCG 43-120 GX 110
CGCG 43-127 GX 90
CGCG 44-7 GX 90
CGCG 44-33 GX 90
CGCG 44-36 GX 90
CGCG 44-67 GX 90
CGCG 44-69 GX 110
CGCG 45-18 GX 90
CGCG 45-26 GX 90, 110
CGCG 45-38 GX 110
CGCG 45-39 GX 110
CGCG 45-74 GX 90
CGCG 45-84 GX 109, 110
CGCG 45-113 GX 109, 110
CGCG 45-123 GX 109
CGCG 45-125 GX 89
CGCG 46-2 GX 89
CGCG 46-5 GX 89
CGCG 46-12 GX 89, 109
CGCG 46-13 GX 89, 109
CGCG 46-17 GX 89
CGCG 46-25 GX 89
CGCG 46-28 GX 109
CGCG 46-33 GX 89, 109
CGCG 46-38 GX 109
CGCG 46-49 GX 109
CGCG 46-56 GX 89, 109
CGCG 46-62 GX 109
CGCG 46-68 GX 109
CGCG 46-81 GX 89
CGCG 46-91 GX 89
CGCG 47-14 GX 89, 109
CGCG 47-15 GX 89, 109
CGCG 47-25 GX 109
CGCG 47-35 GX 109
CGCG 47-36 GX 109
CGCG 47-37 GX 89
CGCG 47-39 GX 89
CGCG 47-46 GX 109
CGCG 47-49 GX 109
CGCG 47-50 GX 89
CGCG 47-56 GX 89, 109
CGCG 47-59 GX 109

CGCG 47-60 GX 89
CGCG 47-94 GX 109
CGCG 47-97 GX 109
CGCG 47-100 GX 109
CGCG 47-116 GX 89, 109
CGCG 47-142 GX 89
CGCG 47-147 GX 109
CGCG 48-8 GX 109
CGCG 48-19 GX 89
CGCG 48-20 GX 109
CGCG 48-22 GX 89
CGCG 48-43 GX 89
CGCG 48-44 GX 109
CGCG 48-55 GX 89
CGCG 48-70 GX 89, 109
CGCG 48-83 GX 108, 109
CGCG 48-100 GX 88
CGCG 48-115 GX 108
CGCG 49-1 GX 88, 108
CGCG 49-3 GX 88, 108
CGCG 49-21 GX 88
CGCG 49-33 GX 88
CGCG 49-39 GX 88, 108
CGCG 49-46 GX 88
CGCG 49-75 GX 88, 108
CGCG 49-115 GX 88, 108
CGCG 49-148 GX 88, 108
CGCG 49-150 GX 88, 108
CGCG 49-155 GX 108
CGCG 49-178 GX 108
CGCG 50-15 GX 88
CGCG 50-22 GX 88
CGCG 50-25 GX 88, 108
CGCG 50-40 GX 108
CGCG 50-43 GX 108
CGCG 50-49 GX 108
CGCG 50-50 GX 108
CGCG 50-59 GX 88, 108
CGCG 50-68 GX 88, 108
CGCG 50-75 GX 108
CGCG 50-83 GX 108
CGCG 50-87 GX 108
CGCG 50-91 GX 108
CGCG 50-113 GX 88
CGCG 50-116 GX 108
CGCG 51-5 GX 88
CGCG 51-17 GX 108
CGCG 51-46 GX 108
CGCG 51-59 GX 107, 108
CGCG 51-66 GX 107, 108
CGCG 51-69 GX 107, 108
CGCG 52-4 GX 87, 107
CGCG 52-19 GX 87
CGCG 52-21 GX 87, 107
CGCG 52-29 GX 87
CGCG 52-35 GX 107
CGCG 52-38 GX 87, 107
CGCG 52-44 GX 87
CGCG 52-51 GX 87
CGCG 53-2 GX 87
CGCG 53-4 GX 87
CGCG 53-30 GX 107
CGCG 54-11 GX 87
CGCG 54-13 GX 87
CGCG 54-27 GX 86, 87
CGCG 55-2 GX 86
CGCG 55-4 GX 86
CGCG 55-20 GX 86, 106
CGCG 56-1 GX 86
CGCG 56-2 GX 86, 106
CGCG 56-4 GX 106
CGCG 56-8 GX 86
CGCG 57-16 GX 95
CGCG 57-19 GX 95
CGCG 58-16 GX 95
CGCG 58-33 GX 94, 95
CGCG 58-43 GX 94, 95
CGCG 58-45 GX 94, 95
CGCG 58-69 GX 94
CGCG 59-1 GX 94
CGCG 59-17 GX 94
CGCG 59-27 GX 94
CGCG 59-32 GX 94
CGCG 59-33 GX 94
CGCG 59-34 GX 94
CGCG 59-36 GX 94
CGCG 59-37 GX 94
CGCG 59-42 GX 94
CGCG 59-45 GX 94
CGCG 59-49 GX 94
CGCG 59-50 GX 94
CGCG 59-57 GX 94
CGCG 60-3 GX 94
CGCG 60-5 GX 94
CGCG 60-10 GX 94
CGCG 60-13 GX 94
CGCG 60-26 GX 94
CGCG 60-30 GX 94
CGCG 61-2 GX 94
CGCG 61-17 GX 94
CGCG 61-31 GX 93, 94
CGCG 61-32 GX 93, 94
CGCG 61-44 GX 93, 94
CGCG 62-22 GX 93
CGCG 62-29 GX 93
CGCG 62-30 GX 93
CGCG 63-6 GX 93
CGCG 63-21 GX 93
CGCG 63-27 GX 93
CGCG 63-32 GX 93
CGCG 63-34 GX 93
CGCG 63-61 GX 93
CGCG 63-62 GX 93
CGCG 63-79 GX 93
CGCG 63-83 GX 93
CGCG 64-9 GX 93
CGCG 64-18 GX 93
CGCG 64-19 GX 93
CGCG 64-20 GX 93
CGCG 64-26 GX 93
CGCG 64-27 GX 93
CGCG 64-28 GX 93
CGCG 64-58 GX 93
CGCG 64-60 GX 93
CGCG 64-85 GX 92, 93
CGCG 64-101 GX 92
CGCG 64-106 GX 92
CGCG 65-18 GX 92
CGCG 65-24 GX 92
CGCG 65-62 GX 92
CGCG 65-69 GX 92
CGCG 65-73 GX 92
CGCG 65-78 GX 92
CGCG 65-82 GX 92
CGCG 66-1 GX 92
CGCG 66-23 GX 92
CGCG 66-28 GX 92
CGCG 66-36 GX 92
CGCG 66-47 GX 92
CGCG 66-62 GX 92
CGCG 66-68 GX 92
CGCG 66-71 GX 92
CGCG 66-94 GX 92
CGCG 66-104 GX 92
CGCG 66-107 GX 92
CGCG 67-7 GX 92
CGCG 67-9 GX 92
CGCG 67-18 GX 92
CGCG 68-76 GX 91
CGCG 68-81 GX 91
CGCG 69-29 GX 91
CGCG 69-120 GX 91
CGCG 71-100 GX 90
CGCG 72-7 GX 90
CGCG 72-13 GX 90
CGCG 72-15 GX 90
CGCG 72-18 GX 90
CGCG 72-23 GX 90
CGCG 72-41 GX 90
CGCG 73-10 GX 90
CGCG 73-29 GX 90
CGCG 73-71 GX 89, 90
CGCG 73-72 GX 89, 90
CGCG 74-13 GX 89
CGCG 74-20 GX 89
CGCG 74-28 GX 89
CGCG 74-30 GX 89
CGCG 74-43 GX 89
CGCG 74-64 GX 89
CGCG 74-76 GX 89
CGCG 74-77 GX 89
CGCG 74-78 GX 89
CGCG 74-94 GX 89
CGCG 74-103 GX 89
CGCG 74-124 GX 89
CGCG 74-126 GX 89
CGCG 74-129 GX 89
CGCG 74-131 GX 89
CGCG 74-135 GX 89
CGCG 74-136 GX 89
CGCG 74-159 GX 89
CGCG 75-34 GX 89
CGCG 75-41 GX 89
CGCG 75-56 GX 89
CGCG 75-74 GX 89
CGCG 75-81 GX 89
CGCG 75-108 GX 89
CGCG 75-110 GX 89
CGCG 76-10 GX 89
CGCG 76-16 GX 89
CGCG 76-34 GX 89
CGCG 76-41 GX 89
CGCG 76-43 GX 89
CGCG 76-58 GX 89
CGCG 76-77 GX 89
CGCG 76-78 GX 89
CGCG 76-80 GX 89
CGCG 76-84 GX 89
CGCG 76-91 GX 89
CGCG 76-92 GX 89
CGCG 76-97 GX 89
CGCG 76-111 GX 88, 89
CGCG 76-133 GX 88, 89
CGCG 77-11 GX 88
CGCG 77-25 GX 88
CGCG 77-26 GX 88
CGCG 77-55 GX 88
CGCG 77-59 GX 88
CGCG 77-63 GX 88
CGCG 77-74 GX 88
CGCG 77-78 GX 88
CGCG 77-125 GX 88
CGCG 77-130 GX 88
CGCG 78-64 GX 88
CGCG 78-80 GX 88
CGCG 79-8 GX 88
CGCG 79-9 GX 88
CGCG 79-18 GX 88
CGCG 79-37 GX 88
CGCG 79-40 GX 88
CGCG 79-47 GX 88
CGCG 79-48 GX 88
CGCG 79-62 GX 87, 88
CGCG 79-85 GX 87, 88
CGCG 79-89 GX 87
CGCG 80-17 GX 87
CGCG 80-19 GX 87
CGCG 80-32 GX 87
CGCG 80-46 GX 87
CGCG 81-18 GX 87
CGCG 82-15 GX 87
CGCG 83-11 GX 86
CGCG 85-11 GX 95
CGCG 85-23 GX 76
CGCG 87-4 GX 75, 95
CGCG 87-21 GX 95
CGCG 87-25 GX 75
CGCG 88-9 GX 94
CGCG 89-16 GX 75, 94
CGCG 90-4 GX 74
CGCG 90-33 GX 74, 94
CGCG 90-67 GX 93
CGCG 91-55 GX 74
CGCG 91-72 GX 93
CGCG 91-91 GX 74
CGCG 92-35 GX 93
CGCG 92-37 GX 93
CGCG 92-45 GX 93
CGCG 92-46 GX 74
CGCG 92-58 GX 74, 93
CGCG 92-59 GX 93
CGCG 92-61 GX 74
CGCG 92-62 GX 74, 93
CGCG 92-72 GX 93
CGCG 93-13 GX 93
CGCG 93-15 GX 93
CGCG 93-18 GX 93
CGCG 93-19 GX 93
CGCG 93-57 GX 92, 93
CGCG 93-59 GX 92, 93
CGCG 94-39 GX 73, 92
CGCG 94-56 GX 92
CGCG 94-98 GX 73
CGCG 94-105 GX 73
CGCG 95-51 GX 73
CGCG 95-70 GX 73, 92
CGCG 95-91 GX 92
CGCG 95-104 GX 73, 92
CGCG 95-106 GX 92
CGCG 96-28 GX 92
CGCG 97-11 GX 72
CGCG 97-35 GX 72, 91
CGCG 97-36 GX 72, A11
CGCG 97-138 GX 72, A11
CGCG 97-167 GX 91
CGCG 97-179 GX 91
CGCG 99-16 GX 72, 91
CGCG 101-31 GX 90
CGCG 101-51 GX 71
CGCG 102-41 GX 71
CGCG 102-58 GX 71, 89, 90
CGCG 102-67 GX 71
CGCG 103-55 GX 89
CGCG 103-65 GX 89
CGCG 103-86 GX 70, 89
CGCG 103-87 GX 89
CGCG 103-120 GX 89
CGCG 104-44 GX 70
CGCG 104-65 GX 89
CGCG 105-34 GX 89
CGCG 105-91 GX 88, 89
CGCG 106-26 GX 69, 70
CGCG 106-33 GX 88
CGCG 107-53 GX 69
CGCG 108-46 GX 69, A9
CGCG 108-59 GX 88, A9
CGCG 108-81 GX 88, A9
CGCG 108-117 GX 88, A9
CGCG 108-138 GX 69, A9
CGCG 108-165 GX 69, A9
CGCG 109-35 GX 68, 69, 87
CGCG 110-9 GX 68
CGCG 110-10 GX 68
CGCG 110-12 GX 87
CGCG 110-18 GX 68
CGCG 112-41 GX 68
CGCG 113-1 GX 86
CGCG 113-24 GX 67
CGCG 115-6 GX 76
CGCG 115-14 GX 76
CGCG 116-2 GX 76
CGCG 116-3 GX 76
CGCG 116-24 GX 76
CGCG 117-6 GX 75, 76
CGCG 117-33 GX 75, 76
CGCG 117-41 GX 75
CGCG 117-65 GX 75
CGCG 118-3 GX 75
CGCG 118-5 GX 75
CGCG 118-14 GX 75
CGCG 118-18 GX 75
CGCG 118-21 GX 75
CGCG 118-40 GX 75
CGCG 119-51 GX 75
CGCG 121-61 GX 74
CGCG 122-6 GX 74
CGCG 122-13 GX 74
CGCG 122-37 GX 74
CGCG 122-56 GX 74
CGCG 122-73 GX 74
CGCG 122-80 GX 74
CGCG 122-85 GX 74
CGCG 123-9 GX 73, 74
CGCG 123-31 GX 73
CGCG 124-54 GX 73
CGCG 125-5 GX 73
CGCG 125-30 GX 73
CGCG 126-14 GX 73
CGCG 126-101 GX 72
CGCG 127-55 GX 72, A11
CGCG 128-16 GX 72, A10
CGCG 131-4 GX 71
CGCG 132-9 GX 71
CGCG 132-36 GX 71
CGCG 132-64 GX 70, 71
CGCG 132-69 GX 70, 71
CGCG 133-4 GX 70
CGCG 134-24 GX 70
CGCG 134-33 GX 70
CGCG 134-38 GX 70
CGCG 134-57 GX 70
CGCG 134-63 GX 70
CGCG 135-13 GX 70
CGCG 135-36 GX 70
CGCG 136-12 GX 69
CGCG 136-36 GX 69
CGCG 136-37 GX 69
CGCG 136-43 GX 69
CGCG 136-50 GX 69
CGCG 136-62 GX 69
CGCG 136-88 GX 69
CGCG 136-99 GX 69
CGCG 137-47 GX 69
CGCG 137-59 GX 69
CGCG 137-60 GX 69
CGCG 138-2 GX 69
CGCG 138-7 GX 69
CGCG 138-12 GX 69
CGCG 138-16 GX 69
CGCG 138-17 GX 69
CGCG 138-50 GX 68, 69
CGCG 138-65 GX 68
CGCG 138-69 GX 68
CGCG 139-22 GX 68
CGCG 140-25 GX 68
CGCG 141-8 GX 68
CGCG 142-19 GX 67
CGCG 143-14 GX 67
CGCG 145-6 GX 58, 76
CGCG 146-3 GX 58
CGCG 146-40 GX 76
CGCG 147-10 GX 57, 58
CGCG 147-13 GX 57, 58, 75, 76
CGCG 147-31 GX 57
CGCG 147-48 GX 57, 75
CGCG 147-51 GX 57, 75
CGCG 148-6 GX 75
CGCG 148-9 GX 57
CGCG 148-19 GX 57, 75
CGCG 148-21 GX 57
CGCG 148-39 GX 75
CGCG 148-115 GX 57, 75
CGCG 148-117 GX 75
CGCG 149-19 GX 57, 75
CGCG 149-34 GX 75
CGCG 149-50 GX 75
CGCG 150-5 GX 74, 75
CGCG 150-14 GX 56, 57, 74, 75
CGCG 150-30 GX 74
CGCG 150-43 GX 56, 74
CGCG 150-48 GX 56, 74
CGCG 150-50 GX 56, 74
CGCG 150-56 GX 56
CGCG 152-10 GX 56
CGCG 152-70 GX 56
CGCG 154-21 GX 73
CGCG 155-46 GX 73
CGCG 155-78 GX 55
CGCG 156-27 GX 55
CGCG 156-31 GX 73
CGCG 156-84 GX 54
CGCG 160-30 GX 53
CGCG 160-40 GX 71, A8
CGCG 160-73 GX 71, A8
CGCG 160-141 GX 71
CGCG 160-151 GX 53, 71
CGCG 160-172 GX 53
CGCG 160-175 GX 53
CGCG 160-185 GX 53
CGCG 160-190 GX 53
CGCG 161-19 GX 53
CGCG 161-85 GX 71
CGCG 161-103 GX 53
CGCG 161-138 GX 71
CGCG 162-29 GX 52, 53, 70, 71
CGCG 163-14 GX 52
CGCG 163-17 GX 70
CGCG 163-32 GX 70
CGCG 163-70 GX 70
CGCG 164-49 GX 70
CGCG 165-13 GX 52
CGCG 165-54 GX 51, 69
CGCG 166-5 GX 69
CGCG 166-23 GX 51, 69
CGCG 166-64 GX 69
CGCG 167-1 GX 51
CGCG 167-55 GX 51, 69
CGCG 167-64 GX 69
CGCG 169-7 GX 50
CGCG 169-8 GX 68
CGCG 169-12 GX 50, 68
CGCG 170-5 GX 50, 68
CGCG 170-10 GX 50
CGCG 172-4 GX 67
CGCG 172-40 GX 49, 67
CGCG 173-21 GX 49
CGCG 178-28 GX 57
CGCG 182-20 GX 56
CGCG 186-30 GX 54
CGCG 186-63 GX 54
CGCG 190-4 GX 53
CGCG 196-50 GX 51, A3
CGCG 198-32 GX 50
CGCG 199-5 GX 50
CGCG 199-26 GX 50
CGCG 202-8 GX 48, 49
CGCG 203-6 GX 41
CGCG 206-29 GX 40
CGCG 211-34 GX 39
CGCG 215-17 GX 54
CGCG 215-50 GX 37, 54
CGCG 221-10 GX 36
CGCG 221-50 GX 35
CGCG 229-24 GX 33
CGCG 230-20 GX 33
CGCG 230-21 GX 33
CGCG 233-2 GX 41
CGCG 233-5 GX 41
CGCG 233-11 GX 41
CGCG 234-31 GX 41
CGCG 234-39 GX 41
CGCG 234-105 GX 40, 41
CGCG 236-13 GX 40
CGCG 237-18 GX 39, 40
CGCG 237-21 GX 39, 40
CGCG 240-22 GX 39
CGCG 241-2 GX 38
CGCG 241-13 GX 38
CGCG 241-15 GX 38
CGCG 241-36 GX 38
CGCG 241-60 GX 38
CGCG 241-74 GX 38
CGCG 244-27 GX 37
CGCG 247-15 GX 36
CGCG 250-18 GX 35
CGCG 251-19 GX 35
CGCG 251-36 GX 35
CGCG 252-16 GX 34
CGCG 253-17 GX 34
CGCG 253-26 GX 34
CGCG 254-21 GX 34
CGCG 257-2 GX 33
CGCG 257-10 GX 33
CGCG 257-16 GX 33
CGCG 257-17 GX 33
CGCG 257-19 GX 33
CGCG 257-29 GX 32, 33
CGCG 257-35 GX 32, 33
CGCG 258-1 GX 27, 42
CGCG 258-9 GX 27, 41, 42
CGCG 258-10 GX 27, 41, 42
CGCG 259-8 GX 27, 41
CGCG 260-20 GX 27
CGCG 260-37 GX 27, 41
CGCG 261-65 GX 26
CGCG 262-21 GX 26
CGCG 262-54 GX 26
CGCG 263-43 GX 26
CGCG 263-70 GX 26
CGCG 264-86 GX 25
CGCG 264-90 GX 25
CGCG 265-17 GX 39
CGCG 265-23 GX 25
CGCG 266-3 GX 25
CGCG 266-9 GX 25
CGCG 266-31 GX 25
CGCG 266-46 GX 25
CGCG 266-48 GX 25
CGCG 267-15 GX 25
CGCG 268-6 GX 24
CGCG 270-24 GX 24
CGCG 272-32 GX 23
CGCG 274-45 GX 22
CGCG 279-3 GX 21
CGCG 282-3 GX 20
CGCG 283-4 GX 16, 27
CGCG 287-31 GX 26
CGCG 292-85 GX 24
CGCG 298-22 GX 22
CGCG 299-48 GX 21, 22
CGCG 300-14 GX 11, 21
CGCG 307-10 GX 16
CGCG 307-15 GX 16
CGCG 307-16 GX 16
CGCG 309-28 GX 15
CGCG 309-35 GX 15
CGCG 312-32 GX 14, 25
CGCG 314-4 GX 13
CGCG 314-19 GX 13
CGCG 315-17 GX 13
CGCG 329-11 GX 16
CGCG 331-49 GX 14
CGCG 332-23 GX 6, 14
CGCG 332-26 GX 14
CGCG 332-33 GX 6, 14
CGCG 333-59 GX 13, 14
CGCG 334-50 GX 13
CGCG 335-28 GX 13
CGCG 337-28 GX 12
CGCG 339-6 GX 11
CGCG 340-27 GX 10, 11
CGCG 341-9 GX 3, 10
CGCG 341-10 GX 3, 10
CGCG 352-36 GX 5
CGCG 368-2 GX 3
CGCG 372-6 GX 104
CGCG 373-40 GX 104
CGCG 374-9 GX 104
CGCG 374-29 GX 104
CGCG 375-3 GX 103
CGCG 375-14 GX 103
CGCG 375-16 GX 103
CGCG 375-30 GX 103
CGCG 375-33 GX 103

CGCG 376-3 GX 103
CGCG 376-7 GX 103
CGCG 376-21 GX 103
CGCG 376-33 GX 103
CGCG 376-41 GX 103
CGCG 380-9 GX 102
CGCG 381-2 GX 101
CGCG 381-57 GX 101
CGCG 382-39 GX 101
CGCG 383-72 GX 120
CGCG 384-74 GX 120
CGCG 384-75 GX 120
CGCG 385-36 GX 120
CGCG 385-82 GX 120, A16
CGCG 385-87 GX 120
CGCG 385-161 GX 120
CGCG 386-17 GX 120
CGCG 386-31 GX 120
CGCG 386-37 GX 119, 120
CGCG 387-20 GX 119
CGCG 388-96 GX 119
CGCG 389-18 GX 119
CGCG 389-27 GX 119
CGCG 389-80 GX 118
CGCG 391-7 GX 118
CGCG 391-8 GX 118
CGCG 391-10 GX 118
CGCG 391-13 GX 118
CGCG 391-15 GX 118
CGCG 391-17 GX 118
CGCG 391-21 GX 118
CGCG 391-26 GX 118
CGCG 391-27 GX 118
CGCG 391-43 GX 118
CGCG 393-7 GX 117
CGCG 393-24 GX 117
CGCG 393-54 GX 117
CGCG 395-16 GX 117
CGCG 398-2 GX 84
CGCG 398-3 GX 84
CGCG 399-3 GX 104
CGCG 399-13 GX 104
CGCG 399-16 GX 104
CGCG 399-26 GX 84
CGCG 399-28 GX 84, 104
CGCG 400-12 GX 83, 84
CGCG 400-17 GX 83, 84
CGCG 400-27 GX 83
CGCG 403-5 GX 83
CGCG 403-19 GX 103
CGCG 404-4 GX 82
CGCG 404-13 GX 82
CGCG 405-33 GX 82
CGCG 406-13 GX 82
CGCG 406-27 GX 102
CGCG 406-34 GX 102
CGCG 406-45 GX 82
CGCG 406-58 GX 82
CGCG 406-122 GX 101
CGCG 407-5 GX 81
CGCG 407-27 GX 81
CGCG 407-40 GX 101
CGCG 407-68 GX 81
CGCG 409-1 GX 81
CGCG 409-37 GX 81, 100, 101, 120
CGCG 410-1 GX 100
CGCG 410-2 GX 100, 120
CGCG 411-14 GX 100, 120
CGCG 411-22 GX 120
CGCG 414-40 GX 99
CGCG 415-4 GX 99, 119
CGCG 415-6 GX 99, 119
CGCG 415-40 GX 98, 99, 118, 119
CGCG 415-50 GX 98, 99
CGCG 417-5 GX 118
CGCG 419-15 GX 97
CGCG 419-17 GX 97
CGCG 419-20 GX 97
CGCG 420-14 GX 97
CGCG 420-15 GX 117
CGCG 420-21 GX 117
CGCG 420-32 GX 117
CGCG 421-30 GX 96, 97, 116, 117
CGCG 423-2 GX 84
CGCG 424-2 GX 84
CGCG 424-27 GX 84
CGCG 424-33 GX 84
CGCG 425-7 GX 84
CGCG 425-36 GX 83, 84
CGCG 426-27 GX 83
CGCG 426-33 GX 83
CGCG 427-7 GX 83
CGCG 427-19 GX 83
CGCG 428-29 GX 83
CGCG 428-63 GX 82, 83
CGCG 428-65 GX 82
CGCG 429-22 GX 82
CGCG 432-2 GX 81
CGCG 435-40 GX 100
CGCG 436-16 GX 100
CGCG 436-34 GX 100
CGCG 438-2 GX 99
CGCG 438-40 GX 99
CGCG 441-2 GX 98
CGCG 447-3 GX 84
CGCG 447-5 GX 66
CGCG 448-33 GX 83, 84
CGCG 449-2 GX 65
CGCG 453-38 GX 64, 82
CGCG 453-65 GX 64
CGCG 454-1 GX 64
CGCG 454-71 GX 63
CGCG 456-5 GX 63
CGCG 456-6 GX 63
CGCG 456-37 GX 63
CGCG 459-10 GX 80
CGCG 459-16 GX 100
CGCG 459-29 GX 80
CGCG 459-40 GX 80
CGCG 459-55 GX 100
CGCG 459-64 GX 100
CGCG 459-66 GX 80
CGCG 460-18 GX 80
CGCG 461-14 GX 79, 80
CGCG 461-15 GX 79, 80
CGCG 461-70 GX 79, 99
CGCG 462-23 GX 79
CGCG 464-14 GX 98
CGCG 465-3 GX 78
CGCG 466-3 GX 78
CGCG 471-4 GX 65
CGCG 475-41 GX 64
CGCG 475-59 GX 63, 64
CGCG 476-30 GX 63
CGCG 476-34 GX 63
CGCG 476-55 GX 63
CGCG 476-66 GX 63
CGCG 476-96 GX 63
CGCG 476-120 GX 63
CGCG 476-122 GX 63
CGCG 478-2 GX 63
CGCG 478-21 GX 63
CGCG 479-54 GX 63, 80
CGCG 482-35 GX 79, 80
CGCG 482-55 GX 79, 80
CGCG 493-8 GX 47, 65
CGCG 493-15 GX 46, 47
CGCG 496-50 GX 46, 64
CGCG 497-7 GX 45, 46
CGCG 497-42 GX 45
CGCG 499-38 GX 45
CGCG 499-40 GX 45
CGCG 499-95 GX 45, 63
CGCG 499-97 GX 63
CGCG 499-98 GX 45, 63
CGCG 500-13 GX 45
CGCG 500-15 GX 45, 63
CGCG 500-32 GX 45
CGCG 500-33 GX 45
CGCG 500-48 GX 45, 63
CGCG 500-65 GX 45, 62
CGCG 500-74 GX 45, 62, 63, 80
CGCG 500-80 GX 62
CGCG 501-15 GX 62
CGCG 501-18 GX 80
CGCG 501-56 GX 62
CGCG 501-58 GX 62
CGCG 502-43 GX 62, A7
CGCG 502-44 GX 62, A7
CGCG 502-84 GX 62, A7
CGCG 504-2 GX 79
CGCG 504-22 GX 61
CGCG 504-69 GX 79
CGCG 504-101 GX 61
CGCG 505-17 GX 61, 79
CGCG 505-19 GX 61
CGCG 513-10 GX 46
CGCG 514-36 GX 46
CGCG 514-79 GX 46
CGCG 520-8 GX 62, A7
CGCG 521-2 GX 62
CGCG 521-74 GX 62, A6
CGCG 522-2 GX 62, A6
CGCG 522-19 GX 62
CGCG 522-28 GX 62
CGCG 523-2 GX 61
CGCG 524-13 GX 61
CGCG 524-41 GX 61
CGCG 524-54 GX 61
CGCG 531-4 GX 31
CGCG 538-41 GX 44
CGCG 538-43 GX 43, 44
CGCG 539-37 GX 43
CGCG 540-19 GX 43
CGCG 540-59 GX 43, A4
CGCG 540-62 GX 43, A4
CGCG 540-79 GX 43, A4
CGCG 541-11 GX 43, 60
CGCG 548-1 GX 30
CGCG 551-15 GX 44
CGCG 552-5 GX 44
CGCG 554-16 GX 43
CGCG 554-17 GX 43
ClvdB 1 OC 115, 116
ClvdB 83 OC 154
ClvdB 85 OC 115
ClvdB 92 OC 135
ClvdB 113 OC 145, A17
ClvdB 130 OC 48, A2
ClvdB 152 OC 9
Cr 21 OC 80
Cr 34 OC 28, 29
Cr 62 OC 42
Cr 69 OC 96
Cr 74 OC 96
Cr 89 OC 76
Cr 91 OC 116
Cr 92 OC 96, 116
Cr 95 OC 96
Cr 96 OC 116
Cr 97 OC 96, 116
Cr 104 OC 115, 116
Cr 106 OC 95, 96, 115, 116
Cr 107 OC 115, 116
Cr 110 OC 115, 116
Cr 111 OC 95, 96
Cr 115 OC 115
Cr 121 OC 154
Cr 132 OC 172
Cr 135 OC 171, 172
Cr 140 OC 171
Cr 185 OC 171
Cr 187 OC 153, 171
Cr 196 OC 170
Cr 197 OC 186, 187
Cr 198 OC 170
Cr 213 OC 186
Cr 220 OC 199
Cr 223 OC 199
Cr 228 OC 199
Cr 232 OC 199
Cr 234 OC 199
Cr 236 OC 198, 199
Cr 240 OC 198
Cr 249 OC 209
Cr 269 OC 208
Cr 271 OC 208
Cr 272 OC 197
Cr 275 OC 197
Cr 277 OC 208
Cr 292 OC 196
Cr 307 OC 182
Cr 316 OC 164, 181, A22
Cr 332 OC 164
Cr 333 OC 164
Cr 338 OC 164
Cr 345 OC 164, A20
Cr 347 OC 146, 164
Cr 350 OC 106
Cr 351 OC 146
Cr 367 OC 145, A17
Cr 371 OC 145
Cr 399 OC 66
Cr 401 OC 105
Cr 416 OC 66
Cr 419 OC 32, 48, A2
Cr 421 OC 32, A2
Cr 427 OC 9
Cr 428 OC 32, A1
Cr 463 OC 17
Cr 464 OC 7, 16
Cr 465 OC 135
Cr 466 OC 135
Cr 469 OC 145
Cr 470 OC 31
Cz 1 OC 18
Cz 2 OC 18
Cz 3 OC 8, 18, 29
Cz 4 OC 29
Cz 5 OC 29
Cz 6 OC 17, 29
Cz 7 OC 17, 29
Cz 8 OC 29
Cz 9 OC 29
Cz 10 OC 29
Cz 12 OC 29
Cz 13 OC 17, 29
Cz 15 OC 28
Cz 17 OC 28
Cz 18 OC 60
Cz 19 OC 77
Cz 20 OC 59
Cz 21 OC 59
Cz 23 OC 77
Cz 24 OC 77
Cz 26 OC 116
Cz 28 OC 115
Cz 29 OC 135
Cz 30 OC 135
Cz 31 OC 153
Cz 32 OC 153, 171
Cz 37 OC 146
Cz 38 OC 105
Cz 40 OC 66
Cz 41 OC 66
Cz 42 OC 19
Cz 43 OC 18
Danks 1 OC 197, 208, 209
Danks 2 OC 197, 208, 209
Djorg 1 GC 164, A20
Do 1 OC 48
Do 2 OC 32
Do 3 OC 48
Do 4 OC 48
Do 5 OC 48, A2
Do 6 OC 32, A2
Do 8 OC 32, A2
Do 9 OC 32, A2
Do 10 OC 32, 48, A2
Do 11 OC 32, A2
Do 13 OC 8
Do 14 OC 78
Do 15 OC 59
Do 16 OC 59
Do 17 OC 96, 97
Do 18 OC 59
Do 19 OC 96, 97
Do 20 OC 59
Do 21 OC 96, 97
Do 22 OC 116
Do 23 OC 115
Do 24 OC 115
Do 25 OC 115
Do 26 OC 95
Do 27 OC 127
Do 28 OC 126
Do 29 OC 126
Do 30 OC 125, 126
Do 31 OC 125, 126
Do 32 OC 105
Do 35 OC 85
Do 36 OC 32, 33
Do 37 OC 48
Do 38 OC 32
Do 39 OC 48, A2
Do 40 OC 48, A2
Do 41 OC 48, A2
Do 42 OC 48, A2
Do 43 OC 48, A2
Do 44 OC 32, A2
Do 45 OC 47
Do 46 OC 18
Do 47 OC 47, 48
DoDz 1 OC 79, 99
DoDz 2 OC 96, 97
DoDz 3 OC 77
DoDz 4 OC 77
DoDz 5 OC 51, A3
DoDz 6 OC 50
DoDz 7 OC 87
DoDz 8 OC 68
DoDz 9 OC 49
DoDz 10 OC 32, 48
DoDz 11 OC 47
DWB 100-5-9 BN 32, A2
E 3 GC 218
ESO 1-6 GX 220
ESO 2-10 GX 214
ESO 2-11 GX 220
ESO 2-12 GX 220
ESO 3-1 GX 220
ESO 3-3 GX 214
ESO 3-4 GX 214, 219
ESO 3-7 GX 214, 219
ESO 3-13 GX 219
ESO 3-14 GX 219
ESO 3-15 GX 219, 220
ESO 4-6 GX 219
ESO 4-10 GX 219, 220
ESO 4-11 GX 219, 220
ESO 4-13 GX 219
ESO 4-17 GX 220
ESO 4-19 GX 220
ESO 4-20 GX 220
ESO 5-4 GX 220
ESO 5-6 GX 218, 220
ESO 5-10 GX 218, 220
ESO 5-11 GX 220
ESO 6-2 GX 218, 220
ESO 6-3 GX 218
ESO 6-6 GX 220
ESO 7-1 GX 220
ESO 7-6 GX 217, 220
ESO 8-1 GX 217
ESO 8-4 GX 216, 217
ESO 8-6 OC 216
ESO 8-7 GX 216
ESO 8-8 GX 220
ESO 9-1 GX 216
ESO 9-10 GX 220
ESO 10-3 GX 215, 220
ESO 10-4 GX 220
ESO 10-5 GX 215
ESO 10-6 GX 215, 220
ESO 11-3 GX 215
ESO 11-4 GX 215
ESO 11-5 GX 214, 215
ESO 12-4 GX 214
ESO 12-10 GX 214
ESO 12-12 GX 214
ESO 12-14 GX 214
ESO 12-15 GX 214
ESO 12-21 GX 214
ESO 13-1 GX 214
ESO 13-9 GX 214
ESO 13-10 GX 214
ESO 13-12 GX 214
ESO 13-16 GX 214
ESO 13-18 GX 214
ESO 13-21 GX 214, 219
ESO 13-22 GX 214, 219
ESO 13-24 GX 214, 219
ESO 13-26 GX 214, 219
ESO 13-28 GX 214, 219
ESO 14-1 GX 219
ESO 14-3 GX 219
ESO 15-1 GX 219
ESO 15-5 GX 219
ESO 15-6 GX 219
ESO 15-8 GX 219
ESO 15-15 GX 219
ESO 15-18 GX 219
ESO 18-2 GX 218
ESO 18-13 GX 218
ESO 18-15 GX 218
ESO 19-3 GX 217
ESO 19-6 GX 217
ESO 21-2 GX 217
ESO 21-3 GX 217
ESO 21-4 GX 217
ESO 21-5 GX 216, 217
ESO 21-6 OC 216, 217
ESO 22-3 GX 216
ESO 22-10 GX 216
ESO 24-8 GX 215, 216
ESO 24-15 GX 215
ESO 24-19 GX 215
ESO 25-2 GX 215
ESO 26-1 GX 215
ESO 26-2 OC 215
ESO 26-5 GX 215
ESO 27-1 GX 214, 215
ESO 27-2 GX 214, 215
ESO 27-3 GX 214, 215
ESO 27-8 GX 214
ESO 27-9 GX 214
ESO 27-12A GX 214
ESO 27-14 GX 214
ESO 27-21 GX 214
ESO 27-24 GX 214
ESO 28-1 GX 204
ESO 28-2 GX 214
ESO 28-3 GX 214
ESO 28-7 GX 204, 214
ESO 28-9 GX 214
ESO 28-12 GX 204, 214
ESO 28-14 GX 214
ESO 30-8 GX 214, 219
ESO 30-9 GX 214, 219
ESO 30-11 GX 213, 214, 219
ESO 30-13 GX 214, 219
ESO 30-14 GX 214, 219
ESO 30-17 GX 219
ESO 30-19 GX 219
ESO 31-4 GX 213
ESO 31-5 GX 219
ESO 31-18 GX 212, 213
ESO 31-20 GX 219
ESO 33-2 GX 219
ESO 33-3 GX 212, 219
ESO 33-4 GX 219
ESO 33-5 GX 219
ESO 33-6 GX 219
ESO 33-7 GX 219
ESO 33-11 GX 212, 219
ESO 33-14 GX 212, 219
ESO 33-22 GX 212, 219
ESO 33-30 GX 212
ESO 34-9 GX 218
ESO 34-11 GX 218
ESO 34-11B GX 218
ESO 34-12 GX 211
ESO 34-13 GX 211, 218
ESO 34-14 GX 218
ESO 35-1 GX 211, 218
ESO 35-4 GX 218
ESO 35-5 GX 218
ESO 35-7 GX 218
ESO 35-9 GX 218
ESO 35-11 GX 218
ESO 35-17 GX 211
ESO 35-18 GX 218
ESO 36-5 GX 210, 218
ESO 36-6 GX 218
ESO 36-10 GX 210, 218
ESO 36-15 GX 210, 218
ESO 36-19 GX 218
ESO 37-2 GX 218
ESO 37-5 GX 217, 218
ESO 37-9 GX 217, 218
ESO 37-10 GX 217, 218
ESO 37-13 GX 210, 217, 218
ESO 38-4 GX 217
ESO 38-6 GX 217
ESO 39-2 GX 217
ESO 40-7 GX 217
ESO 40-12 GX 217
ESO 40-13 GX 217
ESO 41-5 GX 216, 217
ESO 41-6 GX 216
ESO 42-1 GX 216
ESO 42-7 GX 208, 216
ESO 43-13 OC 216
ESO 44-1 GX 207, 216
ESO 44-2 GX 216
ESO 44-9 GX 207, 216
ESO 44-10 GX 216
ESO 44-22 GX 215, 216
ESO 44-25 GX 206, 207, 215, 216
ESO 45-2 GX 215
ESO 45-3 GX 215
ESO 45-4 GX 206
ESO 45-5 GX 215
ESO 47-34 GX 215
ESO 48-2 GX 215
ESO 48-3 GX 215
ESO 48-7 GX 214, 215
ESO 48-10A GX 205
ESO 48-11 GX 205, 214, 215
ESO 48-13 GX 205, 214, 215
ESO 48-14 GX 205, 214, 215
ESO 48-17 GX 205, 214, 215
ESO 48-20 GX 214, 215
ESO 48-21 GX 205, 214, 215
ESO 48-22 GX 205
ESO 50-2 GX 204
ESO 50-6 GX 204
ESO 51-11 GX 204
ESO 52-1 GX 213
ESO 52-4 GX 213
ESO 52-8 GX 213
ESO 52-14 GX 213
ESO 52-17 GX 213
ESO 52-20 GX 213
ESO 52-21 GX 213
ESO 53-2 GX 213
ESO 53-23 GX 213
ESO 54-21 GX 212
ESO 55-18 GX 212
ESO 55-23 GX 212

ESO 55-29 GX 212, A24
ESO 55-33 GX 212, A24
ESO 55-35 GX 212, A24
ESO 56-104 GX 212, A24, A25
ESO 56-115 GX 212, A24, A25
ESO 56-15 GX 212, A24
ESO 56-154 GX 212, A24
ESO 56-19 GX 212, A24
ESO 57-73 GX 211, A24
ESO 57-73A GX 211, A24
ESO 57-80 GX 211
ESO 58-3 GX 211
ESO 58-4 GX 211
ESO 58-5 GX 211
ESO 58-9 GX 211
ESO 58-13 GX 211
ESO 58-14 GX 211
ESO 58-19 GX 211
ESO 58-25 GX 211
ESO 59-1 GX 211
ESO 59-6 GX 211
ESO 59-7 GX 211
ESO 59-9 GX 211
ESO 59-11 GX 211
ESO 59-12 GX 211
ESO 59-16 GX 211
ESO 59-17 GX 211
ESO 59-19 GX 211
ESO 59-23 GX 211
ESO 59-25 GX 211
ESO 59-27 GX 211
ESO 60-4 GX 210, 211
ESO 60-10 GX 210
ESO 60-11 GX 210
ESO 60-12 GX 210
ESO 60-15 GX 210
ESO 60-18 GX 210
ESO 60-19 GX 210
ESO 60-24 GX 210
ESO 60-26 GX 210
ESO 60-27 GX 210
ESO 61-5 GX 210
ESO 61-7 GX 210
ESO 61-8 GX 210
ESO 61-9 GX 210
ESO 61-11 GX 210
ESO 61-13 GX 210
ESO 61-15 GX 210
ESO 61-17 GX 210
ESO 61-20 GX 210
ESO 61-21 GX 210
ESO 62-8 OC 210
ESO 62-11 OC 210
ESO 63-11 GX 209
ESO 64-3 GX 209
ESO 64-5 OC 209
ESO 65-3 OC 209
ESO 65-7 OC 208
ESO 68-2 GX 208
ESO 68-11 GX 207
ESO 68-12 GX 207
ESO 68-13 GX 207
ESO 68-16 GX 207
ESO 69-6 GX 207
ESO 69-9 GX 207
ESO 70-9 GX 207
ESO 70-13 GX 207
ESO 71-4 GX 206
ESO 71-10 GX 206
ESO 71-13 GX 206
ESO 71-18 GX 206
ESO 72-6A GX 206
ESO 72-13 GX 206
ESO 72-17 GX 206
ESO 73-1 GX 206
ESO 73-5 GX 206
ESO 73-6 GX 206
ESO 73-13 GX 206
ESO 73-18 GX 206
ESO 73-19 GX 206
ESO 74-4 GX 205
ESO 74-7 GX 205
ESO 74-9A GX 205
ESO 74-20 GX 205
ESO 74-25 GX 205
ESO 75-5 GX 205
ESO 75-6 GX 205
ESO 75-28 GX 205
ESO 75-32 GX 205
ESO 75-39 GX 205
ESO 75-41 GX 205
ESO 75-43 GX 205
ESO 75-56 GX 205
ESO 76-4 GX 205
ESO 76-5 GX 205
ESO 76-11 GX 205
ESO 76-19 GX 205
ESO 76-21 GX 204, 205
ESO 76-22 GX 204, 205
ESO 76-30 GX 204, 205
ESO 76-31 GX 204, 205
ESO 77-2 GX 204
ESO 77-3 GX 204
ESO 77-4 GX 204
ESO 77-23 GX 204
ESO 77-28 GX 204
ESO 78-10 GX 204
ESO 78-22 GX 204
ESO 78-23 GX 192, 204
ESO 79-1 GX 192, 204
ESO 79-2 GX 204
ESO 79-3 GX 204
ESO 79-5 GX 204
ESO 79-7 GX 204
ESO 79-8 GX 192, 204
ESO 79-10 GX 204
ESO 79-13 GX 204
ESO 79-13A GX 204
ESO 79-15 GX 203, 204, 213
ESO 79-16 GX 204, 213
ESO 80-1 GX 203, 204, 213
ESO 80-3 GX 213
ESO 80-4 GX 203, 213
ESO 80-6 GX 203, 213
ESO 81-1 GX 203, 213
ESO 81-6 GX 213
ESO 81-7 GX 213
ESO 81-8 GX 213
ESO 82-2 GX 203, 213
ESO 82-6 GX 202, 203, 213
ESO 82-10 GX 213
ESO 83-6 GX 202, 212, 213
ESO 83-7 GX 212, 213
ESO 83-10 GX 212
ESO 83-12 GX 212
ESO 84-9 GX 202, 212
ESO 84-11 GX 212
ESO 84-14 GX 212
ESO 84-15 GX 212
ESO 84-18 GX 202, 212
ESO 84-20 GX 202, 212
ESO 84-21 GX 212
ESO 84-212 GX 212
ESO 84-22 GX 212
ESO 84-26 GX 212
ESO 84-27 GX 212
ESO 84-28 GX 202, 212
ESO 84-32 GX 212
ESO 84-33 GX 212
ESO 84-34 GX 212
ESO 84-35 GX 202, 212
ESO 84-40 GX 202, 212
ESO 85-1 GX 202, 212
ESO 85-2 GX 212
ESO 85-4 GX 202, 212
ESO 85-5 GX 212
ESO 85-14 GX 202, 212
ESO 85-24 GX 212
ESO 85-30 GX 212
ESO 85-34 GX 212
ESO 85-38 GX 212
ESO 85-47 GX 201, 212
ESO 85-88 GX 212
ESO 86-10 GX 212
ESO 86-18 GX 212
ESO 86-23 GX 212
ESO 86-52 GX 211, 212
ESO 86-56 GX 211, 212
ESO 86-62 GX 211
ESO 87-3 GX 211
ESO 87-10 GX 211
ESO 87-20 GX 211
ESO 87-28 GX 201, 211
ESO 87-29 GX 211
ESO 87-32 GX 211
ESO 87-34 GX 211
ESO 87-38 GX 211
ESO 87-42 GX 211
ESO 87-46 GX 211
ESO 87-49 GX 211
ESO 87-50 GX 211
ESO 87-54 GX 211
ESO 88-4 GX 211
ESO 88-9 GX 211
ESO 88-15 GX 200, 211
ESO 88-17 GX 211
ESO 88-18 GX 211
ESO 88-21 GX 211
ESO 88-22 GX 211
ESO 88-23 GX 211
ESO 88-26 GX 211
ESO 89-1 GX 211
ESO 89-9 GX 211
ESO 89-12 GX 211
ESO 89-13 GX 211
ESO 90-4 GX 210
ESO 90-9 GX 210
ESO 90-12 GX 210
ESO 90-14 GX 210
ESO 90-15 GX 210
ESO 91-3 GX 210
ESO 91-6 GX 210
ESO 91-7 GX 199, 210
ESO 91-11 GX 210
ESO 91-16 GX 210
ESO 91-18 GX 210
ESO 92-1 GX 210
ESO 92-3 GX 210
ESO 92-4 GX 210
ESO 92-6 GX 210
ESO 92-14 GX 210
ESO 92-15 GX 210
ESO 92-18 OC 210
ESO 92-21 GX 210
ESO 92-22 GX 210
ESO 93-3 GX 209
ESO 93-8 OC 209
ESO 97-12 GX 208
ESO 97-13 GX 208
ESO 99-2 GX 208
ESO 99-5 GX 208
ESO 99-6 OC 207, 208
ESO 99-8 GX 207
ESO 100-14 GX 207
ESO 100-22 GX 196, 207
ESO 100-23 GX 207
ESO 100-25 GX 207
ESO 101-1 GX 207
ESO 101-3 GX 207
ESO 101-4 GX 207
ESO 101-5 GX 207
ESO 101-7 GX 207
ESO 101-8 GX 207
ESO 101-11 GX 207
ESO 101-13 GX 207
ESO 101-14 GX 196, 207
ESO 101-17 GX 207
ESO 101-18 GX 207
ESO 101-19 GX 207
ESO 101-20 GX 195, 196, 207
ESO 101-21 GX 207
ESO 101-23 GX 207
ESO 101-24 GX 207
ESO 102-2 GX 207
ESO 102-4 GX 207
ESO 102-5 GX 207
ESO 102-6 GX 207
ESO 102-7 GX 207
ESO 102-9 GX 195, 207
ESO 102-12 GX 207
ESO 102-18 GX 206, 207
ESO 102-23 GX 206, 207
ESO 103-5 GX 206
ESO 103-19 GX 206
ESO 103-35 GX 206
ESO 103-39 GX 206
ESO 103-46 GX 206
ESO 103-49 GX 206
ESO 103-56 GX 206
ESO 103-59 GX 206
ESO 104-7 GX 206
ESO 104-22 GX 206
ESO 104-44 GX 206
ESO 104-51 GX 194, 206
ESO 105-4 GX 206
ESO 105-7 GX 206
ESO 105-9 GX 206
ESO 105-12 GX 206
ESO 105-20 GX 206
ESO 105-22 GX 206
ESO 105-26 GX 206
ESO 106-2 GX 205, 206
ESO 106-9 GX 205
ESO 107-2 GX 205
ESO 107-4 GX 205
ESO 107-14 GX 205
ESO 107-15 GX 205
ESO 107-16 GX 205
ESO 107-32 GX 205
ESO 107-35 GX 205
ESO 107-37 GX 193, 205
ESO 107-39 GX 205
ESO 107-44 GX 205
ESO 108-16 GX 205
ESO 108-17 GX 205
ESO 108-18 GX 193, 205
ESO 108-21 GX 205
ESO 108-23 GX 205
ESO 109-10 GX 205
ESO 109-23 GX 204, 205
ESO 109-32 GX 192, 193, 204
ESO 109-34 GX 204
ESO 110-8 GX 204
ESO 110-13 GX 204
ESO 110-14 GX 204
ESO 110-16 GX 192, 204
ESO 110-25 GX 204
ESO 110-27 GX 204
ESO 110-28 GX 204
ESO 110-29 GX 192, 204
ESO 111-6 GX 192
ESO 111-7 GX 192
ESO 111-9 GX 192
ESO 111-10 GX 192
ESO 111-14 GX 192
ESO 111-22 GX 192
ESO 111-24 GX 192
ESO 112-4 GX 192
ESO 112-5 GX 192
ESO 112-8 GX 192
ESO 112-9 GX 192
ESO 112-11 GX 192
ESO 113-4 GX 192, 203
ESO 113-6 GX 192, 203
ESO 113-10 GX 203
ESO 113-11 GX 203
ESO 113-14 GX 203
ESO 113-27 GX 203
ESO 113-29 GX 203
ESO 113-30 GX 203
ESO 113-31 GX 203
ESO 113-32 GX 203
ESO 113-35 GX 203
ESO 113-41 GX 203
ESO 113-42 GX 203
ESO 113-43 GX 203
ESO 113-45 GX 203
ESO 113-48 GX 203
ESO 113-50 GX 203
ESO 114-1 GX 203
ESO 114-4 GX 203
ESO 114-7 GX 203
ESO 114-8 GX 203
ESO 114-14 GX 203, 213
ESO 114-16 GX 203
ESO 114-19 GX 203
ESO 114-21 GX 203
ESO 114-23 GX 203
ESO 114-24 GX 203
ESO 114-27 GX 203
ESO 115-8 GX 203
ESO 115-9 GX 203
ESO 115-13 GX 203
ESO 115-15 GX 203
ESO 115-17 GX 203
ESO 115-21 GX 203
ESO 115-22 GX 203
ESO 115-25 GX 203
ESO 116-1 GX 203
ESO 116-5 GX 202, 203
ESO 116-12 GX 202
ESO 116-14 GX 202
ESO 116-15 GX 202
ESO 116-18 GX 202
ESO 117-16 GX 202
ESO 117-18 GX 202
ESO 117-19 GX 202, 212
ESO 118-3 GX 202
ESO 118-12 GX 202
ESO 118-15 GX 202
ESO 118-19 GX 202
ESO 118-21 GX 202
ESO 118-22 GX 202
ESO 118-28 GX 202
ESO 118-30 GX 202
ESO 118-31 GX 202
ESO 118-34 GX 202
ESO 119-4 GX 202
ESO 119-8 GX 202
ESO 119-11 GX 202, 212
ESO 119-12 GX 202
ESO 119-13 GX 202
ESO 119-15 GX 202
ESO 119-16 GX 202
ESO 119-21 GX 202
ESO 119-23 GX 201, 202
ESO 119-25 GX 201, 202
ESO 119-38 GX 201
ESO 119-43 GX 201
ESO 119-44 GX 201
ESO 119-45 GX 201
ESO 119-46 GX 201, 212
ESO 119-47 GX 201, 212
ESO 119-48 GX 201
ESO 119-49 GX 201
ESO 119-52 GX 201
ESO 119-53 GX 201
ESO 119-54 GX 201
ESO 119-55 GX 201
ESO 119-57 GX 201
ESO 119-58 GX 201
ESO 120-8 OC 201
ESO 120-12 GX 201
ESO 120-16 GX 201
ESO 120-21 GX 201
ESO 120-23 GX 201
ESO 120-26 GX 201
ESO 121-6 GX 201
ESO 121-26 GX 201
ESO 121-41 GX 201
ESO 122-1 GX 201
ESO 122-2 GX 201
ESO 122-3 GX 201
ESO 122-6 GX 201
ESO 122-11 GX 200, 201
ESO 122-16 GX 200
ESO 123-4 GX 200
ESO 123-9 GX 200, 211
ESO 123-16 GX 200
ESO 123-17 GX 200, 211
ESO 123-19 GX 200
ESO 123-23 GX 200
ESO 123-26 OC 200
ESO 124-11 GX 200
ESO 124-14 GX 200
ESO 124-15 GX 200
ESO 124-18 GX 200
ESO 124-19 GX 200
ESO 126-1 GX 199
ESO 126-2 GX 199
ESO 126-3 GX 199
ESO 126-4 GX 199
ESO 126-10 GX 199
ESO 126-11 GX 199
ESO 126-13 GX 199
ESO 126-14 GX 199
ESO 126-17 GX 199, 210
ESO 126-19 GX 199
ESO 126-21 GX 199, 210
ESO 126-22 GX 199
ESO 126-23 GX 199, 210
ESO 126-24 GX 199
ESO 126-25 GX 199
ESO 128-16 OC 199
ESO 129-19 OC 198
ESO 129-32 OC 198
ESO 130-12 GX 198
ESO 130-13 OC 198
ESO 131-9 OC 198
ESO 134-12 OC 197
ESO 135-10 GX 196
ESO 136-12 GX 196
ESO 136-15 GX 196
ESO 136-16 GX 196
ESO 137-2 GX 196
ESO 137-3 GX 196
ESO 137-4 GX 196
ESO 137-6 GX 196
ESO 137-7 GX 196
ESO 137-8 GX 196
ESO 137-10 GX 196
ESO 137-11 GX 196
ESO 137-12 GX 196
ESO 137-16 GX 196
ESO 137-18 GX 196
ESO 137-19 GX 196
ESO 137-21 GX 196
ESO 137-23 OC 196
ESO 137-24 GX 196
ESO 137-25 GX 196
ESO 137-26 GX 196
ESO 137-27 GX 196
ESO 137-29 GX 196
ESO 137-32 GX 196, 207
ESO 137-34 GX 196
ESO 137-35 GX 196
ESO 137-36 GX 196
ESO 137-37 GX 196
ESO 137-38 GX 196
ESO 137-39 GX 196
ESO 137-42 GX 196
ESO 137-43 OC 196
ESO 137-44 GX 196
ESO 137-45 GX 196
ESO 138-1 GX 196
ESO 138-5 GX 196
ESO 138-6 GX 196
ESO 138-7 GX 195, 196, 207
ESO 138-9 GX 195, 196
ESO 138-10 GX 195, 196
ESO 138-11 GX 195, 196, 207
ESO 138-12 GX 195, 196
ESO 138-13 GX 195, 196
ESO 138-14 GX 195, 207
ESO 138-15 GX 195
ESO 138-16 GX 195
ESO 138-17 GX 195
ESO 138-18 GX 195
ESO 138-21 GX 195
ESO 138-22 GX 195
ESO 138-24 GX 195
ESO 138-26 GX 195
ESO 138-27 GX 195
ESO 138-29 GX 195, 207
ESO 138-30 GX 195, 207
ESO 139-1 GX 195, 207
ESO 139-4 GX 195
ESO 139-5 GX 195
ESO 139-8 GX 195
ESO 139-9 GX 195
ESO 139-11 GX 195
ESO 139-12 GX 195
ESO 139-13 OC 195
ESO 139-14 GX 195
ESO 139-21 GX 195
ESO 139-24 GX 195
ESO 139-26 GX 195
ESO 139-28 GX 195
ESO 139-32 GX 195
ESO 139-34 GX 195
ESO 139-36 GX 195
ESO 139-37 GX 195
ESO 139-38 GX 195
ESO 139-41 GX 195
ESO 139-45 GX 195
ESO 139-46 GX 195
ESO 139-47 GX 195
ESO 139-48 GX 195
ESO 139-49 GX 195
ESO 139-51 GX 195
ESO 139-53 GX 195
ESO 139-54 OC 195
ESO 139-55 GX 195
ESO 140-2 GX 195
ESO 140-6 GX 195
ESO 140-12 GX 195
ESO 140-18 GX 195
ESO 140-23 GX 195
ESO 140-31 GX 195
ESO 140-38 GX 195
ESO 140-43 GX 195, 206
ESO 141-3 GX 195
ESO 141-17A GX 194, 195, 206
ESO 141-17B GX 194, 195, 206
ESO 141-21 GX 194, 195
ESO 141-24 GX 194
ESO 141-32 GX 194
ESO 141-42 GX 194, 206
ESO 141-47 OC 194
ESO 141-55 GX 194
ESO 142-13 GX 194
ESO 142-19 GX 194
ESO 142-23 GX 194
ESO 142-30 GX 194
ESO 142-34 GX 194
ESO 142-48 GX 194
ESO 142-49 GX 194
ESO 142-51 GX 194
ESO 142-58 GX 194
ESO 142-59 GX 194
ESO 143-5 GX 194
ESO 143-21 GX 194
ESO 143-27 GX 194
ESO 143-35 GX 194
ESO 143-38 GX 194
ESO 145-1 GX 193
ESO 145-4 GX 193
ESO 145-6 GX 193
ESO 145-16 GX 193
ESO 145-21 GX 193
ESO 145-24 GX 193
ESO 145-25 GX 193
ESO 146-1 GX 193
ESO 146-2A GX 193

ESO 146-6 GX 193
ESO 146-11 GX 193
ESO 146-14 GX 193, 205
ESO 146-206 GX 193
ESO 146-28 GX 193
ESO 147-5 GX 193
ESO 147-6 GX 193
ESO 147-10 GX 193
ESO 147-13 GX 193
ESO 147-14 GX 192, 193
ESO 147-20 GX 192, 193
ESO 148-1 GX 192
ESO 148-2 GX 192
ESO 148-4 GX 192
ESO 148-7 GX 192
ESO 148-17 GX 192
ESO 148-18 GX 192
ESO 148-19 GX 192
ESO 148-20 GX 192
ESO 148-21 GX 192
ESO 149-1 GX 192
ESO 149-3 GX 192
ESO 149-11 GX 192
ESO 149-13 GX 192
ESO 149-16 GX 192
ESO 149-18 GX 192
ESO 149-21 GX 192
ESO 150-1 GX 192
ESO 150-5 GX 192
ESO 150-13 GX 192
ESO 150-14 GX 192
ESO 150-15 GX 192
ESO 150-20 GX 192
ESO 150-21 GX 192
ESO 150-22 GX 192
ESO 150-24 GX 192
ESO 151-4 GX 192, 203
ESO 151-5 GX 192, 203
ESO 151-12 GX 192, 203
ESO 151-18 GX 192, 203
ESO 151-19 GX 192, 203
ESO 151-26 GX 203
ESO 151-27 GX 203
ESO 151-32 GX 203
ESO 151-40 GX 203
ESO 151-43 GX 203
ESO 152-25 GX 203
ESO 152-26 GX 203
ESO 152-34 GX 203
ESO 153-1 GX 203
ESO 153-3 GX 203
ESO 153-4 GX 203
ESO 153-4A GX 203
ESO 153-13 GX 203
ESO 153-15 GX 203
ESO 153-16 GX 203
ESO 153-17 GX 203
ESO 153-18 GX 203
ESO 153-19 GX 203
ESO 153-20 GX 203
ESO 153-23 GX 203
ESO 153-24 GX 203
ESO 153-27 GX 203
ESO 153-29 GX 203
ESO 153-30 GX 203
ESO 153-34 GX 203
ESO 154-2 GX 203
ESO 154-9 GX 203
ESO 154-10 GX 203
ESO 154-13 GX 203
ESO 154-22 GX 203
ESO 154-23 GX 202, 203
ESO 154-28 GX 202, 203
ESO 155-14 GX 202
ESO 155-16 GX 202
ESO 155-20 GX 202
ESO 155-35 GX 202
ESO 155-41 GX 202
ESO 155-54 GX 202
ESO 156-8 GX 202
ESO 156-11 GX 202
ESO 156-18 GX 202
ESO 156-29 GX 202
ESO 157-23 GX 202
ESO 157-30 GX 202
ESO 157-36 GX 202
ESO 157-42 GX 202
ESO 157-43 GX 202
ESO 157-44 GX 202
ESO 157-49 GX 202
ESO 157-50 GX 202
ESO 158-2 GX 202
ESO 158-3 GX 202
ESO 158-7 GX 202
ESO 158-15 GX 202
ESO 158-17 GX 201
ESO 158-18 GX 201
ESO 159-2 GX 201
ESO 159-3 GX 201
ESO 159-4 GX 201
ESO 159-12 GX 201
ESO 159-19 GX 201
ESO 159-20 GX 201
ESO 159-23 GX 201
ESO 159-25 GX 201
ESO 160-2 GX 201
ESO 160-11 GX 201
ESO 160-16 GX 201
ESO 160-19 GX 201
ESO 160-20 GX 201
ESO 160-22 GX 201
ESO 160-23 GX 201
ESO 161-1 GX 201
ESO 161-8 GX 201
ESO 161-19 GX 201
ESO 161-23 GX 201
ESO 161-24 GX 201
ESO 161-25 GX 201
ESO 162-15 GX 200
ESO 162-17 GX 200
ESO 163-11 GX 200
ESO 163-14 GX 200
ESO 163-19 GX 200
ESO 164-4 GX 200
ESO 164-10 GX 200
ESO 165-1 GX 200
ESO 165-9 OC 199
ESO 166-4 OC 199
ESO 168-2 GX 199
ESO 170-3 GX 198
ESO 170-4 GX 198
ESO 171-4 GX 198
ESO 171-5 GX 198
ESO 171-8 GX 198
ESO 171-13 GX 198
ESO 172-9 GX 198
ESO 172-12 GX 197, 198
ESO 174-2 GX 197
ESO 174-3 GX 197
ESO 175-1 GX 197
ESO 175-5 GX 197
ESO 175-6 OC 197
ESO 179-13 GX 196
ESO 181-2 GX 195
ESO 182-1 GX 195
ESO 182-4 GX 195
ESO 182-6 GX 195
ESO 182-7 GX 195
ESO 182-8 GX 195
ESO 182-10 GX 195
ESO 182-12 GX 195
ESO 182-13 GX 195
ESO 182-17 GX 195
ESO 183-3 GX 195
ESO 183-5 GX 195
ESO 183-6 GX 195
ESO 183-7 GX 195
ESO 183-8 GX 195
ESO 183-11 GX 195
ESO 183-19 GX 195
ESO 183-30 GX 194, 195
ESO 183-34 GX 194, 195
ESO 184-4 GX 194, 195
ESO 184-5 GX 194, 195
ESO 184-6 GX 194
ESO 184-11 GX 194
ESO 184-15 GX 194
ESO 184-16 GX 194
ESO 184-26 GX 194
ESO 184-32 GX 194
ESO 184-33 GX 194
ESO 184-34 GX 194
ESO 184-41 GX 194
ESO 184-42 GX 194
ESO 184-43 GX 194
ESO 184-51 GX 194
ESO 184-57 GX 194
ESO 184-58 GX 194
ESO 184-59 GX 194
ESO 184-60 GX 194
ESO 184-63 GX 194
ESO 184-71 GX 194
ESO 184-72 GX 194
ESO 184-74 GX 194
ESO 184-75 GX 194
ESO 184-77 GX 194
ESO 184-81 GX 194
ESO 185-13 GX 194
ESO 185-19 GX 194
ESO 185-21 GX 194
ESO 185-25 GX 194
ESO 185-27 GX 194
ESO 185-32 GX 194
ESO 185-34 GX 194
ESO 185-35 GX 194
ESO 185-37 GX 194
ESO 185-53 GX 194
ESO 185-54 GX 194
ESO 185-64 GX 194
ESO 185-68 GX 194
ESO 186-4 GX 194
ESO 186-12 GX 194
ESO 186-21 GX 194
ESO 186-32 GX 194
ESO 186-36 GX 194
ESO 186-39 GX 194
ESO 186-43 GX 194
ESO 186-46 GX 194
ESO 186-47 GX 194
ESO 186-52 GX 194
ESO 186-55 GX 194
ESO 186-57 GX 194
ESO 186-59 GX 194
ESO 186-60 GX 194
ESO 186-62 GX 194
ESO 186-65 GX 194
ESO 186-69 GX 194
ESO 186-75 GX 194
ESO 187-6 GX 194
ESO 187-8 GX 194
ESO 187-20 GX 194
ESO 187-25 GX 194
ESO 187-26 GX 194
ESO 187-27 GX 194
ESO 187-34 GX 193, 194
ESO 187-35 GX 193, 194
ESO 187-36 GX 193, 194
ESO 187-38 GX 193, 194
ESO 187-39 GX 193, 194
ESO 187-42 GX 193, 194
ESO 187-51 GX 193
ESO 187-58 GX 193
ESO 187-59 GX 193
ESO 188-12A GX 193
ESO 188-18 GX 193
ESO 189-5 GX 193
ESO 189-10 GX 193
ESO 189-12 GX 193
ESO 189-16 GX 193
ESO 189-21 GX 193
ESO 189-31 GX 193
ESO 189-32 GX 193
ESO 190-10 GX 193
ESO 190-11 GX 193
ESO 190-12 GX 193
ESO 192-11 GX 192
ESO 192-12 GX 192
ESO 193-6 GX 177
ESO 193-11 GX 177
ESO 193-15 GX 177
ESO 193-17 GX 177
ESO 193-18 GX 177
ESO 193-19 GX 177
ESO 193-22 GX 192
ESO 193-26 GX 177
ESO 193-29 GX 177
ESO 193-36 GX 177
ESO 193-37 GX 177
ESO 193-41 GX 177
ESO 194-4 GX 177, 192
ESO 194-13 GX 177
ESO 194-15 GX 177
ESO 194-21 GX 177, 192
ESO 194-22 GX 177
ESO 194-24 GX 177
ESO 194-32 GX 177, 191
ESO 194-36 GX 177, 191
ESO 194-38 GX 177, 191
ESO 194-39 GX 192
ESO 195-1 GX 177, 191
ESO 195-3 GX 177, 191
ESO 195-17 GX 191
ESO 195-24 GX 191, 192, 203
ESO 195-29 GX 191, 203
ESO 195-35 GX 191
ESO 196-3 GX 191, 203
ESO 196-11 GX 191, 203
ESO 196-19 GX 191
ESO 196-21 GX 191
ESO 197-2 GX 191
ESO 197-9 GX 191
ESO 197-10 GX 191
ESO 197-15 GX 203
ESO 197-16 GX 191
ESO 197-18 GX 191
ESO 197-21 GX 203
ESO 197-24 GX 203
ESO 198-1 GX 191
ESO 198-2 GX 191
ESO 198-6 GX 191
ESO 198-8 GX 190, 191, 203
ESO 198-11 GX 190, 191
ESO 198-13 GX 190
ESO 198-14 GX 190
ESO 198-19 GX 190
ESO 198-20 GX 190
ESO 198-21 GX 190, 203
ESO 198-27 GX 190, 203
ESO 199-1 GX 190
ESO 200-29 GX 190
ESO 200-53 GX 190, 202
ESO 200-54 GX 190, 202
ESO 201-4 GX 190
ESO 201-14 GX 189, 190
ESO 201-16 GX 189, 190, 202
ESO 201-17 GX 189
ESO 201-21 GX 189
ESO 201-22 GX 189
ESO 202-1 GX 189
ESO 202-7 GX 189
ESO 202-15 GX 189
ESO 202-23 GX 189
ESO 202-31 GX 202
ESO 202-35 GX 189
ESO 202-37 GX 189
ESO 202-40 GX 189
ESO 202-41 GX 202
ESO 202-42 GX 189
ESO 202-43 GX 189, 202
ESO 202-48 GX 189
ESO 202-55 GX 189
ESO 202-56 GX 189
ESO 203-2 GX 189
ESO 203-7 GX 189
ESO 203-9 GX 189
ESO 203-15 GX 189
ESO 203-16 GX 189
ESO 203-17 GX 189
ESO 203-19 GX 189
ESO 203-22 GX 201
ESO 203-24 GX 189
ESO 204-3 GX 189
ESO 204-8 GX 189, 201
ESO 204-11 GX 189
ESO 204-13 GX 188, 189
ESO 204-14 GX 188, 189
ESO 204-22 GX 201
ESO 204-32 GX 188
ESO 204-34 GX 188, 201
ESO 204-35 GX 188
ESO 204-36 GX 201
ESO 205-9 GX 201
ESO 205-11 GX 188, 201
ESO 205-12 GX 188
ESO 205-13 GX 188, 201
ESO 205-14 GX 201
ESO 205-18 GX 188
ESO 205-23 GX 188
ESO 205-27 GX 188
ESO 205-27A GX 188
ESO 205-28 GX 188
ESO 205-31 GX 188
ESO 205-34 GX 188, 201
ESO 206-1 GX 188, 201
ESO 206-14 GX 188
ESO 206-17 GX 188, 201
ESO 206-A20 GX 188
ESO 207-7 GX 201
ESO 207-13 GX 188
ESO 207-15 GX 188
ESO 207-18 GX 188
ESO 207-19 GX 188
ESO 207-22 GX 187, 188, 200
ESO 207-25 GX 187, 188, 200
ESO 208-3 GX 187
ESO 208-15 GX 187, 200
ESO 208-21 GX 187
ESO 208-25 GX 187
ESO 208-26 GX 187
ESO 208-31 GX 200
ESO 208-33 GX 187
ESO 209-9 GX 187
ESO 211-1 OC 200
ESO 213-11 GX 186
ESO 214-2 GX 185, 186
ESO 214-13 GX 185
ESO 214-16 GX 185
ESO 214-17 GX 185
ESO 215-5 GX 185
ESO 215-7 GX 185
ESO 215-8 GX 185
ESO 215-10 GX 185
ESO 215-12 GX 185
ESO 215-13 GX 185
ESO 215-15 GX 185
ESO 215-19 GX 185
ESO 215-21 GX 185
ESO 215-23 GX 185
ESO 215-27 GX 185
ESO 215-31 GX 185
ESO 215-32 GX 185
ESO 215-33 GX 185
ESO 215-37 GX 185
ESO 215-39 GX 185
ESO 215-40 GX 185
ESO 216-3 GX 185
ESO 216-5 GX 198
ESO 216-8 GX 185
ESO 216-10 GX 185
ESO 216-21 GX 185
ESO 216-24 GX 185
ESO 216-27 GX 185
ESO 216-28 GX 185
ESO 216-31 GX 185
ESO 216-37 GX 185
ESO 216-38 GX 185, 198
ESO 217-9 GX 185
ESO 217-12 GX 185
ESO 217-14 GX 185
ESO 217-15 GX 184, 185, 198
ESO 217-16 GX 184, 185, 198
ESO 217-17 GX 184, 185
ESO 217-20 GX 184, 185
ESO 217-22 GX 184
ESO 217-30 GX 198
ESO 217-31 GX 198
ESO 218-2 GX 198
ESO 218-8 GX 184, 198
ESO 219-7 GX 184
ESO 219-14 GX 184
ESO 219-16 GX 184
ESO 219-18 GX 184
ESO 219-21 GX 184
ESO 219-22 GX 184
ESO 219-27 GX 184
ESO 219-30 GX 184, 197
ESO 219-34 GX 184
ESO 219-37 GX 184, 197
ESO 219-41 GX 184
ESO 219-43 GX 184
ESO 220-6 GX 197
ESO 220-7 GX 184
ESO 220-8 GX 184
ESO 220-9 GX 184
ESO 220-10 GX 184
ESO 220-12 GX 184
ESO 220-19 GX 183, 184
ESO 220-20 GX 183, 184
ESO 220-21 GX 183, 184
ESO 220-22 GX 183, 184
ESO 220-23 GX 183, 184
ESO 220-26 GX 183, 184
ESO 220-28 GX 183, 197
ESO 220-32 GX 183
ESO 220-35 GX 197
ESO 220-37 GX 183
ESO 221-1 GX 183
ESO 221-2 GX 183
ESO 221-3 GX 183
ESO 221-4 GX 183
ESO 221-5 GX 183
ESO 221-6 GX 183
ESO 221-7 GX 197
ESO 221-8 GX 183
ESO 221-9 GX 183
ESO 221-10 GX 183
ESO 221-12 GX 183
ESO 221-13 GX 183
ESO 221-14 GX 183
ESO 221-18 GX 183
ESO 221-20 GX 183
ESO 221-21 GX 183
ESO 221-22 GX 183
ESO 221-25 GX 183
ESO 221-26 GX 183
ESO 221-27 GX 183
ESO 221-28 GX 183, 197
ESO 221-32 GX 183
ESO 221-35 GX 197
ESO 221-37 GX 183
ESO 222-1 GX 183
ESO 222-4 GX 183
ESO 222-15 GX 183
ESO 223-2 GX 183
ESO 223-7 GX 183
ESO 223-9 GX 183
ESO 223-12 GX 196
ESO 226-6 OC 182, 196
ESO 231-1 GX 180, 195
ESO 231-2 GX 180
ESO 231-3 GX 180
ESO 231-6 GX 180, 195
ESO 231-7 GX 195
ESO 231-9 GX 180
ESO 231-11 GX 180
ESO 231-14 GX 180
ESO 231-14A GX 180
ESO 231-17 GX 180
ESO 231-18 GX 180, 194
ESO 231-23 GX 180, 194
ESO 231-24 GX 180
ESO 231-29 GX 180
ESO 231-30 OC 180, 194
ESO 232-1 GX 180
ESO 232-4 GX 180, 194
ESO 232-5 GX 180
ESO 232-6 GX 180, 194
ESO 232-9 GX 180
ESO 232-19 GX 180
ESO 232-21 GX 180, 194
ESO 232-23 GX 180
ESO 233-14 GX 194
ESO 233-22 GX 179
ESO 233-27 GX 179
ESO 233-35 GX 179
ESO 233-36 GX 179
ESO 233-37 GX 179
ESO 233-44 GX 179, 194
ESO 233-49 GX 179
ESO 233-50 GX 179
ESO 234-3 GX 179, 194
ESO 234-4 GX 179
ESO 234-9 GX 179
ESO 234-10 GX 179, 194
ESO 234-11 GX 179
ESO 234-13 GX 179
ESO 234-14 GX 179
ESO 234-15 GX 194
ESO 234-16 GX 179
ESO 234-19 GX 179
ESO 234-21 GX 179
ESO 234-24 GX 179, 194
ESO 234-28 GX 179, 194
ESO 234-32 GX 179, 194
ESO 234-43 GX 179
ESO 234-44 GX 179
ESO 234-47 GX 179
ESO 234-49 GX 179
ESO 234-50 GX 179
ESO 234-51 GX 179
ESO 234-52 GX 194
ESO 234-53 GX 179
ESO 234-55 GX 179, 194
ESO 234-68 GX 179, 194
ESO 234-69 GX 179, 194
ESO 234-70 GX 179, 194
ESO 235-4 GX 179
ESO 235-9 GX 194
ESO 235-16 GX 179
ESO 235-22 GX 179
ESO 235-23 GX 179
ESO 235-26 GX 193, 194
ESO 235-32 GX 193, 194
ESO 235-33 GX 179
ESO 235-35 GX 193, 194
ESO 235-39 GX 179
ESO 235-42 GX 179
ESO 235-45 GX 179
ESO 235-47 GX 179
ESO 235-49 GX 179
ESO 235-50 GX 179, 193
ESO 235-51 GX 179, 193
ESO 235-53 GX 179
ESO 235-54 GX 179, 193
ESO 235-55 GX 179
ESO 235-57 GX 179
ESO 235-58 GX 179
ESO 235-59 GX 179
ESO 235-60 GX 179
ESO 235-61 GX 179
ESO 235-83 GX 179
ESO 235-84 GX 179
ESO 235-85 GX 179
ESO 236-3 GX 179, 193

ESO 236-6 GX 179
ESO 236-7 OC 179, 193
ESO 236-8 GX 179
ESO 236-11 GX 179
ESO 236-14 GX 179
ESO 236-18 GX 178, 179
ESO 236-19 GX 178, 179
ESO 236-25 GX 178, 179
ESO 236-26 GX 178, 179
ESO 236-34 GX 178, 193
ESO 236-35 GX 178
ESO 236-36 GX 178
ESO 236-37 GX 178
ESO 237-19 GX 178, 193
ESO 237-30 GX 178
ESO 237-35 GX 178
ESO 237-42 GX 178
ESO 237-47 GX 178
ESO 237-48 GX 178
ESO 237-49 GX 178
ESO 237-51 GX 178
ESO 237-52 GX 178
ESO 238-4 GX 178
ESO 238-5 GX 178
ESO 239-17 GX 177, 178
ESO 240-2 GX 177
ESO 240-3 GX 177
ESO 240-4 GX 177, 192
ESO 240-10 GX 177
ESO 240-11 GX 177
ESO 240-12 GX 177, 192
ESO 240-13 GX 177
ESO 240-14 GX 192
ESO 241-6 GX 177
ESO 241-10 GX 177
ESO 241-20 GX 177
ESO 241-21 GX 177
ESO 241-22 GX 177
ESO 241-23 GX 177
ESO 242-7 GX 177
ESO 242-9 GX 177
ESO 242-12 GX 177
ESO 242-14 GX 177
ESO 242-16 GX 177
ESO 242-17 GX 177
ESO 242-18 GX 177
ESO 242-20 GX 177
ESO 242-23 GX 177
ESO 242-24 GX 177
ESO 243-2 GX 177, 191
ESO 243-6 GX 177, 191
ESO 243-11 GX 191
ESO 243-12 GX 191
ESO 243-24 GX 191
ESO 243-29 GX 191
ESO 243-30 GX 191
ESO 243-31 GX 191
ESO 243-35 GX 191
ESO 243-37 GX 191
ESO 243-41 GX 191
ESO 243-43 GX 191
ESO 243-45 GX 191
ESO 243-49 GX 191
ESO 243-51 GX 191
ESO 243-52 GX 191
ESO 244-6 GX 191
ESO 244-8 GX 191
ESO 244-10 GX 191
ESO 244-12 GX 191
ESO 244-16 GX 191
ESO 244-17 GX 191
ESO 244-20 GX 191
ESO 244-21 GX 191
ESO 244-22 GX 191
ESO 244-23 GX 191
ESO 244-30 GX 191
ESO 244-31 GX 191
ESO 244-32 GX 191
ESO 244-36 GX 191
ESO 244-37 GX 191
ESO 244-39 GX 191
ESO 244-44 GX 191
ESO 244-45 GX 191
ESO 244-47 GX 191
ESO 244-48 GX 191
ESO 244-49 GX 191
ESO 245-3 GX 191
ESO 245-5 GX 191
ESO 245-6 GX 191
ESO 245-7 GX 191
ESO 245-9 OC 191
ESO 245-10 GX 191
ESO 245-12 GX 191
ESO 246-8 GX 191
ESO 246-13 GX 190
ESO 246-15 GX 190
ESO 246-21 GX 190
ESO 246-22 GX 190
ESO 246-25 GX 190
ESO 247-8 GX 190
ESO 248-2 GX 190
ESO 248-6 GX 190
ESO 248-12 GX 190
ESO 249-1 GX 190
ESO 249-8 GX 190
ESO 249-25 GX 190
ESO 249-27 GX 190
ESO 249-35 GX 189, 190
ESO 249-36 GX 189, 190
ESO 250-5 GX 189
ESO 250-12 GX 189
ESO 250-21 GX 189
ESO 251-1 GX 189
ESO 251-2 GX 189
ESO 251-4 GX 189
ESO 251-5 GX 189
ESO 251-6 GX 189
ESO 251-7 GX 189
ESO 251-11 GX 189
ESO 251-16 GX 189
ESO 251-21 GX 189
ESO 251-23 GX 189
ESO 251-26 GX 189
ESO 251-28 GX 189
ESO 251-32 GX 189
ESO 251-37 GX 189
ESO 251-39 GX 189
ESO 251-41 GX 189
ESO 252-4 GX 189
ESO 252-7 GX 189
ESO 252-10 GX 189
ESO 252-12 GX 189
ESO 252-18 GX 189
ESO 252-22 GX 189
ESO 253-1 GX 189
ESO 253-2 GX 189
ESO 253-7 GX 189
ESO 253-12 GX 188, 189
ESO 253-15 GX 188, 189
ESO 253-26 GX 188
ESO 253-27 GX 188
ESO 253-29 GX 188
ESO 253-30 GX 188
ESO 254-6 GX 188
ESO 254-17 GX 188
ESO 254-22 GX 188
ESO 254-31 GX 188
ESO 254-35 GX 188
ESO 254-36 GX 188
ESO 254-37 GX 188
ESO 254-41 GX 188
ESO 254-43 GX 188
ESO 255-5 GX 188
ESO 255-7 GX 188
ESO 255-10 GX 188
ESO 255-11 GX 188
ESO 255-13 GX 188
ESO 255-18 GX 188
ESO 255-19 GX 188
ESO 256-7 GX 188
ESO 256-11 GX 188
ESO 256-12 GX 188
ESO 257-19 GX 187
ESO 258-1 OC 187
ESO 260-17 OC 186
ESO 261-3 OC 186
ESO 261-7 OC 186
ESO 262-4 GX 186
ESO 263-3 GX 186
ESO 263-4 GX 186
ESO 263-5 GX 186
ESO 263-6 GX 186
ESO 263-7 GX 186
ESO 263-8 GX 186
ESO 263-12 GX 186
ESO 263-13 GX 186
ESO 263-14 GX 186
ESO 263-15 GX 186
ESO 263-16 GX 186
ESO 263-18 GX 186
ESO 263-19 GX 186
ESO 263-21 GX 186
ESO 263-23 GX 186
ESO 263-24 GX 186
ESO 263-28 GX 185, 186
ESO 263-29 GX 185, 186
ESO 263-30 GX 185, 186
ESO 263-31 GX 185, 186
ESO 263-33 GX 185, 186
ESO 263-35 GX 185, 186
ESO 263-37 GX 185, 186
ESO 263-46 GX 185
ESO 263-47 GX 185
ESO 263-48 GX 185
ESO 263-51 GX 185
ESO 264-4 GX 185
ESO 264-11 GX 185
ESO 264-18 GX 185
ESO 264-21 GX 185
ESO 264-23 GX 185
ESO 264-24 GX 185
ESO 264-25 GX 185
ESO 264-28 GX 185
ESO 264-30 GX 185
ESO 264-31 GX 185
ESO 264-32 GX 185
ESO 264-34 GX 185
ESO 264-35 GX 185
ESO 264-36 GX 185
ESO 264-39 GX 185
ESO 264-41 GX 185
ESO 264-42 GX 185
ESO 264-43 GX 185
ESO 264-46 GX 185
ESO 264-47 GX 185
ESO 264-48 GX 185
ESO 264-49 GX 185
ESO 264-50 GX 185
ESO 264-57 GX 185
ESO 265-3 GX 185
ESO 265-7 GX 185
ESO 265-9 GX 185
ESO 265-11 GX 185
ESO 265-16 GX 185
ESO 265-22 GX 185
ESO 265-29 GX 185
ESO 265-31 GX 185
ESO 265-33 GX 185
ESO 265-35 GX 185
ESO 266-2 GX 185
ESO 266-3 GX 185
ESO 266-8 GX 185
ESO 266-12 GX 185
ESO 266-15 GX 185
ESO 266-19 GX 185
ESO 266-20 GX 185
ESO 266-22 GX 185
ESO 266-23 GX 185
ESO 267-5 GX 184, 185
ESO 267-11 GX 184, 185
ESO 267-13 GX 184
ESO 267-16 GX 184
ESO 267-17 GX 184
ESO 267-27 GX 184
ESO 267-29 GX 184
ESO 267-30 GX 184
ESO 267-34 GX 184
ESO 267-36 GX 184
ESO 267-41 GX 184
ESO 267-43 GX 184
ESO 268-3 GX 184
ESO 268-4 GX 184
ESO 268-5 GX 184
ESO 268-8 GX 184
ESO 268-15 GX 184
ESO 268-22 GX 184
ESO 268-23 GX 184
ESO 268-27 GX 184
ESO 268-30 GX 184
ESO 268-33 GX 184
ESO 268-34 GX 184, A23
ESO 268-35 GX 184
ESO 268-36 GX 184, A23
ESO 268-37 GX 184
ESO 268-38 GX 184
ESO 268-44 GX 184
ESO 268-46 GX 184
ESO 268-47 GX 184
ESO 269-2 GX 184
ESO 269-6 GX 184
ESO 269-8 GX 184
ESO 269-9 GX 184
ESO 269-12 GX 184
ESO 269-13 GX 184, A23
ESO 269-14 GX 184
ESO 269-20 GX 184
ESO 269-22 GX 184
ESO 269-25 GX 184
ESO 269-30 GX 184
ESO 269-31 GX 184
ESO 269-38 GX 184
ESO 269-43 GX 184
ESO 269-48 GX 184
ESO 269-49 GX 184
ESO 269-52 GX 184
ESO 269-53 GX 184
ESO 269-56 GX 184
ESO 269-57 GX 184
ESO 269-58 GX 184
ESO 269-60 GX 184
ESO 269-61 GX 184
ESO 269-66 GX 184
ESO 269-69 GX 184
ESO 269-70 GX 184
ESO 269-72 GX 184
ESO 269-74 GX 184
ESO 269-74A GX 184
ESO 269-75 GX 184
ESO 269-78 GX 184
ESO 269-80 GX 184
ESO 269-81 GX 184
ESO 269-82 GX 184
ESO 269-85 GX 184
ESO 269-90 GX 184
ESO 270-5 GX 184
ESO 270-6 GX 184
ESO 270-7 GX 184
ESO 270-13 GX 184
ESO 270-14 GX 184
ESO 270-15 GX 184
ESO 270-17 GX 183, 184
ESO 270-21 GX 183, 184
ESO 270-26 GX 183
ESO 270-28 GX 183
ESO 271-4 GX 183
ESO 271-5 GX 183
ESO 271-10 GX 183
ESO 271-17 GX 183
ESO 271-18 GX 183
ESO 271-20 GX 183
ESO 271-22 GX 183
ESO 271-26 GX 183
ESO 271-27 GX 183
ESO 272-1 GX 183
ESO 272-4 GX 183
ESO 272-9 GX 183
ESO 272-24 GX 183
ESO 272-25 GX 183
ESO 273-4 GX 183
ESO 273-14 GX 183
ESO 274-1 GX 182, 183
ESO 274-6 GX 182
ESO 274-16 GX 182
ESO 274-19 GX 182
ESO 280-6 OC 181
ESO 280-7 GX 181
ESO 280-13 GX 180, 181
ESO 281-1 GX 180
ESO 281-8 GX 180
ESO 281-24 OC 180
ESO 281-28 GX 180
ESO 281-33 GX 180
ESO 281-38 GX 180
ESO 281-40 GX 180
ESO 282-10 GX 180
ESO 282-12 GX 180
ESO 282-14 GX 180
ESO 282-17 GX 180
ESO 282-18 GX 180
ESO 282-20 GX 180
ESO 282-21 GX 180
ESO 282-24 GX 180
ESO 282-26 OC 180
ESO 282-27 GX 180
ESO 282-28 GX 180
ESO 283-4 GX 180
ESO 283-15 GX 180
ESO 283-19 GX 180
ESO 283-20 GX 180
ESO 284-2 GX 180
ESO 284-4 GX 179, 180
ESO 284-7 GX 179, 180
ESO 284-9 GX 179, 180
ESO 284-11 GX 179, 180
ESO 284-13 GX 179
ESO 284-16 GX 179
ESO 284-17 GX 179
ESO 284-20 GX 179
ESO 284-21 GX 179
ESO 284-26 GX 179
ESO 284-32 GX 179
ESO 284-33 GX 179
ESO 284-37 GX 179
ESO 284-38 GX 179
ESO 284-39 GX 179
ESO 284-40 GX 179
ESO 284-41 GX 179
ESO 284-43 GX 179
ESO 284-44 GX 179
ESO 284-45 GX 179
ESO 284-46 GX 179
ESO 284-47 GX 179
ESO 284-48 GX 179
ESO 284-51 GX 179
ESO 284-53 GX 179
ESO 285-1 GX 179
ESO 285-13 GX 179
ESO 285-14 GX 179
ESO 285-19 GX 179
ESO 285-20 GX 179
ESO 285-23 GX 179
ESO 285-24 GX 179
ESO 285-26 GX 179
ESO 285-28 GX 179
ESO 285-30 GX 179
ESO 285-32 GX 179
ESO 285-35 GX 179
ESO 285-36 GX 179
ESO 285-40 GX 179
ESO 285-42 GX 179
ESO 285-44 GX 179
ESO 285-48 GX 179
ESO 285-49 GX 179
ESO 285-52 GX 179
ESO 285-55 GX 179
ESO 286-10 GX 179
ESO 286-16 GX 179
ESO 286-17 GX 179
ESO 286-18 GX 179
ESO 286-19 GX 179
ESO 286-22 GX 179
ESO 286-26 GX 179
ESO 286-27 GX 179
ESO 286-29 GX 179
ESO 286-31 GX 179
ESO 286-32 GX 179
ESO 286-33 GX 179
ESO 286-35 GX 179
ESO 286-36 GX 179
ESO 286-37 GX 179
ESO 286-41 GX 179
ESO 286-42 GX 179
ESO 286-44 GX 179
ESO 286-46 GX 179
ESO 286-47 GX 179
ESO 286-49 GX 179
ESO 286-50 GX 179
ESO 286-52 GX 179
ESO 286-58 GX 179
ESO 286-59 GX 179
ESO 286-60 GX 179
ESO 286-63 GX 179
ESO 286-69 GX 179
ESO 286-71 GX 179
ESO 286-80 GX 179
ESO 286-82 GX 179
ESO 287-2 GX 179
ESO 287-4 GX 179
ESO 287-7 GX 179
ESO 287-9 GX 179
ESO 287-10 GX 179
ESO 287-13 GX 179
ESO 287-16 GX 179
ESO 287-26 GX 179
ESO 287-30 GX 179
ESO 287-32 GX 179
ESO 287-33 GX 179
ESO 287-35 GX 179
ESO 287-37 GX 178, 179
ESO 287-39 GX 178, 179
ESO 287-40 GX 178, 179
ESO 287-41 GX 178, 179
ESO 287-42 GX 178, 179
ESO 287-43 GX 178, 179
ESO 287-45 GX 178, 179
ESO 287-46 GX 178, 179
ESO 287-55 GX 178
ESO 288-13 GX 178
ESO 288-21 GX 178
ESO 288-25 GX 178
ESO 288-30 GX 178
ESO 288-32 GX 178
ESO 288-40 GX 178
ESO 288-45 GX 178
ESO 288-49 GX 178
ESO 289-4 GX 178
ESO 289-5 GX 178
ESO 289-10 GX 178
ESO 289-11 GX 178
ESO 289-15 GX 178
ESO 289-26 GX 178
ESO 289-32 GX 178
ESO 289-37 GX 178
ESO 289-42 GX 178
ESO 289-44 GX 178
ESO 289-47 GX 178
ESO 289-48 GX 178
ESO 290-1 GX 178
ESO 290-4 GX 178
ESO 290-6 GX 178
ESO 290-7 GX 178
ESO 290-9 GX 178
ESO 290-10 GX 178
ESO 290-20 GX 178
ESO 290-21 GX 178
ESO 290-25 GX 178
ESO 290-35 GX 178
ESO 290-39 GX 178
ESO 290-40 GX 178
ESO 290-42 GX 178
ESO 290-44 GX 178
ESO 290-51 GX 178
ESO 290-52 GX 178
ESO 291-3 GX 177, 178
ESO 291-4 GX 177, 178
ESO 291-6 GX 177, 178
ESO 291-9 GX 177, 178
ESO 291-24 GX 177
ESO 291-32 GX 177
ESO 292-7 GX 177
ESO 292-9 GX 177
ESO 292-13 GX 177
ESO 292-14 GX 177
ESO 292-22 GX 177
ESO 292-24 GX 177
ESO 292-25 GX 177
ESO 293-10 GX 159, 177
ESO 293-14 GX 159, 177
ESO 293-17 GX 159
ESO 293-22 GX 177
ESO 293-27 GX 159, 177
ESO 293-29 GX 159, 177
ESO 293-34 GX 177
ESO 293-37 GX 177
ESO 293-43 GX 159
ESO 293-45 GX 177
ESO 293-49 GX 159
ESO 294-7 GX 159
ESO 294-10 GX 177
ESO 294-16 GX 159, 177
ESO 294-17 GX 177
ESO 294-20 GX 159, 177
ESO 294-21 GX 177
ESO 294-22 GX 177
ESO 294-23 GX 159
ESO 295-2 GX 159, 176
ESO 295-7 GX 176, 177, 191
ESO 295-9 GX 176
ESO 295-10 GX 176
ESO 295-12 GX 177, 191
ESO 295-13 GX 176
ESO 295-22 GX 176
ESO 295-31 GX 176, 191
ESO 295-32 GX 176
ESO 295-38 GX 191
ESO 296-2 GX 176
ESO 296-5 GX 176, 191
ESO 296-6 GX 176
ESO 296-11 GX 191
ESO 296-19 GX 176
ESO 296-28 GX 176
ESO 296-29 GX 176
ESO 296-34 GX 191
ESO 296-35 GX 176
ESO 296-38 GX 176
ESO 297-3 GX 176
ESO 297-8 GX 176
ESO 297-12 GX 176
ESO 297-15 GX 176, 191
ESO 297-16 GX 176, 191
ESO 297-17 GX 176, 191
ESO 297-18 GX 176, 191
ESO 297-20 GX 191
ESO 297-23 GX 176, 191
ESO 297-25 GX 176
ESO 297-27 GX 191
ESO 297-32 GX 176
ESO 297-34 GX 175, 176
ESO 297-36 GX 175, 176
ESO 297-37 GX 175, 176
ESO 298-3 GX 191
ESO 298-7 GX 175
ESO 298-8 GX 191
ESO 298-11 GX 175, 191

ESO 298-15 GX 175, 191
ESO 298-16 GX 175
ESO 298-19 GX 175
ESO 298-21 GX 175
ESO 298-23 GX 175
ESO 298-28 GX 175
ESO 298-30 GX 175
ESO 298-31 GX 190, 191
ESO 298-36 GX 175
ESO 298-37 GX 175
ESO 299-1 GX 175, 190, 191
ESO 299-13 GX 175
ESO 299-18 GX 175, 190
ESO 299-20 GX 175
ESO 300-5 GX 175
ESO 300-10A GX 175
ESO 300-12 GX 175
ESO 300-14 GX 190
ESO 301-7 GX 190
ESO 301-9 GX 190
ESO 301-11 GX 174, 175
ESO 301-14 GX 174
ESO 301-15 GX 190
ESO 301-22 GX 174
ESO 301-25 GX 174
ESO 302-6 GX 174, 190
ESO 302-7 GX 174
ESO 302-8 GX 174
ESO 302-9 GX 174
ESO 302-12 GX 174
ESO 302-13 GX 174
ESO 302-14 GX 174
ESO 302-16 GX 190
ESO 302-23 GX 174, 189, 190
ESO 302-24 GX 189, 190
ESO 303-2 GX 189
ESO 303-3 GX 189
ESO 303-5 GX 174
ESO 303-20 GX 174, 189
ESO 303-21 GX 174
ESO 304-6 GX 173, 174
ESO 304-17 GX 189
ESO 304-19 GX 189
ESO 304-21 GX 189
ESO 305-9 GX 173
ESO 305-14 GX 173
ESO 305-15 GX 189
ESO 305-17 GX 189
ESO 305-21 GX 173
ESO 305-22 GX 173
ESO 305-25 GX 173
ESO 306-3 GX 173
ESO 306-4 GX 189
ESO 306-7 GX 173, 189
ESO 306-9 GX 189
ESO 306-12 GX 188, 189
ESO 306-13 GX 188, 189
ESO 306-16 GX 173, 188, 189
ESO 306-17 GX 173, 188
ESO 306-22 GX 173
ESO 306-25 GX 173
ESO 306-26 GX 188
ESO 306-30 GX 173
ESO 307-5 GX 172, 173
ESO 307-12 GX 188
ESO 307-13 GX 172, 173, 188
ESO 307-15 GX 188
ESO 307-17 GX 172
ESO 307-25 GX 172
ESO 308-16 GX 172
ESO 308-23 GX 172
ESO 308-24 GX 172
ESO 308-25 GX 172
ESO 309-3 OC 188
ESO 309-5 GX 172
ESO 309-8 GX 172
ESO 309-13 GX 172
ESO 309-15 GX 172
ESO 309-17 GX 188
ESO 309-19 GX 172
ESO 310-2 GX 172
ESO 310-6 GX 171, 172
ESO 310-14 GX 171
ESO 311-7 GX 187
ESO 311-12 GX 187
ESO 311-14 OC 187
ESO 312-4 OC 187
ESO 313-3 OC 187
ESO 313-12 OC 170, 171, 187
ESO 314-2 GX 170
ESO 314-14 OC 170
ESO 315-3 GX 170
ESO 315-4 GX 170
ESO 315-6 GX 170
ESO 315-12 GX 170
ESO 315-14 OC 170
ESO 315-17 GX 170
ESO 315-19 GX 170
ESO 315-20 GX 186
ESO 316-2 GX 170
ESO 316-4 GX 169, 170
ESO 316-8 GX 169, 170
ESO 316-13 GX 169, 170
ESO 316-18 GX 186
ESO 316-20 GX 169, 170
ESO 316-21 GX 186
ESO 316-25 GX 186
ESO 316-26 GX 186
ESO 316-28 GX 169
ESO 316-29 GX 186
ESO 316-30 GX 169, 186
ESO 316-31 GX 169, 186
ESO 316-32 GX 169
ESO 316-33 GX 169
ESO 316-34 GX 169
ESO 316-38 GX 169
ESO 316-40 GX 169
ESO 316-42 GX 169
ESO 316-44 GX 169
ESO 316-46 GX 169
ESO 316-47 GX 169
ESO 317-3 GX 169
ESO 317-5 GX 169
ESO 317-6 GX 169
ESO 317-7 GX 186
ESO 317-8 GX 186
ESO 317-9 GX 169
ESO 317-14 GX 169, 185, 186
ESO 317-15 GX 169
ESO 317-16 GX 169
ESO 317-17 GX 169
ESO 317-19 GX 169
ESO 317-20 GX 185, 186
ESO 317-21 GX 169
ESO 317-22 GX 169
ESO 317-23 GX 169
ESO 317-27 GX 169, 185, 186
ESO 317-32 GX 185
ESO 317-36 GX 169
ESO 317-38 GX 169
ESO 317-41 GX 185
ESO 317-42 GX 169
ESO 317-45 GX 169
ESO 317-46 GX 169
ESO 317-54 GX 169
ESO 318-2 GX 169, 185
ESO 318-4 GX 169
ESO 318-6 GX 169, 185
ESO 318-8 GX 169, 185
ESO 318-13 GX 169
ESO 318-14 GX 185
ESO 318-17 GX 169
ESO 318-19 GX 169
ESO 318-21 GX 169, 185
ESO 318-24 GX 169
ESO 319-11 GX 168, 169, 185
ESO 319-16 GX 168, 169
ESO 319-24 GX 185
ESO 319-26 GX 185
ESO 320-2 GX 185
ESO 320-4 GX 168
ESO 320-5 GX 168
ESO 320-7 GX 168
ESO 320-15 GX 185
ESO 320-19 GX 168
ESO 320-20 GX 168
ESO 320-24 GX 168
ESO 320-26 GX 168
ESO 320-27 GX 168
ESO 320-30 GX 168
ESO 320-31 GX 168
ESO 320-32 GX 168, 185
ESO 320-35 GX 168
ESO 321-1 GX 168
ESO 321-2 GX 184, 185
ESO 321-3 GX 184, 185
ESO 321-4 GX 168
ESO 321-5 GX 168
ESO 321-7 GX 168, 184
ESO 321-10 GX 168
ESO 321-12 GX 168, 184
ESO 321-14 GX 168
ESO 321-16 GX 168
ESO 321-17 GX 184
ESO 321-18 GX 168
ESO 321-19 GX 168
ESO 321-20 GX 168
ESO 321-21 GX 168, 184
ESO 321-25 GX 168
ESO 321-26 GX 168
ESO 322-5 GX 168
ESO 322-7 GX 168, 184
ESO 322-9 GX 168
ESO 322-100 GX 184, A23
ESO 322-101 GX 184, A23
ESO 322-102 GX 184, A23
ESO 322-11 GX 168, 184
ESO 322-18 GX 168
ESO 322-19 GX 168, 184
ESO 322-20 GX 168, 184
ESO 322-21 GX 168, 184
ESO 322-25 GX 168, 184
ESO 322-27 GX 168, 184
ESO 322-28 GX 184
ESO 322-31 GX 184
ESO 322-32 GX 167, 168
ESO 322-34 GX 184
ESO 322-37 GX 184
ESO 322-38 GX 184
ESO 322-40 GX 167, 168, 184
ESO 322-42 GX 184
ESO 322-45 GX 184, A23
ESO 322-51 GX 184, A23
ESO 322-54 GX 167, 168, A23
ESO 322-74 GX 184, A23
ESO 322-75 GX 167, 184, A23
ESO 322-76 GX 167, 184, A23
ESO 322-84 GX 167, A23
ESO 322-85 GX 167, 184, A23
ESO 322-89 GX 184, A23
ESO 322-92 GX 167, 184, A23
ESO 322-94 GX 184, A23
ESO 322-96 GX 184, A23
ESO 322-99 GX 184, A23
ESO 323-2 GX 167, 184, A23
ESO 323-4 GX 167
ESO 323-5 GX 184, A23
ESO 323-7 GX 167, 184, A23
ESO 323-8 GX 184, A23
ESO 323-9 GX 184, A23
ESO 323-10 GX 184, A23
ESO 323-11 GX 184, A23
ESO 323-12 GX 167, 184, A23
ESO 323-14 GX 184, A23
ESO 323-15 GX 167
ESO 323-18 GX 184, A23
ESO 323-19 GX 184, A23
ESO 323-20 GX 184, A23
ESO 323-23 GX 167, 184, A23
ESO 323-24 GX 167
ESO 323-25 GX 167
ESO 323-27 GX 167, 184, A23
ESO 323-28 GX 184, A23
ESO 323-32 GX 184, A23
ESO 323-33 GX 184, A23
ESO 323-34 GX 184, A23
ESO 323-38 GX 184, A23
ESO 323-42 GX 167, 184, A23
ESO 323-44 GX 167, A23
ESO 323-49 GX 167, A23
ESO 323-54 GX 167, 184
ESO 323-63 GX 167, 184
ESO 323-66 GX 184
ESO 323-67 GX 184
ESO 323-68 GX 184
ESO 323-71 GX 167
ESO 323-73 GX 167
ESO 323-77 GX 167, 184
ESO 323-79 GX 167
ESO 323-81 GX 167, 184
ESO 323-82 GX 184
ESO 323-85 GX 184
ESO 323-89 GX 167
ESO 323-90 GX 184
ESO 323-92 GX 167
ESO 323-93 GX 184
ESO 323-97 GX 167
ESO 323-99 GX 184
ESO 324-1 GX 184
ESO 324-3 GX 167
ESO 324-5 GX 167
ESO 324-7 GX 167, 184
ESO 324-9 GX 167
ESO 324-11 GX 167
ESO 324-15 OC 184
ESO 324-18 GX 167
ESO 324-21 GX 167, 184
ESO 324-23 GX 167
ESO 324-24 GX 184
ESO 324-26 GX 167
ESO 324-27 GX 167, 184
ESO 324-29 GX 184
ESO 324-33 GX 167
ESO 324-35 GX 167, 183, 184
ESO 324-36 GX 167
ESO 324-40 GX 167, 183, 184
ESO 324-42 GX 183, 184
ESO 324-44 GX 167
ESO 325-1 GX 167
ESO 325-4 GX 167
ESO 325-6 GX 167
ESO 325-11 GX 183
ESO 325-16 GX 167
ESO 325-19 GX 167
ESO 325-25 GX 183
ESO 325-28 GX 167
ESO 325-30 GX 167
ESO 325-38 GX 166, 167, 183
ESO 325-40 GX 166, 167
ESO 325-42 GX 166, 167, 183
ESO 325-43 GX 166, 167
ESO 325-44 GX 166, 167
ESO 325-45 GX 166, 167
ESO 325-52 GX 166
ESO 325-53 GX 166
ESO 325-55 GX 166
ESO 326-6 GX 166, 183
ESO 327-7 GX 166
ESO 327-20 GX 166
ESO 327-23 GX 166
ESO 327-27 GX 166
ESO 327-31 GX 166
ESO 327-32 GX 166
ESO 327-36 GX 166
ESO 327-39 GX 183
ESO 328-5 GX 166
ESO 328-31 GX 182, 183
ESO 328-41 GX 165, 166
ESO 328-43 GX 182
ESO 328-46 GX 165, 166
ESO 329-2 OC 182
ESO 329-7 GX 165
ESO 329-12 GX 165
ESO 329-15 GX 165
ESO 329-16 GX 182
ESO 329-22 GX 165, 182
ESO 332-20 OC 164, A22
ESO 332-22 OC 164, 181, A22
ESO 335-5 OC 181
ESO 335-11 GX 163
ESO 336-3 GX 180
ESO 336-4 GX 180
ESO 336-6 GX 163
ESO 336-8 GX 180
ESO 336-12 GX 180
ESO 336-13 GX 163
ESO 336-16 GX 180
ESO 337-6 GX 180
ESO 338-4 GX 180
ESO 338-12 GX 162, 180
ESO 338-17 GX 162
ESO 338-21 GX 162
ESO 339-1 GX 162, 180
ESO 339-3 GX 162, 180
ESO 339-4 GX 162
ESO 339-6 GX 162
ESO 339-8 GX 162
ESO 339-9 GX 162
ESO 339-11 GX 162
ESO 339-12 GX 162, 179, 180
ESO 339-17 GX 162
ESO 339-20 GX 162
ESO 339-21 GX 179, 180
ESO 339-25 GX 162
ESO 339-26 GX 162
ESO 339-27 GX 162
ESO 339-28 GX 162
ESO 339-31 GX 162
ESO 339-34 GX 162
ESO 339-36 GX 162
ESO 340-1 GX 162, 179
ESO 340-2 GX 162, 179
ESO 340-3 GX 162
ESO 340-5 GX 162, 179
ESO 340-6 GX 162, 179
ESO 340-7 GX 162
ESO 340-8 GX 162, 179
ESO 340-9 GX 162
ESO 340-10 GX 179
ESO 340-12 GX 162
ESO 340-13 GX 179
ESO 340-14 GX 179
ESO 340-15 GX 179
ESO 340-16 GX 162, 179
ESO 340-17 GX 162
ESO 340-21 GX 179
ESO 340-25 GX 162, 179
ESO 340-26 GX 179
ESO 340-28 GX 162, 179
ESO 340-29 GX 162, 179
ESO 340-32 GX 162
ESO 340-36 GX 162
ESO 340-38 GX 179
ESO 340-40 GX 162
ESO 340-42 GX 162, 179
ESO 340-43 GX 179
ESO 341-4 GX 161, 162
ESO 341-6 GX 161, 162
ESO 341-11 GX 161
ESO 341-12 GX 161
ESO 341-13 GX 161
ESO 341-17 GX 161
ESO 341-21 GX 161, 179
ESO 341-23 GX 161
ESO 341-26 GX 161
ESO 341-29 GX 161
ESO 341-32 GX 161
ESO 342-1 GX 161
ESO 342-5 GX 161, 179
ESO 342-6 GX 161
ESO 342-8 GX 161
ESO 342-10 GX 161
ESO 342-12 GX 161, 179
ESO 342-13 GX 161
ESO 342-19 GX 161
ESO 342-22 GX 161
ESO 342-25 GX 161
ESO 342-26 GX 179
ESO 342-27 GX 179
ESO 342-30 GX 179
ESO 342-32 GX 179
ESO 342-34 GX 161
ESO 342-35 GX 161
ESO 342-38 GX 161
ESO 342-42 GX 161, 179
ESO 342-45 GX 161, 179
ESO 342-48 GX 161
ESO 342-50 GX 161
ESO 342-52 GX 179
ESO 343-1 GX 161, 179
ESO 343-3 GX 161
ESO 343-7 GX 161, 178, 179
ESO 343-9 GX 161, 178, 179
ESO 343-11 GX 161
ESO 343-13 GX 161
ESO 343-14 GX 161
ESO 343-15 GX 161
ESO 343-18 GX 161
ESO 343-20 GX 161
ESO 343-21 GX 161
ESO 343-23 GX 161
ESO 343-24 GX 161, 178
ESO 343-26 GX 161
ESO 343-28 GX 161
ESO 343-31 GX 161
ESO 343-34 GX 161
ESO 343-36 GX 161, 178
ESO 345-2 GX 160, 178
ESO 345-11 GX 160
ESO 345-17 GX 160
ESO 345-21 GX 160
ESO 345-28 GX 160
ESO 345-32 GX 160
ESO 345-33 GX 160
ESO 345-42 GX 160
ESO 345-44 GX 178
ESO 345-45A GX 160, 178
ESO 345-46 GX 160
ESO 345-50 GX 160, 178
ESO 346-1 GX 160
ESO 346-3 GX 160
ESO 346-6 GX 160, 178
ESO 346-7 GX 160
ESO 346-9 GX 160, 178
ESO 346-14 GX 160
ESO 346-18 GX 160, 178
ESO 346-24 GX 160, 178
ESO 346-25 GX 178
ESO 346-32 GX 160
ESO 346-33 GX 160
ESO 347-2 GX 160, 177, 178
ESO 347-3 GX 160
ESO 347-8 GX 177
ESO 347-17 GX 159
ESO 347-20 GX 177
ESO 347-21 GX 159
ESO 347-27 GX 159
ESO 347-29 GX 159
ESO 348-9 GX 159
ESO 349-1 GX 159
ESO 349-2 GX 159
ESO 349-3 GX 159
ESO 349-5 GX 159
ESO 349-9 GX 159
ESO 349-10 GX 159
ESO 349-13 GX 159
ESO 349-17 GX 159
ESO 349-18 GX 159
ESO 349-19 GX 159
ESO 349-20 GX 159
ESO 349-22 GX 159
ESO 349-25 GX 159
ESO 349-26 GX 159
ESO 349-27 GX 159
ESO 349-33 GX 159
ESO 349-37 GX 159
ESO 349-38 GX 159
ESO 349-39 GX 159
ESO 350-4 GX 159
ESO 350-7 GX 159
ESO 350-9 GX 159
ESO 350-15 GX 159
ESO 350-19 GX 159
ESO 350-20 GX 159
ESO 350-22 GX 159
ESO 350-27 GX 159
ESO 350-28 GX 159
ESO 350-34 GX 159
ESO 350-37 GX 159, 176
ESO 350-38 GX 159, 176
ESO 351-3 GX 159, 176
ESO 351-16 GX 176
ESO 351-21 GX 176
ESO 351-28 GX 176
ESO 351-30 GX 176
ESO 352-2 GX 176
ESO 352-6 GX 176
ESO 352-7 GX 176
ESO 352-8 GX 176
ESO 352-15 GX 176
ESO 352-18 GX 176
ESO 352-20 GX 176
ESO 352-28 GX 176
ESO 352-38 GX 176
ESO 352-41 GX 176
ESO 352-47 GX 176
ESO 352-49 GX 176
ESO 352-50 GX 176
ESO 352-51 GX 176
ESO 352-54 GX 176
ESO 352-55 GX 176
ESO 352-57 GX 176
ESO 352-61 GX 176
ESO 352-62 GX 176
ESO 352-63 GX 176
ESO 352-64 GX 176
ESO 352-69 GX 176
ESO 352-71 GX 176
ESO 352-73 GX 176
ESO 352-76 GX 176
ESO 353-5 GX 176
ESO 353-7 GX 176
ESO 353-9 GX 176
ESO 353-14 GX 176
ESO 353-20 GX 176
ESO 353-25 GX 176
ESO 353-26 GX 176
ESO 353-29 GX 176
ESO 353-31 GX 176
ESO 353-33 GX 176
ESO 353-38 GX 176
ESO 353-40 GX 176
ESO 353-41 GX 176
ESO 353-45 GX 176
ESO 353-49 GX 176
ESO 354-3 GX 176
ESO 354-4 GX 176
ESO 354-19 GX 175, 176
ESO 354-21 GX 175, 176
ESO 354-25 GX 175, 176
ESO 354-26 GX 175, 176
ESO 354-29 GX 175, 176
ESO 354-34 GX 175
ESO 354-36 GX 175
ESO 354-41 GX 175
ESO 355-4 GX 175
ESO 355-8 GX 175
ESO 355-10 GX 175
ESO 355-26 GX 175
ESO 355-30 GX 175
ESO 355-31 GX 175
ESO 356-2 GX 175
ESO 356-4 GX 175
ESO 356-9 GX 175

ESO 356-13 GX 175
ESO 356-14A GX 175
ESO 356-18 GX 175
ESO 356-20 GX 175
ESO 356-22 GX 175
ESO 356-24 GX 175
ESO 356-26 GX 175
ESO 357-1 GX 175
ESO 357-7 GX 175
ESO 357-10 GX 175
ESO 357-12 GX 174, 175
ESO 357-25 GX 174, 175
ESO 358-5 GX 174
ESO 358-6 GX 174
ESO 358-10 GX 174
ESO 358-20 GX 174
ESO 358-22 GX 174
ESO 358-25 GX 174
ESO 358-42 GX 174
ESO 358-47 GX 174
ESO 358-50 GX 174
ESO 358-51 GX 174
ESO 358-59 GX 174
ESO 358-60 GX 174
ESO 358-61 GX 174
ESO 358-63 GX 174
ESO 358-66 GX 174
ESO 358-67 GX 174
ESO 359-2 GX 174
ESO 359-3 GX 174
ESO 359-13 GX 174
ESO 359-16 GX 174
ESO 359-18 GX 174
ESO 359-29 GX 174
ESO 359-31 GX 174
ESO 360-2 GX 174
ESO 360-4 GX 174
ESO 360-7 GX 174
ESO 360-10 GX 174
ESO 360-14 GX 174
ESO 361-9 GX 173
ESO 361-12 GX 173
ESO 361-15 GX 173
ESO 361-19 GX 173
ESO 361-23 GX 173
ESO 361-25 GX 173
ESO 362-1 GX 173
ESO 362-3 GX 173
ESO 362-8 GX 173
ESO 362-9 GX 173
ESO 362-11 GX 173
ESO 362-12 GX 173
ESO 362-13 GX 173
ESO 362-15 GX 173
ESO 362-18 GX 173
ESO 362-19 GX 173
ESO 363-3 GX 173
ESO 363-15 GX 173
ESO 363-17 GX 173
ESO 363-18 GX 173
ESO 364-2 GX 173
ESO 364-4 GX 173
ESO 364-7 GX 173
ESO 364-8 GX 173
ESO 364-12 GX 173
ESO 364-16 GX 172, 173
ESO 364-17 GX 172, 173
ESO 364-18 GX 172, 173
ESO 364-19 GX 172, 173
ESO 364-23 GX 172, 173
ESO 364-28 GX 172
ESO 364-29 GX 172
ESO 364-33 GX 172
ESO 364-35 GX 172
ESO 364-36 GX 172
ESO 364-39 GX 172
ESO 364-40 GX 172
ESO 364-41 GX 172
ESO 364-43 GX 172
ESO 365-1 GX 172
ESO 365-5 GX 172
ESO 365-7 GX 172
ESO 365-9 GX 172
ESO 365-10 GX 172
ESO 365-11 GX 172
ESO 365-13 GX 172
ESO 365-16 GX 172
ESO 365-20 GX 172
ESO 365-21 GX 172
ESO 365-25 GX 172
ESO 365-26A GX 172
ESO 365-26B GX 172
ESO 365-27 GX 172
ESO 365-28 GX 172
ESO 365-29 GX 172
ESO 365-35 GX 172
ESO 366-4 GX 172
ESO 366-5 GX 172
ESO 366-8 GX 172
ESO 366-9 GX 172
ESO 366-11 GX 172
ESO 366-16 GX 172
ESO 366-28 GX 172
ESO 366-30 GX 172
ESO 367-6 GX 172
ESO 367-7 GX 171, 172
ESO 367-8 GX 171, 172
ESO 367-10 OC 171, 172
ESO 367-17 GX 171, 172
ESO 367-18 GX 171, 172
ESO 367-22 GX 171, 172
ESO 368-14 OC 171
ESO 370-6 GX 171
ESO 371-3 GX 170, 171
ESO 371-16 GX 170
ESO 371-17 GX 170
ESO 371-20 GX 170
ESO 371-24 GX 170
ESO 371-25 OC 170
ESO 371-26 GX 170
ESO 371-28 GX 170
ESO 371-30 GX 170
ESO 372-8 GX 170
ESO 372-9 GX 170
ESO 372-12 GX 170
ESO 372-16 GX 170
ESO 372-18 GX 170
ESO 372-23 GX 170
ESO 373-5 GX 170
ESO 373-8 GX 170
ESO 373-10 GX 170
ESO 373-12 GX 170
ESO 373-13 GX 170
ESO 373-19 GX 170
ESO 373-20 GX 170
ESO 373-21 GX 170
ESO 373-26 GX 170
ESO 373-30 GX 170
ESO 374-3 GX 170
ESO 374-8 GX 170
ESO 374-13 GX 169, 170
ESO 374-14 GX 169, 170
ESO 374-28 GX 169
ESO 374-38 GX 169
ESO 374-44 GX 169
ESO 374-45 GX 169
ESO 374-46 GX 169
ESO 375-3 GX 169
ESO 375-7 GX 169
ESO 375-20 GX 169
ESO 375-22 GX 169
ESO 375-26 GX 169
ESO 375-41 GX 169
ESO 375-47 GX 169
ESO 375-52 GX 169
ESO 375-57 GX 169
ESO 375-62 GX 169
ESO 375-64 GX 169
ESO 375-69 GX 169
ESO 375-70 GX 169
ESO 375-71 GX 169
ESO 375-72 GX 169
ESO 376-6 GX 169
ESO 376-7 GX 169
ESO 376-9 GX 169
ESO 376-11 GX 169
ESO 376-12 GX 169
ESO 376-20 GX 169
ESO 376-22 GX 169
ESO 376-23 GX 169
ESO 376-26 GX 169
ESO 377-3 GX 169
ESO 377-6 GX 169
ESO 377-7 GX 169
ESO 377-10 GX 169
ESO 377-12 GX 169
ESO 377-19 GX 169
ESO 377-21 GX 169
ESO 377-24 GX 169
ESO 377-29 GX 169
ESO 377-31 GX 168, 169
ESO 377-34 GX 168, 169
ESO 377-38 GX 168, 169
ESO 377-41 GX 168
ESO 377-46 GX 168
ESO 378-3 GX 168
ESO 378-5 GX 168
ESO 378-8 GX 168
ESO 378-11 GX 168
ESO 378-12 GX 168
ESO 378-20 GX 168
ESO 378-27 GX 168
ESO 379-1 GX 168
ESO 379-6 GX 168
ESO 379-13 GX 168
ESO 379-19 GX 168
ESO 379-20 GX 168
ESO 379-21 GX 168
ESO 379-22 GX 168
ESO 379-26 GX 168
ESO 379-27 GX 168
ESO 379-30 GX 168
ESO 379-31 GX 168
ESO 379-35 GX 168
ESO 380-1 GX 168
ESO 380-6 GX 168
ESO 380-7 GX 168
ESO 380-8 GX 168
ESO 380-14 GX 168
ESO 380-19 GX 168
ESO 380-25 GX 168
ESO 380-29 GX 168
ESO 380-30 GX 168
ESO 380-33 GX 168
ESO 380-34 GX 168
ESO 380-35 GX 168
ESO 380-40 GX 168
ESO 380-41 GX 168
ESO 380-42 GX 168
ESO 380-43 GX 168
ESO 380-46 GX 168
ESO 381-4 GX 167, 168
ESO 381-5 GX 167, 168
ESO 381-9 GX 167, 168
ESO 381-12 GX 167
ESO 381-14 GX 167
ESO 381-17 GX 167
ESO 381-20 GX 167
ESO 381-23 GX 167
ESO 381-29 GX 167
ESO 381-32 GX 167
ESO 381-38 GX 167, A21
ESO 381-41 GX 167, A21
ESO 381-42 GX 167, A21
ESO 381-44 GX 167, A21
ESO 381-46 GX 167
ESO 381-47 GX 167
ESO 381-48 GX 167
ESO 381-50 GX 167, A21
ESO 381-51 GX 167, A21
ESO 382-2 GX 167, A21
ESO 382-7 GX 167
ESO 382-10 GX 167, A21
ESO 382-12 GX 167
ESO 382-16 GX 167
ESO 382-17 GX 167
ESO 382-23 GX 167
ESO 382-24 GX 167
ESO 382-25 GX 167
ESO 382-27 GX 167
ESO 382-31 GX 167
ESO 382-32 GX 167
ESO 382-33 GX 167
ESO 382-34 GX 167
ESO 382-41 GX 167
ESO 382-43 GX 167
ESO 382-44 GX 167
ESO 382-45 GX 167
ESO 382-51 GX 167
ESO 382-54 GX 167
ESO 382-58 GX 167
ESO 382-60 GX 167
ESO 382-67 GX 167
ESO 383-2 GX 167
ESO 383-5 GX 167
ESO 383-8 GX 167
ESO 383-10 OC 167
ESO 383-12 GX 167
ESO 383-19 GX 167
ESO 383-20 GX 167
ESO 383-25 GX 167
ESO 383-26 GX 167
ESO 383-27 GX 167
ESO 383-30 GX 167
ESO 383-31 GX 167
ESO 383-35 GX 167
ESO 383-38 GX 167
ESO 383-44 GX 167
ESO 383-45 GX 167
ESO 383-47 GX 167
ESO 383-48 GX 167
ESO 383-50 GX 167
ESO 383-52 GX 167
ESO 383-53 GX 167
ESO 383-55 GX 167
ESO 383-60 GX 167
ESO 383-64 GX 167
ESO 383-67 GX 167
ESO 383-71 GX 167
ESO 383-73 GX 167
ESO 383-76 GX 167
ESO 383-81 GX 167
ESO 383-87 GX 167
ESO 383-88 GX 167
ESO 383-91 GX 167
ESO 383-92 GX 167
ESO 383-93 GX 167
ESO 384-2 GX 167
ESO 384-3 GX 167
ESO 384-5 GX 167
ESO 384-6 GX 167
ESO 384-7 GX 167
ESO 384-9 GX 167
ESO 384-11 GX 167
ESO 384-12 GX 167
ESO 384-13 GX 167
ESO 384-14 GX 166, 167
ESO 384-16 GX 166, 167
ESO 384-18 GX 166, 167
ESO 384-19 GX 166, 167
ESO 384-21 GX 166, 167
ESO 384-23 GX 166, 167
ESO 384-25 GX 166, 167
ESO 384-26 GX 166, 167
ESO 384-27 GX 166, 167
ESO 384-29 GX 166, 167
ESO 384-33 GX 166, 167
ESO 384-35 GX 166, 167
ESO 384-37 GX 166, 167
ESO 384-41 GX 166
ESO 384-43 GX 166
ESO 384-47 GX 166
ESO 384-49 GX 166
ESO 384-51 GX 166
ESO 384-53 GX 166
ESO 384-55 GX 166
ESO 384-57 GX 166
ESO 385-2 GX 166
ESO 385-3 GX 166
ESO 385-5 GX 166
ESO 385-8 GX 166
ESO 385-12 GX 166
ESO 385-14 GX 166
ESO 385-15 GX 166
ESO 385-17 GX 166
ESO 385-25 GX 166
ESO 385-28 GX 166
ESO 385-30 GX 166
ESO 385-32 GX 166
ESO 385-33 GX 166
ESO 385-47 GX 166
ESO 385-49 GX 166
ESO 386-2 GX 166
ESO 386-4 GX 166
ESO 386-6 GX 166
ESO 386-9 GX 166
ESO 386-11 GX 166
ESO 386-12 GX 166
ESO 386-14 GX 166
ESO 386-19 GX 166
ESO 386-21 GX 166
ESO 386-31 GX 166
ESO 386-33 GX 166
ESO 386-34 GX 166
ESO 386-38 GX 166
ESO 386-39 GX 166
ESO 386-40 GX 166
ESO 386-41 GX 166
ESO 386-43 GX 166
ESO 386-44 GX 166
ESO 386-47 GX 166
ESO 387-5 GX 166
ESO 387-12 GX 166
ESO 387-16 GX 165, 166
ESO 387-19 GX 165, 166
ESO 387-21 GX 165, 166
ESO 387-26 GX 165, 166
ESO 387-29 GX 165, 166
ESO 387-33 GX 165
ESO 389-5 OC 165
ESO 389-6 GX 165
ESO 392-13 OC 164
ESO 395-2 GX 163
ESO 396-3 GX 163
ESO 396-7 GX 163
ESO 396-16 GX 163
ESO 397-1 OC 163
ESO 397-18 GX 162
ESO 398-12 GX 162
ESO 398-20 GX 162
ESO 398-27 GX 162
ESO 398-29 GX 162
ESO 399-10 GX 162
ESO 399-13 GX 162
ESO 399-14 GX 162
ESO 399-15 GX 162
ESO 399-18 GX 162
ESO 399-19 GX 162
ESO 399-23 GX 162
ESO 399-25 GX 162
ESO 399-26 GX 162
ESO 400-4 GX 162
ESO 400-5 GX 162
ESO 400-6 GX 162
ESO 400-7 GX 162
ESO 400-13 GX 162
ESO 400-15 GX 162
ESO 400-17 GX 162
ESO 400-19 GX 162
ESO 400-20 GX 162
ESO 400-23 GX 162
ESO 400-24 GX 162
ESO 400-25 GX 162
ESO 400-28 GX 162
ESO 400-30 GX 162
ESO 400-33 GX 162
ESO 400-37 GX 162
ESO 400-38 GX 162
ESO 400-40 GX 162
ESO 400-41 GX 162
ESO 400-43 GX 161, 162
ESO 401-7 GX 161
ESO 401-25 GX 161
ESO 401-26 GX 161
ESO 402-3 GX 161
ESO 402-10 GX 161
ESO 402-21 GX 161
ESO 402-26 GX 161
ESO 403-3 GX 161
ESO 403-4 GX 161
ESO 403-6 GX 161
ESO 403-9 GX 161
ESO 403-12 GX 161
ESO 403-17 GX 161
ESO 403-24 GX 161
ESO 403-26 GX 161
ESO 404-3 GX 161
ESO 404-11 GX 160, 161
ESO 404-12 GX 160, 161
ESO 404-15 GX 160, 161
ESO 404-17 GX 160, 161
ESO 404-18 GX 160, 161
ESO 404-19 GX 160, 161
ESO 404-21 GX 160, 161
ESO 404-23 GX 160, 161
ESO 404-27 GX 160, 161
ESO 404-28 GX 160
ESO 404-30 GX 160
ESO 404-31 GX 160
ESO 404-39 GX 160
ESO 404-40 GX 160
ESO 404-43 GX 160
ESO 404-45 GX 160
ESO 405-6 GX 160
ESO 405-9 GX 160
ESO 405-11 GX 160
ESO 405-13 GX 160
ESO 405-15 GX 160
ESO 405-29 GX 160
ESO 405-31 GX 160
ESO 406-4 GX 160
ESO 406-10 GX 160
ESO 406-11 GX 160
ESO 406-17 GX 160
ESO 406-18 GX 160
ESO 406-21 GX 160
ESO 406-22 GX 160
ESO 406-31 GX 160
ESO 406-37 GX 160
ESO 406-42 GX 160
ESO 407-2 GX 160
ESO 407-7 GX 160
ESO 407-9 GX 160
ESO 407-14 GX 159, 160
ESO 407-18 GX 159
ESO 408-8 GX 159
ESO 408-12 GX 159
ESO 408-17 GX 159
ESO 408-21 GX 159
ESO 408-22 GX 159
ESO 408-28 GX 159
ESO 408-37 GX 159
ESO 409-1 GX 141
ESO 409-3 GX 141
ESO 409-21 GX 141
ESO 409-25 GX 141
ESO 410-5 GX 159
ESO 410-18 GX 159
ESO 410-27 GX 141, 158
ESO 411-3 GX 141, 158
ESO 411-6 GX 159, 176
ESO 411-10 GX 176
ESO 411-28 GX 176
ESO 411-29 GX 176
ESO 411-30 GX 176
ESO 411-33 GX 176
ESO 411-34 GX 158
ESO 412-3 GX 158
ESO 412-15 GX 176
ESO 412-21 GX 176
ESO 412-27 GX 158, 176
ESO 413-4 GX 176
ESO 413-16 GX 158
ESO 413-18 GX 158
ESO 413-24 GX 176
ESO 414-8 GX 176
ESO 414-25 GX 157, 158
ESO 414-28 GX 157
ESO 414-32 GX 157
ESO 415-3 GX 175
ESO 415-7 GX 157, 175
ESO 415-10 GX 175
ESO 415-19 GX 175
ESO 415-20 GX 175
ESO 415-22 GX 157
ESO 415-26 GX 175
ESO 415-28 GX 175
ESO 415-31 GX 175
ESO 416-1 GX 157, 175
ESO 416-5 GX 175
ESO 416-8 GX 157
ESO 416-9 GX 175
ESO 416-12 GX 175
ESO 416-18 GX 175
ESO 416-25 GX 175
ESO 416-32 GX 175
ESO 416-37 GX 157
ESO 416-39 GX 175
ESO 416-41 GX 175
ESO 417-1 GX 175
ESO 417-3 GX 157
ESO 417-11 GX 157
ESO 417-18 GX 175
ESO 417-20 GX 175
ESO 418-7 GX 156
ESO 418-7A GX 156
ESO 418-8 GX 174
ESO 418-9 GX 174
ESO 418-11 GX 174
ESO 418-14 GX 174
ESO 419-3 GX 156
ESO 419-12 GX 156
ESO 419-13 GX 174
ESO 420-3 GX 156, 174
ESO 420-5 GX 174
ESO 420-6 GX 174
ESO 420-9 GX 174
ESO 420-10 GX 156, 174
ESO 420-13 GX 174
ESO 420-18 GX 156
ESO 421-19 GX 155, 173
ESO 422-6 GX 173
ESO 422-10 GX 173
ESO 422-16 GX 155
ESO 422-18 GX 173
ESO 422-23 GX 173
ESO 422-27 GX 173
ESO 422-29 GX 155
ESO 422-33 GX 155
ESO 422-40 GX 155, 173
ESO 422-41 GX 173
ESO 423-2 GX 173
ESO 423-16 GX 173
ESO 423-20 GX 155
ESO 423-24 GX 155, 173
ESO 424-1 GX 155, 173
ESO 424-25 OC 173
ESO 424-27 GX 173
ESO 424-33 GX 173
ESO 424-36 GX 155
ESO 425-1 GX 172, 173
ESO 425-2 GX 154, 155
ESO 425-6 OC 154, 172
ESO 425-10 GX 154

ESO 425-11 GX 154, 172
ESO 425-12 GX 154
ESO 425-14 GX 154
ESO 425-15 OC 154, 172
ESO 425-16 GX 172
ESO 425-18 GX 154
ESO 425-19 GX 154
ESO 426-1 GX 154
ESO 426-2 GX 172
ESO 426-7 GX 172
ESO 426-8 GX 154
ESO 426-14 GX 154
ESO 426-18 GX 172
ESO 426-26 OC 172
ESO 427-8 GX 172
ESO 427-13 GX 172
ESO 427-21 GX 172
ESO 427-26 GX 154
ESO 427-29 GX 154, 172
ESO 427-32 OC 172
ESO 427-34 GX 154
ESO 428-2 GX 154
ESO 428-4 GX 154
ESO 428-9 GX 154, 172
ESO 428-11 GX 154, 172
ESO 428-13 GX 153, 154, 171, 172
ESO 428-14 GX 153, 154, 171, 172
ESO 428-16 GX 153, 154
ESO 428-20 GX 153, 154, 171, 172
ESO 428-22 GX 153, 154
ESO 428-23 GX 153, 154, 171, 172
ESO 428-28 GX 171, 172
ESO 428-29 GX 153, 154, 171, 172
ESO 428-31 GX 171
ESO 428-32 GX 171
ESO 428-37 GX 171
ESO 429-2 OC 153
ESO 429-9 GX 171
ESO 429-13 OC 171
ESO 429-19 GX 153, 171
ESO 430-1 GX 153
ESO 430-14 OC 171
ESO 430-18 OC 171
ESO 430-20 GX 153
ESO 430-26 GX 171
ESO 430-28 GX 171
ESO 431-1 GX 153, 171
ESO 431-2 GX 171
ESO 431-18 GX 171
ESO 432-2 GX 170, 171
ESO 432-3 OC 152, 153
ESO 432-12 GX 170
ESO 432-13 GX 170
ESO 433-2 GX 152
ESO 433-4 GX 170
ESO 433-7 GX 170
ESO 433-8 GX 170
ESO 433-10 GX 152
ESO 433-12 GX 170
ESO 433-13 GX 170
ESO 433-15 GX 170
ESO 434-5 GX 170
ESO 434-7 GX 170
ESO 434-9 GX 170
ESO 434-15 GX 170
ESO 434-18 GX 170
ESO 434-21 GX 152
ESO 434-23 GX 152
ESO 434-27 GX 170
ESO 434-28 GX 152
ESO 434-32 GX 152, 170
ESO 434-33 GX 170
ESO 434-34 GX 170
ESO 434-41 GX 170
ESO 435-1 GX 152, 170
ESO 435-3 GX 152, 170
ESO 435-5 GX 170
ESO 435-9 OC 152
ESO 435-10 GX 169, 170
ESO 435-14 GX 151, 152
ESO 435-16 GX 151, 152
ESO 435-17 OC 169, 170
ESO 435-19 GX 169, 170
ESO 435-20 GX 151, 152
ESO 435-29 GX 169, 170
ESO 435-33 OC 169, 170
ESO 435-43 GX 151
ESO 435-50 GX 169
ESO 436-1 GX 151
ESO 436-2 OC 151, 169
ESO 436-19 GX 151, 169
ESO 436-26 GX 169
ESO 436-27 GX 169
ESO 436-29 GX 169
ESO 436-31 GX 151, 169
ESO 436-32 GX 169
ESO 436-34 GX 151, A19
ESO 436-35 GX 169
ESO 436-39 GX 169
ESO 436-44 GX 151, A19
ESO 436-46 GX 151, A19
ESO 437-4 GX 151, A19
ESO 437-9 GX 151, A19
ESO 437-14 GX 169
ESO 437-15 GX 151, A19
ESO 437-18 GX 169
ESO 437-21 GX 151, A19
ESO 437-22 GX 151, A19
ESO 437-30 GX 169
ESO 437-31 GX 151, 169
ESO 437-33 GX 169
ESO 437-35 GX 169
ESO 437-42 GX 169
ESO 437-44 GX 151, A19
ESO 437-45 GX 151, A19
ESO 437-49 GX 169
ESO 437-56 GX 169
ESO 437-65 GX 169
ESO 437-67 GX 169
ESO 437-72 GX 169
ESO 438-5 GX 151
ESO 438-8 GX 169
ESO 438-9 GX 151
ESO 438-10 GX 151
ESO 438-14 GX 151, 169
ESO 438-15 GX 151
ESO 438-17 GX 150, 151
ESO 438-18 GX 150, 151
ESO 438-23 GX 150, 151, 168, 169
ESO 439-9 GX 150, 168
ESO 439-10 GX 150, 168
ESO 439-11 GX 168
ESO 439-20 GX 150, 168
ESO 440-4 GX 150
ESO 440-6 GX 168
ESO 440-11 GX 150
ESO 440-19 GX 168
ESO 440-26 GX 168
ESO 440-27 GX 150
ESO 440-32 GX 168
ESO 440-34 GX 168
ESO 440-37 GX 150
ESO 440-38 GX 168
ESO 440-39 GX 168
ESO 440-41 GX 168
ESO 440-43 GX 150, 168
ESO 440-44 GX 150, 168
ESO 440-46 GX 150
ESO 440-49 GX 168
ESO 440-55 GX 150
ESO 441-1 GX 150, 168
ESO 441-7 GX 168
ESO 441-11 GX 150
ESO 441-12 GX 168
ESO 441-14 GX 168
ESO 441-17 GX 168
ESO 441-22 GX 168
ESO 442-2 GX 150
ESO 442-4 OC 150, 168
ESO 442-6 GX 168
ESO 442-13 GX 149, 150
ESO 442-14 GX 149, 150, 167, 168
ESO 442-15 GX 167, 168
ESO 442-25 GX 149
ESO 442-26 GX 149, 167, A21
ESO 442-28 GX 167, A21
ESO 443-4 GX 149, 167, A21
ESO 443-6 GX 167, A21
ESO 443-11 GX 167, A21
ESO 443-14 GX 167, A21
ESO 443-15 GX 167, A21
ESO 443-17 GX 149, 167, A21
ESO 443-21 GX 149, 167, A21
ESO 443-24 GX 167, A21
ESO 443-28 GX 167, A21
ESO 443-34 GX 167, A21
ESO 443-36 GX 167, A21
ESO 443-37 GX 167, A21
ESO 443-39 GX 167, A21
ESO 443-41 GX 167, A21
ESO 443-42 GX 149, 167, A21
ESO 443-43 GX 167, A21
ESO 443-48 GX 167, A21
ESO 443-50 GX 149, A21
ESO 443-53 GX 167, A21
ESO 443-54 GX 167, A21
ESO 443-55 GX 167, A21
ESO 443-56 GX 167, A21
ESO 443-59 GX 149
ESO 443-66 GX 149, 167, A21
ESO 443-67 GX 167, A21
ESO 443-69 GX 149, A21
ESO 443-72 GX 167, A21
ESO 443-73 GX 167, A21
ESO 443-77 GX 167, A21
ESO 443-79 GX 149
ESO 443-80 GX 149
ESO 443-85 GX 167
ESO 444-2 GX 149
ESO 444-7 GX 167
ESO 444-10 GX 167
ESO 444-16 GX 167
ESO 444-17 GX 167
ESO 444-18 GX 167
ESO 444-21 GX 167
ESO 444-33 GX 167
ESO 444-37 GX 167
ESO 444-46 GX 167
ESO 444-47 GX 167
ESO 444-55 GX 149, 167
ESO 444-71 GX 167
ESO 444-72 GX 167
ESO 444-84 GX 149
ESO 444-86 GX 167
ESO 445-1 GX 167
ESO 445-2 GX 167
ESO 445-8 GX 149
ESO 445-14 GX 167, A18
ESO 445-15 GX 167, A18
ESO 445-16 GX 167, A18
ESO 445-20 GX 149, 167, A18
ESO 445-26 GX 167, A18
ESO 445-33 GX 167, A18
ESO 445-40 GX 167, A18
ESO 445-44 GX 167, A18
ESO 445-49 GX 167, A18
ESO 445-51 GX 149
ESO 445-53 GX 167, A18
ESO 445-54 GX 167, A18
ESO 445-57 GX 167, A18
ESO 445-58 GX 167, A18
ESO 445-59 GX 167, A18
ESO 445-59A GX 167, A18
ESO 445-64 GX 149
ESO 445-65 GX 149, 167, A18
ESO 445-66 GX 167, A18
ESO 445-69 GX 167, A18
ESO 445-73 GX 149, 167, A18
ESO 445-74 OC 167, A18
ESO 445-75 GX 149
ESO 445-76 GX 149, 167, A18
ESO 445-81 GX 166, 167
ESO 445-83 GX 148, 149
ESO 445-85 GX 166, 167
ESO 445-86 GX 166, 167
ESO 445-89 GX 166, 167
ESO 446-1 GX 166, 167
ESO 446-2 GX 166, 167
ESO 446-3 GX 166, 167
ESO 446-7 GX 166
ESO 446-8 GX 166
ESO 446-13 GX 166
ESO 446-17 GX 166
ESO 446-18 GX 148, 166
ESO 446-19 GX 148, 166
ESO 446-23 GX 166
ESO 446-26 GX 166
ESO 446-27 GX 166
ESO 446-29 GX 148
ESO 446-31 GX 148, 166
ESO 446-35 GX 148, 166
ESO 446-36 GX 148, 166
ESO 446-45 GX 166
ESO 446-51 GX 148, 166
ESO 446-53 GX 148, 166
ESO 447-10 GX 148
ESO 447-17 GX 148
ESO 447-18 GX 148, 166
ESO 447-19 GX 148
ESO 447-21 GX 148, 166
ESO 447-22 GX 166
ESO 447-23 GX 148
ESO 447-29 OC 148, 166
ESO 447-30 GX 166
ESO 447-31 GX 166
ESO 447-36 GX 148
ESO 449-4 GX 165, 166
ESO 449-5 GX 165, 166
ESO 450-9 GX 165
ESO 450-11 GX 147
ESO 450-18 GX 147
ESO 451-1 GX 147
ESO 451-8 GX 147, 165
ESO 452-5 GX 147
ESO 452-7 GX 147
ESO 452-8 GX 147, 165
ESO 452-11 GC 146, 147
ESO 453-4 GX 146, 164
ESO 456-9 OC 164, A20
ESO 456-38 GC 145, 146
ESO 457-15 GX 163
ESO 458-10 GX 163
ESO 459-6 GX 163
ESO 459-10 GX 144, 145
ESO 460-4 GX 144, 162
ESO 460-8 GX 162
ESO 460-9 GX 162
ESO 460-13 GX 162
ESO 460-18 GX 144, 162
ESO 460-23 GX 144
ESO 460-25 GX 144
ESO 460-26 GX 144
ESO 460-30 GX 144
ESO 460-31 GX 144
ESO 460-32 GX 144
ESO 460-33 GX 144
ESO 460-34 GX 144
ESO 460-35 GX 144
ESO 461-2 GX 162
ESO 461-3 GX 144
ESO 461-5 GX 162
ESO 461-6 GX 162
ESO 461-7 GX 162
ESO 461-24 GX 162
ESO 461-25 GX 162
ESO 461-29 GX 162
ESO 461-37 GX 144, 162
ESO 461-44 GX 144, 162
ESO 462-8 GX 144
ESO 462-10 GX 144, 162
ESO 462-13 GX 162
ESO 462-15 GX 144
ESO 462-16 GX 144
ESO 462-20 GX 144, 162
ESO 463-10 GX 161, 162
ESO 463-25 GX 161
ESO 463-26 GX 161
ESO 464-1 GX 143
ESO 464-3 GX 161
ESO 464-5 GX 161
ESO 464-16 GX 143
ESO 464-17 GX 161
ESO 464-21 GX 143, 161
ESO 466-1 GX 143, 161
ESO 466-4 GX 143, 161
ESO 466-21 GX 142, 143
ESO 466-24 GX 142, 143
ESO 466-26 GX 142, 143
ESO 466-28 GX 142, 143
ESO 466-36 GX 160, 161
ESO 466-43 GX 160, 161
ESO 466-51 GX 160, 161
ESO 467-3 GX 142
ESO 467-13 GX 142
ESO 467-15 GX 142
ESO 467-16 GX 142
ESO 467-23 GX 142
ESO 467-26 GX 142, 160
ESO 467-27 GX 142
ESO 467-30 GX 142
ESO 467-36 GX 160
ESO 467-37 GX 142
ESO 467-43 GX 160
ESO 467-46 GX 160
ESO 467-51 GX 142
ESO 467-53 GX 142
ESO 467-58 GX 160
ESO 468-6 GX 142
ESO 468-20 GX 160
ESO 469-8 GX 142
ESO 469-11 GX 160
ESO 469-14 GX 142
ESO 469-15 GX 160
ESO 469-17 GX 142, 160
ESO 470-13 GX 159
ESO 470-18 GX 141
ESO 471-2 GX 141
ESO 471-6 GX 159
ESO 471-22 GX 159
ESO 471-24 GX 141
ESO 471-28 GX 159
ESO 471-47 GX 141, 159
ESO 471-49 GX 141, 159
ESO 471-54 GX 159
ESO 472-4 GX 141
ESO 472-10 GX 141
ESO 473-5 GX 141
ESO 473-16 GX 141
ESO 473-18 GX 141
ESO 473-24 GX 141
ESO 473-25 GX 141
ESO 474-25 GX 158
ESO 474-26 GX 158
ESO 475-14 GX 158
ESO 475-15 GX 158
ESO 476-4 GX 158
ESO 476-5 GX 158
ESO 476-8 GX 158
ESO 476-10 GX 158
ESO 476-16 GX 158
ESO 476-25 GX 158
ESO 477-7 GX 158
ESO 477-14 GX 158
ESO 477-16 GX 157, 158
ESO 477-22 GX 157, 158
ESO 478-6 GX 157
ESO 479-1 GX 157
ESO 479-4 GX 157
ESO 479-20 GX 157
ESO 479-25 GX 157
ESO 479-31 GX 157
ESO 479-33 GX 157
ESO 479-34 GX 157
ESO 479-37 GX 157
ESO 479-38 GX 157
ESO 479-40 GX 157
ESO 479-43 GX 157
ESO 480-9 GX 157
ESO 480-22 GX 157
ESO 480-25 GX 157
ESO 481-7 GX 157
ESO 481-11 GX 157
ESO 481-14 GX 157
ESO 481-16 GX 156, 157
ESO 481-17 GX 156, 157
ESO 481-18 GX 156, 157
ESO 481-19 GX 156, 157
ESO 481-21 GX 156, 157
ESO 481-29 GX 156
ESO 481-30 GX 156
ESO 482-1 GX 156
ESO 482-5 GX 156
ESO 482-32 GX 156
ESO 482-35 GX 156
ESO 482-36 GX 156
ESO 482-36A GX 156
ESO 482-46 GX 156
ESO 482-47 GX 156
ESO 483-1 GX 156
ESO 483-6 GX 156
ESO 483-9 GX 156
ESO 483-12 GX 156
ESO 483-13 GX 156
ESO 484-5 GX 156
ESO 485-4 GX 155, 156
ESO 485-12 GX 155
ESO 485-16 GX 155
ESO 485-21 GX 155
ESO 486-3 GX 155
ESO 486-4 GX 155
ESO 486-7 GX 155
ESO 486-19 GX 155
ESO 486-22 GX 155
ESO 486-23 GX 155
ESO 486-29 GX 155
ESO 486-32 GX 155
ESO 486-34 GX 155
ESO 486-37 GX 155
ESO 486-38 GX 155
ESO 486-39 GX 155
ESO 486-41 GX 155
ESO 486-45 OC 155
ESO 486-49 GX 155
ESO 486-52 GX 155
ESO 486-53A/B GX 155
ESO 486-57 GX 155
ESO 487-5 GX 155
ESO 487-17 GX 155
ESO 487-22 GX 155
ESO 487-30 GX 155
ESO 487-35 GX 155
ESO 487-36 GX 155
ESO 487-37 GX 155
ESO 488-1 GX 155
ESO 488-7 GX 155
ESO 488-9 GX 155
ESO 488-12 GX 155
ESO 488-15 GX 155
ESO 488-19 GX 155
ESO 488-27 GX 155
ESO 488-35 GX 155
ESO 488-49 GX 154, 155
ESO 488-51 GX 154, 155
ESO 488-59 GX 154, 155
ESO 488-60 GX 154
ESO 489-1 OC 154
ESO 489-4 GX 154
ESO 489-5 GX 154
ESO 489-6 GX 154
ESO 489-7 GX 154
ESO 489-10 GX 154
ESO 489-11 GX 154
ESO 489-15 GX 154
ESO 489-19 GX 154
ESO 489-22 GX 154
ESO 489-23 GX 154
ESO 489-28 GX 154
ESO 489-29 GX 154
ESO 489-31 GX 154
ESO 489-35 GX 154
ESO 489-37 GX 154
ESO 489-40 GX 154
ESO 489-47 GX 154
ESO 489-50 GX 154
ESO 489-53 GX 154
ESO 489-54 GX 154
ESO 489-57 GX 154
ESO 490-5 GX 154
ESO 490-6 GX 154
ESO 490-7 GX 154
ESO 490-10 GX 154
ESO 490-12 GX 154
ESO 490-14 GX 154
ESO 490-16 GX 154
ESO 490-17 GX 154
ESO 490-18 GX 154
ESO 490-20 GX 154
ESO 490-22 GX 154
ESO 490-26 GX 154
ESO 490-28 GX 154
ESO 490-31 GX 154
ESO 490-36 GX 154
ESO 490-37 GX 154
ESO 490-38 GX 154
ESO 490-41 GX 154
ESO 490-45 GX 154
ESO 491-1 GX 154
ESO 491-6 GX 154
ESO 491-9 GX 154
ESO 491-10 GX 154
ESO 491-12 GX 154
ESO 491-13 GX 154
ESO 491-15 GX 154
ESO 491-20 GX 154
ESO 491-21 GX 154
ESO 491-22 GX 154
ESO 492-2 GX 154
ESO 493-3 OC 153
ESO 494-4 GX 153
ESO 494-7 GX 153
ESO 494-9 OC 153
ESO 494-10 GX 153
ESO 494-12 GX 153
ESO 494-16 GX 153
ESO 494-19 GX 153
ESO 494-21 GX 153
ESO 494-22 GX 153
ESO 494-24 GX 153
ESO 494-25 GX 153
ESO 494-26 GX 153
ESO 494-27 GX 153
ESO 494-29 GX 153
ESO 494-31 GX 153
ESO 494-35 GX 153
ESO 494-36 GX 153
ESO 494-39 GX 153
ESO 494-42 GX 153
ESO 495-5 GX 153
ESO 495-6 GX 153
ESO 495-9 GX 153
ESO 495-11 GX 153
ESO 495-12 GX 153
ESO 495-13 GX 153
ESO 495-17 GX 153
ESO 495-21 GX 152, 153
ESO 496-13 GX 152
ESO 496-19 GX 152

ESO 496-22 GX 152
ESO 497-1 GX 152
ESO 497-2 GX 152
ESO 497-17 GX 152
ESO 497-18 GX 152
ESO 497-22 GX 152
ESO 497-26 GX 152
ESO 497-27 GX 152
ESO 497-28 GX 152
ESO 497-29 GX 152
ESO 497-39 GX 152
ESO 497-42 GX 152
ESO 498-3 GX 152
ESO 498-4 GX 152
ESO 498-5 GX 152
ESO 498-6 GX 152
ESO 498-13 GX 152
ESO 499-2 GX 152
ESO 499-4 GX 152
ESO 499-5 GX 152
ESO 499-7 GX 152
ESO 499-8 GX 152
ESO 499-9 GX 152
ESO 499-11 GX 152
ESO 499-13 GX 152
ESO 499-21 GX 152
ESO 499-22 GX 151, 152
ESO 499-23 GX 151, 152
ESO 499-26 GX 151, 152
ESO 499-28 GX 151, 152
ESO 499-30 GX 151, 152
ESO 499-32 GX 151, 152
ESO 499-34 GX 151, 152
ESO 499-37 GX 151, 152
ESO 500-4 GX 151
ESO 500-6 GX 151
ESO 500-17 GX 151
ESO 500-18 GX 151
ESO 500-31 GX 151
ESO 500-32 GX 151
ESO 501-1 GX 151
ESO 501-3 GX 151, A19
ESO 501-10 GX 151, A19
ESO 501-11 GX 151
ESO 501-12 GX 151
ESO 501-13 GX 151, A19
ESO 501-23 GX 151
ESO 501-24 GX 151
ESO 501-25 GX 151, A19
ESO 501-35 GX 151, A19
ESO 501-51 GX 151, A19
ESO 501-56 GX 151, A19
ESO 501-65 GX 151, A19
ESO 501-66 GX 151, A19
ESO 501-67 GX 151, A19
ESO 501-68 GX 151, A19
ESO 501-75 GX 151, A19
ESO 501-79 GX 151
ESO 501-80 GX 151
ESO 501-82 GX 151, A19
ESO 501-84 GX 151, A19
ESO 501-86 GX 151
ESO 501-88 GX 151
ESO 502-5 GX 151
ESO 502-7 GX 151
ESO 502-11 GX 151
ESO 502-12 GX 151
ESO 502-15 GX 151
ESO 502-16 GX 151
ESO 502-17 GX 151
ESO 502-18 GX 151
ESO 502-20 GX 151
ESO 502-23 GX 151
ESO 503-5 GX 151
ESO 503-7 GX 151
ESO 503-11 GX 150, 151
ESO 503-22 GX 150
ESO 504-1 GX 150
ESO 504-3 GX 150
ESO 504-4 GX 150
ESO 504-5 GX 150
ESO 504-10 GX 150
ESO 504-17 GX 150
ESO 504-19 GX 150
ESO 504-24 GX 150
ESO 504-25 GX 150
ESO 504-27 GX 150
ESO 504-28 GX 150
ESO 504-30 GX 150
ESO 505-2 GX 150
ESO 505-3 GX 150
ESO 505-7 GX 150
ESO 505-8 GX 150
ESO 505-9 GX 150
ESO 505-12 GX 150
ESO 505-13 GX 150
ESO 505-14 GX 150
ESO 505-15 GX 150
ESO 505-23 GX 150
ESO 506-2 GX 150
ESO 506-3 GX 150
ESO 506-4 GX 150
ESO 506-27 GX 149, 150
ESO 506-32 GX 149, 150
ESO 506-33 GX 149, 150
ESO 507-6 GX 149
ESO 507-7 GX 149
ESO 507-8 GX 149
ESO 507-13 GX 149
ESO 507-14 GX 149
ESO 507-17 GX 149
ESO 507-21 GX 149
ESO 507-24 GX 149
ESO 507-25 GX 149
ESO 507-26 GX 149
ESO 507-27 GX 149
ESO 507-28 GX 149
ESO 507-29 GX 149
ESO 507-32 GX 149
ESO 507-35 GX 149
ESO 507-36 GX 149
ESO 507-37 GX 149
ESO 507-43 GX 149
ESO 507-45 GX 149
ESO 507-46 GX 149
ESO 507-62 GX 149
ESO 507-67 GX 149
ESO 508-7 GX 149
ESO 508-8 GX 149
ESO 508-11 GX 149
ESO 508-19 GX 149
ESO 508-24 GX 149
ESO 508-25 GX 149
ESO 508-30 GX 149
ESO 508-33 GX 149
ESO 508-34 GX 149
ESO 508-43 GX 149
ESO 508-44 GX 149
ESO 508-51 GX 149
ESO 508-56 GX 149
ESO 508-62 GX 149
ESO 508-63 GX 149
ESO 508-66 GX 149
ESO 508-78 GX 149
ESO 509-3 GX 149
ESO 509-8 GX 149
ESO 509-9 GX 149
ESO 509-103 GX 149
ESO 509-12 GX 149
ESO 509-19 GX 149
ESO 509-21 GX 149
ESO 509-23 GX 149
ESO 509-26 GX 149
ESO 509-44 GX 149
ESO 509-45 GX 149
ESO 509-53 GX 149
ESO 509-55 GX 149
ESO 509-74 GX 149
ESO 509-95 GX 149
ESO 509-97 GX 149
ESO 509-98 GX 149
ESO 510-3 GX 149
ESO 510-7 GX 149
ESO 510-9 GX 149
ESO 510-10 GX 149
ESO 510-13 GX 149
ESO 510-17 GX 149
ESO 510-26 GX 148, 149
ESO 510-36 GX 148, 149
ESO 510-36A GX 148, 149
ESO 510-43 GX 148, 149
ESO 510-44 GX 148, 149
ESO 510-52 GX 148, 149
ESO 510-54 GX 148
ESO 510-55 GX 148
ESO 510-56 GX 148
ESO 510-58 GX 148
ESO 510-59 GX 148
ESO 510-61 GX 148
ESO 510-65 GX 148
ESO 510-70 GX 148
ESO 510-71 GX 148
ESO 510-74 GX 148
ESO 511-1 GX 148
ESO 511-3 GX 148
ESO 511-6 GX 148
ESO 511-8 GX 148
ESO 511-17 GX 148
ESO 511-18 GX 148
ESO 511-21 GX 148
ESO 511-23 GX 148
ESO 511-26 GX 148
ESO 511-30 GX 148
ESO 511-31 GX 148
ESO 511-32 GX 148
ESO 511-33 GX 148
ESO 511-34 GX 148
ESO 511-35 GX 148
ESO 511-42 GX 148
ESO 511-44 GX 148
ESO 512-5 GX 148
ESO 512-12 GX 148
ESO 512-16 GX 148
ESO 512-18 GX 148
ESO 512-19 GX 148
ESO 512-23 GX 148
ESO 513-4 GX 148
ESO 513-11A GX 148
ESO 513-11B GX 148
ESO 513-15 GX 148
ESO 513-30 GX 148
ESO 514-5 GX 147, 148
ESO 514-6 GX 147, 148
ESO 514-15 GX 147, 148
ESO 514-16 GX 147
ESO 514-17 GX 147
ESO 514-23 GX 147
ESO 514-27 GX 147
ESO 515-3 GX 147
ESO 515-13 GX 147
ESO 516-8 GX 147
ESO 518-7 OC 146
ESO 521-38 OC 145, A17
ESO 522-5 OC 145, A17
ESO 524-1 OC 145
ESO 525-8 OC 144
ESO 526-7 GX 144
ESO 526-11 GX 144
ESO 527-11 GX 144
ESO 527-19 GX 144
ESO 528-8 GX 144
ESO 528-17 GX 144
ESO 528-21 GX 144
ESO 528-23 GX 143, 144
ESO 528-33 GX 143, 144
ESO 528-34 GX 143, 144
ESO 528-36 GX 143, 144
ESO 530-10 GX 143
ESO 530-29 GX 143
ESO 530-30 GX 143
ESO 530-32 GX 143
ESO 530-34 GX 143
ESO 530-48 GX 143
ESO 531-2 GX 143
ESO 531-22 GX 143
ESO 532-11 GX 142, 143
ESO 532-14 GX 142, 143
ESO 532-19 GX 142
ESO 532-21 GX 142
ESO 532-22 GX 142
ESO 532-26 GX 142
ESO 533-4 GX 142
ESO 533-5 GX 142
ESO 533-9 GX 142
ESO 533-10 GX 142
ESO 533-14 GX 142
ESO 533-18 GX 142
ESO 533-20 GX 142
ESO 533-21 GX 142
ESO 533-25 GX 142
ESO 533-28 GX 142
ESO 533-35 GX 142
ESO 533-37 GX 142
ESO 533-45 GX 142
ESO 533-50 GX 142
ESO 534-1 GX 142
ESO 534-2 GX 142
ESO 534-3 GX 142
ESO 534-4 GX 142
ESO 534-9 GX 142
ESO 534-10 GX 142
ESO 534-13 GX 142
ESO 534-24 GX 142
ESO 534-32 GX 142
ESO 535-1 GX 142
ESO 535-15 GX 141, 142
ESO 535-16 GX 141, 142
ESO 536-2 GX 141, 142
ESO 536-14 GX 141
ESO 538-8 GX 141
ESO 538-10 GX 141
ESO 538-22 GX 141
ESO 538-24 GX 141
ESO 539-5 GX 141
ESO 539-7 GX 141
ESO 539-14 GX 141
ESO 540-1 GX 141
ESO 540-2 GX 141
ESO 540-3 GX 141
ESO 540-10 GX 141, 158
ESO 540-14 GX 141, 158
ESO 540-16 GX 141, 158
ESO 540-19 GX 158
ESO 540-31 GX 158
ESO 540-32 GX 158
ESO 541-1 GX 158
ESO 541-4 GX 158
ESO 541-5 GX 158
ESO 541-11 GX 158
ESO 541-13 GX 158
ESO 541-15 GX 158
ESO 541-26 GX 158
ESO 542-6 GX 140, 158
ESO 542-7 GX 140, 158
ESO 542-8 GX 140, 158
ESO 542-30 GX 140, 158
ESO 543-12 GX 158
ESO 543-20 GX 158
ESO 544-2 GX 157, 158
ESO 544-4 GX 157, 158
ESO 544-20 GX 157
ESO 544-30 GX 157
ESO 545-2 GX 157
ESO 545-5 GX 157
ESO 545-13 GX 157
ESO 545-16 GX 157
ESO 545-34 GX 157
ESO 545-40 GX 157
ESO 545-42 GX 157
ESO 546-5 GX 139, 157
ESO 546-6 GX 157
ESO 546-7 GX 157
ESO 546-31 GX 157
ESO 546-34 GX 157
ESO 547-4 GX 157
ESO 547-5 GX 157
ESO 547-9 GX 157
ESO 547-11 GX 138, 157
ESO 547-12 GX 138, 157
ESO 547-15 GX 157
ESO 547-20 GX 138, 157
ESO 548-1 GX 156, 157
ESO 548-5 GX 156, 157
ESO 548-14 GX 138, 156
ESO 548-21 GX 156
ESO 548-23 GX 138, 156
ESO 548-25 GX 156
ESO 548-32 GX 138, 156
ESO 548-33 GX 156
ESO 548-35 GX 156
ESO 548-36 GX 156
ESO 548-40 GX 156
ESO 548-46 GX 138, 156
ESO 548-47 GX 156
ESO 548-65 GX 156
ESO 548-68 GX 156
ESO 548-70 GX 156
ESO 548-75 GX 138, 156
ESO 548-81 GX 156
ESO 549-2 GX 156
ESO 549-6 GX 156
ESO 549-8 GX 156
ESO 549-18 GX 156
ESO 549-21 GX 156
ESO 549-22 GX 156
ESO 549-23 GX 156
ESO 549-36 GX 138, 156
ESO 549-40 GX 156
ESO 550-2 GX 138, 156
ESO 550-5 GX 138, 156
ESO 550-14 GX 156
ESO 550-24 GX 156
ESO 550-25 GX 156
ESO 551-16 GX 156
ESO 551-30 GX 137, 155, 156
ESO 551-31 GX 155, 156
ESO 552-3 GX 155
ESO 552-11 GX 155
ESO 552-12 GX 155
ESO 552-20 GX 155
ESO 552-40 GX 155
ESO 552-43 GX 155
ESO 552-58 GX 155
ESO 552-66 GX 155
ESO 553-2 GX 137, 155
ESO 553-3 GX 137, 155
ESO 553-9 GX 155
ESO 553-10 GX 155
ESO 553-14 GX 155
ESO 553-16 GX 155
ESO 553-20 GX 155
ESO 553-23 GX 155
ESO 553-26 GX 155
ESO 553-33 GX 155
ESO 553-42 GX 155
ESO 553-43 GX 155
ESO 553-44 GX 155
ESO 553-46 GX 155
ESO 554-2 GX 155
ESO 554-4 GX 155
ESO 554-18 GX 155
ESO 554-19 GX 155
ESO 554-24 GX 155
ESO 554-27 GX 155
ESO 554-28 GX 155
ESO 554-29 GX 155
ESO 554-38 GX 155
ESO 555-1 GX 155
ESO 555-2 GX 155
ESO 555-4 GX 155
ESO 555-5 GX 155
ESO 555-6 GX 155
ESO 555-10 GX 155
ESO 555-13 GX 154, 155
ESO 555-14 GX 154, 155
ESO 555-19 GX 154, 155
ESO 555-20 GX 154, 155
ESO 555-22 GX 154, 155
ESO 555-23 GX 154, 155
ESO 555-27 GX 154, 155
ESO 555-28 GX 154
ESO 555-29 GX 154
ESO 555-36 GX 154
ESO 555-39 GX 154
ESO 555-40 GX 154
ESO 556-1 GX 154
ESO 556-2 GX 154
ESO 556-5 GX 154
ESO 556-12 GX 154
ESO 556-15 GX 154
ESO 556-19 GX 154
ESO 556-22 GX 154
ESO 557-2 GX 136, 154
ESO 557-3 GX 154
ESO 557-5 GX 154
ESO 557-6 GX 154
ESO 557-9 GX 154
ESO 558-5 GX 154
ESO 559-2 OC 153, 154
ESO 559-13 OC 153
ESO 560-12 GX 153
ESO 560-13 GX 153
ESO 560-14 GX 153
ESO 561-2 GX 153
ESO 561-3 GX 153
ESO 561-5 OC 153
ESO 561-9 GX 153
ESO 561-23 GX 153
ESO 561-24 GX 153
ESO 561-25 GX 153
ESO 561-27 GX 153
ESO 561-30 GX 153
ESO 561-32 GX 153
ESO 561-33 GX 153
ESO 562-6 GX 153
ESO 562-7 GX 153
ESO 562-13 GX 153
ESO 562-14 GX 134, 153
ESO 562-19 GX 153
ESO 562-23 GX 152, 153
ESO 563-2 GX 152, 153
ESO 563-3 GX 152, 153
ESO 563-11 GX 152, 153
ESO 563-12 GX 152, 153
ESO 563-13 GX 152, 153
ESO 563-14 GX 152, 153
ESO 563-16 GX 152, 153
ESO 563-17 GX 152
ESO 563-21 GX 152
ESO 563-26 GX 152
ESO 563-28 GX 152
ESO 563-31 GX 134, 152
ESO 563-36 GX 152
ESO 564-11 GX 152
ESO 564-27 GX 152
ESO 564-30 GX 152
ESO 564-31 GX 152
ESO 564-32 GX 152
ESO 564-36 GX 152
ESO 565-1 GX 152
ESO 565-7 GX 152
ESO 565-8 GX 152
ESO 565-11 GX 152
ESO 565-19 GX 152
ESO 565-21 GX 152
ESO 565-22 GX 152
ESO 565-24 GX 152
ESO 565-25 GX 152
ESO 565-27 GX 152
ESO 565-29 GX 152
ESO 565-30 GX 152
ESO 565-33 GX 152
ESO 566-2 GX 152
ESO 566-7 GX 152
ESO 566-8 GX 152
ESO 566-10 GX 152
ESO 566-14 GX 152
ESO 566-19 GX 152
ESO 566-24 GX 152
ESO 566-30 GX 151, 152
ESO 567-5 GX 151, 152
ESO 567-6 GX 151, 152
ESO 567-25 GX 151
ESO 567-26 GX 151
ESO 567-29 GX 151
ESO 567-32 GX 151
ESO 567-33 GX 151
ESO 567-37 GX 151
ESO 567-39 GX 151
ESO 567-48 GX 132, 151
ESO 567-52 GX 151
ESO 568-10 GX 151
ESO 568-11 GX 151
ESO 568-12 GX 151
ESO 568-19 GX 151
ESO 569-1 GX 151
ESO 569-9 GX 151
ESO 569-12 GX 151
ESO 569-14 GX 151
ESO 569-16 GX 151
ESO 569-24 GX 151
ESO 570-2 GX 151
ESO 570-4 GX 151
ESO 570-12 OC 151
ESO 570-19 GX 150, 151
ESO 571-12 GX 131, 150
ESO 571-15 GX 150
ESO 571-16 GX 150
ESO 572-7 GX 150
ESO 572-8 GX 150
ESO 572-9 GX 150
ESO 572-18 GX 150
ESO 572-22 GX 150
ESO 572-23 GX 150
ESO 572-24 GX 150
ESO 572-25 GX 150
ESO 572-30 GX 150
ESO 572-32 GX 150
ESO 572-35 GX 150
ESO 572-44 GX 150
ESO 572-49 GX 150
ESO 573-2 GX 150
ESO 573-3 GX 150
ESO 573-6 GX 150
ESO 573-11 GX 150
ESO 573-12 GX 150
ESO 573-14 GX 150
ESO 573-16 GX 150
ESO 573-17 GX 150
ESO 573-22 GX 150
ESO 574-24 GX 149, 150
ESO 574-28 GX 149
ESO 574-29 GX 149
ESO 574-31 GX 149
ESO 574-32 GX 149
ESO 574-33 GX 149
ESO 575-1 GX 149
ESO 575-10 GX 149
ESO 575-13 GX 149
ESO 575-21 GX 149
ESO 575-29 GX 149
ESO 575-32 GX 149
ESO 575-33 GX 149
ESO 575-42 GX 149
ESO 575-43 GX 149
ESO 575-44 GX 149
ESO 575-44A GX 149
ESO 575-47 GX 149
ESO 575-53 GX 149
ESO 575-59 GX 149
ESO 575-61 GX 149
ESO 576-3 GX 149
ESO 576-5 GX 149
ESO 576-8 GX 149

IC 557 GX 93
IC 558 GX 56, 74
IC 559 GX 93
IC 560 GX 113
IC 561 GX 113
IC 562 GX 113
IC 563 GX 113
IC 564 GX 113
IC 565 GX 93
IC 566 GX 113
IC 568 GX 93
IC 569 GX 93
IC 571 GX 93
IC 572 GX 93
IC 574 GX 133
IC 575 GX 133
IC 577 GX 93
IC 578 GX 93
IC 579 GX 133
IC 581 GX 93
IC 582 GX 73, 74, 93
IC 584 GX 93
IC 585 GX 93
IC 586 GX 133
IC 587 GX 113
IC 588 GX 113
IC 590 GX 113
IC 591 GX 93
IC 592 GX 112, 113
IC 593 GX 112, 113
IC 594 GX 112, 113
IC 595 GX 92, 93
IC 596 GX 92, 93
IC 598 GX 39
IC 599 GX 112, 113, 132, 133
IC 600 GX 112
IC 602 GX 92
IC 603 GX 112, 132
IC 605 GX 112
IC 607 GX 92
IC 608 GX 132
IC 609 GX 112
IC 610 GX 73
IC 613 GX 92
IC 614 GX 112
IC 616 GX 92
IC 623 GX 112
IC 624 GX 132
IC 625 GX 151
IC 626 GX 132
IC 627 GX 112
IC 628 GX 92, 112
IC 630 GX 132
IC 632 GX 112
IC 633 GX 112
IC 634 GX 92, 112
IC 635 GX 92
IC 636 GX 112
IC 639 GX 92
IC 642 GX 73
IC 646 GX 25
IC 648 GX 92
IC 649 GX 112
IC 651 GX 112
IC 653 GX 112
IC 654 GX 132
IC 657 GX 112
IC 658 GX 92
IC 659 GX 132
IC 664 GX 92
IC 669 GX 92
IC 670 GX 92
IC 671 GX 112
IC 672 GX 132
IC 673 GX 112
IC 674 GX 38
IC 676 GX 92
IC 677 GX 92
IC 678 GX 92
IC 680 GX 112
IC 681 GX 132
IC 687 GX 38
IC 691 GX 24
IC 692 GX 91, 92
IC 693 GX 111, 112, 131, 132
IC 694 GX 24
IC 696 GX 91
IC 698 GX 91
IC 699 GX 91
IC 700 GX 72
IC 701 GX 72
IC 705 GX 38
IC 706 GX 131
IC 707 GX 72
IC 708 GX 38
IC 709 GX 38
IC 711 GX 38
IC 712 GX 38
IC 716 GX 111
IC 718 GX 91
IC 719 GX 91
IC 720 GX 91
IC 722 GX 91
IC 723 GX 131
IC 724 GX 91
IC 725 GX 111
IC 727 GX 91
IC 728 GX 111
IC 731 GX 38
IC 736 GX 91
IC 739 GX 72, A10
IC 741 GX 111
IC 742 GX 72, A10
IC 743 GX 131
IC 745 GX 111
IC 746 GX 72, A10
IC 748 GX 91
IC 749 GX 37, 38
IC 750 GX 37, 38
IC 751 GX 37, 38
IC 753 GX 111
IC 754 GX 111
IC 755 GX 91
IC 756 GX 111
IC 758 GX 13, 24
IC 760 GX 150, 168
IC 761 GX 131
IC 762 GX 72, A10
IC 764 GX 150, 168
IC 766 GX 131
IC 767 GX 91
IC 768 GX 91
IC 769 GX 91
IC 771 GX 91
IC 773 GX 91, A15
IC 775 GX 91, A13
IC 776 GX 91, A15
IC 777 GX 72
IC 779 GX 54, 72
IC 780 GX 72
IC 781 GX 91, A13
IC 782 GX 91, 111, A15
IC 783 GX 91, A13
IC 783A GX 91, A13
IC 784 GX 111
IC 785 GX 131
IC 786 GX 131
IC 789 GX 91, A15
IC 790 GX 91
IC 791 GX 72
IC 792 GX 91
IC 794 GX 91, A13
IC 796 GX 91
IC 797 GX 91, A13
IC 800 GX 90, 91, A13
IC 801 GX 24
IC 806 GX 130, 149, 150
IC 807 GX 130, 149, 150
IC 810 GX 90
IC 812 GX 110
IC 813 GX 71
IC 816 GX 90
IC 818 GX 53, 71
IC 821 GX 53, 71
IC 826 GX 53
IC 827 GX 90
IC 829 GX 130
IC 830 GX 24
IC 832 GX 71, A8
IC 835 GX 71, A8
IC 836 GX 13
IC 838 GX 71, A8
IC 840 GX 90
IC 842 GX 53, 71, A8
IC 843 GX 53, 71, A8
IC 844 GX 167, A21
IC 847 GX 23
IC 849 GX 110
IC 851 GX 71
IC 852 GX 23
IC 853 GX 23
IC 857 GX 71, 90
IC 858 GX 71, 90
IC 860 GX 71
IC 862 GX 71
IC 863 GX 130, 149
IC 867 GX 71
IC 871 GX 110
IC 874 GX 149
IC 875 GX 23
IC 876 GX 110
IC 879 GX 149
IC 881 GX 90
IC 882 GX 90
IC 883 GX 53
IC 885 GX 71
IC 891 GX 110
IC 892 GX 110
IC 894 GX 71, 90
IC 896 GX 110
IC 900 GX 90
IC 901 GX 90
IC 902 GX 36, 37
IC 903 GX 110
IC 904 GX 110
IC 905 GX 71
IC 908 GX 110
IC 909 GX 71
IC 920 GX 129, 130
IC 933 GX 71
IC 939 GX 109, 110
IC 942 GX 23
IC 943 GX 109, 110
IC 944 GX 89, 90
IC 946 GX 89
IC 947 GX 109
IC 948 GX 89
IC 951 GX 36
IC 952 GX 109
IC 954 GX 12
IC 959 GX 89
IC 960 GX 70, 71, 89
IC 962 GX 89
IC 966 GX 89, 109
IC 967 GX 89
IC 968 GX 109
IC 971 GX 129
IC 972 PN 129, 148
IC 976 GX 109
IC 978 GX 109
IC 979 GX 89
IC 982 GX 70, 89
IC 983 GX 70, 89
IC 984 GX 70
IC 985 GX 109
IC 986 GX 109
IC 988 GX 109
IC 989 GX 109
IC 990 GX 52
IC 991 GX 129
IC 992 GX 109
IC 994 GX 89
IC 995 GX 23
IC 996 GX 23
IC 997 GX 109
IC 999 GX 70, 89
IC 1000 GX 70, 89
IC 1007 GX 109
IC 1010 GX 109
IC 1011 GX 109
IC 1012 GX 52
IC 1014 GX 89
IC 1017 GX 70
IC 1021 GX 70
IC 1024 GX 109
IC 1027 GX 23
IC 1028 GX 36
IC 1029 GX 36
IC 1035 GX 89
IC 1037 GX 70
IC 1041 GX 109
IC 1042 GX 109
IC 1044 GX 89
IC 1046 GX 12
IC 1048 GX 109
IC 1049 GX 12, 23
IC 1050 GX 70
IC 1053 GX 89
IC 1054 GX 109
IC 1055 GX 129
IC 1056 GX 36
IC 1062 GX 70
IC 1063 GX 109
IC 1065 GX 12
IC 1066 GX 109
IC 1067 GX 109
IC 1068 GX 109
IC 1069 GX 23
IC 1071 GX 109
IC 1075 GX 70
IC 1076 GX 70
IC 1077 GX 148
IC 1078 GX 88, 89
IC 1079 GX 88, 89
IC 1080 GX 128, 129
IC 1081 GX 148
IC 1082 GX 88, 89
IC 1084 GX 128, 129
IC 1085 GX 70, 88, 89
IC 1090 GX 36
IC 1091 GX 128
IC 1092 GX 88
IC 1093 GX 88
IC 1096 GX 70
IC 1097 GX 70
IC 1099 GX 22
IC 1100 GX 12, 22
IC 1101 GX 88, 108
IC 1102 GX 108
IC 1105 GX 108
IC 1106 GX 108
IC 1110 GX 12
IC 1113 GX 88
IC 1116 GX 88
IC 1118 GX 88
IC 1124 GX 69
IC 1125 GX 108
IC 1128 GX 108
IC 1129 GX 11, 12
IC 1131 GX 88
IC 1132 GX 69
IC 1133 GX 88
IC 1135 GX 69, 88
IC 1136 GX 108
IC 1141 GX 88
IC 1142 GX 69
IC 1143 GX 4
IC 1144 GX 35
IC 1145 GX 11
IC 1146 GX 11
IC 1147 GX 11
IC 1149 GX 88
IC 1151 GX 69, 88, A9
IC 1152 GX 35
IC 1153 GX 35
IC 1154 GX 11
IC 1155 GX 88, A9
IC 1156 GX 69
IC 1158 GX 108
IC 1165 GX 88, A9
IC 1169 GX 88
IC 1174 GX 88
IC 1178 GX 69, 88, A9
IC 1182 GX 69, 88, A9
IC 1183 GX 69, 88, A9
IC 1196 GX 88
IC 1197 GX 87, 88
IC 1198 GX 87, 88
IC 1199 GX 87, 88
IC 1205 GX 87, 88
IC 1206 GX 87, 88
IC 1209 GX 87
IC 1210 GX 11, 22
IC 1211 GX 22
IC 1214 GX 11
IC 1215 GX 11
IC 1216 GX 11
IC 1218 GX 11
IC 1221 GX 35
IC 1222 GX 35
IC 1223 GX 35
IC 1225 GX 11
IC 1228 GX 11
IC 1231 GX 22
IC 1235 GX 11
IC 1236 GX 68
IC 1237 GX 21, 22
IC 1241 GX 11
IC 1242 GX 107
IC 1244 GX 50
IC 1245 GX 50
IC 1248 GX 21
IC 1249 GX 50
IC 1251 GX 11
IC 1254 GX 11
IC 1255 GX 86, 87
IC 1256 GX 68
IC 1257 GC 126, 127
IC 1258 GX 21
IC 1261 GX 11
IC 1262 GX 34
IC 1263 GX 34
IC 1264 GX 34
IC 1265 GX 34
IC 1266 PN 181
IC 1267 GX 21
IC 1268 GX 68, 86
IC 1269 GX 68
IC 1274 BN 145, A17
IC 1275 BN 145, A17
IC 1276 GC 126
IC 1277 GX 49
IC 1279 GX 49
IC 1283/84 BN 145
IC 1286 GX 21
IC 1287 BN 126
IC 1288 GX 49
IC 1289 GX 49
IC 1291 GX 33
IC 1295 PN 125, A14
IC 1296 GX 49
IC 1297 PN 162, 163
IC 1301 GX 33
IC 1302 GX 48
IC 1303 GX 48
IC 1309 GX 124, 144
IC 1310 OC 48
IC 1311 OC 32, A2
IC 1313 GX 124
IC 1317 GX 104
IC 1318(a) BN 48, A2
IC 1318(b) BN 32, A2
IC 1318(c) BN 32, 48, A2
IC 1318(d) BN 32, 48, A2
IC 1318(e) BN 32, 48, A2
IC 1319 GX 144
IC 1320 GX 104
IC 1324 GX 124
IC 1330 GX 124
IC 1331 GX 124
IC 1332 GX 124
IC 1334 GX 124
IC 1336 GX 143
IC 1337 GX 123, 124
IC 1339 GX 123, 124, 143
IC 1340 BN 47
IC 1341 GX 123, 124
IC 1342 GX 123, 124
IC 1344 GX 123, 124
IC 1346 GX 123, 124
IC 1347 GX 123, 124
IC 1350 GX 123, 124
IC 1356 GX 123, 124
IC 1357 GX 123
IC 1359 GX 83
IC 1361 GX 83, 103
IC 1364 GX 103
IC 1365 GX 103
IC 1368 GX 103
IC 1369 OC 32
IC 1370 GX 103
IC 1371 GX 103
IC 1373 GX 103
IC 1375 GX 103
IC 1377 GX 103
IC 1381 GX 103
IC 1384 GX 103
IC 1385 GX 103
IC 1386 GX 143
IC 1387 GX 103
IC 1389 GX 143
IC 1390 GX 103
IC 1392 GX 47
IC 1393 GX 143
IC 1394 GX 83
IC 1396 BN OC 19
IC 1398 GX 83
IC 1401 GX 103
IC 1405 GX 103
IC 1408 GX 123
IC 1411 GX 103
IC 1412 GX 123, 142, 143
IC 1414 GX 83
IC 1417 GX 123
IC 1418 GX 103
IC 1420 GX 64, 65
IC 1421 GX 123
IC 1423 GX 103
IC 1425 GX 103
IC 1427 GX 83
IC 1431 GX 123
IC 1433 GX 122, 123
IC 1434 OC 19
IC 1435 GX 142
IC 1436 GX 122, 123
IC 1437 GX 102, 103
IC 1438 GX 142
IC 1439 GX 142
IC 1440 GX 122
IC 1442 OC 19
IC 1443 GX 142
IC 1445 GX 122, 142
IC 1447 GX 102, 122
IC 1451 GX 122
IC 1453 GX 122
IC 1454 PN 2
IC 1455 GX 102
IC 1458 GX 122
IC 1459 GX 160
IC 1460 GX 102
IC 1461 GX 82
IC 1464 GX 122
IC 1466 GX 102
IC 1467 GX 102
IC 1468 GX 102
IC 1470 BN 18
IC 1471 GX 122
IC 1472 GX 64, 82
IC 1473 GX 46, 64
IC 1474 GX 82, 102
IC 1478 GX 82
IC 1479 GX 122
IC 1481 GX 82, 102
IC 1482 GX 101, 102
IC 1489 GX 121, 122
IC 1491 GX 121
IC 1492 GX 101
IC 1495 GX 121
IC 1496 GX 101
IC 1498 GX 101, 121
IC 1501 GX 101
IC 1502 GX 2
IC 1503 GX 101
IC 1504 GX 101
IC 1505 GX 101
IC 1506 GX 101
IC 1507 GX 101
IC 1508 GX 81
IC 1509 GX 121
IC 1510 GX 101
IC 1513 GX 81
IC 1515 GX 101
IC 1516 GX 101
IC 1517 GX 101
IC 1520 GX 121
IC 1522 GX 101
IC 1524 GX 101
IC 1525 GX 30
IC 1526 GX 81
IC 1527 GX 101
IC 1529 GX 121
IC 1531 GX 159
IC 1532 GX 204
IC 1533 GX 121
IC 1534 GX 30
IC 1536 GX 30
IC 1540 GX 63
IC 1542 GX 63
IC 1543 GX 63
IC 1544 GX 63
IC 1546 GX 63
IC 1549 GX 81
IC 1550 GX 45
IC 1551 GX 81
IC 1553 GX 141
IC 1554 GX 159
IC 1555 GX 159
IC 1558 GX 141
IC 1559 GX 63, 80
IC 1561 GX 141, 158
IC 1562 GX 141, 158
IC 1563 GX 121, 140
IC 1564 GX 81, 100
IC 1565 GX 81, 100
IC 1566 GX 81, 100
IC 1568 GX 81, 100
IC 1569 GX 100
IC 1571 GX 120
IC 1574 GX 141, 158
IC 1575 GX 120
IC 1576 GX 158
IC 1579 GX 158
IC 1583 GX 80
IC 1584 GX 80
IC 1585 GX 80
IC 1586 GX 80
IC 1587 GX 158
IC 1588 GX 158
IC 1590 OC 18
IC 1592 GX 100, 120
IC 1595 GX 191
IC 1596 GX 80
IC 1597 GX 192
IC 1598 GX 100, 120

IC 4834 GX 206
IC 4835 GX 194
IC 4836 GX 194
IC 4837 GX 194
IC 4837A GX 194
IC 4838 GX 194
IC 4839 GX 194
IC 4840 GX 194
IC 4841 GX 206
IC 4842 GX 194
IC 4843 GX 194
IC 4844 GX 194
IC 4845 GX 194
IC 4846 PN 125
IC 4847 GX 206
IC 4848 GX 194
IC 4849 GX 194, 206
IC 4851 GX 194
IC 4852 GX 194
IC 4853 GX 206
IC 4854 GX 194
IC 4856 GX 194
IC 4857 GX 194
IC 4859 GX 206
IC 4860 GX 206
IC 4862 GX 206
IC 4864 GX 215
IC 4866 GX 194
IC 4867 GX 33
IC 4869 GX 194
IC 4870 GX 206
IC 4872 GX 194
IC 4874 GX 180
IC 4875 GX 194
IC 4876 GX 194
IC 4877 GX 180, 194
IC 4878 GX 194
IC 4879 GX 194
IC 4880 GX 194
IC 4881 GX 194
IC 4882 GX 194
IC 4883 GX 194
IC 4884 GX 194
IC 4885 GX 194
IC 4886 GX 180, 194
IC 4887 GX 206
IC 4888 GX 194
IC 4889 GX 194
IC 4890 GX 194
IC 4892 GX 206
IC 4894 GX 180, 194
IC 4897 GX 180, 194
IC 4899 GX 206
IC 4901 GX 194
IC 4903 GX 206
IC 4904 GX 206
IC 4905 GX 194
IC 4906 GX 194
IC 4908 GX 194
IC 4909 GX 179, 180
IC 4911 GX 179, 180, 194
IC 4913 GX 162
IC 4915 GX 194
IC 4916 GX 179, 180
IC 4919 GX 194
IC 4923 GX 194
IC 4926 GX 162
IC 4929 GX 206
IC 4931 GX 162
IC 4933 GX 194
IC 4934 GX 206
IC 4935 GX 194
IC 4936 GX 194
IC 4937 GX 194
IC 4938 GX 194
IC 4939 GX 194
IC 4941 GX 194
IC 4942 GX 194
IC 4943 GX 179
IC 4944 GX 194
IC 4945 GX 206
IC 4946 GX 179
IC 4951 GX 194
IC 4952 GX 194
IC 4954/55 BN 48, 66
IC 4956 GX 179
IC 4960 GX 206
IC 4961 GX 194
IC 4963 GX 194
IC 4964 GX 205, 206, 215
IC 4965 GX 194
IC 4967 GX 205, 206
IC 4968 GX 206
IC 4970 GX 205, 206
IC 4975 GX 194
IC 4979 GX 194
IC 4980 GX 194
IC 4981 GX 205, 206
IC 4983 GX 194
IC 4985 GX 205, 206
IC 4986 GX 194
IC 4991 GX 179
IC 4992 GX 205, 206
IC 4994 GX 194
IC 4995 GX 194
IC 4996 OC 48, A2
IC 4997 PN 84
IC 4998 GX 162
IC 4999 GX 144
IC 5002 GX 194
IC 5005 GX 144
IC 5009 GX 205
IC 5013 GX 162
IC 5014 GX 205, 215
IC 5019 GX 162
IC 5020 GX 162
IC 5021 GX 194
IC 5023 GX 205
IC 5024 GX 205
IC 5025 GX 215
IC 5026 GX 215
IC 5031 GX 205
IC 5032 GX 205
IC 5034 GX 194
IC 5038 GX 205
IC 5039 GX 143, 144, 161, 162
IC 5041 GX 143, 144, 161, 162
IC 5042 GX 205
IC 5049 GX 161
IC 5050 GX 104, 124
IC 5052 GX 205
IC 5053 GX 205
IC 5054 GX 205
IC 5063 GX 194
IC 5064 GX 194
IC 5065 GX 143, 161
IC 5067 BN 32, A1
IC 5068 BN 32, A1
IC 5069 GX 205
IC 5070 BN 32, A1
IC 5071 GX 205
IC 5073 GX 205
IC 5074 GX 205
IC 5075 GX 205
IC 5076 BN 32
IC 5078 GX 123, 124
IC 5079 GX 193
IC 5084 GX 205
IC 5086 GX 143, 161
IC 5087 GX 205, 215
IC 5090 GX 103
IC 5092 GX 205
IC 5094 GX 205
IC 5096 GX 205
IC 5100 GX 205
IC 5101 GX 205
IC 5102 GX 205, 215
IC 5103 GX 215
IC 5104 GX 65
IC 5105 GX 161, 179
IC 5105A GX 161, 179
IC 5105B GX 161, 179
IC 5106 GX 205
IC 5109 GX 215
IC 5110 GX 193
IC 5111 GX 103
IC 5116 GX 205
IC 5117 PN 31, 32
IC 5120 GX 205
IC 5122 GX 143
IC 5123 GX 205
IC 5125 GX 193
IC 5128 GX 161
IC 5130 GX 205, 214, 215
IC 5131 GX 161
IC 5138 GX 205
IC 5139 GX 161
IC 5140 GX 205
IC 5141 GX 193
IC 5142 GX 205
IC 5145 GX 83
IC 5146 BN 31
IC 5148-50 PN 160, 161
IC 5149 GX 142, 143
IC 5152 GX 178, 193
IC 5154 GX 205
IC 5156 GX 160, 161
IC 5157 GX 160, 161
IC 5162 GX 193
IC 5165 GX 205
IC 5169 GX 160
IC 5170 GX 178
IC 5171 GX 178
IC 5173 GX 205
IC 5174 GX 160
IC 5175 GX 160
IC 5176 GX 205
IC 5177 GX 82, 83
IC 5178 GX 142
IC 5179 GX 160
IC 5180 GX 46
IC 5181 GX 178
IC 5185 GX 205
IC 5186 GX 160
IC 5187 GX 193
IC 5188 GX 193
IC 5190 GX 193
IC 5197 GX 193
IC 5199 GX 160
IC 5201 GX 178
IC 5202 GX 205
IC 5203 GX 193
IC 5207 GX 193
IC 5208 GX 205
IC 5210 GX 142
IC 5211 GX 142
IC 5212 GX 160
IC 5217 PN 31
IC 5220 GX 193
IC 5222 GX 205
IC 5224 GX 178
IC 5225 GX 142
IC 5227 GX 205
IC 5231 GX 64
IC 5233 GX 64
IC 5240 GX 178
IC 5241 GX 102
IC 5242 GX 64
IC 5243 GX 64
IC 5244 GX 204, 205
IC 5246 GX 204, 205
IC 5247 GX 204, 205
IC 5249 GX 204, 205
IC 5250 GX 204, 205
IC 5252 GX 204, 205
IC 5256 GX 204, 205
IC 5258 GX 64
IC 5261 GX 142
IC 5262 GX 160
IC 5263 GX 204
IC 5264 GX 160
IC 5266 GX 204
IC 5267 GX 178
IC 5267A GX 178
IC 5267B GX 178
IC 5269 GX 160
IC 5269A GX 160
IC 5269B GX 160
IC 5269C GX 160
IC 5270 GX 160
IC 5271 GX 160
IC 5272 GX 204
IC 5273 GX 160
IC 5274 GX 64
IC 5279 GX 204
IC 5280 GX 204
IC 5282 GX 64
IC 5284 GX 64
IC 5285 GX 64
IC 5287 GX 102
IC 5288 GX 204
IC 5289 GX 160
IC 5290 GX 142
IC 5298 GX 63, 64
IC 5301 GX 204
IC 5302 GX 204
IC 5304 GX 122
IC 5305 GX 82
IC 5309 GX 82
IC 5312 GX 63, 64
IC 5314 GX 63, 64
IC 5315 GX 63, 64
IC 5317 GX 63, 64
IC 5321 GX 121, 122, 141
IC 5323 GX 204
IC 5324 GX 204
IC 5325 GX 177
IC 5326 GX 141
IC 5328 GX 177
IC 5328A GX 177
IC 5328B GX 177
IC 5329 GX 63
IC 5331 GX 63
IC 5332 GX 159
IC 5334 GX 101
IC 5338 GX 63
IC 5339 GX 204
IC 5341 GX 63
IC 5342 GX 63
IC 5343 GX 141
IC 5350 GX 141
IC 5351 GX 101
IC 5353 GX 141
IC 5354 GX 141
IC 5355 GX 45
IC 5356 GX 101
IC 5357 GX 101
IC 5358 GX 141
IC 5362 GX 141
IC 5365 GX 159
IC 5370 GX 45
IC 5373 GX 45
IC 5374 GX 101
IC 5375 GX 101
IC 5376 GX 45
IC 5378 GX 81
IC 5381 GX 81
IRAS 09371+1212 BN 93
Kemble 2 OC 10
King 1 OC 8
King 2 OC 18
King 4 OC 29
King 5 OC 28
King 6 OC 28
King 7 OC 28, 42, 43
King 8 OC 59
King 9 OC 19
King 10 OC 19
King 11 OC 8
King 12 OC 18
King 13 OC 18
King 14 OC 8
King 15 OC 18
King 16 OC 8
King 17 OC 59
King 18 OC 19
King 19 OC 18
King 20 OC 18
King 21 OC 8, 18
King 25 OC 85
King 26 OC 85
KMHK 211 OC 212, A24
LBN 1036 BN 135
LDN 557 DN 105, 106
LDN 582 DN 105
LDN 617 DN 105
LDN 673 DN 85
LDN 684 DN 85
LDN 889 DN 32, 48, A2
LDN 935 DN 32, A1
LDN 1616 DN 117
LDN 1710 DN 164
LDN 1773 DN 146
Lederman 1 OC 43, 44
Leo I GX 92, 93
Leo II GX 73
Leo III GX 55, 56
LH 120-N070 BN 212, A24, A25
Liller 1 GC 164
Lind 1 OC 204, 214
Lo 1 OC 199
Lo 27 OC 199
Lo 28 OC 199
Lo 46 OC 199
Lo 59 OC 199
Lo 89 OC 199
Lo 112 OC 199
Lo 153 OC 199
Lo 165 OC 199
Lo 172 OC 199
Lo 213 OC 199
Lo 306 OC 198
Lo 372 OC 198
Lo 402 OC 198
Lo 481 OC 198
Lo 565 OC 198
Lo 624 OC 198
Lo 670 OC 209
Lo 682 OC 198
Lo 694 OC 198
Lo 757 OC 208, 209
Lo 807 OC 197, 208
Lo 821 OC 197
Lo 894 OC 208
Lo 915 OC 197
Lo 991 OC 197, 208
Lo 995 OC 208
Lo 1002 OC 208
Lo 1010 OC 197
Lo 1095 OC 197
Lo 1101 OC 197
Lo 1152 OC 197
Lo 1171 OC 197
Lo 1194 OC 197
Lo 1202 OC 197
Lo 1225 OC 197
Lo 1256 OC 197
Lo 1282 OC 197
Lo 1289 OC 197
Lo 1339 OC 197
Lo 1378 OC 197
Lo 1409 OC 197
Lo 2045 OC 196
Lo 2158 OC 196
Lo 2313 OC 196
Lynga 1 OC 197, 208
Lynga 2 OC 197
Lynga 3 OC 196
Lynga 4 OC 196
Lynga 5 OC 196
Lynga 6 OC 182, 196
Lynga 7 GC 196
Lynga 8 OC 182
Lynga 9 OC 182
Lynga 11 OC 182
Lynga 12 OC 181, 182
Lynga 13 OC 181, 182, A22
Lynga 14 OC 181
Maffei I GX 29
Maffei II GX 29
Mayer 1 OC 18
Mayer 2 OC 28
Mayer 3 OC 153
MCG −7-1-8 GX 177
MCG −7-13-1 GX 172, 173
MCG −7-39-5 GX 180
MCG −7-48-23 GX 159
MCG −5-1-2 GX 141
MCG −5-1-5 GX 141
MCG −5-1-6 GX 159
MCG −5-1-9 GX 141, 159
MCG −5-1-10 GX 159
MCG −5-1-13 GX 159
MCG −5-1-14 GX 141, 159
MCG −5-1-15 GX 15933
MCG −5-1-16 GX 141, 159
MCG −5-1-17 GX 159
MCG −5-1-18 GX 141, 159
MCG −5-1-19 GX 141
MCG −5-1-20 GX 159
MCG −5-1-21 GX 141
MCG −5-1-26 GX 159
MCG −5-1-28 GX 159
MCG −5-1-29 GX 141
MCG −5-1-30 GX 159
MCG −5-1-39 GX 141, 159
MCG −5-1-42 GX 141
MCG −5-1-44 GX 159
MCG −5-1-46 GX 141, 159
MCG −5-2-1 GX 141
MCG −5-2-2 GX 141
MCG −5-2-4 GX 159
MCG −5-2-5 GX 141
MCG −5-2-7 GX 159
MCG −5-2-9 GX 141
MCG −5-2-12 GX 141, 159
MCG −5-2-13 GX 141
MCG −5-2-20 GX 141
MCG −5-2-22 GX 141
MCG −5-2-23 GX 141, 158
MCG −5-2-24 GX 141, 158
MCG −5-2-25 GX 141, 158
MCG −5-2-29 GX 141, 158, 159, 176
MCG −5-2-30 GX 141, 158
MCG −5-2-32 GX 141, 158, 159, 176
MCG −5-2-35 GX 159, 176
MCG −5-3-4 GX 158
MCG −5-3-8 GX 176
MCG −5-3-9 GX 176
MCG −5-3-11 GX 158
MCG −5-3-16 GX 176
MCG −5-3-17 GX 176
MCG −5-3-18 GX 176
MCG −5-3-21 GX 176
MCG −5-3-23 GX 158
MCG −5-3-26 GX 158
MCG −5-3-28 GX 158
MCG −5-4-5 GX 158
MCG −5-4-8 GX 176
MCG −5-4-10 GX 176
MCG −5-4-12 GX 176
MCG −5-4-13 GX 176
MCG −5-4-14 GX 176
MCG −5-4-17 GX 176
MCG −5-4-18 GX 176
MCG −5-4-20 GX 176
MCG −5-4-23 GX 176
MCG −5-4-26 GX 176
MCG −5-4-27 GX 176
MCG −5-4-29 GX 176
MCG −5-4-34 GX 176
MCG −5-4-35 GX 158
MCG −5-4-37 GX 176
MCG −5-4-38 GX 176
MCG −5-4-39 GX 158
MCG −5-4-40 GX 176
MCG −5-4-42 GX 158
MCG −5-5-1 GX 176
MCG −5-5-9 GX 176
MCG −5-5-10 GX 158
MCG −5-5-13 GX 158, 176
MCG −5-5-17 GX 158
MCG −5-5-18 GX 158
MCG −5-5-20 GX 158
MCG −5-5-21 GX 158
MCG −5-5-26 GX 175, 176
MCG −5-5-28 GX 175, 176
MCG −5-6-2 GX 175, 176
MCG −5-6-3 GX 157, 175
MCG −5-6-4 GX 175
MCG −5-6-10 GX 157
MCG −5-6-12 GX 157, 175
MCG −5-7-4 GX 157
MCG −5-7-6 GX 157, 175
MCG −5-7-7 GX 157, 175
MCG −5-7-11 GX 157
MCG −5-7-20 GX 157
MCG −5-7-23 GX 157
MCG −5-7-25 GX 157
MCG −5-7-26 GX 175
MCG −5-7-27 GX 157, 175
MCG −5-7-28 GX 157
MCG −5-7-30 GX 175
MCG −5-7-31 GX 157
MCG −5-7-36 GX 175
MCG −5-7-38 GX 175
MCG −5-7-39 GX 175
MCG −5-7-40 GX 175
MCG −5-7-41 GX 157
MCG −5-7-43 GX 157
MCG −5-7-44 GX 175
MCG −5-8-3 GX 157, 175
MCG −5-8-5 GX 157
MCG −5-8-6 GX 175
MCG −5-8-8 GX 157
MCG −5-8-10 GX 175
MCG −5-8-12 GX 157
MCG −5-8-14 GX 157
MCG −5-8-15 GX 157
MCG −5-8-16 GX 157
MCG −5-8-17 GX 157
MCG −5-8-19 GX 157
MCG −5-8-20 GX 175
MCG −5-8-21 GX 175
MCG −5-8-23 GX 175
MCG −5-9-2 GX 174, 175
MCG −5-9-3 GX 174, 175
MCG −5-9-6 GX 156
MCG −5-9-7 GX 156
MCG −5-9-18 GX 174
MCG −5-10-1 GX 156, 174
MCG −5-10-3 GX 156
MCG −5-10-8 GX 174
MCG −5-10-14 GX 156, 174
MCG −5-11-3 GX 174
MCG −5-11-8 GX 174
MCG −5-11-10 GX 156, 174
MCG −5-11-12 GX 156, 174
MCG −5-11-13 GX 156, 174
MCG −5-11-14 GX 174
MCG −5-11-15 GX 156, 174
MCG −5-12-1 GX 155, 156, 173, 174
MCG −5-12-5 GX 155
MCG −5-12-7 GX 173
MCG −5-12-12 GX 155
MCG −5-12-13 GX 155, 173
MCG −5-13-1 GX 155, 173
MCG −5-13-3 GX 155
MCG −5-13-14 GX 155, 173
MCG −5-13-19 GX 173
MCG −5-13-20 GX 173

MCG −5-13-21 GX 173
MCG −5-14-1 GX 155, 173
MCG −5-14-2 GX 155, 173
MCG −5-14-5 GX 173
MCG −5-14-8 GX 173
MCG −5-14-9 GX 155, 173
MCG −5-14-10 GX 155, 173
MCG −5-14-12 GX 155
MCG −5-14-14 GX 155
MCG −5-14-17 GX 173
MCG −5-14-19 GX 173
MCG −5-14-22 GX 173
MCG −5-14-23 GX 173
MCG −5-14-24 GX 173
MCG −5-15-1 GX 154, 155, 172, 173
MCG −5-15-3 GX 172, 173
MCG −5-15-5 GX 172
MCG −5-15-6 GX 154
MCG −5-16-1 GX 154
MCG −5-16-2 GX 172
MCG −5-16-3 GX 154
MCG −5-16-4 GX 154, 172
MCG −5-16-5 GX 172
MCG −5-16-7 GX 172
MCG −5-16-11 GX 154
MCG −5-16-14 GX 154
MCG −5-16-18 GX 154
MCG −5-16-21 GX 172
MCG −5-16-22 GX 154
MCG −5-17-8 GX 154
MCG −5-21-1 GX 171
MCG −5-23-5 GX 170
MCG −5-23-13 GX 152, 170
MCG −5-23-16 GX 170
MCG −5-24-7 GX 169, 170
MCG −5-24-13 GX 169, 170
MCG −5-24-25 GX 151
MCG −5-24-29 GX 151
MCG −5-24-30 GX 151, 169
MCG −5-25-15 GX 169
MCG −5-25-17 GX 151, A19
MCG −5-25-18 GX 151, A19
MCG −5-25-21 GX 151, A19
MCG −5-25-22 GX 151, A19
MCG −5-25-24 GX 151, A19
MCG −5-25-26 GX 151, A19
MCG −5-25-29 GX 151, A19
MCG −5-25-30 GX 169
MCG −5-25-32 GX 151, A19
MCG −5-25-37 GX 151, A19
MCG −5-26-2 GX 169
MCG −5-26-4 GX 169
MCG −5-26-6 GX 151
MCG −5-26-11 GX 151, 169
MCG −5-26-13 GX 169
MCG −5-26-15 GX 169
MCG −5-27-1 GX 151
MCG −5-27-6 GX 168, 169
MCG −5-27-9 GX 150, 151, 168, 169
MCG −5-27-11 GX 168
MCG −5-27-14 GX 168
MCG −5-27-17 GX 168
MCG −5-28-1 GX 168
MCG −5-28-2 GX 168
MCG −5-28-8 GX 168
MCG −5-28-13 GX 168
MCG −5-28-14 GX 168
MCG −5-28-16 GX 168
MCG −5-29-4 GX 150, 168
MCG −5-29-6 GX 168
MCG −5-29-11 GX 150
MCG −5-29-12 GX 150
MCG −5-29-17 GX 168
MCG −5-29-19 GX 168
MCG −5-29-22 GX 168
MCG −5-29-27 GX 168
MCG −5-29-35 GX 168
MCG −5-30-4 GX 168
MCG −5-31-1 GX 149, 167, A21
MCG −5-31-2 GX 149
MCG −5-31-5 GX 167, A21
MCG −5-31-6 GX 149
MCG −5-31-8 GX 149, 167, A21
MCG −5-31-10 GX 167, A21
MCG −5-31-14 GX 167, A21
MCG −5-31-17 GX 167, A21
MCG −5-31-22 GX 167, A21
MCG −5-31-27 GX 167, A21
MCG −5-31-31 GX 167, A21
MCG −5-31-33 GX 149
MCG −5-31-39 GX 167
MCG −5-32-1 GX 149, 167
MCG −5-32-2 GX 149, 167
MCG −5-32-3 GX 167
MCG −5-32-4 GX 167
MCG −5-32-7 GX 167
MCG −5-32-8 GX 167
MCG −5-32-12 GX 167
MCG −5-32-14 GX 149, 167
MCG −5-32-19 GX 149, 167
MCG −5-32-27 GX 149
MCG −5-32-30 GX 149
MCG −5-32-42 GX 167
MCG −5-32-45 GX 167
MCG −5-32-47 GX 167
MCG −5-32-48 GX 167
MCG −5-32-55 GX 167
MCG −5-32-57 GX 167
MCG −5-32-61 GX 167
MCG −5-32-64 GX 167
MCG −5-32-65 GX 167
MCG −5-32-70 GX 167, A18
MCG −5-32-72 GX 149
MCG −5-32-74 GX 167, A18
MCG −5-33-4 GX 149, 167, A18
MCG −5-33-5 GX 167, A18
MCG −5-33-10 GX 167, A18
MCG −5-33-13 GX 167, A18
MCG −5-33-17 GX 167, A18
MCG −5-33-23 GX 167, A18
MCG −5-33-29 GX 149
MCG −5-33-39 GX 166, 167
MCG −5-33-41 GX 166, 167
MCG −5-33-43 GX 166
MCG −5-33-46 GX 148
MCG −5-34-4 GX 166
MCG −5-34-7 GX 148, 166
MCG −5-34-8 GX 166
MCG −5-34-13 GX 166
MCG −5-35-4 GX 166
MCG −5-35-6 GX 166
MCG −5-35-7 GX 166
MCG −5-37-2 GX 147
MCG −5-45-2 GX 163
MCG −5-45-4 GX 145, 163
MCG −5-45-5 GX 163
MCG −5-45-6 GX 162, 163
MCG −5-46-2 GX 144, 162
MCG −5-46-4 GX 144, 162
MCG −5-47-1 GX 162
MCG −5-47-2 GX 162
MCG −5-47-3 GX 162
MCG −5-47-5 GX 144
MCG −5-47-8 GX 162
MCG −5-47-12 GX 162
MCG −5-47-13 GX 162
MCG −5-47-19 GX 162
MCG −5-47-20 GX 144
MCG −5-47-21 GX 144, 162
MCG −5-47-22 GX 162
MCG −5-47-23 GX 144, 162
MCG −5-47-25 GX 144, 162
MCG −5-48-1 GX 162
MCG −5-48-2 GX 144
MCG −5-48-3 GX 162
MCG −5-48-5 GX 144, 162
MCG −5-48-8 GX 162
MCG −5-48-13 GX 162
MCG −5-48-15 GX 162
MCG −5-48-18 GX 162
MCG −5-48-19 GX 162
MCG −5-48-21 GX 162
MCG −5-48-24 GX 143, 144
MCG −5-48-27 GX 143, 144
MCG −5-49-3 GX 143
MCG −5-49-5 GX 161
MCG −5-49-14 GX 143
MCG −5-49-15 GX 143
MCG −5-50-1 GX 143, 161
MCG −5-50-3 GX 143
MCG −5-50-4 GX 143, 161
MCG −5-50-5 GX 161
MCG −5-50-6 GX 161
MCG −5-50-7 GX 161
MCG −5-50-9 GX 143
MCG −5-50-11 GX 161
MCG −5-50-12 GX 161
MCG −5-50-13 GX 161
MCG −5-50-14 GX 161
MCG −5-50-16 GX 143, 161
MCG −5-50-17 GX 143, 161
MCG −5-51-1 GX 143, 161
MCG −5-51-2 GX 143
MCG −5-51-3 GX 143
MCG −5-51-4 GX 143, 161
MCG −5-51-5 GX 143
MCG −5-51-6 GX 143, 161
MCG −5-51-7 GX 143, 161
MCG −5-51-13 GX 161
MCG −5-51-14 GX 143, 161
MCG −5-51-15 GX 143, 161
MCG −5-51-16 GX 161
MCG −5-51-18 GX 143, 161
MCG −5-51-23 GX 143
MCG −5-51-24 GX 142, 143
MCG −5-51-25 GX 142, 143
MCG −5-51-26 GX 142, 143
MCG −5-51-27 GX 142, 143, 160, 161
MCG −5-51-30 GX 160, 161
MCG −5-51-31 GX 160, 161
MCG −5-52-1 GX 160, 161
MCG −5-52-2 GX 142, 143
MCG −5-52-3 GX 142, 143
MCG −5-52-4 GX 142, 143
MCG −5-52-5 GX 142, 143
MCG −5-52-12 GX 160, 161
MCG −5-52-13 GX 142, 143
MCG −5-52-14 GX 160, 161
MCG −5-52-15 GX 142, 143
MCG −5-52-16 GX 142, 143
MCG −5-52-17 GX 142, 143
MCG −5-52-18 GX 142, 143
MCG −5-52-25 GX 142
MCG −5-52-30 GX 142, 160
MCG −5-52-33A GX 142
MCG −5-52-35 GX 142
MCG −5-52-37 GX 142
MCG −5-52-40 GX 142
MCG −5-52-41 GX 142
MCG −5-52-42 GX 142
MCG −5-52-44 GX 142
MCG −5-52-45 GX 142
MCG −5-52-46 GX 142
MCG −5-52-48 GX 142
MCG −5-52-52 GX 142, 160
MCG −5-52-55 GX 142
MCG −5-52-59 GX 142
MCG −5-52-62 GX 142
MCG −5-52-63 GX 142
MCG −5-52-64 GX 142
MCG −5-52-67 GX 142
MCG −5-52-71 GX 142
MCG −5-52-72 GX 160
MCG −5-53-5 GX 160
MCG −5-53-7 GX 160
MCG −5-53-8 GX 142, 160
MCG −5-53-15 GX 142
MCG −5-53-18 GX 160
MCG −5-53-20 GX 160
MCG −5-53-25 GX 142
MCG −5-53-26 GX 142
MCG −5-53-28 GX 142
MCG −5-53-29 GX 160
MCG −5-53-32 GX 142, 160
MCG −5-54-2 GX 142
MCG −5-54-3 GX 142, 160
MCG −5-54-4 GX 142
MCG −5-54-5 GX 160
MCG −5-54-6 GX 160
MCG −5-54-7 GX 160
MCG −5-54-8 GX 160
MCG −5-54-9 GX 160
MCG −5-54-10 GX 142
MCG −5-54-12 GX 160
MCG −5-54-15 GX 142
MCG −5-54-17 GX 142
MCG −5-54-24 GX 142, 160
MCG −5-55-1 GX 141, 142
MCG −5-55-2 GX 141, 142
MCG −5-55-3 GX 141, 142
MCG −5-55-4 GX 141, 142
MCG −5-55-6 GX 141, 142
MCG −5-55-8 GX 141, 159
MCG −5-55-11 GX 141
MCG −5-55-13 GX 141
MCG −5-55-14 GX 141, 159
MCG −5-55-17 GX 159
MCG −5-55-18 GX 141
MCG −5-55-19 GX 141
MCG −5-55-20 GX 141
MCG −5-55-24 GX 141
MCG −5-55-25 GX 159
MCG −5-55-26 GX 141
MCG −5-55-27 GX 141
MCG −5-55-28 GX 141, 159
MCG −5-55-30 GX 159
MCG −5-55-32 GX 141, 159
MCG −5-55-33 GX 141, 159
MCG −5-55-34 GX 141
MCG −5-56-4 GX 141
MCG −5-56-7 GX 141, 159
MCG −5-56-8 GX 141, 159
MCG −5-56-12 GX 141
MCG −5-56-15 GX 141
MCG −5-56-18 GX 141, 159
MCG −5-56-21 GX 141, 159
MCG −5-56-22 GX 141
MCG −5-56-25 GX 141
MCG −5-56-26 GX 141, 159
MCG −5-56-28 GX 159
MCG −5-56-29 GX 141, 159
MCG −5-56-31 GX 159
MCG −5-56-33 GX 141, 159
MCG −4-1-1 GX 141
MCG −4-1-3 GX 141
MCG −4-1-6 GX 141
MCG −4-1-8 GX 141
MCG −4-1-9 GX 141
MCG −4-1-22 GX 141
MCG −4-1-23 GX 141
MCG −4-1-24 GX 141
MCG −4-1-25 GX 141
MCG −4-1-27 GX 141
MCG −4-1-28 GX 141
MCG −4-2-3 GX 141
MCG −4-2-5 GX 141
MCG −4-2-7 GX 141
MCG −4-2-13 GX 141
MCG −4-2-21 GX 141
MCG −4-2-25 GX 141, 158
MCG −4-2-31 GX 141, 158
MCG −4-2-38 GX 141, 158
MCG −4-2-39 GX 141, 158
MCG −4-2-44 GX 141, 158
MCG −4-3-2 GX 158
MCG −4-3-4 GX 158
MCG −4-3-7 GX 158
MCG −4-3-8 GX 158
MCG −4-3-10 GX 158
MCG −4-3-11 GX 158
MCG −4-3-12 GX 158
MCG −4-3-13 GX 158
MCG −4-3-14 GX 158
MCG −4-3-16 GX 158
MCG −4-3-17 GX 158
MCG −4-3-18 GX 158
MCG −4-3-20 GX 158
MCG −4-3-23 GX 158
MCG −4-3-25 GX 158
MCG −4-3-26 GX 158
MCG −4-3-28 GX 158
MCG −4-3-34 GX 158
MCG −4-3-35 GX 158
MCG −4-3-36 GX 158
MCG −4-3-38 GX 158
MCG −4-3-40 GX 158
MCG −4-3-43 GX 158
MCG −4-3-45 GX 158
MCG −4-3-46 GX 158
MCG −4-3-48 GX 158
MCG −4-3-52 GX 158
MCG −4-3-54 GX 158
MCG −4-4-2 GX 158
MCG −4-4-3 GX 158
MCG −4-4-4 GX 158
MCG −4-4-7 GX 158
MCG −4-4-8 GX 158
MCG −4-4-12 GX 158
MCG −4-4-15 GX 158
MCG −4-4-17 GX 158
MCG −4-4-19 GX 158
MCG −4-4-21 GX 158
MCG −4-4-22 GX 158
MCG −4-5-1 GX 158
MCG −4-5-2 GX 158
MCG −4-5-5 GX 158
MCG −4-5-7 GX 158
MCG −4-5-14 GX 158
MCG −4-5-24 GX 157, 158
MCG −4-6-1 GX 157, 158
MCG −4-6-2 GX 157, 158
MCG −4-6-7 GX 157
MCG −4-6-8 GX 157
MCG −4-6-14 GX 157
MCG −4-6-15 GX 157
MCG −4-6-17 GX 157
MCG −4-6-18 GX 157
MCG −4-6-20 GX 157
MCG −4-6-22 GX 157
MCG −4-6-23 GX 157
MCG −4-6-24 GX 157
MCG −4-6-26 GX 157
MCG −4-6-27 GX 157
MCG −4-6-28 GX 157
MCG −4-6-29 GX 157
MCG −4-6-33 GX 157
MCG −4-6-36 GX 157
MCG −4-6-39 GX 157
MCG −4-6-42 GX 157
MCG −4-6-43 GX 157
MCG −4-6-44 GX 157
MCG −4-7-3 GX 157
MCG −4-7-6 GX 157
MCG −4-7-7 GX 157
MCG −4-7-8 GX 157
MCG −4-7-11 GX 157
MCG −4-7-12 GX 157
MCG −4-7-14 GX 157
MCG −4-7-15 GX 157
MCG −4-7-16 GX 157
MCG −4-7-17 GX 157
MCG −4-7-18 GX 157
MCG −4-7-19 GX 157
MCG −4-7-22 GX 157
MCG −4-7-23 GX 157
MCG −4-7-25 GX 157
MCG −4-7-26 GX 157
MCG −4-7-27 GX 157
MCG −4-7-28 GX 157
MCG −4-7-29 GX 157
MCG −4-7-33 GX 157
MCG −4-7-35 GX 157
MCG −4-7-36 GX 157
MCG −4-7-39 GX 157
MCG −4-7-41 GX 157
MCG −4-7-44 GX 157
MCG −4-7-45 GX 157
MCG −4-7-49 GX 157
MCG −4-8-1 GX 157
MCG −4-8-3 GX 157
MCG −4-8-4 GX 157
MCG −4-8-6 GX 157
MCG −4-8-7 GX 157
MCG −4-8-10 GX 157
MCG −4-8-13 GX 157
MCG −4-8-14 GX 157
MCG −4-8-15 GX 157
MCG −4-8-18 GX 157
MCG −4-8-19 GX 157
MCG −4-8-20 GX 157
MCG −4-8-22 GX 157
MCG −4-8-28 GX 157
MCG −4-8-34 GX 157
MCG −4-8-35 GX 157
MCG −4-8-39 GX 157
MCG −4-8-40 GX 157
MCG −4-8-43 GX 157
MCG −4-8-44 GX 157
MCG −4-8-45 GX 157
MCG −4-8-46 GX 157
MCG −4-8-47 GX 157
MCG −4-8-49 GX 157
MCG −4-8-54 GX 156, 157
MCG −4-8-55 GX 156, 157
MCG −4-9-1 GX 156, 157
MCG −4-9-9 GX 156
MCG −4-9-10 GX 156
MCG −4-9-18 GX 156
MCG −4-9-21 GX 156
MCG −4-9-23 GX 156
MCG −4-9-27 GX 156
MCG −4-9-32 GX 156
MCG −4-9-38 GX 156
MCG −4-9-43 GX 156
MCG −4-9-49 GX 156
MCG −4-9-57 GX 156
MCG −4-10-9 GX 156
MCG −4-10-12 GX 156
MCG −4-10-14 GX 156
MCG −4-10-16 GX 156
MCG −4-11-2 GX 156
MCG −4-11-3 GX 156
MCG −4-11-5 GX 156
MCG −4-11-9 GX 156
MCG −4-11-11 GX 156
MCG −4-11-12 GX 156
MCG −4-11-16 GX 156
MCG −4-11-18 GX 156
MCG −4-11-19 GX 156
MCG −4-11-20 GX 156
MCG −4-11-23 GX 156
MCG −4-11-24 GX 155, 156
MCG −4-11-25 GX 155, 156
MCG −4-12-2 GX 155, 156
MCG −4-12-3 GX 155, 156
MCG −4-12-5 GX 155, 156
MCG −4-12-6 GX 155, 156
MCG −4-12-9 GX 155, 156
MCG −4-12-17 GX 155
MCG −4-12-18 GX 155
MCG −4-12-20 GX 155
MCG −4-12-21 GX 155
MCG −4-12-22 GX 155
MCG −4-12-23 GX 155
MCG −4-12-25 GX 155
MCG −4-12-30 GX 155
MCG −4-12-32 GX 155
MCG −4-12-34 GX 155
MCG −4-12-37 GX 155
MCG −4-12-38 GX 155
MCG −4-12-40 GX 155
MCG −4-12-41 GX 155
MCG −4-12-42 GX 155
MCG −4-13-2 GX 155
MCG −4-13-3 GX 155
MCG −4-13-4 GX 155
MCG −4-13-5 GX 155
MCG −4-13-8 GX 155
MCG −4-13-10 GX 155
MCG −4-13-14 GX 155
MCG −4-13-15 GX 155
MCG −4-13-16 GX 155
MCG −4-14-1 GX 155
MCG −4-14-8 GX 155
MCG −4-14-12 GX 155
MCG −4-14-13 GX 155
MCG −4-14-14 GX 155
MCG −4-14-16 GX 155
MCG −4-14-17 GX 155
MCG −4-14-18 GX 155
MCG −4-14-19 GX 155
MCG −4-14-20 GX 155
MCG −4-14-21 GX 155
MCG −4-14-23 GX 155
MCG −4-14-26 GX 155
MCG −4-14-27 GX 155
MCG −4-14-28 GX 155
MCG −4-14-30 GX 155
MCG −4-14-31 GX 155
MCG −4-14-32 GX 155
MCG −4-14-34 GX 155
MCG −4-14-35 GX 155
MCG −4-14-37 GX 155
MCG −4-14-38 GX 155
MCG −4-14-39 GX 155
MCG −4-14-41 GX 155
MCG −4-15-6 GX 154, 155
MCG −4-15-7 GX 154
MCG −4-15-9 GX 154
MCG −4-15-12 GX 154
MCG −4-15-15 GX 154
MCG −4-15-16 GX 154
MCG −4-15-17 GX 154
MCG −4-15-22 GX 154
MCG −4-16-6 GX 154
MCG −4-16-11 GX 154
MCG −4-16-16 GX 154
MCG −4-20-1 GX 153
MCG −4-20-5 GX 153
MCG −4-21-1 GX 153
MCG −4-21-6 GX 152, 153
MCG −4-21-7 GX 152, 153
MCG −4-21-9 GX 152
MCG −4-23-4 GX 152
MCG −4-23-6 GX 152
MCG −4-23-7 GX 152
MCG −4-23-9 GX 152
MCG −4-23-12 GX 152
MCG −4-23-14 GX 152
MCG −4-24-1 GX 152
MCG −4-24-6 GX 152
MCG −4-24-16 GX 151
MCG −4-24-17 GX 151
MCG −4-24-18 GX 151
MCG −4-24-20 GX 151
MCG −4-25-1 GX 151
MCG −4-25-5 GX 151
MCG −4-25-6 GX 151
MCG −4-25-8 GX 151
MCG −4-25-10 GX 151
MCG −4-25-20 GX 151, A19
MCG −4-25-21 GX 151, A19
MCG −4-25-23 GX 151, A19
MCG −4-25-25 GX 151, A19
MCG −4-25-30 GX 151
MCG −4-25-37 GX 151, A19
MCG −4-25-38 GX 151, A19
MCG −4-25-43 GX 151, A19
MCG −4-25-48 GX 151, A19
MCG −4-25-50 GX 151, A19
MCG −4-26-5 GX 151

MCG –4-26-7 GX 151
MCG –4-26-8 GX 151
MCG –4-26-16 GX 151
MCG –4-26-17 GX 151
MCG –4-27-7 GX 150, 151
MCG –4-27-9 GX 150, 151
MCG –4-27-13 GX 150
MCG –4-28-1 GX 150
MCG –4-28-2 GX 150
MCG –4-28-7 GX 150
MCG –4-29-2 GX 150
MCG –4-29-9 GX 150
MCG –4-29-10 GX 150
MCG –4-29-12 GX 150
MCG –4-29-13 GX 150
MCG –4-29-14 GX 150
MCG –4-29-15 GX 150
MCG –4-29-17 GX 150
MCG –4-30-3 GX 150
MCG –4-30-5 GX 150
MCG –4-30-9 GX 149, 150
MCG –4-30-11 GX 149, 150
MCG –4-30-12 GX 149, 150
MCG –4-30-16 GX 149
MCG –4-30-22 GX 149
MCG –4-30-23 GX 149
MCG –4-30-35 GX 149
MCG –4-30-36 GX 149
MCG –4-31-2 GX 149
MCG –4-31-3 GX 149
MCG –4-31-4 GX 149
MCG –4-31-7 GX 149
MCG –4-31-8 GX 149
MCG –4-31-9 GX 149
MCG –4-31-13 GX 149
MCG –4-31-14 GX 149
MCG –4-31-20 GX 149
MCG –4-31-22 GX 149
MCG –4-31-23 GX 149
MCG –4-31-24 GX 149
MCG –4-31-27 GX 149
MCG –4-31-34 GX 149
MCG –4-31-37 GX 149
MCG –4-31-38 GX 149
MCG –4-31-42 GX 149
MCG –4-31-44 GX 149
MCG –4-31-46 GX 149
MCG –4-31-49 GX 149
MCG –4-32-11 GX 149
MCG –4-32-14 GX 149
MCG –4-32-15 GX 149
MCG –4-32-22 GX 149
MCG –4-32-25 GX 149
MCG –4-32-29 GX 149
MCG –4-32-37 GX 149
MCG –4-32-38 GX 149
MCG –4-32-40 GX 149
MCG –4-32-43 GX 149
MCG –4-32-46 GX 149
MCG –4-32-48 GX 149
MCG –4-32-51 GX 149
MCG –4-33-1 GX 149
MCG –4-33-3 GX 149
MCG –4-33-4 GX 149
MCG –4-33-5 GX 149
MCG –4-33-6 GX 149
MCG –4-33-10 GX 149
MCG –4-33-11 GX 149
MCG –4-33-12 GX 149
MCG –4-33-15 GX 148, 149
MCG –4-33-17 GX 148, 149
MCG –4-33-18 GX 148, 149
MCG –4-33-20 GX 148, 149
MCG –4-33-22 GX 148, 149
MCG –4-33-28 GX 148, 149
MCG –4-33-30 GX 148, 149
MCG –4-33-31 GX 148, 149
MCG –4-33-32 GX 148, 149
MCG –4-33-35 GX 148, 149
MCG –4-33-37 GX 148
MCG –4-33-41 GX 148
MCG –4-33-42 GX 148
MCG –4-33-44 GX 148
MCG –4-33-51 GX 148
MCG –4-33-52 GX 148
MCG –4-34-8 GX 148
MCG –4-34-17 GX 148
MCG –4-34-19 GX 148
MCG –4-35-2 GX 148
MCG –4-35-5 GX 148
MCG –4-35-7 GX 148
MCG –4-35-8 GX 148
MCG –4-35-11 GX 148
MCG –4-35-13 GX 148
MCG –4-35-14 GX 148
MCG –4-35-15 GX 148
MCG –4-35-17 GX 148
MCG –4-36-1 GX 148
MCG –4-36-2 GX 148
MCG –4-36-7 GX 147, 148
MCG –4-36-11 GX 147, 148
MCG –4-36-12 GX 147, 148
MCG –4-37-4 GX 147
MCG –4-37-5 GX 147
MCG –4-37-6 GX 147
MCG –4-38-2 GX 147
MCG –4-46-3 GX 144
MCG –4-47-4 GX 144
MCG –4-47-8 GX 144
MCG –4-47-11 GX 144
MCG –4-48-1 GX 144
MCG –4-48-8 GX 144
MCG –4-48-10 GX 144
MCG –4-48-15 GX 144
MCG –4-48-17 GX 144
MCG –4-48-19 GX 144
MCG –4-48-25 GX 143, 144
MCG –4-48-26 GX 143, 144
MCG –4-48-27 GX 143, 144
MCG –4-49-1 GX 143
MCG –4-49-2 GX 143
MCG –4-49-4 GX 143
MCG –4-49-5 GX 143
MCG –4-49-9 GX 143
MCG –4-49-11 GX 143
MCG –4-50-2 GX 143
MCG –4-50-4 GX 143
MCG –4-50-6 GX 143
MCG –4-50-7 GX 143
MCG –4-50-8 GX 143
MCG –4-50-9 GX 143
MCG –4-50-14 GX 143
MCG –4-50-16 GX 143
MCG –4-50-18 GX 143
MCG –4-50-21 GX 143
MCG –4-50-22 GX 143
MCG –4-50-23 GX 143
MCG –4-50-24 GX 143
MCG –4-50-26 GX 143
MCG –4-50-27 GX 143
MCG –4-51-2 GX 143
MCG –4-51-4 GX 143
MCG –4-51-7 GX 143
MCG –4-51-9 GX 143
MCG –4-51-16 GX 142, 143
MCG –4-51-17 GX 142, 143
MCG –4-52-6 GX 142, 143
MCG –4-52-13 GX 142
MCG –4-52-17 GX 142
MCG –4-52-18 GX 142
MCG –4-52-24 GX 142
MCG –4-52-27 GX 142
MCG –4-52-34 GX 142
MCG –4-52-35 GX 142
MCG –4-52-37 GX 142
MCG –4-52-38 GX 142
MCG –4-52-44 GX 142
MCG –4-52-46 GX 142
MCG –4-53-1 GX 142
MCG –4-53-3 GX 142
MCG –4-53-8 GX 142
MCG –4-53-12 GX 142
MCG –4-53-13 GX 142
MCG –4-53-14 GX 142
MCG –4-53-17 GX 142
MCG –4-53-19 GX 142
MCG –4-53-25 GX 142
MCG –4-53-28 GX 142
MCG –4-53-30 GX 142
MCG –4-53-31 GX 142
MCG –4-53-32 GX 142
MCG –4-53-37 GX 142
MCG –4-54-2 GX 142
MCG –4-54-5 GX 142
MCG –4-54-9 GX 142
MCG –4-54-10 GX 142
MCG –4-54-12 GX 142
MCG –4-54-13 GX 142
MCG –4-55-8 GX 141
MCG –4-55-9 GX 141
MCG –4-55-13 GX 141
MCG –4-55-14 GX 141
MCG –4-55-18 GX 141
MCG –4-55-21 GX 141
MCG –4-56-1 GX 141
MCG –3-1-1 GX 141
MCG –3-1-2 GX 121
MCG –3-1-3 GX 121
MCG –3-1-4 GX 121, 141
MCG –3-1-5 GX 141
MCG –3-1-6 GX 141
MCG –3-1-7 GX 141
MCG –3-1-9 GX 121, 141
MCG –3-1-10 GX 121
MCG –3-1-11 GX 121
MCG –3-1-12 GX 141
MCG –3-1-13 GX 141
MCG –3-1-14 GX 121
MCG –3-1-15 GX 121
MCG –3-1-16 GX 121
MCG –3-1-17 GX 121
MCG –3-1-18 GX 121
MCG –3-1-20 GX 141
MCG –3-1-22 GX 141
MCG –3-1-25 GX 141
MCG –3-1-28 GX 121
MCG –3-2-1 GX 121
MCG –3-2-2 GX 141
MCG –3-2-3 GX 121
MCG –3-2-5 GX 121, 141
MCG –3-2-7 GX 141
MCG –3-2-8 GX 141
MCG –3-2-9 GX 121
MCG –3-2-10 GX 121
MCG –3-2-11 GX 141
MCG –3-2-12 GX 121
MCG –3-2-13 GX 121
MCG –3-2-14 GX 121
MCG –3-2-15 GX 121
MCG –3-2-16 GX 121
MCG –3-2-17 GX 121
MCG –3-2-18 GX 121
MCG –3-2-21 GX 141
MCG –3-2-25 GX 121, 140, 141, 158
MCG –3-2-27 GX 121, 140
MCG –3-2-32 GX 141, 158
MCG –3-2-37 GX 140
MCG –3-2-38 GX 140
MCG –3-2-39 GX 140
MCG –3-2-40 GX 140
MCG –3-2-42 GX 140, 141, 158
MCG –3-3-1 GX 140, 141, 158
MCG –3-3-4 GX 140
MCG –3-3-5 GX 140
MCG –3-3-6 GX 140
MCG –3-3-8 GX 140
MCG –3-3-9 GX 140
MCG –3-3-10 GX 140
MCG –3-3-14 GX 140
MCG –3-3-16 GX 158
MCG –3-3-17 GX 140
MCG –3-3-18 GX 140
MCG –3-3-19 GX 140
MCG –3-3-21 GX 140, 158
MCG –3-4-2 GX 140, 158
MCG –3-4-14 GX 140
MCG –3-4-15 GX 158
MCG –3-4-16 GX 158
MCG –3-4-17 GX 158
MCG –3-4-18 GX 140, 158
MCG –3-4-20 GX 158
MCG –3-4-21 GX 158
MCG –3-4-22 GX 140, 158
MCG –3-4-23 GX 140, 158
MCG –3-4-26 GX 140, 158
MCG –3-4-28 GX 158
MCG –3-4-30 GX 158
MCG –3-4-32 GX 140
MCG –3-4-34 GX 140
MCG –3-4-38 GX 140
MCG –3-4-42 GX 158
MCG –3-4-44 GX 140, 158
MCG –3-4-45 GX 158
MCG –3-4-46 GX 140
MCG –3-4-47 GX 140, 158
MCG –3-4-48 GX 140, 158
MCG –3-4-49 GX 140
MCG –3-4-50 GX 158
MCG –3-4-54 GX 140
MCG –3-4-57 GX 140
MCG –3-4-58 GX 140, 158
MCG –3-4-59 GX 140
MCG –3-4-60 GX 140
MCG –3-4-61 GX 140
MCG –3-4-62 GX 140
MCG –3-4-64 GX 158
MCG –3-4-66 GX 140, 158
MCG –3-4-67 GX 140
MCG –3-4-68 GX 140, 158
MCG –3-4-70 GX 158
MCG –3-4-71 GX 140
MCG –3-4-72 GX 158
MCG –3-4-74 GX 140
MCG –3-4-75 GX 140
MCG –3-4-76 GX 140
MCG –3-4-78 GX 140
MCG –3-4-79 GX 140, 158
MCG –3-5-2 GX 140
MCG –3-5-3 GX 140
MCG –3-5-6 GX 140, 158
MCG –3-5-7 GX 140
MCG –3-5-8 GX 140
MCG –3-5-9 GX 140
MCG –3-5-10 GX 140, 158
MCG –3-5-12 GX 140
MCG –3-5-14 GX 140
MCG –3-5-18 GX 140
MCG –3-5-19 GX 139, 140
MCG –3-5-20 GX 139, 140
MCG –3-6-2 GX 139
MCG –3-6-3 GX 139
MCG –3-6-5 GX 139
MCG –3-6-6 GX 139, 157
MCG –3-6-7 GX 139
MCG –3-6-8 GX 139, 157
MCG –3-6-9 GX 139
MCG –3-6-11 GX 157
MCG –3-6-12 GX 157
MCG –3-6-15 GX 139
MCG –3-6-20 GX 139
MCG –3-6-21 GX 157
MCG –3-6-22 GX 139
MCG –3-6-23 GX 139
MCG –3-7-3 GX 157
MCG –3-7-4 GX 157
MCG –3-7-6 GX 139
MCG –3-7-7 GX 139
MCG –3-7-8 GX 157
MCG –3-7-9 GX 157
MCG –3-7-10 GX 139
MCG –3-7-12 GX 157
MCG –3-7-13 GX 139
MCG –3-7-14 GX 139
MCG –3-7-17 GX 157
MCG –3-7-18 GX 139
MCG –3-7-19 GX 139
MCG –3-7-20 GX 157
MCG –3-7-23 GX 139, 157
MCG –3-7-24 GX 139
MCG –3-7-26 GX 157
MCG –3-7-27 GX 139, 157
MCG –3-7-28 GX 139
MCG –3-7-32 GX 139
MCG –3-7-35 GX 139
MCG –3-7-36 GX 157
MCG –3-7-38 GX 139, 157
MCG –3-7-39 GX 139, 157
MCG –3-7-40 GX 139, 157
MCG –3-7-41 GX 139
MCG –3-7-42 GX 139
MCG –3-7-47 GX 139
MCG –3-7-48 GX 157
MCG –3-7-50 GX 139
MCG –3-7-51 GX 139
MCG –3-7-52 GX 139
MCG –3-7-53 GX 139, 157
MCG –3-7-57 GX 139
MCG –3-7-60 GX 157
MCG –3-7-61 GX 139
MCG –3-8-5 GX 139
MCG –3-8-6 GX 139, 157
MCG –3-8-7 GX 139, 157
MCG –3-8-12 GX 139, 157
MCG –3-8-17 GX 139
MCG –3-8-18 GX 139
MCG –3-8-19 GX 139, 157
MCG –3-8-21 GX 139
MCG –3-8-22 GX 157
MCG –3-8-23 GX 157
MCG –3-8-25 GX 139
MCG –3-8-26 GX 139
MCG –3-8-27 GX 157
MCG –3-8-30 GX 139
MCG –3-8-32 GX 139
MCG –3-8-33 GX 139
MCG –3-8-34 GX 139
MCG –3-8-37 GX 139
MCG –3-8-41 GX 139, 157
MCG –3-8-45 GX 139, 157
MCG –3-8-46 GX 138, 139
MCG –3-8-47 GX 157
MCG –3-8-50 GX 138, 139
MCG –3-8-51 GX 138, 139
MCG –3-8-52 GX 138, 139
MCG –3-8-55 GX 138, 139
MCG –3-8-57 GX 138, 139
MCG –3-8-58 GX 138, 139
MCG –3-8-69 GX 138, 139
MCG –3-8-72 GX 138
MCG –3-9-3 GX 138
MCG –3-9-4 GX 138
MCG –3-9-5 GX 138
MCG –3-9-6 GX 138, 157
MCG –3-9-10 GX 138
MCG –3-9-13 GX 157
MCG –3-9-16 GX 138
MCG –3-9-20 GX 138
MCG –3-9-21 GX 138
MCG –3-9-24 GX 138, 156, 157
MCG –3-9-25 GX 138
MCG –3-9-26 GX 138
MCG –3-9-27 GX 138
MCG –3-9-29 GX 138, 156, 157
MCG –3-9-30 GX 138, 156, 157
MCG –3-9-31 GX 138
MCG –3-9-32 GX 138
MCG –3-9-34 GX 138
MCG –3-9-36 GX 138
MCG –3-9-41 GX 138
MCG –3-9-44 GX 138, 156
MCG –3-9-47 GX 138
MCG –3-9-48 GX 138, 156
MCG –3-10-9 GX 138
MCG –3-10-10 GX 156
MCG –3-10-16 GX 138
MCG –3-10-18 GX 156
MCG –3-10-24 GX 138
MCG –3-10-26 GX 156
MCG –3-10-34 GX 156
MCG –3-10-36 GX 156
MCG –3-10-37 GX 156
MCG –3-10-38 GX 138
MCG –3-10-41 GX 138
MCG –3-10-42 GX 138
MCG –3-10-45 GX 138
MCG –3-10-46 GX 156
MCG –3-10-50 GX 138, 156
MCG –3-10-51 GX 138, 156
MCG –3-10-52 GX 156
MCG –3-11-1 GX 156
MCG –3-11-7 GX 138
MCG –3-11-8 GX 156
MCG –3-11-10 GX 138
MCG –3-11-12 GX 138, 156
MCG –3-11-14 GX 137, 138
MCG –3-11-17 GX 137, 138
MCG –3-11-18 GX 137, 138
MCG –3-11-19 GX 137
MCG –3-12-1 GX 137, 156
MCG –3-12-3 GX 137
MCG –3-12-8 GX 137
MCG –3-12-9 GX 156
MCG –3-12-11 GX 137
MCG –3-12-13 GX 137
MCG –3-12-15 GX 156
MCG –3-12-16 GX 155, 156
MCG –3-13-2 GX 137
MCG –3-13-3 GX 137
MCG –3-13-4 GX 137, 155
MCG –3-13-5 GX 137, 155
MCG –3-13-6 GX 137
MCG –3-13-7 GX 155
MCG –3-13-8 GX 137
MCG –3-13-9 GX 137, 155
MCG –3-13-10 GX 137, 155
MCG –3-13-11 GX 137, 155
MCG –3-13-13 GX 137
MCG –3-13-14 GX 155
MCG –3-13-16 GX 137, 155
MCG –3-13-17 GX 137, 155
MCG –3-13-18 GX 137
MCG –3-13-20 GX 137, 155
MCG –3-13-22 GX 155
MCG –3-13-23 GX 155
MCG –3-13-24 GX 137, 155
MCG –3-13-27 GX 137
MCG –3-13-28 GX 137
MCG –3-13-30 GX 137
MCG –3-13-31 GX 137
MCG –3-13-32 GX 155
MCG –3-13-33 GX 137, 155
MCG –3-13-35 GX 137, 155
MCG –3-13-41 GX 155
MCG –3-13-42 GX 137
MCG –3-13-44 GX 137, 155
MCG –3-13-46 GX 137, 155
MCG –3-13-48 GX 137, 155
MCG –3-13-50 GX 155
MCG –3-13-51 GX 137
MCG –3-13-52 GX 137
MCG –3-13-53 GX 137
MCG –3-13-56 GX 155
MCG –3-13-57 GX 155
MCG –3-13-58 GX 137
MCG –3-13-60 GX 155
MCG –3-13-61 GX 155
MCG –3-13-62 GX 137
MCG –3-13-63 GX 137
MCG –3-13-64 GX 137, 155
MCG –3-13-67 GX 137, 155
MCG –3-13-68 GX 155
MCG –3-13-69 GX 137
MCG –3-13-74 GX 155
MCG –3-14-1 GX 137
MCG –3-14-3 GX 137, 155
MCG –3-14-4 GX 137
MCG –3-14-5 GX 155
MCG –3-14-11 GX 137
MCG –3-14-12 GX 155
MCG –3-14-17 GX 136
MCG –3-14-19 GX 155
MCG –3-14-20 GX 155
MCG –3-15-4 GX 136, 155
MCG –3-15-5 GX 136
MCG –3-15-6 GX 136
MCG –3-15-7 GX 136, 155
MCG –3-15-8 GX 136, 155
MCG –3-15-14 GX 136
MCG –3-15-15 GX 155
MCG –3-15-17 GX 155
MCG –3-15-21 GX 155
MCG –3-15-23 GX 155
MCG –3-15-26 GX 155
MCG –3-15-27 GX 136
MCG –3-16-1 GX 154, 155
MCG –3-16-4 GX 136
MCG –3-16-5 GX 154, 155
MCG –3-16-6 GX 136
MCG –3-16-7 GX 154, 155
MCG –3-16-8 GX 136, 154, 155
MCG –3-16-16 GX 154
MCG –3-16-18 GX 136
MCG –3-16-20 GX 154
MCG –3-16-23 GX 136
MCG –3-17-1 GX 136
MCG –3-17-4 GX 136
MCG –3-17-5 GX 136, 154
MCG –3-18-4 GX 154
MCG –3-21-1 GX 153
MCG –3-21-8 GX 134
MCG –3-22-2 GX 134, 153
MCG –3-22-7 GX 152, 153
MCG –3-22-10 GX 134
MCG –3-22-11 GX 152, 153
MCG –3-23-3 GX 134, 152
MCG –3-23-7 GX 152
MCG –3-23-8 GX 152
MCG –3-23-9 GX 134
MCG –3-23-10 GX 134, 152
MCG –3-23-15 GX 152
MCG –3-23-17 GX 152
MCG –3-23-18 GX 152
MCG –3-24-4 GX 133, 152
MCG –3-24-6 GX 133, 152
MCG –3-24-9 GX 152
MCG –3-24-11 GX 133
MCG –3-24-13 GX 133
MCG –3-25-1 GX 133
MCG –3-25-3 GX 133
MCG –3-25-4 GX 133
MCG –3-25-5 GX 133
MCG –3-25-9 GX 133
MCG –3-25-10 GX 133, 152
MCG –3-25-12 GX 152
MCG –3-25-14 GX 133, 152
MCG –3-25-15 GX 133
MCG –3-25-16 GX 133
MCG –3-25-18 GX 152
MCG –3-25-23 GX 133
MCG –3-25-25 GX 133, 152
MCG –3-25-26 GX 152
MCG –3-25-27 GX 133
MCG –3-25-31 GX 152
MCG –3-26-5 GX 151, 152
MCG –3-26-6 GX 151, 152
MCG –3-26-9 GX 151, 152
MCG –3-26-13 GX 133
MCG –3-26-15 GX 151, 152
MCG –3-26-16 GX 133, 151
MCG –3-26-21 GX 133
MCG –3-26-23 GX 151
MCG –3-26-26 GX 151
MCG –3-26-27 GX 151

MCG –3-26-30 GX 132, 133, 151
MCG –3-26-38 GX 132, 151
MCG –3-27-1 GX 151
MCG –3-27-2 GX 151
MCG –3-27-3 GX 151
MCG –3-27-5 GX 132, 151
MCG –3-27-6 GX 151
MCG –3-27-7 GX 151
MCG –3-27-8 GX 132, 151
MCG –3-27-9 GX 132, 151
MCG –3-27-10 GX 132, 151
MCG –3-27-11 GX 151
MCG –3-27-12 GX 151
MCG –3-27-14 GX 151
MCG –3-27-15 GX 151
MCG –3-27-16 GX 132, 151
MCG –3-27-17 GX 132, 151
MCG –3-27-19 GX 151
MCG –3-27-22 GX 132
MCG –3-27-23 GX 132, 151
MCG –3-27-24 GX 132, 151
MCG –3-27-25 GX 151
MCG –3-27-26 GX 132, 151
MCG –3-28-1 GX 132
MCG –3-28-2 GX 132, 151
MCG –3-28-3 GX 132
MCG –3-28-9 GX 151
MCG –3-28-10 GX 132, 151
MCG –3-28-13 GX 132, 151
MCG –3-28-16 GX 132, 151
MCG –3-28-17 GX 132
MCG –3-28-20 GX 132, 151
MCG –3-28-24 GX 132
MCG –3-28-27 GX 151
MCG –3-28-28 GX 151
MCG –3-28-29 GX 151
MCG –3-28-32 GX 132
MCG –3-28-33 GX 132, 151
MCG –3-29-2 GX 132
MCG –3-29-6 GX 131, 132, 150, 151
MCG –3-29-9 GX 150
MCG –3-30-2 GX 131
MCG –3-30-3 GX 131, 150
MCG –3-30-4 GX 131
MCG –3-30-5 GX 150
MCG –3-30-7 GX 131
MCG –3-30-9 GX 150
MCG –3-30-11 GX 150
MCG –3-30-13 GX 131
MCG –3-30-15 GX 150
MCG –3-30-18 GX 150
MCG –3-30-19 GX 131
MCG –3-31-5 GX 150
MCG –3-31-6 GX 150
MCG –3-31-9 GX 150
MCG –3-31-13 GX 150
MCG –3-31-18 GX 131
MCG –3-31-21 GX 150
MCG –3-31-23 GX 131
MCG –3-31-25 GX 150
MCG –3-31-26 GX 150
MCG –3-32-3 GX 131, 150
MCG –3-32-4 GX 131, 150
MCG –3-32-8 GX 150
MCG –3-32-9 GX 150
MCG –3-32-11 GX 150
MCG –3-32-15 GX 149, 150
MCG –3-32-16 GX 149, 150
MCG –3-32-17 GX 149, 150
MCG –3-32-18 GX 149, 150
MCG –3-33-5 GX 130
MCG –3-33-9 GX 149
MCG –3-33-10 GX 149
MCG –3-33-11 GX 149
MCG –3-33-12 GX 130
MCG –3-33-15 GX 130
MCG –3-33-16 GX 149
MCG –3-33-18 GX 149
MCG –3-33-20 GX 149
MCG –3-33-22 GX 149
MCG –3-33-23 GX 130, 149
MCG –3-33-28 GX 130, 149
MCG –3-33-30 GX 130, 149
MCG –3-33-31 GX 130
MCG –3-34-2 GX 130
MCG –3-34-3 GX 149
MCG –3-34-4 GX 130
MCG –3-34-6 GX 149
MCG –3-34-8 GX 130
MCG –3-34-10 GX 130
MCG –3-34-14 GX 130, 149
MCG –3-34-19 GX 130
MCG –3-34-20 GX 130
MCG –3-34-22 GX 130, 149
MCG –3-34-25 GX 130, 149
MCG –3-34-26 GX 149
MCG –3-34-30 GX 130, 149
MCG –3-34-33 GX 130
MCG –3-34-38 GX 130
MCG –3-34-40 GX 130
MCG –3-34-41 GX 130
MCG –3-34-42 GX 130
MCG –3-34-44 GX 130, 149
MCG –3-34-45 GX 149
MCG –3-34-48 GX 130, 149
MCG –3-34-49 GX 130, 149
MCG –3-34-51 GX 130
MCG –3-34-53 GX 130
MCG –3-34-54 GX 130
MCG –3-34-55 GX 130
MCG –3-34-56 GX 130
MCG –3-34-58 GX 130
MCG –3-34-59 GX 130
MCG –3-34-60 GX 130
MCG –3-34-61 GX 130, 149
MCG –3-34-62 GX 130
MCG –3-34-63 GX 130
MCG –3-34-64 GX 130
MCG –3-34-65 GX 130, 149
MCG –3-34-67 GX 130
MCG –3-34-72 GX 149
MCG –3-34-74 GX 149
MCG –3-34-75 GX 149
MCG –3-34-76 GX 130
MCG –3-34-78 GX 149
MCG –3-34-79 GX 130
MCG –3-34-80 GX 149
MCG –3-34-81 GX 149
MCG –3-34-82 GX 130, 149
MCG –3-34-83 GX 149
MCG –3-34-87 GX 130
MCG –3-35-1 GX 149
MCG –3-35-2 GX 149
MCG –3-35-3 GX 130
MCG –3-35-4 GX 130
MCG –3-35-5 GX 130
MCG –3-35-6 GX 130, 149
MCG –3-35-7 GX 130
MCG –3-35-8 GX 149
MCG –3-35-9 GX 130
MCG –3-35-10 GX 130
MCG –3-35-12 GX 149
MCG –3-35-15 GX 149
MCG –3-35-16 GX 149
MCG –3-35-17 GX 129, 130
MCG –3-35-18 GX 149
MCG –3-35-19 GX 149
MCG –3-35-20 GX 129
MCG –3-36-1 GX 129
MCG –3-36-2 GX 148, 149
MCG –3-36-4 GX 129, 148
MCG –3-36-6 GX 129
MCG –3-36-8 GX 129, 148
MCG –3-36-9 GX 129, 148
MCG –3-38-1 GX 148
MCG –3-38-2 GX 148
MCG –3-38-8 GX 129, 148
MCG –3-38-10 GX 148
MCG –3-38-22 GX 129
MCG –3-38-25 GX 129, 148
MCG –3-38-29 GX 128, 129
MCG –3-38-32 GX 148
MCG –3-38-33 GX 148
MCG –3-38-42 GX 128, 129
MCG –3-39-5 GX 128
MCG –3-39-9 GX 128, 147, 148
MCG –3-39-10 GX 147
MCG –3-40-3 GX 128
MCG –3-41-1 GX 147
MCG –3-50-1 GX 125
MCG –3-50-3 GX 125
MCG –3-51-1 GX 124, 144
MCG –3-51-5 GX 124, 144
MCG –3-51-7 GX 144
MCG –3-51-9 GX 144
MCG –3-52-1 GX 144
MCG –3-52-2 GX 124
MCG –3-52-5 GX 124
MCG –3-52-6 GX 144
MCG –3-52-10 GX 144
MCG –3-52-12 GX 144
MCG –3-52-13 GX 144
MCG –3-52-14 GX 124
MCG –3-52-15 GX 144
MCG –3-52-17 GX 144
MCG –3-52-20 GX 124
MCG –3-53-1 GX 124
MCG –3-53-2 GX 143
MCG –3-53-3 GX 143
MCG –3-53-7 GX 124
MCG –3-53-14 GX 143
MCG –3-53-15 GX 123, 124
MCG –3-53-16 GX 123, 124
MCG –3-53-19 GX 123, 124
MCG –3-53-20 GX 143
MCG –3-53-23 GX 123, 124
MCG –3-53-24 GX 123, 124
MCG –3-53-25 GX 123, 124
MCG –3-53-26 GX 123
MCG –3-53-27 GX 123
MCG –3-54-2 GX 143
MCG –3-54-3 GX 143
MCG –3-55-4 GX 123, 143
MCG –3-55-6 GX 123
MCG –3-56-6 GX 123
MCG –3-56-7 GX 142
MCG –3-56-10 GX 122, 123, 142
MCG –3-56-11 GX 122, 123
MCG –3-57-1 GX 122
MCG –3-57-3 GX 142
MCG –3-57-8 GX 122
MCG –3-57-10 GX 142
MCG –3-57-11 GX 142
MCG –3-57-12 GX 122
MCG –3-57-16 GX 122, 142
MCG –3-57-18 GX 122, 142
MCG –3-57-19 GX 122
MCG –3-57-20 GX 142
MCG –3-57-22 GX 122
MCG –3-57-23 GX 122
MCG –3-57-24 GX 122
MCG –3-58-3 GX 122, 142
MCG –3-58-4 GX 122, 142
MCG –3-58-6 GX 122
MCG –3-58-7 GX 142
MCG –3-58-8 GX 122
MCG –3-58-11 GX 122
MCG –3-58-12 GX 122
MCG –3-58-13 GX 122
MCG –3-58-14 GX 122, 142
MCG –3-58-15 GX 122
MCG –3-58-16 GX 122
MCG –3-58-17 GX 122
MCG –3-59-1 GX 122
MCG –3-59-5 GX 121, 122
MCG –3-59-7 GX 121, 122
MCG –3-60-4 GX 141
MCG –3-60-5 GX 141
MCG –3-60-6 GX 121
MCG –3-60-7 GX 141
MCG –3-60-9 GX 141
MCG –3-60-11 GX 121
MCG –3-60-13 GX 141
MCG –3-60-15 GX 141
MCG –3-60-17 GX 141
MCG –3-60-19 GX 121
MCG –3-60-23 GX 121, 141
MCG –2-1-1 GX 121
MCG –2-1-2 GX 121
MCG –2-1-3 GX 121
MCG –2-1-4 GX 121
MCG –2-1-5 GX 121
MCG –2-1-6 GX 121
MCG –2-1-9 GX 121
MCG –2-1-10 GX 121
MCG –2-1-11 GX 121
MCG –2-1-12 GX 121
MCG –2-1-14 GX 121
MCG –2-1-15 GX 121
MCG –2-1-17 GX 121
MCG –2-1-18 GX 121
MCG –2-1-21 GX 121
MCG –2-1-22 GX 121
MCG –2-1-23 GX 121
MCG –2-1-24 GX 121
MCG –2-1-26 GX 121
MCG –2-1-27 GX 121
MCG –2-1-28 GX 121
MCG –2-1-29 GX 121
MCG –2-1-30 GX 121
MCG –2-1-34 GX 121
MCG –2-1-35 GX 121
MCG –2-1-36 GX 121
MCG –2-1-37 GX 121
MCG –2-1-41 GX 121
MCG –2-1-42 GX 121
MCG –2-1-44 GX 121
MCG –2-1-45 GX 121
MCG –2-1-46 GX 121
MCG –2-1-48 GX 121
MCG –2-1-50 GX 121
MCG –2-1-51 GX 121
MCG –2-1-52 GX 121
MCG –2-1-54 GX 121
MCG –2-2-3 GX 121
MCG –2-2-4 GX 121
MCG –2-2-5 GX 121
MCG –2-2-6 GX 121
MCG –2-2-7 GX 121
MCG –2-2-9 GX 121
MCG –2-2-10 GX 121
MCG –2-2-12 GX 121
MCG –2-2-13 GX 121
MCG –2-2-15 GX 121
MCG –2-2-16 GX 121
MCG –2-2-18 GX 121
MCG –2-2-19 GX 121
MCG –2-2-20 GX 121
MCG –2-2-22 GX 121
MCG –2-2-25 GX 121
MCG –2-2-26 GX 121
MCG –2-2-28 GX 121
MCG –2-2-29 GX 121
MCG –2-2-30 GX 121
MCG –2-2-31 GX 121
MCG –2-2-33 GX 121
MCG –2-2-34 GX 121
MCG –2-2-35 GX 121
MCG –2-2-36 GX 121
MCG –2-2-37 GX 121
MCG –2-2-38 GX 121
MCG –2-2-40 GX 121
MCG –2-2-41 GX 121
MCG –2-2-42 GX 121
MCG –2-2-43 GX 121
MCG –2-2-46 GX 121, 140
MCG –2-2-47 GX 121, 140
MCG –2-2-48 GX 121, 140
MCG –2-2-49 GX 121, 140
MCG –2-2-50 GX 121, 140
MCG –2-2-51 GX 121, 140
MCG –2-2-52 GX 121, 140
MCG –2-2-57 GX 121, 140
MCG –2-2-58 GX 121, 140
MCG –2-2-59 GX 121, 140
MCG –2-2-60 GX 121, 140
MCG –2-2-61 GX 121, 140
MCG –2-2-62 GX 121, 140
MCG –2-2-64 GX 121, 140
MCG –2-2-67 GX 121, 140
MCG –2-2-68 GX 121, 140
MCG –2-2-70 GX 121, 140
MCG –2-2-71 GX 121, 140
MCG –2-2-72 GX 121, 140
MCG –2-2-73 GX 121, 140
MCG –2-2-74 GX 121, 140
MCG –2-2-75 GX 121, 140
MCG –2-2-80 GX 140
MCG –2-2-82 GX 140
MCG –2-2-83 GX 140
MCG –2-2-86 GX 140
MCG –2-2-89 GX 140
MCG –2-2-92 GX 140
MCG –2-3-2 GX 140
MCG –2-3-3 GX 140
MCG –2-3-4 GX 140
MCG –2-3-5 GX 140
MCG –2-3-6 GX 140
MCG –2-3-7 GX 140
MCG –2-3-8 GX 140
MCG –2-3-9 GX 140
MCG –2-3-12 GX 140
MCG –2-3-13 GX 140
MCG –2-3-14 GX 140
MCG –2-3-15 GX 140
MCG –2-3-16 GX 140
MCG –2-3-18 GX 140
MCG –2-3-19 GX 140
MCG –2-3-20 GX 140
MCG –2-3-22 GX 140
MCG –2-3-23 GX 140
MCG –2-3-25 GX 140
MCG –2-3-26 GX 140
MCG –2-3-29 GX 140
MCG –2-3-36 GX 140
MCG –2-3-37 GX 140
MCG –2-3-38 GX 140
MCG –2-3-39 GX 140
MCG –2-3-40 GX 140
MCG –2-3-42 GX 140
MCG –2-3-44 GX 140
MCG –2-3-45 GX 140
MCG –2-3-46 GX 140
MCG –2-3-51 GX 140
MCG –2-3-52 GX 140
MCG –2-3-53 GX 140
MCG –2-3-54 GX 140
MCG –2-3-55 GX 140
MCG –2-3-56 GX 140
MCG –2-3-57 GX 140
MCG –2-3-58 GX 140
MCG –2-3-59 GX 140
MCG –2-3-60 GX 140
MCG –2-3-61 GX 140
MCG –2-3-64 GX 140
MCG –2-3-66 GX 140
MCG –2-3-67 GX 140
MCG –2-3-68 GX 140
MCG –2-3-69 GX 140
MCG –2-3-70 GX 140
MCG –2-3-71 GX 140
MCG –2-3-74 GX 140
MCG –2-4-1 GX 140
MCG –2-4-3 GX 140
MCG –2-4-4 GX 140
MCG –2-4-5 GX 140
MCG –2-4-6 GX 140
MCG –2-4-7 GX 140
MCG –2-4-8 GX 140
MCG –2-4-9 GX 140
MCG –2-4-10 GX 140
MCG –2-4-11 GX 140
MCG –2-4-12 GX 140
MCG –2-4-13 GX 140
MCG –2-4-14 GX 140
MCG –2-4-15 GX 140
MCG –2-4-17 GX 140
MCG –2-4-18 GX 140
MCG –2-4-19 GX 140
MCG –2-4-20 GX 140
MCG –2-4-21 GX 140
MCG –2-4-22 GX 140
MCG –2-4-23 GX 140
MCG –2-4-24 GX 140
MCG –2-4-28 GX 140
MCG –2-4-29 GX 140
MCG –2-4-31 GX 140
MCG –2-4-32 GX 140
MCG –2-4-35 GX 140
MCG –2-4-38 GX 140
MCG –2-4-39 GX 140
MCG –2-4-42 GX 140
MCG –2-4-43 GX 140
MCG –2-4-44 GX 140
MCG –2-4-45 GX 140
MCG –2-4-46 GX 140
MCG –2-4-47 GX 140
MCG –2-4-48 GX 140
MCG –2-4-49 GX 140
MCG –2-4-50 GX 140
MCG –2-4-51 GX 140
MCG –2-4-52 GX 140
MCG –2-4-54 GX 140
MCG –2-4-56 GX 140
MCG –2-4-57 GX 140
MCG –2-4-59 GX 140
MCG –2-4-61 GX 140
MCG –2-4-62 GX 140
MCG –2-4-64 GX 140
MCG –2-5-2 GX 140
MCG –2-5-6 GX 140
MCG –2-5-8 GX 140
MCG –2-5-9 GX 140
MCG –2-5-11 GX 140
MCG –2-5-12 GX 140
MCG –2-5-13 GX 140
MCG –2-5-14 GX 140
MCG –2-5-15 GX 140
MCG –2-5-16 GX 140
MCG –2-5-17 GX 140
MCG –2-5-18 GX 140
MCG –2-5-19 GX 140
MCG –2-5-20 GX 140
MCG –2-5-21 GX 140
MCG –2-5-22 GX 140
MCG –2-5-23 GX 140
MCG –2-5-24 GX 140
MCG –2-5-25 GX 140
MCG –2-5-26 GX 140
MCG –2-5-27 GX 140
MCG –2-5-29 GX 140
MCG –2-5-30 GX 140
MCG –2-5-32 GX 140
MCG –2-5-35 GX 140
MCG –2-5-38 GX 140
MCG –2-5-39 GX 139, 140
MCG –2-5-40 GX 139, 140
MCG –2-5-41 GX 139, 140
MCG –2-5-46 GX 139, 140
MCG –2-5-48 GX 139, 140
MCG –2-5-49 GX 139, 140
MCG –2-5-50 GX 139, 140
MCG –2-5-50A GX 139, 140
MCG –2-5-51 GX 139, 140
MCG –2-5-53 GX 139, 140
MCG –2-5-54 GX 139, 140
MCG –2-5-55 GX 139, 140
MCG –2-5-56 GX 139, 140
MCG –2-5-57 GX 139, 140
MCG –2-5-62 GX 139, 140
MCG –2-5-64 GX 139, 140
MCG –2-5-65 GX 139, 140
MCG –2-5-67 GX 139
MCG –2-5-70 GX 139
MCG –2-5-71 GX 139
MCG –2-5-72 GX 139
MCG –2-5-74 GX 139
MCG –2-5-76 GX 139
MCG –2-6-1 GX 139
MCG –2-6-2 GX 139
MCG –2-6-4 GX 139
MCG –2-6-6 GX 139
MCG –2-6-9 GX 139
MCG –2-6-13 GX 139
MCG –2-6-14 GX 139
MCG –2-6-16 GX 139
MCG –2-6-17 GX 139
MCG –2-6-18 GX 139
MCG –2-6-19 GX 139
MCG –2-6-20 GX 139
MCG –2-6-22 GX 139
MCG –2-6-25 GX 139
MCG –2-6-26 GX 139
MCG –2-6-27 GX 139
MCG –2-6-28 GX 139
MCG –2-6-29 GX 139
MCG –2-6-35 GX 139
MCG –2-6-37 GX 139
MCG –2-6-39 GX 139
MCG –2-6-40 GX 139
MCG –2-6-41 GX 139
MCG –2-6-42 GX 139
MCG –2-6-43 GX 139
MCG –2-6-44 GX 139
MCG –2-6-45 GX 139
MCG –2-6-47 GX 139
MCG –2-6-49 GX 139
MCG –2-6-50 GX 139
MCG –2-6-51 GX 139
MCG –2-6-52 GX 139
MCG –2-6-53 GX 139
MCG –2-6-54 GX 139
MCG –2-6-55 GX 139
MCG –2-6-56 GX 139
MCG –2-7-1 GX 139
MCG –2-7-2 GX 139
MCG –2-7-3 GX 139
MCG –2-7-4 GX 139
MCG –2-7-6 GX 139
MCG –2-7-7 GX 139
MCG –2-7-8 GX 139
MCG –2-7-10 GX 139
MCG –2-7-11 GX 139
MCG –2-7-12 GX 139
MCG –2-7-14 GX 139
MCG –2-7-17 GX 139
MCG –2-7-20 GX 139
MCG –2-7-22 GX 139
MCG –2-7-23 GX 139
MCG –2-7-24 GX 139
MCG –2-7-25 GX 139
MCG –2-7-26 GX 139
MCG –2-7-29 GX 139
MCG –2-7-32 GX 139
MCG –2-7-33 GX 139
MCG –2-7-34 GX 139
MCG –2-7-36 GX 139
MCG –2-7-38 GX 139
MCG –2-7-39 GX 139
MCG –2-7-40 GX 139
MCG –2-7-41 GX 139
MCG –2-7-42 GX 139
MCG –2-7-43 GX 139
MCG –2-7-49 GX 139
MCG –2-7-50 GX 139
MCG –2-7-51 GX 139
MCG –2-7-56 GX 139
MCG –2-7-57 GX 139
MCG –2-7-60 GX 139
MCG –2-7-61 GX 139
MCG –2-7-63 GX 139

MCG –2-7-65 GX 139
MCG –2-7-66 GX 139
MCG –2-7-67 GX 139
MCG –2-7-68 GX 139
MCG –2-7-69 GX 139
MCG –2-7-70 GX 139
MCG –2-7-72 GX 139
MCG –2-7-73 GX 139
MCG –2-7-74 GX 139
MCG –2-7-75 GX 139
MCG –2-7-76 GX 139
MCG –2-8-2 GX 139
MCG –2-8-3 GX 139
MCG –2-8-4 GX 139
MCG –2-8-7 GX 139
MCG –2-8-8 GX 139
MCG –2-8-9 GX 139
MCG –2-8-10 GX 139
MCG –2-8-12 GX 139
MCG –2-8-13 GX 139
MCG –2-8-14 GX 139
MCG –2-8-16 GX 139
MCG –2-8-17 GX 139
MCG –2-8-18 GX 139
MCG –2-8-20 GX 139
MCG –2-8-21 GX 139
MCG –2-8-22 GX 139
MCG –2-8-25 GX 139
MCG –2-8-26 GX 139
MCG –2-8-27 GX 139
MCG –2-8-31 GX 138, 139
MCG –2-8-32 GX 138, 139
MCG –2-8-33 GX 138, 139
MCG –2-8-37 GX 138, 139
MCG –2-8-38 GX 138, 139
MCG –2-8-39 GX 138, 139
MCG –2-8-40 GX 138, 139
MCG –2-8-50 GX 138
MCG –2-8-52 GX 138
MCG –2-9-2 GX 138
MCG –2-9-3 GX 138
MCG –2-9-5 GX 138
MCG –2-9-6 GX 138
MCG –2-9-7 GX 138
MCG –2-9-8 GX 138
MCG –2-9-13 GX 138
MCG –2-9-16 GX 138
MCG –2-9-17 GX 138
MCG –2-9-18 GX 138
MCG –2-9-19 GX 138
MCG –2-9-20 GX 138
MCG –2-9-21 GX 138
MCG –2-9-23 GX 138
MCG –2-9-24 GX 138
MCG –2-9-27 GX 138
MCG –2-9-28 GX 138
MCG –2-9-29 GX 138
MCG –2-9-31 GX 138
MCG –2-9-32 GX 138
MCG –2-9-33 GX 138
MCG –2-9-34 GX 138
MCG –2-9-35 GX 138
MCG –2-9-36 GX 138
MCG –2-9-37 GX 138
MCG –2-9-38 GX 138
MCG –2-9-39 GX 138
MCG –2-9-40 GX 138
MCG –2-9-41 GX 138
MCG –2-9-43 GX 138
MCG –2-9-45 GX 138
MCG –2-9-46 GX 138
MCG –2-10-2 GX 138
MCG –2-10-3 GX 138
MCG –2-10-4 GX 138
MCG –2-10-6 GX 138
MCG –2-10-7 GX 138
MCG –2-10-9 GX 138
MCG –2-10-11 GX 138
MCG –2-10-12 GX 138
MCG –2-10-13 GX 138
MCG –2-10-14 GX 138
MCG –2-10-17 GX 138
MCG –2-10-18 GX 138
MCG –2-10-19 GX 138
MCG –2-10-20 GX 138
MCG –2-10-21 GX 138
MCG –2-11-1 GX 138
MCG –2-11-2 GX 138
MCG –2-11-3 GX 138
MCG –2-11-4 GX 138
MCG –2-11-5 GX 138
MCG –2-11-6 GX 138
MCG –2-11-7 GX 138
MCG –2-11-10 GX 138
MCG –2-11-11 GX 138
MCG –2-11-14 GX 138
MCG –2-11-15 GX 138
MCG –2-11-16 GX 138
MCG –2-11-19 GX 137, 138
MCG –2-11-20 GX 137, 138
MCG –2-11-21 GX 137, 138
MCG –2-11-22 GX 137, 138
MCG –2-11-23 GX 137, 138
MCG –2-11-24 GX 137, 138
MCG –2-11-25 GX 137, 138
MCG –2-11-26 GX 137, 138
MCG –2-11-28 GX 137, 138
MCG –2-11-29 GX 137
MCG –2-11-30 GX 137
MCG –2-11-32 GX 137
MCG –2-11-33 GX 137
MCG –2-11-34 GX 137
MCG –2-11-35 GX 137
MCG –2-11-36 GX 137
MCG –2-11-37 GX 137
MCG –2-11-38 GX 137
MCG –2-11-39 GX 137
MCG –2-12-2 GX 137
MCG –2-12-3 GX 137
MCG –2-12-4 GX 137
MCG –2-12-5 GX 137
MCG –2-12-6 GX 137
MCG –2-12-7 GX 137
MCG –2-12-8 GX 137
MCG –2-12-12 GX 137
MCG –2-12-13 GX 137
MCG –2-12-15 GX 137
MCG –2-12-16 GX 137
MCG –2-12-17 GX 137
MCG –2-12-18 GX 137
MCG –2-12-19 GX 137
MCG –2-12-20 GX 137
MCG –2-12-21 GX 137
MCG –2-12-22 GX 137
MCG –2-12-23 GX 137
MCG –2-12-24 GX 137
MCG –2-12-25 GX 137
MCG –2-12-26 GX 137
MCG –2-12-27 GX 137
MCG –2-12-28 GX 137
MCG –2-12-29 GX 137
MCG –2-12-30 GX 137
MCG –2-12-33 GX 137
MCG –2-12-35 GX 137
MCG –2-12-36 GX 137
MCG –2-12-37 GX 137
MCG –2-12-38 GX 137
MCG –2-12-39 GX 137
MCG –2-12-40 GX 137
MCG –2-12-41 GX 137
MCG –2-12-43 GX 137
MCG –2-12-44 GX 137
MCG –2-12-45 GX 137
MCG –2-12-46 GX 137
MCG –2-12-47 GX 137
MCG –2-12-48 GX 137
MCG –2-12-50 GX 137
MCG –2-12-52 GX 137
MCG –2-12-53 GX 137
MCG –2-12-55 GX 137
MCG –2-12-56 GX 137
MCG –2-12-57 GX 137
MCG –2-12-58 GX 137
MCG –2-13-1 GX 137
MCG –2-13-2 GX 137
MCG –2-13-3 GX 137
MCG –2-13-4 GX 137
MCG –2-13-5 GX 137
MCG –2-13-6 GX 137
MCG –2-13-7 GX 137
MCG –2-13-8 GX 137
MCG –2-13-9 GX 137
MCG –2-13-10 GX 137
MCG –2-13-11 GX 137
MCG –2-13-12 GX 137
MCG –2-13-13 GX 137
MCG –2-13-14 GX 137
MCG –2-13-15 GX 137
MCG –2-13-16 GX 137
MCG –2-13-17 GX 137
MCG –2-13-18 GX 137
MCG –2-13-19 GX 137
MCG –2-13-20 GX 137
MCG –2-13-21 GX 137
MCG –2-13-22 GX 137
MCG –2-13-23 GX 137
MCG –2-13-24 GX 137
MCG –2-13-25 GX 137
MCG –2-13-26 GX 137
MCG –2-13-31 GX 137
MCG –2-13-32 GX 137
MCG –2-13-33 GX 137
MCG –2-13-34 GX 137
MCG –2-13-35 GX 137
MCG –2-13-36 GX 137
MCG –2-13-37 GX 137
MCG –2-13-38 GX 137
MCG –2-13-39 GX 137
MCG –2-13-40A GX 137
MCG –2-14-1 GX 137
MCG –2-14-2 GX 137
MCG –2-14-3 GX 137
MCG –2-14-4 GX 137
MCG –2-14-5 GX 137
MCG –2-14-6 GX 137
MCG –2-14-7 GX 137
MCG –2-14-9 GX 137
MCG –2-14-10 GX 137
MCG –2-14-11 GX 137
MCG –2-14-12 GX 137
MCG –2-14-15 GX 136, 137
MCG –2-14-16 GX 136, 137
MCG –2-15-1 GX 136
MCG –2-15-4 GX 136
MCG –2-15-5 GX 136
MCG –2-15-6 GX 136
MCG –2-15-9 GX 136
MCG –2-15-11 GX 136
MCG –2-15-13 GX 136
MCG –2-16-2 GX 136
MCG –2-16-3 GX 136
MCG –2-16-8 GX 136
MCG –2-21-1 GX 134
MCG –2-22-1 GX 134
MCG –2-22-3 GX 134
MCG –2-22-4 GX 134
MCG –2-22-5 GX 134
MCG –2-22-6 GX 134
MCG –2-22-8 GX 134
MCG –2-22-9 GX 134
MCG –2-22-11 GX 134
MCG –2-22-12 GX 134
MCG –2-22-13 GX 134
MCG –2-22-17 GX 134
MCG –2-22-18 GX 134
MCG –2-22-21 GX 134
MCG –2-22-22 GX 134
MCG –2-22-23 GX 134
MCG –2-22-24 GX 134
MCG –2-22-25 GX 134
MCG –2-22-26 GX 134
MCG –2-22-27 GX 134
MCG –2-23-1 GX 134
MCG –2-23-3 GX 134
MCG –2-23-4 GX 134
MCG –2-23-6 GX 134
MCG –2-23-7 GX 133, 134
MCG –2-23-8 GX 133, 134
MCG –2-24-1 GX 133
MCG –2-24-3 GX 133
MCG –2-24-4 GX 133
MCG –2-24-5 GX 133
MCG –2-24-6 GX 133
MCG –2-24-7 GX 133
MCG –2-24-8 GX 133
MCG –2-24-9 GX 133
MCG –2-24-10 GX 133
MCG –2-24-11 GX 133
MCG –2-24-13 GX 133
MCG –2-24-14 GX 133
MCG –2-24-17 GX 133
MCG –2-24-19 GX 133
MCG –2-24-23 GX 133
MCG –2-24-27 GX 133
MCG –2-24-28 GX 133
MCG –2-24-29 GX 133
MCG –2-25-1 GX 133
MCG –2-25-2 GX 133
MCG –2-25-3 GX 133
MCG –2-25-5 GX 133
MCG –2-25-6 GX 133
MCG –2-25-7 GX 133
MCG –2-25-8 GX 133
MCG –2-25-9 GX 133
MCG –2-25-10 GX 133
MCG –2-25-11 GX 133
MCG –2-25-13 GX 133
MCG –2-25-17 GX 133
MCG –2-25-18 GX 133
MCG –2-25-19 GX 133
MCG –2-25-20 GX 133
MCG –2-25-23 GX 133
MCG –2-25-24 GX 133
MCG –2-25-25 GX 133
MCG –2-26-1 GX 133
MCG –2-26-2 GX 133
MCG –2-26-4 GX 133
MCG –2-26-6 GX 133
MCG –2-26-8 GX 133
MCG –2-26-9 GX 133
MCG –2-26-10 GX 133
MCG –2-26-11 GX 133
MCG –2-26-12 GX 133
MCG –2-26-13 GX 133
MCG –2-26-14 GX 133
MCG –2-26-15 GX 133
MCG –2-26-16 GX 133
MCG –2-26-17 GX 133
MCG –2-26-18 GX 133
MCG –2-26-19 GX 133
MCG –2-26-20 GX 133
MCG –2-26-21 GX 133
MCG –2-26-22 GX 133
MCG –2-26-23 GX 133
MCG –2-26-24 GX 133
MCG –2-26-25 GX 133
MCG –2-26-27 GX 133
MCG –2-26-28 GX 133
MCG –2-26-29 GX 133
MCG –2-26-30 GX 133
MCG –2-26-31 GX 132, 133
MCG –2-26-35 GX 132, 133
MCG –2-26-37 GX 132, 133
MCG –2-26-38 GX 132, 133
MCG –2-26-39 GX 132, 133
MCG –2-26-40 GX 132, 133
MCG –2-26-41 GX 132, 133
MCG –2-26-42 GX 132
MCG –2-27-1 GX 132
MCG –2-27-2 GX 132
MCG –2-27-3 GX 132
MCG –2-27-4 GX 132
MCG –2-27-9 GX 132
MCG –2-28-1 GX 132
MCG –2-28-2 GX 132
MCG –2-28-5 GX 132
MCG –2-28-6 GX 132
MCG –2-28-7 GX 132
MCG –2-28-9 GX 132
MCG –2-28-10 GX 132
MCG –2-28-14 GX 132
MCG –2-28-16 GX 132
MCG –2-28-17 GX 132
MCG –2-28-20 GX 132
MCG –2-28-21 GX 132
MCG –2-28-22 GX 132
MCG –2-28-23 GX 132
MCG –2-28-25 GX 132
MCG –2-28-26 GX 132
MCG –2-28-28 GX 132
MCG –2-28-29 GX 132
MCG –2-28-30 GX 132
MCG –2-28-31 GX 132
MCG –2-28-32 GX 132
MCG –2-28-35 GX 132
MCG –2-28-36 GX 132
MCG –2-28-37 GX 132
MCG –2-28-38 GX 132
MCG –2-28-39 GX 132
MCG –2-28-40 GX 132
MCG –2-28-42 GX 132
MCG –2-28-43 GX 132
MCG –2-28-44 GX 132
MCG –2-28-45 GX 132
MCG –2-28-47 GX 132
MCG –2-28-48 GX 132
MCG –2-28-49 GX 132
MCG –2-29-1 GX 132
MCG –2-29-3 GX 132
MCG –2-29-5 GX 132
MCG –2-29-6 GX 132
MCG –2-29-8 GX 132
MCG –2-29-9 GX 132
MCG –2-29-10 GX 132
MCG –2-29-11 GX 132
MCG –2-29-13 GX 132
MCG –2-29-14 GX 132
MCG –2-29-16 GX 132
MCG –2-29-18 GX 132
MCG –2-29-21 GX 131, 132
MCG –2-29-24 GX 131, 132
MCG –2-29-29 GX 131, 132
MCG –2-29-30 GX 131, 132
MCG –2-29-31 GX 131, 132
MCG –2-29-34 GX 131
MCG –2-29-36 GX 131
MCG –2-29-37 GX 131, 132
MCG –2-29-39 GX 131
MCG –2-29-40 GX 131
MCG –2-30-1 GX 131
MCG –2-30-3 GX 131
MCG –2-30-7 GX 131
MCG –2-30-10 GX 131
MCG –2-30-11 GX 131
MCG –2-30-14 GX 131
MCG –2-30-15 GX 131
MCG –2-30-16 GX 131
MCG –2-30-18 GX 131
MCG –2-30-19 GX 131
MCG –2-30-20 GX 131
MCG –2-30-21 GX 131
MCG –2-30-22 GX 131
MCG –2-30-24 GX 131
MCG –2-30-25 GX 131
MCG –2-30-26 GX 131
MCG –2-30-27 GX 131
MCG –2-30-28 GX 131
MCG –2-30-29 GX 131
MCG –2-30-31 GX 131
MCG –2-30-32 GX 131
MCG –2-30-33 GX 131
MCG –2-30-34 GX 131
MCG –2-30-36 GX 131
MCG –2-30-38 GX 131
MCG –2-30-39 GX 131
MCG –2-31-2 GX 131
MCG –2-31-3 GX 131
MCG –2-31-4 GX 131
MCG –2-31-5 GX 131
MCG –2-31-6 GX 131
MCG –2-31-8 GX 131
MCG –2-31-9 GX 131
MCG –2-31-10 GX 131
MCG –2-31-11 GX 131
MCG –2-31-12 GX 131
MCG –2-31-13 GX 131
MCG –2-31-15 GX 131
MCG –2-31-17 GX 131
MCG –2-31-19 GX 131
MCG –2-31-19A GX 131
MCG –2-31-22 GX 131
MCG –2-31-25 GX 131
MCG –2-32-2 GX 131
MCG –2-32-5 GX 131
MCG –2-32-6 GX 131
MCG –2-32-10 GX 131
MCG –2-32-11 GX 131
MCG –2-32-12 GX 131
MCG –2-32-15 GX 130, 131
MCG –2-32-16 GX 130, 131
MCG –2-32-17 GX 130, 131
MCG –2-32-18 GX 130, 131
MCG –2-32-19 GX 130, 131
MCG –2-32-21 GX 130
MCG –2-32-22 GX 130
MCG –2-32-23 GX 130
MCG –2-32-24 GX 130
MCG –2-32-25 GX 130
MCG –2-32-26 GX 130
MCG –2-33-3 GX 130
MCG –2-33-4 GX 130
MCG –2-33-5 GX 130
MCG –2-33-6 GX 130
MCG –2-33-9 GX 130
MCG –2-33-10 GX 130
MCG –2-33-12 GX 130
MCG –2-33-14 GX 130
MCG –2-33-15 GX 130
MCG –2-33-17 GX 130
MCG –2-33-20 GX 130
MCG –2-33-24 GX 130
MCG –2-33-25 GX 130
MCG –2-33-28 GX 130
MCG –2-33-33 GX 130
MCG –2-33-36 GX 130
MCG –2-33-37 GX 130
MCG –2-33-43 GX 130
MCG –2-33-44 GX 130
MCG –2-33-45 GX 130
MCG –2-33-46 GX 130
MCG –2-33-47 GX 130
MCG –2-33-52 GX 130
MCG –2-33-54 GX 130
MCG –2-33-55 GX 130
MCG –2-33-57 GX 130
MCG –2-33-58 GX 130
MCG –2-33-59 GX 130
MCG –2-33-62 GX 130
MCG –2-33-63 GX 130
MCG –2-33-64 GX 130
MCG –2-33-65 GX 130
MCG –2-33-66 GX 130
MCG –2-33-68 GX 130
MCG –2-33-71 GX 130
MCG –2-33-73 GX 130
MCG –2-33-75 GX 130
MCG –2-33-76 GX 130
MCG –2-33-80 GX 130
MCG –2-33-82 GX 130
MCG –2-33-83 GX 130
MCG –2-33-84 GX 130
MCG –2-33-85 GX 130
MCG –2-33-88 GX 130
MCG –2-33-91 GX 130
MCG –2-33-93 GX 130
MCG –2-33-95 GX 130
MCG –2-33-97 GX 130
MCG –2-33-98 GX 130
MCG –2-33-100 GX 130
MCG –2-34-1 GX 130
MCG –2-34-2 GX 130
MCG –2-34-3 GX 130
MCG –2-34-5 GX 130
MCG –2-34-6 GX 130
MCG –2-34-7 GX 130
MCG –2-34-8 GX 130
MCG –2-34-10 GX 130
MCG –2-34-12 GX 130
MCG –2-34-15 GX 130
MCG –2-34-16 GX 130
MCG –2-34-18 GX 130
MCG –2-34-19 GX 130
MCG –2-34-24 GX 130
MCG –2-34-28 GX 130
MCG –2-34-29 GX 130
MCG –2-34-31 GX 130
MCG –2-34-32 GX 130
MCG –2-34-33 GX 130
MCG –2-34-36 GX 130
MCG –2-34-38 GX 130
MCG –2-34-40 GX 130
MCG –2-34-45 GX 130
MCG –2-34-46 GX 130
MCG –2-34-47 GX 130
MCG –2-34-48 GX 130
MCG –2-34-50 GX 130
MCG –2-34-51 GX 130
MCG –2-34-52 GX 130
MCG –2-34-53 GX 130
MCG –2-34-54 GX 130
MCG –2-34-55 GX 130
MCG –2-34-56 GX 130
MCG –2-34-57 GX 130
MCG –2-34-58 GX 130
MCG –2-34-59 GX 130
MCG –2-34-60 GX 130
MCG –2-34-61 GX 130
MCG –2-35-1 GX 130
MCG –2-35-2 GX 130
MCG –2-35-3 GX 130
MCG –2-35-4 GX 130
MCG –2-35-5 GX 130
MCG –2-35-6 GX 130
MCG –2-35-7 GX 130
MCG –2-35-8 GX 130
MCG –2-35-9 GX 130
MCG –2-35-10 GX 130
MCG –2-35-11 GX 130
MCG –2-35-13 GX 130
MCG –2-35-16 GX 129, 130
MCG –2-35-18 GX 129
MCG –2-36-2 GX 129
MCG –2-36-3 GX 129
MCG –2-36-8 GX 129
MCG –2-36-10 GX 129
MCG –2-36-11 GX 129
MCG –2-36-12 GX 129
MCG –2-36-13 GX 129
MCG –2-36-15 GX 129
MCG –2-36-16 GX 129
MCG –2-36-17 GX 129
MCG –2-36-18 GX 129
MCG –2-37-1 GX 129
MCG –2-37-2 GX 129
MCG –2-37-4 GX 129
MCG –2-37-6 GX 129
MCG –2-37-7 GX 129
MCG –2-37-9 GX 129
MCG –2-37-10 GX 129
MCG –2-37-13 GX 129
MCG –2-37-14 GX 129
MCG –2-38-1 GX 129
MCG –2-38-2 GX 129

MCG –1-26-13 GX 133
MCG –1-26-16 GX 133
MCG –1-26-17 GX 133
MCG –1-26-19 GX 113
MCG –1-26-21 GX 133
MCG –1-26-24 GX 113, 133
MCG –1-26-29 GX 132, 133
MCG –1-26-30 GX 112, 113
MCG –1-26-31 GX 132, 133
MCG –1-26-36 GX 132, 133
MCG –1-26-38 GX 112, 132
MCG –1-26-39 GX 132
MCG –1-26-40 GX 112, 132
MCG –1-27-1 GX 132
MCG –1-27-2 GX 132
MCG –1-27-3 GX 132
MCG –1-27-6 GX 132
MCG –1-27-7 GX 112, 132
MCG –1-27-9 GX 112
MCG –1-27-10 GX 112
MCG –1-27-11 GX 132
MCG –1-27-13 GX 132
MCG –1-27-14 GX 112, 132
MCG –1-27-15 GX 132
MCG –1-27-18 GX 132
MCG –1-27-19 GX 132
MCG –1-27-20 GX 132
MCG –1-27-21 GX 132
MCG –1-27-22 GX 132
MCG –1-27-24 GX 112
MCG –1-27-25 GX 132
MCG –1-27-27 GX 132
MCG –1-27-30 GX 132
MCG –1-27-31 GX 112, 132
MCG –1-27-33 GX 132
MCG –1-28-3 GX 132
MCG –1-28-4 GX 132
MCG –1-28-5 GX 112
MCG –1-28-6 GX 132
MCG –1-28-11 GX 112
MCG –1-28-17 GX 132
MCG –1-28-20 GX 132
MCG –1-28-21 GX 132
MCG –1-28-22 GX 132
MCG –1-28-24 GX 132
MCG –1-28-25 GX 112, 132
MCG –1-28-26 GX 112, 132
MCG –1-29-1 GX 132
MCG –1-29-3 GX 132
MCG –1-29-4 GX 112
MCG –1-29-5 GX 112, 132
MCG –1-29-6 GX 132
MCG –1-29-10 GX 111, 112, 131, 132
MCG –1-29-12 GX 111, 112, 131, 132
MCG –1-29-13 GX 131, 132
MCG –1-29-14 GX 131, 132
MCG –1-29-15 GX 131, 132
MCG -1-29-17 GX 111, 112, 131, 132
MCG –1-29-18 GX 111, 112, 131, 132
MCG –1-29-19 GX 131, 132
MCG –1-29-20 GX 111, 112
MCG –1-29-27 GX 131
MCG –1-30-2 GX 131
MCG –1-30-3 GX 131
MCG –1-30-8 GX 131
MCG –1-30-11 GX 131
MCG –1-30-12 GX 131
MCG –1-30-13 GX 131
MCG –1-30-14 GX 131
MCG –1-30-22 GX 131
MCG –1-30-24 GX 131
MCG –1-30-25 GX 131
MCG –1-30-27A GX 111
MCG –1-30-32 GX 111
MCG –1-30-33 GX 111
MCG –1-30-34 GX 111
MCG –1-30-38 GX 131
MCG –1-30-39 GX 131
MCG –1-30-41 GX 111, 131
MCG –1-30-42 GX 131
MCG –1-30-43 GX 111
MCG –1-30-48 GX 131
MCG –1-31-2 GX 111
MCG –1-31-3 GX 111
MCG –1-31-8 GX 111
MCG –1-32-1 GX 131
MCG –1-32-4 GX 131
MCG –1-32-7 GX 131
MCG –1-32-11 GX 131
MCG –1-32-15 GX 111
MCG –1-32-16 GX 131
MCG –1-32-17 GX 111, 131
MCG –1-32-18 GX 111, 131
MCG –1-32-19 GX 131
MCG –1-32-20 GX 131
MCG –1-32-23 GX 110, 111
MCG –1-32-25 GX 110, 111, 130, 131
MCG –1-32-28 GX 130, 131
MCG –1-32-29 GX 110, 111
MCG –1-32-30 GX 130, 131
MCG –1-32-35 GX 130
MCG –1-32-38 GX 110, 130
MCG –1-32-39 GX 110, 130
MCG –1-33-1 GX 110, 130
MCG –1-33-2 GX 130
MCG –1-33-3 GX 130
MCG –1-33-6 GX 130
MCG –1-33-7 GX 110, 130
MCG –1-33-11 GX 110, 130
MCG –1-33-14 GX 110
MCG –1-33-17 GX 130
MCG –1-33-22 GX 130
MCG –1-33-27 GX 130
MCG –1-33-28 GX 130
MCG –1-33-30 GX 110
MCG –1-33-31 GX 130
MCG –1-33-32 GX 130
MCG –1-33-34 GX 130
MCG –1-33-35 GX 130
MCG –1-33-48 GX 130
MCG –1-33-51 GX 130
MCG –1-33-52 GX 130
MCG –1-33-54 GX 130
MCG –1-33-59 GX 110, 130
MCG –1-33-60 GX 130
MCG –1-33-61 GX 130
MCG –1-33-62 GX 130
MCG –1-33-63 GX 130
MCG –1-33-68 GX 130
MCG –1-33-71 GX 130
MCG –1-33-72 GX 130
MCG –1-33-76 GX 110, 130
MCG –1-34-8 GX 130
MCG –1-34-13 GX 110, 130
MCG –1-34-14 GX 130
MCG –1-34-16 GX 110, 130
MCG –1-34-17 GX 130
MCG –1-35-2 GX 110
MCG –1-35-7 GX 130
MCG –1-35-8 GX 130
MCG –1-35-10 GX 109, 110, 129, 130
MCG –1-35-11 GX 129, 130
MCG –1-35-12 GX 129, 130
MCG –1-35-13 GX 129, 130
MCG –1-35-17 GX 129
MCG –1-35-20 GX 109
MCG –1-35-21 GX 109, 129
MCG –1-35-22 GX 109, 129
MCG –1-36-1 GX 109
MCG –1-36-11 GX 129
MCG –1-36-12 GX 129
MCG –1-37-3 GX 109
MCG –1-37-4 GX 109
MCG –1-37-6 GX 109, 129
MCG –1-37-7 GX 109, 129
MCG –1-37-8 GX 129
MCG –1-37-9 GX 129
MCG –1-37-10 GX 129
MCG –1-37-11 GX 129
MCG –1-38-1 GX 109, 129
MCG –1-38-2 GX 109
MCG –1-38-3 GX 109
MCG –1-38-6 GX 129
MCG –1-38-7 GX 109
MCG –1-38-8 GX 129
MCG –1-38-11 GX 128, 129
MCG –1-38-12 GX 128, 129
MCG –1-38-14 GX 128, 129
MCG –1-38-18 GX 108, 109
MCG –1-38-19 GX 108, 109
MCG –1-38-20 GX 128
MCG –1-38-21 GX 108
MCG –1-38-22 GX 128
MCG –1-39-3 GX 128
MCG –1-39-5 GX 108
MCG –1-39-6 GX 108, 128
MCG –1-40-1 GX 128
MCG –1-40-4 GX 128
MCG –1-40-5 GX 128
MCG –1-40-8 GX 128
MCG –1-42-1 GX 127
MCG –1-42-2 GX 107, 127
MCG –1-42-3 GX 107
MCG –1-42-4 GX 107, 127
MCG –1-43-2 GX 107
MCG –1-50-1 GX 125
MCG –1-51-1 GX 104
MCG –1-52-2 GX 104, 124
MCG –1-52-3 GX 104, 124
MCG –1-52-4 GX 124
MCG –1-52-5 GX 124
MCG –1-52-8 GX 124
MCG –1-52-14 GX 104
MCG –1-52-16 GX 104, 124
MCG –1-52-17 GX 124
MCG –1-52-18 GX 104
MCG –1-53-2 GX 104
MCG –1-53-4 GX 104, 124
MCG –1-53-6 GX 104
MCG –1-53-7 GX 104, 124
MCG –1-53-8 GX 104
MCG –1-53-9 GX 104
MCG –1-53-10 GX 104
MCG –1-53-11 GX 104, 124
MCG –1-53-12 GX 124
MCG –1-53-13 GX 124
MCG –1-53-18 GX 103, 104
MCG –1-53-19 GX 103, 104
MCG –1-53-20 GX 123, 124
MCG –1-53-21 GX 103, 104, 123, 124
MCG –1-53-22 GX 123
MCG –1-53-23 GX 123
MCG –1-54-3 GX 103
MCG –1-54-8 GX 123
MCG –1-54-11 GX 123
MCG –1-54-12 GX 103
MCG –1-54-16 GX 103
MCG –1-54-20 GX 103
MCG –1-55-1 GX 103, 123
MCG –1-55-3 GX 103
MCG –1-55-10 GX 123
MCG –1-55-11 GX 103
MCG –1-55-13 GX 123
MCG –1-56-2 GX 122, 123
MCG –1-56-3 GX 102
MCG –1-56-5 GX 102
MCG –1-57-1 GX 122
MCG –1-57-4 GX 102
MCG –1-57-5 GX 102, 122
MCG –1-57-7 GX 102
MCG –1-57-8 GX 102
MCG –1-57-9 GX 102
MCG –1-57-15 GX 102
MCG –1-57-16 GX 102
MCG –1-57-17 GX 122
MCG –1-57-18 GX 102, 122
MCG –1-57-21 GX 102
MCG –1-57-23 GX 102
MCG –1-57-24 GX 102
MCG –1-58-5 GX 122
MCG –1-58-8 GX 102
MCG –1-58-9 GX 102
MCG –1-58-10 GX 102
MCG –1-58-15 GX 122
MCG –1-58-16 GX 102
MCG –1-58-18 GX 122
MCG –1-58-22 GX 102, 122
MCG –1-59-3 GX 102
MCG –1-59-13 GX 102
MCG –1-59-14 GX 102
MCG –1-59-16 GX 102
MCG –1-59-18 GX 102
MCG –1-59-21 GX 101, 102
MCG –1-59-24 GX 101
MCG –1-59-27 GX 101
MCG –1-60-11 GX 101
MCG –1-60-12 GX 101, 121
MCG –1-60-14 GX 101
MCG –1-60-15 GX 101
MCG –1-60-16 GX 101, 121
MCG –1-60-21 GX 101
MCG –1-60-22 GX 101
MCG –1-60-23 GX 101
MCG –1-60-25 GX 121
MCG –1-60-26 GX 121
MCG –1-60-27 GX 101
MCG –1-60-28 GX 121
MCG –1-60-30 GX 121
MCG –1-60-37 GX 121
MCG –1-60-43 GX 101
MCG –1-60-44 GX 121
MCG –1-60-45 GX 101
MCG +0-1-3 GX 101
MCG +0-1-7 GX 101
MCG +0-1-13 GX 101
MCG +0-1-14 GX 101
MCG +0-1-15 GX 101
MCG +0-1-16 GX 101
MCG +0-1-17 GX 101
MCG +0-1-22 GX 101
MCG +0-1-23 GX 101
MCG +0-1-24 GX 101
MCG +0-1-25 GX 101
MCG +0-1-34 GX 101
MCG +0-1-35 GX 101
MCG +0-1-36 GX 101
MCG +0-1-46 GX 101
MCG +0-1-47 GX 101
MCG +0-1-53 GX 101
MCG +0-1-55 GX 101
MCG +0-1-56 GX 101
MCG +0-1-57 GX 101
MCG +0-1-58 GX 101
MCG +0-1-59 GX 101
MCG +0-2-1 GX 101
MCG +0-2-2 GX 101
MCG +0-2-11 GX 101
MCG +0-2-13 GX 101
MCG +0-2-16 GX 101
MCG +0-2-20 GX 101
MCG +0-2-23 GX 101
MCG +0-2-24 GX 101
MCG +0-2-25 GX 101
MCG +0-2-34 GX 101
MCG +0-2-45 GX 101
MCG +0-2-47 GX 101
MCG +0-2-54 GX 101
MCG +0-2-57 GX 101
MCG +0-2-58 GX 101
MCG +0-2-60 GX 101
MCG +0-2-65 GX 101, 120
MCG +0-2-68 GX 101, 120
MCG +0-2-69 GX 101, 120
MCG +0-2-78 GX 101, 120
MCG +0-2-90 GX 101, 120
MCG +0-2-94 GX 101, 120
MCG +0-2-108 GX 101, 120
MCG +0-2-117 GX 101, 120
MCG +0-2-120 GX 120
MCG +0-2-126 GX 120
MCG +0-2-133 GX 120
MCG +0-2-134 GX 120
MCG +0-3-18 GX 120
MCG +0-3-21 GX 120
MCG +0-3-22 GX 120
MCG +0-3-26 GX 120
MCG +0-3-28 GX 120
MCG +0-3-33 GX 120
MCG +0-3-36 GX 120
MCG +0-3-44 GX 120
MCG +0-3-51 GX 120
MCG +0-3-59 GX 120
MCG +0-3-74 GX 120
MCG +0-4-10 GX 120
MCG +0-4-11 GX 120
MCG +0-4-19 GX 120
MCG +0-4-32 GX 120
MCG +0-4-42 GX 120
MCG +0-4-102 GX 120
MCG +0-4-104 GX 120
MCG +0-4-111 GX 120
MCG +0-4-112 GX 120, A16
MCG +0-4-114 GX 120, A16
MCG +0-4-116 GX 120, A16
MCG +0-4-123 GX 120
MCG +0-4-126 GX 120, A16
MCG +0-4-132 GX 120
MCG +0-4-140 GX 120, A16
MCG +0-4-148 GX 120, A16
MCG +0-4-149 GX 120, A16
MCG +0-4-155 GX 120
MCG +0-4-167 GX 120, A16
MCG +0-5-2 GX 120
MCG +0-5-12 GX 120
MCG +0-5-13 GX 120
MCG +0-5-22 GX 120
MCG +0-5-23 GX 120
MCG +0-5-24 GX 120
MCG +0-5-26 GX 120
MCG +0-5-39 GX 119, 120
MCG +0-5-40 GX 119, 120
MCG +0-5-42 GX 119
MCG +0-5-45 GX 119
MCG +0-5-46 GX 119
MCG +0-5-47 GX 119
MCG +0-5-50 GX 119
MCG +0-6-10 GX 119
MCG +0-6-15 GX 119
MCG +0-6-33 GX 119
MCG +0-6-36 GX 119
MCG +0-6-42 GX 119
MCG +0-6-43 GX 119
MCG +0-6-46 GX 119
MCG +0-6-47 GX 119
MCG +0-6-48 GX 119
MCG +0-6-53 GX 119
MCG +0-7-19 GX 119
MCG +0-7-29 GX 119
MCG +0-7-31 GX 119
MCG +0-7-63 GX 119
MCG +0-7-72 GX 119
MCG +0-7-84 GX 119
MCG +0-8-14 GX 119
MCG +0-8-25 GX 119
MCG +0-8-28 GX 119
MCG +0-8-37 GX 119
MCG +0-8-39 GX 119
MCG +0-8-45 GX 119
MCG +0-8-66 GX 118, 119
MCG +0-8-69 GX 118, 119
MCG +0-8-75 GX 118, 119
MCG +0-8-77 GX 118, 119
MCG +0-8-91 GX 118
MCG +0-9-1 GX 118
MCG +0-9-5 GX 118
MCG +0-9-7 GX 118
MCG +0-9-8 GX 118
MCG +0-9-9 GX 118
MCG +0-9-19 GX 118
MCG +0-9-20 GX 118
MCG +0-9-22 GX 118
MCG +0-9-26 GX 118
MCG +0-9-31 GX 118
MCG +0-9-34 GX 118
MCG +0-9-36 GX 118
MCG +0-9-37 GX 118
MCG +0-9-38 GX 118
MCG +0-9-47 GX 118
MCG +0-9-49 GX 118
MCG +0-9-53 GX 118
MCG +0-9-55 GX 118
MCG +0-9-56 GX 118
MCG +0-9-57 GX 118
MCG +0-9-58 GX 118
MCG +0-9-64 GX 118
MCG +0-9-65 GX 118
MCG +0-9-70 GX 118
MCG +0-9-71 GX 118
MCG +0-9-73 GX 118
MCG +0-9-74 GX 118
MCG +0-9-76 GX 118
MCG +0-9-79 GX 118
MCG +0-9-81 GX 118
MCG +0-9-82 GX 118
MCG +0-9-88 GX 118
MCG +0-9-89 GX 118
MCG +0-9-90 GX 118
MCG +0-10-6 GX 118
MCG +0-10-18 GX 118
MCG +0-10-20 GX 118
MCG +0-10-21 GX 118
MCG +0-10-22 GX 118
MCG +0-10-23 GX 118
MCG +0-10-24 GX 118
MCG +0-11-2 GX 118
MCG +0-11-12 GX 118
MCG +0-11-15 GX 117, 118
MCG +0-11-20 GX 117, 118
MCG +0-11-25 GX 117, 118
MCG +0-11-43 GX 117
MCG +0-11-45 GX 117
MCG +0-11-46 GX 117
MCG +0-11-47 GX 117
MCG +0-11-48 GX 117
MCG +0-11-52 GX 117
MCG +0-12-1 GX 117
MCG +0-12-10 GX 117
MCG +0-12-11 GX 117
MCG +0-12-14 GX 117
MCG +0-12-15 GX 117
MCG +0-12-19 GX 117
MCG +0-12-30 GX 117
MCG +0-12-31 GX 117
MCG +0-12-33 GX 117
MCG +0-12-43 GX 117
MCG +0-12-46 GX 117
MCG +0-12-48 GX 117
MCG +0-12-49 GX 117
MCG +0-12-50 GX 117
MCG +0-12-51 GX 117
MCG +0-12-54 GX 117
MCG +0-12-58 GX 117
MCG +0-12-65 GX 117
MCG +0-12-71 GX 117
MCG +0-13-2 GX 117
MCG +0-13-5 GX 117
MCG +0-13-9 GX 117
MCG +0-13-13 GX 117
MCG +0-13-17 GX 117
MCG +0-13-18 GX 117
MCG +0-13-23 GX 117
MCG +0-13-24 GX 117
MCG +0-13-29 GX 117
MCG +0-13-30 GX 117
MCG +0-13-32 GX 117
MCG +0-13-35 GX 117
MCG +0-13-40 GX 117
MCG +0-13-41 GX 117
MCG +0-13-42 GX 117
MCG +0-13-43 GX 117
MCG +0-13-45 GX 117
MCG +0-13-46 GX 117
MCG +0-13-47 GX 117
MCG +0-13-51 GX 117
MCG +0-13-57 GX 117
MCG +0-13-58 GX 117
MCG +0-13-59 GX 117
MCG +0-13-61 GX 117
MCG +0-13-62 GX 117
MCG +0-13-68 GX 117
MCG +0-14-1 GX 117
MCG +0-14-2 GX 117
MCG +0-14-3 GX 117
MCG +0-14-4 GX 117
MCG +0-14-8 GX 117
MCG +0-14-9 GX 117
MCG +0-14-10 GX 117
MCG +0-14-11 GX 117
MCG +0-14-15 GX 117
MCG +0-14-16 GX 117
MCG +0-14-17 GX 117
MCG +0-14-20 GX 116, 117
MCG +0-20-1 GX 115
MCG +0-20-3 GX 115
MCG +0-20-6 GX 114, 115
MCG +0-21-3 GX 114
MCG +0-22-3 GX 114
MCG +0-22-5 GX 114
MCG +0-22-16 GX 114
MCG +0-22-18 GX 114
MCG +0-22-20 GX 114
MCG +0-23-2 GX 114
MCG +0-23-4 GX 114
MCG +0-23-20 GX 113, 114
MCG +0-24-3 GX 113
MCG +0-24-8 GX 113
MCG +0-24-11 GX 113
MCG +0-24-13 GX 113
MCG +0-24-16 GX 113
MCG +0-25-4 GX 113
MCG +0-25-5 GX 113
MCG +0-25-9 GX 113
MCG +0-25-18 GX 113
MCG +0-25-23 GX 113
MCG +0-25-25 GX 113
MCG +0-26-1 GX 113
MCG +0-26-4 GX 113
MCG +0-26-9 GX 113
MCG +0-26-14 GX 113
MCG +0-26-16 GX 113
MCG +0-26-22 GX 112, 113
MCG +0-26-27 GX 112, 113
MCG +0-26-28 GX 112, 113
MCG +0-26-29 GX 112, 113
MCG +0-26-31 GX 112, 113
MCG +0-27-1 GX 112
MCG +0-27-2 GX 112
MCG +0-27-4 GX 112
MCG +0-27-5 GX 112
MCG +0-27-7 GX 112
MCG +0-27-10 GX 112
MCG +0-27-11 GX 112
MCG +0-27-13 GX 112
MCG +0-27-16 GX 112
MCG +0-27-19 GX 112
MCG +0-27-22 GX 112
MCG +0-27-23 GX 112
MCG +0-27-24 GX 112
MCG +0-27-25 GX 112
MCG +0-27-27 GX 112
MCG +0-27-40 GX 112
MCG +0-28-2 GX 112
MCG +0-28-7 GX 112
MCG +0-28-8 GX 112
MCG +0-28-13 GX 112
MCG +0-28-15 GX 112
MCG +0-28-16 GX 112

MCG +0-28-17 GX 112
MCG +0-28-21 GX 112
MCG +0-28-23 GX 112
MCG +0-28-28 GX 112
MCG +0-28-29 GX 112
MCG +0-29-1 GX 112
MCG +0-29-2 GX 112
MCG +0-29-4 GX 112
MCG +0-29-5 GX 112
MCG +0-29-6 GX 112
MCG +0-29-7 GX 112
MCG +0-29-8 GX 112
MCG +0-29-10 GX 112
MCG +0-29-13 GX 112
MCG +0-29-15 GX 112
MCG +0-29-17 GX 111, 112
MCG +0-29-18 GX 111, 112
MCG +0-29-19 GX 111, 112
MCG +0-29-21 GX 111, 112
MCG +0-29-23 GX 111, 112
MCG +0-29-26 GX 111, 112
MCG +0-29-27 GX 111, 112
MCG +0-29-28 GX 111, 112
MCG +0-29-29 GX 111, 112
MCG +0-29-33 GX 111, 112
MCG +0-29-35 GX 111, 112
MCG +0-29-36 GX 111
MCG +0-29-38 GX 111
MCG +0-30-2 GX 111
MCG +0-30-3 GX 111
MCG +0-30-4 GX 111
MCG +0-30-8 GX 111
MCG +0-30-9 GX 111
MCG +0-30-10 GX 111
MCG +0-30-13 GX 111
MCG +0-30-15 GX 111
MCG +0-30-16 GX 111
MCG +0-30-20 GX 111
MCG +0-30-25 GX 111
MCG +0-30-27 GX 111
MCG +0-30-29 GX 111
MCG +0-30-30 GX 111
MCG +0-30-31 GX 111
MCG +0-31-1 GX 111
MCG +0-31-5 GX 111
MCG +0-31-10 GX 111
MCG +0-31-15 GX 111
MCG +0-31-23 GX 111
MCG +0-31-25 GX 111
MCG +0-31-35 GX 111
MCG +0-31-39 GX 111
MCG +0-31-40 GX 111
MCG +0-31-41 GX 111
MCG +0-31-42 GX 111
MCG +0-31-44 GX 111
MCG +0-32-5 GX 111
MCG +0-32-7 GX 111
MCG +0-32-10 GX 111
MCG +0-32-13 GX 111
MCG +0-32-15 GX 111
MCG +0-32-16 GX 111
MCG +0-32-18 GX 111
MCG +0-32-25 GX 110, 111
MCG +0-32-29 GX 110, 111
MCG +0-32-36 GX 110
MCG +0-33-3 GX 110
MCG +0-33-15 GX 110
MCG +0-33-19 GX 110
MCG +0-33-20 GX 110
MCG +0-33-23 GX 110
MCG +0-33-28 GX 110
MCG +0-33-29 GX 110
MCG +0-34-5 GX 110
MCG +0-34-6 GX 110
MCG +0-34-7 GX 110
MCG +0-34-8 GX 110
MCG +0-34-12 GX 110
MCG +0-34-13 GX 110
MCG +0-34-18 GX 110
MCG +0-34-19 GX 110
MCG +0-34-20 GX 110
MCG +0-34-24 GX 110
MCG +0-34-28 GX 110
MCG +0-34-30 GX 110
MCG +0-34-34 GX 110
MCG +0-34-35 GX 110
MCG +0-34-36 GX 110
MCG +0-34-37 GX 110
MCG +0-34-38 GX 110
MCG +0-35-4 GX 110
MCG +0-35-5 GX 110
MCG +0-35-7 GX 110
MCG +0-35-18 GX 110
MCG +0-35-19 GX 109, 110
MCG +0-35-20 GX 109, 110
MCG +0-36-4 GX 109
MCG +0-36-6 GX 109
MCG +0-36-10 GX 109
MCG +0-36-14 GX 109
MCG +0-36-15 GX 109
MCG +0-36-18 GX 109
MCG +0-36-22 GX 109
MCG +0-36-30 GX 109
MCG +0-37-9 GX 109
MCG +0-37-10 GX 109
MCG +0-37-13 GX 109
MCG +0-37-14 GX 109
MCG +0-37-17 GX 109
MCG +0-37-23 GX 109
MCG +0-38-4 GX 109
MCG +0-38-10 GX 109
MCG +0-38-13 GX 108, 109
MCG +0-38-19 GX 108, 109
MCG +0-39-3 GX 108
MCG +0-39-4 GX 108
MCG +0-39-9 GX 108
MCG +0-39-10 GX 108
MCG +0-39-13 GX 108
MCG +0-39-14 GX 108
MCG +0-39-17 GX 108
MCG +0-39-24 GX 108
MCG +0-40-2 GX 108
MCG +0-40-5 GX 108
MCG +0-41-5 GX 107, 108
MCG +0-42-5 GX 107
MCG +0-43-3 GX 107
MCG +0-44-1 GX 107
MCG +0-46-2 GX 106
MCG +0-50-1 GX 105
MCG +0-51-2 GX 104
MCG +0-51-6 GX 104
MCG +0-51-7 GX 104
MCG +0-51-12 GX 104
MCG +0-52-16 GX 104
MCG +0-52-17 GX 104
MCG +0-52-19 GX 104
MCG +0-52-23 GX 104
MCG +0-52-28 GX 104
MCG +0-52-38 GX 104
MCG +0-52-40 GX 104
MCG +0-52-43 GX 104
MCG +0-53-2 GX 104
MCG +0-53-8 GX 104
MCG +0-53-10 GX 103, 104
MCG +0-53-11 GX 103, 104
MCG +0-53-13 GX 103, 104
MCG +0-54-2 GX 103
MCG +0-54-5 GX 103
MCG +0-54-18 GX 103
MCG +0-54-25 GX 103
MCG +0-54-29 GX 103
MCG +0-55-1 GX 103
MCG +0-55-5 GX 103
MCG +0-55-12 GX 103
MCG +0-55-28 GX 103
MCG +0-56-2 GX 103
MCG +0-56-9 GX 103
MCG +0-56-10 GX 102, 103
MCG +0-56-11 GX 102, 103
MCG +0-56-13 GX 102, 103
MCG +0-56-14 GX 102, 103
MCG +0-56-15 GX 102, 103
MCG +0-57-1 GX 102
MCG +0-57-2 GX 102
MCG +0-57-6 GX 102
MCG +0-57-9 GX 102
MCG +0-58-4 GX 102
MCG +0-58-12 GX 102
MCG +0-58-13 GX 102
MCG +0-58-18 GX 102
MCG +0-58-22 GX 102
MCG +0-58-23 GX 102
MCG +0-58-24 GX 102
MCG +0-58-31 GX 102
MCG +0-58-34 GX 102
MCG +0-59-6 GX 102
MCG +0-59-7 GX 102
MCG +0-59-13 GX 102
MCG +0-59-22 GX 102
MCG +0-59-25 GX 101, 102
MCG +0-59-28 GX 101, 102
MCG +0-59-32 GX 101, 102
MCG +0-59-33 GX 101, 102
MCG +0-59-34 GX 101, 102
MCG +0-59-43 GX 101, 102
MCG +0-60-11 GX 101
MCG +0-60-13 GX 101
MCG +0-60-16 GX 101
MCG +0-60-20 GX 101
MCG +0-60-21 GX 101
MCG +0-60-23 GX 101
MCG +0-60-25 GX 101
MCG +0-60-26 GX 101
MCG +0-60-27 GX 101
MCG +0-60-28 GX 101
MCG +0-60-30 GX 101
MCG +0-60-31 GX 101
MCG +0-60-32 GX 101
MCG +0-60-33 GX 101
MCG +0-60-36 GX 101
MCG +0-60-37 GX 101
MCG +0-60-40 GX 101
MCG +0-60-41 GX 101
MCG +0-60-46 GX 101
MCG +0-60-48 GX 101
MCG +0-60-49 GX 101
MCG +0-60-50 GX 101
MCG +0-60-51 GX 101
MCG +0-60-54 GX 101
MCG +0-60-57 GX 101
MCG +1-1-13 GX 81
MCG +1-1-16 GX 81, 101
MCG +1-1-29 GX 81
MCG +1-1-33 GX 101
MCG +1-1-49 GX 81
MCG +1-2-4 GX 81
MCG +1-2-8 GX 81
MCG +1-2-9 GX 81
MCG +1-2-12 GX 101
MCG +1-2-14 GX 81, 101
MCG +1-2-15 GX 81, 101
MCG +1-2-17 GX 101
MCG +1-2-18 GX 81
MCG +1-2-23 GX 81, 100, 101, 120
MCG +1-2-24 GX 81, 100, 101, 120
MCG +1-2-26 GX 81, 100
MCG +1-2-34 GX 81, 100
MCG +1-2-36 GX 81, 100
MCG +1-2-38 GX 101, 120
MCG +1-2-42 GX 81, 100
MCG +1-2-54 GX 100
MCG +1-3-8 GX 100
MCG +1-3-13 GX 100
MCG +1-4-2 GX 100
MCG +1-4-3 GX 120
MCG +1-4-4 GX 100
MCG +1-4-7 GX 100, 120
MCG +1-4-10 GX 100, 120
MCG +1-4-13 GX 100, 120
MCG +1-4-27 GX 100, 120
MCG +1-4-42 GX 100
MCG +1-4-44 GX 100
MCG +1-4-47 GX 100
MCG +1-4-60 GX 100
MCG +1-5-1 GX 120
MCG +1-5-4 GX 100, 120
MCG +1-5-8 GX 100, 120
MCG +1-5-12 GX 100
MCG +1-5-18 GX 100
MCG +1-5-20 GX 100, 120
MCG +1-5-30 GX 119, 120
MCG +1-5-33 GX 119, 120
MCG +1-5-47 GX 99, 119
MCG +1-6-9 GX 99, 119
MCG +1-6-13 GX 119
MCG +1-6-15 GX 119
MCG +1-6-18 GX 99
MCG +1-6-20 GX 99
MCG +1-6-23 GX 99, 119
MCG +1-6-24 GX 99, 119
MCG +1-6-25 GX 99
MCG +1-6-30 GX 99, 119
MCG +1-6-60 GX 99
MCG +1-7-3 GX 99, 119
MCG +1-7-4 GX 99, 119
MCG +1-7-5 GX 99
MCG +1-7-7 GX 119
MCG +1-7-13 GX 99
MCG +1-7-20 GX 99
MCG +1-7-25 GX 99
MCG +1-8-5 GX 99
MCG +1-8-8 GX 119
MCG +1-8-11A GX 99
MCG +1-8-27 GX 98, 99
MCG +1-8-28 GX 98, 99, 118, 119
MCG +1-8-30 GX 98, 99
MCG +1-8-33 GX 98, 99
MCG +1-8-34 GX 98, 99, 118, 119
MCG +1-8-35 GX 98, 99, 118, 119
MCG +1-8-36 GX 98, 99, 118, 119
MCG +1-8-40 GX 98, 118
MCG +1-8-41 GX 98
MCG +1-8-42 GX 118
MCG +1-10-8 GX 98
MCG +1-11-1 GX 98
MCG +1-11-3 GX 98
MCG +1-11-4 GX 98
MCG +1-11-10 GX 118
MCG +1-11-11 GX 98
MCG +1-11-14 GX 97, 98
MCG +1-12-1 GX 117
MCG +1-12-4 GX 97
MCG +1-12-10 GX 97
MCG +1-12-11 GX 97
MCG +1-13-8 GX 117
MCG +1-14-3 GX 97, 117
MCG +1-14-5 GX 97
MCG +1-14-7 GX 97
MCG +1-14-9 GX 97
MCG +1-14-10 GX 97
MCG +1-14-22 GX 97
MCG +1-14-27 GX 116, 117
MCG +1-14-28 GX 116, 117
MCG +1-14-36 GX 96, 97, 116, 117
MCG +1-17-1 GX 96
MCG +1-19-1 GX 95
MCG +1-19-2 GX 95
MCG +1-19-3 GX 95
MCG +1-19-5 GX 95
MCG +1-20-2 GX 95
MCG +1-20-5 GX 94, 95
MCG +1-20-7 GX 94, 95
MCG +1-20-8 GX 94
MCG +1-21-5 GX 94
MCG +1-21-8 GX 114
MCG +1-21-14 GX 114
MCG +1-21-16 GX 94
MCG +1-21-17 GX 94
MCG +1-21-21 GX 114
MCG +1-22-3 GX 114
MCG +1-22-7 GX 114
MCG +1-22-10 GX 114
MCG +1-22-11 GX 114
MCG +1-22-13 GX 114
MCG +1-22-14 GX 94
MCG +1-22-15 GX 94
MCG +1-23-1 GX 114
MCG +1-23-4 GX 94
MCG +1-23-16 GX 93, 94, 113, 114
MCG +1-23-19 GX 113, 114
MCG +1-23-20 GX 113, 114
MCG +1-23-21 GX 113
MCG +1-24-2 GX 113
MCG +1-24-8 GX 93
MCG +1-24-11 GX 93, 113
MCG +1-24-15 GX 113
MCG +1-24-19 GX 93, 113
MCG +1-24-23 GX 113
MCG +1-25-3 GX 93
MCG +1-25-8 GX 93
MCG +1-25-15 GX 93
MCG +1-25-26 GX 113
MCG +1-25-29 GX 113
MCG +1-25-33 GX 113
MCG +1-26-1 GX 113
MCG +1-26-2 GX 113
MCG +1-26-4 GX 93
MCG +1-26-11 GX 113
MCG +1-26-12 GX 113
MCG +1-26-15 GX 113
MCG +1-26-16 GX 92, 93, 112, 113
MCG +1-26-17 GX 92, 93, 112, 113
MCG +1-26-20 GX 112, 113
MCG +1-26-28 GX 92, 93
MCG +1-26-30 GX 112, 113
MCG +1-27-2 GX 92
MCG +1-27-3 GX 92
MCG +1-27-5 GX 112
MCG +1-27-6 GX 112
MCG +1-27-7 GX 112
MCG +1-27-10 GX 112
MCG +1-27-18 GX 112
MCG +1-27-20 GX 92, 112
MCG +1-27-26 GX 92
MCG +1-27-29 GX 92
MCG +1-27-30 GX 112
MCG +1-28-1 GX 112
MCG +1-28-3 GX 112
MCG +1-28-14 GX 92
MCG +1-28-16 GX 92
MCG +1-28-20 GX 112
MCG +1-28-22 GX 92
MCG +1-28-25 GX 92
MCG +1-28-28 GX 92
MCG +1-28-29 GX 112
MCG +1-28-30 GX 112
MCG +1-28-32 GX 112
MCG +1-28-38 GX 112
MCG +1-29-3 GX 112
MCG +1-29-7 GX 112
MCG +1-29-9 GX 112
MCG +1-29-12 GX 92, 112
MCG +1-29-15 GX 92
MCG +1-29-16 GX 92, 112
MCG +1-29-17 GX 92, 112
MCG +1-29-22 GX 112
MCG +1-29-35 GX 91, 92, 111, 112
MCG +1-29-38 GX 111, 112
MCG +1-29-39 GX 111, 112
MCG +1-29-42 GX 91, 92
MCG +1-29-46 GX 111
MCG +1-29-47 GX 111
MCG +1-29-48 GX 91
MCG +1-30-2 GX 91
MCG +1-30-5 GX 111
MCG +1-30-6 GX 111
MCG +1-30-7 GX 91
MCG +1-30-8 GX 111
MCG +1-30-10 GX 111
MCG +1-30-12 GX 91
MCG +1-30-16 GX 91
MCG +1-30-18 GX 91
MCG +1-30-19 GX 111
MCG +1-31-2 GX 91, 111
MCG +1-31-7 GX 91, 111
MCG +1-31-9 GX 91, 111
MCG +1-31-10 GX 91
MCG +1-31-13 GX 91, 111
MCG +1-31-14 GX 111
MCG +1-31-19 GX 91
MCG +1-31-20 GX 91
MCG +1-31-21 GX 111
MCG +1-31-30 GX 91, 111, A15
MCG +1-31-33 GX 111, A15
MCG +1-31-53 GX 91, 111, A15
MCG +1-32-6 GX 91, A15
MCG +1-32-29 GX 111, A15
MCG +1-32-41 GX 91, 111, A15
MCG +1-32-49 GX 111, A15
MCG +1-32-56 GX 91, 111, A15
MCG +1-32-59 GX 111, A15
MCG +1-32-76 GX 91, A15
MCG +1-32-93 GX 110, 111
MCG +1-32-96 GX 110, 111
MCG +1-32-108 GX 90, 91
MCG +1-32-111 GX 90, 91
MCG +1-32-113 GX 110, 111
MCG +1-32-126 GX 110, 111
MCG +1-32-133 GX 90
MCG +1-33-6 GX 110
MCG +1-33-8 GX 110
MCG +1-33-26 GX 110
MCG +1-33-29 GX 90, 110
MCG +1-33-33 GX 90
MCG +1-35-5 GX 110
MCG +1-35-13 GX 110
MCG +1-35-26 GX 110
MCG +1-35-28 GX 110
MCG +1-35-37 GX 89, 90
MCG +1-36-24 GX 89
MCG +1-36-31 GX 109
MCG +1-37-4 GX 109
MCG +1-37-25 GX 109
MCG +1-37-26 GX 109
MCG +1-37-33 GX 89
MCG +1-37-41 GX 109
MCG +1-37-43 GX 109
MCG +1-37-50 GX 89, 109
MCG +1-38-17 GX 109
MCG +1-38-19 GX 109
MCG +1-38-20 GX 108, 109
MCG +1-38-24 GX 88, 89
MCG +1-38-27 GX 108, 109
MCG +1-38-28 GX 88, 108
MCG +1-39-3 GX 88
MCG +1-39-5 GX 108
MCG +1-39-10 GX 88
MCG +1-39-13 GX 88
MCG +1-39-14 GX 88
MCG +1-39-16 GX 88, 108
MCG +1-39-17 GX 108
MCG +1-39-18 GX 108
MCG +1-39-24 GX 108
MCG +1-40-10 GX 88, 108
MCG +1-40-13 GX 108
MCG +1-41-4 GX 108
MCG +1-41-6 GX 88
MCG +1-41-15 GX 107, 108
MCG +1-42-2 GX 87
MCG +1-42-5 GX 87
MCG +1-42-8 GX 107
MCG +1-43-9 GX 87
MCG +1-45-2 GX 86
MCG +1-46-4 GX 86
MCG +1-52-9 GX 104
MCG +1-52-13 GX 84
MCG +1-54-5 GX 83, 103
MCG +1-54-10 GX 83
MCG +1-55-4 GX 83
MCG +1-55-5 GX 83
MCG +1-55-10 GX 83
MCG +1-55-13 GX 83
MCG +1-55-15 GX 103
MCG +1-56-1 GX 83, 103
MCG +1-56-4 GX 83
MCG +1-56-6 GX 83
MCG +1-56-11 GX 83
MCG +1-56-13 GX 83
MCG +1-56-14 GX 83
MCG +1-56-16 GX 82, 83
MCG +1-56-17 GX 82, 83
MCG +1-56-21 GX 82
MCG +1-58-6 GX 82, 102
MCG +1-58-22 GX 82
MCG +1-58-27 GX 82
MCG +1-58-28 GX 82
MCG +1-58-32 GX 82
MCG +1-59-2 GX 82
MCG +1-59-9 GX 102
MCG +1-59-15 GX 82
MCG +1-59-27 GX 82
MCG +1-59-28 GX 82, 102
MCG +1-59-29 GX 82, 102
MCG +1-59-41 GX 82
MCG +1-59-43 GX 82
MCG +1-59-58 GX 81, 82
MCG +1-59-59 GX 81, 82
MCG +1-59-63 GX 81, 82
MCG +1-59-65 GX 81, 82
MCG +1-59-77 GX 81, 82
MCG +1-59-78 GX 81, 82, 101, 102
MCG +1-59-79 GX 81, 82
MCG +1-59-82 GX 81, 82
MCG +1-59-84 GX 81
MCG +1-59-85 GX 81
MCG +1-60-1 GX 81
MCG +1-60-9 GX 101
MCG +1-60-11 GX 101
MCG +1-60-13 GX 81
MCG +1-60-14 GX 81
MCG +1-60-18 GX 81
MCG +1-60-23 GX 81
MCG +1-60-38 GX 81
MCG +1-60-39 GX 81
MCG +1-60-44 GX 81
MCG +2-1-9 GX 81
MCG +2-1-12 GX 81
MCG +2-1-14 GX 81
MCG +2-1-16 GX 81
MCG +2-1-31 GX 81
MCG +2-2-10 GX 81
MCG +2-2-13 GX 81
MCG +2-2-22 GX 81, 100
MCG +2-3-2 GX 100
MCG +2-3-3 GX 100
MCG +2-3-7 GX 100
MCG +2-3-12 GX 100
MCG +2-3-13 GX 100
MCG +2-3-27 GX 100
MCG +2-3-31 GX 100
MCG +2-3-33 GX 100
MCG +2-4-22 GX 100
MCG +2-4-25 GX 100
MCG +2-4-28 GX 100
MCG +2-4-29 GX 100
MCG +2-4-31 GX 100
MCG +2-4-43 GX 100
MCG +2-4-45 GX 100
MCG +2-4-46 GX 100
MCG +2-5-8 GX 100
MCG +2-5-18 GX 99, 100
MCG +2-5-24 GX 99, 100

MCG +2-5-39 GX 99, 100
MCG +2-6-9 GX 99
MCG +2-6-21 GX 99
MCG +2-6-22 GX 99
MCG +2-6-40 GX 99
MCG +2-6-45 GX 99
MCG +2-6-46 GX 99
MCG +2-6-50 GX 99
MCG +2-7-5 GX 99
MCG +2-7-10 GX 99
MCG +2-7-12 GX 99
MCG +2-8-5 GX 99
MCG +2-8-14 GX 99
MCG +2-8-20 GX 99
MCG +2-8-26 GX 99
MCG +2-8-31 GX 99
MCG +2-8-37 GX 98, 99
MCG +2-8-41 GX 98, 99
MCG +2-8-45 GX 98, 99
MCG +2-8-48 GX 98, 99
MCG +2-9-2 GX 98
MCG +2-9-7 GX 98
MCG +2-9-8 GX 98
MCG +2-15-2 GX 96
MCG +2-18-3 GX 95
MCG +2-18-4 GX 95
MCG +2-20-3 GX 95
MCG +2-20-8 GX 94, 95
MCG +2-21-2 GX 94
MCG +2-21-5 GX 94
MCG +2-21-6 GX 94
MCG +2-21-10 GX 94
MCG +2-22-1 GX 94
MCG +2-22-2 GX 94
MCG +2-22-3 GX 94
MCG +2-22-6 GX 94
MCG +2-23-5 GX 94
MCG +2-23-11 GX 93, 94
MCG +2-23-19 GX 93, 94
MCG +2-23-23 GX 93, 94
MCG +2-23-24 GX 93, 94
MCG +2-23-25 GX 93, 94
MCG +2-23-26 GX 93
MCG +2-23-27 GX 93
MCG +2-24-5 GX 93
MCG +2-24-7 GX 93
MCG +2-24-15 GX 93
MCG +2-25-8 GX 93
MCG +2-25-9 GX 93
MCG +2-25-10 GX 93
MCG +2-25-14 GX 93
MCG +2-25-19 GX 93
MCG +2-25-21 GX 93
MCG +2-25-23 GX 93
MCG +2-25-28 GX 93
MCG +2-25-32 GX 93
MCG +2-25-34 GX 93
MCG +2-25-35 GX 93
MCG +2-25-36 GX 93
MCG +2-25-37 GX 93
MCG +2-25-39 GX 93
MCG +2-25-42 GX 93
MCG +2-25-49 GX 93
MCG +2-25-51 GX 93
MCG +2-25-52 GX 93
MCG +2-25-53 GX 93
MCG +2-25-54 GX 93
MCG +2-26-3 GX 93
MCG +2-26-4 GX 93
MCG +2-26-8 GX 93
MCG +2-26-11 GX 93
MCG +2-26-12 GX 93
MCG +2-26-13 GX 93
MCG +2-26-16 GX 93
MCG +2-26-17 GX 93
MCG +2-26-23 GX 93
MCG +2-26-29 GX 92, 93
MCG +2-26-33 GX 92, 93
MCG +2-27-9 GX 92
MCG +2-27-13 GX 92
MCG +2-27-14 GX 92
MCG +2-27-23 GX 92
MCG +2-27-24 GX 92
MCG +2-27-33 GX 92
MCG +2-28-4 GX 92
MCG +2-28-10 GX 92
MCG +2-28-24 GX 92
MCG +2-28-26 GX 92
MCG +2-28-27 GX 92
MCG +2-28-29 GX 92
MCG +2-28-31 GX 92
MCG +2-28-40 GX 92
MCG +2-28-46 GX 92
MCG +2-28-49 GX 92
MCG +2-28-51 GX 92
MCG +2-29-1 GX 92
MCG +2-29-2 GX 92
MCG +2-29-3 GX 92
MCG +2-29-4 GX 92
MCG +2-29-5 GX 92
MCG +2-30-9 GX 91
MCG +2-30-17 GX 91
MCG +2-30-21 GX 91
MCG +2-30-31 GX 91
MCG +2-30-35 GX 91
MCG +2-30-42 GX 91
MCG +2-31-7 GX 91
MCG +2-31-10 GX 91
MCG +2-31-18 GX 91
MCG +2-31-20 GX 91
MCG +2-31-27 GX 91
MCG +2-32-22 GX 91
MCG +2-32-25 GX 91
MCG +2-32-161 GX 90, 91
MCG +2-33-12 GX 90
MCG +2-33-38 GX 90
MCG +2-33-44 GX 90
MCG +2-33-49 GX 90
MCG +2-35-16 GX 90
MCG +2-35-18 GX 89, 90
MCG +2-35-22 GX 89
MCG +2-35-24 GX 89
MCG +2-36-2 GX 89
MCG +2-36-10 GX 89
MCG +2-36-21 GX 89
MCG +2-36-23 GX 89
MCG +2-36-30 GX 89
MCG +2-36-33 GX 89
MCG +2-36-39 GX 89
MCG +2-36-41 GX 89
MCG +2-36-42 GX 89
MCG +2-36-45 GX 89
MCG +2-36-49 GX 89
MCG +2-36-52 GX 89
MCG +2-36-53 GX 89
MCG +2-36-54 GX 89
MCG +2-36-57 GX 89
MCG +2-37-8 GX 89
MCG +2-37-9 GX 89
MCG +2-37-11 GX 89
MCG +2-38-5 GX 89
MCG +2-38-7 GX 89
MCG +2-38-8 GX 89
MCG +2-38-20 GX 89
MCG +2-38-21 GX 89
MCG +2-38-24 GX 88, 89
MCG +2-38-29 GX 88, 89
MCG +2-38-30 GX 88, 89
MCG +2-38-31 GX 88, 89
MCG +2-38-33 GX 88, 89
MCG +2-38-38 GX 88
MCG +2-38-39 GX 88
MCG +2-38-42 GX 88
MCG +2-39-1 GX 88
MCG +2-39-6 GX 88
MCG +2-39-8 GX 88
MCG +2-39-10 GX 88
MCG +2-39-13 GX 88
MCG +2-39-13A GX 88
MCG +2-39-18 GX 88
MCG +2-39-20 GX 88
MCG +2-39-28 GX 88
MCG +2-40-9 GX 88
MCG +2-40-10 GX 88
MCG +2-40-15 GX 88
MCG +2-41-2 GX 88
MCG +2-41-3 GX 88
MCG +2-41-6 GX 88
MCG +2-41-10 GX 87, 88
MCG +2-41-14 GX 87, 88
MCG +2-41-18 GX 87, 88
MCG +2-42-3 GX 87
MCG +2-42-7 GX 87
MCG +2-43-2 GX 87
MCG +2-44-1 GX 87
MCG +2-45-5 GX 86
MCG +2-45-7 GX 86
MCG +2-46-12 GX 86
MCG +2-52-6 GX 84
MCG +2-52-7 GX 84
MCG +2-52-25 GX 84
MCG +2-53-6 GX 83, 84
MCG +2-53-7 GX 83, 84
MCG +2-53-8 GX 83, 84
MCG +2-53-9 GX 83, 84
MCG +2-53-10 GX 83, 84
MCG +2-53-11 GX 83, 84
MCG +2-54-5 GX 83
MCG +2-54-6 GX 83
MCG +2-54-12 GX 83
MCG +2-54-16 GX 83
MCG +2-54-17 GX 83
MCG +2-54-18 GX 83
MCG +2-54-21 GX 83
MCG +2-54-24 GX 83
MCG +2-55-5 GX 83
MCG +2-55-8 GX 83
MCG +2-55-22 GX 83
MCG +2-55-23 GX 83
MCG +2-55-24 GX 83
MCG +2-55-25 GX 83
MCG +2-55-26 GX 83
MCG +2-56-3 GX 83
MCG +2-56-4 GX 83
MCG +2-56-6 GX 83
MCG +2-56-21 GX 82, 83
MCG +2-56-22 GX 82, 83
MCG +2-56-25 GX 82, 83
MCG +2-58-1 GX 82
MCG +2-58-2 GX 82
MCG +2-58-3 GX 82
MCG +2-58-9 GX 82
MCG +2-58-16 GX 82
MCG +2-58-25 GX 82
MCG +2-58-26 GX 82
MCG +2-58-28 GX 82
MCG +2-58-29 GX 82
MCG +2-58-33 GX 82
MCG +2-58-39 GX 82
MCG +2-58-42 GX 82
MCG +2-58-46 GX 82
MCG +2-58-47 GX 82
MCG +2-58-48 GX 82
MCG +2-58-50 GX 82
MCG +2-58-54 GX 82
MCG +2-58-58 GX 82
MCG +2-58-59 GX 82
MCG +2-58-61 GX 82
MCG +2-58-64 GX 82
MCG +2-59-2 GX 82
MCG +2-59-12 GX 82
MCG +2-59-26 GX 81, 82
MCG +2-59-39 GX 81, 82
MCG +2-59-42 GX 81, 82
MCG +2-59-47 GX 81
MCG +2-59-49 GX 81
MCG +2-60-12 GX 81
MCG +2-60-13 GX 81
MCG +2-60-17 GX 81
MCG +2-60-18 GX 81
MCG +2-60-21 GX 81
MCG +2-60-25 GX 81
MCG +3-1-33 GX 63
MCG +3-2-1 GX 63
MCG +3-2-5 GX 63
MCG +3-2-19 GX 63, 80, 81, 100
MCG +3-2-22 GX 63, 80
MCG +3-3-1 GX 80
MCG +3-3-6 GX 80
MCG +3-3-8 GX 100
MCG +3-3-9 GX 80
MCG +3-3-16 GX 80, 100
MCG +3-4-13 GX 100
MCG +3-4-15 GX 80
MCG +3-4-21 GX 80
MCG +3-4-34 GX 80, 100
MCG +3-4-37 GX 80
MCG +3-4-45 GX 80
MCG +3-4-46 GX 80
MCG +3-4-47 GX 80
MCG +3-5-1 GX 80
MCG +3-5-9 GX 80
MCG +3-5-13 GX 80, 100
MCG +3-5-31 GX 80
MCG +3-6-3 GX 79, 80
MCG +3-6-8 GX 99
MCG +3-6-13 GX 99
MCG +3-6-17 GX 79, 80
MCG +3-6-18 GX 79, 80, 99
MCG +3-6-32 GX 79
MCG +3-6-37 GX 79
MCG +3-6-40 GX 79
MCG +3-6-41 GX 79
MCG +3-6-42 GX 79, 99
MCG +3-6-44 GX 79
MCG +3-6-47 GX 79, 99
MCG +3-6-49 GX 79, 99
MCG +3-6-50 GX 79
MCG +3-7-1 GX 99
MCG +3-7-5 GX 79, 99
MCG +3-7-6 GX 79, 99
MCG +3-7-8 GX 99
MCG +3-7-13 GX 79
MCG +3-7-19 GX 79
MCG +3-7-20 GX 79
MCG +3-7-23 GX 79
MCG +3-7-24 GX 79
MCG +3-7-26 GX 79
MCG +3-7-31 GX 79
MCG +3-7-32 GX 79
MCG +3-7-36 GX 79
MCG +3-7-40 GX 79, 99
MCG +3-7-42 GX 79
MCG +3-7-47 GX 79
MCG +3-7-49 GX 99
MCG +3-7-50 GX 99
MCG +3-8-6 GX 99
MCG +3-8-7 GX 99
MCG +3-8-11 GX 79, 99
MCG +3-8-21 GX 79
MCG +3-8-27 GX 99
MCG +3-8-29 GX 99
MCG +3-8-32 GX 98, 99
MCG +3-8-35 GX 98, 99
MCG +3-8-36 GX 98, 99
MCG +3-8-41 GX 79
MCG +3-9-1 GX 79, 98
MCG +3-9-5 GX 79
MCG +3-10-1 GX 78
MCG +3-10-2 GX 78
MCG +3-11-3 GX 78
MCG +3-13-5 GX 68, 87
MCG +3-18-5 GX 76
MCG +3-18-6 GX 76
MCG +3-18-7 GX 76
MCG +3-18-9 GX 76
MCG +3-19-13 GX 75
MCG +3-19-14 GX 75
MCG +3-19-18 GX 75
MCG +3-19-20 GX 75
MCG +3-20-7 GX 75
MCG +3-21-1 GX 94
MCG +3-21-2 GX 75, 94
MCG +3-21-7 GX 94
MCG +3-21-15 GX 75
MCG +3-21-18 GX 94
MCG +3-22-8 GX 75, 94
MCG +3-22-16 GX 75
MCG +3-22-18 GX 75
MCG +3-22-20 GX 74, 75, 94
MCG +3-23-13 GX 74
MCG +3-23-19 GX 74
MCG +3-23-21 GX 74, 93, 94
MCG +3-23-22 GX 74
MCG +3-23-32 GX 74
MCG +3-24-2 GX 74
MCG +3-24-3 GX 74
MCG +3-24-13 GX 93
MCG +3-24-17 GX 93
MCG +3-24-24 GX 74, 93
MCG +3-24-29 GX 74, 93
MCG +3-24-36 GX 93
MCG +3-24-38 GX 74
MCG +3-24-44 GX 93
MCG +3-24-46 GX 74, 93
MCG +3-24-51 GX 74
MCG +3-24-55 GX 74, 93
MCG +3-25-1 GX 93
MCG +3-25-3 GX 93
MCG +3-25-4 GX 74, 93
MCG +3-25-7 GX 74
MCG +3-25-23 GX 93
MCG +3-25-27 GX 93
MCG +3-25-38 GX 93
MCG +3-26-6 GX 93
MCG +3-26-7 GX 93
MCG +3-26-30 GX 92, 93
MCG +3-26-36 GX 92, 93
MCG +3-26-44 GX 92, 93
MCG +3-26-48 GX 92, 93
MCG +3-27-1 GX 73
MCG +3-27-2 GX 92
MCG +3-27-3 GX 92
MCG +3-27-14 GX 73, 92
MCG +3-27-28 GX 73
MCG +3-27-41 GX 92
MCG +3-27-45 GX 92
MCG +3-27-46 GX 73
MCG +3-27-53 GX 92
MCG +3-27-61 GX 73
MCG +3-27-63 GX 73, 92
MCG +3-27-65 GX 92
MCG +3-27-75 GX 92
MCG +3-28-2 GX 73
MCG +3-28-5 GX 92
MCG +3-28-11 GX 92
MCG +3-28-22 GX 92
MCG +3-28-26 GX 92
MCG +3-28-29 GX 73
MCG +3-28-36 GX 92
MCG +3-28-38 GX 92
MCG +3-28-45 GX 73, 92
MCG +3-28-52 GX 73, 92
MCG +3-28-54 GX 73
MCG +3-28-56 GX 73, 92
MCG +3-29-18 GX 72, 73
MCG +3-29-24 GX 72, 73, 92
MCG +3-29-25 GX 72, 73, 92
MCG +3-29-27 GX 72, 73
MCG +3-29-34 GX 72, 73
MCG +3-29-55 GX 72
MCG +3-29-58 GX 72
MCG +3-29-60 GX 72
MCG +3-29-61 GX 72
MCG +3-30-7 GX 91
MCG +3-30-11 GX 91
MCG +3-30-14 GX 72
MCG +3-30-15 GX 91
MCG +3-30-21 GX 72
MCG +3-30-26 GX 72
MCG +3-30-33 GX 91
MCG +3-30-38 GX 72, 91
MCG +3-30-44A GX 72, 91
MCG +3-30-49 GX 72, A11
MCG +3-30-51 GX 72, A11
MCG +3-30-55 GX 72, A11
MCG +3-30-56 GX 91
MCG +3-30-71 GX 72, A11
MCG +3-30-80 GX 91
MCG +3-30-83 GX 72, A11
MCG +3-30-94 GX 72, A11
MCG +3-30-98 GX 72, A11
MCG +3-30-108 GX 72, A11
MCG +3-30-113 GX 72, A11
MCG +3-30-115 GX 72, A11
MCG +3-30-117 GX 91
MCG +3-30-125 GX 91
MCG +3-31-7 GX 91
MCG +3-31-15 GX 72, 91
MCG +3-31-25 GX 91
MCG +3-31-28 GX 72, 91
MCG +3-31-29 GX 72, A10
MCG +3-31-30 GX 91
MCG +3-31-41 GX 72, 91
MCG +3-31-49 GX 72, 91
MCG +3-31-61 GX 72, 91
MCG +3-31-64 GX 91
MCG +3-31-75 GX 91
MCG +3-31-94 GX 72, 91
MCG +3-32-3 GX 72, 91
MCG +3-32-82 GX 90
MCG +3-33-22 GX 90
MCG +3-34-3 GX 71
MCG +3-34-26 GX 90
MCG +3-34-28 GX 71, 90
MCG +3-34-40 GX 90
MCG +3-35-3 GX 71
MCG +3-35-6 GX 90
MCG +3-35-13 GX 71, 90
MCG +3-35-14 GX 71, 90
MCG +3-35-19 GX 71
MCG +3-35-22 GX 89, 90
MCG +3-35-26 GX 89, 90
MCG +3-35-27 GX 71, 89, 90
MCG +3-35-33 GX 89
MCG +3-35-34 GX 89
MCG +3-35-35 GX 71, 89
MCG +3-36-7 GX 89
MCG +3-36-17 GX 89
MCG +3-36-25 GX 89
MCG +3-36-38 GX 89
MCG +3-36-44 GX 89
MCG +3-36-47 GX 89
MCG +3-36-50 GX 89
MCG +3-36-52 GX 89
MCG +3-36-54 GX 89
MCG +3-36-63 GX 89
MCG +3-36-72 GX 89
MCG +3-36-73 GX 89
MCG +3-36-93 GX 70
MCG +3-37-24 GX 70
MCG +3-38-27 GX 70
MCG +3-38-30 GX 89
MCG +3-38-35 GX 89
MCG +3-38-38 GX 70
MCG +3-38-43 GX 89
MCG +3-38-44 GX 89
MCG +3-38-57 GX 89
MCG +3-38-71 GX 88, 89
MCG +3-38-81 GX 70, 88
MCG +3-38-82 GX 70
MCG +3-38-83 GX 70, 88
MCG +3-38-84 GX 88
MCG +3-39-1 GX 88
MCG +3-39-11 GX 70
MCG +3-39-14 GX 70
MCG +3-39-19 GX 88
MCG +3-39-23 GX 88
MCG +3-40-1 GX 88
MCG +3-40-9 GX 88
MCG +3-40-12 GX 69, 88
MCG +3-40-18 GX 69, 88
MCG +3-40-60 GX 69
MCG +3-41-3 GX 88
MCG +3-41-4 GX 69
MCG +3-41-28 GX 88, A9
MCG +3-41-29 GX 88, A9
MCG +3-41-33 GX 69, A9
MCG +3-41-47 GX 69, 88, A9
MCG +3-41-61 GX 88, A9
MCG +3-41-64 GX 88
MCG +3-41-83 GX 88, A9
MCG +3-41-95 GX 88, A9
MCG +3-41-124 GX 88, A9
MCG +3-41-137 GX 69, A9
MCG +3-41-141 GX 69, 87, 88, A9
MCG +3-42-8 GX 87
MCG +3-42-14 GX 69
MCG +3-42-18 GX 69
MCG +3-42-19 GX 69
MCG +3-42-26 GX 87
MCG +3-43-9 GX 68
MCG +3-44-3 GX 87
MCG +3-45-3 GX 68
MCG +3-45-5 GX 68
MCG +3-45-8 GX 68
MCG +3-45-25 GX 68, 86
MCG +3-45-26 GX 68, 86
MCG +3-45-32 GX 68
MCG +3-46-6 GX 86
MCG +3-46-21 GX 86
MCG +3-47-8 GX 67
MCG +3-47-10 GX 67, 85, 86
MCG +3-53-1 GX 65
MCG +3-53-9 GX 83, 84
MCG +3-53-14 GX 65
MCG +3-54-2 GX 65
MCG +3-54-5 GX 65
MCG +3-54-9 GX 65, 83
MCG +3-54-10 GX 65, 83
MCG +3-54-11 GX 83
MCG +3-56-9 GX 64, 83
MCG +3-56-13 GX 64
MCG +3-56-17 GX 82, 83
MCG +3-57-2 GX 64
MCG +3-57-6 GX 82
MCG +3-57-10 GX 64
MCG +3-57-11 GX 82
MCG +3-57-19 GX 64
MCG +3-57-22 GX 64
MCG +3-57-23 GX 64
MCG +3-57-28 GX 64
MCG +3-57-29 GX 64
MCG +3-57-31 GX 64
MCG +3-58-1 GX 64
MCG +3-58-2 GX 82
MCG +3-58-5 GX 64
MCG +3-58-7 GX 64, 82
MCG +3-58-8 GX 64
MCG +3-58-13 GX 64
MCG +3-59-1 GX 64
MCG +3-59-9 GX 64
MCG +3-59-11 GX 64
MCG +3-59-20 GX 82
MCG +3-59-22 GX 82
MCG +3-59-26 GX 63, 64
MCG +3-59-27 GX 63, 64
MCG +3-59-29 GX 82
MCG +3-59-30 GX 82
MCG +3-59-40 GX 63, 64
MCG +3-59-46 GX 63, 64, 81, 82
MCG +3-59-47 GX 63, 64
MCG +3-59-48 GX 63, 64
MCG +3-59-49 GX 63, 64
MCG +3-59-51 GX 63, 64
MCG +3-59-53 GX 81, 82
MCG +3-59-57 GX 63, 81, 82
MCG +3-59-58 GX 63
MCG +3-59-59 GX 63
MCG +3-59-60 GX 63, 81
MCG +3-59-64 GX 63

MCG +3-60-9 GX 63
MCG +3-60-11 GX 81
MCG +3-60-21 GX 63
MCG +3-60-23 GX 63, 81
MCG +3-60-24 GX 63
MCG +3-60-25 GX 81
MCG +3-60-28 GX 63, 81
MCG +3-60-29 GX 63
MCG +3-60-31 GX 63
MCG +3-60-36 GX 63
MCG +4-1-19 GX 63
MCG +4-1-20 GX 63
MCG +4-1-40 GX 63
MCG +4-1-43 GX 63
MCG +4-1-44 GX 63
MCG +4-1-46 GX 63
MCG +4-1-48 GX 63
MCG +4-2-10 GX 63
MCG +4-2-19 GX 63
MCG +4-2-26 GX 63
MCG +4-2-30 GX 63
MCG +4-2-39 GX 63, 80
MCG +4-2-43 GX 63, 80
MCG +4-2-45 GX 63, 80
MCG +4-2-46 GX 63, 80
MCG +4-2-47 GX 63, 80
MCG +4-2-50 GX 63, 80
MCG +4-2-53 GX 80
MCG +4-3-8 GX 80
MCG +4-3-11 GX 80
MCG +4-3-12 GX 80
MCG +4-3-16 GX 80
MCG +4-3-17 GX 80
MCG +4-3-20 GX 80
MCG +4-3-21 GX 80
MCG +4-3-26 GX 80
MCG +4-3-27 GX 80
MCG +4-3-32 GX 80
MCG +4-3-33 GX 80
MCG +4-3-39 GX 80
MCG +4-3-40 GX 80
MCG +4-3-42 GX 80
MCG +4-3-43 GX 80
MCG +4-5-1 GX 80
MCG +4-5-3 GX 80
MCG +4-5-6 GX 80
MCG +4-5-7 GX 80
MCG +4-5-8 GX 80
MCG +4-5-10 GX 80
MCG +4-5-27 GX 79, 80
MCG +4-5-31 GX 79, 80
MCG +4-5-36 GX 79, 80
MCG +4-5-46 GX 79, 80
MCG +4-5-47 GX 79, 80
MCG +4-6-1 GX 79
MCG +4-6-2 GX 79
MCG +4-6-10 GX 79
MCG +4-6-11 GX 79
MCG +4-6-15 GX 79
MCG +4-6-16 GX 79
MCG +4-6-19 GX 79
MCG +4-6-22 GX 79
MCG +4-6-26 GX 79
MCG +4-6-28 GX 79
MCG +4-6-29 GX 79
MCG +4-6-31 GX 79
MCG +4-6-32 GX 79
MCG +4-6-36 GX 79
MCG +4-6-42 GX 79
MCG +4-6-43 GX 79
MCG +4-6-44 GX 79
MCG +4-6-49 GX 79
MCG +4-6-60 GX 79
MCG +4-6-61 GX 79
MCG +4-7-1 GX 79
MCG +4-7-2 GX 79
MCG +4-7-3 GX 79
MCG +4-7-5 GX 79
MCG +4-7-6 GX 79
MCG +4-7-8 GX 79
MCG +4-7-13 GX 79
MCG +4-7-15 GX 79
MCG +4-7-17 GX 79
MCG +4-7-19 GX 79
MCG +4-7-26 GX 79
MCG +4-8-1 GX 79
MCG +4-8-3 GX 79
MCG +4-8-4 GX 79
MCG +4-8-12 GX 79
MCG +4-10-4 GX 78
MCG +4-10-9 GX 78
MCG +4-10-11 GX 78
MCG +4-10-12 GX 78
MCG +4-10-14 GX 78
MCG +4-10-15 GX 78
MCG +4-10-17 GX 78
MCG +4-10-24 GX 78
MCG +4-10-28 GX 78
MCG +4-16-1 GX 76
MCG +4-16-2 GX 76
MCG +4-16-3 GX 76
MCG +4-16-5 GX 76
MCG +4-17-1 GX 76
MCG +4-17-6 GX 76
MCG +4-17-10 GX 76
MCG +4-18-5 GX 75, 76
MCG +4-18-7 GX 75, 76
MCG +4-18-9 GX 75, 76
MCG +4-18-14 GX 75, 76
MCG +4-19-3 GX 75
MCG +4-19-7 GX 75
MCG +4-19-17 GX 75
MCG +4-20-34 GX 75
MCG +4-20-69 GX 75
MCG +4-21-7 GX 74, 75
MCG +4-21-13 GX 74, 75
MCG +4-21-25 GX 74
MCG +4-22-27 GX 74
MCG +4-22-32 GX 74
MCG +4-22-52 GX 74
MCG +4-24-22 GX 73
MCG +4-25-14 GX 73
MCG +4-25-19 GX 73
MCG +4-25-28 GX 73
MCG +4-25-34 GX 73
MCG +4-25-47 GX 73
MCG +4-25-49 GX 73
MCG +4-25-50 GX 73
MCG +4-26-7 GX 73
MCG +4-26-9 GX 73
MCG +4-26-12 GX 73
MCG +4-26-13 GX 73
MCG +4-26-18 GX 73
MCG +4-26-20 GX 73
MCG +4-26-23 GX 73
MCG +4-26-26 GX 73
MCG +4-26-27 GX 73
MCG +4-26-27A GX 73
MCG +4-26-32 GX 73
MCG +4-27-1 GX 73
MCG +4-27-17 GX 72, 73
MCG +4-27-25 GX 72, 73
MCG +4-27-34 GX 72
MCG +4-27-36 GX 72
MCG +4-27-39 GX 72
MCG +4-27-46 GX 72
MCG +4-27-53 GX 72
MCG +4-27-57 GX 72
MCG +4-27-72 GX 72
MCG +4-27-78 GX 72, A10
MCG +4-28-16 GX 72
MCG +4-28-29 GX 72, A11
MCG +4-28-31 GX 72
MCG +4-28-32 GX 72
MCG +4-28-33 GX 72, A11
MCG +4-28-36 GX 72
MCG +4-28-45 GX 72
MCG +4-28-47 GX 72, A11
MCG +4-28-48 GX 72, A11
MCG +4-28-62 GX 72, A10
MCG +4-28-65 GX 72, A10
MCG +4-28-70 GX 72, A10
MCG +4-28-77 GX 72, A10
MCG +4-28-87 GX 72, A10
MCG +4-28-92 GX 72, A10
MCG +4-28-115 GX 72, A10
MCG +4-28-116 GX 72, A10
MCG +4-28-117 GX 72, A10
MCG +4-28-121 GX 63
MCG +4-28-123 GX 72, A10
MCG +4-29-2 GX 72, A10
MCG +4-29-4 GX 72, A10
MCG +4-29-12 GX 72, A10
MCG +4-29-13 GX 72, A10
MCG +4-29-32 GX 72, A10
MCG +4-29-37 GX 72, A10
MCG +4-29-38 GX 72, A10
MCG +4-29-44 GX 72
MCG +4-29-45 GX 72
MCG +4-29-48 GX 72
MCG +4-29-50 GX 72
MCG +4-29-69 GX 72
MCG +4-30-10 GX 71, 72
MCG +4-31-5 GX 71
MCG +4-31-6 GX 71
MCG +4-32-2 GX 71
MCG +4-32-12 GX 71
MCG +4-32-13 GX 71
MCG +4-32-14 GX 71
MCG +4-32-17 GX 71
MCG +4-32-19 GX 71
MCG +4-32-26 GX 71
MCG +4-33-9 GX 71
MCG +4-33-10 GX 71
MCG +4-33-19 GX 71
MCG +4-33-20 GX 71
MCG +4-33-37 GX 70, 71
MCG +4-33-39 GX 70, 71
MCG +4-33-40 GX 70
MCG +4-33-44 GX 70
MCG +4-35-5 GX 70
MCG +4-35-6 GX 70
MCG +4-36-2 GX 70
MCG +4-36-10 GX 70
MCG +4-36-16 GX 70
MCG +4-36-23 GX 70
MCG +4-36-38 GX 69, 70
MCG +4-36-46 GX 69
MCG +4-37-3 GX 69
MCG +4-37-4 GX 69
MCG +4-37-14 GX 69
MCG +4-37-42 GX 69
MCG +4-37-44 GX 69
MCG +4-37-55 GX 69
MCG +4-38-1 GX 69
MCG +4-38-14 GX 69
MCG +4-38-15 GX 69
MCG +4-38-43 GX 69
MCG +4-38-44 GX 69
MCG +4-38-45 GX 69
MCG +4-39-3 GX 69
MCG +4-39-13 GX 69
MCG +4-39-18 GX 68, 69
MCG +4-41-1 GX 68
MCG +4-41-18 GX 68
MCG +4-42-2 GX 68
MCG +4-42-17 GX 67, 68
MCG +4-42-22 GX 67, 68
MCG +4-43-4 GX 67
MCG +4-43-31 GX 67
MCG +4-43-33 GX 67
MCG +4-49-1 GX 65
MCG +4-49-6 GX 65
MCG +4-50-4 GX 65
MCG +4-50-7 GX 65
MCG +4-51-7 GX 65
MCG +4-51-8 GX 65
MCG +4-51-10 GX 65
MCG +4-51-15 GX 64, 65
MCG +4-52-1 GX 64, 65
MCG +4-52-3 GX 64
MCG +4-52-6 GX 64
MCG +4-52-7 GX 64
MCG +4-52-9 GX 64
MCG +4-52-10 GX 64
MCG +4-52-11 GX 64
MCG +4-53-14 GX 64
MCG +4-54-9 GX 64
MCG +4-54-20 GX 64
MCG +4-54-22 GX 64
MCG +4-54-23 GX 64
MCG +4-54-27 GX 64
MCG +4-54-34 GX 64
MCG +4-54-39 GX 63, 64
MCG +4-55-2 GX 63, 64
MCG +4-55-23 GX 63
MCG +4-55-24 GX 63
MCG +4-55-26 GX 63
MCG +4-55-28 GX 63
MCG +4-55-32 GX 63
MCG +4-55-44 GX 63
MCG +4-55-45 GX 63
MCG +4-55-47 GX 63
MCG +4-56-1 GX 63
MCG +4-56-6 GX 63
MCG +4-56-9 GX 63
MCG +4-56-10 GX 63
MCG +4-56-14 GX 63
MCG +4-56-19 GX 63
MCG +5-1-3 GX 45
MCG +5-1-4 GX 45
MCG +5-1-8 GX 45, 63
MCG +5-1-14 GX 63
MCG +5-1-15 GX 63
MCG +5-1-17 GX 63
MCG +5-1-21 GX 63
MCG +5-1-27 GX 45
MCG +5-1-31 GX 45
MCG +5-1-47 GX 63
MCG +5-1-55 GX 45
MCG +5-1-56 GX 63
MCG +5-1-58 GX 45
MCG +5-1-61 GX 45
MCG +5-1-63 GX 45, 63
MCG +5-2-1 GX 45
MCG +5-2-2 GX 45
MCG +5-2-3 GX 45, 63
MCG +5-2-5 GX 45
MCG +5-2-8 GX 45, 63
MCG +5-2-11 GX 45
MCG +5-2-23 GX 45
MCG +5-2-28 GX 45, 62
MCG +5-2-31 GX 45, 62, 63, 80
MCG +5-2-45 GX 62, 80
MCG +5-2-47 GX 62, 80
MCG +5-3-3 GX 62
MCG +5-3-6 GX 62, 80
MCG +5-3-7 GX 62, 80
MCG +5-3-29 GX 62
MCG +5-3-32 GX 62, 80
MCG +5-3-35 GX 62
MCG +5-3-39 GX 62
MCG +5-3-42 GX 62
MCG +5-3-66 GX 62, A7
MCG +5-3-72 GX 62, A7
MCG +5-3-74 GX 62
MCG +5-4-12 GX 62, A7
MCG +5-4-14 GX 62, A7
MCG +5-4-26 GX 62, A7
MCG +5-4-31 GX 80
MCG +5-4-33 GX 80
MCG +5-4-48 GX 62, A7
MCG +5-4-49 GX 80
MCG +5-4-55 GX 62, A7
MCG +5-4-56 GX 62, 80
MCG +5-4-59 GX 62, A7
MCG +5-4-68 GX 62
MCG +5-4-70 GX 62
MCG +5-4-71 GX 62
MCG +5-4-72 GX 62
MCG +5-5-1 GX 80
MCG +5-5-2 GX 62
MCG +5-5-9 GX 62
MCG +5-5-10 GX 62
MCG +5-5-11 GX 80
MCG +5-5-29 GX 61, 62
MCG +5-5-52 GX 61
MCG +5-6-7 GX 61
MCG +5-6-8 GX 61
MCG +5-6-12 GX 61
MCG +5-6-14 GX 61
MCG +5-6-35 GX 61
MCG +5-6-36 GX 61
MCG +5-7-5 GX 61
MCG +5-7-39 GX 61
MCG +5-7-50 GX 79
MCG +5-8-2 GX 61, 79
MCG +5-10-1 GX 60
MCG +5-10-5 GX 60
MCG +5-16-1 GX 58
MCG +5-16-7 GX 58, 76
MCG +5-16-10 GX 76
MCG +5-17-2 GX 76
MCG +5-17-7 GX 76
MCG +5-17-10 GX 76
MCG +5-17-12 GX 58
MCG +5-17-19 GX 75, 76
MCG +5-18-7 GX 57, 58, 75, 76
MCG +5-18-11 GX 75
MCG +5-18-13 GX 57
MCG +5-18-24 GX 75
MCG +5-18-25 GX 75
MCG +5-18-26 GX 57, 75
MCG +5-19-4 GX 75
MCG +5-19-18 GX 57
MCG +5-19-30 GX 57, 75
MCG +5-19-33 GX 75
MCG +5-19-36 GX 57
MCG +5-19-37 GX 57, 75
MCG +5-19-38 GX 57, 75
MCG +5-20-3 GX 57, 75
MCG +5-20-7 GX 75
MCG +5-20-11 GX 75
MCG +5-20-20 GX 57
MCG +5-20-26 GX 57, 75
MCG +5-20-28 GX 57
MCG +5-20-29 GX 56, 57
MCG +5-21-1 GX 74, 75
MCG +5-21-5 GX 74
MCG +5-22-6 GX 74
MCG +5-22-20 GX 56, 74
MCG +5-22-25 GX 56, 74
MCG +5-22-27 GX 56
MCG +5-22-34 GX 74
MCG +5-23-17 GX 56
MCG +5-23-37 GX 56
MCG +5-24-1 GX 56
MCG +5-24-5 GX 55, 56
MCG +5-24-11 GX 73
MCG +5-24-22 GX 55
MCG +5-25-9 GX 73
MCG +5-25-10 GX 55
MCG +5-25-14 GX 73
MCG +5-25-16 GX 73
MCG +5-25-28 GX 73
MCG +5-25-30 GX 55
MCG +5-25-31 GX 73
MCG +5-26-2 GX 55, 73
MCG +5-26-3 GX 73
MCG +5-26-4 GX 73
MCG +5-26-7 GX 55
MCG +5-26-13 GX 73
MCG +5-26-27 GX 73
MCG +5-26-29 GX 73
MCG +5-26-31 GX 73
MCG +5-26-37 GX 55
MCG +5-26-46 GX 55, 73
MCG +5-26-47 GX 55
MCG +5-26-51 GX 55
MCG +5-26-58 GX 73
MCG +5-26-68 GX 73
MCG +5-27-1 GX 73
MCG +5-27-5 GX 55
MCG +5-27-25 GX 73
MCG +5-27-49 GX 54, 55
MCG +5-27-52 GX 72, 73
MCG +5-27-68 GX 72
MCG +5-27-70 GX 72
MCG +5-27-74 GX 72
MCG +5-27-79 GX 72
MCG +5-27-86 GX 54
MCG +5-27-89 GX 72
MCG +5-28-3 GX 72
MCG +5-28-11 GX 72
MCG +5-28-13 GX 72
MCG +5-28-15 GX 72
MCG +5-28-20 GX 72
MCG +5-28-32 GX 54, 72
MCG +5-28-40 GX 72
MCG +5-28-41 GX 72
MCG +5-28-44 GX 72
MCG +5-28-67 GX 72
MCG +5-28-71 GX 72
MCG +5-28-74 GX 54
MCG +5-29-3 GX 72
MCG +5-29-12 GX 54
MCG +5-29-15 GX 72
MCG +5-29-17 GX 72
MCG +5-29-24 GX 54, 72
MCG +5-29-35 GX 72
MCG +5-29-45 GX 72
MCG +5-29-66 GX 54
MCG +5-30-9 GX 54
MCG +5-30-17 GX 54
MCG +5-30-64 GX 53, 54
MCG +5-30-70 GX 71
MCG +5-30-74 GX 53
MCG +5-30-79 GX 71
MCG +5-30-87 GX 53, 71
MCG +5-30-94 GX 71
MCG +5-30-101 GX 71
MCG +5-30-108 GX 71
MCG +5-30-109 GX 71, A8
MCG +5-30-111 GX 71, A8
MCG +5-30-115 GX 53
MCG +5-30-116 GX 71, A8
MCG +5-31-18 GX 71, A8
MCG +5-31-22 GX 53
MCG +5-31-23 GX 71, A8
MCG +5-31-36 GX 71, A8
MCG +5-31-37 GX 71, A8
MCG +5-31-42 GX 53
MCG +5-31-45 GX 53, 71, A8
MCG +5-31-46 GX 71, A8
MCG +5-31-67 GX 53
MCG +5-31-74 GX 71, A8
MCG +5-31-94 GX 71, A8
MCG +5-31-95 GX 71, A8
MCG +5-31-106 GX 53, 71, A8
MCG +5-31-127 GX 53, 71, A8
MCG +5-31-132 GX 71
MCG +5-31-133 GX 71
MCG +5-31-139 GX 53, 71
MCG +5-31-140 GX 71
MCG +5-31-141 GX 71
MCG +5-31-147 GX 53
MCG +5-31-151 GX 53
MCG +5-31-152 GX 71
MCG +5-31-153 GX 71
MCG +5-31-154 GX 71
MCG +5-31-158 GX 53
MCG +5-31-167 GX 53
MCG +5-32-1 GX 53
MCG +5-32-4 GX 53
MCG +5-32-17 GX 53
MCG +5-32-35 GX 53
MCG +5-32-49 GX 53
MCG +5-32-50 GX 53
MCG +5-32-56 GX 71
MCG +5-32-62 GX 71
MCG +5-33-11 GX 71
MCG +5-33-13 GX 71
MCG +5-33-18 GX 53
MCG +5-33-19 GX 71
MCG +5-33-20 GX 71
MCG +5-33-38 GX 70, 71
MCG +5-33-39 GX 52, 53
MCG +5-33-41 GX 52, 53, 70, 71
MCG +5-33-47 GX 52, 70
MCG +5-33-48 GX 52, 70
MCG +5-34-3 GX 52, 70
MCG +5-34-23 GX 70
MCG +5-35-11 GX 52
MCG +5-35-18 GX 70
MCG +5-36-9 GX 70
MCG +5-36-10 GX 52
MCG +5-36-27 GX 69
MCG +5-37-17 GX 69
MCG +5-37-23 GX 51
MCG +5-38-3 GX 69
MCG +5-38-6 GX 69
MCG +5-38-12 GX 51
MCG +5-38-19 GX 69
MCG +5-38-39 GX 51
MCG +5-38-42 GX 69
MCG +5-38-52 GX 51
MCG +5-40-1 GX 68
MCG +5-40-7 GX 68
MCG +5-40-14 GX 50, 68
MCG +5-40-15 GX 68
MCG +5-40-21 GX 50
MCG +5-40-23 GX 68
MCG +5-40-24 GX 68
MCG +5-40-25 GX 50, 68
MCG +5-40-38 GX 50
MCG +5-40-43 GX 50
MCG +5-40-47 GX 50
MCG +5-41-15 GX 68
MCG +5-42-7 GX 50, 68
MCG +5-42-10 GX 68
MCG +5-42-14 GX 68
MCG +5-42-17 GX 49, 50
MCG +5-42-29 GX 49, 50, 67, 68
MCG +5-42-41 GX 49, 50
MCG +5-43-9 GX 67
MCG +5-43-16 GX 49, 67
MCG +5-43-19 GX 49
MCG +5-43-20 GX 49
MCG +5-44-2 GX 49
MCG +5-44-9 GX 49
MCG +5-45-5 GX 67
MCG +5-52-2 GX 46, 64
MCG +5-52-3 GX 46, 64
MCG +5-52-4 GX 46
MCG +5-52-6 GX 64
MCG +5-52-13 GX 46, 64
MCG +5-53-1 GX 46
MCG +5-53-7 GX 46
MCG +5-53-17 GX 46
MCG +5-53-21 GX 46
MCG +5-54-3 GX 46
MCG +5-54-8 GX 46
MCG +5-54-12 GX 46
MCG +5-54-13 GX 46
MCG +5-54-14 GX 46
MCG +5-54-15 GX 46
MCG +5-54-16 GX 46
MCG +5-54-17 GX 46
MCG +5-54-19 GX 46
MCG +5-54-20 GX 64
MCG +5-54-23 GX 46
MCG +5-54-25 GX 46
MCG +5-54-31 GX 64
MCG +5-54-38 GX 64
MCG +5-54-39 GX 46, 64
MCG +5-54-41 GX 46
MCG +5-54-47 GX 46
MCG +5-54-53 GX 64
MCG +5-54-56 GX 45, 46, 63, 64
MCG +5-54-58 GX 63, 64

MCG +5-54-60 GX 45, 46, 63, 64
MCG +5-54-61 GX 45, 46, 63, 64
MCG +5-55-1 GX 45, 46
MCG +5-55-7 GX 45, 46, 63, 64
MCG +5-55-9 GX 45, 46, 63, 64
MCG +5-55-10 GX 45, 46, 63, 64
MCG +5-55-12 GX 45, 63
MCG +5-55-14 GX 63
MCG +5-55-17 GX 45
MCG +5-55-21 GX 45
MCG +5-55-22 GX 45
MCG +5-55-24 GX 45, 63
MCG +5-55-27 GX 45
MCG +5-55-28 GX 45
MCG +5-55-30 GX 45
MCG +5-55-41 GX 45
MCG +5-55-45 GX 45
MCG +5-55-48 GX 45
MCG +5-55-50 GX 45, 63
MCG +5-55-51 GX 45, 63
MCG +5-56-7 GX 63
MCG +5-56-11 GX 45
MCG +5-56-20 GX 45, 63
MCG +6-1-2 GX 45
MCG +6-1-4 GX 45
MCG +6-1-6 GX 45
MCG +6-1-8 GX 45
MCG +6-1-9 GX 45
MCG +6-1-12 GX 45
MCG +6-1-16 GX 45
MCG +6-1-18 GX 45
MCG +6-1-20 GX 45
MCG +6-1-22 GX 45
MCG +6-1-23 GX 45
MCG +6-1-26 GX 45
MCG +6-1-28 GX 45
MCG +6-1-30 GX 45
MCG +6-2-1 GX 45
MCG +6-2-12 GX 45, 62
MCG +6-2-14 GX 45, 62
MCG +6-2-17 GX 62
MCG +6-2-18 GX 62
MCG +6-2-19 GX 62
MCG +6-3-2 GX 62
MCG +6-3-4 GX 62
MCG +6-3-7 GX 62
MCG +6-3-9 GX 62, A7
MCG +6-3-10 GX 62, A7
MCG +6-3-12 GX 62, A7
MCG +6-3-13 GX 62, A7
MCG +6-3-16 GX 62
MCG +6-3-17 GX 62
MCG +6-3-22 GX 62
MCG +6-3-25 GX 62, A7
MCG +6-3-26 GX 62, A7
MCG +6-4-2 GX 62, A7
MCG +6-4-5 GX 62
MCG +6-4-9 GX 62
MCG +6-4-14 GX 62, A7
MCG +6-4-36 GX 62
MCG +6-4-42 GX 62
MCG +6-4-44 GX 62
MCG +6-4-46 GX 62
MCG +6-4-50 GX 62
MCG +6-4-51 GX 62
MCG +6-4-52 GX 62
MCG +6-4-55 GX 62
MCG +6-5-5 GX 62, A6
MCG +6-5-10 GX 62, A6
MCG +6-5-18 GX 62, A6
MCG +6-5-21 GX 62, A6
MCG +6-5-24 GX 62, A6
MCG +6-5-38 GX 62, A6
MCG +6-5-40 GX 62, A6
MCG +6-5-45 GX 62, A6
MCG +6-5-47 GX 62, A6
MCG +6-5-48 GX 62, A6
MCG +6-5-53 GX 62, A6
MCG +6-5-61 GX 61, 62, A6
MCG +6-5-63 GX 61, 62, A6
MCG +6-5-69 GX 61, 62
MCG +6-5-73 GX 61, 62, A6
MCG +6-5-74 GX 61, 62, A6
MCG +6-5-80 GX 61
MCG +6-5-83 GX 61
MCG +6-5-90 GX 61
MCG +6-5-106 GX 61
MCG +6-6-1 GX 61
MCG +6-6-4 GX 61
MCG +6-6-5 GX 61
MCG +6-6-7 GX 61
MCG +6-6-12 GX 61
MCG +6-6-20 GX 61
MCG +6-6-21 GX 61
MCG +6-6-27 GX 61
MCG +6-6-29 GX 61
MCG +6-6-43 GX 61
MCG +6-6-47 GX 61
MCG +6-6-53 GX 61
MCG +6-6-57 GX 61
MCG +6-6-59 GX 61
MCG +6-6-72 GX 61
MCG +6-6-77 GX 61
MCG +6-7-15 GX 61
MCG +6-7-19 GX 61
MCG +6-7-24 GX 61
MCG +6-7-46 GX 61
MCG +6-7-47 GX 61
MCG +6-8-2 GX 61
MCG +6-8-15 GX 60, 61
MCG +6-8-27 GX 60
MCG +6-9-4 GX 60
MCG +6-9-9 GX 60
MCG +6-9-10 GX 60
MCG +6-15-1 GX 58
MCG +6-15-2 GX 58
MCG +6-15-15 GX 58
MCG +6-16-1 GX 58
MCG +6-16-4 GX 58
MCG +6-16-8 GX 58
MCG +6-16-11 GX 58
MCG +6-16-12 GX 58
MCG +6-16-13 GX 58
MCG +6-16-14 GX 58
MCG +6-16-16 GX 58
MCG +6-16-23 GX 58
MCG +6-16-29 GX 58
MCG +6-16-30 GX 58
MCG +6-16-31 GX 58
MCG +6-16-32 GX 57, 58
MCG +6-16-38 GX 57, 58
MCG +6-17-9 GX 57
MCG +6-18-1 GX 57
MCG +6-18-2 GX 57
MCG +6-18-6 GX 57
MCG +6-18-8 GX 57
MCG +6-18-9 GX 57
MCG +6-18-11 GX 57
MCG +6-19-1 GX 57
MCG +6-19-2 GX 57
MCG +6-19-3 GX 57
MCG +6-19-8 GX 57
MCG +6-19-13 GX 56, 57
MCG +6-19-17 GX 56, 57
MCG +6-19-20 GX 56
MCG +6-19-22 GX 56
MCG +6-20-3 GX 56
MCG +6-20-4 GX 56
MCG +6-20-13A GX 56
MCG +6-20-20 GX 56
MCG +6-20-21 GX 56
MCG +6-20-26 GX 56
MCG +6-20-30 GX 56
MCG +6-20-31 GX 56
MCG +6-20-36 GX 56
MCG +6-20-37 GX 56
MCG +6-20-46 GX 56, A5
MCG +6-20-48 GX 56, A5
MCG +6-20-49 GX 56, A5
MCG +6-20-50 GX 56
MCG +6-21-5 GX 56, A5
MCG +6-21-12 GX 56, A5
MCG +6-21-18 GX 56, A5
MCG +6-21-24 GX 56, A5
MCG +6-21-29 GX 56, A5
MCG +6-21-32 GX 56, A5
MCG +6-21-33 GX 56, A5
MCG +6-21-36 GX 56, A5
MCG +6-21-38 GX 56, A5
MCG +6-21-40 GX 56
MCG +6-21-41 GX 56
MCG +6-21-42 GX 56
MCG +6-21-43 GX 56
MCG +6-21-44 GX 56
MCG +6-21-45 GX 56
MCG +6-21-46 GX 56
MCG +6-21-48 GX 56
MCG +6-21-50 GX 56
MCG +6-21-51 GX 56
MCG +6-21-52 GX 56
MCG +6-21-58 GX 56
MCG +6-21-61 GX 56
MCG +6-21-64 GX 56
MCG +6-21-68 GX 56
MCG +6-21-69 GX 56
MCG +6-21-70 GX 56
MCG +6-22-4 GX 56
MCG +6-22-6 GX 56
MCG +6-22-7 GX 56
MCG +6-22-9 GX 56
MCG +6-22-10 GX 56
MCG +6-22-11 GX 56
MCG +6-22-12 GX 56
MCG +6-22-16 GX 56
MCG +6-22-22 GX 56
MCG +6-22-24 GX 56
MCG +6-22-26 GX 56
MCG +6-22-30 GX 56
MCG +6-22-32 GX 56
MCG +6-22-34 GX 55, 56
MCG +6-22-36 GX 55, 56
MCG +6-22-37 GX 55, 56
MCG +6-22-38 GX 55, 56
MCG +6-22-40 GX 55, 56
MCG +6-22-42 GX 55, 56
MCG +6-22-44 GX 55, 56
MCG +6-22-45 GX 55, 56
MCG +6-22-48 GX 55, 56
MCG +6-22-51 GX 55, 56
MCG +6-22-52 GX 55, 56
MCG +6-22-57 GX 55, 56
MCG +6-22-61 GX 55, 56
MCG +6-22-62 GX 55, 56
MCG +6-22-63 GX 55, 56
MCG +6-22-64 GX 55
MCG +6-22-66 GX 55
MCG +6-22-67 GX 55
MCG +6-22-68 GX 55
MCG +6-22-69 GX 55
MCG +6-22-70 GX 55
MCG +6-22-72 GX 55
MCG +6-22-73 GX 55
MCG +6-23-1 GX 55
MCG +6-23-2 GX 55
MCG +6-23-10 GX 55
MCG +6-23-12 GX 55
MCG +6-23-14 GX 55
MCG +6-23-15 GX 55
MCG +6-23-16 GX 55
MCG +6-24-7 GX 55
MCG +6-24-13 GX 55
MCG +6-24-20 GX 55
MCG +6-24-22 GX 55
MCG +6-24-30 GX 55
MCG +6-24-32 GX 55
MCG +6-24-34 GX 55
MCG +6-24-38 GX 55
MCG +6-24-39 GX 55
MCG +6-24-42 GX 55
MCG +6-24-44 GX 55
MCG +6-24-47 GX 55
MCG +6-25-1 GX 55
MCG +6-25-3 GX 55
MCG +6-25-5 GX 55
MCG +6-25-6 GX 55
MCG +6-25-8 GX 55
MCG +6-25-9 GX 55
MCG +6-25-12 GX 55
MCG +6-25-14 GX 55
MCG +6-25-15 GX 55
MCG +6-25-18 GX 55
MCG +6-25-21 GX 55
MCG +6-25-22 GX 55
MCG +6-25-25 GX 55
MCG +6-25-29 GX 55
MCG +6-25-31 GX 55
MCG +6-25-32 GX 54, 55
MCG +6-25-36 GX 54, 55
MCG +6-25-37 GX 54, 55
MCG +6-25-42 GX 54, 55
MCG +6-25-43 GX 54, 55
MCG +6-25-50 GX 54, 55
MCG +6-25-51 GX 54, 55
MCG +6-25-59 GX 54, 55
MCG +6-25-61 GX 54, 55
MCG +6-25-62 GX 54
MCG +6-25-63 GX 54
MCG +6-25-65 GX 54
MCG +6-25-68 GX 54
MCG +6-25-70 GX 54
MCG +6-25-71 GX 54
MCG +6-25-72 GX 54
MCG +6-25-73 GX 54
MCG +6-25-74 GX 54
MCG +6-25-79A GX 54
MCG +6-26-3 GX 54
MCG +6-26-5 GX 54
MCG +6-26-10 GX 54
MCG +6-26-12 GX 54
MCG +6-26-13 GX 54
MCG +6-26-15 GX 54
MCG +6-26-16 GX 54
MCG +6-26-18 GX 54
MCG +6-26-22 GX 54
MCG +6-26-26 GX 54
MCG +6-26-35 GX 54
MCG +6-26-38 GX 54
MCG +6-26-44 GX 54
MCG +6-26-50 GX 54
MCG +6-26-56 GX 54
MCG +6-26-65 GX 54
MCG +6-26-66 GX 54
MCG +6-26-67 GX 54
MCG +6-27-2 GX 54
MCG +6-27-8 GX 54
MCG +6-27-16 GX 54
MCG +6-27-17 GX 54
MCG +6-27-20 GX 54
MCG +6-27-22 GX 54
MCG +6-27-25 GX 54
MCG +6-27-28 GX 54
MCG +6-27-31 GX 54
MCG +6-27-32 GX 54
MCG +6-27-33 GX 54
MCG +6-27-34 GX 54
MCG +6-27-41 GX 54
MCG +6-27-46 GX 54
MCG +6-27-48 GX 54
MCG +6-27-50 GX 54
MCG +6-27-52 GX 54
MCG +6-27-54 GX 54
MCG +6-28-5 GX 54
MCG +6-28-6 GX 54
MCG +6-28-11 GX 54
MCG +6-28-12 GX 53, 54
MCG +6-28-22 GX 53, 54
MCG +6-28-42 GX 53
MCG +6-28-44 GX 53
MCG +6-29-1 GX 53
MCG +6-29-6 GX 53
MCG +6-29-10 GX 53
MCG +6-29-11 GX 53
MCG +6-29-15 GX 53
MCG +6-29-21 GX 53
MCG +6-29-28 GX 53
MCG +6-29-37 GX 53
MCG +6-29-43 GX 53
MCG +6-29-46 GX 53
MCG +6-29-56 GX 53
MCG +6-29-58 GX 53
MCG +6-29-59 GX 53
MCG +6-29-63 GX 53
MCG +6-29-66 GX 53
MCG +6-29-67 GX 53
MCG +6-29-70 GX 53
MCG +6-29-72 GX 53
MCG +6-29-73 GX 53
MCG +6-29-75 GX 53
MCG +6-29-79 GX 53
MCG +6-29-80 GX 53
MCG +6-29-81 GX 53
MCG +6-29-82 GX 53
MCG +6-29-83 GX 53
MCG +6-29-84 GX 53
MCG +6-30-3 GX 53
MCG +6-30-7 GX 53
MCG +6-30-9 GX 53
MCG +6-30-13 GX 53
MCG +6-30-17 GX 53
MCG +6-30-18 GX 53
MCG +6-30-21 GX 53
MCG +6-30-26 GX 53
MCG +6-30-27 GX 53
MCG +6-30-29 GX 53
MCG +6-30-30 GX 53
MCG +6-30-31 GX 53
MCG +6-30-32 GX 53
MCG +6-30-34 GX 53
MCG +6-30-35 GX 53
MCG +6-30-38 GX 53
MCG +6-30-41 GX 53
MCG +6-30-42 GX 53
MCG +6-30-44 GX 53
MCG +6-30-45 GX 53
MCG +6-30-49 GX 53
MCG +6-30-57 GX 53
MCG +6-30-64 GX 53
MCG +6-30-65 GX 53
MCG +6-30-66 GX 53
MCG +6-30-67 GX 53
MCG +6-30-70 GX 53
MCG +6-30-71 GX 53
MCG +6-30-73 GX 53
MCG +6-30-79 GX 53
MCG +6-30-80 GX 53
MCG +6-30-84 GX 53
MCG +6-30-89 GX 53
MCG +6-30-91 GX 53
MCG +6-30-93 GX 53
MCG +6-30-94 GX 53
MCG +6-30-104 GX 53
MCG +6-30-106 GX 53
MCG +6-31-15 GX 53
MCG +6-31-17 GX 53
MCG +6-31-18 GX 53
MCG +6-31-25 GX 53
MCG +6-31-29 GX 52, 53
MCG +6-31-31 GX 52, 53
MCG +6-31-35 GX 52, 53
MCG +6-31-36 GX 52, 53
MCG +6-31-37 GX 52, 53
MCG +6-31-38 GX 52, 53
MCG +6-31-42 GX 52, 53
MCG +6-31-43 GX 52, 53
MCG +6-31-44 GX 52, 53
MCG +6-31-47 GX 52, 53
MCG +6-31-49 GX 52, 53
MCG +6-31-62 GX 52
MCG +6-31-65 GX 52
MCG +6-31-66 GX 52
MCG +6-31-70 GX 52
MCG +6-31-74 GX 52
MCG +6-31-77 GX 52
MCG +6-31-82 GX 52
MCG +6-31-84 GX 52
MCG +6-31-86 GX 52
MCG +6-31-87 GX 52
MCG +6-31-92 GX 52
MCG +6-31-95 GX 52
MCG +6-32-8 GX 52
MCG +6-32-9 GX 52
MCG +6-32-17 GX 52
MCG +6-32-24 GX 52
MCG +6-32-31 GX 52
MCG +6-32-39 GX 52
MCG +6-32-40 GX 52
MCG +6-32-47 GX 52
MCG +6-32-48 GX 52
MCG +6-32-51 GX 52
MCG +6-32-52 GX 52
MCG +6-32-54 GX 52
MCG +6-32-56 GX 52
MCG +6-32-57 GX 52
MCG +6-32-58A GX 52
MCG +6-32-63 GX 52
MCG +6-32-64 GX 52
MCG +6-32-66 GX 52
MCG +6-32-68 GX 52
MCG +6-32-69 GX 52
MCG +6-32-70 GX 52
MCG +6-32-74 GX 52
MCG +6-32-76 GX 52
MCG +6-32-78 GX 52
MCG +6-32-79 GX 52
MCG +6-32-80 GX 52
MCG +6-32-82 GX 52
MCG +6-32-84 GX 52
MCG +6-32-85 GX 52
MCG +6-32-87 GX 52
MCG +6-32-89 GX 52
MCG +6-33-5 GX 52
MCG +6-33-9 GX 52
MCG +6-33-17 GX 52
MCG +6-33-19 GX 52
MCG +6-33-20 GX 52
MCG +6-33-22 GX 52
MCG +6-34-1 GX 51, 52
MCG +6-34-2 GX 51, 52
MCG +6-34-4 GX 51, 52
MCG +6-34-5 GX 51, 52
MCG +6-34-10 GX 51
MCG +6-34-13 GX 51
MCG +6-34-14 GX 51
MCG +6-34-15 GX 51
MCG +6-34-16 GX 51
MCG +6-34-18 GX 51
MCG +6-34-22 GX 51
MCG +6-34-23 GX 51
MCG +6-34-24 GX 51
MCG +6-34-25 GX 51
MCG +6-34-26 GX 51
MCG +6-34-27 GX 51
MCG +6-34-28 GX 51
MCG +6-35-1 GX 51
MCG +6-35-5 GX 51
MCG +6-35-7 GX 51
MCG +6-35-8 GX 51
MCG +6-35-11 GX 51
MCG +6-35-12 GX 51
MCG +6-35-15 GX 51
MCG +6-35-18 GX 51
MCG +6-35-24 GX 51
MCG +6-35-25 GX 51
MCG +6-35-27 GX 51
MCG +6-35-29 GX 51
MCG +6-35-31 GX 51
MCG +6-35-32 GX 51
MCG +6-35-34 GX 51
MCG +6-35-36 GX 51
MCG +6-35-40 GX 51
MCG +6-35-41 GX 51
MCG +6-35-44 GX 51
MCG +6-35-46 GX 51
MCG +6-36-3 GX 51
MCG +6-36-4 GX 51
MCG +6-36-5 GX 51
MCG +6-36-6 GX 51
MCG +6-36-8 GX 51
MCG +6-36-9 GX 51
MCG +6-36-12 GX 51
MCG +6-36-20 GX 51
MCG +6-36-23 GX 51
MCG +6-36-24 GX 51
MCG +6-36-31 GX 51
MCG +6-36-33 GX 51
MCG +6-36-34 GX 51
MCG +6-36-36 GX 51, A3
MCG +6-36-38 GX 51
MCG +6-36-40 GX 51
MCG +6-36-42 GX 51
MCG +6-36-43 GX 51
MCG +6-36-45 GX 51
MCG +6-36-50 GX 51
MCG +6-36-51 GX 51
MCG +6-36-53 GX 51
MCG +6-37-1 GX 50, 51
MCG +6-37-2 GX 50, 51
MCG +6-37-4 GX 50, 51
MCG +6-37-11 GX 50
MCG +6-37-18 GX 50
MCG +6-37-19 GX 50
MCG +6-37-20 GX 50
MCG +6-37-23 GX 50
MCG +6-37-29 GX 50
MCG +6-37-31 GX 50
MCG +6-37-32 GX 50
MCG +6-38-1 GX 50
MCG +6-38-4 GX 50
MCG +6-38-5 GX 50
MCG +6-38-12 GX 50
MCG +6-38-18 GX 50
MCG +6-38-19 GX 50
MCG +6-38-21 GX 50
MCG +6-38-22 GX 50
MCG +6-39-1 GX 50
MCG +6-39-5 GX 50
MCG +6-39-6 GX 50
MCG +6-39-9 GX 50
MCG +6-39-10 GX 50
MCG +6-39-11 GX 50
MCG +6-39-12 GX 50
MCG +6-39-13 GX 50
MCG +6-39-14 GX 50
MCG +6-39-16 GX 50
MCG +6-39-21 GX 50
MCG +6-39-23 GX 50
MCG +6-39-29 GX 49, 50
MCG +6-39-30 GX 49, 50
MCG +6-40-2 GX 49
MCG +6-40-3 GX 49
MCG +6-40-6 GX 49
MCG +6-40-7 GX 49
MCG +6-40-13 GX 49
MCG +6-40-15 GX 49
MCG +6-40-16 GX 49
MCG +6-40-17 GX 49
MCG +6-41-5 GX 49
MCG +6-41-6 GX 49
MCG +6-41-12 GX 49
MCG +6-41-17 GX 49
MCG +6-41-20 GX 49
MCG +6-41-21 GX 49
MCG +6-41-24 GX 49
MCG +6-43-3 GX 48
MCG +6-47-6 GX 47
MCG +6-47-7 GX 47
MCG +6-48-10 GX 46
MCG +6-48-20 GX 46
MCG +6-49-1 GX 46
MCG +6-49-15 GX 46
MCG +6-49-17 GX 46

MCG +8-13-74 GX 40, 41
MCG +8-13-91 GX 40, 41
MCG +8-13-93 GX 40, 41
MCG +8-13-102 GX 40, 41
MCG +8-13-105 GX 40, 41
MCG +8-13-106 GX 40, 41
MCG +8-13-109 GX 40, 41
MCG +8-14-6 GX 40
MCG +8-14-30 GX 40
MCG +8-14-32 GX 40
MCG +8-14-34 GX 40
MCG +8-14-38 GX 40
MCG +8-15-5 GX 40
MCG +8-15-24 GX 40
MCG +8-15-26 GX 40
MCG +8-15-32 GX 40
MCG +8-15-35 GX 40
MCG +8-15-56 GX 40
MCG +8-16-8 GX 40
MCG +8-16-13 GX 40
MCG +8-16-16 GX 40
MCG +8-16-19 GX 40
MCG +8-16-22 GX 40
MCG +8-16-23 GX 40
MCG +8-17-9 GX 39
MCG +8-17-14 GX 39
MCG +8-17-27 GX 39
MCG +8-17-34 GX 39
MCG +8-17-35 GX 39
MCG +8-17-40 GX 39
MCG +8-17-48 GX 39
MCG +8-17-55 GX 39
MCG +8-17-69 GX 39
MCG +8-17-70 GX 39
MCG +8-17-84 GX 39
MCG +8-17-87 GX 39
MCG +8-17-102 GX 39
MCG +8-17-103 GX 39
MCG +8-17-109 GX 39
MCG +8-17-110 GX 39
MCG +8-18-7 GX 39
MCG +8-18-12 GX 39
MCG +8-18-13 GX 39
MCG +8-18-14 GX 39
MCG +8-18-16 GX 39
MCG +8-18-23 GX 39
MCG +8-18-25 GX 39
MCG +8-18-29 GX 39
MCG +8-18-32 GX 39
MCG +8-18-33 GX 39
MCG +8-18-34 GX 39
MCG +8-18-35 GX 39
MCG +8-18-40 GX 39
MCG +8-18-42 GX 39
MCG +8-18-56 GX 39
MCG +8-18-57 GX 39
MCG +8-18-59 GX 39
MCG +8-18-60 GX 39
MCG +8-18-62 GX 39
MCG +8-18-64 GX 39
MCG +8-19-8 GX 39
MCG +8-19-11 GX 39
MCG +8-19-32 GX 38, 39
MCG +8-19-35 GX 38, 39
MCG +8-19-39 GX 38
MCG +8-20-12 GX 38
MCG +8-20-17 GX 38
MCG +8-20-18 GX 38
MCG +8-20-22 GX 38
MCG +8-20-24 GX 38
MCG +8-20-31 GX 38
MCG +8-20-32 GX 38
MCG +8-20-38 GX 38
MCG +8-20-40 GX 38
MCG +8-20-45 GX 38
MCG +8-20-52 GX 38
MCG +8-20-56 GX 38
MCG +8-20-61 GX 38
MCG +8-20-71 GX 38
MCG +8-20-77 GX 38
MCG +8-20-87 GX 38
MCG +8-20-89 GX 38
MCG +8-20-92 GX 38
MCG +8-21-1 GX 38
MCG +8-21-2 GX 38
MCG +8-21-12 GX 38
MCG +8-21-13 GX 38
MCG +8-21-16 GX 38
MCG +8-21-37 GX 38
MCG +8-21-38 GX 38
MCG +8-21-39 GX 38
MCG +8-21-42 GX 38
MCG +8-21-44 GX 38
MCG +8-21-45 GX 38
MCG +8-21-46 GX 38
MCG +8-21-58 GX 38
MCG +8-21-61 GX 38
MCG +8-21-67 GX 38
MCG +8-21-69 GX 38
MCG +8-21-72 GX 38
MCG +8-21-74 GX 38
MCG +8-21-75 GX 38
MCG +8-21-88 GX 38
MCG +8-21-90 GX 38
MCG +8-21-92 GX 38
MCG +8-21-93 GX 38
MCG +8-22-13 GX 38
MCG +8-22-21 GX 38
MCG +8-22-32 GX 38
MCG +8-22-54 GX 37, 38
MCG +8-22-73 GX 37
MCG +8-22-75 GX 37
MCG +8-22-76 GX 37
MCG +8-22-80 GX 37
MCG +8-22-84 GX 37
MCG +8-23-7 GX 37
MCG +8-23-24 GX 37
MCG +8-23-25 GX 37
MCG +8-23-26 GX 37
MCG +8-23-29 GX 37
MCG +8-23-31 GX 37
MCG +8-23-32 GX 37
MCG +8-23-35 GX 37
MCG +8-23-40 GX 37
MCG +8-23-49 GX 37
MCG +8-23-51 GX 37
MCG +8-23-61 GX 37
MCG +8-23-65 GX 37
MCG +8-23-67 GX 37
MCG +8-23-69 GX 37
MCG +8-23-80 GX 37
MCG +8-24-14 GX 37
MCG +8-24-26 GX 37
MCG +8-24-28 GX 37
MCG +8-24-29 GX 37
MCG +8-24-31 GX 37
MCG +8-24-33 GX 37
MCG +8-24-34 GX 37
MCG +8-24-35 GX 37
MCG +8-24-36 GX 37
MCG +8-24-40 GX 37
MCG +8-24-41 GX 37
MCG +8-24-45 GX 37
MCG +8-24-56 GX 37
MCG +8-24-68 GX 37
MCG +8-24-70 GX 37
MCG +8-24-86 GX 37
MCG +8-24-89 GX 37
MCG +8-24-99 GX 37
MCG +8-24-101 GX 37
MCG +8-24-108 GX 37
MCG +8-24-109 GX 37
MCG +8-25-17 GX 37
MCG +8-25-18 GX 36, 37
MCG +8-25-43 GX 36
MCG +8-26-15 GX 36
MCG +8-26-20 GX 36
MCG +8-26-22 GX 36
MCG +8-26-28 GX 36
MCG +8-26-31 GX 36
MCG +8-27-18 GX 36
MCG +8-27-27 GX 36
MCG +8-27-28 GX 36
MCG +8-27-31 GX 36
MCG +8-27-40 GX 36
MCG +8-27-41 GX 36
MCG +8-27-43 GX 36
MCG +8-27-52 GX 36
MCG +8-27-55 GX 36
MCG +8-27-60 GX 36
MCG +8-27-62 GX 36
MCG +8-28-3 GX 35, 36
MCG +8-28-4 GX 35, 36
MCG +8-28-19 GX 35
MCG +8-28-20 GX 35
MCG +8-28-28 GX 35
MCG +8-28-31 GX 35
MCG +8-28-36 GX 35
MCG +8-28-37 GX 35
MCG +8-28-42 GX 35
MCG +8-29-4 GX 35
MCG +8-29-5 GX 35
MCG +8-29-8 GX 35
MCG +8-29-12 GX 35
MCG +8-29-13 GX 35
MCG +8-29-18 GX 35
MCG +8-29-27 GX 35
MCG +8-29-31 GX 35
MCG +8-30-1 GX 35
MCG +8-30-10 GX 35
MCG +8-30-20 GX 35
MCG +8-30-24 GX 35
MCG +8-30-25 GX 35
MCG +8-30-26 GX 35
MCG +8-30-27 GX 35
MCG +8-30-35 GX 35
MCG +8-30-39 GX 35
MCG +8-31-19 GX 34
MCG +8-31-21 GX 34
MCG +8-31-44 GX 34
MCG +8-32-5 GX 34
MCG +8-32-21 GX 34
MCG +8-33-34 GX 34
MCG +8-33-41 GX 33, 34
MCG +8-34-1 GX 33
MCG +8-34-2 GX 33
MCG +8-34-33 GX 33
MCG +8-36-9 GX 32, 33
MCG +8-36-10 GX 32, 33
MCG +8-43-8 GX 30
MCG +9-10-4 GX 27
MCG +9-10-5 GX 27
MCG +9-11-2 GX 27, 41
MCG +9-11-8 GX 27
MCG +9-12-1 GX 27
MCG +9-12-4 GX 27
MCG +9-12-6 GX 27
MCG +9-12-22 GX 26, 27
MCG +9-12-23 GX 26, 27
MCG +9-12-49 GX 26
MCG +9-13-1 GX 26
MCG +9-13-17 GX 26
MCG +9-13-47 GX 26, 40
MCG +9-13-50 GX 26
MCG +9-13-56 GX 26, 40
MCG +9-13-63 GX 26
MCG +9-13-67 GX 26
MCG +9-13-71 GX 26
MCG +9-13-74 GX 26
MCG +9-13-88 GX 26
MCG +9-14-17 GX 26
MCG +9-14-29 GX 26
MCG +9-14-31 GX 26
MCG +9-14-38 GX 26
MCG +9-14-45 GX 26
MCG +9-14-50 GX 26
MCG +9-14-61 GX 26
MCG +9-14-80 GX 26
MCG +9-14-83 GX 26
MCG +9-15-11 GX 26
MCG +9-15-21 GX 26
MCG +9-15-33 GX 26
MCG +9-15-51 GX 25, 26, 39
MCG +9-15-52 GX 25, 26, 39
MCG +9-15-110 GX 25
MCG +9-15-111 GX 25, 39
MCG +9-15-116 GX 25
MCG +9-15-118 GX 25, 39
MCG +9-16-8 GX 39
MCG +9-16-9 GX 25
MCG +9-16-14 GX 25
MCG +9-16-22 GX 39
MCG +9-16-29 GX 25, 39
MCG +9-16-30 GX 25, 39
MCG +9-16-42 GX 25
MCG +9-16-43 GX 25
MCG +9-16-44 GX 25
MCG +9-16-48 GX 25
MCG +9-17-9 GX 25
MCG +9-17-11 GX 25
MCG +9-17-16 GX 25
MCG +9-17-19 GX 25
MCG +9-17-23 GX 25
MCG +9-17-26 GX 25
MCG +9-17-33 GX 25
MCG +9-17-42 GX 25
MCG +9-17-49 GX 25
MCG +9-17-55 GX 25
MCG +9-17-56 GX 25
MCG +9-17-64 GX 25
MCG +9-17-65 GX 25
MCG +9-18-4 GX 25
MCG +9-18-14 GX 25
MCG +9-18-16 GX 25
MCG +9-18-24 GX 25, 38
MCG +9-18-59 GX 24, 25
MCG +9-18-63 GX 24, 25
MCG +9-18-69 GX 38
MCG +9-18-70 GX 24, 25
MCG +9-18-72 GX 24, 25, 38
MCG +9-18-84 GX 24
MCG +9-18-95 GX 24
MCG +9-18-102 GX 24
MCG +9-19-3 GX 24
MCG +9-19-16 GX 24
MCG +9-19-18 GX 24
MCG +9-19-35 GX 24
MCG +9-19-42 GX 24
MCG +9-19-44 GX 24
MCG +9-19-49 GX 24
MCG +9-19-51 GX 38
MCG +9-19-52 GX 24
MCG +9-19-53 GX 24, 38
MCG +9-19-57 GX 24, 38
MCG +9-19-64 GX 24
MCG +9-19-68 GX 24, 38
MCG +9-19-72 GX 24
MCG +9-19-73 GX 24
MCG +9-19-75 GX 24
MCG +9-19-97 GX 24, 38
MCG +9-19-107 GX 24
MCG +9-19-112 GX 24
MCG +9-19-125 GX 24
MCG +9-19-126 GX 24
MCG +9-19-137 GX 24
MCG +9-19-148 GX 24
MCG +9-19-149 GX 24
MCG +9-19-165 GX 24
MCG +9-19-167 GX 24, 38
MCG +9-19-172 GX 24
MCG +9-19-176 GX 24
MCG +9-19-196 GX 24
MCG +9-20-2 GX 38
MCG +9-20-27 GX 24
MCG +9-20-49 GX 24
MCG +9-20-68 GX 24
MCG +9-20-77 GX 24, 37, 38
MCG +9-20-80 GX 24
MCG +9-20-91 GX 24
MCG +9-20-96 GX 24
MCG +9-20-97 GX 24
MCG +9-20-103 GX 24
MCG +9-20-108 GX 24, 37
MCG +9-20-116 GX 37
MCG +9-20-130 GX 24
MCG +9-20-131 GX 24
MCG +9-20-138 GX 24
MCG +9-20-151 GX 24
MCG +9-20-153 GX 24, 37
MCG +9-20-155 GX 24
MCG +9-20-156 GX 24, 37
MCG +9-20-157 GX 24
MCG +9-20-172 GX 24, 37
MCG +9-20-174 GX 24
MCG +9-20-176 GX 24
MCG +9-20-181 GX 24
MCG +9-20-183 GX 24
MCG +9-21-1 GX 24
MCG +9-21-8 GX 24, 37
MCG +9-21-9 GX 24, 37
MCG +9-21-11 GX 24
MCG +9-21-29 GX 24
MCG +9-21-32 GX 24
MCG +9-21-41 GX 24
MCG +9-21-42 GX 37
MCG +9-21-43 GX 37
MCG +9-21-46 GX 37
MCG +9-21-68 GX 23, 24, 37
MCG +9-21-70 GX 23, 24
MCG +9-21-72 GX 23, 24
MCG +9-21-76 GX 23, 24
MCG +9-21-83 GX 23, 24
MCG +9-21-84 GX 23, 24
MCG +9-21-87 GX 23, 24
MCG +9-21-94 GX 23, 24
MCG +9-21-98 GX 23
MCG +9-21-100 GX 23
MCG +9-21-101 GX 23
MCG +9-22-3 GX 23
MCG +9-22-13 GX 23
MCG +9-22-16 GX 23
MCG +9-22-20 GX 23
MCG +9-22-23 GX 23
MCG +9-22-24 GX 37
MCG +9-22-26 GX 23
MCG +9-22-35 GX 37
MCG +9-22-36 GX 23
MCG +9-22-41 GX 23, 37
MCG +9-22-44 GX 23
MCG +9-22-47 GX 23
MCG +9-22-53 GX 23
MCG +9-22-54 GX 23
MCG +9-22-57 GX 23
MCG +9-22-65 GX 23, 37
MCG +9-22-67 GX 23
MCG +9-22-68 GX 23
MCG +9-22-74 GX 23
MCG +9-22-79 GX 23
MCG +9-22-84 GX 23
MCG +9-22-92 GX 23
MCG +9-23-3 GX 23
MCG +9-23-9 GX 23
MCG +9-23-23 GX 23
MCG +9-23-25 GX 23
MCG +9-23-41 GX 23
MCG +9-23-42 GX 23
MCG +9-23-47 GX 23
MCG +9-23-51 GX 23
MCG +9-23-52 GX 23
MCG +9-23-61 GX 23
MCG +9-24-4 GX 23
MCG +9-24-10 GX 23
MCG +9-24-12 GX 23
MCG +9-24-13 GX 23
MCG +9-24-16 GX 23
MCG +9-24-17 GX 23
MCG +9-24-22 GX 23, 36
MCG +9-24-24 GX 23, 36
MCG +9-24-26 GX 23, 36
MCG +9-24-27 GX 23
MCG +9-24-34 GX 23
MCG +9-24-35 GX 23, 36
MCG +9-24-38 GX 23
MCG +9-24-43 GX 23
MCG +9-24-54 GX 23
MCG +9-24-60 GX 22, 23
MCG +9-25-3 GX 22, 23
MCG +9-25-12 GX 22
MCG +9-25-13 GX 22
MCG +9-25-15 GX 22
MCG +9-25-22 GX 22, 36
MCG +9-25-29 GX 22
MCG +9-25-31 GX 22
MCG +9-25-32 GX 22
MCG +9-25-33 GX 35, 36
MCG +9-25-36 GX 22, 35, 36
MCG +9-25-37 GX 22
MCG +9-25-45 GX 35
MCG +9-25-52 GX 22, 35
MCG +9-25-53 GX 22
MCG +9-26-6 GX 22
MCG +9-26-12 GX 22
MCG +9-26-18 GX 22
MCG +9-26-24 GX 22
MCG +9-26-27 GX 22
MCG +9-26-32 GX 22
MCG +9-26-52 GX 22
MCG +9-26-53 GX 22
MCG +9-27-4 GX 22
MCG +9-27-12 GX 22
MCG +9-27-13 GX 22
MCG +9-27-18 GX 22
MCG +9-27-26 GX 22, 35
MCG +9-27-27 GX 22
MCG +9-27-28 GX 22
MCG +9-27-31 GX 22
MCG +9-27-45 GX 22
MCG +9-27-47 GX 22
MCG +9-27-50 GX 22
MCG +9-27-54 GX 22
MCG +9-27-56 GX 22
MCG +9-27-58 GX 22, 35
MCG +9-27-68 GX 22, 34, 35
MCG +9-27-69 GX 22, 34, 35
MCG +9-27-85 GX 22
MCG +9-27-88 GX 22
MCG +9-27-89 GX 22
MCG +9-28-4 GX 22
MCG +9-28-15 GX 21, 22
MCG +9-28-19 GX 21, 22
MCG +9-28-30 GX 21
MCG +9-28-31 GX 21
MCG +9-28-43 GX 21
MCG +9-28-45 GX 21
MCG +9-29-14 GX 21
MCG +9-29-16 GX 21, 34
MCG +9-29-17 GX 21, 34
MCG +9-29-23 GX 21
MCG +9-29-31 GX 21
MCG +9-29-36 GX 21
MCG +9-29-42 GX 21
MCG +9-29-52 GX 21
MCG +9-30-9 GX 21, 34
MCG +9-30-23 GX 21, 33
MCG +9-31-7 GX 21, 33
MCG +9-31-12 GX 20, 21, 33
MCG +9-31-20 GX 20
MCG +9-31-21 GX 20
MCG +9-31-23 GX 20
MCG +9-31-32 GX 20
MCG +10-2-1 GX 18
MCG +10-9-6 GX 27
MCG +10-9-14 GX 27
MCG +10-9-18 GX 27
MCG +10-10-3 GX 27
MCG +10-10-21 GX 15, 26, 27
MCG +10-11-3 GX 26
MCG +10-11-9 GX 26
MCG +10-11-29 GX 26
MCG +10-11-33 GX 26
MCG +10-11-37 GX 26
MCG +10-11-41 GX 15, 26
MCG +10-11-42 GX 26
MCG +10-11-53 GX 26
MCG +10-11-70 GX 26
MCG +10-11-82 GX 26
MCG +10-11-92 GX 15, 26
MCG +10-11-94 GX 26
MCG +10-11-96 GX 15, 26
MCG +10-11-118 GX 26
MCG +10-11-126 GX 26
MCG +10-11-132 GX 26
MCG +10-11-133 GX 26
MCG +10-11-135 GX 26
MCG +10-11-146 GX 26
MCG +10-11-150 GX 15, 26
MCG +10-12-1 GX 26
MCG +10-12-12 GX 26
MCG +10-12-14 GX 26
MCG +10-12-16 GX 26
MCG +10-12-19 GX 14
MCG +10-12-26 GX 26
MCG +10-12-28 GX 15, 26
MCG +10-12-36 GX 26
MCG +10-12-50 GX 26
MCG +10-12-63 GX 26
MCG +10-12-67 GX 26
MCG +10-12-72 GX 26
MCG +10-12-78 GX 26
MCG +10-12-79 GX 26
MCG +10-12-81 GX 26
MCG +10-12-82 GX 26
MCG +10-12-85 GX 26
MCG +10-12-87 GX 26
MCG +10-12-94 GX 26
MCG +10-12-100 GX 26
MCG +10-12-107 GX 26
MCG +10-12-108 GX 26
MCG +10-12-109 GX 26
MCG +10-12-118 GX 26
MCG +10-12-120 GX 26
MCG +10-12-130 GX 26
MCG +10-12-131 GX 14, 15, 26
MCG +10-12-137 GX 26
MCG +10-12-141 GX 26
MCG +10-12-143 GX 26
MCG +10-12-144 GX 26
MCG +10-12-145 GX 26
MCG +10-13-9 GX 26
MCG +10-13-20 GX 26
MCG +10-13-21 GX 26
MCG +10-13-24 GX 26
MCG +10-13-32 GX 26
MCG +10-13-47 GX 25, 26
MCG +10-13-64 GX 14, 25
MCG +10-13-66 GX 25
MCG +10-14-8 GX 14, 25
MCG +10-14-12 GX 25
MCG +10-14-19 GX 14, 25
MCG +10-14-22 GX 25
MCG +10-14-39 GX 25
MCG +10-14-44 GX 25
MCG +10-14-47 GX 25
MCG +10-14-56 GX 25
MCG +10-15-3 GX 25
MCG +10-15-34 GX 25
MCG +10-15-51 GX 25
MCG +10-15-67 GX 25
MCG +10-15-70 GX 25
MCG +10-15-79 GX 25
MCG +10-15-90 GX 25
MCG +10-15-93 GX 14, 25
MCG +10-15-96 GX 25
MCG +10-15-99 GX 25
MCG +10-15-100 GX 25
MCG +10-15-103 GX 25
MCG +10-15-105 GX 25
MCG +10-15-106 GX 25
MCG +10-15-108 GX 25
MCG +10-15-116 GX 25
MCG +10-15-118 GX 25
MCG +10-16-9 GX 25
MCG +10-16-18 GX 25
MCG +10-16-24 GX 25

MCG +10-16-25 GX 25
MCG +10-16-41 GX 24, 25
MCG +10-16-44 GX 24, 25
MCG +10-16-49 GX 24
MCG +10-16-52 GX 24
MCG +10-16-61 GX 24
MCG +10-16-71 GX 24
MCG +10-16-74 GX 24
MCG +10-16-77 GX 13, 24
MCG +10-16-81 GX 24
MCG +10-16-89 GX 24
MCG +10-16-90 GX 13, 24
MCG +10-16-92 GX 24
MCG +10-16-110 GX 24
MCG +10-16-118 GX 24
MCG +10-16-122 GX 24
MCG +10-16-123 GX 24
MCG +10-16-134 GX 24
MCG +10-17-4 GX 24
MCG +10-17-7 GX 24
MCG +10-17-10 GX 24
MCG +10-17-19 GX 24
MCG +10-17-57 GX 24
MCG +10-17-58 GX 24
MCG +10-17-64 GX 24
MCG +10-17-75 GX 24
MCG +10-17-79 GX 24
MCG +10-17-97 GX 24
MCG +10-17-120 GX 24
MCG +10-17-121 GX 24
MCG +10-17-141 GX 24
MCG +10-17-147 GX 24
MCG +10-18-6 GX 24
MCG +10-18-21 GX 24
MCG +10-18-27 GX 24
MCG +10-18-28 GX 24
MCG +10-18-44 GX 24
MCG +10-18-87 GX 24
MCG +10-18-88 GX 24
MCG +10-18-89 GX 13, 24
MCG +10-19-1 GX 24
MCG +10-19-14 GX 23, 24
MCG +10-19-27 GX 12, 13, 23
MCG +10-19-29 GX 23
MCG +10-19-49 GX 23
MCG +10-19-71 GX 23
MCG +10-19-75 GX 23
MCG +10-19-77 GX 23
MCG +10-19-79 GX 23
MCG +10-19-83 GX 23
MCG +10-19-85 GX 23
MCG +10-19-86 GX 23
MCG +10-19-93 GX 23
MCG +10-19-94 GX 23
MCG +10-19-102 GX 23
MCG +10-20-11 GX 23
MCG +10-20-16 GX 23
MCG +10-20-19 GX 23
MCG +10-20-21 GX 23
MCG +10-20-23 GX 23
MCG +10-20-36 GX 23
MCG +10-20-59 GX 23
MCG +10-20-78 GX 23
MCG +10-20-79 GX 23
MCG +10-20-84 GX 23
MCG +10-21-10 GX 23
MCG +10-21-17 GX 12, 23
MCG +10-21-36 GX 23
MCG +10-21-46 GX 22
MCG +10-22-6 GX 22
MCG +10-22-9 GX 22
MCG +10-22-10 GX 22
MCG +10-22-21 GX 22
MCG +10-22-22 GX 22
MCG +10-22-38 GX 22
MCG +10-23-9 GX 22
MCG +10-23-15 GX 22
MCG +10-23-23 GX 22
MCG +10-23-39 GX 22
MCG +10-23-53 GX 22
MCG +10-23-54 GX 22
MCG +10-23-55 GX 22
MCG +10-23-68 GX 22
MCG +10-23-72 GX 22
MCG +10-23-75 GX 11, 22
MCG +10-23-85 GX 22
MCG +10-24-32 GX 22
MCG +10-24-52 GX 11, 22
MCG +10-24-58 GX 22
MCG +10-24-64 GX 22
MCG +10-24-80 GX 21, 22
MCG +10-24-86 GX 21, 22
MCG +10-24-87 GX 21, 22
MCG +10-24-91 GX 21, 22
MCG +10-24-101 GX 11, 21
MCG +10-24-121 GX 21
MCG +10-24-123 GX 21
MCG +10-25-11 GX 21
MCG +10-25-14 GX 21
MCG +10-25-17 GX 21
MCG +10-25-32 GX 21
MCG +10-25-46 GX 21
MCG +10-25-81 GX 11, 21
MCG +10-25-104 GX 21
MCG +10-25-121 GX 10, 11, 21
MCG +10-26-6 GX 21
MCG +10-26-17 GX 21
MCG +10-26-39 GX 10, 21
MCG +10-26-43 GX 21
MCG +10-27-1 GX 21
MCG +10-27-11 GX 20
MCG +10-27-12 GX 20
MCG +10-28-3 GX 10, 20
MCG +10-28-17 GX 20
MCG +10-29-3 GX 20
MCG +11-7-13 GX 16
MCG +11-8-12 GX 15
MCG +11-8-16 GX 15
MCG +11-8-19 GX 15
MCG +11-8-21 GX 15
MCG +11-8-25 GX 15
MCG +11-8-27 GX 15
MCG +11-8-30 GX 15
MCG +11-8-42 GX 15
MCG +11-8-43 GX 15
MCG +11-8-45 GX 15
MCG +11-8-50 GX 15
MCG +11-9-21 GX 15
MCG +11-9-41 GX 15
MCG +11-9-50 GX 15
MCG +11-10-37 GX 15
MCG +11-10-52 GX 15
MCG +11-10-59 GX 15
MCG +11-10-63 GX 15
MCG +11-10-64 GX 15
MCG +11-10-65 GX 15
MCG +11-10-67 GX 15
MCG +11-10-69 GX 15
MCG +11-11-5 GX 14, 15
MCG +11-11-31 GX 14
MCG +11-12-8A GX 14
MCG +11-12-8B GX 14
MCG +11-12-10 GX 14
MCG +11-13-25 GX 14
MCG +11-13-36 GX 14
MCG +11-14-3AB GX 13
MCG +11-14-5 GX 13
MCG +11-14-8 GX 13
MCG +11-14-25A GX 13
MCG +11-14-33 GX 13, 24
MCG +11-15-16 GX 13
MCG +11-15-36 GX 13
MCG +11-15-54 GX 13, 24
MCG +11-18-4 GX 12
MCG +11-19-15 GX 11
MCG +11-19-21 GX 11
MCG +11-19-32 GX 11
MCG +11-20-1 GX 11
MCG +11-20-8 GX 11
MCG +11-20-16 GX 11
MCG +11-20-19 GX 11
MCG +11-20-28 GX 11
MCG +11-25-3 GX 9
MCG +12-3-2 GX 2, 7
MCG +12-5-6 GX 7, 16
MCG +12-5-14 GX 7
MCG +12-6-2 GX 16
MCG +12-6-13 GX 16
MCG +12-8-22 GX 15
MCG +12-8-40 GX 6
MCG +12-9-29 GX 6, 14
MCG +12-9-33 GX 14
MCG +12-9-37 GX 14
MCG +12-9-58 GX 14
MCG +12-10-13 GX 14
MCG +12-10-20 GX 14
MCG +12-10-30 GX 14
MCG +12-10-43 GX 14
MCG +12-10-54 GX 14
MCG +12-10-67 GX 13, 14
MCG +12-11-16 GX 13
MCG +12-11-42 GX 13
MCG +12-12-7 GX 5, 13
MCG +12-12-15 GX 13
MCG +12-12-16 GX 13
MCG +12-13-31 GX 12
MCG +12-13-34 GX 12
MCG +12-14-25 GX 12
MCG +12-15-5 GX 12
MCG +12-15-13 GX 11, 12
MCG +12-15-21 GX 11
MCG +12-15-25 GX 11
MCG +12-15-39 GX 11
MCG +12-15-45 GX 11
MCG +12-15-47 GX 11
MCG +12-15-53 GX 11
MCG +12-16-12A GX 11
MCG +12-16-37 GX 11
MCG +13-6-15 GX 6
MCG +13-6-23 GX 6
MCG +13-7-27 GX 6
MCG +13-7-28 GX 6
MCG +13-7-36 GX 5, 6
MCG +13-7-38 GX 5, 6
MCG +13-7-39 GX 5, 6
MCG +13-8-8 GX 5, 6
MCG +13-8-12 GX 5, 6
MCG +13-8-14 GX 5
MCG +13-8-26 GX 5
MCG +13-8-40 GX 5
MCG +13-8-41 GX 5
MCG +13-8-43 GX 5
MCG +13-8-54 GX 5
MCG +13-8-56 GX 5
MCG +13-8-57 GX 5
MCG +13-8-62 GX 5
MCG +13-9-6 GX 5
MCG +13-9-7 GX 5
MCG +13-9-21 GX 5
MCG +13-9-23 GX 5
MCG +13-9-31 GX 5
MCG +13-9-43 GX 5
MCG +13-11-2 GX 4
MCG +13-11-3 GX 4
MCG +13-11-12 GX 4
MCG +13-12-3 GX 4
MCG +13-12-25 GX 4
MCG +14-2-9 GX 1, 7
MCG +14-4-29 GX 1
MCG +14-4-49 GX 1
MCG +14-6-15 GX 5
MCG +14-6-23 GX 5
MCG +14-6-28 GX 4, 5
MCG +15-1-12 GX 1
Mel 20 OC 43
Mel 31 OC 59
Mel 66 OC 187
Mel 71 OC 135
Mel 72 OC 135
Mel 101 OC 209, 210
Mel 105 OC 209
Mel 111 OC 72
Mel 186 OC 106
Mel 227 OC 215
Mi 92 BN 48, 66
Moffat 1 OC 196
Mrk 6 OC 29
Mrk 18 OC 186
Mrk 38 OC 145
Mrk 50 OC 18
NGC 1 GX 63
NGC 2 GX 63
NGC 3 GX 81
NGC 4 GX 81
NGC 5 GX 45
NGC 7 GX 141, 159
NGC 9 GX 63
NGC 10 GX 159
NGC 11 GX 45
NGC 12 GX 101
NGC 13 GX 45
NGC 14 GX 81
NGC 15 GX 63
NGC 16 GX 63
NGC 19 GX 45
NGC 20 GX 45
NGC 22 GX 63
NGC 23 GX 63
NGC 24 GX 141
NGC 25 GX 192
NGC 26 GX 63
NGC 27 GX 45, 63
NGC 28 GX 192
NGC 29 GX 45
NGC 31 GX 192
NGC 34 GX 121
NGC 35 GX 121
NGC 36 GX 81
NGC 37 GX 192
NGC 38 GX 101, 121
NGC 39 GX 45
NGC 40 PN 8
NGC 41 GX 63
NGC 42 GX 63
NGC 43 GX 45
NGC 45 GX 141
NGC 47 GX 121
NGC 48 GX 30
NGC 49 GX 30
NGC 50 GX 121
NGC 51 GX 30
NGC 52 GX 63
NGC 53 GX 192
NGC 54 GX 121
NGC 55 GX 159
NGC 57 GX 63, 81
NGC 59 GX 141
NGC 60 GX 101
NGC 61A GX 121
NGC 61B GX 121
NGC 62 GX 121
NGC 63 GX 81
NGC 64 GX 121
NGC 65 GX 141
NGC 66 GX 141
NGC 67 GX 45
NGC 68 GX 45
NGC 69 GX 45
NGC 70 GX 45
NGC 71 GX 45
NGC 72 GX 45
NGC 72A GX 45
NGC 73 GX 121
NGC 74 GX 45
NGC 75 GX 81
NGC 76 GX 45, 63
NGC 77 GX 141
NGC 78A GX 101
NGC 78B GX 101
NGC 79 GX 63
NGC 80 GX 63
NGC 81 GX 63
NGC 83 GX 63
NGC 85 GX 63
NGC 86 GX 63
NGC 87 GX 177
NGC 88 GX 177
NGC 89 GX 177
NGC 90 GX 63
NGC 92 GX 177
NGC 93 GX 63
NGC 94 GX 63
NGC 95 GX 81
NGC 96 GX 63
NGC 97 GX 45, 63
NGC 98 GX 177
NGC 99 GX 81
NGC 100 GX 81
NGC 101 GX 159
NGC 102 GX 121
NGC 103 OC 18
NGC 104 GC 204, xvi, xxi, xxii
NGC 105 GX 81
NGC 106 GX 101, 121
NGC 107 GX 121
NGC 108 GX 45, 63
NGC 109 GX 63
NGC 110 OC 8
NGC 112 GX 45
NGC 113 GX 101
NGC 114 GX 101
NGC 115 GX 159
NGC 116 GX 121
NGC 117 GX 101
NGC 118 GX 101
NGC 119 GX 192
NGC 120 GX 101
NGC 121 GC 204
NGC 124 GX 101
NGC 125 GX 101
NGC 126 GX 101
NGC 127 GX 101
NGC 128 GX 101
NGC 129 OC 18
NGC 130 GX 101
NGC 131 GX 159
NGC 132 GX 101
NGC 133 OC 8
NGC 134 GX 159
NGC 135 GX 121, 140
NGC 136 OC 18
NGC 137 GX 81
NGC 138 GX 81, 101
NGC 139 GX 81, 101
NGC 140 GX 45
NGC 141 GX 81, 101
NGC 142 GX 141
NGC 143 GX 141
NGC 144 GX 141
NGC 145 GX 101, 121
NGC 146 OC 8
NGC 147 GX 30
NGC 148 GX 159
NGC 149 GX 45
NGC 150 GX 141
NGC 151 GX 121, 140
NGC 152 OC 204, 214, A26
NGC 154 GX 121, 140
NGC 155 GX 121, 140
NGC 157 GX 121, 140
NGC 159 GX 192
NGC 160 GX 63, 80
NGC 161 GX 101, 120
NGC 162 GX 63
NGC 163 GX 121, 140
NGC 164 GX 101, 120
NGC 165 GX 121, 140
NGC 166 GX 121, 140
NGC 167 GX 141
NGC 168 GX 141, 158
NGC 169 GX 63, 80
NGC 170 GX 101, 120
NGC 172 GX 141, 158
NGC 173 GX 101, 120
NGC 174 GX 141, 158, 159, 176
NGC 175 GX 141, 158
NGC 176 OC 204, 214, A26
NGC 177 GX 141, 158
NGC 178 GX 121, 140
NGC 179 GX 121, 140, 141, 158
NGC 180 GX 81, 100
NGC 181 GX 45, 62, 63, 80
NGC 182 GX 101, 120
NGC 183 GX 45, 62, 63, 80
NGC 184 GX 45, 62, 63, 80
NGC 185 GX 30
NGC 186 GX 101, 120
NGC 187 GX 121, 140
NGC 188 OC 1
NGC 189 OC 18
NGC 190 GX 81, 100
NGC 191 GX 121, 140
NGC 192 GX 101, 120
NGC 193 GX 101, 120
NGC 194 GX 101, 120
NGC 195 GX 121, 140
NGC 196 GX 101, 120
NGC 197 GX 101, 120
NGC 198 GX 101, 120
NGC 199 GX 101, 120
NGC 200 GX 101, 120
NGC 201 GX 101, 120
NGC 202 GX 101, 120
NGC 203 GX 101, 120
NGC 204 GX 101, 120
NGC 205 GX 30
NGC 206 Star Cld 30, 45, 62
NGC 207 GX 121, 140
NGC 208 GX 120
NGC 209 GX 141, 158
NGC 210 GX 140
NGC 212 GX 192
NGC 213 GX 100
NGC 214 GX 63, 80
NGC 215 GX 192
NGC 216 GX 141, 158
NGC 217 GX 140
NGC 218 GX 45, 62
NGC 219 GX 120
NGC 220 OC 204, 214, A26
NGC 221 GX 30, 45, 62
NGC 222 OC 204, 214, A26
NGC 223 GX 120
NGC 224 GX 30
NGC 225 OC 18
NGC 226 GX 45, 62
NGC 227 GX 120
NGC 228 GX 63, 80
NGC 229 GX 63, 80
NGC 230 GX 141, 158
NGC 231 OC 204, 214, A26
NGC 232 GX 141, 158
NGC 233 GX 45, 62
NGC 234 GX 100
NGC 235 GX 141, 158
NGC 235A GX 141, 158
NGC 236 GX 120
NGC 237 GX 120
NGC 238 GX 177
NGC 239 GX 120
NGC 240 GX 100
NGC 241 OC 204, 214, A26
NGC 242 OC 204, 214, A26
NGC 243 GX 62, 80
NGC 244 GX 140
NGC 245 GX 120
NGC 246 PN 140
NGC 247 GX 158
NGC 248 BN OC 204, 214, A26
NGC 249 BN 204, 214, A26
NGC 250 GX 100
NGC 251 GX 80
NGC 252 GX 80
NGC 253 GX 158
NGC 254 GX 176
NGC 255 GX 140
NGC 256 BN OC 204, 214, A26
NGC 257 GX 100
NGC 258 GX 80
NGC 259 GX 120
NGC 260 GX 80
NGC 261 BN OC 204, 214, A26
NGC 262 GX 62
NGC 263 GX 140
NGC 264 GX 176
NGC 265 OC 204, 214, A26
NGC 266 GX 62
NGC 267 BN 204, 214, A26
NGC 268 GX 120, 140
NGC 269 OC 204, 214, A26
NGC 270 GX 140
NGC 271 GX 120
NGC 272 OC 62
NGC 273 GX 140
NGC 274 GX 140
NGC 275 GX 140
NGC 276 GX 158
NGC 277 GX 140
NGC 278 GX 44
NGC 279 GX 120
NGC 280 GX 80
NGC 281 BN 18
NGC 282 GX 62
NGC 283 GX 140
NGC 284 GX 140
NGC 285 GX 140
NGC 286 GX 140
NGC 287 GX 62
NGC 288 GC 158
NGC 289 GX 176
NGC 290 OC 204, 214, A26
NGC 291 GX 140
NGC 292 GX 204, A26
NGC 293 GX 140
NGC 294 BN OC 204, 214, A26
NGC 295 GX 62
NGC 296 GX 62
NGC 298 GX 140
NGC 299 BN OC 204, A26
NGC 300 GX 176
NGC 301 GX 140
NGC 303 GX 140
NGC 304 GX 80
NGC 306 BN OC 204, A26
NGC 307 GX 120
NGC 309 GX 140
NGC 311 GX 62
NGC 312 GX 192, 203
NGC 314 GX 176
NGC 315 GX 62
NGC 317A GX 44
NGC 317B GX 44
NGC 318 GX 62
NGC 319 GX 191
NGC 320 GX 158
NGC 321 GX 120, 140
NGC 322 GX 191
NGC 323 GX 192, 203
NGC 324 GX 176, 191
NGC 325 GX 120, 140
NGC 326 GX 80
NGC 327 GX 120, 140
NGC 328 GX 192, 203
NGC 329 GX 120, 140
NGC 330 OC 204, A26
NGC 331 GX 120
NGC 332 GX 100
NGC 333 GX 140
NGC 333B GX 140
NGC 334 GX 176
NGC 335 GX 158
NGC 337 GX 140
NGC 337A GX 140
NGC 338 GX 62

NGC 339 OC 214, A26
NGC 340 GX 140
NGC 341 GX 140
NGC 342 GX 140
NGC 344 GX 158
NGC 345 GX 140
NGC 346 BN OC 204, A26
NGC 347 GX 140
NGC 348 GX 192, 203
NGC 349 GX 140
NGC 350 GX 140
NGC 351 GX 120
NGC 352 GX 120
NGC 353 GX 120
NGC 354 GX 80
NGC 355 GX 140
NGC 356 GX 140
NGC 357 GX 140
NGC 358 OC 8, 17, 29
NGC 359 GX 120
NGC 360 GX 204
NGC 361 OC 204, A26
NGC 362 GC 204, A26
NGC 363 GX 140
NGC 364 GX 120
NGC 365 GX 176
NGC 366 OC 8, 17, 29
NGC 367 GX 140
NGC 368 GX 191
NGC 369 GX 140, 158
NGC 371 OC 204, A26
NGC 373 GX 62, A7
NGC 374 GX 62, A7
NGC 375 GX 62, A7
NGC 376 OC 204, A26
NGC 377 GX 158
NGC 378 GX 176
NGC 379 GX 62, A7
NGC 380 GX 62, A7
NGC 381 OC 29
NGC 382 GX 62, A7
NGC 383 GX 62, A7
NGC 384 GX 62, A7
NGC 385 GX 62, A7
NGC 386 GX 62, A7
NGC 387 GX 62, A7
NGC 388 GX 62, A7
NGC 389 GX 62
NGC 390 GX 62, A7
NGC 391 GX 120
NGC 392 GX 62, A7
NGC 393 GX 62
NGC 394 GX 62, A7
NGC 395 OC 204, 213, A26
NGC 396 GX 120
NGC 397 GX 62, A7
NGC 398 GX 62, A7
NGC 399 GX 62, A7
NGC 403 GX 62, A7
NGC 404 GX 62
NGC 406 GX 204, 213, A26
NGC 407 GX 62, A7
NGC 409 GX 176
NGC 410 GX 62, A7
NGC 411 OC 204, 213, A26
NGC 413 GX 120
NGC 414 GX 62, A7
NGC 415 GX 176
NGC 416 OC 204, 213, A26
NGC 417 GX 158
NGC 418 GX 176
NGC 419 OC 204, 213, A26
NGC 420 GX 62, A7
NGC 422 OC 204, 213, A26
NGC 423 GX 158, 176
NGC 424 GX 176
NGC 425 GX 62
NGC 426 GX 120
NGC 427 GX 176
NGC 428 GX 120
NGC 429 GX 120
NGC 430 GX 120
NGC 431 GX 62, A7
NGC 432 GX 203
NGC 433 OC 29
NGC 434 GX 203
NGC 434A GX 203
NGC 435 GX 120
NGC 436 OC 29
NGC 437 GX 100, 120
NGC 438 GX 176
NGC 439 GX 176
NGC 440 GX 203
NGC 441 GX 176
NGC 442 GX 120
NGC 443 GX 62, A7
NGC 444 GX 62
NGC 445 GX 120
NGC 446 GX 120
NGC 447 GX 62, A7
NGC 448 GX 120
NGC 449 GX 62, A7
NGC 450 GX 120
NGC 451 GX 62, A7
NGC 452 GX 62
NGC 454 GX 203
NGC 455 GX 100, 120
NGC 456 BN OC 204, 213, 214, A26
NGC 457 OC 29
NGC 458 OC 204, 213, A26
NGC 459 GX 80, 100
NGC 460 OC 204, 213, 214, A26
NGC 461 GX 176
NGC 462 GX 120
NGC 463 GX 100
NGC 465 OC 204, 213, 214, A26
NGC 466 GX 203
NGC 467 GX 120
NGC 468 GX 62, A7
NGC 469 GX 100
NGC 470 GX 120
NGC 471 GX 100
NGC 472 GX 62, A7
NGC 473 GX 100
NGC 474 GX 120
NGC 475 GX 100
NGC 476 GX 100
NGC 477 GX 44, 62
NGC 478 GX 158
NGC 479 GX 120
NGC 480 GX 140
NGC 481 GX 140
NGC 482 GX 176, 191
NGC 483 GX 62, A7
NGC 484 GX 203
NGC 485 GX 100
NGC 486 GX 100, 120
NGC 487 GX 140
NGC 488 GX 100, 120
NGC 489 GX 100
NGC 490 GX 100, 120
NGC 491 GX 176
NGC 491A GX 176
NGC 492 GX 100, 120
NGC 493 GX 120
NGC 494 GX 62, A7
NGC 495 GX 62, A7
NGC 496 GX 62, A7
NGC 497 GX 120, A16
NGC 498 GX 62, A7
NGC 499 GX 62, A7
NGC 500 GX 100, 120
NGC 501 GX 62, A7
NGC 502 GX 100
NGC 503 GX 62, A7
NGC 504 GX 62, A7
NGC 505 GX 100
NGC 507 GX 62, A7
NGC 508 GX 62, A7
NGC 509 GX 100
NGC 511 GX 100
NGC 512 GX 62, A7
NGC 513 GX 62, A7
NGC 514 GX 100
NGC 515 GX 62, A7
NGC 516 GX 100
NGC 517 GX 62, A7
NGC 518 GX 100
NGC 519 GX 120, A16
NGC 520 GX 120
NGC 521 GX 120
NGC 522 GX 100
NGC 523 GX 62, A7
NGC 524 GX 100
NGC 525 GX 100
NGC 526 GX 176
NGC 526B GX 176
NGC 527 GX 176
NGC 527B GX 176
NGC 528 GX 62, A7
NGC 529 GX 62, A7
NGC 530 GX 120, A16
NGC 531 GX 62, A7
NGC 532 GX 100
NGC 533 GX 120
NGC 534 GX 176
NGC 535 GX 120, A16
NGC 536 GX 62, A7
NGC 538 GX 120, A16
NGC 539 GX 158
NGC 540 GX 158
NGC 541 GX 120, A16
NGC 542 GX 62, A7
NGC 543 GX 120, A16
NGC 544 GX 176
NGC 545 GX 120, A16
NGC 546 GX 176
NGC 547 GX 120, A16
NGC 548 GX 120, A16
NGC 549 GX 176
NGC 550 GX 120
NGC 551 GX 62
NGC 554 GX 158
NGC 555 GX 158
NGC 556 GX 158
NGC 557 GX 120, A16
NGC 558 GX 120, A16
NGC 559 OC 17
NGC 560 GX 120, A16
NGC 561 GX 62, A7
NGC 562 GX 44
NGC 563 GX 158
NGC 564 GX 120, A16
NGC 565 GX 120, A16
NGC 566 GX 62, A7
NGC 567 GX 140
NGC 568 GX 176
NGC 569 GX 100
NGC 570 GX 120, A16
NGC 571 GX 62, A7
NGC 572 GX 176
NGC 573 GX 44
NGC 574 GX 176
NGC 575 GX 80
NGC 576 GX 191, 203
NGC 577 GX 120, A16
NGC 578 GX 158
NGC 579 GX 62, A7
NGC 581 OC 29
NGC 582 GX 62, A7
NGC 583 GX 158
NGC 584 GX 140
NGC 585 GX 120, A16
NGC 586 GX 140
NGC 587 GX 62
NGC 588 BN 62
NGC 589 GX 140
NGC 590 GX 44
NGC 591 GX 62
NGC 592 Star Cld 62
NGC 593 GX 140
NGC 594 GX 140
NGC 595 BN 62
NGC 596 GX 140
NGC 597 GX 176
NGC 598 GX 62
NGC 599 GX 140
NGC 600 GX 140
NGC 602 BN 213, 214
NGC 604 BN 62
NGC 605 GX 44
NGC 606 GX 80
NGC 608 GX 62
NGC 609 OC 17
NGC 612 GX 176
NGC 613 GX 158, 176
NGC 614 GX 62
NGC 615 GX 140
NGC 617 GX 140
NGC 619 GX 176
NGC 620 GX 44
NGC 621 GX 62
NGC 622 GX 120
NGC 623 GX 176
NGC 624 GX 140
NGC 625 GX 191
NGC 626 GX 176
NGC 628 GX 100
NGC 630 GX 176
NGC 631 GX 100, 120
NGC 632 GX 100, 120
NGC 633 GX 176
NGC 634 GX 62
NGC 636 GX 140
NGC 637 OC 17
NGC 638 GX 100
NGC 639 GX 158, 176
NGC 640 GX 140
NGC 641 GX 191
NGC 642 GX 158, 176
NGC 643 OC 214
NGC 643B GX 214
NGC 643C GX 214, 219
NGC 644 GX 191
NGC 645 GX 100, 120
NGC 646 GX 213
NGC 647 GX 140
NGC 648 GX 140, 158
NGC 649 GX 140
NGC 650-51 PN 29, 44
NGC 652 GX 100
NGC 653 GX 62
NGC 654 OC 29
NGC 655 GX 140
NGC 656 GX 80
NGC 657 OC 29
NGC 658 GX 100
NGC 659 OC 29
NGC 660 GX 100
NGC 661 GX 80
NGC 662 GX 62, A6
NGC 663 OC 29
NGC 664 GX 120
NGC 665 GX 99, 100
NGC 666 GX 62, A6
NGC 667 GX 158
NGC 668 GX 62, A6
NGC 669 GX 62, A6
NGC 670 GX 80
NGC 671 GX 99, 100
NGC 672 GX 80
NGC 673 GX 99, 100
NGC 675 GX 99, 100
NGC 676 GX 99, 100, 119, 120
NGC 677 GX 99, 100
NGC 678 GX 80
NGC 679 GX 62, A6
NGC 680 GX 80
NGC 681 GX 139, 140
NGC 682 GX 139, 140
NGC 683 GX 99, 100
NGC 684 GX 80
NGC 685 GX 203
NGC 686 GX 158
NGC 687 GX 62, A6
NGC 688 GX 62, A6
NGC 689 GX 158
NGC 690 GX 139, 140
NGC 691 GX 80
NGC 692 GX 191
NGC 693 GX 99, 100
NGC 694 GX 80
NGC 695 GX 80
NGC 696 GX 176
NGC 697 GX 80
NGC 698 GX 176
NGC 699 GX 139, 140
NGC 700 GX 62, A6
NGC 701 GX 139, 140
NGC 702 GX 119, 120
NGC 703 GX 62, A6
NGC 704 GX 62, A6
NGC 705 GX 62, A6
NGC 706 GX 99, 100
NGC 707 GX 139, 140
NGC 708 GX 62, A6
NGC 709 GX 62, A6
NGC 710 GX 62, A6
NGC 711 GX 80, 99
NGC 712 GX 62, A6
NGC 713 GX 139
NGC 714 GX 62, A6
NGC 715 GX 139
NGC 716 GX 99
NGC 717 GX 62, A6
NGC 718 GX 119
NGC 719 GX 80
NGC 720 GX 139
NGC 721 GX 62
NGC 722 GX 80
NGC 723 GX 158
NGC 725 GX 139
NGC 726 GX 139
NGC 727 GX 176
NGC 731 GX 139
NGC 732 GX 61, 62, A6
NGC 733 GX 61, 62
NGC 734 GX 139, 158
NGC 735 GX 61, 62, A6
NGC 736 GX 61, 62
NGC 738 GX 61, 62
NGC 739 GX 61, 62
NGC 740 GX 61, 62
NGC 741 GX 99, 119
NGC 742 GX 99, 119
NGC 743 OC 29
NGC 744 OC 29
NGC 745 GX 203
NGC 746 GX 44
NGC 747 GX 139
NGC 748 GX 119
NGC 749 GX 158, 176
NGC 750 GX 61, 62
NGC 751 GX 61, 62
NGC 752 OC 61, 62, A6
NGC 753 GX 61, 62, A6
NGC 754 GX 203
NGC 755 GX 139
NGC 756 GX 139
NGC 758 GX 119
NGC 759 GX 61, 62, A6
NGC 761 GX 61, 62
NGC 762 GX 119, 139
NGC 765 GX 79, 80
NGC 766 GX 99
NGC 767 GX 139
NGC 768 GX 119
NGC 769 GX 61, 62
NGC 770 GX 79, 80
NGC 772 GX 79, 80
NGC 773 GX 139
NGC 774 GX 99
NGC 775 GX 157, 158
NGC 776 GX 79, 80
NGC 777 GX 61, 62
NGC 778 GX 61, 62
NGC 779 GX 119, 139
NGC 780 GX 79, 80
NGC 781 GX 99
NGC 782 GX 203
NGC 783 GX 61, 62
NGC 784 GX 79, 80
NGC 785 GX 61, 62
NGC 786 GX 99
NGC 787 GX 139
NGC 788 GX 139
NGC 789 GX 61, 62
NGC 790 GX 119, 139
NGC 791 GX 99
NGC 792 GX 99
NGC 794 GX 79, 80
NGC 795 GX 203
NGC 796 BN OC 214, 219
NGC 797 GX 61, 62
NGC 798 GX 61, 62
NGC 799 GX 119
NGC 800 GX 119
NGC 801 GX 61, 62
NGC 802 GX 213
NGC 803 GX 99
NGC 804 GX 61
NGC 805 GX 79
NGC 806 GX 139
NGC 807 GX 79
NGC 808 GX 157, 158
NGC 809 GX 139
NGC 810 GX 99
NGC 811 GX 139
NGC 812 GX 44
NGC 813 GX 213
NGC 814 GX 139
NGC 815 GX 139
NGC 816 GX 61, 79
NGC 817 GX 79, 99
NGC 818 GX 61
NGC 819 GX 61, 79
NGC 820 GX 99
NGC 821 GX 99
NGC 822 GX 191
NGC 823 GX 157
NGC 824 GX 175
NGC 825 GX 99
NGC 826 GX 61
NGC 827 GX 99
NGC 828 GX 61
NGC 829 GX 139
NGC 830 GX 139
NGC 831 GX 99
NGC 833 GX 139
NGC 834 GX 61
NGC 835 GX 139
NGC 836 GX 157
NGC 837 GX 157
NGC 838 GX 139
NGC 839 GX 139
NGC 840 GX 99
NGC 841 GX 61
NGC 842 GX 139
NGC 844 GX 99
NGC 845 GX 61
NGC 846 GX 44
NGC 848 GX 139
NGC 849 GX 157
NGC 850 GX 119
NGC 851 GX 119
NGC 852 GX 203
NGC 853 GX 139
NGC 854 GX 175
NGC 855 GX 79
NGC 856 GX 119
NGC 857 GX 175
NGC 858 GX 157
NGC 860 GX 61
NGC 861 GX 61
NGC 862 GX 191
NGC 863 GX 119
NGC 864 GX 99, 119
NGC 865 GX 79
NGC 868 GX 119
NGC 869 OC 29, i, vi, vii
NGC 870 GX 99
NGC 871 GX 99
NGC 872 GX 139, 157
NGC 873 GX 139
NGC 874 GX 157
NGC 875 GX 119
NGC 876 GX 99
NGC 877 GX 99
NGC 878 GX 157
NGC 879 GX 139
NGC 880 GX 119
NGC 881 GX 139
NGC 882 GX 99
NGC 883 GX 139
NGC 884 OC 29, i, vi, vii
NGC 886 OC 17
NGC 887 GX 139
NGC 888 GX 203
NGC 889 GX 191
NGC 890 GX 61
NGC 891 GX 43, 44
NGC 892 GX 157
NGC 893 GX 191
NGC 895 GX 119, 139
NGC 896 BN 17, 29
NGC 897 GX 175
NGC 898 GX 43, 44
NGC 899 GX 157
NGC 900 GX 79
NGC 901 GX 79
NGC 902 GX 139
NGC 904 GX 79
NGC 905 GX 139
NGC 906 GX 43, 44
NGC 907 GX 157
NGC 908 GX 157
NGC 909 GX 43, 44
NGC 910 GX 43, 44
NGC 911 GX 43, 44
NGC 912 GX 43, 44
NGC 913 GX 43, 44
NGC 914 GX 43, 44
NGC 915 GX 79
NGC 916 GX 79
NGC 918 GX 79
NGC 919 GX 79
NGC 920 GX 43, 44
NGC 921 GX 139
NGC 922 GX 157
NGC 923 GX 43, 44
NGC 924 GX 79
NGC 925 GX 61
NGC 926 GX 119
NGC 927 GX 99
NGC 928 GX 79
NGC 929 GX 139
NGC 931 GX 61
NGC 932 GX 79
NGC 933 GX 43
NGC 934 GX 119
NGC 935 GX 79
NGC 936 GX 119
NGC 937 GX 43
NGC 938 GX 79
NGC 939 GX 190, 191
NGC 940 GX 61
NGC 941 GX 119
NGC 942 GX 139
NGC 943 GX 139
NGC 944 GX 139
NGC 945 GX 139
NGC 946 GX 43
NGC 947 GX 157

NGC 948 GX 139
NGC 949 GX 61
NGC 950 GX 139
NGC 951 GX 157
NGC 953 GX 61, 79
NGC 954 GX 190
NGC 955 GX 119
NGC 956 OC 43
NGC 957 OC 29
NGC 958 GX 119
NGC 959 GX 61
NGC 960 GX 139
NGC 962 GX 79
NGC 963 GX 119
NGC 964 GX 175
NGC 965 GX 157
NGC 966 GX 157
NGC 967 GX 139, 157
NGC 968 GX 61
NGC 969 GX 61
NGC 970 GX 61
NGC 972 GX 61, 79
NGC 973 GX 61
NGC 974 GX 61
NGC 975 GX 99
NGC 976 GX 79
NGC 977 GX 139
NGC 978A GX 61
NGC 978B GX 61
NGC 979 GX 190
NGC 980 GX 43, 61
NGC 981 GX 139
NGC 982 GX 43, 61
NGC 984 GX 79
NGC 985 GX 139
NGC 986 GX 175
NGC 986A GX 175
NGC 987 GX 61
NGC 988 GX 139
NGC 989 GX 139
NGC 990 GX 99
NGC 991 GX 139
NGC 992 GX 79
NGC 993 GX 119
NGC 995 GX 43
NGC 996 GX 43
NGC 997 GX 99
NGC 998 GX 99
NGC 999 GX 43
NGC 1000 GX 43
NGC 1001 GX 43
NGC 1002 GX 61
NGC 1003 GX 43, 61
NGC 1004 GX 119
NGC 1005 GX 43
NGC 1007 GX 119
NGC 1008 GX 119
NGC 1009 GX 119
NGC 1010 GX 139
NGC 1011 GX 139
NGC 1012 GX 61
NGC 1013 GX 139
NGC 1015 GX 119
NGC 1016 GX 119
NGC 1017 GX 139
NGC 1018 GX 139
NGC 1019 GX 119
NGC 1020 GX 119
NGC 1021 GX 119
NGC 1022 GX 139
NGC 1023 GX 61
NGC 1023A GX 61
NGC 1024 GX 99
NGC 1025 GX 203
NGC 1026 GX 99
NGC 1027 OC 29
NGC 1028 GX 99
NGC 1029 GX 99
NGC 1030 GX 79
NGC 1031 GX 203
NGC 1032 GX 119
NGC 1033 GX 139
NGC 1034 GX 139
NGC 1035 GX 139
NGC 1036 GX 79
NGC 1037 GX 119
NGC 1038 GX 119
NGC 1039 OC 43
NGC 1041 GX 119, 139
NGC 1042 GX 139
NGC 1043 GX 119
NGC 1044 GX 99
NGC 1045 GX 139
NGC 1046 GX 99
NGC 1047 GX 139
NGC 1048 GX 139
NGC 1048A GX 139
NGC 1049 GC 175
NGC 1050 GX 61
NGC 1051 GX 139
NGC 1052 GX 139
NGC 1053 GX 43
NGC 1054 GX 79
NGC 1055 GX 119
NGC 1056 GX 79
NGC 1057 GX 61
NGC 1058 GX 61
NGC 1060 GX 61
NGC 1061 GX 61
NGC 1063 GX 119, 139
NGC 1064 GX 139
NGC 1065 GX 139
NGC 1066 GX 61
NGC 1067 GX 61
NGC 1068 GX 119
NGC 1069 GX 139
NGC 1070 GX 119
NGC 1071 GX 139
NGC 1072 GX 119
NGC 1073 GX 119
NGC 1074 GX 139
NGC 1075 GX 139
NGC 1076 GX 139
NGC 1077A GX 43, 61
NGC 1077B GX 43, 61
NGC 1078 GX 139
NGC 1079 GX 157, 175
NGC 1080 GX 119
NGC 1081 GX 139
NGC 1082 GX 139
NGC 1083 GX 139
NGC 1084 GX 139
NGC 1085 GX 119
NGC 1086 GX 43
NGC 1087 GX 119
NGC 1088 GX 99
NGC 1089 GX 139
NGC 1090 GX 119
NGC 1091 GX 139, 157
NGC 1092 GX 139, 157
NGC 1093 GX 61
NGC 1094 GX 119
NGC 1095 GX 119
NGC 1096 GX 203
NGC 1097 GX 175
NGC 1097A GX 175
NGC 1098 GX 139, 157
NGC 1099 GX 139, 157
NGC 1100 GX 139, 157
NGC 1101 GX 119
NGC 1102 GX 157
NGC 1103 GX 139
NGC 1104 GX 119
NGC 1105 GX 139
NGC 1106 GX 43
NGC 1107 GX 99
NGC 1108 GX 139
NGC 1109 GX 99
NGC 1110 GX 139
NGC 1111 GX 99
NGC 1114 GX 139, 157
NGC 1115 GX 99
NGC 1116 GX 99
NGC 1117 GX 99
NGC 1118 GX 139
NGC 1119 GX 139, 157
NGC 1120 GX 139
NGC 1121 GX 119
NGC 1122 GX 43
NGC 1124 GX 157
NGC 1125 GX 139
NGC 1126 GX 119
NGC 1127 GX 99
NGC 1129 GX 43
NGC 1130 GX 43
NGC 1131 GX 43
NGC 1132 GX 119
NGC 1133 GX 139
NGC 1134 GX 99
NGC 1135 GX 203
NGC 1136 GX 203
NGC 1137 GX 119
NGC 1138 GX 43
NGC 1139 GX 139
NGC 1140 GX 139
NGC 1143 GX 119
NGC 1144 GX 119
NGC 1145 GX 157
NGC 1148 GX 138, 139
NGC 1149 GX 118, 119
NGC 1150 GX 138, 139
NGC 1151 GX 138, 139
NGC 1152 GX 138, 139
NGC 1153 GX 118, 119
NGC 1154 GX 138, 139
NGC 1155 GX 138, 139
NGC 1156 GX 79
NGC 1157 GX 138, 139
NGC 1158 GX 138, 139
NGC 1159 GX 43
NGC 1160 GX 43
NGC 1161 GX 43
NGC 1162 GX 138, 139
NGC 1163 GX 138, 139, 157
NGC 1164 GX 43
NGC 1165 GX 175
NGC 1166 GX 98, 99
NGC 1167 GX 61
NGC 1168 GX 98, 99
NGC 1169 GX 43
NGC 1171 GX 43
NGC 1172 GX 138, 139
NGC 1175 GX 43
NGC 1177 GX 43
NGC 1179 GX 157
NGC 1180 GX 138, 139
NGC 1181 GX 138, 139
NGC 1182 GX 138, 139
NGC 1184 GX 7
NGC 1185 GX 138, 139
NGC 1186 GX 43
NGC 1187 GX 157
NGC 1188 GX 138, 139
NGC 1189 GX 138, 139
NGC 1190 GX 138, 139
NGC 1191 GX 138, 139
NGC 1192 GX 138, 139
NGC 1193 OC 43
NGC 1194 GX 118, 119
NGC 1195 GX 138, 139
NGC 1196 GX 138, 139
NGC 1198 GX 43
NGC 1199 GX 138, 139
NGC 1200 GX 138, 139
NGC 1201 GX 157
NGC 1202 GX 138
NGC 1203A GX 138
NGC 1203B GX 138
NGC 1204 GX 138
NGC 1206 GX 138
NGC 1207 GX 61
NGC 1208 GX 138
NGC 1209 GX 138
NGC 1210 GX 157
NGC 1211 GX 118
NGC 1213 GX 61
NGC 1214 GX 138
NGC 1215 GX 138
NGC 1216 GX 138
NGC 1217 GX 175
NGC 1218 GX 118
NGC 1219 GX 118
NGC 1220 OC 28
NGC 1221 GX 118
NGC 1222 GX 118
NGC 1223 GX 118
NGC 1224 GX 43, A4
NGC 1225 GX 118
NGC 1226 GX 61
NGC 1227 GX 61
NGC 1228 GX 157
NGC 1229 GX 157
NGC 1230 GX 157
NGC 1231 GX 138
NGC 1232 GX 157
NGC 1232A GX 157
NGC 1233 GX 61
NGC 1234 GX 138
NGC 1236 GX 98
NGC 1238 GX 138
NGC 1239 GX 118
NGC 1241 GX 138
NGC 1242 GX 138
NGC 1244 GX 213
NGC 1245 OC 43
NGC 1246 GX 213
NGC 1247 GX 138
NGC 1248 GX 118, 138
NGC 1249 GX 202
NGC 1250 GX 43, A4
NGC 1252 OC 202
NGC 1253 GX 118
NGC 1253A GX 118
NGC 1254 GX 118
NGC 1255 GX 157
NGC 1256 GX 157
NGC 1257 GX 43, A4
NGC 1258 GX 157
NGC 1259 GX 43, A4
NGC 1260 GX 43, A4
NGC 1261 GC 202
NGC 1262 GX 138
NGC 1263 GX 138
NGC 1264 GX 43, A4
NGC 1265 GX 43, A4
NGC 1266 GX 118
NGC 1267 GX 43, A4
NGC 1268 GX 43, A4
NGC 1270 GX 43, A4
NGC 1271 GX 43, A4
NGC 1272 GX 43, A4
NGC 1273 GX 43, A4
NGC 1274 GX 43, A4
NGC 1275 GX 43, A4
NGC 1276 GX 43, A4
NGC 1277 GX 43, A4
NGC 1278 GX 43, A4
NGC 1280 GX 118
NGC 1281 GX 43, A4
NGC 1282 GX 43, A4
NGC 1283 GX 43, A4
NGC 1284 GX 138
NGC 1285 GX 138
NGC 1286 GX 138
NGC 1287 GX 118
NGC 1288 GX 174, 175
NGC 1289 GX 118
NGC 1290 GX 138
NGC 1291 GX 190
NGC 1292 GX 156, 157
NGC 1293 GX 43, A4
NGC 1294 GX 43, A4
NGC 1295 GX 138
NGC 1296 GX 138
NGC 1297 GX 156, 157
NGC 1298 GX 118
NGC 1299 GX 138
NGC 1300 GX 156, 157
NGC 1301 GX 156, 157
NGC 1302 GX 156, 157
NGC 1303 GX 138
NGC 1304 GX 118
NGC 1305 GX 118
NGC 1306 GX 156, 157
NGC 1308 GX 118
NGC 1309 GX 138
NGC 1310 GX 174, 175
NGC 1311 GX 202
NGC 1313 GX 213
NGC 1313A GX 213
NGC 1314 GX 118
NGC 1315 GX 156, 157
NGC 1316 GX 174, 175
NGC 1316C GX 174
NGC 1317 GX 174, 175
NGC 1319 GX 156, 157
NGC 1320 GX 118
NGC 1321 GX 118
NGC 1322 GX 118
NGC 1323 GX 118
NGC 1324 GX 118, 138
NGC 1325 GX 156
NGC 1325A GX 156
NGC 1326 GX 174, 175
NGC 1326A GX 174
NGC 1326B GX 174
NGC 1327 GX 156
NGC 1328 GX 118
NGC 1329 GX 138, 156
NGC 1331 GX 156
NGC 1332 GX 156
NGC 1333 BN 60
NGC 1334 GX 43
NGC 1335 GX 43
NGC 1336 GX 174
NGC 1337 GX 138
NGC 1338 GX 138
NGC 1339 GX 174
NGC 1341 GX 174
NGC 1342 OC 60
NGC 1343 GX 16, 17
NGC 1344 GX 174
NGC 1345 GX 138, 156
NGC 1346 GX 118, 138
NGC 1347 GX 156
NGC 1348 OC 28, 43
NGC 1349 GX 118
NGC 1350 GX 174
NGC 1351 GX 174
NGC 1351A GX 174
NGC 1352 GX 156
NGC 1353 GX 156
NGC 1354 GX 138
NGC 1355 GX 118, 138
NGC 1356 GX 190
NGC 1357 GX 138
NGC 1358 GX 118, 138
NGC 1359 GX 156
NGC 1360 PN 156
NGC 1361 GX 138
NGC 1362 GX 156
NGC 1363 GX 138
NGC 1364 GX 138
NGC 1365 GX 174
NGC 1366 GX 174
NGC 1368 GX 138
NGC 1369 GX 174
NGC 1370 GX 156
NGC 1371 GX 156
NGC 1372 GX 138
NGC 1373 GX 174
NGC 1374 GX 174
NGC 1375 GX 174
NGC 1376 GX 118, 138
NGC 1377 GX 156
NGC 1379 GX 174
NGC 1380 GX 174
NGC 1380A GX 174
NGC 1381 GX 174
NGC 1382 GX 174
NGC 1383 GX 156
NGC 1384 GX 98
NGC 1385 GX 156
NGC 1386 GX 174
NGC 1387 GX 174
NGC 1388 GX 138
NGC 1389 GX 174
NGC 1390 GX 156
NGC 1391 GX 156
NGC 1393 GX 156
NGC 1394 GX 156
NGC 1395 GX 156
NGC 1396 GX 174
NGC 1397 GX 118
NGC 1398 GX 156
NGC 1399 GX 174
NGC 1400 GX 156
NGC 1401 GX 156
NGC 1402 GX 156
NGC 1403 GX 156
NGC 1404 GX 174
NGC 1405 GX 138
NGC 1406 GX 174
NGC 1407 GX 156
NGC 1409 GX 118
NGC 1410 GX 118
NGC 1411 GX 190
NGC 1412 GX 156
NGC 1413 GX 138
NGC 1414 GX 156
NGC 1415 GX 156
NGC 1416 GX 156
NGC 1417 GX 118
NGC 1418 GX 118
NGC 1419 GX 174
NGC 1421 GX 138
NGC 1422 GX 156
NGC 1423 GX 138
NGC 1424 GX 118
NGC 1425 GX 156, 174
NGC 1426 GX 156
NGC 1427 GX 174
NGC 1427A GX 174
NGC 1428 GX 174
NGC 1431 GX 118
NGC 1432 BN 78, A12
NGC 1433 GX 190
NGC 1434 GX 138
NGC 1435 BN 78, A12
NGC 1437 GX 174
NGC 1437A GX 174
NGC 1438 GX 156
NGC 1439 GX 156
NGC 1440 GX 156
NGC 1441 GX 118
NGC 1444 OC 28
NGC 1445 GX 138
NGC 1447 GX 138
NGC 1448 GX 190
NGC 1449 GX 118
NGC 1450 GX 138
NGC 1451 GX 118
NGC 1452 GX 156
NGC 1453 GX 118
NGC 1455 GX 156
NGC 1459 GX 156
NGC 1460 GX 174
NGC 1461 GX 138
NGC 1462 GX 98
NGC 1463 GX 202
NGC 1464 GX 138
NGC 1465 GX 60
NGC 1466 GC 212
NGC 1467 GX 138
NGC 1468 GX 138
NGC 1469 GX 16
NGC 1470 GX 138
NGC 1472 GX 138
NGC 1473 GX 212
NGC 1476 GX 190
NGC 1477 GX 138
NGC 1478 GX 138
NGC 1481 GX 156
NGC 1482 GX 156
NGC 1483 GX 190
NGC 1484 GX 174
NGC 1485 GX 16
NGC 1486 GX 156
NGC 1487 GX 190
NGC 1489 GX 156
NGC 1490 GX 212
NGC 1491 BN 28, 42, 43
NGC 1492 GX 174
NGC 1493 GX 189, 190
NGC 1494 GX 189, 190
NGC 1495 GX 189, 190
NGC 1496 OC 28
NGC 1497 GX 78
NGC 1499 BN 60
NGC 1500 GX 202
NGC 1501 PN 28
NGC 1502 OC 16, 28
NGC 1503 GX 212
NGC 1504 GX 138
NGC 1505 GX 138
NGC 1506 GX 202
NGC 1507 GX 118
NGC 1508 GX 78
NGC 1509 GX 138
NGC 1510 GX 189, 190
NGC 1511 GX 212
NGC 1511A GX 212
NGC 1511B GX 212
NGC 1512 GX 189, 190
NGC 1513 OC 42
NGC 1514 PN 60
NGC 1515 GX 202
NGC 1515A GX 202
NGC 1516A GX 137, 138
NGC 1516B GX 137, 138
NGC 1517 GX 97, 98
NGC 1518 GX 156
NGC 1519 GX 137, 138, 156
NGC 1520 OC 219
NGC 1521 GX 156
NGC 1522 GX 202
NGC 1526 GX 212
NGC 1527 GX 189
NGC 1528 OC 28, 42
NGC 1529 GX 202, 212
NGC 1530 GX 7
NGC 1531 GX 174
NGC 1532 GX 174
NGC 1533 GX 202
NGC 1534 GX 202, 212
NGC 1535 PN 137, 138
NGC 1536 GX 202
NGC 1537 GX 174
NGC 1538 GX 137, 138
NGC 1539 GX 78
NGC 1540 GX 156
NGC 1540A GX 156
NGC 1541 GX 117
NGC 1542 GX 117
NGC 1543 GX 202
NGC 1544 GX 1
NGC 1545 OC 42
NGC 1546 GX 202
NGC 1547 GX 137, 156
NGC 1548 OC 60
NGC 1549 GX 202
NGC 1550 GX 117
NGC 1552 GX 117
NGC 1553 GX 202

NGC 1554/55 BN 78
NGC 1556 GX 189
NGC 1557 OC 212
NGC 1558 GX 189
NGC 1559 GX 202, 212
NGC 1560 GX 16
NGC 1561 GX 137
NGC 1562 GX 137
NGC 1563 GX 137
NGC 1564 GX 137
NGC 1565 GX 137
NGC 1566 GX 202
NGC 1567 GX 189
NGC 1568 GX 117
NGC 1569 GX 16
NGC 1571 GX 189
NGC 1572 GX 174, 189
NGC 1573 GX 7, 16
NGC 1574 GX 202
NGC 1576 GX 117
NGC 1577 GX 137
NGC 1578 GX 189, 202
NGC 1579 BN 60
NGC 1580 GX 117, 137
NGC 1581 GX 202
NGC 1582 OC 42
NGC 1583 GX 137, 156
NGC 1584 GX 137, 156
NGC 1585 GX 189
NGC 1586 GX 117
NGC 1587 GX 117
NGC 1588 GX 117
NGC 1589 GX 117
NGC 1590 GX 97
NGC 1591 GX 156
NGC 1592 GX 156
NGC 1594 GX 117, 137
NGC 1595 GX 189
NGC 1596 GX 202
NGC 1597 GX 137
NGC 1598 GX 189
NGC 1599 GX 117
NGC 1600 GX 117, 137
NGC 1601 GX 117, 137
NGC 1602 GX 202
NGC 1603 GX 117, 137
NGC 1604 GX 117, 137
NGC 1605 OC 42
NGC 1606 GX 117, 137
NGC 1607 GX 117
NGC 1608 GX 117
NGC 1609 GX 117
NGC 1610 GX 117
NGC 1611 GX 117
NGC 1612 GX 117
NGC 1613 GX 117
NGC 1614 GX 137
NGC 1615 GX 77, 78
NGC 1616 GX 189
NGC 1617 GX 202
NGC 1618 GX 117
NGC 1620 GX 117
NGC 1621 GX 117
NGC 1622 GX 117
NGC 1623 GX 137
NGC 1624 BN OC 42
NGC 1625 GX 117
NGC 1627 GX 117
NGC 1628 GX 117
NGC 1629 OC 212
NGC 1630 GX 155, 156
NGC 1631 GX 155, 156
NGC 1632 GX 137
NGC 1633 GX 97
NGC 1634 GX 97
NGC 1635 GX 117
NGC 1636 GX 137
NGC 1637 GX 117
NGC 1638 GX 117
NGC 1640 GX 155, 156
NGC 1641 OC 212
NGC 1642 GX 117
NGC 1643 GX 117, 137
NGC 1644 OC 212, A24
NGC 1645 GX 117, 137
NGC 1646 GX 137
NGC 1647 OC 77
NGC 1648 GX 137
NGC 1649 OC 212, A24
NGC 1650 GX 137
NGC 1651 OC 212, A24
NGC 1652 OC 212, A24
NGC 1653 GX 117
NGC 1654 GX 117
NGC 1656 GX 117, 137
NGC 1657 GX 117
NGC 1658 GX 189
NGC 1659 GX 117
NGC 1660 GX 189
NGC 1661 GX 117
NGC 1662 OC 97
NGC 1663 OC 97
NGC 1664 OC 42
NGC 1665 GX 117, 137
NGC 1666 GX 137
NGC 1667 GX 137
NGC 1668 GX 189
NGC 1669 GX 212
NGC 1670 GX 117
NGC 1672 GX 202
NGC 1673 OC 212, A24
NGC 1676 OC 212, A24
NGC 1677 GX 117
NGC 1678 GX 117
NGC 1679 GX 173
NGC 1680 GX 189
NGC 1681 GX 117, 137
NGC 1682 GX 117
NGC 1683 GX 117
NGC 1684 GX 117
NGC 1685 GX 117
NGC 1686 GX 137
NGC 1687 GX 173
NGC 1688 GX 202
NGC 1690 GX 117
NGC 1691 GX 117
NGC 1692 GX 155
NGC 1693 OC 212, A24
NGC 1694 GX 117
NGC 1695 OC 212, A24
NGC 1696 OC 212, A24
NGC 1697 OC 212, A24
NGC 1698 BN OC 212, A24
NGC 1699 GX 117
NGC 1700 GX 117
NGC 1701 GX 155, 173
NGC 1702 OC 212, A24
NGC 1703 GX 202
NGC 1704 OC 212, A24
NGC 1705 GX 202
NGC 1706 GX 202, 212
NGC 1707 OC 97
NGC 1708 OC 27, 28
NGC 1709 GX 117
NGC 1710 GX 137
NGC 1711 OC 212, A24
NGC 1712 BN OC 212, A24
NGC 1713 GX 117
NGC 1714 BN OC 212, A24
NGC 1715 BN 212, A24
NGC 1716 GX 155
NGC 1718 OC 212, A24
NGC 1719 GX 117
NGC 1720 GX 137
NGC 1721 GX 137
NGC 1722 BN OC 212, A24
NGC 1723 GX 137
NGC 1724 OC 42
NGC 1725 GX 137
NGC 1726 GX 137
NGC 1727 BN OC 212, A24
NGC 1728 GX 137
NGC 1729 GX 117
NGC 1730 GX 137
NGC 1731 BN OC 212, A24
NGC 1732 OC 212, A24
NGC 1733 OC 212, A24
NGC 1734 OC 212, A24
NGC 1735 BN OC 212, A24
NGC 1736 BN OC 212, A24
NGC 1737 BN 212, A24
NGC 1738 GX 155
NGC 1739 GX 155
NGC 1740 GX 117
NGC 1741 GX 117
NGC 1743 BN OC 212, A24
NGC 1744 GX 155
NGC 1745 BN 212, A24
NGC 1746 OC 77
NGC 1747 BN 212, A24
NGC 1748 BN 212, A24
NGC 1749 OC 212, A24
NGC 1751 OC 212, A24
NGC 1752 GX 137
NGC 1753 GX 117
NGC 1754 GC 212, A24
NGC 1755 OC 212, A24
NGC 1756 OC 212, A24
NGC 1758 OC 77
NGC 1759 GX 173
NGC 1760 BN 212, A24
NGC 1761 BN OC 212, A24
NGC 1762 GX 117
NGC 1763 BN 212, A24
NGC 1764 OC 212, A24
NGC 1765 GX 201, 202, 212
NGC 1766 OC 212, A24
NGC 1767 BN OC 212, A24
NGC 1768 OC 212, A24
NGC 1769 BN 212, A24
NGC 1770 BN OC 212, A24
NGC 1771 GX 212
NGC 1772 BN OC 212, A24
NGC 1773 BN 212, A24
NGC 1774 OC 212, A24
NGC 1775 OC 212, A24
NGC 1776 OC 212, A24
NGC 1777 OC 219
NGC 1778 OC 59
NGC 1779 GX 137
NGC 1780 GX 155
NGC 1782 BN OC 212, A24
NGC 1783 OC 212
NGC 1784 GX 137
NGC 1785 OC 212, A24
NGC 1786 GC 212, A24
NGC 1787 OC 212
NGC 1788 BN 117
NGC 1789 OC 212, A24
NGC 1790 OC 27
NGC 1791 BN OC 212, A24
NGC 1792 GX 173
NGC 1793 OC 212, A24
NGC 1794 GX 155
NGC 1795 OC 212, A24
NGC 1796 GX 201, 202
NGC 1796A GX 201
NGC 1796B GX 201
NGC 1797 GX 137
NGC 1798 OC 42
NGC 1799 GX 137
NGC 1800 GX 173
NGC 1801 OC 212, A24
NGC 1802 OC 77
NGC 1803 GX 189
NGC 1804 OC 212, A24
NGC 1805 OC 212, A24
NGC 1806 OC 212, A24
NGC 1807 OC 97
NGC 1808 GX 173
NGC 1809 GX 212, A24
NGC 1810 OC 212, A24
NGC 1811 GX 155, 173
NGC 1812 GX 155, 173
NGC 1814 OC 212, A24
NGC 1815 OC 212, A24
NGC 1816 OC 212, A24
NGC 1817 OC 97
NGC 1818 OC 212, A24
NGC 1819 GX 97, 117
NGC 1820 OC 212, A24
NGC 1821 GX 137
NGC 1822 OC 212, A24
NGC 1823 OC 212, A24
NGC 1824 GX 201
NGC 1825 OC 212, A24
NGC 1826 OC 212, A24
NGC 1827 GX 173
NGC 1828 OC 212, A24
NGC 1829 BN OC 212, A24
NGC 1830 OC 212, A24
NGC 1831 OC 212
NGC 1832 GX 137
NGC 1833 BN OC 212, A24
NGC 1836 OC 212, A24
NGC 1837 BN OC 212, A24
NGC 1838 OC 212, A24
NGC 1839 BN 212, A24
NGC 1840 OC 212, A24
NGC 1841 GC 219, 220
NGC 1842 OC 212, A24
NGC 1843 GX 137
NGC 1844 OC 212, A24
NGC 1845 OC 212, A24
NGC 1846 OC 212, A24
NGC 1847 OC 212, A24
NGC 1848 BN 212, A24
NGC 1849 OC 212, A24
NGC 1850 OC 212, A24
NGC 1851 GC 173, 189
NGC 1852 BN OC 212, A24
NGC 1853 GX 201
NGC 1855 OC 212, A24
NGC 1856 OC 212, A24
NGC 1857 OC 59
NGC 1858 BN 212, A24
NGC 1859 OC 212
NGC 1860 OC 212, A24
NGC 1861 OC 212, A24
NGC 1862 OC 212, A24
NGC 1863 OC 212, A24
NGC 1864 OC 212, A24
NGC 1865 OC 212, A24
NGC 1866 OC 212
NGC 1867 OC 212, A24
NGC 1868 OC 212
NGC 1869 BN OC 212, A24
NGC 1870 OC 212, A24
NGC 1871 BN OC 212, A24
NGC 1872 OC 212, A24
NGC 1873 BN OC 212, A24
NGC 1874/76/77 BN 212, A24
NGC 1875 GX 96, 97
NGC 1878 OC 212, A24
NGC 1879 GX 173
NGC 1880 BN 212, A24
NGC 1881 BN 212, A24
NGC 1882 OC 212, A24
NGC 1883 OC 42
NGC 1885 OC 212, A24
NGC 1886 GX 155
NGC 1887 OC 212, A24
NGC 1888 GX 136, 137
NGC 1889 GX 136, 137
NGC 1890 OC 212
NGC 1891 OC 173
NGC 1892 GX 212
NGC 1893 OC 59
NGC 1894 OC 212, A24
NGC 1895 BN 212, A24
NGC 1896 OC 59, 77
NGC 1897 OC 212, A24
NGC 1898 GC 212, A24
NGC 1899 BN 212, A24
NGC 1900 OC 212
NGC 1901 OC 212, A24
NGC 1902 OC 212, A24
NGC 1903 OC 212, A24
NGC 1904 GC 155
NGC 1905 OC 212, A24
NGC 1906 GX 136, 137
NGC 1907 OC 59
NGC 1910 OC 212, A24
NGC 1911 OC 212, A24
NGC 1912 OC 59
NGC 1913 OC 212, A24
NGC 1914 BN OC 212, A24
NGC 1915 OC 212, A24
NGC 1916 GC 212, A24
NGC 1918 BN 212, A24
NGC 1919 BN OC 212, A24
NGC 1920 BN OC 212, A24
NGC 1921 BN 212, A24
NGC 1922 OC 212, A24
NGC 1923 BN OC 212
NGC 1924 GX 116, 136
NGC 1925 BN OC 212
NGC 1926 OC 212, A24, A25
NGC 1928 OC 212, A24, A25
NGC 1929 BN 212, A24, A25
NGC 1930 GX 189
NGC 1931 BN OC 59
NGC 1932 OC 212, A24
NGC 1934 BN 212, A24, A25
NGC 1935 BN 212, A24, A25
NGC 1936 BN 212, A24, A25
NGC 1938 OC 212, A24, A25
NGC 1939 OC 212, A24, A25
NGC 1940 OC 212, A24, A25
NGC 1941 BN OC 212, A24
NGC 1942 OC 212
NGC 1943 BN 212, A24, A25
NGC 1944 OC 212
NGC 1945 BN 212, A24
NGC 1946 OC 212, A24
NGC 1947 GX 212
NGC 1948 BN 212, A24
NGC 1949 BN OC 212, A24, A25
NGC 1950 OC 212, A24, A25
NGC 1951 OC 212, A24
NGC 1952 BN 77
NGC 1954 GX 136
NGC 1955 OC 212, A24, A25
NGC 1956 GX 219
NGC 1957 GX 136
NGC 1958 OC 212, A24, A25
NGC 1959 OC 212, A24, A25
NGC 1960 OC 59
NGC 1961 GX 16
NGC 1962 BN OC 212, A24, A25
NGC 1963 OC 173
NGC 1964 GX 155
NGC 1965 BN 212, A24, A25
NGC 1966 BN 212, A24, A25
NGC 1967 OC 212, A24, A25
NGC 1968 BN OC 212, A24, A25
NGC 1969 OC 212, A24, A25
NGC 1970 BN 212, A24, A25
NGC 1971 OC 212, A24, A25
NGC 1972 OC 212, A24, A25
NGC 1973 BN 116
NGC 1974 BN OC 212, A24, A25
NGC 1975 BN 116
NGC 1976 BN OC 116, 136
NGC 1977 BN OC 116
NGC 1978 OC 212, A24
NGC 1979 GX 155
NGC 1980 BN OC 116, 136
NGC 1981 OC 116
NGC 1982 BN 116, 136
NGC 1983 BN 212, A24, A25
NGC 1984 BN 212, A24, A25
NGC 1985 BN 59
NGC 1986 OC 212, A24, A25
NGC 1987 OC 212, A24, A25
NGC 1989 GX 173
NGC 1992 GX 173
NGC 1993 GX 136, 155
NGC 1994 OC 212, A24, A25
NGC 1996 OC 77
NGC 1997 OC 212
NGC 1998 GX 188, 189
NGC 1999 BN 136
NGC 2000 OC 212, A24
NGC 2001 OC 212, A24, A25
NGC 2002 OC 212, A24
NGC 2003 OC 212, A24
NGC 2004 OC 212, A24, A25
NGC 2005 GC 212, A24, A25
NGC 2006 OC 212, A24
NGC 2007 GX 188, 189
NGC 2008 GX 188, 189
NGC 2009 OC 212, A24, A25
NGC 2010 OC 212, A24, A25
NGC 2011 BN OC 212, A24, A25
NGC 2012 GX 219
NGC 2013 OC 27
NGC 2014 BN OC 212, A24, A25
NGC 2015 BN OC 212, A24, A25
NGC 2016 OC 212, A24, A25
NGC 2017 OC 136, 155
NGC 2018 BN OC 212, A24
NGC 2019 GC 212, A24, A25
NGC 2020 BN 212, A24, A25
NGC 2021 BN OC 212, A24, A25
NGC 2022 PN 96
NGC 2023 BN 116
NGC 2024 BN 116
NGC 2025 OC 212, A24
NGC 2026 OC 77
NGC 2027 OC 212, A24
NGC 2028 OC 212, A24, A25
NGC 2029 BN 212, A24, A25
NGC 2030 BN OC 212, A24
NGC 2031 OC 212, A24, A25
NGC 2032 BN 212, A24, A25
NGC 2033 OC 212, A24, A25
NGC 2034 BN 212, A24
NGC 2035 BN 212, A24, A25
NGC 2036 OC 212, A24, A25
NGC 2037 BN OC 212, A24, A25
NGC 2038 OC 212, A24, A25
NGC 2040 BN OC 212, A24, A25
NGC 2041 OC 212, A24
NGC 2042 OC 212, A24, A25
NGC 2044 OC 212, A24, A25
NGC 2046 OC 212, A24, A25
NGC 2047 OC 212, A24, A25
NGC 2048 BN 212, A24, A25
NGC 2049 GX 173
NGC 2050 BN OC 212, A24, A25
NGC 2051 OC 212, A24
NGC 2052 BN 212, A24, A25
NGC 2053 OC 212, A24, A25
NGC 2055 BN OC 212, A24, A25
NGC 2056 OC 212, A24, A25
NGC 2057 OC 212, A24, A25
NGC 2058 OC 212, A24, A25
NGC 2059 OC 212, A24, A25
NGC 2060 OC 212, A24, A25
NGC 2061 OC 173
NGC 2062 OC 212, A24
NGC 2064 BN 116
NGC 2065 OC 212, A24, A25
NGC 2066 OC 212, A24, A25
NGC 2067 BN 116
NGC 2068 BN 116
NGC 2069 BN 212, A24, A25
NGC 2070 BN OC 212, A24, A25, xix, xx, xxi, xxii
NGC 2071 BN 116
NGC 2072 OC 212, A24, A25
NGC 2073 GX 155
NGC 2074 BN OC 212, A24, A25
NGC 2075 BN OC 212, A24, A25
NGC 2076 GX 136
NGC 2077 BN OC 212, A24, A25
NGC 2078 BN 212, A24, A25
NGC 2079 BN 212, A24, A25
NGC 2080 BN OC 212, A24, A25
NGC 2081 BN 212, A24, A25
NGC 2082 GX 212
NGC 2083 BN 212, A24, A25
NGC 2084 BN 212, A24, A25
NGC 2085 BN OC 212, A24, A25
NGC 2086 BN 212, A24, A25
NGC 2087 GX 201
NGC 2088 OC 212, A24, A25
NGC 2089 GX 136, 155
NGC 2090 GX 173
NGC 2091 OC 212, A24, A25
NGC 2092 OC 212, A24, A25
NGC 2093 BN OC 212, A24, A25
NGC 2094 OC 212, A24, A25
NGC 2095 OC 212, A24, A25
NGC 2096 OC 212, A24, A25
NGC 2097 OC 201, 212
NGC 2098 BN OC 212, A24, A25
NGC 2099 OC 59
NGC 2100 OC 212, A24, A25
NGC 2101 GX 201
NGC 2102 OC 212, A24, A25
NGC 2103 BN 212, A24
NGC 2104 GX 188, 201
NGC 2105 OC 212, A24
NGC 2106 GX 155
NGC 2107 OC 212, A24, A25
NGC 2108 OC 212, A24, A25
NGC 2109 BN OC 212, A24, A25
NGC 2110 GX 136
NGC 2111 OC 212, A24, A25
NGC 2112 OC 116
NGC 2113 BN OC 212, A24, A25
NGC 2114 OC 212, A24, A25
NGC 2115 GX 188
NGC 2115A GX 188
NGC 2116 OC 212, A24, A25
NGC 2117 OC 212, A24, A25
NGC 2118 OC 212, A24, A25
NGC 2119 GX 96
NGC 2120 OC 212
NGC 2121 OC 212, A24
NGC 2122 BN OC 212, A24, A25
NGC 2123 OC 212
NGC 2124 GX 154, 155
NGC 2125 OC 212, A24, A25
NGC 2126 OC 41
NGC 2127 OC 212, A24, A25
NGC 2128 GX 27
NGC 2129 OC 76, 77
NGC 2130 OC 211, 212, A24
NGC 2131 GX 154, 155
NGC 2132 OC 201
NGC 2133 OC 212, A24
NGC 2134 OC 212, A24
NGC 2135 OC 211, 212, A24
NGC 2136 OC 211, 212, A24
NGC 2137 OC 211, 212, A24
NGC 2138 OC 211, 212
NGC 2139 GX 154, 155
NGC 2140 OC 211, 212, A24
NGC 2141 OC 96
NGC 2143 OC 96, 116
NGC 2144 GX 218, 219
NGC 2145 OC 211, 212, A24
NGC 2146 GX 6, 7
NGC 2146A GX 6
NGC 2147 BN OC 211, 212, A24
NGC 2148 GX 201
NGC 2149 BN 136
NGC 2150 GX 211, 212, A24
NGC 2151 OC 211, 212, A24
NGC 2152 GX 188
NGC 2153 OC 211, 212, A24

NGC 2154 OC 211, 212, A24
NGC 2155 OC 211, 212
NGC 2156 OC 211, 212, A24
NGC 2157 OC 211, 212, A24
NGC 2158 OC 76
NGC 2159 OC 211, 212, A24
NGC 2160 OC 211, 212, A24
NGC 2161 OC 218, 219
NGC 2162 OC 211, 212
NGC 2163 BN 76
NGC 2164 OC 211, 212, A24
NGC 2165 OC 27, 41
NGC 2166 OC 211, 212, A24
NGC 2168 OC 76
NGC 2169 OC 96
NGC 2170 BN 136
NGC 2171 OC 211, 212, A24
NGC 2172 OC 211, 212, A24
NGC 2173 OC 211, 212
NGC 2174 BN 76
NGC 2175.1 OC 76
NGC 2176 OC 211, 212, A24
NGC 2177 OC 211, 212, A24
NGC 2178 GX 211, 212
NGC 2179 GX 154
NGC 2180 OC 116
NGC 2181 OC 211, 212
NGC 2182 BN 136
NGC 2183 BN 136
NGC 2184 OC 116
NGC 2185 BN 136
NGC 2186 OC 96, 116
NGC 2187 GX 211, 212, A24
NGC 2187A GX 211, 212, A24
NGC 2188 GX 172
NGC 2190 OC 218, 219
NGC 2191 GX 201
NGC 2192 OC 58
NGC 2193 OC 211, 212
NGC 2194 OC 96
NGC 2196 GX 154
NGC 2197 OC 211, 212, A24
NGC 2199 GX 211, 212, 218, 219
NGC 2200 GX 188
NGC 2201 GX 188
NGC 2202 OC 96, 116
NGC 2203 OC 218, 219
NGC 2204 OC 154
NGC 2205 GX 201, 211
NGC 2206 GX 154
NGC 2207 GX 154
NGC 2208 GX 27, 41
NGC 2209 OC 211, 218, 219
NGC 2210 GC 211, A24
NGC 2211 GX 154
NGC 2212 GX 154
NGC 2213 OC 211, A24
NGC 2214 OC 211
NGC 2215 OC 136
NGC 2216 GX 154
NGC 2217 GX 154
NGC 2218 OC 76
NGC 2219 OC 116
NGC 2220 OC 188
NGC 2221 GX 201
NGC 2222 GX 201
NGC 2223 GX 154
NGC 2224 OC 96
NGC 2225 OC 136
NGC 2226 OC 136
NGC 2227 GX 154
NGC 2228 GX 211
NGC 2229 GX 211
NGC 2230 GX 211
NGC 2231 OC 211
NGC 2232 OC 116
NGC 2233 GX 211
NGC 2234 OC 96
NGC 2235 GX 211
NGC 2236 OC 96
NGC 2237-9,46 BN 95, 96, 115, 116
NGC 2241 OC 211
NGC 2242 PN 41
NGC 2243 OC 172
NGC 2244 OC 115, 116, xiii
NGC 2245 BN 95, 96
NGC 2247 BN 95, 96
NGC 2249 OC 211
NGC 2250 OC 115, 116, 135, 136
NGC 2251 OC 95, 96
NGC 2252 OC 95, 96, 115, 116
NGC 2253 GX 15
NGC 2254 OC 95, 96
NGC 2255 GX 172
NGC 2256 GX 6
NGC 2257 GC 211
NGC 2258 GX 6
NGC 2259 OC 95, 96
NGC 2260 OC 115, 116
NGC 2261 BN 95, 96
NGC 2262 OC 115, 116
NGC 2263 GX 154
NGC 2264 BN OC 95
NGC 2266 OC 76
NGC 2267 GX 172
NGC 2268 GX 1, 6
NGC 2269 OC 115
NGC 2271 GX 154
NGC 2272 GX 154
NGC 2273 GX 27
NGC 2273B GX 27
NGC 2274 GX 58
NGC 2275 GX 58
NGC 2276 GX 1
NGC 2280 GX 154
NGC 2281 OC 41
NGC 2282 BN 115
NGC 2283 GX 154
NGC 2286 OC 115
NGC 2287 OC 154
NGC 2288 GX 58
NGC 2289 GX 58
NGC 2290 GX 58
NGC 2291 GX 58
NGC 2292 GX 154
NGC 2293 GX 154
NGC 2294 GX 58
NGC 2295 GX 154
NGC 2296 BN 135
NGC 2297 GX 211
NGC 2298 GC 172
NGC 2300 GX 1
NGC 2301 OC 115
NGC 2302 OC 135
NGC 2303 GX 41
NGC 2304 OC 76, 95
NGC 2305 GX 211
NGC 2306 OC 135
NGC 2307 GX 211
NGC 2308 GX 41
NGC 2309 OC 135
NGC 2310 GX 172, 188
NGC 2311 OC 115
NGC 2314 GX 6
NGC 2315 GX 41
NGC 2316 BN 135
NGC 2318 OC 135
NGC 2319 OC 115
NGC 2320 GX 41
NGC 2321 GX 41
NGC 2322 GX 41
NGC 2323 OC 135
NGC 2324 OC 115
NGC 2325 GX 154
NGC 2326 GX 40, 41
NGC 2326A GX 40, 41
NGC 2327 BN 135
NGC 2328 GX 188
NGC 2329 GX 40, 41
NGC 2330 GX 40, 41
NGC 2331 OC 76
NGC 2332 GX 40, 41
NGC 2333 GX 58
NGC 2335 OC 135
NGC 2336 GX 6
NGC 2337 GX 40, 41
NGC 2338 OC 115, 135
NGC 2339 GX 76
NGC 2340 GX 40, 41
NGC 2341 GX 76
NGC 2342 GX 76
NGC 2343 OC 135
NGC 2344 GX 40, 41
NGC 2345 OC 135
NGC 2346 PN 115
NGC 2347 GX 15
NGC 2348 OC 211
NGC 2350 GX 95
NGC 2353 OC 135
NGC 2354 OC 154
NGC 2355 OC 95
NGC 2357 GX 75, 76
NGC 2359 BN 135
NGC 2360 OC 135
NGC 2362 OC 153, 154
NGC 2364 OC 135
NGC 2365 GX 75, 76
NGC 2366 GX 15
NGC 2367 OC 153, 154
NGC 2368 OC 135
NGC 2369 GX 200, 211
NGC 2369A GX 200, 211
NGC 2369B GX 200, 211
NGC 2370 GX 75
NGC 2371-72 PN 57, 75
NGC 2373 GX 57
NGC 2374 OC 135
NGC 2375 GX 57
NGC 2376 GX 75
NGC 2377 GX 135
NGC 2379 GX 57
NGC 2380 GX 153, 154
NGC 2381 GX 211
NGC 2383 OC 153
NGC 2384 OC 153
NGC 2385 GX 57
NGC 2387 GX 57
NGC 2388 GX 57
NGC 2389 GX 57
NGC 2392 PN 75
NGC 2393 GX 57
NGC 2395 OC 95
NGC 2396 OC 135
NGC 2397 GX 211
NGC 2397A GX 211
NGC 2397B GX 211
NGC 2398 GX 75
NGC 2401 OC 135
NGC 2402 GX 95
NGC 2403 GX 15
NGC 2405 GX 75
NGC 2406 GX 75
NGC 2407 GX 75
NGC 2410 GX 57
NGC 2411 GX 75
NGC 2414 OC 135
NGC 2415 GX 57
NGC 2416 GX 95
NGC 2417 GX 200, 211
NGC 2418 GX 75, 95
NGC 2419 GC 57
NGC 2420 OC 75
NGC 2421 OC 153
NGC 2422 OC 135
NGC 2423 OC 135
NGC 2424 GX 57
NGC 2425 OC 135
NGC 2426 GX 26
NGC 2427 GX 187
NGC 2428 OC 135
NGC 2429A GX 26
NGC 2429B GX 26
NGC 2430 OC 135
NGC 2431 GX 26
NGC 2432 OC 153
NGC 2433 GX 95
NGC 2434 GX 211
NGC 2435 GX 57
NGC 2437 OC 135
NGC 2438 PN 135
NGC 2439 OC 171
NGC 2440 PN 153
NGC 2441 GX 6, 15
NGC 2442 GX 211
NGC 2444 GX 57
NGC 2445 GX 57
NGC 2446 GX 26
NGC 2447 OC 153
NGC 2449 GX 75
NGC 2450 GX 75
NGC 2451 OC 171
NGC 2452 PN 153
NGC 2453 OC 153
NGC 2454 GX 94, 95
NGC 2455 OC 153
NGC 2456 GX 26
NGC 2457 GX 26
NGC 2458 GX 26
NGC 2460 GX 26
NGC 2462 GX 26
NGC 2463 GX 26
NGC 2466 GX 211
NGC 2467 OC 153
NGC 2468 GX 26
NGC 2469 GX 26
NGC 2470 GX 114
NGC 2472 GX 26
NGC 2474 GX 26
NGC 2475 GX 26
NGC 2476 GX 57
NGC 2477 OC 171
NGC 2479 OC 134, 153
NGC 2480 GX 75
NGC 2481 GX 75
NGC 2482 OC 153
NGC 2483 OC 153
NGC 2484 GX 57
NGC 2485 GX 94
NGC 2486 GX 75
NGC 2487 GX 75
NGC 2488 GX 26
NGC 2489 OC 171
NGC 2490 GX 75
NGC 2491 GX 94
NGC 2492 GX 75
NGC 2493 GX 57
NGC 2494 GX 114
NGC 2495 GX 57
NGC 2496 GX 94
NGC 2497 GX 26
NGC 2498 GX 75
NGC 2499 GX 94
NGC 2500 GX 40
NGC 2501 GX 134
NGC 2502 GX 200
NGC 2503 GX 75
NGC 2504 GX 94, 114
NGC 2505 GX 26
NGC 2506 OC 134
NGC 2507 GX 94
NGC 2508 GX 94
NGC 2509 OC 153
NGC 2510 GX 94
NGC 2511 GX 94
NGC 2512 GX 75
NGC 2513 GX 94
NGC 2514 GX 94
NGC 2516 OC 200, xix, xx, xxii
NGC 2517 GX 134
NGC 2518 GX 26, 40
NGC 2521 GX 26
NGC 2522 GX 75, 94
NGC 2523 GX 6, 15
NGC 2523A GX 6
NGC 2523B GX 6, 15
NGC 2523C GX 6, 14, 15
NGC 2524 GX 57
NGC 2525 GX 134
NGC 2526 GX 94
NGC 2527 OC 153
NGC 2528 GX 57
NGC 2530 GX 75, 94
NGC 2532 GX 57
NGC 2533 OC 153, 171
NGC 2534 GX 26
NGC 2535 GX 75
NGC 2536 GX 75
NGC 2537 GX 40
NGC 2537A GX 40
NGC 2538 GX 114
NGC 2539 OC 134
NGC 2540 GX 75
NGC 2541 GX 40
NGC 2543 GX 57
NGC 2544 GX 6, 14, 15
NGC 2545 GX 75
NGC 2546 OC 171
NGC 2547 OC 187
NGC 2548 OC 114, 134
NGC 2549 GX 26
NGC 2550 GX 6
NGC 2550A GX 6, 14, 15
NGC 2551 GX 6, 14, 15
NGC 2552 GX 40
NGC 2553 GX 75
NGC 2554 GX 75
NGC 2555 GX 114
NGC 2556 GX 75
NGC 2557 GX 75
NGC 2558 GX 75
NGC 2559 GX 153
NGC 2560 GX 75
NGC 2561 GX 114
NGC 2562 GX 75
NGC 2563 GX 75
NGC 2564 GX 153
NGC 2565 GX 75
NGC 2566 GX 153
NGC 2567 OC 171
NGC 2568 OC 171
NGC 2569 GX 75
NGC 2570 GX 75
NGC 2571 OC 153, 171
NGC 2572 GX 75
NGC 2573 GX 220
NGC 2573A GX 220
NGC 2573B GX 220
NGC 2574 GX 134
NGC 2575 GX 75
NGC 2576 GX 75
NGC 2577 GX 75
NGC 2578 GX 134
NGC 2579 OC 171
NGC 2580 OC 171
NGC 2581 GX 75
NGC 2582 GX 75
NGC 2583 GX 114, 134
NGC 2584 GX 114
NGC 2585 GX 114
NGC 2587 OC 153, 171
NGC 2588 OC 171
NGC 2590 GX 114
NGC 2591 GX 6
NGC 2592 GX 75
NGC 2593 GX 75, 94
NGC 2594 GX 75
NGC 2595 GX 75
NGC 2596 GX 75, 94
NGC 2598 GX 75
NGC 2599 GX 75
NGC 2600 GX 26
NGC 2601 GX 210, 211
NGC 2602 GX 26
NGC 2603 GX 26
NGC 2604 GX 57, 75
NGC 2607 GX 75
NGC 2608 GX 75
NGC 2609 OC 200
NGC 2610 PN 134
NGC 2611 GX 75
NGC 2612 GX 134
NGC 2613 GX 153
NGC 2614 GX 14
NGC 2615 GX 114
NGC 2616 GX 114
NGC 2617 GX 114
NGC 2618 GX 114
NGC 2619 GX 74, 75
NGC 2620 GX 74, 75
NGC 2621 GX 74, 75
NGC 2622 GX 74, 75
NGC 2623 GX 74, 75
NGC 2624 GX 74, 75
NGC 2625 GX 74, 75
NGC 2626 BN 171, 187
NGC 2627 OC 152, 153, 170, 171
NGC 2628 GX 74, 75
NGC 2629 GX 14
NGC 2630 GX 6, 14
NGC 2632 OC 74, 75
NGC 2633 GX 6
NGC 2634 GX 6, 14
NGC 2634A GX 6, 14
NGC 2635 OC 170, 171
NGC 2636 GX 6, 14
NGC 2637 GX 74, 75
NGC 2638 GX 56, 57
NGC 2639 GX 40
NGC 2640 GX 200
NGC 2641 GX 14
NGC 2642 GX 114
NGC 2643 GX 74, 75
NGC 2644 GX 114
NGC 2645 OC 187
NGC 2646 GX 6, 14
NGC 2647 GX 74, 75
NGC 2648 GX 94
NGC 2649 GX 56
NGC 2650 GX 14
NGC 2651 GX 94
NGC 2654 GX 26
NGC 2655 GX 6
NGC 2656 GX 26
NGC 2657 GX 94
NGC 2658 OC 170, 171
NGC 2659 OC 187
NGC 2660 OC 187
NGC 2661 GX 94
NGC 2662 GX 134
NGC 2663 GX 170
NGC 2665 GX 152
NGC 2667 GX 74
NGC 2668 GX 56
NGC 2669 OC 200
NGC 2670 OC 186, 187
NGC 2671 OC 186, 187
NGC 2672 GX 74
NGC 2673 GX 74
NGC 2674 GX 134
NGC 2675 GX 26
NGC 2676 GX 39, 40
NGC 2677 GX 74
NGC 2679 GX 56
NGC 2681 GX 26, 39
NGC 2682 OC 94
NGC 2683 GX 56
NGC 2684 GX 39
NGC 2685 GX 26
NGC 2686A GX 39
NGC 2686B GX 39
NGC 2687A GX 39
NGC 2687B GX 39
NGC 2688 GX 39
NGC 2689 GX 39
NGC 2690 GX 114
NGC 2691 GX 56
NGC 2692 GX 25, 26
NGC 2693 GX 25, 26, 39
NGC 2694 GX 25, 26, 39
NGC 2695 GX 114
NGC 2697 GX 114
NGC 2698 GX 114
NGC 2699 GX 114
NGC 2701 GX 25, 26
NGC 2702 GX 114
NGC 2704 GX 56
NGC 2706 GX 113, 114
NGC 2708 GX 113, 114
NGC 2709 GX 113, 114
NGC 2710 GX 25, 26
NGC 2711 GX 74, 93, 94
NGC 2712 GX 39
NGC 2713 GX 113, 114
NGC 2714 GX 200
NGC 2715 GX 6
NGC 2716 GX 113, 114
NGC 2717 GX 152
NGC 2718 GX 93, 94
NGC 2719 GX 56
NGC 2719A GX 56
NGC 2720 GX 93, 94
NGC 2721 GX 113, 114
NGC 2722 GX 113, 114
NGC 2723 GX 113, 114
NGC 2724 GX 56
NGC 2725 GX 93, 94
NGC 2726 GX 25
NGC 2728 GX 93, 94
NGC 2729 GX 113, 114
NGC 2730 GX 93, 94
NGC 2731 GX 93, 94
NGC 2732 GX 6
NGC 2734 GX 93, 94
NGC 2735 GX 74
NGC 2735A GX 74
NGC 2736 BN 186
NGC 2737 GX 74
NGC 2738 GX 74
NGC 2739 GX 25, 39
NGC 2740 GX 25, 39
NGC 2742 GX 25
NGC 2742A GX 14, 25
NGC 2743 GX 74
NGC 2744 GX 74
NGC 2745 GX 74
NGC 2746 GX 56
NGC 2747 GX 74
NGC 2748 GX 6
NGC 2749 GX 74
NGC 2750 GX 74
NGC 2751 GX 74
NGC 2752 GX 74
NGC 2753 GX 74
NGC 2754 GX 152
NGC 2755 GX 39
NGC 2756 GX 25
NGC 2758 GX 152
NGC 2759 GX 56
NGC 2761 GX 74
NGC 2762 GX 39
NGC 2763 GX 133
NGC 2764 GX 74
NGC 2765 GX 113
NGC 2766 GX 56, 74
NGC 2767 GX 39
NGC 2768 GX 25
NGC 2769 GX 39
NGC 2770 GX 56
NGC 2771 GX 39
NGC 2772 GX 152

NGC 2773 GX 93
NGC 2774 GX 74
NGC 2775 GX 93
NGC 2776 GX 39
NGC 2777 GX 93
NGC 2778 GX 56, A5
NGC 2779 GX 56, A5
NGC 2780 GX 56, A5
NGC 2781 GX 133
NGC 2782 GX 39, 56
NGC 2783 GX 56, 74
NGC 2784 GX 152
NGC 2785 GX 39, 56
NGC 2787 GX 14
NGC 2788 GX 210
NGC 2788B GX 210
NGC 2789 GX 56, 74
NGC 2790 GX 74
NGC 2791 GX 74, 93
NGC 2792 PN 186
NGC 2793 GX 56, A5
NGC 2794 GX 74, 93
NGC 2795 GX 74, 93
NGC 2796 GX 56
NGC 2797 GX 74, 93
NGC 2798 GX 39
NGC 2799 GX 39
NGC 2800 GX 25
NGC 2801 GX 74
NGC 2802 GX 74
NGC 2803 GX 74
NGC 2804 GX 74
NGC 2805 GX 14
NGC 2806 GX 74
NGC 2807 GX 74
NGC 2808 GC 210
NGC 2809 GX 74
NGC 2810 GX 14
NGC 2811 GX 133
NGC 2812 GX 74
NGC 2813 GX 74
NGC 2814 GX 14
NGC 2815 GX 152
NGC 2817 GX 113
NGC 2818 OC PN 170
NGC 2819 GX 93
NGC 2820 GX 14
NGC 2821 GX 152
NGC 2822 GX 210
NGC 2823 GX 56, A5
NGC 2824 GX 74
NGC 2825 GX 56, A5
NGC 2826 GX 56, A5
NGC 2827 GX 56, A5
NGC 2828 GX 56, A5
NGC 2829 GX 56, A5
NGC 2830 GX 56, A5
NGC 2831 GX 56, A5
NGC 2832 GX 56, A5
NGC 2833 GX 56, A5
NGC 2834 GX 56, A5
NGC 2835 GX 152
NGC 2836 GX 210
NGC 2838 GX 56
NGC 2839 GX 56, A5
NGC 2840 GX 56, A5
NGC 2841 GX 39
NGC 2842 GX 210
NGC 2843 GX 74
NGC 2844 GX 39, 56
NGC 2845 GX 170
NGC 2848 GX 133
NGC 2849 OC 170, 186
NGC 2850 GX 113
NGC 2851 GX 133
NGC 2852 GX 39, 56
NGC 2853 GX 39, 56
NGC 2854 GX 39
NGC 2855 GX 133
NGC 2856 GX 39
NGC 2857 GX 39
NGC 2858 GX 113
NGC 2859 GX 56, A5
NGC 2860 GX 39
NGC 2861 GX 113
NGC 2862 GX 74
NGC 2863 GX 133
NGC 2864 GX 93, 113
NGC 2865 GX 152
NGC 2866 OC 186, 199
NGC 2867 PN 199
NGC 2868 GX 133
NGC 2870 GX 25
NGC 2872 GX 93
NGC 2873 GX 93
NGC 2874 GX 93
NGC 2876 GX 133
NGC 2877 GX 113
NGC 2878 GX 113
NGC 2880 GX 14, 25
NGC 2881 GX 133
NGC 2882 GX 93
NGC 2883 GX 170
NGC 2884 GX 133
NGC 2885 GX 74
NGC 2887 GX 210
NGC 2888 GX 152
NGC 2889 GX 133
NGC 2890 GX 133
NGC 2891 GX 152
NGC 2892 GX 14
NGC 2893 GX 56, 74
NGC 2894 GX 93
NGC 2895 GX 25
NGC 2896 GX 74
NGC 2897 GX 113
NGC 2898 GX 113
NGC 2899 PN 199
NGC 2900 GX 113
NGC 2902 GX 133
NGC 2903 GX 74
NGC 2904 GX 170
NGC 2906 GX 93
NGC 2907 GX 133
NGC 2908 GX 5, 6
NGC 2909 GX 14
NGC 2910 OC 199
NGC 2911 GX 93
NGC 2912 GX 93
NGC 2913 GX 93
NGC 2914 GX 93
NGC 2915 GX 218
NGC 2916 GX 74
NGC 2917 GX 113
NGC 2918 GX 56
NGC 2919 GX 93
NGC 2920 GX 152
NGC 2921 GX 152
NGC 2922 GX 56
NGC 2923 GX 93
NGC 2924 GX 133
NGC 2925 OC 199
NGC 2926 GX 56
NGC 2927 GX 74
NGC 2928 GX 93
NGC 2929 GX 74
NGC 2930 GX 74
NGC 2931 GX 74
NGC 2933 GX 74, 93
NGC 2935 GX 152
NGC 2936 GX 113
NGC 2937 GX 113
NGC 2938 GX 6
NGC 2939 GX 93
NGC 2940 GX 93
NGC 2941 GX 74, 93
NGC 2942 GX 56
NGC 2943 GX 74, 93
NGC 2944 GX 56
NGC 2945 GX 152
NGC 2946 GX 74, 93
NGC 2947 GX 133
NGC 2948 GX 93
NGC 2949 GX 93
NGC 2950 GX 25
NGC 2951 GX 113
NGC 2952 GX 133
NGC 2954 GX 93
NGC 2955 GX 56
NGC 2956 GX 152
NGC 2957 GX 14
NGC 2957A GX 14
NGC 2958 GX 93
NGC 2959 GX 14
NGC 2960 GX 113
NGC 2961 GX 14
NGC 2962 GX 93, 113
NGC 2963 GX 14
NGC 2964 GX 56
NGC 2965 GX 56
NGC 2966 GX 113
NGC 2967 GX 113
NGC 2968 GX 56
NGC 2969 GX 133
NGC 2970 GX 56
NGC 2971 GX 56
NGC 2972 OC 186
NGC 2973 GX 170
NGC 2974 GX 113
NGC 2975 GX 133
NGC 2976 GX 14
NGC 2977 GX 5, 6
NGC 2978 GX 133
NGC 2979 GX 133
NGC 2980 GX 133
NGC 2981 GX 56
NGC 2982 OC 186
NGC 2983 GX 152
NGC 2984 GX 93
NGC 2985 GX 14
NGC 2986 GX 152
NGC 2987 GX 113
NGC 2988 GX 74
NGC 2989 GX 152
NGC 2990 GX 93, 113
NGC 2991 GX 74
NGC 2992 GX 133
NGC 2993 GX 133
NGC 2994 GX 74
NGC 2996 GX 152
NGC 2997 GX 170
NGC 2998 GX 39
NGC 3001 GX 170
NGC 3003 GX 56
NGC 3005 GX 39
NGC 3006 GX 39
NGC 3007 GX 133
NGC 3008 GX 39
NGC 3009 GX 39
NGC 3010 GX 39
NGC 3011 GX 56
NGC 3012 GX 56
NGC 3013 GX 56
NGC 3014 GX 113
NGC 3015 GX 113
NGC 3016 GX 93
NGC 3017 GX 113
NGC 3018 GX 113
NGC 3019 GX 93
NGC 3020 GX 93
NGC 3021 GX 56
NGC 3022 GX 113, 133
NGC 3023 GX 113
NGC 3024 GX 93
NGC 3025 GX 152
NGC 3026 GX 74
NGC 3027 GX 14
NGC 3028 GX 152
NGC 3029 GX 133
NGC 3030 GX 133
NGC 3031 GX 14
NGC 3032 GX 56, 74
NGC 3033 OC 199
NGC 3034 GX 14
NGC 3035 GX 133
NGC 3036 OC 199, 210
NGC 3037 GX 152
NGC 3038 GX 170
NGC 3039 GX 113
NGC 3040 GX 74
NGC 3041 GX 93
NGC 3042 GX 113
NGC 3043 GX 25
NGC 3044 GX 113
NGC 3045 GX 152
NGC 3047 GX 113
NGC 3047A GX 113
NGC 3048 GX 93
NGC 3049 GX 93
NGC 3051 GX 152
NGC 3052 GX 152
NGC 3053 GX 93
NGC 3054 GX 152
NGC 3055 GX 113
NGC 3056 GX 152
NGC 3057 GX 5, 6
NGC 3058 GX 133
NGC 3059 GX 210, 217, 218
NGC 3060 GX 93
NGC 3061 GX 5, 6
NGC 3062 GX 113
NGC 3064 GX 133
NGC 3065 GX 14
NGC 3066 GX 14
NGC 3067 GX 55, 56
NGC 3068 GX 73, 74
NGC 3069 GX 93
NGC 3070 GX 93
NGC 3071 GX 55, 56
NGC 3072 GX 151, 152
NGC 3073 GX 25
NGC 3074 GX 55, 56
NGC 3075 GX 93
NGC 3076 GX 151, 152
NGC 3077 GX 14
NGC 3078 GX 151, 152
NGC 3079 GX 25
NGC 3080 GX 93
NGC 3081 GX 151, 152
NGC 3082 GX 169, 170
NGC 3083 GX 113
NGC 3084 GX 151, 152
NGC 3085 GX 151, 152
NGC 3086 GX 113
NGC 3087 GX 169, 170
NGC 3088A GX 73, 74
NGC 3088B GX 73, 74
NGC 3089 GX 151, 152
NGC 3090 GX 113
NGC 3091 GX 151, 152
NGC 3092 GX 113
NGC 3093 GX 113
NGC 3094 GX 93
NGC 3095 GX 169, 170
NGC 3096 GX 151, 152
NGC 3098 GX 73, 74
NGC 3099 GX 55, 56
NGC 3100 GX 169, 170
NGC 3101 GX 113
NGC 3102 GX 25
NGC 3104 GX 39, 55, 56
NGC 3105 OC 199
NGC 3106 GX 55
NGC 3107 GX 93
NGC 3108 GX 169, 170
NGC 3109 GX 151, 152
NGC 3110 GX 133
NGC 3111 GX 39
NGC 3112 GX 151, 152
NGC 3113 GX 151
NGC 3114 OC 199
NGC 3115 GX 133
NGC 3116 GX 55
NGC 3117 GX 113
NGC 3118 GX 55
NGC 3119 GX 93
NGC 3120 GX 169
NGC 3121 GX 93
NGC 3124 GX 151
NGC 3125 GX 151, 169
NGC 3126 GX 55
NGC 3127 GX 133
NGC 3128 GX 133
NGC 3130 GX 92, 93
NGC 3131 GX 73
NGC 3132 PN 169, 186
NGC 3133 GX 133
NGC 3134 GX 92, 93
NGC 3135 GX 39
NGC 3136 GX 210
NGC 3136A GX 210
NGC 3136B GX 210
NGC 3137 GX 151, 169
NGC 3138 GX 132, 133
NGC 3139 GX 132, 133
NGC 3140 GX 132, 133
NGC 3141 GX 132, 133
NGC 3142 GX 132, 133
NGC 3143 GX 132, 133
NGC 3144 GX 5, 6
NGC 3145 GX 132, 133
NGC 3146 GX 151
NGC 3147 GX 5, 6, 14
NGC 3149 GX 217, 218
NGC 3150 GX 55
NGC 3151 GX 55
NGC 3152 GX 55
NGC 3153 GX 92, 93
NGC 3154 GX 73, 92, 93
NGC 3155 GX 5, 6
NGC 3156 GX 112, 113
NGC 3157 GX 169
NGC 3158 GX 55
NGC 3159 GX 55
NGC 3160 GX 55
NGC 3161 GX 55
NGC 3162 GX 73
NGC 3163 GX 55
NGC 3164 GX 25
NGC 3165 GX 112, 113
NGC 3166 GX 112, 113
NGC 3168 GX 25
NGC 3169 GX 112, 113
NGC 3171 GX 151
NGC 3172 GX 1
NGC 3173 GX 151
NGC 3175 GX 151
NGC 3177 GX 73
NGC 3178 GX 132
NGC 3179 GX 39
NGC 3182 GX 25
NGC 3183 GX 5
NGC 3184 GX 39
NGC 3185 GX 73
NGC 3186 GX 92
NGC 3187 GX 73
NGC 3188 GX 25
NGC 3188A GX 25
NGC 3190 GX 73
NGC 3191 GX 39
NGC 3193 GX 73
NGC 3195 PN 217, 218
NGC 3196 GX 73
NGC 3197 GX 5, 6
NGC 3198 GX 39
NGC 3199 BN 199
NGC 3200 GX 132, 151
NGC 3201 GC 186
NGC 3202 GX 38, 39
NGC 3203 GX 151
NGC 3204 GX 73
NGC 3205 GX 38, 39
NGC 3206 GX 25
NGC 3207 GX 38, 39
NGC 3208 GX 151
NGC 3209 GX 73
NGC 3211 PN 199, 210
NGC 3212 GX 5
NGC 3213 GX 73
NGC 3214 GX 25
NGC 3215 GX 5
NGC 3216 GX 73
NGC 3217 GX 92
NGC 3220 GX 25
NGC 3221 GX 73
NGC 3222 GX 73
NGC 3223 GX 169
NGC 3224 GX 169
NGC 3225 GX 25
NGC 3226 GX 73
NGC 3227 GX 73
NGC 3228 OC 185, 186, 199
NGC 3230 GX 92
NGC 3231 OC 14
NGC 3232 GX 73
NGC 3233 GX 151
NGC 3235 GX 73
NGC 3236 GX 25
NGC 3237 GX 55
NGC 3238 GX 25
NGC 3239 GX 73, 92
NGC 3240 GX 151
NGC 3241 GX 169
NGC 3242 PN 151
NGC 3243 GX 112
NGC 3244 GX 169
NGC 3245 GX 73
NGC 3245A GX 73
NGC 3246 GX 112
NGC 3247 BN OC 199
NGC 3248 GX 73
NGC 3249 GX 169
NGC 3250 GX 169
NGC 3250A GX 169, 185, 186
NGC 3250B GX 169, 185, 186
NGC 3250C GX 169, 185, 186
NGC 3250D GX 169
NGC 3250E GX 169, 185
NGC 3251 GX 73
NGC 3252 GX 5, 14
NGC 3253 GX 92
NGC 3254 GX 55, 73
NGC 3255 OC 199
NGC 3256 GX 185, 186
NGC 3256A GX 185, 186
NGC 3256B GX 185
NGC 3256C GX 185
NGC 3257 GX 169
NGC 3258 GX 169
NGC 3258A GX 169
NGC 3258C GX 169
NGC 3258D GX 169
NGC 3258E GX 169
NGC 3259 GX 14
NGC 3260 GX 169
NGC 3261 GX 185
NGC 3262 GX 185
NGC 3263 GX 185
NGC 3264 GX 25
NGC 3265 GX 73
NGC 3266 GX 14
NGC 3267 GX 169
NGC 3268 GX 169
NGC 3269 GX 169
NGC 3270 GX 73
NGC 3271 GX 169
NGC 3273 GX 169
NGC 3274 GX 73
NGC 3275 GX 169
NGC 3276 GX 169
NGC 3277 GX 73
NGC 3278 GX 169
NGC 3279 GX 92
NGC 3280 GX 132
NGC 3280A GX 132
NGC 3280B GX 132
NGC 3281 GX 169
NGC 3281A GX 169
NGC 3281C GX 169
NGC 3281D GX 169
NGC 3282 GX 151
NGC 3285 GX 151, A19
NGC 3285A GX 151, A19
NGC 3285B GX 151, A19
NGC 3286 GX 25
NGC 3287 GX 73
NGC 3288 GX 25
NGC 3289 GX 169
NGC 3290 GX 132, 151
NGC 3292 GX 132
NGC 3293 OC 199
NGC 3294 GX 55
NGC 3296 GX 132
NGC 3297 GX 132
NGC 3298 GX 38
NGC 3299 GX 92
NGC 3300 GX 92
NGC 3301 GX 73
NGC 3302 GX 169
NGC 3303 GX 73
NGC 3304 GX 55
NGC 3305 GX 151, A19
NGC 3306 GX 92
NGC 3307 GX 151, A19
NGC 3308 GX 151, A19
NGC 3309 GX 151, A19
NGC 3310 GX 25
NGC 3311 GX 151, A19
NGC 3312 GX 151, A19
NGC 3313 GX 151
NGC 3314A GX 151, A19
NGC 3314B GX 151, A19
NGC 3315 GX 151, A19
NGC 3316 GX 151, A19
NGC 3318 GX 185
NGC 3318A GX 185
NGC 3318B GX 185
NGC 3319 GX 38
NGC 3320 GX 38
NGC 3321 GX 132
NGC 3323 GX 73
NGC 3324 BN OC 199
NGC 3325 GX 112
NGC 3326 GX 92, 112
NGC 3327 GX 73
NGC 3329 GX 5
NGC 3330 OC 199
NGC 3331 GX 151
NGC 3332 GX 92
NGC 3333 GX 169
NGC 3334 GX 55
NGC 3335 GX 151
NGC 3336 GX 151, A19
NGC 3337 GX 112
NGC 3338 GX 92
NGC 3340 GX 112
NGC 3341 GX 92, 112
NGC 3343 GX 5, 13, 14
NGC 3344 GX 73
NGC 3346 GX 92
NGC 3347 GX 169
NGC 3347A GX 169
NGC 3347B GX 169
NGC 3347C GX 169
NGC 3348 GX 13, 14
NGC 3349 GX 92
NGC 3350 GX 55
NGC 3351 GX 92
NGC 3352 GX 73
NGC 3353 GX 25
NGC 3354 GX 169
NGC 3356 GX 92
NGC 3357 GX 92
NGC 3358 GX 169

NGC 3359 GX 13, 14
NGC 3360 GX 132
NGC 3361 GX 132
NGC 3362 GX 92
NGC 3363 GX 73
NGC 3364 GX 13, 14
NGC 3365 GX 112
NGC 3366 GX 185
NGC 3367 GX 92
NGC 3368 GX 92
NGC 3369 GX 151
NGC 3370 GX 73, 92
NGC 3372 BN 199, xviii, xix, xxii
NGC 3374 GX 38
NGC 3375 GX 132
NGC 3376 GX 92
NGC 3377 GX 92
NGC 3377A GX 92
NGC 3378 GX 169, 185
NGC 3379 GX 92
NGC 3380 GX 73
NGC 3381 GX 55
NGC 3383 GX 151
NGC 3384 GX 92
NGC 3385 GX 112
NGC 3386 GX 92, 112
NGC 3389 GX 92
NGC 3390 GX 169
NGC 3391 GX 92
NGC 3392 GX 13, 14
NGC 3393 GX 151
NGC 3394 GX 13, 14
NGC 3395 GX 55
NGC 3396 GX 55
NGC 3398 GX 25
NGC 3399 GX 92
NGC 3400 GX 73
NGC 3402 GX 132
NGC 3403 GX 5, 13, 14
NGC 3404 GX 132
NGC 3405 GX 92
NGC 3406 GX 25, 38
NGC 3407 GX 25
NGC 3408 GX 25
NGC 3409 GX 132, 151
NGC 3410 GX 25, 38
NGC 3411 GX 132
NGC 3412 GX 92
NGC 3413 GX 55
NGC 3414 GX 73
NGC 3415 GX 38
NGC 3416 GX 38
NGC 3417 GX 92
NGC 3418 GX 73
NGC 3419 GX 92
NGC 3419A GX 92
NGC 3420 GX 132, 151
NGC 3421 GX 132
NGC 3422 GX 132
NGC 3423 GX 92, 112
NGC 3424 GX 55
NGC 3425 GX 92
NGC 3426 GX 73
NGC 3427 GX 92
NGC 3428 GX 92
NGC 3429 GX 92
NGC 3430 GX 55
NGC 3431 GX 132, 151
NGC 3432 GX 55
NGC 3433 GX 92
NGC 3434 GX 112
NGC 3435 GX 25
NGC 3437 GX 73
NGC 3438 GX 92
NGC 3439 GX 92
NGC 3440 GX 25
NGC 3441 GX 92
NGC 3442 GX 55
NGC 3443 GX 73, 92
NGC 3444 GX 92
NGC 3445 GX 25
NGC 3446 OC 185
NGC 3447 GX 92
NGC 3447A GX 92
NGC 3448 GX 25
NGC 3449 GX 169
NGC 3450 GX 151
NGC 3451 GX 73
NGC 3452 GX 132
NGC 3453 GX 151
NGC 3454 GX 73, 92
NGC 3455 GX 73, 92
NGC 3456 GX 132
NGC 3457 GX 73, 92
NGC 3458 GX 24, 25
NGC 3459 GX 132, 151
NGC 3462 GX 92
NGC 3463 GX 151
NGC 3464 GX 151
NGC 3465 GX 5
NGC 3466 GX 92
NGC 3467 GX 92
NGC 3468 GX 38, 55
NGC 3469 GX 132
NGC 3470 GX 24, 25
NGC 3471 GX 24, 25
NGC 3473 GX 73, 92
NGC 3474 GX 73, 92
NGC 3475 GX 73
NGC 3476 GX 92
NGC 3477 GX 92
NGC 3478 GX 38
NGC 3479 GX 132
NGC 3481 GX 132
NGC 3482 GX 185
NGC 3483 GX 151
NGC 3485 GX 92
NGC 3486 GX 73
NGC 3487 GX 73, 92
NGC 3488 GX 24, 25
NGC 3489 GX 92
NGC 3490 GX 92
NGC 3491 GX 92
NGC 3492 GX 92
NGC 3493 GX 73
NGC 3495 GX 112
NGC 3496 OC 198, 199
NGC 3499 GX 24, 25
NGC 3500 GX 5
NGC 3501 GX 73, 92
NGC 3502 GX 132
NGC 3503 BN OC 198, 199
NGC 3504 GX 73
NGC 3506 GX 92
NGC 3507 GX 73
NGC 3508 GX 132
NGC 3509 GX 112
NGC 3510 GX 73
NGC 3511 GX 151
NGC 3512 GX 73
NGC 3513 GX 151
NGC 3514 GX 151
NGC 3515 GX 73
NGC 3516 GX 13
NGC 3517 GX 24
NGC 3519 OC 198
NGC 3520 GX 151
NGC 3521 GX 112
NGC 3522 GX 73
NGC 3523 GX 5
NGC 3524 GX 92
NGC 3526 GX 92
NGC 3527 GX 73
NGC 3528 GX 151
NGC 3529 GX 151
NGC 3530 GX 24
NGC 3532 OC 198, xviii, xix, xxii
NGC 3533 GX 169
NGC 3534 GX 73
NGC 3535 GX 112
NGC 3536 GX 73
NGC 3537A GX 132
NGC 3539 GX 73
NGC 3540 GX 55
NGC 3541 GX 132
NGC 3542 GX 55
NGC 3543 GX 24
NGC 3545A GX 55
NGC 3545B GX 55
NGC 3546 GX 132
NGC 3547 GX 92
NGC 3549 GX 24
NGC 3550 GX 73
NGC 3552 GX 73
NGC 3553 GX 73
NGC 3554 GX 73
NGC 3555 GX 73
NGC 3556 GX 24
NGC 3557 GX 169
NGC 3557B GX 169
NGC 3558 GX 73
NGC 3559 GX 92
NGC 3561 GX 73
NGC 3561A GX 73
NGC 3562 GX 13
NGC 3563 GX 73
NGC 3563A GX 73
NGC 3564 GX 169
NGC 3565 GX 151
NGC 3567 GX 92, 112
NGC 3568 GX 169
NGC 3569 GX 55
NGC 3570 GX 73
NGC 3571 GX 151
NGC 3572 OC 198
NGC 3573 GX 169
NGC 3574 GX 73
NGC 3576 BN 198
NGC 3577 GX 38
NGC 3579 BN 198
NGC 3580 GX 112
NGC 3581 BN 198
NGC 3582 BN 198
NGC 3583 GX 38
NGC 3584 BN 198
NGC 3585 GX 151
NGC 3586 BN 198
NGC 3587 PN 24
NGC 3588 GX 73
NGC 3589 GX 24
NGC 3590 OC 198
NGC 3591 GX 132
NGC 3592 GX 73, 92
NGC 3593 GX 92
NGC 3594 GX 24
NGC 3595 GX 38
NGC 3596 GX 92
NGC 3597 GX 151
NGC 3598 GX 73, 92
NGC 3599 GX 73
NGC 3600 GX 38
NGC 3601 GX 92, 112
NGC 3602 GX 73, 92
NGC 3603 OC 198
NGC 3605 GX 72, 73
NGC 3606 GX 168, 169
NGC 3607 GX 72, 73
NGC 3608 GX 72, 73
NGC 3609 GX 72, 73
NGC 3610 GX 24
NGC 3611 GX 112
NGC 3612 GX 72, 73
NGC 3613 GX 24
NGC 3614 GX 38
NGC 3614A GX 38
NGC 3615 GX 72, 73
NGC 3617 GX 150, 151
NGC 3618 GX 72, 73
NGC 3619 GX 24
NGC 3620 GX 217
NGC 3621 GX 168, 169
NGC 3622 GX 13
NGC 3623 GX 92
NGC 3624 GX 92
NGC 3625 GX 24
NGC 3626 GX 72, 73
NGC 3627 GX 91, 92
NGC 3628 GX 91, 92
NGC 3629 GX 72, 73
NGC 3630 GX 111, 112
NGC 3631 GX 24
NGC 3633 GX 111, 112
NGC 3634 GX 131, 132
NGC 3635 GX 131, 132
NGC 3636 GX 131, 132
NGC 3637 GX 131, 132
NGC 3638 GX 131, 132
NGC 3639 GX 72, 73
NGC 3640 GX 111, 112
NGC 3641 GX 111, 112
NGC 3642 GX 24
NGC 3643 GX 111, 112
NGC 3644 GX 111, 112
NGC 3645 GX 111, 112
NGC 3646 GX 72, 73
NGC 3647 GX 111, 112
NGC 3648 GX 54, 55
NGC 3649 GX 72, 73
NGC 3650 GX 72, 73
NGC 3651 GX 72, 73
NGC 3652 GX 54, 55
NGC 3653 GX 72, 73
NGC 3654 GX 13
NGC 3655 GX 91, 92
NGC 3656 GX 24
NGC 3657 GX 24
NGC 3658 GX 54, 55
NGC 3659 GX 72, 73, 91, 92
NGC 3660 GX 131, 132
NGC 3661 GX 131, 132
NGC 3662 GX 111, 112
NGC 3663 GX 131, 132
NGC 3664 GX 111, 112
NGC 3664A GX 111, 112
NGC 3665 GX 54
NGC 3666 GX 91, 92
NGC 3667 GX 131, 132
NGC 3667A GX 131, 132
NGC 3668 GX 13
NGC 3669 GX 24
NGC 3670 GX 72
NGC 3671 GX 24
NGC 3672 GX 131, 132
NGC 3673 GX 150
NGC 3674 GX 24
NGC 3675 GX 38
NGC 3677 GX 38
NGC 3678 GX 72
NGC 3679 GX 111, 112, 131, 132
NGC 3680 OC 185
NGC 3681 GX 91, 92
NGC 3682 GX 13
NGC 3683 GX 24
NGC 3683A GX 24
NGC 3684 GX 72, 91, 92
NGC 3685 GX 111
NGC 3686 GX 72, 91, 92
NGC 3687 GX 54, 72
NGC 3688 GX 131, 132
NGC 3689 GX 72
NGC 3690 GX 24
NGC 3691 GX 91
NGC 3692 GX 91
NGC 3693 GX 131
NGC 3694 GX 54
NGC 3695 GX 54
NGC 3696 GX 131
NGC 3697 GX 72
NGC 3699 PN 198
NGC 3700 GX 54
NGC 3701 GX 72
NGC 3702 GX 131
NGC 3704 GX 131
NGC 3705 GX 91
NGC 3706 GX 168
NGC 3707 GX 131
NGC 3710 GX 72
NGC 3711 GX 131
NGC 3712 GX 72
NGC 3713 GX 72
NGC 3714 GX 72
NGC 3715 GX 131
NGC 3716 GX 111
NGC 3717 GX 168
NGC 3718 GX 24
NGC 3719 GX 111
NGC 3720 GX 111
NGC 3721 GX 131
NGC 3722 GX 131
NGC 3723 GX 131
NGC 3724 GX 131
NGC 3725 GX 24
NGC 3726 GX 38
NGC 3727 GX 131
NGC 3728 GX 72
NGC 3729 GX 24
NGC 3731 GX 91
NGC 3732 GX 131
NGC 3733 GX 24
NGC 3734 GX 131
NGC 3735 GX 13
NGC 3736 GX 5, 13
NGC 3737 GX 24
NGC 3738 GX 24
NGC 3739 GX 72
NGC 3740 GX 24
NGC 3741 GX 38
NGC 3742 GX 168
NGC 3743 GX 72
NGC 3744 GX 72
NGC 3745 GX 72
NGC 3746 GX 72
NGC 3748 GX 72
NGC 3749 GX 168
NGC 3750 GX 72
NGC 3751 GX 72
NGC 3752 GX 5
NGC 3753 GX 72
NGC 3754 GX 72
NGC 3755 GX 54
NGC 3756 GX 24
NGC 3757 GX 24
NGC 3758 GX 72
NGC 3759 GX 24
NGC 3759A GX 24
NGC 3761 GX 72
NGC 3762 GX 24
NGC 3763 GX 131
NGC 3764 GX 72, 91
NGC 3765 GX 72
NGC 3766 OC 198
NGC 3767 GX 91
NGC 3768 GX 72, 91
NGC 3769 GX 38
NGC 3769A GX 38
NGC 3770 GX 24
NGC 3771 GX 131
NGC 3772 GX 72
NGC 3773 GX 91
NGC 3774 GX 131
NGC 3775 GX 131
NGC 3776 GX 111
NGC 3777 GX 131
NGC 3778 GX 185
NGC 3779 GX 131
NGC 3780 GX 24
NGC 3781 GX 72
NGC 3782 GX 38
NGC 3783 GX 168
NGC 3784 GX 72
NGC 3785 GX 72
NGC 3786 GX 54
NGC 3787 GX 72, A11
NGC 3788 GX 54
NGC 3789 GX 131
NGC 3790 GX 72, 91
NGC 3791 GX 131
NGC 3793 GX 54
NGC 3795 GX 24
NGC 3795B GX 24
NGC 3796 GX 24
NGC 3798 GX 72
NGC 3799 GX 91
NGC 3800 GX 91
NGC 3801 GX 72, 91
NGC 3802 GX 72, 91
NGC 3803 GX 72, 91
NGC 3804 GX 24
NGC 3805 GX 72, A11
NGC 3806 GX 72, 91
NGC 3808 GX 72
NGC 3808A GX 72
NGC 3809 GX 24
NGC 3810 GX 91
NGC 3811 GX 38
NGC 3812 GX 72
NGC 3813 GX 54
NGC 3814 GX 72
NGC 3815 GX 72
NGC 3816 GX 72, A11
NGC 3817 GX 91
NGC 3818 GX 131
NGC 3819 GX 91
NGC 3820 GX 91
NGC 3821 GX 72, A11
NGC 3822 GX 91
NGC 3823 GX 131
NGC 3824 GX 24
NGC 3825 GX 91
NGC 3826 GX 72
NGC 3827 GX 72, A11
NGC 3828 GX 91
NGC 3829 GX 24
NGC 3830 GX 72
NGC 3830A GX 72
NGC 3831 GX 131
NGC 3832 GX 72
NGC 3833 GX 91
NGC 3834 GX 72, A11
NGC 3835 GX 24
NGC 3836 GX 131
NGC 3837 GX 72, A11
NGC 3838 GX 24
NGC 3839 GX 91
NGC 3840 GX 72, A11
NGC 3841 GX 72, A11
NGC 3842 GX 72, A11
NGC 3843 GX 91
NGC 3844 GX 72, A11
NGC 3845 GX 72, A11
NGC 3846 GX 24
NGC 3846A GX 24
NGC 3847 GX 54
NGC 3848 GX 91
NGC 3849 GX 111
NGC 3850 GX 24
NGC 3851 GX 72, A11
NGC 3853 GX 91
NGC 3855 GX 54
NGC 3857 GX 72, A11
NGC 3859 GX 72, A11
NGC 3860 GX 72, A11
NGC 3861 GX 72, A11
NGC 3862 GX 72, A11
NGC 3863 GX 91
NGC 3864 GX 72, A11
NGC 3865 GX 131
NGC 3866 GX 131
NGC 3867 GX 72, A11
NGC 3868 GX 72, A11
NGC 3869 GX 91
NGC 3870 GX 38
NGC 3871 GX 54
NGC 3872 GX 91
NGC 3873 GX 72, A11
NGC 3875 GX 72, A11
NGC 3876 GX 91
NGC 3877 GX 38
NGC 3878 GX 54
NGC 3879 GX 13
NGC 3880 GX 54
NGC 3881 GX 54
NGC 3882 GX 198
NGC 3883 GX 72, A11
NGC 3884 GX 72, A11
NGC 3885 GX 150
NGC 3886 GX 72, A11
NGC 3887 GX 131
NGC 3888 GX 24
NGC 3889 GX 24
NGC 3890 GX 5
NGC 3891 GX 54
NGC 3892 GX 131
NGC 3893 GX 38
NGC 3894 GX 24
NGC 3895 GX 24
NGC 3896 GX 38
NGC 3897 GX 54
NGC 3898 GX 24
NGC 3900 GX 72
NGC 3901 GX 5
NGC 3902 GX 72
NGC 3903 GX 168
NGC 3904 GX 150, 168
NGC 3905 GX 131
NGC 3906 GX 38
NGC 3907 GX 111
NGC 3907B GX 111
NGC 3908 GX 91
NGC 3909 OC 185
NGC 3910 GX 72, A10, A11
NGC 3911 GX 72
NGC 3912 GX 72
NGC 3913 GX 24
NGC 3914 GX 91
NGC 3915 GX 111
NGC 3916 GX 24
NGC 3917 GX 24, 38
NGC 3918 PN 198
NGC 3919 GX 72, A10
NGC 3920 GX 72, A10
NGC 3921 GX 24
NGC 3922 GX 38
NGC 3923 GX 150
NGC 3924 GX 38
NGC 3925 GX 72, A10
NGC 3926A GX 72, A10
NGC 3926B GX 72, A10
NGC 3928 GX 38
NGC 3929 GX 72, A10
NGC 3930 GX 54
NGC 3931 GX 24, 38
NGC 3932 GX 38
NGC 3933 GX 91
NGC 3934 GX 91
NGC 3935 GX 54
NGC 3936 GX 150
NGC 3937 GX 72, A10
NGC 3938 GX 38
NGC 3940 GX 72, A10
NGC 3941 GX 54
NGC 3942 GX 131
NGC 3943 GX 72, A10
NGC 3944 GX 72
NGC 3945 GX 24
NGC 3946 GX 72, A10
NGC 3947 GX 72, A10
NGC 3949 GX 38
NGC 3950 GX 38
NGC 3951 GX 72, A10
NGC 3952 GX 111
NGC 3953 GX 24
NGC 3954 GX 72, A10

NGC 3955 GX 150
NGC 3956 GX 150
NGC 3957 GX 150
NGC 3958 GX 24
NGC 3959 GX 131
NGC 3960 OC 198
NGC 3961 GX 13
NGC 3962 GX 131
NGC 3963 GX 24
NGC 3964 GX 72
NGC 3966 GX 54
NGC 3967 GX 131
NGC 3968 GX 91
NGC 3969 GX 150
NGC 3970 GX 131
NGC 3971 GX 54, 72
NGC 3972 GX 24
NGC 3973 GX 91
NGC 3974 GX 131
NGC 3975 GX 24
NGC 3976 GX 91
NGC 3977 GX 24
NGC 3978 GX 24
NGC 3979 GX 111
NGC 3981 GX 150
NGC 3982 GX 24
NGC 3983 GX 72, A10
NGC 3984 GX 54, 72
NGC 3985 GX 37, 38
NGC 3986 GX 54
NGC 3987 GX 72, A10
NGC 3988 GX 72
NGC 3989 GX 72, A10
NGC 3990 GX 24
NGC 3991 GX 54
NGC 3992 GX 24
NGC 3993 GX 72, A10
NGC 3994 GX 54
NGC 3995 GX 54
NGC 3996 GX 91
NGC 3997 GX 72, A10
NGC 3998 GX 24
NGC 3999 GX 72, A10
NGC 4000 GX 72, A10
NGC 4001 GX 37, 38
NGC 4002 GX 72, A10
NGC 4003 GX 72, A10
NGC 4004 GX 72
NGC 4005 GX 72, A10
NGC 4006 GX 111
NGC 4008 GX 72
NGC 4009 GX 72, A10
NGC 4010 GX 37, 38
NGC 4011 GX 72, A10
NGC 4012 GX 91
NGC 4013 GX 37, 38
NGC 4014 GX 91
NGC 4015 GX 72, A10
NGC 4016 GX 72
NGC 4017 GX 72
NGC 4018 GX 72, A10
NGC 4020 GX 54
NGC 4021 GX 72, A10
NGC 4022 GX 72, A10
NGC 4023 GX 72, A10
NGC 4024 GX 150
NGC 4025 GX 54
NGC 4026 GX 37, 38
NGC 4027 GX 150
NGC 4027A GX 150
NGC 4029 GX 91
NGC 4030 GX 111
NGC 4031 GX 54
NGC 4032 GX 72, A10
NGC 4033 GX 131, 150
NGC 4034 GX 13
NGC 4035 GX 131
NGC 4036 GX 24
NGC 4037 GX 91
NGC 4038 GX 150
NGC 4039 GX 150
NGC 4040 GX 72, 91
NGC 4041 GX 13, 24
NGC 4043 GX 111
NGC 4044 GX 111
NGC 4045 GX 111
NGC 4045A GX 111
NGC 4047 GX 37, 38
NGC 4048 GX 72
NGC 4049 GX 72
NGC 4050 GX 131
NGC 4051 GX 37, 38
NGC 4052 OC 209
NGC 4053 GX 72
NGC 4054 GX 24
NGC 4056 GX 72, A10
NGC 4057 GX 72, A10
NGC 4058 GX 111
NGC 4060 GX 72, A10
NGC 4061 GX 72, A10
NGC 4062 GX 54
NGC 4063 GX 111
NGC 4064 GX 72
NGC 4065 GX 72, A10
NGC 4066 GX 72, A10
NGC 4067 GX 91
NGC 4068 GX 24
NGC 4069 GX 72, A10
NGC 4070 GX 72, A10
NGC 4071 PN 209
NGC 4072 GX 72, A10
NGC 4073 GX 111
NGC 4074 GX 72, A10
NGC 4075 GX 111
NGC 4076 GX 72, A10
NGC 4077 GX 111
NGC 4078 GX 91
NGC 4079 GX 111
NGC 4080 GX 72
NGC 4081 GX 13
NGC 4082 GX 91
NGC 4083 GX 91
NGC 4084 GX 72, A10
NGC 4085 GX 37
NGC 4086 GX 72, A10
NGC 4087 GX 150
NGC 4088 GX 37
NGC 4089 GX 72, A10
NGC 4090 GX 72, A10
NGC 4091 GX 72, A10
NGC 4092 GX 72, A10
NGC 4093 GX 72, A10
NGC 4094 GX 131
NGC 4095 GX 72, A10
NGC 4096 GX 37
NGC 4097 GX 54
NGC 4098 GX 72, A10
NGC 4100 GX 37
NGC 4101 GX 72, A10
NGC 4102 GX 24
NGC 4103 OC 198
NGC 4104 GX 72
NGC 4105 GX 150, 168
NGC 4106 GX 150, 168
NGC 4108 GX 13
NGC 4108A GX 13
NGC 4108B GX 13
NGC 4109 GX 37
NGC 4110 GX 72
NGC 4111 GX 37
NGC 4112 GX 168, 184
NGC 4114 GX 131
NGC 4116 GX 111
NGC 4117 GX 37
NGC 4118 GX 37
NGC 4120 GX 13
NGC 4121 GX 13
NGC 4122 GX 54
NGC 4123 GX 111
NGC 4124 GX 91
NGC 4125 GX 13
NGC 4126 GX 91
NGC 4127 GX 5
NGC 4128 GX 13
NGC 4129 GX 131
NGC 4131 GX 54, 72
NGC 4132 GX 54, 72
NGC 4133 GX 5
NGC 4134 GX 54, 72
NGC 4135 GX 37
NGC 4136 GX 54, 72
NGC 4137 GX 37
NGC 4138 GX 37
NGC 4139 GX 111
NGC 4141 GX 24
NGC 4142 GX 24
NGC 4143 GX 37
NGC 4144 GX 37
NGC 4145 GX 54
NGC 4146 GX 72
NGC 4147 GC 72
NGC 4148 GX 54
NGC 4149 GX 24
NGC 4150 GX 54
NGC 4151 GX 54
NGC 4152 GX 91
NGC 4155 GX 72
NGC 4156 GX 54
NGC 4157 GX 37
NGC 4158 GX 72
NGC 4159 GX 5
NGC 4161 GX 24
NGC 4162 GX 72
NGC 4163 GX 54
NGC 4164 GX 91
NGC 4165 GX 91
NGC 4166 GX 72, 91
NGC 4168 GX 91
NGC 4169 GX 54, 72
NGC 4172 GX 24
NGC 4173 GX 54, 72
NGC 4174 GX 54, 72
NGC 4175 GX 54, 72
NGC 4176 GX 131
NGC 4177 GX 131
NGC 4178 GX 91
NGC 4179 GX 111
NGC 4180 GX 91, A15
NGC 4181 GX 24
NGC 4183 GX 37
NGC 4184 OC 198, 209
NGC 4185 GX 72
NGC 4186 GX 91
NGC 4187 GX 37
NGC 4188 GX 131
NGC 4189 GX 91
NGC 4190 GX 54
NGC 4191 GX 91, A15
NGC 4192 GX 91
NGC 4193 GX 91
NGC 4194 GX 24
NGC 4195 GX 24
NGC 4196 GX 72
NGC 4197 GX 91, 111, A15
NGC 4198 GX 24
NGC 4199 GX 24
NGC 4200 GX 91
NGC 4201 GX 131
NGC 4202 GX 111
NGC 4203 GX 54
NGC 4204 GX 72
NGC 4205 GX 13
NGC 4206 GX 91
NGC 4207 GX 91
NGC 4210 GX 13
NGC 4211 GX 72
NGC 4211A GX 72
NGC 4212 GX 91
NGC 4213 GX 72
NGC 4214 GX 54
NGC 4215 GX 91, A15
NGC 4216 GX 91
NGC 4217 GX 37
NGC 4218 GX 37
NGC 4219 GX 184
NGC 4219A GX 184
NGC 4220 GX 37
NGC 4221 GX 13
NGC 4222 GX 91
NGC 4224 GX 91, A15
NGC 4225 GX 131
NGC 4226 GX 37
NGC 4227 GX 54
NGC 4229 GX 54
NGC 4230 OC 198
NGC 4231 GX 37
NGC 4232 GX 37
NGC 4233 GX 91, A15
NGC 4234 GX 111, A15
NGC 4235 GX 91, A15
NGC 4236 GX 13
NGC 4237 GX 91
NGC 4238 GX 13
NGC 4239 GX 91
NGC 4240 GX 131
NGC 4241 GX 91, A15
NGC 4242 GX 37
NGC 4244 GX 54
NGC 4245 GX 54, 72
NGC 4246 GX 91, A15
NGC 4247 GX 91, A15
NGC 4248 GX 37
NGC 4249 GX 91, 111, A15
NGC 4250 GX 13
NGC 4251 GX 72
NGC 4252 GX 91, 111, A15
NGC 4253 GX 54, 72
NGC 4254 GX 91, A13
NGC 4255 GX 111, A15
NGC 4256 GX 13
NGC 4257 GX 91, 111, A15
NGC 4258 GX 37
NGC 4259 GX 91, 111, A15
NGC 4260 GX 91, A15
NGC 4261 GX 91, 111, A15
NGC 4262 GX 91, A13
NGC 4263 GX 131
NGC 4264 GX 91, 111, A15
NGC 4266 GX 91, 111, A15
NGC 4267 GX 91, A13
NGC 4268 GX 91, 111, A15
NGC 4269 GX 91, A15
NGC 4270 GX 91, 111, A15
NGC 4271 GX 24
NGC 4272 GX 54
NGC 4273 GX 91, 111, A15
NGC 4274 GX 54, 72
NGC 4275 GX 72
NGC 4276 GX 91, A15
NGC 4277 GX 91, 111, A15
NGC 4278 GX 54, 72
NGC 4279 GX 131
NGC 4281 GX 91, 111, A15
NGC 4282 GX 91, 111, A15
NGC 4283 GX 54, 72
NGC 4284 GX 24
NGC 4285 GX 131
NGC 4286 GX 54, 72
NGC 4287 GX 91, 111, A15
NGC 4288 GX 37
NGC 4289 GX 111, A15
NGC 4290 GX 24
NGC 4291 GX 5
NGC 4292 GX 111, A15
NGC 4293 GX 72
NGC 4294 GX 91, A13
NGC 4295 GX 72
NGC 4296 GX 91, A15
NGC 4297 GX 91, A15
NGC 4298 GX 91, A13
NGC 4299 GX 91, A13
NGC 4300 GX 91, 111, A15
NGC 4301 GX 111, A15
NGC 4302 GX 91, A13
NGC 4303 GX 111, A15
NGC 4303A GX 111, A15
NGC 4304 GX 168
NGC 4305 GX 91, A13
NGC 4306 GX 91, A13
NGC 4307 GX 91
NGC 4308 GX 54
NGC 4309 GX 91, A15
NGC 4310 GX 54, 72
NGC 4312 GX 91, A13
NGC 4313 GX 91, A13
NGC 4314 GX 54, 72
NGC 4316 GX 91
NGC 4318 GX 91, A15
NGC 4319 GX 5
NGC 4320 GX 91, A13
NGC 4321 GX 91, A13
NGC 4322 GX 91, A13
NGC 4324 GX 91, 111, A15
NGC 4325 GX 91, A13
NGC 4326 GX 91, A15
NGC 4328 GX 91, A13
NGC 4329 GX 131
NGC 4330 GX 91, A13
NGC 4331 GX 5
NGC 4332 GX 13
NGC 4333 GX 91, A15
NGC 4334 GX 91, A15
NGC 4335 GX 24
NGC 4336 GX 72
NGC 4337 OC 198
NGC 4338 GX 72
NGC 4339 GX 91, A15
NGC 4340 GX 91
NGC 4341 GX 91, A15
NGC 4342 GX 91, A15
NGC 4343 GX 91, A15
NGC 4344 GX 72, 91
NGC 4346 GX 37
NGC 4348 GX 111
NGC 4349 OC 198
NGC 4350 GX 91
NGC 4351 GX 91, A13
NGC 4352 GX 91, A13
NGC 4353 GX 91, A15
NGC 4356 GX 91, A15
NGC 4357 GX 37
NGC 4358 GX 24
NGC 4359 GX 54
NGC 4360 GX 91
NGC 4361 PN 150
NGC 4362 GX 24
NGC 4363 GX 5
NGC 4364 GX 24
NGC 4365 GX 91, A15
NGC 4366 GX 91, A15
NGC 4369 GX 54
NGC 4370 GX 91, A15
NGC 4371 GX 91, A13
NGC 4372 GC 209
NGC 4373 GX 168
NGC 4373A GX 168
NGC 4373B GX 168
NGC 4374 GX 91, A13
NGC 4375 GX 72
NGC 4376 GX 91, 111, A15
NGC 4377 GX 91, A13
NGC 4378 GX 111, A15
NGC 4379 GX 91, A13
NGC 4380 GX 91, A13
NGC 4382 GX 72
NGC 4383 GX 91
NGC 4384 GX 24
NGC 4385 GX 111
NGC 4386 GX 5
NGC 4387 GX 91, A13
NGC 4388 GX 91, A13
NGC 4389 GX 37
NGC 4390 GX 91, A13
NGC 4391 GX 13
NGC 4392 GX 37
NGC 4393 GX 72
NGC 4394 GX 72
NGC 4395 GX 54
NGC 4396 GX 91, A13
NGC 4402 GX 91, A13
NGC 4403 GX 131
NGC 4404 GX 131
NGC 4405 GX 91
NGC 4406 GX 91, A13
NGC 4408 GX 72
NGC 4410A GX 91
NGC 4410B GX 91
NGC 4411A GX 91, A15
NGC 4411B GX 91, A15
NGC 4412 GX 111, A15
NGC 4413 GX 91, A13
NGC 4414 GX 54
NGC 4415 GX 91, A15
NGC 4416 GX 91, A15
NGC 4417 GX 91
NGC 4418 GX 111
NGC 4419 GX 91, A13
NGC 4420 GX 111
NGC 4421 GX 91, A13
NGC 4422 GX 111, 131
NGC 4423 GX 91, 111, A15
NGC 4424 GX 91
NGC 4425 GX 91, A13
NGC 4428 GX 131
NGC 4429 GX 91, A13
NGC 4430 GX 91, A15
NGC 4431 GX 91, A13
NGC 4432 GX 91, A15
NGC 4433 GX 131
NGC 4434 GX 91, A15
NGC 4435 GX 91, A13
NGC 4436 GX 91, A13
NGC 4438 GX 91, A13
NGC 4439 OC 198
NGC 4440 GX 91, A13
NGC 4441 GX 13
NGC 4442 GX 91
NGC 4444 GX 184
NGC 4445 GX 91
NGC 4446 GX 91, A13
NGC 4447 GX 91, A13
NGC 4448 GX 72
NGC 4449 GX 37
NGC 4450 GX 72, 91
NGC 4451 GX 91
NGC 4452 GX 91, A13
NGC 4453 GX 91, A15
NGC 4454 GX 111
NGC 4455 GX 72
NGC 4456 GX 168
NGC 4457 GX 111, A15
NGC 4458 GX 91, A13
NGC 4459 GX 91, A13
NGC 4460 GX 37
NGC 4461 GX 91, A13
NGC 4462 GX 150
NGC 4463 OC 209
NGC 4464 GX 91, A15
NGC 4465 GX 91, A15
NGC 4466 GX 91, A15
NGC 4467 GX 91, A15
NGC 4468 GX 91, A13
NGC 4469 GX 91, A15
NGC 4470 GX 91, A15
NGC 4472 GX 91, A15
NGC 4473 GX 91, A13
NGC 4474 GX 91, A13
NGC 4475 GX 72
NGC 4476 GX 91, A13
NGC 4477 GX 91, A13
NGC 4478 GX 91, A13
NGC 4479 GX 91, A13
NGC 4480 GX 111
NGC 4481 GX 13
NGC 4482 GX 91, A13
NGC 4483 GX 91
NGC 4484 GX 131
NGC 4485 GX 37
NGC 4486 GX 91, A13
NGC 4486A GX 91, A13
NGC 4486B GX 91, A13
NGC 4487 GX 131
NGC 4488 GX 91
NGC 4489 GX 91
NGC 4490 GX 37
NGC 4491 GX 91, A13
NGC 4492 GX 91
NGC 4493 GX 111
NGC 4494 GX 72
NGC 4495 GX 54, 72
NGC 4496A GX 111
NGC 4496B GX 111
NGC 4497 GX 91, A13
NGC 4498 GX 91
NGC 4499 GX 168
NGC 4500 GX 24
NGC 4501 GX 90, 91, A13
NGC 4502 GX 90, 91
NGC 4503 GX 90, 91, A13
NGC 4504 GX 130, 131
NGC 4506 GX 90, 91, A13
NGC 4507 GX 168
NGC 4509 GX 54
NGC 4510 GX 13
NGC 4511 GX 24
NGC 4512 GX 13
NGC 4513 GX 13
NGC 4514 GX 54, 72
NGC 4515 GX 90, 91
NGC 4516 GX 90, 91, A13
NGC 4517 GX 110, 111
NGC 4517A GX 110, 111
NGC 4518 GX 90, 91
NGC 4518B GX 90, 91
NGC 4519 GX 90, 91
NGC 4519A GX 90, 91
NGC 4520 GX 130, 131
NGC 4521 GX 13
NGC 4522 GX 90, 91
NGC 4523 GX 90, 91, A13
NGC 4524 GX 130, 131
NGC 4525 GX 54
NGC 4526 GX 90, 91
NGC 4527 GX 110, 111
NGC 4528 GX 90, 91, A13
NGC 4529 GX 72
NGC 4531 GX 90, 91, A13
NGC 4532 GX 90, 91
NGC 4533 GX 110, 111
NGC 4534 GX 54
NGC 4535 GX 90, 91
NGC 4536 GX 110, 111
NGC 4537 GX 37
NGC 4538 GX 110, 111
NGC 4539 GX 72
NGC 4540 GX 90, 91, A13
NGC 4541 GX 110, 111
NGC 4542 GX 37
NGC 4543 GX 90, 91
NGC 4544 GX 110, 111
NGC 4545 GX 13
NGC 4546 GX 110, 111
NGC 4547 GX 24
NGC 4548 GX 90, 91, A13
NGC 4549 GX 24
NGC 4550 GX 90, 91, A13
NGC 4551 GX 90, 91, A13
NGC 4552 GX 90, 91, A13
NGC 4553 GX 167, 168
NGC 4555 GX 72
NGC 4556 GX 72
NGC 4558 GX 72
NGC 4559 GX 72
NGC 4561 GX 71, 72

NGC 4562 GX 72
NGC 4563 GX 71, 72
NGC 4564 GX 90, 91, A13
NGC 4565 GX 71, 72
NGC 4566 GX 24
NGC 4567 GX 90, 91, A13
NGC 4568 GX 90, 91, A13
NGC 4569 GX 90, 91, A13
NGC 4570 GX 90, 91
NGC 4571 GX 90, 91, A13
NGC 4572 GX 5
NGC 4573 GX 184
NGC 4574 GX 167, 168
NGC 4575 GX 167, 168, 184
NGC 4576 GX 110, 111
NGC 4578 GX 90, 91
NGC 4579 GX 90, 91, A13
NGC 4580 GX 90, 91, 110, 111
NGC 4581 GX 110, 111
NGC 4583 GX 53, 54
NGC 4584 GX 90, 91
NGC 4585 GX 71, 72
NGC 4586 GX 110, 111
NGC 4587 GX 110, 111
NGC 4588 GX 90, 91
NGC 4589 GX 5
NGC 4590 GC 149, 150
NGC 4591 GX 90, 91
NGC 4592 GX 110, 111
NGC 4593 GX 110, 111, 130, 131
NGC 4594 GX 130, 131
NGC 4595 GX 90, 91
NGC 4596 GX 90, 91
NGC 4597 GX 110, 130
NGC 4598 GX 90
NGC 4599 GX 110
NGC 4600 GX 110
NGC 4601 GX 167, 168, 184, A23
NGC 4602 GX 110, 130
NGC 4603 GX 167, 168, 184, A23
NGC 4603A GX 167, 168, 184
NGC 4603B GX 184, A23
NGC 4603C GX 167, 168, 184, A23
NGC 4603D GX 167, 168, 184, A23
NGC 4604 GX 110, 130
NGC 4605 GX 24
NGC 4606 GX 90
NGC 4607 GX 90
NGC 4608 GX 90
NGC 4609 OC 198, 209
NGC 4611 GX 90
NGC 4612 GX 90
NGC 4613 GX 71, 72
NGC 4614 GX 71, 72
NGC 4615 GX 71, 72
NGC 4616 GX 167, 168, 184, A23
NGC 4617 GX 37
NGC 4618 GX 37
NGC 4619 GX 53, 54
NGC 4620 GX 90
NGC 4621 GX 90
NGC 4622 GX 167, 168, 184, A23
NGC 4622A GX 167, 168, 184, A23
NGC 4622B GX 167, 168, 184, A23
NGC 4623 GX 90
NGC 4625 GX 37
NGC 4626 GX 130
NGC 4627 GX 53, 54
NGC 4628 GX 130
NGC 4629 GX 110
NGC 4630 GX 110
NGC 4631 GX 53, 54
NGC 4632 GX 110
NGC 4633 GX 90
NGC 4634 GX 90
NGC 4635 GX 71, 72
NGC 4636 GX 110
NGC 4637 GX 90
NGC 4638 GX 90
NGC 4639 GX 90
NGC 4640 GX 90
NGC 4641 GX 90
NGC 4642 GX 110
NGC 4643 GX 110
NGC 4644 GX 24
NGC 4645 GX 184, A23
NGC 4645A GX 184, A23
NGC 4645B GX 184, A23
NGC 4646 GX 24
NGC 4647 GX 90
NGC 4648 GX 5
NGC 4649 GX 90
NGC 4650 GX 167, 184, A23
NGC 4650A GX 167, 184, A23
NGC 4651 GX 90
NGC 4652 GX 24
NGC 4653 GX 110
NGC 4654 GX 90
NGC 4655 GX 37
NGC 4656 GX 53, 54
NGC 4657 GX 53
NGC 4658 GX 130
NGC 4659 GX 90
NGC 4660 GX 90
NGC 4661 GX 167, 184, A23
NGC 4662 GX 53
NGC 4663 GX 130
NGC 4665 GX 110
NGC 4666 GX 110
NGC 4668 GX 110
NGC 4669 GX 24
NGC 4670 GX 71
NGC 4671 GX 130
NGC 4672 GX 184, A23
NGC 4673 GX 71
NGC 4674 GX 130
NGC 4675 GX 24
NGC 4676A GX 53
NGC 4676B GX 53
NGC 4677 GX 184, A23
NGC 4678 GX 110
NGC 4679 GX 167, A23
NGC 4680 GX 130
NGC 4681 GX 184, A23
NGC 4682 GX 130
NGC 4683 GX 184, A23
NGC 4684 GX 110
NGC 4685 GX 71
NGC 4686 GX 24
NGC 4687 GX 53
NGC 4688 GX 110
NGC 4689 GX 90
NGC 4690 GX 110
NGC 4691 GX 110
NGC 4692 GX 71
NGC 4693 GX 13
NGC 4694 GX 90
NGC 4695 GX 24
NGC 4696 GX 184, A23
NGC 4696A GX 184, A23
NGC 4696B GX 184, A23
NGC 4696C GX 167, 184, A23
NGC 4696D GX 184, A23
NGC 4696E GX 167, 184, A23
NGC 4697 GX 110, 130
NGC 4698 GX 90
NGC 4699 GX 130
NGC 4700 GX 130
NGC 4701 GX 110
NGC 4702 GX 71
NGC 4703 GX 130
NGC 4704 GX 37
NGC 4705 GX 110, 130
NGC 4706 GX 184, A23
NGC 4707 GX 24, 37
NGC 4708 GX 130
NGC 4709 GX 184, A23
NGC 4710 GX 90
NGC 4711 GX 53
NGC 4712 GX 71
NGC 4713 GX 90, 110
NGC 4714 GX 130
NGC 4715 GX 71
NGC 4716 GX 130
NGC 4717 GX 130
NGC 4718 GX 110, 130
NGC 4719 GX 53
NGC 4720 GX 110
NGC 4721 GX 71
NGC 4722 GX 130
NGC 4723 GX 130
NGC 4724 GX 130
NGC 4725 GX 71
NGC 4726 GX 130
NGC 4727 GX 130
NGC 4728 GX 71
NGC 4729 GX 184, A23
NGC 4730 GX 184, A23
NGC 4731 GX 130
NGC 4732 GX 24
NGC 4733 GX 90
NGC 4734 GX 110
NGC 4735 GX 71
NGC 4736 GX 37
NGC 4737 GX 53
NGC 4738 GX 71
NGC 4739 GX 130
NGC 4741 GX 37
NGC 4742 GX 130
NGC 4743 GX 184, A23
NGC 4744 GX 184, A23
NGC 4745 GX 71
NGC 4746 GX 90
NGC 4747 GX 71
NGC 4748 GX 130
NGC 4749 GX 13
NGC 4750 GX 13
NGC 4751 GX 184, A23
NGC 4753 GX 110
NGC 4754 GX 90
NGC 4755 OC 198, xviii, xix, xxii
NGC 4756 GX 130
NGC 4757 GX 130
NGC 4758 GX 90
NGC 4760 GX 130
NGC 4761 GX 130
NGC 4762 GX 90
NGC 4763 GX 130, 149
NGC 4764 GX 130
NGC 4765 GX 110
NGC 4766 GX 130
NGC 4767 GX 167, A23
NGC 4767A GX 167, A23
NGC 4767B GX 167, A23
NGC 4770 GX 130
NGC 4771 GX 110
NGC 4772 GX 110
NGC 4773 GX 130
NGC 4774 GX 53
NGC 4775 GX 130
NGC 4776 GX 130
NGC 4777 GX 130
NGC 4778 GX 130
NGC 4779 GX 90
NGC 4780 GX 130
NGC 4781 GX 130
NGC 4782 GX 130
NGC 4783 GX 130
NGC 4784 GX 130
NGC 4785 GX 184
NGC 4786 GX 130
NGC 4787 GX 71, A8
NGC 4788 GX 71, A8
NGC 4789 GX 71, A8
NGC 4789A GX 71, A8
NGC 4790 GX 130
NGC 4792 GX 130
NGC 4793 GX 71, A8
NGC 4794 GX 130
NGC 4795 GX 90
NGC 4798 GX 71, A8
NGC 4799 GX 110
NGC 4800 GX 37
NGC 4801 GX 24
NGC 4802 GX 130
NGC 4803 GX 90
NGC 4805 GX 71, A8
NGC 4806 GX 149, 167, A21
NGC 4807 GX 71, A8
NGC 4807A GX 71, A8
NGC 4808 GX 110
NGC 4809 GX 110
NGC 4810 GX 110
NGC 4811 GX 184, A23
NGC 4812 GX 184, A23
NGC 4813 GX 130
NGC 4814 GX 24
NGC 4815 OC 209
NGC 4816 GX 71, A8
NGC 4818 GX 130
NGC 4819 GX 71, A8
NGC 4820 GX 130
NGC 4821 GX 71, A8
NGC 4822 GX 130
NGC 4823 GX 130
NGC 4824 GX 71, A8
NGC 4825 GX 130
NGC 4826 GX 71
NGC 4827 GX 71, A8
NGC 4828 GX 71, A8
NGC 4829 GX 130
NGC 4830 GX 149
NGC 4831 GX 149
NGC 4832 GX 167, A23
NGC 4833 GC 209
NGC 4834 GX 23, 24
NGC 4835 GX 184
NGC 4835A GX 184
NGC 4836 GX 130
NGC 4837 GX 37
NGC 4838 GX 130
NGC 4839 GX 71, A8
NGC 4840 GX 71, A8
NGC 4841A GX 71, A8
NGC 4841B GX 71, A8
NGC 4842A GX 71, A8
NGC 4842B GX 71, A8
NGC 4843 GX 110
NGC 4845 GX 110
NGC 4846 GX 53
NGC 4847 GX 130
NGC 4848 GX 71, A8
NGC 4849 GX 71, A8
NGC 4850 GX 71, A8
NGC 4851 GX 71, A8
NGC 4852 OC 197, 198
NGC 4853 GX 71, A8
NGC 4854 GX 71, A8
NGC 4855 GX 130
NGC 4856 GX 130
NGC 4857 GX 13
NGC 4858 GX 71, A8
NGC 4859 GX 71, A8
NGC 4860 GX 71, A8
NGC 4861 GX 53
NGC 4862 GX 130
NGC 4863 GX 130
NGC 4864 GX 71, A8
NGC 4865 GX 71, A8
NGC 4866 GX 90
NGC 4867 GX 71, A8
NGC 4868 GX 53
NGC 4869 GX 71, A8
NGC 4870 GX 53
NGC 4871 GX 71, A8
NGC 4872 GX 71, A8
NGC 4873 GX 71, A8
NGC 4874 GX 71, A8
NGC 4875 GX 71, A8
NGC 4876 GX 71, A8
NGC 4877 GX 130
NGC 4878 GX 130
NGC 4880 GX 90
NGC 4881 GX 71, A8
NGC 4883 GX 71, A8
NGC 4885 GX 130
NGC 4886 GX 71, A8
NGC 4887 GX 130
NGC 4888 GX 130
NGC 4889 GX 71, A8
NGC 4890 GX 110
NGC 4892 GX 71, A8
NGC 4893 GX 53
NGC 4894 GX 71, A8
NGC 4895 GX 71, A8
NGC 4895A GX 71, A8
NGC 4896 GX 71, A8
NGC 4897 GX 130
NGC 4898 GX 71, A8
NGC 4899 GX 130
NGC 4900 GX 110
NGC 4901 GX 37
NGC 4902 GX 130
NGC 4903 GX 167, A21
NGC 4904 GX 110
NGC 4905 GX 167, A21
NGC 4906 GX 71, A8
NGC 4907 GX 71, A8
NGC 4908 GX 71, A8
NGC 4909 GX 184
NGC 4911 GX 71, A8
NGC 4914 GX 53
NGC 4915 GX 110
NGC 4916 GX 53
NGC 4917 GX 37
NGC 4918 GX 110
NGC 4919 GX 71, A8
NGC 4920 GX 130
NGC 4921 GX 71, A8
NGC 4922 GX 53, 71, A8
NGC 4923 GX 71, A8
NGC 4924 GX 130
NGC 4925 GX 130
NGC 4926 GX 71, A8
NGC 4926A GX 71, A8
NGC 4927 GX 71, A8
NGC 4928 GX 130
NGC 4929 GX 71, A8
NGC 4930 GX 184
NGC 4931 GX 71, A8
NGC 4932 GX 37
NGC 4933A GX 130
NGC 4933B GX 130
NGC 4933C GX 130
NGC 4934 GX 71, A8
NGC 4935 GX 90
NGC 4936 GX 167, A21
NGC 4938 GX 23, 24, 37
NGC 4939 GX 130
NGC 4940 GX 184
NGC 4941 GX 110, 130
NGC 4942 GX 130
NGC 4943 GX 71, A8
NGC 4944 GX 71, A8
NGC 4945 GX 184
NGC 4945A GX 184
NGC 4946 GX 184
NGC 4947 GX 167
NGC 4947A GX 167
NGC 4948 GX 130
NGC 4948A GX 130
NGC 4949 GX 53, 71, A8
NGC 4950 GX 184
NGC 4951 GX 130
NGC 4952 GX 53, 71, A8
NGC 4953 GX 167
NGC 4954 GX 5
NGC 4955 GX 149, 167, A21
NGC 4956 GX 53
NGC 4957 GX 71, A8
NGC 4958 GX 130
NGC 4959 GX 53
NGC 4960 GX 71, A8
NGC 4961 GX 71, A8
NGC 4963 GX 37
NGC 4964 GX 23
NGC 4965 GX 149
NGC 4966 GX 53, 71
NGC 4967 GX 23
NGC 4968 GX 149
NGC 4969 GX 90
NGC 4970 GX 149
NGC 4971 GX 71
NGC 4973 GX 23
NGC 4974 GX 23
NGC 4975 GX 110, 130
NGC 4976 GX 184
NGC 4977 GX 23
NGC 4978 GX 71
NGC 4979 GX 71
NGC 4980 GX 149, A21
NGC 4981 GX 130
NGC 4983 GX 71
NGC 4984 GX 130
NGC 4985 GX 37
NGC 4986 GX 53
NGC 4987 GX 23, 37
NGC 4988 GX 184
NGC 4989 GX 110, 130
NGC 4990 GX 110, 130
NGC 4991 GX 110
NGC 4992 GX 90
NGC 4993 GX 149
NGC 4994 GX 149
NGC 4995 GX 130
NGC 4996 GX 110
NGC 4997 GX 130
NGC 4998 GX 37
NGC 4999 GX 110
NGC 5000 GX 71
NGC 5001 GX 23
NGC 5002 GX 53
NGC 5003 GX 37
NGC 5004 GX 53, 71
NGC 5004A GX 53, 71
NGC 5005 GX 53
NGC 5006 GX 149
NGC 5007 GX 12, 13, 23
NGC 5009 GX 37
NGC 5010 GX 130
NGC 5011 GX 184
NGC 5011A GX 184
NGC 5011B GX 184
NGC 5011C GX 184
NGC 5012 GX 71
NGC 5013 GX 110
NGC 5014 GX 53
NGC 5015 GX 110
NGC 5016 GX 71
NGC 5017 GX 130
NGC 5018 GX 149
NGC 5019 GX 110
NGC 5020 GX 90
NGC 5021 GX 37
NGC 5022 GX 149
NGC 5023 GX 37
NGC 5024 GC 71
NGC 5025 GX 53
NGC 5026 GX 184
NGC 5027 GX 90
NGC 5028 GX 130
NGC 5029 GX 37
NGC 5030 GX 130
NGC 5031 GX 130
NGC 5032 GX 71
NGC 5033 GX 53
NGC 5034 GX 12, 13
NGC 5035 GX 130
NGC 5036 GX 110
NGC 5037 GX 130
NGC 5038 GX 130
NGC 5039 GX 110
NGC 5040 GX 23, 37
NGC 5041 GX 53
NGC 5042 GX 149
NGC 5043 OC 197
NGC 5044 GX 130
NGC 5045 Star Cld 208, 209
NGC 5046 GX 130
NGC 5047 GX 130
NGC 5048 GX 149
NGC 5049 GX 130
NGC 5050 GX 110
NGC 5051 GX 149
NGC 5052 GX 53, 71
NGC 5053 GC 71, 90
NGC 5054 GX 130
NGC 5055 GX 37
NGC 5056 GX 53
NGC 5057 GX 53
NGC 5058 GX 90
NGC 5059 GX 90
NGC 5060 GX 90
NGC 5061 GX 149
NGC 5062 GX 167
NGC 5063 GX 167
NGC 5064 GX 184
NGC 5065 GX 53
NGC 5066 GX 130
NGC 5068 GX 149
NGC 5070 GX 130
NGC 5071 GX 90
NGC 5072 GX 130
NGC 5073 GX 130
NGC 5074 GX 53
NGC 5075 GX 90
NGC 5076 GX 130
NGC 5077 GX 130
NGC 5078 GX 149
NGC 5079 GX 130
NGC 5080 GX 90
NGC 5081 GX 71
NGC 5082 GX 184
NGC 5083 GX 53
NGC 5084 GX 149
NGC 5085 GX 149
NGC 5086 GX 184
NGC 5087 GX 149
NGC 5088 GX 130
NGC 5089 GX 53
NGC 5090 GX 184
NGC 5090A GX 184
NGC 5090B GX 184
NGC 5091 GX 184
NGC 5092 GX 71
NGC 5093 GX 37, 53
NGC 5094 GX 130
NGC 5095 GX 110
NGC 5096 GX 53
NGC 5097 GX 130
NGC 5098A GX 53
NGC 5098B GX 53
NGC 5099 GX 130
NGC 5100 GX 90
NGC 5101 GX 149
NGC 5102 GX 167
NGC 5103 GX 37
NGC 5104 GX 110
NGC 5105 GX 130
NGC 5107 GX 53
NGC 5108 GX 167
NGC 5109 GX 23
NGC 5110 GX 130
NGC 5111 GX 130
NGC 5112 GX 53

NGC 7510 OC 18
NGC 7511 GX 82
NGC 7512 GX 46
NGC 7513 GX 142
NGC 7514 GX 46
NGC 7515 GX 82
NGC 7516 GX 64
NGC 7517 GX 102
NGC 7518 GX 82
NGC 7519 GX 82
NGC 7521 GX 102
NGC 7523 GX 82
NGC 7524 GX 102
NGC 7525 GX 82
NGC 7527 GX 64
NGC 7528 GX 82
NGC 7529 GX 82
NGC 7530 GX 102
NGC 7531 GX 177, 178
NGC 7532 GX 102
NGC 7533 GX 102
NGC 7534 GX 102
NGC 7535 GX 82
NGC 7536 GX 82
NGC 7537 GX 102
NGC 7538 BN 18
NGC 7539 GX 64
NGC 7540 GX 82
NGC 7541 GX 102
NGC 7542 GX 82
NGC 7543 GX 64
NGC 7544 GX 102
NGC 7545 GX 160
NGC 7546 GX 102
NGC 7547 GX 64
NGC 7548 GX 64
NGC 7549 GX 64
NGC 7550 GX 64
NGC 7552 GX 177
NGC 7553 GX 64
NGC 7554 GX 102
NGC 7556 GX 102
NGC 7557 GX 82
NGC 7558 GX 64
NGC 7559A GX 82
NGC 7559B GX 82
NGC 7562 GX 82
NGC 7562A GX 82
NGC 7563 GX 82
NGC 7564 GX 82
NGC 7566 GX 102
NGC 7567 GX 82
NGC 7568 GX 63, 64
NGC 7570 GX 82
NGC 7572 GX 63, 64
NGC 7573 GX 141, 142
NGC 7576 GX 102
NGC 7577 GX 82
NGC 7578A GX 63, 64
NGC 7578B GX 63, 64
NGC 7579 GX 82
NGC 7580 GX 82
NGC 7582 GX 177
NGC 7583 GX 82
NGC 7584 GX 82
NGC 7585 GX 102
NGC 7586 GX 82
NGC 7587 GX 82
NGC 7588 GX 63, 64
NGC 7589 GX 102
NGC 7590 GX 177
NGC 7591 GX 82
NGC 7592 GX 102
NGC 7593 GX 82
NGC 7594 GX 82
NGC 7596 GX 122
NGC 7597 GX 63, 64
NGC 7598 GX 63, 64
NGC 7599 GX 177
NGC 7600 GX 122
NGC 7601 GX 82
NGC 7602 GX 63, 64
NGC 7603 GX 102
NGC 7604 GX 82
NGC 7606 GX 122
NGC 7608 GX 82
NGC 7609 GX 82
NGC 7610 GX 82
NGC 7611 GX 82
NGC 7612 GX 82
NGC 7615 GX 82
NGC 7616 GX 82
NGC 7617 GX 81, 82
NGC 7618 GX 30
NGC 7619 GX 81, 82
NGC 7620 GX 63, 64
NGC 7621 GX 81, 82
NGC 7622 GX 192, 204
NGC 7623 GX 81, 82
NGC 7624 GX 63, 64
NGC 7625 GX 63, 64, 81, 82
NGC 7626 GX 81, 82
NGC 7628 GX 63, 64
NGC 7629 GX 101, 102
NGC 7630 GX 81, 82
NGC 7631 GX 81, 82
NGC 7632 GX 177
NGC 7633 GX 204
NGC 7634 GX 81, 82
NGC 7635 BN 18
NGC 7636 GX 141, 142, 159, 160
NGC 7637 GX 214
NGC 7638 GX 81, 82
NGC 7639 GX 81, 82
NGC 7640 GX 30, 45, 46
NGC 7641 GX 81, 82
NGC 7642 GX 101, 102
NGC 7643 GX 81, 82
NGC 7644 GX 81, 82
NGC 7645 GX 141, 142, 159, 160
NGC 7646 GX 121, 122
NGC 7647 GX 81, 82
NGC 7648 GX 81, 82
NGC 7649 GX 81, 82
NGC 7650 GX 192
NGC 7651 GX 81, 82
NGC 7652 GX 192
NGC 7653 GX 81, 82
NGC 7654 OC 18
NGC 7655 GX 204
NGC 7656 GX 141
NGC 7657 GX 192
NGC 7658A GX 159
NGC 7658B GX 159
NGC 7659 GX 81, 82
NGC 7660 GX 63
NGC 7661 GX 204
NGC 7662 PN 30
NGC 7663 GX 101, 102
NGC 7664 GX 63
NGC 7665 GX 121, 122
NGC 7667 GX 101, 102
NGC 7671 GX 81, 82
NGC 7672 GX 81, 82
NGC 7673 GX 63
NGC 7674 GX 81, 82
NGC 7675 GX 81
NGC 7676 GX 192
NGC 7677 GX 63
NGC 7678 GX 63
NGC 7679 GX 101
NGC 7680 GX 45
NGC 7681 GX 63, 81
NGC 7682 GX 101
NGC 7683 GX 81
NGC 7684 GX 101
NGC 7685 GX 101
NGC 7686 OC 30
NGC 7687 GX 101
NGC 7688 GX 63
NGC 7689 GX 192
NGC 7690 GX 177, 192
NGC 7691 GX 81
NGC 7692 GX 101, 121
NGC 7693 GX 101
NGC 7694 GX 101
NGC 7695 GX 101
NGC 7696 GX 101
NGC 7697 GX 204
NGC 7698 GX 63
NGC 7699 GX 101
NGC 7700 GX 101
NGC 7701 GX 101
NGC 7702 GX 192
NGC 7703 GX 81
NGC 7704 GX 101
NGC 7705 GX 101
NGC 7706 GX 101
NGC 7707 GX 30
NGC 7708 OC 8
NGC 7709 GX 121
NGC 7710 GX 101
NGC 7711 GX 81
NGC 7712 GX 63
NGC 7713 GX 159
NGC 7713A GX 159
NGC 7714 GX 101
NGC 7715 GX 101
NGC 7716 GX 101
NGC 7717 GX 121
NGC 7718 GX 63
NGC 7719 GX 141
NGC 7720 GX 63
NGC 7721 GX 121
NGC 7722 GX 81
NGC 7723 GX 121
NGC 7724 GX 121
NGC 7725 GX 101
NGC 7726 GX 63
NGC 7727 GX 121
NGC 7728 GX 63
NGC 7729 GX 45, 63
NGC 7730 GX 141
NGC 7731 GX 101
NGC 7732 GX 101
NGC 7733 GX 204
NGC 7734 GX 204
NGC 7735 GX 63
NGC 7736 GX 141
NGC 7737 GX 63
NGC 7738 GX 101
NGC 7739 GX 101
NGC 7740 GX 63
NGC 7741 GX 63
NGC 7742 GX 81
NGC 7743 GX 81
NGC 7744 GX 177
NGC 7745 GX 63
NGC 7746 GX 101
NGC 7747 GX 63
NGC 7749 GX 141, 159
NGC 7750 GX 101
NGC 7751 GX 81
NGC 7752 GX 45, 63
NGC 7753 GX 45, 63
NGC 7754 GX 121
NGC 7755 GX 159
NGC 7756 GX 101
NGC 7757 GX 101
NGC 7758 GX 141
NGC 7759 GX 121
NGC 7760 GX 45
NGC 7761 GX 121
NGC 7762 OC 8
NGC 7763 GX 121
NGC 7764 GX 159, 177
NGC 7765 GX 63
NGC 7766 GX 63
NGC 7767 GX 63
NGC 7768 GX 63
NGC 7769 GX 63
NGC 7770 GX 63
NGC 7771 GX 63
NGC 7772 OC 81
NGC 7773 GX 45
NGC 7774 GX 81
NGC 7775 GX 63
NGC 7776 GX 121
NGC 7777 GX 63
NGC 7778 GX 81
NGC 7779 GX 81
NGC 7780 GX 81
NGC 7781 GX 81
NGC 7782 GX 81
NGC 7783A/B GX 101
NGC 7783C GX 101
NGC 7784 GX 63
NGC 7785 GX 81, 101
NGC 7786 GX 63
NGC 7787 GX 101
NGC 7788 OC 18
NGC 7789 OC 18
NGC 7790 OC 18
NGC 7792 GX 81
NGC 7793 GX 159
NGC 7794 GX 81
NGC 7796 GX 192
NGC 7797 GX 101
NGC 7798 GX 63
NGC 7799 GX 45
NGC 7800 GX 81
NGC 7802 GX 81
NGC 7803 GX 81
NGC 7805 GX 45
NGC 7806 GX 45
NGC 7807 GX 141
NGC 7808 GX 121
NGC 7809 GX 101
NGC 7810 GX 81
NGC 7811 GX 101
NGC 7812 GX 159
NGC 7813 GX 121
NGC 7814 GX 81
NGC 7816 GX 81
NGC 7817 GX 63
NGC 7818 GX 81
NGC 7819 GX 45
NGC 7820 GX 81, 101
NGC 7821 GX 121
NGC 7822 BN 8
NGC 7823 GX 192, 204
NGC 7824 GX 81
NGC 7825 GX 81, 101
NGC 7827 GX 81, 101
NGC 7828 GX 121
NGC 7829 GX 121
NGC 7831 GX 45
NGC 7832 GX 101
NGC 7834 GX 81
NGC 7835 GX 81
NGC 7836 GX 45
NGC 7837 GX 81
NGC 7838 GX 81
Om-2 Cyg OC 32
Pal 1 GC 7
Pal 2 GC 59
Pal 3 GC 113
Pal 4 GC 72
Pal 5 GC 108
Pal 6 GC 146
Pal 8 GC 145
Pal 10 GC 66, 67
Pal 11 GC 124, 125
Pal 12 GC 143
Pal 13 GC 82
Pal 14 GC 87, 88
Pal 15 GC 107
Parsamyan 15 BN 115
PGC 3311 GX 140
PGC 3562 GX 158
PGC 4427 GX 191
PGC 5993 GX 203, 213
PGC 9006 GX 190, 191
PGC 9803 GX 175
PGC 11343 GX 175
PGC 15321 GX 174
PGC 16516 GX 189
PGC 18246 GX 201
PGC 18727 GX 211
PGC 18976 GX 201
PGC 24283 GX 14
PGC 32573 GX 132, 151
PGC 38038 GX 198
PGC 47626 GX 167
PGC 55070 GX 165
PGC 60692 GX 207
PGC 63102 GX 194
PGC 64536 GX 179
PGC 67298 GX 178, 193
PGC 67429 GX 193
PGC 67483 GX 178
Pi 2 OC 187
Pi 3 OC 171
Pi 4 OC 187
Pi 5 OC 170, 171
Pi 7 OC 170, 171
Pi 8 OC 187
Pi 11 OC 186
Pi 12 OC 186
Pi 13 OC 186, 199
Pi 14 OC 199
Pi 15 OC 186
Pi 16 OC 199
Pi 19 OC 197
Pi 20 OC 196
Pi 21 OC 196
Pi 22 OC 182, 196
Pi 23 OC 182
Pi 24 OC 164
PK 80-6.1 PN 47
PMH79 1 OC 9
PN G0.0−6.8 PN 163
PN G0.1−5.6 PN 163
PN G0.1−2.3 PN 146, 164
PN G0.1−1.1 PN 146, 164
PN G0.1+2.6 PN 146
PN G0.1+4.3 PN 146
PN G0.1+17.2 PN 146, 147
PN G0.2−4.6 PN 163
PN G0.2−1.9 PN 146, 164
PN G0.3−4.6 PN 163
PN G0.3−2.8 PN 163, 164
PN G0.3+6.9 PN 146
PN G0.4−2.9 PN 163, 164
PN G0.4−1.9 PN 146, 164
PN G0.5−3.1 PN 163, 164
PN G0.5−1.6 PN 146, 164
PN G0.6−2.3 PN 145, 146, 163, 164
PN G0.6−1.3 PN 146, 164
PN G0.7−7.4 PN 163
PN G0.7−3.7 PN 163, 164
PN G0.7−2.7 PN 145, 146, 163, 164
PN G0.7+3.2 PN 146
PN G0.7+4.7 PN 146
PN G0.8−7.6 PN 163
PN G0.8−1.5 PN 146
PN G0.9−4.8 PN 163
PN G0.9−2.0 PN 145, 146, 163, 164
PN G1.0−2.6 PN 145, 146, 163, 164
PN G1.0+1.9 PN 146
PN G1.1−1.6 PN 146
PN G1.2−3.9 PN 145, 146, 163, 164
PN G1.2−3.0 PN 145, 146, 163, 164
PN G1.2+2.1 PN 146
PN G1.3−1.2 PN 146
PN G1.4−3.4 PN 145, 146, 163, 164
PN G1.4+5.3 PN 146
PN G1.5−6.7 PN 163
PN G1.6−1.3 PN 146
PN G1.7−4.6 PN 145, 163
PN G1.7−4.4 PN 145, 163
PN G1.7−1.6 PN 145, 146
PN G1.7+5.7 PN 146
PN G1.8−3.8 PN 145, 163
PN G2.0−6.2 PN 163
PN G2.0−2.0 PN 145, 146
PN G2.1−4.2 PN 145, 163
PN G2.1−2.2 PN 145, 146
PN G2.1+3.3 PN 146
PN G2.2−9.4 PN 163
PN G2.2−6.3 PN 163
PN G2.2−2.7 PN 145, 146
PN G2.2−2.5 PN 145, 146
PN G2.2+0.5 PN 146
PN G2.3−7.8 PN 163
PN G2.3−3.4 PN 145
PN G2.3+2.2 PN 146
PN G2.4−3.7 PN 145
PN G2.4−3.2 PN 145
PN G2.5−5.4 PN 145, 163
PN G2.5−1.7 PN 145, 146
PN G2.6−3.4 PN 145
PN G2.6+2.1 PN 146
PN G2.6+4.2 PN 146
PN G2.6+8.1 PN 146
PN G2.7−4.8 PN 145
PN G2.8−2.2 PN 145, 146
PN G2.8+1.7 PN 146
PN G2.8+1.8 PN 146
PN G2.9−3.9 PN 145
PN G2.9+6.5 PN 146
PN G3.0−2.6 PN 145, 146
PN G3.1+2.9 PN 146
PN G3.1+3.4 PN 146
PN G3.2−6.2 PN 145, 163
PN G3.2−4.4 PN 145
PN G3.3−7.5 PN 145, 163
PN G3.3−4.6 PN 145
PN G3.4−4.8 PN 145
PN G3.6−2.3 PN 145, 146
PN G3.6+3.1 PN 146
PN G3.7−4.6 PN 145
PN G3.7+7.9 PN 146
PN G3.8−17.1 PN 163
PN G3.8−4.5 PN 145
PN G3.8−4.3 PN 145
PN G3.8+5.3 PN 146
PN G3.9−14.9 PN 163
PN G3.9−3.1 PN 145
PN G3.9−2.3 PN 145, 146
PN G3.9+1.6 PN 146
PN G4.0−11.1 PN 163
PN G4.0−5.8 PN 145
PN G4.0−3.0 PN 145
PN G4.1−3.8 PN 145
PN G4.2−5.9 PN 145
PN G4.2−4.3 PN 145
PN G4.2−3.2 PN 145
PN G4.3−2.6 PN 145, A17
PN G4.3+1.8 PN 146
PN G4.5+6.8 PN 146
PN G4.6+6.0 PN 146
PN G4.7−11.8 PN 163
PN G4.8−22.7 PN 162
PN G4.8−5.0 PN 145
PN G4.8+2.0 PN 146
PN G4.9−4.9 PN 145
PN G4.9+4.9 PN 146
PN G5.0−3.9 PN 145
PN G5.0+3.0 PN 146
PN G5.0+4.4 PN 146
PN G5.1−8.9 PN 145
PN G5.1−3.0 PN 145, A17
PN G5.2−18.6 PN 163
PN G5.2+4.2 PN 146
PN G5.2+5.6 PN 146
PN G5.4−1.9 PN 145, A17
PN G5.5−4.0 PN 145, A17
PN G5.5−2.5 PN 145, A17
PN G5.5+2.7 PN 146
PN G5.5+6.1 PN 146
PN G5.6−4.7 PN 145
PN G5.7−5.3 PN 145
PN G5.7−3.6 PN 145, A17
PN G5.8+5.1 PN 146
PN G5.9−2.6 PN 145, A17
PN G6.0−41.9 PN 161
PN G6.0−3.6 PN 145, A17
PN G6.0+2.8 PN 146
PN G6.0+3.1 PN 146
PN G6.1+8.3 PN 146
PN G6.2−3.7 PN 145
PN G6.2+1.0 PN 146, A17
PN G6.3+3.3 PN 146
PN G6.3+4.4 PN 146
PN G6.4−4.6 PN 145
PN G6.4+2.0 PN 146
PN G6.5−3.1 PN 145, A17
PN G6.7−2.2 PN 145, A17
PN G6.8−19.8 PN 162, 163
PN G6.8−8.6 PN 145
PN G6.8−3.4 PN 145
PN G6.8+2.0 PN 146
PN G6.8+2.3 PN 146
PN G6.8+4.1 PN 146
PN G7.0−6.8 PN 145
PN G7.0−6.0 PN 145
PN G7.0+6.3 PN 146
PN G7.5+4.3 PN 146
PN G7.5+7.4 PN 146
PN G7.6+6.9 PN 146
PN G7.8−4.4 PN 145
PN G7.8−3.7 PN 145
PN G7.9+10.1 PN 126, 127
PN G8.1−4.7 PN 145
PN G8.2−4.8 PN 145
PN G8.2+6.8 PN 146
PN G8.3−1.1 PN 145, A17
PN G8.4−3.6 PN 145
PN G8.6−7.0 PN 145
PN G8.6−2.6 PN 145
PN G8.8+5.2 PN 146
PN G9.0+4.1 PN 146
PN G9.3+2.8 PN 145, 146
PN G9.3+4.1 PN 146
PN G9.4−9.8 PN 145
PN G9.6−10.6 PN 145
PN G9.6+10.5 PN 126
PN G9.8−7.5 PN 145
PN G9.8−4.6 PN 145
PN G10.4+4.5 PN 126, 146
PN G10.6+3.2 PN 145, 146
PN G10.7−6.7 PN 145
PN G10.7+7.4 PN 126
PN G10.8+18.0 PN 127
PN G11.0−5.1 PN 145
PN G11.0+6.2 PN 126
PN G11.1+7.0 PN 126
PN G11.1+11.5 PN 126
PN G11.3−9.4 PN 145
PN G11.3+2.8 PN 126, 145, 146
PN G11.7−6.6 PN 145
PN G11.7−0.0 PN 145
PN G11.9+4.2 PN 126
PN G12.2+4.9 PN 126
PN G12.5−9.8 PN 145
PN G12.6−2.7 PN 145
PN G13.0−4.3 PN 145
PN G13.1+4.1 PN 126
PN G13.3+32.7 PN 107
PN G13.4−3.9 PN 145
PN G13.7−10.6 PN 145
PN G13.8−7.9 PN 145
PN G13.8−2.8 PN 145
PN G14.0−5.5 PN 145

PN G321.3−16.7 PN 207
PN G321.3+2.8 PN 196
PN G321.8+1.9 PN 196
PN G322.1−6.6 PN 196, 207
PN G322.4−2.6 PN 196
PN G322.4−0.1 PN 196
PN G323.1−2.5 PN 196
PN G323.9+2.4 PN 196
PN G324.0+3.5 PN 196
PN G324.1+9.0 PN 183
PN G324.2+2.5 PN 196
PN G324.8−1.1 PN 196
PN G325.0+3.2 PN 196
PN G325.4−4.0 PN 196
PN G325.8−12.8 PN 207
PN G325.8+4.5 PN 182, 196
PN G326.0−6.5 PN 196
PN G326.1−1.9 PN 196
PN G327.1−2.2 PN 196
PN G327.1−1.8 PN 196
PN G327.5+13.3 PN 183
PN G327.7−5.4 PN 196
PN G327.8−7.2 PN 196
PN G327.8−6.1 PN 196
PN G327.8−1.6 PN 196
PN G327.9−4.3 PN 196
PN G328.2+1.3 PN 196
PN G328.9−2.4 PN 196
PN G329.0+1.9 PN 182, 196
PN G329.3−2.8 PN 196
PN G329.4−2.7 PN 196
PN G329.5−2.2 PN 196
PN G329.5+1.7 PN 182, 196
PN G330.2+5.9 PN 182
PN G330.6−3.6 PN 196
PN G330.6−2.1 PN 196
PN G330.7+4.1 PN 182
PN G330.9+4.3 PN 182
PN G331.0−2.7 PN 196
PN G331.1−5.7 PN 196
PN G331.4−3.5 PN 196
PN G331.4+0.5 PN 182, 196
PN G331.5−3.9 PN 196
PN G331.5−2.7 PN 196
PN G331.7−1.0 PN 182, 196
PN G332.0−3.3 PN 196
PN G332.2+3.5 PN 182
PN G332.3−4.2 PN 196
PN G332.8−16.4 PN 195
PN G332.9−9.9 PN 195
PN G333.4−4.0 PN 196
PN G333.4+1.1 PN 182
PN G334.3−1.4 PN 182
PN G334.8−7.4 PN 195, 196
PN G335.2−3.6 PN 181, 182, 196
PN G335.4−1.1 PN 182
PN G335.4+9.2 PN 182
PN G335.5+12.4 PN 165
PN G335.6−4.0 PN 181, 182, 196
PN G335.9−3.6 PN 181, 182
PN G336.2−6.9 PN 195
PN G336.2+1.9 PN 182
PN G336.3−5.6 PN 181, 195, 196
PN G336.8−7.2 PN 195
PN G336.9−11.5 PN 195
PN G336.9+8.3 PN 182
PN G337.4−9.1 PN 195
PN G337.4+1.6 PN 182
PN G337.5−5.1 PN 181
PN G337.6−4.2 PN 181
PN G339.9+88.4 PN 71
PN G340.4−14.1 PN 195
PN G340.8+10.8 PN 165
PN G340.8+12.3 PN 165
PN G340.9−4.6 PN 181
PN G341.2−24.6 PN 194
PN G341.5−9.1 PN 181
PN G342.1+27.5 PN 147, 148
PN G342.5−14.3 PN 181, 195
PN G342.7+0.7 PN 181, A22
PN G342.8−6.6 PN 181
PN G342.9−4.9 PN 181
PN G342.9−2.0 PN 181
PN G343.0−1.7 PN 181, A22
PN G343.3−0.6 PN 181, A22
PN G343.4+11.9 PN 165
PN G343.5−7.8 PN 181
PN G343.6+3.7 PN 164, 181, 182, A22
PN G343.9+0.8 PN 181, A22
PN G344.2−1.2 PN 181, A22
PN G344.2+4.7 PN 164, 165
PN G344.4−6.1 PN 181
PN G344.4+2.8 PN 164, 181, 182, A22
PN G344.8+3.4 PN 164, A22
PN G345.0−4.9 PN 181
PN G345.0+3.4 PN 164, A22
PN G345.0+4.3 PN 164, A22
PN G345.2−1.2 PN 181
PN G345.3−10.2 PN 181
PN G345.5+15.1 PN 165
PN G345.6+6.7 PN 164, 165
PN G345.9−11.2 PN 181
PN G345.9+3.0 PN 164, A22
PN G346.0+8.5 PN 165
PN G346.3−6.8 PN 181
PN G346.9+12.4 PN 165
PN G347.4+5.8 PN 164
PN G347.7+2.0 PN 164
PN G348.0+6.3 PN 164
PN G348.4−4.1 PN 164, 181
PN G348.8−9.0 PN 181
PN G349.2−3.5 PN 164
PN G349.3−4.2 PN 164, 181
PN G349.8+4.4 PN 164
PN G350.1−3.9 PN 164
PN G350.5−5.0 PN 164
PN G350.8−2.4 PN 164
PN G350.9+4.4 PN 164
PN G351.0−10.4 PN 181
PN G351.1+4.8 PN 164
PN G351.2+5.2 PN 164
PN G351.3+7.6 PN 164
PN G351.6−6.2 PN 164
PN G351.7−10.9 PN 181
PN G351.9−1.9 PN 164
PN G351.9+9.0 PN 164
PN G352.0−4.6 PN 164
PN G352.1+5.1 PN 164
PN G352.6+0.1 PN 164
PN G352.6+3.0 PN 164
PN G352.8−0.2 PN 164
PN G352.9−7.5 PN 163, 164
PN G352.9+11.4 PN 146
PN G353.2−5.2 PN 164
PN G353.3+6.3 PN 164
PN G353.5−4.9 PN 164
PN G353.7−12.8 PN 163, 180, 181
PN G353.7+6.3 PN 164
PN G354.2+4.3 PN 164
PN G354.4−7.8 PN 163
PN G354.5+3.3 PN 164
PN G354.9+3.5 PN 164
PN G355.1−6.9 PN 163, 164
PN G355.1−2.9 PN 164, A20
PN G355.1+2.3 PN 164
PN G355.2−2.5 PN 164, A20
PN G355.4−4.0 PN 164, A20
PN G355.4−2.4 PN 164, A20
PN G355.6−2.7 PN 164, A20
PN G355.7−3.5 PN 164, A20
PN G355.7−3.4 PN 164, A20
PN G355.7−3.0 PN 164, A20
PN G355.9−4.2 PN 164, A20
PN G355.9+2.7 PN 164
PN G355.9+3.6 PN 164
PN G356.1−3.3 PN 164, A20
PN G356.1+2.7 PN 164
PN G356.2−4.4 PN 164, A20
PN G356.3−6.2 PN 163, 164
PN G356.5−3.9 PN 164, A20
PN G356.5−3.6 PN 164, A20
PN G356.5−2.3 PN 164, A20
PN G356.5+1.5 PN 164
PN G356.5+5.1 PN 146
PN G356.6−7.8 PN 163
PN G356.7−6.4 PN 163
PN G356.7−4.8 PN 163, 164, A20
PN G356.8−11.7 PN 163
PN G356.8−5.4 PN 163, 164
PN G356.8+3.3 PN 146, 164
PN G356.9−5.8 PN 163, 164
PN G356.9+4.4 PN 146, 164
PN G356.9+4.5 PN 146, 164
PN G357.0+2.4 PN 164
PN G357.1−6.1 PN 163
PN G357.1−4.7 PN 163, 164, A20
PN G357.1+1.9 PN 164
PN G357.1+3.6 PN 146, 164
PN G357.1+4.4 PN 146
PN G357.2−4.5 PN 163, 164, A20
PN G357.2+1.4 PN 164
PN G357.2+2.0 PN 164
PN G357.2+7.4 PN 146
PN G357.3+3.3 PN 146, 164
PN G357.3+4.0 PN 146
PN G357.4−4.6 PN 163, 164, A20
PN G357.4−3.5 PN 164, A20
PN G357.4−3.2 PN 164, A20
PN G357.5+3.1 PN 146, 164
PN G357.5+3.2 PN 146, 164
PN G357.6−3.3 PN 164, A20
PN G357.6+1.0 PN 164
PN G357.6+1.7 PN 146, 164
PN G357.6+2.6 PN 146, 164
PN G357.9−5.1 PN 163, 164
PN G357.9−3.8 PN 163, 164, A20
PN G358.0−5.1 PN 163, 164
PN G358.0+2.6 PN 146, 164
PN G358.0+7.5 PN 146
PN G358.0+9.3 PN 146
PN G358.2−1.1 PN 164, A20
PN G358.2+3.5 PN 146
PN G358.2+3.6 PN 146
PN G358.2+4.2 PN 146
PN G358.3−2.5 PN 164, A20
PN G358.3+1.2 PN 146, 164
PN G358.3+3.0 PN 146
PN G358.4+3.3 PN 146
PN G358.5−4.2 PN 163, 164, A20
PN G358.5−2.5 PN 164, A20
PN G358.5+2.6 PN 146
PN G358.5+2.9 PN 146
PN G358.5+3.7 PN 146
PN G358.5+5.4 PN 146
PN G358.6−5.5 PN 163
PN G358.6+1.8 PN 146, 164
PN G358.6+7.8 PN 146
PN G358.7−5.2 PN 163, 164
PN G358.7−2.7 PN 164, A20
PN G358.7+5.2 PN 146
PN G358.8−0.0 PN 146, 164
PN G358.8+3.0 PN 146
PN G358.8+4.0 PN 146
PN G358.8+4.1 PN 146
PN G358.9−3.7 PN 163, 164, A20
PN G358.9−0.7 PN 164
PN G358.9+3.2 PN 146
PN G358.9+3.4 PN 146
PN G359.0−4.8 PN 163, 164
PN G359.0−4.1 PN 163, 164, A20
PN G359.0+2.8 PN 146
PN G359.1−2.9 PN 164, A20
PN G359.1−2.3 PN 164, A20
PN G359.1−1.7 PN 164, A20
PN G359.1+15.1 PN 146
PN G359.2−33.5 PN 179
PN G359.2+1.2 PN 146
PN G359.2+4.7 PN 146
PN G359.3−3.1 PN 163, 164, A20
PN G359.3−1.8 PN 164
PN G359.3−0.9 PN 146, 164
PN G359.3+1.4 PN 146
PN G359.3+3.6 PN 146
PN G359.4−3.4 PN 163, 164, A20
PN G359.4+2.3 PN 146
PN G359.5+2.6 PN 146
PN G359.6−4.8 PN 163
PN G359.6+2.2 PN 146
PN G359.7−4.4 PN 163, 164
PN G359.7−2.6 PN 164, A20
PN G359.7−1.8 PN 164
PN G359.8−7.2 PN 163
PN G359.8+2.4 PN 146
PN G359.8+3.7 PN 146
PN G359.8+5.2 PN 146
PN G359.8+5.6 PN 146
PN G359.8+6.9 PN 146
PN G359.9−5.4 PN 163
PN G359.9−4.5 PN 163, 164
PN G359.9+5.1 PN 146
Poole J1855.0+1047 OC 85
PWM78 2 GC 106, 126
PWM78 3 OC 18
RCW 47 BN 199
RCW 50 BN 199
RCW 57 BN 198
RCW 58 BN 209
RCW 102 BN 182, 196
RCW 103 BN 182, 196
RCW 126 BN 164
RCW 132 BN 164
Ro 1 OC 66, 84, 85
Ro 2 OC 66
Ro 3 OC 66
Ro 4 OC 48, 66
Ro 5 OC 48
Ro 6 OC 48, A2
Ru 1 OC 135, 136
Ru 2 OC 154, 172
Ru 3 OC 154, 172
Ru 4 OC 135
Ru 5 OC 154
Ru 6 OC 135
Ru 7 OC 135
Ru 8 OC 135
Ru 10 OC 154
Ru 11 OC 154
Ru 12 OC 154
Ru 13 OC 154
Ru 14 OC 172
Ru 15 OC 153, 154
Ru 16 OC 153, 154
Ru 17 OC 153, 154
Ru 18 OC 153
Ru 19 OC 153
Ru 20 OC 153
Ru 21 OC 171
Ru 22 OC 153, 171
Ru 23 OC 153
Ru 24 OC 135
Ru 25 OC 153
Ru 26 OC 135
Ru 27 OC 153
Ru 28 OC 171
Ru 29 OC 153
Ru 30 OC 171
Ru 31 OC 171
Ru 32 OC 153
Ru 33 OC 153
Ru 34 OC 153
Ru 35 OC 171
Ru 36 OC 153
Ru 37 OC 134, 135, 153
Ru 38 OC 153
Ru 39 OC 153
Ru 40 OC 153
Ru 41 OC 153
Ru 42 OC 153
Ru 43 OC 153
Ru 44 OC 153
Ru 45 OC 134
Ru 46 OC 153
Ru 47 OC 171
Ru 48 OC 171
Ru 49 OC 153
Ru 50 OC 171
Ru 51 OC 171
Ru 52 OC 171
Ru 53 OC 153
Ru 54 OC 171
Ru 55 OC 171
Ru 56 OC 171, 187
Ru 57 OC 153
Ru 58 OC 171
Ru 59 OC 171
Ru 60 OC 187
Ru 61 OC 171
Ru 62 OC 153
Ru 63 OC 187
Ru 64 OC 170, 171, 187
Ru 65 OC 187
Ru 66 OC 170, 171
Ru 67 OC 187
Ru 68 OC 170
Ru 69 OC 186, 187
Ru 70 OC 186, 187
Ru 71 OC 186, 187
Ru 72 OC 170
Ru 73 OC 186
Ru 74 OC 170
Ru 75 OC 199
Ru 76 OC 186, 199
Ru 77 OC 199
Ru 78 OC 199
Ru 79 OC 199
Ru 81 OC 186
Ru 82 OC 199
Ru 83 OC 199
Ru 84 OC 210
Ru 85 OC 199
Ru 86 OC 199
Ru 87 OC 186
Ru 88 OC 210
Ru 89 OC 199
Ru 90 OC 199
Ru 91 OC 199
Ru 92 OC 199
Ru 94 OC 209
Ru 95 OC 198
Ru 96 OC 198, 209
Ru 97 OC 198, 209
Ru 98 OC 209
Ru 99 OC 209
Ru 100 OC 198, 209
Ru 101 OC 198, 209
Ru 103 OC 198
Ru 104 OC 198
Ru 105 OC 198
Ru 106 GC 184, 198
Ru 107 OC 208, 209
Ru 108 OC 197
Ru 110 OC 208
Ru 111 OC 197
Ru 112 OC 196, 197, 208
Ru 113 OC 196
Ru 114 OC 196
Ru 115 OC 196
Ru 116 OC 182, 196
Ru 117 OC 182, 196
Ru 119 OC 182, 196
Ru 120 OC 182
Ru 121 OC 182
Ru 122 OC 164, 181, A22
Ru 123 OC 164
Ru 124 OC 164, 181
Ru 126 OC 164
Ru 127 OC 164, A20
Ru 128 OC 164, A20
Ru 129 OC 146, 164
Ru 130 OC 164
Ru 131 OC 146, 164
Ru 133 OC 146
Ru 134 OC 146, 164
Ru 135 OC 126
Ru 136 OC 145, 146, A17
Ru 137 OC 145, 146, A17
Ru 138 OC 145, 146, A17
Ru 139 OC 145, 146, A17
Ru 140 OC 163
Ru 141 OC 126
Ru 142 OC 125, 126
Ru 143 OC 125, 126
Ru 144 OC 125, 126
Ru 145 OC 145
Ru 146 OC 145
Ru 148 OC 42
Ru 149 OC 154
Ru 150 OC 154
Ru 151 OC 135
Ru 152 OC 171
Ru 153 OC 171
Ru 154 OC 187
Ru 155 OC 171
Ru 157 OC 153
Ru 158 OC 170
Ru 159 OC 199
Ru 160 OC 186
Ru 161 OC 199
Ru 162 OC 199, 209, 210
Ru 163 OC 209
Ru 164 OC 198
Ru 165 OC 198
Ru 167 OC 197
Ru 168 OC 146
Ru 169 OC 145, 146, A17
Ru 170 OC 126
Ru 171 OC 125, 126
Ru 172 OC 48
Ru 173 OC 47, 48
Ru 174 OC 47, 48
Ru 175 OC 47
Sch 1 OC 199
Sh 1 OC 198, 199
Sh2-1 BN 147
Sh2-3 BN 164
Sh2-9 BN 147
Sh2-13 BN 164
Sh2-16 BN 146, 164
Sh2-35 BN 145
Sh2-46 BN 126
Sh2-53 BN 126
Sh2-54 BN 126
Sh2-55 BN 125, 126
Sh2-64 BN 106
Sh2-82 BN 66
Sh2-84 BN 66
Sh2-88 BN 66
Sh2-90 BN 66
Sh2-101 BN 48
Sh2-104 BN 48
Sh2-108 BN 48, A2
Sh2-112 BN 32
Sh2-115 BN 32
Sh2-129 BN 19
Sh2-132 BN 19
Sh2-155 BN 8, 18, 19
Sh2-157a BN 18
Sh2-188 PN 29
Sh2-205 BN 28
Sh2-224 BN 42
Sh2-231 BN 59
Sh2-235 BN 59
Sh2-240 BN 77
Sh2-241 BN 58
Sh2-247 BN 76
Sh2-257 BN 76, 96
Sh2-261 BN 96
Sh2-264 BN 96
Sh2-276 BN 116
Sh2-278 BN 116, 117, 136, 137
Sh2-282 BN 115, 116
Sh2-301 BN 154
Sh2-302 BN 135
Sh2-307 BN 153
Sh2-311 BN 153
Skiff J0058.4+6828 OC 8
Skiff J0458.2+4301 OC 42
Skiff J0507.2+3050 OC 59
Skiff J0614.8+1252 OC 96
Skiff J0619.3+1832 OC 76
Skiff J1942.3+3839 OC 48
Skiff J2330.2+6015 OC 18
SL 4 DN 186
SL 7 DN 182
SL 8 DN 182
SL 11 DN 165
SL 17 DN 181, A22
SSWZ94 2 OC 171
SSWZ94 4 GC 209
SSWZ94 6 OC 48
St 2 OC 29
St 3 OC 8, 17, 29
St 4 OC 29
St 5 OC 17
St 8 OC 59
St 10 OC 59
St 11 OC 18
St 12 OC 18
St 13 OC 198
St 14 OC 198, 209
St 15 OC 198
St 16 OC 197, 208, 209
St 17 OC 8, 18
St 18 OC 8
St 19 OC 18
St 20 OC 8, 18
St 21 OC 18
St 23 OC 28
St 24 OC 18
Ste 1 OC 49
Terzan 1 GC 164
Terzan 2 GC 164
Terzan 3 GC 165
Terzan 4 GC 164
Terzan 5 GC 146
Terzan 6 GC 164, A20
Terzan 7 GC 162, 163
Terzan 8 GC 162
Terzan 9 GC 145, 146
Terzan 10 GC 145, 146, A17
Terzan 11 GC 145, A17
Tom 1 OC 154
Tom 2 OC 154
Tom 4 OC 29
Tom 5 OC 28
Ton 2 GC 164
Tr 1 OC 29
Tr 2 OC 29
Tr 3 OC 17
Tr 5 OC 95, 96
Tr 6 OC 153
Tr 7 OC 153
Tr 9 OC 153
Tr 10 OC 186, 187
Tr 11 OC 199
Tr 12 OC 199
Tr 13 OC 199

Tr 14 OC 199
Tr 15 OC 199
Tr 16 OC 199
Tr 17 OC 198, 199
Tr 18 OC 198
Tr 19 OC 198
Tr 20 OC 198
Tr 21 OC 197, 208
Tr 22 OC 197
Tr 23 OC 196
Tr 24 OC 164, 181, A22
Tr 25 OC 164
Tr 26 OC 146, 164
Tr 27 OC 164, A20
Tr 28 OC 164, A20
Tr 29 OC 164, 181
Tr 30 OC 163, 164, A20
Tr 31 OC 145, 146
Tr 32 OC 126
Tr 33 OC 145
Tr 34 OC 125, 126
Tr 35 OC 105, A14
UGC 3 GX 63
UGC 4 GX 101
UGC 5 GX 101
UGC 6 GX 63
UGC 10 GX 81
UGC 11 GX 63
UGC 12 GX 45, 63
UGC 13 GX 63
UGC 14 GX 63
UGC 17 GX 81
UGC 23 GX 81
UGC 27 GX 81, 101
UGC 29 GX 63
UGC 31 GX 63, 81
UGC 33 GX 81, 101
UGC 35 GX 81
UGC 36 GX 81
UGC 40 GX 63
UGC 46 GX 63, 81
UGC 48 GX 30
UGC 50 GX 63
UGC 51 GX 81, 101
UGC 52 GX 81
UGC 53 GX 63
UGC 61 GX 30
UGC 63 GX 45
UGC 64 GX 30, 45
UGC 66 GX 81
UGC 67 GX 81
UGC 68 GX 63
UGC 69 GX 63
UGC 79 GX 63
UGC 81 GX 81
UGC 85 GX 30
UGC 87 GX 63
UGC 88 GX 101
UGC 92 GX 63
UGC 93 GX 45
UGC 95 GX 63
UGC 99 GX 81
UGC 102 GX 45
UGC 105 GX 63
UGC 108 GX 63
UGC 110 GX 63
UGC 112 GX 30
UGC 113 GX 63
UGC 117 GX 45
UGC 119 GX 81
UGC 122 GX 63, 81
UGC 127 GX 63
UGC 128 GX 45
UGC 129 GX 101
UGC 130 GX 45
UGC 132 GX 81
UGC 139 GX 101
UGC 143 GX 81, 101
UGC 146 GX 63
UGC 147 GX 45, 63
UGC 148 GX 81
UGC 151 GX 81
UGC 152 GX 45, 63
UGC 155 GX 81
UGC 156 GX 81
UGC 157 GX 63
UGC 158 GX 30
UGC 160 GX 45
UGC 164 GX 63
UGC 166 GX 45
UGC 169 GX 63
UGC 171 GX 30
UGC 175 GX 30
UGC 178 GX 30
UGC 179 GX 63
UGC 183 GX 30
UGC 191 GX 81
UGC 196 GX 30
UGC 197 GX 63
UGC 202 GX 63
UGC 210 GX 63
UGC 212 GX 101
UGC 215 GX 45, 63
UGC 221 GX 63
UGC 223 GX 63
UGC 224 GX 101
UGC 226 GX 81
UGC 227 GX 63
UGC 228 GX 63
UGC 232 GX 45
UGC 233 GX 81
UGC 236 GX 30
UGC 238 GX 45
UGC 240 GX 81
UGC 242 GX 63
UGC 243 GX 30
UGC 244 GX 63
UGC 248 GX 63
UGC 249 GX 81
UGC 250 GX 81
UGC 253 GX 81
UGC 256 GX 30
UGC 260 GX 81
UGC 261 GX 63
UGC 262 GX 45
UGC 263 GX 30
UGC 265 GX 63
UGC 272 GX 101
UGC 275 GX 101
UGC 277 GX 101
UGC 278 GX 63
UGC 279 GX 45
UGC 282 GX 101
UGC 283 GX 101
UGC 284 GX 45
UGC 287 GX 81
UGC 290 GX 81
UGC 299 GX 45
UGC 303 GX 30
UGC 305 GX 81
UGC 306 GX 30
UGC 310 GX 63
UGC 312 GX 81
UGC 313 GX 81
UGC 314 GX 81
UGC 318 GX 45
UGC 319 GX 45
UGC 321 GX 63
UGC 328 GX 101, 120
UGC 329 GX 101, 120
UGC 330 GX 45
UGC 331 GX 81, 100
UGC 334 GX 45
UGC 335 GX 81, 100
UGC 336 GX 30
UGC 342 GX 101, 120
UGC 346 GX 45
UGC 348 GX 101, 120
UGC 355 GX 45
UGC 358 GX 101, 120
UGC 360 GX 63, 80
UGC 364 GX 63, 80
UGC 367 GX 63, 80
UGC 371 GX 45, 62, 63, 80
UGC 372 GX 30
UGC 375 GX 63, 80
UGC 379 GX 81, 100, 101, 120
UGC 381 GX 45, 62
UGC 384 GX 45, 62
UGC 385 GX 81, 100
UGC 386 GX 81, 100
UGC 388 GX 45, 62
UGC 392 GX 2
UGC 393 GX 63, 80, 81, 100
UGC 394 GX 30
UGC 398 GX 63, 80
UGC 400 GX 45, 62, 63, 80
UGC 402 GX 101, 120
UGC 411 GX 63, 80
UGC 418 GX 81, 100
UGC 422 GX 81, 100
UGC 425 GX 63, 80
UGC 429 GX 100
UGC 431 GX 45, 62
UGC 433 GX 45, 62
UGC 439 GX 120
UGC 442 GX 45, 62
UGC 444 GX 45, 62
UGC 455 GX 120
UGC 460 GX 30
UGC 465 GX 45, 62
UGC 466 GX 120
UGC 469 GX 80
UGC 471 GX 30, 44
UGC 475 GX 18, 30, 44
UGC 477 GX 80
UGC 478 GX 62
UGC 480 GX 62
UGC 484 GX 62
UGC 485 GX 62
UGC 486 GX 30, 44
UGC 488 GX 100
UGC 492 GX 120
UGC 496 GX 120
UGC 500 GX 100
UGC 501 GX 80
UGC 505 GX 120
UGC 506 GX 80
UGC 507 GX 120
UGC 510 GX 80
UGC 511 GX 62
UGC 512 GX 100
UGC 518 GX 62
UGC 521 GX 100
UGC 522 GX 30, 44, 62
UGC 524 GX 62, 80
UGC 525 GX 62, 80
UGC 529 GX 62, 80
UGC 533 GX 100
UGC 539 GX 44
UGC 540 GX 62, 80
UGC 542 GX 62, 80
UGC 544 GX 120
UGC 545 GX 100
UGC 548 GX 62
UGC 556 GX 62, 80
UGC 557 GX 62
UGC 558 GX 100
UGC 563 GX 120
UGC 566 GX 62
UGC 567 GX 62
UGC 571 GX 100
UGC 572 GX 100
UGC 576 GX 44
UGC 578 GX 62
UGC 579 GX 120
UGC 583 GX 120
UGC 587 GX 120
UGC 588 GX 120
UGC 591 GX 80
UGC 595 GX 120
UGC 596 GX 100
UGC 598 GX 62
UGC 599 GX 120
UGC 600 GX 44
UGC 602 GX 62
UGC 603 GX 100
UGC 605 GX 80
UGC 607 GX 100
UGC 608 GX 44
UGC 610 GX 100
UGC 612 GX 80
UGC 614 GX 62
UGC 615 GX 100
UGC 616 GX 80
UGC 617 GX 80
UGC 618 GX 120
UGC 622 GX 44
UGC 626 GX 120
UGC 627 GX 100
UGC 628 GX 80
UGC 629 GX 62, 80
UGC 631 GX 100
UGC 632 GX 62
UGC 633 GX 62, A7
UGC 634 GX 100
UGC 636 GX 80
UGC 640 GX 100
UGC 643 GX 80
UGC 644 GX 100
UGC 646 GX 62, A7
UGC 654 GX 80
UGC 655 GX 44
UGC 661 GX 120
UGC 665 GX 44
UGC 669 GX 62, A7
UGC 670 GX 2
UGC 673 GX 62
UGC 675 GX 120
UGC 677 GX 100
UGC 678 GX 120
UGC 685 GX 100
UGC 690 GX 62
UGC 692 GX 62, A7
UGC 694 GX 120
UGC 695 GX 120
UGC 696 GX 80
UGC 697 GX 62, A7
UGC 701 GX 120
UGC 705 GX 100
UGC 706 GX 100
UGC 708 GX 100
UGC 711 GX 120
UGC 714 GX 62, A7
UGC 717 GX 100
UGC 719 GX 100
UGC 722 GX 100
UGC 723 GX 80
UGC 724 GX 62, A7
UGC 725 GX 44
UGC 726 GX 120
UGC 727 GX 100
UGC 728 GX 44
UGC 729 GX 120
UGC 731 GX 44
UGC 732 GX 62, A7
UGC 733 GX 100
UGC 736 GX 120
UGC 741 GX 100
UGC 742 GX 62, A7
UGC 743 GX 62, A7
UGC 746 GX 44
UGC 748 GX 62, A7
UGC 749 GX 120
UGC 751 GX 80, 100
UGC 755 GX 62
UGC 756 GX 62, A7
UGC 757 GX 120
UGC 761 GX 44
UGC 768 GX 120
UGC 771 GX 120
UGC 772 GX 120
UGC 774 GX 100
UGC 777 GX 44
UGC 783 GX 44
UGC 784 GX 120
UGC 785 GX 100
UGC 790 GX 120
UGC 791 GX 120
UGC 793 GX 120
UGC 797 GX 120
UGC 800 GX 80
UGC 803 GX 100
UGC 808 GX 100
UGC 809 GX 62, A7
UGC 811 GX 62
UGC 813 GX 44
UGC 816 GX 44
UGC 817 GX 120
UGC 822 GX 62
UGC 824 GX 100
UGC 826 GX 44
UGC 830 GX 120
UGC 831 GX 62
UGC 833 GX 100
UGC 834 GX 100, 120
UGC 835 GX 62
UGC 836 GX 44
UGC 838 GX 100
UGC 839 GX 120
UGC 841 GX 62, A7
UGC 842 GX 120
UGC 845 GX 80
UGC 847 GX 120
UGC 849 GX 100
UGC 851 GX 80
UGC 855 GX 100
UGC 860 GX 100
UGC 862 GX 62, A7
UGC 863 GX 2
UGC 871 GX 100, 120
UGC 875 GX 120
UGC 878 GX 62, A7
UGC 880 GX 44
UGC 881 GX 120
UGC 882 GX 100
UGC 883 GX 80, 100
UGC 887 GX 100, 120
UGC 890 GX 120
UGC 891 GX 100
UGC 892 GX 120, A16
UGC 897 GX 100
UGC 899 GX 80
UGC 900 GX 80
UGC 901 GX 62, A7
UGC 902 GX 44
UGC 903 GX 80, 100
UGC 904 GX 80
UGC 909 GX 62
UGC 910 GX 100
UGC 911 GX 62, A7
UGC 913 GX 62, A7
UGC 921 GX 120, A16
UGC 928 GX 120, A16
UGC 929 GX 120, A16
UGC 933 GX 100
UGC 934 GX 62
UGC 937 GX 62, A7
UGC 940 GX 62, A7
UGC 958 GX 100
UGC 959 GX 62, A7
UGC 974 GX 120, A16
UGC 975 GX 62, A7
UGC 984 GX 120, A16
UGC 985 GX 100
UGC 987 GX 62, A7
UGC 989 GX 100
UGC 993 GX 100
UGC 996 GX 120, A16
UGC 1003 GX 120, A16
UGC 1006 GX 120
UGC 1014 GX 100
UGC 1020 GX 80, 100
UGC 1022 GX 62
UGC 1023 GX 100
UGC 1024 GX 100
UGC 1025 GX 80
UGC 1026 GX 100
UGC 1028 GX 120
UGC 1030 GX 120, A16
UGC 1032 GX 80
UGC 1033 GX 62, A7
UGC 1035 GX 44
UGC 1039 GX 1
UGC 1040 GX 120, A16
UGC 1041 GX 100
UGC 1042 GX 44
UGC 1043 GX 120, A16
UGC 1045 GX 62, A7
UGC 1046 GX 62
UGC 1055 GX 120, A16
UGC 1056 GX 100
UGC 1057 GX 100
UGC 1059 GX 62
UGC 1062 GX 120, A16
UGC 1068 GX 44
UGC 1070 GX 44, 62
UGC 1072 GX 120, A16
UGC 1073 GX 80
UGC 1077 GX 80
UGC 1083 GX 100
UGC 1086 GX 62, A7
UGC 1087 GX 100
UGC 1088 GX 62
UGC 1090 GX 62
UGC 1093 GX 80, 100
UGC 1095 GX 62
UGC 1098 GX 80
UGC 1099 GX 80
UGC 1101 GX 44
UGC 1102 GX 120
UGC 1104 GX 80
UGC 1110 GX 100
UGC 1112 GX 120
UGC 1113 GX 80, 100
UGC 1118 GX 120
UGC 1119 GX 80, 100
UGC 1120 GX 120
UGC 1123 GX 120
UGC 1125 GX 62
UGC 1129 GX 100
UGC 1131 GX 62
UGC 1132 GX 44
UGC 1133 GX 120
UGC 1138 GX 120
UGC 1139 GX 100
UGC 1142 GX 44
UGC 1144 GX 100
UGC 1145 GX 62
UGC 1148 GX 2, 7
UGC 1151 GX 120
UGC 1152 GX 62
UGC 1154 GX 80
UGC 1155 GX 120
UGC 1160 GX 62
UGC 1161 GX 62
UGC 1162 GX 44
UGC 1163 GX 120
UGC 1166 GX 62
UGC 1167 GX 100
UGC 1168 GX 44
UGC 1169 GX 120
UGC 1174 GX 120
UGC 1176 GX 100
UGC 1178 GX 62
UGC 1179 GX 44
UGC 1180 GX 120
UGC 1185 GX 44
UGC 1186 GX 44
UGC 1191 GX 100
UGC 1195 GX 100
UGC 1197 GX 80
UGC 1198 GX 1
UGC 1200 GX 100
UGC 1203 GX 100, 120
UGC 1206 GX 100
UGC 1209 GX 100
UGC 1211 GX 100
UGC 1212 GX 62, A6
UGC 1213 GX 62
UGC 1214 GX 120
UGC 1219 GX 80, 99, 100
UGC 1221 GX 62
UGC 1225 GX 119, 120
UGC 1228 GX 80
UGC 1230 GX 80
UGC 1233 GX 80
UGC 1234 GX 62, A6
UGC 1235 GX 119, 120
UGC 1237 GX 99, 100
UGC 1239 GX 99, 100
UGC 1240 GX 119, 120
UGC 1242 GX 99, 100
UGC 1243 GX 80
UGC 1245 GX 99, 100
UGC 1246 GX 99, 100
UGC 1251 GX 62, A6
UGC 1253 GX 99, 100
UGC 1257 GX 62, A6
UGC 1260 GX 99, 100
UGC 1261 GX 99, 100
UGC 1265 GX 80
UGC 1269 GX 62, A6
UGC 1271 GX 99, 100
UGC 1272 GX 62, A6
UGC 1274 GX 99, 100
UGC 1277 GX 62, A6
UGC 1281 GX 62
UGC 1282 GX 99, 100
UGC 1285 GX 1
UGC 1290 GX 99, 100
UGC 1293 GX 119, 120
UGC 1295 GX 62
UGC 1297 GX 119, 120
UGC 1303 GX 44
UGC 1306 GX 62
UGC 1308 GX 62, A6
UGC 1311 GX 62, 80
UGC 1312 GX 99, 100
UGC 1318 GX 62
UGC 1319 GX 62, A6
UGC 1322 GX 99, 100
UGC 1324 GX 80
UGC 1325 GX 99, 100
UGC 1326 GX 99, 100
UGC 1327 GX 62
UGC 1328 GX 80, 99, 100
UGC 1329 GX 80
UGC 1331 GX 44
UGC 1332 GX 44
UGC 1338 GX 62, A6
UGC 1339 GX 62, A6
UGC 1341 GX 62
UGC 1344 GX 62, A6
UGC 1347 GX 62, A6
UGC 1350 GX 62, A6
UGC 1353 GX 62, A6
UGC 1355 GX 44
UGC 1359 GX 62, 80
UGC 1361 GX 62, A6
UGC 1365 GX 119
UGC 1366 GX 62, A6
UGC 1368 GX 99
UGC 1369 GX 80
UGC 1372 GX 80, 99
UGC 1375 GX 80
UGC 1378 GX 2, 7, 17
UGC 1382 GX 119
UGC 1383 GX 99
UGC 1384 GX 80
UGC 1385 GX 62, A6
UGC 1387 GX 62, A6
UGC 1389 GX 44
UGC 1391 GX 99

UGC 1394 GX 44
UGC 1395 GX 99
UGC 1396 GX 80
UGC 1397 GX 62
UGC 1398 GX 61, 62, A6
UGC 1399 GX 80
UGC 1400 GX 61, 62, A6
UGC 1410 GX 119
UGC 1415 GX 61, 62, A6
UGC 1416 GX 61, 62, A6
UGC 1418 GX 44, 61, 62
UGC 1420 GX 99
UGC 1422 GX 61, 62
UGC 1425 GX 99, 119
UGC 1427 GX 99
UGC 1432 GX 79, 80, 99
UGC 1433 GX 79, 80
UGC 1435 GX 99, 119
UGC 1442 GX 119
UGC 1444 GX 119
UGC 1445 GX 79, 80
UGC 1446 GX 119
UGC 1447 GX 44
UGC 1448 GX 119
UGC 1449 GX 119
UGC 1450 GX 99
UGC 1451 GX 79, 80
UGC 1453 GX 79, 80
UGC 1454 GX 119
UGC 1459 GX 61, 62, A6
UGC 1460 GX 61, 62, A6
UGC 1462 GX 79, 80
UGC 1464 GX 119
UGC 1468 GX 99
UGC 1470 GX 61, 62
UGC 1474 GX 61, 62
UGC 1478 GX 79, 80
UGC 1479 GX 79, 80
UGC 1485 GX 79, 80
UGC 1490 GX 79, 80
UGC 1492 GX 44
UGC 1493 GX 61, 62
UGC 1493A GX 44
UGC 1495 GX 79, 80, 99
UGC 1496 GX 99
UGC 1498 GX 99
UGC 1502 GX 61, 62
UGC 1503 GX 61, 62
UGC 1504 GX 44
UGC 1505 GX 99
UGC 1508 GX 44
UGC 1512 GX 99
UGC 1514 GX 79, 80
UGC 1518 GX 79, 80
UGC 1519 GX 79, 80
UGC 1525 GX 119
UGC 1531 GX 79, 80, 99
UGC 1533 GX 79, 80
UGC 1545 GX 99, 119
UGC 1546 GX 79, 80
UGC 1547 GX 79, 80
UGC 1549 GX 79, 80
UGC 1551 GX 79, 80
UGC 1552 GX 44
UGC 1553 GX 119
UGC 1558 GX 99
UGC 1560 GX 79, 80
UGC 1561 GX 79
UGC 1562 GX 44
UGC 1563 GX 44
UGC 1565 GX 79
UGC 1572 GX 99
UGC 1577 GX 61
UGC 1580 GX 99
UGC 1581 GX 61
UGC 1582 GX 61
UGC 1584 GX 99
UGC 1585 GX 44
UGC 1587 GX 99
UGC 1590 GX 61, 79
UGC 1591 GX 61, 79
UGC 1593 GX 99
UGC 1596 GX 61, 79
UGC 1597 GX 119
UGC 1603 GX 119
UGC 1604 GX 61
UGC 1605 GX 99
UGC 1607 GX 44
UGC 1612 GX 44
UGC 1614 GX 99
UGC 1617 GX 119
UGC 1618 GX 119
UGC 1620 GX 119
UGC 1622 GX 99
UGC 1624 GX 119
UGC 1626 GX 44
UGC 1627 GX 119
UGC 1630 GX 99
UGC 1634 GX 44
UGC 1641 GX 61
UGC 1646 GX 99, 119
UGC 1648 GX 79
UGC 1649 GX 99
UGC 1650 GX 61
UGC 1651 GX 61
UGC 1652 GX 79
UGC 1654 GX 61
UGC 1659 GX 99
UGC 1660 GX 61
UGC 1661 GX 44
UGC 1663 GX 99
UGC 1666 GX 61
UGC 1668 GX 61
UGC 1669 GX 99, 119
UGC 1670 GX 99
UGC 1674 GX 61
UGC 1682 GX 61
UGC 1684 GX 99
UGC 1685 GX 61
UGC 1687 GX 99
UGC 1690 GX 61, 79
UGC 1691 GX 61
UGC 1692 GX 44
UGC 1694 GX 99
UGC 1696 GX 61, 79
UGC 1697 GX 119
UGC 1698 GX 119
UGC 1699 GX 29
UGC 1701 GX 61
UGC 1702 GX 99
UGC 1703 GX 61
UGC 1704 GX 44
UGC 1706 GX 79
UGC 1707 GX 119
UGC 1716 GX 119
UGC 1721 GX 61
UGC 1724 GX 99
UGC 1726 GX 61
UGC 1728 GX 44
UGC 1729 GX 61
UGC 1731 GX 79
UGC 1733 GX 79
UGC 1735 GX 61
UGC 1739 GX 79
UGC 1742 GX 99
UGC 1746 GX 119
UGC 1750 GX 61
UGC 1752 GX 79
UGC 1756 GX 119
UGC 1757 GX 61
UGC 1767 GX 61
UGC 1769 GX 61
UGC 1772 GX 61
UGC 1773 GX 99
UGC 1775 GX 99, 119
UGC 1776 GX 61
UGC 1778 GX 61
UGC 1780 GX 44, 61
UGC 1782 GX 44
UGC 1787 GX 61
UGC 1788 GX 61
UGC 1792 GX 61, 79
UGC 1794 GX 119
UGC 1796 GX 43, 44, 61
UGC 1797 GX 119
UGC 1800 GX 61
UGC 1802 GX 43, 44
UGC 1803 GX 99
UGC 1804 GX 61
UGC 1805 GX 61
UGC 1807 GX 43, 44
UGC 1808 GX 79
UGC 1809 GX 119
UGC 1810 GX 61
UGC 1812 GX 79
UGC 1814 GX 99
UGC 1817 GX 99
UGC 1820 GX 61
UGC 1825 GX 61
UGC 1826 GX 61
UGC 1827 GX 43, 44
UGC 1829 GX 61
UGC 1830 GX 43, 44
UGC 1832 GX 43, 44
UGC 1833 GX 79
UGC 1837 GX 43, 44
UGC 1839 GX 119
UGC 1840 GX 43, 44
UGC 1841 GX 43, 44
UGC 1844 GX 79
UGC 1854 GX 43, 44
UGC 1855 GX 43, 44, 61
UGC 1856 GX 61
UGC 1857 GX 61
UGC 1858 GX 43, 44
UGC 1859 GX 43, 44
UGC 1860 GX 79
UGC 1862 GX 119
UGC 1863 GX 119
UGC 1865 GX 61
UGC 1866 GX 43, 44
UGC 1867 GX 43, 44
UGC 1871 GX 79
UGC 1877 GX 61
UGC 1879 GX 99
UGC 1881 GX 79
UGC 1882 GX 61
UGC 1883 GX 99
UGC 1885 GX 79
UGC 1886 GX 61
UGC 1889 GX 61
UGC 1890 GX 61
UGC 1891 GX 79
UGC 1892 GX 79
UGC 1893 GX 43, 44
UGC 1896 GX 61
UGC 1897 GX 99
UGC 1899 GX 79
UGC 1905 GX 119
UGC 1910 GX 61
UGC 1914 GX 43, 44
UGC 1917 GX 79
UGC 1918 GX 79
UGC 1919 GX 61
UGC 1921 GX 79
UGC 1924 GX 61
UGC 1930 GX 29, 43
UGC 1934 GX 119
UGC 1938 GX 79
UGC 1939 GX 79
UGC 1941 GX 61
UGC 1945 GX 119
UGC 1946 GX 99
UGC 1955 GX 79
UGC 1957 GX 43
UGC 1962 GX 119
UGC 1963 GX 61
UGC 1965 GX 79
UGC 1966 GX 99
UGC 1970 GX 79
UGC 1971 GX 79
UGC 1972 GX 61
UGC 1976 GX 61
UGC 1980 GX 61
UGC 1982 GX 99
UGC 1987 GX 43
UGC 1988 GX 43, 61
UGC 1989 GX 43
UGC 1993 GX 61
UGC 1995 GX 119
UGC 1999 GX 79
UGC 2001 GX 43
UGC 2003 GX 61
UGC 2004 GX 119
UGC 2005 GX 119
UGC 2007 GX 99
UGC 2010 GX 119
UGC 2011 GX 61
UGC 2012 GX 79
UGC 2014 GX 61
UGC 2015 GX 61
UGC 2017 GX 79
UGC 2018 GX 119
UGC 2019 GX 119
UGC 2020 GX 79
UGC 2021 GX 99, 119
UGC 2022 GX 61
UGC 2023 GX 61
UGC 2024 GX 119
UGC 2025 GX 79
UGC 2028 GX 79
UGC 2031 GX 79
UGC 2033 GX 61
UGC 2034 GX 43, 61
UGC 2035 GX 43
UGC 2041 GX 99
UGC 2043 GX 43
UGC 2050 GX 43
UGC 2051 GX 119
UGC 2053 GX 61, 79
UGC 2054 GX 61
UGC 2056 GX 119
UGC 2060 GX 43
UGC 2061 GX 119
UGC 2062 GX 119
UGC 2064 GX 79
UGC 2065 GX 61
UGC 2067 GX 61
UGC 2069 GX 61
UGC 2073 GX 43
UGC 2074 GX 43
UGC 2075 GX 99
UGC 2077 GX 61
UGC 2079 GX 79
UGC 2081 GX 119
UGC 2082 GX 79
UGC 2083 GX 61
UGC 2090 GX 61
UGC 2091 GX 119
UGC 2092 GX 99
UGC 2094 GX 61
UGC 2101 GX 43
UGC 2104 GX 79
UGC 2105 GX 61
UGC 2109 GX 61
UGC 2111 GX 43
UGC 2116 GX 61
UGC 2120 GX 119
UGC 2121 GX 119
UGC 2122 GX 61, 79
UGC 2126 GX 43, 61
UGC 2131 GX 61
UGC 2134 GX 79
UGC 2140 GX 79
UGC 2140A GX 79
UGC 2143 GX 61
UGC 2146 GX 43
UGC 2151 GX 79
UGC 2156 GX 61
UGC 2157 GX 61
UGC 2161 GX 43
UGC 2162 GX 119
UGC 2165 GX 61
UGC 2167 GX 99
UGC 2169 GX 61
UGC 2172 GX 43
UGC 2174 GX 61
UGC 2175 GX 43
UGC 2176 GX 99, 119
UGC 2179 GX 61
UGC 2180 GX 61
UGC 2181 GX 119
UGC 2185 GX 43, 61
UGC 2186 GX 43
UGC 2194 GX 43
UGC 2196 GX 43
UGC 2197 GX 61
UGC 2199 GX 119
UGC 2201 GX 61
UGC 2205 GX 61
UGC 2206 GX 61
UGC 2213 GX 61
UGC 2215 GX 43
UGC 2216 GX 119
UGC 2218 GX 61
UGC 2221 GX 61
UGC 2222 GX 61
UGC 2223 GX 61
UGC 2227 GX 43
UGC 2229 GX 119
UGC 2233 GX 43
UGC 2238 GX 99
UGC 2239 GX 61
UGC 2240 GX 43
UGC 2243 GX 61
UGC 2249 GX 43
UGC 2252 GX 99
UGC 2255 GX 99
UGC 2256 GX 43, 61
UGC 2259 GX 61
UGC 2261 GX 43
UGC 2267 GX 79
UGC 2269 GX 43
UGC 2270 GX 43
UGC 2271 GX 119
UGC 2272 GX 79
UGC 2275 GX 119
UGC 2277 GX 43, 61
UGC 2280 GX 43
UGC 2282 GX 99
UGC 2285 GX 99
UGC 2286 GX 79
UGC 2290 GX 79
UGC 2292 GX 119
UGC 2295 GX 119
UGC 2296 GX 79
UGC 2302 GX 119
UGC 2303 GX 79, 99
UGC 2305 GX 61
UGC 2311 GX 119
UGC 2314 GX 43
UGC 2317 GX 43
UGC 2323 GX 99
UGC 2324 GX 119
UGC 2327 GX 79
UGC 2328 GX 61
UGC 2329 GX 99
UGC 2336 GX 99
UGC 2338 GX 119
UGC 2339 GX 99
UGC 2342 GX 43
UGC 2343 GX 119
UGC 2345 GX 119
UGC 2348 GX 99
UGC 2350 GX 43
UGC 2351 GX 43
UGC 2358 GX 7
UGC 2361 GX 43
UGC 2362 GX 99
UGC 2363 GX 61
UGC 2364 GX 99
UGC 2367 GX 99
UGC 2369 GX 99
UGC 2370 GX 43
UGC 2372 GX 99, 119
UGC 2376 GX 61
UGC 2378 GX 99
UGC 2380 GX 29, 43
UGC 2385 GX 119
UGC 2387 GX 99
UGC 2392 GX 61
UGC 2399 GX 99
UGC 2403 GX 119
UGC 2405 GX 99
UGC 2411 GX 7
UGC 2413 GX 98, 99
UGC 2418 GX 118, 119
UGC 2419 GX 98, 99
UGC 2424 GX 79, 98, 99
UGC 2426 GX 98, 99, 118, 119
UGC 2430 GX 98, 99
UGC 2434 GX 98, 99
UGC 2435 GX 61
UGC 2437 GX 98, 99
UGC 2438 GX 98, 99
UGC 2441 GX 118, 119
UGC 2442 GX 79
UGC 2443 GX 118, 119
UGC 2444 GX 98, 99
UGC 2446 GX 118, 119
UGC 2448 GX 61
UGC 2449 GX 43
UGC 2453 GX 98, 99
UGC 2456 GX 61
UGC 2457 GX 79
UGC 2459 GX 43
UGC 2463 GX 43, 61
UGC 2465 GX 61
UGC 2466 GX 61
UGC 2469 GX 98, 99, 118, 119
UGC 2470 GX 43
UGC 2473 GX 43
UGC 2479 GX 118, 119
UGC 2483 GX 61
UGC 2485 GX 7
UGC 2486 GX 79, 98, 99
UGC 2491 GX 61
UGC 2494 GX 61
UGC 2495 GX 43
UGC 2497 GX 61, 79
UGC 2498 GX 79, 98, 99
UGC 2500 GX 43
UGC 2501 GX 118, 119
UGC 2504 GX 43
UGC 2506 GX 79
UGC 2513 GX 118, 119
UGC 2517 GX 118
UGC 2519 GX 7
UGC 2522 GX 118
UGC 2523 GX 118
UGC 2526 GX 61
UGC 2527 GX 118
UGC 2528 GX 43
UGC 2530 GX 79
UGC 2534 GX 43
UGC 2537 GX 43
UGC 2538 GX 43
UGC 2540 GX 61
UGC 2542 GX 17
UGC 2543 GX 61
UGC 2546 GX 61
UGC 2550 GX 61
UGC 2553 GX 79
UGC 2559 GX 43
UGC 2563 GX 79
UGC 2568 GX 43, A4
UGC 2570 GX 79
UGC 2573 GX 43
UGC 2585 GX 118
UGC 2587 GX 118
UGC 2594 GX 118
UGC 2596 GX 43
UGC 2598 GX 43, A4
UGC 2602 GX 98
UGC 2603 GX 7
UGC 2604 GX 61, A4
UGC 2607 GX 118
UGC 2608 GX 43, A4
UGC 2611 GX 118
UGC 2612 GX 43, A4
UGC 2614 GX 43, A4
UGC 2617 GX 43, 60, 61, A4
UGC 2618 GX 43, A4
UGC 2620 GX 7
UGC 2622 GX 98
UGC 2623 GX 60, 61
UGC 2627 GX 60, 61
UGC 2628 GX 118
UGC 2629 GX 60, 61
UGC 2633 GX 60, 61
UGC 2636 GX 60, 61
UGC 2640 GX 43, A4
UGC 2641 GX 118
UGC 2649 GX 118
UGC 2653 GX 60, 61
UGC 2654 GX 43, A4
UGC 2655 GX 43, A4
UGC 2659 GX 43, 60, 61, A4
UGC 2661 GX 60, 61
UGC 2673 GX 43, A4
UGC 2674 GX 98
UGC 2677 GX 118
UGC 2678 GX 60, 61
UGC 2679 GX 118
UGC 2680 GX 118
UGC 2684 GX 78, 79, 98
UGC 2685 GX 60, 61
UGC 2686 GX 43, 60, 61, A4
UGC 2687 GX 118
UGC 2689 GX 43, 60, 61, A4
UGC 2690 GX 118
UGC 2691 GX 118
UGC 2692 GX 118
UGC 2695 GX 98
UGC 2698 GX 43, 60, 61, A4
UGC 2700 GX 43, A4
UGC 2702 GX 60, 61
UGC 2703 GX 98
UGC 2704 GX 118
UGC 2705 GX 118
UGC 2706 GX 60, 61
UGC 2708 GX 43, 60, 61, A4
UGC 2709 GX 60, 61
UGC 2710 GX 60, 61
UGC 2711 GX 118
UGC 2712 GX 98
UGC 2716 GX 78, 98
UGC 2717 GX 43, 60, A4
UGC 2723 GX 43, A4
UGC 2725 GX 43, A4
UGC 2730 GX 43, 60, A4
UGC 2733 GX 43, A4
UGC 2736 GX 43, 60, A4
UGC 2740 GX 98
UGC 2742 GX 43, 60, A4
UGC 2744 GX 118
UGC 2746 GX 60
UGC 2748 GX 118
UGC 2750 GX 60
UGC 2752 GX 43, 60
UGC 2755 GX 60
UGC 2756 GX 43, 60
UGC 2766 GX 43, 60
UGC 2767 GX 7
UGC 2770 GX 60
UGC 2771 GX 60
UGC 2773 GX 43
UGC 2775 GX 43
UGC 2780 GX 60
UGC 2783 GX 60
UGC 2784 GX 60
UGC 2789 GX 16, 17
UGC 2794 GX 43
UGC 2797 GX 78

UGC 2798 GX 43, 60
UGC 2800 GX 16, 17
UGC 2802 GX 118
UGC 2807 GX 60
UGC 2808 GX 60
UGC 2809 GX 78
UGC 2810 GX 60
UGC 2812 GX 118
UGC 2813 GX 16, 17
UGC 2814 GX 118
UGC 2817 GX 60
UGC 2818 GX 60
UGC 2823 GX 98
UGC 2826 GX 16, 17
UGC 2828 GX 60
UGC 2829 GX 98
UGC 2831 GX 118
UGC 2836 GX 60
UGC 2837 GX 43, 60
UGC 2840 GX 78, A12
UGC 2842 GX 118
UGC 2844 GX 43
UGC 2849 GX 43
UGC 2851 GX 43
UGC 2852 GX 98, 118
UGC 2855 GX 16
UGC 2857 GX 60
UGC 2858 GX 43
UGC 2859 GX 43, 60
UGC 2861 GX 60
UGC 2862 GX 98
UGC 2865 GX 7, 16
UGC 2866 GX 16
UGC 2868 GX 60
UGC 2871 GX 98
UGC 2877 GX 60
UGC 2881 GX 60
UGC 2882 GX 60
UGC 2883 GX 118
UGC 2885 GX 60
UGC 2886 GX 60
UGC 2888 GX 60
UGC 2889 GX 60
UGC 2890 GX 16
UGC 2892 GX 78
UGC 2894 GX 98
UGC 2896 GX 7
UGC 2901 GX 60
UGC 2904 GX 98
UGC 2906 GX 7
UGC 2907 GX 7
UGC 2908 GX 42, 43
UGC 2913 GX 118
UGC 2914 GX 98
UGC 2916 GX 16
UGC 2920 GX 60
UGC 2925 GX 118
UGC 2927 GX 78
UGC 2928 GX 78
UGC 2931 GX 78
UGC 2936 GX 118
UGC 2942 GX 78
UGC 2944 GX 60
UGC 2945 GX 60
UGC 2949 GX 78
UGC 2950 GX 60
UGC 2952 GX 60
UGC 2956 GX 60
UGC 2958 GX 78
UGC 2963 GX 118
UGC 2964 GX 78
UGC 2968 GX 78, 97, 98
UGC 2969 GX 117, 118
UGC 2971 GX 60
UGC 2972 GX 60
UGC 2976 GX 78
UGC 2978 GX 60
UGC 2983 GX 117, 118
UGC 2984 GX 97, 98
UGC 2988 GX 78
UGC 2989 GX 60, 78
UGC 2991 GX 60
UGC 2994 GX 117, 118
UGC 2997 GX 97
UGC 2998 GX 117
UGC 3002 GX 117
UGC 3004 GX 117
UGC 3006 GX 117
UGC 3008 GX 117
UGC 3009 GX 78
UGC 3010 GX 97, 117
UGC 3014 GX 117
UGC 3016 GX 60
UGC 3018 GX 117
UGC 3021 GX 60
UGC 3023 GX 117
UGC 3028 GX 60
UGC 3029 GX 117
UGC 3031 GX 117
UGC 3035 GX 97
UGC 3042 GX 16
UGC 3046 GX 16
UGC 3047 GX 60
UGC 3048 GX 16
UGC 3052 GX 60
UGC 3053 GX 78
UGC 3054 GX 117
UGC 3057 GX 7
UGC 3059 GX 117
UGC 3061 GX 97
UGC 3066 GX 97, 117
UGC 3067 GX 97
UGC 3068 GX 97, 117
UGC 3069 GX 7, 16
UGC 3070 GX 117
UGC 3078 GX 60
UGC 3080 GX 117
UGC 3081 GX 117
UGC 3084 GX 97
UGC 3087 GX 97, 117
UGC 3089 GX 97
UGC 3090 GX 16
UGC 3091 GX 117
UGC 3092 GX 16
UGC 3099 GX 77, 78
UGC 3101 GX 7
UGC 3105 GX 117
UGC 3106 GX 117
UGC 3108 GX 42
UGC 3109 GX 117
UGC 3110 GX 7, 16
UGC 3114 GX 16
UGC 3117 GX 117
UGC 3122 GX 97
UGC 3127 GX 117
UGC 3128 GX 117
UGC 3129 GX 77, 78, 97
UGC 3132 GX 7
UGC 3134 GX 117
UGC 3135 GX 77, 78
UGC 3136 GX 117
UGC 3137 GX 7
UGC 3141 GX 117
UGC 3143 GX 16
UGC 3144 GX 7
UGC 3145 GX 117
UGC 3147 GX 16
UGC 3149 GX 16
UGC 3150 GX 7, 16
UGC 3157 GX 77
UGC 3163 GX 16
UGC 3164 GX 117
UGC 3165 GX 77
UGC 3167 GX 16
UGC 3168 GX 117
UGC 3171 GX 117
UGC 3172 GX 97
UGC 3174 GX 117
UGC 3175 GX 7
UGC 3176 GX 16
UGC 3179 GX 117
UGC 3180 GX 97
UGC 3181 GX 97
UGC 3182 GX 16
UGC 3186 GX 117
UGC 3188 GX 97
UGC 3189 GX 16
UGC 3191 GX 117
UGC 3192 GX 117
UGC 3193 GX 117
UGC 3194 GX 117
UGC 3195 GX 117
UGC 3197 GX 7
UGC 3200 GX 117
UGC 3202 GX 117
UGC 3205 GX 59
UGC 3206 GX 117
UGC 3207 GX 117
UGC 3208 GX 117
UGC 3209 GX 117
UGC 3211 GX 117
UGC 3214 GX 117
UGC 3215 GX 117
UGC 3218 GX 16, 27, 28
UGC 3223 GX 117
UGC 3224 GX 97, 117
UGC 3230 GX 7
UGC 3231 GX 117
UGC 3233 GX 117
UGC 3234 GX 97
UGC 3235 GX 16
UGC 3237 GX 117
UGC 3241 GX 7
UGC 3245 GX 16
UGC 3246 GX 117
UGC 3247 GX 97
UGC 3248 GX 117
UGC 3250 GX 16
UGC 3252 GX 16
UGC 3253 GX 1, 7
UGC 3256 GX 117
UGC 3257 GX 1, 7
UGC 3258 GX 117
UGC 3259 GX 16
UGC 3260 GX 27, 42
UGC 3261 GX 77, 97
UGC 3263 GX 117
UGC 3264 GX 117
UGC 3267 GX 16
UGC 3269 GX 97
UGC 3271 GX 117
UGC 3273 GX 27
UGC 3274 GX 97
UGC 3275 GX 97
UGC 3276 GX 7
UGC 3277 GX 16
UGC 3281 GX 16
UGC 3282 GX 97
UGC 3283 GX 117
UGC 3284 GX 16
UGC 3287 GX 117
UGC 3293 GX 96, 97
UGC 3294 GX 116, 117
UGC 3296 GX 116, 117
UGC 3301 GX 116, 117
UGC 3302 GX 7
UGC 3303 GX 116, 117
UGC 3306 GX 116, 117
UGC 3307 GX 16
UGC 3309 GX 16
UGC 3314 GX 27
UGC 3317 GX 7, 16
UGC 3318 GX 7
UGC 3319 GX 16
UGC 3325 GX 41, 42, 59
UGC 3326 GX 7
UGC 3331 GX 116
UGC 3336 GX 1
UGC 3340 GX 6, 7
UGC 3341 GX 77
UGC 3342 GX 16
UGC 3343 GX 16
UGC 3344 GX 16
UGC 3346 GX 27, 41
UGC 3349 GX 16
UGC 3351 GX 27
UGC 3353 GX 6, 7
UGC 3354 GX 27
UGC 3355 GX 41
UGC 3357 GX 6, 7
UGC 3363 GX 77, 96
UGC 3364 GX 6, 7
UGC 3365 GX 15, 16
UGC 3369 GX 41
UGC 3371 GX 6, 7
UGC 3373 GX 6, 7
UGC 3374 GX 41
UGC 3375 GX 27, 41
UGC 3376 GX 96
UGC 3378 GX 6, 7
UGC 3379 GX 15, 16
UGC 3382 GX 15, 16, 27
UGC 3384 GX 6, 7, 15, 16
UGC 3385 GX 6, 7
UGC 3386 GX 15, 16
UGC 3390 GX 58, 59
UGC 3394 GX 27
UGC 3396 GX 6, 7
UGC 3401 GX 6, 7
UGC 3402 GX 15
UGC 3403 GX 15
UGC 3404 GX 6, 7
UGC 3405 GX 6, 7
UGC 3407 GX 41
UGC 3408 GX 6, 7
UGC 3409 GX 15
UGC 3410 GX 6, 7
UGC 3411 GX 15, 27
UGC 3412 GX 6, 7
UGC 3413 GX 6, 7
UGC 3414 GX 15
UGC 3415 GX 15
UGC 3416 GX 15
UGC 3418 GX 41
UGC 3420 GX 6, 7
UGC 3422 GX 15
UGC 3423 GX 6, 7
UGC 3425 GX 15
UGC 3426 GX 15
UGC 3428 GX 15
UGC 3431 GX 6, 7
UGC 3432 GX 27
UGC 3433 GX 116
UGC 3435 GX 6
UGC 3436 GX 15
UGC 3437 GX 15
UGC 3438 GX 15
UGC 3441 GX 6
UGC 3442 GX 1, 6
UGC 3445 GX 27
UGC 3446 GX 27
UGC 3447 GX 76
UGC 3448 GX 15
UGC 3450 GX 76
UGC 3453 GX 6, 15
UGC 3457 GX 116
UGC 3458 GX 15
UGC 3460 GX 6
UGC 3464 GX 6
UGC 3467 GX 41
UGC 3471 GX 6
UGC 3472 GX 6
UGC 3474 GX 15
UGC 3475 GX 58
UGC 3476 GX 58
UGC 3477 GX 41
UGC 3478 GX 15
UGC 3479 GX 58
UGC 3480 GX 27
UGC 3481 GX 41, 58
UGC 3484 GX 27
UGC 3486 GX 6
UGC 3487 GX 41, 58
UGC 3488 GX 27
UGC 3489 GX 76
UGC 3493 GX 41
UGC 3499 GX 58
UGC 3500 GX 1, 6
UGC 3502 GX 15
UGC 3503 GX 76
UGC 3504 GX 27
UGC 3506 GX 41
UGC 3510 GX 41, 58
UGC 3511 GX 15
UGC 3512 GX 41
UGC 3518 GX 76
UGC 3521 GX 1, 6
UGC 3522 GX 1, 6
UGC 3525 GX 41, 58
UGC 3528 GX 1, 6
UGC 3529 GX 58
UGC 3531 GX 76
UGC 3532 GX 41
UGC 3534 GX 76
UGC 3535 GX 41
UGC 3536 GX 58, 76
UGC 3536A GX 1
UGC 3537 GX 58
UGC 3538 GX 41
UGC 3539 GX 15
UGC 3540 GX 6
UGC 3545 GX 27
UGC 3548 GX 6
UGC 3549 GX 6
UGC 3552 GX 76
UGC 3553 GX 76
UGC 3554 GX 41
UGC 3555 GX 76
UGC 3558 GX 76
UGC 3561 GX 41
UGC 3567 GX 41
UGC 3568 GX 41
UGC 3569 GX 27
UGC 3571 GX 58, 76
UGC 3573 GX 76
UGC 3574 GX 27
UGC 3575 GX 15
UGC 3576 GX 41
UGC 3577 GX 15
UGC 3578 GX 95
UGC 3580 GX 15
UGC 3581 GX 6
UGC 3584 GX 76
UGC 3585 GX 76
UGC 3587 GX 76
UGC 3588 GX 41
UGC 3590 GX 58
UGC 3592 GX 41, 58
UGC 3593 GX 41, 58
UGC 3594 GX 76
UGC 3595 GX 26, 27
UGC 3596 GX 58
UGC 3597 GX 41
UGC 3598 GX 26, 27
UGC 3599 GX 76
UGC 3601 GX 41, 58
UGC 3602 GX 95
UGC 3604 GX 6
UGC 3606 GX 15
UGC 3608 GX 41
UGC 3611 GX 76
UGC 3612 GX 58
UGC 3613 GX 95
UGC 3614 GX 41
UGC 3615 GX 58
UGC 3619 GX 58
UGC 3621 GX 95
UGC 3626 GX 15
UGC 3627 GX 26, 27, 41
UGC 3630 GX 115
UGC 3631 GX 58, 76
UGC 3634 GX 95
UGC 3635 GX 76, 95
UGC 3636 GX 6
UGC 3638 GX 41
UGC 3639 GX 76
UGC 3641 GX 95
UGC 3642 GX 15
UGC 3644 GX 15
UGC 3645 GX 41
UGC 3646 GX 26
UGC 3649 GX 58, 76
UGC 3652 GX 76
UGC 3654 GX 1
UGC 3655 GX 41
UGC 3656 GX 76
UGC 3658 GX 76, 95
UGC 3660 GX 15
UGC 3664 GX 58
UGC 3668 GX 6
UGC 3674 GX 76
UGC 3675 GX 6
UGC 3676 GX 76
UGC 3678 GX 76
UGC 3679 GX 41
UGC 3683 GX 40, 41
UGC 3684 GX 26, 40, 41
UGC 3685 GX 26
UGC 3688 GX 95
UGC 3691 GX 95
UGC 3692 GX 58, 76
UGC 3694 GX 58
UGC 3696 GX 40, 41
UGC 3697 GX 15
UGC 3701 GX 15
UGC 3702 GX 76
UGC 3703 GX 58
UGC 3704 GX 26
UGC 3705 GX 6, 15
UGC 3706 GX 40, 41
UGC 3707 GX 95
UGC 3714 GX 15
UGC 3715 GX 1
UGC 3717 GX 6, 15
UGC 3723 GX 58
UGC 3724 GX 40, 41
UGC 3725 GX 40, 41
UGC 3726 GX 76
UGC 3728 GX 58
UGC 3730 GX 6, 15
UGC 3731 GX 58, 76
UGC 3735 GX 58
UGC 3738 GX 95
UGC 3742 GX 58
UGC 3743 GX 58
UGC 3745 GX 76
UGC 3749 GX 15
UGC 3751 GX 76
UGC 3752 GX 58
UGC 3753 GX 76
UGC 3755 GX 95
UGC 3758 GX 40, 41
UGC 3764 GX 15
UGC 3765 GX 26
UGC 3766 GX 95
UGC 3768 GX 95
UGC 3769 GX 115
UGC 3770 GX 76
UGC 3771 GX 15
UGC 3772 GX 95
UGC 3774 GX 57, 58
UGC 3775 GX 95
UGC 3776 GX 57, 58
UGC 3777 GX 57, 58, 75, 76
UGC 3780 GX 57, 58
UGC 3781 GX 40, 57, 58
UGC 3785 GX 95
UGC 3788 GX 57, 58
UGC 3789 GX 26
UGC 3792 GX 26, 40
UGC 3794 GX 6
UGC 3799 GX 26
UGC 3803 GX 75, 76
UGC 3804 GX 15
UGC 3805 GX 75, 76, 95
UGC 3806 GX 75, 76
UGC 3808 GX 75, 76
UGC 3811 GX 57, 58
UGC 3812 GX 40
UGC 3813 GX 95
UGC 3816 GX 26
UGC 3817 GX 40
UGC 3819 GX 95, 115
UGC 3820 GX 75, 76, 95
UGC 3822 GX 57, 58
UGC 3823 GX 75, 76
UGC 3824 GX 75, 76
UGC 3825 GX 40
UGC 3826 GX 26
UGC 3827 GX 75, 76
UGC 3828 GX 26
UGC 3829 GX 57, 58
UGC 3830 GX 115
UGC 3831 GX 40
UGC 3832 GX 26
UGC 3833 GX 57
UGC 3836 GX 15
UGC 3837 GX 57
UGC 3838 GX 15
UGC 3839 GX 95
UGC 3840 GX 75
UGC 3842 GX 75
UGC 3844 GX 40
UGC 3845 GX 40
UGC 3849 GX 57
UGC 3850 GX 15
UGC 3855 GX 26
UGC 3858 GX 6, 15
UGC 3859 GX 6, 15
UGC 3860 GX 40, 57
UGC 3862 GX 75
UGC 3863 GX 40
UGC 3864 GX 15
UGC 3867 GX 26
UGC 3869 GX 75
UGC 3871 GX 40
UGC 3873 GX 75
UGC 3874 GX 75
UGC 3875 GX 26
UGC 3876 GX 75
UGC 3877 GX 95
UGC 3879 GX 57
UGC 3882 GX 26
UGC 3885 GX 26
UGC 3886 GX 15, 26
UGC 3887 GX 57
UGC 3892 GX 95
UGC 3893 GX 15
UGC 3894 GX 15
UGC 3897 GX 26
UGC 3900 GX 95
UGC 3903 GX 75
UGC 3904 GX 57
UGC 3905 GX 15, 26
UGC 3906 GX 6
UGC 3909 GX 6, 15
UGC 3911 GX 75
UGC 3912 GX 115
UGC 3913 GX 57
UGC 3919 GX 15
UGC 3920 GX 75
UGC 3922 GX 26
UGC 3923 GX 26
UGC 3924 GX 95
UGC 3928 GX 26
UGC 3929 GX 6
UGC 3932 GX 75
UGC 3933 GX 40
UGC 3936 GX 95
UGC 3937 GX 57
UGC 3938 GX 95
UGC 3939 GX 75
UGC 3941 GX 95
UGC 3943 GX 26

UGC 5237 GX 39
UGC 5238 GX 113
UGC 5241 GX 25
UGC 5243 GX 25
UGC 5244 GX 14
UGC 5245 GX 113
UGC 5246 GX 74
UGC 5247 GX 14
UGC 5249 GX 113
UGC 5258 GX 93
UGC 5267 GX 93
UGC 5270 GX 93
UGC 5272 GX 56
UGC 5274 GX 93
UGC 5276 GX 56
UGC 5277 GX 14
UGC 5281 GX 56
UGC 5282 GX 56
UGC 5286 GX 93
UGC 5287 GX 56
UGC 5288 GX 93
UGC 5290 GX 39, 56
UGC 5291 GX 93
UGC 5295 GX 39
UGC 5302 GX 14
UGC 5304 GX 93
UGC 5308 GX 93
UGC 5313 GX 74
UGC 5320 GX 74
UGC 5324 GX 93
UGC 5326 GX 56
UGC 5330 GX 74
UGC 5336 GX 14
UGC 5339 GX 73, 74
UGC 5340 GX 73, 74
UGC 5341 GX 73, 74
UGC 5342 GX 93
UGC 5343 GX 93
UGC 5344 GX 93
UGC 5345 GX 39
UGC 5349 GX 55, 56
UGC 5354 GX 39
UGC 5356 GX 39
UGC 5358 GX 93
UGC 5359 GX 73, 74
UGC 5364 GX 55, 56
UGC 5368 GX 39
UGC 5369 GX 25
UGC 5373 GX 93, 113
UGC 5376 GX 113
UGC 5377 GX 113
UGC 5378 GX 113
UGC 5380 GX 113
UGC 5382 GX 55, 56
UGC 5383 GX 113
UGC 5385 GX 93
UGC 5386 GX 14
UGC 5388 GX 113
UGC 5389 GX 55, 56
UGC 5391 GX 55, 56
UGC 5393 GX 55, 56
UGC 5394 GX 55, 56
UGC 5395 GX 93
UGC 5396 GX 93
UGC 5400 GX 93
UGC 5402 GX 5, 6
UGC 5403 GX 73, 74
UGC 5408 GX 25
UGC 5409 GX 93
UGC 5415 GX 14
UGC 5420 GX 73, 74
UGC 5421 GX 25
UGC 5422 GX 25
UGC 5423 GX 14
UGC 5424 GX 113
UGC 5427 GX 55, 73
UGC 5431 GX 73
UGC 5432 GX 93, 113
UGC 5434 GX 73
UGC 5435 GX 25
UGC 5436 GX 73
UGC 5442 GX 14
UGC 5451 GX 39
UGC 5453 GX 93
UGC 5454 GX 93
UGC 5455 GX 14
UGC 5456 GX 93
UGC 5459 GX 25
UGC 5460 GX 25, 39
UGC 5467 GX 73
UGC 5470 GX 92, 93
UGC 5474 GX 55
UGC 5475 GX 25
UGC 5476 GX 39, 55
UGC 5477 GX 92, 93
UGC 5478 GX 55
UGC 5479 GX 25
UGC 5480 GX 25
UGC 5481 GX 55
UGC 5483 GX 112, 113
UGC 5487 GX 112, 113
UGC 5489 GX 73
UGC 5490 GX 55
UGC 5493 GX 112, 113
UGC 5495 GX 92, 93
UGC 5497 GX 14
UGC 5498 GX 73
UGC 5499 GX 73
UGC 5501 GX 112, 113
UGC 5506 GX 112, 113
UGC 5508 GX 14
UGC 5509 GX 73
UGC 5514 GX 73
UGC 5515 GX 112, 113
UGC 5518 GX 55
UGC 5520 GX 14
UGC 5521 GX 112, 113
UGC 5522 GX 92, 93
UGC 5524 GX 73
UGC 5526 GX 92, 93
UGC 5528 GX 112, 113
UGC 5530 GX 14
UGC 5534 GX 25
UGC 5537 GX 92, 93
UGC 5539 GX 112, 113
UGC 5540 GX 55
UGC 5542 GX 25
UGC 5543 GX 112
UGC 5545 GX 39
UGC 5546 GX 25
UGC 5547 GX 92
UGC 5549 GX 25
UGC 5552 GX 73, 92
UGC 5563 GX 55
UGC 5573 GX 92
UGC 5576 GX 14
UGC 5577 GX 55
UGC 5586 GX 112
UGC 5588 GX 73
UGC 5595 GX 92
UGC 5596 GX 5
UGC 5598 GX 73
UGC 5600 GX 5
UGC 5604 GX 38, 39
UGC 5605 GX 38, 39
UGC 5607 GX 112
UGC 5609 GX 5
UGC 5612 GX 14
UGC 5615 GX 25
UGC 5616 GX 92
UGC 5621 GX 73
UGC 5622 GX 55
UGC 5623 GX 55
UGC 5626 GX 25
UGC 5627 GX 92
UGC 5630 GX 14
UGC 5633 GX 92
UGC 5638 GX 73
UGC 5639 GX 73, 92
UGC 5642 GX 92
UGC 5644 GX 92
UGC 5646 GX 92
UGC 5648 GX 112
UGC 5651 GX 73, 92
UGC 5656 GX 55
UGC 5657 GX 38, 39
UGC 5667 GX 112
UGC 5672 GX 73
UGC 5673 GX 25
UGC 5675 GX 73
UGC 5676 GX 25
UGC 5677 GX 112
UGC 5678 GX 112
UGC 5679 GX 73
UGC 5682 GX 5
UGC 5683 GX 73
UGC 5686 GX 5
UGC 5687 GX 92
UGC 5688 GX 14
UGC 5689 GX 5, 14
UGC 5690 GX 73
UGC 5691 GX 73
UGC 5692 GX 14
UGC 5695 GX 92
UGC 5696 GX 73
UGC 5698 GX 38
UGC 5700 GX 14
UGC 5701 GX 5
UGC 5702 GX 92, 112
UGC 5707 GX 38
UGC 5708 GX 112
UGC 5709 GX 73
UGC 5710 GX 73
UGC 5713 GX 73
UGC 5714 GX 38
UGC 5715 GX 112
UGC 5720 GX 25
UGC 5722 GX 25
UGC 5724 GX 14
UGC 5727 GX 14
UGC 5733 GX 25
UGC 5734 GX 25
UGC 5735 GX 92
UGC 5736 GX 112
UGC 5737 GX 92
UGC 5738 GX 55
UGC 5739 GX 92
UGC 5740 GX 38
UGC 5744 GX 38
UGC 5745 GX 112
UGC 5746 GX 38
UGC 5749 GX 73
UGC 5751 GX 73
UGC 5754 GX 38
UGC 5757 GX 5
UGC 5759 GX 55
UGC 5760 GX 92
UGC 5764 GX 55
UGC 5765 GX 14
UGC 5771 GX 38
UGC 5772 GX 112
UGC 5776 GX 14
UGC 5782 GX 5
UGC 5783 GX 112
UGC 5784 GX 112
UGC 5787 GX 112
UGC 5788 GX 92, 112
UGC 5790 GX 112
UGC 5791 GX 38
UGC 5797 GX 112
UGC 5798 GX 38
UGC 5801 GX 73
UGC 5804 GX 55
UGC 5805 GX 73
UGC 5806 GX 55
UGC 5808 GX 92
UGC 5813 GX 55
UGC 5814 GX 5
UGC 5818 GX 92
UGC 5819 GX 55
UGC 5820 GX 5
UGC 5822 GX 73
UGC 5823 GX 112
UGC 5825 GX 73
UGC 5829 GX 55
UGC 5830 GX 73
UGC 5832 GX 92
UGC 5833 GX 73
UGC 5835 GX 13, 14
UGC 5838 GX 38, 55
UGC 5839 GX 55
UGC 5841 GX 5
UGC 5843 GX 13, 14
UGC 5844 GX 73
UGC 5845 GX 38
UGC 5846 GX 25
UGC 5847 GX 112
UGC 5848 GX 25
UGC 5849 GX 112
UGC 5854 GX 5
UGC 5855 GX 73
UGC 5858 GX 92
UGC 5859 GX 38
UGC 5861 GX 55
UGC 5865 GX 112
UGC 5867 GX 112
UGC 5868 GX 55
UGC 5869 GX 92
UGC 5870 GX 55
UGC 5872 GX 38
UGC 5881 GX 73
UGC 5883 GX 25
UGC 5884 GX 73
UGC 5885 GX 55
UGC 5886 GX 112
UGC 5888 GX 25
UGC 5892 GX 92
UGC 5893 GX 55
UGC 5894 GX 73
UGC 5896 GX 112
UGC 5897 GX 92
UGC 5898 GX 55
UGC 5903 GX 73
UGC 5904 GX 13, 14
UGC 5910 GX 55
UGC 5912 GX 73
UGC 5913 GX 112
UGC 5916 GX 73
UGC 5917 GX 38
UGC 5918 GX 13, 14
UGC 5921 GX 73
UGC 5922 GX 112
UGC 5923 GX 92
UGC 5924 GX 73
UGC 5926 GX 5
UGC 5928 GX 25, 38
UGC 5929 GX 112
UGC 5932 GX 13, 14
UGC 5934 GX 55
UGC 5936 GX 55
UGC 5939 GX 5
UGC 5941 GX 38
UGC 5943 GX 112
UGC 5944 GX 92
UGC 5945 GX 73, 92
UGC 5946 GX 5
UGC 5947 GX 73
UGC 5950 GX 92
UGC 5953 GX 38
UGC 5955 GX 13, 14
UGC 5958 GX 73
UGC 5974 GX 112
UGC 5976 GX 25
UGC 5979 GX 13, 14
UGC 5983 GX 55
UGC 5984 GX 55
UGC 5989 GX 73
UGC 5990 GX 55
UGC 5991 GX 38
UGC 5994 GX 92
UGC 5996 GX 38, 55
UGC 5998 GX 38
UGC 5999 GX 92
UGC 6002 GX 55
UGC 6003 GX 112
UGC 6008 GX 38
UGC 6013 GX 38
UGC 6014 GX 92
UGC 6015 GX 38
UGC 6016 GX 25
UGC 6029 GX 38
UGC 6031 GX 55, 73
UGC 6032 GX 13
UGC 6035 GX 73, 92
UGC 6036 GX 55
UGC 6038 GX 38
UGC 6040 GX 112
UGC 6043 GX 92
UGC 6046 GX 92
UGC 6049 GX 92, 112
UGC 6053 GX 92
UGC 6057 GX 112
UGC 6059 GX 24, 25
UGC 6062 GX 92
UGC 6065 GX 5
UGC 6066 GX 92
UGC 6068 GX 92, 112
UGC 6070 GX 55
UGC 6071 GX 38
UGC 6072 GX 92
UGC 6074 GX 38
UGC 6075 GX 38
UGC 6076 GX 38
UGC 6078 GX 92
UGC 6080 GX 24, 25
UGC 6081 GX 92
UGC 6087 GX 112
UGC 6089 GX 55
UGC 6093 GX 92
UGC 6100 GX 38
UGC 6101 GX 38
UGC 6102 GX 73
UGC 6103 GX 38
UGC 6104 GX 92
UGC 6106 GX 38
UGC 6109 GX 38
UGC 6110 GX 24, 25
UGC 6112 GX 92
UGC 6114 GX 38
UGC 6117 GX 38
UGC 6119 GX 112
UGC 6121 GX 55
UGC 6125 GX 38
UGC 6127 GX 38
UGC 6129 GX 38
UGC 6130 GX 92
UGC 6131 GX 38
UGC 6132 GX 55
UGC 6133 GX 13
UGC 6135 GX 38
UGC 6136 GX 38
UGC 6137 GX 92
UGC 6138 GX 73
UGC 6141 GX 92, 112
UGC 6142 GX 112
UGC 6143 GX 55
UGC 6149 GX 38
UGC 6151 GX 73
UGC 6152 GX 55, 73
UGC 6154 GX 5
UGC 6155 GX 112
UGC 6157 GX 73, 92
UGC 6161 GX 38
UGC 6162 GX 24, 38
UGC 6163 GX 73
UGC 6165 GX 38
UGC 6166 GX 73
UGC 6168 GX 92
UGC 6169 GX 92
UGC 6171 GX 73
UGC 6173 GX 73
UGC 6175 GX 73
UGC 6176 GX 73
UGC 6179 GX 13
UGC 6181 GX 73
UGC 6182 GX 24
UGC 6183 GX 55
UGC 6185 GX 92
UGC 6186 GX 24
UGC 6187 GX 38
UGC 6190 GX 73
UGC 6192 GX 24
UGC 6194 GX 73
UGC 6198 GX 55, 73
UGC 6201 GX 38
UGC 6204 GX 73
UGC 6205 GX 38
UGC 6206 GX 92
UGC 6207 GX 73
UGC 6210 GX 92, 112
UGC 6212 GX 112
UGC 6219 GX 73
UGC 6220 GX 55, 73
UGC 6227 GX 38
UGC 6228 GX 24
UGC 6231 GX 92
UGC 6232 GX 38
UGC 6233 GX 92
UGC 6237 GX 13
UGC 6239 GX 112
UGC 6246 GX 73
UGC 6247 GX 73
UGC 6248 GX 92
UGC 6249 GX 24
UGC 6250 GX 73
UGC 6251 GX 24
UGC 6252 GX 73
UGC 6253 GX 73
UGC 6255 GX 38
UGC 6258 GX 73
UGC 6260 GX 112
UGC 6266 GX 38
UGC 6271 GX 55
UGC 6273 GX 55
UGC 6274 GX 55
UGC 6276 GX 55
UGC 6279 GX 55
UGC 6289 GX 112
UGC 6292 GX 54, 55, 72, 73
UGC 6296 GX 72, 73, 92
UGC 6298 GX 54, 55
UGC 6300 GX 92
UGC 6301 GX 72, 73
UGC 6303 GX 54, 55
UGC 6307 GX 54, 55
UGC 6309 GX 24, 38
UGC 6311 GX 112
UGC 6312 GX 92
UGC 6314 GX 54, 55
UGC 6316 GX 13
UGC 6320 GX 72, 73
UGC 6322 GX 72, 73
UGC 6324 GX 72, 73
UGC 6325 GX 72, 73
UGC 6326 GX 54, 55
UGC 6329 GX 112
UGC 6332 GX 72, 73
UGC 6333 GX 72, 73
UGC 6334 GX 72, 73
UGC 6335 GX 24
UGC 6337 GX 54, 55
UGC 6338 GX 54, 55
UGC 6340 GX 112
UGC 6345 GX 111, 112
UGC 6355 GX 54, 55
UGC 6359 GX 111, 112
UGC 6361 GX 111, 112
UGC 6367 GX 54, 55
UGC 6378 GX 13
UGC 6380 GX 38
UGC 6381 GX 13
UGC 6393 GX 54, 55
UGC 6394 GX 54, 55
UGC 6397 GX 54, 55
UGC 6399 GX 38
UGC 6402 GX 111, 112
UGC 6410 GX 38
UGC 6413 GX 111, 112
UGC 6414 GX 72
UGC 6415 GX 72
UGC 6421 GX 72
UGC 6425 GX 72
UGC 6429 GX 13
UGC 6432 GX 111, 112
UGC 6433 GX 54
UGC 6435 GX 111, 112
UGC 6437 GX 72
UGC 6440 GX 111, 112
UGC 6446 GX 24
UGC 6448 GX 13
UGC 6452 GX 24
UGC 6454 GX 54
UGC 6455 GX 38, 54
UGC 6456 GX 5
UGC 6457 GX 111, 112
UGC 6462 GX 91, 92
UGC 6465 GX 72
UGC 6469 GX 111
UGC 6473 GX 5, 13
UGC 6476 GX 72
UGC 6483 GX 72, 91
UGC 6489 GX 38
UGC 6491 GX 54
UGC 6495 GX 72
UGC 6497 GX 54
UGC 6499 GX 54
UGC 6501 GX 24
UGC 6505 GX 24, 38
UGC 6508 GX 72
UGC 6509 GX 72
UGC 6510 GX 111
UGC 6512 GX 54
UGC 6517 GX 54
UGC 6518 GX 24
UGC 6520 GX 13, 24
UGC 6522 GX 72
UGC 6525 GX 72
UGC 6526 GX 54
UGC 6527 GX 24
UGC 6528 GX 24
UGC 6529 GX 38, 54
UGC 6530 GX 13
UGC 6531 GX 54
UGC 6532 GX 13
UGC 6534 GX 13
UGC 6541 GX 38
UGC 6544 GX 72
UGC 6545 GX 54
UGC 6548 GX 72
UGC 6551 GX 54
UGC 6552 GX 13
UGC 6555 GX 24, 38
UGC 6556 GX 91
UGC 6559 GX 91
UGC 6566 GX 24
UGC 6568 GX 111
UGC 6570 GX 54
UGC 6575 GX 24
UGC 6576 GX 38
UGC 6583 GX 72
UGC 6586 GX 91
UGC 6587 GX 111
UGC 6588 GX 91
UGC 6593 GX 72
UGC 6594 GX 91
UGC 6603 GX 54
UGC 6606 GX 38
UGC 6607 GX 72, A11
UGC 6608 GX 111
UGC 6609 GX 72, A11
UGC 6610 GX 54
UGC 6611 GX 38
UGC 6614 GX 72, 91
UGC 6616 GX 24

UGC 6617 GX 91
UGC 6625 GX 72, A11
UGC 6628 GX 38
UGC 6631 GX 72, 91
UGC 6637 GX 72
UGC 6639 GX 54
UGC 6645 GX 72
UGC 6647 GX 91
UGC 6653 GX 91
UGC 6655 GX 91
UGC 6659 GX 54
UGC 6664 GX 54
UGC 6665 GX 111
UGC 6666 GX 91
UGC 6667 GX 24, 38
UGC 6669 GX 91
UGC 6670 GX 72, A11
UGC 6674 GX 72
UGC 6682 GX 24
UGC 6686 GX 91
UGC 6687 GX 72, A11
UGC 6697 GX 72, A11
UGC 6698 GX 13
UGC 6711 GX 13
UGC 6713 GX 38
UGC 6716 GX 54
UGC 6717 GX 91
UGC 6719 GX 72, A11
UGC 6725 GX 72, A11
UGC 6726 GX 38
UGC 6727 GX 24
UGC 6728 GX 5
UGC 6732 GX 24
UGC 6734 GX 91
UGC 6736 GX 111
UGC 6740 GX 91
UGC 6741 GX 72, 91
UGC 6743 GX 72, A11
UGC 6750 GX 111
UGC 6751 GX 72
UGC 6753 GX 91
UGC 6755 GX 38
UGC 6758 GX 91
UGC 6761 GX 54, 72
UGC 6762 GX 24
UGC 6764 GX 13
UGC 6766 GX 24
UGC 6767 GX 24
UGC 6769 GX 111
UGC 6771 GX 111
UGC 6773 GX 38
UGC 6774 GX 24
UGC 6775 GX 91
UGC 6776 GX 38
UGC 6777 GX 54
UGC 6780 GX 111
UGC 6782 GX 72
UGC 6789 GX 5
UGC 6791 GX 72
UGC 6792 GX 54
UGC 6794 GX 91
UGC 6802 GX 24, 38
UGC 6804 GX 91
UGC 6805 GX 38
UGC 6806 GX 72, A10
UGC 6811 GX 38
UGC 6812 GX 38
UGC 6816 GX 24
UGC 6817 GX 54
UGC 6818 GX 38
UGC 6821 GX 72, A10
UGC 6827 GX 54
UGC 6828 GX 24
UGC 6831 GX 91
UGC 6838 GX 111
UGC 6840 GX 24
UGC 6846 GX 72, A10
UGC 6847 GX 72, A10
UGC 6850 GX 111
UGC 6853 GX 54, 72
UGC 6854 GX 111
UGC 6858 GX 24
UGC 6861 GX 72, A10
UGC 6865 GX 38
UGC 6871 GX 91
UGC 6876 GX 72, A10
UGC 6879 GX 111
UGC 6881 GX 72, A10
UGC 6883 GX 72
UGC 6886 GX 91
UGC 6891 GX 72, 91
UGC 6892 GX 54
UGC 6893 GX 54
UGC 6894 GX 24
UGC 6900 GX 54
UGC 6901 GX 38
UGC 6903 GX 111
UGC 6911 GX 91
UGC 6912 GX 24
UGC 6913 GX 72, 91
UGC 6916 GX 54
UGC 6917 GX 37, 38
UGC 6919 GX 24
UGC 6922 GX 37, 38
UGC 6923 GX 24
UGC 6927 GX 54
UGC 6929 GX 54
UGC 6930 GX 37, 38
UGC 6931 GX 24
UGC 6934 GX 111
UGC 6939 GX 24
UGC 6945 GX 54
UGC 6955 GX 54
UGC 6956 GX 37, 38
UGC 6957 GX 24
UGC 6958 GX 111
UGC 6968 GX 72
UGC 6969 GX 24
UGC 6970 GX 111
UGC 6976 GX 72, A10
UGC 6983 GX 24
UGC 6987 GX 54
UGC 6992 GX 37, 38
UGC 6997 GX 54
UGC 6998 GX 111
UGC 7000 GX 111
UGC 7003 GX 91
UGC 7004 GX 111
UGC 7007 GX 54
UGC 7009 GX 13, 24
UGC 7012 GX 54, 72
UGC 7016 GX 91
UGC 7017 GX 54, 72
UGC 7019 GX 13, 24
UGC 7020 GX 37, 38
UGC 7020A GX 13
UGC 7031 GX 54, 72
UGC 7032 GX 91
UGC 7034 GX 111
UGC 7035 GX 111
UGC 7040 GX 72, A10
UGC 7043 GX 91
UGC 7053 GX 111
UGC 7057 GX 111
UGC 7059 GX 5
UGC 7064 GX 54
UGC 7064A GX 24
UGC 7065 GX 111
UGC 7069 GX 37
UGC 7070 GX 24
UGC 7071 GX 54
UGC 7072 GX 72
UGC 7073 GX 72, 91
UGC 7074 GX 72, 91
UGC 7079 GX 13
UGC 7082 GX 24, 37
UGC 7084 GX 54
UGC 7085 GX 91
UGC 7085A GX 54
UGC 7086 GX 5
UGC 7089 GX 37
UGC 7097 GX 5
UGC 7098 GX 54
UGC 7100 GX 72, 91
UGC 7105 GX 54
UGC 7109 GX 91
UGC 7115 GX 72, A10
UGC 7125 GX 54
UGC 7129 GX 37
UGC 7131 GX 54
UGC 7132 GX 54
UGC 7133 GX 72
UGC 7137 GX 72, A10
UGC 7138 GX 72
UGC 7141 GX 72, A10
UGC 7143 GX 72, A10
UGC 7144 GX 24
UGC 7145 GX 54
UGC 7153 GX 13, 24
UGC 7157 GX 72
UGC 7159 GX 54
UGC 7164 GX 13
UGC 7170 GX 72
UGC 7175 GX 54
UGC 7176 GX 37
UGC 7177 GX 111
UGC 7179 GX 13
UGC 7184 GX 111
UGC 7185 GX 111
UGC 7186 GX 72
UGC 7187 GX 54
UGC 7189 GX 5
UGC 7190 GX 54, 72
UGC 7194 GX 91
UGC 7196 GX 91
UGC 7207 GX 54
UGC 7208 GX 54
UGC 7212 GX 54
UGC 7213 GX 37, 54
UGC 7217 GX 72
UGC 7218 GX 24
UGC 7221 GX 72
UGC 7224 GX 72
UGC 7226 GX 5
UGC 7230 GX 91
UGC 7238 GX 5
UGC 7239 GX 91, A15
UGC 7242 GX 13
UGC 7243 GX 54
UGC 7249 GX 91
UGC 7257 GX 54
UGC 7263 GX 72
UGC 7266 GX 72
UGC 7267 GX 24, 37
UGC 7270 GX 72
UGC 7271 GX 37
UGC 7280 GX 111
UGC 7286 GX 72
UGC 7287 GX 72
UGC 7300 GX 72
UGC 7301 GX 37
UGC 7307 GX 91
UGC 7321 GX 72
UGC 7325 GX 37
UGC 7327 GX 91
UGC 7331 GX 72, 91
UGC 7332 GX 111
UGC 7340 GX 37
UGC 7346 GX 72, 91
UGC 7354 GX 111, A15
UGC 7355 GX 91, A13
UGC 7357 GX 72
UGC 7358 GX 37
UGC 7367 GX 37
UGC 7370 GX 111
UGC 7379 GX 24
UGC 7383 GX 91, A15
UGC 7387 GX 111, A15
UGC 7388 GX 54
UGC 7394 GX 111
UGC 7395 GX 54
UGC 7396 GX 111
UGC 7399A GX 72, 91
UGC 7406 GX 24
UGC 7408 GX 37
UGC 7416 GX 37, 54
UGC 7423 GX 91, A15
UGC 7424 GX 91, A15
UGC 7428 GX 54
UGC 7436 GX 91, A13
UGC 7444 GX 37
UGC 7464 GX 111
UGC 7486 GX 37
UGC 7487 GX 111
UGC 7490 GX 13
UGC 7516 GX 111, A15
UGC 7522 GX 111, A15
UGC 7525 GX 37
UGC 7531 GX 111
UGC 7534 GX 24
UGC 7557 GX 91, A15
UGC 7559 GX 54
UGC 7560 GX 37
UGC 7567 GX 91, A15
UGC 7576 GX 72
UGC 7577 GX 37
UGC 7580 GX 91, A15
UGC 7590 GX 91, A15
UGC 7593 GX 37
UGC 7596 GX 91, A15
UGC 7598 GX 54
UGC 7599 GX 54
UGC 7605 GX 54
UGC 7607 GX 111, A15
UGC 7608 GX 37
UGC 7612 GX 111
UGC 7617 GX 37
UGC 7618 GX 24
UGC 7635 GX 24
UGC 7636 GX 91, A15
UGC 7639 GX 37
UGC 7642 GX 111
UGC 7644 GX 111
UGC 7659 GX 24
UGC 7673 GX 54, 72
UGC 7678 GX 54
UGC 7688 GX 90, 91
UGC 7689 GX 54
UGC 7690 GX 37
UGC 7691 GX 24
UGC 7697 GX 72
UGC 7698 GX 54
UGC 7699 GX 54
UGC 7705 GX 24
UGC 7715 GX 110, 111
UGC 7719 GX 54
UGC 7720 GX 110, 111
UGC 7730 GX 13
UGC 7739 GX 90, 91
UGC 7750 GX 54, 72
UGC 7752 GX 110, 111
UGC 7761 GX 13
UGC 7763 GX 110, 111
UGC 7767 GX 5, 13
UGC 7774 GX 37, 53, 54
UGC 7780 GX 110, 111
UGC 7795 GX 90, 91
UGC 7798 GX 110, 111
UGC 7799 GX 53, 54
UGC 7800 GX 110, 111
UGC 7802 GX 90, 91
UGC 7806 GX 110, 111
UGC 7809 GX 13
UGC 7811 GX 53, 54
UGC 7812 GX 53, 54
UGC 7813 GX 110, 111
UGC 7816 GX 53, 54
UGC 7820 GX 110, 111
UGC 7823 GX 37
UGC 7824 GX 110, 111
UGC 7836 GX 53, 54, 71, 72
UGC 7841 GX 110
UGC 7848 GX 13
UGC 7857 GX 90
UGC 7872 GX 5
UGC 7873 GX 110
UGC 7883 GX 110
UGC 7890 GX 71, 72
UGC 7905 GX 24
UGC 7908 GX 5, 13
UGC 7910 GX 37
UGC 7911 GX 110
UGC 7913 GX 110
UGC 7916 GX 53
UGC 7922 GX 24
UGC 7941 GX 13
UGC 7942 GX 90
UGC 7943 GX 90, 110
UGC 7945 GX 110
UGC 7949 GX 53
UGC 7950 GX 24, 37
UGC 7955 GX 71
UGC 7956 GX 5
UGC 7959 GX 71
UGC 7976 GX 110
UGC 7978 GX 53
UGC 7979 GX 90
UGC 7982 GX 110
UGC 7991 GX 110
UGC 7993 GX 24
UGC 7995 GX 5
UGC 8004 GX 53
UGC 8012 GX 24, 37
UGC 8013 GX 71, A8
UGC 8015 GX 90
UGC 8025 GX 53, 71, A8
UGC 8032 GX 90
UGC 8040 GX 24
UGC 8041 GX 110
UGC 8042 GX 90
UGC 8045 GX 90
UGC 8046 GX 24
UGC 8047 GX 24
UGC 8050 GX 24
UGC 8052 GX 5, 13
UGC 8053 GX 110
UGC 8056 GX 90
UGC 8058 GX 23, 24
UGC 8066 GX 110
UGC 8067 GX 110
UGC 8069 GX 53, 71, A8
UGC 8074 GX 110
UGC 8076 GX 53, 71, A8
UGC 8080 GX 71, A8
UGC 8081 GX 90
UGC 8084 GX 110
UGC 8085 GX 90
UGC 8091 GX 90
UGC 8093 GX 90
UGC 8107 GX 23, 24
UGC 8114 GX 90
UGC 8119 GX 37
UGC 8120 GX 5, 13
UGC 8122 GX 71, A8
UGC 8127 GX 110
UGC 8139 GX 53
UGC 8146 GX 23, 24
UGC 8148 GX 5
UGC 8151 GX 23, 24, 37
UGC 8153 GX 110
UGC 8155 GX 90
UGC 8162 GX 37
UGC 8164 GX 5
UGC 8170 GX 90
UGC 8179 GX 53
UGC 8181 GX 53
UGC 8183 GX 5
UGC 8186 GX 110
UGC 8189 GX 37
UGC 8192 GX 90
UGC 8199 GX 53
UGC 8201 GX 12, 13
UGC 8203 GX 53
UGC 8204 GX 90
UGC 8205 GX 23
UGC 8207 GX 53
UGC 8211 GX 23
UGC 8214 GX 12, 13, 23
UGC 8215 GX 37
UGC 8222 GX 23
UGC 8223 GX 110
UGC 8225 GX 37
UGC 8226 GX 23
UGC 8229 GX 71
UGC 8231 GX 23
UGC 8233 GX 90
UGC 8234 GX 12, 13, 23
UGC 8237 GX 12, 13, 23
UGC 8238 GX 110
UGC 8244 GX 71
UGC 8245 GX 5
UGC 8246 GX 53
UGC 8248 GX 71
UGC 8250 GX 53
UGC 8253 GX 90
UGC 8255 GX 90
UGC 8262 GX 110
UGC 8263 GX 110
UGC 8264 GX 1, 5
UGC 8265 GX 110
UGC 8266 GX 53
UGC 8272 GX 37
UGC 8275 GX 110
UGC 8282 GX 23
UGC 8285 GX 90
UGC 8287 GX 5
UGC 8290 GX 71
UGC 8294 GX 53
UGC 8296 GX 90
UGC 8298 GX 90
UGC 8299 GX 53
UGC 8303 GX 53
UGC 8304 GX 37
UGC 8306 GX 90
UGC 8313 GX 37
UGC 8318 GX 53
UGC 8320 GX 37
UGC 8322 GX 90
UGC 8323 GX 53
UGC 8324 GX 110
UGC 8327 GX 37
UGC 8331 GX 37
UGC 8335 GX 12, 13, 23
UGC 8338 GX 53
UGC 8339 GX 23
UGC 8340 GX 110
UGC 8343 GX 71
UGC 8349 GX 90
UGC 8357 GX 110
UGC 8359 GX 71
UGC 8360 GX 110
UGC 8363 GX 71
UGC 8365 GX 37
UGC 8370 GX 90
UGC 8374 GX 12, 13
UGC 8377 GX 53
UGC 8382 GX 90, 110
UGC 8383 GX 90
UGC 8385 GX 90
UGC 8392 GX 53
UGC 8397 GX 53
UGC 8399 GX 53
UGC 8400 GX 37
UGC 8409 GX 71
UGC 8425 GX 90
UGC 8427 GX 90
UGC 8431 GX 53
UGC 8436 GX 12
UGC 8441 GX 23
UGC 8448 GX 71
UGC 8450 GX 90
UGC 8451 GX 53
UGC 8452 GX 90
UGC 8454 GX 1, 5
UGC 8461 GX 53
UGC 8466 GX 53
UGC 8467 GX 12, 23
UGC 8471 GX 53
UGC 8483 GX 53, 71
UGC 8489 GX 37
UGC 8491 GX 12, 23
UGC 8492 GX 53
UGC 8496 GX 53
UGC 8498 GX 53
UGC 8502 GX 53
UGC 8506 GX 37
UGC 8507 GX 71
UGC 8508 GX 23
UGC 8510 GX 53, 71
UGC 8511 GX 12, 23
UGC 8516 GX 71
UGC 8521 GX 110
UGC 8525 GX 12
UGC 8532 GX 90, 110
UGC 8534 GX 110
UGC 8535 GX 71, 90
UGC 8539 GX 53
UGC 8543 GX 110
UGC 8548 GX 53
UGC 8551 GX 23
UGC 8554 GX 53
UGC 8560 GX 53
UGC 8561 GX 53
UGC 8563 GX 90
UGC 8564 GX 53
UGC 8569 GX 90
UGC 8572 GX 90
UGC 8575 GX 90
UGC 8581 GX 110
UGC 8583 GX 53
UGC 8585 GX 90
UGC 8588 GX 36, 37
UGC 8596 GX 90
UGC 8597 GX 36, 37
UGC 8598 GX 71
UGC 8600 GX 53
UGC 8604 GX 12
UGC 8605 GX 53
UGC 8607 GX 110
UGC 8609 GX 53
UGC 8611 GX 36, 37
UGC 8613 GX 90
UGC 8614 GX 90
UGC 8615 GX 1
UGC 8617 GX 90, 110
UGC 8619 GX 53
UGC 8621 GX 53
UGC 8626 GX 90
UGC 8627 GX 53
UGC 8630 GX 53
UGC 8631 GX 110
UGC 8634 GX 110
UGC 8635 GX 110
UGC 8636 GX 71
UGC 8638 GX 71
UGC 8639 GX 23, 36, 37
UGC 8642 GX 36, 37
UGC 8649 GX 23
UGC 8650 GX 110
UGC 8651 GX 36, 37, 53
UGC 8656 GX 36
UGC 8657 GX 90, 110
UGC 8658 GX 23
UGC 8663 GX 90, 110
UGC 8670 GX 36, 53
UGC 8671 GX 23
UGC 8683 GX 53
UGC 8684 GX 23
UGC 8685 GX 53
UGC 8686 GX 110
UGC 8688 GX 36
UGC 8689 GX 89, 90, 109, 110
UGC 8690 GX 109, 110
UGC 8693 GX 53

UGC 8701 GX 71
UGC 8702 GX 36
UGC 8703 GX 71
UGC 8705 GX 71
UGC 8708 GX 89, 90
UGC 8713 GX 53
UGC 8714 GX 23
UGC 8715 GX 53
UGC 8718 GX 53
UGC 8726 GX 36, 53
UGC 8728 GX 89, 90
UGC 8732 GX 12
UGC 8733 GX 36
UGC 8736 GX 53
UGC 8737 GX 12
UGC 8739 GX 53
UGC 8740 GX 109, 110
UGC 8741 GX 23
UGC 8742 GX 53
UGC 8747 GX 4, 5
UGC 8753 GX 71
UGC 8754 GX 53
UGC 8756 GX 36
UGC 8759 GX 71
UGC 8760 GX 53
UGC 8762 GX 71
UGC 8763 GX 71
UGC 8769 GX 89
UGC 8778 GX 53
UGC 8781 GX 71
UGC 8786 GX 109
UGC 8787 GX 109
UGC 8788 GX 71
UGC 8793 GX 53
UGC 8794 GX 71
UGC 8798 GX 36
UGC 8801 GX 109
UGC 8806 GX 53
UGC 8811 GX 12
UGC 8816 GX 109
UGC 8818 GX 89, 109
UGC 8823 GX 12
UGC 8824 GX 53
UGC 8825 GX 53
UGC 8827 GX 89
UGC 8829 GX 53
UGC 8833 GX 53
UGC 8836 GX 23
UGC 8837 GX 23
UGC 8839 GX 71, 89
UGC 8841 GX 36, 53
UGC 8844 GX 109
UGC 8850 GX 70, 71
UGC 8851 GX 53
UGC 8854 GX 53
UGC 8856 GX 52, 53
UGC 8858 GX 52, 53
UGC 8859 GX 23
UGC 8861 GX 89
UGC 8872 GX 89
UGC 8873 GX 70, 71
UGC 8876 GX 36
UGC 8878 GX 89
UGC 8879 GX 70, 71
UGC 8882 GX 23
UGC 8883 GX 89
UGC 8887 GX 70, 71
UGC 8888 GX 70, 71
UGC 8889 GX 70, 71
UGC 8892 GX 23
UGC 8894 GX 12
UGC 8896 GX 89
UGC 8902 GX 89
UGC 8904 GX 70, 71
UGC 8906 GX 89, 109
UGC 8907 GX 89
UGC 8909 GX 23
UGC 8911 GX 70, 71
UGC 8917 GX 36, 52, 53
UGC 8918 GX 89
UGC 8923 GX 52, 53
UGC 8924 GX 109
UGC 8939 GX 109
UGC 8945 GX 52, 53
UGC 8948 GX 89
UGC 8959 GX 12
UGC 8960 GX 52, 53
UGC 8961 GX 70, 71
UGC 8962 GX 52, 53
UGC 8964 GX 4, 5
UGC 8966 GX 89
UGC 8967 GX 89
UGC 8970 GX 23
UGC 8972 GX 89
UGC 8973 GX 89
UGC 8975 GX 52, 53
UGC 8977 GX 89
UGC 8980 GX 52, 53
UGC 8984 GX 52, 53
UGC 8985 GX 23
UGC 8986 GX 109
UGC 8987 GX 89
UGC 8988 GX 23
UGC 8993 GX 109
UGC 8994 GX 109
UGC 8995 GX 89
UGC 8996 GX 89
UGC 8999 GX 52, 70
UGC 9002 GX 89
UGC 9003 GX 52
UGC 9006 GX 109
UGC 9007 GX 89
UGC 9008 GX 89
UGC 9012 GX 52
UGC 9015 GX 89
UGC 9022 GX 52
UGC 9023 GX 89
UGC 9024 GX 70
UGC 9025 GX 89
UGC 9032 GX 23
UGC 9035 GX 52, 70
UGC 9037 GX 89
UGC 9039 GX 12
UGC 9041 GX 89
UGC 9042 GX 52
UGC 9043 GX 89
UGC 9044 GX 89
UGC 9045 GX 52
UGC 9048 GX 52
UGC 9052 GX 4, 5
UGC 9055 GX 89
UGC 9056 GX 36
UGC 9057 GX 109
UGC 9060 GX 23
UGC 9067 GX 89
UGC 9068 GX 36
UGC 9071 GX 23
UGC 9078 GX 70, 89
UGC 9081 GX 52
UGC 9083 GX 36
UGC 9084 GX 89
UGC 9087 GX 70
UGC 9088 GX 52
UGC 9089 GX 89
UGC 9090 GX 70
UGC 9098 GX 36
UGC 9101 GX 70
UGC 9104 GX 89
UGC 9105 GX 36
UGC 9107 GX 52
UGC 9110 GX 89
UGC 9113 GX 52
UGC 9117 GX 89
UGC 9120 GX 109
UGC 9121 GX 89
UGC 9125 GX 36
UGC 9126 GX 89
UGC 9128 GX 70
UGC 9134 GX 89
UGC 9138 GX 70
UGC 9155 GX 70
UGC 9162 GX 89
UGC 9164 GX 70
UGC 9167 GX 70, 89
UGC 9169 GX 89
UGC 9177 GX 89
UGC 9178 GX 23, 36
UGC 9182 GX 70
UGC 9203 GX 52
UGC 9206 GX 89
UGC 9211 GX 36
UGC 9212 GX 70
UGC 9213 GX 52
UGC 9214 GX 52
UGC 9215 GX 109
UGC 9216 GX 36, 52
UGC 9221 GX 52
UGC 9223 GX 36, 52
UGC 9225 GX 89
UGC 9229 GX 109
UGC 9234 GX 70
UGC 9240 GX 36
UGC 9241 GX 52
UGC 9242 GX 52
UGC 9243 GX 52
UGC 9244 GX 89, 109
UGC 9245 GX 23
UGC 9246 GX 89, 109
UGC 9249 GX 89
UGC 9253 GX 52
UGC 9258 GX 109
UGC 9259 GX 89
UGC 9262 GX 52
UGC 9263 GX 109
UGC 9267 GX 89
UGC 9273 GX 89
UGC 9274 GX 70
UGC 9277 GX 109
UGC 9282 GX 70
UGC 9284 GX 52
UGC 9285 GX 109
UGC 9286 GX 89
UGC 9288 GX 89
UGC 9291 GX 52
UGC 9292 GX 109
UGC 9294 GX 70
UGC 9299 GX 109
UGC 9310 GX 109
UGC 9314 GX 36
UGC 9315 GX 52
UGC 9316 GX 70
UGC 9317 GX 70
UGC 9320 GX 52
UGC 9322 GX 70
UGC 9324 GX 36
UGC 9338 GX 89, 109
UGC 9339 GX 89
UGC 9340 GX 70
UGC 9348 GX 109
UGC 9350 GX 52
UGC 9356 GX 89
UGC 9362 GX 109
UGC 9364 GX 89
UGC 9365 GX 109
UGC 9371 GX 109
UGC 9372 GX 52
UGC 9374 GX 89
UGC 9376 GX 36, 52
UGC 9379 GX 36, 52
UGC 9380 GX 109
UGC 9385 GX 89, 109
UGC 9386 GX 36, 52
UGC 9387 GX 52
UGC 9389 GX 89
UGC 9391 GX 23
UGC 9394 GX 89
UGC 9396 GX 70
UGC 9398 GX 109
UGC 9400 GX 89, 109
UGC 9401 GX 70
UGC 9411 GX 89
UGC 9412 GX 23
UGC 9413 GX 4
UGC 9418 GX 70
UGC 9422 GX 36
UGC 9425 GX 52
UGC 9429 GX 36, 52
UGC 9432 GX 109
UGC 9441 GX 36
UGC 9443 GX 89
UGC 9444 GX 70, 89
UGC 9448 GX 23, 36
UGC 9452 GX 23
UGC 9454 GX 89
UGC 9469 GX 109
UGC 9470 GX 109
UGC 9471 GX 89, 109
UGC 9473 GX 52
UGC 9474 GX 89
UGC 9476 GX 36
UGC 9479 GX 109
UGC 9480 GX 70
UGC 9485 GX 109
UGC 9491 GX 109
UGC 9492 GX 89
UGC 9500 GX 89
UGC 9503 GX 70
UGC 9504 GX 52
UGC 9510 GX 52
UGC 9513 GX 89
UGC 9515 GX 89
UGC 9517 GX 89
UGC 9518 GX 52
UGC 9519 GX 52
UGC 9521 GX 89
UGC 9523 GX 89
UGC 9529 GX 12
UGC 9530 GX 89
UGC 9534 GX 89
UGC 9537 GX 52
UGC 9542 GX 36
UGC 9544 GX 70
UGC 9555 GX 89
UGC 9556 GX 23
UGC 9558 GX 70, 89
UGC 9559 GX 36
UGC 9560 GX 52
UGC 9561 GX 89
UGC 9562 GX 52
UGC 9567 GX 36
UGC 9569 GX 36
UGC 9578 GX 70
UGC 9588 GX 52
UGC 9594 GX 70
UGC 9596 GX 70
UGC 9598 GX 36
UGC 9601 GX 108, 109
UGC 9612 GX 36
UGC 9614 GX 88, 89
UGC 9616 GX 88, 89
UGC 9618 GX 70
UGC 9620 GX 70
UGC 9622 GX 70
UGC 9623 GX 52
UGC 9625 GX 88, 89
UGC 9629 GX 22, 23
UGC 9632 GX 22, 23
UGC 9635 GX 70
UGC 9638 GX 22, 23
UGC 9639 GX 36
UGC 9640 GX 88, 89
UGC 9641 GX 36
UGC 9644 GX 70
UGC 9647 GX 52
UGC 9650 GX 4
UGC 9654 GX 88, 89
UGC 9660 GX 36
UGC 9661 GX 108, 109
UGC 9663 GX 22, 23
UGC 9665 GX 36
UGC 9667 GX 88, 89, 108, 109
UGC 9668 GX 4
UGC 9672 GX 70
UGC 9675 GX 88, 89
UGC 9682 GX 108
UGC 9683 GX 4
UGC 9684 GX 36
UGC 9685 GX 88
UGC 9688 GX 22
UGC 9689 GX 88
UGC 9691 GX 36, 52
UGC 9696 GX 88
UGC 9702 GX 22, 36
UGC 9705 GX 70
UGC 9708 GX 108
UGC 9712 GX 88
UGC 9713 GX 70
UGC 9716 GX 52
UGC 9718 GX 70
UGC 9721 GX 88
UGC 9722 GX 22, 36
UGC 9730 GX 4
UGC 9732 GX 108
UGC 9734 GX 12
UGC 9737 GX 22
UGC 9744 GX 108
UGC 9746 GX 108
UGC 9748 GX 4
UGC 9749 GX 12
UGC 9750 GX 4
UGC 9755 GX 88
UGC 9757 GX 88, 108
UGC 9758 GX 88
UGC 9759 GX 22
UGC 9760 GX 108
UGC 9761 GX 35, 36
UGC 9763 GX 70
UGC 9766 GX 22
UGC 9770 GX 70
UGC 9777 GX 70
UGC 9780 GX 35, 36
UGC 9781 GX 88, 108
UGC 9787 GX 108
UGC 9794 GX 88
UGC 9796 GX 35, 36
UGC 9798 GX 88
UGC 9799 GX 88
UGC 9804 GX 108
UGC 9807 GX 88
UGC 9808 GX 88
UGC 9809 GX 51, 52
UGC 9814 GX 88
UGC 9815 GX 35
UGC 9820 GX 69, 70
UGC 9821 GX 88
UGC 9825 GX 69, 70
UGC 9828 GX 69, 70
UGC 9829 GX 108
UGC 9831 GX 51, 52, 69, 70
UGC 9833 GX 69, 70
UGC 9834 GX 51, 52
UGC 9835 GX 35
UGC 9837 GX 22
UGC 9838 GX 88
UGC 9840 GX 35
UGC 9841 GX 69
UGC 9842 GX 51
UGC 9843 GX 69
UGC 9844 GX 88
UGC 9845 GX 88
UGC 9846 GX 88
UGC 9849 GX 22
UGC 9853 GX 22
UGC 9855 GX 12
UGC 9856 GX 35
UGC 9857 GX 35
UGC 9858 GX 35, 51
UGC 9859 GX 69
UGC 9864 GX 88, 108
UGC 9873 GX 35
UGC 9874 GX 4
UGC 9875 GX 69
UGC 9886 GX 108
UGC 9887 GX 108
UGC 9892 GX 35
UGC 9896 GX 11, 12
UGC 9897 GX 88
UGC 9900 GX 88
UGC 9901 GX 88
UGC 9910 GX 51
UGC 9912 GX 88
UGC 9917 GX 69
UGC 9919 GX 88
UGC 9922 GX 51
UGC 9924 GX 35
UGC 9925 GX 88
UGC 9927 GX 69
UGC 9934 GX 22
UGC 9936 GX 35
UGC 9941 GX 88
UGC 9944 GX 4, 11, 12
UGC 9945 GX 108
UGC 9950 GX 4
UGC 9951 GX 88
UGC 9953 GX 108
UGC 9954 GX 69
UGC 9958 GX 69
UGC 9959 GX 35
UGC 9960 GX 108
UGC 9964 GX 88
UGC 9968 GX 108
UGC 9976 GX 88, 108
UGC 9977 GX 108
UGC 9978 GX 88
UGC 9979 GX 108
UGC 9980 GX 108
UGC 9982 GX 11, 12
UGC 9984 GX 69
UGC 9990 GX 108
UGC 9991 GX 88
UGC 9992 GX 11, 12
UGC 9996 GX 88
UGC 9997 GX 35
UGC 9999 GX 69
UGC 10001 GX 35
UGC 10005 GX 108
UGC 10010 GX 35
UGC 10013 GX 22
UGC 10014 GX 88
UGC 10018 GX 11, 12
UGC 10020 GX 69
UGC 10023 GX 88
UGC 10029 GX 88, 108
UGC 10030 GX 108
UGC 10031 GX 22
UGC 10034 GX 51
UGC 10035 GX 69
UGC 10037 GX 88
UGC 10041 GX 88, 108
UGC 10042 GX 88
UGC 10043 GX 69
UGC 10050 GX 69
UGC 10052 GX 69
UGC 10054 GX 4
UGC 10057 GX 11
UGC 10059 GX 69
UGC 10060 GX 69
UGC 10061 GX 88
UGC 10062 GX 69
UGC 10065 GX 51
UGC 10068 GX 88
UGC 10070 GX 35
UGC 10072 GX 11
UGC 10073 GX 69
UGC 10074 GX 69
UGC 10077 GX 88
UGC 10078 GX 11
UGC 10084 GX 69
UGC 10085 GX 69
UGC 10087 GX 35
UGC 10093 GX 69, 88
UGC 10097 GX 35
UGC 10099 GX 35
UGC 10104 GX 51
UGC 10109 GX 35
UGC 10111 GX 88
UGC 10115 GX 11
UGC 10118 GX 35
UGC 10120 GX 51
UGC 10121 GX 69, A9
UGC 10122 GX 51
UGC 10127 GX 69
UGC 10129 GX 22, 35
UGC 10130 GX 88
UGC 10134 GX 88, A9
UGC 10137 GX 88
UGC 10138 GX 69
UGC 10142 GX 11
UGC 10143 GX 88, A9
UGC 10144 GX 88, A9
UGC 10146 GX 88, 108
UGC 10147 GX 88, 108
UGC 10151 GX 69
UGC 10152 GX 35
UGC 10153 GX 69
UGC 10155 GX 51
UGC 10156 GX 35
UGC 10158 GX 88
UGC 10160 GX 69
UGC 10162 GX 11
UGC 10164 GX 88
UGC 10166 GX 51
UGC 10168 GX 35
UGC 10169 GX 88
UGC 10176 GX 88
UGC 10194 GX 11
UGC 10195 GX 69, A9
UGC 10200 GX 35
UGC 10201 GX 88, A9
UGC 10204 GX 88, A9
UGC 10205 GX 51
UGC 10207 GX 51
UGC 10211 GX 69
UGC 10212 GX 4
UGC 10213 GX 88
UGC 10214 GX 22
UGC 10223 GX 69
UGC 10225 GX 87, 88
UGC 10227 GX 51
UGC 10231 GX 11
UGC 10232 GX 69
UGC 10234 GX 51
UGC 10236 GX 69
UGC 10237 GX 4
UGC 10238 GX 87, 88
UGC 10239 GX 87, 88
UGC 10241 GX 35
UGC 10243 GX 69
UGC 10247 GX 22
UGC 10249 GX 87, 88
UGC 10250 GX 69
UGC 10257 GX 51
UGC 10259 GX 51, 69
UGC 10260 GX 69
UGC 10262 GX 51, 69
UGC 10263 GX 1
UGC 10264 GX 107, 108
UGC 10271 GX 22
UGC 10273 GX 69
UGC 10278 GX 35
UGC 10279 GX 22
UGC 10280 GX 4
UGC 10284 GX 22
UGC 10287 GX 87, 88
UGC 10288 GX 107, 108
UGC 10290 GX 107, 108
UGC 10294 GX 11
UGC 10297 GX 69
UGC 10298 GX 11
UGC 10306 GX 107
UGC 10310 GX 35
UGC 10312 GX 51
UGC 10321 GX 69
UGC 10325 GX 35

UGC 10330 GX 35, 51
UGC 10331 GX 22
UGC 10334 GX 11
UGC 10337 GX 87
UGC 10339 GX 107
UGC 10342 GX 51
UGC 10344 GX 51
UGC 10349 GX 35, 51, A3
UGC 10351 GX 69
UGC 10354 GX 35, 51, A3
UGC 10357 GX 35, 51, A3
UGC 10360 GX 87
UGC 10361 GX 22
UGC 10362 GX 51, A3
UGC 10367 GX 51, A3
UGC 10368 GX 4, 11
UGC 10372 GX 51
UGC 10374 GX 35
UGC 10380 GX 87
UGC 10381 GX 51, A3
UGC 10383 GX 11
UGC 10387 GX 87
UGC 10388 GX 87
UGC 10394 GX 107
UGC 10396 GX 22, 35
UGC 10404 GX 51, A3
UGC 10405 GX 69, 87
UGC 10406 GX 107
UGC 10407 GX 35, A3
UGC 10408 GX 22
UGC 10410 GX 69
UGC 10412 GX 87
UGC 10413 GX 69
UGC 10414 GX 87
UGC 10415 GX 35, A3
UGC 10416 GX 87
UGC 10420 GX 51, A3
UGC 10425 GX 11
UGC 10427 GX 35, A3
UGC 10429 GX 51, A3
UGC 10430 GX 35, A3
UGC 10435 GX 69
UGC 10436 GX 35, A3
UGC 10437 GX 35
UGC 10445 GX 69
UGC 10446 GX 4
UGC 10447 GX 4
UGC 10449 GX 11, 22
UGC 10454 GX 22
UGC 10455 GX 69
UGC 10456 GX 22
UGC 10457 GX 35
UGC 10459 GX 35, 51, A3
UGC 10463 GX 87
UGC 10465 GX 107
UGC 10466 GX 4
UGC 10471 GX 4
UGC 10473 GX 50, 51
UGC 10474 GX 107
UGC 10476 GX 11
UGC 10477 GX 50, 51
UGC 10478 GX 11
UGC 10480 GX 35
UGC 10486 GX 35
UGC 10490 GX 68, 69, 87
UGC 10492 GX 107
UGC 10493 GX 22
UGC 10497 GX 11
UGC 10500 GX 22
UGC 10502 GX 11
UGC 10503 GX 11
UGC 10504 GX 50, 51
UGC 10509 GX 4
UGC 10510 GX 22
UGC 10511 GX 22, 35
UGC 10512 GX 50, 51
UGC 10514 GX 68, 69
UGC 10515 GX 22
UGC 10517 GX 22
UGC 10518 GX 11, 22
UGC 10528 GX 68
UGC 10529 GX 50
UGC 10531 GX 34, 35
UGC 10536 GX 11, 22
UGC 10542 GX 22
UGC 10543 GX 68
UGC 10547 GX 50
UGC 10548 GX 22
UGC 10549 GX 68
UGC 10553 GX 34, 35, 50
UGC 10554 GX 107
UGC 10561 GX 11, 22
UGC 10566 GX 50
UGC 10567 GX 50
UGC 10569 GX 87
UGC 10571 GX 34, 35
UGC 10579 GX 22
UGC 10586 GX 34, 35
UGC 10587 GX 11
UGC 10589 GX 22
UGC 10590 GX 22
UGC 10593 GX 22
UGC 10599 GX 50
UGC 10601 GX 22
UGC 10602 GX 34
UGC 10604 GX 4
UGC 10610 GX 34
UGC 10614 GX 11
UGC 10622 GX 11
UGC 10623 GX 107
UGC 10627 GX 34
UGC 10632 GX 4
UGC 10635 GX 50
UGC 10640 GX 107
UGC 10644 GX 21, 22
UGC 10646 GX 21, 22
UGC 10648 GX 21, 22
UGC 10650 GX 68
UGC 10651 GX 34
UGC 10653 GX 68
UGC 10658 GX 68
UGC 10659 GX 87
UGC 10662 GX 50
UGC 10663 GX 50
UGC 10668 GX 50
UGC 10672 GX 68
UGC 10673 GX 50, 68
UGC 10674 GX 87
UGC 10675 GX 50
UGC 10677 GX 50
UGC 10679 GX 50
UGC 10681 GX 34
UGC 10683 GX 107
UGC 10685 GX 87
UGC 10687 GX 21, 22
UGC 10688 GX 50
UGC 10692 GX 68
UGC 10693 GX 34
UGC 10695 GX 34
UGC 10699 GX 87
UGC 10702 GX 68
UGC 10704 GX 4
UGC 10705 GX 87
UGC 10706 GX 50
UGC 10707 GX 34
UGC 10708 GX 50
UGC 10710 GX 34
UGC 10712 GX 50
UGC 10713 GX 11
UGC 10714 GX 50
UGC 10715 GX 50
UGC 10717 GX 68
UGC 10720 GX 87
UGC 10721 GX 68
UGC 10722 GX 34
UGC 10728 GX 68
UGC 10731 GX 11
UGC 10732 GX 50
UGC 10733 GX 50
UGC 10736 GX 11
UGC 10738 GX 87, 107
UGC 10740 GX 1
UGC 10743 GX 87
UGC 10749 GX 50
UGC 10753 GX 50
UGC 10758 GX 21
UGC 10759 GX 34
UGC 10760 GX 34
UGC 10765 GX 34
UGC 10770 GX 21
UGC 10775 GX 87
UGC 10778 GX 87
UGC 10779 GX 87
UGC 10785 GX 50
UGC 10787 GX 68
UGC 10789 GX 87
UGC 10791 GX 11
UGC 10796 GX 21
UGC 10797 GX 87
UGC 10801 GX 50
UGC 10802 GX 87
UGC 10803 GX 4, 11
UGC 10805 GX 87
UGC 10806 GX 34
UGC 10811 GX 21
UGC 10812 GX 34, 50
UGC 10813 GX 68
UGC 10814 GX 34
UGC 10820 GX 34
UGC 10821 GX 21
UGC 10822 GX 21
UGC 10825 GX 106, 107
UGC 10828 GX 68
UGC 10830 GX 21
UGC 10831 GX 68
UGC 10837 GX 68
UGC 10839 GX 21
UGC 10840 GX 68
UGC 10844 GX 86, 87
UGC 10845 GX 34
UGC 10862 GX 86
UGC 10864 GX 86
UGC 10866 GX 68
UGC 10868 GX 86
UGC 10869 GX 21
UGC 10872 GX 86
UGC 10879 GX 68
UGC 10880 GX 21
UGC 10888 GX 21
UGC 10890 GX 50
UGC 10892 GX 4
UGC 10894 GX 68
UGC 10899 GX 68
UGC 10905 GX 68
UGC 10907 GX 4
UGC 10908 GX 34
UGC 10909 GX 68
UGC 10912 GX 11
UGC 10913 GX 86
UGC 10918 GX 86
UGC 10919 GX 68, 86
UGC 10923 GX 1
UGC 10926 GX 68
UGC 10928 GX 68
UGC 10929 GX 50
UGC 10933 GX 50
UGC 10936 GX 21
UGC 10938 GX 34
UGC 10941 GX 68, 86
UGC 10943 GX 106
UGC 10944 GX 68
UGC 10946 GX 34
UGC 10949 GX 4
UGC 10950 GX 86
UGC 10956 GX 106
UGC 10963 GX 11
UGC 10966 GX 68
UGC 10969 GX 50
UGC 10971 GX 21
UGC 10972 GX 68
UGC 10976 GX 50
UGC 10979 GX 68
UGC 10981 GX 86
UGC 10984 GX 21, 34
UGC 10985 GX 86
UGC 10988 GX 21
UGC 10990 GX 50
UGC 10995 GX 11
UGC 10999 GX 68
UGC 11000 GX 50
UGC 11001 GX 86
UGC 11017 GX 50, 68
UGC 11024 GX 68
UGC 11025 GX 68
UGC 11027 GX 68
UGC 11028 GX 21
UGC 11029 GX 68
UGC 11030 GX 106
UGC 11031 GX 50
UGC 11035 GX 50
UGC 11036 GX 21
UGC 11038 GX 11
UGC 11039 GX 68
UGC 11041 GX 50
UGC 11042 GX 68
UGC 11043 GX 68
UGC 11044 GX 68
UGC 11045 GX 50
UGC 11046 GX 68
UGC 11050 GX 50
UGC 11055 GX 86
UGC 11057 GX 86
UGC 11058 GX 49, 50
UGC 11060 GX 67, 68
UGC 11063 GX 67, 68
UGC 11064 GX 67, 68
UGC 11065 GX 34
UGC 11066 GX 10, 11
UGC 11067 GX 86
UGC 11068 GX 67, 68
UGC 11070 GX 67, 68
UGC 11073 GX 86
UGC 11074 GX 86
UGC 11076 GX 49, 50
UGC 11082 GX 67, 68
UGC 11087 GX 49, 50
UGC 11090 GX 67, 68
UGC 11093 GX 86
UGC 11097 GX 67, 68
UGC 11098 GX 49, 50, 67, 68
UGC 11099 GX 10, 11
UGC 11105 GX 67
UGC 11106 GX 10, 11
UGC 11107 GX 67, 86
UGC 11112 GX 34
UGC 11113 GX 67
UGC 11120 GX 67
UGC 11122 GX 86
UGC 11123 GX 67
UGC 11124 GX 49
UGC 11127 GX 67
UGC 11129 GX 49
UGC 11131 GX 106
UGC 11132 GX 49
UGC 11140 GX 49
UGC 11141 GX 86
UGC 11142 GX 67
UGC 11150 GX 67
UGC 11151 GX 49, 67
UGC 11152 GX 67
UGC 11155 GX 67
UGC 11156 GX 67
UGC 11157 GX 49
UGC 11162 GX 49, 67
UGC 11163 GX 49
UGC 11165 GX 10
UGC 11168 GX 86
UGC 11170 GX 67
UGC 11171 GX 49
UGC 11177 GX 86
UGC 11179 GX 67
UGC 11181 GX 34
UGC 11185 GX 34
UGC 11186 GX 34
UGC 11190 GX 34, 49
UGC 11193 GX 10
UGC 11194 GX 67
UGC 11195 GX 49
UGC 11197 GX 67
UGC 11199 GX 49
UGC 11202 GX 34
UGC 11204 GX 34
UGC 11206 GX 34
UGC 11208 GX 10
UGC 11214 GX 86
UGC 11215 GX 21
UGC 11217 GX 33, 34
UGC 11219 GX 49
UGC 11222 GX 10
UGC 11228 GX 33, 34
UGC 11229 GX 67
UGC 11230 GX 10
UGC 11233 GX 49
UGC 11235 GX 33, 34
UGC 11237 GX 67
UGC 11241 GX 21, 33, 34
UGC 11242 GX 86
UGC 11244 GX 33, 34
UGC 11246 GX 67
UGC 11252 GX 33
UGC 11261 GX 67
UGC 11262 GX 33
UGC 11265 GX 49
UGC 11267 GX 1
UGC 11268 GX 49
UGC 11270 GX 3
UGC 11275 GX 49
UGC 11278 GX 49
UGC 11281 GX 49
UGC 11285 GX 67
UGC 11287 GX 33
UGC 11289 GX 67
UGC 11291 GX 49
UGC 11292 GX 21
UGC 11293 GX 85, 86
UGC 11294 GX 67
UGC 11295 GX 3
UGC 11296 GX 49
UGC 11298 GX 21, 33
UGC 11301 GX 67, 85, 86
UGC 11303 GX 49
UGC 11304 GX 49
UGC 11307 GX 67
UGC 11311 GX 49
UGC 11312 GX 49
UGC 11313 GX 49
UGC 11314 GX 67
UGC 11315 GX 67
UGC 11319 GX 33, 49
UGC 11320 GX 67
UGC 11323 GX 67
UGC 11325 GX 49
UGC 11326 GX 85
UGC 11329 GX 33
UGC 11331 GX 3, 10
UGC 11333 GX 49
UGC 11334 GX 3, 10
UGC 11336 GX 49
UGC 11337 GX 67
UGC 11338 GX 49
UGC 11342 GX 10
UGC 11343 GX 21
UGC 11344 GX 67
UGC 11346 GX 67
UGC 11350 GX 67
UGC 11353 GX 67
UGC 11355 GX 67
UGC 11357 GX 33
UGC 11363 GX 10
UGC 11367 GX 49
UGC 11368 GX 67
UGC 11369 GX 67
UGC 11370 GX 67
UGC 11371 GX 67
UGC 11372 GX 49
UGC 11373 GX 21
UGC 11375 GX 67
UGC 11376 GX 33
UGC 11377 GX 3, 10
UGC 11379 GX 67
UGC 11380 GX 49
UGC 11393 GX 67
UGC 11394 GX 67
UGC 11397 GX 49
UGC 11399 GX 49
UGC 11402 GX 3, 10
UGC 11404 GX 49, 67
UGC 11406 GX 33
UGC 11408 GX 33
UGC 11411 GX 10
UGC 11411A GX 20
UGC 11412 GX 20
UGC 11420 GX 33
UGC 11421 GX 33
UGC 11426 GX 48, 49
UGC 11427 GX 10
UGC 11428 GX 48, 49
UGC 11430 GX 33
UGC 11433 GX 48, 49
UGC 11435 GX 20
UGC 11439 GX 33
UGC 11446 GX 33
UGC 11448 GX 48
UGC 11450 GX 33
UGC 11453 GX 20
UGC 11455 GX 10
UGC 11456 GX 48
UGC 11459 GX 33, 48
UGC 11460 GX 33
UGC 11464 GX 20, 33
UGC 11465 GX 33
UGC 11466 GX 33
UGC 11467 GX 33
UGC 11468 GX 33
UGC 11473 GX 33
UGC 11475 GX 20
UGC 11476 GX 10
UGC 11480 GX 33
UGC 11481 GX 104, 105
UGC 11482 GX 84, 85
UGC 11485 GX 33
UGC 11486 GX 33
UGC 11487 GX 10
UGC 11488 GX 104, 105
UGC 11489 GX 104
UGC 11492 GX 20
UGC 11493 GX 84, 104
UGC 11494 GX 33
UGC 11496 GX 10
UGC 11497 GX 104
UGC 11498 GX 84, 104
UGC 11499 GX 32, 33
UGC 11500 GX 32, 33
UGC 11501 GX 104
UGC 11502 GX 20
UGC 11503 GX 32, 33
UGC 11505 GX 104
UGC 11507 GX 32, 33
UGC 11509 GX 20
UGC 11510 GX 20
UGC 11511 GX 84
UGC 11512 GX 84
UGC 11515 GX 10, 20
UGC 11521 GX 104
UGC 11522 GX 84, 104
UGC 11523 GX 84, 104
UGC 11524 GX 84, 104
UGC 11525 GX 104
UGC 11526 GX 84
UGC 11527 GX 104
UGC 11529 GX 10
UGC 11532 GX 84
UGC 11537 GX 104
UGC 11539 GX 9, 10, 20
UGC 11541 GX 104
UGC 11543 GX 84
UGC 11551 GX 84
UGC 11552 GX 84
UGC 11555 GX 84, 104
UGC 11557 GX 20
UGC 11559 GX 104
UGC 11561 GX 104
UGC 11562 GX 104
UGC 11564 GX 84
UGC 11566 GX 104
UGC 11567 GX 104
UGC 11568 GX 84
UGC 11571 GX 84
UGC 11572 GX 84
UGC 11575 GX 104
UGC 11576 GX 104
UGC 11577 GX 104
UGC 11578 GX 84
UGC 11581 GX 104
UGC 11582 GX 66
UGC 11583 GX 20
UGC 11584 GX 104
UGC 11585 GX 104
UGC 11587 GX 84
UGC 11591 GX 104
UGC 11593 GX 84
UGC 11595 GX 104
UGC 11599 GX 84
UGC 11602 GX 84
UGC 11605 GX 84
UGC 11607 GX 104
UGC 11608 GX 65, 66
UGC 11610 GX 84
UGC 11611 GX 84
UGC 11615 GX 65, 66
UGC 11620 GX 84
UGC 11622 GX 104
UGC 11623 GX 84
UGC 11631 GX 104
UGC 11634 GX 84
UGC 11635 GX 3
UGC 11638 GX 84
UGC 11639 GX 84
UGC 11643 GX 65
UGC 11644 GX 84
UGC 11648 GX 9
UGC 11649 GX 104
UGC 11651 GX 65
UGC 11653 GX 65
UGC 11657 GX 103, 104
UGC 11659 GX 83, 84
UGC 11661 GX 83, 84
UGC 11671 GX 83
UGC 11672 GX 83
UGC 11675 GX 83
UGC 11676 GX 65
UGC 11678 GX 9
UGC 11680 GX 103
UGC 11682 GX 65, 83
UGC 11683 GX 65
UGC 11689 GX 9
UGC 11694 GX 83
UGC 11695 GX 103
UGC 11698 GX 83
UGC 11699 GX 83
UGC 11700 GX 83
UGC 11706 GX 83
UGC 11707 GX 65
UGC 11709 GX 65
UGC 11713 GX 83
UGC 11714 GX 103
UGC 11717 GX 65
UGC 11719 GX 83
UGC 11720 GX 83
UGC 11721 GX 83
UGC 11722 GX 65
UGC 11723 GX 103
UGC 11724 GX 103
UGC 11728 GX 65

UGC 11735 GX 83
UGC 11740 GX 83
UGC 11743 GX 47
UGC 11749 GX 65
UGC 11751 GX 83
UGC 11753 GX 47
UGC 11754 GX 65
UGC 11757 GX 32
UGC 11758 GX 83
UGC 11760 GX 103
UGC 11765 GX 83
UGC 11769 GX 65
UGC 11775 GX 47
UGC 11781 GX 47
UGC 11782 GX 83
UGC 11787 GX 65
UGC 11788 GX 65
UGC 11789 GX 103
UGC 11790 GX 103
UGC 11792 GX 83, 103
UGC 11793 GX 65
UGC 11798 GX 31
UGC 11799 GX 31
UGC 11801 GX 31
UGC 11802 GX 31
UGC 11803 GX 83
UGC 11805 GX 31
UGC 11806 GX 31
UGC 11808 GX 31
UGC 11814 GX 103
UGC 11816 GX 103
UGC 11818 GX 9
UGC 11819 GX 31
UGC 11820 GX 83
UGC 11821 GX 83
UGC 11828 GX 103
UGC 11830 GX 65
UGC 11834 GX 65
UGC 11836 GX 31
UGC 11838 GX 65
UGC 11841 GX 47
UGC 11846 GX 83
UGC 11849 GX 65
UGC 11851 GX 83, 103
UGC 11852 GX 64, 65
UGC 11853 GX 103
UGC 11858 GX 31
UGC 11859 GX 103
UGC 11860 GX 64, 65
UGC 11861 GX 2, 3, 9
UGC 11862 GX 46, 47
UGC 11864 GX 31
UGC 11865 GX 83
UGC 11868 GX 64, 65
UGC 11871 GX 83
UGC 11878 GX 64, 65
UGC 11890 GX 46, 47
UGC 11891 GX 31
UGC 11892 GX 46, 47
UGC 11893 GX 46
UGC 11895 GX 46
UGC 11896 GX 83
UGC 11897 GX 31
UGC 11905 GX 64
UGC 11907 GX 103
UGC 11908 GX 83
UGC 11909 GX 31
UGC 11911 GX 31
UGC 11912 GX 46
UGC 11915 GX 102, 103
UGC 11919 GX 31
UGC 11920 GX 31
UGC 11921 GX 82, 83
UGC 11924 GX 64
UGC 11927 GX 31, 46
UGC 11928 GX 82, 83
UGC 11929 GX 46
UGC 11935 GX 31
UGC 11944 GX 64, 82, 83
UGC 11946 GX 31
UGC 11947 GX 82, 83
UGC 11948 GX 82, 83
UGC 11949 GX 46
UGC 11950 GX 46
UGC 11952 GX 82, 83
UGC 11955 GX 46
UGC 11961 GX 31
UGC 11964 GX 64
UGC 11973 GX 31
UGC 11974 GX 46
UGC 11975 GX 46
UGC 11978 GX 64
UGC 11979 GX 31
UGC 11981 GX 46, 64
UGC 11987 GX 82
UGC 11991 GX 31
UGC 11994 GX 46
UGC 11995 GX 46
UGC 12002 GX 31
UGC 12006 GX 46
UGC 12007 GX 46
UGC 12009 GX 46
UGC 12011 GX 46
UGC 12015 GX 64
UGC 12016 GX 31, 46
UGC 12017 GX 82
UGC 12018 GX 46
UGC 12020 GX 46
UGC 12021 GX 82, 102
UGC 12022 GX 46
UGC 12023 GX 82, 102
UGC 12032 GX 31
UGC 12036 GX 64
UGC 12037 GX 46
UGC 12038 GX 64
UGC 12039 GX 46
UGC 12040 GX 46
UGC 12044 GX 46
UGC 12050 GX 64
UGC 12051 GX 46
UGC 12054 GX 82
UGC 12056 GX 46
UGC 12057 GX 31
UGC 12059 GX 64
UGC 12060 GX 46
UGC 12063 GX 46
UGC 12064 GX 46
UGC 12066 GX 64
UGC 12067 GX 64
UGC 12069 GX 2
UGC 12071 GX 46
UGC 12073 GX 46
UGC 12074 GX 82
UGC 12075 GX 46
UGC 12082 GX 46
UGC 12084 GX 64
UGC 12086 GX 31
UGC 12088 GX 31
UGC 12093 GX 64
UGC 12095 GX 31
UGC 12107 GX 64
UGC 12121 GX 46
UGC 12124 GX 64
UGC 12125 GX 31, 46
UGC 12127 GX 46
UGC 12131 GX 46
UGC 12133 GX 82
UGC 12137 GX 46
UGC 12138 GX 82
UGC 12143 GX 46
UGC 12149 GX 46
UGC 12150 GX 46
UGC 12151 GX 102
UGC 12154 GX 8, 9
UGC 12156 GX 46
UGC 12157 GX 46
UGC 12158 GX 64
UGC 12160 GX 2
UGC 12161 GX 46
UGC 12163 GX 46, 64
UGC 12164 GX 46
UGC 12173 GX 46
UGC 12177 GX 46
UGC 12178 GX 82
UGC 12179 GX 46
UGC 12181 GX 46
UGC 12182 GX 2, 8, 9
UGC 12184 GX 82
UGC 12185 GX 46
UGC 12188 GX 46
UGC 12190 GX 64
UGC 12191 GX 64
UGC 12193 GX 64
UGC 12196 GX 82
UGC 12199 GX 31, 46
UGC 12200 GX 64
UGC 12201 GX 46
UGC 12202 GX 82
UGC 12204 GX 31, 46
UGC 12206 GX 46
UGC 12208 GX 102
UGC 12210 GX 46
UGC 12212 GX 46, 64
UGC 12213 GX 82
UGC 12214 GX 46
UGC 12215 GX 46
UGC 12218 GX 46
UGC 12221 GX 2
UGC 12222 GX 82
UGC 12224 GX 82
UGC 12227 GX 62
UGC 12233 GX 64
UGC 12234 GX 46
UGC 12235 GX 46
UGC 12236 GX 46
UGC 12237 GX 82
UGC 12238 GX 46
UGC 12242 GX 46
UGC 12250 GX 82
UGC 12252 GX 46
UGC 12253 GX 82
UGC 12255 GX 82, 102
UGC 12258 GX 64, 82
UGC 12259 GX 64
UGC 12260 GX 46
UGC 12263 GX 8
UGC 12265 GX 64
UGC 12266 GX 82
UGC 12271 GX 102
UGC 12272 GX 64
UGC 12278 GX 64
UGC 12281 GX 82
UGC 12282 GX 31, 46
UGC 12283 GX 64
UGC 12287 GX 18, 19
UGC 12289 GX 64
UGC 12291 GX 64
UGC 12293 GX 64
UGC 12295 GX 102
UGC 12297 GX 46
UGC 12298 GX 46
UGC 12303 GX 64
UGC 12307 GX 82
UGC 12308 GX 82
UGC 12310 GX 64
UGC 12311 GX 46
UGC 12323 GX 46
UGC 12326 GX 64
UGC 12333 GX 64
UGC 12334 GX 64
UGC 12336 GX 102
UGC 12341 GX 31
UGC 12344 GX 64
UGC 12346 GX 102
UGC 12347 GX 64
UGC 12350 GX 82
UGC 12351 GX 64
UGC 12352 GX 64
UGC 12356 GX 46
UGC 12359 GX 82
UGC 12362 GX 46
UGC 12363 GX 102
UGC 12372 GX 46
UGC 12381 GX 31
UGC 12385 GX 46
UGC 12388 GX 82
UGC 12389 GX 30, 31
UGC 12390 GX 64
UGC 12394 GX 46
UGC 12396 GX 30, 31
UGC 12400 GX 64
UGC 12407 GX 82
UGC 12410 GX 46
UGC 12411 GX 30, 31
UGC 12413 GX 64
UGC 12423 GX 82
UGC 12425 GX 64
UGC 12430 GX 46, 64
UGC 12433 GX 30, 31
UGC 12444 GX 46
UGC 12446 GX 102
UGC 12451 GX 82, 102
UGC 12452 GX 102
UGC 12454 GX 82
UGC 12461 GX 82, 102
UGC 12466 GX 102
UGC 12470 GX 63, 64
UGC 12472 GX 82
UGC 12474 GX 45, 46
UGC 12475 GX 102
UGC 12476 GX 45, 46
UGC 12479 GX 102
UGC 12482 GX 45, 46, 63, 64
UGC 12490 GX 63, 64
UGC 12491 GX 30
UGC 12492 GX 102
UGC 12494 GX 82
UGC 12495 GX 82
UGC 12496 GX 30
UGC 12504 GX 2
UGC 12506 GX 82
UGC 12507 GX 30
UGC 12508 GX 102
UGC 12515 GX 63, 64
UGC 12517 GX 30
UGC 12519 GX 81, 82
UGC 12521 GX 101, 102
UGC 12522 GX 81, 82
UGC 12530 GX 45, 46, 63, 64
UGC 12538 GX 45, 46
UGC 12544 GX 81, 82
UGC 12547 GX 81, 82, 101, 102
UGC 12548 GX 81, 82, 101, 102
UGC 12552 GX 81, 82
UGC 12553 GX 81, 82
UGC 12557 GX 45, 46, 63, 64
UGC 12558 GX 30
UGC 12561 GX 81, 82
UGC 12566 GX 45, 46, 63, 64
UGC 12570 GX 45, 46
UGC 12571 GX 81, 82
UGC 12573 GX 30
UGC 12581 GX 81, 82
UGC 12582 GX 81, 82
UGC 12585 GX 81, 82
UGC 12587 GX 63
UGC 12588 GX 30
UGC 12591 GX 63
UGC 12601 GX 81, 82
UGC 12613 GX 81
UGC 12619 GX 63
UGC 12625 GX 45, 63
UGC 12628 GX 101
UGC 12631 GX 63
UGC 12632 GX 30, 45
UGC 12634 GX 30, 45
UGC 12635 GX 101
UGC 12636 GX 63
UGC 12639 GX 45
UGC 12645 GX 45
UGC 12646 GX 63
UGC 12648 GX 101
UGC 12650 GX 45
UGC 12651 GX 45
UGC 12653 GX 81
UGC 12655 GX 63
UGC 12657 GX 45, 63
UGC 12661 GX 101
UGC 12663 GX 63
UGC 12666 GX 45
UGC 12667 GX 45
UGC 12669 GX 30
UGC 12672 GX 45
UGC 12677 GX 81
UGC 12681 GX 63
UGC 12682 GX 63
UGC 12685 GX 101
UGC 12687 GX 81
UGC 12688 GX 81
UGC 12689 GX 81, 101
UGC 12690 GX 101
UGC 12693 GX 45
UGC 12696 GX 63
UGC 12707 GX 63, 81
UGC 12709 GX 101
UGC 12710 GX 63, 81
UGC 12711 GX 45
UGC 12713 GX 45
UGC 12714 GX 45
UGC 12717 GX 81, 101
UGC 12725 GX 63
UGC 12729 GX 101
UGC 12731 GX 63
UGC 12732 GX 63
UGC 12733 GX 63
UGC 12739 GX 101
UGC 12741 GX 45
UGC 12742 GX 30
UGC 12747 GX 63
UGC 12749 GX 81
UGC 12750 GX 30
UGC 12753 GX 81
UGC 12755 GX 63
UGC 12756 GX 81
UGC 12758 GX 101
UGC 12764 GX 63
UGC 12767 GX 81
UGC 12774 GX 101
UGC 12776 GX 45
UGC 12784 GX 63, 81
UGC 12785 GX 63
UGC 12791 GX 63
UGC 12792 GX 63
UGC 12793 GX 81
UGC 12795 GX 30
UGC 12796 GX 30
UGC 12800 GX 81
UGC 12803 GX 45, 63
UGC 12804 GX 63
UGC 12809 GX 30
UGC 12810 GX 101
UGC 12816 GX 101
UGC 12818 GX 81
UGC 12822 GX 81
UGC 12823 GX 63
UGC 12835 GX 63
UGC 12838 GX 101
UGC 12839 GX 63
UGC 12840 GX 63
UGC 12843 GX 63, 81
UGC 12844 GX 63
UGC 12845 GX 45
UGC 12847 GX 101
UGC 12851 GX 30
UGC 12854 GX 81
UGC 12855 GX 63
UGC 12856 GX 81
UGC 12857 GX 101
UGC 12860 GX 81
UGC 12861 GX 45, 63
UGC 12862 GX 30
UGC 12864 GX 45
UGC 12869 GX 45
UGC 12871 GX 81
UGC 12873 GX 63
UGC 12879 GX 63
UGC 12881 GX 101
UGC 12886 GX 63
UGC 12888 GX 30
UGC 12889 GX 30
UGC 12890 GX 81
UGC 12893 GX 63, 81
UGC 12896 GX 63
UGC 12897 GX 63
UGC 12899 GX 63
UGC 12900 GX 63
UGC 12901 GX 63
UGC 12904 GX 45
UGC 12914 GX 63
UGC 12915 GX 63
UGC 12921 GX 2
UGCA 127 GX 136
UKS 1 GC 146, A17
Up 1 OC 54
vdB 1 BN 18
vdB 8 BN 17
vdB 14 BN 28
vdB 15 BN 28
vdB 16 BN 60, 78
vdB 20 BN 78, A12
vdB 23 BN 78, A12
vdB 24 BN 60
vdB 26 BN 97, 98
vdB 29 BN 59, 77
vdB 31 BN 59
vdB 37 BN 97
vdB 38 BN 96, 97
vdB 49 BN 116
vdB 68 BN 136
vdB 69 BN 136
vdB 80 BN 136
vdB 93 BN 135
vdB 96 BN 153, 154
vdB 97 BN 135
vdB 98 BN 153
vdB 111 BN 87
vdB 123 BN 106
vdB 126 BN 66
vdB 128 BN 48
vdB 131 BN 32, A2
vdB 132 BN 32, A2
vdB 133 BN 48
vdB 140 BN 19
vdB 142 BN 19
vdB 143 BN 9
vdB 145 BN 31
vdB 152 BN 9
vdB-Ha 4 OC 171
vdB-Ha 19 OC 171
vdB-Ha 23 OC 171
vdB-Ha 34 OC 187
vdB-Ha 37 OC 187
vdB-Ha 54 OC 186, 187
vdB-Ha 55 OC 170
vdB-Ha 56 OC 186
vdB-Ha 58 OC 199
vdB-Ha 63 OC 186
vdB-Ha 66 OC 199
vdB-Ha 67 OC 186, 199
vdB-Ha 72 OC 199
vdB-Ha 73 OC 186
vdB-Ha 75 OC 199
vdB-Ha 78 OC 199
vdB-Ha 84 OC 199
vdB-Ha 85 OC 186
vdB-Ha 87 OC 199
vdB-Ha 88 OC 186, 199
vdB-Ha 90 OC 199
vdB-Ha 91 OC 199
vdB-Ha 92 OC 199
vdB-Ha 99 OC 199
vdB-Ha 106 OC 199
vdB-Ha 110 OC 198
vdB-Ha 111 OC 209
vdB-Ha 118 OC 198
vdB-Ha 131 OC 209
vdB-Ha 132 OC 209
vdB-Ha 140 OC 209
vdB-Ha 150 OC 208
vdB-Ha 151 OC 197
vdB-Ha 164 OC 208
vdB-Ha 176 GC 182
vdB-Ha 197 OC 181, 182
vdB-Ha 200 OC 181, 182
vdB-Ha 205 OC 164, 181, A22
vdB-Ha 211 OC 181, A22
vdB-Ha 214 OC 164
vdB-Ha 217 OC 164, 181
vdB-Ha 221 OC 164
vdB-Ha 222 OC 164
vdB-Ha 223 OC 164
vdB-Ha 231 OC 164
vdB-Ha 245 OC 146, 164
vdBH 63 BN 208
vdBH 65a BN 208
vdBH 81 BN 181, 195, 196
Wa 3 OC 153
Wa 6 OC 187
Wa 7 OC 135
We 1 OC 181, 182
WG71 1 OC 213, 214, 219